MW00709318

ENCYCLOPEDIA *of* COASTAL SCIENCE

Encyclopedia of Earth Sciences Series

ENCYCLOPEDIA OF COASTAL SCIENCE

Volume Editor

Maurice Schwartz is Professor Emeritus, Department of Geology at Western Washington University, Bellingham, Washington, United States; where he served near the end of his tenure as Dean of the Graduate School and Bureau for Faculty Research. As well as being an author of numerous articles which have appeared in the coastal geology literature he was the editor of *Spits and Bars* (Dowden, Hutchinson & Ross, 1972), *Barrier Islands* (Dowden, Hutchinson & Ross, 1973), *Encyclopedia of Beaches and Coastal Environments* (Hutchinson Ross, 1982), and co-edited with Eric Bird *The World's Coastline* (Van Nostrand Reinhold, 1985).

Advisory Board

Aim of the Series

The Encyclopedia of Earth Sciences Series provides comprehensive and authoritative coverage of all the main areas in the Earth Sciences. Each volume comprises a focused and carefully chosen collection of contributions from leading names in the subject, with copious illustrations and reference lists.

These books represent one of the world's leading resources for the Earth Sciences community. Previous volumes are being updated and new works published so that the volumes will continue to be essential reading for all professional earth scientists, geologists, geophysicists, climatologists, and oceanographers as well as for teachers and students.

See the back of this volume for a current list of titles in the Encyclopedia of Earth Sciences Series. Go to www.eseo.com to visit the "Earth Sciences Encyclopedia Online"—the online version of this Encyclopedia Series.

About the Editors

Professor Rhodes W. Fairbridge has edited more than 30 Encyclopedias in the Earth Sciences Series. During his career he has worked as a petroleum geologist in the Middle East, been a WW II intelligence officer in the SW Pacific and led expeditions to the Sahara, Arctic Canada, Arctic Scandinavia, Brazil and New Guinea. He is currently Emeritus Professor of Geology at Columbia University and is affiliated with the Goddard Institute for Space Studies.

Professor Michael Rampino has published more than 100 papers in professional journals including *Science*, *Nature*, and *Scientific American*. He has worked in such diverse fields as volcanology, planetary science, sedimentology, and climate studies, and has done field work on six continents. He is currently Associate Professor of Earth and Environmental Sciences at New York University and a consultant at NASA's Goddard Institute for Space Studies.

ENCYCLOPEDIA OF EARTH SCIENCES SERIES

ENCYCLOPEDIA *of* COASTAL SCIENCE

edited by

MAURICE L. SCHWARTZ
Department of Geology
Western Washington University, WA, USA

A C.I.P. Catalogue record for this book is available from the Library of Congress

ISBN-13 978-1-4020-1903-6 (HB)
ISBN-13 978-1-4020-3565-4 (e-book)
ISBN-10 1-4020-1903-3 (HB)
ISBN-10 1-4020-3565-9 (e-book)

Published by Springer,
PO Box 17, 3300 AA Dordrecht, The Netherlands

Printed on acid-free paper

Printed in the Netherlands

To Evelyn Kest, my companion and inspiration

Contents

Contributors

John P. Ahrens
6702 Springfield Drive
Mason Neck VA 22079
USA
Phone: 1 703 550 8412
Fax: 1 301 713 0799
e-mail: j.ahrens@worldnet.att.net

Depth of Disturbance
Simple Beach and Surf Zone Models

Marie Aizpuru
Laboratoire d'Ecologie Terrestre (UMR 5552)
Centre de Teledetection
13, avenue de Colonel Roche, BP 4072
31029 Toulouse Cedex 4
France

Mangroves, Ecology
Mangroves, Remote Sensing

Anita de Alava
UDECIMAR, Facultad de Ciencias
Igua 4225,
PO Box 10773
1140 Montevideo
Uruguay

South America, Coastal Ecology

Carl L. Amos
School of Ocean and Earth Science
Southampton Oceanography Centre
Empress Dock, Southampton
Hampshire SO14 3ZH
UK
Phone: 44 023 80 596068
Fax: 44 023 80 593059
e-mail: c.amos@soc.soton.ac.uk

Ripple Marks

Edward J. Anthony
Coastal Geomorphology and Shoreline
Management Unit JE 2208
Universite du Littoral Côte d'Opale
MREN, 32 Avenue Foch, 62930 Wimereux
France
Phone: +33 3 28 26 01 22,
Fax: +33 3 28 23 74 13
e-mail: edward.anthony@univ-littoral.fr

Beach Erosion
Wave and Tide-Dominated Coasts

Rowland J. Atkins
Golder Associates
#204-4430 Chatterton Way
Victoria BC V8X 5J2
Canada
Phone: 1 250 881 7372
Fax: 1 250 881 7470
e-mail: ratkins@golder.com

Sediment Suspension by Waves

Pieter G.E.F. Augustinus
Department of Physical Geography
University of Utrecht
Heidelberglaan 2
PO Box 80.115
Utrecht 3508 TC
The Netherlands
Phone: 31 30 2532777
Fax: 31 30 2531145
e-mail: p.augustinus@geog.uu.nl

Dikes

Jean-Christophe Aznar
Desmid-CNRS
1, rue Parmentier
13200 Arles
France
Phone: 04 90 93 86 86
e-mail: alderv.desmid.wanadoo.fr

Desalination

Martin J. Baptist
Wl/delft hydraulics
Rotterdamseweg 185
Delft
The Netherlands
Phone: 31 15 285 8585
Fax: 31 15 285 8582
e-mail: info@wldelft.nl

Biogeomorphology

André Barreto
Universidade do Vale Itajai
Centro de Ciências Tecnológicas da Terra e do Mar
Rua Uruguai, 458
Itajai SC
Brasil CEP: 88302-202
e-mail: abarreto@cttmar.univali.br

South America, Coastal Ecology

Dilip K. Barua
3370 East 45th Avenue
Vancouver, BC V5R 3E7
Canada
Phone: 604 871 1594
e-mail: dkbarua@shaw.ca

Beaufort Wind Scale
Coastal Upwelling and Downwelling
Wave Hindcasting

Daniel F. Belknap
Department of Earth Sciences
University of Maine
Orono ME 04469
USA
Phone: 1 207 581 2159
Fax: 1 207 581 2202
e-mail: belknap@maine.edu

Alluvial-Plain Coasts
Glaciated Coasts

Carlos Emilio Bemvenuti
Fundação Universidade Federa do Rio Grande (FURG)
Departmento De Oceanografia
CP 474, Rio Grande
Brasil, CEP: 96201-900
e-mail: docbemve@furg.br

South America, Coastal Ecology

Lindino A. Benedet
Coastal Planning and Engineering, Inc.
2481 West Boca Raton Boulevard
Boca Raton FL 33431
USA

Jet Probes

Jérôme Besnehard
TOTAL FINA ELF-DGEP/AEO
2, place de la Coupole-la Defence 6
92078 Paris La Defence cedex
France

Mangrove, Ecology

Eric Bird
Geostudies
343 Beach Road
Black Rock, Victoria 3193
Melbourne
Australia
Phone: 61 3 9589 3738
Fax: 61 3 9589 3738
e-mail: ecfb@unimelb.edu.au

Australia, Coastal Geomorphology Barrier
Coastline Changes
Drift and Swash Alignments
Glossary of Coastal Geomorphology
Eolianite
Mangroves, Coastal Geomorphology
Ria Coasts

William A. Birkemeier
US Army Corps of Engineers
Engineer Research and Development Center
1261 Duck Road, Kitty Hawk NC 27949
USA

Depth of Closure on Sandy Beaches

François Blasco
Laboratoire d'Ecologie Terrestre
UMR 5552-BP4072
31029 Toulouse Cedex 4, France
Phone: 33 5 61 55 85 43
Fax: 33 5 61 55 85 44
e-mail: francois.blasco@cict.fr

Mangroves, Ecology
Mangroves, Remote Sensing

Alan F. Blumberg
HydroQual, Inc.
1200 MacArthur Blvd.
Mahwah NJ 07430,
USA
Phone: 201 529 5151

Cohesive Sediment Transport

Shawn M. Boeser
United Nations
OCHA Humanitarian Information Centre
Liberia
e-mail: smb@drizzle.com

Geographic Information Systems
Global Positioning Systems

Henry Bokuniewicz
Marine Sciences Research Center
State University of New York
Stony Brook NY 11794
USA
Phone: 1 631 632 8674
Fax: 1 631 632 8820
e-mail: hbokuniewicz@notes.cc.sunysb.edu

Endogenic and Exogenic Factors
Ingression, Regression, Transgression
Littoral

Robert W. Brander
School of Biological Earth and Environmental Sciences
University of New South Wales
Sydney NSW 2052
Australia
Phone: 61 2 9385 1558
Fax: 61 2 9385 2899
e-mail: rbrander@unsw.edu.au

Rip Currents

B. Chris Brewster
United States Life Saving Association
3850 Sequoia Street
San Diego, CA 92109–6518
USA
Phone: 858 581 1221
Fax: 720 535 3274
e-mail: brewster@lifesaver1.com

Lifesaving and Beach Safety

Helmut Brückner
Department of Geography
University of Marburg
Deutschhausstr. 10
D-35032 Marburg
Germany
Phone: +49 06421 282 4261
Fax: +49 06421 282 8950
e-mail: h.brueckner@mailer.uni-marburg.de

Geochronology

Michael Bruno
Davidson Laboratory
Stevens Institute of Technology
Castle Point on Hudson
Hoboken NJ 07030
USA
Phone: 1 201 216 5338
Fax: 1 201 216 8214
e-mail: mbruno@stevens-tech.edu

Physical Models

Per Bruun
32 Baynard Cove Road
Hilton Head Island, SC 29928
USA
Phone: 1 843 671 3862
Fax: 1 843 785 1154

Beach Drain
Bypassing at Littoral Drift Barriers
Dredging of Coastal Environments

David M. Bush
Department of Geology
State University of West Georgia
Carrollton GA 30118
USA

Managed Retreat

Ilya V. Buynevich
Department of Earth Sciences
Boston University
Boston MA 02215
USA

Phone: 1 617 353 2530
Fax: 1 617 353 3290
e-mail: valeriy@bu.edu

Ground-Penetrating Radar

Gillian Cambers
PO Box 783, Rincon
Puerto Rico 00677, USA
Phone: 1 787 832 3585
Fax: 1 787 265 2880
e-mail: g_cambers@hotmail.com

Caribbean Islands, Coastal Ecology and Geomorphology

Sarah K. Campbell
Department of Anthropology
Western Washington University
Bellingham WA 98225
USA
Phone: 1 360 650 4793
Fax: 1 360 650 7668
e-mail: campbsk@cc.wwu.edu

Shell Middens

Marie-Claire P. Chaineaur
Institute for Development, Research, and Enquiry in Coastal Zone Studies
Brussels, Belgium

Health Benefits

Roger H. Charlier
Free University of Brussels
Brussels
Belgium B-1050
Phone: 32 2 649 0755
Fax: 32 2 649 0755
e-mail: rhcharlier@hotmail.com
roger.charlier@pop.kpn.be

Health Benefits
History, Coastal Protection
Tidal Power
Tide Mill
Wave Power

Zhongyuan Chen
Department of Geography
East China Normal University
Shanghai 200062
China
Phone: 86 21 62232706/62233303
Fax: 86 21 62232466/62756217
e-mail: z.chen@sklec.ecnu.edu.cn

Coastal Subsidence

Jiyu Chen
Department of Geography
East China Normal University
Shanghai 200062
China
e-mail: chensklec@citiz.net

Dams, Effect on Coasts

Bénédicte Cherbuy
CEFE-CNRS
Route de Mende
34000 Montpellier
France
Phone: 04 67 61 32 74
Fax: 04 67 41 06 16
e-mail: cherbuy@cefe.cnrs-mop.fr

Desalination

Michael J. Chrzastowski
Illinois State Geological Survey
615 East Peabody Drive
Champaign IL 61820-6964
USA
Phone: 1 217 244 2194
Fax: 1 217 244 0029
e-mail: chrzasto@isgs.uiuc.edu

Beach Features

Paolo Ciavola
Dipartimento di Scienze Della Terra
Università di Ferrara
Corso Ercole I d'Este, 32
444100 Ferrara, Italy
Phone: 39 0532 293720
Fax: 39 0532 206468
e-mail: ciavola@libero.it
cv@inife.it

Tracers

H. Edward Clifton
US Geological Survey
345 Middlefield Rd., MS 999
Menlo Park CA 94025
USA
Phone: 1 650 964 1552
e-mail: eclifton@earthlink.net

Coastal Sedimentary Facies
Shoreface

Russell G. Cole
NIWAR
PO Box 893, Nelson
New Zealand
Phone: 64 3 548 1715
Fax: 64 3 548 1716
e-mail: R.Cole@niwa.cri.nz

New Zealand, Coastal Ecology

J. Andrew G. Cooper
Environmental Studies
University of Ulster
Coleraine, Co. Londonderry
BT52 1SA, Northern Ireland
UK
Phone: 44 1265 324429
Fax: 44 1265 324911
e-mail: jag.cooper@ulst.ac.uk

Microtidal Coasts
Carrying Capacity in Coastal Areas

P.J. Cowell
Coastal Studies Unit
Department of Geography
University of Sydney
Sydney NSW 2006
Australia
Phone: 61 2 9351 2189
Fax: 61 2 9351 3644
e-mail: p.cowell@csu.usyd.edu.au

Coastal Changes, Gradual

Stephen J. Craig-Smith
School of Tourism and Leisure Management
UQ Business School
The University of Queensland
Ipswich, Qld 4305
Australia
Phone: 61 7 3381 1007
Fax: 61 7 3381 1012
e-mail: s.craigsmith@uq.edu.au

Cuspate Forelands

Joel C. Creed
Universidade do Estado do Rio de Janeiro
Departmento de Ecologia-IBRAG
PHLC Sala 220
Rua São Francisco Xavier 524
Reo de Janeiro RJ
Brasil
CEP: 20559-900
e-mail: jcreed@openlin.com.br

South America, Coastal Ecology

Mark Crowell
Office of Mitigation
Federal Emergency Management Agency
Washington DC 20742
USA
e-mail: Mark.Crowell@dhs.gov

Erosion: Historical Analysis and Forecasting

Simone Rabelo da Cunha
Universidade do Vale de Itajai
Centro de Ciências Tecnológicas da Terra e do Mar
Rua Uruguai, 458, Itajai, SC
Brasil, CEP: 88302-202
Phone/Fax: 55 407 341 7715
e-mail: simone@cttmar.univali.br

South America, Coastal Ecology

William R. Dally
Surfbreak Engineering Services, Inc
1010 Atlantic St., Suite A-2
Melbourne Beach FL 32951
USA
Phone: 1 407 733 2296
Fax: 1 407 733 2298
e-mail: wdally@surfbreakengineering.com

Surf Zone Processes
Surfing

Robin Davidson-Arnott
Department of Geography
University of Guelph
Guelph ON N1G 2W1
Canada
Phone: 1 519 824 4120 ext 6719
Fax: 1 519 837 9140
e-mail: rdarnott@uoguelph.ca

Beach and Nearshore Instrumentation

Richard A. Davis Jr.
Department of Geology
University of South Florida
Tampa FL 33620
USA
Phone: 1 813 974 2773
Fax: 1 813 974 2654
e-mail: rdavis@chuma.cas.usf.edu

Human Impact on Coasts

Alastair Dawson
Centre for Quaternary Science
William Morros Building
Coventry University, Priory St.
Coventry CV1 5FB
UK
e-mail: gex014@coventry.ac.uk

Tsunamis

Robert G. Dean
Department of Coastal and Ocean Engineering
University of Florida
Gainesville FL 32611-6590
USA
Phone: 1 352 392 2416
Fax: 1 352 392 3466
e-mail: dean@coastal.ufl.edu

Dynamic Equilibrium of Beaches

Omar Defeo
UNDECIMAR, Facultad de Ciencias
Igua 4225
PO Box 10773
11400 Montevideo
Uruguay

South America, Coastal Ecology

Henry F. Diaz
NOAA/OAR/CDC
325 Broadway, Boulder CO 80303
USA
Fax: 1 303 497 7013
e-mail: Henry.F.Diaz@noaa.gov

El Niño–Southern Oscillation

John R. Dingler
US Geological Survey MS-999
345 Middlefield Road
Menlo Park CA 94025
USA
Phone: 1 650 329 5069
Fax: 1 650 329 5190
e-mail: jdingler@octopus.wr.
jdingler@usgs.gov

Beach Processes

Yuri Dolotov
Institute of Water Problems
Laboratory of Hydrodynamics
Russian Academy of Sciences
Lenin Prospect 14, Moscow 117901
Russia
Fax: 7 095 135 54 15

Black and Caspian Seas, Coastal Ecology and Geomorphology

Arthur D. Donovan
British Petroleum
UTG, Room 810
PO Box 4587
Houston TX 77210
USA
Cell: 1 832 725 4028
Phone: 1 281 366 5198
Fax: 1 281 366 7416
e-mail: donovaad@bp.com

Sequence Stratigraphy

J. Pat Doody
National Coastal Consultants
5 Green Lane
Brampton, Huntingdon
Cambs. PE 28 4RE
UK
Phone: 44 1480 392 706
e-mail: pat.doody@ntlworld.com

Europe, Coastal Ecology
History, Coastal Ecology
Monitoring Coastal Ecology

Bruce Douglas
Laboratory for Coastal Research
Florida International University
Miami FL 33199
USA

Erosion: Historical Analysis and Forecasting

Iver W. Duedall
College of Engineering
Florida Institute of Technology
150, West University Boulevard
Melbourne FL 32901-6900
USA
e-mail: duedall@winnie.fit.edu

Demography of Coastal Populations

Marshall D. Earle
Neptune Sciences, Inc.
12310 Pinecrest Rd.
Suite 203, Reston VA 20191
USA
Phone: 1 703 620 5858
Fax: 1 703 620 3900
e-mail: mearle@neptunesci.com

Surf Modeling

Billy L. Edge
Department of Civil Engineering
Texas A&M University
College Station TX 77845
USA
Phone: 1 409 847 8712
e-mail: b-edge@tamu.edu

Sand Rights

Karen Edyvane
School of Geography & Environmental Studies
University of Tasmania
Private Bag 78, Hobart

Tasmania 7001
Australia
Phone: 61 3 6226 2205
Fax: 61 3 6226 2989
e-mail: karen.Edyvane@utas.edu.au

Australia, Coastal Ecology

D. Eisma
Netherlands Institute for Sea Research
PO Box 59
Den Burg, Texel
The Netherlands
Phone: 31 72 58 947 36
Fax: 31 72 58 947 36

Asia, Eastern, Coastal Geomorphology

Lesley C. Ewing
CSPBA, c/o California Coastal Commission
45 Fremont St., Suite 2000
San Francisco CA 94105
USA
Phone: 1 415 904 5291
Fax: 1 415 904 5400
e-mail: lewing@coastal.ca.gov

Sand Rights

Rhodes W. Fairbridge
Center for Climate Systems Research
Goddard Institute for Space Studies/NASA
Columbia University
2880 Broadway
New York NY 10025
USA
Phone: w 1 212 663 8793
Fax: w 1 212 678 5552
Phone: s 1 516 267 8329
Fax: s 1 516 267 8329
e-mail: elicardo@giss.nasa.gov

Asteroid-Impact Coasts
Thalassostatic Terraces
Weathering in the Coastal Zone

Daidu Fan
Marine Geology Laboratory
Tongji University, Shanghai 200092
People's Republic of China
Phone: 86 21 65982208
Fax: 86 21 65025320
e-mail: cxlik@online.sh.cn

Tidal Flats, Open Ocean Coasts

Michael S. Fenster
Environmental Studies Program
Randolph-Macon College
Ashland VA 23005
USA
Phone: 1 804 752 3745
Fax: 1 804 752 4724
e-mail: mfenster@rmc.edu

Offshore Sand Sheets
Setbacks

Beatrice Padovani Ferreira
Universidade Federal de Pernambuco
Departmento de Oceanografia
Av. Arquitetura S/N
Recife PE
Brazil CEP: 50740-550
e-mail: beatrice@ibama.gov.br

South America, Coastal Ecology

Charles W. Finkl
CPE Coastal Geology and Geomatics
2481 N.W. Boca Raton Blvd.
Boca Rton, FL 33431
USA
Phone: 561 391 8102
Fax: 561 391 9116
e-mail: cfinkl@coastalplanning.net

Beach Nourishment
Coastal Soils
Jet Probes
Journals
Nearshore Geomorphological Mapping
Organizations
Vibracore

Duncan M. FitzGerald
Department of Earth Sciences
Boston University
Boston MA 02215
USA
Phone: 1 617 353 2530
Fax: 1 617 353 3290
e-mail: dunc@bu.edu

Ground-Penetrating Radar
Tidal Inlets

B.W. Flemming
Senckenberg Institute
Schleusen street 39A
D-26382 Wilhelmshaven
Germany
Phone: 49 4421 9475 36
Fax: 49 4421 9475 50
e-mail: bw.flemming@sam.terramare.de

Tidal Environments

Donald L. Forbes
Geological Survey of Canada
Atlantic Geoscience Centre
Bedford Institute of Oceanography
PO Box 1006, Dartmouth
Nova Scotia B2Y 4A2
Canada
Phone: 902 426 7737
Fax: 902 425 4104
e-mail: dforbes@nrcan.gc.ca

Paraglacial Coasts

William T. Fox
Department of Geology
Center for Environmental Studies
Williams College
Williamstown MA 01267
USA
Phone: 1 413 597 3927
Fax: 1 413 597 3489
e-mail: william.t.fox@williams.edu

Bioconstruction
Bioerosion
Reefs, Non-Coral

Peter W. French
Department of Geography
Royal Holloway
University of London
Egham, Surrey, TW20 0EX
UK
Phone: 44 1784 44 35 71
Fax: 44 1784 47 28 36
e-mail: p.french@rhbnc.ac.uk

Coastal Zone Management

Patrick L. Friend
School of Ocean and Earth Sciences
Southampton Oceanography Centre
Empress Dock, Southampton
Hampshire SO14 3ZH
UK

Ripple Marks

Francis A. Galgano
Department of Geography and
Environmental Engineering
United States Military Academy

West Point NY 10996
USA
Phone: 1 914 938 4035
Fax: 1 914 938 3339
e-mail: bf6360@usma.edu
Modes and Patterns of Shoreline Change

Cynthia J. Gefvert
South Florida Water Management District
3301 Gun Club Road
West Palm Beach FL 33406
USA
Phone: 1 561 682 6923
Fax: 1 561 682 5923
e-mail: cgefvert@sfwmd.gov
Submarine Groundwater Discharge

Roland Gehrels
Quaternary Environments Research Group
Department of Geographical Sciences
University of Plymouth, Plymouth PL4 8AA
UK
Phone: 44 1752 233079
Fax: 44 1752 233054
e-mail: wrgehrels@plymouth.ac.uk
Sea-Level Changes During the Last Millennium

Guy Gelfenbaum
US Geological Survey
Coastal and Marine Geology Program, MS 999
345 Middlefield Road
Menlo Park CA 94025
USA
Phone: 1 650 329-5483
Fax: 1 650 329 5190
e-mail: ggelfenbaum@usgs.gov
Coastal Currents

Laurel T. Gorman
Information Technology Laboratory
US Army Waterways Experiment Station
3909 Halls Ferry Rd.
Vicksburg MS 39180
USA
Phone: 601 634 4484
e-mail: Laurel.Gorman@erdc.usace.army.mil
Monitoring Coastal Geomorphology

Vivian Gornitz
Center for Climate Systems Research
Columbia University
New York NY 10025
USA
e-mail: vgornitz@giss.nasa.gov
Eustasy
Natural Hazards
Storm Surge

Peter H.F. Graber
138 Altura Way
Greenbrae CA 94904
USA
Phone: 1 415 461 9032
Fax: 1 415 461 8338
e-mail: suzanneeg@prodigy.net
Coastal Boundaries

David Greenland
Department of Geography
Louisiana State University
Baton Rouge LA 70803
USA
e-mail: greenlan@lsu.edu
Coastal Climate

Brian Greenwood
Division of Physical Sciences
Scarborough Campus
University of Toronto
1265 Military Trail
Scarborough ON M1C 1A4
Canada
Phone: 1 416 287 7326
Fax: 1 416 287 7204
e-mail: greenw@scar.utoronto.ca
Bars

Murray R. Gregory
Department of Geology
University of Auckland
Private Bag 92019
Auckland
New Zealand
e-mail: m.gregory@auckland.ac.nz
Marine Debris—Onshore, Offshore, Seafloor Litter

Eric R. Gundlach
E-Tech International Inc.
PO Box 2976
Acton MA 01720-6976
USA
Phone: 1 978 264 9682
Fax: 1 978 264 1926
e-mail: egundlach@compuserve.com
Oil Spills

Peter L. Guth
Department of Oceanography
US Naval Academy
572 Holloway Road
Annapolis, MD 21402-5026
USA
Phone: 1 410 293 6560
Fax: 1 410 293 2137
e-mail: pguth@usna.edu
Coastal Warfare

Nina P. Hadley
Tidal Delta Consulting
5632 1/2 20th Avenue NE
Seattle WA 98105
USA
Phone: 1 206 527 0885
e-mail: ninah@u.washington.edu
Tourism and Coastal Development

Kevin R. Hall
Department of Civil Engineering
Centre for Water and the Environment
Queens University
Kingston ON K7L 3N6
Canada
Phone: 613 533 2127
Fax: 613 533 2128
e-mail: hallk@civil.queeensu.ca
Bioengineered Shore Protection

Jean-François Hamel
Society for the Exploration and Valuing of the Environment (SEVE)
655 rue de la Rivière
Katevale, Quebec J0B 1WO
Canada
Phone: 819 843 3466
Fax: 819 843 3466
e-mail: seve@sympatico.ca
Atlantic Ocean Islands, Coastal Ecology
Arctic, Coastal Ecology

Guoqi Han
Northwest Atlantic Fisheries Centre
Fisheries and Oceans Canada
PO Box 5667
St. John's, NF A1C 5X1
Canada

Phone: 709 772 4326
Fax: 709 772 4105
e-mail: hang@dfo-mpo.gc.ca

Altimeter Surveys, Coastal Tides and Shelf Circulation

Edward B. Hands
2 Bugle Ridge Road
Vicksburg MS 39180-4314
USA
e-mail: e_hands@msn.com

Tidal Datums

Jim D. Hansom
Department of Geography and Topographic Science
University of Glasgow
Glasgow G12 8QQ
Scotland
UK
Phone: 44 0141 330 5406
Fax: 44 0141 330 4894
e-mail: jhansom@geog.gla.ac.uk

Atlantic Ocean Islands, Coastal Geomorphology
Boulder Pavement

Cheryl J. Hapke
USGS, Pacific Science Center
1156, High St.
Santa Cruz CA 95064
USA

Photogrammetry

Miles O. Hayes
Research Planning, Inc.
PO Box 328
1121 Park Street
Columbia SC 29202
USA
Phone: 1 803 256 7322
Fax: 1 803 254 6445
e-mail: miles@researchplanning.com

Barrier Islands
Tide-Dominated Coasts
Wave-Dominated Coasts

Terry R. Healy
Earth Sciences Department
University of Waikato
Hamilton
New Zealand
Phone: 64 7 8384061
Fax: 64 7 8560115
e-mail: trh@waikato.ac.nz

Coastal Wind Effects
Dune Ridges
Estuaries
Littoral Drift Gradient
Muddy Coasts
New Zealand, Coastal Geomorphology
Salt Marsh
Tidal Creeks
Uplift Coasts
Volcanic Coasts
Wave Focusing

Patrick A. Hesp
Department of Geography and Anthroplogy
Louisiana State University
Baton Rouge, LA 70803–4105
USA
Phone: 1 225 578–6244
Fax: 1 225 578–4420
e-mail: pahesp@lsu.edu

Eolian Processes

Frank M.J. Hoozemans
WL/Delft Hydraulics
2600 MH Delft
The Netherlands
Phone: 31 15 285 85 05
Fax: 31 15 285 87 18
e-mail: frank.hoozemans@wldelft.nl

Global Vulnerability Analysis

David Hopley
Coastal and Marine Consultancies Pty Ltd
3 Wingadee Court
Annandale
Townsville QLD 4814
Australia
Phone: 61 7 4725 2832
Fax: 61 7 4779 1400
e-mail: dhopley@austarnet.com.au

Atolls
Cays
Coral Reefs
Coral Reefs, Emerged
Trottoirs

Diane P. Horn
School of Geography
Birbeck College
University of London
Malet Street
London WC1E 7HX
England
UK
Phone: 171 631 6480
Fax: 171 631 6498
e-mail: d.horn@bbk.ac.uk

Hydrology of Coastal Zone

James R. Houston
Coastal Engineering Research Center
USAE Waterways Experiment Station
3909 Halls Ferry Rd.
Vicksburg MS 39180
USA
Phone: 1 601 634 2000
Fax: 1 601 634 2818
e-mail: houstojl@expl.wes.army.mil

Geohydraulic Research Centers

John R.C. Hsu
Department of Marine Environment and Engineering
National Sun Yat-Sen University
70 Lien-Hai Road
Kaohsiung City 80424
Taiwan
Phone: 886 7 525 2000 ext 5076
Fax: 16 886 7 525 5060
e-mail: jrchsu@mail.nsysu.edu.tw

Engineering Applications of Coastal Geomorphology

Terry M. Hume
Water Quality Centre
DSIR-Marine and Freshwater
Department of Scientific and Industrial Research
PO Box 11-155
Hamilton
New Zealand
Phone: 64 7 8561729
Fax: 64 7 8560151
e-mail: t.hume@niwa.cri.nz

Tidal Prism

Hillert Ibbeken
Instutut fur. Geologie, Geophysik und Geoinformatik
Free University of Berlin
Malteser street 74-100, D-12249
Berlin
Germany

Pluvial Lake Shore Deposits

Baban Ingole
Biology Division

National Institute of Oceanography
Dona Paula, Goa 403004
India
e-mail: baban@csnio.ren.nic.in

Indian Ocean Coasts, Coastal Ecology

Douglas L. Inman
Department of Center for Coastal Studies
Scripps Institution of Oceanography
University of California, San Diego
La Jolla CA 92093 0209
USA
Phone: 1 858 534 4334
Fax: 1 858 534 0300
e-mail: dinman@ucsd.edu

Accretion and Erosion Waves on Beaches
Climate Patterns in the Coastal Zone
Energy and Sediment Budgets of the Global Coastal Zone
Littoral Cells
Modeling Platforms, Terraces, and Coastal Evolution
Scour and Burial of Objects in Shallow Water

Nancy L. Jackson
Environment Policy Studies
New Jersy Institute of Technology
University Heights
Newark NJ 071012
USA
Phone: 1 973 596 8467
Fax: 1973 642 4689
e-mail: jacksonn@njit.edu

Developed Coasts

Saskia Jelgersma
Elzenlaan 6
Bergen aan Zee 1865 BM
The Netherlands
Phone: 31 72 5894370
Fax: 31 72 5894370

Sedimentary Basins

Scott A. Jenkins
Center for Coastal Studies
Scripps Institution of Oceanography
University of California, San Diego
La Jolla CA 92093-0209
USA
Phone: 858 534 6480
Fax: 858 534 0300
e-mail: sjenkins@ucsd.edu

Accretion and Erosion Waves on Beaches
Climate Patterns in the Coastal Zone
Energy and Sediment Budgets of the Global Coastal Zone
Modeling Platforms, Terraces, and Coastal Evolution
Scour and Burial of Objects in Shallow Water

Lucille Lewis Johnson
Department of Anthropology
Mail Drop 375
Vassar College
Poughkeepsie NY 12601
USA
Phone: 1 914 (845) 437 5507
Fax: 1 914 (845) 437 7287
e-mail: johnsonl@vassar.edu

Archaeology

Pavel A. Kaplin
Department of Geography
Moscow State University
Moscow 119899
Russia
Phone: 095 9392152
Fax: 095 9392123

Black and Caspian Seas, Coastal Ecology and Geomorphology

Michael S. Kearney
Department of Geography
University of Maryland
College Park MD 20742
USA
Phone: 1 301 405 4057
Fax: 1 301 314 9299
e-mail: mk11@umail.umd.edu

North America, Coastal Ecology

Dieter H. Kelletat
Universität-GH Essen
FB9-Physiogeographie
Universitaetsstrasse 15
Essen 1 D-45117
Germany
Phone: 49 201 183 3162
Fax: 49 201 183 2811
e-mail: dieter.kelletat@uni-essen.de

Dalmatian Coasts
Europe, Coastal Geomorphology
Geographical Coastal Zonality
Holocene Coastal Geomorphology
Notches
Pleistocene Epoch
Sharm Coasts

Michael J. Kennish
Institute of Marine and Coastal Sciences
Rutgers University
New Brunswick NJ 08901-8521
USA
Phone: 1 732 932 8959 ext 240
Fax: 1 732 932 6557
e-mail: kennish@imcs.rutgers.edu

Environmental Quality
Estuaries, Anthropogenic Impacts

Syed Khalil
Coastal Restoration Division
Office of Coastal Restoration and Management
Louisiana Department of Natural Resources
617 North 3rd St., PO Box 44027
Baton Rouge, LA 70804-4027
USA
Phone: 1 225 342 7308
e-mail: syedk@dnr.state.la.us

Vibracore

Robert Kirk
Department of Geography
University of Canterbury
Private Bag 4800
Christchurch
New Zealand
Phone: 64 3 3642 893
Fax: 64 3 3642 2586
e-mail: b.kirk@regy.canterbury.ac.nz

Shore Platforms

Victor Klemas
College of Marine Studies
University of Delaware
Newark DE 19716
USA
Phone: 1 302 831 8256
Fax: 1 302 831 6838
e-mail: klemas@udel.edu

Remote Sensing: Wetlands Classification

Tormod Klemsdal
Department of Geography
University of Oslo
PO Box 1042 Blindern
N-0316 Oslo Norway
Phone: 47 22 856931
Fax: 47 22 857230
e-mail: tormod.klemsdal@geografi.uio.no

Strandflat

Paul D. Komar
College of Oceanic and Atmospheric Sciences
Oregon State University
Corvallis OR 97331 5503
USA
Phone: 1 541 737 5210
Fax: 1 541 737 2064
e-mail: pkomar@oce.orst.edu

Placer Deposits
Rhythmic Patterns

John C. Kraft
Department of Geology
University of Delaware
Newark DE 19716
USA
Phone: 1 610 287 9385
1 302 831 8706
Fax: 1 302 831 4158
e-mail: 00218@udel.edu

Paleocoastlines

Nicholas C. Kraus
U.S. Army, Coastal Hydraulics Laboratory
Vicksburg MS 39180
USA
Phone: 1 601 634 2016
Fax: 1 601 634 3080
e-mail: Nicholas.C.Kraus@erdc.usace.army.mil

Beach Profile
Navigation Structures
Profiling, Beach
Shore Protection Structures

Steven L. Krupa
South Florida Water Discharge Management District
33012 Gun Club Road
West Palm Beach FL 33406
USA
Phone: 1 561 682 6923
Fax: 1 561 682 6442
e-mail: skrupa@sfwmd.gov
krupa3@bellsouth.net

Submarine Groundwater Discharge

Mark A. Kulp
Department of Geology and Geophysics
University of New Orleans
New Orleans LA 70148
USA
Phone: 1 504 280 3119
Fax: 1 504 280 7396
e-mail: mkulp@uno.edu

Deltas

Pravin D. Kunte
National Institute of Oceanography
Dona Paula, Goa 403004
India
Phone: 91 (0)832 221322, 226253 ext 4387
Fax: 91 (0)832 223340, 229102
e-mail: kunte@csnio.ren.nic.in
pdkunte@hotmail.com

Databases

Jacques Laborel
UMR DIMAR, Université de la Méditerranée
Case 901, Faculté des Sciences de Luminy
F13288 Marseille, Cedex 9
France
Phone: 33 91 283 6032
e-mail: rutabaga1@wanadoo.fr

Algal Rims
Bioherms and Biostromes
Sea-Level Indicators, Biologic

Françoise Laborel-Deguen
UMR DIMAR, Université de la Méditerranée
Case 901, Faculté des Sciences de Luminy
F13288 Marseille, Cedex 9
France

Sea-Level Indicators, Biologic

V. Chris Lakhan
Department of Earth Sciences
School of Physical Sciences
University of Windsor
Windsor, Ontario
Canada N9B 3P4
Phone: 1 519 253 4232 ext 2183
Fax: 1 519 253 6214/973 7081
e-mail: lakan@uwindsor.ca

Coastal Modeling and Simulation
Time Series Modeling

Kurt Lambeck
Research School of Earth Sciences
Australian National University
Canberra ACT 0200
Australia
e-mail: kurt.lambeck@anu.edu.au

Isostasy

Magnus Larson
Department of Water Resources Engineering
University of Lund
Box 118
S-221 00 Lund
Sweden
Phone: 46 46 222 8729
Fax: 46 46 222 4435
e-mail: Magnus.Larson@tvrl.lth.se

Numerical Modeling

Stephen P. Leatherman
Laboratory for Coastal Research & International Hurricane Center
Florida International University
10555 West Flagler St.
Miami FL 33199
USA
Phone: 1 305 348 6304
Fax: 1 305 348 1605
e-mail: leatherman@fiu.edu

Airborne Laser Terrain Mapping and Light Detection and Ranging
Erosion: Historical Analysis and Forecasting
Modes and Patterns of Shoreline Change
Rating Beaches

Congxian Li
Marine Geology Laboratory
Tongji University
Shanghai 200092
People's Republic of China
Phone: 86 21 65982208
Fax: 86 21 65025320
e-mail: cxlik@online.sh.cn

Tidal Flats, Open Ocean Coasts

Jeffrey H. List
US Geological Survey
384 Woods Hole Road
Woods Hole MA 02543
USA
Phone: 1 508 457 2343
Fax: 1 508 457 2310
e-mail: jlist@usgs.gov

Sediment Budget

Donald Macintosh
Department of Ecology and Genetics
University of Aarhus
Aarhus C DK-8000
Denmark
e-mail: don.macintosh@biology.au.dk

Asia, Eastern, Coastal Ecology

M. MacLeod
Department of Natural Resource Economics
Scottish Agricultural College
Edinburgh EH9 3JG
Scotland
UK
Phone: 0131 535 4387
Fax: 0131 667 2601
e-mail: m.j.macleod@ed.sac.ac.uk
Carrying Capacity in Coastal Areas

Molly McGraw
Department of Geography and Anthropology
Louisiana State University
Baton Rouge LA 70803
USA
Phone: 225 578 6130
Fax: 225 578 4420
e-mail: walker@lsu.edu
History, Coastal Geomorphology

Orville T. Magoon
Coastal Zone Foundation
PO Box 279
Middletown CA 95461
USA
Phone: 1 707 987 2358 ext 207
Fax: 1 707 987 9351
e-mail: omagoon@guenoc.com
Sand Rights

Mauro Maida
Universidade Federal de Pernambuco
Departmento de Oceanografia
Av. Arquitetura S/N
Recife PE
Brasil
CEP: 50740-550
South America, Coastal Ecology

Gerhard Masselink
Department of Geography
Loughborough University
Loughborough LE11 3TU
UK
Phone: 15 9 222746
Fax: 15 9 223930
e-mail: g.masselink@lboro.ac.uk
Wave Refraction Diagrams
Waves

Patricia M. Masters
University of California, San Diego
Scripps Institution of Oceanography
La Jolla CA 92093-0309
USA
Phone: 1 858 534 4334
Fax: 1 858 534 0300
e-mail: pmasters@ucsd.edu
Modeling Platforms, Terraces, and Coastal Evolution

George A. Maul
College of Engineering
Florida Institute of Technology
150, West University Boulevard
Melbourne FL 32901-6988
USA
Phone: 1 407 674 7453
Fax: 1 407 674 7212
e-mail: gmaul@fit.edu
Coastal Temperature Trends
Demography of Coastal Population
Pressure Gradient Force
Small Islands
Vorticity
Wave Climate

Vincent May
Department of Conservation Sciences
Bournemouth University
Talbot Campus
Fern Barrow
Poole Dorset BH12 5BB
UK
Phone: 44 1202 595178
Fax: 44 1202 595255
e-mail: vmay@bournemouth.ac.uk
Chalk Coasts
Conservation of Coastal Sites
Marine Parks

Randolph A. McBride
Department of Geography and Earth Science
George Mason University
Fairfax VA 22030
USA
Phone: 1 703 993 1642
Fax: 1 703 993 1216
e-mail: rmcbride@gmu.edu
mcbrdra@aol.com
Offshore Sand Banks and Linear Sand Ridges

John McKenna
Coastal Studies Research Group
School of Environmental Studies
University of Ulster
Coleraine BT52 1SA
Northern Ireland
UK
Phone: 44 28 70 324055
Fax: 44 28 70 324911
e-mail: j.mckenna@ulst.ac.uk
Boulder Beaches

Anton McLachlan
College of Science
Sultan Qaboos University
PO Box 36, SQU 123
Oman
Phone: 968 515400
Fax: 968 513415
e-mail: antonmcl@squ.edu.om
Africa, Coastal Ecology

Andrew McMinn
Institute of Antarctic and Southern Ocean Studies
University of Tasmania
GPO Box 252-77, Hobart 7001
Tasmania
Australia
Phone: 61 3 6226 2980
Fax: 61 3 6226 2973
e-mail: andrew.mcminn@utas.edu.au
Antarctica, Coastal Ecology and Geomorphology

Franco S. Medioli
Centre for Marine Geology
Dalhousie University
Halifax
Nova Scotia B3H3J5
Canada
e-mail: medioli@is.dal.ca
Sea-Level Indicators—Biological in Depositional Sequences

Annie Mercier
Society for the Exploration and Valuing of the Environment (SEVE)
655 rue de la Rivière
Katevale, Quebec J0B 1WO
Canada
Phone: 819 843 3466
Fax: 819 843 3466
e-mail: seve@sympatico.ca
Arctic, Coastal Ecology
Atlantic Ocean Islands, Coastal Ecology

Mark D. Merlin
Biology Program
Dean Hall, rm. 2
University of Hawaii at Mãnoa
Honolulu, Hawaii 96822
USA
e-mail: merlin@hawaii.edu

Pacific Ocean Islands, Coastal Ecology

Marc L. Miller
School of Marine Affairs
University of Washington
Seattle WA 98195
USA
e-mail: mlmiller@u.washington.edu

Tourism and Coastal Development

Chizuko Mizobe
Sandy Hook Cooperative Research Institute
74 Magruder Road
Institute of Marine Coastal Sciences
Rutgers University
Highlands NJ 07732
USA

South America, Coastal Geomorphology

Ram K. Mohan
Blasland, Bouck & Lee, Inc.
100 Four Falls Corp Center, Suite 106
W. Conshohocken PA 19428–2950
USA
Phone: 848 530 9119
Fax: 848 530 9118
e-mail: rkm@bbl-inc.com

Capping of Contaminated Coastal Areas
Geotextile Applications

Luis J. Moreno
(Despacho C-276)
Subdirección General de Actuaciones en la Costa
Dirección General de Costas
Secretaría de Estado de Aguas y Costas
Ministerio de Medio Ambiente
Plaza de San Juan de la Cruz, s/n
28071 Madrid
Spain
Phone: 34 91 597 6870
Fax: 34 91 597 5924
e-mail: luis.moreno@m.dgc.mma.es

Headland-Bay Beach

Andrew Morang
CERC, US Army Waterways Experiment Station
3909 Halls Ferry Road
Vicksburg MS 39180
USA
Phone: 1 601 634 2064
Fax: 1 601 634 3080
e-mail: moranga@wes.army.mil

Depth of Closure on Sandy Coasts
Monitoring Coastal Geomorphology

Nils-Axel Mörner
Paleogeophysics and Geodynamics
Stockholm University
S-10691 Stockholm
Sweden
Phone: 46 8 770 6771
Fax: 46 8 790 6777
e-mail: morner@pog.su.se

Changing Sea Levels
Late Quaternary Marine Transgression

Robert A. Morton
US Geological Survey
Center for Coastal Geology
600 Fourth St. South
St. Petersburg FL 33701-4846
USA
Phone: 1 727 803 8747 ext 3080
Fax: 1 727 803 2032
e-mail: rmorton@usgs.gov

Mapping Shores and Coastal Terrain
Mining of Coastal Materials

Robert A. Muller
Coastal Studies Institute
Louisiana State University
Baton Rouge LA 70803
USA

Meteorological Effects on Coasts

Gisèle Muller-Parker
Department of Biology
Western Washington University
Bellingham WA 98225
USA
Phone: 1 360 650 4885
Fax: 1 360 650 7284
e-mail: gisele@biol.wwu.edu

Coral Reef Islands

George Mustoe
Department of Geology, MS 9080
Western Washington University
Bellingham WA 98225
USA
Phone: 1 360 650 3585
Fax: 1 360 650 7302
e-mail: mustoeg@cc.wwu.edu

Honeycomb Weathering
Tafone

C.G.R. Nair
Science, Technology and Environment Department
University of Kerala
Trivandrum -14, Kerala
India
Phone: 11 91 471 323545
e-mail: cgrnair@vsnl.com

Geotextile Applications

G.N. Nayak
Department of Marine Sciences
Goa University
Goa, 403 203
India
Phone: 91 832 221375 ext 262
Fax: 91 832 224184
e-mail: gnnayak@unigoa.ernet.in

Indian Ocean Coasts, Coastal Geomorphology

Daniel Din Ndongo
Departement de Biologie Vegetale - BP 24157
Universite de Douala
Douala, Cameroun

Mangroves, Remote Sensing

William J. Neal
Department of Geology
Grand Valley State University
Allendale MI 49401
USA

Managed Retreat

Robert J. Nicholls
Coastal Geomorphology
Midlesex University
Queensway Enfield EN3 4SF
UK
Phone: 44 181 362 5569
Fax: 44 181 362 6957
e-mail: r.nicholls@mdx.ac.uk

Global Vulnerability Analysis

Alan W. Niedoroda
URS Corporation
3676 Hartsfield Road
Tallahassee FL 32303
USA

Phone: 1 850 574 3197
Fax: 1 850 576 3676
e-mail: alan_niedoroda@urscorp.com

Continental Shelves
Shelf Processes

Yaacov Nir
15 Shimeoni St.
Rehovot 76248, Israel
Phone: 972 8 0468862
Fax: 972 8 0468862
e-mail: yankalenir@hotmail.com
jacobni@netvision.net.il

Coastal Wells

Karl F. Nordstrom
Institute of Marine and Coastal Sciences
Rutgers University
71 Dudley Road
New Brunswick NJ 08901-8521
USA
Phone: 1 732 932 6555 ext 502
e-mail: nordstro@imcs.rutgers.edu

Bay Beaches
Developed Coasts

Patrick D. Nunn
Department of Geography
University of the South Pacific
PO Box 1168
Suva
Fiji
Phone: 679 212542
Fax: 679 301487
e-mail: nunn_p@usp.ac.fj

Pacific Ocean Islands, Coastal Geomorphology

George F. Oertel
Department of Oceanography
Old Dominion University
Norfolk VA 23529
USA
Phone: 1 757 683 4935
Fax: 1 757 683 5303
e-mail: goertel@odu.edu

Coastal Lakes and Lagoons
Coasts, Coastlines, Shores and Shorelines

Julian D. Orford
School of Geography
Queen's University
Belfast BT7 1NN
UK
Phone: 44 2890 335142
Fax: 44 2890 321280
e-mail: j.orford@qub.ac.uk

Gravel Barriers

Antony R. Orme
Department of Geography
University Of California
Los Angeles CA 90095-1524
USA
Phone: 1 310 825 1516
e-mail: orme@geog.ucla.edu

Africa, Coastal Geomorphology

Philip D. Osborne
Pacific International Engineering
PO 1599
Edmonds WA 98020
USA
Phone: 1 425 921 1720
Fax: 1 425 744 1400
e-mail: philo@piengr.com

Cross-Shore Variation of Grain Size on Beaches

W.R. Osterkamp
US Geological Survey
1675 W. Anklam Rd.
Tuscon AZ 85745
USA
Phone: 1 520 670 6821 ext 113
Fax: 1 520 670 6806
e-mail: wroster@usgs.gov

Mining of Coastal Materials

Ervin G. Otvos
Institute of Marine Sciences
703 East Beach Blvd.
PO Box 7000
Ocean Springs MS 39566-7000
USA
Phone: 1 228 872 4235
Fax: 1 228 872 4204
e-mail: ervin.otvos@usm.edu

Beach Ridges
Cheniers

Bruce Parker
Coast Survey Development Laboratory
National Ocean Service, NOAA
1315 East-West Highway, Rm 7806
Silver Spring MD 20910
USA
Phone: 1 301 713 2801
Fax: 1 301 713 4501
e-mail: Bruce.Parker@noaa.gov

Tides

Roland P. Paskoff
10, Square Saint-Florentin
Le Chesnay F78150
France
Phone: 33 1 39556002
Fax: 33 1 39559291
e-mail: paskoff@club-internet.fr

Faulted Coasts
Karst Coasts

Shea Penland
Department of Geology and Geophysics
University of New Orleans
New Orleans LA 70148
USA
Phone: 1 504 280 3119
Fax: 1 504 280 7396
e-mail: spenland@uno.edu

Deltas

Conrad A. Pilditch
Department of Biological Sciences
University of Waikato
PB 3105
Hamilton
New Zealand
Phone: 64 7 838 4466 ext 6132
Fax: 64 7 838 4324
e-mail: c.pilditch@waikato.ac.nz

New Zealand, Coastal Ecology

Orrin H. Pilkey
Division of Earth and Ocean Sciences
Duke University
Durham NC 27708-0227
USA
Phone: 1 919 684 4238
Fax: 1 919 684 5833
e-mail: opilkey@geo.duke.edu

Foreword
Managed Retreat

Paolo A. Pirazzoli
16 rue de la Grange Bateliere

Paris 75009
France
Phone: 33 1 4507 5558
Fax: 33 1 4507 5830
e-mail: pirazzol@cnrs-bellevue.fr

Marine Terraces
Sea-Level Indicators, Geomorphic
Seismic Displacement
Submerging Coasts
Tectonics and Neotectonics
Tide Gauges

Katherine Pond
Robens Centre for Public and Environmental Health
University of Surrey
Guilford, Surrey
UK
Phone: 44 1483 879935
Fax: 44 1483 879971
e-mail: k.pond@surrey.ac.uk

Water Quality

Abel Prieur
FRE 2158-CNRS
Centre de Paleontologie
stratagraphique et Pleoecologie
27-43 boulevard du 11 novembre 1918
F-69622 Villeurbanne Cedex
France
Phone: 33 04 72 444 84 88
e-mail: abel.prieur@univ_lyon1.fr

Asia, Middle East, Coastal Ecology and Geomorphology

Norbert P. Psuty
Sandy Hook Cooperative Research Programs
Institute of Marine and Coastal Sciences
Rutgers University
Highlands NJ 07732
USA
Phone: 1 732 708 1462
e-mail: psuty@imcs.rutgers.edu

South America, Coastal Geomorphology

Michael R. Rampino
Earth and Environmental Science Program
New York University
100 Washington Square East
New York NY 10003
USA
Phone: 1 212 998 3743
Fax: 1 212 995 3820
e-mail: mrr1@nyu.edu
michael.rampino@nyu.edu

Erosion Processes

Elijah W. Ramsey III
US Geological Survey
National Wetlands Research Center
700 Cajundome Blvd.
Lafayette LA 70506
USA
Phone: 1 318 266 8575
Fax: 1 318 266 8616
e-mail: elijah_ramsey@usgs.gov
sijan_sapkota@usgs.gov

Remote Sensing of Coastal Environments

R. Keith Raney
Applied Physics Laboratory
Johns Hopkins University
Laurel MD
USA
Phone: 1 240 228 5384
Fax: 1 240 228 5548
e-mail: keith.raney@jhuapl.edu

Synthetic Aperture Radar Systems

Anto Raukas
Institute of Geology
Estonian Academy of Sciences
7 Estonia Boulevard
Tallinn EE 10143
Estonia
Phone: 372 6454659
Fax: 372 6312074
e-mail: raukas@gi.ee

Klint

Denise J. Reed
Department of Geology and Geophysics
University of New Orleans
New Orleans LA 70148
USA
Phone: 1 504 280 7395
Fax: 1 504 280 7396
e-mail: djreed@uno.edu

Vegetated Coasts
Wetlands

William Ritchie
Aberdeen Institute of Coastal Science &
Management
King's College
Aberdeen AB42 3UE
Scotland,
UK
Phone: 44 1224 274474/272522
Fax: 44 1224 272497
e-mail: w.ritchie@abdn.ac.uk

Machair
Oil Spills, High Energy Coasts
Washover Effects

Neil Roberts
School of Geography
University of Plymouth
Drake Circus, Plymouth
Devon PL4 BAA
Phone: 44 1752 233 051
Fax: 44 1752 233 054
e-mail: cnroberts@plymouth.ac.uk

Holocene Epoch

Rosana Moreira da Rocha
Universidade Federal do Parana
Departmento do Zoologia
CP 19020 Curitiba PR Brasil
CEP: 81.531-990
e-mail: rmrocha@bio.ufpr.br

South America, Coastal Ecology

Peter J. Rosen
Department of Earth and Environmental
Sciences
14 Holmes Hall
Northeastern University
Boston MA 02115
USA
Phone: 1 617 373 4380
Fax: 1 617 373 4378
e-mail: prosen@lynx.neu.edu

Boulder Barricades

John M. Rybczyk
Huxley College of the Environment
Western Washington University
Bellingham WA 98225
USA
Phone: 1 360 650 2081
Fax: 1 360 650 2842
e-mail: john.rybczyk@wwu.edu

Coastal Subsidence
Deltaic Ecology

Yoshiki Saito
MRE, Geological Survey of Japan, AIST
Central 7, Higashi 1-1-1
Tsukuba 305 8567
Japan
Phone: 81 298 61 3895
Fax: 81 298 61 3747
e-mail: yoshiki.saito@aist.go.jp

Beach Stratigraphy

Paul Sanlaville
UMR 5647, CNRS-Université de Lyon 2
Maison de l'Orient Méditerranéen
7 rue Raulin
Lyon 69007
France
e-mail: paul.sanlaville@lyon.mom.fr

Asia, Middle East, Coastal Ecology and Geomorphology

Anja Scheffers
Instutut fur Geographie
Universitat Essen
Universtatssr. 15
D-45117 Essen
Germany
Phone: 0049 201 1833358
Fax: 0049 201 1832811
e-mail: anja.scheffers@uni-essen.de

Europe, Coastal Geomorphology

Gerhard Schellmann
Department of Physical Geography and Landscape Studies
University of Bamberg
Am Kranen 1, PO Box 15 49
D-96045 Bamberg
Germany
Phone: +49 0951 863 2380
Fax: +49 0951 863 1193
e-mail: gerhard.schellmann@ggeo.uni-bamberg.de

Geochronology

Maurice Schwartz
Department of Geology
Western Washington University
Bellingham WA 98225
USA
Phone: 1 360 650 3586
Fax: 1 360 650 7302
e-mail: coastal@cc.wwu.edu

Tors

David B. Scott
Department of Earth Sciences
Dalhousie University
Halifax NS B3H 3J5
Canada
Phone: 902 494 3604
Fax: 902 494 3877
e-mail: dbscott@is.dal.ca

Coastal Changes, Rapid
Sea-Level Indicators—Biological in Depositional Sequences

V. Semeniuk
V.C. Semeniuk Research Group
21 Glenmere Road
Warwick WA 6024
Australia
Phone: 61 8 94473708
Fax: 61 8 94473705
e-mail: vcsrg@iinet.net.com

Tidal Flats

Richard J. Seymour
Scripps Institution of Oceanography
University of California at San Diego
La Jolla CA 92093-0220
USA
Phone: 1 858 534 2561
Fax: 1 858 455 5575
e-mail: rseymour@ucsd.edu

Cross-Shore Sediment Transport
Gross Transport
Longshore Sediment Transport
Net Transport

Ian Shennan
Department of Geography
University of Durham
Durham DH1 3LE
UK
Phone: 44 191 374 2484
Fax: 44 191 374 2456
e-mail: ian.shennan@durham.ac.uk

Peat

Charles Sheppard
Department of Biological Sciences
University of Warwick
Coventry CV4 7AL
UK
Phone: 44 1203 524975
Fax: 44 1203 524619
e-mail: sh@dna.bio.warwick.ac.uk

Coral Reef Coasts

Douglas J. Sherman
Department of Geography
University of Southern California
Los Angeles CA 90089-0663
USA
Phone: 1 213 740 1123
Fax: 1 213 740 0056
e-mail: sherman @usc.edu

Dissipative Beaches
North America, Coastal Geomorphology
Reflective Beaches

Andrew D. Short
Geography, Coastal Studies
University of Sydney
Sydney NSW 2006
Australia
Phone: 61 2 9351 3625
Fax: 61 2 9351 3644
e-mail: a.short@csu.usyd.edu.au

Carbonate Sandy Beaches
Sandy Coasts
Wave Environments

Parmeshwar L. Shrestha
Exponent Inc.
320 Goddard Way, Suite 200
Irvine CA 92618
USA
e-mail: pshrestha@exponent.com

Cohesive Sediment Transport

David P. Simpson
Pacific International Engineering
310 Waterfront Park Building
144 Railroad Ave.
Distribution
Edmonds WA 98020
USA
e-mail: dsimpson@piengr.com

Cross-Shore Variation of Grain Size on Beaches
Gravel Beaches

Rodman E. Snead
Department of Geography
University of New Mexico
Albuquerque NM 87131
USA
Phone: 1 505 247 3105

Desert Coasts

Gordon C. Staples
RADARSAT-2 International
13800 Commerce Parkway
Richmond BC V6V 2J3
Canada
Phone: 1 604 231 4950
Fax: 1 604 231 4940
e-mail: gstaples@rsi.ca

RADARSAT-2

Wayne Stephenson
Department of Geography and Environmental Studies
University of Melbourne
Melbourne
Victoria 3010
Australia

Shore Platforms

J. Court Stevenson
Horn Point Laboratoty
University of Maryland Center for Environmental Science
PO Box 775
Cambridge MD 21613
USA
Phone: 1 410 221 8442
Fax: 1 410 221 8490
e-mail: court@hpl.umces.edu

North America, Coastal Ecology

Michael P. Stewart
School of Spatial Sciences
Curtin University of Technology
Perth 6152 WA
Australia
Phone: 61 8 9266 7600
Fax: 61 8 9266 2703
e-mail: stewart@vesta.curtin.edu.au

Geodesy

Robert R. Stickney
Texas Sea Grant College Program
2700 Earl Rudder Freeway, Suite 1800
College Station, TX 77845
USA
Phone: 1 979 845 3854
Fax: 1 979 845 7525
e-mail: rrstickney@aol.com
stickney@tamu.edu

Aquaculture

Marcel J.F. Stive
NCK/Netherlands Centre for Coastal Research
Delft University of Technology
PO Box 5048
Delft 2600 GA
The Netherlands
Phone: 31 15 2783345
Fax: 31 15 2785124
e-mail: marcel.stive@wldelft.nl

Artificial Islands

Gregory W. Stone
Coastal Studies Institute
Louisiana State University
Baton Rouge LA 70803
USA
Phone: 1 225 388 6188
Fax: 1 225 388 2520
e-mail: gagreg@lsu.edu

Meteorological Effects on Coasts
Nearshore Wave Measurement

Katherine Stone
Meyers, Widders, Gibson, and Long, LLP
5425 Everglades St., Suite 100
Ventura CA 93006
USA
Phone: 1 805 644 7188
Fax: 1 805 644 7390
e-mail: kewstone@aol.com

Sand Rights

William Streever
BP Exploration (Alaska) Inc.
PO Box 196612
Anchorage, AL 99519-6612
Phone: 1 907 564 4383
Fax: 1 907 564 5020
e-mail: streevbj@bp.com

Wetland Restoration

Melanie J. Stright
US Department of the Interior
Minerals Management Service
381 Elden St. (MS 4042)
Herndon VA 20170
USA
Phone: 1 703 787 1736
Fax: 1 703 787 1026
e-mail: melanie.stright@mms.gov

Archaeological Site Location, Effect of Sea-Level Change

William B. Stronge
Florida Atlantic University
777 Glades Rd.
PO Box 3091
Boca Raton FL 33431
USA
Phone: 1 561 297 2833
Fax: 1 561 297 2850
e-mail: strongew@fau.edu

Economic Value of Beaches

Tsuguo Sunamura
Department of Earth and Space Sciences
Osaka University
Osaka 560-0043
Japan
Phone: 81 6 6850 5799
Fax: 81 6 6850 5480
e-mail: sunamura@ess.sci.osaka-u.ac.jp

Cliffs, Erosion Rates
Cliffs, Lithology versus Erosion Rates

Thomas A. Terich
Center for Geography, MS9085
Huxley College
Western Washington University
Bellingham WA 98225
USA
Phone: 1 360 650 3286
Fax: 1 360 650 7702
e-mail: taterich@cc.wwu.edu

Driftwood

Joost H.J. Terwindt
Department of Physical Geography
State University of Utrecht
Heidelberglaan 2
Postbus 80.115
3508 TC Utrecht
The Netherlands
Phone: 31 33 4618039
Fax: 31 33 4618039
e-mail: jenmterw@euronet.nl

Polders

E. Robert Thieler
US Geological Survey
Coastal and Marine Geology Program
384 Woods Hole Road
Woods Hole MA 02543
USA
Phone: 1 508 457 2350
Fax: 1 508 457 2310
e-mail: rthieler@usgs.gov

Photogrammetry

B.G. Thom
Faculty of Built Environment
University of New South Wales
Sydney, NSW, 2052
Australia
Phone: 61 2 9337 1465
Fax: 61 2 9337 1465
email: bthom@mail.usyd.edu.au
Coastal Changes, Gradual

James G. Titus
Mail Code 2174
U.S. Environmental Protection Agency
Washington DC 20460
USA
Phone: office: 202 260 7821
home: 301 464 5423
beach: 609 494 9742
e-mail: titus.jim@epa.gov
Greenhouse Effect and Global Warming
Sea-Level Rise, Effect

Monica M.P. Tognella-De-Rosa
Universidade do Vale de Itajai
Centro de Ciências Tecnológicas da Terra e do Mar
Rua Uruguai, 458, Itajai, SC
Brasil
e-mail: tognella@cttmar.univali.br
South America, Coastal Ecology

Alan S. Trenhaile
Geography Department
University of Windsor
Windsor, Ontario N9B 3P4
Canada
Phone: 519 253 4232 ext 2184
Fax: 519 973 7081
e-mail: tren@uwindsor.ca
Beach Sediment Characteristics
Cliffed Coasts
Mass Wasting
Rock Coasts Processes

David T. Tudor
Bath Spa University College
Newton Park
Bath BA2 9BN
UK
e-mail: dttudor@msn.com
Marine Debris—Onshore, Offshore, Seafloor Litter

Robert J. Turner
Division of Natural Sciences
Southampton College
239 Montauk Highway
Southampton, NY 11968-4198
Phone: 1 516 287 8400
Fax: 1 516 287 8419
e-mail: rturner@southampton.liu.edu
Beachrock

Takaaki Uda
Public Works Research Center
Takara Bldg.,
1–6–4, Taito, Taito-ku
Japan 110–0016
Phone: 81 3–3835–3609
Fax: 81 3–3835–3609
e-mail: uda@pwrc.or.jp
www@pwri.go.jp
Spits

M. Wafar
Biological Oceanography Division
National Institute of Oceanography
Dona Paula, Goa 403 004
India
Phone: 11 713 973 3148
Fax: 11 713 973 3340
e-mail: wafar@csnio.ren.nic.in
Indian Ocean Islands, Coastal Ecology and Geomorphology

Sayeeda Wafar
Biological Oceanography Division
National Institute of Oceanography
Dona Paula, Goa 403 004
India
Indian Ocean Islands, Coastal Ecology and Geomorphology

H. Jesse Walker
Department of Geography and Anthropology
Louisiana State University
Baton Rouge LA 70803-4105
USA
Phone: 1 225 578 6130
Fax: 1 225 578 4420
e-mail: hwalker@lsu.edu
Arctic, Coastal Geomorphology
Beach Nourishment
History, Coastal Geomorphology
Ice-Bordered Coasts

Martyn Waller
School of Geography
Kingston University
Penrhyn Road
Kingston-upon-Thames
Surrey KT1 2EE
UK
Phone: 44 208 547 2000
Fax: 44 208 547 7502
e-mail: m.waller@kingston.ac.uk
Bogs

Ping Wang
Department of Geology, SCA 528
University of South Florida
4202 E. Fowler Ave.
Tampa FL 33620
USA
Phone: 1 813 974 9170
Fax: 1 813 974 2654
e-mail: pwang@chuma1.cas.usf.edu
Nearshore Sediment Transport Measurement
Nearshore Wave Measurement
Tidal Flats, Open Ocean Coasts

Shin Wang
Department of Geography
National Taiwan University
PO Box 23-175
Taipei
Taiwan, ROC
Phone: 886 2 23627652
Fax: 886 2 23627652
e-mail: swang@ntu.edu.tw
Coastal Hoodoos

Detlef A. Warnke
Department of Geological Sciences
California State University
Hayward CA 94542-3088
USA
Phone: 1 510 885 4716
Fax: 1 510 885 2526
e-mail: dwarnke@csuhayward.edu
Pluvial Lake Shore Deposits

Niels West
Department of Marine Affairs
University of Rhode Island
Kingston RI 02881-0817
USA
Phone: 401 874 4046

Fax: 401 874 2156
e-mail: nielsmaf@uri.com

Beach Use and Behaviors
Cleaning Beaches

Dean Whitman
Laboratory for Coastal Research
Florida International University
Miami FL 33199
USA

Airborne Laser Terrain Mapping and Light Detection and Ranging

Patricia L. Wiberg
Department of Environmental Sciences
University of Virginia
Charlottesville VA 22903-4123
USA
Phone: 1 804 924-7546
e-mail: pw3c@virginia.edu

Wave–Current Interaction

Allan T. Williams
Applied Sciences Department
University of Glamorgan
Pontypridd
Mid Glamorgan
Wales
UK
Phone: 44 01656 783928
Fax: 44 91443 480558
e-mail: allan.williams@virgin.net

Marine Debris—Onshore, Offshore, Seafloor Litter

P.P. Wong
Department of Geography
National University of Singapore
10 Kent Ridge Crescent
Singapore 119260
Phone: 65 874 3859
Fax: 65 777 3091
e-mail: geowpp@nus.edu.sg

Reclamation
Tourism, Criteria for Coastal Sites

Guy Woppelmann
Centre Littoral de Geophysique
Universite de La Rochelle
Avenue Michel Crepeau
17042 La Rochelle Cedex 1
France
Phone: 33 5 46458205
Fax: 33 5 46458249
e-mail: gwoppelm@univ-lr.fr

Tide Gauges

Alejandro Yáñez-Arancibia
Instituto de Ecologia A.C.
Division de Recurso Naturales
Jefe Programa de Recursos Costeros
PO Box 63
Xalapa 91000
Km 2.5 Antigua Carretera Xalapa-Coatepec
Xalapa
Ver. Mexico
Phone: 52 28 421800 ext 3413
Fax: 52 28 187809
e-mail: aya@ecologia.edu.mx

Middle America, Coastal Ecology and Geomorphology

Prasanna Yennavar
Biological Oceanography Division
National Institute of Oceanography
Dona Paula, Goa 403 004
India

Indian Ocean Islands, Coastal Ecology and Geomorphology

Wyss W.-S. Yim
Department of Earth Sciences
University of Hong Kong
Pokfulam Road
Hong Kong SAR
China
Phone: 852 2859 2829/2859 1084 Messages
Fax: 852 2517 6912
e-mail: wwsyim@hku.hk

Submerged Coasts

Keqi Zhang
Laboratory for Coastal Research
Florida International University
Miami FL 33199
USA

Airborne Laser Terrain Mapping and Light Detection and Ranging

Foreword

As millions of people move toward the coastline, the rise in sea level is moving the coastline toward them. Anxious for a good view of the sea, property owners crowd the retreating beaches and governments seek the advice of coastal scientists and engineers to find a way to control the erosion problem. As a consequence, the study of sea level change, the greenhouse effect, the coastal processes and ways to respond to retreating coastlines have become paramount priorities. Coastal science has benefited tremendously.

For example, the search for sand used for beach nourishment, currently a favored erosion control approach, is pouring millions of dollars into the hands of those who operate vibracores, side-scan sonar, and shallow seismic devices. The applied science that is carried out to find sand is almost exactly the same as the preferred basic science approach to determine the recent history of shorefaces, coasts, and islands, and to discern the processes that have shaped them. This combined need-to-know and funding bonanza are in significant part responsible for a leap in our comprehension of nearshore processes in the last two decades. So, Maurice Schwartz's enormous effort in bringing us this new *Encyclopedia of Coastal Science* is most timely.

Specialty encyclopedias, especially one concerned with a fast moving science, such as this, provide a contemplative ledge, a place to pause and consolidate recent accomplishments. Inclusion of almost all of the seasoned practitioners of coastal earth science as contributors adds significantly to this volume's usefulness and credibility as well as its potential as a point of consolidation for our science.

This volume is replete with essays on a great number of topics that incorporate the most recent illations, judgments, and discoveries in coastal studies. What are these great advances, especially those that have occurred in coastal science in the 20 years since the Schwartz-edited *Encyclopedia of Beaches and Coastal Environments* hit the book stores? This question, a perennial favorite in Ph.D. exams, will be answered differently according to one's subspecialty. My first choice is the push forward in understanding the shoreface processes, the all-important link between the subaerial beach and the continental shelf. The topic is covered in at least half a dozen articles in the encyclopedia including those by Ed Clifton, Miles Hayes, Robert Dean, Alan Niedoroda, and Victor Goldsmith. There are also articles on the major technological advances that have fueled the advance of coastal science, including photogrammetry, side-scan sonar, synthetic aperture radar systems, and LIDAR among others (R. Thieler, Doug Inman, Keith Raney, and Stephen Leatherman).

A global perspective is furnished in separate articles about the coastal geomorphology of each continent, including the Antarctic, with companion articles on the ecology of the same coasts. A different approach to the big picture is provided through articles on coastline types such as tide- and wave-dominated (Miles Hayes), ice-bordered (Jess Walker), bay beaches (Karl Nordstrom), vegetated coasts (Denise Reed), muddy coasts (Terry Healy), chalk coasts (Vincent May), karst coasts (Roland Paskoff) and uplift coasts (Terry Healy). This small sampling of authors is indicative of the care that editor Schwartz has taken to fill the contributor slots with the "right" people with appropriate specialties.

Maurice Schwartz presents us with his second coastal encyclopedia. The first one, *The Encyclopedia of Beaches and Coastal Environments* (Hutchinson Ross, 1982) became the reference bible for coastal geologists and geomorphologists. At least it was for me. Not surprisingly, in this fast moving specialty, it is now out of date. I believe this new encyclopedia will prove to be equally important, but to a broader spectrum of coastal specialists.

The *Encyclopedia of Coastal Science* follows true to the title and cuts a broad swath. Titles are more sweeping. Individual entries are mostly topical essays rather than the dictionary-like definitions of the earlier encyclopedia. For example, instead of individual short definitions and descriptions of bulkheads, groins, jetties, seawalls, offshore breakwaters and the like, Nick Kraus has contributed two entries: Shore Protection Structures and Navigation Structures. The range of article topics is huge. Some of the outliers include coastal warfare, cleaning beaches, sand rights, oil spills, setbacks, polders, and tourism.

I believe the *Encyclopedia of Coastal Science* will prove to be a priceless resource for both educators and researchers in the coastal sciences as well as for those who use science in managing our nearshore resources. I would even like to see it on the bookshelves of beach cottages, although the volume is not intended for a lay audience. This resource is perfect for those who need succinct and authoritative information on most aspects of coastal science.

Orrin H. Pilkey
Duke University

Preface

Map measurements of the world's coastline length have yielded a figure of 500,000 km. However, when all of the very real and intricate coastal crenulations are considered, the actual length is probably closer to 1,000,000 km. Added to this is the fact that 40% of the 6,000,000,000 people presently inhabiting the earth live within 100 km of a coastline. From these observations, it can be seen that coasts are a very major geomorphic and social feature on the face of the planet. And for this reason, scholars in a multitude of disciplines have long been studying the many facets of the zone where the land meets the sea.

In this collected volume, authorities in many fields expound on certain aspects of their expertise, not so much in a dictionary of terms sense as in a series of essays that may be broken down into such categories as: atmosphere and oceanography, ecology, engineering and technology, geomorphology, and human activities related to the coasts (see Appendix 6: Topic Categories). The reader may not completely agree with some of their views; in fact, some of the authors do not agree entirely with each other. Perusing through professional journals in these fields would show the same variety of opinions on a given topic. For that is the nature of science, holding forth on a subject as interpreted by long and careful study of the evidence. What is then to be found here, between the covers of this volume, are 306 entries that contain a wealth of information on different aspects of the world's coast, which we all hold so dear. If there are any questions of omissions or judgment, the fault then lies entirely with me, the editor.

In a similar vein, one would expect the terminology of a science to have universal acceptance; but, sadly, that is not the case. For example, the ubiquitous term "shoreline" can be employed in the historical geomorphic sense of the line formed by the edge of the water against the land as it rises and falls through tidal cycles or atmospheric changes; or as it has been defined by the US Coast and Geodetic Survey for mapping purposes as the high-tide line, high-water line, or wet–dry boundary. In order for the term to mean the same thing wherever it appears in this volume the geomorphic meaning has been adopted, or clarified where it has deviated from that. For further clarification of this dichotomy, the reader is referred to the entry titled Coasts, Coastlines, Shores, and Shorelines.

Though there have been many trials and tribulations during the four to five years that it has taken to bring this volume to publication, there have also been moments of humor that lightened the load along the way. While explanations for late contributions ran rampant, none was more acceptable than that from the contributor down-under who, while working on a major topic, brought forth two "bubs" (a girl and a boy) to add to an already large family. For sheer inventiveness to a contributor who did not want to repeat a previously published survey, that progressed around a continent in a clockwise fashion, there was your editor's suggestion that he simply proceed in a counter-clockwise direction. In the end it came out only halfway there. Then too, probably the best single line in any entry contained here is the quote to the effect that a certain coastal feature is "... rather like pornography—difficult to define, but you know it when you see it!" That could only be topped by correspondence from another down-under contributor who used wildly colorful expressions that can not even be repeated here.

Of course, I am most appreciative to all of the very many people who have been involved in this project. However, two individuals stand out most significantly. The first and foremost is the editor-in-chief of this earth-science encyclopedia series, Rhodes Fairbridge, who has been my teacher, mentor, and friend for the past 40 years. The second is Peter Binfield, my editor at Kluwer Academic Publishers, who has guided me through this project with expertise, patience, and humor. To both of these gentlemen, I offer my most profound gratitude. Thanks are also due to Russ Burmester, Vicki Critchlow, Gene Hoerauf, Larry Palmer, Kevin Short, and Chris Sutton, at Western Washington University, for considerable technical support.

Sadly, media specialist Kevin Short, who worked his computer magic on many graphics in the volume including the cover photo, passed away suddenly in January of 2004 at the age of 44. Colleagues and friends alike will miss Kevin and remember him for his kindness, humor, and creativity.

Maurice Schwartz

A

ACCRETION AND EROSION WAVES ON BEACHES

An accretion/erosion wave is a local irregularity in beach form that moves along the shore in the direction of net littoral drift. The initial irregularity may be caused by a wide variety of events such as the bulge from an ephemeral stream delta, the material from the collapse of a sea cliff, erosion or accretion associated with convergence and divergence of wave energy over an offshore bar, erosion downdrift of a structure such as a groin, sudden loss of sand by slumping at the head of a submarine canyon, or rapid accretion due to beach nourishment as when dredge spoil is placed on a beach. Given the wide variety of causes leading to local beach irregularities, accretion/erosion waves are common transport modes along beaches.

The wave-like form of an accretion/erosion wave is related to the change in sediment transport rate along the beach (divergence of the drift). Specifically, an irregularity in beach topography along an otherwise straight beach produces wave refraction and diffraction that locally modifies the littoral drift system. Wave convergence at an accretionary bulge reduces the littoral drift passing the bulge, causing downcoast erosion. Consequently, the accretionary bulge moves downdrift with an erosional depression preceding it. An initial erosional depression in beach form, as in the lee of a groin, moves downdrift as a traveling sand deficit because the transport potential of the downdrift side of the depression is always greater.

Accretion and erosion waves are best observed from the air or by comparison of beach profiles with time and distance along the beach. The associated change may be several hundred meters in beach width, but more typically is about 10–20 m over a distance of about 1–2 km and may be masked locally by cusps and other small-scale beach features.

Background

The concept of an accretion/erosion wave was developed to account for the downdrift movement of a sand delta deposited across the beach by an ephemeral stream (Inman and Bagnold, 1963). It was observed that the downdrift movement occurred as an accretionary bulge preceded by an erosional depression (Figure A1). The net littoral drift perturbs deltaic accretion through a series of spit extensions (t_2). Over time, the cumulative spit extensions will progressively displace the accretionary bulge in the downdrift direction, while local wave refraction and refraction-induced divergence of the littoral drift cause erosion downdrift of the bulge (t_3). The areas of accretion and erosion migrate downdrift together in a phase-locked arrangement referred to as an accretion/erosion wave. The movement of the accretion/erosion form was quantified by surveys of the flood delta of the San Lorenzo River in central California (Hicks and Inman, 1987), and further evaluated from the sudden release of sand at the San Onofre power plant in southern California (Inman, 1987).

The propagation rate of the accretion/erosion wave form is slow initially because the on–offshore dimensions and volume of sand to be moved are at a maximum, and a significant fraction of that volume remains outside the region of rapid transport by waves (Figure A1; t_1). Once the material enters the surf zone, the longshore transport rates are much higher and the entire accretion/erosion form moves faster, spreads out along the beach, and decreases in cross-shore amplitude. Measurements near the sand release at San Onofre, CA, showed that the form of the accretional wave initially traveled with a speed of about 0.6–1.1 km/yr in the 1.8 km near the release point (nearfield) and much faster farther from the release (farfield). The delta from the Santa Cruz River floods of 1982/83 was large (800,000 m^3) and extended offshore to depths of over 10 m, and that material moved about 0.5–1.5 km/yr during the first year (Hicks and Inman, 1987). Subsequently, the downdrift erosion and accretion waves from the delta moved with speeds of 2.2–2.8 km/yr (Table A1).

It has been observed that any structure that interrupts the littoral drift of sand along a beach results in an erosional chain reaction traveling downdrift from the structure (Inman and Brush, 1973). The propagation rates of the downdrift erosion wave was evaluated from beach

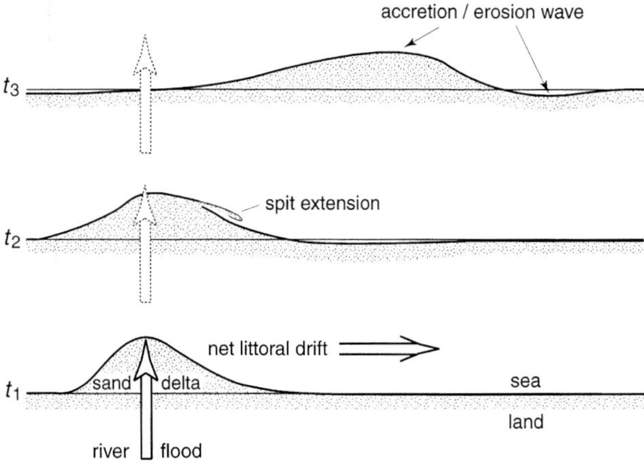

Figure A1 Formation of accretion/erosion wave downdrift from an episodically formed sand delta at time t_1, where $t < t_2 \ll t_3$ (modified from Inman and Bagnold, 1963).

Table A1 Propagation speeds of accretion/erosion waves

Location	Type/cause	References	Net downdrift transport rate 10^3 m^3 yr^{-1}	Speed of accretion/erosion wave (km yr^{-1}) Nearfield[a]	Farfield
Santa Cruz, CA	Accretion from San Lorenzo River Delta	Hicks and Inman (1987)	268	0.5–1.5	
Santa Cruz, CA	Erosion and bypass accretion from Santa Cruz Harbor	Hicks and Inman (1987)	268		2.2–2.8
Santa Barbara, CA	Erosion from harbor and bypass accretion	Inman (1987)	214		2.5–2.8
San Onofre, CA	Accretion from sand release	Inman (1987)	200	0.6–1.1	
Oceanside, CA	Erosion from harbor	Inman and Jenkins (1985)	200		2.2–4.0
Assateague Island, MD	Erosion from jetties at Ocean City Inlet	Leatherman *et al.* (1987)	153	~0.3	
Outer Banks, NC	Migration of Oregon Inlet	Inman and Dolan (1989)	590	0.023	
Nile Delta, Egypt	Accretion wave from onshore migration of sand blanket	Inman *et al.* (1992)	1,000		0.5–1.0

[a] Nearfield is within 1–2 lengths of the perturbing feature such as a sand delta.

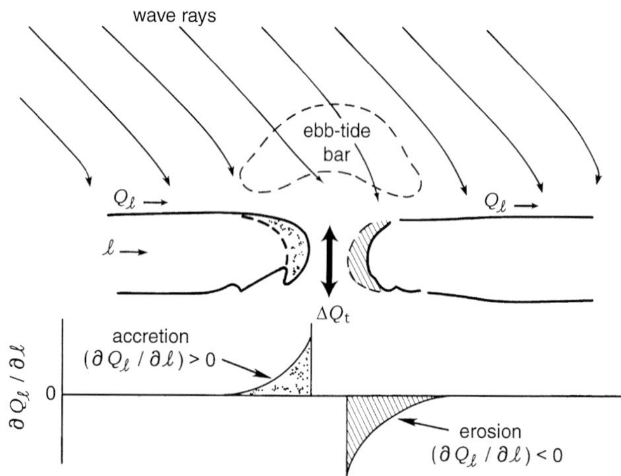

Figure A2 Schematic diagram of the divergence of drift ($\partial Q_\ell/\partial \ell$) at a migrating tidal inlet with net tidal flux of sediment, Q_t (modified from Inman and Dolan, 1989).

surveys following the construction of the harbor jetties at Santa Barbara, CA (Inman, 1987), and the enlargement of the harbor at Oceanside, CA (Inman and Jenkins, 1985). Once in the farfield of the structure, the erosion wave, followed by the accretion wave moved downdrift at 2.5–2.8 km/yr at Santa Barbara and 2.2–4.0 km/yr at Oceanside (Table A1).

Accretion/erosion waves also occur along beaches and barriers downdrift of tidal inlets (e.g., Inman and Dolan, 1989). The erosion wave from the jetties at Ocean City, MD, is a well-known example. The inlet between Fenwick and Assateague barrier islands was stabilized by jetties in 1935. The jetties trapped the littoral drift and caused an erosion wave to travel downdrift along Assateague Island, resulting in a landward recession of the entire barrier island of 460 m in 20 years (Shepard and Wanless, 1971).

The Nile Delta experiences accretion/erosion waves driven by the currents of the east Mediterranean gyre that sweep across the shallow shelf with speeds up to 1 m/s. Divergence of the current downdrift of the Rosetta and Burullus promontories entrains blankets of sand that episodically impinge on the beach. These sand blankets cause shoreline irregularities with average amplitudes of 100 m and wavelengths of about 8 km that travel along the shore at rates of 0.5–1 km/yr as accretion/erosion waves (Inman *et al.*, 1992) (see entry on *Littoral Cells*).

A related example of a traveling accretion/erosion feature occurs when the littoral drift impinges on an inlet causing it to migrate downdrift (Figure A2). The migration proceeds as an accretion of the updrift bank in response to positive fluxes of sediment delivered by the net littoral drift Q_ℓ, while the downdrift bank of the inlet erodes due to a negative divergence of drift across the inlet, $\partial Q_\ell/\partial \ell < 0$. The negative divergence of the drift across the inlet is caused by wave refraction over the ebb-tide bar and by a loss of a portion of the drift to flood-tide entrainment at the inlet, ΔQ_t. Also the offshore tidal bar, maintained by the ebb-tide flow, moves downdrift with the inlet migration. Although the migration rates of the up- and downdrift banks of the inlet and the tidal bar are phase-locked, they are out of phase with the local net sediment changes in the shorezone bordering the inlet. The inlet banks and channel form an accretion/erosion sequence that travels along the beach and surf zone while the ebb-tide bar forms an accretion wave that moves along the shore in deeper water. Their relative on/offshore positions depend on the inlet tidal velocities that are functions of the size of the inlet and the volume of tidal flow through it (Inman and Dolan, 1989; Jenkins and Inman, 1999).

Accretion/erosion waves associated with river deltas and migrating inlets are common site-specific cases that induce net changes in the littoral budget of sediment. However, it appears that accretion/erosion waves in some form are common along all beaches subject to longshore transport of sediment. This is because coastline curvature and bathymetric variability (e.g., shelf geometry and offshore bars) introduce local variability in the longshore transport rate.

Mechanics of migration

An accretion/erosion wave is a wave- and current-generated movement of the shoreline in response to changing sources and sinks in the local balance of sediment flux along a beach. The downdrift propagation of the wave form is driven by advective and diffusive fluxes of sedimenb mass (Figure A3(a)). For convenience, these processes are usually expressed in terms of the longshore flux of sediment volume Q_ℓ into and out of a control cell (Q_{in}, Q_{out}; Figure A3(b)). By convention, fluxes of sediment into the cell are positive and fluxes out are negative. The net change of the volume fluxes between the updrift and downdrift boundaries of the control cell ($Q_{in} - Q_{out}$ = divergence of drift) will result in a net rate of change in the position of the shoreline $\partial x/\partial t$. Shifts in shoreline position will in turn cause the beach profile within the control cell to adjust to new equilibrium positions (Figure A3(b)). The new profile positions alter local wave refraction causing adjustments in the flux of sediment leaving the control cell (Q_{out}). The variation in Q_{out} will alternately accrete and erode the beach downdrift of the control cell. As a consequence, propagation of the accretion/erosion wave involves

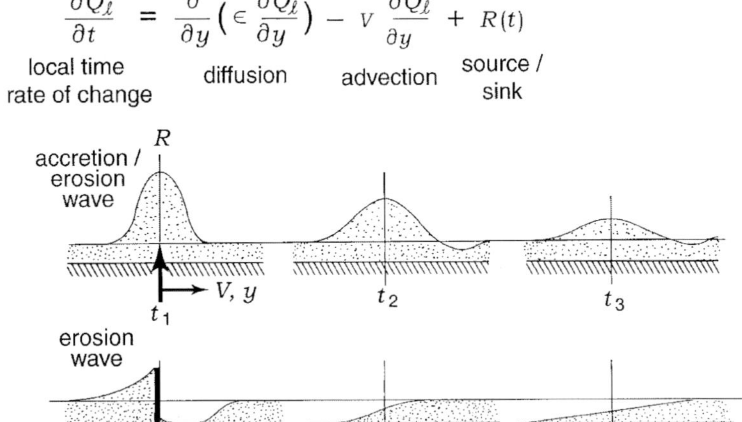

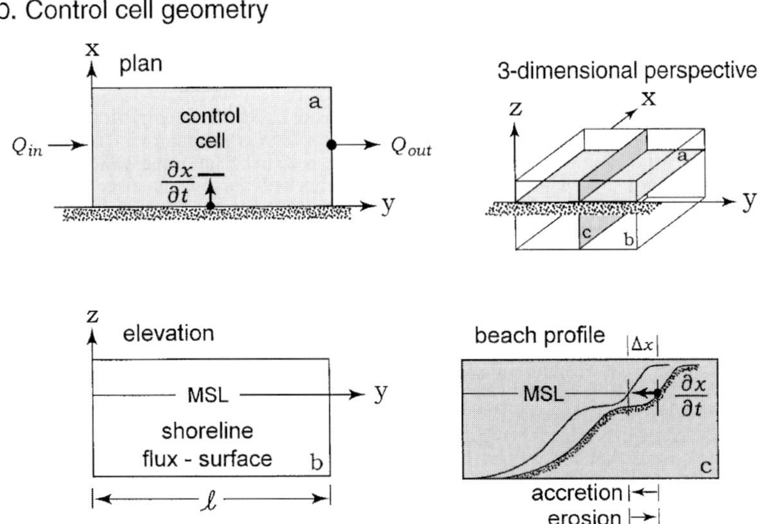

Figure A3 The balance of sediment for a propagating accretion/erosion wave (modified from Inman and Dolan, 1989).

a chain reaction in the local sediment flux balances. The reaction is set off by a disturbance on the updrift side of the control cell that yields a shoreline response on the downdrift side.

At a tidal inlet, these dynamics are impacted by additional fluxes of sediment into or out of a control cell centered at the inlet. When the tidal transport of sediment is ebb-dominated ($\Delta Q_t > 0$), the sediment flux into the control cell builds the ebb-tide bar and increases the rate of sediment that passes over the bar to the downdrift side of the inlet (Figure A2). This stabilizes the inlet position by decreasing deposition on the updrift side and erosion on the downdrift side. Flood-dominated tidal transport ($\Delta Q_t < 0$) has the opposite effect and will cause the inlet to migrate faster (Jenkins and Inman, 1999).

Douglas L. Inman and Scott A. Jenkins

Bibliography

Hicks, D.M., and Inman, D.L., 1987. Sand dispersion from an ephemeral river delta on the central California coast. *Marine Geology*, **77**: 305–318.

Inman, D.L., 1987. Accretion and erosion waves on beaches. *Shore and Beach*, **55**: 61–66.

Inman, D.L., and Bagnold, R.A., 1963. Littoral processes. In Hill, M.N. (ed.), *The Sea, Volume 3, The Earth Beneath The Sea*. New York, London: John Wiley and Sons, pp. 529–553.

Inman, D.L., and Brush, B.M., 1973. The coastal challenge. *Science*, **181**: 20–32.

Inman, D.L., and Dolan, R., 1989. The outer banks of North Carolina: budget of sediment and inlet dynamics along a migrating barrier system. *Journal of Coastal Research*, **5**: 193–237.

Inman, D.L., and Jenkins, S.A., 1985. Erosion and accretion waves from Oceanside harbor. In *Oceans 85: Ocean Engineering and the Environment*, Volume 1. Marine Technological Society and IEEE, Piscataway New Jersey. pp. 591–593.

Inman, D.L., Elwany, M.H.S., Khafagy, A.A., and Golik, A., 1992. Nile Delta profiles and migrating sand blankets. *Proccedings of the 23rd Coastal Engineering Conference*. Billy Edge, (ed.) ASCE, New York: American Society of Civil Engineers, pp. 3273–3284.

Jenkins, S.A., and Inman, D.L., 1999. Sand transport mechanics for equilibrium in tidal inlets. *Shore and Beach*, **67**: 53–58.

Leatherman, S.P., Dean, R.G., Everts, C.E., and Fulford, E., 1987. Shoreline and sediment budget analysis of North Assateague Island, Maryland. *Proceedings of Coastal Sediments '87*. ASCE, pp. 1460–1471.

Shepard, F.P., and Wanless, H.R., 1971. *Our Changing Coastlines*. New York: McGraw-Hill Book Co.

Cross-references

Beach Features
Beach Processes
Coasts, Coastlines, Shores, and Shorelines
Energy and Sediment Budgets of the Global Coastal Zone
Littoral Cells

Longshore Sediment Transport
Scour and Burial of Objects in Shallow Water
Sediment Budget

AFRICA, COASTAL ECOLOGY

Introduction

The continent of Africa straddles the equator and extends to about 35° latitude both north and south. It is, therefore, dominated by warm water regions and its coastline includes environments such as coral reefs and mangroves, except at the northwest and southwestern extremes, where temperate environments occur. The African coastline, excluding Madagascar (4,000 km), totals 35,000 km. A review of the literature dealing with sandy beaches in Africa (Bally, 1986) found the earliest paper to have been published in the 1880s and, of more than 1,000 papers published over 100 years, 19% concerned north Africa, 16% west Africa, 15% east Africa and Madagascar, and 55% southern Africa, particularly South Africa. Early papers were primarily taxonomic but later publications were mainly ecological. English was the language of 59% of the papers, followed by French (29%), German (7%), and Italian (3%). This sketch for literature on sandy beaches is probably representative of all papers dealing with coastal ecology in Africa. If so, it can be concluded that coverage of coastal ecology in Africa is patchy and, whereas the coastal ecology of South Africa has been extensively researched, most other regions are less well known.

The physical environment

Precipitation along the African coast is mostly low, with the exception of Madagascar, central east Africa, and west Africa, where rainfall is high (Figure A4). It is in central to northwest Africa that the largest rivers enter the sea, whereas greatest evaporation occurs in northwest Africa and in the Red Sea. Lowest salinity in coastal waters is therefore in the Gulf of Guinea and highest values are in the Red Sea. The coasts of Africa experience semi-diurnal tides, with the exception of the east coast of Madagascar, the horn and the Mediterranean coast of Egypt, where tides are mixed. Spring tide range mostly approximates 2 m, except in the Mozambique Channel where it is larger and can exceed 4 m, and along the Red Sea and Mediterranean coasts where tides are small (Davies, 1972). The entire east coast experiences temperatures permanently above 20°C, as does the central west coast. Upwelling in the southwest and northwest lowers temperatures regionally. Major currents are the Somali and Mozambique/Agulhas in the east and the Benguella, Guinea, and Canary currents in the west. Wind and wave energy is greatest in the south and lowest in the equatorial regions, the Mediterranean and the Red Sea.

Regional descriptions

West Africa

The northwest coast, from Morocco to Senegal, is sandy and relatively unindented, but from Dakar southeast to Monrovia it becomes very indented and there are numerous offshore islands. Moving further south, the coast becomes low lying and sandy and more deltaic in nature, with large lagoons further east. The eastern area is dominated by the delta of the Niger River. Further south, the Congo River has a major influence and freshwater can extend far offshore. High precipitation and numerous rivers in central west Africa result in warm, low salinity water, known as Guinean waters, circulating in the Gulf of Guinea.

The most notable feature of this coast is the extent of mangrove forests, estimated at 28,000 km^2 or 15% of the world's mangroves. Coastal lagoons range from tidal swamps and seasonal marshlands, associated with the river deltas and estuaries, to extensive coastal lagoons, which are typical of the Guinea coast. Sea grass beds are not well developed and there are no true coral reefs due to the cool waters of the Benguela and the Canary currents. Sandy beaches occur throughout the coast, particularly along the coast of Mauritania and northern Senegal. There are permanent areas of upwelling off Senegal, Zaire, and Namibia, coupled to the Canary and Benguela currents.

The marine resources of the west African region are important in the local and regional economies. Although seaweeds are not diverse, invertebrates (lobsters and shrimps) are exploited and there is a rich icthyofauna with about 250 species and high levels of endemism. Five species of marine turtles nest along the coast and on the islands and there are three species of crocodiles. Millions of migratory birds, especially waders, visit the coast seasonally. The northwest African manatee occurs in suitable habitats from Senegal to Angola.

The coastal human population in west Africa exceeds 50 million and the rate of industrialization and urban population growth is accelerating along the coastal zone. Oil production has increased markedly in Nigeria, Angola, and Gabon. Tourism is an important earner of foreign exchange in the economies of several countries including Gambia and Guinea Bissau. Increased fishing effort and the introduction of more efficient technologies has led to over exploitation of fisheries resources.

East Africa

Oceanic current patterns and monsoon seasons have a major influence on the biogeography and biodiversity of east Africa. The main oceanic currents are the Somali, Madagascar, and Mozambique/ Agulhas currents. The continental shelf is mostly narrow, varying from a few kilometers off Pemba to nearly 145 km in the Bight of Sofala, Mozambique. The shelves and banks are areas of intensive biological activity and productivity; in general, the narrower the shelf, the less productive the sea area. The western Indian Ocean is fairly poor in fisheries compared with other regions. The mainland coast is relatively unindented due to the absence of large rivers and coastal waters are moved by coast parallel currents. Much of the Mozambican coastline consists of low coastal plains forming long stretches of sand beaches and dunes interspersed with muddy rivers. North of the Zambezi estuary, many small coral islands fringe the shore. To the south, the islands of the Bazaruto group and Inhaca are mainly sand.

The total area of mangrove coverage in east Africa is about 10,000 km^2 or 5% of the world mangrove. Diversity of mangrove communities in east Africa is higher than in the west, with 10 species compared with only 7. Sea grass beds are found in all countries where there are low energy environments. The east African coast features fringing and patch reefs along the coastline from Somalia to Mozambique. Sandy beaches are well developed throughout. Small-scale local upwelling occurs seasonally, particularly in the waters off Somalia.

The coast of east Africa supports an enormous diversity of life. More than 350 marine algal species and more than 135 species of coral are known and there are more than 900 fish species, mostly associated with coral reefs. The five marine turtles are abundant. The region has a varied assemblage of seabirds, including frigate birds, tropic birds, boobies, shear-waters, terns, noddies, and gulls. The dugong is present, but its distribution and migration along the mainland coasts is not known. At least 15 species of cetaceans occur along this coast.

The coastal human population in east Africa is about 80 million. Shipping/port development and tourism are the fastest growing industries in the east African coastal zone. These industries are becoming incompatible with conservation in areas where pollution threatens to destroy the scenic beauty of beaches and coral reefs.

Red Sea

The Red Sea, over 2,000 km long and up to 360 km wide, is an important repository of marine biodiversity on a global scale. It features a range of coastal habitats including coral reefs, mangroves, salt marshes/sabkhas, rocky and sandy beaches, dune systems, and sea grass beds. These habitats accommodate more than 500 species of seaweeds, 1,000 fish species, 200 species of stony coral, 130 species of soft corals, more than 200 species of echinoderms, and 200 species of birds. There is a considerable variety of coral reef types with great structural complexity. The diversity and the number of endemic species of corals are extremely high.

The coastal human population is about 5 million. Urbanization, oil and other industries are developing rapidly in the Red Sea and tourism plays a major role in the economy of several countries. Pressures from recreation and tourism are high in the northern part and in the Gulf of Aqaba. Much of the oil produced in the Middle East is transported through the Red Sea and land-based activities (e.g., power and desalination plants, sewage treatment plants, industrial facilities, solid, wastes, and others) are adversely affecting the coastal and marine environment.

North Africa

The north African part of the Mediterranean is warm and arid. While it exhibits a low level of biological productivity, the Mediterranean Sea, as well as the surrounding land, is characterized by a moderate degree of biological diversity. Among the ecosystems that occupy coastal marine areas, the rocky intertidal, estuaries, and sea grass meadows (*Posidonia*

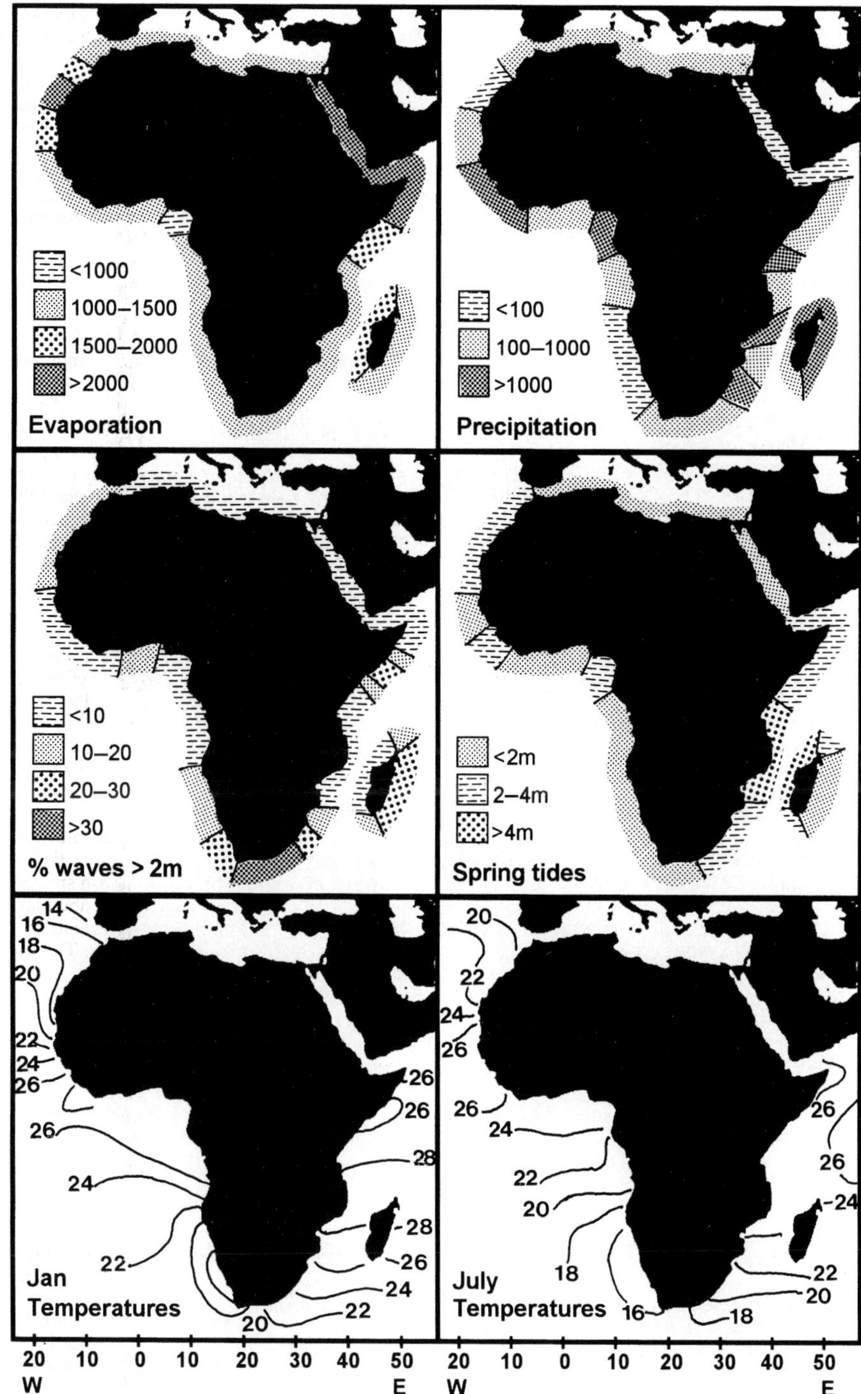

Figure A4 Physical features of African coasts.

beds, especially in Libya and Tunisia) are of significant ecological value. The Nile delta is a major feature of this area. Endangered species include the Mediterranean monk seal, marine turtles, and marine birds. It has been estimated that 500 species from the Red Sea have entered the Mediterranean through Suez, since damming the Nile has reduced the freshwater barrier (Por, 1968). The coastal population is about 40 million and human impacts are considerable. Tourism is of great importance in several areas.

Southern Africa

This region, including South Africa and Namibia, is subtropical to temperate, with cool upwelling waters in the west. It mostly consists of open high-energy coast receiving swell from the Southern Ocean. There are numerous large zetaform bays on the exposed south coast and few sheltered waters. The west is arid and the east is moist. Sandy beaches and rocky shores dominate and there are many small estuaries. Large volumes of sand transport characterize the south, with extensive, coupled dune and beach systems. Rocky shores are notable for their diversity of large limpets. Productive kelp beds occur on the west coast, and salt marshes are typical of estuaries and lagoons in the south and west. Biodiversity is highest in the east and decreases towards the west coast. Fur seals and penguins occur around the islands and two species of turtles breed on the east coast of South Africa.

There are a number of invertebrate fisheries: in addition to abalone in the southwest and subsistence fishing on the east coast, lobsters are exploited in the south and west, and penaeid prawns on the east coast. Demersal and seine fisheries are well developed, especially on the west coast. The coastal population is about 15 million and there are five major harbors. A wide range of activities impinges on the coast: besides recreation, tourism and industry, these include dune mining for heavy minerals, beach mining for diamonds, and damming of rivers.

Biogeography

The "shallow water" or coastal marine biogeography of Africa is dominated by warm water regions (Figure A5), the Indo-Pacific component on the east coast and the Atlantic component on the west coast. The Red Sea and Mediterranean coasts have distinctive, warm water faunas, which may be considered tropical and subtropical, respectively. Their marine boundaries are at Gibraltar and at the entrance to the Red Sea. Southern Africa harbors three provinces in addition to the Indo-Pacific component to the east: a warm temperate region on the extreme south coast and two temperate regions, the Namaqua and Namib provinces on the southwest. The eastern boundary between the southern warm temperate region and the Indo-Pacific is a broad transition region spanning the Eastern Cape and Natal coasts of South Africa. The junction between the Namaqua and Namib provinces, both influenced by upwelling, lies in southern Namibia near Luderitz and the boundary between the Namib and the Guinean is in southern Angola. The boundary between the tropical Guinean province and the north African

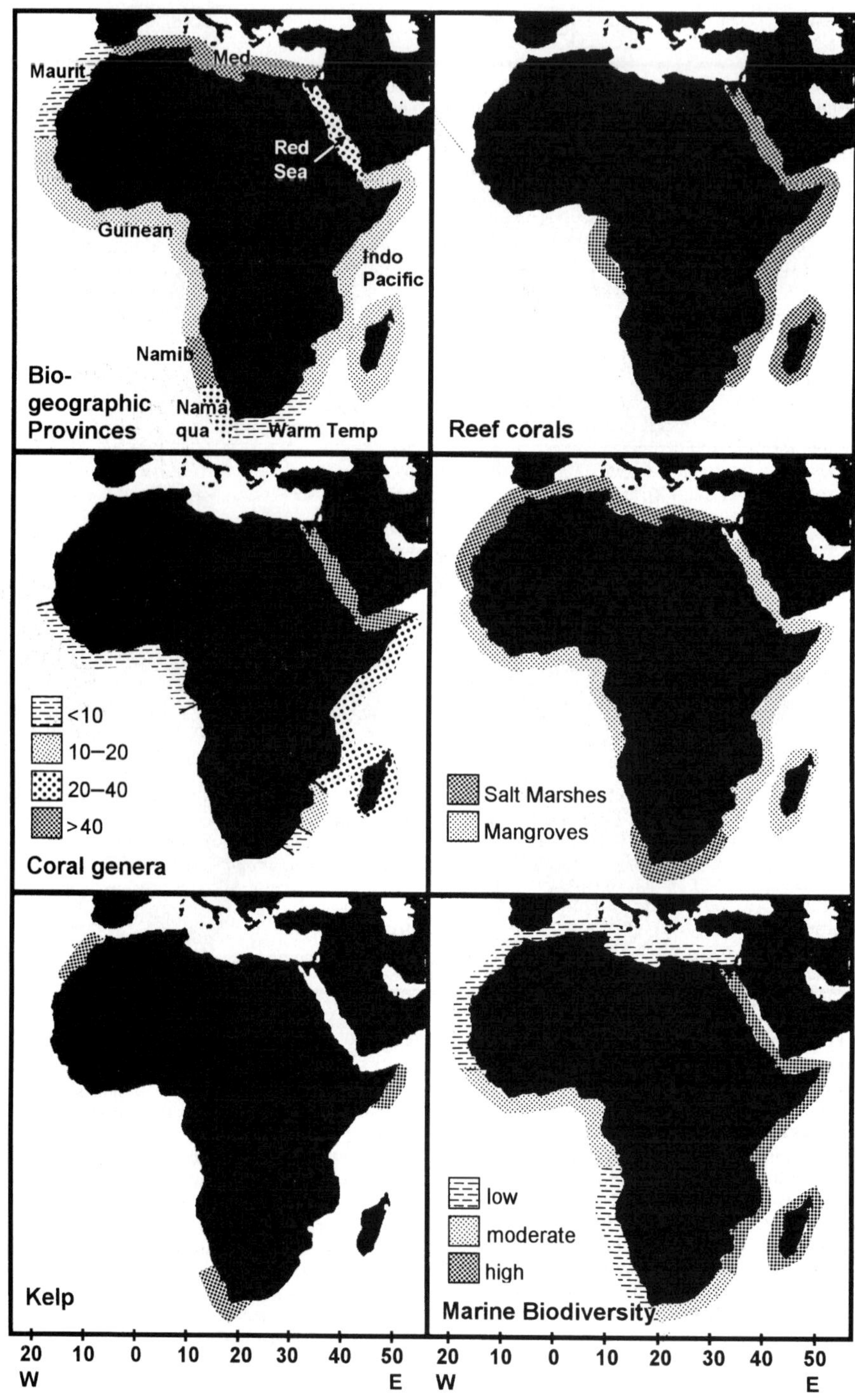

Figure A5 Biogeographic features of African coasts.

Mauritanian province lies around 15°N. Biodiversity is greatest in the Indo-Pacific province and in the Red Sea on the east coast and lowest in the temperate provinces in the southwest and northwest (Namaqua, Namib, and Mauritanian) and the Mediterranean.

Environments

The warm water provinces are characterized by corals, mangroves and, in some cases, sea grass beds, whereas the temperate provinces harbor kelps, where there is upwelling, and salt marshes. Sandy beaches and associated dunes, rocky shores and estuaries occur throughout.

Coral reefs

True coral reefs occur around Madagascar and along the African east coast from the Red Sea to Mozambique with some reef forming (hermatypic) corals extending further south into South Africa. On the west coast, the influence of upwelling restricts corals to a more limited region around the equator and islands off the northwest, but these are not true coral reefs. Those on the east coast are much richer in genera, being a part of the Indo-Pacific region (Figure A5). Coral reefs are mostly of the fringing type and occur predominantly in two provinces, the Red Sea and East Africa/Madagascar (Sheppard *et al.*, 1992). Reef development in between, around the horn and Gulf of Aden, is limited by upwelling. In the northern Red Sea reefs are well developed and drop into deep, clear water, whereas in the south they occur in shallower, more turbid water and are less well developed. Besides fringing reefs, barrier reefs, patch reefs, and even atolls can occur. Coral diversity in the Red Sea is high and there is a clear north/south zonation into 13 communities. Coral reefs also display a depth zonation with maximum coral diversity at 5–30 m depths, depending on exposure and water clarity. Coral cover is mostly less than 50% on slopes but in sheltered areas *Porites* can attain 80% cover. Coral reefs support a diverse associated fauna of cryptic invertebrates and fishes, which play a variety of roles in these complex communities. The Red Sea harbors more than 1,000 species of fishes, many associated with coral reefs. Greatest fish abundance occurs near the reef top, whereas greatest species richness tends to occur at depths of 10–15 m.

Sandy beaches

Sandy beaches, backed by dunes, are the dominant coastal form, comprising as much as 70% of the coastline, and have been well studied in South Africa. Beach form is controlled by the interaction between sediment particle size and wave and tide energy. Conditions of high wave energy, large tide ranges, and fine sand give rise to wide flat beaches referred to as dissipative; whereas, narrow steep beaches, called reflective beaches, develop under conditions of low wave and tide energy and coarse sand. African beaches range from high-energy dissipative systems on the southwest tip of the continent, through intermediate types to microtidal reflective beaches in the Mediterranean and Red Sea. Macrotidal beaches and sand flats occur in the Mozambique Channel. High-energy dissipative systems have been shown to be richest in terms of productivity, with large populations of filter feeders (Brown and McLachlan, 1990). Such beaches, together with macrotidal flats, support diverse benthic faunas. Reflective systems, which are typical of tropical regions with modest tide ranges, generally support lower diversity and lack surf zones. Temperate high-energy beaches with extensive surf zones often develop blooms of surf zone diatoms, which fuel rich food chains. Besides meiofauna and microfauna, beaches support a macrofauna of scavenger/predators, filter feeders and deposit feeders. Among the macrofauna, donacid clams, gastropods of the genus *Bullia*, ghost crabs (*Ocypode*), cirolanid isopods, mysid shrimps, and polychaete worms are typical. The species richness per beach ranges from less than 10 species on reflective beaches to more than 30 on tropical dissipative beaches and tidal flats. Zonation on sandy beaches is less marked than on rocky shores and throughout most of Africa consists of three zones: ghost crabs at the top of the shore, cirolanid isopods on the midshore, and a variety of species lower down. In the temperate areas, sandhoppers (talitrid amphipods) and/or oniscid isopods occur at the top of the shore. Beach fauna is, therefore, controlled primarily by physical factors and beach type, with biogeography and climate playing a lesser role.

Sea grass beds

In the Mediterranean, the Red Sea and throughout the western Indian Ocean, sea grass beds are common in intertidal areas, coastal lagoons, shallow sandy bottoms with good light penetration and sandy areas adjacent to shallow reefs. At least 10 species are common and include members of the genera *Possidonia, Thalassia, Halodule, Syringodium, Halophila, Cymodocea*, and *Thalassodendron*. Sea grasses are not true grasses but are closer to pondweeds. They are aquatic plants with leaves above the surface and rhizomes and stems buried in the sand. Sea grass beds create habitats distinguished by high primary productivity, enhanced by encrusting algae epiphytes. However, there is limited direct herbivory, probably due to distasteful compounds and indigestibility of seagrasses. Instead, seagrasses are consumed via detritus food chains after processing by microorganisms. Sea grasses stabilize the sediment, create habitat, and serve as nursery areas and support a high diversity of associated species, especially molluscs, polychaets, crustaceans, and fishes. Many commercial fishes utilize them as nursery areas. Ten species occur in the Red Sea, *Halodule* and *Halophila*, being the commonest.

Dunes

Dune forms depend on rainfall (vegetation growth) and sediment transport (supply X wind) and are closely coupled to beach type. Simple vegetated foredune ridges and hummocks are the most widespread type. Such dunes typically occur behind low energy and reflective beaches and are characteristic of the moist tropics. Transgressive dune sheets occur in windy areas with large volumes of sand transport, for example, the south coast of South Africa, and parabolic dunes occur in places with predominantly unidirectional winds. Coastal dunes are rapidly colonized by plants, *Ipomoea* and *Scaevola* being typical foredune pioneers. These may be followed by scrub and thicket until climax forest is reached. In the arid southwest and north of Africa coastal dunes are highly mobile and support limited vegetation. Climax dune vegetation may be scrub in arid areas, or even pioneers in hyperarid situations. In the moist tropics, forest may extend right down to the beach. Coupled to landward succession in dune vegetation is a change in animal communities: bird diversity increases as vegetation structure becomes more vertical; insects and small mammals also respond to this gradient. In general, therefore, coastal dune ecology is a function of dune forms and vegetation succession, which in turn are controlled by sand supply and climate.

Rocky shores

Rocky shores form about 30% of the coastline of Africa. Intertidal rocky-shore organisms are exposed to a wide range of physical conditions. Unlike the situation on sandy beaches, organisms on rocky shores cannot burrow to escape adverse conditions: they must simply be tough enough to tolerate the fluctuations. Tides establish a gradient of physical stress, with the high-shore being sun-baked and desiccated, and the low-shore more mild. This gradient leads to a distinctive vertical zonation of organisms, with the top of the shore characterized by a small number of species, low biomass and productivity, and progressive increases in all these variables as one moves down the shore. Physical stress tends to limit the extent to which species can advance up the shore, but biological interactions between species often set limits to how far they extend down the shore. For example, competition from mussels ousts barnacles; grazers, such as limpets and chitons, control the growth of algae; predators, such as whelks, crabs, fish, and birds, limit the zonation of their prey. Moving horizontally along rocky shores, different gradients come into play. The first of these is wave action, which operates on a scale of meters to kilometers. Mobile predators and grazers tend to be inhibited by waves, whereas algae benefit from a reduction in grazing and from an increase in the turnover of nutrients. Filter feeders, such as mussels, are enhanced because waves import the organic particles that comprise their food. The overall effect is that biomass and productivity are highest on wave-beaten shores, such as those occurring on the south coast of Africa (Branch and Griffiths, 1988).

At larger scales of 100s to 1,000s of kilometers, there are gradients of productivity. High nutrient levels associated with upwelling accelerate algal growth, supporting high biomass of grazers and predators but a low diversity of species. On a similarly large scale, there are climatic gradients from temperate to tropical conditions. High temperatures in the tropics lead to greater physical stresses, so that rocky shores there tend to be sparsely occupied, with low biomass and productivity, but high levels of species diversity. Tough coralline turfs become a dominant element among the seaweeds. Grazers and sedentary predators are often confined to shelters during low tide to avoid potentially lethal conditions. On the other hand, mobile predators, such as crabs and fish, are more abundant in the tropics than on temperate shores. In sum, vertical changes in community structure on rocky shores are dictated by gradients of physical stress overlain by biological interactions. Moving horizontally along rocky shores, community patterns and processes controlling them change radically in response to wave action, productivity, and climate.

Kelp beds

Kelps are large and complex in structure, with a root-like holdfast, a long stipe and frond-like blades. Kelps are fast growing plants and are restricted to coastal areas where sunlight is readily available and nutrient levels are high. In Africa, they are confined to zones of upwelling, where cool and nutrient-rich water is brought to the surface by winds that drive surface waters offshore. Upwelling is concentrated on the west coast of southern Africa, off west Africa and in the region of the Somalian horn of Africa, where the world's only tropical upwelling areas can be found. Kelp forests thus occur in the southwest in the Namaqua and Namib provinces (*Ecklonia, Laminaria, Macrocystis*) and in the northwest in the Mauritanian province (*Laminaria*) and off Somalia. Kelp plants are of direct economic significance because they produce alginic acid, widely used in food and other products.

Kelps form dense underwater forests, which break the force of wave action. They are themselves, however, powerfully influenced by waves, which tangle and tear out swathes during storms. Whole kelp plants wash ashore on sandy beaches where they contribute substantially to energy flow. Sandy shores "supplemented" in this way have high levels of biomass and productivity, and their life is concentrated at the top of the shore where the kelp deposits. Surprisingly little kelp is directly eaten by herbivores; most is abraded from the growing plants and contributes to a pool of organic matter that fuels particle-feeders and filter feeders, such as sea cucumbers, mussels, ascidians, and sponges that dominate much of the floor of kelp beds. Large fragments of kelp also break off and support urchins and abalone. Intertidal grazers also benefit: the highest biomasses of grazers ever recorded in the world occur on rocky shores on the west coast of South Africa. Their existence depends on the vital "subsidy" they receive in the form of drift from adjacent subtidal kelp beds. Clearly, kelp beds play significant ecological roles that extend well beyond their confines. They are commercially valuable in their own right, but also sustain other species of commercial importance, such as abalone.

Estuaries

Estuaries, lagoons, and river mouths are highly variable around the coasts of Africa, depending on climate. Other than for the Zambesi in the east, all the major rivers (Kunene, Orange, Congo, Niger, Volta) drain to the west and the Nile empties into the Mediterranean. The Nile (20,000 km^2), Zambesi (700 km^2), Volta (9,000 km^2), Senegal (8,000 km^2), Oueme (1,000 km^2), and Niger (36,000 km^2) have extensive delta systems. Large coastal lagoons occur in west and in southeastern Africa and many wetlands are of considerable conservation value. In the arid areas of southern and southwest Africa, the Red Sea and the Mediterranean coast, most estuaries are ephemeral and close during the dry season. Many estuaries have been impacted by damming of rivers, in some cases the reduction in freshwater supply eliminating normal salinity gradients and/or reducing floodplains (e.g., Niger). Mangroves, salt marshes, and phytoplankton contribute to primary production and support benthic fauna of high abundance but low diversity. Estuaries are important nursery areas for penaeid prawns and a variety of fishes.

The regulation and impoundment of freshwater in river systems is probably the single most important threat to natural functioning of estuarine systems in many parts of Africa. Most of these estuaries are classified as temporary open/closed. Floods, in particular, are important in maintaining functional links between the estuaries and the sea, but because of impoundment schemes, changes in the frequency and intensity of flood events is becoming evident. Many estuaries close more frequently and for longer periods since the removal of accumulated sediment in marsh channels by floods is not as effective as before. Thus, many estuaries are beginning to function differently compared with the natural state. Not only do freshwater abstraction schemes lead to negative downstream impacts, but they also have the capacity to influence the marine nearshore. Major regulations along the Zambesi river (Kariba and Cahora Bassa dams) are having a negative influence in coastal waters, with die back of mangroves and a collapse of the coastal prawn fisheries (Davies *et al.*, 1993).

Salt marshes

Salt marshes are temperate habitats, mostly associated with estuaries. African salt marshes occur in the Mediterranean, on the northwest and on the south and southwest coasts. Most African salt marshes are of limited extent; for example, total salt marsh area in South Africa is 1,700 ha, most of which is confined to five systems. There is clear zonation of the marsh flora from the subtidal to the top of the intertidal zone, *Spartina*, *Zostera*, *Sarcocornia*, and *Limonium* are the most typical genera. These African salt marshes are distinguished from those elsewhere by their warm temperate character and limited size. Associated fauna includes a variety of marsh crabs and shrimps, wading birds and, at high tide, fishes.

Mangroves

Total world mangrove area is 181,000 km^2 and 20% of this is in Africa, 15% on the west and 5% on the east coasts (Spalding *et al.*, 1997). On the east coast, in the Red Sea and Madagascar, they form part of the Indo-Pacific province; there are 10 species and the most widespread are *Avicennia marina, Rhizophora mucronata*, and *Sonneratia alba*. On the west coast of Africa their affinities are with New World tropical regions and, of the seven indigenous species, *Avicennia germinans, Rhizophora racemosa*, and *Laguncularia racemosa* are widespread. Distribution may be as much limited by aridity as temperature. Estuaries may harbor large mangroves systems but they can also occur as narrow strips along the coastline where rivers are absent. Associated fauna includes mudskippers *Periopthalmus*, mangrove snails *Cerithidea*, fiddler crabs, *Uca*, and oysters. Fishes and penaied prawns utilize mangroves as nursery areas. Mangroves are important sources of detritus for estuarine food chains. Wood collection threatens mangroves in several areas (e.g., Mozambique).

Islands

Coastal islands form important rookeries for seals and nesting sites for seabirds. Colonies of Cape fur seals, Jackass penguins, Cape gannets, cormorants and other seabirds occur on numerous coastal islands, especially off South Africa and Namibia. Socotra is also an important breeding area for boobys and terns and also harbors six endemic species of birds and eight endemic reptiles. Red Sea islands support colonies of gulls and terns. Islands off Mauritania are important for migratory birds and support large breeding colonies of gulls and terns.

Utilization

Coastal resources are extensively utilized throughout Africa. These activities range from subsistence gathering of shellfish and mangrove wood to commercial exploitation of abalone, kelp, lobsters, and inshore fishes. Subsistence utilization can cause degradation and damage to coastal habitats. Dynamite fishing also threatens Africa's coastal ecosystems in some localities as it disturbs coral reefs and lagoon systems. Nonliving resources that are exploited include oil (Angola, Nigeria), diamonds in beach and nearshore sediments (Namibia), heavy minerals in dune sands (South Africa), sand, rock, and groundwater. The coast is also a focal point for recreational activities including swimming, surfing, angling, and diving. Indeed, the African coastline is becoming as important as its game reserves and ancient civilizations in attracting tourism. For most recreational activity, sandy beaches are focal points. One third of the coastline is considered to be under threat from developments and other human activities.

Conservation and management

Marine and coastal resources contribute significantly to African economies, especially through fisheries (e.g., Namibia) and tourism (e.g., Mauritius). Current development trends and pressures from increasing urbanization and industrialization are steadily degrading fragile ecosystems. Pollution, mining, and oil exploration are also threatening coastal ecosystems. Oil spills due to well blowouts have caused serious problems in the Niger delta, decimating "black water" biodiversity. In addition, toxic wastes from developed nations have been illegally dumped along the coasts of poor African nations. Many industries dispose of untreated wastes directly into rivers running into the sea. In the Red Sea and in north Africa there is an increasing risk of pollution as over 10^9 tons of oil are transported through the area annually and there are limited maritime traffic regulations. The coastal countries of Africa are also susceptible to the problem of accelerated coastal erosion. This is driven by natural processes that are exacerbated by sea-level rise, upstream construction of dams, other coastal infrastructure, and clearing of mangrove systems.

In recent years, Africa has seen major political, economic, and social changes and has transformed from a rural society to a complex modern region whose ties to natural capital remain strong despite economic development. Population growth (2.8%) is almost twice the global average (1.5%), far in excess of the average rate of economic growth. In many African countries, national parks and conservation areas include parts of the coast as increasing human impact necessitates protection and management. Coastal reserves, which have been established in a number of areas to afford protection or control utilization, need to be expanded, to involve the local populations and to be better managed. Important resources/habitats requiring protection are dunes, mangroves, coral reef, and islands. Integrated coastal zone management, in its infancy or absent in most of Africa, is essential for the future well-being of these spectacular coasts.

Anton McLachlan

Bibliography

Bally, R., 1986. A bibliography of sandy beaches and sandy beach organisms on the African Continent. *South African National Scientific Programmes Report, No 126, Pretoria: CSIR*.

Branch, G.M., and Griffiths, C.L., 1988. The Benguela Ecosystem. Part V. The coastal zone. *Oceanography and Marine Biology Annual Review*, **26**: 395–486.

Brown, A.C., and McLachlan, A., 1990. *Ecology of Sandy Shores*. Amsterdam: Elsevier.

Davies, J.L., 1972. *Geographical Variation in Coastal development*. London: Longman.

Davies, B.R., O'Keefe, J.H., and Sneddon, C.D., 1993. *A Synthesis of the Ecological Functioning, Conservation and Management of South African River Ecosystems*. Pretoria: Water Research Commission.

Por, D.F., 1968. *Lessepsian Migration, Ecological Studies 23*. New York: Springer Verlag.

Sheppard, C., Price, A., and Roberts, C. 1992. *Marine Ecology of the Arabian Region*. London: Academic Press.

Spalding, M., Blasco, F., and Field, C.,1997. *World Mangrove Atlas*. International Society for Mangrove Ecosystems. Yorkshire: Smith Settle, Otley.

Cross-references

Africa, Coastal Geomorphology
Beach Processes
Coral Reefs
Demography of Coastal Populations
Dune Ridges
Estuaries
Human Impact on Coasts
Mangroves, Ecology
Rock Coast Processes
Salt Marsh
Wetlands

AFRICA, COASTAL GEOMORPHOLOGY

The African continent measures 30×10^6 km^2 and its relatively unbroken coastline is 30,000 km long, compared with the 70,000 km coast of Asia, which is only 1.5 times larger than Africa, and the 76,000 km coast of smaller North America (24×10^6 km^2) with its numerous Arctic islands. Over long distances, the African coast is unbroken by sizable inlets, and its major river mouths, except the Congo, are either deltaic or blocked by sand barriers. Excepting Madagascar (587,000 km^2), no large islands lie off the African coast.

Offshore, Africa's continental shelf covers only 1.28×10^6 km^2, compared with 9.39×10^6 km^2 for Asia and 6.74×10^6 km^2 for North America. The shelf averages only 25 km in width, wider off southern Tunisia, Guinea, and major deltas, and reaching 240 km wide across the Agulhas Bank, but narrowing to 5 km off Somalia, northern Mozambique, and Kwa-Zulu. This narrow shelf and paucity of sheltering islands allow deep-water waves and surface ocean currents to approach unmodified close to the mainland shore where they are unusually influential in moving sediment (Orme, 1996).

Explanations for Africa's relatively smooth coastline and narrow continental shelf are to be found in the tectonic processes that triggered the rupture of Gondwana in Mesozoic time and in the geomorphic processes that have shaped the coast more precisely during later Cenozoic time.

Coastal origins

The broad outlines of Africa's coastal margins may be explained in terms of plate-tectonic events over the past 200 Ma (million years before present). Africa's margins were initially blocked out by the rupture of Gondwana in Mesozoic time, and further modified during the progressive opening of the Atlantic and Indian Oceans, the gradual closure of the Tethys Ocean, and the Cenozoic opening of the Red Sea (Figure A6; Orme, 1996; Summerfield, 1996).

Prior to these events, Africa's cratonic nucleus had been fused during the Panafrican–Brasiliano orogeny in later Neoproterozoic time (650–530 Ma) with cratons from other southern continents to form the supercontinent of Gondwana. Later, towards the close of Paleozoic time (330–260 Ma), Gondwana became joined to the supercontinent of Laurussia to form a vast single landmass, Pangea, a suture marked prominently by the Appalachian (Alleghanian) and Atlas collisional orogens of North America and northwest Africa, respectively (Ziegler *et al.*, 1997; Orme, 2002). Compared with Gondwana, however, Pangea did not last long. Following Triassic crustal extension (248–206 Ma), North America began separating from Africa along the Appalachian–Atlas suture and, by mid-Jurassic time (180–160 Ma), an incipient north-central Atlantic Ocean, underlain by oceanic basalts spewing from a spreading center beneath the Canary Islands, lay off northwest Africa (Steiner *et al.*, 1998). The later development of this coastal margin reflects variable rates of sea-floor spreading, epeirogenic flexuring, marginal basin development, and sedimentation.

The continent's western margin farther south was initiated by sequential events, which caused South America to separate from Africa over a period of 100 Ma, from early Jurassic to middle Cretaceous time (Uchupi and Emery, 1991; Milani and Filho, 2000). In the north, following Triassic crustal extension, sea-floor spreading in the early Jurassic (190–180 Ma) introduced ocean waters to the Guinea–Sierra Leone margin. In the south, the Malvinas (Falkland) Plateau, which had formerly wrapped around Africa's southern tip, began shearing westward in the early Jurassic (200–180 Ma), causing the continent's west-facing margin to unzip progressively northward from eastern Brazil. Later, in response to these stresses, right-lateral wrenching along a massive transform system began separating northern Brazil from the south-facing Guinea coast in middle Cretaceous time (120–100 Ma). Throughout this Atlantic margin, initial crustal extension and enhanced magmatism were followed by basin formation along the rifted margin and then, as sea-floor spreading moved Africa farther from the Mid-Atlantic Ridge, by widening oceanic tectonic and sedimentary regimes. Thus, magmatism at a common eruptive center on the Mid-Atlantic Ridge generated early Cretaceous volcanic rocks at Etendeka in northern Namibia and the Paraná flood basalts of southern Brazil, which on subsequent rifting and sea-floor spreading have since separated (Glen *et al.*, 1997). The Liberian coastal margin off Cape Palmas reflects fracture zones that developed around 140 Ma, since when some 8,000 m of sediment have blanketed the shelf and slope. Farther south, the deep Angola and Cape basins developed during Cretaceous time to the north and south of the Walvis Ridge, respectively. South of this ridge, the western continental margin consists of downdropped basement blocks, aligned NNW and overlain by a prograded sediment wedge formed by debris brought down by the Orange River and other streams that have been eroding the Great Escarpment since it was first tilted upward in early Cretaceous time (Dingle and Scrutton, 1974; Summerfield, 1996). Islands such as Ascension and Tristan da Cunha in the central South Atlantic Ocean mark continuing hotspot volcanism on the Mid-Atlantic Ridge at the western edge of the African plate. At Africa's southern tip, the Malvinas Plateau sheared westward from the Hercynian Cape Fold Belt (280–230 Ma) along major transform structures during Jurassic time. Since then, and notably since epeirogenic uplift of the southern Great Escarpment in Cretaceous time, terrigenous debris has crossed the narrow continental shelf to blanket the continental slope and adjacent basins. Farther out, the submarine Agulhas Plateau was probably created by volcanism in fracture zones vacated by the Malvinas Plateau (Barrett, 1977).

Following Permian tensions and Triassic saltwater intrusion, much of Africa's eastern margin was blocked out during earlier Jurassic time (200–160 Ma). In Mozambique, early tensional rifting produced N–S horst and graben that were later buried beneath debris from the Limpopo and Zambezi river systems (Dingle and Scrutton, 1974). The N–S segments of coast south of the Zambezi and Limpopo river mouths reflect these structures, whereas the NE-trending coasts north of these rivers reflect right-lateral offsets of basement cratons, bringing Precambrian rocks close to shore in northern Mozambique. Some 6,000 m of Cretaceous and Cenozoic terrigenous sediment have since accumulated in offshore basins, including the massive Zambezi cone. Madagascar is a large fragment of continental crust that probably lay

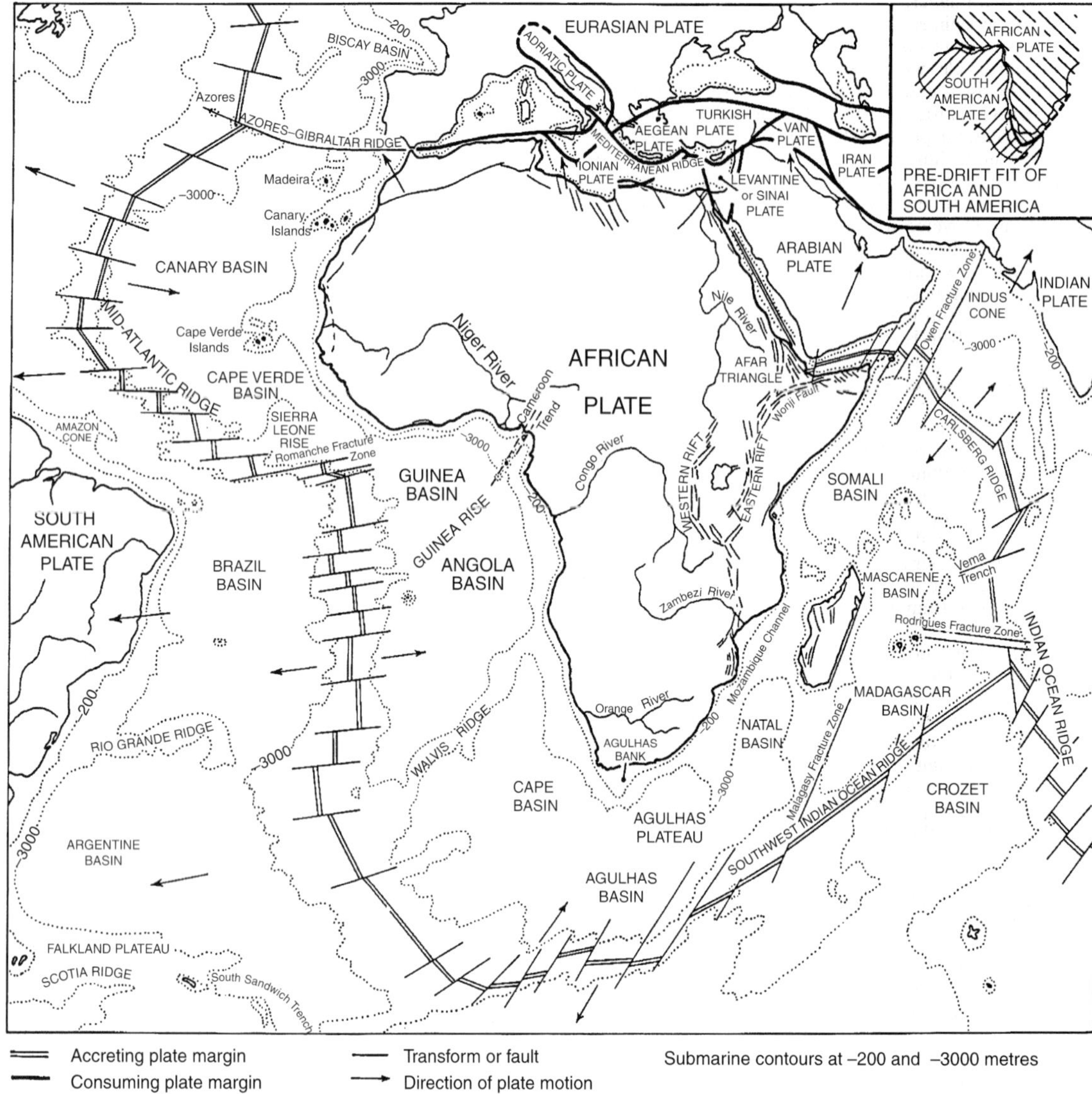

Figure A6 Tectonic relations of the African Plate (from Orme, 1982, with permission of Kluwer Academic Publishers).

alongside western India before the latter began separating from eastern Africa in Jurassic time. Although Madagascar's place in Gondwana has been much debated, it probably translated southward to its present location from Tanzania and Kenya, with the reef-capped Comoro volcanic archipelago extruding along resultant fracture zones (Maugé *et al.*, 1982; Ziegler *et al.*, 1997). Farther north, the southeast margins of Somalia and Arabia also formed from the Jurassic rupture of Gondwana.

The Red Sea and Gulf of Aden coasts reflect relatively late intraplate separation of the African–Arabian portion of Gondwana along extensions of the East African Rift system. In early Cenozoic time, after prolonged stability, the region became the site of intensive magmatic and tectonic activity. By late Eocene time (~35 Ma), vertical uplift had produced a massive Afro-Arabian dome which subsequently broke into Somali, Nubian, and Arabian segments. The Somali segment was bounded to the north and west by crustal flexures and tensional faults along which lateral displacement began in the Miocene, forming the discrete Somali plate. Since then, as the Arabian segment has moved north and east away from the Somali and African plates, new oceanic crust up to 200 km across has formed in the widening Gulf of Aden and Red Sea.

The Mediterranean coastal margin reflects a long and complex relationship with the European portion of the former supercontinent of Laurussia. The rupture of Pangea that began in late Triassic time, sometime before 200 Ma, permitted a large tropical ocean, the Tethys, to intervene between central Europe and north Africa at a time when the African plate lay wholly south of the Equator. As Africa subsequently moved northward and rotated anticlockwise across the Equator, transpressional tectonics reshaped the Tethys into several microplates and convoluted mid-Cenozoic (Atlas–Alpine) collisional orogens, which today form southern Europe and underlie the Mediterranean Sea. The effects of the Atlas orogeny are well revealed along the north African coast between Agadir and Gabes. Farther east, the situation is less clear because regional tectonism involves both transpressional and transtensional interactions between several microplates, activities that yield frequent earthquakes and volcanic events in southern Europe but not in north Africa. In the Aegean subduction zone, for example, the African plate is presently moving under the Aegean microplate at a rate of 2.7 cm yr^{-1}.

In summary, Africa's Atlantic and Indian Ocean coasts reflect divergent or passive plate margins inasmuch as the African plate has diverged from oceanic spreading centers from which the continent is now separated by vast expanses of post-Gondwanan basaltic ocean floor variably mantled with marine and terrigenous sediment. In embryo form, the Red Sea and Gulf of Aden coasts are also divergent margins but continuing crustal adjustments and volcanism belie the term passive. In contrast, the western Mediterranean margin is an active margin, but this categorization is less easily applied farther east.

Coastal and offshore geology

Tectonic origins apart, the present shape of the African coast also reflects the character of the rocks and structures introduced to the coastal zone during and since the breakup of Gondwana. The principal rocks influencing coastal features may be grouped into Precambrian basement, consolidated Phanerozoic platform covers, and poorly consolidated late Cenozoic sediments.

Precambrian rocks outcrop over 57% of Africa's surface, reflecting numerous sedimentary cycles whose deposits were intensely folded and fractured, metamorphosed, and often granitized during at least eight orogenic cycles. Gneiss, schist, quartzite, and migmatite are important but post-orogenic molasse deposits and tabular to strongly folded platform covers of sandstone, limestone, tillite, and volcanics also occur. Today, these basement rocks reach the coastal zone in the Anti-Atlas of Morocco, the Guinea coast between Monrovia and Accra, at intervals from Angola to Cape Province, and again in northern Mozambique, eastern Madagascar, and the Red Sea. Because intense fracturing favors deep weathering, these rocks rarely form high cliffs.

Phanerozoic rocks occur mostly as tabular platform covers occupying large basins between swells in the basement complex, but folded cover rocks form the Cape Fold Belt (Hercynian) of South Africa and the Atlas Fold Belt (Hercynian and Alpine) from Morocco to Tunisia. In Africa as a whole, the great extent of continental deposits and the paucity of marine sediment are noteworthy. The continental rocks include the Nubian Supergroup (Cambrian–Cretaceous) which straddles the Red Sea, and the 7,000 m thick Karroo Supergroup (Carboniferous–Triassic) of southern Africa whose tillites, shales, sandstones, and basalts form bold escarpments inland and sea cliffs at the shore.

In contrast, late Cenozoic sediments within the coastal zone are mostly marine sediments or fluvial deposits reworked by waves, currents, and winds. These materials are often poorly consolidated and thus erodible but, except in the Mozambique coastal plain, rarely extend far inland. More resistant late Cenozoic rocks include coral-reef limestones and aeolianites.

Offshore, there are about a dozen major basins around Africa, formed following the rupture of Gondwana and now partly filled with late Mesozoic and Cenozoic sediment. Where these basins straddle the present coast, both marine and terrigenous sedimentation have occurred, leading to epeirogenic seaward subsidence of the coastal margin which in turn favors more sedimentation (Orme, 1996). The contrast between sedimentation in these basins and denudation in their hinterland has caused significant isostatic responses to loading and unloading, leading, for example, to 600 m of uplift inland from the southwest coast and adjustments to the Orange River below Augrabies Falls (Gilchrist and Summerfield, 1990). Continued deformation and volcanism have complicated coastal evolution, notably in coastal extensions of the 7,000 km long East African Rift system and the Cameroon volcanic line.

Moving clockwise from Ras Asir at the Horn of Africa, the first basin encountered is the 7,000 m deep Somali Basin which merges south with the 8,000 m deep Kenya and 3,000 m deep Dar es Salaam Basins. When the Somali Basin formed along Gondwana's rifted margin, marine transgressions and carbonate deposition followed, but late Jurassic uplift of the Bur basement complex along NE-trending faults separated the interior of this basin from continuing carbonate deposition at the coast. Cenozoic rejuvenation of these faults has largely defined the east coast from Ras Asir to Tanzania (Orme, 1985, 1996). Farther south, the 4,000 m deep Mozambique Basin has also been broken by N–S faults at the south end of the East African Rift system. The 7,000 m thick sedimentary cover of the Madagascar Basin has also been severely faulted.

Along the west coast, a more or less continuous embayment of folded sediments and salt domes is represented by the 4,000 m deep Luanda Basin, the 3,000 m deep Cabinda Basin, and the 8,000 m deep Gabon Basin. The Niger Basin, the seaward extension of the Benue graben, contains up to 10,000 m of Cretaceous and Cenozoic deposits, including important petroleum resources beneath the Niger delta. The narrow Ivory Coast Basin is faulted to depths of 4,000 m and the Senegal Basin reaches depths of 7,000 m at the coast. The Tarfaya Basin farther north, bounded inland by the Zemmour Fault, contains 10,000 m of Mesozoic and Cenozoic deposits beneath Cape Juby.

Early evolution of the Mediterranean coast differed from other coasts because sediments in its developing marginal basins became involved in mid-Cenozoic orogeny, as seen in the mountain arcs and deep sedimentary basins of the Algerian coast. Farther east, the 5,000 m deep Tripoli Basin is separated by NW-trending faults from the extensive 8,000 m deep Sirte Basin of Cyrenaica.

Relative sea-level change

Since the rupture of Gondwana, continuing tectonic, isostatic, and eustatic forces have left their imprint on Africa's emerging coastal margins. For example, sea-level relations have been strongly impacted by Atlas tectonism along the continent's northwest margin during and since mid-Cenozoic time, while late Cenozoic rifting has impacted the Red Sea coast. Isostatic adjustments of Africa's plate margin to denudational unloading and sediment loading have also affected sea levels. Further, the eustatic high sea levels of mid-Cretaceous time caused shallow seas to transgress north Africa and link the Tethys with the Gulf of Guinea. After Paleocene regression, Eocene seas again flooded the Sahara and a shallow gulf persisted into Miocene time (~20 Ma) in Libya and Egypt.

Relative sea-level changes over the past few million years, and especially for the Quaternary, have been the focus of much research, in part because they aid prediction of future coastal behavior. In general terms, tectonism, isostasy, and eustasy may have combined locally to force Quaternary sea-level changes ranging from 300 m above to 200 m below present levels. Along the Moroccan coast, for example, a deformed Pliocene marine surface up to 20 km wide, the Moghrebian rasa, has been raised between 100 and 600 m above sea level, providing a useful measure of Alpine tectonism (Weisrock, 1980). Flights of deformed Pleistocene coastlines also occur along the western Mediterranean coast, notably at Algiers, Bizerte, and Monastir, while raised coral reefs occur along deformed Red Sea and northern Somali coasts. Elsewhere, the stability of much of the African plate relative to high last interglacial seas (~125 ka) is indicated by little deformed coastlines and coral reefs a few meters above present sea level, notably along southern coasts (Orme, 1973; Hobday and Orme, 1975). Offshore, former sea cliffs, beaches, aeolianites, and river valleys, now submerged on the continental shelf, testify to low late Pleistocene sea levels (Orme, 1974, 1976).

The last major global sea-level change, the Flandrian transgression which accompanied the melting of the last Pleistocene ice sheets and culminated around 5 ka (thousand years before present), initiated the present phase of coastal erosion and deposition. Locally, this transgression may have risen a few meters above present sea level, as suggested by the Nouakchottian deposits in Mauritania and Senegal, but such evidence for seemingly high Holocene sea levels may also be explained by local epeirogenic forcing or by climate-induced changes in hydrology and sediment delivery (Ausseil-Badie *et al.*, 1991). Relative sea-level rise during historic times is reflected in submerged Phoenician tombs and Roman ports along the Mediterranean coast, by submerged Arab legacies along the east coast, and more recently by tide-gauge records.

Coastal processes

Climatic factors

Directly or indirectly, climate affects most processes shaping the coast. As elsewhere, Africa's coastal climates reflect the impact of seasonal changes in Earth–Sun relations on the Intertropical Convergence Zone (ITCZ), air masses, and wind regimes. During the southern summer, the ITCZ shifts southward with the Sun to a zone running from the Guinea coast to Madagascar. Excepting the influx of cool moist maritime polar air masses from the Atlantic into the Mediterranean, most of north Africa is covered by warm dry outflowing air associated with continental tropical air masses over the Sahara and southwest Asia, promoting the dust-laden Harmattan over the Guinea coast and the northeast monsoon along the east coast, respectively. During this, the *jilaal* season in Somalia, warm northeasterly winds up to 30 km hr^{-1} drive ocean waters and aeolian sands southward along the coast (Orme, 1985). Farther south, warm rainy conditions and onshore winds linked to inflowing marine air or local thermal convection cells prevail.

During the northern summer, the ITCZ shifts northward to the Sahel. Hot wet conditions prevail over central Africa and the Guinea coast, Saharan aridity reaches the Mediterranean, and, although much of southern Africa is dry, cool maritime air masses and storms from the Southern Ocean impact the southern coast. This, the *hagaa* season in Somalia, sees southwest monsoon winds with mean velocities over 40 km hr^{-1} move aeolian sand and surface waters northward along the east coast, while the hot desiccating dust-laden *kharif* wind invades the Gulf of Aden from northern Somalia.

Oceanic factors

Excepting the Mediterranean and northwest coasts where northwesterly swells predominate, the African coast is dominated by southerly swells generated by storms in the Southern Ocean (Figure A7). From Cape Vert to Cape Agulhas, these swells are mostly southwesterly, decreasing in height northward and from southern winter into southern summer. Along the Guinea coast, these swells are reinforced by southwesterly monsoon winds during the northern summer. Because of the orientation of Africa's west coast, the southwesterly swells generate mainly east-flowing longshore currents in the Gulf of Guinea and mainly north-flowing longshore currents along the west-facing coast farther south. This nearshore wave-driven circulation is reinforced

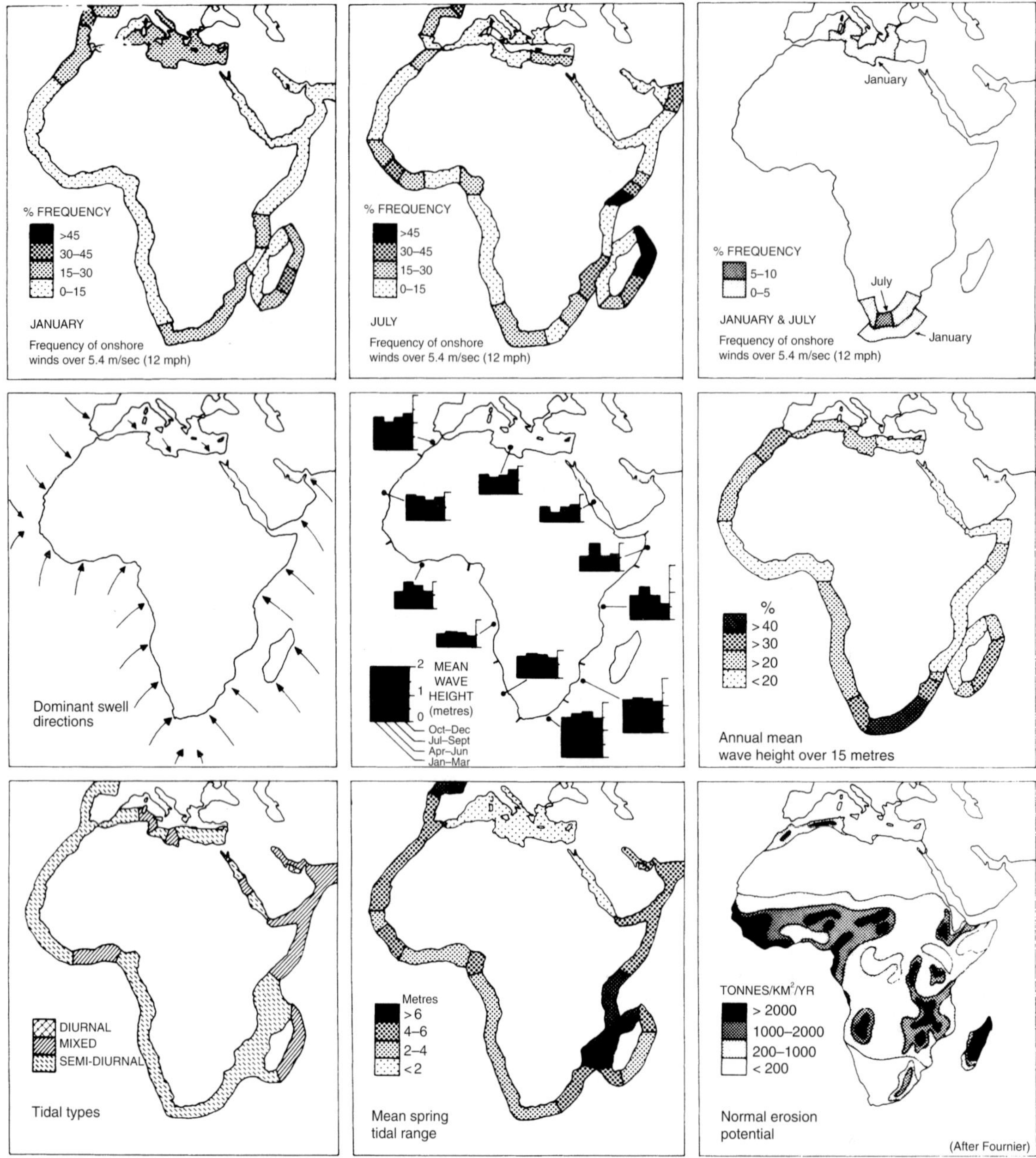

Figure A7 Selected processes affecting the African coast (in part after Hayden *et al.*, 1973; Davies, 1977; from Orme, 1982, with permission of Kluwer Academic Publishers).

offshore by the Guinea and Benguela Currents. Exceptions to these patterns occur in the lee of major promontories such as the Niger delta. Throughout the Atlantic coast, wave erosion and longshore sediment transport are strongest and most frequent when swells are highest—from December to March north of Cape Vert, and from June to September farther south (Davies, 1977).

Along the east coast, from Cape Agulhas to Ras Asir, most swells also originate from the southwest but are strongly refracted to reach the coast from the southeast (Figure A7). This circulation is further complicated by reversing monsoonal winds and currents off Somalia, by westward-moving cyclones and tropical easterlies off Madagascar, and by the south-flowing Agulhas Current farther south. Under the influence of the southwest monsoon, for example, the Somali Current reaches surface velocities of $3.7\ \mathrm{m\ s^{-1}}$ and the uppermost 200 m of its water column transports $60 \times 10^6\ \mathrm{m^3\ s^{-1}}$ northward. Along the KwaZulu coast, littoral transport of coarser terrigenous sediment is commonly northward within the surf zone, but finer sediment that is flushed farther seaward is carried south with the Agulhas Current (Orme, 1973). Onshore winds are stronger and more frequent during the southern winter, when winds exceeding $60\ \mathrm{km\ hr^{-1}}$ may drive powerful surf against the continent's southern tip.

Highest wave energy occurs around the southern tip, from Cape Point to Durnford Point, and moderately high wave energy prevails from the Orange to Limpopo river mouths, along Madagascar's east coast, and along Morocco's Atlantic coast (Figure A7). Lower wave energy occurs in the Red Sea, the Mozambique Channel, the Gulf of Guinea, and eastern Mediterranean.

Semidiurnal tides predominate around Africa, with mixed regimes confined to the central Mediterranean, Red Sea, Guinea, Somali, and eastern Madagascar coasts (Figure A7). Mean tidal range is less than 2 m throughout the Red Sea and Mediterranean, over 6 m along the Mozambique Channel, and 2–6 m elsewhere.

Hydrologic factors

The continent's hydrologic regimes are important because they deliver freshwater and sediment to the coast, which in turn influence coastal ecology and sediment budgets (Orme, 1996; Walling, 1996). Sediment delivery is highest in areas of rugged relief and/or high rainfall, at least seasonally (Figure A7). Thus, the many small rivers that drain steeply from the mountains of Algeria and Morocco to the Mediterranean may have only modest sediment loads but large sediment yields per unit area of their basins, often >1,000 tonnes (t) $km^{-2} yr^{-1}$ (Milliman and Syvitski, 1992). Sediment yields may also be high where open forest or savanna vegetation has been modified or destroyed by human activity, for example, along the coast of Guinea, east Africa, and Madagascar. Sediment yields are lower where luxuriant vegetation inhibits surface erosion or where rivers have long run-out reaches across coastal lowlands, and lowest where heavy rainfall is lacking, notably in desert areas.

Most larger rivers have lower sediment yields because they often cross different climatic regimes and broad floodplains, both of which afford opportunities for floodplain storage. However, because their drainage basins are large, they may deliver substantial sediment loads to the sea annually. Thus, the Congo and Limpopo, respectively, carry average loads of 43×10^6 t and 33×10^6 t to the sea annually (Milliman and Syvitski, 1992). Of particular importance to the present coast are the reduced sediment loads in rivers that have been dammed upstream. Thus the Orange, which formerly transported 89×10^6 t of sediment annually, now moves only 17×10^6 t to the sea. The Zambezi, which once transported 48×10^6 t annually, now carries only 20×10^6 t, largely owing to construction of the Cabora Bassa Dam upstream. Most dramatically, annual sediment loads from the Nile and Volta, formerly 120×10^6 t and 19×10^6 t, respectively, have diminished to virtually zero as a result of construction of the Aswan High Dam (completed 1964, capacity $162 \times 10^6 m^3$) and earlier structures on the Nile, and of the Akosombo Dam (1961) on the Volta. These structures have major consequences for coastal sediment budgets.

Despite high sediment yields from seasonally wet areas and high sediment loads from many larger undammed rivers, Africa's vast internal drainages and the relatively aridity and low gradient of most external basins restrict sediment yields to the coast (Walling, 1996). Thus, compared with other continents, direct wave attack on coastal cliffs and carbonate reefs is a more important source of sediment. Further, the wide extent of seasonally dry climates means that, for much of the year, ocean processes predominate over fluvial processes around river mouths. Thus, many mouths are closed during drought and partially blocked by coastal barriers at other times.

Human factors

Although the African coast has suffered less from human activity than that of Europe or North America, some historic impacts are noteworthy and the rate of human-induced change has accelerated in recent decades. Apart from the construction of major dams inland, noted above, land-use practices inland have often led to extensive soil erosion and increased sediment yields (Stocking, 1996). In Kwa-Zulu, for example, recent sediment yields are two to three times the average for the past 5,000 years, owing in large measure to overgrazing and accelerated erosion in coastal basins (Orme, 1973). Along many parts of the African coast, increased sediment yields have clogged estuaries while sediment plumes have been worked onshore to form new beach-dune ridges. Conversely, along Somalia's east coast, Pleistocene dunes have been reactivated and gullied by overgrazing of vegetation by cattle, goats, and camels, aggravated by tillage practices of an increasingly desperate refugee population (Orme, 1985). Elsewhere, beaches and dunes have been mined for diamonds and heavy minerals, respectively, while beaches and wetlands near major ports have been consumed by harbor and urban development. Pollution and recreation are also beginning seriously to impact coral reefs in east Africa.

The Mediterranean coast

Africa's 5,000-km long north coast is divisible into two contrasting segments (Figure A8). The 3,300 km eastern segment from Sinai to the Gulf of Gabes features low cliffs of little deformed Cenozoic sedimentary rocks interspersed with barrier beaches, lagoons, and dunes. The 1,700 km western segment from Cape Bon to Cape Spartel reflects Alpine tectonism in its rugged coastal mountains and deep coastal basins. Throughout this coast, prevailing northwest winds generate northwesterly swells and wind waves, strongest in winter, which in turn promote net littoral drift eastward. With a mean tidal range of 0.2–0.5 m, but 1.8 m in the Gulf of Gabes, tidal forcing is weak. Wave energy, littoral drift, annual rainfall, and fluvial sediment inputs, diminish eastward.

The 12,500 km^2 Nile delta fronts 280 km of the Egyptian coast. Although fluvial sediment rich in Fe and Al from the Ethiopian Highlands may still be found in beaches east of the delta, water and sediment delivery by the Nile have been much reduced by irrigation diversions and dams built since the 1880s. Of the six or more distributaries used by the Nile some 2,000 years ago, only the Damictta and Rosetta branches still function. Further, because longshore currents continue to move sediment eastward, the delta's seacoast is suffering severe erosion, at rates up to 60 m yr^{-1}, leaving high-density magnetite, ilmenite, zircon, and garnet as black-sand placers while moving lighter minerals downdrift (El-Ashry, 1985; Frihy and Komar, 1991). A 40 km deep littoral zone, the *barari* (barren) fronts the delta, comprising barrier beaches and back-barrier marshes, salt flats, and lagoons of which the largest are Manzala (1,450 km^2) and Burullus (560 km^2). Over the past 8,000 years, delta subsidence has ranged from 1 mm yr^{-1} near Alexandria to 5 mm yr^{-1} near Port Said (Stanley, 1990). During historic time, while delta subsidence was offset by sedimentation, coastline location did not change much but, with sediment delivery now curtailed, several areas have subsided to below sea level. Farther east, the Sinai coast is a broad sandy plain with bold dunes and a 100 km long, 500 m wide barrier beach fronting Bardawil Lagoon along whose shallow hypersaline margins gypsum and halite readily precipitate. The ephemeral Wadi el Arish is the only sizable stream course to reach this coast.

Westward from the delta, dark Nile sediment is replaced by carbonate grains from local sources. Several beach-dune ridges backed by elongate sabkhas parallel the coast, reflecting episodic barrier formation during late Quaternary time. In Cyrenaica, the Jebel el Akdar limestone plateau generates higher rainfall and greater surface runoff, and reaches the coast in rocky cliffs and narrow terraces that leave little room for coastal plain. From Benghazi to Cape Bon, the coast is again dominated by barrier beaches, dunes, sabkhas, and salt marshes. The 150 km^2 Sabkha el Melah in southern Tunisia exemplifies Holocene evaporite sedimentation in which marine detrital limestones, euxinic beds, magnesium carbonates, gypsum, polyhalite, and halite have successively accumulated over the past 6,000 years (Busson and Perthuisot, 1977).

From northern Tunisia westward, coastal character changes as mid-Cenozoic (Atlas) and neotectonic structures form prominent cliffs, marine terraces, and subsiding basins, and as the seasonally wetter, stormier climate mobilizes coastal and fluvial sediment. Quaternary tectonic and eustatic forces are reflected in suites of deformed marine terraces rising to over 200 m above sea level near Monastir and Bizerte, and in 400 m of littoral sediments beneath the subsiding Medjerda delta whose outlet has changed frequently in response to historic floods (Paskoff, 1978). Farther west, along the Algerian Tell and Moroccan Rif, structural basins floored with Quaternary sediment open through swampy lowlands into Skikda, Bejaia, and Arzew Bays. Wadis discharge abundant sediment to the coast in winter, to be partly reworked onshore in aeolian dunes. The Isser and Soummam rivers have a mean yield of 6.1×10^6 and 4.2×10^6 t yr^{-1}, respectively, while sediments carried by the Moulouya from the High Atlas contribute to impressive shingle beaches east of Nador, Morocco. Alpine massifs up to 1,000 m above sea level reach the shore in active sea cliffs involving crystalline basement, Mesozoic limestones and sandstones, and Cenozoic molasse and volcanic rocks, and then descend across a narrow shelf down the 2,000 m Habibas submarine escarpment.

The Atlantic coast

The Atlantic coast of Africa may be divided into three segments: the semiarid to arid coast from Cape Spartel to Cape Vert characterized by the cool Canaries Current, northwesterly swells and south-flowing littoral drift; the humid tropical coast from Cape Vert to beyond the Congo River, dominated by the Guinea Current, southerly swells and east-flowing littoral drift; and the sub-humid to arid coast from Angola south to the Cape characterized by the cool Benguela Current, southwesterly swells and strong north-flowing littoral drift (Figure A9).

From Cape Spartel to Cape Dra, the Moroccan coast is underlain by Hercynian and mid-Cenozoic structures that have been planed off by later marine erosion, forming the Pliocene Moghrebian *rasa* and sequences of Pleistocene marine terraces, all since deformed. The continental margin offshore has seen episodic sedimentation and subsidence since the break-up of Gondwana. Winter floods transport large sediment loads from the Atlas Mountains down the Sebou, Rharb, Sous, and other rivers to feed beach-dune complexes up to 80 m high which, shaped by northwest winds in winter and northeast trade winds in summer, enclose many lagoons and sabkhas.

From Cape Dra to Cape Vert, the coast overlies the Tarfaya and Senegal structural basins whose post-Gondwana cover rocks, planed by episodic marine transgressions, form low coastal cliffs and a relatively smooth continental shelf. Saharan sands blanket this coast along a broad front. Extensive late Cenozoic dune systems show that aeolian processes have long dominated the region. Saharan sands are driven seawards by the

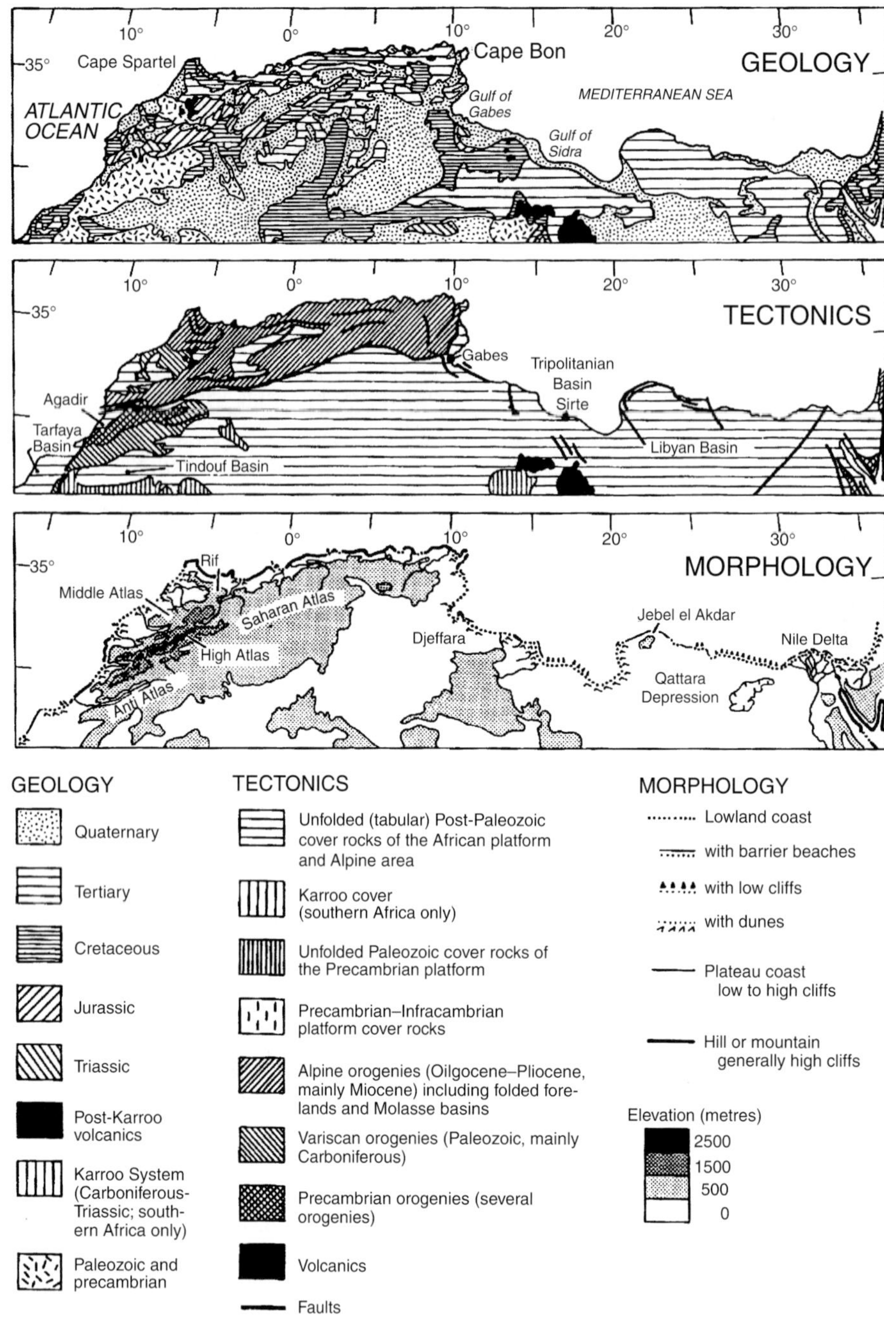

Figure A8 The Mediterranean coast: geology, tectonics, morphology (from Orme, 1982, with permission of Kluwer Academic Publishers).

irifi, a hot easterly wind, to form the great red *draas* and superimposed barchans that trend obliquely across the Mauritanian coast between Cape Blanc and the Senegal River, notably in the Azefal and Akchar ergs (Kocurek *et al.*, 1991). The Sahara now contributes 13×10^6 m^3 of quartz sand to the continental shelf annually, but this amount may have been much larger in the past (Sarnthein and Walger, 1974). Late Pleistocene aeolianites occur to at least −50 m off Cape Vert while Saharan dust is found in deep ocean sediments far beyond the shelf. Coastal sands are reworked onshore by the *alisio*, a persistent northerly wind, to form dunes ridges up to 2,000 m wide and 40 m high, which in turn trap sabkhas (*niayes*) seaward of Saharan draas in Mauritania and Senegal. Lacking rainfall and a mountain backdrop, stream flows and fluvial sediment discharge are now negligible, except during rare rain storms, along the coast between Cape Dra and the Senegal River, but relict fluvial sediments on the continental shelf, including detrital phosphatic placers derived from onshore phosphorites, show that fluvial processes were more important during Pleistocene time.

The 4,250 km^2 Senegal delta forms the outlet for a 196×10^3 km^2 drainage basin whose mostly summer rains average 1,381 mm annually. Despite considerable evaporative loss, annual discharge through the delta averages 870 m^3 s^{-1}, ranging from negligible during prolonged drought to 3,700 m^3 s^{-1} during floods (Guilcher, 1985). The delta formerly entered a bay created by the Flandrian (Nouakchottian) transgression but, as the delta plain accreted, marine processes deflected the river mouth 120 km southward behind a massive barrier spit. The distal end of this barrier, the unstable 30 km long Langue de Barbarie, has been breached 25 times by storm waves or river floods since 1850. Although spring tidal range is only 1.2 m, the low gradient of the Senegal River allows tidal influences to penetrate over 400 km upstream. Southward littoral drift along the open coast ranges from 0.5×10^6 to 1.0×10^6 t yr^{-1}, part of which is lost down Kayor submarine canyon off Cape Vert.

From Cape Vert to the Niger delta, cliffed coasts formed in Precambrian basement and Phanerozoic sedimentary rocks alternate with lowlands underlain by Cretaceous and Cenozoic sediments and

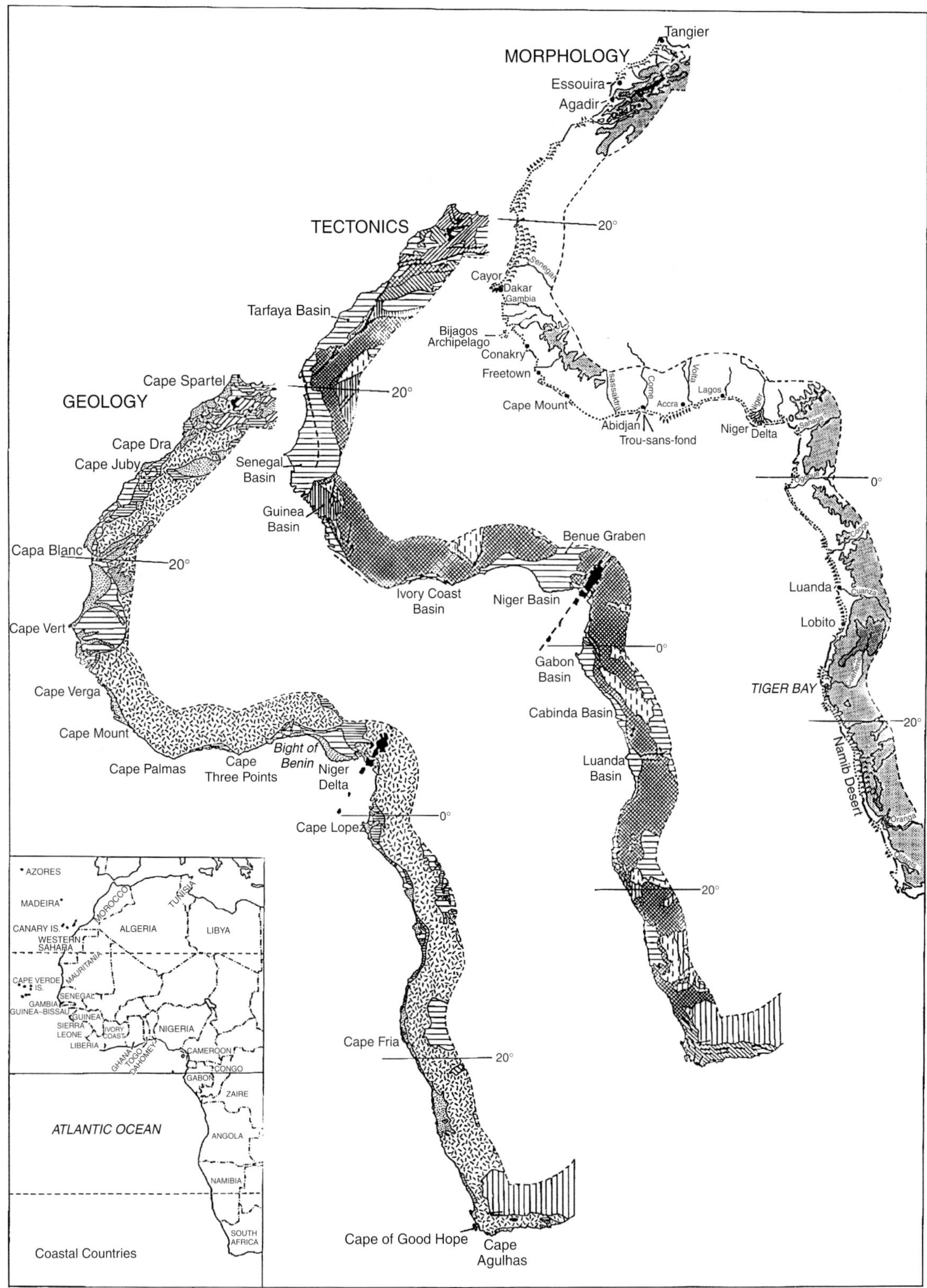

Figure A9 The Atlantic coast: geology, tectonics, morphology (see Figure A8 for key) (from Orme, 1982, with permission of Kluwer Academic Publishers).

fronted by beach ridges and mangrove swamps. Paleozoic rocks emerge along the south coast of Guinea-Bissau and at Cape Verga; basic intrusives form the Kaloum and Freetown peninsulas; Precambrian granite and gneiss underlie the coasts of Liberia and western Ivory Coast, and reappear with Paleozoic sandstone in low cliffs from Cape Three Points nearly to the Volta delta. In contrast, the subsiding and downfaulted Ivory Coast Basin and the larger Niger Basin are underlain by great thicknesses of Cretaceous and Cenozoic sediment overlain by coastal barriers, lagoons, and swamps.

The southwest-facing coast between Cape Vert and Cape Palmas, tropical rains of 2,000–3,000 mm yr^{-1}, enhanced by inland relief and southwest winds drawn onshore during the northern summer, is reflected in abundant discharge of water and sediment from short coastal rivers. Broad deltaic estuaries, such as those of the Casamance, Geba, Konkoure, and Scarcies rivers, reach the sea through intricate networks of tidal creeks, mangrove swamps, and supratidal mudflats or *tannes*. While most estuaries remain open throughout the year, barrier beaches locally front coastal swamps and lagoons in southern Sierra Leone and Liberia. Modern beaches and Holocene beach ridges farther inland, notably on Sherbro Island and the Turner Peninsula testify to abundant sediment supplies, powerful onshore swells, and seasonally reversing littoral drift. Offshore, the continental shelf ranges from 200 km wide on the shallow Great Geba Flat around the Bijagos Archipelago to < 20 km wide at Cape Palmas. This shelf contains many submerged river channels, deltas, coral reefs, and shoreline sequences to depths of −90 m, all evidence of low Pleistocene sea levels (McMaster *et al.*, 1971). Fine sediment is readily swept southward across shelf by the surface Canaries Current and tidal ebb currents, and then returned northward by strong bottom countercurrents.

East from Cape Palmas, the south-facing Guinea coast causes southwesterly swells to promote east-flowing littoral drift, aided for fine sediment by the Guinea Current. Littoral drift reflects sediment availability. Thus, resistant Precambrian rocks east of Cape Palmas and again east of Cape Three Points restrict littoral drift to around 0.2×10^6 m^3 yr^{-1} but, where Cenozoic basinal and recent fluvial sediments reach the coast east of Fresco and again east of Accra, littoral drift reaches 0.8×10^6 m^3 yr^{-1}, forming massive barrier beaches across the mouths of many rivers. Where sediment supplies have been curtailed, notably by construction of the Akosombo Dam (1961) across the Volta River, or where littoral drift is interrupted, for example, by jetties off the Vridi Canal and Lagos Harbor, downdrift beach erosion occurs. Offshore, the continental shelf is relatively narrow (10–70 km) and featureless but exceptions occur in the Trou-sans-Fond, a massive submarine canyon that heads at −10 m off Abidjan but descends to −2,000 m and traps 4,000,000 m^3 yr^{-1} of littoral drift; the large Volta submarine delta; and submerged beach–aeolianite ridges that indicate a late Pleistocene regression to the shelf edge at −120 m.

The Niger delta is the outlet for a 1.11×10^6 km^2 drainage basin whose main artery, the 4,460-km Niger River rises in the Guinea Mountains only 250 km from the Atlantic. Following mid-Cretaceous separation of South America from Africa in this locality, and subsequent crustal adjustments and sea-level changes, the delta began forming in Eocene time when the coastline lay near Onitsha, 250 km inland from the present coast. The basin began acquiring its present form during a later Pleistocene pluvial stage when the NE-flowing Soudanese Niger overflowed into the SE-flowing Nigerian Niger. The Flandrian transgression later drowned the outer 50 km of the delta's late Pleistocene distributary system. The delta now comprises 500,000 km^3 of Eocene–Holocene sediment, up to 8 km thick, and covers 28,827 km^2, of which two-thirds lie above sea level (Hospers, 1971). Present annual discharge of freshwater into the delta is about 200×10^9 m^3 and of sediment about 18×10^6 m^3, mostly as silt and clay that radiate unequally through a dozen major distributaries into the Gulf of Guinea. Depositional environments and sedimentary facies are distributed concentrically within the delta (Allen, 1970). The deltaic plain comprises upper and lower floodplains, tidal creeks, and mangrove swamps fronted by barrier beaches. The submerged delta comprises river-mouth bars, delta-front ramps, pro-delta slope, open shelf, and relict Pleistocene littoral deposits. Prevailing southwesterly swells strike symmetrically against the nose of the delta, promoting divergent littoral drifts east toward the Cross River and west toward the Benin River. Submarine canyons channel about 1×10^6 m^3 of sediment from these drifts into fans beyond the delta foot. The delta mass is broken by arcuate gravity faults downthrown toward the gulf and also affected by compaction and subsidence of its constituent sediments, aggravated in recent years by extensive subsurface oil and gas withdrawal. Subsidence and mangrove clearance have in turn increased erosion to 40 m yr^{-1} along the exposed muddy coast northwest of the Benin River (Ebisemiju, 1987).

The west-facing coast of Africa south of the Mount Cameroun volcanic pile (4,070 m) comprises several post-Gondwanan basins containing Cretaceous and Cenozoic sedimentary and volcanic, rocks separated by Precambrian basement, all variably mantled by recent marine, fluvial, and aeolian deposits. The Niger Basin extends through the Cameroun volcanics into the swampy and deltaic Sanaga coastal plain. The Gabon Basin is fronted by barrier-lagoon complexes and north-oriented spits terminating at Cape Lopez. In contrast, the Cabinda Basin has active sea cliffs north and south of the Congo estuary. South of Precambrian outcrops near Ambriz, the Luanda Basin is fronted in part by 100–150 m sea cliffs cut in Cretaceous and Cenozoic rocks and in part by *restingas*, north-oriented barrier spits enclosing bays and lagoons, notably the Luanda spit (12 km long), Lobito spit (9 km long), and the spit enclosing Tiger Bay (37 km long, before breaching in 1962; Guilcher *et al.*, 1974). These barrier spits form in response to powerful southwesterly swells and predominant north-flowing littoral drift. However, littoral processes are incapable of closing the mouth of the Congo River where powerful discharge annually jets up to 50×10^6 m^3 of sediment seaward into a submarine canyon, denying most of this load to the littoral drift. This submarine canyon descends to −2,700 m before spilling across an abyssal fan with leveed distributaries to −4,900 m.

Farther south, although Precambrian basement and Cretaceous basalts form cliffs locally, the coast from southern Angola through Namibia to northern Cape Province is largely mantled by relict and active coastal dunes, including the Kunene Sand Sea and the 400 km long Namib Sand Sea (Lancaster, 1996), and by relict alluvial fans. The perennial Kunene River along Namibia's northern boundary sheds much terrigenous debris onto the continental shelf. Similarly, the Orange River along the country's southern border, though blocked by sand during the dry season, has long been the source of the beach and dune sand that moves north with the dominant southwesterly winds and swells, prograding the coast and deflecting the mouths of the Swakop and Kuiseb rivers. The diamantiferous marine-terrace deposits of the Sperrgebeit, ranging from 90 m above to −60 m below sea level, originate from switching late Cenozoic outlets of the Orange River. The shelf here varies from <30 km wide off rocky coasts to >160 km wide where terrigenous Orange River sediments have built a broad submarine delta. Where aeolian sand seas block lesser streams, however, shelf sediments are composed mainly of diatomaceous muds, foramineriferal sands, and skeletal clasts.

The southern tip

Between the Olifants and Tugela rivers, strong southerly swells and winter storms have fashioned a rugged cliffed coast interspersed with sandy bays from Precambrian and Paleozoic rocks of the Cape Fold Belt (280–230 Ma) and the thick formations of the Karroo Supergroup. Thus, massive sandstone-on-granite cliffs of the Cape Peninsula are separated from fold mountains farther east by the broad dune-mantled Cape Flats. From Cape Agulhas to Algoa Bay, the coast cuts across the easterly strike of the Cape Fold Belt, causing cliffed anticlinal ridges to alternate with synclinal corridors fronted by barrier beaches and extensive dunefields. In the Alexandria Basin, a transverse dunefield is moving inland along a 50 km wide front driven by WSW winds off Algoa Bay. Sand enters this dunefield at a rate of 375,000 m^3 yr^{-1} and leaves at 45,000 m^3 yr^{-1} for a net gain of 330,000 m^3 yr^{-1} (Illenberger and Rust, 1988). Farther northeast, gently tilted rocks of the Karroo Supergroup are exposed in majestic cliffs fronting coastal plateaus.

Quaternary marine terraces rise at intervals to over 60 m along this coast, while beachrock, related to both high interglacial and Holocene sea levels forms a broken pavement along much of the immediate shore. Beachrock typically consists of quartz sand and shell fragments cemented by micrite or aragonite overlain by laminated calcrete. Relict beachrock and aeolianite are often exposed along the coast and also form submerged ridges offshore (Figure A10). Off Cape Agulhas, the continental shelf forms the 240 km wide Agulhas Bank, relatively featureless except for the shelf edge between −110 and 380 m which is cut by several canyons. The bank is formed from over 6,000 m of post-Gondwanan sediments lying on basement rocks and, during low Pleistocene sea levels, was exposed over 22,500 km^2 (Dingle, 1973).

Compared to Africa's great rivers, the Gourits, Sundays, Great Kei, Mtamvuna, and other rivers are relatively short but, in descending swiftly from the Cape fold mountains and the Great Escarpment, they transport large sediment loads seasonally (Figure A11). In the Mediterranean-type climate of Cape Province, maximum flows occur in the southern winter (May–August) but farther north, in Natal-Kwa-Zulu, summer (December–March) thunderstorms are more important. Rivers draining the bold Drakensberg (3,482 m) enter the sea through Pleistocene channels excavated to −100 m below sea level, now partly filled with Holocene estuarine sediment (Orme, 1974, 1976; Cooper, 1993). Over the past 200 years, sediment yields have been accelerated by soil erosion but diminished by dam construction.

The Indian Ocean coast

Africa's mostly low-lying east coast, from the Tugela estuary to Ras Azir, straddles post-Gondwanan coastal basins filled with several thousand meters of Mesozoic and Cenozoic sediments the 4,000 m deep Mozambique Basin in the south, characterized by north–south faults at the south end of the East African Rift system, and the 3,000–8,000 m

Figure A10 Pleistocene raised beach deposit overlying Pleistocene aeolianite, Durban Bluff, South Africa (from Orme, 1982, with permission of Kluwer Academic Publishers).

Figure A11 Mtamvuna river mouth and wet (summer) season sediment plume, Transkei-Kwa-Zulu border (photo: A.R. Orme).

deep Dar es Salaam–Kenya–Somali Basin continuum farther north (Figure A12).

The onshore portion of the Mozambique Basin forms a coastal plain 300 km wide in central Mozambique but pinches out southward near the Tugela estuary and northward at Mocambo Bay. Its 2,100 km-long shoreline is characterized by long barrier beaches topped by parabolic and lobate dunes up to 180 m high, behind which lie shallow lagoons, extensive swamps, old coastal barriers, and extensive aeolian cover sands (Orme, 1973). In Kwa-Zulu, the massive coastal barrier, although veneered with Holocene sands, is superimposed upon remnants of a major late Pleistocene barrier-lagoon complex, the Port Durnford Formation, whose nearshore, beach, and swamp facies were probably deposited during the last interglacial around 125 ka (Figure A13; Hobday and Orme, 1975). A similar scenario seems likely along much of the Mozambique coast. The

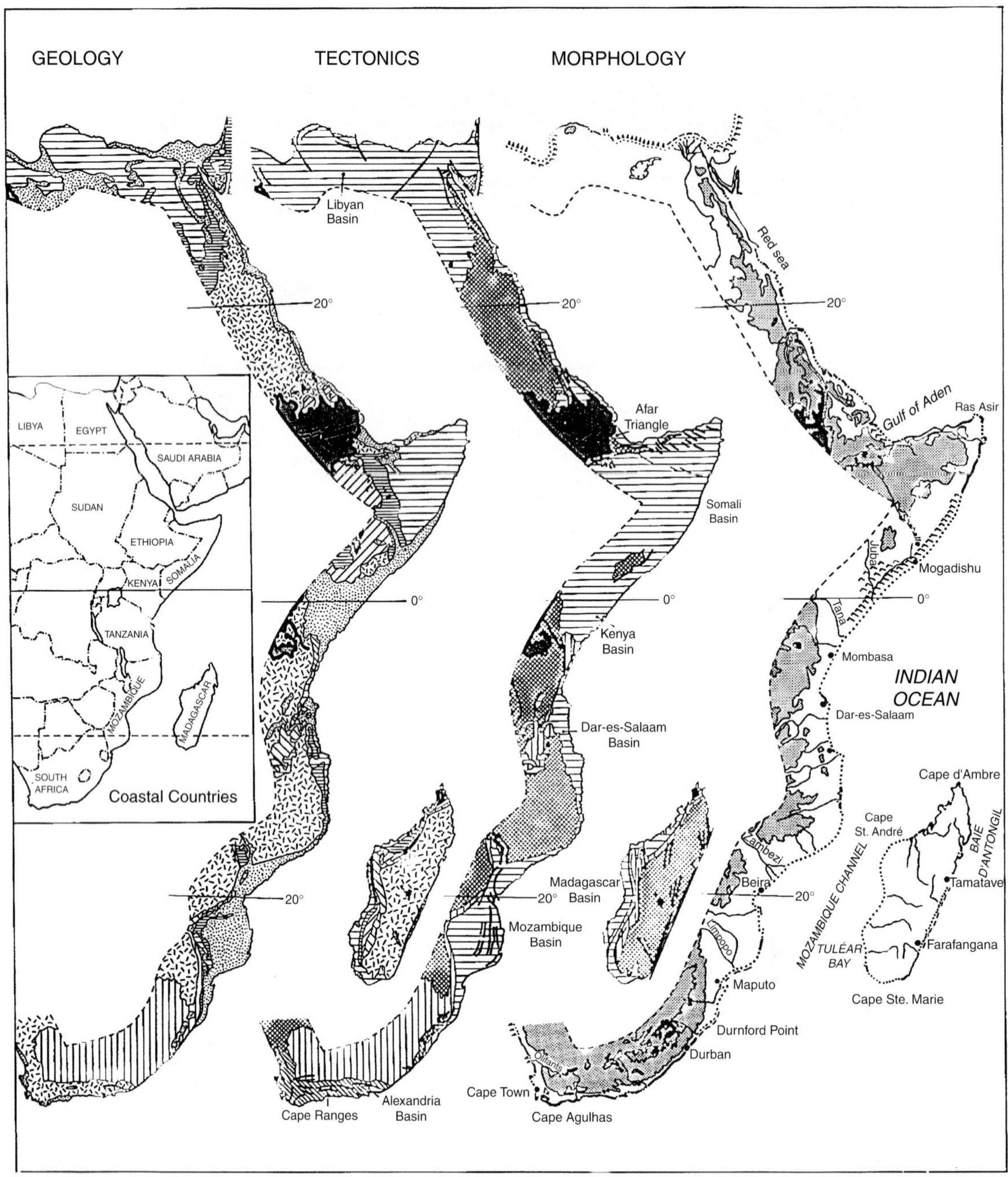

Figure A12 The Indian Ocean and Red Sea coasts: geology, tectonics, morphology (see Figure A8 for key (from Orme, 1982, with permission of Kluwer Academic Publishers).

older coastal barrier was subsequently breached by rivers draining to lowered late Pleistocene sea levels and then invaded by the Flandrian transgression, for example, forming Lake St. Lucia in Kwa-Zulu and Delagoa Bay behind Inhaca Island in Mozambique (Figure A14). Late Pleistocene estuaries excavated to more than −55 m below sea level are now choked by complex sequences of Holocene marine, estuarine, and fluvial sediment (Orme, 1974, 1976). Offshore, beachrock and aeolianites towards the edge of the continental shelf testify to the magnitude of late Pleistocene drawdown.

Shorezone dynamics along the coast of the Mozambique Basin are conditioned by prevailing northeasterly and southwesterly winds, by powerful swells from the Southern Ocean, by longshore currents up to 2 m s^{-1}, by reversing coastal flows related to the Agulhas Current that parallel the shore with surface velocities up to 0.8 m s^{-1}, and by seasonally high discharges from coastal drainage basins. The Tugela River brings 10.5×10^6 t of sediment to the coast annually and its sediment plume, like those of neighboring rivers, may be traced far out into the Agulhas Current. Owing to unwise land use and increased erosion inland, recent sediment delivery rates to the coast have greatly exceeded prehistoric rates, leading, for example, to accelerated beach-ridge and dune formation north of the Tugela estuary (Figure A15). The 7,150 km^2 Zambezi delta forms the outlet for a 1.33×10^6 km^2 savanna drainage basin and a 3,540 km mainstream. Discharge through the delta averages 7,000 m^3 s^{-1}, and sediment that is not trapped within coastal swamps and lagoons, or transported coastwise as littoral drift, is flushed over the continental shelf onto the Zambezi cone. North of the Mozambique Basin, from Mocambo Bay to Kisware Bay, faulted Cretaceous and Cenozoic rocks underlie an embayed cliffed coast fronted by raised coral bluffs and fringing reefs. The continental shelf here is often only a few hundred meters wide, plunging to −2,500 m within 10 km of the shore.

The coastal zone of Tanzania and Kenya comprises numerous, deformed and faulted Pleistocene marine terraces supporting relict

Figure A13 Port Durnford Formation, Kwa-Zulu, showing basal fossiliferous interglacial swamp peat overlain by transgressive beach sands, buried beneath weathered aeolianites (photo: A.R. Orme).

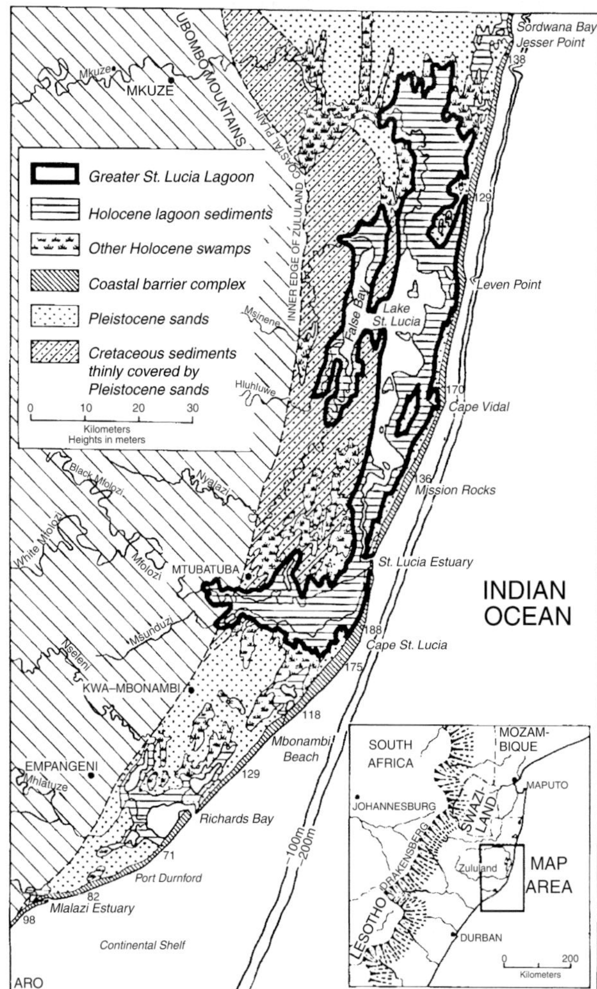

Figure A14 Greater Lake St. Lucia lagoon, Kwa-Zulu, on the site of a late Pleistocene barrier–lagoon complex, was reformed by the Flandrian transgression and has since been largely filled with Holocene lagoonal sediment.

barrier beaches and coral reefs, as well as active Holocene barrier beaches, mangrove swamps, offshore bars, and coral reefs. Three late Quaternary beach ridges surmount the lowest terrace between Dar es Salaam and Tanga, reflecting an abundant sediment supply worked onshore by strong swells, modified by wind action, and then stabilized by vegetal growth (Alexander, 1968). Raised coral reefs rise to 140 m along the Kenya coast but their margins are often obscured where cliffing was inhibited by the presence of wide lagoons and coral growth during marine regressions (Hori, 1970). The offshore islands of Zanzibar (1,660 km^2), Pemba (984 km^2), and Mafia are also relict coralline structures. All three islands are defined eastward by massive faults and Pemba is separated from the mainland by a 700 m deep graben. The Rufiji River, draining a 0.18×10^6 km^2 drainage basin, transports an average sediment load of 17×10^6 t to the sea annually (Milliman and Syvitski, 1992).

From the Tana River for 1,600 km northward to Gifle in Somalia, although fringing reefs persist, the coast is dominated by massive aeolian deposition promoted by semiarid conditions, strong monsoon winds, and episodic supplies of fluvial sediment. The Tana delta is blocked seaward by dunes, while complex Quaternary dune complexes up to 200 m high have deflected the Shebele River for 400 km parallel with the present coast, allowing it to reach the sea through the Juba estuary only after heavy rains (Orme, 1985). Paleodunes along this coast have been widely destabilized and gullied as a result of overgrazing and poor tillage practices, most recently by a population swelled by refugees from strife farther inland. Dune stabilization efforts have so far met with only limited success. From Gifle to Ras Asir, Cenozoic carbonate rocks form rocky marine terraces and sea cliffs 100–300 m high. The 1,100 km long *guban* (sunburnt plain) coast of the Gulf of Aden is partly cliffed, partly backed by alluvial plains, sabkhas, and marine terraces. Thick Plio-Pleistocene marine sequences and coral reefs raised 180–280 m above sea level around Berbera emphasize the recent tectonic emergence of this coastal zone. No perennial rivers reach Africa's east coast north of the Shebele River.

Madagascar

The coast of Madagascar reflects the structural and topographic asymmetry of this large island. The linear, fault-controlled east coast is broken by only one major embayment, the Baie d'Antongil graben (Figure A14). Basement rocks plunge from 2,000 m highlands onto a narrow coastal plain and shelf underlain by Cretaceous sediments. Persistent southeast trade winds, storms in the Southern Ocean, high average wind velocities, and west-moving cyclones from December to March all promote strong wave action. In response, barrier beaches and shallow linear lagoons have formed along 800 km of coast from Fénérive southward to Port Dauphin. These lagoons, separated from one another by low rises or *pangalanes*, are used by coastwise traffic.

In contrast, the west coast fronts a broad alluvial lowland underlain by thick post-Gondwanan sediments and the shelf widens to 150 km off

Figure A15 The Tugela River drains the east face of the Drakensberg and much of Kwa-Zulu. Driven by prevailing southerly swells, its sediment plume provides abundant material for coastal beach and dune formation farther north (photo: A.R. Orme).

Cap St. André and on the Leven Bank. Along the northwest coast between Cap St. André and rugged Cap d'Ambre at the island's northern tip, the Flandrian transgression breached a barrier system to produce complex rias, partly drowned cuestas, deltas, mangrove swamps, and islands. The Isles Glorieuses northwest of Cap d'Ambre, contain coral reefs, now 3 m above sea level, which formed around 150 ka. The semi-arid southwest coast is dominated by active parabolic dunes and oxidized paleodunes shaped by prevailing southerly winds. The shelf is narrower here and in Tuléar Bay is notched by massive submarine canyons, that off the Onilahy River descending to −2,600 m. Coral reefs of many ages and types occur: barrier reefs, fringing reefs, coral banks, and islets. Because muddy detritus inhibits reef growth, these reefs occur between major estuaries. Relative sea-level change since the last interglacial (~125 ka) is reflected in the barrier beach 8–12 m above sea level behind east coast pangalanes, in raised coral reefs at the northern tip, and in submarine coastlines and aeolianites elsewhere (Battistini, 1978).

The Red Sea coast

The Red Sea extends 2,000 km NNW from the narrow Bab el Mandeb, ranges from 150 to 300 km in width, and deepens to over −2,300 m along its central axis (Figure A14). This embryo ocean occupies a complex rift system between the African and Arabian plates. Separation began with left-lateral transtension in Cretaceous–Eocene time but accelerated with a 7° counter-clockwise rotation of the Arabian block after Oligocene time. The Red Sea opened to the Indian Ocean in Miocene time but this link has not been continuous. It remained open to the Mediterranean until the early Quaternary. Then, with closure of the Mediterranean link and eustatic and tectonic changes at Bab el Mandeb, the Red Sea became isolated from time to time during the later Quaternary. With potential evaporation approximating 2,000 mm yr^{-1} but rainfall ranging from only 3 mm yr^{-1} in the north to 150 mm yr^{-1} in the south, episodic closure of the Red Sea from the Indian Ocean has had a dramatic impact on water volume and thus coastal environments.

Owing to the recency of separation, faulted Precambrian basement rocks lie close to the coast and dissected mountains plunge toward a narrow ribbon of coastal plain underlain by later Cenozoic sediments. The northern Red Sea, together with its northern arms in the 10–25 km wide, −1,800 m deep Gulf of Aqaba and shallower Gulf of Suez graben, reflect mainly normal tensional faulting of continental margins. South of Ras Banas, however, the Red Sea is structurally more complex and falls stepwise into a narrow central trough where ascending volcanics have overlapped the continental margins.

The African coast of the Red Sea is characterized by a faulted staircase of coral reefs that form wedge-like plates up to 10 m thick, several kilometers wide, and range from 50 m above to −100 m below present sea level. Onshore, these reefs are mantled by aeolian sand, lagoonal evaporites, estuarine deposits, and alluvial gravels. The coastal plain widens only where major valleys reach the coast, as in the Tokar delta into which the Baraka River flows between June and September, or where structural troughs occur, such as the Dallol Basin, a saline depression that descends to −116 m below sea level where the Red Sea rift turns inland south of Massawa. Offshore faulted paleoreefs form the Dahlak Archipelago at the northern end of the Danakil horst. The paucity of blanketing terrigenous sediment allows modern fringing reefs to thrive in warm transparent waters. Reef margins are dominated to windward by *Acropora* associations, to leeward by *Porites* associations (Braithwaite, 1982). Lateral reef growth of several centimeters annually even occurs in front of now dry early Holocene river deltas and abandoned *marsas* (*sharms*), corridors cut by wadis draining to low Pleistocene sea levels. Barrier reefs and atolls are less common but are found overlying subsiding fault blocks from Port Sudan to Ras Hadarba (Mergner and Schuhmacher, 1985).

Antony R. Orme

Bibliography

Alexander, C.S., 1968. The marine terraces of the northern coast of Tanganyika. *Zeitschrift für Geomorphologie*, **7**: 133–154.

Allen, J.R.L., 1970. Sediments in the modern Niger delta: a summary and review. In Morgan, J.P., and Shaver, R.S. (eds.), *Deltaic Sedimentation, Modern and Ancient*. Tulsa: Society of Economic Paleontologists and Mineralogists, Special Publication No. 15, pp. 138–151.

Ausseil-Badie, J., Barusseau, J.P., Descamps, C., Diop, E.H.S., Giresse, P., and Pazdur, M., 1991. Holocene deltaic sequence in the Saloum estuary, Senegal. *Quaternary Research*, **36**: 178–194.

Barrett, D.M., 1977. Agulhas Plateau off southern Africa: a geophysical study. *Geological Society of America Bulletin*, **88**: 749–763.

Battistini, R., 1978. Observations sur les cordons littoraux Pleistocènes et Holocènes de la côte est de Madagascar. *Madagascar Revue de Géographie*, **35**: 9–37.

Braithwaite, C.J.R., 1982. Patterns of accretion of reefs in the Sudanese Red Sea. *Marine Geology*, **46**: 297–325.

Busson, G., and Perthuisot, J.-P., 1977. Intérêt de la Sebkha el Melah (Sudtunisien) pour l'interprétation des séries evaporatiques anciennes. *Sedimentary Geology*, **19**: 139–164.

Cooper, J.A.G., 1993. Sedimentation in a river-dominated estuary. *Sedimentology*, **40**: 979–1017.
Davies, J.L., 1977. *Geographical Variation in Coastal Development*. London: Longmans.
Dingle, R.V., 1973. Post-Paleozoic stratigraphy of the eastern Agulhas Bank, South Africa continental margin. *Marine Geology*, **15**: 1–24.
Dingle, R.V., and Scrutton, R.A., 1974. Continental breakup and the development of post-Paleozoic sedimentary basins around southern Africa. *Geological Society of America Bulletin*, **85**: 1467–1474.
Ebisemiju, F.S., 1987. An evaluation of factors controlling present rates of shoreline retrogradation in the western Niger delta, Nigeria. *Catena*, **14**: 1–12.
El-Ashry, M., 1985. Egypt. In Bird, E.C.F., and Schwartz, M.L. (eds.), *The World's Coastline*. New York: Van Nostrand Reinhold, pp. 513–517.
Frihy, O.E., and Komar, P.D., 1991. Patterns of beach-sand sorting and shoreline erosion of the Nile Delta. *Journal of Sedimentary Petrology*, **61**: 544–550.
Gilchrist, A.R., and Summerfield, M.A., 1990. Differential denudation and flexural isostasy in formation of rifted-margin upwarps. *Nature*, **346**: 739–742.
Glen, J.M.G., Renne, P.R., Milner, S.C., and Coe, R.S., 1997. Magma flow inferred from anistrophy of magnetic susceptibility in the coastal Paraná–Etendeka igneous province: evidence of rifting before flood volcanism. *Geology*, **25**: 1131–1134.
Guilcher, A., 1985. Senegal and Gambia. In Bird, E.C.F., and Schwartz, M.L. (eds.), *The World's Coastline*. New York: Van Nostrand Reinhold, pp. 555–560.
Guilcher, A., Medeiros, C.A., Matos, J.E., and Oliveira, J.T., 1974. Les restingas (flèches littorales), d'Angola. *Finisterra*, **9**: 171–211.
Hayden, B., Vincent, M., Resio, D., Biscoe, C., and Dolan, R., 1973. *Classification of the Coastal Environments of the World: Part II—Africa*. Washington: Office of Naval Research.
Hobday, D.K., and Orme, A.R., 1975. The Port Durnford Formation: a major Pleistocene barrier–lagoon complex along the Zululand coast. *Transactions of the Geological Society of South Africa*, **77**: 141–149.
Hori, N., 1970. Raised coral reefs along the southeastern coast of Kenya, East Africa. *Geographical Reports, Tokyo Metropolitan University*, **5**: 25–47.
Hospers, J., 1971. The geology of the Niger delta area. In Delaney, F.M. (ed.), *The Geology of the East Atlantic Continental Margin*. London: ICSU/SCOR, pp. 125–142.
Illenberger, W.K., and Rust, I.C., 1988. A sand budget for the Alexandria dunefield, South Africa. *Sedimentology*, **35**: 513–521.
Kocurek, G., Havholm, K.G., Deynoux, M., and Blakey, R.C., 1991. Amalgamated accumulations resulting from climatic and eustatic changes, Akchar Erg, Mauritania. *Sedimentology*, **38**: 751–772.
Lancaster, N., 1996. Desert environments. In Adams, W.M., Goudie, A.S., and Orme, A.R. (eds.), *The Physical Geography of Africa*. Oxford: Oxford University Press, pp. 211–237.
Maugé, L.A., Ségoufin, J., Vernier, E., and Froget, C., 1982. Géomorphologie et origine des bancs du nord-est du Canal de Mozambique—Ocean Indien Occidental. *Marine Geology*, **47**: 37–55.
McMaster, R.L., Lachance, T.P., Ashraf, A., and de Boer, J., 1971. Geomorphology, structure, and sediment of the continental shelf and upper slope off Portuguese Guinea, Guinea, and Sierra Leone. In Delany, F.M. (ed.), *The Geology of the East Atlantic Continental Margin*. London: ICSU/SCOR, pp. 109–119.
Mergner, H., and Schuhmacher, H., 1985. Quantitative Analyse von Korallengmeinschafter des Danganeb Atolls, Mittleres Rotes Meer. *Helgoländer Meersuntersuchungen*, **39**: 375–417.
Milani, E.J., and Filho, A.T., 2000. Sedimentary basins of South America. In Codani, U.G., Milani, E.J., Filho, A.T., and Campos, D.A. (eds.), *Tectonic Evolution of South America*. Rio de Janeiro: International Geological Congress, pp. 389–449.
Milliman, J.D., and Syvitski, J.P.M., 1992. Geomorphic/tectonic control of sediment discharge to the ocean: the importance of small mountainous rivers. *Journal of Geology*, **100**: 525–544.
Orme, A.R., 1973. Barrier and lagoon systems along the Zululand coast, South Africa. In Coates, D.R. (ed.), *Coastal Geomorphology*. Binghamton: State University of New York Press, pp. 181–217.
Orme, A.R., 1974. *Estuarine Sedimentation along the Natal Coast*. Washington: Office of Naval Research.
Orme, A.R., 1976. Late Pleistocene channels and Flandrian sediments beneath Natal estuaries. *Annals of the South African Museum*, **71**: 78–85.
Orme, A.R., 1982. Africa, coastal morphology. In Schwartz, M.L. (ed.), *The Encyclopedia of Beaches and Coastal Environments*. Stroudsburg: Hutchinson Ross, pp. 17–32.
Orme, A.R., 1985. Somalia. In Bird, E.C.F., and Schwartz, M.L. (eds.), *The World's Coastline*, New York: Van Nostrand Reinhold, pp. 703–711.
Orme, A.R., 1996. Coastal environments. In Adams, W.M., Goudie, A.S., and Orme, A.R. (eds.), *The Physical Geography of Africa*. Oxford: Oxford University Press, pp. 238–266.
Orme, A.R., 2002. Tectonism, climate, and landscape. In Orme, A.R. (ed.), *The Physical Geography of North America*. New York: Oxford University Press, pp. 3–35.
Paskoff, R.P., 1978. Evolution de l'embouchure de la Medjerda (Tunisie). *Photo-Interprétation*, **5**: 1–23.
Sarnthein, M., and Walger, E., 1974. Der Aolische Sandstrom aus der W-Sahara zur Atlantikkuste. *Geologische Rundschau*, **63**: 1065–1087.
Stanley, D.J., 1990. Recent subsidence and northeast tilting of the Nile delta, Egypt. *Marine Geology*, **94**: 147–154.
Steiner, C., Hobson, A., Favre, P., and Hernandez, J., 1998. Mesozoic sequence of Fuerteventura (Canary Islands): witness of early Jurassic sea-floor spreading in the central Atlantic. *Geological Society of America Bulletin*, **110**: 1304–1317.
Stocking, M.A., 1996. Soil erosion. In Adams, W.M., Goudie, A.S., and Orme, A.R. (eds.), *The Physical Geography of Africa*. Oxford: Oxford University Press, pp. 326–341.
Summerfield, R.A., 1996. Tectonics, geology, and long-term landscape development. In Adams, W.M., Goudie, A.S., and Orme, A.R. (eds.), *The Physical Geography of Africa*. Oxford: Oxford University Press, pp. 1–17.
Uchupi, E., and Emery, K.O., 1991. Pangaean divergent margins: historical perspective. *Marine Geology*, **102**: 1028.
Walling, D.E., 1996. Hydrology and rivers. In Adams, W.M., Goudie, A.S., and Orme, A.R. (eds.), *The Physical Geography of Africa*. Oxford: Oxford University Press, pp. 103–121.
Weisrock, A., 1980. *Géomorphologie et Paléo-environments de l'Atlas Atlantique*. Rabat, Service Géologique du Maroc.
Ziegler, A.M., Hulver, M.L., and Crowley, D.B., 1997. Permian world topography and climate. In Martini, I.P. (ed.), *Late Glacial and Postglacial Environmental Changes: Quaternary, Carboniferous-Permian, and Proterozoic*. New York: Oxford University Press, pp. 111–146.

Cross-references

Africa, Coastal Ecology
Beachrock
Changing Sea Levels
Coastal Climate
Coral Reef Coasts
Desert Coasts
Human Impact on Coasts
Late Quaternary Marine Transgression
Mangroves, Ecology
Mangroves, Geomorphology
Sharm Coasts

AIRBORNE LASER TERRAIN MAPPING AND LIGHT DETECTION AND RANGING

Introduction

Recent advances in the technology known as airborne laser altimetry or Light Detection and Ranging (LIDAR) allow us to map shoreline (hereinafter the term "Shoreline" generally refers to the "high-water line") position and coastal topography promptly and accurately in order to provide erosion rate updates. Precise measurement of beach and dune dimensions relative to beachfront development can also be obtained in order to determine coastal vulnerability. Overflights can be scheduled after major storm impact to assess quantitatively the amount of beach erosion, dune scarping, and overwash deposition. Considering the U.S. trend of pervasive coastal erosion and ever-increasing beachfront development, punctuated by storm impact, this new technology will fill a major void in terms of providing timely and accurate data for scientific assessments and management programs. In summary, the airborne laser altimetry represents a quantum step forward in mapping coastal erosion and flood hazard zones.

The coastal zone, which represents the interface between the land and sea, is one of the most dynamic areas on Earth. Change is occurring at all time scales from hours (with storm impact) to decades and longer (due to sea-level rise). Therefore, topographic base maps rapidly become outdated as the coastline is continually adjusting to a combination of natural forces and human modifications. At the same time there is continued population growth in the coastal zone and increasing demand for beachfront development (Cullinton *et al.*, 1990). It is estimated that there is already \$3 trillion of development along the U.S. coast vulnerable to erosion and storm flooding. This same trend of coastward migration is manifested worldwide.

Beach erosion is a significant and growing problem. It is estimated that 70% of the world's sandy beaches are eroding (Bird, 1985); in the United States it approaches 90% along the Atlantic coastal plain. The worldwide extent of erosion strongly suggests that eustatic sea-level rise is an important underlying factor, although many other natural processes can locally contribute to the problem. Human intervention alters these natural processes through such actions as dredging of tidal entrances, emplacements of groins and jetties, hardening of shores with seawalls or revetments, and beach nourishment (Leatherman and Dean, 1990). Therefore, erosion rates vary widely on a geographic basis, ranging up to 5 m/yr along the Cape Hatteras, North Carolina shoreline.

Coastal vulnerability

Coastal vulnerability in response to storm impact can be assessed by obtaining accurate and up-to-date measurements of beach and dune dimensions in relationship to the position of buildings. The Federal Emergency Management Agency (FEMA) spent over \$1 billion acquiring topographic data for the coastal and riverine floodplains of the United States in order to generate 100-year flood maps (Flood Insurance Rate Maps or FIRMs). But these data are much out-of-date and often inaccurate for coastal areas where beach and dune dimensions are affected by long-term coastal recession as well as short-duration storm impact. FEMA's data requirements are for 15–20 cm vertical accuracy.

The width of the dry sand beach (or beach backshore), which extends from the berm to the dune, is a measure of coastal vulnerability. In addition to the large seasonal fluctuation in width, most beaches are narrowing through time in response to a long-term erosional trend. As the beach narrows, storm waves can eventually reach the barrier dunes, causing vertical scarping (erosion) while the surge persists. FEMA uses the "540 rule" in combination with a height criteria to determine if a sand dune can withstand a 100-year storm and hence whether the landward-flanking houses are safe from flooding (A-Zone) or subject to attack by waves, requiring (V or velocity) Zone designation. The insurance rate for a particular property varies greatly, depending upon which zone a house is located. The "540 rule" refers to the volume of sand (540 ft^3) per foot of shoreline, but the status of the barrier dunes (e.g., volume and height) is not presently known because of continued storm impact and long-term erosional trends.

Coastal mapping technology

The approach of using classical transit and rod surveys to map shorelines was initiated in the mid-1800s by the National Ocean Service (and its forerunners) of NOAA (Figure A16). While field surveys of this type provide the most accurate information, this approach is time consuming and expensive, making these traditional surveys impractical for large areas (e.g., 10s km of shoreline). These data have been augmented and updated with historic aerial photography since the 1940s, but such hard-copy imagery must be corrected for distortion using stereoplotters or other computer-assisted photogrammetric techniques at considerable cost (Leatherman, 1983). Satellite imagery collected since the early 1970s has provided a synoptic view of coastal processes. Unfortunately, for most applications, this imagery cannot provide the level of detail necessary for precise measurement of shoreline change (Leatherman, 1993). More recently, kinematic global positioning system (GPS) surveys have been made by driving a truck-mounted receiver along the high-water line (HWL) (Morton *et al.*, 1993; French and Leatherman, 1994). While this method allows much more rapid surveying of the shoreline than conventional techniques, it is limited by access restrictions and still allows only kilometers of shoreline to be mapped per day.

Recent advances in microcomputers, laser ranging technology, and GPS have resulted in the development of compact and lightweight airborne laser terrain mapping (ALTM) systems which can inexpensively acquire shoreline topographic data of unprecedented detail and accuracy. These systems, also known as LIDAR represent a tremendous advancement in our capability not only to map the shoreline but also to obtain digital elevations for critical coastal features (e.g., sand dunes) for determining coastal vulnerability.

Modern ALTM systems are able to map an area several hundred meters wide and hundreds of kilometers long with a nominal spacing of 2–3 m and an accuracy of ± 15 cm in a few hours. The existing database

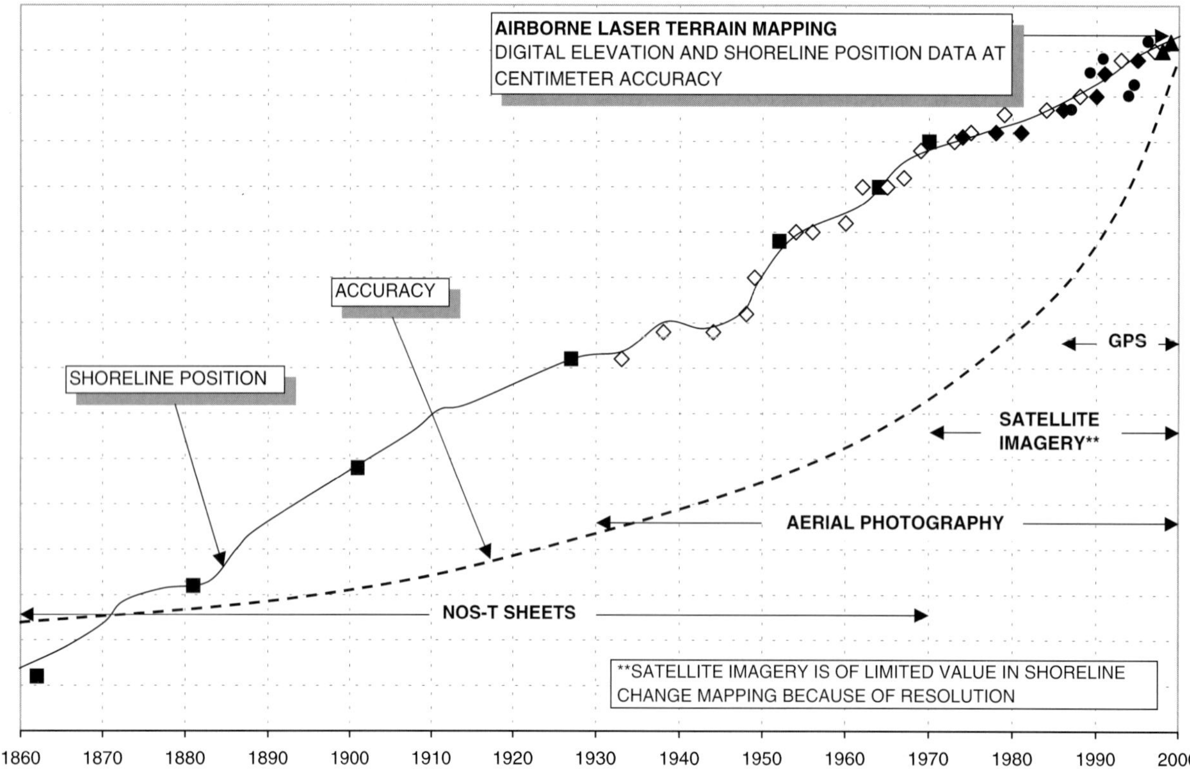

Figure A16 Evolution of shoreline mapping technologies.

of shoreline locations over time can be easily updated by using this new technology. The entirety of Pinellas County Florida of 717 km^2 was mapped in a week (Carter *et al.*, 1998). Comparison of the airborne measurements with surveyed ground control points showed a mean difference of less than 10 cm. Pre- and post-storm airborne laser data can be collected to obtain highly detailed and accurate topographic information on coastal changes (beach erosion, dune destruction, and overwash deposition) resulting from hurricane landfall or winter northeastern impact.

ALTM system

There are several different ALTM systems, but the basic working principles of these systems are similar. Three basic components: the laser scanner, a kinematic GPS, and Inertial Navigation System (INS) are employed simultaneously to determine the ground elevation. The laser scanner detects the range from aircraft to ground by recording the time difference between laser pulses sent out and reflected back. In addition, many systems allow the recording of multiple returns and the return intensity for each laser pulse. Pulse repetition rates of most ALTM systems range between 2 and 25 kHz. A rotating or oscillating mirror mounted in front of the laser causes the laser to scan back and forth, allowing the coverage of a wide swath beneath the flight path. This oscillation of the scanner mirror, in combination with forward motion of the aircraft, typically results in a zigzag scan pattern beneath the flight path (Figure A17).

For most shoreline mapping applications the ALTM system is deployed on a fixed wing aircraft at heights ranging from 350 to 1,000 m above the ground surface (Figure A17). Aircraft positions are recorded continuously by a GPS receiver mounted in the aircraft. A second GPS station situated at a known ground position provides differential corrections for more accurate estimation of the aircraft trajectory. The roll, pitch, and heading of aircraft are continuously recorded by the INS from 10 to 100 times per second.

After the flight, a precise aircraft trajectory is determined by post-processing the aircraft and ground station GPS carrier phase data. This trajectory is then combined with the laser range data, scanner mirror angle, and the INS measurements to determine the precise horizontal coordinates and vertical elevations of each laser reflection. Commonly, the irregularly spaced elevations are interpolated onto a regularly spaced grid to produce a digital elevation model (DEM). Reflection intensity measurements can also be gridded to produce an image similar to a digital photograph.

Applications

Several recent studies have demonstrated the use of LIDAR technology for mapping shoreline morphology and hazard (Carter *et al.*, 1998; Sallenger *et al.*, 1999; Krabill *et al.*, 2000). Studies of coastal vulnerability require coverage of beach and dune dimensions relative to beachfront development. To survey barrier beaches at 1 m resolution using the ALTM system, usually a pair of parallel passes is made along the shore with 20% overlap, yielding an approximately 400 m wide swath of data. In addition, a digital camera system is often installed in the aircraft. Such systems are valuable for comparison of historical shoreline markers such as the HWL or dry/wet boundary with the laser measurements.

Shoreline surveys can be scheduled seasonally or after the passage of major storms such as hurricanes or northeasters to assess quantitatively the amount of beach erosion, dune scarping, and overwash deposition. For example, an ALTM survey was conducted along the central-east coast of Florida immediately before and after the passage of Hurricane Floyd in September 1999 (Figure A18). This airborne laser technology will literally revolutionize our data collection efforts and permit a much better scientific understanding of how beaches change seasonally and annually as well as respond to storm impact.

In summary, this new and important technology represents a quantum step forward in mapping the coastal erosion hazard zones and provides a chance for breakthrough science in terms of understanding shoreline changes relative to the forcing functions, including storm impact.

Stephen P. Leatherman,
Dean Whitman and Keqi Zhang

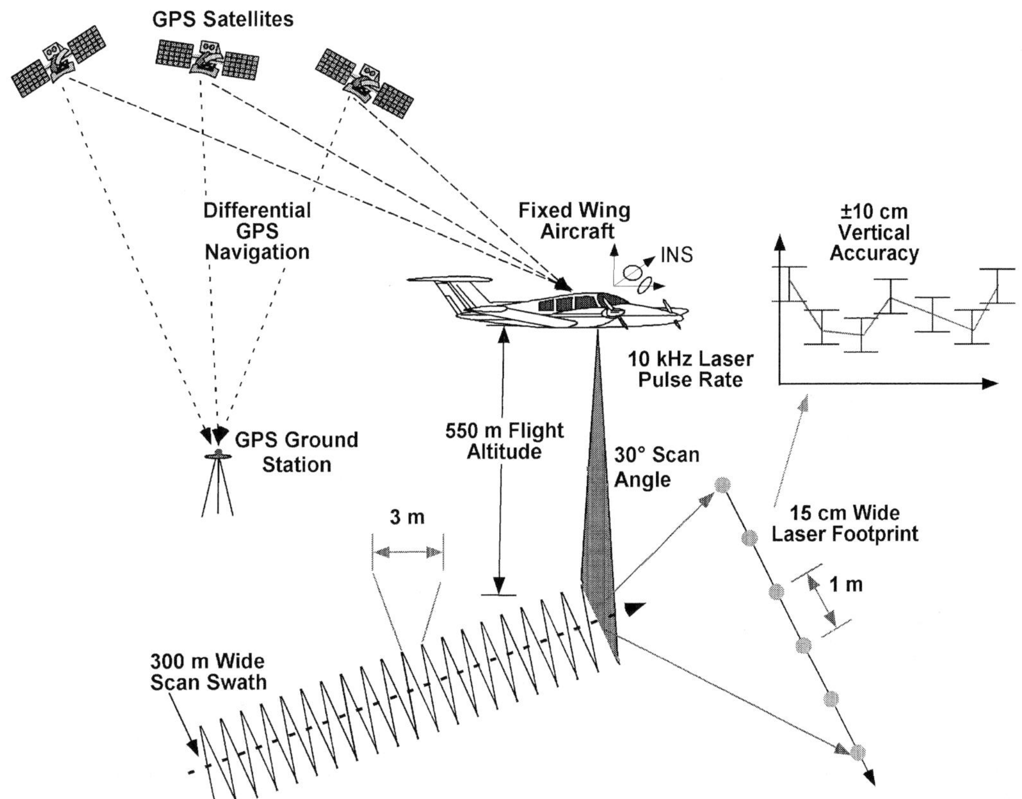

Figure A17 Schematic of ALTM technology.

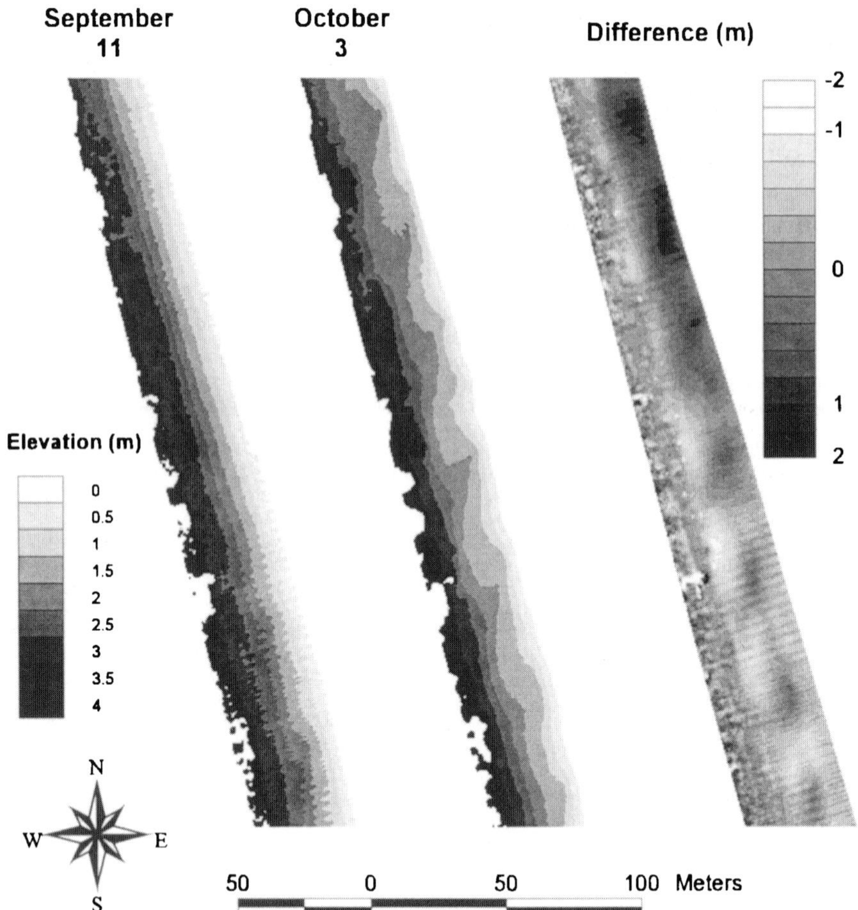

Figure A18 Example of ALTM data for near Vero Beach, Florida.

Bibliography

Bird, E.C.F., 1985. *Coastal Change—A Global Review*. Chichest: John Wiley-Interscience.

Carter, W.E., Shrestha, R.L., and Leatherman, S.P., 1998. Airborne laser swath mapping: application to shoreline mapping. *Proceedings of the INSMAP Conference*. Melbourne, Australia.

Cullinton, T.J., Warren, M.A., Goodspeed, T.R., Remer, D.G., Blackwell, C.M., and McDonough, J.J., III, 1990. *50 Years of Population Change Along the Nation's Coasts, 1960–2010*. Silver Spring, MD: NOAA-National Ocean Services.

French, G.T., and Leatherman, S.P., 1994. The use of global positioning systems in coastal studies. Washington, DC: *Technical Report Prepared for FEMA*.

Krabill, W.B., Thomas, C.W., Swift, R.N., Fredrick, E.B., Manizade, S.S., Yungel, J.K., Martin, C.F., Sonntag, J.G., Duffy, M., Hulslander, W., and Brock, J.C., 2000. Airborne laser mapping of Assateague national seashore beach. *Photogametric Engineering and Remote Sensing*, **66**: 65–71.

Leatherman, S.P., 1983. Shoreline mapping: a comparison of techniques. *Shore and Beach*, **51**: 28–33.

Leatherman, S.P., 1993. In Gurney *et al.* (eds.), *Coastal Change in Atlas of Satellite Observations Related to Global Change*. Cambridge University Press, pp. 327–339.

Leatherman, S.P., and Dean, R.G. 1990. Beach erosion rates and the national flood insurance program. *American Geophysical Union EOS Transactions*, **72**: 9–10.

Morton, R.A., Leach, M.P., Paine, J.G., and Cardoza, M.A., 1993. Monitoring beach change using GPS surveying techniques. *Journal of Coastal Research*, **9**: 702–720.

Sallenger, A.H., Jr., Krabill, W., Brock, J., Swift, R., Jansen, M., Manizade, S., Richmond, B., Hampton, M., and Eslinger, D., 1999. Airborne laser study quantifies El Nino-induced coastal change. *American Geophysical Union EOS Transactions*, **80**: 90–93.

Cross-references

Coastline Changes
Coasts, Coastlines, Shores, and Shorelines
Geographic Information Systems
Mapping Shores and Coastal Terrain
Remote Sensing of Coastal Environments

ALGAL RIMS

History

Since the beginning of coral reef studies algal formations of various sizes and shapes, made of coralline algae and other organisms and generally associated with coral reefs, have been described under various names (algal ridges, crests, mounds, reefs, etc.). They were described as being developed upon the windward edge of reefs, but mention of algal mounds, crusts or rims developed on rock also exist in the early literature. A detailed description of algal rim structure was given by Tracey *et al.* (1948) on Bikini atoll.

Locations were mainly Pacific, but analogous formations were also described in the Atlantic, such as the "boilers" of Bermuda (Agassiz, 1895). Later papers (Kempf and Laborel, 1968; Gessner, 1970; Glynn, 1973; Adey and Burke, 1976; Focke, 1978; Jindrich, 1983; Bosence, 1984) dealt with algal and animal populations associated with Atlantic rims, and demonstrated their identity with Indo-Pacific rims, and their independance from coral reefs proper, since algal rims can develop directly upon littoral rock, or in regions where corals do not thrive.

Definition and morphology

Algal rims are reef-like structures (bioherms or biostromes, following their size, development, and influence on local sedimentation processes) developing on the outer edge of reef-flats as well as on rocky coasts submitted to permanent, strong surf (trade winds). They exist both in Atlantic and Indo-Pacific regions but rim-like formations have also been described in the Mediterranean (Blanc and Molinier, 1955).

They are built mainly by massive or encrusting coralline algae (mostly *Porolithon*), associated with Hydrocorals (*Millepora*), and Vermetid Gastropods (*Dendropoma*), with a few sturdy corals. Specific composition varies with exposure, slope, and nature of substrate (Focke, 1978); coralline algae replacing vermetids when wave energy increases.

Rims are often built-up upon the inner parts of spur-and-groove systems by fusion of algal heads (Tracey *et al.*, 1948). In extreme surf conditions, upward water movement leads to algal terraced pinnacles or blowholes.

Rim populations gradually mingle seaward with populations of the outer slope, rich in coralline algae down to several meters deep but enriched in corals (such as Atlantic *Acropora palmata*).

In weaker surf conditions, the rim gradually passes into a spur-and-groove reef edge or to thin rim-like formations along rocky coasts. Algal rims or equivalent structures may develop in tideless as well as in tidal conditions.

An often overlooked fact is that structures with algal rim or rimmed terrace morphology may be formed not by constructional evolution but by an erosive process. The Bermudian "boilers" may thus be divided into "erosive boilers" generated by erosion of an emerged stack of soft rock, the edge of which is protected by a thin organic cover (original description of a Bermudian boiler) and into "constructive boilers" obtained by the upward growth of a coral patch, covered by a surface layer of vermetids (Ginsburg and Schroeder, 1973).

Relation with past sea levels

Some components of algal rims (*Dendropoma* vermetids, some *Lithophyllum*) have a narrow repartition range around MSL and their presence in cores or on elevated coastlines is a precise indicator of ancient sea levels, with an approximation of less than 1 m (Jindrich, 1983; Laborel, 1986; Laborel *et al.*, 1994). Conversely, corals such as *Acropora palmata*, having a vertical range of 5–10 m are less precise indicators (Lighty *et al.*, 1982) unless a true rim may be recognized in the core.

Further study of algal rims is nevertheless needed to obtain better biological indications of past sea level stands.

Jacques Laborel

Bibliography

Adey, W.H., and Burke, R.B., 1976. Holocene bioherms (algal ridges and bank barrier reefs) of the eastern Caribbean. *Bulletin of the Geological Society of America*, **87**: 95–109.

Agassiz, A., 1895. A visit to the Bermudas in 1894. *Bulletin of the Museum of Comparative Zoology Harvard University*, **26**: 205–281.

Blanc, J.J., and Molinier, R., 1955. Les formations organogènes construites superficielles en Méditerranée occidentale. *Bulletin de l' Institiut Océanographique de Monaco*, **1067**: 1–26.

Bosence, D.W., 1984. Construction and preservation of two recent coralline algal reefs, Sainte Croix, Caribbean. *Paleontology*, **27**: 549–574.

Focke, J.W., 1978. Limestone cliff morphology on Curaçao (Netherlands Antilles), with special attention to the origin of notches and vermetid coralline algal surf benches. *Zeitschrift für Geomorphologie*, **22**: 329–349.

Gessner, F., 1970. *Lithothamnium* terrassen in Karibischen Meer. *Internationale Revue der Gesamten Hydrobiologie*, **55**: 757–762.

Ginsburg, R.N., and Schroeder, J.H., 1973. Growth and submarine fossilisation of algal cup reefs, Bermuda. *Sedimentology*, **20**: 574–614.

Glynn, P.W., 1973. Aspects of the ecology of coral reefs in the Western Atlantic region. In Jones, O.A., and Endean, R. (eds), *Biology and Geology of Coral Reefs*. New York: Academic Press, pp. 271–324.

Jindrich, V., 1983. Structure and diagenesis of recent algal-foraminifer reefs, Fernando de Noronha, Brazil. *Journal of Sedimentary Petrology*, **53**(2): 0449–0459.

Kempf, M., and Laborel, J., 1968. Formations de Vermets et d'Algues calcaires des côtes du Brésil. *Recueil des Travaux de la Station Marine d'Endoume*, **43**: 9–23.

Laborel, J., 1986. Vermetids. In Orson van de Plassche (ed.), *Sea-Level Research: A Manual for the Collection and Evaluation of Data*, Volume 12. Norwich: Geo Books, pp. 281–310.

Laborel, J., Morhange, C., Lafont, R., Le Campion, J., Laborel-Deguen, F., and Sartoretto, S., 1994. Biological evidence of sea-level rise during the last 4500 years on the rocky coasts of continental southwestern France and Corsica , *Marine Geology*, **120**: 203–223.

Lighty, R.G., McIntyre, I., and Stuckenrath, R., 1982. *Acropora palmata* reef framework: a reliable indicator of sea level in the western Atlantic for the past 10 000 years. *Coral Reefs*, **1**: 125–130.

Tracey, J.I., Ladd, J.S., and Hoffmeister, J.E., 1948. Reefs of Bikini, Marshall Islands. *Bulletin of the Geological Society of America*, **59**: 861–878.

Cross-references

Bioherms and Biostromes
Coral Reefs
Sea-Level Indicators, Biologic

ALLUVIAL-PLAIN COASTS

Alluvial-plain coasts are prograding shore systems "formed where the broad alluvial slope at the base of a mountain range is built out into a lake or the sea" and are part of a "neutral" coast (Johnson, 1919, p. 188). This term is little used in modern publications, but is still a useful classification of coastlines built by interaction of river and coastal processes where sediment input and relative sea-level change are relatively balanced, producing a coastline that neither progrades nor retreats rapidly. Alluvial-plain coasts may be closely associated with deltas, estuaries, beach-ridge plains, barrier–lagoon systems, and in some settings glacial outwash plains. Variations in geomorphology and process responses among these coastal environments are determined by the ratio of relative sea-level change to sediment supply, as well as by climate, wave and tidal energy, and underlying geologic controls such as relief, complexity of coastline shape, and tectonic setting.

In sequence stratigraphic terminology, alluvial-plain coasts are expected within a highstand systems tract (Posamentier *et al.*, 1988; Posamentier and Vail, 1988) or falling-stage systems tract (Plint and Nummedal, 2000). These systems tracts form during rising to highstand sea level and highstand to falling sea level, respectively. However, the use of these interpretations clearly requires an understanding of rates of sediment influx and the influence of local basin subsidence or tectonic uplift.

Low-relief alluvial plains are built by fluvial processes such as meandering stream pointbar accretion, deposition on levees, over-bank flood deposition, and a variety of channel avulsion processes that create oxbow lakes, flood chutes, and crevasse-splay deposits. In higher-relief coastal settings fan deltas may develop. Fan deltas are sloping alluvial sediments deposited in a sweeping arc where mountain streams flow out onto lowlands and into a sea or lake, with a substantial subaqueous extent (Holmes, 1965; Prior and Bornhold, 1989). Their subaerial portions are alluvial fans, that is, they are dominated by braided stream and sediment gravity flows in irregular flow conditions. Their subaqueous portions include fan-shaped deltaic and prodelta facies, usually with only a narrow shelf. Alluvial-plain coasts are influenced by littoral processes, either by the reworking by waves and tides of fluvial sediments brought into the coastal setting, or by the gradual infill of pre-existing embayments.

Wave-dominated alluvial-plain coasts occur on relatively simple coastlines, where there is sufficient fetch and wave energy to prevent the outward building of a distinct delta form. The coast may be somewhat tectonically influenced and of moderate relief, such as the Mediterranean Valencia coast of eastern Spain (Figure A19) (Viñals and Fumanal, 1995). The very low relief Texas coast has alluvial-plain coastal segments at High Island, and on the Brazos River delta (Rodriguez *et al.*, 2000, 2001). The Peruvian coast is battered by higher wave energy and has steep cliffs in many sectors. However, valley mouths receive abundant coarse and fine sediment from rivers heading in the foothills of the Andes, modulated by the important climatic variability of the El Nino–Southern Oscillation as well as earthquake destabilization. Fluvial reworking and resurfacing creates alluvial-plain coasts near river mouths, while sand and gravel are reworked into well-developed beach ridge plains northward (downdrift) of the Santa, Piura, and Chira rivers in northwestern Peru (Richardson, 1983; Sandweiss, 1986; Wells, 1996).

Embayment-infill alluvial plains occur on complex coast-lines with higher relief. Sediments are trapped in coastal compartments with relatively limited fetch and lower wave energy. The coastal valleys of western Turkey (Figure A20) and Greece are excellent examples of this type of coast. The coastlines in these embayments have prograded many kilometers in historic times, with the result that famous archaeological sites such as Troy and Thermopylae are now well inland of the coastal settings described in antiquity (Kraft *et al.*, 1980, 1983, 1987). Figure A20 shows

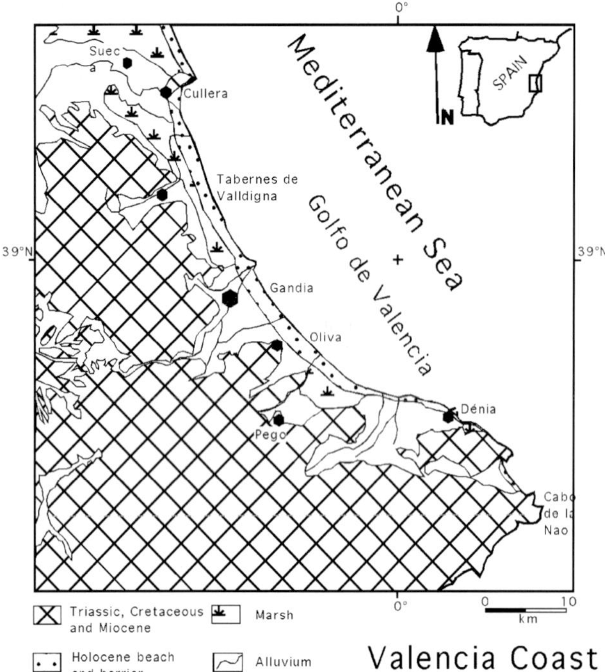

Figure A19 Alluvial-plain coast of eastern Spain, the Valencia Coast (after Viñals and Fumanal, 1995, figure 1).

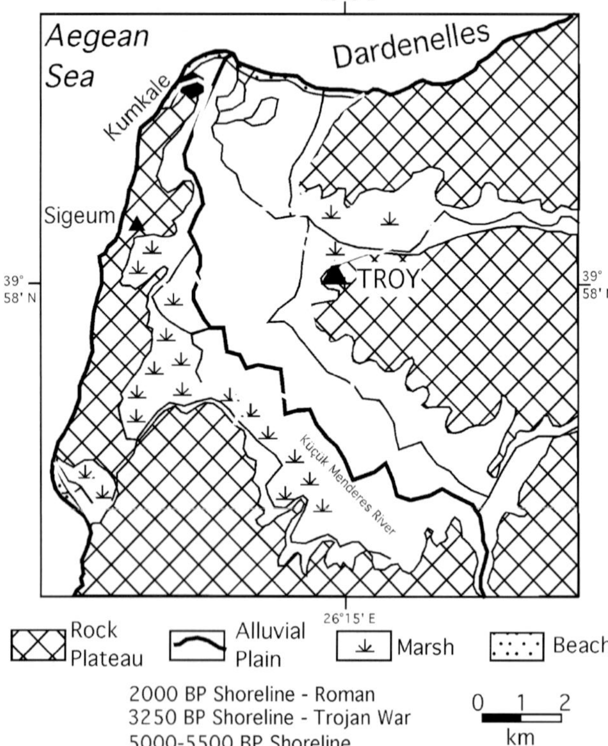

Figure A20 Alluvial-plain coast of northwestern Turkey, the Trojan plain of the Küçük Menderes River (after Kraft *et al.*, (2003, figures 1 and 4–6).

the progradation of the Küçük Menderes alluvial plain 7 km from a coastline within a protected embayment at 5,000–5,500 yr BP, more than 5 km since the Trojan war 3,250 BP, and more than 3 km since Roman times 2,000 BP. These reconstructions are based on interdisciplinary coastal geology, geoarchaeology, and written records (Kraft *et al.*, 2003).

Recognition of the potential for coastline change is crucial to interpretation of archaeological sites on alluvial-plain coasts.

Alluvial-plain coasts associated with large rivers may grade into delta systems, representing an inactive lobe of a delta cycle (Coleman, 1988) or distant from the primary river influx. The Huanghe (Yellow River) entering the Bo Hai Sea of northeastern China has an active modern birdsfoot delta, but there are also broad stretches of alluvial-plain coast on the Bo Hai coast that formed by coastal reworking of former fluvial/deltaic environments (Saito *et al.*, 2001). Marsh and swamp environments, represented by coals in ancient sequences (e.g., Fielding, 1987) are commonly associated with delta plain and alluvial-plain coasts.

Recognition of alluvial-plain coasts in ancient rock sequences may be challenging, particularly to distinguish from delta plain and barrier–lagoon systems. However, the Cretaceous Mesaverde Group littoral facies provide clear examples of alluvial-plain coasts on the margin of the interior seaway of Colorado and Wyoming (Hollenshead and Pritchard, 1961; Kraft and Chrzastowski, 1985). This succession includes transitions from fluvial delta plain (Menefee Fm.) with associated coal and carbonaceous shales, to the Lookout Point Fm. sandstone, in a regressive succession, overlain by the Cliff House Fm. sandstone, in a transgressive succession. Determining the relative influences of eustatic sea-level fluctuations and rates of sediment supply, as well as climate, subsidence, and local process variations, can be even more challenging in the ancient rocks, than in Holocene systems, where geomorphology and geologic settings are more completely understood.

Alluvial-plain coasts may be underrepresented in the literature because they are combined with delta, barrier, or estuarine environments. They represent intermediate conditions of balance between sediment accumulation and sea-level change. They are neither strongly prograding, like deltas, nor transgressed like estuaries. They reflect primary fluvial processes of deposition, but also significant reworking by littoral processes. They may be recognized by the lack of distinct barrier and lagoon environments and close juxtaposition of alluvial and beach systems.

Daniel F. Belknap

Bibliography

Coleman, J.M., 1988. Dynamic changes and processes in the Mississippi River delta. *Geological Society of America Bulletin*, **100**: 999–1015.

Fielding, C.R., 1987. Coal, depositional models for deltaic and alluvial plain sequences. *Geology*, **15**: 661–664.

Hollenshead, C.T., and Pritchard, R.L., 1961. Geometry of producing Mesa Verde sandstones, San Juan basin. In Peterson, J. (ed.), *Geometry of Sandstone Bodies*. Tulsa, OK: American Association of Petroleum Geologists Symposium Volume, pp. 98–118.

Holmes, A., 1965. *Principles of Physical Geology*, 2nd edn., London: Thomas Nelson.

Johnson, D.W., 1919. *Shore Processes and Shoreline Development*. Facsimile edn., 1972. New York: Hafner Publishing Company.

Kraft, J.C., and Chrzastowski, M.J., 1985. Coastal stratigraphic sequences. In Davis, R.A., Jr. (ed.), *Coastal Sedimentary Environments*, 2nd edn. New York: Springer-Verlag, pp. 625–663.

Kraft, J.C., Belknap, D.F., and Kayan, I., 1983. Potentials of discovery of human occupation sites on the continental shelves and nearshore coastal zone. In Masters, P.M., and Flemming, N.C. (eds.), *Quaternary Coastlines and Marine Archaeology*. London: Academic Press, pp. 87–120.

Kraft, J.C., Kayan, I., and Erol, O., 1980. Geomorphic reconstructions in the environs of ancient Troy. *Science*, **209**: 776–782.

Kraft, J.C., Rapp, G., Jr., Szemler, G.J., Tziavos, C., and Kase, E., 1987. The Pass at Thermopylae, Greece. *Journal of Field Archaeology*, **12** (2): 181–198.

Kraft, J.C., Rapp, G., Jr., Kayan, I., and Luce, J.V., 2003. Harbor area at ancient Troy: an interdisciplinary approach. *Geology*, **31**: 163–166.

Plint, A.G., and Nummedal, D., 2000. The falling stage systems tract: reognition and importance in sequence stratigraphic analysis: In: D. Hunt and R.L. Gawthorpe, (eds.), *Sedimentary Responses to Forced Regressions. Geological Society*, London, Special Publications, 172, pp. 1–17.

Posamentier, H.W., Jervey, M.T., and Vail, P.R., 1988. Eustatic controls on clastic deposition I-conceptual framework. In Wilgus, C.K., Hastings, B.S., Kendall, C.G. St. C., Posamentier, H.W., Ross C.A., and Van Wagoner, J.C. (eds.), *Sea-level Changes—An Integrated Approach*. SEPM, Special Publication 42, pp. 109–124.

Posamentier, H.W., and Vail, P.R., 1988. Eustatic controls on clastic deposition II-sequence and systems tract models. In Wilgus, C.K., Hastings, B.S., Kendall, C.G. St. C., Posamentier, H.W., Ross, C.A.,

and Van Wagoner, J.C. (eds.), *Sea-level Changes—An Integrated Approach*. SEPM, Special Publication 42, pp. 125–154.
Prior, D.B., and Bornhold, B.D., 1989. Submarine sedimentation of a developing Holocene fan delta. *Sedimentology*, **36**: 1053–1076.
Richardson, J.B., 1983. The Chira Beach ridges, sea level change, and the origins of maritime economies on the Peruvian coast. *Annals of Carnegie Museum*, **52**(11), 265–275.
Rodriguez, A.B., Fassell, M.L., and Anderson, J.B., 2001. Variations in shoreface progradation and ravinement along the Texas coast, Gulf of Mexico. *Sedimentology*, **48**: 837–853.
Rodriguez, A.B., Hamilton, M.D., and Anderson, J.B., 2000. Facies and evolution of the modern Brazos Delta, Texas; wave versus flood influence. *Journal of Sedimentary Research*, **70**: 283–295.
Saito, Y., Yang, Z.-S., and Hori, K., 2001. The Huanghe (Yellow River) and Changjiang (Yangtze River) deltas; a review on their characteristics, evolution and sediment discharge during the Holocene. *Geomorphology*, **41**: 219–231.
Sandweiss, D.H., 1986. The beach ridges at Santa, Peru: El Nino, uplift and prehistory. *Geoarchaeology*, **1**: 17–28.
Viñals, M.J., and Fumanal, M.P., 1995. Quaternary development and evolution of the sedimentary environments in the central Mediterranean Spanish coast. *Quaternary International*, **29**(30): 119–128.
Wells, L.E., 1996. The Santa beach ridge complex: sea-level and progradational history of an open gravel coast in central Peru. *Journal of Coastal Research*, **12**: 1–17.

Cross-references

Barrier
Beach Ridges
Classification of Coasts
Deltas
Estuaries
Paleocoastlines
Sandy Coasts
Sequence Stratigraphy

ALTIMETER SURVEYS, COASTAL TIDES, AND SHELF CIRCULATION

Satellite altimetry produces unique global measurements of instantaneous sea surface heights relative to a reference surface, and is one of the essential tools for monitoring ocean surface conditions and conducting physical oceanographic research. So far, scientific results from altimetric data have significantly improved our knowledge of global ocean tides and mesoscale circulation variability in the deep ocean. However, the application of altimeter data to coastal waters has been limited (Han, 1995). This article briefly discusses issues related to applications of satellite altimetry to coastal tides and subinertial shelf circulation.

Satellite altimetry

A radar altimeter aboard a satellite is a nadir-looking active microwave sensor. Its signal pulse, transmitted vertically downward, reflects from the ocean surface back to an altimeter antenna. The round-trip time and the propagation speed of the electromagnetic waves are used to compute the range between the antenna and the ocean surface.

From the altimeter-measured range, the instantaneous sea surface relative to a reference surface, such as an ellipsoid, can be determined if a satellite orbit relative to the reference surface is known. With the knowledge of the oceanic geoid, the sea surface topography relative to the geoid due to ocean dynamic circulation including the temporal averages can be mapped. Although the geoid is not well determined yet, repeated observations can provide a measurement of the temporal variability of the sea surface height since the geoid can be treated as time-invariant for oceanographic applications.

Oceanographic applications of satellite altimetry require an accuracy of a few centimeters. This requirement constrains not only the altimeter sensor, but also pertinent atmospheric and oceanographic corrections that have to be made, the satellite orbit, and, in some applications, the reference geoid. One-second data are provided to the science community as the Geophysical Data Records (GDR). So far, the accuracy of altimetric sensors has steadily improved to a few centimeters. The atmospheric corrections consist of ionospheric delay and wet and dry tropospheric delays; the oceanographic effects are composed of the sea state bias, inverse barometric response, the elastic ocean tides, solid Earth tides, and pole tides. The radial uncertainty of the orbit has long been one of the largest error sources in satellite altimetry until the TOPEX/Poseidon (T/P) mission. The root mean square accuracy of the orbit used for producing T/P GDRs is estimated to be 3.5 cm (Fu *et al.*, 1994). At present, the geoid errors occurring at various spatial scales are the largest error source, and the major obstacle to the derivation of the mean current from altimeter data.

Because almost all radar altimeter missions use a repeating orbit, repeat track analysis or collinear analysis becomes a conventional approach for application of altimeter data in physical oceanography. A mean at a location is calculated from data available, and then the mean is removed from the data to produce sea surface height anomalies relative to the mean. Thus, the repeat track analysis eliminates the geoid and its errors, the mean sea surface, and a portion of the orbit error. There are three variants from the standard approach. One is to subtract a selected instantaneous sea surface instead of the mean sea surface; another calculates the along-track slope deviations instead of sea level deviations. For a tidal analysis, collinear differences are usually calculated to bypass the uncertainty associated with the mean sea surface.

Coastal tides

Coastal tides are forced by adjacent deep-ocean tides which themselves are generated by the tide-generating potential associated with the motion of both the moon and the sun. The tidal waves in coastal seas propagate as trapped Kelvin waves, subject to significant amplifications in magnitude and reductions in wavelength.

Several empirical methods can be used to derive tides from altimetric time series. These include but are not limited to harmonic analysis, response analysis, and the inverse method. The first two methods are applied to each geographical location, and the last method considers the spatial correlation as defined by *a priori* spatial covariance in the tidal signal.

Both harmonic and response analyses are purely temporal analyses, common in the analysis of hourly tide-gauge data. The methods are suitable for either coastal seas (Woodworth and Thomas, 1990; Han *et al.*, 1996) or deep oceans (Cartwright and Ray, 1990), not constrained by large satellite track intervals. However, an assimilative model in which altimetric tidal information is optimally blended with a dynamical model would be a great benefit in providing spatial interpolation and extrapolation of tidal fields (Han *et al.*, 2000).

Tides in altimetric data have aliasing periods longer than their physical periods due to the altimeter sub-Nyquist sampling. Consequently, it may be difficult to resolve all leading tidal constituents using the harmonic analysis because of the limited duration of the altimeter data. This difficulty can be mitigated by the response analysis. Once the response weights of a tidal species have been determined, a solution for any tidal constituent in this species can be derived from the response analysis, even though some of these constituents are not resolved at all, given the satellite-sampling period (Cartwright and Ray, 1990).

The inverse method is a joint analysis in time and space, in which the spatial and spectral properties of tides and associated errors are considered using *a priori* covariance explicitly estimated from all the available information for the tides and various altimetric data errors. Several tidal constituents are simultaneously derived by an inversion of the data and the corresponding errors associated with these constituents by computing *a posterior* covariance (Mazzega and Berge, 1994). However, the inverse method is not applicable if the tidal correlation scales are smaller than satellite track intervals.

Subinertial shelf circulation

Another important application of satellite altimetry in physical oceanography is to subinertial variability over a continental shelf. A key component of the geophysical corrections is the removal of oceanic tides.

In coastal seas, tides are often larger than subinertial variability and have horizontal scales of ~100 km owing to the local coastline and bottom topography. Global tidal models may not be adequate for shallow waters; for example, the M_2 amplitudes and phases in Geosat GDRs are significantly different from both tide-gauge and altimeter data over the Scotian Shelf and the Grand Banks (Han *et al.*, 1993) and in the Amazon mouth (Minster *et al.*, 1995). The tide model in T/P GDRs was found not sufficient over the Canadian Pacific shelf (Foreman *et al.*, 1998).

Because the semidiurnal and diurnal constituents are aliased into fluctuations with longer periods, their errors may obscure subinertial fluctuations of interest. For example, the K_1 has an aliasing period of 173 d in the T/P data, close to a semiannual cycle. One way to reduce the tidal correction errors is to use regional tidal models that have higher accuracy (Han *et al.*, 1993). The separation of the low-frequency variability of interest from the aliased tidal constituents can also be achieved using harmonic analysis. The results may be subject to notable statistical uncertainties, depending on the length of data, magnitudes of tidal errors, and error correlation.

The altimeter data can be used to study sea surface height variability directly and to calculate sea surface slopes and thus to derive geostrophic surface currents. The geostrophic currents at depth can be determined when combined with interior density structure. Han and Tang (1999) combined the T/P data with density and wind data to study seasonal variability of current and transport in the Labrador Current. T/P sea level anomalies are being used to examine seasonal and interannual variability of coastal sea levels.

The spatial sampling resolution is critical for shelf circulation. Dynamic models are needed for a space–time interpolation to make the best use of the information. The assimilation of altimeter data into numerical models is one of the most promising approaches to study subinertial shelf circulation.

Discussion: challenges and opportunities

Application of satellite altimetry to coastal tides and shelf circulation is more challenging than to similar problems for the deep ocean (Han, 1995). The reasons are multifold. First, dynamic features in coastal seas are strongly influenced by the complex coastline and bottom topography, and have temporal scales of several days to weeks and spatial scales of tens of kilometers. A satellite altimeter is not good at sampling these small-scale features. Second, an altimeter is subject to tracking problems upon crossing the land–sea boundary; useful data may not be available within tens of kilometres from the land. Third, atmospheric and oceanographic corrections are more uncertain for coastal waters than for the deep ocean.

High temporal and spatial resolutions are very important, particularly in the coastal seas. Unfortunately, they are incompatible, and a reasonable compromise must be made. Consequently, the cross-track spatial resolution is often low relative to the small-scale nature of the coastal processes. Dynamical models with data assimilation are useful for optimally blending the model dynamics with altimetry. Only dynamical models may be able to recover the deep ocean conditions from their sea surface manifestation observed by an altimeter.

A temporal and spatial blending of multi-mission altimeter data can provide a partial solution to the poor spatial resolution and an unprecedented opportunity of studying coastal climate change and its impacts, especially when combined with prognostic models. In particular, the availability of near-real-time satellite altimetry data will be very useful for nowcasts and forecasts of coastal ocean conditions.

Guoqi Han

Bibliography

Cartwright, D.E., and Ray, R.D., 1990. Oceanic tides from geosat altimetry. *Journal of Geophysical Research*, **95**: 3069–3090.

Foreman, M.G.G., Crawford, W.R., Cherniawsky, J.Y., Gower, J.F.R., Cuypers, L., and Ballantyne, V.A., 1998. Tidal correction of TOPEX/POSEIDON altimetry for seasonal sea surface elevation and current determination of the Pacific coast of Canada. *Journal of Geophysical Research*, **103**: 27979–27998.

Fu, L.L., Christensen, E.J., Yamarone, C.A., Jr., Lefebvre, M., Menard, Y., Dorrer, M., and Escudier, P., 1994. TOPEX/POSEIDON mission overview. *Journal of Geophysical Research*, **99**: 24369–24381.

Han, G., 1995. Coastal tides and shelf circulation by altimeter. In Ikeda, M., and Dobson, F.W. (eds.), *Oceanographic Application of Remote Sensing*. Boca Raton: CRC Press, pp. 45–56.

Han, G., and Tang, C.L., 1999. Velocity and transport of the labrador current determined from altimetric, hydrographic, and wind data. *Journal of Geophysical Research*, **104**: 18047–18057.

Han, G., Ikeda, M., and Smith, P.C., 1993. Annual variation of sea surface slopes on the Scotian Shelf and Grand Banks from geosat altimetry. *Atmosphere-Ocean*, **31**: 591–615.

Han, G., Ikeda, M., and Smith, P.C., 1996. Oceanic tides derived from TOPEX/Poseidon altimetry over the Scotian and Newfoundland shelves. *Atmosphere-Ocean*, **34**: 589–604.

Han, G., Hendry, R., and Ikeda, M., 2000. Assimilating TOPEX/Poseidon derived M2 tides in a primitive equation model over the Newfoundland Shelf. *Continental Shelf Research*, **20**: 83–108.

Mazzega, P., and Berge, M., 1994. Ocean tides in the Asian semi-enclosed seas from Topex/Poseidon. *Journal of Geophysical Research*, **99**: 24867–24881.

Minster, J.F., Genco, M.L., and Brossier, C., 1995. Variations of the sea level in the Amazon estuary. *Continental Shelf Research*, **15**: 1287–1302.

Woodworth, P.L., and Thomas, J.P., 1990. Determination of the major semidiurnal tides of the northwest European Continental Shelf from geosat altimetry. *Journal of Geophysical Research*, **95**: 3061–3068.

Cross-references

Geodesy
Remote Sensing of Coastal Environments
Shelf Processes
Sea-Level Change
Tides

ANTARCTICA, COASTAL ECOLOGY AND GEOMORPHOLOGY

Ecology of coastal Antarctica

Antarctica covers an area of 14 million km^2 and has a coastline over 32,000 km long (Figure A21). However, more than 80% of this coastline is comprised of ice cliffs formed by either glaciers or the polar plateau. On average, the continental shelf is both narrower and deeper than other continents, averaging more then 600 m deep.

Marine life in Antarctic waters is controlled by the presence of sea ice. This ice, which can be more than 2 m thick, grows out from Antarctica each year until it covers approximately 20 million km^2; more than doubling the area of the continent. It reaches its maximum extent by late October but by late January it has retreated to the coast. The ice affects Antarctic life in two main ways. Most importantly, it reduces the amount of light reaching the underlying water column and so reduces the primary productivity and secondly, together with icebergs, the sea ice also disturbs and scours much of the shallow coastal zone.

The first explorer to sail south of the Antarctic Circle was secondly Captain James Cook in 1773. He noted the abundant seal and whale life but concluded "that no good would come of it" (Antarctica). His observations on the abundant seal and whale life, however, were to have a profound impact on the ecology of the Southern Ocean as it led to the focus of the whaling and sealing industries moving from the Northern to the Southern Hemisphere. Over the subsequent 150 years these industries effectively removed the higher order predators from the

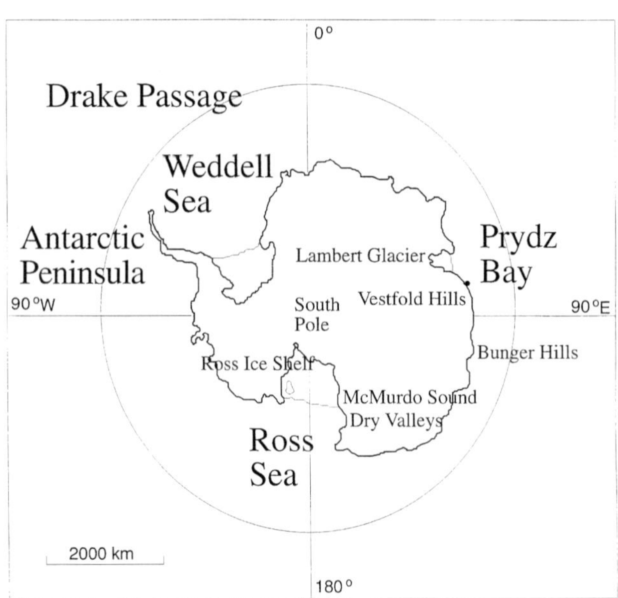

Figure A21 Map of Antarctica.

ecosystem and only now is there evidence that seal and whale populations are beginning to recover from their early savage onslaught. The first record of penguins was much earlier and surprisingly, by Sir Francis Drake in 1644. He commented

> Wee found great store of strange birds, which could not flie at all, nor yet runne so fast as they could escape with their lives; in body they are less than a goose and bigger than a mallard, short and thicke sett together, having no feathers, but insteed thereof a certaine hard and mattered downe; their beaks are not much unlike the bills of crowes, they lodge and breed upon the land, where making earthes, as the conies doe, in the ground, they lay their egges and bring up their young, their feeding and provision to live on is the sea, where they swimm in such sort, as nature may seeme to have granted them no small prerogative of swiftnesse.

Many of the early explorers had biologists aboard and by the beginning of the 20th century most of the important components of the marine ecosystem had been described. Research over more recent years has concentrated on the significance of the microbial ecosystem, specific adaptations to the unique characters of the Antarctic environment, and the effects of human impacts, such as global warming, the ozone hole, over fishing, and pollutants.

The base of all marine food chains is comprised of algae. In the open ocean this is comprised exclusively of phytoplankton. In coastal areas seaweeds and benthic microalgal mats are also important. In Antarctica, a forth and often dominant component is contributed by the sea ice algae.

Phytoplankton in the waters around Antarctica are comprised of the same basic algal divisions as elsewhere. Diatoms are particulary prominant but summer blooms are often comprised of phytoflagellates such as Prymnesiophytes (*Phaeocystis* in particular), Cryptophytes, Prasinophytes, and dinoflagellates.

The largest size class of phytoplankton, that is, the net plankton (20–200 μm), is dominated by diatoms and these comprise most of the large spring algal blooms. The nanoplankton (2–20 μm), unlike most other areas is also dominated by diatoms, *Fragilariopsis cylindrus* in particular, but *Phaeocystis* and Parmales can also form large blooms. Cyanobacteria (i.e., blue green algae) characterize the picoplankton (0.2–2 μm) in most other areas but these are almost completely absent from Antarctic waters.

Each year as the ice melts and the pack ice retreats, it leaves behind it a surface layer of slightly fresher water. Algal cells fall out of the melting ice and contribute to a major annual algal bloom which tracks the retreating ice edge back to the continent. When it reaches the continent a major summer coastal phytoplankton bloom results. This bloom can account for more than 60% of the annual primary productivity and has a similar biomass concentration to phytoplankton blooms elsewhere in the world (up to 10 μg Chla l^{-1}). At other times of the year the phytoplankton biomass is very low (<0.1 μg Chla l^{-1}) but comparable to many other remote ocean sites. In most oceanic areas of the world the availability of nutrients (i.e., nitrate, phosphate, and silicate) limits the growth of algal blooms. In Antarctica, however, this is rarely the case and nutrients are nearly always present in excess. Here phytoplankton growth is usually limited by low light levels and deep mixing. Limited iron (Fe) availability is now also thought to reduce phytoplankton productivity over large areas of the Southern Ocean, but this is unlikely to be the case in coastal areas where iron is available from both the sediments and wind-blown dust. Early explorers often noted the teeming wildlife of Antarctic waters and concluded that it was an area of high productivity but in fact more recently we have found that the reverse is the case. The explorers typically sailed south as the ice was retreating in spring and early summer, a time that coincides with the one time in the year when the melting ice creates the conditions that allow a transient algal bloom. This bloom in turn attracted the abundant zooplankton, birds, whales, etc. The rest of the year the Southern Ocean is remarkably unproductive.

Seaweeds make only a minor contribution to Antarctic primary productivity, first because there is little rocky substrate and second because of the disturbance caused by sea ice and icebergs. Diversity is relatively low and is dominated by red algae. Marine benthic algal mats have not received much attention although it is likely that they contribute up to 50% of the primary production in many shallow marine areas. They are dominantly comprised of diatoms.

When the sea ice reforms in autumn phytoplankton cells are taken up into the ice. The few species that are able to survive do well and grow to be released as an innoculum when the ice again melts the following spring.

The algae form communities either at the top of the ice, within the interior or on the bottom. In late winter and spring in coastal areas the most important communities grow as either encrusting or hanging communities on the bottom of the ice. It is only these bottom communities that are available as food to grazing invertebrates and fish. The algal biomass growing on the bottom of the ice is so high (380 mg Chla m^{-2}) the undersurface of the ice often appears dark brown to black and has a biomass approaching the theoretical maximum for the available sunlight. In coastal areas the dominant species are the diatoms *Nitzschia stellata, Entomeneis kjellmannii, Berkelaya adeliense*, and *Thalassiosira australis*.

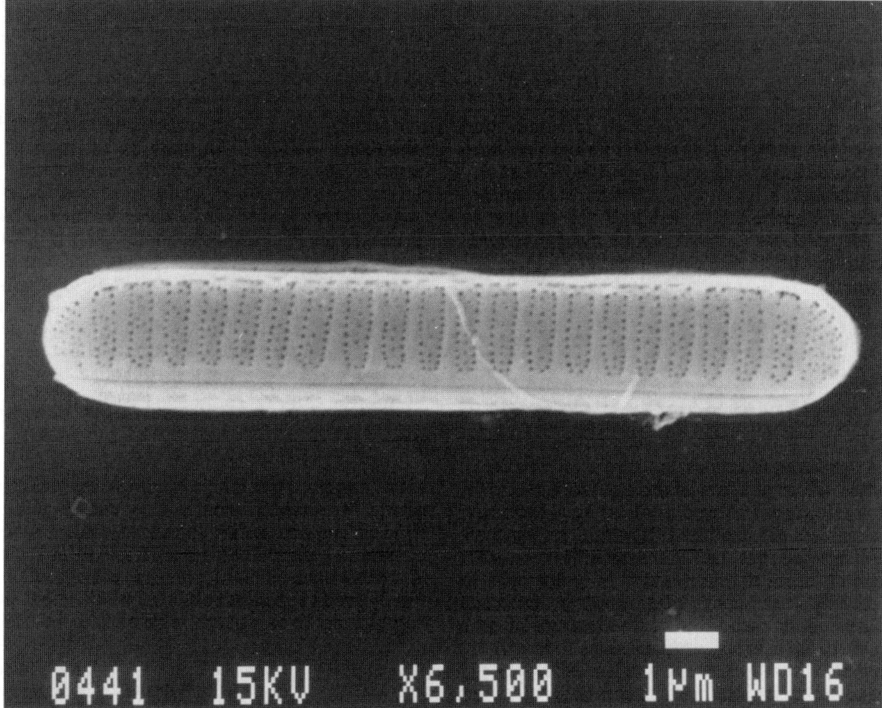

Figure A22 Scanning electron microscope photograph of *F. cylindrus*, an important diatom species in both the phytoplankton and sea ice. It also comprises a major proportion of many bottom sediments.

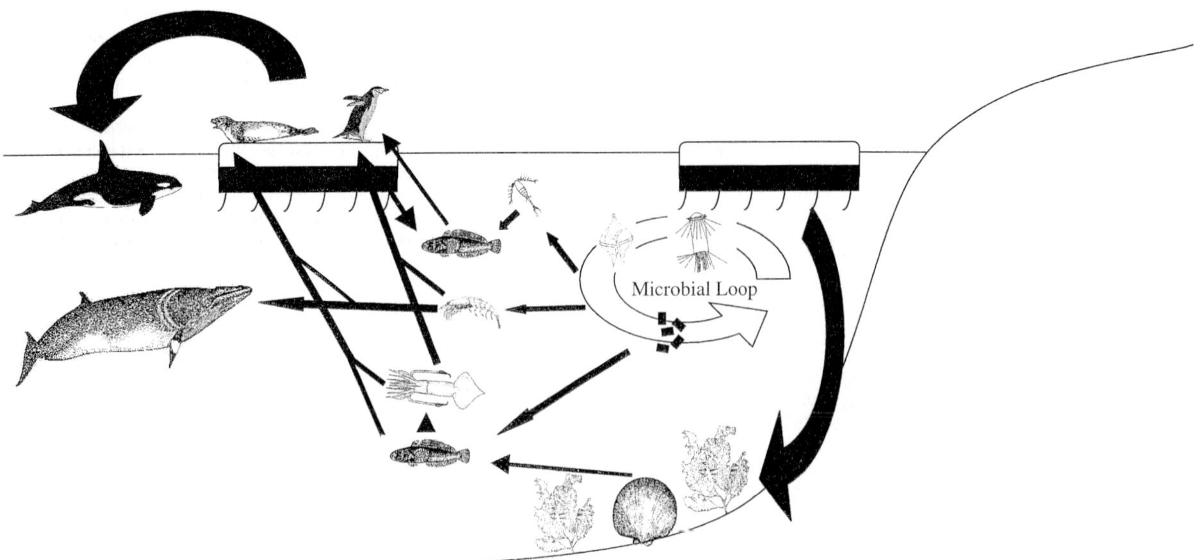

Figure A23 Simplified food web diagram of Antarctic coastal communities. Arrows indicate that an organism is eaten by the animal that is being pointed at. Diagram shows the importance of the microbial loop and sea ice communities to the whole ecosystem. The large arrow from the sea ice to the bottom indicates that most ice algal production falls to the bottom where it is grazed by invertebrates.

Further off shore in the pack ice, the ice communities are dominated by the diatoms *Fragilariopsis curta* and *F. cylindrus* (Figure A22).

With the arrival of warmer temperatures in late spring, melt pools develop on the surface of the ice. These are predominantly colonized by a dinoflagellate, *Polarella*, and chrysophytes. These organisms are released into the water column when the ice finally melts or breaks out in late December or January.

The seafloor, sea ice, and the water column all contain abundant and diverse communities of bacteria and protozoans. These organisms, together with the phytoplankton, comprise the elemensts of the microbial loop (Figure A23). The loop is so named because it describes the route taken by the organic material which is created by the phytoplankton. Phytoplankton use the sun's light energy to synthesize organic matter which they use to grow and reproduce. These cells are gazed on by heterotrophic protists, which in Antarctic coastal waters are mostly dinoflagellates, amoebae, and choanoflagellates. These organisms either die or are attacked by bacteria or viruses, which then decompose to become nutrients for the phytoplankton. It has been estimated that as much as 50% of the energy, which has been harvested from the sun by the phytoplankton, is circled around this loop and thus not made available to multicellular grazers.

The sea ice algae is directly grazed on by small crustaceans such as amphipods, krill, and copepods. These are able to consume up to 2% of the total biomass per day. The other main grazer on sea ice algae is the ice fish, *Pagothenia borchgrevinki*. Much of the primary production of the sea ice falls to the bottom when the ice begins to melt. Here it becomes available as food, together with the benthic algal mats, to the benthic invertebrate community. This community is comprised of a diverse range of sponges, starfish, sea urchins, sea slugs, worms, and shellfish.

The best known invertebrate of the Southern Ocean is the Antarctic krill, *Euphausia superba*. This shrimp-like crustacean, which is up to 20 cm long, forms incredibly large aggregations of up to 100 million individuals and is arguably the most abundant multicellular organism on earth. However, it is not usually a creature of coastal areas; there, it's smaller cousin *Euphausia crystallorophius*, takes over. These creatures together with smaller zooplankton such as copepods and fish larvae comprise the main herbivorous grazers of the phytoplankton.

Zooplankton

The herbivorous zooplankton is composed largely of heterotrophic protists (e.g., dinoflagellates), pelagic crustaceans (e.g., copepods, amphipods, and euphasids), laval fish, and salps. Compared to most other areas the biodiversity of the Antarctic coastal zooplankoton community is relatively low although the biomass is high. Diversity tends to increase with depth. Salps, gelatinous zooplankton, are very significant filter feeders in pelagic waters around Antarctica but their role in coastal communities is unknown.

Biomass and reproduction is strongly seasonal with the peaks coinciding with the development of the maximum phytoplankton bloom soon after the disappearance of the secure. Zooplankton biomass in McMurdo Sound in February has been estimated at 1.5–3.4 g dry weight m^{-3}, a biomass toward the top of the range for coastal communities elsewhere.

Fish and squid

Compared to most other areas the diversity and biomass of Antarctic coastal fish is very low. There is only one common truly pelagic fish, the Antarctic Silverfish(*Pleuragramma antarcticum*), and this commonly comprises more than 90% of the pelagic fish biomass. This species has been found in the diet of all fish that live above the bottom, Weddell Seals, whales, and both Emporer and Adelie Penguins. The diet of *Pleuragramma* itself is comprised largely of copepods, mysids, and other fish.

There is a community of fish associated with the underside of the sea ice. This community is characterized by *Trematomus* spp. and *P. borchgrevinki*. These fish feed directly on the sea ice biota. Other fish species live and feed near the bottom.

Many species of Antarctic fish have developed remarkable adaptations to help them survive in the Antarctic marine environment. The body fluids of most fish have a salinity which would freeze at approximately −0.7°. Since the water temperature is usually around −1.8° they would freeze solid unless they had a mechanism to deal with it. Many, such as *P. borchgrevinki*, have evolved antifreezes comprised of glycoproteins. These depress the freezing point of the fish by inhibiting the growth of ice crystals as contact will cause damage as they get larger. Fish which do not possess antifreezes, such as *P. antarcticum*, must avoid coming into contact with ice crystals as contact will cause snap freezing.

Squid almost certainly comprise a critical component of the Antarctic coastal ecosystem and yet very little is known about them. Approximately, 72 species occur south of the Antarctic convergence but only a relatively small number are thought to live near the coast. They are known to feed on krill but also feed on a variety of other zooplankton and fish species. They themselves comprise an important component of the diets of several whale species (e.g., Sperm Whale, Killer Whale), seal species (Weddell Seals, Ross Seals, and Elephant Seals), and birds (particularly Emperor Penguins and albatrosses).

Benthic communities

Antarctic coastal benthic environments are characterized by both stable low temperatures and the effects of ice. Seasonal changes in water temperature are small compared to most temperate areas with a range of only 2–3°, from −2° in winter to approximately +1° in summer. Melting ice in spring and summer reduces the salinity of surface waters. Conversely, forming ice in autumn and winter expels salt producing water of higher salinity, which then sinks to the bottom.

Sea ice can affect benthic communities down to at least 10 m. The moving ice scours and abrades littoral and sublittoral areas. Icebergs disturb the seafloor down to depths of over 150 m. These bergs, driven by currents and winds can scour rocky surfaces and plough into sediments.

Benthic microalgal mats grow in most shallow nearshore environments. These mats, which are mostly dominated by diatoms, can account for up to 50% of the annual primary productivity. The biomass of these mats is strongly seasonal and is governed by the amount of light reaching the bottom. In winter, when the sun either does not rise or is low on the horizon for only a few hours each day biomass is at a minimum. The amount of light reaching the bottom is usually further reduced by a cover of sea ice up to 2 m thick.

Over 700 species of seaweed are known form the southern Ocean of which approximately 35% are thought to be endemic. Seaweeds are uncommon down to a depth of 5 m due to the scouring action of the sea ice. In McMurdo Sound *Iridaea cordata* is dominant at shallow depths, *Phyllophora antarcica* at intermediate depths and *Leptophyllum coulmanicm* below 20 m. At all locations the diversity increases with depth.

The equable but cold marine environment of Antarctica has led to the development of benthic communities characterized by high biomass but low growth rates. Biodiversity is similar to benthic communities elsewhere but endemism in most major groups, for example, echinoderms, bryzoa, polychaetes, amphipods, etc, is between 50% and 90%

Epibenthic (i.e., those living on the surface rather than within the substrate) invertebrate communities are also strongly influenced by the actions of sea ice. The upper zone, 0–15 m is almost devoid of animals but is briefly colonized by detritus-feeding echinoids (*Sterechinus neumayeri*) and starfish (*Odonaster validus*), isopods, and small fish. An intermediate zone, 15–33 m is characterized by abundant soft corals, sponges, anemones, hydroids, and ascidians. Fish, mostly *Pagothenia bernachii* and *Trematomus centronotus*, are also conspicuous. The lower zone, below 33 m and down to approximately 200 m is characterized by a diverse community of sponges. Infaunal communities (i.e., those living within the sediment) are dominated by polychaetes (worms) and crustaceans; mollusks, unlike temperate areas contribute less than 5% of the total number of individuals. The number of species found is relatively low, usually less than 100, but the number of individuals can be enormous with over 155,000 individuals per meter square found in McMurdo Sound.

A characteristic of Antarctic benthic communities is the high proportion of suspension feeders. This is thought to be a response to the short period of phytoplankton availability during the short summer ice-free period. This has also led to an increasing prevalence of scavenging and opportunistic feeding strategies compared to elsewhere.

Penguins and other sea birds

The abundance of birds in many areas of Antarctica gives a deceptive impression of their biodiversity. Thirty-eight bird species are known to breed south of the Polar Front but only ten actually breed on the Antarctic continent. This level of diversity is very low compared to most other natural areas. Penguins comprise approximately 90% of the Antarctic avian biomass. Four species, the Emperor (*Aptenodytes forsteri*), Adelie (*Pygoscelis adeliae*), Gentoo (*Pygoscelis papuaa*), and Chinstrap (*Pygoscelis antarctica*) breed on the Antarctic continent, but of these the Gentoo and Chintrap Penguins are restricted to the Antarctic Peninsula area. The Emperor Penguin is the largest of the penguins. It alone of all the vertebrates winters on the continent. All other birds and mammals migrate north for the winter. The Adelie Penguin is the most abundant penguin and colonizes areas of bare rock on the coast or on nearby offshore islands. Colonies can be up to several hundred thousand in size. All penguin diets are dominated by krill, although squid and fish supplement the diets to varying degrees depending on location and seasonal abundance.

Most of the remaining birds breeding on the Antarctic continent belong to the petrel family (Procellariiformes). These include the Wilsons Storm Petrel, Snow Petrel, Antarctic Petrel, Cape Petrel, Antarctic Fulmar, and Giant Petrel. The Southern Skua, a large gull-like bird is an aggressive predator at many bird colonies. Although rarely taking adult birds, it feeds on eggs, chicks, and sick individuals. The only other species is the Blue Eyed Cormorant, which is restricted to northern areas of the Antarctic Peninsula.

Bird distribution patterns and breeding cycles are strongly influenced by the short intense pulse of marine primary production between December and February. As this is the only time when food is abundant, the breeding cycle of all birds, except the Emperor Penguin, must be rushed to coincide with maximum food availability. It is essential that the chicks are fledged by the end of summer so they can migrate north before the growing sea ice prevents them access to their food supply.

Seals and whales

Six species of seals are found around Antarctica, the Weddell Seal, the Ross Seal, the Crabeater Seal, the Leopard Seal, the Antarctic Fur Seal, and the Elephant Seal. However, of these only the Weddell Seal actually breeds on the Antarctic continent. This species, which breeds close to the continent on the fast ice and feeds mostly on fish, squid, and krill, forms groups of up to several hundred. Leopard Seals, together with the Killer Whale, are the main top-order predators. These species both feed on penguins and other seals. The Ross and Crabeater seals (the world's most abundant large mammal) spend their entire lives in the pack ice and never venture to shore. The Elephant and Fur seals are mostly creatures of the sub Antarctic and only juveniles or vagrants make appearances on the Antarctic coast.

Twelve species of toothed whale and seven species of baleen whale are present in the Southern Ocean. None are endemic to the region and most have widespread migration patterns. As a result of the whaling impact most of the larger whales are uncommon, with Blue Whale numbers probably down to only a few hundred. However, the smallest of the baleen whales, the Minke Whale, was never seriously harvested and is possibly more abundant than ever. This species is common in coastal areas. All baleen whales feed on zooplankton, predominantly on krill.

Of the toothed whales, the Killer Whale is the most abundant. It is common throughout the pack ice zone and when the ice retreats from the coast in summer, is also a common visitor to coastal areas. It feeds on most larger animals including fish, penguins, and seals. There are no known human deaths attributable to Killer Whales although there are numerous stories of near misses.

The ecosystem of the Antarctic is often seen as pristine and untouched but this is not the case. The whaling and sealing industries of the 19th and early 20th centuries massively depleted the numbers of top-order predators and effectively removed an important component of the ecosystem. Similarly, in the 1970s, over-fishing led to the commercial extinction of most of the more abundant fish species. National bases have degraded areas of the coastline and dumped waste into the nearby ocean. PVCs, DDT, and other toxic compounds have been found in the tissues of most animals. Introduced diseases (e.g., Newcastle disease, canine distemper) have caused widespread deaths among bird and seal populations. The ozone hole and global warming threaten to make even more significant changes. In spite of all these factors, Antarctica is still one of the least impacted ecosystems.

Antarctic geomorphology

Antarctica is simultaneously both the highest and driest continent on earth. It covers an area of approximately 14×10^6 km^2 and has a coastline approximately 32,000 km long. The average elevation is over 2,300 m but this is largely made up of an ice cap that is over 4,000 m thick near its center close to the south Pole. The ice cap contains 90% of the world's freshwater reserves and covers more than 95% of the continent. The remaining ice-free areas are mostly small, scattered, and located near the coast. The Antarctic coastline itself is predominantly ice bound. Over 38% of its length is comprised of ice cliffs where the ice cap terminates abruptly at the sea. A further 44% of the coast is associated with floating ice shelves, large areas where the polar ice cap extends from the continent and floats on the sea. The largest of these, the Ross Ice Shelf, covers an area of 536,000 km^2; an area larger than the U.S. state of Texas or the country of France. About 13% of the coast is occupied by glaciers and only 5% of the coastline is ice-free and even in these areas sea ice covers the sea surface for up to 11 months of the year.

Much of the land surface of Antarctica is currently below sea level, with the ice cap grounding at depths of up to 1,200 m. Were the ice cap to melt, the land surface would rebound by up to several hundred meters but more than 50% of the current land surface would still be under water and the Antarctic Peninsula would become a chain of islands.

Geological investigations of Antarctica have been closely associated with its exploration history. Of particular interest was the position of the South Magnetic Pole. The British explorer, James Clarke Ross, had this objective in mind when he entered what is now known as the Ross Sea in 1841. However, as he sailed southeast he was confronted by the Transantarctic Mountains blocking his way to the Magnetic Pole. Disappointed, he sailed south reaching 76°S and in the process discovering amongst other things the Ross Ice Shelf and Mount Erebus. Most of the subsequent expeditions of the late 19th and early 20th centuries, which often had funding and backing from scientific associations such as the British Royal Society or Royal Geographical Society, took geological and geographical scientific parties along. Both of Robert Falcon Scott's expeditions included geologists as did those of Shackleton, Mawson (himself a geologist and also searching for the South Magnetic Pole), Drygalski, and others. On Scott's fateful return trip from the Pole, the explorers never considered discarding their heavy load of geological specimens even though they probably contributed to their demise. The single most important event in the scientific understanding of Antarctica was the International Geophysical Year in 1957. This led to the establishment of a large number of permanent bases on the Antarctic continent with the specific aim of investigating the structure and evolution of Antarctica.

The present shape of the Antarctica continent is profoundly influenced by its geological past. Antarctica was an integral part of the ancient supercontinent of Gondwana. This continent started to break up in the Jurassic (approximately 200 million years ago) but the last major segments, that is, Australia and South America, did not finally break from Antarctica until the Oligocene, approximately 40 million years ago. For virtually all of this time Antarctica was situated over the South Pole. The northward movement of Australia and South America away from Antarctica led to the development of Drake Passage beneath South America and the Macquarie Passage beneath Australia. This opening ocean led to the progressive thermal isolation of Antarctica and the onset of continental glaciation. As the Southern Ocean widened, Antarctica continued to cool and a permanent ice cap developed, a feature of Antarctica that has been present ever since. The stability of the ice cap, however, has varied and there is now strong evidence for major periods of ice cap shrinkage and expansion through the Pliocene and early Pleistocene (5.1 million years ago).

The erosive and depositional action of ice is almost entirely responsible for the topography, landforms, and continental shelf bathymetry of Antarctica. While melt water streams are seasonally present in some areas they have been inconsequential in either landscape evolution or marine processes. In Antarctica, ice either covers the surface or is responsible for its features. Less than 5% of the surface area of Antarctica is ice-free. These ice-free areas are small and clustered around the coast. The largest of these is the Dry Valleys of the McMurdo Sound area followed by the Bunger Hills and Vestfold Hills of eastern Antarctica. These areas typically show many of the classical features of recently glaciated terrains, including moraines, striated surfaces, drumlins, etc. Many of these areas are also characterized by rich lake provinces. The Vestfold Hills, for instance, an area of only approximately 400 km^2 contains more than 200 lakes, many of which are hyposaline reflecting a marine origin and subsequent stranding by isostatic sea-level rise. Lakes in the Vestfold Hills area are less than 12,000 years old but some lakes in the Dry Valleys may be as old as 500,000 years. Some of these lakes are permanently ice covered (e.g., Lake Hoare and Lake Fryxell in the Dry Valleys), while some are so saline that they never freeze (Deep Lake, Lake Stineer, Vestfold Hills). One of the more unusual features of many of these lakes is that they are permanently stratified (meromictic), that is the water column is layered with zones of increasing salinity with depth. These layers have typically been undisturbed for thousands of years.

Soft sediment shores are uncommon around Antarctica. This is largely because of the absence of significant river action. Small local deltaic features are sometimes present but these are always of very limited extent. Sandy beaches are particularly uncommon in Antarctica as there is rarely sufficient water action to provide the necessary sorting.

The continental shelf of Antarctica is generally narrow, varying in width between 64 and 240 km wide. However, its most distinguishing features are its irregular topography and its great depth, which averages over 500 m; a depth which is more than five times the global average. The outer shelf is typically 400–500 m depth but, unusually, has a landward sloping profile that deepens toward the coast. Depths of 100–1,200 m are common close to the coast in many areas around Antarctica. The reasons for this unusual profile have been speculated on since the late 1920s. It has been shown that glacial isostasy, that is, the effect of the heavy ice cap depressing the crustal surface, can account for up to 150 m of this depth but the remainder is almost certainly due to the effects of glacial erosion at times of low sea-level stand and glacial advance. In some areas during periods of lower sea level, glaciers carved channels far out onto the continental shelf. These drowned fjords are often beneath the effects of modern iceberg scouring and contain basins with restricted circulation. This has led to the preservation of very well-preserved late Quaternary sequences that have a remarkably good time resolution. Such sequences can be found off the Antarctic Peninsula and on the MacRobertsonland Coast. These glacial troughs are usually associated with modern major ice streams and outlet glaciers and they are likely to be the result of multiple glacial advances. Geomorphic landforms associated with these troughs include megascale glacial lineations, striations, and drumlins.

Sedimentation on the continental shelf is dominated by biogenic and glaciomarine processes. Biogenic sediment is comprised of the remains of animals and plants, predominantly from the plankton, while glaciomarine sediment is sediment-derived originally from glaciers but subsequently deposited in the ocean. In offshore areas up to 50% of the sediment can be of biogenic origin. This is overwhelmingly composed of diatom remains. Diatoms are microscopic plants, 2–100 μm in diameter, with siliceous cell walls. Diatom diversity in these sediments is relatively low due to extensive silica dissolution, a factor that becomes more pronounced with increasing water depth. Dominant species include *F. curta, F. cylindrus, Fragilariopsis kerguelenensis, Thalassiosira lentiginosa*, and *Eucampia antarctica*. Other biogenic components include minor quantities of silicoflagellates, Parmales, and foraminera. The remaining sediment is predominantly of glaciogenic origin, although a small component is also of eolian origin.

Around Antarctica, icebergs influence sedimentation up to a distance of several hundred kilometers from the continent. Sediment entrained in a basal debris zone is transported towards the coast. Sediment enters the sea by one of three principal processes; either at ice shelves, ice cliffs, or via glaciers. Ice at the edge of ice shelves calves into icebergs, which drift out to sea and drop their sediment often far from the coast. Most of the sediment, however, is deposited in the grounding zone (i.e., where the ice shelf begins to float). Ice at ice cliffs is eroded and drops its sediment load in close proximity. Most sediment reaches the coast by glaciers, which calve rapidly into small sediment-laden icebergs. It has been estimated that 62% of the ice transported to the coast of Antarctica is calved as icebergs from ice shelves, 22% from glaciers, and 16% from ice cliffs (Drewry and Cooper, 1981).

Most of the continental shelf, at least down to 400 m, is affected by icebergs. These regularly plough into the sediment, disrupting any bedding and causing the sediments to become well mixed and unlaminated. These sediments are sometimes referred to as iceberg turbate. Calcareous sediments are rare on the Antarctic continental shelf but have been recorded on the shelf break in Prydz Bay. Here the dominant component is barnacle shells.

In Antarctica glaciers are by far the most important contributors of terrestrial sediment to the marine environment. Compared to many glaciers in temperate or Arctic areas, they are predominantly sluggish-moving, very cold, and usually extend out to sea as a floating ice tongue. Most Antarctic glaciers are classified as "cold glaciers," that is the temperature of the ice is below the pressure melting point. This results in a glacier that is dry and frozen to the bed. On the Antarctic Peninsula, where the temperatures are higher and the precipitation is much greater they are more dynamic and the glaciers extend to near the mouths of short fjords. These glaciers which often undergo substantial melting in summer are often classified as "subpolar" glaciers. In cold glaciers, where the ice is frozen to the bed, there is comparatively little erosion of bedrock or deposition of sediment. Furthermore, there is rarely any glaciofluvial sediment.

Glaciers enter the sea via fjords, deep usually linear embayments with shallow sills at their entrance. In most temperate areas the glaciers have retreated leaving a deep marine embayment. In Antarctica, however, the glaciers mostly reach the sea and then extend as a floating ice tongue. Minor glaciers are present in areas where the ice cap has retreated such as in the Vestfold Hills and Dry Valleys. The world's largest glacier is the Lambert Glacier in eastern Antarctica. This glacier drains approximately 25% of the eastern Antarctic ice cap and occupies a trough up to 80 km wide and 800 km long. This glacier has retreated in the geological past and would have left the world's largest fjord.

Global warming resulting from the current accumulation of greenhouse gases is already having a major impact on Antarctic glaciers and ice shelves. Temperatures over the Antarctic are predicted to rise faster than most other places on earth. This is leading to an increase in the rate of disintegration of major ice shelves, particularly evident on the Antarctic Peninsula. A major focus of recent research has been to establish the natural limits of Antarctic climate fluctuations in the recent and more distant past. Minor warming and cooling trends of a few degrees have been identified since the end of the last ice age (i.e., over the last 10,000 years) which have led to small scale advances and retreats of the ice cap and glaciers. However, during the Pleistocene (i.e., 1.6 million to 10,000 years) the ice cap at times extended to the edge of the continental shelf. Periods of increased temperature, comparable to those predicted for Antarctica, have been identified in the Pliocene (i.e., 1.6–5 million years ago). During these times there is strong evidence of major ice cap retreat; indicating a possible warm future for Antarctica.

Further recommended ready is given in the following bibliography.

Andrew McMinn

Bibliography

Anderson, J.B., 1999. *Antarctic Marine Geology*. Cambridge: Cambridge University Press.

Drewry, D.J., and Cooper, A.P.R., 1981. Processes and models of Antarctic glaciomarinesedimentation. *Annals of Glaciology*, **2**: 117–122.

Hambrey, M., 1994. *Glacial Environments*. London: UCL Press.

Knox, J., 1994. *Biology of the Southern Ocean*. Cambridge: Cambridge University Press.

Cross-references

Arctic, Coastal Ecology
Arctic, Coastal Geomorphology
Glaciated Coasts
Ice-Bordered Coasts
Paraglacial Coasts

AQUACULTURE

Definition and history

Aquaculture can be most simply defined as underwater agriculture. It is the rearing of aquatic plants and animals through some type of intervention by humans. Aquaculture is conducted in freshwater and saltwater. The term mariculture is often used in relation to marine aquaculture. Aquaculture involves plants and animals produced for human food, as ornamentals, for bait, and in recent years, as sources of nutritional supplements and pharmaceuticals. Some plants and animals are also grown as foods for other aquaculture species. For example, shrimp hatcheries typically grow one or more species of algae to feed brine shrimp (*Artemia salina*), which are fed upon by larval shrimp.

While there is some production of rooted aquatic plants, echinoderms (e.g., sea cucumbers), tunicates (e.g., sea squirts), amphibians (frogs), and reptiles (turtles and alligators), the majority of the world's aquaculture production comes from seaweeds, molluscs (e.g., abalone, clams, oysters, scallops), crustaceans (e.g., shrimp, crawfish), and finfish.

Aquaculture is widely acknowledged to have been born in China, perhaps as long as 4,000 years ago (Avault, 1996) when carp culture was developed. The first written document on aquaculture, a very short book by Fan Li, entitled *Fish Breeding*, appeared in 475 BC in China (Borgese, 1977). Carp culture was also developed in Europe, possibly during the Middle Ages. Tilapia, a type of fish native to north Africa and the Middle East, is depicted on the tombs of the Pharaohs in ancient Egypt. Whether tilapia was being cultured or not is unclear. Oyster culture appears to have been practiced during the days of the Roman Empire, while shrimp were being produced in China by AD 730 (Borgese, 1977). Native Hawaiians constructed fishponds and apparently developed a primitive form of aquaculture at least several hundred years ago. Thus, aquaculture in one form or another has a long history.

By the mid-19th century, trout culture had become a well-developed art in both Europe and North America (Kirk, 1987; Stickney, 1996). In the United States, brook trout were being produced and at least a few commercial fish farms had been established.

Overfishing in the United States was first reported in the 1750s, but it was not until 1871, when Spencer F. Baird convinced Congress to create the US Fish and Fisheries Commission (Stickney, 1996), that any attempt to rebuild wild fisheries was made. Baird enlisted the best fish culturists of the time and put them to work learning how to spawn an array of fishes and invertebrates with the goal of restocking both the inland and marine waters of the nation. In addition, the Commission shipped fish to foreign countries (e.g., establishing populations of rainbow trout in Europe and both rainbow trout and chinook salmon in New Zealand). Fish such as brown trout and common carp were introduced to the United States by the Commission.

While in most cases the stocking of eggs or newly hatched larvae by the Commission did little more than provide food for wild fishes, a significant amount of information and technology were developed, remnants of which can be seen in the most up-to-date of modern hatcheries. Techniques were developed for successfully producing fingerlings of many recreationally and commercially important species, including largemouth bass, channel catfish, walleye, striped bass, and Pacific salmon to name but a few.

The Commission was responsible for producing and distributing the eggs and larvae of hundreds of millions of fishes and invertebrates. Attempts were made to establish Atlantic salmon on the Pacific coast and Pacific salmon in the waters of New England. Both attempts failed, though not because of a lack of effort. Millions of eggs were shipped over a number of years to accommodate the program. Pacific salmon were ultimately established in the Great Lakes, but not until the 1980s.

Ultimately, attempts to enhance marine fisheries through stocking by the Commission and its successor organizations were terminated, though new enhancement programs, largely by the various individual states, have been initiated in recent years. Stocking programs for inland waters, by both state and federal agencies have continued.

Modern commercial aquaculture can arguably be attributed to have its origins in one or more of several countries and decades within the 20th century. For instance, breakthroughs that made possible the development of commercial shrimp culture occurred in Japan during the 1930s, though the industry did not develop significantly until the 1980s after which considerably more information had been developed in Japan, Taiwan, the United States, and other nations; information that was critical to move the technology beyond the research and demonstration phase.

The late 1960s and the 1970s serve as a benchmark period during which commercial aquaculture developed rapidly and began making significant contributions to the world's foodfish supplies. Fish farmers in the southern United States initially produced buffalo (*Ictiobus* sp.), but soon turned to channel catfish (*Ictalurus punctatus*). While the potential for rearing catfish to food size profitably was demonstrated in the 1950s (Swingle, 1956, 1958), commercial activity did not expand from producing and selling catfish fingerlings to growout of foodfish until the mid-1960s (Wellborn and Tucker, 1985).

During the same two decades (1960s and 1970s), commercial crawfish culture developed in Louisiana and rainbow trout production increased dramatically, particularly in the Thousand Springs area of Idaho (Stickney, 1996).

Marine shrimp research with various species in the former genus *Penaeus* (which has recently been split into several genera) was being conducted in both the United States and Taiwan, which led to two different types of larval rearing systems, both of which have been employed extensively. Taiwan and Japan were leading shrimp producing nations in Asia until the 1980s when Thailand, China, Indonesia, the Philippines, India, and others developed active industries. In the Western Hemisphere, the shrimp industry began in the 1980s and was centered in Ecuador, which continues to dominate Latin American production, though many other nations are now involved. The 1970s saw a great deal of interest in the culture of freshwater shrimp, *Macrobrachium rosenbergii*, but for various reasons, there are very few farms currently producing that species.

Also, during the 1980s, the feasibility of producing salmon in net-pens placed in the marine environment was demonstrated (Stickney, 1994). Much of the initial work was conducted on the west coast of the United States with coho (*Oncorhynchus kisutch*) and chinook (*Oncorhynchus tshawytscha*) salmon, and some commercial production occurred in both the United States and Canada. However, it soon became apparent that Atlantic salmon (*Salmo salar*) was more amenable to culture and the industry concentrated on that species, with Norway becoming the dominant salmon-growing nation. Chile became a major salmon-producing nation in the 1990s. Canada continues to produce significant quantities of salmon, and Maine has become the dominant salmon producing state in the United States, though there is still some production in the state of Washington. Large hatchery programs to produce Pacific salmon for release into the wild continue in the Pacific Northwest.

Many other aquatic species have been commercially produced within the past few decades. At present, there are well over 100 species under culture around the world, with more being developed each year. In Europe, sea bass (*Dicentrarchus labrax*) and sea bream (*Sparus aurata*) are being produced in the Mediterranean region, while Atlantic halibut (*Hippoglossus hippoglossus*) are cultured in Norway and Scotland along with the previously mentioned Atlantic salmon. Plaice (*Pleuronectes platessa*) and sole (*Solea solea*) are also among the fishes cultured in Europe.

Popular marine fishes in Japan are salmon (*Oncorhynchus* spp.), red sea bream (*Pagrus major*), and yellowtail (*Seriola quinqueradiata*). Milkfish (*Chanos chanos*) are produced primarily in the Philippines, Thailand, and Indonesia.

Tilapia (*Oreochromis* spp.) are being reared in both salt and freshwater throughout the tropical world and indoors in some temperate areas. Their fast growth and excellent flavor make them excellent aquaculture species.

Among the invertebrates being reared around the world are oysters (e.g., *Crassostrea* spp. and *Ostrea* spp.), including of course, pearl oysters, mussels (primarily *Mytilus* spp.), abalone (e.g., *Haliotis* spp.), and scallops (e.g., *Pecten* spp. and *Patinopecten yessoensis*).

Then there are the plants, dominated by red and brown seaweeds, but including green seaweeds (all of which are forms of algae), but also including higher plants such as water chestnuts, watercress, and such ornamental plants as water lilies, to name but a few. A wide variety of microscopic single-celled algae are produced as food for larval stages of aquacultured animals and, in the case of species such as *Spirulina*, as nutritional supplements for humans (*Spirulina* is also used as a food additive to produce color in fish). Extracts from aquatic plants are also being developed into pharmaceutical products.

Global production

Seaweed aquaculture comprises nearly one-quarter of total world production (New, 1999). Seaweeds play a large role in Asian cuisine, particularly that of Japan. Extracts from seaweeds, including agar and carageenin, are components of products utilized by people around the world. Seaweed extracts are used in automobile tires, toothpaste, ice cream, pharmaceuticals, and a wide variety of other commonly employed products.

By 1996, total world aquaculture production exceeded 34 million metric tons (mmt), with nearly 11 mmt coming from seaweeds (FAO, 1998). A breakdown of production by type of organism is presented in Table A2.

Table A2 World aquaculture production for various groups of organisms in 1996 (data from FAO, 1998)

Organism group	Production (mmt)
Freshwater finfish	14.4
Diadromous finfish[a]	1.7
Marine finfish	0.6
Crustaceans	1.1
Molluscs	8.5
Seaweeds	7.7
Micellaneous[b]	0.1
Total	34.1

[a] Includes fishes that spend part of their life cycle in freshwater and part in saltwater or that can live in either medium throughout their lives. Examples are salmon, eels, and milkfish.

[b] Includes amphibians, reptiles, tunicates, and other minor species.

Table A3 World aquaculture production in 1996 by continent and the top 10 nations in terms of total aquaculture production (data from FAO, 1998)

	Percentage of world total
Continent	
Asia	88.9
Europe	6.0
North America	2.3
South America	1.6
Oceania	0.3
Former USSR	0.4
Africa	0.4
Nation	
China	45.7
India	7.4
Japan	7.0
Republic of Korea	4.5
Indonesia	3.6
United States	3.6
Philippines	3.2
Taiwan	2.8
Spain	2.6
France	2.2

The most widely produced finfish are various species of carp, with the vast majority of that production occurring in China. In fact, China was responsible for 45.7% of the world's aquaculture production, excluding seaweeds, in 1996. When Asia is compared with other regions, the dominance of that continent in terms of aquaculture production is even more impressive (Table A3). Comparison of the top 10 countries in terms of aquaculture production also demonstrates the dominance of China, but also the important contributions of other Asian nations (Table A3).

The world's capture fisheries peaked and now hover around 80 mmt, while demand continues to increase. Aquaculture expansion has been identified as the only means of meeting that demand, and currently available data show that aquaculture production has rapidly increased in recent years. New (1999) developed a comprehensive overview of global aquaculture production that includes a comparison of data between 1987 and 1996. Production over the period covered increased from 13.4 mmt in 1987 to 34.1 mmt in 1996. That rate of expansion cannot be maintained indefinitely, however. Many of the best places to conduct aquaculture are currently being utilized, while various constraints mitigate against aquaculture development in regions where climate might be conducive to good growth of aquatic species. Water supply and quality are constraints in many areas. In the coastal zone, the high cost of land and competition with other users often mitigate against using it for aquaculture.

One approach is to establish fish farms in the open ocean. The costs and logistics associated with that idea present significant problems, but at least some success has been realized. Enhancement stocking, first undertaken by the US Fish and Fisheries Commission over a century ago, is currently being re-evaluated. Revitalization of the red drum (*Sciaenops ocellatus*) fishery in Texas through an aggressive stocking program coupled with strict fishing regulations (including a ban on commercial fishing) provide an indication that such programs have potential and may be cost effective. One approach would be to pay for the programs through license fees, which would apply to commercial and/or recreational fishermen depending upon the species employed.

Much of the increase in aquaculture production over the years, at least in developed nations, can be attributed to increased knowledge that has led to improved technology. One example is the channel catfish industry in the United States. When the industry was in its infancy, a profit could be made if the fish were sold at $1.10/kg. Forty years later, a profit can still be made when catfish prices to the producer are in the range of $1.65–1.80 per kg. The reason that catfish farmers can still realize a profit given the small increase in price per unit weight over nearly a half century of production is related to increases in production per hectare, which is a function of improvements in management of water quality, improvements in the food provided, and to a lesser extent, selective breeding programs.

In the 1950s, when catfish farming was in its infancy, a production level of about 500 kg/ha was considered credible. By the late 1960s that had increased to 1,500 or even 3,000 kg/ha. Currently, some catfish farmers produce as much as 10,000 kg/ha in ponds. Without those increases in production, the catfish industry would have probably collapsed.

Culture systems

The vast majority of aquaculture species are grown in ponds, though a number of other types of culture systems are also in use. They include linear and circular raceways (the latter are also known as tanks), cages, and net-pens. Specialized culture systems for molluscs and seaweeds include bottom culture in bays, pole culture, net culture, basket culture, and raft culture (Stickney, 1994).

Culture systems are often classified as extensive or intensive, though in reality the intensity of culture occurs along a continuum and relates to the complexity of the culture system. An example of a highly extensive system would be the culture of oysters in natural waters by spreading shell (cultch) material on the bottom and allowing larval oysters (spat) to settle and grow. Very little intervention by humans, other than preparation of the bottom and the spreading of shell is involved.

At the other extreme are high-intensity closed-recirculating water systems. Such systems employ raceways as culture chambers. Water is continuously exchanged in each raceway. Exiting water is treated, at a minimum to remove solids (usually by settling or mechanical filtration) and convert ammonia (NH_3 or NH_4^+) and nitrite (NO_2^-) to less toxic nitrate (NO_3^-) in a unit called a biofilter. Additional components that maintain temperature in the desired range, remove foam, convert nitrate to nitrogen gas (N_2), and remove other nutrients, may also be present. Most closed systems, receive supplemental aeration to maintain a high level of dissolved oxygen in the culture chambers. In truly closed systems, new water is added only to replace that lost to splashout, evaporation, or to flush solids from the system. Systems that have the same basic components but in which some percentage of the water is routinely or continuously changed by adding new water, are called semi-closed.

Between the extremes are the other systems mentioned above. Culture ponds, unlike most farm ponds, are most commonly rectangular in shape, are usually no more than 2 m deep, have a drain structure of some type incorporated into them and are, in most cases, located in an area where there is a reliable source of water. While some culture ponds depend on rainwater runoff to fill them, most ponds, at least in the developed world, employ well water or water from a stream, lake, or reservoir for filling and maintenance. Culture ponds come in many sizes, but most used for commercial production range from about 0.1 to 10 ha, with those between about 1 and 5 ha being the most commonly seen. Small ponds are used most often for breeding and rearing of early life stages, while the growout commonly occurs in larger ponds.

If two or more species are reared in the same culture unit—a process known as polyculture—that unit is most likely a pond. The Chinese have a very long history of carp polyculture, which involves different species with different feeding habits. Different species are stocked to consume phytoplankton, zooplankton, benthic invertebrates, and rooted aquatic plants. Benthic feeding carp may specialize on worms and insect larvae or molluscs. Today, the Chinese may also provide agricultural byproducts and prepared feeds. They, like aquaculturists in many other countries, maintain a high level of pond fertility to encourage the growth of natural food organisms. Cattle, swine, and poultry manure are common sources of fertilizer. Livestock are routinely seen being reared adjacent to ponds into which they deposit fresh manure. Rearing livestock adjacent to fishponds is common in many countries, particularly in the developing world.

The majority of raceway systems are of the flow-through variety. That is, water is continuously exchanged. Depending on stocking density and water quality, the exchange rate may range from several times an hour to several times a day.

Cages are relatively small structures with ridged walls, commonly made of welded wire or rubber-coated wire. Cages are used to contain fish in water bodies that are not enclosed (bays, rivers), and in those which are too large, too deep, or have some physical impediment to harvest (such as standing dead timber). Many reservoirs and lakes are good examples. Cages are not usually more than one to a few cubic meters in volume. Cages may be taken to the shoreside to facilitate harvesting.

Net-pens are large cages where nylon netting is employed for the sides and bottom instead of ridged material. The standard design most commonly used in protected marine waters where the net-pens are not exposed to high waves is typically $20 \times 20\ m^2$ on a side and can be 20 or more meters deep. Both cages and net-pens are provided with floats to keep them at the surface. Cages usually sit low in the water and are fitted with tops, while the sides of net-pens typically extend above the water surface sufficiently to prevent fish from jumping over the open top. Walkways around the perimeter of each net-pen allow access for feeding and harvesting.

Recently, there has been a move toward rearing fish in offshore locations. Net-pens designed for the open ocean are usually fully enclosed by netting because they may be subjected to storms. Some are designed so they can be totally submerged throughout the growing cycle or during stormy periods.

Oysters, mussels, clams, and scallops are less susceptible to predation by starfish and other organisms if they are grown off the bottom. Poles, lines, and baskets have been used in conjunction with off-bottom culture. In Spain, for example, ropes to which oysters or mussels are attached are suspended from large rafts.

Enhancement, as previously mentioned, is an option that deserves further consideration. While the culturists associated with the US Fish and Fisheries Commission in the latter half of the 19th century failed in their efforts to increase the numbers of fish in the sea, advances since that time greatly increase the potential for success.

Hatchery technology has developed to the point where the larvae of many species of interest can be reared from egg to appropriate stocking size economically. Enhancement efforts need to be conducted in an environmentally sound manner, however. Fish and invertebrates should not be stocked in such numbers that they overwhelm the available food supply; they should not cause declines in the populations of other valuable species in the community; and their genetic composition should reflect that of wild populations, to name but a few considerations.

Getting started

Several steps should be taken in advance of actually establishing an aquaculture facility. First, the prospective culturist should determine what species will be produced and type of water system to be employed. A suitable location should also be identified. All three factors tend to be co-dependent. Sometimes, the prospective aquaculturist will own property selected as the site of the new facility. If that property is in a subtropical area and there is no suitable source of cold water, it would be undesirable to attempt to produce trout or salmon. Similarly, if the property is distant from the coast and there is no convenient source of saline groundwater, producing marine fish might not be wise. The exception to those examples might be employment of a closed water system within which temperature and salinity could be controlled. Economic analyses of each option under consideration should be conducted to help determine the best course of action.

Once the above matters have been resolved, it is necessary to produce a business plan. Such a plan will be required by lending institutions and venture capitalists. The plan should include a complete financial analysis showing all costs associated with constructing, equipping, and operating the facility over a period of several years, along with an estimate of profit or loss for each year projected. All estimated costs and income should be based on information that can be documented. The culturist should provide a listing of the permits that may be required for the facility as well as a timetable for obtaining those permits and the fees that are associated. Conceptual drawings, or even better, detailed engineering blueprints, should be included with the business plan.

The prospective aquaculturist should determine what, if any, local, state, or national governmental permits are required. In some instances, the permitting process can be long and arduous. It may also be costly, particularly if the applicant is required to battle opponents within the legal system and conduct an extensive environmental impact study prior to establishing a facility. In general, inland aquaculture sites and those located along the coast that employ recirculating technology or other water systems that do not impact the marine or estuarine environment are more easily permitted than those that are established in public waters.

Facility management

On the surface, aquaculture is a biological discipline. In reality, like terrestrial farming, aquaculture involves a variety of sciences as well as engineering, business management, and such skills as plumbing, welding, carpentry, and electrical wiring. In the science arena, while expertise in biology is critical on any fish farm, an understanding of chemistry may be even more important. As the size and intensity of aquaculture operations increase, so does the need for additional and better-trained personnel. A subsistence culture operation in a developing country might be operated by an individual or a family, with perhaps a few additional laborers. A large pond operation that involves a hatchery, larval rearing, and production facilities, or an intensive raceway operation may require a large number of unskilled laborers as well as employees proficient in various specialties. Some of the technical aspects associated with operation and management of an aquaculture facility are discussed in the following subsections.

Water quality

Maintenance of good water quality is critical in any aquaculture facility. There are virtually hundreds of thousands of chemicals that can be found dissolved in water. In addition, there are physical characteristics such as temperature and transparency that can affect performance of organisms in an aquaculture facility. It is neither possible nor necessary to monitor all aspects of water chemistry. In most culture systems, after an initial screening of the water supply for toxicants (e.g., herbicides, pesticides, high levels of trace metals), routine monitoring involves only a few variables. Water temperature and dissolved oxygen, and in marine systems, salinity are typically measured at least several times a week and records of them are maintained. Some culturists employ sensors that constantly monitor those parameters and download the information into computers. Water transparency monitoring is also conducted in pond systems and provides a simple means of determining the state of plankton blooms. This is particularly important in systems that employ fertilization to provide natural food.

In high intensity systems, ammonia may be routinely monitored. Nitrite levels might also be tracked, particularly during colonization of biofilters with bacteria, since the types of bacteria that convert nitrite to nitrate tend to lag behind those that convert ammonia to nitrite. Thus, if only ammonia is measured, the culturist may get the impression that the water is not toxic, when in fact, it may contain a high level of nitrite. Toxic nitrite levels have occurred in heavily stocked catfish ponds in the United States during the late summer, so routine monitoring during that critical period is often undertaken. While ponds are not equipped with biofilters, the bacterial conversion of ammonia to nitrite and nitrate does occur.

Water temperature is of critical importance because it controls metabolic rate in the species being cultured. Aquatic animals can often be categorized as being warmwater or coldwater species. Warmwater animals are those that have an optimum temperature for growth of 25°C or higher, while coldwater species have an optimum below about 15°C. There are some animals, such as walleye (*Stizostedion vitreum vitreum*) and northern pike (*Esox lucius*) that have an optimum between 15°C and 25°C. Such animals are sometimes referred to as midrange species.

Most species have a broad tolerance for environmental temperature. Many species of fish live over a range of temperatures from near 0°C to somewhat above 30°C. Channel catfish and largemouth bass (*Micropterus salmoides*) are but two examples. Many coldwater species, such as trout and salmon, cannot tolerate warm temperatures, while many tropical fishes, such as tilapia, die at temperatures where trout and salmon thrive. Tilapia grow most rapidly when the water temperature is about 30°C, show significantly reduced growth at about 25°C, and stop growing around 20°C. Below the latter temperature tilapia lose their disease resistance and will die when temperatures fall much below 15°C (Avault and Shell, 1968).

Terrestrial animals live in an atmosphere that contains some 20% oxygen, while aquatic species live in a medium that contains only a few parts per million (ppm) of the same essential element. Depending on species, the minimum level of dissolved oxygen in the water that should be present to ensure that the animals have a sufficient supply varies from about 3 ppm (e.g., for catfish) to 5 ppm (e.g., for salmon and trout). From a practical standpoint, it is always desirable to have dissolved oxygen at saturation.

The amount of oxygen that water can hold (the saturation level) varies with temperature, salinity, and altitude. As each of those variables increases, the ability of water to hold oxygen decreases. Thus, warm, saline water at high altitude would hold less oxygen than cool, freshwater at sea level. In virtually all locations where aquatic animals are cultured, dissolved oxygen saturation level reaches or exceeds 5 ppm.

Dissolved oxygen follows a diurnal (daily) cycle. During daylight hours, phytoplankton and other types of vegetation in the aquatic environment produce oxygen as a result of photosynthesis. As a result, the oxygen level in water tends to increase during the daytime. Oxygen is also introduced into water through diffusion from the atmosphere, a process that is enhanced through wind turbulence or mechanical aeration. At night, both plants and animals respire, thereby consuming oxygen during a period when no photosynthesis is occurring (though diffusion is still operating). The result is a cycle in the availability of oxygen in the water. The cycle is also influenced by weather conditions and season of the year. During cloudy days, the rate of photosynthesis is reduced (because of reduced light level), while during clear weather the rate of photosynthesis is accelerated. Recall that temperature plays a role in the saturation level of dissolved oxygen in water; thus the seasonal variation. In most cases, problems associated with low dissolved oxygen occur during warm weather after a period of cloudy days. This often occurs during the late summer or early fall in temperate areas (the time just prior to harvest).

Aquaculturists routinely measure dissolved oxygen in pond systems early in the morning, particularly during the time of year when depletions are likely. Dissolved oxygen level typically peaks in the late afternoon before dusk and gradually falls during the night, and begins to increase after dawn. The lowest dissolved oxygen level will usually occur just prior to dawn, so the aquaculturist needs to check conditions at that time and determine which, if any, ponds may be approaching critically low oxygen levels.

When oxygen levels are low, the culture animals may alert the observant culturist. Fish may swim at the surface with their mouths open. They appear to be gulping for air. Shrimp may climb above the water level on rooted plants to expose their gills to the atmosphere. By the time such signs are noted, the animals may have already been severely stressed. It is much better to routinely measure dissolved oxygen during the night and take action to prevent severe depletions before they occur.

Under moderate stocking regimes only a few ponds on a given aquaculture facility will experience oxygen depletions on the same day and it is not necessary to have each pond provided with emergency aeration equipment. In many cases today, however, stocking densities are so high that oxygen depletions may occur with some frequency in nearly every pond. Such facilities may have aeration equipment permanently assigned to each pond. Paddlewheel aerators are commonly seen in culture ponds. They may be turned on only when the oxygen level falls below a specific level, or they may routinely be run during the night, or in some cases, continuously.

Mechanical aeration is not the only means of aerating ponds. New, oxygen-rich water can be added, though that can be expensive. Mechanical aeration is not inexpensive, however. It requires the equipment and the electricity or fuel required to operate that equipment.

Aquatic animals may have a wide range of tolerance for salinity (euryhaline) or may be restricted to a rather limited salinity range (stenohaline). Salinity is measured in parts per thousand (ppt). The salinity of freshwater is typically about 0.5 ppm, while full strength seawater is usually about 35 ppt. The optimum salinity for an aquatic species may be constant throughout the life cycle or it may change at various life stages. Salmon, for example, spawn and their eggs hatch in freshwater where the larvae live for a few weeks to two years before migrating to the ocean. Fish that spawn in freshwater but grow to adulthood in seawater are called anadromous. Those that spawn in the ocean and migrate to freshwater to grow to adulthood, such as eels, are catadromous.

While most aquatic organisms survive and grow well at a constant optimal salinity, it may be necessary to change the salinity when rearing certain species. It is important to know when those salinity changes should be made and the extent of the changes.

Ammonia is produced by many aquatic species as a metabolic waste product. Ammonia enters the water through the gills. As previously indicated it can be present in the water in two forms: ionized (NH_4^+) or unionized (NH_3). It is the unionized form is highly toxic to aquatic animals. The ratio between the two forms is determined by factors such as water temperature and pH. As either or both temperature and pH increases, the amount of unionized relative to ionized ammonia increases. Under most conditions, total ammonia concentration should not exceed about 1 mg/l. That will ensure that the level of unionized ammonia is not excessive.

Water pH can undergo fairly dramatic changes diurnally, particularly when there is a strong algae bloom present. As plants and animals respire at night in the absence of sufficient light to promote photosynthesis, the increased level of carbon dioxide in the water causes the pH to drop unless the system is well buffered. During the daytime, carbon dioxide is removed from the water by photosynthetic organisms and pH may rise. Carbonate (CO_3^{2-}) and bicarbonate (HCO_3^-) ions make up the foundation of the buffer system. They can absorb hydrogen ions (H^+) or in the case of bicarbonate, act as a source of H^+. If the supply of bicarbonate becomes exhausted, pH can change dramatically. Poorly buffered systems can be as acidic as pH 6 in the early morning and as basic as pH 10 in the afternoon if the buffer system is weak. In recirculating systems, pH will tend to drop over time as organic acids become concentrated in the water. Most aquatic species cannot tolerate such swings in pH. Calcium carbonate in the form of limestone or crushed oyster shell can be employed in the water system to help maintain pH by providing a source of carbonate ions.

Routine measurement of pH and alkalinity (a measure of carbonate and bicarbonate ions) every few days is common on aquaculture facilities. Hardness (a measurement of the calcium plus magnesium level in the water) may also be monitored. Some species, such as the euryhaline red drum that can be found in estuaries as well as in the coastal ocean of the southeastern Atlantic and Gulf of Mexico coasts of the United States, can survive and grow well in hard freshwater, but performance is not as good in soft freshwater.

Genetics and reproduction

Most species being successfully produced by aquaculturists can be maintained throughout their life cycles. Closing the life cycle has been relatively simple for some species, but very difficult for others. Aquatic animals with large eggs, such as salmon, trout, catfish, and tilapia—each of which has eggs 3 mm or larger in diameter—readily spawn in captivity. Their eggs can be incubated relatively easily and the newly hatched fish will consume prepared feeds as first food after yolk sac absorption.

Many aquatic animals of aquaculture interest have very small eggs and larvae so small as to be nearly invisible to the naked eye. The larvae tend to be very fragile and often have extremely limited swimming ability. If exposed to water currents in a hatchery raceway, they may become impinged on screens and killed. For many species, providing a complete feed that will be accepted has proved difficult so it is often necessary to culture living zooplankton as first food.

The most simple approach is to allow the animals to spawn naturally. Providing the proper conditions for spawning is not always possible, and in some cases, culturists still do not know how to duplicate what the animals experience in nature. Hormone injections are sometimes used to induce spawning. In the case of marine shrimp, removal of an eyestalk from females can trigger ovulation.

Fertilized eggs may be allowed to incubate naturally or they may be taken to a hatchery for artificial incubation. Various types of specialized hatching facilities have been developed. Not only is the survival rate generally higher in a hatchery than in a pond, but the culturist can get a good estimate of the number of offspring that have been produced since they are easy to observe in the hatchery as compared with an outdoor pond situation.

Only a few aquatic animals have been truly domesticated through many generations of selective breeding. Most are still only a step or two removed from their wild counterparts. In some cases, for example, in conjunction with hatcheries that produce fish for stocking into nature, a conscious effort is made to maintain the genetic integrity of the wild stock. For animals bred to be reared in captivity only, selection of brood stock for rapid growth, improved body configuration, disease resistance, and other desirable qualities may be employed.

The use of genetic engineering in conjunction with aquaculture species is in its infancy, but there has been some research activity in that controversial arena. Researchers have successfully incorporated selected genes from one species into another, but there do not appear to be any commercially produced transgenic aquatic animals in the market.

Nutrition and feeding

Feed represents the largest variable cost for most types of aquaculture. The exception is mollusc culture where wild phytoplankton is fed upon in most instances (cultured algae is used in mollusc hatcheries and fertilization of ponds is widely used in China and other countries to produce natural food). Many fishes and crustaceans are provided with commercial feeds. When prepared feeds are employed, they often represent nearly half of the variable costs associated with operating the aquaculture facility.

Most fishes and crustaceans will either accept prepared feeds upon first feeding or can be weaned from natural foods within a few weeks after first feeding. Prepared feeds may be a mixture of ground and mixed feedstuffs that are fed directly, or the ingredients may be formed into pellets. Feed pellets can be made in a variety of sizes and shapes; they can be dry, moist, or semi-moist; they may be hard or soft, and they may either sink or float depending upon the species being fed, local ingredient availability, and the type of equipment used in feed manufacture.

For some species, the nutritional requirements have been determined in some detail. Examples are salmon, trout, catfish, tilapia, carp, and to a lesser extent shrimp, striped bass, red drum, and a number of others (Wilson, 1991; D'Abramo *et al.*, 1997). Depending upon the species, requirement information may be available on dietary energy, protein and amino acids, lipids and fatty acids, carbohydrates, vitamins, and minerals.

Feeds may contain animal protein from a few percent (channel catfish) to one-third or more of the diet (some salmonids), or they may be made up entirely of plant feedstuffs with or without added vitamins and minerals. Most of the animal protein in aquaculture feeds comes from fish meal, though poultry byproduct meal, and meat and bone meal are among the other options. Locally available ingredients are typically employed when possible. Fish feeds in Asia, for example, often contain relatively high percentages of rice bran, which has limited nutritional value but is readily available. Soybean meal is a major ingredient in many fish feeds in the United States. Diets may also contain corn meal, cottonseed meal, and/or peanut meal depending on availability and cost. A wide variety of other ingredients are also used.

Aquatic animals on prepared feeds are usually fed once or twice daily during the growout period. The total amount of feed provided on a daily basis is usually not more than 3–4% of the biomass of the animals being fed. Young animals are usually fed at a higher rate (as much as 50% of body weight initially, and declining as the animals become more proficient at finding feed and their relative growth rate slows). Young animals may be fed more frequently. Some carnivorous species may be offered small amounts of feed as often as every few minutes to keep them satiated and reduce cannibalism.

Feeding may be done by hand, though increasingly, automated feeding systems are being employed. Such systems have been developed for virtually every type of water system. Some feed automatically when activated by timers, while others may be activated when animals go to the feeder. The latter, called demand feeders, are used in conjunction with some finfish rearing operations and feature a rod or plate that is suspended in the water which when bumped by the fish causes a few feed pellets to be released.

Diseases

Aquatic animals are susceptible to a wide variety of diseases. Organisms responsible for disease in aquatic species include fungi, bacteria, nematodes, cestodes, trematodes, as well as parasitic protozoans, copepods, and isopods. Some can cause death while others may stress the affected animal to the point that it becomes more susceptible to additional diseases. Not all aquatic animal diseases are caused by other organisms. Some, such as gas bubble disease, are caused by water quality problems (in this case gas supersaturation). Nutritional deficiency diseases also exist. For example, vitamin C deprivation can lead to decalcification of bone, which can lead to spontaneous fracturing of the vertebral column.

While a variety of disease treatments have been developed and are in widespread use around the world, only a very small number of drugs have been approved for use on aquatic organisms in the United States. The US Food and Drug Administration must approve each drug utilized on foodfish. Because the industry is relatively small, the market for drugs often does not justify the financial investment required by the pharmaceutical companies to obtain clearance.

Controversies in aquaculture

In recent years, aquaculture has come under increasing criticism from environmentalists. Particularly heavy criticism has been leveled against aquaculture being conducted in pubic waters. The conduct of aquaculture on private land has received less opposition, except in cases where effluents from such facilities enter public receiving waters.

Salmon net-pen culture in Washington and Maine, and marine shrimp culture in Texas are two examples of aquaculture activities that have been targeted by anti-aquaculture groups. The net-pen industry in Puget Sound, Washington was initially objected to because upland landowners felt the net-pens represented visual pollution. That objection failed to gain traction in the courts, but many other criticisms followed (Stickney, 1990). Included were claims of:

- Waste feed and feces causing dead zones under net-pens
- Nutrients released from decaying feed and feces leading to eutrophication
- Antibiotics in feed entering the food chain leading to the development of resistant bacterial strains
- Net-pens interfering with recreational and commercial fishing
- Net-pens interfering with navigation
- Fish in net-pens transmitting diseases to wild fish
- Escapees from net-pens interbreeding with wild fish, or in the case of nonnative Atlantic salmon successfully colonizing and displacing wild salmon
- Net-pen aquaculture operations being noisy and producing noxious odors.

Many of the claims have at least some basis in reality or hold the potential of becoming problems, though each can be effectively dealt with if the net-pen facility is properly constructed and placed in the proper location (Parametrix, 1990).

If located in areas where tidal exchange and currents are insufficient to carry away waste products, an anoxic zone can occur. In Japan, uncontrolled development of net-pen facilities in several bays led to so-called self-pollution that resulted in heavy mortalities. Strict regulations on the number of net-pens per unit area are now in force in Japan and the problem has been resolved.

Salmon farmers, as other aquaculturists, are dedicated to maintaining the highest possible water quality and other environmental conditions for the animals under their care. Degradation of the culture environment causes stress and can lead to disease. If disease does occur, an approved antibiotic may be employed, but only for 10 days at recommended rates, so large amounts of such pharmaceuticals are not used.

Atlantic salmon have been known to escape but they cannot interbreed with Pacific salmon (the situation is different in Maine where Atlantic salmon are native). Many attempts were made to establish Atlantic salmon in the Pacific Northwest by the US Fish and Fisheries Commission over a century ago. All of those attempts failed. Atlantic salmon that have escaped into the waters of the Pacific Northwest in North America do not seem to compete well with native Pacific salmon.

Proper siting will help avoid problems with nutrient concentration and interference with other users. Excessive noise is not a valid criticism of net-pen operations and there would only be noxious odors in the event of a fish kill that was not immediately cleaned up.

The Texas shrimp farming industry has come under criticism for releasing sediment and nutrient-laden water into adjacent public waters. In one part of the state the suspended solids load from shrimp pond effluents has been blamed for heavy siltation of navigable waters, while nutrients in the effluent are blamed for harmful algae blooms. There has also been concern expressed that diseases in cultured shrimp could spread to wild populations. The industry is dealing with the situation by constructing wetlands through which water exiting the production ponds passes. Solids can settle and nutrients can be taken up by the wetland vegetation. The water may then be recirculated back to the ponds rather than released to the environment. Significant improvements in water quality have been achieved and production costs reduced when constructed wetlands are employed.

The other major criticism of the shrimp industry in Texas and other states involves the use of exotics. Most of the shrimp raised on farms in the United States are exotics from Asia or Latin America. Research on native species was all but halted in the 1970s, but many feel that attempts should be initiated to solve some of the problems so the industry can move away from exotic or nonindigenous species.

The development of extensive pond culture for shrimp has led to destruction of mangrove areas in many tropical countries. Some nations have recognized the importance of mangroves as habitat for marine organisms and wildlife and as buffers that can dampen the effects of storms on upland regions. As a result, mangrove habitat destruction has been outlawed or significantly reduced due to governmental regulation. In addition, mangroves grow in acid soils, which are not as conducive to supporting aquaculture operations as other soil types.

Many countries have brought in nonindigenous species and introduced them into culture. Besides Atlantic salmon and marine shrimp as previously discussed, examples include tilapia in North, Central, and South America as well as throughout much of tropical Asia; Latin American species of marine shrimp in the United States; various species of carp in the United States and Latin America; and walking catfish (*Clarias* spp.) in Europe. For species which have only recently been introduced into aquaculture as exotics, there is concern about escapement and establishment of wild populations that could compete with and perhaps displace native species. The rapid spread of common carp throughout North America beginning in the 19th century and the dispersal and establishment of tilapia in parts of Asia (where tilapia are now considered native though they have been present in Asia only since the 1930s) and Latin America are good examples.

While transgenic fish are not apparently being commercially produced for sale, researchers have been able to incorporate growth hormone producing genes from one species into another and obtained an increased growth rate. Transgenic fish with so-called antifreeze genes that might, for example, allow tilapia to survive at much lower temperatures than normal have been discussed. This type of activity is controversial and there are many stories in the media that have convinced at least a portion of the public that transgenic fish are threats to the environment as well as to the health of consumers. A great deal of debate surrounding this activity will occur before transgenic fishes become commonplace in the market.

Robert R. Stickney

Bibliography

Avault, J.W., Jr., 1996. *Fundamentals of Aquaculture*. Baton Rouge: AVA Press.

Avault, J.W., Jr., and Shell, E.W., 1968. Preliminary studies with the hybrid tilapia *Tilapia nilotica* × *Tilapia mossambica*. *FAO Fishery Report*, **44**: 237–242.

Borgese, E.M., 1977. *Seafarm*. New York: Harry N. Abrams.

D'Abramo, L.R., Conklin, D.E., and Akiyama D.M. (eds.), 1997. *Crustacean nutrition. Advances in World Aquaculture Volume 6.* Baton Rouge: World Aquaculture Society.

FAO, 1998. Aquaculture production statistics 1987–1996. *FAO Fisheries Circular No. 815, Review 10.* Food and Agriculture Organization of the United Nations, Rome, Italy.

Kirk, R.R., 1987. *A History of Marine Fish Culture in Europe and North America*. Farnham, Surrey: Fishing News Books.

New, M., 1999. Global aquaculture: Current trends and challenges for the 21st century. *World Aquaculture*, **30**(1): 8ff.

Parametrix, 1990. *Fish Culture in Floating Net Pens*, Olympia: Washington Department of Fisheries.

Stickney, R.R., 1990. Controversies in salmon aquaculture and projections for the future of the aquaculture industry. In *Proceedings of the Fourth Pacific Congress on Marine Science and Technology, Tokyo, Japan, July 16–20*, Tokyo: PACON, pp. 455–461.

Stickney, R.R., 1994. *Principles of Aquaculture*. New York: John Wiley & Sons.

Stickney, R.R., 1996. *Aquaculture in the United States*. New York: John Wiley & Sons.

Swingle, H.S., 1956. Preliminary results on the commercial production of channel catfish in ponds. *Proceedings, Annual Conference of Southeastern Game and Fish Commissioners*, **10**: 63–72.

Swingle, H.S., 1958. Experiments on growing fingerling channel catfish to marketable size in ponds. *Proceedings, Annual Conference of Southeastern Game and Fish Commissioners*, **12**: 63–72.

Wellborn, T.L., and Tucker, C.S., 1985. An overview of commercial catfish culture. In Tucker C.S. (ed.), *Channel Catfish Culture*. New York: Elsevier, pp. 1–12.

Wilson, R.P. (ed.), 1991. *Handbook of Nutrient Requirements of Finfish*. Boca Raton: CRC Press.

Cross-references

Economic Value of Beaches
Environmental Quality
Estuaries
Human Impact on Coasts
Mangroves
Water Quality

ARCHAEOLOGICAL SITE LOCATION, EFFECT OF SEA-LEVEL CHANGE

In the past, archaeologists often drew conclusions about prehistoric peoples' use of coastal areas and coastal resources from the observable archaeological sites along the existing coastline. However, as archaeologists began to understand the magnitude of sea-level changes associated with the Pleistocene glacial stages, it became apparent that archaeological sites reflecting coastal adaptations along the present coastline were representative of only those periods of time when the prehistoric coastlines coincided with their present positions.

The coastal sites from most prehistoric periods are now either submerged on the continental shelves due to post-glacial rises in eustatic sea level (see *Eustasy*), or located at elevations far above present sea level due to post-glacial isostatic rebound of the earth's lithosphere. In some areas, local tectonic movements have also affected the elevation of former coastal sites in relation to present sea level. Therefore, it is very important for archaeologists studying coastal archaeological sites to understand how the location of the coastline has changed throughout the prehistory of an area. Understanding these changes will help in reconstructing the proper paleoenvironmental setting of the archaeological sites at the present coastline, and in knowing where to look for former coastal archaeological sites.

Sea level changes

Since the mid-Pleistocene (approximately 900,000 years ago) (see *Pleistocene Epoch*), periods of glacial building have reduced the absolute amount of sea water in the ocean basins, lowering global sea levels on the order of 120 m below present levels during periods of maximum glaciation (Peltier, 1999). The most recent glacial stage, the Wisconsinan or Würm, began approximately 100,000 years ago. Data indicate that the early Wisconsinan maximum glacial advance probably occurred between 72,000 and 62,000 years ago (Ruddiman and McIntyre, 1981; Shackleton, 1987; Oldale and Colman, 1992) and resulted in a maximum lowering of global sea level of only 50 m below present sea level (Oldale and Colman, 1992). The late Wisconsinan glacial maximum occurred at approximately 21,000 BP (Fairbanks, 1989; Bard *et al.*, 1990), and resulted in a lowering of global sea level to approximately 120 m below present sea level. These periods of lower sea level resulted in large areas of the continental shelves (see *Continental Shelf*) being exposed as dry land. During the glacial stages, most continental coastlines were far seaward of their present positions (see *Paleocoastlines*), and land bridges formed between some landmasses currently separated by the sea (Masters and Flemming, 1983).

Eustatic sea level refers to the absolute changes in sea level that result from water being removed from the ocean basins during glacial building, or added to the ocean basins during glacial melting (see *Eustasy*). Isostatic adjustments refer to the differential depression or uplift of the earth's lithosphere in response to the weight of glacial ice (see *Isostasy*), the amount of water in the ocean basins, and sediment loading. The lithosphere beneath the areas formerly covered by glacial ice rebounded as the ice melted. The area of uplift around the margins of the ice masses, known as the glacial forebulge area, collapsed as the weight of the ice was released. The ocean basins subsided as a result of water loading due to glacial meltwater and, in some locations, of sediment loading. The continental margins also may have uplifted in response to subsidence of the ocean basins (Walcott, 1972). When these isostatic adjustments of the earth's lithosphere are combined with the effects of eustatic sea-level change and with local tectonic factors, such as vertical movements along fault planes and plate boundaries, a very dynamic picture of global paleocoastline positions emerges. For any given area, the changes in the elevation of the coastline, through time, in relation to the present coastline are referred to as relative sea level.

Relative sea-level change

The graphic depiction of relative sea-level change, through time, is called a relative sea-level curve. Relative sea-level curves are based on radiocarbon-dated samples of *in situ* organic material taken from the seafloor or from shallow geologic cores (see *Sea-Level Indicators, Biologic*), or on geomorphic features indicative of relict coastline environments (see *Sea-Level Indicators, Geomorphic*). The organic materials selected for dating are those known to occur only within former coastline environments (e.g., shells in growth position, certain plant roots, undisturbed brackish-water peat deposits). Relative sea-level curves are constructed by plotting the age in radiocarbon years before present and the elevation (positive or negative) in relation to present sea level of these *in situ* organic materials.

In formerly glaciated areas that have experienced isostatic rebound, paleocoastline positions may be tens to hundreds of meters above present sea level. In areas far removed from the glacial ice masses, the paleocoastline positions may be well below present sea level. In areas not affected by isostatic adjustments or tectonics, the paleocoastline positions will reflect changes in eustatic sea level (Bloom, 1977; Stright, 1995).

Global models of relative sea-level change

Various researchers have been developing and refining global mathematical models to simulate the worldwide glacial isostatic adjustments resulting from the late Wisconsinan glaciation (e.g., Walcott, 1972; Clark *et al.*, 1978; Tushingham and Peltier, 1991, 1992; Peltier, 1994, 1999). For example, in Tushingham and Peltier's ICE-3G and ICE-4G models, unknown parameters are estimated, such as the thickness of the earth's lithosphere, the thickness of the glacial ice, and the timing and history of deglaciation. Then the model is run and compared against the available relative sea-level data around the world. Systematic divergences in the actual relative sea-level data and the model's predictions are examined, and changes are made, if necessary, in the assumed glacial and lithospheric parameters of the model. In more recent studies, even the effect of the earth's altered rotation due to the deformation of the geoid from the weight of glacial ice has been factored into these models (Peltier, 1999).

These global models can be of great value to archaeology. If the relative sea-level history can be accurately predicted for any given geographic area, it can be used to determine where to find coastal archaeological sites for any given prehistoric period, and to understand the systematic biases that have been introduced into archaeological interpretations of human migrations and adaptations. The only factor that cannot be accounted for by such global models is the influence of local tectonic activity on prehistoric site locations.

Lateral changes in the position of the coastline

The position of the coastline may also change laterally through progradation (building seaward) or erosion landward, even in the absence of relative sea-level changes. A prograding coastline would tend to bury and protect archaeological sites (Coastal Environments, Inc., 1977) while an eroding coastline would tend to destroy archaeological sites and redistribute the more durable archaeological materials, such as stone and bone artifacts, along the shore (Stright *et al.*, 1999).

Finding former coastal archaeological sites

To determine where former coastal archaeological sites might exist, it is first necessary, as discussed above, to consult local relative sea-level curves or the results of one of the global glacial isostatic adjustment models to determine which areas were above sea level and available for human habitation. Paleocoastlines that are now above present sea level can be surveyed using standard archaeological methods. However, paleocoastline environments that are now submerged require more specialized archaeological methods (Johnson and Stright, 1992).

Finding submerged archaeological sites

Although the physical and cultural remains of prehistoric people might occur anywhere along these submerged land surfaces, basic human subsistence needs for such things as freshwater, plant and animal food resources, shelter, and raw material for tool manufacture were often determinant in site selection by prehistoric human populations. Landforms such as rivers, lakes, and bays represent locations where one or more of these basic subsistence needs could be met.

Although the archaeological sites themselves are generally not of sufficient size or acoustic contrast to the surrounding sediments to be detected by acoustic remote sensing instruments, the landforms with which they are most likely to be associated can be detected by these instruments (see *Remote Sensing: Coastal Environments*). Those landforms still exposed at the seafloor can be detected either directly by diver surveys or remotely by side-scan sonar, a remote sensing instrument that transmits a high-frequency sound signal at an oblique angle to the seafloor. The returning signal's strength depends on the density of the object encountered at the seafloor. These signals can then be interpreted to infer the presence of relict geomorphic features at the seafloor. The seafloor landforms, thus detected, can then be further investigated by divers or remotely operated vehicles (ROVs).

Archaeological sites buried beneath the seafloor can be located using a high-resolution seismic system such as a subbottom profiler, which images the buried land surface and detects landforms with which sites are likely to be associated. A series of these profiles can then be interpolated into a topographic map of the former land surface. If the landforms are not too deeply buried, they can be investigated by using divers to excavate the overlying sediments. However, complete excavation of underwater sites is a very difficult process; the feasibility of which is dependent on the water depth, the depth of the potential site below the seafloor, the sediment type, and other site conditions. A more realistic approach is to conduct an initial investigation using a series of sediment cores. Standard sedimentary analysis techniques can detect indicators of a buried archaeological site, even in the absence of recognizable artifacts (Coastal Environments, Inc., 1982; Pearson *et al.*, 1986; Stright, 1986).

Although sediment cores provide only a very limited subsurface sample, they can be used to establish the presence or absence of an archaeological site. If it is determined that a site is present at the sampled location, transects of sediment cores can then be used to delineate the aerial extent of the site and depth of the site deposits. Analysis of the sediments in the cores can be used to reconstruct the paleoenvironmental context of the site, determine the subsistence resources that may have been available at the site, and may even provide sufficient organic material to allow radiocarbon dating of the site (Pearson *et al.*, 1986).

Conclusions

Archaeological sites now submerged on the continental shelves are important because they can provide information on prehistoric human adaptations, migrations, and cultural contacts that may not be available from terrestrial sites. Many of the earlier archaeological interpretations, based on data from coastal archaeological sites, need to be carefully reexamined because the major changes in coastline positions caused by glacial advances and retreats may not have been taken into account. Most sites that would demonstrate early prehistoric human adaptations to coastal environments are now submerged. Likewise, land bridges that would have provided easy migration routes between current landmasses are also now submerged. Although raised paleocoastlines do exist, there may be few archaeological sites associated with these features, because the rapid rate of isostatic rebound may have precluded the development of stable coastal ecosystems necessary to support human populations (Oldale, 1985). The bias to our understanding of the archaeological record caused by the shifting positions of paleocoastlines is perhaps that we have thought coastal environments were not as important in the prehistoric past as they likely were, and perhaps we have underestimated prehistoric human migration pressures, opportunities, and contacts.

Melanie J. Stright

Bibliography

Bard, E., Hamelin, B., and Fairbanks, R.G., 1990. U-Th ages obtained by mass spectrometry in corals from Barbados; sea level during the past 130,000 years. *Nature*, **346**: 456–458.

Bloom, A.L., 1977. *Atlas of Sea-Level Curves: International Geological Correlation Programme, Project 61*. Ithica, NY: Cornell University.

Clark, J.A., Farrell, W.F., and Peltier, W.R., 1978. Global changes in postglacial sea level: a numerical calculation. *Quaternary Research*, **9**: 265–287.

Coastal Environments, Inc., 1977. Cultural resources evaluation of the northern Gulf of Mexico continental shelf. Prepared for Interagency Archaeological Services, Office of Archaeology and Historic Preservation, National Park Service, United States Department of the Interior.

Coastal Environments, Inc., 1982. Sedimentary studies of prehistoric archaeological sites. Prepared for Division of State Plans and Grants, National Park Service, United States Department of the Interior.

Fairbanks, R.G., 1989. A 17,000-year glacio-eustatic sea level record: influence of glacial melting rates on the Younger Dryas event and deep-ocean circulation. *Nature*, **342**: 637–642.
Johnson, L.L., and Stright, M.J., 1992. *Paleoshorelines and Prehistory: An Investigation of Method*. Boca Raton, FL: CRC Press.
Masters, P.M., and Flemming, N.C., 1983. *Quaternary Coastlines and Marine Archaeology: Towards the Prehistory of Land Bridges and Continental Shelves*. London: Academic Press.
Oldale, R.N., 1985. Rapid postglacial shoreline changes in the western Gulf of Maine and the Paleo-Indian environment. *American Antiquity*, **50**(1): 145–150.
Oldale, R.N., and Colman, S.M., 1992. On the age of the penultimate full glaciation of New England. In Clark, P.U., and Lea, P.D. (eds.), *The Last Interglacial-Glacial Transition in North America*. Boulder, CO: Geological Society of America, Special Paper 270.
Pearson, C.E., Kelley, D.B., Weinstein, R.A., and Gagliano, S.M., 1986. Archaeological investigations on the outer continental shelf: a study within the Sabine River Valley, offshore Louisiana and Texas. Prepared for the U.S. Department of the Interior, Minerals Management Service, OCS Study MMS 86-0119.
Peltier, W.R., 1994. Ice age paleotopography. *Science*, **265**: 195–201.
Peltier, W.R., 1999. Global sea level rise and glacial isostatic adjustment. *Global and Planetary Change*, **20**: 93–123.
Ruddiman, W.F., and McIntyre, A., 1981. Oceanic mechanisms for amplification of the 23,000-year ice-volume cycle. *Science*, **212** (4495): 617–627.
Shackleton, N.J., 1987. Oxygen isotopes, ice volume and sea level. *Quaternary Science Reviews*, **6**(3/4): 183–190.
Stright, M.J., 1986. Human occupation of the continental self during the late Pleistocene/early Holocene: methods for site location. *Geoarchaeology*, **1**(4): 347–364.
Stright, M.J., 1995. Archaic period sites on the continental shelf of North America: the effect of relative sea-level changes on archaeological site locations and preservation. In Bettis, E.A., III (ed.), *Archaeological Geology of the Archaic Period in North America*. Boulder, CO: Geological Society of America, Special Paper 297.
Stright, M.J., Lear, E.M., and Bennett, J.F., 1999. Spatial data analysis of artifacts redeposited by coastal erosion: a case study of McFaddin Beach, Texas. U.S. Department of the Interior, Minerals Management Service, OCS Study MMS 99-0068.
Tushingham, A.M., and Peltier, W.R., 1991. ICE-3G: a new global model of late Pleistocene deglaciation based upon geophysical predictions of post-glacial relative sea level change. *Journal of Geophysical Research*, **96**(B3): 4497–4523.
Tushingham, A.M., and Peltier, W.R., 1992. Validation of the ICE-3G model of Würm-Wisconsin deglaciation using a global data base of relative sea level histories. *Journal of Geophysical Research*, **97**(B3): 3285–3304.
Walcott, R.I., 1972. Past sea levels, eustasy, and deformation of the earth. *Quaternary Research*, **2**: 1–14.

Cross-references

Archaeology
Beach Erosion
Beach Ridges
Changing Sea Levels
Coastal Subsidence
Coastline Changes
Continental Shelves
Eustasy
Holocene Coastal Geomorphology
Late Quaternary Marine Transgression
Paleocoastlines
Pleistocene Epoch
Remote Sensing of Coastal Environments
Sea-Level Indicators, Biologic
Sea-Level Indicators, Geomorphic
Sea-Level Rise, Effect
Seismic Displacement
Submerged Coasts
Vibracore

ARCHAEOLOGY

Humans have inhabited coastal areas and made use of their resources for hundreds of thousands of years. The richness of the coastal zone, the overlapping of maritime, littoral and inland resources, the ease of transportation along it, have led people to cluster along shores, including the people of today who are destroying much of the record of past inhabitants. However, coastlines are also the most dynamic environments on earth. Tides and currents effect daily and seasonal changes in the shore margin; storms dramatically resculpt the shore; isostatic, eustatic, and tectonic changes (cf. *Isostacy, Eustacy, Tectonics and Neotectonics*) can result in former coastlines being located far inland or far out to sea.

In trying to understand human use of coastlines, it is necessary first to discover where the coast was at the time the site one is interested in was occupied. Due to the ever-changing relationship between the land and the sea, sites of past coastal people may be well inland, well submerged under the sea or still coastal. To discover past coastal sites, it is essential that archaeologists work in cooperation with geologists and other coastal scientists, and take into account both local and global factors which influence relative sea level (cf. *Archaeological Site Location*). Once sites are located, and are determined to be cultural rather than natural features, archaeologists turn to more culturally interesting questions about human adaptation to the coastline and how it changed with time. For example, were these people simply living by the sea, like summer beach residents today or in Pompeii, or were they using marine resources? If they were using marine resources, was their primary focus on industrial or alimentary needs? If the latter, were they littoral hunter-gatherers, who made use of coastal and nearshore resources, or were they maritime hunter-gatherers, who had boats from which to gather creatures of the deep sea?

To understand human adaptation to coastal environments, archaeologists in all parts of the world have looked particularly closely at diet, at the season of occupation of sites, and at relationships between coastal and inland peoples. Changes in adaptation can be forced by changes in the environment, by cultural change and development or, most likely, by a complex interaction between culture and environment. In the short term, archaeologists consider the human response to disaster: for example, if their settlement is destroyed by a *tsunami* (*q.v.*), do people move away or do they reestablish themselves in the same place? One can examine changes through time in coastal occupation, both those forced in some way by environmental changes, for example, rising or falling sea levels, changes in ocean temperatures, which affect littoral or maritime resources, and those caused by social, political, or demographic changes in the human populations of the coast. In the following pages, studies from various parts of the world will be used as exemplars of current examinations of these issues.

Historical development

The archaeology of coastlines has a long history, with some of the earliest studies in archaeology focusing on the Danish *kjokken-möddings*, or shell middens, of the Mesolithic period. These were recognized as cultural by 1837 and shortly thereafter, the Royal Danish Academy of Sciences established a commission to study them, directed by Jens J.A. Worsaae, who was the world's first professional prehistoric archaeologist. In the 1850s, Worsaae and his colleagues published six volumes on their studies, which established that the Danish middens were cultural and had developed through time, that the only domesticated animal the people possessed was the dog, and that these middens were occupied in all seasons of the year but summer. They also identified the plants of the paleo-environment and studied the distribution of cultural features within the mounds. Finally, they initiated experimental archaeological studies: they gave bones to dogs to chew on to determine why there was a preponderance of bird long bone mid-shafts in the middens. This work was highly influential, stimulating scientists to study local shell middens on both coasts of the United States, and in Japan, Brazil, and Southeast Asia. By 1911, shell middens had been examined from the Aleutian Islands to South Africa (Trigger, 1989). The study of coastal sites has not slowed down, since: a 1991 bibliography of coastal and maritime archaeology, admittedly incomplete, contains 2,800 entries (Kerber, 1991).

Finding coastal sites

Finding coastal archaeological sites can be very difficult due to isostatic, eustatic, and tectonic forces that change the location of the sea relative to the land. The west coast of North America is an area in which all of these forces have been active to varying degrees during the period of human occupation. Fedje and Christensen show that, in the Queen Charlotte Islands, coastlines dating from 13,000 to 9,500 BP [uncorrected radiocarbon dates] are deeply drowned while those dating from 9,200 to 3,000 BP are stranded in the rainforest up to 15 meters above modern levels. Coastlines have been approximately coincident with the current position for only the last two to three millennia and for a century or two centered around 9,400 BP." (1999, p. 635). Modeling these *paleocoastlines* (*q.v.*) allowed them to discover a number of coastal archaeological sites in both drowned and uplifted zones and they are confident that more will be discovered. From southern Washington to northern California, on the other hand, late prehistoric coastal sites have been drowned due to earthquake-induced subsidence (cf. *Subsiding Coasts*) along the Cascadia subduction zone, seriously biasing our understanding of Washington's prehistory (Cole *et al.*, 1996). While many sites thus drowned may be lost, the authors have found a number of examples buried under tidal mud along coastal *estuaries* (*q.v.*). In southern California, Waters *et al.* (1999) have identified a complex series of terraces along streams that are caused both by global patterns of sea-level fall and rise and by local climatic factors, possibly in response to an *El Niño-Southern Oscillation* (*q.v.*) event. These events have caused pre-4,000-year-old sites to be deeply buried and consequently difficult to find. The later sites, which are closer to the surface, have an advantage over many southern California sites in being below the bioturbation layer. Waters and his colleagues have found large late marine oriented sites in these channels, which are changing our understanding of late prehistoric cultural development in this area.

Determining site function

Natural/cultural

In many parts of the world, shell midden sites have been quarried out for fertilizer or to use as fill. Thus, many midden deposits may be lost or exist in places far removed from their original deposition. Even in remote places like the Shumagin Islands south of the Alaska peninsula, people have moved midden from one island to another to serve as fertilizer for gardens. Thus, merely finding on shore shell deposits, even when it can be determined that they are of human creation, does not mean that one has found an archaeological site *in situ*.

In Sri Lanka, the lowering of sea level has led to shellfish dying and their shells being washed and blown into windrows on beaches, lagoon and lake bottoms, sand dunes, and headlands. These are often difficult to tell from cultural deposits although artifacts found among the shells suggest that people were responsible for some of the deposition (Katupotha, 1995). In southwestern Louisiana, Henderson *et al.* (1998) also wished to determine whether shell rich deposits on coastal beach ridges were natural or cultural. In this case, the archaeological deposits were found lying on or actually constructed out of reworked *chenier* (*q.v.*) deposits. The authors discovered a suite of taphonomic, sedimentologic, and stratigraphic methods, which served to distinguish the cultural deposits from the natural chenier deposits.

Butler and her colleagues have been working during the past decade to discover how natural deposits of salmon differ from those left by human hunter-gatherers. This is a particularly important problem in the Pacific Northwest where the harvesting and storage of vast quantities of anadromous salmon is thought to underlie the development of the cultural complexity for which the area is famous. Butler has looked at a particular site, the Dalles Roadcut on the Columbia River and concluded that the deposition is probably cultural; has considered the varying natural decay of different salmon bones, and has investigated whether fish bones that have passed through mammalian digestive tracts bear recognizable marks (cf. Butler and Schroeder, 1998). All of these studies aid archaeologists in determining whether salmon deposits are natural or cultural.

An ethnoarchaeological study of fish middens on the coast of Senegal showed that caught and filleted fish mimicked natural fish deposits as the heads were not removed in filleting and, therefore, the entire skeleton remained (Van Neer and Morales Muñez, 1992). On Lake Turkana in Kenya, on the other hand, investigators could distinguish base camps, fish processing camps and fish waste discard sites from each other and from natural death assemblages by the nature and number of bones remaining (Stewart and Gifford-Gonzales, 1994). Both of these ethnoarchaeological studies can aid archaeologists in interpreting prehistoric shell deposits.

Related to the problem of distinguishing natural from cultural deposits is that of determining why species representation and shell size of shellfish found in middens might change through time. In looking at middens on the West Coast of Southern Africa, Jerardino (1997) determined that changes in species were due to humans exploiting different tidal zones, while size changes in individual species were due to a combination of water conditions and human exploitation.

Industrial

The industrial use of coastal resources by prehistoric peoples has not received much attention in recent years, though the attention it has received has been widespread. For example, working in the Sinai, Bar-Yosef Mayer (1997) demonstrates that shells entered sites as raw materials for beads rather than as food remains, while in the Arctic, Savelle (1997) notes that whale bones are often more indicative of housing needs than diet. Most investigators look at shellfish gathering as an enterprise that requires few or no tools aside from a container in which to carry the collected goods. However, Johnson and Bonsall (1999) note that in both the Shumagin Islands in Alaska and in Scotland, tools are found in shell midden contexts which appear to have been used in the collection and processing of shellfish. Wedge-ended rods of stone or sea mammal bone are used for prying limpets from rocks and, possibly, for digging the animal out of its shell, while small splinters of bird bone are used to pry periwinkles and other small snails loose from their shells, developing worn points in the process.

Coastal adaptation

Once coastal sites have been located and identified, archaeologists attempt to discover what people were doing at the sites. This involves figuring out what foods they were eating, by examining both the food remains left at sites and the bones of the humans, which reflect the nature of their diets. In looking at the food remains, it is necessary to figure out what the dietary implications are of, for example, ten bivalve shells versus two salmon bones versus one sea lion femur. Archaeologists are also interested in discovering whether people were using this site year-round or only seasonally, whether they were littoral gatherers or maritime hunters and fishermen, and what their relationship with other groups might be. Finally, archaeologists look both at individual sites occupied for a long period of time and at the pattern of sites within an area to discover how human adaptation to the coastline changed through time.

Diet

The most direct means of discovering what people ate is to examine their bones. Analyses of trace elements and stable isotopes can distinguish between diets rich in marine foods and those rich in terrestrial foods. Low Ba/Sr and Ba/Ca ratios and high $\delta^{13}C$ values in bones indicate a diet rich in marine foods, while the reverse is true for terrestrial diets. High $\delta^{13}C$ values also occur when plants such as maize, which use a C4 metabolic pathway, are eaten, but maize and marine foods in the diet can be distinguished by examining $\delta^{15}N$ and $\delta^{34}S$ values.

Gilbert *et al.* (1994) examined modern plants and animals of the western Cape of South Africa in order to determine marine and terrestrial signatures for Ba/Sr and Ba/Ca ratios, finding distinct differences in both edible tissue and bone. Re-analyzing bones, which had already been tested for $\delta^{13}C$ values, led to the conclusion that both methods worked, but that stable isotopes revealed more subtle differences.

Aufderheide and his colleagues (1994) had the advantage of working with mummified human remains from northern Chile. Here, they were looking at recent immigrants to the coast from the highlands and discovered that the community appeared to have split into two different groups, one right on the coast subsisting almost entirely on marine foods, and the other in the mid-valley eating both marine foods, probably supplied by the coastal community, and valley bottom foods. These results were supported by the presence of external auditory canal extoses, formed due to prolonged cold water diving, in the coastal population.

Human remains are not always available in the sites or levels where archaeologists would like to find them, and the prehistoric people's descendants are often not pleased to have their ancestors subjected to scientific analysis. Cannon *et al.* (1999) have demonstrated at the site of Namu on the British Columbia coast that dog bones can serve as surrogates for human bones in analyzing diets. It is probable that in most prehistoric coastal settlements the diet of dogs was very similar to that of their masters, thus providing archaeologists a much wider sample of bones to analyze for dietary information.

Another approach to dietary reconstruction is to estimate the amount of food represented by the remains present in a site. James Savelle has been particularly active in deriving meat utility indices for marine mammals (cf. Diab, 1998). This also relates to issues of variable preservation, discussed above, and to recovery methods. Fish bones are often underrepresented in archaeological assemblages because of a lack of screening of excavated materials or of using screens with too large a mesh. Ross and Duffy (2000) have demonstrated that a 1 mm mesh screen results in considerably greater recovery of fish bones from an Australian site and have also developed deflocculation and flotation techniques, which make sorting the screened remains practical in the real world: sorting remains from a 100 ml dry screened sample took 20 h, whereas the deflocculated and floated sample was sorted in only 2.5 h. Moreover, the chemical added to water for flotation was sugar and an excellent deflocculant was baking soda, making the procedure inexpensive and safe as well. The result of the 1 mm screening was to verify the reports of southeast Queensland Aborigines that both fishing and shellfishing had been important to them since time immemorial in contrast to prior archaeological investigations in which the paucity of fish remains recovered had led to a hypothesis of a late introduction of fishing to this coast.

Analysis of midden materials also gives rise to problems concerning how to quantify the materials recovered. Faunal analysts now tend to provide both number of identified specimens (NISP) and minimum number of individuals (MNI) data for mammal and fish remains, but the quantification of shellfish remains is a thornier issue. Shells appear in sites in the thousands and are often broken into small fragments. Because shells vary in size and in friability, NISP is not a useful figure. MNI is often used, but ignores a great deal of the data since only the hinge areas of bivalves can be counted. Thus archaeologists have attempted to use the weight of shellfish remains as a proxy for the importance of various species in past diets. This is an area of active discussion at present (cf. Claassen, 2000).

Season of occupation

Coastal settlements can be year-round residences or can represent part of a seasonal round, being occupied for one or two periods during the year while the people live elsewhere, exploiting other resources, during the rest of the year. Determining seasonality from the middens themselves requires some means of discovering when animals were killed, which has proven possible for mammals, fish, and shellfish. Marine mammals are born at particular times during the year, so the age at death of immature animals can indicate when a site was occupied: if all of the young animals found in a site died at the same age, it suggests that the site was occupied for only that season. The analysis of seal bones in South Africa indicated that early in the Holocene people exploited seals periodically throughout the year, while in the late Holocene they restricted this activity to the spring (August to November; Woodborne *et al.*, 1995). A similar pattern of change from year round to seasonal occupation is seen in New Zealand through the analysis of seasonal and annual growth rings on Cod otoliths (earbones; Higham and Horn, 2000). Shellfish, such as oysters and clams, also possess seasonal and annual growth rings, which have been used to analyze site seasonality. In Mexico, oxygen isotope ratios in marsh clams record large-scale salinity fluctuations in the marshes caused by alternating wet and dry seasons. Therefore, the oxygen isotopes in clams from midden sites can be used to determine the season of shellfish harvesting (Kennett and Voorhies, 1996).

Littoral or maritime

In trying to understand Mesolithic coastal sites in western Scotland, Bonsall (1996) makes use of ethnographic data which indicate that the Australian Aborigines had three recognized types of shell middens: processing camps, where shellfish were removed from their shells and prepared for transport elsewhere; dinnertime camps, where people involved in other activities stopped, collected shellfish and enjoyed a meal, and home bases. In Scotland, given the generally inclement weather, remnants of the first two types of camp would probably be found in shelters or other protected locales. Bonsall hypothesizes that the Obanian sites found around the western coast of Scotland are probably the remnants of processing and dinnertime camps occupied by gatherers whose base camps were located inland. The latter had a much richer artifact inventory, including the quintessential Mesolithic microliths, which are not found in the coastal sites.

In the much more complex society of Inka Peru, the fishermen of the Chincha Valley formed a separate community from the valley farmers and overlords whose protein they supplied. Specialized fishing communities seem to have existed in coastal Peru before the Inka took the area over and the Inka probably incorporated these communities into their empire with little change except for skimming the fish off for their own use rather than allowing the valley paramount lord to do so. The fishing here was primarily, if not entirely, littoral netting of small fish such as anchovies and anchovetas, which were then dried for exchange with other members of the valley and imperial communities (Sandweiss, 1992).

In contrast to these littoral adaptations, the residents of the Hoko River site in Washington state possessed a full-scale maritime adaptation. Due to half of the site being waterlogged, there is excellent preservation of organic artifacts, including numerous fishhooks, landing skids for canoes, pack baskets for transporting the fish, primarily halibut, and mats to protect the canoes and keep them from drying out between fishing trips. Around 3,000 years ago, the site was seasonally occupied for a number of years, "mainly in the spring/summer season, when [the people] focused on offshore hook-and-line fishing for bottom fish, and particularly flatfishes" (Croes, 1995, p. 229).

Coastal/inland relationships

Many studies look at the relationships between coastal people and their inland neighbors, or between coastal sites and the inland sites, which formed other parts of their seasonal round. Kennett and Voorhies used the oxygen isotope information from marsh clams, discussed above, to determine changes in the relationship between people and their environments. During the early late Archaic period, people collected clams year round, though they focused on the dry season.

> Through the late Archaic period a general trend occurred toward wet season use of these locations. This culminated at the end of the late Archaic period with the exclusive use of the littoral zone during wet season months. These data indicate a fundamental shift in the way these estuarine locations were being used. We argue that people living in this region altered their overall subsistence strategy during the late Archaic period due to scheduling conflicts that occurred with the adoption of maize agriculture. (Kennett and Voorhies, 1996, p. 689)

A similar process appears to have taken place in the Southeastern United States. Here ranked societies developed which took advantage of both coastal and interior resources. The origins of these societies are not well known, but evidence from Chesapeake Bay suggests that a pattern of springtime collection of oysters for local consumption was replaced, as maize arrived, with intensive spring harvesting and preserving of oysters. Anadromous fish and shellfish were major sources of protein along the coastal plain, where interior mammals such as white tailed deer were rare. Farther south, in Florida, the development of the Calusa cultural pattern is very imperfectly known, but maritime and littoral resources were clearly of major importance (Waselkov, 1997).

On Kodiak Island, Alaska, the permanent settlements are on the shore and sites in the interior tend to be summer salmon fishing camps. The interior Outlet Site, while it was primarily a fishing camp, also showed evidence of fall/winter bird hunting, and boasted substantial houses. Steffian and Saltonstall (2000) hypothesize that, as the population on Kodiak grew, the period of use of interior resources was extended and the coastal people, needing to lay stronger claim to their summer settlements, built substantial houses to serve as signs of ownership when they were absent.

Change in coastal adaptation

The majority of archaeological studies of coastal societies focus on change through time as people adjust to the constantly changing physical and cultural environment. Studies of the changing relationships between people and coasts have taken place in all parts of the world and only a few can be highlighted to give the flavor of current research on this topic.

The classic, and in many ways still unmatched study of human reaction to changing coastlines is Shackleton's study of Franchthi Cave in Greece (cf. 1988). During the 20,000 years the cave was occupied, sea level rose from −115 m to its present level. At varying times in the past the shore in the vicinity of the cave supported different shellfish communities. By comparing the proposed communities to the shellfish remains in the cave, Shackleton was able to hypothesize that the people selectively gathered these resources rather than merely collecting the closest available species.

A short distance to the north of Franchthi Cave, near Volos, Greece, Zangger (1991) has studied changes in settlement along a prograding rather than submerging coastline. At ca. 6000 BP, Holocene sea levels reached their maximum height, and the shore in the vicinity of Volos was 3 km farther inland than it is today. Combined archaeological and geological investigation has shown that during the following 6,000 years coastal cities followed the coastline as it prograded to the east.

In many urban coastal areas, engineers have captured land from the sea, using dikes and landfills to increase human living space. Nowhere did this process begin earlier or proceed further than in Holland. By 1984, Dutch archaeologists were already involved in detailed geoarchaeological studies looking at the changing relationships between people and the land. Brandt, van der Leeuw and van Wijngaarden-Bakker (1984) showed how rising sea level changed the configuration of the land west of Amsterdam and, thus, its use by early settlers. By the early Iron Age, at least, human livestock was also changing the land, simplifying its ecology and creating a human dominated landscape, a process that continued as people began to enclose the land with dikes.

On the Texas Gulf Coast, Archaic Period habitation closely followed relative sea-level rise. When sea-level rise was rapid, estuarine resources were impoverished, and humans abandoned the region. When sea-level rise slowed, "the emergence of highly photosynthetic bay/lagoon shallows support[ed] nutrient-rich shoreline vegetation [which] provided the basis for a rich, exploitable aquatic biomass" (Ricklis and Blum, 1997, p. 287), which the people proceeded to exploit. In the past 10,000 years, this cycle repeated itself three times, with settlements between 7,500 and 6,800 BP, between 5,900 and 4,200 BP and between 3,000 BP and the late Prehistoric Period, and abandonments between 6,800–5,900 BP and 4,200–5,900 BP. During the first two occupations the people focused on shellfish; during the course of the last,

> ongoing sedimentation under conditions of stable sea level led to formation of the continuous modern barrier island chain by ca. 2,500–2,000 BP. As the disconnected barriers coalesced, back-barrier lagoons became very extensive vegetated shallows, providing protective and nutrient-rich habitats for important fish species, with a resultant increase in estuarine fish biomass. A marked increase in fish remains in archaeological contexts, beginning at the time, indicates the emergence of intensive fishing as a major subsistence focus. (Ricklis and Blum, 1997, p. 306)

Short- and long-term responses to environmental disaster

Many coastal changes are gradual and may take generations to make themselves felt. As Brandt *et al.* state, people "perceive slow, continuous changes as unchanged, as similar until so much change has occurred that it is clearly seen. Then, they ... change ... much more drastically than the circumstance requires" (1984, p. 14). Other changes, on the other hand, such as hurricanes, tornadoes, earthquakes, and tsunamis are sudden, dramatic, and disastrous. Human response to sudden disaster varies according to the circumstances and can be looked at on various scales (Johnson, 2002).

Two recent studies of North Pacific coastal peoples come to quite different conclusions concerning human response to catastrophic change. Jordan and Maschner (2000) have studied the western end of the Alaska Peninsula, where isostacy, eustacy, tectonics, and volcanoes have served as a dynamic stage for human settlement. The cultural phases they have identified mark changes in village organization and subsistence orientation in the area and reflect some combination of environmental, social, economic, and political change. The dominant factor in the environmental history of the area has been isostatic recovery, but Jordan and Maschner argue that two large or great earthquakes have had major effects on regional prehistory. The first, in ca. 2,200 BP, led to coseismic subsidence which increased site erosion and submergence. No sites are dated in the area between 2,400 and 2,100 cal yr BP suggesting that sites occupied shortly before the earthquake were destroyed and the area was abandoned for a century following the earthquake. In the following cultural period there is a major change in village organization and subsistence. The most radical change in village organization took place about 850–950 cal yr BP in concert with changes in almost all the eastern Aleutian arc. There does not seem to be any local environmental instability correlated with these changes. However, there was a great earthquake in the Gulf of Alaska between 700 and 900 BP. Environmental recovery probably took a relatively short period of time, but cultural disruption may have lasted considerably longer. Jordan and Maschner argue that the displacement of populations from the region of the epicenter may have triggered the cultural changes in the wider area.

On the south side of the Alaska Peninsula, Johnson does not see as strong a reaction to catastrophe. The initial occupation of the islands appears to have been rapid, following a major uplift episode, which substantially increased terrace living space in the islands. The immigrants may have been fleeing from areas where the effect of the earthquakes was subsidence rather than uplift. Johnson concludes that "for maritime hunter-gatherers like the Shumagin Islanders, earthquakes are terrifying occurrences which may slow cultural development and force populations to move their settlements, but they do not seem to have a major effect on cultural development. Shumagin history, like ours, is more affected by human events than by natural events, however catastrophic" (Johnson, 2002).

Archaeology in service to geology

While archaeologists most often use the results of other scientists to help interpret their sites, occasionally archaeological data are used to help interpret geological data. The majority of these studies concern sea-level change. Since people do not live underwater, sites found beneath the sea are clear evidence of rising sea level or sinking land; sites with clear maritime focus located well away from the shore, on the other hand, indicate lowered sea level or rising land. These sites can often be dated, using either cultural or geophysical means, with more precision and certainty than natural deposits, allowing geologists to pinpoint the changes of interest to them more precisely.

In a major review, Owen Mason (1993) looked at the geoarchaeology of *beach ridges* (*q.v.*) and cheniers on prograding coastlines worldwide. He looked at studies in which archaeologists and geologists working cooperatively achieved a thorough understanding of the chronology of these features as well as those in which a lack of cooperation between the two has led to less satisfactory results for both. In using archaeological sites to date beach ridge formation, one must first carefully evaluate cultural adaptation of the past people to be certain that site location is closely tied to the position of the sea relative to the inhabited ridge.

An example of the study of rising sea levels is that of Morhange *et al.* (1996) which used archaeological, biological, and sedimentary data to model the sea level at Marsielles over the past 4,000 years. In this area, the archaeological indicators of sea level include objects such as wrecks, which were originally deposited below sea level, the walls of quays and wharf pilings found at sea level, and roads and room floors which were originally above sea level. Using these urban features in combination with barnacles and beach indicators led to a detailed and well-supported sea-level record. On a much broader scale, Flemming (1998) reviews drowned sites worldwide that indicate Paleolithic, Neolithic, and Bronze Age sea levels.

Archaeological sites can also provide evidence of falling sea level. In a non-civilized, shell midden context on the Northwest Coast, Cannon (2000) used coring to recover samples from the surface to the base of the cultural deposits in a series of sites. By dating the basal cultural deposits and establishing their relationship to present sea level, he was able to identify steady sea level decline over the past 10,000 years in this region.

Finally, relative sea levels may change in a complex and irregular manner. In the Shumagin Islands, Alaska, where episodic tectonic uplift is coupled with interseismic downwarping and post-glacial sea-level rise over the past 10,000 years, archaeological site locations served to constrain geological modeling, since occupied sites could not be located under water (Winslow and Johnson, 1988).

Intensive studies

North Pacific

Integrated studies of North Pacific prehistory are descendants of 19th century studies of Arctic cultures crystallized in Gjessing's formulation of the Circumpolar Stone Age concept. Early studies of northern maritime adaptations were much too broad and general and have been replaced by more focused studies of particular regions. In the north Pacific, the opening of the "Ice Curtain" led to increased contact between New World and Russian archaeologists interested in the North Pacific. As these studies progressed, links to more southern Pacific cultures became obvious, drawing in Japanese scholars, who have studied both their own and New World cultures. A major focus of these studies has been the inhabitation of the Americas in the terminal *Pleistocene Epoch* (*q.v.*). Additional studies have been concerned with issues of culture contact and influence around the North Pacific Basin, with initial adaptations to the coastline, both littoral and maritime, and with interrelationships between Pacific coastal peoples and their interior neighbors. There has also been concern about the fidelity of representation of cultural components, given post-glacial sea-level changes, but not as much as there probably should have been.

Channel islands

A nearby area in which there has been an explosion of archaeological research in the last few years is the Channel Islands area in southern California. In this region, prehistorically, there was a mainland adaptation, which often, if not always, included both coastal and inland facies, and an island adaptation, which, perforce, was always primarily littoral or maritime, though sometimes provided with inland resources through trade with the mainland people. Archaeologists and geologists working in this area have been trying to determine environmental changes in the area, how these changes affected the marine and shore-based resources, and what effect these changes had on cultural development and change. Also of concern is whether the whole area can be treated as a unit for paleoecological and paleocultural studies or whether subdivisions, between the southern and northern islands or between the islands and the mainland, are significant enough to make conclusions derived from one zone inapplicable in the other. In the process of arguing these various positions, local scholars have significantly advanced our understanding of the factors involved in both ecological and cultural development in highly variable coastal environments (cf. Kennett and Kennett, 2000).

Future work

"Interdisciplinary efforts to reconstruct shoreline history and associated environmental parameters could significantly raise the potential of locating early postglacial archaeological sites" (Fedje and Christensen, 1999, p. 650). More modeling of past environments, both on large- and small-scales, is necessary, as well as additional detailed studies of regions and sites. More bathymetric studies are needed in many areas: in coastal Alaska, for example, bathymetric charts are very general and not useful for detailed coastal studies, even where they do not meander off into dotted lines. Given the maritime focus that the majority of coastal peoples developed, sooner or later, and the freedom of movement boats provided, regional studies are essential (Waselkov, 1997). As detailed understanding of local sea level variation increases, we will be better able to understand how prehistoric people adapted to and used these changes in cultural development.

Conclusions

Absence of evidence is not necessarily evidence of absence, particularly in coastal environments. A dictum all coastal archaeologists should repeat when going to bed and upon arising in the morning: Consider the absent! In modeling past coastal subsistence/settlement systems, archaeologists must take account of sites that are gone! What we don't know will hurt us! Archaeologists should use all methods available, particularly modern tools like *Geographic Information Systems* (*q. v.*), to predict site loss, to get a handle on how much of the archaeological record is actually left in any particular area. Prehistoric coastal peoples saw their settlements disappear in storms just as present coastal people do, and the effect of these disappearances on the archaeological record is something coastal archaeologists ignore at their peril.

Lucille Lewis Johnson

Bibliography

Aufderheide, A.C., Kelley, M.A., Rivera, M., Gray, L., Tieszen, A.L., Iversen, E., Krouse, H.R., and Carevic, A., 1994. Contributions of Chemical Dietary Reconstruction to the Assessment of Adaptation by Ancient Highland Immigrants (Alto Ramirez) to Coastal Conditions at Pisagua, Northern Chile. *Journal of Archaeological Science*, **21**: 515–524.

Bar-Yosef Mayer, D.E., 1997. Neolithic shell bead production in Sinai. *Journal of Archaeological Science*, **24**: 97–111.

Bonsall, C., 1996. The "Obanian Problem:" Coastal Adaptation in the Mesolithic of Western Scotland. In Pollard, T., Morrison, A. (eds.), *The Early Prehistory of Scotland*. Edinburgh: Edinburgh University Press, pp. 183–197.

Brandt, R.W., van der Leeuw, S.E., and van Wijngaarden-Bakker, L.H., 1984. Transformations in a Dutch Estuary: research in a wet landscape. *World Archaeology*, **16**: 1–17.

Butler, V.L., and Schroeder, R.L., 1998. Do digestive processes leave diagnostic traces on fish bone? *Journal of Archaeological Science*, **25**: 957–971.

Cannon, A., 2000. Settlement and Sea-Levels on the Central Coast of British Columbia: Evidence from Shell Midden Cores. *American Antiquity*, **65**: 67–77.

Cannon, A., Schwarcz, H.P., and Knyf, M., 1999. Marine Based Subsistence Trends and the Stable Isotope Analysis of Dog Bones from Namu, British Columbia. *Journal of Archaeological Science*, **26**: 399–407.

Claassen, C., 2000. Quantifying Shell: Comments on Mason, Peterson and Tiffany. *American Antiquity*, **65**: 415–418.

Cole, S.C., Atwater, B.F., McCutcheon, P.T., Stein, J.K., and Hemphill-Haley, E., 1996. Earthquake-Induced Burial of Archaeological Sites along the Southern Washington Coast about A.D. 1700. *Geoarchaeology*, **11**: 165–177.

Croes, D.R., 1995. *The Hoko River Archaeological Site Complex*. Pullman: Washington State University Press.

Diab, Mark C., 1998. Economic Utility of the Ringed Seal (*Phoca hispida*): Implications for Arctic Archaeology. *Journal of Archaeological Science*, **25**: 1–26.

Fedje, D.W., and Christensen, T., 1999. Paleoshorelines and Locating Early Holocene Coastal Sties in Haida Gwaii. *American Antiquity*, **64**: 635–652.

Flemming, N.C., 1998. Archaeological Evidence for Vertical Movement on the Continental Shelf during the Paleolithic, Neolithic and Bronze Age Periods. *Coastal Tectonics*, **146**: 129–146.

Gilbert, C., Sealy, J., and Sillen, A., 1994. An Investigation of Barium, Calcium and Strontium as Paleodietary Indicators in the Southwest Cape, South Africa. *Journal of Archaeological Science*, **21**: 173–184.

Henderson, W.G., Anderson, L.C., and McGimsey, C.M., 1998. Distinguishing Shell-Rich Natural Archaeological Deposits of the Chenier Plain, Southwestern Louisiana. *Geological Society of America, 1998 Annual Meeting, Abstracts with Programs*, **30**, 122.

Higham, T.F.G., and Horn, P.L., 2000. Seasonal Dating Using Fish Otoliths: Results from the Shag River Mouth Site, New Zealand. *Journal of Archaeological Science*, **27**: 439–448.

Jerardino, A., 1997. Changes in Shellfish Species Composition and Mean Shell Size from a Late-Holocene Record of the West Coast of Southern Africa. *Journal of Archaeological Science*, **24**: 1031–1044.

Johnson L.L., 2002. Natural disasters and cultural change in the Shunmugin Islands. In Torrence, R., and Grattan, J. (eds.), *Natural Disasters and Cultural Change*. London: Routledge, pp. 193–203.

Johnson, L.L., and Bonsall, C., 1999. Mesolithic Adaptations on Offshore Islands: the Aleutians and Western Scotland. In Cziesla, E., Kersting, T., and Pratsch, S. (eds.), *Den Goen Spannen … Festschrift für Bernhard Gramsch sum 65. Geburtstag*. Weissbach, Germany: Beier and Beran, pp. 107–115.

Jordan, J.W., and Maschner, H.D.G., 2000. Coastal Paleogeography and Human Occuaption of the Western Alaska Peninsula. *Geoarchaeology*, **15**: 385–414.

Katupotha, J., 1995. Evolution and Geological Significance of Holocene Emerged Shell Beds on the Southern Coastal Zone of Sri Lanka. *Journal of Coastal Research*, **11**: 1042–1061.

Kennett, D.J., and Kennett, J.P., 2000. Competitive and Cooperative Responses to Climatic Instability in Coastal Southern California. *American Antiquity*, **65**: 379–395.

Kennett, D.J., and Voorhies, B., 1996. Oxygen Isotopic Analysis of Archaeological Shells to Detect Seasonal Use of Wetlands on the Southern Pacific Coast of Mexico. *Journal of Archaeological Science*, **231**: 689–704.

Kerber, J.E., 1991. *Coastal and Maritime Archaeology: A Bibliography*. Metuchen, NJ: Scarecrow Press.

Mason, O.K., 1993. The geoarchaeology of beach ridges and cheniers: studies of coastal evolution using archaeological data. *Journal of Coastal Research*, **9**: 126–146.

Morhange, C., Laborel, J., Hesnard, A., and Prone, A., 1996. Variation of relative mean sea level during the last 4,000 years on the northern shores of Lacydon, the ancient harbour of Marseilles (Chantier J. Verne). *Journal of Coastal Research*, **12**: 841–849.

Ricklis, R.A., and Blum, M.D., 1997. Geoarchaeological record of Holocene sea level change and human occupation of the Texas Gulf Coast. *Geoarchaeology*, **12**(4): 278–314.

Ross, A. and Duffy, R., 2000. Fine mesh screening of midden materials and the recovery of fish bone: the development of flotation and deflocculation techniques for an efficient and effective procedure. *Geoarchaeology*, **15**: 21–41.

Sandweiss, D.H., 1992. The archaeology of Chincha fishermen: specialization and status in Inka Peru. *Bulletin of Carnegie Museum of Natural History*, **29**.

Savelle, J.M., 1997. The role of architectural utility in the formation of zooarchaeological whale bone assemblages. *Journal of Archaeological Science*, **24**: 869–885.

Shackleton, J.C., 1988. Reconstructing past shorelines as an approach to determining factors affecting shellfish collecting in the prehistoric past. In Bailey, G., and Parkington, J. (eds.), *The Archaeology of Prehistoric Coastlines*. Cambridge: Cambridge University Press, pp. 11–21.

Steffian, A., and Saltonstall, P., 2000. Outlet Site. Community archaeology, Summer 1999.

Stewart, K.M., and Gifford-Gonzalez, D., 1994. An ethnoarchaeological contribution to identifying hominid fish processing sites. *Journal of Archaeological Science*, **21**: 237–248.
Trigger, B.G., 1989. *A History of Archaeological Thought*. Cambridge: Cambridge University Press.
Van Neer, W., and Morales Muñiz, A., 1992. "Fish Middens": anthropogenic accumulations of fish remains and their bearing on archaeoichthyological analysis. *Journal of Archaeological Science*, **19**: 683–695.
Waselkov, G.A., 1997. Maritime adaptations of ranked societies on the southeastern coast. Presented in the Symposium, "*Late Prehistoric Maritime Adaptations in North America*," organized by L.L. Johnson at the 1997 Annual Meeting of the AAAS, Seattle, Washington.
Waters, M.R., Byrd, B.F., and Reddy, S.N., 1999. Geoarchaeological investigations of San Mateo and Las Flores creeks, California: implications for coastal settlement models. *Geoarchaeology*, **14**: 289–306.
Winslow, M.A., and Johnson, L.L., 1989. Human settlement patterns in a tectonically unstable environment: eastern Aleutian Islands, Alaska. *Geoarchaeology*, **4**: 297–318.
Woodborne, S., Hart, K., and Parkington, J., 1995. Seal bones as indicators of the timing and duration of hunter-gatherer coastal visits. *Journal of Archaeological Science*, **22**: 727–740.
Zangger, E., 1991. Prehistoric coastal environments in Greece: the vanished landscapes of Dimini Bay and Lake Lerna. *Journal of Field Archaeology*, **18**: 1–15.

Cross-references

Beach Erosion
Changing Sea Levels
Coastal Changes, Gradual
Coastal Changes, Rapid
Coastal Subsidence
Coastline Changes
Conservation of Coastal Sites
Continental Shelves
Demography of Coastal Populations
Dune Ridges
Geochronology
Holocene Epoch
Human Impact on Coasts
Late Quaternary Marine Transgression
Sea-Level Rise, Effect
Seismic Displacement
Submerged Coasts
Uplift Coasts

ARCTIC, COASTAL ECOLOGY

Unlike Antarctica, which is an ice-covered continental plateau surrounded by oceans, the Arctic is made up of a central ocean nearly enclosed by land. This entry will mainly focus on the coastal ecology of the Canadian Arctic (Nunavut and Northwest Territories), northern Alaska (United States), Norway, Greenland, and Iceland, as well as the islands of the Barents, Kara, Laptev, and Siberian seas (Figure A24). The intertidal zone, the benthic, and pelagic subtidal communities and the ecosystems associated with the ice itself will be discussed, as well as the ecological importance of polynyas, which are characteristic habitats of the Arctic.

Basic characteristics of the Arctic

The major marine ecosystems of the Arctic include the High Arctic oceanic region, which comprises the deep sea; the High Arctic coastal region between the abyss and the shore; the High Arctic brackish water subregion surrounding the shores of northern Russia, Canada, and Alaska, where the fauna must adjust to seasonal changes in temperature and salinity; and the boreal littoral region of Norway influenced by warmer currents which favor a faunal penetration from the south. Lastly, the Low Arctic shallow region marks the transitional area between the boreal Norwegian fauna and the High Arctic fauna.

The Arctic coastlines of Greenland, northern Iceland, Svalbard, northern Norway, Novaya Zemlya Islands, Murman coast, eastern Siberia, and Arctic archipelago of Canada are generally steep and deeply cut by fjords. However, the northern coast of Alaska, the Canadian Northwest Territories mainland, and most of the Arctic Russian coast are low relief coastal plains. The southern part of the Arctic coasts is characterized by the tundra biome where only grasses, mosses, lichens and about 400 flowering plants grow on a permanently frozen soil during short summers. Most of the biomass is in the root systems. The northernmost portions of Arctic landmasses, consisting mainly of the northeastern islands of Nunavut (Canada) and northern and central Greenland, lie under a permanent ice and snow cover.

Overall, much of the Arctic coastline is ice-covered for about seven months a year and the water temperature generally remains below 0°C. Air temperatures as low as −65 to −68°C have been recorded in Arctic Russia and Canada, and precipitation in the Arctic generally averages less than 250 mm per year. Melting and refreezing processes, wind-chill, runoff from the land and ice movement generate temperature and salinity variations in the upper water layer. The extreme seasonal oscillation in light is an additional factor affecting the biotic components of the Arctic ecosystem.

Due to its particular structure as a semi-enclosed sea (Figure A24), the Arctic Ocean is fairly isolated from the other major oceanic masses, except through the Bering Strait, the Denmark Strait, and the Barents and Norwegian Seas. However, no invertebrate genus is endemic either to the Low or High Arctic. This suggests that most Arctic genera are distributed worldwide or at least found in the Atlantic, the Pacific, or both. Moreover, 77% of the Arctic genera are also common to the Antarctic.

The marine Arctic environment is generally perceived as stressful to the fauna and flora that live there (Menzies, 1975). Low temperature is apparently not the primary factor that has shaped the Arctic marine ecosystem, but rather the marked seasonal oscillation of other physical components. Among the most important characteristics of the Arctic are reduced light, seasonal darkness, prolonged low temperature and icy period, which result in impoverished fauna and flora and low seasonal productivity. Contrary to habitats of lower latitudes, including coral reefs, where biodiversity is, generally, inversely proportional to the respective biomass of each species present, the Arctic is characterized by the dominance of a few species both in abundance and biomass.

From mid-November to late January, the sun is not seen above the horizon in the far North, whereas from mid-May to late July, the sun does not set. However, owing to its geographical position, the Arctic receives an overall low input from the slant illumination of the sun in summer. Moreover, on the polar sea itself, the permanent pack ice reflects 60–90% of the meager incoming radiation. When there is constant heat during the summer, widespread melting of snow and ice occurs, uncovering the seas and land surfaces. The weak organic production is in particular contrast to the high productivity of adjacent subarctic and boreal regions. Production in Arctic waters is probably below an average of 50 g of carbon per square meter of sea surface a year, and mainly concentrated in summer. Phytoplankton peaks in August in northern Canada, while polar zooplankton counts increase by tenfold in July compared with values in the winter dark. Because of this seasonal production, fish feeding on the lower elements of the food chain must glean most of their food in a short period and be able to subsist on small amounts of ingested food and stored energy for long periods. The phases of high productivity and feeding coincide with higher temperatures, which allow higher metabolic rates and levels of activity. Alternatively, when productivity is low, colder temperatures occur and hence, metabolism and energy consumption are low. When and where resource turnover and availability are sufficiently high, the systems may be relatively productive and sustain complex trophic levels, composed of numerous production and consumer organisms (Figure A25).

Supralittoral, littoral, and sublittoral zones

A bar of coarse sediment can form by waves or ice scouring on certain Arctic shores of Alaska and Canada. Pools of brackish water and typical marsh vegetation, including *Carex* spp., *Pucinellia*, mats of mosses, blue-green algae and diatoms, often occur behind these beaches. These supralittoral Arctic marshes can sustain populations of cladocerans, copepods, branchio-pods, dipteran larvae and oligochaete worms, numerous insects, shorebirds, and caribou herds (Broad, 1982).

The intertidal and shallow subtidal areas are relatively barren, at least on the surface. For instance, the Arctic coasts of Alaska and western Canada have no littoral flora and only sparse vegetation in the supralittoral (Broad, 1982). The virtual absence of intertidal fauna, which results in the absence of a fauna in the upper littoral zone, may be attributed to the freezing air and low humidity. However, the barrenness is usually explained, according to many authors, by the mechanical action of ice on the shores and/or to the weak tidal amplitude itself.

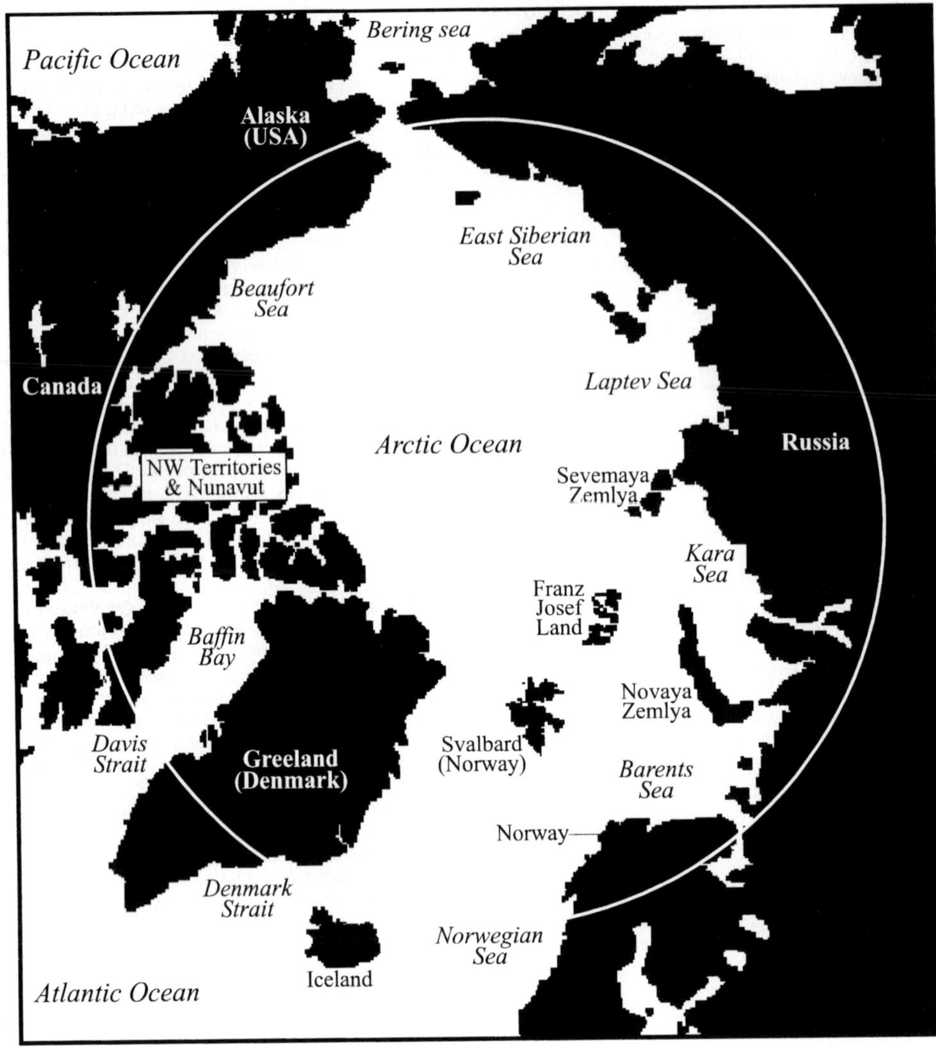

Figure A24 Continental boundaries, major islands and seas of the Arctic. The approximate limits of the Arctic circle (66° 30′N) are shown in white.

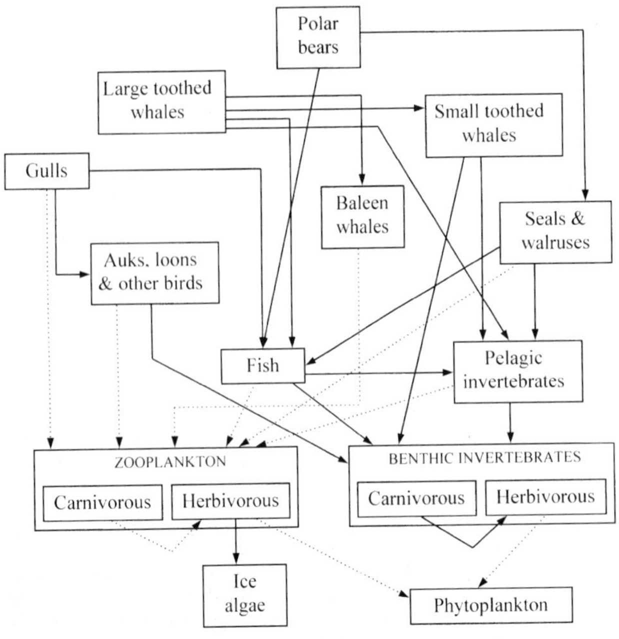

Figure A25 Schematic Arctic food chain.

When wind, current, and tide drive ice floes against the shore, bottom sediments, rocks, and benthic fauna and flora are mixed, overturned, and ground together. The resulting clouds of mud may extinguish the life of nearby organisms under a blanket of mud and smother filter-feeders hundreds of meters away. Consequently, when they are present in the Arctic, the intertidal species found in the Atlantic or Pacific are usually located in the subtidal area.

While sparse, the littoral fauna of the High Arctic is characteristic. In Alaska, the shore to a depth of about 2 m, the approximate depth of seasonal ice, is probably colonized by less than 50 species of macrobenthos, mainly oligochaete worms, midge larvae, amphipods (*Gammarus setosus* and *Onisimus litoralis*) and the isopod *Saduria entomon*, which form nearly the entire biomass (3 ± 5 g m^{-2}; Broad, 1982). The sublittoral fauna beyond the depth reached by ice is more diverse, including species such as mysid shrimps (*Mysis relicta* and *Neomysis rayii*), polychaete worms (*Scolecolepides arcticus, Ampharete vega, Prionospiro cirrifera, Terebellides stroemi* and others), bivalve mollusks (*Cyrtodaria kurriana* and *Lyocyma fluctuosa*), the priapulid *Halicryptus spinulosis*, amphipods (*Pontoporeia affinis* and *Calliopus laevuisculus*), and the four-horned sculpin *Myoxocephalus quadricornis* (Broad, 1982). The boundaries between Arctic and Subarctic shores of northern Canada and Greenland can be determined by the sudden disappearance in the northern parts of three common invertebrates, *Mytilus edulis, Littorina saxatilis* var. *groenlandica*, and *Balanus balanoides*. The last two are the most decisive indicators as their disappearance is more abrupt. In northern Iceland, the littoral zone is characterized by the succession of different species of *Fucus*, from *Fucus spiralis* in the upper section to *Fucus edentatus* in the lower littoral (Munda, 1991).

The sublittoral of Alaskan shores harbors benthic macroalgae, including 3 species of Chlorophyta, which infrequently grow in the littoral zone, 14 species of Phaeophyta and 11 species of Rhodophyta. The intertidal and subtidal zones are feeding grounds for about 30 species of fish, including anadromous salmonids, typically marine gadids, pleuronectids, cottids and a few freshwater species (Broad, 1982).

The Arctic shores of eastern Greenland are similar to those of Alaska. A few algae occur on rocks with *Fucus inflata* in the lower littoral. The fauna of the littoral consists of mostly oligochaete worms (*Enchytreaus albidus* and *Lumbricella lineatus*), mites (*Molgus littoralis* and *Ameronothrus lineatus*) and sublittoral crustaceans (*Gammarus wilkitzkii* and *Mysis oculata*) (Broad, 1982). Warm currents along some coasts of Norway, Iceland, Greenland, and other regions of the Barents and Norwegian Seas promote the occurrence of a boreal biota along Arctic shores. In such environments, the most common species of the mid-littoral and continuing subtidal are the blue mussel *M. edulis*, the gastropods *L. saxatilis*, with some *Buccinum groenlandicum, Margarita helcina* and crustaceans of the genera *Balanus, Gammarus, Caprella* and *Ischyrocerus*. Some species, such as the polychaete worm *Spirorbis spirium* and certain gastropods also colonize floating thallus of *Fucus* and other macroalgae. Jorgensen *et al.* (1999) indicated that 387 species were observed in the Kara Sea, Arctic Russia, with predominance for polychaetes, crustaceans and mollusks. Boreal–Arctic species clearly dominate this area. The sedimentation rate, as well as depth, sediment structure, and salinity apparently influence the faunal distribution.

Studies performed in southern Baffin Island indicated that numerous pelagic and benthic invertebrates were found below ice-covered sea. The fauna and flora in the long lasting ice cover, the small tides and the reduced wave action allow fine sediments to settle and produce low oxygen concentrations near the bottom. The brittle star *Ophiura sarsi*, shrimp *Sclerocrangon boreas*, sea cucumber *Chiridota laevis*, starfish *Crossater papposus* and *Leptasterias groenlandicus* are observed commonly. Rocks deposited by ice and human artifacts provide a scattered hard substrate, which supports a kelp community of *Laminaria* and *Agarum*. Barnacles, ascidians, sea cucumbers, sea anemones, gastropods are also observed abundantly on rocky bottoms.

Bluhm *et al.* (1998) indicated that the sea urchin *Strongylocentrotus pallidus* was considered an important part of the benthic standing stock and carbon flux in the northern Barents Sea. Holte and Gulliksen (1998) noted that macrofaunal dominant species in the sediment of the north Norwegian coast were detritivorous and carnivorous polychaetes, and detritivorous bivalves. Sponges, bivalves, polychaetes, and nematodes are common in the Svalbard waters (Strömberg, 1989).

Primary production (phytoplankton)

The algae that grow on the under-surface of the ice, the phytoplankton and the benthic macroalgae, produce biomass that feeds numerous animals. Together, these plants are called the primary producers. Among them, phytoplankton produces the majority of the edible material, although ice-associated algae are of equal importance in some areas (see Sub-ice flora and fauna section below).

The onset of ice cover and the loss of winter solar radiation have the clearest impact on the primary production in the Arctic. Although, the level of incident radiation in the polar summer is high, much of it is lost by the high reflectivity from the snow and ice. This restricts the production and leads to extreme seasonal oscillation in both phytoplankton and herbivorous zooplankton. The onset of seasonal production is clearly linked to the melting of snow from the ice surface and the subsequent melting of the ice. The annual primary production in polar regions varies from one area to another and ranges from 5 to 290 g C m^{-2} (Sakshaug, 1989). Blooms are formed mainly in spring or summer when nutrient supply is adequate and vertical mixing is restricted to the upper 30–50 m. They are particularly frequent at the ice edge because melting generates water column stability (Alexander, 1980). In the northern waters of Iceland, spring growth of phytoplankton begins in late March and culminates during mid-April (Gislason and Astthorsson, 1998).

According to Zenkevitch (1963), more than 200 species of phytoplankton are present in the Barents Sea. About 80% are diatoms, 20% are dinoflagellates and flagellates. During winter and early spring, phytoplankton biomass is small (<0.5 mg chlorophyll a m^{-3}) and mainly composed of small flagellates. The spring bloom gives place to the growth of larger groups, such as diatoms and dinoflagellates. During summer, two main composition patterns are found. One is dominated by microflagellates in the upper oligotrophic layers and diatoms and other large groups dominate the other, in the sub-surface chlorophyll maximum. Toward autumn, the chlorophyll sub-surface maximum tends to disappear and most of the phytoplankton comprises microflagellates (Loeng, 1989).

Small flagellates dominate the first phase of the phytoplankton bloom in ice-covered water. The main spring bloom occurs in the surface layer close to the ice edge and is usually dominated by diatoms, although flagellates can also be abundant. Toward the end of the diatom bloom, the maximum phytoplankton biomass is found at the nutricline, located below the pycnocline. After the bloom, the major part of the diatoms sink out of the euphotic zone (Loeng, 1989).

Strömberg (1989) indicated that as the ice melts in the Svalbard waters, a very active pelagic primary and secondary production begins, which follows the ice-edge as it retreats. Horsted (1989) noted that there are generally two maxima of primary production from April and from August to September in Greenland waters.

Sub-ice flora and fauna

Wherever a sufficient amount of light penetrates the ice and enough nutrients are present, ice algae develop. These algae make ice-covered seas unique. The ice-edge phenomena are closely linked to the cryopelagic organisms. Thus, one of the short food chains is: primary production (ice-algae, phytoplankton), sea ice amphipods (*G. wilkitzkii, Apherusa glacialis*), polar cod (*Boreogadus saida*), sea birds, or mammals. The biomass of the cryopelagic fauna is high close to the ice-edge and decreases with increasing distance from it.

Daily and seasonal melting and formation of the ice has a thermostatic effect, so that the temperature of the water immediately below the ice is almost constant. It is also often of low salinity, hence the ice biota have evolved as euryhaline and stenothermal organisms. Mel'nikov (1980) distinguishes an endemic or autochthonous flora, consisting of Chlorophyta and Cyanophyta, present year-round and developing each spring in the snow above the ice and gradually leaching downward through the ice; and a non-endemic flora dominated by diatoms. This latter flora develops at the bottom of the ice, is planktonic in origin, and is carried upward as the ice thickens in the fall. He indicated that when diatoms die, they become a source of food for the endemic species of flora. The associated fauna is endemic, *G. wilkitzkii, Mysis polaris, Derjugina tolli, Tisbe furcata* among others, or non-endemic.

The brown zone forming the bottom layer of sea ice is dominated by pennate diatoms, but contains other algae and several species of small invertebrates. Chlorophyll measurements in Arctic Canada show that production of the ice diatoms peaks in mid-June before the phytoplankton blooms in the water below the ice, at a time when the sea ice is still snow-covered (Mansfield, 1975). As soon as the snow melts this ice flora disappears, suggesting that it is adapted to low light intensities. It extends the season of production by preceding the normal phytoplanktonic bloom. These algae also provide a source of food for benthos and fish in coastal waters, especially with the cods *B. saida* and *Arctogadus glacialis* (Mansfield, 1975).

Werner and Arbizu (1999) noted that the lower part of the ice in the Laptev Sea, northern Russia, showed an accumulation of organic matter, mainly composed of organic carbon and chlorophyll, that could provide a food source to the fauna living below the ice in the pelagic habitat, and to the underlying benthos. In fact, they found that large quantities of nauplii, copepods, foraminifers and pteropods were common. Poltermann (1998) found that the Franz Josef Land islands sea ice was inhabited by several amphipods species. Abundance, biomass and small-scale distribution of these cryopelagic amphipods reached 420 ind m^{-2}. Amphipods were concentrated at the edges of the ice floes and were less frequent in areas further away under the ice. Species such as *G. wilkitkzii* were dominant with the less abundant *Apherusa glacialis, Onisimus nanseni*, and *Onisimus glacialis*.

Secondary production (zooplankton)

Water near the surface of the Arctic Ocean supports the greatest summer biomass of zooplankton, the abundance generally decreasing with increasing depth. At least 158 species of zooplankton are known to be found in the central Arctic and adjacent seas including Kara, Laptev, East Siberian, and Beaufort. Crustacea, Ctenophora, Molluska, Annelida, Ostracoda, and Chaetognatha are the most common (Grainger, 1989).

Like phytoplankton, zooplankton shows an extreme annual oscillation in Arctic Canada. This is mainly due to the increase in populations of the herbivorous species, particularly copepods, which are the dominant species. As phytoplankton grows, copepods grow and reproduce, and then as their food supply comes to an end, they gradually die off. In contrast, the primarily carnivorous zooplankton species show relatively little change in numbers throughout the year and no particular period of reproduction (Mansfield, 1975).

Gislason and Astthorsson (1999) indicated that the seasonal variation in biomass and abundance of zooplankton in the waters north of Iceland were low during the winter and peaked once during spring in late May. The principal constituents were *Calanus finmarchicus*, *Pseudocalanus* spp., *Metridia longa*, *C. hyperboreus*, chaetognaths and euphausid larvae. Kosobokova *et al.* (1997) studied the composition and distribution of zooplankton in the Laptev Sea where total biomass ranged between 0.1 and 7.9 g m^{-2}. The dominant species were *Calanus glacialis*, *Calanus hyperboreus*, and *M. longa*. Copepods are also the most important grazers in the Barents Sea with ca. 38 species, *C. glacialis*, *C. finmarchicus*, and *C. hyperboreus* being among the most common (Loeng, 1989). Euphausids as *Thysanoessa* spp. are the most important grazers. Among carnivorous zooplankton species, the chaetognaths *Sagitta elegans* and *Eukrohnia hamata*, together with ctenophores are considered to be important (Loeng, 1989).

Dunbar (1989) and Grainger (1989) pointed out that diel vertical migration of zooplankton in the Arctic Ocean is rare, but seasonal vertical migration is fairly common. In the Barents Sea, larger zooplankton species, such as *C. glacialis*, are thought to overwinter in deeper layers, returning to the surface layer in early spring to mature and spawn. Spawning takes place during the diatom bloom and the new generation feeds on the remnants of this bloom and subsequent production of phytoplankton. Zooplankton in turn is preyed on by capelin, which during summer has a northward feeding migration (Loeng, 1989).

Zenkevitch (1963) characterized the Svalbard water area in the same way as the Barents Sea, with a dominance of copepods *C. finmarchicus* making up 90% of the zooplankton biomass. The plankton biomass was reported to decrease in a northerly and easterly direction. In the area north of Svalbard, chaetognaths, ctenophores, and amphipods comprise most of the remaining biomass (Strömberg, 1989).

Horsted (1989) observed that zooplankton biomass in Greenland waters had its maximum in July at the same time as the second maximum of primary production, while macroplankton had a somewhat longer lasting maximum from June–July onward. Thus, the newly hatched cod larvae seem to depend heavily upon the availability of naupliar and copepodid stages of *C. finmarchicus*.

Polynyas

In the Arctic, there are ice-free areas in the proximity of completely ice-covered regions, so called polynyas. The correlation between areas of open water in ice-covered Arctic and increased biological productivity is clear (Bazely and Jefferies, 1997; Stirling, 1997).

Polynyas form where warm, upwelling sea currents prevent the water surface from freezing. They are either small and temporary or extensive and free of ice all winter. The largest ones appear at the same place every year. The biggest of all polynyas is located at the head of Baffin Bay, Canada, and can reach a surface area as large as Lake Superior. In the desert of floating ice, polynyas represent a rich feeding ground and ideal habitat for many tiny plants and animals at the base of the food web. They also represent a temperate environment that enhances survival of Arctic animals. For instance, the ice around polynyas is ideal for the growth of ice algae. Owing to this large amount of food, countless amphipods and larger crustaceans, squids, fishes and mammals can be observed. Ice algae develop soon in spring, as soon as the sun rises high enough to promote minimum photosynthesis, long before the weather begins to warm up. This early start of primary production and its effect on other levels of the food chain is important for seabirds that need time to build up their strength before they can lay their eggs. Moreover, since seabirds can only hunt for food in open water, they are largely dependent upon polynyas. Among them are fulmars, murres, guillemots, and gulls. Marine mammals such as whales, seals, polar bears, and walruses also converge at polynyas to take opportunity of the abundance of food. If no patches of open water existed in the winter pack ice, many animals could not survive.

Fish

There are no schooling pelagic fish, though the cryopelagic Atlantic cod (*Gadus morhua*) is known to occur in large numbers during the spring period in the Barents Sea, and the polar cod (*B. saida*) appears to form scattering layers. The giant among benthic fish is the Greenland shark (*Somniosus microcephalus*) long known for its scavenging behavior largely oriented on whale carcasses. Where the Bering Sea intrudes into the Beaufort Sea, abundant pelagic herrings (*Clupea harengus*) and capelins (*Mallotus villosus*) occur in the region of the Mackenzie delta. Demersal species such as the Arctic and starry flounders (*Liopsetta glacialis* and *Platichthys stellatus*) and the Greenland and saffron cods (*Gadus ogac* and *Eleginus gracialis*) are also common. Similarities exist in the southern part of Baffin Island where the Atlantic cod and several deepwater forms like the Greenland halibut (*Reinhardtius hippoglossoides*) and the rock and roughhead grenadiers (*Coryphaenoides rupestris* and *Macrourus berglax*) are observed. The Barents Sea is a feeding ground and nursery for large commercially important fish stocks such as Arctic cods, herrings, Greenland halibuts, haddocks, capelins, redfish, and saithes (Loeng, 1989).

Arctic fish must be able to live under conditions of reduced light and in total darkness. Air breathing predators, except for certain seal species, are displaced to gaps, leads and polynyas through interference with respiration. Ice also reduces avian predation to practically zero. In these conditions, ice may be considered advantageous for fish. Moreover, a new biological habitat that might be described as an inverted benthos is provided by ice for fish. A distinct phytoplankton flora occurs immediately below the ice and an abundant crustacean fauna may be tributary to this food source. This diatom-crustaceans community is an important one that supports the Arctic cod, which helps to sustain Arctic char, birds, seals, and belugas.

Birds

About 12 species of birds are known to live all year round in the Arctic, including the raven, gyrfalcon, willow and rock ptarmigans, snowy owl, redpoll, Ross's and ivory gulls, thick-billed murre, dovekie and the black guillemot. Most of them shift south slightly during the Arctic winter. All other birds, about 90 species, are migrants. Numerous species of birds nest in the Arctic, such as the fulmar *Fulmarus glacialis* and the kittiwake *Rissa tridactyla*. In the Greenland Sea, the most abundant species are the fulmar, little auk *Alle alle*, guillemot *Uria lomvia*, and kittiwake (Mehlum, 1997). The abundance of organisms that bloom during summer, including insect larvae, mollusks and crustaceans, are a substantial source of food. Lack of food rather than the cold itself is what makes the Arctic so inhospitable for birds in winter. Nevertheless, considering that the cold is fierce, some of the permanent residents have evolved special adaptation to cope with it.

Mammals

The generally scattered distribution of marine fish in Arctic waters has not prevented mammals from attaining a dominant position among the marine vertebrates, for they depend mainly on the larger invertebrates, both planktonic and benthic, for their food. Thus, most common mammals of the Arctic are either part of or dependent upon coastal ecosystems. They include polar bear, arctic foxes, wolverines, reindeer, musk oxen, lemmings, and many species of marine mammals.

Five species of seals spend at least part of the year in the Arctic. The bearded seal (*Erignathus barbatus*) and ringed seal (*Pusa hispida*) remain all year round. The others are the harp seal, hooded seal, and harbor seal. Except for the harbor seal, all species give birth to their pups on the ice. Most seals are important predators of benthic invertebrates and fish. The most thoroughly investigated species has been the ringed seal, the most truly Arctic of all the mammals. This species is able to survive under the winter fast ice by keeping open breathing holes. In spring, the pup is born in a cavern among rafted ice blocks or on the surface of the ice in a lair excavated in the overlying snow. Predation by the polar bear (*Ursus maritimus*) and Arctic fox (*Alopex lagopus*) can cause very high mortality of young in certain areas. In fact, the polar bears feed almost exclusively on seals. The ability of ringed seals to live under fast ice is not shared by any other Arctic pinnipeds, though the bearded seal is known to keep breathing holes in some areas.

Walrus (*Odobenus rosmarus*) live in the Arctic waters east of Somerset Island, notably in Baffin Bay, Davis Strait and Foxe Basin. Pacific walrus are found along the coast of Alaska and westward along the Arctic shores of Siberia. Walrus give birth on ice floes and, when not feeding, spend most of their time lolling on the ice floes in great numbers. Most of them are bottom feeders grazing on clams and others bivalves in shallow waters. They also eat whelks, sea cucumbers, and other benthic invertebrates.

Four species of whales may be encountered in Arctic waters. Three of them, the bowhead whale (*Balaena mysticetus*), beluga (*Delphinapterus leucas*), and narwhal (*Monodon monoceros*) are permanent residents. Like the walrus, narwhals and belugas are gregarious. Narwhals are found in large herds in the fjords of eastern and northern Baffin Island, Devon Island and Ellesmere Island. These whales occur in the High Arctic waters of the northern hemisphere, where they occupy the most northerly habitat of any cetacean species. The fourth whale species, encountered in summer, is the killer whale (*Orcinus orca*). Other species, such as right, gray, blue, fin, minke, and humpback whales move to the Arctic during their normal annual migration cycles (Gambell, 1989). Toothed whales like sperm, Baird's beaked, northern bottlenose, long-finned pilot, and beaked whales, among others, can also be observed in the Arctic.

The bowhead whale is the only plankton eater, its diet is focused on euphausids. The beluga and narwhal mostly prey on fish and squids as well as pelagic and benthic invertebrates. Killer whales largely prey on other marine mammals.

Human impact

The native people of the Arctic are at the top of the food chain and harvest fish, birds, and marine and terrestrial mammals. Inuits have their homeland stretched from the northern tip of Russia across Alaska and northern Canada to parts of Greenland, but their populations remain sparse.

In spite of its remoteness, the Arctic is threatened by airborne pollutants spreading from industrial areas in the lower latitudes, by growing mineral exploitations (coal, nickel, uranium, tin, gold) and also by natural gas and petroleum industries. As a result, Arctic ecosystems and Inuit populations are seriously endangered.

Moreover, during the course of the past century, there has been an overall increase in global temperature of some 0.5°C. The past decade has been the warmest of the past 100 years with the five warmest individual years having been recorded during it. The recent trends observed are generally consistent with those projected by the global circulation models under increased atmospheric CO_2 concentrations (Maxwell, 1997). Higher temperature is generally associated with less sea ice and snow cover extent. Thus, global warming could be one of the most important factors affecting the biodiversity, distribution, abundance, and seasonal cycles of Arctic species in the future.

Jean-François Hamel and Annie Mercier

Bibliography

Alexander, V., 1980. Interrelationship between the seasonal sea ice and biological regimes. *Cold Regions Science and Tehnology*, **2**: 157–178.

Bazely, D.R., and Jefferies, R.L., 1997. Trophic interactions in arctic ecosystems and the occurrence of a terrestrial trophic cascade. In Woodin, S.J., and Marquiss, M. (eds.), *Ecology of Arctic environments*. Special publication no. 13 of the British Ecological Society, pp. 183–207.

Bluhm, B.A., Piepenburg, D., and Juterzenka , K.V., 1998. Distribution, standing stock, growth, mortality and production of *Strongylocentrotus pallidus* (Echinodermata: Echinoidea) in the northern Barents Sea. *Polar Biology*, **20**: 325–334.

Broad, A.C., 1982. Arctic, coastal ecology. In Schwartz, M.L. (ed.), *The encyclopedia of Beaches and Coastal Environments*. Hutchinson Ross, Stroudsburg, pp. 55–57.

Dunbar, M.J., 1989. The Arctic Ocean as a biological environment. In *Proceedings of the Sixth Conference of the Comité Arctique International*. Leiden: E.J. Brill, pp. 35–47.

Gambell, R., 1989. Status of the cetaceans populations of the Arctic and Subarctic Seas. In *Proceedings of the Sixth Conference of the Comité Arctique International*. Leiden: E.J. Brill, pp. 207–251.

Gislason, A., and Astthorsson, O.S., 1998. Seasonal variations in biomass, abundance and composition of zooplankton in the subarctic waters north of Iceland. *Polar Biology*, **20**: 85–94.

Grainger, E.H., 1989. Vertical distribution of zooplankton in the central Arctic Ocean. In *Proceedings of the Sixth Conference of the Comité Arctique International*. Leiden: E.J. Brill, pp. 48–60.

Holte, B., and Gulliksen, B., 1998. Common macrofaunal dominant species in the sediments of some north Norwegian and Svalbard glacial fjords. *Polar Biology*, **19**: 375–382.

Horsted, S.A., 1989. Some features of oceanographic and biological conditions in Greenland waters. In *Proceedings of the Sixth Conference of the Comité Arctique International*. Leiden: E.J. Brill, pp. 456–476.

Jorgensen, L.L., Pearson, T.H., Anisimova, N.A., Gulliksen, B., Dahle, S., Denisenko, S.G., and Matishov, G.G., 1999. Environmental influences on benthic fauna associations of the Kara Sea (Arctic Russia). *Polar Biology*, **22**: 395–416.

Kosobokova, K.N., Hanssen, H., Hirche, H.-J., and Knickmeier, K., 1997. Composition and distribution of zooplankton in the Laptev Sea and adjacent Nansen Bassin during summer 1993. *Polar Biology*, **19**: 63–76.

Loeng, H., 1989. Ecological features of the Barents Sea. In *Proceedings of the Sixth Conference of the Comité Arctique International*. Leiden: E.J. Brill, pp. 327–365.

Mansfield, A.W., 1975. Marine ecology in Arctic Canada. In *Circumpolar Conference on Northern Ecology*. Ottawa, Canada, pp. 29–47.

Maxwell, B., 1997. Recent climate patterns in the Arctic. In Oechel, W.C., Callaghan, T., Gilmanov, T., Holten, J.I., Maxwell, B., Molau, U., and Sveinbjornsson, B. (eds.), *Global Change and Arctic Terrestrial Ecosystems*. New York: Springer-Verlag, pp. 21–46.

Mehlum, F., 1997. Seabird species associations and affinities to areas covered with sea ice in the northern Greenland and Barents Seas. *Polar Biology*, **18**: 116–127.

Mel'nikov, I.A., 1980. *The ecosystem of Arctic Pack Ice*. Biol. Tsentral'nogo Akticheskogo Basseina, Transl. Dept. Sec. Can. Fisheries and Oceans Canada, Arctic Biological Station, Canada.

Menzies, R.J., 1975. Origin and evolution of the Arctic marine ecosystem. In *Circumpolar Conference on Northern Ecology*. Ottawa, Canada, pp. 15–25.

Munda, I.M., 1991. Shoreline ecology in Iceland, with special emphasis on the benthic algal vegetation. In Mathieson, A.C. and Nienhuis, P.H. (eds.), *Intertidal and Littoral Ecosystems*. Ecosystems of the world 24, Amsterdam: Elsevier, pp. 67–81.

Poltermann, M., 1998. Abundance, biomass and small-scale distribution of cryopelagic amphipods in the Franz Josef Land area (Arctic). *Polar Biology*, **20**: 375–382.

Sakshaug, E., 1989. The physiological ecology of polar phytoplankton. In *Proceedings of the Sixth Conference of the Comité Arctique International*. Leiden: E.J. Brill, pp. 61–89.

Stirling, I., 1997. The importance of polynyas, ice edges, and leads to marine mammals and birds. *Journal of Marine Systems*, **10**: 9–21.

Strömberg, J.-O., 1989. Northern Svalbard waters. In *Proceedings of the Sixth Conference of the Comité Arctique International*. Leiden: E.J. Brill, pp. 402–425.

Werner, I., and Arbizu, P.M., 1999. The sub-ice fauna of the Laptev Sea and the adjacent Arctic Ocean in summer 1995. *Polar Biology*, **21**: 71–79.

Zenkevitch, L., 1963. *Biology of the seas of the USSR*. New York: Wiley Interscience.

Cross-references

Arctic, Coastal Geomorphology
Antarctic, Coastal Ecology and Geomorphology
Green House Effect and Global Warming
Ice-Bordered Coasts
Littoral

ARCTIC, COASTAL GEOMORPHOLOGY

Introduction

The Arctic, long considered a ". . . region of darkness and mists, where sea, land and sky were merged into a congealed mass" (Nansen, 1911, p. 1), has subsequently been defined according to its astronomic, biotic, climatic, cryologic, geomorphic, and hydrologic characteristics (Walker, 1983). From the standpoint of coastal morphology, it is the juncture of whichever category is used with the coastline that is important. The determinant that provides the greatest extent to the coastline is sea ice. In the Northern Hemisphere using sea ice as the limiting boundary results in the inclusion of the coastlines of Hudson Bay, Labrador, and Newfoundland, the Sea of Okhotsk and most of the Baltic and Bering Seas (See *Ice-Bordered Coasts*). By many other criteria, most of these coastlines do not qualify as Arctic.

In this entry, the discussion centers on those coasts that border the Arctic Ocean and surround Greenland and the islands of the Canadian Archipelago. When the lengths of these coastlines are measured on the American Geographical Society 1 : 5,000,000 *Map of the Arctic Region*, they total 82,000 km, Greenland (16,000 km), the Canadian Archipelago (26,000 km) and the other arctic islands (9,000 km) (Figure A26). Other calculations, depending on the detail used, provide larger numbers. For example, Bird (1985) states that the coastline of the islands of the Canadian Archipelago is at least 90,000 km long and Nielsen (1985) reports that that of Greenland is some 40,000 km long.

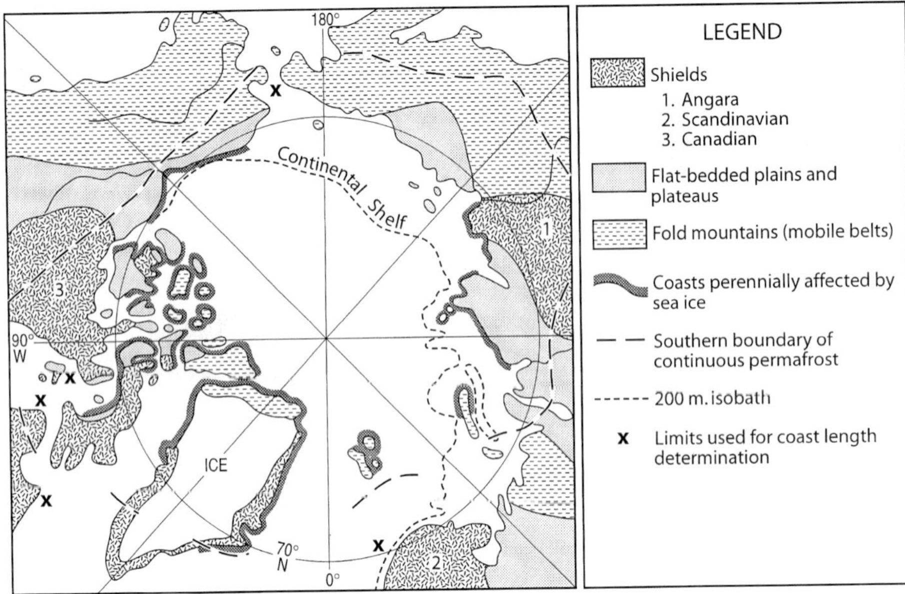

Figure A26 Geologic base, continental shelf and sea ice distribution in the Arctic (compiled from numerous sources including Sater *et al.* (1971); Péwé (1983); Walker (1998)).

The geologic base

Three large, stable shields composed of Pre-Cambrian rocks form the cornerstones of the Arctic, one each in Canada, Scandinavia, and central Siberia (Figure A26). The actual exposed lengths of these Pre-Cambrian rocks along the coast is less than what the size of the shields would lead one to believe because their margins have been buried under eroded shield materials. Between and adjacent to the shields, including their buried margins, are folded mountains (or portions of the so-called "mobile belts" of the Northern Hemisphere) some of which extend to the coast (Figure A26).

Like the embedded coasts surrounding the Atlantic Ocean, Arctic coastal margins are affected by minimal tectonic activity. Earthquakes do occur, as on the north coast of Baffin Island, north of the Mackenzie delta and in the northwest Canadian Archipelago, but they are of low magnitude.

The most recent major event to impact most of the arctic coastal zone is glaciation. The main coastal areas not directly affected by glacial ice during the Pleistocene include the east-central Siberian lowland, a small part of the western Canadian Archipelago and the north coast of Alaska. However, even those coastal zones, though never actually covered by ice sheets, were affected by drainage across them from glacial fronts and by the changes in location of the interface between land and sea due to changes in sea level as the continental glaciers waxed and waned. Today, only a small percentage of the arctic coastline is being modified by glaciers. Major segments are found on Greenland, Ellesmere Island, Novaya Zemblya, and Spitzbergen.

In addition to the direct modifications of the coastlines by glaciers in the form of moraines, drumlins, fiords and strandflats, the rebound that followed deglaciation has (and is) converted formerly submerged coastal belts into subaerial coastal plains. Such occurrences are especially common in northern Canada, the Canadian Archipelago, Scandinavia and the islands north of Siberia. In many locations former coastal features are found today at elevations of as much as 250 m. In the Canadian Archipelago and the Hudson Bay area rebound is continuing at rates of as much as 1 m/century (Andrews, 1970).

The continental shelf and the coastal plain

The two basic contrasting zones of coastal significance in the Arctic are its subaerial and subaqueous portions. These two zones are highly variable in width and in the forms and processes they possess. The continental shelf varies in width from a few kilometers, as off parts of Greenland, to more than 800 km in the East Siberian Sea. With sea-level rise and with coastal erosion, the width of the subaqueous portions of the Arctic Ocean has been increasing. Weber (1989, p. 815), for example, writes that "The Chukchi Shelf was eroded far into the continent and transects the principal mountain ranges of Northern Alaska." Much of the continental shelf of the Arctic Ocean is flat and shallow, cut only by a few submarine valleys such as those off rivers like the Kolyma and Indigirka and off Barrow, Alaska. Most of the smaller features on the shelves are the result of ice gouging and deltaic deposition or are remnants of subaerial erosion that occurred during lower sea levels (Reimnitz *et al.*, 1988; Weber, 1989).

The coastal plain, although not as extensive as the continental shelf, possesses a greater variety of forms and processes than the shelf it borders. Although, the coastal plain of today is subaerial much of it, as in Alaska and northwestern Canada, is the "... landward edge of a continental shelf that has experienced repeated transgressions and withdrawals of the sea ..." (Bird, 1985, p. 243). It is comprised mainly of gravels, sands, and silts that presently are ice-bonded in the permafrost. In places it is low-lying and level. Near the Indigirka River mouth, for example, storm surges have reached as far as 30 km inland (Zenkovich, 1985).

The major forms of the coastal plain that border the ocean include barrier islands and lagoons, sand and gravel beaches, mudflats and marshes, sand dunes, low coastal bluffs, deltas and lengthy rias (*gubas* in Russian).

Among the arctic coast's most conspicuous and extensive features are its barrier islands and sandy spits. They are present along much of the coast of arctic Alaska and northwest Canada and along various parts of the Siberian coastline. Some barrier islands are remnants of the coastal plain whereas most are composed of gravel and sand which originated offshore.

Although, most of the deltas in the Arctic are small, they are sufficiently numerous to occupy a sizeable proportion of the coastline. For example, 135 km (or 9%) of the coastline of Alaska between Cape Lisburne and the Canadian border is deltaic (Wiseman *et al.*, 1973).

Some arctic deltas, like the Mackenzie, Lena, Indigirka, Kolyma, and Colville, face the open ocean, whereas others, such as the Yenisey and Ob, are located at the head of lengthy and narrow gubas (Figure A27). These two deltas, even though confined to the head of their estuaries, are the fourth and fifth largest deltas in Siberia. Among arctic deltas, that of the Lena River (Figure A28) is unique because its structure, shape, and relief appear to be the result of tectonic activity as well as modern hydrological processes (Aré and Reimnitz, 2000).

The deltas of the Arctic, like those elsewhere, are relatively young in that their present-day expression stems only from the time sea-level rise reached a nearly stillstand position about 5000 BP. All of the older deltas in the Arctic contain most of the features, such as distributaries, abandoned channels, lakes, sand bars, mudflats, and sand dunes, that are typical of deltas elsewhere. However, they also possess such cryospheric forms as ice-wedges, ice-wedge polygons, pingos, and thermokarst lakes.

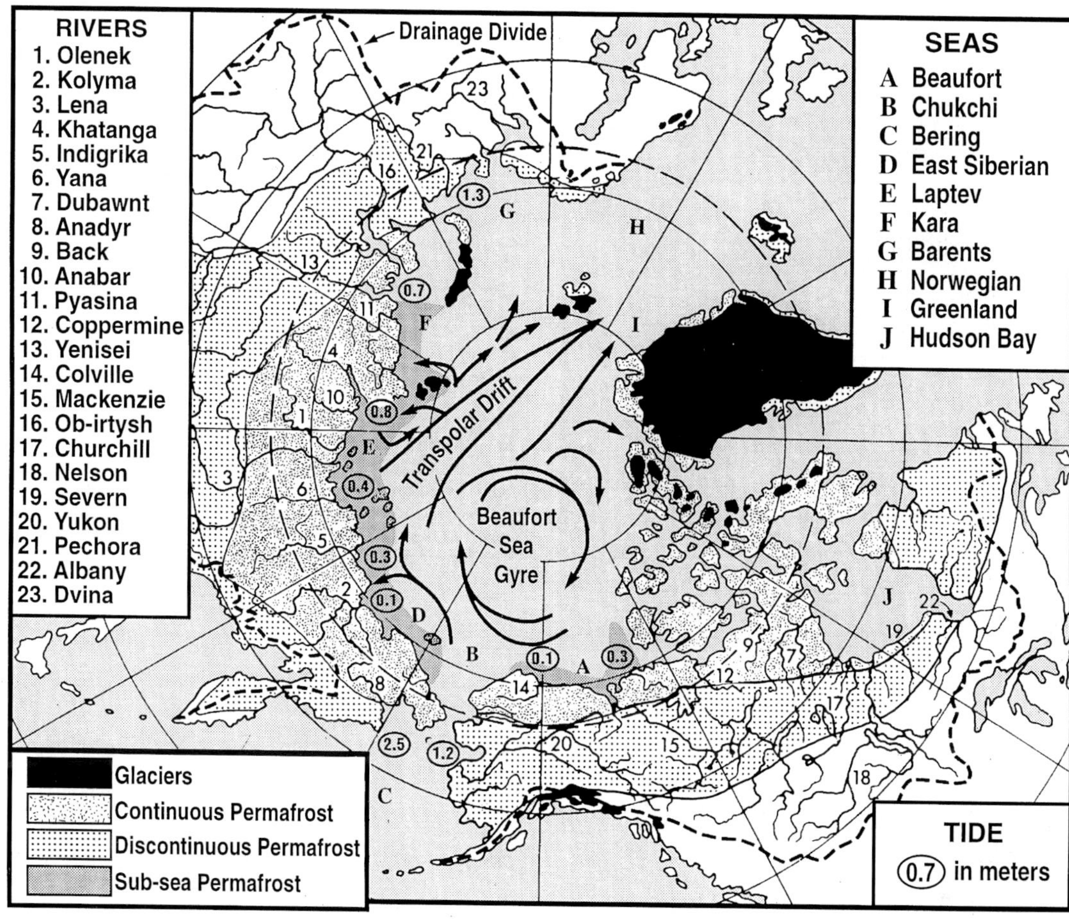

Figure A27 The Arctic illustrating the numerous factors that affect the coast (compiled from numerous sources including NOAA (1981); Lewis (1982); Péwé (1983); Walker (1998)).

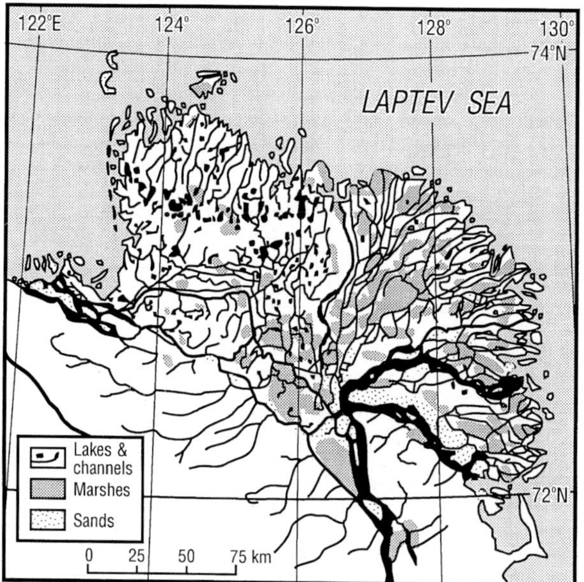

Figure A28 The Lena River delta, the largest at 32,000 km^2 in the Arctic.

Conditions, forms and processes

The present-day appearance of the arctic's coastline, like that of coastlines elsewhere, depends on modifications that have occurred to the geologic base it inherited. Along many coastlines, these modifications include those engineered by humans. In the case of the Arctic, however, human modifications are still minimal, although they do occur at the mouth of some rivers, adjacent to some coastal villages, and where mining operations, including petroleum exploration and production enterprises, have been developed.

Thus, most of the coastal forms in the Arctic, as is true also of the Antarctic, are the result of natural conditions and processes. Included are those conditions and processes associated with cold climates such as low temperature, snow, ice (river and sea) and permafrost with its many forms of ground ice as well as those of more universal occurrence like relief, structure, sediment type, river discharge, offshore gradient, wave action, currents, storm surges, and tides (Figure A27).

Cold climate processes are frequently divided into two types: glacial and periglacial (French, 1989). Although in the arctic of today glacial processes impact only short lengths of shore, periglacial processes are of major importance along the entire coastline (Harper, 1978) including those sections most recently deglaciated.

The *sine qua non* of periglacial processes is the freezing and thawing of the ground, which is mainly temperature dependent. The change from one state to the other may be daily, seasonal or over longer (thousands of years in some cases) periods of time.

Permafrost

Permafrost (a condition in which ground temperature remains below 0°C for more than two years) is present along most of the coastline of the Arctic from which it extends both landward and seaward (Figure A27). Although permafrost is defined by temperature, it is water in the form of ice that is especially critical in coastal modification (Figure A29). The included ice, which serves to bond unconsolidated sediments as long as they remain in a frozen state, makes the mass especially unstable when thawed. Ground ice is highly variable in type and quantity. In the upper few meters of permafrost it may account for more than half of the

Figure A29 Ice complex coast on Muostakh Island facing the Laptev Sea. The cliff is about 20 m high and is retreating at a rate of about 11 m/a (photo courtesy of Feliks Aré).

Figure A30 Thermoerosional niche in deltaic permafrost.

Figure A31 Example of block collapse and ice wedge from E. de K. Leffingwell's classic research monograph on permafrost and coastal morphology (1919).

volume whereas its presence in bedrock is minimal. It can range in form from microscopic pore ice to large segregated masses (Mackay, 1972).

Permafrost cliffs, which border much of the coastal plain of the Arctic, are relatively stable for 8–10 months of the year because they are protected from erosion by being in a frozen state, being covered by snowdrifts, and by facing a frozen, immobile sea. With snowmelt, an active layer (the thin layer that alternately freezes and thaws on the top of permafrost) begins to form and some surface flow is initiated. However, it is after the shore has become sufficiently free of ice that wave action can impact the base of the cliff and major erosion can occur. With wave impact, thawing at the base of the ice-bonded cliff occurs and with sediment removal a thermo-erosional niche (*termoerozionaja niza* in Russian) is created (Gusev, 1952) (Figure A30). The height of thermoerosional niches depends largely on the change in water level during wave attack and the depth of penetration varies with the stability of the sea cliffs.

The permafrost of the Arctic frequently supports other features such as pingos (Mackay, 1972) and ice-wedges which often express themselves at the surface in the form of ice-wedge polygons (Lachenbruch, 1962). Although their size varies, polygons are usually on the order of 10–30 m in diameter. Along the coastline, paralleling wedges serve as lines of weakness and often lead to block collapse (Figures A31 and A32) as thermoerosional niches develop. Those wedges that are perpendicular to the shore can serve as points of more rapid retreat, especially if the matrix is dominated by peat. Such coastlines are highly irregular

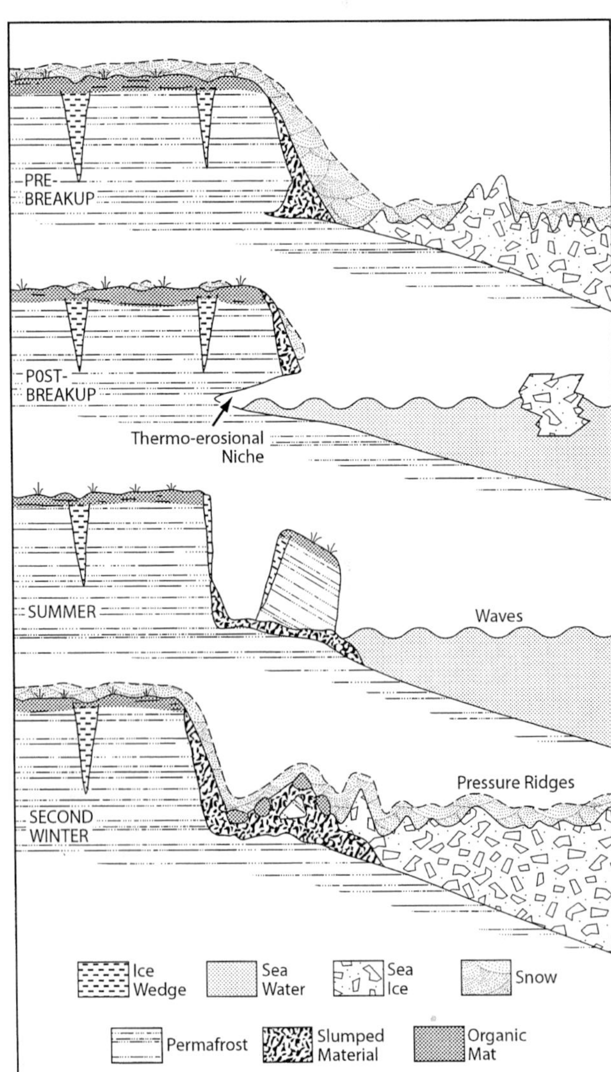

Figure A32 Block collapse and erosion along permafrost coastlines.

Figure A33 Bedrock shore on the Taymyr Peninsula, Siberia (photo courtesy of Feliks Aré).

in detail. In addition to the large block falls that result from thermoerosional niche development, many coastlines exhibit thermokarst forms such as ground-ice slumps (creating what some have called thermocirques) and breached thermokarst lakes. Along hard rock coasts (which are in the minority in the Arctic) erosion of a cliff face appears to be mainly by frost wedging (Figure A33).

Sub-sea permafrost (offshore permafrost, submarine permafrost) occurs beneath the seabed of the continental shelves of northwest Canada, north Alaska, and Siberia (Figure A27). It is relic in that it formed when sea level was low and much of the present-day continental shelf was subaerial. It is of major interest today because of its importance to the petroleum industry and as a store of carbon dioxide and methane (Osterkamp, 2001).

Sea ice

Sea ice which borders virtually all of the arctic coast during winter, is present along some sections even in summer (Figure A26). During the time it is shorefast and bottomfast marine action along the coastline is nil. Also, during summer, much of the coastline is protected from wave attack because the presence of sea ice offshore reduces the fetch available to wind and dampens wave action. However, during these periods, often in the Fall of the year when the ice pack has been removed some distance offshore, storm waves may become powerful eroding agents. In 1986, strong winds over the Chukchi and Beaufort Seas resulted in severe erosion of low coastal cliffs impacting heavily the villages of Wainwright and Barrow, Alaska. Other erosive storms also occurred along these coasts in 1954, 1956, and 1963.

Some coastal specialists consider that the passive nature of sea ice is its most important role *vis-à-vis* the coastline. Nonetheless, sea ice is also an active agent in coastline and offshore modification. Ice push during the periods of freeze up and to a lesser extent during breakup create highly irregular surfaces on arctic beaches. Large pressure ridges formed during freeze up help protect the coastline during breakup because they may last well into the Fall. Although some of the reworked beach forms, including mounds, ridges and kettles, may last for years, depending mainly on how high up on the beach they developed, most tend to be ephemeral because they are removed by summer wave action. Large amounts of beach material (especially sand and gravel but also, in some locations, boulders) can be bulldozed, scoured, rafted, and resuspended by sea ice. In some instances, sea ice rides up on the shore and even overtops low cliffs. If rideup is over a snow bank or an ice foot it will cause minimal modification to the frozen ground beneath.

In addition to the shoreface itself, sea ice causes ice scour out to depths of more than 20 m often several kilometers from the coastline. Deep (2–3 m) gouges or trenches are created by the keels of drifting floes.

Figure A34 Remnant snowdrift along Beaufort Sea in western Canada in August. Note the coarse textured beach and driftwood typical of many arctic beaches.

Most floes eventually become grounded in which case current movement around them can resuspend bottom sediment and transport it elsewhere.

From the standpoint of the coastline, changes occurring to sea ice because of global warming are likely to be very important in the future. Recent data show that, not only is the ice pack thinning, but its seasonal regimen is changing (Parkinson, 2000). If such a trend continues, ice-free periods along coastlines will lengthen and the fetch will increase in length. Thus, both the intensity and duration of waves impacting the shore will also increase.

Snow

Although the amount of snow that falls in the Arctic is limited, partly because of low temperatures, it is an important factor in regard to the coastal zone. It accumulates as snow drifts against irregularities such as coastal bluffs and serves as protective ramps often until late summer (Figure A34). Snow also becomes incorporated in the icefoot in combination with seawater and freshwater ice, and organic and inorganic sediment where it becomes part of the beach (McCann and Taylor, 1975). Possibly the most important role of snow insofar as the arctic coastline is concerned is its service as the main source of water for the rivers that flow with their sediment load into the Arctic Ocean.

Impact of rivers on the Arctic coast

Four of the ten longest rivers of the world (Ob, Yenisei, Lena and Mackenzie) drain into the Arctic Ocean. In addition, there are numerous smaller rivers, some of which are confined to the zone of continuous permafrost. Many of these smaller rivers cease flow completely during winter. Even the larger rivers are highly seasonal with the major discharge of water and sediment occurring during the snowmelt season (Walker, 1998). Thus, the impact of rivers on the coast is confined, like that of most other arctic coastal modifiers, to a few months of the year. Nonetheless, during that restricted season river water impacts the sea ice over and under which it flows.

Floodwater aids in earlier melting of the sea ice at the mouth of the river than otherwise would be the case. Nearshore, much of the flood water flows over the top of the bottom/shore fast ice until it finds holes, as along pressure ridges, down through which it drains. From these discharge holes, which Reimnitz and Bruder (1972) call *strudel*, water continues to flow seaward beneath the ice (Figure A35). As the water pours through these holes in the ice, it creates depressions in the seafloor deposits below.

As sediment-laden floodwater spreads out over the sea ice, its velocity decreases and deposition occurs. Most of the sediment deposited on this bottomfast ice becomes a part of the subaqueous delta as the sea ice melts from under it (Walker, 1974). The sediment that is carried by floodwaters beneath the sea ice may be transported many tens of kilometers seaward and become incorporated in long-shore currents and therefore, lost to the deltas.

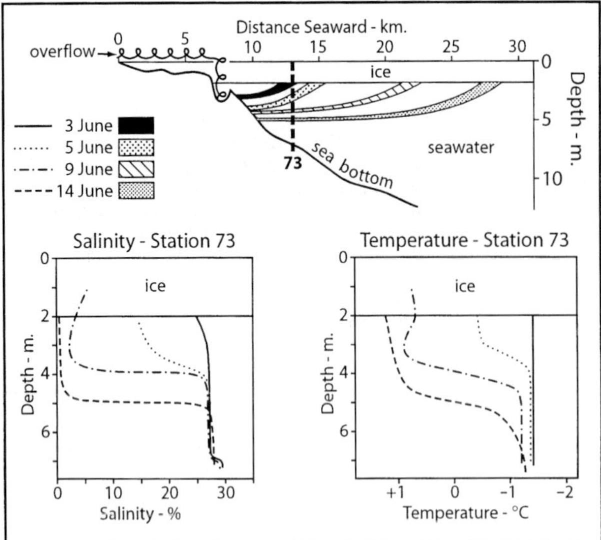

Figure A35 Floodwater overflow and the development of a freshwater wedge beneath sea ice during breakup.

Ocean currents, tides, and waves

"The Arctic Ocean, relatively isolated in terms of worldwide wind systems and partially or wholly covered with sea ice, depending on season, is a body of water in which low tides, slow-moving currents, and low-energy waves predominate" (Walker, 1982, p. 61).

Although there are a few locations in the Arctic where macrotides occur (e.g., off southeast Baffin Island, Canada, and in the Mezen Gulf, Russia, where ranges are more than 10 m), along most of the arctic coast tidal ranges are less than 2 m (Figure A27). Marshes are often associated with the high tide range locations. Despite low astronomic tidal ranges, storm "tides" occasionally result in deep intrusions of water over low coastal plain surfaces. The inner edge of surge lines are often delineated by the presence of driftwood which is a common contribution of the rivers that flow through the taiga of Asia and North America.

The dominant currents in the Arctic Ocean are the Beaufort Sea Gyre and the Transpolar Drift (Figure A27). They are the currents that provide the general direction of flow to pack ice and the occasional ice island that breaks off of Greenland and Ellesmere Island. There are other localized currents that affect the coastal zone by transporting sediments along shore and cause sea ice to impinge the coastline. Such ice movements are affected by wind as well as by ocean currents.

Waves, even though their formation is impossible for most of the year because of the canopy of sea ice that covers the ocean, are, nonetheless, the dominant agent in shore modification along most of the arctic coastline. The length of time, waves are effective, varies greatly ranging from several months along the longest-lasting ice-free zones to only occasionally or even rarely along coasts that are often ice-bound even during summer as exampled by parts of the northwest Canadian Archipelago.

Coastal erosion and climate change

Many of the coasts of the Arctic are retreating, some by tens of meters per year especially along some Siberian coastlines (Aré, 1988). Along the Beaufort Sea coast of Alaska, erosional rates have been such that since the sea reached its present level 4–5,000 BP, the coastline has retreated as much as 25 km (Reimnitz *et al.*, 1988). Rates ranging from 1 to 5 m/a are common in the ice-rich permafrost bluffs bordering much of the Arctic Ocean. In areas north of the Siberian coast, erosion has resulted in the disappearance of some offshore islands.

As most of this retreat is the result of the combined effect of normal marine (wave action) erosion and the periglacial process of thermokarst collapse and thermal erosion, it appears that an increase in the rate of coastal retreat with any rise in sea level and sea-ice degradation is likely.

Such a change would increase the amount of sediment contributed to the shore from cliff erosion which by some calculations already exceeds that contributed by the rivers flowing into the Arctic Ocean (Brown and Solomon, 1999). However, it is likely that the same climate changes that affect sea level and sea-ice degradation will also increase permafrost thaw and river discharge and, therefore, the sediment loads of arctic rivers. Whether the relative proportions of the two sources will change is uncertain.

Conclusion

Although a number of localized programs have been undertaken in the last four decades, some of them out of research stations such as the Naval Arctic Research Laboratory, Barrow, Alaska; the Inuvik Research Laboratory, Inuvik, Canada; and the Permafrost Research Institute, Yakutsk, Russia, Arctic coastal research has generally been neglected.

In 1998, at the International Permafrost Conference, a Working Group on Coastal and Offshore Permafrost was formed. This led to the development of an Arctic Coastal Dynamics (ACD) project which has among its objective the establishment of rates of erosion and accumulation, the development of a network of monitoring sites along the coast and the initiation of research on critical coastal processes in the Arctic. The ACD is an international undertaking that bodes well for the future of arctic coastal research.

H. Jesse Walker

Bibliography

Andrews, J.T., 1970. Present and postglacial rates of uplift for glaciated northern and eastern North America derived from postglacial uplift curves. *Canadian Journal of Earth Sciences*, **7**: 703–715.

Aré, F., 1988. Thermal abrasion of sea coasts. *Polar Geography and Geology*, **12**(1, 2): 157.

Aré, F., and Reimnitz, E., 2000. An overview of the Lena River Delta setting: geology, tectonics, geomorphology, and hydrology. *Journal of Coastal Research*, **16**(4): 1083–1093.

Bird, J.B., 1985. Arctic Canada. In Bird, E.C.F. and Schwartz, M.L. (eds.), *The World's Coastline*. New York: Van Nostand Reinhold Co., pp. 241–251.

Brown, J., and Solomon, S., 1999. *Arctic Coastal Dynamics*. Geological Survey of Canada Open File 3929. Natural Resources, Ottawa, Canada.

French, H.M., 1989. Cold climate processes. In Fulton, R.J. (ed.), *Quaternary Geology of Canada and Greenland*, Vol. K-1. Geological Society of America, pp. 604–611.

Gusev, A.I., 1952. On the methods of surveying the banks at the mouths of rivers of the Polar Basin. *Transactions of the Institute of the Arctic*, **107**: 127–128.

Harper, J.R., 1978. *The Physical Processes Affecting Tundra Cliff Stability*. Unpublished Dissertation, Baton Rouge: Louisiana State University.

Lachenbruch, A., 1962. *Mechanics of Thermal Contraction Cracks and Ice-Wedge Polygons in Permafrost*. New York: Geological Society of America, Special Publication No. 70.

Leffingwell, E. de K., 1919. The Canning River region, Northern Alaska. United States *Geological Survey Professional Paper* 109.

Lewis, E.L., 1982. The Arctic Ocean: water masses and energy exchange. In Rey, L. (ed.) *The Arctic Ocean: The Hydrographic Environment and the Fate of Pollutants*. New York: Wiley, pp. 43–68.

Mackay, J.R., 1972. The world of underground ice. *Association of American Geographers, Annals*, **62**: 1–22.

McCann, S.B., and Taylor, R.B., 1975. Beach freezeup sequence at Radstock Bay, Devon Island, Arctic Canada. *Arctic and Alpine Research*, **7**: 379–386.

Nansen, F., 1911. *In Northern Mists*. Vol. I. New York: Stokes.

Nielsen, N., 1985. Greenland. In Bird, E.C.F. and Schwartz, M.L. (eds.), *The World's Coastline*. New York: Van Nostrand Reinhold Co., pp. 261–265.

NOAA., 1981. *United States Coast Pilot*, Vol. 9. U.S. Department of Commerce.

Osterkamp, T., 2001. Sub-sea permafrost. *Encyclopedia of Ocean Sciences*, San Diego: Academic Press.

Parkinson, C.L., 2000. Variability of arctic sea ice; the view from space, an 18-year record. *Arctic*, **53**(4): 341–358.

Péwé, T., 1983. Alpine permafrost in the contiguous United States. *Arctic and Alpine Research*, **15**: 145–156.

Reimnitz, E., and Bruder, K.F., 1972. River discharge into an ice-covered ocean and related sediment dispersal, Beaufort Sea, coast of Alaska. *Bulletin, Geological Society of America*, **83**(3): 861–866.

Reimnitz, E., Graves, S.M., and Barnes, P.W., 1988. *Beaufort Sea Coastal Erosion, Shoreline Evolution, and the Erosional Shelf Profile.* U.S. Geological Survey. Miscellaneous Investigations Series.
Sater, J., Ronhovde, A., and Van Allen, L., 1971. *Arctic Environment and Resources.* Washington, D.C.: The Arctic Institute of North America.
Walker, H.J., 1974. The Colville River and the Beaufort Sea: some interactions. In Reed, J.C., and Sater, J.E. (eds.), *The Coast and Shelf of the Beaufort Sea.* Washington, DC: The Arctic Institute of North America, pp. 513–540.
Walker, H.J., 1982. Arctic, coastal morphology. In Schwartz, M. (ed.), *The Encyclopedia of Beaches and Coastal Environments.* New York: Van Nostrand Reinhold Co., pp. 57–61.
Walker, H.J., 1983. E Pluribus Unum: small landforms and the Arctic landscape. In Gardner, R. and Scoging, H. (eds.), *Mega-geomorphology.* Oxford: Clarendon Press.
Walker, H.J., 1998. Arctic deltas. *Journal of Coastal Research,* **14**(3): 718–738.
Weber, J.R., 1989. Physiography and bathymetry of the Arctic Ocean seafloor. In Herman, Y. (ed.), *The Arctic Seas.* New York: Van Nostrand Reinhold Co., pp. 797–828.
Wiseman, W.J., Coleman, J.M., Gregory, A.,Hsu, S.A., Short, A.D., Suhayda, J.N., Walters, C.D., and Wright, L.D., 1973. *Alaskan Arctic Coastal Processes and Morphology.* Baton Rouge: Coastal Studies Institute, Technical Report, 149.
Zenkovich, V.P., 1985. Arctic USSR. In Bird, E.C.F., and Schwartz, M.L. (eds.), *The World's Coastline.* New York: Van Nostrand Reinhold Co., pp. 863–869.

Cross-references

Antarctic, Coastal Ecology and Geomorphology
Arctic, Coastal Ecology
Glaciated Coasts
Ice-Bordered Coasts
Paraglacial Coasts

ARTIFICIAL ISLANDS

Introduction

Islands are defined as a relatively small landsurface, surrounded by water. The largest island, Greenland, has a surface, which is still four times smaller than that of the smallest continent, Australia. The total surface of the earth's islands approximates 10 million km^2, which is comparable to Europe's total surface area.

It is common to distinguish between continental and oceanic islands. Continental islands are considered part of a continent when they are located on the associated continental shelf. Oceanic islands, in contrast, do not belong to a continental shelf. The majority of oceanic islands are of volcanic origin. Some oceanic islands (e.g., Madagascar, Greenland) could be considered as subcontinents, which have been separated from larger continents through tectonic processes.

All realizations of artificial or man-made islands, as constructed so far, are close to shores on relatively shallow water either in intertidal zones, in bays, on the shoreface, or on the nearby shallow continental shelf, and most are located in a sheltered environment. Based on the above definition artificial islands belong to the class of continental islands. Because of their general close vicinity to the shore, artificial islands are very similar to land reclamations into the sea, certainly when such reclamation is separated from the existing land by a waterway or channel, which often has a water management purpose.

The history of island and land reclamation into the sea is some 2,000 years old, starting with dwelling mounds in floodprone areas (cf. Kraus, 1996; this reference provides historic context of works in Canada, Germany, Japan, the Netherlands, and Taiwan, described below). In the 16th and 17th century, land reclamations were becoming common practice in western Europe. In the 20th century, this practice has become worldwide. While this continues to be the case, the construction of islands or reclamations further offshore is only of recent times. Islands in support of a larger infrastructure or temporary islands for petroleum exploitation are of the last decades. Many feasibility studies on artificial islands with an intrinsic purpose (such as industry, urbanization, and airports) have been undertaken all over the world since the 1970s. However, it was not until 1994 that the first more offshore-located island was completed at Kansai International Airport in the Bay of Kobe of Japan.

Concepts and applications

Dwelling mounds

Historically, artificial islands and reclamations go back a long way (cf. Van Veen, 1962). The first known historic recordings of artificial islands are by the Romans. Plinius in 47 AD describes how the Friesians along the northern boundaries of the Roman empire lived on artificial mounds (called "terpen" in Dutch) in order to keep "their heads above the water" during spring tides and storm surges. In all they built 1,260 of these mounds in the northeastern part of the Netherlands, which still exist today, although by now situated in reclaimed land. The areas of the mounds vary from 2 to 16 ha, and their surface level may reach as high as 10 m above mean sea level. The volume content of a single mound may be up to 1 million cubic meters (Mm^3).

Poldertype reclamations

This period of family based shelter against high waters reverted into a more collective form of shelter by seawalls and associated reclamations. The start of these constructions is expected to have been shortly after the declaration of the Lex Frisonian in 802. In the 16th and 17th century all over western Europe, including Russia, polder-type reclamations have been realized both on seashores, estuaries and rivers, and lakes. A polder that is used to reclaim part of the sea bottom consists of an endiked continental shelf or intertidal basin area, with its surface, being the original sea bottom, below mean sea level which is maintained dry by the use of pumping stations (see entry on *Polders*).

One of the world's largest polder reclamations (Van de Ven, 1993) concerns that of the Zuiderzee in the Netherlands, which started in 1918 (with the main damming realized in 1932) and is not yet finalized today. The closing off and the partial reclamation of the Zuiderzee have resulted in the gain of 166,000 ha of new land distributed over four polders. This new land is used for agriculture, urban development, recreation, and nature conservation.

Elevated reclamations

The strong economic development over the last decades of Asian countries where the availability of land is scarce, for example, Japan, Singapore, Taiwan, and South Korea, has also resulted there in a series of reclamations. With the exception of South Korea the acceptation of the polder-type reclamation in Asia has shown to be low. Although the execution method initially is similar, viz. endiking the area first and then drying by pumping, there appears to exist a preference in Asia for elevating the reclamation to above mean sea level (see entry on *Reclamation*).

Examples of this practice are the recent realizations (mid-1990s) of the industrial reclamation estates Chang Hua and Yun-Lin on the west coast of Taiwan, comprising 3,000 ha and 10,000 ha, respectively. The fill of these reclamations amounted to 800 Mm^3 of sand, which was dredged by trailing suction hopper dredgers in the nearby offshore area. The reclamation locations were primarily on diluvial substrates formed by pre-Holocene ebb-tidal deltas, which implies that subsidence problems were virtually absent.

Petroleum exploration and exploitation islands

In the petroleum exploration and exploitation industry, there exist several examples of island reclamations. One such example concerns Rincon Island off California (U.S. Army Corps of Engineers, 1984).

Another important example concerns a series of artificial exploration drilling rig islands of temporary nature in the Beaufort Sea – McKenzie Delta region, Canada. Through the construction of a number of such islands in the period 1973 to 1986 the presence of oil and gas reserves was confirmed, but exploitation in this remote, arctic environment was costly. The decline of the oil prices had hampered exploitation. The islands, with an expected lifetime of 3 years, were constructed in water depths varying between 3 and 21 m, just before the frost and ice formation would set in. A total of eight exploration islands of the sacrificial beach type and nine exploration islands of the sandbag-retained type were constructed. In both cases a dense sediment core was placed, which provided the exploration space needed (surface areas of the order of 1 ha). The sacrificial beach type implied that the relatively steep slope of the core was protected with a gently sloping lower beach of less dense sediment, which would be able to withstand erosive forces by water (in summer) and by ice (in winter). The sandbag-retained type resolved the issue of required resistance by protecting the slopes of the core through geotextile sandbags (with volumes of 1.5–3 m^3). The issue of subsidence in this area is virtually absent because of the permafrost conditions.

Infrastructure supporting islands

Over the last decades, a number of islands have been reclaimed to support the construction of a larger infrastructure. An early example is the island Neeltje Jans in the mouth of the Eastern Scheldt estuary. This was created on a subtidal flat, separating the main estuarine channels, with a twofold purpose. The island served both as a working area for construction of the elements that were to form the Eastern Scheldt Storm Surge Barrier (completed in 1988) and to form part of the barrier itself.

Other examples concern the construction of islands in cases where the two banks of a waterway are connected by a combined bridge–tunnel connection. At the transition of the bridge to tunnel and vice versa the island serves as the connection area. One North-American example is the Chesapeake Bay bridge–tunnel connection, a European example is the Øresund bridge–tunnel connection between Denmark and Sweden.

Airport islands

It might be stated that the introduction of offshore artificial islands in larger depths with a ground level above mean sea level was benchmarked by the construction of Kansai International Airport, Japan (1994). Soon, this was followed by Chep Lap Kok International Airport, Hongkong (1998), and Inchon International Airport, South Korea (2001). A common problem to these three airport islands is formed by the foundation. Being international airports demands that important loading forces (400 metric tons for future intercontinental aircraft) need to be sustained. Since all of these islands are constructed in relatively sheltered areas, the local sea bottom commonly consists of alluvial clay, which calls for special measures.

Kansai International Airport, Japan: The main concern in creating the artificial island has been the foundation of the island. Not only was the local water depth some 18 m, but an alluvial clay layer of more than 20 m covered the geotechnically stable diluvial (Pleistocene) clay substrate. The alluvial clay layer was artificially compacted by about 1 million piles that were driven into the layer to drain the water out. Subsequently, a sea defense was constructed of some 11 km circumference, within which 178 Mm^3 of material was dumped. Because of the scarce availability of suitable sand dredge material the majority of the fill was taken from quarries near Osaka. Thus, an island was created which has an elevation of 33 m above the sea bottom. Since the construction, the subsidence has continued and is expected to continue another 30–50 years. In the next phase the surface area of the island will be extended.

Chep Lap Kok International Airport, Hong Kong: A quarter of the artificial island consists of the island of Chep Lap Kok (350 ha), three quarters consists of a sea reclamation. The average local depth was some 6 m. In contrast with the solutions for Kansai and Inchon, the alluvial clay layer of 10–15 m was removed. The reclamation consisted of a lower layer of rubble mound material quarried from Chep Lap Kok and supplemented with sand dredged in the nearby area. To avoid expensive layer gradation transitions between the rubble mound material and the sand geotextiles were applied. The fill quantities amounted to 350 Mm^3, of which two-third was dredged sand.

Inchon International Airport, South Korea: This airport site covers 5,600 ha in total, of which 4,700 ha consists of reclaimed land between the islands of Yongjong and Yongyu along with 900 ha of existing land. To endike the area between the islands, three dams of a total length of 17 km were constructed, varying in crest height between 7.5 and 9.4 m. Their widths at their foundations are 90–120 m and 20 m at the crest. The reclamation of the 4,700 ha required approximately 180 Mm^3 of fill. Around 80% of the fill was dredged from the sea and the remaining 20% was quarried from mountains and hills in the area. The top layer of the reclamation site, which is largely an intertidal area, is composed of a soft alluvial layer of 5 m average thickness. A sand drain technique was applied to achieve higher soil-carrying capacity. Vertical sand pipes of 400 mm diameter were driven into the soft soil every 2.8–3.8 m to drain pore water. The expected additional settlement is less than 0.5 m.

Conclusions

Although the history of artificial islands or island reclamations is long, it may be expected that the increasing pressure of urbanization, industry, and tourism in the densely populated coastal regions is yet to result in a boom of construction of such islands e.g., Palm and Globe islands (Dubai) under construction. The three main technical problems that one will generally face in the design and execution, are those of geotechnical foundation, of connection to the mainland and of availability of fill material. Besides this, one faces the important issue of environmental impact, where the aspects of impact minimization, nature substitution, and working in harmony with natural system forces are the keywords.

Marcel J.F. Stive

Bibliography

Kraus, N.C. (ed.) 1996. History and heritage of coastal engineering: a collection of papers on the history of coastal engineering. International Conference on Coastal Engineering 1950–1996 *American Society of Civil Engineers*, p. 601.

US Army Corps of Engineers, 1984. *Shore Protection Manual*. Vicksburg: Coastal Engineering Research Center.

Van Veen, J., 1962. *Dredge, Drain, Reclaim: The Art of a Nation*. The Hague: Nijhoff Publishers.

Van de Ven, G.P, (ed.) 1993. *Man made Lowlands: History and Water Management and Land Reclamation in the Netherlands*. Utrecht: Matrijs Publishers.

Cross-references

Dredging of Coastal Environments
Geotextile Applications
Polders
Reclamation
Small Islands
Storm Surge

ASIA, EASTERN, COASTAL ECOLOGY

Coastal geography

Extending from Bangladesh in the west, to the Korean Peninsula and Siberia in the northeast (latitudes 89–129°E), Eastern Asia lies within the Indo-West Pacific Biogeographical region described by Ekman (1967). Eastern Asia contains diverse coastal land formations and habitats, ranging from rocky shores and sandy beaches, to coral reefs, seagrass beds, salt marshes, mudflats, and mangrove swamps. The main landmass of Eastern Asia includes long, uninterrupted coastlines, for example, the coast of Rhakine (Arakan) in Myanmar, the coastlines of central Vietnam and China, and the peninsulas of Malaysia and Korea. The region also supports immense archipelagoes, most notably those making up Indonesia and the Philippines, which comprise of more than 13,700 and 7,000 islands, respectively. Innumerable small islands make up the Mergui Archipelago extending from Myanmar to Thailand, the Andaman and Nicobar Islands, and several minor archipelagoes. Huge deltas have also been formed by some of the world's great rivers (the Ganges–Brahmaputra in Bangladesh, the Irrawaddy in Myanmar, the Red River and Mekong River in Vietnam, the Yangtze River and Yellow River in China).

Climate and ecology

Climatically, Eastern Asia features a tropical equatorial subregion consisting of Malaysia, Singapore, Sumatra, Java, and Borneo; and northern subtropical to temperate subregions, which include the Bay of Bengal, the South and East China Seas, and the Yellow Sea (Figure A36). The Malay-Indonesia region has a hot and humid climate and is bathed by shallow, highly productive seas overlying the Sunda Shelf (the submerged peninsula of Pleistocene Sundaland). Extensive coastal areas of the Sunda Shelf are less than 60 m in depth. The Philippines borders the eastern edge of the Sunda Shelf, with much deeper waters of the Pacific Ocean beyond. Although the southern archipelago of Indonesia lies mainly within the humid tropical zone, there are some subhumid to semiarid areas, particularly to the east, including the northeast coast of Java and the islands of Timor, Lombok, and Sumba.

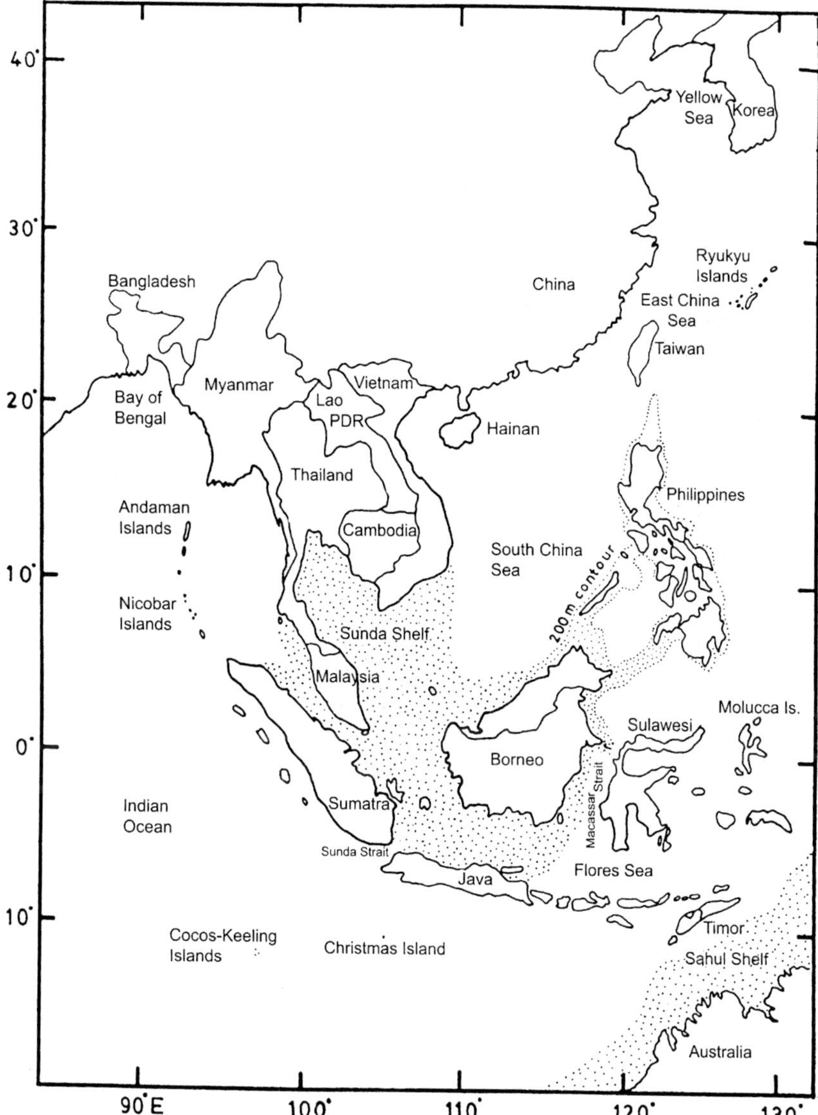

Figure A36 General map of Eastern Asia; the Sunda Shelf and Sahul Shelf regions are shaded.

North of the equatorial region, the NE and SW monsoons dominate the climate and surface ocean circulation. Rainfall becomes more and more seasonal eastwards and northwards from the equatorial subregion, with typhoons and cyclones (defined as storms with wind speeds exceeding 73 mph) associated particularly with the warm SW monsoon season. Typhoons originating in the Pacific pass through the Philippines before reaching the coastline of Vietnam and southern China. There is a "typhoon belt" from central Vietnam to Hainan Island where most of these typhoons strike land, causing wave surges of up to 2.5 m above normal sea level. Similar storms originating in the Bay of Bengal move northwards and commonly strike the coast of Bangladesh as cyclones. A key consequence of these monsoonal weather phenomena is that the most vulnerable coastal regions of Vietnam and Bangladesh are protected by man-made earthen sea dikes that have transformed the coastal ecology of the upper intertidal zone; these coastal defenses have also served to reclaim land for agriculture, salt-making, and aquaculture.

Further north, the coastlines of China and the Korean Peninsula are influenced by the East China Sea and the Yellow Sea. The latter is a productive, semi-enclosed water body with an average depth of only 44 m and a maximum of 100 m. The Yellow Sea is influenced strongly by several major rivers, especially the Huang He (Yellow River) and Yangtze River, which discharge more than 1.6 billion tonnes of sediments annually. Sedimentation and salinity, as well as seasonal temperature changes, control the ecology and productivity of the Yellow Sea ecosystem. The coastline of China consists mainly of mudflats and salt marshes. There are also large areas of intertidal mudflats and salt marshes along the west coast of the Korean Peninsula that provide an important ecological support function to migratory birds (including the very rare black-faced spoonbill *Platelea minor*: world population in 1998 only 613), as well as fisheries stocks (Kellerman and Koh, 1999).

Coastal habitats of Eastern Asia

This section describes the ecology of the main coastal habitats founds in Eastern Asia, particularly the mangrove forests, seagrass beds, and coral reefs of Southeast Asia which are the most productive and diverse in the world. However, Eastern Asia is a region where the natural ecology of the coast shows a complete spectrum of impacts due to human activities, from near pristine beaches and coral reefs on remote islands, to massive coastal land conversion for agriculture, aquaculture, and urban/industrial uses. The pressures from human population and development, particularly in heavily populated countries like Bangladesh, China, Indonesia, and Vietnam, have transformed or degraded the coastal zone at an alarming rate over the past 20 years.

Mangroves

Mangroves, or mangrove swamp forests, consist of trees and shrubs growing in the upper intertidal zone of muddy tropical and subtropical shores, including estuaries and lagoons. Mangroves develop best along sheltered shores where there is high and prolonged rainfall and an average temperature exceeding 20°C (Macnae, 1968). Mangrove forests provide important habitat and other ecological support functions for many terrestrial and aquatic animals, including both resident and visiting species, especially birds, as well as fish and shellfish populations that support substantial coastal fisheries (Macintosh, 1982). For this reason, it is usual today to regard mangroves as an ecosystem rather than simply a community of highly adapted plants and animals. The biology and ecology of mangroves has been reviewed in detail, notably by Macnae (1968), Tomlinson (1986), Robertson and Alongi (1992), and Hogarth (1998), while their traditional exploitation for forestry, fisheries, and aquaculture production is also well documented. An early account by Watson (1928) and a recent update by Gan (1995) describe the mangroves of Matang in Perak, Malaysia, which have been under management for wood production since the 1900s (mainly for charcoal). Matang is widely regarded as the best-managed mangrove forest ecosystem in the world. Schuster (1952) gives an excellent account of the partial conversion of mangroves in Java to integrated mangrove fishponds, or tambaks, for milkfish culture, with marine shrimps (Penaeidae) and mud crabs (*Scylla* spp.) being the secondary crop. The current tambak management system in Indonesia is described by Sukardjo (2000). While these represent good examples of the sustainable economic use of mangroves in Eastern Asia, it is really only within the past decade or so that all the functions, attributes, and values of mangroves (and other tropical coastal ecosystems) have become widely recognized and understood (e.g., White and Cruz-Trinidad, 1998).

According to the World Mangrove Atlas, 41.5% (75,173 sq. km) of the world's total mangrove area (181,077 sq. km) are found in South and Southeast Asia. Indonesia (42,550 sq. km of mangroves) dominates with 23% of the world resource (Spalding *et al.*, 1997). Mangroves occur throughout the tropical and subtropical regions of the world, but the Indo-Pacific region is clearly their center of diversity, with 63 out of the 90 or so genera of mangrove and associated plants occurring here. Mangrove forests extend into the colder temperate subregion of Eastern Asia as far as southern China (Hainan Island and Fujian Province), Taiwan, and the Ryukyu Islands of Japan (Figure A37). The Ryukyus (26–27°N) are the northern limit for mangroves; here temperatures in winter fall to 17–22°C. At these northern extremes, the mangrove vegetation is much more stunted and limited in species compared to the magnificent mangrove forests of tropical Asia. The drop-off in mangrove species with latitude is shown clearly in China's Fujian Province. In southern Fujian, there are six mangrove species (*Aegiceras corniculatum, Avicennia marina, Acanthus ilicifolius, Bruguiera gymnorhiza, Excoecaria agallocha*, and *Kandelia candel*); this falls to four species in central Fujian and to only one species (*K. candel*) in northern Fujian (Peng and Wei, 1983). Further north along the coastline of China, salt marsh vegetation, especially *Salicornia*, replaces mangrove as the dominant intertidal wetland community.

It is in Southeast Asia where mangroves have achieved their greatest development and diversity, especially along the sheltered shores bordering the shallow seas of the Sunda Shelf within the Malay–Indonesia subregion (Figure A38). This includes the coasts of Peninsular Malaysia and Sumatra adjoining the Melaka (Malacca) Straits, the Gulf of Thailand and the coast of Cambodia, the Mekong Delta of southern Vietnam, the northern coast of Java, and almost the entire coastline of Borneo (see Figure A37). The prolific reproductive capacity of the main forest-forming Asian mangrove species is one reason for their great success throughout Southeast Asia (Figure A39).

Under the more exposed conditions of the Bay of Bengal and South China Sea, it is in the great deltas of the Ganges–Brahmaputra

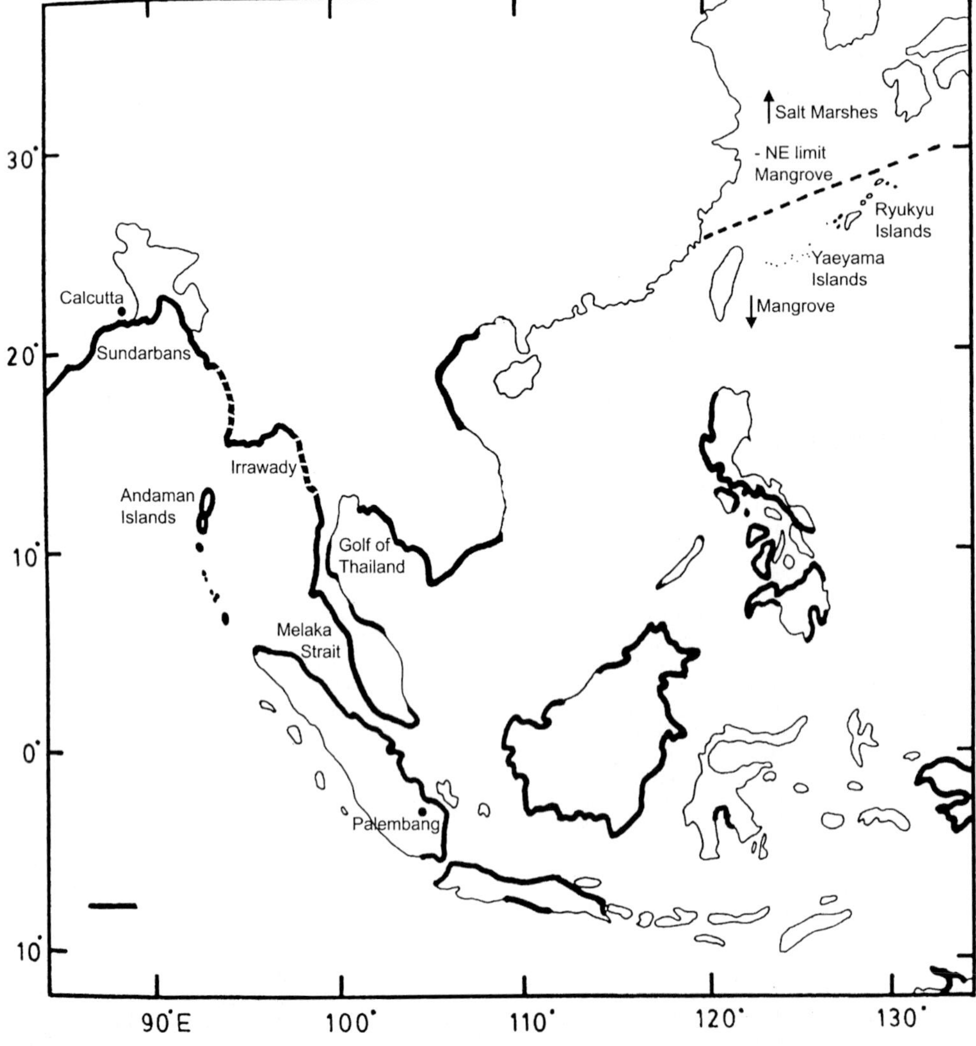

Figure A37 Distribution of mangrove swamp forest in Eastern Asia (heavy coastlines).

Figure A38 Tall, stilt-rooted *Rhizophora* trees dominating a typical estuarine mangrove forest in Southeast Asian (Ranong, southern Thailand).

Figure A39 Spear-like propagules of *Rhizophora mucronata* adapted for penetrating soft muddy sediments. Propagules give mangrove trees of the Family Rhizophoraceae, great capacity for self-generation and colonization.

(Bangladesh), Irrawaddy (Myanmar), and Red River (northern Vietnam) where mangroves are most extensive and luxuriant. In each case, massive deposition of alluvial sediments from the river systems creates mud banks along the shallow coastal zone that enable mangrove forests to colonize and flourish. At the same time, the presence of mangroves affects the hydrological regime by modifying tidal flows and velocities, influencing sedimentation and erosion patterns, and reducing the impact of waves along the coast (Wolanski and Ridd, 1986; Kjerfve, 1990). Thus, mangrove systems are highly dynamic, both in the short-term and over longer time periods. The longer-term dynamic changes in coastal land formation resulting from these processes are well illustrated historically, especially in the Sunda Shelf subregion. Palembang was a thriving port city when Marco Polo visited Sumatra in 1292, but since then coastal accretion has left Palembang more than 50 km inland (Macnae, 1968).

The rates of "land-building" by mangroves are lower on more exposed coastlines due to less favorable current and/or climatic conditions. Mangroves rapidly colonize the mudbanks formed in the Irrawaddy Delta, but some sediments are carried away eastwards and offshore by coastal currents generated during the monsoons. In the Mekong Delta, there is a net flow of sediments from east to west, and consequently some parts of the eastern shore of the Cam Mau Peninsula are actually eroding, despite the great sediment loads being deposited by the Mekong River system. Moreover, Typhoon Linda, which struck southern Vietnam in November 1997, destroyed the last vestiges of the original mixed forest, as well as damaging many of the older mangrove forest plantations in Ca Mau. As in China, the colder climate in northern Vietnam, coupled with frequent typhoons, limit mangroves in the Red River Delta to low, shrubby vegetation (dominated by *Avicennia* spp. and *K. candel*). In contrast, the mangrove forests of southern Vietnam feature taller trees of *Rhizophora, Sonneratia, Brugueira*, and *Ceriops* spp., including large tracts of plantation mangrove consisting of a single commercial species, *Rhizophora apiculata* integrated with canals for aquaculture (Figure A40).

Nipa swamp forest

Along tropical riverbanks, and in low-lying muddy swamplands subjected to greater freshwater inundation, the nipa palm (*Nypa fruticans*) replaces mangrove forest as the characteristic plant community (Figure A41). Nipa palms often intergrade with mangroves in estuarine transition zones near the upper limit of saltwater penetration. Large areas of nipa swamp have been destroyed by land reclamation projects and for shrimp farming, but nipa remains an important wetland community in many Southeast Asian countries, both ecologically and economically. Nipa is still used widely in Southeast Asia as a traditional thatching material for house-building; it can also be processed for sugar and alcohol production (Chan and Salleh, 1987).

Figure A40 A dense monoculture of planted *R. apiculata* trees in Ca Mau, southern Vietnam. Mangrove forestry on the central platform is integrated with aquaculture in the surrounding water canals (foreground).

Figure A41 A dense stand of the palm *N. fruticans,* at the transition zone from mangroves to freshwater vegetation, Mekong Delta, Vietnam.

Salt marshes and mudflats

Salt marshes replace mangroves in the coastal ecology of the temperate subregions of Eastern Asia, especially along the sheltered western coastline of the Korean Peninsula. The southern portion of the Yellow Sea, including the entire west coast of Korea, is dominated by tidal flats up to 10 km wide, with an operating tidal range of 4–10 m. This forms the Yellow Sea ecoregion that includes an estimated 2,850 km^2 of tidal flats (Kellerman and Koh, 1999). Salt marsh vegetation characterized by *Suaeda japonica* once flourished on the upper part of the shore, but much of this has been lost through land conversion for agriculture.

The tidal flats of the Yellow Sea ecoregion contain muddy, sandy, and mixed sediments; these areas are highly productive, with a rich benthic fauna of polychaetes, bivalve and gastropod mollusks, crustaceans, holothurians, and branchiopods (Koh and Shin, 1988). The tidal flats are an extremely important habitat for migratory wading birds on their passage between Indonesia–Australia and Eastern Siberia–Alaska. Ecologically, the coastal habitats of the Yellow Sea have been likened to those in northern Europe, particularly the Wadden Sea (Kellermann and Koh, 1999).

Salt marsh plants also replace mangroves locally in more arid conditions, even within tropical Southeast Asia. On dry muddy soils above the mangrove zone, or where coastal dikes and embankments have been constructed, the salt-tolerant creeper *Sesuvium portulacastrum* often covers such exposed habitats, its fleshy leaves being capable of retaining water even during prolonged droughts. In Vietnam, where virtually all the mangrove zone has been disturbed by sea dike and shrimp pond construction, several other marginal species colonize the supralittoral zone, including the shrubs *Clerodendron inerme* and *Pluchea indica* and the tree *Thespesia populnea*, which produces a hibiscus-like flower. Such high, arid habitats are hostile to most marine animals, but fiddler crabs (*Uca* spp., Family Ocypodidae) and mangrove sesarmid crabs (Family Grapsidae) burrow deeply into the soil for protection. A few mangrove species, especially *E. agallocha* and *Lumnitzera racemosa* also tolerate the dry, acidic conditions on the slopes of coastal dikes. Local farmers in Vietnam often plant these mangrove trees, along with *Thespesia* and *Casuarina*, to provide shade and wind protection for their houses and farmland. Very poor people also gather *Sesuvium* for use as animal fodder.

Sandy beaches

Eastern Asia includes some of the most magnificent sandy beaches in the world, many of which are still unspoiled, particularly those fringing remote coral islands. The world's longest beach is located on the Arakan coast (now Rhakine) of Myanmar, while the powdery white beaches of Boracay Island (Western Visayas, the Philippines) are regarded by many to be the world's finest. Beautiful sandy beaches also make up 90% of the eastern coastline of Peninsular Malaysia. Not surprisingly, the sandy beaches of Indonesia, Malaysia, Philippines, Thailand, and more recently Vietnam, are promoted strongly as a key attraction to tourists. Particularly renowned are the beaches on many tropical islands such as Bali and Lombok (Indonesia), Penang and the Lankawi Islands (Malaysia), Phuket and Koh Samui (Thailand), Boracay (Philippines), and Phu Quoc (Vietnam).

Beach sediments of volcanic origin occur in parts of Indonesia, such as eastern Java, creating beaches of black or gray sand. Beach ridges and sand dunes are not well developed in the humid tropical regions of Asia, but examples occur in parts of southern Java and southwestern Sumatra; here a natural beach ridge and dune vegetation of *Casuarina, Pandanus, Calophyllum, Inophyllum*, and *Barringtonia* occurs, together with planted or self-established coconut palms. A similar sandy ridge and sand dune formation occurs much more extensively in monsoonal Asia, especially along parts of the coastline of Myanmar and Vietnam. Natural and planted *Casuarina equisetifolia* trees are a feature of the low sand ridges behind the beach slope. In many provinces of Vietnam planted *Casuarina* trees act as a wind break in front of coastal villages, as well as providing a traditional, and renewable, source of wood (Figure A42). The slender form of the branches and leaves of this native conifer enables *Casuarina* to withstand the arid conditions and strong winds associated with sea-facing sandy beaches.

Within the upper intertidal zone on exposed shores with strong surf, the sand is too dry and unstable for most marine animals, but burrows of the ghost crab *Ocypode ceratophthalma* are a common site on the upper beach slope. This large ocypodid crab burrows deeply for protection from the sun and desiccation. Ghost crabs emerge at low tide to scavenge from the sand line, or to chase and capture smaller fiddler crabs (*Uca* spp.) or soldier crabs (*Dotilla* spp.) which occupy the lower, less-exposed intertidal zone. As the fastest running land crustacean, *Ocypode* is particularly well adapted to catch its prey and to escape predators.

Of the seven species of sea turtles found in the world, four species visit sandy beaches in Malaysia to lay their eggs. The beaches of Rantau Abang, in Trengannu on the east coast of Peninsular Malaysia, are one of only four main nesting sites worldwide known for the giant leathery turtle, *Dermochelys coriacea*, and the only nesting site in Eastern Asia. This rare and endangered species is the world's largest living reptile, capable of reaching 2 m in length with a weight of more than 900 kg. It is also the only sea turtle without a hard shell; instead it has a leathery, ridged carapace. Although this unique species has a worldwide distribution, being found even in temperate seas, it only breeds in the tropics. After laying many large parchment-like eggs in a nest dug carefully in the sand at the top of the beach slope, the female turtle returns to the sea. As soon as the eggs hatch the baby turtles struggle out of the sand and crawl down the beach slope to enter the sea. Despite efforts to protect its nesting beaches in Malaysia, the giant leathery turtle has declined significantly over the past 30 years, disturbance to its nesting habitat being one of several critical factors.

The other turtles nesting on Malaysia's beaches are also endangered, having suffered severe decline due to egg harvesting and relentless fishing pressure because of their high value. It is also likely that their feeding habitats have declined significantly. The olive ridley turtle (*Lepidochelys olivacea*) is the world's smallest marine turtle; its diet includes crustaceans, mollusks, and jellyfish. The green turtle (*Chelonia mydas*) has a vegetarian diet of seagrasses and algae, though it may also consume sponges. The beautiful hawksbill turtle (*Eretmochelys imbricata*) also feeds on sponges. The hawksbill and ridley turtles are hunted for their shells, while the green turtle is prized for its meat and oil. The same four species of marine turtle occur in the Philippines (Gomez, 1980), but they have declined greatly in numbers, just as in Malaysia, despite conservation measures including closed seasons for turtle fishing and egg-collecting (Palma, 1993). Sea turtles are particularly difficult to conserve because of their precise habitat requirements, as well as their high

Figure A42 Planted *C. equisetifolia* trees serving to stabilize the upper beach zone and provide a windbreak, Nam Dinh Province, northern Vietnam.

Figure A43 A coral reef flat of hard and soft corals in Singapore; heavy coastal industries can be seen in the distance.

value to poachers. The disturbances to their historical breeding beaches, and the great decline in coral reefs and seagrass beds which provide sea turtles with feeding habitats, place these remarkable marine reptiles at great risk, even if hunting them can be controlled.

Coral reefs

There are almost 100,000 sq. km of coral reefs in Southeast Asia, representing about one-third of the world's total coral resource. Like mangroves, corals show their greatest diversity in Southeast Asia, with 600 of the almost 800 reef-building coral species found in this tropical subregion of Eastern Asia. The center of coral diversity is eastern Indonesia, the Philippines, and the Spratly Islands with over 70 genera of hard coral represented. Indonesia and the Philippines, alone contain 77% of the region's coral reefs due to their multitude of islands featuring the sheltered, clear water conditions corals require. Coral reef fishes, crustaceans, and mollusks also show their greatest diversity in this subregion and up to 3,000 animal species may be present on a single coral reef. A recent identification guide to corals of the world (Veron and Stafford-Smith, 2000) provides an excellent reference to the corals of Eastern Asia. Several new coral species are described and there are probably many more species still to be identified.

Notwithstanding their great beauty, diversity, and ecological importance, the world's coral reefs have suffered a dramatic decline in recent years. About 10% may already have been degraded beyond recovery; a further 30% are likely to decline seriously within the next 20 years. Recently, the World Resources Institute estimated the potential threat to coral reefs using standard criteria to calculate a "reefs at risk" indicator (Bryant *et al.*, 1998). The reefs identified as being at greatest risk are in South and Southeast Asia, East Africa, and the Caribbean. Specifically, over 85% of the reefs of Indonesia and Malaysia, and over 90% of those in Cambodia, Singapore, the Philippines, Taiwan, and Vietnam are threatened (Figure A43).

Seagrass beds

Seagrasses are distributed widely through tropical and subtropical Eastern Asia, mainly as low intertidal to subtidal beds in shallow sandy bays, lagoons, and around near shore islands. They often form an associated wetland community below mangrove-fringed shores with sandy sediments, or occur interspersed with coral reefs. Like corals, seagrasses do not tolerate the muddy or turbid habitat conditions typical of estuaries and mudflats. In Vietnam, for example, where more than 5,000 ha of seagrass beds containing 15 species have been identified, the main sites for seagrasses are north of the Red River Delta in Quang Ninh Province, in the central region of Vietnam (Hue and Khanh Hoa-Nha Trang) where there are many magnificent sandy bays and lagoons, and around the southern island of Phu Quoc (Vietnam Environmental Monitor, 2002). Seagrass beds in Vietnam have become heavily degraded due to excessive harvesting for animal food and organic fertilizer, as well as destructive fishing practices (use of dynamite and cyanide) in seagrass and coral habitats.

Coastal reclamation projects have also caused widespread loss of Eastern Asia's seagrass beds and coral reefs, but there is also growing evidence of decline due to natural causes, including coral bleaching associated with warming of the seas and "wasting disease" in seagrasses. Haynes *et al.* (1998) concluded that most seagrass losses, both natural and anthropogenic, stem from reduced light intensity due to sedimentation and/or epiphytic development caused by water nutrient enhancement.

Coastal development activities commonly increase sediment loading in the surrounding waters, causing great harm to seagrass beds and coral reefs. One well-documented example is Hong Kong's Chep Lap Kok Airport (Lee, 1997). Increased sediment levels caused by the airport's construction virtually wiped out the nearby beds of *Zostera japonica* and *Halophilia ovata*. While *Halophilia* has recovered well following the construction period, the *Zostera* beds have almost completely disappeared, although some transplanted patches have survived and enlarged. This difference between seagrass species may be explained by the observation that *Zostera* is mainly subtidal and intolerant of shading, whereas *Halophilia* also occurs in the lower intertidal zone and can survive in relatively low light intensity.

Seagrass beds play various ecological roles, including sediment adsorption and nutrient absorption from seawater; this may serve an important protection function for adjacent coral reefs, since corals are even more sensitive to sediment and nutrient loading. Seagrasses also provide important nursery habitats for many fish species, including groupers fingerlings, for example, which are heavily collected throughout Southeast Asia for aquaculture. Seagrass beds are also critical feeding areas for the green turtle (*C. mydas*) and the rare sea mammal, the dugong (*Dugong dugun*), which feeds almost exclusively and selectively on various seagrasses, especially *Halodule* and *Halophilia*. Dugongs may choose these particular grasses in preference to other genera (*Thalassia* and *Enhalus*) because they are easily digested (*Halophilia*) and rich in available nitrogen (*Halodule*). Dugongs are capable of digesting a large volume (8–15% of body weight per day) of low quality plant material (Reynolds and Odell, 1991), an observation that explains why the great loss of seagrass beds in Eastern Asia has been so damaging to the survival prospects for this unique mammal. It has also been reported that grazing by dugongs and turtles plays an ecological role in seagrass communities by stimulating the growth of new leaves, thereby changing the structure and nutritional status of the seagrass bed (Aragones and Marsh, 2000).

Exploitation of coastal resources

Human pressure

Eastern Asia is a region where the great natural diversity of coastal topography, biological communities, and species has been transformed in many countries by immense human population and development pressures. Although human occupation and economic use of the coastal zone have a long tradition in the region, dating back centuries in the case of coastal farming and aquaculture, the pace of coastal development has accelerated since the 1980s, raising the need for habitat conservation and integrated coastal zone management as never before.

The effects of high population growth in the coastal zone, unsustainable exploitation of coastal resources, and conversion of coastal habitat for agriculture, aquaculture, and urban and industrial development, have now reached crisis levels. In reviewing the threatened status of wetlands in Asia in 1989, Wetlands International (an international NGO) noted that Southern and Eastern Asia (including the Indian subcontinent) had an average population density about eight times that of the rest of the world and was increasing at a rate of 55 million people annually. This led Wetlands International to conclude that, "Most of the threats to wetlands in Asia are a direct consequence of the need to feed and house this massive and ever-increasing number of human beings." Unfortunately, this is a very accurate assessment of the problems facing Eastern Asia's coastal ecosystems in general.

Table A4 Estimated loss of the original mangrove forest area in different countries (based on data from World Resources Institute, 1996)

Country	Loss of mangrove forest (% loss by area)
Bangladesh	73
Brunei	17
Indonesia	45
Malaysia	32
Myanmar	58
Singapore	76
Thailand	87
Vietnam	62
Unweighted average for Asia	61

A good historical example is the Chakaria Sundarbans region of Bangladesh. Located on the eastern border of the Bay of Bengal, this is the eastern limit of the great Sunderbans mangrove forest shared between Bangladesh and India and the largest mangrove ecosystem in the world. One hundred years ago the Sundarbans was reserved forest, but over the next several decades the mangroves were exploited for wood and fishery products and then encroachment followed for human settlement and salt production. Mangrove conversion accelerated greatly during the 1980s due to the development of coastal shrimp farming, until by 2001 all the remaining mangrove had been converted to shrimp ponds and human settlements (Hossain and Kwei-Lin, 2001). Even by the 1980s, the Chakaria Sundarbans were considered too degraded to merit any special conservation efforts.

There has been a similar history of mangrove exploitation in Vietnam, where now less than 30% of the original mangrove area remains and virtually no pristine forest has survived. Only 60 years ago, mangroves in Vietnam covered an area of up to 400,000 ha (Maraund, 1943 cited by Hong and San, 1993), of which about 250,000 ha flourished in the Mekong Delta. The greatest concentration of mangrove (150,000 ha) was in the Minh Hai Peninsula at the southern tip of Vietnam (now divided into Ca Mau and Bac Lieu provinces). The most recent estimate for the total area of mangrove forest remaining in Vietnam is only 110,000 ha (Vietnam Environmental Monitor, 2002). Overall, the Eastern Asia region has lost more than 60% of its mangrove forest in just a few decades (Table A4).

It is not just Eastern Asia's fragile tropical mangrove forests, seagrass beds, and coral reefs that have become severely depleted or degraded. Approximately 600 million people live in the area around the Yellow Sea, including major cities such as Shanghai, Seoul-Inchon, and Pyongyang-Nampo. The Yellow Sea is among the most heavily exploited coastal and marine areas of the world, with more than 100 species of fish, crustaceans, and cephalopods captured by its fisheries. Today, coastal habitats and fishery resources in the Yellow Sea are threatened by both land-based and sea-based infrastructures, by pollution, and by overfishing and other forms of unsustainable resource exploitation. More than 100 million tonnes of domestic sewage and about 530 million tonnes of industrial wastewater are discharged annually into the nearshore areas of the Yellow Sea. These discharges come from large coastal cities, shipping, and oil exploration; they contain

Figure A44 Recreational use of a sandy beach in Johore, Pensinsular Malaysia, with dense surrounding urban infrastructure—a scene typical of many beaches near large cities throughout Eastern Asia.

many pollutants, including heavy metals (cadmium, lead, mercury), oil residues, and nitrogenous compounds.

Other economic uses of the coastal zone in Eastern Asia tend to reflect the stage of development of the countries concerned. Like Bangladesh and Vietnam, the more developed South-east Asian countries of Indonesia, Malaysia, Philippines, and Thailand have heavily exploited sheltered bays, lagoons, estuaries, and other mangrove and mud flat areas for coastal agriculture and aquaculture. There has also been localized industrial and urban development, as well as heavy investment in tourist infrastructure centered on islands and mainland sandy beach sites near major cities; for example, Pattaya and Hua Hin in Thailand, Port Dickson and Johore in Peninsular Malaysia (Figure A44). Estimates for Thailand made by the Royal Thai Forestry Department in 1993 revealed that, of the original 372,448 ha of mangrove forest (measured in 1961), 133,812 ha (35.9%) had been converted into salt pans, mining, agriculture, and infrastructure, 64,991 ha (17.5%) was occupied by coastal shrimp farms, and 4,961 ha (1.35%) had been converted to other uses (leaving only 45% undeveloped).

In the most populated and developed parts of Asia, especially Hong Kong, Singapore, and parts of Peninsular Malaysia, China, and Korea, coastal land is at such a premium that urban and industrial development (e.g., airports, container ports, oil-refineries) have almost completely replaced earlier natural resources use. Such economic developments have overtaken the coastal zone in a remarkably short period of time. Only 50 years ago Schuster (1952) noted "The greater part of down town Singapore is built on *Rhizophora* piles driven into the mud of the harbor region." Today only carefully conserved vestiges of the original coastal habitat remain in Singapore (e.g., Figure A43).

Coastal aquaculture

Of the many human uses of the coastal zone, aquaculture stands out not only in terms of its scale and importance throughout the Eastern Asia region, but also because of its interaction with the ecology of the coastal environment (e.g., Primavera, 1993). Unlike the conversion of coastal habitat to agriculture or urban and industrial uses, the production of human food by aquaculture is still critically dependent on the carrying capacity of the surrounding aquatic ecosystem. Where coastal aquaculture causes environmental degradation, there is a rapid negative feedback on the viability of the aquaculture production system itself, as the experiences with intensive shrimp farming in Eastern Asia show only too well (e.g., Smith, 1999). The term "self-pollution" is sometimes used to describe this problem in coastal aquaculture (Phillips and Macintosh, 1997).

Shrimp farming

Shrimp farming has had a dramatic impact on the coastal ecology of Eastern Asia over the past 20 years or so. Although coastal aquaculture has a long history of operation in Southeast Asia (Schuster, 1952; Macintosh, 1982), it was the commercialization of technology for breeding one particular species, *Penaeus monodon*, or the black tiger shrimp in Taiwan, which enabled shrimp seed (post-larvae) to be produced, then reared intensively in coastal shrimp farms. The world demand for farmed shrimp also rose significantly in the 1980s, led at that time by Japan, because the yields from shrimp capture fisheries had long since reached a ceiling level. From early pioneering research on the breeding of *Marsupenaeus japonicus* in Japan, research then developed in China involving the local species *Fenneropenaeus chinensis* and in Taiwan with *P. monodon*. Cool climatic conditions and high costs had mitigated against shrimp farming in Japan involving the local species *M. japonicus*, but the black tiger shrimp was readily accepted by the Japanese market and Taiwanese shrimp producers rapidly spread their knowledge to other parts of tropical Asia where conditions for farming *P. monodon* were much better. During the 1980s, investors in Thailand, Philippines, Indonesia, and Malaysia rushed to adopt the new technology, converting mangrove forests, traditional shrimp ponds, salt pans, coastal rice fields, and (in the Philippines) sugar plantations, into intensive shrimp farms. From Southeast Asia, "shrimp fever" as it became known (Primavera, 1993) spread to Vietnam and Bangladesh, countries with huge areas of coastal mangroves and mudflats considered more suitable for conversion to extensive, low-cost, shrimp ponds. Since 1995, records show that on average 138 million shrimp juveniles, or "seed" (mainly of *P. monodon* and *Macrobrachium rosenbergii*) have been collected annually from the Sundarbans reserve forest in Bangladesh to support coastal shrimp farming. Shrimp seed collecting is extremely destructive as many other species of fish and crustacean seed are caught and discarded by the shrimp fishers, but it is a form of livelihood for the great majority of families living in the Bangladesh Sundarbans and involve an estimated 225,000 shrimp seed collectors (Rouf and Jensen, 2001).

Shrimp farming was also promoted on a massive scale in China, with shrimp farms occupying 160,000 ha from Hainan Island in the south to Liaoning Province in the northern temperate zone. Production reached a peak of 200,000 tonnes in 1992 before coastal water pollution and

Figure A45 Intense use of the coastal zone for aquaculture; the photograph shows ponds and a complex of water pipes supporting fish and shellfish farming during the aquaculture boom in southern Taiwan in the 1990s.

shrimp diseases had a devastating impact on output (Smith, 1999). Shrimp farming has also developed in Korea, but as with its neighbors China and Taiwan, aquaculture has been badly affected by coastal industrialization and associated problems of water pollution (including self-pollution) and shrimp disease outbreaks (Figure A45). The shrimp farms in Korea are confined almost entirely to the west coast where it is more sheltered. The majority of farms are located within the intertidal zone or near to the coastline. Reflecting the colder operating conditions in Korea compared to Southeast Asia, the main species cultured are the local shrimp *F. chinensis* followed by *M. japonicus*. Annual production levels are rather low (less than 1 tonne/ha), but the environmental limitations on shrimp production in Korea are offset by the high price farmers can obtain for live shrimp delivered to the restaurant trade, including live sales to Japan.

Figure A46 A large mud crab (*Scylla paramamosain*) for sale in a local market in Vietnam. Mud crabs support important artisanal fisheries and coastal aquaculture production throughout Eastern Asia.

Mud crab culture

Poor coastal communities from Bangladesh to the Philippines are still heavily dependent on mangrove-based fisheries and aquaculture to support their livelihoods (Figure A46). This includes crab fishing for *Portunus* (blue swimming crab) and *Scylla* (mud crab), as well as shrimp fisheries and shrimp culture. Mud crab culture has also developed rapidly as a secondary activity to shrimp farming within the mangrove forest ecosystem. It carries much lower investment costs and risks than shrimp farming and is therefore more suitable for poor families (Overton and Macintosh, 1997). In the Lower Mekong Delta of Vietnam, farmers can now rear from 500 to 800 kg of mud crab per hectare in a simple mangrove pond (unpublished data). With a farm gate value of about USD3/kg, even a small crab pond can provide a poor Vietnamese family with an annual net income of USD200–300. This is a significant contribution to the economy of families living in remote coastal areas where the potential daily income from fishing, salt production, or agricultural laboring is only USD1–2.

Other forms of coastal aquaculture

Many other types of aquaculture are also economically and ecologically important in the coastal zone of Eastern Asia. Economically, caged fish culture for valuable species such as groupers (*Epinephelus* spp.) and snappers (*Lutjanus* spp.) appears to be very attractive, but it carries many risks, especially from fish diseases and can be highly polluting due to the use of trash fish as the main type of feed. The environmental damage stemming from large-scale collection of wild fingerlings (especially groupers) from coral reef and seagrass habitats to support aquaculture is another great cause of concern. For large-scale aquaculture

Figure A47 The ark shell, or "blood cockle" (genus *Anadara*), one of the many bivalve mollusks found abundantly on coastal mudflats throughout Eastern Asia and important in aquaculture, especially in Malaysia and China.

Figure A48 Tourists, long-tailed macaques, and mangrove forest in the Can Gio Biosphere Reserve, near Ho Chi Minh City, Vietnam. The reserve has conservation, sustainable development, education, and research functions.

production, and in terms of their ecological impacts, the culture of bivalve mollusks and seaweeds has many advantages over most fish and crustacean species. Mollusks and large seaweeds are cultured in huge quantities by both China and the Republic of Korea. For example, marine aquaculture production in China reached 10.6 million tonnes in 2000, with 93% of this total contributed by bivalve mollusks and seaweeds. The main species of mollusks include oysters (*Crassostrea*), mussels (*Mytilus*), hard clams (*Meretrix meretrix*), short-necked clams (*Mactra* spp.), scallops (*Clamys farreri*), and ark shells of the genus *Anadara* (Figure. A47). Gastropod mollusks of the genus *Haliotis*, better known as abalone, are more difficult to rear, but are in high demand as a delicacy. These snail-like animals occur naturally on subtidal rocks (especially in Korean waters) where they graze on algae. However, overfishing has decimated abalone populations because they are so valuable and easy for divers to harvest.

Seaweeds grow naturally on rocks in the lower intertidal and subtidal zones, especially in the warmer southerly regions of the Yellow Sea. The seaweed *Pelvetia siliquosa* occurs widely in the Shandong and Liadong regions of China, as well as around the Korean Peninsula. It grows most luxuriantly in Korean waters where it has been harvested for hundreds of years and exported to China. *Sargassum pallidum*, another edible seaweed, dominates in the west Yellow Sea, and *Plocamium telfairiae* is also common there. A cold water seaweed, the kelp *Laminaria japonica*, has become the most important seaweed cultured in China. Introduced from Japan, kelp farming now occupies more than 3,000 ha of China's coastal waters (yielding 10,000 tonnes dry weight/year) and is supported by hatcheries producing young plants for transfer to growing frames in the sea once the water temperature drops below 20°C.

Coastal habitat conservation and restoration

Given the already very high human population pressure in many Asian countries, it is difficult to be optimistic about the future for coastal habitats and the communities of plants and animals they support. Coral reefs and seagrass beds are particularly sensitive to destructive fishing methods, pollution, and any increased sedimentation caused by coastal construction or other development activities. Even mangrove forests, which are more robust and relatively easy to manage on a sustainable basis by tree replanting, have disappeared at an alarming rate (Table A4).

There are some positive indications that coastal protected areas (e.g., biosphere reserves, nature parks, and sanctuaries) can preserve habitat and protect species biodiversity. In addition, protected areas can play a valuable role in raising public awareness of the need for such conservation efforts (Figure A48). Habitat restoration, especially of mangroves, is also being supported very actively in several Asian countries. Thousands of hectares of mangrove have already been planted successfully in Bangladesh and in the Red River Delta and Mekong Delta of Vietnam in order to protect the coastal zone from storms, flooding, and erosion, and to accelerate the reclamation of land from the sea. Mangrove planting also leads to a higher diversity of animals and birds, and to improved fisheries production. However, the full ecological impact of mangrove reforestation need careful assessment, both in terms of the plant and animal communities that develop in mangrove plantations and the physical changes that take place in the ecosystem (Macintosh *et al.*, 2002).

In contrast to mangroves, it is very difficult to restore coral reefs and seagrass beds and most efforts to date have been only experimental in scale, involving, for example, the transplanting of healthy coral fragments or individual coral heads onto a degraded reef (e.g., Clark and Edwards, 1995; Rinkevich, 2000). For corals and seagrasses, strict conservation of the coastal habitats supporting these delicate plant and animal communities, including protection from pollution, should be the first priority. This is not simply a plea to protect these diverse coastal communities for biological interest; there are sound economic reasons for doing so. White and Cruz-Trinidad (1998), for example, estimated that the sustainable annual benefit from 1 ha of coral reef in the Philippines is in the range USD31,900 to USD113,000, including its fisheries, tourism, coastal protection, and biodiversity values. This suggests that, conservatively, coral reefs are contributing almost USD1 billion annually to the Philippines' economy, but equally such huge economic benefits will be lost if coral habitat continues to be destroyed.

Donald Macintosh

Bibliography

Aragones, L.V., and Marsh, H., 2000. Impact of dugong grazing and green turtle cropping on tropical seagrass communities. *Pacific Conservation Biology*, **5**: 277–288.

Bryant, D., Burke, L., McManus, J., and Spalding, M. 1998. *Reefs at Risk*. Washington: World Resources Institute Report,

Chan, H.-T., and Salleh, Md. N., 1987. *Traditional uses of the Mangrove Ecosystem in Malaysia*. Mangrove Ecosystem Occasional Papers

No. 1. UNDP/UNESCO Regional Mangroves Project RAS/86/120. New Delhi: UNESCO,

Clark, S., and Edwards, A.J., 1995. Coral transplantation as an aid to reef rehabilitation: evaluation of a case study in the Maldive Islands. *Coral Reefs* **14**: 201–213.

Ekman, S., 1967. *Zoogeography of the Sea*. London: Sidgwick and Jackson.

Gan, B.K., 1995. A *Working Plan for the Matang Mangrove Forest Reserve (Fourth Revision)*. Malaysia: State Forest Department of Perak Darul Ridzuan.

Gomez, E.D., 1980. A review of literature on marine turtles in the Philippines. *Kalikasan, Philippines Journal of Biology*, **9**: 95–99.

Haynes, D.J., Slater, J., Devlin, M, and Make, L. 1998. Great Barrier reef water quality and dugong protection areas. *Reef Research*, **8**: 10–15.

Hogarth, P.J., 1998. *The Biology of Mangroves*. Oxford: Oxford University Press.

Hong, P.N., and San, T.H., 1993. *Mangroves of Vietnam*. Regional Wetlands office: International Union for the Conservation of Nature (IUCN).

Hossain, Md. S., and Kwei-Lin, C., 2001. *Land Use Zoning for Integrated Coastal Zone Management*. ITCZM Monograph No. 3. Bangkok: Asian Institute of Technology.

Kellerman, A., and Koh, C.-H., 1999. Korean tidal flats: the west Pacific mirror of the European Wadden Sea. *Wadden Sea Newsletter*, **1**: 7–8.

Kjerfve, B., 1990. *Manual for Investigation of Hydrological Processes in Mangrove Ecosystems.* UNESCO/UNDP Regional Project "Research and its Application to the management of the Mangroves of Asia and the Pacific" (RAS/86/120). Paris: UNESCO.

Koh, C.-H., and Shin, H.-C., 1988. Environmental characteristics and distribution of macrobenthos in a mudflat of the west coast of Korea (Yellow Sea). *Netherland. Journal of Sea Research*, **22**: 279–290.

Lee, S.-Y., 1997. Annual cycle of biomass of a threatened population of the internal seagrass *Zostera japonica* in Hong Kong. *Marine Biology*, **129**: 183–193.

Macintosh, D. J., 1982. Fisheries and aquaculture significance of mangrove swamps, with special reference to the Indo-West Pacific region. In Muir, J.F., and Roberts, R.J. (eds.), *Recent Advances in Aquaculture*, Volume 1. London: Croom Helm, pp. 3–85.

Macintosh, D.J., Ashton, E.C., and Havanon, S., 2002. Mangrove rehabilitation and intertidal biodiversity: a study in the Ranong mangrove ecosystem, Thailand. *Estuarine, Coastal and Shelf Science*, **55**: 331–345.

Macnae, E., 1968. A general account of the fauna and flora of mangrove swamps and forests in the Indo-West Pacific Region. *Advances in Marine Biology*, **6**: 73–270.

Overton, J.L., and Macintosh, D.J., 1997. Mud crab culture: prospects for the small-scale Asian farmer. *Infofish International*, **5**: 6–32.

Palma, J.A.M., 1993. Marine turtle conservation in the Philippines. In Nacu, A., Trono, R., Palma, J.A., Torres, D., and Agas, F., Jr. (eds.), *Proceedings of the First ASEAN Symposium-Workshop on Marine Turtle Conservation*. Manila: WWF (Worldwide Fund for Nature), pp. 105–122.

Peng, L., and Wei, X.-M., 1983. Ecological notes on the mangroves of Fujian, China. In Teas, H.J. (ed.), *Biology and Ecology of Mangroves. Tasks for Vegetation Science*, Volume 8. The Hague, Netherlands: Dr. W. Junk Publishers, pp. 31–36.

Phillips, M.J., and Macintosh, D.J., 1997. Aquaculture and the environment: challenges and opportunities. In Nambiar, K.P.P., and Singh, T. (eds.), *Sustainable Aquaculture*. Proceedings of Indofish-Aquatech '96 International Conference on Aquaculture. Kuala Lumpur, Malaysia, 25–27 September 1996. Kuala Lumpur, Malaysia: Infofish, pp. 159–170.

Primavera, J.H., 1993. A critical review of shrimp pond culture in the Philippines. *Revues in Fisheries Science*, **1**: 151–201.

Reynolds, J.E., and Odell, D.K., 1991. *Manatees and Dugongs*. New York: Facts on File.

Rinkevich, B., 2000. Steps towards the evaluation of coral reef restoration by using small branch fragments. *Marine Biology*, **136**: 807–812.

Robertson, A.I., and Alongi, D.M., 1992. *Tropical Mangrove Ecosystems. Coastal and Estuarine Studies,* Volume 41. Washington, DC: American Geophysical Union.

Rouf, Md. A., and Jensen, K.R., 2001. *Coastal Fisheries Management and Community Livelihood*. ITCZM Monograph No. 4. Bangkok: Asian Institute of Technology.

Schuster, W.H., 1952. *Fish Culture in Brackish Water Ponds of Java*. Rome: Indo-Pacific Fisheries Council, Special Publication 1.

Smith, P.T. (ed.), 1999. Towards sustainable shrimp culture in Thailand and the region. *Australian Centre for International Agricultural Research (ACIAR) Proceedings No. 90.*

Spalding, M.D., Blasco, F., and Field, C. (eds.), 1997. *World Mangrove Atlas. Okinawa:* The International Society for Mangrove Ecosystems.

Sukardjo, S., 2000. Indonesia: Mangrove-friendly aquaculture. In Primavera, J.H., Garcia, L. Ma. B., Casanos, M.T., and Surtida, M.B. (eds.), *Proceedings. of the Workshop on Mangrove-Friendly Aquaculture*, January 11–15, 1999. Iloilo City, Philippines: Southeast Asian Fisheries Development Center, pp. 105–140.

Tomlinson, P.B., 1986. *The Botany of Mangroves*. Cambridge: University Press.

Veron, J.E.N., and Stafford-Smith, M., 2000. *Corals of the World.* Townsville MC, QLD 4810, Australia: Australian Institute of Marine Science.

Vietnam Environmental Monitor, 2002. Hanoi: The National Environment Agency/The World Bank/The Danish Agency for International Development (DANIDA).

Watson, J.G., 1928. Mangrove forests of the Malay Peninsula. *Malayan Forest Records*, **6**: 1–275.

White, A.T., and Cruz-Trinidad, A., 1998. *The Values of Philippine Coastal Resources: Why Protection and Management are Critical.* Cebu City, Philippines: Coastal Resources Management Project.

Wolanski, E., and Ridd, P., 1986. Tidal mixing and trapping in mangrove swamps. *Estuarine, Coastal and Shelf Science*, **23**: 759–771.

World Resources Institute, 1996. *World Resources 1996–7: A Guide to the Global Environment*. WRI/UNEP/UNDP/WB. New York: Oxford University Press.

Cross-references

Aquaculture
Asia, Eastern, Coastal Geomorphology
Carrying Capacity in Coastal Areas
Coral Reefs
Demography of Coastal Population
Human Impact on Coasts
Mangroves, Ecology
Mangroves, Geomorphology
Salt Marsh
Sandy Coasts
Tidal Flats
Tourism and Coastal Development

ASIA, EASTERN, COASTAL GEOMORPHOLOGY

The shores of eastern Asia, extending from Thailand and Indonesia to eastern Siberia and including the Philippines, Taiwan, and Japan, largely follow the tectonically active zones along the eastern and northern sides of the Eurasian plate (Inman and Nordstrom, 1971). Large stretches of coastal area along Indonesia, the Philippines, Taiwan, Sakhalin, Kamchatka, and the East Asian island chains are therefore tectonically unstable. In these areas the structural trends are generally parallel to the coast, which was distinguished by Suess (1892) as the Pacific type. Outside these areas, away from the tectonically active collision zones, the coastal regions are generally more stable and the structural trends are usually not parallel to the coast, which corresponds to Suess' Atlantic type. This type is present along most of the Asian mainland from Thailand to Siberia.

Comparatively straight coasts, situated along mountain chains, sometimes with river deltas and local alluvial foreland, are found mainly in western Sumatra, southern Java, northern Vietnam, and the Pacific coast of Siberia. A drowned older topography with an irregular coastline is present along parts of southern Vietnam, the mainland coast north of the Red River, on the islands of eastern Indonesia, on northern Kalimantan (Borneo), the Philippines, the Shandong Peninsula in China, Japan, southern Sakhalin, and the eastern end of Siberia. These coasts are usually somewhat altered by the sea, with bays containing beaches, spits, and barriers, sometimes being filled up with sediment and partially surrounded by alluvial foreland. The main exceptions are the large river deltas, the west coast of Kamchatka that consists largely of sand barriers, some beach barriers in Japan and on Sakhalin, and some areas like eastern Taiwan where tectonic uplift matches the Holocene sea-level rise. Elsewhere the coast is predominantly depositional, consisting of beaches, spits, barriers, tombolos, mudflats, marshes, mangrove swamps, and coral reefs.

The general direction of beaches, spits, and barriers is related to the direction of the swell. Between northern Japan and Indonesia the swell comes chiefly from northern directions, and the beaches and spits face largely E–NE. Where the swell is southerly, as in the Indian Ocean and in the Pacific north of 45°N, the beaches face mainly SE–SW. A major distinction can be made between the coasts north and south of about 40°N (Davies, 1973). North of this latitude storm waves are predominant and pebbles form an important part of the beach material. During the winter, sea ice (pack-ice) borders the coast as far south as Vladivostok and northern Japan. Further north, along the northern shore of the Sea of Okhotsk and north of the Kamchatka peninsula, permafrost reaches the sea, and from the eastern Siberian coast ice cliffs have been reported. South of 40°N the ocean swell is more important, and large areas from southern Japan to Malaysia are affected by tropical storms. In this area, especially south of 25°N, sandy beaches are predominant while pebbles are relatively unimportant as beach material.

Coral reefs and beach rock are confined to the tropical and subtropical zones—they are not present north of 35°N. Mangrove grows south of 31°N (on the Asian mainland coast south of 26°N) and dominates large stretches of coast in Indonesia, Malaysia, and Thailand, especially in areas that are sheltered from the ocean swell. An important part of the East Asian coastline consists of several large river deltas (those of the Mekong, Yang-tse-Kiang or Chang Jiang, and the Hoang-Ho or Yellow river) and a large number of smaller deltas often of still appreciable dimensions such as the deltas of the Chao Phraya, the Red River, and the Si Kiang. This preponderance of deltas is due to the high mechanical erosion on the mainland between Thailand and Korea (>60 ton per km^2 per year), which, associated with the presence of large and numerous rivers, has resulted in extensive deltaic and estuarine deposition.

Tidal effects are especially important along the mainland coast from Taiwan to Kamchatka, where the tides have an appreciable range. Usually the tidal range in eastern Asia is less than 4 m (and around Japan mostly less than 2 m) but in the northern Sea of Okhotsk, in the Yellow Sea, and along the Chinese coast it reaches more than 6 m. Tidal marshes have been formed locally, as, for example, along the Gulf of Bohai, in southern Korea, and in Hangzhou Bay.

A striking feature on many coasts from Japan through southeast Asia to Indonesia is the evidence of raised coastlines at 0.5–1 m, about 2 m, and 5–6 m above present sea level. These are indicated by flat, sandy terraces, beach ridges, coral reefs, mollusks and/or barnacles, notches, benches, platforms, and beachrock. Similar raised coastlines are known from many other coasts in Asia, Africa, South America, and Australia (for references, see Tjia, 1975, and Clark *et al.*, 1978). Daly (1934) and many others therefore assumed that during the Holocene, about 6,000–4,000 years ago, sea level stood about 5 m higher than at present. Others have contended that these features are related to temporary high water during storms, or to local uplift. Tjia (1975) has suggested that glacioeustatic uplift of the areas in the Northern Hemisphere that were glaciated during the Pleistocene, may account for the absence of any indications for high sea levels during the Holocene. C14 dates of raised reefs and mollusk shells from the Pacific and Indian Ocean, Southeast Asia, and Australia point to a slight Holocene transgression as well as to local uplift. Clark *et al.* (1978) indicate that regional differences in the Holocene sea-level rise are related to differences in crustal rebound after the Pleistocene through ice cap melting and the increasing water load of the oceans.

The general features of the east Asian coasts are summarized in Figure A49, which is based essentially on the world map given in Valentin (1952), but with emphasis on the coastline characteristics. More detailed knowledge, however, is still lacking for large stretches of coast, especially in eastern Siberia and Southeast Asia.

Siberia

The Siberian coast east of the Taymir Peninsula is flat up to the Kolyma River estuary and shows a drowned topography with only a higher coast at the Laptev Straits, where bedrock is exposed. The flat coast is locally raised a few meters and here small cliffs are formed by intensive thermal abrasion. This is caused by the regular melting of ice and frozen deposits (permafrost) along the cliffs and removal by waves of the loose sediment. The coastal flats are flooded over large distances (up to 30 km) during wind-induced surges. During offshore winds the muddy nearshore seafloor can be exposed over similar distances. Several large rivers reach the sea in this area such as the Lena, Indigirka, and Kolyma rivers, but only the Lena has a vast delta, which is largely a relict feature. The other rivers have long, partly filled-in estuaries and insignificant deltas. The islands off the coast (the New Siberian Islands and Wrangel Island) have a bedrock nucleus surrounded by loose but frozen sedimentary deposits. Thermal erosion is large and can annually remove up to 50 m from the cliffs.

East of the Kolyma River mouth the coast is partially rock, where the mountain ranges between Kolyma and the Bering Strait come near to the sea. There is usually a narrow coastal plain, several kilometers wide, and at Cape Billing a sand barrier extends for about 50 km into the sea with an up to 10 km wide lagoon on the landward side. Transverse sandbars divide the lagoon into several rounded lakes; their formation is related to the development of wind-induced circulation cells in the lagoon. The sand barrier has been formed in an area that is protected against pack-ice (probably by Wrangel Island). In general, longshore sediment drift is limited by the short period that the coast is free of ice. The sediment is mainly derived from reworked glacial deposits. The easternmost part of Siberian coast along the Bering Strait is characterized by high cliffs and fjords where the Chukhotsk range meets the coast (Zenkovich, 1985).

The Pacific coast of Siberia down to Korea extends from the arctic into the temperate zone and has complex tectonics with stable areas alternating with areas of emergence and subsidence. The coastal area is usually mountainous with rather short rivers (except the large Amur River) and narrow coastal plains. Waves and tides are variable: storm waves are up to 12 m in the Bering Strait and in the Sea of Okhotsk, while lower waves (normally up to 2–3 m, and up to 7 m during the winter) dominate in the other areas. Most of the coast is mesotidal, but locally, as in Penzinskaya Bay, becomes macrotidal (up to 12 m). There is a relatively large sediment supply to the coast, because of a wet, cold, or temperate climate in combination with easily weathered rock, such as volcanic deposits and Pleistocene fluvioglacial sediments. Barriers, coastal lagoons, and tombolos are common features, separated by promontories of rock or (unconsolidated) sediment (Kaplin, 1985). The largely drowned topography has for a large part been filled in during the postglacial period. Along Sikhote Alin the coast is relatively straight and uninterrupted because of a low mountain range that lies parallel to the coast.

Japan, Sakhalin

The shores of Sakhalin and Japan consist of large stretches of ria coast, alternating with many small and several large stretches of beaches, spits, barriers, sand dunes, lagoons, and river outlets. Especially on northern Sakhalin, but also on Hokkaido and along Honshu, large parts of the coast are depositional alternating with rocky promontories. Locally, volcanism close to the coast has resulted in cliffs and beaches of volcanic material. Most of the coastal lowland in Japan is intensively used so that a large proportion of the coast is man-made: less than 50% has remained natural. Coastal terraces are widely distributed in northern Japan; in the south they occur mainly on the promontories along the Pacific coast. The terraces are predominantly present on the more easily abraded Tertiary and Quaternary formations, whereas the indented ria coasts have been developed mainly in folded Palaeozoic and Mesozoic rock. The highest terraces have been found in southwestern Hokkaido at +585 m, but usually they reach to +150 m (Kosugi, 1971). Below present sea level submerged flat surfaces have been found to depths of −150 and −190 m.

The terraces as well as raised coral reefs (Yabe and Sugiyama, 1935) indicate that the Quaternary eustatic changes in sea level may have had an important effect on coastal development in Japan, but there is also much evidence for differential uplift, tilting, and local subsidence. Consequently, in an appreciable number of papers by Japanese authors the combined effects of both eustatic sea-level changes and crustal movements have been discussed (Richards and Fairbridge, 1965; Yoshikawa *et al.*, 1965; Richards, 1970). The presence of ria coasts is generally seen as evidence of an overall subsidence since the Pleistocene, but narrow terraces scattered along the rias suggest a more recent emergence. Some terraces (at +40–+60 m) are widely distributed along the Japanese coast and are seen as the result of eustatic sea-level changes. Coastal terraces, formed at the same sea level, have also been found at different altitudes within short distances, even locally disappearing below present sea level, as a result of crustal movements. From this complex situation it has been deduced that the Holocene sea-level rise started around 18,000 BP and that between 6,000 and 3,000 BP sea level was 3–5 m higher than at present, which is indicated by former wave-cut notches, sea caves, benches, small terraces, marine deposits, and neolithic remains (Tada *et al.*, 1952). Clark *et al.* (1978) have shown that these features can also be explained by crustal rebound after the Pleistocene and the increasing water load of the ocean.

A marked feature along the Japanese and Sakhalin coasts is the widespread occurrence of shore platforms (Takahashi, 1974). These are most

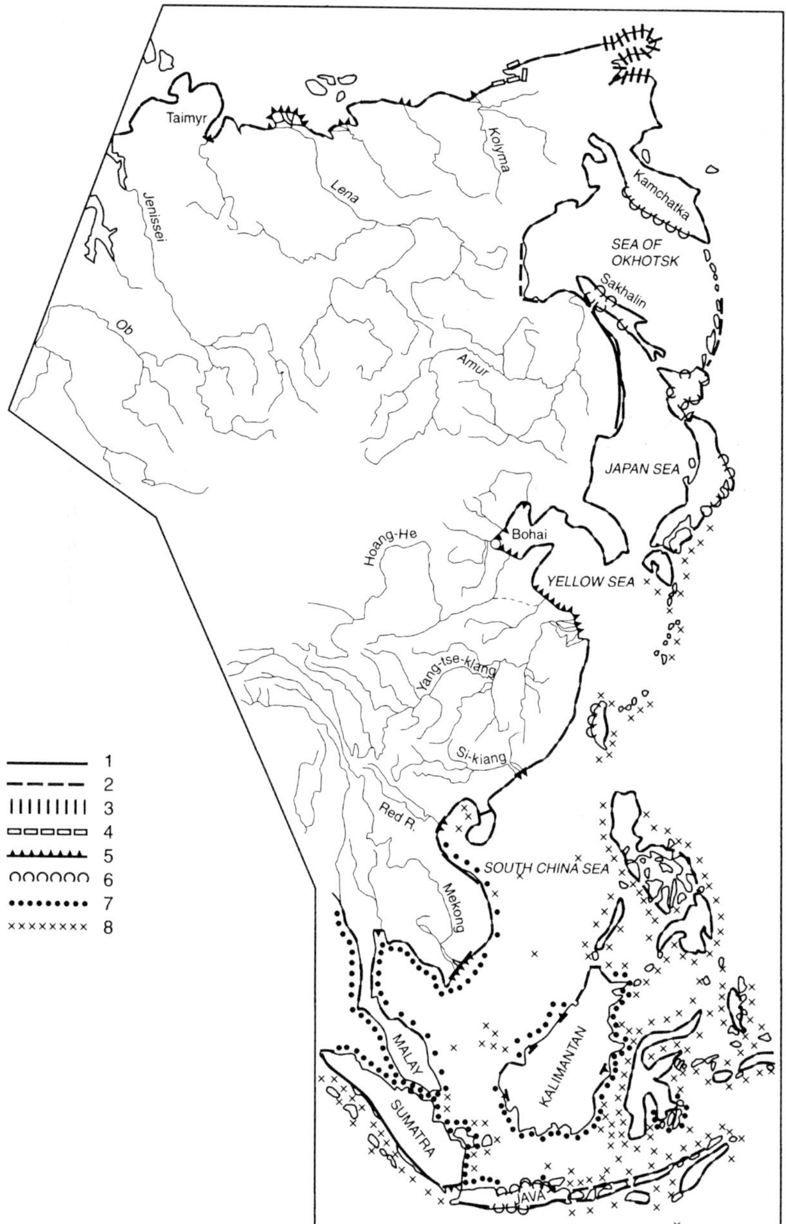

Figure A49 Coastal morphology of eastern Asia. (1) Comparatively straight coasts along mountain chains, sometimes with river deltas and local alluvial foreland; (2) Drowned older topography with an irregular coastline, somewhat altered by the sea; (3) Fjords; (4) Tableland coast Table and coast with cliffs; (5) Large river deltas; (6) Predominantly spits, barriers, mud flats, marshes; (7) Mangrove swamp; (8) Coral reefs.

conspicuous along the coasts of the Pacific and the East China Sea, less on the Sea of Japan coasts, and least along the Seto Inland Sea. Their maximum width varies from 530 m along the Pacific to 80 m on the Sea of Japan coasts, which roughly corresponds to the wave conditions on each coast. The average width is 30–80 m and the maximum length about 4 km; they occur on all types of rock, but tend to be wider on Neogene mudstones, sandstone, and tuff-breccia.

The East Asia mainland, Korea to Malaya

The mainland coast from Korea to the Malayan Peninsula is largely a drowned or ria-type coast, with abrasion at the headlands, accumulation (with sandy beaches, spits, and beach/ ridges) in the bays and, especially in China and southern Korea, numerous islands. Large estuaries and river deltas are present where major rivers reach the coast. In West and South Korea tidal flats have been formed in the bays and in the shallower parts between the islands (Guilcher, 1976), which have been partly reclaimed. Coastal terraces are present along the East Korea coast up to 100–150 m and along the West Korea coast up to 30–60 m, which reflects the emergence of the east coast and the submergence of the west coast. Also from the area around Hong Kong (Williams, 1971) coastal terraces and old beaches have been reported: they occur at +5 and +14 m (Berry, 1961) and are probably also present elsewhere. At higher levels, up to +70 m, benches are found, below present sea level flat surfaces can be distinguished down to −60 m. Especially the terraces at +5 and +14 m are thought to be caused by eustatic sea-level changes. Flat surfaces at +130 and +230 m are probably not of marine origin.

Around the Gulf of Bohai, between the Liaodong and Shandong peninsulas, the coast has been filled with recent alluvium from the Huang He (Yellow River), Luan He, Liao, and a number of smaller rivers. Long stretches of tidal marshland have been formed. Along the northwestern shore, however, up to 80 km of alluvial deposits were eroded again in historical times (von Wismann, 1940), probably because the Huang He then had a southerly course and no new sediment from that river reached the Gulf of Bohai at that time. The present situation dates from 1853, when the Huang He again took a northerly course.

Between 1448 and 1853 it had been reaching the sea south of the Shandong Peninsula after having joined the Huai He River. While at present the silt from the Huang He moves northward anticlockwise through the Gulf of Bohai, the silt flowing out of the southern mouth before 1853 was transported further south as far as the Yang-tse (Chang Jiang) River mouth, which was deflected southward. In 400 years the coastline in this area moved 36–62 km eastward.

South of the Yang-tse-Kiang (Chang Jiang) a ria coast ends at the delta of the Si-Kiang, while further south the coast is interrupted by the deltas of the Red River, the Mekong, the Chao Phraya, and other smaller rivers. Tidal flats of variable widths have been formed along Hangzhou Bay, Wenzhou Bay, and numerous smaller bays and estuaries. Salt-marshes along the upper parts of the flats have mostly been reclaimed. South of Hong Kong mangrove swamps are present and fringing coral reefs become increasingly numerous; in particular around the Gulf of Thailand extensive coastal barriers and sandy beaches have been formed. From Dong Hoi to Cape Varella in Vietnam the coast is formed by a 15–40 km wide belt of barrier complexes, lagoons, and sand dunes; from Cape Varella to Cape Padaran the coast is again a typical ria coast. Terraces have been found in Vietnam and Cambodia between +2 and +80 m (Carbonnel, 1964; Saurin, 1965). The terrace at +2 m and raised reefs, mollusks (oysters), and conglomerates at the same level have been found all along the coasts of Vietnam and Cambodia. Other terraces occur more regionally; the +4 m terrace only in Vietnam, the +15 and +25 m terraces in southern Vietnam and Cambodia, and the terrace at +80 m only near Cape Padaran.

Taiwan, Philippines

Compared to the mainland coast, Taiwan is very different. Situated in the tectonically active zone it has been slowly uplifted during the Quaternary. This uplifting continues at present, although there is some evidence for regional temporary subsidence. Eustatic sea-level changes are considered to have had little effect on the coast, since the changes in sea level have been more than balanced by the amount of uplift (Hsu, 1962). On the Pacific side, the coast is steep and erosion predominates (as also on the southwestern and northern coasts), but on the west coast, which is sheltered from the ocean swell and the tropical storms as well as from the strong northeastern monsoon, a flat, marshy coast is formed with tidal flats. Coral reefs are present on the Pacific side as well as round the Ryukyu Islands further north and along the Pacific coasts of Kyushu and Shikoku.

The Philippine coasts, situated like those of eastern Taiwan in the tectonically active zone, are comparatively steep but intersected by river outlets, broad valleys, and inlets with beaches, surrounded by coral reefs. Although the area has been drowned during the Holocene transgression, the islands have been rising during the Pleistocene and Holocene and there is evidence of differential uplift. Numerous and often extensive marine terraces are present; on Sabtan Island and on Luzon the highest are at +180 m, on Bataan at +275 m, and in northern Mindanao at +360 m (van Bemmelen, 1970), but no correlations have been made.

Malay Peninsula

The Malay Peninsula, together with the eastern part of Sumatra, western and central Kalimantan (Borneo), and the Sunda-Java Sea, belongs to the Sunda landmass, which has probably been tectonically stable since the middle Pleistocene. The presence of beach ridges several meters above present sea level has been interpreted as being the result of recent uplift, but can also be explained by crustal rebound after the Pleistocene and the increasing water load of the oceans (Clark *et al.*, 1978).

The east coast of Malaya is relatively smooth with shallow bays and few indentations. More than 80% of this coast is formed by sandy beaches interrupted by river outlets and small deltas; cliffs form about 10% of the coast (Nossin, 1965; Swan, 1968). Tidal swamps with mangrove are usually found landward of the beaches and sandbars. They are sheltered against the swell of the strong monsoon that comes from the northeast between October and April. During this period there is more erosion than accretion, especially between late October and January, when spring tides are at their highest. Accretion is widespread between February and September, when the tides are less effective and the wind blows mainly offshore. A minor period of erosion occurs from May to July when the spring tides reach secondary maxima.

The west coast of Malaya is for a large part covered with mangrove, which continues northward to the Irrawaddy delta and from there to the mouths of the Ganges. Locally, this mangrove belt is interrupted by cliffs, river outlets, and beaches. The islands around Singapore have been studied in detail by Swan (1971), who has drawn attention to the fact that in this sheltered, low-energy coast the coastal forms are very different from those supposed to be characteristic for such coasts in the humid tropics. Mangrove swamps, tidal flats, beaches, and fringing reefs, are present as well as cliffs, caves, shore platforms, and beach conglomerate (ironstone). This diversity is due to the intense and continuous chemical weathering, which makes possible subaerial and marine erosion by small waves, and to differences in rock type and exposure. Coastal sediment is produced through cliff erosion, which takes place along both sheltered and exposed parts of the coast.

Indonesia

In Indonesia, mangrove swamps are present along the coasts of Sumatra, northern Java, along most of Kalimantan (Borneo), along northern Borneo (Sarawak), and at the southern end of Sulawesi (Celebes). All these areas are relatively sheltered against the ocean swell. Cliffs and beaches are locally present (Wall, 1964), but although the rivers of Sumatra and Kalimantan carry large amounts of sediment to the sea, only the Kapuas and Pawan rivers, the Rajang and Baram rivers on the northern coast, and the Mahakam River in the east have built up sizeable deltas. Elsewhere, where sediment supply is abundant, broad alluvial lowlands have been formed. The coasts of Java are predominantly beaches interrupted by river outlets and deltas and locally by cliffs. However, along eastern Java, western Sumatra, and on most of the eastern Indonesian islands, the coasts are relatively steep. A few river deltas with associated swamps and alluvial foreland are present on the west coast of Sumatra and on the northeast coast of Java.

The main feature in western Indonesia is the large, drowned Sundaland (Molengraaff, 1921). Two large submarine valley systems, being the continuation of the present rivers in Sumatra and Kalimantan, are present on the continental shelf. During the Pleistocene periods of low sea level, one system drained Sumatra and east Kalimantan, discharging into the South China Sea. The other system, draining Java and south Kalimantan, discharged south of Makassar Strait. A divide crossed the Sunda Sea between Sumatra and Kalimantan across Billiton and the Karimata Islands. The Sundaland plain has been dissected at least twice during the Pleistocene as a consequence of the eustatic lowering of sea level (van Overeem, 1960). In the whole area, including Malaya, eustatic terraces are present from +50 to −90 m (Tjia, 1970).

On the former east coast of the Sundaland, a large barrier reef has been formed, stretching from Balikpapan on Kalimantan to the island of Sumbawa in the south. It is interrupted in many places and shows a large gap of about 100 km wide facing a deep embayment in the former Sundaland. The main river outlet was probably situated here. The reef began as a late Pleistocene fringing reef and has grown upward with the gradual rise in sea level. Locally, it reaches the sea surface as separate coral islands. Similar fringing reefs of smaller dimensions are present around Sulawesi and the smaller islands in eastern Indonesia, along the west coast of Sumatra (where they show gaps in front of the river deltas), and further north along the Nicobar and Andaman Islands. In the Sunda Sea, coral growth is restricted to a number of isolated areas away from muddy river outlets (Kuenen, 1933; Umbgrove, 1947). In eastern Indonesia, numerous atolls and barrier reefs have grown upward from gradually subsiding submarine ridges and platforms, rising abruptly and steeply from a depth of 1,000–2,000 m. The effects of winds and waves on the reefs are conspicuous, especially in eastern Indonesia, where the reefs grow more vigorously on the windward side. Sea currents cause erosion and may shape a whole group of reefs. Solution of coral occurs within the tidal range. On the former northern coast of the Sundaland no barrier reef has been formed. Here the former coast was flat and gradually merged into an extensive sandy and probably muddy shelf. The water in this area was presumably too turbid for coral reef growth. This is also the case along the south coast of Irian Jaya (New Guinea), where large alluvial plains have been formed, covered along the coast by mangrove.

D. Eisma

Bibliography

Berry, L., 1961. Erosion surfaces and emerged beaches in Hong Kong. *Geological Society of America Bulletia*, **72**: 1383–1394.

Carbonnel, J.P., 1964. Sur l'existence d'un complexe de terrasses quaternaires dans l'ile de Saracen (Cambodge). *Societe Geologique de France Comptes Rendu*, **9**: 371–373.

Clark, J.A., Farrell, W.E., and Peltier, W.R., 1978. Global change in post glacial sea level: a numerical calculation. *Quaternary Research*, **9**: 265–289.
Daly, R.A., 1934. *The Changing World of the Ice Age*. New Haven, CT: Yale University Press, p. 271.
Davies, J.L., 1973. *Geographical Variations in Coastal Development*. New York: Hafner, p. 204.
Guilcher, A., 1976. Les cotes a rias de Coree et leur evolution morphologique. *Annales Geographie*, **85**: 641–671.
Hsu, T.L., 1962. A study on the coastal geomorphology of Taiwan. *Geological Society of China Proceedings*, **5**: 29–45.
Inman, D.L., and Nordstrom, C.E., 1971. On the tectonic and morphologic classification of coasts. *Journal of Geology*, **79**: 1–21.
Kaplin, P., 1985. Pacific USSR. In . Bird, E.C.F., and Schwartz, M.L. (eds.), *The World's Coastline*, Volume 115. New York: Van Nostrand Reinhold Co, pp. 857–862.
Kosugi, K., 1971. Etude analytique des depots de sable et galets actuels d'origines diverses. Application au processus de formation des surfaces de terrasse littorale dans le Hokkaido (II). *Science Report of the Tokohu University*, 7th series, **20**: 257–326.
Kuenen, Ph.H., 1933. *Geology of Coral Reefs*. Leiden: E.J.Brill, p. 125.
Molengraaff, G.A.F., 1921. Modern deep-sea research in the East Indian Archipelago. *Geographical Journal*, **27**: 95–118.
Nossin, J.J., 1965. Analysis of younger beach ridge deposits in eastern Malaya. *Zeitschrift Geomorphologie*, **9**: 186–208.
Richards, H.G., 1970. Annotated bibliography of Quaternary shorelines. Supplement 1965–1969, VIII INQUA-Congress, Academy of Natural Science of Philadelphia, Special Publication 10, p. 240.
Richards, H.G., and Fairbridge, R.W., 1965. Annotated Bibliography of Quaternary shorelines (1945–1964), VII INQUA-Congress, Academy of Natural Science of Philadelphia, Special Publication 6, p. 280.
Saurin, E., 1965. Terrasses littorales de Son Hai. *Archaeology Geology de Viet-Nam*, **7**: 20–24.
Suess, E., 1892. *Das Anlitz der Erde*, Volume 1. Vienna: F. Tempsky, 778p.
Swan, S.B.St.C., 1968. Coastal classification with reference to the east coast of Malaya. *Zeitschrift Geomorphologie*, **7** (Supplement), 114–132.
Swan, S.B.St.C., 1971. Coastal geomorphology in a humid tropical low energy environment: the islands of Singapore. *Journal of Tropical Geography*, **33**: 43–61.
Tada, F., Nakano, T., and Iseki, H., 1952. Shoreline development of the Pacific coast of Japan in prehistoric time. *Proceedings of the 17th International Congress Geographical Union*, 386–391.
Takahashi, T., 1974. Distribution of shore platforms in southwestern Japan. *Science Report of the Tohoku University*, 7th series, **24**: 33–45.
Tjia, H.D., 1970. Quaternary shorelines of the Sunda Land, Southeast Asia. *Geologie en Mijnbouw*, **49**: 135–144.
Tjia, H.D., 1975. Holocene eustatic sea levels and glacio-isostatic rebound. *Zeitschrift Geomorphologie*, **22**(Supplement), 57–71.
Umbgrove, J.H.F., 1947. Coral reefs of the East Indies. *Geological Society of America Bulletin*, **58**: 729–778.
Valentin, H., 1952. *Die Kusten der Erde*. Petermanns Geographsiche Mitteilung, Erganzungsheft H 246. Gotha, Justus Perthes, 118p.
Van Bemmelen, R.W., 1970. *The Geology of Indonesia*, The Hague: Martinus Nijhoff. Volume 1, 732 p.; Volume 2, p. 265.
Van Overeem, A.J.A., *für* 1960. The geology of the cassiterite placers of Billiton, Indonesia. *Geologie en Mijnbouw*, **39**: 444–457.
Von Wissmann, H., 1940. Sudwest-Kiangsu, der Wuhu-Taihu-Kanal und das Problem des Yangdse-Deltas. Wiss. Veroff. Deutsch. Mus. Landerk. Leipzig. N.F., **8**, 63–105.
Wall, J.R.D., 1964. Topography–soil relationships in lowland Sarawak. *Journal of Tropical Geography*, **18**: 192–199.
Williams, A.T., 1971. Beach morphology and tidal cyclic fluctuations around Hong Kong Island. *Journal of Tropical Geography*, **32**: 62–68.
Yabe, H., and Sugiyama, T., 1935. Geological and geographical distribution of reef corals in Japan. *Journal of Paleontology*, **9**(3): 183–217.
Yoshikawa, T., Kaizuka, S., and Ota, Y., 1965. Coastal development of the Japanese islands. *Proceedings of the 7th Congress Association Quaternary Research*, **8**: 457–465.
Zenkovich, V.P., 1985. Arctic USSR. In Bird, E.C.F., and Schwarz, M.L. (eds.), *The World's Coastline*, Volume 116. New York: Van Nostrand Reinhold Co, pp. 863–871.

Cross-references

Asia, Eastern, Coastal Ecology
Beachrock
Changing Sea Levels
Cliffed Coasts
Coral Reefs
Coral Reefs, Emerged
Deltas
Ice-Bordered Coasts
Mangroves
Sea-Level Indicators, Biologic
Sea-Level Indicators, Geomorphic
Shore Platforms
Uplift Coasts

ASIA, MIDDLE EAST, COASTAL ECOLOGY AND GEOMORPHOLOGY

Between the Bosphorus to the west and the Iran–Pakistan border to the east, the coasts of the Middle East may be divided into two different provinces, the Mediterranean one to the west and the Indian Ocean one to the east. The first province concerns the Aegean and Mediterranean coasts of Turkey on the one hand and the Levantine coast on the other hand. The second province includes three rather different sectors: the western coast of the Red Sea and the coasts of the Persian Gulf, two elongated and narrow inland seas, and the Indian Ocean coasts, comprising the south Arabian coast, the Gulf of Oman, and the Makran coasts. Each of these sectors presents its own particularities together with some common characteristics, whether in tectonics, climatology, ecology, or geomorphology (Figures A50–A58).

The coasts of Anatolia

Climate, vegetation, hydrology

Along the Aegean and Mediterranean coasts of Anatolia, the annual precipitation ranges from 400 mm in the plains, with a minimum in the sheltered areas like the Çukurova plain, to more than 1,000 mm on the hills and flanks of mountains, notably in the Lycian sector. In winter, the Aegean and Mediterranean coastal areas are by far milder than in central Anatolia, with mean January temperature higher than 10°C in sectors bordering the sea along the Mediterranean Sea and higher than 5°C on the lower slopes of the hills and mountains. But everywhere freezing may occur, especially in the Aegean area. Summers are very hot; mean August temperatures exceed 25°C and maximum absolute temperatures reach 40°C or more, particularly in the Cilician plain. Rainfall occurs in winter and the summer is dry for about six months. Vegetation is much more xerophytic than in the West Mediterranean province. The seaward slopes of the coastal mountain ridges are truly Mediterranean in vegetation and in climate. Along the Amanus Mountains, the evergreen maquis zone of *Quercus calliprinos* and *Pistacia palaestina* sometimes starts at sea level and extends to an altitude of about 700 m. In the Lycian–Cilician sector, the eu-Mediterranean belt is confined to a very narrow ribbon with the same plant association of the evergreen maquis. In the Aegean sector, the evergreen maquis mainly comprises the association *Pistacia lentiscus, Olea europaea*, and *Myrtus* (Zohary, 1973).

The ratio of precipitation/evaporation shows a deficit and the eastern Mediterranean Sea receives water coming from the Black Sea through the Bosphorus and Dardanelles Straits, the salinity in the Aegean Sea is notably lower than in the eastern Mediterranean; in August, it is less than 38‰ in the north, less than 39‰ in the south of the Aegean Sea, whereas it is more than 39‰ in the eastern Mediterranean along the Turkish and Levantine coasts. Marine currents are generally north–south in the Aegean Sea and east–west along the Mediterranean coast of Turkey. On the whole, sea water temperatures increase from the north to the south in the Aegean Sea and from the west to the east in the Mediterranean, rising for example in August from 25°C along the south-west coast of Anatolia to over 28°C along the Cilician coast.

Regional features

The Turkish Aegean coastline is very long (3,484 km) and indented. It presents a great variety of lithological units, with alternation of crystalline massifs and Mesozoic geosynclinal belts including mostly limestone, flysch, and ophiolitic formations. This area is tectonically controlled by a highly complex structure of east–west oriented active horsts and grabens. Peninsulas, islands, rocky cliffs, and a narrow continental shelf correspond to the horsts. In the grabens there are wide continental

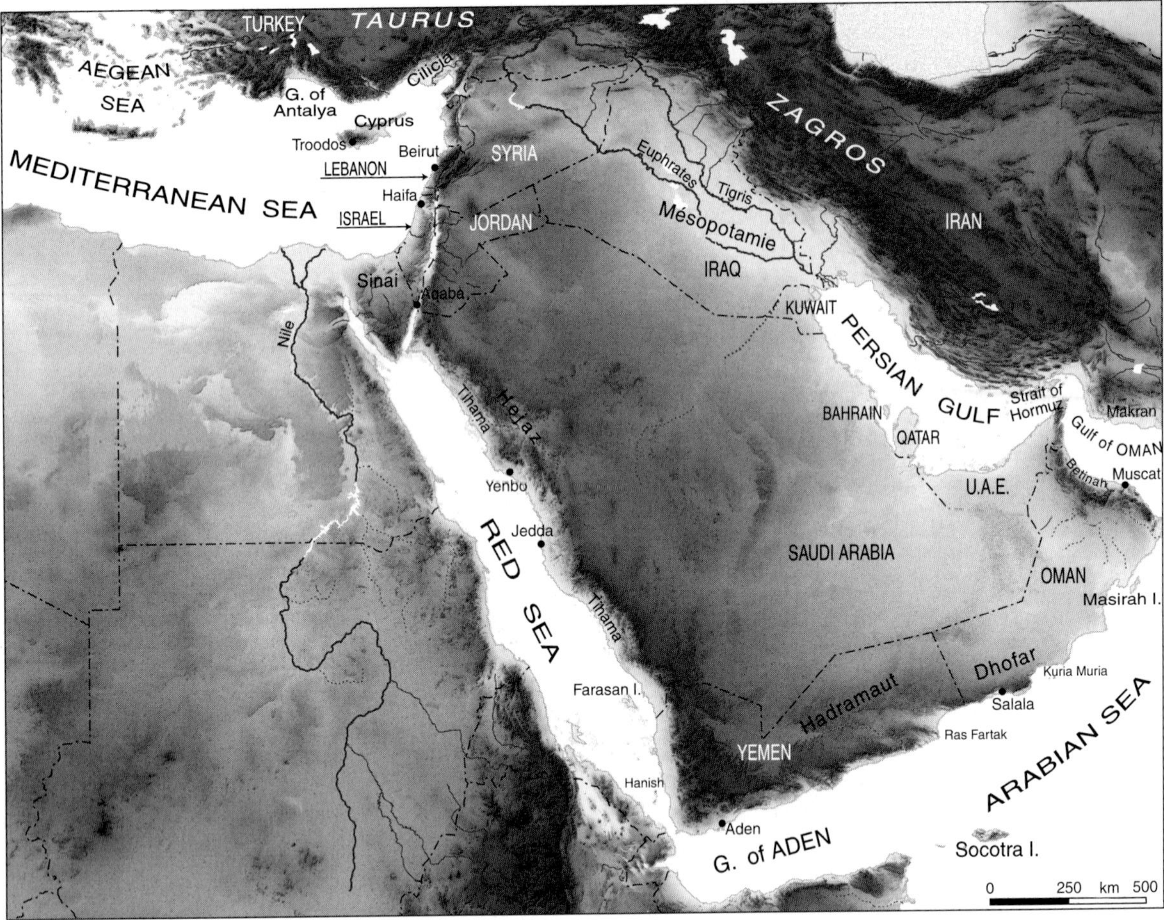

Figure A50 Location map.

shelves prolonging vast deltaic plains bordered with marshes, dunes, and sandy beaches. In the north, the Sea of Marmara, related to the greater Anatolian fault, is very deep (>1,300 m).

During the Middle Miocene severe tectonic phases occurred and grabens formed. These grabens were submerged by marine transgressions during the Quaternary, and during the glacial periods rivers deepened their courses. During the Holocene, when the sea-level rise had reached its present level about 6,000 years ago, major marine embayments occurred in several depressions where the main rivers ended; such as the Gediz, the Küçük Menderes, and the Büyük Menderez. Many important cities of the Hellenistic to Roman times were located on the immediate coastline and closely associated with the sea, such as Ephes, Priene, Milatus, or Heracleia; in the vicinity of Ancient Troy, the marine embayment protruded nearly 10 km south of the site. But, quickly, deltaic progradation and floodplain aggradation, caused by deforestation and severe erosion on the steep slopes of this mountainous country, markedly changed the landscape: the depressions were infilled with prograding river alluvium and deltas. These towns were separated from the sea; Ephes in the plain of the Küçük Menderes and Priene or Milatus in the Büyük Menderez basin, and the site of Troy is now about 6 km from the sea. Sometimes, for instance, in the Gediz Basin and the Izmir area, it seems that the evolution was more complex and mainly controlled by intense tectonics. The progradation was less marked where there is no major drainage system to provide sediment, for instance, in the south as in the Gökova Gulf. Now, progradation has largely halted mainly where the grabens widen rapidly into larger coastal embayments and modern deltas are wave dominated (Kraft *et al.*, 1980; Erol, 1985; Aksu *et al.*, 1987; Hakyemez *et al.*, 1999; Kayan, 1999).

The Mediterranean coastline of Turkey (1,707 km) is more regular and rectilinear than the Aegean coast, because of its longitudinal structure. The coastline is subparallel to the Taurus Mountains consisting of Paleozoic to Tertiary folded formations where carbonate rocks outcrop largely, favoring numerous karstic features. West of Antalya, the Taurus Mountains, S–N oriented, rise sharply and steeply out of the sea with high cliffs and only very small deltaic coastal plains. From Antalya to Alanya, there is a major embayment with a wide coastal plain edged by extensive coastal dunes and sandy beaches with frequent beach-rock formations; the only exception are the travertine cliffs at Antalya. Marine Pleistocene terraces occur to the east of the Antalya Bay at heights from 5 m to 20–30 m or even higher. Here, as in the Hatay area, beaches and bio-erosional notches and benches are frequent, between +0.80 and +2.5 m, testifying to late Holocene uplifts (Pirazzoli, 1986; Kayan, 1994). East of Alanya, once more, the Taurus Mountains rise directly out of the sea with very steep flanks and cliffs in Paleozoic schists and Mesozoic limestones. Scattered sandy beaches exist at the mouth of the rivers but elsewhere the coastline is dominated by plunging cliffs or by pebble beaches at the foot of the cliffs. The main river is the Göksu, characterized by a small lobed delta plain. East of Göksu, between the SW–NE Taurus Mountains and the N–S Amanus mountains (or Nur Mountains), there is the large Çukurova plain (or Cilician plain) which is a progradational plain built by two big rivers, the Seyhan to the west and the Ceyhan to the east. The plain, formerly swampy or marshy, but now largely drained and dry, is edged by a series of broad Holocene accretion beaches. At the foot of the Amanus Mountains, the rectangular-shaped Gulf of Iskenderun is structurally controlled. At the very southern end of the coast of Turkey, the Orontes River crosses the Amanus Mountains in a narrow defile and ends in a small delta near the ancient harbor of Seleucia ad Orontes (Erol, 1985). Almost everywhere, the coastline is now receding owing to the taking of sediments on beaches and dunes, the setting up of dams, and slight raising of the sea level, and slabs of beach rock are more and more frequent on the beaches.

Cyprus Island lies south of Turkey. Its coastline, about 450 km long, is partly steep and rocky in the north (notably in the Karpos Peninsula) and in the SW (at the foot of Troodos massif) and partly beach-fringed, particularly in the southeast, alongside the Mesogea central plain, mainly between Limassol and Famagusta where there are some brackish lagoons.

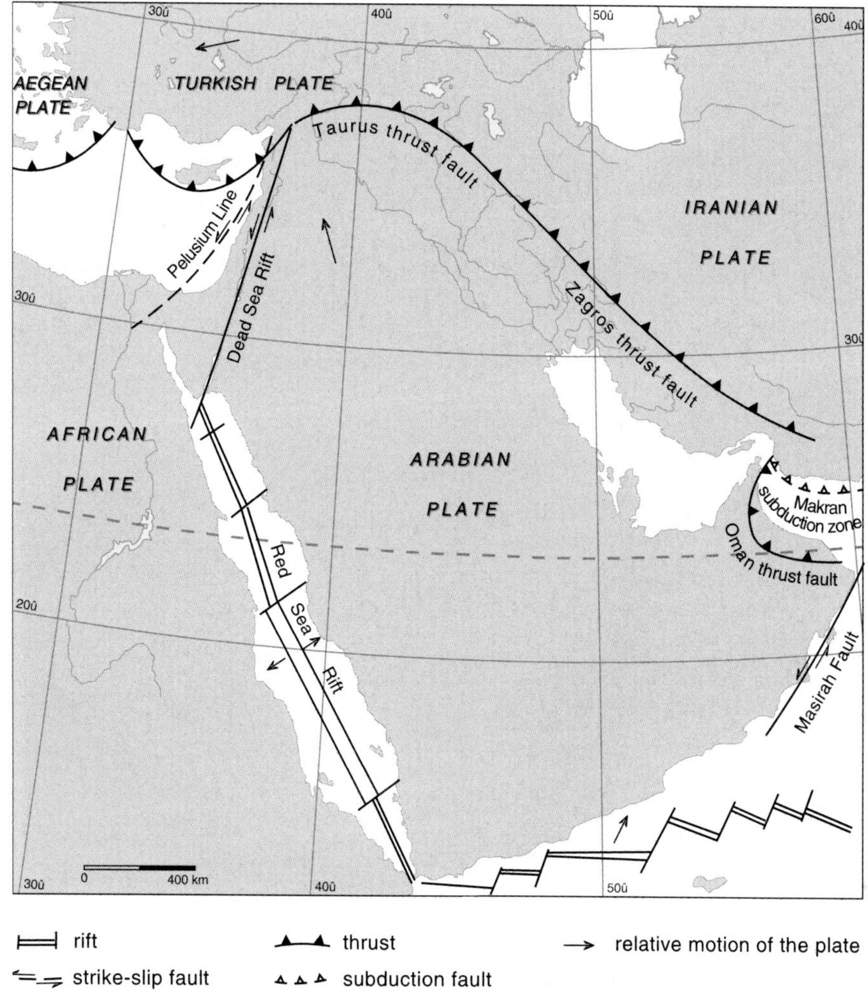

Figure A51 Tectonic map.

The Levantine coast

Climate and marine hydrology

The climate is clearly Mediterranean, with hot and dry summers, and rainy and mild winters and the winds mainly blowing from the west. Average rainfall decreases from the north (ca. 1,000 mm north of Latakia) to the south (313 mm at Gaza). Average air temperature along the coast usually ranges from 11°C to 14°C in January (with minima reaching the freezing point) and from 23°C to 26°C in August with absolute maximum temperatures over 45°C. Tidal range does not exceed 40 cm. There is an anticlockwise east Mediterranean current, which moves south–north along the Levantine coast. Longshore currents occur too, influenced by coastal morphology and wind direction, but the resultant shifting is to the north. All these currents carry the Nile sediments northward up to south Lebanon. Westerly and especially southwest winds are dominant. As the fetch is large, the coast undergoes strong wave action. In winter, storms are sometimes severe: in January 1968, with wind speeds of almost 150 km per hour, wave heights reached 7 m off the Beirut region.

The living world of the eastern coasts of the Mediterranean is impoverished and the littoral fauna and flora are poorly diversified in comparison to that of the western Mediterranean as a result of the extremely low biological productivity of the Levantine basin, mainly since the opening of the Suez Canal and the erection of the Upper Assouan Dam. Since the opening of the Suez Canal, over 300 migrant species have invaded the Mediterranean from the Red Sea at a faster and faster rate; 9 new species of fish arrived between 1902 and 1939, but 22 from 1965 to 1987. Among the "Lessepsian" crustaceans, decapods are particularly representative, such as *Myra fugax* or *Atergatis roseus*, and *Charybdis longicollis* represents up to 70‰ of the sandy–muddy bottoms. Among the Serpulids, 7 Lessepsian species have been identified, belonging to 3 genera, and 33 algae are possibly lessepsian (Por, 1982, 1990).

Coastal morphology

The rectilinear Levantine coast is open to the western winds and swells. The only noteworthy irregularities of the coast are such peninsulas as Ras el-Bassit in Syria, Ras el-Mina (near Tripoli), Ras Chekka, Ras Beirut, and Tyr in Lebanon, and Mount Carmel in Israel. Some of them (Tripoli or Tyr) are ancient rocky islands more or less recently connected to the mainland by a sandy bar (tombolo). Bays are generally small, the most important being the Akkar Bay, on the Syria–Lebanon frontier, and the Haifa Bay in Israel. Transverse faults, often east–west, account for the main features of the coastline. The north Levantine coast runs parallel to the folded coastal mountain range of Jabal Ansarieh in Syria and Mount Lebanon (culminating at 3,083 m) in Lebanon. The mountain range approaches very close to the Mediterranean Sea, mainly in Lebanon where, in the Jounieh Bay, for example, the altitude reaches 600 m less than 1 km from the sea as the crow flies. Nevertheless, high cliffs are uncommon, except in the Jabal Akra in the north of Syria, or at Ras Chekka or Ras Beirut in Lebanon. In the northern Levant, the continental shelf is narrow and broken by very deep and steep canyons, as in the Beirut Bay; in the south, the continental shelf gets wider and wider and its slope is very gentle. The coastal plain is discontinuous and narrow in the north Levant, widening in south Lebanon, between Saida and Tyr, and even more south of the Mount Carmel; getting wider and wider to the south, it reaches a width of 40 km near Gaza. On the other hand, north of Haifa, the coast is

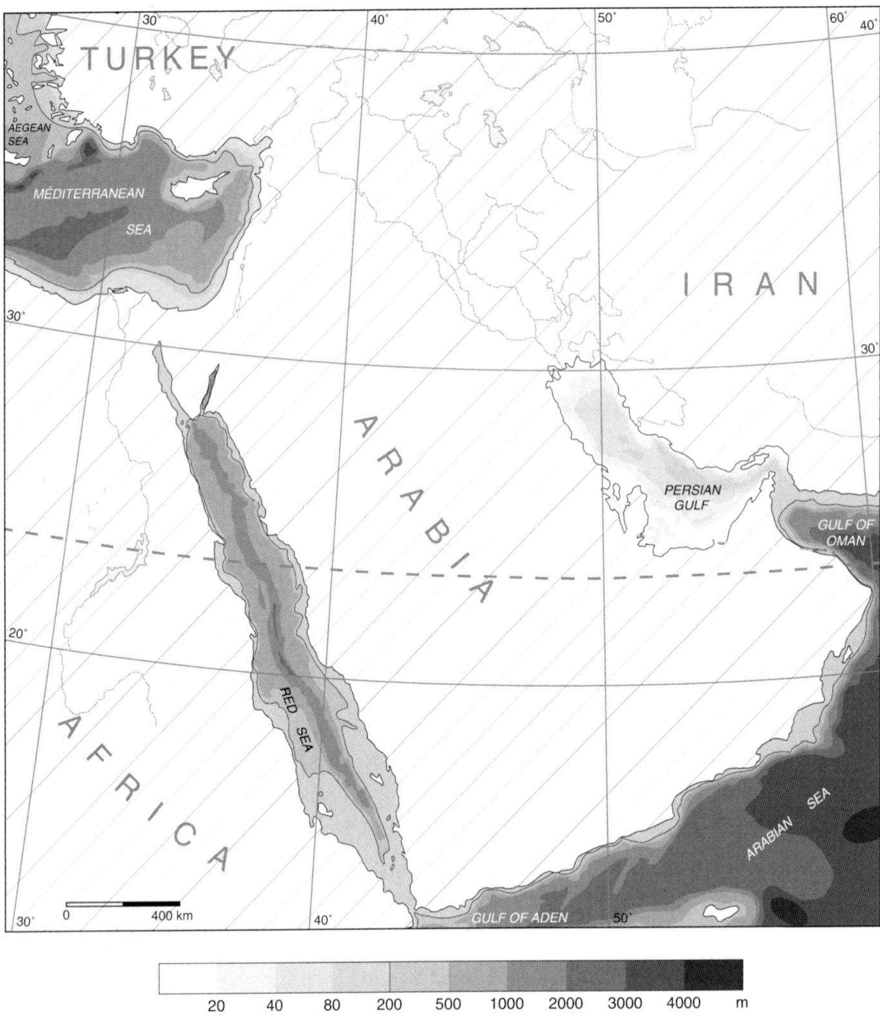

Figure A52 Bathymetry.

frequently rocky, in Cretaceous, Eocene, or Miocene limestones or in Pleistocene eolian or marine sandstones, with an indented outline and numerous creeks or coves.

Cliffs are generally one to some meters high and their base spreads out in a horizontal "trottoir" at mean high sea level. One to three meters wide on limestone, wider on sandstone, this trottoir has been cut into the rock, mainly by biochemical weathering, and is edged with a Vermetid biological construction (the sessil gastropods *Dendropoma petraeum* cemented by the calcareous red alga *Neogoniolithon notarisii* and perforated by Clionid boring sponges and rock-boring pelecypod *Lithophaga lithophaga*), forming an elevated rim buffeted by the fine weather swell. On the surface of the trottoir live *Vermetus triqueter*, a dense cover of *Mytilus galloprovincialis* and *Brachydontes* sp. The intertidal belt mainly shows two species of gastropods, *Patella caerulea* and *Monodonta turbinata*, and, in the upper mid-littoral belt, the barnacle *Chthamalus stellatus* (Safriel, 1974; Por, 1982). On limestone, the present day trottoir is frequently doubled by a fossil one, about +1 m m.s.l. and wider than the present one. This ancient trottoir has been interpreted as representing the occurrence of a rapid uplift movement during the early Byzantine period (Pirazzoli, 1986).

At the creek-heads there are pebble beaches fed by wadis, but the coastal plains are fringed with sandy dunes and beaches. In south Lebanon, the sands, light in color, are mainly of local origin and composed of broken calcium carbonate skeletons of pelecypods (mainly *Glycymeris violacescens*) and calcareous algae. Along the Sharon and Pleshet plains, in Israel, the sands are composed mostly of quartzose grains and heavy minerals mostly coming from the Nile delta. In the Beirut area, the quartzose sands originate in the lower Cretaceous sandstones which outcrop largely in the mountains nearby.

Ancient coastlines

Relicts of Pleistocene *kurkar* (dune calcarenite, called *ramleh* in Arabic) ridges form several parallel alignments in the coastal plain south of Mount Carmel, witnesses of several coastal dunes associated with transgressive coastlines; the waves sometimes attack the westernmost kurkar, cutting it into a cliff preceded by a wide rocky platform (Nir, 1982). The youngest one also occurs as small islands or islets off the coast of Palestine, Lebanon (at Sidon and near Tripoli), and Syria (Arados).

Syria, Lebanon, and Mount Carmel show fine examples of little deformed, but strongly uplifted Pleistocene coastlines up to 300 m above present sea level (Sanlaville, 1977), resembling wide horizontal flat platforms edged with indurated beach sediments, sometimes containing paleolithic artifacts. The last interglacial deposits, between 6 and 12 m, contain a rich Tyrrhenian fauna; at Naame, 15 km south of Beirut, 14 species living on the western coast of Africa but missing today on the Mediterranean coast, were found, such as *Strombus bubonius, Cantharus viverratus, Conus testudinarius, Arca afra, Cymatium costatum*, etc. (Fleisch *et al.*, 1981). Holocene marine deposits are also frequent, indicating recent oscillations of the sea level or tectonic movements up to 2 m, but the aggradation beaches were often clearly in phase with fluviatile discharges controlled by climatic oscillations or anthropic factors (Sanlaville *et al.*, 1997).

Erosion and pollution

Some decades ago, beach rock occurred here and there along the Levantine coast but it is now widespread all along the coast, in relation

Figure A53 Sea surface salinity (‰ in August).

to strong beach erosion which results in the disappearance of the beach sand and the exhumation of an ancient slightly consolidated buried beach, then strongly indurated when exhumed. Everywhere the coast is retreating, sometimes rapidly, and beach sediments and dune sands disappear quickly. This coast is also badly polluted and the landscapes are greatly altered by sea walls, breakwaters, harbors, and all sorts of concrete buildings. Because of the westerly swell, the presence of slabs of beach rock at the lower edge of beaches or the presence of deep and sharp lapiaz on the rocky coastal areas, and severe pollution, this coast is not as attractive as it was, despite good climatic conditions.

The Red Sea coast

Structure, climate, and marine hydrology

The Red Sea occupies a complex rift system between the African and Arabian plates, when separation occurred in the mid-Tertiary times. The Red Sea opened into the Indian Ocean in Miocene times and remained open into the Mediterranean until the early Quaternary. The beginning graben event reactivated erosion and due to the penetration of the sea, carbonates and evaporites were deposited, with coarse–clastic intercalations. With the closure of the Mediterranean connection and tectonic and eustatic changes affecting the Bab al Mandeb narrows, the Red Sea became isolated during the later Quaternary and, because of intense evaporation, greatly diminished in volume with a dramatic impact on shore-zone morphology and ecology.

Connection between the waters of the Red Sea and the Indian Ocean is hindered by a shallow submarine sill in the Bab el Mandeb narrows; the narrows are only 20 km wide and near the Hanish Island, the sill does not exceed −137 m. In the north, the Suez Canal established an artificial but very weak contact with the Mediterranean Sea. The climate of the area is hot and arid. In February, the temperature in the coastal plain (Tihama) varies from 12°C in the north to over 26°C in the south (23.5°C in Jeddah), and in August it is over 30°C (Jeddah, 32.4°C). The Tihama receives less than 50 mm of rain. North of Jeddah, the rainfall does not reach 100 mm in the mountains (occurring in winter and spring), but it is more important in the south: over 400 mm in the Asir and over 600 mm in Yemen (in summer and spring) where mountains are very high. But water is lost before reaching the sea, so the Red Sea does not receive freshwater, since evaporation is very high (more than 2 m per year). The shortage must be supplied by the Indian Ocean through the Bab al Mandeb, where three currents are superposed in summer and two in winter as a result of water density and seasonal winds. So the Red Sea is warm and very salty. Surface sea water temperature stands between 28°C and 31°C in the northern part in August (under 28°C in the Aqaba Gulf) and rises above 31°C in the southern part. In February, the surface sea water temperature rises from the north (less than 20°C in the Aqaba Gulf) to the south (26°C). Salinity is very high in summer, increasing regularly from the south (37‰) to the north (over 41‰ in the Gulf of Suez). Sea water temperature and transparency are very favorable to coral reefs and the mangal vegetation is well developed, dominated by *Avicennia marina*; along the southern Red Sea coast there is also found *Rhizophora mucronata, Bruguiera gymnorrhiza*, and *Ceriops tagal* (Glover, 1998). Winds are predominantly from the NW all over the Red Sea in summer; in winter, NW winds occur in the northern part of the Red Sea but SE winds blow in the southern part. The marine currents are oriented north–south in summer and south–north in winter. Tidal range is insignificant; 0.10 m in Jeddah, but atmospheric pressure and winds can bring about variations of the sea level up to 0.90 m, which is very dangerous for boats due to coral reefs.

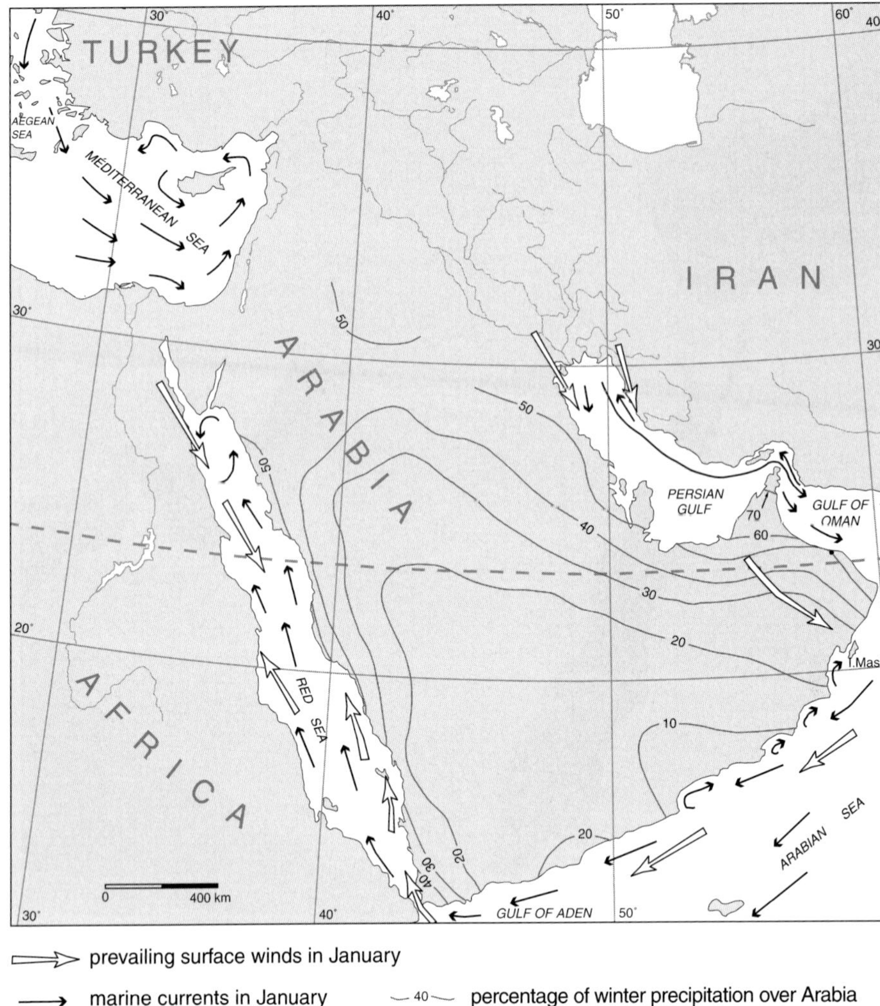

Figure A54 Winds and currents in January.

Coastal morphology

The landscape of the region is predominantly mountainous; the highest peaks are often in excess of 2,500 m, mainly in the south, in Asir or Yemen where mountains reach a height of over 3,000 m and up to 3,760 m at the Jabal Nabi Shuaib, in Yemen. In places, the strongly dissected mountains come right down to the sea. But generally a flat plain, 5–40 km wide, constitutes an almost 2,000 km long marginal corridor, called the Tihama, following the border of the Arabian Mountains. Plutonic, volcanic, and metamorphic rocks outcrop in the mountains. In the Tihama, the basement is generally hidden below Tertiary and Quaternary rocks. Tertiary sediments, which can reach a thickness of over 2,000 m, generally present a division into three series: Oligocene, Lower, and Middle Miocene limestones and conglomerates, Middle and Upper Miocene evaporites, and Upper Miocene–Pliocene limestones and clastic red series. Frequently, Tertiary rocks are exposed only at the edge of the basement and in a few places in the coastal plain where mainly Quaternary deposits outcrop, constituted of gravel terraces, extensive alluvial cones and uplifted coral reefs.

Coming out of the mountains, the alluvial surfaces fall relatively steeply to the coast. They can be recognized by their covering of pebbles and gravels with particularly dark desert varnish. Upstream, near the basement, a terrace consists of a thick sequence of coarse gravels and downtream mainly of fine sediments. There are generally several stepped alluvial terraces, the oldest characterized by coarse pebbles and a very dark patina, the youngest by a sandy to gravelly material and a lighter color. Relatively coarse windblown sands often occur in the upstream part of the coastal plain. Dating the fluvial terraces is rather difficult, but relative dating is sometimes possible. Between Umm Lajj and Yanbu, the basalts from the Jabal Nabah, K/Ar dated 1.4 ± 0.6 My, flowed into wadis that were already cut into the oldest terrace, which might be Lower Quaternary or Upper Pliocene and show evidence of vertical block movements; the formation of the middle terrace antidates the Umm Lajj basalts (0.4 ± 0.2 My). The age of the youngest gravel terrace can be estimated, due thanks to its direct connection to the 6–8 m marine terrace, which belongs to the last interglacial (Hötzl *et al.*, 1984).

Ancient coral reefs outcrop in the lower part of the coastal plain, in places slightly covered with gravels. The main reef body, very flat, is the 6–8 m marine terrace. Toward the sea, the main reef itself outcrops and, behind it, increasingly lagoonal facies. This ancient reef can be attributed to the last interglacial (isotopic stage 5e) and exhibits no real deformation. An older one, faulted, is probably 250 ky old (isotopic stage 7). A Holocene erosion step at 1–2 m above m.s.l., only a few meters wide, is carved into the edges of the 6–8 m reef, even on its landward side. The reef terraces form striking isolated erosional remnants and more or less broad table-like crests. Depressions and channels developed behind the reefs, which, subsequently, were sometimes occupied by bays (Sharm) when the sea level rose later. The best example is Sharm Yanbu with its narrow deep channel and its two branches representing the continuation of two wadi channels. The protective reef plate kept this crest from being eroded, while the loose material behind it was excavated. Such back-coast depressions recur repeatedly. These bays are the result of the last drop in sea level in the latest Pleistocene and the later increase in the Holocene. On the eastern edge of the reef crest, an inter-fingering of coral and beach formations with clastic continental sediments can often be seen. This interfingering points to the existence of a humid climate during the interglacial period. Carbon-14 dated remnants of trees indicate heavier-than-today precipitation during the so-called "Neolithic Pluvial."

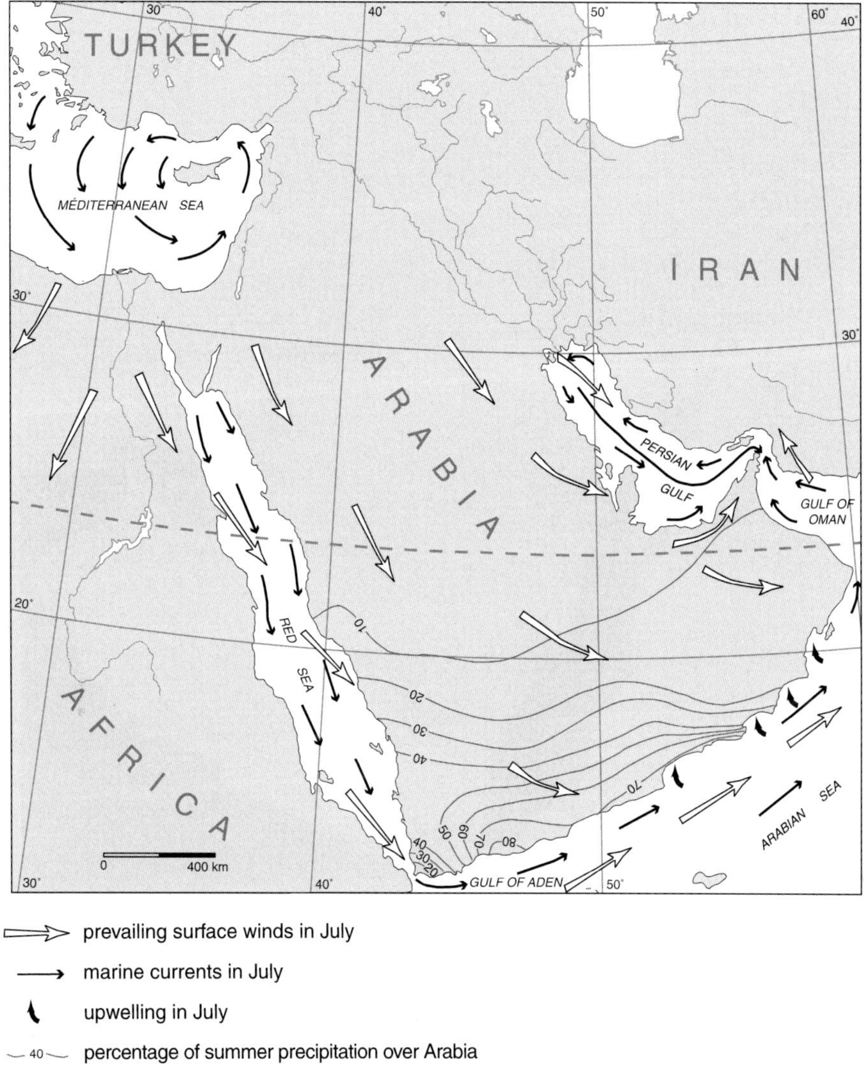

Figure A55 Winds and currents in July.

Coral reefs

Coral reefs are mainly of the fringing type; fringing reefs in a very nearshore position in front of steep cliffs, or fringing reefs with a lagoon within shallow water. They grow close to the mainland and are absent from wadi mouths. Barrier reefs and even atolls also occur, mainly in the central and southern part of the Red Sea. The outlines of the reef crest and the orientation of the foreslopes follow the tectonic pattern. Spectacular drop-offs are widespread and represent predominantly horst and graben structures parallel to the rift.

In the Aqaba Gulf, the fringing reefs exhibit a characteristic biozonation (Dullo and Montaggioni, 1998): (1) a local and poorly developed beach; (2) the back-reef zone, a sandy depression between the beach and the reef-flat, with a maximum width of 50 m and a depth of 1–2 m, the bottom sediment is colonized by scattered coral heads (*Stylophora*, *Seriatopora, Platygyra, Millepora*), alcyonarian colonies (*Lithophyton, Cladiella*, *Sinularia*) and seagrass beds (*Halophila, Halodule*); (3) the reef-flat zone, about 20 m wide, this is a dead coral pavement bordered by a 1 m rear-reef step. Coral communities are composed of small-sized colonies (*Stylophora*, *Seriatopora, Acropora*) associated with hydrocorals (*Millepora*). The reef front forms a nearly vertical drop-off, 2–4 m high, with *Millepora* and branching red algae; (4) the outer slope zone, with loose sedimentary slopes (sandy talus and patches of *Halophila*) or coral-reef slopes, with their upper parts dominated by branching growth forms (*Stylophora*, *Seriatopora*, *Acropora*, *Echinopora*). Along the coast of Arabia, the fringing reef constitutes an almost continuous belt and often has wide back-reef zones.

On the central and southern coast of Arabia, coral reefs also occur as offshore knoll reef platforms. In the vicinity of Jeddah, about 3 km west of the coastline, the offshore reef consists of a shallow platform covering an area of 800 km^2, bounded both shorewards and seawards by a vertical escarpment dropping rapidly to depths of 400–800 m, marking the edge of the Red Sea trough. The platform is occupied by scattered reefs separating sandy bottoms. The present reefs come from Holocene reefs settled on Pleistocene reef limestones. The coral community of the innermost part is dominated by *Stylophora pistillata*. Coralline algae are represented by massive branching and crustose forms (*Lithothamnium, Spongites*). The upper parts of the steeply reef flanks are settled by *Millopora*.

In the southern part of the Red Sea, where the sea reaches its maximum width of 360 km, remarkably flat shoals (called *shab*) occur. The best example is the Farasan bank which extends from 16°N to 22°N. Lat. and reaches a width of up to 120 km. The sea is almost always less than 100 m deep and often only a few meters deep. The two main islands, which were very important for pearling at the beginning of the 20th century, are 60 km long and 8 km wide (al Kabir) and 35 km long and 10 km wide (Sajid). These islands are flat and low (under 20 m), except in a very few places where heights of up to 75 m are to be found, in domed eminences where gypsum and anhydrite outcrop as a result of halokinesis (Dabbagh *et al*., 1984).

Regional features

The Gulf of Aqaba is the continuation of the Red Sea rift, formed as a consequence of transform movements along the Aqaba–Levant structure. This narrow gulf slopes steeply to depths of over 1,800 m. In the north of the Red Sea, the coastline is very straight, free from larger indentations, with a trend of 140/150°, identical to that of the entire

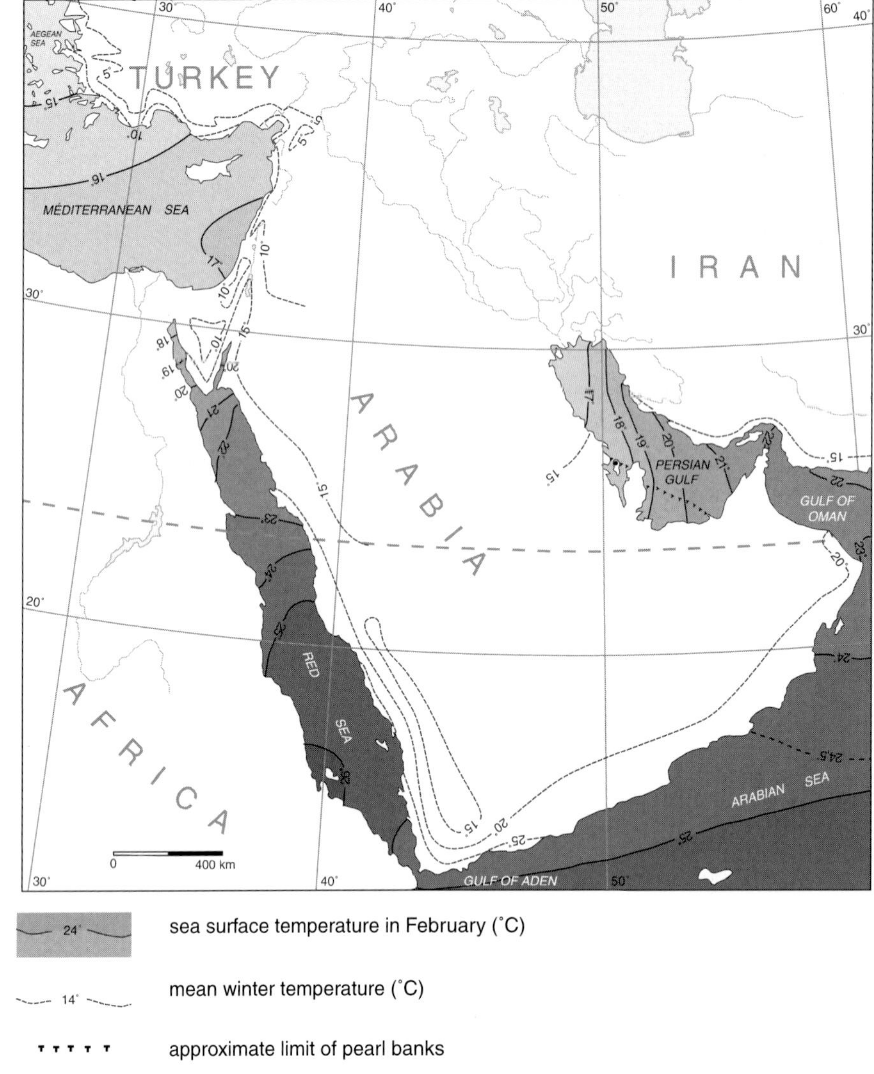

Figure A56 Mean winter temperatures.

graben. Sigmoid curves are typical of the central and southern parts of the Red Sea coast. The 200 m isobath shows significant deviations. It is seldom parallel to the coast, underwater ridges, and troughs repeatedly show the conjugated 70° course indicated by transform faults. The main wadis are generally related to young NE striking faults and block-faults.

The Wadi al Hamdh, with a catchment extending almost to al Madinah, is the most important drainage system in the Hejaz Mountains. Its load of sediment considerably contributed to the creation of a broad aggradation plain south of al Wajh. Even under the present arid conditions, its water, enriched with mud and rubble, regularly reaches the coastal plain, and sometimes the sea. A continuous reef belt has developed some 30 km away from the coast (Al-Sayari *et al.*, 1984).

Because of a wide break in the recent coral reef along the coast, Jeddah is one of the few places on the Red Sea coast where a large harbor could be located. Isobaths show an E–W channel from Wadi Fatimah into the Red Sea. The upper course of Wadi Fatimah extends into the massive basalt flows of the Harrat Rahat and the remaining catchment is made up mainly of Precambrian rocks. The middle course of Wadi Fatimah is to be considered as a fault-bounded graben. The presence of plutonic and metamorphic rocks limits the potential aquifers to the alluvions of the Wadi Fatimah which is an early source of water for the Jeddah city, as the ruins of a qanat(conduit) system show (Hacker *et al.*, 1984).

The plain of Jizan forms a narrow NW-oriented coastal strip. The strongly dissected Asir and Yemen high mountains overlook the whole area. At their base, parallel foothills, consisting of metamorphic schists or volcanic rocks reach heights of between 200 and 700 m. The plain itself averages about 40 km in width and is composed of alluvial deposits, except for some volcanic cones and the salt dome (diapir) of Jizan (Müller, 1984). Several important intermittent wadis flow across the plain, fed by the summer and spring rain falling in the mountains, but only their upper courses are perennial. Many volcanic intrusions, which occurred during the development of the Red Sea rift, are characteristic of the plain. Given the good state of preservation of some cinder cones, volcanic activity has continued up to recent geologic times. The coast is preceded by a reef over almost its entire length. Immediately adjacent to the coast is a sabkhah zone, some kilometers wide, consisting of silty, clayey sediments with high salt content. On its eastern border, this zone passes into the alluvial accumulation of the middle terrace (Purser and Bosence, 1998).

The Persian Gulf coast

Hydrology, climate, and structure

The Persian Gulf has an average depth of 35 m and a maximum depth of 100 m near its narrow entrance, the Strait of Hormuz, 60 km wide. Tidal range varies from 1 m at Dubai, 2.5 m at Bahrain, and 3.5 m at Kuwait to a maximum of 5 m in the Clarence Strait, between Qeshm Island and the mainland, and the tidal range averages ca. 1 m at Basra, 150 km upstream on the Shatt al Arab. Regional tidal currents are oriented approximately parallel to the axis of the Gulf. Tidal velocity may exceed 60 cm/s, their bidirectional movements favoring the development of spectacular oolitic tidal deltas (Purser and Seibold, 1973).

The considerable loss of water in the Gulf through evaporation is not compensated by precipitation and river inflow, the only notable inflow of fresh water coming from the Shatt al Arab, in fact mainly from the Karun (ca. 6,000 m^3/s). As a result, compensatory waters come from

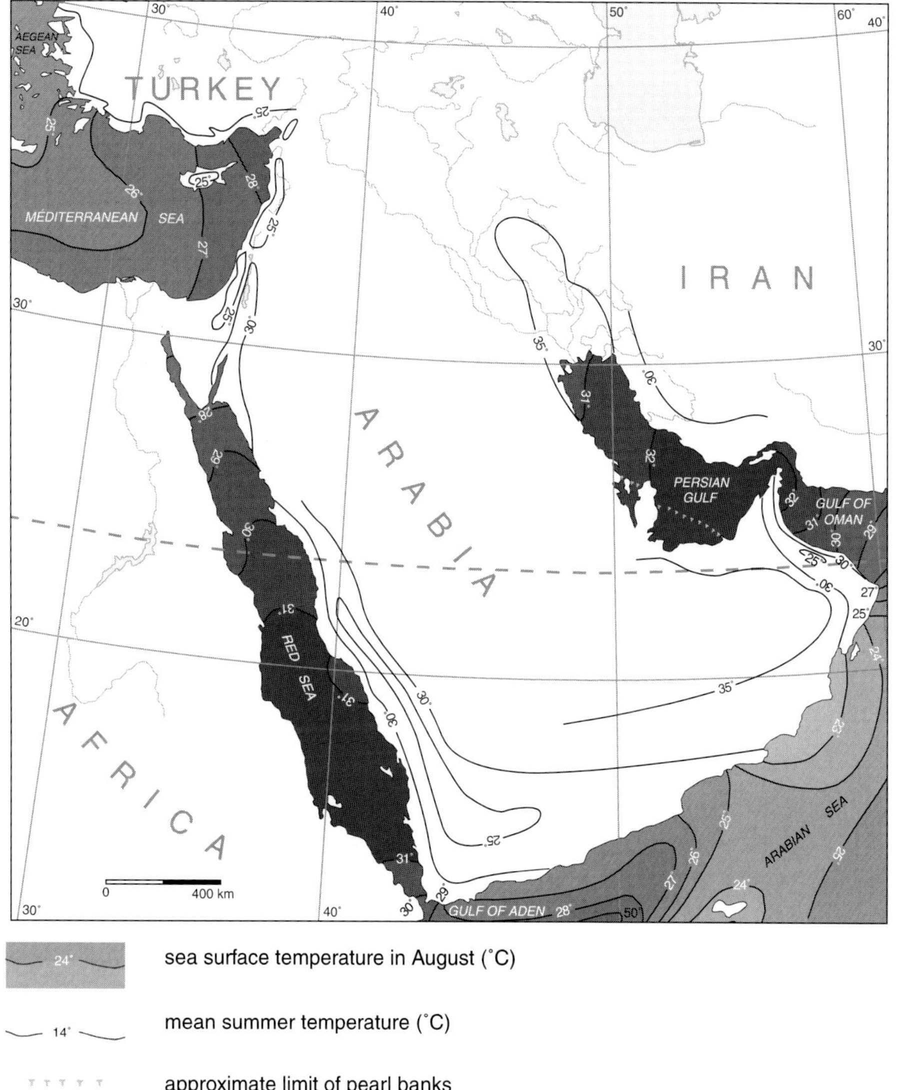

Figure A57 Mean summer temperatures.

the Arabian Sea, with surface current moving anticlockwise along the Iranian coast and coming down the Arabian coast, determining the distribution of salinity, temperatures, and nutrients. In summer, surface water attains temperatures over 32°C in the central part but very much higher in the Arabian lagoons. Temperatures in winter fall below 20°C in the northern part of the Gulf and even under 18°C north of Bahrain, where corals and mangrove cannot live (the absolute minimum is −5°C at Abadan). The range of variation in water temperature tends to increase away from the entrance of the Persian Gulf. The salinity of surface waters is high along the northern Iranian coast (39‰–40‰) but diminishes down to 30‰ west of the mouth of the Shatt al Arab, increasing then to the south along the Arabian coast (more than 38‰ east of Qatar and even 60‰ in the Gulf of Salwa).

The climate of the Persian Gulf area is hot and semiarid to arid. Mean monthly temperatures increase from 14°C at Kuwait to 18°C at Abu Dhabi in January and stand ca. 34°C in July all along the Arabian coast. Extreme temperatures are cool in winter and torrid in summer, the maximum absolute temperature reaches 49.2°C at Sharjah and 52°C at Abqaiq. Mean annual rainfall is less than 100 mm on the Arabian coast, slightly higher on the Lower Mesopotamian Plain and on the Iranian coast (ranging from 100 to 200 mm with a maximum of 230 mm at Bushire), rains occurring mainly in winter with a second peak in spring. The prevailing "Shamal" wind blows down the axis of the Gulf from the NW, then from the north to the south. Summer months are essentially calm, but during the winter the Shamal may be very strong. Mangrove stands are found at intervals along the Persian Gulf and *Avicennia* is the single species because it is the most resistant to low and high temperatures and to an intense evaporation resulting in high soil salinities. The northernmost site is in Bahrain and the greatest areas occur in the United Arab Emirates and in the Clarence Strait, in Iran, but the mangrove environment was richer and more extensive during the Middle Holocene and *Terebralia palustris*, frequent at that time, is now extinct (Glover, 1998).

The small but numerous streams draining the Zagros ranges are characterized by rough flash-floods and deliver a significant amount of terrigenous sediments. In contrast, the Arabian side of the basin is completely lacking in fluviatile influx so almost pure carbonate sediments predominate along the Arabian coastline. Oolitic sediments are forming both in exposed and sheltered environments and are particularly abundant in the southern part of the Gulf.

From the peak of the last glaciation until about 14,000 yr BP, the Gulf was free of marine influence. By 14,000 yr BP the Strait of Hormuz had opened up as a narrow waterway and by about 12,500 yr ago marine incursion into the central basin had started. The present coastline was reached shortly before 6,000 yr BP but the relative sea level then rose 1–2 m above its present level (Lambeck, 1996).

The difference in slope between the two flanks of the Gulf reflects the fundamentally different tectonic histories of Iran and Arabia. Its elongate axis separates two distinct geological provinces, the stable Arabian foreland and the unstable Iranian fold belt. The foundations of the modern Persian Gulf were largely laid in the Plio-Pleistocene Zagros orogeny.

The collision of the Arabian and Euro-Asian plates resulted in the formation both of a broad structural depression, about 2,000 km long, which extended from Syria to Hormuz and which eventually became an extensive sedimentary basin, and of the Zagros Mountains consisting in anticlines hugely folded with steeply dipping flanks, characterized by a

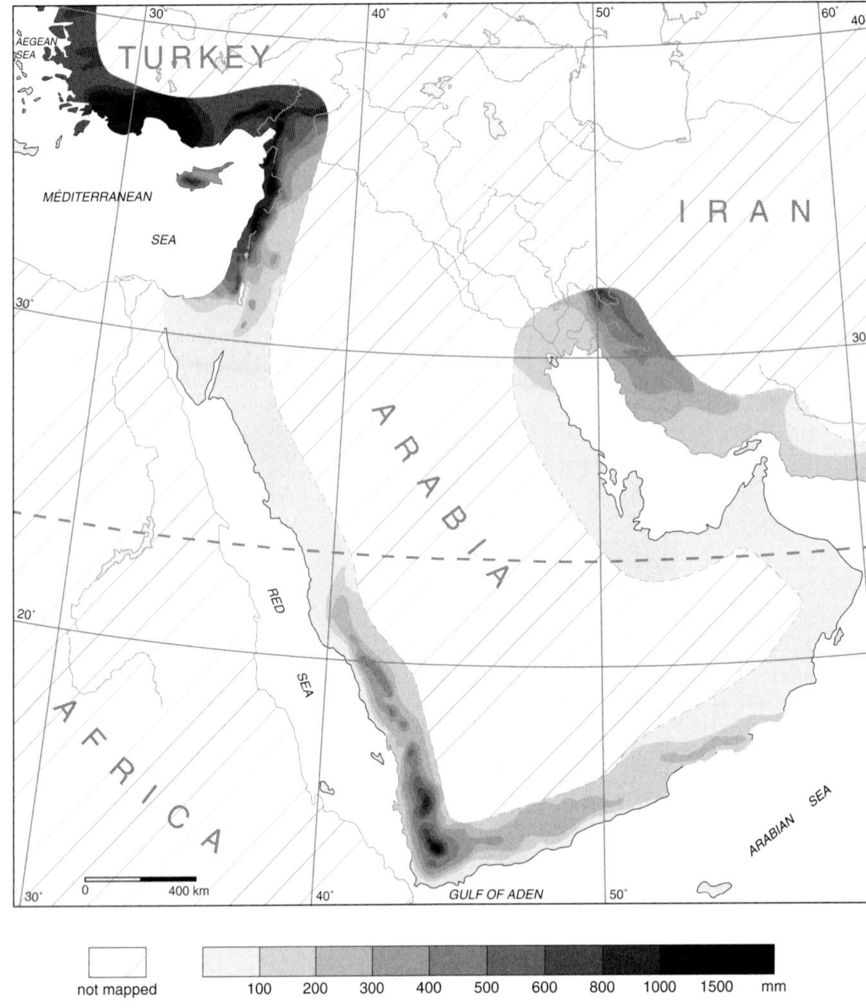

Figure A58 Mean annual precipitation in coastal areas.

NW–SE oriented belt, with mainly Late Cenozoic sedimentary formations in which clays, silts, and carbonates are predominant. The tectonic instability results in frequent earthquakes.

The low-lying Arabian coast and adjacent shallow seafloor are characterized by large, low-dipping anticlines with N–S or NE–SW trends, for example, in Bahrain and Qatar. In the Persian Gulf there are some 20 islands, as well as numerous submarine highs or shoals. Most of the islands and submarine shoals are clearly due to salt diapirism which, affecting the early Palaeozoic Hormuz salt series, was very active during the Plio-Pleistocene period and down to recent times (Kassler, 1973). The NW–SE Arabian coastline and the straight coastline from Abu Dhabi to Ras al-Khaimah may be partly controlled by faulting.

Regional features

The Shatt al Arab deltaic system. The extreme north of the Gulf corresponds with the deltaic system of the Shatt al Arab, the combined delta of the Euphrates, Tigris, and Karun. At the latitude of Basra, the two wide alluvial fans of Wadi Batin (Pleistocene) and Karun River (Holocene) restrict the width of the Mesopotamian plain, being responsible for the upstream development of extensive palustrine environments (Hammar Lake and marshes), where the Euphrates and Tigris deliver most of their water and suspended sediment. After loss of their suspended sediment in the marshes and shallow lakes, the clear waters are drained to the Shatt al Arab (Purser *et al.*, 1982). In fact, much of the fluviatile waters and suspended sediment of the Shatt al Arab come from the Karun River. South of the two big alluvial fans, the Shatt al Arab forms a major deltaic system, dominated by strong tidal conditions, with large sebkhas flanking the waterways, wide tidal areas, sebkhas, marshes, and islands (such as the large Bubiyan Island). During the latter period of the flooding of the Gulf, in the mid-Holocene, much of these areas was likely to have been a shallow marine– lagoonal environment when sea level rose 1 or 2 m above its present level (Sanlaville, 1989, 2000).

The Arabian coast is low-lying, with only a slight relief corresponding to the broad gentle anticlines that trend N–S or NE–SW. South of the large Bay of Kuwait and down to the Gulf of Bahrain, the coast is low and almost rectilinear, with rocky promontories, wide sandy beaches, and extensive silted swamps. The sandy point of Ras Tanurah extends SSE for about 10 km. The Dammam Peninsula is related to an Eocene limestone anticline, as is Bahrain Island, rising to 122 m at Jabal Dukhan. The Gulf of Bahrain between Bahrain and the mainland is very shallow with extensive submerged spits and bars separated by long channels, while the Bahrain archipelago is surrounded by wide, partly coralline tidal flats, which are largely exposed at low tide. Near Bu Ashira, in Bahrain, a mangrove swamp is formed of thick but not very deep mud and small mangrove bushes where the fauna consists mostly of gastropods, especially of the Potamididae and Planaxidae families (Smythe, 1972). In the Qatar Peninsula, Eocene dolomite and limestone form massive and continuous outcrops forming cliffs separated from the sea by a strip of sebkhas, several kilometers wide. At the maximum of the Holocene transgression, the cliffs constituted an irregular coastline. After stabilization of the present sea level, fine sediment began to fill and regularize the embayments, creating hook-shaped spits at their southern ends and wide supratidal flats eventually formed on a complicated tidal channel system. Today, unusually strong easterly winds combine with high spring tides to flood the sebkhas as far inland as the ancient cliff. After a few weeks, evaporation of the flood water produces crusts of salt, gypsum, and dried algal mats which are progressively blown away (Shinn, 1973).

Most of the beaches are composed of carbonate sand and the bulk of this material is of biological origin, made up of small fragments of shells and corals. This beach complex includes subtidal and supratidal zones. On the eastern coast of Saudi Arabia, tidal flats occupy 30–40% in space of the numerous large or small bays along the coastline. They consist of mud and very fine sand deposited in bays and sheltered zones where wave energy is low and also of rock flats composed of mud, sands, and shell fragments cemented into beachrock known locally as *farush* (Basson *et al.*, 1977).

The mud flats generally consist of very fine, silty sediments which can be divided into several zones:

1. The marsh grass zone with plants such as *Phragmites communis, Aeluropus lagopoides, Bienertia cycloptera.*
2. The halophyte zone covered at high tide by only a few centimeters of water. It is the uppermost portion of the true intertidal region covered with *Arthrocnemon macrostachyum* and *Halocnemon strobilaceum*, with a far-stretching root system. In the mud surface there are sometimes to be seen the round holes of the crab *Cleistoma dotilliforme.*
3. The mangrove zone with *A. marina*, with trees 1 or 2 m high, with their characteristic pneumatophores.
4. The algal mat zone composed of cemented sediment with blue-green algae and diatoms. Microscopic animals live in this mat, including gastropods, ostracods, nematods, flatworms, copepods, and oligochaete worms.
5. The *Macrophtalmus* zone between the mangrove zone and the low-tide level. These *crabs (Macrophtalmus depressus* and *Macrophtalmus grandidieri*) live in a burrow in the mud.
6. The Cerithidae zone whose density can reach 2,100 individuals per square meter, with sometimes *Murex kuesterianus.*
7. Tidal creeks or drainage channels, dug by ebb tide, are a characteristic feature of the muddy tidal flats, with the presence of tide pools, about 2 or 3 m wide, 50 cm deep, and 10–30 m long, subject to great fluctuations of temperature and salinity. These tidal creeks are inhabited by a great variety of animals (shrimps, fishes, swimming crabs), which find there a shelter from the predators and rich food.
8. The sand flats frequently have a grayish color due to the presence of organic matter. At low tide, one can observe ripple marks produced by the ebbing water. Often, sands constitute a pellicular layer overlying the rock or a beach rock. The tidal channels are not strongly marked in that zone, where many crabs live, such as *Ocypode saratan, M. depressus, M. grandidieri Scopimera scabricauda.*
9. Rock flats also exist, for example, in the Gulf of Salwah, in bays where beach rock find exceptional conditions of formation. The *farush* consists of broken shells, sand, and mud cemented together. This type of environment presents very great biodiversity with polychaete worms, gastropods (*Thais* sp., *Monilea* sp., *Trochus* sp., *Littorina* sp.) pelecypods (*Pinctada* sp., *Isognomon* sp., *Anomia* sp., *Barabatia* sp.) and decapod Crustacea (*Alpheus* sp., *Pilumnus* sp., *Xantho* sp.), and also algae such as *Enteromorpha* sp., *Rhizoclonium kochianum, Achnanthes* sp.

Middle Holocene beach sediments are to be found 1 or 2 m above the present sea level, indicating a higher sea level than today. Pleistocene marine calcarenites are also known in that area where some paleo-wadis (W. Ar Rimah, W. As Sahba, and W. Ad Dawasir) have built gravel fans of enormous size.

The relatively linear character of the northern Arabian coast is modified by the Qatar Peninsula which strongly influences the marine currents and patterns of sedimentation on the SE side of the Persian Gulf. To the east of the Qatar Peninsula there is a broad, shallow area, 10–20 m deep, studded with numerous shoals and salt dome islands. An irregular bathymetric ridge, the *Great Pearl Bank Barrier*, extends eastwards from Qatar along the central part of the United Arab Emirates coast, whose coastline is characterized mainly by low, evaporitic, supratidal flats, which locally attain 10 km in width, and by storm beaches in more exposed settings.

The western and central parts of the Trucial coast are protected laterally by Qatar Peninsula and the Great Pearl Bank barrier. The coastline is characterized by a complex of islands and peninsulas. Each island is growing by accretion around a Pleistocene rock core. Accretion also occurs as tails of sediments, which locally form tombolos. The inter-island channels end seawards in spectacular oolite deltas. Small swamps, colonized by the black mangrove *A. marina* and other halophytic plants, have developed, favoring the deposition of carbonate muds. The shallowness of the lagoons, together with the protection provided by the coastal barrier complex has led to a very active intertidal flat accretion which has produced the wide sabkhah plain, approximately 5 km wide, during the last 4,000 years. The high salinity and temperature of the water, together with the extreme aridity, have led to the extensive development of dolomite and other evaporite minerals (mainly gypsum and anhydrite) within the supratidal sediments (Evans *et al.*, 1969; Purser and Evans, 1973). As the lagoons become completely filled, aeolian coastal sands may transgress over them.

The northern part of the coast is unprotected and direcly faces the entire length of the Gulf. It suffers from the effects of maximum wave fetch and a strong eastward longshore transport, which resulted in the development of storm beaches backed by coastal dunes mainly composed of skeletal carbonate sands, and the construction of major spit systems; between Sharjah and Ras al Khaimah, a series of subparallel spits has prograded the coastline some 5–10 km seaward, bringing vast lagoons into existence, the biggest being the Umm al Qowayn lagoon, which shelters vast mangroves. Further inland, the extensive quartzose dune fields of the Arabian desert extend up to the alluvial fans which skirt the Oman mountains. The NE section of the Trucial coast end in the cliffs and rocky shorelines of the Musandam Peninsula. They are mainly composed of limestone and are drained by a series of deeply incised wadis which terminate in spectacular alluvial fans, some of which reaching the coast.

The northern Persian Gulf coast is mountainous, with ridges up to 1500 m high formed by hugely folded, anticlines but the intensity of the folding diminishes markedly seawards where Pliocene folding produced regularly spaced elongate synclines and anticlines. In the south many anticlines are cut by salt diapirs. The chain of islands in the Gulf (Kharg, Shuaib, Qeshm, etc.) forms part of the Zagros foothill belt. The Zagros is seismically active, earthquakes are quite frequent and much detrital sediment is deposited in the form of alluvial fans delivered through short, ephemeral streams perpendicular to the main structural axis.

North of Bushehr, the coast is at first swampy, then low and sandy. South of Bushehr, the Iranian shores are essentially linear and rocky, the mountains rising abruptly above the sea, with narrow coastal plains associated with the estuaries of numerous small rivers flowing down the Zagros mountains. The coastal valleys yield evidence of two distinct fills separated by a phase of incision. According to artifacts, the deposition of the older ceased by 6,000 BC. The process of the younger began about 1,250 years ago. Most of the streams are now incising their channels (Vita-Finzi, 1978).

The Mehran River, oriented parallel to the Gulf, is situated along the axis of a syncline and has developed a big marine alluvial-fan delta, prograding into the Clarence Strait, between the mainland and Qeshm Island. Its very flat fan is dominated by silty and sandy deposits covered upstream by scattered vegetation of *Chenopodiaceae* and *Graminae*. Downstream there are mud banks with large desiccation polygons, the shores of the strait and of the smallest tidal creeks being populated by a dense belt of mangroves (*Avicennia*). The mud banks area, 15 km wide, includes a series of extensive, mostly bare tidal flats with a characteristic network of tidal channels (Baltzer and Purser, 1990).

The Persian Gulf area is the foremost producer and exporter of oil in the world and is now experiencing huge transformations. The landscapes have been strongly altered by the creation of harbors, channels, and artificial islands, the filling of lagoons and bays, and the construction of roads and towns. Oil pollution is a great danger and the equilibrium of the sandy shores is threatened.

The Arabian Sea coasts

Climate and hydrology

The coastline of the Arabian Sea, along the southern coast of Arabia, the Gulf of Oman and the Iranian Makran, is very hot and arid. Mean temperature in January is >20°C (22.8°C at Salalah and 25.6°C at Aden) and the maximum absolute temperature in August is >45°C (46.7°C at Mascate and 47.2°C at Masirah). Mean annual rainfall is <100 mm (38 mm at Aden, 94 mm at Masirah) and the rainfall variability is extremely high. Winter is the main rainfall period, except on the coast of Hadramaut and Dhofar where rainfall occurs mainly in summer. Precipitation over the coastal area of Salalah is usually in the form of a fine drizzle with daily totals seldom exceeding 5 mm but the presence of fog (on average 54 days per year) has a marked effect on vegetation growth.

As a whole, the Arabian Sea is very deep with a rather narrow continental shelf. Coral reefs exist intermittently where the continental shelf is wider. The tidal range is generally between 2 and 4 m. All during the year, the coast is buffeted by the heavy Indian Ocean swell. Winds and currents are parallel to the coastline in accordance with the monsoon circulation. The NE monsoon begins in October but is mainly characteristic between November and March, with a maximum in December and January. Winds are moderate and the sea propitious to navigation. The SW monsoon is longer and stronger. It begins in April and is very strong from June to September. Then the sea is very rough and navigation is dangerous so the fishers do not leave the harbors. Strong southwesterly winds blow along the southern coast of the Arabian Peninsula, forcing the ocean surface in a northeasterly direction and causing a compensatory upwelling of cold subsurface water. Along the Dhofar coast, between Ras Fartak and Masirah, sea surface temperature can be up to 5°C cooler than the ambient offshore values. These cold waters are rich in nutrients that trigger phytoplankton blooms and make for a great wealth of fish, 70% of the biological sediment is deposited during the summer monsoon months. Tropical storms and cyclones occur from time to time, almost entirely confined to May–June and October–November and cross the coast of Yemen and mainly Oman, generally between Salalah and Masirah, about once every three years. They occur more frequently along the Makran coast but are fewer in the Gulf of Oman. Well-developed mangrove stands are found at intervals along the Arabian Sea shore, mainly in the bays or lagoons.

Regional features

The southern coast of Arabia is more than 2,600 km long between the Bab el-Mandeb and Rass el-Hadd. It corresponds to a young rift border; its structure is very complex and the coast is often bordered with high mountains. Volcanic rocks outcrop here and there, particularly in the western part where they appear as islands and promontories, for instance in the Perim Island, at the entrance of the Red Sea, in the Aden district where extinct volcanoes are connected to the continent by tombolos, or in the Shuqra area. The Old Aden settled in a crater whose rim reaches the height of 551 m at the Jabal Shamsan. Narrow plains bordered by sandy beaches alternate with high rocky cliffs. At Ras Fartak, vertical cliffs reach up to 580 m high. About 450 km to the south, Socotra is a barren high (1,519 m) island of limestone and granite, 112 km long and 15 km wide with generally steep coasts and reef-fringed promontories. In eastern Dhofar, the Kurya Muria archipelago is edged with fringing reefs. The eastern part of the southern Arabian coast is low, sandy, and barren, notably in the Wahiba Sands area, in Oman. This coastal area belongs to the Sudanian vegetation region, well differentiated from the adjacent Sahara–Arabian region by hundreds of genera and thousands of species that are not to be found in the Holoarctis, especially different species of Acacia, mainly *Acacia tortilis, Capparis decidua, Calotropis procera, Panicum turgidum, Indigofera spinosa*, etc. (Zohary, 1973).

The Arabian coast of the Gulf of Oman, between Ras al-Hadd and the Strait of Hormuz, in the Musendam Peninsula, sweeps in an arc, some 650 km long, dominated by an isolated folded mountain range, the Jabal al-Hajar, whose highest point rises to 3,035 m. At either end of this range the mountains plunge straight into the sea and the coast is rocky and cliffy, with deep water inlets. In the center of the arc the mountains are separated from the sea by the great Batina Plain, some 280 km long and 20–30 km wide. The coast is sandy with a strip of sand dunes. Behind it, the piedmont plain is covered with silts, gravels, and conglomerate laid down in a now semi-fossilized drainage system. During the Quaternary, the deposition of wadi sediments occurred during the interglacial pluvials, whereas erosional processes predominated during the cooler aridials and numerous sea-level terraces were formed during the interglacials; traces of three coastlines dating from the last interglacial can be found and the Holocene sea level was higher than the present one (Hanns, 1998).

The Makran coast is part of an accretionary wedge of late Cretaceous to Holocene sediments which accumulated near an oceanic subduction margin, between two major strike-slip fault zone, the left-lateral Chaman fault zone to the east and, to the west, the right-lateral Zendan fault, which separates the simply folded Zagros Ranges from the turbidites and ophiolitic melanges of the Inner Makran.

The Makran ranges rise irregularly from the coastal plain in a series of ridges underlain by folded sedimentary rocks. Along the coast, outcrop mainly Pliocene thick, neritic, and monotonous sequences of calcareous mudstones (the so-called *Chatti mudstones* of Pakistan). They have been locally affected by gentle folding but the beds are often sufficiently undeformed to produce extensive and almost horizontal platforms. So, the coast is marked by a series of prominent headlands (Konarak, Chah Bahar, Gavater, etc.) separated by low areas. These rocky headlands rise as isolated flat-topped hills formed from marine sandstone, conglomerate, and coquina of the Ormara and Jiwani Plio-Pleistocene formations. The Konarak terrace, one of the best developed, is an 18 km long and 300–900 m wide platform. Most of the scientists who investigated this area were strongly interested in these platforms which exhibit spectacular stepped shore platforms and raised beaches (Page *et al.*, 1979; Snead, 1993; Reyss *et al.*, 1998). On Qeshm Island as many as 18 marine terraces were identified, up to 220 m in elevation, and as many as 19 levels, up to 246 m, near Chah Bahar. The lowest levels, less than 4 m high, were 14C dated as Holocene; for higher levels, up to 26 m in elevation. Uranium series analyses gave apparent ages between 100 and 140 ka BP, and therefore may be ascribed to oxygen-isotopic stage 5e (or 5c). These dates show that uplift rates did not exceed 0.2 mm/year during the late Quaternary, except in areas where salt domes exist, for example, on Qeshm Island where uplift rate was faster (6 mm/year). Deposits of the oldest levels are not suitable for dating because the shells and corals are very recrystallized. These marine terraces give evidence of the complex interrelationships of tectonism, coastal erosion and sedimentation, and eustatic sea-level changes (Reyss *et al.*, 1998).

However, the main features of the coastal area could be the low-lying alluvial plains forming a coastal strip, 5 to more than 20 km wide, between the coastline and the mountains (Page *et al.*, 1979). Several stepped terraces partly rocky and partly alluvial exist at the foot of the Makran ranges. The streams have eroded the siltstone and mudstone bedrock into low-lying pediments and built huge deltas. Locally, the plain is covered with sand dunes. Typical mud volcanoes outcrop as well in Makran as on the coastal plain of the Persian Gulf. In the coastal plain there are many shore deposits, series of prograding accretion beach ridges, isolating lagoons and marshes which shelter vast mangroves, before being filled by silt. In some places, ancient islands have been connected to the mainland by one or two spits which finally form a tombolo (Ras Tang, Konarak). So, the 19.3 km-long Gurdim terrace paralleling the coastline is connected with the coastal plain by a wide 12.8 km tombolo, displaying well-developed beach ridges. The Makran coast is unprotected and during the SW monsoon and especially in tropical storms and cyclones, the ocean and surf are extremely rough. Erosion continues to affect the headlands, but elsewhere progradation seems to have dominated along the flat sandy areas since the maximum of the postglacial transgression. Radiocarbon dates for shell samples collected in the central part of the tombolo of Konarak and elsewhere on the ancient beaches of the Makran coast gave ages between 4870 ± 100 yr BP and 6255 ± 320 yr BP (Page *et al.*, 1979; Vita-Finzi, 1981). The same observations were made on beach deposits of the Pakistan Makran (Sanlaville *et al.*, 1991). So, in Makran the coastline has been prograding since the Middle Holocene owing to both slight uplift and marine or alluvial sedimentation.

Paul Sanlaville and Abel Prieur

Bibliography

Aksu, A.E., Piper, D.J.W., and Konuk, T., 1987. Quaternary growth patterns of Büyük Menderes and Küçük Menderes deltas, Western Turkey. *Sedimentary Geology*, **52**: 227–250.

Al-Sayari, S.S., Hötzl H., Moser, H., Rauert, W., and Zötl, J.G., 1984. Quaternary from Dhuba to Al Wajh. In Jado, A.R., and Zötl, J.G. (eds.), *Quaternary Period in Saudi Arabia*, Volume 2. NewYork: Springer-Verlag, pp. 66–82.

Baltzer, F., and Purser, B.H., 1990. Modern alluvial fan and deltaic sedimentation in a foreland tectonic setting: the lower Mesopotamian Plain and the Arabian Gulf. *Sedimentary Geology*, **67**: 175–197.

Basson, P.W., Burchard, J.E., Hardy, J.T., and Price, A.R.G., 1977. Biotopes of the western Arabian Gulf. Marine life and environments of Saudi Arabia. *Aramco Department of Loss Prevention and Environmental Affairs, Dhahran, Saudi Arabia*, 242.

Dabbagh, A., Hötzl, H., and Schnier, H., 1984. Farasan islands. In Jado A.R., and Zötl J.G. (eds.), *Quaternary Period in Saudi Arabia*, Volume 2. NewYork: Springer, pp. 212–220.

Dullo, W.C., and Montaggioni, L., 1998. Modern Red Sea coral reefs: a review of their morphologies and zonation. In Purser B.H., and Bosence D.W.J. (eds.), *Sedimentation and Tectonics in Rift Basins, Red Sea and Gulf of Suez*. London: Chapman and Hall, pp. 583–594.

Erol, O., 1985. Turkey and Cyprus. In Bird E.C., and Schwartz M.L. (eds.), *The World's Coastline*. NewYork: Van Nostrand Reinhold Company, pp. 491–498.

Evans, G., Schmidt, V., Bush, P., and Nelson, H., 1969. Stratigraphy and geologic history of the Sabkha, Abu Dhabi, Persian Gulf. *Sedimentology*, **12**: 145–159.

Fleisch, H., Comati, J., and Elouard, P., 1981. Poursuite et synthèse des études sur le gisement à *Strombus bubonius* Lmk (Tyrrhénien) de Naamé (Liban). *Quaternaria*, **XXIII**: 51–79.

Glover, E., 1998. Mangroves, molluscs and man. Archaeological evidence for biogeographical changes in mangrove around the Arabian peninsula. *Arabia and its Neighbours*, 63–78.

Hacker *et al.*, 1984. Region around Jeddah: geology, geomorphology and climate. In Jado A.R., and Zötl J.G. (eds.), *Quaternary Period in Saudi Arabia*, Volume 2. NewYork: Springer, pp. 107–122.
Hakyemez, H.Y., Erkal, T., and Götkas, F., 1999. Late Quaternary evolution of the Gediz and Büyük Menderes grabens, Western Anatolia, Turkey. *Quarternary Science Reviews*, **18**: 549–554.
Hanns, C., 1998. Predominant features of the Quaternary relief development seawards of the Oman mountains as reflected in wadi and coastal terraces and other coastal features. In Alsharhan A.S., Glennie K.W., Whittle G.L., and Kendall (eds.), *Quaternary Deserts and Climatic Change*. Rotterdam: A. Balkema, pp. 17–27.
Hötzl, H., Jado, A.R., Lippolt, H.J., and Puchelt, H., 1984. The Quaternary from Umm Lajj to Yanbu al Bahr. In Jado A.R., and Zötl J.G. (eds.), *Quaternary Period in Saudi Arabia*, Volume 2. NewYork: Springer, pp. 82–102.
Kassler, P., 1973. The structural and geomorphic evolution of the Persian Gulf. In Purser B.H. (ed.), *The Persian Gulf: Holocene Carbonate Sedimentation and Diagenesis in a Shallow Epicontinental Sea*. Berlin: Springer-Verlag, pp. 11–32.
Kayan, I., 1994. Geomorphological outlines of Turkey. *Archaeometry*, 365–374.
Kayan, I., 1999. Holocene stratigraphy and geomorphological evolution of the Aegean coastal plains of Anatolia. *Quaternary Science Reviews*, **18** (4/5): 541–548.
Kraft, J.C., Kayan, I., and Erol, O., 1980. Geomorphic reconstructions in the environs of Ancient Troy. *Science*, **209**: 776–782.
Lambeck, K., 1996. Shoreline reconstructions for the Persian Gulf since the last glacial maximum. *Earth Planetary Science Letters*, **142**: 43–57.
Müller, E., 1984. The Quaternary from Umm Lajj to Yanbu al Bahr. In Jado A.R., and Zötl J.G. (eds.), *Quaternary Period in Saudi Arabia*, Volume 2. NewYork: Springer, pp. 141–150.
Nir Y., 1982. Asia, Middle East, coastal morphology: Israel and Sinai. In Schwartz M.L. (ed.), *The Encyclopedia of Beaches and Coastal Environment*. Stroudsburg, PA: Hutchinson Roos Publishes., pp. 86–98.
Page, W.D., Alt J.N., Cluff, L.S., and Plafker, G., 1979. Evidence for the recurrence of large-magnitude earthquakes along the Makran coast of Iran and Pakistan. *Tectonophysics*, **52**: 533–547.
Pirazzoli, P.A., 1986. The Early Byzantine tectonic paroxysm. *Zeitschrift für Geomorphologie*. N.F., Suppl.-Bd., **62**: 31–49.
Por F.D., 1982. Asia, Middle East, coastal ecology. In Schwartz, M.L. (ed.), *The Encyclopedia of Beaches and Coastal Environments*. Stroudsburg, PA: Hutchinson Roos Publishing. Co., pp. 82–84.
Por F.D., 1990. Lesseptian migration. An appraisal and new data. In Godeaux J. (ed.) A propos des migrations lesseptiennes. *Bulletin de l'Institut Oceanographique de Monaco*, no. spécial **7**: pp. 1–10.
Purser, B.H. (ed.), 1973. *The Persian Gulf: Holocene Carbonate Sedimentation and Diagenesis in a Shallow Epicontinental Sea*. Berlin: Springer-Verlag.
Purser, B.H., and Bosence, D.W.J. (eds.), 1998. *Sedimentation and Tectonics in Rift Basins, Red Sea and Gulf of Suez*. London: Chapman and Hall.
Purser, B.H., and Evans, G., 1973. Regional sedimentation along the Trucial coast, SE Persian Gulf. In Purser B.H. (ed.), *The Persian Gulf: Holocene Carbonate Sedimentation and Diagenesis in a Shallow Epicontinental Sea*. Berlin: Springer-Verlag, pp. 211–231.
Purser, B.H., and Seibold, E., 1973. The principal environmental factors influencing Holocene sedimentation and diagenesis in the Persian Gulf. In Purser B.H. (ed.), *The Persian Gulf: Holocene Carbonate Sedimentation and Diagenesis in a Shallow Epicontinental Sea*. Berlin: Springer-Verlag, pp. 1–10.
Purser, B.H., Al Azzawi, M., Al Hassani, N.H., Baltzer, F., Hassan, K.M., Orszag-Sperber, F., Plaziat, J.-C., Yacoub, S.Y., and Younis, W.R., 1982. Caractères et évolution du complexe deltaïque Tigre-Euphrate. *Memoire of the Geological society of France*, **144**: 207–216.
Reyss, J.L., Pirazzoli, P.A., Haghipour, A., Hatté C., and Fontugne, M., 1998. Quaternary marine terraces and tectonic uplift rates on the south coast of Iran. In Stewart I.S., and Vita-Finzi C. (eds.), *Coastal Tectonics*. Geological Society, London, Special Publications **146**: pp. 225–237.
Safriel, U., 1974. Vermetid gastropods and intertidal reefs in Israel and Bermuda. *Science*, **186**: 1113–1115.
Sanlaville, P., 1977. *Etude géomorphologique de la région littorale du Liban*, 2 Volumes. Beyrouth: Publications Université Libanaise, Section des Etudes géographiques.
Sanlaville, P., 1989. Considérations sur l'évolution de la basse Mésopotamie au cours des derniers millénaires. *Paléorient*, **15**(2): 5–27.
Sanlaville, P., 2000. *Le Moyen-Orient arabe. Le milieu et l'homme*. Paris: A. Colin.
Sanlaville, P., Besenval, R., Evin, J., and Prieur, A., 1991. Evolution de la région littorale du Makran à l'Holocène. *Paléorient*, **17**(1): 3–18.
Sanlaville, P., Dalongeville, R., Bernier, P., and Evin, J., 1997. The Syrian coast: a model of Holocene coastal evolution. *Journal of Coastal Research*, **12**(2): 385–396.
Shinn, E.A., 1973. Carbonate coastal accretion in an area of longshore transport, NE Qatar, Persian Gulf. In Purser B.H. (ed.), *The Persian Gulf: Holocene Carbonate Sedimentation and Diagenesis in a Shallow Epicontinental Sea*. Berlin: Springer-Verlag, pp. 179–191.
Smythe, K.R., 1972. Marine mollusca from Bahrain Island, Persian Gulf. *Journal of Conchyliculture*, **27**: 491–496.
Snead, R.J., 1993. Uplifted marine terraces along the Makran coast of Pakistan and Iran. In Shroder J.F., Jr. (ed.), *Himalaya to the sea. Geology, Geomorphology and the Quaternary*. London, New York: Routledge, pp. 327–362.
Vita-Finzi, C., 1978. Recent alluvial history in the catchment of the Arabo-Persian Gulf. In Brice W.C. (ed.), *The Environmental History of the Near and Middle East Since the Last Ice Age*. London: Academic Press, pp. 255–270.
Vita-Finzi, C., 1981. Late Quaternary deformation on the Makran coast of Iran. *Zeitschrift für Geomorphologie*. N.F., Suppl.-Bd. **40**: 213–226.
Zohary, M., 1973. *Geobotanical Foundations of the Middle East*. 2 Volumes. Stuttgart: G. Fisher.

Cross-references

Artificial Islands
Beachrock
Climate Patterns in the Coastal Zone
Coastal Climate
Coral Reef Coasts
Coral Reefs
Desert Coasts
Mangroves, Ecology
Oil Spills
Paleocoastlines
Sharm Coasts
Tidal Creeks
Trottoirs

ASIA, SOUTHERN—See INDIAN OCEAN COASTS

ASTEROID-IMPACT COASTS

A highly unusual category of general coastal morphology is created by asteroid impacts at some time in the geologic past. Asteroid impact creates a circular or ovoid crater beneath which is a brecciated zone that extends many thousand meters below the former surface of the Earth's crust. The eroded relics of these ancient craters have been called "*astroblemes*" (Dietz, 1961).

Three coastal areas in North America are believed to owe their morphology, at least in part, to asteroid impact. They are:

1. *Chesapeake Bay, and adjacent areas of Maryland and Virginia.* The coast is characterized by an unusual pattern of a drowned dendritic drainage system, that is to say, organized like the branches of a well-shaped tree. It is fed by the valleys of the Susquehanna, Potomac, and Rappahanock rivers. The asteroid or "bolide" struck 35.2 (+/−0.3) million years ago, in Late Eocene times in soft coastal plain and shelf sediments which to the impacting object, about 3–5 km in diameter and travelling at about 80,000 km/h, would have had the consistency of warm butter. Molten glass shards (cooled as tektites) were strewn over 9 million km^2. Penetrating about 600 m of youthful sediments it continued to a limited depth in the crystalline basement. The principal crater is about 85 km in diameter, and overlying (postimpact) sediments have slowly sagged into the depression and the surrounding area. In the center of this astrobleme, there is an inner depression with a peak in the center (just as in some other impact sites such as the Miocene Ries crater of Germany). A much larger area extending up to 90 km from the

principal crater rim is marked by a belt of radial fractures spreading out over much of the Salisbury Embayment. A polymictous debris plume reached NE, up the Atlantic coast at least 400 km. The subsurface structures have been proven by extensive seismic profiling and drilling (Poag, 1997). In the 35 Myr since the impact, the site has been the focus of differential compaction and subsidence (with eustatic revivals) that determines the general form of today's coastline. Another but much smaller asteroid fragment at the same time created the 25-km-wide Toms Canyon crater 140 km east of Atlantic City.

2. *Gulf of Mexico (Chicxulub crater)*. This impact occurred near the west side of the Yucatan peninsula of Mexico about 65.2 (+/−0.4) million years ago, generating a crater 180 km across. Its ejected debris has been traced to a semicircular rim extending in an arc through central Texas which suggests an incoming orbit heading NNW. This event was destined to become the Cretaceous/Tertiary boundary, marking a revolution involving major biotic extinctions, notably the dinosaurs, ammonites, and other classes (75% of all marine species). While the western and northern borders of the Gulf of Mexico had already developed as a broad arc in Mesozoic times, its southern border is marked by a meridional fault system that today truncates the western borders of Yucatan, a nearly horizontal platform of Miocene limestones. The impact theory, initiated by W. and L. Alvarez in the late 1970s (from evidence in Italy) triggered an intense interest that has been well reviewed by Marvin (1990) and expanded to events throughout the entire Phanerozoic (Rampino and Haggerty, 1996). It seems likely that anomalous patterns of coastlines worldwide should be explored for possible astrobleme ancestry.

3. *Hudson Bay*. The present form of this embayment is an epicontinental sea bounded particularly in the southeast by an extraordinary arcuate coast that coincides with an Archaean/ Proterozoic boundary injected by lopolithic sills of diabase. An ancient astrobleme is suggested by Dietz (1961).

Rhodes W. Fairbridge

Bibliography

Dietz, R.S., 1961. Astroblemes. *Scientific American*, **205**: 50–58.

Marvin, U.B., 1990. Impact and its revolutionary implications for geology. Geological Society of America, Special Paper, **247**, pp. 147–154.

Poag, C.W., 1997. The Chesapeake Bay bolide impact: a convulsive event in Atlantic coast plain evolution. *Sedimentary Geology*, **108**: 45–90.

Rampino, M., and Haggerty, B., 1996. Impact crises and mass extinctions: a working hypothesis. Geological Society of America, Special Paper, **307**, pp. 11–30.

Cross-references

Estuaries
Sea-Level Rise
Submerging Coasts

ATLANTIC OCEAN ISLANDS, COASTAL ECOLOGY

While many islands lie in Atlantic waters, this entry will focus only on a few of the most significant to illustrate a variety of insular habitats from the Northern and Southern Hemispheres (Figure A59). Except for the Faeroe, Azores, Canaries, Bermudas, Ascension, Gough, and the islands of the Scotia Arc, relatively little is known about the coastal ecology of most of the islands scattered in the Atlantic. When available, data on the general biodiversity, the intertidal and subtidal zones, and the seasonal variation of the plankton will be presented, together with some aspects of the terrestrial ecology relevant to the coast and typical of islands.

The fauna and flora of most islands have several features that distinguish them from those of continents. Many of these are related to difficulties of dispersal. Organisms that can disperse well are more likely to be found on islands than those that cannot. Species of plants and animals on islands show equilibrium between new arrivals and local extinctions. Generally, the number of species is greater on large islands than small islands, varying in function of their isolation from other landmasses (Table A5). Thus, islands usually present an impoverished terrestrial and marine wildlife compared to equal areas of mainland (Williamson, 1981).

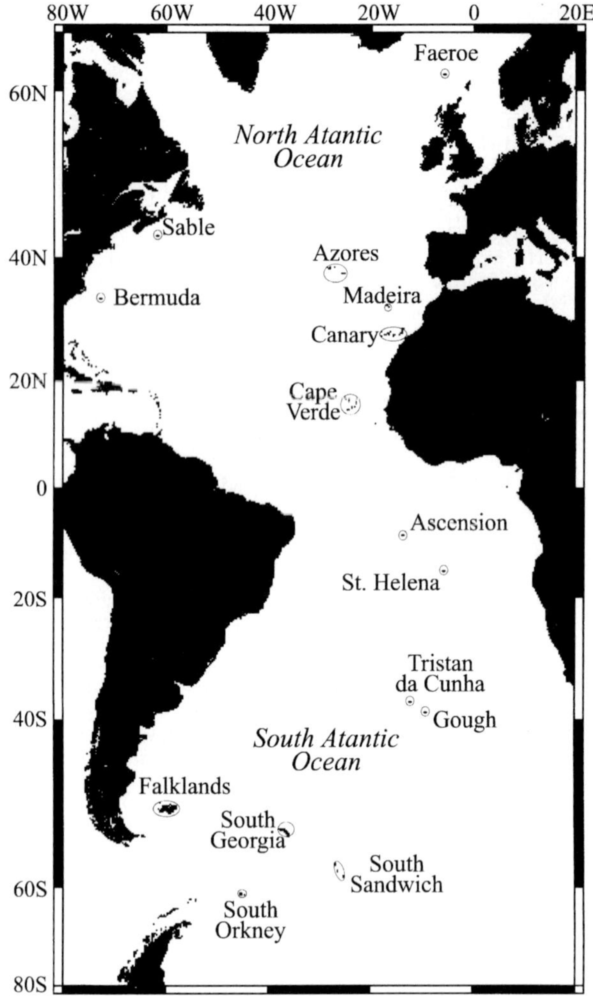

Figure A59 Map of the Atlantic Ocean locating the islands discussed in this entry.

Table A5 Isolation index of different islands or group of islands in the Atlantic Ocean (compiled from the United Nations Environment Programme)

Islands	Isolation index[a]
Bermuda	91
Azores	74–77
Madeira	66
Canary	12–41
St. Helena	113
Ascension	119
Tristan da Cunha	78
Gough	125
Falklands	59
South Georgia	113
South Sandwich	71–76
South Orkney	28–31

[a] The isolation index is the sum of the square roots of the distances to the nearest equivalent or larger island, the nearest island group or archipelago and the nearest continent. A range indicates the isolation variation among the different islands composing an archipelago or group of islands.

Faeroe (Denmark)

The Faeroe group comprises 1,398 km^2 of rugged and mountainous islands located just north of the United Kingdom, on the edge of the Rockall Rise. The coastline is indented, with fjord-like inlets, the marine life is abundant and the climate is mainly temperate oceanic.

As their situation in the transitional zone between the North Atlantic and the Arctic, the Faeroe Islands harbor a flora and fauna made up of both boreal and arctic species (Blindheim, 1989). The phytoplankton productivity of this area is considered to be one of the greatest in the world oceans. The blooms occur in March and consist mainly of diatoms (Braarud *et al.*, 1958). Compared to the spring bloom, the summer abundance of pelagic algae is rather poor, but more diverse (Blindheim, 1989). The zooplankton bloom begins around March–April, in conjunction with the phytoplanktonic bloom. The summer pelagic fauna is dominated by copepods such as *Calanus finmarchicus*, *Pseudocalanus minutus*, and also by the arrow worms *Sagitta elegans* and *Eukrohnia hamata*. The Euphausiacea represent another important component of the zooplankton. Common pelagic fish in the waters off Faeroe are herrings, Greenland halibuts, blue whittings, redfish, cods, and haddocks.

On the vertical slopes of the coast, from the lowest water spring tide mark, every centimeter of the subtidal benthic zone is covered in a profuse carpet of life. Kelp and other algae are gradually replaced by sea anemones and sponges in deeper water.

Marine birds like the puffin *Fatercula arctica* and marine mammals such as the pilot whale *Globicephala melaena* and the white-beaked dolphin *Lagenorhynchus albirostris* are abundant.

Sable (Canada)

Sable Island is located approximately 300 km southeast of the Canadian province of Nova-Scotia. The island itself is about 40 km long and 1 km wide, hence its temperate climate is markedly influenced by the sea. Ocean waters maintain winter air near freezing and keep summer temperatures below what would be considered room temperature. Fierce ocean currents sweep around the island causing the peculiar shifting of sand that constantly changes its contours. Between the dunes are numerous depressions usually filled with freshwater, and supporting a variety of aquatic plants and insects. A total of 179 species of terrestrial plants can be found on Sable Island. Along the coast, the moving sand is stabilized by the marram *Ammphila breviligulata*, the sandwort *Honckenya peploides*, and the beach pea *Lathyrus maritimus*.

Since all the shores of the island are sandy, benthic invertebrate diversity is low, although the sand dollar *Echinarachnius parma*, shrimp *Crangon septemspinosa*, and gastropod *Lunatia heros* are observed commonly. Among fish larvae, the pollock *Pollachius virens* and haddock *Melanogrammus aeglefinus* have been found around Sable Island. The American sand lace, smooth skate, yellowtail flounder, Atlantic cod, Atlantic mackerel, witch flounder, and spiny dogfish are also present.

Sable Island is the only known nesting place for the ipswich sparrow *Passerculus sandwichensis princeps*. Over 2,500 pairs of terns, mostly Arctic terns *Sterna paradisaea*, nest on the island along with over 500 pairs of black-backed gulls *Larus marinus* and 2,000 pairs of herring gulls *Larus argentatus*. A few sandpipers *Calidris* spp. and semipalmated plovers *Charadrius semipalmatus* also nest on the island and broods of black ducks *Anas rubripes* and red-breasted mergansers *Mergus sarrator* are produced near the ponds. During migration and after strong gales, nearly any species of bird in eastern North America may show up on Sable Island. About 300 Sable Island horses, introduced in 1738, are now living wild on the island. Sable Island also supports the largest breeding colonies of gray seals *Halichoerus grypus* and harbor seals *Phoca vitulina* in the western Atlantic Ocean. Harbor seals use Sable Island as a mating ground (Coltman *et al.*, 1999).

Not far from Sable Island is the Sable Gully, which is the largest marine canyon in eastern Canada, and is considered an important habitat for a wide diversity of marine life, including over 200 vulnerable northern bottlenose whales (*Hyperoodon ampullatus*).

Bermuda

The island group of Bermuda is located in the western central region of the North Atlantic oceanic gyre known as the Sargasso Sea, approximately 1,250 km southeast of New York City. Bermuda comprises about 130 islands and islets covering a land of just over 50 km^2. The climate is wet subtropical. These islands are characterized by tidal and nontidal ponds, rocky coasts, sandy beaches, and shallow marine bays. They harbor the most northerly mangroves in the world and the most northerly coral reefs in the Atlantic, including coral/algal and algal/vermetid reefs.

The ocean temperature of the surface waters off Bermuda varies seasonally between 18°C and 28°C. This island group can be considered a semi-tropical area where ocean conditions differ markedly from those encountered further south in the true tropics. In Bermuda, winter cooling and mixing of the surface layers occurs and a seasonal thermocline develops from late spring through fall (Morris *et al.*, 1977). The phytoplankton of the Sargasso Sea that surrounds Bermuda shows low concentrations, especially during spring and summer with about 3,000 cells L^{-1}. The density of phytoplankton reaches a maximum in fall and winter with 50,000 cells L^{-1}. In summer, the phytoplankton is mainly composed of coccolithophores, dinoflagellates, and diatoms. The main species, especially between late fall and spring, is the coccolithophore *Coccolithus huxleyi*. A spring bloom of diatoms, 200,000 cells L^{-1}, occurs around April. Dinoflagellates dominate in late summer concurrent with the lowest productivity of the year.

The zooplankton of the waters surrounding the Bermuda reef platform and of the inshore waters are distinct from the oceanic species. The larvae of numerous benthic invertebrates are a significant component of the reef and inshore zooplankton, which is 5 to 10 times more abundant than in the Sargasso Sea, fluctuating around 2,000 individuals m^{-3}. Maximum zooplankton abundance generally occurs after the phytoplanktonic bloom in late spring-summer and again in winter, following the spring and fall increases of phytoplankton. Both zooplankton peaks are of only short duration and are followed by rapid population declines. Copepods are the most abundant members of the inshore populations and dominate in summer and fall, representing about 70–95% of all species present. In winter, the appendicularian *Oikopleura graciloides* dominate and fish eggs are prevalent in spring. Copepods nauplii and gastropod veligers are abundant in spring and again in early winter. Though less abundant, medusae, cladocerans, decapod and fish larvae are also part of the zooplankton of Bermuda (Morris *et al.*, 1977).

The benthic habitats of Bermuda comprise rocky shores of the intertidal and subtidal areas, gullies, crevices and caves, muddy bottoms, mangroves, sandy-bottoms, and coral reefs. The coral reefs of Bermuda can further be divided into four different types: the ledge flat reef, the patch reefs, the coral knobs, and boiler reefs (Morris *et al.*, 1977). The muddy substrata, mainly located in shallow protected areas, are associated with mangroves but also occur in the deepest bassin depressions. The sublittoral zones around Bermuda, including the surroundings of coral reefs, are mostly sandy. Numerous algae and sea grasses cover the shallows and offer home or shelter to various invertebrates. Burrowing species are the most abundant group of invertebrates present.

The rocky shore, sublittoral, and coral reef habitats shelter the greatest number of species. Overall, 118 common species are known from the rocky-shore subtidal area, including several algae, cnidarians, mollusks, and echinoderms. The rocky intertidal zone harbors 82 common species, dominated by algae, mollusks, and crustaceans. Many of these species live in the lower fringe or directly in tide pools. In the coral reefs, there are about 79 common species, mainly corals, but also algae, sponges, mollusks, and crustaceans. Mangrove swamps sustain about 21 species, dominated by algae, shrubs, mollusks, crustaceans, and bryozoans (Morris *et al.*, 1977).

Macronesia

Macronesia encompasses the islands of Azores, Madeira, Canaries, and Cape Verde, which are located in the northeastern Atlantic.

The Azores consist of 10 Portuguese islands covering 2,244 km^2 and having a highly volcanic topography with many peaks and hollows (Bird, 1985). They lie about 1,500 km off the main coast of Europe and their climate is moist-moderate with abundant rainfall.

Currently, 274 species of algae are listed in the Azores, including 45 Chlorophyta, 52 Phaeophyta, and 177 Rhodophyta, among which are 10 endemic species. Although the Azores share algal biotopes with the Atlantic coast of mainland Europe, intertidal and subtidal seashores in the archipelago mostly lack the functionally important community of fucoids and laminarians widespread in the North Atlantic. Large brown algae such as *Fucus spiralis* are sporadic in the Azores, while *Laminaria ochroleuca* is known from only one location. Other large brown algae (*Cystoseira* spp., *Sargassum* spp.) occur in deep sheltered pools, lagoons, and some subtidal biotopes.

The intertidal areas of the Azores typically present three distinct zones. The highest band is dominated by littorinids (*Littorina striata*, *Melaraphe neritoides*). The upper boundary of the second zone is marked by the presence of barnacles *Chthamalus stellatus* and some small shoots of *Ulva rigida*, *Gelidium pusillum*, *Gelidium microdon*, *Ralfsia verrucosa*, and *Enteromorpha* spp. The third zone comprises an extensive area dominated by *G. pusillum* and *Centroceras capillacea*. In the adjacent subtidal area, the algal community changes along the depth gradient. *Ulva* spp., *Hypnea musciformis*, and *Asparagopsis armata* dominate the shallow depths. The 10 m depth community is characterized by the algae *Dictyota dichotoma*, *Halopteris* spp., and *Stypocaulon scoparia*, while *Zonaria tournefortii* is dominant at 30 m depth. With the exception of mollusks *Bittium* spp., that can be found everywhere, differences in the invertebrate fauna are also observed along the depth gradient. Sponges are an example of this: *Tethya aurantium*, *Halichondria panicea*, *Hymeniacidon sanguinea*, and *Timea unistellata* occur in the intertidal zone, whereas *Petrosia ficiformis*, *Erylus discophorus*, *Myxilla macrosigma*, and *Clathrina coriacea* are found between 10 and 15 m depth (Neto and Avila, 1999).

The southern oyster drill *Stramonita* (*Thais*) *haemastoma* preys intensively on the populations of *Mytilus edulis*, *Patella candei*, *Patella aspera*, *Crassostrea virginica*, *Ostrea equetris*, *Ischadium recuvrum*, *Balanus eburneus*, *Balanus amphitrite*, and *Chthamalus stellatus* which colonize the intertidal zone of the Azores. Sea urchins *Arbacia lixula* are also common on the rocky shores.

Over 175 species of mesopelagic fish are believed to occur around the Azores. The most widely spread families are included in the Stomiidae, Gonostomatidae, Sternoptychidae, and Paralepididae. The yellowmouth barracuda *Sphyraena viridensis* is a common pelagic coastal predator, mostly preying on juvenile horse mackerels *Trachurus picturatus*.

Large numbers of juvenile loggerhead sea turtles *Caretta caretta* thrive off the coast of the Azores. The sperm whale *Physeter macrocephalus* is the most common whale in Azorian waters. A total of 23 species of cetaceans have been observed, including the false killer whale *Pseudorca crassidens*, Risso dolphin *Grampus griseus*, bottlenose dolphin *Tursiops truncatus*, common dolphin *Delphinus delphis*, stripped dolphin *Stenella coeruleoalba* and spotted dolphin *Stenella frontalis*. *Balaneoptera acurostrata* and *Globicephala macrorhynchus* have also been recorded in winter and early spring.

The Portuguese archipelago of Madeira consists of one large and several small islands lying some 600 km off the coast of Morocco and representing a total land area of 810 km^2, mostly made up of extinct volcanoes. The climate is semi-arid subtropical. The slopes are steep, but well vegetated and terminated in high coastal cliffs. Beaches are mainly gravely, but beach borders are also found. The coastal marine fauna of Madeira is a mixture of species with Mediterranean affinities and distinctly tropical species that often reach their northern limit there (Wirtz, 1999).

Little is known of the precise marine ecology of this group. The sea urchin *Diadema antillarum* is an extensive herbivore that preys on algal bed. Thirty-eight species of fish are easily observed in the coastal waters of Madeira. The endangered Madeira soft-plumaged petrel *Pterodroma mollis madeira* occurs there, as well as the Mediterranean monk seal *Monachus monachus*, which breeds on the island. In fact, Madeira is one of the most important remaining localities for this species threatened with extinction.

The Canary group comprises 7,500 km^2 of dry islands spreading close to the coast of northwest Africa. These Spanish islands are mostly surrounded by sandy beaches and rocky shores and their climate is semi-arid subtropical.

In the Canary Islands, the chlorophyll concentrations range from 0.09 mg m^{-3} to 0.25 mg m^{-3}, with deep maxima between 75 and 100 m depth. The biomass in the shallower water corresponds to 2,000–28,000 cells L^{-1} near the coast (Ojeda, 1996). Some coastal lagoons are characterized by the conspicuous presence of Cyanophyta and Chlorophyta up to 10^5 cells ml^{-1}, thus imparting a green coloration to the water.

Primary productivity is high in the water off Canary Islands. The northern portion of the islands supports a high phytoplankton production where the Canary current collides with the coast. The biomass is not large but the turnover is pronounced. Bacteria associated with phytoplankton are abundant at the outer edge of the main phytoplankton concentration. Zooplankton accumulates in the marginal southern areas and is quite important in the upwelling area in the upper 50 m of the water column. The zooplankton community is generally composed of copepods, but mysidacea are usually important in all regions. Zooplankton also includes amphipods, ostracods, and mysids.

The sea urchin *Diadema antillarum* is commonly observed. Falcon *et al.* (1996) found 76 species of coastal fish around the Canary Islands, the most common being *Abudefduf luridus*, *Canthigaster rostrata*, *Chromis limbatus*, *Sparisoma cretense*, and *Thalassoma pavo*. Lorenzo (1995) indicated that many waders (*Charadrius alexandrinus*, *Calidris alba*, *Arenaria interpres*) were found in intertidal lava platforms and in supratidal lagoons of the Canary Islands.

St. Helena and Ascension (United Kingdom)

St. Helena is an isolated volcanic peak rising 4,400 m from the seafloor, slightly east of the mid-Atlantic ridge, about 1,900 km off the coast of Angola. It covers 125 km^2 as a deeply eroded volcanic cone. It has a dry subtropical climate. The bay on the south coast has a beach backed by calcareous dunes, which have been driven far inland by the southeast trade winds. The coastal areas are rugged and barren whereas the higher elevations in the center of the island present a lush vegetation (Bird, 1985).

St. Helena's isolated position in the South Atlantic has given rise to an unusual and remarkable land and marine flora and fauna. There are 60 species of flowering plants, of which 49 are endemic. Thirteen endemic species of ferns are present, almost all on the verge of extinction. Of 1,100 land invertebrates, 400 are unique to St. Helena. At least six unique land birds once occurred on St. Helena, only one, the wirebird *Charadrius sanctae helenae*, survives today. Also, modest seabird populations nest on isolated cliffs and rocks. The island is home to the largest species of earwig, *Labidura herculeana*. Endemic species of shrimps *Typhlatya rogersi* and *Procaris ascensionis* are present in rock pools. Ten shore fishes are found only around the island, and 16 more are found only here and at Ascension. The green turtle *Chelonia mydas*, and the hawksbill turtle *Eretomochelys imbricata* may breed along the coast of St. Helena.

Ascension is a 97 km^2 volcanic island isolated in the south Atlantic. The peak of a dormant volcano rises 3,000 m above the seafloor and a further 859 m above sea level to the summit of Green Mountain. The vegetation is sparse with scattered guano deposits and the island has a dry tropical climate. Imposing sea cliffs line the more exposed eastern coast (Bird, 1985). Beaches, intertidal rock pools, and rocky shores are mainly composed of bare rock, regularly covered by a layer of crustose coralline algae that has cemented all rubble into a uniform surface.

The general biodiversity of Ascension is not high, not only because of its isolation but also due to the absence of lagoons, estuaries, seagrass beds, mangroves, and coral reefs (Manning and Fenner, 1990). Price and John (1978) concluded that the primary facet presented by the intertidal and sublittoral fringe on Ascension was still that of stark bare rock, sometimes abutting or adjoining clean sand. Other factors such as strong tidal surge, supplemented by rollers, enormous waves that assault the island, influence the shore and shallow depth colonization. Finally, the blackfish *Melichthys niger* is another important factor regulating the presence of other species, as it deters colonization by grazing all surfaces that it can reach (Day, 1983), including the crustose algae described earlier. Stands of algae and colonies of sabellariid worms, barnacles, and solitary coral are limited to crevices, blowholes, and isolated tide pools where blackfish cannot penetrate (Manning and Fenner, 1990).

The sandy beaches of Ascension are high-energy areas made up of very coarse sand. Only digging crustaceans *Hippa testudinaria* have been found in the sand and crabs *Grapsus* spp. have been seen roaming over the open beaches at night. Isolated tide pools with rims of protective rock are often inhabited by dense beds of oysters *Saccostrea cucullata* (Manning and Fenner, 1990).

Much of Ascension's global conservation importance comes from the island's remoteness, which has produced one of the most remarkable island floras and faunas in the world. It is of world significance for its 11 species of breeding seabirds, especially the unique Ascension Island frigate bird *Fregata aquila*. There are six unique species of land plants, nine of marine fish and shellfish, and over twenty of land invertebrates.

Some 27 species of amphipods have been described in the pool habitats along the shore, and Manning and Fenner (1990) observed 74 species of decapods on Ascension. Of these, 41 occur in the western Atlantic, and most are common.

The echinoderm fauna of Ascension comprises 25 species, including brittle stars, sea cucumbers, sea urchins, and starfish. Eight of these species are amphi-Atlantic, one species is restricted to Ascension, three species are also known in the western Atlantic, four species are also found in eastern Atlantic, five species are circumtropical and four species are also found in St. Helena. These findings suggest that species dispersion by larvae is mediated by surface and subsurface transport of planktonic stages (Pawson, 1978).

About 81 species of shore fishes are known, including 11 endemic species. Ascension has one of the most important breeding populations of green turtle *C. mydas* in the world. The green turtles mate and lay their eggs on sandy beaches, whereas hawksbill turtles *E. imbricata* are less frequently seen along the shores.

Tristan da Cunha and Gough (United Kingdom)

Tristan da Cunha is a circular volcanic island of 103 km^2 rising 2,062 m above sea level. The island is bleak and barren, but zoned vegetation occurs above 1,350 m. It has a wet temperate climate and is surrounded by rocky shores.

Gough is a deeply dissected volcanic island located 350 km southeast of Tristan da Cunha. High cliffs with hanging valleys and waterfalls surround the surface area of 65 km^2. The luxuriant evergreen scrub forest thins out above the 300 m contour. The climate is a cool-temperate oceanic. The flora of Gough is typical of southern cold-temperate oceanic islands, with relatively low species diversity, and a large preponderance of ferns and cryptogams.

Both Tristan da Cunha and Gough are characterized by their isolation and high level of endemism near the shore. However, with increasing depth, the amount of faunal similarity increases between regions across the South Atlantic.

The marine area can be split into two distinct algal zones. From sea level to 5 m depth, algae consist mainly of bull kelp *Durvillea antarctica*, and beyond 20 m are dominated by *Laminaria pallida* and giant kelp *Macrocystis pyrifera*. Forty species of algae have been recorded, of which two species are endemic to Gough. Most littoral species found at Gough are widespread on other southern ocean islands, and 79 invertebrate species have been recorded. The absence of limpets and bivalves in the littoral and subtidal zones is noted. Sea urchins *Arbacia dufresnii* are abundant in the marine area, as are whelks *Argobuccinum* sp., chitons, starfishes, sea anemones, bryzoans, barnacles, slipper limpets, nudibranchs and sponges. Important marine species include the Tristan rock lobster *Jasus tristani* (from close inshore to 400 m depth around Gough), and octopi. Both are economically exploited by fishermen under close regulation. About 51 species of fish are known to occur in nearshore waters of Tristan da Cunha and Gough.

Gough has been described as a strong contender for the title of "most important seabird colony in the world," with 54 species recorded in total, of which 22 breed on the island and 20 are seabirds. About 48% of the world's population of northern rockhopper penguin *Eudyptes chrysocome moseleyi* breed at Gough. Atlantic petrel *Pterodroma incerta* is endemic to Gough and the Tristan group of islands. Gough also harbors a major colony of the great shearwater *Puffinus gravis* with up to three million pairs breeding on the island. The main southern ocean breeding sites of little shearwater *Puffinus assimilis* are Tristan da Cunha and Gough Island, with breeding pairs numbering several million. The wandering albatross *Diomedea exulans dabbenena* is virtually restricted to Gough, with up to 2,000 breeding pairs. The only survivors of southern giant petrel *Macronectes gigantous* also breed on Gough, with an estimated 100–150 pairs. Gough moorhen *Gallinula comeri* is found in fern bush vegetation areas, and estimates of population size vary from 300–500 pairs to 2,000–3,000 pairs. About 200 pairs of Gough finch *Rowettia goughensis* have been recorded, although recent estimates suggest that there may be 1,000 pairs. Other seabirds breed on these islands, such as the yellow-nosed albatross *Diomedea chlororhynchos chlororhynchos* and the great shearwater *P. gravis*. The penguin *Eudyptes crestata* is present on Tristan.

Subantarctic fur seals *Arctocephalus tropicalis* (200,000 individuals and increasing), and southern elephant seals *Mirounga leonina* (about 100 individuals) are the only two native breeding mammals. The former breed at beaches all round the island, whereas the latter are restricted to the island's sheltered east coast. Two other marine mammals are found in the area, namely the southern right whale *Eubalaena glacialis australis* and the dusky dolphin *Lagenorhynchus obscurus*. Reptiles, amphibians, freshwater fish, and native terrestrial mammals are absent, although the introduced house mouse *Mus musculus* is widespread and abundant.

Falklands (United Kingdom)

The Falklands include two main islands and several hundreds smaller islands and islets that cover about 12,000 km^2. The coastline is lengthy and highly indented with numerous rias at the mouths of winding valleys. The landscape is treeless, with tussocky grasses and extensive peat bog mantles (Bird, 1985). Some rocky coasts possess offshore kelp beds of *Macrocystis*. The climate is dry cool temperate.

The Falklands are exceptionally rich in marine life, including benthic and pelagic forms. Also, about 85% of the world population of the black-browed albatross (*Diomedea melanophris*), and the largest concentration of rockhopper penguins (*E. chrysocome*) are found on the islands. A total of 63 species of sea and land birds occur, including the night heron *Nycticorax nycticorax cyanocephalus*, ashy-headed goose *Chloephaga poliocephala*, kelp goose *Chloephaga hybrida malvinarum*, Garnot's ground tyrant *Muscisaxicola muscisaxicola macloviana*, and the common diving petrel *Pelecanoides urinatrix berard*. The penguins *Pygoscelis papua*, *E. crestata*, *Eudyptes chrysolophus*, and *Spheniscus magellanicus* are present on the Falklands. The islands are breeding grounds for sea lions (*Otaria flavescens*), elephant seals (*Mirounga leonoina*) and fur seals (*Arctocephalus gazella*). Fifteen species of whales and dolphins occur in the surrounding seas. Pelagic invertebrates such as the squid *Ilex argentinus* are abundant near the coast and can be considered the basic diet of birds and mammals.

Scotia Arc Islands (United Kingdom)

South Georgia, South Sandwich, and South Orkney Islands combined represent about 10,000 km^2 of land that peaks over 2,000 m. They are known as the islands of the Scotia Arc. Most of them rise steeply from the sea and are rugged and mountainous. Their coastlines are generally cliffed and rocky, with many inlets and some bay beaches.

South Georgia is a remote island located 1,330 km from any other land. It is largely barren and has steep, glacier-covered mountains. The climate is wet subpolar; it is variable, with mostly westerly winds throughout the year, interspersed with periods of calm; nearly all precipitation falls as snow. The South Georgia fauna is depauperated and consists mainly of direct developers. The intertidal fauna is of geologically recent origin. It is suggested that South Georgia shores stem from colonization by rafting from remote sources. There are abundant rocky shores with *Macrosystis* kelp offshore. Some 24 species of breeding seabirds and 26 visitors, 6 species of seals, and 10 endemic bottom dwelling fish are found around South Georgia. Penguins *Aptenodytes patagonicus*, *E. chrysolophus, Pygocelis antarctica* are present.

The South Sandwich Islands are of volcanic origin with some active volcanoes. They have a wet polar climate, and sparse vegetation of lower plants (lichens and mosses), richer around fumaroles with dense bryophyte communities. The penguins *A. patagonicus*, *E. chrysolophus, Pygocelis papua*, and *Pygocelis adeliae* are present.

The South Orkney Islands are located between the Antarctic Peninsula and South Georgia and they have a subpolar climate. The littoral zone is either lifeless or very poor. Scarce diatoms and a few species of chitons and gastropods are found along the littoral. At low tide, there are green and red algae in small rock pools, and amphipods and planarians under stones. At around 2 or 3 m depth, green and red algae appear with an increasing number of fauna including starfish, urchins, sponges, and ascidians. At around 8–10 m depth, new species of starfish appear and the general animal biomass increases with depth. At 30 m, ascidians, sea urchins, starfish, and brittle stars form vast colonies. The penguins *P. antarctica, P. papua* and *P. adeliae* are present around South Orkney.

Overall, the biota of the intertidal region is poor, even during summer, in the Scotia Arc Islands. Due to the rigorous conditions, most biota is confined to lower levels of the shore (Stephenson and Stephenson, 1972). The midlittoral zone that is most exposed to air is characterized by the presence of the *Porphyra* algae followed by *Ulothrix* and *Urospora* along the center of the zone. There are also algae in the tide pools and crevices. Limpets *Patinigera polaris* occur in crevices, whereas amphipods, nemertines, and flatworms are found clustered under stones and rocks. The tide pools of the midlittoral are home to many Antarctic species of algae such as *Leptosomia, Iridaea, Adenocytis* that can also extend their distribution to the infralittoral fringe area with *Desmarestia, Curdiea, Monostroma*, and *Plocamium*. The infralittoral fringe is characterized by numerous algae, mainly of the genera *Desmarestia* and *Ascosiera*. The huge *Phyllogigas grandifolius* dominates this area and its blades harbor a small fauna comprised, among others, of worms, sponges, hydroids, and mollusks (Stephenson and Stephenson, 1972).

There is no supralittoral fringe in many places, the mid-littoral fringe has a very restricted population but there is an increasing number of species in the infralittoral zone (Stephenson and Stephenson, 1972). Several animals such as sponges, alcyonarians, pycnogonids, amphipods, isopods, polychaetes, and nemertine worms are giant representative of their group (Knox, 1970).

The chlorophyll abundance in the surface waters shows a high degree of variability. The chlorophyll *a* in South Orkney fluctuates around 4 mg m^{-3} in summer. High pigment concentrations have been found especially between South Orkney and the South Sandwich Islands. The main species that compose the phytoplankton biomass are diatoms. *Chaetoceros neglectus, Chaetoceros dichaeta, Chaetoceros atlanticus, Nitzshia* sp., *Corethron criophilum* are the most common reaching densities around 10^5 cells L^{-1} (Zernova, 1970). The standing crop of zooplankton in the subantarctic represents 55 g m^{-3} between 0 and 50 m depth. Around South Georgia, the relationship between the density of phytoplankton and zooplankton is inverse (Knox, 1970). Copepods reach maximum densities at around 600 m depth, decreasing continuously toward the surface. The main species of zooplankton are *Euphausia superba, Euphausia frigida, Thysanoessa* spp., *Parathemisto* spp., and *Salpa fusiformis*. The main krill biomass is observed in areas of the island arc. The complicated underwater relief and coastline of this area leads to the formation of numerous gyres and local eddy currents trapping the krill (Makarov *et al.*, 1970).

Krill has been found to be an important part of the diet of numerous fish living around the islands, including members of the families Rajidae, Paralepidae, Myctophidae, Scopelarchidae, Muraenolepidae, Gadidae, Moridae, Macruridae. The occurrence of various fish in the epipelagic waters of the Scotia Arc Islands, which are described by many authors as a highly productive krill zone, are also profitable to whales, seals, birds, and Antarctic fishes, but also to subantarctic and subtropical epipelagic species of the southern hemisphere. The crabeater, leopard, Ross and some other seals can be found in the Scotia Arc Islands.

Annie Mercier and Jean-François Hamel

Bibliography

Bird, E.C.F., 1985. Atlantic Ocean Islands. In Bird, E.C.F. and Schwartz, M. (eds.), *The World's Coastline*. New York: Van Nostrand Reinhold, pp. 1035–1039.

Blindheim, J., 1989. Ecological features of the Norwegian Sea. In *Proceedings of the sixth conference of the Comité Arctique International*. Leiden: E.J. Brill, pp. 366–401.

Braarud, T., Ringdal Gaarder, K., and Nordli, O., 1958. Seasonal changes in the phytoplankton at various points of the Norwegian west coast. *Fiskeridirektoratet Skrifter Serie Havundersoekelser*, **12**: 77.

Coltman, D.W., Bowen, W.D., and Wright, J.M., 1999. A multivariate analysis of phenotype and paternity in male harbor seals, *Phoca vitulina*, at Sable Island, Nova Scotia. *Behavioral Ecology*, **10**: 169–177.

Day, R.W., 1983. Effects of benthic algae on sessile animals: observational evidence from coral reef habitats. *Bulletin of Marine Science*, **33**: 597–605.

Falcon, J.M., Brito, S.A., and Bundrick, C.M., 1996. Structure of and relationships within and between the littoral, rock-substrate fish communities off four islands in the Canarian Archipelago. *Marine Biology*, **125**: 215–231.

Knox, G.A., 1970. Antarctic marine ecosystems. In Holdgate, M.W. (ed.), *Antarctic Ecology*. London: Academic Press, pp. 69–96.

Lorenzo, J.A., 1995. Habitat use of wintering waders in the coast of El Medano (Tenerife, Canary Islands). *Miscellania Zoologica, Barcelona*, **18**: 153–160.

Makarov, R.R., Naumov, A.G., and Shevtsov, V.V., 1970. The biology and distribution of the Antarctic krill. In Holdgate, M.W. (ed.), *Antarctic Ecology* Vol. I. London: Academic Press, pp. 173–176.

Manning, R.B., and Fenner, A.C., 1990. Decapod and Stomatopod Crustacea from Ascension Island, South Atlantic Ocean. *Smithsonian Contributions to Zoology, Washington: Smithsonian Institution Press*, **503**, p. 91.

Morris, B., Barnes, J., Brown, F., and Markham, J., 1977. *The Bermuda Marine Environment. A Report of the Bermuda Inshore Waters Investigations* 1976–1977. St. George West, Bermuda: Bermuda Biological Station Special Publication No. 15.

Neto, A.I., and Avila, S.P., 1999. Subtidal algal communities and their associated molluscan fauna in Sao Vicente Bay (Sao Miguel, Azores). *Abstracts of the 34th European Marine Biology Symposium, Ponta Delgada, Azores*, p. 7.

Ojeda, A., 1996. Phytoplanktonic biomass and chlorophyll *a* in western Canary Islands. *Oceanographia y Recursos Marinos en El Atlantico Centro-Oriental, ICCM*, pp. 91–121.

Pawson, D.L., 1978. The echinoderm fauna of Ascension Island, South Atlantic Ocean. *Smithsonian Contributions to the Marine Sciences, Washington: Smithsonian Institution Press*.

Price, J.H. and John, D.M., 1978. Subtidal ecology in Antigua and Ascension: A comparaison. *Progress in Underwater Science Report of the Underwater Association, New Series*, **3**: 111–133.

Stephenson, T.A. and Stephenson, A., 1972. *Life between Tidemarks on Rocky Shores*. W.H. Freeman and Company, New York: pp. 81–96.

Williamson, M., 1981. *Island Populations*. Oxford University Press, UK.

Wirtz, P., 1999. New records of coastal marine animals from Madeira and from the Azores, and their zoogeographical interpretation. *Abstracts of the 34th European Marine Biology Symposium, Ponta Delgada, Azores*, p. 9.

Zernova, V.V., 1970. Phytoplankton of the Southern Ocean. In Holdgate, M.W. (ed.), *Antarctic Ecology*. London: Academic Press, pp. 137–142.

Cross-references

Arctic, Coastal Ecology
Antarctic, Coastal Ecology and Geomorphology
Atlantic Ocean Islands, Coastal Geomorphology
Caribbean Islands, Coastal Ecology and Geomorphology

ATLANTIC OCEAN ISLANDS, COASTAL GEOMORPHOLOGY

From south to north, the islands of the Atlantic covered here include the Scotia Arc (South Shetland, South Orkney, South Sandwich, and South Georgia), Bouvet, the Falklands, Tristan da Cunha and Gough, St. Helena and Ascension, Macronesia (Cape Verde, Canaries, Madiera, and Azores), Bermuda, Sable, Faeroes, and Jan Mayen (Figure A60). The Caribbean islands, Iceland, Great Britain, and Ireland are covered elsewhere.

Setting

With only a few exceptions such as Bermuda and Sable, it is clear from the spatial distribution of almost all of the other individual islands or island groups in the Atlantic Ocean that most owe both their location and coastal morphology to a turbulent tectonic and often volcanic past. The tectonic opening of the Atlantic Ocean is the key factor in this history. Commencing in the Late Jurassic some 140 million years ago and fully open some 65 million years ago (Hansom and Gordon, 1998), the Atlantic basin is characterized by the westward movement of both American plates and the eastward movement of the African and Eurasian plates. Basaltic magma upwelling into the crustal gap in the mid-ocean spreading center has over time produced a long submarine ridge composed of a series of fissure volcanoes known as the mid-Atlantic Ridge. In places the volume of upwelling magma has been sufficient to allow the construction of volcanoes that extend to the ocean surface and form the individual volcanic islands and island groups of Bouvet, Gough, Tristan da Cunha, St. Helena, Ascension, Macronesia, Kelard, and Jan Mayen. Elsewhere, where more than two plates and spreading directions are involved, this relatively simple mid-Atlantic geometry is more complex. For example, the development of the Scotia Arc in the South Atlantic was contingent on the opening of the Drake Passage and the eastward movement of several micro-plates along a progressively elongating arc. The boundaries of the Scotia Arc are marked in the south by the South Shetland and South Orkney Islands, in the north by South Georgia and in the east by the still-extending volcanic island arc of the South Sandwich Islands (Hansom and Gordon, 1998). As a result, the rocks of South Georgia, South Shetlands, and South Orkney are mainly composed of fragments of older rocks and intruded volcanics, most of which were created elsewhere and subsequently transported to their present sites. By contrast, 70% of the rocks of the South Sandwich Islands are recently extruded basalts (Baker, 1990).

This tectonic and volcanic past has also ensured that, with exception of parts of the Scotia Arc in the South Atlantic and Bermuda and Sable Island in the North Atlantic, almost all of the remaining mid-ocean islands are of volcanic origin (Bird, 1985a, b). Many remain as active volcanoes that rise from deep water and their coastlines are mainly characterized by steep cliffs cut into lava, ash or hyaloclastite. Since many of the volcanoes themselves are youthful, the coastlines are also young and in a state of continuing adjustment in response to marine activity and ongoing volcanic events. For example, Fogo, in the Cape Verdes, is an active stratovolcano with a symmetrical cone rising out of the ocean (Figure A61). Its slopes are composed mainly of subaerial lava flows, some of which are dated, and all of which end in marine cliffs, sometimes with small cliff-foot beaches composed of basalt boulders and gravels.

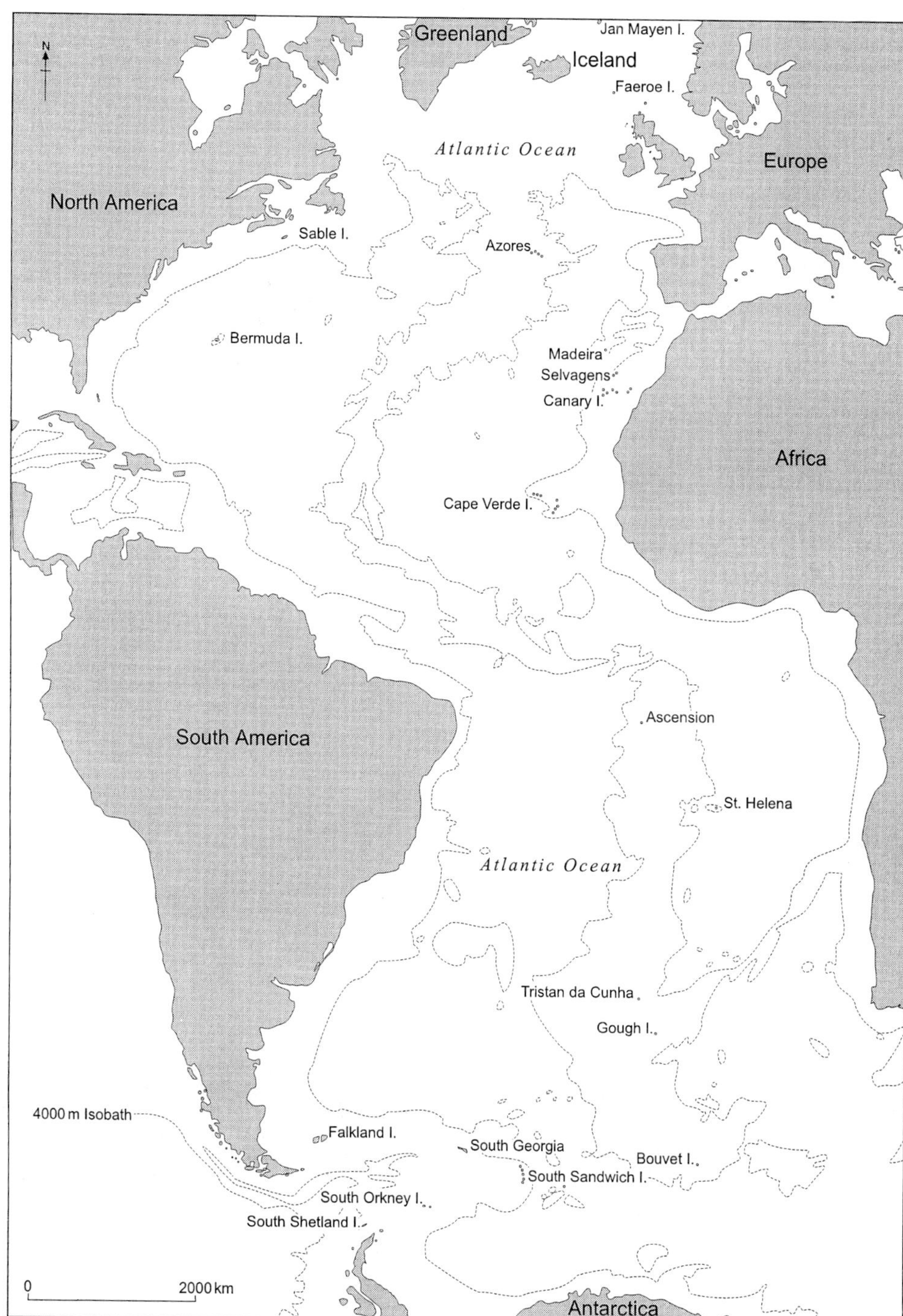

Figure A60 Location of the Atlantic Ocean Islands.

The Scotia Arc

Like most Antarctic coasts, the coasts of the Scotia Arc islands are mainly cliffed and rocky. Many of the cliffs are cut in glacier ice and beaches are regionally rare (Hansom and Gordon, 1998). Rapid cliff retreat rates of 1 cm a^{-1} have been calculated for the cliffs of Fildes Peninsula in the South Shetland Islands and are thought to be the result of efficient frost shattering (Hansom, 1983a). Since all of these coasts are affected by sea ice for up to eight months of the year, the more normal coastal processes associated with wave and tide operate intermittently and often in conjunction with floating ice fragments (Hansom and Kirk, 1989). This means that the relative order associated with wave processes is annually swept away and replaced by the bulldozing and grinding action of floating ice so that beach sediment is moved in a nonselective way to produce disorganized fabrics and landforms. In the South Shetland Islands much of the coast is either rock or ice-cliff, although some glacier termini rest on a low rocky platform at sea level. There are important exceptions, however, and significant beach development occurs on the peninsulas that protrude beyond the ice cover. The Fildes and Byers Peninsulas on King George Island and Livingston Island, respectively, are adorned with extensive emerged beach deposits, some of which connect offshore islets and skerries to the shore platforms of the main island (Figure A62). Prominent shore platforms occur at sea level and at altitudes of 3–8 m, 11–17 m, 28–50 m and 85–100 m above sea level and are cut into a variety of rock types (Hansom, 1983a). Although gravel beaches are found up to 200 m above sea level the most extensive are sited on the lower shore platforms at altitudes of 2.5–3 m, 5.50–6 m, 8–10.5 m, 12–13.5 m, and 16.5–18.5 m above sea level, as well as at present sea level (John and Sugden, 1971; Hansom, 1983a). Since the extensive sub-horizontal shore platforms that occur at present sea level in sheltered locations give way to steeper, ramp-like platforms in more exposed locations, Hansom (1983a) related their development to the interplay between the frequency of floating ice blocks that horizontally abrade the surface and wave activity that produces ramps. The upper part of the sub-horizontal platforms within sheltered bays is often adorned with an undulating layer of boulders that have been smoothed and compacted into extensive boulder pavements by the action of floating ice blocks (see Boulder Pavement) (Hansom, 1983b). Three of the islands, Penguin, Bridgeman, and Deception, are relatively new and active volcanoes and erosion of the friable rocks of the outer coast has produced cliffs of a variety of heights. The inner crater of Deception Island has been inundated by the sea and has several sand and gravel beaches that are frequented by penguins and tourists alike, in sharp contrast to the steep rocky outer coast.

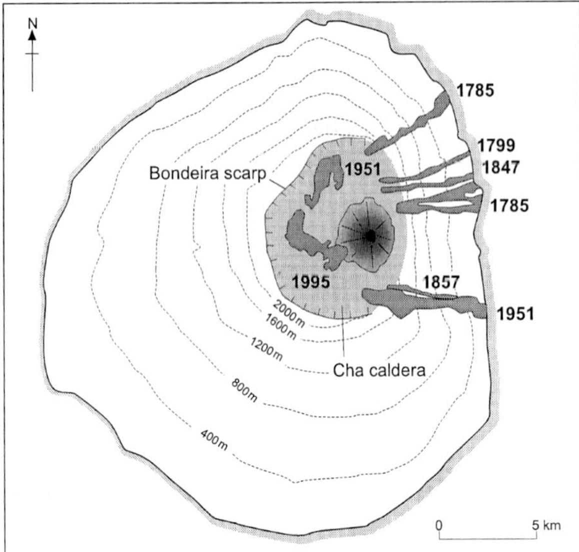

Figure A61 The island of Fogo, Cape Verde group is an active volcano and has a steep coast that is characteristic of many of the Mid-Atlantic Islands. 18–20th century lavaflows indicated as map.

Whereas the extensive beaches of parts of the South Shetland Islands allow for easy landing for both wildlife and boats, the coastline of the South Orkney Islands is more formidable. Three of the four largest islands, Coronation, Powell, and Laurie are dominated by large permanent ice caps spilling down steep rocky cliffs which plunge below sea level. A few bouldery beaches occur, for example, on Laurie Island. The remaining island, Signy is characterized by ancient schists and amphibolites that have been glacially scoured to produce a relatively low and rolling coastal plain with a rocky sloping shore albeit with only a few bouldery beaches. The coast of the South Sandwich Islands is as inhospitable as that of South Orkney, since only a small proportion of land is free of ice and there are few beaches or easy landing points. The South Sandwich group consists of a 400-km-long volcanic island arc caused by the subduction of the South Sandwich plate beneath the South American plate. Rising from an 8,428 m-deep ocean trench, the eleven main islands are all volcanoes made up of 60% lava and 40% tephra (Baker, 1990). The coastline is steep and cliffed but because many of the islands are still active volcanoes, erosion of softer tephras has produced a crenulate coastline with infrequent small bouldery beaches. All of the islands except Zavodovski are characterized by reefs and skerries and these are especially well-developed between smaller island clusters, such as in the extreme south at Thule, Cook and Bellingshausen islands. Low altitude bare coastal plains do occur on some islands, such as Zavodovski and Bellingshausen, but few scientists

Figure A62 Ardley Island, in the South Shetland Islands, is connected to one of the main islands by an intertidal sand tombolo and is adorned with raised beaches whose altitudes are the same as those on the adjacent islands.

Figure A63 Saint Andrews Bay, South Georgia, is fed by glacigenic sediment and is characteristic of the extensive outwash beaches of the north coast of the island.

have investigated these islands and so, the nature of the plains is unknown. In the strong westerly ocean circulation of the Scotia Sea, pumice from South Sandwich is known to travel great distances. The pumice from a submarine eruption close to Zavodovski Island in March 1962 reached the beaches of New Zealand in late 1963, having been transported east within the Antarctic Circumpolar Current at 11–17 cm s^{-1} (Hansom and Gordon, 1998).

South Georgia on the northern limb of the Scotia Arc, is a 170-km-long and 40-km-wide sequence of volcaniclastic sediments folded into a mountain range that rises out of the Atlantic Ocean to heights of over 2,700 m. It is not volcanically active but is heavily glacierized and has all the characteristics of a glacially eroded coastal fold mountain range. Long ridge-like peninsulas separated by deep glacial troughs jut out into the sea. The south coast is more heavily glacierized than the north and is extremely rocky and dominated by steep cliffs cut in ice or rock. The few small cliff-foot gravel or sandy beaches that occur are swept by storm waves from the southwest and occasionally by storm surges (Hansom, 1981). On the north-facing coast, the glacier cover is more restricted and a significant amount of ice-free ground exists, particularly on the peninsulas. The peninsulas are flanked by high and steep cliffs but these often have narrow shore platforms at their base together with numerous offshore islets and skerries. Some fragments of emerged shore platforms are known to occur at altitudes of 2, 5, 6–7, and 20–50 m above sea level but a systematic study of distribution and altitude is lacking (Clapperton, 1971). The lower of the shore platforms are often adorned with emerged beaches at 2–4 m and 6–7.5 m above sea level (Clapperton *et al.*, 1978). Within the northern bays and fjords of South Georgia lie the most extensive beaches of the Antarctic region. Fed by numerous debris-laden glaciers and short glacifluvial streams, substantial areas of the sheltered inner fjord heads have become infilled by glacifluvial outwash plains fringed by extensive sand and gravel beaches (Hansom and Kirk, 1989). The largest of these outwash beaches occur at Salisbury Plain, Fortuna Bay, Cumberland Bay and St. Andrews Bay (Figure A63) (Gordon and Hansom, 1986). Some of the inner reaches of the fjords of South Georgia are affected by sufficiently large quantities of floating ice calved from adjacent glaciers to allow the development of intertidal boulder pavements similar to those produced by sea ice further south (Hansom, 1983b).

Falkland Islands

The coastline of the Falkland Islands, like the islands themselves, is low, flat, and reminiscent of the coastline of the Outer Hebrides of Scotland. Although hundreds of smaller islands exist, the main island group comprises West and East Falkland separated by Falkland Sound. The intricate and crenulate nature of the coastline is probably more related to the submergence and faulting of the underlying Devonian, Carboniferous, and Permian limestones and sandstones, rather than to the efficiency or otherwise of marine erosion and deposition. In spite of the majority of the coastline being rocky, steep and high cliffs are mainly absent except in the extreme west where 100–200 m cliffs occur at Beaver Island. Everywhere and especially within the broad embayments and inlets, there are innumerable small islets and skerries whose detailed morphology and outline appear to be structurally controlled. A notable feature of the rocky coastline is the abundance of dense stands of giant kelp in the nearshore. On the western flank of the fault that controls Falkland Sound, narrow coast-parallel outcrops of hard and softer rock have been eroded to produce an intricate series of small headlands composed of more resistant rock separated by small arcuate bays cut into less resistant beds. The result is an essentially linear coastline stretching the entire length of Falkland Sound, broken by regular and uniform bays where the outer rocks have been breached. The crenulate nature of the coastline and extensive areas of nearshore shallow water means that a wide range of sheltered locations exists for potential beach development. Streams, although abundant and rarely dry, do not appear to contribute significantly to beaches, since there is a lack of beach development adjacent to the mouths of the creeks and streams entering tidal waters. Beaches of sand and gravels tend to be located in either outer coast sites where ocean waves gain access or within inlets where more open aspects allow wind-generated local waves to produce beaches. The high proportion of shell-sand of the beaches points to an important biological input from nearshore shallow waters. Significant beach development on the outer coast occurs at Bull Point, Lively Island, and at Bertha's Beach where sandy spits and tombolos connect small islets to mainland East Falkland. Paloma Beach, Elephant Beach, and Concordia Beach in the north of East Falkland are all examples of large open coast sandy beaches where strong winds have resulted in blown sand spreads over inland areas. In West Falkland, large beaches occur within the inlet of Whitsand Bay. Unfortunately, as a result of indiscriminate mine-laying during the Falklands War, several of the beaches of East Falkland are now unsafe and access is restricted.

Bouvet, Gough, Tristan da Cunha, St. Helena and Ascension

Bouvet Island (Figure A60) stands at the southern end of the mid-Atlantic Ridge, the slopes of the central cone terminating on all sides in precipitous cliffs that descend abruptly to sea level. Probably a complex of volcanic cones, the island rises almost symmetrically to a flat ice-covered dome 935 m high. The ice covers the eastern side of the island where it reaches the sea in an ice cliff 122 m high. The north and west sides are free of ice but are much steeper. Between 1955 and 1958, a new 0.2 km^2 rock platform emerged at 25 m above sea level as a result of either eruption or rock falls associated with tremors (Stonehouse, 1972). Since little low-level ground was previously to be found around the coast of Bouvet, the new addition has become a prime-breeding site for penguins, petrels, and seals. Gough Island (Figure A60) lies 2,300 km to the east of the Cape of Good Hope, South Africa and is the eroded summit of a Tertiary volcano. Although the island is mountainous, rising to 910 m above sea level, the slopes are cut by deep gullies or gulches and the coastline is characterized by steep cliffs that appear to have undergone erosion to produce a variety of narrow boulder beaches at the foot of the cliffs. There are also numerous islets, stacks, and skerries that mostly lie within 100 m of the coast. Tristan da Cunha lies 350 km to the northwest of Gough. It is a circular island of 98 km^2 with four other small islands close by. The base of the central volcanic peak lies

Figure A64 The steep cliffs of Saint Helena are capped by outcrops of trachyte at Great Stone Top on the south east coast. (Photo courtesy of Barry Weaver.)

3,700 m deep on the seabed and the summit rises to almost 2,100 m above sea level. The lower slopes are almost entirely lava and form high cliffs that surround the island. Emerged beaches, platforms, and caves occur at 5 m above sea level on Tristan but 12 m above sea level on the small adjacent islands, indicating differential tectonic uplift in the area (Bird, 1985b). In some places, small and inaccessible beaches occur, some of which are sandy. A small plateau at the foot of the main cliffs on the northwest side at Herald Point provides the site for Edinburgh, the only inhabited part of the island. The cliffs are lower along the Edinburgh coast, allowing access to the sea via small gullies. A narrow gravel beach has developed along this stretch of coast, fed by material eroded from the cliffs. In the months following the 1961 eruption at Big Point, on the updrift section of this coast, rapid recession of the cliff was noted (10 m in 8 weeks). This contributed to accretion of the downdrift gravel beaches which formed a spit enclosing a small lagoon (Bird, 1985b).

St. Helena (Figure A60) is 122 km^2 in area and lies in mid-ocean, 2,900 km from Brazil and 1,900 km from the west coast of Africa. Like Tristan and Gough, the coastline comprises high stepped and sometimes vertical cliffs cut by steep v-sided valleys. Emerged shore platforms stand at 4–6 m above sea level (Bird, 1985b) and the numerous offshore islets and skerries common on the south and west coast, such as at Egg Island and Cockburn Battery, are almost absent from the north and east coast. The highest cliffs occur at Flagstaff Bay in the north and at Man and Horse Cliffs in the southwest where they reach 580 m. The vertical steps in the coastal cliffs often correspond to differences in the composition of the outcropping of lava flows. At Great Stone Top, Turks Cap, and Prosperous Bay on the east coast, thick flows of trachyte form vertical faces, which cascade debris onto stone chutes below (Figure A64). Some of these extend into the sea to form fringing boulder beaches but nowhere on the outer coast are these well developed or extensive. Only in favored locations where the deep gulches reach the coast do small beaches, some of sand, occur, such as at Lemon Valley Bay. At Sandy Cove on the south coast, the sandy beach is backed by calcareous sand dunes. In mid-ocean some 1,300 km north of St. Helena lies Ascension Island, a Holocene stratovolcano with no known historic eruptions (Simkin and Siebert, 1994). Although more arid than St. Helena, the coasts of both islands are similarly high and surrounded by steep cliffs cut into volcanic lava. At Boatswain Bird islet, a natural arch has been eroded into a 98 m high monolithic stack of trachyte (Bird, 1985b). However, Ascension also has superb white sand beaches derived from shell and coral. Outside of the Caribbean area, the Atlantic Ocean is not noted for its corals although small structures are known from the coast of western Africa (Trenhaile, 1997). Although little seems to be known of the nature of the sediment supply to the Ascension beaches, the carbonates can only come from shells and coralline algae (such as *Lithothamnion*) growing in the narrow shallow water fringing the island. Erosion of adjacent rock shores and terrestrial sediment in-washed by infrequent rains probably account for the remainder of the beach sediment.

Macronesia

The island groups comprising Macronesia stretch lie in a long line that extends north between the coast of West Africa and the Portuguese coast (Figure A60). All of the island groups are volcanic in origin and many remain active volcanoes. Four main stratovolcanoes occur in the arid Cape Verde Islands, Fogo, Brava, Santo Antao, and San Vicente. Fogo is 2,829 m high yet only 24 km wide and so the steep seaward slopes end abruptly in retreating cliffs. The volcano has erupted 10 times between 1500 and 1995 each one sending long streams of lava down the flanks into the ocean and altering the coastal geometry. A similar picture characterizes the other islands of the group, each being surrounded by steep cliffs cut by deep gullies at the foot of which occur infrequent gravelly beaches. Emerged features have been noted at six levels up to 100 m above sea level. Some bays contain fringing algal reefs and in some locations boulders have been moved inland by wave activity by up to 200 m (Bird, 1985b). The Canary Islands comprise seven main islands, six of which host volcanoes. Some of these have erupted as recently as 1971. The coastline of the Canaries resembles that of the other mid-ocean volcanic islands in as much as the central volcanic spine of the islands falls steeply to a mainly rocky and cliff coast cut by deeply incised ravines. For example, in the northwest of Tenerife at Acantilado de Los Gigantes, sub-vertical cliffs reach 500 m. The occurrence of such cliffs has constrained some of the tourist-related expansion that many of the coastal towns and villages have undergone in recent years. Some cliffs have been formed in sediments brought down by torrents in the ravines. These deposits have since been tectonically uplifted together with the boulder beaches that once fronted them. The north coast of Gran Canaria has good examples of such cliffs, together with well-developed shore platforms a few meters above sea level. However, the volcanism of the Canaries has in places produced several long and low craggy volcanic peninsulas that have allowed beach development to occur. Fuerteventura, Lanzarote, Tenerife, and Gran Canaria all have sandy beaches some with extensive sand dunes. The long white sand beach at San Andres, on the southeast coast of Tenerife, has been artificially nourished with sand brought from the Sahara Desert. Between Maspalomas and Playa del Ingles, in the south of Gran Canaria, 328 ha of fine shell sand have accumulated at the mouth of the Fataga ravine to enclose a freshwater lagoon. On account of the relative aridity, most dunes in the Canaries are sparsely vegetated and the unfixed dunes at Maspalomas advance east towards the lighthouse at a meter per year. Some of the dunes have succumbed to tourist development, such as at Playa de las Canteras in the north of Gran Canaria. In the easternmost part of the islands, the supply of beach sediment from the nearshore and the inland ravines is augmented from a more exotic source. Fuertoventura and Lanzarote are affected by the summer *scirrocco*, a hot wind from the Sahara 90 km to the east, which carries large quantities of dust and desert sand to the islands. Locally known as the *kalima*, it turns the day into twilight and regularly covers surfaces with a thin layer of sediment. Over the centuries, this sediment has been an important source of sand for the beaches of the easternmost islands. In the south, Fuertoventura narrows at the Pared isthmus where sandy beaches extend eastward to connect the original island of Jandia to the main island by what is now a low and narrow peninsula. On the eastern shore, the sands of the 26 km-long Playa de Sotavento (Sp: leeward) are protected from the dominant westerly waves. On the western side, the Playas of Cafete, Pared and Barlovento (Sp: windward) are more exposed with strong currents and undertows. Nearby Lanzarote is a similarly dramatic landscape of lava fields and steep cliffs with intervening sand beaches, such as the sweeping Puerto del Carmen beach on the east coast. Accretion is common at many of the valley-mouth inlets.

Maderia comprises the main island itself, together with the smaller island of Porto Santo, the nearby Islas Desertas, together with the uninhabited Selvagens to the south. All of these islands are volcanic in origin but have not been active in the last 1.5 million years. As a result the, main volcanic core of Madeira, together with its plateau-like lateral subsidiary vents, has become eroded to produce deeply incised valleys and gorges running down to the sea. Between the mouths of the river valleys are high cliffs of vertical columns of basalt with layers of red and yellow tufa exposed in places. The 580 m plunging cliffs of Cabo Giroa, west of Funchal, are among Europe's highest, but high cliffs are found everywhere on the coast of Madiera. In the north, fragments of shore-platform occur as well as small islets and skerries, particularly near Faial. There are several well-developed stacks near Ponta do Sao Lourenco in the northeast. Small beaches of rounded gray basalt gravel occur at several places, particularly where river mouths occur. The only sandy beach is at Prainha, on the extreme east of the island where the sand is mainly basalt. Shell sand occurs on the low plateau area in the extreme east of the island. The nearby island of Porto Santo is 14 km long and 5 km wide, its generally low volcanic profile veneered by thin deposits of sand, calcareous sandstone and clay. The cliffs of the north coast reach 100 m but the south coast is dominated by an 8 km long white sand beach fronted by a shallow sandy-floored bay backed by low cliffs of cemented sands. Protected from the main force of southwesterly storms by Madeira, from the west by the small island of Baixo and from the north by its own cliffs, the south coast of Porto Santo is relatively sheltered. It appears to have been subject to sediment accumulation over a substantial period of time derived both from a combination of shell sand from shallow nearshore surfaces, aeolian sand blown west on the *scirrocco* from the Sahara to the east and from topsoil erosion caused by early deforestation. The Selvagens are composed of limestone capped by lava and ash and are cliffed to 100 m in the north but with a gentler beach-fringed coast in the south. Emerged beaches and dune calcarenites occur on Pleistocene marine terraces at 3 and 7 m above sea level (Bird, 1985b).

The most northerly of the islands of Macronesia, the nine islands of the Azores are a widely separated group of mountainous but fertile islands which share the steep nature of many of the mid-oceanic islands but also have long beaches and many fishing harbors. The coast of the largest island, Sao Miguel, is a microcosm of the Azores coastline. Where the coast is backed by higher volcanic peaks, it is characterized by steep cliffs fronted by patchy low shore platforms, offshore stacks and islets, such as at Mosteiros. Emerged beaches and platforms occur at various heights up to 60 m (Bird, 1985b). However, the highly indented coast also has short sandy beaches between cliff headlands. Where the hinterland is lower, longer sand beaches occur such as at Praia dos Moinhos and at Populo where small pine-clad sand dunes occur. Several coastal features of note occur on the other islands. Pico Island is 42 km long and dominated by a 2351 m cone-shaped strato-volcano of the same name. Various historical lava flows have extended the coastline and now have been eroded back to form low cliffs. In places sheltered from the dominant westerly waves, smaller volcanic forms survive on the coast, such as at Barca on the island of Graciosa where a volcanic cone has been sectioned by marine erosion to form bays fringed by crumbling cliffs of ash that cascade onto boulder beaches below. The 1957/58 eruption at Capelhino on Faial showered the adjacent coast in ash and contributed to accretion of the beach. At Porto Pim, also on Faial, a substantial sandy spit has developed (Bird, 1985a). Such is the power of the surf reaching the Azores that Pico has been chosen as the site of a pilot plant to produce energy from waves.

Faeroes and Jan Mayen

The rugged outer coast of the 16 main islands that comprise the Faeroes is the result of the deep incision of past glaciers into a thick pile of horizontally bedded flood basalts. These are slowly eroding, especially on the west and north coasts, due to constant exposure to high-energy westerly and northerly North Atlantic waves. The outer coast is dominated by high and sub-vertical cliffs that reach 820 m at Hestmuli in the north of the island of Kunoy. The rate of recession is unknown but is higher on the seaward facing slopes where cliffs have developed that tend to be steeper than immediately adjacent landward-facing glaciated slopes, such as at Tindholmur in the extreme west. The slopes of the inner coast are the sides of over-steepened glacial troughs that have been flooded by Holocene sea-level rise and are more subdued and better vegetated than the cliffs of the outer coast (Figure A65). Small skerries, islets, and stacks (Fr: drangur) are common as are deep clefts and geos (Fr: gjogvs) where dykes and faults have been differentially eroded into the horizontal basalts of the west coast. Small, often structurally controlled, ramped shore platforms are common around most of the coast but tend to be more fragmented and smaller on the more exposed west and south coast. More continuous and better-developed platforms occur near the entrances to the fjords, particularly in the east. For example, well-developed shore platforms occur on the east coast of Vidoy. There are very few beaches in the Faeroes and the ones that do occur are almost exclusively found at the sheltered fjord heads. In such locations, wave approach is unidirectional and the limited amounts of

Figure A65 Glacial troughs have been cut into horizontally bedded basalts and subsequently flooded by Holocene sea level rise to produce cliffs that are characteristic of the inner coast of the Faeroese fjords. The plunging cliffs of the Faeorese outer coast are everywhere exposed to high energy Atlantic waves.

sediment supplied by the small streams accumulates to produce small pocket beaches composed of black basalt sand (such as at Tjornuvik, Funningsfjordur, and Vidareidi) and occasionally backed by low sand dunes (such as at Sandur and Saksun).

The coastline of Jan Mayen is dominated by the volcanic bulk of Beerenberg, a 2,277 m stratovolcano that comprises the northern half of the 54-km-long island (Norsk Polarinstitutt, 1959). The coastline of Nord-Jan is steep and rugged, comprising 200–400-m-high cliffs eroded in ash, lava and tephra as well as into the ice walls of tidewater glaciers that spill down from the central crater. The coastline shows signs of recent advance as a result of lava tongues from eruptions in 1970 and 1985, which have extended into the ocean. These recent eruptions and flows are an extension of the mode of construction of the island over an estimated 700,000 years. Such additions to the coastline have resulted in the base of preexisting cliffs becoming buried by newer lavas and it is onto this platform that more recent glacier terminal moraines have been deposited, such as at Smithbreen in the northeast of the island (Kinsman and Sheard, 1962). At Kroksletta, near the northern cape, 4,000-year-old moraines that rest on top of an emerged shore platform and beach at 8–10 m above sea level have been buried by subsequent lava flows. The protruding seaward edge of the Kroksletta lava is now cliffed to a height of 5–13 m (Kinsman and Sheard, 1962). The south part of Jan Mayen is a hilly ridge of scoria craters and mounds and trachyte domes whose lower elevations and older age has produced a coastline of low gradient rocky platforms connected by gravelly beaches. In several places around the Sud-Jan coastline a prominent rock platform is present upon which is sited emerged beach gravels. This feature is particularly well developed on the north coast of Sud-Jan at Sorbukta, Engelsbukta, and Haugenstranda. Sediment supply to the central section coastline that joins Sud-Jan to Nord-Jan appears to be healthy both from erosion of the lava cliffs to the west and from shallow water sources: the 10 m depth contour lies about 0.7 km offshore. This has resulted in the construction of a substantial barrier beach enclosing a lagoon along an 11-km-long stretch of south-central coast at Lagunevollen, with a smaller barrier and lagoon at Stasjonsbukta on the north coast. The composition of these barrier beaches is not known but assumed to be composed of mixed sands and gravels transported from the west. Such a large beach complex is unusual for a small mid-oceanic island. However, the full impact of Atlantic westerly waves and swell at Sud-Jan is modified by a large circular submarine reef (probably an eroded volcanic cone), which reaches to within 11 m of the sea surface some 10 km south of Sud-Jan. Although the north coast is subject to variable amounts of sea ice for an average of four months, the south coast is mainly free of ice (Gloerson *et al.*, 1992). Elsewhere, the coastline of Sud-Jan is steep and rocky but the cliffs are not as high as in Nord-Jan and small gravelly cliff-foot beaches fed by rockfall and wave erosion of the adjacent lava occur.

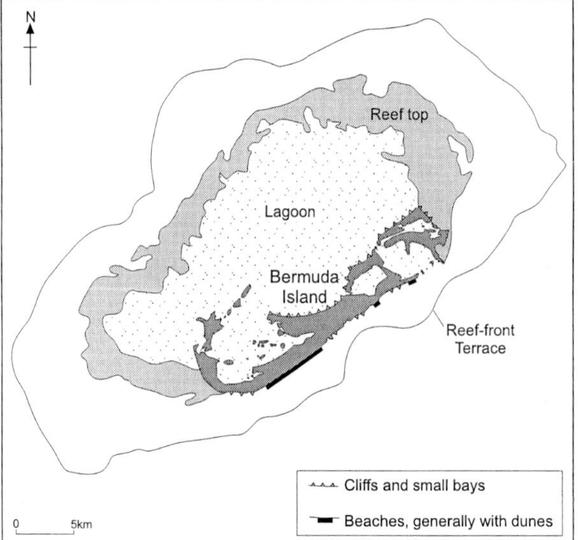

Figure A66 The calcarenite islands of Bermuda are sited atop a lava plateau. The present coastline is mainly composed of low cliffs with long beaches in the south (after Vacher, 1973).

Bermuda and Sable

With two notable exceptions, Bermuda and Sable, all of the Atlantic Ocean Islands have, or have had, a close and recent association with tectonic and volcanic activity. However, Bermuda, lying to the east of the Carolinas coast of the United States is a group of 150 limestone islands located about 1,000 km east of Cape Hatteras (Figure A60). The islands rest on the southern margin of a 650 km^2 platform that is presently submerged to depths of 20 m. However, boreholes at Bermuda have penetrated through a thin capping of calcarenites to meet lava at a variety of depths of 171, 43, 33, and 21–24 m (King, 1972). The platform of volcanic rocks is the site for the northernmost coral reefs in the North Atlantic. According to Vacher (1973) the present-day Bermuda platform consists of four geomorphological-ecological provinces: a reef-front terrace at 20 m depth; a main reef composed of 4 m deep coral-algal reefs everywhere except in the south where algal intergrowths occur; a 16 m deep lagoon in the north and the islands themselves forming a northeast trending chain near the southern edge of the platform (Figure A66). The exposed limestones of the islands are Pleistocene calcarenite, 90% of which is aoelian and the rest is beach and shallow water in origin (Vacher, 1973). The aeolianites were built from calcarenite produced mainly offshore and transported to the shore and so the inference is that during an earlier phase of submergence, the coastline was supplied by abundant reef-derived materials, which then led to beach and dune development before becoming lithified. This supply of reef-derived sediment to beaches and dunes continues today and progradation appears to be healthy (Bird, 1985a), but patchy. Nevertheless the present supply also appears to be more restricted than the former supply, since only on the island's south shore do extensive beaches backed by dune occur. These extensive sandy beaches are pink because of large numbers of fresh *Homotrema* clasts derived from the reefs 1 km from the shoreline (Mackenzie *et al.*, 1965). However, much of the modern south coast is cliff with only limited sources of offshore-derived sediment. Unlike during the earlier development of the calcarenites, the sediments of the lagoon now appear to play little part in beach supply. As a result, the islands' lagoon-facing shore is erosional with a line of cliffs and few pocket beaches (Vacher, 1973).

Sable Island (Figure A60), on the continental shelf to the east of Nova Scotia, Canada, is a low and wind-swept series of sandy islands famous for its sandy shoals and shipwrecks. The development of the Sable Island Bank, on which the islands sit, is related to the proximity of the continental shelf-break and the maximum ice positions of the Late Wisconsin and later readvances. These ice movements deposited a thick suite of glacigenic tills and superficial sandy material, that was subsequently subject to transgression, modification, and transport (King, 2001). The island has spectacular but desolate sandy beaches backed by sand dunes that reach up to 30 m high and cover 40% of the island's surface (Byrne and McCann, 1995). The intertidal zones of the long sandy beaches are characterized by prominent shoreface attached ridges with intervening depressions that reach up to 12 m deep (Dalrymple and Hoogendoorm, 1997). Strong alongshore currents cause eastward migration of these bars alongshore at rates of up to 50 m a^{-1} and at angles of up to 50° to the coastal orientation (Dalrymple and Hoogendoorm, 1997). The dunes of Sable consist of primary dunes that have developed *in situ* together with secondary dunes that have migrated across the island. The resultant coastal morphology represents a mix of both natural processes and anthropogenic disturbance. For example, the constantly changing coastal outline has resulted in the relocation of the lighthouse and is thought to be partly due to dune mobilization and reduced vegetation cover under the enhanced grazing pressure of introduced ponies (Owens and Bowen, 1977). Sable Island has undergone 14.5 km of eastward migration but, in spite of this, its 30 km^2 has been maintained over the past 200 years (Cameron, 1965). Sable Island thus seems to be subject to a regime of deposition that appears roughly balanced by an equivalent amount of erosion.

J.D. Hansom

Bibliography

Baker, P.E., 1990. The South Sandwich Islands. In LeMasurier, W.E., and Thomson, J.W. (eds.), *Volcanoes of the Antarctica Plate and Southern Oceans*. Washington, D.C.: American Geophysical Union, pp. 361–395.

Bird, E.C.F., 1985a. *Coastline Changes A Global Review*. Chichester: Wiley, p. 219.

Bird, E.C.F., 1985b. Atlantic Ocean Islands. In Bird, E.C.F., and Schwartz, M.L. (eds.), *The World's Coastline*. New York: Van Nostrand, pp. 1035–1039.

Byrne, M.L., and McCann, S.B., 1995. Canadian Landform Examples-31. The dunescapes of Sable Island. *Canadian Geographer*, **39**(4): 363–368.
Cameron, H.L., 1965. The shifting sands of Sable Island. *Geographical Review*, **44**: 363–376.
Clapperton, C.M., 1971. Geomorphology of the Stromness Bay-Cumberland Bay area, South Georgia. *British Antarctic Survey Reports*, 70, p. 25.
Clapperton, C.M., Sugden, D.E., Birnie, R.V., Hansom, J.D., and Thom, G., 1978. Glacier fluctuations in South Georgia and comparison with other island groups in the Scotia Sea In Van Zinderen Bakker, E.M., (ed)., *Antarctic Glacial History and World Palaeoenvironments*, Rotterdam: A.A. Balkema, pp. 95–104.
Dalrymple, R.W., and Hoogendoorm, E.L., 1997. Erosion and deposition on migrating shoreface-attached ridges, Sable Island, Eastern Canada. *Geoscience Canada*, **24**(10): 25–36.
Gloerson, P., Campbell, W.J., Cavalieri, D.J., Comiso, J.C., Parkinson, C.L., and Zwally, H.J., 1992. *Arctic and Antarctic Sea Ice, 1978–1987. Satellite passive-microwave observations and analysis (NASA SP-511)*. Washington, D.C.: Scientific and Technical Information Program, NASA.
Gordon, J.E., and Hansom, J.D., 1986. Beach forms and changes associated with retreating glacier ice, South Georgia, *Geografiska Annaler. series A*, **68A**(1–2): 15–24.
Hansom, J.D., 1981. Storm surging in South Georgia. *British Antarctic Survey Bulletin*, **53**: 141–146.
Hansom, J.D., 1983a. Shore platform development in the South Shetland Islands Antarctic. *Marine Geology*, **53**: 211–229.
Hansom, J.D., 1983b. Ice-formed intertidal boulder pavements in the sub-Antarctic. *Journal of Sedimentary petrology* , **53**(1): 0135–0145.
Hansom, J.D., and Gordon, J.E., 1998. *Antarctic Environments and Resources: a Geographical Perspective*. Addison, Wesley, Longman, p. 402.
Hansom, J.D., and Kirk, R.M., 1989. Ice in the intertidal zone: examples from Antarctica. *Essener Geographische Arbeiten*, Bd. **18**: 211–236.
John, B.S., and Sugden, D.E., 1971. Raised marine features and phases of glaciation in the South Shetland Islands. *British Antarctic Survey Bulletin*, **24**: 45–111.
King C.A.M., 1972, *Beaches and Coasts*. New York: St. Martins press.
King, E.L., 2001, A glacial origin for Sable Island: ice and sea-level fluctuations from seismic stratigraphy on Sable Island Bank, Scotian Shelf, offshore Nova Scotia. *Geological Survey of Canada*, Current research. D19, 18.
Kinsman, D.J.J. and Sheard, J.W., 1962. The Glaciers of Jan Mayen. *Journal of Glaciology*, **4**: 439–448.
Mackenzie, F.T., Kulm, L.D., Cooley, R.L., and Barnhart, J.T., 1965, *Homtrema rubrum* (Lamark), a sediment transport indicator. *Journal of Sedimentary petrology,* **35**: 265–272.
Norsk Polarinstitutt, 1959. *Topografisk Kort over Jan Mayen*. Blad 1 & 2, 1;50 000, Oslo.
Owens, E.H., and Bowen, A.T., 1977, Coastal environments of the maritime provinces. *Maritime Sediments*, **13**: 1–32.
Simkin, T. and Siebert, L., 1994. *Volcanoes of the World*. Tucson: Geoscience Press.
Stonehouse, B., 1972. *Animals of the Antarctic: The Ecology of the Far South*. London: Peter Lower/Eurobook.
Trenhaile, A.S., 1997. *Coastal Dynanmics and Landforms*. Oxford: Clarendon Press.
Vacher, L., 1973, Coastal Dunes of Younger Bermuda. In Coates, D.R. (ed.), *Coastal Geomorphology*. London: George Allen & Unwin, pp. 355–391.

Cross-references

Antarctica, Coastal Ecology and Geomorphology
Atlantic Ocean Islands, Coastal Ecology
Boulder Pavements
Cliffs, Erosion Rates
Cliffs, Lithology versus Erosion Rates
Glaciated Coasts
Scour and Burial of Objects in Shallow Water
Shore Platforms
Volcanic Coasts
Wave Power

ATOLLS

Atolls are coral reefs found in the open ocean consisting of an annular rim surrounding a central lagoon. There is a general presumption that atolls have volcanic foundations and some 425 have been recognized around the world (Wiens, 1962; McLean and Woodroffe, 1994). However, because of similar superficial morphology many shallower water coral reefs have been termed "shelf atolls" (Ladd, 1977). In Indonesian waters 55 such reefs have been recognized though many will not have the volcanic foundations of the majority of atolls found in the Pacific and Indian Oceans.

The largest atoll is Kwajalein in the Marshall Islands (120×32 km^2) followed by Rangiroa in the Tuamotus (79×34 km^2), though many smaller atolls are only a few kilometers in diameter. In a sample of 99 atolls, Stoddart (1965) indicated a mean area of 272.5 km^2, and atolls worldwide are considered to have a total area of 115,000 km^2.

The contribution of atoll studies to marine geology

Ever since the historic scientific expedition of Charles Darwin in the *Beagle* (1832–36) atolls have captured the imagination of marine geologists and Darwin's synthesis of his ideas on the formation of coral reefs (1842) have stimulated 160 years of research. This stimulus has come from the facts that:

- reef-building corals grow in shallow (<100 m) water
- many atolls rise from depths of several thousand meters
- there is a similarity of morphology (annular rim and lagoon) wherever atolls are formed.

Darwin's hypothesis of atolls evolving from fringing reefs around a volcanic island via a barrier reef stage was the catalyst for subsequent mainstream investigations into glacio-eustatic sea-level change, vertical and horizontal tectonic movements, and the foundations and internal structure of oceanic atolls. Atoll investigations have subsequently played a major part in research into carbonate diagenesis, isostasy, continental drift, plate tectonics, and seafloor spreading. They have generally confirmed the remarkable insight of Darwin regarding the origin of atolls.

The internal structure and origin of atolls

Darwin recognized that resolution of the problem of atoll development required deep drilling and, after several aborted attempts, T. Edgeworth David of Sydney University in 1896–98 drilled the atoll of Funafuti to a depth of 339.7 m, all in shallow water coral, but so fragmented that those who opposed subsidence argued that the core passed through a detrital slope deposit. It was to be almost 50 years before drilling totally penetrated the coral of an atoll to the basaltic basement at a depth of more than 1,300 m on Enewetak atoll. Subsequent drilling, much of it associated with atomic tests, at Enewetak and Bikini atolls but also at Midway by the US Geological Survey, and at Muroroa and Fangataufa in French Polynesia, by the French, confirmed both the volcanic foundations of atolls and the depth of *in situ* coral far below the photic limits of both modern coral growth and growth at glacially lowered sea levels. Volcanic basement was determined at 1,283 and 1,408 m at Enewetak, 415 and 438 m on Mururoa (though seismic results show basalt as shallow as −170 m, Guille *et al.*, 1996), 153 and 503 m at Midway and 270–400 m at Fangataufa. These figures are typical of those from many other atolls, which have now been investigated by drilling or seismic survey.

While shallow water corals have been recognized to depths greater than 1,000 m, clearly confirming widespread subsidence, considerable lithification, cementation, and diagenesis has taken place in the lower sections of the atoll foundations, which have been determined as up to 30,000,000 years in age. Alteration of the carbonate rocks results from fluctuations of more than 100 m in sea level with vertical migration of phreatic water table favoring the transformation of the original aragonite of the biogenic structure to calcite. At depths greater than 500 m cold permeating sea water dissolves both aragonite and calcite leaving behind only low-Mg calcite. Dolomitization takes place within thick aquifers of mixed fresh and saline waters and is common in the lowest portions of the atoll foundations.

Like other reefs, atolls were most recently exposed during the last glaciation when karstic processes prevailed on the atoll surface, as in previous low sea-level phases, forming a clearly identifiable "solution unconformity" beneath Holocene reef material which is commonly only 10–20 m thick. Karstic features, including the saucer-like shape of atolls, have long been considered to be the major determinant of the gross morphology of atolls (e.g., see Purdy, 1974).

The internal structure and radiometric ages of basalts beneath atolls have been integral parts of the evidence to develop the ideas of plate tectonics and seafloor spreading (e.g., Scott and Rotondo, 1983; Guille *et al.*, 1996). Volcanoes develop at oceanic "hot spots" from which they migrate through tectonic plate movement, acquiring coral reefs if they are in warm enough waters. Subsidence takes place due to both volcanic loading and cooling, and reefs and islands pass through the classic Darwinian sequence of fringing, barrier, and atoll formations producing linear chains along the direction of plate movement, such as the Hawaiian and Society Islands. If they are out of the zone of coral growth, subsidence continues and atolls become submerged as guyots, for example, the Emporer Seamounts to the northwest of the Hawaiian chain.

The surface morphology of atolls

The high energy situation of most atolls means that once they reached sea level after their last period of exposure, they have quickly developed strong zonational patterns. Reef slopes commonly display groove and spur structures, reef crests are surmounted by significant algal ridges and reef flats of about 500 m width are a conduit for sediments which form accumulating aprons of sand within the lagoon. Lagoons, with or without patch reefs are often of similar depth, within the range of 10–30 m with only a thick Holocene sediment veneer over the older Pleistocene foundations. Water circulation within the lagoons is complex, with wave setup raising swell side water levels and strong flow over the rim. Water escapes through leeward passes but in many instances a return bottom current back towards the swell side may be present.

Particularly on the swell and/or windward side of atolls, are linear islands termed "motus" separated by narrow passages or "hoa." The majority are constructed of coral gravel broken from the reef front and deposited a set distance back on the reef flat as the entraining waves lose their transportational capacity. Spectacular episodes of deposition have been described during hurricanes or tropical cyclones but these are not absolutely necessary for motu formation as equatorial atolls such as the Maldives, not experiencing strong, tropical revolving storms, also have significant island development. These islands may be only a metre or so above high tide and are regarded as the coastal landforms most susceptible to sea-level rise associated with global climate change.

David Hopley

Bibliography

Darwin, C., 1842. The structure and distribution of coral reefs. London: Smith, Elder and Co.

Guille, G., Goutiere, G., Sornein, J.F., Buiges, D., Gachon, A., and Guy, C., 1996. *The Atolls of Mururua and Fangataufa (French Polynesia). I, Geology–Pelrology–Hydrogeology, From Volcano to Atoll.* Musée Oceanographique Monaco.

Ladd, H.S., 1977. Types of coral reefs and their distribution. In Jones, O.A., and Endean R. (eds.), *Biology and Geology of Coral Reefs*, Vol. IV, Geology 2, New York: Academic Press, pp. 1–19.

McLean, R.F., and Woodroffe C.D., 1994. Coral atolls. In Carter, R.W.G. and Woodroffe, C.D. (eds.), *Coastal Evolution: Late Quaternary Shoreline Dynamics*. Cambridge: Cambridge University Press, pp. 267–302.

Purdy, E.G., 1974. Reef configurations: cause and effect. *Society of Economic Paleontologists and Mineralogists, Special Publication*, No. 18, 9–76.

Scott, G.A.J., and Rotondo, G., 1983. A model to explain the differences between Pacific plate island-atoll types. *Coral Reefs*, **1**: 139–149.

Stoddart, D.R., 1965. The shape of atolls. *Marine Geology*, **3**: 369–383.

Wiens, H.J., 1962. *Atoll Environment and Ecology*. New Haven: Yale University Press.

Cross-references

Algal Rims
Beachrock
Bioconstruction
Bioherms and Biostromes
Cays
Coral Reef Islands
Coral Reefs
Pacific Ocean Islands, Coastal Geomorphology
Small Islands

AUSTRALIA, COASTAL ECOLOGY

Australia is a large island continent with over 61,700 km of coastline (including near by islands). The coastline stretches from the tropics (10°S) to cool temperate (44°S), north–south, encompassing 30° of longitude (113°E to 153°E), east–west. Australia's ocean territory is one of the largest marine jurisdictions in the world (16.1 million km^2) and extends from the tropical epicenter of global marine biodiversity to the Antarctic (Zann, 1995). Australia's continental inshore environment contains a major slice of the marine biodiversity of the Southern Hemisphere, including a large number of unique species (in the temperate south) and species threatened in neighboring regions (in the tropical north).

Several features distinguish Australian coastal environments from other coastal environments in the world: high climate variability; low runoff; and the influence of the El Niño–Southern Oscillation. Australia has the least runoff, the fewest rivers and the lowest rainfall of any inhabited continent. In addition, high evaporation rates and low continental relief (particularly along the southern coast) combine to markedly reduce the rate of runoff to oceans. Australia's high climate variability is partly associated with the effects of the El Niño–Southern Oscillation, which results in drought followed by heavy rains and flooding—particularly in the east (DEST, 1996). In southern areas, ecosystems are attuned to drought/flood cycles while in northern areas, ecosystems are attuned to wet/dry cycles.

Australia's unique geological and tectonic history have produced a highly diverse and distinctive marine biota. Over geological timescales, Australia's coastal environment and biota has evolved and been greatly influenced by geological, tectonic, and climatic stability, and along the southern coast, by geographical and climatic isolation (Poore, 1995). Australia has had a relatively stable climate due to the northward movement of the Australian tectonic plate (which has compensated for global cooling). The rate of change of relative sea level along the coastlines has been slow and largely controlled by changes in global sea level and eustatic processes, rather than localized tectonic movements. This has allowed the adaptation and migration of the marine biota. In addition, Australia's variable climate, range of coastal environments, and the low nutrient status of its coastal waters, have contributed significantly to the diversity and endemism of Australia's marine environments, particularly among some marine taxon (e.g., fish, mangroves, seagrasses, macroalgae, sponges). Low nutrient regimes generally promote biological diversity and coevolutionary strategies to rapidly harvest, utilize, and recycle limited nutrient resources.

The position, size, and shape of the Australian continent have also contributed to the high level of marine biodiversity. Australia straddles the Tropic of Capricorn, so over one-third of the continent is in the tropics with the remainder in the temperate latitudes. The large continental landmass of Australia (and extensive continental shelf) and the shape of the continent provide extensive east–west coasts, with few contiguous coasts in the world possessing such extensive east–west coastlines. Australia's northern coastline adjoins tropical water masses and the southern coastline (the longest in the Southern Hemisphere) adjoins the Southern Ocean, resulting in distinct tropical and temperate biotas. The El Niño–Southern Oscillation also has a major influence on water temperatures and marine life.

In southern Australia, endemism among the marine fauna and flora has been enhanced by the isolation of the coastal shelf biota from other southern continents (by ocean basins) since the Paleocene or Eocene and also, from a latitudinal temperate gradient to the north (Poore, 1995). Further, geological stability has allowed relatively undisturbed sequences, compared with many other parts of the world, and the survival of many ancient and relict species of Pangaean and Gondwanan origin, particularly for many invertebrate species (Durham, 1979; Fleming, 1979). The long east–west, ice-free, extent of the southern coastline (i.e., the longest stretch of south facing coastline in the Southern Hemisphere) has also contributed to the diversity among the temperate biota. Together, these factors have contributed to globally significant levels of biodiversity and endemism in the marine biota, particularly among the marine algae (Womersley, 1990; Phillips, 2001) and some invertebrate taxon (i.e., bryozoans, ascidians, sponges) (Edyvane, 1996). In addition, many of the marine species, which inhabit the temperate reefs of Australia, are characterized by short larval periods and localized dispersal. For these reasons, there is a great tendency for local and regional rarity and endemism in temperate waters, with species distributions characterized by small, isolated, localized populations.

Australia's "State of the Environment Report" (Commonwealth of Australia 2001) concluded that Australia's marine environments and habitats are generally in good condition, to the extent that this can be measured. Many of Australia's remote areas are assumed to be in good health because they are thought to face few pressures, but information is lacking to confirm this. Only a few areas, however, can be regarded as pristine, because pressures such as nutrient loading, pollution with persistent chemicals, and fishing, have affected nearly all parts of Australia's marine and estuarine ecosystems. Near many cities and other parts of the coastline the condition of some habitats (particularly seagrasses, saltmarshes, and mangroves), is poor. There is increasing concern at the state of estuaries, with assessments of declining water quality, threats to estuarine fisheries and other impacts in many areas. Most rivers, estuaries, and coastal waters near Australia's large population centers show signs of eutrophication. (Commonwealth of Australia 2001) Blooms of toxic marine algae, some species of which may have been introduced from other countries in ships' ballast waters, are periodic, serious problems in parts of southern Australia.

Physical setting

Oceanography

The waters surrounding Australia (and New Zealand) are linked to three of the large ocean basins of the Southern Hemisphere, the Pacific, the Indian, and Southern Oceans. The southwest Pacific Ocean, which borders the eastern margins of Australia, is dominated by the large anti-clockwise circulation of the South Pacific. The elements of this gyre consists of the western-flowing South Equatorial Current, which after crossing the tropical Pacific, deflects southwards to form the East Australian Current, and then turns eastward. South of Australia is the massive Antarctic Circumpolar Current or "West Wind drift". On the western margins of Australia, the Indian Ocean South Equatorial Current, driven by trade winds and westerlies, has a similar anti-clockwise circulation, flowing towards the east coast of Africa and merging with the Agulhas Current, the strongest western boundary current in the Southern Hemisphere. The subtropical convergence and the subpolar convergence zones isolate tropical and subtropical waters of Australasia from the nutrient-rich waters of the Southern Ocean, although some mixing of subantarctic waters does occur at depth in the south Tasman Sea (see Figure A67). For the most part, the southern coast faces the full force of the Southern Ocean and as such, experience some of the highest wave energies in Australia. The El Nino–Southern Oscillation (and El Nino/La Nina cycles) has a major influence on prevailing current systems around Australia, dramatically affecting water temperatures and marine life, including major pelagic fisheries.

Both sides of the island continent of Australia experience poleward-going warm currents, which act as conduits to bring tropical ocean waters with their entrained plants and animals to southern Australian waters. On the east is the East Australian Current (EAC) (a western boundary current) and on the west is the Leeuwin Current, which flows southward along the outer shelf to the southwest corner of the continent, and then runs eastward along the southern margin (Cresswell, 1991). The EAC flows south from the Coral Sea (Church, 1987) close to the continental slope, until central New South Wales (NSW) where it runs further offshore (Cresswell *et al.*, 1983). Once or twice a year the EAC turns into loops in the Tasman Sea off the NSW coast. These loops become detached from the current and form EAC eddies, warm disc-shaped water parcels having a diameter of 200–300 km, and extending 1,500–2,000 m into the ocean (Nilsson and Cresswell, 1981).

On the west coast of Australia, the Leeuwin Current is unique. Unlike other eastern boundary currents (such as the California, Benguela, and Peru Currents and the currents off North Africa), the Leeuwin Current flows southward (rather than equatorial) and is associated with warm water incursions and downwelling (rather than cold water and upwelling) (Cresswell and Golding, 1980; Bye, 1983). The Leeuwin Current is primarily driven by thermosteric effects associated with the flow of extremely warm waters from the Pacific to Indian oceans (via the Indonesian Archipelago). This strongly affects shelf circulation along the entire west coast (and to a lesser extent, south coast) and has profound effects on marine life and also, fisheries (i.e., Southern Bluefin tuna, western rock lobster, mackerel, Horse mackerel, Australian salmon, Australian herring) (Lenanton *et al.*, 1991). The

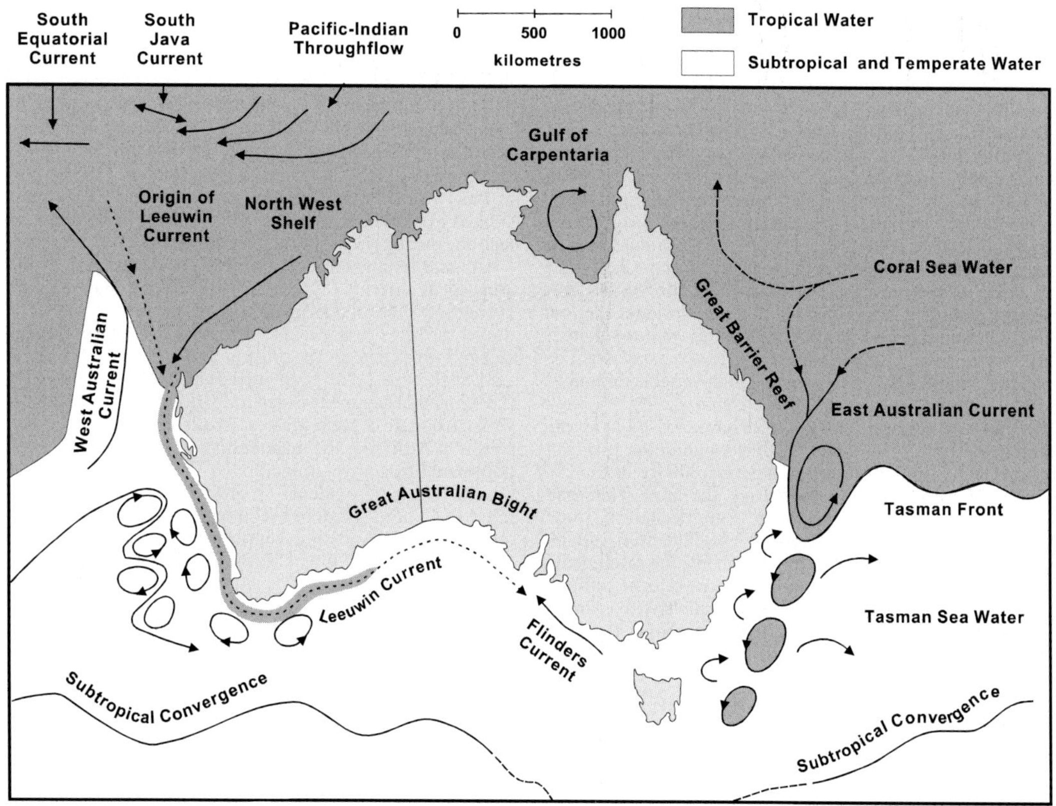

Figure A67 Major oceanographic features of Australia.

warm low-salinity waters of the Leeuwin Current are thought to be responsible for the distribution, migration, and patterns of dispersal of many pelagic marine organisms from the warm waters of the northwest of Australia to the southern seaboard. This is evidenced by a unique "tropical" or Indo-Pacific element, both in the demersal and pelagic fauna of the southern coast (Maxwell and Cresswell, 1981).

Wind-driven continental shelf waves have been identified as a primary source of current variability in practically all Australian continental shelf waters, such as the Great Barrier Reef (Cahill and Middleton, 1993), the New South Wales shelf, the Great Australian Bight (Provis and Lennon, 1981), Bass Strait (Middleton and Viera, 1991) and the North West Shelf (Webster, 1985). With topographic scales of several hundred kilometers the flows are strongly controlled by the coastline and somewhat insulated from the open ocean, and can be extremely complex (e.g. Gulf of Carpentaria, Bass Strait). Between November and May cyclones are a common feature of Australian tropical waters, and can generate strong currents and high mean sea levels, as well as high seas.

Isolation from open ocean waters allow some water masses of distinctive properties to form, particularly in embayments and gulfs. Inverse estuaries (i.e. estuaries which are more saline at the head, rather at the mouth of the estuary) are a characteristic feature of Australia's coasts, particularly in semi-arid regions, due to Australia's low rainfall and variable climate. In the upper reaches of Spencer Gulf and Gulf St. Vincent (in South Australia), low runoff coupled with high evaporation rates result in salinities exceeding 48 psu and 42 psu, respectively (Nunes and Lennon, 1986; de Silva Samarasinghe and Lennon, 1987). Similarly, the limited depth of Bass Strait (approximately 100 m), results in greater cooling of waters in winter (compared with the deeper adjacent ocean), and subsequent cascade down the continental slope into the Tasman Sea (Godfrey *et al.*, 1980; Villanoy and Tomczak, 1991).

Nutrients

Australian coastal waters are generally low in nutrients (especially nitrate and phosphate), and hence, are low in phytoplankton and productivity. The oligotrophic nature of Australia's coastal waters is due to both, continental and oceanographic features. Oceanographic factors that contribute to the low nutrient status include: the isolation of Australia's coastal waters (i.e., Great Australian Bight) from rich subantarctic waters to the south; the lack of a nutrient-rich eastern boundary current (and associated upwelling) in the Indian Ocean (originating in high latitudes)—but rather, warm water, poleward currents that bring tropical, nutrient-depleted waters along the continent's east and west coasts (i.e., the East Australian Current and Leeuwin Current); the dominance of large areas by subtropical waters with limited nutrient reserves down to 100–200 m; and the lack of major sources of upwelling of nutrients along the Australian coastline. In addition, wind-induced upwellings, which give enhanced productivity to Equatorial Pacific and Equatorial Atlantic waters, are largely ineffective in the tropical waters of Australia, because of the confused array of closely positioned land masses and islands and their confined seas to the north. Several continental factors also contribute to general nutrient impoverishment. These include: the low nutrient status (particularly in phosphates) of the well-weathered, coastal-drained soils in many areas of Australia; low coastal relief or topography; and low run-off (due to the low and high variable rainfall in Australia).

Australia's marine and estuarine communities have evolved in a generally low ambient nutrient environment, but with in some places a highly variable, nutrient supply, due to the great variability in the climate—particularly the variability in runoff from the land. This may account for the apparent sensitivity and large-scale decline of some Australian biota (such as seagrasses) to nutrient enrichment of estuarine and coastal waters. Following European settlement, the catchments of Australian rivers are now, poorly vegetated, and low rates of infiltration of rainfall result. Thus, runoff from Australian rivers is more highly pulsed with long droughts interspersed with major flooding events. This flooding typically leads to turbidity and nutrient plumes into coastal waters.

Much of Australia's continental shelf and coastal marine environment lies within the tropics. At higher latitudes, water temperature, levels of wind mixing and inputs of light necessary for photosynthesis go through pronounced seasonal cycles (Harris *et al.*, 1987, 1991; Clementson *et al.*, 1989). Because of the annual winter replenishment of nutrients, temperate oceanic and continental shelf regions tend to support larger seasonal blooms of algae and greater production at higher trophic levels than do (non-upwelling) tropical ocean and shelf systems. However, there are no large seasonal blooms producing surpluses of organic matter. As a result, Australia lacks the large demersal fisheries that characterize Northern Hemisphere continental shelf systems. Lacking a well-defined seasonal cycle, biological variability in tropical ecosystems is more closely related to disturbance events and oceanographic processes such as upwelling, floods, tidal mixing, and cyclones.

Nutrient enrichment occurs in several isolated areas around Australia. These include persistent upwellings along much of Australia's eastern seaboard (along the New South Wales coast between Port Stephens and the Queensland border) and at sites along the southern coast of Victoria (Gippsland coast) and South Australia (southern Eyre Peninsula and between Port Macdonnell and Robe), bottom coastal enrichment off New South Wales, cyclonic eddies off eastern Tasmania (southwest Tasman Sea), convective overturn (off Tasmania), North West Shelf enrichment, and tidally induced enrichments of the Great Barrier Reef waters (Wenju *et al.*, 1990; Furnas, 1995). Inshore southern habitats under the influence of localized nutrient-rich, cold water, seasonal upwellings, have resulted in very high levels of reef biodiversity (particularly among the macroalgae and invertebrates) and productivity. Off southern Eyre Peninsula, the region is also a key area of pilchard abundance (Ward *et al.*, 1998) and one of the most important sites for seabirds and marine mammals in temperate Australia (Edyvane, 2000).

Continental shelf and shelf processes

Australia's continental shelf covers approximately 2 million km^2, an area equal to about 25% of the continent's land surface. It is shallow, generally less than 200 m in water depth, and in vertical profile is a smooth, flat surface, which dips gently seawards, with some parts rimmed by shelf edge barrier reef systems (e.g., the Great Barrier Reef shelf). In a global context, the depth of the shelf break (20–550 m water depth and defined as 200 m by international convention) and the width (2–1,500 km) exhibit a wide variability. The shelf break is located about 10 km offshore from Fraser Island (east coast) and North West Cape at Exmouth (west coast), but over 500 km distance offshore on the Arafura Shelf in the north.

The continental shelf is comprised of granitic crustal material—unlike the deep ocean bed which is underlain by basalt. During Pleistocene periods of lower relative sea levels, the shelf was exposed subaerially and the Australian mainland was "joined" by dry land to several of the adjacent large islands such as Tasmania and New Guinea. Such "land bridges" are considered to have facilitated the migration of animals and humans in the late Pleistocene "Ice Age". The shallow seas comprising the Sahul Shelf, Gulf of Carpentaria, Torres Strait, and Bass Strait were thus subjected to erosion and sedimentation by rivers and wind on several occasions during the past 150,000 years. The basins in Bass Strait, Bonaparte Gulf, and the Gulf of Carpentaria are considered to have been the sites of large fresh to brackish water lakes and lagoons during these periods of emergence.

Coastal sedimentation processes are dominated by: (1) swell waves and storm currents (80% of the world's continental shelves); (2) tidal currents (17%); or (3) intruding ocean currents (3%). Currents produced during storm events—either as tropical cyclones or temperate storms—dominate in the erosion and transportation of sediment over 82% of the Australian shelf surface area. Detrital and organic sediment (i.e. biogenic) sediment is generally dominant on the continental shelf, with carbonate content greatest in offshore sediments (Harris, 1995). In southern Australia, the inshore non-calcareous sediments are generally dominated by coarse-grained relict (i.e., sediment deposited in a previous low sea environment) or reworked quartz sand. In contrast, the inner shelf, non-calcareous sediments of tropical regions are commonly terrigenous silts and clays supplied by the large river systems of northern Australia (e.g., Great Barrier Reef region) (Belperio and Searle, 1988).

Sediments on Australia's southern continental shelf are arranged in zones parallel to the coast (reflecting the dominance of ocean swell and storm generated currents and long period swell waves) and represent some of the largest modern, cool-water, open shelf accumulations of carbonate sediments in the world (Connolly and von der Borch, 1967; Wass *et al.*, 1970; Gostin *et al.*, 1988; James and von der Borch, 1991). These include the Otway region of southeastern South Australia (i.e., Lacepede and Bonney Shelf) (James *et al.*, 1992) and western Victoria (Otway Shelf) and Great Australian Bight (i.e., Eucla Platform) (James and Bone, 1992), the Rottnest Shelf, and the New South Wales Shelf (Davies, 1979). While hermatypic corals and the calcareous green alga

Halimeda characterize tropical shelf carbonate sediments (north of 24°S), bryozoa or "lace corals" and coralline algae dominate the cooler water sediments along the southern margins of Australia (south of 38°S) (Gostin *et al.*, 1988).

The sediment loads supplied to the coastal zone by many rivers are trapped in estuaries, hence, little or no sediment may reach the deeper waters of the continental shelf (particularly along the southern shelf). River catchments in the seasonally wet northern parts of Australia generally have a higher sediment yield (on the order of 100–300 t/km^2/yr) than those in the south (on the order of 10–30 t/km^2/yr) (Harris, 1995). Export of continental terrigenous sediment to the wide southern shelf is low because of the low continental relief and lack of significant drainage to the sea, particularly in the Great Australian Bight (i.e., from southwestern Australia, to the mouth of Murray River in central South Australia). Within sheltered areas of low coastal relief, redistribution of coastal sediments result in shallow coastal embayments, dominated by tidal salt marshes, mangroves, and sea grass (e.g., Port Phillip Bay, Spencer Gulf, Gulf St. Vincent), while in exposed coastal areas of high wave energy (e.g., the Great Australian Bight), these reworked sediments supply sediment for extensive beach and dunal systems which dominate these regions (Gostin *et al.*, 1988).

Tropical cyclones, associated with atmospheric low-pressure systems, are the cause of storm events in much of northern Australia. The sections of the Australian shelf most frequently affected by cyclones are the North West Shelf with up to 25 cyclones per decade and the Great Barrier Reef with up to 15 cyclones per decade. Tropical cyclones induce strong currents that erode and transport sediment over a wide area. Modeling studies on the North West Shelf (Hearn and Holloway, 1990) have shown that, under the influence of tropical cyclones, strong westward flowing coastal and inner shelf currents are established between the eye of the cyclone and the coast. Such cyclone-induced currents are clearly a significant factor affecting sediment movement on cyclone-dominated shelves, but they may also influence the long-term (net) sediment movement on some otherwise tidally dominated sections of the shelf.

Tidally dominated shelves occur generally where the mean spring tidal range measured along the coast exceeds 4 m (macrotidal). Around Australia, tidal ranges greater than 4 m occur along the northwestern coastline between Port Hedland and Darwin and reach a maximum range of about 9.2 m in King Sound. The southern Great Barrier Reef is also macrotidal, with a tidal range of 8.2 m in Broad Sound. Tidal currents are also an important sand-transporting agent in mesotidal (2–4 m tidal ranges) areas, such as Torres Strait, Moreton Bay (Queensland), and Bass Strait. Tidal currents are able to dominate sand transport in microtidal areas (tidal range less than 2 m) in restricted cases where coastal geometry affords shelter from ocean generated swell and wind-driven currents. Such is the case in many bays (e.g., Shark Bay), the approaches to some major ports (e.g., Port Phillip Bay) and in partly enclosed gulfs (e.g., Spencer Gulf, Gulf St. Vincent, and the Gulf of Carpentaria). Sediment transport and dispersal controlled by intrusive ocean currents affects only a little more than 1% of the Australian continental shelf. The only location where such currents are known to dominate sediment movement around Australia is the shelf offshore from Fraser Island, where the southward-flowing East Australian Current intrudes.

Coastal marine ecosystems and habitats

Marine ecosystems

The marine ecosystems of Australia has been the subject of several biogeographic studies and classifications in recent decades (Womersley and Edmonds, 1958; Womersley, 1959, 1981a,b, 1990; Bennett and Pope, 1960; Knox, 1963; Markina, 1976; Edgar, 1984). The biogeographical regions or provinces identified are traditionally regarded as coastal regions characterized by a relatively distinct and homogeneous flora and fauna, with only a small percentage of species common to adjacent regions, and usually differing in sea temperatures from adjacent provinces by more than 5°C. Many of these classifications have, until recently, been based on a knowledge of the distribution of intertidal biota (rather than on assessments of whole floras and faunas). Traditionally, the tropical and subtropical marine environments of Australia, i.e. >20°C, have been classified into three major biogeographical regions or provinces: the Damperian Province (from the Houtman Abrolhos, 28°50′S, off Western Australia, around northern Australia to Torres Strait); the Solanderian (or Banksian) Province (the mainland coast from Torres Strait, south to about 25°S in southern Queensland; and the Great Barrier Reef Province (comprising the Great

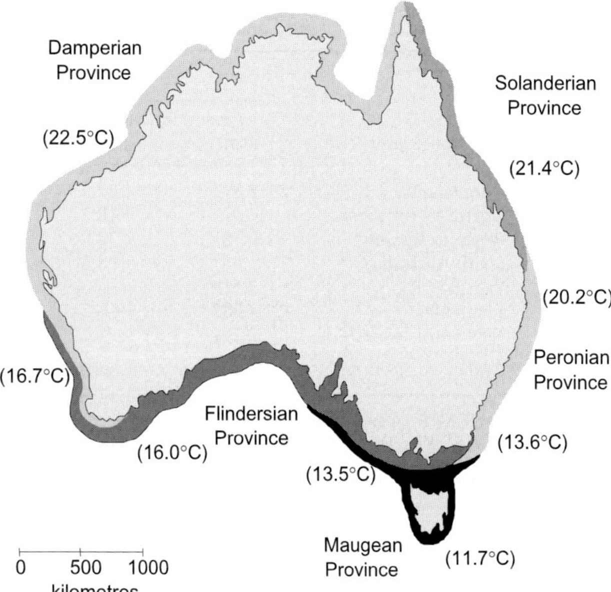

Figure A68 Marine biogeographical provinces of Australia, and average surface sea temperatures (from Edyvane, 1999; courtesy of SARDI).

Barrier Reef and associated islands) (see Figure A68). However, the widespread distribution of most pan-tropical Indo-Pacific algae suggests that the whole of the tropical Australian coast is probably best regarded as one province, with possibly subprovincial regions (Womersley, 1990).

The temperate marine environments of Australia have traditionally been classified into three provinces: the Flindersian Province, which encompasses the warm–cool temperate (i.e., 12–20°C) waters of southwest Western Australia, along the southern coast of Australia, to southern New South Wales, including the waters of South Australia, Victoria and Tasmania; the Maugean Province (or Subprovince of the Flindersian), which encompasses the cold temperate (i.e., 10–15°C) biota of Tasmania and Victoria and, to a lesser extent, South Australia (east of Robe) and New South Wales (south of Eden); and the Peronian Province, which encompasses the warm temperate (15–20°C) biota of the southern coast of Queensland, down the New South Wales coast to the Victorian border (Bennett and Pope, 1960; Knox, 1963; Womersley, 1981a,b; Edgar, 1984). The biota of Tasmania comprise elements of all three southern Australian marine biogeographic provinces (i.e., Peronian, Maugean, and Flindersian) (Edgar, 1984).

In recent years, biogeographic or regional ecosystem classifications have been developed for many coastal and nearshore marine environments around Australia (Muldoon, 1995). In several states, this has involved utilizing analytical multivariate procedures to classify patterns in nearshore ecosystem diversity. Many of these regionalizations have now been integrated as part of a national effort to develop an "*Interim Marine and Coastal Regionalisation of Australia*" (IMCRA) to assist in the development of a representative system of Marine Protected Areas (IMCRA, 1998). To-date, a total of two pelagic and nine demersal provinces (and biotones) are recognized for the continental shelf (out to the limit of the Australian EEZ), based on physical oceanography, seafloor topography and the distribution of pelagic and demersal fish (CSIRO, 1996a,b; IMCRA, 1998). At a finer scale, a total of 60 mesoscale bioregions have been identified for the nearshore marine environments of Australia, based on a wide range of physical and biological descriptors, such as climate, oceanography (water temperature, wave energy), tidal range, coastal geomorphology, and biology (habitats, marine mammals, endemic species) (IMCRA, 1998).

Coastal and marine habitats

The coastal habitats of Australia are largely influenced by the interplay of physical factors, particularly oceanography, climate, and coastal geology. As such the inshore environments of Australia range from low-energy muddy tidal deltas, which dominate the tropical north, to the swell-dominated rocky coasts and sandy shores of the temperate south. Muddy sediments dominate the tropical environments of Australia,

Table A6 Coastline and coastal types as a percentage of length of coastline around Australia. Percentages remaining from the total is coral reefs and estuary mouths (adapted from Fairweather and Quinn, 1995)

	Coastline		Coast type				
State/territory	(km)	Percentage of total	Rock %	Sand %	Mud %	Mud/mangrove %	Total
Queensland	13,347	22.3	4	36	1	19	59
Northern Territory	10,953	18.3	9	40	6	48	97
Western Australia	20,781	34.8	19	49	2	19	87
South Australia	5,067	8.5	33	59	1	8	100
Victoria	2,512	4.2	26	64	7	9	99
Tasmania	4,882	8.2	29	68	2	2	99
New South Wales[a]	2,194	3.7	33	65	1	2	100
Total	59,736	100	18	47	2	19	84

[a] Includes Jervis Bay Territory (57 km).
Source: Australian Land Information Group (AUSLIG) 1 : 100,000 Coastline Database (1993).

Table A7 Coastal habitats around Australia (adapted from Galloway, 1982; Saenger, 1995; Kirkman, 1997; Edyvane, 1999)

	Coastal habitat		
State/territory	Mangrove (km^2)	Seagrass (km^2)	Number of estuaries
Queensland	4,602 (39.8%)	22,300[a] (40.1%)	307
Northern Territory	4,119 (35.6%)	900[b] (1.6%)	137
Western Australia	2,517 (21.8%)	22,000 (39.6%)	145
South Australia	211 (1.8%)	9,620 (17.3%)	15
Victoria	12 (0.1%)	100 (0.2%)	35
Tasmania	0	500 (0.9%)	63
New South Wales	99 (0.9%)	155 (0.3%)	81
Total	11,558	55,575	783

[a] Includes Torres Strait (18,000 km^2).
[b] Comprises Gulf of Carpentaria (900 km^2).

either under mangrove forests or as wide, near-flat shores on macrotidal coasts. While rocky shores are more widespread in temperate marine environments (particularly in the Great Australian Bight and Tasmania), occupying approximately 30% of the coastline (cf. approximately 11% in tropical areas), sandy beaches dominate the temperate coasts of Australia (see Table A6). While sandy beaches are common all around Australia, they occupy approximately 50–70% of the temperate coastline, particularly along the eastern and western coast of Australia—regions swept by Eastern Australian Current and the Leeuwin Current, respectively (Fairweather and Quinn, 1995). Coral reefs occupy 14% of Australia's coastal habitat and are essentially restricted to tropical regions.

Australian estuaries occur over a wide range of geological and climatic conditions and consequently display a great variety of form (Saenger, 1995). Australia has 783 major estuaries; 415 in the tropics, 170 in the subtropics, and 198 in temperate areas (Table A7). Estuaries are most common in the humid tropics. There are few estuaries on the long arid coastline in the southwest and west. In the semi-arid and arid regions of Australia, low rainfall and variable climate result in irregular freshwater inputs and prevalence of "inverse estuaries", that is estuaries that generally flow (and flood) only after local rains have fallen. Gulf St. Vincent and Spencer Gulf represent the largest temperate inverse estuaries in Australia, while Shark Bay and Exmouth Gulf (in Western Australia) are the largest tropical estuaries. Naturally rich in nutrients, estuaries are ecologically highly productive and are important habitats for adult marine fauna as well as critical nursery habitats for the juveniles of many species.

Coastal saltmarshes. Coastal saltmarshes in Australia are intertidal plant communities dominated by herbs and low shrubs, which are often associated with estuaries. Although there is a high degree of endemism at the species level in the saltmarsh flora of Australia, at generic and family level there is a strong similarity (in structure and composition) between Australian saltmarshes and those elsewhere in the Southern Hemisphere and also in the Northern Hemisphere (Adam, 1990). When viewed in a world context, Australian saltmarshes are not as distinctly "Australian" as are the various terrestrial communities of the continent (Adam, 1995).

Australia has 13,595 km^2 of coastal saltmarshes, with the most extensive areas occurring in the tropics (Bucher and Saenger, 1991). On the continental scale, there is a striking trend of increased species richness and community complexity with increasing latitude (Saenger *et al.*, 1977; Specht, 1981; Adam, 1990). Saltmarshes in northern Australia, although extensive, are very species poor (frequently *S. virginicus*—dominated), with frequently less than 10 vascular plant species in their total flora. Compared to northern Australia, saltmarshes in southern Australia (i.e., Victoria and Tasmania), which cover much smaller areas, frequently support at least three times as many species (Adam, 1995). Not only is the flora of individual sites much greater at higher latitudes, but the regional saltmarsh flora is also much higher.

In northern Australia both mangrove and saltmarsh may be restricted to the lower, more frequently flooded intertidal zone while the upper intertidal takes the form of extensive hypersaline flats with only very sparse and localized vegetation. In southern Australia a biogeographic distinction can be drawn between saltmarshes on arid or seasonally arid (Mediterranean climate) coasts and those on temperate coasts with relatively high rainfall (Bridgewater, 1982; Adam, 1990). On drier coasts saltmarsh vegetation is characterized by a diversity of succulent, shrubby chenopods with a tendency to more open vegetation towards the upper tidal limit, while on wetter shores vegetation is denser with more grassland and sedgeland communities. Much of the primary production of saltmarshes is utilized in detrital pathways, which include microbially mediated decomposition, and take place either within salt marshes or in adjacent waters. The assumption of high productivity is linked with the "outwelling hypothesis", by which coastal wetlands act as a source of detritus and nutrients exported to offshore. Data on the linkages of saltmarshes to adjacent waters in Australia are lacking (Adam, 1995).

Saltmarshes provide habitat for numerous organisms of both terrestrial and marine origin. The most important fish habitats in intertidal wetlands are seagrasses and mangroves: Australian saltmarshes, unlike those overseas, generally have few permanent creeks and pans and so provide few fish habitats. Nevertheless, studies indicate that at least some saltmarshes may provide habitats utilized by fish (Morton *et al.*, 1987; Connolly *et al.*, 1997). Coastal wetlands are an important habitat for birds. The number of species breeding in saltmarshes is small, but the upper marsh vegetation provides nest sites for some species. A large part of the population of one of the rarest species in Australia, the orange-bellied parrot (*Neophema chrysogaster*), overwinters on saltmarshes in Victoria where it feeds on the seeds of chenopods. Migratory waders feed largely on invertebrates in intertidal sand and mudflats but saltmarshes may provide secure high tide roosts. Little is known about the utilization of saltmarsh by other faunal groups.

Mangrove forests. Australia has the third largest area of mangroves in the world (approximately 11,500 km^2) and some of the most diverse communities (Galloway, 1982). The mangrove flora of Australasia (the area including New Guinea, New Caledonia, Australia and

New Zealand) is one of the richest in the world, having approximately five times the species richness of all other regions excepting the neighboring region of Indo-Malesia (Duke, 1992). Generally, the greatest concentration of mangrove species in Australasia is found in southern New Guinea and northeastern Australia, where 45 taxa of mangrove plants are shared (Robertson and Alongi, 1995). The most recent review of mangrove species in Australia (Duke, 1992) recognized 39 taxa of mangrove plants belonging to 21 genera and 19 families. The taxa include at least four rare hybrids of more common species. Only one species, the newly discovered *Avicennia integra*, appears to be endemic to Australia. All other species are widely distributed on the island of New Guinea or in Southeast Asia.

Mangroves are generally associated with low energy, muddy coastlines, particularly tropical deltas. However, they can grow on a wide variety of substrates including sand, volcanic lava, or carbonate sediments (Chapman, 1976). Mangrove environments dominate the tropical coastal habitats of northern Australia, occupying approximately 11,228 km^2 (cf. 333 km^2 in southern temperate Australia) (see Table A7). Mangrove forests show their greatest development on humid, tropical coastlines where there are extensive intertidal zones (as found on low gradient or macrotidal coasts), and abundant fine-grained sediment, high rainfall and freshwater runoff from catchments. However, there are large local forests in the subtropics and as far south as Corner Inlet in Victoria at 38°S. The temperate marine environments of Australia are generally characterized by a lack of mud, tidal delta, and mangrove habitats, apart from large sheltered embayments (such as Port Phillip Bay, Spencer Gulf and Gulf St. Vincent). The largest areas of temperate mangrove forest occur in South Australia (approximately 332 km^2), largely within the sheltered gulfs of Spencer Gulf and Gulf St. Vincent (Galloway, 1982). Similarly, species richness in mangrove communities is greatest in the northeast humid tropics and there is a gradual decrease in the number of species with increase in latitude (Robertson and Alongi, 1995). Some estuarine systems on Cape York have up to 35 species of mangrove plants while to the south (Victoria and South Australia) only one species, *Avicennia marina*, occurs.

Variations in air temperature, rainfall, river runoff, sediment type (and deposition rate), tidal amplitude, and geomorphology along the Australian coast produce a variety of regional coastal settings for mangroves to grow (Robertson and Alongi, 1995). Mangrove communities vary from tall (up to 40 m), highly productive, closed canopy forests dominated by species of the Family Rhizophoraceae (particularly species of *Rhizophora* and *Bruguiera*) in the high rainfall, humid tropics of northeastern Queensland and the Kimberley coast of Western Australia—to the open woodlands of *A. marina*—which dominate the mangrove habitats at latitudes greater than 30°S. As conditions become more arid within the tropics (such as the Pilbara coast in Western Australia), water and salinity stress on mangroves increase in the intertidal zone and species richness decreases, and open canopy woodlands or short (1–5 m), and low productivity open shrublands develop (Semeniuk *et al.*, 1978).

Understanding of mangrove food webs in Australia currently only refers to tidally driven (non-estuarine) tropical mangrove forests (of north Queensland). These studies suggest that the carbon budget does not follow the classic overseas models of mangrove food chains—with high sediment bacterial biomass and production (sediment bacteria are a significant sink for carbon fixed by mangrove vegetation) greater than the sum of litter fall and dead wood turnover (Alongi, 1990). Meiofaunal and macrofaunal densities are low in mangrove forests, possible due to low food quality (i.e., very low nitrogen content) of mangrove detritus. Elements of the macrofauna, particularly decapod crustaceans (such as grapsid crabs of the subfamily Sesarminae), play major roles in food chains and nutrient transformations (i.e. retention of nutrient litter and hence, total tidal export of materials from forests) in intertidal tropical forests (Robertson, 1991). Little work has been done on other mangrove forest types and current models are unlikely to apply to the range of forest types in the different regions of Australia.

Mangrove ecosystems are important habitats for nekton. Fish species of the families Ambassidae, Clupeidae, Engraulididae, Gobiidae, and Leiognathidae are numerically dominant in mangrove waters, while species of Sparidae, Haemulidae, Lutjanidae, Carcharhinidae, Centropomidae, and Carangidae contribute significantly to the biomass of mangrove fish communities. Many of the dominant fish families occur in mangroves only as juveniles and recruitment from offshore waters is dependent on a number of factors, not least of which is the strength of the wet season and thus the prevailing inshore seawater salinities (Robertson and Duke, 1990). While recruitment factors are important, the variation in forest types and the availability of food and microhabitats are important factors in controlling spatial variation in fish densities and community structure (Robertson and Blaber, 1992).

There can be significant export of detritus from mangrove habitats to nearshore receiving waters. The importance of mangrove detritus in supporting higher consumers (such as fish and prawns) in estuarine and coastal areas of Australia has been confirmed. For instance, the zoea larvae of sesarmid and ocypodid crabs from mangrove forests are a major food source for juvenile fish in north Queensland estuaries. The juveniles of some penaeid prawns feed on mangrove detritus or on meiofauna that is mangrove dependent.

Some of Australia's most important single species commercial and recreational fisheries are directly linked to mangroves during at least part of their life cycles, such as the banana prawn (*Penaeus merguiensis*), bream (*Acanthopagrus australis* and *Acanthopagrus berda*), grunter (*Pomadasys kakaan*), barramundi and mangrove jack (*Lutjanus argentimaculatus*) (Robertson and Alongi, 1995). Other commercial fisheries have more indirect but equally important connections to mangroves. In the wet tropics of northeastern Australia, the seagrass meadows on which juvenile tiger prawns (*Penaeus esculentus*) depend often occur immediately seaward of mangrove forests (Robertson and Lee Long, 1991). Mangroves in Australia are also important nursery grounds for a variety of small non-commercial fish species, which are an important food source for carnivorous estuarine fishes such as barramundi and trevallies and jacks (Carangidae), and offshore billfish and mackerel.

As elsewhere in the world, mangroves play a major role in coastal protection in Australia. Mangroves in northern Australia are likely to play a major role in ameliorating the effects of cyclones and storm surges on nearby urban centers and low-lying agricultural lands. Mangrove forests are also major sites of trapping and stabilization of sediments derived from river catchments.

Seagrass beds. Seagrasses are productive, widespread, and ecologically significant features of Australia's tropical and temperate nearshore environments (Larkum *et al.*, 1989). Australia has the highest biodiversity of seagrasses in the world, the largest areas of temperate seagrass and one of the largest areas of tropical seagrass. Australian seagrasses are characterized by high speciation and also, endemism, especially in temperate regions. Australia has over 30 species of seagrass, which can be broadly categorized into tropical and temperate groups (Kuo and McComb 1989; Kirkman, 1997). On the basis of species composition and distribution a transition zone occurs at approximately 30°S (Moreton Bay) on the east coast and 25°S (Shark Bay) on the west coast (Larkum and den Hartog, 1989).

Of the 55,500 km^2 of seagrass meadows in Australia, the largest meadows occur in tropical Australia, in Torres Strait (17,500 km^2 or 31.5%) and Shark Bay (13,000 km^2 or 23.4%). Within temperate waters, seagrass meadows are largest and most diverse along the southern coast of Western Australia (9,000 km^2 or 16.2%) (Poiner and Peterken, 1995), and Spencer Gulf (5,520 km^2) and Gulf St. Vincent (2,440 km^2) in South Australia (Edyvane, 1999). In contrast, seagrass abundance (and diversity) is low in southeastern Australia, where the high-energy coastline restricts seagrass to estuaries and protected bays. For instance, seagrass occupies approximately 500 km^2 (or 0.9%) in coastal Tasmanian waters, 150 km^2 (or 0.3%) in the waters of New South Wales and 100 km^2 (or 0.2%) in Victoria waters (see Table A7).

Australia's seagrasses can be divided into those with temperate and those with tropical distributions, with Shark Bay on the west coast and Moreton Bay on the east coast located at the center of the overlap zones. Temperate species are distributed across the southern half of the continent, extending northwards on both the east and west coasts. These species have been studied most extensively, particularly the large genera *Amphibolis*, *Posidonia*, and *Zostera*, although species remain that have been little studied. The highest biomasses, and highest regional species diversity, occur in southwestern Australia, where seagrasses occur inside fringing coastal limestone reefs, or in semi-enclosed embayments. Large seagrass meadows are present in protected areas across the Great Australian Bight, and into South Australia and Tasmania. Along the New South Wales coastline, seagrasses are confined to estuaries such as Botany Bay, except for the large sheltered embayment of Jervis Bay (Kirkman, 1997).

In northern Australia, seagrass species possess tropical affinities, for example, *Thalassia* and *Cymodocea*. Tropical beds can be of high diversity, but generally possess lower biomasses than in temperate zones. While large areas of seagrasses occupy embayments such as Hervey Bay, Queensland, tropical seagrasses are generally confined to intertidal environments, or to deep water (Lee Long *et al.*, 1993). The genera *Halophila* and *Halodule* extend beyond the tropics into cooler waters. In the tropics these genera are heavily grazed by turtles and dugongs (Lanyon *et al.*, 1989). Grazing of tropical seagrasses by dugongs and sea turtles provides another unique aspect of tropical Australian seagrass meadows. Caribbean and Indo-Pacific seagrass meadows are characterized by the absence or depletion of macrograzers (i.e., manatees, sea turtles, dugongs), due to overfishing. Hence, tropical Australian seagrass meadows are probably more representative of the natural conditions in which tropical seagrasses evolved.

Seagrass growth is confined to estuaries and protected embayments in southeastern Australia, due to the high-energy coastline. Although a generalized north-to-south distribution pattern can be given, the distribution and dominance of the three sea grass genera (*Zostera, Halophila,* and *Posidonia*) is dictated by the occurrence and nature of those coastal features, which offer a suitable growth habitat for them.

Unlike temperate seagrass species, tropical seagrasses tend to occur in mixed-species stands. Thirteen species from seven genera are found across northern Australia. The region has a greater diversity of seagrass species and communities than elsewhere in the Indo-Pacific. Torres Strait supports one of the largest seagrass areas in Australia. A total of 17,500 km^2 of seagrass supporting habitat associated with 295 km of coastline or reef has been identified and mapped. The Gulf of Carpentaria, a shallow marine embayment, supports 906 km^2 of seagrass habitat along 671 km of coastline. The northwestern coast of Australia is the least well-known areas of northern Australia. Tropical seagrass meadows directly support dugong (*Dugong dugon*) and green sea turtles (*Chelonia mydas*) (Lanyon *et al.*, 1989).

Seagrass communities are of considerable importance in the processes of coastal ecosystems because of their high rates of primary production and their ability to trap (and stabilize) coastal sediments and organic nutrients. Their importance to commercial and recreational fisheries is well documented (Cappo *et al.*, 1998; Butler and Jernakoff, 1999). Additionally, seagrasses are important in substrate stabilization, supply and fixation of biogenic calcium carbonate, detrital food chains, nutrient cycling and as substrate for epibiota, and as critical habitats for many species (Larkum *et al.*, 1989).

Rocky reefs. Rocky shores are the main hard substrata of headlands along the open coasts and sometimes within estuaries. Rocky reefs occur along the entire length of the Australian coastline, mostly within shallow ocean waters. However, they are more widespread in temperate marine environments (particularly in the Great Australian Bight and Tasmania), occupying approximately 30% of the coastline (cf. approximately 11% in tropical areas) (Table A6). Prominent areas of reef occur seaward of most headlands and rock platforms. Significant amounts of subtidal rocky reef are also found in deeper waters offshore, around islands, and in estuaries; among the latter, in drowned river valleys.

Species diversity is particularly high in temperate rocky reef habitats (Keough and Butler, 1995). The southern coastline has the world's highest diversity of red and brown algae (around 1,155 species), bryozoans (lace corals), crustaceans, and ascidians (sea squirts). Diverse assemblages of brown, red, and green macroalgae, along with sponges, ascidians and other sessile invertebrates enhance habitat complexity and provide many opportunities for specialization (e.g., Jones and Andrew, 1990; Lincoln Smith and Jones, 1995). Furthermore, the large macroalgae (such as kelp), that partially cover most rocky reefs, enhance overall species diversity by providing patches of shaded habitat favored by distinct assemblages of organisms (Kennelly, 1995).

The intertidal ecology of Australian rocky shores (particularly temperate coasts) has been relatively well described (Womersley and Edmonds, 1958; Womersley and King, 1990). In contrast, very few studies have been conducted on tropical rocky shores. Similarly, the subtidal hard substrata or reefs of Australia are little known. It has only been since the advent of SCUBA apparatus in the 1960s, that detailed subtidal reef studies have been conducted and these have largely been restricted to temperate reefs (e.g., Shepherd and Womersely, 1970, 1971, 1976, 1981; Shepherd and Sprigg, 1976; Shepherd, 1983; Edgar, 1984; Jones and Andrew, 1990; Underwood and Kennelly, 1990).

The ecology of subtidal temperate reefs in Australia is characterized by the structural dominance and diversity of large macroalgae and an abundance of sessile and mobile invertebrate assemblages (i.e., sponges, bryozoans, ascidians, hydroids, echinoderms, molluscs, crustaceans) (for recent views, see Underwood and Chapman, 1995; Keough and Butler, 1995). In the cold temperate waters of Australia (east of Robe, South Australia), shallow, wave-exposed nearshore reefs are dominated by large canopy-forming species of brown macroalgae, such as the Giant Kelp (*Macrocystis*), *Phyllospora comosa* and the just subtidal, Bull Kelp (*Durvillea potatorum*) (Womersley, 1981a,b; Womersley and King, 1990; Underwood and Kennelly, 1990; Underwood *et al.*, 1991). In warm temperate waters (west of Robe), wave-exposed nearshore reefs are dominated by smaller (i.e., up to 2 m high), canopy-forming brown macroalgae, such as *Ecklonia radiata*, and fucoid algae (such as *Seirococcus axillaris, Scytothalia dorycarpa*, and species of *Cystophora* and *Sargassum*). In deeper waters, these large algae are replaced by smaller turfing red algae and sessile invertebrates (particularly sponges, bryozoans, ascidians, hydroids, and soft corals).

Urchins are important algal grazers on temperate reefs (Keough and Butler, 1995). Dominant genera on open coasts range from *Centrostephanus* and *Heliocidaris* in New South Wales, to *Heliocidaris* alone in South Australia, to a mixture of *Heliocidaris, Tripneustes*, and *Echinometra* in southwestern Western Australia. In the warm temperate waters of eastern Australia, heavy grazing by the common eastern Australian urchin (*Centrostephanus rodgersii*), typically results in distinct "urchin barrens" (Jones and Andrew, 1990). While this habitat (and urchin) has, until recently, been rare in southern Australia, it has become more common with the introduction (and increasing dominance) of *Centrostephanus* in southern waters over the past several decades. This is probably due to the increased southern penetration of the warm water, Eastern Australian Current along the eastern seaboard (due possibly to climate change), which has resulted in a significant increase in warm temperate components of the marine biota in southern Australia (Edgar, 1999).

Rocky reef provides refuge and feeding opportunities for a wide variety of fish and mobile invertebrates (Jones and Andrew, 1990; Lincoln Smith and Jones, 1995). Small fish and invertebrates can escape predators by hiding in cracks and crevices. Larger fish, along with octopus and cuttlefish appear to use rocky reef as cover from which they can ambush passing prey. Pelagic fish are also attracted to rocky reef areas by aggregations of small baitfish. Many species of invertebrates and algae live on or within rocky habitat and provide a diverse range of foods. Some fish along with abalone and sea urchins eat drift and/or attached algae associated with rocky reefs (Jones and Andrew, 1990). Many commercially and/or recreationally important fish and invertebrates depend on rocky reef habitat for some or all of their life (i.e., rock lobster, abalone). However, very little is known of the habitat linkages of these species, and their trophic interactions with other reefal organisms.

The species and size composition of fish communities using rocky reefs is likely to vary between (superficially) similar reefs, even within a relatively small area (e.g., a particular embayment, estuary or length of coastline), because of factors such as water quality, larval supply and proximity to other habitats such as sea grasses and mangroves (Bell *et al.*, 1988; Hannan and Williams, 1998). Thus, many reefs are likely to have a degree of uniqueness in terms of their ecological function. Estuarine reefs, though relatively species-poor, are likely to represent an important "stepping stone" for larger juvenile fishes moving between nursery habitats (such as sea grass and mangroves) and habitats used by adults.

In areas of high wave exposure and localized, cold-water oceanic upwellings, very high levels of invertebrate and macroalgal diversity (i.e., 95–125 dominant species) occur on nearshore reefs, particularly for species of Rhodophyta (i.e., red algae) (Shepherd, 1981; Edyvane and Baker, 1996). Species of articulate and crustose coralline red algae are particularly prevalent in these mixed-red algal communities, and are known to form key microhabitats for the larvae of economically important invertebrates (such as the abalone, *Haliotis rubra*) (Shepherd and Turner, 1985). In areas of very high current or tidal movement, rocky reefs are dominated by suspension feeders (i.e., sponges, echinoderms, bryozoans), which form diverse sponge gardens, even in relatively shallow waters (<15 m).

Some macroalgal-dominated rocky reefs, by virtue of their latitude and/or offshore location, support distinct communities of tropical species. Diverse assemblages of reef-building corals along with associated tropical fish, algae, and invertebrates can be found as far south as the Solitary Islands (New South Wales) and Rottnest Island (Western Australia). Offshore islands are often host to tropical species well beyond their normal (inshore) ranges as they are often subject to warmer water temperatures than those experienced inshore (Kuiter, 1993), due to greater influence by the warm East Australian Current (Kailola *et al.*,1993).

Sandy shores. Sandy and soft-sediment shores and their biotic assemblages are unlike any other marine benthic habitat. In contrast to rocky shores, sandy and soft-sediment habitats are characterized by their three-dimensional nature (i.e., depth is an important variable); range of grain size, depth and chemistry of the sediments (which exerts a profound influence on the types of organisms living within it); large size range of organisms which include some that ingest the sediment matrix, as well as living on or within it; and the lack of large attached plants and dominance of microscopic primary producers (see review by Fairweather and Quinn, 1995). Importantly, soft-sediment habitats provide the contiguity of habitat with other types, such as sea grasses, mangroves, salt marsh and open ocean, which facilitates the movement of organisms among them—both between and within different stages of their life cycles.

As elsewhere in the world, unvegetated, soft-sediment shores remain one of the most under-researched marine benthic habitats in Australia (Fairweather, 1990; Fairweather and Quinn, 1995). In Australia soft-sediment research has been neglected in favor of studies on coral reefs, mangroves, rocky shores, and vegetated habitats within estuaries (Fairweather and Quinn, 1995). In a review of research papers published between 1980 and 1987, Fairweather (1990) reported that only 10% of the 729 papers examined dealt with sandy bottoms, and only 6% with muddy bottoms. This is despite the fact that sandy shores represent on average, approximately 47% (or 36–68%) of Australia's coastline, while muddy shores (particularly adjacent to mangrove forests) comprise approximately 21% (or 1–48%) of the coastline (Fairweather and Quinn, 1995).

Soft-sediment research studies in Australia are very limited and have largely been descriptive (i.e., species lists and zonation schemes) with a few quantitative or semi-quantitative studies (i.e., community descriptions linked to depth zonation), and include tidal flat studies (Rainer, 1981), subtidal soft-sediment studies on detailed examinations of spatial and temporal variation (Morrissey *et al.*, 1992a,b) and quantitative surveys of benthos in Port Phillip Bay (Poore and Rainer, 1979). Significantly, soft-sediment research in Australia has benefited considerably from foreign scientists who have established research programs in Australia, particularly on tidal flats in Western Australia (i.e., Black and Peterson, 1987; Peterson and Black, 1987) and sandy beaches in NSW and Western Australia (i.e., Dexter 1983a,b, 1984; MacLachlan and Hesp, 1984).

Due to the paucity of experimental studies, there are presently no cohesive theories about any aspect of the ecology of soft-sediment intertidal assemblages in Australia (Fairweather and Quinn, 1995). Similarly, understanding of trophic interactions on sandy beaches and hence the flow of energy in soft-sediment ecosystems, is limited.

Coral reefs. The central Indo-Pacific is the world's center of coral reef diversity, with at least 50% of the world's coral reefs, covering an area greater than 300,000 km^2. Australia has the largest area of coral reefs of any nation and the largest coral reef complex, the Great Barrier Reef (GBR). The GBR is the largest single coral reef in the world. There are also large areas of coral reefs in Torres Strait, the Coral Sea Territories and central and northern Western Australia. The Tasman Sea reefs (Elizabeth and Middleton Reefs and Lord Howe Island fringing reef) are the highest-latitude coral reefs in the world, thriving in conditions that are marginal for coral growth elsewhere. Australia's coral reefs form seven distinctive groups: high-latitude reefs of eastern Australia (Solitary Islands, Lord Howe Island, Elizabeth and Middleton Reefs), the GBR, reefs of the Coral Sea (e.g., Ashmore Reef, south to Cato and Wreck reefs), reefs of northern Australia and shallow, turbid waters of the eastern Arafura Sea, Cocos (Keeling) Atoll and Christmas Island, reefs of the Northwest Shelf, reefs of coastal Western Australia (e.g., Ningaloo and Houtman Abrolhos).

The central Indo-Pacific region of northern Australia and Southeast Asia supports by far the richest invertebrate fauna associated with coral reefs in the world (i.e., more than 100 coral genera and some 500 coral species, compared with only 25 genera in the Western Atlantic, and about 60 species from the Caribbean). Australia, because of its geographic position within the world's center of marine biodiversity, is critical to the conservation of corals because of the level of degradation in the region's equatorial reefs. Approximately 70% of all central Indo-Pacific coral reefs have been significantly degraded (Vernon, 1995). This degradation has occurred primarily through over-fishing (which has effectively removed the top of the food pyramid in most Southeast Asian and Japanese areas of reef), eutrophication and increased sedimentation (from urban outfall, deforestation, agricultural runoff and coastal development), and direct intrusive activities (principally through subsistence food gathering—particularly mining practices, shell collecting, and unregulated tourism).

The GBR is not the most diverse in terms of species (Indonesian and Philippine reefs have greater numbers of coral species) but it is extremely diverse in terms of reef types, habitats, and environmental regimes. This is because the GBR is large enough to extend from the low latitude tropics to subtropical zones, to have regions with very different climates (wind patterns and rainfall), tidal regimes, water qualities, bathymetry, island types, substrata, and even geological histories.

Specific threats to Australia's coral reefs include, elevated nutrients in the inner GBR, coral bleaching events, outbreaks of crown-of-thorns (*Acanthaster*) starfish in the outer central and northern GBR and Tasman reefs, damage from the passage of tropical cyclones and outbreaks of coral-eating *Drupella* snails in Ningaloo Reef, Western Australia. Coral reefs are well represented in marine protected areas in Australia. Australia's reefs are less affected by human activities than those in other countries, mainly due to low to moderate levels of use and their remoteness, but elevated nutrients and sediments resulting from inland soil erosion are a threat in non-arid regions.

Research on Australia's coral reefs has largely excluded inter-reefal regions (these comprise more than 90% of the area of the GBR and are heavily trawled for fish and prawns); has inadequately considered biology and processes at a total-systems level; has largely excluded the far northern GBR, the reefs of the Northern Territory and, most importantly, the reefs of Western Australia; and has failed to consider microbial processes (Veron, 1995). Research has also failed to adequately address the needs of long-term conservation, in particular, the effects of sediments and nutrients on reefs, the root cause of *Acanthaster* outbreaks, and the long-term effects of commercial and recreational fishing.

Evolution of marine biodiversity

Australia's coastal environments range from tropical to temperate latitudes and consequently, encompass a wide range of coastal habitats, such as coral reefs, algal reefs, seagrass beds, mangroves, saltmarshes and also, non-vegetated habitats (sandy beaches, mudflats). All major groups of marine organisms are represented, and many of the highly diverse groups and species are endemic—particularly in the southern temperate Australian waters. Australia has the world's largest areas and highest species diversity of tropical and temperate seagrasses, largest area of coral reefs, highest mangrove species diversity and third-largest area of mangroves (Zann, 1995).

The distribution patterns today are the result of contributions from two different early Tertiary biotas:

1. The pan-Pacific Tethyan biota and its successor, the Indo-West Pacific biota, have dominated the northern coasts of Australia since the beginning of the Tertiary and also contribute to temperate biotas, especially in the southwest. To the north, barriers to interchange of shelf and coastal biotas with Southeast Asia are only slight. There are, therefore, many widespread tropical elements in the northern Australian biota and a low percentage of endemicity. At its southern limit the Tethyan element is limited by the latitudinal temperature gradient.
2. The temperate Paleoaustral fauna has dominated southeastern Australian coasts also from the early Tertiary and is now the major element of the biota of the entire southern coast.

Australia's marine biological diversity, like that of the land, is notable for its high proportion of endemic species, particularly in southern Australia. While the marine flora and fauna of tropical Australia and the Indo-Pacific mixed some 20 million years ago (when the continental plates of Australia and South East Asia collided), the marine biota of southern temperate Australia has remained geologically and climatically isolated for over 65 million years—resulting in some of the highest levels of endemism in the world (Poore, 1995). In southern Australia about 80–90% of the species in most marine groups (e.g., fish, echinoderms, molluscs) are considered to be endemic (Poore, 1995) (see Table A8). In northern Australia, only about 10% of the species in most groups are endemic. Approximately 85% of fish species, 95% of molluscs, and 90% of echinoderms are endemic. In contrast, approximately 13%, 10%, and 13% of fish, mollusc and echinoderms, respectively, are endemic in the tropical regions of Australia (Wilson and Allen, 1987). The species in northern waters are mostly shared with the Indonesian archipelago, the epicenter of global marine biological diversity, but a region where many marine ecosystems are under threat.

Australia's southern marine platform, one of the largest in the world, is of considerable evolutionary significance. Due to its tectonic stability since the late Eocene (approximately 40 million years ago), sequences deposited have been little disturbed compared with many other parts of the world. As such, it provides a unique glimpse of the direct ancestral lineages for many invertebrate species found there today. These include the bivalve genus *Bassina* (family Veneridae), which has a continual fossil record from the Oligocene, and *Neotrigonia*, the only extant member of the family Trigoniidae, which was widespread and abundant in the Mesozoic (Durham, 1979; Fleming, 1979). Many of the endemic genera, such as the cowries *Notocypraaea* and the volutes *Ternivoluta*, have fossil lineages dating back to the early Tertiary and may be survivors from an ancient paleoaustral fauna. Other endemic genera, such as the cowries *Zoila*, the bivalves *Miltha* (family Lucinidae) and gastropods of

Table A8 Marine biodiversity in Australia (from Edyvane, 1999; courtesy of SARDI)

Taxon/group	Temperate		Tropical	
	Number of species	Degree of endemism	Number of species	Degree of endemism
Macroalgae	1,155	75% (red algae)	400	low
Sea grass	22	95%?	15	low
Fish	600	85%	1,900	13%
Echinoderms	220	~90%	high	13%
Asteroids	50	high	?	–
Crinoids	7	high	high	–
Ophiuroids	74	high	?	–
Echinoids	49	high	?	–
Holothurians	40	80%	?	–
Ascidians	210	73%	?	?
Mollusks	?	95%	?	10%
Cnidaria	–	–	–	–
Anthozoa	200	?	high	?
Hydrozoa	200	high	low	?
Scyphozoa	10	?	high	?
Sponges	~1,000	high	?	?
Bryozoa	~500	high	low	?
Pycnogonids	50	73%	low	?

the genus *Diastoma* (family Diastomatidae) are relicts of once widespread Tethyan groups that have become extinct elsewhere (Wilson and Allen, 1987).

Australia has a very long coastline (both latitudinally and longitudinally) but trends in species diversity are poorly documented. While saltmarshes, mangroves, fishes, corals, molluscs, echinoderms, and decapod crustaceans (crabs, hermit crabs, lobsters, shrimps, prawns) decrease in diversity from north to south, the same trends are not apparent in other marine taxa, that is macroalgae, seagrasses, sponges, bryozoans, nudibranchs, ascidians, and peracarid crustaceans (amphipods, isopods, and others) (Table A8). In marine amphipods (Barnard, 1991), the reverse is true. In isopod crustaceans a diversity gradient overall is not yet apparent but relative dominance of families certainly changes with latitude as it does in many other taxa. These biogeographic distributions have repercussions in functional ecology—for example, the taxonomic composition of crustacean scavenger guilds. Cypridinid ostracodes, cirolanid isopods, and lysianassoid amphipods play this role but last of these become relatively more important as one moves from tropical to temperate waters.

A noteworthy aspect of the temperate fauna is the occurrence in many genera of eastern and western species pairs that do not interbreed because of geographic isolation (allopatric species). Among the fish fauna there are some 18 of these pairs of sister species, including the wrasse, *Achoerodus viridis* (Eastern Blue Groper) and *Achoerodus gouldii* (Western Blue Groper), *Pelates octolineatus* (western species) and *Pelates sexlineatus* (eastern species) of the grunter family (Terapontidae), and *Vicentia* species (western species) and *Vicentia conspersa* (eastern species) of the cardinal fish family (Apogonidae) (Wilson and Allen, 1987). As a result, warm temperate fish faunas of the Pacific and Indian Oceans are distinct from each other. East–west species pairs have also been recorded in brachyuran crabs, mollusks, and asteroids in Bass Strait (Dartnall, 1974).

A number of Australia's marine species, although found elsewhere, find the greatest security in Australian waters. These include the green turtle (*C. mydas*), hawksbill turtle (*Eretmochelys imbricata*), and loggerhead turtle (*Caretta caretta*), now rare except in Australian waters (Marsh *et al.*, 1995). The dugong (*D. dugon*), which is suffering from population decline in many parts of its range, is found in greater numbers in Australian waters than anywhere else in the world.

Trends in marine biota

The seas around Australia are globally significant in terms of many taxonomic groups. However, little is known of the status of most of the marine species. Scientific interest to date has largely centered on the higher vertebrates such as turtles, seabirds, seals, dugongs, and whales. Generally, microscopic organisms (algae, fungi, and bacteria), invertebrates and fish have been neglected. Many marine species remain undescribed and relatively little is known about most of the described species. The conservation status of very few invertebrate species is known. A number of species appear vulnerable because they are rare and have quite restricted habitats. An enormous taxonomic and monitoring effort would be required in Australia to describe all species and to determine their status. While regulations governing many of the fished species have long existed in Australia, marine fish conservation is a relatively new field and the conservation status of most species is poorly known.

Australia's marine phytoplankton includes representatives of all 13 algal classes, including diatoms (5,000 species) and dinoflagellates (2,000 species). The marine plant diversity of Australia is of global significance, particularly among the macroalgae of southern Australia, which represents one of the richest and most unique marine flora's in the world (Womersley, 1990; Huisman *et al.*, 1998; Phillips, 2001). The richness of the temperate macroalgal flora (i.e., 1,155 species) is approximately 50–80% greater than for other comparable regions around the world, with an estimated 800 species and over 75% endemism recorded in the Rhodophyta (red algae) alone (Womersley, 1990). This is largely due to the length of the southerly facing rocky coastline (i.e., the longest, ice-free, temperate coastline in the world) and the long period of geological isolation (Edyvane, 1996). While other regions around the world, such as Japan and the Pacific North America, have recorded a higher number of macroalgal species (i.e., 1,452 and 1,254 species, respectively), the coastal waters of these regions encompass a wide range of climatic conditions, from arctic to tropical. Similarly, the level of temperate species biodiversity in macroalgae is approximately three times the level recorded in the tropical regions of Australia, where approximately 200–400 species of macroalgae have been described (Womersley, 1990).

The angiosperms (flowering plants) are also very well represented. Australia has 11 of the world's 12 genera of seagrasses and over half the total number of species (Larkum and den Hartog, 1989). Australia's waters also contain the highest level of species diversity and endemism for seagrasses in the world, with the greatest levels of speciation and endemism in temperate waters, where 22 species have been recorded (cf. 15 species in tropical waters) (Poiner and Peterken, 1995). Australia has 16 of the world's 20 families of mangroves and 40 of the world's 55 species of mangroves.

Among the marine fauna, Australia has one of the largest fish faunas in the world, due to the wide latitudinal range of Australian marine habitats, the presence of the Great Barrier Reef (the world's largest coral reef) and the radiation of a number of temperate fish families (Paxton *et al.*, 1989). The warm East Australian Current and Leeuwin Current, flowing down the east and west coasts, respectively, also seasonally bring tropical species into more southerly latitudes. Australia has an estimated 4,000–4,500 species of fish, of which around 3,600 have been described. About a quarter of the species are endemic; most of these are found in southern Australia.

Diversity of several major marine taxa is highest in the tropical regions of Australia. This includes molluscs, echinoderms, and fish.

Among the molluscs, most families are represented in the Australian fauna and have more species in the north of the country than in the south. The level of endemicity in the north is low (about 10%) with most species distributed widely in the Indo-West Pacific region. In contrast, endemicity of the southern Australian fauna is about 95% and several endemic genera occur. Some endemic genera are relicts of the once widespread Tethyan fauna (as are the endemic families, Trigoniidae, Campanilidae, and Diastomatidae, each with a sole living representative). Other endemic genera are relicts of the ancient Paleoaustral fauna.

Echinoderms, with only a few hundred Australian species, repeat the biogeographic patterns of the fishes and molluscs with only 13% of species endemic in the tropical region and 90% in the temperate region. Many echinoderms have a long larval life which may explain why 22% of Tasmanian species occur also in New Zealand where they are transported by the West Wind Drift.

Among the fish fauna, approximately 3,400 species of marine fishes are pelagic or oceanic and wide-ranging in tropical and temperate seas (Paxton *et al.*, 1989). About three-quarters occur on the shelf and nearshore. The greatest number of species are in the tropics where approximately 120 families, 600 genera and more than half of the species (1,900) are found; and most of these are common to the Indo-West Pacific region. Although most species have pelagic eggs and larvae a moderate level of endemicity (13%) is maintained with the help of southerly flowing currents on both the east and west Australian coasts.

The fish fauna of temperate Australia comprises about 600 species of which 85% is endemic and 11% is shared with New Zealand (Wilson and Allen, 1987; Kuiter, 1993). Endemism at the genus level is about 38% (of an estimated 290 genera). One contribution to this high level of endemicity is radiation in a few families with low dispersability; viviparous clinids, brooding syngnathids, and nesting gobiescocids and gobiids. In some 8 families of fish that have more than 10 species, more than half of the total world species occur in Australian waters (e.g., Cheilodactylidae). The five marine fish families Brachio-nichthydae, Pataecidae, Enoplosidae, Gnathanacanthidae and Dinolestidae are endemic to Australia and a number of these families are monotypic. Several families of fish are noted for their explosive radiations in southern Australia. They include pipefishes and sea horses (Syngnathidae), leatherjackets (Monacanthidae), and anglerfishes (Antennariidae). The Australian pipefish fauna is regarded as uniquely diverse. Nearly 80% of the 47 genera in the Indo-Pacific region are found in Australia and 14 of these genera (38%) are endemic. Ten of these genera are monotypic (Wilson and Allen, 1987).

Australia also has an extremely rich chondrichthyan (shark, chimeras or ghost sharks, and rays) fish fauna. It is estimated that at least 296 species, comprising 166 sharks, 117 rays, and 13 chimeras, live in Australian waters. The richness of this fauna is apparent through comparisons with other areas: 182 species have been recorded from southern Africa; 174 from the Japanese Archipelago; and 130 from the eastern North Atlantic and Mediterranean Seas (Last and Stevens, 1994). More than half of the entire Chondrichthyan fauna, some 54%, is endemic to Australia. Of the shark fauna, 48% of the species are endemic, with most of the endemic species being found in tropical or warm temperate areas. More than half of the Australian chimeras appear to be regional endemics, and some 73% of Australian rays are endemic.

Among the reptiles, Australia has about 30 of the world's 50 species of sea snakes, around half of which are endemic (Marsh *et al.*, 1995). The family of aipysurids live in coral reef waters and the family of hydrophiids live in inter-reefal waters of Australia's tropics. Six of the seven world's turtle species are found in Australian waters. One, the flatback turtle, is endemic. Breeding migrations may cover hundreds to thousands of kilometers and many turtles breeding in Australia may live around the island of Papua New Guinea, the southwestern Pacific Islands and Indonesia, making species management difficult. Turtles are slow to reach maturity and breed perhaps only five times in their lives, making them extremely vulnerable to overexploitation.

About 110 species of seabirds belonging to 12 families are found in Australia and its external territories. Of these, 76 species breed and spend their entire lives in the region. The remaining 34 species are regular or occasional visitors. Some 14 species or subspecies of Australia's seabirds (13% of the total) are considered to be threatened, largely because their colonies on oceanic islands are few in number and are vulnerable to harvesting and natural disasters. The wandering albatross on Macquarie Island, Abbot's booby on Christmas Island and the Australian subspecies of the little tern are classified as "Endangered" under IUCN criteria. Lord Howe's Kermadec petrel and white-bellied storm-petrel and Christmas Island's Christmas frigatebird are considered "Vulnerable."

In contrast to temperate biota, reef-building scleractinian corals are essentially tropical and their distribution is substantially different from that of echinoderms, molluscs, and fishes. The Australian fauna is a subset of Indo-West Pacific corals and is largely confined to the Great Barrier Reef and smaller reefs on the west coast (Vernon, 1995). There is not a gradual reduction in the number of species with increasing latitude but rather an abrupt depletion of species at the termination of the Great Barrier Reef and at the Houtman Abrolhos. Coral reefs do occur further south—for example, at Lord Howe Island and Solitary Islands—but the number of species is few. Few species of coral occur on the southern coast of Australia and reefs are not formed there.

The tropical dugong is the only fully herbivorous marine mammal and the only Sirenian (sea cow) to occur in Australia (Marsh *et al.*, 1995). It is extinct or near extinct in most of its former range which extended from East Africa to South East Asia and the Western Pacific. Northern Australia has the last significant populations (estimated to be over 80,000) in the world. Large, long-lived mammals, dugongs become sexually mature at around 10 years and calve every 3–5 years, making them vulnerable to over-hunting. Major concerns are possible over-hunting of Torres Strait populations, mortalities in fish gillnets and shark nets and loss of sea grass habitat.

Three species of eared seals or pinnipeds breed in mainland Australian waters: the rare and endemic Australian sea lion (*Neophoca cinerea*), the Australian fur seal (*Arctocephalus pusillus doriferus*) and New Zealand fur seal (*Arctocephalus forsteri*) (Shaughnessy, 1999). The Australian populations of the New Zealand Fur Seal are limited in their distribution to southern Tasmania and the Great Australian Bight, and are found on the islands of Recherche Archipelago (WA), eastwards to Kangaroo Island (SA). The Australian Sea Lion is one Australia's most endangered marine mammals and one of the rarest and most endangered pinnipeds in the world and is endemic to Australia (Gales *et al.*, 1994). Prior to seal-hunting, this species occurred along the whole of the southern coastline, but is now limited to the offshore islands of Western and South Australia, from the Houtman Abrolhos to the islands of Recherche Archipelago (WA), and from Nuyts Archipelago to Kangaroo Island (SA). Breeding populations of the Australian Fur Seal, which also breeds in South Australia, are confined to southeastern Australia, including Tasmania (Shaughnessy, 1999). Major human threats include entanglement in fishing nets and ocean litter, incidental killing, oil pollution, and disturbances by visitors.

Eight species of baleen whales and 35 species of toothed whales, porpoises, and dolphins are found in Australian waters, although none is endemic (Bannister *et al.*, 1996). This is almost 60% of the world's total cetacean species. Australian breeding populations of southern right whales were seriously depleted by hunting as early as 1845. Their population has slowly increased from small remnants totaling a hundred or so, to between 500 and 800. Australian breeding populations of humpback whales were depleted by 1963. Numbers are recovering and there are now thought to be up to 4,000 breeding in Australian waters. Gillnets, shark nets set off bathing beaches, discarded fishing nets and ingestion of plastic litter are considered threats to cetaceans in Australian waters.

Marine and coastal research

While marine and coastal research in Australia is relatively young, Australia has gained preeminence in a number of research areas, including coral reef/tropical marine studies, marine geosciences, physical oceanography, fisheries research, coastal hydrodynamics and numerical modeling, and remote sensing techniques (DITAC, 1989; Zann, 1996). While some subjects and geographic areas are well known, there are serious gaps in knowledge and understanding of Australia's marine environment, particularly in relation to coastal and reef ecosystems, ocean/atmosphere processes, marine chemistry, chemical oceanography and importantly, baseline information on coastal and marine ecosystems, habitats and processes (to facilitate sustainable use).

Research on Australia's tropical marine environments and habitats (such as mangroves and coral reefs), particularly in the Great Barrier Reef region, is relatively advanced. This is largely due to the location and research missions of national marine research and management agencies (Fairweather and Quinn, 1995; Edyvane, 1996). In 1991, around 40% of marine research projects in Australia were undertaken in the tropics. Most research was undertaken in the Great Barrier Reef region (31% of total), followed by the New South Wales coast (16%), Bass Strait (12%), the southwest coast (5%) and the northwest (3%) (Zann, 1996).

Despite the high concentration of Australia's population (approximately 80% of the total population) in the coastal temperate areas of the continent, the habitats and biodiversity of temperate marine ecosystems (particularly subtidal habitats), have largely been under-researched and hence, are poorly documented and understood (Fairweather and Quinn, 1995). This has resulted in the lack of habitat conservation (and integrated management) in many areas of southern Australia, despite the high level of threats and the high level of biodiversity and endemism among a range of marine faunal and floral groups.

Temperate habitat studies in Australia have largely focused on isolated studies of intertidal rocky shores and estuaries and sea grass areas—habitats recognized as important in the management of nearshore fisheries (Fairweather and Quinn, 1995). Research on temperate rocky shores has principally focused on intertidal habitats or the biology of individual reef species or assemblages and particularly, population studies of commercially important species (Fairweather and Quinn, 1995). Regional habitat studies are generally rare and descriptions of macrofloral (or faunal) assemblages of subtidal habitats are geographically very patchy. For some regions (e.g., South Australia, Tasmania) detailed information is known based on isolated subtidal field surveys conducted in the 1970s and 1980s (i.e., Shepherd and Womersley, 1970, 1971, 1976, 1981; Shepherd and Sprigg, 1976; Shepherd, 1983; Edgar, 1984). It is only recently that studies involving broadscale coastal mapping and documentation of temperate subtidal communities, including, reef, sea grass, and soft-bottom communities, have commenced (Edyvane, 1999; Edgar *et al.*, 1997). Elsewhere (such as in the Great Australian Bight and offshore waters), there are almost no published inventories or maps on subtidal benthic habitats (Edyvane, 1996, 2000). This reductionist approach to the study of coastal nearshore ecosystems has principally hindered understanding of these ecosystems and the ecological linkages between reefal and non-reefal habitats (such as sea grass, mud, sandy coasts).

Habitat inventories are particularly required for rocky coasts and reefs and the continental shelf and large gulfs—the two habitats supporting the most valuable commercial fisheries (Cappo *et al.*, 1998). The need for a better understanding of habitat processes and dynamics was highlighted for all major habitat types—as testable paradigms about fisheries production exist only for temperate seagrass. Relative to their fisheries value, overall gaps in knowledge are particularly outstanding for the continental shelf (and slope), followed by rocky coasts, and reefs. Perhaps least studied in Australia are the nursery role and biological processes of sandy shores.

Much of the marine ecological research conducted in temperate Australia has focused on the study of ecological processes, particularly on rocky intertidal shores, rather than a description of the patterns of biodiversity and the distribution and abundance of marine assemblages. These studies have particularly focused on the role of intra- and interspecific competition and predation and their influence on the structure of rocky intertidal assemblages (Underwood and Kennelly, 1990), and also, patterns and variability in recruitment of marine invertebrates, both on small and large spatial scales (see review by Fairweather and Quinn, 1995; Keough and Butler, 1995).

Marine ecological research in Australia requires more effort into baseline mapping of marine biodiversity and the impacts of human use (at a range of temporal and spatial scales), long time-series datasets for trend analysis, and the modeling of coastal processes, and also a greater focus on the linkages between: catchment practices and effects on nearshore ecosystems; reefal habitats and non-reefal habitats; offshore and nearshore ecosystems; and the relationship between coastal physiography, physical processes and scaled patterns of biodiversity, that is ecosystem geography (Edyvane, 1996).

In recent years, a national audit of catchments, rivers and estuaries has been undertaken (Commonwealth of Australia, 2002). Several large, ecosystem-scale modeling studies of Australia's marine environments have also been undertaken, primarily for pollution management. These include Port Phillip Bay in Victoria (CSIRO, 1996), and Cockburn Sound and Albany Harbor (Simpson and Masini, 1990) in Western Australia, and the Huon River estuary in Tasmania, (CSIRO 2000) and Hervey Bay in Queensland (Tibbetts *et al.*, 1998, Dennison of Abal 1999). A major ecosystem study of the North West Shelf (in Western Australia) has been undertaken for tropical fisheries and multiple-use, ecosystem based management (NWSJES 2002).

Summary

Australia's marine environment extends from the coast to the boundary of it's 200 nautical mile exclusive economic zone and covers million square kilometers of seas (an area 16% larger than the land). As the world's largest island, Australia has wide range of coastal and marine environments, which stretch approximately 61,700 km, from the tropical northern regions to temperate southern latitudes. Along this extensive coastline occur a wide range of habitats and biological communities including rocky shores, sandy beaches, algal reefs, kelp forests, which dominate the temperate south, and coral reefs, estuaries, bays, seagrasses beds, mangrove forests and coastal salt marshes, which dominate the tropical north and also, the less understood mid-water, outer-shelf, and deepwater habitats. Australia's marine environments also include external territories in the Indian Ocean, South Pacific, Southern Ocean, and Antarctica.

The extent and diversity of Australia's marine and coastal environments has resulted in some of most diverse, unique and spectacular marine life in the world—supporting some of the greatest number of marine species in the world. For instance, Australia has the world's largest areas and highest species diversity of tropical and temperate seagrasses, the highest diversity of marine macroalgae, the largest area of coral reefs, the highest mangrove species diversity and global levels of biodiversity for a range of marine invertebrates (e.g., bryozoans, ascidians, nudibranchs) (Zann, 1995). Approximately 4,000 species of fish, 43 species of whales and dolphins, and 6 of the 7 world species of marine turtles are recorded from Australia.

All major groups of marine organisms are represented in Australian waters, and many of the highly diverse species are endemic or unique to Australia's waters. The nature of the temperate Australian marine biota, particularly the high level of endemism, is principally a result of the unique geological and tectonic history of Australia, particularly the long period of geological and climatic isolation. In temperate southern Australian waters, which have been geographically and climatically isolated for around 65 million years, most known species (i.e., 90–95%) are endemic or restricted to the area (Poore, 1995). In contrast, in the waters of tropical northern Australia, which are connected by currents to the Indian and Pacific Ocean tropics, most species (i.e., 85–90%), are shared with the Asia-Pacific region.

Karen S. Edyvane
University of Tasmania

Bibliography

Adam, P., 1990. *Saltmarsh Ecology*. Cambridge: Cambridge University Press.

Adam, P., 1995. Saltmarsh. In Zann, L.P. and Kailola, P. (eds.), *The State of the Marine Environment Report for Australia*. Technical Annex: 1. The Marine Environment. Australia: Great Barrier Reef Marine Park Authority, pp. 97–105.

Alongi, D.M., 1990. The ecology of tropical soft-bottom benthic ecosystems. *Annual Review of Oceanography and Marine Biology*, **28**: 381–496.

Andrew, N. (ed.), 1999. *Under Southern Seas: The Ecology of Australia's Rocky Reefs*. Sydney: University of New South Wales Press.

Bannister, J.L., Kemper, C.M., and Warneke, R.M., 1996. *The Action Plan for Australian Cetaceans*. Canberra: Australian Nature Conservation Agency.

Barnard, J.L., 1991. Amphipodological agreement with Platnick. *Journal of Natural History*, **25**: 1675–1676.

Bell, J.D., Steffe, A.S. and Westoby, M., 1988. Location of seagrass beds in estuaries: effect on associated fish and decapods. *Journal of Experimental Marine Biology and Ecology*, **122**: 127–146.

Belperio, A.P. and Searle, D.E., 1988. Terrigenous and carbonate sedimentation in the Great Barrier Reef province. In Doyle, L.J. and Roberts, H.H. (eds), *Carbonate—clastic transitions*. Amsterdam: Elsevier, pp. 143–174.

Bennett, I. and Pope, E.C., 1960. Intertidal zonation of the exposed rocky shores of Tasmania and its relationship with the rest of Australia. *Australian Journal of Marine and Freshwater Research*, **11**: 182–221.

Black, R. and Peterson, C.H., 1987. Biological versus physical explanations for the non-random pattern of host occupation by a macroalga attaching to infaunal bivalve molluscs. *Oecologia*, **73**: 213–221.

Bridgewater, P.B., 1982. Phytosociology of coastal saltmarshes in the mediterranean climatic region of Australia. *Phytocoenologia*, **10**: 257–296.

Bucher, D. and Saenger, P., 1991. An inventory of Australian estuaries and enclosed marine waters: an overview of results. *Australian Geographical Studies*, **29**: 370–381.

Butler, A. and Jernakoff, P., 1999. *Seagrass in Australia. Strategic Review and Development of an R & D Plan*. Victoria: CSIRO Publishing.

Bye, J.A.T., 1983. The general circulation in a dissipative ocean with longshore wind stresses. *Journal of Physical Oceanography*, **13**: 1553–1563.
Cahill, M.L. and Middleton, J.H., 1993. Wind-forced motion on the Northern Great Barrier Reef. *Journal of Physical Oceanography*, **23**: 1176–1191.
Cappo, M., Alongi, D.M., Williams, D. McB., and Duke, N., 1998. *A Review and Synthesis of Australian Fisheries Habitat Research. Major Threats, Issues and Gaps in Knowledge of Coastal and Marine Fisheries Habitats. Volume 1: A Prospectus of Opportunities for the FRDC Ecosystem Protection Program*. Australian Institute of Marine Science. Townsville, Australia.
Chapman, V.J., 1976. *Mangrove Vegetation*. Vaduz, Germany: J.Cramer.
Church, J.A., 1987. The East Australia Current adjacent to the Great Barrier Reef. *Australian Journal of Marine and Freshwater Research*, **38**: 671–683.
Clementson, L.A., Harris, G.P., Griffiths, F.B., and Rimmer, D.W., 1989. Seasonal and inter-annual parameters in Storm Bay, Tasmania. I. Physics, chemistry and biomass of components of the food chain. *Australian Journal of Marine and Freshwater Research*, **40**: 25–39.
Commonwealth of Australia. 2001. *Australia State of the Environment 2001*. Independent Report to the Commenwealth Minister for the Environment and Heritage. Australian State of the Environment Committee. CSIRO Publishing on behalf of the Department of the Environment and Heritage.
Commonwealth of Australia. 2002. *Australian Catchment, River and Estuary Assessment 2002*. Volumes 1 and 2. National Land and Water Resources Audit, Land & Water Australia.
Connolly, J.R. and von der Borch, C.C., 1967. Sedimentation and physiography of the sea floor south of Australia. *Sedimentary Geology*, **1**: 181–220.
Connolly, R.M., Dalton, A., and Bass, D.A., 1997. Fish use of an inundated saltmarsh flat in a temperate Australian estuary. *Australian Journal of Ecology*, **22**: 222–226.
Cresswell, G.R., 1991. The Leeuwin Current—observations and recent models. *Journal of the Royal Society of Western Australia*, **74**: 1–14.
Cresswell, G.R. and Golding, T.J., 1980. Observations of a south-flowing current in the south-eastern Indian Ocean. *Deep-Sea Research*, **27A**: 449–466.
Cresswell, G.R., Ellyett, C., Legeckis, R., and Pearce, A.F., 1983. Nearshore features of the East Australian Current system. *Australian Journal of Marine and Freshwater Research*, **34**: 105–114.
CSIRO, 1996a. *Interim Marine Bioregionalisation of Australia: Towards a National System of Marine Protected Areas*. Australia: CSIRO Divisions of Fisheries, Oceanography.
CSIRO, 1996b. *Port Phillip Bay Environmental Study. Final Report*. Australia: CSIRO.
CSIRO Huon Estuary Study Team. 2000. Huon Estuary Study—environmental research for integrated catchment management and aquaculture. Final report to Fisheries Research and Development Corporation. Project number 96/284, June 2000. CSIRO Division of Marine Research. Marine Laboratories, Hobart.
Dartnall, A.J., 1974. Littoral biogeography. In Williams, W.D. (ed.), *Biogeography and Ecology in Tasmania*. The Hague: Junk, pp. 171–194.
Davies, P.J., 1979. Marine geology of the continental shelf off southeast Australia. *Bulletin of the Bureau of Mineral Resources*, **195**: 51.
de Silva Samarasinghe, J.R. and Lennon, G.W., 1987. Hypersalinity, flushing and transient, salt-wedges in a tidal gulf—an inverse estuary. *Estuarine, Coastal and Shelf Science*, **24**: 483–498.
Dennison, WC & EG Abal. 1999. *Moreton Bay Study: A Scientific Basis for the Healthy Waterways Campaign*. South East Queensland Water Quality Strategy, Brisbane.
Department of the Environment, Sport and Territories, 1996. (DEST) *Australia State of the Environment 1996*. Australia: CSIRO Publishing.
Dexter, D.M., 1983a. A guide to sandy beach fauna of New South Wales. *Wetlands*, (Australia), **3**: 94–104.
Dexter, D.M., 1983b. Community structure of intertidal sandy beaches of NSW, Australia. In McLachlan, A., and Erasmus, T. (eds.), *Sandy Beaches as Ecosystems*. The Hague: W.Junk Publishers, pp. 451–472.
Dexter, D.M., 1984. Temporal and spatial variability in the community structure of the fauna of four sandy beaches in south-eastern NSW. *Australian Journal of Marine and Freshwater Research*, **35**: 663–672.
DITAC, 1989. *Oceans of Wealth?* Department of Industry, Technology and Commerce. Canberra: Australian Government Publishing Service.
Duke, N.C., 1992. Mangrove floristics and biogeography. In Robertson, A.I. and Alongi, D.M. (eds.), *Tropical Mangrove Ecosystems*. Washington: AGU Press, pp. 63–100.
Durham, J.W. 1979. The fossil record, plate tectonics, and the development of the Southern Hemisphere. In *Proceedings of the International Symposium on Marine Biogeography and Evolution in the Southern Hemisphere Volume 1*, New Zealand Department of Scientific and Industrial Research, New Zealand, pp. 165–186.
Edgar, G.J., 1984. General features of the ecology and biogeography of Tasmanian rocky reef communities. *Papers and Proceedings of the Royal Society of Tasmania*, **118**: 173–186.
Edgar, G.J., 1999. Tasmania. In Andrew, N. (ed.), *Under Southern Seas: the Ecology of Australia's Rocky Reefs*. Sydney: University of New South Wales Press, pp. 30–39.
Edgar, G.J., Moverley, J.S., Barrett, N.S., Peters, D., and Reed, C., 1997. The conservation-related benefits of a systematic marine biological sampling program: the Tasmanian bioregionalisation as a case study. *Biological Conservation*, **79**: 227–240.
Edyvane, K.S., 1996. The role of Marine Protected Areas in temperate ecosystem management. In Thackway, R. (ed.), *Developing Australia's Representative System of Marine Protected Areas. Criteria and Guidelines for Identification and Selection. Proceedings of a technical meeting held at the South Australian Aquatic Sciences Centre, West Beach, Adelaide, 22–23 April 1996*. Commonwealth of Australia.
Edyvane, K.S., 1999. *Conserving Marine Biodiversity in South Australia. Part 2. Identification of Areas of High Conservation Value in South Australia*. South Australia: South Australian Research and Development Institute.
Edyvane, K.S., 2000. The Great Australian Bight. In Sheppard, C.R.C. (ed.), *Seas at the Millennium: an Environmental Evaluation*. Oxford: Elsevier Science.
Edyvane, K.S. and Baker, J., 1995. The South Australian regionalisation project. In Muldoon, J. (ed.), *Towards a Marine Regionalisation for Australia*. Australia: Great Barrier Reef Marine Park Authority, pp. 121–140.
Edyvane, K.S. and Baker, J.L., 1996. Mapping biodiversity studies in South Australia: research for integrated coastal-marine management? In Harvey, N. (ed.), *Proceedings of the Australian Coastal Management Conference, Glenelg, South Australia, 17–19 April, 1996*. South Australia: University of Adelaide, pp. 172–184.
Fairweather, P.G., 1990. Ecological changes due to our use of the coast: research needs versus effort. *Proceedings of the Ecological Society of Australia*, **16**: 71–77.
Fairweather, P.G. and Quinn, G.P., 1995. Marine ecosystems: hard and soft shores. In Zann, L.P. and Kailola, P. (eds), *The State of the Marine Environment Report for Australia*. Technical Annex: 1. The Marine Environment. Australia: Great Barrier Reef Marine Park Authority, pp. 25–36.
Fleming, C.A. 1979. Evolution of the southern pacific marine biota: the expanding fossil record. *Proceedings of the International Symposium on Marine Biogeography and Evolution in the Southern Hemisphere Volume 1*, New Zealand Department of Scientific and Industrial Research, New Zealand, pp. 5–26.
Furnas, M.J., 1995. Land–sea interactions and oceanographic processes affecting the nutrient dynamics and productivity of Australian marine ecosystems. In Zann, L.P. and Kailola, P. (eds), *The State of the Marine Environment Report for Australia*. Technical Annex: 1. The Marine Environment. Australia: Great Barrier Reef Marine Park Authority, pp. 61–73.
Gales, N.J., Shaughnessy, P.D., and Dennis, T.E., 1994. Distribution, abundance and breeding cycle of the Australian sea lion *Neophoca cinerea* (Mammalia: Pinnipedia). *Journal of Zoology, London*, **234**: 353–370.
Galloway, R.W., 1982. Distribution and physiographic patterns of Australian mangroves. In Clough, B.F. (ed.), *Mangrove Ecosystems in Australia: Structure, Function and Management*. Canberra: Australian National University Press.
Godfrey, J.S., Jones, I.S.F., Maxwell, J.G., and Scott, B.D., 1980. On the winter cascade from Bass Strait into the Tasman Sea. *Australian Journal of Marine and Freshwater Research*, **30**: 275–286.
Gomon, M.F., Glover, J.C.M., and Kuiter, R.H. (eds.), 1994. *The Fishes of Australia's South Coast*. Adelaide: State Print.
Gostin, V.A., Belperio, A.P., and Cann, J.H., 1988. The Holocene non-tropical coastal and shelf carbonate province of southern Australia. *Sedimentary Geology*, **60**: 51–70.
Hannan, J.C. and Williams, R.J., 1998. Recruitment of juvenile marine fishes to seagrass habitat in a temperate Australian estuary. *Estuaries*, **21**: 29–51.

Harris, G., Nilsson, C., Clementson, L., and Thomas, D., 1987. The water masses of the east coast of Tasmania: seasonal and inter-annual variability and the influence on phytoplankton biomass and productivity. *Australian Journal of Marine and Freshwater Research*, **38**: 569–590.

Harris, G.P., Griffiths, F.B., Clementson, L.A., Lyne, V., and van der Doe, H., 1991. Seasonal and inter-annual variability in physical processes, nutrient cycling and the structure of the food chain in Tasmanian shelf waters. *Journal of Plankton Research*, **13**(suppl.): 69–82.

Harris, P.T., 1995. Marine geology and sedimentology of the Australian continental shelf. In Zann, L.P. and Kailola, P. (eds), *The State of the Marine Environment Report for Australia*. Technical Annex: 1. The Marine Environment. Australia: Great Barrier Reef Marine Park Authority, pp. 11–23.

Hearn, C.J. and Holloway, P.E., 1990. A three-dimensional barotropic model of the response of the Australian north west shelf to tropical cyclones. *Journal of Physical Oceanography*, **20**: 60–80.

Huisman, J.M., Cowan, R.A., and Entwisle, T.J., 1998. Biodiversity of Australian marine macroalgae—a progress report. *Botanica Marina*, **41**: 89–93.

Interim Marine and Coastal Regionalisation of Australia Technical Group, 1998. *Interim Marine and Coastal Regionalisation of Australia: an Ecosystem Classification for Marine and Coastal Environments. Version 3.3*. Canberra: Environment Australia.

James, N.P. and Bone, Y., 1992. Synsedimentary cemented calcarenite layers in Oligo-Miocene shelf limestones, Eucla Platform, southern Australia. *Journal of Sedimentary Petrology*, **62**: 860–872.

James, N.P. and von der Borch, C.C., 1991. Carbonate shelf edge off southern Australia: a prograding open-platform margin. *Marine Geology*, **19**: 1005–1008.

James, N.P., Bone, Y, von der Borch, C.C., and Gostin, V.A., 1992. Modern carbonate and terrigenous clastic sediments on a cool-water, high-energy, mid-latitude shelf; Lacepede Shelf, southern Australia. *Sedimentology*, **34**: 877–904.

James, N.P., Boreen, T.D., Bone, Y., and Feary, D.A., 1994. Holocene carbonate sedimentation on the west Eucla Shelf, Great Australian Bight: a shaved shelf. *Sedimentary Geology*, **90**: 161–177.

Jones, G.P. and Andrew, N.L., 1990. Herbivory and patch dynamics on rocky reefs in temperate Australasia: the roles of fish and sea urchins. *Australian Journal of Ecology*, **15**: 505–520.

Kailola, P.J., Williams, M.J., Stewart, P.C., Reichelt, R.E., McNee, A., and Grieve, C., 1993. *Australian Fisheries Resources*. Bureau of Resource Sciences and the Fisheries Research and Development Corporation. Commonwealth of Australia.

Kennelly, S.J., 1995. Kelp beds. In Underwood, A.J. and Chapman, M.G.(eds.), *Coastal Marine Ecology of Temperate Australia*. Sydney: University of New South Wales Press, pp. 106–120.

Keough, M.J., and Butler, A.J., 1995. Temperate subtidal hard substrata. In Zann, L.P. and Kailola, P. (eds), *The State of the Marine Environment Report for Australia. Technical Annex: 1. The Marine Environment*. Australia: Great Barrier Reef Marine Park Authority, pp. 37–52.

Kirkman, H., 1997. *Seagrasses of Australia*. Canberra: State of the Environment Technical Paper Series (Estuaries and the Sea), Department of the Environment.

Knox, G.A., 1963. The biogeography and intertidal ecology of the Australasian coasts. *Annual Review of Oceanography and Marine Biology*, **1**: 341–404.

Kuiter, R., 1993. *Coastal Fishes of South-Eastern Australia*. Bathurst: Crawford House Press.

Kuo, J. and McComb, A.J., 1989. Seagrass taxonomy, structure and development. In Larkum, A.W.D., McComb, A.J. and Shepherd, S.A. (eds.), *Biology of Seagrasses*. Amsterdam: Elsevier, pp. 6–73.

Lanyon, J., Limpus, C.J., and Marsh, H., 1989. Dugongs and turtles: grazers in the seagrass system. In Larkum, A.W.D., McComb, A.J., and Shepherd, S.A. (eds.), *Biology of Seagrasses*. Amsterdam: Elsevier, pp. 610–634.

Larkum, A.W.D., and den Hartog, C., 1989. Ecology and biogeography of seagrasses. In Larkum, A.W.D., McComb, A.J., and Shepherd, S.A. (eds.), *Biology of Seagrasses*. Amsterdam: Elsevier, pp.112–156.

Larkum, A.W.D., McComb, A.J., and Shepherd, S.A. (eds.), 1989. *Biology of Seagrasses*. Elsevier, Amsterdam.

Last, P.R. and Stevens, J.D. 1994. *Sharks and Rays of Australia*. Melbourne: CSIRO Publishing.

Lee Long, W.J., Mellors, J.E., and Coles, R.G., 1993. Seagrasses between Cape York and Hervey Bay, Queensland, Australia. *Australian Journal of Marine and Freshwater Research*, **44**: 19–32.

Lenanton, R.C., Joll, L., Penn, J., and Jones, G.K., 1991. The influence of the Leeuwin Current on coastal fisheries of Western Australia. *Journal of the Royal Society of Western Australia*, **74**: 101–114.

Lincoln Smith, M.P. and Jones, G.P., 1995. Fishes of shallow coastal habitats. In Underwood, A.J. and Chapman, M.G., *Coastal Marine Ecology of Temperate Australia*. Sydney: University of New South Wales Press, pp. 240–253.

Maclachlan, A. and Hesp, P., 1984. Faunal response to morphology and water circulation of a sandy beach with cusps. *Marine Ecology Progress Series*, **19**: 133–144.

Markina, N.P., 1976. Biogeographic regionalisation of Australian waters of the Indian Ocean. *Oceanology*, **15**: 602–604.

Marsh, H., Corkeron, P.J., Limpus, C.J., Shaughnessy, P.D., and Ward, T.M., 1995. The reptiles and mammals in Australian seas: their status and management. In Zann, L. (ed.), *The State of the Marine Environment Report for Australia. Technical Annex 1: The Marine Environment*. Canberra: Department of Environment, Sport and Territories, pp. 151–166.

Maxwell, J.G.H. and Cresswell, G.R., 1981. Dispersal of tropical marine fauna to the Great Australian Bight by the Leeuwin Current. *Australian Journal of Marine and Freshwater Research*, **32**: 493–500.

Middleton, J.H., 1995. The oceanography of Australian seas. In Zann, L.P. and Kailola, P. (eds), *The State of the Marine Environment Report for Australia. Technical Annex: 1. The Marine Environment*. Australia: Great Barrier Reef Marine Park Authority, pp. 1–10.

Middleton, J.F. and Viera, F., 1991. The forcing of low frequency motions in Bass Strait. *Journal of Physical Oceanography*, **21**: 695–708.

Morrissey, D.J., Howitt, L., Underwood, A.J., and Stark, J.S., 1992a. Spatial variation in soft-sediment benthos. *Marine Ecology Progress Series*, **81**: 197–204.

Morrissey, D.J., Underwood, A.J., Howitt, L., and Stark, J.S., 1992b. Spatial variation in soft-sediment benthos. *Marine Ecology Progress Series*, **164**: 233–245.

Morton, R.M., Pollock, R.B., and Beumer, J.P., 1987. The occurrence and diet of fishes in a tidal inlet to a saltmarsh in southern Moreton Bay, Queensland. *Australian Journal of Ecology*, **12**: 217–237.

Muldoon, J. (ed.), 1995. *Towards a Marine Regionalisation for Australia*. Ocean Rescue 2000 Workshop Series, Publication Number 1. Great Barrier Reef Marine Park Authority.

Nilsson, C.S. and Cresswell, G.R., 1981. The East Australian Current and it's eddies. *Progress in Oceanography*, **9**: 133–184.

North West Shelf Joint Environmental Study Interim Report. 2002. CSIRO Division of Marine Research. Marine Laboratories, Hobart.

Nunes, R.A. and Lennon, G.W., 1986. Physical property distributions and seasonal trends in Spencer Gulf, South Australia: an inverse estuary. *Australian Journal of Marine and Freshwater Research*, **37**: 39–53.

Paxton, J.R., Hoese, D.F., Allen, G.R., and Hanley, J.E., 1989. *Zoological Catalogue of Australia Vol. 7: Pisces (Petromyzontidae to Carangidae)*. Canberra: Australian Government Publishing Service/Australian Bological Resources Survey.

Peterson, C.H. and Black, R., 1987. Resource depletion by active suspension feeders on tidal flats: influence of local density and tidal elevation. *Limnology and Oceanography*, **32**:143–166.

Phillips, J.A., 2001. Marine macroalgal biodiversity hotspots: why is there high species richness and endemism in southern Australian marine benthic flora? *Biodiversity and Conservation*, **10**: 1555–1577, 2001.

Platnick, N.I., 1991. Patterns of biodiversity: tropical vs temperate. *Journal of Natural History*, **25**: 1083–1088.

Poiner, I.R. and Peterken, C., 1995. Seagrasses. In Zann, L. (ed.), *The State of the Marine Environment Report for Australia. Technical Annex 1: The Marine Environment*. Canberra: Department of Environment, Sport and Territories, pp. 107–117.

Poore, G.C.B., 1995. Biogeography and diversity of Australia's marine biota. In Zann, L. (ed.), *The State of the Marine Environment Report for Australia. Technical Annex 1: The Marine Environment*. Canberra: Department of Environment, Sport and Territories, pp. 75–84.

Poore, G.C.B. and Rainer, S., 1979. A three-year study of benthos of muddy environments in Port Phillip Bay, Victoria. *Estuarine and Coastal Shelf Science*, **9**: 477–497.

Provis, D.G. and Lennon, G.W., 1981. Some oceanographic meas-urements in the Great Australian Bight. In *Proceedings of the Fifth Australian Conference on Coastal and Ocean Engineering, 1881. Offshore Structures*. Australia: the Institution of Engineers, pp. 272–277.

Rainer, S., 1981. Temporal patterns in the structure of macrobenthic communities of an Australian estuary. *Estuarine and Coastal Shelf Science*, **13**: 597–620.

Resource Assessment Commission, 1993. *Coastal Zone Enquiry. Final Report*. Canberra: Australian Government Publishing Service.

Robertson, A.I., 1991. Plant–animal interactions and the structure and function of tropical mangrove forest ecosystems. *Australian Journal of Ecology*, **16**: 433–443.

Robertson, A.I. and Alongi, D.M. (eds.) 1992. *Tropical Mangrove Ecosystems*. Washington, AGU Press.

Robertson, A.I. and Alongi, D.M., 1995. Mangrove systems in Australia: structure, function and status. In Zann, L.P. and Kailola, P. (eds), *The State of the Marine Environment Report for Australia. Technical Annex: 1. The Marine Environment*. Australia: Great Barrier Reef Marine Park Authority, pp. 119–133.

Robertson, A.I. and Blaber, S.J.M., 1992. Plankton, epibenthos and fish communities. In Robertson, A.I. and Alongi, D.M. (eds.), *Tropical Mangrove Ecosystems*. Washington, AGU Press, pp. 173–224.

Robertson, A.I. and Duke, N.C., 1990. Recruitment, growth and residence times of fishes in a tropical Australian mangrove swamp. *Estuarine and Coastal Shelf Science*, **31**: 723–743.

Robertson, A.I. and Lee Long, W.J., 1991. The influence of nutrient and sediment loads on tropical mangrove and seagrass ecosystems. In Yellowlees, D. (ed.), *Proceedings of the workshop on land use patterns and nutrient loading of the Great Barrier Reef region*. Townsville: James Cook University of North Queensland, pp. 197–208.

Rochford, D.J., 1979. *Nutrient Status of the Oceans Around Australia. Report 1977–1979*. Hobart: CSIRO Division of Fisheries and Oceanography.

Rochford, D.J., 1986. Seasonal changes in the distribution of Leeuwin Current waters off southern Australia. *Australian Journal for Marine and Freshwater Research*, **37**: 1–10.

Saenger, P., 1995. The status of Australian estuaries and enclosed marine waters. In Zann, L.P., and Kailola, P. (eds), *The State of the Marine Environment Report for Australia. Technical Annex: 1. The Marine Environment*. Australia: Great Barrier Reef Marine Park Authority, pp. 53–59.

Saenger, P., Specht, M.M., Specht, R.L., and Chapman, V.J., 1977. Mangal and coastal salt-marsh communities in Australasia. In Chapman, V.J. (ed.), *Wet Coastal Ecosystems*. Amsterdam: Elsevier, pp. 293–345.

Sanderson, J.C. and Thomas, D.P., 1987. Subtidal macroalgal communities in the D'Entrecasteaux Channel, Tasmania. *Australian Journal of Ecology*, **12**: 41–51.

Semeniuk, V., Kenneally, K.F., and Wilson, P.G., 1978. *Mangroves of Western Australia*. Perth Western Australia, No. 12: Western Australian Naturalist Club.

Shaughnessy, P.D., 1999. *The Action Plan for Australian Seals*. Canberra: Environment Australia.

Shaughnessy, P.D., Gales, N.J., Dennis, T.E., and Goldsworthy, S.D., 1994. Distribution and abundance of New Zealand fur seals, *Arctocephalus forsteri*, in South Australian and Western Australia. *Wildlife Research*, **21**: 667–695.

Shepherd, S.A., 1981. Ecological strategies of a deep water red algal community. *Botanica Marina*, 24: 457–463.

Shepherd, S.A., 1983. Benthic communities of upper Spencer Gulf, South Australia. *Transactions of the Royal Society of South Australia*, **107**: 69–85.

Shepherd, S.A. and Sprigg, R.C., 1976. Substrate, sediments and subtidal ecology of Gulf St.Vincent and Investigator Strait. In Twidale C.R., Tyler, M.J. and Webb, B.P. (eds.), *Natural History of the Adelaide Region*. Royal Society of South Australia, pp. 161–174.

Shepherd, S.A. and Turner, J.A., 1985. Studies on South Australian abalone (genus *Haliotis*). VI. Habitat preference, abundance and predators of juveniles. *Journal of Experimental Marine Biology and Ecology*, **93**: 285–298.

Shepherd, S.A. and Womersley, H.B.S., 1970. The sublittoral ecology of West Island, South Australia. I. Environmental features and algal ecology. *Transactions of the Royal Society of South Australia*, **94**: 105–138.

Shepherd, S.A. and Womersley, H.B.S., 1971. Pearson Island expedition 1969. I. The subtidal ecology of benthic algae. *Transactions of the Royal Society of South Australia*, **95**: 155–167.

Shepherd, S.A. and Womersley, H.B.S., 1976. The subtidal algal and seagrass ecology of St.Francis Island, South Australia. *Transactions of the Royal Society of South Australia*, **100**: 177–191.

Shepherd, S.A. and Womersley, H.B.S., 1981. The algal and seagrass ecology of Waterloo Bay, South Australia. *Aquatic Botany*, **11**: 305–371.

Simpson, C.J. and Masini, R.J., (eds.). 1990. *Albany Harbours Environmental Study 1988–1989*. Environmental Protection Authority Bulletin, No. 412, Perth, Western Australia.

Specht, R.R., 1981. Biogeography of halophytic angiosperms (saltmarsh, mangrove and seagrass). In Keast, A. (ed.), *Ecological Biogeography of Australia*. The Hague: Junk, pp.575–589.

Tibbetts, IR, NJ Hall, WC Dennison (eds). 1998. *Moreton Bay and Catchment*. School of Marine Science, The University of Queensland, Brisbane.

Underwood, A.J. and Chapman, M.G., 1995. *Coastal Marine Ecology of Temperate Australia*. Sydney: University of New South Wales Press.

Underwood, A.J. and Kennelly, S., 1990. Ecology of marine algae on rocky shores and subtidal reefs in temperate Australia. *Hydrobiologia*, **192**: 3–20.

Underwood, A.J., Kingsford, M.J., and Andrew, N.L., 1991. Patterns in the shallow subtidal marine assemblages along the coast of New South Wales. *Australian Journal of Ecology*, **16**: 231–250.

Vernon, J.E.N., 1995. Coral reefs—an overview. In Zann, L.P., and Kailola, P. (eds), *The State of the Marine Environment Report for Australia. Technical Annex: 1. The Marine Environment*. Australia: Great Barrier Reef Marine Park Authority, pp. 135–144.

Villanoy, C.A., and Tomczak, M., 1991. Influence of Bass Strait water on the Tasman Sea thermocline. *Australian Journal of Marine and Freshwater Research*, **42**: 451–464.

Ward, T.M., Jones, G.K., and Kinloch, M., 1998. *Baitfisheries of Southeastern Australia*. Final report to FRDC.

Wass, R.E., Connolly, J.R., and MacIntyre, R.J., 1970. Bryozoan carbonate sand continuous along southern Australia. *Marine Geology*, **9**: 63–73.

Watkins, D., 1993. *A National Plan for Shorebird Conservation in Australia*. Australasian Wader Studies Group, Royal Australasian Ornithologists Union and World Wide Fund for Nature. RAOU Report No. 90.

Webster, I.T., 1985. Frictional continental shelf waves and the circulation response of a continental shelf to wind-forcing. *Journal of Physical Oceanography*, **15**: 855–864.

Wenju, C., Schahinger, R.B. and Lennon, G.W. 1990. Layered models of coastal upwelling: a case study of the South Australian region. In Davies A.M.(ed.), *Modelling Marine Systems*, Vol. 1 CRC Press, Boca Raton, Florida, USA, pp. 73–91.

Wilson, B.R. and Allen, GR. 1987. Major components and distribution of marine fauna. In Dyne, G.W. (ed.), *Fauna of Australia. Volume 1A. General Articles*. Canberra: Australian Government Publishing Service, pp. 43–68.

Womersley H.B.S., 1959. The marine algae of Australia. *Botanical Review*, **25**: 545–614.

Womersley H.B.S., 1981a. Biogeography of marine macroalgae. In Clayton, M.N., and King, R.J. (eds.), *Marine Botany: an Australasian Perspective*. Melbourne: Longman Cheshire, pp. 293–307.

Womersley, H.B.S., 1981b. Marine ecology and zonation of temperate coasts. In Clayton, M.N. and King, R.J. (eds.), *Marine Botany: an Australasian Perspective*. Melbourne: Longman Cheshire, pp. 211–240.

Womersley, H.B.S 1990. Biogeography of Australasian marine macroalgae. In Clayton, M.N., and King, R.J. (eds.), *Biology of Marine Plants*. Melbourne: Longman Cheshire, pp. 349–366.

Womersley, H.B.S. and Edmonds, S.J., 1958. A general account of the intertidal ecology of South Australian coasts. *Australian Journal of Marine and Freshwater Research*, **9**: 217–260.

Womersley, H.B.S. and King, R.J., 1990. Ecology of temperate rocky shores. In Clayton, M.N. and King, R.J. (eds.), *Biology of Marine Plants*. Melbourne: Longman Cheshire, pp. 266–295.

Zann, L.P., 1995. *Our Sea, Our Future: Major Findings of the State of the Marine Environment Report for Australia*. Australia: Great Barrier Reef Marine Park Authority.

Zann, L.P., 1996. *State of the Marine Environment Report for Australia: Technical Summary*. Great Barrier Reef Marine Park Authority for the Department of the Environment, Sport and Territories, Ocean Rescue 2000 program. Commonwealth of Australia.

Cross-references

Australia, Coastal Geomorphology
Coral Reefs
History, Coastal Ecology
Indian Ocean Coasts, Coastal Ecology
Mangroves, Coastal Ecology
New Zealand, Coastal Ecology
Rock Coast processes
Salt Marsh
Sandy Coasts
Wetlands

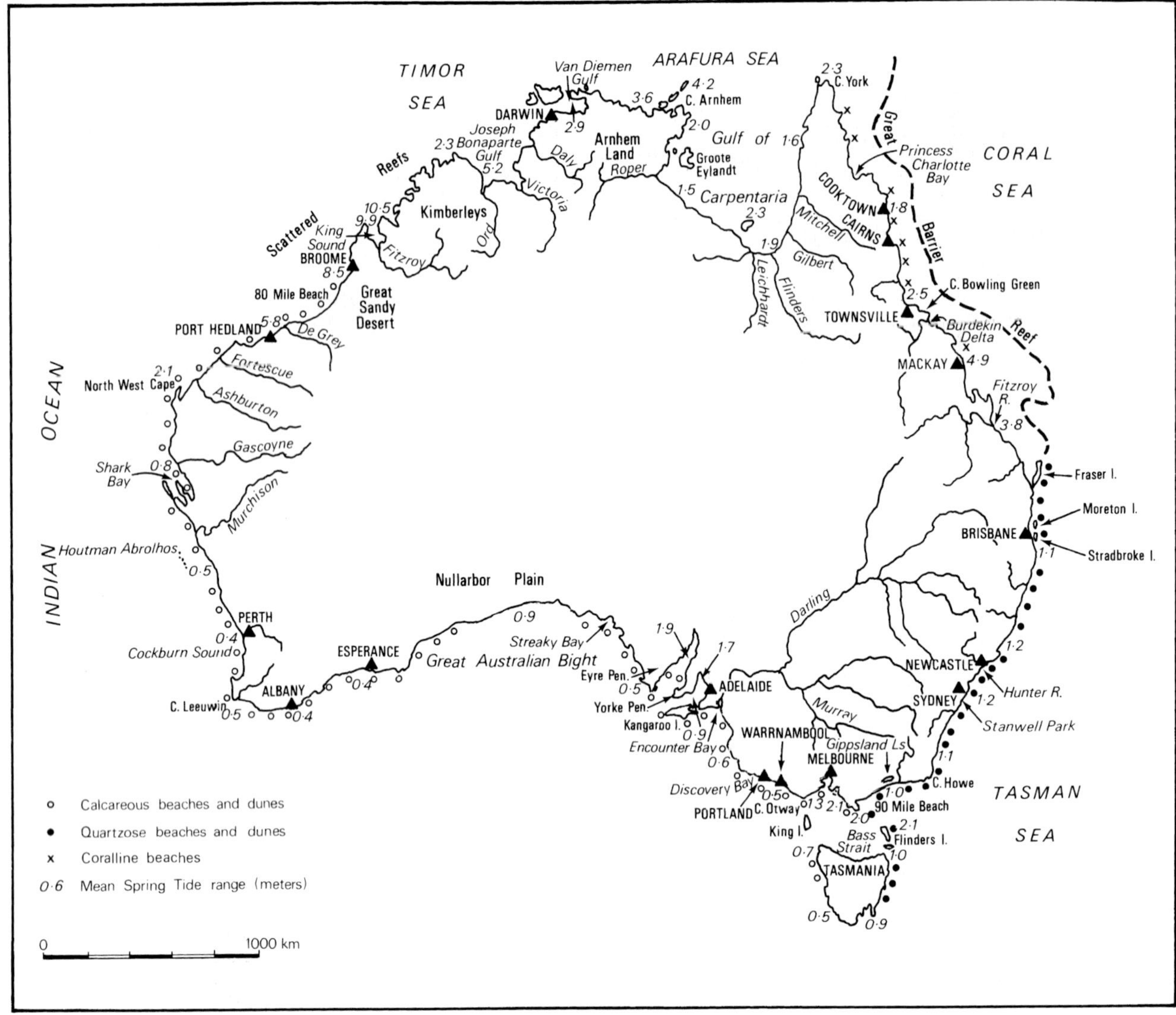

Figure A69 Location and feature map.

AUSTRALIA, COASTAL GEOMORPHOLOGY

The Australian coastline is about 20,000 km long (Figure A69), and has many long gently curving sandy beaches interspersed with sectors of cliffs and steep coast. The beaches on the western and southern coasts, from Broome in the northwest to Wilsons Promontory in Victoria and the western coasts of the Bass Strait islands and Tasmania are predominantly calcareous, whereas those of the eastern and northern coasts are generally quartzose, except in the vicinity of fringing coral reefs. Many of the beaches form the seaward margins of depositional sand barriers bearing multiple beach ridge or dune topography (Bird, 1973; Thom, 1984), and on the western and southern coasts the calcareous sands of the older (Pleistocene) dunes have been partially lithified by secondary carbonate precipitation to form dune calcarenites (calcareous aeolianites). These are extensive between Broome and Cape Leeuwin on the Western Australian coast, between Streaky Bay in South Australia and Cape Otway in Victoria, and locally around the shores of Bass Strait, including the west coasts of King Island and Flinders Island and the northwest coast of Tasmania. In places these Pleistocene dune calcarenites have been exposed to marine erosion and cut back to form cliffs and shore platforms (Figure A70).

Cliffs and rocky shores have also developed where older geological formations reach the coast: in Tertiary formations near Cape Cuvier, in granites and older shield formations between Cape Leeuwin and Albany, and in Tertiary formations at the head of the Great Australian Bight. Pre-Cambrian rocks reach the coast on the Eyre and Yorke peninsulas, on the Fleurieu Peninsula and on Kangaroo Island in South Australia. In Victoria there are cliffs cut in various formations, including Cenozoic basalts, Tertiary and Mesozoic sediments, Paleozoic rocks, and intruded granites. Tasmania has extensive rocky and steep coasts developed on pre-Carboniferous rocks and intruded granites, with Mesozoic formations, including Jurassic dolerites, on its southeastern coast. In New South Wales, there are cliffed and rocky headlands on Mesozoic and older rock formations, and in Queensland steep coasts, with limited cliffing, occur on similar formations inshore from the Great Barrier Reef (Figure A71). On the north coast of Australia cliffed and rocky sectors are extensive in Arnhemland and where the Kimberley Ranges reach the sea, both areas of Paleozoic and older rock formations.

In southwestern and southeastern Australia most of the rivers flow into estuarine lagoons with outlets constricted by sandy barrier formations. These lagoons show stages of infilling, largely by fluvial sediments, and some have been reclaimed by river deposition as broad swampy alluvial plains. In northern Australia, river mouths are less encumbered by coastal sand deposition and form estuarine gulfs and deltaic regions. The shores of estuaries, lagoons, and sheltered embayments are often bordered by marshes and swamps, which have advanced on to tidal sandflats or mudflats. The latter become more extensive on sectors of the northern coast where tide range is large.

Coralline reef formations are extensive off the northern half of Australia, particularly in the Great Barrier Reef off Queensland. Numerous outlying patch reefs and coastal fringing reefs occur off the

Figure A70 Cliffs and shore platforms cut in Pleistocene dune calcarenite on the Nepean Ocean coast near Melbourne, Victoria (photo: E.C.F. Bird, Copyright-Geostudies).

Figure A71 Steep coast of Macalister Range, north of Cairns, Queensland, with an absence of cliffing on a moderate wave energy coast inshore from the Great Barrier Reef (photo: E.C.F. Bird, Copyright-Geostudies).

north and northwest coasts, around to Houtman Abrolhos off Western Australia (Fairbridge, 1967).

Variations in morphology around the Australian coast are related to geological, climatic, oceanic, and biological factors that vary regionally around the margins of the continent (Davies, 1980), taking account of changes in climate and sea level that have occurred on and around the Australian land mass (7.8 million km^2) and its bordering continental shelves (2.5 million km^2).

Geological factors

The western half of the Australian continent is a shield region dominated by Paleozoic and older formations that produce the hard, rocky coasts of the Kimberley region and southwest Western Australia. However, on much of the western coast these formations are obscured by Pleistocene dune calcarenites and associated Holocene dunes. On the Eyre Peninsula in South Australia the dune calcarenite mantle has been cut back by marine erosion to expose underlying harder Pre-Cambrian rocks, which become promontories as bordering cliffs in the less resistant dune calcarenites recede.

The eastern margin of Australia consists of folded Paleozoic rock formations with structural trends roughly parallel with the eastern coastline. These formations have been uplifted and dissected to form the Eastern Highlands, with a watershed varying from 20 to 400 km inland, and a succession of relatively small drainage basins feeding rivers that flow down to the Pacific coast. The structural pattern is complicated by the presence of sedimentary basins occupied by Mesozoic formations: the Sydney Basin, for example, contains the major sandstone formations prominent in cliffs along the New South Wales coast between Jervis Bay and Newcastle.

Between the Western Shield and the Eastern Highlands the continent is dominated by Cenozoic rocks, which occupy a number of basins, several of which extend to and beyond the present coastline. In eastern Victoria the Latrobe Valley syncline passes beneath a depression occupied by the Gippsland Lakes region, where complex sand barriers of Pleistocene and Holocene age are fringed by the Ninety Mile Beach. In western Victoria Tertiary rocks dip gently off the uplifted Mesozoic formations of the Otway Ranges and pass beneath an extensive volcanic province, locally active into the early Holocene: the sequence is clearly displayed in cliff sections westward from Cape Otway to Discovery Bay. Bass Strait is underlain by another sedimentary basin, and on a smaller scale Port Phillip and Western Port Bays in Victoria occupy fault-bounded sunklands (Bird, 1993). Western Tasmania is dominated by Lower Paleozoic rock formations, but in the southeast these are overlain by Mesozoic rocks in an area where Cenozoic faulting has influenced the evolution of ridges and valleys now partly drowned as rias.

In South Australia, a sedimentary basin underlies the Lower Murray valley, and the river flows across Tertiary formations to enter coastal lagoons (Lakes Albert and Alexandrina) behind the Pleistocene dune calcarenite barriers of Encounter Bay. At the head of the Great Australian Bight the Eucla Basin contains Tertiary formations that outcrop on a long stretch of even-crested cliffs bordering the riverless Nullarbor Plain. Similar cliffed formations mark the Carnarvon Basin at Cape Cuvier, but the Tertiary rocks in the Canning Basin, in northwest Australia, vanish beneath the Great Sandy Desert and the dune calcarenites of the Eighty Mile Beach, and the Joseph Bonaparte Gulf Basin is largely a shelf structure. The Carpentaria Basin runs northward under the Gulf of Carpentaria, a graben with Cenozoic formations declining from the western side of Cape York Peninsula, and the much smaller Laura Basin occupies the hinterland of Princess Charlotte Bay on the eastern side.

Despite the existence of these several Cenozoic basins Australia has been generally a rather stable continent, and there are extensive planation surfaces, Cenozoic and older, that show little tectonic deformation. In consequence the continent is notably flat and low-lying, with 80% of the land area below the 500 m contour. Many of the rivers (e.g., the Murray) have low channel gradients as they approach the sea, and carry mainly fine-grained sediments (silt and clay) to their mouths. Steeper gradients exists in rivers draining the Pacific slopes of the Eastern Highlands, and some of these carry sand and gravel in the loads delivered to the sea. Fluvial sediment yields are also influenced by outcrop lithology and past as well as present weathering regimes. Most of the granite outcrops in the Eastern Highlands have been deeply weathered, and rivers draining them derive quartz-rich sands; weathered basalt supply silt and clay, and river gravels are produced by the disintegration of Paleozoic metasediments. On the coast marine erosion has generally removed weathered mantles, so that the granites, for example, usually protrude as rocky headlands, as at Wilsons Promontory in Victoria.

Figure A72 The River Don, in north Queensland, carries sand and gravel down to the coast, where it is spread along the shore by wave action (photo: E.C.F. Bird, Copyright-Geostudies).

Although, the general picture is one of prolonged stability in Australia, marginal neotectonic activity has been recorded locally, especially in Victoria, and the possibility that slight coastal warping has continued into Holocene times cannot be ruled out.

Climatic factors

The prevailing aridity of the interior and much of western and southern Australia has resulted in few rivers draining to the coast: those few flow intermittently, with occasional brief episodes of flooding. In the southwest of Western Australia and the southeast of South Australia winter rainfall nourishes seasonal runoff, but Australia's largest river system, the Murray-Darling, draining a catchment of about a million square kilometers, has only a limited discharge at its mouth compared with those of similar scale in more humid regions. In consequence of hinterland aridity, terrigenous sediment supply to the coasts of much of western and southern Australia has been meager, and coastal deposition nourished primarily from seafloor sources is dominated by carbonate sands of biogenic origin. The presence of such material in shelf sediments is a further consequence of the lack of terrigenous supply, hinterland aridity having evidently prevailed also during Pleistocene phases of lowered sea level.

By contrast, the Eastern Highlands have a higher and more regular rainfall, and the steeper catchments draining to the Pacific coast nourish river systems that have delivered a more substantial terrigenous sediment supply. Coastal sediments are dominated by quartzose sands, some of which are supplied directly by such rivers as the Shoalhaven, the Hunter, the Burdekin, and the Don (Figure A72), particularly during episodes of flooding. Much of the sand deposited in beaches and barriers along the Victorian and New South Wales coasts has, however, been swept in from the seafloor during successive marine transgressions in Pleistocene and Holocene times by the reworking of shelf deposits of quartzose sand derived partly from fluvial supply during low sea-level phases and partly from weathered rock outcrops now submerged. The contrast between the carbonate sands and dune calcarenites of southern and western Australia and the quartzose sands of the southeastern and eastern coast is thus related to a contrast in shelf sedimentation, due in turn to a contrast in hinterland topography and climate through Quaternary times.

Along the Queensland coast terrigenous sediment supplied by rivers includes quartzose sands and lithic gravels derived from relatively steep catchments as well as muddy sediment derived from tropical weathering mantles on catchment outcrops. In addition, coastal sedimentation is locally complicated by the presence of fringing and nearshore reefs built by corals and associated organisms in tropical waters. Reef clastics derived from these are added to adjacent beaches, but despite the proximity of a rich coralline province the beaches of Queensland are generally poor in carbonates (Bird, 1978).

In Northern Australia, summer rainfall (some from occasional cyclones) results in substantial runoff from river systems, but catchments are generally larger and less steep than those of the east coast, and fluvial loads are dominated by fine-grained sediment derived from prolonged weathering of surface outcrops, sandy yields being limited and localized. Muddy sedimentation is extensive on the shores of the Gulf of Carpentaria and in northern embayments and inlets, particularly around Van Diemen Gulf, Joseph Bonaparte Gulf, and King Sound, each of which receives major inputs of river water and sediment.

Other features related to climate include the nature and rate of rock weathering in the shore zone, most active under humid tropical conditions in northern Australia, and the vigor of wind action, which is much stronger on temperate coasts than in the humid tropical north, a contrast that helps to explain the more extensive development and greater mobility of coastal dunes in western, southern, and southeastern Australia (Figure A73) than in the north and northeast. Davies (1980) elucidated such contrasts as these within a climatically based sector classification, clockwise around the continent, from warm temperate humid (Fraser Island to Portland), warm temperate arid (Portland to North West Cape), tropical arid (North West Cape to Broome), and tropical humid (Broome to Fraser Island) coastal sectors.

Oceanic factors

Swell generated by storms in the Southern Ocean moves in to the western, southern, and southeastern coasts of Australia, and its refracted patterns are effective in shaping the long, curved outlines typical of sandy beaches. In New South Wales, for example, refraction of the

Figure A73 Parabolic dunes spilling inland from the sandy beach on the coast of Encounter Bay, South Australia (photo: E.C.F. Bird, Copyright-Geostudies).

dominant southeasterly swell is responsible for the asymmetrically curved outlines of beaches between bedrock headlands (Chapman *et al.*, 1982).

In the southern half of Australia coasts are frequently exposed to storm wave action generated in coastal waters as cyclonic depressions move through from west to east. With relatively deep water close inshore, these are typically high wave energy coasts with bold cliffs and headlands and smooth depositional outlines. Extensive sand deposition has constricted the valley mouth inlets formed by marine submergence during the Late Quaternary marine transgression and formed estuarine thresholds of inwashed sand at their entrances (Roy, 1984).

On the east coast north of Fraser Island ocean swell is interrupted by reef structures, and in the lee of the Great Barrier Reef it is largely excluded from coastal waters (Hopley, 1982). Within these waters the prevailing southeasterly winds generate only moderate wave energy, and consequently the coast has but limited sectors of cliffing, poor development of shore platforms and irregular depositional outlines, with many spits, cuspate forelands and some protruding deltas, notably the Burdekin; sectors of mudflat and mangrove swamp occur on shores in the lee of headlands and islands. Sandy barriers and dune systems have been deposited on a much smaller scale than on the oceanic coasts, and take the form of beach-ridge plains, often built as cheniers on emergent mudflats. River mouths are less encumbered by sand deposition, and coastal lagoons do not occur north of Fraser Island. In the absence of ocean swell, shoreward drifting of sea floor sand has made only a limited contribution to coastal sedimentation on the Queensland coast.

On the north coast of Australia swell from the Indian Ocean is attenuated by the exceptionally broad continental shelf, and its effects are further diminished by the large tide ranges. In the Timor and Arafura Seas and within the Gulf of Carpentaria local wind action produces generally low to moderate wave energy on bordering coasts. Steep sectors show little cliffing and retain the intricate outlines of submerged landscapes of fluvial dissection, while depositional sectors are typically irregular and varied in detail. Storm surges accompanying tropical cyclones accomplish rapid short-term changes, and may be responsible for the emplacement of sandy cheniers on coastal plains, for example, on the depositional lowland south of Van Diemen Gulf and on the low-lying coasts of the Gulf of Carpentaria, where surges of up to 7 m have been reported.

Although, Pacific tsunamis have generated only minor surges on the Australian coast in the past two centuries it is possible that they occurred at earlier stages on a larger scale, and may have shaped some apparently emerged features (Bryant *et al.*, 1996).

The broad patterns of longshore drifting on the Australian coast are northward and eastward from Cape Leeuwin and northward along the east coast, but where coastal outlines are adjusted to the dominant swell patterns little net drifting occurs (Davies, 1980).

In addition to variations in wave energy there are marked contrasts in tide range around the Australian coast (Figure A69). In the southern part of the continent, from North West Cape around to Fraser Island, mean spring tide ranges are generally microtidal (<2 m), rising to mesotidal (2–4 m) in Spencer Gulf and Gulf St. Vincent in South Australia and in Western Port Bay and Corner Inlet in Victoria. In southwestern Australia the spring tide range is less than 0.5 m, and it can be difficult to detect tidal oscillations because wind stress and barometric pressure variations raise and lower the sea surface by larger amounts. On the Queensland coast and in the southeastern part of the Gulf of Carpentaria tide ranges are typically mesotidal, but the northwestern coast between Port Hedland and Darwin is macrotidal (>4 m), with a maximum of 10.5 m in Collier Bay, near Yampi Sound.

Where tide range is, small wave energy is concentrated within a narrow vertical range, and where it is large, wave energy is more dispersed. Thus, the tidal variation emphasizes the contrast between high wave energy conditions in the southern part of Australia and low wave energy conditions along the northern coast. Large tide ranges also generate strong ebb and flow currents, which impede the deposition of wave-shaped structures, such as spits and barriers, and lead to the formation of wide funnel-shaped estuaries, such as that of the Fitzroy, opening into King Sound. Broad sandflats and mudflats are exposed at low tide and, with diminished and intermittent wave action above mid-tide level, salt marshes and mangrove swamps become more extensive than on microtidal coasts, where the small tide range compresses the biological zonations.

Figure A74 Mangroves advancing seaward on the coast of Cairns Bay, north Queensland. An outer zone of *Avicennia* is backed by *Rhizophora* and then a varied mangrove community (photo: E.C.F. Bird, Copyright-Geostudies).

Figure A75 The intertidal zone in King Sound, Western Australia, has branching tidal creeks fringed by mangroves, desiccated sandy mudflats, and a serrated high tide shoreline where desert dune ridges and swales have been submerged by the sea (photo: E.C.F. Bird, Copyright-Geostudies).

Biological factors

The effects of organisms on coastal processes are limited by their geographical range, conditioned mainly by climatic and oceanic factors. Reef-building corals live in Australian coastal waters from Houtman Abrolhos in the west around to the Capricorn Group at the southern end of the Great Barrier Reef. They are most prolific off the northeast coast, while off the northwest they give place to bryozoan and oyster banks and off the west coast the reefs are submerged dune calcarenite ridges with only a veneer of corals (Fairbridge, 1967).

Mangroves extend as far south as Corner Inlet (38°51′45″S) adjacent to Wilsons Promontory in Victoria. In the humid tropical sector of northeast Queensland there are up to 27 mangrove species in wide swamps backed by rain forest vegetation (Figure A74), but in drier sectors there are fewer species and the seaward mangrove fringe is backed by salt marshes and hypersaline mudflats. The effects of mangroves on sedimentation vary from one species to another, but the white mangrove (*Avicennia marina*), the most widespread of the Australian mangrove species and the only one extending as far south as Victoria, has a pneumatophore network that acts as a sediment filter and is consequently an agent of land-building alongside estuaries and embayments sheltered from strong wave action (Bird, 1993). Introduced *Spartina anglica* has re-shaped the intertidal profiles of Andersons Inlet, a Victorian estuary, and the Tamar estuary in Tasmania, by developing a depositional marshland terrace, but it does not grow in warmer environments farther north.

Shelly beaches (as distinct from beaches of shelf-derived biogenic carbonate sand) are confined to coastal sectors where shell-bearing organisms live abundantly, notably on rocky shores and reefs and within estuaries and lagoons. These are extensive towards the heads of the South Australian Gulfs and on the western shores of Port Phillip Bay, where their dominance reflects the local absence of other types of beach material on a coast formed by the marginal submergence of a basaltic plain (Bird, 1993).

Inherited features

Australian coastal landforms include various features inherited from past phases of contrasted environments. Some of these are related to changes of sea level during Quaternary times: the submergence of valley mouths and the shoreward drifting of sand from the sea floor to form beaches, barriers, and coastal dunes have already been noted. In addition, there are coastal features formed during past phases of higher sea level, some of which have persisted from Pleistocene times while others may be refered to a higher Holocene sea-level episode.

Figure A76 Cliffs cut in soft Miocene limestone on the Port Campbell coast in Victoria, with some of the outlying stacks known as the Twelve Apostles (photo: E.C.F. Bird, Copyright-Geostudies).

These features include emerged shore platforms, some with associated fossil beach deposits backed by former cliffs now degraded to bluffs; emerged coralline reefs; emerged sandy thresholds near the mouths of estuaries and lagoons; and beach ridge and barrier features standing higher, in relation to sea level, than they did at the time of their formation.

The interpretation of some of these features has been questioned. It is necessary to distinguish between emerged shelly beaches and the shelly aboriginal kitchen middens frequently encountered above high-tide level, and to be sure that apparently emerged features were not formed when storm surges or tsumanis carried wave action briefly to higher levels. Coastal features that have undoubtedly emerged may be the outcome of tectonic uplift of the land margin, a fall in sea level, or some combination of the two. Emerged features dating from Pleistocene times often show evidence of dissection by fluvial incision or wind scour extending below present sea level during the last phase of lowered sea level, as on the inner barrier of the Gippsland Lakes (Bird, 1993), whereas emerged features of Holocene age do not show such dissection.

On parts of the Australian coast the landforms bear the imprint of weathering under past phases of climate that were warmer, cooler, wetter, or drier than at present. These include lateritic weathering profiles in cliff exposures of Tertiary rock in Victoria, inherited from a Pliocene phase of warmer and wetter climate, and the presence of periglacial deposits on cliffs in southernmost Tasmania, inherited from a Pleistocene phase of colder climate. On the east coast of the Cape York Peninsula near Cape Flattery extensive systems of elongated parabolic dunes, now largely covered by scrub and forest, originated under drier and perhaps windier conditions in the past. Patterns of coastal rock weathering northward and southward of the wettest sector of humid tropical coast in northeast Queensland indicate that the warm and wet environment has been at times more extensive than it is now. The high parabolic dunes on Fraser Island must also have formed under different conditions in the past, for they now carry a cover of rain forest. It is possible that sand mobilization here was also partly related to phases of rising sea level, when vegetated dunes deposited during earlier phases of stable or falling sea level were trimmed back by marine erosion and exposed to the effects of onshore wind action.

The shores of King Sound include narrow sandy peninsula ridges and intervening corridor embayments that are features inherited from an arid landscape of sand ridges and swales that has been partly submerged by the Late Quaternary marine transgression (Figure A75).

Shore platforms

Shore platforms on the Australian coast show a variety of forms, some sloping seaward, others almost horizontal. Some are related to coastal structures, especially where they coincide with exhumed bedding planes on resistant strata, but many transgress shore structures. Seaward sloping platforms are essentially abrasion ramps, but subhorizontal platforms have been influenced by shore weathering processes which have tended to flatten and maintain platforms originally cut by abrasion. Subhorizontal platforms are well developed at about high-tide level on cliffed coasts cut in basalt and fine-grained sedimentary formations, such as sandstones and shales, which are subject to disintegration by wetting and drying processes (and possibly also by salt crystallization) down to a level of saturation by sea water. These are well developed on sandstones and mudstones on the New South Wales coast south of Sydney.

On dune calcarenite coasts similar platforms have developed at a lower level, equivalent to the level attained by fringing reefs built upward by corals. Their flatness is evidently the outcome of solution processes effective down to the level at which precipitation of carbonates becomes dominant, indurating the platform surface, a process that has been demonstrated on the Nepean Peninsula in Victoria (Figure A70). Where waves are supplied with sand or gravel derived from cliff erosion, fluvial outflow or sea floor sources, abrasion becomes more effective, and the subhorizontal benches give place to platforms that slope seaward.

Modern dynamics

Erosion is predominant on Australian beaches, even on the shores of barriers and beach-ridge plains that prograded earlier in Holocene times, and most beaches are backed by cliffed dunes. Exceptions are found on growing sectors of spits and cuspate forelands, on deltas still receiving fluvial sediment (notably in northern Australia), and on sectors where beach sand has accumulated alongside protruding structures, such as harbor breakwaters. There are also some sectors, notably on the western coast of Western Australia, in Streaky Bay in South Australia, off northwest Tasmania and near Corner Inlet in Victoria where sand is still moving in from nearshore shoals to maintain or prograde beaches (Sanderson *et al.*, 1998).

The prevalence of beach erosion could be due to a slight rise in sea level in recent decades or to increased storminess in coastal waters, but the most likely explanation is a general reduction in the supply of sand from the sea floor, which had previously nourished beach and barrier progradation. Other factors have contributed locally, such as the reduction of fluvial sand supply following dam construction in the Barron and Burdekin Rivers in North Queensland and the diminution of sand supply where the longshore sand supply has been intercepted by breakwaters, as at Sandringham on the coast of Port Phillip Bay (Bird, 1996).

Cliffed coasts have continued to recede, especially on sectors where soft rock formations confront stormy seas, as on the Tertiary sands and clays of the Port Campbell district in Victoria (Figure A76). Minor progradation has been detected on some mangrove-fringed coasts, as in Cairns Bay, Queensland, and on salt marshes, particularly where *Spartina* grass has been introduced. Elsewhere there is evidence of reduction of mangrove and salt marsh fringes, as in Western Port Bay, Victoria, where there has been drainage of swampland and construction of boat harbors.

Human impacts

Coastal morphology in Australia has been modified in various ways by human activities, particularly in the vicinity of urban centers, where ports have been constructed, seawalls built to halt coastline recession, cliffs stabilized and groins inserted in an attempt to retain a beach fringe. A substantial part of the formerly cliffed northeast coast of Port Phillip Bay, on Melbourne's Bayside, now consists of artificial bluffs lined by such structures, and depletion of the natural beach fringe has been offset by artificial beach nourishment. Similar works have been introduced to the Adelaide coast, where the beach is now maintained by periodic renourishment using sand dredged from a prograded beach alongside a breakwater at the Outer Harbor.

Many coastal dune areas show evidence of vegetation loss and initiation or acceleration of blowouts as a consequence of clearing, burning, trampling, or vehicle damage to the dune vegetation. Attempts have been made to restore stability by fencing dune areas and planting vegetation, including the introduced European marram grass, *Ammophila arenaria*. Such activities as beach nourishment and dune stabilization mark a trend toward management of coastal morphology in Australia, with the aims of conserving many natural systems and their biota and providing recreational opportunities in the coastal environment.

Eric Bird

Bibliography

Bird, E.C.F., 1973. Australian coastal barriers. In Schwartz, M.L. (ed.), *Barrier Islands*, Stroudsburg: Dowden, Hutchinson & Ross, pp. 410–426.

Bird, E.C.F., 1978. Variations in the nature and source of beach materials on the Australian coast. In Davies, J.L., and Williams, M.A.J. (eds.), *Landform Evolution in Australasia*. Canberra: Australian National University Press, pp. 144–157.

Bird, E.C.F., 1993. *The Coast of Victoria*. Melbourne University Press.

Bird, E.C.F., 1996. *Beach Management*. Chichester: Wiley.

Bryant, E.A., Young, R.W., and Price, D.M., 1996. Tsunamis as a major control on coastal evolution, Southeastern Australia. *Journal of Coastal Research*, **12**: 831–840.

Chapman, D.M., Geary, M., Roy, P.S., and Thom, B.G., 1982. *Coastal Evolution and Coastal Erosion in New South Wales*. Sydney: Coastal Council of New South Wales.

Davies, J.L., 1980. *Geographical Variation in Coastal Development*. 2nd edn. London: Longman.

Fairbridge, R.W., 1967. Coral reefs of the Australian region. In Jennings, J.N., and Mabbutt, J.A. (eds.), *Landforms Studies from Australia and New Guinea*. Canberra: Australian National University Press, pp. 386–417.

Hopley, D., 1982. *The Geomorphology of the Great Barrier Reef*. Chichester: Wiley.

Roy, P.S., 1984. New South Wales estuaries: their origin and evolution. In Thom, B.G. (ed.), *Coastal Geomorphology in Australia*. Sydney: Academic Press, pp. 99–121.

Sanderson, P.G., Eliot, I., and Fuller, M., 1998. Historical development of a foredune plain at Desperate Bay, Western Australia. *Journal of Coastal Research*, **14**: 1187–1201.

Thom, B.G., 1984. Sand barriers of Eastern Australia: Gippsland—a case study. In Thom, B.G. (ed.), *Coastal Geomorphology in Australia*. Sydney: Academic Press, pp. 233–261.

Cross-references

Changing Sea Levels
Coastal Changes, Gradual
Coastal Changes, Rapid
Cliffs, Erosion Rates
Coastal Subsidence
Deltas
Glaciated Coasts
Greenhouse Effect and Global Warming
Mapping Shores and Coastal Terrain
Volcanic Coasts

B

BARRIER ISLANDS

Definition and occurrence

Barrier islands are elongate, shore-parallel accumulations of unconsolidated sediment (primarily sand), some parts of which are supratidal, that are separated from the mainland by bays, lagoons, or wetland complexes. They are most abundant along the coastlines of the trailing edges of continental plates and of epicontinental seas and lakes (e.g., Caspian and Black Seas). They do not occur on coasts with tidal ranges greater than around 4 m, because their primary mechanism wave action is not focused long enough at a single level during the tidal cycle to form the island and the strong tidal currents associated with such large tides transport the available sand to offshore regions. Barrier islands do occur, primarily as spit forms, on leading edge and glaciated coasts, but they are a minority coastline type in those areas.

Barrier islands are the dominant coastline type along the Atlantic and Gulf coasts of the United States, most of them having been formed within the last 4,000–5,000 years during a near stillstand of sea level.

Barrier islands on depositional coasts

Depositional, coastal plain shorelines typically have barrier islands located between major river deltas and estuaries. Two types of barrier islands may be present, those that consistently migrate landward (retrograding) and those that build seaward (prograding). The island type depends upon the ratio of relative sea-level change to sediment supply. Diminished or low sediment supply and/or rapid sea-level rise promotes the development of retrograding barrier islands and *vice versa* for prograding barrier islands.

Retrograding barrier islands are composed of coalescing washover fans and terraces that are overtopped at high tides, usually several times a year (Figure B1). Stratigraphically, a relatively thin wedge of sand and shell of the washover terrace overlies backbarrier sediments, which are typically composed of muddy sediments deposited in the lagoons or wetlands behind the islands. These islands are impractical sites for human development because of their constant landward migration.

Prograding barrier islands are composed of multiple beach ridges. Many have a drumstick configuration because of selective sand accumulation on the updrift end of the island (the direction sand comes from). The most notable changes on prograding islands occur when the adjacent tidal inlets migrate or when the inlets expand dramatically during hurricane storm surges.

Stratigraphically, prograding barrier islands are composed of sand 8–10 m thick (thickness depends upon wave size), which has prograded over offshore muds (Figure B1). When human development occurs on these islands, buildings are usually secure from all but the most extreme hurricanes, if they have been set back an adequate distance from the front-line dunes and tidal inlets. But that security will vanish if a major rise in sea level occurs (Hayes, 1996).

The morphology of prograding barrier islands is controlled by a combination of wave and tidal forces (Hayes, 1979; Davis, 1994a). Under *wave-dominated* conditions, which most commonly occur in microtidal areas (tidal range ≤2 m; Davies, 1964), the barriers are long, typically tens of kilometers, with widely spaced inlets that have large flood-tidal deltas and small ebb-tidal deltas. Washover fans are common and the islands are flanked on the landward side by bays and/or lagoons (Figure B2). Barrier islands along *mixed-energy* coasts (Hayes, 1979), which typically occur in mesotidal areas (tidal range = 2–4 m), are stunted and short (usually <10 km) with abundant tidal inlets that contain large ebb-tidal deltas and small to nonexistent flood-tidal deltas. These islands are flanked on the landward side by complex tidal channels, tidal flats, and wetlands (Figure B3). Barrier islands do not occur on *tide-dominated* coasts.

Origin of barrier islands

Barrier islands are thought to most commonly originate in one of the three possible ways: (1) by spit elongation (Fisher, 1967); (2) retreating transgressive barrier islands (Swift, 1975); and (3) a process termed *transgressive–regressive interfluve hypothesis* by Hayes (1994). In many parts of the world, it is clear that the source of sand for the existing barrier islands originated from an updrift headland, and as a spit extended away from the headland it was cut into segments during storms, creating tidal inlets that eventually attained permanence. Swift (1975) stated that barrier islands originated at a lower stand of sea level and migrated over the drowning coastal plain as sea level rose during the early Holocene. The "primary barrier" of Pierce and Colquhoun (1970), the nucleus for many prograding barrier islands, no doubt originated in this way. According to the transgressive–regressive interfluve hypothesis, as sea level rose, the transgressive barrier eventually perched on the topographic high of an interfluve located between major alluvial valleys that were carved during the last Pleistocene lowstand (Figure B4). Once sea level stabilized around 4500 years BP, prograding barrier islands developed in areas with adequate sediment supply. As the island grew, beach ridges prograded away from the interfluve, with major tidal inlets forming at both ends of the island over the antecedent lowstand valleys (Moslow, 1980). This mechanism explains the origin of many of the major prograding barrier islands along the coast of the southeastern United States.

Barrier islands on leading edge and glaciated coasts

The somewhat rare barrier islands on the leading edge, west coast of the United States are, for the most part, relatively short spits that have built

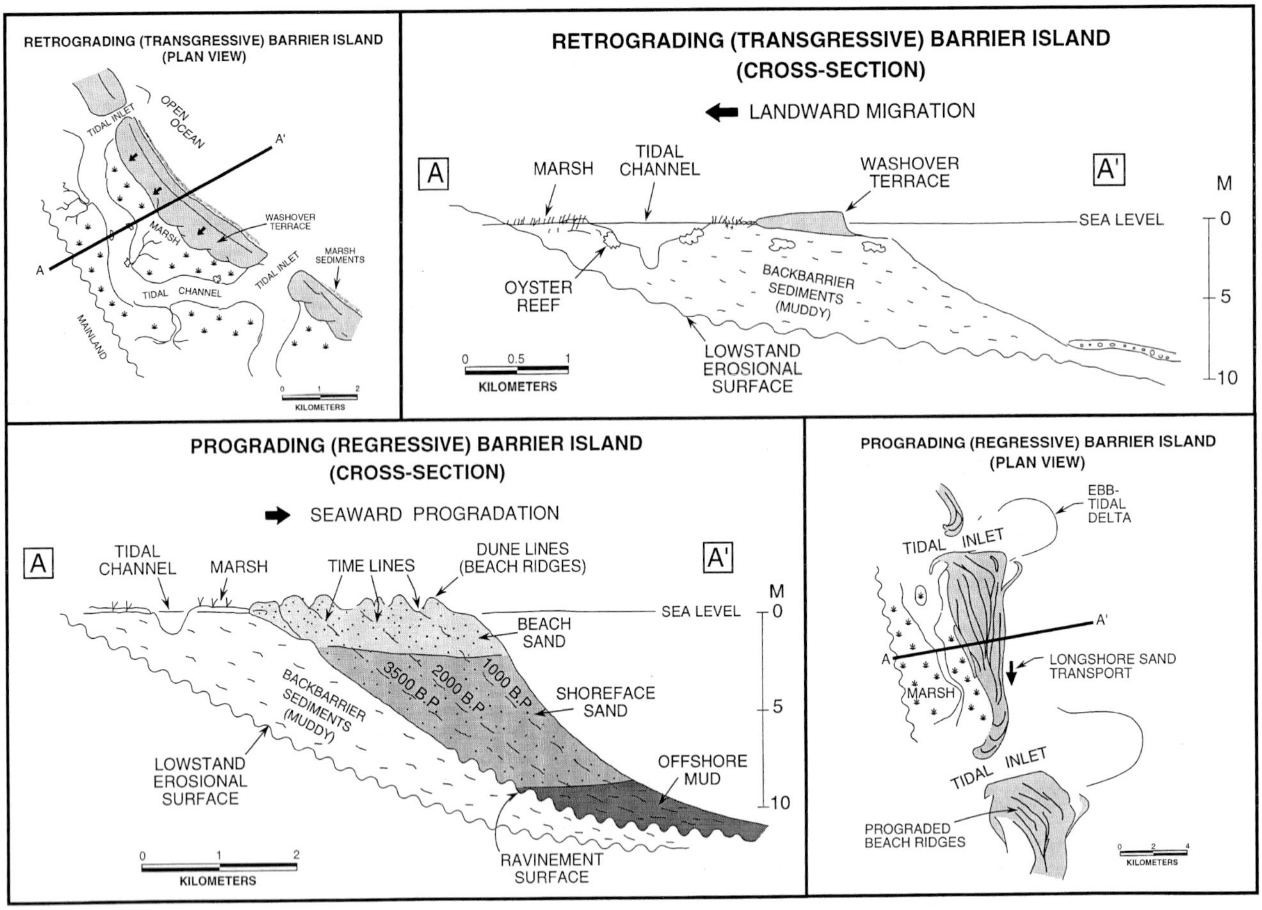

Figure B1 Morphology and stratigraphy of prograding and retrograding barrier islands.

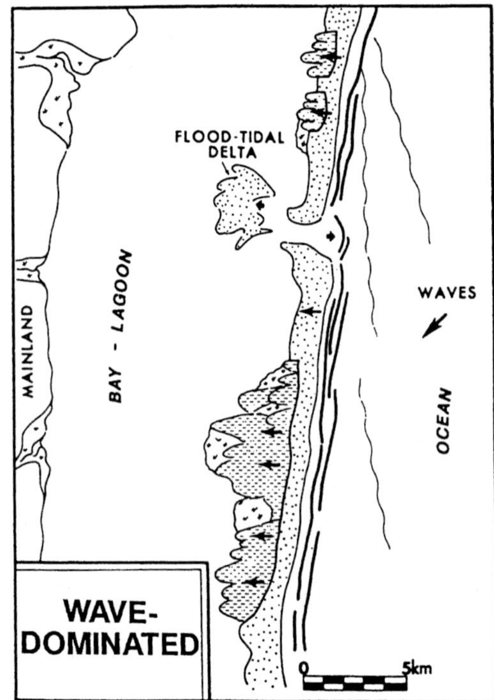

Figure B2 Typical morphology of a prograding, wave-dominated barrier-island shoreline.

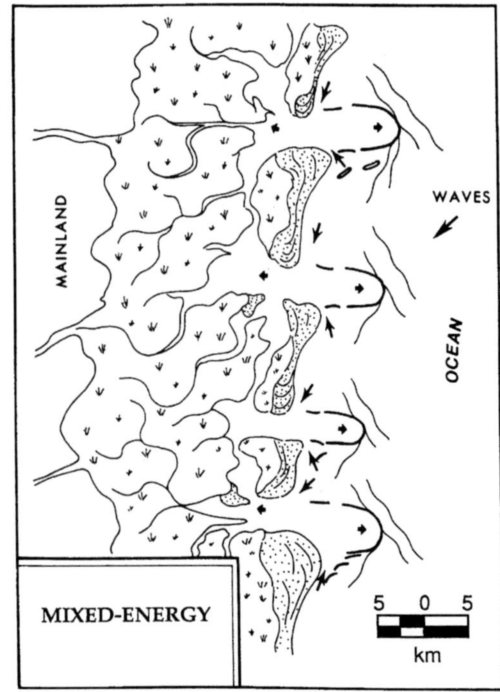

Figure B3 Typical morphology of a prograding, mixed-energy barrier-island shoreline.

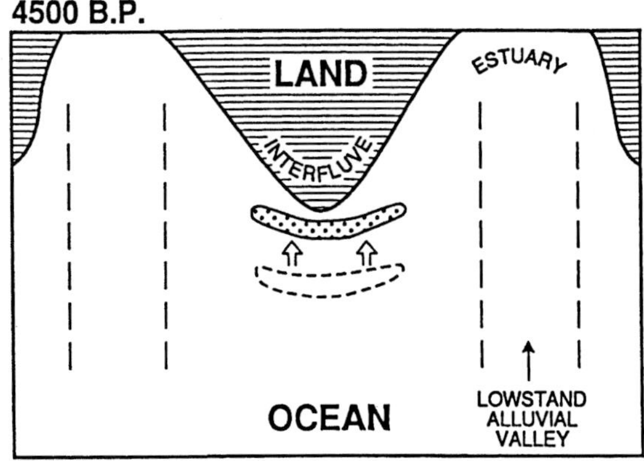

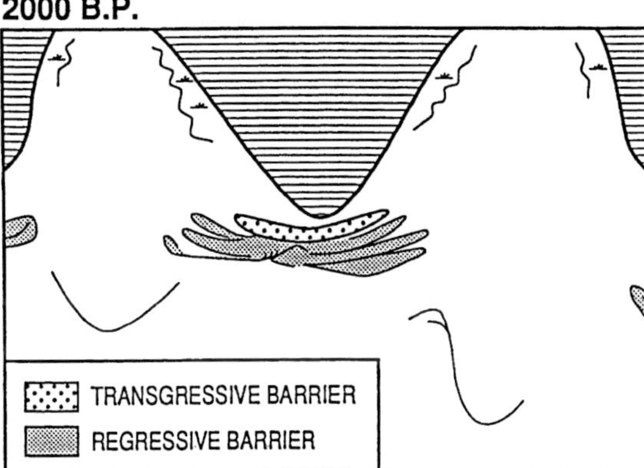

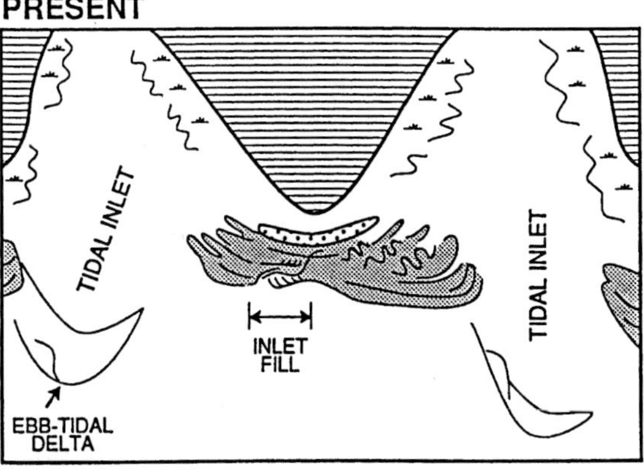

Figure B4 Model for the origin of a prograding, mixed-energy barrier-island along the southeastern US coastline–transgressive–regressive interfluve hypothesis. Based on Pierce and Colquhoun (1970) and Moslow (1980).

away from rocky headlands or river mouths. River discharge controls the shape of the spit during high discharge, and waves control it during low discharge (Dingler and Clifton, 1994; Smith, *et al.*, 1999).

Although occurring in a wide variety of types, which were classified into six major categories by Fitzgerald and Van Heteren (1999), barrier islands make up <25% of the glaciated coastline of New England. Most of these islands originate as spits, which are transformed into a variety of forms by tidal and wave action. Antecedent topography and geology also play important roles in shaping the morphology of the barrier islands along this complex coastline.

Further reading

For further information on the subject of barrier islands the reader is referred to: Schwartz (1973), Siringan and Anderson (1993), Davis (1994b), Moslow and Heron (1994), and Sexton and Hayes (1996).

Miles O. Hayes

Bibliography

Davies, J.L., 1964. A morphogenic approach to world shorelines. *Zeitschrift für Geomorphologie*, **8**: 27–42.

Davis, R.A., Jr., 1994a. Barrier Island systems—a geologic overview. In Davis, R.A., Jr. (ed.), *Geology of Holocene Barrier Island Systems*. New York: Springer-Verlag, pp. 1–46.

Davis, R.A., Jr. (ed.), 1994b. *Geology of Holocene Barrier Island Systems*. New York: Spring-Verlag.

Dingler, J.R., and Clifton, H.E., 1994. Barrier systems of California, Oregon, and Washington. In Davis, R.A., Jr. (ed.), *Geology of Holocene Barrier Island Systems*. New York: Springer-Verlag, pp. 115–205.

Fitzgerald, D.M., and Van Heteren, S., 1999. Classification of Paraglacial Barrier Systems: coastal New England, USA. *Sedimentology*, **46**: 1083–1108.

Fisher, J.J., 1967. Origin of Barrier Island Chain Shoreline. *Geological Society of America Special Paper 115*, pp. 66–67.

Hayes, M.O., 1979. Barrier Island morphology as a function of tidal and wave regime. In Leatherman, S.P. (ed.), *Barrier Islands from the Gulf of St. Lawrence to the Gulf of Mexico*, New York: Academic Press, pp. 1–27.

Hayes, M.O., 1994. The Georgia Bight Barrier System. In Davis, R.A., Jr. (ed.), *Geology of Holocene Barrier Island Systems*. New York: Springer-Verlag, pp. 233–304.

Hayes, M.O., 1996. Barrier Island Basics. *Geotimes*, **41**: 17.

Leatherman, S.P. (ed.), 1979. *Barrier Islands from the Gulf of St. Lawrence to the Gulf of Mexico*. New York: Academic Press.

Moslow, T.F., 1980. *Stratigraphy of Mesotidal Barrier Islands*. PhD Thesis, Depart Geology, University of South Carolina.

Moslow, T.F., and Heron, S.D., Jr., 1994. The outer banks of North Carolina. In Davis, R.A., Jr. (ed.), *Geology of Holocene Barrier Island Systems*. New York: Springer-Verlag, pp. 47–74.

Pierce, J.W., and Colquhoun, D.J., 1970. Holocene evolution of a portion of the North Carolina Coast. *Geological Society of America. Bulletin*, **81**: 3697–3714.

Schwartz, M.L. (ed.), 1973. *Barrier Islands*. Stroudsburg: Dowden, Hutchinson & Ross.

Sexton, W.J., and Hayes, M.O., 1996. Holocene deposits of reservoir-quality sand along the central south Carolina coastline. *American Association of Petroleum Geologists Bulletin*, **80**: 831–855.

Siringan, F.P., and Anderson, J.B., 1993. Seismic facies, architecture, and evolution of the Bolivar roads tidal/delta complex, East Texas Gulf Coast. *Journal of Sedimentary Petrology*, **63**: 794–808.

Smith, D.G., Meyers, R.A., and Jol, H.M., 1999. Sedimentology of an Upper-Mesotidal (3.7 m) Holocene Barrier, Willapa Bay, SW Washington, U.S.A.. *Journal of Sedimentary Research*, **69**: 1290–1296.

Swift, D.J.P., 1975. Barrier-island genesis; evidence from the Central Atlantic Shelf, Eastern USA. *Sedimentary Geology*, **14**: 1–43.

Cross-references

Barrier
Changing Sea Levels
Spits
Tidal Inlets
Tide-Dominated Coasts
Wave-Dominated Coasts

BARRIER

A barrier (coastal barrier) is an elongated coastal ridge of deposited sediment built-up by wave action above high tide level offshore or across the mouths of inlets or embayments. It is usually backed by a lagoon or swamp, which separates it from the mainland or from earlier barriers. A barrier, thus defined, is distinct from a bar, which is submerged at least at high tide (Shepard, 1952), and from reefs of biogenic origin (see Coral Reefs). Most barriers consist of sand, but some contain gravel as well as sand, and others consist entirely of gravel (shingle): see Gravel Barriers. Chesil Beach, on the south coast of England, is a well-known shingle barrier, and similar features are seen on the southeast coast of Iceland, and on the east and south coasts of South Island, New Zealand. Commonly the gravel has been derived from glacial or periglacial deposits, as on the north coast of Alaska and the southern shores of the Baltic Sea.

Barriers are said to occupy about 13% of the world's coastline (Leontyev and Nikiforov, 1965). They are most extensive on the Gulf and Atlantic coasts of North America and the ocean coasts of Australia, South Africa, and eastern South America, but they also occur on a smaller scale elsewhere, notably in Sri Lanka and New Zealand. Some barriers are transgressive, migrating landward across lagoon and swamp deposits; others remain in position, or are widened seaward by progradation, usually indicated by successively formed beach or dune ridges. Transgressive barriers occur where sediment is washed or blown over into backing lagoons or swamps, particularly during storms. Low sectors of a barrier through which storm waves or surges flow are called swashways, and sediment swept over a barrier through these is deposited as a washover fan on the inner shore.

On some coasts there are multiple barriers, the inner and older barriers being of Pleistocene age bordered (and overlain) by outer and younger barriers of Holocene age. Thus the Gippsland Lakes in southeastern Australia are enclosed by an Inner Barrier and an Outer Barrier (the Ninety Mile Beach), separated by lagoons and swamps, and with relics of an earlier, Prior Barrier that predates the enclosure of the existing Lakes (Bird, 1973). In this case, barriers have developed seaward of earlier marine coastlines, but evidence of preceding exposure to the open sea is not always present, particularly in lagoons where the enclosing barriers have been transgressing landward during a phase of rising sea level, as on the Siberian coast (Zenkovich, 1967).

The term barrier beach indicates a single, narrow elongated ridge (usually less than 200 m wide) built parallel to the coast, without surmounting dunes. A barrier island is bordered by transverse gaps (tidal inlets, lagoon entrances, river outlets), which may be migratory and subject to closure; it usually bears beach ridges, dunes and associated swamps, and minor lagoons, and may incorporate recurves (Schwartz, 1973). Examples include Scolt Head Island on the east coast of England, and several along the Atlantic coast of the United States. A barrier with many interruptions becomes a barrier island chain. A barrier attached to the mainland at one end can be termed a barrier spit, as on the coast north of the Columbia River in Washington State (where Long Beach is a barrier spit partly enclosing Willapa Bay); one built across the mouth of an embayment a bay (baymouth) barrier. There are also mid-bay barriers and bay-head barriers, defined by their position.

In general, barriers are found on coasts where the tide range is small (as on the southern Baltic coast), and become chains of barrier islands where currents produced by larger tides maintain transverse gaps (as on the Danish, German, and Dutch North Sea coasts).

Barriers can form in various ways (Schwartz, 1971), multiple causality being related to the nature and supply of sediment, the transverse profile of the coast, tide range, wave conditions, and relative sea-level change. Some barriers may have formed by the emergence of nearshore swash bars as sea-level fell (e.g., Knotten, on the Danish island of Laesø), but many developed during and after the Late Quaternary marine transgression by submergence of pre-existing sand ridges and the shoreward drifting of sea floor sediment. Of these, some are still transgressive (as on parts of the Atlantic coast of the United States) while others have become anchored, and widened by progradation (as on parts of the southeast Australian coast). Growth and landward transgression of barriers has been demonstrated in recent years on parts of the coast of the Caspian Sea, which has risen about 2 m since 1977 (Kaplin and Selivanov, 1995).

Barrier spits may show features indicative of longshore growth, such as former recurved terminations on the landward side (as on Orfordness in England or the Langue de Barbarie in West Africa), but others have been built and widened by wave-deposited sediment from the sea floor (as on Clatsop Spit in Oregon), and many result from combinations of onshore and longshore sediment drifting.

The shaping of barriers can be traced with reference to patterns of beach and dune ridges indicating stages in their growth, and from their stratigraphy, which may indicate phases of upward growth, landward movement, and seaward progradation, as in Van Straaten's (1965) classic study of barriers on the Netherlands coast and Thom's (1984) study of sand barriers in eastern Australia. Barriers of unconsolidated sand are readily re-shaped by wave and wind action, but where barrier sediments have become lithified (e.g., the Pleistocene dune calcarenites in Australia and elsewhere) they are more durable, and may show cliffing (as in the inner barriers of the Coorong in South Australia). Some barriers incorporate segments of pre-existing terrain, such as the glacial moraines on Long Island in New York, Walney Island in northwest England and Sylt on the German North Sea coast.

Eric Bird

Bibliography

Bird, E.C.F., 1973. Australian coastal barriers. In Schwartz, M.L. (ed.), *Barrier Islands*. Stroudsburg: Dowden, Hutchinson & Ross, pp. 410–426.

Kaplin, P., and Selivanov, O.A., 1995. Recent coastal evolution of the Caspian Sea as a model for coastal responses to the possible acceleration of global sea level rise. *Marine Geology*, **124**: 161–175.

Leontyev, O.K., and Nikiforov, L.G., 1965. Reasons for the world-wide occurrence of barrier beaches. *Oceanology*, **5**: 61–67.

Schwartz, M.L., 1971. The multiple causality of barrier islands. *Journal of Geology*, **79**: 91–94.

Schwartz, M.L. (ed.), 1973. *Barrier Islands*. Stroudsburg: Dowden, Hutchinson & Ross.

Shepard, F.P., 1952. Revised nomenclature for coastal depositional features. *American Association of Petroleum Geologists*, **36**: 1902–1912.

Thom, B.G., 1984. Sand barriers of eastern Australia: Gippsland—a case study. In Thom, B.G. (ed.), *Coastal Geomorphology in Australia*. Sydney: Academic Press, pp. 233–261.

Van Straaten, L.M.J.U., 1965. Coastal barrier deposits in south and north Holland, *Meded. Geol. Stichting*, **17**: 41–87.

Zenkovich, V.P., 1967. *Processes of Coastal Development* (trans. O.G. Fry, ed. J.A. Steers). Edinburgh: Oliver and Boyd.

Cross-references

Barrier Islands
Bars
Coral Reefs
Drift and Swash Alignments
Gravel Barriers
Spits

BARRIERS, GRAVEL—See GRAVEL BARRIERS

BARS

Sedimentary ridges, both symmetric and asymmetric, and generally larger than bedforms that characterize the upper shoreface of coastal zones dominated by waves are called *wave-formed bars*. They were recognized as early as 1845 on the marine coasts of Europe (Elie de Beaumont), by 1851 in the Great Lakes of North America (Desor), and subsequently on marine and lacustrine coasts worldwide (see Schwartz, 1982, pp. 135–139). However, confusion still surrounds this term because of its use for ridges with a wide range of size, morphology, location, and orientation relative to the shoreline. Also, the term *bar* has been used in a variety of environments, from subaerial to those dominated by tidal currents or river currents. Furthermore, the present understanding of the origin(s) and dynamics of *wave-formed bars* is still incomplete.

Shepard (1950) called shore-parallel ridges and troughs *longshore bars* and *troughs*, equating them with the terms *ball* and *low* of Evans (1940), and associated them with plunging breakers. He emphasized the *seasonality* of such bars on the west coast of the United States, and subsequently terms such as *winter* and *summer, storm* and *normal*, and *storm* and *swell* have been applied to denote the presence or absence of bars. Although a correlation between profile form and storm waves or season may exist in some localities (e.g., Inman *et al.*, 1993), it is not universal. Both *barred* and *non-barred* profiles occur at times in some areas, while in others only one profile type may persist throughout the year. There is usually a distinct *relaxation time* between the forcing conditions and bar adjustment; thus in the short-term, bars are generally in a transient state. In the longer term (years to decades), wave-formed bars represent the *equilibrium morphology* for many coastal environments.

Bar morphology

Wave-formed bars are most clearly identified as near-symmetrical or asymmetrical undulations in the upper shoreface profile (Figure B5). They occur intertidally and subtidally, and may range in number from one to more than thirty, this number often varying through time. Short and Aagaard (1993) introduced a bar parameter, $B^* = x_s/gT^2 \tan\beta$, to

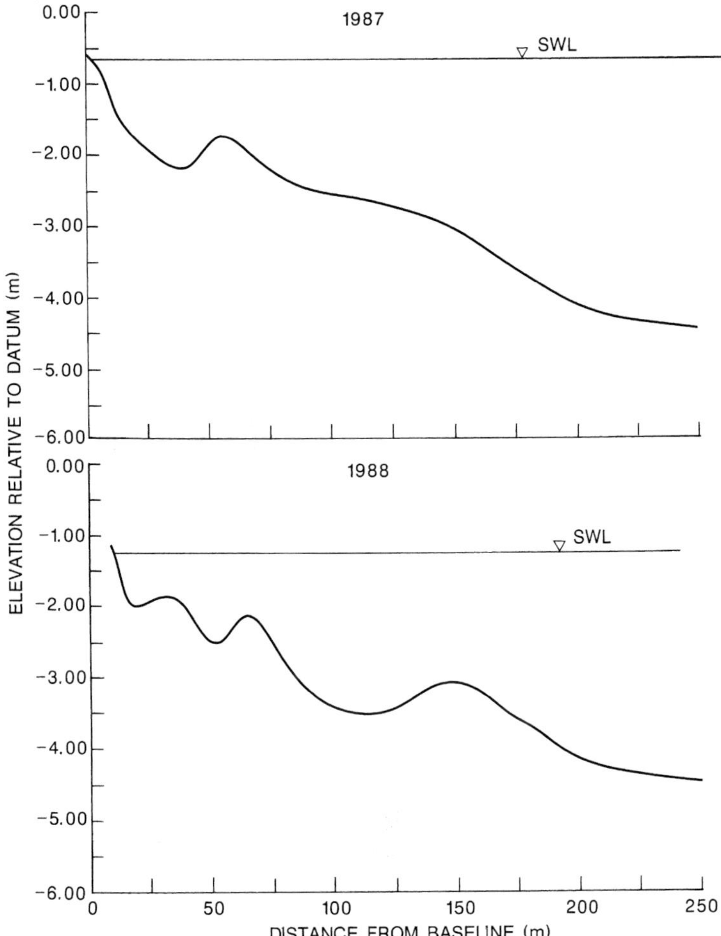

Figure B5 Typical barred profiles from a sandy nearshore environment in the Canadian Great Lakes. The profiles were surveyed in successive years along the same transect. Note the differing number and position of the wave-formed bars at the same location, even though the mean beach slope is the same.

identify the number of bars on a linear sloping shoreface (tan β) terminating at a constant depth at a distance offshore, x_s. When $B^* < 20$, the profile is non-barred, for $B^* = 20$–50, 1 bar occurs; for $B^* = 50$–100, 2 bars, for $B^* = 100$–400, 3 bars; and for $B^* > 400$ there are 4 bars. Crest heights above the adjacent trough can range from less than a decimeter (Carter, 1978) to more than 4.75 m (Greenwood and Mittler, 1979). In plan view, they form continuous or compartmentalized, linear, sinuous, or crescentic patterns, and range from shore-parallel to shore-normal in orientation, often producing periodic or *rhythmic* topography both alongshore and cross-shore. The morphometry of bars has been studied by Greenwood and Davidson-Arnott (1975), Hands (1976), and Reussink *et al.* (2000) in order to define the equilibrium form and dynamics induced by a specific set of environmental constraints.

Bar classification

A *universal* classification of *wave-formed bars* does not yet exist, and indeed it may never be possible to define perfectly mutually exclusive classes. A simple descriptive classification based on morphology and the associated environmental constraints is illustrated in Table B1 (Greenwood and Davidson-Arnott, 1979). The group names are those in common use, and the definitive paper describing each type is cited. Other classifications are based on the concept that bars are part of a temporal sequence of beach profile evolution and that they are scaled to that of the controlling wave process. The morphological sequence is controlled by incident wave energy (high and low frequency) and was identified either through aerial photographs or more recently through video-imagery (e.g., Short, 1979; Lippmann and Holman, 1990; see Figure B6). Many coastal environments do not experience such sequential behavior.

Ridge and runnel topography (*Type I*) is found on low-angle, macro- to meso-tidal foreshore slopes dominated by surf action and foreshore drainage during the tidal cycle. Although low in amplitude these bars are usually stable in form and position or migrate only slowly. In contrast, the *cusp-* or *bar-type sand waves* (*Type II*) are extremely dynamic, often destroyed during storms and regenerated as the storm wanes and smaller amplitude, longer period waves propagate shoreward. These bars result from surf bores and swash action near the toe of the swash slope (an alternative name is *swash* bar). Furthermore, they may develop from *Type VI* bars as they migrate relatively rapidly both alongshore and onshore, and in the latter case may *weld* to the foreshore (Davis *et al.*, 1972; Aagaard *et al.*, 1998). Note that there is confusion with respect to the term *ridge and runnel* as used in northwest Europe and North America (Orford and Wright, 1978; Orme and Orme, 1988). Here, the term *ridge and runnel* is restricted to its initial definition by King and Williams (1949); the forms described by Hayes and subsequent workers (Hayes and Boothroyd, 1969) are classified here as *Type II* bars.

Type III multiple parallel bars (e.g., Nilsson, 1973; Exon, 1975) and *Type IV transverse bars* (e.g., Niedoroda and Tanner, 1970; Carter, 1978; Dolan and Dean, 1985) tend to be limited to low-angle shorefaces and small to moderate wave heights, coupled with limited water level shifts. However, they have been identified on more energetic shorelines (Konicki and Holman, 2000).

The number of bars increases with decreasing beach slope (Davidson-Arnott, 1988). The height and spacing of the multiple bars increases in the offshore direction, and bar form is near symmetrical in contrast to the *Type II* group. Transverse bars run normal or obliquely to the shoreline and can range in length from 3 m up to 4 km, with heights from less than 0.05 m up to 2 m and alongshore spacing of the

Table B1 Bar morphologies and environmental constraints

			Morphology					Environmental constraints			
Class	Name	Definitive description	Size (m)	Planform	Profile	Number seaward	Location	Wave energy	Breaking processes	Tidal range	Slope
Type I	Ridge and runnel	King and Williams (1949)	$h \sim 0.2$–1.5 $l \sim 10^3$	Straight, shore parallel	Asymmetric landward	1–4	Intertidal	L–M	Surf-swash, beach drainage	Ma–Me	0.007 0.024
Type II	Cusp- or bar-type sand wave	Sonu (1973)	$h \sim 0.2$–1.5 $l \sim 10^2$	Straight to spit shaped, shore parallel	Asymmetric landward	1–2	Intertidal and low tide terrace	M	Breakers, surf-swash	Me–Mi	>0.01
Type III	Multiple parallel	Zenkovitch (1967)	$h \sim 0.2$–0.75 $l \sim 10^3$	Straight to sinuous, shore parellel	Near symmetric	4–30 or more	Nearshore & intertidal	L–M	Spilling	Mi–Me	<0.01
Type IV	Transverse	Niederoda and Tanner (1970)	$h \sim 0.2$–0.75 $l \sim 10^2$	Straight, shore normal	Symmetric and asymmetric landward	1	Nearshore and intertidal	L	Surf-swash,	Mi spilling	<0.0045
Type V	Nearshore I	Shepard (1950)	$h \sim 0.15$–1.0 $l \sim 10^2$ (?)	Straight, shore parallel	Asymmetric landward	1–2	Nearshore	M–H	Plunging	Mi–Me	<0.1
Type VI	Nearshore II	Evans (1940)	$h \sim 0.25$–3.0 $l \sim 10^3$	Straight, sinuous to crescentic, shore parallel	Asymmetric landward	1–4	Nearshore	M	Spilling	Mi	<0.01

Note: h = bar height; l = bar length; Ma = macro; Me = meso; Mi = micro; L = low; M = moderate; H = high.
Source: Modified from Greenwood and Davidson-Arnott, 1979.

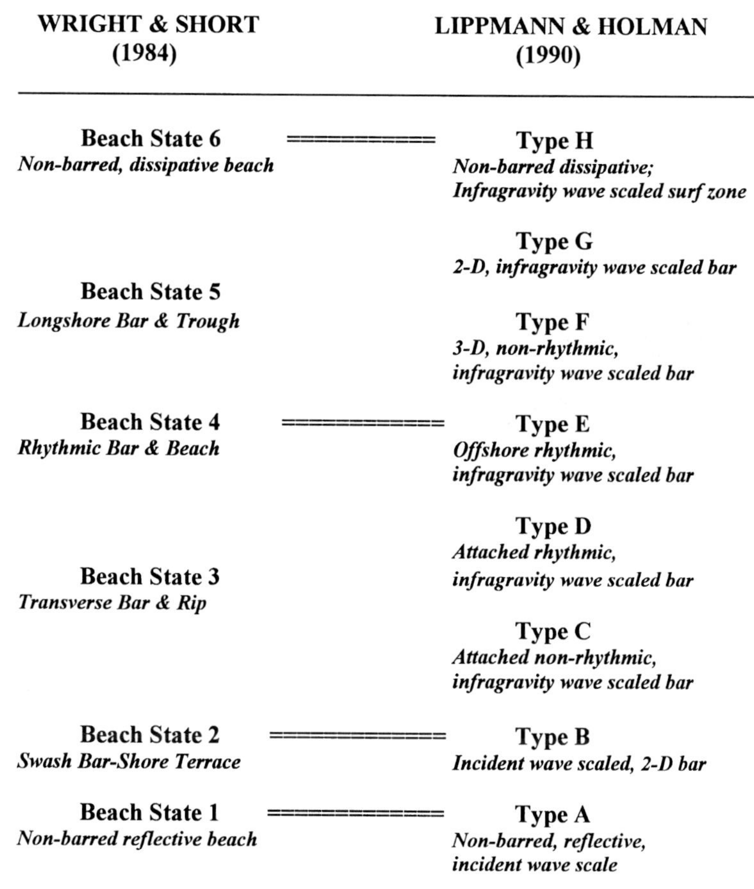

Figure B6 Classification and scaling of sequential upper shoreface morphologies. The equivalence between the contrasting sequences of Wright and Short and Lippmann and Holman is indicated (modified after Lippmann and Holman, 1990).

order of 10^0–10^2 m (Carter, 1978; Gelfenbaum and Brooks, 1997). The larger forms may migrate alongshore at rates up to 8 m a^{-1}. Usually, transverse bars are anchored to the shoreline (indeed they appear as an extension of a shoreline protuberance), but Konicki and Holman (2000) recorded the unusual case of transverse bars running offshore from a *Type VI* bar.

The division of *nearshore bars* into two groups is based upon size, stability, and the controlling waveform. *Type V bars* are associated with large plunging breakers, which produce narrow, low amplitude ridges on relatively steep slopes: they lack a well-defined asymmetry and are essentially unstable modifications of non-barred nearshore profiles. *Type VI bars*, in contrast, are relatively large configurations formed seaward of the low water level. Where there is more than one bar, the distance offshore, depth-of-water over the crest, and bar height all usually increase offshore in a regular manner, although in some cases the height decreases after some offshore distance (Lippmann *et al.*, 1993; Ruessink and Kroon, 1994). The volume of sediment in each bar form usually increases consistently offshore. *Type VI* bars may be three-dimensional, sinuous-to-crescentic, and the alongshore length scales may range from 10^2 to 10^3 m (Greenwood and Davidson-Arnott, 1975; Bowman and Goldsmith, 1983). Where more than one bar is rhythmic, the alongshore wavelength decreases shoreward.

Bar genesis

The boundary conditions necessary for bar formation depend upon the longer term evolution of the coast, which dictates the nature of the bed materials (grain mineralogy, size, sorting, etc.), the bathymetric setting (slope, exposure, etc.), and the geographic location (wave climate, tidal regime, etc.). Local forcing conditions for bars have been studied both theoretically and empirically, and by experiments in the laboratory and the field (see van Rijn 1998). In general, barred profiles are associated with large values of both wave steepness and wave height-to-grain size ratios, and are associated with the final stages of shoaling and dissipation of wave energy through breaking, and the complex hydrodynamics, which accompany these processes (Wright *et al.*, 1979). Furthermore, the size of wave-formed bars induces a very strong feedback to the shoaling and breaking process. Although cause and effect are far from clear, it is evident that equilibrium bar profiles can exist only where the time-averaged sediment transport (suspended and bedload) is zero everywhere on that profile.

A large number of specific hypotheses have been proposed for bar formation over the last 50 years and all involve mechanisms for convergence of sediment transport; these hypotheses were primarily related to *Type V and VI* bars and fall into three major groups:

(1) *break point hypotheses* relate bars directly to wave breaking and result from: (i) a seaward transport of sediment entrained by roller or helical vortices under plunging or spilling breakers, respectively (Miller, 1976; Zhang, 1994; Figure B7); (ii) convergence of sediment at the breakpoint through onshore transport associated with increasing asymmetry and skewness of the high-frequency incident waves and offshore transport through set up induced undertow (Dally and Dean, 1984; Dally, 1987; Thornton *et al.*, 1996). However, Sallenger and Howd (1989) concluded that bars are not necessarily coupled to the breakpoint, but can grow and migrate, while within the inner surf zone, landward to the point of initial breaking.

(2) *infragravity wave hypotheses* propose that low frequency waves generated within the surf zone (surf beat) or offshore and reflected produce a convergent pattern of drift velocities, which interact with the large incident short wave oscillatory velocities to induce a range of bar forms from two- to three-dimensional crescentic forms (e.g., Bowen and Inman, 1971; Short 1975; Bowen 1980; Holman and Bowen, 1982; Bowen and Huntley, 1984). These waves can be standing or progressive and can be produced in a number of different ways as a result of energy dissipation during breaking and are frequently related to amplitude modulation of the incident wave field (*groupiness*; Roelvink and Broker, 1993; Reussink, 1998). Alternating scour and deposition by mass

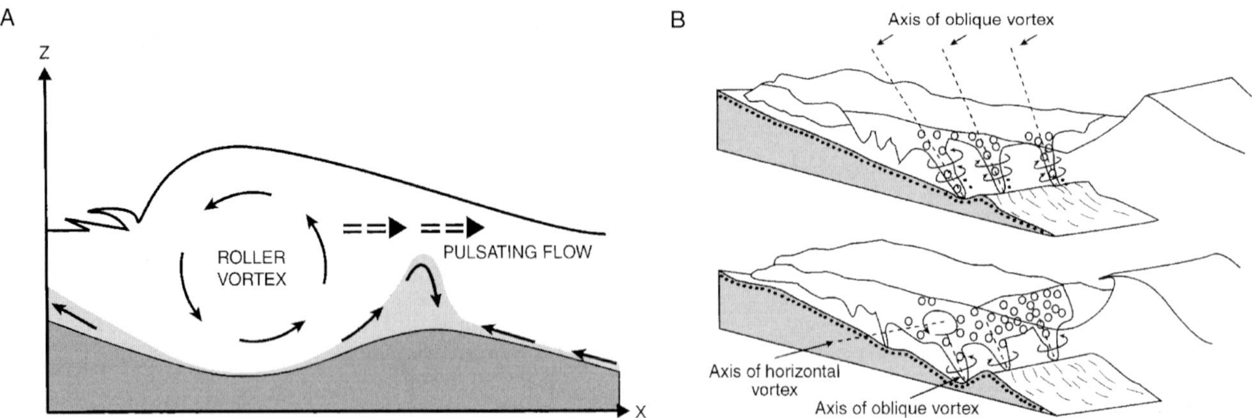

Figure B7 Bar formation by breaking waves: (A) trough scouring by a roller vortex under plunging breakers and offshore sediment transport converging with sediment driven onshore by shoaling waves (modified after Miller, 1976); (B) trough scouring by oblique vortices generated under spilling breakers (modified after Zhang, 1994).

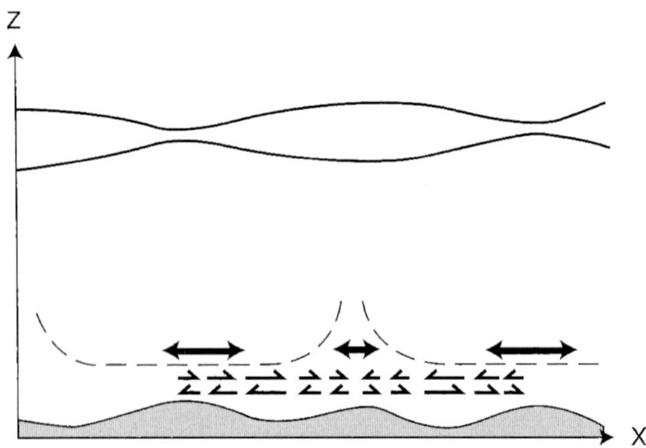

Figure B8 Bar formation as a result of mass transport in the boundary layer of a strongly reflected incident wave. The surface wave envelope is shown as well as the circulation within the bottom boundary layer. Bed load will converge at nodes of the surface elevation and suspended load at antinodes. Note: the boundary layer flow is indicated by single-headed arrows; the mean flow is indicated by double-headed arrows.

transport velocities in the bottom boundary layer generated by standing waves, resulting from the interaction of reflected and incident waves, was shown to occur in the laboratory by Carter *et al.* (1973; Figure B8). The boundary layer was actually segregated. At the bed, drift velocities converge at nodes, while at some distance above the drift velocities converge at antinodes. Under large waves when bars are most active, suspension transport is dominant and therefore sediment will converge and bars will form at the antinodal position of standing waves (e.g., Bowen, 1980). Reflection of waves in the infragravity range was clearly demonstrated by Suhayda (1974) and shown to relate to bar forms by Short (1975) and Katoh (1984). Sediment moves to null positions in the drift velocity field of low-frequency standing (Figure B9(A)) or progressive edge waves (Figure B9(B)), which are periodic both alongshore and offshore. Recent field measurements have clearly shown the importance of group-bound long waves to suspended sediment transport in barred surf zones (e.g., Osborne and Greenwood, 1992), but isolating the drift velocities associated with these secondary waves is difficult. This second-order drift velocity hypothesis requires one dominant wave frequency, which is not common (see Bauer and Greenwood, 1990 for an exception). However, there are a number of suggestions to overcome this inadequacy of the edge wave hypothesis. Aagaard (1990) has argued for the excitation of cutoff mode edge waves (limited by the beach slope) and selection of the dominant mode as that mode which is closest to the wave group period. A phase coupling between the primary orbital motion of a partially standing long wave and groupy short waves was also proposed by O'Hare (1994) to avoid this requirement of narrow bandedness in the infragravity spectrum. Other mechanisms producing a limited number of edge wave frequencies and modes are topographic control (Kirby *et al.*, 1981; Bryan and Bowen, 1996) and interaction of edge waves and the longshore current (Howd *et al.*, 1992). O'Hare and Huntley (1994) propose a leaky wave origin for an inner surf zone bar, which is relatively insensitive to the group period, incident wave height, and the width of the infragravity spectrum.

(3) *self-organization hypotheses* propose that processes associated with the complex, nonlinear feedback between the sand bed and the hydrodynamics give rise to a range of topographic forms. For example, alongshore and offshore sediment movement was proposed under meandering or cellular nearshore circulations produced by (i) instability of longshore flows (Figure B10; Barcilon and Lau, 1973; Hino, 1974; Falques, 1991; Damgaard Christiansen *et al.*, 1994); (ii) coupling between morphodynamic instability and mean flows (Deigaard *et al.*, 1999; Vittorio *et al.*, 1999; Falques *et al.*, 2000); and (iii) Bragg scattering from periodic topography (Heathershaw and Davies, 1985; O'Hare and Davies, 1993; Rey *et al.*, 1995; Yu and Mei, 2000). These mechanisms cannot produce bars directly, but require some initial perturbation of the profile. However, it has been shown that some bar characteristics are not well predicted by these models (e.g., the cross-shore/ alongshore spacing—see Konicki and Holman, 2000). The nonlinear action between shoaling waves and the bed (Boczar-Karakiewicz and Davidson-Arnott, 1987) was also proposed as a mechanism for generating periodic patterns of sediment transport which matched the spacing and general shape of multi-barred shorelines.

The horizontal roller vortex mechanism is most applicable to single *Type V* bars of the US west coast, and justifies the early correlation of bar formation with wave steepness. Multibarred profiles reflect either multiple breakpoints (Dally, 1980; Davidson-Arnott, 1981) or bar formation by distinct differences in wave energy; for example, an outer bar may be produced under storm waves and an inner bar by less energetic conditions (King and Williams, 1949). Water level shifts and coincident shifts in breaker location could also produce a multiple barred system. Oblique, helical vortices were produced under spilling rather than plunging breakers in the laboratory and could account for both single and multiple barred profiles (Zhang, 1994). However, the mass transport velocities under reflected standing waves would perhaps best explain the formation of *Type III* multiple parallel bars; simple reflection of the incident waves could not be the cause, since the length scales of the bars would require much longer periods. The theoretical convergence patterns of drift velocities under standing edge waves provide strong support for their role in forming crescentic *Type VI* bars. Progressive edge waves may be responsible for linear bars of the same group (Huntley, 1980). Further, the edge wave periods necessary to produce the length scales found in nature is of the same order as the well-known *surf beat*. However, the generation of these trapped modes of oscillation still remains ill-defined, even though field observations of low-frequency peaks in the nearshore energy spectrum have been made on barred coasts and related to the presence of edge waves (Huntley, 1980; Bauer and Greenwood, 1990).

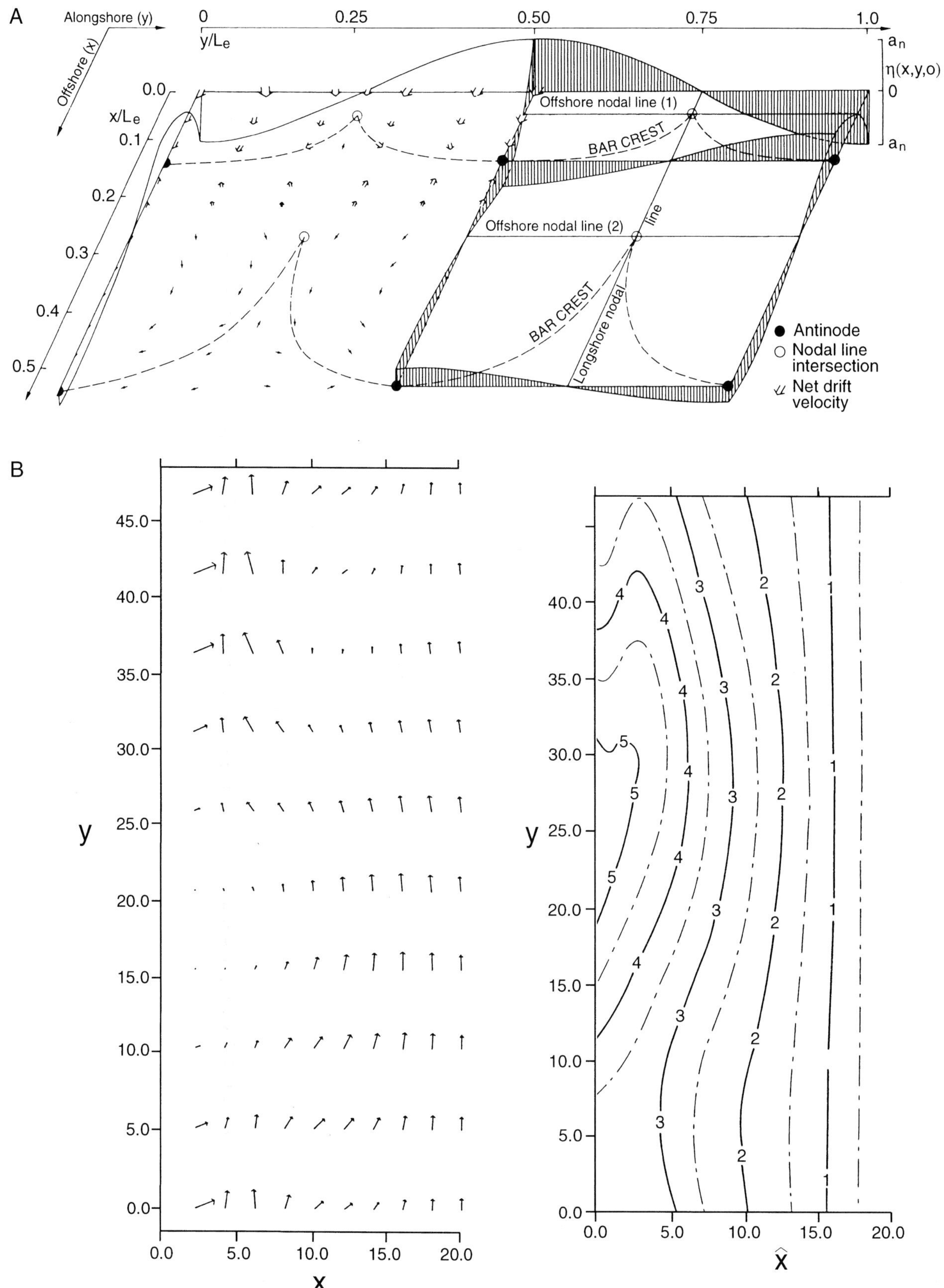

Figure B9 Bar formation by infragravity waves: (A) net drift velocities associated with standing edge waves and the creation of crescentic bars (modified after van Beek, 1974). (B) dimensionless drift velocities and equilibrium nearshore bathymetry associated with the propagation of two edge wave modes (1 and 2) of the same frequency in the same direction (modified after Holman and Bowen, 1982). Note: *y* represents the alongshore direction and *x* the across-shore direction.

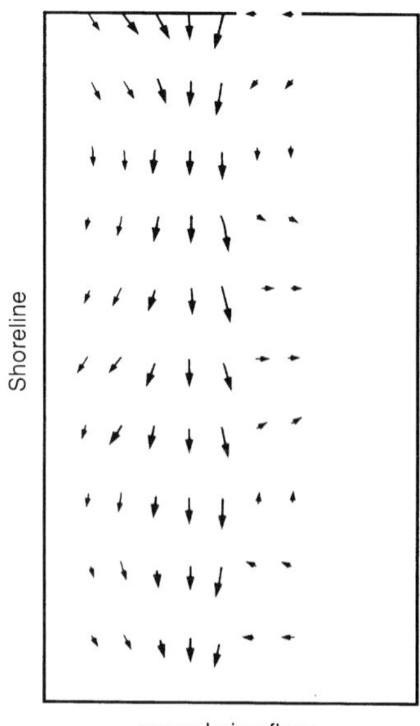

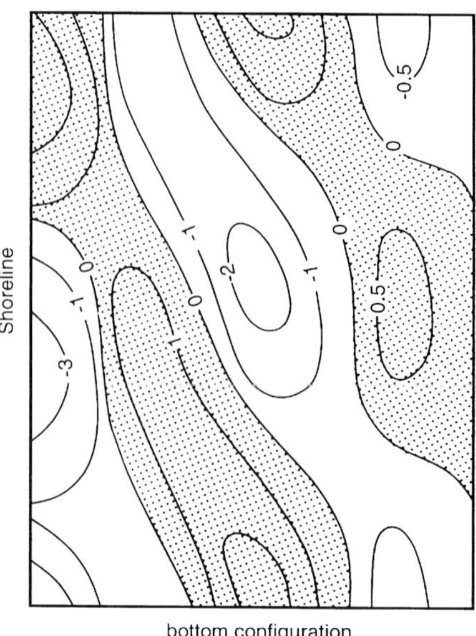

Figure B10 Bar formation due to hydrodynamic instability between longshore currents and the sand bed (modified after Hino, 1974). Note the meandering nature of the longshore flow and the sinuous bar topography that is produced.

Barred topography has long been associated with the occurrence of cellular nearshore circulations (Shepard *et al.*, 1941), and Hino (1974) proposed that an instability of the fluid sediment interface would generate variations in sediment transport resulting in sinuous or crescentic undulations of the surf-zone bed (Figure B9). Certainly the role of rip-cell circulation in bar dynamics has been well documented for *bar-* and *cusp-type sand waves* (Bowen and Inman, 1969; Davis and Fox, 1972; Sonu, 1973; Greenwood and Davidson-Arnott, 1975; Wright and Short, 1984), for *transverse bars* (Niedoroda, 1973), and for *Type VI* bars, both crescentic and straight (Greenwood and Davidson-Arnott, 1979). However, it is also possible that the regularity in nearshore circulations is in fact controlled by the presence of edge waves (Holman and Bowen, 1982). Whichever mechanism initiates bars, there will be feedback between the topography and the hydrodynamics, perhaps giving rise to some "hybrid" model of formation (Holman and Sallenger, 1993).

Bar morphodynamics

In general, the smaller the *wave-formed bar* the more dynamic it is, as there is less sediment involved in morphological changes (Sunamura and Takeda, 1984). However, there is considerable variability in morphodynamic behavior, depending upon bar type, the general environmental constraints, and indeed the antecedent state of the bar (i.e., whether or not it is close to its equilibrium position). Bars also tend to migrate at lower rates as the tidal range increases, since at some stage the bars are being exposed subaerially and remain static at this time. Bar dynamics have generally been related to behavior under specific storm events. However, the magnitude, frequency, and sequencing (*chronology*) of such events may be important in the nearshore, which as a nonlinear dynamical system, is extremely sensitive to feedback processes (see Moller and Southgate, 1997; Southgate and Moller, 2000; see Elgar, 2001 for an alternative view). There now exist at least two long time series of morphological change: (1) thirty years of annual profiling along 100 km of the Dutch coast (Ruessink and Kroon, 1994); (2) sixteen years of bathy metry recorded at Duck, NC (Plant *et al.*, 1999). Extensive measurements of the cross-shore location, and alongshore bar shape, are now being made successfully on a near continual basis at a number of locations worldwide using video-imagery (e.g., Lippmann and Holman, 1990; van Enckvort and Ruessink, 2001).

Type I bars are relatively stable in general, although landward migration rates of ~10 m per month have been recorded. Under low energy conditions the ridges have been observed to be: (1) destroyed by storms and regenerated in the post-storm period (Mulrennan, 1992); and (2) formed by storms (Hale and McCann, 1982). *Type II* bars have been shown to migrate at relatively rapid rates, both onshore and alongshore, and *Type III* bars migrate also at a relatively rapid rate. *Type II, IV*, and *VI* bars have been shown to occur as part of a temporal sequence of beach evolution by Wright *et al.*, (1979), Wright and Short (1984), Sunamura (1988), and Lippmann and Holman (1990). This sequence ranges from fully dissipative (barred profile) to fully reflective (non-barred profile) wave conditions, and therefore, is related to the surf similarity parameter ($\varepsilon = a_b\omega^2/g \tan^2\beta$; where a_b = breaker amplitude, ω = incident radian wave frequency, g = the gravitational constant, β = beach slope). In the Australian Model, the two-dimensional shore-parallel longshore bar and trough occurs at the fully dissipative beach stage, the rhythmic bar and beach at an intermediate stage, and the non-barred profile occurs at the fully reflective stage. In regions where a more limited range of waves exist, the beach may simply change between one or two stages, and where the environmental constraints are more restrictive still, then the bars may assume only one characteristic morphology. Further refinement of the stage model used the *Dean Parameter* ($\Omega = H_b/\omega_s T$; where H_b = breaker height in meters; ω_s = sediment fall velocity in meters per second; T = wave period in seconds). Barred profiles occurred when $\Omega > 0.85$ and non-barred profiles occurred when $\Omega < 0.85$ (Wright *et al.*, 1985). Sunamura (1988) used the dimensionless parameter $K^* = H_b^2/gT^2d$, where g is the gravitational constant and d is the grain size, to classify sequences dependent upon erosional or accretional beach stages. Erosion is characterized by $K^* \geq 20$ and is associated with offshore bar migration, slope decreases, and a dissipative state; while $5 \leq K^* \leq 20$ indicates onshore migration and beach accretion. Yet, a further parameter was introduced by Kraus and Larson (1988) to separate barred and non-barred profiles, $P = gH_0^2/\omega_s^3 T$, where H_o = offshore wave height. A value of 9,000 separates barred (greater values) from non-barred profiles (Dalrymple, 1992).

Type VI nearshore bars have been found to migrate onshore, offshore, and alongshore, with offshore rates reaching 2.5 m h^{-1} during storms and erosion/accretion rates of 0.05 m h^{-1} (Sallenger *et al.*, 1985; Aagaard and Greenwood, 1995). Onshore migration rates are generally smaller, but may still reach 1 m h^{-1}. When the *Type VI* bars are three-dimensional, they may migrate alongshore at rates up to 10 m per month (Greenwood and Davidson-Arnott, 1975). Ruessink *et al.* (2000) examined the relative rates of across-shore and alongshore migration using complex empirical orthogonal functions applied to profile data. The alongshore migration rate ranged up to 150 m per day and was strongly related to the alongshore component of the offshore wave energy flux. Short-term variability in bar crest position was shown to be due to changes in the quasi-regular topography, and not to alongshore uniform on–offshore migration. While offshore migration under storms has been clearly related to hydrodynamic forcing, especially the setup-driven undertow (Gallagher *et al.*, 1998) or mean currents modulated by infragravity waves (Aagaard and Greenwood, 1995), the onshore migration of Type VI bars is poorly known. Generally the motion is attributed to skewed fluid velocities and accelerations (Elgar *et al.*, 2001).

On the Dutch coast a multiple bar *Type VI* system exhibited characteristics of a feedback-dominated system, producing cyclic changes over either 4 or 15–18 years (Wijnberg and Terwindt, 1995). Plant and Holman (1997) showed that bars on the east coast of the United States exhibited unpredictable behavior in relation to wave height changes and yet still moved through a sequential pattern of form changes. This paradoxical behavior they related to feedback effects. The forcing for these transitions is as controversial as bar genesis, since direct hydrodynamic forcing has been proposed as well as a self-organization mechanism.

Little work has been done specifically upon bar decay, other than the welding process associated with *Type II* bars (e.g., Davis *et al.*, 1972; Aagaard *et al.*, 1998). However, the one major exception is the study of the multiple bar system along the Dutch coast. Here, the bar system shifts progressively offshore over time and the outermost bar decays. This has been attributed to the action of highly asymmetric, nonbreaking waves (Larson and Kraus, 1992; Reussink and Kroon, 1994; Wijnberg, 1997). Plant *et al.* (2001) suggest that a morphologic feedback mechanism can lead to bar decay. As bars move onshore under nonbreaking conditions they are also reduced in height; thus they move further away from wave breaking, allowing further bar decay. This has been observed at Duck, NC (Lippmann *et al.*, 1993).

Predictive models of bar genesis and dynamics

Because of the relatively poor knowledge of long-term bar behavior and the inadequacy of local sediment transport models for the complex nearshore environment, predictive models for the genesis and dynamics of wave-formed bars are still far from complete. In general models proposed are either (1) process-based models (e.g., Bowen, 1980), or (2) behavior-based models (de Vriend *et al.*, 1993). The latter range from: (1) highly parameterized models to predict summer–winter (bar–berm) profiles (e.g., Aubrey, 1979) or sequential bar evolution (Wright and Short, 1984; Sunamura, 1988) to (2) statistically based models for predicting bar dynamics (Aubrey *et al.*, 1980) to (3) morphological models to simulate large-scale beach changes (e.g., Cowell *et al.*, 1995).

Brian Greenwood

Bibliography

Aagaard, T., 1990. Infragravity waves and nearshore bars in protected, storm-dominated coastal environments. *Marine Geology*, **94**: 181–203.

Aagaard, T., and Greenwood, B., 1995. Suspended sediment transport and morphological response on a dissipative beach. *Continental Shelf Research*, **15**: 1061–1086.

Aagaard, T., Nielsen, J., and Greenwood, B., 1998. Suspended sediment transport and nearshore bar formation on shallow intermediate-state beach. *Marine Geology*, **148**: 203–225.

Aubrey, D.G., 1979. Seasonal patterns of onshore/offshore sediment movement. *Journal of Geophysical Research*, **84**: 6347–6354.

Aubrey, D.G., Inman, D.L., and Winant C.D., 1980. The statistical prediction of beach changes in southern California. *Journal of Geophysical Research*, **85**: 3264–3276.

Barcilon, A.L., and Lau, J.P., 1973. A model for the formation of transverse bars. *Journal of Geophysical Research*, **78**: 2656–2664.

Bauer, B.O., and Greenwood, B., 1990. Modification of a linear bar-trough system by a standing edge wave. *Marine Geology*, **92**: 177–204.

Boczar-Karakiewicz, B., and Davidson-Arnott, R.G.D., 1987. Nearshore bar formation by non-linear wave processes—a comparison of model results and field data. *Marine Geology*, **77**: 287–304.

Bowen, A.J., 1980. Simple models of nearshore sedimentation: beach profiles and longshore bars. In McCann, S.B. (ed.), *The Coastline of Canada: littoral processes and shore morphology*. Geological Survey of Canada, Special Paper: 80–10, pp. 1–11.

Bowen, A.J., and Huntley, D.A., 1984. Waves, long waves and nearshore morphology. *Marine Geology*, **60**: 1–13.

Bowen, A.J. and Inman, D.L., 1969. Rip currents, 2: laboratory and field observations. *Journal of Geophysical Research*, **74**(23): 5479, 5490.

Bowen, A.J., and Inman, D.L., 1971. Edge waves and crescentic bars. *Journal Geophysical Research*, **76**: 8662–8671.

Bowman, D., and Goldsmith, V., 1983. Bar morphology of dissipative beaches: an empirical model. *Marine Geology*, **51**: 15–33.

Bryan, K., and Bowen, A.J., 1996. Edge wave trapping and amplification on barred beaches. *Journal of Geophysical Research*, **101**: 6543–6552.

Carter, R.W.G., 1978. Small-scale transverse bars in Lough Neagh, Northern Ireland. *Journal Earth Science, Royal Society of Dublin*, **1**: 205–209.

Carter, T.G., Liu, P.L. and Mei, C.C., 1973. Mass transport by waves and offshore sand bedforms. *American Journal of Waterways, Harbors and Coastal Engineering Division*, ASCE **99**: 165–184.

Cowell, P.J., Roy, P.S. and Jones, R.A., 1995. Simulation of LSCB using a morphological behaviour model. *Marine Geology*, **126**: 45–61.

Dally, W.R., 1987. Longshore bar formation—surf beat or undertow? Coastal Sediments '87, ASCE, New York, pp. 71–86.

Dally, W.R., 1980. Numerical model for beach profile evolution. M.S. thesis, Department Civil Engineering, Newark: University of Delaware.

Dally, W.R., and Dean, R.G., 1984. Suspended sediment transport and beach profile evolution. J. Waterway, Port, Coastal and Ocean Eng., **110**: 15–33.

Dalrymple, R.A., 1992. Prediction of storm/normal beach profiles. *Journal of Waterway, Port, Coastal and Ocean Engineering*, **118**: 193–200.

Damgaard Christiansen, E., Deigaard, R., and Fredsøe, J., 1994. Sea bed stability on a long straight coast. In *Proceeding 24th International Conference on Coastal Engineering*, American Society Civil Engineers, New York, pp. 1865–1879.

Davidson-Arnott, R.G.D., 1981. Computer simulation of nearshore bar formation. *Earth Surface Processes and Landforms*, **6**: 23–34.

Davidson-Arnott, R.G.D., 1988. Controls on formation of barred nearshore profiles. *Geographical Review*, **78**: 185–193.

Davis, R.A., Jr., and Fox, W.T., 1972. Coastal processes and nearshore sand bars. *Journal of Sedimentary Petrology*, **42**: 401–412.

Davis, R.A., Jr., Fox, W.T., Hayes, M.O., and Boothroyd, J.C., 1972. Comparison of ridge and runnel systems in tidal and non-tidal environments. *Journal of Sedimentary Petrology*, **42**: 413–421.

De Vriend, H.J., Capobianco, M., Chesher, T., De Swart, H.E., Latteux, B., and Stive, M.J.F., 1993. Approaches to long term modelling of coastal morphology: a review. *Coastal Engineering*, **21**: 225–269.

Deigaard, R., Dronen, N., Fredsoe, J., Jensen, J.H., and Jorgensen, M.P., 1999. A morphological stability analysis for a long straight barred coast. *Coastal Engineering*, **36**: 171–195.

Dolan, T.J., and Dean, R.G., 1985. Multiple longshore sand bars in the upper Chesapeake Bay. *Estuarine, Coastal and Shelf Science*, **5**(21): 727–743.

Elgar, S., 2001. Coastal profile evolution at Duck, North Carolina: a precautionary note. *Journal of Geophysical Research*, **106**: 4625–4627.

Elgar, S., Gallagher, E.L., and Guza, R.T., 2001. Nearshore sandbar migration. *Journal of Geophysical Research*, **106**: 11623–11627.

Evans, O.F., 1940. The low and ball of the east shore of Lake Michigan, *Journal of Geology*, **48**: 467–511.

Exon, N.F., 1975. An extensive offshore sand bar field in the western BalticSea. *Marine Geology*, **18**: 197–212.

Falques, A., 1991. A note on the Barcilon and Lau model for transverse bars. *Revista de Geofisica*, **47**: 191–95.

Falques, A., Coco, G., and Huntley, D.A., 2000. A mechanism for the generation of wave-driven rhythmic patterns in the surf zone. *Journal Geophysical Research*, **105**: 24071–24087.

Gallagher, E., Guza, R.T., and Elgar, S., 1998. Observations of sand bar behaviour on a natural beach. *Journal of Geophysical Research*, **103**: 3203–3215.

Gelfenbaum, G., and Brooks, G.R., 1997. Long term observations of migrating shore-normal bars. *Proceedings of Coastal Dynamics* '97, Plymouth, ASCE, New York, pp. 654–663.

Greenwood, B., and Davidson-Arnott, R.G.D., 1975. Marine bars and nearshore sedimentary processes, Kouchibouguac Bay, New Brunswick, Canada. In Hails, J. and Carr, A. (eds.), *Nearshore Sediment Dynamics and Sedimentation*. London: John Wiley & Sons Ltd., pp. 123–150.

Greenwood, B., and Davidson-Arnott, R.G.D., 1979. Sedimentation and equilibrium in wave-formed bars: A review and case study. *Canadian Journal of Earth Science*, **16**: 313–332.

Greenwood, B., and Mittler, P.R., 1979. Structural indices of sediment transport in a straight, wave-formed, nearshore bar. *Marine Geology*, **32**: 191–203.

Hale, P.B., and McCann, S.B., 1982. Rhythmic topography in a mesotidal low-wave-energy environment. *Journal of Sedimentary Petrology*, **52**: 415–429.

Hands, E.H., 1976. Observations of barred coastal profiles under the influence of rising water levels, Eastern Lake Michigan, 1967–71. U.S. Army, Corps of Engineers, *CERC Technical Report*, 76-1, 113 pp.

Hayes, M.O., and Boothroyd, J.C., 1969. Storms as modifying agents in the coastal environment. In Hayes, M.O. (ed.), *Coastal Environments: NE Massachusetts*. Amherst: Department Geology, University Massachusetts.

Heathershaw, A.D., and Davies, A.G., 1985. Resonant wave reflection by transverse bedforms and its relation to beaches and offshore bars. *Marine Geology*, **62**: 321–338.

Hino, M., 1974. Theory on formation of rip current and cuspidal coast. *Coastal Engineering Japan*, **17**: 23–38.

Holman, R.A., and Bowen, A.J., 1982. Bars, bumps and holes: models for the generation of complex beach topography. *Journal of Geophys. Research*, **87**: 457–468.

Holman, R.A., and Sallenger Jr., A.B., 1993. Sand bar generation: a discussion of the Duck Experiment Series. *Journal of Coastal Research*, **15** (Special Issue): 76–92.

Howd, P.A., Bowen, A.J., and Holman, R.A., 1992. Edge waves in the presence of strong longshore currents. *Journal of Geophys. Research*, **97**: 11357–11371.

Huntley, D.A., 1980. Edge waves in a crescentic bar system. In McCann, S.B. (ed.), *Coastline of Canada: Littoral Processes and Shore Morphology*. Geological Survey of Canada, Paper 80-10, Report 2, pp. 111–121.

Inman, D.L., Elwany, M.H.S., and Jenkins, S.A., 1993. Shorerise and bar–berm profiles on ocean beaches. *Journal of Geophysical Research*, **98**: 18181–18199.

Katoh, K., 1984. Multiple longshore bars formed by long period standing waves. *Report of the Port and Harbour Research Institute*, **23**: 3–46.

King, C.A.M., and Williams, W.W., 1949. The formation and movement of sand bars by wave action. *Geographical Journal*, **112**: 70–85.

Kirby, J.T., Dalrymple, R.A., and Liu, P.L.-F., 1981. Modification of edge waves by barred beach topography. *Coastal Engineering*, **5**: 35–49.

Konicki, K.M., and Holman, R.A., 2000. Transverse bars at Duck, North Carolina. Proceedings of the 25th International. Conference of Coastal Engineering, ASCE, New York, pp. 3588–3599.

Kraus, N.C., and Larson, M., 1988. Beach profile changes measured in the tank for large waves, 1956–57 and 1962. U.S. Army Corps of Engineers, Coastal Engineering Research Center, *Technical Report*. CERC-89-9.

Larson, M., and Kraus, N.C., 1992. Analysis of cross-shore movement of natural longshore bars and material to create longshore bars. U.S. Army Corps of Engineers, CERC Technical Report, DRP-92-5, 89 pp.

Lippmann, T.L., and Holman, R.A., 1990. The spatial and temporal variability of sand bar morphology. *Journal of Geophysical Research*, **95**: 11575–11590.

Lippmann, T.L., Holman, R.A., and Hathaway, K.K., 1993. Episodic, non-stationary behaviour of a two sand bar system at Duck, NC, USA. *Journal of Coastal Research*, **15**(Special Issue): 49–75.

Miller, R.L., 1976. Role of vortices in surface zone prediction: sedimentation and wave forces. In Davis, R.A., Jr. and Ethington, R.L. (eds.), *Beach and Nearshore Sedimentation*. Society of Economic Paleontologists and Mineralogists Special Publication 24, pp. 92–114.

Moller, I., and Southgate, H.N., 1997. Fractal properties of beach profile evolution at Duck, North Carolina. *Journal of Geophysical Research*, **105**: 11489–11507.

Mulrennan, M.E., 1992. Ridge and runnel beach morphodynamics: an example from the central east coast of Ireland. *Journal of Coastal Research*, **8**: 906–918.

Niedoroda, A.W., 1973. Sand bars along low energy beaches. Part 2: Transverse bars, In Coates, D.R. (ed.), *Coastal Geomorphology*. Binghamton: State University of New York, pp. 103–113.

Niedoroda, A.W., and Tanner, W.F., 1970. Preliminary study of transverse bars. *Marine Geology*, **9**: 41–62.

Nilsson, H.D., 1973. Sand bars along low energy beaches Part 1: multiple parallel sand bars of southeastern Cape Cod. In Coates, D.R. (ed.), *Coastal Geomorphology*, Publication In Geomorphology, New York; State University of New York, pp. 99–102.

O'Hare, T., 1994. The role of long waves in sand bar formation: a model exploration. In *Proceedings of Coastal Dynamics, '94*. Barcelona, ASCE, New York, pp. 74–78.

O'Hare, T.J., and Davies, A.G., 1993. Sand bar formation beneath partially standing waves: laboratory experiments and model simulations. *Continental Shelf Research*, **13**: 1149–1181.

O'Hare, T.J., and Huntley, D.A., 1994. Bar formation due to wave groups and associated long waves. *Marine Geology*, **116**: 313–325.

Orford, J.D., and Wright, P., 1978. What's in a name? Descriptive or genetic implications of "ridge and runnel" topography. *Marine Geology*, **28**: M1–M8.

Orme, A.R., and Orme, A.J., 1988. Ridge and runnel enigma. *Geographical Review*, **78**: 169–184.

Osborne, P.D., and Greenwood, B., 1992. Frequency dependent cross-shore suspended sediment transport 2: a barred shoreface, Bluewater Beach, Ontario, Canada. *Marine Geology*, **106**: 25–51.

Plant, N.G., and Holman, R.A., 1997. Strange kinematics of sandbars. *Proceedings of Coastal Dynamics '97*, Plymouth, ASCE, New York, pp. 355–364.

Plant, N.G., Holman, R.A., Freilich, M.H., and Birkemeier, W.A., 1999. A simple model for interannual sandbar behaviour. *Journal of Geophysical Research*, **104**: 15755–15776.

Plant, N.G., Freilich, M.H., and Holman, R.A., 2001. Role of morphologic feedback in surf zone sandbar response. *Journal of Geophysical Research*, **106**: 973–989.

Rey, V., Davies, A.G., and Belzons, M., 1995. On the formation of bars by the action of waves on an erodible bed: a laboratory study. *Journal of Coastal Research*, **11**: 1180–1194.

Roelvink, J.A., and Broker, I., 1993. Cross-shore profile models. *Coastal Engineering*, **21**: 163–191.

Ruessink, B.G., 1998. Infragravity waves in a dissipative multiple bar system. Ph.D. thesis, University of Utrecht, 243 pp.

Ruessink, B.G. and Kroon, A., 1994. The behaviour of a multiple bar system in the nearshore zone of Terschelling, the Netherlands, 1965-1993. *Marine Geology*, **121**: 187–197.

Ruessink, B.G., van Enckevort, I.M.J., Kingston, K.S., and Davidson, M.A., 2000. Analysis of observed two- and three-dimensional nearshore bar behaviour. *Marine Geology*, **64**: 237–257.

Sallenger, A.H., Jr., and Howd, P.A., 1989. Nearshore bars and the breakpoint hypothesis. *Coastal Engineering*, **12**: 301–313.

Sallenger, A.H., Jr., Holman, R.A. and Birkemeier, W.A., 1985. Storm-induced response of a nearshore bar system. *Marine Geology*, **64**: 237–258.

Schwartz, M., 1982. *Encyclopaedia of Beaches and Coastal Environments*. Stroudsburg, PA: Hutchinson Ross Publishing Company.

Shepard, F.P., 1950. Longshore bars and longshore troughs. Beach Erosion Board, Technical *Memo* 15, p. 32.

Shepard, F.P., Emery, K.O., and LaFond, E.C., 1941. Rip currents: a process of geological importance. *Journal of Geology*, **49**: 337–369.

Short, A.D., 1979. Three-dimensional beach stage model. *Journal of Geology*, **87**: 553–571.

Short, A.D., 1975. Multiple offshore bars and standing waves. *Journal of Geophysical Research*, **80**: 3838–3840.

Short, A.D., and Aagaard, T., 1993. Single and multi-bar beach change models. *Journal of Coastal Research*, **15**(Special Issue): 141–157.

Southgate, H.N., and Moller, I., 2000. Fractal properties of coastal profile evolution at Duck, North Carolina. *Journal of Geophysical Research*, **105**: 11489–11507.

Sonu, C.J., 1973. Three-dimensional beach changes. *Journal Geology*, **81**: 42–64.

Suhayda, J.N., 1974. Standing waves on beaches. *Journal of Geophysical Research* **79**: 3065–3071.

Sunamura, T., 1988. Beach morphologies and their change. Chapter 2, In Horikawa, K. (ed), Nearshore dynamics and coastal processes: theory, measurement and predictive models. Tokyo: University of Tokyo Press, pp. 136–166.

Sunamura, T., and Takeda, I., 1984. Landward migration of inner bars. *Marine Geology*, **60**: 63–78.

Thornton, E.B., Humiston, R.T., and Birkemeier, W., 1996. Bar/trough generation on a natural beach. *Journal of Geophysical Research*, **101**: 12097–12110.

Van Beek, J.L., 1974. Rhythmic patterns of beach topography. *Technical Report*, 187, 46 pp., Coastal Studies, Baton Rouge: Louisiana State University.

Van Enckevort, I.M.J., and Ruessink, B.G., 2001. Effect of hydrodynamics and bathymetry on video estimates of nearshore sand bar position. *Journal Geophysical Research*, **106**: 16969–16979.

Van Rijn, L.C., 1998. *Principles of Coastal Morphology*. Amsterdam: Aqua Publications.

Vittorio, G., de Swart, H.E., and Blondeaux, P., 1999. Crescentic bedforms in the nearshore region. *Journal of Fluid Mechanics*, **381**: 271–303.

Wijnberg, K.M., 1997. On the systematic offshore decay of breaker bars. *Proceedings 25th International Conference Coastal Engineering*, New York, pp. 3600–3613.

Wijnberg, K.M., and Terwindt, J.H.J., 1995. Quantification of decadal morphological behaviour of the central Dutch coast. *Marine Geology* **126**: 301–330.

Wright, L.D., and Short, A.D., 1984. Morphodynamic variability of surf zones and beaches: a synthesis. *Marine Geology*, **56**: 93–118.

Wright, L.D., Short, A.D., and Green, M.O., 1985. Short-term changes in the morphodynamic states of beaches and surf zones: an empirical predictive model. Marine Geology, 62: 339–364.

Wright, L.D., Chappell, J., Thom, B.G., Bradshaw, M.P., and Cowell, P., 1979. Morphodynamics of reflective and dissipative beach and inshore systems, southern Australia, *Marine Geology*, **32**: 105–140.

Yu, J., and Mei, C.C., 2000. Formation of sand bars under surface waves. *Journal of Fluid Mechanics*, **416**: 315–348.

Zhang, D.P., 1994. Wave flume experiments on the formation of longshore bars produced by breaking waves. *Science Report, Institute Geoscience, University Tsukuba*, **15**: 47–105.

Zenkovith, V.P., 1967. *Processes of Coastal Development*. London: Oliver and Boyd.

Cross-references

Beach Features
Beach Processes
Net Transport
Profiling, Beach
Rhythmic Patterns
Surf Zone Processes
Wave–Current Interaction

BAY BEACHES

The length of shore in bays, sounds, lagoons, and estuaries (here termed bays) greatly exceeds the length of ocean shore in many countries. Beaches are common in these bays, but they are often so small and isolated that they escape attention, except in populated locations. The definition of bay in relation to the open coast is somewhat subjective, and bays such as Monterey Bay, California may have wave-energy levels that are among the highest in the world. This discussion is confined to low-energy beaches that occur in mostly enclosed bays where the fetch distances for local wave generation are generally less than 50 km. The principal factors affecting the morphodynamics of these beaches are locally generated waves and wave-induced currents, but wind-induced and tidal currents play a role in morphologic change. Fluvial processes may become dominant at estuarine shores in narrow basins or tributaries. At the low end of the wave-energy continuum, other terms, such as stream bank, intertidal marsh margin, or bay bottom may be more appropriate than beach.

Shore processes

Waves generated by local winds in bays have low heights (usually mean heights <0.2 m and storm wave heights <1.0 m) and short periods (2.0–4.5 s) (Nordstrom, 1992). Ocean waves entering bays play a limited role in beach change where shores do not face ocean entrances (Jackson, 1995). Tidal range affects the vertical distribution of wave energy over the profile, determining the width of the beach and the duration that waves break at any elevation. Bay beaches are usually characterized by a steep upper foreshore with a broad, flat fronting terrace. On tidal beaches, spilling waves break in a broad surf zone across the gently sloping terrace at low tide, but the energy in the waves is low. At high tide, waves reach the upper foreshore with little loss of energy and usually break as plunging waves.

Longshore currents are predominantly generated by the breaking of local wind-waves but refracted ocean waves, tidal flows, and wind drift are locally important and may result in flows bayward of the breaking waves that are opposite flows generated by local wind-waves. Tidal currents are important near channels, projecting headlands, and constrictions in the bay, and they may be the dominant agent of sediment transport on the terrace bayward of the foreshore. Ice forms faster and has a greater influence on mid- and high-latitude bay beaches than on ocean beaches because bay waters are colder in winter, shallower, and less saline; ice lasts longer because low wave energies are slow to remove it. Ship and boat wakes are higher on bay beaches than on ocean beaches because vessels can pass close to the shore, but the average energy in the wakes is usually only a small percentage of the average energy of wind waves in all but the smallest bays.

Water level changes can be locally induced by winds blowing across the bay or they can be induced by flow of water through inlets from surges generated on the open coast. Winds can increase water levels on the downwind side of the bay while lowering water levels on the upwind side, but a large opening to the sea on the downwind side of the bay can result in lower water levels downwind.

Beach and shore characteristics

Beaches comprise a large proportion of the shore in many bays (Nordstrom and Roman, 1996). Important examples include Delaware Bay (Jackson, 1995), Chesapeake Bay (Rosen, 1980; Ward *et al.*, 1989), Puget Sound (Downing, 1983; Terich, 1987). Beaches in smaller bays, with limited availability of sand and gravel may be small, highly localized, or confined to ocean entrances. Many beaches have been created in urbanized estuaries where none would occur naturally because wave energies are too low. These artificial beaches are often wider than natural beaches in undeveloped areas. Some new beaches are accidental by-products of landfill operations; some are created intentionally as new beach recreation areas (Nordstrom, 1992).

Bay beaches may be unvegetated or partially vegetated and composed of sand, gravel, or shell. Surface sediments are often coarser on bay beaches than on ocean beaches with a similar source. Lag gravel is common on the beach surface, formed from particles exhumed by swash or by preferential elimination of fines by low-energy waves. Individual pebbles move readily over the sand surface, and swash excursions create bands of gravel on the upper foreshore.

The depth of mobilization of sediments on the upper foreshore is shallow (e.g., <0.2 m under storm conditions), and the active beach may be only a thin veneer of unconsolidated material overlying an immobile layer of coarse sediments, clay, peat, or a shore platform. Mobilization of sediments on the low tide terrace by waves may occur only to depths of 10–30 mm, and biological activity may play a greater role than wave processes in altering the characteristics of the surface and subsurface (Nordstrom, 1992).

Vegetation plays a greater role in influencing morphologic change on bay beaches than on ocean beaches because of greater abundance of vegetation in bays and the reduced ability of the low-energy waves to move it. Vegetation helps bind bottom sediment and attenuate wave energies; vegetation flotsam in the breaker and surf zones alters the wave and current characteristics and the likelihood of entrainment of beach sediment; vegetation litter in the wrack line forms barriers to waves, currents, and swash uprush.

Bay shorelines are often composed of numerous isolated beaches with different orientations. They have high variability in morphology and rate of erosion over small areas resulting from local differences in fetch, wind direction, stratigraphy, inherited topography, resistant outcrops on the foreshore, variations in submergence rates, and amounts of sediment in eroding formations (Phillips, 1986; Rosen, 1980). Beach compartments are isolated into longshore drift cells defined by deep coves or headlands formed by resistant rock, marsh, or human structures.

The net rate of longshore transport on estuarine beaches varies with orientation, fetch distance, and size of each drift cell and ranges from tens of cubic meters to tens of thousands of cubic meters (Wallace, 1988). Although rates of transport are low, the magnitude of erosion can be high because the quantities of sediment in transport represent a sizable fraction of the total unconsolidated sediment in the active beach. Many bay shores are eroding at greater rates than nearby ocean shores.

Beach change

The upper foreshores of most bay beaches are modally reflective. Conspicuous cyclic morphologic change is confined to the immediate vicinity of the foreshore. Sediment removed from the upper foreshore during high-wave-energy events is deposited on the lower foreshore with a change to a concave upward profile shape. Sediments moved farther offshore onto the terrace form only a thin veneer over the surface instead of forming the break point bar that is prominent on many ocean beaches. Landward and bayward displacement of the entire foreshore profile may also occur while the profile slope is maintained. This parallel-slope retreat and advance is common when sediment exchange is due primarily to longshore transport and is most pronounced near the ends of drift compartments (Nordstrom, 1992).

Resource values of bay beaches

The fronting terrace of a low-energy bay beach has a relatively stable substrate that allows macroscopic plants and fauna to thrive. The upper foreshore is more energetic and may have less species diversity and abundance. Infauna and macroalgae provide prey to juveniles of commercially valuable fish, and the intertidal area provides habitat for recreationally important clams and numerous species of epifauna and infauna. The upper foreshore may be an important spawning area for horseshoe crabs. Fish and invertebrates are prey for foraging birds, especially in the regularly exposed intertidal zone. Wrack from plant litter is inhabited by numerous amphipods and insects. The swash zone and dry upper foreshore are also foraging areas for birds, including upland species.

Bay beaches are not as intensively used for recreation as ocean beaches, but they have important complementary values. They provide convenient surfaces for launching and landing boats and boards for wind surfing. They are favored by parents with children because they provide a safer environment than on the ocean. Many bay beaches are underutilized for recreation because of the unclean appearance of the beaches or lack of awareness of their existence or unique attributes, but ease of access causes bay beaches close to urban areas to have relatively high rates of use.

Shore protection and management

Erosion control strategies for bay beaches may differ from strategies for ocean beaches because of differences in the scale of erosional forces and in the value of resources. Protection programs are facilitated because beach segments are small, isolated drift cells, often under jurisdiction of only one management agency and because small-scale, low-cost protection may be utilized. Low wave energies and gentle offshore gradients make construction of fixed offshore engineering works more practical than on high-energy beaches. Shore-parallel walls are often successful because they can withstand direct attack of local waves; they take up minimal space on the beach and adjacent upland; and they limit the loss of biological resources on the fronting terrace or bay bottom. Projects funded by national or state/ provincial governments are often not economically feasible, resulting in a fragmented approach to protection by individual property owners. Simple engineering principles are often ignored in constructing small-scale protection structures, including lack of filter cloth or weep holes in bulkheads, failure to build structures deep enough to prevent toe scour or high enough to prevent overtopping, weak fastenings, and failure to use adequate sized armor stones or perform maintenance. As a result, there is much evidence of structural failure. Beach fill is increasingly used for protection or recreation, but fill can cover benthic habitat and eliminate shallow-water areas for aquatic plants. Bayside nourishment projects can be inexpensive because only small quantities of fill are required. Fill materials brought in from outside the region may retain their exotic appearance because of limited mixing by low-energy waves.

There has been considerable federal and state intervention in decisions on developing bay shores, especially in productive estuaries, but this intervention is rarely conducted to maintain beach resources. Alternative human uses such as transportation, industrial developments, residences, and boating are compatible with a coastal location according to most policies, and actions to enhance these uses may eliminate beaches. The number and value of bay beaches can be enhanced by implementing beach nourishment operations, altering vegetation, constructing appropriate protection structures, acquiring key sites for public use, and enhancing access. The ease of constructing and maintaining bay beaches and the paucity of quality recreation space in many urban areas make creation of new beaches as surrogates for ocean beaches an attractive option.

Karl F. Nordstrom

Bibliography

Downing, J., 1983. *The Coast of Puget Sound: Its Processes and Development*. Seattle, WA: Puget Sound Books.

Jackson, N.L., 1995. Wind and waves: influence of local and non-local waves on mesoscale beach behavior in estuarine environments. *Annals of the Association of American Geographers*, **85**: 21–37.

Nordstrom, K.F., 1992. *Estuarine Beaches*. London: Elsevier Applied Science.

Nordstrom, K.F. and Roman, C.T. (ed.), 1992. *Estuarine Shores: Evolution, Environments and Human Alterations*. London: John Wiley & Sons.

Phillips, J.D., 1986. Spatial analysis of shoreline erosion: Delaware Bay, NJ. *Annals of the Association of American Geographers*, **76**: 50–62.

Rosen, P.S., 1980. Erosion susceptibility of the Virginia Chesapeake Bay shoreline. *Marine Geology*, **34**: 45–59.

Terich, T.A., 1987. *Living with the Shore of Puget Sound and the Georgia Strait*. Durham, NC: Duke University Press.

Wallace, R.S., 1988. Quantification of net shore-drift rates in Puget Sound and the Straight of Juan de Fuca, Washington. *Journal of Coastal Research*, **4**: 395–403.

Ward, L.G., Rosen, P.S., Neal, W.J., Pilkey, O.H., Jr., Pilkey, Sr., O.H., Anderson, G.L., and Howie, S.J., 1989. *Living with Chesapeake Bay and Virginia's Ocean Shores*. Durham, NC: Duke University Press.

Cross-references

Beach Erosion
Beach Nourishment
Beach Processes
Dissipative Beaches
Estuaries
Human Impacts on Coasts
Reflective Beaches
Sediment Transport
Shore Protection Structures

BEACH AND NEARSHORE INSTRUMENTATION

Instrumentation in studies of the coast generally, and of the beach and nearshore zone in particular is designed to measure attributes of form and changes in the form (bed) over time, including bedforms; fluid processes related to waves, water level and currents in the water and wind on the beach; and sediment concentration and mass transport rate in the water and on the beach. These measurements may be made at a variety of temporal scales ranging from fractions of a second to months and years and spatial scales ranging from a few square millimeters to hundreds of square kilometers. Some attributes are measured individually, but much of the focus today, and over the past three decades, has been on measurements of morphodynamics, in which the objective is to measure fluid and sediment transport processes and the resulting change in morphology at a temporal scale of minutes to days and occasionally months. Much activity is focused on sandy and to a lesser extent muddy coasts and much of the instrumentation described here is devoted to these, but some work also takes place on the erosion of cohesive clay and bedrock coasts. The highly dynamic nature of the nearshore and swash zone in particular poses many problems for the design of instruments for measuring fluid and sediment transport processes. In addition to the need for very rugged instruments and supports for mounting them, there are difficulties posed by the lack of access to much of the nearshore during storm conditions, and by the presence of bubbles and organic matter in the water column. Ultimately, the instrumentation is designed and deployed to measure particular properties and processes of the beach and nearshore zones, and therefore, this review is organized by the measurement objective rather than particular instrument types.

Measurement of form and changes in form (erosion and deposition)

Erosion of cohesive and bedrock coasts

Where the coastline is developed in bedrock, till, and cohesive muds the focus of attention is usually on the measurement of rates of erosion in relation to the strength attributes of the material and the erosional or forcing processes (Sunamura, 1992). On a small scale, erosion by weathering and abrasion of rock coasts generally takes place so slowly that measurements are made at point locations on a timescale of months to years. The micro erosion meter originally used to measure solution of limestone was adapted for use on intertidal shore platforms (Trudgill *et al.*, 1981). It consists of a pointer attached to a micrometer gauge on a mount that can be placed on pins drilled into the rock platform. The mount swivels to allow measurement to be taken at several points around the station so that an average value can be obtained. Measurements are commonly taken at intervals of months or years because of the relatively slow rate of downcutting (Kirk, 1977; Viles and Trudgill, 1984). A cruder version of the instrument has been adapted for measurements of erodability of tills and clays underwater (Askin and Davidson-Arnott, 1981; Davidson-Arnott and Langham, 2000). Because the erosion rates are typically up to several centimeters per year, measurements can be made on a weekly to monthly basis.

Measurements of the erodability of fine-grained cohesive muds commonly found in a variety of marine and estuarine environments, have commonly been made with a variety of benthic flumes—essentially inverted channels of various configurations which can be deployed either on the exposed tidal flat or underwater (Amos *et al.*, 1992; Maa *et al.*, 1993; Houwing, 1999). Water is circulated through the channel at increasing speeds until the shear on the bed induces erosion and the erosion rate is measured either directly, or indirectly by measurement of suspended sediment concentrations. Recent experiments have also been made with the Cohesive Strength Meter, an automated device which employs a carefully regulated vertical jet of water and monitors the rate of erosion with respect to the impact force (Tolhurst *et al.*, 1999).

Erosion and deposition of sediments

While large-scale changes in form can be measured by a variety of techniques described below, these techniques are usually carried out at finite intervals of days, weeks, or months. Measurement of changes in the bed at particular locations on the timescale of dynamic measurements of fluid flow and sediment transport, typically on the order of minutes to hours, has proved to be surprisingly difficult to do in shallow water. One major problem in the nearshore during storms on sandy coasts is the difficulty of distinguishing the bed from the material immediately above it, which is being transported as bed load or suspended load close to the bed. Techniques for measuring changes in elevation at points in the nearshore range from simple erosion rods to optical and acoustical instruments.

Simple measurements of change in bed elevation and the total depth of activation can be made with rods emplaced along a profile or on a grid which are surveyed before and after a storm (Greenwood *et al.*, 1979). The rods can be emplaced by wading and diving. The maximum scour depth can be resolved by placing a washer on the sand surface and then measuring the depth of burial following the storm. Results from a grid of these can be used to measure volume change in the nearshore (Greenwood and Mittler, 1984). Similarly, thin rods can be used on the subaerial beach to measure erosion by wind and thus provide a comparison volume to measurements of sand transport or deposition. Other approaches involving this simple technology in coastal applications include the use of a bedframe device to measure rates of sediment deposition in foredunes (Davidson-Arnott and Law, 1990, 1996) and the use of Surface Elevation Tablet (SET) stations in measuring net change in saltmarshes (Cahoon *et al.*, 2000). Recently, automated devices which act in a similar fashion have been developed. One approach uses a vertical array of photocells spaced at a small increment, usually on the order of 1 cm, with the bed being distinguished by either a change in the voltage output or a circuit which can detect where the break is between exposed and buried cells (e.g., Lawler, 1992). An alternative method uses the difference in conductivity between sediments and seawater to distinguish the bed level (Ridd, 1992). The value of these instruments is that they are relatively low cost and therefore provide the potential for deployment of sufficient sensors to give reasonable spatial coverage across the surf zone.

It should be possible to detect the bed using a small echo sounder mounted on a support above the bed, though this has proved notoriously difficult when there are large amounts of sediment moving over the bed and in suspension. One adaptation of this approach is to mount the transducer on a frame with a sealed stepper motor that permits it to traverse a section of the bed, thus permitting determination of two-dimensional bedform properties and migration rates (Greenwood *et al.*, 1993). Transducers and miniature versions of sidescan sonar have been used more successfully in deeper water where sediment concentrations are much lower. Recent developments in acoustic doppler technology give a much better definition of the bed. Several versions of acoustic doppler velocity profilers (ADCPs) are available which enable the speed of currents to be detected at incremental distances from the sensor. When pointed downward, these are able to distinguish the bed more precisely than simple sonar devices, because the doppler shift is absent from sediments that are not moving (see section below on Sediment concentration, mass transport rate, and deposition for information relating to the ADCP). Small, relatively cheap acoustic sounders are available for use in air and can be used on the beach to measure changes in elevation of the bed or the water surface in wells installed to measure the water table. These devices can also be mounted on tracks to give a profile of changes in elevation during a transport event and the dimensions of any bedforms that develop.

Measurement of form and form change

Measurement of the dune, beach, and nearshore form on a scale of meters to kilometers has traditionally been done using standard survey and hydrographic techniques. Surveys out to the limit of wading have been carried out with levels and theodolites, and the use of a total station incorporating an electronic distance measurement (EDM) unit and electronic data storage is now standard. These permit rapid surveys over a range of elevations and the output is readily incorporated into a wide range of contouring and geographic information system (GIS) software packages which can produce digital elevation models and permit easy extraction of volume change through repetitive surveys (see Figure B11). In shallow water, depth has traditionally been determined using standard echo sounders mounted on a boat (Gorman *et al.*, 1998). Digital recording has now replaced the standard paper trace and positional data can be recorded simultaneously using a global positioning system (GPS). Towed arrays or acoustic multibeam transducers can be used to give simultaneous mapping of a wide swath, including information on large bedforms (Morang *et al.*, 1997). Better definition of the seafloor and three-dimensional bedform features can be attained with sidescan sonar, which utilizes a towed transducer that emits a signal at right angles to the tow direction and records returns from a swath either side of the transducer (Morang *et al.*, 1997).

The use of GPS which integrates signals from three or more satellites to determine location and elevation for a variety of surveying tasks, is now becoming standard in measuring beach form and change as it is in so many other fields. Simple systems can give positional accuracy of a few meters and elevation to about 10 m. Much greater accuracy can be obtained through the use of differential systems, which simultaneously capture the signal from the satellites and from a land-based station whose position and elevation is known precisely (see Figure B11). Moderate priced differential systems make use of Coast Guard beacons, which are set up along the coast for navigational purposes. These can give positional accuracy of ±2–3 cm and vertical resolution of about double that, though the accuracy decreases with distance from the beacon. More expensive differential systems use a base station set up over a known position and a rover station for the actual survey. The systems can be used to measure the height and position of particular points but they can also be put in a backpack or on a vehicle allowing continuous recording of a traverse. This permits the mapping of linear features such as the waterline, thalweg of tidal creeks, bar crest, and top and bottom of cliffs, thus permitting much better delineation of these features and permitting more accurate delineation of change through repetitive surveys.

A major problem for morphodynamic experiments in the nearshore and surf zone is to obtain measurements of form change during intense storm events. While measurements of sediment transport and nearshore water motion can be obtained throughout an event, most measurements of form change have been obtained through standard surveys carried out during low wave conditions before and after the event. Some data during storms can be obtained from jetties and from specially constructed platforms that span the surf zone. However, some specialized equipment makes data collection during quite high wave conditions possible. These include various sled devices, which can have either a mast with a prism for measurement by a total station or a GPS station to enable position and elevation to be determined. The sleds may be towed by boat beyond the surf zone and winched onshore or a pulley system may be attached to an anchor seaward of the surf zone, enabling the sled to be pulled offshore without recourse to a boat. One highly specialized instrument is the CRAB used extensively at the CERC facility at Duck, North Carolina to carry out a variety of tasks in the water, including surveys in waves up to 3 m (e.g., Plant *et al.*, 1999).

Production of topographic maps from stereo pairs of aerial photographs has been a standard procedure for five decades but photo rectification and automated contouring have required expensive equipment and are rarely used for small-scale beach studies. However, new developments in video technology and digital photogrammetry are making remote measurements of form change much more practical. Video technology has been applied for more than a decade to measure waves and swash run-up (see below) but it has also been applied to measurement of the position of nearshore bars through time exposure of wave breaking (Konicki and Holman, 2000; Ruessink *et al.*, 2000; Alport *et al.*, 2001). The intensity of wave breaking is captured by creating time exposure images over a period on the order of 10 min and the resultant smooth white bands outline the zones of wave breaking on shallow bar crests and at the beach. Video cameras can also be used to monitor changes in dynamic features such as tidal inlets and associated ebb and flood tidal deltas (Morris *et al.*, 2001).

The use of digital images from still and video cameras to produce digital elevation models (DEMs) through a variety of computer software packages is of especial interest in mapping coastline changes and changes in the morphology of the beach and foredune area (Chandler, 1999). The technique makes use of overlapping pairs of photographs produced either in the traditional way through the movement of a camera installed in a plane, helicopter, or a land-based vehicle, or through the use of images taken from two fixed positions. In the case of aerial photography or moving vehicles, the position of each digital image can be linked to real time positional data provided by DGPS. Where fixed cameras are used on the beach, control points whose position and

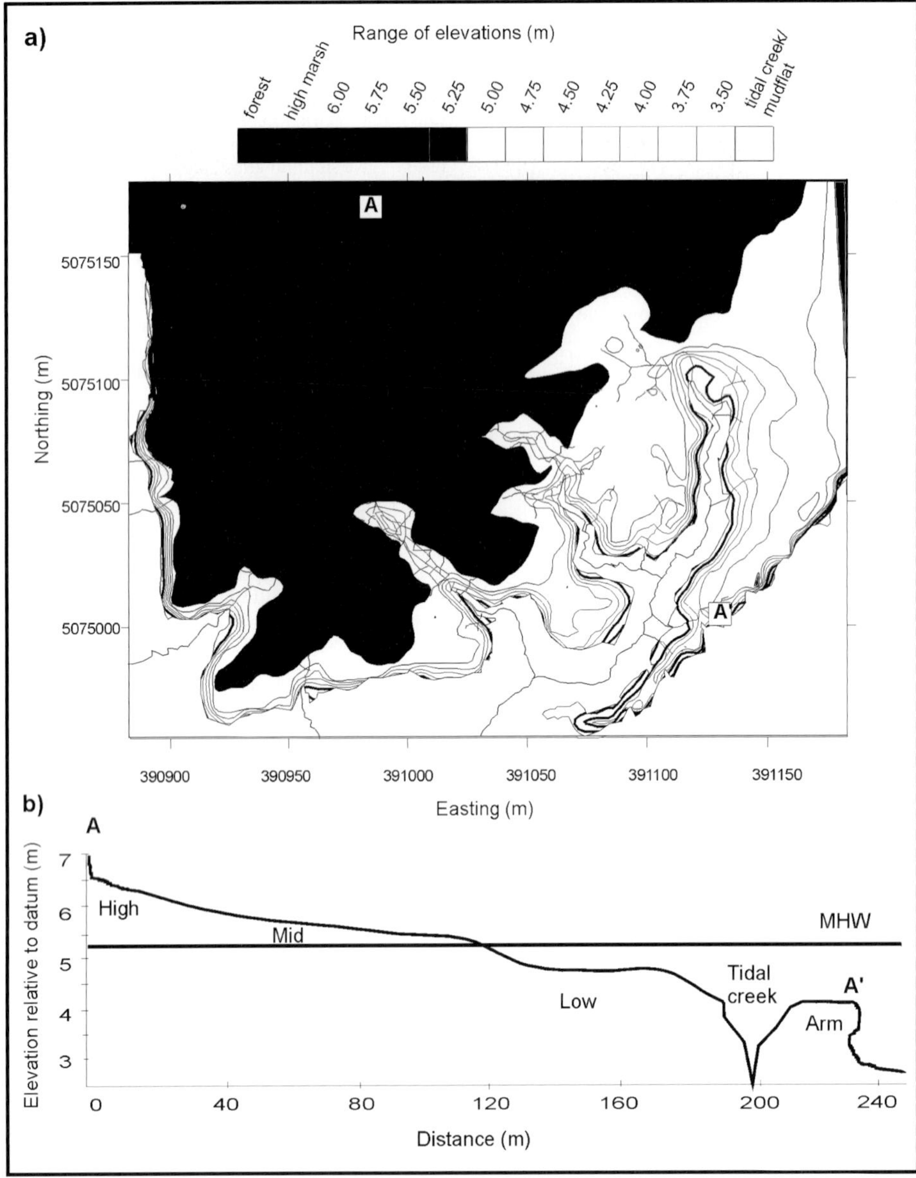

Figure B11 Digital elevation model of a saltmarsh and tidal creeks, Bay of Fundy, Canada, produced from measurements made with a total station and with a DGPS system.

elevation have been surveyed precisely are used to aid in rectification (Hancock and Willgoose, 2001). The advantage of these automated photogrammetric systems is that they can provide a very large number of data points for construction of the DEMs and much of the processing can be automated, thus allowing the evolution of topography over days, weeks, or months to be captured.

On a larger scale, the development of light detection and ranging (LIDAR) technology combined with DGPS permits topographic mapping of both the land surface and the nearshore bed to depths of 10–15 m (Irish and White, 1998; Sallenger *et al.*, 2001). The technology makes use of a laser transmitter/receiver, which transmits laser pulses toward the surface and records the traveling time of the reflected pulse. The pulse is reflected from the land surface and, over water, the return from the bottom can also be detected down to depths that depend on the degree of absorption, scattering, and refraction in the water; these in turn depend on sun angle and intensity and on the degree of turbidity in the water. The system can be deployed in a helicopter or fixed wing aircraft. Apart from the unique ability to map both the land and shallow nearshore, the technique offers a relatively low-cost method for determining topographic changes due to major storms and hurricanes (Sallenger *et al.*, 2001), and for surveying changes in areas such as salt marshes and tidal mud flats which are difficult to access with standard surveying approaches.

Winds, waves, water levels, and currents

Much of the focus in field studies of the beach and nearshore zone is on measuring the morphodynamics of sandy coasts, and to a lesser extent that of muddy coasts. These experiments require measurement of fluid and sediment dynamics over a range of timescales from fractions of a second to hours, days, and months and over spatial scales ranging from a few centimeters to hundreds of meters. Until recently, different instrumentation has been required to measure the fluid dynamics from that measuring sediment dynamics. Over the past three decades, mechanical devices for measuring fluid dynamics have been increasingly replaced by solid-state electronics involving the application of a range of direct and remote technologies. Because of the broad range of forcing variables and the spatial scales involved in determination of sediment transport, a wide range and large number of instruments is typically employed. Instrumentation typically involves measurement of wind speed and direction (both for aeolian transport on the beach and for the dynamics of the water surface), water surface elevation, wave form and direction, and water motion.

Wind speed and direction

Wind speed and direction have traditionally been measured by some form of mechanical cup or propellor-type anemometer and resistance

Figure B12 Array of cup anemometers and wind vanes mounted on towers to measure wind flow over the beach foreshore at Innisfree Beach, Ireland. Two versions of integrating sediment traps are seen on the left—the smaller traps are cylindrical traps after Leatherman (1978) and the other traps are wedge traps (Nickling and McKenna-Neuman, 1997).

wind vane mounted on a mast above the water or land surface. These give good resolution of the horizontal wind velocity at a particular elevation. Typically, in studies of aeolian transport on the beach several vertical arrays of anemometers will be deployed in order to obtain measurements of internal boundary layer development and to estimate the bed shear velocity u_* (Greely *et al.*, 1996; see Figure B12). The vertical flow can be obtained with systems of three propellor type anemometers or with two mounted at 45° angles as in the K-Gill anemometer (Atakturk and Katsaros, 1989). Sonic anemometers now offer the ability to measure fluid motion in all three dimensions, though their size is still large enough to make measurements close the bed (and thus the saltation layer) difficult. Recent modification of the Irwin sensor, a vertical pitot tube that can be mounted flush with the bed (Irwin, 1980), offers the ability to obtain direct measurements of wind stress near the bed with little disturbance to the flow.

Mean water level

Measurements of water surface elevation are collected routinely to measure changes due to tides, storm surge, and wave set-up and set-down across the breaker and surf zones. Traditional mechanical floats have now been replaced by optical and acoustic sensors installed in stilling wells. Most studies of surf zone dynamics have made use of mean values of the surface elevation measured from wave staffs or pressure transducers, though the accuracy of these measurements is on the order of +/− several centimeters. More precise measurements that can be used for investigating detailed mechanisms of nearshore circulation can be obtained through the use of manometer tubes deployed into the surf zone from the shore (Nielsen and Dunn, 1998).

Waves

Field measurements of waves in the inner nearshore and surf zones can be obtained by some form of surface piercing wave staff, which has the advantage of providing a direct measure of the wave form. These systems make use of the conduction of electricity by water, particularly seawater, and record either the change in electrical resistance or capacitance of the system as water rises and falls over a length of uninsulated cable which is part of an electrical circuit (Ribe and Russin, 1974; Timpy and Ludwick, 1985). The change in resistance or capacitance can be conditioned to produce a variation in an output signal which may be a DC current or a frequency. The sensor itself may be fixed to a support jetted into the sand in the nearshore (see Figure B13) or attached to some physical structure such as a jetty or platform. The advantage of the surface staff is that it provides a direct measure of the water surface form and they have been used extensively in many studies, particularly in fetch limited areas where installation can be accomplished during calm conditions (Davidson-Arnott and Randall, 1984; Greenwood and Sherman, 1984).

The disadvantage of wave staffs is that they are subject to high wave forces when deployed in shallow water and they have largely been replaced with some form of pressure transducer housed in a watertight case. These can be deployed some distance below the surface, or on the bed and they are often colocated with other sensors such as electromagnetic current meters and nepelometers (see below). They sense the change in pressure associated with the passage of individual waves. The pressure variation with depth can be predicted from wave theory and thus it is possible to develop a transform function that will relate the recorded variations in pressure to the surface wave form (Lee and Wang, 1984). Since there is usually a spectrum of frequencies present in the pressure transducer record, the transform should be performed for all of the frequencies present. This is not a trivial task, though it can be done routinely in a data analysis program. There is some loss of information on the true form of the surface wave as well as the loss of the higher frequencies with increasing depth of deployment, but this is offset in studies in and close to the breaker and surf zones by the ease of deployment and the reduced exposure to breaking wave forces.

The water surface can also be measured remotely using a video camera to measure the change in surface elevation against a graduated pole or screen. This gives a good measure of the wave form without interference and it enables determination of whether the wave is broken or not. The record can be digitized manually or a computer software algorithm can be used to extract the position of the surface automatically. Video cameras have also been used extensively to extract data on run-up frequencies on the beach (Holman and Sallenger, 1985). Recent developments in LIDAR technology may also permit application to measuring waves (Irish *et al.*, 2001).

Individual wave staffs or pressure transducers provide a picture of the variations in water surface elevation through time—that is, they give

Figure B13 Resistance wave staffs (left) deployed over an intertidal ridge and runnel, Nova Scotia, Canada. A grid of depth of disturbance rods is deployed across the bar (center) and frames supporting OBS and electromagnetic current meters can be seen in the far right.

information on wave height and period but not on the direction of travel. This requires either the deployment of several instruments in an array, which permits determination of the wave direction through a comparison of travel time between various sensors (Bodge and Dean, 1984; Howell, 1992), or the measurement of both the horizontal and vertical components of water motion using a pressure transducer and bidirectional electromagnetic current meter or an acoustic doppler current meter (see following section).

Because of the rapid oscillatory motion associated with wave action in shallow water, mechanical current meters are generally not useful in studies of fluid processes in the inner nearshore and surf zone, though miniature-ducted impellor current meters have proved useful in some locations (Wright *et al.*, 1982; Masselink and Hegge, 1995). Measurements of hydrodynamics in the nearshore and surf zone were revolutionized in the 1970s by the development of electromagnetic current meters (EMCMs) and they have been used extensively in almost all field experiments as well as in the laboratory (Huntley and Bowen, 1975; Cushing, 1976). Examples of their use can be found in numerous experiments including the Nearshore Sediment Transport Study (NSTS, Seymour, 1989), the Canadian Coastal Sediment Study (C^2 S^2; Willis, 1987) and in the various experiments carried out at the CERC at Duck, North Carolina (Birkemeier *et al.*, 1997). The current meters produce a fluctuating magnetic field around the sensor head and measure the voltage generated by fluid flow in the field using Farraday's law. Typically, the instruments have four sensors mounted orthogonally so as to detect flow along two orthogonal axes. The current meter is usually mounted so as to detect horizontal flow, but it is possible to mount it with one axis vertically. The sensors have a fast response time, permitting sampling at frequencies >5 Hz and are able to detect very small mean flows in a highly fluctuating environment (see Figure B14). The majority of EMCMs used in the field have been made by Marsh-McBirney Inc. of Maryland, USA. Large models have a 10.5 cm diameter head and the small ones, which have been used extensively in the surf zone have a 4 cm head. These current meters have been evaluated extensively (Aubrey and Trowbridge, 1985, 1988; Guza, 1988) and their widespread use allows for ease of comparison between different studies. While EMCMs are now being replaced by various forms of acoustical instrument, they are still useful in the breaker and surf zone because of their smaller sensitivity to the presence of air bubbles.

In the past decade, various forms of acoustical doppler instruments have been developed which have been used in laboratory and field experiments. Essentially, they emit an acoustic signal which is reflected by fine material in the water column from a focal point and received by orthogonally mounted transducers. The relative motion in each axis is then determined by the doppler shift of the signal. The acoustic doppler velocimeter (ADV) is the simplest of the instruments and measures velocity at a single point on the order of a few centimeters from the emitting transducer.

Sediment concentration, mass transport rate, and deposition

Obtaining measurements of sediment transport is the third key element in morphodynamic experiments in the coastal zone. There are a large number of instruments available and a variety of approaches have been taken, but much work remains to be done to obtain reliable measurements over a reasonable spatial and temporal scale (White, 1998). Estimates of net longshore transport over periods of weeks, months, or years at a location can be obtained from measurements of the amount trapped at a total barrier, either over a short period at a purpose-built groin (Wang and Kraus, 1999) or at a large jetty. A number of studies have also used fluorescent or radioactive tracers for measurement of longshore sediment transport or of transport pathways within the surf zone. However, the focus here is on instrumentation for instantaneous measurement of the transport rate either directly, or indirectly through the combination of measurements of sediment concentration and net water motion. Direct measurement techniques include various traps and acoustic doppler instruments that measure concentration and velocity simultaneously through some portion of the water column. Indirect techniques for measuring concentration include optical and conductivity devices.

Figure B14 Electromagnetic current meter (left) and three OBS probes mounted on a "goalpost" in the intertidal zone, Skallingen, Denmark. Electronics for the current meter are installed in the waterproof housing secured to a pole jetted into the sand to the left of the goalpost. A profile line of large depth of disturbance rods is visible at the right.

Traps and pumps

A number of devices have been used in attempts to trap sediment suspended in the water column of the net transport in the swash or surf zone, but the oscillatory motion associated with wave action makes this task much more difficult than, for example, under unidirectional flow in a river. Pump samplers have been used with varying degrees of success and various bottles for capturing the suspended sediment load. These all require considerable effort and the logistical difficulties coupled with doubts as to the accuracy of the sampling process have limited their further use. Recently, arrays of streamer traps consisting of long bags of fine mesh fixed to a rigid rectangular inlet have been used to measure transport where there is a net current present. These traps are able to capture large amounts of sediment, but it is still not clear that they can provide reliable estimates of the net transport or that they provide accurate results over a range of conditions.

Optical devices

Optical devices emit light and give a measure of the sediment concentration in the water column at a point either through the degree of attenuation of the light beam or through the amount of light reflected from particles in suspension. They therefore do not provide a direct measure of sediment transport and thus must be collocated with a device such as an EMCM which measures the fluid flow. The light transmitted is generally in a narrow wave band in the infrared range in order to minimize the effects of natural light in the water.

Transmissometers measure the degree of attenuation of the light over a fixed distance separating the emitter from the receiver. They tend to be relatively bulky instruments best suited for work some distance seaward of the breaker zone in depths greater than 10 m where suspended sediment concentrations are relatively low. A Sea Tech transmissometer with a 5 cm path length was developed for use in shallow water (Huntley, 1983) but the much smaller probes associated with instruments measuring reflected light proved more suitable for the inner nearshore and surf zones.

Much of our understanding of the dynamics of suspended sediment transport in the nearshore over the past three decades has come from the use of the optical backscatterance sensor (OBS) originally developed at the University of Washington (Downing *et al.*, 1981) and now produced commercially by D & A instruments. The OBS is a miniature nephelometer which measures the backscatterance of light by sediments suspended in the fluid. It utilizes a narrow infrared beam which has the advantage of minimizing interference by sunlight and confining the sampling volume to a short distance from the probe. The sensor is compact (2.1 cm diameter) with transmitter and receiver mounted next to each other at the end of the probe, thus minimizing flow interference and enabling sensors to be mounted in close proximity to other probes or to electromagnetic current meters with which they are often collocated (see Figures B14, B15(a)). The OBS probe is very rugged, enabling it to be deployed in areas of strong currents and breaking wave impacts, and it is clearly superior to other nephelometers for work in the nearshore marine environment (Greenwood *et al.*, 1990). They are designed to measure suspended sediment concentrations in areas where concentrations may be high and/or may vary rapidly over short time periods (i.e., on the order of 0.25 Hz). They have been used in a wide range of marine environments, including the shoreface and continental shelf (Wright *et al.*, 1991; Kineke and Sternberg, 1992), the breaker and surf zones (Black and Rosenberg, 1994), and estuaries (Kineke *et al.*, 1991).

Provided they are not deployed too close to the bed (interference with the bed itself) or too close to the surface (effects of ambient light) both transmissometers and OBS probes work well. They are linear over a wide range of grain sizes from clay to sand and there has been extensive testing and calibration of both types of instruments in the laboratory and field (Downing and Beach, 1989; Osborne *et al.*, 1993; Greenwood and Jagger, 1995; Bunt *et al.*, 1999; Sutherland *et al.*, 2000). However, they perform best with a narrow range of grain size and calibration, where there is a wide range of grain size they are subject to considerable error (Bunt *et al.*, 1999).

Routine field calibration of the instruments is difficult and laboratory calibration requires quite complex facilities and involves the difficulties of obtaining representative samples of suspended sediment to be returned to the lab for testing. Recent testing of a laser *in situ* scattering and transmissometry (LISST) instrument produced by Sequoia Scientific (Traykovski *et al.*, 1999; Gartner *et al.*, 2001; Mikkelsen and Pejrup, 2001) offers the potential to measure the complete particle size distribution and concentration simultaneously. Initially, this may offer a means of calibrating cheaper, less bulky sensors but further developments may lead to smaller versions which could be deployed close to the bed.

Acoustic Doppler velocity profilers

Acoustical instruments offer the possibility of measuring both particle concentration and velocity simultaneously and over some appreciable portion of the water column, and thus providing a direct measure of the transport rate. Acoustic Doppler velocity profilers (ADVP) use the same basic technology as the ADV described above. However, they measure the return signal in very small increments of time, thus allowing the determination of velocity and concentration in discrete "bins." They can be used from a boat with position fixed by DGPS and can thus give a complete picture of flow over bedforms in estuaries and tidal channels (Best *et al.*, 2001).

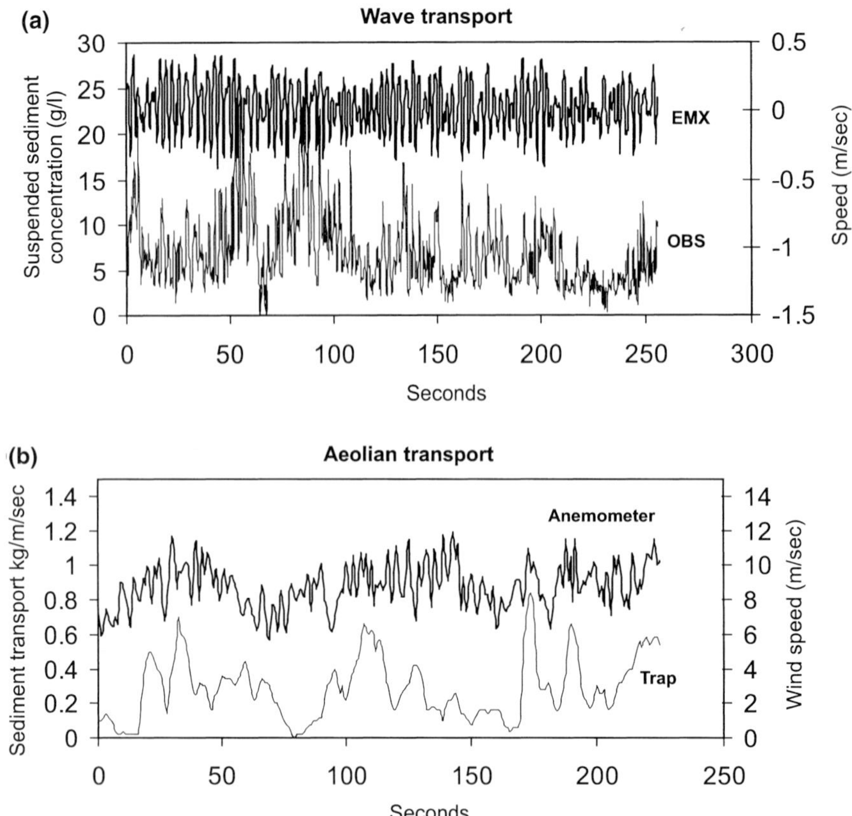

Figure B15 Comparison of flow and sediment transport in the nearshore and on the beach. (a) On-offshore flow speed measured with an electromagnetic current meter and suspended sediment concentration measured with an OBS probe at an elevation of 0.15 m above the bed on the crest of an intertidal sand bar (see Figure B13). The instruments were sampled at 4 Hz. (b) Horizontal wind speed measured with a cup anemometer and sediment transport rate measured with a wedge trap fitted with an electronic balance. The instruments were sampled at 1 Hz and smoothed with a 5 s running mean.

Aeolian sand transport

The design of field instrumentation for measuring sand transport by wind has tended to lag behind that available for measuring transport in the water. Sediment transport from the beach over a period of days or weeks can be measured indirectly by measuring accumulation in vegetated sand dunes by profiling or by the use of a bedframe device (Davidson-Arnott and Law, 1990, 1996).

Most direct measurements of sediment transport have been made with simple vertical traps which are oriented into the wind and which allow the sand captured to collect in the base. The sediment collected over a period of time is weighed to give an average transport rate for the collection period. A major problem is to design the trap so that it is isokinetic, otherwise sand in transport is diverted away from the trap opening by the pressure buildup. Simple vertical traps, which have been used widely may have an efficiency <30%. Wedge-shaped traps have improved aerodynamics and are likely much closer to isokinetic (Nickling and McKenna-Neuman, 1997—see Figure B12). However, these traps are more sensitive to changes in wind direction and will undersample when the wind angle exceeds 5°; thus they can only be used for periods of 15–30 min without attention. A variety of other trap designs are available (Goossen *et al.*, 2000) but all have a number of problems with accuracy.

Horizontal traps offer the opportunity to sample all of the transport load and provide a means for calibrating vertical traps. They require a large pit several meters across and again will integrate total transport over periods of tens of minutes to hours (Greely *et al.*, 1996). Use of a wet horizontal trap can reduce some of the logistics (Wang and Kraus, 1999) because the trap need only be a few centimeters deep.

A number of trap designs are now being used to obtain measurements of the instantaneous mass transport rate, thus permitting comparison of the transport rate with measurements of the wind flow. The trap design of Nickling and McKenna-Neuman has been modified to incorporate a continuous weighing electronic balance (Mckenna-Neuman *et al.*, 2000; see Figure B15(b)). Bauer and Namikas (1998) used the same trap but designed a combination tipping bucket and strain gage to weigh the sand collected over long time periods. The design of Jackson (1996) uses a similar weighing mechanism to that used by Bauer and Namikas, but the trap itself is a circular collection funnel that is mounted flush with the surface. This avoids the problem of isokinetic sampling associated with vertical traps and has the advantage of omni directional collection. However, it measures flux to the surface rather than the total transport rate.

Impact measurement

The drawbacks of trap designs and the need for high speed, continuous sampling of the transport rate have led to the development of several instruments that measure the impact of saltating grains and then attempt to calibrate this to the transport rate. The saltiphone (Arens, 1996) uses a microphone to record impacts and the intensity is then recorded as a voltage signal. While this gives a measure of the relative transport rate, it has proved difficult to calibrate and is sensitive to variations in grain size. The SENSIT (Stockton and Gillette, 1990) responds to the impact of saltating grains on a piezoelectric crystal and counts the number of impacts per second. It has also proved difficult to calibrate to give a measure of the mass transport rate and seems to offer only a relative measure of the grains in saltation. Because of the small sampling area a vertical array of the sensors must be deployed in order to measure the total mass transport rate.

Robin Davidson-Arnott

Bibliography

Alport, M.J., Basson, J., Mocke, G., Naicker, J., and Saltau, C., 2001. Discrimination and analysis of video imaged shorelines and nearshore processes. In *Proceedings Coastal Dynamics '01*. American Society of Civil Engineers, New York, pp. 989–997.

Amos, C.L., Daborn, G.R., Christian, H.A., Atkinson, A., and Robertson, A., 1992. In situ erosion measurements on fine-grained sediments from the Bay of Fundy. *Marine Geology*, **108**: 175–196.

Arens, B., 1996. Rates of aeolian sand transport on a beach in a humid temperate climate. *Geomorphology*, **7**: 3–18

Askin, R.W., and Davidson-Arnott, R.G.D., 1981. Micro erosion meter modified for use underwater. *Marine Geology*, **40**: M45–M48.

Atakturk, S.S., and Katsaros, K.B., 1989. The K-Gill: a twin propellor-vane anemometer for measurements of atmospheric turbulence. *Journal of Atmospheric and Oceanic Technology*, **6**: 509–515.

Aubrey, D.G., and Trowbridge, J.H., 1985. Kinematic and dynamic estimates from electromagnetic current meter data. *Journal of Geophysical Research*, **90**(C5): 9137–9146.

Aubrey, D.G., and Trowbridge, J.H., 1988. Reply (to comments by Guza, 1988). *Journal of Geophysical Research*, **93**(C2): 1344–1346.

Bauer, B.O., and Namikas, S.L., 1998. Design and field test of a continuously weighing tipping-bucket assembly for aeolian sand traps. *Earth Surface Processes and Landforms*, **23**: 1171–1183.

Best, J.L., Kostaschuk, R.A., and Villard, P., 2001. Quantitative visualization of flow fields associated with alluvial sand dunes. *Journal of Flow Visualization*, **4**: 373–381.

Birkemeier, W.A., Long, C.E., and Hathaway, K.K., 1997. DELILAH, DUCK94 and Sandy Duck: three nearshore field experiments. In *Proceedings Coastal Engineering 1996*. American Society of Civil Engineers, New York, pp. 4052–4065.

Black, K.P., and Rosenberg, M.A., 1994. Suspended sand measurements in a turbulent environment: field comparison of optical and pump sampling techniques. *Coastal Engineering*, **24**: 137–150.

Bodge, K.R., and Dean, R.G., 1984. Wave measurement with differential pressure gages. In *Proceedings Coastal Engineering 1984*. American Society of Civil Engineers, pp. 755–769.

Bunt, J.A.C., Larcombe, P., and Jago, C.F., 1999. Quantifying the response of optical backscatter devices and transmissometers to variations in suspended particulate matter. *Continental Shelf Research*, **19**: 1199–1220.

Cahoon, D.R., French, J.R., Spencer, T., Reed, D., and Moller, I., 2000. Vertical accretion versus elevational adjustment in UK saltmarshes: an evaluation of alternative methodologies. In Pye, K., and Allen, J.R.L. (eds.), *Coastal and Estuarine Environments. Journal of the Geological Society*, Special Publication **175**: 223–238.

Chandler, J., 1999. Effective application of automated digital photogrammetry for geomorphological research. *Earth Surface Processes and Landforms*, **24**: 51–63.

Cushing, V., 1976. Electromagnetic water current meter. In *Proceedings of Oceans '76. Proceedings Coastal Engineering 1992*. Institute of Electrical and Electronic Engineers, pp. 298–301.

Davidson-Arnott, R.G.D., and Langham, D.R.J., 2000. The effects of softening on nearshore erosion of a cohesive shoreline. *Marine Geology*, **166**: 145–162.

Davidson-Arnott, R.G.D., and Law, M.N., 1990. Seasonal patterns and controls on sediment supply to coastal foredunes, Long Point, Lake Erie. In Nordstrom, K.F., Psuty, N.P., and Carter, R.W.G. (eds.), *Coastal Dunes: Form and Process*. Chichester: John Wiley & Sons, pp. 177–200.

Davidson-Arnott, R.G.D., and Law, M.N., 1996. Measurement and prediction of long-term sediment supply to coastal foredunes. *Journal of Coastal Research*, **12**: 654–663.

Davidson-Arnott, R.G.D., and Randall, D.C., 1984. Spatial and temporal variations in spectra of storm waves across a barred nearshore. *Marine Geology*, **60**: 15–30.

Downing, J.P., and Beach, R.A., 1989. Laboratory apparatus for calibrating optical suspended solids sensors. *Marine Geology*, **86**: 243–249.

Downing, J.P., Sternberg, R.W., and Lister, C.R.B., 1981. New instrumentation for the investigation of sediment suspension processes in the shallow marine environment. *Marine Geology*, **42**: 19–34.

Gartner, J.W., Cheng, R.T., Wang, P.-F., and Richter, K., 2001. Laboratory and field evaluation of the LISST-100 instrument for suspended particle size determinations. *Marine Geology*, **175**: 199–219.

Goossens, D., Offer, Z., and London, G., 2000. Wind tunnel and field calibration of five aeolian sand traps. *Geomorphology*, **35**: 233–252.

Gorman, L., Morang, A., and Larson, R., 1998. Monitoring the coastal environment; Part IV: mapping, shoreline changes, and bathymetric analysis. *Journal of Coastal Research*, **14**: 61–92.

Greeley, R., Blumberg, D.G., and Williams, S.H. 1996. Field measurements of the flux and speed of wind-blown sand. *Sedimentology*, **43**: 41–52.

Greenwood, B., and Jagger, K., 1995. Sensitivity of optical sensors to grain size variations in the sand mode: implications for transport measurements. In *Proceedings Canadian Coastal Conference*. Canadian Coastal Science and Engineering Association, pp. 383–398.

Greenwood, B., and Mittler, P.R., 1984. Sediment flux and equilibrium slopes in a barred nearshore. *Marine Geology*, **60**: 79–98.

Greenwood, B., and Sherman, D.J., 1984. Waves, currents sediment flux and morphological response in a barred nearshore. *Marine Geology*, **60**: 31–61.

Greenwood, B., Hale, P.B., and Mittler, P.R., 1979. Sediment flux determination in the nearshore zone. In *Proceedings Workshop on Instrumentation for Currents and Sediments in the Nearshore Zone*. National Research Council of Canada, pp. 99–115.

Greenwood, B., Osborne, P.D., Bowen, A.J., Hazen, D.G., and Hay, A.E., 1990. C-COAST: The Canadian Coastal Sediment Transport Programme—suspended sediment transport in the nearshore zone. In *Proceedings Canadian Coastal Conference*. National Research Council of Canada, pp. 319–336.

Greenwood, B., Richards, R.G., and Brander, R.W., 1993. Acoustic imaging of sea-bed geometry: a high resolution remote tracking sonar (HERTSII). *Marine Geology*, **112**: 207–218.

Guza, R.T., 1988. Comments on "Kinematic and dynamic estimates from electromagnetic current meter data" by D.G. Aubrey and J.H. Trowbridge. *Journal of Geophysical Research*, **93**(C2): 1337–1343.

Hancock, G., and Willgoose, G., 2001. The production of digital elevation models for experimental model landscapes. *Earth Surface Processes and Landforms*, **26**: 475–490.

Holman, R.A., and Sallenger, A.H., Jr., 1985. Set up and swash on a natural beach. *Journal of Geophysical Research*, **90**: 945–953.

Houwing, E.-J., 1999. Determination of the critical erosion threshold of cohesive sediments on intertidal mudflats along the Dutch Wadden Sea coast. *Estuarine, Coastal and Shelf Science*, **49**: 345–355.

Howell, G.L., 1992. A new nearshore directional wave gage. In *Proceedings Coastal Engineering 1992*. American Society of Civil Engineers, pp. 295–307.

Huntley, D.A., 1983. In situ sediment monitoring techniques: a survey of the state of the art in the USA. In *Proceedings Canadian Coastal Conference*. National Research Council of Canada, pp. 151–165.

Huntley, D.A., and Bowen, A.J., 1975. Comparison of the hydrodynamics of steep and shallow beaches. In Hails, J., and Carr, A. (eds.), *Nearshore Sediment Dynamics and Sedimentation*. London: John Wiley & Sons, pp. 69–110.

Irish, J.L., and White, T.E., 1998. Coastal engineering applications of high resolution lidar bathymetry. *Coastal Engineering*, **35**: 47–71.

Irish, J.L., Wozencraft, J.M., and Cunningham, A.G., 2001. Water wave measurement with lidar from a fixed platform. In *Proceedings Coastal Dynamics '01*. American Society of Civil Engineers, pp. 998–1006.

Irwin, H.P.A.H., 1980. A simple omnidirectional sensor for wind-tunnel studies of pedestrian level winds. *Journal of Wind Engineering and Industrial Aerodynamics*, **7**: 219–239.

Jackson, D.W.T., 1996. A new, instantaneous aeolian sand trap design for field use. *Sedimentology*, **43**: 791–796.

Kineke, G.C., and Sternberg, R.W., 1992. Measurements of high concentration suspended sediments using the optical backscatterance sensor. *Marine Geology*, **108**: 253–258.

Kineke, G.C., Sternberg, R.W., Cacchione, D.A., Krank, K., and Drake, D.E, 1991. Distribution and characteristics of suspended sediment on the Amazon shelf. *Oceanography*, **4**: 21–26.

Kirk, R.M., 1977. Rates and forms of erosion on intertidal platforms at Kaikoura Peninsula, South Island, New Zealand. *New Zealand Journal of Geology and Geophysics*, **20**: 571–613.

Konicki, K.M., and Holman, R.A., 2000. The statistics and kinematics of transverse sand bars on an open coast. *Marine Geology*, **169**: 69–101.

Lawler, D.M., 1992. Design and installation of a novel automatic erosion monitoring system. *Earth Surface Processes and Landforms*, **17**: 455–463.

Leatherman, S.P., 1978. A new aeolian sand trap design. Sedimentology, **25**: 303–306.

Lee, D.-Y., and Wang, H., 1984. Measurement of surface waves from subsurface gage. In *Proceedings Coastal Engineering*. American Society of Civil Engineers, pp. 271–286.

Maa, J.P.-Y., Wright, L.D., Lee, C.-H., and Shannon, T.W., 1993. VIMS sea carousel: a field instrument for studying sediment transport. *Marine Geology*, **115**: 271–287.

Masselink, G., and Hegge, B., 1995. Morphodynamics of meso- and macrotidal beaches: examples from central Queensland, Australia. *Marine Geology*, **129**: 1–23.

McKenna Neuman, C., Lancaster, N., and Nickling, W.G., 2000. Effect of unsteady winds on sediment transport intermittency along the stoss slope of a reversing dune. *Sedimentology*, **47**: 211–226.

Mikkelsen, O.A., and Pejrup, M., 2001. The use of a LISST-100 laser particle sizer for in-situ estimates of floc size, density and settling velocity. *Geo-Marine Letters*, **20**: 187–195.

Morang, A., Larson, R., and Gorman, L., 1997a. Monitoring the coastal environment; Part 1: waves and currents. *Journal of Coastal Research*, **13**: 111–133.

Morang, A., Larson, R., and Gorman, L., 1997. Monitoring the coastal environment; Part 111: geophysical and research methods. *Journal of Coastal Research*, **13**: 1064–1085.
Morris, B.A., Davidson, M.A., and Huntley, D.A., 2001. Measurements of the response of a coastal inlet using video moniotoring techniques. *Marine Geology*, **175**: 251–272.
Nickling, W.G., and McKenna Neuman, C., 1997. Wind tunnel evaluation of a wedge-shaped aeolian transport trap. *Geomorphology*, **18**: 333–345.
Nielsen, P., and Dunn, S.L., 1998. Manometer tubes for coastal hydrodynamics investigations. *Coastal Engineering*, **35**: 73–84.
Osborne, P.D., Vincent, C.E., and Greenwood, B., 1993. Measurement of suspended sediment concentrations in the nearshore: intercomparison of optical and acoustic backscatter sensors. *Continental Shelf Research*, **14**: 159–174.
Plant, N.G., Holman, R.A., Freilich, M.H., and Birkemeir, W.A., 1999. A simple model for interannual bar behaviour. *Journal of Geophysical Research*, **104**(C7): 15,755–15,776.
Ribe, R.L., and Russin, E.M., 1974. Ocean wave measuring instrumentation. In *Proceedings International Symposium on Ocean Wave Measurement and Analysis*. American Society of Civil Engineers, pp. 396–416.
Ridd, P.V., 1992. A sediment level sensor for erosion and siltation detection. *Estuarine, Coastal and Shelf Science*, **35**: 355–362.
Ruessink, B.G., van Enckvort, I.M.J., Kingston, K.S., and Davidson, M.A., 2000. Analysis of two- and three-dimensional nearshore bar behaviour. *Marine Geology*, **169**: 161–183.
Sallenger, A.H., Krabill, W., Swift, R., and Brock, J., 2001. Quantifying hurricane-induced coastal changes using topographic lidar. In *Proceedings Coastal Dynamics '01*. American Society of Civil Engineers, pp. 1007–1018.
Seymour, R.J. (ed.), 1989. Nearshore Sediment Transport. New York: Plenum.
Stockton, P., and Gillette, D.A., 1990. Field measurements of the sheltering effect of vegetation on erodible land surfaces. *Land Degradation and Rehabilitation*, **2**: 77–85.
Sunamura, T., 1992. *The Geomorphology of Rocky Coasts*. Chichester: John Wiley & Sons.
Sutherland, T.F., Lane, P.M., Amos, C.L., and Downing, J., 2000. The calibration of optical backscatter sensors for suspended sediment of varying darkness levels. *Marine Geology*, **162**: 587–597.
Timpy, D.L., and Ludwick, J.C., 1985. Bore height measurement with improved wave staff. *Journal of Waterways, Port, Coastal and Ocean Engineering*, **111**: 495–510.
Tolhurst, T.J., Black, K.S., Shayler, S.A., Mather, S., Black, I., Baker, K., and Paterson, D.M., 1999. Measuring the *in situ* erosion shear stress of intertidal sediments with the cohesive strength M (CSM). *Estuarine, Coastal and Shelf Science*, **49**: 281–294.
Traykovski, P., Latter, R.J., and Irish, J.D., 1999. A laboratory evaluation of the laser in situ scattering and transmissometry instrument using natural sediments. *Marine Geology*, **159**: 355–367.
Trudgill, S.T., High, C.J., and Hanna, F.K., 1981. Improvements to the micro erosion meter. British Geomorphology Research Group, *Technical Bulletin* **29**, p. 17.
Viles, H.A., and Trudgill, S.T., 1984. Long term remeasurements of micro-erosion meter rates, Aldabra Atoll, Indian Ocean. *Earth Surface Processes and Landforms*, **9**: 89–94.
Wang, P., and Kraus, N., 1999. Horizontal water trap for measurement of aeolian sand transport. *Earth Surface Processes and Landforms*, **24**: 65–70.
White, T.E., 1998. Status of measurement techniques for coastal sediment transport. *Coastal Engineering*, **35**: 17–45.
Willis, D.H., 1987. The Canadian coastal sediment study: an overview. In *Coastal Sediments '87*. American Society of Civil Engineers, pp. 682–693.
Wright, L.D., Nielsen, P., Short, A.D., and Green, M.O., 1982. Morphodynamics of a macrotidal beach. *Marine Geology*, **50**: 97–128.
Wright, L.D., Boon, J.D., Kim, S.C., and List, J.H., 1991. Modes of cross-shore sediment transport on the shorefaceof the Middle Atlantic Bight. *Marine Geology*, **96**: 19–51.

Cross-references

Airborne Laser Terrain Mapping
Erosion Processes
Geographic Information Systems
Global Positioning Systems
Monitoring, Coastal Geomorphology
Muddy Coasts
Photogrammetry
Sandy Coasts

BEACH CUSPS—See RHYTHMIC PATTERNS

BEACH DRAIN

Introduction

For over half a century, reporters have suggested a link between the elevation of beach groundwater and erosional or accretional trends of the beach face. Beach dewatering (the artificial lowering of the water table within beaches by a system of drains and pumps) is suggested by its proponents as a practical alternative to more traditional methods of coast-stabilization. Within the last 15–20 years several tests have been installed, and to date seven to eight commercial dewatering systems have operated. The following is a review of the origins and development of the dewatering concept from early work on beach face permeability and beach groundwater dynamics, to recent field and laboratory studies that have explicitly examined the effect of artificial groundwater manipulation on beach face accretion and erosion.

The origin of the beach drain

The beach drain (Figure B16) is not a new concept, but was revived in the last 20 years due to commercial interests (Turner and Leatherman, 1997).

The origins of the beach drain concept can be traced back 50 years to early work in two parallel fields of coastal research: the role of beach face permeability in controlling erosion or accretion (e.g., Bagnold, 1940); and the tidal dynamics of beach groundwater (e.g., Grant, 1948). The installation within the last 10 years of prototype beach dewatering systems in Europe (Vesterby, 1994) and the United States (Lenz, 1994) signified the transition of the beach dewatering concept from the hypothetical to the practical. The potential use of beach drain technology is beginning to be noted within the mainstream coastal engineering community (e.g., Abbott and Price, 1994, pp. 334–336), and in the last five years a limited number of journal articles (e.g., Weisman *et al.*, 1995; Li *et al.*, 1995) and more frequent papers presented at the coastal engineering conferences (e.g., Davis and Hanslow, 1991; Ogden and Weisman, 1991; Davis *et al.*, 1992, 1993; Oh and Dean, 1994) have served to raise the awareness of the beach dewatering concept. However, if beach dewatering technology is to meet the promise that its proponents claim, the answers to a number of fundamental questions must be addressed.

Counter to the impression that may be gained from publications and other materials produced by commercial players in the beach dewatering industry, the underlying physical mechanisms that may contribute to the success of the beach drain concept are *not* yet fully elucidated.

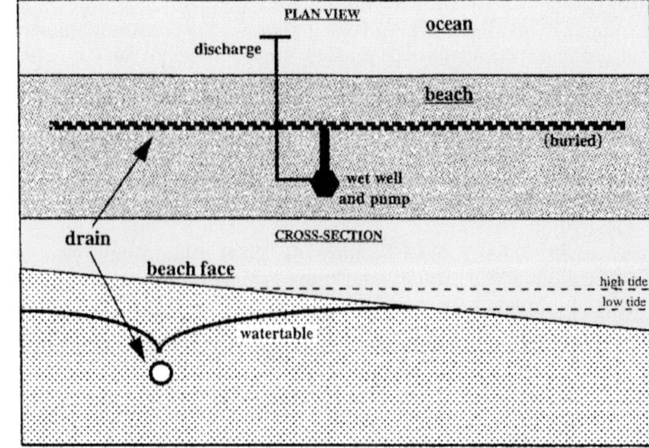

Figure B16 Schematic diagram of a beach dewatering system. Length of the system may vary from a few hundred meters to several hundred meters (from Turner and Leatherman, 1997, reprinted by permission of the Journal of Coastal Research).

Dewatering is a well-established practice in the excavation industry, but the inclusion of a highly dynamic land–ocean boundary where sediment motion is a function of both static inter-granular forces, and surf and swash zone hydrodynamics, makes the description of transport mechanisms across a dewatered beach face unique. On the more practical side to many coastal scientists and engineers, the field evidence from operating dewatered sites remains inconclusive. A comprehensive and independent assessment of the mid-to-long-term operation of a prototype installation is yet to be reported in the scientific literature, and until such a study is completed it is unlikely that the prevailing mood of healthy skepticism (e.g., Bruun, 1989) will be either validated or changed.

History and development

Bagnold's seminal laboratory investigations (Emery and Foster, 1948) undertook the first published study describing the dynamics of the water table in sandy beaches. They referred to prior unpublished work of Zinn (1942).

Several laboratory and field tests are described by Turner and Leatherman (1997) including Bagnold (1940), and Emery and Foster (1948). Ogden and Weisman (1991) undertook two-dimensional (2-D) tests using irregular waves ranging from erosive to accretive and concluded that for the range of conditions tested, the beach drain had no significant effect on the rate of erosion or accretion at the still water line, but did promote berm development and hence overall beach face steepening. A more recent study by the same researchers (Weisman *et al.*, 1995) examined the effectiveness of beach dewatering under the influence of the tides, and concluded that water table lowering maintains its effectiveness in promoting berm growth and beach face steepening for both tidal and nontidal cases. Heaton (1992) undertook a series of single and multiple wave experiments, and quantified a general trend that increasing water table elevation resulted in an increasing volume of sediment eroded from the beach face. Oh and Dean (1994) reported a set of three experiments where the water table was alternatively elevated, lowered, and equal to mean sea level, and concluded that an elevated water table resulted in the overall destabilization and erosion of previously marginally stable regions of the beach face. A simple seepage model (Oh and Dean, 1994) demonstrated that outflow across the beach face may act to reduce the effective weight (and hence stability) of surficial sediment.

Considerable activity has taken place in Australia. Davis *et al.* (1992, 1993) made field tests and found that for fair conditions the drain increases beach stability while storm conditions had the opposite effects. Nielsen (1990, 1992) did considerable testing including the infiltration effects on sediment mobility under recorded stabilizing as well as destabilizing forces and their relation to fluidization. Nielson's latest (2001) is an attempt to determine infiltration on effects on sediment mobility. Many field tests are described by Turner and Leatherman (1997) providing results or conclusions from each separate test programs.

First full-scale test—Thorsminde, Denmark

A test site at Hirtshals was not considered a success by the Danish Geotechnical Institute and was subsequently dismantled, but the results from the Hirtshals West site were deemed encouraging, and it was decided to undertake the first large-scale test of the dewatering concept at Thorsminde on the west coast of Denmark. Hansen (1986) provides details of the beach and installation, which are summarized to varying degrees by Ovesen and Schuldt (1992) and Vesterby (1991, 1994). The test site is located on the exposed North Sea coast, where the shoreline fluctuated seasonally by ± 15 m with a reported average erosion rate of 2–4 m/year.

The conclusion after year of operation was:

1. The usual seasonal fluctuation in shoreline position was halted and net recession ceased.
2. The southern drained region prograded seaward approximately 10 m and stabilized at a distance of 20–25 m in front of the drain line, while the northern drained region, after an initial period of recession, also stabilized at a distance of 20–25 m in front of the drain.
3. End effects appeared to extend the effective drain length by 100–200 m, particularly on the southern down-drift side of the dewatering system.

Continued tests with independent observers were not very successful. The report by the Coastal Directorate (Bruun, 1989) has the following conclusion:

1. Under mild wave conditions the coastal drain system stabilizes beach profiles and provides a wider, higher high-tide beach. The coastal drain system is useful under certain specific conditions as described.
2. The coastal drain does not stop beach or dune erosion during storms. It is in no way a substitute for artificial nourishment. Its effectiveness on an eroding shore will decrease with time.

First installation in the USA—Sailfish Point, Florida

In 1988, Coastal Stabilization, Inc. (a subsidiary to Moretrench American Corporation) installed a 180 m-long-beach dewatering system at Sailfish Point, near the southern end of Hutchinson Island, on the Atlantic coast of Florida, USA. The beach is composed of fine-grained, well-sorted sand; the most notable feature along this otherwise open Atlantic coast is the natural coastline protection provided by a rock reef located approximately 100–150 m offshore. It has been suggested that despite the presence of the reef, between 1972 and 1986 recession of the high-tide shoreline exceeded 2 m/year. It is important to note that this erosional trend is reported to have reversed and become accretionary prior to the installation of the beach drains in 1988 (Terchunian, 1989; Dean, 1989).

The dewatering system installed at Sailfish Point (referred to as "Stabeach" by Coastal Stabilization, Inc.) consisted of a 0.3–0.5 m diameter PVC pipe buried at an elevation of approximately -2.5 m, providing a collection drain for numerous 1.5 m long horizontal well points attached at approximately 3 m intervals along its length. Collected water traveled via a suction pipe to a pumping station located landward of the dune line (Lenz, 1994). An independent report prepared for Coastal Stabilization, Inc., by Dean (1989) after 11 months of monitoring concluded that it was not possible to separate natural beach changes from those induced by the dewatering system; but a second report by the same author (Dean, 1990) after approximately 20 months of operation provided the first independent evidence that the dewatering system was having a positive effect on the beach. From a straightforward analysis of time series of sand volumes and the position of the high-tide shoreline, Dean concluded that, while it remained difficult to separate natural beach changes and those caused by water table lowering:

1. The dewatering system appeared to have resulted in local moderate accretion, in contrast to a general erosional trend to the north and a relatively small accretionary trend to the south.
2. The system appeared to result in a considerably more stable high-tide shoreline relative to both control segments north and south.

Recent installations

Some beach drains have been installed in the United Kingdom, United States, and Denmark. The results, however, were generally nonconclusive (Turner and Leatherman, 1997).

Conclusions

This brief report provides an overview of the history and current status of beach dewatering as a potential practical alternative to more traditional methods of coastal stabilization. The specific findings are as follows:

1. A link between the elevation of coastal groundwater and erosion or accretion trends at the shore has been reported in the coastal literature for over 50 years. The origins of this work can be traced to parallel but initially unrelated strands of beach research in the 1940s that were simultaneously providing new insight into the role of swash infiltration in determining erosion and accretion at the beach face, and the dynamics of beach groundwater in controlling the saturation characteristics of the foreshore.

2. In the mid-1970s, the first laboratory investigations were reported that examined the artificial lowering of beach groundwater as a method to promote shore accretion and stability, and the results proved encouraging. By the late 1970s the results of the first field investigation of this approach were reported, but the results of this work were less conclusive.

3. Commercial interest in beach dewatering as a practical alternative to more traditional methods of shore stabilization was initiated in the early 1980s as the result of an unrelated engineering project on the Danish coast. The decreasing efficiency of a buried seawater filtration system was observed to correspond to the rapid build up of sediment in front of intake pipes.

4. A full-scale test of the dewatering concept on the open North Sea coast of Denmark was undertaken during the period 1985–91. Initial results proved encouraging, and for the first two and half years of the system's operation published data suggest that, relative to untreated control sites, the dewatered beach stabilized and showed a positive trend of shore accretion. During the ensuing four years, the published

monitoring results were less conclusive, and it was interpreted that the beach drain was having no discernible positive effect on enhancing net beach width. Relative to the eroding control sections of beach, it was tentatively concluded that the dewatering system reduced the rate at which the coastline was eroding.

New dewatering sites should at present be regarded as experimental, rather than a proven solution to erosion management. The main problem with the drain is that it does not produce sand. It only takes some sand away from adjoining beaches. Compared to artificial nourishment the drain is uneconomical. Coastal researchers must investigate further both the dynamics of coastal groundwater determining the time-varying saturation characteristics of the beach face; and the modification of sediment transport mechanisms at the beach face induced by groundwater infiltration and seepage. Only when a physical understanding of these processes is gained, will the mechanisms determining the success or failure of the dewatering concept be understood (Turner and Leatherman, 1997).

Per Bruun

Bibliography

Abbott, M.B., and Price, W.A. (eds.), 1994. *Coastal Estuarial and Harbor Engineers' Reference Book*. New York: E & FN Spon.

Bagnold, R.A., 1940. Beach formation by waves; some model experiments in a wave tank. *Journal of the Institute of Civil Engineers*, **15**: 27–54.

Bruun, P., 1989. The coastal drain: what can it do or not do? *Journal of Coastal Research*, **8**(1): 123–125.

Davis, G.A., and Hanslow, D.J., 1991. Beach stabilization by drainage of the watertable. *Proceedings of the 10th Australasian Conference on Coastal and Ocean Engineering.*

Davis G.A., Hansow, D.J., Hibbert, K., and Nielsen, P., 1992. Gravity drainage: a new method of beach stabilization through drainage of the watertable. *Proceedings of the International Conference on Coastal Engineering*. pp. 1129–1141

Davis, G.A., Hanslow, D.J., Hibbert, K., and Nielsen, P., 1993. Beach stabilization by gravity drainage. In Bruun, P. (ed.), *Proceedings of the Hilton Head Island South Carolina, USA International Symposium*, Volume 2, p. 707.

Dean, R.G., 1989. Independent analysis of beach changes in the vicinity of the Stabeach System at Sailfish Point, Florida. Report prepared for Coastal Stabilization Inc., 15 p.

Emery, K.O., and Foster, J.F., 1948. Water tables in marine beaches. *Journal of Marine Research*, **7**: 644–654.

Grant, U.S., 1948. Influence of the water table on beach aggradation and degradation. *Journal of Marine Research*, **7**: 655–660.

Hansen, H.K., 1986. Coastal drain system: full-scale test at Thorminditangen. *Summary Report* No. 170–83322, Lyngby, Denmark: Danish Geotechnical Institute.

Heaton, K.L., 1992. Local effects of groundwater table variation on on-shore off-shore sediment transport. Unpublished BE (Hons) thesis. Perth: University of Western Australia, 134 p.

Lenz, R.G., 1994. Beachface dewatering—a tool for coastal stabilization. Alternative technologies in beach preservation. *Proceedings of the 1994 National Conference on Beach Preservation Technology*, pp. 27–52.

Li, L., Barry, D.A., and Pattiaratchi, C., 1995. Numerical modelling of beach dewatering through drainage. *Journal of Waterways, Port, Coastal, and Ocean Engineering*.

Nielsen, P., 1990. Tidal dynamics of the water table in beaches. *Water Resources Research*, **26**(9): 2127–2134.

Nielsen, P., 1992. *Coastal Bottom Boundary Layers and Sediment Transport*. River Edge, NJ: World Scientific Co. Pte. Ltd.

Nielsen, P. *et al.*, 2001. Infiltration effects on sediment mobility under waves. *Coastal Engineering*, **42**(2): 105–114.

Ogden, M.R., and Weisman, R.N., 1991. Beach stabilization using drains; an experimental model study. In *Procedings of the Coastal Sediments '91*. American Society of Civil Engineers, pp. 1955–1969.

Oh, T.M., and Dean, R.G., 1994. Effects of controlled water table on beach profile dynamics. *Procedings of the 24th International Conference on Coastal Engineering*, pp. 234–235.

Ovesen, N.K., and Schuldt, J.C., 1992. Beach Management System-Documentation. Summary Report. Danish Geotechnical Institute.

Terchunian, A.V., 1989. Performance of the Stabeach System at Hutchinson Island, Florida. *Proceedings of the 2nd National Conference on Beach Preservation Technology, Florida Shore and Beach Preservation Association*, pp. 229–238.

Turner, I.L., and Leatherman, P.L., 1997. Beach Dewatering as a "soft" engineering solution to coastal erosion—a history and critical review. *Journal of Coastal Research*, **13**(4): 1050–1063.

Vesterby, H., 1991. Coastal drain system: a new approach to coastal restoration. *Proceedings of the GEO-Coast '91*, pp. 651–654.

Vesterby, H., 1994. Beach face dewatering—the European experience. Alternative Technologies in Beach Preservation: *Proceedings of the 1994 National Conference on Beach Preservation Technology*, pp: 53–68.

Weisman, R.N., Seidel, G.S., and Ogden, M.R., 1995. The effect of water table manipulation on beach profiles. *Journal of Waterways, Port, Coastal and Ocean Engineering*, **121**: 134–142.

Zinn, D.J., 1942. An ecological study of the interstitial microfauna of some marine sandy beaches, with special reference to Copepoda. Unpublished Ph.D. dissertation. New Haven: Yale University.

Cross-references

Beach Erosion
Cross-Shore Sediment Transport
Depth of Disturbance
Hydrology of the Coastal Zone

BEACH EROSION

Introduction

Beaches are loose accumulations of sand, gravel, or a mixture of the two that bound an estimated 30% of the world's coasts (Bird, 1996). Because they consist of more or less loosely packed noncohesive sediments, beaches act as buffers that absorb, reflect, and dissipate energy delivered to the shore by waves. By doing so, they shelter areas behind the beach, especially during storms, from wave attack and flooding. Such back-beach zones may be cliffs, dunes, or low-lying marshes and lagoons. On many coasts of the world, the beaches and these associated back-beach environments have been taken up by development (Nordstrom, 1994). A lot of this development has occurred over the last three to four decades, thriving on the worldwide growth in domestic and international tourism, and largely favored by the diversification of beach recreational activities. The boom in coastal development, especially on low-lying sandy coasts, has been matched by an increasing awareness that the beaches that form the foundations of prosperity of many communities are eroding in many places. An estimated more than 70% of the world's beaches are now eroding (Bird, 1996). Lack of foresight in construction and development planning has, in many cases, led to massive and irreversible urbanization of the coast that renders many communities vulnerable to the insidious effects of beach erosion. Erosion impairs the capacity of a beach to act as a buffer against storms. This means that beach erosion may have serious negative repercussions for low-lying island states, for shorefront communities, and for beach-based leisure activities on which depend many jobs and from which many coastal communities draw income.

Beach studies started mainly in connection with military activities, notably during the preparation of the World War II landings on the French coast. Since then, they have increased dramatically, especially over the last three decades, as beach erosion has become a critical issue in coastal zone management in many countries. Evaluating the implications of beach erosion necessitates a clear definition of what beach erosion is and how it is measured, notwithstanding the fact that the physical processes involved in the dynamics of erosion are still not well understood.

Perception of beach erosion and measurement of erosion rates

Proper coastal management requires a clear definition of beach erosion and accurate quantification of erosion rates. Although beach erosion has received great attention from coastal scientists, government agencies, local authorities, and beachfront owners, its perception and exact definition are controversial issues, mainly as a result of the diverse interests of the different parties involved in beaches and/or their management (Esteves and Finkl, 1998). This statement, made in reference to beaches in Florida, holds true for beaches in many developed countries. Beach erosion is a process whereby a beach loses its sediment, resulting in a depletion of its sediment budget. This process occurs where the beach can no longer balance energy produced by waves and by water piling up against it, leading to net sediment loss and lowering and retreat of the beach. Basically therefore, beach erosion may be viewed as resulting from an imbalance between, on the one hand, the energy inputs and, on the other, the resistance of the beach bed and sediment liable to be mobilized by the fluid forces. The erosion process itself is thus a way of eventually reestablishing balance through dissipation of energy. However, this is a scientific and objective view of beach erosion. Perception of the problem generally tends to be associated with developed shores in urban areas mainly where sandy beaches are important to the economy (Finkl and Esteves, 1998). As these authors have shown for the beaches of Florida, which account for about 25% of the total sandy shores in the United States, this bias of the erosion perception is shown in discrepancies in the delimitation of both erosion problem areas (EPA) and critically eroded areas (CEA) among different surveys. There are no common standards for objectively classifying beach erosion. Each party perceives beach erosion in its own way. Furthermore, beach erosion is not commonly perceived as a problem on undeveloped shores. In an effort at objective standardization, Esteves and Finkl (1998) and Finkl and Esteves (1998) propose a useful, comprehensive beach erosion classificatory scheme covering developed and undeveloped coasts.

There are also no common standards for quantifying rates of beach change (Moore, 2000) and for determining high-tide shoreline position (Galgano *et al.*, 1998; Morton and Speed, 1998; Douglas and Crowell, 2000). Beach erosion is generally quantified through some statistical treatment of retreat rates and volumetric losses (e.g., Leatherman, 1983). The input data comes either from field surveys that have gained in accuracy with the advent of electronic stations and differential global positioning systems (GPS), or from numerically rectified aerial photographs, maps, and land-use documents. Other new methods include digital video imagery near ground level or from low-flying aircraft, and airborne scanning laser altimetry or light detection and ranging (LIDAR) (Mason *et al.*, 2000; see also entry on *Mapping Beaches and Coastal Terrain*). Beach erosion rates and volumetric losses may also be estimated from the depletion of beach nourishment material where such nourishment is regularly carried out (e.g., Finkl, 1996, see entry on *Beach Nourishment*).

Rates of beach erosion may range from a net moderate loss of less than a meter a year to several meters following just one storm event. Such rates may also vary alongshore, decreasing from a maximum in "hot spots" or high-tide shoreline areas subject to the most severe perturbations, to "cold spots" where the effects of such perturbations are no longer felt and the high-tide shoreline is stable. Extreme rates of beach retreat in isolated "hot spots" along the southeast barrier-island coast of the United States approach 4 m yr^{-1}, causing substantial loss of land and oceanfront property (Finkl, 1993). Reliable determination of rates of beach retreat is important in coastal planning, especially as regards construction setbacks.

Beach erosion processes within the profile

The beach is a three-dimensional (3-D) sediment body that extends alongshore from the upper limits of wave run-up to the outer limits of wave action, the so-called closure depth, in the nearshore zone. However, while the upper limits may be relatively easy to identify using geomorphic features (Morton and Speed, 1998), the offshore limits are not, for obvious reasons. Beach erosion may be a short-term (order of hours to seasons) process that reflects adjustment to wave energy changes, or a longer-term (order of years) one that reflects an increasingly deficient beach sediment budget (Figure B17). On sandy beaches, short-term changes involving erosion are commonly part of a so-called morphodynamic cycle of adjustment of the beach profile to seasonal or nonseasonal changes in wave energy (Short, 1999). Seasonal changes commonly correspond to the classic winter profile flattened by storms and the summer profile that accretes under fair weather conditions. Beach profile adjustment generally leads to better absorption of the nearshore and incident wave energy, leading over a more or less long period of time (hours to months), to an equilibrium situation and shoreline stability. The period of adjustment depends on the wave energy inputs, the beach morphology, and the sediment volume. Rapid beach recovery is quite common, but recovery may sometimes take several years following major storms (Morton *et al.*, 1994; Galgano *et al.*, 1998). Sandy beach morphology and sediment volume are intricately related, defining profiles that are either short, steep, and reflective, commonly associated with coarse sediment, or wide, flat, and dissipative, commonly with fine sand (see entries on *Reflective Beaches* and *Dissipative Beaches*). In the former, much of the sand is locked up in the intertidal beach, forming especially a voluminous subaerial beach sometimes comprising an upper beach terrace called a berm. In the latter, much of the sand is stored in shallow intertidal to subtidal bars. High-energy waves impinging on steep reflective sandy beaches result in the fastest response times, resulting in erosion of the upper beach and berm, and seaward removal of the sand to form barred dissipative beaches. In such situations, erosion of the upper beach is therefore compensated by accumulation on the lower beach, without there being necessarily a net loss of sediment. However, coastal managers are sensitive to changes in subaerial beach volume, so that such short-term upper beach and berm erosion, especially where severe, may raise anxiety.

Beaches characterized dominantly by gravel differ in their behavior. Such coarse-grained beaches have generally steep, narrow reflective profiles that are commonly inert and unresponsive, to a certain degree, to increases in wave energy. This is either a result of the spatial organization of the constituent clasts and/or because of the capacity of these coarse-grained beaches to absorb wave energy through high percolation rates (Carter, 1988; Forbes *et al.*, 1995; Orford *et al.*, 1996). Macrotidal beaches, found in areas with large tidal ranges (>4 m at spring tides), also commonly show slow or moderate response to high-energy events, compared to their more common microtidal counterparts, because of their wide, dissipative profiles and the rapidity of migration of the wave domains that goes with the important tidal excursion.

On any sandy or gravelly beach profile, short-term morphodynamic changes may be embedded in longer-term changes involving net sediments gains or losses, the latter being synonymous with overall beach erosion throughout the profile. Whatever its origins, a net loss of beach sediment results in durable changes in beach morphology as the beach seeks to adjust to this situation of sediment deficit. Durable erosion generally results in a permanently scarped beach profile exhibiting an upper beach scarp (Figure B18). In some cases, erosion can lead

Figure B17 The erosion and retreat of a gravel beach in Picardy, France, has left a World War II blockhouse stranded on the beach.

Figure B18 Durable beach erosion is commonly manifested by an upper beach scarp. The erosion here, which threatens coastal settlements and has led to inland relocations of several villages and an international highway, has occurred downdrift of the port of Cotonou in Benin, West Africa on a coast subject to strong longshore drift.

to the total disappearance of a beach. The sediment lost accumulates elsewhere, either further alongshore, in other beaches, in estuarine and lagoonal sinks, or in offshore sinks.

In spite of a considerable amount of beach research over the past three decades, the sediment transport processes operating on beaches and involved in beach erosion are still poorly known (Butt and Russell, 2000; also see entries on *Beach Processes* and *Surf Zone Processes*). "Normal" waves and wave shoaling processes in the nearshore zone lead to net shoreward flows that may result in sediment drifting alongshore or from offshore working its way toward the beach. In essentially reflective beach systems, these shoreward flows are balanced by gravity-driven seaward return flows down the beach face. The net sediment budget of the beach would then depend on its ability to diminish the seaward return flow volume through processes such as infiltration and grain size and bedform adaptations to flow strength. These processes depend on beach slope and grain size, both interrelated, and on the water table on the beach. Low, dry season water tables on microtidal beaches composed of medium to coarse sand in West Africa favor exceptionally steep beach slopes (up to 18°) that result from berm buildup through sand deposition by swash infiltration. Higher rainy season water tables have the opposite effect of encouraging little infiltration, thus favoring sediment transport down the beach. Erosion processes on the more complicated dissipative beaches depend on the complex interplay of various modes of fluid motion near the beach (Komar, 1998). These include incident gravity waves, infragravity or low-frequency waves generated by transfer of energy from the former, alternations of high and low waves or wave grouepiness, wave-induced currents, tidal currents, currents due to wind forcing, wave–current interactions, and patterns of energy concentration related to the way these various fluid forces interact with the morphology. On a few of the world's coasts subject to strong, sustained winds, eolian processes may also contribute in removing sand from the beach.

Gravel beaches show specific behavioral modes because of their coarse grain sizes. The dynamics of these beaches are fully discussed elsewhere (Forbes *et al.*, 1995; Orford *et al.*, 1996; see entry on *Gravel Barriers*). On sandy beaches subject to episodic high-energy erosive events, a variety of related meteorological and hydrodynamic factors combine to enhance beach stripping. These are setup of the water level close to the beach, due to waves, onshore wind forcing, and sometimes the direct exposure of the beach to the low pressure system that generates the storm waves, and durable saturation of the beach face through both rainfalls that commonly accompany stormy weather and enhanced swash run-up. Water pileup on the seaward-sloping beach must be balanced by seaward return flows, or in low-lying sand barrier systems, by overwash. In these high-energy events, liquefaction and removal of the beach sand may be accompanied by deposition of these sediments in offshore areas where the energy of seaward flows peters out or is balanced by shoreward-directed energy. The beach profile retreat is also a short-term mechanism of creating accommodation space for the temporary water pileup. The sediment transported seaward may initially travel via major rip current pathways generated by incident and infragravity waves. Sand may flow alongshore from these pathways, because of the commonly oblique incidence of waves and wind setup, feeding strong longshore currents. Subsequently, as larger waves and strong winds lead to more pileup of water on the beach, the seaward return flows may no longer be simply canalized in rip channels and mass balancing seaward flows may occur, resulting in generalized beach stripping and offshore sediment loss. The intensity of beach erosion depends on various factors such as the wave energy level, the antecedent beach morphology, orientation of winds relative to the coast, their strength, beach grain size, tidal range and tidal state, and the duration of high-energy conditions.

Longshore manifestations of beach erosion

A beach may comprise one or several sediment cells with bounding limits to longshore drift. Swash and drift-aligned beaches (see entry on *Drift and Swash Alignment*), respectively, designate beaches associated with weak and strong rates of longshore drift (Davies, 1980). In many cases of beach erosion, the process is a subtle, insidious one that does not require the high-energy events described above (although these may spectacularly enhance erosion rates) other than seasonal increases in wave energy to which the beach is generally well adapted. This is particularly the case on coasts subject to strong longshore drift rates on which depend the overall stability of the beach. Erosion functions essentially where major engineering structures block the sediment load drifting alongshore. Continuity of sediment transport downdrift by the longshore current is assured by beach erosion. The strongest drift rates, sometimes exceeding 1 million $m^3 a^{-1}$ of sand or gravel, are found where large swell waves impinge with marked obliquity on long, open beaches, as on the Gulf of Guinea coast in West Africa, in New Zealand, and the Kerala coast of India.

The longshore manifestations of beach erosion have received attention in the literature, both in terms of the plan shape of freestanding beaches (as opposed to short, headland-bound bay beaches) and of the effects of major engineering structures. The plan shape of freestanding beaches may change rapidly in response to sediment depletion. These changes basically reflect sediment cell divisions (Carter, 1988) that may also involve switches from drift to swash alignment, in an attempt by the beach to adjust to sediment deficit by diminishing longshore transport. Examples have been described from both sandy (e.g., Anthony, 1991) and gravelly beaches (e.g., Forbes *et al.*, 1995; Orford *et al.*, 1996; Anthony and Dolique, 2001). The large-scale changes in beach plan shape are also accompanied by beach textural and profile reorganizations. Downdrift of jetties, a major cause of beach erosion (see next section), the high-tide shoreline morphology in plan commonly defines a log-spiral curve (see entry on *Headland Bay Beach*) or a half-heart bay (Silvester and Hsu, 1993) that may extend several kilometers. This shape illustrates the more severe retreat that affects the beach just downdrift of such structures. Continuity of sediment transport by the longshore current after the jetty is assured by sometimes rapid and significant beach erosion. Erosion diminishes downdrift of this "hot spot" as the longshore current becomes increasingly charged with sediment, leading to a more linear high-tide shoreline. At some distance downdrift, erosion becomes nil and the high-tide shoreline may even show advance from the accumulation of sediment eroded from the beach updrift. These longshore changes are sometimes manifested by a fast "erosion front" and a slow "erosion front" separated by a salient, or "bump," that may exacerbate erosion downdrift (Bruun, 1995). The existence of such two fronts along any eroding beach probably reflects two sediment cells on either side of a central downdrift accumulation terminus fed by beach erosion within the more updrift cell. The accumulation "bump," or salient, influences wave incidence angles in such a way as to minimize drift and capture sediment, thus aggravating erosion within the following longshore cell, as examples from gravel barrier beaches have shown (e.g., Orford *et al.*, 1996; Anthony and Dolique, 2001). According to Bruun (1995), the distance of downdrift migration of erosion fronts on the Atlantic shoreline of Florida is of the order of 30–40 km, the fronts migrating essentially from inlet to inlet. This distance is similar to that downdrift of the seaport of Lomé, in West Africa (Anthony and Blivi, 1999).

On some beaches, especially headland-bound bay beaches, it is not uncommon for seasonal or longer-term changes in the predominant direction of wave approach to induce changes in longshore drift. This process results in "beach rotation" (Short, 1999), which is the periodic lateral movement of sand towards alternating ends of the embayed beach. It results in erosion at one end of the beach, while the other accretes. In rare cases, beach rotation is due to short- to medium-term (order of a few years) changes in nearshore bathymetry that affect wave refraction and dissipation patterns. The massive mud banks delivered by the Amazon river to the muddy coast of South America migrate westward toward the Orinoco delta, inducing changes in incident wave energy levels by strongly modulating wave refraction and diffraction patterns. These generate lateral movement of sand in embayed beaches between bedrock headlands in Cayenne, French Guiana, resulting in alternations in erosion (Figure B19) and accretion over time, without net sediment loss (other than through illicit sand extraction). Similar effects on beaches elsewhere may be generated by changing sand bank configurations offshore, as in the case of the sandy beaches of northern France bounding the English Channel and the southern North Sea.

The causes of beach erosion

The sediment that accumulates on the shore to form a beach may come from various sources. Any poorly consolidated material on which waves and currents impinge may be a source. Such material may be an initial coastal and nearshore deposit of diverse origin such as glacial till or fluvial sediments, or may be delivered to the shore through landslides or by volcanoes. These sources are usually cut into coastal cliffs and underwater slopes that recede as they deliver sediments to the shore for beach construction. Dunes may also deliver sand to the beach but the beach and dunes, especially those immediately bounding the beach, should be considered as an interrelated system with sand interchanges. Some infilled estuaries and many sand- or gravel-rich deltas also supply sediment to beaches, especially at times of high river discharge.

Any natural or human action that affects the supply capacity of a given source and the cross-shore and longshore sediment transport processes on beaches may result in erosion. In many cases, especially on long open beaches, several factors, whose specific roles are difficult to

Figure B19 An example of a beach in Cayenne, French Guiana, affected by periodic rotation due to mudbanks migrating alongshore. The erosion presently affecting this end of the beach (concomitant with accretion at the opposite end) has been aggravated by illicit sand extraction. Note the massive rock protection on the upper beach.

Figure B20 Beach accumulation and erosion, respectively, updrift and downdrift of a jetty in Upper Normandy, France. Erosion on this coast has been exacerbated by the stabilization of cliffs that hitherto supplied flint clasts to the beaches.

disentangle, may jointly cause beach erosion. The most readily discernible causes of beach erosion are where identified human actions and activities perturb the beach sediment budget and the morphodynamic functioning of the beach. This cause of beach erosion dramatically developed in the 20th century with the multiplication of dams across rivers and large-scale urbanization of the coastal zone worldwide. On many coasts of the world, the construction of dams has, over the long run, led to coastal sediment starvation and beach erosion. The effects of artificial structures on the shore (see *Shore Protection Structures*), and especially beaches, have received considerable attention in the literature (Walker, 1988; Silvester and Hsu, 1993; Bird, 1996; Charlier and De Meyer, 1998). One important cause of beach erosion worldwide is the construction of jetties and ports (Figure B20). In many coastal communities, as along the eastern United States, the lagoons behind barrier islands are important economic waterways whose inlets need to be deepened by dredging and kept open permanently by groins and breakwaters. These impede the longshore drift of sand that is vital in nourishing beaches downdrift. Esteves and Finkl (1998) estimate that 90% of beach erosion in southeast Florida has been caused by human action, mostly the construction of deepened inlets with protective jetties. Deepwater ports constructed on open beach coasts subject to strong longshore drift have similar negative effects, as in the Bight of Benin in West Africa (Anthony and Blivi, 1999). Here, national seaports in Togo, Benin, and Nigeria have resulted in dramatic beach erosion downdrift of the port breakwaters, and in equally spectacular beach accretion updrift. The continual beach erosion on this coast (Figure B18) has led

Figure B21 Chronic beach and dune erosion in Wissant, a tourist and recreational resort in northern France.

to successive inland relocations of coastal communities and of the major international highway linking the three countries, at great cost to their already beleaguered economies.

Another source of perturbation of beach sediment budgets and a cause of beach erosion is coastal urbanization, which involves the development of urban fronts with high-rise condominiums and hotels on the upper beach. Some of the best examples include the US Atlantic and Gulf coasts (Nordstrom, 1994; Esteves and Finkl, 1998), and the Mediterranean rivieras (Anthony, 1997). In many cases, related dune systems have been flattened or severely degraded, and this has had a dramatic effect on beach stability. Dunes tend to be overlooked as the "savings account" of the shore while the beaches act basically as a "checking account." The dunes store important volumes of sand that help in balancing the beach budget. In the past, uncontrolled shorefront urbanization has commonly entailed narrowing of many beaches, diminishing in time their wave-energy buffering capacity, and leading to beach erosion (Figure B21). Beachfront urbanization also often requires defense structures, notably walls and revetments emplaced on the upper beach. Depending on their design, these structures may act as static barriers that reflect wave energy offshore, thus aggravating beach erosion, although some doubt has been recently cast on this negative effect of sea walls (Kraus and McDougal, 1996). In the past, urbanization and the development of road and rail networks has sometimes involved the direct quarrying of sand or gravel from beaches with fragile sediment budgets. This practice is still frequent in developing countries that lack awareness of the environmental consequences of such beach sediment depletion.

To stabilize already eroding beaches, groins are sometimes built across the beach with the aim of trapping sediment drifting past. Breakwaters are also sometimes built off the beach to dissipate some of the wave energy that erodes the beach. In playing an efficient role sometimes in alleviating beach erosion, these structures may be instrumental in simply transferring the erosion problem further downdrift, often to the detriment of another community. It is not uncommon to see groin fields sprouting downdrift, increasing in numbers as the problem goes from one community to the next. The gravel barrier beach in Upper Normandy (Figure B20) and Picardy, France, is a clear illustration of this downdrift "march" of erosion and of the attendant groin field. In Picardy, a groin field emplaced to stabilize an eroding gravel beach grew from 6 groins in 1976 to 96 groins in 2000 over a distance of 10 km (Anthony and Dolique, 2001). The initial erosion of this beach started with the construction of several jetties updrift in the 19th and 20th centuries, and was aggravated by the artificial consolidation and stabilization of several sectors of cliffs that hitherto liberated gravel flint clasts to the beach longshore drift cell (Figure B20). This practice of cliff stabilization has, in some cases like this one, led to beach sediment starvation and erosion.

Natural sediment depletion is considered as a major cause of worldwide beach erosion (Bird, 1996). On many of the world's coasts, especially in areas where sea level over the past 5–6,000 years has been relatively stable, the sand forming the beaches was derived from sediments on the inner continental shelf. These drowned nearshore deposits have been reworked by waves and driven onshore to form successive beach ridges and dunes sometimes several kilometers wide, as along large stretches of the Australian, West African, and Brazilian coasts.

This process, called progradation, has stopped in most areas as the nearshore sediment supply has petered out. The beaches bounding these prograded coasts are sensitive to any long-term changes in wave energy, resulting, for instance, from greater storminess or sea-level rise. Although exhaustion of nearshore sediment stocks is commonly invoked as a cause of beach erosion, it is hard to substantiate because of the lack of records of long-term beach and nearshore profile changes.

Beaches, as mentioned earlier, may show short-term changes in profile in response to storms and fair weather conditions. Apart from the various causes of sediment depletion evoked above, changes in the state of the sea also lead to durable beach erosion. These changes include greater storminess, short-term variations in sea level related to major changes in sea surface temperatures such as, those associated with El Nino events, and secular sea-level rise, commonly imputed to global warming. Changes in offshore wave energy are due to storms, such as the northeasters in the eastern United States, and cyclones, or may reflect more subtle increases in wave energy due to greater storminess and sea-level rise. Exceptional waves generated by submarine landslides or earthquakes may also lead to significant beach erosion. These events generate destructive high-energy waves that remove the beach sediments offshore. The seaward return flows may lead to losses of sand beyond the offshore limits of the beach profile, such that the sand cannot be returned to the beach during the following fair-weather wave conditions. On beaches bounding low-lying coasts, permanent losses of sediment may also occur inland as waves wash over the shore. Greater storminess implies more frequent episodes of higher incident wave energy often accompanied by strong wind setup of water level inshore. Many beaches do not have the available sediment stocks to adapt to such increases in wave power and to the currents resulting from wind and wave forcing. Sea-level rise, either on a short-term basis, due to short-term events such as El Nino, or of a secular nature due to global warming, would similarly favor wave energy impingement higher up the beach face (see entry on *Sea-Level Rise, Effect*). New sediment commonly does not move in from alongshore to balance the increase in wave energy, and the beach erodes as its sediment stocks are transferred seaward. Depending on the wave energy regime and the rate of sea-level rise, such sediment may be permanently trapped offshore as the base of wave action moves upward through sea-level rise. This pattern of beach erosion resulting from sea-level rise has been extensively debated in terms of what has become known as the Bruun rule (e.g., SCOR Working Group, 1991; Thieler *et al.*, 2000).

Managing beach erosion

Good beach management requires both accurate bookkeeping on rates and patterns of beach change and implementation of the right strategies in the face of erosion. Beaches are a multiresource asset in many ways, involving huge sums of money in developed economies, both in terms of revenue and for design and implementation of management policies. As a result, the number of parties involved in beach management may be considerable, ranging from state legislators and engineering bodies, through recreational and tourist agencies, to scientists, beachfront home owners individually or as associations, and environmental and ecological pressure groups. Beaches are, as such, objects of conflicting interests. In many developed economies, beach erosion has become the fundamental coastal zone management problem, and a national issue in several countries bordered by long stretches of densely developed low-lying shores, such as the Netherlands and the United States. It has also become a cause of major concern for low-lying island states subject to sea-level rise (Leatherman, 1997). In the face of beach erosion, the management options are very few indeed. These include the determination of development setback lines in order to accommodate future erosion without endangering constructions. In the absence of precise determination of beach erosion rates, this strategy may fail, as on the Bight of Benin coast in West Africa (Figure B18). A second strategy is that of letting erosion take its course, generally in undeveloped areas where the process does not constitute a hazard. A third strategy is that of fighting beach erosion at all cost, especially where vital national, economic, or military interests are at stake. The finest example of such a policy is that of the Netherlands (Hillen and Roelse, 1995). On some developed shores such as parts of south Florida, the value of beach real estate and the revenue from beaches are such that the high-tide shoreline position has to be maintained, generally through the implementation of costly solutions such as efficient bypassing of inlets (see entry on *Bypassing at Littoral Drift Barriers*) and, especially, regular beach nourishment (Finkl, 1996). These are often, and increasingly, the only efficient ways of restoring the beach sediment volume. These "soft engineering" techniques have been discussed in numerous papers in scientific journals, especially the *Journal of Coastal Research* and *Shore and Beach*, as well as in regular newspaper commentaries in many countries. A specific comment needs to be made here on engineering practice in managing beach erosion. In many countries, including the United States, beach erosion has been managed using various assumed empirical relationships. Some of these relationships have been reviewed recently by Thieler *et al.* (2000) who draw attention to their oversimplified assumptions relative to the complex reality of beaches. In Europe, engineering practice in some countries, notably the Netherlands, has treated coastal management within a geomorphic systems approach, rather than simply in terms of deterministic engineering models. While the dependence on engineering models is still well entrenched in France, the tendency in Britain has shifted in recent years toward considering beach and, more generally, shore erosion management, in terms of a geomorphic systems approach (Hooke, 1999) that integrates local experience (Brunsden and Moore, 1999). The complexity of beach erosion and the large number of parties involved in its management should call for a sensible and balanced mix of science with a systems approach, engineering expertise, past and present experiences, and the specificities of the local context in which erosion occurs.

Many developed countries are today faced with minor to critical beach erosion problems, largely because of lack of foresight in coastal development patterns. While they may have the resources to combat beach erosion, the same is not true for developing countries which cannot divert much needed money toward beach management, often considered as a "low priority" area. These countries are increasingly subject to the pressures of an often rapid pace of economic development, and of beach-based tourist activities, while facing the threats of sea-level rise from global warming. It is perhaps reassuring that because of the still moderate level of coastal development in many of these countries, they have the opportunity of avoiding the mistakes made in the past by the developed countries by planning such development in a way as to ensure sustenance of the beach resource. This opportunity can be seized through active transfer of knowledge from the developed to the developing countries.

Edward J. Anthony

Bibliography

Anthony, E.J., 1991. Beach-ridge plain development: Sherbro Island, Sierra Leone. *Zeitschrift fur Geomorphologie, Suppl.-Bd*, **81**: 85–98.

Anthony, E.J., 1997. The status of beaches and shoreline development options on the French Riviera: a perspective and a prognosis. *Journal of Coastal Conservation*, **3**: 169–178.

Anthony, E.J., and Blivi, A.B., 1999. Morphosedimentary evolution of a delta-sourced, drift-aligned sand barrier–lagoon complex, western Bight of Benin. *Marine Geology*, **158**: 161–176.

Anthony, E.J., and Dolique, F., 2001. Natural and human influences on the contemporary evolution of gravel shorelines between the Seine estuary and Belgium. In Packham, J.R, Randall, R.E., Barnes, R.S.K and Neal, A. (eds.), *The Ecology and Geomorphology of Coastal Shingle*. Otley, West Yorkshire: Westbury Academic and Scientific Publishers, pp. 132–148.

Bird, E.C.F., 1996. *Beach Management*. Chichester: Wiley and Sons.

Brunsden, D., and Moore, R., 1999. Engineering geomorphology on the coast: lessons from west Dorset. *Geomorphology*, **31**: 391–409.

Bruun, P., 1995. The development of downdrift erosion. *Journal of Coastal Research*, **11**: 1242–1257.

Butt, T., and Russell, P., 2000. Hydrodynamics and cross-shore sediment transport in the swash zone. *Journal of Coastal Research*, **16**: 255–268.

Carter, R.W.G., 1988. *Coastal Environments*. London: Academic Press.

Charlier, R.H., and De Meyer, C.P., 1998. *Coastal Erosion—Response and Management*. Berlin: Springer.

Davies, J.L., 1980. *Geographical Variation in Coastal Development*, 2nd edn. London: Longman.

Douglas, B.C., and Crowell, M., 2000. Long-term shoreline position prediction and error propagation. *Journal of Coastal Research*, **16**: 145–152.

Esteves, L.S., and Finkl, C., Jr., 1998. The problem of critically eroded areas (CEA): an evaluation of Florida beaches. *Journal of Coastal Research*, **26** (Special issue): 11–18.

Finkl, C., Jr., 1993. Pre-emptive strategies for enhanced sand bypassing and beach replenishment activities in southeast Florida: a geological perspective. *Journal of Coastal Research*, **18**(Special issue): 59–89.

Finkl, C., Jr., 1996. What might happen to America's shoreline if artificial beach nourishment is curtailed: A prognosis for southeastern

Florida and other sandy regions along regressive coasts. *Journal of Coastal Research*, **12**: iii–ix.

Finkl, C., Jr., and Esteves, L.S., 1998. The state of our shores: A critical evaluation of the distribution, extension, and characterization of beach erosion and protection in Florida. *Proceedings of the 11th National Conference on Beach Preservation Technology*. Tallahassee, FL, pp. 302–318.

Forbes, D.L., Orford, J.D., Carter, R.W.G., Shaw, J., and Jennings, S.C., 1995. Morphodynamic evolution, self-organization and instability of coarse-clastic barriers on paraglacial coasts. *Marine Geology*, **126**: 63–86.

Galgano, F.A., Douglas, B.C., and Leatherman, S.P., 1998. Trends and variability of shoreline position. *Journal of Coastal Research*, **26** (Special issue): 282–291.

Hillen, R., and Roelse, P., 1995. Dynamic preservation of the coastline in the Netherlands. *Journal of Coastal Conservation*, **1**: 17–28.

Hooke, J.M., 1999. Decades of change: contributions of geomorphology to fluvial and coastal engineering and management. *Geomorphology*, **31**: 373–389.

Komar, P.D., 1998. *Beach Processes and Sedimentation*. Upper Saddle River, NJ: Prentice Hall.

Kraus, N.C., and McDougal, W.G., 1996. The effects of seawalls on the beach: Part I, an updated literature review. *Journal of Coastal Research*, **12**: 691–701.

Leatherman, S.P., 1983. Shoreline mapping: a comparison of techniques. *Shore and Beach*, **51**: 28–33.

Leatherman, S.P. (ed.), 1997. Island states at risk. *Journal of Coastal Research*, **24**(Special issue).

Mason, D.C., Gurney, C., and Kennett, M., 2000. Beach topography mapping—a comparison of techniques. *Journal of Coastal Conservation*, **6**: 113–124.

Moore, L.J., 2000. Shoreline mapping techniques. *Journal of Coastal Research*, **16**: 111–124.

Morton, R.A., and Speed, F.M., 1998. Evaluation of shorelines and legal boundaries controlled by water levels on sandy beaches. *Journal of Coastal Research*, **14**: 1373–1384.

Morton, R.A., Paine, J.G., and Gibeaut, J.C., 1994. Stages and durations of post-storm beach recovery, southeastern Texas, USA. *Journal of Coastal Research*, **10**: 884–908.

Nordstrom, K.F., 1994. Developed coasts. In Carter, R.W.G., and Woodroffe, C. (eds.), *Coastal Evolution: Late Quaternary Shoreline Morphodynamics*. Cambridge University Press, pp. 477–509.

Orford, J.D., Carter, R.W.G., and Jennings, S.C., 1996. Control domains and morphological phases in gravel-dominated coastal barriers of Nova Scotia. *Journal of Coastal Research*, **12**: 589–604.

S.C.O.R. Working Group 89, 1991. The response of beaches to sea-level changes: a review of predictive models. *Journal of Coastal Research*, **7**: 895–921.

Short, A.D. (ed.), 1999. *Beach and Shoreface Morphodynamics*. Chichester: John Wiley & Sons.

Silvester, R., and Hsu, J.R.C., 1993. *Coastal Stabilization: Innovative Concepts*. Englewood Cliffs, NJ: Prentice Hall.

Thieler, E.R., Pilkey, O.H., Jr., Young, R.S., Bush, D.M., and Chai, F., 2000. The use of mathematical models to predict beach behavior for U.S. coastal engineering: a critical review. *Journal of Coastal Research*, **16**: 48–70.

Walker, H.J. (ed.), 1988. *Artificial Structures and Shorelines*. Dordrecht: Kluwer Academic.

Cross-references

Beach Nourishment
Beach Processes
Bypassing at Littoral Drift Barriers
Dissipative Beaches
Drift and Swash Alignments
Gravel Barriers
Mapping Shores and Coastal Terrain
Modes and Patterns of Shoreline Change
Reflective Beaches
Sea-Level Rise, Effect
Shore Protection Structures
Surf Zone Processes

BEACH FEATURES

Limits and formation of beach features

When discussing the types of features that can be observed along a beach, it is important to first consider the boundaries in the coastal zone that define the limits of a beach. In everyday usage and in the scientific literature, there are some differences in defining these limits, primarily with regard to the seaward limit. Recreational beach users will often consider the beach to extend no farther seaward than the shoreline, and thus limit the beach to an entirely emergent feature, having a width that varies with changing water level. Scientific usage typically extends the beach out to the maximum limit of low water regardless of the water level at any particular time. In some scientific usage, such as in the discussion of coastal sediment dynamics, the seaward limit of the beach may be considered to extend out to the breaker zone (Figure B22) well beyond the low-water shoreline. The most useful definition, and the one used here, is that the beach refers to the zone containing unconsolidated material that extends from the limit of ordinary low-water (or mean low-tide level) on its seaward side to the limit of influence by storm waves on its landward side (Figure B22) (Hunt and Groves, 1965; Baker *et al.*, 1966; Coastal Engineering Research Center, 1984).

Based on morphology, the beach is divisible into two zones. The *backshore* is the more landward and higher part of the beach and is typically a near-horizontal to gently landward-sloping surface. The backshore is not affected by the run-up of waves except during storm events, and so this is the typically dry part of the beach. The landward limit of the beach, which is the limit of influence of storm waves, generally is marked by a change in material, a change in morphology, or a change to a zone of permanent vegetation. Examples of such a landward limit include dunes, cliffs or bluffs, or even engineered structures such as bulkheads or revetments. The *foreshore*, also called the *beachface*, is the more seaward part of the beach. The foreshore has an overall seaward slope, but may include one or more ridges and troughs on its lower slope. Because the foreshore extends to the limit of ordinary low-water, at times of high-water the lower part of the foreshore is submerged.

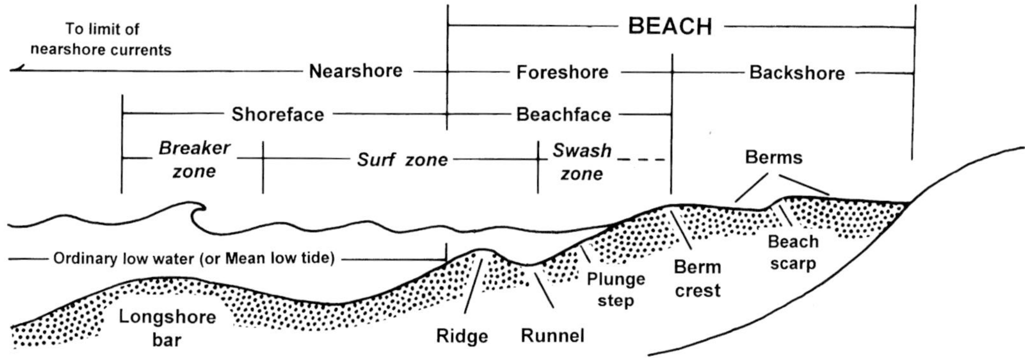

Figure B22 Generalized beach and nearshore profile showing names of major beach features and zones.

Critical in the definition of a beach is the presence of unconsolidated materials. These unconsolidated materials are what make a beach, and it is the erosion, transport, and deposition of these materials that results in beach features. Worldwide, the most common beach material is sand-size sediments composed of mineral, shell, or rock fragments. Coarser beach materials include gravel, cobbles, shingle, and even boulders. The beach will be made from whatever is locally available for the waves to rework. Along shores impacted by commercial or industrial activity, it is not uncommon to find beaches composed in part or completely of bricks, broken concrete, demolition debris, or any other material that may have been dumped along the shore and subjected to movement and redistribution by wave action.

The types of features that may occur along and across a beach vary in time, scale, and relative position. The primary agent in forming beach features is wave action (Davis, 1985). Other important agents are the rise and fall of water level, currents, wind, and ice in settings where coastal ice can form. Some beach features can form and persist indefinitely with minimal change in shape or location, but these are the exception. Because the beach is a dynamic setting, most beach features are ephemeral. Once formed, most beach features will only persist until new wave, water level, current, or wind conditions destroy them and replace them with new features.

Beach slope

Beach slope, which refers to the slope of the foreshore or beachface, deserves special mention as a beach feature because it is one of the characteristics used to distinguish different beaches. Slope is a dynamic feature that changes with changes in wave conditions as well as the gain or loss of different sediment sizes on the beach face. In general, the slope angle, measured relative to a horizontal plane, increases as the grain size increases; thus beaches composed of material such as pebbles or cobbles will tend to have a steeper beach face than ones made of sand. This slope difference relates to the greater permeability of the larger materials (Bagnold, 1940; Bascom, 1951; King, 1972). The wave run-up (or swash) can percolate downward through the interstitial spaces of the larger materials, and this minimizes the erosional influence of the run-back (or backwash). Storm conditions will flatten the beach slope as beach sediment is eroded and transported seaward. In the calmer wave conditions following the storm, beach slope recovers to a steeper slope as material is accreted to the beach.

Major beach features

Major beach features are here defined as those having large topographic expression or large areal extent. One of the beach features that can have the largest vertical expression on a beach is a *beach scarp*. Beach scarps are erosional features that occur when the slope of the beachface is lowered during storm events, and the beachface migrates landward by cutting into the backshore. The result is a near-vertical slope along the limit of this erosion (Figure B23). The height of a beach scarp may be just a few centimeters or a meter or more depending on the degree of wave action and the type of beach material. Beach scarps are commonly observed in areas where a new supply of sediment (i.e., beach nourishment) has recently been applied in an effort to replenish and build up the beach and wave action has cut into this nourishment and redistributed the sediment in the process of reestablishing an equilibrium beach profile.

Figure B23 Beach scarp resulting from recent storm erosion along a sand beach on the Illinois shore of Lake Michigan at Illinois Beach State Park. (photo by Michael Chrzastowski, Illinois State Geological Survey.)

One of the reasons that beach scarps are prominent beach features is that these near-vertical erosional features are cut into a near-horizontal area of the upper beach. This upper beach, in many cases, is a broad, near-horizontal to gently landward-sloping area called a *beach berm*, or simply a *berm* (Figure B22). Berms are depositional features formed from the wave-induced onshore accumulation of sediment. Local coastal conditions may preclude formation of a berm along some beach segments, while other beach segments may have two or more berms at different elevations. When more than one berm occurs, the lower berm(s) (sometimes called the *ordinary berm*) is a result of average or more typical waves, and the higher berm(s) (sometimes called the *storm berm*) is a result of the less frequent larger waves. A beach scarp may exist between two berms having different elevations. The seaward margin of the berm is typically defined by a rather abrupt change in slope from the near horizontal surface of the berm to the inclined surface of the beachface. The line defined by this change in slope is called the *berm crest* or *berm edge*. The berm crest is the distinguishing beach feature that divides the beach into the foreshore and backshore zones (Figure B22).

When low-water occurs, large-scale beach features are exposed that formed underwater and have a morphology influenced by waves, water-level changes, and associated currents. *Ridges and runnels* are the most common of these features. Ridges are elongate low mounds of beach material that are parallel or subparallel to the shore; runnels are the low areas or troughs that occur between the ridges and on the landward side of the shoremost ridge. A single ridge–runnel set may occur with the runnel on the landward side of the ridge or, if the lower foreshore is a broad, low-slope area, multiple sets of ridges and runnels may extend across this area. Such multiple sets of ridges and runnels form a washboard or corrugated topography across the lower beachface that contrasts with the smoother surface across the upper beachface. Another term for these features is "*ball*" referring to the ridge, and "*low*" referring to the runnel. The term "trough" is also sometimes applied to the runnel.

Ridges and runnels, when present in multiple sets, are one example of the types of repetitive patterns that can be observed in beach features. Another major beach feature with a repetitive nature is the *beach cusp*. Beach cusps are low mounds of beach material, separated by crescent-shaped troughs, occurring in a series along the shore. Yet another repetitive feature, and one of potentially large vertical scale, is the *beach ridge*. Beach ridges are depositional features, formed by the mounding of beach material by wave action, usually during storm events. Beach ridges are formed in the backshore zone, in some places at the most landward limit of wave influence, and they can extend as a nearly continuous linear feature for many kilometers along the shore. Wind transport may contribute sand to the tops of these ridges and form superimposed dunes (i.e., dune ridges). A single beach ridge may develop, persist for some time, and then be destroyed by renewed storm action. Sequential beach ridges may form through a series of depositional events and, with time, the juxtaposition of these ridges will contribute to the progradation of the coast.

A major beach feature common to barrier island beaches is the *washover fan*. During storm events, elevated water levels and large wave run-up can combine to transport large volumes of beach material across the beach to be deposited in broad, lobate accumulations on the landward margin of the beach. These deposits, called *washover fans*, essentially extend the landward limit of the beach and play an important role in the landward and upward migration of the beach during conditions of rising sea level (Kraft and Chrzastowski, 1985). On barrier island beaches, the formation of washover fans results in the burial of back-barrier marshes and filling along the barrier margin of lagoons.

Coasts that are subject to seasonal ice formation, such as the Great Lakes coasts of North America, can have various beach features that are related to the presence of coastal ice. A hummocky topography may develop along the upper foreshore as a result of sediment pushed into ridges and mounds by wave-thrusted ice. Shallow depressions may also develop across the upper foreshore where wave run-up may remove thin slabs of ice-cemented sand. Once an ice complex forms along the shore, a hummocky topography may develop in the lower foreshore by the action of grounded ice and the scour and fill by waves and currents around the edges of the ice. On the outer edge of the ice, an erosional trough may develop caused by the downward deflection of wave energy as waves impact the ice face. This trough can be a half-meter deep and 2–3 m wide (Barnes *et al.*, 1994). The trough location will shift toward or away from shore as the ice margin shifts in these directions. These troughs and all other ice-related beach features are relatively short-lived. Once the ice conditions cease, any ice-induced modifications to the beach morphology are quickly eliminated by ice-free wave conditions.

Minor beach features

Minor beach features are defined here as those with minimal topographic expression or small areal extent. Although limited in height and area, some of these beach features can be visually prominent because of contrast in color, texture, or materials compared to the surrounding beach. Prime examples of such prominent, small-scale beach features are the *tidemark* which is the high-water mark left by tidal water, and *swash marks* formed along the landward limit of wave swash on the beach face. The tide mark is generally a nearly continuous wavy line defined by an accumulation of driftwood, seaweed, and other floatable debris collectively called *flotsam*, left on the beach by the previous highest tide level. Swash marks are a series of superimposed scalloped or fan-shaped patterns defined by fine sand, mica flakes, or bits of seaweed deposited along the most landward reach of the swash. Swash marks are beach features that are in a nearly continuous state of formation and destruction with each new swash event. So too are *backwash patterns* which are diagonal patterns formed on the beach by the dispersion of backwash flowing around small obstacles such as a shell or pebble. A falling tide or the lowering of water level after a storm can contribute to the formation of *rill marks* which are small, erosional furrows or channels across the beachface caused by the seaward flow of water as the water table in the beach lowers and water percolates out onto the beachface in a spring-like manner. *Air holes* may also occur on the beachface as water percolates and forces air from the pore spaces up to the surface.

Near the ridge and runnel on the lower foreshore slope, a subtle linear feature may occur called the *step* or *plunge step*. This is a subtle decline in the foreshore profile that is caused by the final plunge of waves before running up the beachface. The plunge step is best developed in settings of low tidal range and steep foreshore slope (Davis, 1985).

Because one or more of the berms of a beach are elevated above the influence of the swash, fine sand across these berms is typically dry and can be influenced by wind action. Although the berms are located on the beach, features can develop here that are common to dunes and desert settings. Sand ripples may develop, small dunes may form in wind shadows such as behind logs or other driftwood, and wind deflation areas may occur where the fine sand has been removed to lower the surface and leave a concentration of coarser particles similar to a desert pavement.

Michael J. Chrzastowski

Bibliography

Bagnold, R.A., 1940. Beach formation by waves; some model experiments in a wave tank. *Journal of the Institution of Civil Engineering*, **15**: 27–52.

Baker, B.B., Jr., Deebel, W.R., and Geisenderfer, R.D. (eds.), 1966. *Glossary of Oceanographic Terms*. Washington, DC: U.S. Naval Oceanographic Office, Special Publication 35.

Barnes, P.W., Kempema, E.W., Reimnitz, E., and McCormick, M., 1994. The influence of ice on southern Lake Michigan coastal erosion. *Journal of Great Lakes Research*, **20**: 179–195.

Bascom, W., 1951. The relationships between sand size and beachface slope. *American Geophysical Union Transactions*, **32**: 866–874.

Coastal Engineering Research Center, 1984. *Shore Protection Manual*, Volume 1. Washington, DC: U.S. Army Corps of Engineers.

Davis, R.A., Jr., 1985. Beach and nearshore zone. In Davis, R.A., Jr. (ed.), *Coastal Sedimentary Environments*. New York: Springer-Verlag, pp. 379–444.

Hunt, L.M., and Groves, D.G., 1965. *A Glossary of Ocean Science and Undersea Technology Terms*. Arlington, VA: Compass Publications.

King, C.A.M., 1972. *Beaches and Coasts*. London: Edward Arnold.

Kraft, J.C., and Chrzastowski, M.J., 1985. Coastal stratigraphic sequences. In Davis, R.A., Jr. (ed.), *Coastal Sedimentary Environments*. New York: Springer-Verlag, pp. 625–663.

Otvos, E.G., 1999. Rain-induced beach processes; landforms of ground water sapping and surface runoff. *Journal of Coastal Research*, **15**: 1040–1054.

Cross-references

Beach Nourishment
Beach Processes
Beach Ridges
Drift and Swash Alignments
Profiling, Beach
Rhythmic Patterns
Ripple Marks
Scour and Burial of Objects in Shallow Water

BEACH NOURISHMENT

Introduction

Beaches occur where there is sufficient sediment for wave deposition above water level along lakes, open ocean coasts, embayments, and estuaries. Beach nourishment most commonly takes place along marine beaches, which are among the most dynamic environments on earth. On a global scale, estimates of marine sandy beaches (see entry on *Sandy Coasts*) range from about 34% (170,000 km) (Hardisty, 1990) to 40% of the world's coastline (Bird, 1996). Beaches form essentially 100% of the coast of The Netherlands, 60% in Australia, and 33% in the United States (Short, 1999). Comprising a significant proportion of the world's coastline, beaches are important considerations for coastal recreation and storm protection, while others are used for residential, commercial, and industrial purposes. Although they serve as natural barriers to storm surge (*q.v.*) and waves (*q.v.*), today about 75% of the world's beaches are subject to erosion (Bird, 1985). In the United States, the percentage of eroded beachfront is somewhat greater than the world average and is estimated by some coastal researchers to approach 90% (e.g., Leatherman, 1988). During the last century, many erosion-control techniques were developed to mitigate the unwanted impacts of erosional events, especially those associated with accelerated rates of erosion in the vicinity of groins, seawalls, or jetties along developed shores. Traditionally, coastal armoring structures such as seawalls, breakwaters, and groins were relied upon to reduce wave energy approaching the shore or to catch sediment moving across or along the shore, and thus provide protection from coastline retreat. Engineering works, however, provide only partial protection and in some cases actually exacerbate the problem they were designed to cure. During the last century, beach nourishment was recognized as an environmentally friendly method of shore protection, especially along the coasts of the western world. Today, artificial beach nourishment is the method of choice for shore protection along many developed coasts with eroding beaches (Figure B24).

Despite the fact that beach nourishment has been used in many hundreds of locations under a wide variety of environmental conditions (e.g., Psuty and Moreira, 1990; Silvester and Hsu, 1993), and frequently integrated with hard structures as part of strategic shore protection efforts, there is much debate about whether the procedure is the best solution to problems of coastline retreat. Although there are many arguments against beach nourishment, artificial supply of beach-sand remains the most practical method of protecting against coastal flooding from storm surges, for advancing the shoreline seaward, and for widening recreational beaches.

Definitions, terminology, and concepts

The term *beach nourishment* came into general use after the first renourishment project in the United States at Coney Island in 1922 (Dornhelm, 1995). In engineering parlance, the terms *beach (re)nourishment*, *beach replenishment*, and *beach restoration* are often used more or less interchangeably in reference to the artificial (mechanical) placement of sand along an eroded stretch of coast where only a small beach, or no beach, previously existed. There are, however, subtle connotations in the application of each term. Sediments that accumulate along the shore in the form of beaches are naturally derived from a variety of sources such as fluvial transport in rivers to deposition of sediments in deltas, from preexisting sediments on the offshore seabed, from chemical precipitates (e.g., oolites on carbonate banks in tropical and subtropical environments), or from organisms living along the shore (e.g., shells and exoskeletons from marine organisms). When the natural sediment supply is interrupted, beaches become sediment-starved and the shoreline retreats landward due to volume loss. Efforts to artificially maintain beaches that are deprived of natural sediment supply thus attempt to proxy nature and (re)nourish the beach by mechanical placement of sand. The beach sediment is thus *replenished* by artificial means. *Beach restoration* implies an attempt to restore the beach to some desired previous condition. A nourished or *constructed beach*

Figure B24 a.

could be placed along a previously beach-less shore, whereas a restored beach is revitalized by the mechanical placement of sediment.

Beach nourishment projects involve placement of sand on beaches to form a *designed structure* so that an appropriate level of protection from storms is achieved. The placement of sand is commonly by methods such as dredging sand from borrow areas on the seafloor (e.g., Finkl *et al.*, 1997), bypassing sand around deepwater inlets or other obstructions (e.g., groins) along the coast that interrupt the littoral drift, or overland delivery of sand from inland quarries to the coast. Although pumping of sand from offshore is the most widespread method of application, due to the large volumes of sediment that are required for most projects, other developments feature placement of sand by trucking or barging from quarries or construction sites, as well as removal of sand from dunes, or relocation of sediment on the berm via *beach scraping* (e.g., Bird, 1990; Healy *et al.*, 1990; McLellan, 1990; Verhagen, 1996). It is now known, however, that removal of sand from dunes is not an appropriate option for sand supply because dune and beach sediment budgets are inextricably interlinked (Psuty, 1988). Although most beach nourishment projects deal with sand-sized particles on low to moderate energy coasts, shore protection efforts are also undertaken in very high-energy environments where gravel beaches are featured (e.g., Zenkovich and Schwartz, 1987; Peshkov, 1993).

Beach nourishment, which entails the construction of new beaches where none existed before or restoration of degraded beaches, usually takes place on developed shores. Along undeveloped shores, beaches provide natural habitat (e.g., nesting grounds for sea turtles and shorebirds) but on developed shores beaches additionally protect coastal infrastructure from storms and are important recreational sites for a globally expanding tourist industry. When beaches are degraded by decreased width and lowering of berms (see entry on *Beach Features*), many communities choose to replace lost sediment by pumping beach-quality sand from offshore to selected renourishment sites onshore.

Arising from reviews of replenishment activities in the United States, a new terminology was developed to describe the ruggedness or persistence of sand placed along the shore to form artificial beaches. Beach durability defines how well the beach performed under a variety of conditions. The definition *of beach durability* by Leonard *et al.* (1990a) states that "... the time between placement and loss of at least 50% of

Figure B24 b.

the fill volume ..." represents the half-life of a beachfill. The identification of *profile evolution*, performance of the beachfill after placement, is an important consideration of durability and longevity.

Historical background to the deployment of beach nourishment for shore protection

The beach at Coney Island, New York, was the first to benefit from a concerted effort at beach nourishment when, in 1922, more than 1×10^6 m^3 of material were dredged from New York Harbor and transported to Coney Island (Hall, 1953; Dornhelm, 1995). Based on the apparent success of this new shore-protection measure, there soon followed a number of other projects along eroding shores elsewhere in New York and along coastlines in New Jersey and southern California. As in the case of Coney Island, beachfills in these early artificial renourishment projects were dredged from sediments in harbors and ship channels. In 1939, Waikiki Beach, Hawaii, was artificially nourished as a recreational beach.

Some other early beach nourishment projects included efforts in Africa and Europe. The construction of jetties for the Durban, South Africa, harbor entrance in 1850 initiated erosion of an adjacent downdrift beach. Groins were built along the shore but they did not stop the aggressive beach erosion. Based on the recommendations of a Belgian engineer, additional long, low groins, in combination with sand bypassing, were added. This was the first attempt at *beach stabilization* using renourishment techniques in South Africa (Swart, 1996). Around 1850, seawalls and groins were deployed along the North Sea coast of Norderney in an effort to stabilize this eroding German barrier island. Although these engineering structures prevented dune erosion, they did not stop the loss of sediments from the beach. In an attempt to alleviate problems associated with coastline retreat due to beach erosion, the first large-scale beach nourishment project in Europe was initiated on Norderney in 1951. By 1989 the beach had been renourished an additional six times (Kunz, 1993).

Beach nourishment projects have been carried out in many other counties including Australia, Belgium, Brazil, Cuba, Denmark, France, Great Britain, Japan, New Zealand, Portugal, Russia (see discussion in

C

Figure B24 c. Beach renourishment on the Gulf Coast of western Florida. (A) Renourishment on Longboat Key, west-central Florida coast near Sarasota showing an eroded beach partly protected by rock revetments and seawalls. In the central foreground there is no beach at high tide. Placement of beachfill is advancing from north to south, as shown in the top of the photo. (B) The restored beach, now 100 m wide, provides a degree of protection from storm surge while providing a much enlarged recreational beach area. (C) Small coastal suction, cutterhead dredger obtaining sand from an offshore borrow near Captiva Key, southwestern Florida. Beach-quality sand is pumped ashore in a slurry via a floating pipeline. About 75% of the renourished beaches on the Florida west coast have a half-life longer than 5 years (Dixon and Pilkey, 1991) (photos: Courtesy of Coastal Planning and Engineering, Inc., Boca Raton, FL).

Walker and Finkl, 2002). Even though the basic goal of beach nourishment is to elevate the beach and advance the shoreline in order to realize all of the consequent benefits of multiple use but especially increased storm protection, the techniques of sand transfer to the shore and design parameters differ among national approaches. In the United States, the State of Florida has a long and distinguished record of beach nourishment along the southeast coast that involves such notable achievements as: (1) the first sand bypassing weir jetty in the world (Hillsboro Inlet, Broward County), (2) the longest continuously operating fixed sand bypass plant (South Lake Worth Inlet, Palm Beach County) in the world, (3) longest half-life of any renourished beach in the United States (Miami Beach, Miami-Dade County), and (4) the first successful groin-aragonite beachfill project in the United States (Fischer Island, Miami-Dade County) (Finkl, 1993; Balsillie, 1996).

Needs for beach nourishment

Although beach erosion (see entry on *Erosion Processes*) is common along most coastlines, it is often difficult to recognize in the field unless there are obvious indications of sediment removal. The development of beach scarps in the berm, dune breaches with overwash, presence of tree stumps or marsh muds on the beachface, and location or damage of buildings precariously close to uprush levels are all signs that beaches are moving landward due to sediment loss. Young *et al.* (1996) describe these features as *geoindicators* that are helpful for evaluating coastline change along beaches. Such indications of beach erosion are important parameters for estimating the sensitivity and extent of the beach-erosion problem and remediation. The removal of beach materials is by wave action, tidal or littoral currents, or wind. Ranges of countermeasures provide protection from beach erosion, foremost among them during the last quarter of the 20th century, being *artificial nourishment* (i.e., the mechanical placement of sand on the beach).

Beach protection measures are necessary because beaches are important natural resources that support multipurpose activities. When well maintained, beaches provide storm surge protection, flood control, recreational activities, and habitat for numerous species of plants and animals (Wiegel, 1988). Lack of proper coastal maintenance may allow beach erosion to reduce dunes and other natural upland protection, increase loss of natural habitats, degrade a major source of tourist revenue, and shrink the overall economy (Strong, 1994; Finkl, 1996). Beaches thus need to be protected because they reduce vulnerability to coastal development in high-velocity areas (see entry on *Global Vulnerability Analysis*).

Although beaches provide a measure of protection to the shore from damage by coastal storms and hurricanes (typhoons and tropical cyclones), their effectiveness as natural barriers against surge flooding depends on their size and shape and on the duration and severity of storms. Beaches are also highly valued as recreational resources that contribute to the economic well-being of many coastal regions in the world. The trend of increasing beachfront development since World War II has resulted in the replacement of dune systems with buildings. This practice has increased exposure of buildings to damage from natural forces (Figure B25), especially high-energy secular events. The presence of buildings close to an eroding coastline enhances the reduction in beach width because the stabilized shore cannot move landward as it would under natural conditions (see entry on *Human Impact on Coasts*). Fixation of the coastline by construction in turn adversely impacts both natural storm protection and recreational quality of affected beaches (NRC, 1995).

The deterioration or degradation of beaches is regarded as undesirable because beaches provide natural protection from storms and have economic value. The unwanted effects of beach erosion commonly place life and property at risk, usually from flooding, and decrease a community's ability to maintain a viable tourist-based economy. Commercial and residential development on upland areas behind beaches and in close proximity to eroding beachfronts are mainly jeopardized by decreasing (eroding) beach widths (e.g., Wiegel, 1988, 1994). Increased potential for economic loss and safety concerns for human life thus drive desires to remediate beach erosion by artificially replacing sand that is lost to erosion.

In addition to the use of beach nourishment for combating coastal erosion, the procedure has been advocated because it: (1) tends to be less expensive and easier to construct than hard structures, (2) is aesthetically desirable, "user friendly" (e.g., Nelson, 1993) and "environmentally green" (Finkl and Kerwin, 1997), (3) provides a source of sand for wind-created or artificially created dunes which add to the protection of inshore areas (e.g., Psuty, 1988; Malherbe and LaHousse, 1998), (4) utilizes "waste products" from dredging or construction projects (e.g., Hillyer *et al.*, 1997), (5) contributes to the littoral sediment budget and may benefit down-drift locations (e.g., Lin *et al.*, 1996), (6) capitalizes on natural processes (e.g., Charlier and De Meyer, 2000) and thus is more acceptable to society, and (7) restores habitat for biota (e.g., Finkl, 1993; Verhagen, 1996).

Causes of beach erosion

Artificial beach nourishment became necessary only when beachfronts were developed for recreational, urban, industrial, and military uses. It is often difficult to understand at once the causes of shore erosion because they can be natural or introduced by human activity along the shore. When induced or accelerated by engineering structures the process is sometimes referred to as *structural erosion* (Pilarczyk, 1990). Beach erosion is influenced by numerous factors such as uplift (e.g., neotectonism) or subsidence (e.g., groundwater withdrawal, compaction of sediments) of the land surface, change in climate patterns (especially storm frequency, deviation of prevailing wind direction), interruption of sediment supply to the coast, eustatic fluctuations of the sea surface, blockage of littoral drift, and construction on the coast. An increase in relative sea level (i.e., drowning of the coast and landward retreat of coastlines) is, however, often cited as the primary geophysical cause of beach erosion (e.g., Leatherman, 1988; Douglas *et al.*, 2000) but many other factors are involved. Construction of dams on major exorheic rivers withholds delivery of sediment to the coast. In the case of the Mississippi River, the sediment load today is about half what it was in pre-dam construction days. Further, sediments that bypass a dam are usually fine-grained and therefore more likely to be carried out to sea and lost to coastal accumulation. Dredging of deep inlets, navigational entrances (see entry on *Tidal Inlets; Navigation Structures*), and the construction of shore protection structures such as jetties are other

Figure B25 Example of an overdeveloped coastal segment along Balneário Camboriú Beach (Santa Catarina State, southern Brazil), where construction of condominiums and tourist facilities restricts the natural dune–beach interaction with phases of seasonal storminess. During perigean spring tides, and storm surges resulting from the passage of atmospheric cyclonic fronts, the beach is under water and the beachfront road and adjacent shops are flooded. Periodic beach replenishment is required to widen the dry-beach width and add height to the berm. Without beach nourishment, developed coastal segments such as this one lose socioeconomic amenities associated with a wide recreational beach and become increasingly vulnerable to flooding.

interrelated factors that contribute to the degradation of natural beach systems. This list is by no means comprehensive and yet it must be concluded that the causes of beach erosion are manifestly complicated and often interrelated.

Although a range of natural processes contributes to the landward retreat of coastlines, urban development along the shore necessitates placement of sand on eroding beaches in an effort to protect infrastructure. The essence of the problem is not the dynamic adjustment of coastlines to fluctuating ambient conditions but construction too close to the water. Most coastline development is deliberate for reasons of access, proximity, or aesthetics. Whatever the initial impetus for developing coastlines, the result has been an expensive exercise in what is usually nationally funded coastal protection. Coastline retreat in Australia, for example, is not as problematical as it is in the United States because the Commonwealth government established Crown lands along most of the coast keeping urban development some distance inland. In other countries such as The Netherlands where land has been reclaimed from the sea, large dikes and other engineering structures (see Walker, 1988) such as dunes and renourished beaches are part of efforts to hold back the sea. Although these areas have multipurpose uses (e.g., storm protection, conservation, recreation, water catchment), they are not open to intensive urban uses (Pilarczyk, 1990).

Coastal development often includes the dredging of ports and harbors and the navigational channels that serve them. Jetties that provide protection from waves in the channel are used to stabilize the geographic migration or orientation of many entrances. Jetties and deep inlets, as well as other shore protection structures such as groins, interrupt the natural littoral drift by impounding sediment or causing it to be jetted offshore beyond the longshore sediment transport system. It is now widely recognized that the interruption of sediment transport along the shore by artificial structures causes downdrift shores to become sediment-starved, which in turn results in shore erosion and retreat of the coastline (Figure B26). Large littoral drift blockers (e.g., deep navigational entrances, groin fields) can initiate downdrift erosion that propagates for several tens of kilometers (Bruun, 1995). In a study of 1,238 km of Florida coastline, distributed among 25 coastal counties and covering about 95% of the state's beaches, Finkl and Esteves (1998) concluded that littoral drift blockers on Florida's Atlantic and Gulf coasts accounted for 72% of the statewide beach erosion. Along the southeastern coast where there are numerous stabilized inlets, and the volume of sediment transported in the littoral drift is relatively small (e.g., $<50{,}000$–$100{,}000\ m^3\ a^{-1}$), littoral drift blockers appear to cause about 90% of the beach erosion. Most of the beach erosion here is thus anthropogenically induced and is, at least theoretically, quite preventable from a technical point of view. In practice, however, remediation of the beach erosion problem is politically recalcitrant.

The causes of beach erosion are complicated interacting processes and it is emphasized that beach nourishment only treats erosional symptoms and does not eliminate the causes. Beachfills are sacrificial in the sense that they are not permanent solutions to the beach erosion problem; they thus provide only temporary protection and it is anticipated that replenishment will be repetitive.

Design of beach nourishment projects

Beach nourishment projects are designed to: (1) increase dune and berm dimensions (i.e., height, length, and width), (2) advance the coastline

Figure B26 Hillsboro Inlet, Broward County, southeast Florida. Erosion of the downdrift coast, seen in the coastal offset (photo center), was caused by stabilization of this inlet by jetties. Subsequent construction of a weir jetty on the updrift side of the inlet permitted sand to overwash into an interior sand trap on the landward side of the jetty. A floating dredge sucks the sand from the trap and pumps it to the downdrift side of the inlet via a submerged pipeline. Dredging is mostly conducted during storms when sediment is overwashed through the weir because the trap quickly fills with sediment. If the sand trap is not cleared as it fills, excess sediment spills into the navigation channel where it becomes a hazard to boaters. This sand bypassing arrangement moves 100% of the estimated net littoral drift to the downdrift beach which is thus maintained in a healthy state.

seaward, (3) reduce storm damage from flooding and wave action, and (4) widen the recreational beach area. Beach nourishment projects are complicated technical procedures that require careful preparation for successful execution of site-specific engineering design (Finkl and Walker, 2002). The scale of mechanical sediment supply is quite variable, ranging from large enterprises that involve federal and local partnerships where sometimes tens of millions of cubic meters of sand are involved to small-beach restorations that may require less than 50,000 m^3 of sand. Equipment for obtaining, transporting, and placement of sand on the beach varies with the scale of the project. Large projects require robust equipment that can handle large volumes under high-energy conditions for open ocean dredging (Figure B27). Placement of sand on small, protected beaches (e.g., pocket beaches, embayed shores) (see also entry on *Bay Beaches*) can often be achieved with small dredges during fair weather conditions (Figure B28). In either case, sediment is often redistributed along the shore by front-end loaders, graders, and tractors to achieve the design profile (see also Figures B24 and B30).

Emergency repair of *erosional hot spots* (localized coastal segments where there is increased erosion and rapid shoreline retreat that dramatically exceeds background rates of erosion, as described by Finkl and Kerwin, 1997) that develop during storms may require only a few thousand cubic meters of sediment until more thorough corrective action can be initiated. Beach nourishment the world over is based on the application of natural sediments, mostly beach-quality sands derived from offshore dredging. Many developed nations now recycle glass products and volumes of recycled glass cullet are increasing yearly. The State of Florida, for example, annually produces in excess of 130,000 tonnes of surplus glass cullet that could be made available for beach nourishment (Finkl and Kerwin, 1997). Glass cullet, a chemically inert form of silica, can be graded (mechanically ground) to desired colors and grain sizes to perfectly match native beach sands. For small projects, costs per cubic meter of placed cullet are usually less than beachfill sands.

The design process specifies the quantity, configuration, and timing of sediment distribution along a specific coastal segment to emulate natural storm protection or recreational area, or both. The design must consider rates of long-term (background) erosion as well as temporal impacts of storms and wave climate to adequately address variables associated with the quantity, quality, and placement of beachfill along the shore. As a general rule, sediment comprising the nourished beach is anticipated to erode at least as fast as the background rate of the pre-nourished coastline. It is usually observed in practice that sediment volume loss rates and coastline retreat for artificial beaches are significantly greater than historical rates for the natural beach (e.g., Dean, 2000), even when differences in grain size and sorting are taken into account (e.g., Ashley *et al.*, 1987). Although an allowance for continued erosion of beachfill is part of the design assessment, the purpose of beachfill design is to maximize the longevity of artificial beaches. The designs can only be optimized by changing the morphological configuration of the beachfill or by the choice of the fill material. The grain-size of borrow material was traditionally considered to be the most important factor for optimizing beachfill. Studies by Eitner (1996), however, indicate that grain-size has little effect on beachfill longevity because grain-size influences the critical threshold stress to a lesser extent than does the grain

Figure B27 Large ocean-going dredge operating off the coast of southeast Florida. The *Illinois*, owned and operated by the Great Lakes Dock and Dredge Company (Oak Brook, IL, USA), is one of the largest suction cutterhead dredges in the United States. The 98-m long dredge, which digs to a nominal depth of 32 m, has a 760 mm discharge diameter and can pump sand 7,635 m without booster pumps. With total installed power of 8,400 kW, and 1,400 kW cutter power coupled with 662,000 L fuel capacity, the *Illinois* is ideal for working offshore for long periods in rough weather conditions. The dredge is not self-propelled and must be moved to projects by tugboats. When on site, the dredge has a large swing radius using a system of anchors and cables to maneuver within about a 300 m range without resetting the anchors. This dredge can work in swell up to 2 m high and usually operates 24 h a day when possible. Beach-quality sand from the borrow area is pumped in a pipeline to the beach. Because the dredge might operate up to 2 km offshore, the pipeline is floated on the surface near the dredge for ease of maneuverability and submerged near the shore for safety of boaters and beach users.

A

B

Figure B28 Small-scale beach renourishment at Alegre Beach, Santa Catarina, Brazil. Local erosion of a beach on the downdrift side of an inlet required sand renourishment for shore protection, recreation, and marine fisheries (beach launching sites for local fisherman). (A) The small coastal dredge pumps sand from the seabed offshore to the beach via a submerged pipeline. (B) Sediment is pumped up onto the upper beachface in a slurry, as shown in the subaerial plume in the photo center. Extra pipe is stored on the berm for future use as the dredge moves about in the offshore borrow area.

density. Only a coarser material such as gravel, which also has a significantly higher critical threshold stress, may effectively extend beachfill longevity. In general, however, most researchers agree that coarser grain sizes produce steeper, more stable, and longer-lived fills (e.g., Bruun, 1990; Pilkey, 1990; Smith, 1990; Dean, 1991). Controversy does, however, surround the kinds of methods used to estimate erosion rates. Houston (1990), for example, emphasizes that designer extrapolations or predictions of replenished beach life of one to several decades is not advisable because beach conditions are too variable and especially vulnerable to cycles of storminess.

There are various methods of *beach nourishment design* that are complimentary in the overall process of optimizing *project performance*. Potential designs are initially evaluated at a preliminary level in which the anticipated project performance is predicted using simple and relatively inexpensive methods. After the performance characteristics are compared with project objectives, the design is refined until the performance predictions confirm an optimal design. For sites without complex boundaries (i.e., straight beaches without terminal groins, inlets, or headlands), prediction tools correctly estimate the time required for renourishment to within approximately 30% of actual project performance (NRC, 1995). Subsequent to establishing the preliminary design, more sophisticated predictive methods are used to optimize the design. This bimodal approach checks preliminary and advanced methods of design, facilitates a rapid and efficient convergence to final design, and provides a clear perspective of how well the design parameters fit project requirements. If the predicted volumetric losses, based on preliminary and advanced methods, differ by more than 50%, the essential elements of the design procedure are reviewed and revised, where necessary.

The design beach

The *design profile* is the shore-normal cross section that an *equilibrated beach* is anticipated to assume. The best estimate of this profile is obtained by the seaward transfer of the natural beach profile by the amount of beach widening that is required (USACE, 1992). Estimates of beachfill volumes are generally increased if the borrow material is finer grained than the native sand. The *construction profile* is the cross section that the contractor is required to achieve. Because the constructed beach, which contains design fill and the advanced-fill volumes, is often steeper than the design cross section due to construction limitations, it is also usually significantly wider than the design profile. Wave action adjusts the construction cross section to a flatter dynamically equilibrated slope within the first few months to a year after placement of the beachfill (cf. Figure B31). Because the dynamically adjusted profile contains *design and advanced fills*, it is wider than the design width during the *nourishment interval* (the time elapsed between nourishment episodes). At the time of renourishment, the design and equilibrium profile are theoretically equal (NRC, 1995).

Mechanical deposition of sediment along a beach nourishment site, during initial construction or renourishment, may not correspond to the natural profile of the beach at the time of placement (Figure B29). In the United States, use of a construction profile rather than a natural profile is the normal placement practice. It is customary for nourishment designs in the United States to establish uniform beach width along a project's length. It is also standard practice to provide sufficient sand to nourish the entire profile from the dune to the depth of significant sand removal, the so-called *depth of closure* (DoC). The DoC is a term used by engineers to define the *depth of active sediment* movement on the seabed (see also entry on *Depth of Closure*). Other terms that are related to this critical concept include profile pinch-off depth, critical depth, depth of active profile, maximum depth of beach erosion, seaward limit of nearshore eroding wave processes, and seaward limit of constructive wave processes. The DoC in beachfill design is defined as "The depth of closure for a given or characteristic time interval is the most landward depth seaward of which there is no significant change in bottom elevation and no significant net sediment transport between the nearshore and the offshore" (Kraus *et al.*, 1998). This definition applies to the open coast where nearshore waves and wave-induced currents are

Figure B29 Emergency beach renourishment at Gravatá Beach, Santa Catarina, Brazil. A coastal highway and commercial infrastructure was threatened by beach erosion during a strong southeaster that brought heavy surf conditions to the coast during the Southern Hemisphere winter of 1999. Removal of the beach by wave and current action, undermined part of the coastal highway interrupting coastal access. Although of finer grain size than the native beach sand, emergency fill was trucked to the site for immediate shore protection.

the dominant sediment-transporting mechanisms. The definition of the DoC infers or stipulates that: (1) the landward water depth at which no sediment change occurs can be reliably identified, (2) there is an estimate of no significant change in bottom elevation and no significant net cross-shore sediment transport, (3) a time frame is related to the renourishment interval or design life of the project, and (4) at the DoC cross-shore transport processes are effectively decoupled from transport processes occurring farther offshore.

Estimates of fill requirements are based on the geometric transfer of the active cross-shore profile seaward by the design amount. If the beachfill grain size matches the native sand and there are no rock outcrops, seawalls, or groins, the design profile (shore-normal cross section) at each alongshore range marker (permanent locations of cross-shore survey sites are typically spaced every 330 m along Atlantic and Gulf coast beaches in Florida) should ideally match the dynamically stable shape of the native beach profile. Cross sections may be more closely spaced in beach nourishment project areas for better engineering control. Enough sediment is included in the design to nourish the entire profile (Hanson and Lillycrop, 1988).

The total sediment volume is independent of the cross-shore profile because the shape of *the renourished profile* is parallel and similar to the existing natural profile. Extra fill is required, however, in front of seawalls in order to achieve the proposed berm elevation. After these seawall volumes are calculated, estimates of nourishment fill volumes are based on seaward translocation of the entire profile. It is emphasized that sand is usually needed along the entire profile, both above and below the water because the beach, by definition, retains subaerial (berm) and submarine (beachface) sections. Placement of the required extra fill volume in front of seawalls typically moves the high-tide shoreline farther seaward than adjoining non-seawall segments. This design requirement, however, causes alongshore gradients in littoral drift that tend to become erosional hot spots. An alternative to providing the additional seawall volumes is to build narrower berms in front of seawalls. Narrower berms are advantageous because they reduce littoral drift gradients that are set up by overly wide sections of nourished beaches in front of seawalls. Similar levels of storm protection (for uplands) are provided by narrower berms when they are backed by seawalls compared to wider berms without them. In many instances, however, beach nourishment in front of seawalls can become problematic, especially where coastline retreat extends landward along coastal segments adjacent to the seawall and where there is deep wave scour in front of the wall (see discussions in Kraus and Pilkey, 1988).

Coastal engineers attempting to predict beach washout and profile response seaward of seawalls often employ beach and dune recession models. Commonly used approaches include EDUNE (Kriebel, 1986), SBEACH (Larson and Kraus, 1990), and GENESIS (Hanson, 1989; Hanson and Kraus, 1989). The numerical models predict the evolution of the cross-shore beach profile toward the so-called equilibrium storm profile (NRC, 1995). Both models are based on principles related to the disparity between actual and equilibrium (theoretical) wave-energy dissipation per unit volume of water within the surf zone. For convenience of calculations, the models assume that sand eroded from the upper beach deposits offshore, with no loss or gain of material to the profile. It is well-known, however, that beach sediment is often transported offshore and is lost from the littoral drift system (e.g., Pilarczyk, 1990). Estimates of storm surge used in coastline recession models, and for calculating run up, are based on USACE (1984, 1986, 1989) engineering manuals. Storm-surge frequencies and extents of coastal flooding are also deployed by the Federal Emergency Management Agency (FEMA) and the National Oceanographic and Atmospheric Administration (NOAA). Storm hydrographs are thus obtained from FEMA, NOAA, and universities to generate probabilities of storm-induced shoreline recession. Wave statistics can be obtained from wave gauge records, published summaries of observations, or wave hindcast estimates such as the Wave Information Study (USACE, 1989). Other methods such as the Empirical Simulation Technique (EST) (Borgman *et al.*, 1992) develop joint probability relationships between various multiple parameters contained in historical data records. Using historical storm records as input, the EST statistically develops a much larger storm-response database while maintaining statistical similarities to original data and it is thus possible to achieve estimates of storm-induced beach recession (Howard and Creed, 2000).

Protocols for overfill on design beaches

Beachfill is usually dispersed out of the nourishment area (i.e., away from the replenished or artificial beach) to adjacent shores or deeper water. The process leading to a decrease in beachfill volume is referred to as "loss," although this sand still temporarily contributes to the stability of the shore in general, but not at the original location. From the point of view of sediment transport, the sand is not "lost" because it is partly retained in the littoral system. From the perspective of the beach manager, however, migration of sediment away from the beachfill represents a tangible decrease of dry beach area.

Erosion of nourished beaches feature two distinct components: (1) the linear regression of the volume of sand in the coastal profile and (2) additional erosion arising from the newly nourished shoreline which becomes more exposed (lying more seaward) than adjacent shore up- and downdrift (Verhagen, 1996). Sediment losses alongshore as well as adjustment or equilibration of the constructed cross-shore profile are responsible for the so-called "additional erosion." In cases where the volume loss associated with the coastal erosion is large compared to the quantity of the beachfill and where the previous rate of erosion is known, a multiplier is used to compensate for all sand loss. Verhagen (1996) suggests a value of 40% extra fill. The presence of structures such as seawalls, due to their interaction with coastal processes, may also require additional fill.

The term *advanced fill* refers to the eroded part of the beach profile before nourishment becomes necessary. The volume and areal distribution of advanced fill is estimated from analysis of the historical rate of erosion and shoreline change. The potential impact of *project fill* on coastal processes is an additional consideration that is taken in account. Procedures used to make these estimates include the historical coastline change method (USACE, 1991) or numerical methods (Hanson and Kraus, 1989). The *historical shoreline change method* assumes that the nourished beach will erode at the same rate as the prenourished beach. This method is commonly employed by beach designers (based on survey results) but can yield a significant underestimate of nourishment requirements (NRC, 1995).

Most long-term erosion (as opposed to episodic storm erosion or development of erosional hot spots) of a nourished beach is initiated by increased gradients of littoral drift along the project length. Major littoral drift gradients affecting the stability of nourished beaches are the preexisting background (regional) rates or historical erosion of the pre-nourished shore and stresses associated with the high-tide shoreline salient that was advanced seaward by the project fill. These perturbations of normal coastal processes are the cause of *end losses* and *spreading* of the fill. All of these littoral drift gradients interact with the nourished beach causing a progressive loss of fill. Exclusive consideration of the background erosion rate neglects end (and spreading) losses, which causes an underestimate of nourishment volume and overestimate of project life. Although losses from the project due to spreading cause accretion on adjacent beaches, they must be included in the advanced-fill design in order to achieve performance objectives (NRC, 1995).

Nourishment profiles

Models of beachfill placement depend on renourishment design schemes that are selected by considering static and dynamic peculiarities of the site, fill requirements, temporal and spatial distribution of natural habitats, and costs. Some of the more common approaches include: (1) placing all of the sand in a dune behind the active beach, (2) using the nourished sand to build a wider and higher berm above mean water level, (3) distributing fill material over the entire beach profile (above and below water), or (4) placing sand offshore in an artificial bar (NRC, 1995). The approach taken partly depends on the location of the source material and the method of delivery to the beach. If the borrow site is a quarry on land and the sand is transported by trucks to the beach (cf. Figure B28), placement on the berm or in a dune is generally most economical. If the material is pumped shoreward from offshore ocean-going dredges (cf. Figure B26), it is usually more practical to place the sand directly on the beach, in the nearshore zone, or to build an artificial bar. If pumped onshore in a sand-water slurry, the sand is subsequently redistributed by grader or bulldozer across the shore to form a more natural beach profile (Figure B30).

The use of large dunes (i.e., man-made dikes) fronted by renourished beaches as an effective coastal protection measure has long been recognized in The Netherlands (Verhagen, 1990; Watson and Finkl, 1990). These constructed dune–beach systems are designed to withstand the 1-in-10,000-year condition of wave intensity and storm surge flooding. This extreme level of protection is justified because entire cities lie behind the coastal defenses.

Bruun (1988) advocates nourishing the entire beach profile, which he terms *profile nourishment*. The main advantage of this approach is that the sand is placed in approximately the same configuration as the existing profile, so that drastic initial adjustments are mostly avoided, especially

Figure B30 Restoration of Captiva Beach in 1996 on Sanibel Island, southwestern Florida, showing beach-quality sand pumped from offshore via a submerged pipeline to the shore. The sand–water slurry debauches from the pipe end (lower center foreground) to form an alluvial fan that builds up along the shore. Shell hunters often congregate about the proximal part of the sediment fan hoping to find collector's specimens. Shore birds search the distal portions of the fan for crustaceans brought up from the seafloor in the offshore borrow area. Road graders and bulldozers redistribute the pumped sediment into the beachfill design shape that often incorporates overfill that will be sacrificed to the littoral system as the beach equilibrates to the ambient wave climate. The beachfill is sometimes initially somewhat darker than the native beach sand, as shown here, due to a small percentage of organic content. Organic matter in the fill bleaches out within a few weeks and the native-colored sands imperceptibly grade into the fill material. Pipe extensions lie at the foot of the foredunes (photo, lower right) and mark the approximate position of the back berm (photo: Courtesy of Coastal Planning and Engineering, Inc., Boca Raton, FL).

the rapid erosion of the nourished berm. When wave action undermines the newly constructed berm, a beach scarp frequently forms along the length of the project fill. These scarps can pose hazards to beachgoers trying to gain access to the water from the berm (Figure B31). In some cases, foot traffic across the scarp tramples the steep slope to a flatter one that cuts into the beachfill. These cuts or beach tracks can provide ingressive pathways for surge and run-up which in turn can accelerate erosion of the project fill volume.

Artificial nourishment of eroded beaches can be indirectly achieved by placing dredged sand in the offshore zone (McLellan, 1990). Dredged material is deposited in shallow water, typically using a split-hull barge either as a mound or shaped as a long liner ridge that simulates a shore-parallel sand bar. It is anticipated that the sand deposited in the offshore mound or artificial bar will migrate onto the beach. Prior to welding onto the beachface, the bar causes waves to break farther offshore, a process that reduces the wave energy on the beach in the lee of the bar. The disposal depth of the offshore nourishment should be shallower than the seaward boundary of active sediment transport (as defined by normal to moderately elevated energy conditions) so that sediment quickly moves onto the subaerial beach.

Mechanical bypass systems

Shore protection structures (*q.v.*) or other coastal construction works can interrupt littoral drift flow patterns and trap sediment near structures, within navigational entrances to port and harbors, and in flood- or ebb-tidal deltas. Sediment trapping by littoral drift blockers causes downdrift beach erosion (e.g., Finkl, 1993; Bruun, 1995). In order to mitigate the downdrift effects of sand starvation along the coast, it is necessary to move sand around barriers in order to supply beaches with sediment. Due to losses of sediment offshore (see previous discussion), the quantity needed for downdrift beach nourishment may be greater than the trapped sediment volume. Some bypassing systems that are geared for normal use may be overwhelmed during large storms while others function best during or immediately after storms when sediment is brought to the *sand trap* (a dredge pit that collects sediment).

Fixed bypassing systems generally are less effective and more expensive to run than mobile floating systems (Bruun, 1993). Most bypassing plants work at less than 50% efficiency, and some at 30%, which means that less than half of the drift is bypassed to the downdrift beaches. The combination of periodic beach replenishment and innovative bypassing techniques is an option that can restore longshore sediment transport and greatly reduce beach erosion (Bruun, 1996). Suggested new alternatives include mobilization of the bypass intakes on rails or cranes, implementation of jet pumps, or seabed fluidizers (Bruun and Willekes, 1992).

Several different kinds of mechanical bypassing systems are used effectively in a variety of coastal settings: (1) mobile dredges in the harbor and or entrance (e.g., Santa Cruz, CA), (2) movable dredge in the lee of a detached breakwater that forms an updrift sand trap (e.g., Channel Islands and Port Hueneme, CA), (3) floating dredge within an entrance using a weir jetty on the updrift side (e.g., Hillsboro Inlet, FL; Boca Raton, FL; Masonboro Inlet, NC; Perdido Pass, AL), (4) fixed pump with dredge mounted on a movable boom (Lake Worth Entrance, FL; South Lake Worth Inlet, FL), (5) jet pumps (eductor) mounted on a movable crane, with main water supply and booster pumps in a fixed building (e.g., Indian River Inlet, Delaware) (NRC, 1995). These, and other installations, and their operational performances are described in engineering and design manuals (e.g., USACE, 1991) which provide guidance for the design and evaluation of sand bypassing systems.

Veneer beachfills

Veneer fills are beach-quality sands that are placed over a relatively large volume of material that is generally not suitable for beach nourishment. The unsatisfactory materials, which may be either grossly coarser or finer

Figure B31 Example of a large beach-replenishment project involving more than 6 million m^3 of sand that was pumped from offshore in southeast Florida near Port Everglades (Fort Lauderdale). As shown in the photo, the beachfill has been eroded back to the native sand beach (left) and the built beach is perched about 2 m above the eroded beach. The beach scarp that developed during erosion of the fill was graded by dragging a heavy I-beam across the sand. The more gradual slope facilitated access to the water by reducing the hazardous vertical face of the scarp.

than normal beach sand, remain as an underlayer beneath the thin surface sand veneer. Veneer beachfills are thus used in situations where beach-quality sand is not available in sufficient quantities to economically undertake a nourishment project. The usual reason for placing a veneer fill is based in economics where the cost is prohibitive if a cross section is totally built of beach-quality sand. Veneer fills are of two basic types: (1) fills where the underlying materials are coarser than typical beach sands (e.g., boulders, coral, rocks) and (2) fills where the underlying materials are finer than typical beach sands (e.g., silts or silty sands where the median grain size is much smaller than native sand). In the United States, veneer beachfills have been used in Corpus Christi, TX; Key West, FL; and Grand Isle, LA (NRC, 1995). A fundamental design problem associated with veneer fills involves selection of a veneer that is thick-enough, so that it will not erode away and expose the underlayer during storms or before scheduled replenishment. Although variable, depending on the local conditions, the thickness of the veneer must provide a sedimentary envelope that incorporates profile variations without compromise.

Pros and cons

In the United States, there are two schools of thought regarding beach nourishment; the larger group favors artificial placement of sand on eroding beaches as part of shore protection measures while the other smaller group discourages the practice on the grounds of environmental, economic, sociological, or political grounds. A recent study by the US National Research Council (NRC) examined the diversity of viewpoints about the success or failure of nourishment projects (Table B2). It was concluded that the factors involved include a large number of interest groups who have different "... viewpoints, objectives, needs, and ideas ..." (NRC, 1995, p. 41).

In the 1980s and early 1990s, there was much debate about the pros and cons of beach nourishment with many illuminating facts coming from both sides of the issue (Pilkey, 1990). As persuasive as many arguments were, the end result was that federal support for new beach nourishment was largely withdrawn when the US Congress removed the USACE from many new projects by reducing or eliminating funding (e.g., Finkl, 1996). If beach renourishment is to continue as a shore-protection measure in the United States, local funds will have to support the practice along many stretches of the coast.

Proponents of beach nourishment favor continuance of the engineering practice for many reasons, the most important of which feature shore protection (mainly flood control against storm surge) and economic value in terms of income from recreational use. The arguments for beach nourishment are legion and include those factors already indicated as part of the needs for shore protection.

Antagonistic to views of beach nourishment are concepts that focus on environmental impacts of offshore dredging (e.g., Nelson, 1993), especially near sensitive environments such as coral reefs and sea-grass beds, and placement of sand along the shore which buries meiofauna and infauna, or which may adversely impact rare species such as sabellariid worm reefs. Divergent opinions also focus on expenditures of public funds for coastal segments that do not provide public access to the beach or interpretations of coastal management practices that call for retreat from the coast. Other issues that are sometimes raised concern beachfill performance (i.e., durability, longevity, half-lives of replenishments) which is keyed into the design life of renourished beaches (e.g., Houston, 1991; Davis *et al.*, 1993; Farrell, 1995). The USACE, for example, often estimated and advertised life spans of a decade or so for many proposed projects. Studies subsequently found that, on average, the life spans of renourished beaches were usually less than anticipated. Durabilities of renourished Atlantic beaches, for example, were found to have a half-life of about 4–5 years. (Pilkey and Clayton, 1989). The percentage of renourished beaches lasting more than 5 years along Atlantic coastal barriers averaged about 65% while those on Gulf and Pacific coasts, respectively, averaged about 75% and 65% (Leonard *et al.*, 1990b; Dixon and Pilkey, 1991; Trembanis and Pilkey, 1998).

Conclusions

Among the different approaches to beach nourishment the world over, are various techniques that essentially boil down to methods of placing suitable sediment along the shore to: (1) maintain an existing but eroding beach, (2) create a new beach where none existed before, or (3) to improve a seriously degraded former beach. No matter the actual method of beach nourishment design that involved acquisition of suitable sediment and placement along the shore, this soft engineering approach to erosion mitigation must be regarded as a temporary solution to a chronic problem. In spite of the fact that in the United States, for example, there has been more than a half-century of experience, beach nourishment remains a procedure with unclear universal application. Experience has shown that there are no simple rules that work everywhere because it is now widely appreciated that local site characteristics must be important criteria in successful design. Peculiarities of local conditions related to bathymetry, sediment grain size or shape and composition, exposure and orientation of the beach to prevailing and storm wind patterns, wave climate, and para- and diabathic sediment

Table B2 Evaluation of beach nourishment projects, based on objectives, interpretation of criteria for success, and various measures of performance (modified from NRC, 1995)

Objective	Criteria for success	Measures of performance
Create, improve, or maintain a recreational beach	A viable recreational asset (acceptable dry-beach width and carrying capacity) during the tourist season	Periodic surveys of beach width using quantitative techniques. Assessment of beach visits and carrying capacity via such methods as aerial photography
Protect coastal infrastructure from wave attack and flooding by storm surge	Sufficient sand, gravel, or cobbles remaining in a configuration that blocks or dissipates wave energy and surge. Hard structures may be included in the solution	Evaluation of structural flooding damage following storms that do not exceed the project design
Maintain an intact dune seawall system	No overtopping during a storm that does not exceed the design water level and wave height limits	Verification of stabilized shore line position
Create, restore, or maintain beach habitat	Episodic erosional extremes do not exceed the design profile. Structures, if allowed, remain intact. Postfill erosion rates comparable to historical values	Profile surveys showing that the sedimentary volume and configuration meet or exceed the design profile
Protect the environment	Sediment volume, areal extent, and condition plus vegetation of the back beach or dune meeting environmental needs	Observation and survey of habitat characteristics and conditions
Avoid long-term ecological changes in affected habitats	Return to pre-nourishment (native beach) conditions within an acceptable time frame	Periodic monitoring of faunal assemblages of critical concern

flux pathways can all affect shore erosion and beach stability. Intricacies of shore processes and their interactions with engineering works such as jetties, dredged channels, groins, and breakwaters, for example, can exacerbate natural shore erosion. Fortunately, it is now realized that many shore protection structures are themselves the main causes of accelerated beach erosion. In southeast Florida, for example, stabilized navigational entrances (i.e., jettied tidal inlets) are responsible for about 90% of the beach erosion problem (e.g., Finkl and Esteves, 1998). As formidable as this figure may seem, it is now evident that improved sediment bypassing at littoral drift blockers, as described by Bruun (1995), can significantly enhance beach nourishment efforts by prolonging what are relatively short life spans of placed sediments. Using a combination weir jetty, interior sand trap, and floating suction cutterhead dredge, the Hillsboro Inlet in Broward County (southeast Atlantic coast of Florida), is able to annually bypass 100% of the estimated net littoral drift (Finkl, 1993). Thus, in many instances, improved sand bypassing at inlets can maintain sediment transport alongshore to supply downdrift beaches with incremental sand.

Shore protection via beach nourishment comes, however, with a high price tag but there often are few options that are practical. Retreat from the shore in highly developed coastal infrastructures is not possible nor is a passive approach where structures or facilities are threatened by beach erosion or coastal flooding. The Dutch, for example, have taken an aggressive approach by actually reclaiming land from the seabed by diking and poldering (see Walker, 1988). Elsewhere, most of the world's developed shores thus face the promise of attempting to maintain present coastlines via beach nourishment.

Even though beach nourishment is the shore protection measure of choice for many coastal managers, the future of the procedure in the short-term (less than 50 years beyond today) may seem bright but in the long term (more than 100 years from today) the prognosis would be poor. If the natural rise in mean sea level (see entry on *Sea-Level Rise*) continues to be exacerbated by human action to the point that relative sea level continues to increase, many coastal areas will experience inability to locate suitable beachfill materials in sufficient quantity to support artificial nourishment. As beachfill materials become scarcer due to increased demands, project costs will escalate, but cost/benefit ratios will probably be favorably maintained because of higher property values per length of coastal segment. Further, if the general rise in eustatic sea level accelerates as some researchers predict, renourished and constructed beaches will be no match for increased vulnerabilities from erosion and storm surges. The problem is, unfortunately, growing as more and more stretches of shore are developed.

Recommendations to improve performance (i.e., life span, durability of beach nourishment projects) include mapping of the shore zone (both subaerial and submarine features) to better understand the topographic features and sediment transport pathways that are related to coastal stability in the local area. Post-project monitoring is another important step that can help assimilate factors that are related to the degradation of beachfill project life. For now, beach nourishment projects meet the needs of many coastal communities that require protection of beaches.

Charles W. Finkl and H. Jesse Walker

Bibliography

Ashley, G.M., Halsey, S.D., and Farrell, S.C., 1987. A study of beachfill longevity: Long Beach Island, NJ. In Kraus, N.C. (ed.), *Coastal Sediments '87*. New York: American Society of Civil Engineers, pp. 1188–1202.

Balsillie, J.H., 1996. Florida's history of beach nourishment and coastal preservation: the early years, 1910–1974. *Proceedings of the 1996 National Conference on Beach Preservation Technology*. Tallahassee, FL: American Shore and Beach Preservation Association, pp. 350–368.

Bird, E.C.F., 1985. *Coastline Changes*. Chichester: Wiley-Interscience.

Bird, E.C.F., 1990. Artificial beach nourishment on the shores of Port Philip Bay, Australia. *Journal of Coastal Research*, **6**(Special issue): 55–68.

Bird, E.C.F., 1996. *Beach Management*. Chichester: Wiley.

Borgman, L.E., Miller, M.C., Butler, H.L., and Reinhard, R.D., 1992. Empirical simulation of future hurricane histories as a tool in engineering and economic analysis. College Station, TX: *ASCE Proceedings, Civil Engineering in the Oceans V*.

Bruun, P., 1988. Profile nourishment: its background and economic advantages. *Journal of Coastal Research*, **4**(2): 219–228.

Bruun, P., 1990. Discussion of: Leonard, L., Clayton, T., and Pilkey, O.H. An analysis of replenished beach design parameters on U.S. East Coast barrier islands. *Journal of Coastal Research*, **6**(1): 15–36. *Journal of Coastal Research*, **6**(4): 1037–1039.

Bruun, P., 1993. An update on sand bypassing procedures and prices. *Journal of Coastal Research*, **18**(Special issue): 277–284.

Bruun, P., 1995. The development of downdrift erosion. *Journal of Coastal Research*, **11**(4): 1242–1257.

Bruun, P., 1996. Navigation and sand bypassing procedures at inlets: Technical management and cost aspects. *Journal of Coastal Research*, **23**(Special issue): 113–119.

Bruun, P., and Willekes, G., 1992. Bypassing and backpassing at harbors, navigation channels, and tidal entrances: Use of shallow-draft hopper dredges with pump-out capabilities. *Journal of Coastal Research*, **4**(4): 687–701.

Charlier, R.H., and De Meyer, C.P., 2000. Ask nature to protect and build-up beaches. *Journal of Coastal Research*, **16**(2): 385–390.

Davis, R.A., Jr., Inglin, D.C., Gibeaut, J.C., Creaser, G.J., Haney, R.L., and Terry, J.B., 1993. Performance of three adjacent but different nourishment projects, Pinellas County, Florida. In Magoon, O. T. (ed.), *Coastal Zone '93*. New York: American Society of Civil Engineers, pp. 379–389.

Dean, R.G., 1991. Equilibrium beach profiles: characteristics and applications. *Journal of Coastal Research*, **7**(1): 53–84.

Dean, R.G., 2000. Beach nourishment design: consideration of sediment characteristics. *Proceedings of the 2000 National Conference on Beach Preservation Technology*. Tallahassee, FL: Florida Shore and Beach Preservation Association, pp. 41–63.

Dixon, K.L., and Pilkey, O.H., 1991. Summary of beach replenishment on the U.S. Gulf of Mexico shoreline. *Journal of Coastal Research*, **7**(1): 249–256.

Dornhelm, R.B., 1995. The Coney Island public beach and boardwalk improvement of 1923. *Shore and Beach*, **63**(1): 7–11.

Douglas, B.C., Kearney, M.S., and Leatherman, S.P., 2000. *Sea Level Rise: History and Consequences*. New York: Academic.

Eitner, V., 1996. The effect of sedimentary texture on beach fill longevity. *Journal of Coastal Research*, **12**(2): 447–461.

Farrell, S. 1995. Beach nourishment at Avalon, New Jersey: A comparison of fill performance with and without submerged breakwaters. In Tait, L.S. (ed.), *Sand Wars, Sand Shortages and Sand-Holding Structures*. Tallahassee, FL: Florida Shore and Beach Preservation Association, pp. 149–164.

Finkl, C.W., 1993. Pre-emptive strategies for enhanced sand bypassing and beach replenishment activities in southeast Florida: A geological perspective. *Journal of Coastal Research*, **18**(Special issue): 59–89.

Finkl, C.W., 1996. What might happen to America's shorelines if artificial beach replenishment is curtailed: a prognosis for southeastern Florida and other sandy regions along regressive coasts. *Journal of Coastal Research*, **12**(l): iii–ix.

Finkl, C.W., and Esteves, L.S., 1998. The state of our shores: a critical evaluation of the distribution, extension, and characterization of beach erosion and protection in Florida. Tallahassee, FL: Florida Shore and Beach Association, pp. 302–318.

Finkl, C.W., and Kerwin, L., 1997. Emergency beach fill from glass cullet: an environmentally green management technique for mitigating erosional 'hot spots' in Florida. *Proceedings of the 10th National Conference on Beach Preservation Technology*. Tallahassee, FL: Florida Shore and Beach Association, pp. 304–319.

Finkl, C.W., and Walker, H.J., 2002. Beach nourishment. In Chen, J., Eisma, D., and Walker, J. (eds.), *Engineered Coasts*. Dordrecht, The Netherlands: Kluwer.

Finkl, C.W., Khalil, S.M., and Andrews, J.L., 1997. Offshore sand sources for beach replenishment: Potential borrows on the continental shelf of the eastern Gulf of Mexico. *Marine Georesources and Geotechnology*, **15**: 155–173.

Hall, J.V., Jr., 1953. Artificially nourished and constructed beaches. *Proceedings, of the 3rd Conference on Coastal Engineering*. Cambridge, MA: pp. 119–136.

Hanson, H., 1989. Genesis—a generalized shoreline change numerical model. *Journal of Coastal Research*, **5**(1): 1–27.

Hanson, H., and Kraus, N.C., 1989. *GENESIS: Generalized Model for Simulating Shoreline Change. Report 1: Reference Manual and Users Guide*. Vicksburg, MS: Coastal Engineering Research Center.

Hanson, M.E., and Lillycrop, W.J., 1988. Evaluation of closure depth and its role in estimating beach fill volume. *Proceedings of Beach Preservation Technology '88*. Tallahassee, FL: Florida Shore and Beach Preservation Association, pp. 107–114.

Hardisty, J., 1990. *Beaches, Form and Process*. London: Unwin Hyman.

Healy, T.R., Kirk, R.M., and deLange, W.P. 1990. Beach renourishment in New Zealand. *Journal of Coastal Research*, **6**(Special issue): 77–90.

Hillyer, T.M., Stakhiv, E.Z., and Sudar, R.A., 1997. An evaluation of the economic performance of the U.S. Army Corps of Engineers Shore Protection Program. *Journal of Coastal Research*, **13**(l): 8–22.

Houston, J.R., 1990. Discussion of: Pilkey, O.H. A time to look back at beach nourishment (editorial) and Leonard, L., Clayton, T., and Pilkey, O.H. An analysis of beach design parameters on U.S. East Coast barrier islands. *Journal of Coastal Research*, **6** (4): 1023–1036.

Houston, J.R., 1991. Beachfill performance. *Shore and Beach*, **59**(3): 15–24.

Howard, S.C., and Creed, C.G., 2000. The empirical simulation technique in predicting storm-induced beach damage. *Proceedings of the 2000 National Conference on Beach Preservation Technology*. Tallahassee, FL: Florida Shore and Beach Preservation Association, pp. 147–157.

Kraus, N.C., and Pilkey, O.H. (eds.), 1988. The effects of seawalls on the beach. *Journal of Coastal Research*, **4**(Special issue): 146p.

Kraus, N.C., Larson, M., and Wise, R., 1998. Depth of closure in beach-fill design. Vicksburg, MS: US Army Engineer Waterways Experiment Station, Coastal Hydraulics Laboratory, CETN II–40 (3/98).

Kriebel, D.L., 1986. Verification study of a dune erosion model. *Shore and Beach*, **54**(3).

Kunz, H., 1993. Sand losses from an artificially nourished beach stabilized by groynes. In Stauble, D.K. and Kraus, N.C. (eds.), *Beach Nourishment Engineering and Management Considerations*. New York: American Society of Civil Engineers, pp. 191–205.

Larson, M., and Kraus, N.C., 1990. *SBEACH: Numerical Model for Simulating Storm-Induced Beach Change. Report 2: Numerical Foundation and Model Tests*. Vicksburg, MS: Coastal Engineering Research Center, *Technical Report* CERC-89-9.

Leatherman, S.P., 1988. Beach response strategies to accelerated sea-level rise. *Proceedings of the 2nd North American Conference on Preparing for Climate Change*. Washington, DC: The Climate Institute, pp. 353–358.

Leonard, L.A., Clayton, T., and Pilkey, O.H., 1990a. An analysis of beach design parameters on U.S. East Coast barrier islands. *Journal of Coastal Research*, **6**(1): 15–36.

Leonard, L.A., Dixon, K.L., and Pilkey, O.H., 1990b. A comparison of beach replenishment activities on U.S. Atlantic, Pacific, and Gulf of Mexico coasts. *Journal of Coastal Research*, **6**(Special issue): 127–140.

Lin, P.C.-P., Hansen, I., and Sasso, R.H., 1996. Combined sand bypassing and navigation improvements at Hillsboro Inlet, Broward County, Florida: The importance of a regional approach. In Tait, L.S. (ed.), *The Future of Beach Nourishment*. Tallahassee, FL: Florida Shore and Beach Preservation Association, pp. 43–59.

Malherbe, B., and LaHousse, B., 1998. Building coastal protection with sand in Belgium. *Journal of Coastal Research*, **26**(Special issue): 101–107.

McLellan, T.N., 1990. Nearshore mound construction using dredged material. *Journal of Coastal Research*, **7**(Special issue): 99–107.

National Research Council (NRC), Committee on Beach Nourishment and Protection, 1995. *Beach Nourishment and Protection*. Washington, DC: National Academy Press.

Nelson, W.G., 1993. Beach restoration in the southeastern US: environmental effects and biological monitoring. *Ocean and Coastal Management*, **19**(2): 157–182.

Peshkov, V.M., 1993. Artificial gravel beaches in the coastal protection. In Kosiyan, R., and Magoon, O.T. (eds.), *Coastlines of the Black Sea. Coastal Zone '93*. New York: American Society of Civil Engineers, pp. 82–102.

Pilarczyk, K.W., 1990. *Coastal Protection*. Rotterdam, The Netherlands: Balkema.

Pilkey, O.H., 1990. A time to look back at beach renourishment. *Journal of Coastal Research*, **6**(1): iii–vii.

Pilkey, O.H., and Clayton, T.D., 1989. Summary of beach replenishment experience on U.S. East Coast barrier islands. *Journal of Coastal Research*, **5**(1): 147–159.

Psuty, N.P., 1988. Sediment budget and dune/beach interaction. *Journal of Coastal Research*, **3**(Special issue): 1–4.

Psuty, N.P., and Moreira, M.E.S.A., 1990. Nourishment of a cliffed coastline, Praia da Rocha, the Algarve, Portugal. *Journal of Coastal Research*, **6**(Special issue): 21–32.

Short, A.D. (ed.), 1999. *Handbook of Beach and Shoreface Morphodynamics*. Chichester: Wiley.

Silvester, R., and Hsu, J.R.C., 1993. *Coastal Stabilization: Innovative Concepts*. Englewood Cliffs, NJ: Prentice-Hall, 578 p.

Smith, A.W.S., 1990. Discussion of: Pilkey, O.H. A time to look back at beach renourishment. *Journal of Coastal Research*, **6**(1): iii–vii; and Leonard, L., Clayton, T., and Pilkey, O.H. An analysis of beach design parameters on U.S. East Coast barrier islands. *Journal of Coastal Research*, **6**(1): 15–36. *Journal of Coastal Research*, **6**(4): 1041–1045.

Strong, W.B., 1994. Beaches, tourism and economic development. *Shore and Beach*, **62**(2): 6–8.

Swart, D.H., 1996. The history of coastal engineering in South Africa. In Kraus, N.C. (ed.), *History and Heritage of Coastal Engineering*. New York: American Society of Civil Engineers, pp. 429–464.

Trembanis, A.C., and Pilkey, O.H., 1998. Summary of beach nourishment along the U.S. Gulf of Mexico shoreline. *Journal of Coastal Research*, **14**(2): 407–417.

U.S. Army Corps of Engineers (USACE), 1984. *Shore Protection Manual*. Washington, DC: U.S. Government Printing Office, U.S. Army Corps of Engineers Publication No. 008-002-00218-9.

U.S. Army Corps of Engineers (USACE), 1986. *Storm Surge Analysis*. Washington, DC: U.S. Government Printing Office. Engineer Manual No. EM 1110-2-1412.

U.S. Army Corps of Engineers (USACE), 1989. *Water Level and Wave Heights for Coastal Engineering Design*. Washington, DC: U.S. Government Printing Office. Engineering Manual 1110-2-1414.

U.S. Army Corps of Engineers (USACE), 1991. *Sand Bypassing System, Engineering and Design Manual*. Washington, DC: U.S. Government Printing Office. Engineer Manual No. EM 1110-2-1616.

U.S. Army Corps of Engineers (USACE), 1992. *Monitoring Coastal Projects*. Washington, DC: U.S. Government Printing Office, Engineer Regulation ER 1110-2-8151.

Verhagen, H.J., 1990. Coastal protection and dune management in The Netherlands. *Journal of Coastal Research*, **6**(1): 169–180.

Verhagen, H.J., 1996. Analysis of beach nourishment schemes. *Journal of Coastal Research*, **12**(l): 179–185.

Walker, H.J. (ed.), 1988. *Artificial Structures and Shorelines*. Dordrecht, The Netherlands: Kluwer Academic Publishers.

Walker, H.J., and Finkl, C.W., 2002. Beach nourishment: case studies. In Chen, J., Eisma, D., and Walker, J. (eds.), *Engineered Coasts*. Dordrecht, The Netherlands: Kluwer Academic Publishers.

Watson, I., and Finkl, C.W., 1990. State of the art in storm-surge protection: The Netherlands Delta Project. *Journal of Coastal Research*, **6**(3): 739–764.

Wiegel, R.L., 1988. Keynote address: some notes on beach nourishment, problems and advancement in beach nourishment. *Proceedings of Beach Preservation Technology '88*. Tallahassee, FL: Florida Shore and Beach Preservation Association, pp. 1–18.

Wiegel, R.L., 1994. Ocean nourishment on the USA Pacific coast. *Shore and Beach*, **62**(1): 11–36.

Young, R.S., Bush, D.M., Pilkey, O.H., and Neal, W.J., 1996. Evaluating shoreline change and associated risk from coastal hazards: an inexpensive qualitative approach. In Berger, A.R., and Iams, W.J. (eds.), *Geoindicators: Assessing Rapid Environmental Changes in Earth Systems*. Rotterdam: Balkema, pp. 193–206.

Zenkovich, V.P., and Schwartz, M.L., 1987. Protecting the Black Sea-Georgian S.S.R. gravel coast. *Journal of Coastal Research*, **3**(2): 201–210.

Cross-references

Beach Erosion
Beach Processes
Bypassing at Littoral Drift Barriers
Cross-Shore Sediment Transport
Dredging of Coastal Environments
Erosion Processes
Management (see Coastal Zone Management)
Natural Hazards
Net Transport
Sandy Coasts
Sediment Budget
Shore Protection Structures
Storm Surge

BEACH PROCESSES

The continuous changes taking place in the coastal zone constitute *beach processes*. A *beach* is one part of the *coastal zone*, which is the transitional area between terrestrial and marine environments. The coastal zone comprises the beach; an underwater region that extends seaward to the depth where waves no longer effect the sea bottom; and inland to features such as *sea cliffs*, *dune fields*, and *estuaries* (*q.v.*). Beach studies focus on understanding spatial and temporal changes in alongshore and cross-shore geomorphic features of the beach (*Beach Features: q.v.*) and in the size and composition of beach sediment. Over time, the coastal zone changes in character in response to changes in wave climate and other physical processes.

Comprehensive knowledge of beach processes is crucial to society because the majority of the world's coastlines are eroding (Thornton *et al.*, 2000). Moreover, *sea-level rise* (*q.v.*) from *global warming* could accelerate coastal land loss, resulting in an increasing rate of loss of coastal habitat and structures. Coastal land loss has a large negative societal impact because approximately two-thirds of the world's population lives adjacent to the ocean or large inland bodies of water. In the United States, more than half of the population lives within 50 miles of the shore, while about 85% of the sandy shore is eroding from a combination of damming of rivers, inlet improvements, sand mining in the coastal zone, sea-level rise, and large storms (*q.v.*). Understanding the processes that cause land loss on sandy coasts is particularly important because beaches are a popular recreational area; are essential to commerce; and protect coastal cliffs, dunes, and structures.

Beaches continually change in response to changes in wave conditions. The changes occur over both short- and long-terms, reflecting both subtle changes associated with daily or weekly variations in tidal level or wave climate and gross changes associated with seasonal variations in wave climate. Beaches usually shift back and forth between a wide, built-up berm with a barless nearshore zone and a small-to-absent berm with one or more well-developed bars in the nearshore. These fluctuations often occur on top of an equilibrium profile that exhibits no net long-term change.

Textbooks that discuss beaches, the coastal zone, and the processes that affect them include Johnson (1919), Guilcher (1958), Shepard (1963), Shepard and Wanless (1971), Bascom (1980), Komar (1983, 1998), Bird (1984), Carter (1988), Carter and Woodroffe (1994), and van Rijn (1998). A report by Thorton *et al.*, (2000) summarizes the state of coastal processes research as of 1999.

Beaches

Many classifications exist to describe different types of coast. Van Rijn's (1998) classification consists of mud, sand, gravel/ shingle, and rocky coasts. In this classification, approximately 10% of the coasts are mud and 15% are terrigenous sand and carbonate sand coasts. Because of their large societal importance, most studies of coastal change have been conducted on sand coasts, which typically are wave-dominated environments (*q.v.*).

The beach is the most prominent visual feature of sand coasts. It is the area of unconsolidated material that extends landward from the low-water line to the place where there is a definite change in material or physiographic form (e.g., a cliff or dune), or to the line of permanent vegetation (*Coastal Boundaries: q.v.*). Beach material can be any combination of sand (grain size = 0.0,625–2 mm), granules (2–4 mm), pebbles (4–64 mm), cobbles (64–256 mm), and boulders (>256 mm) (*Sediment Classification: q.v.*). The grain size on most beaches is in the sand range.

Grain size and composition

In temperate climates, beaches typically consist of quartz and feldspar grains derived from the weathering of terrestrial rocks. Commonly, denser minerals (*heavy minerals*) that are specific to also occur in small percentages. On volcanic islands, the sand frequently includes lava fragments and associated minerals. Worldwide, many beaches have a calcium carbonate fraction from the breakup of shells, concentrations of foraminiferans, and nearby coral reefs (*q.v.*). For example, on the island of Hawaii, beach sand can be black, green, or white. The black sand comes from the erosion of lava beds and decomposition of hot lava flowing into the ocean; the green sand comes from the mineral olivine, which crystallizes when magma cools; and the white sand consists of calcium carbonate. Although calcium carbonate beaches are usually associated with the tropics, beaches in other climes also can have a large shell fraction.

Because the heavy-mineral fraction in beach sand is indicative of the provenance of that sand, heavy minerals can be used to trace the movement of sand along the beach. For example, Trask (1952) showed that the sand reaching the harbor at Santa Barbara, CA comes from the area around Morro Bay, CA, more than 160 km up coast. He established this by using the mineral augite as a tracer. Along the stretch of coast between the two cities, augite occurs in all beach sands, but the only source area is in the vicinity of Morro Bay. The concentration of augite decreases between Morro Bay and Santa Barbara as non-augite sources (e.g., local streams and cliffs) add to the sand moving along the coast.

Beach profile

Figure B32 shows a typical shore-normal cross section, or *profile*, of the coastal zone. The coastal zone can be divided into four major subzones: *backshore*, *foreshore*, *inshore*, and *offshore*. The locations of the beach subzones depend on whether the beach has accreted or eroded (*Beach Erosion: q.v.*). The offshore and inshore are seaward of the low-water line. The former is a relatively flat part of the profile seaward of the breaker zone. The latter includes the breaker and surf zones (*q.v.*). The foreshore is the sloping portion of the profile (typically 2–10°) and encompasses the normal intertidal part of the beach. On an eroding beach, the foreshore could cover the entire beach. The *beach face* is the upper portion of the foreshore that is normally exposed to wave *uprush*; it is often synonymous with the foreshore. On an accreting beach, the *berm crest* is the transitional area between the foreshore and backshore; often it is a striking feature several meters above low water, especially on cobble and boulder beaches (Guilcher, 1958). The backshore extends landward from the berm crest to the edge of the beach; the nearly horizontal part is the *berm*, which forms from the deposition of lower-foreshore sediment that is transported over the berm crest. There can be more than one berm on a backshore. A nearly vertical escarpment (*scarp*) caused by erosion of an earlier berm crest can separate berms on multi-berm beaches.

The loss of beach sand usually corresponds to a gain of sand in the nearshore, and *vice versa.* Beach sand also can be lost to sanddunes, estuaries, and submarme canyons Human activities can result in both beach loss and gain. *Pocket beaches* can be an exception where the sand often shifts alongshore with changes in wave climate. Visually, the result is a shift in the profile of the beach in both the horizontal and vertical. Although such changes are commonly associated with winter

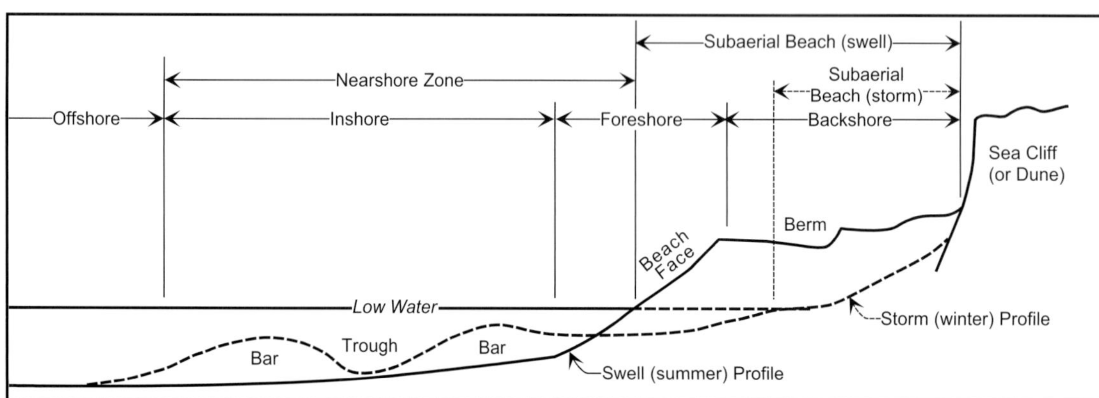

Figure B32 Terminology for the coastal zone along a shore-normal profile.

(erosion) and summer (accretion), in reality they occur any time of year in response to stormy and fair weather. *Storm waves* erode the berm and shift the shoreline (defined as the high-water line or wet–dry boundary) landward. *Swell waves* build up the berm and shift the shoreline seaward.

For sandy beaches, Wiegel (1964) developed a relationship between median grain size, slope of the beach face, and wave climate using data from many US beaches. Grain size and beach slope were measured at mean tide level where the correlation was best (Bascom, 1951). The relationship shows that that the slope depends on two factors—grain size and wave exposure (Figure B33). For any given wave climate, slope increases with increasing grain size. Correspondingly for a given grain size, the smaller the waves, the steeper the beach face. Thus, a beach face becomes flatter when eroding (larger, storm waves) and steeper when accreting (smaller, swell waves).

Beach processes

A complete understanding of beach change requires an understanding of the processes active throughout the coastal zone. Accordingly, beach processes is a subset of *coastal processes*, or *coastal morphodynamics* (morpho-: form, structure; dynamics: motivating or driving forces). Thus, coastal processes involve investigations of the interactions of coastal-zone features and hydrologic, meteorologic, and fluvial forces by means of sediment transport (*q.v.*). Coastal geomorphology and fluid dynamics couple at a continuum of temporal and spatial scales such that the fluid dynamics produces sediment transport, which produces geomorphic change. Progressive modification of the geomorphic features in turn alters boundary conditions for the fluid dynamics, which evolve to produce further changes in sediment-transport patterns and consequently the geomorphic features (Cowell and Thom, 1994).

Coastal processes happen over a wide range of spatial and temporal scales; the upper limits are generally set at 10 km and 1 yr, respectively. For example, the properties of waves entering the coastal zone from deep water and interacting with the nearshore profile determine the overall characteristics of nearshore waves and flows. However, small-scale processes control the turbulent dissipation of breaking waves, bottom boundary layer, and bedform processes that determine the local sediment flux. Cross- and longshore variations in waves, currents, and bottom slope cause spatial gradients in sediment fluxes resulting in changes due to erosion or accretion. Traditionally, the study of coastal processes has been restricted to small and intermediate scales (Thornton *et al.*, 2000); making it but one of several influences on the coastal zone (Figure B34).

The rate of response of geomorphic features to the fluid dynamics depends on scale; larger features take relatively longer to change (temporal scale in Figure B35). Hence, equilibrium is almost instantaneous for small-scale processes, and quasi-equilibrium becomes more noticeable as the geomorphic scale increases. For example, bedform scale and forcing history link the rate of transition between bedform types inside and just outside of the surf zone. Under large waves, significant changes in small-scale bedforms can occur within a single wave cycle, but changes in large-scale bedforms can exhibit significant hysteresis.

Wind, waves, tides, storms, and stream discharge are important driving forces in the coastal zone. Streams transport sediment from the

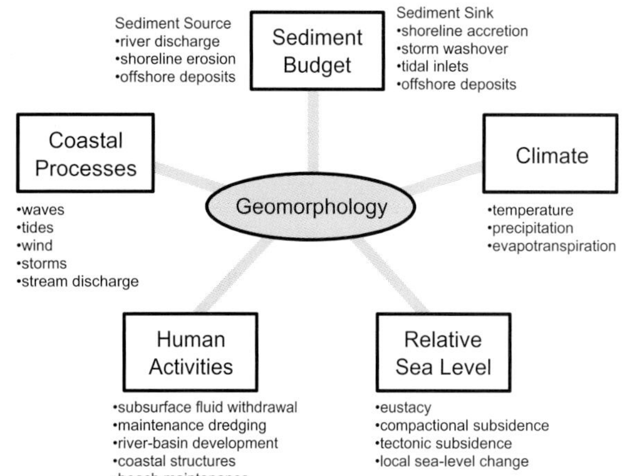

Figure B34 Processes that influence the geomorphology of the coastal zone (Thornton *et al.*, 2000).

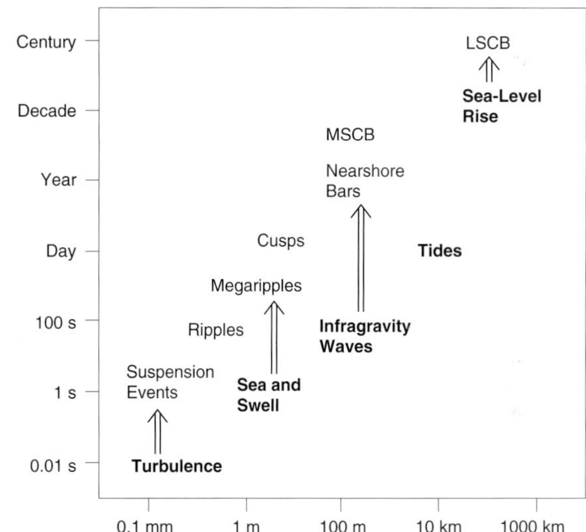

Figure B35 Temporal and spatial scales for coastal processes (Ruessink, 1998). Fluid forces are boldfaced.

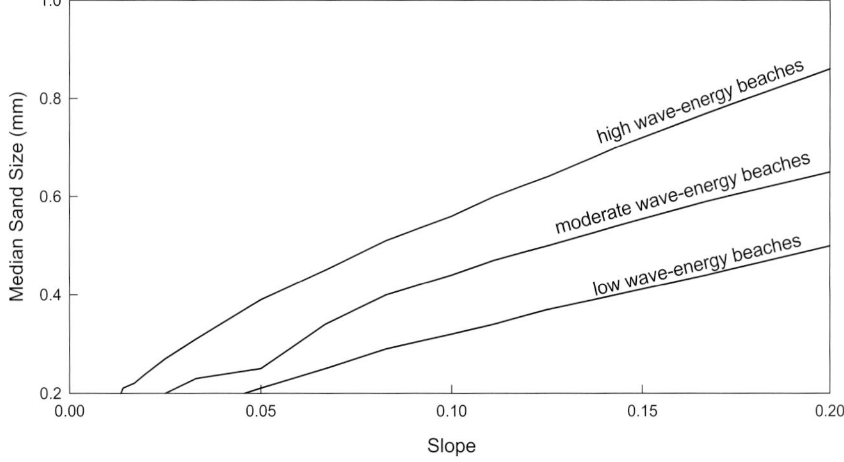

Figure B33 Beach slope as a function of sand size and wave energy.

hinterland to the coastal zone for the other forces to distribute. Wind directly transports beach sand and generates waves (*Meteorological Effects: q.v.*). Waves produce currents that transport sand cross-shore and longshore. Tides play a supporting role by exposing different parts of the beach face to waves and currents. Storms produce strong wind, waves, and elevated sea level, which can cause extensive coastal erosion; movement of sand to the nearshore or across barrier islands (*q.v.*) and spits (*q.v.*); and coastal flooding, as well as intensifying small- and intermediate-scale processes.

Waves are the major source of energy driving beach change in the nearshore. As a wave approaches the coast, it reaches a water depth where it begins to interact with the bottom (*shoal*). That depth occurs when the ratio of water depth to deep-water wavelength is less than 0.5, which is commonly in the depth range of 10–20 m. *Wave base* is the term Bradley (1958) gave to the depth at which normal wave erosion begins (also see Dietz, 1963). *Depth of Closure* (*q.v.*) is the maximum depth of extreme bottom changes; it is a function of the nearshore storm-wave height exceeded 12 h per year and the associated wave period (Hallermeier, 1981). Shoreward, a wave undergoes a systematic transformation (*shoaling*) where wavelength (L) decreases and height (H) increases (H/L is the wave *steepness*). When the ratio of the wave height to water depth reaches 0.73–1.03 (Galvin, 1972), the wave breaks, producing a beachward flow (*bore*). The result is an upward (*run-up*) and downward (*backwash*) flow of water on the beach face (the *swash zone*). If run-up reaches the back of the beach, it can erode cliffs, dunes, and structures.

Sand transport begins soon after waves begin to shoal. Transport volume and velocity increase shoreward in proportion to increasing intensity of the wave-bottom interaction. Whether there is beach erosion or beach accretion depends on wave height and period. Storms raise sea level by piling up water against the coast (*storm surge*) (*q.v.*), and greatly increase wave height and steepness (*storm* waves). Such waves tend to produce beach face erosion with the sediment being moved to the nearshore. Lower, less-steep waves (*swell* waves) produce accretion by moving nearshore sand onto the beach face. Storm and swell waves can occur any time of the year, although the former are more common in the winter and the latter in the summer. Consequently, the belief that erosion is a winter phenomenon (season of storm waves) and accretion a summer phenomenon (season of swell waves) is not completely correct. Basically, the wave climate continuously varies, and the beach face never reaches an equilibrium state where the volume of sand moving up the beach face equals that moving down.

Breaking waves create a circulation system where the water driven shoreward across the surf zone returns to the offshore via a strong, narrow flow called a *rip current* (Figure B36) whose spacing ranges from tens to hundreds of meters. Velocities in rip currents are often strong enough to carry sediment and swimmers bathers alongshore through the breakers; in those cases, a distinct watercolor demarcates the rip current. Often the velocity is too strong to swim against, so people caught in a rip current must swim parallel to shore to escape.

If a wave enters the coastal zone at an angle to the bathymetric contours, its crest bends to align with those contours (*refraction*). If the wave breaks at an angle to the beach, a longshore current develops. Because sand grains move with the flow, longshore sediment transport (*q.v.*) occurs (*littoral drift*). Because the sand grains are subject to both run-up and littoral drift, they follow a zigzag path along the beach face.

Geomorphic features

Based on the temporal and spatial scales of sediment transport and geomorphic change, the coastal zone can be divided into two cross-shore subzones—the upper and middle shoreface—and three longshore subzones—micro, meso, and macro cell. In the upper shoreface, breaking waves and bores generate active sand transport and rapid geomorphic response. In the middle shoreface, slow sand transport rates result in slow geomorphic change. Micro cells include smaller geomorphic features such as ripples and small beach-face features that change in times of a day or less. Meso cells include geomorphic features such as sand bars (*q.v.*), beach profiles, and beach cusps that change in a year or less. Macro cells extend kilometers and include large coastal geomorphic features (Figure B35).

Micro-scale beach features

When waves begin to shoal, there is a back-and-forth water motion at the seafloor with onshore and offshore excursions being equal. The sediment moves the same way, and symmetric *oscillation ripples* form normal to the direction of wave advance. Nearer to shore, the water motion at the seafloor becomes asymmetric because the onshore component of the wave orbits is larger than the offshore one. These *current ripples* have a gentle seaward facing slope and a steep onshore one. Near the breaker zone and in much of surf zone, the bed is flat because of intense water motion. However, current ripples can form in the seaward flowing *rip currents*.

At the upper swash limit, deposition of debris forms scalloped, marginal lines known as *swash marks*. Backwash flowing around small obstacles form seaward opening "V"s, and in some cases, rhombic patterns develop as a result of the minor currents generated in the backwash. Seepage of interstitial water down the foreshore slope at low tide cuts miniature channels termed *rill marks*.

Meso- and macro-scale beach features

Generally, sediment eroded from the foreshore and transported offshore forms one or more longshore bas with a trough shoreward of each one. The bars can extend alongshore for kilometers except for breaks caused by rip currents. Studies by Evans (1940), Keulegan (1948), King and Williams (1949), and Shepard (1950) concluded that there is a strong relationship between the bars and breaking waves. Keulegan, who studied bars formed in a laboratory wave channel, found that wave height and steepness govern the bar position. An increase in wave height moves the bar seaward (deeper water). Holding the wave height constant and reducing the wave period moves the bar shoreward. Starting with a smooth profile, the bar initially forms just shoreward of the breaker position. As it grows, both the bar and breakers migrate landward. When the bar is fully developed, it modifies energy transfer in the surf

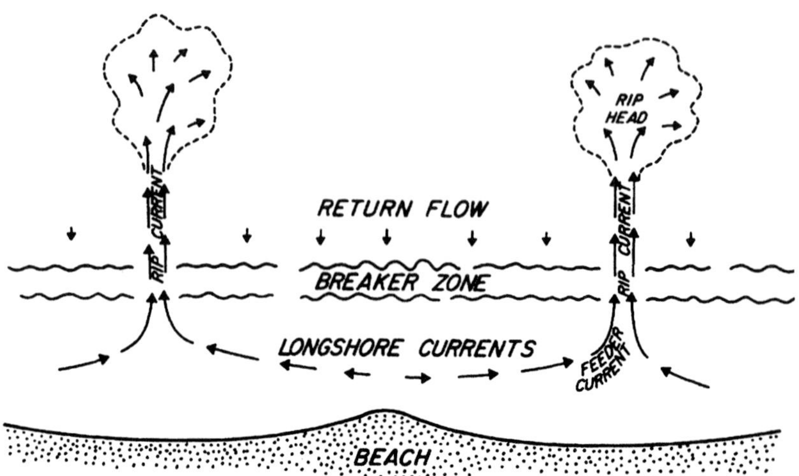

Figure B36 The nearshore circulation cell. Breaking waves create a circulation system where the water-driven shoreward across the surf zone travels alongshore (*longshore current*) and returns to the offshore via a strong, narrow flow (*rip current*) (after Komar, 1998).

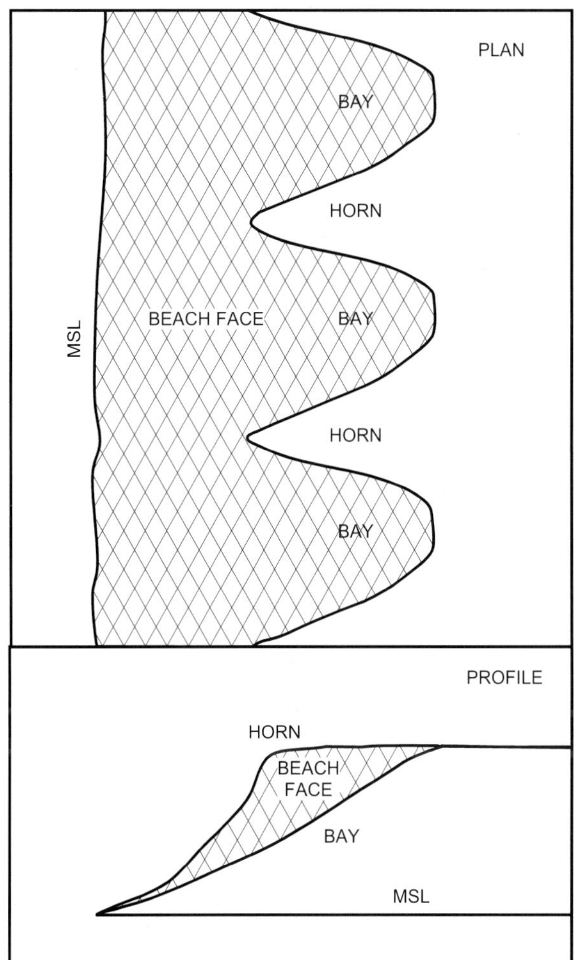

Figure B37 Plan and profile views of a cuspate beach.

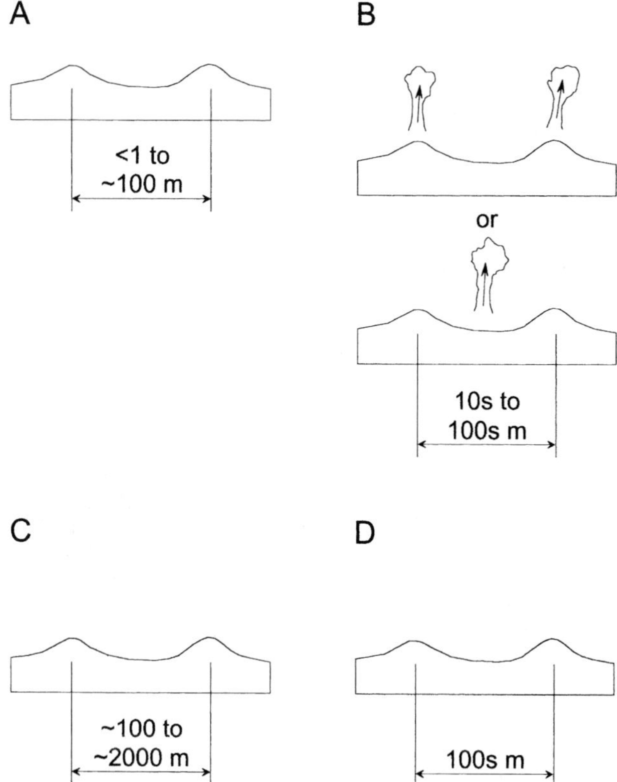

Figure B38 Classification of rhythmic shoreline forms: (A) reflective beach cusps; (B) rip current embayment cusps; (C) crescentic bay cusps; (D) transverse and oblique bars (after Komar, 1998).

zone by reducing the amount of energy that the breaking waves can impart to the waves that reform in the surf zone. Shepard (1950) found that there can be no bar and trough for short-period storm waves because there is not a well-defined breaker zone.

Beaches can be two- or three-dimensional based on the linearity of the berm and beach face. On two-dimensional beaches, the berm crest and foreshore contours are straight and parallel. On three-dimensional beaches, the berm crest and foreshore have rhythmic, crescentic features (*Beach Cusps: q.v.*) that vary greatly in height and length. All cusps are characterized by seaward facing ridges or *horns* separated by *embayments* or *bays* (Figure B37). Attempts to classify these features by size (e.g., Dolan and Ferm, 1968; Dolan *et al.*, 1974) have been unsuccessful because there is a large overlap in sizes of cusps formed by different mechanisms. Komar (1983) developed a classification scheme with four types of cusps whose origins can be attributed to different processes of waves and currents within the nearshore (Figure B38). These are reflective beach cusps, rip current embayment-cusp systems, crescentic bar-cusp systems, and transverse and oblique bars (see Figure B38 for wavelengths).

There has been much disagreement as to the origin of reflective beach cusps. One recent theory is that *edge waves*, waves trapped at the shore with net motion in the longshore direction and decreasing amplitude offshore (Holman, 1983), play a role in the formation of beach cusps. In a laboratory wave tank, Guza and Inman (1975) found that beach cusps developed in response to edge waves. In the field, Huntley and Bowen (1978) observed the formation of cusps in the presence of edge waves. Werner and Fink (1993) proposed that beach cusps form through strong, positive feedback between wave run-up and beachface topography.

A rip current embayment-cusp system develops when rip currents erode the beach face creating embayments. The cusps are midway between the embayments (Bowen and Inman, 1969). The cusps correspond to positions of zero longshore sediment transport produced by waves combining with the feeder currents that flow alongshore toward the rip currents (Figure B36). In some cases, rip current erosion can be so extensive that the embayments cut across the beach, exposing foredunes and cliffs to wave attack (Komar and Rea, 1976; Komar and McDougal, 1988). Beach sediment can be deposited in the lee of a rip current so that the cusps correspond to the rip locations. This appears to occur most commonly on steep beaches under relatively low wave conditions (Komar, 1971).

Crescentic bars are rhythmic lunate features with uniform spacing primarily found underwater, commonly on long straight beaches (Shepard, 1952; Hom-ma and Sonu, 1963). They appear to be confined to regions of small- to medium-tidal range (Bowen and Inman, 1971) and form best where the beach slope is low. The generally accepted mechanism for their formation is by edge waves in the infragravity range (Bowen and Inman, 1971).

Transverse and oblique sandbars are nearshore features that do not parallel the beach, and they are welded to the beach face. There are various explanations for their formation that include the rotation of rip-current segmented bars (Sonu, 1972), processes akin to the migration of river bars or the development of river meanders (Bruun, 1954; Sonu, 1969; Dolan, 1971), and the superposition of edge waves (Holman and Bowen, 1982).

Beach change

Beach cycles

Various cycles affect the beach, depending essentially on changes in wave steepness and effective sea level. In regions of tidal action, the swash–backwash zone and breaker zone shift landward and seaward with the flood and ebb tide. The range of the tides and the slope of the foreshore determine the distance over which the shift takes place. With other parameters remaining constant, breaker height will be greater at high than at low tide. At high tide, the prevailing waves approach less hindered over the relatively deeper water of the nearshore, whereas at low tide, approaching waves are modified by the shoaler depths of the gently sloping nearshore bottom. Under this changing regime, scour may be slightly increased at high tide.

At approximately 7.5-day intervals, the tidal range changes from minimum (neap) to maximum (spring). The change from neap to spring tides can produce upper-foreshore erosion with lower-foreshore and nearshore deposition (LaFond, 1938; Inman and Filloux, 1960). LaFond and Rao (1954) postulate the cumulative result of higher effective sea level, higher waves, and a time lag in recovery during the low spring tide cause the redistribution of sand. The opposite sand movement can occur during neap tides.

Effects of beach erosion

Erosion strips sediment from the beach face and moves it to the nearshore or redistributes it alongshore. When enough sand is removed, there is no longer a high, wide beach, and waves can attack coastal features such as cliffs, dunes, and anthropogenic structures. In some places, storms transport beach sand across spits and barrier islands and deposit the sand in adjacent lagoons. If the sand returned to the beach is less than the volume eroded, the beach narrows, and if possible, the shoreline shifts landward.

Shoreline retreat is a natural process that is of little or no concern in unpopulated areas. However, in populated areas, shoreline retreat is a major issue. Several years can pass between storms severe enough to cause significant damage to a stretch of coast. Consequently, many people build or buy homes and other facilities on the coast with the idea that the adjacent beach is permanent. Later they watch storm waves remove the beach sand and directly attack their property or the coastal cliffs and dunes that protect them. Then, affected communities quickly want to know how to save their beaches and protect their homes and facilities. Although beaches usually rebuild after storms, a beach does not always return to its pre-storm position, and the community must take remedial measures to reverse long-term shoreline retreat.

Because storm waves threaten coastal facilities when fronting beaches lack sand, post-storm beach accretion is essential to minimize economic loss in the coastal zone. In areas where there is insufficient beach to protect coastal structures, there are several procedures to prevent or mitigate shoreline retreat. Traditionally, these procedures require building a protective structure (*q.v.*) on the beach or at the landward edge of the backshore. These include *seawalls*, *revetments*, *groins*, and *breakwaters*. While these structures often protect the property behind them, the fronting beach typically does not return because increased water turbulence at the structure prevents sand deposition during swell conditions (see Dean, 1999 for examples). The result is a section of coast with no beach, and if longshore sand transport is not properly taken into account, the shoreline downdrift of a structure also can lose its sand and retreat. Furthermore, structures often fail if improperly designed, allowing coastal retreat to resume.

Beach nourishment (*q.v.*) is another technique used to prevent shoreline retreat by augmenting the native beach sand with sand imported from other areas. Although beach nourishment creates wide beaches, this technique may not provide a long-term solution to beach loss especially where erosion rates are high or there is a persistent problem. A major problem is that the cost of importing sand can be high, especially since the sand should be similar in character to the native sand and because more sand is frequently needed after a storm season. When this technique is successful, there will be a year-around beach for public use and shoreline protection.

Other techniques include relocating coastal structures to allow for shoreline retreat and defining setback lines for coastal development. Shoreline retreat permits nature to take its course, but often is infeasible in populated areas. Setback lines, which are based on historical shoreline retreat rates, need to be implemented before coastal development begins.

Sediment budget and littoral cell

The budget of littoral sediments is simply an application of the principle of conservation of mass to the littoral sediments—the time rate of change of sand within the system depends upon the rate at which sand enters the system versus the rate at which it leaves. An analysis, therefore, involves evaluations of the relative importance of various sediment sources and losses to the nearshore zone, and a comparison of the net gain or loss with the observed rate of beach erosion or accretion.

The budget of littoral sediment involves making assessments of the sedimentary contributions (credits) and losses (debits) and equating these to the net gain or loss (balance of sediments) in a given sedimentary compartment or littoral cell (Bowen and Inman, 1966; Komar, 1996). The balance of sediments between the credits and debits should be approximately equivalent to the local beach erosion or accretion.

Table B3 The budget of littoral sediments

Credit	Debit
Longshore transport into the area	Longshore transport out of the area
River transport	Wind transport away from the beach
Sea cliff erosion	Offshore transport
Biogenous deposition	Solution and abrasion
Hydrogenous deposition	Mining
Wind transport onto beach	
Beach nourishment	

Table B3 summarizes the possible credits and debits of sand for a littoral sedimentary budget, while some of the more important components are diagrammed in Figure B39. In general, the longshore movement of sand into a littoral compartment, river transport, and seacliff erosion provide the major natural credits; longshore movement out of the compartment, offshore transport (especially through submarine canyons), transport into estuaries, and wind transport shoreward to form sand dunes are the major debits. Included in Table B3 are the major human-induced credits and debits, including beach nourishment, which is increasingly used to rebuild lost beaches, and mining (*q.v.*), which directly removes sediment from the nearshore.

Research techniques

More is known about the geomorphology of the coastal zone than about the processes that modify the geomorphology. For example, it has only been during the past decade that the coupling between waves, circulation, and changes in nearshore bathymetry has begun to be observed and modeled. In addition, research is now focusing on fluid velocities and particle flux profiles in the bottom boundary layer and in the surface boundary layer under breaking waves. Studies of velocity and sediment concentration measurements in the swash zone are moving forward with the development of new instruments.

At present, it is not possible to forecast the effect of an upcoming storm season on a section of coast. However, it is possible to ascertain both the ultimate storm profile and the rate at which a beach returns to its original profile or shifts to a new equilibrium profile. The principle method for obtaining quantitative data on beach change is to repeatedly survey a beach. Comparing the resulting profiles will give erosion and accretion rates for the time encompassed by the surveys.

Geomorphology

Various surveying techniques are available to determine the geomorphic character of the coastal zone. Offshore, a boat-mounted depth sounder can be used to measure the bottom profile (*Bathymetric Surveys: q.v.*), and a side-scan sonar can be used to collect oblique views of the bottom. Both of these techniques are limited to depths where boats can safely operate, which does not include the breaker and surf zones. Furthermore, their accuracy is limited because of boat movement by waves and currents.

Vehicles have been developed to measure beach profiles across the nearshore. These include remotely controlled tractors (Seymour *et al.*, 1978; Dally *et al.*, 1994), sleds (Sallenger *et al.*, 1983), and an 11-m high motorized tripod that can drive across the beach and into the nearshore (Birkemeier and Mason, 1978). MacMahan (2001) mounted an echosounder and a *global positioning system* (GPS) (*q.v.*) unit onto a waverider to profile from deep water through the surf zone.

Surveying with a rod and transit is the most common method used to obtain beach profiles. Techniques range from the "Emery Board" procedure that uses two wooden rods separated by a rope, to sophisticated instruments that use light beams to measure the distance to a prism. The vertical accuracy of the latter instruments can be less to than 1 cm. With all these instruments, however, a closely spaced grid of points can be difficult to achieve except in small beach areas. When a beach is two-dimensional, a single cross-shore survey is sufficient to characterize it, but a beach with cusps and other three-dimensional features requires multiple cross-shore and, in some cases, alongshore surveys. With temporal surveys along a fixed shore-normal line, the existence of cusps can cause errors in comparing beach volume and beach-face location. At present, kinematic GPS units mounted on survey rods and various kinds of vehicles are being used to rapidly survey large sections of beach (Kaminsky *et al.*, 1998).

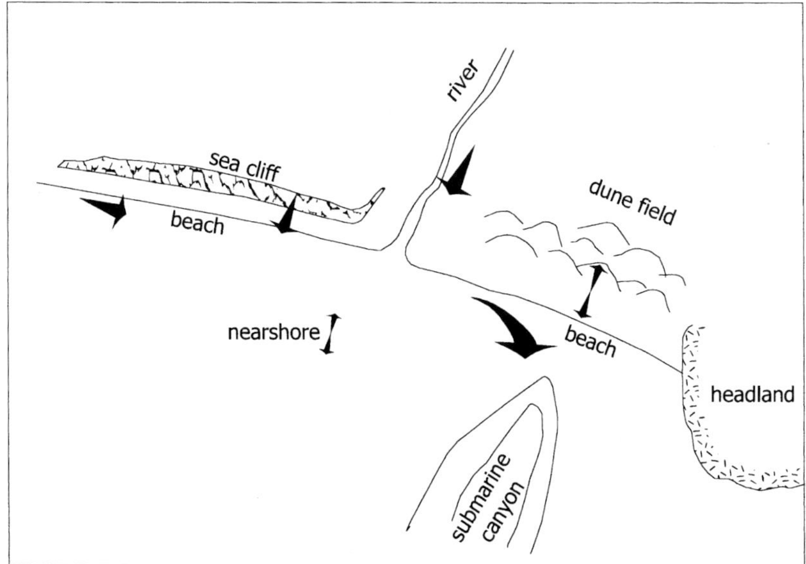

Figure B39 Principle constituents of a littoral zone sediment budget in a natural setting. Size of arrows is a rough indication of relative importance, though they can vary for a given situation. Not shown are anthropogenic impacts such as sand mining (after Komar, 1998).

Remote sensing techniques can be used to study large sections of the subaerial part of the coastal zone. Some of the techniques also can be used to look at geomorphic features in the nearshore. Air photos provide a qualitative look at the geomorphology, and quantitative measurements are possible with images that can be orthorectified. Such images are especially useful for measuring cliff retreat. Plant and Holman (1997) measured intertidal beach shape using a combination of time exposures and differential GPS. Lippmann and Holman (1989) investigated the locations and forms of offshore bars by taking several-minute time exposures from a camera mounted high above the beach and looking longshore. Recently, LIDAR (light detection and ranging) (*q.v.*), an airborne scanning instrument, has been used to map the subaerial part of the coastal zone (Brock *et al.*, 1999) and shallow parts of the nearshore (Irish and Lillycrop, 1999). These instruments are capable of rapidly estimating elevations to within a couple of centimeters approximately every 3 m^2 over regional scales.

Physical processes

Many instruments are available to measure water and sediment motion at all scales throughout the coastal zone. Many of them are sturdy enough to withstand the forces generated in the breaker and surf zones. Pressure sensors and bidirectional current meters have a sampling rate fast enough to measure water depth and wave-generated currents, respectively. Optical sensors measure sediment concentration. Sonic devices measure rapid changes in bottom elevation and, in some cases, the height of sediment suspension.

Future research in coastal processes

The goal of future research in coastal processes includes developing predictive models for:

- bedload and suspended sediment transport under combined wave and current forcing;
- turbulent wave/current boundary layers over 3-D small-scale morphology;
- effects of moving sediment on boundary layer;
- contribution to sediment transport by bedform migration;
- effects of grain-size distribution on sediment transport (Thornton *et al.*, 2000).

Morphology and its variability are important end products of predictive models. However, because sediment transport is not well understood, prediction of morphological change is inadequate at all scales. For example, at smaller scales, ripples and megaripples are observed to be ubiquitous, but have not been incorporated into models even though their effect on the flow field (as roughness elements) and sediment transport may be significant. Complex patterns in long-term, large-scale morphology have also been observed. However, models for morphology change have predictive skill only over the short term, whereas long-term, large-scale predictions are not yet possible. Research issues include:

- predicting morphology across the spectrum of length scales;
- free versus forced large-scale morphology models;
- understanding feedback between morphology and the flow field;
- coupling between length scales.

John R. Dingler

Bibliography

Bascom, W.N., 1951. The relationship between sand size and beach face slope. *American Geophysical Union Transactions*, **32**: 866–874.

Bascom, W.N., 1980. *Waves and Beaches*. New York: Doubleday & Company.

Bird, E.C.F., 1984. *Coasts: An Introduction to Coastal Geomorphology*. Oxford, England: Basil Blackwell.

Birkemeier, W.A., and Mason, C., 1978. The CRAB: A unique nearshore surveying vehicle. *Journal of Survey Engineering*, **110**(1): 1–7.

Bowen, A.J., and Inman, D.L., 1966. Budget of littoral sands in the vicinity of Point Arquello, California. U.S. Army Coastal Engineering Research Center, *Technical Memorandum* No. 19.

Bowen, A.J., and Inman, D.L., 1969. Rip currents, 2: laboratory and field observations. *Journal of Geophysical Research*, **74**: 5479–5490.

Bowen, A.J., and Inman, D.L., 1971. Edge waves and crescentic bars. *Journal of Geophysical Research*, **76**(36): 8662–8671.

Bradley, W.C., 1958. Submarine abrasion and wave-cut platforms, *Geological Society of America Bulletin*, **69**: 957–974.

Brock, J., Sallenger, A., Krabill, W., Swift, R., Manizade, S., Meredith, A., Jansen, M., and Eslinger, D., 1999. Aircraft laser altimetry for coastal processes studies. *Coastal Sediments '99*, 2414–2428.

Bruun, P.M., 1954. Migrating sand waves and sand humps, with special reference to investigations carried out on the Danish North Sea coast. *Proceedings of the 5th Coastal Engineering Conference, on American Society of Civil Engineers*, pp. 269–295.

Carter, R.W.G., 1988. *Coastal Environments*. San Diego, CA: Academic Press.

Carter, R.W.G., and Woodroffe, C.D., 1994. *Coastal Evolution*. New York: Cambridge University Press.

Cowell, P.J., and Thom, B.G., 1994. Morphodynamics of coastal evolution. In Carter, R.W.G., and Woodroffe, C.D. (eds.), *Coastal Evolution*. New York: Cambridge University Press, pp. 33–86.

Dally, W.R., Johnson, M.A., and Osiecki, D.A., 1994. Initial development of an amphibious ROV for use in big surf. *Marine Technology Society Journal*, **28**(1): 3–10.

Dean, C., 1999. *Against the Tide*. New York: Columbia University Press.
Dietz, R.S., 1963. Wave base, marine profile of equilibrium, and wave built terraces: a critical appraisal. *Geological Society of America Bulletin*, **74**: 971–990.
Dolan, R., 1971. Coastal landforms: crescentic and rhythmic. *Geological Society of America Bulletin*, **82**: 177–180.
Dolan, R., and Ferm, J.C., 1968. Crescentric landforms along the Atlantic coast of the United States. *Science*, **159**: 627–629.
Dolan, R., Vincent, L., and Hayden, B., 1974. Crescentic coastal landforms. *Zeitschrift für Geomorphologie*, **18**: 1–12.
Evans, O.F., 1940. The low and ball of the East Shore of Lake Michigan. *Journal of Geology*, **48**: 476–511.
Galvin, C.J., Jr., 1972. Waves breaking in shallow water. In Meyer, R.E. (ed.), *Waves on Beaches*. New York: Academic Press, pp. 413–456.
Guilcher, A., 1958. *Coastal and Submarine Morphology*. New York: John Wiley & Sons.
Guza, R.T., and Inman, D.L., 1975. Edge waves and beach cusps. *Journal of Geophysical Research*, **80**(21): 2997–3012.
Hallermeier, R.J., 1981. A profile zonation for seasonal sand beaches from wave climate. *Coastal Engineering*, **4**: 253–277.
Holman, R.A., 1983. Edge waves and the configuration of the shoreline. In Komar, P.D. (ed.), *CRC Handbook of Coastal Processes and Erosion*. Boca Raton, FL: CRC Press, pp. 21–34.
Holman, R.A., and Bowen, A.J., 1982. Bars, bumps, and holes: models for the generation of complex beach topography. *Journal of Geophysical Research*, **87**(C1): 457–468.
Hom-ma, M., and Sonu, C.J., 1963. Rhythmic patterns of longshore bars related to sediment characteristics. *Proceedings of the 8th Coastal Engineering Conference on American Society of Civil Engineers*, pp. 248–278.
Huntley, D.A., and Bowen, A.J., 1978. Beach cusps and edge waves. *Proceedings of the 16th Coastal Engineering Conference on American Society of Civil Engineers*, pp. 1378–1393.
Inman, D.L., and Filloux, J., 1960. Beach cycles related to tide and local wind wave regimes. *Journal of Geology*, **68**: 225–231.
Irish, J.L., and Lillycrop, W.J., 1999. Scanning laser mapping of the coastal zone: the SHOALS system. *ISPRS Journal of Photogrammetry and Remote Sensing*, **54**(2/3): 123–129.
Johnson, D.W., 1919. *Shore Processes and Shoreline Development*. New York: John Wiley & Sons.
Kaminsky, G., Ruggiero, P., and Gelfenbaum, G., 1998. Monitoring coastal change in southwest Washington and northwest Oregon during the 1997–98 El Nino. *Shore and Beach*, **66**(3): 42–51.
Keulegan, G.H., 1948. An experimental study of submarine sand bars. U.S. Army Corps of Engineers. *Beach Erosion Board Technical Report No. 3*.
King, C.A.M., and Williams, W.W., 1949. The formation and movement of sand bars by wave action. *Geographical Journal*, **107**: 70–84.
Komar, P.D., 1971. Nearshore cell circulation and the formation of giant cusps. *Geological Society of America Bulletin*, **82**: 3593–3600.
Komar, P.D., 1983. Rhythmic shoreline features and their origins. In Gardner, R., and Scoging, H. (eds.), *Mega-Geomorphology*. Oxford, England: Clarendon Press, pp. 92–112.
Komar, P.D., 1998. *Beach Processes and Sedimentation*, 2nd edn. Upper Saddle River, NJ: Prentice-Hall.
Komar, P.D., and McDougal, W.G., 1988. Coastal erosion and engineering structures. The Oregon experience. *Journal of Coastal Research*, **4**: 77–92.
Komar, P.D., and Rea, C.C., 1976. Erosion of Siletz Spit, Oregon. *Shore and Beach*, **44**: 9–15.
Komar, P.D., 1996. The budget of littoral sediments—concepts and applications. *Shore and Beach*, **64**: 18–26.
La Fond, E.C., 1938. Relationship between mean sea level and sand movements. *Science*, **88**: 112–113.
LaFond, E.C., and Rao, R.P., 1954. Beach erosion cycles near Waltair on Bay of Bengal. *Andhrea University Mem. Oceanography*, **1**: 63–77.
Lippmann, T.C., and Holman, R.A., 1989. Quantification of sand bar morphology: a video technique based on wave dissipation. *Journal of Geophysical Research*, **95**(C7), 11,575–11,590.
MacMahan, J., 2001. Hydrographic surveying from a personal watercraft. *Journal of Surveying Engineering*, **127**(1): 12–24.
Plant, N.G., and Holman, R.A., 1997. Intertidal beach profile estimation using video images. *Marine Geology*, **140**(1–2): 1–24.
Ruessink, B.G., 1998. Infragravity waves in a dissipative multiple bar system. Doctoral Thesis, Department of Physical Geography, University of Utrecht, The Netherlands.
Sallenger, A.H., Jr., Howard, P.C., Fletcher, C.H., and Howd, P.A., 1983. A system for measuring bottom profile, waves and currents in the high-energy nearshore environment. *Marine Geology*, **51**: 63–76.
Seymour, R.J., Higgins, A.L., and Bothman, D.P., 1978. Tracked vehicle for continuous nearshore profiles. *Proceedings of the 16th Coastal Engineering Conference American Society of Civil Engineers*, pp. 1542–1554.
Shepard, F.P., 1950. Beach cycles in Southern California. U.S. Army Corps of Engineers. *Beach Erosion Board Technical Memorandum* No. 20.
Shepard, F.P., 1952. Revised nomenclature for depositional coastal features. *Bulletin of American Association of Petroleum Geologists*, **36**: 1902–1912.
Shepard, F.P., 1963. *Submarine Geology*, 2nd edn. New York: Harper & Row.
Shepard, F.P., and Wanless, H.R., 1971. *Our Changing Coastlines*. New York: McGraw–Hill.
Sonu, C.J., 1969. Collective movement of sediment in littoral environment. *Proceedings of the 11th Coastal Engineering Conference on American. Society of Civil Engineers*, pp. 373–400.
Sonu, C.J., 1972. Field observation of nearshore circulation and meandering currents. *Journal of Geophysical Research*, **77**(18): 3232–3247.
Thornton, E., Dalrymple, R.A., Drake, T.G., Elgar, S., Gallagher, E.L., Guza, R.T., Hay, A.E., Holman, R.A., Kaihatu, J.M., Lippmann, T.C., and Ozkan-Haller, H.T., 2000. State of nearshore processes research, II. *Naval Postgraduate School Technical Report* NPS-OC-00-001.
Trask, P.D., 1952. Sources of beach sand at Santa Barbara, California, as indicated by mineral grain studies. U.S. Army Corps of Engineers. *Beach Erosion Board Technical Memorandum* No. 28.
van Rijn, L.C.,1998. *Principles of Coastal Morphology*. Amsterdam, The Netherlands: Aqua Publications.
Werner, B.T., and Fink, T.M., 1993. Beach cusps as self-organized patterns. *Science*, **260**: 968–971.
Wiegel, R.L., 1964. *Oceanographical Engineering*. Englewood Cliffs, NJ: Prentice-Hall.

Cross-references

Barrier Islands
Bars
Beach Cusps (see Rhythmic Patterns)
Beach Erosion
Beach Features
Beach Nourishment
Cliffed Coasts
Coastal Boundaries
Coral Reefs
Cross-Shore Sediment Transport
Depth of Closure on Sandy Coasts
Dune Ridges
Erosion Processes
Global Positioning Systems
Greenhouse Effect and Global Warming
Longshore Sediment Transport
Meteorologic Effects on Coasts
Mining of Coastal Materials
Muddy Coasts
Nearshore Geomorphological Mapping
Profiling, Beach
Rhythmic Patterns
Ripple Marks
Rock Coast Processes
Sandy Coasts
Sea-Level Rise, Effect
Sediment Budget
Shore Protection Structures
Spits
Surf Zone Processes
Tides
Wave-Dominated Coasts
Waves

BEACH PROFILE

The beach profile is one the most studied features of coastal morphology. The shape of the beach profile determines the vulnerability of the coast to storms, the extent of usable beach for habitat and recreation, and the legal boundary distinguishing public and private ownership of land (Shalowitz, 1962, 1964; Anders and Byrnes, 1991). The first modern studies of the beach profile were motivated to understand its shape and variability in support of amphibious operations during World War II, when personnel and supply boats had to cross the beach profile from offshore to the dry beach (Bascom, 1980).

Beach profile terminology

The term "beach profile" refers to a cross-sectional trace of the beach perpendicular to the high-tide shoreline and extends from the backshore cliff or dune to the inner continental shelf or a location where waves and currents do not transport sediment to and from the beach. The profile shape is variable, depending on the time of year within the annual beach cycle and, also, the elapsed time after a storm. Waves, water level, and sediment grain size are the main controlling factors of beach profile shape.

Terminology associated with the beach profile is shown in Figure B40. The backshore runs from the seaward-most dune or the cliff to the land and water intersection. One or more berms may appear on a beach, depending on seasonal changes in water level. Berms are flat areas created during times of accretionary wave conditions, typically during summer. The beach intersects the water at the foreshore, and the foreshore is typically a plane slope that extends over a water level range from low tide to high tide. During a storm, a vertical step or scarp may form on the berm. The inshore covers the surf zone from the seaward end of the foreshore to past the seaward-most longshore sand bar, joining to the offshore. Several bars and associated troughs may appear on the beach profile.

Approximations of beach profile shape

As a first approximation, it is often possible to represent the profile of a gravel, pebble, or sandy beach (Here, the profile is assumed to have an unlimited supply of sand and that no "hard bottom" is present such as limestone reefs, coral reefs, and other non-erodable (hard) substances.) by a straight line of constant slope as,

$$h = x \tan \beta, \qquad \text{(Eq. 1)}$$

where h is the still-water depth, x the distance from shoreline, and $\tan \beta$ is the beach slope. This expression, defining a "plane beach" or planar beach slope has convenience for making simple calculations. Typically, however, the foreshore is the only area of the beach profile well represented by a straight line.

A more realistic representative profile shape was introduced by Bruun (1954) and studied extensively by Dean (1977, 1991). This profile is called the "equilibrium" or "x to the two-thirds profile" and is given as,

$$h = Ax^{2/3}, \qquad \text{(Eq. 2)}$$

where A is the shape parameter and will be discussed below. For water with temperature of about 20°C and typical sand sizes with sediment fall speed varying between about 1 and 10 cm s^{-1}, Kriebel *et al.* (1991) found that A could be related to fall speed w by,

$$A = 2.25\left(\frac{w^2}{g}\right)^{1/3}, \qquad \text{(Eq. 3)}$$

where g is the acceleration due to gravity. Moore (1982) was the first to study the functional dependence of A and found it to be an increasing function of the median grain size d_{50} for a wide range of materials. The empirical curve can be approximated by,

$$\begin{aligned} A &= 0.41(d_{50})^{0.94}, & d_{50} &< 0.4 \\ A &= 0.23(d_{50})^{0.32}, & 0.4 \le d_{50} &\le 10 \\ A &= 0.23(d_{50})^{0.28}, & 10 \le d_{50} &\le 40 \\ A &= 0.46(d_{50})^{0.11}, & 40 &\le d_{50} \end{aligned} \qquad \text{(Eq. 4)}$$

for which d_{50} is expressed in millimeters. The equilibrium profile can encompass a large range in grain size, as seen by the values in equation 4. Because the shape parameter A increases with increasing sediment grain size or fall speed, finer-grained beaches have gentle slopes and coarser-grained beaches are steeper, in accord with observations (Bascom, 1980).

Bars and troughs

Bars and troughs, also called longshore bars and longshore troughs, are the major perturbations from the equilibrium profile. Typically, areas with small tide range possess the most prominent bars because the wave breaker line remains in one position longer. Multiple-barred beaches are common on the Great Lakes, bays, and the Gulf of Mexico coast of the United States, for example, where the tide range is small. Some of these bars are formed by various predominant waves, such as typical waves and storm waves. Likewise, if the wave conditions occupy a relatively narrow range of height and period, such as on the north shore of Long Island, New York (facing the Long Island Sound), bars tend to be more prominent as compared to bars on the south shore of Long Island, because the Atlantic Ocean has a much more variable wave climate and smears out such bottom features.

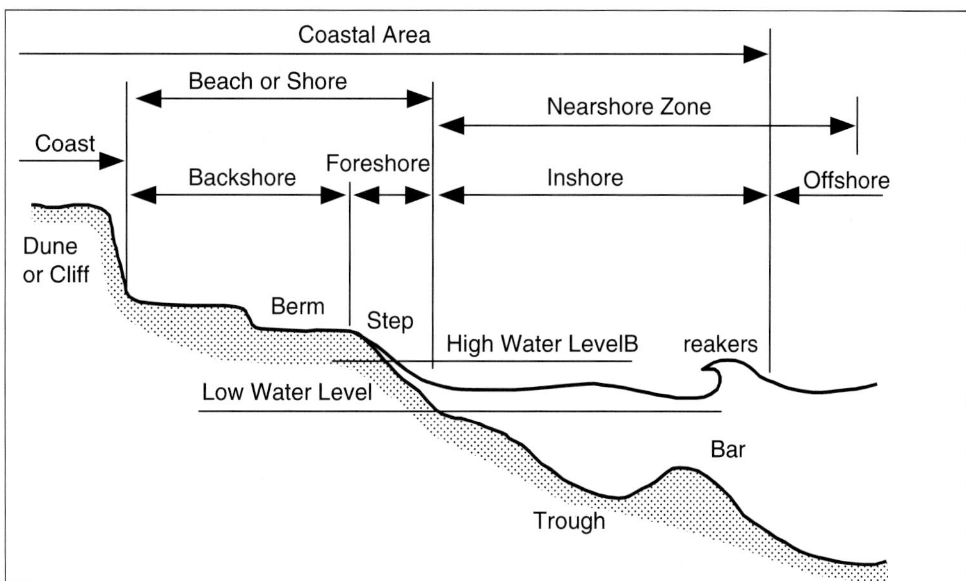

Figure B40 Terminology associated with the beach profile.

Larson and Kraus (1989) analyzed large-scale laboratory data for breaking waves and sandy beach beaches and found that the depth over the crest of a bar h_c was related to the breaking wave height H_b as

$$h_c = 0.66\,H_b. \quad \text{(Eq. 5)}$$

Conversely, if the depth over the crest of a well-established bar is measured, the breaking wave height that created the bar may be inferred from equation 5 to be $H_b = 1.5\,h_c$.

The beach profile at North Padre Island, TX, located along the Gulf of Mexico, was surveyed in the mid-1970s in one of the earliest applications of a sea sled, and then again in the mid-1990s with a sled, assuring high accuracy (see Profiling, Beach). Figure B41 plots a time series of surveys made at the same location on the beach. In both eras, profile elevation was referenced to mean sea level (MSL) at a local ocean tide gauge. Although the beach may have advanced or receded during the two decades, the shape of the profile can be compared because of the common vertical datum.

Figure B41 indicates that one to three (occasionally four) bars can appear on the profile at North Padre Island. It can be estimated through equation 5 that these bars are related to different classes of waves as outer bar—severe storm waves; middle bar—typical storm waves; and inner bar—waves under normal Gulf of Mexico conditions. In Figure B42, the average of 18 beach profile surveys taken alongshore at North Padre Island in 1996 is plotted together with the equilibrium ($x^{2/3}$) profile, equation 2, determined by the median grain size in the surf zone (0.18 mm). Sediment sampling performed during the surveys demonstrated a decrease in grain size with distance offshore, and such a decrease in size of surficial sediments is typical along the beach profile, with the coarsest material located at the beach face and bars, and the finer material located offshore. Coarse material is also found at the landward sides of longshore bars, because the finer material is transported away from this hydrodynamically energetic area.

Seasonal characteristics of the beach profile

During winter and the occurrence of seasonal storms (periodic northeasters, tropical storms, hurricanes), waves and cross-shore currents remove material from the beach and deposit it in bars far offshore. Large volumes, for example, 30–100 $m^3\,m^{-1}$ width of beach, can be removed from the beach berm and dune in a single large storm. Whether a beach will erode or accrete and the bar move onshore or offshore can be estimated with a dimensionless parameter called the Dean number formed as $N = H/wT$, where H is the wave height in deeper water, w is the beach sediment fall speed, and T is the wave period. For $N > 3.2$, erosion is

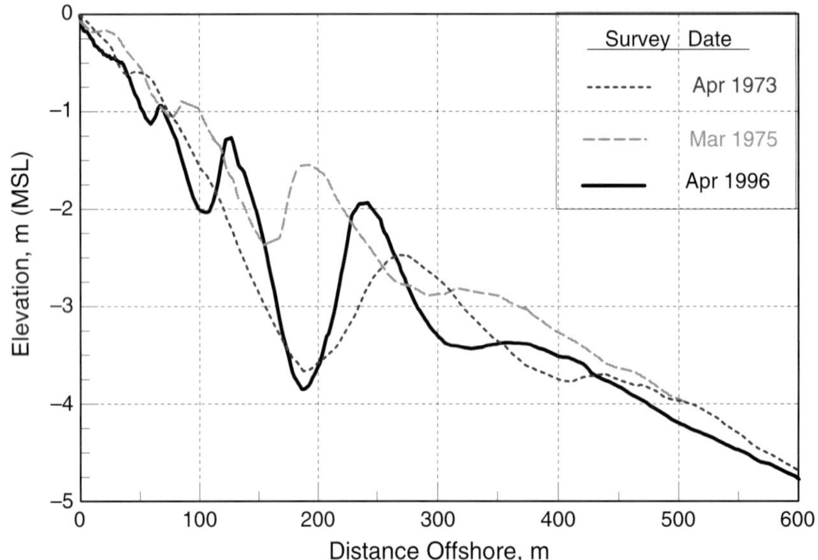

Figure B41 Beach profile surveys taken by sled two decades apart, North Padre Island, TX.

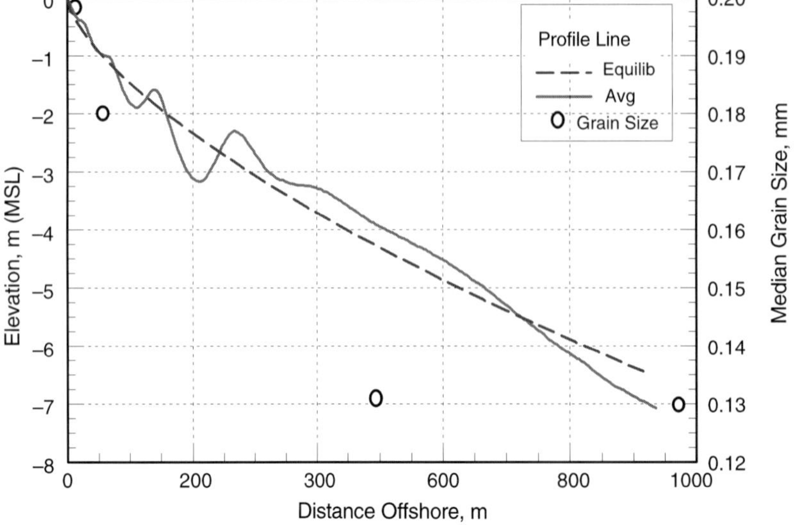

Figure B42 Average beach profile and equilibrium profile, North Padre Island, TX, 1996.

probable, whereas for $N < 3.2$, accretion is probable (Kraus *et al.*, 1991). Large values of the Dean number tend to occur during storms, when wave heights are large, and small values tend to occur in the summer, when wave heights are small or after storms, when the wave period of the swell waves becomes long. During the following milder accretionary waves of summer, material gradually moves onshore and returns to the beach, deflating the bar(s) and building the berm. Wind-blown sand then gradually rebuilds the dunes.

Sometimes, storms can be so severe that the beach does not recover on the timescale of human lifetime or engineering projects. Such is probably the case on the south shore of Long Island for the "Great New England Hurricane" of September 1938, which weakened the barrier islands and caused many breachings or cutting of temporary inlets. Sand taken offshore by such strong storms lies is such deep water that summer or "recovery" waves cannot readily transport it back to the beach.

The seasonal averages of a large number of surveys made on the same cross-shore transect (Line 62) are plotted in Figure B43. The surveys were made at the US Army Corps of Engineers' Field Research Facility (FRF), in Duck, NC, located on the "Outer Banks" barrier island chain. The profile is surveyed every two weeks as routine monitoring or more frequently for specific research goals by means of a large motorized tripod, estimated to have a vertical accuracy of ±2 cm. The National Geodetic Vertical Datum (NGVD) is close to MSL at the FRF. Bars are absent from the plots because the average is taken over a large number of surveys. Sand moves offshore in winter (arbitrarily defined as the interval January to March) and returns in summer (June–August). A broad hump in the winter average at about 2–4 m depth indicates the presence of storm bars during that season. During summer months, the steep profile in shallower water created by the winter waves is gradually replenished and becomes shallower.

One property of the beach profile observed in Figure B43 is that the spring (April–June) and fall (October–December) average profiles almost plot on top of one another, and in between the two terminal states of summer and winter (Larson and Kraus, 1994). The regularity in profile response to waves indicates that the processes should be predictable with relatively simple techniques. Although not shown, the average of all profiles corresponds well with the equilibrium profile with a median grain size of 0.2 mm.

The seasonal response of the profile is shown in another way in Figure B44, which plots the average change in depth irrespective of sign (absolute value) as a function of average depth for the winter and summer profiles. The change in depth is less in summer than in winter, which is intuitively reasonable because waves are smaller in summer. The average maximum depth change occurs in winter, near the shoreline, and is

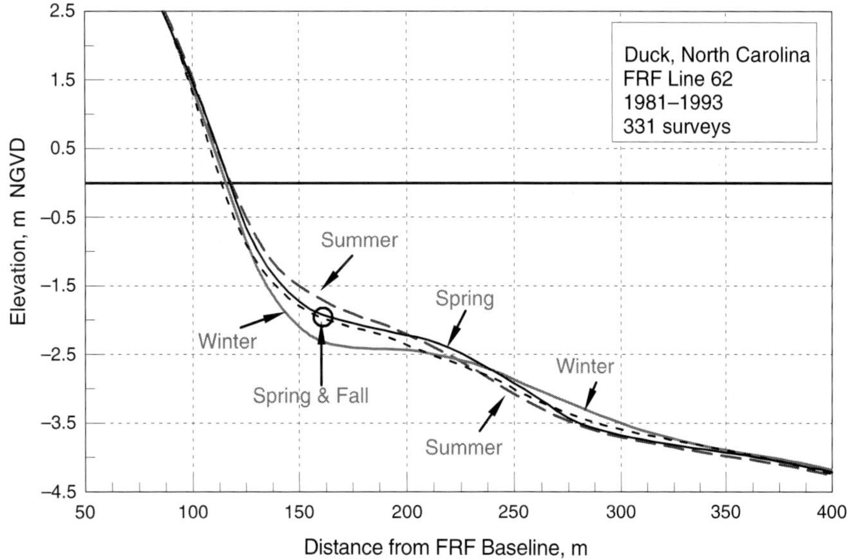

Figure B43 Average seasonal profiles, Duck, NC.

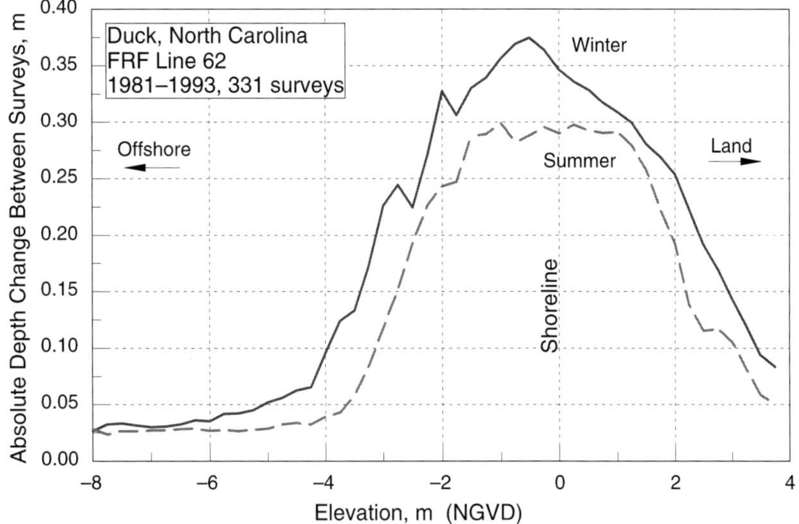

Figure B44 Average absolute depth change for summer and winter surveys, Duck, NC.

about 0.5 m (1.5 ft). The location where the profile changes little through time, delineating the area where sediment in the offshore is no longer exchanged with the beach, is called of the depth of closure (Kraus *et al.*, 1999). Knowledge of how much the profile elevation changes is required for modeling coastal processes, for designing beach fills, for placing pipes, outfalls, and cables, across the surf zone, and for placement of instruments so that they do not become buried.

Nicholas C. Kraus

Bibliography

Anders, F.J., and Byrnes, M.R., 1991. Accuracy of shoreline change rates as determined from maps and aerial photographs. *Shore and Beach*, **59**(1): 17–26.

Bascom, W., 1980. *Waves and Beaches, The Dynamics of the Ocean Surface*, Revised and Updated Edition. Garden City, NY: Anchor Books.

Bruun, P., 1954. Coast erosion and development of beach profiles. U.S. Army Corps of Engineers. *Beach Erosion Board, Technical Memorandum* No. 44.

Dean, R.G., 1977. Equilibrium beach profiles: U.S. Atlantic and Gulf coasts. Department of Civil Engineering, *Ocean Engineering Report* No. 12. Newark, DE: University of Delaware.

Dean, R.G., 1991. Equilibrium beach profiles: principles and applications. *Journal of Coastal Research*, **7**(1): 53–84.

Kraus, N.C., Larson, M., and Kriebel, D.L., 1991. Evaluation of beach erosion and accretion predictors. *Proceedings of Coastal Sediments '91*. American Society of Civil Engineers, pp. 572–587.

Kraus, N.C., Larson, M., and Wise, R.A., 1999. Depth of closure in beach-fill design. *Proceedings of the 12th National Conference on Beach Preservation Technology*. Tallahassee, FL: Florida Shore and Beach Preservation Association, pp. 271–286.

Kriebel, D.L., Kraus, N.C., and Larson, M., 1991. Engineering methods for predicting beach profile response. *Proceedings of Coastal Sediments '91*. American Society of Civil Engineers, pp. 557–571.

Larson, M., and Kraus, N.C., 1989. SBEACH: numerical model for simulating storm-induced beach change. Report 1: empirical foundation and model development, Technical Report CERC-89–9, U.S. Army Engineer Waterways Experiment Station, Vicksburg, MS: Coastal Engineering Research Center.

Larson, M., and Kraus, N.C., 1994. Temporal and spatial scales of beach profile change, Duck, North Carolina. *Marine Geology*, **117**: 75–94.

Moore, B.D., 1982. Beach profile evolution in response to changes in water level and wave height. MCE thesis, Department of Civil Engineering. Newark: University of Delaware.

Shalowitz, A.L., 1962. *Shore and Sea Boundaries. Volume 1: Boundary Problems Associated with the Submerged Lands Cases and the Submerged Lands Acts*. Publication 10–1, U.S. Department of Commerce, Coast and Geodetic Survey, 420 pp.

Shalowite, A.L., 1964. *Shore and Sea Boundaries. Volume 2: Interpretation and Use of Coast and Geodetic Survey Data*. Publication 10–1, U.S. Department of Commerce, Coast and Geodetic Survey.

Cross-references

Accretion and Erosion Waves on Beaches
Bars
Coastal Warfare
Depth of Closure on Sandy Beaches
Monitoring, Coastal Geomorphology
Meteorologic Effects on Coasts
Storm Surge
Surf Zone Processes
Tidal Datums

BEACH RATING—See RATING BEACHES

BEACH RIDGES

Definitions

Johnson (1919) defined beach ridges as depositional features constructed by waves at successive shore positions. Reineck and Singh (1975) characterized beach ridges as having been formed at high-tide level of "rather coarse sediment" and related to storms or exceptionally high water stages. Bates and Jackson (1980) designated beach ridges as low mounds of beach and beach-and-dune material, heaped up by waves over the backshore beyond the present limit of storm waves or ordinary tides, but there is a risk of confusing wave-deposited and wind-deposited ridges. Referring to relict strandplain dunes as beach ridges, Carter (1986) used the term very broadly to cover "all large constructional forms of the upper beach, capable of preservation," applying it also to landward-shifting offshore bars, already welded swash bars, and prograded berm ridges.

Davis *et al.* (1972), Fraser and Hester (1977), Carter (1986, 1988) and others referred to onshore-migrating swash bars and/or to the stranded end products as "beach ridges." For some time after welding to the shore, the prograding sandy berm ridges may continue to be impacted by daily beach processes. Most North American and Australian authors considered stabilized onshore beach ridges as either of predominantly wave-built or of wave and wind-constructed, composite origin (e.g., Price, 1982). Hesp (1984, 1985), Mason (1990), and Mason *et al.* (1997), thus distinguishing between low profile "smooth, terrestrial," squat berm ridges and steep dune ridges that often overlie and bury them.

Beach ridge presently is defined as a relict shore ridge that is more or less parallel with the coastline and with other landward-adjacent ridges. It is built by wave swash (a berm ridge) that may be surmounted by wind-deposited sediment (a foredune). Once such a ridge becomes isolated from daily active beach processes by coastal progradation, which may lead to the construction of one or more new ridges to seaward, it becomes a beach ridge. On wide eolian backshore plains, shore-parallel dune ridges may also form behind active foredunes. Regardless of their dimensions, shapes, and origin, active beach/shore ridges impacted and modified almost daily by shore processes are excluded from the designation.

Associated landforms

A berm was originally defined as a narrow, scarp-backed, and wave-cut horizontal surface in the beach foreshore (e.g., Komar, 1976). Subsequently, it came to mean a wedge-shaped ridge, between an upper foreshore slope and a landward-inclined berm top surface (King, 1972). Its base is the horizontal plan that intersects the foreshore slope at the level of the backshore plain. Hine (1979) defined a berm as a shore-parallel linear body of triangular cross section with a horizontal to slightly landward-dipping surface (berm top) and a steeper seaward-dipping slope (beach face). Berms are short-lived and frequently reforming landforms, often absent from a beach.

Swash currents deposit sediment that builds the landward-sloping high-tidal berm above the level of the adjacent backshore. The ephemeral high-tidal sand berms are of aggradational origin, with secondary indications of erosional scarping. Increased onshore winds during falling tides briefly stabilize the water level. Intermediate-level berms with vertical scarplets may form during these stillstands.

Berm ridges, occasionally sizable and more permanent than berm surfaces are wave-built intertidal-supratidal landforms. The lithosomes are composed of intertidal and high tidal (swash-overwash) deposit, bounded by the backshore plane and the berm surface along its foreshore margin (Figure B45). After becoming isolated by progradation from the daily effects of beach processes, these inactive landforms attain the status of wave-constructed beach ridges. Formed on mainland or island beaches or on shore-parallel sand spits, berm ridges are bracketed between the foreshore and the landward (or lagoonward) margin of the backshore. On the landward side they may follow the shoreline of an elongated lagoon or beach pond ("cat's eye pond"; Coastal Research Group, 1969), enclosed by a sand spit (Figure B46). Shore-parallel inter-ridge swales bracket each ridge. Sets of prograding berm ridges form on beaches of limited sand supply. Short *et al.* (1989) reported on a nearly exclusively swash-built beach ridge plain in Australia, and in Egypt, Goodfriend and Stanley (1999) described a shelly sandridge plain, composed of 20–30 cm high, wave-built ridges without eolian cover.

Several authors have regarded high-tidal sand berms as incipient (wave-built) beach ridges during the Australian "berm debate" (Davies, 1957; Bird, 1960; Hails, 1969). Later, Bird and Jones (1988) proposed that if a berm survived a 15-day tidal cycle, it becomes a beach ridge. Berms have also been credited with providing a foundation for the development of embryonic foredunes that develop into full-sized ones (Davies, 1957; Bird and Jones, 1988). However, foredunes often form along the seaward margin of the backshore plain as well. The presence of berms, often absent from beaches, especially from dissipative and high-energy beaches (e.g., Short, 1984), is not an indispensable precondition for foredune development (Hesp, 1984, 1985).

Berm formation by wave action on the Tabasco shore of the Gulf of Mexico has been attributed to alternating "cut-and-fill" cycles of erosion and aggradation (Psuty, 1966). This berm-shaping process,

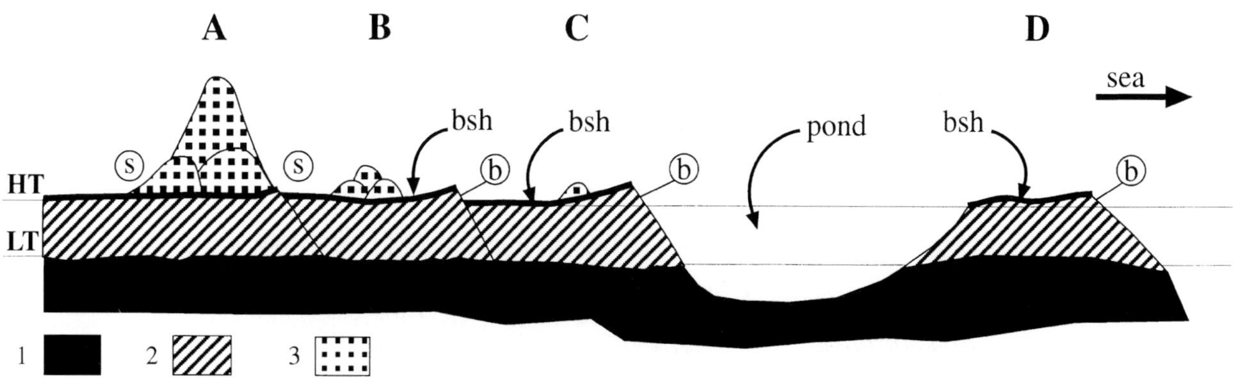

Figure B45 Beach ridge-associated depositional facies and landforms on a prograding strandplain. Depositional facies: (1) subtidal; (2) wave-built, intertidal-to-supra-high tidal; (3) eolian. Landforms: b, berm; bsh, backshore plain; s, swale; A, foredune; B, accreting embryonic dunes on backshore plain in the early foreshore stage; C, single embryonic dune on berm ridge (backshore-berm) surface; pond: spit-growth-enclosed beach pond; D, pond-isolated berm ridge (intertidal-supratidal sand spit interval), without embryonic dunes. HT, high-tide level; LT, low-tide level. (Otvos, 2000) Reprinted with permission from Elsevier Science.

however, appears to be a localized and ephemeral phenomenon without bearing on long-term strandplain development. The adjacent, 20-km wide Tabasco strandplain, for example, has been receiving abundant eolian sand supply and represents a foredune ridge plain, underlain by backshore and berm ridge deposits.

Truncation lines that separate mainland and island beach ridge sets

Where beach ridge progradation has been abruptly terminated by shore erosion, then followed by renewed beach growth, there are cross-cutting truncation lines that separate generations of beach ridges in mainland and island barriers of Quaternary age. This process was historically documented on several Mississippi coast strandplain islands (Otvos, 1981). The St. Joseph Bay-area barrier spit and small mainland strandplains in NW Florida provide good illustrations (Figure B47).

Gravel-boulder ("storm") ridges. Storm-associated high tides and waves build gravel ramparts as high as 6 m (Clapperton, 1990). Gravel-boulder ridges, associated with storm surges therefore rise well over their associated sea (lake) levels. Coarse clastic sediments, including shelly material resist backwash erosion and become stranded on these shore ridges. Permanent shingle emplacement at superelevated tide levels is aided by backwash percolation into the permeable gravelly substrate (Carter, 1988). For a given still-water level, the height of wave-built ridges built during winter storms may vary by as much as 2–2.6 m (Adams and Wesnousky, 1998). These "storm" ridges are common on glaciated and bedrock shores; also on tectonically or isostatically raised marine and lacustrine terraces. Examples abound in Canada's Maritime Provinces, New England, and on high Pacific shores between Alaska and Mexico.

Coarse clastic beach ridges or bedrock terrace veneers accompany raised strandlines of pluvial and glacial lakes in the North American interior basins (Fulton, 1989; Morrison, 1991). Coarse clastic sediments were delivered by high-gradient streams, alluvial fans, mass wasting, occasionally even fluvioglacial processes. Adjacent bedrock areas that served as sediment sources have undergone intensive physical weathering under periglacial and cold-temperate conditions. Pluvial Lake Bonneville and its successor, early stage Great Salt Lake in Utah and Nevada; as well as Lake Lahontan in Nevada and California provide the best examples (Morrison, 1965, 1991; Currey, 1980; Adams and Wesnousky, 1998, 1999). Gravel-boulder ridges, deposited on wave-cut bedrock terraces form discontinuous tabular and tabular cross-stratified bodies, several meters thick (Adams and Wesnousky, 1998). Playa beach ridges that contain carbonate nodules, include paleosols and incorporate secondarily calcreted and gypsum-creted grit in the arid Lake Eyre basin, Australia, one of the world's largest internally drained regions (Nanson *et al.*, 1998).

Bouldery-gravelly coarse clastic sediments that often dominate ice-dammed glacial lake shorelines along the fluctuating glacial margin in North America have originated from reworked moraine till, fluvioglacial delta, periglacial colluvium, and in particular ice-contact (esker, kame, outwash delta) deposits. Major examples include relict shore features on glacial Lakes Agassiz, Algonquin, and Ojibway (Fulton, 1989, pp. 144–145, 257, 343, 362–364). Due to sand scarcity, the short life span of a given strandline, and erosive wave regimes, instead of regular beach ridges gravelly-bouldery shore zone lags frequently veneer wave-cut lake terrace surfaces.

Sediment types and ridge forms of sandy beach ridges. Depending on the wide range of wave and current conditions and source sediments on a given marine or lake shore sector, the upper beach deposits (intertidal–supratidal intervals) may be represented by fine-to-coarse, even gravelly sands in the berm ridges. Whereas, sorting tends to be good in uniformly sandy beaches, it becomes moderately sorted when coarser clastic fractions are also included (e.g., Thompson, 1992). Chappell and Grindrod (1984) described a rare transition between sandy ridges of a regular beach ridge plain and a small adjacent chenier plain, composed of shell-rich ridges. Reflecting the low relief and gentle seaward and landward slopes of the berm ridges, basinward-dipping parallel laminae and low-angle (3–5°) cross lamination tend to characterize the upper foreshore slope.

Subhorizontal or gently landward-dipping laminations occur on the landward beach ridge surfaces. The highly variable beach ridge dimensions, the height above still water level and slope angles depend on wave conditions, local tidal, or lake level ranges, including wind-induced rise in sea and lake-levels along given shore sectors.

"Pebble-armored ridges." Pebble sheets plastered onto sand dunes during storms were designated as gravelly ramp barriers (Orford and Carter, 1982; Mason, 1990). The hydraulic ratios and shapes of shell bioclasts result in higher transport and dispersal rates than with regard to larger, denser silicate rock clasts. Whereas, gravel/boulder ridges accumulate during direct storms, their impact tends to flatten and disperse already existing ridges, composed of sand and lighter, platy shell clasts (Greensmith and Tucker, 1969; Rhodes, 1982, p. 217). Higher-energy events enable accumulation even of bioclastic rudites (Woodroffe *et al.*, 1983; Meldahl, 1993).

Beach ridges, strandplain versus terrace development

Strandplain formation may be a continuous process with grain-by-grain addition of sand to the widening foreshore. Continuous progradation of the neap berm at mid-to-high level results in a gently undulating, almost level beach plain. On mesotidal foreshores where neap high tide remains below the highest foreshore levels, continuously accreting neap berms are uninterrupted by inter-berm swales (Hine, 1979, Figure 17(A)). Increased sand supply along the low-microtidal Gulf of Mexico beaches leads to steady outbuilding of the foreshore, accompanied by consequent progradation of narrow, closely spaced beach ridges. Discontinuous beach ridge progradation involved either the stranding or remolding of landward-migrated mesotidal swash bars on the foreshore (Hine, 1979; Carter, 1986) or spit growth from and downdrift reattachment (Figure B46) to the beach in microtidal settings (Otvos, 1981). Both processes isolate elongated ponds. Fronted seaward (lakeward) by newly formed active foredunes, and slowly filled by eolian and washover sands, such ponds may gradually become wide supratidal inter-ridge swales (Figure B46).

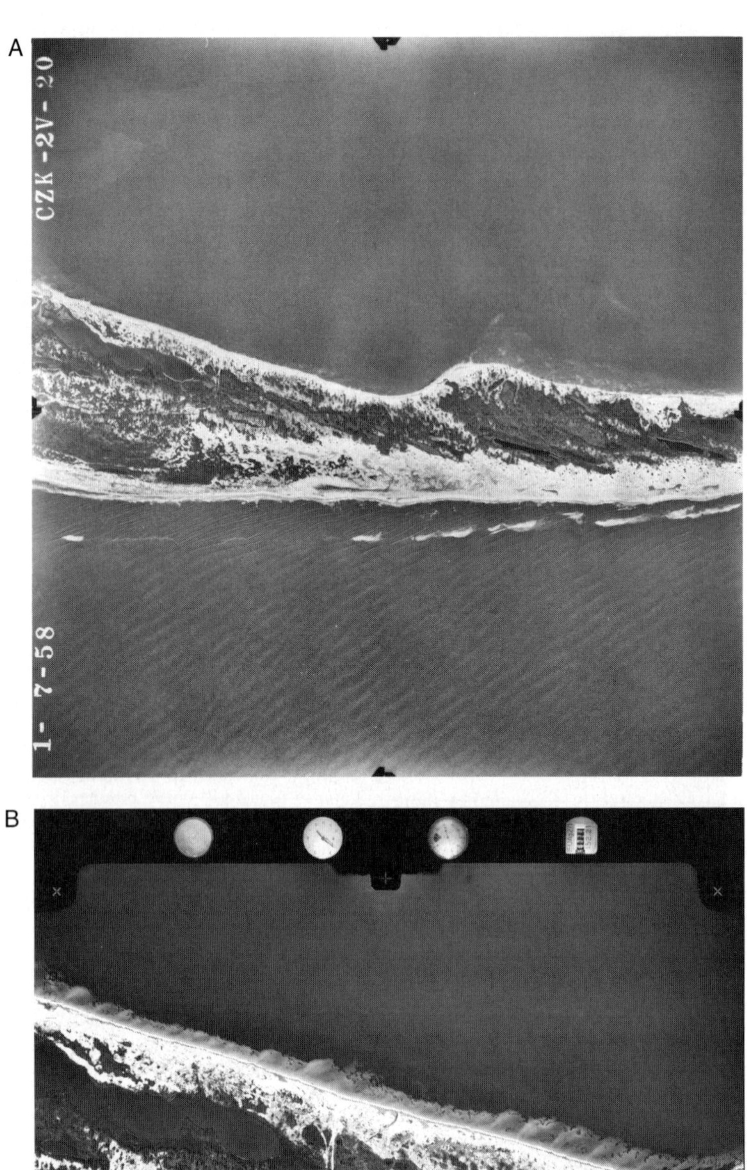

Figure B46 Prograding strandplain, southeastern Horn Island, MS. (A) Inter-swale ponds between old beach ridge sets in wooded interior. Elongated shore-parallel ponds of different orientation isolated from Gulf of Mexico (south) by barren narrow strip of backshore and berm ridges (USDA aerial photo, January, 1958. Width of image: 4.56 km). (B) Eleven years later: western half of previous image. Shore-parallel westward spit and spit-platform growth about to form new cat's eye pond (bottom). Already isolated beach pond to east. Eroding (white) and forested (dark) old interior beach ridge sets, with swale ponds in the island interior (USCGS aerial photo, October, 1969. Width of image: 2.3 km).

A

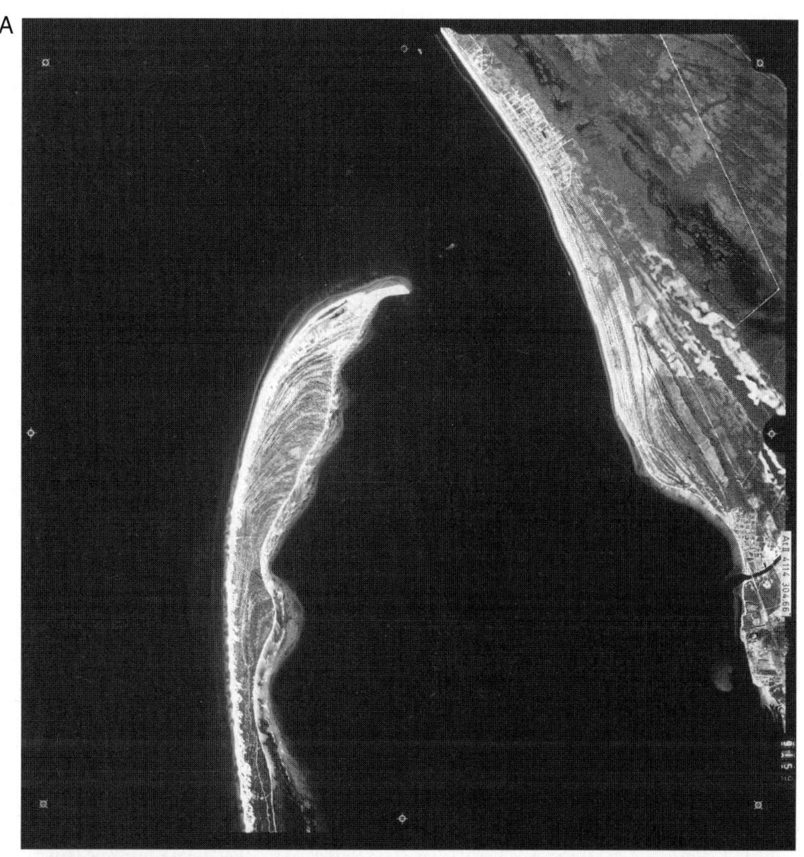

B

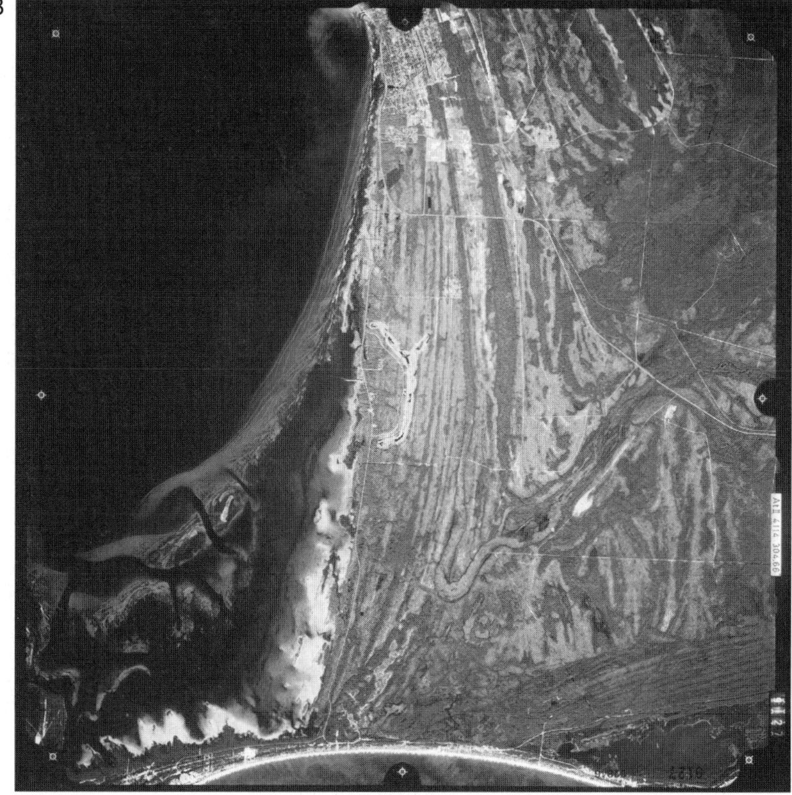

Figure B47 (A) Generations of narrow Late Holocene strandplain fans, northern St. Joseph Bay, NW Florida. Left: north tip of St. Joseph Peninsula (barrier spit). Right: small Holocene Palm Point mainland strandplain. Upper right corner: wide Late Pleistocene (Sangamon) beach ridges and partially filled swales represent erosion-impacted strandplain sectors. (B) St. Joseph Bay, NW Florida. Quaternary strandplains. Wide beach ridges of N–W-trending Late Pleistocene (Sangamon) strandplain (center field of photo) again contrast with narrow, crisply outlined Late Holocene strandplain ridges (lower right corner) (USGS aerial photo, October, 1978. Width of image: ca. 15 km). (Otvos, in press).

A comparison of Late Pleistocene strandplain ridges with the sharply outlined, narrow Late Holocene beach ridges illustrates the fact that prolonged infilling and erosional modification of Pleistocene strandplain ridges result in more subdued, more gently sloping beach ridges, separated by wider swales (Figure B47).

Instead of strandplains, gently undulating, nearly level eolian sand terraces form when sand supply and beach progradation does not keep pace with rapidly growing and sand-trapping beach vegetation. Beach progradation and/or eolian sand supply rates under these conditions are relatively low (e.g., Ruz and Allard, 1994).

Rates of beach ridge development

Depending on ridge dimensions, sediment supply rates, hydrodynamic and vegetative conditions, beach ridge development may proceed slowly or quickly. Thus, Nanson *et al.* (1998) report on a beach ridge along an Australian playa lake that during a high lake stage, formed in less than one year. Development rates in a number of other calculated examples from worldwide locations ranged between 1.8 and 3.3 yr/ridge, at other sites the rates were as low as 30–150 yr/ ridge (in: Otvos, 2000, pp. 90–91).

Beach ridges as ancient sea/lake levels markers

Former sea (lake) levels may be identified when a hortizontal interface is recognizable between the wave-built foreshore and the overlying eolian lithosome in a given ridge. This was the case in Lake Michigan strandplain ridges (Fraser and Hester, 1977; Thompson, 1992) where low-angle sand and gravel cross beds and trough-cross-bedded lacustrine sands of wave-built origin underlie land snail-bearing, cross-bedded, in part massive, structureless dune sands. On pure sand beaches that lack granule and pebble clasts due to the very short transport distance from the immediately adjacent source of sediment, distinctions between wave- and wind-deposited lithosomes often are difficult or impossible to make on granulometric grounds alone.

At times of superelevated lake and sea levels, associated with storm-related temporary rise of the water levels wave-built sandy, shelly, or gravelly beach ridges may aggrade significantly above normal high-tide levels (e.g., Mason, 1990; Mason and Jordan, 1993). Precise reconstruction of former sea levels from beach ridges therefore may be problematical.

Similar to beach ridge summits, coarse clasts are useful markers of reference surfaces to document postdepositional tectonic and isostatic changes, including vertical displacement, tilting, and warping of the land surface. In their absence, wave-cut bedrock terraces and lag clast-veneered strandlines may also serve this purpose. Strandline stairsteps were documented on isostatically uplifted marine and glacial lake shores in subarctic North America and Scandinavia (e.g., Hudson Bay, Tyrell Sea; Fulton, 1989). Flights of raised beaches characterize former pluvial/playa lake shores in western North America, Australia, and other presently arid and semiarid regions. Bounding surfaces of lower foreshore Lepidopthalmus (formerly, Callianassa) ghost shrimp-burrowed barrier ridge deposits and correlative adjacent lagoonal-saltmarsh surfaces in six Late Pliocene-Pleistocene coastal terrace sequences were similarly utilized on the Georgia-northeast Florida seaboard (Hoyt, 1969).

Ervin G. Otvos

Bibliography

Adams, K.D., and Wesnousky, S.G., 1998. Shoreline processes and the age of the Lake Lahontan highstand in the Jessup embayment. *Geological Society of America Bulletin*, **110**: 1318–1332.

Adams, K.D., and Wesnousky, S.G., 1999. The Lake Lahontan highstand: age, surficial characteristics, soil development, and regional shoreline correlation. *Geomorphology*, **30**: 357–392.

Bates, R.L., and Jackson, J.A., 1980. *Glossary of Geology*, 2nd edn. Falls Church, VA: American Geological Institute.

Bird, E.C.F., 1960. The formation of beach ridges. *Australian Journal of Science*, **22**: 349–350.

Bird, E.C.F., and Jones, D.J.B., 1988. The origin of foredunes on the coast of Victoria, Australia. *Journal of Coastal Research*, **4**: 181–192.

Carter, R.W.G., 1986. The morphodynamics of beach-ridge formation: Magilligan, Northern Ireland. *Marine Geology*, **73**: 191–214.

Carter, R.W.G., 1988. *Coastal Environments*. New York: Academic Press.

Chappell, J.C., and Grindrod, J., 1984. Chenier plain formation in northern Australia. In Thom, B. G. (ed.), *Coastal Geomorphology in Australia*. New York: Academic Press, pp. 197–232.

Clapperton, C.M., 1990. Quaternary glaciations in the southern ocean and Antarctic Peninsula area. *Quaternary Science Reviews*, **9**: 229–252.

Coastal Research Group, 1969. *Coastal Environments, NE Massachusetts and New Hampshire, Guidebook, SEPM Field Trip*. Boston: University of Massachusetts.

Currey, D.R., 1980. Coastal geomorphology of Great Salt Lake and vicinity. Great Salt Lake—an overview. *Utah Geological and Mineral Survey Bulletin*, **116**: 69—82.

Davies, J.L., 1957. The importance of cut and fill in the development of sand beach ridges. *Australian Journal of Science*, **20**: 105–111.

Davis, R.A., Jr., Fox, H.T., Hayes, M.O., and Boothroyd, J.C., 1972. Comparison of ridge-and-runnel systems in tidal and non-tidal environments. *Journal of Sedimentary Petrology*, **42**: 401–412.

Fraser, G.S., and Hester, N.C., 1977. Sediments and sedimentary structures of a beach-ridge complex, southwestern shore of Lake Michigan. *Journal of Sedimentary Petrology*, **47**: 1187–1200.

Fulton, R.J. (ed.), 1989. Quaternary geology of Canada and Greenland. Geology of Canada, No.1, The Geological Society of America, Geology of North America Series (DNAG), K–1.

Goodfriend, G.A., and Stanley, D.J., 1999. Rapid strand-plain accretion in the northeastern Nile Delta in the 9th century AD and the demise of the port of Pelusium. *Geology*, **27**: 147–150.

Greensmith, J.T., and Tucker, E.V., 1969. The origin of Holocene shell deposits in the chenier plain facies of Essex (Great Britain). *Marine Geology*, **7**: 403–425.

Hails, J.R., 1969. The origin and development of the Umina-Woy ridge system, Broken Bay, NSW. *The Australian Geographer*, **11**: 1–12.

Hesp, P.A., 1984. Foredune formation in southeast Australia. In Thom, B.G. (ed.), *Coastal Geomorphology in Australia*. Sydney: Academic Press, pp. 69–97.

Hesp, P.A., 1985. The formation of sand "beach ridges" and foredunes. *Search*, **15**: 289–291.

Hine, A.L., 1979. Mechanism of berm development and resulting beach growth along a barrier spit complex. *Sedimentology*, **26**: 333–351.

Hoyt, J.H., 1969. Late Cenozoic structural movements, northern Florida. *Gulf Coast Association of Geological Societies Transactions*, **19**: 1–9.

Johnson, D.W., 1919. *Shore Processes and Shoreline Development*. New York: Hafner Publishing.

King, C.A.M., 1972. *Beaches and Coasts*, 2nd edn. London: Arnold.

Komar, P.D., 1976. *Beach Processes and Sedimentation*. Englewood Cliffs, NJ: Prentice-Hall.

Mason, O.K., 1990. Beach ridge geomorphology of Kotzebue sound: implications. Paleoecology and archeology. Ph.D. dissertation. Anchorage: University of Alaska.

Mason, O.K., and Jordan, J.W., 1993. Heightened North Pacific storminess during synchronous Late Holocene erosion of northwest Alaska beach ridges. *Quaternary Research*, **40**: 55–69.

Mason, O.K., Hopkins, D.M., and Plug, L., 1997. Chronology and paleoclimate of storm-induced erosion and episodic dune growth across Cape Espenberg Spit, Alaska. *Journal of Coastal Research*, **18**: 770–798.

Meldahl, K.H., 1993. Geographic gradients in the formation of shell concentrations. Plio-Pleistocene marine deposits. Gulf of California. *Paleogeography, Paleoclimatology, Paleoecology*, **101**: 1–25.

Morrison, R.B., 1965. New evidence on Lake Bonneville stratigraphy and history from southern Promontory Point, Utah. *U.S. Geological Survey Professional Paper* 525-C, pp. C110–C119.

Morrison, R.B., 1991. Quaternary stratigraphic, hydrologic, and climatic history of the Great Basin, with emphasis of Lake Lahontan, Bonneville, and Tecopa. Geological Society of America, The Geology of North America, K-2, pp. 283–320.

Nanson, G.C., Callen, R.A., and Price, D.M., 1998. Hydroclimatic interpretation of Quaternary shorelines on South Australian playas. *Paleogeography, Paleoclimatology, Paleoecology*, **144**: 281–305.

Orford , J.D., and Carter, R.W.G., 1982. Crestal, overtop and washover sedimentation on a fringing sandy gravel barrier coast. *Journal of Sedimentary Petrology*, **52**: 265–278.

Otvos, E.G., 1981. Barrier island formation through nearshore aggradation-stratigraphic and field evidence. *Marine Geology*, **43**: 195–243.

Otvos, E.G., 2000. Beach ridges—definitions and significance. *Geomorphology*, **32**: 83–108.

Otvos, E.G., in press. Holocene coastal barriers, Gulf of Mexico. Chronology, evolution, and sea-level relationships. *Journal of Coastal Research*.

Price, W.A., 1982. Beach ridge plain. In Schwartz, M.L. (ed.), *Encyclopedia of Beaches and Coastal Environments*. Stroudsburg, PA: Hutchinson Ross Publishing, pp. 159–160.
Psuty, N.P., 1966. The geomorphology of beach ridges in Tabasco. *Coastal Studies Institute Technical Report*, 30. Baton Rouge, LA: Louisiana State University.
Reineck, H.E., and Singh, I.B., 1975. *Depositional Sedimentary Environments*. Berlin, New York: Springer Verlag.
Rhodes, E.G., 1982. Depositional model for a chenier plain, Gulf of Carpentaria, Australia. *Sedimentology*, **29**: 201–222.
Ruz, M.-H., and Allard, M., 1994. Foredune development along a subarctic emerging coastline, eastern Hudson Bay, Canada. *Sedimentology*, **42**: 725–734.
Short, A.D., 1984. Beach and nearshore facies: southeast Australia. *Marine Geology*, **60**: 261–282.
Short, A.D., Buckley, R.C., and Fotheringham, D.G., 1989. Preliminary investigations of beach ridge progradation on Eyre Peninsula and Kangaroo Island. *Marine Geology*, **90**: 289–296.
Thompson, T.A., 1992. Beach-ridge development and lake-level variation in southern Lake Michigan. *Sedimentary Geology*, **80**: 305–318.
Woodroffe, C.D., Curtis, R.J., and McLean, R.F., 1983. Development of a chenier plain, Firth of Thames, New Zealand. *Marine Geology*, **53**: 1–22.

Cross-references

Barrier
Barrier Islands
Beach Processes
Cheniers
Dunes and Dune Ridges
Meteorological Effects on Coasts
Rock Coast Processes
Sea-Level Indicators, Geomorphic
Spits
Tectonics and Neotectonics

BEACH SAFETY—See LIFESAVING AND BEACH SAFETY

BEACH SEDIMENT CHARACTERISTICS

Beach sediments are derived from a wide variety of sources, including cliff erosion, rivers, glaciers, volcanoes, coral reefs, sea shells, the Holocene rise in sea level, and the cannibalization of ancient coastal deposits. The nature of the source and the type and intensity of the erosional, transportational, and depositional processes in a coastal region determine the type of material that makes up a beach. In turn, the characteristics of the sediments strongly influence beach morphology and the processes that operate on it (Trenhaile, 1997).

Grain size

The grain size of pebbles and other large clastic material can be measured with callipers, and sieves are used for sand and other coarse beach sediments. A number of techniques are used to determine the size of finer sediments including Coulter Counters, pipettes, hydrometers, optical settling instruments, and electron microscopes. The grain size can be expressed using the Wentworth scale, which is based on classes that are separated by factors of two, so that each is twice the size of the one below. A $\log_2$ transform can be used to provide integers for each of the Wentworth class limits:

$$D_\phi = -\log_2 (D_{\mathrm{mm}})$$

where D_ϕ is the grain diameter in phi units (ϕ) and D_{mm} is the corresponding diameter in millimeters (Table B4). Unfortunately, the term "grain diameter" can refer to several different things (Sleath, 1984):

(1) the mesh size of the sieve through which the grains are just able to pass;
(2) the diameter of a sphere of the same volume;
(3) the length of the long, short, or intermediate axes of the grain, or some combination of these lengths; or
(4) the diameter of a smooth sphere of the same density and settling velocity as the grains.

Table B4 Sediment grain size classification

Type	ϕ units	Wentworth (mm)
Boulder	>−8	>256
Cobble	−8 to −6	256–64
Pebble	−6 to −2	64–4
Granule	−2 to −1	4–2
Sand		
Very coarse sand	−1 to 0	2–1
Coarse sand	0–1	1–0.5
Medium sand	1–2	0.5–0.25
Fine sand	2–3	0.25–0.125
Very fine sand	3–4	0.125–0.0625
Silt		
Coarse silt	4–5	0.0625–0.0312
Medium silt	5–6	0.0312–0.0156
Fine silt	6–7	0.0156–0.0078
Very fine silt	7–8	0.0078–0.0039
Clay		
Coarse clay	8–9	0.0039–0.00195
Medium clay	9–10	0.00195–0.00098

The weight-percentages of the sediment can be plotted against the diameter in phi units in the form of histograms or frequency curves. Grain-size distributions are most frequently represented, however, by plotting the grain size data on a probability, cumulative percentage ordinate, and the phi scale on an arithmetic abscissa. The percentiles on the cumulative size distribution can be used to estimate the mean, standard deviation, and other simple descriptive statistical measures, although the calculations can also be made by computer. For comparative purposes, sediment samples can be represented by the mean or median grain size, or by the size of the grain that is coarser than some percentage of the sample.

There have been many attempts to identify the transportational processes and the depositional origin of sediments based on their sediment-size distributions. The grain-size distributions of beach sediments often consist of three straight-line segments, rather than the single straight line of a normal distribution plotted on a Gaussian probability axis. The three segments have been variously interpreted as representing: coarse bed load, fine suspended load, and intermediate-sized grains that move in intermittent suspension; the effect of packing controls on a grain matrix, the larger grains being a lag deposit, with the finest grains resting in the spaces between grains of median size; and different laminae in the beach, representing several depositional episodes. A further possible explanation is that the segmentation of grain-size distributions on log-normal cumulative probability paper may reflect the use of an inappropriate probability model. The log-normal model poorly represents the extremes of natural grain-size distributions, which may conform much better to a hyperbolic probability function (Trenhaile, 1997). Some workers believe that the four parameters of a logarithmic hyperbolic distribution are more sensitive to sedimentary environments and dynamics than the statistical moments of the normal probability function, but others have found that there is little difference (Sutherland and Lee, 1994). Grain sizes may also be fitted to a skew log-Laplace model, a limiting form of the log-hyperbolic distribution which is essentially described by two straight lines, and is defined by three parameters (Fieller *et al*., 1984).

Grain shape

The shape of beach grains can be expressed in various ways. The roundness of a grain, which refers to the smoothness of its surface, has been defined as the ratio of the radius of curvature at its corners to the radius of curvature of the largest inscribed circle. Grain sphericity describes the degree to which its shape approaches that of a sphere with three equal orthogonal axes. The shape of a grain can range from spherical, to plate, to rod-like forms, according to the relationship between the three axes, which can be depicted in the form of a ternary diagram. Grain shape can be measured and defined using a variety of indices. They include the E shape factor (ESF):

$$\mathrm{ESF} = D_s \left[\frac{D_s^2 + D_i^2 + D_l^2}{3} \right]^{-0.5}$$

and the Corey shape factor (CSF):

$$\mathrm{CSF} = \frac{D_s}{(D_i D_l)^{0.5}}$$

where D_l, D_s, and D_i are the long, short, and intermediate axes of the grain, respectively.

The shape of coarse clasts can be determined fairly easily by direct measurement, but this is usually impossible or too time-consuming for sand and other small grains. Therefore, the roundness and sphericity of sand grains has often been estimated by visual comparison with a set of standard grain images of known roundness, although Fourier analysis is increasingly being used (Powers, 1953; Thomas *et al.*, 1995). Winkelmolen's (1971) "rollability" index, the time taken for a grain to roll down the inside of a revolving, slightly inclined cylinder, is easier to measure than other shape parameters, and the shape distribution factors, obtained by plotting grain rollability against grain size, may be more characteristic and indicative of the mode of origin of coastal sediments.

Grain density

The density of a grain is determined by its mineralogy (Table B5). In temperate regions, most beach sediment originated from the granitic rocks of continents, and they largely consist of quartz and, to a much lesser extent, feldspar grains, but carbonates may dominate in the tropics, especially where there are coral reefs. The sediments in pocket beaches enclosed between prominent headlands, and in beaches derived from other restricted source areas, however, can be strongly influenced by the mineralogy of the local geological outcrops, or by the accumulation of shelly carbonate material. Beaches can consist almost entirely of heavy minerals in volcanic areas, and the usually small amounts of heavy minerals in continental beach sediments, such as magnetite, hornblende, and garnet, help to identify the source rocks, their relative importance, and the direction of longshore transport.

Bulk density and packing

Bulk density reflects the way the grains are arranged or packed together. Spherical grains of uniform size can be packed in four ways. The centers of grains in unstable cubic packing describe the corners of a cube, whereas a tetragonal arrangement is formed by moving the upper layer of grains so that they occupy the hollows between the grains below. With orthorhombic packing, the centers of the lower layer of grains form a diamond pattern, with the centers of the grains in the upper layer directly above. A rhombohedral arrangement is created by moving the upper layer of grains into the hollows created by the lower layer. The porosity of the sediments is 48%, 30%, 40%, and 26% with cubic, tetragonal, orthorhombic, and rhombohedral packing of spherical grains, respectively.

The shape of the grains exerts an important influence on the bulk properties of a sediment, including its packing geometry, stability, porosity, and permeability. Small cavities are created in a deposit by shell fragments and other flat, flaky, or plate-like particles, which greatly increase its porosity. Differences in the size of the grains also affect packing density and porosity. Smaller grains occupy the spaces between larger grains, increasing the packing density and decreasing the porosity. Grains that are less than about one-seventh the size of the larger grains can pass down through the voids between the larger grains. Packing is also influenced by deposition rates. Cubic arrangements develop when there are high depositional rates and grain collisions, and rhombohedral packing, when slow deposition allows grains to settle into their optimum positions. Grains settling onto the bed with high fall velocities jostle and vibrate the underlying layers, increasing the packing density and reducing the porosity. Suspended grains settling out in still water are also less densely packed that those that are deposited by waves and currents.

Table B5 The mean density of some minerals found in beach sands

Mineral	Density (kg m^{-3})
Aragonite	2,930
Augite	3,400
Calcite	2,710
Foraminifera shells	1,500
Garnet	3,950
Hornblende	3,200
Magnetite	5,200
Microcline	2,560
Muscovite	2,850
Orthoclase	2,550
Plagioclase	2,690
Quartz	2,650
Rutile	4,400
Zircon	4,600

Grain sorting

Grains are sorted or separated according to their shape, size, and density (Table B6). Beach sediments are generally better sorted than river sediments, but less well than dunes. Beach grain-size distributions are occasionally positively skewed, but the skew is generally negative. Although the presence of a tail of coarse grains has been attributed to the removal of fine grains, or the addition of coarse clasts or shells, skewness can also arise from a single sedimentary event, and it is not necessarily symptomatic of the mixing of two or more sediment populations (McLaren, 1981).

Cross-shore and longshore changes in beach sediment characteristics can result from mechanical and chemical breakdown, differential transport of grains according to their size, longshore variations in wave energy, the addition or loss of sediment, or the mixing of two or more distinct sediment populations. Sorting occurs through selection, breaking, and mixing (Carter, 1988). Rejection and acceptance phenomena play an important role in the selection process, and in perpetuating sorted grain distributions on beaches. Rejection accelerates the transport of coarse grains over finer grains, whereas shielding impedes the movement of fine grains over coarser grains. Grains moving over material of similar size have a high probability of being assimilated or accepted by the underlying material.

Erosion of a source material produces a lag deposit that is coarser, better sorted, and more positively skewed than the original sediment. If all the transported sediment is deposited, the deposit will be finer, better sorted, and more negatively skewed than the source. If the transported sediment is only selectively deposited, the deposit will be better sorted and more positively skewed than the source. The deposit will be finer than the source if only material finer than the mean size of the source is eroded, but it may be coarser if sediment larger than the mean size is removed from the original deposit (McLaren, 1981).

The mean grain size of beach sediments depends on the characteristics of the source and the nature of the sedimentary processes. Mean grain size varies according to differences in wave energy along beaches and on the exposed and sheltered sides of islands, and it also changes through time as gently sloping, storm-eroded beaches recover to their steeper, fully accreted states. In the cross-shore direction, the coarsest sediments are generally found on a beach at the plunge point of the breaking waves, and the grains tend to become finer seawards and

Table B6 Factors controlling sediment sorting (Steidtmann, 1982; with permission of Blackwell Science)

Rate of sediment accumulation
- Slow—allows reworking of grains
- Rapid—allows little or no reworking of grains
- None—scour

Nature of the sediment surface
- Size distribution of grains
- Packing and arrangement of grains
- Type of bedforms present

Style of grain motion
- Traction, including sliding and rolling
- Saltation
- Suspension

Fluid characteristics
- Velocity or shear velocity
- Turbulence
- Depth

Grain characteristics
- Size
- Shape
- Density

landwards of this point. There are often coarser sediments on the upper part of the beach, however, which could either have been stranded over the berm crest by large swash events, or it could be a deflation lag deposit, resulting from the aolian removal of finer grains to form dunes. It is not known whether larger or smaller grains move most easily alongshore, and therefore whether examples of beach sediments becoming coarser downdrift represent anomalous or normal situations. In any case, whereas there is often longshore grading on beaches in essentially closed embayments, it is generally lacking or poorly developed where there are large amounts of sediment moving alongshore, or where active sediment throughput does not allow enough time for it to develop (Carter, 1988). The degree of grain-size sorting normal to the beach is also a contentious issue. Some workers have found that the poorest sorting occurs in the breaker and surf zones and the best in the swash zone, whereas others have found that the degree of sorting declines on either side of the breaker zone.

Beach sediments are also sorted according to grain density, and particularly to the abundance and mineralogy of the heavy mineral component. Small heavy mineral grains occupy the spaces between the larger and less dense quartz and feldspar grains, shielding them from the flow so that they are less easily entrained. The lighter quartz grains are transported alongshore more rapidly than the heavy minerals, even when both types of grains have the same settling velocity—presumably because the smaller size of the heavy mineral grains inhibits entrainment during each brief suspension episode. Selective longshore transport of quartz grains may therefore result in heavy minerals becoming concentrated in erosive lag deposits.

There are often concentrations of heavy minerals on beaches in the form of bands or streaks near the high tide or upper swash zones, in the troughs of ripples, or where there are shells, coarse clasts, or other flow obstructions. The upper swash zone may consist of dark layers of fine, heavy mineral grains grading upwards into light colored layers of coarser, quartz–feldspar grains. The alternating layers are between about 1 and 25 mm in thickness, and they typically extend along the beach for a few tens of meters (Clifton, 1969). The formation of swash laminae has been attributed to shear sorting in the downrush, which causes the coarser grains to migrate upwards into the zone of lower shear, while the finer and heavier grains move downwards, into the zone of maximum shear at the bed. An alternate explanation is that the smaller particles tend to fall into the spaces between the larger grains, thereby displacing coarser grains toward the surface.

Heavy mineral concentrations in the cross-shore direction have either been attributed to wave asymmetry, the heavy minerals being carried onshore by high current velocities, but not by the weaker offshore flows, or to beach erosion and offshore transport of the quartz–feldspar grains. There may be poor separation under vigorous wave conditions, however, and the heavy and light minerals can be entrained and transported together.

There has been little research on the effect of sand grain shape on longshore and cross-shore sorting patterns. The proportion of angular grains increases in the direction of longshore transport between Delaware and Chesapeake Bays, possible because their lower settling velocities allow them to remain in suspension longer, so that they are carried further and at higher rates than more rounded grains. On Long Island, however, grains become rounder with longshore transport. In a laboratory and field study, grains of similar size and mineralogy (quartz) were differentially transported and sorted within the swash zone, with the more rounded grains being deposited near the top of the uprush (Trenhaile *et al.*, 1996).

Alan S. Trenhaile

Bibliography

Carter, R.W.G., 1988. *Coastal Environments*. London: Academic Press.

Clifton, H.E., 1969. Beach lamination: nature and origin. *Marine Geology*, **7**: 553–559.

Fieller, N.R.J., Gilbertson, D.D., and Olbricht, W., 1984. A new method for environmental analysis of particle size distribution data from shoreline sediments. *Nature*, **311**: 648–651.

McLaren, P., 1981. An interpretation of trends in grain size measures. *Journal of Sedimentary Petrology*, **51**: 611–624.

Powers, M.C., 1953. A new roundness scale for sedimentary particles. *Journal of Sedimentary Petrology*, **23**: 117–119.

Sleath, J.F.A., 1984. *Sea Bed Mechanics*. Chichester: Wiley.

Steidtmann, J.R., 1982. Size-density sorting of sand-size spheres during deposition from bedload transport and implications concerning hydraulic equivalence. *Sedimentology*, **29**: 877–883.

Sutherland, R.A., and Lee, C.-T., 1994. Application of the log-hyperbolic distribution to Hawaiian beach sands. *Journal of Coastal Research*, **10**: 251–262.

Thomas, M.C., Wiltshire, R.J., and Williams, A.T., 1995. The use of Fourier descriptors in the classification of particle shape. *Sedimentology*, **42**: 635–645.

Trenhaile, A.S., 1997. *Coastal Dynamics and Landforms*. Oxford: Oxford University Press.

Trenhaile, A.S., Van der Nol, L.V., and LaValle, P.D., 1996. Grain roundness and beach sorting. *Journal of Coastal Research*, **12**: 1017–1023.

Winkelmolen, A.M., 1971. Rollability, a functional shape property of sand grains. *Journal of Sedimentary Petrology*, **41**: 703–714.

Cross-references

Cross-Shore Sediment Transport
Cross-Shore Variation of Grain Size on Beaches
Longshore Sediment Transport
Sediment Suspension by Waves
Surf Modeling
Surf Zone Processes

BEACH STRATIGRAPHY

A beach is the boundary between the land and water bodies such as oceans and lakes that develops on wave-dominated coasts. It is defined as a shore consisting mainly of unconsolidated materials extending from the low-water line to where marked changes in physiographic form and/or materials are observed or to the permanent vegetation line. The zone between the low-water and high-water levels, which has a concave topography and slopes gradually seaward, is known as the foreshore or beach face. The area landward from the crest of the most seaward berm of a beach is called the backshore.

The slope gradient of the beach face varies according to material, particularly the grain size, and wave intensity (Carter, 1988; Hardisty, 1990). In general, beaches consisting of coarse-grained materials and high-energy beaches have steeper slopes. Waves and currents continuously change the slope gradient and materials of beaches, resulting in the formation of characteristic sediment facies (Harms *et al.*, 1975; McCubbin, 1981).

Succession of coastal sediments

At accumulating or progradational beaches, the succession of coastal sediments consists of lower shoreface, upper shoreface, foreshore, backshore, and dunes in ascending order. This is a typical succession on a wave- or storm-dominated sandy coast. The shoreface, located in the nearshore zone, has a concave topography formed by waves. The upper shoreface, also called the inshore, is a zone with bar and trough topography constantly influenced by waves and wave-induced currents. The migration of bars landward or seaward and rip currents result in the tabular cross-stratification and trough cross-stratification that characterize the upper shoreface sediments. Two-dimensional (2-D) and three-dimensional (3-D) wave ripple structures are also commonly found. These sedimentary facies reflect mostly fair-weather wave conditions. The upper shoreface sediments overlie the lower shoreface sediments, which are characterized by swaley cross-stratification (SCS) or hummocky cross-stratification (HCS). HCS is characterized by low-angle ($<15°$) erosional lower set boundaries with subparallel and undulatory laminae that systematically thicken laterally and by scattered lamina-dip directions (Harms *et al.*, 1975). SCS is amalgamated HCS with abundant swaley erosional features. These sedimentary structures are thought to be formed by the oscillatory currents of storm waves with offshore-directed currents. During storms, beaches are eroded and longshore bars migrate seaward. Strong oscillatory currents caused by storm waves agitate sea-bottom sediments at the shoreface. Some of the sediments are transported offshore by bottom currents caused by coastal set-up and gravity currents. Oscillatory currents related to calming storm waves produce HCS/SCS in the shoreface to inner shelf region overlain by wave ripple lamination. HCS/SCS is found only in sediments of coarse silt to fine sand. Because lower shoreface sediments are mainly deposited during storms, there is a sharp boundary formed by bar migration between upper and lower

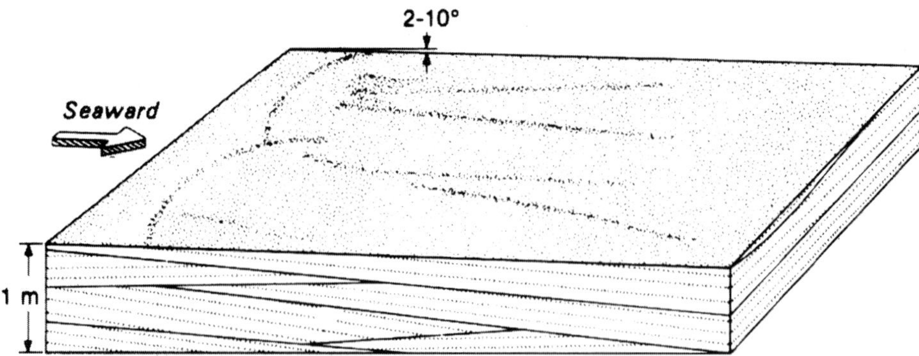

Figure B48 Swash cross-stratification. Stratification and set boundaries are formed parallel to changing slope of beachface and dip generally seaward (after McCubbin, 1981).

shoreface sediments. The lower shoreface topography depends on inner-shelf topography. Because typical shoreface topography can form only on a gentle/flat basal surface, no clear shoreface topography can form in the steep shelf regions of active plate margins. Thus, sometimes only the upper shoreface is referred to as the shoreface.

The uppermost part of the upper shoreface sediments is a step zone sediment characterized by slightly coarser materials, which are overlain by foreshore sediments. The foreshore sediments are characterized by gently seaward-dipping (2–10°) parallel lamination and wedge-shaped set boundaries. This structure is called swash cross-stratification or wedge-shaped cross-stratification (Figure B48). The essential characteristics of this stratification are 1–30 cm-thick bedsets, low-angle dips of laminae and set contacts, an average dip direction toward the sea or lake, mostly erosional set contacts, and laminae lying parallel to set contacts.

Backshore sediments overlie foreshore sediments with a gradual contact and are characterized by low-angle landward-dipping parallel lamination, current ripples, plant remains such as rootlets, and heavy mineral concentrations. Light minerals are removed by winds and form eolian coastal dunes behind the backshore. Heavy mineral concentrations are also a characteristic feature of erosional beaches, where they are residuals of the eroded beach sediments.

The coastal succession and sedimentary facies reflect the intensity of current velocity under fair-weather and storm conditions and seaward-decreasing energy conditions. Under fair-weather conditions, from the foreshore to the upper and lower shoreface, the bedforms (sedimentary structures) found are upper plane beds (parallel lamination), 3-D and 2-D subaqueous dunes (trough and tabular cross-bedding), and 3-D and 2-D ripples (ripple lamination), respectively. On the other hand, under storm conditions, beaches are eroded and the lower shoreface resembles an upper flow regime resulting in the formation of HCS. Ripples are formed in shelf regions.

Foreshore sediments

There are three hierarchies of foreshore sediments: lamination, tide-controlled structures, and storm wave/current-controlled structures.

Foreshore sediments are characterized by parallel lamination formed by the combined processes of wave swash (uprush) and backwash. Each lamina shows reverse grading from fine to coarse with thicknesses of a few millimeters to 2 cm related to each swash and backwash event as a result of either downward filtering of fine particles, or Bagnoldian dispersive pressure resulting from shear between the grains in the flow (Clifton, 1969; Allen, 1984). The fabric of the foreshore sediments shows elongated grains that orient themselves normal to the shoreline, and both landward-imbricated and seaward-imbricated grains are reported. However, these imbricated structures are influenced by the combination of waves and tides.

Reversals of the imbrication dip are thought to result from a predominance of swash transport during the flood stage and backwash transport during the ebb stage. The tidal pattern also influences the depositional thickness of the foreshore sediments. The thick layers are deposited during cycles of higher tidal range, and the thin layers are deposited during cycles of smaller tidal range (Yokokawa and Masuda, 1991). Grain size is also influenced by tides. Water-level changes by tides cause the breaker zone of waves and swash/backwash to shift. Allen (1984) showed that coarser sediments are deposited during flood stages, and finer sediments are deposited during ebb stages.

Storm waves and storm-induced currents have an erosional impact on beaches. During subsequent waning and fair-weather conditions, beaches recover as a result of sediment accretion by waves. This cycle results in an upward-fining succession from a basal erosional surface with coarse-grained materials to finer sandy materials. The coarse deposits formed under high wave energy just after the storm show a remarkable dominance of seaward-dipping imbrication, independent of tidal cycles (Yokokawa and Masuda, 1991). By regarding major erosional surfaces in beach sediments as sequence boundaries according to the sequence stratigraphic model, the depositional zone of foreshore sediments and their stacking pattern can be analyzed. A bedset with a thickness of tens of centimeters bounded by major erosional surfaces is regarded as a depositional sequence, and a lamina set with a thickness of several centimeters to ca. 20 cm is regarded as a parasequence. The depositional pattern of lamina sets shows a landward shift of the depositional zone (onlap) in the lower part of the bedset and a seaward shift of the depositional zone (downlap/progradation) in the upper part. Moreover, bedsets also form a higher order sequence (Masuda *et al.*, 1995).

Changes in waves from seasonal changes in wind direction and wave strength and type produce seasonal beaches. For example, high, strong waves may create high-level beaches with coarse sediments and a steep beachface, or occasionally erosional beaches with residual coarse sediments and heavy minerals, at the high-water level in winter; and calm waves may make gentle, accretional beaches in summer, depending on the location of the beach.

Beaches are distributed not only along wave-dominated coasts but also along tide-dominated coasts influenced by waves. In general, tide-dominated coasts have muddy or sandy tidal flats in the intertidal zone. However, waves create narrow beaches in the upper part of the intertidal zone, occasionally with beach ridges landward from the beach. A typical example is the coast of the Mekong River delta, which is a meso-tidal coast with waves. Beaches and well-developed beach ridges are found in the upper part of the intertidal zone to the supratidal zone (Ta *et al.*, 2002).

Yoshiki Saito

Bibliography

Allen, J.R.L., 1984. *Sedimentary Structures: Their Character and Physical Basis*. Amsterdam: Elsevier.

Carter, R.W.G., 1988. *Coastal Environments*. London: Academic Press.

Clifton, H.E., 1969. Beach lamination: nature and origin. *Marine Geology*, **7**: 553–559.

Hardisty, J., 1990. *Beaches: Form and Process*. London: Unwin Hyman.

Harms, J.C., Southard, J.B., Spearing, D.R., and Walker, R.G., 1975. *Depositional environments as interpreted from primary sedimentary structures and stratification sequences*. SEPM, Short Course No. 2, 161p.

Masuda, F., Okazaki, H., Yokokawa, M., Murakoshi, N., and Sakai, T., 1995. Mini-sequence stratigraphy for Pleistocene beach deposits. In Saito, Y., Hoyanagi, K., and Ito, M. (eds.), *Sequence Stratigraphy,*

Towards a New Dynamic Stratigraphy. Memoir of the Geological Society of Japan, No. 45, pp. 101–109.
McCubbin, D.G., 1981. Barrier-island and strand-plain facies. In Scholle, P.A., and Spearing, D. (eds.), *Sandstone Depositional Environments*. AAPG Memoir 31, pp. 247–279.
Ta, T.K.O., Nguyen, V.L., Tateishi, M., Kobayashi, I., Saito, Y., and Nakamura, T., 2002. Sediment facies and late Holocene progradation of the Mekong River delta in Bentre Province, southern Vietnam: an example of evolution from a tide-dominated to a tide- and wave-dominated delta. *Sedimentary Geology*, **152**: 313–325.
Yokokawa, M., and Masuda, F., 1991. Tidal influence on foreshore deposits, Pacific coast of Japan. In Smith, D.G., Reinson, G.E., Zaitlin, B.A., and Rahmani, R.A. (eds.), *Clastic Tidal Sedimentology*. Canadian Society of Petroleum Geologists. Memoir 16, pp. 315–320.

Cross-references

Beach Erosion
Beach Features
Beach Processes
Rhythmic Patterns
Shelf Processes

BEACH USE AND BEHAVIORS

Introduction

Beaches comprise only 9% of the total conterminous coastline in the United States (Ozmore, 1976). Unfortunately, while no national census of beach visits exists, several studies rank beach recreation as one of the most popular outdoor recreational activities in the United States. It is, therefore, surprising that so little research has been undertaken that addresses the socioeconomic aspects of these activities. This anomaly is particularly noticeable when contrasting the volume of physical and biological research undertaken dealing with beaches and the nearshore environment. Historically, beach recreational activities have centered on the following three activities: bathing, shore-based fishing, and beachcombing. During the past 20 years, many new activities have emerged, several of which use beaches primarily as a staging area. Such activities include surfing, windsurfing, boogie boarding, and a host of shallow-water boating activities including kayaking, canoeing, personal water crafts (PWC), and surfboarding. Many of these activities are incompatible with the more traditional uses of the beach, resulting in user conflicts. Some of these have been managed through the introduction of local, state, and federal legislation, while others have been adjudicated in the courts. Finally, the absolute increase in the number of users of beaches, as well as the diversity of activities occurring, have resulted in a growing demand for both access and accessibility to the nation's beaches.

This entry begins with a historical overview of beach uses, followed by a discussion of three related concerns: increased beach density; the demands this has placed on beach access and accessibility; and how this problem has been addressed. The entry concludes with a discussion about the increasing threat to beachgoers from pollutants in coastal waters.

Overview of beach uses and the factors affecting beach activities

It is likely that beach recreation owes its origin to the perceived value of beaches as healthy environments capable of relieving serious medical conditions (Goodhead and Johnson, 1996). In Britain, during the early part of the 1800s, many people visited beaches with the belief that immersion in, and the drinking of, seawater was healthy and would result in the relief of a number of physical ailments (Meyer-Arendt, 1986). Half a century later, these activities had evolved into resorts generally located within a day's travel of major European and North American cities. Newport, and to a lesser extent Narragansett, Rhode Island, became well-known resorts in New England and were connected by rail to both Boston and New York. In England, Brighton served the same function. More recently, the Hamptons on Long Island have become important beach destinations point for the wealthy. However, nearly all of the research dealing with the early history of beach recreation is descriptive.

One of the few examples where geographers have sought to move from purely descriptive studies to nomothetic research can be found in the extensive writings of Meyer-Arendt, who built on the early work of British geographers with an interest in beach recreation. These studies centered on the concomitant urbanization of coastal areas. Meyer-Arendt studied a number of beach resorts along the northeast Gulf coast, and, based on these efforts, developed the Coastal Resort Morphology Mode (Meyer-Arendt, 1986). This is a spatiotemporal, five-stage recreational land use model. The initial stage is characterized by easy beach access that has attracted limited residential developments, which, in turn, support a small recreational business district (RBD). Toward the end of this first stage, increased day visitation takes place. This second phase is referred to as the "Take Off" stage and is characterized by increased recreational development, extending outward from the RBD. Most of this development is along the coastline on both sides of the RBD. Sometimes a recreational fishing pier is constructed, usually at the foot of the RBD. The third phase is dominated by further development and urban expansion. True central business district (CBD) land uses characterize the area immediately surrounding the RBD. Residential developments continue to expand outward, and most of the early structures located closest to the CBD undergo rapid demolition or conversion to more up-scale recreational developments. Most of the structures still cater to a seasonal clientele, but with a small core of year-round residents. If the resort is located on a barrier beach, developments will have reached the bay-shore. As a result, much of the wetlands located there will have been destroyed by canalization, or filled in, resulting in significant environmental impacts. The fourth stage is consolidating the developments characterized in the previous stage, except that condominium developments now cater to those who no longer can afford to buy (let alone build) single-family homes. The final phase is characterized by complete saturation, where lower income residents and those on fixed incomes are forced to sell out, in part because of high property values and property taxes. Dolan and his co-workers analyzed the rise and decline of religious sea camps during the 19th century, only vestiges of which exist today.

Following the end of World War II, beach visitation became one of the most popular outdoor recreational activities that cut across all socioeconomic groups, although significant social and ethnic discrimination was still in evidence. Furthermore, the popularity of beach visitation continued to increase in concert with a general population migration from inland to coastal areas (Kimmelman *et al.*, 1974). Perhaps a contributing factor to the popularity of beach recreation related to the low cost associated with bathing, where transportation often represented the only cost of engaging in the activity. During this period, the predominant activities were bathing, sunning, and socializing, attractions that are as popular today as they were then. The only difference is that today many beachgoers are experiencing significant competition due to other outdoor recreational activities that utilize the beach as a staging area for other water-based activities. These include shore-based fishing and the launching of a variety of light vessels that can be trailered or car-topped, including kayaks, canoes, surf and sailboards, and PWCs.

For many beachgoers, the beach represents a place on which a host of activities can be undertaken, including bathing, sunbathing, ball playing, and socializing. While some degree of specialization appears to take place on certain beaches, most beach visitors tend to participate in several different activities during a day-on-the-beach.

Most early studies concerned with beach uses sought to describe and classify beach users and beaches based on perceived preferences. Several of these studies emerged from the Chicago School of Geography; under the direction of Gilbert White (1973) and his students, where the resource users' perceptions of the environment were seen as the primary factors influencing behavior. The initial research thrust dealt with perceived flooding risks, but these studies soon expanded to include all kinds of perceived environmental factors influencing behavior, including those affecting beach visitation. Few studies have analyzed the activities and social interactions occurring on the beach (Gerlach, 1987). Examples of these include Hecock who concluded that beachgoers were attracted to certain beaches based on their physical characteristics (Hecock, 1966). This study suggested that younger beachgoers preferred beaches with a stronger wave environment where bodysurfing could be undertaken. Conversely, families with small children preferred beaches where the wave environment was more gentle and where the beach slope was less steep, allowing children to play safely in the shallows (Jubenville, 1976).

One area that most coastal recreational planners and resource managers have addressed concerns the number of beach visitors that a given beach can accommodate. While no overall accepted standard exists on the number a given site can accommodate before the perceived value of a visit begins to decline, some efforts were made to address this issue nationally. The Outdoor Recreational Resources Review Commission suggested that 2,000 bathers could be accommodated per one mile of beach (Rockefeller, 1962). One problem with this measure is that no distinction is being made on the basis of the width of the beach. Jackson (1972), citing a California study, suggested that each bather in lakes required a minimum of 50 square feet of water. Other factors play a role in the decision-making process leading to a person, family, or group deciding to visit a given beach. In a study conducted in the New York–New Jersey Metropolitan area, West (1973) found that access and especially accessibility were considered more important factors compared to water quality.

By far, most of the social studies conducted on beaches have dealt with density and crowding (Boots, 1979). In this context, several authors identified "crowding" as a factor influencing beach use (Sowman, 1987). De Ruyck *et al.* (1987) identified two types of densities, one of which defined overall density as a number of visitors per unit area. He also defined "patch densities," a term he referred to as "social carrying capacity" on three beaches in South Africa. These researchers found that density tolerance was influenced by the size of the beach (the smaller the beach, the greater the willingness to accept more people (greater densities). He also found that "crowd attracting beach activities," such as impromptu ball games and other sports events, resulted in higher crowding tolerance by the visitors.

In an unrelated study, West (1974) also found that beachgoers' perception of beach density varied depending upon the respondent's residence. Those beach visitors living in urban areas were willing to tolerate greater beach crowding compared with those living in suburban areas.

Beach access and accessibility

Access and accessibility are terms often used interchangeably, however, in this entry access refers to the ability to move from an existing "right-of-way," such as a road or public parking lot, to a public beach. Accessibility refers to the obstacles that a beach visitor may encounter in traveling from his or her home to the beach. Such obstacles may include a lack of parking facilities, high entry fees, or in an urban context, a lack of public transportation to the beach.

Physical access to the shore is governed by two sets of law, one related to common law, the other by legislation. The common law principles concerning beach use originate from old Roman Law, which held that beach resources (seaweed, fish, and shellfish) were held by the sovereign, who then allowed the citizens to fish and collect seaweed from the shore. This principle was adopted in Britain during the Roman reign and eventually transferred to North America during the Colonial Period where, following independence, the concept of the "sovereign" was replaced with the general public. This meant that the government held the submerged lands seaward of the mean high water line (MHWL) in trust for the general public. In most US states, the legal definition of the public domain is seaward of the MHWL (Anon, 1988). The MHWL, in turn, is defined on the basis of the location of the average high-tide shoreline during a full metonic cycle.

A legislative approach to increasing public access was initially implied in the Coastal Zone Management Act (1972), and in the subsequent amendments. The 1986 amendments were identified as an area of special interest. Most coastal states have made some efforts to increase public access to the nation's beaches, although accomplishments vary widely. One of the aims of California's and Oregon's, and to a lesser extent Washington State's Coastal Management Program has been to increase physical access at certain intervals along their respective coastlines. Along California's rural coast, the aim is to provide coastal access every three miles. This goal is comparable to those formulated in Oregon and Washington. The objective of providing access to the shore at regular intervals has been more problematic along the Eastern Seaboard, in large part because of much higher population densities, less land in public ownership, and overall higher land prices. Together these factors have made eminent domain acquisition much more difficult and costly. Some states have attempted to increase coastal access using the principle of perfecting public right-of-ways. Rhode Island has undertaken a statewide search to identify existing and abandoned right-of-ways, largely through legal research. This effort has significantly increased public access to the state's coastal areas.

The absence of physical access to the beach is only one of the many constraints that a potential beach visitor is likely to encounter. Lack of accessibility may at times be a greater hindrance to visiting the beach. Such factors may be deliberate attempts by local cities and towns to limit or outright prohibit out-of-town visitors on local beaches. In other instances, impeded accessibility is unintentional or unavoidable (Heatwole and West, 1980).

Limiting or prohibiting beach access to out-of-town citizens on facilities owned and operated by local municipalities may vary from outright prohibition to charging unreasonably high entry or parking fees. Many of these instances have been adjudicated in the courts, which have generally ruled that where higher entrance fees have been levied against nonresidents, such fees may be permitted as long as the increased fees cover the additional costs resulting from accommodating the increased number of nonresidents. The courts have generally assumed that a portion of a resident's property tax is designated to the operation of recreational facilities, including beaches, and that opening such beaches to nonresidents often means increased expenditures to insure the health and welfare of the visitors. This may mean higher costs to cover the costs of additional guards, beach patrols, cleanups, and other services. The courts have generally felt that such additional expenditures could be recovered by charging the nonresidents a higher fee compared with those levied on residents (Neptune v Burrough of the City of Avon, 1972).

The popularity of beaches and beach uses has increased significantly during the latter part of the 20th century, a development that is likely to increase for the foreseeable future. This increased demand has raised two concerns: use conflict and water quality declines.

Conflict resolution

As mentioned in the introduction, many additional beach uses now exist. Some of these are incompatible with traditional recreational beach activities. Examples include shore-based fishing, various boating activities, including water skiing, use of personal water crafts and surfing. Most of these conflicts have been dealt with on the local level, while a few have been adjudicated in a court of law. Of the management procedures that have been introduced on the local level, zoning procedures are probably the most common. Zoning, as it was first conceptualized in New York City in 1916 (Haar, 1977), was originally intended to control building height. Zoning maps later followed with zoning ordinances specifying restricted or prohibited uses.

Recreational applications of zoning have been attempted both on land and on the water in an attempt to reduce conflicts between and among different recreational pursuits. On the water, zoning has been used by a number of municipalities to segregate swimmers and bathers from boaters—especially powerboaters, surfers, and PWCs. Two versions of zoning have been used: permanent zones and space/time zoning ordinances. In the case of permanent zones, a protected activity (e.g., swimming or bathing) is protected from all other activities by prohibiting those from entering the designated area. A less common practice is sometimes referred to as time zoning. In this instance, the competing uses are assigned different periods when each activity can take place, thereby eliminating any conflicts between competing uses. If a given beach is sought by both swimmers and surfers, the beach may be restricted to one user group while the other use may be permitted during different periods. A coastal municipality may allow surfers access to the beach during the early morning and again in the late afternoon. Swimmers and bathers may have exclusive use of the beach and adjacent nearshore during the period from mid-morning to late afternoon.

The same procedures may be utilized on land in areas where users compete for the same area. Sunbathing and shore-based fishing are both legitimate recreational activities that sometimes may compete for the same stretch of beach area. Shore-based fishing may be restricted to the early morning and late afternoon, while sunbathing may be permitted from mid-morning to late afternoon.

Environmental impacts on beach use

Socioeconomic factors are not the only variables influencing beach recreation quantitatively as well as qualitatively. There are at least two additional variables that increasingly have played a role in this nation's beach recreational activities. One concerns the increased outbreaks of algal blooms, in particular, those classified as harmful. The other concerns the impact that beach activities may have on endangered species and the restrictions imposed on beach visitors to protect threatened and endangered biological resources.

Algal blooms have occurred along the nation's coasts at least since the Spanish first settled Florida. However, there is growing evidence that these incidents are increasing quantitatively and qualitatively. The number of harmful algal bloom (HAB) incidents have increased significantly in recent years as have the impacts on marine life, swimmers, bathers, and people handling fish and shellfish affected by these incidents. While the cause for these events has not yet been determined, there is growing evidence that land-based pollution is partially responsible (Anon, 2000). The effects of HAB events range from the discoloration of large patches of waters to fish kills, die-offs of manatees in Florida, and possibly the deaths of small marine mammals in the United States, Scandinavia, and the Mediterranean. So-called "red tide" incidents in Florida, have significantly affected swimmers and bathers. Toxins released from these HABs can also become airborne, resulting in respiratory irritation, coughing, and sneezing by people who are not even in direct contact with the affected waters (Luttenberg, 2001).

The second factor that has influenced swimming and bathing in recent years is the potential conflict between the Endangered Species Act and all types of beach recreational activities. During the 1980s and 1990s, large stretches of barrier beaches on Cape Cod were closed to fishing, overland vehicular traffic, and bathing in an effort to protect the Piping Plover nests and fledglings from being trampled. In 1988, it was estimated that only 20 pairs of piping plovers were nesting within the Cape Cod National Seashore (Lopez, 1998). Largely because of the severe restrictions placed on beach traffic (both pedestrian and vehicular), a substantial increase in nesting pairs has been noted throughout the seashore (Lopez, 1998). These accomplishments, however, have not been made without impacts on beach recreation in general, shore-based fishing or bathing. Within the Cape Cod National Seashore, less than 10 miles of the Atlantic shore are now open to ORV traffic during the nesting season (from March through July). Similar restrictions have been imposed on bathing and beachcombing in piping plover nesting areas.

Conclusions

Beaching and bathing continue to be two of the most popular outdoor recreational activities both here and abroad, yet with few exceptions, not many studies have addressed the behavior and motivation of the beach-going public. Studies conducted on or adjacent to beaches generally fall into two groups. The first has sought to analyze the reaction of the beach visiting public to deteriorating water quality. The second group of studies has concentrated on infrastructure changes that have taken place in the areas immediately inland from many popular bathing beaches. While these studies are important both socially and economically, it is suggested that many additional findings would enhance our understanding of the factors motivating the beach visitor and the management of beaches. Answers that are needed include studies dealing with crowding and density tolerances and better understanding of beach preferences by different user groups. Are some beaches attracting certain population groups simply because they are more accessible, or because the amenities found on the beach attract specific user groups interested in participating in activities (e.g., surfing) that may not be readily available on all beaches? The role of physical access is still an issue in many communities, notwithstanding that access along the shore is recognized by most states as a public right, Public beaches constitute less than 10% of all the beaches in the United States. This increasingly scarce resource may be better managed if we had a better understanding of the factors that attract and detract the public to certain beaches.

Niels West

Bibliography

Anon, 1988. *Managing the Shoreline for Water Dependent Uses: A Handbook of Legal Tools*, Marine Law Institute, University of Maine School of Law, Portland, ME.

Anon, 2000. National Assessment of Harmful Algal Blooms in U.S. Waters, Washington: National Science and Technology Council Committee on the Environment and Natural Resources.

Boots, B.N., 1979. Population Density, Crowding and Human Behavior. *Progress in Human Geography*, **3**(1): 13–63.

De Ruyck, M.C., Soares, A.G., and McLachlan, A. 1987. Social carrying capacity as a management tool for sandy beaches. *Journal of Coastal Management*, **13**(2): 822–830.

Gerlach, J., 1987. Spring Break at Padre Island: A New Type of Tourism. *Paper presented at the Annual Meeting of the AAG*, Portland, OR.

Goodhead, T. and Johnson, D., 1996. Maritime Leisure. In Goodhead, T., and Johnson, D. (eds.), *Coastal Recreation Management*. London: E & FN Spoon Publishers.

Haar, C.M., 1977. *Land-Use Planning: A Casebook on the Use, Misuse and Re-use of Urban Land*, Boston, Little Brown and Company.

Heatwole, C.A., and West, N. 1980. Mass transit and beach access in New York City. *Geographical Review*, **70**: 210–217.

Hecock, R.D., 1966. Public beach recreation opportunities and patterns of consumption on Cape Cod. unpublished Ph.D. dissertation, Worcester: Graduate School of Geography, Clark University.

Jackson, R., 1972. 'Recreational zoning and lake planning'. *Town Planning Review*, **43**(1): 41–55.

Jubenville, A., 1976. *Outdoor Recreation Planning*, Philadelphia: W.B. Saunders.

Kimmelman, B. *et al.*, 1974. *Studies in Environment: Outdoor Recreation and the Environment*, Vol. V. Washington: Office of Research and Development, EPA, Publication EPA 600/5/73-012.

Lopez, J.J. 1998. *The use of negotiated rulemaking in piping plover/off-road vehicle management at Cape Cod national seashore*. unpublished Masters of Arts thesis, Kingston: Department of Marine Affairs, University of Rhode Island.

Luttenberg, D. 2001. Managing harmful algal blooms in U.S. waters: the evolution and implementation of monitoring and management. unpublished Paper, Kingston: Department of Marine Affairs, University of Rhode Island.

Meyer-Arendt, K.J., 1986. 'Barrier island settlement and landuse evolution: a gulf coast model'. *Proceedings* of *The Coastal Society's 10th National Conference*, New Orleans, LA.

Neptune City v Borough of Avon-By-the-Sea 1972:294 A.2d47 {N.J.1972}

Ozmore, K., 1976. 'Texas—A pioneer in open beaches, In Susan H. Anderson (ed.), *Recreation—Marine Promise: Proceedings, Discussion, and Overview of the National Conference on Marine Recreation*. Newport Beach, California, October 2–4, 1975.

Rockefeller, L. 1962, outdoor Recreation for America: A Report to the President & to Congress by the outdoor Recreation Resources Review Commission, U.S. Gov. Printing Office, Washington DC.

Sowman, M.R. 1987. A Procedure for assessing recreational carrying capacity of coastal resort areas. *Landscape and Urban Planning*, **14**.

West, N., 1973. An analysis of factors influencing beach activities in the New York-New Jersey metropolitan estuarial region. unpublished Ph.D. dissertation, New Brunswick: Department of Geography, Rutgers, the State University, New Brunswick.

West, N., 1974 Perceived density as a factor in selecting beaches. unpublished New Brunswick: Study, Department of Geography, Rutgers, the State University.

White, G.F., 1973. *Natural Hazard Perception and Choice*, Cambridge: Oxford University Press.

Cross-references

Beach Processes
Cleaning Beaches
Coastal Boundaries
Coastal Zone Management
Developed Coasts
Environmental Quality
Human Impact on Coasts
Lifesaving and Beach Safety
Rating Beaches
Tourism and Coastal Development
Tourism, Criteria for Coastal Sites

BEACHROCK

Formation and distribution of beachrock

Beachrock is defined by Scoffin and Stoddart (1987, p. 401) as "the consolidated deposit that results from lithification by calcium carbonate of sediment in the intertidal and spray zones of mainly tropical coasts." Beachrock units form under a thin cover of sediment and generally overlie unconsolidated sand, although they may rest on any type of foundation. Maximum rates of subsurface beachrock cementation are thought to occur in the area of the beach that experiences the most wetting *and* drying—below the foreshore in the area of water table excursion between the neap low and high tide levels (Amieux *et al.*, 1989; Higgins, 1994). Figure B49 shows a beachrock formation displaying typical attributes.

There are a number of theories regarding the process of beach sand cementation. Different mechanisms of cementation appear to be responsible at different localities. The primary mechanisms proposed for the origin of beachrock cements are as follows:

(1) physicochemical precipitation of high-Mg calcite and aragonite from seawater as a result of high temperatures, $CaCO_3$

Figure B49 Multiple unit beachrock exposure at barrio Rio Grande de Aguada, Puerto Rico. The sculpted morphology, development of a nearly vertical landward edge, and dark staining of outer surface by cyanobacteria indicate that this beachrock has experienced extended exposure. Landward relief and imbricate morphology of beachrock units define shore-parallel runnels that impound seawater (photo: R. Turner).

supersaturation, and/or evaporation (Ginsburg, 1953; Stoddart and Cann, 1965);

(2) physicochemical precipitation of low-Mg calcite and aragonite by mixing of meteoric and fresh groundwater with seawater (Schmalz, 1971);

(3) physicochemical precipitation of high-Mg calcite and aragonite by degassing of CO_2 from beach sediment pore water (Thorstensen *et al.*, 1972; Hanor, 1978); and

(4) precipitation of micritic calcium carbonate as a byproduct of microbiological activity (Taylor and Illing, 1969; Krumbein, 1979; Strasser *et al.*, 1989; Molenaar and Venmans, 1993; Bernier *et al.*, 1997).

Although most beachrock cement morphologies suggest an inorganic origin, physicochemical mechanisms operating alone do not adequately account for the discontinuous distribution of beachrock formations. As Kaye (1959, p. 73) put it, "the problem hinges more on an adequate explanation for the absence of beachrock from many beaches than on its presence in others." The discontinuity of beach cementation, along with the complex assemblage of cement types found in adjacent samples of beachrock led Taylor and Illing (1971) to propose that the microenvironment exerts a greater influence on the cementation process than does the macroenvironment.

Several beachrock researchers concur with this assessment and support the theory that *initial* cementation in beach sands is controlled by the distribution and metabolic activity of bacteria because: (1) dark, organic-rich micritic rims have been identified around cemented grains in most petrographic studies of beachrock (Krumbein, 1979; Beier, 1985); (2) microbially mediated precipitation of carbonates has been repeatedly demonstrated in both marine and terrestrial environments (Buczynski and Chafetz, 1993); and (3) bacterial populations are particularly large and productive in the intertidal zone of water table fluctuation where beach lithification occurs. Once biologically mediated cryptocrystalline cements are established as nucleation sites, larger crystals precipitated via physicochemical processes can grow and bridge the sediment grains.

Rates of beachrock formation are undoubtedly variable but are generally believed to be quite rapid, on the scale of months to years (Frankel, 1968). For example, Hopley (1986) reported that beachrock formed within six months on Magnetic Island near Townsville, Australia, while Moresby (1835) wrote that Indian Ocean natives made an annual harvest of beachrock for building stone and within a year they had a new lithified crop.

Several Pleistocene and older beachrock formations have been identified. However, the dynamic nature of sandy coastlines and a historically fluctuating sea level necessitate that most occurrences of *intertidal* beachrock are less than 2,000 years old. This is commonly supported by the incorporation of modern man-made artifacts in beachrock formations rather than by ^{14}C dates, as beachrock is poorly suited for radiocarbon dating.

The majority of recent beachrock is formed on beaches in the same regions that favor coral reef formation. This is generally below 25° latitude where there is a well-defined dry season and "the temperature of ground water at a depth of about 76 cm in beaches remains above 21°C for at least 8 months of the year" (Russell, 1971, p. 2343). However, beachrock can also form at higher latitudes. For example, beachrock exposures are common throughout the Mediterranean and have been reported along portions of the coasts of Norway, Denmark, Poland, Japan, New Zealand, South Africa, the Black Sea, and the northern Gulf of Mexico. Beachrock formations have also been reported on lakeshores in Pennsylvania, Michigan, Africa, New Zealand, southeast Australia, and the Sinai Peninsula.

Subaerially exposed beachrock units constitute only a small proportion of the cemented sediments in the intertidal zone. For example, Emery and Cox (1956) found beachrock *exposures* on only 24% of the predominantly calcareous beaches of Oahu, Kauai, and Maui, whereas jet-probing conducted by Moberly (1968, p. 32) revealed that "exposed or covered beachrock appears to be present at all calcareous beaches in the state" of Hawaii. In the event of continued sea-level rise and human activities that exacerbate coastal erosion, much more beachrock will be exhumed.

Morphology of beachrock formations

Beachrock formations typically consist of multiple units, representing multiple episodes of cementation and exposure. Beachrock that forms below the foreshore has an upper surface slope that tends to mimic that of the seaward dipping (4–10°) internal beach bedding. However, beach sand cementation has also been found to occur below the berm and foredune of a beach (Russell, 1971; Hopley and MacKay, 1978). Those authors found that the beachrock forming below the backshore had a nearly horizontal upper surface that corresponded to the groundwater table and truncated the original beach bedding.

Most intertidal beachrock formations are detached from subaerial and subtidal cemented sediments. Beachrock is laterally discontinuous as well, usually exposed for only short distances before disappearing under loose sand or ending entirely. It is likely that the formation and preservation of beachrock on a given section of beach is negatively correlated to alongshore increase in wave energy and frequency of beach erosion.

The reported thickness of beachrock formations ranges from a few centimeters up to 5 m, with approximately 2 m being most common. Variations in degree of cementation within a beachrock unit can be controlled by variability in porosity, permeability, and composition of different sand layers (Molenaar and Venmans, 1993). Generally, precipitation of cements is most rapid near the top of a beachrock unit. Accordingly, young beachrock units are better cemented at the top and noticeably less so near the base. This attribute makes them more susceptible to scour at their base upon exposure, commonly resulting in undercutting and slumping. It is this undercutting that fosters the development of nearly vertical landward edges on beachrock units. In areas where a chronic deficit in sand supply or erosive conditions have exhumed the seaward edge of a beachrock formation, it is frequently observed to be steep as well.

Long-term exposure of beachrock will radically change the ecology of a sandy shoreline by providing a hard substrate that can support an increased diversity of animal and plant life. The reader is referred to the papers of McLean (1974), Jones and Goodbody (1984), and Miller and Mason (1994) to learn more about the ecology and biophysical modification of intertidal beachrock exposures.

Beachrock and coastal evolution

Although beachrock, as defined, forms in the intertidal zone, it does not always remain there. On prograding coasts, a series of beachrock units may form at depth, leaving older units stranded well behind the active beach. On retreating coasts, outcrops of beachrock may be evident offshore where they may serve as a hard substrate for coralgal reef growth. If the strike of the beach changes over time, then the strike of the beachrock units will reflect that change.

Armed with the knowledge that beachrock is formed in the intertidal zone, many geologists have related beachrock outcrops to changes in sea level for particular coasts. Semeniuk and Searle (1987) demonstrated that beachrock formation can keep pace with slow shore recession, resulting in a wide, continuous band of beachrock, but that rapid shore recessions (or periods of high wave energy and foreshore instability) are represented by gaps (unconsolidated sediment) in a sequence of beachrock units. Assuming a nearly constant rate of sea-level rise, these gaps may indicate that beachrock can temporarily stabilize the position of the shore under erosive conditions until sea level has risen enough to cause the shore to jump back (Cooper, 1991). Many other researchers have asserted that beachrock outcrops will protect a beach from erosion, as well as control the plan configuration of a coastline.

Research by Turner (1999) has demonstrated that the influence beachrock has on beach processes will largely depend on the extent and morphology of the exposure, both of which evolve over time. Cumulative exposure and erosion of a beachrock formation over a period of years to decades can foster a gradual increase in the landward and seaward relief of the beachrock units and the development of shore-parallel runnels and shore-perpendicular breaches in the beachrock. The high seaward relief of such a beachrock unit effectively attenuates incident wave energy and retards onshore sediment transport. The high landward relief of the beachrock unit can act as the seaward wall of a runnel that blocks offshore return of backwash and forces impounded seawater and entrained sand to flow laterally on the foreshore to low spots and shore-normal breaches in the beachrock formation. Beachrock breaches and runnels are erosionally enlarged over time, locally increasing onshore inputs of wave energy and longshore sediment transport rates on the foreshore.

On a beach on Puerto Rico's west coast, beach width and volume were found to be least stable where the seaward beachrock unit was breached and most stable away from the breaches behind high relief beachrock. Sections of foreshore most protected by a high relief beachrock ridge exhibited the lowest volumes of subaerial sand storage, unusually narrow beach widths, and the slowest beach erosion recovery rates. In short, a beach with a high relief intertidal beachrock exposure is more likely to be sediment deficient and out of synch with the wave regime. This puts the backshore of a beachrock beach at risk of catastrophic retreat following the development of a breach in the beachrock or in the event of a high energy wave event coupled with a storm surge or spring high tide.

Conclusions

The transformation of sandy beaches to rocky beachrock beaches is increasingly common in the tropics and subtropics. Where beachrock is exposed by erosion, it acts as a natural breakwater or revetment, decelerating further shoreline and backshore retreat. However, it also tends to retard beach buildup and is poorly suited to recreational use, both major issues in the tropics where tourism is often the primary source of income. The potential for beachrock to significantly alter the evolution of a coast justifies additional research on its influence on beach processes. In particular, the characteristics and effects of beachrock on dissipative beaches have received little attention and are likely to be significantly different than those observed on more reflective beaches.

Despite many petrographic investigations of beachrock cements, the processes responsible for beachrock formation are still poorly understood. Given the likelihood of cement diagenesis in the beach environment, there is a need to pursue other research methods. For example, the subsurface formation of beachrock should be tracked on a variety of beaches over an extended period. The processes affecting beach sand cementation should also be examined under controlled conditions in a laboratory setting. Preliminary experiments conducted by Turner (1995) indicate that the addition of dissolved nitrate or organic carbon to beach sand microcosms stimulates bacterial growth and the precipitation of intergranular calcium carbonate. This leads to the question as to whether coastal discharges of groundwater contaminated with fertilizers or human wastes are increasing the rate and geographic range of beachrock formation.

Robert J. Turner

Bibliography

Amieux, P., Bernier, P., Dalongeville, R., and Medwecki, V., 1989. Cathodoluminescence of carbonate-cemented Holocene beachrock from the Togo coastline (West Africa): an approach to early diagenesis. *Sedimentary Geology*, **65**: 261–272.

Beier, J.A., 1985. Diagenesis of Quaternary Bahamian beachrock: petrographic and isotopic evidence. *Journal of Sedimentary Petrology*, **55**: 755–761.

Bernier, P., Guidi, J.B., and Bottcher, M.E., 1997. Coastal progradation and very early diagenesis of ulramafic sands as a result of rubble discharge from asbestos excavations (northern Corsica, western Mediterranean). *Marine Geology*, **144**: 163–175.

Buczynski, C., and Chafetz, H.S., 1993. Habit of bacterially induced precipitates of calcium carbonate: examples from laboratory experiments and Recent sediments. In Rezak, R. and Lavoie, D.L. (eds.), *Carbonate Microfabrics*. New York: Springer-Verlag, pp. 105–116.

Cooper, J.A.G., 1991. Beachrock formation in low latitudes: implications for coastal evolutionary models. *Marine Geology*, **98**: 145–154.

Emery, K.O., and Cox, D.C., 1956. Beachrock in the Hawaiian Islands. *Pacific Science*, **10**: 382–402.

Frankel, E., 1968. Rate of formation of beachrock. *Earth and Planetary Science Letters*, **4**: 439–440.

Ginsburg, R.N., 1953. Beachrock in south Florida. *Journal of Sedimentary Petrology*, **23**: 85–92.

Hanor, J.S., 1978. Precipitation of beachrock cements: mixing of marine and meteoric waters vs. CO_2-degassing. *Journal of Sedimentary Petrology*, **48**: 489–501.

Higgins, C.G., 1994. Subsurface environment of beaches—temperature and salinity. Geologic Society of America, Abstracts with Programs, Seattle Meeting, A–364.

Hopley, D., 1986. Beachrock as a sea-level indicator. In van de Plassche, O., (ed.), *Sea-level Research: A Manual for the Collection and Evaluation of Data*. Norwich: Geo Books, Regency House, pp. 157–173.

Hopley, D., and Mackay, M.G., 1978. An investigation of morphological zonation of beach rock erosional features. *Earth Surface Processes*, **3**: 363–377.

Jones, B. and Goodbody, Q.H., 1984. Biological alteration of beachrock on Grand Cayman Island, British West Indies. *Bulletin of Canadian Petroleum Geology*, **32**: 201–215.

Kaye, C.A., 1959. Shoreline features and Quaternary shoreline changes, Puerto Rico. *U.S.G.S. Professional Paper*, **317**(B): 49–140.

Krumbein, W.E., 1979. Photolithotropic and chemoorganotrophic activity of bacteria and algae as related to beachrock formation and degradation (Gulf of Aqaba, Sinai). *Geomicrobiology*, **1**: 139–203.

McLean, R.F., 1974. Geologic significance of bioerosion of beachrock. *Proceedings of the 2nd International Coral Reef Symposium, Brisbane*, **2**: 401–408.

Miller, W.R., and Mason, T.R., 1994. Erosional features of coastal beachrock and aeolianite outcrops in Natal and Zululand, South Africa. *Journal of Coastal Research*, **10**(2): 374–394.

Moberly, R., 1968. Loss of Hawaiian littoral sand. *Journal of Sedimentary Petrology*, **38**(1): 17–34.

Molenaar, N., and Venmans, A.A.M., 1993. Calcium carbonate cementation of sand: a method for producing artificially cemented samples for geotechnical testing and a comparison with natural cementation processes. *Engineering Geology*, **35**: 103–122.

Moresby, R.M., 1835. Extracts from Commander Moresbys' report on the northern atolls of the Maldives. *Journal of the Royal Geography Society of London*, **5**: 398–404.

Russell, R.J., 1971. Water-table effects on seacoasts. *Geology Society of America Bulletin*, **82**: 2343–2348.

Schmalz, R.F., 1971. Formation of beach rock at Eniwetok Atoll. In Bricker, O.P., (ed.), *Carbonate Cements*. Baltimore: Johns Hopkins University Press, pp. 17–24.

Scoffin, T.P., and Stoddart, D.R., 1987. Beachrock and intertidal cements. In Scoffin, T.P. (ed.), *An Introduction to Carbonate Sediments and Rocks*. Glasgow: Blackie Publishing Company, pp. 401–425.

Semeniuk, V. and Searle, D.J., 1987. Beach rock ridges/bands along a high-energy coast in southwestern Australia—their significance and use in coastal history. *Journal of Coastal Research*, **3**(3): 331–342.

Stoddart, D.R. and Cann, J.R., 1965. Nature and origin of beach rock. *Journal of Sedimentary Petrology*, **35**(1): 243–273.

Strasser, A., Davaud, E., and Jedoui, Y., 1989. Carbonate cements in Holocene beachrock: example from Bahiret el Biban, southeastern Tunisia. *Sedimentary Geology*, **62**: 89–100.

Taylor, J.C.M., and Illing, L.V., 1969. Holocene intertidal calcium carbonate cementation, Qatar, Persian Gulf. *Sedimentology*, **12**: 69–107.

Thorstenson, D.C., Mackenzie, F.T., and Ristvet, B.L., 1972. Experimental vadose and phreatic cementation of skeletal carbonate sand. *Journal of Sedimentary Petrology*, **42**(1): 162–167.

Turner, R.J., 1995. Bacteria and algae-mediated precipitation of calcium carbonate in Puerto Rico sand and seawater suggests that nutrient-rich ground water discharges enhance coastal sand cementation. Geologic Society of America, Abstracts with Programs, New Orleans Meeting, **27**(6): A-346.

Turner, R.J., 1999. Morphodynamic relationship between beachrock exposure and littoral zone processes on the west coast of Puerto Rico. Unpublished dissertation, University of North Carolina, Chapel Hill, 395 p.

Cross-references

Beach Features
Coral Reef Coasts
Eolianite
Rock Coast Processes
Sea-Level Indicators, Geomorphic

BEAUFORT WIND SCALE

The Beaufort scale of wind velocity relates wind speed to the physical appearance of the sea surface by considering such factors as apparent wave height and the prominence of breakers, whitecaps, foam and spray. It is the oldest method of judging wind force. Originally devised by Admiral Sir Francis Beaufort of the British Navy in 1805 to simplify the signaling of wind and weather conditions between sailing vessels, it has since been repeatedly modified to make it more relevant to modern navigation. Table B7 gives an updated modern version of the Beaufort scale, adapted from British Admiralty (1952), Thomson (1981), and US Army Coastal Engineering Research Center (1984). Meyers *et al.* (1969) presented an elaborate version of the wind scale based on British Admiralty (1952), McEwen and Lewis (1953), and Pierson *et al.* (1953). Wind speed measured at 11 m (36 feet) above sea surface is usually applied to use the scale. The wave heights are approximate and represent fully arisen sea state. As with any subjective judgment method, the Beaufort Scale is far from perfect. Similar subjective scales have been proposed to assess tornado and hurricane damages. The Fujita scale (F-scale) was proposed in 1951 by Tetsuya Fujita for rating the severity of tornadoes as a measure of the damage. The Saffir–Simpson scale is used for rating the severity of damages by a hurricane.

Dilip K. Barua

Bibliography

British Admiralty, 1952. *Manual of Seamanship*. London: H.M. Stationary Office.

McEwen, W.A. and Lewis, A.H., 1953. Encyclopedia of nautical knowledge. Cambridge, MD: Cornell Maritime Press.

Meyers, J.J., Holm, C.H., and McAllister, R.F., 1969. Handbook of ocean and underwater engineering. New York: McGraw-Hill Book Company.

Pierson, W.J., Neuman, G., and James, R.W., 1953. Practical methods for observing and forecasting ocean waves. New York: New York University, College of Engineering.

Thomson, R.E., 1981. *Oceanography of British Columbia Coast*. Sidney, B.C.: Department of Fisheries and Oceans.

US Army Coastal Engineering Research Center, 1984. *Shore Protection Manual*, 4th edn., Washington, D.C.: US Government Printing Office.

Cross-references

Climate Patterns in the Coastal Zone
Coastal Climate
Meteorologic Coastal Wind Effects on Coasts
Nearshore Wave Measurement
Wave Hindcasting

BIOCONSTRUCTION

The term, *bioconstruction*, usually refers to a bioconstructed limestone that has been built-up by colonial and sediment-binding organisms including algae, corals, bryozoans, and stromatoporoids. The term, bioconstructed limestone, was introduced by Carozzi and Zadnick (1959) in their study of the Silurian Wabash reef in southern Indiana. The word, bioconstructed, was used to distinguish the limestones and dolomites which were found in a reef from the dolomitic calcarinites preserved in the reef flanks and the dolomitic shales in the country rock (Carozzi and Zadnick, 1959). The term, bioconstruction, was next applied to Devonian stromatoporoid reefs in the Beaverhill Lake Formation, Upper Devonian, Alberta Canada (Carozzi, 1961).

European use of the word bioconstruction

The word, *bioconstruction*, was widely accepted and used in European geologic journals, but has not appeared in any North American journals since 1961. The European use of the term, *bioconstruction*, includes what the North American geologists would refer to as reefs, bioherms, and biostromes. Based on living coral reefs, Ladd (1944) defined a reef as a rigid, wave-resistant framework constructed by large skeletal organisms. A broader definition of a reef as "a discrete carbonate structure formed by in-situ organic components that develops topographic relief upon the seafloor" has been proposed by Wood (1999, p. 5). Cumings (1930) defined a bioherm as a mound-like, dome-like, lens-like, or reef-like mass of rock built-up by sedentary organisms (such as corals, algae, foraminifera, mollusks, gastropods, and stromatoporoids), composed almost exclusively of their calcareous remains and enclosed or surrounded by rock of different lithology. A biostrome is defined as a distinctively bedded and widely extensive lenticular, blanket-like mass of rock built by and composed mainly of the remains of sedentary organisms and not swelling into a mound-like or lens-like form; an organic layer, such as a bed of shells, crinoids, or corals, or a modern reef in the course of formation, or even a coal seam (Cumings, 1930).

Types of bioconstructions

Examples of several different types of bioconstructions, which would fall into the categories of reef, bioherms, and biostromes, are included to show how the term bioconstruction is used in the European literature. In Spain, rugose corals and calcareous algae bioconstructions are also called biostromes (Rodrigues and Sanchez, 1994). In Jurassic and Cretaceous strata in Germany, Rehfeld (1996) describes different forms of sponge bioconstructions which comprise bioherms, biostromes, and sponge meadows. The wave resistant calcisponge and algal reefs of the Capitan reef facies, partially wave resistant reef mounds and non-wave resistant skeletal mounds in the Guadalupe Mountains of New Mexico, are described as Permian bioconstructions (Noe, 1996). Therefore, bioconstruction is a general term for limestone and dolomite deposits formed by colonial and sediment binding organisms which include reefs, bioherms, and biostromes.

Table B7 Beaufort wind scale

Beaufort No.	Name	Wind speed knot	m/s	Effect of wind at sea surface	Significant wave height (m)	Effect of wind on land
0	Calm	<1	0.0–0.2	Like a mirror	0	Still, smoke rises vertically
1	Light air	1–3	0.3–1.5	Ripples form with the appearance of scales, but without foam crests	0.1–0.2	Smoke drifts, vanes remain motionless
2	Light breeze	4–6	1.6–3.3	Small wavelets, crests appear glossy but no breaking	0.3–0.5	Leaves rustle, vanes move, wind can be felt on face
3	Gentle breeze	7–10	3.4–5.4	Larger wavelets begin to break, some scattered white horses	0.6–1.0	Constant movement of leaves and small twigs, flags begin to stream
4	Moderate breeze	11–16	5.5–7.9	Small waves predominant but fairly frequent white horses	1.5	Dust and loose paper are lifted, thin branches move
5	Fresh breeze	17–21	8.0–10.7	Moderate waves, distinctly elongated, many white horses, chance of spray	2.0	Small trees in leaf begin to sway
6	Strong breeze	22–27	10.8–13.8	Long waves with extensive white foam, breaking crests, spray likely	3.5	Large branches move, power lines whistle, stop lights sway, umbrellas difficult to control
7	Moderate gale	28–33	13.9–17.1	Sea heaps up and white foam from breaking waves begins to be blown in streaks, spindrift begins to be seen	5.0	Entire trees sway, some resistance to walkers, car feels force of wind
8	Fresh gale	34–40	17.2–20.7	Moderately high waves of greater lengths, edges of crests break into spindrift, foam is blown into well-marked streaks	7.5	Twigs break off trees, difficult walking against wind
9	Strong gale	41–47	20.8–24.4	High waves, rolling sea, dense streaks of foam, spray may affect visibility	9.5	Roof tiles lifted off, windows may be blown in, trees may topple
10	Whole gale	48–55	24.5–28.4	Very high waves with long overhanging crests, foam in great patches blown in dense white streaks downwind, heavy rolling sea causes ships to slam, visibility reduced by spray, sea surface takes on whitish appearance	12.0	Trees uprooted considerable structural damage to some buildings
11	Storm	56–66	28.5–32.7	Exceptionally high waves, sea covered with long white patches of foam blown downwind, wave crests blown into froth everywhere, visibility impeded by spray	15.0	Widespread damage, extensive flooding in low lying areas if wind is directed onshore
12	Hurricane	>66	>32.7	Air filled with foam and spray, sea completely white, visibility seriously impaired	>15.0	Severe structural damage to buildings, widespread devastation and flooding

Conclusions

Bioconstruction is distinctly a European term for a limestone which has been built-up by colonial and sediment binding organisms such as algae, corals, bryozoans, and stromatoporoids. It combines what North American geologists would refer to as reefs, bioherms, and biostromes.

William T. Fox

Bibliography

Carozzi, A.V., 1961. Reef petrography in the Beaverhill Lake Formation, Upper Devonian, Swan Hills area, Alberta, Canada. *Journal of Sedimentary Petrology*, **31**: 497–513.

Carozzi, A.V. and Zadnik, V.E., 1959. Microfacies of Wabash reef, Wabash, Indiana. *Journal of Sedimentary Petrology*, **29**: 164–171.

Cumings, E.R., 1930, List of species from the New Corydon, Kokomo, and Kenneth Formations of Indiana, and from reefs in the Mississineaw and Liston Creek formations. *Proceedings of the Indiana Academy of Science*, **39**: 204–212.

Ladd, H.S., 1944, Reefs and other bioherms. National Research Council, Division of Geology and Geography, *Annual Report 4*, Appendix K, pp. 26–29.

Noe, S.U., 1996, Late-stage reef evolution of the Permian Reef Complex; shelf margin and outer-shelf development of the Tansill Formation (Late Permian), northern Guadalupe Mountains, New Mexico, USA. In Reitner, J., Neuweiler, F., and Gunkel, F. (eds.), *Global and Regional Controls on Biogenic Sedimentation; 1, Reef Evolution, Research Reports*. Goettinger Arbeiten zur Geologie und Palaeontologie. Sonderband. SB2, pp. 249–258.

Rehfeld, U., 1996, Mediating and limiting processes during the development of spongiolitic bioconstructions in Jurassic and Cretaceous strata; a paleontological, facial and geochemical analysis. In Reitner, J., Neuweiler, F., and Gunkel, F. (eds.), *Global and Regional Controls on Biogenic Sedimentation; 1, Reef Evolution, Research Reports*. Goettinger Arbeiten zur Geologie und Palaeontologie. Sonderband. SB2, pp. 249–258.

Rodriguez, S. and Sanchez, C.F., 1994. Rugose corals and calcareous algae bioconstructions in the Torreon section, Visean, Badajoz. *Coloquios de Paleontologia*, **6**: 61–75.

Wood, R.A., 1999, *Reef Evolution*. Oxford: Oxford University Press.

Cross-references

Atolls
Bioerosion
Bioherms and Biostromes
Coral Reefs
Reefs, Non-Coral
Tidal Environments

BIOENGINEERED SHORE PROTECTION

In an effort to arrest shore erosion at many coastal locations and to provide protection to marinas and harbors, it may be necessary to construct structures in high wave energy zones. Current practice involves utilization of structures constructed using large armor stones, concrete and steel walls, and a variety of other "hard" engineering techniques. Quite often, these structures do not add to the aesthetic and recreational attributes of a site and may impact significantly on the local environment. Integration of bioengineered components into the design of breakwaters and shore protection systems can be utilized, in certain cases, to enhance the project by providing better biological habitat and ancillary water quality improvement. Thus the goal of a project changes to include not only the stabilization of the eroding area or the provision of "quiet" waters, but to increase the quantity and quality of habitat available to fish and waterfowl communities, while providing an effective and aesthetic control of natural environment.

Background

In both the engineered and natural environment, the flow of water often causes erosion. The causes must be understood before the problem can be addressed. In the coastal zone, the flow of water results from wave action, the associated runup and backwash, wave breaking, alongshore currents, and the natural flow of water along side and overtop of the high-tide shoreline. In addition to the interaction of the high-tide shoreline or lakeward structure with water, a considerable amount of animal and human activity create additional stresses on the high-tide shoreline. Bioengineering methods of shore protection offer a practical solution that can also create an aesthetically pleasing and environmentally beneficial "buffer zone." Bioengineering, in this context, is the utilization of vegetation, either by itself or in combination with other defense mechanisms, depending upon the local environment. The other defense mechanisms may include the use of rock lining, offshore islands, wave screens, and submerged shoals that limit the wave energy reaching a site. Quite often, these defense structures can be designed to provide significant enhancement to the environment, particularly in providing suitable fish habitat for spawning, feeding, and hiding from predators.

The value of vegetation for protecting the soil depends on the combined effects of roots, stems, and foliage. Roots and rhizomes reinforce the soil. Immersed foliage elements absorb and dissipate energy and may cause sufficient interference with the flow to prevent scour. In a sediment-laden environment, they may also promote deposition.

A coastline requiring protection can be considered as two separate areas and thus habitat enhancement can be geared toward two communities; the high energy nearshore environment and the onshore environment, which can be suitably modified to ensure low wave energy levels. Enhancement of the nearshore zone can include construction of rock revetments as reef habitat, inclusion of submerged offshore structures to reduce wave energy levels reaching the shore and primary wave defense structures which provide habitat enhancement potential by the nature of their design (Figure B50). Selection of stone and design of its placement is developed in a manner to provide a reef like habitat beyond minimum stone placement required for the minimal shore

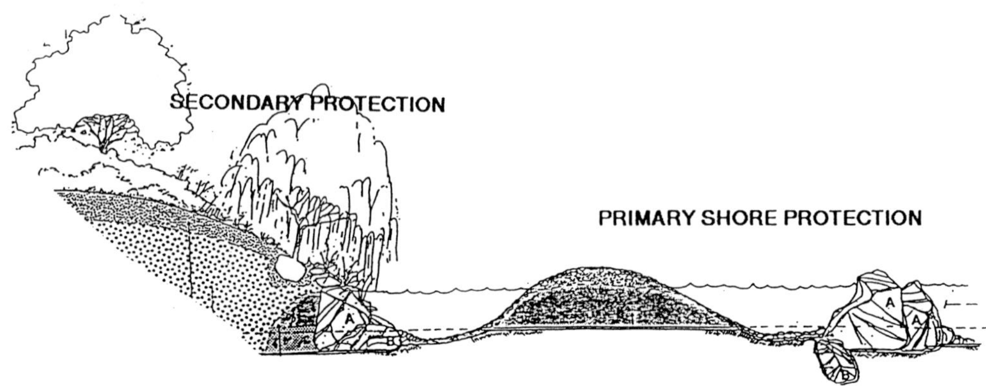

Figure B50 Utilization of shoals.

stabilization. The effectiveness of both natural and artificial reef like habitats as fish community habitats has been well documented. Proper design and installation of rock will provide protective cover and feeding areas and will supply the needs for small aquatic and benthic organisms by providing protection from the high energy wave action and from larger predators.

Low wave energy areas can be created behind a primary defense such as an offshore rock structure or wave screen, or through the creation of lagoons behind stable control structures (Figure B51). The development of constructed wetland pockets and areas for other shallow water plants can occur in these lagoons. This can be promoted by the establishment of "biological" riprap in the form of brush and woody plant debris. These materials will provide a setting that will foster the accumulation of shore plants from wind blown seed banks. As the brush decomposes, it provides a limited release of nutrients to the developing plant community and is eventually replaced by living plants. The establishment of new habitat will provide the opportunity for colonization by wetland plant and animal species that require quiescent waters. Transient use of the habitats by a variety of aquatic and migratory waterfowl is an additional potential for these environments.

Goals and objectives

The designer is encouraged to consult specialists in the fields of coastal hydraulics, fisheries, geomorphology, biology, landscape architecture, or any field that could make the project a success. The design of bioengineered breakwaters and shore protection that functions environmentally requires a multidisciplined approach. Usually, no individual has all the expertise required to ensure successful implementation.

The following geomorphologic, hydraulic, and biological changes may occur as a result of modification of the shore, which would occur from the creation of a marina or harbor, or from local erosion protection schemes:

- Loss or elimination of aquatic vegetation
- Loss or elimination of backshore vegetation
- Removal of specific nearshore bathymetrical features
- Modified substrate conditions
- Modified hydrodynamic, flow, sediment, and water quality regimes
- Changes in nutrient conditions and reductions in food organisms
- Aesthetic degradation
- Reductions in habitat diversity and environmental stability
- Increased water temperatures.

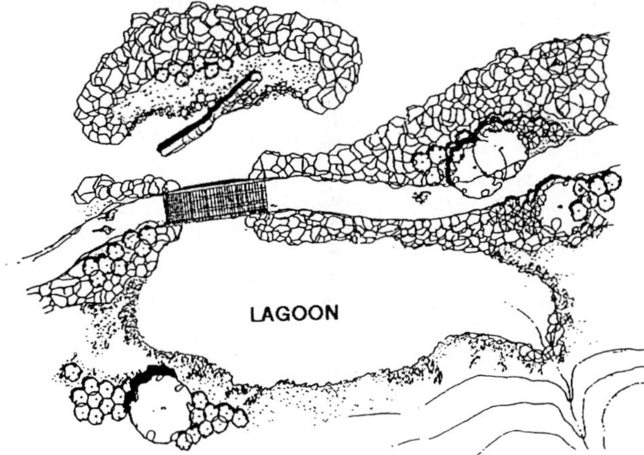

Figure B51 Development of a lagoon cell.

The shore is a dynamic system where impacts are difficult to predict. Engineered structures, when properly designed and constructed, can provide both species and habitat diversity and thereby mitigate potential adverse changes. However, the goals and objectives of the shore protection design must be correctly identified early in the design process. The designer must be aware of the design goals and objectives to correctly identify, size, and locate the various functional elements within the system. Biodiversity within and adjacent to the shore is interrelated with the quality in updrift and downdrift areas. Changes to any one of these components may adversely impact on others.

Habitat requirements

Aquatic life generally requires a habit that contains the following:

1. Sufficient water depth and volume for each life stage.
2. Adequate water quality with preferred ranges of temperature, dissolved oxygen, PH, etc.
3. A variety of continuous hydrodynamic conditions varying from deep water to shallow water for breeding and cover. Also flow conditions that sort bed load materials to provide a good environment for bottom dwelling organisms are advantageous.
4. Adequate cover to provide shade, concealment, and orientation.
5. Adequate food to maintain metabolic processes, growth, and reproduction.

Shore improvements should be designed for the individual fish species. Specific requirements for reproduction, juvenile rearing, and adult rearing with regard to feeding location, concealment from predators and competitors, and sanctuary from flow extremes and ice formation varies between species. Loss of the natural bathymetric features, which are utilized by particular species as a result of implementation of shore protection, could eliminate many of the requirements necessary to sustain significant biodiversity along the nearshore area. In addition, removal of existing shore vegetation, in either the emergent of submergent zones would significantly reduce or eliminate the potential to sustain a fish population.

Utilizing vegetation

In certain low wave energy environments, vegetation may be used by itself to provide suitable protection to an eroding shore. Reeds and

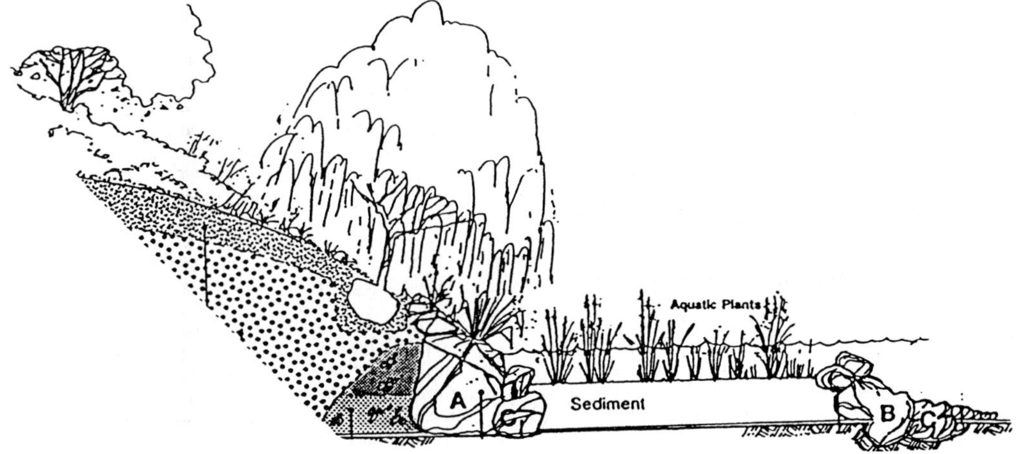

Figure B52 Emergent vegetation used in conjunction with stone.

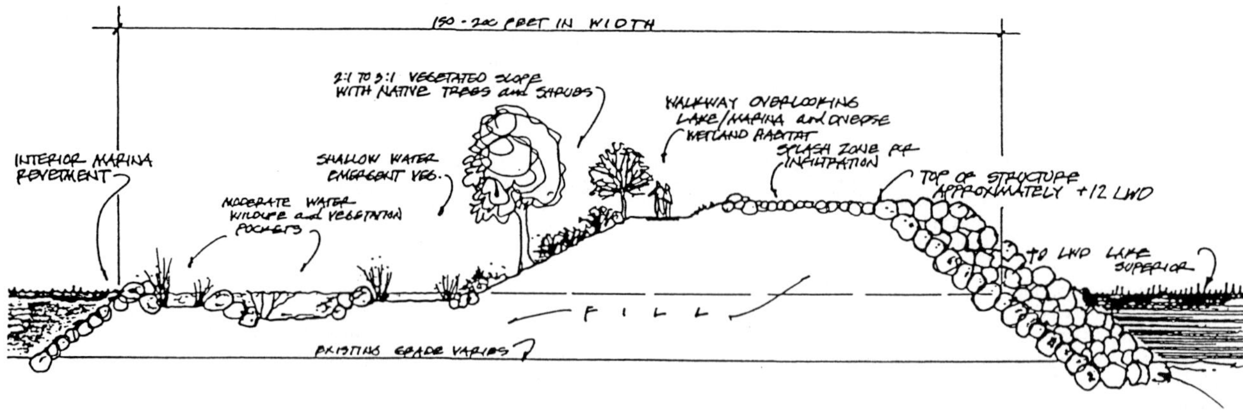

Figure B53 Control structure/lagoon.

other marginal plants can form an effective buffer zone by absorbing wave energy and restricting the alongshore flow velocity adjacent to the shore. They therefore have a protective value. Specific functions that they can perform include:

1. Absorbing and dissipating wave-wash energy.
2. Interference and protection of the shoreline bank from the flow.
3. Reinforcement of the surface soil through the root mat and prevention of scour of the bank material.
4. Sediment accumulation brought about by the dense plant stems.

Marginal plants require very wet ground and generally will not survive in water that is more than 0.5 m deep for long period of time. They flourish in conditions of low flow velocity and their integrity is weakened by wave action in excess of 0.5–0.75 m. Different species offer different levels of protection with regard to wave energy dissipation. For incident wave conditions under 0.5 m, reed beds having a width of 2–2.5 m may dissipate 60–80% of the incoming wave energy. In areas with higher levels of wave energy, riprap and geotextiles may be used in conjunction with vegetation to provide effective bank protection (Figure B52). In areas of high incident wave energy, an area of low wave energy can be created behind a primary defense such as an offshore rock structure or wave screen, or through the creation of lagoons behind stable control structures (Figure B53), as described above.

Natural methods of protection generally have low capital cost in comparison with conventional engineering methods. However, they may well have higher recurrent cost due to regular inspection, trimming and cutting, and repair. In areas where a combination of conventional and bioengineered structures are required, recent experience at several sites on the Great Lakes has established that these techniques may cost 20–30% more than conventional techniques alone.

Possible disadvantages are that natural protection schemes take time to mature and to become fully effective. Depending on the type, natural protection may take several growing seasons to reach the desired standard of protection.

Bioengineering differs from other conventional forms of engineering in two key respects, which strongly influence the design approach:

1. Bioengineering involves considerable practical experience and judgment, as opposed to the application of quantitative design theory or rules.
2. Careful management is required not only in the establishment of vegetation, but also in its aftercare over the initial growing seasons.

Use of vegetation requires the following points to be considered

The principal plant groups that can be used are aquatic plants, grasses, shrubs, and trees. Selection is based on consideration of the different roles to be performed by the vegetation, taking into account the physical and chemical properties of the soil, the climatic conditions, and the soil/water regime under which the plant must survive. Vegetation establishment may take several growing seasons and is a seasonal activity that must be managed and maintained. The engineer must prepare and agree to specific management objectives and a management program with the owner/client. This is in order to ensure that the vegetation is maintained in a fit condition to perform its intended roles.

Zones and horizons of natural protection

With natural methods of protection, and particularly methods involving the use of live material, the effectiveness of different materials is strongly dependent on their location in relation both to the dominant external water level and to the subsoil soil/water regime. To achieve effective protection using natural materials, the designer will almost inevitably need to use different methods of protection in different zones and horizons of the shore (Coppin and Richards, 1989).

Use of reeds

The emergent and marginal types of aquatic plants, such as the common reed, bulrush, and great pond sedge, are frequently used for interference and protection purposes to form a protective margin along the shore at the waterline. They also encourage siltation by absorbing current flow energy, and thus reducing the sediment-carrying capacity of the flow. Reeds can be easily weakened by erosion and loosening of the soil around the rhizomes due to wave energy. It is therefore necessary to protect the zone containing roots from high-velocity flow or significant wave attack. Provided this is done, the stems and leaves will protect the shore bank above.

Uses of shrubs and trees

A limited range of trees are water-tolerant and can be used in bioengineering structures for bank protection in both the aquatic and damp zones. The willow, alder, and black poplar are the principal water-tolerant species. In particular, a dense root structure is able to provide some protection as well as substantial reinforcement effect to enhance the stability of the shore both above and below the mean water level. The willow and poplar are particularly useful for bioengineering because they can be propagated from cut limbs. The cut limbs can be placed such that secondary root growth develops and shoots sprout from dormant buds. Trees, which are not water-tolerant, do not have any major direct function in shore stabilization, although they may provide shade to control the growth of aquatic life as discussed earlier.

Use of grasses

Grass is used very extensively in bank protection in the zones above the high water level. Grass roots cannot tolerate prolonged submergence periods. A wide variety of grass species and mixtures therefore are appropriate to satisfy the functional, environmental, and management requirements for a protection scheme. The principal functions which grass fulfills are those of interference, protection, root reinforcement, and soil restraint. The surface root structure forms a composite soil/root mat, which enhances the erosion resistance of the bare subsoil,

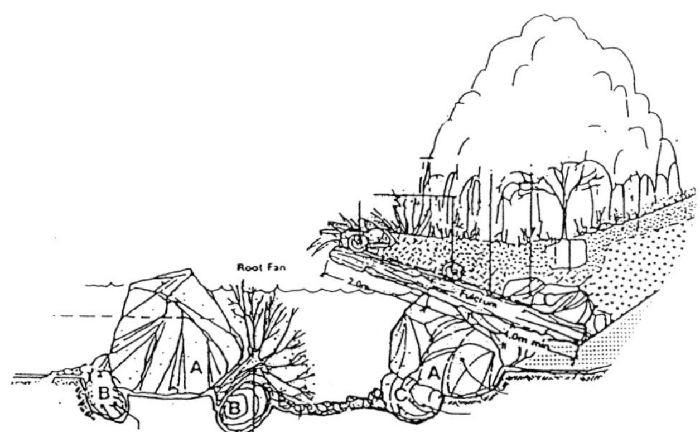

Figure B54 Timber (CANTILEVER) to provide habitat and shade.

and which is anchored into the subsoil by deeper roots. The engineering function of grass may be augmented by the use of geotextile or cellular concrete reinforcement to form composite protection. With both types of reinforcement, the visual effect of grass is retained. Erosion of grass cover by wave runup generally occurs by the scouring of soil from around the roots of a plant, thereby weakening its anchorage until the plant itself is removed by the drag of the flowing water. The effectiveness of grass protection can also be seriously reduced by any localized patches of bare soil or poor grass cover.

The rate of growth of different grasses varies considerably. Complete grass cover should normally be achieved by the middle of the first growing season while full protective strength of the sward is reached during the second season. Provision should be made for aftercare including mowing, fertilizing, and weed control.

Use of timber and woody material

A variety of timber and other dead woody materials can be used in the shore protection scheme usually fulfilling reinforcement, protection, and sometimes drainage functions (see Figure B54). Natural hardwoods will retain their integrity for 5–10 years if built into the bottom of a bank below the water level. Out of the water they can last longer but the worst environment for timber is the alternately wet and dry zone around mean water level.

Monitoring

As part of the project design for the shore stabilization enhancements, a monitoring program is required. The purpose of the monitoring program is to measure the success and applicability of the enhancement methods to other shore projects.

Baseline habitat conditions should be assessed by observation and characterization of existing conditions. A plant survey and macroinvertebrate sampling of the nearshore benthic environment and a terrestrial plant survey should be performed to document existing plant and animal populations. Incidental observations of birds should be made as part of fieldwork. Sampling of nearshore fish populations should be coordinated with local regulatory agencies. Post-construction monitoring of the establishment of biological communities should be completed to evaluate the success of a particular scheme.

Conclusions

Shore protection enhancements similar to those described in this entry have been successfully implemented at numerous sites on the Great Lakes, most notably in Canada at Red Rock Marina, Lake Superior; Thunder Bay Harbor, Lake Superior; Kingston, Lake Ontario; and various reaches of the St. Lawrence Seaway, and at Bender Park, Lake Michigan; Silver Bay, Lake Superior; and in Louisiana (Gulf of Mexico) in the United States. The range of design wave conditions range from 0.75 to 4.5 m at these various sites. Many other projects are in the process of implementation.

Utilization of bioengineered shore protection, in concert with virtually transparent offshore protection (submerged breakwaters, wave screens, etc.) can provide for significant levels of protection while maintaining the natural beauty of an area, and, in most circumstances, providing significant opportunities for habitat enhancement and increased biodiversity.

Further suggested reading may be found below.

Kevin R. Hall

Bibliography

Coppin, N.J. and Richards, I.G., 1989. *The Use of Vegetation in Civil Engineering*. CIRIA / London: Butterworths.

Hemphill, R.W. and Bramley, M.E., 1989. *Protection of River Canal Banks. CIRIA* / London: Butterworths.

Fisheries and Aquatic Biota, 1986. Volume 3F, Environmental Reference Book. Toronto: Ontario Ministry of Transportation.

Stream Enhancement Guide, 1980. Victoria: Canada Fisheries and Oceans and Province of British Columbia.

Cross-references

Beach Erosion
Coastal Zone Management
Geotextile Applications
Monitoring, Coastal Ecology
Shore Protection Structures
Vegetated Coasts
Wetlands Restoration

BIOEROSION

In his study of the erosion of steep cliffs around Huntington Sound in Bermuda by excavating sponges, Neumann (1966) defined the term bioerosion as the removal of consolidated material or lithic substrate by direct action of organisms. Soon after the term, bioerosion, was introduced, geologists and biologists described many different types of bioeroding organisms including algae, bacteria, foraminifera, sponges, bryozoa, annelid worms, barnacles, gastropods, bivalves, echinoderms, fish, and mammals. The process of bioerosion was also reported from many different marine and non-marine environments ranging from mountain slopes to the tops of deep sea knolls, and from rocky intertidal zones and coral reefs to the flanks of continental shelves. Bioerosion has also been reported from climatic zones extending from tropical and subtropical to the subarctic and arctic. Several different types of experiments have been devised for studying the rates of bioerosion by different types of organisms and in different environments. Although bioerosion was first recorded for living sponges in the intertidal zone, evidence for bioerosion was also found in the ancient rocks extending back at least to the Silurian.

Bioeroding organisms

Several different types of microbial borers have been described from modern environments and ancient rocks. Microbial borers including cyanobacteria and chlorophytes were found in modern reef environments at depths between 0 and 230 m, and boring hetertrophs are present between 100 and 300 m (Vogel *et al.*, 1996). Evidence for boring algae (cyanophyta) has been preserved in Silurian bivalves and may be responsible for the silicification of their shells (Liljedahl, 1986). Twenty species of foraminifera, ranging age from Jurassic (Callovian) to Recent, are known to make cavities in hard substrates (Venec, 1996). The bioeroding foraminifera were found in turbulent, warm, shallow-water environments.

A wide variety of living and fossil invertebrates have been identified as bioeroders. Several species of boring sponges have been reported from reef areas in Bermuda (Neumann, 1966) and Grand Cayman Island in the British West Indies (Acker and Risk, 1985), and on a deep sea knoll at depths of 1,600 to 1,800 m (Boerboom, 1996). The bioeroding mollusks include chitons, gastropods, and bivalves. *Chiton pelliserpentis* removed hardened mudrock during feeding at Mudrock Bay in Kaikoura, New Zealand (Horn, 1984). The spawn of the gastropod, Nerita, settled on the sea bottom and eroded carbonate rocks at Cathedral Point in Costa Rica (Fischer, 1980). The bivalve genus *Lithophaga* was an active chemical borer in reefs from the Carboniferous through the Eocene (Krumm, 1992). Rock-boring echinoids excavated large cavities in reefs in the South Florida keys (Kues and Siemers, 1974) and on Enewetak Atoll in the Marshall Islands (Russo, 1980). The rock-boring barnacle, *Lithotyra*, eroded the rock face while grazing in the intertidal zone (Ahr and Stanton, 1973). The polychaete annelid, *Eunice*, burrowed into the carbonate rocks along the shore of the Gulf of California.

Vertebrates including fish and mammals play an important role in bioerosion on reefs and mountain slopes. Parrotfish have been observed feeding on coral in reef environments and their rates of bioerosion were measured (Frydl and Stearn, 1978). Recolonization experiments on coral reef communities near Aquaba on the Red Sea demonstrated that herbivorous fish were a major factor in structuring coral reef communities (van Treeck *et al.*, 1996). In the Pyrenees of Spain, the indirect effect of digging by small mammals was considered more significant than the direct detachment of soil cover (Martines and Pardo, 1990).

Rates of bioerosion

Several different field experiments have been used to estimate the rates of bioerosion by different organisms and in different environments. On the carbonate coastline of Bermuda, experiments show that the sponge *Cliona lampa* is capable of removing 6–7 kg of material from 1 sq. m of carbonate substrata in 100 days, corresponding to an erosion rate of calcarenite of more than 1 cm per year (Neumann, 1966). In Kaikoura, New Zealand, *Chiton pelliserpentis* removed mudrock from the surface at a rate of 47.3 g/sq. m on the high shore and 173 g/sq. m on the low shore (Horn, 1984). This was equivalent to about 2% of total on the high shore and 5.5% on the low shore. On Moorea reef barrier flat in French Polynesia, bioerosion rates for echinoids was estimated at 4.5 kg/sq. m per year and for scarid fish at 1.7 kg/sq. m per year (Peyrot *et al.*, 1996).

Environments of bioerosion

Although most examples of bioerosion have been studied from tropical reefs and intertidal zones, bioerosion also has been reported from high-latitude environments, the outer continental shelf and deep-sea knolls. Algae borings were found in gastropod shells and echinoderm tests in the high-latitude, low-energy environments in the firths of Clyde and Lorne, Scotland (Akpan and Farrow, 1985). Boring sponges were dredged up from Newfoundland from depths of approximately 1,600–1,800 m on top of Orphan Knoll, 550 km northeast of Saint John's (Boerboom, 1996). Evidence for bioerosion was also found in the clastic sediments on the outer continental shelf around the Hudson Canyon off the eastern coast of the United States (Twichell *et al.*, 1984). A workshop on bioerosion convened by Bromley (1999) has reviewed several different aspects of bioerosion ranging from the style of bioerosion in Late Jurassic reefs to the role of bioerosion in carbonate budgets in Indo-Pacific reefs.

Conclusions

Bioerosion by microorganisms, invertebrates, and vertebrates is widespread throughout many different carbonate environments from the early Paleozoic to the Recent. Bioeroding organisms have been reported from mountain slopes to deep-sea knolls, from the rocky intertidal zone to coral reefs and from the tropics to the arctic circle. The rates of bioerosion vary from a few grams per square meter to several kilograms per square meter depending on the organisms involved and the depositional environments.

William T. Fox

Bibliography

Acker, K.L. and Risk, M.J., 1985. Substrate destruction and sediment production the boring sponge Cliona caribbaea on Grand Cayman Island. *Journal of Sedimentary Petrology*, **55**: 701–711.

Ahr, W, M. and Stanton, R.J. Jr., 1973. The sedimentologic and paleoecologic significance of Lithotrya, a rock-boring barnacle. *Journal of Sedimentary Petrology*, **43**: 20–23.

Akpan, E. and Farrow, G., 1985. Shell bioerosion in high-latitude low-energy environments; firths of Clyde and Lorne, Scotland. *Marine Geology*, **67**: 139–150.

Bromley, R.G., 1999. Bioerosion; a collection of workshop papers. *Bulletin of the Geological Society of Denmark Marine Geology*, **44**: 129–180.

Boerboom, C.M., 1996. Bioerosion and micritization in the deep sea; a look at the coral *Desmophyllum cristagalli*. Bachelor's thesis, McMaster University. Hamilton, ON.

Fischer, R., 1980. Bioerosion by spawn of *Nerita funiculate* and *Nerita scabricosta*, Gastrapoda. *Neues Jahrbuch fuer Geologie und Palaeontologie*, Monatshefte, **5**: 287–292.

Frydl, P. and Stearn, C.W., 1978. Rate of bioerosion by parrotfish in Barbados reef environments. *Journal of Sedimentary Petrology*, **48**: 1149–1157.

Horn, P.L., 1984. Beachrock erosion due to feeding by Chiton (*Sypharochiton pelliserpentis*) at Mudstone Bay, Kaikoura, New Zealand. *Mauri Ora*, **11**: 57–61.

Krumm, D.K., 1992. The paleoecology of the chemically boring bivalve genus Lithophaga. *Annual Report*—Western Society of Malacologists, 24, p. 16.

Neumann, A.C., 1966. Observations on coastal erosion in Bermuda and measurements of the boring rate of the sponge, *Cliona lampa*. *Limnology and Oceanography*, **11**: 92–108.

Kues, B.S., and Siemers, C.T., 1974. The geobiological significance of the modern echinoid Echinometra as an agent of bioerosion, southern Florida keys. *Abstracts with Programs—Geological Society of America*, **6**: 834.

Liljedahl, L., 1986. Endolithic microorganisms and silicification of a bivalve fauna from the Silurian of Gotland. *Lethaia*, **19**: 267–278.

Martines, R., and Pardo, A.M.P, 1990. First data on bioerosion due to small Mammalia in Spanish Pyrenees. *Ekologiya*, **1**: 27.

Neumann, A.C., 1966. Observations on coastal erosion in Bermuda and measurements of the boring rate of the sponge, *Cliona lampa*. *Limnology and Oceanography*, **11**: 92–108.

Peyrot, C.M., Le-Campion, A.T., Harmelin, V.M, Romano, J.C., Chazottes, V., Pari, N., and Le-Campion, J.,1996. Bioerosion in the carbonate cycle; quantification of the process in French Polynesia. Memoires de la Societe 36. at Mudstone Bay, Kaikoura, New Zealand. *Mauri Ora*, **11**: 57–61.

Russo, A., 1980. Bioerosion by two rock boring echinoids (*Echinometra mathaei* and *Echinostrephus aciculatus*) on Enewetak Atoll, Marshall Islands. *Journal of Marine Research*, **38**: 99–110.

Twichell, D.C., Grimes, C.B., Jones R.S., and Able, K.W., 1984. The role of bioerosion in shaping the outer continental shelf around Hudson Canyon, offshore Eastern United States. *Abstracts with Programs—Geological Society of America*, p. 16.

van Treeck, P., Schuhmacher, H., and Paster, M., 1996. Grazing and bioerosion by herbivorous fishes; key processes structuring coral reef communities. *Goettinger Arbeiten zur Geologie und Palaeontologie*, Sonderband. SB2, pp. 133–137.

Venec, P.M.T., 1996. Bioeroding foraminifera; a review. *Marine Micropaleontology*, **28**: 19–30.

Vogel, K., Kiene, W., Gektidis, M., and Radtke, 1996. Scientific results from investigations of microbial borers and bioerosion in reef environments. *Goettinger Arbeiten zur Geologie und Palaeontologie*, Sonderband. SB2, pp. 139–143.

Cross-references

Atolls
Bioconstruction
Cliffs, Erosion Rates
Coral Reefs
Erosion Processes
Karst Coasts
Tidal Environments

BIOGENOUS COASTS—See VEGETATED COASTS

BIOGEOMORPHOLOGY

Biogeomorphology is a discipline that combines ecology and geomorphology. Geomorphology is the study of landforms and their formation. Ecology is the study of the relationships between biota and their environment. The environment is defined as factors that affect biota. These factors can be abiotic (physical, chemical), biotic (other organisms), or anthropogenic (humans). Abiotic geomorphological processes may affect biota and biota may in turn affect geomorphological processes. The interaction between both defines the discipline of biogeomorphology. *Biogeomorphology is the study of the interaction between geomorphological processes and biota.*

Essential concepts

The term biogeomorphology was first used in the 1980s (Viles, 1988), although earlier studies have been conducted that were focused on biogeomorphology without using this term. Biogeomorphology is studied in terrestrial as well as in aquatic systems. In coastal systems biogeomorphological interactions are clearly demonstrated in the shallow, productive waters, and in various sedimentary environments. Examples of biogeomorphological interrelationships include sand dune development, tidal flats, salt marshes, mangrove systems, and coral reefs.

Relevant geomorphological factors in coastal systems are bathymetry, bed composition (rock, gravel, sand, silt), and the transport of sediment. It also includes factors that drive morphological processes, such as water flow and wave energy. The biota involved in coastal biogeomorphology include plants and animals, ranging from very small (algae) to very large (whales).

The geomorphological influence on biota is in its most direct form the influence on habitats (living environments) of flora and fauna. The coastal morphology and geomorphological processes define the gradients between high and low, between wet and dry, and between sedimentation and erosion. These gradients and the processes that cause them are determinative for gradients in grain size of the sediment, nutrient levels, organic matter levels, and moisture. Plants and animals are tuned to specific conditions and will therefore be abundant in specific locations.

The biological influence on geomorphological processes is the influence of biota to create, maintain, or transform their own geomorphological surroundings. This is demonstrated by the influence of vegetation on the hydraulic resistance, erodability and sedimentation, or by the influence of fauna on sediment characteristics through bioturbation and biostabilization.

In some cases morphological processes are dominant over biological processes and therefore the biota have to adjust to their environment. In other cases biological processes are dominant. The most interesting are those cases where there is a mutual interaction that leads to feedback coupling of processes. When looking for these cases, it is important to examine the temporal and spatial scales of the mutually interacting processes. Biogeomorphological interrelationships can be found in several coastal environments, for both hard and soft substrates.

Biogeomorphology for hard substrates

On rocky shores and coral reefs a typical community of organisms thrives that affects the erosion rates of its substrate. Influenced by abiotic factors such as wave energy, splash water, inundation frequency and -period, depth, desiccation and substrate type, a clear zonation can be found of various cyanobacteria, (macro-)algae, fungi, lichens, molluscs, sponges, worms, sea urchins, fish, etc. Some of these organisms dwell on the surface of the substrate, while others live within the substrate. Their effect on erosion of the substrate is divided in "biological corrosion," processes that modify the substrate but provides no erosion product, and "biological abrasion" (see *Bioerosion*), processes that do generate an erosion product. Grazing, burrowing and boring on or in the substrate carries out biological abrasion, and is most significantly found in coral reef systems.

Biogeomorphology for soft substrates

In soft coastal systems, the interrelationships between geomorphological factors and biota can mainly be noticed for benthic fauna and flora. The presence of benthic species is affected by hydraulic and morphologic conditions, such as depth, current velocity, salinity, and grain size. The effect of soft substrate communities on geomorphology is divided into biostabilization and biodestabilization. Biostabilization leads to an increase in soil resistance, preventing erosion, while biodestabilization leads to an increased erodability.

Biostabilization by plants

On tidal flats, small algae (diatoms) are capable of affecting the geomorphology. These diatoms can form extensive algal mats and excrete EPS mucus, which is a sticky substance made of polysaccharides that glues the sediment together and therefore protects the sediment against erosion. Sea grass is dependent on clear water, it needs sunlight to grow. A sea grass meadow slows down the current velocity near the bed and therefore sand and silt will not resuspend in the water, which otherwise would lead to turbid water. Furthermore, their root system binds the substrate. Ultimately, deposition of suspended sediment is encouraged in a sea grass meadow, which leads to the supply of organic material with nutrients, needed for growth.

Seaweeds are also capable of adjusting their physical environment by damping down wave energy; and salt marshes also play an important role in stabilizing sediments. Salt marsh vegetation makes fine sediment settle down resulting in a continuous heightening of the marsh. The higher the marsh gets, the more vegetation can grow and the better the marsh is protected against erosion. Other stabilizing effects result from cementation of beachrock by cyanobacteria and stromatolite formation by algae.

Biostabilization by animals

Some macrozoobenthos can actively catch sediment particles from the water column and bring them to the bed. The presence of a mussel bank, for example, will alter the bed in different ways. Mussels slow down the water flow and they protect the bed against erosion. Mussels also actively catch small particles from the water column by filterfeeding and subsequently excrete these as pseudofeces. This results in a change in the soil composition to finer sediments.

Animal tube fields are also believed to stabilize the sediment, because there is a clear accumulation of fine particles and organic matter between the tubes. The tube itself may affect small-scale turbulence and therefore have a stabilizing effect, however, a great deal may be attributed to the community of microorganisms between the tubes that excrete mucus. Other stabilizing effects result from large banks of dead shells and mucus binding by meio- and macrofauna.

Biodestabilization

Benthic fauna may destabilize the substrate by their digging and feeding activities (bioturbation). The constant mixing and recycling of sediment in the top centimeters of the bed results in a characteristic vertical particle-size profile. The selective uptake and excretion of preferred particle sizes results in sorting and pelletizing sediments. Together with the digging of burrows and the constant movement within the substrate, these activities lead to the generation of a surface micro-relief that has a higher hydraulic roughness and is more prone to erosion. Furthermore, bioturbation also affects the sediment water content, porosity, and sediment cohesion.

Scale interactions in biogeomorphology

Different physical and biological processes can have dynamic interactions when they operate on the same spatial and temporal scales. Processes that act on a very small scale may appear as noise in the interactions with processes on larger scales. Their effect can be accounted for by proper averaging procedures (e.g., for turbulence). Processes that act on a large-scale may be treated as slowly varying or even constant boundary conditions when studying their effects on processes on smaller scales (e.g., sea-level rise due to climate change). Techniques for scale interactions are reasonably well established in geomorphology (De Vriend, 1991) and are based on scale linkage via sediment transport. In biology, however, population and community dynamics give rise to spatial and temporal structures that are not easily linked. In recent years, the importance of scale has been increasingly recognized (Legendre *et al.*, 1997) as an essential aspect of understanding the biotic and abiotic processes that affect the biogeomorphology of coastal systems.

Martin J. Baptist

Bibliography

De Vriend, H.J., 1991. Mathematical modelling and large-scale coastal behaviour, Part 1: Physical processes. *Journal of Hydraulic Research*, **29**(6): 727–740.

Legendre, P., Thrush, S.F., Cummings, V.J., Dayton, P.K., Grant, J., Hewitt, J.E., Hines, A.H., McArdle, B.H., Pridmore, R.D., Schneider, D.C., Turner, S.J., Whitlatch, R.B., and Wilkinson, M.R., 1997. Spatial structure of bivalves in a sand flat: Scale and generating processes. *Journal of Experimental Marine Biology and Ecology*, **216**: 99–128.

Viles, H.A. (ed.), 1988. *Biogeomorphology*. Oxford: Basil Blackwell Ltd.

Cross-references

Algal Rims
Beachrock
Bioconstruction
Bioerosion

Bioengineered Shore Protection
Bioherms and Biostromes
Coral Reefs
Reefs, Non-Coral
Rock Coast Processes
Shore Platforms
Vegetated Coasts

BIOHERMS AND BIOSTROMES

History

Originally coined by Cumings (1932), the word bioherm along with its brother term biostrome have been widely used in reef literature, but their proper stratigraphic definition is often misunderstood.

In the original meaning (Chevalier, 1961) a bioherm was defined as a mound or lens-shaped organic build-up, edified by the skeletons of various organisms and lying unconformably inside a stratigraphic series of different lithology. Conversely, a biostrome was a flat, layered reef structure, wide or narrow in shape and causing no stratigraphic disturbance inside its sedimentary environment.

Discussion

Both words "bioherm" and "biostrome" were obviously coined for fossil build-ups, whose stratigraphic position in the sedimentary sequence can be studied; and they were also commonly used for the description of living or subfossil structures, whether the sedimentary environment of the latter is accessible or not to study.

Definitions vary according to authors: In the Encyclopaedia Britannica a bioherm is defined as an "ancient organic reef of moundlike form built by a variety of marine invertebrates ... (and coralline algae). A structure built by similar organisms that is bedded but not moundlike is called a biostrome."

Many geologists, however, extend these definitions to gravity deposited mounds or layers of skeletal remains, such as shells or broken coral, including reworked or transported material, as illustrated by Roger Suthren in his on-line lectures in Sedimentology, a second year Geology module at Oxford Brookes University: "Bioherms: (are) mound or lens-shaped (biological build-ups). Some are in-place organic structures (reefs), others are banks of loose, transported carbonate sediment consisting largely of shells or skeletons. Biostromes: (are) laterally extensive beds, sheets or ribbons of carbonate material. Some have grown in-place (reefs); others consist of transported shells and skeletons."

For Battistini *et al.* (1975) a bioherm is a: "lens shaped organic reef ... embedded *in situ* inside sedimentary layers of different lithological nature ... it may be surrounded by a peripheral talus of biodetrital sediments," whereas a biostrome is a "layered, bank like organic reef of variable extension, creating no discontinuity inside the embedding sedimentary layers."

There is, therefore, no general agreement upon a complete definition taking into account at one and the same time such different characters as: age, stratigraphic conformity or unconformity, along with the autochtonous or allochtonous nature of deposited organisms.

Furthermore, many authors (notably among biologists and geographers) tend to use "bioherm" as a general term not only for major biological build-ups such as extensive algal rims or coral reefs (e.g., see Adey and Burke, 1976) but also for small-scale organic build-ups, for which the word "biostrome" would better fit. Bosence and Pedley, who had first used "bioherm" in a preliminary publication (1979) dealing with Miocene layers of calcareous algae in Malta, appropriately dropped it for "biostrome" in their final paper (1982).

It is, therefore, difficult for an actualist (whether geologist or not) to find criteria sufficiently precise and reliable to distinguish between the alternate notions of bioherm and biostrome. For example, an algal rim growing on the outer edge of a coral reef is indeed a bioherm, or a part of a bioherm since it takes an active part in the sedimentary processes of the latter, but the same kind of formation thinly coating a limestone or a volcanic shore, or on a vertical cliff, without altering sedimentation should be called a biostrome even if both formations are in continuity with one another.

Further difficulty lies in the fact that, for actualists, detrital accumulations of dead shells and broken skeletal material (generally mud-supported) are considered as something very different from a true build-up or reef, since the latter is fundamentally made of an *in situ* developed formation, resulting in boundstone or framestone lithologies *sensu* Bathurst (1971).

Conclusions

Unless bio-accumulated detrital mounds and layers are taken out of the definition of bioherms and biostromes (a revision that only geologists can decide), and the status of small-scale build-ups is settled, the use of the latter words should preferably be restricted to the stratigraphic study of the fossil formations for which they were first coined (their associated detrital facies, and other types of detrital formations being included or not). Students of living reefs are conversely encouraged to prefer more general terms (such as "biological build-up," "reef-like structure," or "biogenic construction") instead.

Jacques Laborel

Bibliography

Adey, W.H. and Burke, R.B., 1976. Holocene bioherms (algal ridges and bank barrier reefs) of the eastern Caribbean. *Geological Society of America Bulletin*, **87**: 95–109.

Bathurst, R.G.C., 1971. *Carbonate Sediments and their Diagenesis. Developments in Sedimentology*, **12**. Amsterdam: Elsevier.

Battistini, R., Bourrouilh, F., Chevalier, J.P., Coudray, J., Denizot, M., Faure, G., Fischer, J.C., Guilcher, A., Harmelin-Vivien, M., Jaubert, J., Laborel, J., Montaggioni, L., Masse, J.P., Mauge, L.A., Peyrot-Clausade, M., Pichon, M., Plante, R., Plaziat, J.C., Plessi, Y.B., Richard, G., Salvat, B., Thomassin, B.A., Vasseur, J., and Weydert, P., 1975. Eléments de terminologie récifale indo pacifique. *Téthys* **7**: 1–111.

Bosence, D.W. and Pedley, H.M., 1979. Palaeoecology of a Miocene coralline algal bioherm, Malta. *Bulletin du Centre de Recherches d'Exploration et de Production d'Elf-Aquitaine*, **3**: 463–470.

Bosence, D.W. and Pedley, H.M., 1982. Sedimentology and palaeoecology of a miocene coralline algal biostrome from the Maltese Islands. *Paleogeography, Paleoclimatology, Paleoecology*, **38**: 9–43.

Chevalier, J.P., 1961. *Recherches sur les Madréporaires et les formations récifales du Miocène de la Méditerranée occidentale*. Thèse, Paris, Société Géologique de France, vol. 2, 562 pp.

Cumings, 1932. Reefs or bioherms? *Geological Society of America Bulletin*, **43**: 331–352.

Cross-references

Algal Rims
Coral Reefs
Reefs, Non-Coral
Sea-Level Indicators, Biologic

BLACK AND CASPIAN SEAS, COASTAL ECOLOGY AND GEOMORPHOLOGY

Coastal zone of the Black Sea

The coasts of the Black Sea are rather uniform and slightly embayed. The Crimea is the only large peninsula protruding offshore. The wide opened bays facing the sea (Odesskii, Kalamitskii, Tendrovskii, Karkynitskii, Yarylgachskii, Burgasskii) as well as the above mentioned Crimean Peninsula are located in the northern part of the region. The southern, eastern, and western coasts are smooth and uniform with small bays. The total extent of the coastline exceeds 4,000 km (Figure B55).

Zenkovich (1958, 1959) contributed much to the study of the Black Sea coasts. In the two-volume monograph, he described coasts of the former Soviet Union and analyzed dynamics and morphology of certain regions. Diverse coastal areas were described by investigators from different countries (Nevesskii, 1967; Shuiskii, 1974; Simeonova, 1976; Kiknadze, 1977; Zenkovich and Schwartz, 1987; Shuiskii and Schwartz, 1988; Kaplin *et al.*, 1991, 1993). The American Society of Civil Engineers has recently published a collection of articles concerning the Black Sea coasts (Kos'yan, 1993).

The environmental problems of the coasts have been discussed in many publications. The most complete summaries were given in the monographs of Sapozhnikov (1992) and Kuksa (1994).

Large-scale investigations were carried out in the frame of the international INEP program "Black Sea Environmental Program." Due to these activities about 2,000 analytical maps of the Black Sea natural environment were compiled, among them the map of the main sources

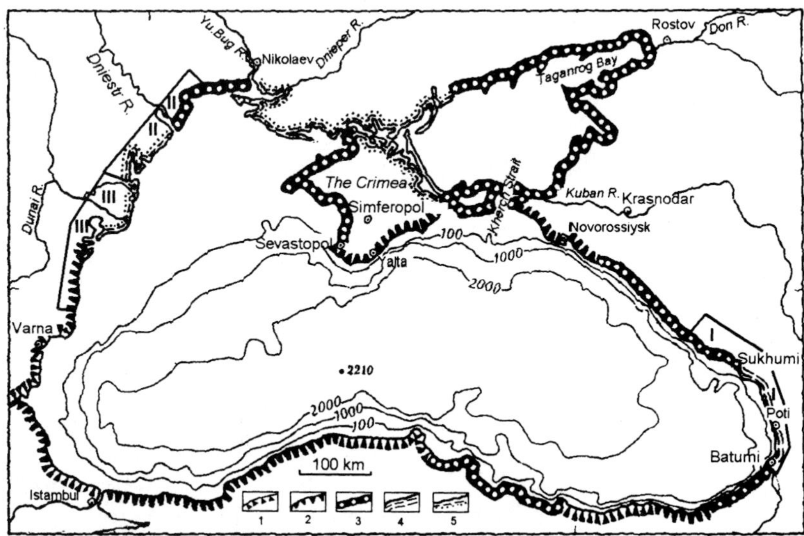

Figure B55 Types of coasts of the Black Sea and the Sea of Azov. 1, straight faulted; 2, erosional bight; 3, graded erosional and depositional; 4, graded depositional; 5, liman and lagoon, erosional and depositional. Key study areas are shown by numbers (I–III).

of pollution in the nearshore zone with subsequent entry to the geoecological information system (Berlyant *et al.*, 1999). The Geographic Information System (GIS) was processed at the Geographical Faculty of the Moscow State University. The users of this GIS may receive not only maps, but also the tables with the data on the amount of pollutants and other information concerning the sources of pollution and natural reserves of the Black Sea. A compact-disc "Black Sea GIS" was published by INOPS/ENVP in 1998.

Environmental problems of the coast

Two main problems could be outlined among the environmental problems of the coasts: (1) influence of the rising sea level upon coastal processes and intensification of erosion related to it; (2) increasing anthropogenic impact.

Anthropogenic impact is mainly manifested by water pollution. Water contamination by pesticides leading to degradation of bottom vegetation was revealed in shallow bays (Kuksa, 1994). It is the result of disposal of freshwater from irrigation systems of Southern Ukraine. Water pollution caused a 3-fold decrease in the phytoplankton biomass in the nearshore zone and a 1.5-fold decrease in the zooplankton and zoobenthos biomass. Considerable pollution of the sea and especially its nearshore zone is determined by the influx of freshwater from the largest river of the region—the Danube. Its influence is noticed along the coasts of Ukraine, Romania, Bulgaria, and even Turkey. The Danube discharges enormous amount of oil-products, heavy metals, pesticides, and other pollutants. Pollutants are mainly accumulated in bottom sediments and biota. For instance, water plants of the Danube coast contain 0.007–0.020 mg/kg of mercury.

The concentration of pollutants discharged by the Danube decreases eastward (near Odessa and Sevastopol) and southward (in Romania and Bulgaria). Other rivers, the Dnieper, Inguri, Rioni, Chorokh and others, contribute much to the contamination of the nearshore waters.

Due to pollution of nearshore waters the role of biogenic sediments (mainly shells) in coastal dynamics decreases. At the end of the 1940s shelly sediments constituted 40–50% of coastal accumulative forms on the northwestern coast (Zenkovich, 1982), while in the 1980s its contribution was less than 10% (Shuiskii, 1974).

Another important ecological factor of anthropogenic origin is the influence of economic activity on the sediment budget in the coastal zone. Regulation of the rivers causes a sharp decrease in the solid river runoff and, hence, less sediments are supplied to beaches. Mass removal of sediments (sand, pebbles, gravel) directly from beaches, quarries, the nearshore zone, and river mouths considerably damaged the coastal zone. In the Caucasian coastal region this process started at the end of the last century when beach sediments were taken for construction of railroads. Mass sediment removal continued in the 1950s–1960s, when ports and other economic objects were built. During 1945–55, 100 million m^3 of beach pebbles were removed from the Tuapse-Adler coast (Kiknadze, 1977). As a result of this action, many beaches of the Caucasian coast became one-half smaller during two or three decades. This caused intensive coastal erosion. Of the 312-km-long Georgian coastline, 220 km were subjected to coastal erosion due to its retreat at a rate of 1–3 m/year. Active coastal erosion manifested by beach destruction was also recorded in the Crimea (Zenkovich, 1982).

During the last few decades many countries have been taking efforts to protect their shores. However, many hydrotechnical constructions such as seawalls, groins, breakwaters, and others have intensified an adverse effect of the sea on the coast. Construction of artificial beaches appeared to be the most effective method. During 1981–86 in Georgia, about 8 million m^3 of sediment was taken from subaerial quarries that facilitated creating artificial beaches with a total area of about 60 ha. As a result, a recreation zone was formed and the problem of shore protection in Georgia was practically solved (Kiknadze, 1977; Zenkovich and Schwartz, 1987). Creation of artificial beaches or additional sediment supply to existing natural ones was undertaken in other regions as well (Odessa, Crimea, Bulgaria).

Coastal geomorphology

In general, erosional coasts predominate along the Black Sea. Elevated mountainous coasts predominate in the eastern and southern parts of the Black Sea. This is a zone of young Alpine orogenesis. Graded and erosional accumulative coasts are typical of the western and northern parts of the sea. Geologically this zone is dominated by hard blocks protruding from the ancient Russian platform and remains of the Baikalian orogenesis. In the Eastern Black Sea erosional processes are especially active due to an extremely narrow continental shelf which sometimes nearly coincides with the coastline as in the Caucasus. Thus, the submarine slope has steep gradients allowing large storm waves to attack the coast.

Slopes of the Great Caucasian Ridge form the largest part of the Caucasian coast, since the axis of the ridge is subparallel to the coastline. This is the reason why cliffy coasts up to 200 m high prevail between Anapa and Sukhumi. The cliffs are cut in the steeply sloping flysch beds and its ridges are noticed in the submarine bench. In the southern part of the Caucasian coast, the Batumi region, foothills of the Little Caucasian Ridge reach the shore. The Colchis Lowland lies between the Great and Little Caucasian Ridges. It follows the large Alpine flexure. The lowland is swamped and its flanks are only slightly higher than the sea level. The lowland experiences a prolonged tectonic submergence. Many rivers flowing from the slopes of both ridges drain onto the Colchis Lowland. Despite this, sandy coasts do not migrate seaward. The heads of submarine canyons are located close to the mouths of the large rivers such as the Inguri, Rioni, Supsa, and others. The alluvial material is removed to the canyons instead of being accumulated on the beaches. Moreover, in many places the shores of the Colchis Lowland are eroded (up to 3 m/year).

The presence of large promontories near Adler, Pitsunda, Sukhumi, Burup-Talii are typical of the Caucasian coast. They are located near

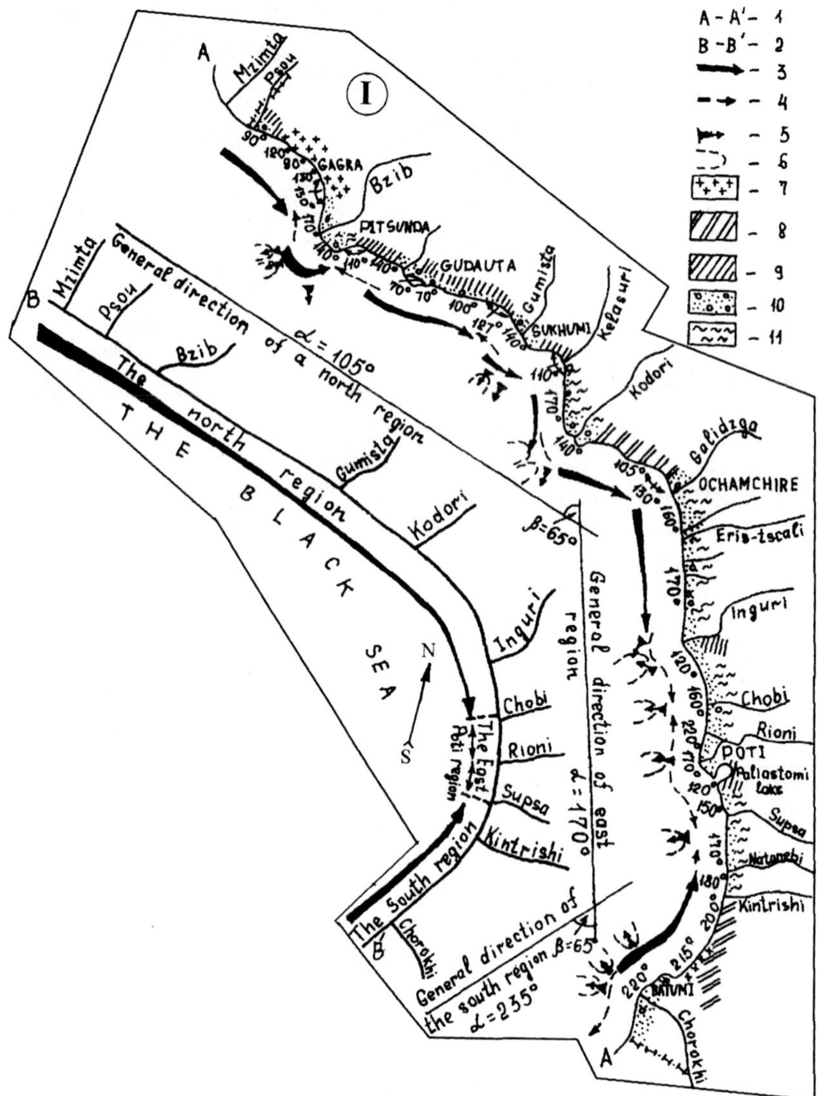

Figure B56 Schematic map of morpho- and litho-dynamics of the Black Sea coast in Georgia (after Kos'yan, 1993). 1, modern coastline; 2, coastline during the period of the drop in sea-level rise 6–5 ka; 3, longshore sediment streams, their direction and relative actual capacity; 4, direction of migration and transport of finer sediment; 5, partial loss of sediments at considerable depths; 6, canyon heads and steep falls; 7, cliffed rocks with erosional relief; 8, semi-cliffed rocks (conglomerates, marl, schists, etc.) with erosional relief; 9, related rocks (marine and lagoonal clays) with plain relief; 10, loose deposits (pebbles, gravel, sands of terraces, dunes and beaches); 11, bog and lacustrine deposits.

the river mouths and consist of the Holocene alluvium (Figure B56). These promontories protrude far offshore and overlie a significant portion of the continental shelf. No other large accumulative landforms are present on the Caucasian coast. The beaches are associated either with numerous river mouths or places of longshore sediment drift discharge. They are mainly composed of gravel and pebbles.

As shown above, accumulative forms are subjected to active erosion. Its intensification is caused by natural reasons: sea-level rise at a rate of 1–2 mm/year and decrease in river discharge due to regulation of rivers and removal of beach sediments. Heads of submarine canyons contribute to coastal dynamics since part of the material transported by alongshore drift is accumulated there. For example, the Akula submarine canyon near Pitsunda accumulates about 80 thousand m^3 of sediments per year (Kiknadze, 1977; Kos'yan, 1993).

Within the Georgian coastal zone alongshore drift is directed to the southeast (Figure B56). Each sediment stream represents a dynamic system with its own source of sediment supply and areas of sediment loss (submarine canyons and steep slopes) or final discharge. The capacity of alongshore sediment streams ranges from 3–15 to 150–220 thousand m^3/year. A small alongshore sediment stream is directed to the north from the Chorokh river mouth to the Colchis Lowland.

High erosional shores are typical of the mountainous coasts of the Crimea. They are subjected to active erosion since they are affected by severe winds (and waves) blowing from the southwest and southeast. Shore destruction is accelerated by landslides occurring in clays. Sometimes the landslides have an area of hundreds of square meters. For instance, the town of Alupka is located on six large landslides and its stability is conditioned by several factors. Of these are influence of underground and surface waters, abrasion, load of buildings and other construction.

Many shores of the Southern Crimea are formed by the slopes of ancient volcanoes (Karadag region) and tectonic faults. Outcrops of volcanic rocks and limestones form capes separated by shores represented by soft shales, clays, and sandstones. Ria-coasts occur near Sevastopol and Balaklava.

Beaches of the Southern Crimea are formed of pebbles, because finer sediments (more than 0.03) are transported down the steep submarine slopes. Removal of pebbles for building purposes caused the disappearance of beaches. However, some of them have been recently restored.

Many shores of the Southern Crimea are artificially protected. Dynamic interaction between different regions is weak due to the absence of large rivers supplying sufficient amounts of alluvial sediments to the coastal zone. Thus, local shore protection is successful and has no negative influence on adjacent coasts. Different, usually complex, engineering structures are used that protect coasts from both landslides and abrasion. Of these are embankments with seawalls, traverses, breakwaters, groins with artificial sediment filling between them, etc. (Zenkovich, 1982; Kos'yan, 1993).

Steep coasts of the Southern Black Sea are formed by densely forested northern slopes of the high Eastern and Western Pontus mountains stretching subparallel to the coastline. The mountains gradually lower westward and near the Bosphorus Strait their height does not exceed 300 m. Erosional and denudation coasts with steep rocky cliffs are widespread in Turkey. Only in separated small bays do the sandy-pebbly "pocket" beaches occur. Areas of sediment accumulation are associated with mouths of such large rivers as the Kizil-Irmak, Sakarja, and Eshil-Irmak. These rivers form rather large deltas prograding far offshore and nearly reaching the edge of the narrow continental shelf. Violent storms produced by severe northwesterly winds deflect the pathways of alluvium to the east thus forming flanked barriers (Kos'yan, 1993).

The largest curves of the Turkish coast correspond to the lowland peninsulas of Bafri and Djiva, related to river deltas and the mountainous Injeburun Peninsula.

Western and northwestern coasts of the Black Sea are rather low with hilly plains of different origin (alluvial, marine, and alluvial–marine) facing the sea. The delta of the Danube, the largest river of Western Europe, is located here. It has a complicated structure. Besides common channel bars there are a series of cheniers (local name "grindu") marking the stages of delta progradation. The river mainly discharges through its northern Kiliiskii channel. Thus, the southern part of the delta is smaller and is being slightly eroded. Active utilization of the Danube water for irrigation by five countries reinforces erosion of the southern part of the delta. There are numerous water reservoirs in the delta: lakes-limans (northern part), complexes of lakes and lagoons (southern part), lakes (inner part). From the north the delta is bounded by the Budzhak plateau, and from the southwest by the lake–lagoon complex of Rozelm-Synop. Abundance of warm water and high fertility of soils favor plant and animal life.

Coasts to the northeast from the delta are represented by plains and low plateaus. The only exception is the anticline of the Tarkhankut Peninsula. Its steep slopes are mainly composed of easily eroded loesses and clays. The rate of erosion ranges from 7 to 20 m/year (Shuiskii, 1974).

These erosional coasts alternate with lagoons and limans. Limans represent the lower parts of river valleys that have been flooded during the Holocene transgression. Most of them are separated from the sea by sandy–shelly accumulative forms (spits or baymouth barriers). Specific environmental conditions exist in limans since their waters are warmer and less salty. As a result, productivity of waters is higher.

A considerable part of the coast is subjected to landslides. Both landslides and coastal abrasion destroy valuable territories of the Ukrainian steppes. At present, the accumulative forms in the mouths of limans are eroding. They are composed of sand layers overlying lagoonal clays, thus giving evidence for migration of the accumulative forms toward the lagoons (Shuiskii and Schwartz, 1988).

Two opposite longshore drifts exist in the region stretching to the northeast from the Danube River to Odessa (Figure B57).

Jagged coasts are characteristic of the region to the east from Odessa including the western Crimea. Adjacent lowlands experience relative submergence at a maximal rate of 30 cm/ 100 years as recorded in the inner part of the Karkinit Bay. Large accumulative forms are the most interesting elements of the coastal relief, that is, the Kinburn spit and the system of the Tendra-Dyarylgach spits related to it.

The Kinburn spit and Odessa shoal (to the west of it) originated in the place of the Dnieper and South Bug deltas junction. Under the

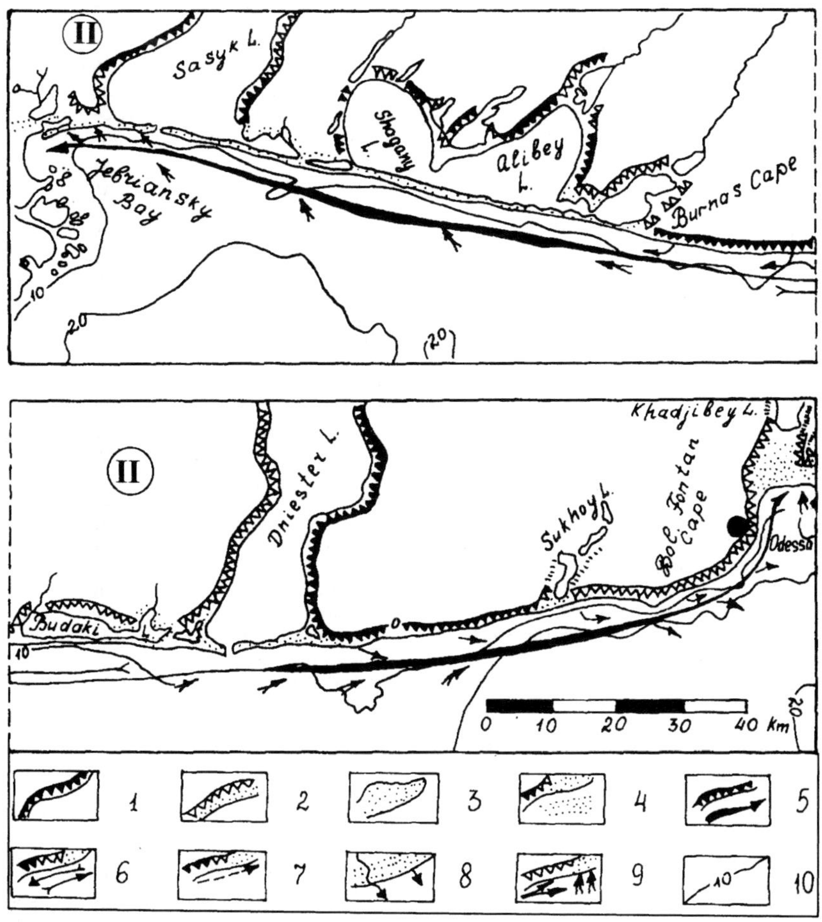

Figure B57 Geomorphological map of the northwestern Black Sea coast between the Danube River delta and Odessa (after Zenkovich, 1958). 1, active erosional scarps; 2, passive erosional scarps; 3, emerged coastal accretion bodies and coastal ridges; 4, emerged coastal accretion bodies and coastal ridges; 5, longshore sediment streams (thickness of arrows proportional to the capacity of a stream); 6, longshore sediment drift; 7, prevailing sediment drift; 8, offshore sediment drift; 9, onshore sediment drift; 10, depths in meters.

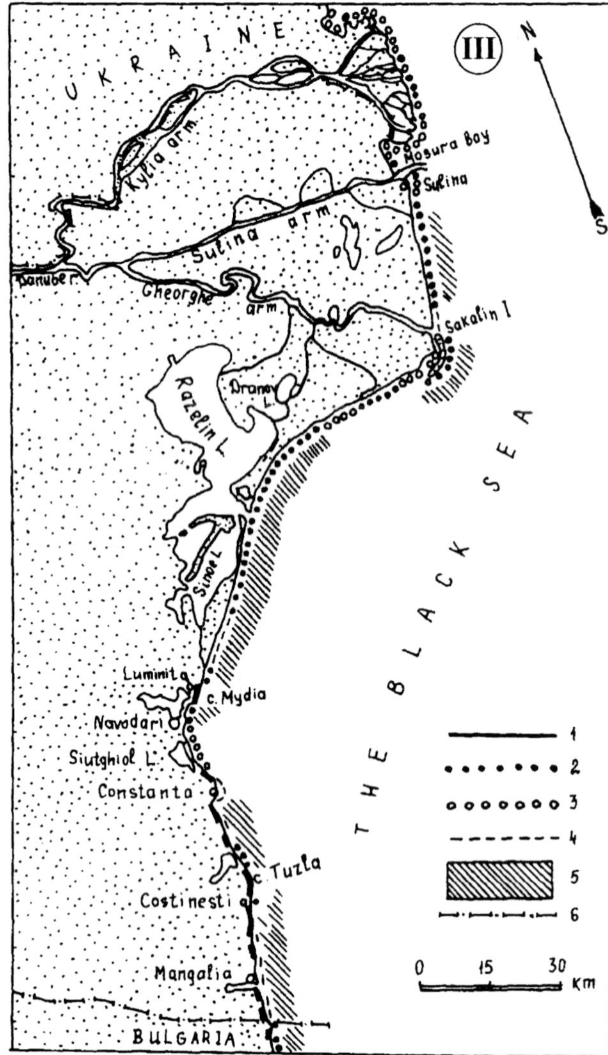

Figure B58 Morphology of the Romanian coast of the Black Sea (after Kos'yan, 1993). 1, active cliffs; 2, retreating coastline of accumulative coasts; 3, prograding coastline of accumulative coasts; 4, stable coastlines; 5, coastal sections with predominance of erosional processes in nearshore zone; 6, state frontier.

Holocene sea-level rise deltaic sediments were reworked and a system of subaerial and submarine sand bars was formed. Dunes and salt lakes located on the Kinburn spit are parallel to the coastline.

The Tendra and Yarylgach spits represent a joined accumulative form that continues to grow. However, landward migration also takes place. As a result, the central part of the accumulative form became attached to the continental shore, and its distal end formed two separate spits.

The eastern part of the Crimean Peninsula, together with the Taman Peninsula, form a single coastal region divided into two parts by the Kerch' Strait. The territory is covered by limans and lagoons associated with the ancient and modern delta of the Kuban' River. However, erosional shores are dominant. The coastline represents a series of arcs where clays of Maikopian age intercalate with solid rocks of Neogene age that form headlands.

Along the Kerch'–Taman' coast relics of the ancient accumulative forms and lagoonal silts were reported that allowed for reconstructing the Holocene history of the coastal area.

The western coast of the Black Sea lies in Romania and Bulgaria. The Romanian coast is subdivided into two parts. Its western part corresponds to the Danube delta that equals 78% of the delta surface. As mentioned above this part of the delta is now eroding at the rate of up to 7 m/year (Kos'yan, 1993).

Southward from the delta the coast is graded. It is composed of loesses, clays, and limestones of Neogene and Pleistocene age. A considerable part of the southern Romanian coast (51 km of the total 101 km) is abraded and consists of active cliffs 2–40 m high. The rate of abrasion averages 1–2 m/year, sometimes reaching 7 m/year. Maximal rates are characteristic for cliffs composed of loesses and clays. Capes are usually formed of Neogene limestones. Submarine benches are typical of the Romanian erosional coasts. Their width sometimes reach 1,100 m (Figure B58).

Coastal accumulative forms are represented by sandy beaches resting against cliffs and, near the river mouths, and by barrier forms separating lakes and lagoons. Sintghiol is the largest sandy barrier. Sedimentary material is supplied to the sea by the Danube delta and active cliffs.

The Romanian coast is actively used for recreation. To protect the coast from destruction certain efforts have been taken: construction of seawalls, cobble filling, etc.

The coasts of Bulgaria are mainly erosional. In southern Bulgaria erosional forms are restricted to the small bays of the zone of Alpine orogenesis. Cliffs of eight different types are distinguished in this area: from 15 to 20 m high cliffs with even surfaces composed of uniform loess and clayey deposits to high cliffs (up to 60–90 m) with uneven surfaces and a series of landslide steps on the slopes. Such cliffs are widespread in the region between Kavarna and Balchik and Kranevo-Zlaty Pyaski. Earthquakes facilitate landslides thus considerably accelerating retreat of the coast. The average rates of erosion vary from 0.005 to 1 m/year. Maximal rates reach 30 m/year. The estimated amount of material released due to abrasion of cliffs is 1,344,100 m^3/year.

Material produced by coastal abrasion and alluvium forms the beaches that occupy 28% of the Bulgarian coast (Simeonova, 1976). Some of the beaches, like that at Varna, prograde at a rate of 0.75 m/year. Similar process operate near the mouth of the Kamchiya River and in the region of the popular resorts of Albena and Zlaty Pyaski. However, most of the accumulative coasts retreat. For instance, between Cherny Nos Peninsula and the Albena resort the rate of retreat is 0.12–0.63 m/year. Generally the rate of retreat grows in the northward direction. Erosion often results from the negative influence of human activities and underestimation of the role of coastal processes. To protect coasts from erosion Bulgarian engineers fill up tetrapods with stones and construct dams separating the bays from the sea and straighten the coastline.

The coasts are also protected by groins (often short and without any filling between them), seawalls and other less effective structures. The most effective means, such as creation of artificial beaches, are not implemented in Bulgaria.

Sea-level oscillations played an important role in the recent evolution of the Black Sea coasts. From the available data it follows that during the 20th century sea level was steadily rising at the average rate of 2.1 ± 1.3 mm/year (Nikonov *et al.*, 1997). This estimation is based on the data collected at 70 points on the Russian, Ukrainian, Georgian, and Bulgarian coasts. The values exceed the average rate of the global sea-level rise, probably due to tectonic submergence of the Black Sea depression. Different estimations of the sea-level oscillations could be definitely attributed to different tectonic movements. The highest rates of sea-level rise were recorded in the Colchis Lowland, while the lowest ones were in the northeastern Black Sea.

Thus, it might be concluded that submergence of the coasts has been the main trend in their recent evolution. This process leads to abrasion of the cliffed coasts and erosion of the accumulative ones. However, under predicted conditions of more rapid sea-level rise, erosion will be intensified and many of the unique accumulative forms will be destroyed. First of all this affects the accumulative forms of the limans in the northwestern coastal area (Tendra, Binburn, Yagyrlach spits). Their destruction may have a severe impact on the ecology of limans that are the zones of extremely high bioproductivity. Sea-level rise will accelerate destruction of the Holocene accumulative forms on the Georgian and Turkish coasts. In this connection, all countries of the region must plan enhancing protective activities and carry out a long-term policy of coastal management.

Coastal zone of the Caspian Sea

The coasts of the Caspian Sea display a great variety of natural environments being located in different landscape zones. Recently, the problems associated with the rapid rise of its sea level have generated particular interest, especially in the context of the expected accelerated rise of the global sea level (Dolotov and Kaplin, 1996).

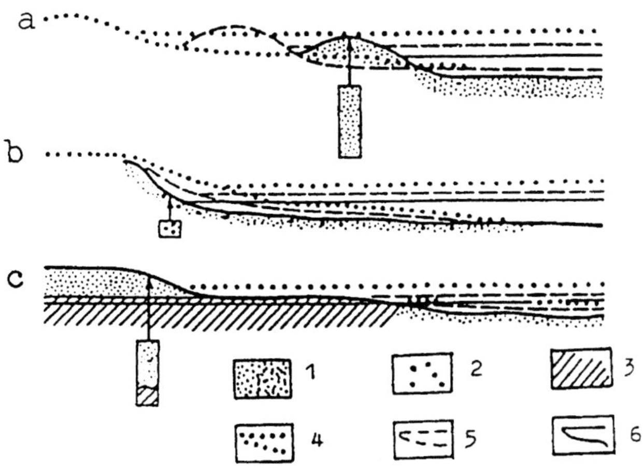

Figure B59 Processes of relief formation and sediment accumulation in shallow-water neashore zone under sea-level fall (after Dolotov, 1996). a, continuous sediment accumulation; b, erosion of external edges of accumulative coasts; c, preservation of sandy accumulative body; 1, sand; 2, pebble; 3, bedrock; 4, 5 and 6, successive positions of sea level.

The history of investigations on the Caspian Sea coasts was discussed in detail in the monograph of Leont'ev and Khalilov (1965). Further generalization was given in the monograph written by Leont'ev with co-authors (Leont'ev *et al.*, 1977). Present environmental problems of this area were outlined in the monographs of Kuksa (1994) and Zonn (1999).

Environmental problems of the Caspian Sea coastal zone arise from active economic development of not only the Caspian Sea itself, but its drainage area and adjacent territories (Kaplin and Ignatov, 1997). This region is distinguished by repeated sea-level changes, both seasonal and multi-annual. That is why the Caspian Sea is a natural laboratory for studying evolution of coasts under different sea-level oscillations.

In the modern historical period a rapid sea-level fall (of nearly 1.7 m) occurred from 1929 until the early 1940s. Dynamic changes in the coastal zone mainly depended upon the rate of sea-level fall (or decrease in depth in the nearshore zone) and the amount of sediments in the coastal area (Dolotov, 1961).

On the shallow-water sand coasts that are typical of the Caspian Sea the relief-forming processes are controlled by sediment budget, gradient of the coast, and configuration of the coastline under sea-level fall going on at different rates. Three types of the coasts with different patterns of relief changes have been identified.

Continuous accumulation of sediments and progradation of coasts (Figure B59(a)) takes place in case sufficient amounts of sediments are supplied to the coastal zone from adjacent land and shores (positive sediment budget over prolonged time period). If a positive sediment budget is replaced by a negative one progradation of the coast changes to landward retreat of the coastline (Figure B59(b)) (irregular sediment supply). Under insufficient sediment supply erosion is replaced by preservation of sandy accumulative bodies that occurs when direct wave attacks over the former coastline have ceased (Figure B59(c)).

Under sea-level fall the evolution of the Caspian Sea coasts went on in the following manner: continuous accumulation of sediments, emergence of the seafloor, and continuous seaward advance of the coastline. The area of coasts enlarged, and economic activity occupied new territories where settlements, roads, oil- and gas-pipelines, and resorts were constructed (Dolotov, 1996). Taking into consideration the predicted future sea-level fall the Soviet Government decided to create a vast recreation zone in the coastal regions of Dagestan and Azerbaidzhan (Molchanova, 1989). The Caspian coast offers several advantages over the Caucasian coast of the Black Sea. Sandy beaches are wider and longer here, solar radiation is more active and the number of sunny days in summer is higher (Veliev *et al.*, 1987). Part of these constructions has already been built.

In 1978, an unexpected and sharp sea-level rise occurred. The average rate of sea-level rise was 14–15 cm/year, but in some years it was as high as 30 cm/year and even more. The direct influence of the sea-level rise was flooding of coastal lowlands and acceleration of coastal erosion. The indirect impact included the rise of groundwater, swamping of the coastal area, salinization of soils, and expansion of surge areas. Environmental conditions of both the nearshore shallow zone and adjacent land sharply changed. Sea-level rise favored erosional processes and general landward migration of the coastline.

At the same time, contrary to the general trend of sea-level rise, in some patches of the western coast accumulation of sediments went on and the coastline migrated seaward. This happens, when sediment input exceeds the amount of unconsolidated sediments flooded by the advancing sea. These coastal regions are of special interest since they give a unique opportunity for future economic development (primarily recreation) even under the ongoing sea-level rise.

Generally, the character of relief changes under sea-level rise depends upon the rate of this rise, relief-forming environmental processes (hydrodynamics), and sediment balance. Coastal morphology determines the character and rate of the natural catastrophic processes together with their impact on economy and population (Dolotov, 1996).

Coastal geomorphology

Based on differences in relief, the Caspian Sea coast is subdivided into four regions (Leont'ev *et al.*, 1977).

The western coast receives about 50% of the total solid river runoff, while the material produced by abrasion is considerably less abundant. As a result of a positive sediment balance coastal erosion is suppressed. Under dominant northwesterly and southeasterly winds longshore drift streams are formed that smooth the coast and create accumulative forms.

River runoff is practically absent in the eastern regions. Active erosion in the recent past has not produced sufficient amounts of sedimentary material, since the cliffs are mainly composed of clays and limestones. Biogenic and chemogenic sedimentation went on slowly. As a whole, the coast is more embayed than in the west. In places where the coastline remains relatively straight over long distances, prevailing waves that are transversal to the coast form large accumulative barriers.

The northern coast is distinguished by extreme shallowness. Southeasterly and northerly winds predominate here. Abundant fine-grained alluvial material supplied by rivers is transported in the form of suspension. Winds and on- and offshore currents form the coasts characterized by frequent and significant displacement of the coastline.

Southern coasts are represented (Voropaev *et al.*, 1998) by coastal lowlands ranging in width from 1 (in the central part) to 60 km (near the large deltas of the Sefidrud and Gyurgyan Rivers). More than 40 small rivers discharge into the Caspian Sea in this region. During the past 50 years, solid river runoff was considerably reduced due to construction of reservoirs on all the large rivers. Granulometric composition of the beaches changes from gravel–pebble (western Mazenderan) to sand (Gilidzhan, central Mazenderan), and silt (eastern Mazenderan).

Several types of accumulative coasts occur in the Caspian Sea: lagoonal shores, coasts with terraces and other accumulative forms, erosional and deltaic shores, mudflats, coastal lowlands formed by onshore winds and waves (Figure B60).

Deltaic coasts occupy considerable parts of the coast—these are the deltas of the Volga and the Ural in the north, Terek and Sulak in the northwest, and Kura in the southwest. Even, quite recently, these rivers discharged considerable amounts of sediments into the sea thus supporting progradation of deltas. This process intensified during sea-level falls. In the northern Caspian Sea, where the Volga and Ural deltas are located, coasts have retreated seaward by dozens and hundreds of kilometers since 1929. In the late 1950s and 1960s large-scale hydro-engineering projects were launched. This caused dramatic decrease in river discharge of such rivers as the Kura, Terek, and Sulak. The rise of the Caspian Sea level caused erosion of deltas. Coastal lowlands of the Northern Caspian region became inundated, and the previously accumulated coastal landforms were destroyed.

Large portions of the Caspian Sea coasts are flat. They were formed in course of regression that took place from the 1930s to the 1970s. Mudflats occur in the northern Caspian region, around Kirov Bay in southern Azerbaidzhan, in the region surrounding Krasnovodsk Bay and to the south from it. During regression aggradation of the northwestern coasts of the sea proceeded at a rate of 60–100 m/year, in the Kizlyar Bay it was faster and reached 150–200 m/year, and northward of it-even 700–800 m/year. In the Kirov Bay wind-induced mudflats were up to 1.5 km wide (Kaplin, 1997). The sea-level fall resulted in considerable advance of land in the eastern coastal regions. For instance, southward from the Cheleken Peninsula the average rate of land advance during the period from 1929 to 1957 was 34–36 m/year (Leont'ev *et al.*, 1977). Naturally, shallow bays of the northern coast were dried up.

The change from regressive to transgressive regime has dramatically affected the drained shores with mudflats. Gentle gradients (close to 0.0001) of submarine slopes caused passive flooding of the coasts.

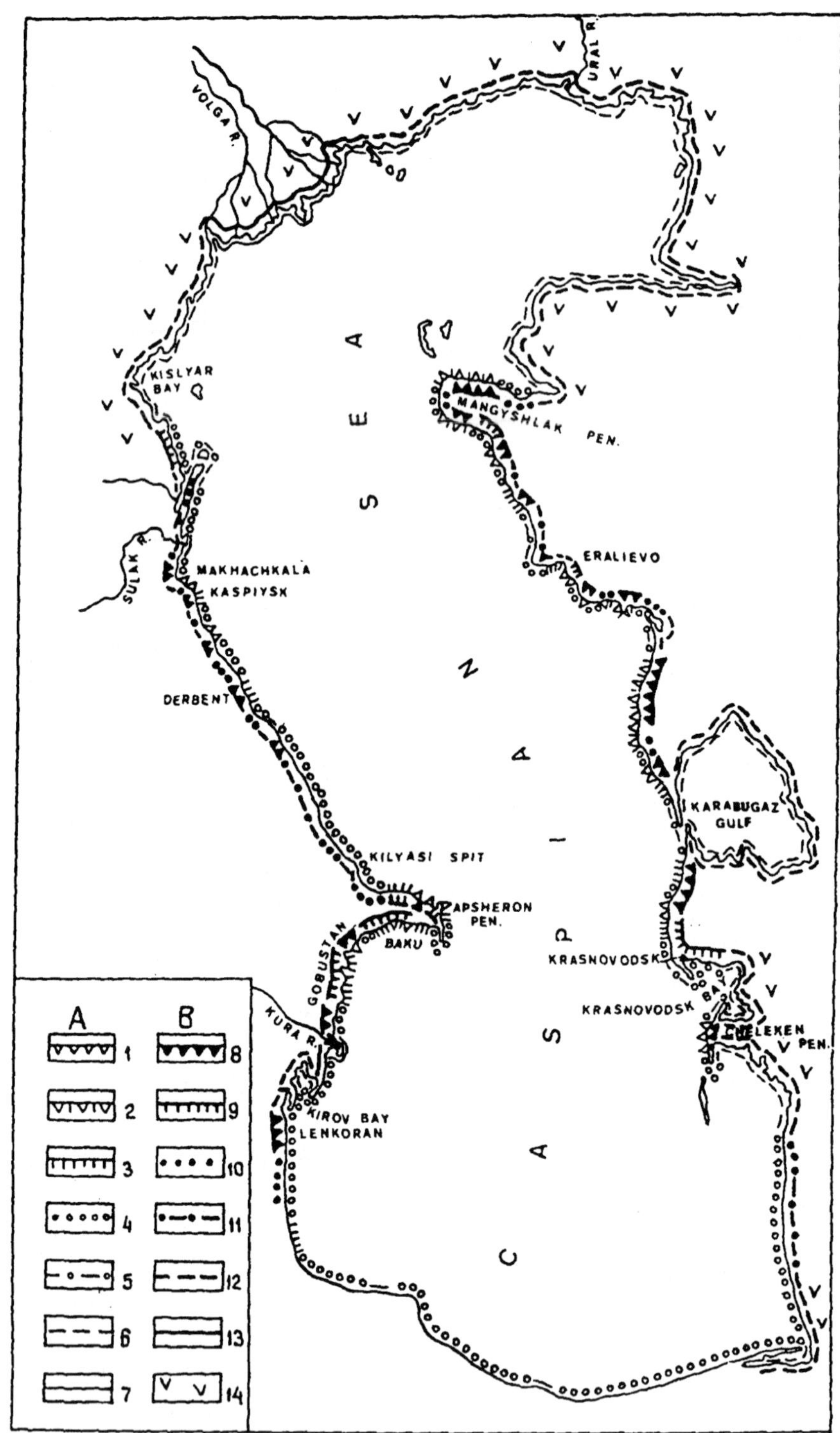

Figure B60 Types of the Caspian Sea coasts (after Ignatov *et al.*, 1993).
A, regressive stage (prior to 1977). 1, erosional shore; 2, erosional shore with passive cliff; 3, erosional–accumulative shore; 4, progradingbeach; 5, accumulative lagoonal shore; 6, mudflats formed by wind-induced surges; 7, deltaic coast. B, transgressive stage (after 1978). 8, erosional shore; 9, erosional–accumulative shore; 10, prograding beach; 11, accumulative lagoonal shore; 12, mud flats; 13, deltaic coast; 14, areas affected by transgressive flooding.

Sea-level rise resulted not only in submergence of land, but rise of the groundwater table and, hence, salinization of groundwater and swamping of the adjacent lowlands. The Kirov Bay was filled with water again. In the vicinity of the Kilyazinskaya spit (northern Azerbaidzhan), a flat coastal terrace that formed in 1940 due to sea-level fall has been partly flooded and swamped. Considerable parts of coastal lowlands in the northern near Caspian region on both sides of the Volga River have been flooded. Accumulation of sediments went on in such bays as the Komsomolets, Kaidak, Mertvyi Kultuk, Bol'shoi Sor in the Bugaz Peninsula, and in depressions between Baer knolls that are widespread in the northern and northeastern Caspian lowlands. Wind-induced surges up to 1.5–2 m high caused passive flooding of this vast area. These phenomena resulted from a longshore drift and caused episodic abrasion of the coasts and erosion of the seafloor, producing furrows and ridge-and-runnel erosional forms. Wind-induced currents re-suspended sands and silts previously accumulated on the seafloor. Suspended material was removed by currents and precipitates near the shoreline to form saltings (mudflats).

Sea-level rise changes accumulative coasts as well. As mentioned above, accumulation that took place during the regressive stage affected nearly the entire coastline, but the pattern was somewhat different. As sea level was falling, cliffs became passive, longshore drift ceased, and spits were eroded, especially at their base. We can exemplify the Kura and Agrakhan spits on the western shore, the accumulative form of Cape Rakushechnyi, Kenderli, and Krasnovodsk spits on the eastern one. With the drop in sea level, the middle and lower parts of submarine slope became eroded. The produced clastic material is gradually transported upward the slope and accumulated near the shore. Intensive landward transportation of sediments inhibited the longshore drift.

Due to the sea-level rise former cliffs became active again and abrasion intensified. Transgression changed the profile of submarine slopes, especially their upper parts. These changes are accompanied by erosion of the frontal part of accumulative forms or creation of bars near the water edge that later turn into beach barriers separating the sea from lagoons. Many lagoonal shores were formed in Dagestan and northern Azerbaidzhan, where gradients of the submarine slope equal 0.005. The present sea-level rise is responsible for accumulation of beach barriers separating lagoons. The beach barriers are overlapping lagoons thus giving an impression that the coastline retreats. However, lagoons keep expanding despite the landward migration of beach barriers since flooding of drained territories and the rise of the groundwater table favor further expansion and deepening of lagoons.

The coasts of Dagestan and Northern Azerbaidzhan, with steeper gradients (up to 0.01), have shore-attached bars formed of coquina. These bars have an asymmetrical profile giving evidence of their landward migration towards young terraces behind them. No lagoons are formed because the coasts are steep and lie above sea level.

Finally, slopes with gradients exceeding 0.01 are subjected to active erosion of both Holocene and recent accumulative forms. This trend leads to significant landward retreat of the coastline (Kaplin, 1997). Evolution of such coasts follows the well known "Bruun's rule" (Bruun, 1962).

The transgression has also affected erosional coasts. The latter are typical of the Eastern (Mangyshlak Peninsula, regions northward from the Kara-Bogaz-Gol Bay, Cheleken Peninsula) and, partially, Western (Dagestan, Lenkoran', Apsheron Peninsula) Caspian Sea. The share of erosional shores on the Dagestan coast increased from 10% to 40%, on the Azerbaidzhan coast—from 20% to 55%, Kazakhstan—from 8% to 13%, and Turkmenistan—from 7% to 22%. Until recently, some places on the Dagestan coast have been protected from erosion by offshore submarine ridges composed of limestone. Benches and ridges have been flooded by the transgression, during storms (especially surges) waves reach the cliffs. If the cliffs used to be protected by accumulative terraces, clastic material has been actively reworked; part of it being transported into the longshore drift and the rest removed down the submarine slope. In general, the above-water parts of the slopes were more actively abraded by sea waves, and the coastline rapidly migrated landward. In some parts of the Dagestan and Lenkoran', the rate of the process reached 20–25 m/year.

Therefore, the recent rise of the Caspian Sea level has significantly modified the dynamics of all identified coastal types. Evolution of the coasts subjected to relative submergence depends upon the gradient of the submarine slope (Kaplin, 1997). Different ways in evolution of accumulative coasts under the sea-level rise (Figure B61) have been discussed in several publications on the Caspian Sea (Kaplin, 1989, 1990; Ignatov *et al.*, 1992, 1993). The coastal dynamics of the Caspian Sea display a certain discrepancy between transgressive and regressive regimes on the one hand, and cycles of the coastal evolution on the other hand. A regressive regime usually corresponds to the cycle of accumulation. However, erosion of coastal accumulative forms started in the 1960s when sea level was still falling. Certainly, one of the reasons why erosion became active is economic activity, such as construction of barrages and irrigation networks that reduced the solid river runoff and led to the deficiency of sediments in the coastal zone. On the other hand, it is natural that an erosional cycle succeeds the accumulative one. The reason is that both a drop in the sea level and its rise under transgression are associated with reformation of the submarine slope and with its erosional cutback. Yet, during regression the zone of the shore slope erosion shifts seaward, not landward, involving parts of the outer slope where benthonic material is of a finer grain size. With time the deposits of fractions that can be transported up the slope and that can built the accretion forms are depleted. Although the rates of the Caspian coastal processes are much higher than of those in the world ocean, these processes are essentially similar. Therefore, research into the Caspian Sea coastal dynamics has importance beyond its regional significance. It may be of great use for simulating the formative laws applying to the coastal dynamics of the world ocean, particularly as the ocean level is rising and is likely to keep rising in the future as a result of global warming (Kaplin, 1997).

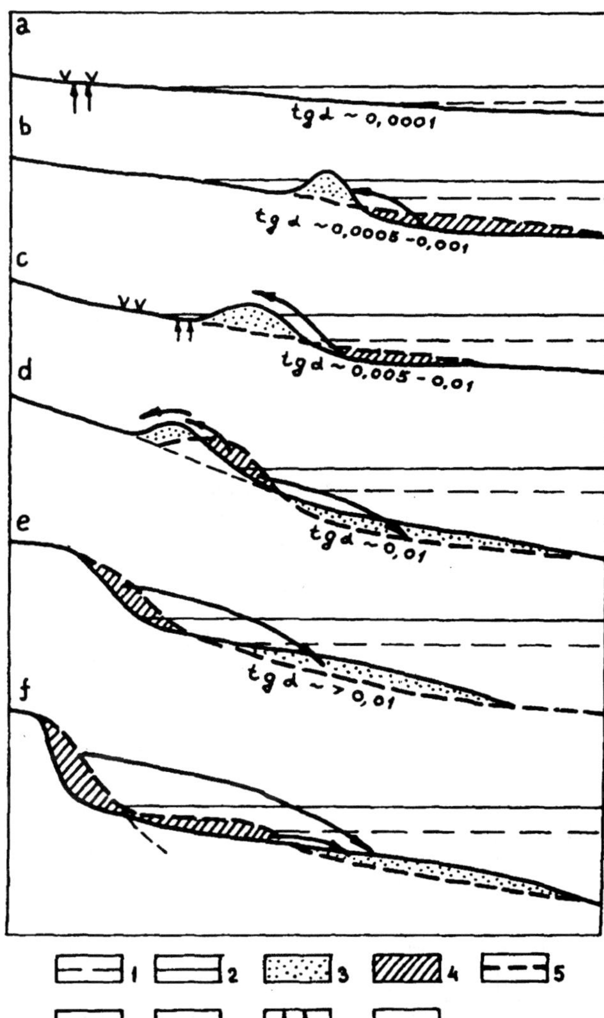

Figure B61 Schematic representation of the Caspian Sea transgressive coasts in function of the offshore gradient (after Kaplin, 1989). 1, regressive sea level; 2, transgressive sea level; 3, sediment accretions; 4, erosion lens; 5, former profile of coastal zone; 6, present-day profile; 7, sediment drift; 8, groundwater rises; 9, bogs.

Environmental problems of the coasts

The present environmental conditions of the Caspian Sea are determined by the effect of rising sea level and increasing anthropogenic impact (Kaplin, 1997).

Considerable sea-level rise has already adversely affected the economy of the coastal states of the Caspian region including the Russian Federation that occupies its northwestern part—the Dagestan, Kalmykia, and Astrakhan' region. Many industrial and habitable buildings, recreational structures and other objects are now in the zone of flooding or in the zone of subsoil water penetration (Dolotov, 1996).

The rising sea level poses threats as follows: flooding and underflooding of coastal areas earlier occupied by communication facilities, livestock farms, grazing lands, fish hatcheries, piers, fish spawning grounds, wildlife and nesting bird habitats. It also prevents cattle from accessing fertile pastures. Incursions of seawaters into areas occupied by human settlements or farms generally lacking treatment facilities resulted in capture and retransportation of technogenic products, municipal effluents and wastes toward the Caspian Sea increasing the supply of pollutants. Flooding causes malfunctioning of irrigation channels and transverse drains in the irrigated areas. Intensive shore erosion has caused losses of considerable land areas.

Special environmental problem is the underflooding of running and suspended wells in the oil and gas fields and subsequent propagation of oil products. Since oil production has or had been carried out over many decades it is evident that significant amounts of oil have been accumulating in the soil, finding their way into, and heavily polluting the sea. The oil film formed over large areas, dramatically reducing water evaporation and affecting the Caspian water balance (Kaplin, 1997).

Starting up of the first turn of the Astrakhan' gas-condensate complex together with exploitation of oil and gas-deposits present a potential threat to existence of unique ecosystem of the river mouths in the Northern Caspian region (Kuksa, 1994). Development of shelf oil and gas fields by the Caspian states strongly affects natural processes by pollution of water and bottom sediments, destruction of plant cover, and considerable reduction of fish resources (Kurbatova, 1994).

In course of the Caspian sea-level rise some of the productive oil and gas fields on the coast will be flooded. The flooded area will also include prospective sites for deep exploratory well-drilling.

The major sources of pollution of the Caspian Sea water consist of: river (surface) runoff; untreated effluents discharged from enterprises, farms, or human settlements in coastal areas or in the river mouths; navigation accidents and technical processes in industries operating directly in the Caspian water area; surface washout during surges; and flooding of producing oil and gas fields, industrial sites, agricultural lands, and human settlements.

River runoff is the largest source of pollutants to the Caspian Sea accounting for 90% or even more of the total influx of pollutant. It is attributed to the fact that the Volga, Ural, Kura, and Terek rivers receive polluted effluents from various industrial facilities and farms along their entire courses. Concentration of various pollutants in river water at river mouths exceeds the maximum permissible levels, frequently by a very wide margin (up to 10-fold or more).

It should be noted that pollutants transport with river runoff is a rather constant process, varying only slightly in different years. The most important consequences are as follows: large amounts of pollutants carried to the sea by rivers under the influence of hydrometeorological factors (wind, currents, waves) penetrate the entire water column and bottom sediments, that later act as the sources of secondary pollution of seawater. Self-purification processes are unable to neutralize the water continuously impacted by chemical inputs. Environmental conditions in river deltas are seriously threatened, mainly in deltas and river mouth beaches. The latter represent the most valuable natural water complexes that act as the principal fish spawning and foraging grounds as well as waterfowl nesting places and rest areas.

The second important source of pollutants transported to the sea are effluents discharged from enterprises, farms, or human settlements situated directly on the coast. An extremely adverse effect on the marine environment is produced by sudden discharge of pollutants resulting from accidents at enterprises of treatment facilities as well as various failures of sewage systems (Kaplin, 1997). All maritime cities, first of all, Astrakhan', Baku, Makhachkala, Turkmenbashi, are sources of pollution that drop to the sea sewage waters (Zonn, 1999).

The third major source of seawater contamination are accidents (oil and oil products spills) occurring during navigation and exploitation of offshore oil- and gas-fields, as well as, owing to the sea-level rise, flooded coastal oil fields. Accidental spills of oil and oil products are a source of significant damage to marine ecosystems, because concentration of pollutants may be extremely high exceeding the permissible level by hundreds or by even thousands of times.

Biological resources and fish reserves are affected as early as at the stage of seismic exploration of oil and gas fields. Special damage was caused by blasting operations that were responsible for the large-scale mortality of sturgeons. During drilling operations on the continental shelf, a special threat is posed by sustained discharges of liquid and solid wastes that are associated with the drilling process. Environmental consequences in the areas of offshore oil- and gas-field development are observable 5–12 km away from the drilling site and are manifested as high levels of oil pollution of water, bottom sediments, aquatic and benthic fauna and flora, and reduced species diversity of benthic communities and degradation of their structure (Kaplin, 1997).

The sea receives water wastes from many sources. From 2 to 5 tons of heavy metals, 60,000–200,000 tons of petroleum products, and more than 5 million tons of organic pollutants are dumped into deltas every year. The sediments are, therefore, contaminated by heavy metals, especially lead and cadmium, and their concentrations exceed many times those of the natural background. The composition of organic matter within the limits of deltas and their flood-plains causes deterioration of the oxygen regime and leads to hydrogen sulfide pollution of the beaches.

The flooding and associated contamination of land initiates specific biogeochemical processes whereby anaerobic gas generation is stimulated. The waters then become contaminated by metal compounds, heavy hydrocarbons, carbon dioxide, and nitrogen compounds, as well as bituminous substances and aromatic (benzene) polycyclic hydrocarbons. Generation of hydrogen sulfide poses the greatest danger (Kaplin, 1995).

Transformations of ecosystems in all rivers of the Caspian Sea basin occurred due to hydraulic engineering and hydro-energetic projects, exploration of oil- and gas-fields, oil-chemical production, irrigation of nearshore territories, and increase of industrial and domestic water supply (Zonn, 1999).

Yuri Dolotov and Pavel Kaplin

Bibliography

Berlyant, A.M., Mamaev, V.O., and Musina, O.P. (eds.), 1999. GIS. The Black Sea. Moscow, Izd. Moskovskogo universiteta (in Russian).

Bruun, P., 1962. Sea-level rise as a cause of shore erosion. *Journal of Waterways and Harbors Division*, **88**(WWI): 117–130.

Dolotov, Yu.S., 1961. Formation and classification of marine depositional terraces on the coasts subjected to the affect of tectonic emergence. *Transactions of Institute of Oceanology of the Soviet Academy Science*, **48**: 172–192 (in Russian).

Dolotov, Yu.S., 1996. On coastal relief change conditioned by sea level oscillations and its influence on economic activity (Caspian Sea as an example). In *Coastal Change*. Proceedings of the International Conference BORDOMER and IOC-UNESCO, **1**: 496–506.

Dolotov, Yu.S. and Kaplin, P.A., 1996. Present shore zone of the Caspian Sea—natural model of coastal evolution under conditions of the global sea-level rise. *Oceanology*, **36**(6): 926–930 (in Russian).

Ignatov, E.I., Kaplin, P.A., Luk'yanova, S.A., and Solov'eva, G.D., 1992. The effect of the recent transgression of the Caspian Sea on its coastal dynamics. *Geomorphology*, **1**: 12–21 (in Russian).

Ignatov, E.I., Kaplin, P.A., Luk'yanova, S.A., and Solov'eva, G.D., 1993. Evolution of the Caspian Sea coasts under conditions of sea-level rise: model for coastal change under increasing "greenhouse effect". *Journal Coastal Research*, **9**(1): 104–111.

Kaplin, P.A., 1989. Shoreline evolution during the 20th century. In Ayala-Costanares, W., Wooster, A. and Yanes-Arencibia, A. (eds.), *Oceanography-1988*. Mexico D.F.: UNAM Press, pp. 59–64.

Kaplin, P.A., 1990. Practical problems for coastal submergences in the light of secular trends. In Paepe, P. *et al.* (eds.), *Greenhouse Effect, Sea Level and Drought*. Ddorecht: Kluwer, pp. 385–393.

Kaplin, P.A., 1995. The Caspian: its past, present and future. In Mandych, A.F. (ed.), *Enclosed Seas and Large Lakes of Eastern Europe and Middle Asia*. Amsterdam: SPB Acad. Publ., pp. 71–117.

Kaplin, P.A., 1997. Implications of climatic change and water-level rise in the Caspian Sea Region. Regional review—1995. UN Environment Program, Geneva, UNEP.

Kaplin, P.A., and Ignatov, E.I. (eds.), 1997. Geoecology of the Caspian region. Issue 1: Geoecological changes under conditions of the Caspian Sea level oscillations. Moscow, Geogr. Faculty of the Moscow State University (in Russian).

Kaplin, P.A., Leont'ev, O.K., Luk'yanova, S.A. and Nikiforov, L.G., 1991. The shores. Moscow, Mysl' (in Russian).

Kaplin, P.A., Porotov, A.V., Selivanov, A.O. and Esin, N.N., 1993. The North Black Sea and the Sea of Azov coasts under a possible greenhouse-induced global sea-level rise. In Kos'yan, R. (ed.), *Coastlines of the Black Sea*. New York: American Society of Civil Engineers Publication, pp. 316–354.

Kiknadze, A.G., 1977. The dynamic system and sediment budget of the Georgian coast. In *Man and Natural Environment*. Sukhumi: Alashara Publication, pp. 5–11 (in Russian).

Kos'yan, R. (ed.), 1993. *Coastlines of the Black Sea*. New York: American Society of Civil Engineers Publications.

Kuksa, V.I., 1994. *Southern Seas* (Aral Sea, Caspian Sea, Sea of Azov and Black Sea) *Under Conditions of Anthropogenic Stress*. St. Petersburg: Gidrometeoizdat.

Kurbatova, I.E., 1994. Study of anthropogenic disturbance of natural environment in the north-eastern coastal areas of the Caspian Sea using space images. *Water Resources*, **21**(4): 487–491 (in Russian).

Leont'ev, O.K., and Khalilov, A.I., 1965. *Natural Conditions of Coastal Formation in the Caspian Sea*. Baku: Academy of Science Azerbaijan SSR (in Russian).

Leont'ev, O.K., Maev, E.G., and Rychagov, G.I., 1977. The geomorphology of the shores and floor of the Caspian Sea. Moscow, Izd. Moskovskogo Universiteta (in Russian).
Molchanova, L.V., 1989. Recreational potential of the Caspian Sea coasts. In Korzun, V.I. (ed.) *Recreational Resources of the Caspian Sea: The Problems of Use and Defense*. Moscow: Nauka, pp. 4–13 (in Russian).
Nevesskii, E.N., 1967. *Depositional Processes in the Nearshore Zone*. Moscow: Nauka, (in Russian).
Nikonov, A.A., Ekman, S.V., and Mishin, A.V., 1997. Modern vertical movements of the Earth's crust in the coasts of the Black Sea and the Sea of Azov (according to the data on sea level change). *Doklady RAS*, **357**(6): 818–822 (in Russian).
Sapozhnikov, V.V., 1992. Ecological state of the Black Sea nearshore zone. In Sapozhnikov, V.V. (ed.) *Ecology of the Black Sea Nearshore Zone*. Moscow: VNIRO, pp. 5–17 (in Russian).
Shuiskii, Yu.D., 1974. Erosional processes and their rates on the Ukrainian shores of the Black Sea and the Sea of Azov. Izvestiya AN SSSR, ser. *Geografiya*, **6**: 10–16 (in Russian).
Shuiskii, Yu.D., and Schwartz, M.L., 1988. Human impact and rates of shore retreat along the Black sea coast. *Journal of Coastal Research*, **4**(3): 405–416.
Simeonova, G.O., 1976. Onshore processes of the Black Sea within Bulgaria. In XXIII International Geographical Congress, Dynamics of sea shores, Tbilisi, pp. 32–35 (in Russian).
Veliev, Kh.A., Mamedov, S.B., and Aliev, Yu.G., 1987. Perspectives of recreational assimilation of the Azerbaidzhan coast of the Caspian Sea. In *Geographical and Economic Problems of investigations and Assimilation of the Southern Seas of the USSR*. Leningrad: USSR Geographical Society (in Russian).
Voropaev, G.V., Krasnozhen, G.F., and Lachojani, H.K., 1998. River drainage and steadiness of the Iranian coast of the Caspian sea. *Water Resources*, **25**(6): 747–758 (in Russian).
Zenkovich, V.P., 1958. *Dynamics and Morphology of the USSR Shores of the Black Sea*. Moscow: USSR Academic Science Publication, vol. 1 (in Russian).
Zenkovich, V.P., 1959. *Dynamics and Morphology of the USSR Shores of the Black Sea*. Moscow: USSR Academic Science Publication, vol. 2 (in Russian).
Zenkovich, V.P., 1982. The Black Sea and the Sea of Azov. In Aseev, A.A., and Korzhnev, S.S. (eds.). *The Far East Shores and Other Shores of the USSR*. Moscow: Nauka, pp. 199–205 (in Russian).
Zenkovich, V.P., and Schwartz, M.L. 1987. Protecting the Black Sea-Georgian SSR gravel coast. *Journal of Coastal Research*, **3**(2): 201–209.
Zonn, I.S., 1999. *The Caspian: Myths and Reality*. Moscow: TOO "Korkis" (in Russian).

Cross-references

Barrier
Beach Processes
Deltas
Marine Debris—Onshore, Offshore, Seafloor Litter
Oil Spills
Sea-Level Rise, Effect
Shore Protection Structures
Water Quality

BLUFFS—See CLIFFED COASTS

BOGS

Terminology

The term bog is used to describe certain forms of wet terrestrial vegetation. Unfortunately, in common with the words employed for many other categories of wetland, there are variations and inconsistencies in usage, regionally (particularly within Europe) as well as globally. Bog has been broadly defined so as to encompass all types of peat forming vegetation (see entry on *Peat*) or narrowly defined to denote only plant communities which are dependent upon precipitation and dust for supplies of water and nutrients. The term peatland is more appropriate for the former. The latter "ombrotrophic" condition may be an absolute state, however, this is by no means always the case and it is perhaps not surprisingly, therefore, that such communities have floristic affinities with other wetland vegetation types. Consequently, recent authors (Wheeler and Proctor, 2000) have preferred to use the term bog to describe a type of vegetation, that is, one which is usually dominated by either (or a combination of) sphagna (mosses), ericoids (dwarf shrubs), or Cyperaceae (sedges). Bogs are characteristically base-poor (with a pH < 5.0) and generally, though not exclusively, occur over a substratum of peat. Rather than peatland, an additional term, mire is preferred here to describe all forms of wet terrestrial vegetation. Bogs frequently grade into base-rich mires, which in Europe are referred to as fens. Fens may be herbaceous or wooded (fen carr in Europe), for which in the United States the terms marsh and swamp are, respectively, commonly used.

Classification and distribution

Mire vegetation is strongly influenced by hydrology and topography and hydro-topographical relationships have been widely used in mire classification schemes. Fundamental is the division made between the ombrotrophic (atmospheric) and minerotrophic (via surface runoff or percolating groundwater) supply of water and nutrients. Waterlogged peat can accumulate above the groundwater level (to a maximum depth of about 10 m) leading to the formation of raised mires. Such mires are entirely ombrotrophic and therefore base-poor, supporting only bog vegetation. Other mires are to some extent influenced by a mixture of ombrotrophic and minerotrophic sources, local scale variations in which are expressed through hydrochemical and floristic gradients. Such mires are by no means always base-rich and bog vegetation will occur either where the soils of the catchment are poor in soluble minerals (minerotrophic bogs) or where precipitation/evaporation ratios are particularly high. The latter situation is not uncommon along the Atlantic seaboards of North America and northwestern Europe. Here extensive, primarily ombrogenous, "blanket" bogs can cover the landscape spreading over relatively steep slopes and descending down to sea level.

On a regional and continental scale, the geographic distribution of mire types reflects the climatic regime. The limit of ombrogenous bog development in southeastern Labrador, for example, coincides with the 1,100 mm precipitation isopleth (Foster and Glaser, 1986). Globally mires are more widely distributed at high and low latitudes and at high altitude. Nevertheless, extensive lowland mires, such as the Everglades, occur even in the subtropics and tropics where in coastal districts they merge into brackish marshes and mangrove swamps. Regional and continental level reviews of bog and related vegetation types can be found in Gore (1983).

A variety of mire types are found in coastal situations (see entry on *Wetlands*). Some North American and northwestern European sites show complete spatial gradations from salt marsh, through fen and minerotrophic bog to raised bog. Similar temporal gradations can be reconstructed from peats deposited within coastal sedimentary sequences. Species composition in coastal mires is likely to be additionally influenced by the input of sodium and chloride via salt-spray, brackish groundwater, or as a result of flooding episodes.

Origins and development

Mire communities can develop from waterbodies (a process referred to as hydroseral succession) or over formerly dry surfaces. Both circumstances apply in coastal situations where vegetation changes in stratigraphic sequences are frequently used to infer waterlevel movements resulting from fluctuations in relative sea level. However, it should be noted that any environmental process influencing waterlevel elevation or nutrient status is capable of producing vegetation change. Such change can also result from internal (termed autogenic) processes, most obviously through the accumulation of sediment. Therefore, while the growth of mire over a dry surface indicates rising waterlevels, mires can develop over marine/brackish sediments as a result of falling, stable, or even slowly rising waterlevels (if exceeded by the rate of sediment accumulation).

Vegetation sequences and the processes influencing the development of coastal mires are reviewed in Waller *et al.* (1999) with particular reference to stratigraphic information collected from the Romney Marsh depositional complex in southeastern England. *Alnus glutinosa* (alder) dominated fen carr (swamp) vegetation developed, above salt marsh clays, and prevailed at sites close to the upland edge and in neighboring river valleys, from ca. 6000 to 2400 yr BP. This community appears to have been sustained both by inflowing base-rich water and rising relative sea-level (preventing vertical isolation from groundwater). At sites immediately behind a coastal barrier peat formation began later. Here salt marsh clays are followed by a sequence of herbaceous fen, minerogenic bog, and ombrotrophic bog. Bog development appears to have required both vertical and spatial isolation from base-rich water sources. The former was induced by a decline in the rate of relative sea-level rise and by climate change. An additional factor appears to be the mobility of the peat matrix. Peat formed from herbaceous vegetation is more mobile than woody peat and acidiphilous vegetation developing on such surfaces is therefore less likely to be flooded with base-rich water. Spatial isolation seems to have been achieved by the extensive landward accumulation of peat and the presence of the barrier. The importance of the latter is demonstrated by the widespread occurrence of bog in back-barrier environments in the Low Countries during the Holocene epoch. Having achieved independence from groundwater the ombrotrophic vegetation of Romney Marsh was able to continue growing for a further 1000 C14 years after other mire types within the depositional complex were subject to renewed marine inundation.

Paleoenvironmental reconstructions using bog sediments

Sediments derived from bog, in common with other organic deposits, comprise an important paleoecological archive. Preserved plant material (seeds, pollen, and vegetative remains) and faunal remains such as Rhizopods (testate amoebae) and Coleoptera (beetles) can be used to elucidate *in situ* environmental changes and in some cases changes occurring in adjacent habitats. Analysis of the pollen preserved in organic sediments has proved a particularly powerful tool for understanding long-term vegetation trends. Mires may also contain archaeological artifacts (see entry on *Archaeology*) and in northwestern Europe a number of exceptionally well-preserved human remains, referred to as Bog bodies, have been recovered. Ombrotrophic bogs, being dependent upon precipitation for their growth, have additionally been an important source of information on climate change during the Holocene epoch. In particular, stratigraphic changes from darker more decomposed peat to lighter fresh *Sphagnum* peat have been taken to indicate periods of faster peat growth and therefore wetter climatic conditions. Changes in bog stratigraphy at many locations across northwestern Europe (including coastal locations) indicate such a climate shift occurred around 2650 yr BP (van Geel *et al.*, 1996). Unfortunately, changes in bog stratigraphy are not always synchronous between, or even within, bogs. Growth rates vary not only geographically but also in response to local hydrological features.

Organic sediments derived from coastal situations are commonly employed in reconstructions of former sea level as they can be radiocarbon dated and certain plant communities can be related to a specific ("reference") waterlevel range (see entries on *Peat* and *Sea-Level Indicators, Biological in Depositional Sequences*). The latter is clearly not the case with sediments derived from ombrotrophic bog. Given the difficulties distinguishing between ombrotrophic and minerotrophic bogs on the basis of floristic composition, and the gradations possible between these conditions, organic sediments derived from acidophilious vegetation should be avoided when collecting material for this purpose.

Human exploitation

Mires are exploited as a source of peat for fuel and horticulture and, following drainage, for cultivation. Ombrotrophic sediments are best suited for the former purpose, minerogenic for the latter. The extensive exploitation of peat for fuel can be traced back to the medieval period in northwestern Europe. For example, large quantities were removed from a series of valleys on the edge of the coast of Norfolk in eastern England. Subsequent flooding created a series of shallow lakes referred to as Broads. Peat continues to be a major energy resource in a number of countries (Russia, Ireland). Reclamation of the lowland mire complexes in Europe occurred from the 17th century onwards through the construction of effective drainage channels subsequently aided by pumping (see entry on *Reclamation*). Such activities result in the lowering of the land surface both as a result of sediment compaction (the loss of interstitial fluid) and erosion (as the surface organic sediments decompose). At Holme Fen, in Eastern England, where *Sphagnum* is an important peat constituent, the ground surface fell by 3.87 m between 1848 and 1978 (Hutchinson, 1980). The large-scale exploitation of bogs has increasingly led to calls for their conservation. Along with other forms of wetland they are included within the RAMSAR Convention (see entry on *Wetlands*).

Martyn Waller

Bibliography

Foster, D.R., and Glaser, P.H., 1986. The raised bogs of south-eastern Labrador, Canada: classification, distribution, vegetation and recent dynamics. *Journal of Ecology*, **74**: 47–72.

Gore, A.J.P., 1983. *Ecosystems of the World, 4B: Mires, Swamp, Bog, Fen and Moor. Regional Studies*. Amsterdam: Elsevier.

Hutchinson, J.N., 1980. The record of peat wastage in the East Anglian Fenlands at Holme Post, 1848–1978 AD. *Journal of Ecology*, **68**: 229–249.

van Geel, B., Buurman, J., and Waterbolk, H.T., 1996. Archaeological and palaeoecological indications of an adrupt climate change in The Netherlands, and evidence for climatological teleconnections around 2650 BP. *Journal of Quaternary Science*, **11**: 451–460.

Waller, M.P., Long, A.J., Long, D., and Innes, J.B., 1999. Patterns and processes in the development of coastal mire vegetation: Multi-site investigations from Walland Marsh, Southeast England. *Quaternary Science Reviews*, **18**: 1419–1444.

Wheeler, B.D., and Proctor, M.C.F., 2000. Ecological gradients, subdivisions and terminology of north-west European mires. *Journal of Ecology*, **88**: 187–203.

Cross-references

Archaeology
Coastal Climate
Hydrology of Coastal Zone
Peat
Reclamation
Salt Marsh
Sea-Level Indicators, Biological in Depositional Sequences
Wetlands
Wetlands Restoration

BOULDER BARRICADES

Definition, distribution and historical development

Boulder barricades are elongate rows of boulders that flank the coastline, separated from the shore by an intertidal flat (Figure B62). They are the result of ice transport and therefore are found only in Arctic and sub-Arctic regions. They are formed by the grounding of boulder-laden ice rafts in nearshore zones during spring ice break-up.

In North America, boulder barricades have been reported in Labrador (Daly, 1902; Rosen, 1979, 1980); Hudson Strait, east Foxe Basin, Baffin Island (Bird, 1964); and the St. Lawrence River (Brochu, 1961; Dionne, 1972). In other areas they have been reported in the Baltic Sea and Fennoscandia (Lyell, 1854; Tanner, 1939). Løken (1962) utilized uplifted boulder barricades in northern Labrador as an accurate sea-level indicator to delineate the Holocene regression. While Tanner (1939) observed a decrease in barricade development corresponding to a reduction in tide range from 1.3 to 0 m in Labrador, the features do occur in other nontidal (i.e., Baltic Sea) areas.

Lyell (1854) first recognized boulder barricades as an ice-deposited landform. Daly (1902) introduced the term, but believed that they were an accumulation of boulders at the seaward limit of wave backwash, with ice playing a secondary role. Tanner (1939) concluded that the features were the result of boulder-laden ice cakes piling up against a fixed shore icefoot. Conversely, Brochu (1961) hypothesized that intertidal ice-cakes were moved seaward during ice breakup, pushing boulders to the low water line.

Boulder barricade formation

Monitoring of coastal ice in central Labrador during both winter and spring breakup are the basis for a model for the entrainment of boulders into ice and the transportation during spring breakup. In tidal regions, such as Labrador, intertidal ice freezes downward with increased freezing at each high tide. Boulders are frozen in the ice and become lifted from the intertidal bottom. Observations on a broad intertidal flat indicate that more boulders are lifted from the upper intertidal zone, so apparently the less-frequent lifting of the ice during spring tides was more effective at encasing boulders than the diurnal lifting from the lower intertidal zone. High melting rates occur from the ice surface in the late winter, so the continued freeze-down and surface-melt result in the transportation of boulders up through the ice.

Figure B62 Boulder barricade in Makkovik Bay, Labrador.

Figure B63 A boulder adrift on an ice cake, Makkovik Bay, Labrador.

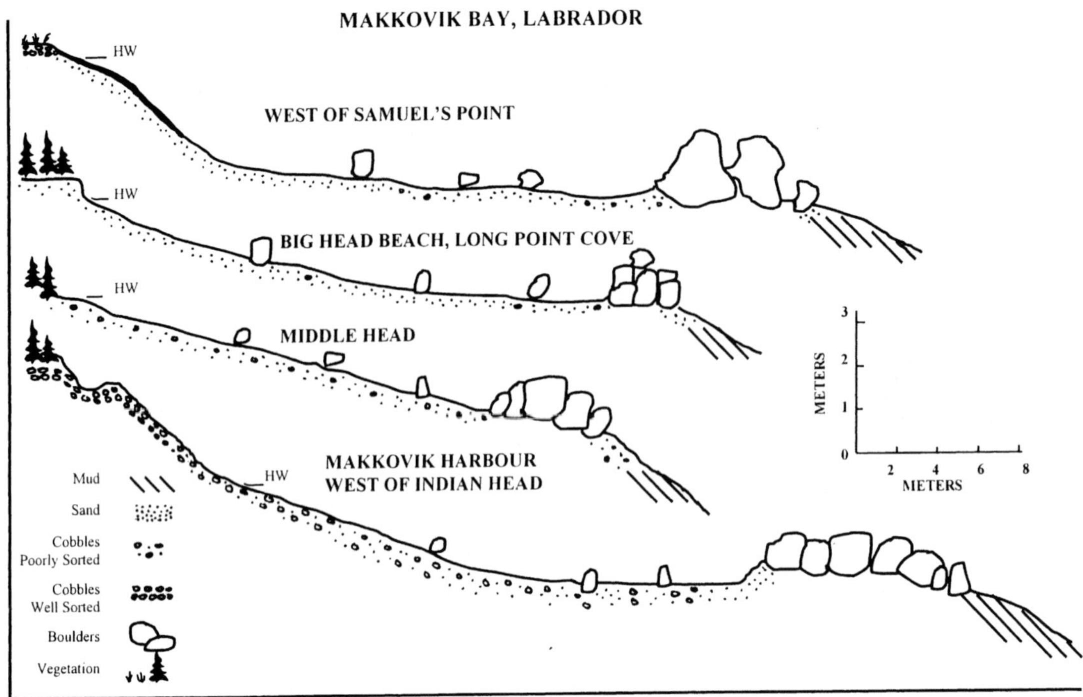

Figure B64 Nearshore profiles at selected sites in Makkovik Bay, Labrador. Random boulders are common in the intertidal zone, while most accumulate as a boulder barricade landward of a steep drop off to deeper water (from Rosen, 1982, with permission of Kluwer Academic Publishers).

In the Baltic Sea off Tallinn, Estonia, which is nearly non-tidal, the infrequent lifting of ice for freeze-down and encasement, and floating for spring transportation may be due to meteorological tides that are a major cause of sea-level fluctuations in the region (Maurice Schwartz, personal communication).

In spring when the snow cover has melted, boulders have been observed sitting on intertidal ice pans (Figure B63). The intertidal zone breaks up before offshore areas because of numerous tidal cracks and the decreased albedo of the mud-laden nearshore ice. Shore leads up to 1 km wide serve as thoroughfares for these wind-transported ice rafts. The boulders may be randomly deposited in the nearshore as *boulder flats*, as commonly occurs on the deltaic flats at the heads of embayments. However, in central Labrador many of the intertidal zones consist of uplifted marine clays with the top surface planed-off by contemporary wave and ice processes. This results in a slope-break near the low water line (Figure B64). Since the ice thickness is comparable to the tide range, there is a high probability for ice-rafts to ground at this position. Accumulation of boulders over successive seasons results in an intermittent barricade, which further serves to trap ice rafts during breakup. Landward of the slope break/boulder barricade position, random boulders, or boulder flats are also common.

At Tallinn, Estonia, the boulder barricades form in a similar setting. In this area, the nearshore is a rock-cut bench and the barricades accumulate at the seaward limit of this bench. (M. Schwartz, personal communication). Conversely, in the St. Lawrence Estuary, there is a range of nearshore boulder forms, including boulder flats, mounds, ridges, and pavements (Dionne, 1972), which corresponds with no evidence of a nearshore slope break.

Summary

Boulder barricades are the result of grounding of boulder-laden ice rafts in the nearshore zone during spring ice breakup. Wind and tides are the major transport mechanisms. The requisite conditions for the formation of boulder barricades are: a rocky coastal setting, sufficient winter ice, and water-level fluctuations to entrain boulders in ice rafts; a distinct slope break in the nearshore zone. Without the third condition, boulders will be desposited as boulder flats.

Boulder barricades are distinctly different from *boulder ramparts* and *ice-push ridges*, which are common on coastal and lake shorelines in that they are located seaward of the strandline, rather along the water line, due to the conditions discussed above.

Peter S. Rosen

Bibliography

Bird, J.B., 1964. *The Physiography of Arctic Canada*. Baltimore, MD: Johns Hopkins Press.

Brochu, M., 1961. Déplacements de Blocs par la Glace le long du Saint Laurent. Étude Géographique No. 30, Ministere des Mines et des Relevés techniques, Ottawa, Canada.

Daly, R., 1902. Geology of the northeast coast of Labrador: Bull. Harvard Mus. Comp. Zool., no. 38, pp. 270–302.

Dionne, J.C., 1972. Charactéristiques des blocs erratiques des rives de l'estuaire du Saint Laurent. *Rev. Géogr.*, Montreal, vol. XXVI, 2, pp. 125–152.

Lϕken, O., 1962. The late-glacial and post glacial emergence and the deglaciation of northern most Labrador. *Geography Bulletin*, **17**, p. 23–56.

Lyell, C., 1854. *Principles of Geology*, New York: D. Appleton and Co.

Rosen, P.S., 1979. Boulder Barricades in Central Labrador. *Journal of Sedimentary Petrology*, **49**: 1113–1124.

Rosen, P.S., and McCann, S.B., 1980. Coastal environments of the Makkovik region Labrador. *Geologic Survey of Canada*, **80–10**: 267–280.

Rosen, P.S., 1982. Boulder Barricades. In Schwartz, M.L. (ed.), Encyclopedia of Beaches and Coastal Environments, Stroudsburg: Hutchinson Ross, pp. 176–178.

Tanner, V., 1939. Om de blockrika strandgördlarna (Boulder barricades) vid subarktiska oceankustar, förekomstsätt og uplomst. *Terra*, **51**: 157–165.

Cross-references

Arctic, Coastal Geomorphology
Ice-Bordered Coasts
Paraglacial Coasts

BOULDER BEACHES

Strictly speaking a boulder beach is one where the mean clast size meets the formal definition of "boulder" in the terms of the Wentworth grade scale, that is with a mean particle dia-meter of >256 mm (−8 *phi*) (Wentworth, 1922). However, the term is also sometimes used in a general sense to describe beaches where the sediment is a mixture of boulders and large cobbles.

Boulder beaches are found in high wave-energy environments where clasts of these large dimensions are released directly by erosion of bedrock, or where material is delivered to the shore zone by slope movements such as rockfall. In both cases sediment size is a function of joint spacing. Preformed boulders may also be supplied by erosion of Quaternary deposits such as glacial till, and by infrequent high-magnitude river floods.

While a considerable body of literature exists on gravel beaches, (usually concerned with pebbles and small cobbles), beaches of large cobbles and boulders have been neglected. This is partly because morphological response times are too long for consideration in the normal time-frame of academic field programs, and also because very large clasts are difficult to characterize. Some publications purporting to be general overviews of coarse sediments completely disregard boulders. Others set arbitrary upper limits to their area of interest in that, even where boulders form part of the beach sediment, sampling, experimental work, and subsequent analyses are confined to, at maximum, the cobble sizes. Most workers are reasonably precise about the lower size limit of the sediment studied, but the upper limit often remains vague: this may be partly due to the prevalent use of the *phi* scale, which becomes increasingly generalized in the coarser sizes. Terminology is often imprecise, for example, the description "gravel" is commonplace, although it covers all clasts coarser than 2 mm diameter (−1 *phi*), and so does not distinguish among pebbles, cobbles, and boulders.

Sedimentology

The few studies that have been carried out on boulder beaches have tended to look at details rather than broad patterns of sedimentation and morphology, for example, Bartrum (1947), Shelley (1968), and Hills (1970). The neglect of large clast beaches has led to attempts to apply sedimentation models derived from studies of pebble and cobble beaches to the boulder beach environment. Doubts concerning the validity of such extrapolations were fully confirmed by a comprehensive study of boulder beaches by H.L. Oak along the coast of New South Wales in Australia (Oak, 1984). Oak proposed that boulder beaches demonstrate certain unique sedimentary characteristics that distinguish them as fundamentally different from pebble and cobble beaches. Hence, relationships established in the many studies of gravel beaches are often inapplicable.

The dominant characteristics of boulder beaches listed by Oak are:

1. A high wave-energy environment, competent to move large clasts.
2. Upbeach fining of sediment.
3. Abundant breakage of sediment.
4. Positively skewed size distributions.
5. Upbeach decrease in roundness.
6. No shape zonation.
7. No sphericity grading.
8. Low foreshore slopes, decreasing as particle size increases.

Of these characteristics numbers 2–4 and 6–8 contrast strongly with the known sedimentary characteristics of pebble and cobble beaches.

Sediment size

Mean clast size is the most significant parameter determining the sedimentary character and behavior of a boulder beach. The pattern of general upbeach coarsening typical of smaller grade gravels is a function of two interlinked processes: (1) storm swash can carry most of the range of available clast sizes upslope, and (2) backwash is competent to carry a major part of this range at least some distance seawards. Neither of these is true of a boulder beach, and so clast size is the primary determinant of movement. Clast size decreases upbeach because the dominant boulders are so large that they can only be moved by traction as bedload, and even then perhaps only in very infrequent intense storms. Only the sub-population derived from breakage of the larger clasts can be suspended. As wave uprush moves up the beach face, permeability reduces its volume, and gravity effects and turbulence reduce its velocity so quickly, that only increasingly finer material can be transported. Backwash effects are negligible, so the smaller clasts, including breakage products, remain where swash deposits them. Waves can move the larger boulders to the trim line at the base of the beach, but cannot move them any distance upslope.

As high-energy marine processes act on predominantly boulder-sized sediment winnowing of the fines produces a sediment assemblage dominated by a relatively small number of well-sorted large clasts, with a very minor subordinate population of smaller fragments, most of which have survived in the high-energy area only because of entrapment. The distinctive positively skewed size distribution of a boulder beach is attributed to the presence among the beach sediments of this tail of fines derived as breakage products of the dominant boulder population. Since large clasts resist continuous movement, spasmodic breakage during storms is the dominant size-reducing process (Bluck, 1969; Matthews, 1983). Abrasion is limited to the effects of passive sandblasting or the small movements of *in situ* abrasion, both of which are largely confined to a limited area at the base of the beach. In contrast, on pebble/cobble beaches breakage is minimal and most size reduction is achieved by attrition. The very fine products of this process will be removed in suspension, unlike pebble-sized breakage products, which may be retained on the beach.

Sediment shape

Shape sorting (and the related characteristic of sphericity sorting) is poorly developed on a boulder beach. This forms a contrast with the characteristic shape zoning of a pebble beach, which results from selective clast transportation in which backwash plays an important role. On a boulder beach the sedimentation process is fundamentally different because the morphology is purely swash-formed. Selective shape sorting becomes increasingly ineffective where clasts are large and where wave conditions are turbulent. On a high-energy boulder beach shape sorting is insignificant because shape is only a dominant influence when entrainment forces are at critical thresholds for selective transport. When forces are not marginal, mass rather than shape, is the dominant control on net up- and down-beach transport potential.

Shape-controlled sorting processes are inoperative because, (1) even when storm waves are competent the prevailing bedload transport mechanism is basically insensitive to clast shape, and size will remain the dominant factor, (2) the large clast sizes and the high porosity of the beach means that backwash does not have the energy potential to create shape sorting, and (3) the rugosity of the beach surface militates against the gravity-induced downslope movements of pebble grades, which are instead trapped in the voids between boulders. Only storm swash leaves its fingerprint and as a result the only primary structure imprinted on a boulder beach is swash-controlled upbeach fining. Size, therefore, exercises not only an initial control on upbeach sorting, but it is also the terminal control. As mean clast size decreases the size control typical of boulder sedimentation gradually gives way to the shape control associated with pebble and cobble sedimentation.

Sediment roundness

Beach boulders are typically smoothed and rounded. On high-energy coasts clast transport, given free movement conditions, is very rapid. This casts considerable doubt on whether the angular, rough-textured blocks produced by wave quarrying and rockfall could possibly acquire such a degree of rounding and smoothing on a short (both spatially and temporally) unimpeded journey from source to boulder beach. Active and passive abrasion and rounding processes continue to take place in the beach environment, but their effects are largely confined to the base of the beach. Clasts at higher elevations on the beach face may have been emplaced by one high-magnitude storm. Marine influences will rarely reach these elevations, and only slow weathering processes can contribute to further rounding. It is doubtful whether the sum of these beach-face processes can entirely account for the evolution of clast shapes from an initially angular form controlled by geology, to the rounded, marine form characteristic of boulder beaches. It seems more likely that the majority of boulder beach clasts have spent some time in the intensely turbulent and abrasive hydrodynamic environment represented by traps such as potholes, gullies, and channels. Eventually a storm liberates the clasts to continue their journey to the beach.

Roundness is most pronounced toward the base of a boulder beach because the large clasts in this area experience marine action over longer time periods, and also because angular breakage products will be transported upbeach by wave action. Rounding cannot be taken as evidence of clast movements within the beach deposit, as rounded profiles can be attributed to pre-emplacement history (see above), and can also be created and maintained by *in situ* processes such as breakage, mutual attrition, and "water load abrasion". The large well-sorted clasts found in the high-energy zone near the seaward margin of the beach characteristically demonstrate rounded, flattened ellipsoidal profiles. The remarkable stability of the lower part of a boulder beach in the face of high-energy wave action is probably due to a combination of large clast size, imbrication caused by strong unidirectional flows, and a variety of other fitting and interlocking processes acting on the beach fabric (Shelley, 1968; Hills, 1970; Bishop and Hughes, 1989).

With distance upbeach angularity tends to increase, especially in the finer grades with more compact and platy shapes. This is a general comment, as the size- rather than shape-controlled swash transport mechanism will occasionally carry clasts exhibiting the full range of roundness well up the beach face. The major reason for the upbeach increase in angularity is the influence of breakage during infrequent storms. The products of breakage will remain as a component of the beach sediments because at higher positions on the beach face wave action that might winnow small-grade material is infrequent, backwash is ineffective, and the only agency acting to increase roundness is the relatively slow process of spheroidal weathering. On the boulders near the landward margin of the larger beaches, surface soundness deteriorates as weathering processes produce a rougher texture, and lichen colonization is common. In a general sense it is probably valid to consider the development of weathering rinds and lichen cover as indices of decreasing marine influence and movement. However, recent rockfall blocks on any part of the beach face carry a weathering/lichen signature from a subaerial environment, not a beach environment.

Sediment orientation

On gravel beaches pebbles transported as bedload generally tend to be oriented with the long axis parallel to the shore, that is; transverse to the direction of swash movement. Such preferred orientation patterns are weakly developed on boulder beaches because a high velocity turbulent flow on a coarse bed leads to decreased regularity in orientation patterns. Clast collisions can change orientations, disturbing or perhaps even completely obscuring the pattern imprinted by the transport process. Clasts can also be oriented by the prevalent waves without undergoing net transport. Thus, while a bedload transport mechanism does generate preferred orientations, the generally weak development of this characteristic on boulder beaches is probably due to the interplay of high-energy wave action with the particularly rough surface of the beach deposit.

Beach profile

On all types of beaches relationships involving slope are regarded as particularly significant because slope is usually considered the primary index of morphological response to wave action. There have been so few published studies on boulder beaches that discussion on their profiles must be tentative.

The most notable feature of the typical boulder beach profile is a lack of variation over time. Even during relatively severe storms changes involve only individual clasts, with the beach slope itself remaining unchanged. Some boulder beaches exhibit obvious concave upwards profiles. In detail, many have one basically rectilinear main facet extending down to mean high-water mark or below, with overall concavity produced by narrow and rudimentary low angle facets in the intertidal area. The beach profile can be conceptualized as providing a "fingerprint" of the resultant of earlier swash/backwash interaction. On boulder beaches backwash is minimal so beach material is pushed shorewards to rest at an angle controlled by the balance of gravity and the dominant swash forces. For this reason, coarse beaches are more likely to be concave upwards than fine beaches.

On boulder beaches mean beach slope decreases as mean size increases. This characteristic appears to be in direct conflict with one of the basic tenets of coastal studies, that is, that coarser sediments produce steeper slopes, partly because the angle of rest is higher, but mainly because high percolation reduces backwash, which would tend to draw down material and lessen the slope. Shepard (1963) published a table in which the predicted average beach face slope for clasts in the 64–256 mm size range (−6 to −8 *phi*) was 24°. However, the slopes of boulder beaches are considerably below this predicted value, an illustration of the dangers inherent in extrapolating from work on finer grade material. Most studies of boulder beaches record slopes in the range 6°–14° with the mean lying around 12°.

Oak (1984) formulated an explanation for the finding that the slopes of boulder beaches are gentler than predicted. On all beaches of whatever mean sediment size, storms produce an equilibrium profile that is flatter than the pre-storm profile. The accepted principle that a beach must adjust to wave energy by flattening its profile holds true then, even for boulder-sized clasts. Beach angle does indeed increase with clast size, but only if waves can move all sediment. In practice only high-energy storm waves are competent to move large clasts, so the profile of a boulder beach is in fact a "lag" storm profile, adjusted to and formed by storm waves. On sand and pebble beaches lag times are short, and in the days after a storm infill will steepen the beach face. However, this does not happen on a boulder beach, as normal wave action cannot bring about a steep fairweather profile, so the beach typically exhibits a relatively low-angle storm profile. Therefore, the relatively gentle slope of a boulder beach, with its concave upwards profile, can be considered indicative of high swash velocities and minimum sediment storage, that is, a storm profile. The persistence of the profile simply reflects the fact that competent storms are infrequent.

John McKenna

Bibliography

Bartrum, J.A., 1947. The rate of rounding of beach boulders. *Journal of Geology*, **55**: 514–515.

Bishop, P., and Hughes, M., 1989. Imbricate and fitted fabrics in coastal boulder deposits on the Australian east coast. *Geology*, **17**: 544–547.

Bluck, B.J., 1967. Sedimentation of beach gravels, examples from south Wales. *Journal of Sedimentary Petrology*, **37**(1): 128–156.

Bluck B.J., 1969. Particle rounding in beach gravels. *Geological Magazine*, **106**(1): 1–14.

Hills, E.S., 1970. Fitting, fretting and imprisoned boulders. *Nature*, **226**: 345–347.

Matthews, E.R., 1983. Measurements of beach pebble attrition in Palliser Bay, southern North Island, New Zealand. *Sedimentology*, **30**(6): 787–799.

Oak, H.L., 1984. The boulder beach: a fundamentally distinct sedimentary assemblage. *Annals Association American Geographers*, **74**(1): 71–82.

Shelley, D., 1968. Fitting boulders: the result of an important shore process. *Nature*, **220**: 1020–1021.

Shepard, F.P., 1973. *Submarine Geology*. New York: Harper and Row.

Wentworth, C.K., 1922. A scale of grade and class terms for clastic sediments. *Journal of Geology*, **30**: 377–392.

Cross-references

Beach Sediment Characteristics
Boulder Barricades
Boulder Pavements
Cliffed Coasts
Cliffs, Lithology versus Erosion Rates
Gravel Barriers
Gravel Beaches
Rock Coast Processes
Shore Platforms

BOULDER PAVEMENTS

Striated boulder pavements can form either on intertidal surfaces in areas affected by floating ice (Martini, 1981; Hansom, 1983, 1986) or at the base of glaciers or on grounded ice sheets (A.G.I., 1974; Boulton, 1978; Visser and Hall, 1984). Pavements have also been described from fluvial environments (Mackay and Mackay, 1977). Their distinctive nature also allows them to be used in the sedimentary record to assist in the reconstruction of past ice-affected environments (Eyles, 1988). Pavements deposited subglacially are argued to be the result of accretion of boulders around an obstacle and to carry striations that are largely unidirectional. Although there are no detailed descriptions of such pavements forming in present glacial environments, they have also been described from the top surface of Quaternary deposits as well as buried within such deposits (Hansom, 1983; Eyles, 1988). Pavements formed on present cold-climate intertidal surfaces are thought to be the result of abrasion and bulldozing of boulder-lag surfaces by floating ice and small icebergs (Martini, 1981; Hansom, 1983, 1986, Gilbert *et al.*, 1984; Forbes and Taylor, 1994). The striations that the boulder surfaces carry are then controlled by the direction of movement of blocks of floating ice together with the rotational striations imparted when such blocks become stranded. Prerequisites for the development of intertidal boulder pavements are held by Hansom (1983) to be: (1) a boulder source; (2) frequent onshore movement of floating ice; and (3) a low-gradient intertidal zone. Given such conditions, the degree of development of the pavement seems to be controlled by the frequency of onshore ice movement, because the best formed pavements occur in areas subject to the highest frequencies of freely moving ice rather than areas that remain frozen for substantial parts of the year.

Figure B65 A well-developed boulder pavement in the South Shetland Islands, Antarctica. The polygonal depressions are caused by tidally grounded ice blocks which smooth and striate the boulders.

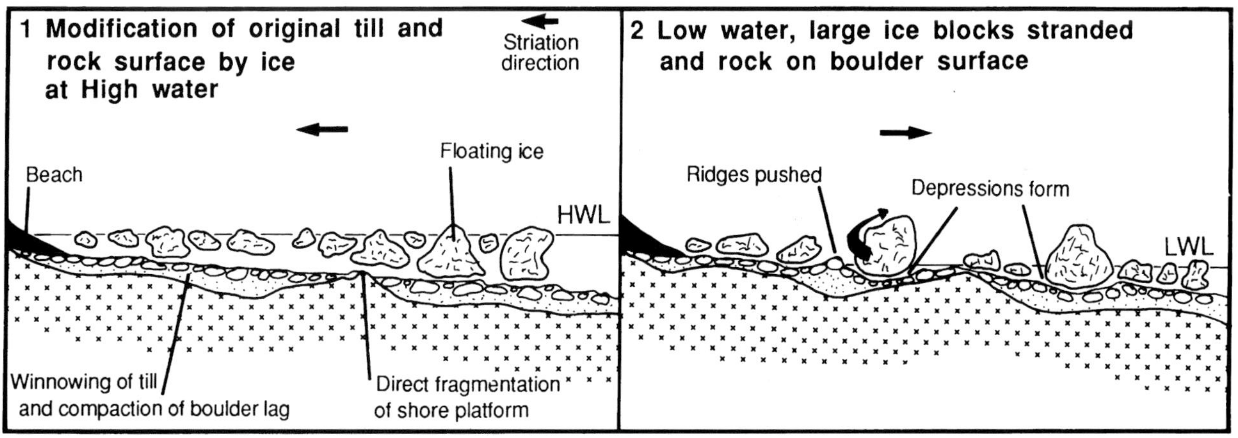

Figure B66 A model showing the development of a boulder pavement at the extremes of the tidal cycle. Progressive stranding of ice blocks causes compaction, polishing, and striation of the boulders as well as forming depressions in the pavement surface.

Marine boulder pavements are composed of smoothed boulders, often of up to 1 m in diameter, that are tightly packed together in the intertidal zone, the pavement surface appearing as a smooth, highly polished, and striated mosaic. The pavement surface is often interrupted by outcropping bedrock together with shore-normal furrows and polygonal depressions that can be up to 5 m across (Figure B65). In the South Shetland Islands (see *Atlantic Ocean Islands*) they have been described as comprising a single layer of boulders underlain by a layer of clay containing locally derived lithologies, whereas in South Georgia, pavements are underlain by glacial till into which the boulders have been packed (Hansom, 1983). The main processes involved in pavement development are summarized in Figure B66. Floating ice blocks coming ashore onto a low gradient boulder-strewn shore bulldoze and pack loose boulders in the zone of grounding, initially in the upper intertidal but increasingly at the seaward edge. Some boulders may come from direct fragmentation of rock outcrops of any underlying shore platform that may exist and some may come from ice-rafted exotics. Polishing of the boulder surface is achieved by rock-shod floating ice abrading and striating the surface of the boulders (Hansom, 1983). The orientations of the striations also inform the development processes of the surface polygonal depressions since the spread of striations on boulder ridges parallel to the shore can only be achieved by partially stranded ice blocks rotating on the pavement surface (Figure B66).

In the Antarctic, boulder pavements are found in varying degrees of development across 10° of latitude from South Georgia to the Antarctic Peninsula and in Victoria Land, pavements of tightly packed and smoothed boulders locally veneer the shallow subtidal shore platforms of the Adare and Hallett Peninsulas (Hansom and Kirk, 1989). The distribution and development of the Antarctic pavements show clear relationships between the frequency of floating ice grounding and wave processes. Where the frequency of ice grounding is high then the pavements are well developed. In the South Shetlands, the probability of floating ice and the percentage of ice concentration are both high. This limits the wave processes that destroy the pavement surface while ensuring frequent ice smoothing and packing. The result is a morphogenetic environment with a mix of ice/wave processes that is optimal for pavement development. Moving south away from this optimal ice/wave zone the incidence of grounding ice is reduced and so in the Antarctic Peninsula, wave processes are negligible, the incidence of fast ice is high and the frequency of ice grounding is low. Pavements here are poorly developed and embryonic. On the open coasts of South Georgia, well to the north of the optimal ice/wave zone, wave processes dominate, the frequency of grounding ice is low and so pavements are again poorly developed. However, within the sheltered inner fjords of South Georgia, wave processes are restricted, the frequency of floating ice is higher on account of glaciers calving into tidewater and, as a result, the boulder pavements are better developed (Hansom and Kirk, 1989). In the fjords of Vestfirdir in Iceland, similar forms also exist but are poorly developed as a result of the juxtaposition of a very limited ice-climate and a very low energy wave environment (Hansom, 1986).

Martini (1981) suggests that the incidence of boulder pavements can be taken as a reliable indicator of intertidal ice action. Thus, the occurrence of emerged pavements at 5, 9, 12.5, and 17 m above sea level in the South Shetland Islands is convincing evidence of unchanged morphogenetic conditions in the area at least since the uppermost of the pavements was formed some 9000 years BP (Hansom, 1983). Eyles (1988) uses boulder pavements in a similar way to reconstruct fluctuations in ice environments in the Gulf of Alaska during the early Pleistocene.

J.D. Hansom

Bibliography

A.G.I., 1974. *American Geophysical Union Glossary of Geology*. Washington, D.C.: American Geophysical Union.

Boulton, G.S., 1978. Boulder shapes and grain size distribution of debris as indicators of transport path through a glacier and till genesis. *Sedimentology*, **25**: 773–779.

Eyles, C.H., 1988, A model for striated boulder pavement formation on glaciated shallow-marine shelves: an example from the Yakataga Formation, Alaska. *Journal of Sedimentary Petrology*, **58**(1): 62–71.

Forbes, D.L., and Taylor, R.B., 1994. Ice in the shore zone and the geomorphology of cold coasts. *Progress in Physical Geography*. **18**(1): 59–89.

Gilbert, R., Aitken, A., and McLaughlan, B., 1984. A survey of the coastal environments in the vicinity of Nain, Labrador. *Maritime Sediments*, **20**: 143–155.

Hansom, J.D., 1983. Ice-formed intertidal boulder pavements in the sub-Antarctic. *Journal Sedimentary Petrology*, **53**(1): 1035–1045.

Hansom, J.D., 1986. Intertidal forms produced by floating ice in Vestfirdir, Iceland. *Marine Geology*, **71**: 289–298.

Hansom, J.D., and Kirk, R.M., 1989. Ice in the intertidal zone: examples form Antarctica. *Essener Geographische Arbeiten*, **18**: 211–236.

Mackay, J.R., and Mackay, D.R., 1977. The stability of ice-push features, Mackenzie River, Canada. *Canadian Journal Earth Science* **14**: 2213–2225.

Martini, I.P., 1981. Ice effect on erosion and sedimentation on the Ontario shores of James Bay, Canada. *Zeitschrift fur Geomorphologie*. N. F. **25**(1): 1–7.

Visser, J.N.J., and Hall, K.J., 1984. Boulder beds in the glaciogenic Permo-Carboniferous Dwyka Formation in South Africa. *Sedimentology*, **32**: 281–294.

Cross-references

Antarctica, Coastal Ecology and Geomorphology
Arctic, Coastal Geomorphology
Boulder Barricades
Boulder Beaches
Glaciated Coasts
Ice-Bordered Coasts
Shore Platforms

BYPASSING AT LITTORAL DRIFT BARRIERS

Definition

A littoral drift barrier is an obstacle against the littoral drift or migration of material along the shore. Such barriers may be natural, for example, major headlands on the shore, or man-made such as jetties, breakwaters, or dredged channels, which established a hindrance for the normal drift of material along the shore.

Natural barriers may be responsible for major changes in the natural uninterrupted shore. The California saw-toothed headland shore is a large example of that. Bypassing is transportation of materials across the barrier, breaking the barrier-effect.

Bypassing by nature

Bypassing is the way that material, after a short interruption caused by an inlet, channel, jetty, or other kind of littoral barrier, is given back to the normal littoral drift zone a distance downdrift from the littoral barriers. If nature did not bypass sand across inlets, passes, and channels on seashores, many *marine forelands* including barriers, spits, and entire peninsulas would not exist. A typical example of this is Florida, which was built of sand washed down by rivers and streams from the Appalachian highland, and carried southward, for final deposition in the huge barrier and ridge systems.

Bar bypassing—limited tidal action

Figure B67 shows a barrier with an inlet. Littoral drift material passes along the barrier. At the downdrift end, it continues on its way across the inlet on a submerged bar, the extent and depth of which depends on the amount and character of the material which bypasses and the intensity of wave and current action. By increasing amounts of littoral material, the bar area increases and depth decreases.

In most cases, migration of tidal channels takes place in the direction of the littoral drift. Sand is transported over the bar under the influence

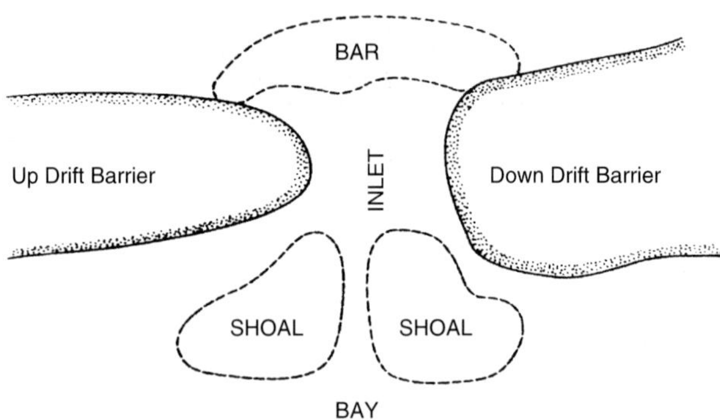

Figure B67 Coastal inlet with predominant bar bypassing (from Bruun, 1990). (Reproduced by permission. From Port Engineering V2 4E copyright (c) 1990, Gulf Publishing Company, Houston, Texas, 800-231-6275. All rights reserved.)

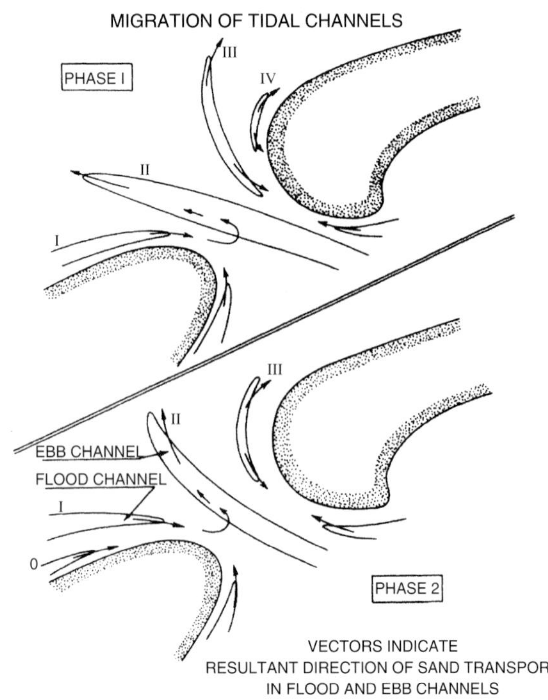

Figure B68 Migration of tidal channels (from Bruun, 1961). (Reproduced by permission of the publisher, ASCE.)

of waves and deposited on the updrift bank of the channels, thus forcing the shifting.

In the vicinity of tidal inlets, the generally strong tidal currents in the inlet change the littoral drift pattern entirely. Along the uninterrupted coastline, wave action is generally the predominant cause for the transportation of material. In the vicinity of tidal inlets, however, transport of material takes place under the combined effect of waves and tidal currents.

In tidal rivers, estuaries and inlets, tidal channels can usually be identified as either *flood* or *ebb* channels. Flood channels carry predominantly flood flow, causing a resultant sand transport in a bayward direction; they usually have a shoal at the end. Ebb channels carry predominantly ebb flow and have resultant material transport seaward and a bar or shoal at the end (Figure B67).

Principles involved in bypassing by tidal flow action—unimproved inlets

In general, sand transfer by tidal flow takes place in two different ways, namely by migration of channels and bars and by transport of sand by tidal flow in the channel. Tidal channels in inlets, particularly those running between the gorge and the ocean, are subject to migration. This means that they change location continuously, moving from one side of the inlet to the other. In Figure B68, this principle is demonstrated by Phases 1 and 2 of a tidal channel system. Channels in Phase1 are numbered, I, II, III, and IV. In Phase 2 the locations of these channels have changed compared with Phase 1, and a new channel, 0, has developed. In this example, the channels move from left to right, and bars or shoals between the channels move in the same direction with the result that a bar occasionally joins the downdrift coast.

One may distinguish between inlets that are mainly bar-bypassers (Figure B67) and inlets that are rather tidal flow bypassers by considering the ratio between tidal prism during spring tide (Ωm^3) and the total amount of material carried to the inlet entrance by the littoral drift (M_{tot} in cubic meters per year). A great many cases were analyzed and showed that inlets with $\Omega/M_{tot} < 50$ were mainly bar-bypassers, while inlets with $\Omega/M_{tot} > 150$ were mainly tidal flow bypassers. Inlets with $50 < M_{tot} < 150$ combined the two modes. Inlets or harbors without or with only little tidal prism have only one bypassing style-man-made bypassing or dredging.

Man-made littoral drift barriers

Human intervention of coastal processes started when they erected shore-perpendicular or parallel breakwaters for protecting ports against waves and sediments and groins for coastal protection on open littoral drift shores. This type of construction began in the 19th century in the Mediterranean and on the shores of the British Isles (Bruun, 1990). The problem of man-induced erosion was magnified when the Dutch invented dredging in an effort to provide greater channel depths for navigation. It was by hard and very expensive experience that they learned that when they put something out in the sea, "something is going to happen." Commonly, shoaling occurs on one side and erosion on the other side of an obstruction. In most instances, this probably came as a surprise and often initiated "desperate efforts" in order to maintain depths at an entrance (e.g., by extending updrift breakwaters or jetties or by dredging operations with available equipment or by both). This provided only a temporary relief for navigation and usually the greater the efforts to maintain depths, the more severe the erosion on the downdrift side.

The first technical counter-measures were the construction of groins and/or seawalls. While both mitigated the nearshore or onshore erosion problem, they also aggravated the downdrift erosion. Not until the late 1930s was it realized that the only practical solution to the problem was the elimination of the barrier effect. This was done by establishing sand bypassing whereby material is pumped or trucked across the barrier to the downdrift beaches.

The need for bypassing was supported by legislation such as the Florida law (1987), which reads as follows (Section 161.142, Declaration of Public Policy Relating to Improved Navigation inlets):

"(1) All construction and maintenance dredging of beach-quality sand should be placed on the downdrift beaches; or, if placed elsewhere, an equivalent quality and quantity of sand from an alternate location should be placed on the downdrift beaches.

(2) On an average annual basis, a quantity of sand should be placed on the downdrift beaches equal to the natural net annual longshore sediment transport."

Quantitative considerations

The quantitative aspect of longshore drift blocking by barriers is very simple. If the barrier causes the loss of a certain quantity of material

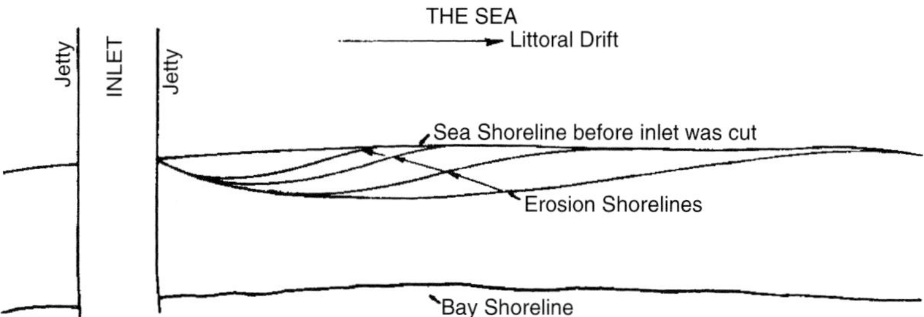

Figure B69 Shoreline development downdrift of the Fort Pierce Inlet, Florida, schematics (from Bruun, 1990). (Reproduced by permission. From Port Engineering V2 4E copyright (c) 1990, Gulf Publishing Company, Houston, Texas, 800-231-6275. All rights reserved.)

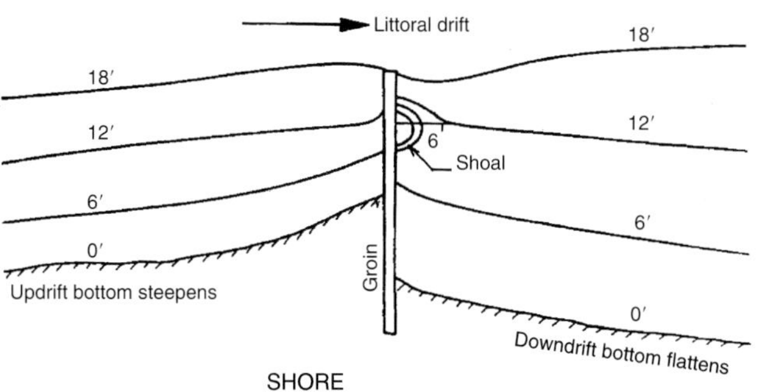

Figure B70 The development of bottom configuration downdrift of a littoral drift barrier (from Bruun, 1990). (Reproduced by permission. From Port Engineering V2 4E copyright (c) 1990, Gulf Pubhlishing Company, Houston, Texas, 800-231-6275. All rights reserved.)

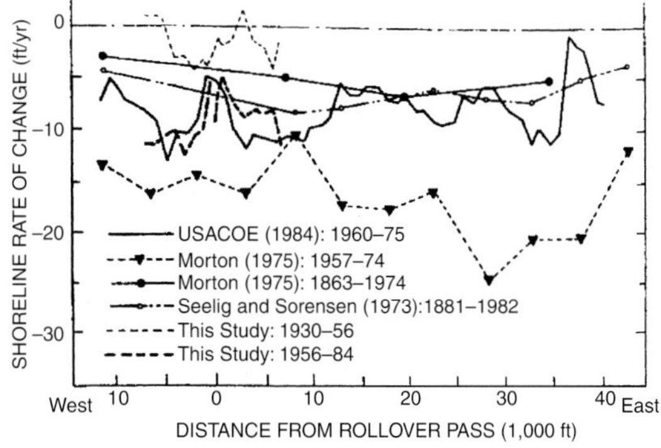

Figure B71 Comparison of shoreline rates of change near Rollover Pass, Texas (from Bruun, 1995, reprinted by permission of the Journal of Coastal Research).

which was "locked up" by the barrier, this quantity is unavailable to downdrift beaches, which consequently will suffer erosion of that magnitude. The more difficult question is: how is erosion, due to loss of sand, distributed downdrift as a function of time.

Coastal geomorphological considerations

Three parameters are important in this context: the length of the adversely affected shore, the cross-sectional retreat of the erosion cut and the rate of expansion of erosion, and its dis-tribution downdrift as functions of time. Length and cross-sectional evolution of the erosion cut give the geometric development as a function of time. The corresponding development in the offshore bottom follows the same general pattern, but there is usually a material change in the configuration of the offshore profiles, which tends to flatten in the downdrift areas (Bruun, 1990, chapters 7 and 8). Figure B69 shows a typical longshore shoreline development trend, Figure B70 (Bruun, 1990) shows the offshore development as well.

The dominant sediment bypassing mechanism at a tidal inlet affects the extent and magnitude of the downdrift high-tide shoreline response. The updrift coastline response to the introduction of a jetty is fairly localized and little dependent upon the sediment bypassing mechanisms active at the tidal inlet. Tidal inlets which are predominantly tidal-flow bypassers have more severe, downdrift effects on high-tide shoreline response than tidal inlets which bypass sediment through bar bypassing. Tidal inlets which are combined tidal flow and bar bypassers have relatively constant downdrift effects (magnitude and rate of change) through time. Significantly deepening the channel through the bar can alter the dominant sediment bypassing mechanism.

Bodge (1992, 1999), Rosati and Ebersole (1996), and Bruun (1995) made efforts to quantify the response of adjacent shores by tidal inlets. Rosati and Krauss (1999) have continued these efforts. Bodges (1999) paper is universally applicable. Bruun (1995) gives about 20 examples of which two are mentioned below.

The literature only mentions few examples where the downdrift long distance development was recorded as function of time to obtain a rate. Examples in the literature include Fields *et al.* (1989) and Bodge (1993, 1999). Theoretical approaches are available, but they concentrate on immediate downdrift reactions. Although admittedly, the effects continue to expand downdrift "infinitely" as indicated by Pelnard-Considere (1956).

Obviously, the migration rate of the downdrift erosion depends upon the quantitative magnitude of the *barrier effect*, for example, the loss of material to inlet shoals. A large loss will expand faster than a small loss.

"Beach/Inlet Processes and Management", Special Issue No. 18, of the *Journal of Coastal Research* (1993, A.J. Mehta, Editor) has a great number of examples on the influence of coastal inlets on the littoral drift system. The short distance effect of the inlet on the downdrift erosion is shown in several figures, but the development downdrift is cut short by only examining the development for a limited distance downdrift.

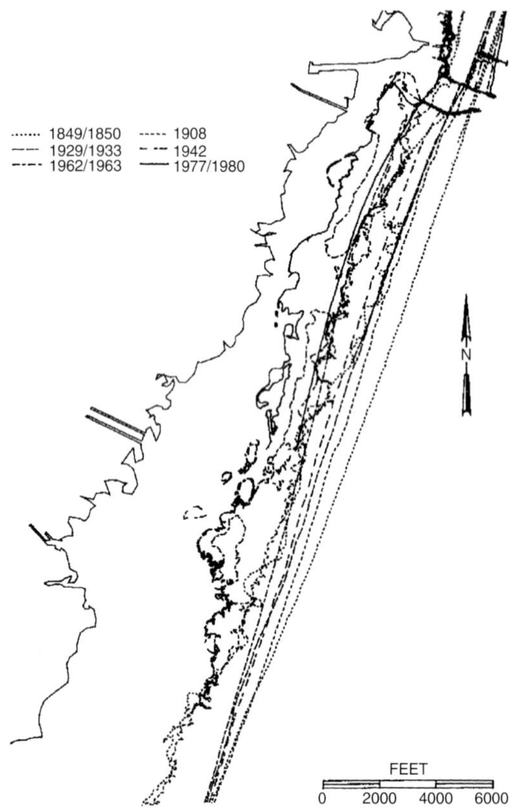

Figure B72 Barrier Island migration showing landward displacement of both Ocean and Bay high tide shorelines of Assateague Island (1850–1980). The Barrier Island maintained its width, 120–210 m, with this rapid translocation. The mainland bayshore has remained essentially stable (from Bruun, 1995, reprinted by permission of the Journal of Coastal Reasearch).

Rollover Pass, Texas

Conditions at Rollover Pass on the Texas Gulf, 31 km northeast of the Galveston Inlet, are described by Bales and Holley (1989) and by Bruun (1995). The pass is a man-made artificially stabilized inlet on the Bolivar Peninsula. Improvements of the pass were completed in 1959. Figure B71 compares shoreline rates of change near Rollover Pass. Referring to the period 1957–74 in the figure, it may be seen that the downdrift effect extended at least 40,000 ft. or 13 km and probably more. Based on 1957–74, one arrives at a migration rate of erosion of 13/17 = 0.8 km/year. The rate is 0.9 km/year if 1959, the completion year for the improvements, is used.

Ocean City Inlet, Maryland

Shoreline evolution on either side of the inlet is dealt with by Leatherman (1984). Historical shoreline changes on the downdrift side of the inlet on Assateague Island are shown in Figure B72. Construction of the Ocean City Inlet jetties, in combination with a net southerly longshore littoral drift, has resulted in severe erosion along north Assateague Island. It appears that none of the 120,000 m^3 of sand that annually flows south along this coastal sector reaches Assateague Island. Since the jetties, built 1934–35 are filled to capacity (1984), the material is largely moving offshore to build a huge ebb tidal delta that is detectable from space through analysis of Landsat imagery. A comparison of the 1942 and 1962 coastlines clearly shows the trend since the

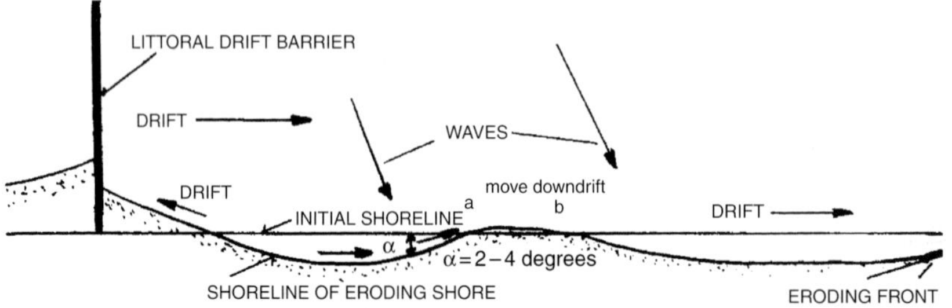

Figure B73 Schematics. The development of downdrift erosion at littoral drift barrier (from Bruun 1995, reprinted by permission of the Journal of Coastal Research).

jetty construction. The arc of erosion south of the inlet is clearly evident when considering historical changes 1950–80 in Figure B72. The historical high-tide shoreline's high-tide changes tend to converge further downdrift. This artificially induced erosion continues to impinge further downdrift through time.

Figure B72 does not extend far enough downdrift to indicate the front of the jetty-induced erosion. By a slight extrapolation, it was found that it is most likely that in 1962 it had reached 10 km south. That is, a front movement of 10 km/20 years— 0.5 km/year.

This was further confirmed by Rosati and Ebersole's (1996) quantitative research, which demonstrated that the downdrift erosion extended at least 14 km downshore (Bruun, 1995, 2000).

The zero or slow down area

The peculiar "zero-area" which sometimes appears downdrift at a rather short distance from the barrier (Figure B73) is a coastal geomorphological feature which *does not necessarily indicate the extreme limit of leeside erosion*. Obviously, the high-tide shoreline has to resume its initial direction following its change of direction on the downdrift side of the barrier. This can only be accomplished by an S-curve, which in turn develops a kind of a "corner" at point "a" as seen in Figure B73. This makes the local following section of the shore resemble "a groin" with some (minor) stabilizing effects updrift, but at the same time aggravating the large-scale erosion downdrift caused by the littoral drift barrier. A consequence of that is that the bump is most developed when the drift is very predominant and less visible for a more neutral situation.

Conclusion

The downdrift high-tide shoreline development at a littoral drift barrier may in some cases, but not always, be described by a short (local) as well as a long distance effect which both move downdrift at various rates; the long distance movement being two to three times faster than the short distance, or about ~0.5 km/ year versus ~1–1.5 km/year. These figures may be subject to considerable variances depending upon wave intensities, barrier morphologies and littoral drift magnitudes as well as upon the relative predominance of the drift. The short distance effect is a coastal geomorphological feature, the long distance a materials deficit feature. Quantitative research is making progress (Bodge, 1999; Rosati and Kraus, 1999).

Bypassing technologies

Figure B74 (Bruun, 1990) is a review of bypassing plants and principles distinguishing between non-scouring and scouring conditions (tidal entrances). Bypassing may be undertaken by bypassing plants or by bypassing arrangements.

The most simple examples of bypassing are found at breakwaters, single or double, perpendicular or parallel to shore provided with dredged entrance channels for navigation (Tables 9–16 of Bruun, 1990, is a comprehensive overview of bypassing plants and arrangements). Table B8 summarizes the situation as of 1990 (Bruun, 1990).

The main difference between bypassing at harbors and at tidal entrances lies in the action of the tidal currents. It may therefore be said that while in the case of harbors, bypassing arrangements may be

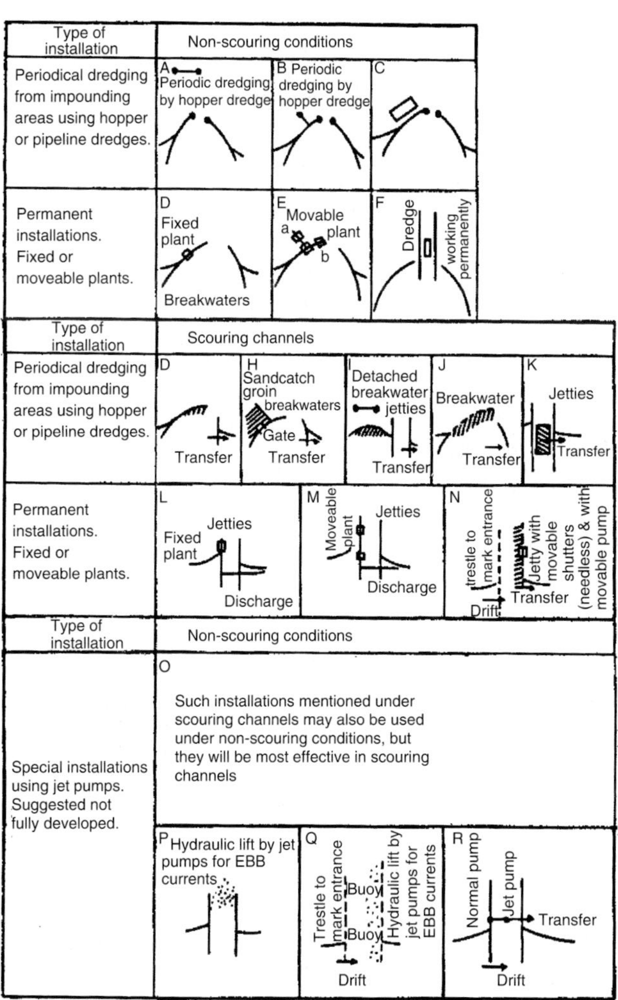

Figure B74 Various principles of bypassing (from Bruun, 1990). (Reproduced by permission. From Port Engineering V2 4E copyright (c) 1990, Gulf Publishing Company, Houston, Texas, 800-231-6275. All rights reserved.)

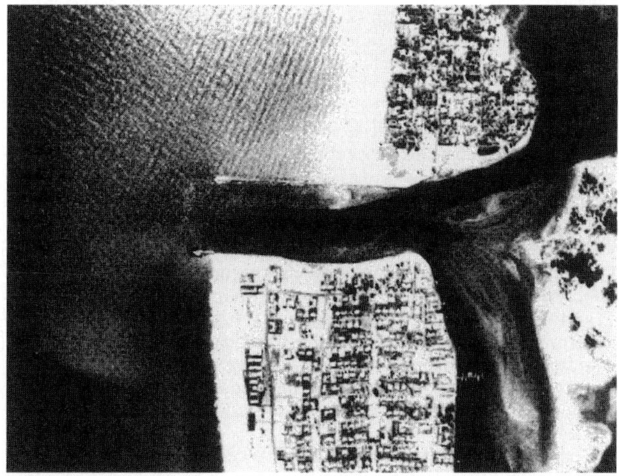

Figure B75 The Palm Beach Inlet, Florida (from Bruun, 1990). (Reproduced by permission. From Port Engineering V2 4E copyright (c) 1990, Gulf Publishing Company, Houston, Texas, 800-231-6275. All rights reserved.)

Table B8 Number of bypassing arrangements established or under Construction (Bruun, 1990, with updates)

	Fixed plants (including jet pumps)	Movable plants	Detached breakwaters	Weir jetties	Sand catch or dredged trap updrift in channel or bay
Built	7	3	3	6	20
Suggested	2	—	1	—	—

This table gives the right order of magnitude of current projects.

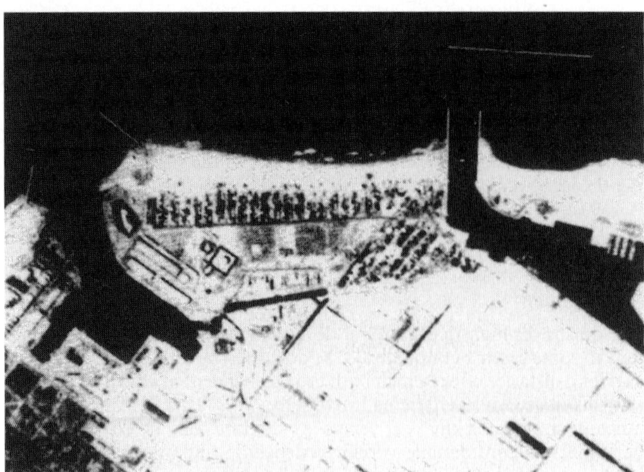

Figure B76 Detached Breakwater, Ventura Harbor, California (from Bruun, 1990). (Reproduced by permission. From Port Engineering V2 4E copyright (c) 1990, Gulf Publishing Company, Houston, Texas, 800-231-6275. All rights reserved).

designed solely on wave mechanics principles, the design at tidal inlets also includes current mechanics.

At Paradeep, State of Orissa, Bay of Bengal, India, a large movable plant which included a 750 hp pump producing 500 t sand/h was installed on a 370-m steel trestle running perpendicular to the updrift breakwater, in the middle sixties (Figure B74(E)). The specifications required that the dredge pump combined with a booster pump should be capable of handling this quantity of slurry through an 46 cm reducing to 41 cm pipeline about 2,200 m long. The plant was supposed to work fairly regularly throughout the year in most weather conditions. The trap capacity, however, proved to be too small to handle the strong deposits during the monsoon and sand bypasses the trap when it is filled, some of it in suspension even before it is filled. The result is that it has become necessary also to operate a hydraulic pipeline dredge in the entrance to remove the sand, which escaped the trap. It is probably rather doubtful that more fixed plants, although proposed or under discussion, will be built as the most recent experiences are not very promising. The 191,000 m^3/yr plant at Palm Beach Inlet in Florida, Figure B75, has not been satisfactory either and has seldom operated at the planned full capacity. It is going to be replaced by a more effective arrangement.

Future development of bypassing

The most reliable or effective trap arrangement is undoubtedly the detached breakwater built offshore on the updrift side (Figures B74(A), (I) and Figure B76). But it is an expensive solution requiring a large offshore area, a usually rather expensive breakwater and an effective suction dredge of a seagoing, therefore also expensive, type plus a rather long and therefore also costly pipeline, possibly with one or more booster stations to push the material all the way to the downdrift side beaches.

Generally, it may be said that developments that are taking place favor the most flexible arrangement of traps to be dredged by floating equipment, which bypass the material across the littoral drift barrier. The success of such arrangements, however, depends partly upon the correct placement of the trap from a sedimentation as well as a practical viewpoint in regard to transfer of material and partly upon the equipment available for transfer and the economics involved.

Trap arrangements at or on the updrift side of a sand catch breakwater like Figures B74(B), (C), (H), however, leave the dredging equipment somewhat exposed to wave action. The submerged weir (Figure B74(J)) was introduced to alleviate this drawback but it has not been fully satisfactory in all cases. While the Hillsboro Inlet, Florida, the "old timer" in the group, must be classified as a success, the arrangement at the Masonboro Inlet, North Carolina and the arrangement at East Pass, Florida, have experienced some difficulties due to weir operation.

The Hillsboro Inlet

The Hillsboro Inlet in southeast Florida, is a natural inlet to the Atlantic Ocean, connecting the Atlantic Intracoastal Waterway to the ocean. It provides free access for commercial and recreational boats, and storm water drainage for a large interior land area.

There was about a 1.2 m depth over a rock bar at the entrance in its natural condition. The inlet channel was improved in the 1960s by an excellent design developed at the University of Florida, confirmed by a hydraulic model study. The channel was cut to a depth of 3 m. A 61 m jetty on the north side for the predominant littoral drift and a 122 m jetty on the south side, were constructed.

The north side jetty has a natural weir (low section) for material transfer across the jetty, and a sand deposit basin (trap) inside for storage of material. The stored material is later transferred by a hydraulic dredge to beaches south of the inlet.

A worn-out dredge was replaced in 1983 by a 41 cm dredge reduced to 30 cm. The year-round dredging operation is very successful due to the weir design. The channel is able to be kept at an operating depth of 2.4 m 92,000 ± 15,000 m^3 of beach quality material is bypassed each year, essentially all the littoral drift sand deposited in the inlet basin and channel.

When sand is dredged promptly from the channel after a storm, less material is lost to the ocean by ebb tides or deposited in the interior channel by flood tides. The beach south of the inlet has accreted yearly and has not had to be renourished since 1983. It appears to be reaching equilibrium. A planned project will increase the outer channel depth to 6 m, improve the geo-metry of the entrance, increase the material captured for bypassing, and improve navigation safety and drainage.

Alternative bypassing systems using jet-pumps

Jet pumps submerged in the entrance for transfer by normal pumping power (Figure B74(R)) may also prove a useful procedure but it only covers a rather local area, although its influence may be expanded for some distance to either side by several pumps and pipelines (Boyce and Polvi, 1972).

The application of jet pumps to stir up material, Figures B74(P), (Q), (Bruun, 1990) using ebb currents as the main flushing or carrying agent, may prove to be a very practical arrangement, but it only helps to carry the material away from a certain local area like agitation dredging and does not transfer the material.

A jet pump system was built at the Nerang River entrance in Queensland, Australia, as described by Bruun (1990). It is based on an updrift array of jet pumps. The pumps have a large capacity, but there are problems with clogging of the pumps by debris.

Use of submerged pumps combined with fluidization

Weisman *et al.* (1982) describe improvement of channel and bypassing stabilities by perforated hydraulic pressure pipes placed below the bottom. A few examples are mentioned here.

Case one, inlet with a dredged, otherwise unprotected channel

It may be improved by lift pipes placed across the bar, at the same time improving bypassing by combined wave and (ebb) current action. A trap may also be placed in the channel to accumulate materials carried to the trap by ebb as well as by flood currents. This trap has a "lift system" in the middle that may be emptied whenever needed, for example, by a fluidization pump.

Case two, inlet entrance improved by special geometry jetties for channel stability and bypassing

Lift pipes are used to obtain optimum stability of the channel across an entrance bar or shoal (almost standard). This also improves bypassing by combined ebb currents and wave action. Channel stability is further improved by a trap in the channel operated continually for lift during ebb flows, so that the channel always stays clear. The trap may be emptied intermittently for transfer, by fluidization. Outside the updrift jetty a large trap is established for continuous transfer of material carried to the trap by littoral currents and onshore bottom creep due to wave action. This transfer may also be undertaken by fluidization using the same pump as for the bar lift.

Advantages of using hydraulic lift for channel stability

The advantages of using hydraulic lifts to increase flushing abilities are well demonstrated in nature by the influence of wave action in "opening up" a cross section. It may also be observed at places where nature delivers—free of charge—the hydraulic pressure. Some natural tidal inlets placed are accordingly all over the world. The lift may be operated according to needs and particularly during and after heavy storms. The lift is able to direct the sediment transport oceanward. Such a lift system may be used in connection with a submerged pump like the Punaise which is an underwater pump of Dutch origin. It has the shape of a thumb-tack. The pointed part digs itself down in the bottom by hydraulic pressure pumps. The fluidized material is carried from the cone-shaped pit through pipeline to the discharge area. The Punaise has been tested in the Netherlands and is being tested in the United States by the US Army Corps of Engineers.

The Punaise

The shape of the dredge gave it the name "The Punaise" (Dutch word for thumb-tack). The first "PinPoint" dredge, Punaise PN250, was commissioned in 1990, the second dredge Punaise PN400 was commissioned in late 1993.

The Punaise works on a very simple principle. A pump and suction pipe are connected to ballast tanks and then the entire structure is submerged to come to rest in the bottom where dredging is to be carried out.

The link to a small shore-based control unit and energy supply is provided by cables and hoses while a pipeline is used to transport the dredged material. The submerged pump excavates by hydraulic erosion. It creates an unstable slope upon which sediment flows to the suction intake. The unique support system makes use of the suction pipe that is embedded into the bottom to a level below the dredging depth (1–3 m) thus providing both horizontal and vertical stability.

Conclusion

1. Fixed bypassing plants will be replaced by more flexible plants. For major projects large floating plants like the trailing hopper dredger which discharges downdrift is common.
2. For medium size projects bypassing most likely will be by proper size, but smaller, hopper, or pipeline dredgers.
3. For projects of more modest size bypassing will be by shallow water hopper dredgers in some cases combined with underwater pumps and fluidized on arrangements (Visser and Bruun, 1997).

Per Bruun

Bibliography

Bales, D.J., and Holley, Ed.R., 1989. Sand transport in Texas tidal inlet, In *Proceedings ASCE Waterway, Port, Coastal and Ocean Engineering*, **115**(4) 427–443.

Bodge, K.R., 1992. *Port Canaveral Inlet Management Plan*. Port Canaveral: Olsen Associates, Inc.

Bodge, K.R., 1993. Gross transport effects and sand management strategy at inlets. *Journal of Coastal Research*, **18**(Special Issue,) 111–124.

Bodge, K.R., 1999. Inlet Impacts and Family of Solutions, *Coastal Sediments 1999*, Reston: American Society of Civil Engineers.

Boyce, A.R., and Polvi, R.D., 1972. Jet pump application in estuary maintenance. Corvallis: *Civ. Eng. Dept., Oregon State University*.

Bruun, P., 1961. Natural by-passing of sand at coastal inlets. *ASCE Translations*, **126**(4): 823–854.

Bruun, P., 1990. *Port Engineering*, Houston: *The Gulf Publishing Co.*

Bruun, P., 1995. The development of downdrift erosion. *Journal Coastal Research* **11**(4): 1242–1257.

Bruun, P., and Gerritsen, F., 1959. Natural bypassing of sand at coastal inlets. *Transactions of the American Society Covil Engineers*, Paper #3293, 823–854.

Fields, J.L., Marino, J.N., and Weishar, L.L., 1989. Effect of Florida tidal inlets on adjacent shorelines, *Florida Shore and Beach Preservation Association Proceedings*, pp. 383–391.

Leatherman, S.P., 1984, Shoreline evolution of North Assateague Island, Maryland, *Shore and Beach*, **52**(3): 3–10.

Mehta, A.J. (ed.), 1993, Beach inlet processes and management. *Journal of Coastal Research*, Special issue No. 18.

Pelnard-Considere, R., 1956. Prediction of Beach Planforms with Essai de Theorie de l'Evolution des Formes de Rivage an Plage de Sable et de Galets, *4th Journees de l'Hydraulique: Les Energies de la Mer*, Question III, No. 1, 239–300.

Rosati, J., and Ebersole, B.A., 1996. Littoral impact of Ocean City Inlet, Maryland, USA, *Chapter 23 of Proceeding of the ASCE Conference*, pp. 2779–2791.

Rosati, J.D., and Kraus, N.C., 1999, Advances in coastal sediment budget methodology. *Shore and Beach*, **67**(2–3): 56–65.

Visser, K., and Bruun, P., 1997, The Punaise underwater dredger. *Journal of Coastal Research* **13**(4): 1329–1333.

Weisman, R.N., Collins, C.C., and Pars, J.M., 1982. Maintaining tidal inlets channels by fluidization. *Journal of Waterway, Port Coastal Ocean Division*, **108**(4): 526–538.

Cross-references

Barrier
Dredging of Coastal Environments
Littoral Cells
Longshore Sediment Transport
Navigation Structures
Tidal Prism

C

CAPPING OF CONTAMINATED COASTAL AREAS

The rapid and under-regulated development of industries in the past has left major contamination problems in several coastal and estuarine areas. Such contamination from historical chemical releases (industrial discharges, storm sewers, wastewater, landfill runoff, and leachate) severely impairs the ecological and recreational functions of the estuary/ocean. The discharged contaminants typically adhere to the fine sediments and settle in low-energy zones, which are conducive to deposition. Once settled, such sediments can exert significant oxygen demand, reduce benthic diversity, and result in poor water quality.

Contaminated sediment removal can be expensive because a high degree of efficiency and reliability is required in such operations. Capping is an attractive, nonintrusive and cost-effective method of remediating contaminated sediments in rivers and harbors, where draft restriction is not a major concern. The same physicochemical properties and hydraulic conditions that favored the initial adsorption to and deposition of the contaminated sediments typically favor successful containment by capping. Since capping is an *in situ* technique, it can often be accomplished at approximately 20–30% of the cost of a remedial dredging project.

Cap definition and considerations

Capping involves the controlled placement of clean sediment (usually sand) over the contaminated sediments in order to isolate them from the environment. The cap thickness is a function of several factors including desired breakthrough time, bioturbation, and contaminant flux. At high-energy sites, the cap is often armored by a layer of stones. Where the size of the armor significantly exceeds that of the base material, a layer of intermediate size stones (filter layer) is provided to prevent hydraulic washout of the base material. Another special case is the use of a geotextile layer on top of the native sediment layer to provide added bearing capacity and load distribution, especially when the sediments are very soft and the cap is multi-layered.

To ensure sufficient protectiveness and isolation of the contaminants, the cap should withstand the worst design case physical and chemical events at the site. Major physical destabilizing forces include extreme water flows, storm waves, and tidal currents. Molecular diffusion, ground water-induced advection, hydrodynamic dispersion, and pore water flux due to consolidation are the major chemical destabilizing forces. A base sand isolation layer provides protection from chemical events and an armor/filter layer (usually riprap and gravel/cobbles) provides protection from physical forces. Bioturbation (sediment processing by benthic organisms) is another major destabilizing force causing mixing of the water/sediment interface, thereby releasing contaminants to the water column. Therefore, the design thickness of the cap should account for physical and chemical isolation, as well as bioturbation. Since the likelihood and type of organisms that may colonize the cap would vary from site to site, a site-specific survey of benthic organisms is often required. This could also be estimated by chemical and isotopic analysis of thinly sectioned sediment cores for sediment mixing depths.

The success of a capping operation depends on the proper selection of construction techniques for the cap and armor materials. Mathematical modeling of cap performance using design equations is frequently used since: (1) it can evaluate the effectiveness of the cap as a chemical barrier for various design parameters, and (2) it can quantify potential contaminant release rates to assess long-term effectiveness of the cap. A brief review of the cap design criteria and guidance for the selection of cap construction equipment are presented here.

Design criteria

Special factors of interest to cap design/evaluation include cap behavior during placement, settlement of the cap, and native sediments following placement, potential contaminant release mechanisms, and stability of the cap in the long term (Mohan, 2000). Design considerations for an underwater cap can be classified into the following six categories (Palermo, 1991; Zeeman, 1993; Mohan, 2000):

Site investigations: A good field database is an essential starting point for an effective cap design. Investigations should include evaluation of water depth, currents, wave forces, flow velocities, extreme events (storms, floods), ice effects and forces, soil and foundation properties (strength, consistency, compressibility, and permeability), contaminant types and levels, and water column/biota contamination data.

Cap area: The cap should ideally cover the entire horizontal extent of the contaminated sediments. Mapping the horizontal distribution of the contaminants is essential to identifying this parameter. In special cases where sediment contamination is vast and widely distributed spatially within the waterbody, the extent of the cap should be determined based on human health exposure risks and isolation of the bulk of the contaminant mass to the food chain.

Cap movement during placement: Underwater caps may be placed using several techniques such as surface placement from barges, clamshell, or pipelines, placement using spreader boxes, tremie pipes, and underwater placement using clamshells. Several mechanisms are of interest as the cap is being placed. One of the main concerns during placement of the cap is the movement of the slug (velocity and properties) as it reaches the ocean or river bed. Theoretical considerations should include impact velocity of the sand, thickness and horizontal dispersion of the sand slug, and required number of lifts for obtaining the design thickness.

Table C1 Comparison of cap construction techniques (Mohan, 1997, Proceedings Coastal Zone '97)

Equipment	Water depth	Material	Accuracy	Resuspension	Mixing	Cost
Barge discharge	Deep	Sand	Medium	Medium	Low	Low
Submerged discharge	Both[a]	All[b]	High	Low	Low	Medium
Skip box	Shallow	All[b]	High	Low	Low	Medium
Conveyor belts	Deep	Sand/rock	Medium	Medium	Medium	Medium
Dry/wet pumping	Both[a]	Sand/silt/clay	Medium	Low	Low	Low
Surface/underwater clamshell	Both[a]	Sand/rock	Medium	High	High	Medium
Modified dredges	Deep	Sand/rock	Medium	Medium	Medium	High
Thin-layer disposal	Deep	Sand	High	Medium	Low	Medium

[a] Deep and shallow water application.
[b] Material types (sand, silt, clay, and rock).

Geotechnical stability: Construction of a sand cap over soft native sediments may cause stability concerns due to the increased stresses in those layers. The primary stability concern during cap placement is the potential for the sediments to fail under the increased stresses, and to form "mudwaves" which could be trapped in subsequent layers of sand cap material. This concern is particularly valid since contaminated sediments typically have very high water contents and low shear strengths, resulting from the lack of sediment consolidation. Cap stability analysis involves consideration of three major factors: (1) settlement of the cap and foundation material, (2) bearing capacity of the cap, and (3) slope stability of the cap.

Chemical stability: Chemical analysis is one of the major design elements since it defines the effectiveness of the cap in containing (and isolating) the contaminants from the water column. There are four major mechanisms that could potentially release contaminants through the cap system: (1) diffusion, in the absence of groundwater flow, (2) advection and dispersion, if groundwater flow through the cap is present, (3) release of pore water due to sediment compression, and (4) erosion of contaminated sediments due to pore water release during sediment compression.

Physical stability: An underwater cap must be designed for a specific level of protection. This implies that the cap/armor should be able to withstand the impact forces resulting from a certain return period event (typically the 100-year event). An armor layer is provided when necessary to protect the sand isolation layer from being eroded away due to hydraulic and environmental forces. In general, bottom velocities exceeding about 0.3 m/s (1 fps) can initiate sediment erosion. Considerations include protection from river flow, wave action, propeller wash forces, ice loads, and forces resulting from pore water release mechanisms.

Hydraulic filtering: The cap/armor should also be designed to withstand the impacts of the hydraulic filtering forces. There are two ways of providing this protection: (1) by installing a separate filter layer between the armor and cap layers, and (2) by installing a "graded" armor layer (i.e., with a wider grain size range) above the cap layer.

Construction techniques

Table C1 lists the factors influencing the selection of cap construction equipment based on site- and material-specific properties.

Summary

Since the early 1980s, several pilot-scale and full-scale projects have been conducted to field-test the effectiveness of capping under a variety of site conditions (Strugis and Gunnison, 1985; Palermo, 1991; Wang *et al.*, 1991; Fredette *et al.*, 1992; Zeeman, 1993; Averett and Francingues, 1994; Nelson *et al.*, 1994; Hull *et al.*, 1998; Laboyrie and Flach, 1998; Lillycrop and Clausner, 1998; Shaw *et al.*, 1998; and Mohan *et al.*, 1999). These studies have established the validity of the environmental isolation provided by the capping process, thereby making capping a viable tool to be screened during coastal clean-up projects. Note that a final decision on the preferred remedy for a specific site should be based on technical, environmental, and economic analysis of the various alternatives (capping, dredging/capping, dredging/disposal, and no action) for the site.

Ram K. Mohan

Bibliography

Averett, D.E., and Francingues, N.R., 1994. Sediment remediation: an international review. In *Proceedings Dredging '94 Conference*, American Society of Civil Engineers, pp. 596–605.

Fredette, T.J., Germano, J.D., Carey, D.A., Murray, P.M., and Kullberg, P.G., 1992. Chemical stability of capped dredged material disposal mounds in Long Island Sound. *Chemistry and Ecology*, **7**: 173–194.

Hull, J., Jersaka, J., Pochop, P., and Cummings, J., 1998. Evaluating a new in-situ capping technology for mitigating contaminated sediments. In *Proceedings Wodcon—XV Conference*. World Dredging Association, pp. 555–576.

Laboyrie, H., and Flach, B., 1998. The handling of contaminated dredged material in the Netherlands. In *Proceedings Wodcon—XV Conference*. World Dredging Association, pp. 513–526.

Lillycrop, L., and Clausner, J., 1998. Numerical design of the 1997 capping project at the Mud Dump Site. In *Proceedings Wodcon—XV Conference*. World Dredging Association, pp. 937–952.

Mohan, R.K., 1997. Design and construction of subaqueous caps for restoring contaminated coastal areas. In *Proceedings, Coastal Zone '97 Conference*. University of Massachusetts, pp. 763–765.

Mohan, R.K., D'Hollander, R.D., Johnson, A.N., Brozowski, P., D'Ambrosiio, K.T., and Jerome, J., 1999. Remediation of contaminated sediments by in-place capping—a case study of Rahway River, NJ. *Journal of Marine Environmental Engineering*, **5**: 1–34.

Mohan, R.K., 2000. Modeling the physical and chemical stability of underwater caps in rivers and harbors. In Herbich, J.B. (ed.), *Handbook of Ocean and Coastal Engineering*. New York: McGraw Hill, Inc., pp. 14.1–14.27.

Nelson, E.E., Vanderheiden, A.L., and Schuldt, A.D., 1994. Eagle harbor superfund project. In *Proceedings Dredging '94 Conference*, American Society of Civil Engineers, pp. 1122–1131.

Palermo, M.R., 1991. *Design Requirements for Capping*. Vicksburg: US Army Waterways Experiment Station.

Shaw, J., Whiteside, P., and Ng, K., 1998. Contaminated mud in Hong Kong: a case study of contained seabed disposal. In *Proceedings Wodcon—XV Conference*. World Dredging Association, pp. 799–810.

Strugis, T.C., and Gunnison, D., 1985. *New Bedford Harbor Superfund Report No. 6*. Vicksburg: US Army Waterways Experiment Station.

Wang, X.Q., Thibodeaux, L.J., Valsaraj, K.T., and Reible, D.D., 1991. Efficiency of capping contaminated sediments in-situ: laboratory-scale experiments on diffusion–adsorption in the capping layer. *Environmental Science and Technology*, **25**: 1578–1584.

Zeeman, A.J., 1993. Subaqueous capping of contaminated sediments. *Canadian Geotechnical Journal*, **31**: 570–577.

Cross-references

Environmental Quality
Geotextile Applications
Human Impact on Coasts
Marine Debris—Onshore, Offshore, Seafloor Litter
Numerical Modeling
Water Quality

CARBONATE SANDY BEACHES

Introduction

Carbonate beaches have a significant proportion of the sediment fabric of biogenic origin, and carbonate in composition. Carbonate beaches are therefore wave deposited accumulations of sediment (sand to boulder in size) deposited on shores where a nearshore supply of biogenic debris is available. The carbonate detritus can originate in a range of environments and from a number of sources including coral and algal reef debris, mixed skeletal material including broken shells, foraminifera, bryozoa, gastropods and ostracods, some terrestrial animals, and ooids.

In the tropics, they are associated with three separate environments. They are most commonly found adjacent to/or nearby coralline-algae reefs. They also occur in lee of tropical intertidal sand and mud flats where shells are winnowed to form shelly beach ridges–cheniers; and they may occur adjacent to carbonate banks in shallow tropical seas. In temperate regions, there are three separate sources of carbonate material. They are found adjacent to carbonate encrusted rocky shores and reefs. Temperate sea grass meadows along lower energy shores supply molluscs and other debris to sand flats and lower energy beaches; and are most extensive adjacent to some high-energy carbonate-producing, temperate continental shelves. Carbonate-dominated beaches also tend to exist in areas with the above conditions, where there is insufficient material of terrigenous origin to significantly dilute the carbonate proportion, particularly in arid regions.

An additional characteristic of many carbonate beaches is the post-depositional modification (diagenesis) of the sediment to form beach calcarenite (beachrock) in warm tropical waters, and aeolian calcarenite (dunerock) in more arid tropical to temperate environments. Once formed, the lithified sediments can have a dramatic impact on the nature and future evolution of the coastal system, particularly when the beach and/or dunerock is exposed to wave-wind processes. Even beaches with little carbonate may experience similar diagenetic processes particularly in warm tropical waters. Given this an additional definition of carbonate beaches are those whose processes and morphology are significantly impacted by the presence of carbonate grains and cements, particularly the formation of beach and dune calcarenite.

This entry will briefly examine the location of carbonate beaches, the major sources of carbonate sands, and the impact of the carbonate sands on beach morphodynamics.

Location

Carbonate beaches exist in tropical and temperate locations, including some in relatively high latitudes. The main prerequisite for a carbonate beach is a source of carbonate-producing detritus, and a mechanism to erode and/or transport it to the shore. Bird and Schwartz (1985) recorded 32 tropical and 14 temperate countries where carbonate beaches and beach and/or dune calcarenite have been recorded. Scholle *et al.* (1983) also provide a list of locations and references and where carbonate beach and dune systems have been recorded, while Robbins and Magoon (2001) is a more recent overview of primarily tropical carbonate beach systems.

Table C2 provides a summary of major beach carbonate sources and materials in tropical and temperate regions. The table and even the locations recorded in Bird and Schwartz is by no means exhaustive, as sediment characteristics are only available in the literature from a small minority of the world's beaches. This overview is based on the published data and locations at present. Future work will no doubt expand the range and perhaps types of carbonate beaches.

Table C2 Location, sources, and form of carbonate beach systems

	Source	Material	Beach morphology
Tropical			
Open coast	coral-algae reefs	coral-algae fragments	reflective
Deltaic coasts	tidal sand/ mud flats	molluscs	cheniers, beach ridges
Shallow tropical seas	carbonate banks	ooids, skeletal, peloidal, pellets, aggregate sands	beach ridges
Temperate			
rocky coast	encrusting organisms	rocky shore biota	variable
low energy	sea grass meadows, tidal sand flats	molluscs, red algae foraminifera, bivalve fragments,	reflective/sand flats
high energy	inner shelf	molluscs, red algae, encrusting bryozoans, echinoids	low to high energy + dune calcarenite

Tropical locations and sources

Tropical carbonate beaches generally occur between 25°N and S in association with three major carbonate-producing environments: coralline-algal reefs, tidal flats, and carbonate banks.

Coral reefs

Tropical carbonate beaches are predominantly located in association with coralline-algae reef systems, particularly fringing reefs, which lie in close proximity to the shore, and all beaches associated with atolls and cays, where the eroded reef debris is the sole source of material. Tropical reef beaches include windward boulder ramparts formed in lee of exposed high-energy reefs, sand and coral cays formed on coral islands, and atolls and mainland beaches composed of sand through boulder debris (Scoffin and Stoddart, 1983). Richmond (2001) provides an overview of tropical Pacific island carbonate beach systems including sediment sources, rates of supply, and physiographic settings.

Presence of an offshore reef is insufficient to guarantee carbonate beaches, as there must also be a mechanism to erode and transport the carbonate detritus from the reef to the shoreline. The world's longest reef system, the 2,000 km long Great Barrier Reef, lies between 20 and 200 km from the Australian coast. Most of the carbonate detritus is deposited tens of kilometers offshore close to the source reefs and does not reach the shore. The mainland coast is therefore dominated by terrigenous sediments, low in carbonate (<20%). Only scattered fringing reefs in the northern reef system actively supply carbonate sediments to the adjacent mainland beaches forming carbonate rich beaches (70–90 %) (Short, 2001).

Tidal flats

Tidal sand and mud flats are extensive in tropical regions in association with river mouth deltas and protected embayments. These surfaces and adjacent subtidal sediments provide a habitat for a range of organisms, particularly molluscs. When exposed to episodic wave erosion and reworking the coarse carbonate detritus, together with any coarse terrigenous material, is eroded and reworked shoreward and deposited as shelly beach ridges, on sand flats, and cheniers, on mud flats (e.g., see Rhodes, 1982).

Carbonate banks

Carbonate banks occur in shallow tropical seas and are arranged perpendicular to shore as tidal-bar belts or shore parallel as marine sand belts. They may be composed of ooids, and/or mixtures of skeletal, peloidal, pellets and aggregate sands. Low tidal deltas and barrier islands surrounded by low-energy carbonate beaches can form in association with the banks. They have been recorded in south Florida and the Bahamas, and in Western Australia's Shark Bay (Scholle *et al.*, 1983).

Temperate locations and sources

Temperate carbonate beaches occur outside the tropics in a range of temperate environments, from cool humid climates in Scotland and Ireland, to Mediterranean climates in Greece and the Black Sea, but are most commonly associated with arid to semiarid temperate coasts in North and South Africa and southern and western Australia. The three major sources of temperate carbonate sands are rocky shores, temperate sea grass meadows, and shelf biota.

Rocky shores

Rocky shores play host to a range of encrusting carbonate organisms (Calhoun and Field, 2001) whose death assemblages can be reworked onshore usually into carbonate enriched embayed or pocket beaches, with source headlands to either side and rock reefs offshore. Knuuti and Knuuti (2001) report on a Maine Beach (43°N) composed of 80% carbonate material consisting of barnacles, echinoids, mussels, and other material. Higher latitude carbonate beaches and dunes have also been reported in Ireland and Scotland (54 and 59°N) (Scholle *et al.*, 1983).

Sea grass meadows-sand flats

Sea grass meadows of *Posidonia* and *Zostera* provide a habitat for a range of molluscs, together with red algae, foraminifera and bivalve fragments, as well as sea grass roots and detritus. This material is reworked shoreward to build intertidal sand flats backed by low-energy shelly beach ridges. Such systems occur commonly in the southern Australian gulfs and Great Australian Bight (Short *et al.*, 1989), and throughout Shark Bay on the central west coast.

Shelf biota

Temperate continental shelves across southern and western Australia, around South Africa and along the western and north African and southern and eastern Mediterranean coast contribute carbonate detritus to the most extensive temperate carbonate beach and dune systems. On the southern Australian shelf Boreen *et al.* (1993) describe a shallow (30–70 m) inner shelf containing a range of hard-bodied organisms particularly molluscs, red algae, encrusting bryozoans, echinoids and soft kelps, and sponges. The shelf is dominated by high-energy wave action, which erodes and transports carbonate detritus shoreward. Along the south and western Australian coast it contributes to a 9,000 km coastal system containing 3,000 beaches occupying 70% of the coast, with regional beach systems composed of between 58% and 80% carbonate, and individual beaches reaching 100% carbonate (Short, 2001) (Figure C1).

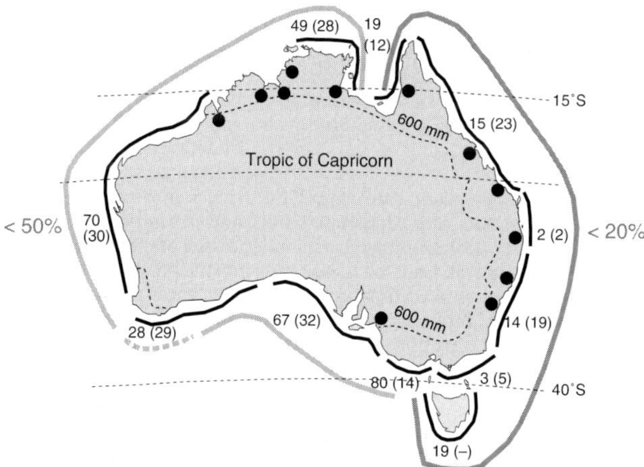

Figure C1 Australian carbonate beach sands. Numbers indicate mean regional percentage of carbonate beach sand, with standard deviation in brackets. Note the dominance of carbonate sands across the southern and western coast derived from continental shelf biota, and in low-energy areas, from sea grass meadows, while in the northwest fringing reefs are a major source. Dots indicate major river systems and 600 mm isohyet delineates Australia's humid fringe (from Short, 2001, with permission of the ASCE).

Beach systems

The beach systems associated with carbonate beaches, behave similar to quartz beaches, and are dependent on the sediment and wave–tide characteristics. They may be composed of boulders through fine sand, may range from wave to tide dominated, and from reflective to dissipative. Their sediment characteristics and grain size is largely related to the sources and level of wave energy. Boulder coral ramparts and cays are associated with windward reef systems (Scoffin and Stoddart, 1983). Most tropical carbonate beaches tend to be of low energy and composed of coarser sands and gravel including those in lee of reefs and sand–mud flats, the latter may also contain a high proportion of whole and shell fragments, a consequence of which is their description also as "shell ridges."

Temperate beaches can also range from coarse to fine. Coarser sands and some gravel are usually associated with the lower energy beach ridges form in lee of sea grass meadows and their low-energy nearshore and/or intertidal sand flats. Shell Beach in Shark Bay, Western Australia, a very low-energy system formed in lee of shallow sand banks and sea grass meadows consists of 100% shells (Figure C2). Local carbonate enrichment of energetic beaches by *Donax* species (commonly known as cockles, coquina, pippis) derived from adjacent foreshore habitats occurs through tropical and temperate locations.

The finest and best-sorted carbonate sediments are associated with shelf-derived carbonates on high-energy coasts. The erosion, transport, and energetic deposition process leads to a diminution of the source material into some of the finest (<0.1 mm), well to very well sorted, beach material. While there has been a limited amount of work in the sources of such material (e.g., Boreen *et al.*, 1993; James *et al.*, 1999) there is no published work in the actual sources (and age) of the beach sands. This is an area of considerable scope, particularly as it would provide information on the rates of contemporary carbonate sediment supply.

Diagenesis

A characteristic of carbonate beaches in all locations is the post-depositional modification of the carbonate fabric through geochemical processes, which commences soon after deposition. This is manifested as the commonly termed beachrock and dunerock. Beachrock, or more correctly beach calcarenite, refers to the precipitation from seawater of carbonate cements in the intertidal zone of tropical beaches, leading to the cementation of the intertidal beach zone. The beach and swash zone is an ideal environment for the precipitation of marine cements. Groundwater supersaturated with $CaCO_3$ is pumped by both wave and tidal activity through the usually coarse, highly porous, permeable sediments. The cement is derived from precipitation from the seaward matrix, and not from solution of the *in situ* sand grains. Cements are typically aragonite or high Mg-calcite (10–14%). Cements precipitated from freshwater contain low Mg-calcite (5%). As a consequence the rock normally has extremely well-preserved structures and is typically exposed as tabular, seaward sloping slabs of beachrock (Figure C3).

Figure C2 Detail (left) and view (right) of Shell Beach, Shark Bay, Western Australia, which is composed of 100% molluscs derived from adjacent sea grass meadows (A.D. Short).

Figure C3 Beachrock detail showing well preserved, cemented beach laminations, Maceio, Brazil (left); and tabular, seaward dipping beachrock, Heron Island, Queensland (A.D. Short).

Figure C4 Beachrock rich in coquina shells, backed by eroded rock forming a beachrock breccia, Marineland, South Carolina (A.D. Short).

This process needs only a few years to decades to occur, though most would have occurred during the late Holocene.

Beachrock formation requires that sediments remain close to the surface and shoreline, the source of wave and tidal pumping. For this reason, they occur on stationary rather than prograding shores.

Once the beachrock is formed it may remain buried and have little impact on contemporary beach processes. However, as the unconsolidated beach narrows the underlying beachrock will influence swash permeability and the groundwater table, leading to a super-elevation of the groundwater and exit point, both of which will enhance beach erosion.

When the rock is exposed at the shore it behaves as an impermeable sloping seawall (Figure C3), which will decrease permeability, enhance erosion and retard sediment deposition and beach recovery. The rock itself may be eroded as tabular boulders, which are deposited to the lee as a boulder breccia (Figure C4).

If the beachrock becomes stranded off the shore, as a shore-parallel or curvilinear reef, it will act as an attached or detached shore-parallel breakwater, with usually discontinuous form longshore. This affects wave breaking, attenuation, and refraction. Wave breaking on the reef may result in zero ocean wave energy in lee of the reef, either permanently and/or at low tide (Figure C5). Wave attenuation over lower reef sections and/or at high tide will reduce waves in lee of the reef; while wave refraction over and or around detached reefs will redirect the wave train. The impact of the above is first to decrease breaker wave energy at the shore, and second to realign the backing sedimentary shore parallel to the refracted wave crests. As a consequence, shores in lee of beach rock reefs have lower energy beach systems and are more crenulate (Figure C5).

In temperate and some more arid-tropical locations, longer term pedogenic process in carbonate rich beach and dune deposits leads to the formation of calcrete (caliche). It consists of usually a loose thin top soil, underlain by laminar, breccia and massive calcrete consisting of recrystalized carbonate sands devoid of original structures, followed by mottled calcrete including lithocasts and lithoskels associated with vegetation roots and insect casts, and finally the loosely consolidated parent material, usually carbonate-rich dune sands (Scholle *et al.*, 1983) (Figure C6). Dune calcarenite occurs across North Africa to the eastern Mediterranean, in South Africa and southern and western Australia.

Dune calcarenite will have a similar impact to beachrock on beach morphodynamics. In addition, because calcrete will affect an entire beach–dune–barrier system, it is usually far more massive than beachrock, and consequently have far greater impact on the adjacent or backing coastline (Short, 2001).

Summary

Carbonate beaches occur throughout the tropics in association with coral reefs, tidal flats, and carbonate banks, and in temperate latitudes adjacent to rocky shores and reefs, and in lee of sea grass meadows and carbonate-rich shelves. Each of these environments produces a range of carbonate detritus sourced from coral fragments, encrusting algae, bryozoa, formanifera, molluscs, echinoids, and other animals. Depending on the erosion and transport processes, the detritus may be deposited at the beach as whole skeletons/shells, through to fine carbonate sands. Once deposited it may contribute to backing coastal dune systems, and be subject to cementation as beachrock in warmer tropical waters, and dunerock in arid to semi-arid carbonate rich beach–dune systems. Once formed the calcarenite can be exposed or submerged as beachrock or calcarenite barrier reefs, causing wave breaking, attenuation and refraction, which control the form of the lee coastline.

Andrew D. Short

Bibliography

Bird, E.C.F., and Schwartz, M.L. (eds.), 1985. *The World's Coastline*. New York: Van Nostrand Reinhold, 1071 pp.

Boreen, T., James, N., Wilson, C., and Heggie, D., 1993. Surficial cool-water carbonate sediments on the Otway continental margin, southeastern Australia. *Marine Geology*, **112**: 35–56.

Calhoun, R.S., and Field, M.E., 2001. Beach and reef-flat sediments along the south shore of Molakai, Hawaii. In Robbins, L.L., and Magoon, O.T. (eds.), *Carbonate Beaches 2000*, USGS and ASCE, Key Largo. Reston, VA: ASCE, pp. 163–171.

James, N.P., Collins, L.B., Bone, Y., and Hallock, P., 1999. Subtropical carbonates in a temperate realm: modern sediments on the southwest Australian shelf. *Journal of Sedimentary Petrology*, **69**: 1297–1321.

Knuuti, K., and Knuuti, K., 2001. A high-latitude carbonate beach in midcoastal Maine. In Robbins, L.L., and Magoon, O.T. (eds.), *Carbonate Beaches 2000*, USGS and ASCE, Key Largo. Reston, VA: ASCE, pp. 172–180.

Rhodes, E.G., 1982. Depositional model for a chenier plain, Gulf of Carpentaria, Australia. *Sedimentology*, **29**: 201–221.

Richmond, B.M., 2001. Overview of Pacific Island carbonate beach systems. In Robbins, L.L., and Magoon, O.T. (eds.), *Carbonate Beaches 2000*, USGS and ASCE, Key Largo. Reston, VA: ASCE, pp. 218–228.

Figure C5 Beachrock reef at Suape, Brazil blocking all wave energy at low tide (left). Beachrock reef at Port Gregory, Western Australia, backed by a Holocene barrier. Note the crenulate lee shoreline, and deep tidal, reef-controlled, channel (right) (A.D. Short).

Figure C6 Detail (left) of laminar, breccia, and massive calcrete (left) and a 20 m high section of aeolian calcarenite containing two buried palaesols (right), Robe, South Australia (A.D. Short).

Robbins, L.L., and Magoon, O.T. (eds.), 2001. *Carbonate Beaches 2000*, USGS and ASCE, Key Largo. Reston. VA: ASCE.

Scholle, P.A., Bebout, D.G., and Moore, C.H., 1983. *Carbonate Depositional Environments*. Tulsa: AAPF Memior 33.

Scoffin, T.P., and Stoddart, D.R., 1983. Beachrock and intertidal cements. In Goudie, A.S., and Pye, K. (eds.), *Chemical Sediments and Geomorphology*. London: Academic Press, pp. 401–425.

Short, A.D., 2001. Distribution and impact of carbonate sands on southern Australian beach-dune systems. In Robbins, L.L., and Magoon, O.T. (eds.), *Carbonate Beaches 2000*, USGS and ASCE, Key Largo. Reston, VA: ASCE, pp. 236–250.

Short, A.D., Buckley, R.C., and Fotheringham, D.G., 1989. Preliminary investigations of beach ridge progradation on Eyre Peninsula and Kangaroo Island. *Transactions of the Royal Society of South Australia*, **113**: 145–161.

Cross-references

Beachrock
Beach Sediment Characteristics
Cays
Cheniers
Coral Reefs
Coral Reef Coasts
Dune Calcarenite (see Eolianite)
Sandy Coasts
Tidal Flats

CARIBBEAN ISLANDS, COASTAL ECOLOGY AND GEOMORPHOLOGY

Geography

Location and topography

The main line of Caribbean Islands lie in an arc extending north from the coast of Venezuela and then trending westwards towards the southern tip of Florida; while other small group of islands lie off the northwestern coast of Venezuela and off the coasts of Central America. Most of the islands are located in the Tropics, and lie north of the 10°N line of Latitude and south of the 23°N Tropic of Cancer, although some of the islands in the Bahamas chain extend north of this line.

The Greater Antilles, consisting of four major islands—Cuba, Jamaica, Hispaniola and Puerto Rico—are, with the exception of Cuba, fairly mountainous with narrow coastal plains, while Cuba is dominated by extensive lowland plains. The Lesser Antilles consist of two chains, an outer chain from Barbados to Anguilla and continuing north to the Bahamas chain, consists of low-lying islands; while the inner chain from Grenada northwards are mountainous volcanic islands.

Geology

The Caribbean islands are quite young in geological terms, the Caribbean basin began forming less than 200 million years ago. The

West Indian Island arcs have been formed along a stress zone between the North and South American crustal plates. The inner chain of the Lesser Antilles developed along a line of weakness in the earth's crust. A series of submarine volcanoes formed, which rose above sea level to form the islands. Many of these islands still have signs of active volcanism such as, hot springs and sulfur emissions. Indeed there have been major volcanic eruptions in recent years, in particular Chances Peak in Montserrat, which started erupting in 1995 creating a volcanic emergency on the island and causing most of the population to flee to the northern third of the island or to other islands. This emergency continued until 1998, and now at the beginning of the 21st century, only the northern third of Montserrat is habitable and the dome on the volcano continues to grow at a significant rate. Other islands have also experienced eruptions in recent times, for example, Soufrière in St. Vincent and the Grenadines erupted in 1979.

The outer chain of the Lesser Antilles, and the Greater Antilles are older and have a more complicated history, consisting of volcanic episodes, alternating with periods of subsidence and sedimentation (Blume, 1974). The outer chain of the Lesser Antilles consists of islands with volcanic bases capped by thick limestone. Islands such as, Puerto Rico have a central mountainous spine consisting of igneous rocks with coastal plains composed of sedimentary rocks on the north and south coast (Bush *et al.*, 1995).

In more recent geological time there have been major changes in sea level, which have affected the geography of the islands. About 14,000 years ago, the sea level was 68 m below the present level and many associated islands, such as Puerto Rico and the Virgin Islands were joined by dry land. The sea reached near its present level about 5,500 years ago at that point the separation of most islands was completed. Trinidad became an island only in recent geological time, when erosion flooding of the Orinoco River separated it from the mainland across the Serpent's Mouth (Bacon, 1978).

Climate

Lying in the Tropics, the islands have an equable climate with little change in temperature from month to month. Average temperatures in the Bahamas are 25.1°C, and rise toward the equator, for example, 26.7°C in Curaçao. Temperatures vary with altitude and aspect. Precipitation amounts vary from island to island and within islands depending on topography and aspect. Most islands have a wet and a dry season, with the main wet season being from May/June to October/November, some islands may have two rainfall maxima. The dry season is generally from December to April.

The Northeast Trade Winds determine the weather in the Caribbean throughout the year. These blow with great constancy from directions between northeast and southeast. The Trade Winds are strongest between June and July and from December through March. During the first half of the year they usually blow from directions between northeast and east, veering more toward the southeast during the mid and latter parts of the year. The trailing ends of frontal systems moving from west to east over the southern part of the United States sometimes affect the Greater Antilles, particularly between October and April.

The Caribbean Islands lie in the hurricane zone. Many hurricanes originate as tropical waves off the west coast of Africa and travel across the Atlantic Ocean gaining energy from the warm ocean waters. As tropical waves strengthen, they pass through several stages, including tropical depression and tropical storm (see Table C3) before reaching hurricane strength.

A hurricane is an intense low-pressure tropical weather system with maximum surface wind speeds that exceed 118 km/h (74 m/h). Hurricanes are graded according to their sustained wind speed (wind gusts may be significantly higher) using the Saffir–Simpson scale of 1–5, see Table C3. The hurricane season lasts from June 1 to November 30, although rarely hurricanes can occur outside of these limits. September is the month when most hurricanes are experienced. Hurricanes generally move in an east to west/northwest direction across the Caribbean region, although exceptions have been recorded, particularly later on in the hurricane season, for example, Hurricane Lenny, a category 4 hurricane, formed just south of Jamaica in November 1999 and moved in an easterly direction and eventually hit Anguilla and St. Maarten.

Oceanography

Water temperatures are generally above 25°C. The region is affected by the North Equatorial Current, which flows generally from east to west across the Caribbean Sea. The surface current then turns northwards through the Yucatan Channel, where a part flows into the Gulf of Mexico and the other part forms the Florida Current, which eventually becomes part of the Gulf Stream.

Table C3 Categories of tropical low-pressure systems

Category	Sustained wind speed km/h	Sustained wind speed m/h
Tropical wave	Up to 36	Up to 22
Tropical depression	37–60	23–37
Tropical storm	61–117	38–73
Hurricane category 1	118–152	74–95
Hurricane category 2	153–176	96–110
Hurricane category 3	177–208	111–130
Hurricane category 4	209–248	131–155
Hurricane category 5	249+	156+

The average tidal range in the Caribbean is small, 0.3 m (1 ft), and 0.5 m (1.5 ft) at Spring Tides. Most islands experience a semi-diurnal tidal pattern. There are two very deep trenches in the Caribbean Sea. The Puerto Rico Trench follows a west–east trend and extends from north of the Dominican Republic to the northern islands of the Lesser Antilles, the maximum depth is 9,200 m. The Cayman Trench lies between Jamaica and the Cayman Islands.

Waves generally approach the region from the east, varying from northeast to southeast, depending on the Trade Wind regime. As a result, in the Lesser Antilles, where the islands generally trend north to south, there is a very clear distinction between the windward, exposed, east coasts with high wave energy, and the leeward, sheltered, west coasts. This distinction is not always so clear in the Greater Antilles, although here it is often the north and east coasts which are the most exposed to high wave energy.

During the Northern Hemisphere winter months (October to April), swell waves from North Atlantic low-pressure systems are often experienced in the Caribbean Islands. These swell waves travel southwards and impact north, east, and west facing coasts in particular. Also during the winter months, waves between the northwest and northeast may be experienced as frontal systems move across the southern United States and sometimes also affect the northern shores of the Greater Antilles.

Very high-energy waves from "unusual" directions may also be experienced during tropical storms and hurricanes. During such events, hurricane waves may directly impact the normally sheltered west coasts of the Lesser Antilles.

Biogeography

Small islands can be expected to have less diversity of species and habitats than continents. Newly formed islands may be isolated from land areas populated by plants and animals by large expanses of sea. And these islands can only be colonized by those species capable of making the sea crossing, as a result they will have fewer species than neighboring mainland areas (Bacon, 1978). The degree of isolation within an archipelago is another factor. Trinidad was separated from the South American mainland in relatively recent times, thus its flora and fauna are very similar to the neighboring mainland. However, the "oceanic" islands of the rest of the Caribbean have no such obvious relationship with the mainland. Distance from the mainland is another factor, for instance the number of species of amphibians and reptiles decreases northwards throughout the Lesser Antilles.

Coastal characteristics

It has been argued that islands do not have coastal zones, instead the whole island is a coastal system, extending from the highest peak to the offshore reefs. Everything that takes place inland, be it vegetation clearing or pesticide application, also affects the coast. The concept is obviously very relevant in some of the very small islands, such as the low-lying cays of the Virgin Islands and the Bahamas, and also some of the small volcanic islands where there is just one central peak or mountain range, such as Saba and Nevis. And while it also has merit for the mid- to large-sized islands, such as Dominica and Jamaica, for example, it is more difficult to apply here because of the overall complexity of those islands. It has further been suggested (Nichols and Corbin, 1997) that an Island System Management framework represents a way to achieve the objectives of Integrated Coastal Management.

Beaches

Differences in the structure and character of an island's rocks exert a strong influence on the coastline, whether it consists of cliffs, low rocky platforms, or some other form such as beaches or wetlands. With tourism playing such a major economic role in the Caribbean region, beaches are economically the most important part of the coastal zone. With the region's dependence on "sun, sea, sand" tourism, the very economic fabric of some islands is determined by its beaches and their characteristics. This is not to say that beaches are unimportant for island residents, indeed the opposite is the case, since islanders value their beaches highly as places for recreation, sports, and simple enjoyment. Furthermore, island beaches provide natural protection for very valuable real estate located along the coastline.

Morphology. Beaches in the Caribbean Islands range from long straight beaches extending for several kilometers, such as on the east coast of Barbados from what is known as the East Coast Road, just north of Bathsheba, to Long Pond and Greenland; to tiny pocket, bayhead beaches, extending just a few hundred meters, and enclosed by cliffed or rocky headlands, such as Lime Kiln Bay on the west coast of Montserrat. There is endless variety, not only in shape and geomorphology, but also in sediment size and type, and the presence or absence of rivers. Incoming wave energy and the presence or absence of coral reefs, are other important factors controlling coastal morphology.

Beach material. The word "beach" in the Caribbean Islands is usually taken to imply sand beach. However, geologically, beaches may consist of other sizes of particles (clay, silt, sand, gravel, cobbles, or boulders). In this entry, however, and given the regional importance of sand beaches and tourism, the word "beach" implies sand beach, unless otherwise qualified. The color of the beach sand usually reflects its source. In Anguilla, for instance, the beaches are for the most part composed of white calcareous sand, with patches of shells mixed in, this sand is derived from the offshore reefs and the breakdown of the limestone of which the island is composed. In Dominica, many of the beaches are composed of black sand with grains of olivine and magnetite, brought down to the coast by rivers, and derived from the erosion of volcanic rocks. The yellow silica sand along Trinidad's north coast is derived from the erosion of the Northern Range's sedimentary rocks.

Not all beaches consist of sand, larger size ranges from gravel to boulders, make up many beaches in the volcanic islands, for instance at Argyle on the east coast of St. Vincent the beach is composed of large stones and boulders. Some of the more exposed beaches on the south coast of Tortola in the British Virgin Islands are composed of large coral fragments and boulders.

The geology and the wave energy determine the type of material on the beach. Strong wave action washes out the sand particles leaving a steep beach profile as is often found on the Atlantic-facing beaches. While calmer conditions often result in fine sand being deposited and a gently sloping beach. Some beaches have different types of material according to the season, for instance some leeward beaches in the Lesser Antilles have stone beaches during the months with higher wave energy, October to April, while during the calmer summer months, sand is moved onshore to cover the stones.

A rock formation found on many Caribbean beaches, either on the beach or seaward of the beach is called beachrock. This consists of sand grains and other beach material, including stones, cemented together by calcium carbonate. Beachrock forms within the body of the beach, beneath the sand surface and near the water table. Once the covering sand has been eroded, the beachrock formation hardens into rock. Exposures of beachrock are indicative of erosion (Cambers, 1998).

Beach flora and fauna. Conditions above high water mark in the spray zone are very unfavorable for plants because of the dry sand and lack of water. Plants that grow here must store rainwater in succulent tissues, and have small-sized leaves to reduce water loss. Two plants often found above the high water mark are the Beach morning glory (*Ipomoea pescaprae*) and the grass Seashore dropseed (*Sporobolus virginicus*). They stabilize the sand surface and make conditions more favorable for other species such as Beach bean (*Canavalia maritime*), and Sea purslane (*Sesuvium portulacastrum*). However, high waves during storms frequently destroy this temporary beach vegetation, which then has to begin again establishing a foothold on the sand.

Inland from the spray zone there is a belt of woodland trees. These must withstand salt spray, strong winds, and saline soils. Common species include Sea grape (*Coccoloba uvifera*), Seaside mahoe (*Thespesia populnea*) and the West Indian almond (*Terminalia catappa*). These all have deep roots, which play an important role in stabilizing the sand. They also act as a windbreak. Another important tree is the Manchineel (*Hippomane mancinella*), this deep rooting tree plays an important role in slowing down beach erosion, however, the milky juice, which exudes from the branches when cut, may blister the skin, and the green fruits or small "apples" are poisonous. Another tree, the Casuarina (*Casuarina equisetifolia*), is an introduced species, which grows very rapidly in the coastal environment. However, it casts dense shade and produces a "carpet" of thick needles, which reduce the growth of ground cover. The coconut palm (*Cocos nucifera*), while not a native tree (it was introduced from the Indo-Pacific region) is very common in the islands and is often planted by tourism developers. However, it is very shallow rooting and is often the first tree to fall during high winds associated with tropical storms and hurricanes.

The beaches provide a habitat for a variety of crustaceans, mollusks, and other animals. Common species include the ghost crab (*Ocypode*), the Chip-chip (*Donax*), the Sand dollar (*Mellita quinquiesperforata*). Jellyfish sometimes washed up on the beach include the Portuguese man-of-war (*Physalia*), and the Tennis ball jellyfish (*Stomolophus*). Birds seen at Caribbean beaches include the Brown pelican (*Pelicanus occidentalis*), Brown booby (*Sula leucogaster*), and Frigate bird (*Fregata magnificens*).

Sea turtles nest on beaches in the Caribbean Islands, the most common species are the Leatherback (*Dermochelys coraicea*), Green (*Chelonia mydas*), Hawksbill (*Eretmochelys imbricata*), and Loggerhead (*Caretta caretta*). Most of these species are declining as a result of over-harvesting, loss of habitat, etc. Many islands have laws defining closed seasons or total bans on the catching of sea turtles or poaching of their eggs.

Beach changes. Coastlines are areas of continuous change where the natural forces of wind and water interact with the land. These changes have been taking place for millennia. History provides the example of Cockburn Town in the Turks and Caicos Islands, where Back Street had to be renamed Front Street at the beginning of the 20th century as erosion took its toll.

Beach changes have been measured in many of the Caribbean Islands since the 1980s as part of a regional program sponsored by the United Nations Educational, Scientific and Cultural Organization (UNESCO) and the University of Puerto Rico Sea Grant College Program. (This monitoring program is known by the acronym COSALC, Coast and Beach Stability in the Caribbean Islands Project, or as it has been recently renamed "Managing Beach Resources and Planning for Coastline Change, Caribbean Islands".)

Beaches change seasonally in the Caribbean Islands, in many islands a winter/summer pattern can be distinguished, with beaches eroding during the winter months, October to April, as a result of higher Trade Wind waves and winter swells from the North Atlantic storms, and the same beaches accreting in the summer months when wave energy is lower. However, the pattern often varies between different coasts, so it is unwise to generalize.

An initial analysis of the beach monitoring results (Cambers, 1997a) showed that 70% of the beaches had eroded and 30% had accreted. Mean annual erosion rates varied between 0.27 and 1.06 m/yr. Natural and anthropogenic factors are involved. The natural factors include winter swells, tropical storms and hurricanes, and sea-level rise. Among the most important anthropogenic factors are beach and dune mining, building too close to the beach zone, badly placed sea defenses, and deterioration of nearshore coral reefs.

Hurricanes appeared to be the major factor controlling the erosion. For instance, in Dominica (Cambers and James, 1994) there was significant erosion after Hurricane Hugo in 1989, followed by accretion in the years 1990–92. However, the beaches never returned to their pre-hurricane size, before further hurricanes impacted Dominica in 1995.

During 1995, three tropical storms/hurricanes hit the eastern Caribbean Islands within a three-week period: Tropical Storm Iris, Hurricane Luis, and Hurricane Marilyn. There was extensive damage, both to the islands' infrastructure and environment especially erosion of the beaches. The COSALC database was used to determine the extent of erosion of the land behind the beach (dunes or coastal lands) as a result of these events. Dunes on the north coast of Anguilla retreated as much as 30 m as a result of Hurricane Luis. Erosion rates were much lower further away from the hurricane center.

During the last part of the 1990s, the islands in the Lesser Antilles, from Dominica to Anguilla, experienced several tropical storms and hurricanes, leading to serious erosion. These numerous high-energy events appeared to introduce a certain vulnerability to the coastal systems making recovery slower and less sustained.

Beach protection. The increasing coastal development in the Caribbean Islands, particularly over the last two decades, coupled with the ongoing coastal erosion, has resulted in many property owners and sometimes government agencies building structures to slow down or stop the erosion. These structures include revetments, seawalls, groins, and offshore breakwaters. They have had varying degrees of success. Some islands, for example, St. Maarten and Anguilla, have attempted to restore their beaches by nourishing them with sand from offshore. Again, these measures have had varying degrees of success, at Maunday's Bay in Anguilla, the beach was nourished three times during the period 1995–99, only to have the sand washed away by a high-energy or hurricane event in 1997, 1998, and 1999.

There are no easy answers to this dilemma, especially since every beach is individual and must be treated as such. The use of adequate coastal development setbacks, which ensure that new coastal development is placed a "safe" distance away from the active beach zone has been proposed. A methodology has been developed, based on using existing data on erosion, hurricanes, and sea-level rise to determine a specific setback distance for each beach (Cambers, 1997b). Thus, beaches experiencing more erosion have higher setback distances. This methodology is at present being applied in Anguilla and Nevis.

Undoubtedly erosion is going to continue in the Caribbean Islands, and the problems of how to protect beachfront property and to maintain beaches, will likely intensify.

Cliffs and rocky shores

Limestone and volcanic cliffs abound in the Caribbean Islands. Caves and "blowholes" exist in limestone cliffs in many islands, for example, Barbados, Bahamas and Turks and Caicos Islands. Steep rock slopes and cliffs are the most predominant coastline type in some of the volcanic islands, for example, Montserrat and Dominica.

Rocky shores are characteristically areas of rapid erosion, and they provide a great variety of habitats for marine plants and animals (Bacon, 1978). Animals include barnacles such as *Tetraclita squamosa* and *Chthamalus fissus*, snails of the genus *Littorina*, mussels such as *Brachidontes exustus*, and crabs such as the Rock crab (*Grapsus grapsus*). Sea eggs, for example, *Lytechinus*, may be found on limestone and other hard rocks. Green algae are abundant and include *Chaetomorpha* and *Cladophora*, and some species of red and brown species may also occur. Boulders may be seen encrusted with calcareous algae.

Sand dunes

Sand dunes are mounds of sand that often lie behind the active part of the beach. In the Caribbean Islands, they range from very low formations, 0.3–0.6 m in height to large hills of sand up to 6 m high. There may be several parallel rows of dunes. Dunes form when sand is carried by the wind from the dry beach area to the land behind the beach. When the wind encounters an obstacle such as a clump of vegetation it slows and the sand is deposited. Significant sand movement will take place when the wind speed, measured at a height 1 m above the ground, exceeds 6 m/s (Bagnold, 1954). In the Caribbean Islands, average wind speeds equal or exceed this value in the months from June to July and from December through March. During storms/hurricanes, dunes are often eroded and the sand may be deposited offshore, after the storm/hurricane, the sand is returned to the beach and the slow process of dune building may recommence. As a result of Hurricane Luis, dunes in Anguilla showed an average retreat of 9 m, recovery is expected to take decades, if it occurs at all (Cambers, 1996).

Dune vegetation promotes the trapping of sand, however, as with beach vegetation, dune plants have to adapt to harsh conditions including high temperatures, dryness, salt spray, and the deposition of sand (Craig, 1984). Dune vegetation in the Caribbean Islands includes the same plants as those described colonizing beaches. In addition, Sea lavender (*Tournefortia gnaphalodes*) has been observed to be a very effective shrub in stabilizing dunes.

Unfortunately sand dunes are often seen as prime sites for sand mining activities in the Caribbean Islands. Many dune areas have completely disappeared as a result of mining for construction material, for example, at Josiah's Bay in Tortola in the British Virgin Islands, and at Diamond Bay on the east coast of St. Vincent, where extensive black sand dunes, more than 6 m high have now gone. At Isabela on the north coast of Puerto Rico, the mining of an extensive system of dunes has taken place in the last two decades, leaving just a narrow strip of residual dunes, which are sometimes breached during high wave events.

Efforts have been made in some island to restore dunes using sand fences, for example, at Arecibo in Puerto Rico, and Shoal Bay North in Anguilla. Research has shown that a sand fence with a 50% porosity (ratio of open space to total area) can result in a 1.2 m high dune forming in 12–24 months (Clark, 1995). Work with sand fences in the Caribbean islands and elsewhere has shown that one of the critical factors in dune growth is the width of dry sand beach in front of the fence.

Mangroves

In estuaries, lagoons and coastal mudflats, characterized by mud deposits and sheltered wave environments, mangrove ecosystems have developed characterized by trees of the red mangrove (*Rhizophora mangle*). Red mangroves have prop roots and can grow on newly deposited sediments and on sand bars and coral reef crests (Bacon, 1978). Large mangrove swamps can be seen on the Gulf Coasts of Trinidad, and the south coast of Jamaica in the Portland Bight area. As sediment gradually accumulates around the roots of the red mangrove, conditions become suitable for the establishment of the black mangrove (*Avicennia germinans*). This species has a horizontal root system from which arise "breathing tubes" pneumatophores. A third mangrove species grows among both red and black mangroves, the white mangrove (*Laguncularia racemosa*). A fourth mangrove species, Buttonwood (*Conccarpus erectus*) is considered a transitional species between true mangroves and terrestrial vegetation.

Coastal mangroves are very important to fisheries and marine productivity. They provide a habitat, which is a breeding ground and nursery area for many species. Juvenile fishes live in the mangrove swamps until they are mature enough to migrate to the open sea, these include young butterfly fish, angel fish, tarpon, snappers, and barracudas, in addition there are juvenile spiny lobsters and shrimps. Other animals include worms, small crustaceans, and sea cucumbers. Clusters of oysters cling to the red mangrove roots, there are two common species, the mangrove oyster (*Crassostrea rhizophorae*) and the flat tree oyster (*Isognomon alatus*) (Jones and Sefton, 1978). There are also sponges and sea anemones. Numerous insects, lizards, tree snails, and birds also live in the mangrove swamps.

There is a complex food web within the mangroves, starting with the mangrove leaves which fall and decay and are then eaten by shellfish, shrimps, crabs and fishes, which in turn become the prey of other juvenile fishes.

Because of their intricate root systems, mangroves also provide good protection to the coastline during storms and hurricanes. As a result, mangrove lagoons, such as Paraquita Bay in Tortola in the British Virgin Islands, are used as a safe mooring area during storms and hurricanes (Clarke and Lettsome, 1989).

Coral reefs

Among the most ecologically diverse systems in nature, coral reefs play an important role in the protection and formation of many Caribbean beaches. A coral reef is made up of many millions of coral polyps, tiny animals that secrete limestone to form themselves a stony skeleton. A healthy coral reef is home to many different plants and animals. Stony or hard corals are the main reef builders, while soft corals for example, the sea fan (*Gorgonia flabellum*) are flexible.

The main types of reefs found in the Caribbean are fringing reefs, which border a coastline, for example, there are several fringing reefs along the west coast of Barbados; patch reefs, which are isolated clumps of coral sometimes measuring only a few meters in diameter; and barrier reefs, which are separated from the coastline by a deep lagoon or channel, for example, the barrier reef off the east coast of Andros Island in the Bahamas.

Coral reefs are especially important to beaches because they protect the shore from high waves. Fringing reefs and barrier reefs often grow very close to the sea surface. Incoming waves break and expend their energy on the reef, and while secondary waves may reform in the lagoon and ultimately affect the beach, they are much lower in energy. Thus in many ways reefs are also natural breakwaters.

Coral reefs also act as a sand source for many beaches. Many fishes, for example, the Parrot fish (*Scarus*), Butterfly fish (*Chaetodon*), and Trigger fish (*Balistes*), actively feed on the coral, biting off chunks of coral and excreting coral sand, which is then moved towards the beach by wave action.

Coral reefs are vulnerable to natural forces, such as hurricanes, increased seawater temperatures, human activities such as pollution, offshore dredging, and ship anchoring. Hurricane Luis, a category 4 hurricane, passed over Anguilla in 1995. A quantitative offshore benthic survey had been conducted the previous year (1994), and comparing this with a post-hurricane survey in 1996 showed that 61% of the intact live reefs, either hard coral or soft corals, were degraded to rubble or, bare rock. Previously intact, though dead, elkhorn coral reefs

(*Acropora palmata*) were largely reduced to rubble. Damage was most severe at the inshore reefs (Bythell and Buchan, 1996).

It may take some years before the damage to a coral reef is manifested at the beach itself, such as through a reduced breakwater effect. And then, it may be difficult to separate out this cause from other causes of beach erosion.

Sea grass beds

Sea grass beds are important habitats found in shallow nearshore waters. They are often found between the beach and coral reefs. There are three main types of sea grasses: Turtle grass (*Thalassia testudinum*), Manatee grass (*Cymodocea filiforme*), and Shoal grass (*Diplanthera wightii*). Sea grasses are flowering plants, not algae, and they spread mainly through a system of underground creeping stems. They are an important marine habitat for many marine animals and their larval stages. They also provide foraging grounds for commercially important species such as grunts and conch. Sea grass beds maintain clear coastal waters by filtering sediments from the water and holding them down with their complex root systems. Calcareous algae living among the sea grass beds produce carbonate sediment.

Sea grass beds are very vulnerable to hurricanes, and often after a hurricane, beaches may be covered with thick mats of dead sea grass, more than 1 m thick. Hurricane Luis in 1995, reduced the sea grass bed cover by an average 45% in Anguilla (Bythell and Buchan, 1996).

Problems facing Caribbean Islands' coasts

Tourism development. Tourism growth is taking place in most of the Caribbean Islands, indeed the tourism business has been termed "the engine of growth" (Patullo, 1996). The Caribbean's share of world tourist arrivals is less than 2%, but this is triple that of South Asia's share. The insular Caribbean ranked ninth in the world on the basis of its tourism receipts in 1990, and in 1994 the Caribbean Tourism Organization estimated that its 34 member states grossed US$ 12 billion from tourism. Tourism is a big business in the Caribbean Islands.

However, besides bringing economic growth and employment for many, tourism also brings environmental problems, partly because in its early days, in the 1970s and 1980s, tourism growth was largely unregulated. This created a precedent and by the time environmental controls were being brought into place, around the early 1980s, much of the damage had already been done. Islands such as Barbados were already facing serious pollution and degradation of their nearshore fringing reefs, Jamaica was dealing with beach exclusivity such that local residents could no longer gain access to certain prime beach sites, and even small islands, such as Montserrat, were seeing a construction boom and increased mining of their beaches.

Efforts to try and reverse some of these trends dominated the 1990s and are continuing today. However, it is difficult to find many success stories, since environmental concerns usually come second to economic growth priorities. And with so much of the Caribbean's tourism industry, from the hotel rooms to the airlines to the tour operators, being controlled by foreign interests (Patullo, 1996) it is sometimes difficult for the islands to take control of this vitally important industry. Yet, in order for the islands to truly benefit from tourism in the long term, this is what must happen.

Climate variation. Predictions about global climate change as a result of global warming still have a considerable degree of uncertainty. However, climate variation can be traced back over the last century at least, and can be used to predict climate trends in the near future. For instance, hurricanes in the North Atlantic Basin have been shown to have a cyclical pattern (Sheets and Williams, 2001). There are periods of two to three decades when there are many hurricanes—active periods, followed by periods of similar duration when there are fewer hurricanes—quiet periods. Records show that the 1900s to the 1920s was a quiet period, followed by an active period from the 1930s to the 1960s. Recently, there has been a quiet period, which extended from 1971 to 1994. However, since 1995, the Atlantic Basin has entered an active period, with the last five years of the 1990s showing an increased number of hurricanes and in particular an increased number of very intense hurricanes (category 3+). For instance, the years 1995–99 showed an average of four category 3+ hurricanes per year, compared with the previous five years, 1990–94, when the average was one category 3+ hurricane per year.

While these are average figures referring to the entire Atlantic Basin, this trend can be seen in many of the Caribbean Islands, particularly the northern Lesser Antilles, from Dominica to Anguilla, where many islands have experienced at least one hurricane per year since 1995. This climatic variation, while not yet fully understood, appears to be related to several climatic factors such as the rainfall in West Africa, upper level winds and ocean currents. With the massive tourism-related coastal development, which has taken place in coastal areas in the Caribbean Islands since the 1970s, this trend is of considerable concern to the islands.

While not directly related to climate variation, another infrequent phenomenon, about which the Caribbean Islands must become increasingly concerned, is the threat of a tsunami. Given its tectonic instability, the region is long overdue for a tsunami event. The last major tsunami in the Caribbean was on October 11, 1918, when a tsunami generated by an underwater earthquake, hit northwestern Puerto Rico and caused considerable damage as well as more than 100 deaths. The increased coastal development that has taken place in recent decades makes the Caribbean Islands very vulnerable to a tsunami threat. Added to this, is the fact, that as yet, there is no tsunami warning system in place in the Caribbean region.

The challenges for integrated coastal management in the Caribbean Islands. In seeking sustainable development, the Caribbean Islands face many problems. These include their small size, relative isolation, vulnerability to natural disasters, limited economic diversification and access to external capital (Commonwealth Secretariat, 2000). They will have to develop their own solutions to these problems and limitations.

Integrated coastal management is one approach that can be adapted to their needs. This will involve adding an intersectoral, horizontal dimension, which includes civil society, to what is essentially a top–down sectoral system of management in most islands. This will not be easy given the dichotomy of timescales between the political agenda (4–5 years) and the environmental agenda (decades). Lying at the center of all the ongoing efforts to achieve sustainable development and integrated coastal management is the need for effective and efficient communication.

Gillian Cambers

Bibliography

Bacon, P.R., 1978. *Flora and Fauna of the Caribbean. An Introduction to the Ecology of the West Indies.* Trinidad: Key Caribbean Publications.

Bagnold, R.A., 1954. *The Physics of Blown Sand and Desert Dunes.* New York: William Morrow and Co.

Blume, H., 1974. *The Caribbean Islands.* London: Longman Group Ltd.

Bush, D.M., Webb, R.M.T., González Liboy, J., Hyman, L., and Neal, W.J., 1995. *Living with the Puerto Rico Shore.* Durham: Duke University Press.

Bythell, J.C., and Buchan, K.C., 1996. *Impact of Hurricane Luis on the Coastal and Marine Resources of Anguilla: Marine Ecological Survey.* Barbados: British Development Division in the Caribbean.

Cambers, G., 1996. *The Impact of Hurricane Luis on the Coastal and Marine Resources of Anguilla, Beach Resources Survey.* Barbados: British Development Division in the Caribbean.

Cambers, G., 1997a. Beach changes in the Eastern Caribbean Islands: Hurricane impacts and implications for climate change, In Leatherman, S.P. (ed.), *Island States at Risk: Global Climate Change, Development and Population. Journal of Coastal Research,* Special Issue No. 24, pp. 29–48.

Cambers, G., 1997b. *Planning for Coastline Change: Guidelines for Construction Setbacks in the Eastern Caribbean Islands.* CSI *info* 4, Paris: UNESCO.

Cambers, G., 1998. *Coping with Beach Erosion.* Paris: UNESCO Publishing.

Cambers, G., and James, A., 1994. Sandy coast monitoring: the Dominica example (1987–1992). *UNESCO Reports in Marine Sciences,* **63**: 91 pp.

Clark, J., 1995. *Coastal Zone Management Handbook.* Boca Raton: CRC Press.

Clarke, N., and Lettsome, B., 1989. *Natural Resources of the British Virgin Islands.* Caribbean Studies, Laurel Publications International.

Commonwealth Secretariat, 2000. *Small States: Meeting Challenges in the Global Economy.* Report of the Commonwealth Secretariat/World Bank Joint Task Force on Small States.

Craig, R.M., 1984. *Plants for Coastal Dunes of the Gulf and South Atlantic Coasts and Puerto Rico.* Agriculture Information Bulletin, No. 460, Washington, DC: US Department of Agriculture.

Jones, A., and Sefton, N., 1978. *Marine Life of the Caribbean.* New York: Macmillan Education Ltd.

Nichols, K.E., and Corbin, C., 1997. Community Based Approaches to Beach Management in St. Lucia, In Cambers, G. (ed.), *Managing*

Beach Resources in the Smaller Caribbean Islands, Papers presented at a UNESCO-University of Puerto Rico Workshop, 21–25 October 1996, Mayaguez, Puerto Rico. Coastal regions and small islands papers #1, UNESCO, pp. 182–190.
Patullo, P. 1996. *Last Resorts: The Cost of Tourism in the Caribbean.* Jamaica: Ian Randlr Publishers.
Sheets, B., and Williams, J., 2001. *Hurricane Watch.* New York: Vintage Books, Random House Inc.

Cross references

Coastal Climate
Coral Reefs
Mangroves, Coastal Ecology
Mangroves, Coastal Geomorphology
Meterorologic Effects on Coasts
Mining of Coastal Materials
Small Islands
Tourism and Coastal Development
Tourism, Criteria for Coastal Sites

CARRYING CAPACITY IN COASTAL AREAS

The concept of capacity has received considerable attention as a result of increasing anthropogenic pressure in certain natural environments. Much consideration has recently been given to increases in coastal populations, with the implication that the carrying capacity of the world's coast is finite and such considerations form part of several coastal management initiatives (UNEP, 1996).

Johnson and Thomas (1996) argue that present interest in tourism capacity is due to growth in tourism combined with increasing awareness of environmental issues. The concept is particularly important in the coastal zone which is undergoing rapid change as a result of demographic changes and industrialization (see Kay and Alder, 1999, p. 21) in the context of global climate and sea-level change. In its broadest sense, carrying capacity refers to the ability of a system to support an activity or feature at a given level. In the coastal zone, these systems can vary greatly in both scale and type, and range from small salt marshes through large beach resorts to entire continental coasts. The activities or features that they support are also varied and include, for example, beach recreation or species abundance. The term "carrying capacity" does not therefore have a single precise definition. Rather, it is a broad term that covers a range of different concepts. These concepts are related by the idea that systems such as beaches have certain limits or thresholds. For example, a maximum number of animals can be grazed on any given dune system. Attempting to determine the actual limits is often problematic and raises some fundamental questions. In the case of dune grazing, various criteria could be used to define the carrying capacity. This could involve assessing the effects of grazing on, for example, the physical integrity of the site, its ecological status, or its recreational value. In practice, these features may be interrelated.

The situation is further complicated by the subjective nature of certain limits. For example, the point at which the aesthetic impact of grazing becomes unacceptable is difficult to define and may vary from one location or cultural setting to another. In recognition of the diverse nature of carrying capacity as a concept, a variety of types of carrying capacity have been identified. Most of these fall into the following categories: physical, ecological, social, and economic.

Physical carrying capacity: This is a measure of the spatial limitations of an area and is often expressed as the number of units that an area can physically accommodate, for example, the number of berths in a marina. Determining the physical capacity for certain activities can, however, become problematic when subjective elements are introduced. For example, the maximum number of people that can safely swim in a bay depends on human perceptions and tolerance of risk.

Ecological carrying capacity: At its simplest, this is a measure of the population that an ecosystem can sustain, defined by the population density beyond which the mortality rate for the species becomes greater than the birth rate. The approach is widely adopted in fisheries science (e.g., Busby *et al.*, 1996). In practice, species interactions are complex and the birth and mortality rates can balance over a range of population densities. In a recreational context, ecological carrying capacity can also be defined as the stress that an ecosystem can withstand, in terms of changing visitor numbers or activities, before its ecological value is unacceptably affected. This approach raises the difficult question of defining ecological value and what constitutes an unacceptable change in it.

Social carrying capacity: This is essentially a measure of crowding tolerance. It has been defined as "... the maximum visitor density at which recreationists still feel comfortable and uncrowded" (De Ruyck *et al.*, 1997, p. 822). In the absence of additional changes, beyond this density visitor numbers start to decline. The social carrying capacity can, however, be influenced by factors such as the recreational infrastructure, visitor attitudes, and sociocultural norms.

Economic carrying capacity: This seeks to define the extent to which an area can be altered before the economic activities that occur in the area are affected adversely. It therefore attempts to measure changes in economic terms (Rees, 1992). Examples from the coastal zone might include examining the effect of increased numbers of trailer parks on agricultural activity in dune systems.

In addition to these single discipline assessments, there are a number of composite measures such as recreational and tourist carrying capacity. These attempt to define the threshold of an area for tourism or recreation by combining a range of indicators (Sowman, 1987). The actual carrying capacity of a coastal area assessed according to any of the above approaches depends largely on the nature of the area. Carter (1989, p. 357) noted that "Coastal environments vary considerably in their ability to absorb anthropogenic pressure. The carrying capacity of dune grassland is many orders of magnitude below that of rock cliffs." While this may be true, at least in some views of carrying capacity, it should be borne in mind that capacities are not necessarily fixed in time. They can often be altered by management practices for example, the provision of recreational facilities can increase the social carrying capacity of an area. They can also alter in response to wider environmental changes, whereby a change in mean sea temperature could affect the ecological carrying capacity of an area for a range of species, or a shift in social attitudes could alter what was considered acceptable degradation. As Arrow *et al.* (1995, p. 520) have noted: "Carrying capacities in nature are not fixed, static or simple relations. They are contingent on technology, preferences, and the structure of production and consumption. They are also contingent on the ever-changing state of interactions between the physical and biotic environment."

M. MacLeod and J.A.G. Cooper

Bibliography

Arrow, K., Bolin, B., Costanza, R., Dasgupta, P., Folke, C., Holling, C.S., Jansson, B-O., Levin, S., Maler, K.-G., Perrings, C., and Pimentel, D., 1995. Economic growth, carrying capacity, and the Environment. *Science*, **268**: 520–521.
Busby, P.J., Wainwright, T.C., Bryant, G.J., Lierheimer, L.J., Waples, R.S., Waknitz, F.W., and Lagomarsino, I.V., 1996. Status review of West Coast Steelhead from Washington, Idaho, Oregon, and California. NOAA Technical Memorandum NMFS-NWFSC-27.
Carter, R.W.G., 1989. *Coastal Environments: An Introduction to the Physical, Ecological and Cultural Systems of Coastlines.* London: Academic Press.
De Ruyck, M.C., Soares, A.G., and MacLachlan, A., 1997. Social carrying capacity as a management tool for sandy beaches. *Journal of Coastal Research*, **13**: 822–830.
Johnson, P., and Thomas, B., 1996. Tourism capacity: a critique. In Briguglio, L. (ed.), *Sustainable Tourism in Islands and Small States: Issues and Policies*, pp. 118–136.
Kay, R., and Alder, J., 1999. *Coastal Planning and Management.* London: E and FN Spon.
Rees, W., 1992. Ecological footprints and appropriated carrying capacity: what urban economics leaves out. *Environment and Urbanization*, **4**: 121–130.
Sowman, M.R., 1987. A Procedure for assessing recreational carrying capacity of coastal resort areas. *Landscape and Urban Planning*, **14**: 331–344.
UNEP, 1996. Guidelines for Integrated Planning and Management of Coastal and Marine Areas in the Wider Caribbean Region. UNEP Caribbean Environment Programme, Kingston, Jamaica, 1996.

Cross-references

Coastal Zone Management
Economic Value of Beaches
Environmental Quality
Human Impact on Coasts
Tourism and Coastal Development
Tourism, Criteria for Coastal Sites

CAYS

Cays are islands that form on coral reefs as the result of wave action accumulating reef-derived sediment in one location. Cays may not rise above the highest tide level, but those that do can quickly acquire a vegetative cover but because they are formed of unconsolidated materials, may remain mobile and unstable.

Accumulation of sediment on a reef top results from the centripetal action of refracted waves around a coral reef. The shape of the reef may mean that even under different wind directions the focal point may remain more or less the same, though a strongly prevalent wind or swell wave direction will favor cay formation. Shape of the reef may also influence the shape of the cay because of the wave refraction pattern it produces. Oval reefs are more likely to produce oval cays, while more elongated reefs produce linear cays. Because relatively small changes in wind direction can produce much greater changes in wave refraction patterns on the elongated reefs, these linear cays may be more unstable. The size of the reef is to some extent irrelevant as cays can be found on reef tops only a hectare in area, or on the largest of reefs. However, on larger reefs the competency of the transmitted wave (the reformed wave after its initial break on the reef crest) to carry sediment may be lost over the wide reef flat before the centripetal focus point is reached, and sediment is more widely deposited as a veneer over the reef flat.

Cay sediments

Because of their microstructure, biogenic sediments are generally bimodal, either coarse (at least gravel size) or sand size. Coarser sediments are almost entirely coral fragments, most commonly clasts of *Acropora* sp. and other branching corals. Considerable energy is required to move such deposits and accumulation of gravel banks is normally found on the windward or high-energy side of a reef no more than 100 m or so from the reef edge. Major episodes of deposition have been described at times of tropical storms. Atoll motus, linear largely gravel islands which are set back from and are parallel to the reef edge are good examples.

The sediments of sand cays are more variable and include fragmented coral, coralline algae, molluscs, *Halimeda* flakes and complete tests of smaller foraminifera (including the "star" sands of some cay beaches). Their size range, however, is remarkably uniform worldwide, commonly in the range of 0–1.5ϕ and sorting is generally less than 1ϕ.

Hydrogeology and cementation

A relatively simple explanation of cay hydrology, the Ghyben–Herzberg model, has generally been applied. It consists of a freshwater lens derived from rainwater, "floating" over the denser seawater. A minimum size of cay is required for it to form and under larger cays the freshwater lens may rise up to a meter above the adjacent sea level and extend several meters down. General flow of water is outwards from the cay center. However, a layered aquifer model (e.g., Oberdorfer and Buddemeier, 1988) that more realistically presumes a Holocene reef overlying Pleistocene reef with contrasting permeabilities is now accepted.

Outflow at the island margins is still present, and is the major process in the cementation of cay sediments beneath the beach as beachrock. Percolating meteoric water is aggressive to the carbonate sediments especially with increasing partial pressure of CO_2 within the cay. However, as this pressure reduces as the groundwater flows out beneath the beach, this process is reversed and cementation takes place.

Cays may also be important nesting sites for birds and their droppings percolate to the water table as guano, where they may accumulate as phosphatic cay sandstone. Both beachrock and cay sandstone are important in providing some stability to coral cays.

Vegetation

Coral cays provide a classic example of vegetation succession, many of the plants being pan-tropical. Seeds either float to the island or reach it through the digestive tract of birds. In time, decaying vegetation adds organic matter to the soil, which allows more advanced plants to become established.

Initial colonizers are robust creepers and vines such as *Ipomoea* sp. and *Canavalia* sp. or salt tolerant succulents such as *Sesuvium* sp. They are followed by grasses (e.g., *Lepturus repens*) then shrubs including *Argusia argentea* or (on coarser or cemented deposits) *Pemphis acidula*. Higher trees, such as *Casuarina equisetifolia* may rise above the shrub layer. As soils become more fertile, an interior forest may become the dominant vegetation with *Pisonia grandis* being the most common on atoll motus and cays of the Pacific.

Because cays are constantly changing and being added to, all stages of the vegetational succession may be found as a concentric pattern on the island.

Classification of coral cays

Cays are classified on the basis of sediment type (gravel or sand), location on the reef (windward or leeward), shape (linear or compact), and stage of vegetation cover (Stoddart and Steers, 1977; Hopley, 1982, 1997). Most mobile and ephemeral are unvegetated sand cays. Vegetated cays are obviously more stable, but linear sand cays may still exhibit a degree of mobility especially the sand spits, which occur at each end of the island. As sand cays become more stable with cemented deposits and vegetation, occasional storms and wind can add to the island height and large stable well-vegetated sand cays may rise as much as 3 m over high water mask (HWM). However, as such cays have been dated to as much as 5,000 years old, in some regions a higher mid-Holocene sea level may have been implicated in their early evolution.

Coarser gravel (shingle) cays develop from linear rubble banks parallel to the reef front, or hammerhead spits behind promontories on the reef front. Though more stable they are more inhospitable to vegetation and their colonization may be slower.

In some parts of the world both windward gravel deposits and leeward sand cays may develop on the same reef top, with the intervening reef flat being subsequently colonized by mangroves. The term "low wooded island" has been applied to these reef islands, which have been described especially from Australia's Great Barrier Reef (Hopley, 1982, 1997) and Indonesia (Tomascik *et al.*, 1997). Both of these areas have the greatest diversity of coral cays in the world.

David Hopley

Bibliography

Hopley, D., 1982. *Geomorphology of the Great Barrier Reef: Quaternary Development of Coral Reefs*. New York: John Wiley—Interscience.

Hopley, D., 1997. Geology of reef islands of the Great Barrier Reef, Australia. In Vacher, H.L. and Quinn, T.M. (eds.), *Geology and Hydrogeology of Carbonate Islands, Developments in Sedimentology* Vol. 54. Amsterdam: Elsevier, pp. 835–866. (See other papers on coral islands in this volume.)

Oderdorfer, J.A., and Buddemeier, R.W., 1988. Climate change: effect on reef island resources. *Proceedings of the 6th International Coral Reef Symposium*, **3**: 523–528.

Stoddart, D.R., and Steers, J.A., 1977. The nature and origin of coral reef islands. In Jones, O.A. and Endean, R. (eds.), *Biology and Geology of Coral Reefs* Vol. IV, Geology 2. New York: Academic Press, pp. 60–105.

Tomascik, T., Mah, A.J., Nontji, A., and Moosa, M.K., 1997. *The Ecology of the Indonesian Seas*. 2 vols. Hong Kong: Periplus, NY: Academic Press.

Cross-references

Atolls
Beachrock
Coral Reef Islands
Coral Reefs
Hydrology of Coastal Zone

CENTRAL AMERICA—See MIDDLE AMERICA

CHALK COASTS

The extraordinary whiteness of chalk has given rise to distinct cliffs visible over large distances. From the Fourth Century BC Greek navigator, Pytheas of Massilia, through Julius Caesar and Shakespeare, to modern writers, chalk coasts were given a unique character. They are most commonly known for their place in Shakespeare's "King Lear," the literature of Maupassant and the paintings of Monet and the wartime nostalgic song, "White Cliffs of Dover." The Romans called England "Albion" because of its white cliffs. The earliest regional study of a chalk coast is Girard's (1907) description of the cliffs of the English Channel. Chalk cliffs occur, however, more widely than these classic locations. Chalk is a $CaCO_3$ limestone found in the Upper Cretaceous outcrops on the European coast, in northern Ireland, Yorkshire and Norfolk, along the southern coast of England from the Isle of Thanet in the east to Beer in the west (Steers, 1946), along the northern coast of France from Calais to the mouth of the Seine (Girard, 1907; Briquet, 1930; Precheur, 1960) and on the islands of Møn, Denmark, and Rügen, Germany.

Chalk coasts are distinctive for four reasons: the relative purity of the chalk itself, their distinctive landscape, their often vertical rapidly retreating cliffs and their place in the geomorphological literature as type locations for particular landforms. The chalk contains veins of flint, a biogenic silica, which are typically nodular in shape, vary in size (long axis) from 2 cm to over 200 cm and can weigh more than 8 kg. They are the main source of beach clasts on these coasts and may exist in their original form, break into smaller angular particles (quickly rounded at their edges), or as rounded pebbles. These pebbles are often the result of erosion of re-worked landslides and fluvio-glacial deposits. Chalk cliffs have been important, but not unique, suppliers of flint shingle to such large and important features as the ness, spit, and tombolo structures of southeast England, including Orfordness, Dungeness, Hurst Castle spit, and Chesil beach (May and Hansom, 2003), and northern France.

Chalk is notable for its whiteness, its purity, and its friability (Hancock, 1976) and is formed mainly of calcareous fossil debris. The white chalk is dominated by two main groups of particle sizes (0.5–4 μm and 10–100 μm). The lower 25–60 m of the chalk is often dominated by chalk-marl facies in which calcareous-rich chalk alternates with marls containing up to 29% clay (Perrin, 1971). The dolomitic chalk outcrops of the Normandy coast, the white chalk of Yorkshire and the White Limestone of Antrim are harder than other chalks because of diagenesis by burial and locally high heat flows. Typically, these harder chalks retreat less rapidly than the other chalk coastlines. Chalk porosity ranges between 35% and 47% (Hancock, 1976), but matrix permeability is exceptionally low. Mass permeability is controlled by joints, fractures, and passages enlarged by solution, all of which are very common in the chalk and have a very important role in the development of some of the more complex chalk coasts.

Cliff forms (Table C4) include very steep slopes, slopes which are strongly faceted, and shallower slopes which are associated with large coastal landslides such as Folkestone Warren (Hutchinson *et al.*, 1980). Many chalk cliffs with low mean retreat rates have small slope facets with mean angles between 30° and 42° that are vegetated by chalk downland grasses and flowering plants. On these slopes, there may be skeletal calcareous soils, but occasionally red-brown sandy clay deposits ("*clay with flints*") on the cliff-top are washed down the cliffs and form local pockets of soil.

Some chalk coasts include very intricate cave and inlet and stack and arch features. Caves and inlets usually develop along discontinuities in the rock and give rise to large numbers of narrow inlets. At Flamborough Head, for example, there are more than 55 narrow inlets or geos and around the Isle of Thanet headlands were riddled with inlets and caves, some of which were filled in during coast protection works in the 1960s. The classic erosional cave-arch-stack-stump model found in many school textbooks owes much to the example of differential erosion of the sea along intersecting faults and joints north of Swanage. Stacks are rare features on chalk coasts, usually because the friability of the upper cliffs or the angles of dip are too great to allow classic pinnacle stacks to survive. In contrast, the English cliffs between Beachy Head and the mouth of the Cuckmere River (the Seven Sisters) are one of the best examples worldwide of a cliffed coast where, despite the local effects of jointing, cliff falls and pocket beaches and the larger

Table C4 Major categories of chalk coast—England and France

Category	Main features	Cliff height (m)	Typical rates of cliff-top retreat ($m\ a^{-1}$)	Length (km) England	Length (km) France
I	Simple plan and profile, former usually similar to beach plan form. Slope angles exceed 75°. Extensive platforms.	Less than 31[a]	0.13–1.33	48	56
II	As I, with many small less steep slope Facets. Platforms narrower usually less than 200 m.	More than 31	0.05–0.51	28	15
III	Cliff with basal rocky pedestal, many Discontinuities, simple upper profile, but complex plan including caves, stacks arches. Platforms common but may be submerged.	Less than 31	0.09–0.14	5	3
IV	High cliffs with complex profile. Plan Usually simple, but upper and lower cliffs exceed 75°. Middle slope typically at mean angles 35°–42° degrees. Boulder fields at cliff foot may affect plan in detail and provide protection for periods of several decades.	More than 31	0.11–0.36	9	22
V	High with extensive and complex Undercliff typically with rotational slides. Upper plan related to mass movements, shoreline plan to debris and boulder fields.	More than 31	0.09–0.36	8	18
VI	Lower steep chalk cliffs with upper concave slope affected by landslides. Simple plan.	Less than 31	0.36	1	10
VII	Steep lower cliff with basal rocky pedestal, many discontinuities. Concave upper profile in overlying sands or clays. Lower cliff has very complex plan including geos, caves, stacks, arches.	Less than 31	0.12–0.95	4	0
Total length of chalk cliffs				103	124

Categories I to VI based on Precheur (1960) and May and Heeps (1985), and category VII based on May *in* May and Hansom (2003).
[a] 31 m = critical height for vertical chalk cliffs (Hutchinson, 1971).

effects of cutting across a series of valleys, the coastline retains a remarkably straight form. It is a prime example of the efficiency of marine processes in maintaining erosion and being able to remove the supply of sediment from erosion. The strength of the chalk is sufficient to maintain a vertical face unlike many other comparable materials where slumping is more typical.

The average annual retreat of the cliff-top edge of chalk coasts ranges from about 0.05 m a^{-1} to over 1 m a^{-1} (May, 1971; May and Heeps, 1985), but many chalk coasts are marked by intermittent cliff changes including some very large falls. Before these falls, the base of the cliff may be undercut in a distinct notch (Hutchinson, 1972) or blocks may be quarried from it. The resultant talus may act as a groin as these large falls can take several decades to be removed. Typical recorded losses include cliff-top area in excess of 7,000 m^2 (May, 1971), 0.5 million tonnes in volume or linear retreat of over 15 m in a single event. In January 1999 a very large fall at Beachy Head on the southern English coast affected a 200 m stretch of a 90 m-high cliff and produced a debris fan over 5 m in height covering some 50,000 m^2. As cliff height and slide volume increase, a "degree of flow" may develop in chalk debris (Hutchinson, 1983). In falls from cliffs between 70 and 150 m in height, a "chalk flow" may carry debris across the shore platforms for distances more than four times the cliff height (Hutchinson, 1980, 1983). These flow slides may occur because high pore-water pressures are generated through the crushing impact of relatively weak blocks of high porosity, near-saturated chalk (Hutchinson, 1984) or as a result of steady-state flows resulting from rapid or undrained loading of the chalk (Leddra and Jones, 1990).

The debris produced by falls varies from clay-sized fines (0.5–100 μm) to boulders of over 1–4 m. Wave action and longshore currents disperse the fines rapidly. Shingle-sized debris is commonly rounded within a few days and is worn down to sand-sized fragments in a few months. The larger material may, however, remain *in situ* for very long periods of time and become colonized by algae and molluscs. The boulders may gradually become smaller, but the platforms often retain a substantial debris cover, with boulder arcs marking the outer edges of the chalk flow debris several decades after the original failure. Intertidal chalk platforms can exceed 1 km in width but are more generally up to about 200 m wide. Their form and development (Wood, 1968) has been attributed to the effects of the geological structures, storm wave action (So, 1965), the presence of flint strata (which may have a very significant armoring effect: May and Hansom, 2003), the tractive effects of beach material (Girard, 1907) and the role of boring organisms (Jehu, 1918). Recorded rates of platform lowering are about 25 mm a^{-1}, but Foote *et al.* (2000) show that micro-erosion rates are very variable and are strongly influenced by platform dwelling organisms. Bioerosion may play a significant role in chalk coast erosion, but further work is needed.

Apart from parts of the Sussex, Kent, and Normandy coasts where walls have been constructed to reduce coastal erosion and protect towns, chalk coasts are little modified by the effects of coastal engineering works and so remain one of the better studied natural cliffed coasts.

Vincent May

Bibliography

Briquet, A., 1930. *Le littoral du nord de la France, et son évolution morphologique*. Orleans: Tessier.

Foote, Y., Moses, C., Robinson, D., Saddleton, P., and Williams, R., 2000. European shore platform erosion dynamics. *Oceanology International 2000*, Brighton Conference.

Girard, J., 1907. *Les falaises de la Manche*. Paris: Leroux.

Hancock, J.M., 1976. The petrology of the chalk. *Proceedings of the Geological Association*, **86**: 499–535.

Hutchinson, J.N., 1972. Field and laboratory studies of a fall in Chalk cliffs at Joss Bay, Isle of Thanet. *Proceedings of the Roscoe Memorial Symposium, Cambridge*, Henley-on-Thames, pp. 692–706.

Hutchinson, J.N., 1980. Various forms of cliff instability arising from coast erosion in the UK. *Fjellsprengningsteknikk-Bergmekanikk-Geoteknikk 1979*, pp. 19.1–19.32.

Hutchinson, J.N., 1983. *Engineering in a landscape*. Inaugural lecture, 9th October 1979. Imperial College of Science and Technology, University of London.

Hutchinson, J.N., 1984. Landslides in Britain and their countermeasures. *Journal of Japan Landslide Society*, **21**(1): 1–24.

Hutchinson, J.N., Bromhead, E.N., and Lupini, J.F., 1980. Additional observations on the landslides at Folkestone Warren. *Quarterly Journal of Engineering Geology*, **13**: 1–31.

Jehu, T.J., 1918. Rock-boring organisms as agents in coast erosion. *Scottish Geographical Magazine*, **34**: 1–11.

Leddra, M.J., and Jones, M.E., 1990. Steady-state flow during undrained loading of chalk. In *Chalk—Proceedings International Symposium, Brighton 1989*. London, pp. 245–252.

May, V.J., 1971. The retreat of chalk cliffs. *Geographical Journal*, **137**: 203–206.

May, V., and Hansom, J., 2003. *Coastal Geomorphology of Great Britain*. Peterborough: Joint Nature Conservation Committee.

May, V.J., and Heeps, C., 1985. The nature and rates of change on chalk coastlines. *Zeitschrift für Geomorphologie NF, Supplementband*, **57**: 81–94.

Perrin, R.M.S., 1971. *The Clay Mineralogy of British Sediments*. London: Mineralogical Society.

Precheur, C., 1960. *Le littoral de la Manche de Ste Adresse à Ault*. Poitiers: S.F.I.L..

So, C.L., 1965. Coastal platforms of the Isle of Thanet, Kent. *Transactions of the Institute of British Geographers*, **37**: 147–156.

Steers, J.A., 1946. *The Coastline of England and Wales*. Cambridge: Cambridge University Press.

Wood, A., 1968. Beach platforms in the Chalk of Kent. *Zeitschrift für Geomorphologie NF*, **12**: 107–113.

Cross-references

Bioerosion
Cliffed Coasts
Cliffs, Erosion Rates
Cliffs, Lithology versus Erosion Rates
Coastline Changes
Europe, Coastal Geomorphology
Shore Platforms

CHANGING SEA LEVELS

The geodetic sea level—better known as the geoid—is the equipotential surface between the attraction and repelling forces. The actual sea level—known as the dynamic sea surface—closely approximates the geoid surface. The deviations of the dynamic surface from the geoid surface are caused by different oceanographic, hydrographic, and meteorological forces acting upon the distribution of the water masses. Neither the dynamic sea level nor the geoid remains undeformed with time. Sea-level changes refer to changes in mean sea level. Eustatic changes denote changes in mean sea level over time units larger than daily and annual cyclic changes.

Our only true benchmarks are the former sea-level positions in the field. The present level of those data points are the combined function both of past changes in sea level and past changes in crustal level. In opposite to all types of crustal deformations, we term all different types of sea-level changes "eustatic" (Mörner, 1986). In the old concept of eustasy, all sea-level changes were parallel over the globe (e.g., Fairbridge, 1961). This is no longer tenable; sea-level changes can never be fully parallel over the globe and are often of compensational, even opposed, character (Mörner, 1976). The old and new concepts of eustasy are illustrated in Figure C7.

Water distribution and sea-level changes

The distribution over the globe of the ocean water masses is defined by the Earth's rotational ellipsoid, the geoid topography, and various dynamic forces (air pressure, current bet, coastal runoff, etc.). Any changes in any of those parameters will induce sea-level changes on the local, regional, or global dimension.

Sea level can only change within limits determined by its physical causes. The main sea-level variables are listed and quantified in Table C5. Corresponding rates and amplitudes are given by Mörner (1996a, b).

The global water volume is controlled by glacial eustasy, evaporation/precipitation, loss/gain of water, and steric expansion/ contraction.

The volume of the ocean basins is controlled by tectonic deformation (ocean floor subsidence, mid-oceanic ridge growth, and coastal tectonics).

The distribution of the water over the globe is controlled by Earth's rate of rotation, geoid deformation, and various dynamic forces (air pressure, current bet, coastal runoff).

Actual changes in the sea level represent an intricate interaction of multiple factors requiring understanding of various subjects and

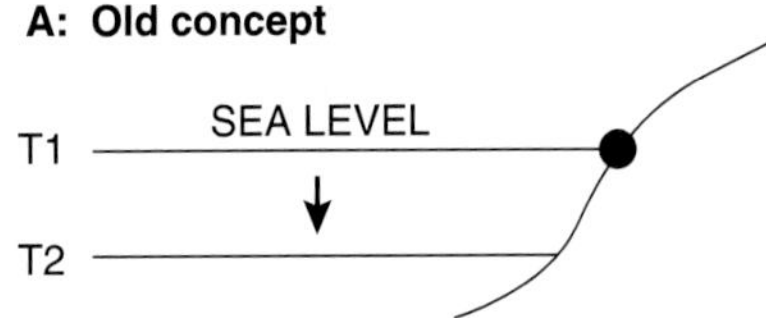

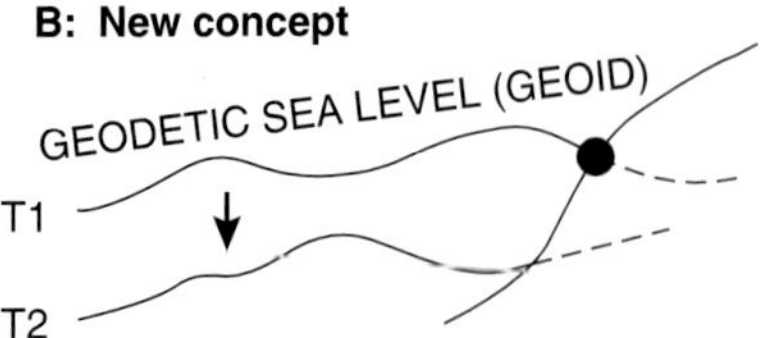

Figure C7 The old and new concept of "eustasy" (after Mörner, 1987) illustrated by the change (arrow) from one sea-level position (black dot and T1) to a new, lower, position (T2). In the old concept (A), any eustatic change in sea level was parallel all over the globe. In the new concept (B), the real sea-level surface is an irregular surface that never remains parallel at a change in sea level.

processes. The dominant controlling factor differs through time and with timescale considered.

On the million-year timescale, sea level is mainly determined by ocean basin volume, mass distribution, and rate of rotation.

On the thousand-year timescale, sea level varies mainly due to glacial eustasy and related glacio-isostasy, earth rotation and geoid shape as determined by lithospheric mass change.

Sea-level changes *on the hundred-year timescale*, seem primarily to reflect changes in water-mass distribution related to changes in the oceanic circulation caused by variations in climate and Earth's rotation rate.

On a yearly scale, sea level varies locally because of dynamic factors and El Niño–Southern Oscillation (ENSO) effects.

Table C5 Sea-level variables and quantities (in meters), for rates, see Mörner (1996b)

Present tides	
Geoid tides	0.78
Ocean tides	up to max. 18 m
Present geoid relief	
Maximum difference	180
Present dynamic sea surface	
Low harmonics	up to max. ~2
Major currents	up to ~5
Earth radial differences	
Equator/polar plane	21,385
Presently stored glacial volume (in meters sea level)	
Antarctica	~60
Greenland	~5
All alpine glaciers	0.5
Glacial eustasy	
Ice Age amplitudes	100–180
Geoid deformations	
On Ma-scale	~50–250
Last 150 Ka (20 Ka cycles)	30–90
Last 20 Ka	~60
Last 8 Ka	5–8
Holocene oscillations	1–2
Global loading models	
Last 20 Ka	some 10–30
Last 5 Ka	some 5 m
Differential rotation	
Holocene Gulf Stream pulses	1.0–0.1
Holocene equatorial current	~1.0
El Niño/ENSO events	0.3
Major current topography	max. 5
Rotation rate and sea level	
Meters sea level per 15 ms rotation	1.0
Air pressure and sea level	
Meters sea level per 1 mb	0.01
Ocean thermal expansion	
Meters sea level per °C per 1,000 m	~0.2

The sea-level debate

Sea-level changes were recorded and debated from the early 18th century (Mörner, 1987). The effects of changes in rotation, in ocean basin subsidence, in deflection of the plume-line and in glaciation were advocated. With the recording of multiple Ice Ages, it became natural to consider cyclic changes in glacial eustatic volume of the oceans (e.g., Daly, 1934).

In the 1960s the debate was intensive whether sea level, after the last glaciation maximum, rose with oscillations in Mid-Holocene time reaching even above the present level (Fairbridge, 1961) or as a smooth line continually rising to its present level (Shepard, 1963). The curves of Jelgersma (1961), Scholl (1964), Mörner (1969), and Tooley (1974) added detailed records from single regions backed up by numerous dates ruling out the occurrence of high-amplitude sea-level oscillations. Assuming that eustatic sea level could be expressed in one globally valid eustatic curve, all differences in observed levels had to be explained in terms of changes in the crustal or sedimentary level. In 1971 Mörner called attention to the fact that most sea-level curves, despite very large differences in level in Mid-Holocene time, converged fairly closely at about 8,000 radiocarbon years BP. This seemed to indicate a major cyclic redistribution of the ocean masses in Holocene time. Later, the explanation became obvious; the geoid relief is subjected to deformation (Mörner, 1976). Consequently, each region has to define its own regional eustatic curve (a global curve became an illusion).

Clark (1980) and Peltier (1998) presented their global glacial isostatic loading models, which called for major sea-level irregularities over the globe.

The "World Atlas of Holocene Sea-Level Changes" (Pirazzoli, 1991) gives a good and comprehensive view of the changes in sea level. Besides global differentiation in the eustatic ocean components, the recorded relative sea-level changes are strongly affected by differential tectonics and sedimentary compaction. Despite this complexity, one may distinguish some general regularities. This is illustrated in Figure C8 where five sea-level curves from quite different parts of the globe are compared. Despite significantly different shapes, trends, and levels, they all show a general rise up to some 6,000–5,000 BP followed an out-of-phase to sometimes opposed oscillatory trend for the last 5,000–6,000 years around the present zero level. The first part records a general glacial eustatic rise in sea level prior to 6,000–5,000 BP plus a local to regional differentiation due to tectonics, geoid deformation, and global isostatic compensation. The second part (the last 5,000–6,000 years) lacks any significant glacial eustatic component and is instead dominated by the redistribution of water masses around the present mean sea level.

The redistribution of water masses during the last 5,000 years seems primarily to be the function of changes in the main oceanic current systems in response to the interchange of angular momentum between the "solid" Earth and the hydrosphere (e.g., Mörner, 1995).

This implies that the Earth was in a totally different mode in the last 5,000 years as compared with the previous 12–13,000 years with different dominant driving forces for the changes in sea level.

To reconstruct past sea-level changes

Field observations of past sea-level positions are the only means of reconstructing sea-level changes in pre-instrumental time. The observational object may be a stratigraphic layer, a morphologic feature or single objects referring to sea-level changes. This object must be fixed in altitude with respect to present sea level (this may range from an absolute level to a relative level). The next step is to interpret the observed field object with respect to the sea level at the time of formation. This implies the study of sea-level relation, sedimentary characteristics, paleoecological inference, paleoenvironmental interpretation, and coastal dynamics. Finally, the object needs proper dating. After all these steps, we have a past sea-level datum or a paleo-sea-level-indicator, which can be used in sea-level curves and sea-level reconstructions. The quality of this datum point depends upon the quality of each individual step in its establishment. The dating of a continuous catching-up coral reef growth may provide an exceptionally continuous record (e.g., Fairbanks, 1989).

We have to understand that there is a long chain of uncertainties concerning sea-level changes as they are interpreted by the investigator, the

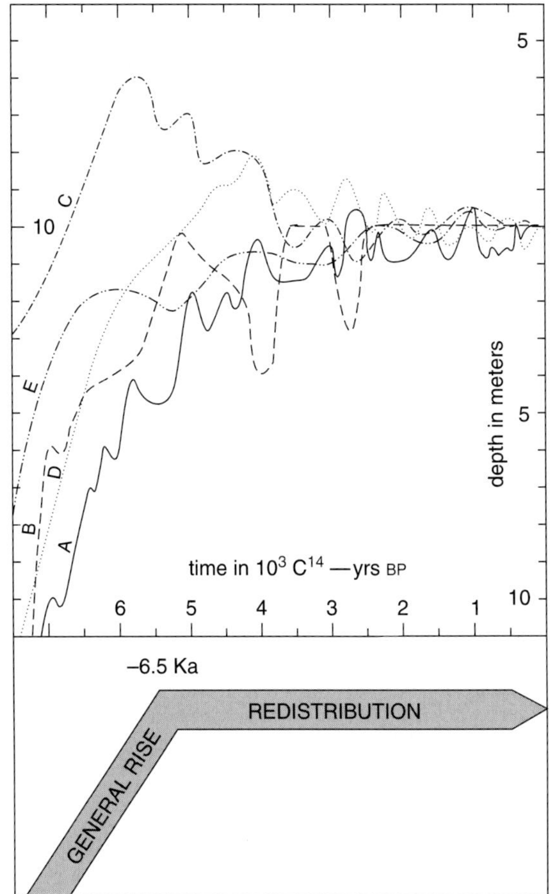

Figure C8 Five sea-level curves (A–E) from different parts of the world (from Mörner, 1996a). Despite tectonic differentiation, all five curves show a general rise up to about 6,000–5,000 BP, recording the end of the glacial eustatic rise in sea level. Thereafter, the curves are characterized by an out-of-phase variability (even opposed) around a base level close to present zero, recording an irregular redistribution of the oceanic water masses over the globe. The lower graph illustrates these two different stages, or modes, of sea-level changes. Reproduced with permission of Zeitschrift fur Geomorphologie.

sea-level changes as they are recorded in the field by fragmentary traces, and the sea-level changes as they actually happened in the past.

Deep-sea records

The deep-sea oxygen isotope records (e.g., Shackleton and Opdyke, 1973) have a predominant control from the changes in ice volume and glacial eustatic changes in oceanic water volume. Therefore, they provide a reasonable reconstruction of the Quaternary glacial eustatic cycles (Figure C9). For the last 140,000 years, there seems to be a good agreement between oxygen isotope variations and actual dates of coral reef levels (e.g., Chappell *et al.*, 1996).

Tide-gauge records

Sea level can be measured by water-marks, mareographs, and tide-gauges. The oldest tide-gauge is the one in Amsterdam from 1682. From a compilation of global tide-gauge station, Gutenberg (1941) tried to reconstruct the mean global eustatic component. He arrived at a value of 1.1 mm/yr rise in sea level. Table C6 includes a number of later estimates, all at about 1 mm/yr.

Tide-gauges record the interplay of changes both in sea level and in land level. Because most tide-gauges are located on harbor constructions, often resting on unconsolidated sediments, subsidence due to compaction is a serious problem.

Satellite altimetry

With the improvement of satellite altimetry (TOPEX/POSEIDON) it has become possible to record changes in the oceanic sea-level position.

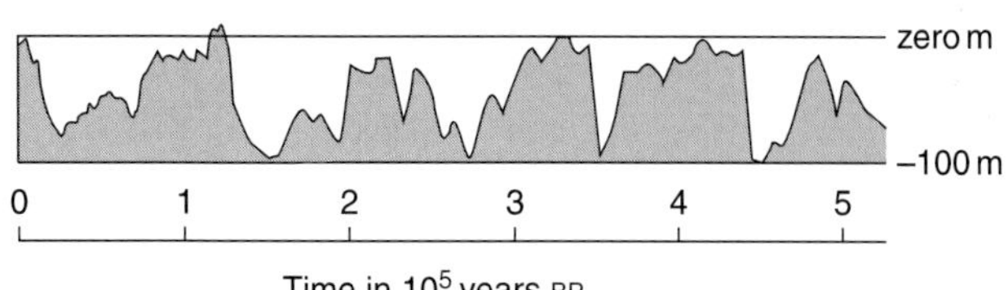

Figure C9 Main Quaternary glacial eustatic cycles as derived from deep-sea oxygen isotope records after Shackleton and Opdyke (1973). A glacial/interglacial sea-level amplitude in the order of 100 m is recorded. It should be noted, however, that improved data sets indicate that the last glaciation maximum was in the order of −120–130 m.

Table C6 Recent, present, and possible future changes in sea level as recorded or calculated from different observational records

Time period	Rates (mm/yr)	Source of information	References
Last 150 years	1.1	Mean of tide-gauges	1
	1.2	Mean of tide-gauges	2
1830–1930	1.1	NW Europe tide gauge data	3
	1.1	Past uplift versus present uplift and eustasy	3
	max. 1.1	Earth's rotation versus tide-gauge	4
Last 100 years	1.0	UK—North Sea tide-gauges	5
	1.1	Fennoscandian tide-gauges	6
1992–96	0.0	Satellite altimetry	7
1997–98	ENSO	Satellite altimetry	7
1999–2000	<0.5	Satellite altimetry	7
2000–2100	0.8	Estimates of all water sources	8

References: (1) Gutenberg, 1941, (2) Fairbridge and Krebs, 1962, (3) Mörner, 1973, (4) Mörner, 1996a, (5) Shennan and Woodworth, 1992, (6) Lambeck *et al.*, 1998, (7) Figure C10, (8) IPCC, 2001.

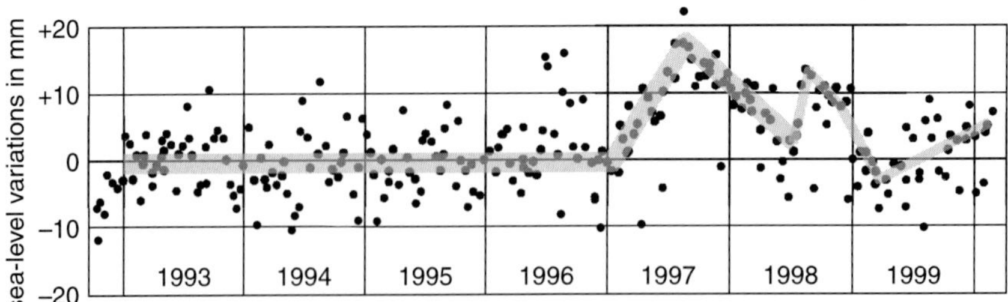

Figure C10 Sea-level changes in millimeters as recorded by TOPEX/POSEIDON between October 1992 and April 2000. The variability is high, in the order of ±5–10 mm. From 1993 to 1996, no trend is recorded, just a noisy record around zero. In 1997 something happens. High-amplitude oscillations are recorded; a rapid rise in early 1997 at a rate in the order of 2.5 mm/yr, followed by a rapid fall in late 1997 and early 1998 at a rate in the order of 1.5 mm/yr, and finally, in late 1998 and 1999, a noisy record with unclear trends. The new factor introduced in 1997 and responsible for the high-amplitude oscillations, no doubt, is the global ENSO event, implying rapid redistribution of oceanic water masses (just as recorded in Figure C8; Mörner, 1995). This means that this data set does not record any general trend (rising or falling) in sea level, just variability around zero plus the temporary ENSO perturbations.

This has opened new means of recording sea-level changes. Previously, we were restricted to changes at the sea/land interface (the shoreline). Now we can record the changes over the entire ocean surface covered by satellites.

Figure C10 gives a satellite record for the time period 1992–2000. From 1993 to 1996, there is hardly any detectable trend, just a noisy variability in the order of ±5–10 mm. From 1997 to 1999 (especially 1997–98), there are rapid high-amplitude perturbations; a rapid rise in early 1997 at a rate of ~2.5 mm/yr, followed by a rapid fall in late 1997 and early 1998 at a rate of ~1.5 mm/yr. These rapid high-amplitude perturbations cannot represent real ocean volume changes, but imply that ocean masses were rapidly redistributed over the globe (including the arctic regions that are not covered by the satellite records), just like what we record for the Late Holocene sea-level changes (Figure C8; Mörner, 1995).

Present sea-level changes

For the last 5,000 years global sea-level records do not concur with a simple trend but rather with site-specific redistribution of ocean masses (Figure C8). The same applies for the present. Despite this, it seems to be possible to detect a slight general rise in sea level since about AD 1850. This is illustrated in Table C6. No acceleration in this sea-level rise can be recorded in the last decades.

Future sea-level changes

Predictions of sea-level changes to be expected in the future must, by necessity, be based on our understanding of the past and present sea-level trends. Considering the past driving forces, the present trend and rates, and possible future variability, there are no reasons to assume any disastrous rise in sea level as often claimed (IPCC, 2001) but rather a value in the order of a moderate rise of 10 cm ± 10 cm in a century.

Nils-Axel Mörner

Bibliography

Chappell, J., Omura, A., Esat, T., McMulloch, M., Pandelfi, J., Ota, Y., and Pillans, B., 1996. Reconciliation of Late Quaternary sea levels derived from coral terraces at Huon Peninsula with deep sea oxygen isotope records. *Earth Planetary, Science Letters*, **141**: 227–236.

Clark, J.A., 1980. A numerical model of worldwide sea-level changes on a viscoelastic Earth. In N.-A. Mörner (ed.), *Earth Rheology, Isostasy, and Eustasy*. New York: John Wiley & Sons, pp. 525–534.

Daly, R.A., 1934. *The Changing World of the Ice Age*. New Haven: Yale University Press.

Fairbanks, R.G., 1989. A 17,000-year glacio-eustatic sea-level record: influence of glacial melting rates on the Younger Dryas event and deep-ocean circulation. *Nature*, **342**: 637–642.

Fairbridge, R.W., 1961. Eustatic changes in sea level. *Physics and Chemistry of the Earth*, **4**: 99–185.

Fairbridge, R.W., and Krebs, O.A., 1962. Sea level and the southern oscillation. *Geophysical Journal*, **6**: 532–545.

Gutenberg, B., 1941. Changes in sea level, postglacial uplift and mobility of the Earth's interior. *Bulletin Geological Society of America*, **52**: 721–772.

IPCC, 2001. Chapter 11. Sea level changes. *IPCC WG1 Third Assessment Report*, 78 pp.

Jergersma, S., 1961. Holocene sea-level changes in the Netherlands. *Mededelingen van de Geologische Stichting*, **C4**: 1–101.

Lambeck, K., Smither, C., and Ekman, M., 1998. Tests of glacial rebound models for Fennoscandinavia based on instrumented sea- and lake-level records. *Geophysical Journal International*, **135**: 375–387.

Mörner, N.-A., 1969. The Late Quaternary history of the Kattegatt Sea and the Swedish West Coast: deglaciation, shorelevel displacement, chronology, isostasy and eustasy. *Sveruges Geologiska Undersökning*, **C640**: 1–487.

Mörner, N.-A., 1973. Eustatic changes during the last 300 years. *Palaeogeography Palaeoclimatology Palaeoecology*, **13**: 1–14.

Mörner, N.-A., 1976. Eustasy and geoid changes. *Journal of Geology*, **84**: 123–151.

Mörner, N.-A., 1986. The concept of eustasy. A redefinition. *Journal of Coastal Research* (Special Issue), **1**: 49–52.

Mörner, N.-A., 1987. Models of global sea level changes. In Tooley, M.J. and Shennan, I. (ed.), *Sea level Changes*. Oxford: Blackwell, pp. 333–355.

Mörner, N.-A., 1995. Earth rotation, ocean circulation and paleoclimate. *GeoJournal*, **37**: 419–430.

Mörner, N.-A., 1996a. Sea level variability. *Zeitschrift fur Geomorphology N.F.*, Suppl-Bd. **102**: 223–232.

Mörner, N.-A., 1996b. Rapid changes in coastal sea level. *Journal of Coastal Research*, **12**: 797–800.

Peltier, 1998. Postglacial variations in the level of the sea: implications for climate dynamics and soild-earth geophysics. *Reviews of Geophysics*, **36**: 603–689.

Pirazzoli, P.A., 1991. *World Atlas of Holocene Sea-Level Changes*. Elsevier Oceanographic Series **58**: 300 pp.

Scholl, D.W., 1964. Recent sedimentary record in mangrove swamps and rise in sea level over the southwestern coast of Florida. Part 1. *Marine Geology*, **1**: 344–366.

Shackleton, N.J., and Opdyke, N.D., 1973. Oxygen isotope and paleomagnetis stratigraphy of equatorial Pacific core V28-238: oxygen isotope temperatures and ice volumes on a 10^5 year and 10^6 year scale. *Quaternary Research*, **3**: 39–55.

Shennan, I., and Woodworth, P.L., 1992. A comparison of late Holocene and twentieth-century sea-level trends from the UK and North Sea region. *Geophysical Journal International*, **109**: 96–105.

Shepard, F.P., 1963. Thirty-five thousand years of sea level. In Clements, T. (ed.), *Essays in Marine Geology in Honor of K.O. Emery*, 1–10. Los Angeles: Univ. S. Cal. Press.

Tooley, M., 1974. Sea-level changes during the last 9000 years in north-west England. *Geographical Journal*, **140**: 18–42.

Cross-references

Coastal Changes, Gradual
El Niño–Southern Oscillation

Eustasy
Geodesy
Late Quaternary Marine Transgression
Sea-Level Changes During the Last Millennium
Sea-Level Datum
Tide-Gauges

CHENIERS

Definition and morphology

Cheniers are stabilized wave-built ridges, usually of sand, but sometimes shelly or gravelly, that have been deposited on an alluvial plain of silt, clay, or peat on prograding deltas or coastal plains. They often occur in sub-parallel sets as elongated, narrow ridges (typically up to 3 m high and 40–400 m wide), standing at or slightly above the highest high tide level. The term chenier (Russell and Howe, 1935a, b) comes from the French native word (chêne) for the "live oak" tree (*Quercus virginiana*) which dominates the vegetation on chenier ridges in Louisiana. They differ from beach ridges (*q.v.*) in that they have been built on an alluvial foundation.

Multiple cheniers are separated by relatively wide intertidal or shallow subtidal swamps or mudflats, which are occasionally submerged by high tides or river floodwaters to form temporary lagoons. These intervening areas may either be barren of vegetation or covered at least in part by grasses, salt marsh, or mangroves. The term chenier plain includes these intervening areas as well as the chenier ridges, where at least two successively formed ridges are present (Otvos and Price, 1979). Transgressive cheniers are generally emplaced during storm surges, when large waves on a temporarily raised sea level sweep beach sediment inland. Successively formed cheniers usually occur in a sub-parallel seaward sequence as the coastline progrades. Even where there is a substantial shell content the slopes bordering cheniers tend to be gentle (Byrne *et al.*, 1959). In the fine sandy, shell-free Suriname cheniers the slope is <1 (Augustinus, 1980). Two of the world's largest chenier plains are found along the coasts of Louisiana and Guiana (Figure C11), but smaller chenier plains occur on many coastal plains, often behind embayments.

Chenier coasts are associated with a broad spectrum of tidal amplitudes that range from microtidal (e.g., SW Louisiana) and mesotidal to macrotidal (e.g., Van Diemen Gulf in northern Australia, Gulf of California). Most cheniers are of mid- to late-Holocene age, but there are cheniers of Pleistocene age in the Colorado Delta region of the Gulf of California (Meldahl, 1993).

Due to constructive overwash from storm surges the chenier crests may slightly exceed highest high tide level. As traced laterally along given chenier ridge sets in Louisiana (Figure C12), locally variable sand supply and wave conditions resulted in highly variable ridge crest elevations. Given an adequate coastal sand supply and sufficiently strong onshore winds, cheniers may be capped by small foredunes, or at least

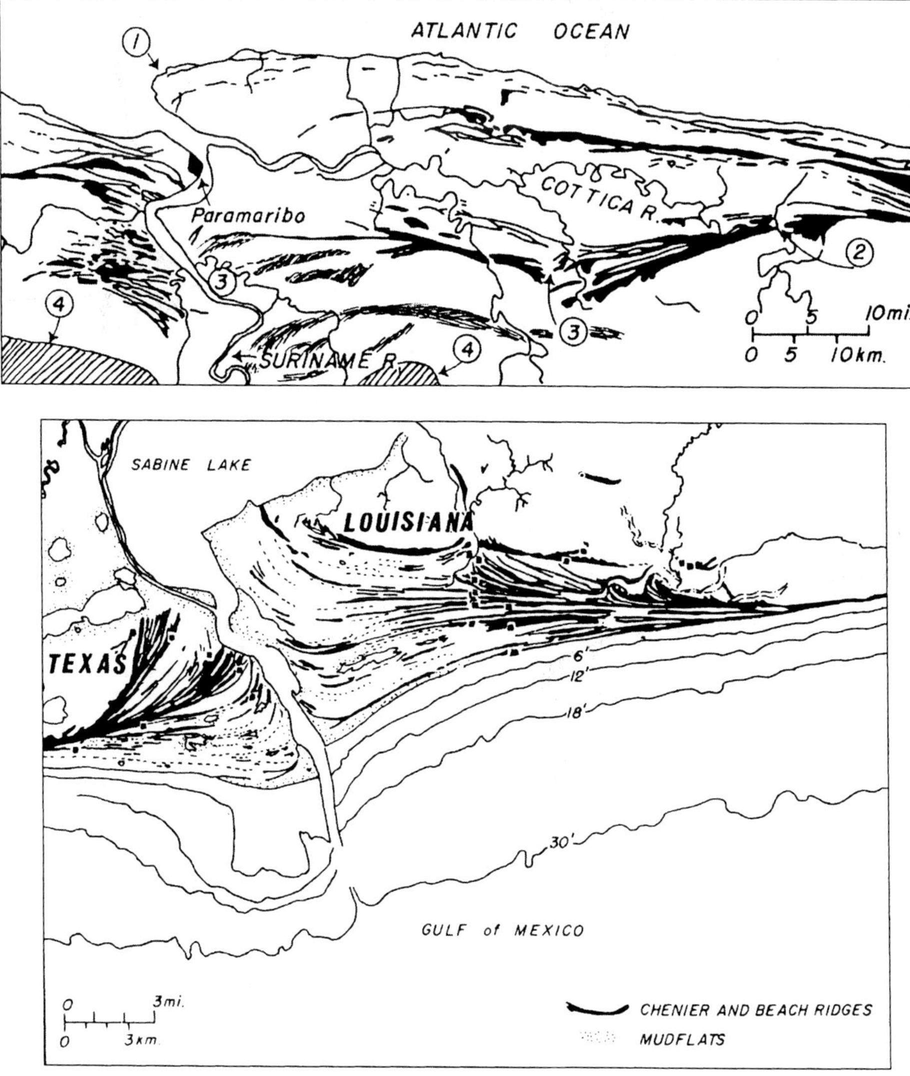

Figure C11 SW Louisiana and Guiana chenier trends and chenier plains (Louisiana after Byrne *et al.*, 1959, Guiana after Vann, 1959; from Otvos and Price, 1979). Reprinted with permission from Elsevier Science.

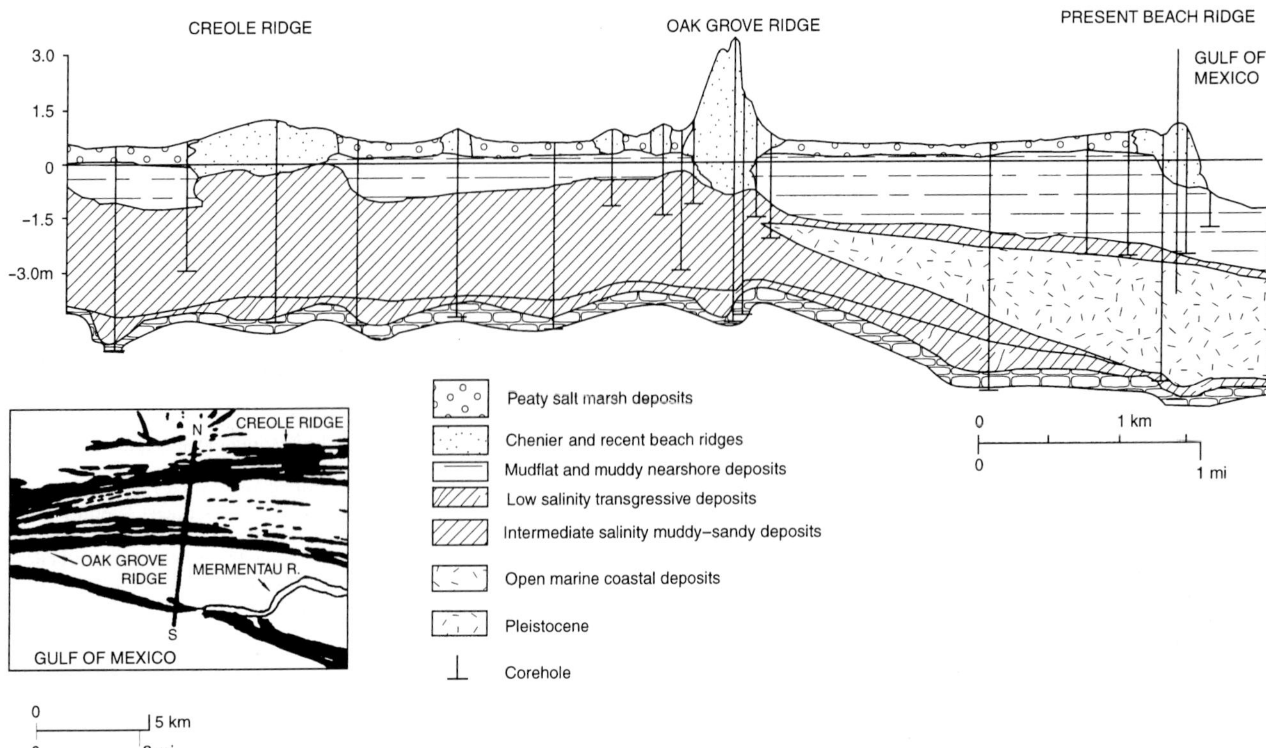

Figure C12 Cross section across Creole and Oak Grove Ridge cheniers, SW Louisiana (after Byrne *et al.*, 1959; from Otvos and Price, 1979). Reprinted with permission from Elsevier Science.

veneered by eolian sand (e.g., Suriname, Augustinus, 1978; SW Louisiana, Byrne *et al.*, 1959). Chenier ridge heights, including those not enhanced by eolian sand accumulation are not related to, or indicative of the elevation of former sea levels.

In small bays mudflat progradation, alternating with sand and shell accumulation and incorporation into developing chenier ridges occur simultaneously (Woodroffe and Grime, 1999). Side-by-side formation of a chenier passing laterally into a sandy beach ridge was described from the northeastern coast of Australia (Chappell and Grindrod, 1984).

Regressive and transgressive cheniers

Regressive cheniers that formed successively along the shore of a prograding delta or coastal plains overlie shallow nearshore marine and intertidal deposits (Figure C13). On tropical shores, these cheniers may extend 1 m or more below the surrounding high-tidal mudflats, down to the buried top of mangrove deposits (Cook and Polach, 1973). Except beneath landward-inclined slopes, formed by overwash, seaward-dipping low angle (<15) lamination and cross-stratification are typical in most of the cheniers. Sandy shell ridges, driven landward over intertidal flats and through mangrove thickets are transported by intermittent wave swash during high tides. Eventually they become stranded and stabilized as transgressive chenier ridges (Figure C13) (Thompson, 1968; Jennings and Coventry, 1973; Woodroffe *et al.*, 1983; Chappell and Grindrod, 1984; Woodroffe and Grime, 1999).

Following the original definition of cheniers as transgressive features (Russell and Howe, 1935a, pp. 27–34; 1935b) several authors defined "true" chenier ridges as landward-driven landforms (Otvos, 2000). However, stratigraphic and morphological data indicate that most Louisiana cheniers are not transgressive. With their shallow ridge bases, transgressive chenier ridges display washover-induced, landward-inclined stratification. They overlie intertidal, low-supertidal mud/marsh surfaces at very shallow depths. Powerful tropical storms often erode and even flatten landward migrating ridges. Even if lesser events effectively translated them landward before and after the storm, the 1974 hurricane surprisingly had no appreciable impact on the landward migration of sand and shell ridges in Shoal Bay, Australia (Woodroffe and Grime, 1999).

Ervin G. Otvos

Bibliography

Augustinus, P.G.E.F., 1978. The changing shoreline of Surinam (South America), Foundation for Scientific Research in Suriname and Netherlands Antilles. Utrecht: The Netherlands.

Augustinus, P.G.E.F., 1980. Actual development of the chenier coast of Suriname (South America). *Sedimentary Geology*, **26**: 91–113.

Byrne, J.V., LeRoy, D.O., and Riley, C.M., 1959. The chenier plain and its stratigraphy, southwestern Louisiana. *Transactions Gulf Coast Association of Geological Societies*, **9**: 237–260.

Chappell, J., and Grindrod, J., 1984. Chenier plain formation in northern Australia. In Thom, B.G. (ed.), *Coastal Geomorphology in Australia*. Sydney: Academic Press, pp. 197–231.

Cook, P.J., and Polach, H.A., 1973. Sedimentology and Holocene history of a tropical estuary (Broad Sound, Queensland). *Bureau of Mineral Resource Bulletin*, **170**: 260.

Jennings, J.N., and Coventry, R.J., 1973. Structures and texture of a gravelly barrier island in the Fitzroy Estuary, Western Australia, and the role of mangroves in the shore dynamics. *Marine Geology*, **15**: 145–167.

Meldahl, K.H., 1993. Geographic gradients in the formation of shell concentrations: Plio–Pleistocene marine deposits, Gulf of California. *Paleogeography, Paleoclimatology, Paleoecology*, **101**: 1–25.

Otvos, E.G., 2000. Beach ridges—definition and significance. *Geomorphology*, **32**: 83–108.

Otvos, E.G., and Price, W.A., 1979. Problems of chenier genesis and terminology—an overview. *Marine Geology*, **31**: 251–263.

Russell, R.J., and Howe, H.V., 1935a. *Geology of Cameron and Vermillion Parishes*. Baton Rouge: Louisiana Geological Survey Bulletin, **6**.

Russell, R.J., and Howe, H.V., 1935b. Cheniers of southwestern Louisiana. *Geographical Review*, **25**: 449–461.

Thomson, R.W., 1968. Tidal flat sedimentation on the Colorado River Delta, northwest Gulf of California. *Geological Society of America Memoirs*, **107**: 133.

Vann, J.H., 1959. The geomorphology of the Guiana coast. Second Coastal Geography Conference. Coastal Studies Institute, Louisiana State University, Washington, D.C., pp. 153–187.

Woodroffe, C.D., Curtis, R.J., and McLean, R.F., 1983. Development of a chenier plain, Firth of Thames, New Zealand. *Marine Geology*, **53**: 1–22.

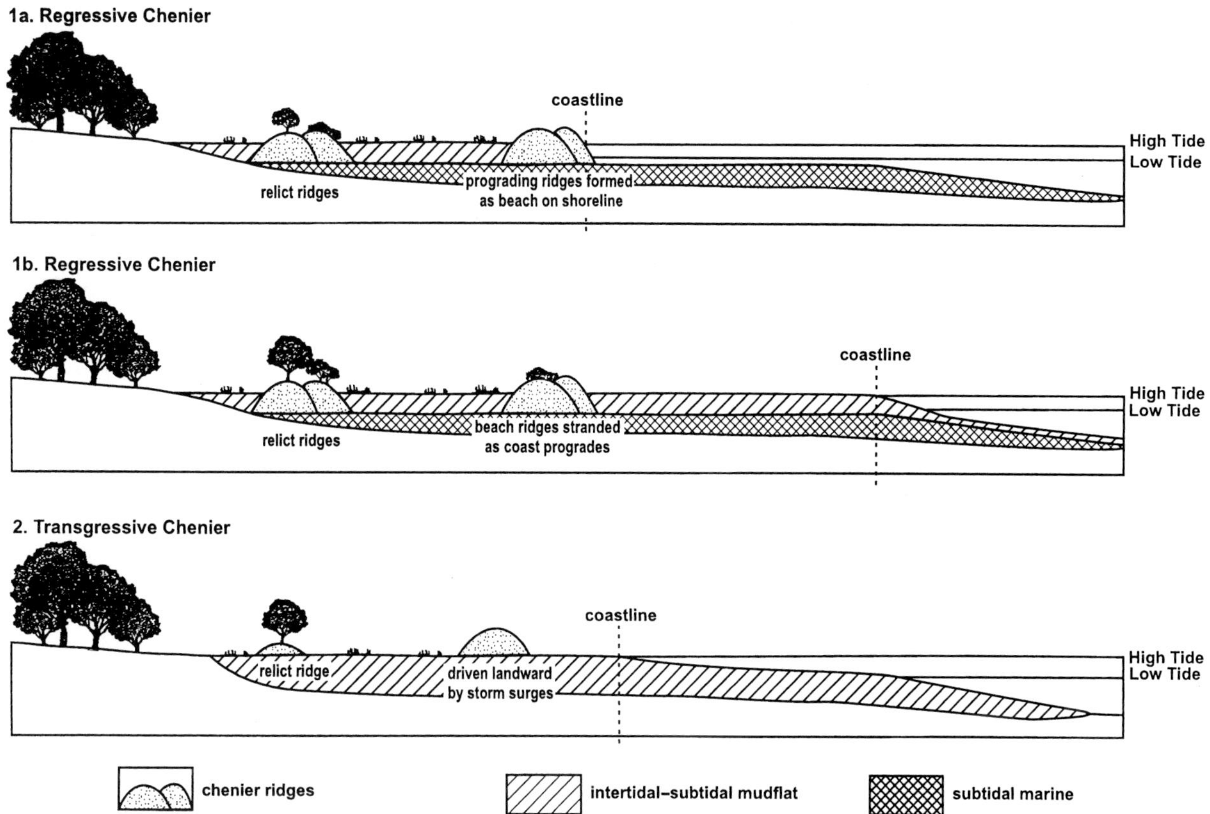

Figure C13 Schematic development of regressive and transgressive cheniers.

Woodroffe, C.D., and Grime, D., 1999. Storm impact and evolution of a mangrove-fringed chenier plain, Shoal Bay, Darwin, Australia. *Marine Geology*, **159**: 303–321.

Cross-references

Beach Features
Beach Ridges
Cross-Shore Sediment Transport
Deltas
Muddy Coasts
Sandy Coasts
Storm Surge

CLASSIFICATION OF COASTS—See HOLOCENE COASTAL GEOMORPHOLOGY

CLEANING BEACHES

Beach clean up is yet another environmental grassroot initiative organized by non-governmental organizations (NGOs), later adopted by government on all levels. As is often the case, the research addressing this and similar environmental initiatives is sparse in large part caused by a belated interest on the part of Federal and State agencies.

Beach degradation can be classified in several ways; based on the origin of the waste (*in situ*, oceanic, or land-based); by the way it is conveyed to the beach environment (point-source and non-point source); by the manner in which it appears on the beach (liquid and solid); and by the manner in which it was generated in the first place (natural and anthropogenic). The waste materials that are washed up on beaches include those falling into most but not all of these classifications. Material excluded from beach clean-up initiatives is liquid waste, except in those instances where it has been transformed while on the beach, or during the time it traveled on or in the water column. Examples of such materials include certain hydrocarbon materials that have metamorphosed into "tarballs." Organized beach clean up also excludes organic waste washed up on beaches such as dead vegetative matter and fish and shellfish carcasses, except in those instances where the accumulation of such is the result of fish kills caused by some anthropogenic activity.

Notwithstanding, that solid waste on the nation's beaches has contributed significantly to the public's concern about beach and nearshore water quality (Stratton, 1969), virtually no governmental initiatives were directed specifically at cleaning up solid waste washed up on the nation's beaches prior to 1980. Such efforts were left to local initiatives. One of these was directed by a few NGO's in which the Ocean Conservancy (TOC) formerly known as the Center for Marine Conservation, took an early lead. The other was done by local municipalities, towns, and counties with jurisdiction over beaches. Each of these initiatives is discussed below.

Governmental involvement in beach clean up can be divided into three different approaches. In nearly all instances, beach clean up is done by positioning waste containers on the beach with the expectation that the patrons will deposit their waste in those containers. In addition, many municipalities with few or small beaches will supplement these activities with manual clean up, usually conducted early in the morning before the beach becomes too crowded. On days when the beach is busy, more garbage will be deposited either in the containers or left on the beach, which may attract flies, wasps, gulls, and other animals; resulting in complaints levied against the Parks and Recreation Department. In jurisdictions where beaches are larger and the waste problem more extensive, the physical clean up is done using machinery in the form of a roto-tiller device and a sifting device that separate the sand from any waste materials buried by the wind or human activities. These activities also take place during the early morning.

A less common means of addressing beach waste and beach clean up is the program sometimes referred to as "Carry In, Carry Out." The State of Rhode Island introduced this program on all State beaches during the late 1980s. Each patron is issued a plastic bag together with an entry ticket and is expected to collect all waste material created during the visit and carry it back home at the end of the trip. This program was quite controversial when first introduced as many administrators, especially in communities where these beaches were located, were afraid that the waste generated would become a local problem by people dumping their waste locally. According to at least one Park's administrator, this has not been a major problem for the town. The successful implementation of this program has had several advantages including, reduced clean-up cost for the State, fewer problems with animals being attracted to waste bins, significant reduction in both visual and olfactory impacts eliminated by the absence of overflowing trash cans.

This absence of governmental interests, especially on the Federal level, has contributed to a near dearth of research on the distribution, volume, source of solid waste on beaches, and its impact on the physical and biological environment. This entry is divided into the following sections: origin and extent of the beach clean-up movement; sources, characteristics of and categories of solid waste, concluding with some observations on the status of beach clean up overall.

Origin of beach clean-up initiatives

The first known organized beach clean up took place in 1986 in Texas and was an initiative sponsored by TOC in which 2,800 volunteers collected a total of 124 tons of waste covering 122 miles of coastline. Two years later, TOC's efforts were nationalized in which 47,500 volunteers participated. In 1989, the effort had spread to Canada and Mexico marking the internationalization of the event. Today, more than 90 countries participate on an annual basis. In an effort to disseminate information on beach trash and beach clean up, and share the costs of implementing the program, TOC entered into partnerships with both public and private organizations and businesses. The beach clean up takes place once-a-year, usually the third Saturday in September during which nearly 8.5 million items are collected, sorted, counted, and analyzed.

More than 20 years after beach pollution was first recognized as a serious environmental problem, Congress passed the Marine Plastic Pollution Research and Control Act (PL 100-220) in 1988 that required the National Ocean and Atmospheric Administration (NOAA) to develop a public outreach program as well as implementing Annex V of the International Convention for the Prevention of Pollution from Ships (ICPPS) (see below). The public's interest in marine pollution soon outstripped NOAA's capacity to provide the materials and services needed, with the result that NOAA contracted the Ocean Conservancy to provide these services. This partnership lasted until 1996 when funding ceased. Prior to 1988, the 1972 reauthorized Coastal Zone Management Act (CZMA) included language on marine debris as one of nine objectives for the Coastal Zone Enhancement Grants program by amending section 309 of the Act (P<104–150, §1456b). NOAA's involvement with ocean pollution and beach clean up has continued through the Sea Grant Program where seven Sea Grant Programs have developed statewide beach clean-up programs. In addition 55 state, territories, commonwealths, and dependencies have established similar programs. Some of these are operated as part of the states' coastal management programs while others operate under the auspices of the Environmental Protection Agency (EPA) or private organizations. Some private organizations, including the Professional Association of Diving Instructors (PADI), have also established beach clean-up partnerships with the Ocean Conservancy. Finally, during the late 1980s and early 1990s beach debris was sampled quarterly on eight National Seashores and National Parks (Ribic *et al.*, 1997).

The International Marine Organization (IMO) sponsored ICPPS sought to address ship-borne solid waste disposal. This Convention was ratified in 1988 and consists of a series of sector-defined annexes of which Annex 5 deals specifically with the disposal of garbage and plastics. This Annex has 88 participating nations including the United States. The US Coast Guard is the agency that enforces both the ICPPS and the Marine Plastic Pollution Research and Control Act.

Sources, characteristics and categories of beach trash

In the United States, it often takes a focusing event before action is initiated. In the case of beach clean up, the focusing event was the medical waste wash-ups and subsequent beach closures that took place on both Long Island and New Jersey beaches during the period 1989–90. The cost of these events in terms of reduced tourism revenues to coastal communities in both New Jersey and New York has been estimated somewhere between two and five billion dollars (Stewart, 1989). The origin of these materials appears to have been the several dumping sites located in the New York Bight some 12 miles offshore. While the origin of this material was marine, approximately 70% of the waste materials washed up on the nation's beaches came from land and most of that originated from inland sources as opposed to trash left on the beach by beach visitors. The latter has been especially difficult to control from a management and legal perspective (Nollkaemper, 1997). While the beach clean up is a reactive as opposed to proactive program, most of the land-based sources of trash are addressed through existing programs and legislation. Numerous communities, companies, and manufacturers have by now implemented recycling programs enabling a large proportion of the total waste stream from being disposed of in landfills. Much of the land-based trash that eventually finds its way to the beach represents discarded materials on roads, parking lots that are either blown and/or washed into streams (Liffman *et al.*, 1997). Of these materials glass and plastic containers, plastic film products (bags, sheets, and other packing materials) represent the most problematic from a marine environmental perspective. Plastic materials are often consumed by fish, marine turtles, and marine mammals, some of which are mistaking these products as their favorite foods like various species of jellyfish including the Portuguese Man-of-War. The ingestion of plastic film has been a major problem for the endangered and threatened sea turtles many of which slowly starve to death following the consumption of plastic bags floating on the surface (Balazs, 1985). One study estimates that more than one hundred thousand marine mammals and more than one million seabirds die each year from ingestion of plastic materials of one kind or another (OCRM, 1991).

Another major plastic component is made up of abandoned fishing gear in the form of nets or rope, most of which consists of non-biodegradable plastic materials. These have resulted in significant impacts in terms of entanglements of both large and small cetacean, a problem Congress recognized by enacting the Marine Mammal Health and Stranding Response Act (1992). This Act established the Marine Mammal Health and Stranding Response Program that enables recognized stranding organizations to collect and disseminate reference data on the health of marine mammals. A related problem impacting sea mammals is the so-called ghost fishing which traps these animals in abandoned or discharged nets and traps which prevents them from surfacing to breathe. Again, Congress stepped in and enacted the Marine Mammal Rescue Assistance Act, 2000 (PL 106-555). This Act provides Federal funding for those organizations authorized to help the release of marine mammals entangled with fishing gear and other anthropogenic materials.

The composition of the debris washing up on the shore represents yet another dimension of the problem and varies from region to region depending upon the major use of the coast, and nearshore and offshore uses. In areas where fishing is important, as in Alaska and the Northeast, lost or abandoned fishing gear represents a much higher proportion of the beach debris compared with regions where fishing is less important. Another activity-based source of beach debris is recreational boating. At least one source suggests that this activity accounts for nearly 60% of all marine generated waste (Eastern Research Group, Inc., 1989). Given that recreational boating continues to grow in popularity, it is likely that this beach debris component will continue to increase relative to other activity-based sources at least until programs and/or legislation have successfully curtailed these contributions. While some attempts have been made to describe regional differences in the types of beach debris (Coe *et al.*, 1997; Golik, 1997; Matsumura *et al.*, 1997) no uniform classification system has yet been developed. The National Park Service classifies beach debris into seven different groups while TOC has defined a total of 81 different categories. For purposes of illustration the International Coastal Cleanup data for 1996 is summarized in Table C7.

Table C7 Beach debris by type (from Coe *et al.*, 1997, with permission of Springer-Verlag)

Debris category	Percent of total
Plastics	61.2
Metal	11.1
Paper	10.8
Glass	10.4
Wood	2.9
Rubber	2.5
Cloth	1.4
Total	100.3

Future trends

As with so many other environmental issues, the beach clean-up initiative is a reactive rather than a proactive program. Unclean beaches represent an aesthetic and potential health hazard as well as a more serious environmental problem that threatens many of our most endangered marine species. Although much has been accomplished in cleaning up the marine environment both here and abroad, much still needs to be done. Remedial work related to beach clean up and marine pollution may be best addressed in two ways: through legal instruments, education and outreach. Each will be discussed briefly below. Several international conventions have been passed in recent years that directly and indirectly seek to limit marine pollution by ships, oil platforms, and other floating or submerged platforms that may emit solid waste to the marine environment. In the United States, the Coast Guard requires each vessel to maintain a log of waste materials discharged. In addition, most ports now have facilities enabling commercial vessels to offload both liquid and solid waste materials while in harbor. These requirements cover all vessels, including recreational and commercial fishing vessels. The issue here has been one of enforcement, which requires actual witnesses of any discharge before the offending vessel can be charged. Perhaps one solution to this problem is to require all vessels, including commercial fishing and recreational boats, to maintain logs on the location, type and quantity of materials discharged whether on or offshore.

While arguments have been made that recreational vessels account for most of the solid waste problems discharged at sea, it is possible that most of these estimates are based on older data, collected and analyzed prior to the implementation of zero discharge zones, and the various campaigns that most states have initiated to make recreational boating more environmentally amenable. Even industry has played a role in disseminating information intended to make boating and boaters more environmentally sensitive. These include the National Clean Boating Campaign (NCBC) which is sponsored by the Marina Environmental Education Foundation. The NCBC is made up of marinas, marine trade associations, governmental agencies, the USCG Auxiliary, several environmental NGOs. While the "Zero Discharge Zone" program initiated in Rhode Island is not directly related to beach clean up, there is overwhelming support within the Rhode Island boating community to support a more aggressive environmental initiative on the part of all levels of government. This suggests that further efforts to communicate the importance of proper bagging and discharge of solid waste will be met with unqualified approval.

Niels West

Bibliography

Balazs, G.H., 1985. Impact of ocean debris on marine turtles: entanglement and ingestion. In Shomura, R.S., and Yoshida, H.O. (eds.), *Proceedings of the Workshop on the Fate and Impact of Marine Debris, 1985*. National Marine Fisheries Service, Washington, DC as cited in OCRM, 1991.

Coe, J.M., Andersson, S., and Rogers, D.B., 1997. Marine debris in the Caribbean Region. In Coe, J.M., and Rogers, D.B. (eds.), *Marine Debris: Sources, Impacts and Solutions*, New York: Springer Verlag.

Eastern Research Group, Inc., 1989. Development of estimates of garbage disposal in the maritime sectors. In *Transportations Systems Center, Research and Special Programs Administration*, U.S. Department of Transportation, Washington, DC, in OCRM, 1991.

Golik, A., 1997. Debris in the Mediterranean Sea: types, quantities and behavior. In Coe, J.M., and Rogers, D.B. (eds.), *Marine Debris: Sources, Impacts and Solutions*.

Liffmann, M., Howard, B., O'Hara, K, and Coe, J.M., 1997. Strategies to reduce, control and minimize land-source marine debris. In Coe, J.M., and Rogers, D.B. (eds.), *Marine Debris: Sources, Impacts and Solutions*.

Matsumura, S., and Nasu, Keiichi, 1997. Distribution of floating debris in the North Pacific Ocean: sighting surveys 1986–1991. In Coe, J.M., and Rogers, D.B. (eds.), *Marine Debris: Sources, Impacts and Solutions*.

Nollkaemper, A., 1997. Legal regulation of upland discharges of marine debris: from local to global controls and back. In Coe *et al.* (eds.), OCRM, 1991. *Marine Debris: Status Report and Bibliography*, Technical Assistance Bulletin # 104, NOAA, Washington, DC.

PL 100-220, 33 USC, Section 1901 et.seq.

PL 102-587, 16 USC, Section 1421.

PL 106-555, 15 USC, Section 1361 note.

Ribic, C.A., Johnson, S.W., and Cole, C.A., 1997. Distribution, type, accumulation, and source of marine debris in the United States, 1989–1993. In Coe, J.M., and Rogers, D.B. (eds.), *Marine Debris: Sources, Impacts and Solutions*.

Stewart, T.R., 1989. Perspectives on Medical Waste. The Rockefeller Institute of Government, Albany, NY as quoted in OCRM, (1991).

Stratton, J.A. *et al.*, 1969. *Science and Environment, Volume 1*, Panel Reports of the Commission on Marine Science, Engineering and Resources, U.S.G.P.O., Washington, DC.

Cross-references

Beach Use and Behaviors
Conservation of Coastal Sites
Environmental Quality
Human Impact on Coasts
Lifesaving and Beach Safety
Marine Debris—Onshore, Offshore, Seafloor Litter
Rating Beaches
Tourism and Coastal Development

CLIFFED COASTS

It has been estimated that there are cliffs, consisting of rock or cohesive clays, around about 80% of the world's oceanic coasts (Emery and Kuhn, 1982). Sea cliffs are found in all types of morphogenic and tectonic settings, although they are generally absent from low-lying, sediment-abundant, Amero-trailing coasts, which have plate collision coasts and high mountains on the opposite side of the continent. Despite the morphological importance of marine cliffs, and the aesthetic contribution they make to the littoral environment, there are only a few systematic descriptions of their form and development in the literature (Trenhaile, 1987, 1997; Sunamura, 1992).

Cliff morphology

The type of scenery that develops on a cliff coast is the product of a number of factors, including the morphology of the hinterlands, present and past climates, wave and tidal environments, changes in relative sea level, and the structure and lithology of the rocks. Although cliff morphology is difficult to classify on the basis of climate or wave regime, some types of cliff are more characteristic of particular morphogenic regions than others (Figure C14).

Steep cliffs usually develop in marine-dominated environments and convex cliffs where subaerial processes dominate. Where both process suites are effective, subaerial processes produce a convex slope in the upper portion of cliffs, while marine processes cut steep cliff faces at their base. Coastal slopes therefore tend to be greater in the vigorous, storm wave environments of the mid latitudes than in high latitudes or in the tropics, where frost and chemical weathering are important. Wave action is generally weak in high latitudes because of sea ice, and in some places coastal configuration. Most steep slopes on these coasts are the product of glacial erosion and there are few true marine cliffs. In the humid tropics, the formation of steep marine cliffs is inhibited by fairly weak wave action, protective coral or algal reefs, and extensive coastal plains. Chemical weathering plays an important role in cliff development, and the weathered material is very susceptible to slumping, mudflows, and other mass movements. Coastal slopes are often covered in vegetation for all but the lowest few meters, and steep cliffs may be restricted to headlands and other exposed areas. Steep, bare cliffs are much more common in the arid tropics, particularly in exposed areas where spray, high evaporation and salt crystallization, alternate wetting and drying, corrosion, and other weathering processes help to steepen coastal slopes and inhibit the growth of vegetation.

There are numerous exceptions to the morphogenic classification of marine cliffs. For example, despite strong waves, there has been only minor modification of glacially sculptured granites in Maine and eastern Canada. On the other hand, despite weak wave action, steep, bare cliffs are common in the limestones of the Canadian Arctic, and they occur throughout the wet and dry tropics where strong waves can be generated by onshore monsoons, tropical cyclones, and Trade Winds.

Geological factors are at least as important as climate and wave regime in determining the relative efficacy of marine and subaerial

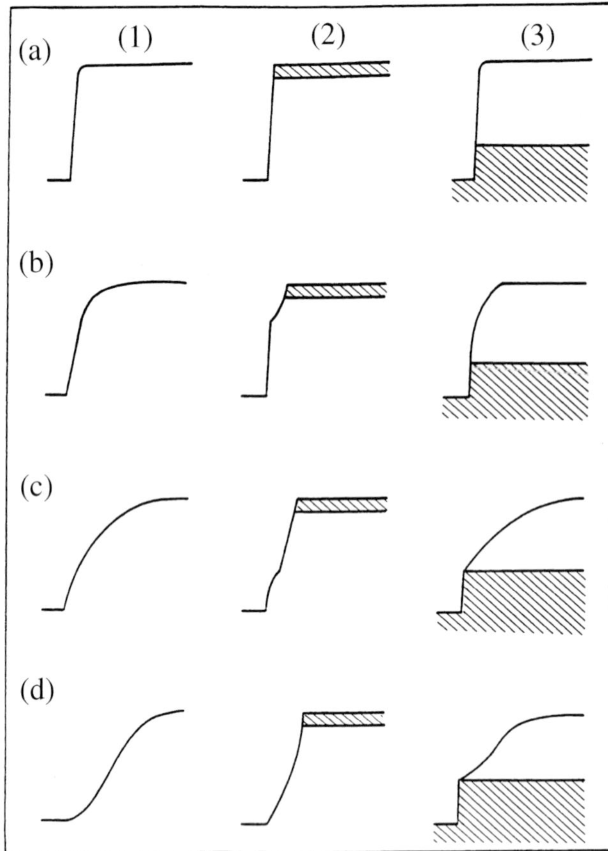

Figure C14 The effect of marine and subaerial processes and variable rock hardness on cliff profiles. Columns (1), (2), and (3) represent homogeneous rocks, harder rocks (shaded) at the top of the cliff, and harder rocks at the cliff base, respectively. Rows (a), (b), (c), and (d) represent environments in which marine erosion is much greater than, greater than, equal to, and less than weathering, respectively. (After Emery and Kuhn, 1982.)

processes, and therefore the shape of cliff profiles (Emery and Kuhn, 1982). For example, weaker material in the upper part of a cliff increases the effects of subaerial weathering, whereas weaker material in the lower part of a cliff facilitates wave erosion. Cliff profiles are also strongly influenced by structural weaknesses, stratigraphic variations, and the attitude or orientation of the bedding (Emery and Kuhn, 1980; Trenhaile, 1987, 1997). Very steep cliffs generally develop in rocks that are either horizontally or vertically bedded, and more moderate slopes in seaward or landward dipping rocks. The relationship between cliff profiles and the dip of bedding and/ or joint planes is very complex, however, and vertical and seaward-sloping cliffs can be formed in horizontal, or seaward or landward dipping rocks.

Composite cliffs have at least two major slope elements. Multistoried cliffs have two or more steep faces separated by more gentle slopes, whereas bevelled (hog's back, slope over wall) cliffs consist of convex or straight seaward-facing slopes above steep, wave-cut faces. Bevelled cliffs can develop in homogeneous rocks where marine and subaerial processes are of comparable importance, or where weaker rocks overlie more resistant materials (Figure C14). High composite cliffs in ancient, resistant massifs developed over very long periods of time, however, and they experienced marked changes in climate and sea level during the Quaternary. Wave-cut cliffs were abandoned during the last glacial stage, when frost and periglacial mass movements were probably dominant processes in the ice-free middle latitudes. The cliffs were gradually replaced by convex slopes developing beneath the accumulating talus, and when the sea rose to its present position marine processes removed the debris and, depending on the relative strength of the waves and the rock, either trimmed the base of the convex slopes to form composite cliffs, or completely removed it to form steep, wave-cut cliffs. Modeling suggests that bevelled profiles develop if the debris reached the top of the cliff during the last glacial stage, and multistoried profiles if the debris only rose part of the way up the cliff face (Griggs and Trenhaile, 1994).

There are usually shore platforms or beaches at the foot of cliffs, but some plunge directly into deep water. Plunging cliffs developed where the postglacial rise in relative sea level was much greater than the rate of sediment accumulation at the cliff foot. Once sea level had become stable, erosion was inhibited by the lack of abrasives at the water level, and reflection of the incoming, nonbreaking waves. Plunging cliffs are particularly common around basaltic islands in the Southern Hemisphere, where there is slow erosion and little longshore sediment transport.

Cliff erosion

The erosion of sea cliffs is episodic, site-specific, and closely related to prevailing meteorological conditions. Although recession rates have been measured in many parts of the world, most are little more than rough estimates. Reported rates vary from virtually nothing up to 100 m yr^{-1} (Kirk, 1977; Sunamura, 1992), and it is difficult to identify patterns that can be correlated with variations in rock type or wave and tidal conditions. A variety of techniques have been used to determine rates of erosion, but the most accurate ones have usually been used over only very short periods of time. Japanese workers have studied the short-term effects of coastal dynamics on cliff erosion. Variations in recession rates were partly explained by differences in the frequency of waves that exceed the minimum height capable of causing erosion, and they increased as the compressive strength of the rocks decreased (Sunamura, 1992); however, many other factors need to be considered. In south Central England, for example, the compressive strength of the rocks decreases toward the east, but this is countered by changes in the fracture pattern, which favors increasing rates of cliff erosion to the west (Allison, 1989).

Other features of cliffed coasts

Small bays, narrow inlets, caves, arches, and stacks are usually the result of erosion along structural weaknesses, particularly bedding, joint, and fault planes, and in the fractured and crushed rock produced by faulting. These features form in rocks which have well defined and well spaced planes of weakness, yet are strong enough to stand as high, near-vertical slopes, and as the roofs of caves, tunnels, and arches. They are therefore uncommon in weak or thinly bedded rocks with dense joint systems (Trenhaile, 1987).

Narrow, steep-sided inlets (geos) often develop along steeply dipping joints or faults in horizontal or gently dipping rocks, and from the erosion of dykes and the collapse of lava tunnels in igneous rocks. The occurrence and morphology of caves are usually controlled by weaknesses associated with joints, faults, breccias, schistosity planes, unconformities, irregular sedimentation, and the internal structure of lava flows. Marine caves are usually quite small, but large karstic caves have been inherited by the sea following cliff recession. It has been assumed that arches and longer tunnels develop through the coalescence of caves driven into the opposite sides of headlands. However, arches near La Jolla, California appear to be associated with coastal indentations and surge channels, that pile up the approaching waves on one side of a point, causing penetration of the weaker rocks at the cliff base (Shepard and Kuhn, 1983). Arches can also be formed by the marine invasion of karstic caves. Fountains of spray are emitted through blowholes when large breakers surge into tunnel-like caves connected to the surface along joint or fault-controlled shafts. Marine inheritance of karstic sinks, sinkholes, and tunnels has produced spectacular blowhole systems in some areas.

Most stacks are the result of the dissection of the coast along joints and other planes of weakness, although some are associated with folding, submergence of tropical tower karst, and differences in resistance associated with solution pipes, induration, or variable lithology. Some stacks develop from the collapse of the roofs of arches, but many form directly from erosion of the cliff face. Spectacular stacks have developed from the dissection of coarse, poorly sorted conglomerates and arkosic sandstones along prominent, well-spaced joint planes in the Bay of Fundy, which has the largest tidal range in the world. Old photographs suggest that these stacks may have life spans, ranging from their initial separation from the cliff to their eventual collapse, of between 100 and 250 years. Notches are ubiquitous at the cliff foot, and they are responsible for the characteristic mushroom-shape appearance of the stacks. Although there is no consistent relationship between the depth of notches on the seaward and landward sides of the stacks, the notches are at higher elevations on the seaward side. The deepest part of most notches is a little below the mean high tidal level, although several are up to a meter or two below it, especially on the landward side of the stacks. Stack morphology and notch depth change in a fairly predictable manner through time, as the stacks become increasingly isolated from the cliff (Trenhaile *et al.*, 1998).

Although it is generally assumed that headlands consist of more resistant materials than those in the adjacent bays, there have been few detailed investigations of variations in rock structure or strength. The occurrence of small bays and headlands often reflects the rather subtle influence of rock structure, including variations in joint density, the orientation of discontinuities, and the thickness, strike, and dip of the beds. Large bays and prominent headlands are more likely to reflect differences in the lithology of the rocks, however, although they could also develop in fairly homogeneous rocks if, as theory suggests, the low cliffs around stream outlets retreat more rapidly than the higher cliffs on the adjacent interfluves.

The plan shape of crenulated coasts may eventually attain an equilibrium state. This would occur when, because of the increasing effect of wave refraction as coasts become more indented, the more resistant rocks on the exposed headlands are eroded by higher waves at the same rate as the weaker rocks in the increasingly sheltered bays are eroded by lower waves. The plan shape would then be maintained through time as the coast retreated landward. As much of the erosive work could have been accomplished in previous interglacial stages, when sea level was similar to the present day, only minor modification of inherited coasts may have been required to attain equilibrium at present sea level; however, we lack reliable, long-term data on cliff erosion rates that are needed to determine whether the plan shape of rock coasts has attained a state of equilibrium.

Inheritance

In many areas, erosional processes are modifying elements of cliffed coasts that were inherited or partly inherited from the past (Bryant *et al.*, 1990; Trenhaile *et al.*, 1999). For example, raised beaches in southwestern Britain contain middle and upper Pleistocene material which suggests that the composite cliffs behind are very old, and dated sediments in marine caves in southwestern Britain, northwestern Spain, off northern France, and in southern Australia have shown that the caves are at least as old as the last interglacial stage (Davies, 1983; Trenhaile *et al.*, 1999). The occurrence of these ancient features is consistent with the paleo sea-level record, which shows that interglacial sea levels were similar to the present level on a number of occasions during the middle and late Pleistocene. Inheritance is most likely to have occurred on tectonically stable coasts, and in resistant, slowly eroding rocks. It is much more difficult to determine whether it has occurred in weaker rock areas, where any till or other sedimentary covers, raised beaches, structural remnants, and other evidence which may have existed would have been removed by fairly rapid wave erosion at the present level of the sea.

Cohesive clay coasts

Cohesive clay coasts have many of the characteristics of weak rock coasts, and they are often fronted by wide shore platforms. Fine-grained sediments lose their cohesion when they are eroded, however, and the debris is generally transported offshore as suspended load. Therefore no permanent or continuous beach develops, and as the eroded material provides no protection to the cliff, the height of the cliff has no effect on the rate of erosion. Variations in groundwater level, which change the strength and stability of clay materials, is a crucial factor in the failure of cohesive coastal slopes. In temperate regions, erosion of cohesive coasts by marine and subaerial processes is often markedly seasonal in nature, reflecting variations in wave energy, temperature, precipitation, and groundwater pressures. There is a strong relationship between landslide and flow activity in cohesive sediments and the occurrence of clays with a high proportion of montmorillonite or other swelling minerals. These minerals increase the frequency and mobility of slope movements, and allow them to occur on more gentle slopes. Shallow sliding occurs as mudslides or, if the water content is very high, as mudflows. Clay cliffs are also susceptible to deep-seated rotational landslides, which occur where basal erosion is rapid enough to remove the mudflow debris and steepen the underlying coastal slopes.

Hutchinson's (1973) form/process classification of cliffs in the London Clay of southeastern England is considered to be generally representative of slopes in stiff fissured clays (see *Mass Wasting*). Three types were distinguished, based on the relative rates of basal marine erosion and subaerial weathering:

1. Type 1 occurs where the rate of basal erosion is broadly in balance with weathering and sediment is supplied to the toe of the slope by shallow mudsliding. Removal of the mudslide debris stimulates further sliding, and unloading, softening, and weathering of the exposed clays.
2. Type 2 is found where basal erosion is more rapid than weathering. Waves remove all the material supplied by mudslides and erosion, and undercutting of the *in situ* clay steepens the profile. This eventually causes a deep-seated mass movement, which is characteristically a rotational slide involving basal failure. The sea removes the slump debris and toe erosion then steepens the slope until another failure occurs.
3. Type 3 coastal slopes develop where there is no basal erosion. When a cliff is abandoned by the sea or coastal defenses are constructed at its base, debris is carried to its foot by shallow rotational slides. Therefore, abandoned cliffs often have a steeper upper slope on which landsliding is initiated, and a flatter lower slope where colluvium accumulates. The slides eventually become quiescent when slope gradients have been reduced to the ultimate angle of stability against landsliding, and hill wash and soil creep gradually convert the bluffs into a series of undulations and then into smooth slopes.

Hutchinson's (1973) three cliff types also develop, in response to long-term changes in lake level, along the northern shore of Lake Erie. Variations in wave intensity and the type of glacial sediments in the cliffs, however, also account for the simultaneous distribution of the three cliff types along this coast (Quigley and Gelinas, 1976). Hutchinson's classification is also generally applicable to cohesive cliffs in Denmark, although there are some significant differences resulting from glacio-isostatic recovery and changes in relative sea level (Prior and Renwick, 1980). Despite the general applicability of Hutchinson's model to a wide range of cohesive sediments, however, alternate models of cliff retreat are relevant to specific areas.

Alan S. Trenhaile

Bibliography

Allison, R.J., 1989. Rates and mechanisms of change in hard rock coastal cliffs. *Zeitschrift fur Geomorphologie*, **73**(Supplement Band): 125–138.

Bryant, E.A., Young, R.W., Price, D.M., and Short, S.A., 1990. Thermo-luminescence and uranium–thorium chronologies of Pleistocene coastal landforms of the Illawarra Region, New South Wales. *Australian Geographer*, **21**: 101–111.

Davies, K.H., 1983. Amino acid analysis of Pleistocene marine molluscs from the Gower Peninsula. *Nature*, **302**: 137–139.

Emery, K.O., and Kuhn, G.G., 1980. Erosion of rock coasts at La Jolla, California. *Marine Geology*, **37**: 197–208.

Emery, K.O., and Kuhn, G.G., 1982. Sea cliffs: their processes, profiles, and classification. *Geological Society of America Bulletin*, **93**: 644–654.

Griggs, G.B., and Trenhaile, A.S., 1994. Coastal cliffs and platforms. In Carter, R.W.G., and Woodroffe, C.D. (eds.), *Coastal Evolution*. Cambridge: Cambridge University Press, pp. 425–450.

Hutchinson, J.N., 1973. The response of London Clay cliffs to differing rates of toe erosion. *Geologia Applicata e Idrogeologia*, **8**: 221–239. Reprinted in 1975 as Building Research Establishment (U.K.) Current Paper CP 27/75.

Kirk, R.M., 1977. Rates and forms of erosion on intertidal platforms at Kaikoura Peninsula, South Island, New Zealand. *New Zealand Journal of Geology and Geophysics*, **20**: 571–613.

Prior, D.B., and Renwick, W.H., 1980. Landslide morphology and processes on some coastal slopes in Denmark and France. *Zeitschrift fur Geomorphologie*, **34** (Supplement Band): 63–86.

Quigley, R.M., and Gelinas, P.J., 1976. Soil mechanics aspects of shoreline erosion. *Geoscience Canada*, **3**: 169–173.

Shepard, F.P., and Kuhn, G.G., 1983. History of sea arches and remnant stacks of La Jolla, California, and their bearing on similar features elsewhere. *Marine Geology*, **51**: 139–161.

Sunamura, T., 1992. *Geomorphology of Rocky Coasts*. Chichester: John Wiley.

Trenhaile, A.S., 1987. *The Geomorphology of Rock Coasts*. Oxford: Oxford University Press.

Trenhaile, A.S., 1997. *Coastal Dynamics and Landforms*. Oxford: Oxford University Press.

Trenhaile, A.S., Pepper, D.A., Trenhaile, R.W., and Dalimonte, M., 1998. Stacks and notches at Hopewell Rocks, New Brunswick, Canada. *Earth Surface Processes and Landforms*, **23**: 975–988.

Trenhaile, A.S., Pérez Alberti, A., Martínez Cortizas, A., Costa Casais, M., and Blanco Chao, R., 1999. Rock coast inheritance: an example from Galicia, northwestern Spain. *Earth Surface Processes and Landforms*, **24**: 1–17.

Cross-references

Cliffs, Erosion Rates
Cliffs, Lithology versus Erosion Rates
Faulted Coasts
Mass Wasting
Notches
Paraglacial Coasts
Rock Coast Processes
Shore Platforms

CLIFFS, EROSION RATES

Cliff erosion occurs when the assailing force of waves acting on the cliff face is greater than the resisting force of the cliff-forming rocks. Sunamura (1977) has indicated that a fundamental relation governing cliff erosion may be expressed by:

$$dX/dt = K \ln (F_w/F_r), \qquad \text{(Eq. 1)}$$

where X is the erosion distance, t is the time, F_w is the wave assailing force, F_r is the rock resisting force, and K is a constant with dimensions of $[LT^{-1}]$. This equation indicates that erosion takes place if $F_w > F_r$, whereas no erosion occurs if $F_w \leq F_r$. The erosion rate, dX/dt, is an instantaneous rate which is expressed in terms of the slope of the tangent line to a curve showing the temporal change in erosion distance. Accurate and continuous monitoring of erosion distance in the field is extremely difficult, so that instantaneous erosion rates have not been obtained. Instead, the average rate, $\Delta X/\Delta t$, has been widely employed, in which ΔX is the erosion distance during a certain period of time, Δt. Depending on the length of Δt, erosion rates are highly fluctuated. Erosion distances have been documented by a variety of techniques (Sunamura, 1992) such as sequential aerial photographs, historic maps, and *in situ* measurements including surveys, micro-erosion meters, and pegs.

Short-term versus long-term erosion rates

On the north shore of Long Island, New York, a hurricane of September 14, 1944 cut back the cliff composed of glacial deposits a horizontal distance of over 12 m in a single day (Davies *et al.*, 1972). This provides a striking contrast to an 80-year average of 0.5 m/year for this coast (Bokuniewicz and Tanski, 1980). Similar examples can be taken from some locations along the California coast (Griggs and Savoy, 1985). During January storms of 1983, waves removed about 14 m of cliff top in one section of Miocene mudstone–siltstone cliffs at Santa Cruz, which had receded at an average rate of 0.2 m/year from 1931 to 1982. Near Capitola, 10 km east of Santa Cruz, a sea cliff of the same geology was cut back 1.5–3 m overnight by the storms, a long-term average recession rate including this site being only 0.3 m/year. Thus, a short-term erosion rate based on the time interval including such a storm event makes the long-term average rate very unreliable.

Temporal variations in erosion rates

On the coast where the resisting force of rocks can be assumed to be time-independent, fluctuations of the wave assailing force are responsible for temporal changes in erosion rates. The wave force is obviously dependent on wave energy level in the offshore region. As seen from equation 1, erosion occurs only when the assailing force exceeds a threshold value. The occurrence frequency of waves that actually cause erosion is, therefore, important when considering erosion rates: the more frequently large waves attack a coast during a certain period of time, the more severely the cliff recedes to produce a higher erosion rate. A study by Sunamura (1982) on two Tertiary cliffs in Japan indicates that short-term erosion rates are closely related to the frequency of erosion-causing waves.

The wave assailing force is greatly controlled by water level, nearshore topography, and talus or ice cover at the cliff base. Fluctuations of these controlling factors produce temporal variations in erosion rates. The most important factor in dramatic erosion is an abnormal rise in water level, known as storm surges. A remarkable water level rise may occur on tidal coasts when storm surges superimpose on high tides. The well-known 1953 North Sea storm surge caused drastic erosion of glacial-sand cliffs at Covehithe, Suffolk, England (Williams, 1960). The storm surge with a maximum height of 2.3 m occurred at high tide yielding a total sea-level rise of 3.5 m above mean sea level. A glacial-sand cliff was cut back 9 m due to the action of storm waves during about 2 h at the peak of the surge.

Beaches working as a buffer diminish drastically depending on wave and tidal conditions as seen at Big Lagoon, north of Eureka on the northern California coast (Savoy and Rust, 1985). A sandy cliff backing a beach has receded at a long-term average rate of 0.6 m/year. During the winter of 1982–83, combinations of storm waves and high tides removed the beach, and the cliff was cut back more than 9 m in this time period.

Fronting beaches frequently experience change due to sediment imbalance caused by human activities such as harbor works, dam construction on rivers, and beach mining or filling. A reduction in beach levels stemming from human action has produced unintentional results of rapid erosion of sea cliffs in some areas. An example of this sort of erosional events can be taken from the Half Moon Bay area of central California (Lajoie and Mathieson, 1985). The construction of a breakwater at Princeton Harbor in the northern part of Half Moon Bay was made between 1956 and 1960. The breakwater interrupted southward littoral drift, resulting in depletion of beaches in front of sea cliffs composed of unconsolidated materials at El Granada adjacent to the south of the breakwater. This caused remarkable changes in erosion rates before and after the engineering work, from 0.1 to 2 m/year. Some other studies dealing with cliff erosion intensified by anthropogenic factors have been reviewed in Sunamura (1992).

The influence of underwater morphological change on temporal variations in long-term erosion rates has been reported from Dunwich in Suffolk, England (Robinson, 1980). Glacial–sand cliffs were cutting back at an annual rate of: 1.6 m/year from 1589 to 1753, 0.85 m/year from 1753 to 1824, 1.5 m/year from 1824 to 1884, 1.15 m/year from 1884 to 1925, and 0.15 m/year from 1925 to 1977, the last being extremely low. This drop in the erosion rate during the recent 50 years can be attributed to the reduction in wave energy arriving at the coast. The energy reduction was due to a significant divergence in wave refraction, produced by a changing bottom configuration associated with the steady northward growth of Sizewell Bank, which took place during the past century.

Talus and ice cover protect the foot of cliffs from wave attack until they are removed by the wave and current action, giving rise to no erosion. Such a protective effect has been observed on the Ohio shore of Lake Erie (Carter and Guy, 1988).

Spatial variations in erosion rates

The long-term erosion rate masks the variation in short-term erosion rates and shows a site-specific value. This value fluctuates from place to place depending on the local variation in the wave assailing force and the rock resisting force. Suppose that several cliffs with rocks of different resistance are exposed to a similar wave climate; then the long-term erosion rate, R^*, may be described from equation 1 as:

$$R^* = K^* \ln (F_w^*/F_r^*) = K^* (A - \ln F_r^*), \qquad \text{(Eq. 2)}$$

where F_w^* is the long-term averaged assailing force and F_r^* is the resisting force which is considered here to be time-independent, and K^* and A $(= \ln F_w^*)$ are constants. This equation indicates that so-called "harder" cliffs have lower erosion rates. The relationship between R^* and F_r^* for cliffs on the Pacific coast of eastern Japan (Sunamura, 1992) and on the Mediterranean coast of southern Italy (Budetta *et al.*, 2000) can be well expressed in terms of equation 2 (Sunamura, 1992).

Assuming that cliffs with similar lithology are subjected to the different wave climate, the long-term rate should be higher where wave conditions are more severe, if other controlling factors are constant. Since F_r^* is constant this time, one may obtain from equation 2.

$$R^* = K^* (B + \ln F_w^*), \qquad \text{(Eq. 3)}$$

where $B = -\ln F_r^* =$ constant. Sunamura (1992) discussed the R^* versus F_w^* relationship using data of the long-term erosion rate in the central part of the north shore of Lake Erie, where cliffs cut in glacial deposits have developed with no marked alongshore variations in their geotechnical properties, but different exposures to waves.

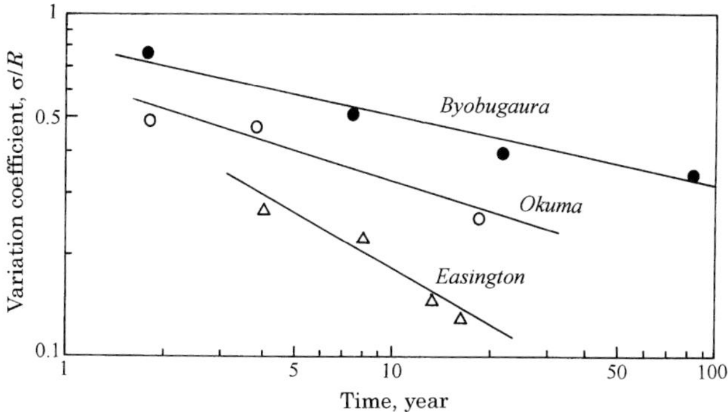

Figure C15 The variation coefficient of the erosion rate, σ/R, plotted against the length of time for the three cliffs. Data from Sunamura (1973) for Byobugaura and Okuma, and from Richards and Lorriman (1987) for Easington.

Alongshore changes in beach width give rise to spatial variations in erosion rates, if other factors are uniform. An example has been reported from the south shore of Lake Erie (Carter *et al.*, 1986); cliffs composed of glaciolacustrine clay were receding from 1876 to 1973 at an annual rate of 1.3–2 m/year where the beach had a width of 0–6 m, whereas only a few to 10 cm/year on the shore with wide beaches (>24.3 m).

The presence or absence of beach sediments at the cliff base, which can work as an abrasive tool to intensify the wave assailing force, yields spatial variations in the erosion rate. Robinson (1977) reported that on the northeast Yorkshire coast of England, erosion rates of Upper Lias shale cliffs fronted by beaches were 15–18.5 times higher than those with no beaches.

Water-level changes induced by local tectonic movement may control of the wave assailing force, which results in regional variations in long-term cliff erosion rates. The influence of the relative sea-level change caused by an earthquake on cliff erosion has been reported from the coast of Oregon in the Pacific northwest of the United States (Komar and Shih, 1993).

On a coastal strip where the respective alongshore variations in the wave assailing force and the cliff resisting force are considered to be uniform, it is anticipated from equation 1 that any location along the coast has receded at the same rate, that is, there is no spatial variation in erosion rate. Actually, however, there are considerable alongshore fluctuations when considering erosion rates in a short-time period. Figure C15 is a plot of the variation coefficient, σ/R, against the length of time, where R is the space-average value of erosion rates and σ is the standard deviation of R; the value of σ/R increases with increasing variability. Larger fluctuations in shorter-term erosion rates reflect sporadic and localized occurrence of mass movement, which results in talus formation at the cliff base, increasing alongshore variability in the wave assailing force. The degree of erosion-rate variation becomes smaller as the timescale lengthens (Figure C15), suggesting that parallel recession of the coastline will take place over a long period.

Tsuguo Sunamura

Bibliography

Bokuniewicz, H., and Tanski, J., 1980. Managing localized erosion of coastal bluffs. In *Proceedings Coastal Zone '80*. American Society of Civil Engineers, 3, pp. 1883–1898.

Budetta, P., Galietta, G., and Santo, A., 2000. A methodology for the study of the relation between coastal cliff erosion and the mechanical strength of soil and rock masses. *Engineering Geology*, **56**: 243–256.

Carter, C.H., and Guy, D.E., Jr., 1988. Coastal erosion: processes, timing and magnitudes at the bluff toe. *Marine Geology*, **84**: 1–17.

Carter, C.H., Monroe, C.B., and Guy, D.E., Jr., 1986. Lake Erie shore erosion: the effect of beach width and shore protection structures. *Journal of Coastal Research*, **2**: 17–23.

Davies, D.S., Axelrod, E.W., and O'Conner, J.S., 1972. Erosion of the north shore of Long Island, Marine Science Research Center, State University of New York, Stony Brook, *Technical Report*, 18.

Griggs, G., and Savoy, L., 1985. Sea cliff erosion. In Griggs, G., and Savoy, L. (eds.), *Living with the California Coast*. Durham: Duke University Press, pp. 27–34.

Komar, P.D., and Shih, S.-M., 1993. Cliff erosion along the Oregon coast: a tectonic-sea level imprint plus local controls by beach processes. *Journal of Coastal Research*, **9**: 747–765.

Lajoie, K.R., and Mathieson, S.A., 1985. San Francisco to Año Nuevo. In Griggs, G., and Savoy, L. (eds.), *Living with the California Coast*. Durham: Duke University Press, pp. 140–177.

Richards, K.S., and Lorriman, N.R., 1987. Basal erosion and mass movement. In Anderson, M.G., and Richards, K.S. (eds.), *Slope Stability*. Chichester: John Wiley & Sons, pp. 331–357.

Robinson, A.H.W., 1980. Erosion and accretion along part of the Suffolk coast of East Anglia, England. *Marine Geology*, **37**: 133–146.

Robinson, L.A., 1977. Marine erosive processes at the cliff foot. *Marine Geology*, **23**: 257–271.

Savoy, L., and Rust, D., 1985. The northern California coast: Oregon border to Shelter Cove. In Griggs, G., and Savoy, L. (eds.), *Living with the California Coast*. Durham: Duke University Press, pp. 81–105.

Sunamura, T., 1973. Coastal cliff erosion due to waves: field investigations and laboratory experiments. *Journal of Faculty of Engineering, University of Tokyo*, **32**: 1–86.

Sunamura, T., 1977. A relationship between wave-induced cliff erosion and erosive force of waves. *Journal of Geology*, **85**: 613–618.

Sunamura, T., 1982. A predictive model for wave-induced cliff erosion, with application to Pacific coasts of Japan. *Journal of Geology*, **90**: 167–178.

Sunamura, T., 1992. *Geomorphology of Rocky Coasts*. Chichester: John Wiley & Sons.

Williams, W.W., 1960. *Coastal Changes*. London: Routledge and Kegan Paul.

Cross-references

Cliffed Coasts
Cliff, Lithology versus Erosion Rate
Coastline Changes
Dams, Effect on Coasts
Human Impact
Rock Coast Processes on Coasts
Storm Surges

CLIFFS, LITHOLOGY VERSUS EROSION RATES

Lithological control of erosion rates

When the assailing force of waves acting on the cliff base is greater than the resisting force of rocks forming there erosion occurs, resulting in augumentation of the instability of an overall cliff profile, eventually leading to intermittent mass movement. Depending mainly on lithological factors such as geological structures, stratigraphic features, and geotechnical properties, the mode of mass movement is generally classified into four types: falls (e.g., Jones and Williams, 1991), topples (e.g., Davies *et al.*, 1998), flows (e.g., Grainger and Kalaugher, 1987), and slides (e.g., Hutchinson, 1986). Cliffs recede with one or hybrid types of these failure modes. Cliff recession has been documented by a variety of techniques (e.g., Sunamura, 1992), of which the most popular one is to measure distances on sequential aerial photographs and/or historic maps. Average erosion rates are determined by dividing the measured distances by the length of time interval. Orders of erosion rates summarized on a lithological basis are generally: 10^{-3} m/year for granitic rocks; 10^{-3}–10^{-2} m/year for limestone; 10^{-2} m/year for flysch and shale; 10^{-1}–10^{0} m/year for chalk and Tertiary sedimentary rocks; 10^{0}–10^{1} m/year for Quaternary deposits; and 10^{1} m/year for unconsolidated volcanic ejecta (Sunamura, 1983). These results show that lithology and cohesiveness of the cliff material greatly control the erosion rate. It should be noted, however, that toe erosion by waves originally causes cliff retreat.

Instantaneous erosion rates can be described by the following equation (see *Cliffs, Erosion Rates*):

$$dX/dt = K \ln (F_w/F_r) \qquad \text{(Eq. 1)}$$

where X is the erosion distance, t is the time, F_w is the wave assailing force, F_r is the rock resisting force, and K is a constant with dimensions of $[LT^{-1}]$. This equation holds when $F_w \geq F_r$; otherwise, $dX/dt = 0$. A short-term erosion rate is highly variable with time (see *Cliffs, Erosion*

Rates), so let us consider a long-term erosion rate, a site-specific value reflecting relative magnitude of wave action to rock strength.

Parameters for rock-mass strength against wave erosion

To find the most suitable parameter for expressing rock resistance to wave erosion, various strength measures have been tested in cliff erosion studies: compressive strength (e.g., Tsujimoto, 1987; Carter and Guy, 1988), tensile strength (e.g., Tsujimoto, 1987), cohesive strength (McGreal, 1979), shear strength (Zeman, 1986; Kamphuis, 1987), penetration strength (Yamanouch, 1977), and Schmidt hammer rebound value (Jones and Williams, 1991), and compressive strength converted from Schmidt hammer readings (Budetta *et al.*, 2000; Stephenson and Kirk, 2000a). Of these parameters compressive strength is found to be positively related to other measures and is a widely used parameter with testing criteria having been well established, so it is suitable to use this measure as a representative strength parameter when considering the intact part of cliff-forming materials (Sunamura, 1992).

Resistance of the cliff material against wave force is quite different according to the presence of discontinuities such as cracks, cleavages, joints, faults, and bedding planes, some being inherent in lithology and others being of tectonic origin; cracks and joints especially can also be brought about by weathering if rocks are prone to weather. When waves hit a cliff with joint- or fault-associated openings, the air in the interstices is suddenly compressed, which results in pressure increase. Measurements of dynamic pressure inside a crevice in a vertical cliff set up in a wave flume indicate that a considerable increase in pressure can be produced depending on the size of opening and input wave height (Sunamura, 1994). Such pressure increase exerts an excessive outward stress on the sidewall of the opening as well as on its roof. As the opening grows, the wave assailing force inside the opening begins to increase its intensity, which in turn gradually widens and deepens the opening, which causes a further increase in the assailing force, finally leading to removal of blocks bounded by joints and faults.

Thus, discontinuities in the cliff material play a significant role in the erosion process. To assess the influence of discontinuities on the overall cliff resistance against wave action, Tsujimoto (1987) employed a nondimensional parameter, V_f/V_l, where V_f and V_l are the sonic velocities measured, respectively, in the field on a rock mass with discontinuities and in the laboratory on an intact rock specimen. With increasing degree of fracturing, V_f decreases, so that the value of V_f/V_l reduces from unity and approaches to zero. Tsujimoto attempted to evaluate F_r multiplying this parameter by compressive strength of an intact rock specimen, S_c:

$$F_r = B\,(V_f/V_l)\,S_c = B\,S_c^*, \qquad \text{(Eq. 2)}$$

where S_c^* [$= (V_f/V_l)\,S_c$] is the overall strength of a rock mass and B is a constant. Budetta *et al.* (2000) proposed a similar strength measure: $F_r = J_p S_c$, in which J_p depends on the volume of a rock mass and on joint condition factors such as the roughness and alteration of joint walls and the size of joints. The measure was tested in the Cilento area of southern Italy.

Long-term erosion rates versus rock-mass strength

Suppose that cliffs with rocks of different resistance are exposed to a similar wave climate; then one obtains from equations 1 and 2:

$$R^* = K^*\,(\ln C^* - \ln S_c^*), \qquad \text{(Eq. 3)}$$

where R^* is the long-term erosion rate, K^* is a constant, and $C^* = \ln (F_w/B) = $ const. Using data collected by Tsujimoto (1987) from open coasts of Japan, the R^* versus S_c^* relationship is plotted in Figure C16, where R^* values denote erosion rates averaged over a long period of time, more than 30 years. The erosion rates are around 1 m/year for Quaternary deposits and Pliocene sedimentary rocks which have strengths in a range of 0.1–3 MPa; they are on the order of 10^{-3}–10^{-2} m/year for Miocene sedimentary rocks with a strength of about 10 MPa; and they are almost 0 m/year for hard rocks (>20 MPa) such as andesite and basalt, the data of these being plotted on the x-axis. The curve in Figure C16 is drawn as a result of a best fit of equation 3 with the data except the four data on the x-axis. The general trend is well described by equation 3 with $K^* = 0.2$ m/year and $C^* = 10$ MPa, although large scatter in the data is seen. Such scatter could be attributed to: (1) cliff resistance to wave erosion cannot be properly evaluated by S_c^*, only for some lithology, and (2) F_w does not always take on a constant value.

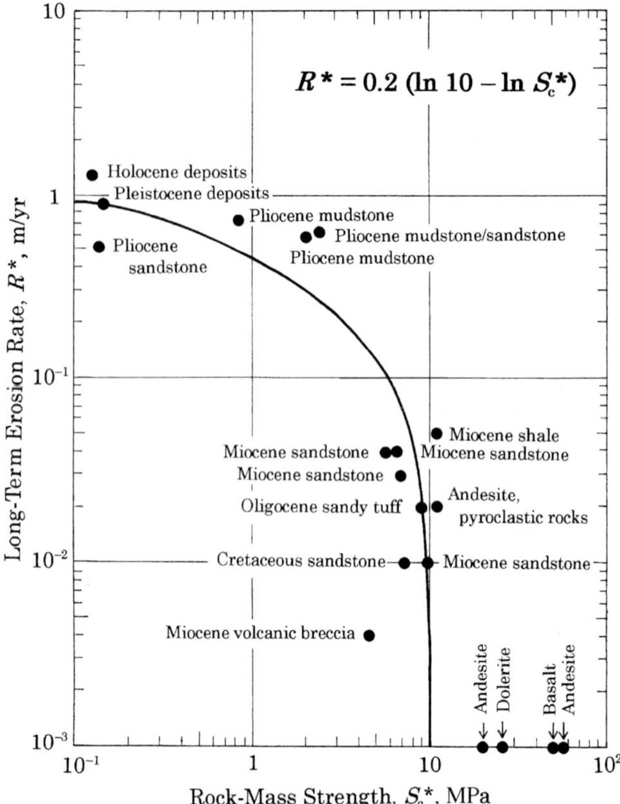

Figure C16 Relationship between long-term cliff erosion rate and cliff-forming rock-mass strength of cliffs on open coasts of Japan. The arrowed data denote that erosion rates are almost null. Data from Tsujimoto (1987).

One of major factors not considered in equation 3 is deterioration of the superficial part of a rock mass due to various types of weathering: alternate wet/dry weathering, frost action, salt crystallization, and the combined action of these. Mottershead (1994) examined this type of deterioration through strength testings on Lower Cretaceous sandstone used in sea walls at Weston-super-Mare, Avon, UK. The result indicated that the strength of weathered (honeycombed) rock is approximately 20% of that of fresh rock. Stephenson and Kirk (2000b) reported that considerable reduction in compressive strength (up to 50%) due to wet/dry weathering has occurred on limestone forming shore platforms at Kaikoura, New Zealand. Other factors responsible for strength reduction are fatigue effect by repetitive action of waves and biological activity, the former having still been unquantified, while the latter giving a significant influence, in some instances, as demonstrated by the fact that compressive strengths of sandstone and siltstone riddled by boring mollusks are found to have only 20–25% of the strength of solid (intact) rocks (Healy, 1968).

Tsuguo Sunamura

Bibliography

Budetta, P., Galietta, G., and Santo, A., 2000. A methodology for the study of the relation between coastal cliff erosion and the mechanical strength of soil and rock masses. *Engineering Geology*, **56**: 243–256.

Carter, C.H., and Guy, D.E., Jr., 1988. Coastal erosion: processes, timing and magnitudes at the bluff toe. *Marine Geology*, **84**: 1–17.

Davies, P., Williams, A.T., and Bomboe, P., 1998. Numerical analysis of coastal cliff failure along the Pembrokeshire Coast National Park, Wales, UK. *Earth Surface Processes and Landforms*, **23**: 1123–1134.

Grainger, P., and Kalaugher, P.G., 1987. Intermittent surging movements of a coastal landslide. *Earth Surface Processes and Landforms*, **12**: 597–603.

Healy, T.R., 1968. Bioerosion on shore platforms in the Waitemata formation, Auckland. *Earth Science Journal*, **2**: 26–37.

Hutchinson, J.N., 1986. Cliffs and shores in cohesive materials: geotechnical and engineering geological aspect. In *Proceedings Symposium on Cohesive Shores*, National Research Council, Canada, pp. 1–44.

Jones, D.G., and Williams, A.T., 1991. Statistical analysis of factors influencing cliff erosion along a section of the West Wales coast, UK. *Earth Surface Processes and Landforms*, **16**: 95–111.

Kamphuis, J.W., 1987. Recession rate of glacial till bluffs. *Journal of Waterway, Port, Coastal and Ocean Engineering*, **113**: 60–73.

McGreal, W.S., 1979. Marine erosion of glacial sediments from a low-energy cliffline environment near Kilkeel, Northern Ireland. *Marine Geology*, **32**: 89–103.

Mottershead, D.N., 1994. Spatial variations in intensity of alveolar weathering of a dated sandstone structure in a coastal environment, Westonn-spur-Mare, UK. In Robinson, D.A., and Williams, R.B.G. (eds.), *Rock Weathering and Landform Evolution*. Chichester: John Wiley & Sons, pp. 151–174.

Stephenson, W.J., and Kirk, R.M., 2000a. Development of shore platforms on Kaikoura Peninsula, South Island, New Zealand. Part One: The role of waves. *Geomorphology*, **32**: 21–41.

Stephenson, W.J., and Kirk, R.M., 2000b. Development of shore platforms on Kaikoura Peninsula, South Island, New Zealand. Part II: The role of subaerial weathering. *Geomorphology*, **32**: 43–56.

Sunamura, T., 1983. Processes of sea cliff and platform erosion. In Komar, P.D. (ed.), *Handbook of Coastal Processes and Erosion*. Boca Raton: CRC Press, pp. 233–265.

Sunamura, T., 1992. *Geomorphology of Rocky Coasts*. Chichester: John Wiley & Sons.

Sunamura, T., 1994. Rock control in coastal geomorphic processes. *Transactions, Japanese Geomorphological Union*, **15**: 253–272.

Tsujimoto, H., 1987. Dynamic conditions for shore platform initiation. *Science Report, Institute of Geoscience, University of Tsukuba*, **8A**: 45–93.

Yamanouchi, H., 1977. A geomorphological study about the coastal cliff retreat along the southwest coast of the Atsumi peninsula, central Japan. *Science Report, Faculty of Education, Gunma University*, **26**: 95–128.

Zeman, A.J., 1986. Erodibility of Lake Erie undisturbed tills. In *Proceedings Symposium on Cohesive Shores*, National Research Council, Canada, pp. 150–169.

Cross-references

Bioerosion
Cliffed Coasts
Cliffs, Erosion Rates
Coastline Changes
Rock Coast Processes
Shore Platforms
Weathering in the Coastal Zone

CLIMATE PATTERNS IN THE COASTAL ZONE

Decadal climate changes are associated with the systematic variations in global patterns of atmospheric pressure, temperature, and sea surface temperature that persist from a year or two to multiple decades. These climate changes have major impacts on coastal processes, modifying wave climate, rainfall, river sediment flux, sand transport paths, coastal erosion, sea level, and inundation. Understanding of the types and causes of climate change is in its infancy. *El Niño/Southern Oscillation* (ENSO) patterns that generate storms comparable to extreme seasonal events may occur every 3–7 years and appear to be associated with other longer period changes. The *Pacific/North American* (PNA) pattern of pressure anomalies and its sea surface temperature equivalent, the *Pacific Decadal Oscillation* (PDO), are associated with multidecadal periods of more intense El Niño type weather, alternating with decades of more intense *La Niña* weather, the complementary phase to El Niño. Similarly, a *North Atlantic Oscillation* (NAO) is associated with decadal changes in atmospheric pressure fields between Iceland and Portugal that alter North Atlantic wave climate and Caribbean hurricane tracks.

Background

The worldwide connection of weather patterns (teleconnections) was first addressed in detail by Sir Gilbert Walker (1928). He identified weather patterns with three global reversals in atmospheric pressure and temperature, the Southern Oscillation, the NAO, and the North Pacific Oscillation. Nile River floods, monsoon reversals in India, rice crop failures in Japan, and droughts in Australia were associated with these global reversals. Although there were little data at that time, Walker emphasized that ocean temperatures play a most important part in world weather. The advent of detailed global databases and rapid computer modeling in recent times has fueled a renewed understanding of the extent and duration of these climate events. Proxy records obtained from Greenland ice cores, tree rings, coral reefs, fish scales, and ocean sediments show that abrupt fluctuations in climate have occurred throughout the Holocene. These shifts may occur within a few years and extend for decades to millennia (e.g., Soutar and Isaacs, 1974; Meko, 1992; Cole *et al.*, 1993, 2000; Heusser and Sirocko, 1997; Gupta *et al.*, 2003). These past events and the presently occurring climate changes are generally thought to be related to the extreme phases of the ENSO phenomenon.

Hypotheses explaining the causes of unusual ENSO events are numerous and range from the extrusion of hot lava on the sea floor to variations in the intensity of the sun's radiations. Most climate modelers favor a complex set of changes in the interaction of the atmosphere and ocean that cause fluctuations in trade winds, monsoon intensity, and sea surface temperature (e.g., Somerville, 1996; Pierce *et al.*, 2000). Although the causes of climate change remain obscure, a clear pattern of atmosphere–ocean events associated with ENSO phenomena has emerged. It is this pattern that gives rise to three-month forecasts of ENSO events. However, ongoing ENSO events are superimposed on a progressive rise in mean global temperature. Consequently, their effects are likely to be less predictable and perhaps more intense than previous events (Inman and Jenkins, 1997; Timmermann *et al.*, 1999; Fedorov and Philander, 2000).

Short-term climate patterns

The seasonal variations in the exposure of the hemispheres to the sun produce changes in the duration of daylight and the angle of the sun's irradiance. These effects modulate solar heating, resulting in variation of the earth's atmospheric pressure field which in turn induces seasonal climatic effects. Seasonal variations are enhanced by the higher convective effects of land and the greater concentration of landmass relative to water in the temperate latitudes of the Northern Hemisphere (Figures C17(a), (b)).

On occasion, the typical seasonal weather cycles are abruptly and severely modified on a global scale. These intense global modifications are signaled by anomalies in the pressure fields between the tropical eastern Pacific and Malaysia known as the ENSO (e.g., Diaz and Markgraf, 1992). The intensity of the oscillation is often measured in terms of the *Southern Oscillation Index* (SOI), defined as the monthly mean sea-level pressure anomaly in millibars normalized by the standard deviation of the monthly means for the period 1951–80 at Tahiti minus than at Darwin, Australia (Figures C17(a), (b), lower). A negative SOI (lower pressure at Tahiti, higher pressure at Darwin) is known as an *El Niño* or warm ENSO event, because of the arrival of unusually warm surface water off the coast of Peru at the time of Christmas, hence the term El Niño. Warm water also occurs along the coast of California, and both regions experience unusually heavy rainfall. A positive SOI is known as *La Niña* and it signals the occurrence of colder than normal surface water in the eastern Pacific, but stronger trade winds in the Pacific and southwest monsoons in the Indian Ocean with heavy rainfall in India and on the Ethiopian plateau (Inman *et al.*, 1996).

During El Niño events the combination of low pressure over the Pacific Ocean and the thermal expansion of warm water cause an increase in water level along the eastern Pacific Ocean. The water level increases by 20–30 cm during strong El Niño events and is depressed an equivalent amount during strong La Niña events. The water level increases lead to significant coastal inundation and sea cliff erosion during the intense storms that are common to El Niño events.

Decadal climate patterns

ENSO events occur with dominant spectral peaks at about 3 and 6 plus years. However, these ENSO events are modified by climate changes that occur on decadal timescales of one quarter to one half century. These changes are often discussed in terms of two atmospheric patterns, the PNA and the NAO, and a sea surface temperature pattern, the PDO. Both PNA and PDO are long period, multidecadal analogs of the seasonal variations of global pressure and temperature and are related to Walker's (1928) North Pacific Oscillation. NAO is a periodic intensification or relaxation of the January phase of the pressure variation (Figure C17(a)). These decadal climate patterns are disguised (aliased) by the interannual changes because they have the same structure and appear as extreme cases of the interannual patterns. This aliasing has delayed the general understanding and acceptance of these concepts.

The PNA pattern is associated with an atmospheric dipole in pressure anomaly over the Pacific Ocean/North America region whose polarity reversals lead to wet and dry climate along the Pacific coast of North America (Wallace and Gutzler, 1981). High-pressure anomaly over the North Pacific Ocean and low-pressure anomaly over the North American continent result in dry (La Niña dominated) climate along the coast of central and southern California, while the opposite polarity in these dipole patterns leads to wet (El Niño dominated) climate (Figures C18(a), (b)). Inman and Jenkins (1999) show that the twenty coastal rivers of central and southern California have streamflow and sediment fluxes during the wet phase of PNA (ca. 1969–98) that exceed those during the preceding dry phase (ca. 1944–68) by factors of 3 and 5, respectively. The sediment flux during the three major El Niño events of the wet phase averaged 27 times greater than the annual flux during the dry phase.

The PDO is a long-lived sea surface temperature pattern with cool and warm phases similar in structure to the La Niña/El Niño phases of ENSO cycles (Goddard and Graham, 1997; Mantua *et al.*, 1997). Although the phase transitions of PNA and PDO may differ slightly in space and time, both events persist for decades and produce the most visible climate signatures in the North Pacific (Francis and Hare, 1994; Minobe, 1997). The El Niño dominated phase of the PDO cycle is characterized by a weakening of the trade winds that results in an eastward movement (slosh) of the warm pool of equatorial water normally contained in the western Pacific by the trades during La Niña conditions (Cole *et al.*, 2000). The sequence of events leading to and associated with the warm (El Niño) phase of the PDO is illustrated in Figure C19. It has been suggested that the breakdown from a prevailing La Niña to an El Niño occurs when brief bursts of westerly winds near the dateline begin the easterly transport of warm water. The positive feedback from the warm water further decreases the trade winds, beginning the transformation to an El Niño (e.g., Fedorov and Philander, 2000).

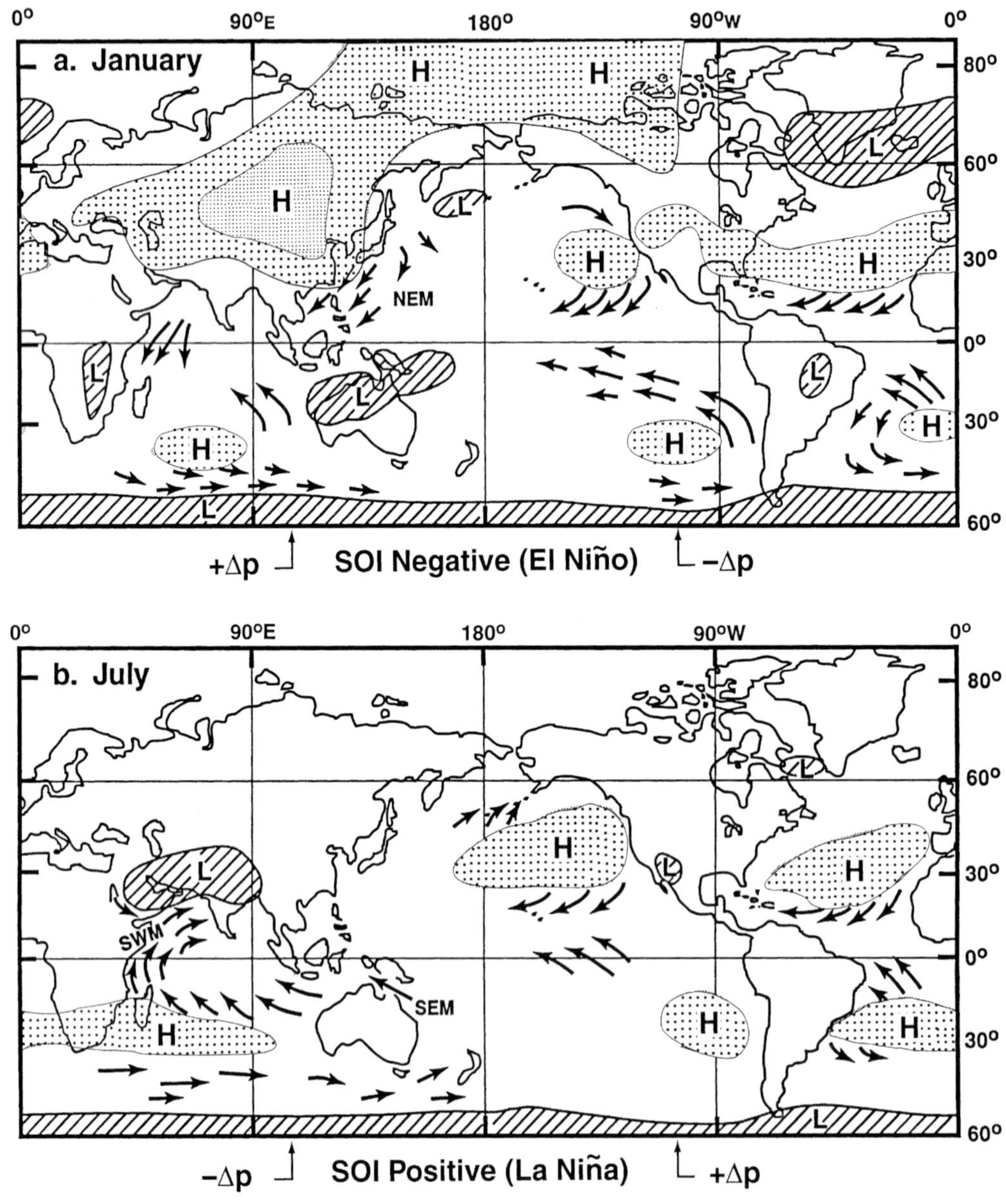

Figure C17 Seasonal pressure and winds for (a) January and (b) July. Contours of 1,020 mb and 1,000 mb are shown around areas of high (H) and low (L) pressure, respectively. Prevailing winds are indicated by arrows: NEM, SEM, SWM designate northeast, southeast, and southwest monsoons. The pressure anomaly Δp is centered around longitudes 105°E and 105°W for negative and positive Southern Oscillation Indices (SOI, modified from Inman *et al.*, 1996).

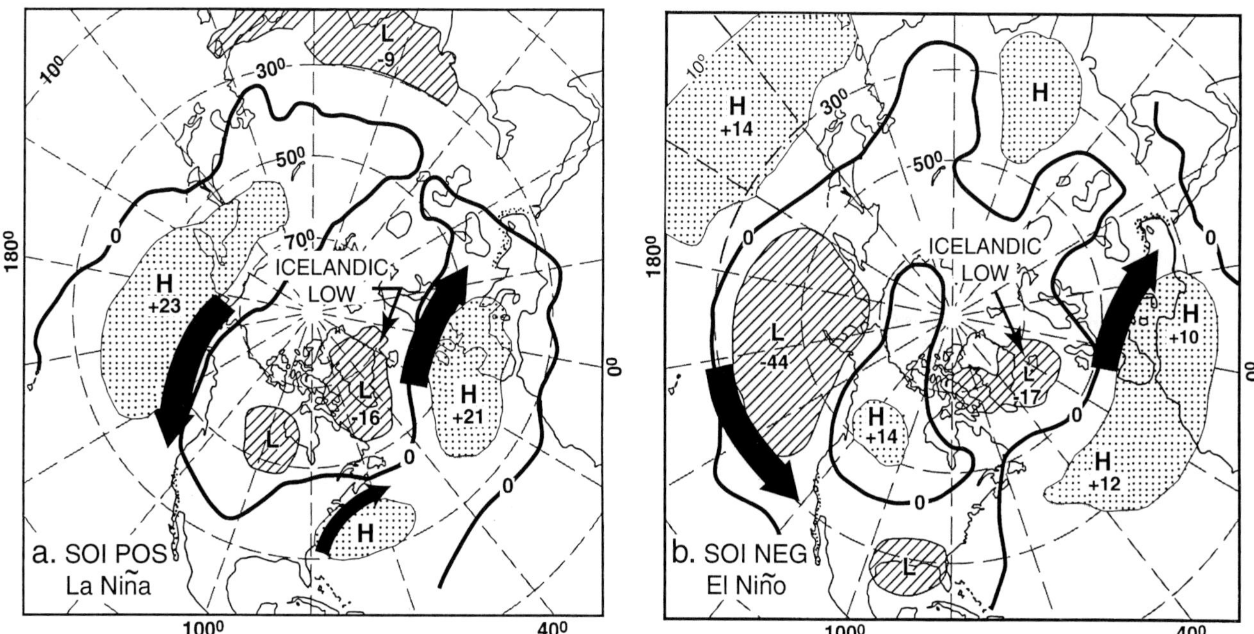

Figure C18 Storm track enhancement (arrows) associated with 700 mb atmospheric pressure anomalies during (a) La Niña and (b) El Niño dominated climate patterns (modified from Inman *et al.*, 1996; pressure data from Redmond and Cayan, 1994).

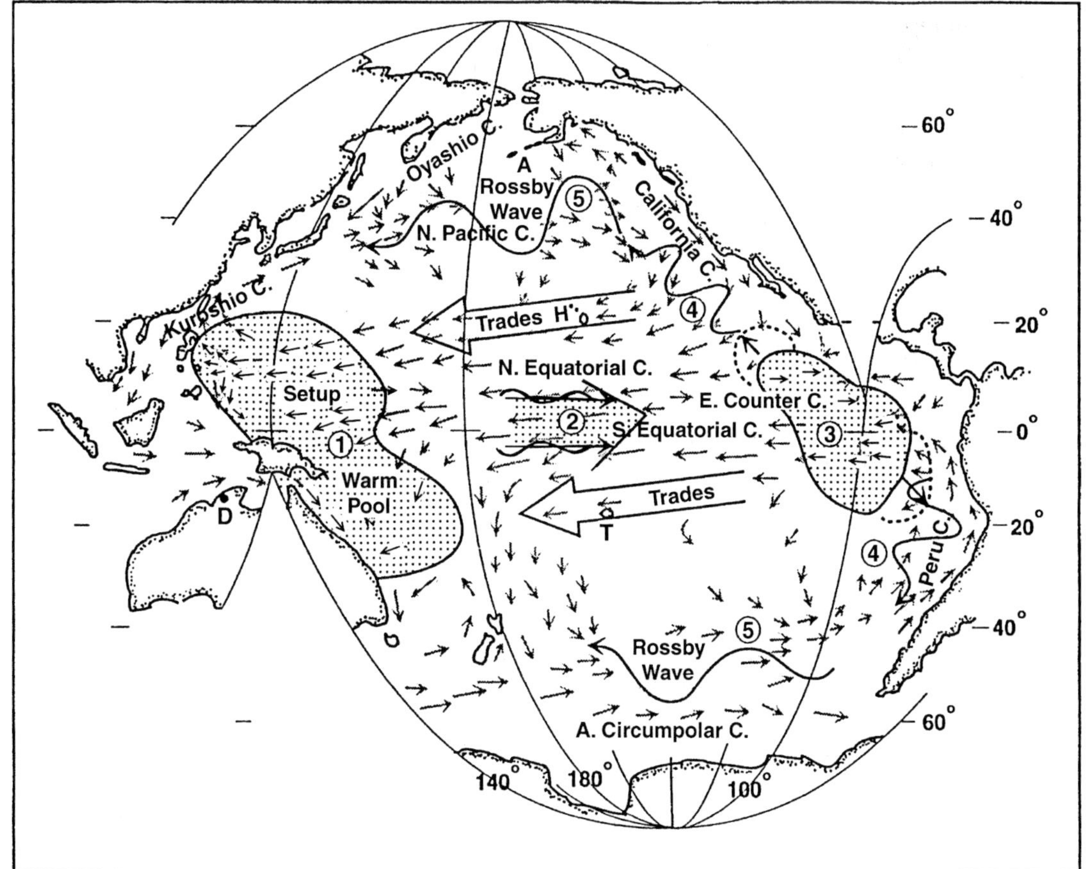

Figure C19 Schematic illustration of sequences leading to a fully developed El Niño event. Numbers refer to (1) warm pool of water setup by trade winds, (2) eastward "slosh" (baroclinic Kelvin wave) of warm water released by relaxation of trade winds, (3) warm pool travels to eastern Pacific and (4) spreads poleward along continental margins (trapped barotropic Kelvin waves), and excites (5) westernly traveling planetary (Rossby) waves. D, T, and H designate Darwin, Tahiti, and Hawaii (modified from Inman *et al.*, 1996).

The stronger trade wind systems during the cool (La Niña) phase of PDO are part of a general spin-up of the atmospheric circulation, which causes the North and South Pacific Gyres to rotate faster. Both effects (wind and current) induce upwelling that maintains cold-water masses along the west coast of the Americas, which sustains the typically cool dry coastal climate of these regions during the La Niña dominated periods of the PDO and PNA. In contrast, the strengthened trade winds of the La Niña period are associated with higher rainfall on the windward sides of Hawaii and other high islands in the trade wind belt. Higher than average rainfall occurred at Hilo and Lihue in the Hawaiian Islands during the La Niña dominated period 1943–68, and lower average rainfall during the subsequent El Niño dominated period 1969–98. However, the El Niño period included episodic Kona storms and the occurrence of hurricane Iniki.

The NAO is an atmospheric dipole that retains its polarity with a low over Iceland and a midlatitude high that enlarges and contracts (Wallace and Gutzler, 1981; Hurrell, 1995). The Icelandic low remains more or less in place while the midlatitude high-pressure anomaly changes in size, strength, and latitude (Figure C18). During La Niña the high-pressure anomaly is stronger and centered over the Bay of Biscay and Spain; during El Niño the high expands but weakens and is centered around 30° North Latitude over the eastern North Atlantic and North Africa. Figure C18 shows that during La Niña events, storm tracks end near the Black Sea causing droughts in Israel; during El Niño the storm tracks end in Israel bringing rain and high waves. The biblical 7-year floods on the Nile River are associated with La Niña intensification of the southwest monsoon over the Indian Ocean (Figure C17). These monsoons bring heavy rainfall over the high Ethiopian plateau that induces flooding on the Nile River.

The dipole systems of the PNA and NAO redirect the tracks of traveling storm fronts, causing changes in coastal wave climate. The strongest pressure gradients occur along the boundaries of the pressure anomalies of the PNA and NAO patterns and these boundaries indicate the prevailing storm paths. Figure C18(a) (La Niña) shows large areas of high pressure over the North Pacific and North Atlantic. These high-pressure areas enhance storm tracks as shown by the arrows for waves approaching North America and Europe, with the prevailing North American storm tracts directed at the Pacific northwest and along the Atlantic east coast. In the eastern Pacific, the principal wave energy during La Niña dominated climate (1943–77) was from Aleutian lows having storm tracks, which usually did not reach southern California. Summers were mild and dry with the largest summer swells coming from very distant Southern Hemisphere storms. The wave climate in southern California changed with the El Niño dominated weather of 1978–98. The prevailing northwesterly winter waves were replaced by high-energy waves approaching from the west or southwest (Figure C18(b)), and the previous Southern Hemisphere swells of summer were replaced by shorter period tropical storm waves during late summer months from the more immediate waters off Central America. The east coast of North America is out of phase with the west coast for severe weather. Heavy rainfall and numerous, powerful Caribbean hurricanes and northeasters occur typically during La Niña years (Figure C18(a)) (see entry on *Energy and Sediment Budgets of the Global Coastal Zone*).

Douglas L. Inman and Scott A. Jenkins

Bibliography

Cole, J.E., Fairbanks, R.G., and Shen, G.T., 1993. Recent variability in Southern Oscillation: isotopic results from a Tarawa Atoll coral. *Science*, **260**: 1790–1793.

Cole, J.E., Dunbar, R.B., McClanahan, T.R., and Muthiga, N.A., 2000. Tropical Pacific forcing of decadal SST variability in the western Indian Ocean over the past two centuries. *Science*, **287**: 617–619.

Diaz, H.F., and Markgraf, V. (eds.), 1992. *El Niño, Historical and Paleoclimatic Aspects of the Southern Oscillation*. Cambridge, UK: Cambridge University Press.

Fedorov, A.V., and Philander, S.G., 2000. Is El Niño changing. *Science*, **288**: 1997–2002.

Francis, R.C., and Hare, S.R., 1994. Decadal-scale regime shifts in the large marine ecosystems of the Northeast Pacific: a case for historical science. *Fisheries Oceanography*, **3**: 279–291.

Goddard, L., and Graham, N.E., 1997. El Niño in the 1990's. *Journal of Geophysical Research*, **102**: 10,423–10,436.

Gupta, A.K., Anderson, D.M., and Overpeck, J.T., 2003. Abrupt changes in the Asian southwest monsoon during the Holocene and their links to the North Atlantic Ocean. *Nature*, **421**: 354–357.

Heusser, L.E., and Sirocko, F., 1997. Millennial pulsing of environmental change in southern California from the past 24 k.y.: a record of Indo-Pacific ENSO events. *Geology*, **25**: 243–246.

Hurrell, J.W., 1995. Decadal trends in the North Atlantic Oscillation: regional temperatures and precipitation. *Science*, **269**: 676–679.

Inman, D.L., Jenkins, S.A., and Elwany, M.H.S., 1996. Wave climate cycles and coastal engineering practice. *Coastal Engineers, 1996, Proceedings of the 25th International Conference (Orlando)*, New York: American. Society Civil Engineers, **25**: 314–327.

Inman, D.L., and Jenkins, S.A., 1997. Changing wave climate and littoral drift along the California coast. In Magoon, O.T. *et al.* (eds.), *California and the World Ocean '97*. Reston, VA: American. Society. Civil Engineers, pp. 538–549.

Inman, D.L., and Jenkins, S.A., 1999. Climate change and the episodicity of sediment flux of small California rivers. *Journal of Geology*, **107**: 251–270.

Mantua, N.J., Hare, S.R., Zhang, Y., Wallace, J.M., and Francis, R.C., 1997. A Pacific interdecadal climate oscillation with impacts on salmon production. *Bulletin of American Meteorological Society*, **78**: 1069–1079.

Meko, D.M., 1992. *Spectral properties of tree-ring data in the United States southwest as related to El Niño, Historical and Paleoclimatic Aspects of the Southern Oscillation*, Cambridge, UK: Cambridge University Press.

Minobe, S., 1997. A 50–70 year climatic oscillation over the North Pacific and North America. *Geophysical Research Letters*, **24**: 683–686.

Pierce, D.W., Barnett, T.P., and Latif, M., 2000. Connections between the Pacific Ocean tropics and midlatitudes on decadal time scales. *Journal of Climate*, **13**: 1173–1194.

Redmond, K.T., and Cayan, D.R., 1994. El Niño/Southern Oscillation and western climate variability. Nashville, TN: 6th AMS Conference on Climate Variations.

Somerville, R.C.J., 1996. *The Forgiving Air, Understanding Environmental Change*. Berkeley, CA: University of California Press.

Soutar, A., and Isaacs, J.D., 1974. Abundance of pelagic fish during the 19th and 20th centuries as recorded in anaerobic sediment off the Californias. *Fishery Bulletin*, **72**: 257–294.

Timmermann, A., Oberhuber, J., Bacher, A., Esch, M., Latif, M., and Roeckner, E., 1999. Increased El Niño frequency in a climate model forced by future greenhouse warming. *Nature*, **398**: 694–697.

Walker, G.T., 1928. World weather. *Quarterly Journal of Royal Meteorological Society*, **54**: 79–88.

Wallace, J.M., and Gutzler, D.L., 1981. Teleconnections in the geopotential height field during Northern Hemisphere winter. *Monthly Weather Review*, **109**: 784–812.

Cross-references

Coastal Climate
Coastal Currents
Coastal Temperature Trends
Coastal Upwelling and Downwelling
Databases (see Appendix 4)
El Niño–Southern Oscillation
Global Vulnerability Analysis
Meteorologic Effects on Coasts

COASTAL BOUNDARIES

Introduction

Worldwide, coastal boundaries range from limits of international jurisdiction in the sea to local property lines along or near the shore.

A full examination of the multitude of shore and sea boundaries in all coastal nations is, of course, beyond the scope of this entry. Instead, the primary focus will be on the definition and demarcation of the legal boundaries between public and private coastal property in the United States.

The entry will also discuss (1) the American federal-state submerged lands boundary, as well as (2) offshore boundaries delimiting the extent of the territorial sea, the contiguous zone and the exclusive economic zone.

In examining coastal boundary principles, it is helpful to agree upon definitions of various legal and technical terms. Although recognizing that different terms are, or have been, used elsewhere, this entry will use certain words and phrases in a specific manner.

For example, in this entry the main categories of coastal property (Shalowitz, 1962; Graber, 1980) will be referred to as uplands, tidelands, and submerged lands, defined as follows:

Uplands are lands above (landward of) the mean high-water line (a technical term to be defined later). They are sometimes referred to as littoral lands or fastlands. (The owners of these lands are littoral owners; they are sometimes referred to, imprecisely, as riparian owners. Technically, "riparian" refers to the banks of a river.)

Tidelands are lands between the mean high-water and mean low-water lines. They are sometimes described as lands over which the tide ebbs and flows. Tidelands are adjacent to uplands on the landward side and to submerged lands on the seaward side. Tidelands are sometimes called the foreshore, shore, or wet-sand area.

Submerged lands are lands below (seaward of) the mean low-water line.

For consistency, this entry will use the word "define" to refer to the legal definition of a coastal boundary, and "demarcate" to refer to the process of locating a boundary on the ground.

In the United States, no single uniform rule governs the determination of the public–private coastal property boundary; under the federal system, each state generally can fashion its own property laws. Under certain circumstances, however, the federal rule of coastal boundaries controls over conflicting state law. The mean high-water line—the modern, technically precise equivalent of the vague English common-law term "ordinary high-water mark"—defines the public–private property boundary under the federal rule and in most states, but some states use the mean low-water line or other lines (Shalowitz, 1962, 1964; Maloney and Ausness, 1974; Graber, 1980).

Before discussing the definition of, and methods of demarcating, the public–private boundary, this entry will summarize (1) pertinent scientific and technical aspects of the tide and tidal datums, and (2) the relevant historic and legal background.

The tide and tidal datums

Most US public–private coastal property boundaries are defined by the course of the tide, and tidal datums generally are used in demarcating the location of the boundary on the ground.

The *tide* is the periodic rise and fall of the surface of the sea resulting from the gravitational attraction of the moon and sun acting upon the rotating earth. A lunar (or tidal) day is 24 h and 50 min, and a lunar month is $29\frac{1}{2}$ days. The *range of the tide* is the difference in the height between the successive high and low tides. The range varies from day to day and from place to place.

Spring tides, or tides of increased range, occur twice during each lunar month as a result of the sun and moon being in line with each other and earth (in conjunction) at new moon and full moon. *Neap tides*, or tides of decreased range, occur twice during each lunar month as a result of the sun and moon being at right angles to earth (in opposition) at the moon's first and third quarters (the moon in quadrature). In most places, however, spring and neap tides occur a day or two after the corresponding phase of the moon because of a lag known as phase age (Shalowitz, 1962; Maloney and Ausness, 1974; Hicks, 1975).

Tides can be classified by the frequency and uniformity of the high and low waters. Along American coasts, there are three principal types of tide: the diurnal, semidiurnal, and mixed. The *diurnal* (daily) type of tide, in which there is only one high water and one low water in a tidal day, is common along most of the Gulf of Mexico coast. The *semidiurnal* (semidaily) type of tide, with two high waters and two low waters each tidal day, predominates along the Atlantic coast (and throughout the world). The *mixed* type of tide, in which there are two high waters and two low waters in a tidal day but with significant differences in the heights of the two highs or the two lows, or both, is found along the Pacific coast (Shalowitz, 1962; Hicks, 1975).

Tidal datums are vertical datums defined by phases of the tide, such as high water. They are used as references for heights on land and depths in the sea, and in demarcating certain US coastal property boundaries, and the limits of the territorial sea, the contiguous zone and the exclusive economic zone (Shalowitz, 1962, 1964; Hicks, 1975).

The National Ocean Survey (NOS), the American federal agency responsible for maintaining tide gauges and preparing nautical charts, determines the elevations of various tidal datums at control tide stations based on 19 years of data. The 19-year period, known as a *tidal epoch*, is deemed to constitute a full cycle because the most significant tidal variations complete their cycles during 19 years. Tidal datums are local datums, accurate only for the location of the tide station from which the tidal information is obtained, and should not be extended into areas having different topographic features without substantiating measurements (Shalowitz, 1962; Maloney and Ausness, 1974).

The standard tidal datums determined by NOS are:

Mean higher high water (MHHW): The arithmetic mean of the higher-high-water heights of a mixed tide over a specific 19-year tidal epoch; only the higher high water of each pair of high waters of a tidal day is included in the mean.

Mean high water (MHW): The arithmetic mean of the high-water heights over a specific 19-year tidal epoch; for a semidiurnal or mixed tide, the two high waters of each tidal day are included in the mean.

Mean sea level (MSL): The arithmetic mean of hourly water elevations over a specific 19-year tidal epoch.

Mean low water (MLW): The arithmetic mean of the low-water heights over a specific 19-year tidal epoch; for a semidiurnal or mixed tide, the two low waters of each tidal day are included in the mean.

Mean lower low water (MLLW): The arithmetic mean of the lower-low-water heights of a mixed tide over a specific 19-year tidal epoch; only the lower low water of each pair of low waters of a tidal day is included in the mean.

The MHW and MLW tidal datums are the principal tidal datums used in U.S. coastal property boundary demarcation, and the MLLW tidal datum is used in determining the offshore boundaries along America's Pacific coast (Shalowitz, 1962, 1964; Hicks, 1975; Hull and Thurlow, 1975).

Historic and legal background

The English common law and, to a lesser extent, the civil law of Continental Europe provide the backdrop for the development of American rules defining coastal property boundaries and governing public and private rights in the seashore.

The civil law of Continental Europe originated in early Roman law, which proclaimed the sea and seashore *res communes*, or "common to all," and not subject to private ownership. Private littoral ownership generally extended only to the line of the highest winter waves (Maloney and Ausness, 1974; Graber, 1980).

In contrast, early English kings appear to have granted favored lords title to and exclusive private rights of fishery in many tidal areas. As the English common law evolved over many centuries, however, it became the general rule that the Crown has *prima facie* title to, and the public certain rights to use, tidelands and navigable rivers. The *jus publicum*, as the public rights are termed, is similar to the early Roman public rights to use the seashore. It is these English common-law rights, as expanded and refined, that became what is referred to in the United States today as the public trust doctrine (Shalowitz, 1962; Maloney and Ausness, 1974; Graber, 1980).

Circa 1568–69, during the reign of Queen Elizabeth I, Thomas Digges, a lawyer, engineer and surveyor, wrote a treatise asserting the Crown's ownership of the lands beneath tidal waters. Although the courts did not embrace his theory immediately, the doctrine of the Crown's *prima facie* title to tidelands became well established under English common law by the end of the 17th century. Sir Matthew Hale, an influential jurist who was to become chief justice, espoused Digges' theory in the treatise *De Juris Maris*, written ca. 1666–67 (Maloney and Ausness, 1974; Graber, 1980).

In the early days of the American colonies, there were no specific restrictions on conveyances of tidelands by colonial proprietors, provided there was no impairment of navigation. For example, a 1641–47 Massachusetts Bay Colony ordinance (general law) authorized the extension of littoral owners' titles seaward to the low-water mark or to 100 rods (1,650 feet) beyond the high-water mark, whichever was more landward. Later, however, application of the evolving English common law became more widespread in the colonies, partly because of the English authorities' desire to unify the colonies for commercial gain (Maloney and Ausness, 1974; Graber, 1980, 1982a).

With the American Revolution, the former colonies, by virtue of their new sovereignty, succeeded to the rights of the English Crown and Parliament in colonial tidelands. Absolute title to all tidelands was vested in the original states, except for those lands that already had been validly granted into private ownership. In 1789, the original states surrendered to the federal government some of their rights in the tidelands by adopting the U.S. Constitution, which provides the bases of the federal government's commerce clause powers and its admiralty jurisdiction. The term "federal navigational servitude" refers to the federal government's paramount authority to control and regulate the navigable waters of the United States under the commerce clause (Graber, 1980, 1981a).

In 1845, the U.S. Supreme Court declared that as new states are subsequently admitted to the Union, they are deemed to have the same sovereignty and property rights as the original 13 states. This concept is known as the equal-footing doctrine. As the United States acquired additional territory, title to all lands beneath tidal and other navigable waters vested in the nation, subject to valid grants by prior governments, in trust for future states. Upon creation of a new sovereign state from such acquired areas, or from the lands formerly within an older state, the new coastal state became vested with title to all lands underlying tidal waters. The after-admitted states' sovereign title to tidelands, except for those lands already validly granted into private ownership, is absolute, although subject to the public trust easement and the federal government's paramount navigational servitude and admiralty jurisdiction (Shalowitz, 1964; Graber, 1980).

In England, during the 18th and early 19th centuries, the common law used the term "ordinary high-water mark" to describe the property boundary between the sovereign's foreshore, or tidelands, and the adjoining private littoral lands. The imprecision of this term may be traced to Lord Hale's 1666–67 treatise. He classified the shoreline on the basis of what he perceived to be three types of tide: "(1st.) The high spring tides, which are the fluxes of the sea at those tides that happen at the two equinoxials. (2d) The spring tides which happen twice every month at full and change of the moon. (3d) Ordinary tides, or nepe [sic] tides, which happen between the full and change of the moon." (Shalowitz, 1962; Graber, 1980).

Apparently, Lord Hale introduced the concept that what he termed "nepe tides" should be considered the "ordinary tides" for property boundary purposes. In his treatise, he concluded that lands subject to inundation by tides of the first two of his three classes can be privately owned, but that the foreshore owned by the Crown extends landward as far as it is covered by the "ordinary ... or nepe [sic] tides." (Shalowitz, 1962).

As pointed out by Aaron L. Shalowitz, the author of the landmark treatise, *Shore and Sea Boundaries*, and a lawyer and engineer for the U.S. Coast and Geodetic Survey, NOS's predecessor agency:

> "... Lord Hale's designation of 'neap tides' shows that it is susceptible of two interpretations: (1) all the tides that occur between the full and change of the moon, and (2) only those tides that occur twice a month at the time of the first and third quarters when the moon is in quadrature" (Shalowitz, 1962, p. 91). Based on today's knowledge of the types of tide, Lord Hale was both unscientific and ambiguous in equating what he termed "nepe tides" with "ordinary tides."

It was not until 1854, long after the American Revolution, that the English high court clarified the definition of the common-law term "ordinary high-water mark" in the *Chambers* decision (*Attorney-General* v. *Chambers*, 1854). The lord chancellor ruled that the boundary was to be determined by using "the line of the medium high tides between the springs and the neaps," excluding both the neap and spring tides.

Public–private coastal property boundary rules

The United States has a wide variety of federal and state legal rules governing (1) the definition of the property boundary between public and private coastal lands, and (2) the legal effect of such physical shoreline (see *Coasts, Coastlines, Shores and Shorelines*) changes as accretion and erosion (terms to be defined later). In most situations, the applicable state law governs.

In many states, however, the federal government originally owned littoral lands that it later granted into private ownership. If there is a conflict between the federal and state rules on the definition of the boundary in those states, or on the legal effect of physical shoreline changes, the federal law controls when title to an upland parcel is derived from a federal grant (Maloney and Ausness, 1974).

Federal rule

The federal rule defines the property boundary as the intersection of the MHW tidal datum with the shore. The rule was first stated by the U.S. Supreme Court in the 1935 *Borax* decision (*Borax, Ltd.* v. *City of Los Angeles*, 1935). At issue was the location of the boundary between upland owned by Borax, as successor to a federal grantee, and the adjoining tideland held by the City of Los Angeles under a California legislative grant.

Borax argued that the boundary should be determined under what it asserted to be the California rule: by using only the high "neap tides." Borax would have gained land thereby because a boundary using a tidal datum based on only the high "neap tides" would be more seaward, or lower, than one derived from a mean of all the high tides. The Supreme Court rejected the "neap tide" argument, holding that the extent of land under a federal grant is a federal question, and thus the boundary between the upland and the adjoining tideland is determined under federal law.

Before clarifying the federal upland–tideland boundary rule, the *Borax* Court reviewed Lord Hale's 1666–67 treatise on the types of shorelines, which had referred to "nepe tides" in an unscientific manner, and the 1854 English *Chambers* decision that the "ordinary high-water mark" was to be determined by excluding both the neap and spring tides and using only the intermediate, or medium, tides. In *Borax*, the Court equated the common-law term "ordinary high-water mark" with the technically precise phrase "line of mean high water." (The boundary is also referred to as the "mean high-water line" and "mean high-tide line.") The Court held that when federal law applies, the public–private coastal property boundary is located by using the MHW tidal datum, as determined by the agency then known as the Coast and Geodetic Survey (now NOS) based on 18.6 years (now rounded to 19 years) of data.

State rules

State property boundary rules are the result of various state constitutional and statutory provisions as well as case law. The majority of the states follow the English common-law concept that the boundary is the "ordinary high-water mark," or its updated, technically precise counterpart, the mean high-water line.

In general, subject to many qualifications, 16 coastal states use the mean *high*-water line as the boundary. They are: Alabama, Alaska, California (subject to the "neap tide" language discussed below), Connecticut, Florida, Georgia, Maryland, Mississippi, New Jersey, New York, North Carolina, Oregon, Rhode Island, South Carolina, Texas (as to lands under common-law grants but not as to lands derived from Spanish or Mexican grants) and Washington. (Maloney and Ausness, 1974; Graber, 1980, 1981b, c, d, 1982b, c, 1983a, b, c, 1984a, b, d, 1985b, 1986a, b, 1988, 1989.)

California's so-called "neap tide" rule originated in the state Supreme Court's 1861 *Teschemacher* decision (*Teschemacher* v. *Thompson*, 1861). The court echoed Lord Hale's 17th-century treatise equating "nepe" or "neap tides" with "ordinary tides" for boundary purposes. Despite an 1872 statute providing that the "ordinary high-water mark" is the seaward limit of upland, California courts continued to refer to "neap tides," although generally in a nontechnical manner, in defining the boundary. Finally, in 1966, in the *Kent Estate* decision, the appellate court held that the boundary is to be determined by using the 19-year mean of the high "neap tides." (*People* v. *Wm. Kent Estate Co.*, 1966). The court apparently attempted to define "neap tides" in a technical manner, but overlooked the fact that there usually is a lag of a day or two between quadrature of the moon and the minimum or neap range of the tide. Although the decision has not been overruled, other modern California decisions, as well as recent statutes, refer to the "line of mean high tide" in defining tidelands, so there is some uncertainty as to the state's boundary rule (Graber, 1981b).

Six Atlantic coast states have departed from the English common-law rule and utilize the mean *low*-water line, or variations of it, as the public–private coastal property boundary. They are: Delaware, Maine, Massachusetts, New Hampshire, Pennsylvania, and Virginia (Maloney and Ausness, 1974; Graber, 1980, 1982a, 1984c, 1985a, c, 1987a, b).

Louisiana, with its French and Spanish heritage, accepts the line of the highest winter tide, a version of the civil-law rule. In Texas, if the original source of upland title was a prestatehood Spanish or Mexican grant, the line of MHHW, another variation in the civil-law rule, is the boundary. Hawaii adheres to its aboriginal, customary concept that the public–private boundary follows the upper reaches of the wash of the waves (Maloney and Ausness, 1974; Graber, 1980, 1981d, 1982d, 1983d).

Despite these generally applicable state rules, the boundary in a specific situation may differ. For example, even in states with a high-water boundary, certain tidelands, or both tide and submerged lands, may have been sold into private ownership by the state, and still are privately held.

Fixed or ambulatory boundaries

The location of the public–private coastal property boundary sometimes is permanently fixed, due to, for example, litigation or a boundary line agreement. In addition, on a rocky coast, for all practical purposes, the boundary may amount to a fixed line (though it is possible that the tidal datum could change or that the coast might erode in the future).

However, in most situations, particularly on sandy shores, the boundary is an ambulatory, or moving, line as the result of natural and/or artificial causes. The main types of physical changes (Maloney and Ausness, 1974; Graber, 1980, 1982c) are as follows:

Accretion: Addition to land along the shoreline resulting from the gradual, imperceptible deposit by the water of sand, sediment or other material; the word *alluvion* refers to the deposit itself, but sometimes is used synonymously with accretion. The test of what is "gradual and imperceptible" is: "Though the witnesses may see, from time to time, that progress has been made, they could not perceive it while the progress was going on." (*County of St. Clair* v. *Lovingston*, 1874).

Reliction: Land formerly covered by the water that has become dry by the gradual, imperceptible recession of water.

Erosion: Gradual, imperceptible action wearing away land along the shore by currents, waves, wind, and other elements.

Submergence: Disappearance of soil under the water and formation of a body of water under it.

Avulsion: Sudden change in the shoreline.

Generally, subject to qualification, the littoral owner benefits as the boundary moves seaward with accretion and reliction, but loses as the boundary shifts landward with erosion and submergence. Avulsion, however, does not result in a change in the legal boundary (Maloney and Ausness, 1974; Graber, 1980, 1982c).

In allowing littoral owners to benefit from accretion, the federal rule and most coastal states' rules do not distinguish natural accretion from accretion due to artificial, or manmade, causes (e.g., a breakwater or pier). Generally, however, a private littoral owner is not entitled to benefit from accretion caused by his or her own intentional acts, such as filling adjoining tide-covered lands (Maloney and Ausness, 1974; Graber, 1980).

In its 1967 *Hughes* decision, the U.S. Supreme Court ruled that the federal rule controlled over Washington state law with regard to accretions to private upland property derived from a federal grant. Washington claimed the state as tideland owner was entitled to the accreted land under an 1889 state constitutional provision denying littoral owners any right to accretions between their property and the ocean. The Court, citing its 1935 *Borax* principle that federal law governs the question of ownership of property within a federal grant, expanded the rule so it also determines the issue of ownership of accretion to that land (*Hughes* v. *Washington*, 1967).

In its 1982 *California* decision, the U.S. Supreme Court held that federal rather than state law controlled in a dispute over accreted land along the Pacific Ocean adjoining a Coast Guard facility. It was undisputed that the upland parcel had been continuously owned by the federal government since California's admission to the Union in 1850, and that the accretion had been artificially caused by construction and maintenance of nearby U.S. jetties. Under the state rule, California as tideland owner would have been entitled to such artificially accreted land (*California* v. *United States*, 1982).

When state rules of accretion apply, the pertinent state's laws must be examined. As previously mentioned, in most states, the littoral owner benefits from accretion, whether it results from natural or artificial causes. California, however, denies the upland owner any artificially caused accreted land; the boundary is to be determined when the shoreline was in its "last natural position," before it was changed by the artificial condition (Graber, 1981b, 1982c).

Demarcation of coastal property boundaries

Over time, the methods of demarcating (locating on the ground) and delineating (showing on a map) public–private coastal property boundaries have evolved from relatively primitive and inexact techniques to extremely sophisticated and precise procedures.

Early methods

The ocean's edge is, of course, a natural boundary, and is depicted on the oldest maps of coastal areas. But until seashore property values escalated in the 20th century, the location of the public–private property boundary was only an approximation. This has been called the first generation of coastal boundary determination by Paul T. O'Hargan, a coastal surveying authority who wrote that it "took the form of an educated guess." This resulted from a lack of data and the difficulties of surveying these boundaries. In many sections of the United States, there were insufficient tidal datums or the means to establish them (O'Hargan, 1975, p. 6).

Current procedures

After the appropriate legal rule defining the coastal property boundary has been identified, a land surveyor or engineer must apply that rule to the site in question to locate the boundary on the ground.

Boundaries defined in terms of the tide involve two engineering components: (1) a vertical one, based on the height of the tide and constituting a tidal datum, such as MHW, and (2) a horizontal one, related to the line where the tidal datum intersects the shore to form the tidal boundary, such as the MHW line (Shalowitz, 1962, 1964).

With the availability of tidal datum elevations published by the federal government, more precise boundary demarcation became possible in the United States (O'Hargan, 1975). The tidal datum elevations published by NOS, however, are accurate only for the locality of the relatively few control tide stations where tidal observations have been made during a full 19-year tidal epoch. There can be widespread differences in the elevations of datums from point to point along the coastline or even in the same bay or inlet (Hicks, 1975).

Unless the site of the boundary to be demarcated is in the immediate vicinity of a control tide station, it is usually necessary to establish one or more subordinate short-term tide stations where tidal observations can be taken. To derive the mean 19-year values needed for determining boundaries, the records from the local, or secondary, stations are simultaneously compared with the control station's records (Hull and Thurlow, 1975; O'Hargan, 1975; Cole, 1983).

The second, or land, component of the property boundary may consist of rock, sand, marsh, mangroves, or another type of vegetation. The nature of the coast where the boundary is being demarcated and delineated may, of course, affect the choice of the procedure used by the surveyor or engineer.

In what O'Hargan termed the second generation of coastal boundary determination, the technique of using contours became common (O'Hargan, 1975). Under this procedure, the surveyor or engineer uses leveling and conventional horizontal surveying techniques on the assumption that the datum line (derived from simultaneous comparisons of the control and the local, or secondary, tide stations) is a level topographic contour in the vicinity of the datum determination (Hull and Thurlow, 1975).

In 1974, however, this contour procedure was criticized at a convention of the American Society of Photogrammetry because the underlying assumption is sometimes incorrect: the variations in tidal datum elevation even at points quite near one another can be extensive (Guth, 1974). In what O'Hargan called the third generation of coastal boundary determination, several new approaches were developed as an outgrowth of that criticism, such as the extrapolated water elevation (EWE). "In this method, the tides are used to make a 'mark' on tide staffs set along the mean high water line. In other words, the datum is extrapolated from a single tide station to various places along the line to be located" (O'Hargan, 1975, p. 24). Florida almost immediately adopted the new procedure as the most accurate method of surveying these boundaries (Guth, 1976).

Various other procedures for determining the property boundary are in use, including biological interpretation by remote sensing imagery and photogrammetric analysis of aerial photographs (Hull and Thurlow, 1975).

Locating historic boundaries

In some situations, the public–private coastal property boundary must be determined as of a date in the past. In California, for example, it frequently is necessary to locate the "last natural position" of the shoreline because of the state rule that the upland owner is not entitled to artificial, or manmade, accretions (Graber, 1981b).

When called upon to demarcate and delineate the "last natural position" or some other historic location of the line in question, a surveyor or engineer must be thorough and innovative. He or she must search for and analyze all available pertinent historic materials, such as old U.S. Coast and Geodetic Survey topographic surveys and hydrographic and navigational charts, and the accompanying field notes and descriptive reports. Other professionals, such as geologists or oceanographers, may be needed to assist in the investigation of the "last natural" or other historic location of the line in question.

Other American coastal boundaries

Although this entry has focused on American public–private coastal property boundaries, there are many other categories of U.S. regulatory, political, and jurisdictional boundaries near or offshore from the coast. This section briefly discusses some of them.

Federal-state boundary

Before World War II, it generally was assumed that states owned the submerged lands offshore from their tidelands. In the postwar years, however, the United States challenged coastal state assertions of title. In 1947, in *United States* v. *California*, the U.S. Supreme Court held that California did not own the marginal belt extending three geographic (nautical) miles from its coast and that the federal government, not the state, has paramount rights in, and power over, that belt, an incident of which embraces control over the resources in the underlying soil, including oil (*United States* v. *California*, 1947). Similar decisions were handed down in 1950 concerning Louisiana and Texas (Shalowitz, 1962; Graber, 1981a).

In 1953, Congress responded to these decisions by enacting the Submerged Lands Act. This law (1) quitclaimed to the coastal states U.S. title claims to certain submerged lands, and (2) defined the submerged lands confirmed to the states in terms of state boundaries as they existed when the state became a member of the Union or as previously approved by Congress, but not extending seaward from the coastline of any state more than three geographic (nautical) miles (one marine league) in the Atlantic and Pacific Oceans or more than nine geographic (nautical) miles (three marine leagues) in the Gulf of Mexico (Submerged Lands Act, 1953; Graber, 1981a). (For simplicity, the text will use only the word "miles," omitting the qualifying words "geographic (nautical)." A geographic mile equals 1.151 statute miles and 1 statute mile equals 1.609 km.)

The Submerged Lands Act has had the general effect of resolving the federal-state boundary at either 3 or 9 miles offshore. The act specifies the "coast line," for purposes of locating the seaward boundaries of the states, as "the line of ordinary low water … ." (Submerged Lands Act, 1953, §3). There has been considerable subsequent litigation between the United States and various coastal states involving specific areas. The U.S. Supreme Court accepted assertions by Florida and Texas that their maritime boundaries extend three leagues (9 miles) into the Gulf of Mexico, but rejected similar claims by Louisiana, Mississippi, and Alabama (Shalowitz, 1962; Graber, 1980, 1981c, d, 1982d, 1986a, 1988).

Federal outer continental shelf

Legally, the outer continental shelf consists of submerged lands subject to federal jurisdiction immediately seaward of the 3- or 9-mile belt of submerged lands owned by the states. In 1953, Congress passed the Outer Continental Shelf Lands Act authorizing the federal government to lease oil, gas, and other mineral resources within the area of submerged lands between the seaward limit of the states' submerged lands and the edge of the continental shelf (Outer Continental Shelf Lands Act, 1973).

Limits under international law

The United States, like other coastal nations, exercises maritime jurisdiction over various zones in the seas adjoining its coasts. Under international law, these national zones include not only the continental shelf but also the territorial sea, the contiguous zone, and the exclusive economic zone. The breadth of the territorial sea and other zones is measured from a baseline, which usually, but not always, is "the low-water line along the coast as marked on large scale charts officially recognized by the coastal [nation]." (Convention on the Territorial Sea and the Contiguous Zone, 1958, art. 3). Along America's Pacific coast, which has a mixed tide, the NOS's charts delineate the MLLW line, instead of the MLW line (Shalowitz, 1962, 1964).

Under international law, each coastal nation is entitled to assert a territorial sea immediately seaward of its inland waters as wide as 12 miles from the baseline. In the territorial sea, the nation has sovereign rights but must allow free passage to foreign vessels. Traditionally, the United States had claimed a territorial sea only 3 miles wide, but on December 27, 1988, President Reagan asserted an American territorial sea with a 12-mile breadth (Proclamation No. 5928, 1988).

The contiguous zone is a zone of the high seas seaward of, and adjacent to, the territorial sea. In this zone, the coastal nation is authorized to prevent infringement of its customs, fiscal, immigration, and sanitary regulations (Convention on the Territorial Sea and the Contiguous Zone, 1958). Under international law, the contiguous zone may not extend more than 24 miles beyond the baseline used for measuring the width of the territorial sea (Law of the Sea Convention, 1982).

Until the end of World War II, waters seaward of the territorial sea and the contiguous zone generally were deemed to be the high seas and not subject to any coastal nation's jurisdiction. However, some nations unilaterally asserted their exclusive rights to offshore fisheries and other resources. The Third United Nations Conference on the Law of the Sea authorized coastal nations to establish 200 mile-wide exclusive economic zones in which they could explore, exploit and manage living and nonliving resources. The national jurisdiction within the zone extends 200 miles seaward from the same baseline used to measure the width of the territorial sea (Law of the Sea Convention, 1982).

The United States asserted a 200 mile-wide fishery zone in 1976 (Fishery Conservation and Management Act, 1976). President Reagan declared the establishment of the nation's 200-mile-wide exclusive economic zone in 1983 (Proclamation No. 5030, 1983).

Conclusion

Over the last 65 years, the American legal system, with some notable exceptions, has made remarkable progress in clarifying the old English common-law definition of the boundary between public and private coastal property.

This welcome improvement may be traced to 1935, when the U.S. Supreme Court handed down a landmark decision equating the vague common-law phrase "ordinary high-water mark" with the technically precise term "line of mean high water." By doing so, the Court provided a scientifically and technically sound basis for the federal rule defining the legal boundary. In addition, the decision inspired some state legislatures and courts to follow suit by clarifying their state boundary rules. It also enabled surveyors and engineers, using procedures that have become more sophisticated over time, to accurately locate the boundary on the ground under the modern rules.

Some issues concerning the definition and demarcation of coastal boundaries are still unresolved, but the progress made since 1935 suggests that the legal system will be able to work with the scientific and engineering communities to resolve those problems.

Although the process of clarifying the public–private coastal property boundary has been a major development, there have also been significant trends concerning offshore boundaries, especially since the end of World War II. The bitter dispute between the federal government and coastal states over control of submerged lands was largely resolved by Congress in 1953 and in several later court decisions. The political solution: dividing ownership of the submerged lands between the United States and the states. And in international law, developments since 1976 have led to an expanded American territorial sea and new 200-mile-wide U.S. fishery and exclusive economic zones.

While the law relating to coastal boundaries, like property law generally, is rooted in the past, it can no longer be assumed that the pertinent legal rules are immutable. Like the shoreline of a sandy beach, change seems inevitable.

Peter H.F. Graber

Bibliography

Attorney-General v. *Chambers*, 4 De G.M. &G. 206, 43 Eng. Rep. 486 (1854).

Borax, Ltd. v. *Los Angeles*, 296 U.S. 10 (1935).

California v. *United States*, 457 U.S. 273 (1982).

Cole, G.M., 1983. *Water Boundaries*. Rancho Cordova, CA: Landmark Enterprises.

Convention on the Territorial Sea and the Contiguous Zone, 15 U.S.T. 1606, T.I.A.S. 5639 (1958).

County of St. Clair v. *Lovingston*, 90 U.S. (23 Wall.) 46, 66–69 (1874).

Fishery Conservation and Management Act of 1976, 90 Stat. 331 (1976).

Graber, P.H.F., 1980. The Law of the Coast in a Clamshell (overview and first in series of 26 articles). *Shore and Beach*, **48**(4): 14–20.

Graber, P.H.F., 1981a. The Law of the Coast in a Clamshell (federal), *Shore and Beach*, **49**(1): 16–20.

Graber, P.H.F., 1981b through 1989. The Law of the Coast in a Clamshell, *Shore and Beach*. Remaining articles in series discuss state laws, including coastal boundaries, on state-by-state basis. They are as follows, by states, with years of publication, volume numbers, issue numbers and page numbers: Alabama, 1988, **56**(2): 12–17; Alaska, 1985b, **53**(2): 3–8; California, 1981b, **49**(2): 20–25; Connecticut, 1984d, **52**(4): 15–18; Delaware, 1985c, **53**(3): 9–14; Florida, 1981c, **49**(3): 13–20; Georgia, 1986b, **54**(3): 3–7; Hawaii, 1983d, **51**(4): 9–18; Louisiana, 1982d, **50**(4): 16–23; Maine, 1984c, **52**(3): 17–20; Maryland, 1984a, **52**(1): 3–10; Massachusetts, 1982a, **50**(1): 13–18; Mississippi, 1986a, **54**(1): 3–7; New Hampshire, 1987a, **55**(1): 12–17; New Jersey, 1982b, **50**(2): 9–14; New York, 1983c,

51(3): 10–16; North Carolina, 1983a, **51**(1): 18–23; Oregon, 1982c, **50**(3): 16–23; Pennsylvania, 1987b, **55**(2): 9–11; Rhode Island, 1989, **57**(2): 20–23; South Carolina, 1984b, **52**(2): 18–25; Texas, 1981d, **49**(4): 24–31; Virginia, 1985a, **53**(1): 8–14; Washington, 1983b, **51**(2): 16–21.
Guth, J.E., 1974. Will the real mean high water line please stand up. In *Proceedings, American Congress of Surveying and Mapping, and American Society of Photogrammetry*.
Guth, J.E., 1976. Survey procedures for determining mean high water. In *Proceedings, American Congress on Surveying and Mapping*.
Hicks, S.D., 1975. *Tide and Current Glossary*. Washington: National Ocean Survey.
Hughes v. *Washington*, 389 U.S. 290 (1967).
Hull, W.V., and Thurlow, C.I., 1975. *Tidal Datums and Mapping Tidal Boundaries*. Washington: National Ocean Survey.
Law of the Sea Convention, 21 I.L.M. 1261 (1982); U.N. Pub. E.83.V.5 (1983).
Maloney, F.E., and Ausness, R.C., 1974. The use and legal significance of the mean high water line in coastal boundary mapping. *North Carolina L. Review*, **53**: 185–273.
O'Hargan, P.T., 1975. The Three Generations of Sovereign Boundary Line Location. Coastal Zone Surveying and Mapping Symposium, American Congress on Surveying and Mapping.
Outer Continental Shelf Lands Act, 67 Stat. 462 (1953).
People v. *Wm. Kent Estate Co.*, 242 Cal.App.2d 156, 51 Cal.Rptr. 215 (1966).
Proclamation No. 5030, 1983.
Proclamation No. 5928, 1988.
Shalowitz, A., 1962. *Shore and Sea Boundaries*, Vol. 1. Washington: U.S. Government Printing Office.
Shalowitz, A., 1964. *Shore and Sea Boundaries*, Vol. 2. Washington: U.S. Government Printing Office.
Submerged Lands Act, 67 Stat. 29 (1953).
Teschemacher v. *Thompson*, 18 Cal. 11 (1861).
United States v. *California*, 332 U.S. 19 (1947).

Cross-references

Beach Erosion
Beach Processes
Coastal Changes, Gradual
Coastal Changes, Rapid
Coastline Changes
Coasts, Coastlines, Shores and Shorelines
Continental Shelves
Erosion Processes
Mapping Shores and Coastal Terrain
Tidal Datums
Tide Gauges
Tides

COASTAL CHANGES, GRADUAL

The concept of gradual coastal change can be seen as embracing different contexts, both spatial and temporal. It is conceivable that gradual change may include those depositional and erosional events that have taken millions of years to produce present-day coastal landforms at one end of the spectrum, while at the other end slow and imperceptible accretion to beaches after storms may represent another form of gradual change. Therefore, a wide-ranging time frame may be used covering geological, engineering, and arguably event scales as depicted in Figure C20.

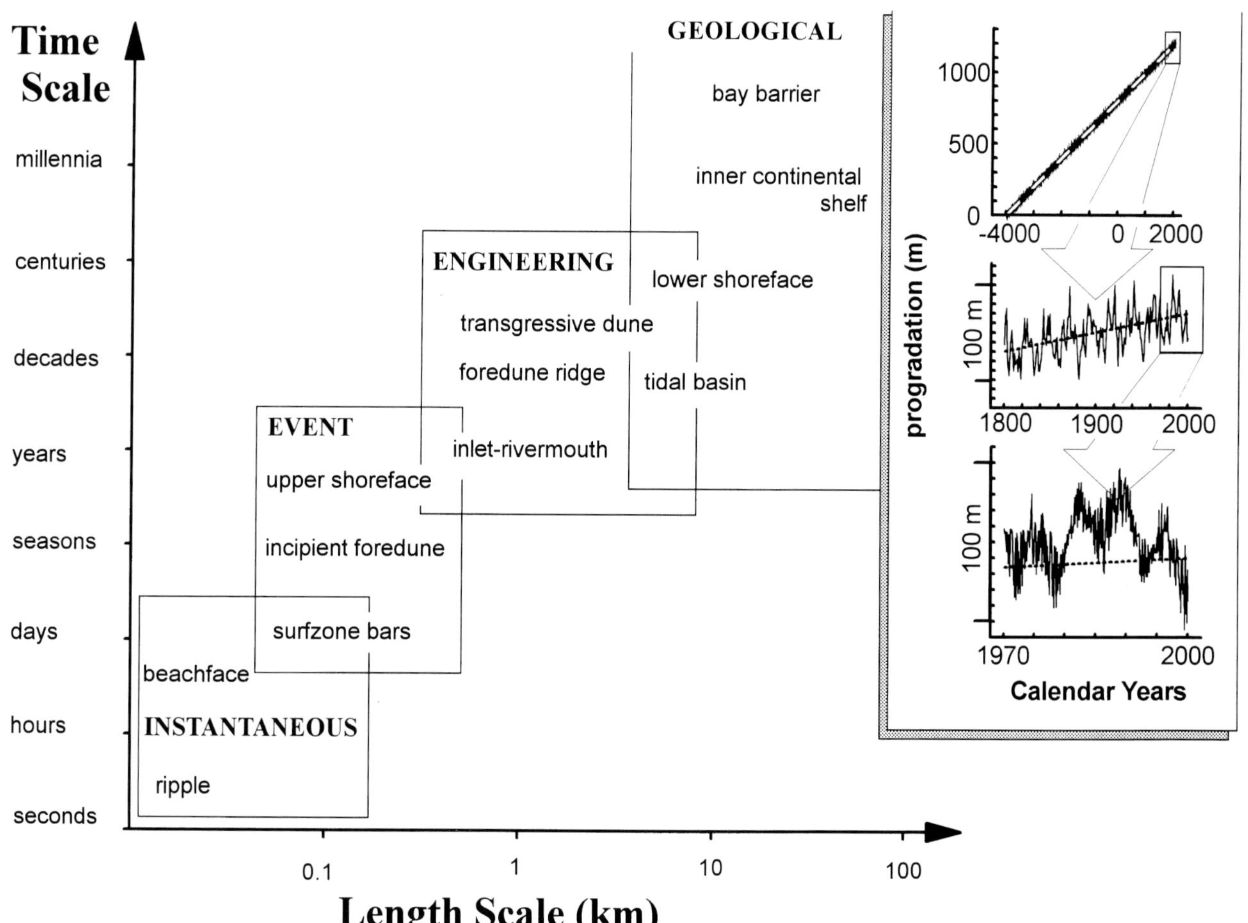

Figure C20 Spatial and temporal scales of coastal geomorphic change depicting examples of features formed at different scales and how oscillations in magnitude and frequency of processes may be superimposed on longer-term trends (e.g., progradation of a strandplain with beach cut and fill cycles superimposed as at Moruya, New South Wales, Australia, see Thom and Bowman, 1981).

This figure demonstrates how high frequency, low magnitude processes at shorter time scales can be superimposed on longer-term oscillations and trends at scales of millennia and longer term as reflected in depositional and erosional land forms.

Cenozoic geologic scale

Coastal geomorphologists have a long-standing interest in the evolution of coastal landforms from Tertiary to Quaternary times. Many coastal regions show an historical imprint or inheritance of depositional or erosional processes reflecting changes in shoreline position as sediments accumulate or cliffs erode. In glaciated areas, the impact of gradual coastal accretion or erosion from Tertiary times is rarely apparent. Similarly, where tectonic and volcanic activities prevail there is little evidence of shoreline features of pre-Quaternary age. However, tectonic uplift as seen, for instance, along the Huon Peninsula in Papua New Guinea, can provide a mechanism for preserving the remnants of gradual coral reef accumulation over hundreds of thousands of years which correlate very well with glacial to interglacial climatic and eustatic cycles (Chappell and Shackleton, 1986).

Classic thinking in coastal geomorphology, epitomized by the work of Douglas Johnson (1919), dramatically demonstrated the nature of gradual change following the application to coasts of the Davisian cyclic model in geomorphology. Assumptions built into this model may be queried, but the basic notion of change driven by the interaction of marine forces (especially waves) and the land still underpins any conceptual framework for understanding how coasts evolve over time. The importance of sea-level change during the Quaternary is now seen as a major factor in coastal geomorphology, a phenomenon of relatively limited significance in the Davis/Johnson model. Shepard, Russell, Fairbridge, Armstrong Price, Steers, and Guilcher (and others) are of a generation that assessed coastal landforms in different regions of the world from a Quaternary perspective. They showed how erosional, and especially depositional landforms, over time, have changed their character under different sea-level conditions. Many coastal regions experienced post-depositional modification and transformation after primary wave, tidal and wind processes have performed their work (Thom, 1975).

The Australian coast contains many examples of gradual change in coastal environments during the Cenozoic. These features have formed since dislocating tectonic forces that ripped the continent apart from Gondwana remnants essentially stopped operating about 60 million years ago. On the east coast, the narrow continental shelf has been subjected to very slow marginal subsidence. This has facilitated two geomorphic outcomes: shelf-edge accretion of a sediment wedge ca. 500 m in thickness, and an inner-shelf abrasion surface up to 10 km wide and ca. 1,000 km long (the East Australian Marine Abrasion Surface—EAMAS). The surface is now veneered with late Quaternary sediments and is entrenched by infilled paleo-valleys carved by rivers draining to lower sea levels. The surface culminates on the landward side with discontinuous marine cliffs cut into Mesozoic or Paleozoic rocks often over 100 m in height and representing truncated spurs and ridges of drowned valleys and plateaux (as seen around Sydney). We are uncertain of the precise age of initiation of the surface and cliffs, but it is easiest to conceive of these features as time-transgressive evolving progressively over millions of years during periods of relatively higher sea level.

Another large-scale Australian example that reflects gradual accumulation of sediments are the barrier-lagoonal deposits formed over millions of years within the lower Murray-Darling basin. In this region, relative sea level since the early Tertiary has fallen. Sands, including heavy minerals such as ilmenite, form horizontally stacked barriers which in Quaternary times have been partially reworked into terrestrial dune fields. Similar "stepped" or extensive strand plains initiated in the Tertiary and extending through the Pleistocene into the Holocene have formed in southwest Western Australia near Perth, in parts of southeast Brazil, and along the Atlantic coastal plain of the United States.

Holocene geologic scale

At the other end of the geologic scale are coastal phenomena which are the result of processes operating at or close to present sea level over the last 6,000 years or so. Within this time frame, erosion and/or deposition can initiate and reconstitute sand barriers, tidal flats, river mouths, tidal deltas, coastal dunes, and rock platforms at the macro scale, and a myriad of landforms at a micro scale. Use of techniques of drilling and dating have enabled age structures and sedimentary histories to be reconstructed from Holocene deposits in coastal areas of many parts of the world.

In areas of postglacial rebound beach deposits dated 6,000 years BP may occur 100–150 m above present sea level (e.g., Gulf of St Lawrence), whereas for areas marginal to the great deltas, like the Mississippi, deposits of similar age may have gradually subsided below sea level and be buried by younger sediments (see Carter and Woodroffe, 1994). In contrast, those coasts experiencing a much more passive neotectonic regime, such as eastern Australia, witness beach/foredune ridge accumulation at about present sea level not only in the Holocene, but also in the last and penultimate interglacials (Roy *et al.*, 1994; see Hopley, 1994, for a review of gradual coral reef accumulations in similar tectonic settings).

One conclusion which can be drawn from an analysis of Holocene sand barriers is how gradual processes of accretion are periodically interrupted by erosion or reduced rates of accretion. The result is a strand plain with distinct linear ridges (beach or foredune) parallel or sub-parallel to the present shoreline. Changes in sediment budgets are critical in the gradual evolution of strand plains. As pointed out by Cowell *et al.* (1999) sand may be supplied from the shoreface as well as littoral sources in order for accretionary processes to work. Closed sediment compartments highlight the effectiveness of gradual onshore transport of sediment from the lower shoreface in certain areas (see Figure C21; Roy *et al.*, 1994). These areas are in contrast to those "leaky" compartments dominated by littoral sand transport associated with river mouth (inlet) or headland bypassing.

Engineering scale

Even closer to the present are those features which result from erosion or deposition over periods of two years or more. Increasingly, coastal

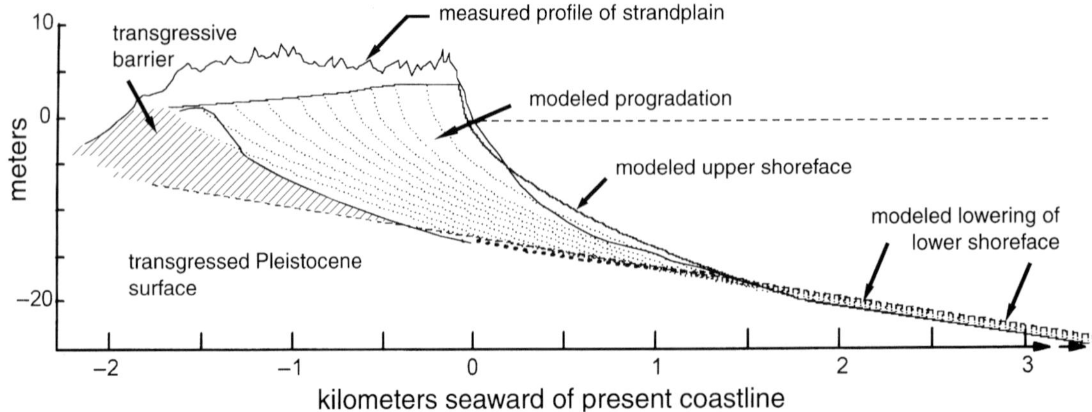

Figure C21 Simulated strandplain development during the Holocene at Tuncurry, New South Wales, Australia (after Cowell *et al.*, 1999). The simulation involved a constant sea level for 6,000 years and an across-shore sand supply from lower to upper shoreface averaging 1.1 m^3 per meter of shoreline together with a steady increase in depth of the lower shoreface at a rate of 0.002 m/y decreasing linearly with distance to 10 km seawards of the present coastline forming 80% of sand now in the strandplain.

geomorphologists are recognizing the impact of oscillations in climatic conditions leading to changes in beach volume, beach alignment, inlet openings/closures, river-mouth bar migration, dune vegetation recovery, etc. However, some of these changes may reflect "self-organization" of morphodynamic conditions involving positive and negative feedback effects independent of climatic forces (see Cowell and Thom, 1994, citing field studies in the United States, Netherlands, and Australia).

Recovery of the subaerial beachface in width and thickness following a period of sudden storm wave erosion is well documented in coastal literature (Komar, 1998). Storm impacts in winter followed by gradual summer accretion are characteristic of many coasts in North America (e.g., California) and Europe. The severity of impact may be enhanced during El Niño years as in Oregon, and that gradual beach recovery may take longer after such severe winter events (see Komar, 1998 for description of El Niño storm impacts). Contrasting situations occur in the southwestern Pacific where beaches may accrete through many years (winter and summer) influenced by El Niño only to erode drastically in La Niña or storm-dominated periods (Thom and Hall, 1991). The role of the Pacific Decadel Oscillation (PDO) as a factor in driving beach cycles of erosion and accretion is now seen as a significant phenomenon which may explain the timing of events that do not easily fit ENSO patterns. El Niño and La Niña can be seen as phenomena lying on top of the large-scale temperature and sea-level height distributions determined by the PDO.

Understanding processes of gradual beach accretion is of significance in beach management. Monitoring beach change by photogrammetric and/or field survey methods yields valuable information on how quickly all or part of a beach may recover following a stormy period. Beaches and foredunes in New South Wales took 5–10 years to reach pre-storm positions following the last major episode of erosion in 1974 (Thom and Hall, 1991). During phases of "gradual and imperceptible" accretion, landowners with land title tied to an "ambulatory" boundary (such as mean high water mark) may seek a redetermination of their land title to claim accreted land. Under the English "doctrine of accretion" exported to many other countries, the boundary remains fixed if sudden erosion cuts back into this new land. A consequence is for landowners to seek to defend their properties with seawalls and bulkheads against erosion. Beach amenity can suffer as a result of these actions (Titus, 1998).

Gradual change in coastline position and form, in tidal flat morphology, in the nature of inlet openings, and in the vegetative cover of coastal dunes are phenomena that affect the lives of millions of people. Understanding the dynamics of gradual changes in coastal geomorphology requires monitoring and historical reconstruction using a variety of techniques and analytical procedures. The use of simulated modeling, as seen for instance with the Shoreface Translation Model developed by Cowell *et al.* (1995), highlight a way of achieving such an understanding when coupled with a knowledge of geomorphic history and processes (see Figure C21).

B.G. Thom and P.J. Cowell

Bibliography

Carter, R.W.G., and Woodroffe, C.D. (eds.), 1994. *Coastal Evolution. Late Quaternary Shoreline Morphodynamics*. Cambridge: Cambridge University Press.

Chappell, J., and Shackleton, N., 1986. Oxygen isotopes and sea level. *Nature*, **324**: 137–140.

Cowell, P.J., and Thom, B.G., 1994. Morphodynamics of coastal evolution. In Carter, R.W.G., and Woodroffe, C.D. (eds.), *Coastal Evolution*. Cambridge: Cambridge University Press, pp. 33–86.

Cowell, P.J., Roy, P.S., and Jones, R.A., 1995. Simulation of large-scale coastal change using a morphological behaviour model. *Marine Geology*, **126**: 45–61.

Cowell, P.J., Roy, P.S., Cleveringa, J., and de Boer, P.L., 1999. Simulating coastal systems tracts using the shoreface translation model. *SEPM Special Publication No.*, **62**: 165–175.

Hopley, D., 1994. Continental shelf reef systems. In Carter R.W.G., and Woodroffe, C.D. (eds.), *Coastal Evolution*. Cambridge: Cambridge University Press, pp. 303–340.

Johnson, D.W., 1919. *Shore Processes and Shoreline Development*. New York: John Wiley and Sons [facsimile edition; New York: Hafner, 1965].

Komar, P.D., 1998. *Beach Processes and Sedimentation*. New Jersey: Prentice Hall.

Roy, P.S., Cowell, P.J., Ferland, M.A., and Thom, B.G., 1994. Wave-dominated coasts. In Carter, R.W.G., and Woodroofe, C.W. (eds.), *Coastal Evolution*. Cambridge: Cambridge University Press, pp. 121–186.

Thom, B.G., 1975. Modification of coastal and deltaic terrain subsequent to deposition. *Zeitschrift für Geomorphologie*, **22**(Supp.): 145–170.

Thom, B.G., and Bowman, G.M., 1981. Beach erosion–accretion at two time scales. *Proceedings 17th International Conference Coastal Engineering*, Sydney, pp. 934–945.

Thom, B.G., and Hall, W., 1991. Behaviour of beach profiles during accretion and erosion dominated periods. *Earth Surface Processes and Landforms*, **16**: 113–127.

Titus, J.G., 1998. Rising seas, coastal erosion, and the takings clause: how to save wetlands and beaches without hurting property owners. *Maryland Law Review*, **57**: 1279–1399.

Cross-references

Changing Sea Levels
Coastal Changes, Rapid
Coral Reefs, Emerged
El Niño-Southern Oscillation
Holocene Coastal Geomorphology
Sea-Level Changes During the Last Millennium
Submerging Coasts
Uplift Coasts

COASTAL CHANGES, RAPID

Introduction

There are many definitions of what might be called rapid events occurring on coastlines from sea-level movements caused by global warming and/or cooling (10's to 1,000's of years) to tsunamis that occur in seconds. In this entry the types of rapid changes discussed will be of the more instantaneous nature from *tsunamis* (*q.v.*) to *storm surges* (*q.v.*), the type of damage done and what the events leave as a trace in the geologic record. These phenomena were deemed sufficiently important by UNESCO/IGCP (International Geological Correlation Program) to form an IGCP project (IGCP 367) that lasted for 5 years (1994–98), where over 500 people from 70 countries were involved working on almost every aspect of rapid change in the coastal zone. Much of this entry will be drawn from the five volumes produced from that project (Scott and Ortlieb, 1996; Shennan and Gehrels, 1996; Murray-Wallace and Scott, 1999; Scott, 1999; Stathis *et al.*, 2000).

Climatic events

Storms

By far, the most dramatic of climatic events are hurricanes or typhoons as they are called in the Pacific. These events produce profound changes on coastlines and particularly man-made structures in the coastal zone. The effects are of two types, wind damage and water damage from elevated tides and torrential rains causing coastal flooding. At the center of a hurricane winds can reach over 200 km/h and storm tides can be as much as 8 m above normal. The amplitude of the storm tide depends on the barometric pressure (the lower the pressure, the higher the tide), the wind speed which causes the water to pile up onshore and the natural state of the tide (i.e., high versus low tide stage). The worst scenario for high storm tides is low pressure and high winds coinciding with a normal high spring tide.

The type of damage done by wind consists of knocking down of forests as occurred in hurricane *Hugo* in South Carolina (USA) in September of 1989 when the eyewall swept through a national forest and simultaneously knocked down most of the forest in a few seconds. Together with that it also destroyed many coastal man-made structures. Probably the most damage and deaths were caused by the storm surge, which at the eyewall location was 6 m above normal. However, even 50 km from the direct hit, the storm surge was almost 3 m. A purely human problem was sorting out which phenomenon caused the damage, the wind or the water, because many people did not have government flood insurance, which meant that their losses were not covered by their commercial hurricane insurance.

Maybe not too surprisingly the traces left in the naturally occurring geologic record are much less profound. At the eyewall hit of *Hugo*, there was a 10 cm sand layer but at a location 50 km north there was no sedimentological trace and in the extensive marsh areas there was not even a thin sand layer (Collins *et al.*, 1999), the natural systems can act as a buffer to storm events, perhaps a lesson to the human population.

The hurricanes do leave a trace in nontidal *coastal inlets* (*q.v.*) of either small sand layers or reworked microfossils, which can be used to build a chronology of storms on some coastlines (Liu and Fearn, 1993; Collins *et al.*, 1999). Also strong surges and sustained winds (either from hurricanes or long lasting winter storms such as "nor'easters" on the Atlantic coast of North America) can create washovers and change inlet positions in *barrier island* (*q.v.*) systems.

Global warming

This effect causes a general sea-level rise on scales from 10's of years to 1,000's of years both by melting of glacial ice caps and steric expansion of the ocean, which may amount to 30–40 cm. Depending on the coastal gradient this may or may not be significant.

It is difficult to measure these changes (Pirazzoli, 1991) because they are usually on the order of a few centimeters. Mostly they have been detected by *tidal gauges* (*q.v.*) for the last 50 years (Pirazzoli, 1991) but these measurements are sometimes questionable. Some *biological markers* are now capable of measuring very small changes (± 5 cm) in sediment records (Scott and Medioli, 1986) such that it is possible to measure rapid sea-level changes in the late Holocene (e.g., Scott and Collins, 1996). It is these mid-Holocene measurements that may help us to understand the suggested *sea-level rise* (*q.v.*) to be caused by anthropogenic *global warming* (*q.v.*).

Seismic events

Earthquakes

Sudden land movements on coastlines can either instantly emerge or submerge coastlines and any structure that might be in the way. This is particularly a problem around earthquake prone coasts such as the Pacific Rim and most of the northeastern Mediterranean.

The type of records preserved are often in coastal marshes because they remain depositional and their strong vertical zonations help to show sharp vertical surface changes, that is, sudden change from forest to marsh if the land drops or vice versa if the land emerges (Atwater, 1987; Atwater *et al.*, 1995; Shennan *et al.*, 1999). There are outstanding records of these phenomena all up and down the west coasts of both North and South America. In rocky coastline areas, *biological markers* maybe uplifted far above their natural range, providing a good indication of rapid change and the magnitude of that change (Ortlieb *et al.*, 1996).

Extended records of these events have been documented in the northwestern part of the United States and southern British Columbia, Canada that show a 300–500 year periodicity for magnitude 8 or 9 (on the Richter Scale) earthquakes (Atwater, 1987) over the last several thousand years. Until recently, it was difficult to plan based on a 300-year return time but Shennan *et al.* (1998, 1999) found in two separate places that there appeared to be a precursor event between one and three years prior to these events that could be detected in coastal forest sequences that submerged during a major seismic disruption. The most compelling evidence is from the traces left by the Great Alaska Earthquake of 1964 that was the second largest seismic event ever recorded at 9.2 on the Richter Scale. The evidence from this quake is convincing because unlike the prehistoric quakes in Oregon (Shennan *et al.*, 1998), in Alaska, we know exactly when the event occurred and we know absolutely that the forest layer was above tidal influence before the March 26, 1964 quake. What is seen is a small change in both diatom and foraminiferal assemblages in the forest *before* the quake that indicate a slight subsidence, maybe 30 cm, that was not detected by anyone living there or by macro-vegetation (Scott *et al.*, 1998; Shennan *et al.*, 1999). So it is possible that there may be a technique now that we can use as an early warning system for these mega-quakes.

Tsunamis

Tsunamis (*q.v.*) are perhaps the most spectacular and frightening of the coastal effects caused by earthquakes. They are large wavelength, fast moving waves resulting from the rapid land slumps or uplifts caused by earthquakes in offshore areas. When they are in the deep water of the open ocean they are hardly noticeable but when they reach the shallow coastal regions the energy is converted to wave height, which in many cases can reach as high as 20 m. The most recently documented cases at this writing were the 1994 Java (Dawson *et al.*, 1996) and the 1998 Papua New Guinea tsunamis where several thousand people were killed (McSaveney *et al.*, 2000).

Tsunamis also leave highly detectable traces in coastal deposits in the form of sand layers contained in marsh or nontidal *coastal pond* sediments. However, these traces cannot be reliably separated from hurricane layers, so if both phenomena occur in the same area it could be confusing to sort out which is which. As with storm traces, the best evidence for offshore transport is the presence of marine microfossils. Many times the tsunami will leave no trace if it travels over hard substrates.

The occurrence of tsunamis is limited largely to the Pacific Rim where they can occur either very close or very far from the actual seismic event; an example of a distant catastrophic tsunami was the killing of several people and destruction caused by a tsunami in Crescent City, California (USA) resulting from the 1964 Alaska earthquake whose epicenter was several thousand kilometers to the north.

Sedimentation/erosion events

Natural changes

There are rapid events in coastal areas caused simply by rapid sediment events that may be a result of flood deposits, rapid deforestation caused by forest fires, or any type of event that changes the sediment budget in a dramatic sense to cause a shift in morphology of the coastline. In the case of a forest fire, large amounts of erosion as a result of deforestation would place sediment into *coastal inlets* (*q.v.*) causing rapid infilling or conversely, if a sediment source is reduced it can cause the shoreline to retreat rapidly in the absence of any other effects (e.g., sea level or storms).

Anthropogenic causes

By far the most spectacular rapid sedimentation events are caused by human intervention. They are almost too numerous to mention but almost every marine *delta* (*q.v.*) in the world is being influenced by sediment starvation which causes rapid land loss as the delta continues to subside with no replenishment of sediment (e.g., the Nile and Mississippi Deltas). The starvation results from dams either that trap the sediment upstream (the Nile) or diversions of the rivers, which do not allow the river to switch back and forth (the Mississippi).

Other less spectacular but damaging rapid sedimentation events are caused by careless placement of structures on the coast that impede *sediment transport* (*q.v.*) rapidly changing the balance of erosion/sedimentation.

Summary

Rapid coastal changes can take many forms; only a few of the most obvious examples are discussed here. Often the geologic record of these events is difficult to detect, especially in dynamic coastal environments, where tidal activity reworks most of the record. The records are best in nontidal *coastal ponds and inlets* where an extreme event brings in allochtonous sediment that is then sandwiched in quiet water, often organic deposits and are relatively simple to detect and date.

David B. Scott

Bibliography

Atwater, B.F., 1987. Evidence for great Holocene earthquakes along the outer coast of Washington State. *Science*, **236**: 942–944.

Atwater, B.F., Nelson, A.R., Clague, J.J., Carver, G.A., Yamaguchi, D.K., Bobrowsky, P.T., Borgois, J., Darienzo, M.E., Grant, W.C., Hemphill-Haley, E., Kelsey, H.M., Jacoby, G.C., Nishenko, S.P., Palmer, S.P., Peterson, C.D., and Reinhardt, M.A., 1995. Summary of coastal geologic evidence for past great earthquakes at the Cascadia subduction zone. *Earthquake Spectra*, **11**: 1–18.

Collins, E.S., Scott, D.B., and Gayes, P.T., 1999. Hurricane records on the South Carolina coast: can they be detected in the sediment record. *Quaternary International*, **56**: 15–26.

Dawson, A.G., Shi, S., Dawson, S., Takahashi, T., and Shuto, N., 1996. Coastal sedimentation associated with the June 2nd and 3rd, 1994 Tsunami in Rajegwesi, Java. *Quaternary Science Reviews*, **15**(8–9): 901–912.

Liu, K.-B., and Fearn, M.L., 1993. Lake-sediment record of late Holocene hurricane activities from coastal Alabama. *Geology*, **21**: 793–796.

McSaveney, M.J., Goff, J.R., Darby, D.J., Goldsmith, P., Barnett, A., Elliot, S., and Nongkas, M., 2000. The 17th July 1998 Tsunami, Papua New Guinea: evidence and initial interpretation. *Marine Geology*, **170**: 81–92.

Murray-Wallace, C.V., and Scott, D.B. (eds.), 1999. Late Quaternary coastal records of rapid change: applications to present and future conditions (IGCP Project 367). *Quaternary International*, **56**: 1–154.

Ortlieb, L., Barrientos, S., Guzman, N., 1996. Coseismic Coastal Uplift and Coralline Algae record in Northern Chile; the 1995 Antofagasta Earth quake case: Quaternary Science Reviews, **15**, nos. 8–9, pp. 949–960.

Pirazzoli, P.A., 1991. *World Atlas of Sea-level Changes*. Amsterdam: Elsevier Oceanographic series 58.

Scott, D.B. (ed.), 1999. Records of Rapid Change in the Late Quaternary (IGCP 367). *Quaternary International*, **60**: 1–126.

Scott, D.B., and Medioli, F.S., 1986. Foraminifera as sea-level indicators. In Plassche O. van de (ed.), *Sea-level Research: A Manual for the Collection and Evaluation of Data*. Norwich, U.K., Geo Books, pp. 435–456.

Scott, D.B., and Ortlieb, L. (eds.), 1996. Records of rapid events in Late Quaternary shorelines. *Quaternary Science Reviews*, **15**(8,9): 761–960.

Scott, D.B., Shennan, I.A., and Combellick, R.A., 1998. Evidence for pre-cursor events prior to the 1964 Great Alaska earthquake from buried forest deposits in Girdwood, Alaska: abstract in *Annual Geological Society of America Meeting*, Toronto, Canada, pp. A-226.

Shennan, I.A., and Gehrels, R. (eds.), 1996. An introduction to IGCP Project 367 Late Quaternary coastal records of rapid change: Application to present and future conditions. *Journal of Coastal Research*, **12**(4): 795–797.

Shennan, I.A., Long, A.J., Rutherford, M.M., Innes, J.B., Green, F.M., and Walker, K.J., 1998. Tidal marsh stratigraphy, sea-level change and large earthquakes II. submergence events during the last 3500 years at Netarts Bay, Oregon, USA. *Quaternary Science Reviews*, **17**: 365–393.

Shennan, I.A., Scott, D.B., Rutherford, M.R., and Zong, Y., 1999. Microfossil analysis of sediments representing the 1964 earthquake, exposed at Girdwood Flats, Alaska, USA. *Quaternary International*, **60**: 55–74.

Stathis, C., Stiros, S., and Scott, D.B. (eds.), 2000. Records of Rapid coastal change in Late Quaternary coastal sediments and landforms. *Marine Geology* (special issue), **170**: 249 p., thematic issue.

Cross-references

Barrier Islands
Changing Sea Levels
Coastal Changes, Gradual
Coastal Lakes and Lagoons
Deltas
Greenhouse Effect and Global Warming
Meteorological Effects on Coasts
Sediment transport (see Cross-Shore Sediment Transport and Longshore Sediment Transport)
Storm Surge
Tide Gauges
Tsunami

COASTAL CLIMATE

Three factors are important in an explanation of coastal climate. The first is the juxtaposition of two surfaces, land (or in some cases, ice) and water, of very different properties. The second is the geographic location of the coast with respect to continental masses and global-scale atmospheric and oceanic currents of air and water, respectively. The third is the topography of the coast and its hinterland. We must also take into account temporal and spatial scale issues so as to order the different aspects and to maximize our understanding of coastal climates. Scale will be used as an organizing framework in the following account.

Several useful data and information sources exist for an entrance into the topic of Coastal Climatology. One is an encyclopedia entry under the same topic name by Walker *et al.* (1987) in *The Encyclopedia of Climatology*. A second is the Master Environmental Library that is an online data source for environmental data for coastal modeling applications (Allard and Siquig, 1998).

Time and space scales

Climatologists recognize different types of climate according to their time and space scales. The local wind currents that redistribute sand around a beach sand dune and the different temperatures on the sunny and shaded side of a stand of beach grass would be in the realm of *microclimatology*. Different climates formed by the different surfaces and local climatic interactions as one might traverse the sand of a barrier island, the water of its related lagoon, the inland coastal marshes, and the forest or agricultural land of the immediate hinterland would constitute a study of *topoclimatology*. A sea breeze might penetrate inland some 75 km from the shore on a given day and would represent a phenomenon studied in *mesoclimatology*. Climatic variable value gradients parallel to the coastline are usually much steeper than those that are perpendicular to the coastline. The geography of the US Pacific Northwest (PNW), for example, gives rise to well-marked gradients in values of climatic variables both in the latitudinal (north–south) direction but also in the longitudinal (west–east) direction. This is not uncommon for coastlines that have some north–south alignment. The location of a coast in relation to global latitude, continental geography, hemispheric-scale winds and storm tracks in the atmosphere, and surface water currents in the ocean, and the state of teleconnective indices such as the Southern Oscillation and North Atlantic Oscillation, would help explain its *macroscale climatology*. The passage from micro to macroscale defines a scale hierarchy. Most scale hierarchies in Earth Sciences, including coastal climates, have at least three characteristics. First, events on the larger scales tend to set the context for those on the scale below. Second, explanation of events on one scale is best given by reference to events on the scale immediately above and below the scale and phenomenon in question. Third, at least at the lower end of the hierarchy, there tends to be a direct relationship between time and space scales. For example, ephemeral events, like an eddy in the wind, happen on small time and space scales while some other events, like a hurricane, sometimes take many days to run their course and have the potential to affect a relatively large geographic area.

Microscale coastal climates

Wind is the principal climatic variable at the microscale in coastal climates and its interaction with topographic and biologic factors has long been studied. Three factors are important at this scale. First, wind at coastlines tends to come from a predominant direction often from offshore. This has the effects of distributing sand and sand dunes in particular ways, and of sometimes shaping the vegetation canopies. The actual type of sand dune formed has been suggested to be a function of the supply of sand, the density of vegetative cover, and the average wind speed (Strahler and Strahler, 1992, p. 406). Second, the capacity of wind to carry sand is an exponential function of its velocity. Thus when wind speeds are decreased, by, for example, the presence of vegetation or a fence on a beach, then sand and other particles being transported are rapidly deposited. Vegetation and wind thus have a synergistic interaction in stabilizing sand movement. Third, typically wind velocity increases logarithmically with height. This fact needs to be taken into account when designing high-rise buildings in coastal environments. Pilkey *et al.* (1998, p. 254) have noted that at any given height pressure from the wind is greater near the shore than it is inland.

Topoclimatology

The coast, interpreted broadly, may be formed of many different surfaces including water, sand, soil, rock, and a variety of vegetated and urban surfaces. Each one of these surfaces has a different energy and moisture balance, and depending on its roughness, a different degree of momentum transfer with the atmosphere. These differences might not be too important on the topoclimate scale on a coastline with a steep cliff topped with relatively homogeneous vegetation because the climatic differences would be overwhelmed by both smaller- and larger-scale processes. However, the differences are important on a coastline typified by a barrier island, lagoon, and hinterland series on days when the large-scale wind is not too strong. The differences are important because they will form a set of distinctly different microclimates that will relate, at least, to the flora and fauna of the area.

Salt transport by wind is also important at this scale. The concentration of salt particles near the surface in coastal climates is directly related to the wind speed (Bigg, 1996). A high number of salt particles makes the coastal air very hazy. The salt particles attract water vapor and play an important role in raindrop formation. Damaging salt spray can sometimes have a pruning effect on coastal vegetation. How far inland or how high up a cliff this effect can occur is determined by wind flow and its interaction with topography on the scale in question. Soils near coasts are typically high in salts transported by wind from the ocean.

Mesoclimatology—the land–sea breeze circulation

It is the different properties of the water and land surface that ultimately lead to the land–sea breeze phenomenon—one of the most studied features of all aspects of coastal climates. The circulation is developed at mid and low latitude coasts on days when larger-scale wind systems are not predominant. Anticyclonic weather in summer time provides an ideal context. The circulation is found both on maritime and freshwater coasts. At the surface of the Earth, the land–sea breeze circulation takes the form of winds coming from water surface during the day and from the land during the night. The surface of a water body usually manifests only a small daily range of temperature. Oke (1987) has outlined the four reasons for this. Solar radiation can easily penetrate a water surface and thus heats a deep layer and not just the surface itself. Convection currents and mass transport in the water permits any heat gain or loss to be spread throughout a large volume of water. Evaporation from the water surface cools the water surface and helps promote vertical mixing within the water itself. Water has a high thermal capacity compared with soil and thus takes about three times the amount of heat to raise its temperature by 1°C as for the same volume of soil. Thus there is a reduced flow of sensible heat from the water surface to the air compared with that over the land during the day. This leads to a reduced warming rate of the air over the water compared with that over the land surface. As described by Oke (1987), greater sensible heat flow over the land in the morning heats the air columns more rapidly and to greater heights than over the water. The atmospheric pressure well above the surface of the land is higher than at the same level over the water. The resulting horizontal pressure gradient results in airflow at upper levels toward and out over the water surface. This flow ultimately produces a higher atmospheric pressure at the surface over the water that, in turn, forms a cross-coastline flow of air from the water to the land. This surface level flow is known as the *sea* or *lake breeze*. The *land breeze* starts in the evening because of the greater contraction and cooling of the air columns over the land compared with the sea.

Oke reports that the daytime sea breeze circulation has greater vertical and horizontal extents and larger wind speeds than the land breeze because it is driven by solar forcing. Typically, the sea breeze might have a velocity of 2–5 m/s, a depth of 1–2 km, and extend inland as far as 30 km. The land breeze characteristically has a velocity of 1–2 m/s and a smaller extent than the see breeze. If the sea breeze is forming in a location where there is cold upwelling water, then the air formed over this water may be cool enough to continue the sea breeze circulation all through the night and not have a land breeze develop at all.

The sea breeze brings with it air that is cooler and more humid than the air found over the land. This cool air plows underneath the warmer air over the land and forms a localized cold front known as a *sea breeze front*. The sea breeze front gradually advances inland during the day and is associated with the development of sea breeze cumulus clouds. These clouds are often taken by the counter (seawards) flow above the surface and dissipate because they are decoupled from their moisture source. Purdom (1976) showed that some parts of the front displayed higher concentrations of convection than in others. He found that the shape of the coastline in some places created stronger upward vertical velocities. Kingsmill (1995) documented for summer months the collision of a sea breeze front traveling from the east side of the Florida peninsular with a gust front traveling from the west. Interestingly, he found that convective activity was most prominent just prior to the collision of the two fronts rather than at the time when the two collided. Oke (1987) has suggested that if the sea breeze travels far enough inland then the influence of the Coriolis force will affect its direction and the breeze will end up traveling parallel to the coast. A fairly deep penetration of the sea breeze front inland was observed in April 1999 at Wokingham, which is 75 km inland from the southern coast of England. The passage of the front was accompanied by an increase of the dew point temperature of 4°C and a change of wind direction from 70° to 210° (Burton, 2000).

The climatology of sea (lake) and land breezes is highly variable depending on the particular location being considered. A 15-year study of Lake Michigan found that lake breezes tend to occur more frequently along the eastern shore of Lake Michigan than along the western shore (Laird and Kristovich, 1998). In addition, there was a gradual increase in the number of lake breeze events from May through August, then a slight decrease into September. The distribution of the frequency of land breezes is similar to that of the lake breezes from May through September, however, they occur about 10% less often than the daytime lake breeze circulation. Typically, sea breeze activity is observed on the south coast of England on 75 days during the year (Burton, 2000).

The land–sea breeze circulation has many practical implications. Oke (1987) quotes one study in which tobacco leaves near the coastline of Lake Erie were damaged by ozone that had been formed over the lake from lakeshore city pollutants and then transported back over the shore and tobacco crop by the lake breeze. In another study, quoted by the same source, blister rust disease impacted pine trees 15–20 km from the Lake Superior lakeshore. This impact was thought to be due to spores from bushes located between the diseased trees and the lake and transported out over the lake at night, back aloft by the counter airflow of the land breeze, and then descending upon the pine trees. Michael *et al.* (1998) showed that during spring and summer, sea breeze circulations can strongly influence airport operations, air-quality, energy utilization, marine activities, and infrastructure in the New York city region. Here, the investigators found the geographic configuration of region presents a special challenge to atmospheric prediction and analysis. The New Jersey and Long Island coasts are approximately at right angles to each other; additionally Long Island Sound separates Long Island from the mainland of Connecticut. The various bodies of water in the region (Atlantic Ocean, Long Island Sound, New York Harbor, Jamaica Bay, etc.) have different surface temperatures. In addition the urbanization of the New York area can modify atmospheric flows.

Apart from land–sea breeze phenomena mesoscale wind fields in general are very important in influencing the coastal and coastal ocean processes. Considerable attention is being given, for example, to the effect of wind on driving and directing freshwater plumes from rivers entering the ocean. This has many implications for cross continental shelf transport (Lohrenz, 1998).

Macroscale climatology

The location of the coast in the world determines in large part its overall climate. A first order classification of coasts on this scale would include tropical windward coasts, tropical leeward coasts, mid- and high-latitude windward coasts, mid- and high-latitude leeward coasts, and ice-influenced coasts. Each of these categories would be modified according to whether or not the hinterland of the coast contained mountains. Examples of climate data from such categories are given in Table C8. Walker *et al.* (1987) quote other ways of classifying coastal climates.

Under the classification suggested here the extremes for high precipitation are seen in the tropical and mid-latitude windward coasts with mountains such as, respectively, on the island of Maui and other islands in Hawaii, and the PNW of the United States. In the case of the north coast of Maui, Hawaii, onshore moist tropical air is lifted up and cooled against the mountains giving rise to some of the world's highest precipitation values. Similar orographic effects occur at the coast of the PNW but here it is moist, mid-latitude Pacific air that is uplifted often during mid-latitude storms, which provides intense precipitation. The extreme for low precipitation is found on tropical coasts on the western sides of continents, as for example, on the coast of Peru. Air flows down the mountain sides of the Andes mountains in Peru and northern Chile and thus is warm and dry when it reaches the coast. More importantly this coastline, and others in this category, has a cool ocean current offshore. Upwelling of cold water gives rise to fog that is often the only source of precipitation in this climate. Ice-ocean boundaries are not considered to be coastlines by some (Walker *et al.*, 1987) but since they form a distinctive interface on the Earth's surface they will be briefly mentioned here. Thus the "coasts" with the lowest temperatures are those formed by the meeting of ice masses and the sea such as the Greenland and Antarctic ice cap coasts. Such coasts are often marked by strong, katabatic winds blowing from the ice cap. Katabatic winds are winds formed from cold, relatively dense air flowing down a topographic feature. The climate of the other categories of coasts is intermediate between the extremes described here.

One of the most important features of coastal macroclimate is the storm. Both tropical storms (hurricanes) and mid-latitude storms or cyclones heavily impact the geologic, biologic, and human dimensions of a coast. Sometimes it is the presence of the coast itself that helps cause or enhance the storms. Cyclogenesis is enhanced, for example, at the east coast of the United States as upper-level disturbances cross the Appalachian mountains and then cross the strong baroclinic zone of the atmosphere near the coast. At the coast, the cyclones flow parallel to the landmass in winter and a strong cooling of air on the landward side leads to the formation of macroscale fronts (Rotunno, 1994).

Considerable time is spent researching the frequency and magnitude of coastal storms. Waves generated by mid-latitude storms, for example,

Table C8 Climate data for selected coastal areas

Tropical	Location	Latitude (Deg Min)	Longitude (Deg Min)	Elevation (m asl[a])	Mean annual temperature (°C)	Mean monthly temperature Range (°C)	Annual precipitation (mm)
Windward							
Mountains	Hana, Hawaii	20 48 N	156 13 W	2	23.4	21.7–24.9	2060
No Mountains	Belem, Brazil	1 23 S	48 29 W	14	25.9	25.4–26.5	2770
Leeward							
Mountains	Arica, Chile	18 28 S	70 22 W	29	18.5	15.4–22.2	1
No Mountains	Bonthe, Sierra Leone	7 32 N	12 30 W	8	26.8	25.1–28.2	3718
Mid-latitude							
Windward							
Mountains	Hokitika, New Zealand	42 43 S	170 57 E	5	11	6.4–15.1	2721
No Mountains	Bordeaux, France	44 51 N	00 42 W	48	12.3	5.2–19.6	900
Leeward							
Mountains	Tokyo, Japan	35 41 N	139 46 E	36	14.7	3.7–26.4	1563
No Mountains	Boston, USA	42 22 N	71 01 W	9	10.8	−0.9–23.2	1086
High-latitude							
Windward							
Mountains	Tromso, Norway	69 39 N	18 57 E	114	3.3	−2.7–12	1119
No Mountains	Spitsbergen, Norway	78 04 N	13 38 E	9	−3.8	−11.9 to 5.0	354
Leeward							
Mountains	Cape Tobin[b] Greenland	70 24 N	21 58 W	42	−7.4	−16.3–2.7	not available
No Mountains	Resolute, Canada	74 43 N	94 59 W	64	−16.2	−31.8 to 4.6	131
Ice-influenced	Nord, Greenland	81 36 N	16 40 W	35	−16.5	−31.0–3.9	not available

Source: Data are from Liljequist (1970) except for Hana data which are from the Western Region Climate Center.
[a] asl, above sea level.
[b] Note that Cape Tobin also represents an ice-influenced coast.

are responsible for much of the coastal erosion that occurs along the Atlantic coast of the United States. Dolan *et al.* (1988) found that at this location, the five-month period December through April had 63% of all the storms for the area. They also found a systematic increase between 1942 and 1984 in the number of storms occurring in May and October resulted in the lengthening of the storm season and a more abrupt transition between the winter and summer storm regime. The authors were able to construct a map of the mean cyclone track and frequency of all the mid-latitude storms in their data set that produced significant waves at Cape Hatteras. The storm track runs Southwest to Northeast following the coast of the Carolinas and continuing out to sea at Cape Hatteras. Locally the storms following this track are called Nor'easters and they often do considerable damage to the coast. The secular variation of such storms is very important. Hayden (1981) discovered that in the early years of the 20th century there was a trend toward increased cyclone frequency over marine, as opposed to continental, areas of the Atlantic US coast. He also found decadal changes in coastal cyclogenesis with a maximum in the record up to 1978 occurring in the 1950s.

It is possible that decadal changes in the frequency of storms whose predominant wind directions is different can lead to marked geographic effects. At Hog Island, Virginia there is a clear dynamic coastline change pattern: erosion occurs on the north when accretion occurs on the south, or *vice versa*. This kind of cyclic process of coastline changes may repeat every 190 years on Hog Island. Currently, the southern part of Hog Island just ended the erosion processes, while the northern part just ended the accretion processes. It is hypothesized that the changing geography of the island on this timescale may be related to the frequency of storms that either have winds with a predominant northeasterly or southeasterly component.

Damage to coasts is also done by Tropical Cyclones that go by different names in different parts of the world—hurricanes, for example, in the Atlantic and typhoons in Southeast Asia. For a tropical cyclone to be called a hurricane there must be a sustained wind of 33.5 m/s (75 mph) or more. Such systems can have diameters exceeding 300 km. Very large typhoons in the western North Pacific can have diameters up to 3,000 km and are sometimes called super-typhoons. Hurricanes are placed into five categories in the United States. Category 1 has wind speeds of 119–153 kph; category 2, 154–177 kph; category 3, 178–209 kph; category 4, 210–249 kph; and category 5, has wind speeds in excess of 249 kph. In the following discussion, I will use the term hurricane as being synonymous to tropical cyclone.

Hurricanes are features of coastal climate mainly during the late summer and autumn seasons since they depend upon high Sea Surface Temperatures (SSTs) as one factor in their initial formation. Hurricanes impact the tropical, subtropical, and mid-latitude coastlines that are mainly on the east coasts of continents or coasts that face south in the Northern Hemisphere and *vice versa*. Hurricanes usually weaken after passing the coastline and moving inland because they depend upon the warm moist ocean surface for their energy. Besides high winds the storms are noteworthy for bringing large quantities of rain to the coast. Sometimes tornadoes are spawned as well. Hurricanes represent an interesting interaction between meso- and macro-climatology since their formation and track once formed is often a function of the macroscale features such as upper atmosphere wind fields. Much coastal damage is done by the interaction of the hurricane with the ocean. The resulting formation of storm surges brings extra high flows of ocean water over the coastline. The storm surge of a category 5 hurricane could be 5.5 m (18 ft) above normal sea level.

As with mid-latitude cyclones, interdecadal variability in frequency is well marked in hurricanes. Hurricane frequency in the Atlantic decreased from the 1890s to the 1920s and 1930s, increased during the 1940s to 1970s, and then decreased in the last part of the 20th century

(Riehl, 1987). The latter decrease coincided with the development of much coastal property in the Southeastern coast of the United States.

Ice-influenced coastlines have their own interesting climatological features besides the katabatic winds already mentioned. An interaction between macro and mesoscale events gives rise to strips of open coastal water surrounded by ice-covered ocean. The open water areas are called polynyas and they typically open and close over the time period of a few days. They are partly formed by coastal winds pushing the ice offshore and are associated with a complex set of water and atmosphere density and flow interactions (Chapman, 1999). The polynyas, sometimes called leads, are large sources of sensible heat to the atmosphere in winter with heat fluxes locally exceeding 500 W/m^2. In summer, the comparatively low albedo of the open water compared with that of nearby ice allows the absorption of solar radiation that warms the water. The fractional lead coverage in the Arctic Ocean as a whole is 1% in winter and greater than 20% in late summer (Curry and Webster, 1999).

There are also phenomena of dynamic climatology existing where coastlines are backed by mountain areas. Rotunno (1994) describes how Kelvin waves may propagate along the basin-wall-like sides of an ocean basin such as at the Pacific coast of North America. Here also strong alongshore jet streams have been documented as a result of the mountains. Phenomena that appear similar to flow separation in classical fluid dynamics have been found to occur in the lee of capes and other coastline salients.

Temporal climate variability in coastal climates

All coastal climates are subject to climate variability on a variety of timescales. The presence of the ocean water has a moderating effect on the climate of the nearby land at all the timescales. Climate variability at coastlines is therefore usually less than that in interior continental landmasses but can still be significant especially at the longer timescales. We may examine an example from the PNW of the United States to demonstrate the kinds of climatic variability that can occur at, or near, a coast over a variety of timescales.

Major changes in the climate of the PNW from glacial to interglacial have been described by Worana and Whitlock (1995). The extreme glacial climate, a little inland from the coast, had a mean annual temperature that was likely to have been 7°C lower than today with the January temperature 14°C lower. Conditions were also probably drier than today with the Polar Front jet stream flowing well to the south. After the last retreat of the ice the climate was warmer. Between 10,000 and 5,600 BP (Early Holocene) temperature is believed to have increased and effective moisture decreased as summer droughts became a climatic feature associated with an expansion of the subtropical high pressure in the eastern Pacific Ocean. On the next timescale down, some cyclic coastal climatic behavior has been suggested. Sediment cores from anoxic basins reveal a 60–70-year cycle of warming and cooling over the past 1,500 years (US GLOBEC, 1994). The last cycle in this pattern, which ecologically is associated with sardine dominance over anchovy and *vice versa*, is seen with warming around 1925, cooling at 1948, and a return to warming about 1977. Ware and Thompson (1991) regard the 60-year cycle in Southern California data as being consistent with their belief that pelagic fishes are responding to a 40–60-year oscillation in primary and secondary production that, in turn, is being forced by a long-period oscillation in wind-induced upwelling. Trenberth (1993) reviews coupled General Circulation Model (GCM) output and suggests, under current projections of CO_2 increase, the western coast of North America is expected to warm by about 1.0–1.5°C by the year 2030 and that northern regions will warm more rapidly than southerly regions. The hydrologic cycle is expected to intensify by about 10% leading to greater amounts of evaporation and precipitation. Trenberth (1995) has also suggested that the relative frequency of the El Niño mode in the past decade "may well be one of the primary manifestations of anthropogenic global warming." There are also interdecadal scale variations in PNW coastal climate. Mantua *et al.* (1996) described a Pacific Decadal Oscillation (PDO) in SSTs that has many implications for salmon production. It has been shown that the PDO manifests itself in coastal climate by giving rise to four distinct climatic periods. The periods are 1896–1915, wet and cool, 1915–46, dry and warm; 1947–75, wet and cool; and 1976–94, dry and warm. The dominant source of variation on the quasi-quintennial scale is related to the El Niño–Southern Oscillation (ENSO). Low Southern Oscillation Index (SOI) values (El Niños) are correlated with reduced precipitation (Redmond and Koch, 1991), snowpack and streamflow in the PNW (Cayan and Webb, 1992). The PNW also tends to display warmer winter temperatures and drier than usual winters under El Niños conditions (Redmond and Koch, 1991). At the opposite end of the oscillation Kahya and Dracup (1993) are among several investigators to demonstrate that high precipitation and high streamflow values in the coastal PNW are often associated with La Niña years. Superimposed on ENSO-scale climatic variability is the variability on the interannual and shorter timescales such as the passage of individual mid-latitude cyclones and anticyclones.

Future research

There is an unlimited amount of work to do on coastal climates. Working through the scales we may suggest the following areas of important research. First, it would be interesting to document the differences in the surface energy budget values across the different surfaces of the subcomponents of the shoreline—barrier island, marsh, lagoon, etc. Second, it is important to find out what are the interactions between the mesoscale and macroscale events with respect to coastal climates. How, for example, do sea breeze fronts affect other weather systems occurring inland? Much more has to be found out about the effect of varying SSTs on weather systems passing over them. This is particularly the case where SSTs show a large range of different values over a short distance. The Gulf Stream ocean current and areas of coastal upwelling water are cases in point. There are many local dynamic features that need investigation such as the trapping of atmospheric waves under marine temperature inversions and low-level coastal atmospheric jet streams. Interdecadal variation in the climatic variables of coastal climate must be established and, if possible, explained. This is important because of the huge amount of coastal development that has happened and will continue to occur worldwide. Finally, the affect of potential global climate change on coastal climates over the next 100 years must be studied. This topic has not received too much attention so far because the global climate models tend to operate at a coarser spatial scale than one that allows a coast to be specifically considered.

David Greenland

Bibliography

Allard, R.A., and Siquig, R.A., 1998. The master environmental library: a source for environmental data for coastal modeling applications. In *Second Conference on Coastal Atmospheric and Oceanic Prediction and Processes*. Boston: American Meteorological Society.

Bigg, G.R., 1996. *The Oceans and Climate*. Cambridge: Cambridge University Press.

Burton, B.J., 2000. Passage of a well-marked sea-breeze front at Wokinham. *Weather*, **55**: 281–286.

Cayan, D.R., and Webb, R.H., 1992. El Niño/Southern Oscillation and streamflow in the western United States. In Diaz, H.F., and Markgraf, V. (eds.), *El Niño: Historical and paleoclimatic aspects of the Southern Oscillation*. Cambridge: Cambridge University Press. pp. 29–68.

Chapman, D.C., 1999. Dense water formation beneath a time-dependent coastal polynyas. *Journal of Physical Oceanography*, **29**(4): 807–820.

Curry, J.A., and Webster, P.J., 1999. *Thermodynamics of Atmospheres and Oceans*. San Diego: Academic Press.

Dolan, R., Lins, H., and Hayden, B., 1988. Mid-Atlantic coastal storms. *Journal of Coastal Research*, **4**(3): 417–433.

Hayden, B.P., 1981. Secular variation in Atlantic Coast extratropical cyclones. *Monthly Weather Review*, **109**(1): 159–167.

Kahya, E., and Dracup, J.A., 1993. Streamflow and La Niña event relationships in the ENSO-Streamflow core areas. In Redmond, K.T., and Tharp, V.L. (eds.), *Proceedings of the Ninth Annual Pacific Climate (PACLIM) Workshop*, April 21–24, 1992. California Department of Water Resources, Interagency Ecological Studies Program, Technical Report 34, pp. 89–96.

Kingsmill, D.E., 1995. Convection initiation associated with a sea-breeze front, a gust front, and their collision. *Monthly weather Review*, **123**: 2913–2933.

Laird, N.F., and Kristovich, D.A.R., 1998. Climatology of lake and land breezes for the coastal region of Lake Michigan. In *Second Conference on Coastal Atmospheric and Oceanic Prediction and Processes*. Boston, MA: American Meteorological Society.

Liljequist, G.H., 1970. *Klimatologi*. Stockholm: Generalstabens Litografiska Anstalt.

Lohrenz, S.,1998. Wind effects on freshwater plumes. In Henrichs, S. *et al.* (eds.), *Coastal Ocean Processes (CoOP): Transport and Transformation Processes Over Continental Shelves with Substantial*

Freshwater Inflows. University of Maryland Center for Environmental Science Technical Report TS-237-00, pp. 122–127.
Mantua, N.J., Hare, S.R., Zhang, Y., Wallace, J.M., and Francis, R.C., 1996. A Pacific interdecadal oscillation with impacts on salmon production. *Bulletin of the American Meteorological Society*. **78**: 1069–1079.
Michael, P., Miller, M., and Tongue, J.S., 1998. Sea breeze regimes in the New York City region—Modeling and Radar observations. *Second Conference on Coastal Oceanic and Atmospheric Prediction*. Phoenix, Arizona: American Meteorological Society.
Oke, T.R., 1987. *Boundary Layer Climates*, 2nd edn. New York: Routledge.
Pilkey, O.H., Neal, W.J., Riggs, S.R., Webb, C.A., Bush, D.M., Pilkey, D.F., Bullock, J., and Cowan, B.A., 1998. *The North Carolina Shore and its Barrier Islands*. Durham: Duke University Press.
Purdom, J.F.W., 1976. Some uses of high-resolution GOES imagery in the mesoscale forecasting of convection and its behavior. *Monthly Weather Review*, **104**: 1474–1483.
Redmond, K., and R. Koch., 1991. ENSO v surface climate variability in the western United States. *Water Resources Research*. **27**: 2381–2399.
Riehl, H., 1987. Hurricanes. In Oliver, J.E., and Fairbridge, R.W. (eds.), *The Encyclopedia of Climatology*. New York: Van Nostrand Reinhold. pp. 483–496.
Rotunno, R., 1994. Coastal meteorology. In *Coastal Ocean Processes: Wind-driven transport processes on the U.S. West Coast*. Woods Hole Oceanographic Institute Technical Report WHOI-94-20, pp. 87–90.
Strahler, A.H., and Strahler, A.N., 1992. *Modern Physical Geography*. 4th ed., New York: Wiley.
Trenberth, K.E., 1993. Northern hemisphere climate change: physical processes and observed changes. In Mooney, H.A., Fuentes, E.R., and Kronberg, B. (eds.), *Earth System Response to Global Change: Contrasts between North and South America*. New York: Academic Press, pp. 35–59.
Trenberth, K.E., 1995. Atmospheric circulation climate changes. *Climatic Change*, **31**: 427–453.
U.S.GLOBEC, 1994. Eastern Boundary Current Program: A science plan for the California Current. U.S. Global Ecosystems Dynamics. Report Number 11, 134 pp.
Walker, H.J., Hsu, S.A., and Muller, R.A., 1987. Coastal Climate. In Oliver, J.E., and Fairbridge, R.W. (eds.), *The Encyclopedia of Climatology*. New York: Van Nostrand Reinhold, pp. 356–362.
Ware, D.M., and R.E. Thompson, 1991. Link between long-term variability in upwelling and fish production in the Northeast Pacific Ocean. *Canadian Journal of Fisheries and Aquatic Science*, **49**: 2296–2306.
Warona, M.A., and Whitlock, C., 1995. Late Quaternary vegetation and climate history near Little Lake, central Coast Range, Oregon. *Geological Society of America Bulletin*, **107**(7): 867–876.

Cross-references

Climate Patterns in the Coastal Zone
Coastal Temperature Trends
Desert Coasts
El Niño–Southern Oscillation
Geographical Coastal Zonality
Ice-Bordered Coasts
Meteorological Effects on Coasts
Sea Breeze Effects
Storm Surge

COASTAL CURRENTS

Introduction

Coastal currents are coherent water masses in motion that are found in the region between the coastline and the edge of the continental shelf. Coastal currents are important because the coastal zone is the place where most nutrients, pollutants, and sediments are introduced into the ocean, and where most larvae are generated and dispersed. Coastal currents are responsible for the transport and dispersal of these biological, chemical, and geological tracers in the water. For example, predictions of the advection and diffusion of spilled oil is dependent upon knowledge of the currents near the coast. Coastal currents often are considered as being made up of two components, alongshore, or parallel to the coast, and cross-shore, or perpendicular to the coast. Away from the influence of inlets or river mouths, cross-shore flows are typically weaker than alongshore flows, yet the larger gradients in cross-shore properties make the cross-shore flows better at redistributing those tracers. In addition, if a current is periodic, such as a tidal current, even though it may be quite strong, it may not result in significant net transport. A non-periodic event-driven current such as from a storm may result in a significant net transport. Some currents that are persistent and result in significant transport of water are named after an adjacent landmass, such as the California Current and the Kuroshio Current. In general, currents are classified by the processes that form them (Wright, 1995). Some of the most important currents that exist near the coast outside of the surf zone include (1) wave-driven currents, (2) tidal currents, (3) wind-driven currents, and (4) buoyant plumes.

Types of coastal currents

Wave-driven currents

Periodic water particle movements are associated with the passage of wind-generated surface gravity waves. The back and forth currents due to waves have a relatively short period, up to about 20 s. The horizontal orbital velocity of water motion is a function of wave amplitude, wave period, wavelength, and water depth (Komar, 1998). According to linear wave theory, the water motion under waves in deep water is circular, with horizontal motions being comparable to vertical motions. The diameter of the circular motion decreases exponentially with water depth, so that in deep water, the influence of the surface wave can no longer be felt. In intermediate water depths the motions are elliptical, becoming smaller and flatter as the bottom is approached, and in shallow water the water motions are back and forth. In deep water, where linear wave theory is best applied, there is no net transport of water due to the passage of a wave. In shallow water, however, linear wave theory breaks down. According to nonlinear Stokes theory, the velocity magnitude is increased and the duration shortened under the crest of the wave, and the velocity magnitude is decreased and the duration lengthened under the trough. The result of this asymmetry in orbital motion is a non-periodic current, or mass transport in the direction of wave propagation, which is called Stokes drift. This mean current can be large for large amplitude and short period waves, but is a small fraction of the instantaneous wave speed, only a few centimeters per second for long period swell.

Tidal currents

Tides and tidal currents in the ocean basins are produced by the interaction of a rotating earth with the gravitational forces of the moon and the sun (see *tides*) (Mofjeld, 1976; Bowden, 1983). These oceanic tides impinge on the continental shelf and produce sea-level oscillations and currents in the coastal zone at tidal frequencies. The dominant fluctuations are either semi-diurnal (twice a day), diurnal (once a day), or a combination of the two. For example, on the East Coast of the United States the tides are semi-diurnal, whereas on the Gulf Coast they are diurnal. The difference is due to the interaction between the forcing and the natural frequencies of the tidal basin. Most coastal environments also have longer variations in the tides and tidal currents with periods of about two weeks. These fortnightly variations result in stronger spring tides and weaker neap tides. Tidal heights (up to a few meters) and tidal currents (up to several tens of centimeters per second) are periodic, thus they result in little net transport, despite excursions of several kilometers per tide. The horizontal component of tidal currents on the continental shelf tends to change direction and amplitude during each tidal cycle. An arrow denoting the current's speed and direction would trace an ellipse with each tidal period. If the direction of the current is restricted by a boundary, as near a coast, then the ellipse will become more elongated, with the long axis aligned parallel to the coast. The speed of a tidal current may increase when the current goes around a headland, passes through a restriction, or passes from deeper to shallower water.

Wind-driven currents

The drag of the wind on the sea surface results in the formation of waves and in the generation of a surface current. The wind-generated current is strongest at the surface and decreases with depth. The magnitude of the surface current depends on the wind speed and on the roughness of the sea surface due to waves, which effects the transfer of energy from the wind to the water. Based on field observations, the magnitude of the surface current is approximately 3% of the wind speed and the direction of the wind-driven current is nearly aligned with the direction of the wind, but not exactly. Due to the rotation of the earth, the Coriolis force causes the wind-generated current to veer to the right of the wind in the Northern Hemisphere and to the left of the wind in the Southern Hemisphere. Various theories predict that the surface current will be directed between 10° and 45° to the right (left) of the wind in the Northern (Southern) Hemisphere, and that the current vector will rotate further to the right (left) with increasing depth below the surface. This rotating current direction and decreasing speed with depth is known as the Ekman spiral. The total transport averaged over the entire Ekman spiral is at right angles to the direction of the wind and is known as Ekman transport. The turning of the upper layer is reduced, however, in shallow water where the water depth is less than the Ekman depth.

Upwelling and downwelling circulation

When the wind blows parallel to the coast and the wind-driven Ekman transport is away from the coast, then the water near the coast will set down causing a lowering of sea level. The sloped sea level will induce an onshore-directed pressure gradient, which when balanced against the Coriolis force will lead to a shore-parallel geostrophic flow in the same direction as the wind beneath the surface Ekman layer. In addition, water near the coast will rise, or upwell, to replace the surface water that is transported offshore. Upwelling circulation is very important because it brings cold, nutrient-rich water from depth to the surface, which can enhance biological productivity. Many of the largest producing fisheries in the world occur in regions of coastal upwelling. Downwelling flows result when the wind is in the opposite direction, causing onshore-directed surface Ekman transport. In the downwelling case, sea level rises against the coast as surface waters are brought from offshore to onshore. In some cases, there can be an interaction between the shore-parallel wind-driven flow and the upwelling or downwelling waters that enhances the coastal flow. As an example, Korso *et al.* (1997) described how cold upwelled water along the coast of Oregon and southward-directed wind-driven surface currents interact resulting in an increased coastal jet occurring along the upwelled front.

Buoyant plumes

Currents arise in the coastal ocean from the flow of freshwater out of rivers and estuaries. Currents result from both the difference in the height of the water between the river and the ocean and from the density difference between the fresh water and the salty ocean water. Upon entering the ocean, the positively buoyant plume will spread laterally on top of the ocean water and thin. After leaving the confines of the river or estuary, the plume will also be acted on by the Coriolis force. The degree of turning of the plume due to the earth's rotation depends on the width of the river mouth, the thickness of the plume, and the Coriolis parameter, which varies with latitude (Garvine, 1987). The boundary of the plume where the water density changes rapidly is known as a front. Fronts are important because they mark zones of flow convergence. Plume currents can be affected by existing tidal currents and by the wind. For example, Stumpf *et al.* (1993) observed a buoyant plume move at speeds of up to 11% of the wind speed. The enhanced speed of the plume relative to normal surface drift is due to reduced friction between the buoyant plume and the underlying ocean water.

Coastal currents and boundaries

Regardless of their origin, coastal currents are effected by their proximity to a boundary, either the seafloor or the atmosphere. Boundaries tend to modify ocean currents through the action of friction. The regions, or layers, in which currents are modified are called boundary layers. Boundary layers near the bed and near the surface can be 5–20 m thick, so on the inner and middle continental shelf, the dynamics of the boundary layers can make up a significant percentage of the water column. In the bottom-boundary layer the flow depends not only on the near-bottom current, but also on the presence of surface gravity waves. This *wave–current interaction* (*q.v.*) tends to modify the near-bed flow. Currents in the bottom-boundary layer can also be complicated by the presence of density stratification and bed morphology (see *Shelf Processes*).

In a narrow zone immediately adjacent to the coast, nearshore currents result when wave asymmetry and breaking cause significant mass transport of water in the direction of the shore. Important nearshore currents include undertow, rip currents, and longshore currents (Komar, 1998) (see *Surf Zone Processes*).

At the outer boundary of the coastal zone, open-ocean currents flowing along the continental slope can at times influence coastal currents. For example, ocean currents such as the California Undercurrent off the US west coast can influence shelf currents by impinging onto the shelf, by entraining water from the shelf, or by inducing flow on the shelf through pressure gradients between the two water masses.

Measurement techniques

Ocean currents are studied experimentally, through field measurements, and theoretically, through the use of mathematical models. Currents are measured with a variety of instruments, including mechanical rotor-type, electro-magnetic, and acoustic current meters. The choice of which current meter to use depends on among other things the environment in which the measurements are made. In the surf zone, currents are typically measured by mounting electro-magnetic current meters to fixed poles jetted into the bottom. Divers deploy the current meters and the data are logged in real-time through a cable to shore. Shelf and slope currents are measured using current meters on a mooring with a weight on the bottom to anchor the mooring in place and a float at or near the surface to hold the mooring vertical. Alternatively, current meters are placed on a tripod, a pyramid-shaped frame that sits on the bottom. Current meters deployed on the shelf and in deeper water are self-contained, requiring an underwater power supply and data acquisition system to store the data until the instrument is recovered. In the past decade acoustic current meters that provide a profile of currents through the water column, as opposed to at a single point, have been used extensively to provide detailed information on coastal currents.

Guy Gelfenbaum

Bibliography

Bowden, K.F., 1983. *Physical Oceanography of Coastal Waters*. Chichester: Ellis Horwood Limited.

Garvine, R.W., 1987. Estuarine plumes and fronts in shelf waters: a layer model. *Journal of Physical Oceanography*, **17**: 1877–1896.

Komar, P.D, 1998. *Beach Processes and Sedimentation*, 2nd edn. Upper Saddle River, NJ: Prentice-Hall.

Korso, P.M., Barth, J.A., and Strub, P.T., 1997. The coastal jet: observations of surface currents over the Oregon continental shelf from HF Radar. *Oceanography*, **10**(2): 53–56.

Mofjeld, H.O., 1976. Tidal currents. In Stanley, D.J., and Swift, D.J.P. (eds.), *Marine Sediment Transport and Environmental Management*. New York: Wiley.

Stumpf, R.P., Gelfenbaum, G., and Pennock, J.R., 1993. Wind and Tidal Forcing of a Buoyant Plume, Mobile Bay, Alabama. *Continental Shelf Research*, **13**(11): 1281–1301.

Wright, L.D., 1995. *Morphodynamics of Inner Continental Shelves*. Boca Raton, FL: CRC Press, Inc.

Cross-references

Pressure Gradient Force
Rip Currents
Shelf Processes
Surf Zone Processes
Tides
Vorticity
Wave–Current Interaction
Waves

COASTAL ENGINEERING—See SHORE PROTECTION STRUCTURES; NAVIGATION STRUCTURES

COASTAL HOODOOS

Introduction

Hoodoo is a descriptive and collective term used by Thornbury (1969) to describe such topographic forms as columns, pillars, and toadstool rocks of bizarre shapes. He recognized that differential weathering helped to develop and modify such outstanding topographic forms. In Bryan's study (1925), he also concluded that those he studied in the southwestern United States were chiefly the work of rainwash plus effects of mechanical and chemical weathering. Basically, the term bears no genetic meaning, but was used as a descriptive term to describe those small landform features of odd shapes. Hoodoos occur in various environments and are associated with many different types of rocks. Previous studies of such hoodoos indicated that differential weathering almost always played an important role in sculpturing their forms. Hoodoos often became a scenic attraction rather than a geologic subject to be appreciated by visitors. Near Tuscon, USA, on the way to Mt. Lemmon, from Windy Point to Geology Vista, highly weathered granitic rocks display hoodoos by the roadside. The place became a vista point and the hoodoos became an attraction for visitors to stop and take pictures. Actually, wherever hoodoos were found, whether they occurred in deserts, on coastal platforms, at the edge of highly dissected terrain or at riverbanks, they often became a site for visitors' appreciation. They are almost always recognized as earth's miracles and definitely as earth's heritage sites. In this regard, hoodoos are more of an artistic and aesthetic subject than a geological phenomenon.

Coastal hoodoos are relatively well studied. Trenhaile (1987) and Ollier (1984) described in detail the processes that caused the formation of hoodoos on coastal platforms and other coastal environments. Trenhaile pointed out that salt weathering which includes hydration, thermal expansion, and crystallization of various salts are important to form such features as tafoni and honeycombs.

He also pointed out that honeycombs are best developed in the zone above the level of normal wave action, but that periodically experiences heavy spray and splash during stormy weather. Bartrum and Turner (1928) first drew attention to the lowering, smoothening, and leveling of inter-tidal rock platforms by weathering processes apparently associated with pools of standing water. Water layering weathering includes alternate wetting and drying (slaking), salt crystallization, and many other chemical weathering processes. Collectively, they serve as a delicate process to sculpture hoodoos. Case hardening which often accompanies water layer leveling on platform surface often causes the formation of boxwork or frame-like patterns and weathering pits, which have raised rims. Some pits resemble miniature volcanoes with water-filled craters (Ollier, 1984). He described elevated pits from Cape Patterson, Victoria as follows: "Small pits are filled by the tide or by rainwater and this keeps the immediately surrounding rock wet, while the platform around is reduced in height by wetting and drying. Eventually the pits comes to be like a miniature volcano with the crater filled with water and so kept wet, rising above a miniature plain which occasionally dries out." Other papers which described water layer weathering and salt weathering include Evans (1970), Gill *et al.* (1981), Goudi (1993), Mustoe (1982), Wellman and Wilson (1965). Pedestal rock is another type of hoodoo often seen in the coastal environment. Pedestal rocks are mushroom-like rocks, which are projections of solid bedrock formed by flaking and hydration in a desert environment (Ollier, 1974). Ollier also described pedestals from Cape Patterson, Victoria, and indicated that they were formed by blocks of arkose resting on a coastal mudstone platform. Hills (1971) described similar features formed by a fallen block lying on a platform and others formed by large calcareous concretions that now stand on pedestals above a water-layer lowered platform surface.

Hoodoos in Yehliu, Taiwan

Such hoodoos, including elevated pits, pedestals, tafoni, honeycombs, and others are well developed in the Yehliu Scenic Area in Taiwan. These hoodoos were studied by Hsu (1964), Lu (1978), Wang and Lee (1984, 1994), Hsu (1988), Hong and Huang (2001).

The most abundant hoodoos in Yehliu are pedestals, which are locally called mushroom rocks. One of the pedestals looks like a western queen's head and was called the Queen's Head (Figure C22). Next to the pedestals is a layer with many Ginger Rocks, which are stand-up calcareous concretions originally occurring in a thin bed (Figure C23). It is quite significant for so many concretions, widely distributed, but limited to a bed of less than 1 m. This thin bed is the key bed that can be used for tracing and correlation of sedimentary strata. Candle Rocks are lower in the sequence stratigraphically (Figure C24). They belong to another sandstone bed with randomly distributed concretions. Concretions exposed often led to the formation of elevated pits after uplifting and erosion. Furrows developed around the concretions, which were filled with water frequently. Surrounding the furrows is a zone of hardening which became relatively raised. After further uplifting, the Candle Rock becomes higher and an apron-like fringe area developed gradually. As a whole they are elevated pits as described by Ollier (1984). The size of the Candle Rocks found in Yehliu may reach a height of 1.5–2.0 m, and a diameter of 2 m.

Figure C22 Queen's Head, a pedestal rock.

Figure C23 Ginger rocks, which are concretions stand out after lowering of adjacent ground.

Geology of the pedestal rocks in Yehliu, Taiwan

The Yehliu coastal scenic area is located approximately 15 km to the northwest of Keelung City, Taiwan. The area is a promontory, approximately 1.5 km long and only about 400 m wide. Part of the promontory is so narrow that it is actually penetrated by sea caves or narrow openings along joints.

Strata exposed in Yehliu area are mainly bedded sandstone of Miocene age. These tilted strata strike N40E and dip at an angle of

Figure C24 Candle rocks, which are elevated pits as described by Ollier (1984). At the very center is a concretion.

18°–22° to the SE. The strata can be subdivided into eight units based on their lithological and morphological characteristics. Many sandstone beds are calcareous and rich in concretions. Pedestal rocks are mushroom-shaped, with a resistant head rock supported by a neck rock made of an underlying, less resistant rock. In general, a pedestal rock is 1.5–3 m high and composed of a head, a neck, and a base. The spheroidal head is 1.5–2 m in diameter and honeycombed. The neck, instead, is smooth-surfaced and thinner than the head. The distribution of the pedestals is clearly lithologically and stratigraphically controlled. They form a parallel and linear array regularly arranged on the tilted shore platform. The thin, calcareous sandstone bed in which the pedestals occur is tilted and protected from direct wave attack by another set of sandstone beds, which form small ridges. From top to bottom the upper rows of pedestals are about 2–4 m above sea level, and the lower pedestals are just above the sea level. The same sandstone bed exhibits no signs of pedestal development under the sea surface. A closer look at the pedestals clearly reveals that these pedestals start to develop once they left the sea surface. Gradually they became separated blocks to form pedestals with no neck, then with a thick neck, and then with a thin neck. Finally the neck broke, the head fell and the pedestal's life ended. The plane which connects the top of the pedestals inclines at an angle of 17°–20°, whereas the surface (actually the shore platform) which connects the bottom of the pedestals has an inclination of about 11°–13°. This is another indication that pedestals grew taller and higher as the shore platform was weathered lower. Those parts that rose above the sea earlier were weathered more. The pedestal rocks, which are higher in elevation, are taller and usually have a thinner neck. Those that stand lower are short and have a massive neck. The height of the pedestals are 1.5–3 m. The tallest is about 3.5 m. The diameter of the head may reach 1.6 m. The spaces between the pedestals average to a distance of 2–3 m. The total number of pedestals is about 180.

Owing to weathering of the calcareous sandstone, the surface crust of the head is dark colored compared to other parts of the pedestal and to those parts of the same bed below the sea level. Gauri (1978) gave a good explanation of the weathering process which formed such a dark-colored crust on a lime-rich rock.

Physical properties of the head and the neck of the pedestals are studied both in the field and in the laboratory. As a result of various means of mechanical tests, such as the Schmid Hammer Test, Slaking Durability Test, and Point Load Test, it is known that there is a difference in the physical strength of the different parts of the pedestals. The head of the pedestals is of much greater strength than the neck.

Water absorption by the head is also slightly lower than the neck, which indicates that the neck rock may have more open crevices that allow water to penetrate.

Microscopic study of thin sections of rock specimens from different parts of the pedestal also reveal important information. The head rock is composed of well-sorted quartz grains cemented mainly by calcite. The neck rock is composed of less well-sorted quartz grains but cemented by calcite and iron minerals. Thin-section study of the surface crust of the head rock, which is darker in color and honeycombed, indicates an enrichment of calcite and clay minerals as cemented materials (Wang and Lee, 1994; Hong and Huang, 2001). The result of microscopic study again supported the interpretation by Gauri (1978).

Wang and Lee (1984, 1994) in their study, concluded that differential weathering, including alternate wetting and drying, salt crystallization and other chemical processes accompanied by uplifting and erosion are responsible for the formation of the pedestals. Based on coral reef dating, estimated uplift rate of the area (about 2 mm/year), elevating of the red oil-painted safety markers on the platform (which is about 1–4 cm above the adjacent ground, and the first time the park placed the paint is about 50 years ago), and other evidences such as the height and the spacing between the pedestals, Wang and Lee (1984, 1994) estimated that the age of the pedestals is not more than 4,000 years since their first appearance above the sea level.

Shin Wang

Bibliography

Bartrum, J.A., and Turner, F.J., 1928. Pillow lavas, peridotites and associated rocks from northern most New Zealand. *Transactions of New Zealand Institute*, **59**: 98–138.

Bryan, K., 1925. The Papago Country, Arizona. *USGS Water Supply Paper*, **499**: 90–93.

Evans, I.S., 1970. Salt crystallization and rock weathering. *Revue de Geomorphologie Dynamique*, **19**: 153–177.

Gill, E.D., Segnit, E.R., and McNeill, N.H., 1981. Rate of formation of honeycomb weathering features (small scale tafoni) on the Otway coast, SE Australia. *Proceedings of the Royal Society of Victoria*, **92**: 149–154.

Gauri, K.L., 1978. The preservation of stone. *Scientific American*, **238** (6): 126–136.

Goudi, A.S., 1993. Salt weathering simulation using a single-immersion technique. *Earth Surface Processes and Landforms*, **18**(4): 369–376.

Hills, E.S., 1971. A study of cliffy coastal profiles based on examples in Victoria, Australia. *Zeitschrift fur Geomorphologie*, **15**: 137–180.

Hong, E., and Huang, E., 2001. Formation of the pedestal rocks in the Taliao formation, northern coast of Taiwan. *Western Pacific Earth Sciences*, **1**(1): 99–106.

Hsu, M.L., 1988. A study of the cavernous rocks in Yehliu (in Chinese). Department of Geography, National Taiwan University, *Science Report* 13, pp. 141–155.

Hsu, T.L., 1964. Hoodoo rocks at Yehliu, northern coast of Taiwan (in Chinese). *Bulletin Geological Survey of Taiwan*, **5**: 29–45.

Lu, S.R., 1978. Re-survey of the genesis of mushroom rocks from Yehliu, northern Taiwan (in Chinese). *Bulletin Geographical Society of China* (Taiwan), **6**: 40–45.

Mustoe, G.E., 1982. The origin of honeycomb weathering. *Geological Society of America, Bulletin*, **93**: 108–115.

Ollier, C., 1974. *Weathering and Landforms*. London: Macmillan Education.

Ollier, C., 1984. *Weathering*, 2nd ed. London and New York: Longman.

Thornbury, W.D., 1969. *Principles of Geomorphology*, 2nd ed. New York: John Wiley and Sons, Inc.

Trenhaile, A.S., 1987. *The Geomorphology of Rock Coast*. Oxford: Oxford University Press.

Wang, S., and Lee, G.H., 1984. Origin of the pedestal rocks in the Yehliu area of northern coast of Taiwan (in Chinese). *Special Publication of the Central Geological Survey* No. 3, pp. 257–272.

Wang, S., and Lee, G.H., 1994. Origin of the pedestal rocks in the Yehliu area of the northern coast of Taiwan. *Chinese Environment and Development*, **5**(1): 47–60.

Wellman, H.W., and Wilson, A.T., Wilson, 1965. Salt weathering, a neglected erosive agent in coastal and arid environments. *Nature*, **204**: 1097–1098.

Cross-references

Honeycomb Weathering
Rock Coast Processes
Shore Platforms
Tafoni
Weathering in the Coastal Zone

COASTAL LAKES AND LAGOONS

Description

Coastal lakes and lagoons are bodies of water occurring in topographic depressions (basins) that are separated from the sea by narrow barriers of land. A range of coastal water bodies exists from areas totally encompassed by land to areas primarily exposed to the sea. The enclosed features, which are primarily shielded from the sea, are called coastal ponds and lakes. Bodies of water with outlets to the sea are called coastal bays and lagoons depending on their shapes. Oertel (1985) called these features *barrier* lagoons because of their association with coastal barriers. In general, coastal lakes and lagoons form secondary coastlines behind the main ocean coastline. The interior shores of these secondary coasts are shielded from direct exposure to ocean waves.

Formation

Coastal lakes and lagoons are a natural result of barrier formation. The formation of coastal barriers causes separation of the coastal ocean from an inland body of water. The diversity of water-surface configurations of "coastal barrier water bodies" is caused by variations in basin and barrier formation. The formation of basins and barriers in the coastal zone is influenced by both marine and terrestrial processes. Sea-level rise is a third factor influencing the formation of basins and barriers. The three main processes of barrier formation that produce coastal water bodies are upbuilding of shoals and bars (de Beaumont, 1845), embayment of interior shores by spit progradation (Gilbert, 1885), and inundation of lowland areas by sea-level rise (McGee, 1890).

Drainage basins and interfluves form irregular topography with contrasting elevations. Rising sea level inundates the lowland areas and leaves the interfluve areas emergent. The distal ends of interfluve areas (necks) are called headlands. Headland shores are often seaward of an inundated shore forming an irregular coastline. Gilbert (1885) illustrated that sediment eroded from headlands moves laterally along the shore. The direction and nature of migration determines the type of secondary landforms that may be developed. The lateral transport and deposition of beach materials onto primary landforms may simply result in the development of broader beaches. Aggrading beaches may advance across smaller headlands or small stream valleys. These features are called barrier beaches and when they block stream drainage to the sea, the resulting water bodies are coastal ponds or coastal lakes.

The creation of coastal ponds is totally dependent on the relationship between the aggradation rate by littoral processes and the flushing rate by stream discharge. Along a given reach of shore the littoral drift of material may vary considerably. Aggradation is related to sediment supply and wave-driven longshore current discharge. Low littoral sand-transport loads may be flushed easily from small channels. However, high-load currents may fill even relatively large channels. The flushing ability of a channel is related to the volume of water that is moved through the channel. At tidal inlets, the discharge through a channel is controlled by basin size and tidal prism. Thus, inlet flushing and inlet filling are governed by characteristics of both the terrestrial and marine environments.

Coastal lakes

There are two principal groups of coastal lakes: coastal barrier lakes and "cats-eye" ponds. *Coastal barrier lakes* occur in small drainage basins that have been dammed-up by littoral drift. The supply of sediment for littoral drift on the Atlantic Coastal Plain of the United States is variable, but the main sources are from reaches of shore between major rivers and estuaries (not the rivers and estuaries). Although these reaches of shore may not extend out into the ocean, they are often called headlands because of their distal position at interfluves between river systems. Barrier lakes are most common where small drainage basins occur adjacent to broad, high headland areas where there is abundant littoral drift to block the drainage to the sea.

The shapes of lakes and orientations of lake shorelines are principally controlled by antecedent topography. Since coastal lakes are water bodies occupying inundated drainage basins, their shoreline orientations and lake shapes are controlled by the hypsometric relationships of v-shaped drainage basins and sea level. Since most basins slope seaward, the deeper and broader parts of the lakes are at the seaward sides of lakes. In plan view, this gives lakes a generally triangular or delta (Δ) shape, with the apex of the triangle pointing landward and the base of the triangle along the seaward shore. During flooding of some basins, only the steep "thalweg" sections of streams are initially flooded, and therefore these barrier lakes are linear in shape.

Since coastal lakes have no access to the coastal ocean, the only source of salt-water intrusion into coastal lakes is through the porous sediment along the seaward shore of the lake. However, lake water levels are generally above sea level and hydraulic-head differences cause groundwater seepage from the lakes toward the coastal sea rather than a flux from the ocean to the lake. Therefore, coastal barrier lakes are generally fresh, although occasional overwash or tidal pumping can produce ephemeral periods of low salt concentrations.

Cats-eye ponds are another type of coastal lake. They are given their name because of their lenticular shape like a cat's eye. Cats-eye ponds are generally smaller than coastal barrier lakes. They occur in the immediate proximity of the shore in beach and dune environments. They are bodies of water trapped in relatively young secondary landforms. The dynamic nature of the shore makes these features relatively ephemeral.

Cats-eye ponds have two general modes of formation. One mode involves beach spits and the other involves dune ridges. Beach spits occur at curves in high-water shorelines that produce small beach headlands. Refraction of littoral sediment transport around beach headlands produces sand bodies (spits) that prograde oblique to the original high-water shoreline. Gaps between spits and the original shore form linear depressions (runnels). If a shore continues to accrete seaward of a runnel, then the swale becomes a semi-permanent linear depression in the backshore. The accumulation of water in these swales produces cats-eye ponds. Since the bases of these systems are very close to sea level, tidal pumping and salt intrusion are common.

A second means of cats-eye pond formation occurs at shores that have prograded seaward leaving a field of secondary dune ridges on their path. This is most common at inlets near the ends of barrier spits and islands. Secondary dune ridges and intradune swales are rapidly colonized by dune plants. As the landscape and plant succession matures, two processes occur that contribute to pond formation. After years of precipitation and evaporation, older landscapes on the backshore develop shallow groundwater layers. Because of irregularities in topography the water table tends to be closer to swale surfaces than the surfaces of the dune ridges. During wet seasons the water table may actually intercept the swale surface creating ponds. However, water-table depths in backshore areas are notoriously variable and swales may dry-out during periods of low precipitation–evaporation relationships. The other main factor effecting pond development in swales is the succession of vegetation on the swale surface. At mature landscapes, the density of plants in a swale is high and the swale surface is enriched with living and dead organic material. The organic material is less permeable than the original sand surface. Rainwater percolates into the groundwater at much slower rates in the organic-enriched zones and some surface water eventually gets perched above the swale surfaces forming ponds. Again, these ponds tend to be ephemeral and although occasional overwash or tidal pumping can produce low salt concentrations, cats-eye ponds tend to be fresh.

Coastal lagoons

Coastal lagoons are generally much larger than coastal lakes. There are two main modes of coastal lagoon formation as described above (Gilbert, 1885; McGee, 1890). Under the inundation model of formation (McGee, 1890), lagoons commonly occur at moderate-sized drainage basins that feed into the coastal ocean (Figure C25). Lagoons form when basins are flooded by the coastal sea and a small conduit is maintained that allows a free exchange of water between the lagoon and the sea. Since the inlet is at the shore, the drift of littoral sands tends to fill the inlet channel. The maintenance of the inlet is dependent on the ability of tidal currents to remove incoming sands at a greater rate than littoral input. However, both of these parameters can be quite variable along a reach of shore and numerous combinations resulting in different rates. In general, the lagoonal tidal prism is the main source of flood and ebb-discharges that flush the inlet channels. The magnitudes of littoral drift are principally governed by sand sources and wave climates. When there is sufficient wave-energy, large headland areas may provide large discharges of sand to the littoral system. In areas of low-littoral drift, small tidal currents may flush a sufficient volume of sediment to allow flushing of the entrance area. Inlets in areas of larger littoral drift require large tidal prisms and fluid discharges to flush sands from inlet channels. At relatively large drainage basins influenced by low-littoral drift rates, inlets are broad and are generally called estuary entrances.

Coastal lagoons in flooded drainage basins inherit the shapes of the basin contours. This generally produces triangular or delta-shaped (Δ) water bodies with coast-parallel seaward margins and v-shaped landward margins. Rehoboth Bay, Indian River Bay and Assawoman Bay

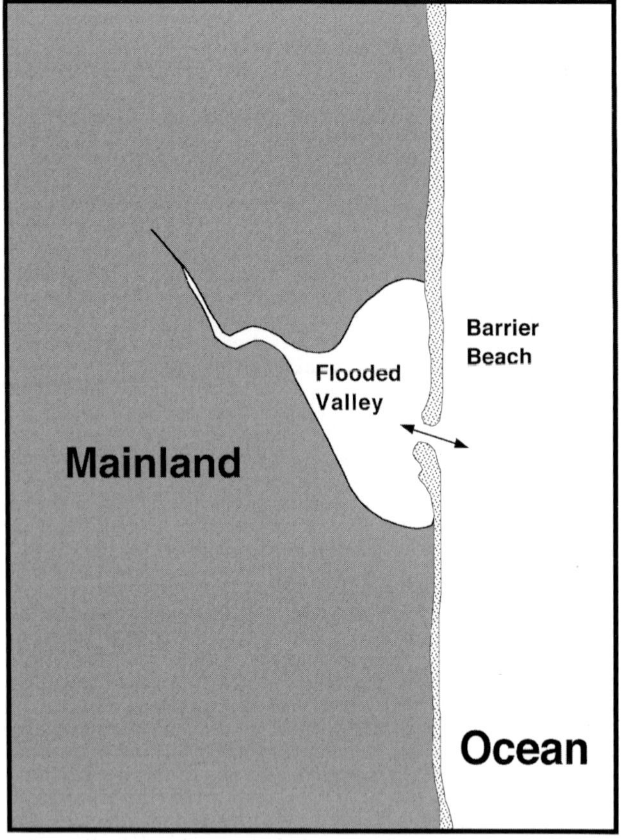

Figure C25 Sketch map illustrating an *inundation model* for coastal barrier lagoon formation. Sea-level rise causes the flooding of a coastal valley, while lateral beach migration creates a partial barrier between the flooded valley and the coastal sea.

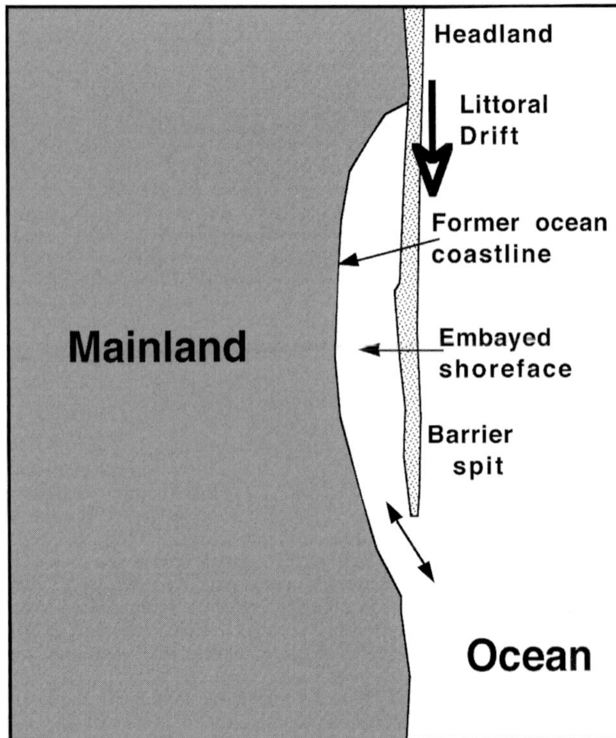

Figure C26 Sketch map illustrating a *spit embayment model* for coastal barrier lagoon formation. Lateral spit migration creates a partial barrier between the coastal sea, and a recessed section of the former ocean coastline. The barrier spit embays the recessed coastline and the former shoreface.

along the Delaware headland are good examples of this *delta bay* type of coastal lagoon. With prolonged sea-level rise, inundation may submerge the interfluve between two adjacent basins causing adjacent delta bays to coalesce. The landward high-water shoreline of a coalesced system takes on the irregular contour shapes of antecedent topography (Figure C25).

The spit-embayment model of lagoon formation (Gilbert, 1885) involves the use of littoral processes to build sand bodies seaward of a previously smoothed and straightened coastline (Figure C26). This type of lagoon formation is probable along wave-dominated coasts with regressive seas or very slow shifts in sea level. In order to build a second offset coastline, a large headland area is generally required. Spit growth from the headland engulfs a portion of the recessed mainland shore and shoreface forming a coastal lagoon. In the spit-migration model, the ocean-spit generally develops parallel to the mainland high-water shoreline and the water-surface area takes on a rectangular shape.

While the mainland and ocean high-water shorelines are generally near parallel, the actual high-water shoreline along the backbarrier of a spit tends to be wavy because of washover processes. Spits are generally narrow, low-profile sand bodies that are subject to overwash during storms. During overwash, sand is dispersed landward across the spit and deposited in fans. Multiple washover fans often coalesce along the backbarrier forming a wavy or scalloped shore.

During the Holocene, transgressing seas have been rising over the much of the world's coasts and mainland shores of lagoons are shifting landward over subaerially sculptured topography. Thus, during the present transgressive stage of Earth history it is improbable that high-water shorelines along the mainland and backbarrier margins in coastal lagoons are parallel. Oertel *et al.* (1992) illustrated the relationships among sea-level rise, antecedent topography, and lagoonal landscapes. They found that during transgression, lagoons evolve from marsh-choked basins to open-water basins. Oertel and Kraft (1994) illustrated the relationships among rising sea level, antecedent topography and several different types of marine barriers along the middle Atlantic Coast of the United States. They used the elements of coastal compartments described by Fisher (1967) as a means of linking different barrier and lagoon types with antecedent topography.

Coastal lagoon circulation

Coastal lagoons are inherently different from estuaries because of their circulation characteristics. Circulation in estuaries is largely governed by the mixing of large volumes of freshwater with seawater that enters through broad inlets. Coastal lagoons generally have relatively small amounts of freshwater and mixing is driven by tidal exchanges through narrow inlets. The percentage of the tidal prism to the total capacity of the lagoon is therefore very important.

The total capacity (volume) of the basin is the product of the wetted-surface area and the average depth of the basin. Coastal Plain lagoons generally have depths ranging from 2 to 3 m, however maximum depths of almost 30 m have been measured in tidal channels of many coastal lagoons. If lagoon floors are relatively smooth, average depths of 2–3 m can be used to determine basin capacity. However, the floors of most coastal lagoons have been shaped by the flooding of antecedent topography during the Holocene transgression. Thus, a reliable average depth can only be gained via a complete hypsographic analysis of the basin. Oertel *et al.* (2000) introduced the term hydro-hypsography to describe the surface area to depth relationships in a flooded basin. An analysis involves determining the area of slabs of water (w_{slab}) at increments of depth from the deepest part of the basin to the surface. From these slabs an accurate volume of the lagoon (Ω_L) can be determined.

$$\Omega_L = \Sigma(w_{slab1} + w_{slab2} + w_{slab3} + \ldots + w_{slabn}).$$

An average hydrologic depth (HD) can be obtained by dividing the lagoon volume (Ω_L) by the lagoon surface area (A_L).

$$\text{HD} = \Omega_L / A_L.$$

In relatively small lagoons with large entrance areas, the tidal phase lag between the inlet and the mainland is small and the tide is essentially a standing wave. The tidal prism can be estimated by obtaining the product of the tidal range and the area. The comparison of tidal prism

(Ω_T) with lagoon capacity is a first-order estimate of hydraulic turnover time (HTT).

$$\mathrm{HTT} = \Omega_L / \Omega_T.$$

In shallow lagoons with large tidal ranges, HTT may be as short as 1–2 tidal cycles. Large lagoons with small tidal ranges may have very large HTTs.

Mixing in coastal lagoons is governed by both tidal exchange and wave stirring. The relatively shallow depths of most coastal lagoons are above the wave bases of locally generated waves. At lagoonal depths of 2–3 m, most of the lagoonal water column can be readily stirred by wave energy.

Multiple-shore transgression

Coastal lakes and lagoons form a coastal boundary zone with three shores. A mainland shore is located along the landward side of a coastal lake or lagoon. A second shore is located along the outer margin of the lagoon on the backbarrier side of an island or spit. A third shore is not in the lagoon but is located along the seaward side of a barrier island or spit and is exposed to ocean waves. Only the mainland and backbarrier shores are in the lagoon; however, ocean shore dynamics have a strong influence on the dynamics of the backbarrier shore.

The high-water shorelines at the mainland and backbarrier shores both respond to Holocene sea-level rise in the same way. As water levels rise, the lagoon inundates the mainland and backbarrier surfaces. Thus, the rate of landward shift of the mainland high-water shoreline ($-\Delta Sl_m$) is coupled with the slope of the antecedent surface. Inundation over gently sloping surfaces is rapid, whereas inundation of steep slopes is slower. At Coastal Plain slopes of 0.05–0.10° and sea-level rise rates of 15 cm per century, high-water shorelines along the mainland can shift landward 86–172 m per century. Since the slope of the mainland surface changes with distance from the coast, rates of recession may increase or decrease as slope changes. Increases and decreases in the rate of landward shift would tend to cause an increase and decrease in lagoon capacity, respectively. The actual capacity of the basin is dependent on the rates of movement of both the inner and outer shorelines of the lagoon.

The high-water shoreline along the backbarrier of the lagoon is more dynamic than the mainland high-water shoreline because the land surface is more dynamic. While inundation of the backbarrier by sea-level rise tends to expand the lagoon capacity, sedimentation along the backbarrier shore tends to shrink lagoon width. The backbarrier margin of a lagoon generally remains stable for long periods of time. However, during storms, sand may wash across a barrier spit or island and be deposited at the backbarrier shore in washover fans. The intermittent overwash events cause the high-water shoreline to prograde into the lagoon ($-\Delta Sl_{bb}$). The relative rates of both the inner and outer high-water shorelines of a lagoon are major factors effecting the lagoon capacity.

$$\Omega_L \propto (-\Delta Sl_m) - (-\Delta Sl_{bb})$$

The migration rate of the mainland shoreline ($-\Delta Sl_m$) is a passive process caused by inundation of the land surface during sea-level rise (Figure C27). However, the landward migration of the high-water shoreline along the backbarrier ($-\Delta Sl_{bb}$) is an active process requiring a wave-energy source to transport sand across a barrier and deposit it as washover fans along a backbarrier shore. The temporal and spatial patterns of backbarrier high-water shoreline migration by overwash are controlled by the wave climate and the elevation of barrier beaches, respectively. In general, the backbarrier high-water shorelines of narrow, low-profile barriers migrate landward at faster rates than at wide, high profile barriers.

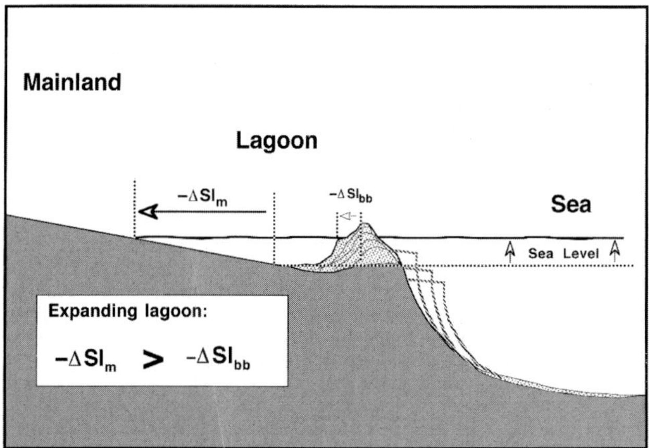

Figure C27 Cross-sectional sketch of a coastal barrier lagoon illustrating a scenario for lagoon expansion during sea-level rise. In this scenario the landward shift of the mainland high-water shoreline ($-\Delta Sl_m$) in response to sea-level rise exceeds the landward shift of the backbarrier high-water shoreline ($-\Delta Sl_{bb}$) in response to prograding washover fans.

Coastal lagoon sedimentation

The main sources of sediment contributing to basin fill come from denudation of the mainland, material flooding through tidal inlets and aggradation of wash-over fans along the backbarrier of islands and spits. Fine-grained sediments are generally associated with the shallower boundaries of the lagoon where marsh vegetation enhances sedimentation of silts and clays. During the Holocene transgression, fringe-marsh facies along lagoonal mainland shores are the basal deposits accreting above the antecedent surface (Oertel *et al.*, 1989). In the northern temperate environments of North America, the main source of fine-grained sediment is from the reworking of older sediments of lagoon floors. In the southern temperate zones, a larger contribution is made from fluvial sources and marsh plants. Coarse-grained sediments occur at flood deltas, the floors of shallow wave-stirred bays, subaqueous washover fans and deep current scoured channels. The flood delta and washover fans are new sources of sediment to coastal lagoons, whereas the sandy floors of wave-stirred bays and deep scour channels are lag deposits.

The relationship between basin capacity and the relative amount of sedimentation in a basin was described by Nichols (1989). When relatively minor amounts of sediment fill only small volumes of a lagoonal basin, then the lagoons are described as *deficit*. When large amounts of sediment fill major amounts of a lagoonal basin, then the lagoons are described as *surplus*. Lucke (1934) presented a model for surplus conditions where tidal inlets were the major source of sediment to coastal lagoons. Relatively coarse-grained littoral sediments are swept into tidal inlets by wave and tidal currents. Most of these sediments are deposited in the inlet areas in the form of flood deltas. Lucke (1934) assumed that coastal lagoons have an initial "open-water" state, and that the littoral sands flooding through inlets would eventually fill the basins forming marsh-choked lagoons. Coarse sediments can also enter lagoons during storm surges when portions of the island are overwashed, dispersing sand to the lagoon floor along the backbarrier of a barrier island. However, the relative amount of sediment accumulation in a basin is related to both sediment input and basin capacity. As noted above, basin capacity changes in response to sea level rise and barrier morphodynamics. Thus, when the rate of lagoon constriction is very rapid, then moderate volumes of sediment input may fill and create surplus conditions.

When the rate of lagoon expansion is very rapid, then even moderate volumes of sediment input may be smaller than the volume increase created by lagoon expansion. These deficit conditions were described by Oertel *et al.* (1992). During the Holocene transgression, lagoons along the mid-Atlantic region of the United States evolved from shallow to deep-water basins. In this region, sea-level rise has maintained deficit lagoons with large open-water bays rather than marsh-choked lagoons. The deficit nature of the lagoons can also be recognized in the stratigraphic record. Transgressive lagoonal sediments of Hog Bay, Virginia, USA have a coarsening upward sequence indicative of lagoonal inundation (Oertel *et al.*, 1989). Basal facies representing the initial inundation of the antecedent soil surface were mottled, poorly sorted sands enriched with fine-grained sediments in flasers. Muddy sediments with sand lenses indicative of fringe marshes occurred above the basal facies. Fringe marsh facies graded into clean muddy laminae with low percentages of sand that are characteristic of low marsh sediments. When the sea level rose above the low marshes, the upper portion of the sequence was reworked by lagoonal waves. The waves winnowed the fines from the area leaving a lag of sandy sediments that characterize open bays in a coastal lagoon.

George F. Oertel

Bibliography

de Beaumont, E., 1845. Lecons de geologie pratique. Paris: Bertrand, P.

Fisher, J.J., 1967. Origin of barrier island chain shoreline, middle Atlantic States. *Geolological Society of American*, **115**(Special Paper): 66–67.

Gilbert, G.K., 1885. Lake Bonneville. *U.S. Geological Survey*, Monograph, 1, 438 pp.

Lucke, J.B., 1934. A theory of lagoon deposits on shoreline of emergence. *Journal of Geology*, **42**: 561–584.

McGee, W.D. 1890. Encroachments of the sea. *Forum*, **9**: 437–449.

Nichols, M.M., 1989. Sediment accumulation rates and relative sea-level rise in lagoons. In Ward, L.G. and Ashley, G.M. (eds.), Physical processes and sedimentology of siliciclastic-dominated lagoonal systems. *Marine Geology*, **63**: 201–220.

Oertel, G.F., 1985. The barrier island system. In Oertel, G.F. and Leatherman, S.P. (eds.), Barrier Islands. *Marine Geology*, **63**: 1–18.

Oertel, G.F., and Kraft, J.C., 1994. New Jersey and Delmarva barrier islands. In Davis, R.A. (ed.), *Geology of Holocene Barrier Island Systems*. pp. 207–232.

Oertel, G.F., Kearney, M.S., Leatherman, S.P., and Woo, H.J., 1989. Anatomy of a barrier island platform, outer barrier lagoon, southern Delmarva Peninsula, Virginia. *Marine Geology*, **88**: 303–318.

Oertel, G.F., Kraft, J.C., Kearney, M.S., and Woo, H.J., 1992. A rational theory for barrier lagoon development. In Fletcher, C.H. and Wehmiller, J.F. (eds.), Quaternary coasts of the United States: marine and lacustrine systems. *Soc. Econ. Paleont. and Min.*, Special Publication **48**: 77–87.

Oertel, G.F., Overman, K., Allen, T.R., Carlson, C.R., and Porter, J.H. 2000. Hypsographic analysis of coastal bay environments using integrated remote sensing techniques, Great Machipongo Bay, Virginia, U.S.A. *Proceeding Volume, Sixth International Conference on Remote Sensing for Marine and Coastal Environments*, Charleston, SC.

Cross-references

Barrier
Beach Processes
Changing Sea Levels
Holocene Epoch
Longshore Sediment Transport
Spits
Tidal Prism

COASTAL MODELING AND SIMULATION

General concepts and terminology

Modeling and simulation can be considered as the complex of activities associated with constructing models of real-world systems, like the coast, and simulating them on the computer. Fundamental to simulation is the process of formulating a computerized model of the coast, and conducting experiments with this model on a digital computer for the purpose of understanding the dynamics at work in the system. This undertaking of modeling and computer simulation of a dynamic system may lead to a new understanding that would not follow directly from the original knowledge about the system (Bossel, 1994).

Using a computer to simulate the dynamics of the coastal system requires that a set of assumptions taking the form of logical or mathematical relationships be developed and formulated into a model. Models can be considered as imitations or approximations of prototypes. Models are not reality, and no model, however complex, can be more than a representation of reality (Bekey, 1977). Depending on the degree of complexity, models can be homomorphic or isomorphic. Homomorphic models are the result of both simplification and imperfect representation of reality. Isomorphic models show a one-to-one correspondence between the elements of the model and the system being represented, and they also have exact relationships or interactions between the elements (Shannon, 1975). While models can be classified in several ways based on their operational parameters, it must be emphasized that computer modeling represents the integrated development of mathematical equations, logical rules, and a computer program embodying the equations, the logical rules and the solutions to them.

Why simulation of the coastal system

Conventional techniques of analytical modeling and analysis are inadequate for understanding the coast, which can be viewed as a complex, dynamic large-scale system or super- system with an integrated arrangement of separate units or component systems, which vary in morphological form, pattern and configuration (Lakhan, 1989). An increasing number of researchers (e.g., Lakhan and Jopling, 1987; Briand and Kamphuis, 1990; Lakhan and LaValle, 1990; Larson *et al.*, 1990; Lakhan, 1991; de Vriend, 1991; Oelerich, 1991; Win-Juinn and Ching-Ton, 1991; Abdel-Aal, 1992; Horn, 1992; Lakhan and LaValle, 1992; Lakhan *et al.*, 1993; Warren, 1993; Black, 1995; Stive and de Vriend, 1995; de Vriend and Ribberink, 1996; Holmes and Samarawickrama, 1997; LaValle and Lakhan, 1997; Sierra *et al.*, 1997; Black *et al.*, 1999) have, therefore, concentrated on the use of computer-based models to gain insights on the complex interactions among interdependent entities (e.g., waves, winds, and currents), which have cumulative impacts on the coastal system. Simulation models are preferred over analytic models because analytic models require many simplifying assumptions to make them mathematically tractable, whereas simulation models have no such restrictions. In brief, a simulation model has several advantages (Adkins and Pooch, 1977). They include the following:

1. It permits controlled experimentation. A simulation experiment can be run a number of times with varying input parameters to test the behavior of the system under a variety of situations and conditions.
2. It permits time compression. Operation of the system over extended periods of time can be simulated in only minutes with ultra fast computers.
3. It permits sensitivity analysis by manipulation of input variables.
4. It does not disturb the real system.

One of the most attractive features of coastal simulation models is based on the fact that their use permits manipulation in terms of real-time of minutes which would take years for the prototype. Of the numerous modeling types (see Lakhan and Trenhaile, 1989; McHaney, 1991) there are two broad types of simulation modeling: continuous simulation and discrete-event simulation. The distinction is based on whether the state can change continuously or at discrete points in time (Miller *et al.*, 1999).

The process of simulating the coastal system

The theoretical and applied literature on modeling and simulation provide the fundamental rules, which can be used to design simulation models of the coastal system. While variations exist in the terminology used and the emphasis placed by different authors (e.g., Law and Kelton, 1991; Pooch and Wall, 1993; Bossel, 1994; Cloud and Rainey, 1998; Zeigler *et al.*, 2000) on each simulation rule and procedure, a realistic and credible simulation model of the coastal system can be formulated by carefully considering each of the following:

(1) Objective Specification, and System Definition;
(2) Model Conceptualization and Formulation;
(3) Data Collection and Model Parameterization;
(4) Model Translation;
(5) Model Verification;
(6) Model Validation; and
(7) Sensitivity Analysis and Results Interpretation.

(1) Objective specification, and system definition

With a dynamic simulation model, it is possible to understand how the coastal system changes over time and also gain an understanding of the dynamics at work in the system. Once a decision has been made to develop a dynamic simulation model, the problem to be investigated and the objectives to be attained must be specified. The components to be included in the model and the abstractions made of them are governed by the objectives (Zeigler, 1984).

After the objective of the simulation model is outlined, the system to be modeled must be defined. In brief, the coastal system is a complex decomposable large-scale open system whose states change in response to both exogenous and endogenous forces. As a real-world system, it is a composition of interacting component systems and associated sub systems. For modeling purposes, caution must, therefore, be taken in demarcating and specifying not only the boundaries of the large-scale coastal system, but also each of the sub systems. Since it is unlikely that the whole coastal system will be modeled the boundary conditions must be portioned into the segment of desired interest. To do this, one must understand the system environment and states. "It must be recognized that any given point in time, a large number of coastal entities interact with a large variety of macro and micro morphological states to produce an apparently random spatial and temporal pattern" (Lakhan, 1989, p. 20). Changes in the state of the coastal system are also affected by a set of poorly understood entities—the components of the system—the processes that are interacting over time, in this case the waves, tides, winds, currents, etc. In defining the coastal system the simulationist must contend with the fact that the coast with its interacting component systems can undergo a sequence of ill-defined transitory states or display cyclic or periodic behavior patterns. It must be realized, therefore, that in order to define the boundaries of the coastal system of interest, it is necessary to identify and evaluate the key system entities and attributes, and their interrelationships over a sufficiently long span of time.

(2) Model conceptualization and formulation

Before the model is developed the simulationist must take cognizance of the fact that like most natural systems the coastal system is governed by a large number of interdependent entities and their attributes, and is complicated by the presence of several feedback mechanisms, some positive and some negative in complex feedback loops (Lakhan, 1989). Hence, it is difficult to identify and separate the independent and dependent entities, which control a portion or the whole coastal system. Here it should be noted that, although complex systems and their environments are objective (i.e., they exist), they are also subjective in that the particular selection of the elements (entities) to be included or excluded, and their configuration are dictated by the researcher (Shannon, 1975). The aim of most simulationists to build a model that has high face validity whereby the model meets specified performance requirements.

For explanation purposes, a simple conceptual model, which can be eventually developed into a highly detailed model to simulate spatio-temporal nearshore behavior, is presented in Figure C28. More entities and attributes have not been included in the model because of the notions of either Occam's razor (see Checkland, 1981) or Bonini's paradox (see Dutton and Starbuck, 1971), which both stress the need for the development of parsimonious models which are simple and straightforward. Essentially, the assumptions and functional relations expressed in the model must be kept as simple and realistic as possible because if the assumptions incorporated in the model are complex, and their mutual interdependencies are obscure, the simulation program will be no easier to understand than the real operating processes (Dutton and Starbuck, 1971). With this knowledge in mind, the model presented in Figure C28 has only the important functional entities and attributes of the coastal system. This conceptual model assumes that the nearshore system is an open system, which is influenced by nearshore waves, with the antecedent nearshore morphology regulated by the energies of wave refraction and reflection. The model also includes the currents and tides that affect the nearshore system in different locations over time. Long-term spatial and temporal changes in nearshore morphology are influenced most by the impacts of longshore currents and sediment supply. The conceptual model emphasizes the entities of waves and currents because they are by far the largest contributors of energy to the nearshore system (US Army Engineer Waterways Experiment Station, Coastal Engineering Research Center, 1984).

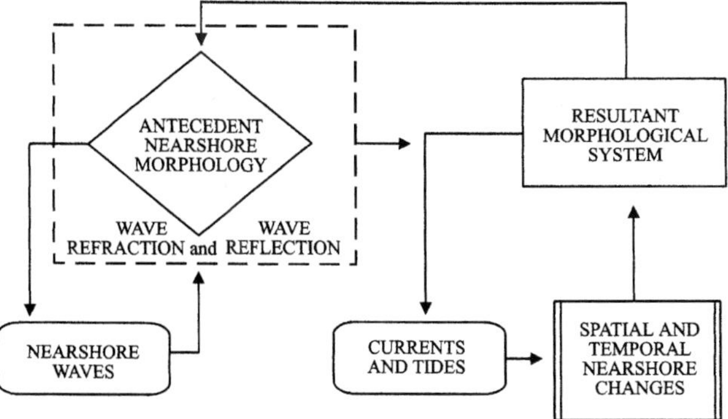

Figure C28 A simple conceptual framework for modeling nearshore changes.

Processes such as tides and local currents also have significant influences on the overall behavior of the nearshore system.

(3) Data collection and model parameterization

One of the principal tasks in the coastal simulation process is to gain adequate information and data on the entities and attributes which govern the coast. The data are necessary for the parameterization of the simulation model. The data must be collected over a continuous and long period of time, so that the time series provide accurate insights on the nature of the entities, which impact on the coastal system. The data must also be homogeneous and free from measurement errors. Since the use of raw empirical data in a simulation model implies that all one is doing is simulating the past (Shannon, 1975), it is imperative to analyze empirical data and identify the probability distributions of those attributes (e.g., wave heights, wave periods, and longshore current velocities) which are to be incorporated in a stochastic simulation model. The model will then use the data indirectly by drawing variates from the selected statistical distributions. The specification of a statistical distribution from which values of the variable (e.g., wave height or wave period) are selected randomly, implies the parameterization of a stochastic model. For a stochastic model, multiple runs must be conducted for each combination of the independent variable values. The inherent randomness of a stochastic model is not required for a deterministic model, which does not use probability distributions. The deterministic model can be parameterized for every combination of independent variable values or for all possible combinations of input variable values.

Selection of the appropriate theoretical distributions for parameterizing a stochastic coastal simulation model can only be done after empirical data are analyzed to see whether they conform to at least one of the known theoretical distributions. To select a theoretical distribution, it is necessary to establish a null hypothesis, and test the deviation of the empirical distribution from the selected theoretical distribution. This is done by testing the distributional assumption and the associated para-meter estimates for goodness-of-fit. The goodness-of-fit that is chosen to test how well the selected distribution fits the observed data depends on several considerations; among them sample size, type of data, and nature of distributional assumptions. Computer programs with the capabilities of running chi-square, Kolmogorov–Smirnov, Cramer–von Mises, and Moments goodness-of-fit tests can be used to check whether a set of empirical data follow or conform to any of the known theoretical distributions. Lakhan (1984) found that with the chi-square goodness-of-fit test, long-term mean wave height and wave period data are Rayleigh distributed; tidal range is normally distributed; longshore current velocity is exponentially distributed; and sediment size is lognormally distributed. Hence, a model to simulate spatial and temporal coastal behavior can be parameterized with these distributions.

If it is found that at a set level of significance, the empirical coastal data cannot be fitted to any of the known continuous or discrete distributions, then it is possible that the data are not independent and uncorrelated. To determine if this is the case the simulationist must use the data to compute the autocorrelation functions. If the analyzed coastal data are found to be autocorrelated, then the model can be parameterized with autocorrelated variates. When the coastal data fit a known theoretical distribution at a specified level of significance, and also are autocorrelated, then the model can benefit from being parameterized with autocorrelated distributed variates. An easily implemented and computationally efficient method for the generation of autocorrelated pseudo random numbers with specific continuous distributions can be found in Lakhan (1981).

The decision to use a particular parameterization procedure depends principally on the length of the simulation run and the nature and characteristics of the entities and their corresponding attributes, which govern any particular coastal system. If the collected coastal data are poorly correlated, and also cannot be "fitted" to any of the known theoretical distributions, then one option will be to use the empirical form of the distribution function. It should be noted, however, that the use of an empirical distribution function is generally not desirable in a stochastic simulation model of the coastal system. In the absence of coastal data, theoretical principles governing the coastal entities can be used for employing appropriate theoretical distributions in the stochastic coastal simulation model.

(4) Model translation

Once the coastal system model is developed with the appropriate equations, theoretical distributions, data specifications, logical rules and constraints, a computer program must be written to implement the model.

The computer program will embody the necessary logical rules, the equations and their solutions, etc. The model must be programmed for a digital computer using an appropriate computer language. The language can be either a general purpose programming language or dedicated simulators and simulation languages. For various aspects of coastal simulation researchers (e.g., LaValle and Lakhan, 1997; Rodriguez-Molina and Garcia-Martinez, 1998; Yu *et al.*, 1998) have demonstrated the applicability of the FORTRAN general purpose programming language.

There are now numerous simulators, discrete simulation languages, continuous simulation languages, and combined simulation languages (Dessouky and Roberts, 1997). The simulators are application-specific modeling tools while the simulation languages have features that facilitate a broad range of modeling applications. While the selection of a dedicated simulation language is a matter of convenience (Nikoukaran and Paul, 1999) it is, nevertheless, worthwhile to consider the specific features and flexibility of simulation programming languages for coastal simulations. In brief, the use of a special purpose simulation language can shorten the development of the computer program, and also allow for faster coding and verification. In addition, a special purpose simulation language may incorporate several simulation functions, which permit the advancement of simulation time, generation of random variates, and execution of simulation events at discrete and continuous time periods.

(5) Model verification

Model verification is essentially the process of substantiating that the model is transformed from one form into another, as intended, with sufficient accuracy. The accuracy of converting a model representation in flowchart form to an executable computer program is evaluated in model verification (Balci, 1997, p. 3). From a software engineering point of view, a computer program must work according to its specifications. The manner in which the program is designed and written is as important as whether the program works. Good programming practice can involve modular programming, and checking intermediate simulation outputs through tracing and statistical testing per module (Kleijnen, 1995). In software testing the program errors must be detected. Debugging is part of the testing process (Weiner and Sincovec, 1984). Static and dynamic approaches can be used for testing the accuracy of computer programs (Sargent, 1982, 1992). In static testing, the computer program of the formulated model is analyzed to determine if it is correct by using such techniques as correctness proofs, syntactic decomposition, and examination of the structural properties of the program. Dynamic testing, on the other hand, consists of different strategies, which include bottom-up testing, top-down and mixed testing (a combination of top-down and bottom-up testing). In dynamic testing, the techniques which are often applied are internal consistency checks, traces, and investigations of input–output relations.

The verification technique of stress testing is based on the notion that while bugs in the program are both embarrassing and difficult to locate they do not manifest themselves immediately or directly, they may, however, show up under stress. With stress testing the parameters of the model are set to unusually high or low values, and then the model is checked to see if it "blows up" in an understandable manner (Bratley *et al.*, 1987). For reliable programming results from the study of large, complex systems, such as the coastal system, software developers can also adopt a system called the Capability Maturity Model, established by the United States military, and currently overseen by the Software Engineering Institute in Pittsburgh, PA.

(6) Model validation

The major task of validating the computer simulation model must be accomplished, once it is ascertained that the computer simulation program is performing exactly the way it is intended. Without discussing the various validation approaches (see Banks *et al.*, 1986a, b; Balci and Sargent, 1984; Sargent, 1988; Balci, 1997) put forward in the simulation literature, it should be stressed that validating computer simulation models is difficult because simulation models contain both simplifications and abstractions of the real-world system. The philosophy of science teaches that models cannot be validated, because doing so implies that the model represents truth (see Popper, 1968). For practical purposes, it makes sense to require that the model faithfully captures the system behavior only to the extent demanded by the objectives of the simulation study. An experimental frame of interest is first determined, and replicative validity can be affirmed if, for all the experiments possible within the experimental frame, the behavior of the model and system agree within acceptable tolerance (Zeigler *et al.*, 2000).

The coastal simulationist can only develop a valid model by making certain that there is very close correspondence between the model and the coastal system. This is done by building a model that has high face validity. The model will be valid if it accurately predicts the performance of the coastal system. If this is not achieved then it is necessary to examine, and change if necessary, the structural and data assumptions of the model. Another form of validity is structural validity, which requires that the model not only is capable of replicating the data observed from the system, but also mimics in step-by-step, component-by-component fashion the way in which the system does its transitions (see Zeigler *et al.*, 2000).

(7) Sensitivity analysis and results interpretation

Sensitivity analysis is an unavoidable step in model validation. The goal of performing sensitivity analysis of a simulation is to determine the effect of input variation and the effect of input uncertainty on the output data (Fürbringer and Roulet, 1999). In any coastal simulation, sensitivity analysis will involve changing the model's input by a small amount and checking the corresponding effect on the model's output. If the output varies greatly for a small change in an input parameter that input parameter may require reevaluation (see Law and Kelton, 1982). It is best to run the model several times with variation in the parameter values. The desired goal is to determine if the basic pattern of the results is sensitive to changes in the uncertain parameters. In execution of the model with different input parameters, it is necessary to check whether the reference mode is obtained with each sensitivity test. If the simulation results are in agreement with the reference mode then the model can be considered robust.

The robustness of the simulation results can be tested by performing additional simulations with different initialization parameters. This comprehensive sensitivity analysis can provide insights on whether the obtained simulation results conform to reality. To interpret the simulation outputs and make inferences on how the coast or any type of system operates, it is necessary to have both theoretical and practical knowledge of the statistical aspects of simulation experimentation (initial conditions, data translation, replications, runlength, etc.) and data outputs. The same principles that apply to the analysis of empirical data also apply to the analysis of simulated data (LaValle and Lakhan, 1993). Statistical tests will vary depending on whether the model is stochastic or deterministic. Lakhan (1986) and Lakhan and LaValle (1987) found that a stochastic simulation model on nearshore profile development will produce output data which are subject not only to random characteristics, but also to various degrees of correlation. Interpreting these results requires the utilization of a statistical methodology which will highlight the stochastic and other properties of the simulation outputs.

V. Chris Lakhan

Bibliography

Abdel-Aal, F.M., 1992. Shoreline change modeling. In Partridge, P.W. (ed.), *Computer Modelling of Seas and Coastal Regions*. Boston, MA: Computational Mechanics Publications, pp. 379–396.

Adkins, G., and Pooch, U.W., 1977. Computer simulation: a tutorial. *Computer*, **10**: 12–17.

Balci, O., 1997. Principles of simulation model validation, verification, and testing. *Transactions of the Society for Computer Simulation International*, **14**(1): 3–12.

Balci, O., and Sargent, R.G., 1984. A bibliography on the credibility assessment and validation of simulation and mathematical models. *Simuletter*, **15**: 15–27.

Banks, J., Gerstein, D.M., and Searles, S.P., 1986a. The verification and validation of simulation models: a methodology. *Technical Report*, School of Industrial and Systems Engineering Atlanta, GA: Georgia Tech.

Banks, J., Gerstein, D.M., and Searles, S.P., 1986b. The verification and validation of simulation models: unresolved issues. *Technical Report*, School of Industrial and Systems Engineering Atlanta, GA: Georgia Tech.

Bekey, G.A., 1977. Models and reality: some reflections on the art and science of simulation. *Simulation*, **29**: 161–164.

Black, K.P., 1995. The numerical hydrodynamic model 3DD and support software. *Occasional Report No. 19*, Waikato, NZ: Department of Earth Sciences, University of Waikato, 69 p.

Black, K., Green, M., Healy, T., Bell, R., Oldman, J., and Hume, T., 1999. Lagrangian modeling techniques simulating wave and sediment dynamics determining sand-body equilibria. In Harff, J.,

Lemke, W., and Stattegger, K. (eds.), *Computerized Modeling of Sedimentary Systems*. New York: Springer-Verlag, pp. 3–22.
Bossel, H., 1994. *Modeling and Simulation*. Wellesley, MA: A.K. Peters, Ltd.
Bratley, P., Fox, B.L., and Schrage, L.E., 1987. *A Guide to Simulation*, 3rd edn. New York: Springer-Verlag.
Briand, M.H.G., and Kamphuis, J.W., 1990. A micro-computer based QUASI 3-D sediment transport model. In *Proceedings of the 22nd Coastal Engineering Conference*. American Society of Civil Engineers, pp. 2159–2172.
Checkland, P., 1981. *Systems Thinking, Systems Practice*. New York: John Wiley and Sons.
Cloud, D., and Rainey, L. (eds.), 1998. *Applied Modeling and Simulation: An Integrated Approach to Development and Operation*. New York: McGraw-Hill.
Dessouky, Y., and Roberts, C.A., 1997. A review and classification of combined simulation. *Computers and Industrial Engineering*, **32**(2): 251–264.
de Vriend, H.J., 1991. Modelling in marine morphodynamics. In Arcilla, A.S., Pastor, M., Zienkiewicz, O.C., and Schrefler, B.A. (eds.), *Computer Modelling in Ocean Engineering 91*, Rotterdam, The Netherlands: A.A. Balkema, pp. 247–260.
de Vriend, H.J., and Ribberink, J.S., 1996. Mathematical modelling of meso-tidal barrier island coasts. Part II: Process-based simulation models. In Lui, P.L.-F. (ed.), *Advances in Coastal and Ocean Engineering*, Vol. 2. Singapore: World Scientific Publishing Co., pp. 151–197.
Dutton, J.M., and Starbuck, W.H., 1971. *Computer Simulation of Human Behavior*. New York: Academic Press, Inc.
Fürbringer, J.-M., and Roulet, C.-A., 1999. Confidence of simulation results: put a sensitivity analysis module in your model. The IEA-ECBCS annex 23 experience of model evaluation. *Energy and Buildings*, **30**: 61–71.
Holmes, P., and Samarawickrama, S.P., 1997. Tide and wave-induced changes at a coastal lagoon entrance. In Arcinas, J.R., and Brebbia, C.A. (eds.), *Computer Modelling of Seas and Coastal Regions III. Coastal 97*. Boston, MA: Computational Mechanics Publications, pp. 75–84.
Horn, D.P., 1992. A numerical model for shore-normal sediment size variation on a macrotidal beach. *Earth Surface Processes and Landforms*, **17**: 755–773.
Kleijnen, J.P.C., 1995. Verification and validation of simulation models. *European Journal of Operational Research*, **82**(1): 145–162.
Lakhan, V.C., 1981. Generating autocorrelated pseudo-random numbers with specific distributions. *Journal of Statistical Computation and Simulation*, **12**: 303–309.
Lakhan, V.C., 1984. A FORTRAN '77 Goodness-of-Fit Program: testing the goodness-of-fit of probability distribution functions to frequency distributions. *Technical Publication No. 2*, Toronto, ON: International Computing Labs, Inc.
Lakhan, V.C., 1986. Modelling and simulating the morphological variability of the coastal system. Presented at the *International Congress on Applied Systems Research and Cybernetics* on August 18, 1986 in Baden-Baden, West Germany.
Lakhan, V.C., 1989. Modeling and simulation of the coastal system. In Lakhan, V.C., and Trenhaile, A.S. (eds.), *Applications in Coastal Modeling*. Amsterdam, The Netherlands: Elsevier Science Publishers, pp. 17–42.
Lakhan, V.C., 1991. Simulating the interactions of changing nearshore water levels, morphology and vegetation growth on Guyana's coastal environment. In McLeod, J. (ed.), *Toward Understanding Our Environment*. Simulation Councils, Inc, USA, pp. 13–20.
Lakhan, V.C., and Jopling, A., 1987. Simulating the effects of random waves on concave-shaped nearshore profiles. *Geografiska Annaler*, **69A**: 251–269.
Lakhan, V.C., and LaValle, P.D., 1987. Development and testing of a simulation model for nearshore profile changes. Presented at the *Canadian Association of Geographers, Marine Studies Group*, May 28, 1987, Hamilton, Ontario.
Lakhan, V.C., and LaValle, P.D., 1990. Development and testing of a simulation model for nearshore profile changes. In Ricketts, P.J. (ed.), *Studies in Marine and Coastal Geography*, Halifax, NS: St. Mary's University, pp. 61–73.
Lakhan, V.C., and LaValle, P.D., 1992. Simulating the onshore–offshore movement of sediment in the coastal environment. *Environmental Software. An International Journal*, **7**: 165–173.
Lakhan, V.C., and Trenhaile, A.S., 1989. Models and the coastal system. In Lakhan, V.C., and Trenhaile, A.S. (eds), *Applications in Coastal Modeling*. Amsterdam, The Netherlands: Elsevier Science Publishers, pp. 1–16.
Lakhan, V.C., Trenhaile, A.S., and LaValle, P.D., 1993. Modelling and simulating morphological changes in the nearshore system. *Third International Geomorphological Conference. Program with Abstracts*. The International Association of Geomorphologists, p. 176.
Larson, M., Kraus, N.C., and Hanson, H., 1990. Decoupled numerical model of three-dimensional beach changes. *Proceedings of the 22nd Coastal Engineering Conference*, American Society of Civil Engineers, pp. 2173–2185.
LaValle, P.D., and Lakhan, V.C., 1993. Comparison between simulated and empirical results on nearshore morphological changes. *Third International Geomorphological Conference. Program with Abstracts*. The International Association of Geomorphologists, p. 175.
LaValle, P.D., and Lakhan, V.C., 1997. Utilizing microcomputer-based models to simulate changes in the nearshore environment. *Environment Modelling & Software*, **12**(1): 19–26.
Law, A.M., and Kelton, W.D., 1982. *Simulation Modeling and Analysis*. New York: McGraw-Hill Co.
Law, A.M., and Kelton, W.D., 1991. *Simulation Modeling and Analysis*, 2nd edn., Singapore: McGraw-Hill.
McHaney, R., 1991. *Computer Simulation*. New York: Academic Press, Inc.
Miller, J.A., Sheth, A.P., and Kochut, K.J., 1999. Perspectives in modeling: Simulation, database, and workflow. In Chen, P.P., Akoka, J., Kangassalo, H., and Thalheim, B. (eds.), *Conceptual Modeling. Current Issues and Future Directions*. Berlin and New York: Springer-Verlag, pp. 154–167.
Nikoukaran, J., and Paul, R.J., 1999. Software selection for simulation in manufacturing: A review. *Simulation Practice and Theory*, **7**: 1–14.
Oelerich, J., 1991. Modelling of time dependent longshore sediment transport. In Arcilla, A.S., Pastor, M., Zienkiewicz, O.C., and Schrefler, B.A. (eds.), *Computer Modelling in Ocean Engineering 91*. Rotterdam, The Netherlands: A.A. Balkema, pp. 271–282.
Pooch, U.W., and Wall, J.A., 1993. *Discrete Event Simulation*. Boca Raton, LA: CRC Press.
Popper, K.R., 1968. *The Logic of Scientific Discovery*. New York: Harper Torchbooks.
Rodriguez-Molina, J.J., and Garcia-Martinez, R., 1998. HydroTrack: A graphical software system for the simulation of pollutant discharges in water. *Environmental Modelling and Software*, **13**(2): 211–223.
Sargent, R.G., 1982. Verification and validation of simulation models. In Cellier, F.E. (ed.), *Progress in Modelling and Simulation*. London: Academic Press Inc., pp. 159–172.
Sargent, R.G., 1988. A tutorial on validation and verification of simulation models. In *Proceedings of the 1988 Winter Simulation Conference*. CA: San Diego, pp. 33–39.
Sargent, R.G., 1992. Validation and verification of simulation models. In *Proceedings of the 1992 Winter Simulation Conference*, Piscataway, NJ: IEEE, December, pp. 104–114.
Shannon, R.E., 1975. *Systems Simulation: The Art and the Science*. Englewood Cliffs, NJ: Prentice-Hall, Inc.
Sierra, J.P., Azuz, I., Rivero, F., Sánchez-Arcilla, A., and Rodriguez, A., 1997. Morphodynamic modelling in the nearshore area. In Arcinas, J.R., and Brebbia C.A. (eds.), *Computer Modelling of Seas and Coastal Regions III. Coastal 97*. Boston, MA: Computational Mechanics Publications, pp. 433–442.
Stive, M.J.F., and de Vriend, H.J., 1995. Modelling shoreface profile evolution. *Marine Geology*, **126**: 235–248.
US Army Engineer Waterways Experiment Station, Coastal Engineering Research Center, 1984. *Shore Protection Manual*, Vol. 1, Washington, DC: US Government Printing Office.
Warren, R., 1993. Development and application of a generic modelling system for coastal processes. In List, J.H. (ed.), *Large-Scale Coastal Behavior '93*. US Geological Survey Open-File Report 93–381, pp. 22–223.
Weiner, R., and Sincovec, R., 1984. *Software Engineering with Modula-2 and Ada*. New York: John Wiley and Sons.
Win-Juinn, C., and Ching-Ton, K., 1991. Numerical model of beach profile changes. In Arcilla, A.S., Pastor, M., Zienkiewicz, O.C., and Schrefler B.A. (eds.), *Computer Modelling in Ocean Engineering 91*. Rotterdam, The Netherlands: A.A. Balkema, pp. 219–225.
Yu, C.S., Berlamont, J., Embrechts, H., and Roose, D., 1998. Modelling coastal sediment transport on a parallel computer. *Physics and Chemistry of the Earth*, **23**(6): 497–504.
Zeigler, B.P., 1984. *Mulifacetted Modelling and Discrete Event Simulation*. London: Academic Press Inc.
Zeigler, B.P., Praehofer, H., and Kim, T.G., 2000. *Theory of Modeling and Simulation*. 2nd edn. New York: Academic Press.

Cross-references

Numerical Modeling
Surf Modeling
Time Series Modeling
Wave Climate Modeling
Wave Refraction Diagrams

COASTAL PROCESSES—See BEACH PROCESSES

COASTAL SEDIMENTARY FACIES

Introduction

Coastal sedimentary environments include a variety of physical and biological processes, which act on coastal sediment to produce associations of composition, texture, primary sedimentary structures, and fossils. These associations constitute sedimentary facies, lithologically distinct, genetically related components of a sedimentary system, or environmental facies, whereby a set of specific environmental processes imparts a distinctive character to a sediment. Coastal sedimentary facies provide the signature of specific types of coastal deposits and facilitate their identification in the geologic record.

This entry describes some of the more common clastic sedimentary facies associated with open coasts and coastal embayments. For a review of carbonate coastal facies, the reader is referred to Demicco and Hardie (1995). The discussion here also does not address facies of high-latitude coasts or low-latitude coasts dominated by mangrove swamps (see Hill *et al.*, 1995; and Cobb and Cecil, 1993, respectively). Many of the facies described here form in both marine and lacustrine settings, although tidally influenced facies are restricted to the marine environment

For more detailed information on coastal sedimentary facies, the reader is referred to the excellent summaries provided by Reineck and Singh (1973), Davis (1985), Walker and James (1992), Galloway and Hobday (1996), and Reading (1996).

Coastal sedimentary processes

A complex array of processes influence coastal sedimentary facies (Figure C29). Of these the most important are waves, tides, and biogenic processes. Sediment input is also critical to the facies character (grain size) and to the nature of the preserved deposit (sedimentation rate).

Waves may exist as "seas," driven by local winds, or as "swell," generated by distant storms. Swell tends to have longer period and to influence the seabed to greater depths than do local sea waves. "High-energy" coasts are likely to be dominated by swell. "Low-energy" coasts receive smaller everyday waves, but can experience very large waves during storms.

Waves move sediment by two mechanisms. The passage of waves induces a back and forth orbital motion at the bottom. In water that is deep relative to the size of the wave, this motion is symmetrical in velocity and duration of flow. In shallower water, the motion becomes asymmetric, with short relatively strong landward flow under the crest of a wave and a more prolonged, weaker, seaward flow under the trough (for the conditions under which this asymmetric flow occurs, see the "Stokes wave" field in Komar, 1976, Figure 3–17). The orbital velocity asymmetry is important for promoting the onshore transport of coarser material and bed load in general.

Waves also, as they encounter the shore, generate sustained flow in the form of shore-parallel longshore currents and seaward-directed rip currents. These currents, which form nearshore circulation cells, not only transport sediment, but also shape the nearshore morphology into bars and troughs.

Tides influence sedimentary facies in two ways. The rhythmic rise and fall of the sea changes water depth and locally exposes extensive intertidal flats. It also generates tidal currents that flow alternately landward (flood tides) and seaward (ebb tides). In morphologically complex bays, either the flood or ebb tide is likely to predominate at any one location.

The response of sediment to these fluid motions depends in large part on its textural nature, which is determined partly by local dynamic processes and partly by what is available to the system. Grain size is also a major influence on the rate and style of bioturbation, which combined with the frequency and intensity of physical processes determine the degree of preservation of physical structures. Bays typically contain abundant silt and clay, which are borne in suspension or move in small aggregates (particularly as fecal pellets) along the bottom. Between the tidal ebb and flood, at high and low water, the flow diminishes, allowing the settling of suspended sediment. This sediment may immediately accumulate on the bottom or it may be concentrated into a dense, highly concentrated fluid that hugs the bottom before deposition.

Biogenic processes are import in the development of coastal facies. Biogenic effects include bioturbation, the generation of distinctive traces, contribution of shell detritus, the fixing of suspended fine sediment by filter feeding and vegetative baffling, and the binding and bioerosion of the substrate. These processes depend on many factors, but faunal activities are particularly sensitive to variations in salinity. Faunal traces in brackish environments tend to be simpler, less diverse,

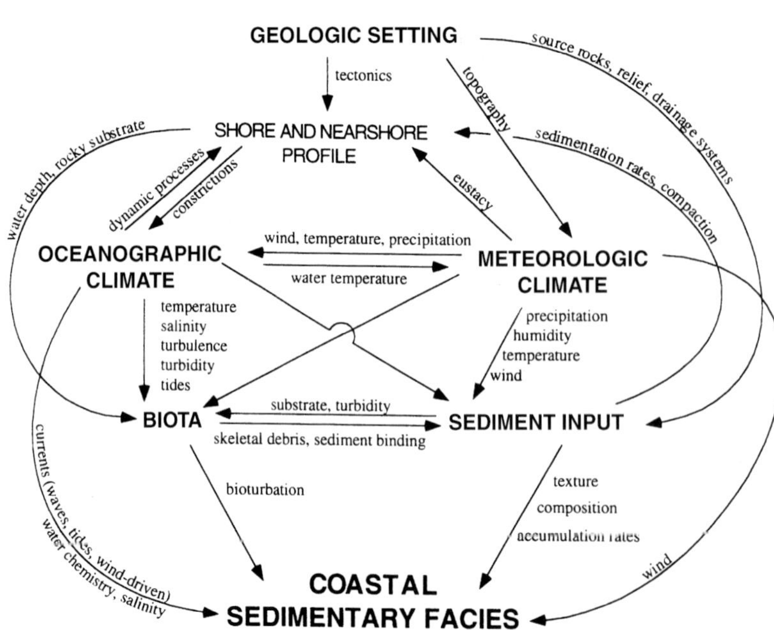

Figure C29 Interacting environmental factors and resulting processes that control the development of coastal facies (modified from Clifton and Hunter, 1982).

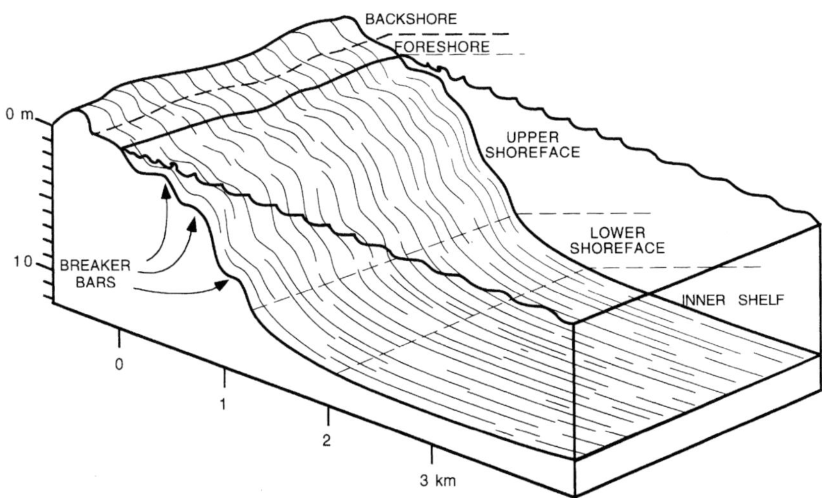

Figure C30 Typical morphologic zonation of a wave-dominated open coast. Considerable vertical exaggeration.

and more likely to be dominated by a single ichnogenus (Pemberton and Wightman, 1992) than those made in oceanic salinities.

Sedimentary facies of wave-dominated open coasts

Waves are most important on open coasts, where the tides mostly only raise and lower the water level. A fundamental aspect of most open clastic coasts is the *shoreface* (Figure C30), a relatively steeply inclined (1 : 200) ramp that extends offshore from the intertidal *foreshore* zone and merges with a nearly flat (1 : 2,000) inner shelf or basin floor. Its origin is discussed by Niederoda and Swift (1981). The shoreface-shelf transition lies at depths that range from 2 to 30 m. On prograding coasts, where the sedimentation rate is commonly sufficient to allow for a wave-induced equilibrium profile, the base of the shoreface rarely exceeds a water depth of ten or so meters.

The *upper shoreface* is equivalent to the nearshore zone, where wave-generated currents predominate. Upper shoreface facies are generally the coarsest part of the coastal system and are characterized by high-angle trough cross-bedding, that is directed parallel to the shoreline (in response to longshore currents) or offshore (in response to rip currents). Modern medium- to coarse-grained nearshore zones are dominated during fair-weather by medium-scale lunate megaripples (dunes) that face and migrate in an onshore direction. Cross-bedding produced by these structures seems rarely preserved, however, in ancient upper shoreface deposits, which seem dominated by storm effects. If gravel is present, it typically is well sorted into discrete beds and layers, and gravel and shells commonly form erosional lags within the deposit. Small, shore-normal gravel-filled gutter casts exist at the base of some gravel beds. Burrowing is common and is dominated by the *Skolithos* ichnofacies, characterized by mostly vertical cylindrical and U-shaped burrows, the dwelling structures of suspension feeders or passive carnivores (Pemberton *et al.*, 1992). The trace *Macaronichnus* is common in fine sand of high-energy coastal deposits and feeding pits of rays or other vertebrates may occur. The thickness of upper shoreface deposits generally ranges from 1 to 8 m, depending on the intensity of wave energy.

A *lower shoreface* facies is identifiable in most pebbly shoreface deposits. This facies, 1–3 m thick, consists of small pebbles of relatively uniform size in a matrix of well sorted fine to very fine-grained sand. The pebbles occur as scattered clasts, in small clusters, or as thin discontinuous stringers 1–2 pebbles thick. The associated sand may be laminated or structureless, and is indistinguishable from inner shelf sand. The facies results from a combination of offshore transport from the upper shoreface during storms and onshore transport of most of this material by the shoaling waves in the aftermath of a storm. Winnowed by the waves, the coarser material moves landward more slowly and some is trapped in burrows, swales, or other depressions on the seafloor. In the absence of pebbles, the upper shoreface facies may be the only recognizable shoreface component.

On transgressive coasts, *shoreface-attached ridges* extend obliquely into the sea. These ridges, composed of fine to coarse sand, are tens of kilometers long, several kilometers wide, and 5–15 m high. Cores through shoreface-attached ridges show that they consist of an upward-coarsening accumulation of storm-event beds (Rine *et al.*, 1991). Some shoreface-attached ridges show evidence of both storm and tidal influences in their internal structures.

The *beach foreshore* facies lie within an intertidal zone subject to the swash and backwash of the waves. Foreshore sands are typically very well sorted and characterized by planar lamination. Foreshore deposits are generally 1–2 m thick, and if the grain size permits, show an upward-fining. Individual lamina may show inverse grading. Placers of heavy minerals define the upper part of some foreshore deposits. Small vertical burrows and the trace *Macaronichnus* typify the ichnologic signature of the foreshore.

Nonmarine coastal facies of the open coast include eolian dunes, characterized by large-scale landward dipping foresets and backshore facies of wind/tidal flats. The character of the *backshore* facies depends on the specific setting and the amount of vegetative growth. Backshore sand is commonly fine and well sorted; layers of pebbles and coarser sand are introduced during floods, and layers of muddy and/or peaty sediment can develop under wetter conditions. Structures such as climbing adhesion ripples or vertebrate footprints document subaerial exposure. Roots are common, and substantial coals may directly overlie the foreshore facies in some ancient deposits.

Mudflats dominate some open coasts proximal to, or down-current from, the mouths of large mud-bearing streams (see Augustinius, 1989). In settings such as the coast of Surinam, inshore fluid mud maintains its presence by frictionally reducing the energy of incoming waves. Episodically, mud deposition ceases, either during storms or because the mud moves along the coast in discrete waves, and sandy beaches develop. A return to muddy deposition isolates the sandy sediment in shore- parallel ridges or *cheniers*. Cheniers on the coast of Surinam are about 2 m high and contain a combination of steeply and gently landward dipping cross-beds or, where fine-grained, gently landward and seaward-dipping cross-beds. Cheniers adjacent to the Mississippi River contain shell aggregates concentrated by storm waves. Mud in the associated flats may be laminated to structureless; mudflats covered by salt marshes will contain root or rhizome structures.

Sedimentary facies of deltaic coasts

Deltas form where sediment discharged at the mouth of rivers exceeds the capacity of marine processes to redistribute it fully. As sites of substantial sediment accumulation, deltas form a volumetrically important part of the sedimentary record. Deltas fall into three end-member types depending on whether they are dominated by rivers, waves, or tides. River-dominated deltas are lobate and commonly quite muddy; the modern Mississippi River delta is a prime example. Wave-dominated deltas have a classic deltoid shape and are more likely to be sandy; the

Niger River delta is a commonly cited example. Tide-dominated deltas are less common, and tend to be muddy or sandy and highly embayed. The Fly River delta of Papua New Guinea is a primary example.

Distributary channels, which develop on delta plains where a primary fluvial channel diverges into a series of straight or sinuous distributaries, are common to all types of delta. A balance of fluvial discharge and tidal range determines the extent of salt water intrusion into the lower reaches of the distributary and the degree of tidal influence. Distributary channel floors are erosional and typically littered with lags of mudclasts and/or plant detritus. Three-dimensional subaqueous sand dunes in the thalweg produce trough cross-bedding. Higher on the channel banks the sediment becomes finer and is likely to be dominated by ripple bedding. Reversals of cross-bedding direction will occur where tidal range is sufficiently great. The nature of burrowing depends on salinity ranges in the distributary. If salinity is low or highly variable, the burrow types are likely to be greatly restricted.

The deposits of the distributary channels typically display an upward coarsening, and may pass upward into muddy levee deposits where complete. Clinoforms may mark the upper part of a distributary channel fill. Cross-sections of the deposits formed by relatively straight distributary channels show a deep central channel flanked by broader lateral extensions ("wings") near the top.

The banks of the distributary channels are capped by *levees*, ridges composed of fine-grained sediment at either side of a channel. Levee deposition occurs during submergence associated with major floods. Internal structures of levees include ripple lamination and small-scale cross-bedding, roots, and in some cases bioturbation.

Deltaic bays are associated with lobate (typically river-dominated) deltas and represent settings of low energy in which the prevailing sedimentation is from suspension, including much organic matter. Episodic influxes of silt and fine sand during floods produce coarser laminae, which may be well preserved in brackish settings where faunal reworking is diminished. The shallower parts of deltaic bays commonly are somewhat sandier and may display evidence of wave reworking in the form of hummocky cross-stratification. Plant growth along the edge of a bay may inhibit the development of normal beaches.

Crevasse splays and deltas accumulate where floodwaters breach a levee and carry fine sand and silt into an adjacent deltaic lake or bay. The resulting lobate deposits become progressively thinner and finer with distance into the bay. Internal sedimentary structures include ripple- and planar-lamination, climbing ripples, trough cross-bedding, and mud drapes. Graded beds, and, near the breach, erosional surfaces, define the deposits of separate floods. Dewatering or other deformational structures are fairly common in these rapidly deposited sands. The upper part of individual flood deposits may be bioturbated, where salinities permit. Splays may amalgamate into coarsening-up crevasse deltas, which resemble other small deltas.

Distributary mouth bars develop where distributary channels of a river-dominated delta debouch into a basin (Figure C31). They consist of coarsening-up accumulations of fine-grained or silty sand. Deposition occurs mostly during flood stage, but the resulting deposit can be partly reworked by waves or tidal currents. Internal structures include ripple, climbing ripple, and planar-lamination. Flaser bedding may be present, as well concentrations of wood or finely divided organic detritus.

Beyond the actual mouth of a distributary lies the *delta front*, an area dominated by flood deposition and possible reworking by waves. Delta front deposits of river-dominated deltas typically consist of interbedded fine to very fine grained sand and mud. Ripple lamination is a common internal structure of the sands, although storm waves can also generate hummocky cross-bedding in this environment. Some graded sand beds appear to be deposited by sediment gravity flows, and deformational structures are common. Bioturbation may be intense to limited depending on the volume of freshwater discharged into this setting. Delta front deposits grade down into *prodelta* muds that form the deeper-water foundation of a delta. Facies on the fronts of wave-dominated deltas may be indistinguishable from those of an open-coast wave-dominated shore.

Tide-dominated deltas have received less study than river-dominated deltas. Here, tidal influence on river flow may extend up-river well beyond the distributary system. Many of the facies (tidal channels and flats) associated with tide-dominated deltas are similar to those discussed in the section on embayment facies. Tidal *sand ridges*, 10 km or more long, several kilometers wide and as much as 10 m high, are common in the funnel-shaped mouths of tide-dominated deltas. Where studied, these ridges display an upward coarsening from interbedded sand and mud to cross-bedded sand. Mud drapes and flaser bedding are common.

Fan deltas are a special type of delta that exists where an alluvial fan builds into a marine or lacustrine basin. Alluvial fans are constructed by a combination of fluvial processes and episodic debris flows; because they typically lie proximal to highlands, alluvial fans are generally composed of coarse sediment. Where waves are too small to redistribute this sediment, it accumulates on the shoreline in the form of a *Gilbert delta*, with steep, nearly planar, foresets that extend to the floor of the adjacent

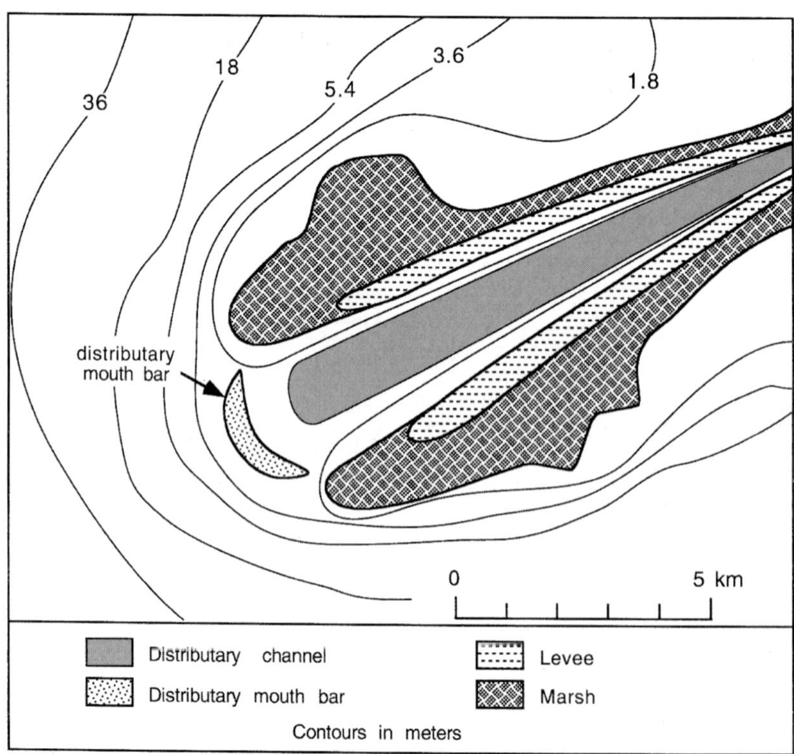

Figure C31 Morphologic elements and facies at the mouth of a river-dominated delta lobe.

basin. Such foresets may define a deltaic unit tens of meters, thick. Sediment slides down the foresets in grain flows, in which the coarser material outruns the finer component. As a consequence, Gilbert deltas are likely to show an overall upward fining from their bases. Individual foreset layers, however, tend to display an upward coarsening normal to the stratification in response to the grain flow process that created them.

Gilbert deltas form in settings other than fan deltas, wherever coarse sediment is introduced into a basin with relatively low wave energy. Braided streams carrying a coarse sediment load commonly create Gilbert deltas where they encounter a shore. Alluvial fans differ from braid plains by containing debris flow deposits composed of unsorted layers of sand, mud, and gravel with little internal structure. Subaerial debris flows on fan deltas flows can extend across the strand line and compose much of the subaqueous part of the delta.

Sedimentary facies of coastal embayments

Coastal embayments are bodies of water somewhat restricted from the open sea. They include river mouths, drowned river valleys, and lagoons or other bodies of saline water isolated from the open sea by a barrier island or spit. Estuaries are semi-restricted bodies of water in which river discharge measurably dilutes the salinity. Lagoons may have no significant freshwater input.

Embayments are subject to a complex interplay of waves, tides, and river discharge. Of these processes, tides typically predominate. Generally, the limited wind fetch in embayments precludes the development of large waves. River discharge, an important factor in many embayments, need not be present. Dalrymple *et al.* (1992) provide a useful model of estuarine facies that recognizes an inner area dominated by fluvial processes, a central area of mixed energy in which both river flow and tidal currents are significant, and an outer area dominated by waves and tides.

Within embayments, specific sedimentary environments produce distinctive facies. In addition to the factors noted above, bay facies depend on the degree of subaerial exposure provided by the tides. Subtidal facies are never exposed, intertidal facies are exposed by the astronomical tides, and supratidal facies are inundated only during coincidence of high astronomical and high meteorological (storm surge or river flood) tides. This section addresses the most common sedimentary facies formed in coast embayments.

Tidal basins as used here are subtidal parts of a bay, which lack discrete tidal channels. As a consequence, tidal currents tend to be feeble, and the sediment largely undisturbed by physical processes. The substrate here is typically fine-grained and dominated by biogenic processes. Commonly the sediment is totally bioturbated. Mollusk shells are common, either as clasts or *in situ*. During periods of minimal sedimentation, articulated and disarticulated shells can form a pavement on the bay floor leading to discrete shell layers in the resulting deposit.

Tidal channels provide a focus for tidal flow within a bay. Their size ranges from tens of meters to kilometers across and meters to tens of meters deep. Their substrate ranges from gravel to mud depending on the nature of river influence and the texture of sediment available to reworking by tidal currents. Typically, tidal channels cut into older sediment or bedrock, and their floors are littered with coarse detritus: shells, pieces of wood, mud clasts, and, today, human debris. Where tidal channels erode into a relatively consolidated muddy substrate, organisms, such as decapods may bore into the older material. The filling of these borings by channel-floor detritus generates a *Glossifungites* ichnofacies (MacEachern *et al.*, 1992). Less cohesive substrates are likely to become colonized by bivalves, producing a thin interval of *in situ* shells below the base of a channel. Where sand is abundant, two- and/or three- dimensional dunes typically occupy a channel floor, producing trough or tabular cross-bedding. Otherwise, channel-floor deposits are likely to be thoroughly bioturbated, owing to slow rates of sediment accumulation.

Because tidal channels tend to migrate, their banks are either erosional or accretionary (Figure C32). Accretionary tidal channel banks form a volumetrically important part of many ancient bay fills. Because accretionary banks are sites of deposition, internal physical structures are likely to be preserved. The nature of accretionary bank stratification depends greatly on sediment texture. Sandy banks are likely to preserve cross- bedding or ripple lamination. With increasing mud, the bedding style changes progressively from wavy (scattered clay drapes) to mixed (roughly equal quantities of sand beds and clay drapes) to flaser (mud encasing sand ripples) to laminated (mud with scattered sand laminae). Rhythmic stratification is common, owing to the regularity of tides or seasonal processes. Tidal channel bank deposits generally (but not invariably) show an upward fining. The migration of tidal channel banks produces large clinoforms that are evident in good exposures.

Tidal sand bars are ridges of sand within tidal channels. Such bars may be entirely subtidal or include both subtidal and intertidal components. Subtidal bars are 1–15 km long, have length to width ratios greater than three, and have a relief of as much as 20 m (Dalrymple and Rhodes, 1995). The larger ones approach the size of tidal sand ridges of large open embayments and seaways (see Houbolt, 1968). Sand waves (dunes) typically mantle the bar surfaces and internal structure consists of trough and tabular cross-bedding. Where bars are less active, their internal structure is likely to consist of sets of cross-strata separated by bioturbated intervals, owing to low rates of aggradation and bedform migration. Larger bars commonly separate areas of flood- and ebb-dominance. Like the tidal sand ridges in the North Sea, these ridges commonly are asymmetric whereby the steeper side faces in the direction of dominant sand transport. As a consequence the dip cross-bedding foresets is unidirectional in the direction of subordinate sand transport. Where tidal sand bars extend into the intertidal range, the vertical sequence likely to result if the bar were fully preserved consists of a lower subtidal and intertidal cross-bedded interval overlain by tide flat and salt marsh deposits (Dalrymple *et al.*, 1990).

Intertidal flats are sandy or muddy areas of low relief that are alternately exposed and inundated by astronomical tides. Sedimentation rates are low and bioturbation predominates unless physical or chemical

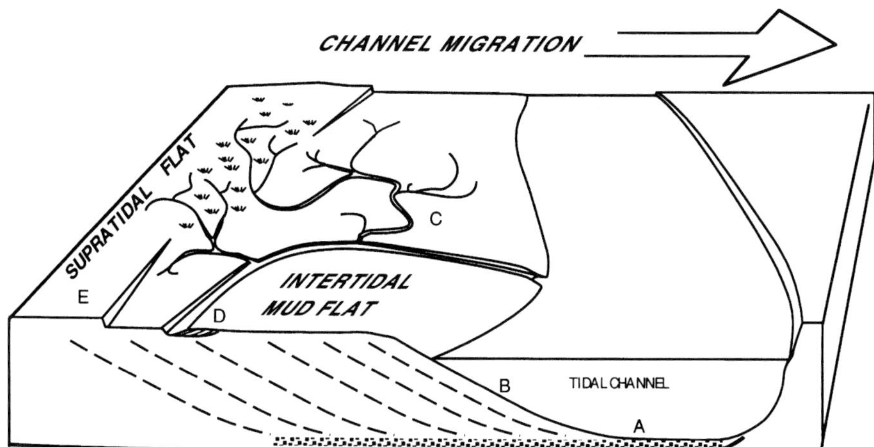

Figure C32 Components and stacking process of a mesotidal muddy tidal channel system.

A. Tidal channel floor composed of bioturbated shell and other lag; Numerous burrows/boring into firm ground beneath the channel floor form a *Glossifungites* ichnofacies; B. Tidal channel accretionary bank, composed of stratified mud and sand; C. Tide flat, underlain by bioturbated mud/or sand; D. Accretionary bank of runoff channel, composed of finely interlaminated sand and mud rich in organic detritus. Low-angle mud across-stratification; E. Supratidal flat, inundated only during combinations of high spring tides and meterologic tides, such as storm surges. Underlain by laminated very fine sand/silt and mud. Numerous root or rhyzome structures.

factors inhibit faunal colonization. An example is on flats exposed to very high tidal range (macrotidal flats). Here, the difference between high tide under neap and spring tidal conditions is sufficient to provide an upper flat that is only inundated during high-spring tides. This intermittent exposure precludes the development of a mature infauna, and stratification is preserved in the form of flaser, mixed, wavy, or lenticular bedding, depending on the proportion of sand and mud. Where mesotidal conditions (tidal range 2–4 m) obtain, the difference between high spring and neap tides is less and bioturbation prevails in nearly all of the flat sediment.

Most muddy tidal flats are crossed by narrow, dendritic *runoff channels* (tidal creeks), typically 1–2 m deep. Most runoff channels have an erosional and an accretionary bank. Sedimentation on the accretionary banks is very rapid and the sediment commonly displays laminae that reflect individual tides as well as neap and spring tidal cycles. Despite their meandering appearance, the rate of channel migration is slow, owing to the erosional resistance of the muddy substrate. Sediment that accumulates rapidly on accretionary banks is prone to slumping into the channel, where it is flushed out by the tidal currents. The limited channel migration produces muddy clinoforms or cross-strata, 1–2 m thick, characterized by finely interlaminated sand, mud and organic detritus, and sporadic slump structures. Runoff channels on sandy tidal flats generally lack well-formed accretionary banks, and the associated distinctive facies.

Salt marshes are common on the upper part of many intertidal flats or marine delta plains (see Frey and Basan, 1985). The substrate may be completely bioturbated or display faint rhythmic lamination. The organic content is typically high, and root or rhizome structures are typical. *Supratidal flats* lie higher still and are inundated only under combinations of high- astronomical and high-meteorological (river flood or storm surge) tides. Stratification, in the form of rhythmic alternations of silt and organic-rich mud, is typically crossed by numerous roots or rhizomes. Commonly the laminae form sets of several distinctive layers, probably deposited at successive high tides during a single meteorological event.

Bayhead deltas are common in estuarine bays. These may take the form of a small river delta, displaying a transition from muddy sediment in the most basinward, deeper parts to sandy sediment in the upper, river-dominated part. Fluvial or distributary channels cap the upper delta surface. The resulting deposit shows an upward coarsening. Evidence of tidal influences, such as tidal bundling or reversals in ripple-lamination or cross-bedding, can serve to distinguish this type of small delta from those produced by crevassing.

In estuaries with a large tidal range, such as San Francisco Bay, the bay head delta comprises a complex of anastomosing channels without a clear deltaic form. In this setting, the bayhead delta consists of tidal channels that gradually transform landward into fluvial channels. The interface between salt and fresh water (head of the salt-water wedge) is typically very turbid and the site of rapid deposition from suspension. As with tide-dominated deltas on an open coast, the tidal influence on river flow may be expressed many kilometers upriver from the head of the salt-water wedge.

Washover fans are lobate accumulations of sand on the seaward side of lagoons, formed where storm waves breach a barrier and sweep sand into the adjacent lagoon (Leatherman *et al.*, 1977). Bases are erosional, commonly marked by a shell lag. Internal structures include inversely graded planar lamination, ripple lamination, and climbing ripples, antidune stratification and, particularly at landward margins, landward-dipping cross-bedding. Tops of individual overwash accumulations can be bioturbated or contain rootlets.

Sedimentary facies of inlet areas

Tidal inlets form a fluid interface between bays and the open coast (see Boothroyd, 1985). Inlets may range from simple channels a few meters deep and tens of meters wide to complexes of channels and shoals more than 10 km or more across wide at the mouths of large embayments. Individual channels in this setting may be tens of meters deep and hundreds of meters wide. Inlet sediment is mostly sand, and occurs in a setting where tidal flow is likely to be at its highest. As a consequence, inlet channels are typically floored by two- and three-dimensional dunes. Individual channels may be dominated by either ebb or flood tide. Lag deposits of shells and or pebbles (where the adjacent shoreface is gravelly) are generally present, and may contain species representative of both open coast and bay. Trough and tabular cross-bedding and abundant burrows characterizes the resulting deposit.

Small inlets tend to migrate laterally along the coast, particularly where wave-driven longshore sediment transport is significant. Larger inlets may be more stationary, but interior inlet channels may migrate

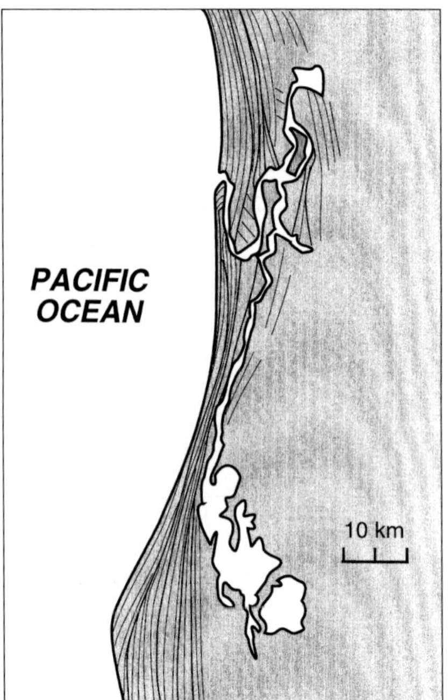

Figure C33 Beach Ridges indicating progradation on the Nayarit coast of Mexico (after Curray *et al.*, 1969). Each ridge marks a previous position of the coastline.

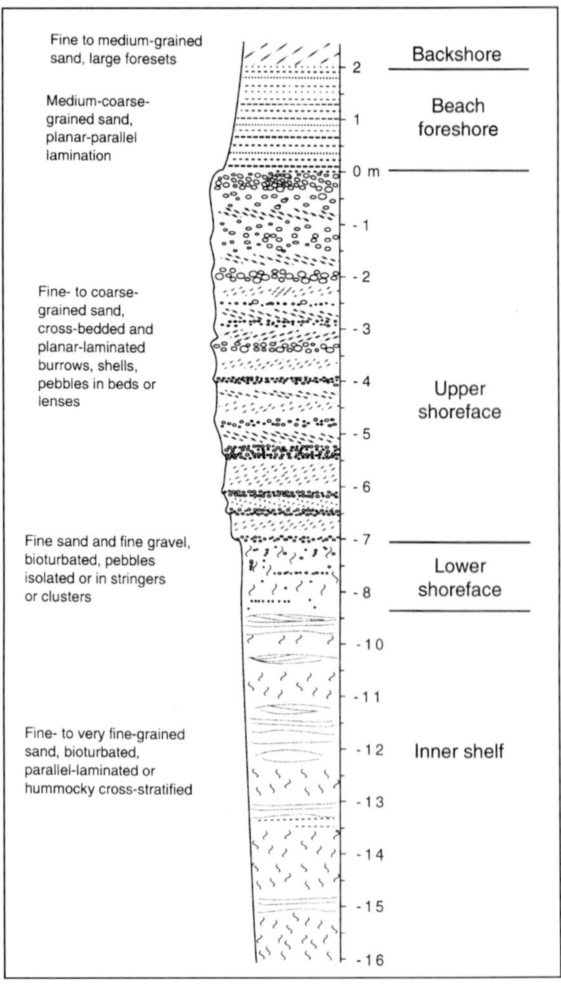

Figure C34 Typical shoreface succession on a prograding pebbly, high-energy coast.

rapidly. In Willapa Bay, Washington, for example, a primary inlet channel, 25 m deep and 600–700 m wide, has moved laterally more than a kilometer in the past 50 years. Inlet facies in such a setting are likely to consist of a lower part composed of cross-bedded sand produced in the primary channel overlain by the deposits of smaller tidal channels and wave- and tide-influenced shoals. Channel migration can produce clinoforms, the thickness of which depends on the depth of the channel. The lateral extent of inlet deposits depends on the degree of inlet migration. Extensive migration will generate laterally extensive inlet facies within a coastal succession.

Lobate bodies of sand, *flood tidal deltas*, form sandy ramps on the bayward side of many inlets. Where the tidal inlet migrates, flood tidal deltas may compose laterally extensive sand bodies along the seaward side of a lagoon enclosed by a barrier island. Initially these deltas form as a set of lobes building into the bay, covered by straight to sinuous two-dimensional landward-facing dunes (Hine, 1975). As they mature, tidal flow is concentrated into channels, ultimately producing a low, seaward-facing ramp dissected by broad flood-dominated tidal channels (Boothroyd, 1985). The resulting deposit is dominated by landward-directed trough and tabular cross-bedding and contains internal erosional surfaces capped by shell lags.

Ebb tidal deltas form on the seaward side of inlets, where they are subject to the influence of both waves and tides. Off large embayments, these deltas may be quite large, exceeding10 km in width and extending 5 km or more into the adjacent sea. Off smaller inlets on a transgressive coast, the deltas may form the nucleus of shoreface attached ridges (McBride and Moslow, 1991). Flood- and ebb-dominated tidal channels cross the shoal constituting the delta: typically the ebb flow is concentrated in the deeper channels. The combination of flood tidal flow and wave swash can produce a generally landward orientation to bedforms on the delta platform (Boothroyd, 1985). Upward-fining storm beds with erosional bases are common, particularly in the shallower parts of an ebb tidal delta.

Preservation of coastal sedimentary facies and resulting facies successions

Thick accumulations of coastal facies typically reflect a record of fluctuating sea level superimposed on a setting of overall tectonic subsidence. The shore response to sea-level change depends on the rate of sedimentation. In general, the influx of sediment at the shore promotes progradation, whereby the additional sediment forces the shore to shift basin-ward. Today, most prograding open coasts are characterized by a series of each ridges, each marking an earlier position of the shore (Figure C33) The resulting accumulation conforms to Walther's Law, which holds that a vertical arrangement of related facies conforms to their lateral distribution at the time of deposition. Progradational successions show an upward-shallowing, produced as shallower water facies encroach over their deeper water counterparts. On deltaic or other open coasts, the result typically generates an upward-coarsening interval. Progradation also occurs on a smaller scale where a bayhead delta builds into an estuary, or a crevasse delta builds into a deltaic bay. Figures C34–C36 show examples of progradational successions.

Fining-up facies successions are generated within bays and on deltas by the lateral migration of tidal or delta distributary channels. In these cases, the shallower parts of the system consist of intertidal flats or salt marshes, or, with distributary channels, levees. Figures C36 and C37 show examples of fining-up successions produced by migrating channels.

Although progradation is the most common circumstance of preservation of coastal sedimentary facies, not all facies are preserved. Those

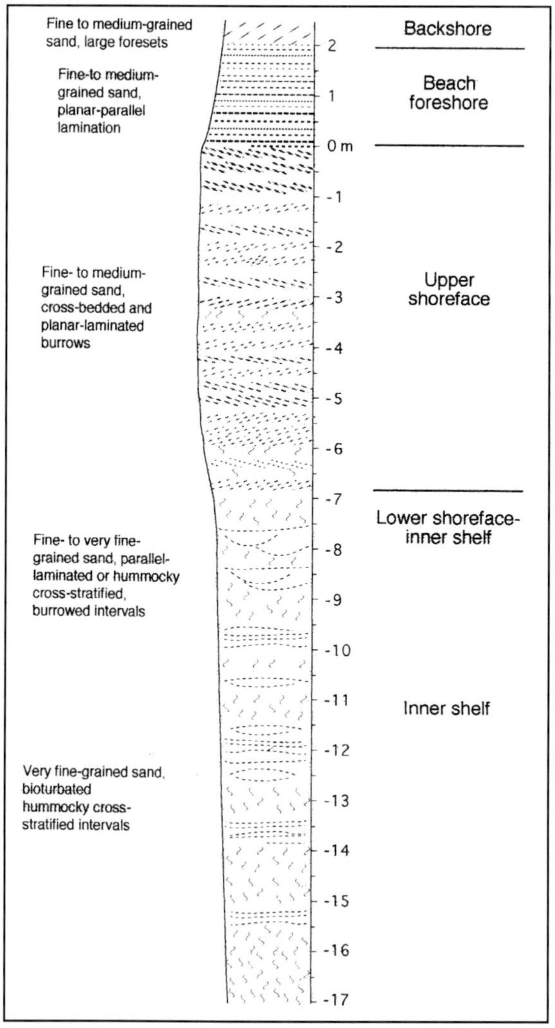

Figure C35 Typical shoreface succession on a prograding finer-grained, non-pebbly, high-energy coast.

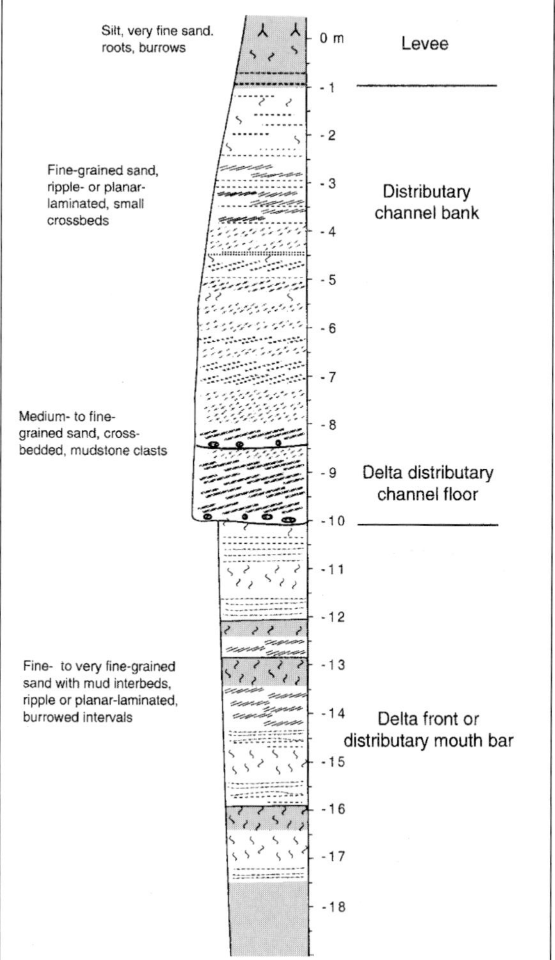

Figure C36 Stratigraphic succession produced in the upper part of a prograding river-dominated delta.

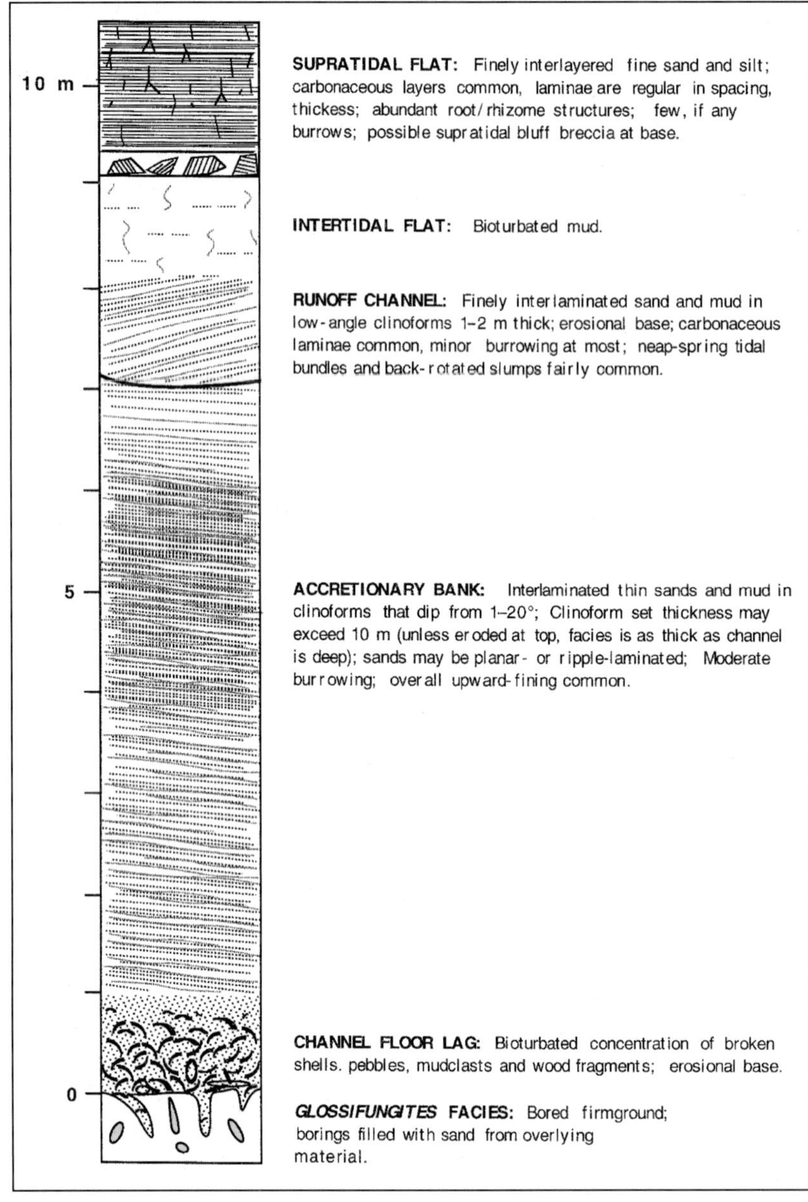

Figure C37 Vertical facies succession produced by migrating muddy tidal channel system. Thickness of sequence depends on depth of tidal channel and degree of subsequent erosion of upper part of succession.

associated with features that stand in relief above other, more landward, features are likely to be destroyed as progradation progresses. Breaker bars, for example, are common features of many upper shorefaces (Figure C30). During progradation, the seaward migration of the bar troughs that lie to landward erases much, if not all, of the bar. Similarly, distributary mouth bars have a limited potential for preservation. During the seaward migration of a distributary mouth system, distributary mouth bars are likely to be eroded and cannibalized by the distributary channel that created them. If that channel avulses, however, the associated bar may be preserved as a feature with limited lateral extent.

Where relative sea level falls during progradation a "forced regression" ensues. Wave erosion associated with the readjustment to a new, lower base level can create a "regressive surface of marine erosion". The amount of sediment removed by this erosion depends on the rate and degree of base level fall and prevailing wave energy. Commonly the surface marks an abrupt change from shelf to shoreface deposits.

Progradation may occur during relative sea-level rise, if the rate of sedimentation is sufficiently high. But commonly, a rise in base level is associated with deposition in the feeder systems. With the resulting reduction of sediment contributed to the coast, marine transgression occurs, whereby the shoreline shifts landward.

During transgression, some facies are likely to be preserved and others lost. During lowstands that commonly precede transgression, coastal rivers incise valleys near the coastline. Estuarine facies deposited in these valleys at the lowstand or during the initial rise tend to be largely preserved through a transgression.

A different situation prevails on the open coast. The natural product of waves and coastal winds is a shoreface and beach capped by a foredune ridge. In most wave-dominated settings, enough sand exists that a small rise in relative sea level is met by an upbuilding of the shore to a topographically higher position. Where a sea-level rise causes a reduction in sediment input to the coast, a basin forms behind the beach ridge and a barrier island is created. Most present-day coasts fronted by barrier islands are in a state of transgression.

Barrier islands formed during transgression have a limited potential for preservation. The landward translation of a shoreface erases the deposits that lie landward from its base, leaving an erosional (*ravinement*) surface overlain by shelf facies (Figure C38). Commonly this surface is the only preserved expression of a transgression.

Under conditions of rapid accommodation and sufficient sediment supply, however, lagoonal facies are preserved (Figure C39). These may occur in inverted succession, whereby the landward-most facies (lagoonal marshes and back-bay muds) are overlain by sandier deposits

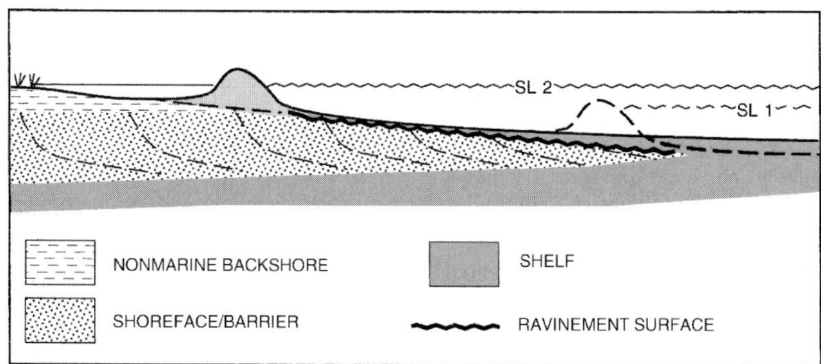

Figure C38 Transgressive erosion caused by the lateral translation of the shoreface (see Brunn, 1962). None of the transgressive coastal facies are preserved, and the transgression is recorded by an erosional ("ravinement") surface overlain by deeper water typically shelf) deposition.

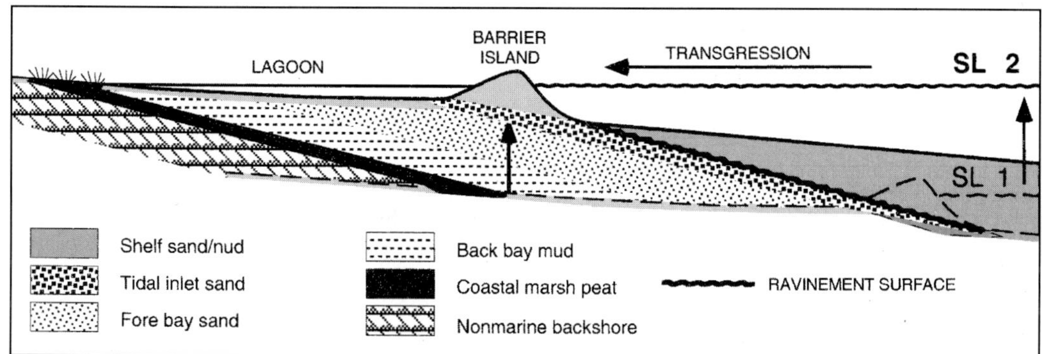

Figure C39 Transgressive succession produced in settings with rapid accommodation and adequate sedimentation. A lagoonal succession develops as the lagoonal facies are buried by the landward-migrating barrier. It is capped by a "ravinement" surface overlain by shelf facies.

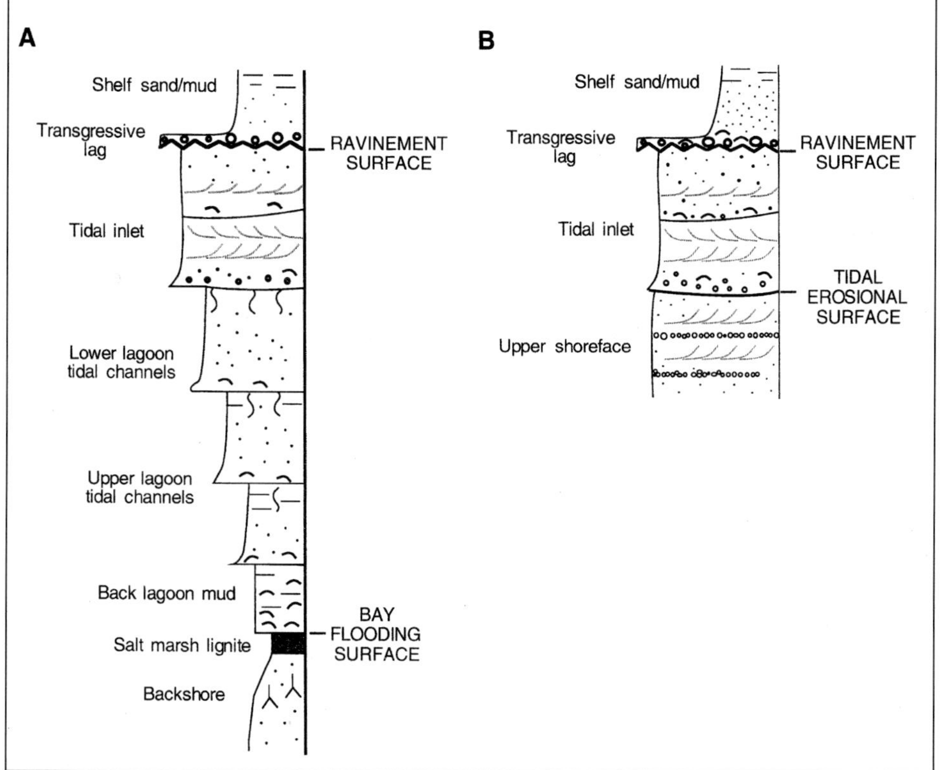

Figure C40 Vertical successions produced by preservation of transgressing lagoons. (A) High rate of accommodation. (B) Modest rate of accommodation.

associated with the approach of the transgressing barrier and capped by tidal inlet facies just below the ravinement surface (Figure C40A). The thickness and organization of these transgressive lagoonal deposits differs depending on the balance between accommodation and sedimentation. Although most are but a few meters thick, some can exceed 20 m. Most display a basal "bay-flooding surface" and a capping ravinement surface, although vigorous tidal currents and low rates of accommodation may in some places lead to the erosion of the lower part of the succession (Figure 40B).

H. Edward Clifton

Bibliography

Augustinius, P.G.E.F., 1989. Cheniers and chenier plains: a general introduction. *Marine Geology*, **90**: 219–229.

Boothroyd, J.C., 1985. Tidal inlets and tidal deltas. In R.A. Davis, Jr. (ed.), *Coastal Sedimentary Environments*, 2nd edn. New York, Springer-Verlag, pp. 445–532.

Bruun, P., 1962. Sea level rise as a cause of shore erosion. *A.S.C.E. Proceedings, Journal Waterways and Harbors Div.*, **88**: 117–130.

Clifton, H.E., and Hunter, R.E., 1982. Coastal sedimentary facies. In M.L. Schwartz (ed.), *The encyclopedia of Beaches and Coastal Environments*. Stroudsburg: Hutchinson Ross Publishing Company, pp. 314–322.

Cobb, J.C., and Cecil, C.B., 1993. *Modern and Ancient Coal-forming Environments*. Denver: Geological Society of America Special Publication 286.

Curray, J.R., Emmel, F.J., and Crampton, P.J.S., 1969. Holocene history of a strand plain, lagoonal coast, Nayarit, Mexico. In A.A. Castanares, and F.B. Phleger (eds.), *Coastal Lagoons, A Sym-posium*. Mexico City, Universidad Nacional Autonoma, pp. 63–100.

Dalrymple, R.W., and Rhodes, R.N., 1995. Estuarine dunes and bars. In G.M.E. Perillo (ed.), *Geomorphology and Sedimentology of Estuaries, Developments in Sedimentology*, Vol. 53, Elsevier, pp. 359–422.

Dalrymple, R.W., Knight, J.R., Zaitlin, B.A., and Middleton, G.V., 1990. Dynamics and facies model of a macrotidal sand-bar complex, Cobequid Bay-Salmon River estuary (Bay of Fundy). *Sedimentology*, **37**: 577–612.

Dalrymple, R.W., Zaitlin, B.A., and Boyd, R., 1992. Estuarine facies models: conceptual basis and stratigraphic implications. *Journal of Sedimentary Petrology*, **62**: 1130–1146.

Davis, R.A., Jr., 1985. Coastal sedimentary environments, 2nd edn. New York: Springer-Verlag.

Demicco, R.V., and Hardie, L.A., 1995. Sedimentary structures and early diagenetic features of shallow marine carbonate deposits Atlas Series No. 1. Tulsa: SEPM (Society for Sedimentary Geology).

Frey, R.W., and Basan, P.B., 1985. Coastal salt marshes. in R.A. Davis, Jr. (ed.), *Coastal sedimentary environments*, 2nd edn. New York: Springer-Verlag, pp. 225–301.

Galloway, W.E., and Hobday, D.K., 1996. Terrigenous clastic depositional systems: applications to fossil fuel and groundwater resources, 2nd edn. New York: Springer-Verlag.

Hill, P.R., Barnes, P.W., Héquette, A., and Ruz, M.-H., 1995. Arctic coastal plain shorelines. In R.W.G. Carter, and C.D. Woodroffe (eds.), *Coastal Evolution: Late Quaternary Shoreline Morphodynamics*, Cambridge: Cambridge University Press.

Hine, A.C., 1975. Bedform distribution and migration patterns on tidal deltas in the Chatham Harbor estuary, Cape Cod, Massachusetts. In L.E. Cronin (ed.), *Estuarine Research*, Vol. II, *Geology and Engineering*. London: Academic Press.

Houbolt, J.J.H.C., 1968. Recent sediments in the Southern Bight of the North Sea. *Geologie Mijnbouw*, **47**: 245–273.

Komar, P.D., 1976. *Beach Processes and Sedimentation*. Englewood Cliffs: Prentice-Hall Inc.

Leatherman, S.P., Williams, A.T., and Fisher, J.S., 1977. Overwash sedimentation associated with a large-scale northeaster. *Marine Geology*, **24**: 109–121.

MacEachern, J.A., Raychandhuri, R., and S.G. Pemberton, 1992. Stratigraphic applications of the *Glossifungites* ichnofacies: delineating discontinuities in the rock record. In S.G. Pemberton (ed.), *Applications of Ichnology to Petroleum Exploration*. SEPM Core Workshop No. 17, pp. 169–198.

McBride, R.A., and Moslow, T.F., 1991. Origin, evolution, and distribution of shore face sand ridges, Atlantic inner shelf, U.S.A. *Marine Geology*, **97**: 57–85.

Niederoda, A.W., and Swift, D.J.P., 1981. Maintenance of the shoreface by wave orbital currents and mean flow. *Geophysical Research Letters*, **8**: 337–348.

Pemberton, S.G., and Wightman, D.M., 1992. Ichnologic characteristics of brackish water deposits. In S.G. Pemberton (ed.), *Applications of Ichnology to Petroleum Exploration*, SEPM Core Workshop No. 17, pp. 141–167.

Pemberton, S.G., Van Waggoner, J.C., and Wach, G.D., 1992. Ichnofacies of a wave-dominated shoreline. In S.G. Pemberton (ed.), *Applications of Ichnology to Petroleum Exploration*, SEPMx Core Workshop No. 17, pp. 339–382.

Reading, H.G., 1996. Sedimentary environments: processes, facies, and stratigraphy (third edition). Oxford: Blackwell Scientific Ltd.

Reineck, H.-E., and Singh, I.B, 1973. Depositional sedimentary environments—with reference to terrigenous clastics. Berlin: Springer-Verlag.

Rine, J.M., Tillman, R.W., Stubblefeld, W.L., and Swift, D.J.P., 1991. Lithostratigraphy of Holocene sand ridges from the nearshore and middle continental shelf of New Jersey, USA. In T.F. Moslow, and E.G. Rhodes (eds.), *Modern and Ancient Shelf Clastics: a core Workshop*. SEPM Core Workshop No. 9, pp. 1–72.

Walker, R.G., and James, N.P., 1992. *Facies Models Response to Sea Level Change*. St. John's: Geological Association of Canada.

Cross-references

Barrier Islands
Bars
Beach Sediment Characteristics
Beach Stratigraphy
Cheniers
Deltas
Dynamic Equilibrium of Beaches
Estuary
Offshore Sand Banks and Linear Sand Ridges
Offshore Sand Sheets
Surf Zone Processes
Tidal Flats
Tidal Flats, Open Ocean Coasts
Tides
Waves

COASTAL SOILS

Introduction

The occurrence of soil in coastal zones is not a simple matter to resolve. Summarized coastal zones form a global perspective that only briefly indicates potential occurrence of soil in relation to coastal landforms, maritime climates, drainage, vegetation, and time available for formation. Increased understanding of coastal soils is, however, only possible from the purview of specific examples that illustrate soil distribution patterns in terms of sequences based on soil-forming factors of topography (toposequences), time (chronosequences), climate change (climosequence), and biological factors (biosequence). The conceptual frameworks for the models were initially postulated by Jenny (1941) and subsequently enhanced by various other researchers (e.g., Butler, 1959, 1967; Runge, 1973; Huggett, 1975). The value of these sequential paradigms is that they provide visualization of complex interrelationships in a framework that clarifies and elucidates coastal-soil transition inland from the shore.

The models are relatively uncomplicated in young (Holocene), more or less uniform parent materials such as those which may occur in small deltaic areas, mangrove swamps, coastal alluvium, and dunes. Complications arise along most coasts because coastal fringes reflect polygenetic development where exceedingly old parent materials and associated soils may occur juxtaposed to incipient pedogenesis. *Environmental complexity* in coastal areas where there is rise and fall of relative sea level, stillstands and diastems, increase and decrease of sediment supply, delta progradation and retrogradation, alluvial channel avulsion, development of deltaic superlobes, chenier formation, climatic change, and so forth, is manifested in weathering zones and soils that provide clues to evolutionary sequences as does the stratigraphy and nature of the included materials. Examples of depositional sequences are reported by Davis (1978), Milliman and Meade (1983), Liu and Walker (1989), Less and Lu, (1992), Roberts and Coleman

(1996), Saito *et al.* (2000), and Yatsko (2000), among others, for a variety of coastal settings. Dune systems, including large eolianites, which can be just as complicated as deltaic sequences, often provide stratigraphy with intercalated Paleosols (e.g., Fairbridge, 1950; Cooper, 1966; Denny and Owens, 1979; Hearty and Kindler, 1997). Pedogenic inertia and persistence (e.g., Huggett, 1975; Finkl, 1980) is another problem where older soils (which in turn may be initial parent material for soil forming under new environmental conditions) may occur in geomorphically young coastal landscapes.

Thus, coastal soils *sensu stricto* are often regarded as those soils that are derived form marine or estuarine parent materials that have recently been exposed subaerially to pedogenesis viz. acid sulfate soils, etc. Others might be related to coastal dunes, beach ridges, chenier plains, salt marsh, or upland freshwater marsh. The soils mostly range in age from late Holocene to early Pleistocene, reflecting a chronosequence that is geologically very young. This restricted point of view excludes many coastal sectors where soils of great age may be found, especially in the intratropical zones where coastal erosion may expose intensely weathered deposits as coastal cliffs in laterite (e.g., Hays, 1967; Petit, 1985; Retallack, 1990; Wang, 2003). Coastal soils *sensu lato* include a very wide range of possibilities as almost any kind of soil may occur in coastal zones. Fortunately, the possibilities are not endless and there is some order to apparent chaos of haphazard occurrence. Some clues as to potentialities for soil distribution patterns, at myriametric scales, on coasts may be gleaned from the morphological classification of coasts on the basis of plate tectonics (e.g., Inman and Nordstrom, 1971) where tectono-physiographic settings limit the scope of pedogenesis. Trailing continental margins provide large expanses of coastal plain sediments, as occurring along the eastern seaboard of the United States, south of Washington, DC, where there are thick weathering sequences composed of residual oxidized materials and soils developed in alluvial, estuarine, and marine deposits (e.g., Markewich *et al.*, 1986). In the vicinity of major rivers, especially on the southern coastal plain of the United States, major floodplains and deltas give rise to organic-rich soils. The coastal plains of southern Florida give rise to Holocene calcareous soils on slightly higher elevations whereas, lower poorly drained sites contain a suite of organic and muck soils related to Everglades, sloughs, swales, and other karst depressional areas that became infilled with fine-grained materials mixed with organic materials following the Holocene transgression (e.g., Gleason *et al.*, 1984; Finkl, 1994, 1995; Lodge, 1994).

Leading continental margins, such as that occurring along the mountainous and (strike-slip) faulted North American west coast where an oceanic plate is subducted below an overriding continental plate, provide a completely different set of tectono-physiographic conditions for soil development. Uplifted marine terraces contain soils of different ages, depending on the rates of uplift and number of terraces along the coast. Tsunami deposits, which are preserved in protected areas, provide allochthonous arenaceous parent materials for soils (Clague *et al.*, 1999) that date back to the penultimate great Cascadia earthquake 1,000 years ago. Coastal valleys in Oregon and Washington often retain interesting stratigraphy, where earthquake-induced subsidence leads to the burial of alluvial and marsh soil profiles and the development of younger soils in the new parent materials above (e.g., Nelson and Kashima, 1993). Pedogenic environments along these kinds of tectonic coasts where there are active seismotectonic regimes, as in Greece (Maroukian *et al.*, 2000) and throughout much of the Mediterranean region and elsewhere, are much more dynamic than the quiescent trailing margins.

Finally, there is the ultimately confounding situation where former coastlines are now stranded long distances from present shores. Many of these rocky shores (e.g., Johnson, 1992; Libbey and Johnson, 1997) provide parent materials for subsequent soil development (neopedogenesis of soil materials), but these occurrences can hardly be considered "coastal" in that pedogenesis is now unrelated to coastal factors of soil formation, other than coastal–marine parent material. Coastal Paleosols are, however, usually associated with modern as well as ancient coastlines whether at present sea level, emergent, or submerged. Emergent examples are best known and have been extensively studied along with Quaternary geomorphology and stratigraphy, for example, in Papau New Guinea (e.g., Bloom and Yonekura, 1985), Barbados (e.g., Radtke, 1989; Schellmann and Radtke, 2001), Bermuda (e.g., Hearty *et al.*, 1992), Bahama Islands (e.g., Hearty and Kindler, 1997), Point Peron in Western Australia (Fairbridge, 1950), Chile and Argentina in South America (e.g., Rutter *et al.*, 1989), and many other coastal areas too numerous to mention. Notable within the present purview of coastal soils, however, is the study by Schellmann and Radtke (2003) who employ morphologic, pedostratigraphic, and chronostratigraphic assessments of marine terraces along the Patagonian Atlantic coast. Many of these studies were used for reconstruction of local sea-level history, including evidence in southwestern Western Australia for a higher relative mean sea level (MSL) during the Holocene (Fairbridge, 1961; Playford and Leech, 1977), but as discussed by Searle and Woods (1986) evidence for the ± 3 m oscillations proposed by Fairbridge can be elusive. Still other ancient shorelines, instead of experiencing terrestrial sequestration from the sea, are found as drowned relicts. Examples are reported form Hawaii (Fletcher and Sherman, 1995), but there is no mention of included drowned soils.

Deltaic plains, such as the Mississippi River delta deposits, provide similar examples where constructive (alluvial accumulation, build up and soil formation) and destructive cycles (e.g., subsidence and drowning) present complex pedogenic environments that respond to fluvial-deltaic processes and longer-term base-level changes (Roberts and Coleman, 1996). Callaway *et al.* (1997), studying sedimentary accretion rates on low tidal-amplitude sites along the Gulf of Mexico, found evidence that contravened previous hypotheses concerning the relationship between tidal range and marsh vertical accretion rates. They found that there was little correlation between mineral and organic matter accumulation rates, with average organic matter accumulation rates showing less variation compared to mineral matter accumulation. There thus may be limits to annual rates of organic matter accumulation. Sediment addition is, however, clearly required for maintenance of coastal marsh soils (e.g., DeLaune *et al.*, 1987, 1990).

There are many different soil classifications in use today throughout the world (see discussions in Finkl, 1981), but none feature-specific sections on coastal soils *per se*. Rather than being perceived as a disadvantage, this situation is advantageous because soils occurring in coastal environments are hierarchically related throughout comprehensive soil classification systems and placed within continental or international schemes that facilitate reference and comprehension. Prominent among the major soil classifications are efforts by American, Australian, Canadian, French, German, Brazilian, and Russian soil researchers but the system employed here is the one adopted by the Food and Agricultural Organization FAO of UNESCO. The FAO system (FAO–UNESCO–ISRIC, 1988; FAO, 1991) is employed because it is specifically designed as an international system and because it provides a comprehensive inventory of world soil resources in the form of electronic databanks and maps. Although this is the primary organizing or categorical basis for discussing coastal soils, classes in other systems are sometimes equally valid (Spaargaren, 2000) and they are sometimes referred to for ease of reference and also to avoid unnecessary complications that would arise from correlation with other systems. Local soil classification schemes often reflect regional or cultural bias in their geography and therefore international efforts are encouraged (e.g., Finkl, 1982a, b). As already appreciated by many coastal researchers, they must be ambidextrous when dealing with coastal soil morphology, genesis, and classification. Unless otherwise stated, classificatory units referred to here belong to the international FAO system of soil classification, but similar and related terminology is also employed from the US soil classification system as applied in *Soil Taxonomy* (Soil Survey Staff, 1975) and keys (Soil Survey Staff, 1992). The following descriptions are based on FAO major soil groups.

Histosols

The major soil groups of Histosols (from Gr. *Histos*, tissue) include a wide variety of peat and muck soils that range from moss peats of the boreal tundra, moss peats, reeds/sedge peats, forest peats of the temperate zone, and mangrove and swamp forest peats of the humid tropics (Spaargaren, 1994; Inubushi *et al.*, 2003). Histosols are unlike all other soils in that they are formed in and by "organic soil material" with physical, chemical, and mechanical properties that differ strongly from those of mineral soil materials. They develop in conditions where organic materials are produced by an adapted (climax) vegetation, and where biochemical decomposition of plant debris is retarded by low temperatures, persistent waterlogging, extreme acidity, oligotrophy and/or the presence of high levels of electrolytes from organic toxins. Organic soil material contains more than 12% organic carbon by weight (20% OM). The diagnostic horizons for Histosols are the histic and folic horizons. Other diagnostic horizons used to separate soil units in Histosols are termed sulfuric, sulfidic, and salic. The combination of specific environmental conditions, composition of the organic soil material, and degree of decomposition produces six soil units that include: Gelic Histosols (permafrost occurs within 200 cm of the soil surface), Thionic Histosols (there is a sulfidic or sulfuric horizon starting within 125 cm of the surface), Salic Histosols (there is a Salic horizon starting within 50 cm of the surface), Folic Histosols (folic horizon present), Fibric Histosols (more than two-thirds of the organic soil material consists of recognizable plant material), and Haplic Histosols (other units).

Histosols form where the rate of accumulation of plant debris exceeds the rate of its decay (e.g., Nyman *et al.*, 1993). In practice, Histosols occur where microbial decomposition of plant debris is impaired by: cold temperatures, excessive wetness, salts or other toxins, and severe aridity or oligotrophy. Histosols under the permanent influence of groundwater, unless artificially drained (e.g., "low moor peats") occur in low-lying positions in fluvial, lacustrine, and marine landscapes, mainly in temperate regions but also to limited extents in the tropics. Other soils occurring in the same environment include Fluvisols, Gleysols and, in coastal regions, Solonchaks (e.g., adjacent to coastal mangrove peats). In lacustrine landforms and in coastal marsh embayments, Histosols may be associated with Vertisols, such as that occurring on the Texas Gulf Coastal Plain (e.g., Stiles *et al.*, 2003). Oligotrophy and prolonged wetness are primarily accountable for the low decay of organic debris. In the wet tropics (mainly the Indonesian region surrounding the Sunda Flat), their formation is conditioned by high rates of organic matter production from the climax forest vegetation. Change in use of tropical peatlands can affect the decay rates of organic matter and production of methane and nitrous oxide gases viz. conversion from secondary forest peatland to paddy fields, as reported for Kalimantan in Indonesia (Inubushi *et al.*, 2003). In addition to land-use change as a forcing function for pedogenic succession, similar migration of soil properties are associated with tidal inundation along transgressive coastal areas such as around the Chesapeake Bay on the eastern US coastal plain. Hussein and Rabenhorst (2001) report, for example, pedogenic transformation of Ultisols to Alfisols and eventually to Histosols in response to sea-level rise, based on changes in sodium adsorption ratios and exchangeable sodium percentages. Salinization and alkalization processes as a function of tidal inundation frequency across low-lying coastal landscapes produce lateral changes in pedogenic properties that ultimately relate to different profile characteristics and classification of soils. Lateral pedological linkages exist with a variety of other soil groups, including Podzols, Fluvisols, Gleysols, Cambisol, and Regosols.

The global extent of Histosols is about 275 million ha (FAO, 1991) (Table C9), roughly half of which are located in the Arctic zone, one-third in temperate lowlands, and one-sixth in tropical lowlands (Spaargaren, 1994). An example of the latter case is the subtropical Florida Everglades, which constitute the largest single body of Histosols in the world (Stephens, 1956, cited in Amador and Jones, 1995). A large area of cold region organic soils occurs on the southern and southwestern subArctic margins of Hudson Bay (Manitoba, Ontario, Quebec) and around Great Bear Lake (Northwest Territories) in Canada. Here, Cryic Fibrisols (with frozen subsurface) occupy about 171,715 km^2 of coastal plain (Clayton *et al.*, 1977).

In lower latitude coastal areas, continually rising sea level has caused brackish or saline waters to engulf drowned river valleys (rias) or to extend over formerly upland soils. This has led to the formation of coastal marsh Histosols in several geomorphic settings (Darmody and Foss, 1979). As sea levels continue to rise, marsh margins are pushed landward and the organic materials continue to thicken so that the older and deeper Histosols generally exist nearer the open water (e.g., Nyman *et al.*, 1993; Rabenhorst, 1997). Although distant from high-latitude glacial activity, coastal Histosols at lower latitudes were impacted by glacio-eustatism (e.g., Rabenhorst and Swanson, 2000). During the glacial maximum (~20,000 yr BP) when large quantities of water were bound in glacial ice, eustatic sea level was at least 150 m below present MSL. Deglaciation cycles caused sea level to rise at such a rapid rate (~10–20 mm a^{-1}) that initially vegetation could not colonize tidal regions. Approximately 3,000–5,000 years ago, sea-level rise slowed so that marsh vegetation became established and organic parent materials began to accumulate (Bloom and Stuvier, 1963; Redfield, 1972). As sea level rose, organic materials accumulated in Histosols so that coastal marshes and mangroves generally accreted at about the same rate. Peat accretion in coastal areas presently ranges from 3 to 8 mm a^{-1}, which is much higher than in non-coastal regions. The highest rates of sea-level rise may, however, be too great for marsh systems to maintain and some areas suffer marsh loss and degradation (e.g., Huiskes, 1990; Nyman *et al.*, 1993; Boesch *et al.*, 1994) where Histosols are eroded, degraded (e.g., Mendelssohn and McKee, 1988) or drowned. Loss may also occur in response to subsidence (primarily as microbial mineralization of soil organic matter), such as in areas of the Florida Everglades where drainage of coastal wetlands (Everglades marsh) (Finkl, 1995) has exposed organic soils subaerially (Stevens *et al.*, 1984) and their surface elevations may decrease by as much as -2.5 cm a^{-1}.

The Florida Everglades is the largest continuous sedge moor outside of the Pleistocene glaciated area of the United States (Gleason *et al.*, 1984). The expansive coastal wetlands and freshwater marsh of southern Florida are a result of the very slow relative rise of sea level during the past 3,200 years (average rate of 4 cm per century) (Spackman *et al.*, 1966; Wanless *et al.*, 1994). A large number of coastal sediment bodies in southern Florida were initiated or stabilized between 3,300 and 3,000 yr BP and again approximately between 2,300 and 2,500 yr BP (Wanless *et al.*, 1994). The origin and development of those coastal and shallow-marine carbonate and clastic sediment bodies is dependent on the interactive roles and timing of sea-level changes, climatic fluctuations, storm influences, vegetative colonization, preexisting topography, and provision of detrital (input or recycled) siliciclastic and carbonate sediment. Of these, the rate of sea-level rise was the primary control on the clastic, carbonate, and organic sediment bodies of southern Florida.

As one of the largest marshes in the world, the Everglades peat deposits are commonly differentiated into subtypes based on compositional variation, thickness, and degree of weathering. Water is required for the formation of peat soil because it restricts atmospheric oxygen from the soil; without oxygen, microorganisms cannot decompose dead marsh plants as fast as they accumulate. Ever-thicker deposits of peat are deposited until an elevation is reached where the surface dries enough that oxygen-dependent decay (the opposing aerobic process), or fire, prevents accumulation (Lodge, 1994). Freshwater peats of the Everglades merge southwards with mangrove peats on the tidal plain fronting Florida Bay. Occupying more than 400 ha of marshland, the Everglades Peat forms from degraded saw grass in long-hydroperiod habits. It is the most abundant peat occurring in the Everglades Basin. The Loxahatchee Peat, covering about 29,000 ha, occurs in deeper marsh areas and contains less saw grass remains and more evidence of slough vegetation such as the roots, rhizomes, leaves, and seeds of water lilies. Very poorly drained organic soils of the Everglades usually have a surface layer of black muck (sapric material—organic material that is too finely divided for identification of plant remains) that overlies black or reddish-brown muck subsoils that in turn rest on sand, limestone bedrock, or other materials (McCollum *et al.*, 1978). The term *muck* refers to highly disintegrated peat soils that are composed of more mineral than organic matter (see discussion in Brown *et al.*, 1990). *Marl* is the product of periphyton. During the dry season, the organic material (dead algae) in the periphyton mass oxidizes, leaving the calcium carbonate particles as a light-colored soil. Marl is descriptively called *calcitic mud* and is the main soil of the short-hydroperiod wet prairie habitats near the margins of the southern Everglades where bedrock lies close to the surface (Lodge, 1994). Marl and peat soils are opposites, requiring aerobic and anaerobic soil conditions, respectively. Peat cannot accumulate in shorter hydroperiod marshes where marl persists, and acid conditions within a peat soil dissolve marl and prevent its accumulation. However, there are large areas in the northern Everglades where a thick layer of marl underlies peat. This occurrence is evidence that the hydroperiod increased substantially in those areas as the Everglades ecosystem evolved. Much of the Everglades region is flooded during the summer rainy season, which is a natural process essential to the health of this wetland. Many areas of urban development, especially on the coastal plain of southeast Florida, occur on former Everglades (wetlands that were drained as part of reclamation programs) (Finkl, 1995) where Histosols have been stripped away and building sites 'demucked' and filled with dredged materials. The region is, however, prone to returning to its natural wetland condition during high rainfall events making localized sectors of the region susceptible to flooding (Finkl, 2000).

Acid sulfate soils

Acid sulfate soils occupy an area of some 24 million ha worldwide where the topsoil is severely acid or will become so if drained (Ritsema *et al.*, 2000). An equal area of these soils may be thinly covered by peat and nonsulfidic alluvium (van Mensvoort and Dent, 1997). Acid sulfate soils are typically associated with mangrove forests that once covered 75% of the coastlines in tropic and subtropical countries (Quarto and Cissna, 1997). Estimates of the extent and distribution of acid sulfate soils suffer, however, from the lack of field surveys, few reliable laboratory data, and from variable definition. More significant than their areal extent is their location where they are concentrated in otherwise densely settled coastal and floodplains, mostly in the tropics, where development pressures are intense with few suitable alternatives for expansion of farming, or urban or industrial development. Two-thirds of the known extent is in Vietnam (e.g., Mekong River Delta), Thailand, Indonesia, Malaysia, western Africa (Senegal, Guinea Bissau), northern Australia, and on coastal plains in southern China (e.g., Pearl River Delta). By far, the largest use of acid sulfate soils is for rice cultivation

Table C9 Main kinds of soils occurring in coastal areas, based on a world reference base for soil resources, indicating the different types of soil cover in relation to landscape features and associated linkages with related soils

Soil group (FAO)[a]	Area ($\times 10^6$ ha)[b]	Major subdivisions[c]	Associated soils[d]	Lateral linkages or intergrades[e]	Landscape position, climate, drainage, topography, or parent materials
Histosols	275	Gelic, Thionic, Salic, Folic, Fibric, Haplic Histosols	Fluvisols, Gleysols, Solonchaks, Vertisols	Podzols, Fluvisols, Gleysols, Cambisols, Regosols	Low-lying positions in fluvial, lacustrine, and marine landscapes. Solonchaks often adjacent to coastal mangrove peats
Anthrosols	NA	Hydragric, Irragric, Cumulic, Hortic Anthrosols	Border most of the major soil groups where soils have been strongly influenced by man	Includes coastal archaeological sites	Old cultivates sites, excluding paddy soils, on coastal plains, river valleys, deltas, etc.
Leptosols	1,655	Lithic, Cryic, Skletic, Rendic, Mollic, Umbric, Dystric, Eutric Leptosols	Regosol, Cambisol, Podzol, Calcisol, Luvisol, Histosol, Anthrosol	Wide range of intergrades	Tropics to polar tundra, coastal lowlands to mountains
Cryosols	1,800	Histic, Thixotropic Cryosols	Histosols, Gleysols, Stagnosols, Cambisols	Gelic or cryic units of Histosols, Gleysols, Podzols, Planosols, Glossisols, Stagnosols, Cambisols	Arctic coastal plains of NE Eurasia, Alaska, Yukon, and NW territories of Canada; islands of Siberian, Beringian, and North American sectors of the Arctic Ocean; coastal Greenland
Fluvisols	350	Thionic, Salic, Vertic, Mollic, Calcaric, Umbric, Dystric, Eutric	Cambisols, Regosols, Arenosols, Leptosols, Gleysols, Solonchaks	Arenosols, Cambisols, Fluvic Gleysols, Salic Fluvisols (intergrade to Solonchaks)	Large deltas and fans viz., Ganges, Mekong, Mississippi, Niger, Po, Rhine; major river floodplains; coastal barriers; tidal flats
Solonchaks	260–340	Gleyic, Stagnic, Mollic, Gypsic, Calcic, Sodic, Haplic Solonchaks	Chlorida and sulfate soils	Salic Fluvisols, Thionic Fluvisol—Solonchak intergrades	Coastal salt lakes, lagoons, pans; chloride facies—soils of marine origin under mangrove; neutral chlorido-sulphate soils—of marine origin but enriched by sedimentary gypsum deposited by rivers; acid sulphate soils—soils of marine origin under mangrove as in Gambia, Senegal, Guinee Bissau
Gleysols	720	Cryic, Thionic, Plinthic, Arenic, Mollic, Umbric, Fluvic, Calcic, Haplic Gleysols	Leptosols, Vertisols, Fluvisols, Solonchaks, Stagnosols, Histosols, Regosols	Histi-Mollic, Histi-Umbric Gleysols, Fluvic Gleysols (intergrades to Fluvisols)	Sandy to loamy (sometimes clayey) textural classes of parent material on alluvial, fluvial, fluvioglacial deposits in riverine or coastal areas. Common in depressional and organic-rich coastal areas; for example, Arctic coastal plains, coastal Alaska; Florida Everglades
Podzols	485	Gelic, Gleyic, Stagnic, Humic, Duric, Umbric, Cambric, Haplic	Histosols, Gleysols, Cryosols, Cambisols, Ferralsols, Planosols, Glossisols	Podzol–Histosol–Gleysol sequences, Cambisol–Podzol sequence	Sandy deposits affected by illuviation and cheluviation, boreal climatic zone, coniferous vegetal cover, temperate and tropical wet climatic areas. For example, NE and NW coastal Canada, Scandinavia, NW coastal Russia; coastal plain of Florida; southwestern Australia; coastal Indonesia
Sesquisols	60	Petric, Aeric, Albic, Stagnic, Humic, Eutric, Haplic Sesquisols	Ferralsols, Alisols, Acrisols, Lixisols	Found in association with Gleysols and Stagnosols in areas conditioned by hydromorphy; with Feralsols, Alisols, Acrisols, and Lixisols in better drained positions in the landscape	Usually ancient landsurfaces truncated by modern coastal erosion or in close proximity to the sea; older coastal plains; for example, western Africa, western India, Mekong catchment, northern and southwestern Australia, eastern Amazon region

Table C9 *(Continued)*

Soil group (FAO)[a]	Area ($\times 10^6$ ha)[b]	Major subdivisions[c]	Associated soils[d]	Lateral linkages or intergrades[e]	Landscape position, climate, drainage, topography, or parent materials
Ferralsols	750	Humic, Geric, Gibbsic, Lixic, Rhodic, Eutric, Plinthic, Gleyic, Haplic Ferralsols	Cambisols, Acrisols, Nitisols, Gleysols, Planosols, Arenosols, Sesquisols	Intergrades to Cambisols, Acrisols, Nitisols, Gleysols, Planosols, Arenosols, Sesquisols	Mid- to end-Tertiary peneplains; for example, Brazilian Shield, French Guyana, Senegal, Madagascar, northern and SW Australia; extreme tropical and subtropical weathering
Planosols	130	Gelic, Vertic, Histic, Mollic, Umbric, Dystric, Eutric Planosols	Vertisols, Acrisols, Luvisols	Stagnic Solonetz, Albic Sesquisols, Stagnic Gleysols; related soils with abrupt textural change but insufficiently wet to show reduction by surface water would fall into Luvisols, Lixisols, Alisols, Acrisols	Low-lying, nearly level fluvial or marine terraces, depressional lands on coastal plains, including wetland rice paddies; most extensive in climes with a strongly seasonal variation in rainfall that results in intermittent saturation and reduction of upper soil horizons
Solonetz	135	Gleyic, Stagnic, Salic, Albic, Mollic, Gypsic, Calcic, Haplic Solonetz	Linked through sodic units to Vertisols, Solonchaks, Gleysols, Calcisols	Salic Solonetz, Sodic Solonchaks; through sodic properties in Vertisols, Solonchaks, Gleysols, and Calcisols; with Gleysols and Stagnosols through the Gleynic and Stagnic Solonetz soil unit	Solonetz landscapes are governed by microrelief, waterlogging at the surface, and salinity in the profile; lake terraces, river deltas, depressions along liman coasts; Rio de la Plata deltaic plain in Argentina; coastal South Australia and coastal southeastern Western Australia
Calcisols	800	Petric, Luvic, Sodic, Cambic, and Haplic Calcisols	Vertisols, Solonchaks, Gleysols, Solonetz, Luvisols	Calcic Gleysols, Calsic Gypsicsols, Calcic Luvisol	Arid coastal regions viz. Baja, California; Patagonia and Argentine Pampa; coastal Atacama; Namibia; coastal northern Africa; Nullarbor Plain in Western Australia and coastal plain in South Australia
Alisols	>100	Plinthis, Gleyic, Humic, Vertic, Ferric, Chromic, Luvic, Haplic Alisols	Ferralsols, Nitisols, Acrisols, Lixisols, Vertisols, Cambisols	Vertic Alisols, Alic Nitisols, Alic Acrisols; facies of other highly weathered soil units in intertropical regions	Coastal plain in southern USA; parts of intertropical coastal India, Africa, and northern Australia
Acrisols	>900	Plinthic, Gleyic, Humic, Arenic, Albic, Ferric, Haplic	Lixisols, Luvisols, Alisols, Nitisols, Glossisols	Planosols, Lixic Ferralsols, Arenosols, Regosols, Cambisols	Ancient shield landscapes, in tropical regions, that are truncated by coastal plains or marine erosion; coastal plain is southeastern USA, Caribbean Central America coastal lowlands, Orinoco deltaic plain, western Africa, Southeastern Asia, Indonesia
Luvisols	650	Gleyic, Albic, Vertic, Calcic, Ferric, Chromic, Dystric, and Haplic	Alisols, Lixisols, Acrisols, Nitisols, Glossisols, Solonetz	Luvic Chernozems, Luvic Gypsicsols, Luvic Phaeozems, Luvic Calcisols, Luvic Arenosols	Humid to subhumid temperate regions of western Europe (Belgium, The Netherlands, coastal Germany), parts of the Mediterranean region, southern Australia (extreme southwestern Western Australia, central South Australia), eastern Indian coastal regions, Sri Lanka
Lixisols	435	Plinthis, Gleyic, Humic, Arenic, Albic, Ferric, Haplic	Nitisols, Alisols, Acrisols, Luvisols, Glossisols	Ferralsols, Regosols, Cambisols, Luvisols	On stable shield landscapes in tropical regions; in seasonally dry tropical, subtropical, and warm temperate regions, on Pleistocene and older surfaces (including. Alluvial fans); coastal savanna regions

Cambisols	1,500	Gelic, Gleyic, Vertic, Fluvic, Mollic, Calcaric, Ferralic, Dystric, Chromic, Eutric Cambisols	Luvisols, Alisols, Acrisols, Lixisols, Nitisols, Ferralsols	Regosols, Leptosols, Gleysols, Fluvisols, Acrisols, Sesquisols	Islands (Puerto Rico, Madagascar, New Zealand, Papua-New Guinea); Atlantic Spain and Portugal; Mediterranean Spain, France, western Italy, Turkey; southeastern Australia
Arenosols	900	Leptic, Albic, Protic, Gypsiric, Calcaric, Luvic, Ferralic, Cambic, Haplic	Regosols, Leptosols, Cambisols, Gleysols, Fluvisols	Acrisols, Lixisols, Ferralsol, Planosols, Histosols, Anthrosols, Solonchaks, Regosols, Leptosols, Calcisols	Dry tropical and temperate interior regions, but some coastal areas in western Africa, western Madagascar, and northwestern Australia
Regosols	260	Gelic, Anthropic, Tephric, Gypsiric, Calcaric, Dystric, Eutric	Arenosols, Leptosols, Gypsisols, Ferralsols, Cryosols	Found in all landscapes of the world; intergrades and extragrades may develop into many other soils, depending on the most important soil forming factor(s)	Shallow, weakly developed mineral soils in the initial stages of soil formation; all coastal landscapes

Notes: Compiled from Spaargaren (1994).

[a] Based on the FAO World Reference Base (WRB) major soil groups, compiled and revised from the FAO/UNESCO International Reference Base for Soil Classification and the Legend of the Soil Map of the World (FAO-UNESCO, ISRIC, 1988).

[b] Total area (in millions of hectares) is based on estimated WRB worldwide occurrence of the soil group; coastal occurrences are a smaller subset of unknown areal extent.

[c] Major subdivisions that are proposed to separate the soil groups on the basis of diagnostic properties. Not all of the soil units listed occur in coastal areas as major soils, but they are listed for completeness, where they may be found in specialized environmental situations.

[d] Other soil groups that occur in the same general area or environment; soil groups were not differentiated on the basis of climate in order to keep the number of units manageable. Differences in topography, drainage, and other diagnostic properties and materials result in a range of different soil groups that occur in close geographic proximity to each other.

[e] Lateral linkages are commonly intergrades that merge soil units in time and space. These units also reflect the spatial integration of soil properties in the coastal landscape where differences in the factors of soil formation (i.e., topography or relief, time, drainage or hydrology, organics, parent or initial materials) may lead to rapid lateral changes in soil units. Relief, drainage (or lack thereof, salt spray, or influence of tidal waters), parent materials (terrestrial versus marine sediments), and time (especially in regard to young soils forming in marine sediments) are particularly relevant to coastal soil distribution patterns.

Figure C41 Initiation of mangrove foothold on intertidal marine platform on Eva's Cay, west (bank or leeward side) of Salt Pond on Long Island, Bahamas. Carbonate muds and micritic sediments show initial phases of biochemical weathering as mangroves colonize the area. Accumulation of organic debris derived from the mangroves eventually results in dark coloration of chemically reduced muds. Mangrove soils often show light (thin bands) and dark (thick bands) lamination in the profile because pedogenesis is interrupted by storm waves that deposit new unweathered materials on top of pedogenically altered materials. This mangrove mud is minimal as the stand is only just taking a foothold; older, well-developed mangrove forests such as that occurring in the world's largest estuarine delta (2,600 km^2) along the Indian coast southeast of Calcutta in West Bengal as Sunderbans (beautiful forests) and many other tropical coastal locations feature thick organic-rich profiles.

viz. the Balanta system in western Africa (van Ghent and Ukkerman, 1993) where land is ridged annually to a height commensurate with the expected freshwater flood of the rainy season. The ridging and annual turning speedup removal of soluble toxins from the surface soil, but the system is declining because of the heavy labor demand and low financial returns.

Acid sulfate soils exhibit enormous spatial variability that is tied to the dynamic estuarine, deltaic, and flood plain environments in which they occur. Figure C41 shows typical colonization by mangroves into carbonate intertidal flats that will eventually become sites for accumulation of organic debris derived from the mangroves. Organic-rich muds will eventually develop and be accompanied by pedogenesis. Significant spatial variability occurs at myriametric scales from pyrite-plugged pores to the fine-grained soil matrix to infield, local, and regional soil patterns (Dent, 1986). Remote sensing is widely used in tropical, tidal environments because surveys are arduous and difficult due to mosquitoes and crocodiles. Acid sulfate soils also exhibit significant temporal variability that is associated with their defining characteristics of acidity and related toxicities. Almost uniquely, acid sulfate soils export their problems in drainage and floodwaters.

Evident in acid sulfate soils are patterns related to sedimentary history viz. the common distinction between highly sulfidic, unripe clays or peat in backswamps and riper, coarser textured soils, or even calcareous soils on levees and creek fillings. Soil patterns are not always visible at the surface because present landforms and vegetation may be only partly and indirectly related to the environment of sulfide accumulation. The area may, for example, have been buried by peat or nonsulfidic alluvium, fresh waters may have succeeded brackish, and new plant communities have replaced prior ones.

Sulfidic soils that oxidize and generate H_2SO_4 when drained are referred to as *potential acid sulfate soils* (Wang and Luo, 2002). If they have been drained and are generating H_2SO_4, they are called *raw acid sulfate soils*, but if they have passed through the acid-generating phase but remain severely acid, they are referred to as *ripe acid sulfate soils*. Sulfidic materials include marine and estuarine sands and clays, gytja in brackish lakes and lagoons, and peats that originally formed in freshwater but which have been inundated subsequently by brackish water. The common factors are: (1) supply of organic matter, (2) severely reducing conditions caused by continuous water logging, and (3) a supply of SO_4^{2-}, usually from tidewater that is reduced to sulfides by bacteria decomposing organic matter, and (4) a supply of Fe from the sediment for the accumulation of iron sulfides which make up the bulk of reduced S compounds. These conditions are fulfilled in tidal swamps and salt marshes where thick deposits of sulfidic clay have accumulated in concert with Holocene sea-level rise (e.g., Pons and van Breeman, 1982; Dent and Pons, 1995). As sedimentation raises the soil surface above MSL, topsoil accumulates under better drainage as a nearly ripe, mottled layer that contains little or no sulfide. Sulfidic clays may be buried by peat or nonsulfidic alluvium where freshwater conditions succeed brackish water.

Reclamation and drainage of sulfidic soils causes dramatic changes. When insufficient carbonate is present to neutralize the H_2SO_4 generated by oxidation of sulfides, extreme acidity develops within weeks or months. Raw acid sulfate soils, characterized by a pH < 3.5, contain jarosite but frequently display a black subsoil as some of the SO_4^{2-} generated by drainage is reduced to FeS deeper in the profile. Experiments on acid sulfate soils in the southern coastal regions of Guangdong, China, for example, show that the sulfur content of acid sulfate soils is 2–6 times higher than that of coastal Solonchaks and 3–9 times higher than that of paddy soil (Wang and Luo, 2002). The potential acidity is 2–3 times higher than the actual acidity and the amount of lime required to correct the potential acid ranges from 143.5 t ha^{-1} to 2687.5 t ha^{-1}.

Acid sulfate soils cause on-site problems and also in adjacent areas with the latter attracting attention (Dent, 1986). On-site problems include Al toxicity in drained soils, Fe and H_2S toxicity in flooded soils, salinity, and nutrient deficiencies. Engineering problems include: (1) corrosion of steel and concrete, (2) uneven subsidence, low bearing strength, and fissuring leading to excessive permeability of unripe soils, (3) blockage of drains and filters by ochre, and (4) difficulties of establishing vegetation cover on earthworks and restored land. Off-site problems stem from drainage effluents, earthworks, excavations, and mines.

The acid drainage waters carry Al released by acid weathering of soil minerals and heavy metals that are released by oxidation of sulfide minerals. Episodic release of toxic drainage waters may occur, for example, at the onset of the wet season after a period of low water table during which oxidation has taken place.

Gleysols

Gleysols (from Russian local name *gley*, mucky soil mass) or soils with gleyic properties are permanently wet and reduced in the subsoil (e.g., Schlichting, 1973), and periodically to permanently wet in the topsoil (Spaargaren, 1994). Mainly found in periodically poorly drained depressions, valleys, and low-lying coastal areas, the solum is either mottled (in case of temporary aeration) or has colors reflecting reduction. These redoximorphic features form under the influence of poorly to well-drained groundwater regimes. Typical gleyic soils are classified as Gleys or Groundwatergleys in many European systems of soil classification (e.g., Austria, Germany, France, Switzerland, United Kingdom), although in the German classification the periodically flooded soils of coastal regions are separated as *Marschböden* from normal Gleysols. Gleysols are found in nearly all climates, from perhumid to arid conditions, and cover an area of almost 720 million ha (FAO, 1991) (Table C9). In regions with permafrost (e.g., Arctic Canada, coastal plain of northern Siberia), Cryic Gleysols occur in association with Cryosols and Gelic Histosols. In the lowlands of the temperate latitudes, they are formed in alluvial sediments associated with Fluvisols near riverbeds in coastal areas, and with Histosols. In the humid tropics, they are found in valleys associated with Acrisols, Lixisols, Nitisols, Alisols, or Ferralsols.

Gleysols (Eutric Gleysols, FAO) are a dominant soil of Canada, with major areas occurring in the subArctic and Hudson Bay Lowland (Clayton *et al.*, 1977). Most Gleysols occur on nearly level to undulating topography, particularly where associated with lacustrine, alluvial, or marine deposits that are moderately calcareous and loamy in texture. Cryic Gleysols occur throughout the entire Arctic region of northern Canada, including the Arctic islands extending from the Labrador coast to the Yukon–Alaska border. About 37,698 km^2 or 0.4% of Canada is mapped as dominantly Cryic Gleysols; these areas occur in the Northwest Territories and Yukon. They are found within the Mackenzie Delta and the Arctic Coastal Plain. Gleysols are often spatially associated with Podzols and Luvisols.

Anthrosols

These soils result from human activities that produce profound modification or burial of the original soil horizons. In the 1988 Revised Legend of the Soil Map of the World (FAO–UNESCO–ISRIC, 1988), these are "... soils in which human activities have resulted in profound modification or burial of the original soil horizons, through removal or disturbance of surface horizons, cuts and fills, secular additions of organic materials, long-continued irrigation, etc...." Implicit in this general definition is the recognition of a distinct set of anthropedogenic processes that are associated with characteristic anthric diagnostic horizons. Anthropogenic soil materials that result from anthropogeomorphic processes are better considered as "non-soil," unless they are subjected to a sustained period of pedogenesis (Spaargaren, 1994). Many different soil classification systems from around the world make provisions for these kinds of soils. Continuous cultivation of paddy soils often produces a hydragric horizon that also includes redoximorphic and illuviation features in the subsoil. Many Anthrosols occur in coastal deltaic environments, especially in Southeast Asia and Indonesia. The Mississippi Delta region in the United States contains extensive areas of Anthrosols that are now subject to drowning by relative sea-level rise and contribute to the land-loss problem in coastal Louisiana (Boesch *et al.*, 1994).

Leptosols—Entosols, Regosols, Lithosols, etc.

The name Leptosols (from Gr. *leptos*, thin) is used in the FAO *World Reference Base for Soil Resources* (Spaargaren, 1994) for shallow soils overlying hard rock or highly calcareous material, and soils that have, by weight, less than 10% fine earth material. Leptosols occupy more than 1,655 million ha from the tropics to polar tundra (Table C9). Different names have been used in reference to these kinds of shallow, rocky soils. The term *Lithosol* was applied by Thorp and Smith (1949) to denote azonal soils having an incomplete solum or no clearly expressed soil morphology and consisting of freshly or incompletely weathered mass of hard rock or hard rock fragments. *Rankers* were defined as soils that have a dark-colored surface horizon, rich in organic matter with a low base status that is overlying unconsolidated siliceous material exclusive of alluvial deposits or hard rock. The name *Rendzina*, as first used by Sibirtzev (1901), was originally assigned to intrazonal soil on calcareous rocks. *Regosols* were first defined by Thorp and Smith (1949) as an azonal group of soils consisting of deep unconsolidated rock (soft mineral deposits) in which few or no clearly expressed soil characteristics developed.

Leptosols represent the initial phases of soil formation and as such, they are of great significance because they are the forerunners of the young or weakly developed soils of other units. These soils occur in landscapes where deposition or erosion, resistant bedrock, inert parent material, saturation, or human activity was limited normal pedological development. These soils commonly form on mountains, sand dunes, flood plains, and riverine lunettes, and coastal plains. In northern Canada, for example, Cryic Orthic Regosols in Arctic Lowlands commonly occupy gently undulating coastal plains developed on stony or sandy glacial till and marine deposits and occupy upwards of 1,305,423 km^2 (Clayton *et al.*, 1977), but not all of this area is coastal. Parent material lithology is a factor that may inhibit pedogenesis where resistant bedrock, for example, increases runoff and erosion, and thereby decreases weathering. These soils also form in sandy parent materials that have an abundance of rock fragments. When parent materials consist of quartzitic sands without coarse fragments, such as in many sand dunes, profile development is also retarded.

Coastal dunes are built of beach sand blown onshore. Calcareous sand may give rise to calcareous eolianite, as on the coast of western Victoria (Australia), southwestern Western Australia, Bermuda, and the Bahamas. Quartz sand dunes are common in eastern Australia and grew particularly large during the last Ice Age, when sea levels were low and wide sandy flats were exposed (Ollier and Pain, 1996; Short, 2003). The high dunes of Queensland such as Fraser Island are the highest dunes in the world. Leptosols are also associated with beach ridges (Tanner, 1995) that are formed as beaches with broad curved shapes and lengths often of many tens of kilometers, as seen in this aerial photograph of a pocket beach in the Bahama Islands where the beach ridges mirror successive storm deposit build up (Figure C42). Cheniers on the Yangtze Delta (China) are also tens of kilometers in length (Yan *et al.*, 1989), up to 500 m wide, and similar to bight coast cheniers (Otvos and Price, 1979) along the Louisiana—Texas coast. As wind blows sand from the beach to make ridges, these accumulations eventually become stabilized to support soil formation. Beach ridges often join together to make a chenier or beach ridge plain (cf. Figure C42) where the ridges usually consist of coarser materials than the intervening swales. The more inland ridges are older than nearshore ones and composition of the ridges may be quite variable. A recent study by Rink and Forrest (2003) regarding the accretion history of beach ridges on Cape Canaveral and Merritt Island, FL, estimated ridges to be 4000 ± 150 years old on the peninsula and about 43,000 years old on the more landward Merritt Island. Average ridge spacing was 109 ± 20 m, giving an average ridge accumulation rate of 80 ± 8 years per ridge on the cuspate foreland. The ridge accumulation rate is related to storm frequency (tropical storms or hurricanes); soil formation proceeded on the ridges subsequent to deposition. On a coastal plain in southeastern Australia, for example, the inland beach ridges are quartzitic in composition and date back to the Miocene, whereas the nearshore sets are calcareous and formed throughout the Quaternary to the late Holocene (Cook *et al.*, 1977). Many Tertiary marine beaches are now found high and dry in Western Australia, at places like Eneabba, Yoganup, and Minninup (Lissman and Oxenford, 1975) that are now some distance inland from the present shore of the Indian Ocean. More recent beach ridge systems are described for the inner Rottnest Shelf coast near Perth in southwestern Australia by Searle *et al.* (1988) where the main geomorphic components in the area are shore-parallel Pleistocene eolianite ridges and their associated intervening depressions, and the Holocene units in basins, slopes, banks—nearshore platforms, beaches, and beach–dune complexes.

Anaerobiosis may slow pedogenesis where there are persistent high water tables that prevent leaching of soil constituents, inhibit redox cycles, or preclude subsurface horizon development. Poorly drained floodplains and coastal marshes host these kinds of soils.

Cryosols

Cryo-hydromorphic soils occur in Arctic, subArctic, and boreal regions under cold continental, (sub)humid, or semiarid climatic conditions. They support a characteristic vegetation cover of dominantly taiga, forest-tundra, or tundra with a ground cover of lichen moss. These soils have permanently frozen subsoil, often at a shallow depth of <1 m, and occupy enormous areas (1,800 million ha) typically on Arctic coastal

Figure C42 Beach ridge plain near the southern tip of Eleuthera Island, Bahamas. The succession of calcareous beach ridges represents a developmental sequence of deposition, often associated with phases of storminess, where younger dunes and soils developed on them occur closer to the shore and older dunes and associated soils occur farther inland. Stages of soil development are usually referred to as chronosequences, where greater profile differentiation occurs on the older dunes. Minimal profile developed associated with weak phases of pedogenesis produce Leptosols (Lithosols), Arenosols, and Cambisols on the seawardmost ridges. Other types of calcimorphic soils occur inland in association with longer times for pedogenesis to precede.

Figure C43 Organic-rich soil developed in fluvial-deltaic sediments flanking the Ijsselmeer. These soils, reclaimed by diking and drainage, are susceptible to flooding and find use as pasture lands and reserves. Many soils developed in parent materials derived from marine sediments become extremely acid when drained due to oxidation of sulfur compounds. The low pH values of acid sulfate soils preclude many uses.

plains (Table C9). Some show a significant accumulation of organ matter (dominantly weakly decomposed tissues of oligotrophic plants) at the surface, others are mainly mineral (or organo-mineral) in nature. They are saturated with water during the entire period of thawing, but lack features associated with reduction or redistribution of iron. The more mineral soils often show the peculiar phenomenon of thixotropy. Thixotropic properties are defined as the ability of relatively compact fine-textured soil materials to transform under pressure into an apparent liquid and unstable state when saturated with loosely bound water (quicksand-like property). When pressure weakens, the material gradually regains its initial compactness. Cryosols have been previously described in Russia as Peaty Frozen, Taiga Frozen, Gleyic Frozen, and Destructive Taiga Frozen soils and more recently as Homogeneous (or Peaty-Duff) and Thixotropic Cryozems. In Canada they are known as Organic and Turbic Cryosols. In *Soil Taxonomy* (Soil Survey Staff, 1975, 1992), they correspond with soils having a pergelic soil temperature regime.

Fluvisols

The central concept of this group relates to soil developed in fluvial (alluvial) sediments. Fluvisols were previously recognized by a variety of names including but not limited to swamp soils, Lithomorphic soils, Brown soils, Groundwater Gley soils, and Vaaggronden ("Vague soils," Netherlands). Whatever the name, these soils are deposited in aqueous sedimentary environments viz. (1) inland fluvial and lacustrine environments, (2) marine environments, and (3) coastal saltings or brackish marsh environments, of which deltas are a special case (Dreissen and Dudal, 1989). Fluvisols occupy significant areas on deltas and fans, coastal barriers, and tidal flats (350 million ha) (Table C9). On aggrading coastlines, combinations of sand or shingle spits and offshore bars allow dunes to accumulate, giving shelter to a nearshore zone where silts and organic muds accumulate in low-energy or quiescent environments. Tidal inundation brings additional supplies of new suspended sediment that flocculates in the water column but which eventually settles to the bottom to help accretion of marsh soils. Aggradation of marsh surfaces takes place relatively slowly and is often associated with strictly zoned vegetation according to the length of time the marsh is inundated. Creeks give tidal waters access to the marsh at flood tide and drain the marsh at ebb tide. Deltaic systems develop where rivers debouching in the sea bring significant amounts of material, providing coastal conditions are satisfactory for sediment accumulation, as seen here in Figure C43 along the Netherlands coast. In the intertropical regions, particularly, the conditions are such that reducing conditions may be developed in which the sulfates in the seawater are reduced to pyrites. Sediments containing pyrites may become parent materials for acid sulfate soils.

Fluvisols show clear sedimentary layering or stratification in a major part of the profile. Soil formation may have affected the upper part of the profile to produce dark-colored surface horizons that are high in organic matter and which form in comparatively little time (Figure C44). Also, the material in the upper part of the profile below the surface horizon may have lost evidence of sedimentary layering due to weathering and have acquired the color and structure of stronger developed soils. These juvenile soils tend to show little evidence of weathering and soil formation below 25 cm depth, except possible gleying. Permanent or seasonal saturation with water, causing recurrent anaerobic conditions and low biological activity, tends to preserve the original stratified nature of the original deposits. Consequently, the more important linkages of Fluvisols are with other weakly developed soils viz. Cambisols, Regosols, Arenosols, Leptosols, Gleysols, and Solonchaks.

Figure C44 Core from a polder soil in The Netherlands. Depending on the nature of sediment grain-size distributions in the original marine sediments, soil texture can range from sand to clay. The flights making up this continuous core show minimal effects of pedogenesis due to the high groundwater table and reducing conditions. The upper part of the soil profile may show redoximorphic properties (e.g., mottles) due to fluctuating water tables and alternation of reducing and oxidizing conditions. The length of the hydroperiod in the upper part of the profile largely determines surface soil characteristics.

Solonchaks

Solonchaks (from Russian *sol*, salt, and *chak*) are widespread in arid and semiarid regions, occurring widely (covering 260– 340 million ha) in many parts of Asia, Australia, South and North America, China, and the Middle East (Table C9). Salts responsible for salinity have many origins in parent materials that are related to marine deposition; volcanic, hydrothermal, and eolian processes; and lithologic properties of initial materials. Mangrove deforestation contributes to salinization of coastal soils, erosion, land subsidence, and release of carbon dioxide to the atmosphere. The presence of these salts, the amount of osmotic pressure of the soil solution, or the toxicity of a given ion leads to special landscapes, either occupied by salt-tolerant (halophyte) vegetation or characterized by the complete absence of vegetation (salt lakes, salt lagoons, salt pans, etc.), depending on the degree of salinity.

Two chemical criteria are used to define these soils: (1) the solubility product of the accumulated salts or of salts that may form and (2) the ion concentration in the soil solution. To be considered salt-affected, the soils must contain an important quantity of salts that are more soluble than gypsum ($CaSO_4 \cdot 2H_2O$: log Ks is −4.85 at 25°C), thus also more soluble than calcium carbonate, jarosite, and iron sulfide. For this reason, gypsiferous soils, calcareous soils, and soils with jarosite are excluded from the Solonchaks and have a home in the Gypsisols, Calcisols, and Thionic Fluvisols, respectively (Spaargaren, 1994). These soils must also have a total salt concentration of the soil extract, expressed as the electrical conductivity (EC) within a depth of 125 cm of the surface. This minimum value, measured in a saturated paste extract (or in a 1 : 1 extract for very sandy soils) must be: (1) >15 dS m^{-1} (or mS cm^{-1}) at 25°C, if the pH is less than 8.5 (neutral salts) or (2) more than 8 dS m^{-1} if the pH is more than 8.5 (for alkaline carbonate soils) or less than 3.5 (for acid sulfate soils other than Thionic Fluvisols).

Chloride soils are of marine origin but influenced by the potential acidity of the soils under mangrove. The pH is very acid to acid (3.5–5) in reaction, calcium is almost lacking and sulfates, entirely derived from seawater, are present in only very small quantities compared to chloride. Sodium is dominant in the soil solution (CL $\gg SO_4 > HCO_3$; Na $\gg$ Ca). White salt deposits (NaCl, $CaCl_2$, $MgCl_2$) may occur at the soil surface as crusts or efflorescence, which seasonally can turn into brown powdery surface layers. Examples are noteworthy in Gambia and in the Casamance region of Senegal.

Neutral chloride–sulfate soils are mostly of marine origin, but can be enriched by sedimentary gypsum deposited by rivers. When the pH is near neutral, sulfates and chlorides occur, and carbonates, sodium, calcium, and magnesium are in variable proportions and exceed bicarbonates present in variable quantities. Salt deposits are mostly NaCl, $CaCl_2$, and $MgCl_2$ associated with gypsum in the profiles. These soils are widespread and examples can be found in the alluvial plains, chotts, and sebkhas of North Africa, the Mexican Lagunas or playas, along the Euphrates River in Syria and Iraq, and in salt flats elsewhere.

Acid sulfate soils of marine origin are formed in potentially acid environments of mangrove vegetation (mangals) where acidification is promoted by dry periods in some tropical regions. When the environment becomes extremely acid (pH < 3.5), the result is acidolyse of clay and liberation of aluminum, iron, and magnesium; there is little calcium and chloride in these systems. The lack of bases and very low pH prevents the formation of jarosite. Examples occur in Gambia, the Casamance region of Senegal, and in Guinee Bissau. In reduced condition, these soils classify as Thionic Fluvisols as they have a sulfidic horizon; in oxidized and acidified states, these soils may classify as Salic Fluvisols if the EC is sufficiently high.

Carbonate soils include alkaline bicarbonate–sulfate soils and strongly alkaline carbonate soils. They form in aerobic environments with the production of sulfates from sulfides present in the original marine muds. The liberated acidity is neutralized by an excess of sodium and calcium carbonates in the biogeochemical system. Variations of these soils are found in the polders of Tchad, New Caledonia, the Balkan region, and Australia. The Solonchaks occupy the lower parts of landscapes where runoff, seepage, and shallow groundwater can accumulate; they are separated from most other soils by having a high salt content at or near the ground surface.

Podzols

The term *Podzol*, meaning "white earth" is derived from the Russian terms *pod* (beneath) and *zol* (ash). These soils are widely known as acid ashy-gray sands over dark sandy loams (Buol *et al.*, 1980). These two contrasting albic and spodic horizons, with an abrupt boundary between them, make these soils among the most dramatic, eye-catching, and photogenic soils in the world (Figure C45). The setting in which such a soil is produced requires the combination of soil-formation factors that yield necessary conditions where there is accumulation of iron, aluminum, and/or organic matter in a subsoil horizon. Podzols thus occupy large areas where there are sandy parent materials from the tropics to the boral climatic zone (485 million ha, Table C9). Many kinds of vegetation, including, grasses, can yield certain organic compounds that enhance podzolization, but litter from a certain few species of plants in particular foster such accumulations viz. the hemlock tree (*Tsuga canadensis*) in forests of northern latitudes, the kauri tree (*Agathis australis*) of New Zealand, and heath (*Calluna vulgaris* and *Erica* sp.) in northern Europe. Podzsols additionally occur under a wide variety of trees (e.g., *Picea, Pinus, Larix, Thuya, Populus, Quercus, Betula*) and understory plants in moist climatic zones that are cool (e.g., Humid Meso- and Microthermal climates). Humic Podzols (Cryohumods in *Soil Taxonomy*) occur as dominant soils in Canada, for example, in the Maritime Provinces (e.g., Cape Breton and Newfoundland), in the coastal forests of the British Columbia mainland, and on coastal islands (Clayton *et al.*, 1977). Extensive areas of these soils occur on the humid subtropical coastal plain of the southeastern United States. Impressive are the 9 m thick Bh horizons in Spodosols on the North Carolina Coastal Plain (Buol *et al.*, 1980). Ages of these soils are extremely variable, some Spodosols being able to form relatively quickly. Burges and Drover (1953), for example, reported evidence that 200 years were required to leach calcite from beach sand in New South Wales, Australia, but 2,000 years to produce an iron-Podzol, and 3,000 years to produce an iron humus Podzol with pH as low as 4.5. An iron-Podzol in coastal California has been estimated at about 1 million years (Jenny *et al.*, 1969). On coastal dunes in cooler climates, Podzols may develop within a few centuries, but in warmer or less-humid climates Podzols require several millennia or longer periods to form (Sevink, 1991). Dune soils are easily leached because of their high permeability and low water-storage capacity. Although sometimes high in carbonates, the acid buffering capacity of the non-calcareous fraction (e.g., quartz) is generally very low. As a consequence, acidification proceeds rapidly with concurrent changes in soil-forming processes and profile characteristics. Rates of leaching and acidification are high in climates with a large precipitation surplus and low decomposition rates (because of low temperatures). In arid climates, leaching will be strongly reduced and mobile components such as carbonates and gypsum may accumulate within the solum. Accumulations of secondary carbonates (e.g., caliche, kunkar) are frequently cited in studies of Holocene dyer systems in drier coastal zones.

Figure C45 Spodozol (Podzol) developed in siliceous, Pleistocene-age Pamlico Sands overlying a portion of the limestone Atlantic Coastal Ridge along the subtropical southeast coast of Florida. Here, prominent albic tongues (eluvial zone) extend at least 1.5 m into the organic- and iron-rich spodic horizon (illuvial zone). These coastal Spodosols are extremely photogenic due to the extreme biochemical differentiation within the profile, dramatic contrasts in soil color, abrupt wavy horizon boundaries, and thick profiles.

Podzolization is thus a bundle of processes that bring about translocation, under the influence of the hydrogen ion, organic matter, iron, and aluminum (and a small amount of phosphorus) from the upper part of the mineral solum to the lower part (Buol *et al.*, 1980). Cheluviation, the dominant process of soil formation of Podzols, involves the combination of two processes: (1) weathering of primary minerals (especially phyllosilicates) by organic complexing acids and (2) translocation of organic matter both in the weathering of primary minerals and in the translocation of Al and Fe complexes. These processes lead to special morphological and analytical characteristics that are used to identify a soil material as Podzolic. Cheluviation affects large areas of soils in the boreal zone. The process is also active in humid regions, especially in the temperate zone but also in the equatorial realm where many examples of Podzols have been described.

Podzols are soils characterized by the presence of an illuvial spodic horizon. In this horizon, amorphous compounds have accumulated consisting of organic matter and aluminum, with or without iron or other cations. The process of translocation and accumulation, known as cheluviation, is usually shown by the occurrence of an albic (bleached) horizon underlain by a spodic horizon. The illuviation of organic compounds can often be demonstrated by the presence of thick cracked organic coatings on the skeleton grains within the spodic horizon. In coarse sandy materials, Podzol morphology is well expressed and strong contrasts occur between eluvial (albic) and illuvial (spodic) horizons (cf. Figure C45). In most Podzols, where clay percentage is often less than 10%, coarse texture ranges from sand to sandy loam. Accordingly, water retention capacity is low, less than 50 mm and, although Podzols develop in humid climates, they often show moisture stress. Podzols are very acid soils, with a pH ranging from 3.5 to 4.5 in surface horizons; the pH may increase up to 5.5 in lower horizons. The cation exchange capacity (CEC) is mostly due to the organic compounds present and base saturation is always very low. Organic matter has a high C/N ratio especially in the surface horizons (C/N = 25 or more) and in the spodic (C/N = 20 or more), indicative of low biological activity and slow degradation of organic matter.

The limits of the Podzols are determined by the minimum expression of the spodic horizon and the minimal contrast between the eluvial and illuvial horizons. Soils showing evidence of illuvial organo Al/Fe complexes but lacking sufficient amounts of it to qualify for Podzols, form intergrades with Cambisols, Arenosols, and Gleysols. Podzols and intergrades are typical of a soil mantle on sandy coastal plains with poor quartzitic materials affected by a shallow water table.

Sesquisols, Ferralsols, Alisols, Acrisols, Lixisols

Sesquisols are soils either containing at shallow depth a layer indurated by iron (petroplinthite), or at some depth mottled material that irreversibly hardens after repeated drying and wetting (plinthite). *Plinthite* is an Fe-rich, humus-poor mixture of clay with quartz and other diluents that occurs as redoximorphic concentrations in platy, polygonal, or reticulate patterns (Shaw *et al.*, 1997). These kinds of soils occur mainly in the tropics but examples are common to old landscapes in (sub)tropical and temperate regions. Those with a shallow petroplinthite horizon are known as (high level) laterites, ironstone soils, or Sols ferrugineux tropicaux a cuirasse. They have widespread occurrence in western Africa, especially in the Sudano-Sahelian region where they cap structural tablelands; they are also common in central-southern India, the upper Mekong catchment, northern Australia, and the eastern part of the Amazon region (Table C9).

The Sesquisols with plinthite are known as Plinthosols, groundwater laterite, sols gris lateritiques, or Plinthaquox (Soil Survey Staff, 1992). They are found in extensively flat terrains with poor internal drainage, such as the late Pleistocene or early Holocene sedimentary basins of eastern and central Amazonia and the central Congo basin, but also on coastal plains such as in southwestern Western Australia (e.g., Swan Coastal Plain, Ridge Hill Shelf, Donnybrook Sunkland). Well-drained soils with loose ironstone concretions (sesquiskeletal material) are frequent nearly everywhere in the tropics and subtropics, in many landscape situations. The material is the result of former plinthite formation, subsequent hardening, and transport or re-weathering.

The central concept of Sesquisols is one of a soil affected, at present or in the past, by groundwater or stagnating surface water in which iron has been segregated to such an extent that a mottled (redoximorphic) layer has been formed which irreversibly hardens when exposed to air and sunshine. Included in the concept are those soils that have a hardened layer at shallow depth. The most important characteristic of Sesquisols therefore is the presence of plinthite and petroplinthite. Plinthite is an iron-rich, humus-poor mixture of kaolinitic clay with quartz and other diluents. It commonly occurs as red mottles in platy, polygonal, or reticulate patterns, and changes irreversibly on exposure to a hardpan or to irregular aggregates on exposure and repeated wetting and drying with free access of oxygen. In a perennial wet soil, plinthite is usually firm but can be cut with a spade. A plinthic horizon is at least 15 cm thick in which plinthite takes up 10% or more by volume. Petroplinthite is a continuous layer of indurated material in which iron oxides are an important cement and in which organic matter is absent, ore present only in traces. The Fe_2O_3 content is generally greater than 30%. The continuous ferruginous duricrust layer may be either massive, reticulate, interconnected platy, or columnar pattern that encloses non-indurated material. The indurated layer may be fractured, but then the average lateral distances between fractures is 10 cm or more and the fractures themselves do not occupy more than 20% by volume within the layer.

Sesquisols dominantly occur in intertropical regions and have linkages with Ferralsols, Alisols, Acrisols, and Lixisols. Sesquisols may occur in distinctly different landscape positions. Petric Sesquisols occupy the higher landscape positions, often as a result of landscape inversion due to lowering of the erosion base (McFarlane, 1976). They now form tablelands and are usually freely drained. The other Sesquisols are found mainly in depressions or other areas with impeded drainage. On the Coastal Plain of Georgia (USA), for example, most soils have sandy surface horizons with high infiltration rates; many, however, have slowly permeable subsoil horizons with high concentrations of plinthite (Shaw *et al.*, 1997) where only 10% platy plinthite is necessary to perch water (Daniels *et al.*, 1978). These soils have developed from Miocene aged sediments and are classified in *Soil Taxonomy* in fine-loamy, siliceous, thermic families of Plinthaquic, Aquic, Arenic Plinthic, Plinthic, and Typic Kandiudults. In Georgia, more than 1.5 million ha of soils containing plinthite have been mapped. Sesquisols are found in association with Gleysols and Stagnosols in areas conditioned by hydromorphy, and with Ferralsols, Alisols, Acrisols, and Lixisols, which occupy the better-drained positions in the landscape.

Ferralsols

Tropical weathering and pedogenesis produces Ferralsols that are characterized by a colloidal fraction dominated by low-activity clays (1 : 1 alumino-silicate minerals) and sesquioxides. These highly weathered soils lack rock fragments with weatherable minerals and are characterized by the presence of low-activity clays and a uniform morphology that lacks distinct horizonation. If there is sufficient iron in the original material, these soils are reddish in color and contain a weak pedal structure; there are few marks of other soil-forming processes such as clay accumulation through translocation. Secondary accumulation of stable minerals such as gibbsite or iron oxyhydrates may be present in concretionary forms (e.g., pisolites) as part of the fine earth fraction. Typical soils are situated on old geomorphic surfaces, which formed through erosion and deposition, and occur today as relict landscapes over large areas (750 million ha, Table C9). Having formed in transported and reworked colluvial materials, these strongly weathered soils have little relationship with underlying geological strata. As a result of transport and deposition, the soils may be stratified (e.g., McFarlane, 1976; Finkl, 1980, 1984). In extreme stages of weathering in a free leaching environment, there is relative accumulation of sesquioxides and small amounts of 1 : 1 alumino-silicate minerals (e.g., kaolinite) as well as resistant minerals such as anatase and rutile. The sesquioxide minerals include goethite, hematite, and gibbsite; the alumino-silicate minerals are generally degraded kaolinites and some muscovite. In this weathering environment, neogenesis of alumino-silicates is rare with the possible exception of aluminum chlorite.

Occurring on geomorphically old surfaces, these soils are associated with Cambisols in areas with rock outcrops or where rock comes near to the surface. On the stable surface they occur together with Acrisols, which seem often to be related to the presence of more acidic parent materials (e.g., gneiss). On more basic rocks they occur associated with Nitisols, with no apparent relation to underlying rocks or topographic positions. Near valleys, Ferralsols merge into Gleysols.

Alisols (from L. *alumen*, alum), occurring in humid (sub) tropical and warm temperate regions, are acid soils with a dense layer of illuviated clay in the subsoil; they cover more than 100 million ha worldwide (Table C9). The intense weathering processes in these soils degrade 2 : 2 and 2 : 1 clay minerals and release large amounts of aluminum to create a very acid environment. Alisols correlate partly with Aqults, Humults, and Udults in *Soil Taxonomy* (Soil Survey Staff, 1975, 1992), the Ferialsols or sols fersiallitiques tres lessive in French systems (e.g., AFES, 1992; CPCS, 1967), and Red Yellow Podzolic soils with a high clay activity in the Brazilian soil classification.

Acrisols (from L. *acris*, very acid) are characterized by a subsurface accumulation of activive clays, a distinct clay increase with depth, and a base saturation (by 1M NH_4Oac) of less than 50%. The soils have been named Red–Yellow Podzolic soils, Podzó licos vermelho-amarelo distróficos a argila de atividade baixa, sols ferrallitiques fortement ou moyennement désatures (CPCS, 1967), Red–Yellow Earths, Latosols, and oxic subgroups of Alfisols and Ultisols. The later have been redefined as kand- and kanhapl- great groups in *Soil Taxonomy* (Soil Survey Staff, 1975, 1992). These soils are common in tropical, subtropical, and warm climatic regions, on Pleistocene and older surfaces that take up more than 900 million ha worldwide (Table C9).

The dominant characteristic of Luvisols (from L. *luere*, to wash) is textural differentiation in the profile that forms a surface horizon depleted in clay and an accumulation of clay in the subsurface argic horizon. These soils are further characterized by moderate to high-activity clays and low aluminum saturation. These soils are known as sols lessivés in France, Parabraunerde in Germany, pseudo-Podzolic soils in Russia, Gray–Brown Podzolic soils in earlier US terminology or as Alfisols in *Soil Taxonomy*.

Lixisols (from L. *lix*, lye) are characterized by a subsurface accumulation of low-activity clays (cation exchange capacity is less than 24 cmol(+)/kg) and moderate to high base saturation. They show a distinct increase in clay content with depth. These soils have been named Red–Yellow Podzolic soils, Podzó licos vermelho-amarelo eutróficos a argila de atividade baixa, sols ferrugineux tropicaux lessives, and sols ferrallitiques faiblement désaturés appauvris (CPCS, 1967), Red and Yellow Earths, Latosols, and oxic subgroups in *Soil Taxonomy* (Soil Survey Staff, 1992). Limits and linkages of Lixisols, Acrisols, Alisols, and Luvisols is entirely based on analytical properties. Therefore, in areas where these kinds of soils have CECs close to this value, they will merge into each other.

These soils, which are relict (out of pedogenic phase with the present environment) and not related to contemporary soil-forming processes in coastal zones, are nevertheless important pedogenic and geomorphic features along many coasts; they occur about 435 million ha worldwide (Table C9). Although commonly associated with truncated landsurfaces that produce coastal cliffs such as those seen on marine sediments along the eastern boundary of the Guiana Shield in northeastern South America (Valeton, 1981); the coastal Trivandrum district in southern Kerala (India) (Karunakaran and Roy, 1981); Long Reef Beach near Sydney, New South Wales (Retallack, 1990); Darwin, Northern Territory (Australia) (McNally *et al.*, 2000); and as lateritic spurs and benches along the Darling Fault Scarp in southwestern Australia (e.g., McArthur and Bettenay, 1960), these weathering profiles also occur as reefs on seaward sloping surfaces that extend below present MSL (Short, 2003) in the Northern Territory and into the Gulf of Carpentaria (Hays, 1967). Interestingly, these Tertiary and older weathering profiles extend from northern Australia to the southern boundary of Papua New Guinea but are now drowned by the eastern Arafura Sea and Torres Strait. Depending on the thickness of the weathering profile and its intersection with present MSL, a variety of materials may occur along the shore with beaches or marine terraces developed in saprolite, pallid zones, mottled materials, or ferruginous duricrust.

In the case of the Darwin porcellanite (Darwin Member of the Bathurst Island Formation), a siliceous duricrust that was produced by leaching and replacement of preexisting rocks, the degree of alteration is less than that found in silcrete (McNally *et al.*, 2000). Pisolithic ferricrete (lateritic or ironstone "pea gravel") occurs in close association with the porcellanitic duricrust throughout the Darwin area. A karst-like relationship between the two types of duricrust (porcellanite versus ferricrete) is exposed along road cuts in Darwin, as described by McFarlane *et al.* (1995) where the ferricrete fills cavities in the porcellanite. This ferricrete is pedogenetically unrelated to the old lateritic pallid zone that developed in the Cretaceous sulfide-bearing marine muds following the retreat of the early Cretaceous sea. Porcellanization around Darwin is mostly confined to lower Cretaceous rocks that weathered to form the pallid zone of a laterite profile 50–70 million years ago. This deep weathering profile was later silicified in a number of cycles that were most intense during the Miocene (Milnes and Thierry, 1992). Both silicification and ferruginization continue to the present day, as evidenced by cliff-base 1940s junk that has been cemented together in a kind of "bottlecrete" by iron-charged spring waters. Silicified and ferruginous "beachrock" occurs at Fanny Bay (Darwin) between the present high and low tide lines. McNally *et al.* (2000) report that these beachrocks occur in a groundwater discharge zone, indicating that some duricrusts have developed since the postglacial sea-level stabilization around 6,000 BP. Marine erosion along the Darwin coast cut marine benches and eroded cliffs into deep mantles of chemical weathering that retain lateritic and siliceous Paleosols. Slumps of pallid zone materials at the cliff base provide residual lithogenic quartz and secondary (epidiagenetic) agglomerated porcellanized particles as beach materials. Parts of the beachface are silicified and ferruginized to form an unusual kind of "beachrock." Younger coastal soils have developed in the older deep weathering profiles that are exposed along the shore. Complicated polygenetic laterite profiles are also reported from Hong Kong (e.g., Wang, 2003) where pedogenic and ground-water origins represent paleosurfaces that require prolonged subaerial exposure, quiescent tectonism, and a certain level of climatic stability.

Planosols

Planosols (from L. *planus*, flat) have a bleached, light-colored eluvial horizon that abruptly passes into dense subsoil with significantly higher clay content. These soils typically occur in seasonally or periodically wet, level areas, often above normal flood levels of nearby rivers or estuaries. Planosols occur on nearly level river or marine terraces in Southeast Asia from Bangladesh to Vietnam, southern Brazil, southern and eastern Africa, and southern Australia. On a global scale, Planosols occupy about 130 million ha (Table C9). Melanization, the addition of dark-colored organic matter, is the process by which a dark surface horizon extends down into the profile. This bundle of specific processes includes: (1) extension of roots from grasses in the profile, (2) partial decay of organic materials in the solum, (3) reworking of the soil and organic matter (bioturbation) by earthworms, ants, rodents, with the development of krotovinas (folled burrows), (4) eluviation and illuviation of organic colloids along with some mineral colloids, and (5) formation of resistant "ligno-protein" residues that give a dark color to these soils (Buol *et al.*, 1980).

Cambisols

Cambisols (from L. *cambiare*, to change) are moderately developed soils that are characterized by slight to moderate weathering of parent materials. They do not contain appreciable amounts of illuviated clay, organic matter, aluminum and/ or iron compounds. These soils are particularly common in temperate and boreal regions that were influenced by ice during recent glacial periods. The poorly developed profiles result because soil parent materials are young and because soil formation is slow under low temperatures (or even permafrost) of high latitudes in northern Russia and Canada. Cambisols also develop in pre-weathered, old parent materials where they occur in association with highly developed soils such as Acrisols and Ferralsols. The largest continuous surface of Cambisols in warm regions is found on young alluvial plains, terraces, and deltas of the Ganges–Brahmaputra river system. Cambisols are also quite common in arid climates, where they are closely associated with Calcisols. Cambisols cover an enormous are of about 1.5 billion ha worldwide (Table C9) and constitute the third largest major soil grouping in the FAO Revised Legend (FAO, 1991). The minimum degree of soil development present in Cambisols separates them from Regosols. Leptosols are not permitted to have a cambic horizon and the shallowness of Leptosols separates them from Cambisols. When the cambic horizon reaches its maximum expression, Cambisols border soils defined by an argic horizon (e.g., Luvisols, Alisols, Acrisols, Lixisols, Nitisols), a spodic horizon (Podzols), or ferralic horizon (Ferralsols). Definition of these horizons is such that the cambic characteristics are excluded.

Arenosols

Arenosols (from L. *arena*, sand) are sandy soils with slight to moderate profile development. These soils are widely distributed and form one of the more extensive major soil groups in the world. FAO (1991) estimates Arenosols to cover about 900 million ha (Table C9), or 7% of the land surfaces; if shifting sand and active dunes are included, the coverage is about 10%. Sandy coastal plains and coastal dune areas are, of course, of smaller extent but ecologically very important. Although Arenosols occur predominantly in arid and semiarid regions, they are typical azonal soils that occur in the widest possible range of climates from very arid to very humid and from hot to cold. Arenosols occur predominantly on eolian sands, either dunes (Figure C46) or flats, but have also formed on marine, littoral, and lacustrine sands of beach ridges, lagoons, deltas, lakes, etc. Other names for Arenosols include Psamments (Soil Survey Staff, 1975, 1992), sols mineraux bruts (France), siliceous, earthy and calcareous sands, and various Podzolic soils (Australia), red and yellow sands (Brazil), and raw sands (Britain and Germany), among others. Arenosols thus have linkages with most other major soil groups. Only a few groups, such as Vertisols, have no linkages. The limits are determined by either textural requirements or the occurrence of certain diagnostic horizons within specified depths. Of most importance in their development are factors that limit soil horizon development in wetlands, alluvial lands, sand lands (Figure C47), rocky areas, and various unconsolidated materials. The spatial distribution of Arenosols indicates that many factors are involved, which operate in various combinations to limit profile development: (1) dry but hot to warm to cold climates and microclimates that parse the amount and duration of water movement in the soils (Arctic, subArctic, Antarctic, and temperate and intertropical desert zones provide these conditions), (2) mass wasting and solifluction remove surficial material faster than

Figure C46 Primary coastal dune on the Atlantic coast of southern Eleuthera Island, Bahamas. These Holocene carbonate dune sands are stabilized by vegetation and anchored to a rocky headland (photo center). Beachrock outcropping on the beachface helps to protect the dune foot from erosion.

most pedogenic horizons can form, (3) cumulization adds new material to the surface of the soil faster than new material can be assimilated into a pedogenic horizon, (4) immobilization of the soil plasma in inert materials, in carbonate-rich flocculated materials, and in some highly siliceous sediments to inhibit profile differentiation by illuviation, (5) exceptional resistance to weathering (pedologic inertia) of some initial materials, (6) infertility and toxicity of some initial materials to plant growth that inhibits biogenetic differentiation (e.g., on serpentine barrens), (7) saturation with water or even submergence for long periods, (8) short exposure of initial material to active factors of soil formation (e.g., marine flats newly exposed by uplift), and (9) a recent drastic change in the biotic factor that initiates formation of a new pedogenic regfime in an old one, which in turn serves as initial material (Buol *et al.*, 1980).

Regosols

Regosols (from Gr. *rhegos*, blanket) are well-drained, medium-textured, deep mineral soils that are derived from unconsolidated materials and which have been traditionally separated from shallow soils (Lithosols, Leptosols, etc.) and from those of sandy or coarser texture (e.g., Arenosols). A Regosol does not have assemblages of diagnostic horizons, properties, or materials that are definitive for any other group of soils. Most properties of Regosols are associated with the materials themselves and not with genetically developed soil features. Conceptually, Regosols are the initial state for pedogenesis representing recently deposited or recently exposed, earthy materials on a groundsurface; these weakly developed soils contain poorly defined horizons. Some soil materials, which represent a late stage of weathering with few distinguishing characteristics, are recognized as Regosols. For example, thin ferralitic materials that cannot go into Cambisols and may be classified with Regosols (Spaargaren, 1994). Although there are examples of initial stages of soil development in all landscapes throughout the world, the areal extent of Regosols is often limited and these soils are often inclusions in other soil units. Although of generally limited extent on a global scale (Table C9), Regosols can be important soils in specialized environments (e.g., tundra), such as Arctic plains in northern Canada where Cryic Regosols account for about 23.5% of Canada's land area (Clayton *et al.*, 1977), extending from Labrador and Baffin Island on the east to the Mackenzie Delta in the west. Nearly 3,000 km^2 of Cumulic Regosols (Dystric Fluvisols in *Soil Taxonomy*) occur in tidal marshes adjacent to the Bay of Fundy in the Maritime Provinces as well as on coastal floodplains adjacent to the Yukon and Mackenzie deltaic river systems. In the terrain, Regosols are mostly associated with degrading or eroding areas, while other soils occur on aggrading, depositional, or stable areas. As time passes and soil formation proceeds, Regosols may develop into many other soils depending on the most important soil forming factor(s). Sand dunes on the central Delmarva Peninsula of Maryland and Delaware, for example, contain weathering zones that are represented by Regosols (Entisols in *Soil Taxonomy*) with profiles that are only a meter or two in thickness (Denny and Owens, 1979). Basal horizons in the dunes constitute an ancient weathering profile, truncated at the top and overlain by a younger deposit of sand. These so-called two-cycle dunes thus contain an older basal dune that is Wisconsinan in age and the early episode of weathering may have been early Holocene, a comparatively warm Hypsithermal interval. The surface sand may be no more than a few hundred years old, the result of

Figure C47 Exposure of fine-grained calcareous dune sands on the landward side of dunes shown in Figure C. The imprint of pedogenesis is minimal and basically associated with accumulation of organic matter on the dune surface. Translocation of organic compounds down into the profile is limited to the upper half-meter or so.

deforestation and cultivation in colonial times (Denny and Owens, 1979). The Sharon Coastal Plain in Israel is predominantly comprised by sand dune fields and eolianite ridges. Cliffs along this plain contain three major sand accumulation periods between 65 and 50 ka, from 7 to 5 ka, and from 5 to 0.2 ka (Frechen *et al.*, 2002), that in turn seem to correlate with periods of sapropel formation in the Mediterranean Sea and so with periods of strongly increased African monsoon activity. The uppermost eolianite (calcareous sandstone) seems to correlate with a sapropel formed at about 7.8 ± 4.0 ka BP. Major phases of soil formation occurred between 35 and 25 ka resulting in a Rhodoxeralf soil (Grumusolic Dark Brown soil) and between 15 and 12 ka forming a Rhodoxeralf (hamra) soil. Frechen *et al.*, (2002) identify at least seven weak pedogenic phases resulting in weakly developed horizons of Regosol-type soils that are intercalated in the eolianite sequences. The oldest eolianite layer contains kurkur (caliche) and seems to be about 55 ka BP in age.

Coastal plain soils

As a general rule, soils of younger landscapes are less differentiated (i.e., they contain simple sequences of horizonation and plasmic soil materials are unorganized) than those of older surfaces where weathering profiles tend to contain highly organized soil materials with complex horizonation (e.g., Butler, 1959; Huggett, 1975; Buol *et al.*, 1980; Gerrard, 1981; Ollier and Paine, 1996). Because considerable time is needed to form differentiated soils where soil horizons and materials are highly organized, it is possible to sequentially arrange soils in chronological order with soils developed in young deposits placed before those formed later. Soils arranged in this way represent a sequence of development, referred to as a *chronosequence* (Jenny, 1941). If prior conditions of weathering continue, the young soils in time develop characteristics seen in the oldest (Gerrard, 1981; Ward and McArthur, 1983). This is a simplistic interpretation of uniform and uninterrupted development, but other more complicated (interrupted) sequences are possible, such as apparent trends due to secular changes in climate, drainage, or biota. Investigation of elemental mass balance trends, based on element translocation patterns in response to climate forcing, show promise for estimating the weathering flux under drier or wetter mean annual precipitation regimes (e.g., Stiles *et al.*, 2003). Optical dating of young coastal dunes (e.g., Ballarini *et al.*, 2003) provides a tool for reconstructing coastal evolution on a timescale of decades to a few hundred years. Quartz optically stimulated luminescence (OSL) dating is based on the fact that OSL dates of less than 10 years are very well-zeroed prior to deposition and burial, as demonstrated by Ballarini *et al.*, (2003) for an accretionary coastal segment near Texel, The Netherlands. Other luminescence sediment dating methods, more amenable to timescales of a very few thousand years, have found application in the study of the activation and stabilization of coastal dunes on a North Carolina barrier island (Berger *et al.*, 2003). Here, the quartz SAR (single-aliquot regenerative-dose method) is preferred for photonic dating in this region because of the relative abundance of quartz. These kinds of dating methods provide tight controls for estimates of pedogenic phases and the estimation of time intervals required to produce specific horizonation sequences in coastal dunes.

Radiocarbon dating is limited by the availability of datable materials and the technique can be applied only to younger soils (<50,000 yr BP) that range from latest Pleistocene to Holocene in age (e.g., Vogel, 1980; Bowman, 1990). The prospect of dating and correlating coastal and bottom deposits as well as terrestrial–marine weathering sequences, as seen in coastal soils, is best achieved along the littoral where sedimentation and subsequent soil development are closely associated with changes in sea level (e.g., Karpytchev, 1993). Lake-level histories as well can sometimes be gleaned from radiocarbon dates for coastal dune fields, as demonstrated by Arbogast and Loope (1999) for Holocene coastal dunes on the shores of Lake Michigan. When marine deposits are correlated with terrestrial weathering sequences by stratigraphic linkage at the coast, pedogenesis can be related to sea-level change and erosion–deposition cycles (Kukla, 1977). If dates related to sea-level changes can be assigned to coastal landforms, then the chronology can be extended inland by using local markers (Ward and McArthur, 1983). At the coast, abundant shells can supplement wood and peat as a basis for radiocarbon dating (Mook and van de Plassche, 1986).

If there is uncrystallized coral present, then dates back to 200,000 years can be obtained from uranium decay methods, and on volcanic shores potassium/argon dating provides useful information. For coasts where these methods are not applicable, the principal means of dating refers to the correlation of stranded shorelines. If the former shores in one locality cannot be dated directly, it is sometimes possible to relate them to the shores in other localities where dates exist, assuming that changes in sea level are instantaneous and global (ward and McArthur, 1983).

Because the earth's crust is in dynamic equilibrium with the asthenosphere and tectonic adjustments are always taking place, no coastal landmass can be regarded as completely stable, no matter how free it is from seismic activity or neotectonism. Few coasts are so mobile, however, that they contain no part of the eustatic record. Landforms in the coastal zone are complex and not all are only the product of terrestrial processes or marine action.

The apparent simplicity of most coastal plains is suggested by their lack of relief, which disguises these plains from their true nature as terminal geomorphic surfaces of complex transgressive–regressive depositional units laid down during glacioeustatic fluctuations. Bounded at the surface by terrestrial and marine erosional, transportational, and depositional landforms, the stratigraphic units (including paleosols) are limited below by unconformities of terrestrial and marine origin. Sedimentary facies between these surfaces reflect coastal plain (alluvial, palustrine, and eolian), estuarine, and marine open-shelf environments (e.g., Curray, 1978; Davis, 1978). Their evaluation requires a complete stratigraphic examination, as offered by Haq (1995) in summaries of the principles and concepts of sequence stratigraphy (the study of rock relationships within a chronostratigraphic framework wherein the succession of rocks is cyclic and is composed of genetically related stratal units viz. sequences and systems tracts).

Ideally, two main sedimentary sequences occur within a transgressive–regressive unit (e.g., Blum *et al.*, 2001). The terrestrial sequence, which develops as sea level regresses seaward through lowering, results from extension of alluvial erosion and sedimentation, and of soil formation, into the coastal zone. The marine sequence forms during subsequent landward transgressions or rise of sea level when marsh, estuarine, beach, dune, and deltaic sediments are introduced. The relative thickness, significance, and spatial extent of these sequences depend on the dynamics of coastal processes and on sediment volumes brought to the shore by rivers. When major rivers deposit large volumes of alluvium at the shore, the marine contribution to the sedimentary pile is limited. Coastal landforms thus result predominantly from alluvial distribution.

Coastal strand plains with no alluvial contribution, such as those forming in the Holocene landscape of Gippsland, Victoria, in Australia (Ward, 1977), lie at the other extreme of the scale. Relative placement of terrestrial and marine sediments here is determined by gradients at the coast, and by the history of eustatic fluctuations relative to the land. If, for example, sea level rises above a previous high stand, the earlier deposits will be truncated, buried, or partially submerged. Conversely, a eustatic fall will leave the earlier deposits stranded as a terrace.

Almost all the soils on the Swan Coastal Plain, southwestern Australia, are formed by material deposited by rivers (fluvial, alluvial soils) and wind (eolian soils). The Yilgarn Block (part of the West Australian craton), east of the Darling Scarp bordering the Swan Coastal Plain, was epeirogenically uplifted as part of cymatogenic movement and warping about 40–50 million years ago. This tectonic uplift increased the differential between base level and the plateau surface causing erosion of the craton by rivers and streams. Detritus from deep chemical weathering (e.g., laterite profiles) or saprolite was either deposited onto the Swan Coastal Plain or carried to the sea (Fairbridge and Finkl, 1980). The eroded material formed new parent materials for soil development on the Plain.

Broadly there are thus two major groups of soils on the Swan Coastal Plain. The first is a series of dune systems near the coast formed as a result of deposits from the sea, and including material originally derived from erosion of the Yilgarn block mixed with marine sediments. Once formed, the dunes can be eroded by strong winds. The sand is mostly eroded from the dunes nearest the coast, and is redeposited on the dunes further inland. The second major soil types are a series of soils formed by deposits directly eroded from the Yilgarn block and which comprise soils on an alluvial (Pinjarra) plain, occurring between the dunes and the scarp. There is a third, very narrow strip of soil, called the Ridge Hill Shelf, next to the scarp, formed from material eroded from the scarp.

Although different conditions prevail in different locations, some general principles emerge from regional comparisons. It is impossible to review the development of all coasts, but some cases illustrate the salient principles that are involved.

Soils of coastal dune systems

Several distinct features characterize coastal sand dune systems (e.g., Cooper, 1966; Nordstrom *et al.*, 1990). The dunes themselves often show a regular succession from the more active and unstable foredunes at the back of the beach to older, more stable vegetated dunes inland (cf. Figures C46 and C47). The other major element is the system of slacks or swales (low-lying, narrow coastal wetlands) that occur between the main dune or beach-ridge trends (cf. Figure C42). Dune slacks are low-lying areas within dune systems that are seasonally flooded and where nutrient levels are low. They occur primarily on the larger dune systems in the United Kingdom, for example, especially in the west and north where the wetter climate favors their development when compared with the generally warmer and/or drier dune systems of continental Europe. The range of plant communities found in slacks is considerable and depends on the structure of the dune system, the successional stage of the dune slack, the chemical composition of the dune sand, and the prevailing climatic conditions. When a coastal dune forms, dune sand accumulates seaward if there is an abundant sediment supply or the dune grows to its maximum height and then moves landwards (Gerrard, 1981). If the dunes are carbonaceous, leaching progresses with increasing age and when coupled with an increase in organic matter, there is a transition in soil reaction from alkaline to acid conditions. The rate of leaching is initially rapid but declines as the amount of carbonate, rather than the amount of rainfall, becomes a limiting factor. The rate of leaching in the early stages also depends on the nature of the shell fragments and matrix grain size, being slower if the fragments are large. The process is temporarily halted if material is added to the dune system by wind action because leached and organic-rich layers are buried.

Differences in soil moisture between the younger and older dunes, and between dunes and swales direct soil formation in regard to oxidation of organic matter (Gerrard, 1981). Young swales are initiated with a considerable advantage because nutrients and other minerals accumulate in the wetlands, a situation that is ultimately reflected in the soil profiles and pH values. The rate of increase of organic content depends on the initial lime content of the dunes. On lime-deficient dunes, early colonization by vegetation takes place with a concomitant increase in the rate of litter accumulation. As more acid conditions prevail, litter breakdown is inhibited and organic matter build up is promoted.

Swan Coastal Plain, southwestern Australia

The tectonically stable limestone coastline of southwestern Australia is well suited to preservation signals of former sea levels. The eustatic record is limited by the time frame when lithified dunes formed in middle to late Pleistocene times. Fairbridge (1961) interpreted the field evidence in calcarenite sequences and Paleosols as fluctuation of sea-level against the backdrop of a stable landmass, incorporating the curve in a proposed global standard. The sequence of former sea level stands was based on wave cut benches, submerged and stranded beach cobble deposits, karst, and marine bands associated with the Pleistocene limestone.

The Swan Coastal Plain in southwestern Western Australia presents a complex chronosequence of mid-Pleistocene to Holocene dune soils, the oldest interior sets (Bassendean dunes of siliceous sands with Podzol morphology) merging with a series of laterite-covered spurs of the Ridge Hill Shelf that form the foothills of the Darling Scarp (McArthur and Bettenay, 1979). The Bassendean Dune System is the oldest of the dune systems of marine origin and today consists of a series of low hills varying from 20 m to almost flat, which rise to 110 m above sea level. They were formed as a belt of coastal dunes and associated shoreline deposits that accumulated mostly during a period of high sea level, about 115,000 years ago during the Riss–Würm interglacial period (McArthur and Bettenay, 1979). Secondary or epidiagenetic calcrete commonly occurs in an interdunal area (between the Spearwood and Quindalup dunes), which is marine or estuarine in origin, as a cap on top of fossiliferous limestone that in turn is covered by a shallow sandy soil. The Swan Coastal Plain contains alluvium and dune sands (McArthur and Bettenay, 1960). The fluviatile deposits, displayed as coalesced alluvial fans, are of different ages and limited to the west by sequences of coastal dunes. The oldest Bassendean dune sequence is siliceous and Podzols have developed on them. The next Spearwood sequence has a core of eolianite beneath the weakly podzolized siliceous sand (Figure C48). The Quindalup dunes are calcareous and sequentially the youngest. Properties of the dune soils change in an orderly fashion with age. Assuming that the oldest was originally calcareous, there is progressive loss of carbonate, a lowering of the pH, and formation of Podzols. Between the two younger dune sequences is a zone that is marine in origin.

In the southern extension of the Swan Coastal Plain, the Quindalup dune systems extends along the coast and are backed by soils of the Stirling Swale, a marsh environment, and soils developed on older dunes farther inland (McArthur and Bettenay, 1956). Alluvial and swamp soils occur still inland on the coastal plain. This sequence of dune systems, often calcareous near the coast and siliceous on the landward side of the coastal plain, separated by intervening swales is typical of many coastal plains around the world. Alluvial soils truncate these sequences as rives make their way to the coast and swamps often prevail in older degraded dune systems.

Gippsland Area, Victoria, Australia

The coastal zone south of the Dividing Range in Victoria is separated from the sea by a narrow sand barrier (Ward, 1977). Generally characterized as extensive lowland with broad, low-lying sand plains and swamps around lagoons, 14 marine terraces elsewhere rise in shallow steps to the 130 m contour. Dune sands form low ridges on the marine terraces and alluvial terraces beside the main rivers terminate at former shorelines defined by stranded marine terraces, except for the youngest one that passes beneath present sea level. Three groups of soils are defined whether the parent sediment is eolian sand, alluvium, or stranded marine terraces. The soils on the eolian sands show progressive development of Spodosol characteristics when viewed in order of age. In the other groups, the most striking feature is the progressive development with increasing age of a contrast in texture between solum and subsoil. Changes in clay mineral suites and chemical composition accompany these changes in morphology.

Figure C48 Exposure of the Spearwood Dune System on the Swan Coastal Plain, southwestern Western Australia, near Perth. This dune system was formed during the Würm I interstadial about 80,000 years ago. The surface soils are leached and the carbonate has been precipitated below to form layers and columns of hard, compact limestone deposited from solution by percolating waters. Infilled solution pipes are clearly evident in these calcareous dunes that show multiple phases of solution and precipitation, erosion and deposition, and soil formation. Many solution pipes contain a thin rind of travertine in addition to being infilled with Terra Rosa type soil materials.

Southeast South Australia, Australia

The most widespread soils of the Kingston—Mt. Gambier region in southeast South Australia are derived mainly from calcareous beach sands (Podzols and Terra Rossa profiles) or from estuarine and lacustrine deposits (Groundwater Rendzina and Solodized Solonetz profiles) (Blackburn *et al.*, 1965). The calcareous beach sands occur in about 30 stranded Quaternary beach ridges that comprise a soil chronosequence with the oldest soils occurring more than 80 km inland from the present coast. Variation in morphological properties of the main kinds of soils shows gradual transition in age sequence that is represented by depth of red soil, thickness of kunkar, and occurrence of ferruginous concretions (iron-rich pisolites). The time span for soil formation for some of the older soils appears to be on the order of several hundred thousand years (Blackburn *et al.*, 1965). Most of the ridges are consolidated, immobile structures on which soil profiles show little evidence of truncation. In some places there is evidence of truncation in the form of secondary siliceous sand dunes and exposed kunkar. Most of the older landsurfaces are represented by non-truncated soil profiles of sand Podzolics.

Soils of Polar Coastal Plains

Cold soils occupy roughly one-third of the world's land area (Rieger, 1983). Maritime and continental climatic types are recognized, but each has varying degrees of expression. In maritime types, precipitation is high and uniformly distributed throughout the year, and temperature differences between winter and summer are minimal. Many sediments and soils in Polar regions have a complex history, particularly as they relate to cyclic deposits, glacial activity, wind action, solifluction, as well as cryogenic processes (Tedrow, 1977; Allard, 2001). The coastal plain on the northern tip of Alaska is flat and the drainage pattern is poorly integrated. Barrow is a low-relief emergent region where lakes, marshes, or former lake basins occupy 50–75% of the region. The remainder of the coastal plain consists of initial surface residuals, areas that do not show evidence of thaw-lake activity. Black (1964) reports that the complexity of the surficial materials near Barrow makes it difficult to subdivide the geologic deposits. Beaches, bars, distributary deposits, and deltaic deposits are interspersed with shallow-water marine sediments and lagoonal, lacustrine, and fluvial deposits. The extremely low ground and air temperatures subdue chemical and biological activity that inhibits pedogenesis. Studying weathering in northeastern Greenland, Washburn (1969, 1973) defines the major weathering processes in terms of oxidation, formation of desert varnish, case hardening, rillenstein formation, carbonate coating, salt wedging, granular disintegration, differential weathering, exfoliation, cavernous weathering, frost cracking, and frost wedging. More recently, Rieger (1983) pointed out the morphological impacts of freeze-thaw cycles, hydrostatic mounding and fissuring, and cryoturbation (viz. turbation accompanied by ice-wedge formation), on soil formation in cold regions. Study of cryoturbated Paleosols associated with past ice-wedge activity on late-Holocene sandy fluvial terraces in northern Québec (e.g., Allard, 2001), for example, corresponds with paleoclimatic information obtained from existing Arctic-wide and regional proxy records. The Little Ice Age stands out as a period of intense ice-wedge activity, followed by a warm thawing-out interval during the first half of the 20th century. From AD 1946 to 1991, climatic cooling reactivated ice-wedge formation (Allard, 2001). In some situations, especially in the main coastal tundra zone, some of the physical and chemical weathering processes are inseparable from incipient soil-forming processes (Tedrow, 1977). Due to the adverse and specialized conditions in these cold lands, researchers recommend different kinds of soil classification systems, which are reviewed in Tedrow (1977). Modern soil classification systems tend to incorporate soil on cold coastal plains as part of more comprehensive approaches, rather than have separate systems for specialized soil environments. In the Canadian system (Clayton *et al.*, 1977), for example, those cold region coastal soils that show weak profile development fall into the Podzolic

(Humo-Ferric Podzol) great group. Mostly Spodic Cryopsamments (in *Soil Taxonomy*) and the Brunisolic order (Cambisols in FAO/UNESCO; Inceptisols, excluding Aquepts in *Soil Taxonomy*) provide for the more normal soils. The Regosolic order (equates to the concept of Entisol, excluding Aquents, in *Soil Taxonomy*) provides for various soils with little or no profile development. Gleysolic orders (Aquepts, Aquents, and some Fluvaquents in *Soil Taxonomy*) include the gamut of poorly drained mineral soils with an organic layer up to 30 cm thick. Within the Canadian system, *cry* is attached to the names of 10 subgroups. The names Cryic Regosol and Cryic Gleysol indicate similarity to Regosol and Gleysol but also to those soils that are underlain by permafrost within a depth of 100 cm from the surface. The major genetic soils (excluding Ranker) of the tundra zone can be arranged in caternary form as described by Tedrow (1977) viz. Arctic Brown, Upland Tundra, Meadow Tundra, and Bog soil. The Arctic Brown represents zonal soils whereas the Tundra soil is indicative of hydromorphic conditions. Organic soils occur in low-lying, poorly drained coastal areas referred to as bogs, swamps, marshes, or muskegs. Since many coastal landscapes have been free of glacial ice within the Holocene (past 10,000 years) there are many depressions in the landscape, water-impoundment areas, and flat plains with sluggish surface drainage. Recently deglaciated landscapes tend to exhibit poorly developed surface drainage patterns that result in water-saturated environments and the formation of Bog soils. Russian investigators commonly refer to the presence of organic deposits in sectors that have "sculpture-accumulative" relief (Tedrow, 1977). Where coastlines are elevated, the organic deposits are usually intermixed with windblown sand. South of Barrow in the Meade River basin, surficial deposits have a sandy character with numerous blowouts, sand dunes, and vegetative mats that are buried by blowing sand. The coastal plain west of the Colville Delta contains geologic materials that incorporate a marine facies with dunes and loess. Because floodplains and sandbars are exposed to the wind action in summer, soils in the vicinity of coastal rivers and deltas contain admixtures of windblown sediments. The sand and silt is blown inland by onshore winds and deposited along the banks. Soil properties are characteristic for a xeric, willow-vegetated dune system along the river but where the dunes merge with tundra inland, the terrain is poorly drained and wetland soils develop.

Interesting organic-rich soils of cold regions are the ornithogenic soils of coastal continental Antarctica (Casey Station, Wilkes Land) where penguin guano plays an important role in nutrient cycles in the ecosystem (Beyer *et al.*, 1997). During the breeding season, seabirds bring huge amounts of organic matter from the sea to the land and this organic matter has accumulated to considerable thickness and consists of a friable layer of droppings, feathers, shells, bones, and dead penguins (Ugolini, 1972; Tedrow, 1977). On Signy Island (maritime Antarctica), 5% of the total snow-free surface is made up of ornithogenic soils from penguin guano. The chemical properties of the ornithogenic soils changes rapidly after the abandonment of the rookeries by the penguins, mainly because of CO_2 evolution and N and P release in the sea (Tatur, 1989). Other processes include the formation of oxalic acids in the first few centimeters and simultaneous concentration of recalcitrant soil compounds such as chitin, urates, and phosphate minerals. Leaching of iron and organic P fractions and the accumulation into the subsoil in maritime Antarctica produces a spodic Bh dark brownish (5 YR 3/1.5) horizon under a pale gray (10 YR 5/2) AE horizon, which are typical of Podzols in temperate regions (Beyer *et al.*, 1997).

Coastal barriers and soils as transitional biophysical systems

Barrier systems are widespread throughout coastal areas (Jelgersma *et al.*, 1970; Leatherman, 1979) and form the parent materials of soils. The barrier usually forms offshore, and a lagoon or backbarrier flat develops between the main barrier and the mainland; barrier–lagoon systems usually show transgressive, stationary, and regressive facies assemblages as seen along the coast of China (Li and Wang, 1991) and elsewhere. Distinctive soil suites similar in age develop on the barrier and the backbarrier flats (Daniels and Hammer, 1992). Some barriers form on the mainland, so only the soils seaward and those in the geomorphically equivalent estuaries or lagoons are the same age as the barrier. Goodwin (1987) describing coastal soil–landform relationships, reports that on the Coastal Plain of the Carolinas in the United States, soils on the landward Talbot Plain are older than the soils on the seaward Arapahoe Ridge or the Pamlico Plain. The Arapahoe Ridge and its association with the Talbot Plain is an example of why stratigraphic relations must be proven because the form alone often leads to incorrect conclusions.

Barrier systems may be wide, such as those in South Carolina, or very narrow as the Arapahoe Ridge in North Carolina (Goodwin, 1987) or the Atlantic Coastal Ridge in southeastern Florida (Finkl, 1994). These systems may be very smooth, broad flats with little local relief, or irregular as in prograding beach ridge systems (e.g., Tanner, 1995; Rink and Forrest, 2003). Most barrier systems are a series of barrier and backbarrier flats where the seaward soils tend to develop on sandy parent materials.

Soils formed in the barrier sediments typically are 80–95% sand, with fine and medium sands dominating (Markewich *et al.*, 1986). Spodic or humate horizons are common in the sandy barriers; in the Netherlands, peat and gytia interfinger with older dune systems and on top of the Older Dunes there is a strongly developed Podzolic soil (Jelgersma *et al.*, 1970). The Younger Dunes began to form in the 12th century and are weakly podzolized. The spodic horizons may be very few centimeters thick beneath dryer sites, or 6 or 7 m thick beneath broad flats with deep-water movement. Most soils on the Arapahoe Ridge have spodic horizons, as do soils on the beach ridge complex. The soils on the beach ridge are complex sands. The A1 and spodic horizons on the ridge are often discontinuous and thin. The soils in the swales have thick organic-rich A horizons and distinct spodic horizons. Soils and vegetation change abruptly in response to differing hydrological gradients across the ridge and swale sequence.

Soils on the backbarrier flats can have a wide range of textures. Many backbarrier systems are extensive, have a nearly level surface, and have finer textures than the adjacent barrier.

Coastal Paleosols

Paleosols are soils that formed on landscapes of the past (Yaalon, 1971). They are identified on the basis of soil-forming processes, which produce the soil profile morphology, that no longer operate due to changes in climate (Figure C49), local environmental conditions, or because of burial by younger (terrestrial or marine) sediments (cf. Figure C49), lava, or inundation (drowning) by the sea. Ruhe (1965) recognized three main types of Paleosols: buried, exhumed, and relict. *Buried Paleosols* form at the landsurface but are subsequently covered by sediment or other materials, a process that separates and removes the soil from the contemporary soil-forming zone (Figure C50). The depth of burial or integrity of the covering layer (e.g., lava) must suspend the soil-forming processes and yet be rapid-enough to prevent development of a Cumulic soil profile. Common examples occur in coastal dune fields, on overwashed barrier flats or spits, or on chenier plains where the rate of episodic shifting sands is slow-enough for soil formation to proceed and the new material to be incorporated into the soil profile. Epidiagenesis is common and resulting changes in original soil properties and horizons (formed prior to burial) must be considered in interpretation of Paleosol composition. *Exhumed Paleosols* occur at the land surface when a buried Paleosol is uncovered by erosion of the blanketing material. These soils may be out-of-phase with existing environmental conditions, as they occur at the time of reexposure. Exhumed Paleosols are commonly found in calcarenite sequences or active (including periodically reactivated) dune environments. *Relict Paleosols* formed on preexisting landscapes but were never buried all the while environmental conditions changed to a different regime, leaving behind weathering products that characterized prior landscape conditions. Common examples occur in a variety of landscapes, but those associated with ancient tropical and subtropical land surfaces are exemplars viz. Seaquisols, Ferralsols, Alisols, Acrisols, Luvisols, and Lixisols.

A common example of buried coastal Paleosols occurs in horizontally bedded sequences, as in tropical southern Florida where there are five distinct marine units punctuated by episodes of subaerial weathering (Enos and Perkins, 1977). Exposure to the atmosphere and chemical weathering of the exposed rock by dissolution resulted in the formation of discontinuous bands of dense caliche-type crusts (pedogenic calcrete as also reported in the Bahamas by Brown, 1986), Paleosols, freshwater limestone, and laminated crusts (Enos and Perkins, 1977; Beach, 1982). Identification of subaerial exposure surfaces is an accepted means of identifying boundaries (unconformities) between different stratigraphic sub-divisions of the Quaternary units. Interestingly, some of these weathering surfaces are radioactive and referred to as Secondary Depositional Crusts (SDC) (Krupa, 1999). These SDCs are typically associated with increased natural gamma radiation immediately below subaerial exposure surfaces. These SDCs seem to be related to the paleo groundwater interface (Krupa, 1999; Harvey *et al.*, 2002), which was formed by residual downward percolation of rainfall through the vadose zone. The SDCs exhibit elemental abundances (e.g., Ca, Mg, Al, Sr, Fe, U, Th, and K) above the natural background level for the area.

Figure C49 Weakly developed phases of pedogenesis in the disturbed upper part of a glacial till associated with the Ronkonkoma Moraine on Long Island, New York, USA. Two phases of soil formation (layers A and B), separated by intervening unweathered reworked sediments, occur near the upper part of the moraine, which advanced about 21,000 years ago as part of the latest Wisconsin glaciation cycle. Soil layer (A) is a buried soil or paleosol. Soil layer (B) is a contemporary surface soil. This most recent phase of soil formation is represented by an organic-rich accumulation at the surface and melanization of underlying materials. The translocation of organic compounds and oxidation of iron imparts a darker color to the subsoil. View is from Long Island Sound toward an erosional exposure that provides morainic sediments and weathered materials to the littoral system.

Figure C50 Paleosol bands (arrows) within a calcarenite sequence on the Atlantic coast of northern Eleuthera Island, Bahamas. These weak phases of soil formation took place during stable phases with the Wisconsin glacial cycle. Reactivation of dune formation imitated additional cycles of deposition followed by stabilization and soil formation. The dark color of the soil bands is due to oxidative weathering of iron-rich minerals that accumulated in the profile by eolian accession. Note that the seaward portion of the dunes and intercalated soils have been eroded, leaving soil layers and dune stratification projecting outward to where the dunes once stood.

The Bermuda record is a complex mosaic of carbonate eolianites, shoreline–marine limestones, calcarenite *protosols* ("accretionary Paleosols"), and Terra Rossa Paleosols (Vacher and Hearty, 1989; Herwitz *et al.*, 1996). Sayles (1931) was among the first to recognize that alternation of marine limestones, eolianites, and Paleosols in local sections provides a record of Pleistocene sea-level fluctuations. Bretz (1960), drawing diverse observations into a comprehensive history, established the correlation that eolian and marine limestones were deposited during Pleistocene interglacials when the shallow platform (−20 m) surrounding Bermuda was submerged and that Terra Rossa Paleosols formed during glacial-age platform emergences. Solutional depressions often formed in the eolianites subsequent to dune stabilization and weathering to produce solution holes or pipes. These depressions were sometimes infilled with fine-grained sediment (Fe-rich dust) that weathered to red soils or Terra Rossa (Herwitz and Muhs, 1995) (cf. Figure C48). The recurrent stratigraphic pattern, as reported by Vacher and Hearty (1989), features an erosional coastline marine deposit overlain by a regressive eolianite. Where the interglacial complex is entirely exposed above present sea level, high-stand erosional coastline deposits can be traced downdip into laterally extensive beach deposits of a depositional coastline. The beach deposits thus grade upward and laterally into eolianite, which in turn extends landward over the marine deposit and oversteps the erosional coastline deposits.

Bahamian paleosols, by way of another example of buried and exhumed Paleosols (Figures C51 and C52), are predominantly calcium carbonate and contain minor amounts of mineral-insoluble residue (quartz, plagioclase, and clay minerals) (Boardman *et al.*, 1995). There are at least four megascopically distinct Paleosols: laminated crusts, homogeneous crusts, breccia/conglomerate, and homogeneous matrix. All four types of Bahamian Paleosols and all mineralogic varieties of paleosols are found on stratigraphically equivalent rock units (Boardman *et al.*, 1995). The number and character of Paleosol-bound parasequences on New Providence Island (Nassau), Bahamas, suggest that at least five interglacial cycles are distinguishable. Whole-rock amino acid racemization (AAR) ratios confirm the stratigraphic sequence and indicate that isotope Stages 1 through 11 are likely included among eight depositional packages (Hearty and Kindler, 1997). The Bahamian stratigraphic and pedogenic sequences, not unlike other calcarenite environments elsewhere, are complicated by the intricate stacking of similar materials in complex geomorphological settings that respond to fluctuating climates and sea levels (e.g., Carew and Mylroie, 1994). Similar late Pleistocene– Holocene coastal dune sequences and buried soils are described for the southeast Queensland coast by Ward and Grimes (1987), who suggest that Holocene sea level soon after 3,900 radio-carbon years BP was lower than present.

Figure C51 Buried paleosol in the supratidal zone on the northeastern Atlantic coast of Eleuthera Island, Bahamas. This buried Paleosol is now partially exposed along the landward margin of an abrasion platform, indicating active erosion of dune cover sands.

Figure C52 Close-up shot giving details of the buried soil layer seen in panoramic view in Figure C51. The abrupt stratigraphic boundaries above and below the soil layers (arrows mark boundaries) and contrasting nature of all three materials verifies the independence of the soil layer, now partially truncated leaving just the stump of a former profile. The hammer and Kubiena box (for collection of undisturbed, oriented samples) are for scale.

In addition to calcareous precipitates, other kinds of indurated materials related to ferruginous crusts and intensely weatherized materials are also common on coastal plains. Delaney (1966), for example, describes interesting mixed and contra-genetic sequences of caliche and laterite (Fe-duricrust) on the Rio Grande do Sul of southeast Brazil and northern Uruguay. Here, Quaternary units consist of a perched blanket of sand (the Itapoã Formation), arkosic sands and gravels (Graxaim Formation), the Serra de Tapes Laterite (a soil- stratigraphic unit), and a marine quartzose sand (Chuí Formation). The relict Serra de Tapes Laterite is not forming today and must have formed during a climate with high temperature and moderately high precipitation, conditions that do not exist on the coastal plain today. Pleistocene caliche was formed in a climate almost opposite to that of the Serra de Tapes Laterite.

Similar kinds of relationships are found in cold regions where ice-permafrost sequences on the Siberian coastal plain, with large polygonal ice wedges, represent excellent archives for paleoenvironmental reconstruction. Such deposits contain numerous well-preserved records (ground ice, Paleosols, peat beds, fossils), which permit characterization of environmental conditions during a clearly defined period of the past 60 ka (Schirrmeister *et al.*, 2002). Lake plains along the eastern shore of Lake Michigan provide analogous setups where buried soils occur in dune fields with cyclical development after relatively long (>100 year) periods of low sediment supply (Loope and Arbogast, 2002). Because soil formation takes place during stable landscape conditions, the presence of buried soils facilitated construction of a general hypothesis for late Holocene interaction of foredune and perched dune models on this lake plain, as it relates to multicentury lake-level fluctuations. It seems, on the basis of this study of 32 dune fields and included Paleosols, that the modern landscape originated after the postglacial peak in lake level (i.e., after the Nipissing transgression, about 5,500–5,000 cal year BP) (Loope and Arbogast, 2002).

The aforementioned descriptions feature Quaternary soils in contemporary coastal landscapes, but there are numerous other Paleosolic developments along ancient shores where old weathering sequences persist in the form of deep weathering profiles of the lateritic or bauxitic type in senile landscapes. Salients among these examples are found in the formation of lateritic bauxites in Cretaceous and Tertiary coastal plains. Bauxites commonly form elongate belts hundreds of kilometers long, parallel to lower Tertiary shorelines in India and South America (Valeton, 1983). Typical sediment associations are found in India, Africa, South and North America, that are characterized by: (1) red beds rich in detrital and dissolved material of reworked laterites, (2) lacustrine sediments and hypersaline precipitates, (3) lignites intercalated with marine clays, layers of siderite, pyrite, maracasite, and jarosite, and (4) marine chemical sediments rich in oolitic iron ores or glauconite. The geographic relationship between *in situ* lateritic bauxites and the shoreline at the time of bauxite formation are interesting, if not some what controversial. Occurrences of bauxites residing parallel to ancient coastlines is documented for the karst bauxites of the Mediterranean (Southern France, Greece, Yugoslavia). Most of these types of lateritic bauxites, which are soft at the time of formation, belong to the Aquox subgroup of the group of Oxisols in *Soil Taxonomy*. The Fe-rich parts eventually form hard ferricretes whereas the Al-rich parts form hard alucretes (Goudie, 1973).

Alternating sequences of landscape stability and instability, as related to phases of soil formation in shield areas, are comprehended in the *cratonic regime* (Fairbridge and Finkl, 1980). This model recognizes a thalassocratic–biostatic condition (a long-term stable phase) during which prolonged deep chemical weathering takes place. In contrast, the epeirocratic–rhexistatic condition is a shorter-term unstable phase when there is general ecological stress and erosion of weathering profiles. The dual-phased, long-term–short-term regime was developed from experience on the West Australian craton (Yilgarn Block) but is applicable to other cratonic margins around the world. The significance to coastal soils relates to occurrences of laterites, bauxites, and associated materials along contemporary coastlines and ancient shores. Phases of soil formation are related to landscape stability and high sea levels; erosional phases where there is stripping of weathered materials from the land surface is associated with low sea levels. This sequence of events along cratonic margins produces a wide range of distinctly contrasting coastal Paleosols that include and which are derived from residual weathering that juxtaposes Quaternary coastal plain soils with relict Tertiary and older weathering profiles.

Epeirocratic–rhexistatic phases on the Yilgarn Block (West Australian craton) have, for example, resulted in the stripping of regolith materials form the craton surface to produce a series of etchplains that represent degrees of soil removal (Finkl and Churchward, 1973; Finkl, 1979). The least amount of stripping is associated with incipient etchplains, which occur on the seaward-most margins of the craton, along the north–south trending Indian Ocean coastal boundary and along the east–west trending Southern Ocean coastal fringe. Buried etchplains occur on the southern margin of the craton where deep chemical weathering profiles are intersected by present sea level and overrun by coastal dune systems.

Conclusion

Coastal soils are, in a certain sense, almost enigmatic because most soils occurring in coastal zones have complex histories and are related to environmental conditions that no longer predominate. This fact is perhaps not surprising because coasts themselves are extremely complex, resulting from combinations of land movements and fluctuations of sea level. Some coasts thus have histories of ingression where relative sea-level rise drowns preexisting landscapes (Kelletat, 1995). The accompanying soils become drowned or are buried by younger coastal sediments and new phases of soil formation. In other cases, the seabed becomes emergent by sea-level fall or neotectonic uplift providing new parent materials for soil formation. In other situations, the land–sea boundary remains more or less stable long enough for soils to develop in coastal deposits such as mangrove muds. Whatever the mechanisms involved, there exists a diverse range of soils in coastal zones. Some are clearly "coastal" in the sense that they are closely related to present shoreline materials and conditions; they are therefore younger Holocene in age. Other soils occurring in coastal zones are not strictly "coastal" because they can form independently of maritime conditions. The complex array of coastal zone soils is not easily comprehended without careful study and some background in pedological concepts, principles, practices, and field experience. Complex terminologies of different soil classification systems also tend to confound all but the most intent coastal researchers who make the effort to ferret out the details. Coastal soils are an interesting topic that offers scope for studying coastal change or evolution as recorded in the pedological record.

It is seen that coastal zones contain the youngest soils on earth as well as some of the oldest. Newly exposed seafloor muds form parent materials for acid sulfate soils and Solonchaks, recently stabilized coastal dunes host youngest Holocene Regosols whereas some older dunes feature strongly differentiated Pleistocene Podzols. Holocene Gleysols and Histosols form in juxtaposed swales and slacks where drainage is poor. Planosols and Solonetz occur on marine terraces and deltaic regions along with Histosols, whereas Cryosols and Leptosols characterize cold region coastal lowlands. Developed on older coastal plains in the low to middle humid latitudes there is a range of pedogenic formation that includes Podzols, Planosols, Aliosls, Luvisols, and Arenosols and Calcisols in drier realms. Deltaic plains often feature Fluvisols, Planosols, Solonetz, and Histosols. Ancient surfaces truncated by coastal erosion, especially in intertropical regions, show Sesquisols, Ferralsols, Alisols, Acrisols, and Lixisols. Last but not least there are Anthrosols that characterize pedogenic regimes in coastal areas where there has been a long record of human habitation and cultivation of field crops. This description of coastal soils may seem like a long digression, but it is actually a light vignette because the situation is far more complex than what has been alluded to here.

This overview nevertheless provides a fair indication of the range and complexity of pedogenesis in coastal zones as well as indicating the close interrelationships between coastal processes, landforms, vegetation, hydrology, maritime climates, sediments, and the imprint of biochemical weathering on coastal landscapes and seascapes. More importantly, however, is the fact that coastal soils provide much biophysical information that is useful in attempts to unravel some of the major events in coastal evolution, especially phases of soil formation that correspond to sea-level stillstands and landscape stability.

Charles W. Finkl

Bibliography

AFES (Association Française pour l'Etude du sol), 1992. *Référentiel pédologique, principaux sols d'Europe*. Paris: INRA.

Allard, M., 2001. Late-Holocene climatic changes as detected by the growth and decay of ice wedges on the southern shore of Hudson Strait, northern Québec, Canada. *Holocene*, **11**(5): 563–578.

Amador, J.A., and Jones, R.D., 1995. Carbon mineralization in pristine and phosphorus enriched peat soils of the Florida Everglades. *Soil Science*, **159**(2): 129–135.

Arbogast, A.F., and Loope, W.L., 1999. Maximum-limiting ages of Lake-Michigan coastal dunes: their correlation with Holocene lake level history. *The Journal of Great Lakes Research*, **29**: 372–382.

Ballarini, M., Wallinga, J., Murray, A.S., van Heteren, S., Oost, A.P., Bos, A.J.J., and van Eijk, C.W.E., 2003. Optical dating of young

coastal dunes on a decadal timescale. *Quaternary Science Reviews*, **22**(10–13): 1011–1018.
Beach, D.K., 1982. *Depositional and Diagenetic History of Pliocene–Pleistocene Carbonates, Northwestern Great Bahama Bank: Evolution of a Carbonate Platform*. Doctoral dissertation, Miami, FL: University of Miami.
Berger, G.W., Murray, A.S., and Havholm, K.G., 2003. Photonic dating of Holocene back-barrier coastal dunes, northern North Carolina, USA. *Quaternary Science Reviews*, **22**(10–13): 1043–1051.
Beyer, L., Knicker, H., Blume, H-P., Bölter, M., Vogt, B., and Schneider, D., 1997. Soil organic matter of suggested spodic horizons in relic ornithogenic soils of coastal continental Antarctica (Casey Station, Wilkes Land) in comparison with that of spodic soil horizons in Germany. *Soil Science*, **162**(7): 518–527.
Black, R.F., 1964. Gubik formation of Quaternary age in northern Alaska. *U.S. Geological Survey Professional Paper*, 302-C, pp. 59–91.
Blackburn, G., Bond, R.D., and Clarke, A.R.P., 1965. *Soil Development Associated with Stranded Beach Ridges in South-East South Australia*. Melbourne, Victoria: CSIRO, Soil Publication No. 22.
Bloom, A.L., and Stuvier, M., 1963. Submergence of the Connecticut coast. *Science*, **139**: 333–334.
Bloom, A.L., and Yonekura, N., 1985. Coastal terraces generated by sea-level change and tectonic uplift. In Woldenberg, M.J. (ed.), *Models in Geomorphology*. Boston: Allen & Unwin: Symposium on Geomorphology International Series No. 14, Binghamton pp. 139–154.
Blum, M.D., Carter, A.E., Zayac, T., and Goble, R., 2001. Middle Holocene sea-level and evolution of the Gulf of Mexico Coast (USA). *Journal of Coastal Research*, **36**(Special issue): pp. 65–80.
Boardman, M.R., McCartney, R.F., and Eaton, M.R., 1995. Bahamian paleosols: origin, relation to paleoclimate, and stratigraphic significance. In Curran, A., and White, B. (eds.), *Terrestrial and Shallow Marine Geology of the Bahamas and Bermuda*. Boulder, CO: Geological Society of America Special Paper 300, pp. 33–49.
Boesch, D.F., Josselyn, M.N., Mehta, A.J., Morris, J.T., Nuttle, W.K. Simenstad, C.A., and Swift, D.J.P., 1994. Scientific assessment of coastal wetland loss, restoration and management in Louisiana. *Journal of Coastal Research*, **20**(Special issue).
Bowman, S., 1990. *Radiocarbon Dating*. Berkeley, CA: University of California Press and the British Museum.
Bretz, J.H., 1960. Bermuda: a partially drowned late mature Pleistocene karst. *Geological Society of America Bulletin*, **71**: 1729–1754.
Brown, R.B., Stone, E.L., and Carlisle, V.W., 1990. Soils. In Myers, R.L., and Ewel, J.J. (eds.), *Ecosystems of Florida*. Orlando: University of Central Florida Press, pp. 35–69.
Brown, T.W., 1986. *The Formation of Pedogenic Calcrete: Its Stratigraphic and Diagenetic Significance in the Quaternary Limestones on San Salvador Island, Bahamas*. Master's thesis, Bloomington, IN: Indiana University.
Buol, S.W., Hole, F.D., and McCracken, R.J., 1980. *Soil Genesis and Classification*. Ames: University of Iowa Press.
Burges, A., and Drover, D.P., 1953. The rate of Podzol development in sands of the Woy Woy District, N.S.W. *Australian Journal of Botany*, **1**: 83–94.
Butler, B.E., 1959. Periodic phenomena in landscapes as a basis for soil studies. C.S.I.R.O. (Australia) Soil Publication No. 14.
Butler, B.E., 1967. Soil periodicity in relation to landform development in southeastern Australia. In Jennings, J.N., and Mabbutt, J.A. (eds.), *Landform Studies from Australia and New Guinea*. Canberra: Australian National University Press, pp. 231–255.
Callaway, J.C., DeLaune, R.D., and Patrick, W.H., Jr., 1997. Sediment accretion rates from four coastal wetlands along the Gulf of Mexico. *Journal of Coastal Research*, **13**(1): 181–191.
Carew, J.L., and Mylroie, J.E., 1994. Discussion of Hearty, P.J., and Kindler, P., 1993. New perspectives on Bahamian geology: San Salvador Island, Bahamas. *Journal of Coastal Research*, **10**(4): 1087–1094.
Clague, J.J., Hutchinson, I., Mathews, R.W., and Patterson, R.T., 1999. Evidence for late Holocene tsunamis at Catala Lake, British Columbia. *Journal of Coastal Research*, **15**(1): 45–60.
Clayton, J.S., Ehrlich, W.A., Cann, D.B., Day, J.H., and Marshall, I.B., 1977. *Soils of Canada, Volume 1: Soil Report*; *Volume II: Soil Inventory*. Ottawa: Supply and Services Canada.
Cook, P.J., Colwell, J.B., Firman, J.B., Lindsay, J.M., Schwebel, D.A., and Von Der Borsch, C.C., 1977. Late Cainozoic sequence of South East Australia and sea level changes. *BMR Journal of Geology and Geophysics*, **2**: 81–88.
Cooper, W.S., 1966. *Coastal Sand Dunes of Oregon and Washington*. Boulder, CO: Geological Society of America Memoir, 72.
CPCS, 1967. *Classification des sols*. Grignon, France: ENSA, 87p.
Curray, J.R., 1978. Transgressions and regressions. Reprint of Original 1964 paper. In Swift, J.P., and Palmer, H.D. (eds.), *Coastal Sedimentation*. Stroudsburg, PA: Dowden, Hutchinson & Ross, Benchmark Papers in Geology No. 42, pp. 97–203.
Daniels, R.B., and Hammer, R.D., 1992. *Soil Geomorphology*. New York: Wiley.
Daniels, R.B., Perkins, H.F., Hajek, B.F., and Gamble, E.E., 1978. Morphology of discontinuous phase plinthite and criteria for its field identification in the southeastern United States. *Soil Science Society of America Journal*, **42**: 944–949.
Darmody, R.G., and Foss, J.E., 1979. Soil–landscape relationships of the tidal marshes of Maryland. *Soil Science Society of America Journal*, **43**: 534–541.
Davis, R.A., 1978. *Coastal Sedimentary Environments*. New York: Springer-Verlag.
Delaney, P.J.V., 1966. *Geology and Geomorphology of the Coastal Plain of Rio Grande do Sul, Brazil and Northern Uruguay*. Baton Rouge, LA: Louisiana State University Press, Coastal Studies Series No. 15.
DeLaune, R.D., Smith, C.J., Patrick, W.H., and Roberts, H.H., 1987. Rejuvenated marsh and bay-bottom accretion on the rapidly subsiding coastal plain of the U.A. Gulf Coast: a second-order effect of the emerging Atchafalaya Delta. *Estuarine, Coastal and Shelf Science*, **25**(4): 381–389.
DeLaune, R.D., Pezeshki, S.R., Pardue, J.H., Whitcomb, J.H., and Patrick, W.H., 1990. Some influences of sediment addition to deteriorating marshes in the Mississippi River deltaic plain: a pilot study. *Journal of Coastal Research*, **6**(1): 181–188.
Dent, D.L., 1986. Acid sulfate soils: a baseline for research and development. Wageningen, The Netherlands: International Institute for Land Reclamation and Improvement (ILRI), ILRI Publication No. 39.
Dent, D.L., and Pons, L.J., 1995. Acid sulphate soils: a world view. *Geoderma*, **67**: 263–276.
Denny, C., and Owens, J.P., 1979. Sand dunes on the central Delmarva Peninsula, Maryland and Delaware. Washington, DC: U.S. Government Printing Office, *Geological Survey Professional Paper* 1067-C.
Dreissen, P.M., and Dudal, R. (eds.), 1989. *The Major Soils of the World. Lecture Notes on their Geography, Formation, Properties and Use*. Wageningen: Agricultural University and Belgium; Katholieke Universiteit Leuven.
Enos, P., and Perkins, R.D., 1977. *Quaternary Sedimentation in South Florida*. Boulder, CO: Geological Society of America Memoir, 147.
Fairbridge, R.W., 1950. The geology and geomorphology of Point Peron, Western Australia. *Journal of the Royal Society of Western Australia*, **34**: 35–72.
Fairbridge, R.W., 1961. Eustatic changes in sea level. In Ahrens, L.H., Press, L., Rankema, K., and Runcorn, S.K. (eds.), *Physics and Chemistry of the Earth*, Volume 4. Oxford: Pergamon, pp. 99–185.
Fairbridge, R.W., and Finkl, C.W., 1980. Cratonic erosional unconformities and peneplains. *Journal of Geology*, **88**: 69–86.
FAO, 1991. *World Soil Resources. An Explanatory Note on the FAO World Soil Resources Map at 1:25,000,000 Scale*. Rome: Food and Agriculture Organization of the United Nations, Report No. 66.
FAO-UNESCO-ISRIC, 1988. Soil Map of the World. Revised Legend. Rome: Food and Agricultural Organization of the United Nations, World Soil Resources, Report No. 60.
Finkl, C.W., 1979. Stripped (etched) landsurfaces in southern Western Australia. *Australian Geographical Studies*, **17**(1): 33–52.
Finkl, C.W., 1980. Stratigraphic principles and practices as related to soil mantles. *Catena*, **7**(2/3): 169–194.
Finkl, C.W. (ed.), 1981. *Soil Classification*, Volume 1. Stroudsburg, PA: Hutchinson Ross, Benchmark Papers in Soil Science.
Finkl, C.W., 1982a. The geography of soil classification. *Quaestiones Geographicae*, **8**: 55–59.
Finkl, C.W., 1982b. Toward a comprehensive soil classification system. *Geographical Bulletin*, **21**: 41–47.
Finkl, C.W., 1984. Chronology of weathered materials and soil age determinations in pedostratigraphic sequences. *Chemical Geology*, **44**(1/3): 311–335.
Finkl, C.W., Jr. (ed.), 1994. Disaster mitigation in the South Atlantic Coastal Zone (SACZ): A prodrome for mapping hazards and coastal land systems using the example of urban subtropical southeastern. In Finkl, C.W. (ed.), *Coastal Hazards: Perception, Susceptibility and Mitigation*. Charlottesville, VA: Coastal Education and Research Foundation, pp. 339–366.
Finkl, C.W., 1995. Water resources management in the Florida Everglades: are 'lessons from experience' a prognosis for

conservation in the future? *Journal of Soil and Water Conservation*, **50**: 592–600.
Finkl, C.W., 2000. Identification of unseen flood hazard impacts in southeast Florida through interpretation of remote sensing and geographic information system techniques. *Environmental Geosciences*, **7**(3): 119–136.
Finkl, C.W., and Churchward, H.M., 1973. The etched landsurfaces of southwestern Australia. *Journal of the Geological Society of Australia*, **20**(3): 295–307.
Fletcher, C.H., and Sherman, C.E., 1995. Submerged shorelines on Oahu, Hawaii: Archive of episodic transgression during the deglaciation? In Finkl, C.W. (ed.), *Holocene Cycles: Climate, Sea Level, and Sedimentation*. West Palm Beach: Coastal Education and Research Foundation, *Journal of Coastal Research*, **17**(Special issue): 141–152.
Frechen, M., Neber, A., Dermann, B., Tsatskin, A., Boenigk, W., and Ronen, A., 2002. Chronostratigraphy of aeolianites from the Sharon Coastal Plain of Israel. *Quaternary International*, **89**(1): 31–45.
Gerrard, A.J., 1981. *Soils and Landforms: An Integration of Geomorphology and Pedology*. London: Allen and Unwin.
Gleason, P.J., Cohen, A.D., Smith, W.G., Brooks, H.K., Stone, P.A., Goodrick, R.L., and Spackman, W., Jr., 1984. The environmental significance of Holocene sediments from the Everglades and saline tidal plain. In Gleason, P.J. (ed.), *Environments of South Florida: Past and Present*, Volume II. Coral Gables: Miami Geological Society, pp. 297–351.
Goodwin, R.A., 1987. *Soil Survey of Pamlico County, North Carolina*. Washington, DC: U.S. Department of Agriculture, Soil Conservation Service.
Goudie, A., 1973. Duricrusts *in Tropical and Subtropical Landscapes*. Oxford: Clarendon.
Haq, B.U. (ed.), 1995. *Sequence Stratigraphy and Depositional Response to Eustatic, Tectonic and Climatic Forcing*. Dordrecht: Kluwer.
Harvey, J.W., Krupa, S.L., Gefvert, C., Mooney, R.H., Choi, J., King, S.A., and Giddings, J.B., 2002. Interactions between surface water and ground water and effects on mercury transport in the north-central Everglades. Reston, VA: U.S. Geological Survey, Water-Resources Investigations Report 02-4050.
Hays, J., 1967. Land surfaces and laterite sin the north of the Northern Territory. In Jennings, J.N., and Mabbutt, J.A. (eds.), *Landform Studies from Australia and New Guinea*. Canberra: Australian National University Press, pp. 182–210.
Hearty, P.J., and Kindler, P., 1997. The stratigraphy and surficial geology of New Providence Island and surrounding islands, Bahamas. *Journal of Coastal Research*, **13**(3): 798–812.
Hearty, P.J., Vacher, H.L., and Mitterer, R.M., 1992. Aminostratigraphy and ages of Pleistocene limestones of Bermuda. *Geological Society of America Bulletin*, **104**: 471–480.
Herwitz, S.R., and Muhs, D., 1995. Bermuda solution pipe soils: a geochemical evaluation of eolian parent materials. In Curran, H.A., and White, B. (eds.), *Terrestrial and Shallow Marine Geology of the Bahamas and Bermuda*. Geological Society of America, Bahamas-Bermuda Special Paper 300, pp. 311–323.
Herwitz, S.R., Muhs, D., Prospero, J., and Vaughn, B., 1996. Origin of Bermuda's clay-rich paleosols and their climatic significance. *Journal of Geophysical Research–Atmospheres*, **101**: 23, 389–23,400.
Huggett, R.J., 1975. Soil landscape systems: a model of soil genesis. *Geoderma*, **13**(1): 1–22.
Huiskes, A.H.L., 1990. Possible effects of sea level changes on saltmarsh vegetation. In Beukema, J.J. *et al.* (eds.), *Expected Effects of Climatic Change on Marine Coastal Ecosystems*. Dordrecht: Kluwer, pp. 167–172.
Hussein, A.H., and Rabenhorst, M.C., 2001. Tidal inundation of transgressive coastal areas: Pedogenesis of salinization and alkalization. *Soil Science Society of America Journal*, **65**(2): 536–545.
Inman, D.L., and Nordstrom, C.E., 1971. On the tectonic and morphologic classification of coasts. *Journal of Geology*, **79**(1): 1–21.
Inubushi, K., Furukawa, Y., Hadi, A., Purnomo, E., and Tsuruta, H., 2003. Seasonal changes of CO_2, CH_4 and N_2O fluxes in relation to land-use change in tropical peatlands located in coastal area of South Kalimantan. *Chemosphere*, **52**(3): 603–609.
Jelgersma, S., De Jong, J., Zagwijn, W.H., and Van Regteren Altena, J.F., 1970. The coastal dunes of the western Netherlands; geology, vegetational history and archaeology. *Mededelingen Rijks Geologische Dienst*, Nieuwe Serie No. 21.
Jenny, H., 1941. *Factors of Soil Formation. A System of Quantitative Pedology*. New York: McGraw Hill.
Jenny, H., Arkley, R.J., and Schultz, A.M., 1969. The pygmy forest-Podzol ecosystem and its dune associates in the Mendocina coast. *Madrono*, **20**: 60–74.
Johnson, M.E., 1992. Ancient rocky shores: a brief history and annotated bibliography. *Journal of Coastal Research*, **8**: 797–812.
Karpytchev, Yu, A., 1993. Reconstruction of Caspian sea-level fluctuations: Radiocarbon dating coastal and bottom deposits. *Radiocarbon*, **35**(3): 400–420.
Karunakaran, C., and Sinha Roy, S., 1981. Laterite profile development linked with polycyclic geomorphic surfaces in south Kerala. In Krishnaswamy, V.S. (ed.), *Lateritisation Processes* (*Proceedings of the International Seminar on Lateritisation Processes*, 11–14 December 1979, Trivandrum, India). Rotterdam: Balkema, pp. 221–231.
Kelletat, D.H., 1995. *Atlas of Coastal Geomorphology and Zonality*. West Palm Beach, FL: Coastal Education and Research Foundation.
Kukla, G.J., 1977. Pleistocene land—sea correlations. I. Europe. *Earth Science Reviews*, **13**: 307–374.
Krupa, S.L., 1999. *Recognition and Analysis of Secondary Depositional Crusts in the Surficial Aquifer System of Southeast Florida*. Master's thesis, Boca Raton: Florida Atlantic University.
Leatherman, S.P., 1979. *Barrier Islands: From the Gulf of St. Lawrence to the Gulf of Mexico*. New York: Academic.
Less, B.G., and Lu, Y., 1992. A preliminary study on formation of the sand dune systems in the northern Australian coastal zone. *Chinese Science Bulletin*, **37**(7): 587–592.
Li, C., and Wang, P., 1991. Stratigraphy of the late Quaternary barrier—lagoon depositional systems along the coast of China. *Sedimentary Geology*, **72**: 189–200.
Libbey, L.K., and Johnson, M.E., 1997. Upper Pleistocene rocky shores and intertidal biotas at Playa La Palmita (Baja California Sur, Mexico). *Journal of Coastal Research*, **13**(1): 216–225.
Lissman, J.C., and Oxenford, R.J., 1975. Eneabba rutile-zircon-ilmenite sand deposit, WA. In Knight, C.L. (ed.), *Economic Geology of Australia and Papua New Guinea, Volume 1 Metals*. Melbourne, Victoria: Australian Institute of Mining and Metallurgy, Monograph No. 5, pp. 1062–1088.
Liu, C., and Walker, H.J., 1989. Sedimentary characteristics of cheniers and the formation of the chenier plains of east China. *Journal of Coastal Research*, **5**: 353–368.
Lodge, T.E., 1994. *The Everglades Handbook: Understanding the Ecosystem*. Delray Beach, FL: St. Lucie Press.
Loope, W.L., and Arbogast, A.F., 2002. Dominance of an ~150-year cycle of sand-supply change in late Holocene dune-building along the Eastern Shore of Lake Michigan. *Quaternary Research*, **54**: 414–422.
Markewich, H.W. *et al.*, 1986. *Soil Development and its Relation to the Ages of Morphostratigraphic Units in Horry County, South Carolina*. Reston, VA: U.S. Geological Survey Bulletin No. 1589.
Maroukian, H., Gaki-Papanastassiou, K., Papanastassiou, D., and Palyvos, N., 2000. Geomorphological observations in the coastal zone of Kyllini Peninsula, NW Peloponnesus-Greece, and their relation to the seismotectonic regime of the area. *Journal of Coastal Research*, **16**(3): 853–863.
McArthur, W.M., and Bettenay, E., 1956. *The Soils and Irrigation Potential of the Capel-Boyanup Area, Western Australia*. Melbourne, Victoria: CSIRO Soils and Land Use Series No. 16.
McArthur, W.M., and Bettenay, E., 1960. The development and distribution of the soils of the Swan coastal plain, Western Australia. CSIRO Australia Soil Publication No. 16.
McArthur, W.M., and Bettenay, E., 1979. The land. In O'Brien, B.J. (ed.), *Environment and Science*. Nedlands: University of Western Australia Press, pp. 22–52.
McCollum, S.H., Cruz, O.E., Stem, L.T., Wittstruck, W.H., Ford, R.D., and Watts, F.C., 1978. *Soil Survey of Palm Beach County Area, Florida*. Washington, DC: U.S. Department of Agriculture, Soil Conservation Service.
McFarlane, M.J., 1976. *Laterite and Landscape*. London: Academic.
McFarlane, M.J., Ringrose, S., Giusti, L., and Shaw, P.A., 1995. The origin and age of karstic depressions in the Darwin-Koolpinyah area, N.T. In Brown, A.G. (ed.), *Geomorphology and Groundwater*. New York: Wiley, pp. 93–120.
McNally, G.H., Clarke, G., and Weber, B.W., 2000. Porcellanite and the urban geology of Darwin, Northern Territory. *Australian Journal of Earth Sciences*, **47**: 35–44.
Mendelssohn, I.A., and McKee, K.L., 1988. *Spartina alterniflora* dieback in Louisiana: time-course investigation of soil waterlogging effects. *Journal of Ecology*, **76**: 509–521.
Milliman, J.D., and Meade, R.H., 1983. World-wide delivery of river sediment to the oceans. *Journal of Geology*, **91**: 1–21.

Milnes, A.R., and Thierry, M., 1992. Silcretes. In Martin, I.P., and Chesworth, W. (eds.), *Soils and Paleosols*. Amsterdam: Elsevier, pp. 349–377.

Mook, W.G., and van de Plassche, 1986. Radiocarbon dating. In van de Plassche, O. (ed.), *Sea-Level Research: A Manual for the Collection and Evaluation of Data*. Norwich: Geo Books, pp. 525–560.

Nelson, A.R., and Kashima, K., 1993. Diatom zonation in southern Oregon tidal marshes relative to vascular plants, foraminifera, and sea level. *Journal of Coastal Research*, **9**(3): 673–697.

Nordstrom, K., Psuty, N., and Carter, W., 1990. *Coastal Dunes: Form and Process*. Chichester: Wiley.

Nyman, J.A., Delaune, R.D., Roberts, H.H., and Patrick, W.H., Jr., 1993. Relationship between vegetation and soil formation in a rapidly submerging coastal marsh. *Marine Ecology Progress Series*, **96**: 269–279.

Ollier, C., and Pain, C., 1996. *Regolith, Soils and Landforms*. Chichester: Wiley.

Otvos, E.G., and Price, W.A., 1979. Problems of chenier genesis and terminology—an overview. *Marine Geology*, **31**: 251–263.

Petit, M., 1985. A provisional world map of duricrust. In Douglas, I., and Spencer, T. (eds.), *Environmental Change and Tropical Geomorphology*. London: Allen & Unwin, pp. 269–279.

Playford, P.E., and Leech, R.E.J., 1977. Geology and hydrology of Rottnest Island. *Geological Survey of Western Australia Report* No. 6.

Pons, L.J., and van Breeman, N., 1982. Factors influencing the formation of potential acidity in tidal swmaps. In Dost, H., and van Breeman, N. (eds.), *Proceedings of the Bangkok Symposium on Acid Sulphate Soils*. Wageningen, The Netherlands: International Institute for Land Reclamation and Improvement (ILRI), ILRI Publication No. 31, pp. 37–51.

Quarto, A., and Cissna, K., 1997. The mangrove action project. In Peck, D. (ed.), *Ramsar Convention on Wetlands*, 2p. [ramsar@ramsar.org; Ramsar Convention Bureau, Rue Mauverney 28, CH-1196 Gland, Switzerland].

Rabenhorst, M.C., 1997. The chrono-continuum: an approach to modeling pedogenesis in marsh soils along transgressive coastlines. *Soil Science*, **167**: 2–9.

Rabenhorst, M.C., and Swanson, D., 2000. Histosols. In Sumner, M.E. (ed.), *Handbook of Soil Science*. Boca Raton, FL: CRC Press, pp. E183–E209.

Radtke, U., 1989. Marine Terrassen und Korallenriffe—Das Problem der quartären Meerespiegelschwankungen erläutert an Fallstudien aus Chile, Argentinien und Barbados. *Düsseldorfer Geographische Schriften*, **27** (Düsseldorf).

Redfield, A.C., 1972. Development of a New England salt marsh. *Ecological Monographs*, **42**: 201–237.

Retallack, G.J., 1990. *Soils of Past: An Introduction to Paleopedology*. London: Harper Collins Academic.

Rieger, S., 1983. *The Genesis and Classification of Cold Soils*. New York: Academic.

Rink, W.J., and Forrest, B., 2004. Dating evidence for the accretion history of beach ridges on Cape Canaveral and Merritt Island, Florida, USA. *Journal of Coastal Research*, **20**(3).

Ritsema, C.J., van Memsvoort, M.E.F., Dent, D.L., van den Bosch, H., and van Wijk, A.L.M., 2000. Acid sulfate soils. In Sumner, M.E. (ed.), *Handbook of Soil Science*. Boca Raton, FL: CRC Press, pp. G121–G154.

Roberts, H.H., and Coleman, J.M., 1996. Holocene evolution of the deltaic plain: a perspective—from Fisk to present. *Engineering Geology*, **45**(1996): 113–138.

Ruhe, R.V., 1965. Quaternary paleopedology. In Wright, H.E., and Frey, D.E. (eds.), *The Quaternary of the United States*. Princeton, NJ: Princeton University Press, pp. 755–764.

Runge, E.C.A., 1973. Soil development sequences and energy models. *Soil Science*, **115**: 183–193.

Rutter, H., Schnack, E.J., Fasano, J.L., Isla, F.I., Del Rio, J., and Radtke, U., 1989. Correlation and dating of Quaternary littoral zones along the coast of Patagonia and Tierra del Fuego. *Quaternary Science Reviews*, **8**: 213–234.

Saito, Y., Wei, H., Zhou, Y., Nishimura, A., Sato, Y., and Yokota, S., 2000. Delta progradation and chenier formation in the Huanghe (Yellow River) delta, China. *Journal of Asian Earth Sciences*, **18** (2000): 489–497.

Sayles, R.W., 1931. Bermuda during the Ice Age. *American Academy of Arts and Sciences*, **66**: 183–190.

Schellmann, G., and Radtke, U., 2001. Neue Ergebnisse zur Verbreitung und Altersstellung gehobener Korallenriffterrassen im Süden von Barbados. In Schellmann, G. (ed.), *Von de Nordseeküste bis Neuseeland—Beiträge zur 19 Jahrestagung des Arbeitskreises "Geographie der Meer und Küsten"* vom 24–27 Mai 2001 in Bamberg. *Bamberger Geographische Schriften*, **20**: 201–224.

Schellmann, G., and Radtke, U., 2003. Coastal terraces and Holocene sea-level changes along the Patagonian Atlantic coast. *Journal of Coastal Research*, **19**(4): 983–1010.

Schirrmeister, L., Siegert, C., Kunitzky, V.V., Grootes, P.M., and Erlenkeuser, H., 2002. Late Quaternary ice-rich permafrost sequences as a paleoenvironmental archive for the Laptev Sea Region in northern Siberia. *International Journal of Earth Science*, **91**: 154–167.

Schlichting, E., 1973. Pseudogleye und Gleye—Genese und Nutzung hydromorpher Böden. In Schlichting, E. and Schwertmann, U. (eds.), *Pseudogley and Gley* (*Transactions of Commissions V and VI of the International Society of Soil Science*). Weinheim, Germany: Verlag Chemie, pp. 1–6.

Searle, D.J., and Woods, P., 1986. Detailed documentation of Holocene sea-level record in the Perth region, South Western Australia. *Quaternary Research*, **26**: 299–308.

Searle, D.J., Semeniuk, V., and Woods, P.J., 1988. Geomorphology, stratigraphy and Holocene history of the Rockingham-Becher Plain, South-western Australia. *Journal of the Royal Society of Western Australia*, **70**(4): 89–109.

Sevink, J., 1991. Soil development in the coastal dunes and its relation to climate. *Landscape Ecology*, **6**(1/2): 49–56.

Shaw, J.N., West, L.T., Truman, C.C., and Radcliffe, D.E., 1997. Morphologic and hydraulic properties of soil with water restrictive horizons in the Georgia Coastal Plain. *Soil Science*, **162**(12): 875–885.

Short, A.D., 2003. Cross-shore sediment transport around Australia—sources, mechanisms, rates, and barrier forms. *ASCE International Conference on Coastal Sediments* (Clearwater Beach, FL, 18–23 May 2003), *Coastal Sediments Book of Abstracts*, pp. 216–217.

Sibirtzev, N.M., 1901. Russian soil investigations. In Finkl, C.W. (ed.), *Benchmark Papers in Soil Science, Volume 1. Soil Classification*. Stroudsburg, PA: Hutchinson Ross Publishing Company, pp. 15–35.

Soil Survey Staff, 1975. *Soil Taxonomy: A Basic System of Soil Classification for Making and Interpreting Soil Surveys*. Washington, DC: U.S. Department of Agriculture, Agriculture Handbook No. 436.

Soil Survey Staff, 1992. *Keys to Soil Taxonomy*. U.S. Department of Agriculture, Soil Management Support Services, *Technical Monograph* No. 19.

Spaargaren, O.C. (compiler and ed.), 1994. *World Reference Base for Soil Resources*. Rome: FAO, Land and Water Development Division, 161p.

Spaargaren, O.C., 2000. Other systems of soil classification. In Sumner, M.E. (ed.), *Handbook of Soil Science*. Boca Raton, FL: CRC Press, pp. E137–E174.

Spackman, W., Dolsen, C.P., and Riegel, W., 1966. Phytogenic organic sediments and sedimentary environments in the Everglades—Mangrove complex. Part I: evidence of a transgressing sea and its effects on environments of the Shark River area of Southeastern Florida. *Paleontographica*, **117**(**B**), 135–152.

Stevens, J.C., Allen, L.H., Jr., and Chen, E.C., 1984. Organic soil subsidence. *Geological Society of America Reviews in Engineering*, **6**: 107–122.

Stiles, C.A., Mora, C.I., Driese, S.G., and Robinson, A.C., 2003. Distinguishing climate and time in the soil record: mass-balance trends in Vertisols from the Texas coastal prairie. *Geology*, **31**(4): 331–335.

Tanner, W.F., 1995. Origin of beach ridges and swales. *Marine Geology*, **129**: 149–161.

Tatur, A., 1989. Ornithogenic soils of maritime Antacrtica. *Polar Research*, **10**: 481–532.

Tedrow, J.C.F., 1977. *Soils of the Polar Landscapes*. New Brunswick, NJ: Rutgers University Press.

Thorp, J., and Smith, G., 1949. Higher categories of soil classification: order, suborder and great group. *Soil Science*, **67**: 117–126.

Ugolini, F.C., 1972. Ornithogenic soils of Antacrtica. In Liano, G.A. (ed.), Antarctic terrestrial biology. *American Geophysical Union Antarctic Research Series*, **20**: 181–193.

Vacher, H.L., and Hearty, P., 1989. History of Stage 5 sea level in Bermuda: review with new evidence of a brief rise to present sea level during Substage 5A. *Quaternary Science Reviews*, **8**: 159–168.

Valeton, I., 1981. Bauxites on peneplained metamorphic and magmatic rocks, on detrital sediments and on karst topography, their similarities and contrasts of genesis. In Krishnaswamy, V.S. (ed.), *Lateritisation Processes* (*Proceedings of the International Seminar on Lateritisation Processes*, 11–14 December 1979, Trivandrum, India). Rotterdam: Balkema, pp. 15–23.

Valeton, I., 1983. Palaeoenvironment of lateritic bauxites with vertical and lateral differentiation. In Wilson, R.C.L. (ed.), *Residual*

Deposits: Surface Related Weathering Processes and Materials. London: Blackwell, pp. 77–90.
van Ghent, P.A.M., and Ukkerman, R., 1993. The Balanta rice farming system in Guinea Bissau. In Dent, D.L., and van Mensvoort, M.E.F. (eds.), *Selected Papers of the Ho Chi Minh City Symposium on Acid Sulfate Soils*. Wageningen, The Netherlands: ILRI Publication No. 53, pp. 103–112.
van Mensvoort, M.E.F., and Dent, D.L., 1997. Acid sulfate soils. In Lal, R. *et al.* (eds.), *Methods for Assessment of Soil Degradation*. Boca Raton, FL: CRC Press, pp. 301–333.
Vogel, J.C., 1980. Accuracy of the radiocarbon time-scale beyond 15000 B.P. *Radiocarbon*, **22**(2): 210–218.
Wang, Y., 2003. Coastal laterite profiles at Po Chue Tam, Lantau Island, Hong Kong: the origin and implication. *Geomorphology*, **52** (3/4): 335–347.
Wang, J., and Luo, S., 2002. Sulfur and its acidity in acid sulfate soil of Taishan coastal plain in southern China. *Communications in Soil Science and Plant Analysis*, **33**(3/4): 579–586.
Wanless, H.R., Parkinson, R.W., and Tedesco, L.P., 1994. Sea level control on stability of wetlands. In Davis, S.M., and Ogden, J.C. (eds.), *Everglades: The Ecosystem and its Restoration*. Delray beach, FL: St. Lucie Press, pp. 199–223.
Ward, W.T, 1977. *Geomorphology and Soils of the Stratford-Bairnsdale Area, East Gippsland, Victoria*. CSIRO Australia Division of Soils, Soils and Land Use Series No. 57.
Ward, W.T., and Grimes, K.G., 1987. History of coastal dunes at Triangle Cliff, Fraser Island, Queensland. *Australian Journal of Earth Sciences*, **34**: 325–333.
Ward, W.T., and McArthur, W.M., 1983. Soil formation on coastal lands and the effects of sea-level changes. In *Division of Soils, CSIRO, Soils: An Australian Viewpoint*. London: Academic, pp. 101–105.
Washburn, A.L., 1969. Weathering, frost action, and patterned Ground in the Mesters Vig District, northeast Greenland. *Meddel. Om Gronland*, **176**(4): 303 p.
Washburn, A.L., 1973. *Periglacial Processes and Environments*. London: Arnold.
Yaalon, D.H., 1971. Soil-forming processes in time and space. In Yaalon, D.H. (ed.), *Paleopedology*. Jerusalem: Israel University Press, pp. 29–40.
Yan, Q. Xu, S., and Shao, X., 1989. Holocene cheniers in the Yangtze Delta, China. *Marine Geology*, **90**: 337–343.
Yatsko, A., 2000. Of marine terraces and sand dunes: the landscape of San Clemente Island. *Pacific Coast Archaeological Society Quarterly*, **36**(1): 26–30.

Cross-references

Alluvial Plain Coasts
Barrier Islands
Beach Ridges
Beachrock
Bogs
Changing Sea Levels
Cheniers
Coastal Sedimentary Facies
Deltaic Ecology
Dune Ridges
Hydrology of Coastal Zone
Mangroves, Ecology
Mangroves, Geomorphology
Muddy Coasts
Peat
Salt Marsh
Sandy Coasts
Submerged Coasts
Vegetated Coasts
Weathering in the Coastal Zone
Wetlands

COASTAL SUBSIDENCE

Introduction and terminology

Coastal subsidence, which can be described as the downward displacement of the land relative to sea level, often occurs in deltaic regions associated with riverine and estuarine sedimentation. These regions are commonly sites of thick deposits of Tertiary, Pleistocene, and Holocene sediments, exceeding 12,000 m in the Mississippi River delta, for example (Penland *et al.*, 1988). Coastal subsidence in these depositional zones are a function of five processes, each described in detail below: (1) downwarping, (2) tectonic activity, (3) consolidation of Tertiary, Pleistocene, and Holocene deposits, (4) shallow subsidence, and (5) underground water and gas extraction (Figure C53). When only natural processes are considered, 1 + 2 + 3 define deep subsidence. Shallow and deep subsidence combined equal total subsidence (Cahoon *et al.*, 1995).

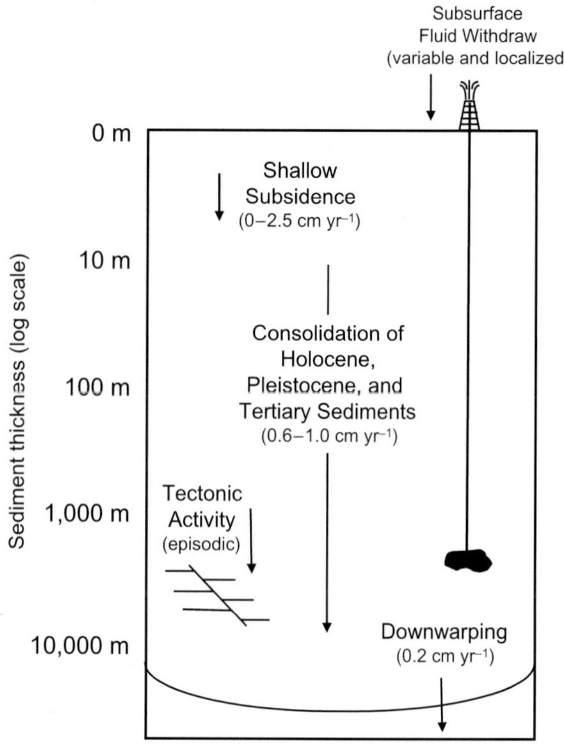

Figure C53 Processes contributing to total subsidence in coastal regions. Comments and numbers in parentheses indicate the relative magnitude of each process in the Mississippi River delta, a region where coastal subsidence has been intensely studied (modified and redrawn from Penland *et al.*, 1988).

Eustatic sea-level rise (the global sea-level rise due to changes in the volumes of glaciers and ice caps, and due to density-temperature relationships) can also be viewed as a "relative" land sinking or subsidence process. On a global scale, eustatic sea levels have risen approximately 125 m since the last glacial maximum 18,000 yr BP. Current rates of eustatic sea-level rise are estimated at 0.15 cm yr^{-1} (Church *et al.*, 2001). The term "relative sea-level rise" is used to describe deep subsidence plus eustatic sea-level rise because it is often difficult to differentiate deep subsidence rates from eustatic sea-level rise along the world's coasts.

Downwarping

The weight of thick sediment sequences deposited during the Quaternary by major rivers that have discharged considerable sediment load, such as the Amazon, Yangtze, Mississippi, Yellow, and the Ganges/Brahmaputra has caused downwarping of the underlying crust (Figure C54) (Coleman, 1982; Chen *et al.*, 2000). For example, in the Mississippi River delta, the combined weight of Cenozoic deposits accumulating in the geosyncline off the river mouth has resulted in the slow downwarping the Mesozoic basement. However, this process, estimated to be occurring at an average rate of 0.2 cm yr^{-1}, accounts for only a small fraction of total subsidence in the region (Penland *et al.*, 1988).

Tectonic activity

The tectonic-fault framework that exists in coastal regions can sometimes control the geomorphic configuration of the coast-delta extension (Coleman, 1982; Chen and Stanley, 1995) and the intensity of the fault motion plays an important role in influencing the overall subsidence rate (Stanley, 1988; Chen and Stanley, 1995). On a large-scale, fault tilting can trigger the diversion of large rivers from one region of a coast to

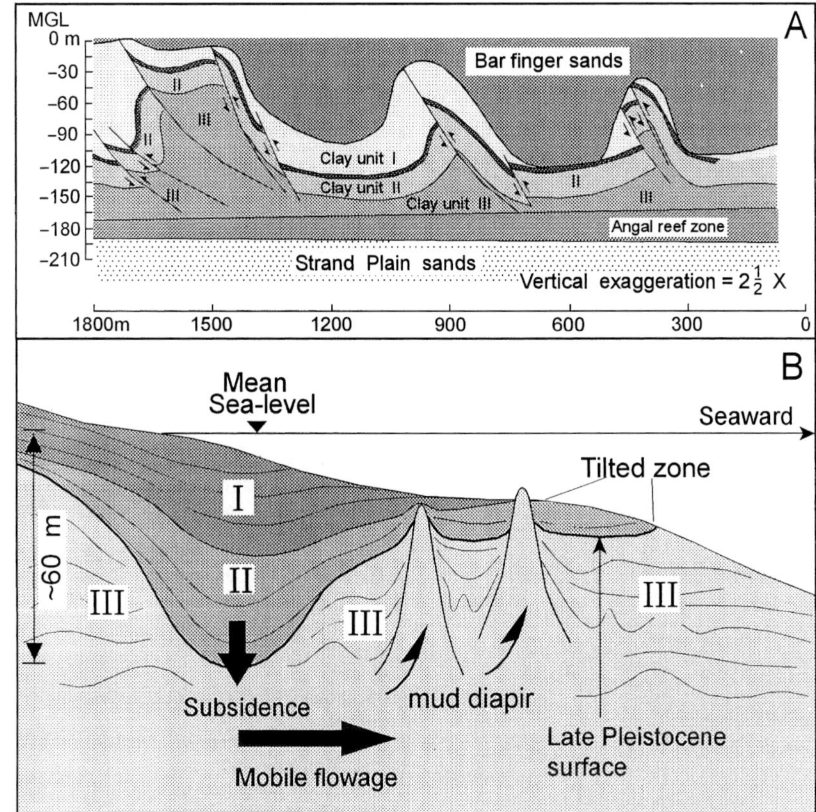

Figure C54 Diagram showing the gravitation subsidence both in the Mississippi and Yangtze estuarine depocenters. (A) The Mississippi model indicating the more seaward displacement than the downward, resulting in tremendous reverse mud slump in front of the river mouth region (modified after Coleman; 1982), and (B) The Yangtze model indicating downward and seaward sediment displacement.

another. Consequently, the estuarine depocenter marked by the thickest sediment sequences, and the most active subsidence, switches laterally. The subsidence triggered by tectonism is a long-term and episodic, geologic processes that is difficult to distinguish from the subsidence caused by the long-term compaction of Cenozoic sediments. A regional estimate of tectonic subsidence along the coast can be made by evaluating subsurface data including petrologic, litho- and chronostratigraphic descriptions from a series of boreholes through sedimentary sections.

Consolidation of Tertiary, Pleistocene, and Holocene deposits

The consolidation of Tertiary and Pleistocene marine and riverine deposits can be a significant component of total subsidence. However, it is the consolidation of Recent sediments that is considered the primary cause of background subsidence in coastal regions. Large amounts of riverine sediments deposited over the past 15,000 years have accumulated along coastal depocenters resulting in high rates of subsidence, regionally. This consolidation, or compaction, is caused by the dewatering of sediments (primary consolidation), the rearrangement of sediment structure (secondary consolidation), and the decomposition of organic matter. In some parts of the Mississippi River delta, subsidence rates due to consolidation exceeds 0.6 cm yr^{-} (Penland *et al.*, 1988).

Shallow subsidence

The compaction in the upper 5–10 m of the sediment contributes a shallow component that is often overlooked when calculating total subsidence in coastal regions (Cahoon *et al.*, 1995). Shallow subsidence is the result of primary consolidation and the decomposition of organic matter. It is an especially important process in coastal systems under stress (from flooding or salt, for example) where below ground plant structures, such as roots and rhizomes, die and decompose, leading to rapid subsurface collapse. In the Mississippi River delta, for example, high rates of relative sea-level rise in the coastal marshes have lead to increasingly long periods of flooding, plant mortality, and rates of shallow subsidence that exceed 2.5 cm yr^{-1}.

Underground water and gas extraction

In some coastal regions, large amounts of fresh groundwater have been extracted along the coast for industrial, agricultural, and domestic use. Rates of subsidence due to this process can be alarmingly high, although the effects are usually localized. For example, subsidence rates within the Po delta of Italy and the Yangtze delta of China have, in the some past years, exceeded 30 cm yr^{-1}, due largely to groundwater withdrawal (Cencini, 1998; Chen, 1998). The reinjection of salt water into the substrata, to replace extracted fluids, has been successfully used to slow down or even halt, but not reverse, extraction-induced subsidence. Hydrocarbon extraction along the coast has the same potential for causing land subsidence, but has been less studied.

Methods of measurement

Estimates of deep subsidence are usually based on long-term records from tide gauges that are mounted on stable piers, bridges, and pilings that extend through the shallow subsidence zone (and thus do not include shallow subsidence). A tidal gauge record spanning at least 18.6 years is required to factor out variations due to the moon's nodal cycle. Typically, mean annual or monthly water levels are regressed against time to yield a rate of relative sea-level rise (Emery and Aubrey, 1991). To estimate the deep subsidence component of relative sea-level rise, the eustatic component (currently 0.15 cm yr^{-1}) is then subtracted. Eustatic sea-level rise is derived from the analysis of tide gauge data from coasts worldwide that are assumed to be geologically stable (Penland *et al.*, 1988).

In addition, a radiocarbon-dated stratigraphy can help to calculate the net subsidence rate along the coast (Stanely, 1988). Stratigraphic analysis of sediment borings can serve to define the subsurface configuration of Recent deposits. Isopachous maps of the Holocene sediment sections can also reveal thickened sediment sequences along the coast. Moreover, high- resolution seismic profiles can provide evidence of Recent subsidence. Deformed late Quaternary strata, mud slumps, diapirs, tilted bedding, and gas-deformed structures in Holocene sequence imply remobilization of underconsolidated sediment within and adjacent to the estuarine depocenter (Chen *et al.*, 2000). Deformation types and their specific position relative to the depocenter record the displacement of sediments as a result of compaction due to sediment overburden.

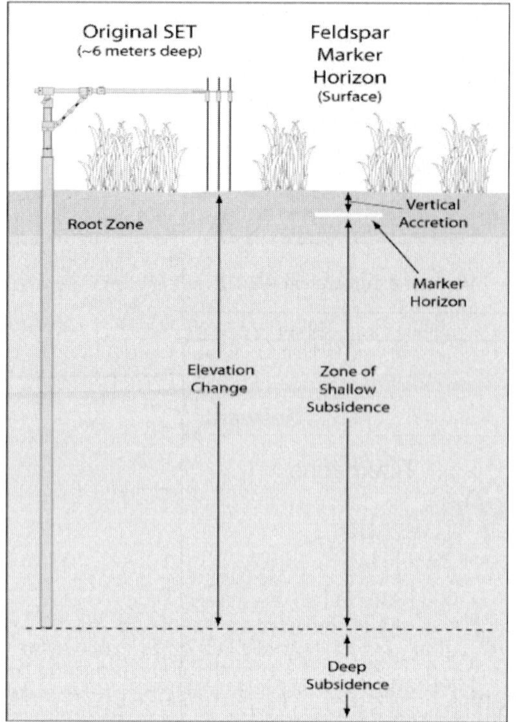

Figure C55 A surface elevation table (SET), shown in use with a feldspar marker horizon, used to estimate rates of shallow subsidence in coastal systems (Figure courtesy of Jim Lynch, and Don Cahoon, United States Geological Survey).

The tidal gauge technique for estimating subsidence (described above), and most estimates of subsidence reported in the literature, measure only deep subsidence, even though rates of shallow subsidence can greatly exceed the rates of either of these processes in some marshes (Cahoon *et al.*, 1995). A recently developed field technique that uses a sediment marker horizon (such as feldspar clay) in conjunction with a surface elevation table (Figure C55), an instrument that measures changes in elevation relative to a shallow subsurface datum, has made it possible to measure shallow subsidence (Cahoon *et al.*, 1995).

Subsidence and its implications

In stable or prograding coastal systems, relative sea-level rise is usually balanced by the accretion of fluvial or reworked marine sediments. However, extensive hydrologic alterations such as dams and levees have sharply reduced the amounts of fluvial sediments transported to coastal regions. As a result, saltwater intrusion, flooding, and coastal erosion has become much more prevalent in the past 100 years. In the Mississippi River delta plain, for example, land loss rates exceed 65 km^2 yr^{-1}, due largely to a subsidence/accretion imbalance. To add to this problem, the eustatic sea-level rise component of relative sea-level rise is expected to accelerate over the next 100 years due to global warming (Church *et al.*, 2001). Therefore, accurate measurements of coastal subsidence are vital for planning coastal protection in densely populated and low-lying region vulnerable to marine inundation.

Zhongyuan Chen and John Rybczyk

Bibliography

Cahoon, D.R., Reed, D.J., and Day, J.W., 1995. Estimating shallow subsidence in microtidal salt marshes of the southeastern United States: Kaye and Barghoorn revisited. *Marine Geology*, **128**: 1–9.

Cencini, D., 1998. Physical processes and human activities. *Journal of Coastal Research*, **14**: 774–793.

Chen, X.Q., 1998. Changjiang (Yangtze) River delta, China. *Journal of Coastal Research*, **14**: 838–858.

Chen, Z., and Stanley, D.J., 1995. Quaternary subsidence and river channel migration in the Yangtze Delta plain, Eastern China. *Journal of Coastal Research*, **11**: 927–945.

Chen, Z., Song, B.P., Wang, Z., and Cai, Y.L., 2000. Late Quaternary evolution of the subaqueous Yangtze Delta, China: stratigraphy, sedimentation, palynology and deformation. *Marine Geology*, **162**: 423–441.

Church, J.A., Gregory, J.M., Huybrechts, P., Kuhn, M., Lambeck, K., Nhuan, M.T., Qin, D., and Woodworth, P.L., 2001. Changes in sea level. In Houghton, J.T., Ding, Y., Griggs, D.J., Noguer, M., van der Linden, P.J., Dai, X., Maskell, K., and Johnson, C.A. (eds.), *Climate Change 2001: The Scientific Basis*. Cambridge: Cambridge University Press, Contribution of Working Group I to the Third Assessment Report of the Intergovernmental Panel on Climate Change, pp. 639–693.

Coleman, J.M., 1982. *Deltas: Processes of Deposition and Models for exploration*, 2nd edn. Boston: International Human Resources Development Corp.

Emery, K.O., and Aubrey, D.G., 1991. *Sea Levels, Land Levels and Tide Gauges*. New York: Springer.

Penland, S., Ramsey, K.E., McBride, R.A., Mestayer, J.T., and Westphal, K.A., 1988. *Relative sea level rise and delta-plain development in the Terrebonne Parish region. Coastal Geology Technical Report*. Baton Rouge: Louisiana Geological Survey.

Stanely, D.J., 1988. Subsidence in the Northeastern Nile Delta: rapid rates, possible causes, and consequences. *Science*, **240**: 497–500.

Cross-references

Coastal Sedimentary Facies
Dams, Effect on Coasts
Deltas
Eustasy
Greenhouse Effect and Global Warming
Isostasy
Sea-Level Rise, Effect
Sedimentary Basins
Submerged Coasts
Submerging Coasts
Tide Gauges

COASTAL TEMPERATURE TRENDS

Coastal temperature trends—temperature change per unit time—are an indicator of global climate change. Few long-term records of water temperatures are extant, but from many tide gauge sites of the US Coast and Geodetic Survey (USC&GS—now incorporated into the NOAA National Ocean Service), there are hand-written observations covering much of the 20th century. Herein a comparison is made between trends calculated from USC&GS water temperatures at selected tide gauge sites and air temperature trends from the Historical Climatology Network (HCN) of the United States.

Prior to development of acoustic tide gauges and digital data-logging and transmission, trained Tidal Observers would visit USC&GS primary tide gauge sites, typically on a daily basis. Their duties included comparisons between the marigram and the tide staff, measuring sea surface water temperature and density, and routine maintenance. The sea surface temperature (SST) values were recorded on standard forms and mean monthly and annual values were calculated by hand. These data forms were then archived, and largely ignored. Only 20 or so USC&GS tide gauge station temperature records of significant multi-decadal length have been recovered, and are analyzed in detail by Maul *et al.* (2001).

Tidal Observers determined SST by lowering a bucket from the pier or other structure supporting the tide gauge, retrieving a sample of water, and measuring the temperature with a standard mercury-in-glass thermometer. Seawater density was measured with a hydrometer (USC&GS, 1929). In typical Coast and Geodetic Survey tradition, all instruments were numbered and routinely calibrated, with emphasis on data quality dating back to 1807 when President Thomas Jefferson founded the agency.

The HCN was established in 1984 to provide a dataset suitable for detecting and monitoring secular changes of regional rather than local climate. There are over 1,200 stations in the HCN, with observations dating to the early 19th century, including air temperature, humidity, barometric pressure, and winds. The NOAA National Weather Service trains weather observers of the HCN—many of whom are volunteers. HCN air temperatures (AirTs) are mostly free from urbanization effects (the warming due to changing the environment from rural to urban). As

Table C10 SST trends from NOAA tide gauge sites and AirT trends from the HCN

Station	Years	N	SST trend*	AirT trend*	SST−AirT*
Boston	1921–94	69	3.6	0.8	2.8
New York	1926–94	61	1.8	1.1	0.7
Atlantic City	1911–91	54	0.9	1.2	−0.3
Baltimore	1914–93	65	0.9	2.1	−1.2
Charleston	1921–92	70	−0.1	−0.3	0.2
Mayport	1944–93	46	0.2	−2.0	2.2
Key West	1926–94	36	−0.0	0.6	−0.6
Galveston	1921–92	36	−0.1	−0.0	−0.1
La Jolla	1916–96	79	0.7	2.9	−2.2
Los Angeles	1923–91	40	0.8	1.3	−0.5
San Francisco	1921–94	57	0.5	0.4	0.1
Seattle	1922–94	41	0.0	1.1	−1.1
Neah Bay	1935–94	53	1.1	0.3	0.8
Summary			0.8 ± 1.0	0.7 ± 1.2	$r^2 = 0.06$

* °C per century.

Figure C56 Map of the contimerous United States showing the location of the 13 SST and AirT sites summarized in Table C10. Only sites with at least 35 years of data have been selected for analysis.

with the USC&GS tide gauge SSTs, HCN data are well regarded as the benchmark for high-quality observations in the United States.

In Table C10, trends in SST and AirT are summarized. Stations were chosen to reflect long-term SST trends at coastal locations somewhat evenly distributed around the United States (Figure C56), but the HCN and SST sites are not exactly coincident geographically. The HCN dataset is generally more complete than the SST dataset, and the controlling dates were driven by the availability of tide gauge temperature observations (N is the number of annual mean water temperature values available to be used in each trend calculation; due to missing data, N is not equal to starting year minus ending year). In all cases, the same years are used for the SST/HCN comparison, with the early 1990s limiting the SST data due to the change from human Tidal Observers to digital stand-alone instrumentation.

Trends in annual mean SST and AirT are calculated by standard linear least-squares methods where an equation of the form $T = a{\cdot}t + b$ (where T is temperature, a is temperature trend ($\partial T/\partial t$) in °C per century, t is time, and b is the linear intercept) is fitted to the data. The standard error in the trend a is typically ±0.03–±0.05°C/century. As can be seen in the summary row at the bottom of Table C10, both SST and AirT seem to be increasing in general, 0.8 ± 1.0 and 0.7 ± 1.2°C/century, respectively, where ±1.0 and ±1.2 is the standard deviation of the average of the corresponding trends.

There is a significant degree of variation from locale to locale. The largest SST trend is at Boston, MA. This unusually large trend, a = 3.6°C/century, may be caused by the geography of Boston Harbor, by a general warming of the Gulf of Maine, and/or by urbanization of the marine environment. Nearby tide gauge sites at New York, NY, Atlantic City, NJ, and Baltimore, MD have trends that are half or less that of Boston. For the northeast as a general value, the trend at the open coastal site of Atlantic City, a = 0.9°C/century, appears to be most representative of the region.

The southeastern United States, including the Gulf of Mexico, on the other hand, shows no statistically significant warming or cooling. SST trends at Charleston, SC are slightly negative, but nearby Mayport, FL are slightly positive. At Key West, FL, the southeastern region's most oceanic site, a = −0.0°C/century, a slight cooling trend.

Along the US west coast, the SST trends amongst the California sites are most similar, averaging about +0.7°C/century. There is an important difference between Seattle, WA and Neah Bay, WA. Seattle is well within Puget Sound with no large rivers to flush the estuary, and Neah Bay is at the western entrance to the Strait of Juan de Fuca near Cape Flattery—an oceanic environment. The zero SST trend at Seattle is unusual among the west coast stations, but serves to illustrate the variability in seawater temperature trends from place to place.

The HCN air temperature trends in Table C10 show a similar range of values compared to the SSTs. In general, the northeastern United States has warming air, the southeastern United States has cooling, and

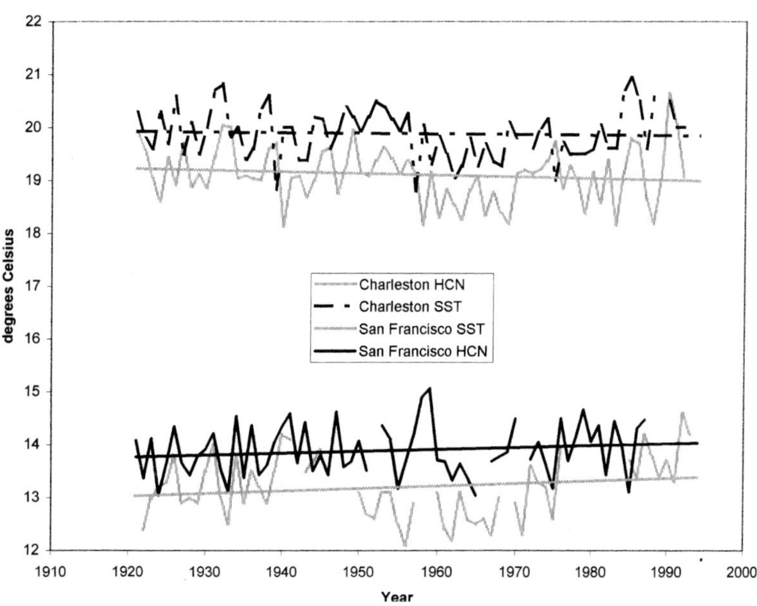

Figure C57 Time series plots of SST and AirT from Charleston, SC and San Francisco, CA. Note the missing data, which is typical of many such geophysical records, and the interannual variability.

the west coast again has warming. Many differences (SST minus AirT—column 6 in Table C10) exist between the tide gauge SST trends and the HCN air temperature trends from juxtaposed sites, the average difference being 0.1 ± 1.4°C/century. The very large difference at La Jolla, CA, for instance, may be more of a reflection that the SSTs are from an open ocean site (the Scripps Pier) and the HCN from an inland site a few kilometers away. A similar explanation is possible for Mayport, FL, but such differences illustrate the vagaries of climate monitoring.

Perhaps the most telling statistic is the value of the coefficient of determination (r^2) between the air and water trends, shown in the summary row in Table C10. A value $r^2 = 0.06$ is a statement that only 6% of the SST trends are explained by the AirT trends, and *vice versa*. The statistical relationship between these two important climate indicators suggests the relationship is nearly random.

To illustrate two of the datasets used to construct Table C10, the mean annual SST and AirT from Charleston, SC and from San Francisco, CA are plotted in Figure C57. Charleston is an example of cooling temperatures and San Francisco shows warming temperatures. It can be seen that temperatures have substantial variability about trend lines—a statement of inter-annual variability caused by climate cycles such as the El Niño–Southern Oscillation or the North Atlantic Oscillation. Figure C57 also shows that there typically are differences in the mean temperatures; in Charleston the mean SST is higher than the AirT, and in San Francisco the mean AirT is higher than the SST (San Francisco's famous fogs are in part caused by the air being cooled by the water).

This entry demonstrates the complex nature of climate change in the coastal United States using a homogenous dataset. As with sea-level change and other indicators of global change, SST and AirT are seen to be site-specific, and that it is poor scientific practice to extrapolate to other regions of Earth. It also shows the challenge and value of data archeology. Many climatic records are not yet digitized and many are deteriorating or are already lost. While these data do not improve clarity in the climate change debate, they offer an important opportunity to learn more about the evolving nature of our planet's geophysical history.

George A. Maul

Bibliography

Maul, G.A., Davis, A.M., and Simmons, J.W., 2001. Seawater temperature trends at USA tide gauge sites. *Geophysical Research Letters*, **28** (20): 3935–3937.

U.S. Coast and Geodetic Survey, 1929. *Instructions, Primary Tide Stations*. Special Publication No. 154, Washington: United States Government Printing Office.

Cross-references

Climate Patterns in the Coastal Zone
Coastal Climate
Environmental Quality
Greenhouse Effect and Global Warming
Meteorological Effects on Coasts
Water Quality

COASTAL UPWELLING AND DOWNWELLING

Definition

An interruption causes a horizontal motion of water to turn into a vertical flux to satisfy the mass balance. This vertical flux of water at the coast is known as coastal upwelling or downwelling, depending on upward or downward movement, respectively. Among others, wind-generated water surface current and its interruption by the coast is a major cause of coastal upwelling and downwelling. Therefore, most of these phenomena have seasonal timescales, while others are event-dependent. Coastal upwelling is more noticeable because it brings cooler subsurface water upward to the surface in the case of a thermally stratified water column, and is accompanied by primary productivity and abundance of fish resources. The phenomenon has considerable economic significance because it affects fisheries, weather, and oceanic currents in many parts of the world. This has prompted much interest in research on upwelling systems around the world. Some executed in the 1970s, including those under the umbrella of Coastal Upwelling Ecosystems Analysis (CUEA), were compiled through an International Symposium on Coastal Upwelling (Richards, 1981). The upwelled water often forms a plume with fronts that can be identified by remote sensing technologies. Upwelling and downwelling scales vary from localized episodes to large ocean areas. Upwelling speeds are very low and vary from 1 to 10 m/day, but upwelling regions may cover thousands of square kilometers.

The following sections describe the physics and types of coastal upwelling and downwelling systems, major places of their occurrence, environmental consequences, mathematical models of upwelling motion resulting from longshore windstress, and measurements.

Physics and types

The upwelling and downwelling processes can be discussed under four fundamentally different initiation mechanisms.

Longshore win

The most important episodes of coastal upwelling and downwelling, in terms of size, duration, and economic importance, occur due to longshore wind. When wind blows parallel to a long coastline for a considerable period of time (for a steady state to develop), a dynamic balance is established between the wind stress on the water surface and other hydrodynamic terms in the equation of motion. Figure C58 illustrates schematically the mechanism of coastal upwelling development due to longshore wind. Two simplifications of the hydrodynamic equation can best explain the cross-shore motion and the resultant episodes of upwelling or downwelling. The first is popularly known as Ekman transport that results from considering a dynamic balance between the

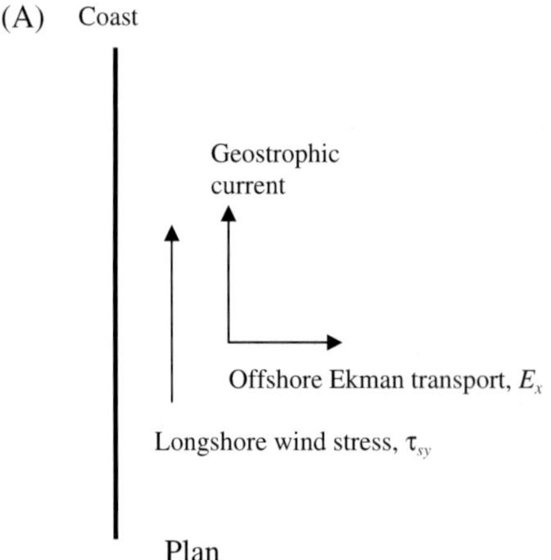

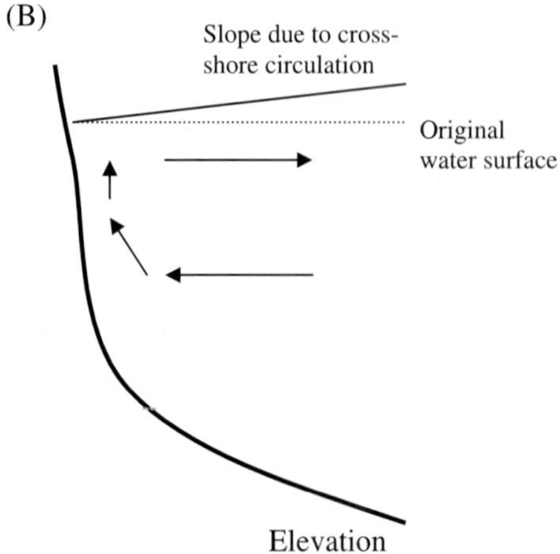

Figure C58 Coastal upwelling mechanism due to longshore wind.

wind-induced frictional force at the water surface and the Coriolis force. Swedish oceanographer Ekman (1905) first introduced this concept to explain quantitatively the ice-drift observations of Nansen in 1893. Due to a balance between these forces, a mass transport of water takes place in the surface layer, which is directed to the right of the blowing wind in the Northern Hemisphere (to the left of the blowing wind in the Southern Hemisphere). Depending on the wind direction relative to the coastline, water level is lowered or elevated along the shoreline, and mass conservation requirement causes vertical motions of upwelling or downwelling, respectively. For the northwesterlies along California and Oregon coast, this creates an offshore transport of water mass, and underwater upwells to the surface (McEwen, 1912). The second simplified explanation of the same motion is based on a geostrophic balance between the pressure gradient force induced by cross-shore Ekman transport and the Coriolis force. The pressure gradient force arises due to the development of a slope in the water surface (Figure C58). Further discussion of this mechanism is given in the section on mathematical models of coastal upwelling and downwelling.

Offshore and onshore wind

On a more localized scale, upwelling and downwelling also occur due to offshore and onshore winds. Onshore winds pileup water along the coast and downwelling occurs to satisfy the mass conservation requirement. The opposite happens, when an offshore wind lowers the water surface at the coast, and the underwater upwells to the surface. Thomson (1981) and Henry and Murty (1972) reported occurrence of upwelling due to an offshore wind in the Departure Bay of the Strait of Georgia. They found that due to upwelling, surface water temperature dropped by 4°C within less than 24 h of offshore wind blowing. Field *et al.* (1981) measured upwelling and downwelling episodes in response to offshore and onshore windstorms, respectively, in a kelp bed in southern Benguela off Cape Peninsula in South Africa.

Underwater ridges, shoals, and large-scale bedforms

When interrupted by underwater ridges, shoals, and large-scale bedforms, the horizontal flow upwells to the surface in the form of a surface boil. Such local upwelling, especially when associated with large horizontal current, may occur in many places in response to a large discontinuity of the seabed. Thomson (1981) reported several upwelling episodes in many passes of the Juan de Fuca Strait and Strait of Georgia along the British Columbia coast.

Open ocean divergence and convergence

From June to October, southeast trade winds cause upwelling in the equatorial Pacific (Thomson, 1981). The Coriolis force deflects the trade-wind generated South Equatorial Current to northward in the north of the equator and to southward in the south of the equator. A belt of divergence, known as *Equatorial Divergence* is thus created that prompts upwelling of cold underwater to the surface. A similar divergence is caused at about 10°N latitude in response to northeast trade winds. Divergence also occurs due to a moving pressure field associated with cyclones. A cyclonic low-pressure system causes surface-water divergence and upwelling occurs around the eye of the cyclone.

Geography

The following areas are some of the notable large-scale upwelling regions in the world (Bowden, 1983; Smith, 1983).

1. In the North Pacific, between 25°N and 45°N, along the California and Oregon coast in response to the equatorward California Current during summer northwesterlies.
2. In the South Pacific, between 5°S and 30°S, along the Peruvian coast in response to the equatorward Peru Current.
3. In the North Atlantic, between 10°N and 40°N, along the northwest coast of Africa in response to the equatorward Canary Current.
4. In the South Atlantic between 5°S and 30°S, along the southwest coast of Africa in response to the equatorward Benguela Current.
5. Along the east coast of Africa and south coast of the Arabian Peninsula in response to the poleward Somali Current during the southwest monsoon.

Consequences

Coastal upwelling is a very well-known phenomenon because of its practical interest. The important consequences of upwelling are abundance of fish resources in the upwelling region, influence on coastal climate, development of sea fogs, and intensification of existing coastal currents.

Very often, the upwelled water brings upward with it a higher concentration of nutrient than that at the resident surface water. The reason for a lower concentration in resident surface water is that overtime the available nutrients are used up by the growth of phytoplankton. The nutrients are usually nitrate, phosphate, and silicate salts. The enhanced nutrient content of surface water helps in the growth of phytoplankton that supports a greater zooplankton concentration. This primary productivity maintains a high fish population and the whole interdependence of lives in the region is known as an upwelling ecosystem. Most of the important marine fisheries of the world are found in regions of upwelling. Although coastal upwelling areas comprise only 1% of the total area of the ocean, they are estimated to produce more than 50% of the total marine fish harvest.

In an upwelling region, the cold upwelled water causes a drop in air temperature of the region, and sea fog develops locally in the presence of moist air. Thomson (1981) mentioned such effects along the west coast of Oregon and the British Columbia coast. Investigations off Oregon and Washington coast showed that the broad and slow southward flowing California Current is intensified into a narrow jet of stream with speeds exceeding 100 cm/s in the upwelling region.

Mathematical models

In this section, some mathematical models are described that deal with large-scale coastal upwelling due to longshore wind. In most models of large vertical motions in the ocean, the Coriolis force is balanced against different forces such as pressure gradient force, friction force, etc. In the simplest form of this model, known as the Ekman–Sverdrup model of upwelling, the Coriolis force is balanced against the steady wind forcing in homogeneous water. Together with continuity equation, the solution of the system of equations (taking x-axis across and y-axis longshore) gives the Ekman transport (Pond and Pickard, 1978; Bowden, 1983; Smith, 1983),

$$E_x = \rho_w wL = \rho_w uD = \frac{\tau_{sy}}{f}, \quad \text{(Eq. 1)}$$

where E_x is offshore Ekman mass transport, ρ_w is density of water, w is vertical velocity of upwelling, L is length across the shore of upwelling influence, u is cross-shore horizontal velocity, D is Ekman layer depth, τ_{sy} is wind stress on the water surface in the longshore direction, f is Coriolis parameter, and is given by $2\Omega \sin \phi$, where Ω is angular speed of the earth (7.29×10^{-5} rad/s) and ϕ is latitude. The wind stress can be estimated using quadratic friction law as,

$$\tau_{sy} = \alpha\rho_a U_{10y}^2 \quad \text{(Eq. 2)}$$

where α is air friction coefficient (~ 0.0015), ρ_a is air density ($= 1.25$ kg/m^3), and U_{10y} is the wind velocity at 10 m above mean sea level in the longshore direction.

The offshore Ekman transport is expressed in terms of Upwelling Index (UI) (Bakun, 1973),

$$\mathrm{UI} = \frac{E_x}{\rho_w}. \quad \text{(Eq. 3)}$$

The UI is also referred to as Bakun Index. The National Marine Fisheries Service of NOAA publishes monthly summaries of Upwelling Indices for major upwelling regions along the US coast.

Hidaka (1954) obtained an analytical solution to the system of hydrodynamic equations that considered a balance between the Coriolis forces and the diffusive transports in the vertical and horizontal directions. From his findings, some scales of upwelling motion could be estimated,

$$S_v = \sqrt{\frac{N_v}{f}}, \quad \text{(Eq. 4)}$$

$$S_h = \sqrt{\frac{N_h}{f}}, \quad \text{(Eq. 5)}$$

and

$$\frac{w}{u} = \frac{S_v}{S_h} = \sqrt{\frac{N_v}{N_h}}. \qquad \text{(Eq. 6)}$$

In these relations, S_v is vertical length scale, S_h is horizontal length scale analogous to mixing length, N_v is vertical eddy viscosity coefficient, and N_h is horizontal eddy viscosity coefficient. Garvine (1971) made an analysis by including the longshore water slope and showed that beyond the narrow coastal boundary layer frictional effects could be neglected.

In reality, ocean water is stratified, and a complete solution can only be obtained by taking into account all the forces. Some early contributions in this direction were given by Yoshida (1967), Allen (1973), and Hamilton and Rattray (1978). Their solutions are dependent on defining some characteristic baroclinic terms:

$$E_v = \frac{N_v}{fh^2}, \qquad \text{(Eq. 7)}$$

and

$$R_b = \frac{Nh}{f}. \qquad \text{(Eq. 8)}$$

where N is Brunt-Väisälä or buoyancy frequency and is given by

$$N = \sqrt{-\frac{g}{p_w}\frac{\partial \rho_w}{\partial z}} \qquad \text{(Eq. 9)}$$

In these relations, E_v is Ekman number in the vertical direction, R_b is baroclinic Rossby radius of deformation, z is vertical coordinate and is positive upward, h is water depth, and g is acceleration due to gravity. Note that Ekman number is a ratio between the friction force and the Coriolis force, and R_b defines a horizontal offshore scale in which upwelling occurs.

Measurements

Field measurements of surface temperature and conductivity in horizontal and vertical directions, and anomalies of these parameters from the expected variation provide a quantitative indication of upwelling and downwelling. Some examples of the stable expected variation in the vertical direction (measured positive upward) are: (1) a positive temperature gradient, (2) a negative salinity gradient, and (3) a negative density gradient. Visible and infrared satellite remote sensing of color and temperature are also used to identify and show the dynamics of upwelled water (Chase, 1981; Petrie *et al.*, 1987).

The best reference texts for coastal upwelling are *Introductory Dynamic Oceanography* by Pond and Pickard (1978), *Coastal Upwelling* by Richards (1981), *Physical Oceanography of Coastal Waters* by Bowden (1983), and a technical report *Physical Features of Coastal Upwelling Systems* by Smith (1983).

Dilip K. Barua

Bibliography

Allen, J.S., 1973. Upwelling and coastal jets in a continuously stratified ocean. *Journal of Physical Oceanography*, **3**: 245–257.

Bakun, A., 1973. *NOAA Technical Report NMFS SSRF-G71.*

Bowden, K.F., 1983. *Physical Oceanography of Coastal Waters.* Chichester: Ellis Horwood Ltd.

Chase, R.R.P., 1981. NASA's potential remote sensing capabilities that could be applied to upwelling studies. In Richards, F.A. (ed.), *Coastal Upwelling.* Washington, DC: American Geophysical Union.

Ekman, V.W., 1905. On the influence of the earth's rotation on ocean currents. *Arkiv foer Matematik Astronomi Och Fysik* (Swedish), **2**: 1–52.

Field, J.G., Griffiths, C.L., Linley, E.A.S., Zoutendyk, P., and Carter, R.A., 1981. Wind-induced water movements in a Benguela kelp bed. In Richards, F.A. (ed.), *Coastal Upwelling.* Washington, DC: American Geophysical Union, pp. 507–513.

Garvine, R.W., 1971. A simple model of coastal upwelling systems. *Journal of Physical Oceanography*, **1**: 169–179.

Hamilton, P., and Rattray, M., 1978. A numerical model of the depth dependent wind-driven upwelling circulation on a continental shelf. *Journal of Physical Oceanography*, **8**: 437–457.

Henry, R.F., and Murty, T.S., 1972. Three-dimensional circulation in a stratified bay under variable wind stress. In Nihoul, J.C.J. (ed.), *4th Liege Colloguim on Ocean Hydrodynamics, Mémoires de la société Royale des Science de Liège.*

Hidaka, K., 1954. A contribution to the theory of upwelling and coastal currents. *Transactions of the American Geophysical Union*, **35**: 431–444.

McEwen, G.F., 1912. The distribution of ocean temperature along the west coast of North America deduced from Ekman's theory of the upwelling of cold water from adjacent ocean depths. *Internationale Revue der gesamten Hydrobiologie und Hydrographie*, **5**: 243–286.

Petrie, B., Topliss, B.J., and Wright, D.G., 1987. Coastal upwelling and eddy development off Nova Scotia. *Journal of Geophysical Research*, **92**(12): 12979–12991.

Pond, S., and Pickard, G., 1978. *Introductory Dynamic Oceanography.* Oxford: Pergamon Press.

Richards, F.A. (ed.), 1981. *Coastal upwelling.* Washington, DC: American Geophysical Union.

Smith, R.L., 1983. *Physical Features of Coastal Upwelling Systems.* Seattle: Washington Sea Grant Communications WSG 83-2.

Thomson, R.E., 1981. *Oceanography of British Columbia Coast.* Sidney, BC: Department of Fisheries and Oceans.

Yoshida, K., 1967. Circulation in the eastern tropical oceans with special reference to upwelling and undercurrents. *Japanese Journal of Geophysics*, **4**: 1–75.

Cross-references

Coastal Climate
Coastal Currents
Coastal Modeling and Simulation
Coastal Wind Effects
El Niño–Southern Oscillation
Meteorologic Effects on Coasts
Numerical Modeling
Shelf Processes
Time Series Modeling
Vorticity

COASTAL WARFARE

Most people today live near the coast, and that concentration has prevailed far back into prehistory. As a result, military history includes many chapters on coastal warfare. Invading armies attack from the sea, and defenders build fortifications to attempt to keep them back. This recorded history starts with Homer's story of the Trojan War, and most recently includes the British recapture of the Falkland Islands and the US Marines night landing in Somalia under the glare of television cameras.

Coastal warfare involves all the elements of warfare: knowledge and use of the environment (terrain, weather, and tides), fortifications and obstacles, deception and surprise, concentration of force at the decisive point, mobility, logistics, leadership, and technology. Coastal operations include naval battles, land battles in coastal plains, and sieges of coastal fortresses. Table C11 list some important coastal battles.

Many naval battles have been fought in coastal waters, both because of strategic considerations and the need to support operations on land. Two classical battles exemplify this category of operation. In 480 BC the Greek fleet defeated the Persians at Salamis. The Greeks used constricted coastal waters to negate the size advantage of the Persian and Phoenician ships, and the naval victory ended the Persian invasion of Greece. In 31 BC, Octavius blockaded Marc Anthony's fleet at Actium on the western coast of Greece. Anthony's fleet finally sortied, with his 220 heavy ships fighting 260 lighter ships. A day of bloody but indecisive fighting ended when Cleopatra fled with 60 ships. When Anthony followed, his remaining ships despaired and Octavius won the battle that propelled him to sole leadership of Rome.

Coastal plains often provide key avenues of approach. In 490 BC, the Spartans and their allies defended the pass at Thermopylae to stop the Persian advance along the coast. Betrayed when the Persians learned of a path around their positions, the Spartans conducted a heroic and doomed fight to buy more time for the Greek forces. As at Marathon earlier that year, topography along the coast dictated the location of the battle.

Coastal fortresses control land and sea avenues of approach, increasing their importance for both attacker and defender. King Edward I of England created a series of "Edwardian" castles in Wales. Of these six castles, perhaps the finest examples of medieval castle design, five had coastal locations. Resupply and reinforcement by sea provided an important

Table C11 Selected coastal battles

490 BC	Marathon	Athens defeated landed Persian force on constricted coastal plain
490 BC	Thermopylae	Persians annihilate Spartans defending a choke point along the coastal plain
480 BC	Salamis	Greek naval forces defeat Persians in coastal water where the Persians cannot take advantage of larger ships
31 BC	Actium	March Anthony's naval force defeated while trying to break out of blockade by forces of Octavius
1274, 1281	Mongol invasions of Japan	Two invasions of Japan defeated by typhoons
1453	Fall of Constantinople	Ottoman siege of Byzantine coastal capital, cutoff both by land and sea
1759	Quebec	British assault force landed at night, scaled the Abraham Heights, and captured the city
1915	Gallipoli	British land at multiple locations along the Gallipoli peninsula but cannot advance inland and eventually withdraw
1943	Tarawa	US forces capture a small and heavily defended Pacific atoll
1944	D-Day invasion of Normandy	Allied forces unleash the largest amphibious assault ever on German defenders
1950	Inchon	UN forces broke out of the Pusan perimeter with an amphibious landing far behind enemy line
1982	Falkland Islands	British recaptured the islands using naval and amphibious forces without a land base anywhere near the islands

Table C12 Selected amphibian vehicles of World War II

Acronym	Name	Capabilities	Number produced
DUKW	Amphibious truck "Duck"	2.5 tons, 25 soldiers 5 knots at sea, or 80 km/h on land	20,000
LVT	Landing vehicle tracked—"amtracs," "amphtracs," or Alligator	8 m long, 6.5–7.5 knots additional 2850 armored LVT(A)	15,501
LCVP	Landing craft vehicle, personnel	3.6 tons of cargo or 36 soliders	23,398
LCM	Landing craft, mechanized	15 m, 1 medium tank	11,392
LCI	Landing craft, infantry (large)	48 m, 200 solidiers 2 gangways instead of bow ramp	1,051
LCT	Landing craft, tank	6 medium tanks	1,435
LST	Landing ship, tank	100 m, 20 Sherman tanks max. speed 11.5 knots 1,490 tons 26 lost in action and 13 in accidents	1,051
LSD	Landing ship dock	4,490–4,546 tons	18

Figure C59 LST (left) and LCVP (right) in the Philippines during World War II (Photo: M. Schwartz).

aspect of their military importance. Perhaps the greatest siege in history ended the millennium-long rule of the Byzantine empire, when the Ottoman Turks cutoff Constantinople by land and sea. Until the very end, the defenders hoped for relief from the sea. Improvements in technology have decreased the value of fortresses, and such sieges have become much less frequent. Stalingrad was not a coastal city, but the fighting there in World War II shows that the potential still exists for urban siege warfare in a coastal setting.

Amphibious assaults

The amphibious assault in the face of a defended beach marks the culmination of coastal warfare. It requires cooperation among army, navy, and air force units, and reached its highest development during World War II. Previous conflicts had moved armies by sea and landed them to begin combat operations, including invasions of Britain by Julius Caesar and William the Conqueror, but as late as World War I landings at Gallipoli used standard ships and small boats to land troops. In the early 1930s, the Japanese developed landing craft with a bow ramp. Copied by the allies, World War II saw the development and fielding of specialized landing craft and amphibious vehicles specifically for amphibious assaults. (Table C12). Four different theaters saw amphibious landings: leapfrogging along New Guinea to the Philippines (Figure C59) in the Southwest Pacific under General Douglas MacArthur, island hopping in the Central Pacific under Admiral Chester Nimitz, the Mediterranean theater movements from North African to Sicily and finally Italy, and the largest single amphibious operation with the landing in Normandy under General Dwight Eisenhower.

In 1950, UN forces in Korea under General MacArthur staged a repeat of the World War II Pacific landings, with old landing ship tanks (LST) and many Marine veterans recalled to active duty. This landing exemplifies many characteristics of an amphibious assault. On environmental grounds Inchon presented an unlikely site. A narrow river channel and wide mudflats restricted ship operations. A 10 m tidal range, and the necessity for two high tides during daylight, restricted the landing to a few days each month. Surprise and impeccable execution overcame the environmental drawbacks to make the Inchon landings a turning point in the Korean War.

Following the success at Inchon, technology improvements have greatly increased the ability of large navies to successfully land forces on a hostile beach. Helicopters, and specially configured aircraft carriers to support them, allow airlifting of troops ashore and bypassing of coastal defenses. Parachute and glider units filled this role in World War II landings like Normandy, but the helicopters can return to the ship and bring in additional troops and supplies or move troops ashore as needed. The squared-off bow ramps have been replaced with extendable ramps that allow a hydrodynamic bow and greatly increase the cruising speed of the amphibious ships while still allowing roll-off of the assault vehicles. Large landing ship docks (LSDs, over 3 times the size of the 18 built during World War II) feature a large docking pool that can be flooded, allowing the smaller amphibious vessels to rapidly float away to begin the assault. LCACs—landing craft air cushioned—use jet engines to float on a air cushion of air, riding rapidly over the waves and obstacles up to 1.2 m high, and up onto the beach. Only one conflict has tested the new technology, the British recapture of the Falkland Islands in 1982. Due to time constraints when the Argentines attacked without warning, and because budgetary cuts meant the British did not have a full complement of new amphibious equipment, the British success relied as much on rapidly requestioned and modified commercial vessels as on the specialized naval vessels. In the First Persian Gulf War, the American amphibious forces had a diversionary role without actually performing a landing, and the new equipment and doctrine did not receive a battle test.

Military principles of amphibious assault mirror those of any offensive action. The attacker must mass forces at the critical location, and then rapidly exploit the advantage. In many cases this proves relatively easy to achieve, because the defender must spread his forces over a large area and many beaches. Until the defender knows where the main attack will take place—and the danger of mistaking a diversionary landing for the main attack always dulls the reaction time of defenders—the attacker can exploit numerical superiority on the landing beaches. Surprise and deception keep the defender guessing about the location of the assault, and keep forces spread over a large area. Throughout history, many defenders have decided that beaches cannot be defended, and have kept most of their forces in reserve prepared to fight a mobile defense once the situation becomes clear. German and Japanese attempts to fortify the beaches and stop the attacker at the water's edge never succeeded in World War II.

The key to the offense becomes rapid exploitation of the initial numerical advantage—failure here doomed the landings at Gallipoli in World War I, and led to the stalemate after the Anzio landings during World War II. In contrast, the 1759 British capture of Quebec resulted from a nighttime landing, and an immediate scaling of the cliffs the French had considered impassible. Rapid exploitation requires good planning, forceful leadership on the beach, and the ability to rapidly reinforce the beach. In almost all opposed landings, the units in the first wave become militarily ineffective. Despite planning and rehearsals, boats land units at the wrong place and split up unit integrity, and casualties due to enemy action and the surf decimate units. Junior leaders and individual initiative can carry the assault well inland and secure the beachhead, but the strategic goals of the invasion require that additional units be rapidly landed with unit integrity and supporting armor, artillery, and air defense support. These units in the follow-on waves must seize the initiative and expand the beachhead before the defenders can respond and mass their units. Then the initial assault units can regroup for later action.

Concurrently with the follow-on combat units, the attacker must land supplies to keep the units on the beach moving inland. At Gallipoli, the British had trouble keeping units supplied with water which had to come in over the beach, and despite multiple successful landings at different locations and times they never exploited the initial success to expand the beachheads. At Inchon, the large tidal range required that LSTs be beached overnight, because they had to come in at high tide to get over the impassible mudflats, and it was impossible to unload them before the tide began to ebb. Beached LSTs provide a tempting target, a necessary risk because the assault troops needed heavy vehicles and the supplies they carried. Logistics remain important until ports can be captured. Allied efforts to create artificial harbors in Normandy failed when a major storm damaged the structures, which meant supplies had to come over the beach for much longer than the logisticians wanted.

Environmental considerations

The requirements for logistical support and combat reinforcement often dictate the time of a landing. A significant tidal range usually restricts landings to a few days each month, because planners want two tides during daylight hours. The first tide near sunrise brings in the assault waves, and the second tide twelve and a half hours later brings supplies and a second wave. Depending on conditions, planners may prefer to assault with a low or high tide. At Normandy for D-Day, the assault came at low tide. This exposed the largest number of obstacles in the surf zone, but did not make the job of combat engineers easy as they tried to neutralize obstacles—the tide rose faster than they could deal with the obstacles. At Inchon, planners chose the high spring tide so the assault vehicles could get to the seawall without getting mired in the mudflats. Neap tides might be selected to avoid the extremes in tidal range, and insure there is enough water to get over obstacles even at low water. Poor knowledge of the tides at Tarawa led to problems because the reef grounded many of the assault vessels, and there were not enough amphibious tractors which could drive over the reef and up out of the water, but a strategic need for speed led planners to accept risks as militarily necessary. At Tarawa, the actual tide encountered did not correspond to either the predictions or later hindcasts, demonstrating the large effects that meteorological conditions can have when the normal tidal range is only about 1.2–1.9 m. Depending on the tides involved, favorable windows can be either two weeks (Normandy) or four weeks apart (Inchon). Tidal currents can also influence the assault timetable, since at Inchon the maximum current speed equaled the speed of many of the amphibious assault vessels.

Geologic conditions for an assault beach generally require two factors: low slope and firm sediment. On a reflective–dissipative continuum, this generally places assault beaches somewhere in the middle of the spectrum. Assaults cannot target beaches backed by cliffs because there these lack easy egress, although the landings at Quebec in 1759 and the American Rangers scaling the cliffs at Normandy to capture key coast artillery positions show that even this rule can be violated to achieve complete surprise or to provide supporting assaults to the main attack. Moderate slopes offshore allow amphibious vehicles easy egress from the water, and moderate slopes ashore allow rapid expansion of the beachhead. The landing beaches at Gallipoli were all chosen because they were small sandy pocket beaches along a coast that generally provided steep terrain right up to the water. Extreme low slopes generally must be avoided, because they come with small sediment sizes that will not allow heavy military vehicles to operate and rapidly exit the beach. The landing at Inchon needed to come at high tide to avoid the extensive mudflats; the selected "beaches" allowed the landing craft to get to the base of a seawall and minimized the problems that would have been encountered on mudflats. Favorable geologic conditions accompany favorable wave conditions, as low slopes correlate with smaller wave heights. Following the experience at Tarawa, assaults also generally try to avoid offshore reefs. At Tarawa, the landing craft approached from inside the lagoon, but grounded on the reef and dropped Marines into the water where many drowned in the deeper water between the reef and the shore. Other dropped gear to get ashore, and then had no weapons or supplies.

Weather always affects amphibious operations. Two Mongol attempts to invade Japan in 1274 and 1281 ended in defeat in large part due to intense cyclones that destroyed much of the invasion fleet. The Japanese called these Kami-kaze, or "divine winds," and used the term during World War II for suicide attacks, many at ships massed to support Allied amphibious landings. The D-Day landings had to wait one day, and nearly had to wait two weeks for the next favorable tide conditions because of a severe storm over the Atlantic. The Inchon fleet suffered minor damage from one typhoon while preparing for the landing in Japan, and put to sea a day early to avoid a second typhoon. The next favorable date for the Inchon landings was a month away, the Pusan perimeter required relief immediately, and the North Koreans were likely to begin mining the channel into Pusan if given more time. For both D-Day and Inchon landings, correct favorable weather predictions by the staff meteorologist allowed the invasions to proceed without having to wait weeks for the next available date with favorable tides.

While defenders have put obstacles in the water and along the beach to slow or stop the defenders, and dug and reinforced positions for their weapons and themselves, these obstacles and fortifications defeated no major landings in World War II. Conversely, despite massive naval and air bombardments preceding most of the landings in World War II, the attackers never dislodged and rarely seriously disrupted the defenders until the assault boats hit the beaches. This paradox has led many military thinkers to assert that a mobile defense back from the beaches will work better than attempting to build the Maginot line at the shore, while for the attackers, a relatively short bombardment maintains surprise while achieving almost the same effect as a much longer bombardment.

Supporting landing operations can be among the costliest naval actions. During the campaign for the Gilbert Islands, the Marines lost 1,115 dead and missing on Tarawa. The sinking of the escort carrier *Liscome Bay*, supporting the army landing on Makin Island, led to the loss of 644 men. The largest naval battle ever fought, the battle of Leyte Gulf, occurred when the Japanese attempted to destroy the forces landing in the Phillipines and the ships supporting them. US losses included two escort carriers, including one lost to kamikaze. During the invasion of Okinawa in April 1945, kamikaze attacks sank 36 ships and landing craft. In the Falklands campaign, the British saw 9 of their 23 frigates and destroyers either sunk or seriously damaged.

The SEAL (Sea, Air, and Land) Teams date only to 1970, but have become respected both in the military and Hollywood. Their primary missions involve military special operations along coastlines.

Coastal warfare promises to remain a major focus of the world's militaries well into the 21st century. Further reading on this subject may be found in the following bibliographic listing.

Peter L. Guth

Bibliography

Bartlett, M.L. (ed.), 1983. *Assault From the Sea: Essays on the History of Amphibious Warfare*. Annapolis, MD: United States Naval Institute.

Clapp, M., and Southby-Tailyour, E., 1996. *Amphibious Assault Falklands*. Annapolis, MD: United States Naval Institute.

Foster, S., 1995. *Hit the Beach: Amphibious Warfare from the Plains of Abraham to San Carlos Water*. London: Arms & Armor Press, Cassell Group.

Gatchel, T.L., 1996. *At the Water's Edge: Defending Against the Modern Amphibious Assault*. Annapolis, MD: United States Naval Institute.

United States Navy, 1993. *... From the Sea: Preparing the Naval Service for the 21st Century*. Washington, DC: Navy White Paper.

Winters, H.A., and Galloway, G.E., Jr., Reynolds, W.J., and Rhyne, D.W., 1998. *Battling the Elements: Weather and Terrain in the Conduct of War*. Baltimore, MD: Johns Hopkins University Press.

Cross-references

Coastal Climate
Coral Reefs
Human Impact on Coasts
Pacific Ocean Islands, Coastal Geomorphology
Scour and Burial of Objects in Shallow Water
Tidal Environments
Tides
Wave Environments

COASTAL WELLS

"And he looked, and saw a well in the field, and behold, three flocks of sheep were lying there beside it, for from the well they watered the flocks. Now the stone on the mouth of the well was large."

Genesis 29, 2

"When all the flocks were gathered there, they would then roll the stone from the mouth of the well..."

Genesis 29, 3

Historical background

Water in arid and semiarid land is one of, if not the most important and vital factor, for the existence of people, livestock, and farming all along the history of mankind. Therefore, in many regions where freshwater was not available, occupation was very difficult and much sporadic if at all. In general, people were dependent on permanent water sources such as springs or rivers that supplied water all year round, and temporary sources such as seasonal rivers, natural water cisterns, etc. Later, when agricultural and farming irrigation systems were developed, an immense need for larger water quantities existed. Throughout history, when there has been a necessity and demand for expansion of settlements, either as a result of political pressure or due to the growth in population, people moved (to their sorrow), from the prolific and rich water regions into much poorer areas.

Coastal areas in general, and more specifically the near-littoral zone, reached an ever-growing importance when the sea became another nutrient source to the communities of the gatherers and hunters. As a result, they looked for new water resources, preferably natural, and if these did not exist, they searched and created artificial sources.

In arid and semiarid regions where natural water sources are very scarce, settling was totally dependent on their talent to either transport water to these regions, or locate and discover other sources, mostly underground. The existence of ground-water at a relatively shallow depth below the surface in the nearshore region permitted water exploitation through well digging at least as early as the Pre-Pottery Neolithic Period (ca. 7,900 years BP; Galili and Nir, 1993). This period parallels the beginning of the development of canal irrigation systems in China and Anatolia (Hübener, 1999).

There is no doubt that during the early periods of humanity, people experienced seasonal living along the coast (Oren, 1992), in regions where mountains reach the coastal zone, in most cases where there had been no water problem. Springs or river water easily covered the daily needs of people and their livestock. While in shallow relief or hilly coastal plains, in those semiarid lands where no springs or rivers exist, the water problem was very crucial to their survival.

Coastal water wells

At a certain stage of development, people discovered the fact that fresh or semi-fresh water may be reached very easily by digging shallow holes on a sandy beach quite close to the coastline (perhaps in imitation of animal behavior). It is assumed, although there is very little chance of finding any evidence at all, that this method was practiced for a very long period at much earlier stages before digging of water wells commenced.

The oldest known well in Israel is found at 10–12m. water depth in an Atlit Pre-Ceramic Neolithic village, located offshore, opposite the southern flank of Mount Carmel (Galili *et al.*, 1988; Galili and Nir, 1993; Nir, 1997). This well is about 5 m deep, mostly built of stones. It was found in a relatively large prehistoric site with many dwellings and other village installations. The well, dug in a clayey soil reached, at its lower part, a permeable carbonate cemented quartz–sandstone layer, the Pleistocene "kurkar" rock, which is known to contain a very prolific high-quality aquifer.

The majority of the historical wells are dug as a vertical pit, having a cylindrical shape, with inner diameters varying from 80 to 200 cm (Forbes, 1955; Nir and Eldar-Nir, 1987, 1988). Wells of this type are technically constructed from top to bottom, layer after layer, strengthening its

Figure C60 A typical Antillian coastal well at Tell Ashqelon, located on the southern coast of Israel (photo: Y. Nir).

walls with local stones, undressed in the old wells, but nicely dressed and fit with the well's round shape in the more "modern" wells from the Hellenistic and Roman periods and on to the present.

The water sources of the coastal wells depend on the ground water table, water flowing from inland higher topography to the drainage basin—the sea. This water table fluctuates with changes in sea level that fully controls its level.

In the coastal plain of Israel, and no doubt in similar geologic terrain elsewhere around the Mediterranean and other coastal regions, groundwater has an average slope of about 1/1000, and its surface is almost linear (Kafri and Arad, 1978). Many sub-recent wells were dug along the coastal plain of Israel, most of which are in the nearshore strip. One of the most famous sites, where a high concentration of wells is found, is at Tell Ashqelon, located on the southern coast (Nir, 1997). More than 50 water wells of ages from Roman to the 20th century were mapped there (Figure C60).

Water wells of the dug-pit method have been in operation up to the present time, mainly in underdeveloped regions. Wells of different shapes, construction materials, depth, etc., exist along most of the world coasts, located in all regions where the ground water table is close to the surface. Their importance in the developed countries, has dropped, while in the developing countries where there are no other natural sources they are still of high importance to the coastal population.

Yaacov Nir

Bibliography

Forbes, R.J., 1955. Water supply. In Forbes, R.J. (ed.), *Studies in Ancient Technology*. Leiden: E.J. Brill Publishing House, pp. 145–148.

Galili, E., Kaufman, D., and Weinstein-Evron, M., 1988. Eight thousand years under the sea. *Archaeology*, **41**: 66–67.

Galili, E., and Nir, Y., 1993. The submerged Pre-Pottery Neolithic water-well of Atlit-Yam, northern Israel, and its palaeoenvironmental implications. *The Holocene*, **3**(3): 265–270.

Hübener, R., 1999. The evolution of irrigated farming. *Natural Resources and Development*, **49**(50): 143–150.

Kafri, U., and Arad, A., 1978. Paleohydrology and migration of the ground-water divide in regions of tectonic stability in Israel. *Geological Society of America Bulletin*, **89**: 1723–1732.

Nir, Y., 1997. Middle and Late Holocene sea-levels along the Israel Mediterranean coast—evidence from ancient water wells. *Journal of Quaternary Science*, **12**: 143–151.

Nir, Y., and Eldar-Nir, I., 1987. Ancient wells and their geoarchaeological significance in detecting tectonics of the Israel Mediterranean coastal region. *Geology*, **15**: 3–6.

Nir, Y., and Eldar-Nir, I., 1988. Construction techniques and building materials used in the ancient water wells along the coastal plain of Israel. In Marinos, P.G., and Koukis, G.C. (eds.), *Engineering Geology of Ancient Works, Monuments and Historical Sites*. Rotterdam: Balkema, pp. 1765–1771.

Oren, E., 1992. Sinai. In *The New Encyclopedia of Archaeological Excavations in the Holy Land*. Published for The Society for Antiquities Research in Israel, by Carta, and The Ministry of Defence Publishing House (in Hebrew), pp. 1104–1116.

Cross-references

Archaeology
Archaeological Site Location, Effect of Sea-Level Changes
Changing Sea Levels
Desert Coasts
Hydrology of Coastal Zone

COASTAL WIND EFFECTS

Wind can have several effects on the processes influencing coastal geomorphology. These include wind stress on the water surface in major storms, such as hurricanes or typhoons, inducing short-term above normal sea elevations or storm surge, short "choppy" waves in estuaries and fetch-limited harbors, downwelling and upwelling processes in the coastal oceans, and sea breeze effects.

Storm surge is caused by the frictional wind stress on the water surface inducing a current in the surface water. Empirically, the resulting current in the top few centimeters of the sea surface will be about 2% of the wind velocity (Komar, 1976). For Class 5 storm (hurricane) winds, a considerable mass transport is advected along the sea surface waters in the wind direction, so that onshore storm winds force the seawater up against the land. Of course this occurs concomitantly with large, wind-forced, storm waves. The total wind effect is therefore a combination of both the direct wind stress effect and the indirect wave run-up effect. Storm wind stress effects are maximized over broad shallow continental shelves and semi-enclosed seas, such as the Gulf of Mexico, the North Sea, and the Baltic Sea, where extreme surge effects can raise the still water level by ~5 m and flood over the low-lying barriers, estuarine wetlands, and coastal land. The storm surge effect enhances the beach erosion process along open sandy duned coastlines by raising the water table in the beach and lowering intergranular friction between the sand grains, so that the pore water pressure from the raised water table induces the sand from the lower beach face to flow out into the surf zone. But wind-generated surge also piles up tidal waters inside estuaries, tidal inlets, and river mouths, often leading to saline flooding of adjacent low-lying coastal land. As a result of the onshore winds and storm surge, the tidal cycles within the estuaries are altered to a much higher than normal tides and reduced low tide.

Short steep, "choppy" waves are generated by fetch-limited winds inside estuaries and semi-enclosed seas. These "choppy" waves have a number of effects including stirring of the bottom muddy sediments in the shallow nearshore creating the characteristic high muddy suspended load "turbid fringe" observed around shallow estuaries on windy days. These waves may even generate "fluid" mud in muddy estuarine environments from the constant bottom agitation of the loosely cohesive bottom mud sediments. Wind-forced choppy waves in enclosed bays, on top of a storm surge, result in erosion cutback of low terraces and marshland above the normal high tide levels, and the lagoonal shore of barrier islands and spits—this is the prime mechanism for shore erosion within sheltered harbors and estuarine lagoons.

Downwelling and upwelling in the coastal zone are largely wind-generated processes (Figure C61). These processes have marked, but often overlooked, influence on the diabathic (cross-shore) sediment transport. For the case of significant onshore winds blowing for some hours, surface waters are forced up against the beach and dunes, raising the still water level at the beach. As a requirement of continuity, the onshore surface current in turn generates a bottom return current (downwelling). Thus sandy sediment grains initially eroded off the beach in a storm, when uplifted by the bottom orbital motion of the waves, are acted upon by the downwelling current as they fall back to the seafloor, and thus become transported offshore, typically to "closure depth." Conversely, with consistent significant offshore winds, wind stress on the sea surface forces the surface water out to sea, and lowers the still water level. This generates an onshore-directed bottom return current (upwelling), and as the longer period (non-storm) swell waves uplift the bottom sediment, it is transported onshore to the beach, facilitating beach face accretion, and rapid rebuilding of the beach berm. Of the two processes, the latter likely transports bottom sandy sediment at a greater rate because the longer period waves have greater bottom orbital

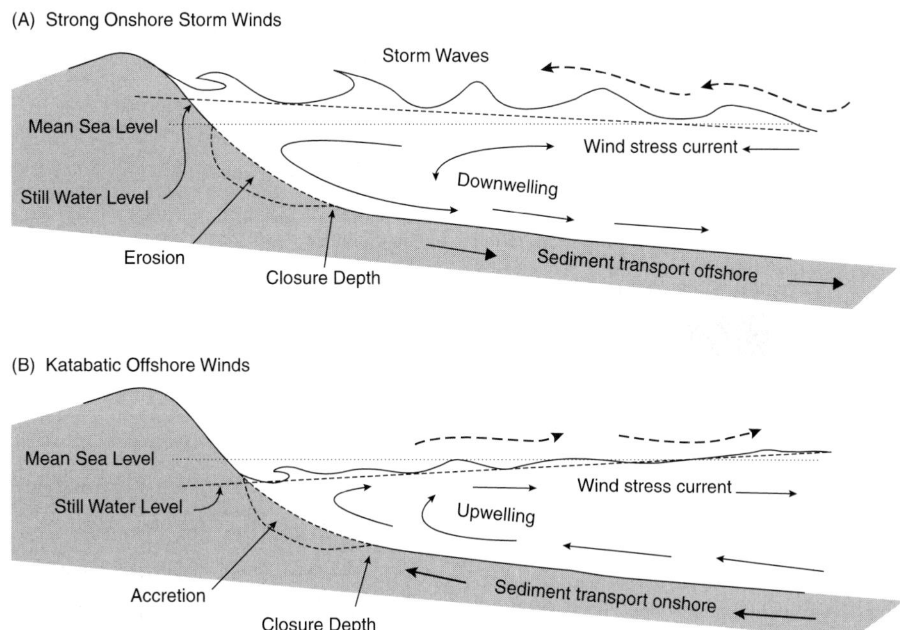

Figure C61 Effect of wind generation inducing coastal upwelling and downwelling and associated diabathic sediment transport.

velocities which uplift more sediment. It is often observed that beach recovery at the end of a storm (when winds reduce in strength and change to an offshore direction) is more rapid than the loss of sand from the beach during the erosion phase.

Sea breeze effects arise from the thermal heating of the land generating a steady onshore wind. The effect of sea breeze on beach morphodynamics has been studied on the Western Australian coast, where sea breezes are some of the strongest in the world. Here the sea breeze cycle is an important influence on beach processes, with onshore winds reaching magnitudes >8 m/s. Masselink and Pattiaratchi (1998) researched the effects of the sea breeze on the nearshore processes, demonstrating how with the onset of a strong onshore sea breeze of 10 m/s, the root mean square (rms) wave height increased from 0.3 to 0.5 m, the wave period decreased from 8 to 4 s, and mean cross-shore flows reached velocities of 0.2 m/s directed offshore. The short period sea breeze-generated waves became superimposed on the background swell. During the sea breeze event, nearshore sediment resuspension became almost continuous, and cross-shore transport was directed offshore. But in addition, the sea breeze affected the beach longshore transport rates when it approached the coast obliquely, and the longshore-suspended sediment transport rate increased by a factor of 100. These sea breeze-induced hydrodynamic changes were associated with erosion of the beach face, and deposition in the surf zone, with a consequent flattening of the profile. The effect was similar to a storm of medium intensity.

Terry R. Healy

Bibliography

Komar, P.D., 1976. *Beach Processes and Sedimentation*. Englewood Cliffs: Prentice-Hall.

Masselink, G., and Pattiaratchi, C.B., 1998. The effect of sea breeze on beach morphology, surf zone hydrodynamics and sediment resuspension. *Marine Geology*, **146**: 115–135.

Cross-references

Beach Erosion
Coastal Upwelling and Downwelling
Cross-Shore Sediment Transport
Storm Surge
Tides
Waves

COASTAL ZONE MANAGEMENT

Coastlines around the world vary greatly in morphology, natural processes, and human usage. From those which experience intensive tourism, to those which are overdeveloped and polluted, or those where property and important natural habitats are threatened by erosion, all need some degree of management intervention to prevent further deterioration. It is for these reasons that the technique of *Coastal Zone Management* (normally abbreviated to *CZM*, and also sometimes referred to as *Integrated Coastal Zone Management—ICZM*) has increased in popularity as a method with which nations can protect and manage their coastal resource. By definition, CZM can be said to be:

> An integrated process which manages all areas of coastal activity which occur along a stretch of coastline, in an holistic manner, so that minimal impacts occur which may be detrimental to the coast itself.

It is a potentially powerful tool, by which the multivariate uses of coasts, such as leisure, residential, industrial, conservation, protection, etc., can be integrated into a management scheme so as not to conflict. The outcome of CZM should be a management plan which identifies coastal problems and outlines solutions which should be appropriate to both the resource value and the natural process "value" for each coastal issue (see Arthurton, 1995).

The historical development of CZM

Ever since humans started to build structures along the coast or to utilize coastal resources, they have been interfering with natural coastal processes. A seawall may protect a coastal development, but it will also stop an important supply of sediment to the coastal sediment budget and may cause wave-induced scour that removes beach sediment. Similarly, a harbor jetty or yacht marina will prevent the longshore drift of sediment, thus starving beaches further along the coast. Both of these examples, will lead to the initiation of new erosion problems on adjacent sections of the coastlines. In a similar manner, a factory or harbor development may lead to increased coastal pollution, and a leisure development to increased public pressure on natural habitats, such as sand dunes. From these types of problem, management issues arise. Historically, such problems have been made worse by the attitudes of society that pervaded at the time. During the 19th century in particular, the tendency to view all natural systems as: (1) "resources" to be exploited for the benefit of humans; and (2) something over which humans should have total control; laid the foundations for many of our contemporary coastal management problems. Planners and local people just did not consider the coastline as something that could possibly be changed because of their impacts on ecology or natural processes.

While in the current climate of increased knowledge and environmental awareness, it is easy to criticize such attitudes, it must be remembered that many of these historic actions were carried out in good faith, but without the relevant knowledge needed to understand their full implications. We now understand that the coastline is not a feature that is fixed in time, but is something which is transitory and liable to considerable temporal and spatial variation in response to natural processes, often with short notice. Similarly, we also know that many coastal habitats are very vulnerable to change, and have experienced great changes due to human interference. As a result, many contemporary coastlines exist with human-induced habitat degradation and/or large areas of inappropriate development, which are now producing serious management issues which need to be addressed. For example, work by Boorman (1976) found that only 40 passes a month across a sand dune (i.e., only 20 people walking to the beach and back) were sufficient to reduce vegetation height by 75%, while 150 passes (75 return journeys) could lead to 50% bare ground. Frequently, these same areas also contain leisure development and infrastructure.

During the 1970s, there was a general worldwide increase in environmental awareness, both in official sectors, such as government and planning, but also within the general public. This was very much a transitory phase for coastal management because while people became more concerned for environmental welfare and protection, civil engineering still pervaded as the generally accepted solution to coastal problems and the general worldwide increase in disposable wealth in the developed nations increased the demand for, and subsequent development of, coastal tourism. The result of this was the continued development of the world's coastlines, which thus allowed the continuation of many of the impacts associated with erosion, sediment starvation, and user conflict. By the 1990s, the development of environmental awareness and understanding had progressed sufficiently to produce a switch to greater emphasis on the preservation of natural processes and restoration of habitats. The result of this has been the increased acceptance of soft defense techniques, such as beach feeding and managed realignment or even abandonment of defenses, over the construction of solid structures (see also *Shore Protection Structures*). The one main driving force to come out of the 1990s was the Earth Summit in Rio de Janeiro, Brazil in 1992 (Quarry, 1992). This led to the production of Agenda 21 as a blueprint for environmental action. Within this document, chapter 17 deals exclusively with oceans and coastlines, and probably represents the strongest international commitment to CZM that currently exists. Agenda 21 commits coastal nations to the implementation of integrated coastal zone management initiatives, and to the sustainable development of coastal and marine areas under their jurisdiction. Within this, a number of objectives commit signatory nations to more specific aspects of coastal management.

1. To provide for an integrated coastal policy and decision-making process so as to promote compatibility and balance of coastal uses. This also stipulates the inclusion and cooperation of government departments, ministries, and agencies which have control over specific aspects of the coast (i.e., to prevent conflicting usage and non-sustainable visitor pressure).
2. To apply preventive and precautionary approaches in respect to coastal development, including prior assessment and systematic observation of the impacts of major projects (i.e., limiting coastal development in unsuitable areas).
3. To promote the development and application of techniques (such as Cost Benefit Analysis and Environmental Impact Assessment) which reflect changes in value resulting from uses in coastal areas. Such changes include pollution; loss of property and land value due to erosion; loss of natural resources and habitat destruction; and, conversely, the increase in hinterland value due to development (i.e., costing benefits and impacts both financially and environmentally).
4. When managing a coastline, to liase with all interested groups, to provide access to relevant information, and to provide opportunities for consultation and participation in planning and decision-making processes associated with the development of management plans (i.e., consider all views, including those of local residents).

These policies have also been supported from other sources, such as the Intergovernmental Panel on Climate Change (IPCC, 1990) which has recognized the need for contemporary management, but also management which allows for the continued protection of the coast in the light of sea-level rise (see also *Changing Sea Levels* and *Sea-Level Rise, Effect*).

Some common coastal resource conflicts—key issues for CZM

The coastal zone is a finite "resource" in that it can only support a certain amount of activity, before its limitations are realized. This process is often termed the "carrying capacity" of the coast, but can mean many different things to different people. For example, an ecologist might argue that the carrying capacity is an ecological term, referring to the degree of human impact an environment (such as a sand dune community) can take before its ecological structure is impaired. Similarly, a tourist might argue that the carrying capacity of a beach refers to the number of people who can use it before personal enjoyment suffers. All such definitions of carrying capacity, regardless of how scientifically valid they may be, are important in CZM because they all reflect impacts on the coastal zone in different ways. It is important that CZM identifies these varying carrying constraints and provides a strategy to prevent them from being reached and exceeded. If this were to occur, then the outcome would be competition and environmental degradation. The key issues that need to be addressed are:

1. Coasts are used for recreation by many different user groups for many different things, producing conflicts of interest and locally intense usage.
2. Coasts are used by developers for many forms of structure, from urban to industrial. This may attract more people to the area (increased visitor pressure), or put people off coming (declining tourism); may cause pollution; or may interfere with natural processes.
3. Coasts experience a series of natural processes which can be modified by, or impinge on, human activity. This relationship of people to process is, therefore, critical because it will govern how the coastal environment undergoes change. This change can be unpredictable in both its rates and style.
4. Activities inland (catchments) and out to sea can also affect the coast. For example, dam building may stop sediment reaching the coast, or oil spillages may cause coastal pollution. Completion of the Aswan Dam, for example, reduced sediment supply to the Nile Delta from 124 to just 50 tonnes a year, leading to extensive erosion (Stanley and Warne, 1993), while the Akosombo Dam reduced sediment supply to the Volta Delta by 99.5% (Ly, 1980). Also, oil spillages can cause slicks that wash up on the coast, with impacts on coastal fauna and flora. See Figure C62 for an overall summary.

CZM needs to reduce impacts to a level that does not induce any further environmental modifications. However, doing this also represents one of the more subjective issues of CZM, in that assessing each form of use or conflict includes some subjective assessment which needs to be balanced against other views when composing the plan. For example, does land claim mean increased farmland/cheap land for development, or does it mean the loss of intertidal habitat, coastal squeeze, and the increased need for coastal defenses in the future?

In carrying out CZM it will be necessary to side with one particular viewpoint, with full justification in the final report for the reasoning for choice selection. Such a decision may well depend on the nature of the coastline and the precise circumstances in question. Using the above example, land claim may be considered beneficial for increased farmland in a country where fertile land is scarce, with management having a role in protecting this resource, but may be considered as detrimental along a coastline with high conservation potential in a country with surplus agricultural production.

While accepting that there will always be some areas of ambiguity involved in making management decisions, there will also be certain factors which are always going to feature highly in any CZM scheme.

Protecting the coast from storms

Many coastal areas are in danger of damage from natural processes. This danger can take many forms, ranging from the rare event, such as earthquakes and tsunami generation; to those which occur more frequently, such as storms, hurricanes, and storm surges; and to those that occur over the long-term, like sea-level rise. The impacts of earthquakes are difficult to provide adequate protection for, with management largely restricted to emergency relief planning. However, CZM can make recommendations which could be important for damage limitation, such as proper coastal zoning where development is not permitted

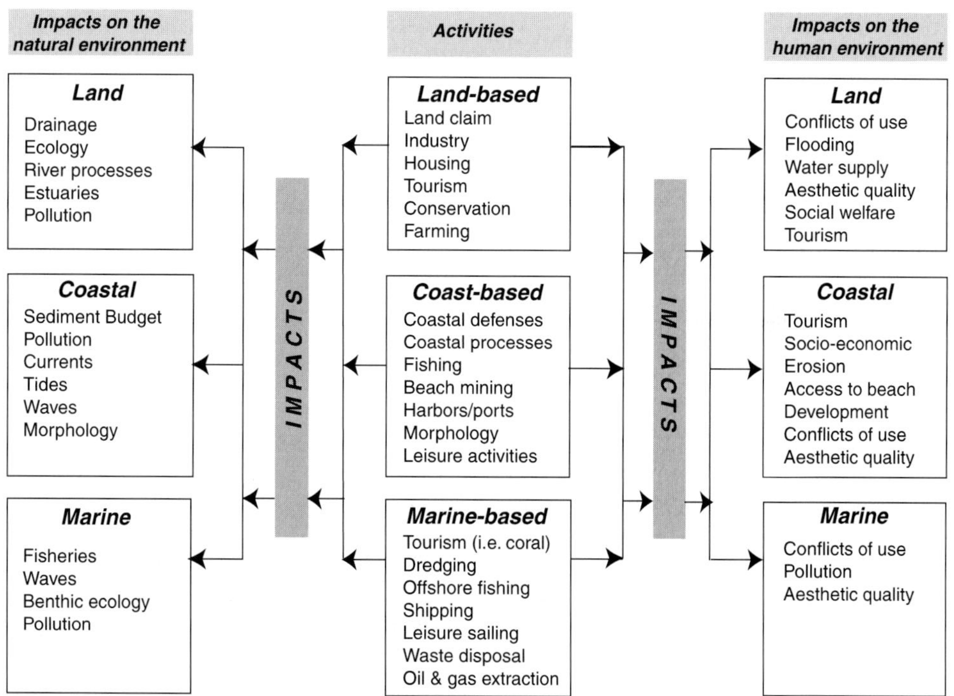

Figure C62 The link between the diverse range of coastal activities and their impacts on both the human and natural environment.

on reclaimed sediments which may undergo liquefaction during earth movements. Storms and hurricanes occur more frequently and can cause sudden and dramatic effects on the coastline, potentially involving the loss of life and property, and also with the potential of altering the coastal morphology by large-scale erosion and the breaching of sea defenses. CZM may decide to develop increased defense provision to cope with these eventualities or, alternatively, to opt for an early warning system which will ensure public safety through evacuation procedures, but not protect property.

In order to protect coasts from storms and hurricanes, there are a series of measures that could be implemented by CZM:

1. In some storm-vulnerable areas, the protection of property by increased defense provision may not be financially viable. In this situation, the protection of human life is of prime concern and so it may be sufficient to arrange evacuation procedures in response to early warning systems.
2. In situations where protection is feasible, CZM may adopt a policy of reinforcing or increasing the height of seawalls.
3. One main concern of storms is the possibility of a storm surge (meteorological tide). Estuaries which are prone to such features, or which contain significant areas of high-value hinterland, need to be protected by tide surge barriers.
4. If hinterland values are low, sufficient protection to property could be provided through the provision of flood lands to take excess water.
5. One of the main causes of problems during storms is the increase in the size of waves. By adopting a policy of artificially building up beaches by beach feeding, sufficient wave attenuation can reduce the impacts of storms at the coast, and thus remove the need for increased hard defenses. Such solutions may be preferable in tourist areas where aesthetics and access considerations are important (see also *Beach Nourishment*).
6. The best form of protection is prevention. Hence, one of the prime management recommendations of CZM should be to avoid building in high-risk areas. Such planning controls can remove the necessity for any of the other methods mentioned here.
7. On coasts where people want to live but which are subject to severe storms, CZM can recommend planning regulations to ensure the construction of "storm-proof" buildings which can withstand the impacts of storm activity. However, it should perhaps be considered that if, after information is made available, people still insist on developing unsuitable areas, then they alone are responsible for the outcome, and should expect no protection to be forthcoming.
8. The impact of storms is also likely to become greater as a result of sea-level rise (see also *Sea-Level Change, Sea-Level Rise, Effect, Waves*). As sea level rises, wave base could deepen, thus increasing the force of a coastal storm. This makes coastal planning even more important, and indicates the need to identify areas susceptible to storm surge inundation.

Managing coastline erosion

One of the main dilemmas facing CZM is whether or not to defend an eroding coast. All such decisions should be taken on the basis of a detailed cost benefit analysis (CBA) and environmental impact assessment (EIA). On this basis, an area that is undeveloped is likely to be left to erode in order to maintain sediment supply to the intertidal zone. When there are homes, places of employment and infrastructure at risk, the decision becomes more complicated as it depends on the value put on the land as to whether it is defended or not. This is determined by a simple CBA, where the value of the land (in terms of property value and income from any resource based upon it) is determined over a fixed time period (normally the life-expectancy of a preferred coastal defense scheme), and compared to the cost of constructing the preferred defense.

An additional consideration in this CBA is the sediment budget (see also *Sediment Budget*). As the successful functioning of the coast as a natural system, and hence its management, is largely based on the movement of sediment from its point of input to its site of deposition, any structure which interferes with this balance will cause a knock-on effect for other parts of the coast, and the setting up of a whole new set of management problems. CZM has to reflect this dynamism, while protecting important areas. In order to assess the importance of sediment input, therefore, it is also common to undertake an EIA to investigate the impacts which any preferred defense method may have on the local environment.

In general terms, in order to maintain a balance between sediment supply (erosion) and deposition (accretion), CZM should opt for defense where necessary, but allow other areas to erode to maintain sediment inputs, even if this means the loss of isolated houses, small villages, or farmland. Only by making such sacrifices can greater protection and stability be given to other, more highly valued coastal areas.

Protection from polluting activities

Many coastlines are subject to a variety of pollutants from industrial, urban, or agricultural sources which can produce physical, chemical, or aesthetic impacts on the coastal environment. Ultimately, all aquatic life forms and much water-based leisure activity depends on water quality for survival and success. However, pressure on the water resource has traditionally been great, with the historical tendency being to treat the seas as a large waste-disposal unit. In addition, pollution is also multi-sourced, and includes industrial pollutants, storm drains, and domestic drainage. In addition to land-based sources, many industries actively dump at sea, both legally and illegally, along the principal of out of sight, out of mind. Often this dumped material is highly toxic or hazardous (see also *Marine Debris*).

CZM should recommend strategies for dealing with such issues. This should cover both routine discharges as well as the risk of accident caused by marine accident. Increased adoption of the "precautionary approach" which arose out of the 1972 Stockholm conference on the Human Environment has allowed easier management of this problem. Subsequently, this has been further supported in legislation by the 1972 London Dumping Convention, 1974 Paris Convention, the 1974 Helsinki Convention, and the 1980 Athens Protocol (Barston, 1994). Many countries deal with these issues directly by instigating methods of emission reduction. In the United States, for example, the use of Best Management Practices (BMPs) are common as a means of input reduction. For example, BMPs for agriculture include contour ploughing, crop rotation, filter strips, and animal waste control as methods to reduce the impacts of agricultural wastes on river and coastal areas. BMPs for urban areas include storm water collection and treatment, and the use of porous surfaces to reduce runoff and even out the storm hydrograph.

Other issues relate to the direct discharge of waste into the marine environment through pipelines. Modern pipelines discharge well beyond the low-water mark, and so the material readily becomes dispersed by marine currents and does not recontaminate the beach. They do, however, have the potential to affect marine organisms and fish stocks. Many countries, however, have old sewage systems and storm water discharge outlets. Typically, these tend to discharge into the intertidal zone, with the result that tidal currents can bring material back inshore. Solid material is often easily detected and CZM plans should have strategy recommendations in place to deal with any situations as they arrive. More seriously, problems occur with water-soluble pollutants which are not as readily detected and can go un-noticed until major environmental problems result. As part of standard CZM recommendations, new waste discharge pipes have to be built to beyond mean low water of the lowest tides.

In estuaries, where dredging may be common as a means of maintaining access for shipping, the problems of disposal of dredged spoil may be particularly acute. One main problem is that these sediments can be highly contaminated due to industrial activity. One possibility is to identify licensed offshore dump sites where dredged material can be placed. However, this material also represents a loss of sediment to the sediment budget and so could lead to issues of sediment starvation elsewhere along the coast. By adopting a wider perspective, CZM can recommend other uses for suitably unpolluted dredgings, such as beach feeding (see also *Beach Nourishment*).

Maintaining biodiversity and protecting habitats

The prime use of many coastal areas is for recreation and tourism. This can lead to some of the greatest conflicts which CZM has to deal with because of the diverse range of uses which a tourist coast has to support. Habitat fragility can be highlighted by problems such as trampling at sensitive sites, such as dune and coral reef areas; while beach resorts may experience widespread conflicts of usage, such as those caused by incompatible activities, such as bird watching and wildfowling, or jet-skiing and swimming. To minimize such conflicts, CZM needs to study which activities occur along the coast under study, to identify which can coexist harmoniously, and methods by which to accommodate others, such as by zoning usage. While conflict and zoning issues can be overcome with good management practice, the protection of habitat may be more difficult. In 1979, the Council of Europe instigated the Convention on the Conservation of European Wildlife and Natural Habitats (more commonly known as the Bern Convention). This has been frequently utilized in European CZM as a means of protecting coastal habitats at risk from development or over-usage. Similarly, in other parts of the world, fragile reef areas have received protection under different legislation, such as by designation as marine conservation areas or marine parks (see also *Marine Parks*).

Threats to biodiversity from tourism, however, is not just a question of people being by the coast. Tourist activities demand support services, infrastructure, accommodation, and land-based entertainment. All such development threatens the habitats already present on the hinterland. Many coastal areas support a wide biodiversity that can be severely reduced by tourism-induced habitat loss. While the Bern convention is applied in Europe, in the United States, Habitat Conservation Plans (HCPs) can be used to balance development and conservation interests. Within these plans, states actively seek land areas to purchase for conservation and habitat creation in order to balance areas of land lost due to development. While this may appear ideal, it does not solve the problem of replacing like with like. In this respect, it is important to consider that in many cases, created habitats do not function or behave in the same way as natural ones.

Access to the coastline

As well as protecting the natural environment, CZM also needs to consider that people need to be considered as part of the management plan. Any development of habitat protection measures, or physical coastal defense structures must protect visitor access to the shore. The coast is an immense recreational resource, earning huge amounts of tourist dollars. As development increases, however, it is possible for the levels of access to parts of the coastline to be reduced. To protect such interests, planning needs to consider the length of coastline and to allow access steps or ramps accordingly. Eroding coasts may, for example, experience loss of steps as cliffs erode, effectively isolating beaches from public access. While this could be seen as a management tool in itself (i.e., make access difficult as a means of protecting vulnerable areas), it will concentrate people into others, thus increasing visitor pressure proportionately.

The protection of coastal development

CZM has to consider the role of hinterland development in two ways. First, and perhaps most obvious, is the protection of infrastructure from erosion and flooding. In this respect, CZM needs to focus primarily on the methods available for coastal defense, and to recommend an holistic defense strategy for use along the coastline in question. When deciding on which defense policy to recommend, a series of points need to be considered:

1. What is the existing situation with respect to defense provision? The coastline can be mapped, marking seawalls, groins, breakwaters, undefended, etc.
2. What state of repair are these structures in? If they are in a poor state of repair, and likely to fail within the lifetime of the CZM plan, further points need to be raised, such as:
   - Is the current defense type still the best form of defense? If yes—plan reconstruction.
   - Is another type of defense better (see *Shore Protection Structures*)? If so what type? Typically, CZM plans should always consider the possibility of replacing hard defenses with soft, or removing defenses altogether.
   - The CZM recommendations need to fit with the holistic strategy for the whole coastline?
   - What will be the emergency response should any existing defenses fail? (rebuild, strengthen or change?)
3. If defenses are still sound, do they fit in to the longer-term defense strategy for the coastline?
4. For un-defended parts of the coastline, is this situation still tenable, or is development likely to become threatened within the lifetime of the plan?

Protecting the "identity" of coastal communities

A second aspect of tourist coast development concerns the social cost of tourism. When coasts experience rapid expansion, such as happened around the coast of the Mediterranean during the 1970s and 1980s and more recently in the Caribbean, it can often lead to small towns and villages losing their identity when they are swallowed up in development. This can, therefore, destroy the social structure of the local inhabitants. Many tourist areas are guilty of this form of identity loss, with coastal villages expanding so that they merge with neighboring villages undergoing similar development. As well as losing all identity and uniqueness, the result is a continuous development along much of a popular tourist coast which also removes any possibility for natural habitat management. CZM needs to recommend strict guidelines in such areas, so as to preserve uniqueness. For example, Canon Beach, in Oregon, has a ban on large fast-food outlets to protect it from mass tourism and day-trippers. Other methods include tight building regulations with regards to style of building and the materials used, limits on how many floors a development can have, and tight rules concerning the provision of open spaces.

Defining limits

From what has already been said, it can be seen that CZM needs to encompass a range of coastal uses, some of which will be marine-based, and some land-based. Clearly, such a distribution of use is going to vary from coast to coast and so the coverage and content of CZM will also need to vary accordingly. Hence, in developing a CZM plan, it is necessary to determine what the main activities are, what part of the coastline they impact, and thus to define the area of land and sea which CZM needs to cover. On coasts where the land-based activity is important, such as tourist-dominated coasts, there is a need to focus CZM more on the land. Conversely, on coasts where marine activities are important, such as for fisheries, there will need to greater CZM emphasis on the sea.

Defining the landward boundary

One of the key problems in deciding how far inland the landward limit of a CZM plan should be is whether to select one based purely on environmental grounds (perhaps the best considering the environmental rationale to the whole process), or one which coincides with an existing administrative boundary (convenient when it comes to actually running the process). In reality, it is very rare that the two should coincide. Possibilities include:

1. The landward limit of maritime climate influence, such as the limit of salt spray, or onshore sediment transport. This limit would allow CZM to protect sensitive supratidal habitats, such as cliff-top grassland communities and sand dunes.
2. Mean high-water mark may be a preferred choice where the main CZM concern is with intertidal habitats. Such a limit would, however, omit any land-based habitats or development or landward vegetation transitional communities.
3. Selecting the boundary of watersheds would include the whole of a river catchment and would encompass any activity which could impact on the coast. For example, the construction of a dam could impact dramatically on sediment loads being brought down river, and thus affect the sediment budget (see also *Dams, Effect on Coasts*). Similarly, land management practices, such as deforestation, could mobilize large volumes of sediment which could swamp sensitive coastal habitats, such as some coral communities, such as occurred in Kanehoe Bay, Hamaii (Jokiel *et al.*, 1993).
4. An arbitrary limit determined with respect to local conditions, such as topography, land-use, or habitats.
5. Limit of coastal development, to allow the protection of tourist interests.
6. The administrative boundary of the maritime authority in question.

Clearly, each possibility has relevance for particular coastal scenarios. Possibility (1) would be relevant on dune coasts, while (2) would be of importance in salt marsh or mangrove situations. However, in reality, the landward limit selected may well be a compromise between several of the above possibilities. Possibility (3), for example, involves huge areas of land and may, in many situations, be unmanageable. In many cases, the distance adopted by CZM tends to be arbitrary, in that a typical distance of ca. 2 km inland from the mean high-water mark of ordinary spring tides (MHWOST) is used, modified according to the geography and issues relevant to a particular coastline.

Defining the seaward boundary

Designating the seaward extent of CZM is often more difficult than landward, because there are no clearly identifiable morphological limits. In reality, therefore, the limit of a country's national jurisdiction may be an obvious choice but could also be problematical where processes which influence the coast, such as shipping, fishing, or oil/gas extraction, occur seaward of this limit. In other circumstances, where no significant marine-based activity occurs, this limit may be too large, and so an alternative may be needed. Such possibilities include:

1. Using the mean low-water mark will include the intertidal zone but ignore all areas seaward of this. Clearly, such a distinction will be only useful where the marine-based activities are of only minor importance.
2. Using the edge of the continental shelf will include more of the sea area, and so cover some fisheries and sediment stores. This latter point may be important where sediment is frequently dredged for coastal sediment recharge or where dredged sediment is disposed of. Clearly, this is a geological limit which will vary between nations. Furthermore, some nations, such as Germany, Holland, and Denmark, do not have a clearly defined shelf of their own.
3. Using a seaward limit determined by some predetermined water depth is useful for the protection of marine flora or fauna, but can lead to a mobile limit in areas of high sediment mobility, or lead to several where water depth becomes deeper and then shallow again.
4. Using one of the existing internationally recognized administrative boundaries makes a useful administrative boundary. In the United Kingdom, for example, there are three such boundaries:
   (a) 3 nautical miles—limit of controlled waters.
   (b) 6 nautical miles—limit of exclusive fishing rights.
   (c) 12 nautical miles—limit of territorial waters.

The CZM process

Many nations now realize, and may have also experienced, many of the impacts associated with the diverse range of activities and interests that occur around their coastlines (see Figure C62). As a result of this and Agenda 21, many have put in place a range of management initiatives to correct and manage them. This growing realization of the problem of both user conflict and historic lack of environmental awareness on our coastlines has brought CZM from an environmental "movement" into the international arena. The advantage of doing this is the development of improved information systems and international standards, yet its drawback is that increasingly, the process has been taken away from coastal managers and put into the hands of politicians.

With the key background issues and coastal problems identified, we are now in a position to look at developing a CZM plan to combat them. First, in order to successfully carry out CZM, it is essential to get away from the traditional piecemeal approaches to management which tend to be sectorially orientated and fragmented, and instead, go for the holistic management based on the coastal cell. In this way, the problems that need managing are dealt with in the context of the coastal process "unit" and thus, the two are carried out relative to each other. However, while the rationale for developing a CZM scheme may be clearly laid out it does raise some important questions. Most important in this are questions such as who does what? Who formulates the plans? Who enforces them? Who pays? This brings us onto the next important consideration, actually turning the concept into a workable framework and plan. This represents perhaps the prime area in which the above ideology has faltered in many cases.

The construction of a management plan

It is important to remember that CZM aims to protect a stretch of coast or an estuary by allowing multi-usage by industry, members of the pubic, and developers, in such a way that impacts on the natural functioning of the coastline is minimized. There is, therefore, a great risk of conflict between these groups if the whole thing is not done well. As the ultimate purpose is to allow the citizen to enjoy the coast in a sustainable way, whether for recreation, work, or as a place to live, it becomes vital that such people have a say in the eventual management plan. The wrong approach (often used) is where a panel of experts, including ecologists, sociologists, engineers, coastal scientists, has developed CZM, with no local consultation. These plans tend to act as a decree, issuing rules and regulations, and dictating to user groups what they can and cannot do. The net outcome of many of these plans is that they meet great opposition by people who resent being told what they can and cannot do in their own neighborhood, with the result that they can sometimes be completely disregarded and so fail to work. A second important consideration is that nothing in the plan is irreversible. The CZM document is advisory and subject to frequent updating and modification to reflect changes in usage patterns, the relative impact of these uses, and the changing natural environment.

Procedure for CZM

There are a series of steps necessary for successful plan implementation:

1. Necessary preliminary decisions and investigations.
   - The coverage of the plan (defining the coastal zone: how much land, how much sea).
   - The main coastal issues for area to be studied (users, conflicts).
   - The main user groups and stakeholders.

- The main environmental processes, erosion/accretion trends, sediment budget, etc.
- Any existing plans and whether or not they are working (development plans, defense strategies, local area plans, etc.). If they are, then they can form the basis of the relevant sections of the CZM plan.
- Amount of monitoring data for the coastline that is already available. Material should aim to cover all areas of concern, such as coastal defense, fisheries, heritage and landscape, industry, transport and development, land management, pollution, recreation and tourism, and wildlife. Data for each of these areas should come from as many sources as possible, including the environment itself, historical records, scientific literature, unpublished records, and members of the public/user groups/societies, etc.

2. Issues during CZM plan constructing and implementation.
   - Investigation of coastal usage and the nature/severity of conflicts (these can be between different user groups, or between activities and natural processes).
   - Incorporation of all user groups and the public in discussions.
   - Classification of the coastline on the basis of any issues of concern and their severity (developed/undeveloped; defended/undefended, etc.). Within each of the issues categories used, it is necessary to make management decisions covering both the current issues, and the future, that is, how will future infrastructure development be managed? How will recreation be controlled? What areas will be defended?
   - On the basis of the above, recommendations should be made relating to planning guidance, use zonation, conservation designation, etc.
   - The design and construction of any structure which CZM has recommended.
   - The design and implementation of regulatory measures (set-back lines, development zones, etc.).
   - The education and persuasion of local people to adopt and apply the plan.
   - Identification of funding sources.
   - Setting up and instigation of a detailed monitoring program.
   - Identification of future coastal defense strategies and the definition of any replacement strategies.
   - Formulation of hazard warning plans (if necessary for threats of flooding, tsunami, hurricanes, etc.).
   - Notification of a date for CZM plan revision (i.e., after 5 years) when all objectives and recommendations are reviewed, any new information assessed, and the report updated and reissued as a second version.
3. Post implementation.
   - Monitor successfulness of recommendations.
   - Obtain data on natural coastal functioning.
   - Make remedial changes if plan is failing in some areas.
   - Prepare for revision and update after designated period.

If all of these are satisfactorily completed, then the final plan will be a document which will:

(a) integrate user-group needs and actions harmoniously between groups;
(b) improve the understanding of how the coast is "working;"
(c) predict how the coast is likely to react to a range of likely environmental scenarios;
(d) provide various emergency response plans to protect local interests;
(e) provide coastal defense and identify future holistic strategies for this;
(f) identify sites, and plan for future development;
(g) identify ecologically vulnerable sites for special designation and protection.

Because CZM is responding to a series of issues which have arisen due to conflicts and problems from existing activity, potential problems arise in dealing with the range of interest groups involved. Few countries operate CZM which has any back-up in law. Most plans are advisory and need the good will of the people who use the coast if they are to succeed. It is for this reason that it is critical to incorporate these people in all stages of the planning and implementation of CZM. In support of plans, however, governments tend to make it a strong recommendation (and possibly a condition of funding) that any coastal development is made with reference to any CZM document.

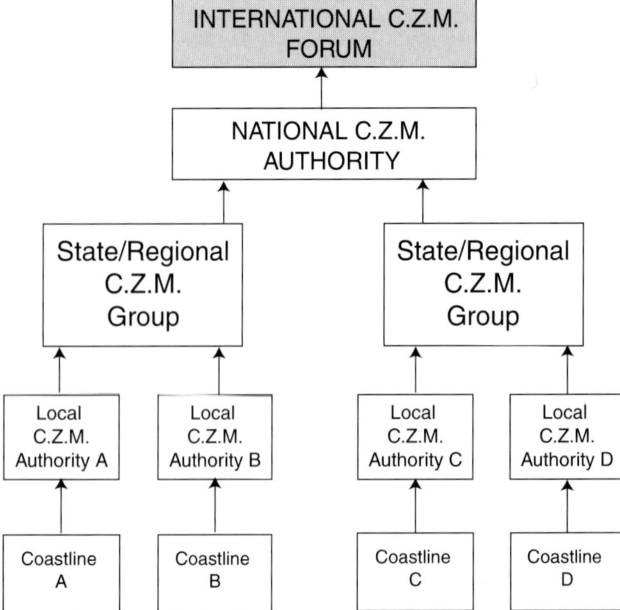

Figure C63 Groups responsible for the production and management of CZM plans.

Running the CZM process

As can be seen from the discussion above there are, potentially, a whole series of groups who have a vested interest in the coastal zone and could, therefore, claim responsibility for running CZM. At a local level, one group needs to oversee the formulation of a plan for a particular stretch of coastline from a neutral standpoint. In the United States, for example, the passing of the CZM Act in 1972, allowed state authorities to develop CZM at the local level. This process involves the appointment of local groups to develop CZM for parts of the coastline, with the relevant state acting as lead authority for all schemes under its jurisdiction (see Figure C63). In other countries, the nature of the lead group may be different but the same principals apply. Typical lead authorities and areas include regional government administrative authority boundaries, or the regional boundaries of an environmental protection agency.

In a similar manner, at a national level a single group needs to oversee the operation of state or regional groups (see Figure C63). Again this is partly to maintain both consistency between local groups and the integrity of the process; but also to liase with similar national bodies in adjacent countries (after all, a coastal cell boundary may not occur conveniently at a country's borders). A further key factor here is that many countries have complicated ownership rights in the coastal zone. In the United Kingdom, for example, the Crown owns the intertidal zone, meaning that it is common for government to get involved at this level.

By using CZM ideology, a successful coastal management authority should ...

1. ... Provide advice and elements of good practice in CZM matters.
2. ... Facilitate contact and discussion between different CZM holders.
3. ... Provide a national framework for CZM planning.
4. ... Ensure comprehensive and standardized coverage of all relevant issues.
5. ... Co-ordinate internationally with other similar national bodies.
6. ... Allocate and seek funding.
7. ... Implement CZM recommendations by making sure that they are known about and followed.

Regions for CZM

One of the main recommendations for CZM is that it should be an holistic study of a piece of coastline. This suggests that it is also necessary to link natural processes with human activities, and so the geographical area over which CZM plans are constructed needs to be based on the behavior of coastal processes. We know that sediment movements around the coast occur in set patterns, with erosion from one area leading to deposition in another and that such processes create a balance within clearly defined areas of the coast, known as coastal cells (see also *Beach Processes* and *Longshore Sediment Transport*). It is important, therefore, that CZM plans take account of these cells, as any management schemes will have to take note of sediment movement, sources, and stores so as to avoid complications with the sediment budget. For example, if one CZM plan recommended the protection of a cliff from erosion, and this supplied sediment to a beach in another CZM authority, this will cut off a major source of sediment supply. If the whole process was covered by one plan, such situations would not arise.

CZM in the longer-term

CZM represents a framework which can be used as a basis for all management plans. By using effective CZM plans, it is possible to effectively manage a coast with sensible development, reduce user conflicts, and to determine the most suitable defense types. Furthermore, the coast can be maintained in a natural form, permitting unimpeded movement of sediment and promoting wildlife habitats. This is fine for the management of human influence. Natural processes, in contrast, are harder to manage because they are less predictable and cannot read CZM plans! Regardless, however, CZM still needs to incorporate natural processes because they will, undoubtedly, impinge upon human activities and so cause conflicts to occur. Typical natural variation in coastal behavior that needs to be considered includes those associated with global warming, such as sea-level rise, increased storminess, increased frequency of extreme events, and hydrological and ecological changes in catchments.

It has been argued by some authorities that this is a rather idealist approach to CZM and could never be achieved. Certainly no one would ever suggest that all the world's coasts have to be naturalized, but it is clear that many parts can. What is perhaps the most frustrating factor in all of this is that huge amounts of money are being spent on outdated forms of defense, when less money could be spent using newer ideas and methods to protect the coast more naturally, more effectively, and for longer, before new investment is needed. The underlying problem lies in lack of education. Governments and national CZM bodies need to inform and convince their populations that to give land back to the sea is not such a bad thing, and in many cases, can be a most efficient use of tax-payers' money.

Further reading sources

As well as cross-referencing to other sections in this book, other key texts may also be of use to the reader as a way of expanding on the information given here. The book by Kay and Alder (1999) covers the process of CZM in some detail and takes the reader through the process, using examples. Similarly, the United Nations Environment Program produced a set of guidelines for integrated management of coastal zones in 1995. A summary of the CZM process is given by French (1997) while details on the types of coastal defense structures, including arguments for and against their usage, is given in French (2001).

Peter W. French

Bibliography

Arthurton, R., 1995, Implications of physical environmental change for coastal zone management. *Coastal Change '95. Proceedings of the International Conference on Coastal Change*. Bordeaux, pp. 761–764.

Barston, R.P., 1994, International dimensions of coastal zone management. *Ocean and Coastal Management*, **23**: 93–116.

Boorman, L.A., 1976, *Dune management: a progress report*. Cambridge: Institute of Terrestrial Ecology, *Unpublished Report*.

French, P.W., 1997. *Coastal and Estuarine Management*. New York & London: Routledge.

French, P.W., 2001. *Coastal Defences: Processes, Problems and Solutions*. New York & London: Routledge.

IPCC, 1990, *Climate Change: The IPCC Scientific Assessment*. Intergrovernmental Panel on Climate Change. Cambridge: Cambridge University Press.

Jokiel, P.L, Hunter, C.L., Taguchi, S, and Watorai, L., 1993. Ecological impact of a fresh water 'reef-kill' in Kanehoe Bay, Oahu, Hawaii. *Coral Reefs*, **12**: 177–184.

Kay, R., and Alder, J., 1999. *Coastal Planning and Management*. New York: Routledge.

Ly, C.K., 1980. The role of the Akosombo Dam in the Volta River in causing coastal erosion in central and eastern Ghana (west Africa). *Marine Geology*, **37**: 323–332.

Quarry, J., (ed.) 1992, *Earth Summit 1992*. The UN Conference on Environment and Development. Rio de Janeiro: The Regency Press Corporation.

Stanley, D.J., and Warne, A.G., 1993, Nile delta: recent geological evolution and human impact. *Science*, **260**: 628–634.

U.N.E.P., 1995, Guidelines for integrated management of coastal and marine areas. U.N.E.P. Regional Seas reports and Studies No. 161. United Nations Environment Programme.

Cross-references

Beach Erosion
Beach Nourishment
Changing Sea Levels
Coastal Boundaries
Coastal Changes, Gradual
Coastal Changes, Rapid
Conservation of Coastal Sites
Cross-Shore Sediment Transport
Dams, Effect on Coasts
Dredging of Coastal Environments
Economic Value of Beaches
Estuary
Geographical Coastal Zonality
Human Impact on Coasts
Longshore Sediment Transport
Mapping Shores and Coastal Terrain
Marine Debris
Marine Parks
Monitoring, Coastal Ecology
Monitoring, Coastal Geomorphology
Rating Beaches
Sea Level Change (see Changing Sea Levels)
Sea-Level Rise, Effect
Sediment Budget
Shore Protection Structures
Tourism and Coastal Development
Waves

COASTLINE CHANGES

Coastline changes (here defined as changes on the high-tide shoreline) can be mapped and measured on various coastal sectors over time scales ranging from a few hours or days (related to tidal cycles, weather events, earthquakes or tsunamis) to long-term trends over decades or centuries. Evidence of changes can be obtained from comparisons of dated historical maps, charts, air photographs, and satellite imagery with the present coastal configuration. Such evidence is available on many coasts over the past century, but in some countries, such as Britain and Denmark, maps and charts surveyed around the beginning of the 19th century permit the determination of changes over longer periods. In the Mediterranean region, and locally elsewhere, there is sporadic evidence of coastline changes over periods of at least 2,000 years, using historical descriptions and dated archaeological sites. On some coasts there is geological evidence of changes over the past 6,000 years, since the Late Quaternary marine transgression brought the world's oceans to approximately their present levels: cliffed coasts have receded and depositional coasts have advanced, and some coasts show evidence of alternations of advance and retreat during this period.

Advance or retreat of a coastline may result from combinations of erosion and deposition, emergence and submergence (Figure C64). Coastline changes are usually expressed in linear terms, as the extent of advance or retreat measured at right angles to the land margin, but they can also be considered in areal terms, as the extent of land gained or lost on a coastal sector, or volumetric terms, as the quantity of material added to, or lost from, the coast. Measurements made over varying periods are commonly averaged as gains or losses per year, but coastline changes may be sudden or intermittent rather than steady and gradual. Attempts have been made to document the advance or recession of coastlines in several countries: the Royal Commission on Coast Erosion in Britain (1907–11), the National "Shoreline" Study in the United States (US Army Corps of Engineers, 1971), and the survey of the Netherlands coastline by Edelman (1977); and in 1984 the International Geographical Union's Commission

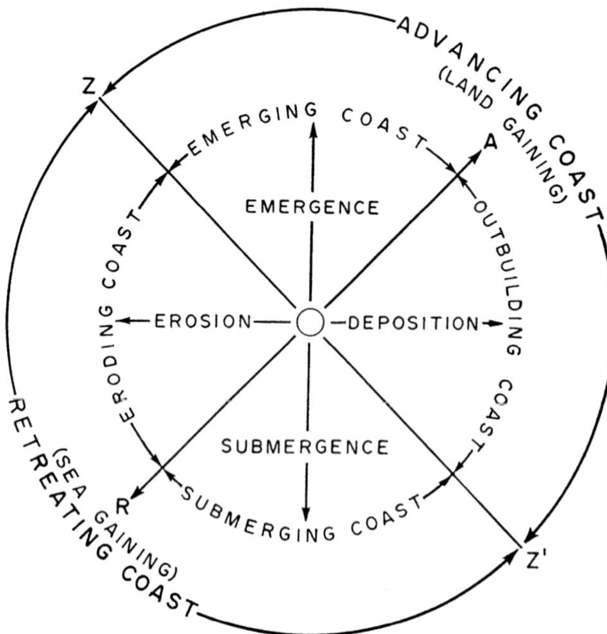

Figure C64 Analysis of coastline changes, based on Valentin (1952), ZZ′ indicates a stable coastline where erosion has been offset by emergence or deposition by submergence, or where no changes have taken place (O).

on the Coastal Environment completed a global review of coastline changes over the preceding century (Bird, 1985). More recently there has been monitoring of changes on particular coastlines using remote sensing and photogrammetry, but this has not yet been pursued systematically around the world's coastline.

Some coastlines have changed very rapidly (advance or retreat >100 m/year), but changes of more than 10 m/year are unusual, and few have changed by more than +/−1 m/year. If, as is predicted, a worldwide sea-level rise develops and accelerates as a consequence of global climatic warming resulting from the enhanced greenhouse effect, submergence and erosion will become widespread, and coastline changes are likely to intensify (Bird, 1993). Coastline changes can be considered in the following categories.

Cliffed coastlines

Cliffs in hard, massive rocks have changed little over the past century, even where they are exposed to stormy seas, but cliffs in soft materials can be cut back very rapidly. Sample rates of cliff retreat were tabulated by Sunamura (1992: Appendix 2). Recession of up to a meter a day has been measured on cliffs cut in unconsolidated volcanic ash, as on the Isla San Benedicto, off the Pacific coast of Mexico, and cliffs cut in alluvium and peat weakened by seasonal freeze–thaw oscillations have retreated up to 100 m in a few summer months before a winter cover of snow and ice briefly stabilized them. On parts of the Polish coast, cliffs cut in glacial drift deposits have shown an average recession of a meter a year, but have lost up to 5 m in a single storm, and similar rates have been documented on the east coast of England, notably on the Holderness cliffs. Recession of cliffs cut in soft formations is often accelerated by slumping and erosion by runoff after heavy rain or from melting snow.

Glaciated coastlines

Where glaciers and ice sheets reach the sea, as in Icy Bay, Alaska, southernmost Chile, and sectors around the Antarctic continent, the coastline may advance with forward movement of the ice, or retreat as the result of melting. Ice cliffs have receded at rates of up to a kilometer per year in Antarctica as the result of melting and calving induced by recent climatic warming. In the Gulf of Alaska, however, rivers augmented by glacifluvial outwash from melting ice have advanced parts of the coastline by more than a kilometer in recent decades by building deltas and prograding beach ridge plains. Rapid progradation occurred on the southeast coast of Iceland when melting of ice, hastened by volcanic eruptions, generated torrential outwash, delivering large quantities of sand and gravel to prograde beaches and build barriers and spits.

Emerging and submerging coastlines

Where land uplift has occurred the coastline has advanced. Isostatic uplift of formerly glaciated areas in Scandinavia and northern Canada has led to the emergence of former nearshore sea floors, as in the Gulf of Bothnia, where the coastlines of Sweden and Finland have advanced more than 100 m over the past century. On some sectors progradation has been augmented by shoreward drifting of sea floor sediment, notably sand from submerged eskers, to build up beaches. Sudden emergence during earthquakes has advanced coastlines on Montague Island, Alaska (1964), Talcahuano in Chile (observed by Charles Darwin during the voyage of the *Beagle* in 1835), the Orontes delta in Turkey (1958), the Makran coast in Pakistan (1945), Cheduba Island in Burma (1762), Tokyo Bay in Japan (1923), and Wellington (1855) and Napier (1931) in New Zealand. Emergence occurred during the phase of sea-level lowering between 1929 and 1977 on the coasts of the Caspian Sea, when islands expanded and became attached to the mainland, while shoals emerged as islands.

Submergence results in coastline recession, except on vertical cliffs or where it is compensated by coastal deposition. Land subsidence in the vicinity of the Mississippi has led to the submergence and erosion of former subdelta coastlines, and land subsidence in the Netherlands, augmenting the effects of a rising sea level, would have submerged a large part of the country had the coastline not been maintained by seawalls and reclamation works. Subsidence in the Venice region, due partly to subsurface compaction and the extraction of groundwater, has accelerated beach retreat on the Lidi barriers and diminished the extent of salt marshes in the Venice Lagoon. Increasing depth and frequency of storm surge flooding here indicates that submergence is partly due to a rising sea level in the Adriatic. Since 1977 the Caspian Sea has been rising and the coastline receding, much of the coastal land that emerged previously having been inundated. Sandy barriers have formed and are migrating landward (Ignatov *et al.*, 1993).

Sudden submergence during earthquakes led to coastline recession at Anchorage in Alaska (1964), in southern Colombia (1979), at Concepcion in Chile (1960), and on the Indus delta (1819, 1845). Tsunamis following sea floor mass movements have caused sudden recession of the Valdez delta in Alaska and along the beach-fringed Ivory Coast.

A correlation has been noted between alternations of water level in the Great Lakes and the advance or retreat of bordering coastlines. During phases of falling lake level the coastline advances as the result of emergence and shoreward drifting of sand to accreting beaches, while during phases of rising lake level the coastline retreats as the result of submergence, beach erosion, and increased cliff recession.

Volcanic coastlines

Volcanic eruptions prograde the coastlines where lava and ash are deposited, as on the southeast coast of Hawaii. They also augment river sediment loads, leading to the advance of deltaic coasts, as on the Rio Samala in Guatemala, and the progradation of beaches, as in the Bay of Plenty, New Zealand after the Tarawere eruption in 1886 and the south coast of Java after recurrent eruptions of the Merapi volcano.

Some oceanic islands are entirely of volcanic origin. Surtsey, off the south coast of Iceland, was built by volcanic eruptions in 1966–67, and has since been reshaped by the rapid recession of cliffs cut in lava and ash on the more exposed southern coastline, generating sand and gravel that have drifted round to accumulate on a triangular northern foreland. The explosive eruption of Krakatau Island, Indonesia, in 1883 left caldera-side cliffs and a surrounding mantle of ash which initially advanced the coastline, but has since been cut back as cliffs receding at up to 6.6 m/year.

Mass movements

Coastal landslides and rock falls temporarily prograde the coastline, forming lobes and fans that are subsequently cliffed and cut back by marine erosion. This sequence has been documented at Tillamook Head in Oregon, Pacific Palisades, California, Black Ven and Fairy Dell in Lyme Bay, Dorset, Rosnaes, and Helgenaes in Denmark, and on the Bulgarian and Crimean coasts of the Black Sea. On the Dorset coast erosion pin measurements have shown intermittent mudslide movements totaling 8 cm in 30 h, graded movements of about 3.5 m in 17 h and surge movements of up to 3 m in 20 min (Allison and Brunsden,

1990). Some landslide lobes have persisted long-enough to act as natural breakwaters, intercepting drifting beach sediment and causing local progradation, as at Golden Cap in Dorset.

Deltaic coastlines

The arrival of large quantities of fluvial sediment at the mouth of a river has led to rapid accretion of sectors of deltaic shoreline, with rates of up to 80 m/year at the south-west Pass of the Mississippi delta and up to 70 m/year on the Irrawaddy delta. On the north coast of Java deltas fed by sediment-laden floodwaters have prograded by up to 180 m/year. Progradation results from deposition of fluvial sediment on river banks to form natural levees, prolonged seaward as sedimentary jetties, and the delivery of sand and gravel to prograde bordering beaches. In addition, there is often a seaward advance of bordering salt marshes, as on the Fraser delta in Canada, or mangrove swamps, as on most tropical deltas, and sedimentation within these eventually builds up new land at high tide level.

Even where deltas have not formed, sediment supplied by rivers can prograde adjacent coastlines. Sand from the Columbia River on the Pacific coast of the United States was distributed northward and southward to prograde Long Beach and the Clatsop Plains. Fluvially nourished beach progradation has also occurred on the west coast of South Island, New Zealand and in north-east Queensland.

Where the mouth of a river is diverted during a major flood, the coastline that had formerly prograded begins to recede, while new progradation develops at the diverted outlet. This has happened on the Huanghe in China (1852), the Rio Sinu in Colombia (1942), the Ceyhan in Turkey (1935), the Medjerda in Tunisia (1973), and the Cimanuk in Java (1947). Artificial diversion of a river mouth by canal cutting has had similar effects on the Brazos delta in Texas (1938), the Rioni in the eastern Black Sea (1939), the Shinano in Japan (1922), and the Cidurian in Java (1927).

Progradation of deltas and coastal plains has accelerated where river sediment yields have increased as the result of soil erosion in their catchments, due to deforestation, overgrazing, or cultivation of steep hinterlands. In Java this has led to rapid progradation of deltas in Jakarta Bay and the Segara Anakan on the south coast is shrinking rapidly as the result of increased sediment input from the Citanduy river. On some coasts, progradation has followed an increase in fluvial sediment yield due to hinterland mining. Hydraulic sluicing in the Sierra Nevada a century ago led to the growth of the Sacramento delta into San Francisco Bay, and open-cast mining has resulted in beach progradation at Par and Pentewan in Cornwall, Chanaral in Chile, Jaba on Bougainville, and around New Caledonia, especially on the Thio and Népoui deltas.

Reduction in fluvial sediment yield to the coast following the building of dams on rivers has led to the onset or acceleration of erosion on many deltas. The best-known example is the Nile delta coastline, which

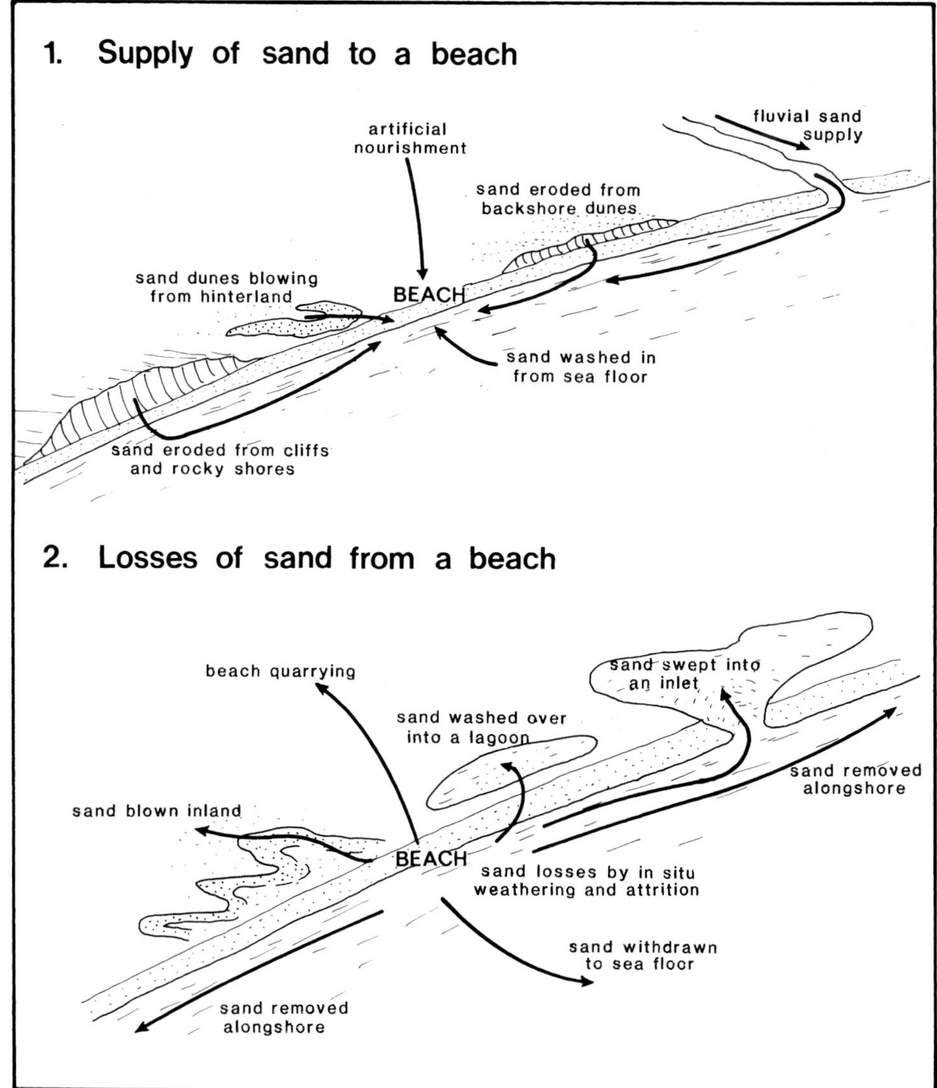

Figure C65 The various ways in which sand is gained by, or lost from, a beach. When gains have exceeded losses beach progradation is likely to occur, while losses exceeding gains leads to the lowering, flattening, and cutting back of the beach profile. Gravel (shingle) beaches show similar patterns of gain and loss, without the wind-blown component (Copyright—Geostudies).

Table C13 The causes of beach erosion

1. Submergence and increased wave attack as the result of a rise in sea level or coastal or nearshore land subsidence: the "Bruun Rule."
2. Diminution of fluvial sand and shingle supply to the coast as a result of reduced runoff or sediment yield from a river catchment (e.g., because of a lower rainfall, or dam construction leading to sand entrapment in reservoirs, or successful revegetation and soil conservation works), or the natural or artificial diversion of a river outlet, or because of the removal of much or all of the weathered mantle from slopes in the river catchments, exposing bare rock that yields little or no sediment to runoff.
3. Reduction in sand and shingle supply eroded from cliffs or shore outcrops (e.g., because of diminished runoff, stabilization of landslides, a decline in the strength and frequency of wave attack, or the building of seawalls to halt cliff recession), or the exposure by cliff recession of formations (such as massive resistant rock, or soft silts and clays) that do not yield beach-forming sediment.
4. Reduction of sand supply to the shore where dunes that had been moving from inland are stabilized, either by natural vegetation colonization or by conservation works, or where the sand supply from this source has run out.
5. Diminution of sand and shingle supply washed in by waves and currents from the adjacent seafloor, either because the supply has declined (e.g., where ecological changes have reduced the production of shelly or other biogenic material), because the transverse profile has attained a concave or steeply declining form which no longer permits shoreward drifting, or because such drifting has been impeded by increased growth of seagrasses or other marine vegetation.
6. Removal of sand and shingle from the beach by quarrying or the extraction of mineral deposits, or losses from intensively used recreational beaches, for example, as the result of beach cleaning operations.
7. Increased wave energy reaching the shore because of deepening of nearshore water (e.g., where a bar or shoal has drifted away, where seagrass vegetation has disappeared, where dredging has taken place, or where the seafloor has subsided).
8. Reduction in sand and shingle supply from alongshore sources as the result of interception (e.g., by a constructed breakwater or groins) or interruption by the growth of a fringing coral reef or some other depositional feature.
9. Increased losses of sand and shingle alongshore as a result of a change in the angle of incidence of waves (e.g., as the result of the growth or removal of a shoal, reef, island, or foreland, or because of breakwater construction or a change in wind regime).
10. Intensification of obliquely incident wave attack as the result of the lowering of the beach face on an adjacent sector (e.g., as the result of dredging, or scour due to wave reflection induced by seawall or boulder rampart construction or beach mining).
11. Increased losses of sand from the beach to the backshore and hinterland areas by onshore wind action, notably where backshore dunes have lost their retaining vegetation cover and drifted inland, lowering the terrain immediately behind the beach and thus reducing the volume of sand to be removed to achieve coastline recession.
12. Increased wave attack due to a climatic change that has produced a higher frequency, duration, or severity of storms in coastal waters.
13. Diminution in the caliber of beach and nearshore material as the result of attrition by wave agitation, leading to winnowing and losses of increasingly fine sediment from the beach, either landward into backshore dunes or seaward to bars and bottom deposits.
14. Diminution of beach volume by weathering, solution, attrition, or impaction (e.g., by heavy vehicle traffic), resulting in lowering of the beach face and coastline recession.
15. Increased scour by waves reflected from backshore structures, such as seawalls or boulder ramparts, leading to reduction of the beach that fronted them.
16. Migration of beach lobes or forelands as the result of longshore drifting—there is progradation as these features arrive at a point on the beach, followed by erosion as they move away downdrift.
17. A rise in the water table within the beach, due to increased rainfall or local drainage modification, rendering the beach sand wetter and thus more readily eroded.
18. Increased losses of beach material due to outwashing by streams or from drains carrying an augmented discharge of water, either because of increased rainfall or from melting snow or ice, or because of natural or artificial modifications (such as urbanization) in the catchments that have increased runoff through beaches.
19. Intensification of wave action where tide range has diminished, as on the shores of a bay or lagoon that has been partly enclosed by natural or artificial structures, impeding tidal ventilation.
20. Erosion where driftwood deposited on the beach is jostled by wave action, leading to erosion of sand and gravel from the beach face.
21. On Arctic coast beaches the removal of a protective sea ice fringe by melting, so that waves reach the beach (e.g., for a longer summer period).

has shown recession of up to 40 m/year since the completion of the Aswan High Dam upstream in 1964. Similar erosion has been documented on the Rhône delta in France. Deltaic coastline erosion has also followed extraction of sand and gravel from river channels or the reduction of runoff by irrigation schemes, soil conservation works, and afforestation in catchments, notably in Italy, Greece, and Turkey.

Beach-fringed coastlines

Changes on beach-fringed coastlines result from variations in the rate of supply of sediment from various sources or in losses from the beach system, as shown in Figure C65. Progradation of beaches and growth of spits supplied with sand and gravel from the erosion of nearby cliffs is well illustrated on coasts dominated by glacial drift deposits, as on Puget Sound on the west coast of North America, New England and the Canadian Maritime Provinces on the east coast of North America, and the coasts of the North Sea and the southern Baltic. Beach erosion can occur when the sediment supply from eroding cliffs diminishes as the result of seawall construction or stabilization, or a reduction in incident wave energy, as at Bournemouth and Brighton in southern England, the Melbourne Bayside coast on Port Phillip Bay and Byobugaura in Japan.

The spilling of waste from coastal quarries has prograded beaches, as at Hoed in Denmark, Porthoustock in Cornwall, and Rapid Bay in South Australia. At Workington in Cumbria, the dumping of waste from coal mines and steel works has locally augmented beaches. On the other hand, quarrying of sand, gravel, or shelly material from beaches has not only depleted them, but has led to accelerated erosion on nearby cliffs, as at West Bay and Seatown in Dorset, Gunwalloe in Cornwall, Hallsands in Devon, and St. Ouen's Bay in Jersey.

Information supplied to the International Geographical Union's Commission on the Coastal Environment in 1972–84 indicated that about 20% of the world's coastline consists of sandy beaches backed by beach ridges, dunes, or other low-lying sandy terrain. Of this, more than 70% had shown net erosion over the preceding few decades and less than 10% sustained pro- gradation, the remaining 20–30% having been stable, or shown no measurable change. The initiation or acceleration of beach erosion was attributed to several factors, listed in Table C13. The relative importance of these factors differs from one coastal sector to another, but analysis of particular eroding beaches have shown that usually more than one has contributed. A detailed discussion is provided in Bird (1996).

Swampy coasts

The seaward margins of salt marshes and mangrove swamps (generally close to the mid-tide shoreline) have advanced seaward where the sediment supply has been abundant, particularly on coasts sheltered from strong wave action and on the shores of prograding deltas. Sedimentation within the widening salt marshes and swamps eventually builds them up to high-tide level, so that the coastline advances. Extensive salt marsh progradation has occurred around Chesapeake Bay during the past 350 years, and has been attributed to increased sediment yields from inflowing rivers as the result of soil erosion in their catchments. Mangrove swamps have prograded on some tropical coasts, particularly near growing deltas, as on the shores of southern and eastern Kalimantan in Indonesia. On the west coast of Peninsular Malaysia, mangroves spread seaward by up to 54 m/year between 1914 and 1929 when sediment yields from agriculture and mining were high, but after dams were built on rivers and tin mining activity reduced erosion became prevalent (Kamaludin, 1993).

On many coasts the seaward margins of salt marshes and mangrove swamps have retreated in recent decades, so that they terminate in small cliffs, fronted by tidal mudflats or sandflats. In southern England and north-western France, and in New England, seaward cliffing is very extensive, and similar features are seen on mangrove swamps on the west coast of Peninsular Malaysia, in Thailand, and the Philippines, possibly in response to a relative sea-level rise in recent decades.

Artificial coastlines

Many coastlines have become artificial, especially during the past century, as the result of the construction of seawalls, boulder ramparts, and other structures intended to halt erosion on cliffs, beaches, and deltas. Spits such as Ediz Hook in Washington and Sandy Hook in New Jersey, previously variable in configuration, have been armoured with seawalls or boulder ramparts. Breakwaters built to shelter harbors or form marinas are now widespread on the coasts of North America, Europe, and the Mediterranean, and some coastal areas have been embanked and reclaimed for ports, industry, and urban development, notably in the Netherlands, China, Singapore, and Tokyo Bay. Reclaimed areas are typically bordered seaward by earth banks or concrete walls, which have become extensive on the coasts of Britain, Europe, and Southeast Asia, particularly Japan, where about a quarter of the coastline is armoured with seawalls and tetrapods (Walker and Mossa, 1986).

The building of artificial structures is still a widely preferred response to erosion and submergence of coastlines, but increasing use is being made of beach nourishment as a means of defending coastlines against erosion which also provides a recreational resource of scenic value.

Eric Bird

Bibliography

Allison, R.J., and Brunsden, D., 1990. Some mudslide movement patterns. *Earth Surface Processes and Landforms*, **15**: 297–312.
Bird, E.C.F., 1985. *Coastline Changes*. Chichester: Wiley.
Bird, E.C.F., 1993. *Submerging Coasts*. Chichester: Wiley.
Bird, E.C.F., 1996. *Beach Management*. Chichester: Wiley.
Edelman, T., 1977. Systematic measurements along the Dutch coast. *Proceedings of the 10th Conference, Coastal Engineering*, pp. 489–501.
Ignatov, Y.I., Kaplin, P.A., Lukyanova, S.A., and Solovieva, G.D., 1993. Evolution of Caspian Sea coasts under conditions of sea level rise. *Journal of Coastal Research*, **9**: 104–111.
Kamaludin, B.H., 1993. The changing mangrove shorelines in Kuala Kuran, Peninsular Malaysia. *Sedimentary Geology*, **83**: 187–193.
Royal Commission on Coastal Erosion, 1907–11. *Reports* (3 Volumes) London: His Majesty's Stationery Office.
Sunamura, T., 1992. *Geomorphology of Rocky Coasts*. Chichester: Wiley.
U.S. Corps of Engineers, 1971. *National Shoreline Survey*. Washington, DC.
Valentin, H., 1952. *Die Küsten der Erde*. Petermanns Geogr. Mitteilungen.
Walker, H.J., and Mossa, J., 1986. Human modification of the shoreline of Japan. *Physical Geography*, **7**: 116–139.

Cross-references

Changing Sea Levels
Coastal Changes, Gradual
Coastal Changes, Rapid
Cliffs, Erosion Rates
Coastal Subsidence
Deltas
Glaciated Coasts
Greenhouse Effect and Global Warming
Mapping Shores and Coastal Terrain
Volcanic Coasts

COASTS, COASTLINES, SHORES, AND SHORELINES

Origins of terms

The word shore comes from the middle English word *Schöre* that came from the middle low German word Schöre. Shore is also an archaic past tense and past participle of shear. In the German language, the verb shear is scheren (Betteridge, 1965). Schere (or shear) means shearing, cutting, or a point of division. Thus, the middle English and low German schöre translates as the sheared line (or shoreline) between the ocean and the land. In the German language, schöre is a line and not a region. However, the English usage of shore (from schöre) became a description of the *area* adjacent to a body of water. Webster's unabridged dictionary defines shore as "land at or near the edge of a body of water" (McKechnie, 1979). The term shoreline is used to describe the *line* marking the edge of a body of water.

The German language actually uses a different word to describe the strip of land adjacent to a body of water. The German word *ufer* is used for bank, shore, beach, or edge. Thus, ufer is more akin to the English shore, and schöre is more akin to the English shoreline.

The word coast comes from the middle English and old French word *coste* meaning rib, hill, or shore. In Webster's unabridged dictionary, coast, is the edge or margin of land next to the sea. The close similarities between the words shore, shoreline, coast, and coastline have lead to many ambiguous and inaccurate usage that are unacceptable for scientific dialogue. Scientific usage of these terms are very specific and require explanations related to their description and use.

Geometric attributes

Coasts, coastlines, shores, and shorelines have geometric characteristics that have precise denotations in scientific nomenclature. Casual usage of the words line, surface, and prism is not appropriate because they are too imprecise or ambiguous. Below are formal definitions of these terms as they should be used in scientific dialogue.

Shorelines and coastlines are lines designating the boundaries between fluid and solid media. A *line* is a one-dimensional construct used to signify the sheared edge between two different materials. A line has length but not breadth. The term coastline and shoreline are both boundary lines between water and land. The term coastline is generally used to describe the approximate boundaries at relatively large spatial scales. Shoreline is used to describe the precise location of the boundary between land and water.

Strips of land adjacent to shorelines and coastlines are called shores and coasts, respectively. Shores and coasts are bounded on one side by a body of water and on the other by a change in terrain to an upland landscape. The *surfaces* of these land strips are two-dimensional features having length and breadth but no thickness. While shore surfaces may have relief, the relief does not give the surface a thickness that implies a volume. The relief only describes elevation changes on a two-dimensional surface. The main difference between a shore and a coast is one of scale. Shores are relatively narrow strips of land adjacent to water bodies, whereas, coasts generally depict relatively broad bands of land adjacent to water bodies.

In a geologic context, processes of erosion, transportation, and deposition sculpt the relief on land surfaces. These processes involve the removal, transport, or deposition of a volume or mass of material, and imply a three-dimensional quality of the shore. The shore must be considered a three-dimensional prism when considering the processes that build or destroy it.

Shore surfaces are two-dimensional areas capping the three-dimensional shore prism. Shorelines are one-dimensional boundaries at the edge of the shore. Generally, the terms shore and coast are used to describe the prism of material below the elongate land surfaces at the edge of a body of water.

Coasts and coastlines

The terms *coast* and *coastline* are used for describing the boundary between the land and the water at broad regional scales. The coast is a strip of land of variable width that extends from a body of water inland to a regional break in terrain features. The coastline is the line that forms the boundary between the coast and a major body of water.

At regional to global scales, the term coastline (rather than shoreline) is used to differentiate the boundaries between the land and the sea. At these scales, changes in margin shapes and boundary orientations are used to describe continents, oceans, and marginal seas. Even at smaller regional scales, coastlines are used to describe the marginal shapes and boundary orientations of estuaries, bays, deltas, barrier islands, and spits. Coastlines may be accurately portrayed on maps having global or regional scales, but the widths of lines constructed at these scales are too large to locate the precise physical position of the sheared boundary between land and water. For example, a volcanic island on a very broad-scale map may look elliptical in shape, but upon closer inspection the border between the land and the sea may illustrate a serrated boundary created by headlands and pocket bays (Figure C66). The same effect would occur at regions of long straight coasts where the precise water boundaries of small estuaries and bays are scarcely discernible at large scales. The broad-scale maps "smooth-out" small coastal irregularities. Thus, the coastline is generally not as precise as the associated shoreline that traces the land–sea boundary at much finer detail. The coastline is a line connecting the most seaward promontories into the sea. Consequently, a length of shoreline is generally much longer than the associated length of coastline (Figure C66).

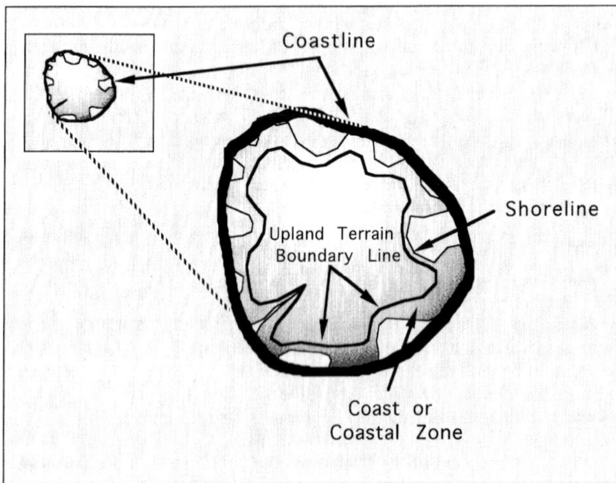

Figure C66 Hypothetical planview of a volcanic island showing the apparent shapes of the island from a broad-scale view and a close-scale view. The coastline is oval-shaped, whereas the shoreline illustrates all of the shore irregularities.

Since the coast is a strip of land adjacent to the coastline its precise boundaries are indefinite and general. The coast is the strip of land between the coastline and a change in terrain to upland landscapes. The coast includes specific coastal environments or landscapes that are distinct from the upland (e.g., bays, estuaries, coastal dunes, maritime forests, maritime climates, tidal waters, etc.).

At early periods in the Earth's history, there were no images or surveyed maps to precisely locate land–water boundaries. Boundaries could only be deducted from other indirect sources of data. Coastlines from these data sources have imprecise locations with respect to their spatial and temporal setting. For example, at periods of 10^4–10^5 years, water volumes in the ocean fluctuate in response to the expansion or contraction of continental glaciers. Glacial waning increases the volume of the sea causing water to transgress over the land surface, while glacial expansion causes water to regress off the land surface. The vertical shift in water level produces horizontal shifts in the position of the boundary between the two media. A paleogeographic depiction of the coastline during these periods may be reconstructed using terrain slope and sea-level projections. These are generalized locations of the boundaries, rather than precise paleo-shoreline locations. A time series of coastline positions on a map may be used to illustrate the regression of water off a land surface, or the transgression of water over a land surface. Thus, temporal scaling also influences our interpretation of coasts and other large-scale areas. Since short-term events generally produce small changes to coastal landscapes, the record of these events cannot be observed at large coastal scales. At broad spatial scales, only large changes produced over long temporal intervals can be observed.

As the sea transgresses over, or regresses off a land surface, the process necessitates a lateral shift in the position of the coastline. At large spatial and temporal scales, the antecedent topography might be considered a passive surface being inundated by rising water. The process is "morphostatic," meaning that the coastline shift is driven by flooding of a surface rather than erosional sculpting of the surface.

Shore and shorelines

The terms *shore* and *shoreline* are used for describing the boundary between the land and the water at very fine local scales. The shore is a strip of land of indefinite width that extends inland from a body of water. The terrain of the shore is primarily composed of forms generated by active or recent marine and marine-related processes. The landscape of the shore generally has a unique plant community draped over the terrain. The landward limit of the shore is located at a boundary separating these terrains and plant communities from upland landscapes. The shore is a dynamic area that has many ephemeral forms sculptured by a variety of marine processes. The landward part of the shore that is primarily molded by aeolian processes is called the

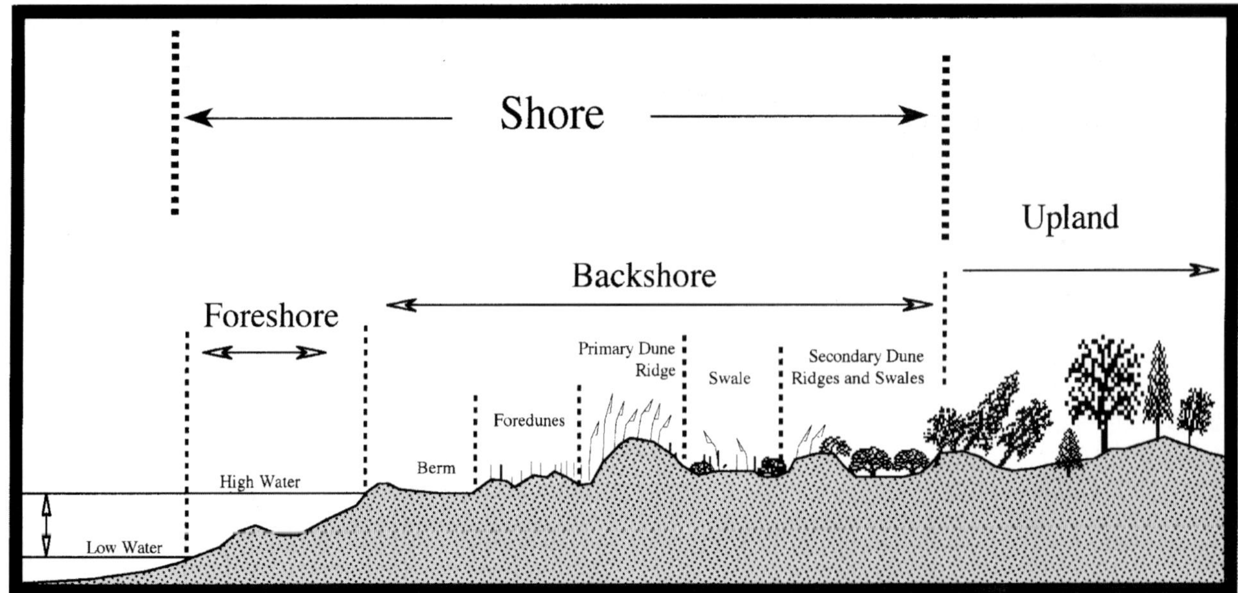

Figure C67 Hypothetical cross section of a shore showing the classification into backshore and foreshore based on topographic features and water levels.

backshore (Figure C67). The surface of the backshore often has a succession of plants draped over a chronosequence of dune ridges between the water's edge and the upland areas. The berm and foredune areas of the backshore are located on shifting sandy surfaces that are sparsely vegetated with specialized herbaceous plants and grasses that tolerate intense light, heat, drought, burial, and salt spray. Dune ridges and swales are located on relatively stable surfaces that have dense concentrations of stress-tolerant grasses and shrubs.

Common grasses that colonize dune surfaces in warm temperate zones include sea oats (*Uniola paniculata*), American beach grass (*Ammophilia breviligulata*), and running beach grass (*Panicum amarum*). Common shrubs in the swales between dune ridges include wax myrtle (*Myrica cerifera*), bayberry (*Myrica pensylvanica*), and silverling (*Baccharis haliminfolia*).

The seaward part of the shore is called the foreshore, which is defined by tidal range (Figure C67). The high-water elevation defines the upper limits of the foreshore, and the low-water elevation defines the lower limit of the foreshore (a more comprehensive description of foreshore zones is described below). The foreshore is the portion of the shore where wave-energy is dissipated on the land. The foreshore is sculptured by the dispersion and shifting of granular particles by the energy of waves and littoral currents. The resultant patterns of erosion and accretion make the foreshore one of the most dynamic areas of the shore.

Shorelines

The shoreline is a line that demarks the precise boundary between the shore and the water. Since the position of the water shifts with the elevation of the tide, then the position of the line at high tide is different from its position at low tide. On Figure C67 it is apparent that there is a high-water shoreline at high tide and a low-water shoreline at low tide. Sea level is generally defined as the elevation halfway between the high- and low-water elevations, and the generic "shoreline" is meant to describe the line associated with sea level, rather with high or low tide. However, during the rise and fall of the tide, there are an infinite number of shorelines between high tide and low tide.

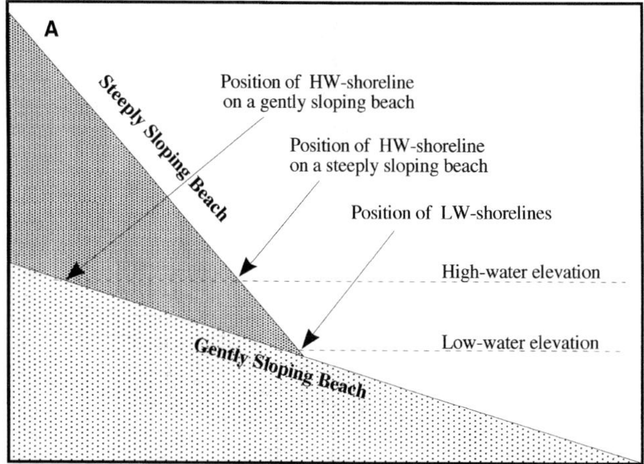

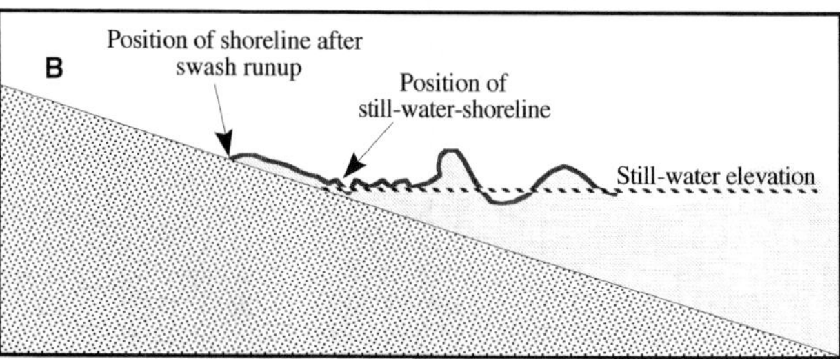

Figure C68 (A) Cross-sectional sketch of a steeply and gently sloping shore. Given a constant vertical tidal range, the horizontal spreads of water across a gently sloping surface is considerably greater than across a steeply sloping surface. (B) After a wave breaks, the rush of water up the beachface (swash) shifts the location of an observed shoreline landward of the projected still-water-shoreline.

The elevations of high tide and low tide are not constant. They are dependent on changes in the magnitudes of gravitational and centrifugal forces pulling on the water surface. Changes in the gravitational forces result from the motion and changes in the relative alignments of the Sun, the moon, and the Earth. When the positions of the Earth, moon, and Sun have additive gravitational effects, then the greatest tidal bulges occur (spring tides). A favorable alignment for gravitational addition occurs twice a month producing spring tides. However, the gravitational forces offset each other twice a month producing, small tidal ranges (neap tides). The elevations of each progressive high and low tide are different with the incremental changes in the resultant gravitational forces of the three celestial bodies. In fact, there are over 140 factors involved in the generation and alteration of the tides. The gravitational force of the moon plays a major role in tidal variations. However, the moons orbit around the Earth is slightly irregular and it takes about 9.3 years for the plane of the moon's orbit to shift above and then below the plane of the solar ecliptic. The cycle is completed in about 18.6 years. These factors can be used to mathematically predict the tidal ranges. An average of all of the high tides and low tides over 18.6 years provides the mean-high and mean-low tide elevations for those predictions. The relationships also can be used to predict the elevations of water level at any position in the 18.6-year cycle, but it assumes no elevation changes are caused by atmospheric conditions. The 18.6-year mean of all high-tide elevations is called mean high water (MHW), the 18.6-year mean of all low-tide elevations is called mean low water (MLW), and the 18.6-year mean of all sea-level elevations is called mean sea level (MSL). There are also means for spring high (MHWS) and spring low tides (MLWS), as well as neap high (MHWN) and neap low tides (MLWN). These temporally averaged elevations can also be affixed to associated sheared lines (shorelines) between the ocean and the land. To standardize the use of the term shoreline, it is appropriate to affix it to a specific water level, such as mean high-water shoreline (MHW-shoreline), mean sea-level shoreline (MSL-shoreline), mean low-water shoreline (MLW-shoreline).

Atmospheric conditions also have influences above and beyond the gravitational effects described above. Low-pressure systems can elevate water surfaces, winds can pile water on the shore, and waves may runup on a shore above the still-water elevation of the sea. Daily effects of these atmospheric events on water level result in "actual" water levels that are generally different from predicted values.

The slope of the shore profile also has a very important influence on the how far water levels may shift up the shore surface (Figure C68). Typically, beaches have slopes between 1 and 3°. At gentle slopes, a 1 m rise in water level related to atmospheric conditions may cause the position of a shoreline to shift landward 188 m. This type of shift can be accomplished without any addition or removal of sediment from the shore ("morphostatics").

Breaking waves are other processes that influence the position of the wet and dry boundary at a shore (Figure C68(B)). As waves break at the toe of the shore, the wave-bores surge forward and strike the beachface. The momentum of the water continues forward and the water (called swash) rushes up the beach profile. Swash often rises 1.5 times the wave height above the still water level (SWL). Since wave climate is spatially and temporally variable, then the average elevation of the swash limit is seasonally different, and different along dissimilar areas of the coast. If an average wave height can be estimated for a region's wave climate, then a mean swash elevation (MSE) can be estimated for any reach of shore. Similarly, a standardized mean swash shoreline would represent the wet/dry boundary of a shore influenced by its associated wave climate. Thus, on any given day the predicted water levels from gravitational tides are rarely accurate because of the variability in atmospheric phenomena. Field and remote observations of shoreline locations are temporally very specific, and the sheared line between the water and the land is generally not related to a specific standardized elevation.

Geologists and oceanographers do not agree on the placement of shorelines on maps and charts. Since geologists are primarily interested in depicting land surfaces, the high-water shoreline is generally used as the datum for these maps. Topographic quadrangles published by the US Geological Survey (USGS) place the shoreline at mean high water or at the North American Geodetic Datum (NAVD 29) which is generally very close to mean high water. On the other hand, the National Oceanographic and Atmospheric Administration (NOAA) publishes charts that are primarily for navigational purposes. Oceanographic cartographers consider any surface that is subaerially exposed as land whether it is always exposed or only exposed periodically by tidal inundation. The datum (and shoreline) for NOAA charts is positioned at MLW or mean lowest low water. It is for this reason that the intertidal zones on maps and charts are often not surveyed, and are simply described as flats and marsh without elevation data.

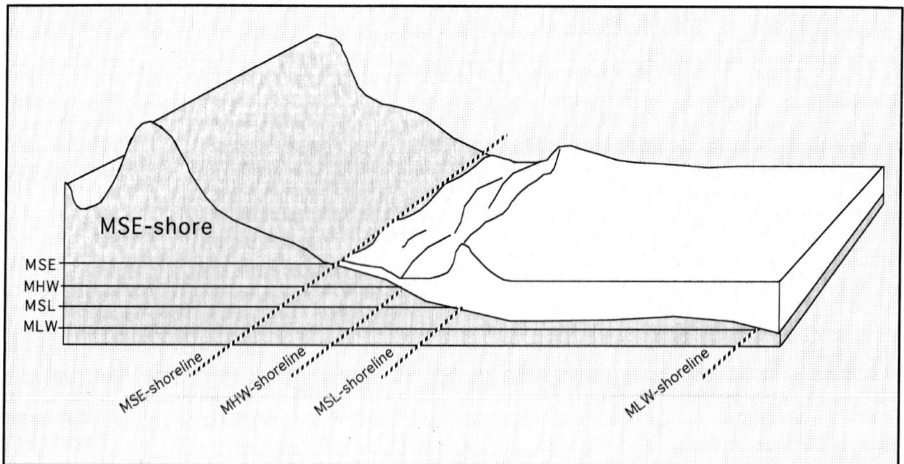

Figure C69 A block-diagram sketch illustrating the relationships among four "mean" water levels, four mean shorelines, and four shore prisms located above each horizon.

Shoreline dynamics

Short- and long-term processes govern the positions of shorelines. A change in shore form created by shifting volumes of sediment is called morphodynamics. The processes of sediment erosion, transportation, and deposition along the shore often result in shifts in shoreline positions. Magnitudes of shoreline shift are expressed as linear units (e.g., meters or feet) where negative values often indicate landward retreat and positive values indicate seaward shifts. During short-term events, shorelines at different elevations on the beachface respond differently. During storms, material is often eroded from the upper part of the shore and deposited at the toe of the shore. As a result, the MHW-shoreline may shift landward, while MLW-shoreline shifts seaward. In this case, the directions of shift for the MHW and MLW-shoreline are opposite, while the MSL-shoreline may have remained stationary. Thus, the use of only one shoreline is insufficient to study the short-term morphodynamic events. However, over long periods, the shift direction of the upper, middle, and lower shorelines tend to be the same, and a specific shoreline might be reliable for studying long-term changes.

Rates of shoreline shift may be used to describe how fast a shoreline is shifting. The rates of shoreline shift are expressed in linear units per time (meter/year or feet/year). During short-term events, consistency in the use of a specific shoreline is important since the upper and lower parts of the beachface respond differently to storms. During long-term intervals (annual to decadal), the gross shift rate and the net shift rates may be significantly different since the directions of long-term shift are bipolar. For example, if a shoreline retreated landward 5 m in 5 years and then advanced 5 m seaward for the next 5 years, the gross shift rate would be 1.0 m/yr, while the net shift rate would be 0.0 m/yr. At very long-term rates (centuries to millennium), shoreline shift rates are impacted by the chronic Holocene rate of sea-level rise.

Shores and shore prisms

Since a shore is a strip of land associated with a shoreline, a standardized subdivision of a shore is logically defined by the series of mean shoreline described above. The mean shorelines are associated with specific elevations on the shore surface. These elevations are associated with a horizontal planar surface that extends under the land surface (Figure C69). The material above each planar surface represents the sand prism of a specific portion of the shore. For example, the mean low-water shore (MLW-shore) is the prism of sand located above the mean low water elevation (and between the MLW-shoreline and the landward limit of the backshore). Each of the standardized shorelines has an associated standardized shore. The mean sea level shore (MSL-shore) is the prism of sediment located above the elevation of mean sea level. The mean high-water shore (MHW-shore) is the prism of sediment located above the elevation of mean high water. The mean swash elevation shore (MSE-shore) is the prism of dry sand located above the mean limit of swash runup on the shore (Figure C68B). These shore prisms are valuable units for describing specific changes to the shore. Since each prism is composed of a specific volume of material, then removal of a portion of that volume is by "geological" definition called erosion. The addition of a volume of material by definition is called accretion. Note that erosion and accretion requires the addition or removal of a mass or volume of material.

Shore accounting

Beach profiles are often made perpendicular to the shoreline to study beach processes. The area between the profile and MLW elevation represents a two-dimensional section of the shore. When combined with a one-dimensional length of shoreline, the resultant prism is a three-dimensional section of shore. Thus, pairs of profiles can be used to estimate the volume of sand stored in reaches of shore. Standardized elevations can also be used to determine the volume of material stored in each of the standardized sections of a shore prism. Thus, an accounting of material can be made for each shore zone that provides information on the erosional and accretional characteristics of the shore. Using the technique of profile comparisons along a stretch of shoreline, rate scales can be devised to normalize the change rate per unit length of shoreline and/or period of time (Oertel *et al.*, 1989).

During storms, material is often eroded from the upper portions of the shore and deposited on the low portions of the shore. The process may result in no net gain to the total shore volume although there was a shift in the location of a large amount of material. An accounting of material in the total MLW-shore might illustrate a stable volume suggesting a passive condition. An accounting of material in the MSE-shore would illustrate erosion, while the accounting of material in toe of the shore between MLW and MSL would show accretion. The dynamic nature of the shore can only be realized by an accounting of material in vertically stratified zones. The shore zones (related to mean water levels) provide a valuable framework for relating patterns of shore-volume change and shoreline shift.

Common misusage

The most common misuse of shoreline is the expression "shoreline erosion" and "shoreline erosion rate." A shoreline is a one-dimensional construct on a planar surface. It has length but no width or depth. In this case, it is a line that denotes the boundary between a solid (land) and a fluid (water). By definition erosion is the removal of a mass or volume of material. Since, the line is one-dimensional, it has no volume or mass. Eroding the line would be equivalent to erasing the boundary between land and water. In strict scientific terminology the expressions "shoreline erosion" or "shoreline erosion rate" are nonsensical. Shorelines shift landward or seaward across a surface, but do not erode. The landward shift of a shoreline should be described as shoreline retreat, while the seaward shift of a shoreline is called shoreline advance. A rise or fall in water level (that does not require the processes of erosion or accretion) may induce shoreline shifts, or it may be a response to erosion or accretion of material from parts of the shore. The advance or retreat of a shoreline is expressed in linear units (meters or feet). The terms erosion and erosion rates must apply to the change in the size of a volume or mass of material.

Since erosion is the removal of a mass or volume of material, shore erosion is used to describe the removal of material from all or part of a section of shore. It is a three-dimensional process requiring a reduction in the size of the shore sediment prism. An increase in the size of a shore sediment prism is described as shore accretion. A decrease in the size of a shore sediment prism is described as shore erosion. Erosion and accretion are expressed in units of volume (generally cubic meters or cubic feet). Erosion and accretion rates are as expressed in volume units per time (e.g., cubic meters/year of cubic feet/year).

The gross amount of material transported in the littoral system is often used as a surrogate for shore erosion. This is another incorrect use of terminology. Material transported in the littoral zone is described as littoral drift. The rate of littoral drift is generally expressed in units of volumetric units per time (e.g., cubic meters of sand per year). This is a transport rate, and is not a substitute for an erosion rate. The rate is a gross amount of material being transported in the littoral zone. Since, erosion is a measure of the removal of material (not transport) the two may not be equivalent. For example, the gross amount of sediment entrained in a littoral current (in a section of shore) may be constant at about 15,000 m^3 in a month. During the transport of this material, the littoral current can remove and add material to the shore. If the amount removed is equal to the amount added then there is no change to the shore sediment prism. However, if the rate of littoral drift is not constant along adjacent sections of shore, then erosion or accretion may result. Gradients in the rates of littoral drift along a section of shoreline may be used to imply erosion or accretion rates. That is, if more material is being transported into a section of shore than is being transported out of a section of shore, then the rate of accretion will be the net difference between the input and output rates.

The use of remotely sensed data is a popular tool for studying rates of shoreline shift. By comparing shoreline locations at two different times, the amount of shift can be estimated over a specific duration. The resultant rates are often used to predict rates of change. Typically, the shoreline is defined by the wet/dry boundary found on an image. Since, each image represents a "snap shot" in time and not one of the standardized elevations (described above), then it is difficult to discern whether any shift was caused by changes in the volume of sediment at the shore (morphodynamics), or changes in water level (morphostatics). Changes in water levels could be a response to different combinations of tidal stage, barometric pressure changes, waves runup, and wind stresses. On a flat beach, these forces could combine to produce several meters of water-level rise resulting in over 200 m of shoreline shift (without a change to the position of a MSL-shoreline or any change in the size of the shore sand prism). This is the +/− error that should be attached to shoreline shift measurements at low-profile beaches. Therefore, it is very difficult to use remotely sensed data to analyze rate of shoreline change without vertical control.

George F. Oertel

Bibliography

Betteridge, H.T., 1965. *The New Cassell's German Dictionary*. New York: Funk and Wagnalls.

McKechnie, J.L., (ed.), 1979. *Webster's New Unabridged Dictionary*. New York: Simon and Schuster.

Oertel, G.F., Ludwick, J.C., and Oertel, D.L.S., 1989. Sand accounting methodology for barrier island sediment budget analysis. *Proceedings of the Sixth Symposium on Coastal and Ocean Management*. Charleston, SC, pp. 43–61.

Cross-references

Beach Processes
Coastal Changes, Rapia
Changing Sea Levels
Littoral Cells
Mapping Shores and Coastal Terrain
Remote Sensing of Coastal Environments
Sediment Transport (see Cross-Shore Sediment Transport and Longshore Sediment Transport)
Tides

COHESIVE SEDIMENT TRANSPORT

Introduction

Sediments are a major component of the suspended load and sediment bed in coastal water bodies (estuaries, lagoons, river mouths, tidal flats, etc.). These sediments can be classified as cohesive and noncohesive. Cohesive sediments are comprised primarily of clay-sized (<2 μm) and silt-sized (<75 μm) particles, mixed with organic matter, and, sometimes, quantities of very fine sand. Noncohesive sediments are primarily sand and gravel-sized material (>75 μm). The underlying property that distinguishes cohesive sediments from noncohesive sediments is that they are cohesive in nature. Cohesion is the predominance of attractive forces over repulsive forces, such that particles in close proximity can bind together to form flocs (aggregates). This can be attributed to the presence of clay and colloidal particles with surface physico-chemical forces that are significant. A clay fraction greater than about 10% is generally sufficient for the sediment to exhibit cohesive properties (van Rijn, 1993).

Cohesive sediments are transported to coastal waters from alluvial and marine sources. Alluvial sediments are derived from various terrestial processes and human activities and are comprised of both cohesive and noncohesive material. The fine sediment load is often termed as wash load in river mechanics. The coarser sediments are deposited in the upper reaches of the river/estuary while the fine sediment fractions are carried further downstream into the estuarine/coastal waters where salinity enhances aggregation, forming flocs that possess higher settling velocities. These aggregates then deposit in areas where velocities are sufficiently small. Marine sediments originate from the nearshore bottom. They are first suspended into the water column by the action of waves and currents and are subsequently transported upstream by the net upstream near-bed flow to zones of low velocities where they eventually deposit. Understanding cohesive sediment transport thus requires knowledge of both sediment behavior and water movement.

Because of the affinity of pollutants to cohesive sediments as a result of adsorption, cohesive sediment transport plays an important role in characterizing the transport and fate of pollutants (e.g., heavy metals, radionuclides, pesticides, and nutrients) input to coastal waters. These pollutants originate from various land-use and marine activities (e.g., urbanization, deforestation, industrialization, mining, agricultural practices, hydrosystem construction works, and dredging/disposal activities) posing a potential threat to receiving waters. Ways to mitigate critical water quality problems are receiving the concentrated attention of environmental engineers, biologists, and ecologists. As a precursor to assessing the impact of pollutants input to a water body, a precise understanding of cohesive sediment dynamics is first necessary.

Dynamics

Cohesive sediments in coastal waters come under the influence of various complex processes. In addition to the transport processes of advection and dispersion in the water column, the principal processes influencing cohesive sediment dynamics are aggregation, settling, deposition, consolidation, and resuspension. These mechanisms are not only dependent on the inherent physico-chemical characteristics of the sediment but also on their interaction with the flow field. It is this interdependence that makes the problem of cohesive sediment transport so complex.

Aggregation

Aggregation refers to the formation of flocs comprised of primary sediment particles. The process of aggregation results from destabilization and collision. When the double layer around each sediment particle is compressed by divalent ions, the particles are destabilized and van der Waals attractive forces predominate, creating a condition conducive to aggregation. Collision occurs as a result of three primary mechanisms: Brownian motion, internal shear, and differential settling. Brownian motion occurs when the sediment particles are bombarded by fluid particles moving in random motion resulting from thermal gradients. Internal shear is due to the presence of velocity gradients in the suspending medium. Differential settling is caused by larger aggregates colliding with smaller aggregates. Brownian motion is responsible for aggregation in stationary or quasi-stationary waters forming aggregates that are weakly bonded. Internal shear predominates in dynamic aquatic systems and the aggregates produced are more dense and durable. Differential settling is important during near-slack periods where concentrations are high. The resulting aggregates are weak and have low density. Aggregation transfers mass through the particle size spectrum toward large-sized aggregates that can be characterized by their higher porosity, increases irregularity and fragility, and higher settling rate (Krone, 1963). Aggregation is influenced by sodium adsorption ratio, pH, salinity, sediment size, shaper, gradation, density, turbulence, temperature, and the efficiency of collision between particles.

Settling and deposition

The settling velocity of cohesive sediments is dependent upon the aggregation process, the suspended sediment concentration, and the ionic concentration of the suspending medium. For low concentrations of sediments in the water column (0.1–0.3 kg m^{-3}), the settling velocity is independent of the concentration. With increasing concentration (1–10 kg m^{-3}), the settling velocity increases due to the formation of stronger, denser, and larger flocs. At high concentrations (>10 kg m^{-3}), a sharp lutocline is formed and the underlying sediment forms a nearly continuous matrix. Interstitial water cannot escape upwards as easily so hindered settling occurs. A high-density suspension characterized by hindered settling is referred to as fluid mud. Reported values of settling velocity in estuarine and coastal waters range from 10^{-7} to 10^{-3} m s^{-1} (Mehta *et al.*, 1989).

As the aggregates settle toward the bed, the near-bed turbulence controls whether the particles bond with particles on the bed or are broken up and reentrained into the water column. The stochastic nature of the near-bed turbulence is characterized by a "probability of deposition," which is defined as the probability that particles reaching the bed actually stick to the bed. This is conveniently expressed as a function of the bed shear stress and a critical shear stress for deposition. The critical shear stress is a function of the type of sediment in suspension and is generally determined from laboratory flume studies. If the critical shear stress for deposition is equal to or less than the bed shear stress, then deposition will cease provided the sediment has uniform properties (Mehta *et al.*, 1989). The deposition rate is thus a function of the aggregate settling velocity, near-bed concentration, and the probability of deposition.

Consolidation

Once the sediment aggregates reach the bed, the flocculated structure breaks down and the aggregates will move close together. The sediments then consolidate under their own weight when particle-to-particle contact is made. The effective stress (i.e., the difference between the total hydrostatic pressure and the pore water pressure) of the soil increases with the release of pore-water pressure. The void ratio of the bed decreases with depth of sediment (i.e., the density and shear strength both increase with depth). Consideration of the consolidation of cohesive sediment beds is crucial for determining whether the bed is susceptible to erosion and how much material can be eroded. The shear strength of consolidated sediments has been estimated from empirical relationships between shear strength and dry density. Experimental results indicate that the shear strength increases with an increase in the clay content, organic matter, salinity, sodium adsorption ratio, and cation exchange capacity. The shear strength decreases with an increase in temperature, pH-value, and the sand concentration in the sediment bed.

Resuspension

Sediment resuspension depends upon the bed shear stress induced by the flow and the resistance of the bed to erosion. In coastal waters, bed shear stresses are computed for the combined effect of waves and currents. These shear stresses can be quite large compared to shear stresses due to currents alone. The resistance to erosion depends upon the sediment type and mineralogy, pore and eroding fluid concentrations, the time history of deposition (i.e., whether the sediments are recently deposited, partially consolidated, or part of a more dense bed), and chemical and biological processes. To account for the multiple-dependence between the above factors and their spatial variability, investigators have relied on laboratory and field experiments to assimilate and analyze sediment cores from various locations. The results of these studies have indicated the need to obtain system-specific data if accurate predictions of erosion are to be realized. The nature of the deposit (i.e., the bed properties) is generally characterized by a critical shear stress for erosion. Investigations on sediment erosion have indicated that when the bed shear stress at the sediment water interface is greater than the critical shear stress of the surface in contact, then the material will erode. The amount of material that can erode is described as a function of the stress in excess of the critical shear stress.

There are two primary modes of erosion, namely mass erosion and surface erosion. Mass erosion refers to the resuspension *en masse* of sediments when the bed shear stress exceeds the critical shear strength of the bed at some depth below the sediment–water interface. Clumps of sediment are eroded. This erosion mechanism occurs when the flow is rapidly accelerating as in zones of strong tidal currents and under storm conditions. Surface erosion, on the other hand, occurs at low to moderate values of the excess shear stress where currents are low to moderate (Mehta *et al.*, 1989). Surface bed sediments are eroded particle by particle until the bed shear stress is less than the critical shear stress for resuspension. The amount of material that can erode from the bed is limited to a finite amount. Laboratory experiments (Parchure and Mehta, 1985; Tsai and Lick, 1987) and field studies (Amos *et al.*, 1992) have revealed that only a finite amount of sediment can be resuspended from a cohesive sediment bed exposed to a constant shear stress as a result of the increase in shear strength with depth of sediment.

Mathematical modeling

To effectively assess the impact of sediment input to coastal waters, it is necessary to predict their spatial and temporal distributions in the water column and in the sediment substrate. In order to do this, it is imperative to consider the interaction between sediment transport processes and the flow field. The advent of sophisticated mathematical models coupled with inexpensive computing power has spearheaded major advances in developing simulation models. These models now strive to incorporate more complex sediment processes into tractable computational techniques. Since the accuracy of cohesive sediment transport predictions is contingent upon the accuracy of the flow field, an understanding of the hydrodynamic circulation processes is essential. In lieu of this, models for hydrodynamics and waves are generally coupled to a sediment transport model so that the various models operate within the same computational framework, in conjunction with one another, with output from one serving as input to another.

The simplest model for cohesive sediment transport is based on zero-dimensional modeling (e.g., Krone, 1985). These models are based on the sediment conservation equation, which represents the mass balance between sediment inflows, outflows, and storage (i.e., sediment deposition). The spatial variability of the sediment properties is ignored. Other models of cohe-sive sediment transport range from simple one-dimensional to sophisticated three-dimensional models. Some of the available models are described in Martin and McCutcheon (1998). The governing equation is based on the mass conservation principle in one-, two-, or three-dimensions, with boundary conditions of no-sediment flux at the free surface, and net erosional or depositional flux at the sediment–water interface. Odd and Owen (1972) used a two-layered one-dimensional model to simulate cohesive sediment transport in the Thames Estuary. Other one-dimensional models (e.g., Lin *et al.*, 1983) do not describe the near-bed concentrations which are required for determination of deposition rates. Two-dimensional models may be depth-averaged or laterally averaged. The assumption is that concentrations are well-mixed over the depth or in the lateral direction. Depth-averaged models (Ariathurai and Krone, 1976; Onishi, 1981; Cole and Miles, 1983; Lick *et al.*, 1994; Shrestha, 1996; Shrestha and Orlob, 1996) compute the sediment deposition rates from mean sediment concentrations averaged over the vertical. Laterally averaged models (Ariathurai *et al.*, 1977; Onishi and Wise, 1982) describe the distribution of sediment concentrations in the longitudinal and vertical directions. Quasi-three-dimensional models account for the vertical distribution of sediments using approximations (e.g., calculates the vertical concentration profile based on the vertical velocity distribution) (Lou *et al.*, 2000). The current state of the art is to utilize three-dimensional finite difference or finite element models for sediment transport (Hayter and Pakala, 1989; Sheng, 1991; Onishi *et al.*, 1993; Shrestha *et al.*, 2000). These models are suitable for applications to systems where the three-dimensionality of the flow, the salinity and temperature structure, and the associated suspended sediment distribution are important.

Sediment processes embedded in the model generally include aggregation, deposition, consolidation, and resuspension of sediments. In addition to the flow field, model inputs include bed properties (i.e., bed type, grain-size distribution, dry density, and sediment resuspension characteristics), fixed and/or time-varying sediment loading from various sources, and sediment settling parameters. Sediment aggregation is generally embedded in the settling velocity formulation and is described as a function of the sediment concentration. In most models, sediment consolidation is accounted for by discretizing the bed into arbitrary layers. Each layer is assigned a certain thickness, density, and a critical shear stress for erosion based on laboratory or field estimates of consolidation. In other models, consolidation is simulated by using a vertically segmented bed model (Shrestha *et al.*, 2000), where the sediment layers depend upon the time after deposition. The assumption is that a recently deposited bed is more susceptible to erosion than an aged bed. The resuspension from the bed is then based not only on the excess shear stress but also on the age of layer. Output from the sediment transport model includes the spatial and temporal distribution of suspended sediment concentrations, the mass of sediment eroded or deposited, and subsequent change in bed elevations.

Model performance is generally assessed through model-data comparison of suspended sediment concentrations and sedimentation rates. Sensitivity of the model input parameters are also assessed. Experience has shown that sediment-loading rates need to be estimated correctly if realistic predictions are to be realized.

The hydrodynamic model is essentially the driver for the sediment transport model. It provides the necessary transport information. Martin and McCutcheon (1998) review a variety of hydrodynamic models ranging from one-dimensional to three-dimensional. The dimensionality of the model is selected based on the intended use of the model. The governing equations include the continuity, momentum, and constituent (i.e., temperature and salinity) transport equations, with an equation of state relating density to temperature and salinity. Inputs to the hydrodynamic model include hydrographical (steady-state and/or time-varying flow rates from rivers and non-point land-based runoffs), astronomical (tides), meteorological (winds and heat fluxes), and internal density gradients (temperature and salinity distributions). Hydrodynamic models should possess the capability to predict different time scale events, ranging from semidiurnal tide scale, diurnal scale, meteorological scale (few days), spring and neap tidal scale (15 days), in addition to monthly and seasonal time scales. Model output includes water surface elevations, currents, horizontal and vertical diffusivities, and distributions of temperature and salinity. The predictive capabilities of the model are assessed via extensive comparisons with data and a confidence is established that the model realistically reproduces the predominant flow dynamics. As an example of a particular three-dimensional model application and its comparison with field data, the reader is referred to Blumberg *et al.* (1999).

Bed shear stresses due to currents and waves are crucial for calculating sediment resuspension and deposition fluxes. Wind waves are high frequency short waves of periods 3–20 s. These waves not only increase mass and momentum fluxes, but also increase bed shear stresses by inducing large velocities near the bottom of the water column. The development of wind waves in coastal waters is based on a balance between wind energy input, wave energy, and wave energy dissipation. The wave climate is predicted from a wave model such as SWAN (Holthuijsen *et al.*, 1993), HISWA (Booij and Holthuijsen, 1995), the Great Lakes Environmental Research Laboratory (GLERL) wave model (Schwab *et al.*, 1984), SMB (USACE, 1984), WAVD (Resio and Perrie, 1989), or ACES (Leenknecht *et al.*, 1992). Inputs to the wave model are bathymetric depths and a two-dimensional, time-dependent surface wind field. Linear wave theory is used to translate the wave height and period into a near-bed peak orbital velocity and peak orbital amplitude. A wave–current model can be used to calculate the bed shear velocity. The bed shear stress is then calculated as the product of the fluid density and the square of the shear velocity.

One of the critical features in the application of the above models is the type and quality of data available for creating model inputs and for calibration and validation of the model. In this regard, data assimilation and analyses represents a core component of any modeling effort. Because of the widespread use of such modeling tools, sediment transport studies are often supplemented by field sampling and monitoring programs. For many studies involving the fate and transport of sediment-associated pollutants, hydrodynamic, wave and sediment transport computations are linked to one or more water quality models.

Summary

This entry emphasizes the key mechanisms influencing cohesive sediment transport in coastal waters. These processes are complex because of the interaction between the physico-chemical properties of the sediment and flow parameters. Mathematical modeling provides a tool to assess the impact of the predominant processes on the transport and fate of sediments discharged to coastal water bodies.

Parmeshwar L. Shrestha and Alan F. Blumberg

Bibliography

Amos, C.L., Grant, J., Daborn, G.R., and Black, K., 1992. Sea carousel—a benthic, annular flume. *Estuarine, Coastal and Shelf Science*, **34**: 557–577.

Ariathurai, C.R., Macarthur, R.C., and Krone, R.B., 1977. Mathematical model of estuarine sediment transport. Dredged Material Research Program, *Technical Report* D-77-12.

Ariathurai, R., and Krone, R.B., 1976. Finite element model for cohesive sediment transport. ASCE, *Journal of Hydraulic Engineering*, **102**: 323–338.

Blumberg, A.F., Khan, L.A., and St. John, J.P., 1999. Three-dimensional hydrodynamic model of New York Harbor region. ASCE, *Journal of Hydraulic Engineering*, **125**(8): 799–816.

Booij, N., and Holthuijsen, L.H., 1995. *HISWA User Manual, Prediction of Stationary, Short-Crested Waves in Shallow Water with Ambient Currents*. Department of Civil Engineering, Delft, The Netherlands: Delft University of Technology.

Cole, P., and Miles, G.V., 1983. Two-dimensional model of mud transport. ASCE, *Journal of Hydraulic Engineering*, **109**(1): 1–12.

Hayter, E.J., and Pakala, C.V., 1989. Transport of inorganic contaminants in estuarial waters. *Journal of Coastal Research*, **5**: 217–230.

Holthuijsen, L.H., Booij N., and Ris, R.C., 1993, A spectral wave model for the coastal zone. In *Proceedings of the 2nd International Conference on Ocean Wave Measurement and Analysis*, pp. 630–641.

Krone, R.B., 1963. *A Study of Rheologic Properties of Estuarial Sediments*. Hydraulic Engineering Laboratory and Sanitary Engineering Research Laboratory, University of California, Berkeley, *SERL Report* No. 63-8.

Krone, R.B., 1985. Simulation of marsh growth under rising sea levels. In Waldrop, W.R. (ed.), *Proceedings of the Conference on Hydraulics and Hydrology in the Small Computer Age*, ASCE, 1, pp. 106–115.

Leenknecht, D.A., Szuwalski, A., and Sherlock, A.R., 1992. Automated coastal engineering system. Coastal Engineering Research Center, Vicksburg: U.S. Army Engineer Waterways Experiment Station.

Lick, W., Lick, J., and Ziegler, C.K., 1994. The resuspension and transport of fine-grained sediments in lake erie. *Journal of Great Lakes Research*, **20**(4): 599–612.

Lin, P., Huan, J., and Li, X., 1983. Unsteady transport of suspended load at small concentrations. ASCE, *Journal of Hydraulic Engineering*, **109**(1): 86–98.

Lou, Jing, Schwab, D.J., Beletsky, D., and Hawley, N., 2000. A model of sediment resuspension and transport dynamics in southern lake michigan. *Journal of Geophysical Research*, **105**(C3): 6591–6610.

Martin, J.L., and McCutcheon, S.C., 1998. *Hydrodynamics and Transport for Water Quality Modeling*. Boca Raton: Lewis Publishers.

Mehta, A.J., Hayter, E.J., Parker, W.R., Krone, R.B., and Teeter, A.M., 1989. Cohesive Sediment Transport. I: Process Description. ASCE, *Journal of Hydraulic Engineering*, **115**(8): 1076–1093.

Odd, N.V.M., and Owen, M.W., 1972. A two-layer model for mud transport in the Thames estuary. In *Proceedings of the Institution of Civil Engineers*. London, England, Supplementary Paper 75175.

Onishi, Y., 1981. Sediment-contaminant transport model. ASCE, *Journal of Hydraulic Engineering*, **107**(9): 1089–1107.

Onishi, Y., and Wise, S.E., 1982. Mathematical model, SERATRA, for sediment-contaminant transport in rivers and its application to pesticide transport in four mile and Wolf Creeks in Iowa. Athens, GA: *USEPA Report* No. EPA/600/3-82-045.

Onishi, Y., Graber, H.C., and Trent, D.S., 1993. Preliminary modeling of wave-enhanced sediment and contaminant transport in new bedford harbor. In Metha, A.J. (ed.), *Nearshore and Estuarine Cohesive Sediment Transport*, pp. 541–557.

Parchure, T.M. and Mehta, A.J., 1985. Erosion of soft cohesive sediment deposits. ASCE *Journal Hydraulic Engineering*, **111**(10): 1308–1326.

Resio, D.T., and Perrie, W., 1989. Implication of an f^{-4} equilibrium range for wind-generated waves. *Journal Physical Oceanography*, **19**: 193–204.

Schwab, D.J., Bennett, J.R., Liu, P.C., and Donelan, M.A., 1984. Application of a simple numerical wave prediction model to Lake Erie. *Journal of Geophysical Research*, **89**(C3): 3586–3592.

Sheng, Y.P., 1991. Three-dimensional modeling of transport of fine sediments and contaminant in lakes and estuaries. In *Proceedings of the ASCE Conference, Hydraulic Engineering*, Nashville, TN, July 29–August 2, pp. 166–171.

Shrestha, P.L., 1996. An integrated model suite for sediment and pollutant transport in shallow lakes. *Advances in Engineering Software*, **27**: 201–212.

Shrestha, P.L., and Orlob, G.T., 1996. Multiphase distribution of cohesive sediments and heavy metals in estuarine systems. ASCE, *Journal of Environmental Engineering*, **122**(8): 730–740.

Shrestha, P.L., Blumberg, A.F., DiToro, D.M., and Hellweger, F.L., 2000. A three-dimensional model for cohesive sediment transport in shallow bays. Minneapolis: *ASCE Proceedings, Joint Conference on Water Resources Engineering and Water Resources Planning and Management*.

Tsai, C.H., and Lick, W., 1987. Resuspension of sediments from Long Island Sound. *Water Science Technology*, **21**(6/7): 155–184.

United States Army Corps of Engineers (USACE), 1984, *Shore Protection Manual*, Volume I. Coastal Engineering Research Center. Vicksburg, MS: Waterways Experiment Station.

van Rijn, L.C., 1993. *Principles of Sediment Transport in Rivers, Estuaries and Coastal Seas*. Amsterdam: Aqua Publications.

Cross-references

Beach Sediment Characteristics
Cross-Shore Sediment Transport
Erosion Processes
Estuaries
Longshore Sediment Transport
Nearshore Sediment Transport Measurement
Numerical Modeling
Sediment Budget
Tidal Environments
Wave Climate
Wave–Current Interaction

COMPUTER SIMULATION—See COASTAL MODELING AND SIMULATION

CONSERVATION OF COASTAL SITES

Conservation of coastal resources has been practiced throughout human history as communities have endeavored to manage the food, water, and other resources of the coast in ways which ensure that each season or year will bring at least comparable bounty from the sea and the land. Protection of coastal sites in the historic past usually meant protection of areas to be used for elite recreation or as locations where coastal resources were to be conserved. Only in recent centuries, however, have communities begun to take specific action to conserve coastal sites for their natural flora and fauna. This has meant that some areas of coastal land and water have been designated as more important for conservation than others. This may have led to distortions in the behavior of coastal ecosystems which are just as important as the distortions which conservation is intended to prevent.

What does "conservation" mean? Usually, it means management in ways which protect the inherent natural and human features and processes of an area. Conservation is often confused with "preservation," but natural features change. Change is normal, but it is usually a means by which sites retain their resilience.

There are many different types of coastal sites which are conserved, including archaeological, biological, cultural, geological, geomorphological, and zoological sites. There is also a wide variety of approaches to their identification, designation, and protection (Table C14).

Conservation for biological reasons

Ehrenfeld (1976) argued that protection of ecosystems was usually justified by two purposes: scientific and human. Typically, these include research and education, the need to protect genetic information, and as parts of the worldwide and interconnected pattern of ecosystems. Often, however, they have been protected because of the perceived importance of particular species. Usually, these are rare and, commonly, higher plants and animals. Conservation was often a response to particular lobby or pressure groups, sometimes even individuals, whose concern about losses of species and habitats led them to establish groups to provide protection to pressure governments into protecting vulnerable species. This was particularly the approach in 19th century North America and Europe. Today, the responsibility lies more with government bodies or conservation nongovernmental organizations (NGOs).

Conservation for geological or geomorphological reasons

Geological and geomorphological sites are usually protected so that stratotypes, fossil assemblages, or structures can be observed and studied or so that specific important features are preserved and protected. The coast is a particularly good location for the observation and investigation of geological features because erosion constantly exposes new strata and because large sections can be observed without excavation. In many countries, the coast provides an accessible location for earth sciences research and education. Sites may also be important because they demonstrate the processes which mold the landscape both now and in the past. Geological sites commonly fall into two groups: "process" sites

Table C14 Typology of coastal conservation sites

Level of designation	Features of importance for designation	Examplar categories	Examplar sites
Global	Having features of ecological, biological, geological, cultural, archaeological significance at international level	World Heritage Site Biosphere Reserve Ramsar Site	Great Barrier Reef (Australia) Danube Delta (Rumania) Karavasta Lagoon (Albania)
Regional	Habitats of regional significance Sites of importance for migrating birds	European SACs European SPAs	Chesil and the Fleet (UK) Wadden See (NL, D, and DK)
National	Key habitats or species	NNRs Biological Reserve National Wildlife Refuge	Studland (UK) Isla del Caño (Costa Rica) Kodiak (USA)
	Protection of fauna and flora with extensive recreational activities	National Parks	Fuji-Hakone-Izu (Japan)
	Protection of fauna and flora with limited recreational activities	National Monument Marine Parks	Fort Jefferson (USA) Oshimoto (Japan)
	Protection of marine areas	Marine Sanctuary MPA Marine Life Conservation District	Farallon Islands (USA) Milieuzone Noordzee (NL) Waikiki (USA)
	Protection of sensitive estuarine areas	National Estuarine Research Reserve	South Slough (Oregon, USA)
	Protection of site of scientific importance (geological or biological)	SSSI Site	Dungeness (UK) Anse du Gris-Nez (France)
	Recreational site with ecological significance	National Recreation Area	Oregon Dunes (USA)
	Protection of archaeological, anthropological, or historic site	Area of Archaeological Importance (AAI) Anthropological Reserve National Historic Site	Hengistbury Head (UK) Cota Brus (Costa Rica) San Juan Island (USA)
Local	Protection of locally important habitats or species	Local Nature Reserve (LNR) Ecological Reserve	Lundy Island (UK) Tomales Bay (USA)
	Protection of fauna and flora and provision for recreational activities	State Park	Humboldt Lagoons (USA)
	Local protection of marine habitats with extensive recreational activities	Municipal Marine Park State Seashore	Balicasag Island (Philippines) Del Norte (USA)

Table C15 Selected geomorphological coastal conservation sites

	Dungeness	Chesil beach and the fleet	Seven Sisters and Beachy Head	North Norfolk coast
Feature	Cuspate shingle foreland	Shingle tombolo and lagoon	Chalk cliffs	Barrier beaches and spits
Key geomorphological importance	Shingle ridges in sequence from c 5,500 years BP	Large coastal fringing tombolo Sediment grading	Rapidly retreating chalk cliffs Intertidal platforms	Barrier islands Complex spits and recurves
Archaeological importance	Reclamation from Roman period to present		Cliff-top Iron Age settlements	Reclamation from mediaeval times
Biological importance	Holly (*Ilex aquifolium*) wood Invertebrates	Little Terns (*Sterna albifrons*) Intertidal and subtidal organisms	Rock-boring organisms Intertidal flora and fauna	Salt marshes
Conservation status	SSSI Geological Conservation Review Site (GCR) SPA	SSSI GCR SAC SPA	SSSI GCR SAC	SSSI GCR SAC SPA

and sites which display features of geological significance. The British Geological Conservation Review which describes all designated geological and geo- morphological sites of national and international importance in Great Britain refers to "process" and "integrity" sites (Table C15). Most fossil locations or stratotypes are integrity sites. At Dungeness, for example, the ridge sequence provides a time sequence which is an integral part of recent coastline history. If it disappears that part of the history goes with it, hence it is an "integrity" site. In contrast, chalk cliffs of the Seven Sisters exemplify well the processes which occur at the coast and so this is predominantly a "process" site.

Archaeological and historical sites

Archaeological and historical sites include major buildings and structures located at the coast as well as sites where artifacts have been identified and excavated. Some sites are now well removed from their original coastal location by the effects of emergence, whereas others have been submerged. In shallow coastal waters, there are often very important submarine archaeological sites associated with wrecks, especially in the more active sea lanes. Some archaeological sites are at risk of disappearance as erosion cuts into them. Unless the site has considerable architectural or historical importance, erosional sites are usually not protected.

Conservation legislation

Although the purpose of conservation laws is usually the same, that is, the protection of features or processes of high scientific or cultural value, the detailed legislation depends upon the legal framework of each nation or state. France, the United States, the United Kingdom, and Japan each have well-developed systems of conservation legislation which lay down specific procedures for the identification, designation, and management responsibility for scientifically important sites. These include coastal sites. The way in which management takes place also depends upon local tradition and legislation. Thus in Britain, coastal conservation is carried out both by statutory bodies such as English Nature (a government body) and by NGOs such as the National Trust and the Royal Society for the Protection of Birds (RSPB). In addition, smaller local bodies such as Wildlife Trusts also own or lease land so that they can conserve and manage areas in which rare or endangered habitats or species occur. However, the statutory bodies often are advisory organizations rather than owners of the sites. Site management depends upon the landowner or occupier of the site. This means that there may be many different organizations involved on conservation of coastal sites within short distances of each other. Their conservation policies and practices may conflict.

In many countries, there is little or no tradition of nongovernmental conservation bodies practicing coastal conservation and so responsibilities lie with government. Thus in France, the government sponsored body, the Conservatoire de l'Espace Littoral et des Rivages Lacustres (CELRL) was established to make coastal lands more accessible at the same time as protecting landscapes and properties of national importance. Modeled in part on the UK's National Trust, it is a governmental body rather than a nongovernmental membership body such as the National Trust. In some locations, First Peoples have traditionally regarded coastal lands as sacrosanct, but have no tradition of legislation to protect these lands. As a result, there are conflicts in conservation practice between traditional views of the value of coastal sites and their legal protection. Examples include the conflicts between European-based legislation and traditional approaches of the Maori in New Zealand. Legislation derived from British law conflicts with traditional law.

There may also be differences between local, national, or international legislation. Thus in the United States, both federal and state legislatures can define sites of coastal conservation and in Europe, the system of Directives which are mandatory on all member states may supplement or even override national conservation practices. The European Union's Birds (European Commission, 1979) and Habitats (European Commission, 1997) Directives have established Europe-wide networks of sites including many at the coast. These are known as Special Protection Areas (SPAs) and Special Areas of Conservation (SACs). In Britain, the seaward boundary of coastal sites such as National Nature Reserves (NNRs) and Sites of Special Scientific Interest (SSSIs) established under the national legislation lies at Low Water Mark, but the boundaries of SACs and SPAs can extend to the territorial limit. There is a very wide range of types and levels of conservation designations worldwide which includes World Heritage Sites, Biosphere Reserves, and Ramsar Sites designated at international level, National Parks, and Marine Sanctuaries at national level and local nature reserves (Table C14).

Conservation measures in different countries

In Europe, there are considerable differences in the history of coastal conservation, usually as a result of the relative importance of government and nongovernmental involvement and action at the coast. Furthermore, there has been considerable conservation action in the past decade as a result of initiatives by the European Community.

France has the longest established nature conservation legislation in Europe. Protection of sites is the responsibility of the Ministère des Affaires Culturelles, with advice provided by the Commission Supérieure des Sites, Perspectives et Paysages and the Commissions Départmentales des Monuments Naturels et des Sites. Decisions at local level must be approved by the national Commission Supérieure. The designation of sites followed the Laws of May 2, 1930 and July 10, 1976, which identified the following criteria for classification of sites as nature reserves:

(1) the preservation of animal and plant species and habitats which are either in danger of extinction in all or part of the national territory or have outstanding characteristics;
(2) the restoration of animal or plant populations or habitats;
(3) the conservation of botanical gardens and arboreta forming reserves for plant species which are rare, unusual, or nearing extinction;
(4) preservation of biotopes and outstanding geological, geomorphological, or speleological features;
(5) preservation or establishment of staging points on the great migration routes;
(6) scientific and technical studies necessary for the development of human knowledge;

(7) preservation of sites of particular interest for studies of the evolution of life and the activities of mankind.

This approach is reflected within the European Union Directives on Birds and Habitats both of which include important coastal areas.

The largest of the French coastal nature reserves is the Camargue (13,171 ha) on the Rhone delta. In contrast, when it was designated as a Site in 1930, Port Cros, a small island off the coast of Provence, had an area of only 140 ha. The initial designation was based on landscape protection, as more was known about the newly protected site, especially its flora and fauna, the significance of the surrounding seas was acknowledged. The site was extended in 1963 to include the surrounding sea, and became the first Marine Protected Area in the Mediterranean. The inclusion of the surrounding sea recognized the importance of such species as Neptune Grass (*Posidonia oceanica*) and Grouper (*Epinephelus marginatus*). This example typifies the development of coastal conservation as sites are first identified to protect a particular species or habitat and then expanded to account for improved understanding of the site and its surroundings. However, conservation can be very slow in practice. For example, at Port Cros the conservation objectives have been restricted by the impact of over half a million visitors annually and local resistance to conservation measures, such as a ban on angling from the shore which restricts traditional coastline activities.

The 62 coastal Départements may designate for protection the habitats of protected species that are threatened and acquire land financed through the Taxe Départmentale des Espaces Naturels Sensibles (TDENS: Departmental Tax for Sensitive Natural Areas), often in association with the CELRL (Meur-Férec, 1997). The Law of May 2, 1930 allows for designation of either Site Classé (classified site) or Site Inscrit (listed site) for sites of special natural and built importance and has been used extensively in the protection of coastal sites (Meur-Férec, 1997). The 1986 Loi Littoral (Coastal Law) provided a statutory framework which recognized the economic, social, and heritage importance of the coast and the unique nature of the coastal zone (Cicin-Sain and Knecht, 1998). This provided a context for individual sites, including, for example, a setback zone of 100 m wide for the whole coast, in which development is prohibited. Marine areas can be declared protected under the 1968 Law of Maritime Hunting, essentially to protect wildfowl! In many countries with long-established traditions of hunting, conservation of these species has often meant controlled management of habitats. Although there are some strongly held objections to game conservation, it is important to emphasize that without these conservation measures many sites which are valued today for their biodiversity would not exist. The 1960 Loi Relative à la Création de Parcs Nationaux (Law for the Creation of National Parks) includes Marine Protected Areas (MPAs) by allowing the terrestrial parks to extend into the maritime region (IUCN, 1994). European Union (EU) regulations and directives further supplement the national policies. Under the Birds (1979) and Habitats (1992) Directives, 133 sites have been proposed for inclusion within the European Community-wide Natura 2000 network of sites.

Two government bodies also have a role in French coastal conservation. The most important is the CELRL established in 1975. It is a public body charged with acquiring coastal and lakeside land, maintaining its ecological character and allowing public access to its sites (Cicin-Sain and Knecht, 1998; European Union for Coastal Conservation, 1995). Once purchased, the land is managed by the CELRL on behalf of the state in perpetuity. However, any acquisition has to take account of the area development plans and prior consultation with the Départements and Communes. This is because the day-to-day management and costs of an acquired site are handed back to the commune(s) or relevant landowner after purchase (EUCC, 1995). It currently owns over 49,000 ha along 650 km of coastline (Meur-Férec, 1997), including the D-Day beaches in Normandy and the Domaine de Certes at Arcachon. The Office National des Forêts (ONF, the National Forestry Office) manages 100,000 ha along 500 km of coastline, mainly forested dunes, such as the Landes. It works with the CELRL and other related administrative bodies to produce management plans for the forests. In the coastal region, ONF's work is more social and ecological than its predominantly economic role on the inland forests (Favennec, 2000). In particular, it is responsible for dune stabilization work, where it often has to balance the needs of conservation with tourism and public access. Thus at Pointe d'Arçay it has implemented a management plan devised with public participation and at Cap-Ferrat has successfully reintroduced marram grass (*Ammophila arenaria*) to the dunes.

Coastal conservation in Britain started, as so often there, with voluntary bodies such as the National Trust and the Royal Society for the Protection of Birds. Previously, conservation measures were largely in the hands of individual land-owners or the Crown and were mainly concerned with management of game or food resources. In 1949, the National Parks and Access to the Countryside Act established the Nature Conservancy as well as the national parks. The Nature Conservancy and its successors (three country agencies: English Nature, the Countryside Council for Wales (CCW), and Scottish Natural Heritage) were concerned with the geological and biological importance of sites for research and education. National Nature Reserves were established together with areas designated as SSSIs. The Royal Commission on Historical Monuments describes and studies sites of historical and archaeological importance, including many coastal sites. The Ancient Monuments and Archaeological Areas Act 1979 provides for the protection of archaeological sites and monuments and does appear to extend into territorial waters, although there is no record of designations at sea. These Acts and bodies are generally empowered only to deal with sites above Low Water Mark. However, the European Directive on Habitats specifies Marine SACs and so specific attention now has to be given to subtidal and submarine sites around the UK coastline. The Directive has force for the coastal waters of all European member states, although the detailed legislation is made at national level.

Submarine and marine sites are dealt with under different legislation. The Merchant Shipping Act 1995 deals with wrecks and salvage, requiring reporting of finds and the Protection of Wrecks Act 1973 provides some protection for important or dangerous wrecks or their sites. War graves may be designated under the Protection of Military Remains Act 1986: diving on and interference with a designated war grave is an offence. International agreement to protect graves in civilian shipwrecks in international waters was given effect in the Merchant Shipping and Maritime Security Act 1997.

The biological sites were reviewed in detail by Ratcliffe (1977) and coastal geomorphological and geological sites have been the subject of a major national review (the Geological Conservation Review (GCR) which has identified almost 100 coastal geomorphological sites (May and Hansom, 2003), plus a number of coastal landslide sites and several hundred geological sites. The Nature Conservation Review (Ratcliffe, 1977) described three stages in site classification:

(1) recording the features of each site to describe the range of ecosystem variation in terms of environmental and biological characteristics;
(2) comparing the quality of these sites; and
(3) choosing a series of nationally important sites.

The criteria used in both assessment and selection of key sites were

(1) size of the site;
(2) diversity;
(3) rarity of the species;
(4) naturalness, that is, areas which have been modified least by humans in so far as they are identifiable;
(5) fragility: habitats which are highly fragmented, rapidly decreasing in number and/or area, difficult to recreate or threatened with extinction;
(6) typicalness;
(7) the recorded history of the site;
(8) the position within an ecological or geographical unit;
(9) the potential natural value of a site;
(10) the intrinsic appeal of a site.

The Nature Conservation Review identified 140 coastal SSSIs with a mean area of 1,957 ha; 8 sites exceeded 10,000 ha in area. At the same time, France had 131 sites of average size 456 ha. 78% French sites were less than 100 ha in area, whereas 42% British sites exceeded 1,000 ha. For coastal geomorphological sites, criteria have included not only the specific importance of the site, but also the processes affecting particular rock-types or structures. As a result, sites with comparable geological conditions have been selected with micro- and meso-tidal ranges and different wave climates. There are 100 geomorphological sites which include large shingle structures such as Dungeness, major landslides such as the west Dorset coast, major dunes such as Morrich More on the east coast of Scotland, and world-renowned features such as Lulworth Cove and the stack of the Old Man of Hoy.

Although their coastal conservation is based upon different traditions both in coastal use and in their laws, France and Great Britain show very well how complex the legislation and administration of coastal conservation has become. Countries with shorter experience have been able to adopt different, and sometimes more efficient, approaches. Furthermore, the European Community has adopted its own framework for conservation in order to establish a community-wide network of related sites.

Funding of coastal conservation is similarly complex and depends upon receiving levels of funding which allow management to be carried out. The British model relies heavily upon voluntary support, membership, industrial support, bequests, and fundraising (Figure C70).

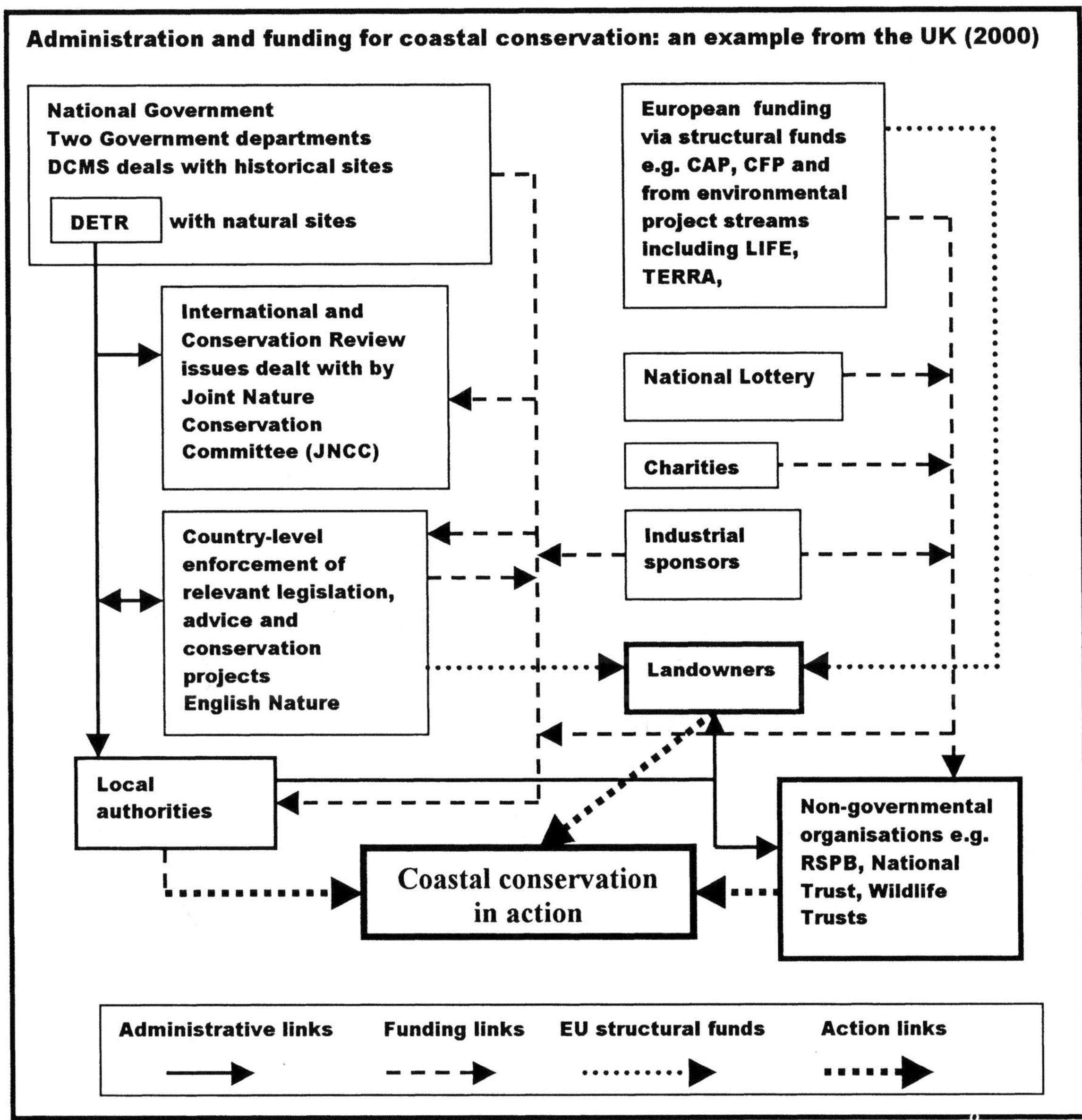

Figure C70 An example of funding for conservation of coastal sites. Key: DCMS = Department of Culture, Media and Sport; DETR = Department of the Environment, Transport and the Regions; CAP = Common Agricultural Policy; CFP = Common Fisheries Policy; RSPB = Royal Society for the Protection of Birds.

Typically, the largest voluntary bodies have strong marketing and sales departments. Funding from national and local government is limited in amount and has to be competed for. Such funding is often restricted in time. In contrast, in France or the United States, for example, conservation is predominantly funded by official sources, with the national and local governments largely responsible for both funding and operation of the conservation sites. The manner and volume of funding may change with a change of government. Thus, although the arrangements shown in Figure C70 for government funding applied in 2000, a reorganization of government departments following a General Election in June 2001 altered some of the responsibilities. In Britain, local authorities have only limited inputs into conservation funding, whereas the French Départements have considerable fiscal and legal tools at their disposal. They may, for example, acquire land financed through the TDENS (Departmental Tax on Sensitive Sites). Often these acquisitions are made in conjunction with the CELRL, and all the 62 coastal Départements have used these powers to purchase land (Meur-Férec, 1997). The day-to-day management and costs of an acquired site are handed back to the commune(s) or relevant landowner after purchase (EUCC, 1995). The majority of funding comes from the state but legacies and public/corporate donations also help with the acquisition and management of sites. Apart from the State, the CELRL is France's second largest landowner but interestingly, its budget is one-fifth of the National Trust's, a British NGO that operates a similar protection by acquisition policy (Meur-Férec, 1997).

At the time of the French Law of 1976, many European countries had hardly established relevant legislation. Greece, for example, had very good legislation concerning its archaeological sites, but was only just beginning to identify natural sites (May and Schwartz, 1981). Similarly, Norway's Nature Conservation Act of 1970 initiated action to preserve and protect plants, animals, and birds and areas of importance to nature conservation which, within a decade, established a network of 13 national parks, almost 200 nature reserves and over 200 locations which were of natural scientific importance. Many are at the coast. In contrast, Poland created its first National Park inland at Bialowieza in 1932 and there are now eight protected coastal areas including two National Parks, the Slowinski (founded 1967: present area 18,619 ha) to the west of Gdansk and the Wolinski Park near the German border (founded 1960: present area 10,937 ha). The Polish Environmental Protection Bill 1992 provides for three "levels" of designation: Strict (usually no admission to the public: 22% of sites), Partial (where management controls are flexible: 59%), and Landscape

Protection (where so-called "soft laws" apply: 19%). This recognizes that although some areas must have strict controls over access in order to protect sensitive areas and species other areas are more resilient and more appropriate for human activities such as recreation. National ecological policy requires that 30% of Polish territory (including coastline) should eventually be designated as "protected," 22% of the area had been designated by 1997. Coastal conservation in Poland is thus a national responsibility rather than a regional or local one. This is a typical pattern of many countries which have relatively recently begun to conserve coastal sites. The strength of national legislation often, but not always, ensures that designation and appropriate funding occur.

Spain combines national and local approaches to conservation. It has a long history of environmental legislation but the first specific conservation (in its modern sense) legislation was the 1975 Law on Espacios Naturales Protegidos (ENPs: Protected Natural Areas). This proposed four categories of protected area: national park, natural park, natural area of national importance or interest, and reserve of scientific interest. Other later laws have established over 25 different categories (WCMC, 1992). The national body responsible for administration of protected areas is the Instituto Nacional para la Conservación de la Naturaleza (ICONA: National Institute for the Conservation of Nature). Set up in 1971, its functions include the encouragement of renewable resource use and maintenance of ecological balance; the creation and administration of national parks and sites of national interest; and the development and exploitation of inland fishing and hunting assets. With decentralization of the government in 1978, responsibility for many conservation issues passed to Spain's 17 autonomous communities (Comunidades Autonomas) which are responsible inside territorial waters for the protection of marine and coastal areas from human activities. The 1989 Law on the Conservation of Natural Areas, Flora, and Wildlife strengthened this responsibility further. The protection of coastal areas is covered by the Ley de Costas (the Shore Act: July 28, 1988) based on the French model (Barragan, 1997). The Shore Act requires protection of the sea up to 100 m offshore and restricts the number of road and pedestrian routes to the coast. It also defines a zone extending 500 m inland which affects land and urban planning as well as prohibiting the disposal of waste in this area (WCMC, 1992). As in France, this provides a national framework within which conservation of coastal sites may take place. Laws concerning sustainable use and conservation of living marine resources have been adapted from the Directives of the European Community since these require member states to bring them into force.

The Waddensee—an example of international cooperation in coastal conservation

International cooperation is sometimes essential if the integrity of coastal ecosystems is to be protected, one of the best examples being the Waddensee in the southern North Sea. The Wadden Sea is the largest unbroken stretch of intertidal mudflats in the world and is extremely productive. With a biomass considered equivalent to rainforests, it is recognized as one of the most important wetland areas in the world. Up to 30 km wide, the Waddensee extends from the northeast Netherlands through Germany to the southwest coast of Denmark and includes low barrier islands, tidal flats, salt marshes, sand dunes, woodland and grassland, and low barrier. It is a very important breeding and overwintering area for a high proportion of the coastal birds of the North Sea. It is also, however, an area of reclamation, oil and gas extraction, commercial fisheries, and resort development. The three countries have cooperated in its conservation since 1978 using a joint memorandum which provides the basis for planning, conservation, and management of the area. In 1997, they adopted the Waddensee Plan, which was developed by involving all stakeholders and has as its focus the conservation, restoration, and development of all the region's ecological features. The interests of the local population, traditional uses of the area, and habitat conservation must be balanced with the other needs of sustainable management and preservation. Seven principles have been used to bring about regulation of activities that threaten its environment and to provide for active intervention as necessary, namely (de Jong *et al.*, 1999):

- Careful decision making—based on using the best available information
- Avoidance—avoiding potentially damaging activities
- Precautionary principle—avoiding damaging activities, even if there is insufficient scientific evidence to link them to specific impacts
- Translocation—moving damaging activities to areas where they will cause a reduced impact
- Compensation—using compensatory measures to offset unavoidably damaging activities
- Restoration—restoring areas which have been damaged
- Best available techniques and best environmental practice should be adopted.

The most important lessons are that international cooperation is essential for the conservation of some vital sites, that there must be agreed targets and procedures, and that local people must be involved especially when their traditional activities form part of the ecosystem as a whole.

In contrast, although some of the less-developed countries of eastern Europe have taken few steps to conserve coastal sites, the role of international conventions such as the Ramsar Convention on Wetlands of International Importance, 1971 and international cooperation in species protection, especially for marine turtles, have had an impact. In some respects, the significance of tourism as an economic driver for conservation has been overlooked although tourism provides a very strong incentive for conservation and sustainable activities. In Albania, for example, the isolation and centralization of the economy and society over the last 50 years affected community behavior as local communities were excluded from the planning process and decision making about natural resources upon which they depended. Not surprisingly, they consider that conservation and habitat protection have ignored their needs and imposed restrictions without providing benefits.

Albania designated its first Ramsar site in 1994 (RAMSAR, 1996, 2001), comprising the Karavasta Lagoon and adjacent Divjaka National Park. In the 1960s, much of the wetland was drained for agriculture, affecting the biodiversity of the lagoon, as well as the surrounding catchment, the Divjaka forest and hills, and its human communities. The total area of the site is 20,000 ha, made up of native forest of *Pinus pinea*, a species unique to the Albanian coast, and the lagoon itself (5,800 ha) which is a brackish water system connected to the sea by artificially maintained channels. The freshwater input to the lagoon comes from runoff from the surrounding agricultural land and Divjaka hills. Karavasta Lagoon has the only coastal breeding population of the Dalmatian pelican (*Pelecanus crispus*) along the Adriatic and Ionian coasts, supporting up to 5% (700–1,000 individuals) of the total world population. A significant proportion of the European breeding populations of Little Tern (*Sterna albifrons*) and Collared Pratincole (*Glareola pratincola*) use the Karavasta wetlands during migration, as well as large numbers of waterfowl, gulls, and terns, including the globally threatened pygmy cormorant, and white-headed duck.

Despite its designation as a Ramsar site, the lagoon has been seriously threatened by inappropriate tourism development which has often damaged fragile coastal habitats upon which its wildlife depend. Pollution and overfishing have added to the stresses on the lagoonal ecosystem. Some of the reclaimed wetlands are affected by high levels of salinity and so are of limited agricultural value. The area is state-owned and managed jointly by the General Directorates of Forestry and Fisheries (both part of the Ministry of Food and Agriculture). In practice, management depends upon a small number of rangers with limited experience of specialized wetland management, training, or equipment. Future strategies include improved planning and management of the lagoon to both protect the ecosystem and its biodiversity (including the pelicans) and derive economic benefits for local communities. A management plan for the site was prepared with international, technical, and financial cooperation through the European Union's PHARE program (PHARE, 1996). The Karavasta Lagoon is an example of where there is a need to involve local communities in biodiversity conservation activities and demonstrate their benefits. Karavasta exemplifies the problem that conservation of coastal sites depends upon management of the human systems which decide how the area will be sustained.

The United States of America

A very wide range of different designations are used in the United States, many within the framework of the National Parks legislation. They range from areas such as National Parks where they exist for both preservation of natural and cultural features and for recreation to specialized research areas, such as the National Estuarine Research Areas that are devoted to research (Table C14). There is a wide network of sites at state and federal level, but the initial establishment of important sites depended upon very strong individual lobbying of the federal government and the role of individuals such as John Muir in getting presidential support. Coastal sites were generally established later than inland sites.

Table C16 Coastal conservation in the United States—California examples

Site	Designation	Responsible authority	Key features	Area (ha)
Southeast Farallon Island	Bird Reservation (1909)	Presidential order following pressure initiated by Director of California Academy of Sciences	Major seabird nesting site Protection to prevent large-scale damage to birdlife by commercial egg companies	26
Southeast Farallon Island	Farallon Reserve (1911)	Re-designation. First statutory protection	Public concern over wildlife Endemic Farallon weed Major seabird nesting site	26
Farallon Islands	National Wildlife Refuge (1972)	US Fish and Wildlife Service and Point Reyes Bird Observatory	300,000 birds in 12 species nest annually. Over 350 species of birds migrate via the islands World's largest nesting populations of Ashy storm petrels, Brandt's cormorants and Western gulls Northernmost breeding site for Northern elephant seals	40
Farallon Islands and surrounding waters up to 6 nautical miles offshore between Bodega Head and Rocky Point and waters within 12 nm of the islands)	Gulf of the Farallones National Marine Sanctuary (1982)	Gulf of the Farallones National Marine Sanctuary (Federal Agency)	Diverse productive marine ecosystem based on upwelling	
Point Reyes National Seashore	National Seashore	National Parks Service	San Andreas Fault Very diverse ecosystems >360 bird species recorded	Almost 29,000
Limantour Estero	Reserve (within National Seashore)	National Parks Service	Estuarine habitats: removal of all life forms prohibited without permit	202
Bodega Dunes	Sonoma Statem Beach	State Department of Parks and Recreation	Dunes up to 50 m high	364
Bodega	Marine Reserve	University of California statewide Natural Reserve System	Undisturbed habitat for research	132
	Marine Life Refuge	State Department of Fish and Game	Marine area up to 330 m offshore	
Duxbury Reef	Marine Reserve	State Department of Fish and Game	Largest exposed shale reef in state Endemic acorn worm and one of only two sites for burrowing anenome (*Halcampa crypta*)	ca. 16
Tomales Bay	Ecological Reserve	State Wildlife Conservation Board	Wetlands and salt marsh	>200
Tomales Bay	State Park	State Department of Parks and Recreation	Sand beaches	
Bolinas Lagoon	Nature preserve	Marin County	Marsh habitats and species	485
Olema Marsh	Not designated	Audubon Canyon Ranch	Privately owned research and wetland restoration project Freshwater marsh	16

The origins and variety of the coastal sites is well illustrated by the coastline north of San Francisco (Table C16). From very humble beginnings, some sites such as the Farallones Islands have grown to major national coastal and marine sites. Designations overlap so that within the Point Reyes National Seashore, for example, there are not only marine and terrestrial reserves, but also historic ranches within the Philip Burton Wilderness Area (California Coastal Commission, 1987). State, county, and private parks and reserves also occur. However, there are few sites owned or managed by private organizations. Estuaries, such as San Francisco Bay, have been extensively reclaimed with the accompanying loss of wetland habitat. This is a worldwide feature of coasts and serious concerns about the loss of biodiversity and productivity have been expressed within the United States for many decades. Since 1850, reclamation has cut off over half of the bay's tidal marshes and open water from tidal action and 95% of the original wetland habitat has been lost (California Coastal Commission, 1987). Nevertheless, San Francisco Bay contains 90% of the coastal wetlands of California and remains a major part of the Pacific coast bird migration route. Responsibility for the wetlands is divided between the Bay Conservation and Development Commission (tidal wetlands) and the US Corps of Engineers (nontidal wetlands), a division which does not recognize the natural hydrological links between them.

In many less-developed countries, funding for conservation is largely provided through tourism receipts, or by international aid programs and funding from international bodies such as the World Wide Fund (WWF) for Nature. Pressure for conservation action and designation typically comes from pressure groups. The implications of conservation action and site or species protection are not always well considered. For example, a ban on turtle fishing along the Pacific coast of Mexico had immediate impacts upon the local economy and knock-on effects to other parts of the coastal ecosystem. The contrast between the approach in many less-developed, but biologically rich, regions is well demonstrated by two examples, Costa Rica and the Mexican state of Oaxaca.

Table C17 Examples of the variety of coastal conservation sites in Costa Rica

Location	Cabo Blanco	Isla del Caño	Manuel Antonio	Ostional
Status	Strict Nature Reserve	Biological Reserve	National Park	National Wildlife Refuge
Area	1,172 ha	300 ha land 5,800 ha marine	682 ha	162 ha land 587 ha marine
Key faunal features	Largest Brown Booby (*Sula leucogaster*) colony in Costa Rica	Coral reefs	Major nesting site for Brown Booby 109 species of mammal, 184 bird species, 10 species of sponge, 19 coral, 24 crustacean, 17 algae, 78 fish	One of the world's most important nesting sites for Olive Ridley Turtle (*Lepidochelys olivacaea*) on beach only 900 m in length Also important for Pacific Green (*Chelonia mydas*) and Leatherback turtles (*Dermochelys coricea*) 102 bird species
Key plant features	Mainly evergreen forest with especially spiny cedar (*Bombacopsis quinatum*), chicle tree (*Manilkara chicle*), and Espave (*Anacardium excelsum*)	Tall evergreen forest dominated by Cow Tree (*Brosimum utile*) Coral reefs	Range of habitats: primary and secondary forest, mangrove swamp, marsh and littoral woodland and lagoons Major nesting site for Brown Booby	Forest includes deciduous trees (especially Frangipani: *Plumeria rubra*), mangrove, cacti and succulents. *Acanthocerus pentagonus* cactus grows on beaches and provides shelter for ctenosaurs (*Ctenosaura similes*)
Archaeological importance		Pre-Columbian cemetery	Pre-Columbian turtle trap	
Geological and geomorphological importance		Plate tectonics: subduction of Cocos Plate beneath Caribbean Plate along Middle American Trench	Sand tombolo	

Costa Rica

The Central American country of Costa Rica has a well-established system of designation and protection of its ecosystems and biodiversity, including coastal sites. As the examples in Table C17 show, these sites include very large areas of land and sea which protect different habitats and some small sites which are most important for particular species. These sites are all nationally designated and form part of a network of sites designed to represent and preserve Costa Rica's wide biodiversity.

Mazunte—an example of conservation in action

Mazunte on the Oaxacan coast of Mexico is, unlike the Waddensee, in one of the poorest areas of the world. It shows how changes in conservation policy can be accompanied by local community involvement in coastal conservation especially when this is focused upon sustainable development. However, it also demonstrates the tenuous nature of much coastal conservation in less-developed areas.

For more than 30 years, Mazunte (the name comes from a Nahuatl word meaning "please lay eggs") was dependent upon turtle catching and canning. Most people in the village were either fishermen or worked in the cannery. A Presidential decree in 1990 banned all turtle fishing following significant lobbying by WWF. The immediate effect was to remove the community's main source of income, as the local community express it, throwing "our people into misery!" Protection of one species (turtles) was enacted at the cost of another. Some local people abandoned the area and moved inland to occupy land within the coastal dry forest, the associated clearance having potentially damaging effects on coastal lagoons as sedimentation increased. The turtle populations were recognized as being under stress from poaching, predators, and disease. The Mexican Center for Marine Turtle Research (CMT) was established in late 1991, using the buildings of the former Society of Industrial Fisheries of Oaxaca, and now has a unique role as the only center which studies seven out of the eight species of marine turtle. In addition, a community-based project known as Ecosolar and a Reserva Ecologica Campesina were established. Within Mazunte itself, a small ecotourism project was based upon sustainable principles and designed to provide accommodation for mainly foreign visitors to the area. Further to the south, the large-scale resort development of Huatulco had a major impact on the unusual dry coastal forest and the coastal beaches and lagoons.

The attempts to conserve not only the turtles and their nesting beaches but also to build a sustainable community have not proved easy, for both human and natural intervention have conspired to threaten the coastal ecosystem. In the aftermath of an uprising in 1994 in Chiapas, there was considerable local unrest which affected Huatulco in August 1996. This had a serious impact on the turtle ecology because, when police were sent to the resort, up to 200 poachers went to local beaches where Olive Ridley turtles were arriving to lay eggs. It is estimated that up to 1 million eggs were lost and large numbers of turtles killed. In October 1997, the area was struck by Hurricane Pauline which caused serious damage to both the CMT and other buildings but also devastated the nesting beaches. 806,000 nests were affected, with half all the nests on Escobilla beach, 4 million eggs and up to 10 million hatchlings lost. Nevertheless, the turtles continue to nest on these beaches. The CMT carries out detailed studies of turtle behavior and ecology both in the Center and on the beaches and the growth of the resort of Huatulco has brought increasing numbers of visitors to Mazunte. The concept of the Reserva Ecologica Campesina is unusual in that it links deliberately the human community with the coastal ecosystem. Knowledge which once served the commercial interest of turtle canning has been harnessed to develop an approach to sustainable development which has shown itself to be very resilient to significant human and natural disruption. The financial support for the CMT is very small, less than US$70,000 annually, and receipts from the ecotourism project are also small. Conservation here is not about conservation of a site, so much as conservation of a whole ecosystem (including its human occupants). For many communities, this will be the future for coastal conservation for it is firmly based in local knowledge, the needs of the local community, and their own initiatives.

Conclusion

The conservation of coastal sites takes many different approaches, often with detailed legal and administrative frameworks for designation and protection, but also in relatively informal ways with the involvement of NGOs and private ownership. In some countries, it has only been through the effective active involvement of individuals and NGOs that conservation has been put into practice. The example of Mazunte is a modern, but more community-based, expression of people's wish to protect and conserve features of their environment which they value which was expressed in different words and actions by the founders of the National Trust or John Muir and the Sierra Club. Once established, such initial routes to coastal conservation need the support of the national and international communities to make sure that they are truly sustainable and not threatened by external political, economic, and social pressures. Although coastal conservation is well established in practice, there are nevertheless enormous threats to many sites from urbanization, resource extraction, tourism, pollution, large-scale coastal projects such as barrages, and reclamation and neglect. Sea-level change and climate change also threaten many sites, although their history shows that most sites are resilient to change provided that human activities allow natural resilience.

Note

The International Union for the Conservation of Nature (IUCN) World Conservation Monitoring Centre (WCMC) in Cambridge, England, is the most important source of information regarding protected sites.

Vincent May

Bibliography

Barragan, J.M., 1997, *Medio ambiente y desarrollo en las areas litorales*, Barcelona: Oikos-tau.

California Coastal Commission, 1987. *California Coastal Resource Guide*. Berkeley: University of California Press.

Cicin-Sain, B., and Knecht, R.W., 1998. *Integrated Coastal and Ocean Management: Concepts and Practices*. Washington, DC: Island Press.

de Jong, F. *et al*., 1999. *Wadden Sea Quality Status Report. Wadden Sea Ecosystem No. 9*. Germany: Petra Potel.

Ehrenfeld, D.W., 1976. The conservation of non-resources, *American Scientist*, **64**: 648–656.

European Commission, 1979. Council Directive on the conservation of wild birds 79/409/EEC.

European Commission, 1997. Council Directive 92/43/EEC on the conservation of natural habitats and of wild fauna and flora as amended by the Act of Accession to the European Union of Austria, Finland and Sweden and by Council Directive 97/62/EEC.

European Union for Coastal Conservation (EUCC), 1995. Conservatoire de l'Espace Littoral et des Rivages Lacustres. *Coastline*, **4**(1): 13–20.

Favennec, J., 2000. The Office National des Forêts, actively involved in managing and protecting the French coast. *Coastline*, **9**(2): 7–19.

IUCN 1994 *Guidelines for Protected Area Management Categories*, Gland: IUCN Commission on National Parks and Protected Areas with the assistance of the World Conservation Monitoring Centre, Gland.

May, V., and Hansom, J., 2003. *Coastal Geomorphology of Great Britain*, Peterborough: Joint Nature Conservation Committee Geological Conservation Review.

May, V.J., and Schwartz, M.L., 1981. Worldwide sites of special scientific interest. In Bird, E.C.F., and Koike, K. (eds.), *Coastal Dynamics and Scientific sites*, Tokyo: Komazawa University, pp. 91–118.

Meur-Férec, C., 1997. A comparative review of coastal conservation policies in France, and England and Wales. *Journal of Coastal Conservation*, **3**: 121–132.

PHARE, 1996. *Karavasta Lagoon-Wetland Management Project: Ramsar Site Management Plan*, Volume 32. France: Station Biologique de la Tour du Valat.

RAMSAR, 1996. *Albania's Karavasta Lagoon*. Ramsar Archives.

RAMSAR, 2001. *The List of Wetlands of International Importance*. Ramsar Archives.

Ratcliffe, D.W., 1977. *A Nature Conservation Review*. Cambridge: Cambridge University Press.

World Conservation Monitoring Centre (WCMC), 1992. *Protected Areas of the World: A Review of National Systems. Barcelona, Spain.*

Cross-references

Archaeology
Environmental Quality
Human Impact on Coasts
Marine Parks
Organizations
Rating Beaches
Tourism and Coastal Development
Tourism, Criteria for Coastal Sites

CONTINENTAL SHELVES

Introduction

Continental shelves are the submarine platforms that surround continents. They are found in the open ocean and in many of the coastal seas. Similar features surround subcontinents, such as Greenland, and many oceanic islands. These features exhibit a wide variety of physical dimensions and characteristic relief. Most commonly, they are relatively flat with slopes varying between $1:10^2$ and $1:10^4$. Depths at the seaward edges of continental shelves are generally on the order of 80–100 m, although there are places where depths are less than 50 m or greater than 200 m. Beyond the shelf edge is the continental slope, which is typically inclined on the order of $1:10^2$ to $1:10^3$ and frequently has much more local relief than the continental shelf.

General morphology

In most places the continental shelf has distinct subzones. Although the nomenclature is not standardized, it is customary to divide the zone between the surf zone and the shelf edge into four major reaches. These are: (1) the *shoreface*, (2) the *inner shelf*, (3) *the mid-shelf*, and (4) the *outer shelf*. On most open-ocean shelves the shoreface, with a slope on the order of $1:10^2$, occurs between depths of 5–25 m. It has a convex upward slope which is steeper at its top. This zone is also occasionally called the shore-rise and is considered by some to be part of the inner continental shelf. The inner continental shelf is usually taken to extend seaward to the point that the costline no longer sig-nificantly influences wind-driven circulation. This distance is strongly affected by the strength and duration of the wind at any given time. It is usually on the order of 5–15 km wide.

The outer-shelf and mid-shelf environments are also defined by their dynamic processes. The outer shelf is where there is significant interaction between deep ocean and shelf circulation patterns. On the mid-shelf, circulation responds to the restricted depths but is not dominated by interaction with ocean currents or near-shore circulation. Because continental shelves have widths in the range of 10^0–10^2 km, the characteristics of these subzones are quite variable.

Large-scale characteristics

Continental shelves can also be divided according to their modes of origin, dominant processes, and geographic locations. In most places around the world where the coastline has been tectonically stable the shelves are broad and of low relief. Often these are simply submerged continuations of adjacent coastal plains, which themselves are extensive. Conversely, where continental margins are overriding adjacent seafloor as a result of continental drift, the shelves tend to be narrow and steep. These are often adjacent to coastal mountain ranges. The inner part of continental shelves at low latitudes is a common site for coral reefs and other forms of carbonate platforms. Shelves around the Antarctic continent, the Arctic Ocean, and certain other high-latitude places are strongly influenced by processes related to sea ice. On some high-latitude shelves, permafrost can be found a few tens of meters below the sea bottom. This attests to the combined effects of severe cold and the relatively recent rise of sea level.

The sedimentary environments and deposits on continental shelves are quite varied. In some places rock platforms extend beneath the shelves and sediments are distributed as thin sheets or in patches. These conditions are encountered where the underlying rock is relatively young (such as in volcanic areas), where the supplying of sediments is limited (e.g., west Florida Shelf), or where recently drowned shelves had been scoured by glaciers. Deposits greater than 10^3 m occur where prolific sources of sediments exist (Mississippi Delta foreshore). In some places, the predictable patterns of sandy nearshore sediments grading to silt and mud near the shelf edge are found. However, there are few general rules. In some locations muddy sediments occur all the way from the shore to the shelf edge. Elsewhere, the entire shelf is covered with sandy sediments, possibly with local patches of silt and mud.

Many broad continental shelves, especially those in stable tectonic areas, are sediment-starved today. In these areas the coastline is characterized by broad estuaries. Sediments, transported in river bedload accumulate at the estuary head. The sandy adjacent continental shelves are the result of reworking of previous deposits. In many cases, the previous deposits represent ancient shore processes from times of lower sea level. At the opposite extreme, there are cases where river deltas extend across the modern shelves so that the seaward face blends into the continental slope. The delta of the Mississippi River is an especially good example.

Shelf features

Although continental shelves tend to have broad areas of low relief, there are many morphological features that characterize this environment. These can be conveniently divided between erosional features, depositional features, and modified relic features. The most common erosive features originated during periods of low sea level when rivers extended entirely across to the shelf edge. Because there were several major sea-level excursions during the Pleistocene, there have been several extended periods where river and river canyon development have been promoted. In some cases, there are no remnants of these ancient river courses evident in the bathymetric shape of the bottom. However, subbottom profiles show channels now filled with sand. In other cases, quite obvious river valleys extend across the continental shelves to submarine canyons that extend down the continental slope. In some places drowned river terraces and bluffs can be found.

Depositional features dynamically coupled to present sedimentary processes and features representing the reworking of previous deposits are both found. The dynamic sedimentary features include sand ridges, sand waves, and a host of lesser bedforms. *Sand ridges* are long features with characteristic heights of 1–10 m, spacings (i.e., wavelengths) of 10^3 m and lengths up to several kilometers. These may be either shoreface-connected or freestanding sets on the lower shoreface and mid-shelf regions. The *shoreface-connected sand ridges* tend to be oriented obliquely to the shore with the acute angle pointing in the direction of dominant longshore sediment transport (Duane *et al.*, 1972; Swift and Field, 1981; Caballeria *et al.*, 2001; Calvete *et al.*, 2001). These features are generally thought to migrate slowly down the coast. The freestanding sand ridges occur in sets in a wide range of shelf environments (Nemeth *et al.*, 2001). They may occur on the tops of shallow sand banks close to the shore. They are also common features in mid-shelf environments. These too are prone to slow migration, although shelf sand ridges in some locations occupy remarkably stable locations.

Sand waves are geometrically similar features that are smaller than sand ridges. They have heights on the order of a few meters, are spaced at distances (i.e., wavelengths) on the order of 10^2 m, and extend 10^2–10^3 m in length (Nemeth *et al.*, 2001). These may appear in extensive fields over broad sandy areas or as secondary features on the crests of sand waves and sand banks. Recent research has shown that sand waves are dynamically stable bottom forms that are brought about by shelf currents through a complex feedback mechanism that includes systematic changes in the structure of turbulence. These large bed forms are stable as either fixed or migrating features.

Shoreface-connected and open-shelf sand ridges as well as sand waves tend to occur in parallel sets with quasi-regular spacing. Many other forms of bottom relief are less regular in shape and spacing. These include sand ribbons and irregular bars. The reasons for these features are generally poorly understood, although many represent the results of reworking of sand deposited during an epoch of lower sea level.

In order to explain many of the morphological features that have been developed on continental shelves through the reworking of previous features, it is necessary to briefly consider sea-level history. Many of the shelf features developed as a result of the migration of the shore across the shelf since the last glacial maximum approximately 18,000 years ago. Worldwide sea level rose approximately 110 m, but the rate was not uniform over this period. The rate of sea-level rise was greatest between 12,000 and 7,000 yr BP. However, throughout the entire recent period the rate of rise has varied within shorter intervals (1,500–4,000 yr). More details are provided in Donoghue *et al.* (2003). Variations in the rate of sea-level rise have caused the shore migration rate to change considerably. At times of slow rise, or rise-reversal, the shore position remains nearly constant. Shore features become well-established and considerable deposits of sand develop. Subsequent episodes of rapid sea-level rise can cause overstepping of these features. They are reworked, but not entirely obliterated. As the availability of dense bathymetric data sets has increased it has become easier to detect and explain these features.

The concept that barrier islands retreat landward in the face of rising sea level is well established. This is brought about by processes that carry sand over and behind the islands while the seaward face erodes. Features formed off the seaward side of these islands can be preserved if their initial volumes are large. In several locations along the northern Gulf of Mexico coast, the huge volumes of sand in ebb tide delta are too great to be disbursed as they are drowned by the ongoing rise of sea level. This results in leaving elongated sand bodies stretching many kilometers onto the shelf. In most cases these features correlate with a position of present estuaries, although they may not be aligned with present tidal inlets.

Because the rate of sea-level rise since the last glacial maximum has been unsteady, with millennia-scale fluctuations, the migration of shore features has also been episodic. In areas where there are plentiful supplies of sand, barrier island complexes extending from the shoreface, across the island, and including the washover deposits in the lagoon, have tended to form during periods when the rate of sea-level change is low. Again, these huge sand deposits resist obliteration when they are drowned during episodes of rapidly rising sea level. Much of the upper portion of these *shore-complexes* is reworked and lost. However, the general morphology can be distinguished where there are sufficiently detailed bathymetric data. In these places subbottom profiles reveal dipping beds representing lower shoreface and back bay deposits (Donoghue *et al.*, 2003).

Many ancient coastal river deltas have also been large enough to avoid being obliterated during periods of rapid sea-level rise. These are now located on the continental shelf. The shelf edge is where many of these are found because the last sea low-stand was below this elevation. Additionally, some major sand shoals on the continental shelf originated as deltas during the transgressive phase of the past sea-level cycle.

Subaqueous deltas are also found on continental shelves. These are sediment clinoforms with nearly flat tops and seaward-dipping foreset beds. However, unlike coastline deltas, these features formed with their tops on the order of 10 m below sea level. Examples of these can be found off the mouth of the Amazon River (Nittrouer and DeMaster, 1996), and in the Italian Adriatic (Trincardi *et al.*, 1996). Unlike the drowned deltas described above, these subaqueous deltas are actively forming at the present time.

There are many extensive morphological features, consisting largely of sand, which occur as isolated *banks* or *shoal complexes*. Although many of these have complex histories and origins, there are others that are members of feature classes with common explanations. Swift (1973) has introduced the term *massif* in conjunction with these features. Generally, a massif is a very large volume of sand that has originated near the coastline and has subsequently been drowned and reworked due to sea-level rise. *Shoal retreat massifs* originated from the complex of sand shoals that form across the mouth of many broad estuaries. The ancestral location of these estuaries may be far out on the present shelf. As sea level rose, both the estuary mouth and the related shoals migrated landward leaving huge volumes of sand behind. Although these are now being reworked by the action of waves and currents, the very size of these deposits protect them against obliteration. However, there must be many cases where smaller features have not been so persistent.

Cape shoal retreat massifs have also been described (swift *et al.*, 1972, 1973). These are thought to originate at the apex of coastline capes in barrier island systems that existed far out on the shelf. The system was subsequently drowned by sea-level rise and traces of the barrier islands obliterated. However, the large collection of sand at the point where the littoral transport systems converged produced a linear shoal mass that extends far onto the present shelf. These are often still connected to capes in barrier island systems.

Shoal retreat and cape retreat massifs, along with many other sand banks that originated from the reworking of previous massive sand deposits, often provided the sites for extensive sand ridge and sand wave fields. This is an excellent example of how a portion of many shelf features represents the result of ongoing processes while other aspects of the same features represent more ancient causes.

The work that resulted in this entry has been assisted by ongoing research in the ONR EuroStrataform Project (Contract Number N00014-03-C-0134).

A.W. Niedoroda

Bibliography

Caballeria, M., Falques, A., Coco, G., and Huntley, D., 2001. A morphodynamic mechanism for transverse bars in the nearshore. In *Procedings of the Coastal Dynamic '01*. American Society of Civil Engineers, pp. 1058–1067.

Calvete, D., Falques, A., deSwart, H.E., and Walgreen, M., 2001. Modeling the formation of shoreface-connected sand ridges on the storm-dominated inner shelves. *Fluid Mechanics*, **441**: 169–193.

Donoghue, J.F., Niedoroda, A.W., Hatchett, L., Locker, S., Clark, R., Koch, J., and Butler, K., 2003. Model for identifying and characterizing offshore sand resources using an interactive GIS. In *Proceedings of the Coastal Sediments '03*. American Society of Civil Engineers.

Duane, D.B., Field, M.E., Meisberger, E.P., Swift, D.J.P., and Williams, S.J., 1972. Linear shoals on the Atlantic continental shelf, Florida to Long Island. In Swift, D.J.P., Duane, D.B., and Pilkey, O.H. (eds.), *Shelf Sediment Transport: Process and Pattern*. Stroudsburg: Dowden Hutchinson & Ross, pp. 447–498.

Nemeth, A., Hulscher, S.J.M.H., and van Damme, R.M.J., 2001. Numerical simulation of sand wave evolution in shallow shelf seas. In *Proceedings of the Coastal Dynamics '01*, American Society of Civil Engineers, pp. 1048–1057.

Nittrouer, C.A., and DeMaster, D.J., 1996. The Amazon shelf setting-tropical, energetic and influenced by a large river. *Continental Shelf Research*, **16**: 553–574.

Swift, D.J.P., 1973. Delaware shelf valley: estuary retreat path, not drowned river valley. *Geological Society American Bulletin*, **84**: 2743–2748.

Swift, D.J.P., and Field, M.E., 1981. Evolution of a classic ridge field, Maryland sector, North American inner shelf. *Sedimentology*, **28**: 461–482.

Swift, D.J.P., Koefoed, J.W., Saulsbury, F.P., and Sears, P., 1972. Holocene evolution of the shelf surface, central and southern Atlantic shelf of North America. In Swift, D.J.P., Duane, D.B., and Pilkey, O.H. (eds.), *Shelf Sediment Transport: Process and Pattern*. Stroudsburg: Dowden Hutchinson & Ross, pp. 100–148.

Swift, D.J.P., Duane, D.B., and McKenny, T.F., 1973. Ridge and swale topography of the middle Atlantic Bight: secular response to the Holocene hydraulic regime. *Marine Geology*, **15**: 227–247.

Trincardi, F., Asioli, A., Cattaneo, A., Correggiari, A., Vigliotti, L., and Accorsi, C.A., 1996. Transgressive offshore deposits on the Central Adriatic shelf: architecture complexity and the record of the Younger Dryas short-term event. *Il Quaternario*, **9**(2): 753–762.

Cross-references

Barrier Islands
Coral Reefs
Deltas
Offshore Sand Banks and Linear Sand Ridges
Offshore Sand Sheets
Reefs, Non-Coral
Shelf Processes

CONVERSION TABLES—See APPENDIX 1

CORAL REEF COASTS

Coral reef coasts occur almost entirely between the Tropics of Cancer and Capricorn, though functional exceptions involving fossil reefs may extend beyond these limits. The reef component of a coral reef coast is constrained by the physiological requirements of the corals themselves, which include clear, warm water, neither too fresh from river outflows (>30 ppt) nor too saline from high evaporation (<40 ppt), with a temperature range of between about 15–30°C, and not heavily sedimented. Exposure levels are rarely limiting; corals thrive from very sheltered to extremely exposed conditions.

Because reefs grow vertically to the low tide level, they provide very effective shelter to the coast, which in turn greatly affects the habitats which lie in their lee. The more closely a reef is located to the coastline, the greater the shelter will be. Extremely sheltered coasts lie behind fringing reefs with broad reef flat components, over which water always remains <1–2 m deep. Reef development may take place some distance offshore, in which case greater fetch and depth between reef and shore leads to higher energies impinging on the coast. In numerous cases, the series of coastline reefs may lie far offshore (east Africa, Great Barrier Reef), possibly even on a line along the edge of the continental shelf, so that coastlines are little protected, added to which the presence of the reef itself may funnel longshore water movement which may add to net coastline erosion and dynamics. In numerous locations, fringing reefs and more distant barriers of reefs, or series of patch reefs, combine to provide complex patterns.

Reefs are biogenic limestone structures, sometimes called bioherms. Thus, over time, reef development shapes the coastline itself, and the shape and morphology of the system decreasingly depends on, or even reflects, the inherited landscape form. While it is usually the case that the limestone-depositing organisms usually can only settle on preexisting

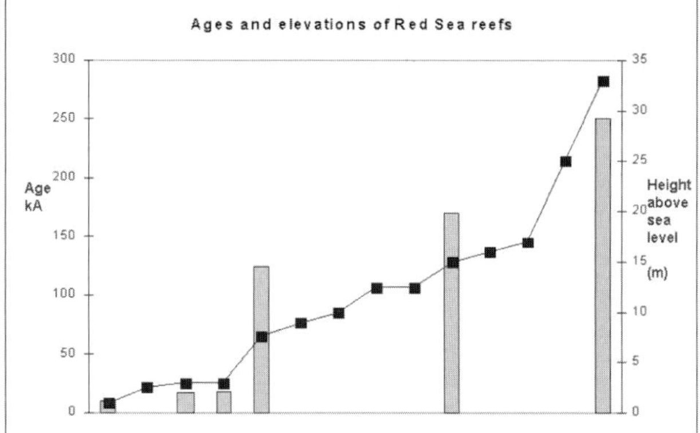

Figure C71 (Top) A central Red Sea fringing reef, whose reef flat is at low tide level. The edge of the reef flat is seen <100 m to seaward, following which the slope dips steeply to hundreds of meters. The person is lying on a tectonically raised fossil reef whose elevation is about 4 m. (Below) Plots of ages of reefs and their heights above sea level. [Data taken from review material in Sheppard *et al.* (1992), table 1.1.]

Figure C72 Fringing reef backed by "hard substrate" mangroves and, on the shallow reef flat, dark patches denote areas of rich seagrasses. The three habitats of reef, mangrove, and seagrasses are closely linked. (Location is in the Sinai peninsula and these mangroves are amongst the most northerly in the world.)

Figure C73 The spur and groove structure along the seaward edge of an oceanic atoll. Formed from growth of calcareous algae (*Porolithon* spp.) it is the primary defence against erosion for the reef and the shore behind (Chagos atolls, Indian Ocean).

hard substrates, they also may develop on unconsolidated substrate where the latter exists in sufficiently sheltered conditions. Good examples of this are found in the Red Sea, a region famous for both its general aridity and excellent fringing reefs lying close to and contiguous with the coast. Here, cores have shown layers of solid reef alternating with layers of unconsolidated sand and rubble of mainly terrestrial origin. The formation of these series comes from alternating phases of reef growth, followed by reef obliteration when heavy rains and sediment deposits carried by flash floods deposit terrestrial materials in thick blankets over the reefs. Alluvial fans develop, prograding to seaward, which then support vast areas of productive shallow water between their leading edge and the shore. Sheltered conditions then allow a new cycle of reef growth to develop on the fans. Depending on current patterns (and repeat episodes of heavy rain) these fans may then infill, extending the land to seaward until greater depth and steep slopes at the seaward edge precludes further progradation.

Uplift of reefs also affects coastline morphology. Clear examples again occur in the Red Sea (Figure C71) where series of tectonically uplifted reefs extend inland on the coastal plain. The Red Sea is an example of a spreading plate with current seismic uplift but, equally, substantial examples of uplifted old coral reef coastlines occur on most continents, notably east Africa and Australia.

Biological systems

Biological systems supported along reef shores are usually rich and diverse. Apart from the productivity and biodiversity of the reefs themselves, any areas of hard substrate between the reef and shore are likely to be colonized extensively by corals. Sand accumulates along these shores from wave pumping over the reefs—mostly the sand being limestone, derived from eroding reef material, and fragments of coral and coralline algae. Soft substrates commonly dominate the shoreward coastal marine environment. If stable enough, seagrasses may thrive, themselves being a rich habitat supporting high productivity and a considerably greater biodiversity than adjacent, uncolonized sand. Once these angiosperms take root, they stabilize the habitat further.

Along the coastline itself in such areas, mangroves may dominate the intertidal habitat, particularly if estuarine or at least slightly lowered salinity conditions prevail. These forests may be extensive and grade inland into true terrestrial forests, or may be very shallow-rooted in very limited sediment (Figure C72) and are known as "hard substrate mangroves." Most mangrove trees have roots and shoots which trap and stabilize sediments. This can also lead to progradation of the coastline to seaward as well as laterally alongshore. The existence of the mangroves, as with seagrasses, commonly is the direct result of the shelter provided by the reef lying to seaward. Conversely, the existence of the large, shore-colonizing stands of rooted plants, traps and consolidates substantial amounts of sediments that otherwise would reach and adversely affect the reefs, which are extremely intolerant of sedimentation. Thus, complex feedback processes determine both the physical and biological nature of the coral reef coastlines, as well as their physical extent. Seaward progradation may attain a balance, depending on the extent of shallow substrate on which the reef exists and over which it may extend, but rates of progradation are fairly slow; considerably slower than may be the case, for example, where mangroves extend the land seaward near heavily sedimented estuaries in the absence of reefs.

Figure C74 The capital city and island of Malé, Republic of Maldives; a coral reef island city. The reef flat is almost entirely infilled. A breakwater (on left) was built by Japanese aid following devastating surges and loss of land. Water abstraction has left the water table damaged, now with saline and sulfurous water.

Coral island shores

These are a special case in which reefs themselves are the only component of the coastline. The modern component may be only a veneer on pre-Holocene reefs, or may themselves develop into thick deposits. Their shallow existence traps sediments, then limited vegetation, birds and further vegetation, in well-documented patterns of biological succession. Known as cays, keys, or islets of atolls, globally their total coastline is considerable, especially in deep oceanic regions distant from continental coasts. In these, all hard substrate is limestone, and all soft substrate is derived limestone sand. Most of the latter will undoubtedly be modern, though where hard substrate exists, outcrops of pre-Holocene material may be expected.

Typically, these coastlines have classical profiles of a low island, followed by a reef flat at low tide level, and a reef slope which dips to seaward, smoothly or in cascades. These reef slopes may be very narrow, less than 20 m wide, this distance in extreme cases being all that separates an island shore from water which rapidly attains depths of several hundred meters. More commonly, reef flats extend for upwards of 100 m, and may be dry or become extremely shallow for up to about a quarter

Figure C75 Beach rock formation. Where freshwater seeps into the intertidal, microbial growth leads to solidification of the fine sands. These appear to be extremely important in the creation of dipped limestone slabs in some areas (Chagos atolls, Indian Ocean).

of the tidal cycle. In exposed areas these may be bordered by an algal ridge; a construction of calcareous red algae which faces the oncoming waves (Figure C73). Usually algal spurs extend more or less regularly to seaward from the ridges for a further 100 m, dipping to 5–10 m depth. These structures have a tremendous wave-baffling role, in that their length and spacing appears to be "tuned" in such a way that incoming waves meet the outgoing swash from the preceding wave, their energies partly cancelling each other out. It appears that the very existence of many coral reef shores depends to some degree on this algal, biogenic coastline structure.

Human habitation on coral reef coastlines

Coral reef coastlines have always been populated, and because reefs and the other rich habitats along these sheltered shores are rich in food sources, notably fish protein, limitations on human numbers have commonly been due more to the availability of freshwater rather than to the supply of food (Sheppard, 2000). In rainy tropical areas, human populations have been considerable for long periods of time, while in arid locations, such as the Middle East, there usually have been far fewer—and more nomadic—people despite the richness of marine foods.

As well as the limiting effects of little or highly seasonal rainfall, the permeable nature of these fossil coral and modern coral island coastlines have long affected the amount of water which may be stored and extracted, and this also has greatly affected the ability of people to live on them. Freshwater lenses on modern coral islands are small and clearly change rapidly in size; they track the quantity of rainfall and they are clearly tidally linked, rising and falling diurnally, often with phase lags of only a very few minutes. They are thus very sensitive to over-abstraction. The available freshwater is also sensitive to increased runoff to sea which takes place following development of human habitation; here significant proportions of the land become covered with built-up areas which are impermeable, causing water to runoff rather than enter the water table. Both the latter and over-abstraction readily cause a collapse of the water table—one marked example is seen in the current sulfurous and saline nature of the water table on Malé, the very built-up and now artificially shaped capital town (and island) of the Republic of the Maldives (Figure C74). Interestingly, it has also been suggested that where the water table seeps laterally to sea in the region of the intertidal zone in carbonate shores, chemical changes, coupled with microbiological growth, causes solidification of "beachrock" a durable form of limestone which also contributes to coast line stability (Figure C75).

Where freshwater and biological systems are not overstretched, human populations may continue to live. The productivity of all three major habitats (reefs, seagrasses, and mangroves) are amongst the most productive in the world (Sheppard, 1995) and, in addition, offshore production in these areas may also yield substantial additional fish such as tuna.

However, numerous factors currently combine to create unsustainable conditions which have led to numerous system collapses and abandonment of coastal villages throughout tropical Asia, Africa, and Central and South America. Pollution from sewage and industries, unsustainable methods of fishing (e.g., dynamite and DDT "bombs"), unsustainable amounts of previously sustainable types of fishing, extraction of coral reef material for building, and infilling of the shallows between reefs and shore to create more land, have all caused ecosystem collapses in coral reef coastlines and have spelled disaster for the local inhabitants. Currently, nearly two-thirds of reefs have been damaged (Bryant *et al.*, 1998) though this figure is nearer 80% in areas such as Southeast Asia. Similarly, removal of the mangroves for timber or to clear space for unsustainable shrimp farming, and whose existence protected the seaward reefs by trapping terrestrial sediments, permits previously stabilized sediments to blanket and kill the reefs leading, eventually, to coastline erosion. Even poor agricultural practices far inland have resulted in massive movement of topsoil into these coasts via river flow, again causing mass mortality of habitats and starvation of people (Sheppard, 2000). Compounding the damage is the fact that such reef-sheltered coastlines tend to be dis- proportionately rich in their provision of breeding, spawning, and nursery areas for marine species.

Effects of climate change and sea-level rise

Although reefs grow vertically to the low tide level, and did so during the rapid and substantial Holocene transgression (the sea-level rise of about 150 m in about 8,000 yr), their ability to continue to do so under present conditions of rising sea level and temperature is not at all certain. First, sea surface temperature increases and El Niño events in recent decades have caused mass bleaching and mortality of corals. In 1998 in the Indian Ocean, for example, coral mortality in shallow reef areas exceeded 80% in many and probably most atolls, and in many patch reefs and fringing coral reefs in the Arabian Gulf, southern Red Sea and many continental shores. As of mid-2000, recovery appears poor. It is known that coral erosion continues, and since growth and erosion in a healthy reef are closely balanced (in favor of growth) net erosion may now be predicted with concomitant reduction of breakwater effect. The importance of the breakwater effect on reefs is marked, and can be seen when reefs have been removed previously by other reasons: for example, by coral "mining" on reefs around islands which have lowered the surface of the reef by 0.5 m, or from mortality of corals through disease which resulted in similar loss; in such cases the loss of the reef has led to severe erosion of the coastline, sometimes exceeding 1 m per year until a new stable point is reached. Given the attractiveness of development on "reef front" locations, infrastructure damage has often been substantial. Both of these forms of reef loss are likely to mimic the loss following recent episodes of temperature-induced mortality.

It is likewise known that coral growth and reef growth become "decoupled" as environmental conditions become increasingly severe (Sheppard *et al.*, 1992), and that even continued coral growth does not necessarily result in continued growth of the essential hard, durable reef. Further, while it is generally understood that reef growth results in deposition of vast quantities of carbonate, a point important in the consideration of possible carbon sinks to combat rising global CO_2 levels, in fact a rising atmospheric CO_2 level also alters water chemistry dynamics in the direction of *reducing* deposition of carbonate; estimates of a 21% reduction of carbonate deposition by 2065 have been postulated given Intergovernmental Panel on Climate Change estimates of CO_2 rise (Leclercq *et al.*, 2000). While the latter work has primarily been directed at the question of the global carbon sink, clearly reduced carbonate deposition on reefs which protect coastlines will become increasingly important too, as the fine balance between reef accretion and erosion becomes disturbed in favor of erosion.

Thus, while healthy reefs may continue to keep pace with a rise in sea level, as has been the case before, the increasingly stressed and damaged reefs appear to be decreasingly capable of fixing CO_2, have a decreasing ability to maintain rich coral- and reef-forming (limestone-depositing) communities, and possibly even a decreased ability to maintain food production for human populations. Evidence is accumulating to show that pressures on reefs, including over-fishing, lead to changes to their ecological stable states which cause them to be dominated by different species groups, even if or when the pressure initially causing the change is removed. The replacement biotic groups may be dominated by non-calcareous species, and the trend appears to be one-way, like a ratchet rather than a wheel. If confirmed further, substantial changes to coastlines currently protected by reefs are likely to occur, the direction of the changes being characterized by increased erosion, reduced productivity, and reduced ability of these coasts to support the human populations currently living there and dependent upon them.

Charles R.C. Sheppard

Bibliography

Bryant, D., Burke, L., McManus, J., and Spalding, M., 1998. *Reefs at Risk. A Map-Based Indicator of Threats to the World's Coral Reefs*. Washington, DC: World Resources Institute.

Leclercq, N., Gattuso, J-P., and Jaubert, J., 2000. CO_2 partial pressure controls the calcification rate of a coral community. *Global Change Biology*, **6**: 329–334.

Sheppard, C.R.C., 1995. Biological communities of tropical oceans. In Nierenberg, W.A. (ed.), *Environmental Biology*. London: Academic Press, pp. 277–289.

Sheppard, C.R.C., 2000. Coral Reefs of the western Indian Ocean: an overview. In McClanahan, T., Sheppard, C.R.C., and Obura, D. (eds.), *Coral Reefs of the Western Indian Ocean: Ecology and Conservation*. Oxford: Oxford University Press.

Sheppard, C.R.C., Price, A.R.G., and Roberts, C.J., 1992. *Marine Ecology of the Arabian Area. Patterns and Processes in Extreme Tropical Environments*. London: Academic Press.

Cross-references

Algal Rims
Beachrock
Biohermes and Biostromes
Coral Reef Islands
Coral Reefs
Coral Reefs, Emerged
Demography of Coastal Populations
Environmental Quality
Human Impact on Coasts
Mangrove, Ecology
Water Quality

CORAL REEF ISLANDS

A coral reef island is composed of rocks from coral skeletons, that is, biologically formed calcium carbonate materials derived from the adjacent coral reef and raised above sea level. Coral reef island sizes range from a few square meters to many square kilometers, and they come in all shapes and proportions. Their soils consist of coral fragments, calcareous algae and other limestone detritus, varied amounts of humus, guano from sea birds, volcanic ash, and drifted pumice (Fosberg, 1976).

Coral reef islands can develop only where suitable conditions sustain coral growth over time. These conditions include favorable physical factors (high temperature, high salinity, good light penetration, and low nutrients) and biological factors (especially a supply of coral larvae) in tropical regions that provide a firm substrate in the shallow photic zone (Birkeland, 1997). Coral islands are created in three ways, which are not mutually exclusive: (1) accumulation of dead coral reef rubble and sediments on top of a reef flat through wave and storm action, resulting in a low island (several meters in height) that may be ephemeral or stable; (2) emergence of a coral reef due to a drop in relative sea level; and (3) uplift of the ocean floor beneath the reef. Coral reef islands resulting from tectonic uplift of coral substrate may range from several to hundreds of meters in height.

The life span of coral reef islands is determined by the balance of erosional and depositional processes, and by sea-level rise. Coral islands drown when their corals are submerged below a critical depth for coral growth (Grigg and Epp, 1989). During their existence, these nonvolcanic islands may become vegetated and support a terrestrial fauna and flora. Coral reef islands are fundamentally different from so-called "coral islands" that consist of fringing coral reefs growing on a foundation provided by an island of volcanic or continental origin. Coral reef islands may be associated with volcanic or continental land masses if they form as the result of depositional processes arising from fringing or barrier reefs. These islands may lie on barrier reefs separated by a lagoon of varying widths from high land, and their proximity to this land may result in some terrigenous sediment input to coral reef islands.

Darwin's theory of reef formation asserts that the three main reef classes, fringing, barrier, and atoll, are each different stages of development of a coral reef (Darwin, 1842). Darwin did not define a coral reef island, simply referring to the islands on atolls as islets of various sizes and shapes. Dana (1872) defined "coral islands" as reefs that stand isolated in the ocean (= atolls), away from the shores of high islands and continents which harbor "coral reefs." His definition is clearly too restrictive, although he clarified it by stating that a coral island has a lagoon in the middle instead of a mountainous island and consists of a very low ratio of habitable land to area. Other terms associated with coral islands also have different local meanings. Extremely small coral islands on atolls are called islets or cays, while others refer to small coastal islands as cays ("keys" in Florida), some of which are not derived from coral reefs. "Coral cay" defines a small low coral reef island. "High" and "low" islands are terms often applied to coral reef islands. High islands are generally not coral reef islands since most consist of volcanic or continental materials with surrounding fringing or barrier reefs. However, uplifted coral reefs may attain a height similar to some high islands. Low islands are true coral reef islands if their sedimentary origin is coral reef material.

Classification of coral reef islands

Stoddart and Steers (1977) placed coral islands into seven different classes based on morphology and sediment characteristics, with vegetation used as a secondary diagnostic. Vegetation patterns depend on elevation, surface features, area, distance from sea, soil development, and human activities (Fosberg, 1976). Stoddart and Steers (1977) stress that these islands do not necessarily evolve from one class to another; any change results from external influences, primarily storms. The first class includes unvegetated and vegetated sand cays, which are low-lying accumulations of coral rubble and sand deposited by wave refraction on the reef. Sand cays are generally associated with barrier reefs and patch reefs in protected areas of reefs, for example, in lagoons behind barrier reefs. They occur on reef patches in shallow, wide lagoons and assume a variety of shapes depending on sea level, wave energy, and tidal ranges (Hopley, 1982). The windward margins of the cays tend to be composed of loose coarse rubble and the leeward sides accumulate sandy sediments. These terrestrial piles of coral material may eventually develop vegetation, mostly of plants from seeds deposited by birds or by ocean currents. A lagoon may form in between, with mangroves helping to stabilize the sediments (Guilcher, 1988). These islands range in maximum dimension from several meters to greater than 1,000 m, with a height of several meters. Unvegetated islands are lower in height and ephemeral in nature. The second class of coral island is the *motu*, which is the most common island in the Indo-Pacific. Motus have a seaward (windward) shingle ridge 3–5 m high and a leeward sandy shore and are generally vegetated. These islands occur most frequently on the exposed windward sides of atolls on the inner reef flat on Pacific atolls where sediments and rubble are deposited by strong unidirectional waves. Motus vary in number, size, and shape on a given reef or atoll. Islands in the Tuamotu atolls range from 250 to 300 per atoll, covering 30–35% of the atoll rim. Most atolls in the Marshalls, Gilberts, and Ellice groups contain more than 20 islands, while the Carolines support fewer than 10 islands per atoll (Stoddart and Steers, 1977). The number of islands on an atoll depends on the number of erosional channels ("*hoa*") cut through formerly continuous atoll rims. The vegetation on motus is strongly influenced by elevation above sea level and island size. A freshwater lens can develop on larger islands which will support a more diverse flora and fauna. The third and fourth classes of coral islands are the shingle cay and mangrove cay. The shingle cay is a storm-related event not common on coral reefs, while mangrove cays are formed by colonization of shoal areas by mangroves. Mangrove cays with a windward sand ridge form the fifth class, and low-wooded islands are the sixth class. The seventh class consists of emerged coral reefs (limestone islands). Examples of emerged coral islands include the northern Florida Keys, the Bahamas, and Bermuda in the Atlantic ocean and Aldabra atoll in the Pacific ocean. These are often dissected into karst topography with steep and strongly undercut shores, sharp peaks, and sinkholes.

Distribution of coral reef islands

Most coral reef islands occur in the Indo-Pacific region. There are over 300 atolls and extensive barrier reefs in the Pacific ocean and only 10 atolls and 2 barrier reefs in the Caribbean region (Milliman, 1973). The total number of coral reef islands is unknown, and varies according to change in sea level and storm activity.

Summary

Coral reef islands occur throughout the world at tropical latitudes. They are best viewed as products of the reef and as an extension of the reef ecosystem (Heatwole, 1976). Ecological studies are needed to determine the physical and biological linkages between the coral reef and its islands. Although most coral islands occur in isolated areas, human influences on the vegetation and fauna are highly evident and easily quantified. These land forms are excellent geological and ecological systems for studying interactions between marine and terrestrial ecosystems.

Gisèle Muller-Parker

Bibliography

Birkeland, C.R., 1997. Introduction. In Birkeland, C.R. (ed.), *Life and Death of Coral Reefs*. New York: Chapman and Hall, pp. 1–12.

Dana, J.D., 1872. *Corals and Coral Islands*. New York: Dodd and Mead.

Darwin, C.R., 1842. *The Structure and Distribution of Coral Reefs*. Reprinted 1984 by the University of Arizona Press, Tucson: Arizona.

Fosberg, F.R., 1976. Coral island vegetation. In Jones, O.A., and Endean, R. (eds.), *Biology and Geology of Coral Reefs*, Volume 3. New York: Academic Press, pp. 255–277.

Grigg, R.W., and Epp, D., 1989. Critical depth for the survival of coral islands: effects on the Hawaiian archipelago. *Science*, **243**: 638–641.

Guilcher, A., 1988. *Coral Reef Geomorphology*. New York: Wiley.

Heatwole, H., 1976. The ecology and biogeography of coral cays. In Jones, O.A., and Endean, R. (eds.), *Biology and Geology of Coral Reefs*, Volume 3. New York: Academic Press, pp. 369–387.

Hopley, D., 1982. *The Geomorphology of the Great Barrier Reef*. New York: Wiley.

Milliman, J.D., 1973. Caribbean coral reefs. In Jones, O.A., and Endean, R. (eds.), *Biology and Geology of Coral Reefs*, Volume 1. New York: Academic Press, pp. 1–50.

Stoddart, D.R., and Steers, J.A., 1977. The nature and origin of coral reef islands. In Jones, O.A., and Endean, R. (eds.), *Biology and Geology of Coral Reefs*, Volume 4. New York: Academic Press, pp. 59–105.

Cross-references

Atolls
Cays
Coral Reefs
Coral Reefs, Emerged
Small islands

CORAL REEFS

Coral reefs are the largest structures built solely by plants and animals and many are clearly visible from space. Gross distribution is limited by a requirement for warm water with an average winter maximum temperature of 18°C generally regarded as providing the poleward margin for coral reefs which are thus found only in a few areas outside the tropics. High temperatures above 30°C may also be lethal to corals. Other requirements include good water circulation and a generally low nutrient status for ambient waters. In the open ocean where nutrient availability is very low, coral reefs exhibit a very efficient nutrient recycling, the marine equivalent of tropical rainforests. Inshore reefs can tolerate higher nutrient levels and may even respond with more rapid growth rates, though this may be countered by the skeleton laid down being more fragile. Turbid waters have a detrimental effect, limiting the penetration of light in the water column and compressing the vertical zonation of the coral reef. In clear open waters corals may grow to depths exceeding 110 m. In turbid water this may be limited to 10 m or less.

The importance of plants on the reef

Many of the restrictions on coral growth are imposed not by the corals themselves but by plants. Within the endodermal tissue of the coral polyps are unicellular algae, zooxanthellae, which have a close symbiotic relationship with the coral host. The zooxanthellae not only provide some of the nutrition for the coral but also aid as important primary producers of oxygen and carbon during daylight hours through the process of photosynthesis. They are also nutrient conservators, satisfying their own requirements from the host metabolic waste which they thus assist in removing. Most importantly, they are direct promoters of inorganic growth of coral skeletons. In daylight, the algae produce more oxygen than can be used by polyp respiration and some of the carbon dioxide produced by the respiratory process is refixed by the algae into new organic matter. As photosynthesis takes up C-12 slightly faster than C-13 the organic matter synthesized by the zooxanthellae has a relative preponderance of C-12 and a pool of carbon compounds enriched in C-13 is left behind. It is from these that the calcium carbonate skeleton is built, aided by pH variations associated with the diurnal photosynthetic process. Corals grow 2–14 times more rapidly in light as opposed to darkness and twice as fast on a sunny day as opposed to a cloudy day.

Other algae also participate in the growth and maintenance of the coral reef. Crustose coralline algae infill and cement the coral structure especially on the high energy margins of the reef where they build distinctive pavements and ridges. *Halimeda* producing delicate plates of calcium carbonate can be extremely prolific in producing sediment and, given appropriate nutrient pulsing, can build bioherms as significant as the coral reefs (as for example in the northern Great Barrier Reef, and the Makasser Straits of Indonesia). At their poleward limits competition from macro algae which can overshade the corals, is a factor in coral reef distribution.

Coral reef diversity and ecology

Coral reefs are amongst the most complex and diverse ecosystems on earth (Dubinsky, 1990). A variety of marine invertebrates and plants contribute to the accumulation of the limestone which are the most massive bioconstructional features in the geological record. Symbiosis with zooxanthellae occurs not only in the corals but in other major calcifying organisms including giant clams and foraminifera which are a major producer of reef sediment. Other important calcifers include bryozoans, sclerosponges, echinoderms, and a wide variety of molluscs. They contribute to both the reef framework and the infilling sediments.

At the scale of the total reef ecosystem gross primary organic production is very high but conversely, because of the amount of recycling which occurs, net productivity may be close to zero. In contrast, calcification is the most active of any ecosystem on earth. With an estimated total area of 617,000 km^2 the estimated total annual production is estimated at 929 million tonnes $CaCO_3$.

The stony scleractinian corals are the most important of the reef biota and their diversity typifies that of many other reef-dwelling species including coral reef fish. It is currently estimated that there are about 220 coral genera and at least 1,300 species, though many are rare and/or azooxanthellate. However, there are two distinct centers of diversity. The most diverse is the Indo-Pacific centered on the south-east Asian archipelago, most significantly the Molucca and Banda Seas of eastern Indonesia. Here there are about 450 species belonging to 80 genera. Individual sites may have as many as 140 species. The other center is in the western Atlantic and is much less rich with only 20 genera and 50 species.

The reasons for this lie in the evolutionary history of the scleractinia which evolved from the earlier rugose corals of the Paleozoic (Veron, 1995). The first scleractinians appeared in the middle Triassic about 250 million yr ago in the western Tethys Sea approximating to southern Europe and the Mediterranean. Corals occupied a circumglobal seaway which included the Tethys Sea and the gap between the two American continents. During the Mesozoic and Tertiary, as the African, Indian, and Eurasian plates collided the Tethys Sea was closed from the west, its diverse coral fauna pushed eastwards into what is now the final relict consisting of the islands and peninsulas of south-east Asia. The Atlantic fauna was isolated first from the east, then in the Pliocene from the west as the Isthmus of Panama isolated the Atlantic from the Pacific.

The gross pattern of diversity is thus a result of geological history. However, local diversity is also dependent on the range of ecological niches available (particularly wide-ranging in south-east Asia) and the frequency of disturbances such as hurricanes, earthquakes, volcanic eruptions, or crown of thorns starfish outbreaks (Connell, 1997). As coral reproduce mainly through a single annual spawning event, a medium frequency of disturbance can allow planula larval recruitment from a wide range of ecological niches because of the longevity of the larval phase which can be several weeks, allowing dispersal over hundreds of kilometers.

Rates of growth and accretion and the importance of bioerosion

The majority of corals are colonial organisms and annual extension rates for single colonies may be more than 10 cm/yr for branching species and 1 cm/yr for massive corals. Each year's increment consists of a wider high-and narrower low-density band which can be clearly identified by x-radiography. The high-density band is normally laid down during a period of stress imposed by either ambient environmental conditions or reproduction.

However, as the corals and other organisms are laying down calcium carbonate, many others are bioeroding the reef substrate. Some of the sediment produced may help to infill the reef framework but much is lost from the immediate vicinity of the reef. Bioeroding organisms include rock grazers in search of epilithic and endolithic blue-green algae on the reef surface, rock browsers which consume reef fabric and sediments such as parrot fish and holothurians, and a wide range of burrowers and borers. Rate of breakdown can match or exceed the rate of calcium carbonate production. For example, individual chitons can consume up to 13 cm^3/yr, grazing fish in Barbados reefs produce 1–2 tonnes/ha/yr of sediment, clionid boring sponges 5–10 tonnes/ha/yr, and grazing echinoids up to 97 tonnes ha/yr, 65% of which is lost from the reef. Boring worms such as Sipunculids and Polychaetes produce up to 20% of the porosity in the reef framework. On Rongelap atoll in the Pacific it is estimated that up to 100,000 tonnes of lagoonal sediment passes through the acid gut of grazing holothurians each year.

Coral reefs which are under stress may have bioerosion rates which exceed the current rate of limestone deposition. However, most healthy reefs are actively accreting. Rates of reef growth are commonly calculated in two different but complementary ways. Biogeochemical measurements of changes to water chemistry over a coral reef can give instantaneous calcification rates related to total reef metabolism (Kinsey, 1985). Such measurements have been made worldwide and when converted to annual rates give remarkably uniform results for the different reef environments. All reefs accrete through the mix of three absolute modes of calcification:

100%	coral cover	10 kg $CaCO_3$ m^2/yr
100%	algal pavement	4 kg $CaCO_3$ m^2/yr
100%	sand and rubble	1 kg $CaCO_3$ m^2/yr

Different zones of the reef are thus accreting at different rates. Sediment is produced from the highly active zones such as the reef front, to the less-active parts of the reef such as lagoons. Taking into consideration the density of the calcium carbonate produced (2.9 for aragonite) and the porosity of coral and reef frameworks (about 50%) these results give an upper limit of about 7–9 mm/yr for reef accretion.

Using the geological record through drilling and radiocarbon dating of reef cores provides an alternative long-term method for calculating coral reef accretion rates, which are nonetheless very compatible with those from reef metabolism. The synthesized results from the Great Barrier Reef (Davies and Hopley, 1983) have been reproduced from many other reef-drilling projects. Framework accretion rates ranged from 1 to 16 mm/yr according to the density of the fabric (lowest for algal pavement, highest for open branching coral frameworks) though the modal rate of 7–8 mm/yr is identical to that from water chemistry. Drilling has also produced rates for detrital sedimentation. In lagoons and sand aprons it is 1–4 mm/yr for sand flats, 7–9 mm/yr for steady trade wind accumulation and 13–18 mm/yr for high-energy low-frequency events.

Coral reefs and sea-level change

Coral reef growth is highly constrained by sea level. Uppermost growth may exceed the mean low water mark in moated pools in reef flats and will die off with any fall in relative sea level. Single tectonic events causing uplift of the land can cause mortality across an entire reef flat and in areas of constant uplift in the tropics, such as at tectonic plate margins, suites of raised reef terraces may occur to several hundreds of meters of elevation. Radiometric dating of such terraces in locations such as the Huon Peninsula in Papua New Guinea and Barbados has determined not only the rates of uplift locally but also global eustatic sea-level records.

As coral reefs have existed for such a long period, global changes in sea level have had a major impact. Global cooling during the late Tertiary and Quaternary, resulting in waxing and waning of ice sheets at the poles and in the northern hemisphere have resulted in falls in sea level of up to 150 m, exposing all extant coral reefs and removing active growth to deeper banks and slopes perhaps severely restricting the area of coral growth at the low sea level times. Fringing reefs were exposed as limestone terraces, shelf barrier reefs as chains of limestone hills, and atolls and other oceanic reefs as limestone islands. The records of sea-level change suggests that over the last million years coral reefs have been exposed for far greater a period than they have been inundated. Reef accretion has been limited to the relatively short high sea-level phases of interglacials.

During periods of exposure, which may have been as much as 100,000 yr at a time, the reefs were subjected to karstic erosion processes. Indeed a series of models for the development of coral reef morphology at both the gross and detailed scale, have been hypothesized based on the karst land forms which were etched at low sea-level phases (Purdy, 1974; Hopley, 1982). This antecedent karst hypothesis has been used to explain the annular shape of atolls and the rim structures of barrier reefs as well as the labrynthine morphology of many reef surfaces (Figure C76) which have been likened to the complex ridges and enclosed depressions of tropical karst regions. Some features such as caves with speleothems, "blue holes" up to 500 m wide and 100 m deep are clearly derived from preexisting karst land forms. Blue holes on the Great Barrier Reef for example (Figure C77) are almost certainly collapse dolines over which a veneer of 10–20 m of Holocene reef growth has occurred (Hopley, 1982).

A problem with the antecedent karst hypothesis is the time available during low stands of sea level for major karst landforms to develop (see Hopley, 1982, chapter 7 for discussion). Recent studies have shown that some morphology previously attributed to karst interaction has in fact been derived from reef growth in the Holocene. However, it is also possible that the growth which occurs during an interglacial high sea level may enhance that produced by erosion in the previous period of exposure.

Figure C76 Complex lagoonal relief in the large reefs of the Pompey Complex, Great Barrier Reef Australia.

Figure C77 A blue hole approximately 300 m wide and 33 m deep in Molar Reef, Great Barrier Reef, thought to have originated from a collapsed doline.

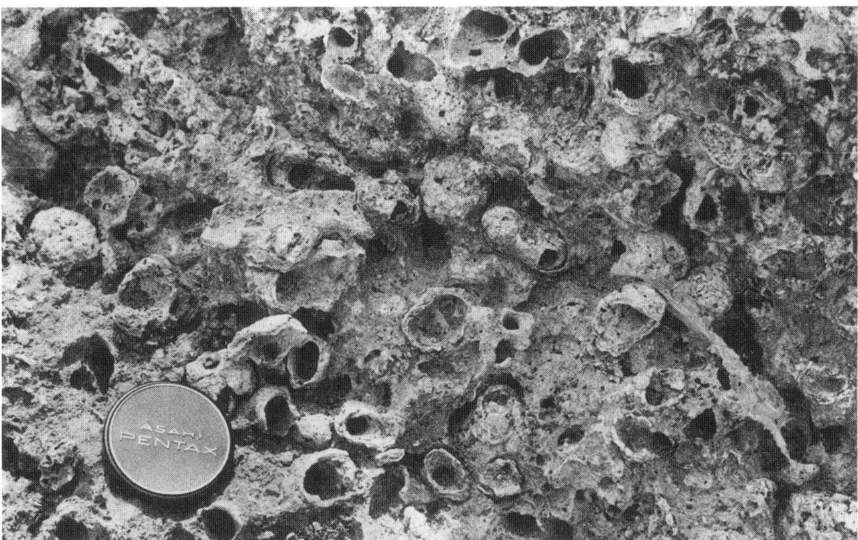

Figure C78 Diagenetically altered *Acropora cervicornis* in the Pleistocene reef tract, Barbados. The aragonite of the original corals has largely disappeared or altered into calcite. Low Mg-calcite molds result from the more rapid replacement of encrusting coralline algae.

Figure C79 A brecciated caliche profile with numerous brown micritic stringers developed on Pleistocene reef limestones ca. 84,000 yr BP on Barbados.

This is then further enhanced by the patterns of erosion during the next low sea-level stand. In all probability both growth and erosion produce the complex relief of coral reef systems.

In addition to the formation of karst landforms during low sea-level exposure, coral reefs also experience diagenetic changes. The original skeletal limestone is largely aragonite (corals, molluscs, *Halimeda*, sclerosponges) or high Mg-calcite (crustose coralline algae) which in subaerial and vadose environments is unstable and may be replaced by a coarser crystalline low Mg-calcite (Figure C78). Caliche crusts (Figure C79) may form on the surface and soils can be retained over parts of the surface even after reinundation. During the subsequent high sea-level phase the older reef is recolonized and new framework deposited over the older surface. However, it remains as a clearly identifiable discontinuity termed a "solution unconformity." Beneath the Holocene reef framework the solution unconformity formed during the last glaciation is typically 10–20 m beneath the surface. Where coral reefs have been drilled to greater depth 10 or more solution unconformities have been identified especially in atolls and shelf edge barrier reefs which suggests continuous subsidence at these sites during the period of reef accretion.

Reef growth in the holocene

Modern coral reefs have developed entirely during the Holocene period, the majority forming a relatively thin veneer over older Pleistocene foundations which are no more than 10–20 m below modern sea level (Figure C80). Very few radiocarbon ages greater than 9,000 yr have been obtained for this Holocene veneer. Only where reefs were in deep water, such as open ocean atolls, were corals able to merely migrate to the flanks of the reef during the last glacial and thus recolonization was simple and immediate. This is also the situation for some of the mid-oceanic volcanic islands, which explains why the deepest and longest postglacial record comes from Tahiti. Here the postglacial reef is more than 90 m thick and basal dates older than 13,500 yr have been obtained. Australia's Great Barrier Reef in contrast rises from a relatively shallow continental shelf generally less than 50 m deep. Even the lowest parts of the older Pleistocene reefs were inundated less than 11,000 yr ago (Hopley, 1995). Delays in re-colonization occurred due to the need for larval recruitment from more distant sources (probably Coral Sea Plateau reefs), and as the result of initially poor water quality as regolith was removed from the Pleistocene foundations and as a

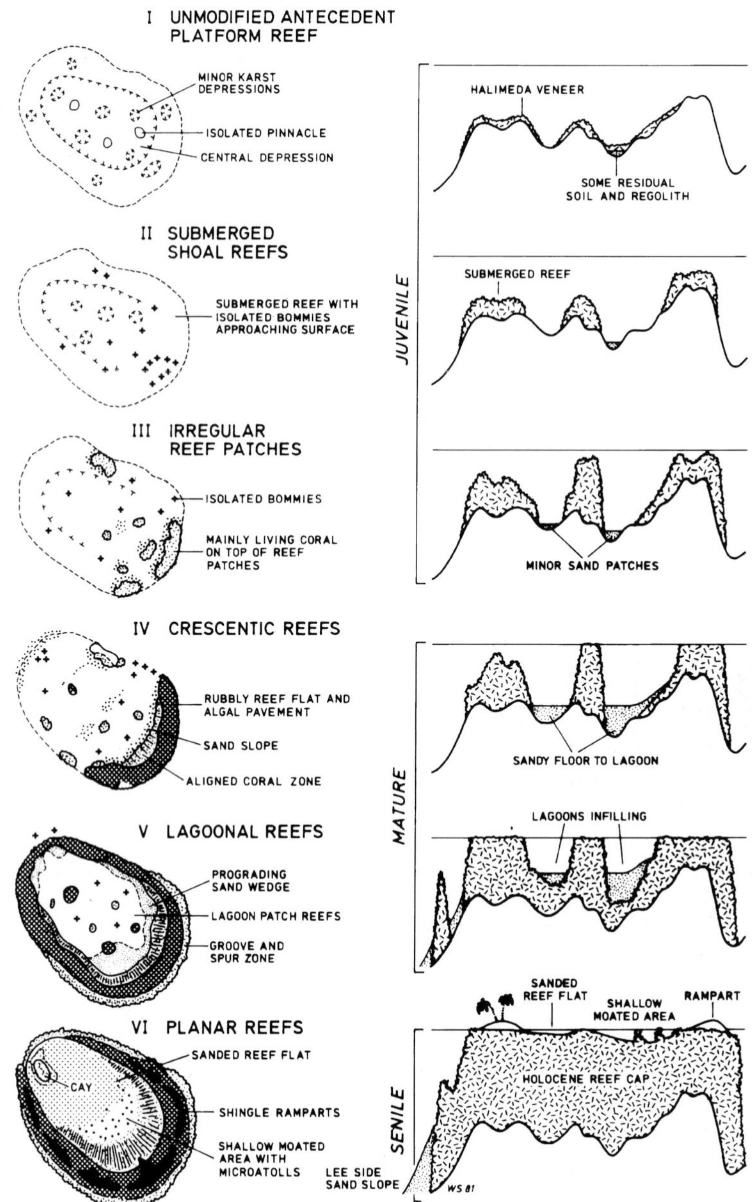

Figure C80 An evolutionary classification of Holocene coral reefs.

result of very close proximity of the mainland of the time to the main reef tract. The major part of the Holocene growth took place only after 8,500 yr BP by which time the mainland coastline was close to its present position and well removed from the outer reef.

On the Great Barrier Reef, and generally many other coral reefs, the major period of Holocene accretion took place between about 8,500 and 6,000 yr ago, when average vertical accretion rates of ca. 5 m per thousand years were achieved. Largely for hydroisostatic reasons (see Hopley, 1982, chapter 13 for discussion) the pattern of Holocene sea-level rise over this period varied in different parts of the world. Over much of the Indo-West Pacific modern sea-level had been achieved by as early as 6,500 yr BP while in the Atlantic relative sea level may have continued to rise right up to the present time, though at a considerably slower rate after 6,000 yr BP. Reefs have responded in different ways with three models recognized (Neumann and MacIntyre, 1985; Guilcher, 1988):

- Keep-up, in which the reef was able to accrete vertically at about the same rate as the rise in sea level and was never more than a few meters below sea level.
- Catch-up, in which the reef lagged behind the sea-level rise due to delays in recolonization and/or slow growth. At times these reefs may have been more than 10 m below sea level.
- Give-up, reefs which after becoming established in the early Holocene died off and remain only as drowned carbonate banks. Various reasons have been put forward. That they became too deep because of the rapid rise in sea level is unlikely in the majority of examples as most would have remained within the photic zone of active reef growth, unless there were also an increase in turbidity (which would have limited the depth of reef growth). Many "give-up" reefs are on shelf edges and sediment shed from upslope reefs may have been one cause of their demise. In the Caribbean, in particular, it has been hypothesized that as sea level first flushed the shelf, a new wave of sedimentation and eutrophication was the cause of shelf edge reef decline (MacIntyre, 1988). In addition, at high latitudes shallow shelf waters may also have killed off reef growth at the shelf edge.

Once coral reefs caught up with sea level then their upward growth phase was terminated and carbonate productivity has been directed into lateral rather than vertical growth.

An evolutionary classification of Holocene reefs

It is possible to examine the growth of modern reefs by way of an evolutionary classification (Hopley, 1982), based on the presumption that the majority of reefs are located over pre-Holocene reefal foundations at no great depth and with an irregular morphology resulting from both the original growth form of the last period of reef growth and the erosion which has taken place during subsequent exposure. This presumes transgression by the sea over the foundations and a sea-level rise which most reefs or most parts of reefs cannot match by upward growth (Figure C80).

Three phases of Holocene reef development are recognized in the classification. In the juvenile phases, initial colonization takes place on the antecedent foundation and upward growth of the reef lags behind sea level rise, enhancing the original relief of the foundation. In the mature phases, reefs reach modern sea level and develop reef flats over the highs in the antecedent surface and especially around the windward reef margins where reef growth appears to be most rapid. One or more lagoons typify this phase but in the later stages of maturity, lateral transport of sediment from the productive margins leads to initial infilling of lagoons, masking of the inherited relief forms, and widening of reef flats. The senile phase is reached when lagoons are infilled completely, and lateral movement of sediment from the windward margins leads to progradation to leeward.

The basic classification is applicable to medium-size reefs growing from antecedent platforms with single depressions. Smaller reefs, where antecedent surfaces may lack a proto-lagoonal relief may develop along similar lines but because the reef flat initially developed around the reefal margins may occupy such a high proportion of the total reef area, any lagoon which forms will be shallow and quickly filled and the mature phases will be either bypassed or be present for only a very short period. In contrast, the largest reefs will be at the mature stages for the longest period because of the low ratio of productive margin to lagoon area. These reefs may also differ from the basic classification during the mature phase by having more than one lagoon.

A major subgroup of reefs are ribbon reefs which form distinctive barriers. They grow from linear foundations which are either discrete antecedent platforms or higher rims on much larger and usually deeper platforms. The basic feature is depth of water behind the narrow reef flat which develops from the foundations. Although sediment moves to leeward, accretion is very slow and widening of the reef flat is retarded. However, the stages in the development of a ribbon reef can be equated with those in the basic classification.

Figure C81 The windward margin of a mid-shelf reef, Great Barrier Reef showing outer living coral zone with buttresses, algal pavement with isolated reef blocks, aligned and nonaligned coral zones.

Figure C82 Reef front dominated by *Acropora* sp. close to wave base. Mainly corymbose are found on the higher crests of the spurs with arborescent and plate-like forms within the deeper channel.

Development of coral reef zonation

Distinctive morphological and ecological zonation of coral reefs is a characteristic which develops only after the reef has reached sea level and commenced horizontal growth of a reef flat (Hopley, 1982, chapter 10). During "catch-up" or even "keep-up" phases, the reef may be too far beneath effective wave base to allow the significant energy gradient from windward to leeward to exist. The reef may be a more homogenous environment of continuous coral cover with the only variable being the attenuation of light on the flanks and the development of more foliacious light capturing colonial forms. As the reef reaches sea level, however, the contrast between windward and leeward sides is enhanced by sediment movement, accumulation of detrital features, and the colonization of the various ecological niches by distinctive life forms. Zones develop parallel to the reef front, their width and distinctiveness determined by the length of time the reef has been at sea level and the strength of the energy gradient (Figure C81).

While the species which inhabit the various zones may differ geographically, the morphology of coral reef zones is similar worldwide and typically may consist of:

Fore-reef slope: Increasing light and wave energy are the controlling parameters on the deeper fore reef. At greater depth are flat, platey, and encrusting light capturing corals. Higher up the slope where light is not limiting branching forms may dominate. Moving into the higher energy wave zone at about 10 or 15 m, more compact or corymbose forms become dominant and coral cover is dense (Figure C82).

Reef crest: Particularly on high-energy reefs the crest develops massive spur and groove morphology. Originally thought to be inherited from karst rillen type features developed at low sea levels, most drilling investigations have indicated a growth origin for spur and grooves. They have developed as a baffle against the wave energy and are self perpetuating as the grooves act as surge channels along which sediment is swept.

Outer reef flat: The coral covered tops of the spurs are continued onto the outermost reef flat as a living coral zone of encrusting and resistant growth forms up to about mean low water mark.

Algal crest: Particularly on very high energy mid-oceanic atoll reefs a rim built by crustose coralline algae may rise to almost mean sea level. On lower energy reefs the feature may be far less prominent consisting of a rubble pavement cemented by the coralline algae. Under even lower energy conditions turf-like algae may dominate in this zone.

Detrital zone: Beyond the algal pavement, and often deposited over it may be a morphologically diverse zone characterized by deposition of coarse material derived from the reef front. This zone can consist of imbricated clasts of branching corals forming low irregular banks parallel to the reef front. Where tropical storms are relatively frequent a more prominent ridge or rampart may develop and, if stable for sufficient time, its lower sections will become cemented. Also in this zone, and on the reef flat to seawards may be much larger boulders and reef blocks the largest of which may be more than 10 m in diameter. These are sections of the reef front broken off in major storms or by tsunami waves.

The inner reef flat: Behind the energy-absorbing outer zones, the inner reef flat may be another zone of dense coral growth. Where wave set up provides a steady flow of water over the reef crest this zone may display an alignment of corals normal to the reef front. Elsewhere, at low tide pools may be moated by the rubble ridges or the algal crest and massive corals will be in the form of micro atolls.

Lagoons and back reef zones: In the deeper water of lagoons or back-reef, finer sand size sediment, which is swept across the reef flat, forms prograding sand sheets which may slowly bury lagoonal or leeward side patch reefs. This sand sheet may be up to 500 m wide and 10 m in thickness and is where much of the carbonate productivity of the outer reef

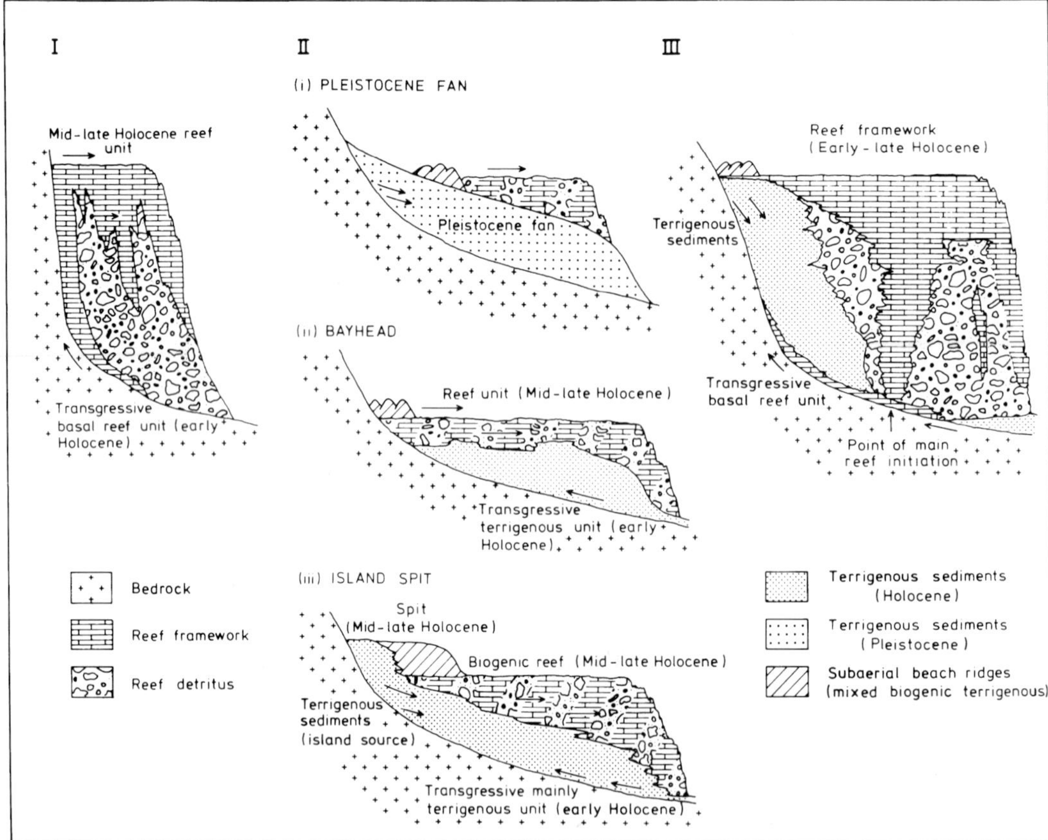

Figure C83 A structural classification of fringing reefs.

is finally deposited. Beyond this patch reefs of various forms may rise from the sandy lagoonal or back reef floor. Each patch reef may have its own "reef flat" surrounded by an algal rim. In shallower lagoons reticulate reefs forming a network of anastomosing ridges may form an intricate pattern of reef growth.

Leeward reef slope: The high energy wave zone is lacking from the leeward side of the reef and coral growth even near the surface may be in the form of branching thickets. If the leeward slope descends into deep enough water, at greater depth more light seeking platey forms may again dominate.

Fringing reefs

Fringing reefs, attached to the mainland or to a high island, may have the weakest zonational development, especially where sheltered by offshore reefs or other islands. They are the commonest reef type with an area exceeding 320,000 km^2 or more than 50% of total reef area. Many grow in turbid waters but can still flourish as long as sediment does not settle and smother the corals. However, because of the greater attenuation of light the vertical zonation of the reef front may be compressed into less than 10 m water depth.

Fringing reefs were formerly considered to be the simplest of reef forms, growing outwards from the mainland or island shoreline as originally advocated by Charles Darwin over 150 yr ago. However, over the last 20 yr a considerable amount of drilling and dating of fringing reefs has been undertaken and they have been shown to have a complexity which matches that of other reef types.

Because they have a foundation on which to grow immediately sea level stabilizes (the island or mainland shore) many fringing reefs have reef flats older than those of nearby offshore reefs. On the Great Barrier Reef micro atolls older than 6,000 yr BP are found on many fringing reef flats. However, reef establishment, growth and internal structure can show considerable variation. The structural classification (Figure C83) is based on the results from Australian fringing reefs, but records from fringing reefs elsewhere in the world suggests that it has a wider application. Three basic types are recognized:

1. Simple reefs formed from the foundation on the lowest portion of the rocky foreshore during the transgression. These reefs were developed while the sea level was still rising. Most of their development has gone into upward growth over the rocky slopes of the island, following the transgression. After the still-stand period and when the reef had reached sea level a small amount of outward growth may have been possible over the reef's own fore-reef talus slope. On the whole, such reefs are growing from relatively deep clear water and reef flat development is therefore limited due to the great vertical extent of these reefs. The structure is thus one of a basal framework unit immediately over rock and then a small biogenic detrital frontal unit with thin reef flat veneer.

2. Reefs developed over preexisting positive sedimentary structures. Reefs have long been regarded as requiring hard substrate for initiation. However, there is increasing evidence that the presence of even a muddy sedimentary structure with positive relief may greatly enhance or speed-up reef flat development. Such sedimentary structures may be in the form of terrigenous mud–sand banks or barriers, lee side sand spits attached to islands, boulder beaches, deltaic bar gravels, and low angle Pleistocene alluvial fans. During the transgressive phase, even though a bank may have existed previously, no reef development is possible because of the inhospitable nature of the substrate. However, once the rocky shores of the adjacent island or mainland are inundated reef colonization takes place rapidly on these shores at shallow depth. Progradation of the reef is then rapid over the preexisting structure with hard substrate now being provided over the sedimentary base by the fore-reef talus from the prograding reef front. The structure of such reefs is thus one of a basal terrigenous sedimentary unit, an inner framework with a prograding carbonate detrital unit extending over the terrigenous base, and an upper, thin, generally less than 4 m, reef flat framework veneer.

3. Reefs developed over more gently sloping substrate, particularly where older foundations of Pleistocene reefs may be present. In these instances, the reef foundation is initiated offshore from the present coastline, although would probably have started as a fringing reef as sea level would have been lower at this time. The rising sea level however, isolated this initial reef which continued to grow upwards during the transgressive phase as an offshore barrier. Possibly, because of poor

circulation and terrigenous input, growth behind this barrier was very slow. After the still-stand and after the reef reached sea level, the outer reef became attached to the island by lagoonal infilling, this infill coming from both the land as a terrigenous unit and from the outer reef as a biogenic carbonate unit. Following the still-stand, some small outward growth may also have taken place. The structure is therefore one of a framework unit offshore from the present shore and an interdigitated terrigenous and biogenic detrital fill behind the framework. A thin, reef flat framework veneer may be present over the entire reef and a small reef front biogenic talus may also be present.

Davies and Hopley (1983) indicated that fringing reefs grew at rates comparable to middle and outer shelf reefs, though usually at the lower end of the growth scale, that is, for framework construction in the range of 1–4 mm/yr. This may be in part a reflection of the greater proportion of massive as opposed to branching framework. As with outer reefs, the fringing reefs showed a bimodal detrital accretion rate, a lower range of 1–5 mm/yr representing accumulation under normal weather conditions, but with higher rates up to 15 mm/yr indicative of infrequent high energy cyclonic events. Fringing reefs generally show very low growth rates in shallow water, particularly when compared to the mid-shelf reefs which have both a protected situation and clear water. The lower rate for fringing reefs is probably a reflection of the turbid water conditions and periodic decline in salinities. However, at depths between 4 and 7 m, fringing reef growth seems to be at least equal to the rate of growth of both mid- and outer-shelf reefs. Below this depth there appears to be a rapid decline in accretion rate. This is interpreted as the result of the high turbidity of inshore waters and rapid decline in light levels at these depths. Equivalent decline in growth rates for outer reefs takes place below approximately 15 m.

The future of the world's coral reefs

While the store of knowledge about the functioning and evolution of coral reefs has expanded dramatically in the last 50 yr, great concerns are currently being expressed about the health and status of the worlds reefs. It is estimated that 10% of coral reefs of the world have already been degraded beyond recognition; 30% are in a critical state such that they will be lost in the next 10–20 yr; a further 30% are so threatened that they may disappear as living ecosystems in 20–40 yr; and only 30% are healthy and stable (Wilkinson, 1993).

The major reasons for this decline are from the anthropogenic stresses of organic and inorganic pollution, sedimentation, and over-exploitation all of which are increasing with the exponential growth of human populations. Unfortunately, the majority of the most rapidly growing populations are in the tropics adjacent to coral reefs with subsistence economies at least partly dependent upon the reefs. However, superimposed upon these direct human pressures are the effects of global climate change with coral reefs being regarded as one of the most sensitive ecosystems to the predicted changes (Buddemeier, 1993).

Many believe that coral reefs are already showing detrimental effects of climate change with a significant increase in coral bleaching as a response to unusually high temperatures detected globally over the last 25 yr (Brown, 1997; Hoegh-Guldberg, 1999). Bleaching involves the expulsion of the symbiotic zooxanthellae (which gives the coral its color) from the coral tissue. Bleaching may be temporary and may be an adaptation mechanism that allows more resistant strains of zooxanthellae to recolonize affected hosts. However, as the zooxanthellae are so important to the metabolic functioning of the coral, severe bleaching leads to mortality of whole colonies. Bleaching events appear to coincide with El Niño episodes with significant occurrences in 1979–80, 1982–3, 1987–8, 1991–5, and 1998. As El Niño events are predicted to become more regular and severe in a "Greenhouse" world, the associated short-term higher temperatures will be superimposed on a more general global warming, leading to increased stresses for coral reefs.

There are other harmful effects, including increased sedimentation from more runoff as tropical areas receive more rainfall, possible greater damage from increasing frequency and severity of tropical storms, and inhibition of calcification with CO_2 enrichment of ocean water. Planktonic and larval stages of many reef organisms, including corals, may be particularly susceptible to increasing ultra violet B (UVB) and shorter ultra violet A (UVA) radiation associated with decreasing levels of ozone in the upper atmosphere.

Rising sea levels can produce some beneficial effects. Inundation of reef flats which have been at sea level for 5,000 yr or more may regenerate coral growth with reef flats contributing more significantly to carbonate production. Wave efficiency across reef flats may increase and sediments stored there may add to reef island construction. Some reef islands, contrary to popular opinion, can actually grow rather than erode as the result of rising sea level, but with reorientation of many islands also probable, the amount of inhabitable area with older cultivatable soils may decrease. Coral reef islands may not disappear, but their attractiveness for human habitation may decrease. Similarly, the structures which are coral reefs will not disappear as the result of anthropogenic stresses, but their dominance by animals (corals) may be overtaken by algae and other plants for which the Greenhouse scenario is more beneficial.

David Hopley

Bibliography

Brown, B.E., 1997, Coral bleaching: causes and consequences. *Coral Reefs*, **16**: 129–138.

Buddemeier, R.W., 1993. Corals, climate and conservation. Proceedings of the *7th International Coral Reef Symposium*, **1**: 3–10.

Connell, J.H., 1997. Disturbance and recovery of coral assemblages. *Coral Reefs*, **16**: 101–113.

Davies, P.J., and Hopley, D., 1983. Growth fabrics and growth rates of Holocene reefs in the Great Barrier Reef. *B.M.R. Journal Australian Geology and Geophysics*, **8**: 237–251.

Dubinsky, Z. (ed.), 1990. *Coral Reefs, Volume 25. Ecosystems of the World*. Amsterdam: Elsevier.

Guilcher, A., 1988. *Coral Reef Geomorphology*. New York John Wiley.

Hoegh-Guldberg, O., 1999, Climate change, coral bleaching and the future of the world's coral reefs. *Journal of Marine and Freshwater Research*, **50**: 839–966.

Hopley, D., 1982. *Geomorphology of the Great Barrier Reef: Quaternary Development of Coral reefs*. New York: John Wiley-Interscience.

Hopley, D., 1995. Continental shelf reef systems. In Carter R.W.G., and Woodroffe, C.D. (eds.), *Coastal Evolution: Late Quaternary Shoreline Morphodynamics*, Cambridge: Cambridge University Press, pp. 303–340.

Kinsey, D.W., 1985. Metabolism, calcification and carbon production I. Systems level studies. *Proceedings of the 5th International Coral Reef Congress*, **4**: 505–526.

Macintyre, I.G., 1988. Modern coral reefs of the western Atlantic: new geological perspectives. *American Association of Petroleum Geologists Bulletin*, **72**: 1360–1369.

Neumann, C., and Macintyre I.G., 1985. Reef response to sea level rise: keep-up, catch-up or give-up. *Proceedings of the 5th International Coral Reef Congress*, **3**: 105–110.

Purdy, E.G., 1974. Reef configurations, cause and effect. In Laporte, L.F. (ed.), *Reefs in Time and Space*, Tulsa, SEPM Special Publication, **18**: pp. 9–76.

Veron, J.E.N., 1995. *Corals in Space and Time: The Biogeography and Evolution of the Scleractinia*. Sydney: University of New South Wales Press.

Wilkinson, C.R., 1993. Coral reefs of the world are facing widespread devastation: can we prevent this through sustainable management practices? *Proceedings of the 7th International Coral Reef Symposium*, **1**: 11–21.

Cross-references

Atolls
Cays
Coral Reef Coasts
Coral Reef Islands
Coral Reefs, Emerged
Karst Coasts

CORAL REEFS, EMERGED

Corals and other reef-building organisms such as algae live within the photic zone, rarely deeper than 100 m and are limited upwards by intertidal exposure. However, well-preserved reefs emerged up to hundreds of meters above modern sea level are found in many parts of the world. All types of reefs are found emerged although the most common are fringing reefs occurring as terraces built over volcanic or other non-carbonate foundations (Figure C84). Emerged barrier reefs appear to be rare, even in Indonesia which has both extensive, modern barrier reefs and numerous emerged fringing reefs (Tomascik *et al.*, 1997). However, a section of the very long (2,000 km) Papua New Guinea barrier reef is emerged in Sabari Island (Figure C85) to the north of the Louisiade Islands, and in the Trobriand and Lusancay Islands to the north. Many emerged atolls occur in the Pacific (e.g., Henderson, Makatea, Mataiva, Nauru, Walpole) and in Indonesia (Maratua, Kakaban). Detailed descriptions of the geology and hydrogeology of many emerged reef examples can be found in Vacher and Quinn (1997).

Causes of emergence

It is generally accepted that world sea levels were last ubiquitously above present during the last interglacial about 125,000 years ago and reefs of this age up to 10 m above present sea level have been reported from many quasi stable areas of the tropical world (e.g., Western Australia, parts of Tonga, Aldabra, Curacao). However, most of the world's reefs were exposed during the periods of glacially lowered sea levels (up to 150 m below present). Landforms produced during exposure have been a major influence on modern reefs (see *Coral Reefs* entry). Emergence during the Holocene is more limited and for isostatic reasons is largely restricted to the Southern Hemisphere (Hopley, 1982, 1987). For example, in the Great Barrier Reef region of north-east Australia emerged reefs of about 1 m and with an age of about 5,000 years are found only on the inner shelf islands and mainland and consist of coalescing micro atolls rather than fully developed fringing reefs as the period at the high level was short and followed a rapid post-glacial rise (Chappell *et al.*, 1983).

Figure C84 Tectonically uplifted emerged reef terraces rising to 270 m on Misima Island, Louisiade Archipelago, Papua New Guinea.

Figure C85 Notch and visor cut into the emerged barrier reef of Sabari Island, Louisiade Archipelago, Papua New Guinea. The reef is probably last interglacial in age.

Figure C86 The back of the 83,000-year reef terrace with sea stack displaying fossil notch and visor cut into older Pleistocene reef limestone, north-east coast of Barbados.

The most spectacular suites of emerged reefs occur in areas of strong uplift at the margins of the world's tectonic plates (e.g., Ryukyu Islands, Barbados (Figure C86), Papua New Guinea (Figure C84), Indonesia). In mid-ocean areas uplift may also occur around hot spot swells or over asthenospheric bumps and is especially responsible for emerged atolls such as Ocean, Nauru, and Marcus Islands (Scott and Rotondo, 1983). Long-term average rates of uplift may be as high as 5 m/ka with the highest reefs being of Tertiary age. For example, on Gunung Dirun Timur, Indonesia, terraces at 1,293 m are of Pliocene age (Tomascik *et al.*, 1997). The majority of the best preserved reefs, however, are Pleistocene and because of the rate of uplift represent not only interglacial high stands but also reef-building periods at stable sea levels which were well below present. Emergence in these unstable areas continues with individual earthquakes having the potential to uplift living reef flats >2 m above sea level (e.g., Cortes, 1993).

Modifications after emergence

After emergence, coral reefs undergo changes to both the geochemistry of the reef fabric and to the morphology of the surface features. The original limestone deposited beneath the sea by coral and other plants and animals is largely aragonite or high magnesium calcite. Once uplifted, however, the limestone is unstable in the freshwater vadose and phreatic environments and diagenesis through solution and reprecipitation as stable phase low magnesium calcite takes place (Matthews, 1974). The open framework of the original reef becomes denser, infilled, and with greater cementation (Figure C78, *Coral Reefs* entry). A coarse calcite crystalline structure is produced and in the uppermost layers a caliche horizon may be formed consisting of a distinctive brown crust with numerous micritic stringers (e.g., Hopley, 1982), (Figure C79, *Coral Reefs* entry). These processes also operated on reefs exposed during major sea level falls of the Pleistocene. Subsequent recolonization during post-glacial transgressions results in distinctive unconformities (see entry *Coral Reefs*), several of which may be found beneath modern reefs.

Erosion of emerged reefs may be relatively slow as the permeable limestone does not support surface drainage. Karst processes dominate but whether or not they have been rapid enough to produce large-scale relief features during the limited exposure of reefs emerged only by sea level fall as suggested by Purdy (1974) is debatable (Hopley, 1982, chapter 7). Bloom (1974) has suggested a facies control over rates of solution which would emphasize the constructional relief differences between reef rim and lagoon floor. Even in ancient reefs, structure may remain evident through facies guidance of solution. For example, the Devonian reefs of Western Australia were buried beneath Permian sandstones then truncated by a Mesozoic–Tertiary planation surface, yet uplift and rejuvenation has resulted in emerged reefs (the Limestone Ranges of the Canning Basin) which faithfully mimic the original morphology. Reefs which are permanently emerged by tectonic uplift do have sufficient time to produce a range of karst landforms including cave systems, solution holes, and gorges. Their surfaces are an intricate network of holes and pinnacles either completely exposed or buried under soil and in some instances, commercially valuable phosphate deposits. Numerous local terms have been applied to this rocky relief (*makatea, feo, ironshore, champignon, platin, pavé*).

Emerged reef erosion may also take place horizontally with processes of bioerosion aided by some solution and physical abrasion creating an intertidal notch with overhanging visor (Figure C85) at rates of up to 7 m/ka (Hopley, 1982). Undercuts of several meters may occur and remain etched into the sides of emerged reefs as they are uplifted, providing erosional evidence of former relative sea levels, in addition to the reefs themselves (Figure C86).

The value of emerged reefs for scientific research

The accessibility of an ecological community and landform which is normally underwater, the relatively tight constraint in the relationship of a raised reef surface to a former sea level, and the accuracy of radiometric dating of constituent corals make emerged reefs highly valuable to several branches of scientific research. For example, they have been used in the development, calibration, and refining of dating techniques (e.g., Grün *et al.*, 1992) but they have made particularly prominent contributions to the understanding of past eustatic sea levels, including those of the last glacial period. In some areas, at least, average uplift rates have remained constant over the latter part of the Pleistocene and are calculated from the present level of reefs for which the original eustatic level is well constrained (e.g., the mid-Holocene or the last interglacial). The uplift rate is then applied to other terrace levels which are dated by radiocarbon or uranium series methods to obtain a past sea level. These techniques were developed originally in Barbados (Broeker *et al.*, 1968; Mesollela *et al.*, 1969) and the Huon Peninsula in Papua New Guinea (Bloom *et al.*, 1974; Chappell, 1974). Results are continually being refined (e.g., Ota *et al.*, 1993) and the degree of agreement between results from different areas justifies the basic assumptions.

Evidence from emerged reefs has also been used in geophysical modeling to develop theories of seafloor spreading, and processes associated with hot spots or melting anomalies, including lithosphere loading, cooling subsidence, and asthenospheric bumps (McNutt and Menard, 1978; Scott and Rotondo, 1983). They have also aided in the understanding of upper earth rheology, flexure of the ocean lithosphere, and global isostatic responses.

Accessibility allows complete species inventories to be made and corals in particular may remain well preserved in even the oldest elevated reefs. From the data collected, the evolution of genera and species can be followed (Veron, 1995) or the changes in ecological communities deduced from timescales ranging through millions of years to the Holocene (Veron and Kelley, 1988; Budd, 2000).

David Hopley

Bibliography

Bloom, A.L., 1974. Geomorphology of reef complexes. In Laporte, L.F. (ed.), *Reefs in Time and Space: Selected Examples from the Recent and Ancient*. Society of Economic Paleontologists and Mineralogists, Special Publication 18, pp. 1–8.

Bloom, A.L., Broeker, W.S., Chappell, J., Matthews, R.K., and Mesollela, K.J., 1974. Quaternary sea level fluctuations on a tectonic coast: new 230Th/234U dates from the Huon Peninsula, New Guinea. *Quaternary Research*, **4**: 185–205.

Broeker, W.S., Thurber, D.L., Goddard, J., Ku, T.-L., Matthews, R.K., and Mesolella, K.J., 1968. Milankovitch hypothesis supported by precise dating of coral reefs and deep-sea sediments. *Science*, **159**: 297–30.

Budd, A.F., 2000. Diversity and extinction in the Cenozoic history of Caribbean reefs. *Coral Reefs*, **19**: 25–35.

Chappell, J., 1974. Geology of coral terraces, Huon Peninsula, New Guinea: a study of Quaternary tectonic movements and sea level changes. *Geological Society of America, Bulletin*, **85**: 553–570.

Chappell, J., Chivas, A., Wallensky, E., Polach, H.A., and Aharon, P., 1983. Holocene palaeoenvironmental changes, central to north Great Barrier Reef, inner zone. *Bureau of Mineral Resources Journal of Australian and Geological Geophysics*, **8**: 223–236.

Cortes, J., 1993. Recently uplifted Caribbean reef. *Coral Reefs*, **12**: 76.

Grün, R., Radkte, U., and Omura, A., 1992. ESR and U-series analyses on corals from Huon Peninsula, New Guinea. *Quaternary Science Reviews*, **11**: 197–202.

Hopley, D., 1982 *Geomorphology of the Great Barrier Reef: Quaternary Development of Coral Reefs*. New York: John Wiley - Interscience.

Hopley, D., 1987. Holocene sea level changes in Australia and the southern Pacific. In Devoy, R.J.N. (ed.), *Sea Surface Studies*, London: Croom Helm.

Matthews, R.K., 1974. A process approach to diagenesis in reefs and reef associated limestones. In Laporte, L.F. (ed.), *Reefs in Time and Space Selected Examples from the Recent and Ancient*. Society of Economic Paleontologists and Mineralogists, Special Publication 18, pp. 234–256.

McNutt, M., and Menard, H.W., 1978. Lithosphere flexure and uplifted atolls. *Journal of Geophysical Research*, **83**: 1206–1212.

Mesollela, K.J., Matthews, R.K., Broeker, W.S., and Thurber, D.L., 1969. The astronomical theory of climatic change: the Barbados data. *Journal of Geology*, **77**: 250–274.

Ota, Y., Chappell, J., Kelley, R., Yonekura, N., Matsumoto, E., and Nishimura, T. 1993. Holocene coral reef terraces and coseismic uplift of Huon Peninsula, New Guinea. *Quaternary Research*, **40**: 177–188.

Purdy, E.G., 1974. Reef configurations, cause and effect. In Laporte, L.F. (ed.), *Reefs in Time and Space: Selected Examples from the Recent and Ancient*. Society of Economic Paleontologists and Mineralogists, Special Publication 18, pp. 9–76.

Scott, A.J., and Rotondo, G.M., 1983. A model to explain the differences between Pacific plate island-atoll types. *Coral Reefs*, **1**: 139–150.

Tomascik, T., Mah, A.J., Nontji, A., and Moosa, M.K., 1997. *The Ecology of the Indonesian Seas*. Hong Kong: Periplus.

Vacher, H.L., and Quinn, T.M., 1997. *Geology and Hydrogeology of Carbonate Islands*. Amsterdam: Elsevier.

Veron, J.E.N., 1995. *Corals in Space and Time: The Biography and Evolution of the Scleractinia*. Sydney: UNSW Press.
Veron, J.E.N., and Kelley, R., 1988. Species stability in reef corals of Papua New Guinea and the Indo Pacific. *Memoir Association of Australasian Palaeontologists*, **6**: 69 pp.

Cross-references

Atolls
Coral Reefs
Notches
Seismic Displacement
Tectonics and Neotectonics
Uplift Coasts

CROSS-SHORE SEDIMENT TRANSPORT

Cross-shore transport

Cross-shore transport refers to the cumulative movement of beach and nearshore sand perpendicular to the shore by the combined action of tides, wind and waves, and the shore-perpendicular currents produced by them. These forces usually result in an almost continuous movement of sand either in suspension in the water column or in flows at the surface of the seafloor. This occurs in a complex, three-dimensional pattern, varying rapidly with time. At any moment, some sand in the area of interest will have an onshore component while other sand is moving generally offshore. The separation of the total transport into components parallel and perpendicular to the shore is artificial and is done as a convenience leading to a simpler understanding of a very complex environment (see entry on *Longshore Sediment Transport* for shore-parallel transport).

Cross-shore transport has been well studied because of its importance to beach erosion. Unlike longshore transport, which is difficult to observe, cross-shore transport can result in large and highly visible changes in the beach configuration over intervals as short as one day. Because of the complexities associated with cross-shore transport, its magnitude and direction is usually inferred from changes in the elevation of the beach and the adjacent ocean bottom. Elevation profiles across the beach and through the surf zone are obtained with conventional surveying techniques, augmented as necessary with water depth measurements from a boat or other floating platform. A reduction in the amount of sand above mean sea level and a corresponding increase in the volume below this reference, for example, would be interpreted as offshore transport. If the beach volume increases at the expense of submerged sand levels, this implies shoreward transport. If the measurements were error-free and the transport was entirely in the cross-shore mode, the amount of sand in the system should remain constant. Gradients in longshore transport, however, can result in local accumulations or losses that make the precise assessment of cross-shore transport difficult.

Small cross-shore adjustments occur during each tide cycle along coasts with significant tide ranges. Much larger cross-shore transport events are associated with major storms, which may last for several days. These changes can be expressed in substantial beach or dune erosion or in rapid formation or migration of bars or other submerged formations. Intervals of smaller waves, and particularly swell of low steepness, can

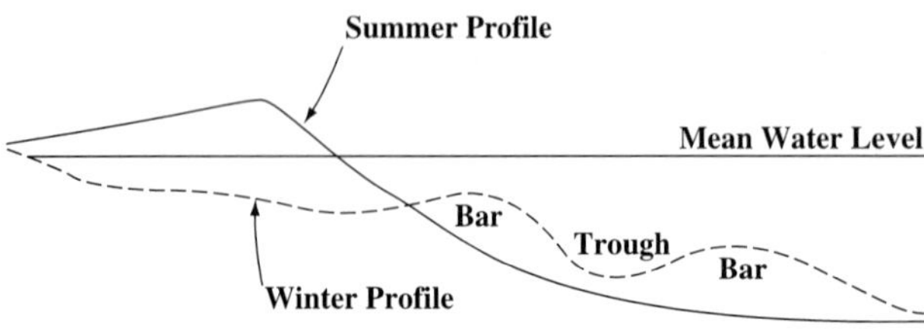

Figure C87 Cross-shore transport. The figure shows schematically how seasonal changes in wave climate affect beach profiles. During winter storms, the dry beach is eroded and seaward cross-shore sediment transport results in the formation of one or more offshore bars. Typically, the envelope within which the shoreline ranges retreats seaward in winter and shoreward under the milder wave climates of summer.

reverse the effects of storms and repair eroded beaches or move bars shoreward. These reversals have much longer timescales, perhaps as long as weeks or months. The accumulated effects of many storms during winter months and long intervals of low waves in summer, for example, can produce seasonal effects through cross-shore transport, as illustrated in Figure C87. Winter beaches may be characteristically lower and narrower with pronounced bars near the location of the largest breakers, while summer beaches are wider and with smaller, less-distinct bars closer to shore. Local wave climates, tide ranges, and sediment size influence these seasonal types, but the changes are a direct result of cross-shore transport.

A number of investigations have been made in an attempt to predict what wave conditions would result in a shift between offshore and onshore transport. Much of the early work was done in laboratory wave tanks, using simple waves (Johnson, 1949). Johnson associated the shift between the formation of summer and winter waves with a critical incident wave steepness in deep water. Other investigators found that the absolute size of the wave changed the critical steepness ratio. A second dimensionless parameter, the ratio of the wave height to the mean sediment diameter, was added by Iwagaki and Noda (1963) to account for scale. Seymour and Castel (1989) review the performance of six simple predictors for the direction of cross-shore transport, which involve some nondimensional combination from among wave height and period, sediment size and fall speed, and beach slope. The best of these was successful only about two-thirds of the time. Bailard and Inman (1981) introduced physics into the study of cross-shore transport with a predictive model that was based upon energetics, the balance between gravitation tending to move sediment downslope and the upslope stress produced by wave asymmetry. More recently, numerical models have been developed to predict storm wave erosion (Larson and Kraus, 1990).

One of the earliest concepts about cross-shore transport concerned the tendency for waves to selectively move coarser sediment sizes shoreward (Cornish, 1898). This results from the asymmetry of waves in shallow water (see entry on *Waves*). As waves approach the beach, the crests become steeper and the troughs longer. The shoreward flows under the crests are higher and for shorter duration than the offshore-directed flows under the troughs. Because larger sediment sizes require higher flow speeds to initiate movement, they tend to be preferentially moved by the shoreward flows. This was later refined to the null point hypothesis in which there is a stable offshore distance for a given sediment size, with a continuous gradation from coarse on the beach to fine in deep water (Ippen and Eagleson, 1955). A number of field investigations have shown that sediment sizes are sorted in the cross-shore direction in nature, although bars and other features tend to produce locally mixed sizes as exceptions to regular gradation.

The study of cross-shore transport has resulted in the concept of an equilibrium underwater profile (Dean, 1991). The shape of the offshore profile is approximated by

$$h(y) = Ay^{0.67},$$

where $h(y)$ is the depth at distance (y) and A is a scale factor related to the grain size distribution of the sand forming the beach. This is an engineering simplification that assumes, under the constant application of a given incident wave, a particular beach will evolve toward a certain profile shape. The incident wave can be further generalized to include all of the seasonal variation, resulting in a single equilibrium profile configuration for that beach. The actual profile may never achieve this contour because the incident waves are constantly changing. However, it is useful to coastal engineers and scientists concerned with understanding the natural variability of beaches. The simple exponential model for the equilibrium beach does not include any provision for bars or other contour complexities such as rock outcroppings.

Richard J. Seymour

Bibliography

Bailard, J.A., and Inman, D.L., 1981. An energetics bedload model for a plane sloping beach: local transport. *Journal of Geophysical Research*, **86**(C3): 2035–2043.
Cornish, V., 1898. On sea beaches and sand banks. *Geography Journal*, **11**: 528–647.
Dean, R.G., 1991. Equilibrium beach profiles: characteristics and applications. *Journal of Coastal Research*, **7**(1): 53–84.
Ippen, A.T., and Eagleson, P.S., 1955. A study of sediment sorting by waves shoaling on a plane beach. U.S. Army Corps of Engineers, *Beach Erosion Board Technical Memorandum*, **63**: p. 83.

Iwagaki, Y., and Noda, H., 1963. Laboratory study of scale effects in two-dimensional beach processes. *Proceedings of the 8th Conference on Coastal Engineering*, **8**: 194–210.

Johnson, J.W., 1949. Scale effects in hydraulic models involving wave motion. *Transactions of the American Geophysical Union*, **30**: 517–525.

Larson, M., and Kraus, N.C., 1990. SBEACH: numerical model for simulated storm-induced beach change. U.S. Army Corps of Engineers, *Waterways Experiment Station, Technical Report* CERC-89-9.

Seymour, R.J., and Castel, D., 1989. Modeling cross-shore transport. In Seymour, R.J. (ed.), *Nearshore Sediment Tranport*, New York: Plenum Press, pp. 387–401.

Cross-references

Beach Erosion
Coastal Processes (see Beach Processes)
Cross-Shore Variation in Sediment Size Distribution
Depth of Closure on Sandy Coasts
Dynamic Equilibrium of Beaches
Erosion Processes
Numerical Modeling
Profiling, Beach
Waves

CROSS-SHORE VARIATION OF GRAIN SIZE ON BEACHES

The distinctive sorting and cross-shore distribution of clastic beach particles is often evident even to the casual observer. The position at which a sediment particle comes to rest on a beach profile and the net patterns of particle size variation resulting from sediment transport processes have long been of interest to scientists and engineers attempting to understand the fundamental relationships governing process and form on beaches.

Beach particle equilibrium and the null-point hypothesis

The null-point concept was introduced to express stability of a beach profile and the associated distribution of sediment on the profile as a reflection of the equilibrium between asymmetric fluid forces (including wave asymmetry, steady currents, and the effects of percolation), bottom slope (gravity force), and the size and weight of sediment (Cornaglia, 1889). Details of the original hypothesis as expressed by Cornaglia did not predict particle and beach morphological behavior as observed, however. Bowen (1980) and Bailard (1981) improved the null-point hypothesis by applying Bagnold's (1966) sediment transport model. According to Bowen, sediment coarser than the equilibrium grain size on the local slope has a smaller value of β/W_s (ratio of local slope and particle settling speed, descriptive of suspended load) and $\beta/\tan\phi$ (ratio of local bottom slope and grain friction angle, descriptive of bedload). Therefore, assuming a shoreward asymmetry in the velocity field, the term involving gravity is reduced, and coarser particles move shoreward to a position having the limiting slope. Similarly, finer particles should move offshore, as observed (e.g., Zenkovitch, 1946).

Sediment transport and profile formation are more complex than are described by the theoretical models of Bowen, Bailard, and others. The concept of a sediment grain being in equilibrium with forcing requires the choice of appropriate temporal and spatial scales for analysis. Time scales as short as a single or even half wave cycle may be involved in the sorting and concentration of mineral grains according to size and density within beach laminae in the swash zone (Clifton, 1969). At somewhat longer timescales, antecedent conditions, such as recent storms or periods of swell, spring/neap variations in tide level, water table variations, or a locally significant change in sediment supply or exposure can alter the shape and location of beach profile features and influence the distribution of sediments comprising the beach. At still longer timescales the relative sea level, sediment sources, prevailing climatology, and offshore bottom slope comprise the framework on which beaches and their particle size distributions develop.

Processes of selective sediment transport

The equilibrium profile depends on the sorting (standard deviation) and shape of the sediment as well as its size. Poor sorting results in less percolation and flatter profiles compared with well-sorted sediments of the same size (Krumbein and Graybill, 1965; McLean and Kirk, 1969). The sorting processes themselves are dependent on the particle properties as well as the hydrodynamics. Sand-sized particles are sorted according to differences in densities, sizes, and shapes of the sediment particles. Four basic processes of dynamic equivalence or selective sorting act on beach sediments (Komar, 1989; Hughes *et al.*, 2000). Settling equivalence, hydraulic equivalence (Rubey, 1933), or suspension sorting is the principle that permits a larger grain of lower density to be deposited simultaneously with smaller grains of higher density. Entrainment equivalence or selective entrainment (Komar and Wang, 1984) refers to the difficulty of entraining smaller and heavier particles within a matrix of comparatively larger and lighter particles. Smaller particle size can be a direct factor in reducing particle mobility in terms of both relative exposure to flow and larger size of pivoting angle. Selective transport refers to the differential rate at which particles are carried by a flowing fluid by virtue of their position in the flow. Dispersive pressure equivalence or shear sorting refers to the process in which coarser grains tend to migrate upward to zones of lower shear while finer and denser grains move downwards to the zone of maximum shear (Bagnold, 1954). The operation of such selective processes in the swash zone may lead to heavy mineral enrichment particularly in erosional episodes.

Part of the cross-shore selective sorting of sediment can be attributed to the patterns of orbital motions under waves as they shoal. Orbital motions under a Stokes-type wave, for example, consist of a forward orbital motion of short duration but high velocity under the crest while the return flow under the trough is slower but of longer duration. Cornish (1898) hypothesized that Stokes asymmetry might explain the shoreward movement of coarse particles and the seaward movement of fine particles. The idea has been verified in experiments (Bagnold, 1940) and modeled with some success (Horn, 1992). A grain may move incrementally shoreward by falling back to the bed before a wave near-bottom orbital trajectory has completed its half-cycle. A grain that is thrown into suspension and falls to the bed in a longer time than one-half the wave period may be deposited seaward of its initial location (Dean, 1973; Nielsen, 1983). Ahrens and Hands (2000) compared sediment stability values with the maximum near-bottom velocity under the breaking wave trough, U_t, calculated with Stream Function Wave Theory to reliably discriminate onshore and offshore sediment transport. Large absolute values of U_t are associated with high suspended load and bedload transport offshore by undertow, and small absolute values of U_t with net onshore bedload transport. Although most of the attention has been focused on the dominant incident waves that arrive at the outer edge of the surf zone, the relative roles of waves of different lengths or frequencies in the selective sorting process in the nearshore zone has not been investigated in detail.

Recent field and laboratory experiments have begun to examine the vertical and temporal coupling between the sediment in transport and the near bed fluid motion to gain further insights regarding the fractionation process (see review by Greenwood and Xu, 2001). Osborne and Vincent (1996) demonstrate the importance of the phase coupling between the unsteady flow at different elevations and the vertically varying instantaneous suspended sediment concentrations in determining the direction and magnitude of sediment transport in combined waves and currents over rippled beds. Prototype scale wave flume measurements document distinct horizontal fractionation of sediment by size (Greenwood and Xu, 2001). Sediment became progressively coarser onshore and progressively finer offshore in the wave flume from the same homogeneous source suggesting that the sorting is explained by suspension transport alone and is controlled by: (1) an increase in the proportion of fines and a decrease in mean size with increasing elevation above the bed; and (2) vertically stratified sediment transport with net onshore transport near the bed and net offshore transport at higher elevations. Direct field measurements are needed from a range of nearshore environments and wave height and period combinations to confirm the relative importance of the mechanisms involved.

Cross-shore variation in grain size and sorting

Mean grain size may decrease both shoreward and seaward from the plunge point at the base of the swash zone at sandy shores. A secondary coarsening of sediments has been observed on the offshore bar, where wave breaking is initiated (e.g., Fox *et al.*, 1966; Greenwood and Davidson-Arnott, 1972). Gravel and coarse sand lags as well as heavy mineral placers can also develop in the troughs or embayments between nearshore bars due to the interaction of waves with strong longshore currents and rips (Komar, 1989). Mean grain size decreases up the foreshore, as a reflection of the diminishing swash intensity. Sediment sorting is poorest (standard deviation of sizes is largest) where energy dissipation is greatest. Skewness is most positive (or least negative) also at the plunge point, indicating a mixing of an over-abundance of fines with coarser sediments, relative to a lognormal distribution of sizes. Bascom (1951) observed a similar pattern, with the additional finding of the second coarsest beach material on the summer berm, and hypothesized that deposition occurred by wave uprush that had overtopped the berm crest and subsequent winnowing of finer fractions by wind.

Beach sand sampled at transects across the active profile at Terschelling (North Sea barrier island of The Netherlands) showed consistent cross-shore pattern of grain size distribution over a period of 19 months (Guillén and Hoekstra, 1996). The proportion of sample in each grain size fraction (50 micron increments) showed a distinctive cross-shore pattern, unique for each size fraction and termed the equilibrium distribution curve. The distribution of the fraction across the profile, relative to the total presence of the fraction, expresses the probability concept in cross-shore sorting.

Application

Engineers and planners need to reliably project structural and beach fill performance, in terms of offshore and longshore losses of beach material, or the change in beach profile or character of the beach face, after the material is sorted by natural processes acting over a period from years to decades. Studies of grain sorting also have relevance to understanding the formation of placers, which may have economic or intrinsic value. Understanding the processes that cause size variation of sediments across a beach profile, and the impact of the material composition on shore morphology is key to wise management of coastal resources.

Philip D. Osborne and David P. Simpson

Bibliography

Ahrens, J.P., and Hands, E.B., 2000. A simple perspective on cross-shore sediment movement. *Shore and Beach*, **68**(4): 3–14.

Bagnold, R.A., 1940. Beach formation by waves: some model experiments in a wave tank. *Journal of Institution of Civil Engineering*, **15**: 27–52.

Bagnold, R.A., 1954. Experiments on a gravity-free dispersion of large solid spheres in a Newtonian fluid under shear. Proceedings of the Royal Society, *London, Series A*, **225**: 49.

Bagnold, R.A., 1966. An approach to the sediment transport problem from general physics. *U.S. Geological Survey Professional Paper* 422-I.

Bailard, J.A., 1981. An energetics total load sediment transport model for a plane sloping beach. *Journal of Geophysical Research*, **86** (C11): 10,398–10,954.

Bascom, W.N., 1951. The relationship between sand-size and beach-face slope. *Transactions of the American Geophysical Union*, **32**: 866–874.

Bowen, A.J., 1980. Simple models of nearshore sedimentation; beach profiles and longshore bars. In McCann, S.B. (ed.), *The Coastline of Canada*. Geological Survey of Canada, pp. 1–11.

Clifton, H.E., 1969. Beach lamination: nature and origin. *Marine Geology*, **7**: 553.

Cornaglia, P., 1889. On beaches, Accademia Nazionale dei Lincei, Atti. Classe di Scienze Fisiche, Matematiche e Naturali, Mem. 5, ser. 4:284–304. Reproduced in translation. In Fisher, J.S., and Dolan, R. (eds.), *Beach Processes and Coastal Hydrodynamics*, Volume 39. Stroudsberg, PA: Benchmark Papers in Geology, pp. 11–26.

Cornish, V., 1898. On sea beaches and sand banks. *Geography Journal* **11**: 528–559 (see also pages 628–647).

Dean, R.G., 1973. Heuristic models of sand transport in the surf zone. *Proceedings of the 1st Australian Conference on Coastal Engineering*, Conference on Engineering Dynamics of the Coastal Zone, Sydney, Australia, pp. 209–214.

Fox, W.T., Ladd, J.W., and Martin, M.K., 1966. A profile of the four moment measures perpendicular to a shore line, South Haven, Michigan. *Journal of Sedimentary Petrology*, **36**: 1126–1130.

Greenwood, B., and Davidson-Arnott, R.G., 1972. Textural variation in the sub environments of the shallow water wave zone Kouchibouguac Bay, New Brunswick, *Canadian Journal of Earth Science*, **9**(9): 679–688.

Greenwood, B., and Xu, Z., 2001. Size fractionation by suspension transport: a large scale flume experiment with shoaling waves. *Marine Geology*, **176**: 157–174.

Guillén, J., and Hoekstra, P., 1996. The "equilibrium" distribution of grain size fractions and its implications for cross-shore sediment transport: a conceptual model. *Marine Geology*, **135**: 15–33.

Horn, D.P., 1992. A numerical model for shore-normal sediment size variation on a macro-tidal beach. *Earth Surface Processes and Landforms*, **17**: 755–773.

Hughes, M.G., Keene, J.B., and Joseph, R.G., 2000. Hydraulic sorting of heavy-mineral grains by swash on a medium-sand beach. *Journal of Sedimentary Research*, **70**(5): 994–1004.

Komar, P.D., 1989. Physical processes of waves and currents and the formation of marine placers. *CRC Critical Reviews in Aquatic Sciences*. **1**(3): 393–423.

Komar, P.D., and Wang, C., 1984. Processes of selective grain transport and the formation of placers on beaches. *Journal of Geology*, **92**: 637.

Krumbein, W.C., and Graybill, F.A., 1965. *An Introduction to Statistical Models in Geology*. New York: McGraw-Hill.

McLean, R.F., and Kirk, R.M., 1969. Relationship between grain size, sorting, foreshore slope on mixed sand shingle beaches. *New Zealand Journal of Geology and Geophysics*, **12**: 138–155.

Nielsen, P., 1983. Entrainment and distribution of different sand sizes under waves. *Journal of Sedimentary Petrology*, **53**: 423–428.

Osborne, P.D., and Vincent, C.E., 1996. Vertical and horizontal structure in suspended sand concentrations and wave-induced fluxes over bedforms. *Marine Geology*, **131**: 195–208.

Rubey, W.W., 1933. The size distribution of heavy minerals within a waterlaid sandstone. *Journal of Sedimentary Petrology*, **3**: 3–29.

Zenkovitch, V.P., 1946. *On the Study of Shore Dynamics*. Volume 1. Akademia nauk SSSR, Institut Okeandogi, Trudy, pp. 99–112.

Cross-references

Beach Processes
Cross-Shore Sediment Transport
Dynamic Equilibrium of Beaches
Sandy Coasts
Sediment Suspension by Waves
Surf Zone Processes

CUSPATE FORELANDS

Cuspate Forelands are accretionary features, which occur on many coastlines of the world; Cape Kennedy (Cape Canaveral) on the east coast of the United States and Dungeness on the south coast of Britain being but two of the more well-known examples (King, 1972). While they vary considerably in shape and size they are generally triangular in outline with the base attached to the coastline and the apex facing the open sea or ocean. Although similar in overall shape, cuspate forelands must not be confused with cuspate spits; the former are cuspate forms which have grown by progradation, usually with nested shore-parallel beach ridges. Cuspate forelands are related to beaches, tombolos, spits, and barriers but whereas the related accretionary features tend to smooth out the coastal outline cuspate forelands accentuate coastal irregularities.

Formation

A number of theories have been forwarded to account for their origin. They may develop in the lee of an offshore island that shelters a segment of the coast in the shadow of the predominant oncoming wave front (Bird, 1984). Waves are refracted to either side of the offshore island and concentrated wave energy affects the mainland shore to either side. Immediately behind the protective island diminished wave energy allows longshore moving sediment to accumulate. If the offshore island is sufficiently near the mainland the cuspate foreland may extend to the island itself in which case a tombolo develops. This has been observed in South Australia at Gabo Island. Here a cuspate foreland developed, it then joined the offshore island to form a tombolo and subsequent storm breaching made it a cuspate foreland once more. Formation in the lee of a protective offshore island however is only one way in which they can form.

More difficult to explain are cuspate forelands which develop on an open coastline with no protective offshore islands (Steers, 1964). The east coast of England affords many examples of this type. Lowestoft Ness (the most easterly point of the British mainland) is one such feature. Whereas the cuspate forelands formed in the lee of protective islands remain relatively stationary, forelands with no such protection can, and very often do, migrate along the coast (Russell, 1958). Surveys in the British Isles have tracked such features migrate along a coastline many kilometers. Banacre Ness in Suffolk has moved northwards by some 2 km over the last 100 years in spite of the dominant longshore drift being to the south (Robinson, 1965). In some cases, the longshore migration has oscillated with periods where a foreland has migrated in one direction followed by periods where reverse migrations have occurred (Hardy, 1966). Occasionally two adjacent cuspate forelands on an open stretch of coast have migrated in opposite directions. Such variations and contrary movement defy simplistic explanations relating foreland development merely to longshore drift.

While beach drifting and dominant waves are important, offshore current residuals and changing seafloor topography are significant contributory factors. Offshore from the British east coast examples can be found of bank and channel complexes developed by the ebb and flood of the tide. Ebb- and flood-dominated channels tend to be separated by sand banks and adjacent to the apex of each foreland can be found an offshore sandbank trending toward the coast (Robinson, 1964). Over time any change in the position of the offshore bank has resulted in a change in the position of the foreland, albeit after a short time lag. The supply of sediment arriving on the beach from the offshore bank is greater than the removal of the sediment along the beach and hence the local build-up of sediment at the landward termination of the adjacent sand bank. The movement of sand onshore from the adjacent sandbank together with wave refraction caused by the offshore sand bank itself accounts for the creation of the foreland and can account for its movement along the coast, which may be in a contrary direction to the main longshore drift.

Surface gravel, which is often present on the British cuspate forelands, is usually found in ridges—it is the tracing of these shingle ridges, which gives a clue to both their formation and migration, (Lewis and Balchin, 1940). The shingle moves along the surface of the beach and accumulates in the area of lower wave energy. While the offshore sandbanks offer less protection than adjacent offshore islands the lesser influence on wave refraction is made up for by their contribution of sediment to the adjacent coastline.

Stephen J. Craig-Smith

Bibliography

Bird, E.C., 1984. *Coasts An Introduction To Coastal Geomorphology*, Canberra: ANU Press.

Hardy, J.R., 1966. An edd-flood channel system and coastal changes near winterton, norfolk. *East Midland Geographer* **4**(1): 24–30.

King, C.M., 1972. *Beaches and Coasts*, London: Edward Arnold.

Lewis, W.V., and Balchin, W.G., 1940. *Post Sea Level at Dungeness. Geography Journal*, **96**: 258–310.

Robinson, A.H.W., 1964. The inshore waters, sediment supply and coastal changes of Port of Lincolnshire. *East Midland Geographer*, **3**(6): 307–321.

Robinson, A.H.W., 1965. *Residual currents in relation to shoreline evolution to the East Anglican Coast. Marine Geology*, **4**: 57–84.

Russell, R.C., 1958. Long straight beaches, *Eclogae Geologicae Helvetiae*, **51**: 591–598.

Steers, J.A., 1964. *The Coastline of England and Wales*, Cambridge: Cambridge University Press.

Cross-references

Barrier
Beach Ridges
Gravel Barriers
Gravel Beaches
Offshore Sand Banks and Linear Sand Ridges
Spits

D

DALMATIAN COASTS

A Dalmatian coast is a prototype of a primary coast by ingression of the rising postglacial sea into a relief of coast-parallel anticlines and synclines from a young orogenesis (Holmes, 1965; Kelletat, 1995; Jackson, 1997), and is named after the landscape of Dalmatia (Croatia, former Yugoslavia, Adriatic Sea, Mediterranean). This type of coastline is very rare, a major example may be the Island of Sumatra with the Mentawai Islands strongly parallel in the Indian Ocean. More resistant elongated cuestas may build the island chains instead of anticlines. The narrow channels between these long islands are called "vallone" or "canale" (from the Italian word for channel or valley), and the Dalmatian coast therefore is named a canale- or vallone-coast, as well (Figure D1). The coastlines of the central part of Croatia, built up mostly by Mesozoic limestones, show only very little forming by true littoral processes. Beaches are missing as well as extended cliffs. The first is because of the small discharge or limited discharge area in a limestone environment with strong karstification, the second because of a very weak surf energy along the Adriatic Sea or inside of the canale and vallone with small fetch. Therefore, the coastlines nearly exactly follow a contour along an older relief, formed during longer periods of terrestrial

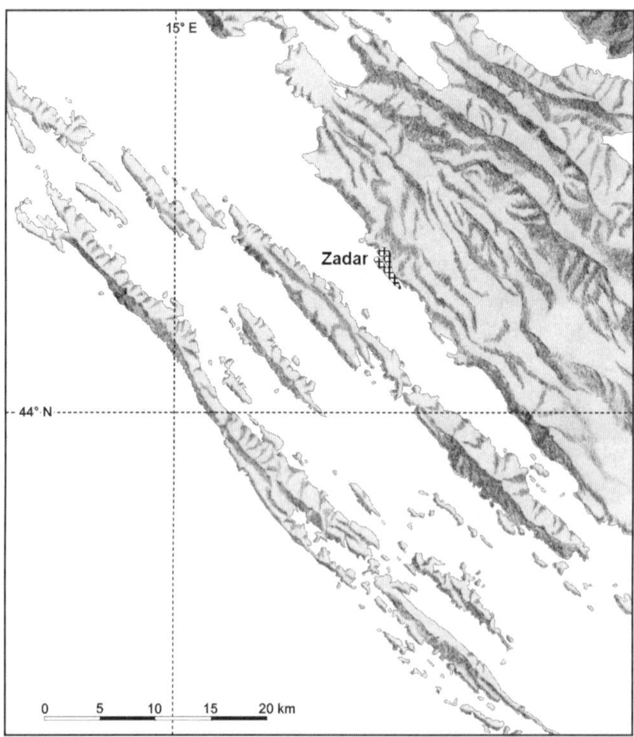

Figure D1 The central part of the Dalmatian coast with many parallel vallone or canale as waterways, and anticlines or cuestas as island chains (from Kelletat, 1995; with permission of the *Journal of Coastal Research*).

denudation. Partly drowned karst features (dolines, uvala, poljes) decorate some of the coastlines, in particular along the mainland coast, where freshwater karst springs occurring at the coast or even under the sea are numerous. The overall picture of the Dalmatian coast gives the impression of submergence and down warping. Indeed higher coastal terraces or deposits from earlier interglacial sea-level highstands are missing. This picture will change very little during a further sea-level rise or sea-level fall, because in the first case the ingression will drown the next depression, while the outermost islands will disappear under a rising sea level, and in the second case new island chains will emerge, while canale or vallone will be transformed into inland depressions.

Dieter Kelletat

Bibliography

Holmes, A., 1965. *Principles of Physical Geology*. New York: Ronald Press.

Jackson, J.A., 1997. *Glossary of Geology*. Alexandria, VA: American Geological Institute.

Kelletat, D., 1995. Atlas of Coastal Geomorphology and Zonality. *Journal of Coastal Research* (Special Issue No. 13.).

Cross-references

Asia, Eastern, Coastal Geomorphology
Endogenic and Exogenic Factors
Europe, Coastal Geomorphology
Holocene Coastal Geomorphology
Karst Coasts
Sea-Level Changes During the Last Millenium
Submerging Coasts

DAMS, EFFECT ON COASTS

People have constructed dams, weirs, or barriers on rivers to utilize and manage water resources for thousands of years. The utilization of water energy also has a long history, such as waterwheels and watermills, etc. The initial dams were used to control water resources and to supply water for irrigation and domestic purposes. Later, the energy of rivers was harnessed behind dams to power primary industries directly, and more recently still, hydroelectricity has been generated using water held in reservoirs, allowing the impoundment and regulation of river flow. The modern era of major dams dates from the 1930s and began in the United States with the construction of the 221 m-high Hoover Dam on the Colorado River. But in the last 50 years there has been a marked escalation in the rate and scale of construction of large dams all over the world, made possible by advances in science and technology. Some major rivers in the world have been intensively manipulated in this way.

Today, approximately one-third of the countries of the world rely on hydropower for more than half of their electricity supply. The hydropower provides 20% of all electricity globally. If this electricity were produced by the combustion of fossil fuel, carbon dioxide emissions would increase by 2 billions tons a year. The construction of dams on rivers, as an infrastructure of national economy, plays an important role in controlling flood, hydroelectricity generation, irrigation, navigation, water supply of industry and agriculture, etc. There is no doubt that the construction of large dams has been successful in achieving their socioeconomic benefits. However, dams alter river hydrological regimes, accordingly induce structure and function changes of ecosystems, and inevitably have marked effects on nature and ecological environment of river drainage basins. An estuary is the only pathway of materials from the drainage basin into the sea. The coastal area far away from dams is the sink of riverine freshwater, sediments, and nutrients. Dams on rivers alter their downstream material discharge into the sea, accordingly induce impacts on estuaries and the coastal zone, such as estuarine channel evolution, the actions between runoff and tide, coastline changes, saltwater intrusion in estuarine and land salinization in delta areas, and ecosystem variations in estuaries and seas. Thereby, damming has positive and negative socioeconomic effects, but in the coastal area it has more negative effects.

Status of dam construction

The International Commission on Large Dams (ICOLD) defines *large dams* as usually >15 m from foundation to crest. Dams of 10–15 m can also be defined as large dams if they meet the following criteria: crest length of 500 m or more; reservoir capacity of at least 1 million m^3; maximum flood discharge of at least 2,000 m^3/s; "specially difficult" foundation problems, or "unusual design." *Major dams* meet one or more of the following criteria: at least 150 m-high; having a volume of at least 15 million m^3; reservoir capacity of at least 25 km^3; or generation capacity of at least one gigawatt.

Table D1 Watersheds with more than five major dams

Watershed	Number of dams
Paranà (South America)	14
Columbia (North America)	13
Colorado (USA)	12
Mississippi (USA)	9
Volga (Russian)	9
Tigris and Euphrates (Mid-East)	7
Nelson (Canada)	7
Danube (Mid-Europe)	7
Yenisey (Russian)	6
Yangtze (China)	6

Based on the definition above, there are some 45,000 large dams, most of which were built in the past 50 years, that now obstruct the world's rivers. There are 306 major dams in the world today and 57 are planned in the near future (Revenga *et al.*, 1998). The most active phase of large dam construction was in the period between 1950 and the mid-1980s, when 885 dams were completed on average each year. At the end of 1960, there were 7,408 such dams registered. By 1986, the total had reached 36,562. Of those constructed, 64% were in Asia, with no less than half of the world total in China. Excluding China, most of the new structures have been built in the temperate zone, but tropical and subtropical countries such as Brazil, Mexico, India, Thailand, Indonesia, Zimbabwe, Nigeria, Cote d'Ivoire, and Venezuela have also become prominent in dam construction. Of these dams 79% are less than 30 m in height, and only 4% exceed 60 m. In total, the world's major reservoirs were estimated to control about 10% of the total runoff from the land in the 1970s. That figure had risen to 13.5% by the early 1990s (Postel *et al.*, 1996) and must be nearing 15% today. The world's largest impoundment, the 8,500 km^2 Volta Reservoir behind Ghana's Akasombo Dam, flooded 4% of that nation's land area. There are 10 watersheds with more than 5 major dams (Table D1).

To date, China has built almost half (22,000–24,000) of the world's estimated 45,000 large dams, and remains one of the most active dam-building countries today. Over 22,000 of the estimated 85,000 significant reservoirs and dams in China are considered as large dams, though there are perhaps a few million small-scale and localized water diversions, check dams and weirs built by local collectives or individual farmers that are unrecorded. China has thus built more than three times the number of large dams in the United States and over five times the number in India. Virtually, all these dams were built since the founding of the People's Republic of China in 1949, as only 22 large dams existed before that time. The water resources controlled about 100 billion m^3, risen to 565 billion m^3 today. At least 90 dams over 60 m are under construction, including the Three Gorges dam on the Yangtze (175 m) and Xiaolangdi on the Yellow River. The utilization rate of river water resources increased to 19.9% in 1997, while the Yellow, Huaihe, and Haihe River are all over 50%, and the Haihe River is nearing 90%. In the United States, 5,500 large dams make it the second most dammed country in the world.

Effects of dams on coast

Alternation of material discharge into the sea

Dam construction on a river changes the flow regime of almost the entire river, both changing its seasonal variations and/or reducing its overall volume, and further change and/or reduce sediment and nutrient outflow. Consequently, almost all parts of the estuarine and coastal environment can be impacted by changes to flow, sediment, and nutrients.

First, damming and reservoir construction change water discharge into the sea. The nature of the impacts depends on the design, purpose, and operation of the dam, among other things. For instance, following construction of the Aswan High Dam on the Nile River, Egypt and Lake Powell River dam on the Colorado River, USA not only altered the seasonal variations of natural flow but also reduced its overall water volume into the sea. One well-documented example of additional stresses created by dams pushing an ecosystem toward collapse has been documented in San Francisco Bay. Prior to 1850, approximately 34 km^3 was discharged on average into San Francisco Bay from the San Joaquin and Sacramento River systems. Today, there exists more than 20 km^3 of water storage capacity in the Central Valley. Approximately 40% of the flow in the these Central Valley rivers are removed for local consumption and 24% is pumped from the San Francisco Bay Delta and exported to Southern California and the Central Valley. Less than 40% of the river flows are now discharged to the Bay (Nichols *et al.*, 1986). In the extreme drought of 1977, releases dropped to 100 m^3/s, from the customary dry season discharges of 400 m^3/s.

Second, because of damming, sediments are intercepted in reservoirs behind dams and reduce sediment discharge into the sea. Before the Aswan High Dam, the Nile River carried about 124 million tons of sediment to the sea each year, depositing nearly 10 million tons on the floodplain and delta. Today, 98% of that sediment remains behind the dam. The construction of the Three Gorges Dam on the Yangtze River, China, will not change total water discharge into the sea, just change seasonal distribution of runoff, but will reduce sediment outflow by approximately 20%. Otherwise, besides reducing sediment discharge, dams may also change the grain size of sediments, large amounts of wash load are intercepted in reservoirs, reducing the supply of fine-grained sediments to estuaries.

Dams not only alter the pattern of downstream flow and sediment (i.e., intensity, timing, and frequency) they also change nutrient regimes. Nutrients carried by runoff and sediments discharge changes with water and sediment outflows. The Colorado, Nile rivers have weak nutrient levels, probably because of high abstraction rates from large reservoirs, leaving little discharge into the sea.

Dams alter the conditions of water, sediment, and nutrient discharges into sea, leading to huge impacts on the environment and ecosystems in estuaries and on coasts (Figure D2), such as coastal erosion, saltwater intrusion, and position changes of mouth bars in estuaries, nursery ground for many valuable fishes. Dam construction on an upstream river, although it does not alter overall volume of outflow, changes its seasonal variations and reduces sediment discharge, which all have profound effects on a river mouth and its adjacent coast.

Coastal erosion

A coastline is dynamic and constantly changing due to coastal erosion and accretion, and its change rate is determined by natural processes, such as rough seas, sea-level rise, high tides, nearshore currents, runoff, as well as human developments that can restrict or accelerate the volume of sediment available for shores. A constant supply of sediment is necessary for depositional coasts to form and be maintained along this coastline. Many human activities reduce the supply of sediment that reaches the sea and, in turn, deprive coasts of replenishment. These activities include dam construction and other developments. Lack of sediment replenishment creates greater vulnerability for coastlines that have always been subject to varying levels of erosion. A principal coastal concern today and in the foreseeable future is coastal erosion. It is estimated that 70% of the world's sandy shoreline are eroding (Bird, 1985). In the United States, the percentage may approach 90% (Leatherman, 1988). This worldwide extent of erosion suggests, that eustatic sea-level rise is an important underlying factor, while many other processes, especially reduced sediment discharge into the sea, are important causes. Dam construction on rivers which intercepts sediments in reservoirs is one of the causes for reduced sediment discharge into sea. Because of dams, the sediment supply to an estuary and its adjacent coast is reduced, thus accelerating coastal erosion. Even in a coastal delta of rich sediment supply, such as the Yangtze Delta, dams on the river may slow its accretional rate and gradually transfer to erosion.

Reduced sediment discharge directly affects coastal delta progradation, leading to loss of beach. For example, the slow accretion of the Nile delta was reversed with the construction of the delta barrage in 1868. Today, other dams on the Nile, including the Aswan High Dam, have further reduced the amount of sediment reaching the delta. As a result, much of the delta coastline is eroding by up to 125–175 m/yr. Deltas below impoundments tend to shrink, reducing habitats, because of the capture of sediments by impoundments. The Nile Delta, Egypt, has shrunk at a rate of 125–175 m/yr (Rozengurt and Haydock, 1993). The consequence of reduced sediments also extends to long stretches of coastline where the erosive effect of waves is no longer sustained by sediment inputs from rivers. Another example of this problem is along the mouth of the Volta River in Ghana. Akosombo Dam has cut off the supply of sediment to the Volta Estuary, affecting also neighboring Togo and Benin, whose coasts are now being eaten away at a rate of 10–15 m/yr. The break-off water discharge in the lower Yellow River, China results in a sharp decrease of sediment discharge into the sea and its delta erosion. The major causes are damming and water transfer. The delta erosion of the Yellow River is an issue that needs to be urgently solved due to water and soil preservation and construction of Xiaolangdi Reservoir.

Accelerated coastal erosion has been detected as a result of regulated flows and dams reducing sediment discharge to the littoral cells. For example, local coastal regions along the Egyptian Mediterranean coast have experienced erosion rates that were three times greater than in the period prior to the Aswan High Dam (Smith and Abdel-Kader, 1988). The construction of the Yangtze Three Gorges Dam will reduce sediment discharge and increase sediment transport capacity of flow in its estuary. Although it is of benefit to the sedimentation of navigation channel, it increases the action of bed sediments and leads to channel instability. Furthermore, it will increase coastline erosion and slow tidal flat accretion seaward (Chen and Xu, 1995).

Saltwater intrusion

Estuarine salinity varies with runoff, wind wave, and tidal current, and has complex spatial and temporal changes. Generally, estuarine salinity is higher during a dry season and is lower during flood season. There is a negative correlation with water discharge. Therefore, the water discharge is an important cause of saltwater intrusion. The change of water discharge leads to the alternation of the saltwater intrusion pattern: distance and intensity. Decreased discharge rates can result in saltwater intrusion along estuaries, and pollute estuarine water and groundwater. It will affect the development of freshwater resources, especially domestic and agricultural water. In the delta area, because of impounding upstream, the aquifers underground becomes saline, and leads to a shortage of fresh groundwater. Besides, the alternation of seasonal distribution of runoff also affects seasonal saltwater intrusion, thus altering the position of estuarine balance zone.

Reduced water discharge due to dams undoubtedly intensifies saltwater intrusion in estuaries. However, seasonal regulation of dams may mitigate saltwater intrusion due to increased water outflow during the dry season. For instance, salinity intrusion is a problem in the Yangtze Estuary because of the vast area using water from the Yangtze. The water stored by Three Gorges at the end of the flood season will be released to increase the dry season flow and may help to alleviate the salinity problem in the estuary.

The intrusion of seawater into delta areas when river flows have declined due to reservoir impoundment is another common downstream effect. Construction of the Farakka Barrage on the River Ganges in India in 1975 has significantly increased the amount of cropland plagued by salinity problems in Bangladesh, as dry-season discharge has declined. Nationally, nearly 350,000 ha of land in Bangladesh was affected by

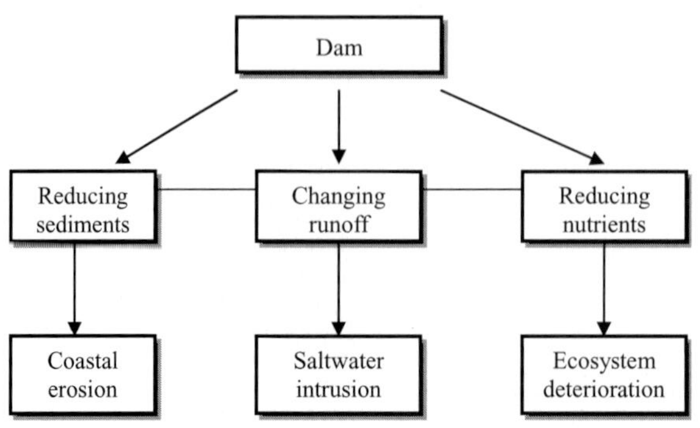

Figure D2 Effects of dam construction of rivers on coast.

salinity problems before the barrage was built, but by the mid-1990s this total had more than doubled to 890,000 ha. Downstream changes in salinity due to construction of the Cahora Bassa Dam in Mozambique are also threatening mangrove forests at the mouth of the Zambezi. Besides, reduced sediment discharge due to dams not only affects delta erosion but also intensifies land salinization.

Effects of ecosystem and biodiversity

A river's estuary, where freshwater meets the sea, is a particularly rich ecosystem. Some 80% of the world's fish catch comes from these habitats, which depend on the volume and timing of nutrients and freshwater. The riverine nutrients exist in water flow and sediments, carried and transported by flow and sediments, increase the nutrient level in estuaries and the adjacent sea. The decrease and alteration of the flows and nutrients reaching estuaries have marked effects on coastal and marine ecosystem and biodiversity, such as the nursery ground of fishes in estuaries and marine fish catch. The alteration of the flows reaching estuaries because of dams is a major cause of the precipitous decline of sea fisheries in many estuaries of the world.

The influence of diminished freshwater outflow to bays, estuaries, and coastal wetlands has become generally accepted during the past decade. One well-documented example of additional stresses created by dams pushing an ecosystem toward collapse has been documented in San Francisco Bay. This reduction in freshwater inflow contributed to a drop in phytoplankton biomass to less than 20% of normal, and zooplankton was also significantly reduced. These conditions resulted in striped bass, one of the key indicator species of the health of the ecosystem, being reduced to the lowest recorded levels (Nichols *et al.*, 1986). Water withdrawal on the North Caspian had the following effects (Rozengurt and Hedgpeth, 1989): (1) the mean salinity increased from 8 to 11 ppt; (2) the estuarine mixing zone was compressed and moved up to the delta; (3) the nutrient yield, especially phosphorus, and sediment load were reduced by as much as 2.5 and 3 times, respectively; (4) biomass of phytoplankton, zooplankton, and benthic organisms were decreased by as much as 2.5 times; and (5) a substantial part of the Volga flood plain that served as a nursery ground for many valuable fishes was transformed into drying swamps or deserts. The regulation of the Volta River in Ghana by the Akasombo and Kpong dams has led to the disappearance of the once-thriving clam industry at the river's estuary, as well as the serious decline of barracuda and other sport fish. The effects of increased salinity on fishes of the Nile Delta have been documented that out of 47 commercial fish species in the Nile prior to the construction of the Aswan High Dam, only 17 were still harvested a decade after its completion. The annual sardine harvest in the eastern Mediterranean has dropped by 83%, probably the effect of a reduction in nutrient-rich silt entering that part of the sea. The effect of lowered nutrient input is generally the greatest in the first year of life of the fishes.

Decreased discharge rates can result in an increase in salinity in estuaries and change the composition of species in this zone. The Danube Delta, central Europe, coastline is receding at a rate of up to 17 m/yr, threatening benefits from tourism down to bird life. The delta supports large populations of bird species that are generally widespread over Europe; some 170 species of birds breed in the delta, including pelicans, herons, ibises, and terns. Impoundments upstream, including seven major dams, on the 2,860-km-long river, channelization and the loss of the nutrient absorption capacity of upstream floodlands has meant that nutrients and other pollutants are affecting delta water quality. Bird populations are at a fraction of their historical numbers. So, although reservoirs may provide new habitat upstream their impacts on birds may be negative in the long-term.

The effect of dams on the ocean might be expected to be significant. The Three Gorges Dam on the Yangtze River in China is expected to reduce the productivity of the East China Sea, one of the largest fishing grounds in the world. It has been estimated that if the Yangtze River outflow is cut back by 10%, the cross-shelf water exchange will be reduced by 9% with a similar reduction in the onshore nutrient supply. Primary production and fish catch in the East China Sea is expected to diminish by a similar proportion (Chen, 2000).

Conclusion

Dam construction on rivers is a great act that human conquers nature and modifies nature. It has twofold effects: positive and negative. The positive roles are irrigation, power, flood control, and water supply, etc. In the meantime, it also leads to negative effects on environment, involving entire river basin, estuary, coast, and sea. The effects of dams on coast include coastal erosion, saltwater intrusion, ecosystem, and biodiversity, etc., and these effects are profound and mostly negative.

Jiyu Chen

Bibliography

Bird, E.C.F., 1985. *Coastline Changes—A Global Review*. Chichester, England: John Wiley-Interscience.

Chen, C.A, 2000. The Three Gorges Dam: in the East China Sea. *Geophysical Research Letters*, **27**(3): 381

Chen, J., and Xu, H., 1995. Impacts of the Yangtze River Three-Gorge hydro-engineering works on the Yangtze Estuary. *Resources and Environment in the Yangtze Valley*, **4**(3): 242–246.

Leatherman, S.P., 1988. Beach response strategies to accelerated sea-level rise. In *Proceedings of the 2nd North American Conference on Preparing for Climate Change*. Washington, DC: The Climate Institute, pp. 353–358

Nichols, F.H., Cloern, J.E., and Luoma, S.N., 1986. The modification of an estuary. *Science*, **231**: 567–573.

Postel, S., 1996. Forging a sustainable water strategy. In Starke, L. (ed.), *State of the World 1996. A Worldwatch Institute Report on Progress Towards a Sustainable Society*. New York: W.W. Norton and Company, pp. 40–59.

Revenga, C., Murray, S., Abramovitz, J., and Hammond, A., 1998. *Watersheds of the World: Ecological Value and Vulnerability*. Washington, DC: World Watch Institute.

Rozengurt, M., and Haydock, I., 1993. Freshwater flow diversion and its implications for coastal zone ecosystems. *Transactions of the 58th North American Wildlife and Natural Resources Conference*. Wildlife Management Institute, pp. 287–293.

Rozengurt, M.A., and Hedgpeth, J.W., 1989. The impact of altered river flow on the ecosystem of the Caspian Sea. *Reviews in Aquatic Science*, **1**: 337–362.

Smith, S.E., and Abdel-Kader, A., 1988. Coastal erosion along the Egyptian delta. *Journal of Coastal Research*, **4**(2): 245–255.

Cross-references

Asia, Eastern, Coastal Geomorphology
Beach Erosion
Deltas
Erosion Processes
Estuary
Hydrology of Coastal Zone
Submerging Coasts

DATABASES—See APPENDIX 4

DATING—See GEOCHRONOLOGY

DATUM—See SEA-LEVEL DATUMS

DEBRIS—See MARINE DEBRIS—ONSHORE, OFFSHORE, SEAFLOOR LITTER

DELTAIC ECOLOGY

A system overview

Deltas are depositional areas where rivers meet the sea, often forming extensive, coastal, alluvial fans. Although there is much variability, deltas tend to develop where high-energy rivers deposit their sediment loads into shallow, low-energy marine systems or, more generally, where river energies exceed marine energies. For example, the Mississippi River discharges an average of 6.2×10^{11} kg of sediments per year into the shallow, low-energy waters of the Gulf of Mexico, allowing for the development of an extensive delta complex spanning $30,000\,\text{km}^2$. Although deltas can dominate the coastal landscape, they are by no means static; characteristically undergoing a cycle of growth, channel

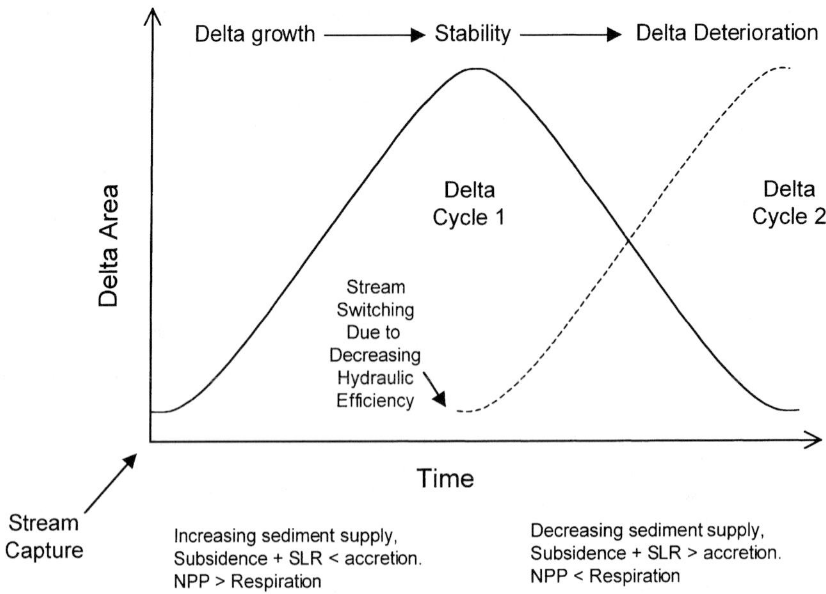

Figure D3 The delta cycle (modified from Roberts, 1997).

abandonment, and deterioration (Figure D3). Thus, the ecological populations and communities associated with deltas are unique in that they are subject to powerful allogenic riverine and marine forces, and to the overarching pattern of the deltaic cycle itself. For this reason, an introductory discussion of deltaic ecology is best approached at an ecosystem level, which considers the interactions between the biota and the abiotic forces that simultaneously drive, stress, and subsidize them.

Ecosystem types

Since an entire deltaic complex can span hundreds of kilo-meters, a wide variety of ecosystem types can be associated with them. Although natural upland habitats can be found along the landward edge of the delta and on the levees that form alongside stream channels, the systems that characterize and dominate deltas are wetlands. Wetlands are distinguished by; (1) the presence of water, either above the surface, or in the rooting zone, (2) hydric soils that develop under anaerobic conditions, and (3) vegetation that is adapted to wet conditions. Since deltas are ecotones where rivers meet the sea, concentrations of salts in the waters associated with these wetlands can range from hypersaline (>35 ppt salt) to fresh (0 ppt salt). In general, the hydroperiod (the pattern of water levels and flooding in a wetland) and the salinity control the types of wetland communities that can be found within a given delta.

In temperate zones, salt and brackish water marshes (marshes are wetlands dominated by herbaceous vegetation) characterize the saline habitats. Freshwater tidal marshes and swamps (swamps are wetlands dominated by trees) characterize either the inland regions of the delta or those areas dominated by freshwater inputs from the active delta lobe. Between latitudes 25°N and 25°S, forests called mangrove swamps characterize the saline habitats. The term "mangrove" is not a taxonomic term, but rather describes a group of 12 genera of woody halophytes adapted to growing in intertidal regions of the tropics. Mangroves show remarkable adaptations to salt and flooding stress, but not to freezing temperatures, therefore they are replaced by salt marshes in the colder temperate zones.

Primary producers and net primary productivity

For several reasons, wetlands associated with deltaic systems are among the most productive in the world. First, several types of primary producers (organisms that assimilate the energy from light to synthesize organic compounds) can be found within one wetland including; photosynthetic bacteria, algae (e.g., phytoplankton, periphyton, and macro algae), bryophytes, and vascular macrophytes ranging from sea grasses to trees. Second, delta systems receive multiple subsidies such as riverine inputs that supply nutrients, sediments, and freshwater, and tides that flush plant toxins from the sediments and maintain moderately aerobic conditions. Finally, although the vascular plants that dominate most wetlands are subject to the stress of periodic flooding and salt, they are still capable of relatively high rates of production because they have evolved numerous strategies that allow them to adapt to, or avoid entirely, anaerobic stress and high concentrations of salt. These strategies include the development of specialized tissue (aerenchyma) that transport oxygen down to the roots, and root-level micro-filters that prevent salt from entering a plant.

Ecologists typically quantify the productivity of a system by measuring its net primary productivity (NPP). NPP is defined as the rate at which biomass is accumulated by photosynthesizers, after accounting for losses due to respiration. NPP is most often expressed in units of carbon, or dry weight, per some unit of time. For example, primary producers in deltaic systems can have rates of NPP exceeding 2.5 $kg\,m^{-2}yr^{-1}$ (above ground dry weight). This is equivalent to rates of NPP measured in tropical rain forests (Whittaker and Likens, 1973). Ecologically, the concept of NPP is relevant because it represents the total amount of energy available to heterotrophs, or secondary producers, in the system, assuming no other inputs from outside the system. In deltaic wetlands, rates of NPP are controlled by (among other things), solar radiation, temperature, nutrients, salinity, and the availability of oxygen.

Secondary production

The biomass accumulated by primary producers in a wetland can be consumed directly by either herbivores, organisms that feed on dissolved organic matter, or detritus feeders (detritivores), and indirectly by carnivores. Secondary production is the rate at which biomass is produced by these consumers of primary production. Due to the inefficiencies of energy transfer, secondary production within a marsh system is much less than primary production, unless there are significant sources of organic matter coming into the wetland from outside the system. This can be the case in some riparian wetlands, for example.

It is estimated that almost 75% of primary production in marsh ecosystems is broken down by bacteria and fungi. Early studies suggested that secondary production in saltmarsh wetland systems was driven, in large part, by these detritivores feeding upon the vascular plants growing within a marsh (Odum, 1980). However, more recent research suggests that a large fraction of secondary production in salt marshes may be directly linked to fauna that consume phytoplankton and benthic algae (Kreeger and Newell, 2000).

Ecological forcing functions

Definition

Forcing functions are the exogenous variables that drive an ecosystem. For example, the sun is the ultimate forcing function that drives primary production, but of course, it is not unique to deltaic ecosystems. As

Table D2 Hierarchy of pulsing events in deltaic systems (modified from Day *et al.*, 1995)

Pulsing event	Time scale	Impacts
Avulsion	1000+ years	New delta lobe formation
Major river floods	50–100 years	Major sediment deposition Enhanced productivity
Hurricanes/ tropical storms	5–20 years	Major sediment deposition
Seasonal river floods	Annual	Sediment deposition Nutrient inputs Decrease salinity Enhance production Organism transport
Spring tides	Bimonthly	Organism transport
Storm events (frontal passages)	Weekly	Sediment deposition Organism transport
Daily tides	1–2 cycles per day	Drainage Enhanced marsh production Flushes accumulating salt and sulfides Organism transport

mentioned earlier, deltaic ecosystems are unique because they are driven by powerful forcing functions that can either subsidize or stress them. For example, river flooding contributes to anoxic conditions in a wetland (a stress), but also brings nutrients to the system (a subsidy). The following section outlines several types of forcing functions that characterize and define deltaic ecosystems.

Pulsed forcing functions

Perhaps more than most systems, deltas ecosystems are subject to pulses of energy, ranging from daily tides to once-a-century category 5 hurricanes, that subsidize the ecosystem. These energetic events, which occur at varying timescales (Table D2) affect primary productivity, material fluxes, and the evolution of the delta itself (Day *et al.*, 2000). Daily tides flush accumulated salts and sulfides that inhibit plant growth, thus enhancing primary productivity. Higher tides, which occur during full and new moons, allow fish to utilize interior marshes for feeding. The sustained winds associated with cold fronts can profoundly affect the hydrodynamics and associated sediment resuspension and deposition in deltaic regions. Annual river floods distribute sediments, nutrients, and freshwater to deltaic systems and have been shown to be positively correlated with primary and secondary productivity. On a longer term, major storms such as hurricanes can deposit several centimeters of sediment onto deltaic marshes and can be a critical process for maintaining a balance between rates of sea-level rise and rates of marsh accretion. Major river floods, which occur only once or twice a century are also associated with substantial sediment deposition and even the channel switching that starts a new delta cycle.

River switching, the ultimate pulsing event

The initial growth of a deltaic lobe starts as river sediments fill in areas along the coast and marine shelf. Eventually, as the delta extends seaward, the existing river becomes longer, the channel slope to the sea decreases and becomes hydraulically inefficient, and the entire river switches to a new channel. This switching event (an avulsion) often occurs abruptly, during a major flood, for example, when an upstream levee is breached and the river captures a new, shorter, and steeper course to the sea. In the final phase of the delta cycle, water and associated sediments are directed to the new channel, and the old deltaic lobe deteriorates as rates of subsidence become greater than the rate that sediments accumulate (Figure D3). Of course while the old delta is deteriorating, a new delta, associated with the new river channel, is forming. The rate of channel switching is governed by the characteristics of the river and the receiving basin. For the Mississippi River delta, perhaps the most studied in the world, the river changes course approximately every 1,000 years. Although pulsing events like the ones described above control ecosystem dynamics on the short term, it is the ecosystem's temporal position within the delta cycle that ultimately controls it.

Subsidence and sea-level rise

Natural rates of subsidence in many deltaic systems can be quite high due to the compaction, consolidation, and downwarping associated with the rapid deposition of alluvial sediments. This subsidence, plus the global eustatic sea-level rise (currently 0.15 cm yr^{-1}), defines the rate of relative sea-level rise (RSLR). Some of the highest rates of RSLR (1.1–1.3 cm yr^{-1}) have been recorded in the Mississippi River delta of the United States. In wetlands, high rates of RSLR can lead to increasingly long periods of inundation that are associated with decreased primary productivity and plant mortality. Therefore, RSLR can be thought of as a forcing function that stresses, rather than subsidizes a wetland ecosystem.

Deltaic wetlands can persist in the face of RSLR when vertical accretion equals or exceeds the rate of subsidence. The rate at which a wetland accretes depends upon sediment and nutrient supply, tidal range, vegetation, and climate. Current models suggest that coastal wetlands can remain stable against a RSLR rate as high as 1.2 cm yr^{-1} (Morris *et al.*, 2002). Since this rate is equal to or higher than the current rate of RSLR in many deltaic systems, this would suggest that these systems are sustainable. However, best estimates suggests that, due to the impacts of global warming, the eustatic sea levels could rise 48 cm by the year 2100 (Church *et al.*, 2001). Recent studies have suggested that few deltaic wetlands would be able to keep pace with this predicted increase in sea-level rise (Day *et al.*, 1999). Exacerbating this problem, reductions in sediment deli-very have led to rapid wetland loss in many deltaic regions worldwide. This is discussed further in the next section.

Anthropogenic disturbances

Although it is technically correct to position humans within an ecosystem, it is not out of order to consider some anthropogenic actions as allogenic forcing functions. For example, as previously discussed, seasonal river flooding deposits sediments onto wetlands that contribute directly to elevation gain, and counterbalance the effects of RSLR.

However, many of the world's major rivers that are responsible for the creation of deltas in the first place have now been dammed. This has not only trapped large amounts of sediments in the reservoirs behind the dams, but has also prevented the seasonal flooding that once distributed sediments onto the delta. For example, due to upstream dams, the sediment load has been reduced by 95% in Nile, Indus, and Ebro rivers, by 75% in the Po and Mississippi rivers, and by over 50% in the Rhone river (Day *et al.*, 2000). Additionally, because of the creation of extensive levee systems along these rivers, the remaining sediment that does reach the coast often discharges directly into deeper waters of the receiving basin, rather than accreting on the deltaic plain. Finally, many wetlands in deltaic regions have become hydrologically isolated by dense networks of canals and spoil banks constructed during the past several centuries. Spoil banks impede drainage and often physically impound wetlands, thus preventing the overland flow of any remaining sediments into coastal wetlands. As an example, because of the combined effect of high rates of RSLR and decreased sediment delivery, wetlands in the Mississippi River delta complex are currently being lost at a rate of 100 km^2 yr^{-1}.

While the hydrologic alternations that reduce pulses are perhaps the greatest anthropogenic forcing function threat to deltaic ecosystems, they are by no means the only ones. Other anthropogenic disturbances include eutrophication, organic enrichment, thermal pollution, toxic inputs, over harvest, and exotic species introduction. Also, because of their position in the landscape, deltas are, and always have been, desirable places to live, farm, and conduct commerce, and humans continue to encroach on these systems.

Toward the conservation and restoration of deltaic ecosystems

Deltaic wetlands provide numerous valuable services such as fisheries production, disturbance regulation, and water quality improvement to a global population that is increasingly concentrated near the coast and dependent upon its resources (Nuttle *et al.*, 1997). Scientists and resources managers have become aware that, to prevent the further loss of these systems, landscape scale solutions directed toward the restoration of sediment supplies are required. Some of the more successful projects have included: (1) creating gaps in the artificial levees that line most distributory streams, (2) removing sea-side dikes to restore tidal flushing, and (3) creating large-scale diversions of river water to re-introduce nutrients, freshwater, and sediments to wetlands that have been hydrologically isolated.

The Caernarvon Freshwater River Diversion project in coastal Louisiana, USA, which has been operating since 1991, is one of the best examples of this third type of restoration. The project consists of a structure containing 5 m^2-gated culverts that divert water from the

Mississippi River into a part of the delta that was losing approximately 400 ha yr^{-1} due to high rates of RSLR. Construction was completed in February 1991 at a cost of US $26.1 million, and the estimated net benefit for fish, wildlife, and recreation is over US $9 million per year. In a 1998 project summary, the Army Corps of Engineers reported a net increase in marshland of 165 ha within a monitored area that originally contained 930 ha of marsh. Oyster and fish productivity, and waterfowl usage has increased dramatically as well.

John M. Rybczyk

Bibliography

Church, J.A., Gregory, J.M., Huybrechts, P., Kuhn, M., Lambeck, K., Nhuan, M.T., Qin, D., and Woodworth, P.L., 2001. Changes in sea level. In Houghton, J.T., Ding, Y., Griggs, D.J., Noguer, M., van der Linden, P.J., Dai, X., Maskell, K., and Johnson, C.A. (eds.), *Climate Change 2001: The Scientific Basis. Contribution of Working Group I to the Third Assessment Report of the Intergovernmental Panel on Climate Change*. Cambridge University Press, pp. 639–693.

Day, J.W., Pont, D., Hensel, P., and Ibanez, C., 1995. Impacts of sea level rise on deltas in the Gulf of Mexico and the Mediterranean: the importance of pulsing events to sustainability. *Estuaries*, **18**: 636–647.

Day, J.W., Rybczyk, J.M., Scarton, F., Rismondo, A., Are, D., and Cecconi, G., 1999. Soil accretionary dynamics, sea level rise and the survival of wetlands in the Venice lagoon: a field and modeling approach. *Estuarine, Coastal and Shelf Science*, **49**: 607–628.

Day, J.W., Psuty, N.P., and Perez, B.C., 2000. The role of pulsing events in the functioning of coastal barriers and wetlands: implications for human impact, management and the response to sea level rise. In Weinstein, M.P., and Kreeger, D.A. (eds.), *Concepts and Controversies in Tidal Marsh Ecology*. Dordrecht: Kluwer Academic Publishers, pp. 633–660.

Kreeger, D.A., and Newell, R.I.E., 2000. Trophic complexity between producers and invertebrate consumers in salt marshes. In Weinstein, M.P., and Kreeger, D.A. (eds.), *Concepts and Controversies in Tidal Marsh Ecology*. Dordrecht: Kluwer Academic Publishers, pp. 187–200.

Morris, J.T., Sundareshwar, P.V., Nietch, C.T., Kjerfve, B., and Cahoon, D.R., 2002. Response of coastal wetlands to rising sea levels. *Ecology*, **83**: 2869–2877.

Nuttle, W.K., Brinson, M.M., Cahoon, D., Callaway, J.C., Christian, R.R., Chmura, G.L., Conner, W.H., Day, R.H., Ford, M., Grace, J., Lynch, J., Orson, R.A., Parkinson, R.W., Reed, D., Rybczyk, J.M., Smith T.J., III, Stumpf, R.P., and Williams, K., 1997. Conserving coastal wetlands despite sea level rise. *EOS*, **78**: 257–261.

Odum, E.P., 1980. The status of three ecosystem-level hypothesis regarding salt marsh estuaries: tidal subsidy, outwelling, and detritus-based food chains. In Kennedy, V.S. (ed.), *Estuarine Perspectives*. New York: Academic Press, pp. 485–495.

Roberts, H.H., 1997. Dynamic changes of the Holocene Mississippi River delta plain: the delta cycle. *Journal of Coastal Research*, **13**: 605–627.

Whittaker, R.H., and Likens G.E., 1973. Primary production: the biosphere and man. *Human Ecology*, **1**: 357–369.

Cross-references

Coastal Subsidence
Dams, Effect on Coasts
Deltas
Dikes
Estuaries
Eustasy
Greenhouse Effect and Global Warming
Mangroves, Ecology
Salt Marsh
Sea-Level Rise, Effect
Wetlands

DELTAS

Deltas are coastal landforms comprised of subaerial and subaqueous packages of fluvial-transported sediments that have formed an alluvial landscape by deposition at the mouth of a river. Deltas form at the coastal interface where riverine sediment supplied to the coastline is not removed by tides or waves. The term delta derives from Herodotus who, in the 5th century BC, noted a geometric similarity between the tract of land at the mouth of the Nile River and the Greek letter "Δ" with its apex directed landward (Moore and Asquith, 1971). Although this distinctive morphology is absent in many river-mouth landscapes, the term has nonetheless been accepted to describe the geographical region near a river mouth and the sedimentary package that develops at a fluvial entrance into a depositional receiving basin.

Globally, deltas can be found on all continents and in all climates (Figure D4). In a general sense, the locations of deltas are similar; at the terminus of a catchment basin that provides sediment load into an ocean, gulf, lagoon, estuary, or lake. Although a delta may form regardless of the size of the fluvial system or receiving basin, some tectonic settings are more conducive than others to the development of major deltaic landscapes. Using the Inman and Nordstrom (1971) tectonic classification of coasts, trailing-edge coasts typically have the largest drainage basins followed successively by marginal-seacoasts and leading-edge coasts. Of the 58 major river systems in the world with drainage areas greater than 10^5 km^2, 56.9% are on trailing edge coasts, 34.5% along marginal-seacoasts, and 8.6% on leading-edge coasts.

Trailing-edge coasts provide a geologic setting favorable to delta development because of their tectonic stability and geologically old, low-relict terrains with extensive river systems that provide an abundant supply of sediment. Moreover, trailing-edge coasts often border broad continental shelves, which provide ideal shallow-water platforms for the formation of deltas. Marginal-seacoasts typically provide a low-energy setting with many trailing coast characteristics that are conducive to delta development. However, the mountainous and immature drainage systems of tectonically unstable, leading-edge coasts with small catchment basins and low sediment supply generally limit the likelihood of extensive deltaic deposition. Table D3 lists the sizes and locations of the some of the world's major deltaic systems.

Delta types and formation

Previous summaries of the world's deltas have primarily focused on the major large deltas (e.g., Coleman and Wright, 1975). A review of the recent literature on late Quaternary deltaic systems reveals that deltas can be classified into four basic types depending on sea-level stage (highstand or lowstand) and/or sediment supply (high or low) (Kindinger, 1988; Boyd *et al*., 1989; Nichol *et al*., 1996). The four basic types of deltas are lacustrine, bayhead (or lagoonal), continental shelf, and continental margin. In an evolutionary sense, these four types of deltas can represent a developmental continuum. Thus, during a sufficiently long sea-level stillstand a fluvial system with adequate sediment supply could result in a lacustrine deltaic system that evolves into a continental-shelf margin deltaic system through progradational outbuilding.

Lacustrine deltas

A lacustrine delta consists of deposits located landward of the coast or paralic zone within an inland lake or sea. The environmental setting in such cases is typically a shallow-water basin with a low-energy regime. Along the Gulf of Mexico, the best example of a lacustrine delta is the Grand Lake delta within the Atchafalaya drainage basin of south-central Louisiana. At the start of the 19th century Grand Lake within the Atchafalaya Basin was a shallow, low-energy lake (Tye and Coleman, 1989). However, by the 1950s a large lacustrine delta had filled this lake and started bypassing into the Atchafalaya Bay as a result of the progressively increasing Mississippi River flow into the Atchafalaya basin (van Heerden and Roberts, 1988). Historically, if left uncontrolled, the Mississippi River would have avulsed from its modern course into the Gulf of Mexico past New Orleans to the Atchafalaya course located farther west. Other lacustrine delta examples are present within Lake Geneva and Lake Constance (Switzerland) (Müller, 1966; Reineck and Singh, 1973).

Bayhead/lagoonal deltas

Bayhead/lagoonal deltas are located in the upper reaches of the coastal zone. In many cases they represent an evolutionary step from a lacustrine delta during a sea-level stillstand or withdrawal, which allows the delta to build into the coastal or paralic zone. Bayhead-delta settings are typically protected by barrier-island systems or remnants of preexisting antecedent topography, resulting in a low-energy environment. Examples of bayhead deltas are found within the northern part of Mobile Bay Alabama (Kindinger *et al*., 1994), Trinity Bay, and San

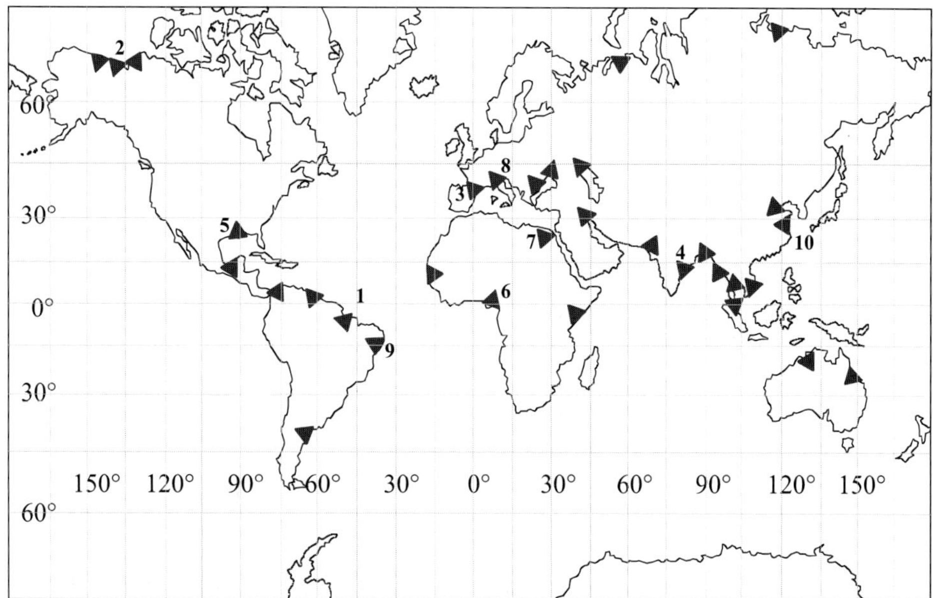

Figure D4 Map showing the global distribution of major deltas. Note that most of the major deltaic systems are located along trailing-edge coastlines close to the equator (modified from Coleman and Prior, 1980). Numbered locations refer to deltas listed in Table D3.

Table D3 Names, locations, and sizes of major deltas (from Wright, 1974, data compiled from Colemon & Wright, 1975).

River	Location	Receiving body of water	Deltaic area ($km^2 \times 10^3$)
Amazon (1)	South America	Atlantic Ocean	467.1
Burdekin	Australia	Coral Sea	2.1
Chao Phraya	Asia	Gulf of Siam	11.3
Colville (2)	North America	Beaufort Sea	1.7
Danube	Europe	Black Sea	2.7
Dneiper	Asia	Black Sea	NA
Ebro (3)	Europe	Mediterranean Sea	0.6
Ganges–Brahmaputra (4)	Asia	Bay of Bengal	105.6
Grijalva	North America	Gulf of Mexico	17.0
Hwang Ho	Asia	Yellow Sea	36.3
Indus	Asia	Arabian Sea	29.5
Irrawaddy	Asia	Bay of Bengal	20.6
Klang	Asia	Straits of Malacca	1.8
Lena	Asia	Laptev Sea	43.6
Mackenzie	North America	Beaufort Sea	8.5
Magdalena	South America	Caribbean Sea	1.7
Mekong	Asia	South China Sea	93.8
Mississippi (5)	North America	Gulf of Mexico	28.6
Niger (6)	Africa	Gulf of Guinea	19.1
Nile (7)	Africa	Mediterranean Sea	12.5
Ord	Australia	Timor Sea	3.9
Orinoco	South America	Atlantic Ocean	20.6
Paraná	South America	Atlantic Ocean	5.4
Pechora	Europe	Barents Sea	NA
Po (8)	Europe	Adriatic Sea	13.4
Red	Asia	Gulf of Tonkin	11.9
Sagavanirktok	North America	Beaufort Sea	1.2
São Francisco (9)	South America	Atlantic Ocean	0.7
Senegal	Africa	Atlantic Ocean	4.3
Shatt-al-Arab	Asia	Persian Gulf	18.5
Tana	Africa	Indian Ocean	3.7
Volga	Europe	Caspian Sea	27.2
Yangtze-Kiang (10)	Asia	East China Sea	66.7

Antonio Bay Texas; (McEwen, 1969; Donaldson *et al.*, 1970; van Heerden and Roberts, 1988).

Continental-shelf deltas

Continental-shelf deltas develop during prolonged sea-level still-stands and/or a sea-level fall. These deltas build out onto the continental shelf. Trailing edge or marginal sea continental-shelf deltas typically are characterized by an abundant sediment supply and their geomorphology reflects the wave, riverine, and tidal conditions in which they develop. Examples of continental-shelf deltas include the Mississippi (US, Louisiana), the Ebro (Spain), and the Colville (US, Alaska) (Frazier, 1967; Maldonado, 1975; Naidu and Mowatt, 1975; Alonso *et al.*, 1990).

Continental-shelf margin deltas

Continental-shelf margin deltas are typically most well-developed during sea-level lowstands or formed by rivers with very large sediment loads during prolonged sea-level stillstands. The majority of shelf-margin deltas in the recent geologic record have developed during falling sea-level or sea-level lowstand conditions as river valleys lengthen, incise across subaerially exposed continental-shelf sediments, and discharge their sedimentary load at the shelf margin (Suter, 1994). With sufficient sediment load and/or during lowstand conditions these deltas continue to build seaward, creating an apron of deltaic sediments along the shelf edge that extends the preexisting continental margin into deeper basinal waters. Numerous lowstand shelf-margin deltas, formed during the late Wisconsin sea-level fall and lowstand that culminated at approximately 18,000 yr BP, are found seaward of the Rio Grand and the Mississippi River deltas (Morton and Price, 1987; Suter *et al.*, 1987; Suter, 1994). Sufficiently large rivers carrying substantial sediment loads may however, construct shelf-margin delta deposits during sea-level highstand conditions. An example of a shelf-margin highstand delta is the depocenter of the main stem of the modern Mississippi River delta, the representative "Bird Foot" delta (Boyd *et al.*, 1989).

Delta processes

Deltas result when sediment-laden fluvial water decelerates upon entering a receiving basin and sediment transport competence of the river is lost. Consequently, fluvial characteristics and river-mouth processes constitute an important controlling factor in the nature of deltaic deposition (Bates, 1953). A consideration of fluvial processes is particularly important in deltaic environments where receiving basin processes such as wave and tide regimes are minimal. In such environments the primary fluvial factors, influencing the nature of deltaic deposition, include the volume of riverine discharge and transported sediment, as well as the textural character of the sediment load. Generally, the forces associated with these factors that dictate the character of fluvial-margin deposition are: (1) the riverine effluent inertia and diffusion, (2) friction between the sediment load and the floor of the receiving basin, and (3) density contrasts between the effluent and the receiving basin waters (Bates, 1953; Wright, 1977; Orton and Reading, 1993).

Homopycnal conditions

Homopycnal flow describes river discharges where the density contrast between riverine and basinal waters is small. Discharge patterns are most heavily influenced by inertia of the riverine water, which allows the effluent to radially spread into the receiving basin. As a result of deltaic progradation, fine-grained bottomset sediments are overlain by relatively coarser grained topset beds and river mouth-bar deposits, creating the classic coarsening upward depositional units common to deltaic systems. Gilbert (1884) first described this type of deltaic system in lakes of the western United States; hence, they are termed Gilbert-type deltas.

Hyperpycnal conditions

Hyperpycnal discharge conditions occur when the riverine inflow is denser than the waters of the receiving basin. Consequently, the sediment-laden effluent moves along the receiving basin bottom as a density current. This type of discharge process is rare and typically restricted to deltaic systems rich in silt and coarse-grained sediments such as in the Bella Coola fjord delta of British Columbia (Kostashcuk, 1985).

Hypopycnal conditions

Hypopycnal discharge conditions occur when the riverine inflow is less dense than the waters of the receiving basin, a condition commonly present in lakes and inland seas. Most of the world's deltas form under hypopycnal conditions. In hypopycnal settings, as the riverine inflow enters the receiving basin it undergoes turbulent mixing resulting in the deposition of the sediment load. Finer-grained sediments are transported well into the receiving basin as the freshwater sediment plume disperses above relatively more dense basinal waters, whereas the relatively coarser grained sediments are deposited in close proximity to the river mouth. In shallow-water hypopycnal discharge conditions, friction between the receiving basin bottom and the riverine leads to multiple bifurcations of the distributary system. Conversely, in deep water during hypopycnal discharge conditions, buoyancy dominates during the dispersal of the effluent, producing elongated distributaries such as present at the modern Mississippi River depocenter (Suter, 1994).

Delta components

The delta plain is the subaerial and subaqeous zone where sediments are discharged from the alluvial valley and accumulate in the receiving basin. Typically, a delta plain is subdivided into an upper delta plain and a lower delta plain, each of which are characterized by different vegetation, morphology, and depositional processes (Figure D5) (Reineck and Singh, 1973; Coleman, 1981; Elliot, 1986). Many of the features of the upper delta plain are inherited from its alluvial valley. In the upper delta plain deposition is primarily fluvial and as the main stem of the river leaves the alluvial valley it discharges onto the upper delta plain and undergoes channel bifurcation. The vegetative landscape of the upper delta plain is a freshwater environment, generally beyond the influence of marine incursions created by the tidal regime of the receiving basin. The major landform includes distributary channels with their subaerial natural levees, point bars, and crevasse splays, as well as interdistributary environments characterized by low-lying swampy areas.

Seaward of the upper delta plain is the relatively lower relief of the lower delta plain that is within the realm of tidal incursion. The lower delta plain is the site of more active deposition through overbank flooding processes. Because the elevations of the natural levees are lower, channel crevassing and distributary shifting is a common and very dynamic process characterizing the lower delta plain. These processes disperse sediment from areas within the upper delta plain into brackish to saline environments of the lower delta plain. Distributary shifting common in the lower delta plain results in a complex distribution of multiaged distributary channels in all stages of evolution (Figure D5). The lower delta plain contains active and abandoned distributary channels in all stages of regressive and transgressive evolution (Penland and Suter, 1989). Because the lower delta plain is a transitional environment influenced by fresh-water fluvial and saline marine waters the vegetative landscape generally contains species tolerant of both brackish and saline water.

Deltaic sediments

In general, deltaic sedimentary packages contain a coarsening upward vertical stratigraphy that reflects seaward progradation of the delta into its receiving basin. At any given time there is typically a well-developed gradient of grain sizes extending seaward from the subaerial to subaqueous portions of a delta. The relatively coarsest grained material is deposited proximal to the river mouth where transport competence is highest, whereas the finest-grained material is carried farther seaward and deposited in more distal locations. Thus, as sediments are delivered to the receiving basin from the river mouth they accumulate in a subaqeous, fine-grained depositional zone referred to as the *prodelta*. The prodelta forms the platform across which deltas prograde and subsequently aggrade. The prodelta sedimentary package is a widespread laterally continuous interval primarily composed of the finest sediment fraction transported by the fluvial system. Thus, the overall finest sediment is found at the base of the prodelta sequence and the whole sequence coarsens upwards. Moreover, the initial prodelta deposits are the thinnest at the base and the beds thicken upward as a result of deltaic progradation. Bioturbation of prodelta sediments is limited by the rate of deposition.

The delta front is located between the zone of fine-grained prodelta deposition and the more landward located, coarser-grained distributary mouth deposits of a progradational deltaic system. The relatively coarser-grained sediments of the delta front generally consist of interbedded clays, silts, and sands. This zone is dominated by the interaction of fluvial and marine processes, resulting in sand-rich accumulations landward of the advancing prodelta. The delta front also represents a zone of transition between deposits representative of progradation and aggradation.

In a progradational deltaic sequence, distributary channel deposits overlie the relatively finer grained delta front and prodelta deposits. The framework of distributary channels consists of distributary mouth bars overlain by natural levee deposits. Distributary mouth bar deposits are primarily subaqueous deposits that grade laterally into relatively finer grained deposits; locally, mouth-bar deposits may contain fine-grained beds within the generally sandy matrix of the mouth bar. The presence of fine-grained deposits primarily represents deposition during low-flow conditions. During flood conditions, natural levee ridges can be over-topped and breached creating crevasse splays. These splays within the distributary network provide conduits for sediment dispersal into interdistributary bays and the origin of crevasse-splay deposits.

Interdistributary bay and marsh sediments are generally as volumetrically significant as prodelta sediments. The interdistributary area is a low-energy depositional environment and less dynamic than areas of the delta plain characterized by multiple channels, channel bifurcations, and channel avulsions. Interdistributary deposits characteristically consist of fining upward deposits consisting of clay-rich bay sediments overlain by organic-rich marsh deposits. This fine-grained depositional environment is punctuated by sandy depositional events associated with overbank flooding and crevasse splay events.

Delta morphology

Although sediment delivery to a delta is by a fluvial system, it is ultimately the dynamic interaction of riverine and basinal processes that control the morphologic, stratigraphic, and sedimentologic variability of a deltaic accumulation. Water depth and basinal configuration, tidal range, wave climate, and coastal currents are the primary basinal processes that control delta morphology. The dynamics of these processes and complex interaction between them leads to a highly variable array of deltaic configurations (Coleman and Wright, 1975).

One of the first and simplest attempts to describe deltaic variability as a function of processes involved depicts the morphology of river deltas as a function of sediment influx and marine processes such as waves and longshore currents (Figure D6). The tidal regime of the receiving basin was not considered in formulation of this model. Fundamentally, this model describes the processes leading to deltaic

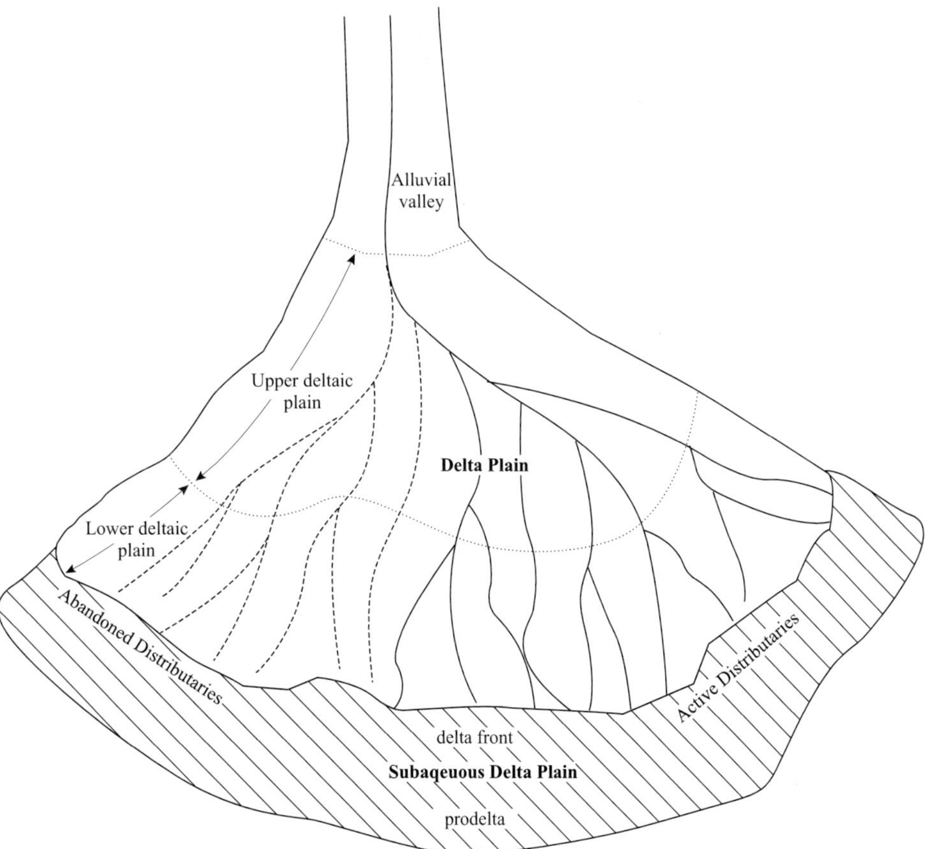

Figure D5 Schematic diagram showing the primary physiographic components and depositional environment of a delta (modified from Coleman and Prior, 1980).

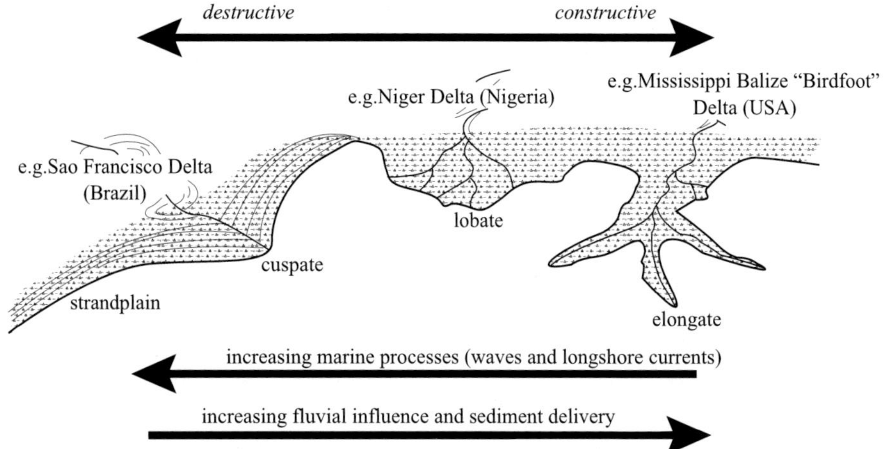

Figure D6 Schematic diagram showing the variation in deltaic geomorphology as a function of the relative influences of fluvial and marine processes (Fisher *et al.*, 1969, with permission of University of Texas, Bureau of Economic Geology).

deposition as either constructive or destructive (Davis, 1983). Deltas dominantly influenced by the nearshore wave climate are termed destructional, characteristically cuspate shaped with poorly developed distributaries and well-developed strand plains such as in the Sao Francisco delta of Brazil. Intermediate between destructional and constructional forms, is the lobate-shaped delta, typified by the Niger delta of Africa. Lobate deltas possess well-developed distributary networks and a smooth coastal outline. Deltas that are strongly influenced by sediment influx are essentially constructional and are characterized by distinctive elongate distributaries. An example of this type of delta is the modern "Birdfoot" depocenter of the Mississippi River.

In the 1970s, new process-response approaches to understanding morphologic variability of deltas incorporated features of the constructional/destructional approach but also considered the effects of the receiving basin tidal regime. In the Galloway (1975) scheme a variety of previously documented delta morphologies (e.g., Fisher *et al.*, 1969; Wright and Coleman, 1973) were classified within the framework of a ternary diagram that effectively differentiated delta morphology as a function of sediment input, wave-energy flux, and tidal energy flux (Figure D7).

Deltas dominated by sediment input are supplied by a large well-developed drainage system, capable of transporting large volumes of sediment to the deltaic coastline. Distributary switching is an important process, resulting in a highly constructive elongate to lobate deltas with straight to sinuous active and abandoned channels. The bulk composition of the sediment supply is muddy to mixed and the framework facies include distributary mouth bar, channel-fill sands, and delta-margin sheet sands (Table D4).

Deltas shaped primarily by waves are usually characterized by a small catchment basin with little sediment influx. The high-energy wave environment leads to winnowing of fine-grained sediments and leaves behind a relatively coarser-grained sediment fraction. Consequently, the bulk composition of wave-dominated deltas is typically sandy. The primary distributary channels are meandering and the framework facies are coastal barrier and beach-ridge sands (Table D4).

Deltas shaped primarily by tides are similar to wave-dominated deltas in that tidal currents winnow away much of the fine-grained sediments delivered to the coast. However, a fundamental difference is that tidally formed sand bodies are oriented shore normal. The distributary channels flare seaward and in some cases are straight and oriented parallel with the tides. The bulk composition of tide-dominated deltas is highly variable. The primary framework facies include fine-grained estuarine fill and coarse-grained sand ridges (Table D4).

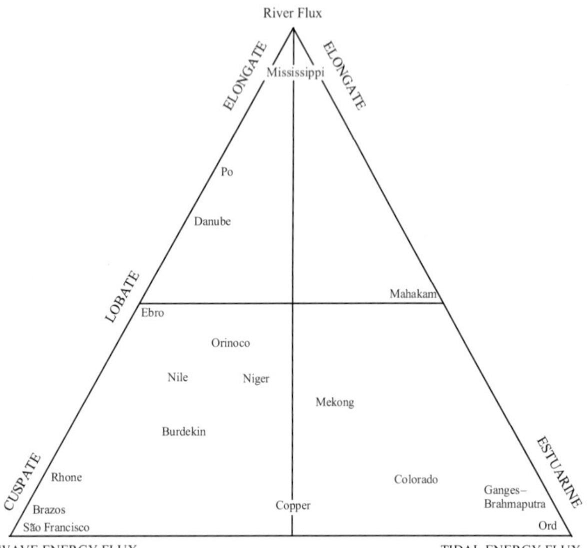

Figure D7 Ternary classification of deltas on the basis of dominate processes and the resulting morphology (after Galloway, 1975).

Modern deltas

Globally, modern deltaic systems are geologically young features that have formed since the initiation of late Wisconsin deglaciation at approximately 18,000 yr BP. Although the development, existence, and evolution of deltas along modern coastlines is contingent upon sea-level change, sediment delivery, and marine processes, the timing of formation for modern deltas suggests that sea level has been a primary factor controlling delta development and stability (Stanley and Warne, 1994).

At the height of the late Wisconsin glacial period approximately 18,000 yr BP, sea level was approximately 120 m lower than present sea level (Fairbanks, 1989). During this interval of lower sea level rivers crossed subaerially exposed continental shelves within incised channel systems. Large volumes of sediment were debouched along bordering continental margins and locally, thick and extensive shelf-margin deltas developed (Morton and Price, 1987; Suter *et al.*, 1987). Subsequently, as the late Wisconsin glaciers melted and sea level rose, exposed shelves were flooded and transgressed with marine waters. Initially, sea level rose so rapidly that the global paralic zone was forced tens of kilometers landward. Marine processes at the leading edge of the transgressing marine waters reworked the formerly exposed continental shelf leaving a marine truncation surface and transgressive coastal-zone deposits in the form of ravinement surfaces and transgressive coastal marine deposits, inner shelf shoals, sand sheets, and shoreface ridges.

Following the initially high eustatic rise rates associated with deglaciation the rate of sea-level rise dropped below a critical threshold rise rate allowing global landward migration of the paralic zone. Beginning approximately 7,000 yr BP, the rate of sea-level rise appears to have slowed enough to allow for deltaic progradation and a relatively synchronous development of the world's deltas (Figure D8). The temporal uniformity in delta development suggests a decline in the rate of sea-level rise, following the melting of late Wisconsin ice sheets, strongly influenced Holocene deltaic formation (Stanley and Warne, 1997).

Deltas and society

Deltaic environments have played an important role in our global society. Deltaic environments served as the culture hearth for early civilizations throughout the world because of their tremendous variety of food resources such as fish and wildlife. Moreover, the rich alluvial soils of deltaic landscapes allowed for the establishment of bountiful crops vital to the establishment and expansion of early cultures (Stanley and Warne, 1997). Today, deltaic wetlands that were more recently viewed as uninhabitable wasteland are once again considered to be of utmost ecological value. Additionally, ancient deltaic deposits are actively explored as they often contain a wealth of fossil fuels such as hydrocarbons and coal, resources that helped spur the industrial revolution and continue to be critically important to our modern technologically driven society.

Table D4 Morphologic, stratigraphic, and sedimentologic characteristics of delta depositional systems that result in fluvial, wave, and tide-dominated systems (modified from Galloway, 1975).

Characteristics	Fluvial-dominated	Wave-dominated	Tide-dominated
Geomorphology	Elongate to lobate	Arcuate	Estuarine to irregular
Channels	Straight to sinuous distributaries	Meandering distributaries	Flaring straight to sinuous distributaries
Sediments	Muddy to mixed	Sandy	Variable
Framework facies	Distributary mouth bar and channel fill sands, delta margin sand sheet	Coastal barrier and beach-ridge sands	Estuary fill and tidal sand ridges
Framework geometry	Parallels depositional slope	Parallels depositional strike	Parallels depositional slope

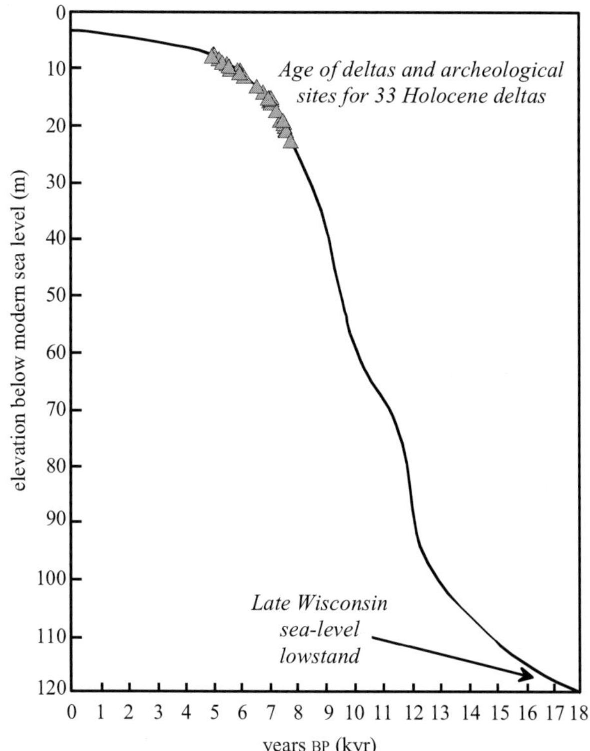

Figure D8 Graph showing the time of formation for latest Quaternary deltaic systems and ages of deltaic archeological sites in relation to deceleration in the rate of sea-level rise following the late Wisconsin sea-level lowstand (ages from Stanley and Warne, 1997; sea-level curve from Fairbanks, 1989).

However, as humans have populated the world's deltas, harnessed the rivers that built them by flood-control structures, and altered natural deltaic coastlines the delicate balance of these dynamic systems has been compromised. Deprived of water and sediment, most of the world's deltas are undergoing dramatic coastal land loss and habitat change as a result of anthropogenic induced as well as naturally occurring subsidence and rising sea level along coastal zones. As a result, a new multidisciplinary field of coastal restoration science is emerging as humans embark to rescue and effectively manage deltaic landscapes. The future of modern deltas will be dependent on our ability to balance environmental protection and economic development, while developing new restoration technologies for their maintenance and restoration.

Shea Penland and Mark A. Kulp

Bibliography

Alonso, B., Field, M.E., Gardner, J.V., and Maldonado, A., 1990. Sedimentary evolution of the Pliocene and Pleistocene Ebro margin, northeastern Spain. *Marine Geology*, **95**: 313–331.

Bates, C.C., 1953. Rational theory of delta formation. *American Association of Petroleum Geologists Bulletin*, **37**: 2119–2161.

Boyd, R., Suter, J.R., and Penland, S., 1989. Sequence stratigraphy of the Mississippi delta. *Gulf Coast Association of Geological Societies Transactions*, **39**: 331–340.

Coleman, J.M., 1981. *Deltas: Processes of Deposition and Models for Exploration*. Minneapolis: Burgess.

Coleman, J.M., and Prior, D.B., 1980. Deltaic sand bodies. In *American Association of Petroleum Geologists Continuing Education Course* 15, Tulsa, OK, p. 171

Coleman, J.M., and Wright, L.D., 1975. Modern river deltas: variability of processes and sand bodies. In Broussard, M.L. (ed.), *Deltas: Models for Exploration*. Houston, TX: Houston Geological Society, pp. 99–146.

Davis, R.A., Jr., 1983. *Depositional Systems: A Genetic Approach to Sedimentary Geology*. Englewood Cliffs: Prentice Hall.

Davis, R.A., Jr., 1994. *The Evolving Coast*. New York: Scientific American Library.

Donaldson, A.C., Martin, R.H., and Kanes, W.H., 1970. Holocene Guadalupe delta of Texas gulf coast. In Morgan, J.P. (ed.), *Deltaic Sedimentation: Modern and Ancient*. Tulsa, OK: Society of Economic Paleontologists and Mineralogists, Special Publication 15, pp. 107–137.

Elliot, T., 1986. Deltas. In Reading, H.G. (ed.), *Sedimentary Environments*. Oxford: Blackwell Scientific Publications, pp. 113–154.

Fairbanks, R.G., 1989. A 17,000-year glacio-eustatic sea level record: influence of glacial melting rates on the Younger Dryas event and deep-ocean circulation. *Nature*, **342**: 637–642.

Fisher, W.L., Brown, L.F., Jr., Scott, A.J., and McGowen, J.H., 1969. Deltas systems in the exploration for oil and gas: a research colloquium. Austin: University of Texas, Bureau of Economic Geology.

Frazier, D.E., 1967. Recent deltaic deposits of the Mississippi River: their development and chronology. *Gulf Coast Association of Geological Societies Transactions*, **27**: 287–315.

Galloway, William E., 1975. Process framework for describing the morphologic and stratigraphic evolution of deltaic depositional systems. In Broussard, M.L. (ed.), *Deltas: Models for Exploration*. Houston, TX: Houston Geological Society, pp. 87–96.

Gilbert, G.K., 1884. The topographical features of lake shores. *United States Geological Survey Annual Report*, **5**: 104–108.

Inman, D.L., and Nordstrom, C.E., 1971. On the tectonic and morphologic classification of coasts. *Journal of Geology*, **97**: 1–21.

Kindinger, J.L., 1988. Seismic stratigraphy of the Mississippi–Alabama shelf and upper continental slope. *Marine Geology*, **83**: 79–94.

Kindinger, J.L., Balson, P.S., and Flocks, J.G., 1994. Stratigraphy of the Mississippi–Alabama shelf and the mobile river incised-valley system. In Dalrymple, R.W., Boyd, R., and Zaitlin, B.A. (eds.), *Society of Economic Paleontologists and Mineralogists, Special Publication* 51, pp. 83–95.

Kostaschuk, R.A., 1985. River mouth processes in a fjord delta, British Columbia Canada. *Marine Geology*, **69**: 1–23.

Maldonado, A., 1975. Sedimentation, stratigraphy and development of the Ebro Delta, Spain. In Broussard, M. L. (ed.), *Deltas: Models for Exploration*. Houston: Houston Geological Society, pp. 311–338.

McEwen, M.C., 1969. Sedimentary facies of the modern Trinity delta. In Lanlford, R.R., and Rogers, J.J.W. (eds.), *Holocene Geology of the Galveston Bay Area*. Houston Geological Society, pp. 53–77.

Moore, G.T., and Asquith, D.O., 1971. Delta: term and concept. *Geological Society of America Bulletin*, **82**: 2563–2568.

Morton, R.A., and Price, W.A., 1987. Late Quaternary sea-level fluctuations and sedimentary phases of the Texas coastal plain and shelf. In Nummedal, D., Pilkey, O.H., and Howard, J.D. (eds.), *Sea-Level Fluctuation and Coastal Evolution*. Society of Economic Paleontologists and Mineralogists, Special Publication 41, pp. 181–198.

Müller, G., 1966. The new Rhine delta in Lake Constance. In Broussard, M.L. (ed.), *Deltas: Models for Exploration*. Houston, TX: Houston Geological Society, pp. 107–124.

Naidu, A.S., and Mowatt, T.C., 1975. Depositional environments and sediment characteristics of the Colville and adjacent deltas, Northern Artic Alaska. In Broussard, M.L. (ed.), *Deltas: Models for Exploration*. Houston, TX: Houston Geological Society, pp. 283–307.

Nichol, S.L., Boyd, R., and Penland, S., 1996. Sequence stratigraphy of a coastal-plain incised valley estuary: Lake Calcasieu, LA. *Journal of Sedimentary Research*, **66**(4): 847–857.

Orton, G.J., and Reading, H.G., 1993. Variability of deltaic processes in terms of sediment supply, with particular emphasis on grain size. *Sedimentology*, **40**: 475–512.

Penland, S., and Suter, J.R. 1989. The geomorphology of the Mississippi River chenier plain. *Marine Geology*, **90**: 231–258.

Reineck, H.E., and Singh, I.B., 1973. *Depositional Sedimentary Environments with Reference to Terrigenous Clastics*. New York: Springer Verlag.

Stanley, D.J., and Warne, A.G., 1994. Worldwide initiation of Holocene marine deltas by deceleration of sea-level rise. *Science*, **265**: 228–231.

Stanley, D.J., and Warne, A.G., 1997. Holocene sea-level change and early human utilization of deltas. *Geological Society of American Today*, **7**(12): 1–7.

Suter, J.R., 1994. Deltaic coasts. In Carter, R.W.G., and Woodroffe, C.D. (eds.), *Coastal Evolution: Late Quaternary Shoreline Morphodynamics*, pp. 87–114.

Suter, J.R., Berryhill, H.L., Jr., and Penland, S., 1987. Late Quaternary sea-level fluctuations and depositional sequences, southwest Louisiana continental shelf. In Nummedal, D., Pilkey, O. H., and

Howard, J. D. (eds.), *Sea-Level Fluctuation and Coastal Evolution*. Society of Economic Paleontologists and Mineralogists, Special Publication 41, pp. 199–219.
Tye, R.S., and Coleman, J.M., 1989. Depositional processes and stratigraphy of fluvially dominated lacustrine deltas: Mississippi Delta plain. *Journal of Sedimentary Petrology*, **59**(6): 973–996.
van Heerden, I., and Roberts, H.H., 1988. Facies development Atchafalaya delta, Louisiana: a modern bayhead delta. *American Association of Petroleum Geologists*, **72**(4): 439–453.
Wright, L.D., 1977. Sediment transport and deposition at river mouths: a synthesis. *Geological Society of America Bulletin*, **88**: 857–868.
Wright, L.D., and Coleman, J.M., 1973. Variations in morphology of major river deltas as functions of ocean wave and river discharge regimes. *American Association of Petroleum Geologists, Bulletin*, **57**: 370–398.

Cross-references

Changing Sea Levels
Coastal Sedimentary Facies
Deltaic Ecology
Ingression, Regression, and Transgression
Late Quaternary Marine Transgression
Sea-Level Changes During the Last Millenium
Sea-Level Rise, Effect
Sedimentary Basins
Sequence Stratigraphy
Submerging Coasts
Wetlands

DEMOGRAPHY OF COASTAL POPULATIONS

Introduction

To most observers who live along a coast it seems clear that the resident population is growing rapidly. But how fast is population growing? Will future growth be concentrated in the coastal zone? Do such demographics portend well for sustainable development?

To answer these questions we have combined the recent work of Burke *et al.* (2001), who reported the fraction of a country's population living within 100 km of the coast, with United Nations (1996) population figures to calculate the 2000 and 2025 coastal populations for 122 countries. Burke *et al.* relied on the work of the Center for International Earth Science Information Network (CIESIN, see reference in Burke *et al.*) at Columbia University, to obtain the coastal population fraction. In our work, we made no attempt to reanalyze the work of CIESIN. Rather, for country-by-country, we multiplied the fractional coastal population reported in Burke *et al.* by United Nations 2000 and 2025 population values (United Nations, 1996).

The United Nations based its calculated demographic projections for each country on birth, death, fertility rates, and age structure of the population. We assumed that the 2025 coastal population fractions would remain unchanged from the 2000 values reported by Burke *et al.* (2001). It is likely, however, that for some countries, such as the United States, coastal populations will grow faster than non-coastal areas (Culliton *et al.*, 1990), as aging populations will live longer and will seek to live near marine environments. Thus, the 2025 populations reported herein should be considered minimum values.

Coastal properties and demographics by country and region

Table D5 lists the results for the 123 countries (column 1) *in thousands of people*. The countries are grouped within eight geographic regions (Burke *et al.*, 2001), with each country listed beneath the region. Coastal lengths and claimed Exclusive Economic Zone are from Burke *et al.* (2001), and the total population values (column 4) are from United Nations (1996). Each country's demographics are broken down into the 2000 and 2025 total populations, the 2000 and 2025 coastal populations (persons living within 100 km of the coast), the difference between 2000 and 2025 coastal populations (column 9), and the relative increase or decrease over the 2000 population (%). The footnotes at the base of Table D5 specify the calculations.

Not surprisingly, China has the largest coastal population, 306 and 355 million people for 2000 and 2025, respectively, or 16% growth in 25 years. Belgium and Denmark will experience 1%, or less, growth in their coastal population from 2000 through 2025. Fifteen other countries in Europe, plus Japan, will experience decreasing coastal populations in 2025. All in all, only 16 countries are projected to have declining coastal populations in the next quarter century or so.

Figures D9–D12 show the overall 2000 and 2025 demographics trends for the eight geographic regions. Figure D9 shows the total *coastal population* estimated for the year 2000, and that projected for 2025 in millions of people. Figures D10 and D11 compare overall total population, by region, with the corresponding total *coastal population*, for 2000 and 2025, respectively. In Figure D12 we present the *population increase per kilometer of coastline* for the eight regions summarized in Table D5.

Middle East and North Africa and sub-Saharan Africa coastal populations will grow by 52% and 81%, respectively, from 2000 to 2025 (Figure D9). Combined, these changes represent an addition of approximately 209 million people living in the coastal area by 2025 over the 2000 coastal population of approximately 330 million people.

The Asian region (which excludes the Middle East), which includes heavily populated China, India, and Indonesia, will experience relatively smaller percentage growth: 25%, by 2025. However, in absolute terms Asian coastal region will be home to 1.6 billion people in 2025, an increase of 325 million over 2000.

North America, Central America and the Caribbean Islands, South America, and Oceania will experience increased growth by 20, 39, 32, and 32%, respectively, or an increase of 121 million people by 2025; the calculated total coastal population in 2025 for these four regions is 528 billion (Table D5).

Europe is the only region were coastal populations may decrease rather than increase. In 2000, the European coastal population is estimated at 287 million people (Table D5); by 2025 we calculate that the coastal population to be 280 million, a drop of 7 million people. The reason for the decrease in coastal population is due to population stability (zero growth rate) or in some cases negative population growth rates.

Figure D12 shows the 2000 and 2025 coastal populations normalized to coastal length, a factor that can indicate "environmental stress" (Culliton *et al.*, 1990). By 2025, the regions of Asia, Middle East/North Africa, and sub-Saharan Africa will contain more than 4,000 persons per kilometer of coastal length. The impacts of such a high density of people along a coast will include marine pollution, over-fishing, and loss of natural habitat.

Consequence of increasing coastal populations

Much of humanity lives in the coastal zone, and probably has done so for millennia. That a 32% increase in coastal population, compared with a 30% increase overall by 2025, should be no surprise, given the uncertainty of such calculations. We estimate then that globally, the coastal populations will grow slightly faster than the overall population, but that the difference is not statistically significant. Clearly, however, sub-Saharan Africa coastal populations will grow much more rapidly than the global average (~81%), while Europe (−2 or −3%) will be much less.

In terms of gross numbers however, it is Asia, with a projected 25% increase translating to 325 million more persons living in the coastal zone by 2025 that is of greatest potential environmental impact. Sub-Saharan Africa's 81% increase equates to 115 million more people in their coastal zones. These two regions, totaling 450 million more coastal dwellers by 2025, account for 69%—more than two-thirds—of all population growth. Socioeconomically, it might well be argued that these two regions also represent the lowest *per capita* income of all Earth's people.

Rapid coastal population growth coupled with limited financial and natural resources is a well-documented recipe arguing against sustainable development (see Maul—*Small Islands*—this volume). The weathered mantra "Give a man a fish—feed him for a day; teach a man to fish—feed him forever" will not withstand the scrutiny of the population statistics in Table D5. Humankind requires resources to survive, infrastructure to provide protection from natural hazards, and a society to provide essentials of public health and governance. Coastal resources are finite, as is Earth's ability to provide in extra-coastal regions. Overfishing is perhaps only the first sign that coastal-dwelling humankind is growing beyond sustainable limits

While feeding burgeoning populations is an essential enterprise, "sustainable development" must also include some other aspects of quality of life. In many tropical regions, the great storms variously called hurricanes, typhoons, or cyclones, have been the cause of more loss of life

Table D5 Coastal properties and demographics by country and region

	Coastal length[a] (km)	Claimed exclusive economic zone[b] (000 km^2)	Total 2000 population[c] (000 people)	Population within 100 km from coast[d] (%)	2000 Population within 100 km coast[e] (000 people)	Total 2025 population[f] (000 people)	2025 Population within 100 km coast[g] (000 people)	Population change, 2000–2025, within 100 km from coast[h] (000 people)	Increase (%)[i]
Asia									
Azerbaijan	871		7828	55.7	4360	9714	5411	1051	24
Bangladesh	3306	39.9	128310	54.8	70314	179980	98629	28315	40
Cambodia	1127		11207	23.8	2667	16990	4044	1376	52
China	30017		1276301	24	306312	1480430	355303	48991	16
Georgia	376	18.9	5418	38.8	2102	5762	2236	133	6
India	17181	2103.4	1006770	26.3	264781	1330201	349843	85062	32
Indonesia	95181	2915	212565	95.9	203850	275245	263960	60110	29
Japan	29020	3648.4	126428	96.3	121750	121348	116858	−4892	−4
Kazakhstan	4528		16928	3.6	609	20047	722	112	18
Korea, Dem People's Reb	4009	72.8	23913	92	22000	30046	27642	5642	26
Korea, Rep	12478	202.6	46883	100	46883	52533	52533	5650	12
Malaysia	9323	198.2	22299	98	21853	31577	30945	9092	42
Myanmar (Burma)	14708	358.5	49342	49	24178	67643	33145	8967	37
Pakistan	2599	201.5	156007	9.1	14197	268904	24470	10274	72
Philippines	33900	293.8	75037	100	75037	105194	105194	30157	40
Singapore	268		3587	100	3587	4912	4912	1325	37
Sri Lanka	2825	500.8	18821	100	18821	23934	23934	5113	27
Thailand	7066	176.5	60495	38.7	23412	69089	26737	3326	14
Turkmenistan	1289		4479	8.1	363	6470	524	161	44
Uzbekistan	1707		25018	2.6	650	36500	949	299	46
Viet Nam	11409	237.8	80549	82.8	66695	110107	91169	24474	37
Sums and Increase (%)[i]			*3358185*		*1294420*	*4246626*	*1619160*	*324740*	*25*
Europe									
Albania	649		3493	97.1	3392	4295	4170	779	23
Belgium	76		10257	83	8513	10271	8525	12	0
Bosnia and Herzegovina	23		4338	46.6	2022	4303	2005	−16	−1
Bulgaria	457	25.7	8306	29.2	2425	7453	2176	−249	−10
Croatia	5663		4485	37.9	1700	4243	1608	−92	−5
Denmark	5316	80.4	5274	100	5274	5324	5324	50	1
Estonia	2956	11.6	1418	85.9	1218	1256	1079	−139	−11
Finland	31119		5179	72.8	3770	5294	3854	84	2
France	7330	706.4	59061	39.6	23388	60393	23916	527	2
Germany	3624	37.4	82688	14.6	12072	80877	11808	−264	−2
Greece	15147		10597	99.2	10512	10074	9993	−519	−5
Iceland	8506	678.7	282	99.9	282	336	336	54	19
Ireland	6437		3574	99.9	3570	3723	3719	149	4
Italy	9226		57194	79.1	45240	51744	40930	−4311	−10
Latvia	565	15.6	2397	75.2	1803	2108	1585	−217	−12
Lithuania	258	3.6	3690	22.9	845	3521	806	−39	−5
Netherlands	1914		15871	93.4	14824	16141	15076	252	2
Norway	53199	1095.1	4407	95.4	4204	4662	4448	243	6
Poland	1032	19.4	38727	13.5	5228	39973	5396	168	3
Portugal	2830	1656.4	9788	92.7	9073	9438	8749	−324	−4
Romania	696	18	22505	6.3	1418	21098	1329	−89	−6
Russian Federation	110310	6255.8	146196	14.9	21783	131395	19578	−2205	−10
Slovenia	41		1914	60.6	1160	1738	1053	−107	−9
Spain	7268	683.2	39801	67.9	27025	37500	25463	−1562	−6

Table D5 *(Continued)*

	Coastal length[a] (km)	Claimed exclusive economic zone[b] (000 km^2)	Total 2000 population[c] (000 people)	Population within 100 km from coast[d] (%)	2000 Population within 100 km coast[e] (000 people)	Total 2025 population[f] (000 people)	2025 Population within 100 km coast[g] (000 people)	Population change, 2000–2025, within 100 km from coast[h] (000 people)	Increase (%)[I]
Sweden	26384	73.2	8898	87.7	7804	9511	8341	538	7
Ukraine	4953	86.4	50801	20.9	10617	45979	9610	−1008	−9
United Kingdom	19717		58336	98.6	57519	59353	58522	1003	2
Yugoslavia			10502	8.1	851	10679	865	14	2
Sums and Increase (%)[I]			*669979*		*287533*	*642682*	*280264*	*−7269*	*−2.53*
Middle East and North Africa									
Algeria	1557		31599	68.8	21740	47322	32558	10817	50
Egypt	5898	185.3	68119	53.1	36171	95766	50852	14681	41
Iran	5890	129.7	76429	23.9	18267	128251	30652	12385	68
Iraq	105	0.7	23109	5.7	1317	41600	2371	1054	80
Israel	205		6077	96.6	5870	7977	7706	1835	31
Jordan	27		6330	29	1836	11894	3449	1614	88
Kuwait	756		1966	100	1966	2904	2904	938	48
Lebanon	294		3289	100	3289	4424	4424	1135	35
Libya Arab Jamahiriva	2025	222.4	6387	78.7	5027	12885	10140	5114	102
Morocco	2008	328.4	28984	65.1	18869	39925	25991	7123	38
Oman	2809	487.4	2717	88.5	2405	6538	5786	3382	141
Saudi Arabia	7572		21661	30.2	6542	42368	12795	6254	96
Syrian Arab Republic	212		16126	34.5	5563	26303	9075	3511	63
Tunisia	1027		9837	84	8263	13524	11360	3097	37
Turkey	8140	176.6	65732	57.5	37796	85791	49330	11534	31
United Arab Emirates	2871	21.2	2444	84.9	2075	3297	2799	724	35
Yemen	3149	465	18118	63.5	11505	39589	25139	13634	119
Sums and Increase (%)[i]			*388924*		*188500*	*610,358*	*287331*	*98831*	*52*
Sub-Saharan Africa									
Angola	2252		12781	29.4	3758	25547	7511	3753	100
Benin	153		6222	62.4	3883	12276	7660	3778	97
Cameroon	1799	10.9	15129	21.9	3313	28521	6246	2933	89
Congo	205		2982	24.5	731	5740	1406	676	92
Congo, Dem Rep	177		51749	2.7	1397	105925	2860	1463	105
Cote d'Ivoire	797	157.4	15144	39.7	6012	24397	9686	3673	61
Equatorial Guinea	603	291.4	452	72.3	327	798	577	250	77
Eritrea (Red Sea)	3446		3809	73.5	2800	6504	4780	1981	71
Gabon	2019	180.7	1235	62.8	776	2118	1330	555	71
Gambia	503	20.5	1244	90.8	1130	1984	1801	672	59
Ghana	758	216.9	19928	42.5	8469	36341	15445	6976	82
Guinea	1614	97	7861	40.9	3215	15286	6252	3037	94
Guinea-Bissau	3176	86.7	1180	94.6	1116	1921	1817	701	63
Kenya	1586	104.1	30340	7.6	2306	50202	3815	1510	65
Liberia	842		3256	57.9	1885	6573	3806	1921	102
Madagascar	9935	1079.7	17395	55.1	9585	34476	18996	9412	98
Mauritania	1268	141.3	2580	39.6	1022	4443	1759	738	72
Mozambique	6942	493.7	19556	59	11538	35444	20912	9374	81
Namibia	1754	536.8	1733	4.7	81	2999	141	60	73
Nigeria	3122	164.1	127868	25.7	32862	238397	61268	28406	86
Senegal	1327	147.2	9495	83.2	7900	16896	14057	6158	78
Sierra Leone	1677		4866	54.7	2662	8200	4485	1824	69

Somalia	3898		11530	54.8	6318	23669	12971	6652	105
South Africa	3751		46257	38.9	17994	71621	27861	9867	55
Sudan	2245		29823	2.8	835	46850	1312	477	57
Tanzania, United Rep	3461	204.3	33687	21.1	7108	62436	13174	6066	85
Togo	53	10.8	4676	44.6	2085	8762	3908	1822	87
Sums and Increase (%)[i]			*482778*		*141107*	*878326*	*255838*	*114730*	*81*
North America									
Canada	265523	3006.2	30679	23.9	7332	36385	8696	1364	19
United States	133312	8078.2	277825	43.3	120298	332481	143964	23666	20
Sums and Increase (%)[i]			*308504*		*127631*	*368866*	*152660*	*25030*	*20*
Central America and Caribbean									
Belize	1996	12.8	242	100	242	375	375	133	55
Costa Rica	2069	542.1	3798	100	3798	5608	5608	1810	48
Cuba	14519	222.2	11201	100	11201	11798	11798	597	5
Dominican Rep	1612	246.5	8498	100	8498	11164	11164	2666	31
El Salvador	756		6319	98.8	6243	9221	9110	2867	46
Guatemala	445	104.5	12222	61.2	7480	21668	13261	5781	77
Haiti	1977	86.4	7817	99.6	7786	12513	12463	4677	60
Honduras	1878	201.2	6485	65.5	4248	10565	6920	2672	63
Jamaica	895	234.8	2587	100	2587	3370	3370	783	30
Mexico	23761	2997.7	98881	28.7	28379	130196	37366	8987	32
Nicaragua	1916		4694	71.6	3361	7639	5470	2109	63
Panama	5637	274.6	2856	100	2856	3779	3779	923	32
Trinidad and Tobago	704	60.7	1341	100	1341	1692	1692	351	26
Sums and Increase (%)[i]			*166941*		*88019*	*229588*	*122376*	*34357*	*39*
South America									
Argentina	8397	925.4	37032	45.1	16701	48896	22052	5351	32
Brazil	33379	344.5	169202	48.6	82232	216596	105266	23033	28
Chile	78563	3415.9	15211	81.5	12397	19548	15932	3535	29
Colombia	5874	706.1	38905	29.9	11633	52668	15748	4115	35
Ecuador	4597		12646	60.5	7651	17796	10767	3116	41
Guyana	1154	122	874	76.6	669	1114	853	184	27
Peru	3362		25662	57.2	14679	35518	20316	5638	38
Suriname	620	119.1	452	87	393	605	526	133	34
Uruguay	1096	110.5	3274	78.5	2570	3692	2898	328	13
Venezuela	6762	385.7	24170	73.1	17668	34775	25421	7752	44
Sums and Increase (%)[i]			*327428*		*166594*	*431208*	*219778*	*53185*	*32*
Oceania									
Australia	66530	6664.1	18838	89.8	16917	23931	21490	4574	27
Fiji	4637	1055	848	99.9	847	1170	1169	322	38
New Zealand	17209	3887.4	3760	100	3760	4878	4878	1118	30
Papua New Guinea	20197	1613.8	4811	61.2	2944	7546	4618	1674	57
Solomon Islands	9880	1377.1	444	100	444	844	844	400	90
Sums and Increase (%)[i]			*28701*		*24912*	*38369*	*32999*	*8087*	*32*
Grand Totals[j]			*5731400*		*2318716*	*7446023*	*2970407*	*651691*	

[a] From Burke *et al.* (2001).
[b] From Burke *et al.* (2001).
[c] From United Nations (1996).
[d] From Burke *et al.* (2001).
[e] Total 2000 population × 0.01 × % population within 100 km from coast.
[f] From United Nations (1996).
[g] Total 2000 population × 0.01 × % population within 100 km from coast.
[h] 2025 population within 100 km from the coast minus 2000 population within 100 km from the coast.
[i] ((2025 coastal population divided by 2000 coastal population) − 1 × 100).
[j] Grand Totals, sums for all countries.

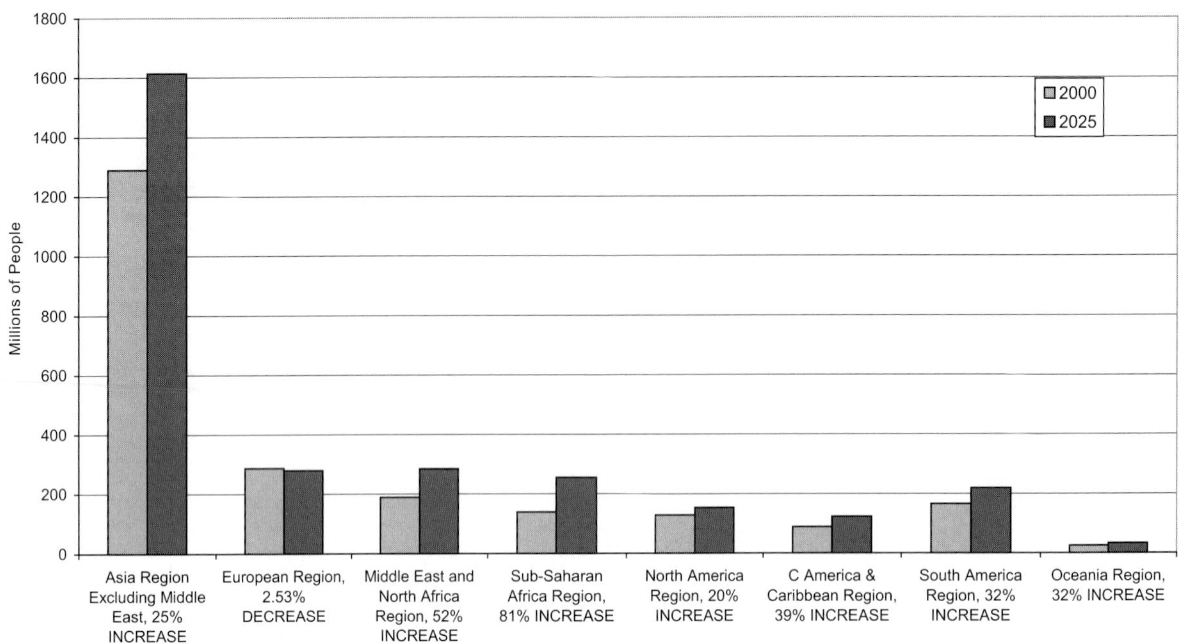

Figure D9 Coastal populations within 100 km of coast by region and relative (%) increase or decrease by region, 2000 and 2025.

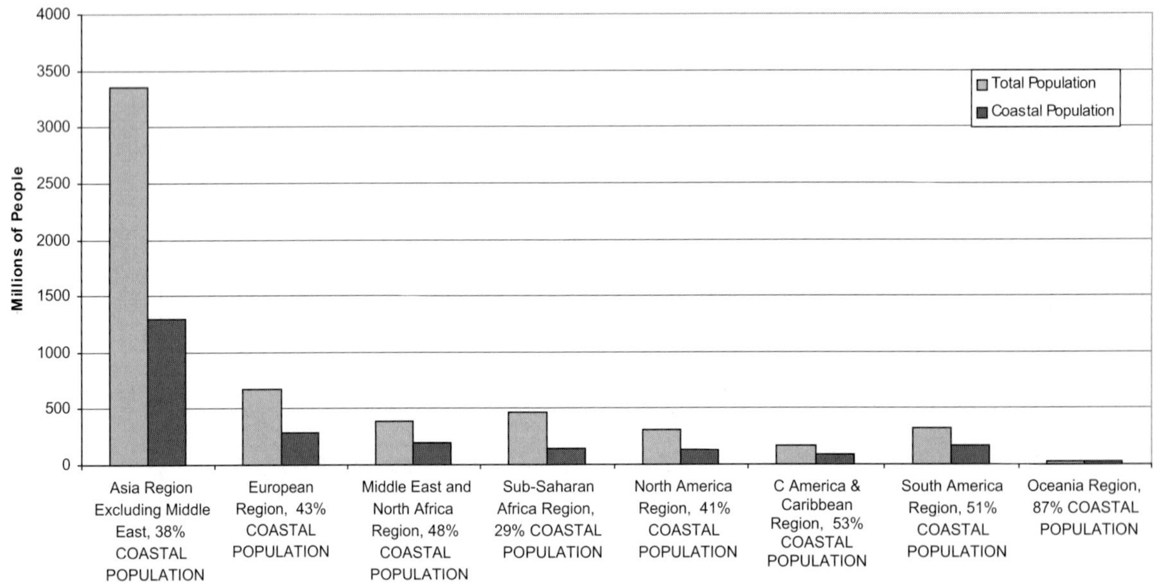

Figure D10 Total population, 2000 (countries with a coastal zone) and coastal population living within 100 km of the coast, by region, 2000.

than any other natural hazard. In the 20th century, an estimated 1,000,000 souls have been lost to this one hazard. With a 69% increase in coastal populations in Asia and Sub-Saharan Africa by 2025, the risk is substantially increased. In extratropical environments, these storms are still extremely destructive, not to mention the intense winter storms of upper mid-latitude flailing against coastal populations. Modern weather forecasting with satellite and radar observations in developed countries clearly is mitigating the risk to coastal communities of such storms.

Another coastal natural hazard, tsunami waves, knows no latitudinal boundaries, as do tropical storms. Although often thought of as a natural hazard of the Pacific Basin, tsunamis are in fact a source of great risk to coastal dwellers in all oceans, especially the Mediterranean and the Caribbean seas. The great Lisbon earthquake and tsunami of 1755 would cause orders-of-magnitude more loss of life and socioeconomic disruption today if repeated in the North Atlantic Ocean.

Beyond the natural hazards, those of geophysical origin, are the biogeographical hazards of intense coastal population growth and development. Humankind produces domestic and industrial wastes. These wastes can be hazardous to others of our species—producing numerous and well-known toxicities or communicable diseases. Given that the greatest growth of our species is in two coastal regions—Asia and Africa—it is reasonable to project increased reports of waterborne public heath outbreaks. Implantation of acceptable waste management practices, for example, alone, is potentially of a magnitude to overcome all financial resources of intergovernmental banking agencies in the next quarter-century.

Finally, we briefly compare future coastal population with future tourist population. According to Goldberg (1994) world tourism in 1970 grew from 160 million to 341 million people in 1985, or an annual growth rate of 5%. If one projects this growth rate exponentially from 1985 through 2025, total world tourism will grow to 2.5 billion, or

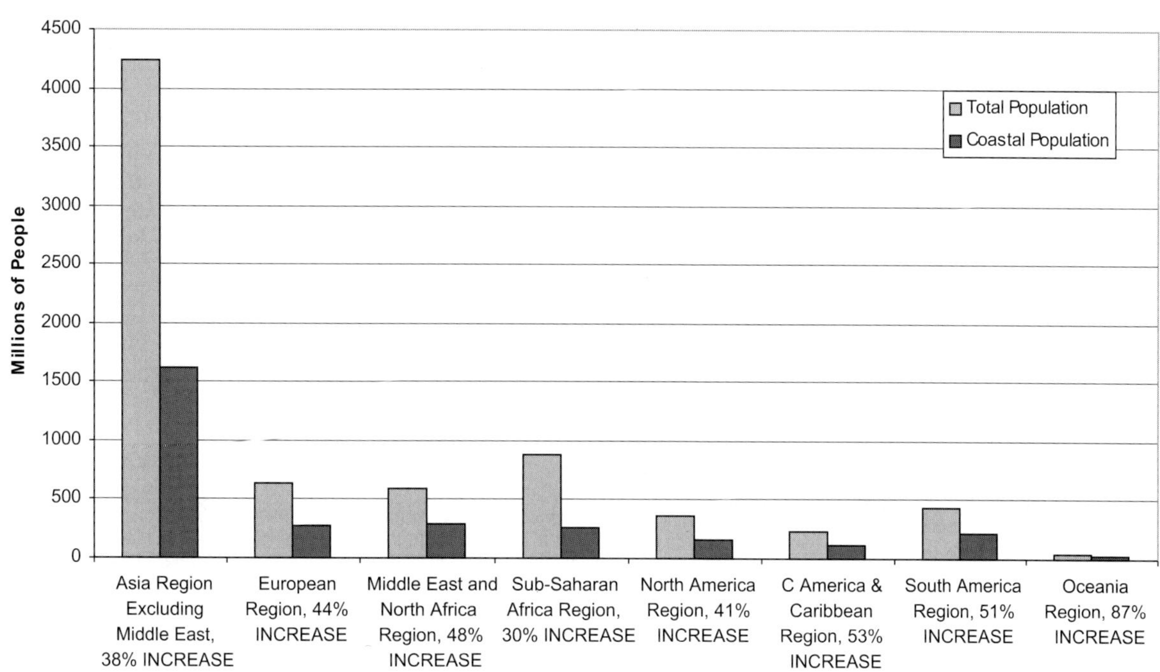

Figure D11 Total population, 2025 (countries with a coastal zone) and coastal population living within 100 km of the coast, by region, 2025.

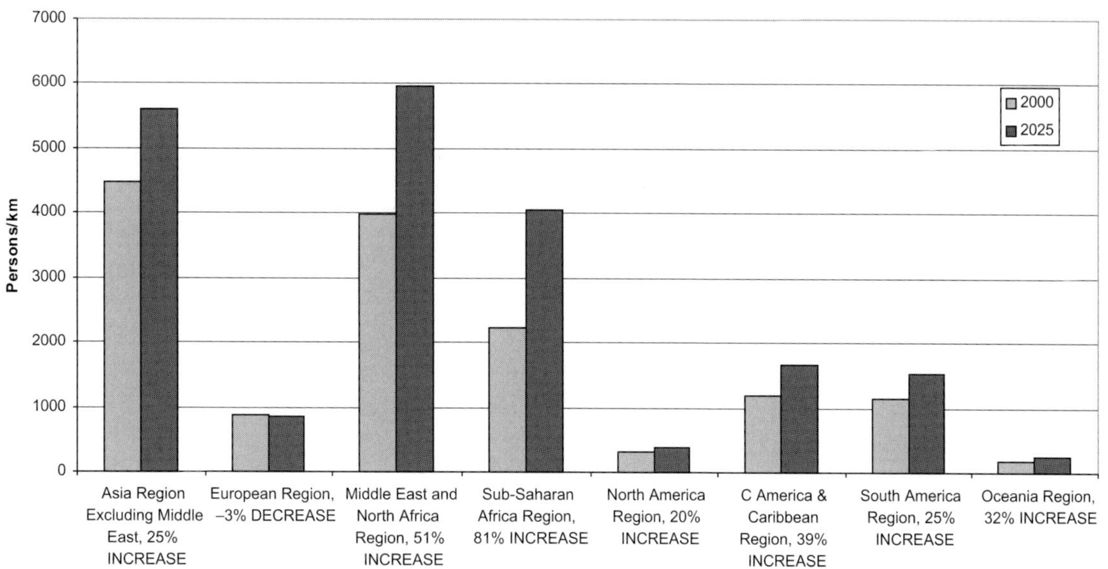

Figure D12 Population per 1 km of coastline, 2000 and 2025, living within 100 km of the coast, by region, relative (%) increase or decrease.

approximately four times the increase in the world resident coastal population calculated (Table D5). While it is problematic whether world tourism can sustain such a large growth, and surely not all tourists seek only the coastal zone, the calculation does demonstrate that some coastal countries may need to either limit tourism or develop new or more efficient infrastructure to accommodate the growing numbers of people living and/or recreating on beaches or in water. For example, the economies for many of the countries in Central America, Caribbean Region, and Oceania are driven by tourism. Thus, such countries will need to carefully consider their options in the areas of waste disposal, habitat, and pollution in order to optimize conditions leading to clean and natural environments to sustain both the local and national economy.

Conclusions

It is not our intention to paint a bleak picture of our coastal future—but of a realistic scenario of population growth and environmental and socioeconomic consequences. Earth is remarkably resilient to humankind's activities. Our planet's ability to dilute pollution is not infinite, and our coastal impacts will exacerbate the global problem. Surprisingly, the global population growth is not significantly different from what is reasonable to expect in the coastal zone—that is within 100 km of the shore. The environmental and socioeconomic impact, however, may be more. The essence of this increased potential impact is simply gravity.

The rivers and streams that bring mid-continent discharges to the coastal waters are essentially pressure gradient forces forced by gravity. Communities that discharge their wastes into coastal waters, whether by gravity or pumping, inject their wastes into littoral ecosystems. The 651,691,000 (±) more of us living within 100 km of the coast by 2025 will impact the very small fraction of Earth's surface area we consider home to 40% of humankind. Surviving that percentage growth is a taxing but not insurmountable challenge.

No other challenge will tax the intellect of humankind more than population growth. It is comforting perhaps that the challenge in the coastal zone is not more daunting than in other regions. However, oceanographers have long known that coastal ecosystems are particularly at risk from human activity. These fragile natural communities are increasingly seen as endangered, not only because of those living within 100 km of the coast, but also because of all others whose effluent reaches the coast.

Iver W. Duedall and George A. Maul

Bibliography

Burke, L., Kura, Y., Kassem, K., Revenga, C., Spalding, M., and McAllister, D., 2001. *Coastal Ecosystems*. Washington, DC: World Resources Institute.

Culliton, T.J., Warren, M.A., Goodspeed, T.R., Remer, D.G., Blackwell, C.M., and MacDonough, J.J., 1990. *50 Years of Population Change along the Nation's Coasts, 1960–2010*. Rockville, MD: National Oceanic and Atmospheric Administration.

Goldberg, E.D. 1994. *Coastal Zone Space, Prelude to Conflicts*. Paris: UNESCO Publishing.

United Nations, 1996. *World Population Prospects*. The 1996 Revision, Annex I: Demographic Indicators, 24 October. New York: UN Department of Economic and Social Affairs, Population Division.

Cross-references

Beach Use and Behaviors
Carrying Capacity in Coastal Areas
Coastal Zone Management
Conservation of Coastal Sites
Economic Value of Beaches
Environmental Quality
Global Vulnerability Analysis
Human Impact on Coasts
Meteorologic Effects on Coasts
Natural Hazards
Small Islands
Tourism and Coastal Development
Water Quality

DEPTH OF CLOSURE ON SANDY COASTS

Kraus *et al.* (1999, p. 272) proposed the following definition of the *depth of closure* (DoC):

> The depth of closure for a given or characteristic time interval is the most landward depth seaward of which there is no significant change in bottom elevation and no significant net sediment exchange between the nearshore and the offshore.

Figure D13 illustrates the concept by showing the change in the nearshore profile resulting from a single storm. In this case, the DoC is 6.45 m as there is little change in the bottom deeper than this. Thus, the DoC separates the active nearshore from a less-active offshore and therefore is an important parameter in many coastal engineering projects. For example, in order to build out the entire beach profile, nourishment quantities are often computed by multiplying the desired added beach width by the DoC. In another application, numerical models of beach change use the DoC as an offshore limit to their computations. Because of its importance, the DoC is a topic of considerable research on how to predict and measure it, and on what it represents with respect to sediment transport.

Analytical prediction

Hallermeier (1977, 1978, 1981a, b, c) developed the first predictive formula for the DoC. Using laboratory tests and limited field data, he hypothesized two limit depths, d_l and d_i (Figure D14). He defined the inshore limit depth, d_l, as the seaward limit of the *littoral zone*, where "intense bed activity (is) caused by extreme near-breaking waves and breaker-related currents" (Hallermeier, 1981c, p. 258). Seaward of d_i, the deeper limit, only insignificant onshore–offshore transport by waves occurs. Between these two limits, is the shoal zone, a buffer region zone where expected waves have neither a strong nor a negligible effect on the sandy bed during a typical annual cycle of wave action.

Hallermeier (1978) suggested an analytical approximation, using linear wave theory for shoaling waves, to predict an *annual* value of d_l:

$$d_l = 2.28H_e - 68.5\left(\frac{H_e^2}{gT_e^2}\right) \qquad \text{(Eq. 1)}$$

where d_l is the annual DoC below mean low water (MLW), H_e is the nonbreaking significant wave height that is exceeded 12 h per year (0.137% of the time), T_e is the associated wave period, and g the acceleration due to gravity.

According to equation 1, d_l is primarily dependent on wave height with an adjustment for wave steepness. Hallermeier (1978, p. 1501) proposed using the 12 h exceeded wave height because it allowed sufficient duration for "moderate adjustment towards profile equilibrium." Equation 1 is based on quartz sand in salt water with a submerged density of $\gamma' = 1.6$ and a median diameter between 0.16 and 0.42 mm, which typifies conditions in the nearshore for many beaches. Because d_l was derived from linear wave theory for shoaling waves, d_l must be seaward of the influence of intense wave-induced nearshore circulation. However, because of various factors, Hallermeier (1978, p. 1502) "proposed that the calculated d_l be used as a minimum estimate of profile close-out depth with respect to low(er) tide level." Because tidal or wind-induced currents may increase wave-induced near-bed flow velocities, Hallermeier suggested using MLW as a reference level to

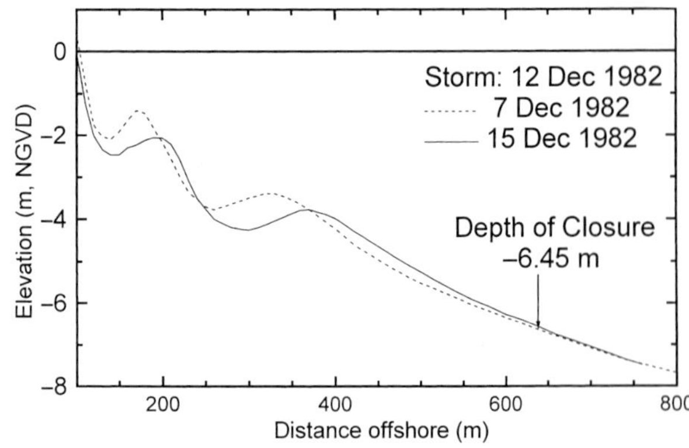

Figure D13 Example nearshore profile change caused by a single storm event with the DoC located at the deepest point of significant bottom change (in this case, <6 cm). Data from the Corps of Engineers' Field Research Facility at Duck, NC, USA.

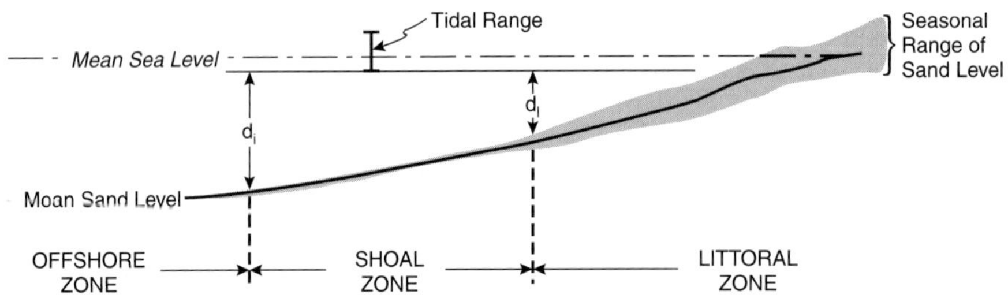

Figure D14 Definition sketch showing limits d_i and d_l, where d_l is the maximum water depth for nearshore erosion by extreme (12 h per year) wave condition (from Hallermeier, 1981c).

obtain a conservative DoC. Although Hallermeier considered only an annual period, Stive *et al.* (1992) extended equation 1 to other time periods by defining H_e as the 12-h exceeded wave height over any time period.

Birkemeier (1985) used two years of profile data from Duck, North Carolina, to examine equation 1. He reasoned that since any storm could be the annual event, an event-dependent d_l could be computed for each storm. Taking this approach, he showed that the functional relationship proposed by Hallermeier appeared reasonable, but equation 1 overpredicted the observed DoC by about 25%. Nicholls *et al.* (1998) expanded on this analysis using 12 years of profile data to examine the parameters in equation 1. Using a depth change criteria of <6 cm between surveys as an indicator of the DoC, they confirmed that using the 12-h exceeded wave height and a datum of MLW provided a conservative estimate of the observed DoC (Figure D15). They also examined the variation of the DoC with time and found that equation 1 became less capable at prediction as the time interval was increased beyond 4 years. They hypothesized that this indicates that, as the time interval increases, the DoC will be governed by other processes beside the highest wave height.

On eastern Lake Michigan, Hands (1983), using 9 years of profile surveys, determined that the closure depth was equal to twice the height of the 5-year return period wave height (H_5):

$$Z \simeq 2H_5 \qquad \text{(Eq. 2)}$$

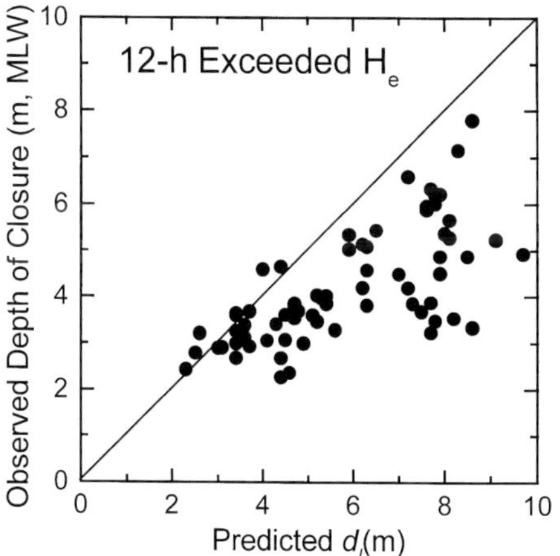

Figure D15 Observed versus predicted DoC using equation 1 for erosional events measured in Duck, NC, USA (modified from Nichols *et al.*, 1998). Except for a few cases of $d_l < 4$ m, equation 1 overpredicts, providing a conservative estimate of the actual observations.

This relationship is similar to equation 1, being based on wave height. One requirement of both equations is that they assume sand-rich profiles and require knowledge of the near-breaking wave climate.

Empirical prediction

The DoC can also be determined empirically by examining seafloor elevation changes measured by repetitive beach and nearshore surveys and identifying the depth at which changes become insignificant (see also Stauble *et al.*, 1993; Grosskopf and Kraus, 1994; Nicholls *et al.*, 1998; Kraus *et al.*, 1999). Since surveys integrate the impact of all processes acting on the profile, they inherently include the influence of storm and calm conditions along with important parameters not included in equation 1 such as grain size, sediment transport, and underlying geology.

In order to identify at what depth changes become insignificant, Kraus and Harikai (1983) examined the standard deviation in depth change computed from a series of surveys, and defined closure as the depth where the standard deviation decreased markedly to a near-constant value. They interpreted the landward region of this depth to be the active profile, where the seafloor was influenced by gravity waves and storm-driven water level changes. The offshore region, of smaller and nearly constant standard deviation, was primarily influenced by lower frequency sediment-transporting processes such as shelf and oceanic currents (Stauble *et al.*, 1993).

An example of how closure was determined empirically at Ocean City, MD, is shown in Figure D16. A clear reduction in the standard deviation occurs at a depth of about 5.5–6 m below the National Geodetic Vertical Datum (NGVD) for the 13 surveys conducted over a period of 5 years. For the 5.6 km of shoreline surveyed at Ocean City, the DoC ranged between 5.5 and 7.6 m. This longshore variation may reflect complex nearshore bathymetry resulting from shore-attached linear shoals. A similar longshore variation in the DoC is shown in Figure D17 for 110 south shore Long Island profile lines. Generally, the depth increases toward the east, with Rockaway Beach averaging 5 m below NGVD and the Montauk zone averaging 7.6 m. These values were based on four high-quality surveys collected between 1995 and 1996 (Morang *et al.*, 1999). The depth increase toward the east corresponds to an eastward increase in wave energy and further reflects the wave height dependency of equation 1. These data also demonstrate that there can be considerable variation in DoC along an 80 km, relatively straight stretch of shore.

While survey data can be used to determine the DoC, surveys have several limitations that require careful consideration. Most importantly, a repetitive survey program should cover at least the time interval of interest with enough frequency to resolve the natural variation which occurs. In general, the longer the program, the more different conditions will be monitored and the more robust the closure depth determination will be. To resolve small changes in depth and in standard deviation, highly accurate surveys (± 2–3 cm) are required and profiles must extend offshore far enough to include the entire active shoreface. In the United States east coast, surveys to 10 m water depth are normally sufficient.

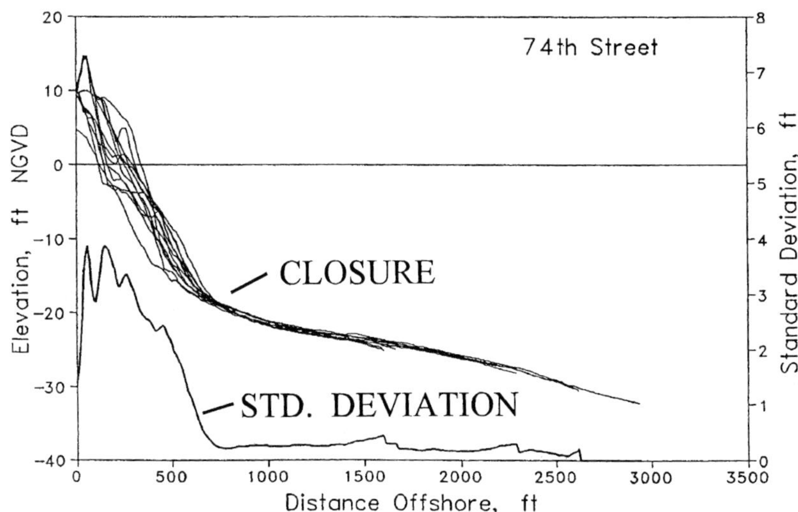

Figure D16 Profile surveys and standard deviation of seafloor elevation at 74th Street, Ocean City, MD (adapted from Stauble *et al.*, 1993). Surveys conducted from 1988 to 1992. Large changes above the datum were caused by beach fill placement and storm erosion.

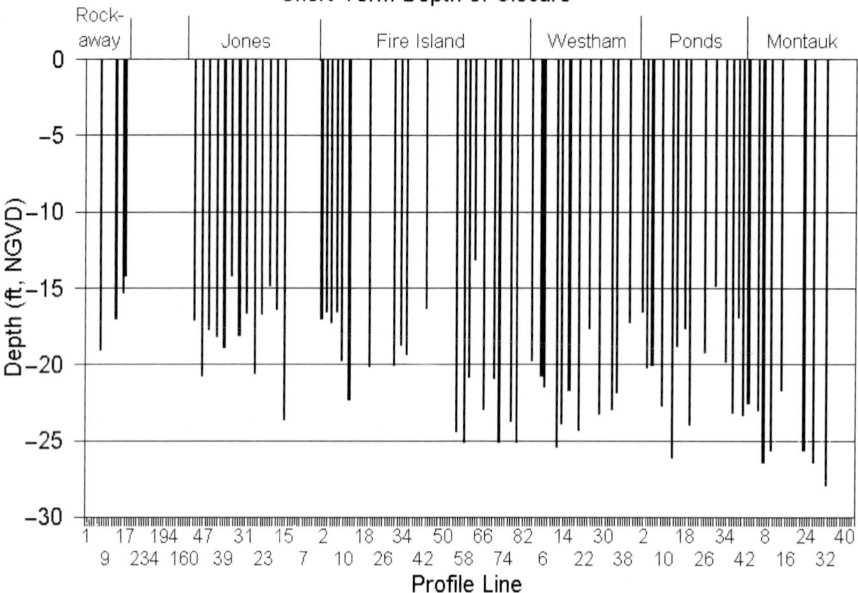

Figure D17 Variation in short-term closure depth along the south shore of Long Island, NY, computed for four survey dates in 1995 and 1996 (adapted from Morang *et al.*, 1999). Zones indicate south shore barriers or regions.

Application

Closure depth is used in a number of coastal engineering applications such as the placement of mounds of dredged material, beach fill, placement of ocean outfalls, and the calculation of sediment budgets. The time dependent nature of the active portion of the shoreface requires a coastal engineer to consider return period. The closure depth that accommodates the 100-year storm will be deeper than one for a 10-year storm. Therefore, a closure depth must be chosen in light of a project's engineering requirements and design life. For example, if a berm is to be built in deep water, where it will be immune from wave resuspension, what is the minimum depth at which it should be placed? This is an important question because of the high costs of transporting and disposing material at sea. It would be tempting to use a safe criterion such as the 100- or 500-year storm, but excessive costs may force the project engineer to consider a shallower site that may be stable only for shorter return period events. The prediction techniques described above allow for both the determination of a closure depth and for the consequences of picking a shallower or deeper depth to be quantified.

Most research into the closure depth have been carried out on sandy beaches with a thick sand layer in wave-dominated environments. Therefore, assumptions behind the concept hold best on open coasts far from structures and inlets and may be less valid for beaches where the morphology is largely controlled by underlying geology, for example, where the active sand is only a veneer above rock or clay. Also, little is known about the DoC for coasts where most sediment is moved by unidirectional currents, such as rivers and tidal channels or where there is long-term net loss of sediment.

Andrew Morang and William A. Birkemeier

Bibliography

Birkemeier, W.A., 1985. Field data on seaward limit of profile change. *Journal of Waterway, Port, Coastal and Ocean Engineering*, **111**(3): 598–602.

Grosskopf, W.G., and Kraus, N.C., 1994. Guidelines for surveying beach nourishment projects. *Shore and Beach*, **62**(2): 9–16.

Hallermeier, R.J., 1977. Calculating a yearly limit depth to the active beach profile. Vicksburg: U.S. Army Engineer Waterways Experiment Station, Coastal Engineering Research Center. *Technical Paper* TP 77–9.

Hallermeier R.J., 1978. Uses for a calculated limit depth to beach erosion. *Proceedings of the Sixteenth Coastal Engineering Conference*. American Society of Civil Engineers, New York: A-S-O-C-E, Ch. 88, pp. 1493–1512.

Hallermeier, R.J., 1981a. A profile zonation for seasonal sand beaches from wave climate. *Coastal Engineering*, **4**: 253–277.

Hallermeier, R.J., 1981b. Terminal settling velocity of commonly occurring sand grains. *Sedimentology*, **28**: 859–865.

Hallermeier, R.J., 1981c. Seaward limit of significant sand transport by waves: an annual zonation for seasonal profiles. Vicksburg: U.S. Army Engineer Waterways Experiment Station, Coastal Engineering Research Center. *Coastal Engineering Technical Aide* CETA 81-2.

Hands, E.B., 1983. The Great Lakes as a test model for profile response to sea level changes. In Komar, P. D. (ed.), *Handbook of Coastal Processes and Erosion*,, Boca Raton, FL: CRC Press, pp. 167–189. (Also reprinted as Miscellaneous Paper CERC-84-14. Vicksburg: U.S. Army Engineer Waterways Experiment Station, Coastal Engineering Research Center.)

Kraus, N.C., and Harikai, S., 1983. Numerical model of the shoreline change at Oarai Beach. *Coastal Engineering*, **7**: 1–28.

Kraus, N.C., Larson, M., and Wise, R. 1999. Depth of closure in beach-fill design. *Proceedings of the 12th National Conference on Beach Preservation Technology*. Florida Shore and Beach Preservation Association, pp. 271–286.

Morang, A., Rahoy, D.S., and Grosskopf, W.M., 1999. Regional geologic characteristics along the south shore of Long Island, New York. *Proceedings of Coastal Sediments '99*. American Society of Civil Engineers, pp. 1568–1583.

Nicholls, R.J., Birkemeier, W.A., and Lee, Guan-hong, 1998. Evaluation of depth of closure using data from Duck, NC, USA. *Marine Geology*, **148**: 179–201.

Stauble, D.K., Garcia, A.W., Kraus, N.C., Grosskopf, W.G., and Bass, G.P., 1993. Beach nourishment project response and design evaluation, Ocean City, Maryland. Vicksburg: U.S. Army Engineer Waterways Experiment Station. *Technical Report* CERC-93-13.

Stive, M.J.F., DeVriend, H.J., Nicholls, R.J., and Capobianco, M., 1992. Shore nourishment and the active zone; a timescale dependent view. *Proceedings of the 23rd Coastal Engineering Conference*. American Society of Civil Engineers, New York, pp. 2464–2473.

Cross-references

Cross-Shore Sediment Transport
Cross-Shore Variation of Grain Size on Beaches
Dynamic Equilibrium of Beaches
Littoral
Net Transport
Numerical Modeling

Profiling, Beach
Sandy Coasts

DEPTH OF DISTURBANCE

Definition

Depth of disturbance, is the depth to which wave action can be expected to disturb the underlying bottom sediment. Sources of disturbance other than wind waves are not discussed. The sediment is considered to be unconsolidated particles in the sand size range.

Background

The approach adopted is to develop analytic expressions to estimate the depth of disturbance through the use of the velocity ratio,

$$U = u_{-h\,\max}/u_{\text{crit}}, \quad \text{(Eq. 1)}$$

where $u_{-h\,\max}$ is the maximum near-bottom orbital velocity of the wave and u_{crit} is the critical velocity required to initiate sediment movement under the wave. If U is greater than 1.0, then movement of the sediment would be expected; if U is less than 1.0, sediment movement would not be expected. Ahrens and Hands (1998) found the parameter U very effective for distinguishing between beach erosion/accretion events. For shallow water conditions, $u_{-h\,\max}$ will be estimated using Stream Function Wave Theory (SFWT), Dean (1974). For deep water, $u_{-h\max}$ will be estimated using linear wave theory. The critical velocity will be estimated using the relation,

$$u_{\text{crit}} = \sqrt{(8\Delta g d)}, \quad \text{for } 0.0625 \leq d \leq 2.0 \text{ mm}, \quad \text{(Eq. 2)}$$

where g is the acceleration of gravity, d is the sediment grain diameter, and $\Delta = (\rho_s - \rho)/\rho$, where ρ_s is the density of the sediment and ρ is the density of water; equation 2 is based on research by Hallermeier (1980).

Concepts and development

It is convenient to approximate the dimensionless near-bottom velocity as a function of relative depth, that is, $u_{-h\,\max}T/H \approx f(h/L_0)$, where H is the wave height, h is the water depth, and L_0 is the deep water wave length given by $L_0 = gT^2/2\pi$, where T is the wave period. The general functional form used is:

$$f_i(h/L_0) = \exp[C_0 + C_1(h/L_0)](h/L_0)^{C_2}, \quad \text{(Eq. 3)}$$

where C_0, C_1, and C_2 are dimensionless coefficients developed from regression analysis. Table D6 gives the values of the coefficients for the two wave theories and ranges of relative depth.

The breaker height, H_b, consistent with SFWT, can be estimated quite accurately as a function of relative depth using the relation,

$$H_b = 0.171 L_0 \tan h[0.73(2\pi h/L_0)], \quad \text{(Eq. 4)}$$

Ahrens and Hands (1998). In general, the wave height can scaled as a fraction of H_b, that is, $H = H_b/a$, where $a = 1.0$ or 4.0 in this article. When the above relations are brought together, the following equation is obtained,

$$U = 0.171 \tan h[0.73(2\pi h/L_0)] f_i(h/L_0)/a\,[16\pi\Delta(d/L_0)]^{1/2}, \quad \text{(Eq. 5)}$$

where $f_i(h/L_0)$ can be any of the three functions indicated by Equation 3 and Table D6. Equation 5 shows that U is a function of three variables,

Table D6 Coefficients used to estimate dimensionless bottom velocities

Function	Wave theory	Wave height	Relative depth range	Coefficients		
				C_0	C_1	C_2
f_1	SFWT	H_b	$0.002 \leq h/L_0 \leq 0.2$	−0.227	−2.078	−0.600
f_2	SFWT	$H_b/4$	$0.002 \leq h/L_0 \leq 0.2$	0.0835	−2.819	−0.601
f_3	Linear	All	$0.2 \leq h/L_0 \leq 2.0$	1.930	−6.368	0.103

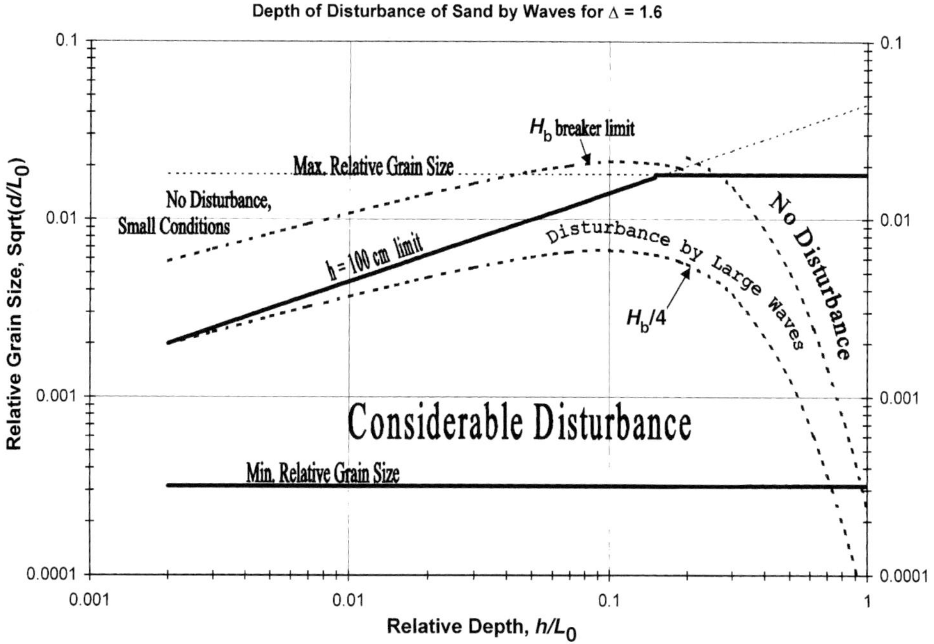

Figure D18 Depth of disturbance plane, relative grain size, $\sqrt{(d/L_0)}$, versus relative depth, h/L_0.

h/L_0, d/L_0, and Δ, however, typically Δ only varies over a small range. To obtain a visual impression of the influence h/L_0, d/L_0, on the depth of disturbance, U is set equal to 1.0 and $\Delta = 1.6$ to produce Figure D18.

Two roughly parallel curves arch across Figure D18 the upper curve is associated with H_b and the lower curve with $H_b/4$. There is a slight discontinuity in the H_b curve because linear wave theory and SFWT are not quite consistent at the transition depth, $h/L_0 = 0.2$. In Figure D18 the area of primary interest is outlined by three heavy, straight lines. The maximum and minimum relative grain sizes are based on the grain size range of $0.0625 \leq d \leq 2.0$ mm and the range of wave periods considered, $2.0 \leq T \leq 20.0$ s. The "h = 100 cm limit" line indicates where the upper limit is 100 cm deep; above this line depths are less than 100 cm and the scale of the wave action is small.

Summary

Figure D18 is a map of the depth of disturbance plane for two wave heights; one equal to the breaker limit, H_b, and the other one quarter the breaker limit. On the right side of the figure, in deep water, is a region of no disturbance. Next to the no disturbance region is a region which is disturbed by large waves, that is, $H_b/4 \leq H \leq H_b$. Beneath this region is a region considerably disturbed by large waves. Above the h = 100 cm limit is a region disturbed by relatively large waves and surprisingly a region of no disturbance. However, above the h = 100 cm limit wave conditions are small since the water depth is less than 100 cm.

John P. Ahrens

Bibliography

Ahrens, J.P., and Hands, E.B., 1998. Parameterizing beach erosion/accretion conditions. *Proceedings of the 26th Coastal Engineering Conference*, American Society of Civil Engineers, **2**, pp. 2382–2394.

Dean, R.G., 1974. Evaluation and development of water wave theories for engineering application; Volume I—Presentation of research results; Volume II—Tabulation of dimensionless stream-function variables. Fort Belvoir, VA: Coastal Engineering Research Center Special Report No. 1.

Hallermeier, R.J., 1980. Sand motion initiation by water waves: two asymptotes. *Journal of Waterway, Port, Coastal, and Ocean Engineering*, ASCE **106**(WW3): 299–318.

Cross-references

Beach Erosion
Beach Processes
Beach Sediment Characteristics
Depth of Closure on Sandy Coasts
Waves

DESALINATION

Water is a valuable natural resource, necessary for domestic, industrial, and agricultural activities. Because of the worldwide population growth, economic development, and erroneous management choices, humanity is nowadays facing an increasing scarcity of satisfactory quality water. This threat is stronger in the arid and semiarid areas (Kulga and Adanali, 1990) but could extend to other regions in the context of a global climatic change which could modify the precipitation patterns at a worldwide scale (Pearce, 1996). The possibility of utilizing saline underground and seawaters, by removing soluble salts from water by desalination technology, is therefore considered today with an increasing attention.

Major processes of desalination

Important research programs had been put in place by the 1980s. Taking into account industrial requirements, three major processes have emerged. Distillation is the older method. Saline water is evaporated partially, leaving the salts behind. The desalinated vapor is then condensed in a separate container. Electrodialyse is based on the mobility of ions in an electrolyte submitted to an electric field. Membranes are placed between a pair of electrodes. Negative ions (anions) migrate toward the anode and positive ions (cations) toward the cathode, creating concentrated and diluted solutions between the alternating membranes. Reverse osmosis is the more recent, and today the most common, technology. The natural osmotic flow is reversed by applying pressure that forces the salt water through a membrane, permeable only for pure water.

A number of other methods may prove valuable under special circumstances or with further development. Freezing is based on the principle that water excludes ions when it crystallizes to ice. In the ion-exchange process, water passes through a bed of specially treated synthetic resin which extracts ions from the solution and replaces them with ions that form water.

Costs of desalination

Total costs of desalination include initial technological investment and energy costs. The minimum energy requirement represents the energy necessary for obtaining freshwater from saline water in the case of an ideal reversible thermodynamic transformation (Nebbia, 1968). In practice, the existing desalination methods consume an amount of energy which is at least 70 times greater than this theoretical minimum (Pearce, 1996). This energy consumption is one of the causes for the high costs of desalination. The selection of an appropriate method depends thus on these economical considerations. Distillation is the more expensive process because of the great quantities of fuel required to vaporize salt water. Electrodialysis and reverse osmosis, which can desalt water much more economically, have therefore replaced distillation for the desalination of brackish water. Because of the high cost of membrane replacement however, these two processes are less suitable for desalinating seawater. Technological advancement has also allowed some reduction in the total desalination cost, by improving energy efficiency (multiflash distillation or hybrid systems), by facilitating transfer processes, by energy recycling (process of cogeneration) or by using solar energy.

Ecological impacts of desalination

Wastewater effluents may have potentially adverse effects on the environment. They present excessive salt concentrations, increased temperatures, and increased turbidity levels and can be charged with chemicals, organics, and metals. In addition, distillation entails an oxygen depletion. These problems can be avoided by treating the effluents before releasing them in the environment but this constraint again increases the cost of desalination.

Conclusion

Desalination offers a nearly unlimited potential of freshwater, particularly in coastal areas. However, this technology is still not extensively developed, because the cost of freshwater production remains a problem. Some countries, which have opted for large-scale desalination, are now searching for more economical solutions (Nativ, 1988). Desalination is essentially utilized to solve survival problems and development of arid and semiarid areas. The use of desalinated water in agriculture is by far less important, particularly because irrigation does not need high water quality (Rhoades *et al.*, 1992). The limits of salinity are generally two to five times those tolerated for industrial and domestic water (Nebbia, 1965). In the case of salinity sensitive crops such as citrus, however, partial desalination by mixing a part of desalinated water to saline underground water, is an interesting solution (Matz, 1965). Because of industrial research and because of the rise in cost for conventionally treated water, desalination turns out to be more and more feasible, economic, and reliable and its importance is expected to increase in the future years. However, even if the problem of freshwater rarefaction can be, in the more or less long-term, technologically solved, we must wonder if an increasing demand for freshwater is compatible with the current will of preserving ecosystems and biodiversity. Excessive water utilization has serious consequences on the function of precarious ecosystems like wetlands (Pearce, 1996). Better management options, centered on conservation and economy of resources constitutes a cheaper, more durable, and more ecological solution in face of water scarcity.

Bénédicte Cherbuy and Jean-Christophe Aznar

Bibliography

Kulga, D., and Adanali, K., 1990. Country report on water resources development in Turkey. Paper to regional seminar on water management strategies in Mediterranean countries, Algiers. CEE, Brussels, Belgium.

Matz, R., 1965. Desalination of sea and brackish water. The present status of the art in Israël. In Tamburini, S. (ed.), *Aqua dolce dal mare*. Milano, Italy.

Nativ, R., 1988. Problems of an over-developed water system, the Israeli case. *Water Quality Bulletin*, **13**: 126–131.

Nebbia, G., 1968. Economics of the conversion of saline waters to fresh water for irrigation. In Boyko, H. (ed.), *Saline Irrigation for Agriculture and Forestry, World Academy of Art and Science*, Volume 4. The Hague: Junk, W.N.V., pp. 267–288.

Pearce, F., 1996. L'enjeu de l'eau. In Skinner, J., and Krivelli, A.J. (eds.), *Conservation des zones humides méditerranéennes*, Volume 5. Arles, France: Station Biologique de la Tour du Valat.

Rhoadhes, J.D., Kandiah, A., and Mashali, A.M., 1992. The use of saline waters for crop production. *Irrigation and drainage paper No 48*. Rome: Food and Agriculture Organization of the United Nations.

Cross-references

Demography of Coastal Populations
Environmental Quality
Human Impact on Coasts
Water Quality

DESERT COASTS

Very great environmental differences exist among the coastal deserts of the world. On an overall worldwide basis climate is the best criteria for subdivions. Climatically, the coastal deserts fall into four principal types which can be called hot, warm, cool, and cold (Figures D19 and D20).

Hot desert coasts

The hot coastal deserts cover the largest areas (Figure D21). Their approximate length is 10,000 miles (16,000 km). This truely tropical hot climate exists only along the southern half of the Red Sea and adjacent coasts of Arabia and the Somaliland horn of Africa (Figures D22–D25). The equatorward sections are warm even in winter with winter temperatures getting above 80°F (25°C) in the heat of the day and only dropping to 70°F (21°C) in the coolest part of the night. The poleward or subtropical type prevails along the Persian Gulf and the Indian Ocean margins of northwestern India, Pakistan, and Iran the northern Red Sea; the Gulf of California, and northwestern Australia.

Most of the hot coastal deserts have relatively little cloud or fog but very high solar radiation. Owing to shallow sill restrictions at the mouths of the Red Sea and the Persian Gulf the water does not mix freely with the colder water of the open ocean, and as a result the water temperature is higher by several degrees. In summer the surface temperature in the southern part of the Red Sea is above 86°F (30°C) and in winter over 77°F (25°C). The Persian Gulf is hotter than the Red Sea in summer, but much cooler than the Red Sea in winter.

Warm desert coasts

The second type of coastal desert can be called warm rather than hot because summers are less hot, with mean temperatures, even in the warmest month being less than 86°F (30°C).The longest single stretch of this climate, about 1,600 miles (2,600 km), is along the Mediterranean Sea, where the Sahara Desert reaches the coast (Figures D26 and D27). Though occasional short rainstorms occur during the winter season, most of the days are bright, mild, and sunny. It might be noted that the climate plus the nearness to densely populated Europe, makes this one of the important winter seaside resort areas of all coastal deserts.

Warm coastal desert climates along exist along the west and south end of Baja California, Mexico (Figure D30), with winter rainfall in the north, summer showers in the south, and non-seasonal very slight rainfall in the center. All the other west-coast deserts of the world also have warm climates in their equatorward positions, such as Angola, Maruritania, Peru, and Australia (Figures D28, D29, D31, and D34). Like the southern end of Baja California (Figure D30), these coastal areas all have their rainy season, such as that occurring, in summer, which is in harmony with their marginal tropical locations. A small but distinct area of this type of warm desert occurs at the southwestern coast of Madagascar (Figure D29). Tourism is developing in some of these areas, especially the southern end of California where visitors from the

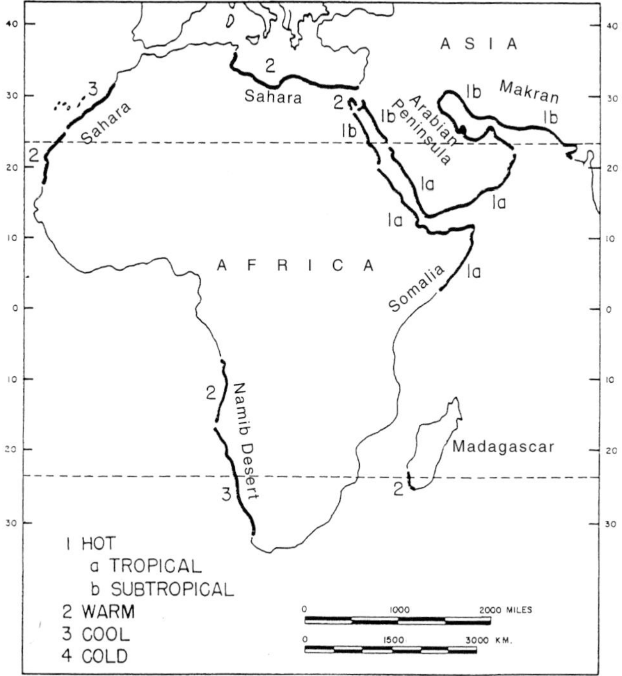

Figure D19 Coastal desert types in Asia and Africa (adapted from Meigs, 1966).

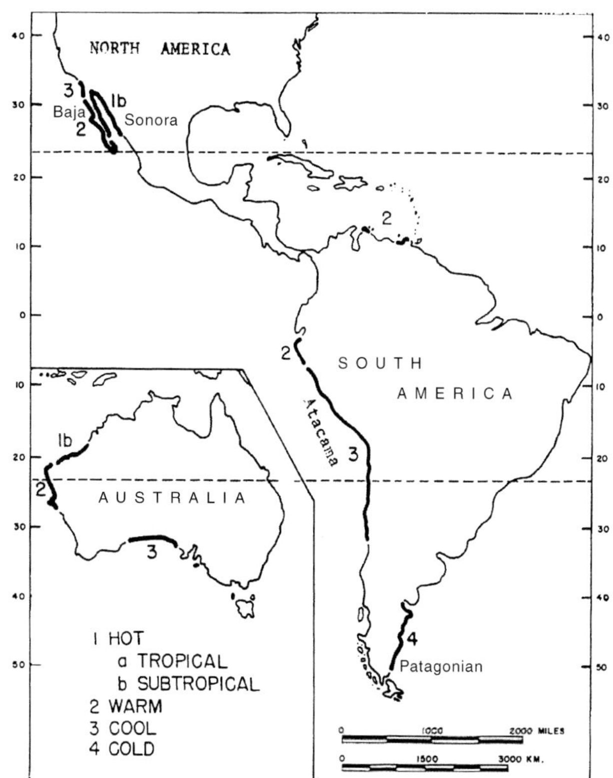

Figure D20 Coastal desert climatic types (adapted from Meigs, 1966).

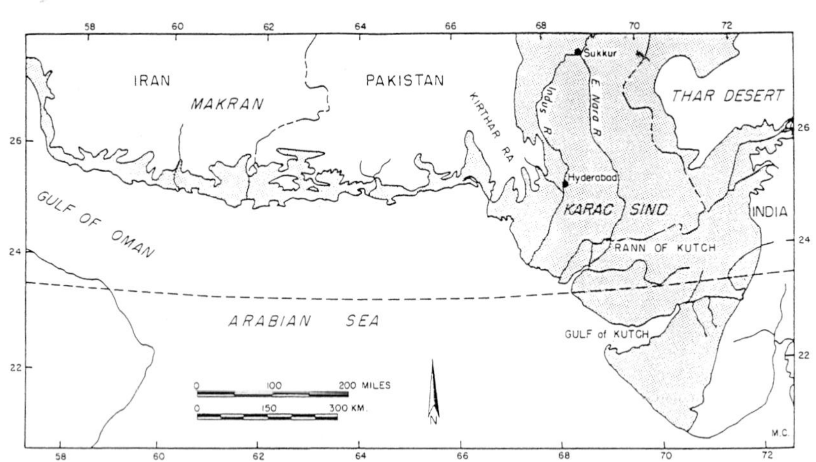

Figure D21 Kathiawar–Makran coastal deserts (adapted from Meigs, 1966).

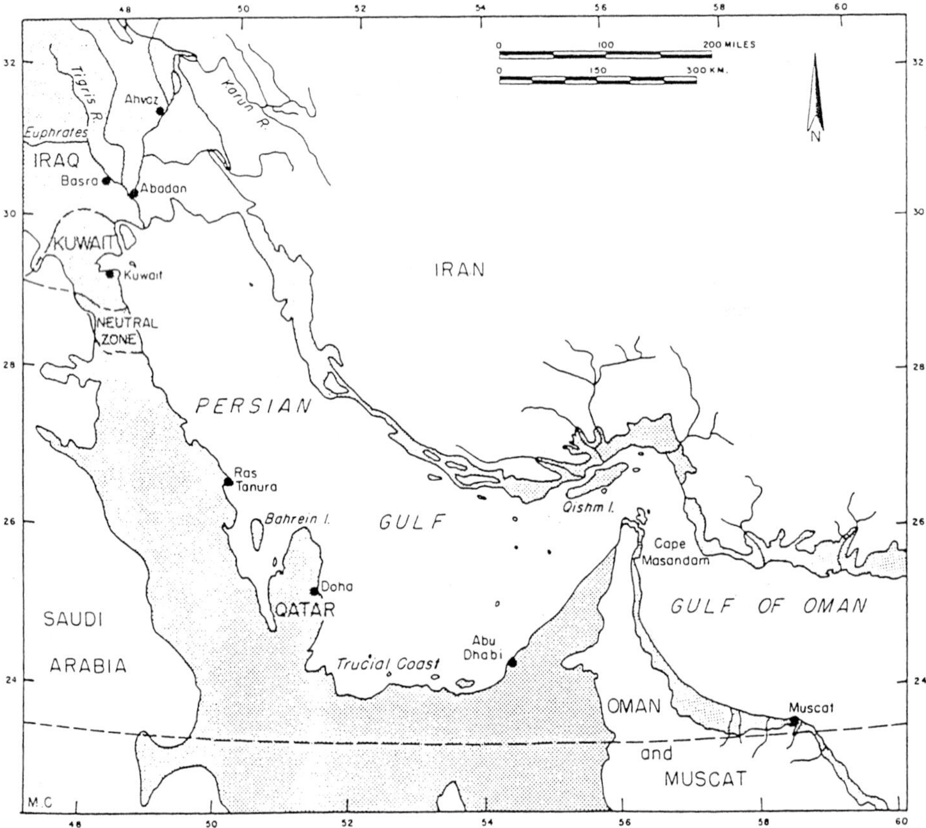

Figure D22 Coastal deserts of the Persian Gulf area (adapted from Meigs, 1966).

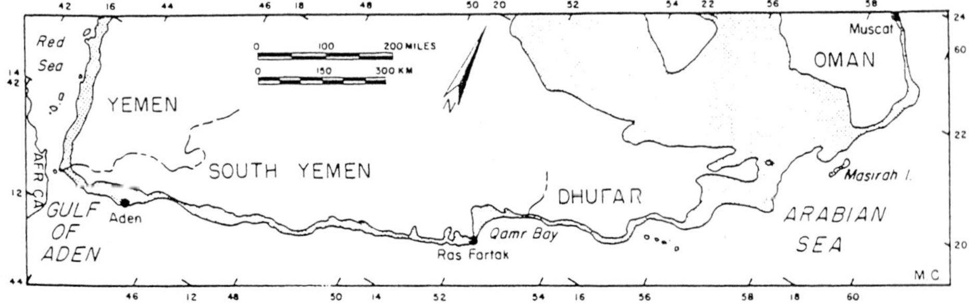

Figure D23 Coastal deserts of southeastern Arabia (adapted from Meigs, 1966).

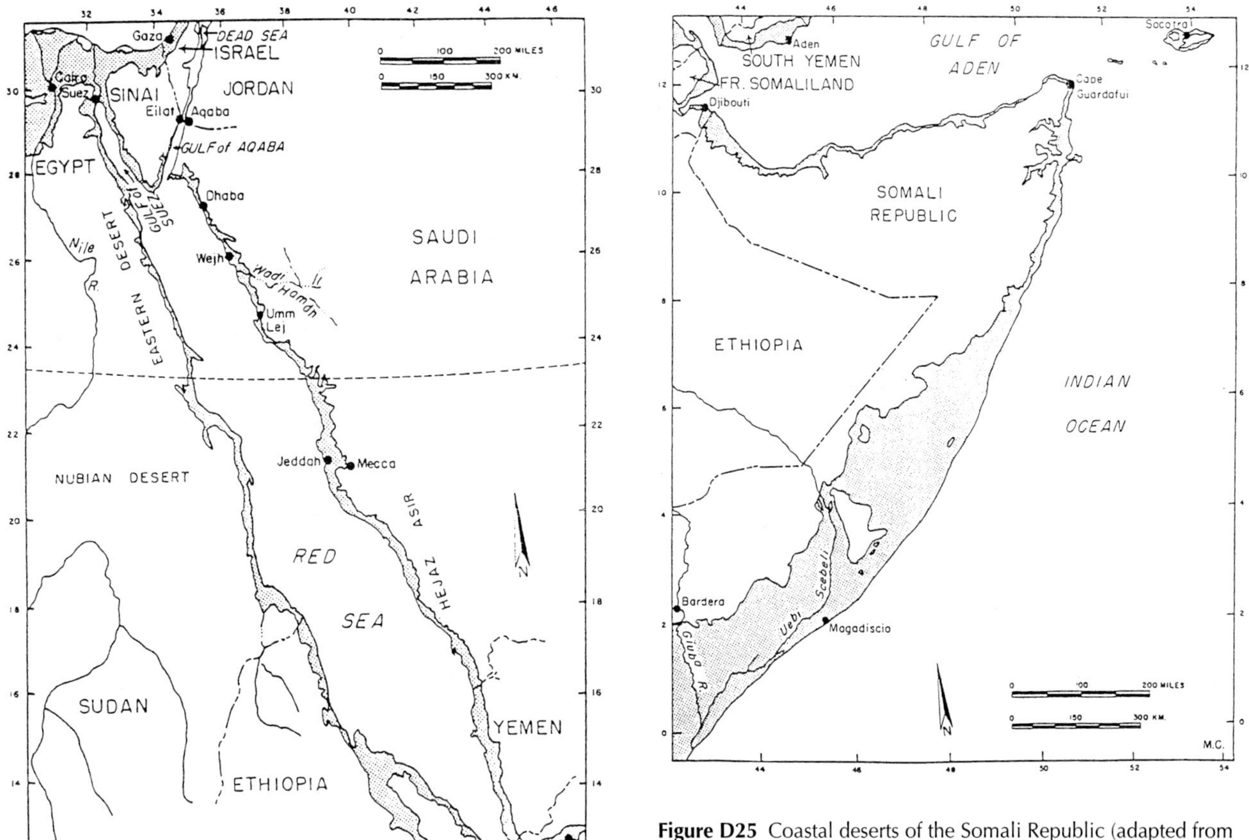

Figure D24 Coastal deserts around the Red Sea (adapted from Meigs, 1966).

Figure D25 Coastal deserts of the Somali Republic (adapted from Meigs, 1966).

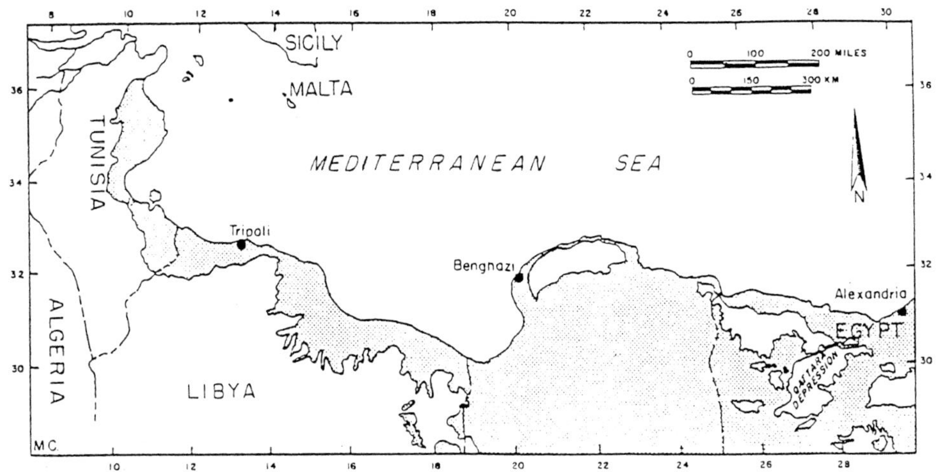

Figure D26 Coastal deserts along the Mediterranean Sea (adapted from Meigs, 1966).

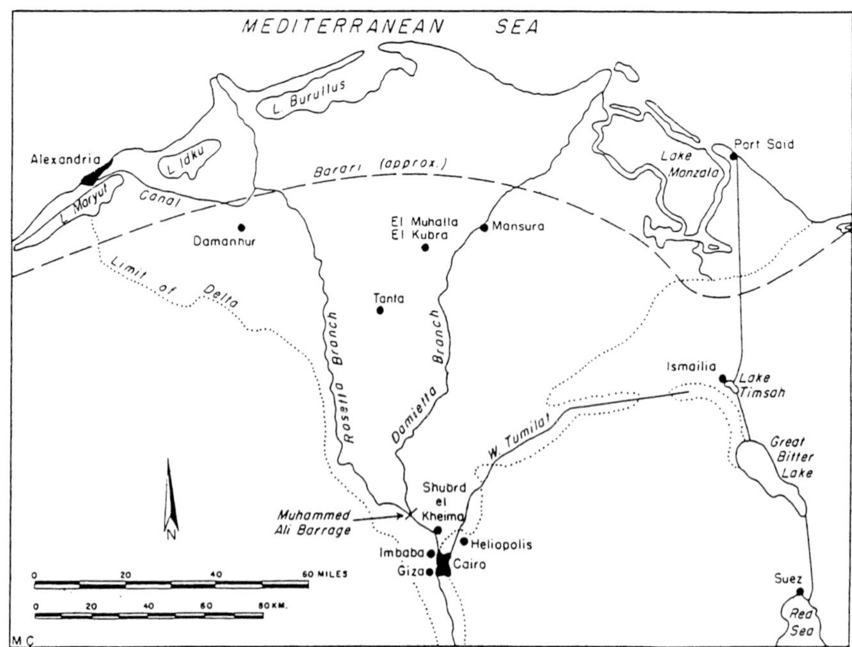

Figure D27 Coastal deserts of the Nile Delta (adapted from Meigs, 1966).

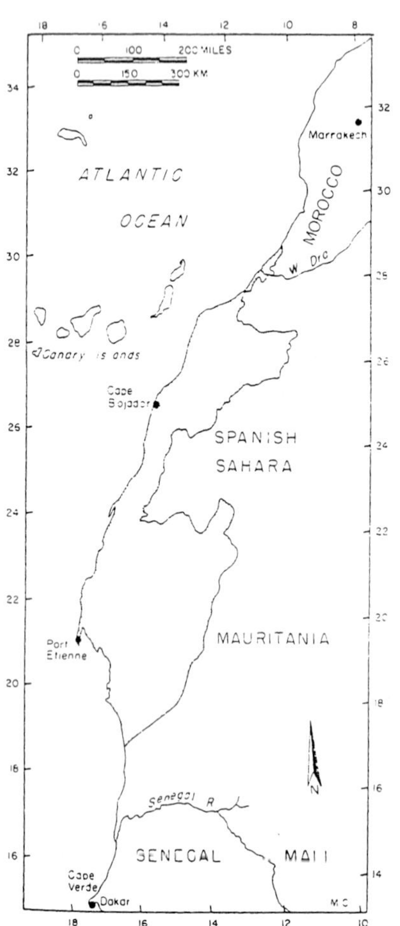

Figure D28 Atlantic-Sahara coastal desert (adapted from Meigs, 1966).

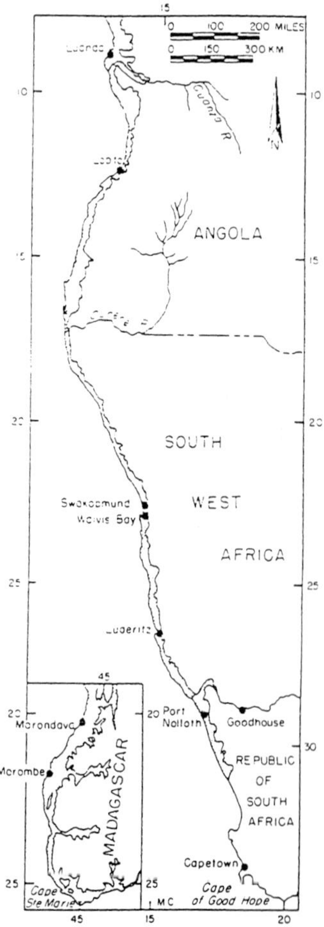

Figure D29 Coastal deserts of southwestern Africa (adapted from Meigs, 1966).

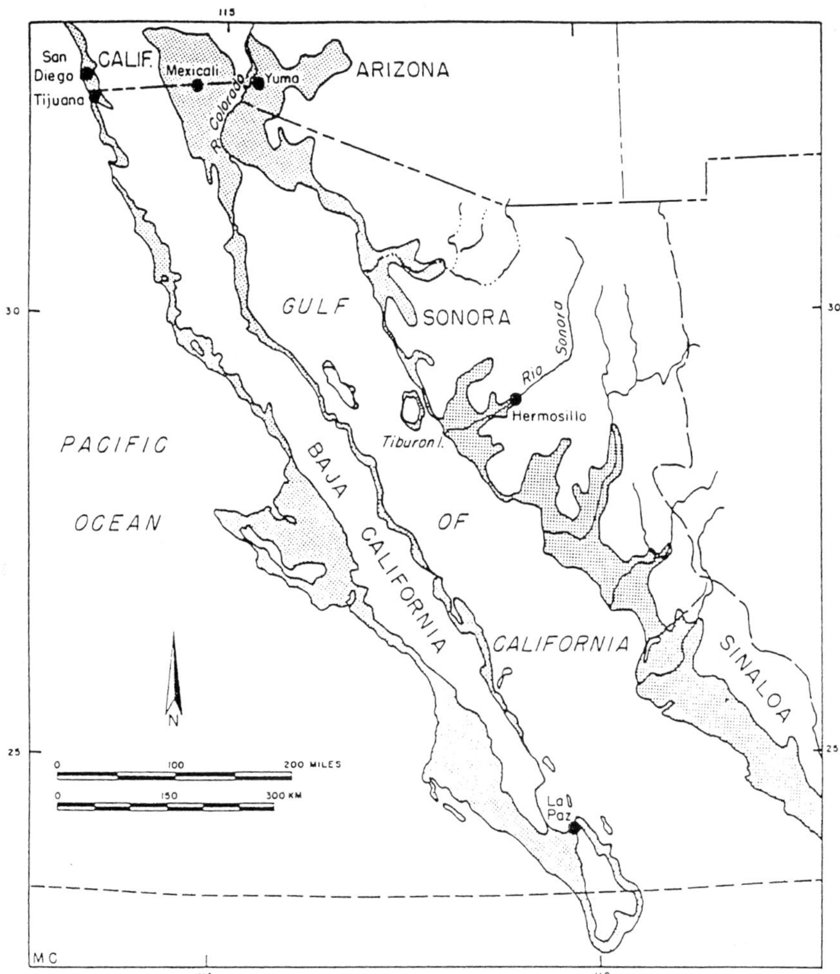

Figure D30 Coastal deserts of western Mexico and Baja California (adapted from Meigs, 1966).

United States and mainland Mexico come by air, sea, or some by road. Near their equatorial ends, these west-coast warm deserts are relatively free from cloud or fog. Toward their polar ends, especially in Peru, southwestern Africa, and Baja California (Figure D29–D31), there is frequent low cloud or fog is summer, as the warm deserts merge into the cool deserts.

A particular tourist and biological resource of this part of Baja California is the gray whale, which travels thousands of miles, from the North Pacific to breed in Scammon's Lagoon, at 27°45′N latitude.

Cool desert coasts

Type 3, the cool desert coasts from the poleward portion of all five west-desert coasts. They occur nowhere else if the shores of the Great Bight of Australia (Figure D35) is considered an extension of the west-coast desert. The longest stretches of this cool desert coast are along the Peru-Chile coast, 2,100 miles (3,400 km) (Figure D32).

Nowhere else on earth is there a coastal desert of such intense aridity which spans so many degrees of latitude as the Chilean-Peruvian desert. The extreme dryness is all the more interesting when one considers that no part of this desert is far from a tropical ocean, for at a maximum it is never more than a few score miles (kilometers) in width, and the fact that the atmospheric humidity is high is attested to by the presence of a great deal of low stratus cloud, fog, drizzle, and heavy dew. Still it seldom rains. Obviously, the dynamic controls making for drought are amazingly well-developed.

Among the climatic anomalies of the Chilean-Peruvian desert two are particularly notable, (1) the unusual intensity of the dryness in conjunction with its latitudal extensiveness and the abruptness with which it ends on both the north and south, so that a transitional steppe climate is only meagerly developed, and (2) the brief periods of heavy rainfall and high temperatures which vary occasionally affect mainly the northern parts of this desert area, this temporarily revolutionizing its weather and changing it to wet tropical conditions. In addition, the Peruvian-Chilean desert has uncommonly well developed those special features of climate which are peculiar to dry western littorals bordered by cool currents. Such as relatively low annual temperatures, small annual and daily ranges of temperature, and the existance of much low stratus fog, cloud, and even drizzle.

Other locations where Type 3, the cool desert coasts, occur are the west coast of Namibia and South Africa, 1,200 miles in length. (2,000 km) (Figure D29), while smaller sections include northern western Sahara and southern Morocco (Figure D28). Baja California has a short segment just south of the California border (Figure D30). These are foggy deserts with fog forming over the cool waters offshore and moving onto the adjacent land either as "high fog or low cloud," or at times directly along the ground. To the cooling effect of the cold water onshore can be added the reduction of solar radiation by the fog blanket. It can also be added that there are regional differences in the season of maximum fog. Along the Chile-Peru coast the season of great fog is in winter; while along the Baja California and northwest Africa coasts there is a summer maximim of the fog. Solar radiations is also affected by high clouds associated with the seasonal rainfall regimes of the neighboring human-occupied areas: winter in most of the cool desert climates, while summer occurs in the main part of the Peruvian coast. However, it can be said, the deserts of the coast provide attractive relief for people from the hot interior valleys or deserts. It might be added

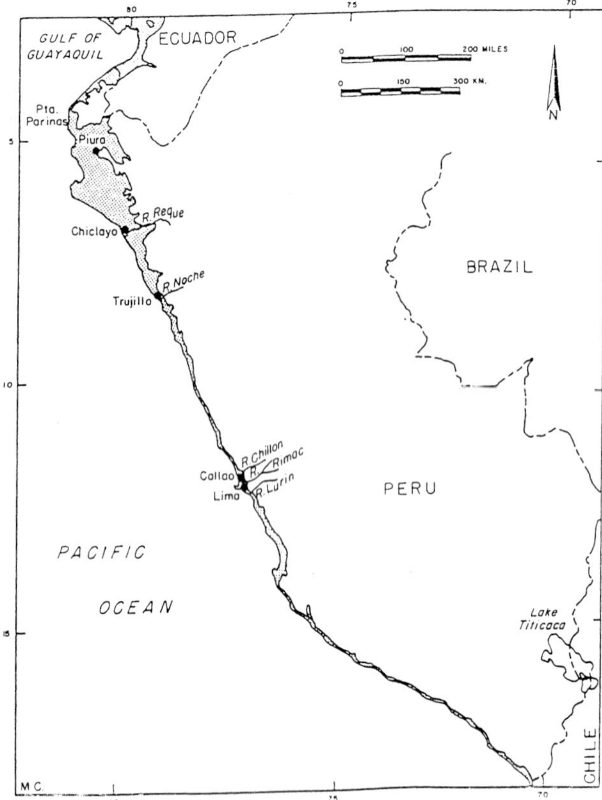

Figure D31 Peruvian coastal deserts (adapted from Meigs, 1966).

that the special circumstances of cool to cold ocean waters offshore has resulted in great fishing grounds.

Cold desert coasts

The coldest coastal desert of all, Type 4 of the classification, occurs along the southeast coast of Patagonian Argentina (Figure D33). Here a cool to cold ocean current reinforces the dry rain-shadow effect of the Andes on the westerly storms that normally prevail in these latitudes. This desert extends to more than 50°S latitude, which is farther poleward than any other coastal desert in the world. The mean temperature of the coldest month is below 50°F (10°C), and the temperature of the warmest month is between 50° and 72°F (10° and 22°C). With such limited heat only the hardiest vegetation and crops grow in this region. Other resources must provide the principle economic basis for settlement.

This account considers only the connecting, not even the oceans. The inland lakes and seas are not considered, not even the large Caspian Sea. This sea has a climate quite different from those of other coastal deserts. Most of its climate (except along the southwest borders) is continental: colder in winter than the coldest of the coastal deserts (Type 4), with averages below freezing; and ranging with the warmest deserts (Type 2) (Figures D34 and D35) in summer. The sea affects the temperature of the land immediately adjacent to it, as occurs with other coastal deserts.

Coastal desert environments

A great variety of landscapes and landforms occur along desert coasts, in fact nearly every type of landform. Desert and semidesert coastal lands make up about 20–25% of all such world regions. Coastal dunes and desert plains predominate but long stretches have steep escarpments or plateaus which rise directly from the sea. In places such as the Makran coast of Pakistan, quite recent uplifted terraces predominate.

The source of sand for desert coasts comes mainly from the sand left on beaches which is moved inland by changing tides and waves plus coastal winds. Another major way for sand to reach desert coasts is by intermittent rivers and streams which carry the sand in short flash

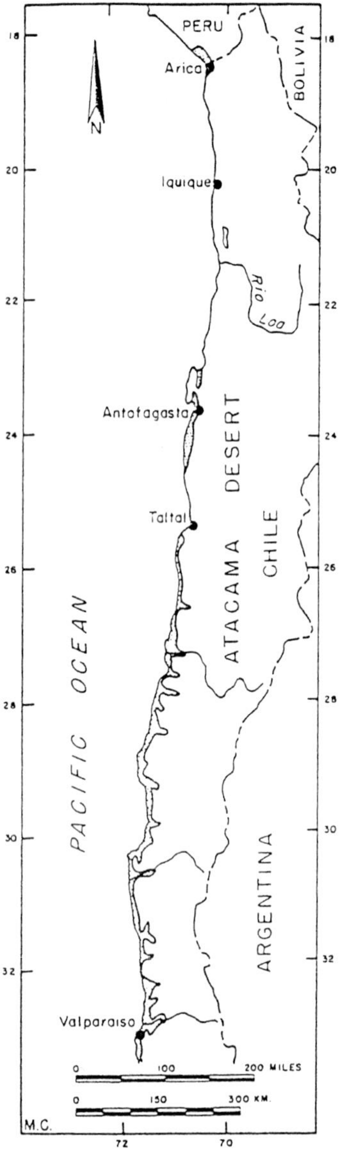

Figure D32 The Atacama desert (adapted from Meigs, 1966).

floods. The sand is either carried into the sea and then moved by waves and currents along the beaches or picked up by winds in river beds and moved to coastal dunes. There is a changing balance between wind transport of sand and by running water in river bads.

Most aeolian bodies occur in basins in the desert. The Pleistocene probably saw periods of more intense fluvial erosion than the present. Many coastal ergs are fashioned out of deposits of Pleistocene or late Tertiary alluvium.

Clay dunes are perhaps surprisingly better authenticated than silt dunes and are more obviously dune like. They have been reported from the Gulf of Texas, northern Algeria, Australia, and the Senegal Delta.

Although it is not difficult to recognize a desert, opinions vary as to the physical features of the environment that should be used as criteria for defining a desert climate. Rainfall must obviously be taken into account, and desert conditions usually prevail in tropical regions when the average rainfall is under 5 in. (127 mm), with semidesert conditions occurring when the average rainfall is over 5 in. (127 mm) but under 15 in. (381 mm). But such figures can be only approximate, because what matters is not only the amount but the effectiveness of rain; and the effectiveness of rain depends on its seasonal occurrence, its rate of evaporation, and the nature of the soil (Cloudsley-Thompson, 1975, p. 8).

The aridity of coastal deserts that results mainly from a lack of rain is often due to the way in which the atmosphere circulates, particularly

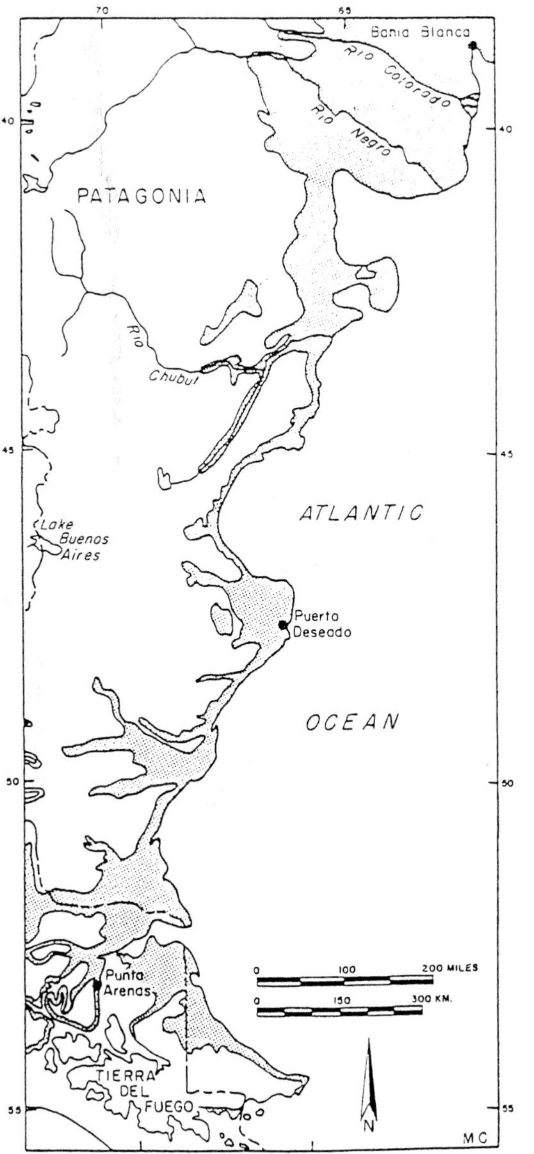

Figure D33 Coastal deserts of Patagonia (adapted from Meigs, 1966).

in its lower layers, and this partly explains the geographical distribution of the world's coastal deserts: near latitudes 30°N and S, the direction of the winds tend to vary, and the air sinks. When this happens, it is warmed by the hot ground, and clouds are dispersed. At the same time, high-pressure belts that separate the polar westerly winds and the tropical easterlies occur near these latitudes. Consequently, the disturbances associated with low pressure, which usually cause rain, are rare. Then, too, there are some local factors that quite often result in low rainfall. These include the rain-shadow effects of mountain ranges, which cause the air to drop all its moisture as it crosses them. For example, the Patagonian desert including the Argentina coast, is the result of both the rain-shadow effect of the Andes Mountains, which block the path of the westerlies, and the cool Faulklands current alongshore.

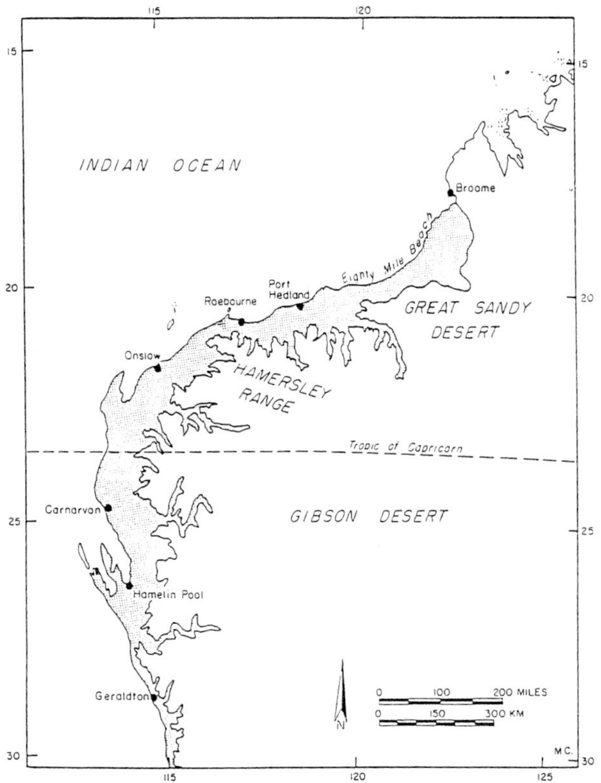

Figure D34 Coastal deserts of western Australia (adapted from Meigs, 1966).

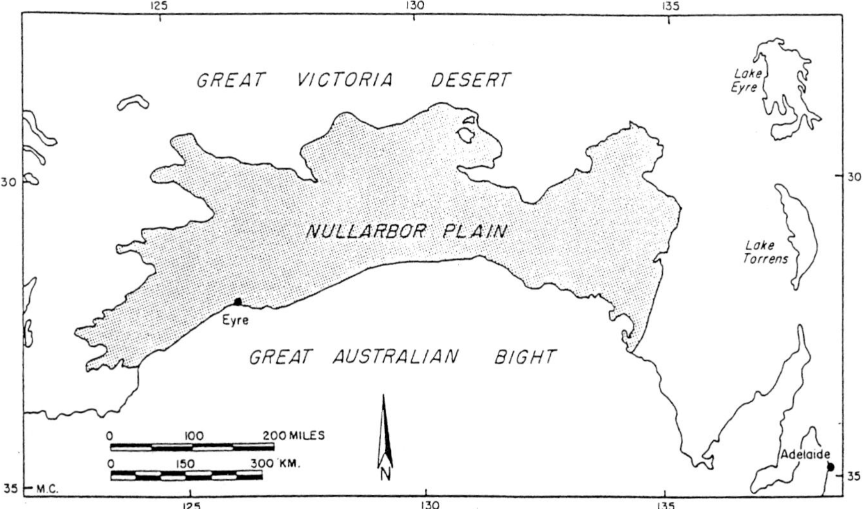

Figure D35 Coastal deserts of southern Australia (adapted from Meigs, 1966).

Desert coastal vegetation

Most desert coasts support a *xerophytic* type of scrub vegetation which grows sparsely in most areas. A few scattered trees, mostly the acacia type in tropical coastal deserts, can be found. In years with more than 8 in. (200 mm) of rainfall, a thick layer of grasses, herbs, and shrubs covers the ground. Climate is the major determinant of vegetation types, and rainfall is much more significant than temperature, however the amount of salt comes into play. In general, it can be said that along most desert coasts many of the shrubs are woody and stunted being about 12 in. (300 mm) high with round cushion-like out-lines. With bleached stems and few leaves many look like skeletons of plants. A few annuals begin sprouting when warmer days occur. Although the number of families is large in most cases the number of species is small.

Again, speaking generally, the first dunes behind the back-shore differ in vegetation from the inland dunes. The difference is due mainly to high salinity and a high water table which often is only 0.5–4.0 ft (0.15–1.2 m) below the surface.

Natural and human factors disturb desert coastal vegetation and lead to temporary changes in plant cover. Disturbed areas result from such natural causes as varying moisture, wind and soil conditions, and destruction by sudden floods. Human activities along the coast which disturb plant successions include the clearing or destruction of the original vegetation for purposes of cultivation, fuel, fodder, and building material. Animal grazing is the most destructive factor.

Several desert coasts have unusual vegetation types. For example, the Namib coast has vegetation that lives off of fog; both the Baja desert coast and the Galapagos Islands have unusual vegetation probably because, for a long period of time, they were so isolated, new species of plants developed.

Animals and plants

It is not easy for plants and animals to survive in desert coastal regions where extremes of drought and flood, of erratic rainfall, or no rainfall at all, only fog and dew, can occur. Therefore, it is not surprising that animals and plants have adaptations that are correspondingly extreme. Desert animals usually run faster than their relatives in temperate regions. Venomous forms are more poisonous than their allies beyond the desert's fringe. Spiny plants are more prickly, and distasteful ones more unpleasant, than anywhere else.

Most desert and semidesert costal regions support some degree of vegetation. The smallest amount of plant life exists in areas of rocky *hammada* and cracking clay; but plants can grow even in such soil whenever it is traversed by wadis. And sandy desert usually supports a flora of some kind, except on the dunes, which tend to stay quite bare. Many plants, such as annual grasses and other small herbs, evade the more extreme desert conditions by completing their life cycle during the short rainy season and passing the remainder of the year lying dormant as fruits or seeds in the soil. Dry seeds can often survive high temperatures without losing their ability to germinate later under more favorable conditions (Cloudsley-Thompson, 1975, p. 26).

Most plants of arid regions have evolved mechanisms not so much of drought tolerance as of resistance to drought. They fight the battle against water loss in a number of ways, often combining different protective devices within a single plant. One means of retaining moisture is to have leaf surfaces coated with fatty substances and resins; these making it harder for water vapor to escape. Other structural adaptations to drought include the presence of dense, hairy coverings (which again keep moisture from escaping) or of pores sunk deep into the leaf tissue so that the dry wind cannot blow directly into them. Smaller and fewer leaves also help resistance to water loss. And— perhaps most important of all devices—a number of desert plants have developed internal tissues for storing water over long periods of time and for giving increased mechanical support to the bloated, water-filled organism after a heavy rainfall. Such plants (among them the cacti) are known as *succulents* (Cloudsley-Thompson, 1975, p. 29).

Some plants can survive on moisture from the air; and at night, with the drop in temperature, dew condenses on their leaves and stems and runs down into the soil around the roots. The welwitschia of the Namib Desert in southwestern Africa is such a plant. It bears two curling leaves, split longitudinally into strips, on which dew condenses, providing water for the shallow roots. The ability of vegetation such as the welwitschia to absorb moisture from the night air, even when dew does not form, can make vitally needed water available for snails, insects, and other small animals of the desert.

Desert vegetation is not entirely ephemeral. There are many plants that habitually grow where evaporation stress is high and water supply low, and that do not need to evade extreme coastal desert conditions. Such plants, all of which show characteristic adaptations to their environment are termed *xerophytes* (from the Greek words meaning "dry" and "plant"). Among them are trees, perennial grasses, and species that develop bulbs, corms, or tubers. Some xerophytes are characterized by the ability to survive long periods of drought and dehydration of the tissues without suffering injury. They are extremely resistant to wilting, and a few species can lose up to 25% of their water content before they begin to wilt.

Another characteristic feature of most desert coastal plants is that they posses painfully sharp thorns and spikes. The probable function of these is to afford protection against browsing and grazing animals. Many contain poisonous or irritant latex. Other secrete resins or tannins in the bark or levels; the pods of some contain strong purgatives. All such devices tend to make the desert plants unpalatable to animals (Cloudsley-Thompson, 1975, pp. 27–30).

Case study of the Pakistan–Iran climate regime

The Pakistan part of this desert coastal type is interesting because it is an area of transition between the Indian summer monsoon system to the east, and the winter cyclonic system of southwest Asia to the west. As a transition area, it receives scanty, unreliable rainfall, averaging less than 10 in. (250 mm) per year from several storm types. Six main weather patterns cross the region, the large subtropical anticyclonic high-pressure cell which predominates most of the year; western depressions originating over the Mediterranean Sea; a modified monsoon pattern; eastern depressions originating over the Bay of Bengal and central India and local thunderstorms and dust storms.

The seasonal division of rainfall for southern Pakistan is shown in Figure D21. Nearly all of Sind and the eastern part of Baluchistan, particularly the Las Bela area, have a pronounced summer monsoon maximum coming in June, July, and August. This line conforms to the eastern slopes of the Kirthar and Sulaiman mountain ranges which act as a barrier to storms from the east. Iran and the western part of Baluchistan have a winter maximum coming from November to March with very little summer precipitation. Central Baluchistan is a region having both winter and summer rainfall with a slight winter maximum, while a thin band in eastern Baluchistan is an area having both winter and summer rainfall, but with a slight summer maximum.

Further reading on this subject may be found in the following bibliography.

Rodman E. Snead

Bibliography

Amirman, D.H.K., and Wilson, A.W., 1973. *Coastal Deserts, Their Natural and Human Environments*. Tucson, AZ: The University of Arizona Press.

Berkofsky, L., and Wurtele, M.G., 1987. *Progress in Desert Research*. Totowa, NJ: Roman and Littlefield.

Cloudsley-Thompson, J., 1975. *The Living Earth. Desert Life*. London: The Danbury Press, Aldus Books Limited.

Logan, R.F., 1960. *The Central Namib Desert*. National Academy of Sciences. National Research Council, Publication 758, 162 p.

Logan, R.F., 1969. Geography of the Central Namib Desert. In McGinnies, W.G., and Goldman, B.J. (eds.), *Arid Lands in Perspective*, Washington, DC, The American Association for the Advancement of Science, and Tucson, AZ: University of Arizona Press.

Mabbutt, J.A., 1977. *Desert Landforms, An Introduction* to *Systematic Geomorphology*, Volume 2, Cambride, MA: The MIT Press.

Mares, M.A. (ed.), 1999, *Encyclopedia of Deserts*, Norman: University of Oklahoma Press.

McGinnies, W.G., and Goldman, B.J., 1968. *Deserts of the World*. Tucson: The University of Arizona Press.

Meigs, P., 1966. *Geography of Coastal Deserts*. Liege Begique Vaillant Carmanne S.A.: UNESCO.

Nordstrom, K., Psuty, N., and Carter, B., 1990. *Coastal Dunes Form and Process*. New York: John Wiley and Sons.

Schneider, S.H. 1996. *Encyclopedia of Climate and Weather* Volume 2, New York: Oxford University Press.

Snead, R.E., 1966. *Physical Geography Reconnaissance Las Bela Coastal Plain. West Pakistan*. Baton Rouge: Louisiana State University Coastal, Series No. 13.

Snead, R.E., 1968. Weather patterns in southern West Pakistan. *Series B*, **16**: 316 *Archives for Meteorology Geophysics and Bioclimatology* 346.

Snead, R.E., 1970. Physical Geography of the Makran Coastal Plain of Iran, *National Technical Information Service*, Springfield, VA: U. S. Department of Commerce, Report No. AD 707745V.

Snead, R.E., 1982. *Coastal Landforms and Surface Features A Photographic Atlas and Glossary*. Stroudsburg, PA: Hutchinson Ross Publishing Company.
Thom, B.G. (ed.), 1984. *Coastal Geomorphology in Australia*. North Ryde, NSW Australia: Academic Press.
Trenhaile, A.S., 1987. *The Geomorphology of Rocky Coasts*. Oxford, England: Oxford University Press.
Trewartha, G.T., 1966. *The Earth's Problems Climates*. London: Methuen and Co., Ltd.
Young, E.D., 1999. Innovations in Earth Sciences, Santa Barbara: ABC-CLIO.

Cross-references

Africa, Coastal Ecology
Africa, Coastal Geomorphology
Asia, Middle East, Coastal Ecology and Geomorphology
Australia, Coastal Ecology
Australia, Coastal Gemorphology
Coastal Climate
Indian Ocean Coasts, Coastal Ecology
Indian Ocean Coasts, Coastal Geomorphology
Meteorologic Effects on Coasts
Middle America, Coastal Ecology and Geomorphology
North America, Coastal Ecology
North America, Coastal Geomorphology

DEWATERING—See BEACH DRAIN

DEVELOPED COASTS

The term "developed coast" often implies a natural system altered by human action. The type, frequency, and magnitude of alterations are influenced by the social and economic value humans place on the resource. Today, developed coasts are modified and physically maintained by humans to enhance navigation, recreation, and protection of settlements. Human modification of the coast has resulted in changes to the dimensions and locations of coastal landforms and the spatial and temporal scales of cycles of erosion and accretion. Globally, human presence on the coast dates back to tens or even hundreds of thousands of years, with some of the earliest documentation on the Mediterranean coast (Nordstrom, 2000). Deforestation and overgrazing were perhaps the earliest human actions that significantly altered the coast by increasing the quantity of sediment delivered (Walker, 1985). Grazing initiated dune migration when practiced directly on the coast. Channelization, diversion, and impoundment of water resources had the opposite effect on the coastal sediment budget; levees and dams decreased sediment delivery from river sources, with the most conspicuous effects occurring in the 20th century.

Direct human alterations to the coast increased in scale with increased occupancy and technological advances associated with the availability of steam power (Marsh, 1885) that enabled the creation and removal of landforms and redistribution of sediment. Pronounced changes to the landscape occurred about the mid-19th century. Harbors were constructed and inlets were regulated through sediment dredging and jetty construction.

Coastal tourism and resort development, beginning in earnest during the second-half of the 19th century, resulted in further alteration of the coast. Seasonal house construction and infrastructure development (transport, utilities, marinas) along the coast resulted in elimination of dunes and elimination of marsh by filling to increase upland width or by dredging to construct marinas. Development occurred near the coastline, increasing vulnerability to wave attack and flooding during storms and requiring shore protection. Large-scale shore protection efforts, primarily groins and seawalls, were initiated in the late 18th century in Europe and in the mid-19th century in the United States (Nordstrom, 2000). Use of these structures, referred to as hard stabilization, increased with increased construction of buildings and support infrastructure. There is considerable debate on the use of these structures in long-term coastline management (Kraus and Pilkey, 1988). Beach nourishment (placement of sediment on the beach and nearshore profile from another source) was first used in the early 20th century and became popular in the second-half of the 20th century. Identified as soft stabilization, nourishment is now the preferred alternative to protect development from storm erosion and flooding (National Research Council, 1995). Introduction of large volumes of sediment to developed shorelines that previously had a sediment deficit has created opportunities for the restoration of many geomorphic features lost to human development.

Characteristic landforms

Many of the characteristic landforms found on developed coasts differ from their natural counterparts in size, location, and spatial and temporal scales of change but are identified using similar terminology. The attributes of landforms found on developed coasts can be the intended or unintended outcome of their modification for human use. Over time, assemblages of human-altered landforms may be accepted as natural and require further human modification to maintain their resource value (van der Meulen *et al.*, 1989).

Beaches on developed coasts are generally narrow, due to the reduction of sediment supply, but nourishment can displace the high-tide shoreline seaward and create a beach that is initially wider than a natural beach. Nourished beaches, when designed for shore protection, are often higher in elevation than their natural counterparts, but beaches downdrift of shore protection structures such as groins will be lower in elevation. Most direct human modifications to beaches are intended to reduce high-tide shoreline mobility over the long-term, although mobility may be higher than expected immediately after the modification such as inlet creation, closure and stabilization, or beach nourishment. The characteristics of sediment found on nourished beaches is dependent on the source, which may be derived from offshore, including inlets.

The location and size of dunes are controlled by direct human actions, such as bulldozing and placement of sand fences, and indirect actions through the passive influence of shore protection structures that trap sediment transported by wind. Human modification by fencing, vegetation planting, or mechanical grading creates forms designed to optimize human values such as shore protection, views of the sea, or ecological enhancement. Small incipient dunes are often lacking because beach litter is eliminated by beach cleaning operations. Secondary dunes are often lacking because that zone is occupied by human structures. Sedimentary characteristics and internal structure of dunes that form around human structures by aeolian processes are similar to natural dunes. Dunes created by bulldozing have poorly defined internal structure and contain sediments that may be too coarse or too fine to be transported by local winds. Dunes created by humans are often less mobile than those created by aeolian processes alone because fencing or buildings constructed close to the shore restrict migration.

Assessing change

Landforms on developed coasts respond to energy inputs according to the same laws as natural landforms, but the spatial and temporal scale of change can differ because of human action (Nordstrom, 2000). Assessment of post-storm recovery of landforms in developed and undeveloped coastal areas reveal that recovery of the backbeach and dunes by natural processes is impeded by the presence of houses in developed areas (Morton *et al.*, 1994), but recovery by human modifications can occur at a far more rapid rate. Closure of inlets created during storms, bulldozing of sediments onto the backbeach from overwash deposits on streets, and reconstruction of dunes using sand fences or earth-moving equipment rapidly restore the coast to the pre-storm landscape to achieve social and economic resource values. Human alterations to the coast may be more dramatic in the future in response to sea-level rise (Titus, 1990).

Assessing landform change on developed coasts requires identification of both human and physical processes that govern change as well as the interactions between them. Human action can be considered either external to or a part of the natural system (Phillips, 1991). As an external influence, human action can be considered an overlay to a natural coastal system. As a part of the natural system, human action must be considered an integral process to assessing change, enabling a better understanding of the functioning of the developed coast, and ways to manage it to achieve both natural and human resource values (Nicholls and Branson, 1998).

Nancy L. Jackson and Karl F. Nordstrom

Bibiliography

Kraus, N., and Pilkey, O. (eds.), 1988. The effects of seawalls on the beach. *Journal of Coastal Research*, **SI4**, p. 146.
Marsh, G.P., 1885. *Earth as Modified by Human Action*. New York: Charles Scribner's Sons.

Morton, R.A., Paine, J.G., and Gibeaut, J.C., 1994. Stages and durations of post-storm beach recovery, southeastern Texas coast. *Journal of Coastal Research*, **10**: 884–908.
National Research Council, 1995. *Beach Nourishment and Protection*. Washington, DC: National Academy Press.
Nicholls, R.J., and Branson, J. (eds.), 1998. *The Geographical Journal*, **164**, part 3, pp. 255–278.
Nordstrom, K.F., 2000. *Beaches and Dunes of Developed Coasts*. New York: Cambridge University Press.
Phillips, J.D., 1991. The human role in earth surface systems: some theoretical considerations. *Geographical Analysis*, **23**: 316–331.
van der Muelen, F., Jungerius, P.D., and Visser, J.H. (eds.), 1989. *Perspectives in Coastal Dune Management*. The Hague: SPB Academic Publishing.
Titus, J.G., 1990. Greenhouse effect, sea level rise, and barrier islands: a case study of Long Beach Island, NJ. *Coastal Management*, **18**: 65–90.
Walker, H.J., 1985. The shoreline: realities and perspectives. In Vallega, A., DaPozzo, C. and Fabbri, P. (eds.) *Coastal Planning: Realities and Perspectives*, Genoa: Commune di Genova, pp. 59–90.

Cross-references

Beach Nourishment
Beach Use and Behaviors
Coastal Zone Management
Demography of Coastal Populations
Economic Value of Beaches
Human Impact on Coasts
Sea-Level Rise, Effect
Shore Protection Structures
Small Islands
Tourism and Coastal Development

DIKES

Introduction

Dikes, especially sea dikes, are coastal constructions build to avoid flooding. The risk of flooding is detrimental to the safety of people and economic, cultural, and ecological values. This aspect has been of great importance since people first thought about defending their dwellings against flood hazards. In the distant past, dwelling mounts were built to protect families or small communities from the sea. They are known from several low-lying coastal areas in the world, for example, from the North coast of Germany and the adjacent Dutch coast, where they have been occupied ever since 2500 BP, as well as from the chenier coast of Suriname, where they date back to 1800 BP.

Population increase urged to a more active method of flood prevention. The people started to construct dikes to keep the water out of whole regions, thus protecting lives and properties against the sea. In the Netherlands, dikes have been built as a community activity from about AD 1100. Ever since, coastal defense has become more and more an engineering activity. Presently, the design criteria for sea dike construction are determined by three important groups of aspects: hydraulic, subsurface, and construction. The first two groups mainly determine the required height and strength of the dike. The last group of aspects is particularly important for the dike's duration of life. Details about the various aspects are given by U.S. Army Corps of Engineers (1984).

Hydraulic aspects

Hydraulic factors influencing the marine water level at small temporal scales are the tidal range, wave height, wave run-up, and the setup of the water level due to wind conditions. Because dikes have to be calculated for extreme water levels, especially a coincidence of extreme conditions of these factors are at stake, for example, high-water spring tide and wave characteristics and setup of the water level during storms. For the establishment of extreme water levels, high-water exceedence frequency curves are used. This method is based on the finding of a systematic relation between the height of a flood-tide level and the number of times this specific level occurs in a century. Extrapolation of the high-water exceedence frequency line enables a probability calculation for the chance that an extreme high-water level, not observed before, may occur. Moreover, the effects of long-term changes in sea level and storm frequency must be taken into consideration.

The final height of a sea dike is determined by the degree of risk of flooding, within the applied economic constraints, the community is willing to accept. In the Netherlands, for instance, the sea defense system is at "Delta height." However, for the distinguished coastal sections of the Dutch coast this "Delta height" has different values. For the uninterrupted West-coast, which protects the most densely populated and economic heart of the country, the minimum height of the dunes and dikes is determined by a water level exceedence chance of 1 in 10,000 years. For most of the barrier islands in the north of the country the "Delta height" corresponds with a high-water exeedence chance of 1 in 2,000 years.

Subsurface aspects

Subsurface aspects like subsidence and tectonic movements codetermine the behavior of the sea level. The Netherlands, for instance, are suffering from a relative rise in sea level, which is the combination of the eustatic sea-level rise and the subsidence of the land (Jelgersma, 1961). Subsidence in this case is mainly due to the isostatic rebound after the last glaciation and the dewatering of the peaty subsoil in the coastal area for agricultural reasons, resulting in compaction. Especially in the past, this drainage has contributed to the general subsidence. At the same time, the central part of the Dutch coast is influenced by a graben system.

In the decision-making of the final height of a sea dike, the compaction of the subsoil due to the weight of the overlying dike body has also to be considered, as well as some compaction in the dike body itself.

The composition of the subsoil is relevant with regard to groundwater flow. A dike is meant to be impermeable for water. However, groundwater can be pressed to flow through the subsoil underneath a dike, due to a difference in hydrostatic pressure on both sides of the dike. The flow velocity depends on the volume of the overhead and on the permeability of the subsoil. This seepage causes saltwater intrusion landward of the dike. In case the currents are sufficiently strong to erode the underlying sediment, this process affects the stability of the dike and ultimately may result in the collapse of the dike.

Construction aspects

The construction of a dike requires building material. In the past, the preferably clayey material was locally dug. Nowadays, however, dikes usually have a sand nucleus, covered by clay to make it impermeable. Suitable clay usually is scarce, whereas sand occurs in large quantities. The advantage of this method is that sand not necessarily needs to be transported on a truck, but can be supplied in suspension by pipeline.

Because of the difference in hydrographic pressure on both sides of a dike, a groundwater table is formed in the sandy dike body. Although the clay cover has to be impermeable for seawater, it must simultaneously be able to let an excess of water from inside the dike through.

Generally, dikes have a delta form. To reduce the effect of wave run-up, the slope of the seaward bank of a dike has to be small. On the contrary, the width of a dike increases in that case. A balance has to be found between cost and safety.

In the flood disaster in the Netherlands in 1953, the effect of overtopping water on the landward side of the dikes appeared to be the most important cause of dike failure. In order to reduce the effects of overtopping water, for example, the penetration of water into the center of the dike body, various degrees of unevenness can be applied.

To avoid erosion of the banks and undermining of the dike body revetments are applied. Usually these revetments are provided with a filter layer, to prevent erosion of the underlying material. Permeable as well as impermeable revetments are applied. It is necessary that the material used is flexible and tight to follow the compaction of the underlying sediment.

Pieter G.E.F. Augustinus

Bibliography

Jelgersma, S., 1961. Holocene sea level changes in the Netherlands. Ph.D. thesis, State University Leiden, Mededelingen Geologische Stichting, CVI-7.
U.S. Army Corps of Engineers, 1984. Shore Protection Manual, Part 1 and 2. Washington DC: Government Printing Office. U.S.

Cross-references

Cheniers
Coastal Subsidence

Geotextile Applications
Hydrology of Coastal Zone
Meteorologic Effects
Polders
Sea-Level Rise Effect
Shore Protection Structures
Storm Surge
Tides
Waves

DISSIPATIVE BEACHES

Definition and classification

Dissipative beaches are systems where most wave energy is expended through the process of breaking. Guza (1974) was apparently the first to use the term dissipative beach. He indicated that the wave-energy status of a nearshore system could be determined using the surf-scaling parameter, ε:

$$\varepsilon = \frac{\alpha\omega^2}{g\tan^2\beta}$$

where α is the wave amplitude at breaking, ω is the wave radian frequency ($\omega = 2\pi/L$, where L is wave length), g is the gravity constant, and β is the beach slope in degrees. The proportion of incident wave energy that is dissipated by breaking increases as ε increases. When ε is less than about 2.5, most wave energy is reflected off the foreshore. For beaches where ε is larger than 20, most energy is dissipated by the turbulence associated with wave breaking. Guza (1974) designated these latter beaches as dissipative. Thus, the relative degree of dissipation or reflection of incident wave energy in a nearshore system may be used as a rationale for the classification of beaches. This is usually accomplished under the rubric of nearshore morphodynamics.

Morphodynamics

The concept of nearshore morphodynamics was developed to characterize systems where form and process are closely coupled through feedback mechanisms. On beaches, waves (*q.v.*) interact strongly with sediments and morphology, and the form of wave breaking is one manifestation of these interactions. For constant wave steepness, H/L (where H is wave height), the breaker type will change as the nearshore slope changes. On a very low gradient slope, spilling breakers would be expected. As the gradient increases, there should be a progression through plunging and collapsing breakers. Finally, on very steep beaches, surging breakers should occur (Galvin, 1968). For a constant nearshore slope, the same sequence of breaker types will occur as wave steepness decreases. Breaker type is closely associated with the expenditure of wave energy in the nearshore (e.g., dissipation or reflection) and the development of nearshore morphology. The morphology, in turn, controls breaker type. These relationships are the underlying bases for the concept of nearshore morphodynamics (see summary by Wright and Short, 1984). The recognition of characteristic sets of dynamic relationships provides the basis for using morphodynamic regimes (or states) as a means for classifying beach types. For example, spilling breakers tend to occur on dissipative beaches. This contrasts with reflective beaches (*q.v.*), where collapsing or surging breakers are common. Plunging waves tend to occur on the intermediate beach states of the morphodynamic model, where neither dissipation nor reflection dominate the response of nearshore morphology.

Characteristics of dissipative beaches

In cross section, morphodynamically dissipative beaches display the classic forms of "storm" or "winter" beach profiles (e.g., Sonu and Van Beek, 1971). According to Wright and Short (1984), other distinguishing characteristics include low gradient nearshore and beach slopes—about 0.01 and 0.03, respectively, and with relatively fine sediment sizes. Dissipative beaches tend to have a substantial sediment volume and at least one linear nearshore bar, although multiple bar–trough systems are common. In dissipative systems, the beach and nearshore zone will exhibit minimal alongshore variability—the system is approximately two-dimensional. Incident wave energy is maximum at the break point, and decreases shoreward. The classic dissipative system displays several coincident sets of spilling breakers (Figure D36), and there is minimal energy remaining at the landward extremity of uprush. Infragravity motion dominates the inner surf zone, and frequently causes linear scarping of the foreshore (e.g., Short and Hesp, 1982). On meso- and macrotidal beaches, the nearshore may be dissipative only at lower tidal stages and reflective at high tide (Short, 1991; Masselink and Hegge, 1995). Short and Hesp (1982) have also linked the dissipative beach state to the formation of large-scale transgressive dune sheets—at least in the Australian context. This linkage was a key development for the derivation of later models of beach–dune interaction (e.g., Sherman and Bauer, 1993).

Douglas J. Sherman

Figure D36 The dissipative beach system at Rossnowlagh, Co. Donegal, Ireland. Note the low-gradient beach slope and multiple lines of spilling breakers. Maximum breaker height is approximately 1.5 m, and period is about 7 s.

Bibliography

Galvin, C.J., 1968. Breaker type classification on three laboratory beaches. *Journal of Geophysical Research*, **73**: 3651–3659.

Guza, R.T., 1974. *Excitation of Edge Waves and Their Role in the Formation of Beach Cusps*. Unpublished Ph.D. dissertation, University of California, San Diego.

Masselink, G., and Hegge, B., 1995. Morphodynamics of meso- and macrotidal beaches: examples from central Queensland, Australia. *Marine Geology*, **129**, 1–23.

Sherman, D.J., and Bauer, B.O., 1993. Dynamics of beach–dune systems. *Progress in Physical Geography*, **17**: 413–447.

Short, A.D., 1991. Macro–meso tidal beach morphodynamics—an overview. *Journal of Coastal Research*, **7**: 417–436.

Short, A.D., and Hesp, P.A., 1982. Wave, beach and dune interactions in southeastern Australia. *Marine Geology*, **48**: 259–284.

Sonu, C.J., and Van Beek, J.L., 1971. Systematic beach changes on the outer banks, North Carolina. *Journal of Geology*, **79**: 416–425.

Wright, L.D., and Short, A.D., 1984. Morphodynamic variability of surf zones and beaches: a synthesis. *Marine Geology*, **56**: 93–118.

Cross-references

Bars
Beach Features
Beach Processes
Reflective Beaches
Sandy Coasts
Surf Zone Processes
Waves

DREDGING OF COASTAL ENVIRONMENTS

Definition of dredging

Dredging is moving material submerged in water from one place to another in water or out of water with dredging equipment. Dredging is applied in projects of navigation, for maintenance of beaches, for reclamation, to secure construction materials, and for environmental dredging.

Dredging for navigation

From the very beginning of civilization, people, equipment, materials, and commodities have been transported by water. Ongoing technological developments and the need to improve cost-effectiveness have resulted in larger, more efficient ships. This, in turn, has resulted in the need to enlarge or deepen many of the rivers and canals, our "aquatic highways," in order to provide adequate access to ports and harbors. Nearly all the major ports in the world have at some time required new dredging works—known as capital dredging—to enlarge and deepen access channels, provide turning basins and achieve appropriate water depths along waterside facilities. Many of these channels have later required maintenance dredging, that is, the removal of sediments which have accumulated in the bottom of the dredged channel, to ensure that they continue to provide adequate dimensions for the large vessels engaged in domestic and international commerce.

Dredging for Construction, Reclamation, and Mining

Dredging is an important way of providing sands and gravels for construction and reclamation projects. In the last two decades, the demand, and the associated extraction rates, for such offshore aggregates have significantly increased. Dredged aggregates have a wide range of uses.

Dredging for the environment

Dredging can be undertaken to benefit the environment in several ways.

Dredged materials are frequently used to create or restore habitats. Recent decades have also seen the increasing use of dredged materials for beach replenishment.

Another environmental use of dredging has been in initiatives designed to remove contaminated sediments, thus improving water quality and restoring the health of aquatic ecosystems. This so-called "remediation" or "cleanup" dredging is used in waterways, lakes, ports, and harbors in highly industrialized or urbanized areas.

Types of dredgers

Dredgers come in two different classes: mechanical and hydraulic.

Mechanical types include the *clamshell* or *grapple dredges* (Figure D37). Larger dredges of that type are no longer favored. The endless *chain bucket* (bucket-ladder) dredge (Figure D38) was used widely earlier in Europe but not in the United States, although they were employed on a few projects like the Panama Canal. For environmental reasons they may still be placed back on the market.

The *mechanical dipper dredge* with its heavy bucket (Figure D39) moved by a very strong arm and boom is still used for dredging relatively loose (usually not solid) rock. The bucket may be provided with special cast iron teeth. Its volume may be several cubic meters.

The *hydraulic dredge* is the most important piece of dredging equipment. The *plain suction dredge* (Figure D40) has no cutter but sucks material off the bottom and discharges it through a stern connected pipe leading to a spoil disposal area.

The *cutterhead pipeline dredge* (Figure D41) has a rotating cutter on the end of the ladder and excavates the material from *in situ* condition and discharges it through the stern to pontoon and shore pipe. The dredge is controlled on stern-mounted spuds and is swung from one side of the channel to the other by means of a swing gear.

The *self-propelled hopper dredge* (Figure D42) has a large hopper in which the dredged material is loaded for later dumping through doors in the bottom. This type of dredge is normally employed where the water is too deep for a pipeline dredge or where spoil areas for such a dredge are not available within economic distances.

The hydraulic dredge has, without any doubt, become the most important piece of equipment in the entire harbor engineering field. Without

Figure D37 The clamshell or grapple dredge (adapted from Bruun, 1990b).

Figure D38 The bucket-ladder dredge (adapted from Bruun, 1990b).

Figure D39 The dipper dredge (adapted from Bruun, 1990b).

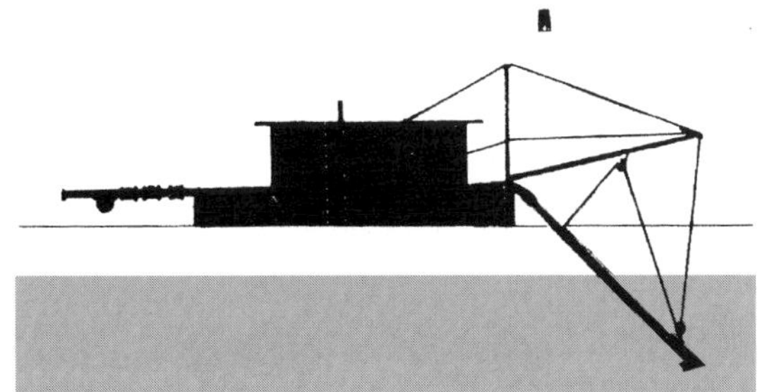

Figure D40 The plain suction dredge (adapted from Bruun, 1990b).

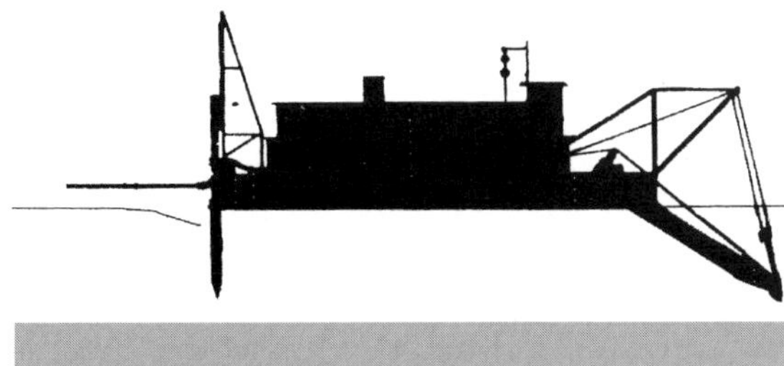

Figure D41 The cutterhead pipeline dredge (adapted from Bruun, 1990b).

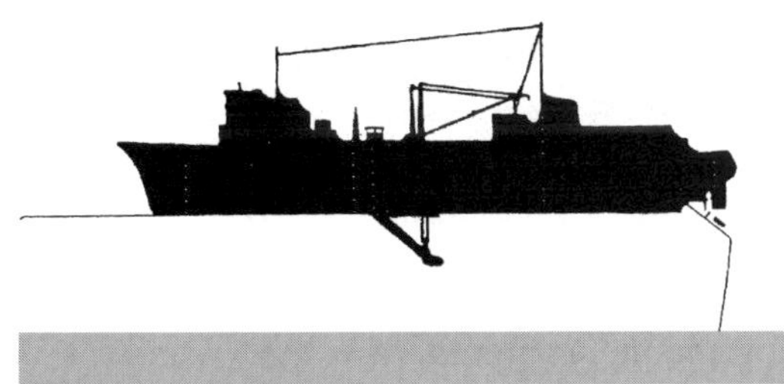

Figure D42 The self-propelled hopper dredge (adapted from Bruun, 1990b).

the dredge, commercial navigation of waterways and rivers would be ended. Waterborne industry would collapse. Ocean shipping as it is known would be nonexistent.

Hydraulic dredges dig canals, ports, and harbors, do maintenance dredging in rivers, canals, and waterways, and excavate for construction of piers, wharves, docks, dams, and underwater foundations. They provide spoil for the reclamation of swamps and marshes; they construct dikes and levees, and dredge sand, gravel, and shell, as well as coal, gold, diamonds, and many other minerals for commercial purposes. The dredge's scope of operation is broad. Most of them have cutterheads.

Small hydraulic dredges, 20–40 cm diameter pipe, operate in water only a few meters deep. Large dredges, 65–90 cm diameter pipe, require more draft but can dig to greater depths. With the aid of booster pumps the distances solids can be pumped are unlimited.

Although the dredge's output is understandably greater in soft materials than in hard, it can excavate almost anything. It digs mud, silt, loam, clay, sand, hardpan, gravel, coral, and even rock. Boulders weighing 500 kg and more have been excavated and transported by dredges.

The self-propelled hopper dredge moves freely without any kind of mooring system. It can operate by pumping holes in the bottom side by side through a usually circular intake which is the reinforced end of the suction pipe, or it can trail along with its trailing suction head attached to the end of the suction pipe "vacuum cleaning" the bottom. This method is more practical and effective than the plain pipe sucking procedure where deep holes may become sediment traps for material which otherwise may have been flushed away. Various organizations like the IADC (International Association of Dredging Companies), the CEDA (Central Dredging Association), the WEDA (Western Dredging Association), and the WODA (World Organization of Dredging Association) organize conferences and publish magazines. (See *Acronyms*).

Special dredgers

Wheel dredgers

The IHC's (The Netherlands) Scorpio Dredger has a dredging wheel that consists of bottomless buckets to which sticky soil does not adhere (Figure D43). They are arranged to form a tunnel from which dredged material cannot easily escape. This effectively prevents the ingress of debris. The result is higher solids concentration, no spillage or blockage, and reduced downtime. Several such dredgers are now in operation all over the world.

The Water Injection Dredger

The Water Injection Dredge (WID) was developed in Europe in the mid-1980s and has been used regularly there since 1987 and recently tested in the United States by the US Army Corps of Engineers (USACE). The principle of WID is to fluidize the shoaled material to a condition that a gravity-driven density current is formed, which transports the sediment down a slope into deeper water where it is no longer an obstruction to navigation, and natural sediment transport processes can take over. This density current remains close to the bottom, minimizing turbidity. Fluidization is accomplished by injecting water directly into the sediment voids under relatively low pressure, through a series of nozzles evenly spaced on a horizontal pipe. A pump mounted on the dredge, barge, or other vessel forces water into the horizontal pipe or head. The pipe pivots from the dredge, allowing the head to be lowered with the aid of a winch. During operation, the head with the injection nozzles is as close to the bottom as possible, while the dredge proceeds across the cut area (Bruun, 1990a, 1996).

Sidecasters

Sidecasters sidecast the dredged materials. They come in various sizes. Small sidecasters are excellent for emergency and shallow water operations, for example, in channels.

The McFarland (Figure D44) also has a hopper. It was put in commission in 1967 and assigned to maintenance of Gulf Coast inlets

Figure D43 The Wheel dredger SCORPIO (adapted from Bruun, 1990b).

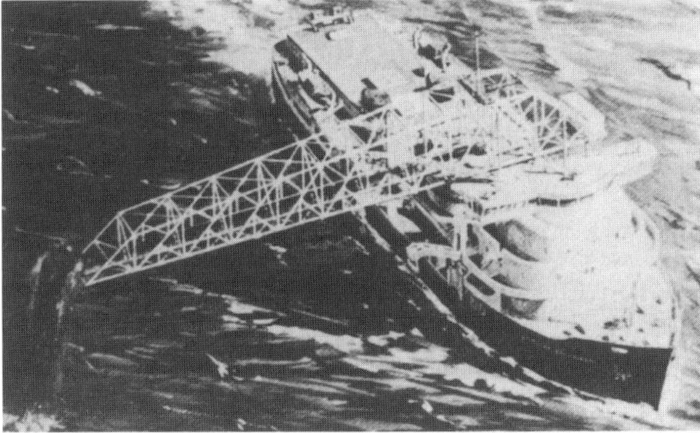

Figure D44 The sidecaster and hopper dredge McFarland (adapted from Bruun, 1990b).

(USACE). It is 91 m long, 22 m wide, with a loaded draft of about 7 m to permit efficient operation in relatively shallow water.

Other types of dredgers are mentioned in the following section on the latest development of dredging equipment.

The latest development in dredging equipment

The latest decades and years have produced dredging equipment on a scale never seen before including hopper dredgers of capacities up to 33,000 m^3, with suction pipes of 1.4 m diameter reaching down to 60 m depth, with deep dredging suction up to 131 m depth, and outputs exceeding to 10,000 m^3/hr. Large projects in the Far East (Hong Kong, Singapore, Thailand, China, Persian Gulf) and also in South America (Argentina, Uruguay) utilize these dredges. Large fills for industrial development including airports, deepening of navigation channels and basins all over the world are in progress or planned. This includes many projects in the United States (New York, Charleston, Savannah, Galveston, Los Angeles, Columbia River, etc.). Materials are used for fills or dumped offshore in designated areas.

The dredging companies involved are mainly Dutch or Belgian. They compete in producing the most effective equipment provided with the latest inventions in survey technique, controls, and maneuvering. They are applied in projects with quantities of tens of millions of cubic meters capital dredging. Smaller size equipment includes hopper dredgers of a few thousand cubic meters hopper capacities for relatively smaller projects, capital as well as maintenance.

Some larger cutterhead hydraulic pipeline dredgers are still being built, in particular for work on harder bottoms and in narrow or protected environments. Some of these are ocean-going with discharge pump capacities of many kilometers. An example is the Bos Kalis 89 cm "ORANJE."

Amphibious "crawl cats" and "crawl dogs" (Bruun, 1996) and underwater pumps like the PUNAISE (Visser and Bruun, 1997) have been built and tested in practice.

One of the most recent additions to the international fleet is the "LANGE WAPPER," a 13,700 m^3 capacity trailing suction hopper dredger which entered service in May 1999. The 20,000 dwt new vessel boasts the extremely shallow draft of 9 m fully laden, setting it in competition with vessels in the 8,000–10,000 m^3 range—yet with substantially higher production rates.

Built by IHC Holland, the Lange Wapper is 120 m long, with a beam of 26.8 m. The total installed power is 13,400k W and dredging depth is in the range 30–50 m. For a draught of 8 m, the vessel can carry a deadweight of 15,500 tonnes.

The Lange Wapper features state-of-the-art computer software to integrate all dredging, navigation, and surveying systems. It is thought to be the first software of its kind in the industry.

Specially designed by DEME, together with IHC Systems (the electronics subsidiary of the IHC group), the software provides a single program capable of handling all data relating to dredging, navigation, and surveying. Previously, software systems could only communicate with each other from standalone packages. Furthermore, up to 10 extra systems were required onboard and extra personnel needed to operate them. The software onboard the Lange Wapper will be available via a number of networked computer terminals and will also be able to access additional data by satellite.

Enhanced controls and accuracies are the main advantages of this new system. Tight dredging parameters will be possible, which will enhance performance in environmentally sensitive conditions and offshore trenching or pipe covering operations.

All dredging projects, however, depend upon hydrographic surveys. As ports and channel systems throughout the world continue to involve expansions and extensions, the need for effective survey programs becomes essential. Throughout the past year there have been some major developments within this sector of the port construction industry.

Dredging for nourishment of beaches

Artificial nourishment of beaches is the most practical and economical measure of coastal protection. Suitable material must be available. Figure D45 shows split-hull dumping of dredged material. Figure D46 is a trailing suction hopper dredge pumping material on the beach and in the nearshore bottom, or through a buoy connected to shore by a (submerged) pipeline. In the United States, nourishment is often done by a hydraulic pipeline dredge connected directly from the borrow area to the beach by (submerged) pipeline.

Figure D45 Direct dumping in the nearshore zone with hopper dredger (from Brunn, 1990b).

Figure D46 Trailing hopper dredger pumping through nozzle placed in bow of the vessel (from Brunn, 1990b).

Profile nourishment

The normal nourishment profile includes a wide horizontal berm with a front slope of about 1 in 10. Because beach nourishment is generally paid for by a public authority, public opinion is important in acquiring sufficient funds. Therefore, it is wise to design beach nourishment in such a way that the public sees that the beach is somewhat wider after the nourishment, but that there is no major adaption in the beach shape during the first storms in autumn. If the purpose of the nourishment is to make a wide recreational beach, this is very important. If the purpose is to prevent flooding, the best place is as high as possible on the beach. If the purpose is to combat chronic erosion, the best place is in the breaker zone.

Profile nourishment (Bruun, 1990a, b) attempts to follow the natural profile in the nearshore as much as this is practically possible. The advantage of profile nourishment is that it decreases the introductory losses of fill producing a more stable condition. OVER THE BOW PUMPING (Figure D46) is most suitable for such operations undertaken in countries like Holland, Denmark, Germany—and at its start in the United States. The Shallow Water Hopper dredgers, in short supply in the United States, are well suited for such operations. They may pump over the bow, if bottom is steep enough, or through a short pipeline (Bruun and Willekes, 1992).

Dredging and the environment

Introduction

The development of navigation requires still deeper and wider channels and basins, as well as fills for industrial harbor and other developments. This, in turn, carries with it the needs for major disposal areas for dredged materials.

Maintenance of depths requires removal of deposited materials which are sometimes—or often—polluted.

Beaches and shores are eroding and need protective nourishment by clean material which will then have to be relocated. Dredging may also be a useful tool for remedying past environmental interference. However, by its very nature, the act of dredging and relocating dredged material has an environmental impact. It is, therefore, of the utmost importance that we should be able to determine whether any planned dredging will have a positive or negative impact on our environment. Evaluation of environmental impact should examine both short- and long-term effects, as well as the sustainability of the altered environment.

While spoiling of material from dredging of tidal inlets on sandy shores has involved little difficulty because the material was clean and ample offshore dumping areas were available, dredging of harbors estuaries, bays, lagoons, and waterways without tidal flushing, has gradually caused many problems with respect to the location of spoil areas.

In either case, a severe disturbance of the area's biology has resulted. This has necessitated a great number of regulations on the extent of such operations.

When using dredging techniques, we must be aware of the environmental effects of the changes we are trying to achieve, as well as the effects of the dredging activity itself which may include: alterations to coastal or river morphology; addition or reduction of wildlife habitat; changes to water currents and wave climates, which might affect navigation, coastal defense, and other coastal matters; reduction or improvement of water quality, affecting benthic fauna, fish spawning, and the like; improvement of employment conditions owing to industrial development; and removal of polluted materials and their relocation to safe, contained areas.

The marine environment is a complex combination of natural features and phenomena, supporting a diverse but largely concealed population. Because of this complexity it is extremely difficult to predict the effects of human-induced changes and short-term operations.

Past environmental ignorance has resulted in many rivers, ports, and harbors in the industrialized nations containing soils that have been contaminated by undesirable levels of metals and chemical compounds. When dredging in these soils, contaminants may be released into the water column and thence into the food chain. Thus, the environmental effects of dredging and relocation of the dredged material may be more severe and will require more detailed analysis. In certain cases, it is the very existence of the polluted soils that has led to dredging: by removing the polluted soils and relocating them to a more secure situation, the environment is improved. Long-term improvement does, of course, depend ultimately on preventing pollution at its source. The treatment and storage of polluted soils is a highly complex subject and requires detailed study.

Many rules and regulations have been issued regarding the handling of polluted materials. International conventions, for example, the London 1972 meeting (Bruun, 1990), have resulted in published advices. The PIANC (Permanent International Association of Navigation Congresses) has issued a number of guidelines including:

- Report of the International Commission. Study of Environmental Effects of Dredging and Disposal of Dredged Materials. Supplement to Bulletin No. 27 (1977)
- Report of the International Commission. Classification of Soils and Rocks to be Dredged. Supplement to Bulletin No. 47 (1984)
- Report. Disposal of Dredged Material at Sea. Supplement to Bulletin No. 52 (1986)
- Beneficial Uses of Dredged Material (1992)
- Dredged Material Management Guide (1997)
- Handling and Treatment of Contaminated Dredged Material, Vol. 1 (1997) and Vol. 2 (1998)
- Management of the Aquatic Disposal of Dredged Material (1998)
- Confined Disposal Facilities for Contaminated Dredged Material

The ADC (Association of Dredging Companies) and the CEDA have printed several reports, here particularly noted and referred to is the 1999 report on "Environmental Aspects of Dredging." In the United States, the USACE, Coastal Engineering Research Center in Vicksburg, MS, through its Dredging Research Division has published a great number of reports as mentioned later.

Management

Management alternatives for dredged material can be grouped into the following four main categories (listed in order of significance): beneficial use, open-water disposal, confined disposal, and treatment.

Beneficial use

The definition of "beneficial use" according to the Dutch ADC and CEDA Guide #5, 1999, is "any use which does not regard the material as a waste."

Dredged material is increasingly regarded as a resource rather than as a waste. The London Convention's Dredged Material Assessment Framework (DMAF) recognizes this by requiring possible beneficial use of the material to be considered before a license for sea disposal may be granted.

Open-water disposal

Open-water disposal is the placement of dredged material at designated open-water sites in oceans, estuaries, rivers, and lakes such that the dredged material is not isolated from the adjacent waters during placement. Open-water disposal generally involves placement of clean or mildly contaminated material.

Open-water disposal of highly contaminated material can also be considered with appropriate control measures. This category includes unrestricted placement on flat or gently sloping waterbeds in the form of mounds or placement with lateral containment (e.g., depressions). For contaminated material, a cap of clean material can provide isolation from the benthic environment. If capping is applied over the mound formed by unrestricted placement, it is called level-bottom capping (LBC). If the capping is applied with lateral containment, it is called contained aquatic disposal (CAD).

Confined disposal

Confined disposal is the placement of dredged material in an engineered containment structure (e.g., dikes, natural, or constructed pits) enclosing the disposal area and isolating the dredged material from surrounding waters or soils during and after placement.

Treatment

Treatment is defined as a way of processing contaminated dredged material with the aim of reducing the amount of contaminated material or reducing the contamination to meet regulatory standards and guidelines. Each project may raise environmental concerns and not only when the material in question is contaminated. For instance, the "beneficial" use of clean, beach-compatible sand for beach nourishment may damage well-established sensitive marine habitats or species. Or the unrestricted open-water placement of clean material may have unacceptable direct physical impacts, such as smothering of bottom-dwelling organisms or local increases in suspended solids concentrations.

Assessments

The environmental assessment of management alternatives (with controls) should consider all potential impacts. Many countries require Environmental Impact Assessment (EIA) studies to be undertaken and the results must be documented in formal Environmental Impact Statements. (EIS.) Even if the law does not require such studies, it is advisable to include them in the project planning as EIS may prove to be very useful in the permitting procedure and in gaining public acceptance for projects.

For decision making, the following eight steps are identified in the Dutch Guide #5, by the IADC and WEDA (1999): establishing the need for dredging, characterization of dredged material, assessment of beneficial use options, preliminary screening of potential disposal alternatives, detailed assessment of disposal alternatives, selection for final design and implementation, permit application and processing, monitoring program design.

The World Bank (1990) distinguishes five main types of dredged sediments:

1. Material from maintenance dredging of ports, harbors, and navigation channels usually consist of fine-grained silt and clays and is often rich in organic matter and contains contaminants.
2. Material from maintenance dredging of sand bars at the entrances to harbors or channels usually consist of fine- to coarse-grained sand and is much less contaminated than (1).
3. Material from capital dredging within a port may vary considerably along the vertical profile. The upper layer is usually fine-grained, organic-rich, and contaminated, and the underlying layer is coarse-grained and uncontaminated. In some circumstances, even the deeper layers might contain elevated levels of contaminants from historical discharges.
4. Material from capital dredging of channels or outer harbor areas is likely to be coarse-grained and uncontaminated.
5. Material derived from remediation dredging is usually fine-grained, organic-rich and, by definition, highly contaminated. Its relocation requires special control measures.

Physical properties

The basic physical characteristics are form and composition, grain size, specific gravity, plasticity, water content, shear strength, water-retention characteristics, permeability, settling behavior, consolidation behavior, compaction, and organic content.

Chemical properties

They include pH-values, calcium carbonate equivalent, cation exchange capacity, salinity, redex potential (elective activity), dissolved oxygen, biochemical oxygen, organic carbon, nutrients, carbon/nitrogen ratio, potassium, and contaminants of any kind.

Biological properties

Biological characterization of sediments may involve testing for the presence of microbial constituents and testing for toxicology.

Microbial constituents of concern include pathogens, viruses, yeasts, and parasites. Sediments should be tested for these constituents whenever dredging sites are close to sewage discharges and placement sites are close to sensitive areas such as drinking water intakes, recreation beaches, or shellfish beds.

Acute toxicity bioassays test the effects of short-term exposure. Toxicity is expressed as median lethal concentration (LC_{50}), that is, concentration which theoretically would kill 50% of the test organisms within a specified time span. The duration varies from hours to days.

Chronic toxicity bioassays assess sublethal effects which can result from prolonged exposure to relatively low concentrations. Such effects include physiological, pathological, immunological, terratological, mutagenic, or carcinogenic effects. Chronic toxicity bioassays may take several weeks. Toxicity bioassays may be used to evaluate: potential impacts of dissolved and/or suspended sediments on water-column organisms (elutriate bioassays); and potential impacts of deposited sediments on benthic organisms (benthic bioassays).

Metals

Metals may exist in the aquatic environment in four basic, but interactive forms. These dissolved forms are: free metal ions; complexed molecules; particulate forms absorbed to the surface of solid particles; and precipates (mostly sulfides).

In free mode (the most toxic), metals are transported with the water and easily taken up by organisms. The particulate associated forms are relatively inactive. In sediments under natural conditions (anoxic, reduced, near neutral pH) only a small fraction of heavy metals is dissolved; the major portion is bound by sulfides or by structurally complex, large organic compounds. Release of these may be induced by increased salinity, reduced pH, and increased redox potential.

Sediment properties versus dredging placement methods

In addition to the original properties at the dredging site, the behavior of sediments during and after placement will also largely depend on the specific characteristics of the dredging and placement techniques.

Dredging techniques may be subdivided into two categories: mechanical and hydraulic dredging.

Mechanical dredgers (e.g., bucket or clamshell) use mechanical force to dislodge, excavate, and lift the sediments. The material is generally placed and transported to the site of discharge by barges. Barges may have bottom doors or split-hulls and the material can be released within seconds as an instantaneous discharge. Cohesive sediment dredged and placed this way remains intact, close to the *in situ* density through the whole dredging and placement process.

Hydraulic dredgers remove and lift the sediments as a slurry (mixture of sediment solids and large amounts of dredging site water). The slurry is transported to the discharge site via pipeline or in hoppers and released with large amounts of entrained water. Hoppers may release the material through bottom doors or through pump-out operations. Hydraulic dredging and transport break down the original structure of sediments.

Beneficial use of dredged materials

The main application is for fills. Recent years have also introduced a number of practical structural usages.

Publications providing guidance on beneficial use include: Beneficial Uses of Dredged Material; Engineer Manual (USACE, 1986); Beneficial Uses of Dredged Material, A Practical Guide (PIANC, 1992); Handling and Treatment of Contaminated Dredged Material (PIANC, 1997); Guidelines for the Beneficial Use of Dredged Material (HR Wallingford, 1996).

Engineering usages

For practical use of dredged materials, one may distinguish between rock, gravel, sand, silt, mud, and clay, differing greatly in grain sizes and shape. Sand and silts may become "quick." Clay is cohesive.

Gravel and sand are dredged for construction purposes. Different kinds of heavy minerals are extracted from sands in sometimes major operations, including from the deep sea (Bruun, 1990).

Coastal protection

Dredged material is used for artificial nourishments of beaches which today is the most common coastal protection measure. The material dredged offshore or in entrances and navigation channels is pumped ashore if suitable for widening and raising the beach elevation.

Materials may also be used to produce protective offshore berms which may be designed as feeder berms, hard berms, and soft berms.

Feeder berms

Feeder berms produced by dumping of materials nearshore serve partly as source of materials for a possible transfer of materials to the beach.

Hard berms

The construction of hard berms involves the placement of suitable material, in depths up to 13 m, to create a permanent feature on the seabed approximately parallel to the shoreline with gentle side slopes that will intercept the troughs of incoming storm waves and decrease erosion of the shoreline. If the berm is made of sand, it will be modified in profile and some of the material will be lost, implying that some maintenance will be necessary. This technique may still be cheaper and more convenient (i.e., causing less disturbance to recreation) than direct beach nourishment.

Soft berms

Naturally occurring underwater mudbanks are known to absorb water wave energy and thereby attenuate waves that pass over them. Energy reductions of 30–90% are not uncommon even in the absence of wave breaking. Attempts have been made to replicate this action by creating underwater mud berms from dredged material.

Open-water disposal

Material dredged may be disposed of in deeper waters or in shallow designated waters. Bypassing at tidal entrances is a common usage. Offshore disposals may be capped to protect the material from being eroded away. Polluted materials may be placed inside dikes. Figure D47 shows the large "Slufter" used by EUROPORT for disposal of dredged materials which are polluted. Land disposals may also use dikes. An impermeable membrane is then placed to avoid penetration of pollutants to the surrounding areas. Material may also be placed in dredged pits in the bottom provided with a proper protective cover.

The dredging research program in the United States by the USACE

The USACE is involved in virtually every navigation dredging operation performed in the United States. The Corps' navigation mission entails maintenance and improvement of about 40,000 km of navigable channels serving about 400 ports, including 130 of the nations 150

Figure D47 The "Slufter" offshore diked disposal are used by EUROPORT (Rotterdam) for polluted materials (adapted from Brunn, 1990b).

largest cities. Dredging is a significant method for achieving the Corps' navigation mission. The Corps dredges an average annual 230 million m^3 of sedimentary material at an annual cost of about US $400 million. The Corps also supports the US Navy's dredging program in both maintenance and new work.

The Corps will continually be challenged in pursuing optimal means of performing its dredging activities. Implementation of an applied research and development (R&D) program to meet demands of changing conditions and generation of significant technology, adopted by all dredging interests, are means of reducing the cost of dredging the nation's waterways and harbors.

The Shallow Water Hopper dredger has always been on their agenda. There are a couple of additions to the fleet financed by private industry. They are of medium-size with about 5,000 m^3 hopper capacity.

Reports on the development of dredging and on the Environmental Effects of Dredging, Technical Notes, are published regularly, the latter category covering the subjects of Aquatic Disposal, Upland Disposal, Wetland Estuary Disposal, Regulatory Management, Beneficial Uses, Equipment, and Miscellaneous as described in numerous reports. Copies may be acquired from the USACE, Waterways Experiment Station, Vicksburg, MS.

The Long-Term Effects of Dredging Operations (LEDO) program by the Dredging Research Division of the USACE, Vicksburg, MS, "focus on cost-effective, environmentally responsible techniques for dredging and dredged material disposal in aquatic, wetland, and upland environments. Current research emphasizes risk-based procedures for effects assessment, exposure assessment, and risk characterization. The Program objective is to provide the latest proven technologies for identifying, quantifying, and managing contaminated sediments in support of cost-effective, environmentally responsible navigation."

"The primary benefit is a more timely complete, and cost-effective execution of the Corps' responsibilities under the Clean Water Act (CWA), Marine Protection, Research, and Sanctuaries Act (MPRSA), Resource Conservation and Recovery Act (RCRA), and Water Resources Development Act (WRDA). This R&D program provides dramatic cost avoidances annually for achieving restoration goals for contaminated dredged material disposal as mandated by law.

Recent major products include:

1. A chronic/sublethal and genotoxic assay that meets national regulatory requirements.
2. National guidance for confined disposal facility contaminant loss assessment procedures. The guidance document was established as Corps policy to meet regulatory mandates.
3. A new analytical screening method that measures bioavailable dioxin and similar compounds with a 90% reduction cost and time required by previous methods (co-sponsored with another research program).
4. The Environmental Residue Effects Database (ERED), the first WWW-accessible database of bioaccumulation effects to improve the accuracy and defensibility of environmental impact predictions while significantly reducing evaluation costs.

LEDO research benefits/value-added to district projects can be demonstrated on a recent example. Since the bioaccumulation interpretation database is Web-accessible, it is immediately and cost-effectively available to all district projects in need of contaminated material assessment."

Dredging and disposal problems in the New York and New Jersey Harbors are mentioned in reports by WEDA (Mohan *et al.*, 1999; Wakeman, 1999). The World Bank, 1990, has also issued its guidelines.

Conclusion

1. A strong development of the Dredging Industry is taking place with still larger and more efficient equipment. This development is most predominant in the low countries in Europe which virtually have taken over the world market.
2. Major projects on improvements of navigation channels and harbors remain. Artificial nourishment of beaches is still increasing.
3. Disposal areas for dredged materials are often running short due to environmental concerns. Diked areas, onshore and offshore, are becoming practices.
4. Environmental dredging is an important factor which requires much attention, planning, and monitoring. Beneficial usages of dredged polluted materials are developing.

Per Bruun

Acronyms

The following abbreviations are used in the text:

IADC	International Association of Dredging Companies (The Netherlands)
CEDA	Central Dredging Association (The Netherlands)
D&PC	Dredging and Port Construction (UK)
PE	Port Engineering Management Magazine (UK)
WEDA	Western Dredging Association (USA)
WODA	World Organization of Dredging Associations (The Netherlands)

Bibliography

Bruun, P., 1990a. Beach nourishment. Improved economy through better profiling and backpassing from offshore sources, *Journal of Coastal Research*, **6**(2): 265–277.

Bruun, P., 1990b. *Port Engineering*. Houston, Gulf Publishing Company.

Bruun, P., 1991. Optimum dredging for artificial nourishment of beaches. *Proceedings ASCE Conference on Water Resources*, pp. 303–307.

Bruun, P., 1996. New Principles and methods in maintaining beaches and depths in channels and entrances. Proceedings of the Bahia Blanca International Symposium, *Journal of Coastal Research*, 1–17.

Bruun, P. and Willekes, G., 1992. Bypassing and backpassing at harbors, navigation channels and tidal entrances: use of shallow-water draft hopper dredgers with pump-out capabilities *Journal of Coastal Research*, **8**(4): 972–977.

International Association of Dredging Companies (IADC) and the Central Dredging Association (CEDA), 1999. *Environmental Aspects of Dredging, Reuse, Recycle or Relocate*. The Hague: International Association of Dredging companies.

Mohan, Ram K., Urso, Dennis C., and Steele, Peter R., 1999. *Optimization of Dredged Material Placement Using the Sub-channel Placement Cell Concept*. Western Dredging Association.

Visser, K., and Bruun, P., 1997. The Punaise underwater dredger, *Journal of Coastal Research*, **13**(4): 1929–1999.

Wakeman, Th.H., 1999. *Using a Materials Assessment Process for Managing Placement of Contaminated Dredged Materials*. Western Dredging Association.

World Bank, 1990. Environmental considerations for port and harbor developments. *World Bank Technical Paper* 126. Washington, DC: The World Bank.

Cross-references

Beach Nourishment
Bypassing at Littoral Drift Barriers
Capping of Contaminated Coastal Areas
Engineering Applications of Coastal Geomorphology
Environmental Quality
Human Impact on Coasts
Mining of Coastal Materials
Reclamation
Water Quality

DRIFT AND SWASH ALIGNMENTS

Drift alignments are found on beach-fringed coasts where the dominant waves arrive obliquely to the shore and (with accompanying currents) maintain a beach parallel to the direction of the resulting longshore drift. They are typically found on straight or saliented coasts where the obliquely arriving waves move sediment alongshore. Swash alignments develop where beaches have been shaped by waves arriving parallel to the shore, usually in curved patterns resulting from wave refraction. They are typically found in embayments where longshore drifting is limited and beach outlines run parallel to the crests of incoming waves. Figure D48 shows how an incoming south-westerly swell has produced beaches with drift alignments in Western Port Bay, Australia, and beaches with swash alignments on adjacent Phillip Island. Drift and swash alignments can be developed experimentally on beaches in a wave tank (Davies, 1980, Figure 90): they are sometimes described as drift-dominated and swash-dominated beaches.

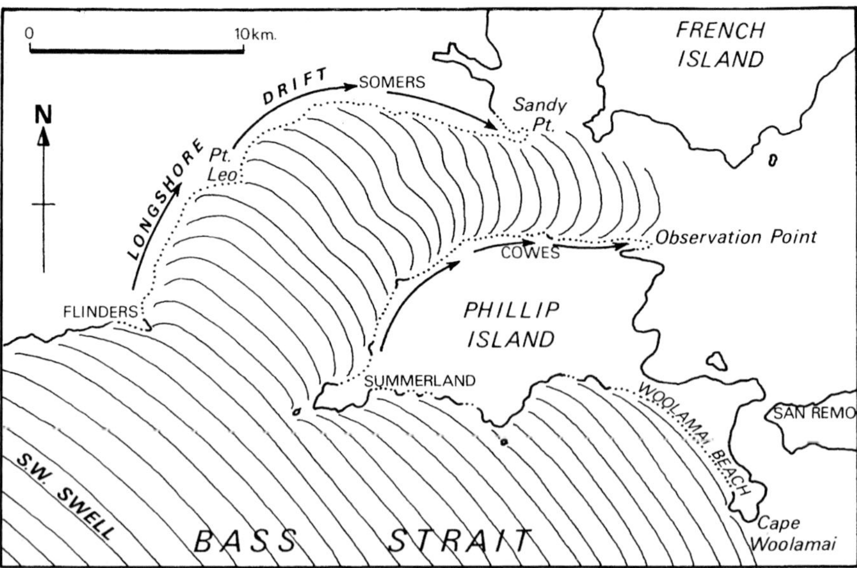

Figure D48 South-westerly swell entering Western Port Bay, Victoria, Australia, produces drift alignments on beaches to Sandy Point and along the north coast of Phillip Island, and swash alignments in bays on the south coast of Phillip Island, notably at Woolamai Beach (Copyright-Geostudies).

Drift alignments

Drift alignments are best developed where waves arrive at an angle of 40°–50° to the coastline. They are typically sinuous in detail, with intermittent lobes and cusps that migrate downdrift, and longshore spits and bars that diverge slightly alongshore. Variations in transverse profile occur as these features pass along the beach. A cyclic equilibrium may be attained where the beach alignment becomes adjusted in such a way that oblique waves impinging on the shore provide sufficient energy to transport the sediment arriving at one end of a drift-dominated beach through to the other, the configuration being maintained. Most beaches on drift alignments have an entirely updrift source of sediment (eroding cliffs or a fluvial input), but some incorporate sediment derived from eroding cliffs or river mouths along the shores, as on the north coast of Hawke Bay, New Zealand (Bird, 1996).

If the angle of wave incidence becomes greater or less than the optimum 40°–50° there is a slackening of longshore rifting and consequent deposition of sediment, but the drift alignment is soon restored when the oblique waves return to the optimum angle. A beach alignment can also be changed by variations in the rate of updrift sediment supply, and by erosion or accretion on updrift sectors of the coast.

Some drift-aligned beaches terminate in spits, such as the Langue de Barbarie on the Atlantic coast of Senegal in West Africa, which has grown southward in response to the dominant north-westerly swell and ends in recurves at the mouth of the Senegal River. Cuspate forelands and recurved (angled) spits may include both drift-aligned and swash-aligned beaches, and as sediment eroded from the drift-aligned shore is carried round the point to be deposited on the swash-aligned shore they migrate downdrift. Dungeness, on the south-east coast of England, is a major shingle foreland that has migrated in this way. Such structures were termed traveling forelands by Escoffier (1954).

Alternatively, beaches on drift alignments may end in sectors of accretion alongside headlands, river mouths, lagoon outlets, or protruding breakwaters.

Swash alignments

Beaches on swash alignments are smoother in outline than those on drift alignments, but are subject to alternations of profile erosion and losses of sediment offshore during storms and profile restoration by onshore sediment flow during calmer phases, a sequence known as "cut-and-fill."

Swash alignments are well developed on beaches in embayments where intervening headlands exclude waves from all but one direction. The curving beach behind Disaster Bay, in southern New South Wales, a narrow embayment between high parallel sandstone cliffs, has a swash alignment with an outline shaped by incoming refracted ocean swell, but too limited in length for oblique waves to generate much longshore drifting. Numerous parallel dune ridges indicate that progradation has maintained this curved plan.

Swash alignments are also well developed on beaches on open coasts where the incoming waves arrive consistently from one direction. In these situations, there is typically a simple relationship between beaches with curving swash alignments and the pattern of incoming refracted waves. In southern Australia, there are long, gently curving beaches which receive south-westerly ocean swell from the Southern Ocean, refracted by nearshore shallowing: Encounter Bay is a good example. South-easterly swell entering Jervis Bay on the east coast produces is refracted into wave patterns that fit the curving swash-dominated outlines of bordering bay eaches that face in various directions, and a similar pattern has been documented in Storm Bay, Tasmania (Davies, 1980). On the New South Wales coast a series of long asymmetrical beaches has developed on swash alignments, where incoming south-easterly waves are refracted round intervening headlands. Swash alignments occur on beaches on the Atlantic coasts of Europe, as on the coast of South Uist in the Hebrides, Rhossili Bay in Wales, and the Bay of Audierne in France.

Once established, gently curving beaches on swash alignments maintain their outlines through cycles of cut-and-fill when beach material is withdrawn seaward from them by storm waves, then swept back onshore in by a gentler swell. Any temporary divergence between incoming wave crests and the coastline initiates local longshore drifting which quickly restores the swash alignment.

Intermediate cases

Few beaches are entirely swash- or drift-dominated. Chesil Beach in Dorset is a shingle beach which has a swash alignment shaped largely by south-westerly swell and storm waves, but longshore drifting occurs on such beaches when westerly waves move sediment south-eastward and when south-easterly waves move it back westward. This alternation has led to lateral grading of this beach from granules at the western end to pebbles and cobbles toward the south-eastern end, near Portland Bill (Bird, 1996).

On the other hand, beaches on drift alignments such as Sandy Hook in New Jersey or the beaches north of the Columbia River in Washington State, are occasionally modified by waves coming in parallel to the shore, and sometimes show alternations of cut-and-fill with onshore offshore (swash–backwash) sediment movement. The existence of parallel beach and dune ridges formed by progradation behind beaches that appear to be drift-dominated, as on the North Beach Peninsula in Washington and Orford Ness in East Anglia, indicates occasional significant swash domination.

Changes in beach alignments

A lateral transition from a drift alignment to a swash alignment occurs on drift-dominated beaches that curve into sectors of accretion alongside headlands or breakwaters.

Where a beach outline is more sharply curved than the approaching waves it is drift-dominated, with a convergence of longshore drifting of beach sediment toward the center of the bay, where the beach progrades until it fits the outline of the arriving swell and so becomes swash aligned. Convergent longshore drifting in Byobugaura Bay, near Tokyo, has prograded the central sector at Katakai in this way, with sediment derived from the erosion of the northern and southern parts, and a similar evolution from drift alignments to a swash alignment has taken place at Guilianova on the east coast of Italy.

Where the beach outline is less-sharply curved than the approaching waves, there is divergent longshore drifting from the center until the beach outline fits the wave pattern and assumes a swash alignment. This has occurred on the shores of the Andalusian Bight, in southern Spain, where erosion of the central sector has been balanced by progradation at Matalascanas to the south and Mazagon to the north.

The beach on the northern flank of Rheban Spit, in south-eastern Tasmania, was built and prograded on a drift alignment as long as it received a longshore sediment supply from the north, but when this diminished the coastline was reshaped by updrift erosion and downdrift accretion until it developed a swash alignment adjusted to waves approaching from the north-east.

On some coasts, paired or opposed spits bordering embayments have grown on drift alignments, as the southern entrance to Menai Strait and in Portmadoc Bay in North Wales. Convergent growth of such spits can lead to the development of a curving beach which becomes swash aligned. Beach ridge patterns on the Swina barrier in Poland indicate such an evolution on the coast of the Pomeranian Gulf.

A change from a swash alignment to a drift alignment may occur where incoming wave patterns are modified by an artificial structure such as a breakwater or by shoal deposition or the growth of a coral reef. At Kunduchi in Tanzania, the sandy coastline had prograded on a swash alignment under the dominance of easterly swell, but the growth of reefs and shoals reduced the effectiveness of these ocean waves, and allowed locally generated and previously subdominant south-easterly waves to supervene, resulting in erosion as the beach assumed a drift alignment.

Eric Bird

Bibliography

Bird, E.C.F., 1996. Lateral grading of beach sediments: a commentary. *Journal of Coastal Research*, **12**: 774–785.

Davies, J.L., 1980. *Geographical Variation in Coastal Development*, 2nd edn. London: Longman.

Escoffier, F.F., 1954. Travelling forelands and the shoreline processes associated with them. *Bulletin Beach Erosion Board, U.S.*, **9**: 11–14.

Cross-references

Barrier Islands
Barrier
Bay Beaches
Cross-Shore Sediment Transport
Cuspate Forelands
Longshore Sediment Transport
Spits

DRIFTWOOD

One of the pleasures of beachcombing is viewing and collecting driftwood of innumerable shapes and sizes which have washed up on a beach. Driftwood is the collective term for all types of freely floating wood on a water body or lodged upon a beach. Driftwood can be found on most marine shores of the world, but is generally in highest concentrations along the shores of the middle and higher latitudes in the boreal forest and arctic regions of the world.

Trees along forested coastlines undermined by wave erosion are a source of driftwood and drift logs, but streams and rivers emptying into the sea are the primary contributors. Limbs, root masses, and whole trees are eroded from the stream banks, or are products of logging practices or log storage and rafting. Trees and logs enter river channels to be transported down river during high flows or are lost within estuaries during log storage and log rafting. The drifting wood can become temporarily lodged in estuaries or flushed directly into the sea. One estimate is that over 30 million board meters of logs enters the sea annually from coastal logging operations from Alaska to California (Sedell and Duval, 1985).

Driftwood movements

The fate of driftwood discharged into the sea is determined by coastal winds, waves, and currents. Some driftwood can transit entire oceans ultimately washing up on some foreign shore. The majority of the wood afloat is more likely to be carried by nearshore waves and currents. Along the Pacific Northwest of the United States, nearshore waves and currents seasonally reverse. In the winter, southwesterly winds drive waves and nearshore currrents to the north. Driftwood floating in these currents tend to end up on beaches far to the north from their river or estuary of origin. During the summer, northwesterly waves and currents carry driftwood southward along the Pacific coast. The volume of driftwood movement during the summer months is usually less as river flows are lower carrying fewer logs to the sea and waves are less energetic, dislodging less wood from beaches.

Large concentrations of driftwood are found along arctic beaches of Siberia, Canada, and Greenland. Rivers of northern Siberia and Canada's Mackenzie River carry driftwood into the Arctic Ocean during the summer. The wood is conveyed by the clockwise Pacific Gyral between the pole and Arctic Canada and the Transpolar Drift Stream from the New Siberian Islands toward the Svalbard and east coast of Greenland (Haggblom, 1982).

Driftwood on beaches

Driftwood and floating logs are driven ashore by waves and rising tides. Smaller pieces of wood can be thrown high on a beach by waves. Large logs, 20–30 m long and 1–2 m in diameter are shoved up the beach face by the repetitive thrusting of oncoming waves. Strong storm-driven waves cram logs and driftwood into dense piles high on the beach face and backshore. Driftwood deposits differ in quantity along any given stretch of shore from a few scattered logs and smaller pieces of wood to huge matted piles up to 3 m high and tens of meters wide. The largest concentrations of driftwood are trapped in coastal embayments such as river mouths and pocket beaches. But large deposits can also be found on the backshore and berms of wide sand beaches and steep rocky high-energy beaches. The wood is driven high up on the shore during storm conditions. There the wood may ultimately become buried by beach sediment. But for much of the wood, the beach is a brief stop off until it is again dislodged and refloated by the next series of storm-driven waves. The unrelenting pounding by waves and rolling in the surf chips away at the bark, branches, and root masses sculpting driftwood into smooth and artistic shapes.

Ecology of driftwood

Driftwood is capable of remaining afloat for long periods of time. Large and small pieces attract growths of epifauna and flora as well as selected insects. It is well-known that both commercial and recreational fish are attracted to floating wood at sea. Tunas and dolphin-fish are routinely found under or near masses of driftwood (Gooding and Magnuson, 1967). The reason is not clear, but it is thought that these fish are simply seeking shade or smaller fish attracted to the driftwood upon which they can prey (Maser *et al.*, 1988).

Once lodged on beaches, accumulations of driftwood tend to trap wind-blown beach sand and coarser swash carried sediments. They create an organic fabric that helps beaches grow or accrete. This is probably best seen near river mouths where river-borne sediments and wood accumulate along the seashore forming sand spits mantled with driftwood and logs. If there is a sufficient volume of sediment delivered to the shore, driftwood helps to capture wind-blown sediment. The wood not only traps sediment but provides decaying organic matter for pioneering plant and dune grass species to take root. The buried wood also creates an internal structural fabric within the building beach or sand spit. It is not uncommon to see large drift logs buried many years ago protruding from the face of an eroding beach scarp. Carbon-14 dating of uncovered driftwood is useful in determining the ages of beach deposits and other environmental conditions. Thus accumulations of driftwood along beaches are thought to help stabilize and contribute to beach accretion processes. Conversely, accumulations of driftwood at the base of beach cliffs can become powerful forces of erosion by battering and abrading cliffs especially when driven by storm waves.

Numerous species of birds are attracted to accumulations of driftwood on beaches and within estuaries. Bald eagles, herons, and pelicans use the logs as perches to rest upon and capture nearby aquatic animals in the surf or mudflats of estuaries.

Driftwood hazards

Driftwood clearly has some ecological benefits, but it also presents some liabilities especially for boaters and beachfront home owners. Floating driftwood, especially large logs, create a navigation hazard for ships and smaller pleasure boats. Driftwood is found in highest concentrations near the shore, in bays, and estuaries generally in same location as ships and boats. Ships' hulls are usually strong-enough to withstand the impact of a floating logs, but damage can occur to rudders and propellers. Smaller pleasure boats traveling at high rates of speed are especially vulnerable to impact damage. Large pieces of driftwood can crack and even pierce hulls as well as inflict serious damage to rudders and propellers. In some waterways around the United States, driftwood is "tagged" with small flags for better visibility by boaters or hauled out of the water.

Shorefront property owners who build their homes closer to the shore than is wise are not only exposing themselves to wave erosion and flooding, but to the intrusion of their homes by wave-driven logs. Such incidents are relatively infrequent, but there are documented cases of large drift logs thrown into the living rooms of shorefront homes in Washington State's Puget Sound.

Beachcombers are attracted to the aesthetic beauty of beaches laden with driftwood. Unfortunately, the wood that is so attractive can also be hazardous. It is not uncommon to learn of a beachcomber who has been killed or injured by a drifting log thrown into the unknowing beach stroller. The State of Washington has posted signs along the shore warning the public of large driftwood.

Driftwood management

Driftwood has long been harvested for firewood and even for minor building projects. For years the practice of harvesting driftwood from beaches has been accepted. However, some have questioned the advisability of continuing such practices, especially in light of the better understood ecological role driftwood plays in estuaries and on beaches. Recognizing the benefits of driftwood to natural ecological systems, the State of Oregon has adopted a policy prohibiting the removal of drift logs. This policy does not preclude the beachcomber from collecting pieces of driftwood, but it is intended to prohibit the large-scale removal of large logs for firewood or other uses. The State of Washington licenses individuals who wish to retrieve drift logs afloat or on beaches. Beyond licensing, Washington does not appear to have any driftwood removal policy applied to the general public.

Thomas A. Terich

Bibliography

Gooding, R.M., and Magnuson, J.J., 1967. Ecological significance of drifting objects to pelagic fishes. *Pacific Science*, **21**: 486–497.

Haggblom, A., 1982. Driftwood in Svalbard as an indicator of sea ice conditions. *Geografiska Annaler*, **64A**: 81–94.

Maser, C. *et al.*, 1988. *From the Forest to the Sea: A Story of Fall Trees.* Gen. Tech. Rep. PNW-GTR-229. U.S. Department of Agriculture, Forest Service, pp. 83–149.

Sedell, J.R., and Duval, W.S., 1985. Water transportation and storage of logs. In Meehan, W.R. (eds.) *Influence of forest and rangeland management on anadromous fish habitat in western North America.* Gen. Tech. Rep. PNW-186. U.S. Department of Agriculture, Forest Service. Pacific Northwest Forest and Range Experiment Station, pp. 1–68.

Cross-references

Beach Erosion
Beach Features
Debris, Onshore and Offshore
Human Impact on Coasts
Natural Hazards

DUNE CALCARENITE—See EOLIANITE

DUNE RIDGES

Dune ridges are known variously as "beach ridges" (Goldsmith, 1985; Komar, 1976), "parallel dunes" (Bird, 1972) or "low foredunes" (Davies, 1977). They are a product of a prograding coast with plentiful sediment supply, low tidal range, shallow nearshore slopes, and relatively low-energy refracted swell wave conditions (Figure D49). Continuing progradation is associated with plentiful diabathic onshore sweep of beach sediment, alternating with episodes of cut and fill, leading to the formation of a series of beach ridges (Davies, 1957). Each ridge is thought to represent the previous location of a beach berm, upon which over time a low foredune develops from "aeolian capping" (Bird, 1972). The height and spacing of the dune ridges is also a function of the effectiveness of the dune colonizing vegetation in binding the sand.

Geomorphically dune ridges may be as low as 0.5 m in elevation above mean high water springs, but more typically 4–8 m high. In width they vary from 50 to 200 m, and individual crests may extend shore-parallel for many kilometers (Figure D49). The ridge alignment parallel to the shore (hence their name "parallel dunes") is suggestive of the swell wave environment breaker alignment. The ridges are normally composed of sand but may also contain gravel. The parallel dune fields are comprised typically of Holocene sands and may be some kilometers wide. For example, Komar (1998) shows an illustration of the dune ridge field at Nayarit, Mexico, which contain some 250 individual ridges which could have evolved in as short a time as 12–20 years, although Pethick (1984) suggests that time intervals between dune ridges more typically varies from 70 to 200 years.

Separating the dune ridges are low swales characterized by impeded drainage and sometimes containing swampy vegetation. The variation in drainage characteristics leads to a tendency for variation in soil development on the well-drained ridge where oxidizing conditions prevail compared with the impeded drainage swale where reducing conditions and the podsolization process lead to dune podsol soils.

Dune ridges are widely associated with Holocene sandy barrier systems. They occur particularly along the Atlantic and Gulf coasts, Mexico, and southeast Australia, and New Zealand. Stratigraphically the dune ridges consist of aeolian low-angle cross-bedding in fine sands overlying low-angle dipping laminae of the beach face and berm, which overly marine deposits of the shoreface. These dune ridge systems evidently formed relatively rapidly (within ~1,000–2,000 years) after the sea reached its approximate present level from the postglacial transgression, presumably as a result of an abundance of sand being swept up with the postglacial transgression.

Terry R. Healy

Figure D49 Parallel dune ridges and swales on the coast near the Nassau River mouth, Queensland (Photo: J. Mabbutt).

Bibliography

Bird, E.C.F., 1972. *Coasts*. Canberra: Australian National University Press.

Davies, J.L., 1957. The importance of cut and fill in the development of beach ridges. *Australian Journal of Science*, **20**: 105–111.

Davies, J.L., 1977. *Geographical Variation in Coastal Development*. London: Longman.

Goldsmith, V., 1985. Coastal dunes. In Davis, R.A. (ed.), *Coastal Sedimentary Environments*. New York: Springer Verlag, pp. 303–378.

Komar, P.D., 1976. *Beach Processes and Sedimentation*. Englewood Cliffs, NJ: Prentice Hall.

Pethick, J., 1984. *An Introduction to Coastal Geomorphology*. London: Edward Arnold.

Cross-references

Beach Features
Beach Ridges
Cheniers
Cross-Shore Sediment Transport
Drift and Swash Alignments
Eolianite
Eolian Processes
Sandy Coasts

DYNAMIC EQUILIBRIUM OF BEACHES

Dynamic equilibrium of beaches describes the tendency for beach geometry to fluctuate about an equilibrium which also changes with time, but much more slowly. Beaches, as discussed here, refers to the visible beach including the shore and its underwater extension to a depth that is nearly static over the long-term. Beaches respond on a wide range of spatial and temporal scales to natural and anthropogenic forcing. Natural agents include waves, tides, currents, winds, and other elements whereas anthropogenic agents include interruption of sediment supply and induced subsidence through withdrawal of ground fluids including water and hydrocarbons. Changes of beach position can occur with or without corresponding changes in the beach profile volume. Over a long time period, coastline position can be considered as the superposition of a long-term trend about which substantial fluctuations occur. Some of these fluctuations are quasi-periodic and others appear random. The paragraphs below discuss the coastline trends followed by the fluctuations.

Long-term changes

Worldwide, the most dominant cause of long-term coastline change is sea-level rise. The global rate of sea-level rise is approximately 12 cm/century, although there is concern that global climate warming may increase this rate substantially in the future through melting of ice and thermal expansion of oceanic waters. The concept of an equilibrium beach profile is useful in considering the effect of sea-level rise on coastline position (Dean, 1991). This has been formalized by Bruun (1962) as the so-called "Bruun Rule." A sea-level rise will cause the profile to be out of equilibrium since the depth at a distance from the coastline is now greater than before and equilibrium can only be restored through a decrease in depth which, in the absence of external sediment addition, must be provided by volume eroded by coastline retreat. The Bruun Rule is a simple equation which depends on a number of factors including wave climate, and sediment characteristics and is expressed as the coastline retreat being a multiple of the sea-level rise, with the multiple usually ranging from 50 to 100. Thus in one century, the expected average worldwide coastline retreat would range from 6 to 12 m; however, the coastline trend at a particular location can deviate from this value by orders of magnitudes ranging from coastline retreat to coastline advancement. In addition, the Bruun Rule can be applied to *relative* sea-level rise resulting from natural or anthropogenically induced subsidence. An example is near the Mississippi River outlet where the relative sea-level rise rate is on the order of 10 times the worldwide rate noted earlier. Entire islands vanish in periods of decades.

Short-term changes

The most rapid short-term changes in coastline position occur during severe storms when the beach may retreat on the order of 50 m in several hours. The primary elements are the elevated water levels (storm tides) and high waves which can cause significant modifications to the nearshore geometry including shore and dune retreat. As discussed for sea-level rise, the storm tide causes the profile to be in temporary disequilibrium and the waves provide the energy to accomplish the profile adjustment. Averaged over long sectors, the primary sediment transport during storms is seaward where a new deposit called an "offshore bar" may be formed or additional sand stored in an existing bar which may be displaced farther seaward during the storm. Elevated water levels and accompanying large wave heights can cause large coastline changes without corresponding changes in the sand volume in the profile. An example illustrating this is a wave tank in which the sand volume is constant; however, the sand can shift offshore resulting in coastline retreat. Coastline retreat associated with a storm and for which there is no volume change will generally be followed by coastline advancement and complete or near complete recovery of the pre-storm coastline position. The timescales of the recession and recovery (advancement) phases differ markedly with the recession occurring in hours to days and the recovery requiring months to years. The modes of profile recovery also differ. The profile changes during erosion as the sediment is transported seaward tend to occur with the profile "hinging" about a subaqueous point with the elevations decreasing landward of the hinge point and increasing seaward of the hinge point. The recovery phase is much more complex with the landward transport of sand occurring in individual sand "packets," each of which, upon reaching the shore, advances the coastline seaward. During long periods of milder wave conditions following a major storm, a number of recovery packets may occur in order to complete the recovery phase. The interest in anticipating and designing for storm-induced beach and dune erosion has led to the development of numerical models (Kriebel and Dean, 1985; Larson and Kraus, 1989). These models are much more effective in representing the erosional phase than the more complex recovery phase. Seasonal coastline changes, described as the average annual coastline fluctuations are poorly understood and vary substantially for various coastlines. The seasonal coastline change range in Florida is approximately 10 m whereas along the south shore of Long Island, NY, the range is about 30 m. The range in southern California is relatively large and is evident by the presence of sandy beaches during summer and rocky beaches left as a lag deposit in the winter when the sand removed has been stored in the offshore bar. Studies have documented that these seasonal coastline changes are related to the wave characteristics with the seaward transport associated with higher waves of shorter period which generally occur in the winter and the landward transport occurring during the milder summer conditions. Some beaches experience somewhat organized beach fluctuations which vary substantially along the shoreline. These appear as "sand waves" (Bruun, 1954; Bakker, 1968; Verhagen, 1989; Thevenot and Kraus, 1995) which propagate along the shore and may have amplitudes of tens of meters sometimes causing distress to property owners with homes located near the high-tide shoreline. The understanding of the causes and relative longevity of sand waves is poorly understood. Proximity to structures (natural or constructed) and/or inlets can increase the coastline fluctuations significantly relative to those along a long unobstructed coastline. The amplified fluctuations near structures is a result of changes in wave direction which causes accumulation of sand against and removal of sand from the beach adjacent to the structure. Similarly, high-tide shorelines in pocket beaches change their orientation in response to changes in wave direction causing large fluctuations near the ends of the beaches (Thompson, 1987). The changes near inlets appears to be primarily due to the influence of and the interaction with the deposit of sand, termed the "ebb tidal shoal." This feature can shelter the adjacent shores from waves causing accumulation, can shift in position thereby exposing the shore to increased wave energy and can release large amounts of sand to the shore in packets. Dredging of inlets can also cause large fluctuations (usually erosional) along the adjacent coastlines. Studies of the long-term coastline position data base in Florida has shown that the standard deviations around the coastline change trend line are approximately 10 m along long coastlines and are on the order of 50 m near natural inlets (Dean *et al.*, 1998).

Summary

In summary, beaches are dynamic, generally exhibiting a trend of coastline change which can be either advancement or retreat, with fluctuations about this trend. Although the underlying processes governing beach change are well-understood qualitatively, our ability to predict quantitatively these changes is likely to remain poor for the coming decades.

Robert G. Dean

Bibliography

Bakker, W.T., 1968. Mathematical theory about sand waves and its application on the Dutch Wadden Isle of Vlieland. *Shore and Beach*, **36**: 4–14.

Bruun, P., 1954. Migrating sand waves or sand humps, with special reference to investigations carried out on the Danish North Sea coast. *Proceedings of the 5th International Conference on Coastal Engineering, ASCE*. New York, pp. 269–295.

Bruun, P., 1962. Sea level rise as a cause of shore erosion. *Journal of Waterway, Port, Coast, and Ocean Engineering*, ASCE, **88**: 117–130.

Dean, R.G. 1991. Equilibrium beach profiles: characteristics and applications, *Journal of Coastal Research*, 7, **1**: 53–84.

Dean, R.G., Cheng, J., and Malakar, S., 1998. Characteristics of shoreline change along the sandy beaches of the state of Florida: an atlas. *Report No*. UFL/COEL-98/015, Gainesville, FL: Department of Coastal and Oceanographic Engineering, University of Florida.

Kriebel, D.L., and Dean, R.G., 1985. Numerical simulation of time-dependent beach and dune response. *Coastal Engineering*, **9**(3): 221–246.

Larson, M., and Kraus, N.C. 1989. SBEACH: numerical model for simulating storm-induced beach change, Report 1: empirical foundation and model development. U.S. Army Coastal Engineering Research Center, Waterways Experiment Station, *Technical Report* CERC-89-9.

Thevenot, M.M., and Kraus, N.C., 1995. Longshore sand waves at Southampton Beach, New York, observations and numerical simulation of their movement. *Marine Geology*, **126**: 249–269.

Thompson, W.C., 1987. Seasonal orientation of California beaches, *Shore and Beach*, **55**(3–4): 67–70.

Verhagen, H.J., 1989. Sand waves along the Dutch coast, *Coastal Engineering*, **13**: 129–147.

Cross-references

Beach Erosion
Beach Processes
Changing Sea Levels
Coastal Changes, Gradual
Coastal Changes, Rapid
Coastal Subsidence
Depth of Closure on Sandy Coasts
Profiling, Beach
Strom Surge

E

ECONOMIC VALUE OF BEACHES

Beaches are economic as well as natural resources. As natural resources, they add beauty to the coast and provide habitats for many creatures including birds and sea turtles. As economic resources, they provide services to people and property that have an economic value. They also generate impacts on the economy and tax base.

Economic services provided by beaches

An important service provided by the beach at the coast is reduced storm damage to upland properties (US Army Corps of Engineers 1996, chapter 5). Beaches reduce storm damage by moving the water line further from upland property. During storm events, water travels less far inland as a result of the beach and so damage to upland property tends to be less. Of course, a beach does not eliminate storm damage to upland properties and in severe storms it may provide little or no protection. For example, the coastline impacted by the center of a hurricane may receive little protection from its beaches, but the upland properties impacted by the outskirts of the hurricane may receive very considerable storm damage reduction from their adjacent beaches. The second major service provided by a beach at the coastline consists of the opportunity for recreation provided to beach users. Although there is no direct fee charged in most places for recreational access to the beach, such recreational services have economic value just like playing a round of golf or a visit to the movies.

Beaches also generate impacts away from the coastline. Chief among these are the impacts on the local economy of the expenditures made by tourists attracted to the beach as a recreational resource (*tourism and coastal development, q.v.*) (Schofield, 1986, chapter 14). There will also be benefits at the regional (e.g., state or province) level to the extent that tourists are attracted from outside the region, and benefits at the national level from the spending by international visitors. The expenditures of beach tourists increase sales and production by impacted business resulting in increased business profits and increased earnings of employees. The well-known multiplier process, popularly called the "ripple effect," will cause these economic benefits to be passed in a chain from front line businesses to their suppliers, and from employees of those businesses to businesses which provide goods and services to those employees. Of course, such positive economic impacts vary depending on the size of the economy, which is impacted. In small economies, many of these impacts are lost to imports. There may also be disbenefits from beaches. For example, the tourists attracted to the beaches may contribute to traffic congestion. The congestion in turn increases travel time in the local area and the involuntary consumption of this time represents an economic loss to the travelers, which can be measured by their wage rates. Additionally, the congestion may increase the frequency of automobile accidents with associated property losses.

Government tax revenues will also be expanded as a result of beaches (Stronge, 1998). The particular impacts, of course, will depend on the particular tax and fee structure in operation in the beach area. In many local communities in the United States, local communities obtain revenues by taxing property values. Beachfront property values tend to be higher than property values away from the beach partly because land use tends to be more intensive at the beach. That is, there are more hotels and high-rise apartment buildings at the beach. But, even beachfront single-family homes tend to have high values relative to homes elsewhere in the community. Other things being equal, land values on the beachfront tend to be high relative to values elsewhere in the community because properties on the beach provide easier access to the recreational and amenity values of the beach. Structural values on properties tend to be positively correlated with land values, and so the high beachfront land values stimulate the construction of relatively expensive beachfront structures. It is not unusual to see small, inexpensive beachfront structures that are in good condition being demolished and replaced by larger, more expensive structures, because sharply increased demand for beachfront land has raised land values to levels where beachfront structures are out of equilibrium with underlying land values. The expenditures of tourists are usually subject to sales taxes, including accommodation taxes, general sales taxes, and taxes on gasoline. These revenue sources are favorably impacted by beach tourism. Finally, income taxes will tend to rise as a result of the increased earnings of workers in the tourist industry. Offsetting these positive impacts on governments may be the required increases in government expenditures resulting from beach tourism. Increased traffic congestion may result in increased expenditures for law enforcement personnel, for example. There will be increased expenditures to clean trash off the beaches and, perhaps, for lifeguards. The fiscal impact of a beach, therefore, is the net impact of the positive increases in government revenues less the required increases in the government expenditures.

Government policies on the beach

Governments regulate the activities of beachfront property owners because actions taken by one property owner may adversely affect nearby property owners or negatively impact the society at large (Fischer, 1990). Many beaches are subject to *beach erosion (q.v.)*, steadily diminishing in size on an annual basis. Beaches are often dynamic systems losing sand during some seasons and gaining sand during others. Along an unbroken stretch of beach, with no headlands jutting into the sea, sand tends to travel along the coastline and it may travel in one direction during one season and reverse direction in another. The direction traveled tends to reflect the weather patterns and, in particular, the direction from which storms come. On an unbroken coastline with no headlands, storms move sand around on the beach but net losses may be low. In some cases, storms may push sand upland away from the beach and, in some cases, sand may be lost offshore. But in many cases, storms move sand away from the coastline and then natural processes of wave action will return most of the sand to its original position after the storm is over.

Inlets, including gaps between coastal islands and river mouths, represent holes in the coastline, which interrupt the flow of sand along the beach. When the tide is coming in sand is pushed into the inlet and when the tide is ebbing sand is deposited offshore. These movements often lead to an offset in the coastline as beaches on one side of the inlet lose more sand than beaches on the other. The asymmetric losses reflect the fact that storms from one direction may be stronger than storms from another. When new inlets are constructed, or old inlets are improved, the equilibrium relationship between adjacent beaches is disturbed and properties on one side of the inlet may be threatened by erosion. Such activities are usually undertaken to improve navigation, but they may also have desirable environmental consequences. In some cases, governments can remove the erosion threat to an adjacent beach by bypassing sand mechanically from one side of the inlet to the other or by dredging the inlet and depositing the sand on the eroding beach. Beachfront properties may also be threatened by erosion if they were originally constructed too close to the high water line. Such structures tend to be relatively old, dating from a time when beaches were regarded as static and, perhaps, when storm events had widened a beach beyond its long run equilibrium size. Older structures may also have been constructed at a time when regulatory authorities were unaware of the dynamic properties of the beach, or when enforcement of regulations was inadequate. In any case, many beaches have properties that are in danger of damage from erosion. Under these circumstances, property owners will often construct structures such as seawalls or revetments to protect their properties from storm damage. These structures may interrupt the flow of sand to adjacent properties or accentuate the force of the waves experienced by neighbors. As a result, such structures tend to stimulate the construction of similar structures along entire segments of the coastline. The actions of the first property owners have negatively impacted their neighbors, and because the structures protrude into the beach area the actions reduce the beach available to the public for recreational purposes and to birds and other creatures, which use the beach as a habitat. To avoid such negative externalities, governments regulate activities of beachfront property owners including, requiring building setbacks and requiring erosion control structures, to pass through a permitting process. In the United States, all three levels of government require permits for beachfront construction. This occurs because beach systems cross local government boundaries and, in some cases, state boundaries, and because there is a national interest in protecting wildlife and other environmental resources.

Beach nourishment

In some cases, erosion control may be accomplished by placing sand on the beach to widen it, often called *beach nourishment* (*q.v.*) (National Research Council, 1995). The conditions that would make beach nourishment successful will depend on the strength of the forces causing erosion. Additionally, it must be possible to obtain a source of sand that is cheap enough to make the project economically feasible. In other cases, it may be appropriate to relocate properties further away from the high water line. This solution is most appropriate where there is available land to accommodate the relocated structure, and where the structure is relatively small and easy to move. Such a structure might be a one or two story second home on a relatively large lot, for example.

When beach nourishment is an appropriate remedy for erosion (i.e., it would be successful), and when it is economically feasible, it has the added advantages of retaining the recreational and environmental values of the "natural" beach. Structural remedies, on the other hand, often reduce the available recreational beach and the available habitats for birds, turtles, and other creatures that use the beach. Additionally, off-beach benefits such as economic and fiscal impacts are also retained by beach nourishment as opposed to structural solutions. It is rarely economically feasible for one property owner, or even a small group of property owners, to undertake a beach nourishment project unless they own a relatively long stretch of coastline. This is because beach nourishment projects have large fixed costs largely unrelated to the volume of sand being placed on the beach. These include costs of sand searches, permitting costs, and mobilization and demobilization of equipment charges. For example, sand searches are expensive because not only must sand be found, but it must be compatible with the sand already on the beach. An archaeological survey may be required to ensure that there are no archaeological artifacts (ship wrecks) in the sand source. Various environmental studies may also be required as part of the permitting process. Additionally, equipment including a dredge may have to be brought from hundreds of miles away. Once at the site, there are setup charges and charges for dismantling the setup at the end of the project. On rare occasions, it may be possible to "piggyback" a small project on a large project that is occurring nearby, thereby greatly reducing the fixed costs incurred by the small project. But beach nourishment projects are highly complex and it is difficult to group projects together because all projects may not be ready for construction at about the same time. Even when a relatively large group of property owners is interested in undertaking a beach nourishment project, they may not be able to agree on a formula for cost sharing. Beach nourishment is a public good in the sense that it is usually not possible to prevent consumers of the services provided by the beach from benefiting from the project unless they have paid for them. This is because beach dynamics will redistribute new sand along the entire section of the beach, and property owners in the middle of the project will gain sand even if they have not paid. This is the well-known "free rider" problem associated with public goods. As a result, beach nourishment, when feasible, may not be selected by owners of properties threatened by erosion even when appropriate and, in principle, economically feasible. But because sand placement retains the recreational and environmental values of the beach, as well as the positive economic and fiscal impacts, society as a whole may wish to encourage beach nourishment when it is an appropriate remedy. Under such circumstances, governments may wish to encourage beach nourishment as a tool for erosion control (Stronge, 1995). At a minimum, government may take a leadership role, financing initial studies and educating property owners about beach nourishment. In many cases, however, governments also subsidize construction and maintenance costs. Such subsidies, of course, will be part of a regulatory regime designed to discourage the use of structures to the extent legally possible.

Measurement issues

Most US beach nourishment projects receive some federal subsidies through the Army Corps of Engineers budget. In order to qualify for such subsidies, beach projects must be economically justified, that is, they must yield benefits in excess of their costs. This requires an economic study to be undertaken in advance of project construction. The economic study projects the benefits and costs of the proposed project. A methodology has been developed over many years for projecting benefits.

A benefit projection methodology is, in effect, a model of how the project's benefits are generated. Like all models, a beach projection methodology involves a simplification of reality. In particular, the model identifies a limited number of key benefits and does not attempt to include all the benefits of the project. Additionally, there are key parameters imposed exogenously on the model, such as the length of the planning horizon and the discount rate used to convert future dollars to present worth or value. Given the potential for inaccuracy introduced by the modeling process, it is surprising that few follow-up studies have been undertaken that would evaluate the accuracy of the model in projecting benefits.

Because benefit projections are based on a model there are a variety of issues that have generated some debate. First, most beach projects are justified on the basis of a very limited number of benefits, namely, property protection and recreational values. At the federal level, off-site benefits such as economic and fiscal impacts are not included, although these benefits may be of the most importance at the local level. The value of environmental benefits is not usually measured, although projects can be prohibited or forced into redesign if they have significant negative environmental impacts. Finally, there may be externalities from beach projects. A beach is a highly visible piece of infrastructure and failure to maintain it may be used by potential migrants to a local community as an indication that the quality of invisible infrastructure (e.g., water, sewer) is poor. Restoration of the beach may lead to a general rise in property values over and beyond what would be predicted by projecting direct benefits.

Storm protection benefits are measured as the discounted value of expected dollar losses assuming the project is constructed, less the discounted expected dollar losses if the project is not constructed. The losses from storms are applied to structure values based on rules of thumb, such as assuming the proportionate dollar loss is twice the proportion of the structure that is damaged by a storm. Thus, for example, when 50% of a building is lost, all of its structure value is lost. Expected losses are calculated as the sum of the probabilities of storm events times the dollar losses resulting from those events. Storm protection benefits are discounted using an exogenously given real interest rate. The basis for the original interest rate is obscure, although it is adjusted regularly using a nominal interest rate on government securities. A lower interest rate makes it easier to justify projects and a higher interest rate makes it more difficult. Opponents of Corps of Engineers projects favor a high interest rate, while supporters of such project favor a low rate. Some people favor basing the interest rate on the government bond yield, but others argue that this will favor public over private projects since government borrowing costs are lower than private borrowing costs. An alternative interest rate could be based on the home equity loan market, under the assumption that taxpayers had to borrow in order to pay their taxes. Beach benefits are discounted over a 50-year period, assuming that the project is maintained on a projected schedule. The 50-year planning horizon is somewhat arbitrary, although if the discount rate is high future benefits and costs will be reduced to zero when discount factors are applied to values generated far into the future. Follow-up studies of the accuracy of projected maintenance intervals are needed.

Recreational benefits are measured based on willingness to pay by beach users. Two approaches have been taken. One is based on travel costs borne by the beach user, that is, out of pocket expenses and time expended by the beach user in traveling to the beach. This approach requires a variety of assumptions to be made about travelers. An alternative approach is to use surveys of beach users to collect information on willingness to pay. It is more common to conduct surveys as personal interviews on the beach rather than by telephone or mail in order to reduce nonresponse bias. There is a large literature in economics relating to contingent valuation surveys, including recommendations for wording questions (Portney, 1994). One issue relates to whether the interviewee realizes that dollars expended on a beach visit are unavailable for alternative products. A second issue relates to allowing an interviewee to select from a list of willingness to pay values, as opposed to giving a yes or no to a single value. Interviewees may have a tendency to pick a middle value on a list. Collecting information by obtaining a response to a single value, the so-called referendum method, requires a larger sample size of interviewees. In the case of beach surveys, there are also unique problems. A certain number of people on the beach strongly oppose user charges for entry to the beach and report little or no willingness to pay values in an effort to discourage the imposition of such charges. There may also be a group who say that the value of the beach is infinite, since it is a beautiful natural resource. Finally, there are difficulties with estimating willingness to pay for beach improvements by interviewing users of the unimproved beach. The number of new users and the increase in average willingness to pay may have to be projected using very limited data.

Issues relating to funding beach improvements

Many beach projects take place on stretches of coastline where the upland properties are at least partly privately owned. Moreover, the average income level of beachfront property owners tends to be high since, as noted above, beachfront property values tend to be high. As a result, benefits from beach improvements may be skewed in the direction of relatively affluent beachfront property owners. In many cases, the public perception that benefits are skewed to beachfront property owners leads to the use of specialized funding sources as a means of financing government subsidies. Thus, for example, a special taxing district may be established to collect additional property taxes from beachfront property owners. In another case, a tax on accommodations may be used, particularly if such accommodations have large clienteles of beach users. In some cases, local funding is obtained by charging beachfront property owners for their storm protection benefits and by charging non-beachfront property owners for their recreation benefits.

In the United States, there are also issues about funding by level of government. Under current law, federally approved beach nourishments may receive subsidies in amounts up to 65% of the total cost for initial projects and up to 50% of the cost of maintenance projects. There are some peculiarities in the federal program for subsidizing beach nourishments. The federal government requires that projects seeking subsidies be economically justifiable based on storm protection benefits alone and this is most easily achieved by projects with highly intense upland land uses, such as high-rise apartment buildings and hotels. On the other hand, the size of the federal government subsidy depends on the extent of public access and parking to the improved beach, that is, the size of the recreational benefits resulting from the beach improvement.

There have been efforts by the executive branch of the federal government to reduce the size of federal subsidies for beach nourishment projects. Most of these efforts have failed to obtain support in Congress. The Clinton administration attempted to limit the number of beaches that could qualify for federal subsidies by distinguishing between beaches of national interest and beaches of state and local interest in much the same way as highways are divided between national roads, state roads, and local roads. The same administration became concerned about the large subsidies that would be required to maintain existing projects, in an era of large budget deficits, and successfully reduced maintenance subsidies from a maximum of 65% of costs to 50%. Both the Clinton and Bush administrations believe that most of the benefits from beach projects accrue at the local level and, therefore, they argue that most of the funding should be obtained from that level. They argue that, from a national viewpoint, economic and fiscal benefits represent a redistribution across regions rather a substantial increase in the national totals. Recreational benefits also represent a redistribution, since beach users may use other beaches or alternative recreational facilities if a beach is not improved. The Bush administration has proposed lowering the maximum federal share to 35%. On the other hand, the federal government maintains control of local beaches as habitats for wildlife and at least some erosion has been caused by navigation improvements to federally maintained inlets. The federal government is arbitrarily adjusting cost sharing proportions without undertaking any study of the national benefits of beaches. Until such a study is undertaken, it is unlikely that an appropriate federal cost share will be obtained.

William B. Stronge

Bibliography

Fischer, David W., 1990. Public policy aspects of beach erosion control. *American Journal of Economics and Sociology*, **49**(2): 185–197.

National Research Council, Marine Board, Commission on Engineering and Technical Systems, 1995. *Beach Nourishment and Protection*, Washington, D.C.: National Academy Press.

Portney, Paul R., 1994. The contingent valuation debate: why economists should care. *Journal of Economic Perspectives*, **8**(4): 3–17.

Schofield, J.A., 1986. *Cost–Benefit Analysis in Urban and Regional Planning*. London, England: Allen and Unwin.

Stronge, William B., 1995. The economics of government funding for beach projects: the Florida case. *Shore and Beach*, **83**(3): 4–6.

Stronge, William B., 1998 The economic benefits of Florida's beaches. In *Proceedings of the 1998. National Conference on Beach Preservation Technology*, Tallahassee, Florida: Florida Shore and Beach Preservation Association, pp. 319–330.

US Army Corps of Engineers, 1995. *Shoreline Protection and Beach Erosion Control Study*. Alexandria, Virginia: Water Resources Support Center, Institute for Water Resources.

Cross-references

Beach Erosion
Beach Nourishment
Managed Retreat
Setbacks
Tourism and Coastal Development

EL NIÑO–SOUTHERN OSCILLATION (ENSO)*

Local fisheries (Enfield, 1988, 1999)

The term El Niño (or "the child") was originally used by Peruvian fishermen in the 19th century to refer to a Christmastime warming of coastal sea surface temperature (SST), often associated with an abrupt decrease in productivity of the local fisheries. The Southern Oscillation (SO) portion of El Niño–Southern Oscillation (ENSO) describes the global-scale surface pressure oscillation documented by workers around the turn of the century and first studied in detail by Sir Gilbert Walker (for historical reviews, see Rasmusson and Carpenter, 1982; various chapters in Glantz *et al.*, 1991; Diaz and Markgraf, 1992, 2000; Diaz and Kiladis, 1995; Allan *et al.*, 1996; Diaz *et al.*, 2001). It was not until the 1960s that Jacob Bjerknes linked the two processes and began to describe the complex

* Updated from the Encyclopedia of Environmental Science (1999) by the author and printed with permission of Kluwer Academic Publishers.

interplay between the ocean and atmosphere, which comprises ENSO and can lead to dramatic perturbations of the global climate system. Here, we will be concerned with the implications of climatic change on ENSO, as well as considering the possible importance of ENSO itself for low-frequency climatic variability.

The root of what is now commonly referred to as ENSO lies in the tropical Pacific, although its influences eventually spread far beyond that ocean basin. To begin to understand the ENSO cycle and its role in climate variability, the mean oceanic and atmospheric conditions over the eastern Indian and Pacific sectors are first considered, at least in general terms. We then examine the principal climatic anomalies worldwide that result from the development of the tropical anomalies. For further reading, see Streten and Zillman (1984) and Kiladis and Diaz (1989).

Mean conditions

The South Pacific high-pressure system is the primary atmospheric circulation feature of the eastern South Pacific Ocean. This semipermanent surface high is associated with equatorward flow along the coast of South America, and strong southeasterly trade winds over the tropical eastern Pacific. These southeasterlies cross the equator and merge with the northeasterly trades of the north Pacific subtropical high at a latitude of about 80°N, giving rise to a zone of surface convergence and heavy rainfall known as the *inter-tropical convergence zone* (ITCZ). Further west, the southeasterly trades weaken and recurve into northeasterlies in the tropical southwest Pacific, merging with southeasterlies from surface high-pressure systems moving eastward from the region of Australia. This produces another northwest–southeast oriented rainfall maximum called the south Pacific convergence zone (SPCZ; Trenberth, 1976).

The distribution of rainfall in the tropical Pacific is closely related to the SST pattern, which in turn can be understood to be driven by the surface wind stress and associated ocean currents (see Pickard and Emery, 1982). One result of the Earth's rotation is to cause surface water to flow to the right of the wind in the Northern Hemisphere and to the left of the wind in the Southern Hemisphere. Thus, easterly trade winds along the equator result in a divergence of water away from the equator, which leads to the "upwelling" of relatively cooler water from depth to conserve mass. Similarly, equatorward wind stress along the western coast of South America leads to divergence of water away from the coast and strong upwelling along the steep subsurface continental margin. Regions of upwelling comprise productive fisheries areas, as the subsurface waters are often rich in nutrients. This effect, along with the northward transport of cold water in the Peru Current, gives rise to the rich fisheries and relatively cold SST for its latitude along the coast of Peru and the Galapagos Islands.

A direct result of the coastal and equatorial upwelling is to increase the stability of the atmospheric boundary layer by the cooling effect of the SST. This, along with the reduced evaporation over colder water, results in the strong suppression of rainfall in these regions. As a result, the coasts of southern Peru and northern Chile are the driest deserts on earth. Anomalously dry climates are also experienced along the so-called "equatorial dry zone" in the central and eastern Pacific as a result of the strong upwelling there. As one moves westward, equatorial SST increases gradually as the trades and upwelling weaken, eventually giving way to the "warm pool" west of the dateline. This region is the largest area of high SST in the world's oceans, averaging greater than 28°C. The warm pool extends west-ward from the western Pacific through the seas surrounding Indonesia and into the eastern Indian Ocean, and is associated with high precipitation over these regions.

The ENSO cycle

ENSO fluctuations are associated with marked deviations from the mean atmospheric and oceanic conditions described above. In the eastern Pacific Ocean, its most obvious manifestation is an increase in SST along the equator and coast of South America, with an associated increase in sea level and depth of the thermocline. This results from a decrease in upwelling due locally to a weakening of the South Pacific high and associated trade wind flow, as well as the eastward propagation of equatorially trapped large-scale waves in the ocean, which suppress the thermocline (Philander, 1990).

The weakening of the South Pacific high is the eastern component of Walker's Southern Oscillation and is associated with an increase in surface pressure over Australasia. Figure E1 illustrates how the surface pressure varies inversely between Tahiti (Papeete) and Darwin, Australia, on timescales of several months and greater. Various indices of the SO have been constructed utilizing normalized pressure anomalies between stations near the poles of the oscillation such as Darwin and Tahiti (see Trenberth, 1984). However, the association between the SO itself and the eastern Pacific SST is not a simple one (Deser and Wallace, 1987). The SO is most closely coupled to SST variability along the equator from about 160°W to the Galapagos, and only loosely related to SSTs near the dateline and in the traditional El Niño region of coastal Peru. Figure E2 gives the temporal evolution of normalized

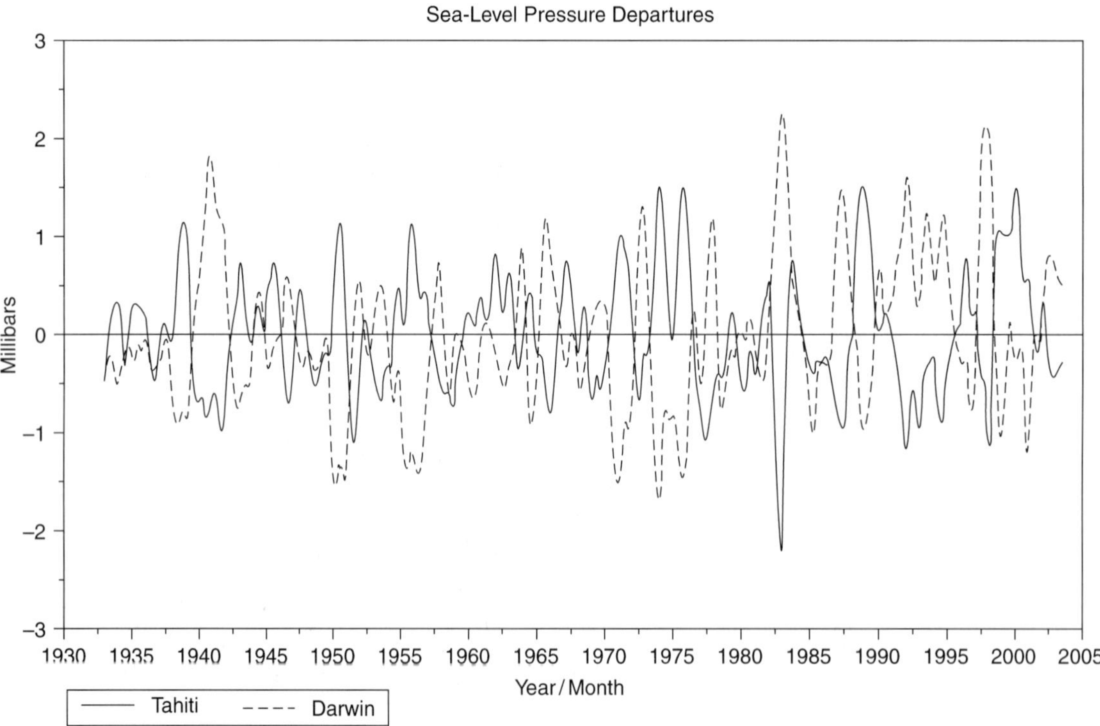

Figure E1 Sea-level pressure anomalies at Tahiti and Darwin, Australia. Monthly departures have been smoothed with a five-month running mean, applied twice.

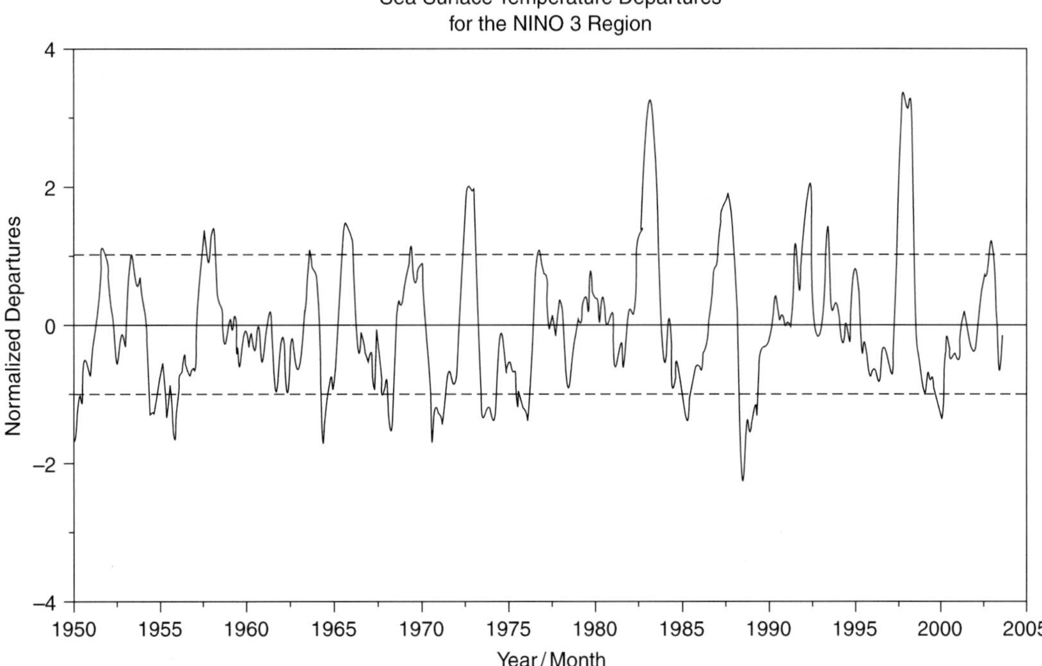

Figure E2 Standardized SST anomalies in the El Niño-3 region for the period 1951 through July 2003. Reference period is the full period of record; monthly data have been smoothed with a three-month running mean, applied twice.

SST variations in the equatorial belt from 6°N to 6°S, 150°W to 90°W, which is known as the Niño-3 region (Rasmusson and Carpenter, 1982). The correlation between the Tahiti minus Darwin sea-level pressure (SLP) difference and the Niño-3 SST is quite high (with about 64% of the variation in one series explained by variations in the other). The five most recent warm events can be seen on these plots, namely 1982–83, 1986–87, and 1991–92, 1997–98, a weak event in 2002–03, along with cold events in 1988–89 and 1998–2000.

Knowledge of the workings of the ENSO phenomenon can be obtained from an examination of its composite behavior (Rasmusson and Carpenter, 1982), along with study of individual cases, which deviate markedly from this mean view (Fu *et al.*, 1986). One interesting aspect of ENSO is the tendency for phase locking to the annual cycle. ENSO extremes often first develop during the northern spring, when the trade winds are weakest and SST is highest on average over the eastern equatorial Pacific. The fundamental instability involved concerns a relationship between SST anomalies and atmospheric convection. In the majority of events, a positive equatorial SST anomaly develops initially near the dateline. Since SST in this region is on average about 28°C, any positive anomaly in this region will create conditions conducive to anomalously active atmospheric convection Barnett *et al.*, 1991).

The development of ENSO events tends to peak during the late summer and early fall of the Northern Hemisphere. Notable exceptions occurred in the warm events of 1982, 1986, and 1991, which developed late in the calendar year and peaked during the northern winter and spring. The feedback leading to a reversal of ENSO conditions is often mediated by the seasonal development of the strong trade wind regime of northern fall, as well as the depletion of heat energy in the warm pool. Much of this energy is apparently transferred to the atmosphere through anomalously high rates of evaporation, and portions of it may also be transferred to higher latitudes by the oceanic meridional circulation. Once the heat content of the ocean is depleted, the stage is set for the transition to the opposite phase of ENSO, the so-called "La Niña" or cold event (e.g., van Loon and Shea, 1985; Philander, 1985, 1990). Cold events also involve an unstable interaction, as described above, for warm events, except that it occurs in reverse, as cold SST in the eastern Pacific help maintain a strong westward pressure gradient and trade winds, in turn favoring the continuance of the SST pattern itself.

ENSO teleconnections

Much of Gilbert Walker's early work was geared towards prediction of Indian rainfall, and the SO was of great interest, as it was known that the strength of the monsoon was inversely proportional to the surface pressure over southern Asia. In his investigations, Walker (1923) established that high pressure over the Australasian region was accompanied by drought over India and Australia and cool, wet winters over the southeastern United States. Surprisingly, little work on atmospheric teleconnections was undertaken until nearly a half a century later, when Bjerknes (1966) uncovered evidence that El Niño conditions were associated with a strengthening of the storm track over the North Pacific. Since then, a wealth of work has been done on the temperature and precipitation signals associated with ENSO. Much of the synopsis that follows has been taken from the work on large-scale ENSO signals of Rasmusson and Carpenter (1982), Ropelewski and Halpert (1986, 1987), Lau and Sheu (1988), Kiladis and van Loon (1988), Kiladis and Diaz (1989), Diaz and Kiladis (1995), and Diaz *et al.* (2001).

Although the term "El Niño" was first used to describe the annual warming of the waters along the Peruvian coast, in the 1960s it became synonymous with the concurrence of abnormally high SST and heavy rains in the usually hyperarid Peruvian coastal plains. Bjerknes (1969) noted that anomalously heavy rains during El Niño also occurred at Pacific island stations such as Canton and Christmas in the equatorial dry zone. As mentioned above, this signal is one manifestation of the equatorward shift of the convergence zones during equatorial warming of the SST associated with ENSO. As a result, the regions normally under the influence of these convergence zones, such as the Caroline and Marshall Islands, Fiji, and New Caledonia, experience drier than normal conditions.

The most pronounced signals occur during the year that an ENSO extreme first develops (here called "Year 0") into the following year (Year + 1), following the convention of Rasmusson and Carpenter (1982). ENSO appears closely tied to the Asian and Australian monsoon circulations, which are notably weaker during warm events (see Webster and Yang, 1992). While precipitation increases in the central and equatorial Pacific during warm events, it becomes markedly drier over regions bordering the western Pacific and eastern Indian Ocean, such as Indonesia and Australia (Allan, 1988; Nicholls, 1992), India (Rasmusson and Carpenter, 1983), and southeast Asia. Failure of monsoon rains during warm events can have catastrophic impacts (Kiladis and Sinha, 1991). Conversely, cold events are often associated with flooding events in India (Parthasarathy and Pant, 1985). The Australasian signal is more evident during the northern fall of year 0, and persists over the monsoon region of northern Australia into the December through February rainy season. However, during strong ENSO events, the entire continent of Australia can be affected by severe drought conditions (Nicholls, 1992). See also the reviews in the special issue of the *Journal of Geophysical Research* [Anderson *et al.* (eds.), 1998].

It should be emphasized that, while warm events can result in devastating precipitation deficits over monsoon regions, these areas generally still receive a large amount of precipitation in all but the most extreme cases. Over certain regions of the data-sparse Indian Ocean, there is some suggestion that warm events actually see above-normal precipitation in phase with that over the central Pacific. Sri Lanka, the Seychelles Islands, and coastal stations of equatorial Africa are certainly wetter than normal during their September–November rainy season (Kiladis and Diaz, 1989), and satellite data from three warm events (1982, 1986, and 1991) support the occurrence of enhanced rainfall over the southern Indian Ocean during the northern winter of those events. If so, this means that the main focus of precipitation over Australasia is actually not so much shifted eastward, but distributed more evenly across the equatorial Pacific and Indian Oceans.

As warm events evolve toward their "mature" phase during the northern winter season (DJF + 1), drier than normal conditions continue in the region of the western Pacific ITCZ from the Philippines eastward (Kiladis and Diaz, 1989), and east of Australia in the normal position of the SPCZ (Kiladis and van Loon, 1988). Similarly, a large region of southeastern Africa, including parts of Zimbabwe, Mozambique, and South Africa, has a marked tendency for drought during DJF + 1 of warm events. Precipitation in this region is highly seasonal, so this signal can have an especially large impact since it occurs during the normal southern summer rainy season (Nicholson and Entekhabi, 1986). Farther north in Africa, there are indications that warm events favor drought conditions from the Sahel eastward to the highlands of Ethiopia during the normal summer rainy season of JJA 0 (see Janowiak, 1988; Lamb and Peppler, 1991).

While coastal Ecuador and northern Peru often experience flooding during warm events, other regions of the Americas often also register large rainfall anomalies. A consistent dry signal is found from JJA 0 through DJF + 1 over northern South America (Rogers, 1988). Over northeast Brazil, the periodic occurrence of severe drought in the agriculturally rich Nordeste region in connection with El Niño events has resulted in severe economic hardship and occasional famines in this region (Hastenrath and Heller, 1977; Chu, 1991). Widespread and severe famine was reported during the great El Niño of 1877–78, while other severe El Niño episodes, including that of 1982–83, have led to great suffering. Although data are sparse, the northern Amazon basin also appears to be affected. In contrast, much of southern South America is wet during JJA 0, and this signal persists into SON 0 in Uruguay, and central Chile and Argentina (see Kiladis and Diaz, 1989; Diaz and Kiladis, 1995; and Diaz *et al.*, 2001). A remnant of this signal is still present in DJF + 1, when the southeastern United States and northern Mexico also shows above normal precipitation. In the heavy rainfall areas of the upper Paraná and Paraguay River Basins, MAM + 1 precipitation tends to be above normal, and this contrasts with the relatively dry summer wet season over Central America.

Although the best correlations between ENSO and temperature are observed in the tropics and subtropics, the Americas can experience large mid latitude temperature anomalies during ENSO events. Strong and relatively mild westerly flow from the Pacific into the North America during warm events is responsible for a large region of positive temperature departures from southern California northward along the west coast to Alaska, then inland across western and central Canada. This signal over northwest North America is one of the most reliable in the extratropics from event to event (Kiladis and Diaz, 1989; Diaz and Kiladis, 1992). Cold events are equally reliable in being associated with anomalously cold winter seasons in this region. In those years, a tendency toward a weak jet stream over the central Pacific leads to atmospheric "blocking" patterns in the Gulf of Alaska, which in turn are associated with anomalous northerly flow over northwestern North America. The increased warm-event storminess over the southeast United States discussed above is also accompanied by cooler temperatures in that region. Similar enhanced zonal flow over sub-tropical South America leads to above-normal temperatures along the west coast; as in North America, this signal is most extensive during its winter.

ENSO and climate variations

It is important to recognize that the ENSO teleconnections represent statistical averages over the last half-century, and the strength of these waxes and wanes on decadal timescales for several reasons. For example, because tropical Pacific SSTs exert only a modest control on seasonal climate variations, random fluctuations of the teleconnections due to sampling alone can occur on multi-decadal timescales (Allan, 2000; Diaz *et al.*, 2001).

It has been noted elsewhere that tropical Pacific SST variations were weak during 1920–40 compared with 1980–2000 (Trenberth and Shea, 1987; Allan *et al.*, 1996; Allan, 2000). Given that the strength of teleconnections increases almost linearly with the amplitude of tropical Pacific SST anomalies (e.g., Kumar and Hoerling, 1998), a reasonable assumption is that stronger teleconnections will occur during decades when tropical Pacific SSTs exhibit larger interannual variance. An additional factor requires no change in ENSO temporal behavior *per se*, but involves slow, multi-decadal, changes in the climatological mean SSTs themselves. These can alter the atmospheric sensitivity to interannual variations in tropical Pacific SSTs because the teleconnections respond to the total, rather than to the anomalous, SSTs.

One plausible factor in this strengthening of ENSO teleconnections is a multi-decadal change in the life cycle of tropical Pacific SST anomalies during warm events (Trenberth, 1984; Mantua *et al.*, 1997). The strength of teleconnection patterns emanating from changes in tropical convection associated with ENSO variability differed considerably before and after 1976 (Diaz *et al.*, 2001). The origin for this change is not known, and may merely reflect the stochastic behavior of the equatorial coupled system itself. Nonetheless, the relevancy for the teleconnection process is large because the warm-event life cycle since 1976 has become closely phased with the climatological annual cycle, and in particular the prolonged anomalous warming in spring of year + 1 now coincides with the peak annual cycle SST warming.

In many areas where a significant ENSO modulation of precipitation is present, the strength of that teleconnection has varied over the past century. It is unclear to what degree the temporal variations of ENSO teleconnections in different regions are related. To the extent that there is a dynamical root to the changes observed in the past century, for example, due to changes in the ENSO SST forcing, then it is reasonable that changes in one region would be accompanied by changes in another.

Decadal scale patterns in tropical and north Pacific SSTs have been shown to correlate to decadal scale changes in the extra-tropical circulation, and associated temperature and precipitation patterns, particularly over North America. These changes have been statistically linked to different phases of a decadal-scale pattern of SST anomaly in the North Pacific known as the Pacific Decadal Oscillation (PDO) (Mantua *et al.*, 1997); however, it is not clear what is cause and what is effect. One interpretation of recent modeling work on the connection between low-frequency changes in tropical SST, ENSO, and decadal scale changes in the general atmospheric circulation suggest a complex interplay between the canonical ENSO system, the slow changes in SST in the Indo-Pacific over the last century, and long-term changes in the atmospheric circulation itself.

Henry F. Diaz

Bibliography

Allan, R.J., 1988. El Niño–Southern Oscillation influences in the Australasian region. *Progress in Physical Geography*, **12**: 4–40.

Allan, R.J., 2000. ENSO and climatic variability in the past 150 years. In Diaz, H.F., and Markgraf, V. (eds.), *El Niño and the Southern Oscillation, Multiscale Variability and Global and Regional Impacts*. New York and Cambridge: Cambridge University Press, pp. 3–55.

Allan, R., Lindesay, J., and Parker, D., 1996. *El Niño Southern Oscillation and Climatic Variability*. CSIRO Publishing, Collingwood, VIC, Austr., 405 pp.

Anderson, D.L.T., Sarachik, E.S., Webster, P.J., and Rothstein, L.M. (eds.), 1998. The TOGA Decade. Reviewing the progress of El Niño research and prediction. *Journal of Geophysical Research*, **103**: 14,167–14,510.

Barnett, T.P., Latif, M., Kirk, E., and Roeckner, E., 1991. On ENSO physics. *Journal of Climate*, **4**: 487–515.

Bjerknes, J., 1966. A possible response of the atmospheric Hadley circulation to equatorial anomalies of ocean temperature. *Tellus*, **18**: 820–829.

Bjerknes, J., 1969. Atmospheric teleconnections from the equatorial Pacific. *Monthly Weather Review*, **97**: 163–172.

Chu, P.-S., 1991. Brazil's climate anomalies and ENSO. In Glantz, M.H., Katz, R.W., and Nicholls, N. (eds.), *Teleconnections Linking Worldwide Climate Anomalies*. Cambridge: Cambridge University Press, pp. 43–71.

Deser, C., and Wallace, J.M., 1987. El Niño events and their relation to the Southern Oscillation: 1925–1986. *Journal of Geophysical Research*, **92**(14): 189–196.

Diaz, H.F., and Kiladis, G.N., 1992. Atmospheric teleconnections associated with the extreme phases of the Southern Oscillation. In Diaz, H.F., and Markgraf, V. (eds.), *El Niño: Historical and Paleoclimatic Aspects of the Southern Oscillation*. Cambridge: Cambridge University Press, pp. 7–28.

Diaz, H.F., and Kiladis, G.N., 1995. Climatic variability on decadal to century timescales. In Henderson-Sellers, A. (ed.), *Future Climates of the World: A Modelling Perspective*. World Survey of Climatology, Amsterdam: Elsevier Publ. Co., pp. 191–244.
Diaz, H.F., and Markgraf, V. (eds.), 1992. *El Niño: Historical and Paleoclimatic Aspects of the Southern Oscillation*. Cambridge: Cambridge University Press, 476 pp.
Diaz, H.F., and Markgraf V. (eds.), 2000. *El Niño and the Southern Oscillation, Multiscale Variability and Global and Regional Impacts*. Cambridge: Cambridge University Press, 496 pp.
Diaz, H.F., Hoerling, M.P., and Eischeid, J.K., 2001. ENSO variability, teleconnections, and climate change. *International Journal of Climatology*, **21**: 1845–1862.
Enfield, D.B., 1988. Is El Niño becoming more common? *Oceanography*, **1**: 23–27.
Enfield, D.B., 1989. El Niño, past and present. *Review of Geophysics*, **27**: 159–187.
Fu, C., Diaz, H.F., and Fletcher, J.O., 1986. Characteristics of the response of sea surface temperature in the central Pacific associated with warm episodes of the Southern Oscillation. *Monthly Weather Review*, **114**: 1716–1738.
Glantz, M.H., Katz, R.W., and Nicholls, N. (eds.), 1991. *Teleconnections Linking Worldwide Climate Anomalies*. Cambridge: Cambridge University Press.
Hastenrath, S., and Heller, L., 1977. Dynamics of climatic hazards in northeast Brazil. *Quarterly Journal of Royal Meteorological Society*, **103**: 77–92.
Janowiak, J.E., 1988. An investigation of interannual rainfall variability in Africa. *Journal of Climate*, **1**: 240–255.
Kiladis, G.N., and Diaz, H.F., 1989. Global climatic anomalies associated with extremes of the Southern Oscillation. *Journal of Climate*, **2**: 1069–1090.
Kiladis, G.N., and Sinha, S.K., 1991. ENSO, monsoon and drought in India. In Glantz, M.H., Katz, R.W., and Nicholls, N. (eds.), *Teleconnections: Linking Worldwide Climate Anomalies*. Cambridge: Cambridge University Press, pp. 431–458.
Kiladis, G.N., and van Loon, H., 1988. The Southern Oscillation. Part VIII: Meteorological anomalies over the Indian and Pacific sectors associated with the extremes of the oscillation. *Monthly Weather Review*, **110**: 120–136.
Kumar, A., and Hoerling, M.P., 1998. Annual cycle of Pacific/North American seasonal predictability associated with different phases of ENSO. *Journal of Climate*, **11**: 3295–3308.
Lamb, P.J., and Peppler, R.A., 1991. West Africa. In Glantz, M.H., Katz, R.W., and Nicholls, N. (eds.), *Teleconnections: Linking Worldwide Climate Anomalies*. Cambridge: Cambridge University Press, pp. 121–189.
Lau, K.-M., and Sheu, P.J., 1988. Annual cycle, quasi-biennial oscillation, and Southern Oscillation in global precipitation. *Journal of Geophysical Research*, **93**(10): 975–988.
Mantua, N.J., Hare, S.R., Zhang, Y., Wallace, J.M., and Francis, R.C., 1997. A Pacific interdecadal climate oscillation with impacts on salmon production. *Bulletin of American Meteorological Society*, **78**: 1069–1079.
Nicholls, N., 1992. Historical El Niño/Southern Oscillation variability in the Australasian region. In Diaz, H.F., and Markgraf, V. (eds.), *El Niño: Historical and Paleoclimatic Aspects of the Southern Oscillation*. Cambridge: Cambridge University Press, pp. 151–73.
Nicholson, S.E., and Entekhabi, D., 1986. The quasi-periodic behavior of rainfall variability in Africa and its relationship to the Southern Oscillation. *Archiv Meteorologic, Geophysik und Bioklimatologic*, **A34**: 311–348.
Parthasarathy, B., and Pant, G.B., 1985. Seasonal relationships between Indian summer monsoon rainfall and the Southern Oscillation. *Journal of Climatology*, **5**: 369–78.
Philander, S.G.H., 1985. El Niño and La Nina. *Journal of the Atmospheric Sciences*, **42**: 2652–2662.
Philander, S.G.H., 1990. El Niño, La Niña, and the Southern Oscillation. Orlando, FL: Academic Press.
Pickard, G.L., and Emery, W.J., 1982. Descriptive Physical Oceanography. New York: Pergamon Press.
Rasmusson, R.M., and Carpenter, T.H., 1982. Variations in tropical sea surface temperature and surface wind fields associated with the Southern Oscillation. *Monthly Weather Review*, **110**: 354–384.
Rasmusson, R.M., and Carpenter, T.H., 1983. The relationship between eastern equatorial Pacific sea surface temperatures and rainfall over India and Sri Lanka. *Monthly Weather Review*, **111**: 517–528.
Rogers, J.C., 1988. Precipitation variability over the Caribbean and tropical Americas associated with the Southern Oscillation. *Journal of Climate*, **1**: 172–182.
Ropelewski, C.F., and Halpert, M.S., 1986. North American precipitation and temperature patterns associated with the El Niño/Southern Oscillation (ENSO). *Monthly Weather Review*, **114**: 2352–2362.
Ropelewski, C.F., and Halpert, M.S., 1987. Global and regional scale precipitation patterns associated with El Niño/Southern Oscillation. *Monthly Weather Review Journal*, **115**: 1606–1626.
Streten, N.A., and Zillman, J.W., 1984. Climate of the South Pacific Ocean. In H. van Loon (ed.) *World Survey of Climatology, Volume 15: The Oceans*. New York: Elsevier.
Trenberth, K., 1976. Spatial and temporal variations of the Southern Oscillation. *Quarterly Journal of Royal Meteorological Society*, **102**: 639–653.
Trenberth, K.E., 1984. Signal versus noise in the Southern Oscillation. *Monthly Weather Review*, **112**: 326–332.
Trenberth, K.E., and Shea, D.J., 1987. On the evolution of the Southern Oscillation. *Monthly Weather Review*, **115**: 3078–3096.
van Loon, H., and Shea, D.J., 1985. The Southern Oscillation. Part IV: The precursors south of 15°S to the extremes of the oscillation. *Monthly Weather Review*, **113**: 2063–2074.
Walker, G.T., 1923. Correlations in seasonal variations of weather. Part VIII: A preliminary study of world weather. *Memoirs Indian Meteorological Department*, **24**: 75–131.
Webster, P.J., and Yang, S., 1992. Monsoon and ENSO: selectively interactive systems. *Quarterly Journal of Royal Meteorological Society*, **118**: 877–926.

Cross-references

Climate Patterns in the Coastal Zone
Coastal Climate
Coastal Temperature Trends
Coastal Upwelling and Downwelling
Coastal Wind Effects
Desert Coasts
Meteorologic Effects on Coasts
Natural Hazards

ENDOGENIC AND EXOGENIC FACTORS

Endogenic (or endogenetic) factors are agents supplying energy for actions that are located within the earth. Endogenic factors have origins located well below the earth's surface. The term is applied, for example, to volcanic origins of landforms, but it is also applied to the original chemical precipitates. Exogenic (or exogenetic) factors are agents supplying energy for actions that are located at or near the earth's surface. Exogenic factors are usually driven by gravity or atmospheric forces. The term is commonly applied to various processes such as weathering, denudation, mass wasting, etc. In coastal science, these factors may be illustrated in two significant applications. One is the classification of coastlines and the other is the discussion of sea-level variations.

Factors in coastline classification

As early as 1885, Suess (as cited in Kennet, 1982) described the fundamental difference between the coastlines of the Atlantic and those of the Pacific Ocean. This distinction is now the basis for classifying the world's coasts into three categories based on the endogenic processes of plate tectonics (Inman and Nordstrom, 1971). Collision coasts are found at convergent plate boundaries such as along the west coasts of North and South America. They tend to be straight and mountainous with narrow continental shelves. Marginal sea coasts, the second category, are protected from the open ocean by island arcs. An example would be the coast of Korea. Trailing-edge coasts are found on the interior areas of tectonic plates. When newly created, like the coast of the Red Sea, trailing-edge coasts appear similar to collision coasts; they are steep, lined by sea cliffs, and have a normal shelf. With age, however, they tend to have broad coastal plains, wide shelves, and low relief like the east coast of the United States.

The evolution of trailing-edge coasts corresponds to early concepts in geomorphology on the aging, or maturing, of continental margins by exogenic forces (Dietz, 1952). Simply put, the continental crust is attacked by wave action and the continental margin shaped by the redistribution of marine sediment by coastal currents. Young, steep coasts are thereby replaced by the gently sloping surface of a thick wedge of coastal plain sediments. Turbidity currents and other elements of gravity-driven mass movements are then instrumental in the basic formation of the continental slope and rise. On a smaller scale, endogenic processes of faulting, volcanic, or other igneous structures provide other coastline types (Shepard, 1963). These are typically erosional, rocky coastlines in which the underlying geologic structure imposes the coastline pattern.

Coastal features formed by exogenic factors may be inherited or active. The surficial processes to which these coastal landforms owe their origin are no longer active. Coastlines formed by glacial action, such as fjords or drumlins, or drowned river valleys, are examples of inherited characteristics. Deltaic coasts, barrier islands, cuspate forelands, tombolos, etc. are all features shaped by exogenic factors since they owe their origin to the active redistribution of sediment by waves and currents. Collectively such features are characteristics of accretional coasts, notwithstanding the fact that such coasts may have serious erosion problems.

Factors in sea-level changes

Endogenic processes also control long-term eustatic sea-level changes. Deep-seated isostasy and the thermal structure of the oceanic lithosphere influence the relief of the ocean basin. Newly created seafloor stands in high relief being thermally expanded. With the passage of time, it cools, contracts, and subsides. As a result, changes in the rate of seafloor spreading and the total length of mid-ocean ridges changes the elevation of the oceanic platform, that is to say, the average depth of the ocean basin, and, consequently, the position of sea-level relative to the continents. A global drop in sea level of some 350 m at the end of the Cretaceous is attributed to the contraction of the mid-ocean ridge system, in part by the disappearance of spreading centers in the Pacific beneath North America.

Exogenic factors that control sea-level variations include seasonal cycles due to temperature variations, storm surges, or geostrophic adjustment of sea surface slopes by variations in the speed of coastal ocean currents. Anticipated sea-level changes, and other coastal impacts, due to global warming have exogenic origins. The current rate of eustatic sea-level rise is due predominantly to the combined effects of the liberation of water from the melting of glaciers and the thermal expansion of the surficial layer of the ocean.

Extraterrestrial factors

The term "exogenic" is also applied to agents having their origins outside the earth such as solar radiation, comets, or meteorites. A collection of features that owe their origin to the impact of meteorites, such as coastal deposits of a tsunami resulting from the Cretaceous impact event, could be ascribed to this class of exogenic factors.

Henry Bokuniewicz

Bibliography

Dietz, R.S., 1952. Geomorphic evolution of continental terrace (continental shelf and slope). *Bulletin of the American Association of Petroleum Geologists*, **36**: 1802–1819.

Inman, D.L., and Nordstrom, C.E., 1971. On the tectonic and morphological classification of coasts. *Journal of Geology*, **79**: 1–21.

Kennet, J.P., 1982. *Marine Geology*. Englewood Cliffs, NJ: Prentice-Hall.

Shepard, F.P., 1963. *Submarine Geology*, 2nd edn. New York: Harper and Row.

Suess, E., 1885. Das Antlitz der Erde, I. Prague, F. Tempsky.

Cross-references

Barrier Islands
Changing Sea Levels
Faulted Coasts
Holocene Coastal Geomorphology
Volcanic Coasts

ENERGY AND SEDIMENT BUDGETS OF THE GLOBAL COASTAL ZONE

The energy for the global coastal zone is primarily powered by solar irradiance and tides. The solar irradiance warms the earth's surface unevenly, driving wind and current systems that redistribute heat and generate storms with rainfall and snow that erode landmasses. The erosion products of sediment and dissolved solids are carried to the sea by rivers and glaciers. Winds, waves, tides, and currents transport and redistribute sediment and sculpt coastal landforms.

Solar irradiance provides an energy flux of about 1.8×10^{14} kilowatt (kW) to the earth. About 2.5×10^{9} kW of mechanical energy in wind-generated waves is incident on the world's 440,000 km coastlines. Another 2.2×10^{9} kW of tidal energy is expended in shallow seas. The total flux of mechanical energy of all kinds in shallow waters of the coastal zone is about 5.5×10^{9} kW. Some of this energy is expended in coastal erosion and transport of the erosion products from the land, and the remainder drives frictional processes and is dissipated as heat. The erosion rate of land is about 6 cm/thousand years resulting in a flux to the sea of 5×10^{9} ton/yr dissolved solids and 20×10^{9} ton/yr particulate solids (1 ton = 1,000 kg).

Introduction

The oceans respond to many different kinds of steady and impulsive forces applied at their boundaries with the atmosphere and seafloor. Although most energy in the nearshore waters comes from wind-generated waves and tidal currents, energy from a number of other sources such as internal waves and ocean currents may also be important locally.

One approach to understanding the relative importance of nearshore processes is to compare the sea's potential to erode the land with the land's potential to supply terrestrial erosion products. Such a comparison ultimately resolves itself into the balance between the budget of power in waves and currents and the budget of sediments available for transport. Of course, this balance varies widely from place to place, and even in the best-studied areas is but poorly understood. However, order of magnitude estimates can be attempted on a worldwide basis, and their consideration here gives some overall perspective for the relative importance of the driving forces that are operative in nearshore waters.

In a practical sense, the waters of the coastal zone can be considered as including the shallow seas and the waters covering the continental shelves of the world, an area of 29×10^{6} km^2, or about 5.5% of the surface area of the world and about 8% of the area of the world oceans (Table E1). The nearshore waters are bounded on the landward side by coastlines that total about 440,000 km in length, and to the seaward by the break in slope at the shelf edge (shelf break) marking the change from the relatively horizontal continental shelf to the steeper continental slope. The continental slopes are one of the striking geographic features of the earth and have a combined length of about 150,000 km.

Table E1 Dimensions of major topographic features

Feature	Area (10^6 km^2)	Length of coastline (10^3 km)
Continents	138.8	210
Large islands (larger than 2,500 km^2)	9.3	136
Small islands (25–2 500 km^2)	0.9	94
Total land	149.0[a]	440
		Length of shelf break (10^3 km)
Oceans (depths >200 m)	332.1	
Continental shelf (0–200 m)	29.0	149
Total marine water	361.1	
Total area earth	510.0	

[a] United Nations (UNEP, 1987) gives total land area of 145×10^6 km^2.
Source: Inman and Nordstrom (1971) for features with length scales greater than 10 km.

Conventionally, the depth of the shelf break is taken as 200 m or 100 fathom, although in some localities the depth may be as great as 400 m. On a worldwide basis, the average depth of the shelf break is about 130 m. The width of the shelf ranges from zero to over 1,300 km and averages about 74 km (Shepard, 1963). Although having little geomorphic basis, international law defines *territorial seas* as usually extending seaward 12 nautical miles (22.2 km) beyond the coastline and an *exclusive economic zone* as usually extending 200 nautical miles (371 km) seaward.

Budget of energy in shallow water

The current systems of the world's atmosphere and ocean are solar powered. The sun irradiates the earth with a total of about 1.8×10^{14} kW, or an intensity of 1.37 kW/m^2, the so-called solar constant. In terms of the rotating, spherical earth, this averages to about 342 W/m^2 of earth's surface. Of this amount about 30% is reflected and backscattered, 19% is absorbed by the atmosphere, and 51% is absorbed at the earth's surface (e.g., Gill, 1982). Solar energy heats the earth's surface unevenly and generates winds. The oceans are relatively opaque to solar irradiance, so that the effect on the oceans is to stratify and stabilize the surface water. As a consequence, the ocean responds much more to wind, even though the intensity of thermal energy from the sun greatly exceeds the mechanical energy from the wind at the water's surface. Averaged over the earth's surface, the intensity of thermal energy absorbed is about 175 W/m^2, while the sea surface stresses and pressures caused by a wind blowing 10 m/s expend about 1.3 W/m^2. In contrast, the mean heat flow to the earth's surface from the interior is approximately 0.1 W/m^2, and occurs mostly at the mid-ocean spreading centers (Pollack *et al.*, 1993). Thus, it is the solar powered atmosphere and ocean currents together that maintain the earth's heat budget by accounting for a poleward transport of energy at the rate of about 8×10^9 kW.

Studies of the energy budgets of the ocean and atmosphere (e.g., Webster, 1994) show that the total energy content of the oceans ($\sim 160 \times 10^{25}$ joule (J)) is 1,000 times larger than that of the atmosphere ($\sim 125 \times 10^{22}$ J). Because of greater mass and heat capacity, the oceans play a critical role in controlling and moderating weather and climate. However, in terms of the kinetic energy available for interacting with the oceans, the atmospheric kinetic energy ($\sim 6 \times 10^{20}$ J) is 200 times larger than that of the oceans.

The wind systems are the direct link between atmosphere and ocean and, through momentum exchange at the ocean's surface, expend kinetic energy at the rate of about 10^{11} kW (Malkus, 1962; Newell *et al.*, 1992). The prevailing wind systems cause all large bodies of water to have windward and leeward shores. The Pacific coasts of the Americas are in general windward coasts while the Atlantic seaboards are leeward coasts. The general wind flow is clockwise in the Northern Hemisphere around semipermanent, midlatitude areas of high pressure, for example, the north Pacific high and the Bermuda high in the Atlantic Ocean. Circulation patterns in the southern oceans are essentially counterclockwise flowing mirror images of those in the Northern Hemisphere (Figure E3).

Sea and swell

The most intense wave action in all oceans results from cyclogenesis over the poleward-flowing, western boundary currents. These warm water jets carry equatorial heat to higher latitudes and produce strong temperature gradients that spin-up intense storms (Figure E3). These boundary cyclonic events generate large, high-frequency *sea waves* along leeward shores and long waves that cross the ocean and appear as low-frequency *swell* on windward coasts. In the north Pacific, dry Siberian winds flowing over the warmer Kuroshio Current produce a series of cyclonic cold fronts that collectively have long fetches and generate the high waves for the Pacific coast of North America during the winter. These storms produce swell when they are distant and sea waves as their tracks near the coast.

Long, low-frequency swell is typically absent from leeward coasts. Rather they are subject to periodic, intense sea waves from the poleward and easterly traveling storm fronts. The tracks for tropical (summer/fall) storms for the middle Atlantic seaboard of the United States essentially

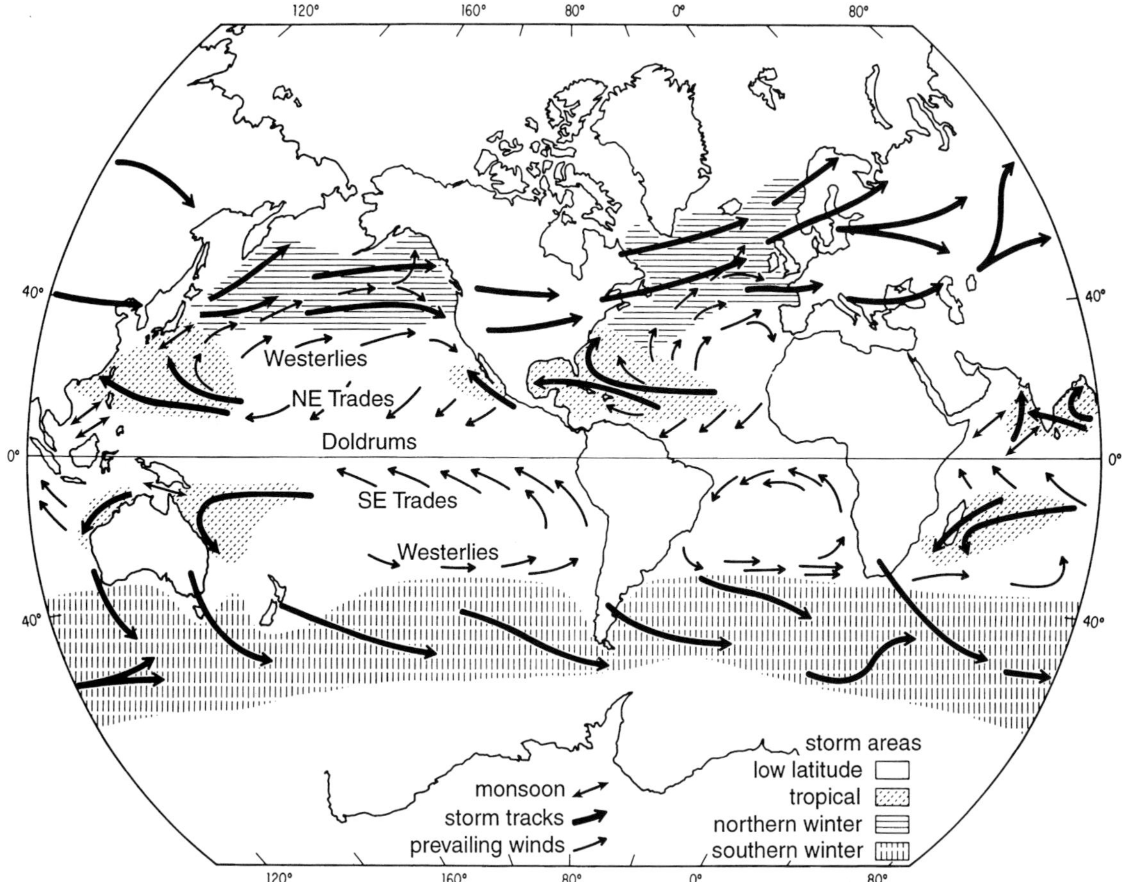

Figure E3 Prevailing winds and storm tracks for the world oceans.

follow the Gulf Stream, while the extratropical (winter/spring) tracks take a more easterly course. In midlatitudes the tracks are channeled between the coast and the perimeter of the Bermuda high. This channel is sometimes known as the Atlantic seaboard cyclone track (e.g., Klein and Winston, 1958). In the vicinity of Cape Hatteras, the highest waves from both extra- and tropical storms are from the northeast. The storm tracks are offshore and generally parallel the coast, so that the counter-clockwise flowing winds in the northwest quadrant of the cyclone (winds blowing from the northeast) are the strongest and have the longest fetch (e.g., Dolan *et al.*, 1990). Thus, prevailing waves along the windward coasts and the strongest cyclonic sea waves along the leeward coasts both have net components toward the equator. As a consequence, there is a worldwide tendency in midlatitudes (20°–45°) for the net wave-induced littoral transport of sand to be toward the equator.

A three meter-high surface wave at sea transmits 100 kW for every meter of wave crest, or 100 MW/km. Since wave energy increases as the square of wave height, most coastal energy is from high waves during storms, particularly storms of long duration. For example, the northeast storm of March 1989 along the Outer Banks of North Carolina had average deepwater wave heights of 3 m (Dolan *et al.*, 1990). These were not unusually high waves, but they persisted for 115 h and had a power expenditure of 11,500 kilowatt hours per meter (kWh/m) of coastline. The average energy flux of waves on the Outer Banks is about 2 kW/m, which sums to a yearly total of 17,500 kWh/m (Inman and Dolan, 1989). Thus this one storm had a power expenditure (and sand transport potential) equivalent to 66% of that for the average year.

Data of instrumented buoys and light vessels in the North Atlantic and North Pacific all show that wave heights have increased significantly and continuously during the past 25 years (e.g., Graham and Diaz, 2001). The North Atlantic data show that the average annual (root-mean square) wave height has increased about 2% per year between 1962 and 1986. Over the 24-year period (1962–86) the range in average annual *significant wave* ($\sqrt{2}$ times root-mean square) height at Seven Stones Light Vessel off Lands End, southwest England, ranged from 2.2 to 2.9 m with an annual increase of 3 cm/yr. The "50-year return value" ranged from 12 to 18 m with an increase of 20 cm/yr (Carter and Draper, 1998). Data from six buoys between latitude 34° and 56° in the eastern North Pacific show similar increasing trends over the period 1975–99. The annual average significant wave height increased on average about 2 cm/yr, ranging for the six buoys from 0.4 to 4.2 cm/yr (Allen and Komar, 2000).

A preliminary estimate of global wave energy can be attempted from the data obtained from instrumented buoys and weather stations in the North Pacific and North Atlantic. The measurements for the eastern North Pacific cover the 25-year time interval from 1975 to 1999 (Allen and Komar, 2000). Over this period, the average annual (root-mean square) wave height increased progressively from about 1.6 to 1.9 m, generating a landward wave energy flux that ranged for these average waves from about 93×10^6 to 131×10^6 kW over the same period. Thus, the wave height increased about 20% while the corresponding energy flux increased about 40%. Similar wave conditions and wave height increases over about the same period were observed for the eastern North Atlantic Ocean (Carter and Draper, 1998). If it is assumed that the energy partitioning between periods of calm and storm results in an energy flux twice that estimated from the average annual wave height (i.e., the significant wave assumption), then the total annual energy flux over the 25-year period in the North Pacific becomes about 190 and 260×10^6 kW with an average for the period of about 230×10^6 kW for this 30°–60° latitude portion of the windward coast of North America.

When the energy flux of 260×10^6 kW for the North Pacific Ocean is extrapolated to the world oceans, the total wave energy flux to the global coastline becomes 2.5×10^9 kW (Table E2, footnote a). This is in agreement with earlier estimates (e.g., Inman, 1973) and gives an average energy flux of 5.7 kW/m to the 440,000 km of world coastline.

What is known about wave climate intensity suggests that windward coasts of oceans in the Northern Hemisphere are in phase with each other during El Niño and La Niña events and out of phase with lee coasts. In terms of decadal climate patterns, which are alternately El Niño or La Niña dominated, the prevailing wind stress and swell directions in the Northern and Southern Hemisphere appear to be out of phase (Trenberth and Hurrell, 1994; Chao *et al.*, 2000). For example, during the La Niña dominated climate of the third-quarter of the 20th century, waves from the southern ocean appeared as "southern swell" along the California coast. Southern swell was absent from the California coast during the previous quarter of the century and again disappeared during the last quarter century (Inman and Jenkins, 1997). This is because of a global realignment in predominant storm tracks with changes in the phase of Pacific Decadal Oscillation (PDO), where El Niño storm tracks tend to be more zonal than those of La Niña storms.

Table E2 Estimates of the flux of mechanical energy in the shallow waters of the world in units of 10^9 kW

Source	10^9 kW
Wind-generated waves breaking against the coastline[a]	2.5
Tidal currents due to surface tides on shelves and in shallow seas[b]	2.2
Currents due to internal waves and tides over shelves[c]	0.5
Large-scale ocean currents in shallow seas[b]	0.2
All other sources	0.1
Total	5.5

[a] Assuming the average energy flux for the six instrumented NOAA buoys in the eastern North Pacific (Allen and Komar, 2000), proportioned over 30°–60° north latitude, is 230×10^6 kW (see text) and that for the 0°–30° portion of the eastern North Pacific is one-half, and that for the leeward ocean coasts is one third that of the windward coasts, giving a total for the coasts bordering the North Pacific Ocean of 460×10^6 kW. The global flux is assumed to be equivalent to that from five and one-half oceans like the North Pacific Ocean (i.e., North and South Pacific Oceans, North and South Atlantic Oceans, Indian Ocean, and one-half "ocean" equivalence for all marginal seas).

[b] Inman and Brush (1973).

[c] Conservative estimate from Wunsch and Hendry (1972).

Consequently, the long waves of El Niño storms do not generally propagate out of the hemisphere in which they were generated. Conversely, the frontal circulation of La Niña storms is more meridional, generating waves that traverse latitude. Therefore, the Northern Hemisphere beaches are usually affected by Southern Hemisphere swells only when the Southern Hemisphere is in a La Niña dominated climate pattern.

Studies of windfields and pressure gradients indicate that the increase in wave heights in the eastern North Atlantic during the last quarter of the 20th century resulted from intensification of the pressure gradients between the Iceland low and the Azores high which could be related to climate change associated with the North Atlantic Oscillation (NAO) (Kushnir *et al.*, 1997). It is likely that some of the increase in wave intensity in the eastern North Pacific Ocean was associated with the PDO and increased sea surface temperatures during the last quarter of the century. However, it is unclear what portion of these increased wave heights in the North Atlantic and North Pacific oceans are also associated with global warming (Graham and Diaz, 2001). If the increases are mainly due to decadal oscillations, they suggest that decadal climate lead to ±30% changes in wave climate (see entry on *Climate Patterns in the Coastal Zone*).

Tsunamis

The most impressive waves in the sea are those generated by submarine earthquakes, landslides, and volcanic explosions (e.g., Van Dorn, 1987). These waves, called *tsunamis*, are well known because of the loss of life and great damage to coastal structures associated with their passage. Of the many types of surface waves, tsunamis contain the highest instantaneous power surges. They are reported to have caused runup to heights as great as 50 m above sea level in modern times and to greater heights in ancient times. The explosion and collapse of the volcanic island of Thera (Santorini) about 1400 BC is thought to have caused a tsunami, about 11 m high at Crete (Yokoyama, 1978), that may have ended Minoan maritime supremacy. The largest, well-documented tsunami was associated with the explosion and collapse of Krakatoa volcano in Sunda Strait between the Indonesian Islands of Java and Sumatra in August, 1883. The energy of the explosion is estimated to have been 2×10^{18} J (500 megatons of TNT) and it ejected 18 km^3 of material into the atmosphere (e.g., Simkin and Fiske, 1983). The resulting tsunami was estimated to be nearly 40 m in height; it drowned 36,000 people and destroyed 5,000 ships. One ship, the *Berouw* was carried 2 km inland on Sumatra where it remains today. The tsunami was observed on tide gauges throughout the world, and dust from the explosion covered the entire world causing brilliant sunsets for many years.

Detectable tsunamis have occurred about once a year. However, truly large tsunamis with energy content as high as 5×10^{15} J occur only about five times per century. Averaged over long periods, tsunamis would generate an energy flux of about 1×10^5 kW. Thus, because of their relative infrequency, these catastrophic waves contribute little to the worldwide budget of power.

Tides

The gravitational forces of moon and sun provide the driving forces for the ocean tides, with the moon's tide generating force being about twice that of the sun. The main lunar (semidiurnal, M_2) tide with a period of 12.42 h is the principal world tide; however, the interaction between lunar and solar forces provides a variety of tide-generating constituents. Locally, tidal amplitudes are determined in part by the interactions of the propagating oceanic tidal bulge with the shape of the ocean basin, favoring oscillations that are harmonics of the tidal forcing and the basin dimensions. The dominant tidal oscillations are usually referred to as *semidiurnal* (twice daily), *diurnal* (once daily), and *mixed* tides that are combinations of the two, such as those along the Pacific coast of the Americas.

The ocean surface tides, having wavelengths that are very long compared with the depth of the ocean, produce some motion even in the deepest water. However, this motion is usually only a few centimeters per second so that the energy dissipation by tides against the deep-sea bottom is slight. The principal tidal dissipation results from the flow of strong tidal currents in shallow areas, such as the Bering Sea and the Argentine shelf. One of the more impressive tidal coastlines occurs in Argentina. The tidal wave travels eastward through Drake Passage and then northward where it shoals over the broad Argentine shelf, attaining spring ranges of nearly 15 m at Rio Gallegos. From the southeastern tip of Tierra del Fuego to Bahia Blanca, a distance of 2,800 km, the tidal range is mostly in excess of 6 m.

"Tides are unique among natural physical processes in that one can predict their motions well into the future with acceptable accuracy without learning anything about their physical mechanism" (Cartwright, 1977). Their prediction from tide gauge records has been possible since Darwin formulated the rules of harmonic tide prediction in 1883. But, prediction from harmonic constituents can be in error when changing wind fields and warm water associated with El Niño events raise and lower sea level by 30 cm and more.

The energy dissipated by various kinds of tidal currents against the shallow sea bottom is one of the major factors that causes the rotational deceleration of the earth. Much of our present knowledge of this dissipation results from investigations by geophysicists into the possible causes of the deceleration in the earth's rotation (Jeffreys, 1920, 1970; Munk and MacDonald, 1960). Tidal dissipation causes a lag in the tidal bulge of the Earth relative to the rotation of the earth–moon system. This bulge exerts a torque that results in a decrease in both the earth's rotation rate and in the lunar orbital velocity.

From telescopic observations of sun and moon over the previous two and one-half centuries, Spencer Jones (Munk and MacDonald, 1960) determined that the moon's deceleration was 22.4 arcsec/century^2. This deceleration would be associated with an energy dissipation in the atmosphere, oceans, and solid earth of about 2.7×10^9 kW. More recently, measurements of the change in the moon's semimajor axis using laser ranging techniques show that the frictional energy dissipation due to the M_2 tides is 3.1×10^9 kW, giving a dissipation rate due to the solar and lunar tides of about 4.0×10^9 kW (e.g., Dickey *et al.*, 1994). The allocation by process for these revised dissipations of tidal energy is not yet understood. The best estimates for the lunar tidal dissipation in shallow seas is that of Miller (1966). He used the method of energy flux into shallow seas, rather than the less accurate method of summing the energy loss due to frictional drag on unit area of the bottom. His value of 1.7×10^9 kW for lunar tidal energy dissipation, when corrected for the solar tidal flux, gives a total energy dissipation for shallow seas of 2.2×10^9 kW (Table E2).

Internal waves

Internal waves are water oscillations that occur within the various density layers beneath the ocean surface. It appears that far more energy is transformed into internal (baroclinic) waves than was previously thought. The vertically integrated energy over the deep sea is typically about 4×10^{-3} J/m^2 (e.g., Garrett and Munk, 1979). Mechanisms generating internal waves include direct coupling between ocean surface waves and the density layers beneath the surface, scattering due to seafloor roughness (Cox and Sandstrom, 1962), and generation by impingement of deep-sea tides on the slopes of the continental shelves and submarine canyons (Rattray *et al.*, 1969).

Cox and Sandstrom (1962) found that internal waves were coupled to surface waves and estimate that the rate of conversion of tidal energy into internal wave energy in the Atlantic Ocean is roughly between 6 and 69% of the power of the surface tide. From this, Munk and MacDonald (1960) estimate that the conversion of energy from surface to internal modes on a global scale may be about 1.5×10^9 kW. Most internal wave energy is contained in the areas of deep ocean, but some of the energy travels into shallow water and is dissipated against the sea bottom. Winant (1974) estimated that the dissipation due to runup of breaking internal waves and surges on the shelves of the world may range between 2×10^5 and 2×10^7 kW.

Wunsch and Hendry (1972) measured currents from an array placed on the continental slope off Long Island, New York, and concluded that the onshore energy flux of internal waves was 6 W/m. Assuming this to be a typical value, they estimated that the worldwide onshore flux of energy would be about 0.7×10^9 kW. A value of about 0.5×10^9 kW for the energy dissipation by internal waves over the world's shelves appears to be a reasonable estimate (Table E2).

Ocean currents, large and small

The kinetic energy associated with large-scale ocean currents is very large compared with the energy in surface waves and tidal currents. Stommel (1958) estimates that the average kinetic energy of currents in the North Atlantic Ocean is about 2×10^{17} J. Flows such as the equatorial and Antarctic Circumpolar currents transport water at rates of 50–200 Sv (Sverdrup = million cubic meter per second). Even the relatively small California current transports about 12 Sv, a discharge that is greater than that of all the rivers and streams in the world. In addition to the surface currents, the world ocean has a surprisingly large circulation of deep and bottom water estimated to be 50 Sv (e.g., Schmitz, 1995).

Intense western boundary currents, such as the Gulf Stream, develop instabilities with horizontal waves or meanders. These meanders form large loops that may close or pinch-off, forming circular eddies or rings. These *mesoscale eddies* are typically 100 km and more in diameter, and since they are in cycloidal balance, conservation of angular momentum causes them to persist for many months to a year or so, as they move thousands of kilometers away from the main current. Eddies that breakoff from the poleward side of the currents have warm-cores and cyclonic rotation, whereas those shed from the equatorial side have cold-cores and anticyclonic motion (e.g., Gill, 1982). The details of the global distribution of the energy in mesoscale eddies is not well known except for satellite observations of western boundary currents. The effects of mesoscale eddies on the continental shelves is poorly understood, although it is known that shelf eddies are associated with wind changes and with coastal promontories (e.g., Davis, 1985).

While the flow of major ocean currents is predominantly in deep-water, their boundaries overlap the continental shelves, and the currents dissipate some of their energy against the shallow sea bottoms. For example, the Guiana (North Brazil) Current, with a volume discharge of about 29 Sv, flows with velocities up to 1 m/s over the 0.5×10^6 km^2 of shelf off the coastlines of the Guianas and northern Brazil (e.g., Fratantoni *et al.*, 1995). The Falkland Current flows with velocities of 25–50 cm/s across the 1×10^6 km^2 of shallow shelf off Patagonia. If the Guiana Current averages 50 cm/s and the Falkland Current averages 25 cm/s, they would dissipate energy at the rate of 0.13×10^9 and 0.03×10^9 kW, respectively. Therefore, it seems likely that the worldwide dissipation of energy from the flow of ocean currents in shallow water may be about 0.2×10^9 kW (Table E2).

Other sources

In addition to generating the ocean's waves, atmospheric winds, and pressures have important local effects nearshore. Pressure fronts generate shelf seiche, and the combination of pressure and wind stress on the water's surface produces local anomalies in sea level that induce circulation of water over the shelf and in submarine canyons. Shelf seiche up to a meter in height or more are common along the Argentine coast (Inman *et al.*, 1962), while the sea level anomalies associated with storm surges may exceed 4 m for severe storms (Pugh, 1987). Also, winds blowing over the coastal zone dissipate energy on beach and sand dunes. On a worldwide basis, this dissipation may be about 1×10^7 kW, and because of the prevalence of sea breeze and other onshore winds, results in a net inland transport of beach sand.

From the foregoing discussion, it appears that the total flux of mechanical energy in the shallow waters of the world is about 5.5×10^9 kW (Table E2). Dissipation and work by one process or another occurs over the entire shelf. The wind-generated surface waves expend their energy primarily nearshore, especially in the breaker zone, while tidal and other ocean currents and internal waves dissipate energy mostly over the outer portions of the shelf. Internal waves, edge waves, shelf seiche, and local winds may produce water motion over the continental shelf and submarine canyons.

Budget of sediment

The coastline is the junction of the realms of terrestrial erosion and marine deposition. Thus, it is the unique singularity between air, land, and sea, and the coastline is the critical datum for dynamical geology.

In 1955, James Gilluly calculated that sediment is now being carried across this boundary at a rate great enough, in the absence of mountain building, to "erase all the topography above sea level in less than 10 million years . . . ," a very short time on the geological timescale. Thus, the supply of sediment associated with runoff from the continents is large, and it has increased during the past 10 million years due to climate variability, especially during the Pleistocene and more recently due to human intervention. The principal sources of beach and nearshore sediments are the rivers that bring large quantities of sand directly to the ocean, the sea cliffs of unconsolidated material that are eroded by waves, and material of biological origin such as shells, coral fragments, and skeletons of other small marine organisms (Figure E4). Also, waves may transport onshore relict sands that were deposited along the inner shelf during lower stands of the sea (e.g., Inman and Dolan, 1989).

Sediment yield increases with the relief of the drainage basin and with decrease in basin size, factors that enhance the sediment flux of small rivers along mountainous coasts (Inman and Jenkins, 1999). Climate, rainfall, and type of geologic formation are also well-known factors in determining sediment yield. Maximum yield occurs for basins in temperate climates that receive an annual rainfall of about 25–50 cm, and decreases for precipitation rates either above or below this maximum (Langbein and Schumm, 1958). In arid climates, the occasional flash floods transport large volumes of sand (Figure E5).

Man's intervention in natural processes is accelerating erosion rates, perhaps by a factor of about two on a global scale (Milliman and Syvitski, 1992; Vitousek *et al.*, 1997). Significant anthropogenic effects began as early as 9,000 years ago with deforestation and the spread of agriculture from the "Fertile Crescent" (e.g., Heun *et al.*, 1997). Man's intervention has accelerated in this century with increased deforestation, expansion of mechanized agriculture, proliferation of dams (Meade, 1996), and since World War II with extensive urbanization of coastal lands particularly in central and southern California. Urbaniza-tion decreases erosion locally, but accelerates streambed erosion in response to the greater frequency and magnitude of peak streamflow due to runoff from impervious urban surfaces (e.g., Trimble, 1997).

The effects of changing climate and sea-level rise have been studied extensively on a glacial/interglacial timescale (e.g., Hay, 1994) and during the Holocene on a millennial scale (e.g., Bond *et al.*, 1997).The effect of decadal scale climate changes on the sediment flux of the 20 largest streams entering the Pacific Ocean along the central and southern California coast was studied by Inman and Jenkins (1999). The annual streamflow ranged from zero during dry years to a maximum of 1.1×10^9 m^3/yr for the Santa Clara River in 1969 with an associated suspended sediment flux of 46×10^6 ton. Trend analyses showed that El Niño/Southern Oscillation (ENSO) induced climate changes recur on a multidecadal timescale in general agreement with the Pacific/North American (PNA) climate pattern: a dry climate extending from 1944 to about 1968 and a wet climate extending from about 1969 to 1998. The dry period is characterized by consistently low annual river sediment flux. The wet period had a mean annual suspended sediment flux about five times greater, caused by strong El Niño events that produce floods with an average recurrence of about five years. The sediment flux of the rivers during the three major flood years averaged 27 times greater than the annual flux during the previous dry climate (see entry *Climate Patterns in the Coastal Zone*).

The effects of climate change are superimposed on erodibility associated with basin geology. The sediment yield of the faulted, overturned Cenozoic sediments of the transverse ranges of California is many times greater than that of the coast ranges and peninsular ranges. Thus the abrupt transition from dry to wet climate in 1969 brought a suspended sediment flux of 100 million tons to the ocean edge of the Santa Barbara Channel from the rivers of the Transverse Range, an amount greater than their total flux during the preceding 25-year dry period. These alternating dry to wet decadal scale changes in climate are natural cycles that have profound effects on fluvial morphology, engineering structures, and the supply of sediment.

Cliff erosion probably does not account for more than about 5% of the material on most beaches. Wave erosion of rocky coasts is usually slow, even where the rocks are relatively soft shales. On the other hand, erosion rates greater than 1 m/yr are not uncommon in unconsolidated sea cliffs and dunes as found on trailing-edge coasts such as the east coast of the United States.

Following initial deposition at the mouths of streams entering the ocean, much of the sand-size portion of terrestrial sediments is carried along the coast by waves and wave-induced longshore currents. The sand carried by these longshore currents may be intercepted by submarine

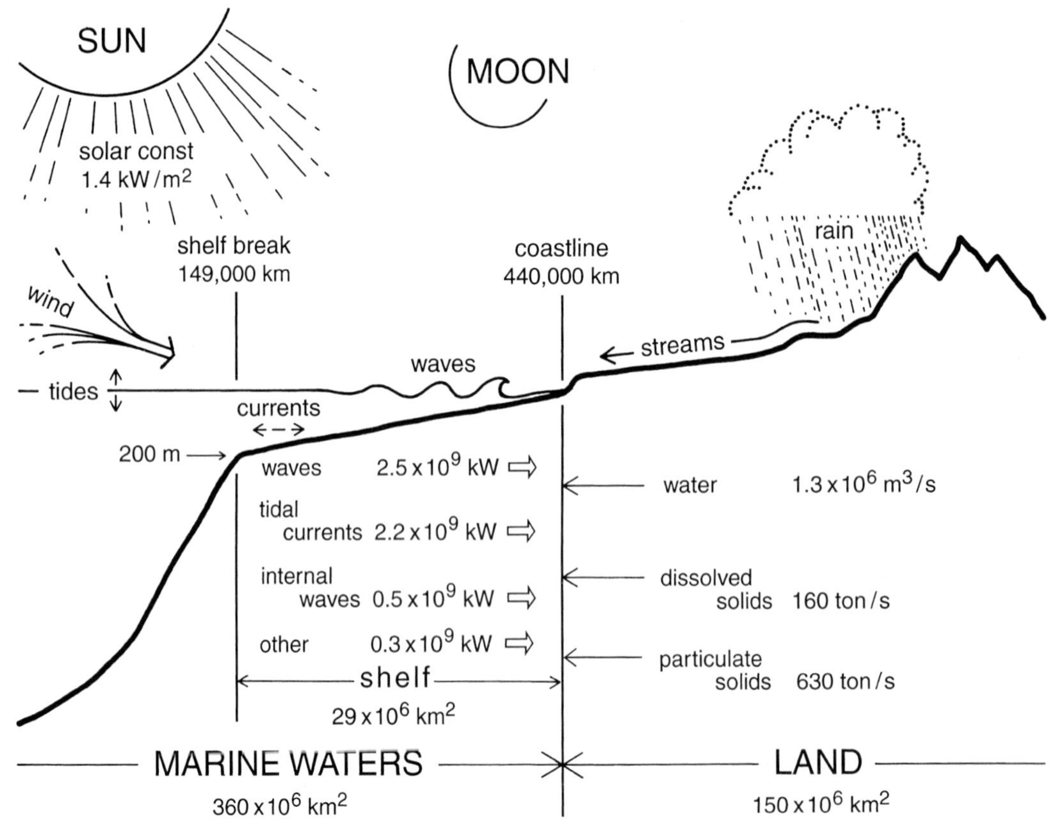

Figure E4 Schematic diagram of budget of energy dissipation in nearshore waters, and the runoff of water and sediment from the land (after Inman, 1973). Data from Tables E1–E3.

Figure E5 Streams carry erosion products from the land to the sea; El Moreno, Gulf of California, Mexico.

canyons that cut into the continental shelf and divert the sand into deeper water offshore. Since submarine canyons occur along most coasts, except those bordered by shallow seas, they probably account for most of the sediment lost into deepwater from the nearshore. Notable among the major sediment transporting submarine canyons of the world are Cap Breton Submarine Canyon on the Atlantic coast of France; Cayar and Trou Sans Fond, and Congo Canyons on the Atlantic coast of Africa; the Indus and the Ganges Submarine Canyons in the Indian Ocean; and the Monterey Submarine Canyon on the Pacific coast of North America. Observations by H.W. Menard (1960) suggest that most of the deep sediments on the abyssal plains along a 400 km section of the California coastline are derived from two submarine canyons: the Delgada Submarine Canyon in Northern California and Monterey Submarine Canyon in central California. These canyons cut the continental shelf almost to the shoreline, and consequently they trap and divert an estimated sediment volume of about 1×10^6 m^3/yr (see entry on *Littoral Cells*).

Traditionally, geologists have estimated long-term erosion and deposition rates from the amount of material deposited during geological time. Such estimates give erosion rates varying from about 1 to 4 cm per thousand years (cm/kyr) for drainage basins of moderate relief to as high as 21–100 cm/kyr for the Himalaya Mountains (e.g., Gilluly, 1955). Assessment of the volume of sedimentary material on the continental United States and in its adjacent seafloors indicates that the average erosion rate of the United States during the past 600 million years was 3–6 cm/kyr (Gilluly *et al.*, 1970).

An increasing number of measurements of river discharge have provided an independent estimate of contemporary erosion rates of the land. The importance of large rivers in transporting the denudation products of the continents to the sea has been known since Lyell (1873) described the flux of sediment into the Bay of Bengal from the Ganges and Brahmaputra Rivers. Since then the contributions from large rivers have been updated and summarized in many studies (e.g., Garrels and Mackenzie, 1971; Inman and Brush, 1973; Meade, 1996) with estimates of the total flux of particulate solids to the ocean of about 16×10^9 ton/yr. The importance of small rivers to the global budget of sediment was first documented by Milliman and Syvitski (1992). They showed that small rivers (drainage basin <10,000 km^2) cover only 20% of the land area but their large number results in their collectively contributing much more sediment than previously estimated, increasing the total flux of particulate solids to about 20×10^9 ton/yr (630 ton/s) by streams and rivers (Table E3).

Table E3 Estimate of the natural runoff of fresh water and solids from the continents into the oceans and coastal waters of the world

Discharge of water into the oceans from all rivers[a]	1.3×10^6 m^3/s
Flux of dissolved solids[b]	160 ton/s
Flux of particulate solids[b]	630 ton/s
Flux of particulate solids from sea cliff erosion[c]	>2 ton/s
Flux of eolian particulate solids[d]	14 ton/s
Average erosion rate of land[b]	6 cm/kyr

[a] Average value (1.1×10^6 m^3/s) from Turekian (1971), Alekin and Brazhnikov (1961), and Holeman (1968) updated in proportion to increase in particulate solids from 530 to 630 ton/s.
[b] Assuming that continental rock of density 2.7 ton/m^3 erodes to form 20% dissolved load and 80% solid particulate load. Assuming flux of particulate solids is 630 ton/s (Milliman and Syvitski, 1992).
[c] Emery and Milliman, 1978.
[d] Windblown dust, volcanic and forest fire ash (Prospero and Carlson, 1981; Rea, 1994; Prospero *et al.*, 2002).

Balance of the budgets of power and sediment

The global flux of erosion products from the land to the sea include 160 ton/s dissolved solids and 630 ton/s particulate solids (Table E3). The dissolved solids are incorporated into the 1.3×10^6 m^3/s flow of "fresh" water as added mass. Most of the particulate solids carried by streamflow are fine silt and clay size material that move as suspended load and washload. The remaining solids are sand and coarser material that move mostly as bed and suspended load. Once the stream discharges into the sea, there are significant differences in the transport paths of the suspended and bedload sediment. Most of the fine suspended sediment moves with the river water and flows out over the sea as a spreading turbid plume, and the sediment, which is subject to flocculation and ingestion by organisms, is eventually deposited in deeper water. The coarser bedload material remains near the river mouth as a submerged sand delta and is later transported along the shore by waves and currents to nourish the downcoast beaches. The few reliable measurements of bedload indicate that it is ca. ≤10% of the total load in large rivers but, in smaller, mountainous streams, may be considerably more (Richards, 1982; Inman and Jenkins, 1999).

It is of interest to compare the global fluxes of energy and sediments into the coastal waters of the world (Figure E4). Since energy provides the capacity for doing work, the flux of energy associated with waves and currents is a measure of their potential to transport sediment. About one-half (2.9×10^9 kW) of the total flux of mechanical energy is associated with currents of various types, internal waves, and turbulence, and this flux acts primarily upon the fine particulate solids that flow out over the ocean surface. The remaining energy (2.5×10^9 kW) is from ocean surface waves and works primarily on sand size solids that are deposited in shallow water. About 85% (535 ton/s) of the particulate solids are fine material, while about 15% (95 ton/s) are sand and coarser.

Waves move sand on, off, and along the shore. Once an equilibrium beach profile is established, the principal transport is along the coast. Theory and field measurements show that the littoral transport rate of sand along a beach is proportional to the energy flux of the incident waves, where the proportionality is a function of the direction of wave approach relative to the shoreline. If 10% of the world's estimated annual wave power at the coast (0.25×10^9 kW) is available for transporting sediment (i.e., the wave breaker angle is about 5°), the waves have the potential to transport the entire sand size load of solids in the world's streams (95 ton/s), a distance of 750 km along the coast each year. Thus, even under the inefficient coupling that exists in nature, it is apparent that the potential for sediment transport in nearshore waters is large.

Douglas L. Inman and Scott A. Jenkins

Bibliography

Alekin, O.A., and Brazhnikova, L.V., 1961. The discharge of soluble matter from dry land of the earth. *Gidrokhimicheskie Materialy*, **32**: 12–24.

Allen, J., and Komar, P., 2000. Are ocean wave heights increasing in the eastern North Pacific. *EOS*, **81**: 561–567.

Bond, G., Showers, W., Cheseby, M., Lotti, R., Almasi, P., deMenocal, P., Priore, P., Cullen, H., Hajds, I., and Bonani, G., 1997. A pervasive millennial-scale cycle in North Atlantic Holocene and glacial climates. *Science*, **278**: 1257–1266.

Carter, D.J.T., and Draper, L., 1988. Has the north-east Atlantic become rougher? *Nature*, **332**: 494.

Cartwright, D.E., 1977. Oceanic tides. *Reports on Progress in Physics*, **40**: 665–708.

Chao, Y., Ghil, M., and McWilliams, J.C., 2000. Pacific interdecadal variability in this century's sea surface temperatures. *Geophysical Research Letters*, **27**: 2261–2264.

Cox, C., and Sandstrom, H., 1962. Coupling of internal and surface waves in water of variable depth. *Journal of Oceanographical Society of Japan*, 20th Anniversary Volume, 499–513.

Davis, R.E., 1985. Drifter observations of coastal surface currents during CODE: the method and descriptive view. *Journal Geophysical Research*, **90**: 4741–4755.

Dickey, J.O. *et al.*, 1994. Lunar laser ranging: a continuing legacy of the Apollo Program. *Science*, **265**: 482–490.

Dolan, R., Inman, D.L., and Hayden, B.P., 1990. The Atlantic coast storm of March 1989. *Journal of Coastal Research*, **6**: 721–725.

Emery, K.O., and Milliman, J.D., 1978. Suspended matter in surface waters: influence of river discharge and upwelling. *Sedimentology*, **25**: 125–140.

Fratantoni, D.M., Johns, W.E., and Townsend, T.L., 1995. Rings of the North Brazil Current: Their structure and behavior inferred from observations and a numerical simulation. *Journal of Geophysical Research*, **100**: 10-633–10-654.

Garrels, R.M., and MacKenzie, F.T., 1971. *Evolution of Sedimentary Rocks*. New York: W.W. Norton & Co.

Garrett, C., and Munk, W., 1979. Internal waves in the ocean. In Van Dyke, M., Wehausen, J.V., and Lumley, J.L., (eds.), *Annual Review of Fluid Mechanics*, Vol. 11. Palo Alto, CA: Annual Reviews Inc., pp. 339–369.

Gill, A.E., 1982. *Atmosphere–Ocean Dynamics*. Orlando, FL: Academic Press.

Gilluly, J., 1955. Geologic contrasts between continents and ocean basins. In Poldervaart, A. (ed.), *Crust of the Earth*. Geological Society America, Special Paper 62, pp. 7–18.

Gilluly, J., Reed, J.C., Jr., and Cady, W.M., 1970. Sedimentary volumes and their significance. *Geological Society of American Bulletin*, **81**: 353.

Graham, N.E., and Diaz, H.F., 2001. Evidence for intensification of North Pacific winter cyclones since 1948. *Bulletin of American Meteorological Society*, **82**: 1869–1893.

Hay, W.W., 1994. Pleistocene–Holocene fluxes are not the earth's norm. In *Material Fluxes on the Surface of the Earth*. Washington, DC: National Academy Press, pp. 15–24.

Heun, M., Schäfer-Pregl, R., Klawan, D., Castagna, R., Accerbi, M., Borghi, B., and Salamini, F., 1997. Site of einkorn wheat domestication identified by DNA fingerprinting. *Science*, **278**: 1312–1314.

Holeman, J.N., 1968. The sediment yield of major rivers of the world. *Water Resources Research*, **4**: 737–741.

Inman, D.L., 1973. Shore processes. In Vetter, R.C. (ed.), *Oceanography: The Last Frontier*. New York: Basic Books, Inc., pp. 317–338.

Inman, D.L., and Brush, B.M., 1973. The coastal challenge. *Science*, **181**: 20–32.

Inman, D.L., and Dolan, R., 1989. The Outer Banks of North Carolina: Budget of sediment and inlet dynamics along a migrating barrier system. *Journal of Coastal Research*, **5**: 193–237.

Inman, D.L., and Jenkins, S.A., 1997. Changing wave climate and littoral drift along the California coast. In Magoon, O.T. *et al.* (eds.), *California and the World Ocean '97*, Reston, VA: ASCE, pp. 538–549.

Inman, D.L., and Jenkins, S.A., 1999. Climate change and the episodicity of sediment flux of small California rivers. *Journal Geology*, **107**: 251–270.

Inman, D.L., and Nordstrom, C.E., 1971. On the tectonic and morphologic classification of coasts. *Journal of Geology*, **79**: 1–21.

Inman, D.L., Munk, W.H., and Balay, M., 1962. Spectra of low frequency ocean waves along the Argentine shelf. *Deep-Sea Research*, **8**: 155–164.

Jeffreys, H., 1920. Tidal friction in shallow seas. *Philosophical Transactions. Royal Society*, London, **221**: 239–264.

Jeffreys, H., 1970. *The Earth; Its Origin, History and Physical Constitution*. Cambridge: Cambridge University Press.

Klein, W.H., and Winston, J.S., 1958. Geographical frequency of troughs and ridges on mean 700 mb charts. *Monthly Weather Review*, **86**: 344–358.

Kushnir, Y., Cardone, V.J., Greenwood, J.G., and Cane, M.A., 1997. The recent increase in North Atlantic wave heights. *Journal of Climate*, **10**: 2107–2113.

Langbein, W.B., and Schumm, S.A., 1958. Yield of sediment in relation to mean annual precipitation. *Transaction of American Geophysical. Union*, **39**: 1076–1084.

Lyell, C., 1873. *Principles of Geology*, Vol. 1. New York: D. Appleton and Co.

Malkus, J.S., 1962. Interchange of properties between sea and air: Large-scale interactions. In Hill, M.N. (ed.), *The Sea, Ideas and Observations*, Vol. 1. NewYork: Interscience Publ., pp. 88–294.

Meade, R.H., 1996. River-sediment inputs to major deltas. In Milliman, J.D., and Haq, B.U. (eds.), *Sea-level Rise and Coastal Subsidence*. Dordrecht: Kluwer, pp. 63–85.

Menard, H.W., 1960. Possible pre-Pleistocene deep-sea fans off central California. *Geological Society America*, **71**: 1271–1278.

Miller, G.R., 1966. The flux of tidal energy out of the deep oceans. *Journal of Geophysical Research*, **71**: 2785–2789.

Milliman, J.D., and Syvitski, J.P.M., 1992. Geomorphic/tectonic control of sediment discharge to the ocean: The importance of small mountainous rivers. *Journal of Geology*, **100**: 525–544.

Munk, W.H., and MacDonald, G.J.F., 1960. *The Rotation of the Earth; a Geophysical Discussion*. London: Cambridge University Press.

Newell, R.E., Newell, N.E., Zhu, Y., and Scott, C., 1992. Tropospheric rivers?—A pilot study. *Geophysical Research Letters*, **12**: 2401–2404.

Pollack, H.N., Hurter, S.J., and Johnson, J.R., 1993. Heat flow from the earth's interior: Analysis of global data set. *Reviews of Geophysics*, **31**: 267–280.

Prospero, J.M., and Carlson, T.N., 1981. Saharan air outbreaks over the tropical North Atlantic. *Pure and Applied Geophysics*, **119**: 677–691.

Prospero, J.M., Ginoux, P., Torres, O., Nicholson, S.E., and Gill, T.E., 2002. Environmental characterization of global sources of atmospheric soil dust identified with the Nimbus 7 Total Ozone Mapping Spectrometer (TOMS) absorbing aerosol product. *Reviews of Geophysics*, **40**: 31 pp.

Pugh, D.T., 1987. *Tides, Surges and Mean Sea Level*. New York: John Wiley and Sons.

Rasmussen, D., 1996. State Mussel Watch Program, 1993–1995, Data Report 96-2WQ. Sacramento: State Water Resources Control Board, California Environmental Protection Agency.

Rattray, M., Dworski, J.G., and Kovala, P.E., 1969. Generation of long internal waves at the continental slope. *Deep-Sea Research*, **16**: 179–196.

Rea, D.K., 1994. The paleoclimatic record provided by eolian deposition in the deep sea: The geologic history of wind. *Reviews of Geophysics*, **32**: 159–195.
Richards, K., 1982. *Rivers, Form and Processes in Alluvial Channels*. London: Methuen.
Schmitz, W.J., Jr., 1995. On the interbasin-scale thermohaline circulation. *Reviews of Geophysics*, **33**: 151–173.
Shepard, F.P., 1963. *Submarine Geology*, 2nd edn, New York: Harper and Row.
Simkin, T., and Fiske, R.S., 1983. *Krakatau 1883: The Volcanic Eruption and Its Effects*. Washington, DC: Smithsonian Institute. Press.
Stommel, H., 1958. *The Gulf Stream, a Physical and Dynamical Description*. Berkeley: University of California Press.
Trenberth, K.E., and Hurrell, J.W., 1994. Decadal atmosphere–ocean variations in the Pacific. *Climate Dynamics*, **9**: 303–319.
Trimble, S.W., 1997. Contribution of stream channel erosion to sediment yield from an urbanizing watershed. *Science*, **278**: 1442–1444.
Turekian, K.K., 1971. Rivers, tributaries, and estuaries. In Hood, D.W., (ed.), *Impingement of Man on the Oceans*. New York: Wiley-Interscience, pp. 9–72.
UNEP, 1987. The state of the world environment. *United Nations Environmental Programme*, UNEP/GC.14/6, 76 pp.
Van Dorn, W.G., 1987. Tide gauge response to tsunamis. Part II: Other oceans and smaller seas. *Journal of Physical Oceanography*, **17**: 1507–1516.
Vitousek, P.M., Mooney, H.A., Lubchenco, J., and Melillo, M.M., 1997. Human domination of Earth's ecosystems. *Science*, **277**: 494–499.
Webster, P.J., 1994. The role of hydrological processes in ocean–atmosphere interactions. *Review of Geophysics*, **32**: 427–476.
Winant, C.D., 1974. Internal surges in coastal waters. *Journal of Geophysical Research*, **79**: 4523–4526.
Wunsch, C., and Hendry, R., 1972. Array measurements of the bottom boundary layer and internal wave field on the continental slope. *Geophysical Fluid Dynamics*, **4**: 101–145.
Yokoyama, I., 1978. The tsunami caused by the prehistoric eruption of Thera. In Doumas, C. (ed.), *Thera and the Aegean World I*. London: Thera and the Aegean World, pp. 277–283.

Cross-references

Accretion and Erosion Waves on Beaches
Beach Processes
Climate Patterns in the Coastal Zone
Coastal Currents
Databases
Erosion Processes
Global Warming (see Greenhouse Effect and Global Warming)
History, Coastal Geomorphology
Human Impact on Coasts
Littoral Cells
Longshore Sediment Transport
Meteorologic Effects on Coasts
Sediment Budget
Tsunami
Waves
Weathering in the Coastal Zone

ENGINEERING APPLICATIONS OF COASTAL GEOMORPHOLOGY

Coastal geomorphology

Literally, the title "geomorphology" comprises earth, shape and science with global aspects in that order. The study of earth science, from the strictly biblical concept of evolution to modern applied geomorphology, has been a long process in itself. In a generic term, geomorphology is indisputably a part of geology, but in institutional affiliation it has been a course in geography in almost all countries around the world. Coastal engineers have also contributed to its development in more recent time. Traditional geomorphology is concerned with the local landforms on the earth's surface and their processes over time. Coastal geomorphology, as a branch within the traditional geomorphology, is a relatively new discipline. It explores the relationship between coastal landforms and their processes affected by factors associated with climatology, oceanography, fluid mechanics, sedimentation, and geophysics.

Before the World War II, geomorphology—including coastal geomorphology—was largely Davisian, based on the concept of cyclic development and erosion, and of Johnsonian, in which the details in short-term coastal changes were ignored. Both approaches considered only the changes in vast stretches of time and space, and were once regarded as a "rather academic and, literally useless approach to coastal geomorphology" (e.g., Pethick, 1986, p. 235). Although these concepts had suffered a major change during the War, their influence can still be traced.

Beginning in the 1940s, detailed work and examples of applied coastal geomorphology have emerged. By the 1950s, the approach had become process-oriented. This advanced further, especially in the 1970s and 1980s. Since then, numerous reports have become available resulting from the work carried out by coastal engineers on various coastal problems around the world.

Between the 1950s and 1980s, another almost universal approach taken by coastal geomorphologists was to attempt a classification of coastal landforms—to recognize and describe coastal features. During this period, some major classifications on coasts were: (1) *submerged*, *emerged*, and *neutral* coasts due to past sea-level variations, (2) four regional types based on plate tectonic theory, (3) *primary* and *secondary* coasts based on the influence due to terrestrial agency or marine processes, (4) *high*, *moderate*, and *low-energy* beaches, (5) *east coast* and *west coast* environment depending on the relative location of the landmass, and (6) *storm* versus *swell* wave environments based on the level of wave energy inputs. However, applications of coastal geomorphology to stabilize eroding beaches were scarce.

Despite the diversity of these technical classifications, it is obvious that many physical features, in terms of the material and the types of physiographic units, can readily be identified along the ocean's edge. These include cliffs, rocky coasts, gravel beaches, sandy beaches, estuaries, deltas, lagoons, coral reefs, salt marshes, muddy fields, and mangrove swamps. For the sandy beaches, notable forms are beach cusps, salients and tombolos, spits, barrier islands, river mouth bars, straight beaches, and especially, curved or headland-bay beaches. These distinct landforms are produced by variable interactions due to the prevailing oceanographical and meteorological factors, as well as the geology of the physiographic units. Davies (1980) has demonstrated a global viewpoint between the orientation of a sandy beach and the direction of the persistent swell waves, resulting from the global wind system. Unfortunately, many basic textbooks in coastal geomorphology and engineering have continued to handle the beach processes at a local level without mentioning its global ramifications. Coastline curvature has been cited in some texts, but no further studies were made on its usage.

Geographers have been attracted by the similarity in the shapes and profiles of various landforms in order to find an appropriate mathematical expression to fit these shapes. This may be seen in the classical examples found in hill slopes, glacial valleys, and the longitudinal profiles of rivers. Researches on coastal landforms have produced a parabolic equation for equilibrium beach profiles (Dean, 1991) and a logarithmic spiral equation for the planform of headland-bay beaches (Yasso, 1965). The latter was widely accepted by geographers and engineers since its inception, until Hsu and his co-workers critically assessed it in the late 1980s. Summary of the development in bay beaches and its applications for shore protection can be found in Silvester and Hsu (1993, 1997).

Headland-bay beaches

The coastline along the boundary between the land and the sea may be either straight over a large length or curved in plan separated by headlands, natural or man-made. The curved beaches are aesthetically beautiful and much more stable, compared with their straight counterparts. They may be fully exposed to waves, partially sheltered or fully protected by natural headlands or engineering structures.

General characters of headland-bay beaches

Of the curved beaches, some may be curved gently, others more indented, or with secondary features along its length. This unique geomorphic feature is ubiquitous not only on oceanic margins, but also on coasts of enclosed seas, lakes, and river foreshores. The bay beaches formed between natural headlands on the coasts can be easily identified on nautical charts, maps, aerial photographs, remote-sensing images, and even in travel brochures. They also appear downcoast of groins, and as salient or tombolo in the lee of protruding headlands, detached (offshore) breakwaters, and breakwaters for harbors. These curved beaches are found in various shapes and sizes, in relation to the prevailing waves and natural outcrops or headlands. The curved planform is produced by the persistent swell waves diffracted from the tip of a headland, combined with wave refraction and a nearshore current circulation system in the lee of the headland.

The curved or embayed beaches have attracted numerous names (Silvester and Hsu, 1993), such as zeta curved bays, half-heart shaped bays, crenulate shaped beaches, logarithmic spiral beaches, curved or hooked beaches, headland-bay beaches, pocket beaches, and offset coasts. Despite their richness in existence and variety in names, stability of bay beaches was not fully tested by geographers and coastal engineers.

Since sandy beaches are constructed by persistent swell waves, their high-tide shoreline orientations are expected to align with the crest lines of these waves. Consequently, a series of asymmetrically curved bays can be found in many places. For example, they can be observed from Sydney to Fraser Island in eastern Australia, from Ilha de Santa Catarina to Rio de Janeiro in Brazil, and in other places. In these regions, the zeta shaped beaches have curved southern ends where they display a lower berm with finer sand particle and less steep beach faces than their straighter and more exposed northern ends. The consistency of coastline indentation and orientation can also be found in other continents around the world. It has taken some decades for this early recognition of bay shapes to become useful for stabilizing beaches in erosive conditions (Silvester, 1979).

Silvester (1960) has questioned whether a bay beach is a stable physiographic feature. He believed that the orientation of the bay beaches is nature's method of balancing wave energy and load of sediment transport. In this manner, beaches have been maintained in position for tens, or even hundreds of years. The orientation of these bays also serves as an excellent indicator of the direction of net sediment movement along the coast. This macroscopic view of the coast imposed by nature is an integrated part of the coastal geomorphology, in addition to the microscopic view of wave kinematics and sediment movements in the marine environment.

Before the stable bay shapes can be applied for useful purpose, its planform has to be quantified and its stability assessed. In geomorphic term, the stability of a bay beach (Figure E6) may be in *static equilibrium* or *dynamic equilibrium*, or even *unstable*. A bay beach is: (1) stable and in *static equilibrium*, if the littoral drift is negligible or supply from upcoast is nonexistent, or (2) stable but in *dynamic equilibrium*, when the littoral drift is still being supplied from upcoast and/or from sources within the bay, or (3) *unstable*, implying potential retreat due to imbalance of sediment input and output. While sand supplied to the system is balanced by that which can be removed by oblique waves, a bay beach can remain in place for decades in dynamic equilibrium. However, this is a rare case today. On the other hand, should supply decrease, due to various reasons to be discussed later in "Causes of man-made erosion," the high-tide shoreline will recede toward the static equilibrium shape, which is the final limit without the impediment of shoreline protection structures.

Three major mathematical expressions have been proposed to fit the planform of bay beaches since 1940s. They are: (1) the logarithmic spiral equation suggested by Krumbein (1944) and quantified by Yasso (1965), (2) the parabolic equation for bays in static equilibrium given by Hsu and Evans (1989) and later reassessed by Silvester and Hsu (1993), and (3) the hyperbolic-tangent equation suggested by Moreno and Kraus (1999). These empirically derived equations have different coordinates systems and origins, and controlling parameters related to bay orientations.

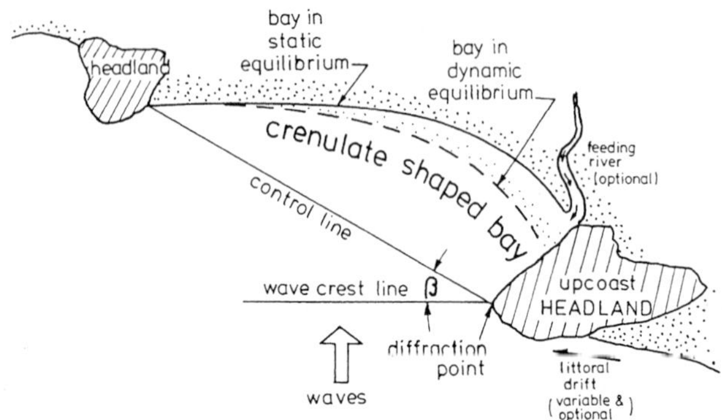

Figure E6 Headland bay beaches in dynamic and static equilibrium.

With a computer, all three methods can be applied to fit the planform of any bay beach (Moreno and Kraus, 1999). But there are some important differences, especially, in the recognition of the point of wave diffraction, the location of the coordinates origin, and beach stability requirement. The log-spiral and hyperbolic-tangent shapes have some drawbacks. First, their coordinate origins do not coincide with the tip of the headland (e.g., the point of wave diffraction). Second, these methods do not address the issue of beach stability or their equations were not produced from bays known in static equilibrium. Third, these two methods do not provide the resultant beach parameters, such as radii and angles, to ease potential applications. As such the log-spiral and hyperbolic-tangent forms cannot determine the effect of sheltering and future extension of any existing headland, or to predict the environmental effect of a new headland or structure on a bay beach.

Parabolic equation for bay beaches in static equilibrium

Hsu and Evans (1989) developed a parabolic equation to fit the peripheries of 27 cases of bay beaches believed to be in static equilibrium in prototype and model conditions. This equation is more preferred than the log-spiral and hyperbolic-tangent equations for the reasons mentioned above. The equation is reproduced here for further discussion:

$$R/R_0 = C_0 + C_1(\beta/\theta) + C_2(\beta/\theta)^2 \qquad \text{(Eq. 1)}$$

There are two independent parameters in equation 1. The first is the *reference wave obliquity angle* β, or the angle between the incident wave crest (assumed linear) and the *control line*, joining the upcoast diffraction point to the near straight downcoast beach. The second basic parameter is the *control line* of length R_0, which is also angled β to the tangent at this downcoast beach end. The radius R to any beach point around the bay periphery is angled at θ from the same wave crest line radiating out from the point of wave diffraction. The three C constants that vary with angle β were generated by regression analysis to fit the peripheries of the 27 data bays used. Examples of bay shape verification can be found in Hsu and Evans (1989) and Silvester and Hsu (1993, 1997). Flexibility in choosing the downcoast control point is a merit of this empirical approach.

Despite having variable dimensions and indentations between the tip of the headland and the downcoast sandy boundary, the stability of a bay beach can now be determined using this parabolic bay shape equation. Values of the non-dimensional ratios of R/R_0 versus increments of 1° or 2° of β from 20° to 80° can be tabulated. This facilitates potential applications to locate the pairs of (R, θ) manually without the need of a computer, once β and R_0 are determined from maps or aerial photographs. When points of (R, θ) are drawn on the plan of the existing beach, the nearness of the existing bay shape to that predicted for static equilibrium can be compared and the beach stability assessed. If the predicted curve, for a hypothetical static equilibrium, is landward of the existing beach, the bay beach is said to be in dynamic equilibrium, or it may become unstable, for diminishing sediment supply.

Beach erosion and conventional protection methods

Continuous conflict has taken place between humans and nature over the narrow belt directly landward of the coastline of the world, where human pressure has been extensive. Within the flat ribbons of sand, between the sea and the hinterland, there are numerous problems of coastal flooding and beach erosion.

Causes of human-induced erosion

The phenomenon of coastline changes are often complicated by oceanographic and meteorological effects on the inherent geological factors (form and material). With swell waves, nature helps move sediment landward to build up the berm and dune, while storm waves transport a large quantity of material to form a protective bar, causing beach disappearance. If waves approach a beach obliquely, a longshore current is induced, which promotes longshore sediment transport, or *littoral drift*. Compared with other landforms on the earth, which have evolved over a vast stretch of time and space, changes in high-tide shoreline positions and variations in beach profile may take place in a matter of hours. It is within this relatively rapid timescale in topographical change that has attracted the attention of many geographers and coastal engineers.

Coastline changes may be produced naturally or human-induced. Natural causes to beach erosion may include wave obliquity, storm attach, wind blown sand, existence of natural headlands, imbalance of sediment transport due to various topographical features, and sea-level rise. Although beach erosion has occurred naturally in many places, the problem is more serious due to the activities of people on the narrow coastal strips locally or elsewhere. Overall, the major causes of human-induced erosion may due to: (1) disturbance to continuity of littoral drift, (2) wave sheltering by coastal structures, (3) offshore movement of sediment, (4) reduction in sediment supply from upcoast, (5) dredging or extraction of sand and gravel for construction, and (6) land subsidence (Uda, 1997; Hsu *et al.*, 2000). Among these, human intervention in and interception of sediment supply have been considered two of the most detrimental factors causing beach erosion in many countries around the world. However, some sources of erosion may be originated from a distant place away from the destination—the beach in erosion. For example, damming a river (for flood control, water supply, or power generation) reduces sediment supply to the beach. More recently, breakwater construction for fishing harbors and commercial ports have also caused beach erosion downcoast, with harbor planners and many consulting engineers responsible for the problems they have caused without considering the geomorphic effect on the beach.

Conventional coastline protection methods

The family of seawalls (including bulkheads, revetments, sea dikes, and normal seawalls) were the very first structures ever constructed on a coast for protection against beach erosion and coastal flooding. They are built parallel to the beach at variable locations in relation to the local beach and surf. Their faces fronting the wave action may be vertical or inclined, with or without armor protection. They may even incorporate wave dissipation blocks near and in front of their toes, to reduce wave energy that causes scouring to the bottom at the toe. Beach flanking immediately downcoast of a long seawall is also common, for which some protection is still required.

Later groins were used to intercept littoral drift at places where erosion had occurred. They usually run normal to the beach and may reach out to the limit of the surf zone. Their length and height vary depending on the purpose they meant to serve. This type of structure has been widely used in many countries. Modern groins may be slightly curved, inclined to the local shore, or even in L-, T-, and Y-shaped or fishtailed. A bay beach may form downcoast of a protruding groin or between two units, where the tip of the upcoast unit serves as the point of diffraction, or upcoast control point for the bay beach so formed.

After groins, construction of detached breakwaters has become popular worldwide. They are either single or mostly in groups, with sandy salient or tombolo formed in their lee. In Japan alone, there were 7,371 units of such structures around the country by 1996, totaling 837 km (MOC, 1997; Hsu *et al.*, 2000). However, a new trend in Japan has been to replace them with submerged structures or artificial reefs, for aesthetic reasons.

More recently, beach nourishment has become the most "cost-effective" alternative for coastal protection and creating recreational beaches, especially in the United States (Bird, 1996). Often the beach fills were not protected by suitable structures, so replenishments were required at relatively frequent intervals. It has been envisaged that the cost of maintaining a wide beach would increase due to transporting the material from a distant source. In some cases, beach compartmentation was implemented.

Engineering applications of coastal geomorphology

A trend has recently emerged in applying the principle of coastal geomorphology, especially the concept of stable bay shape, for beach protection. Examples showing a range of other engineering applications can also be found in the literature (Silvester and Hsu, 1997).

Stability verification for existing bay beaches

With the parabolic bay shape equation (Hsu and Evans, 1989), it is now possible to verify the stability of an existing sandy bay beach. The general procedure is as follows (refer to equation 1, with Figures E6 and E7):

1. Draw a tangent from a point (referred to as the *downcoast control point*) on the straight downcoast section of a bay beach, and relocate this tangent to the point of wave diffraction (the *upcoast control point*), thus representing the *wave crest line* in the definition sketch (Figure E7),
2. Denote the *control line*, by joining the upcoast and downcoast control points, assign the distance between them as R_0; also measure the reference obliquity angle β,

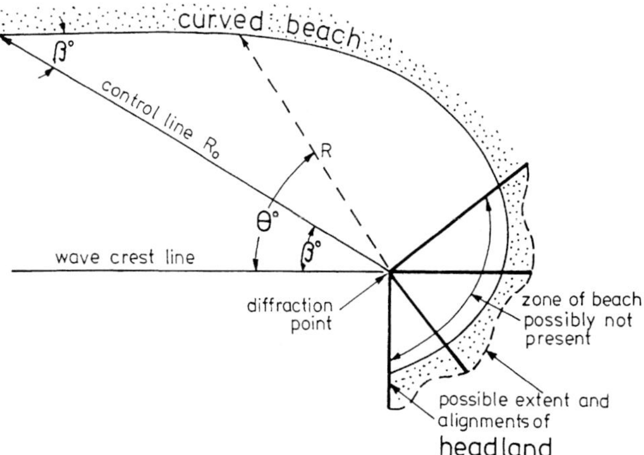

Figure E7 Definition sketch for a parabolic bay shape in static equilibrium.

3. Calculate the values of radii R for a range of corresponding angle θ (for $\theta > \beta$), using equation 1 or a prepared table of R/R_0, so to give sufficient pairs of (R, θ) on the beach,
4. Sketch the predicted bay shape in static equilibrium by joining the points (R, θ) drawn on the existing beach planform,
5. Determine the stability of the existing bay beach by its nearness to the curve given by the prediction.

Should the existing bay shape conform well to the predicted bay shape in static equilibrium, the beach is said to be in *static equilibrium* under the prevailing swell condition. On the other hand, if the existing shape is seaward of the predicted curve, the beach is in dynamic condition, implying either in *dynamic equilibrium*, when sediment balance is maintained, or *unstable*, should sediment supply decrease. However, it must be remembered that the stable high-tide shoreline being predicted applies to swell condition or calm weather. When storm waves arrive, the beach berm will be eroded and material transported offshore to form a submerged bar, which may take some days or longer to return. In the case of static equilibrium, the same high-tide shoreline may result later, but all infrastructures should be kept landward of the receded storm limit.

The stage of bay beach stability may be switched between static and dynamic equilibrium. It can be either from a dynamic to another dynamic condition, or from a dynamic to static, or from static to a dynamic condition, depending on the initial condition of the bay beaches and variability in sediment balance and construction or extension of breakwater (Hsu and Silvester, 1996). This concept of interchange ability for beach stability is extremely useful for practical applications. For example, it can be applied to: (1) the environmental impact study upon constructing or extending a jetty or breakwater within a bay beach in static equilibrium, and (2) the stabilization of a bay beach in eroding condition by locating the tip of a new headland to be installed at a proper location. In these situations, a stable bay beach in static equilibrium is created, thus minimizing the need for protective structures to be constructed downcoast later on.

Any new attempt to construct a jetty or breakwater for a marina or harbor at or within a bay beach already in static equilibrium will result in accretion in the lee at the expense of erosion downcoast, unless certain preventive measures are in place. This natural reshaping will take place within the entire embayment upon the construction of a new headland, despite some modest attempts with seawalls, groins, and detached breakwaters to prevent its fruition. On the other hand, should an engineering task be to construct a breakwater within a bay in dynamic equilibrium; the beach can conveniently be converted into static equilibrium by choosing a proper location for the tip of the new structure. This geomorphic approach of preventing beach erosion should benefit the community as a whole, in addition to the traditional engineering need for providing calm waves within a harbor.

Stabilizing eroding beaches

To combat widespread beach erosion, conventional methods have been used as mentioned in the previous section. Instead of constructing hard structures to defend the eroding beaches, it would be better to create a coastal environment in which the beach can protect itself by the formation of bays to meet the condition of diminishing littoral drift. This requires the reorientation of local high-tide shorelines to produce minimal or null littoral drift within the embayment. The concept is termed *headland control*, in which artificial headlands are used to help reorient the high-tide shoreline parallel to the crest lines of the prevailing swell waves. The concept can be applied to stabilize all shapes of eroding high-tide shorelines, be they straight, bayed, or convex in plan (Silvester and Hsu, 1993, 1997). The artificial headlands may be either one of the conventional protective structures of groins and detached breakwaters with large spacing, or in combination. This can also be used to help retain beach fill staying longer on a renourished beach.

On a straight coast, a system of groins and/or detached breakwaters at close intervals has been used in many countries around the world. The concept of headland control can be applied to modify the conventional protection scheme already in place by removing some units thus widening the distance between them, or to install new headlands if protection did not exist. However, the most important difference, in the headlands used in the conventional defense and the present *headland control*, is in the distance between two adjacent units and their relative orientation to the coast. For a stable beach to form in the vicinity of these artificial headlands the space between two units must be sufficiently large, often to several wave lengths, thus allowing a beach so formed to dissipate wave energy naturally under swell wave condition. The new beach should also have sufficient width and volume of material to satisfy the need of a design storm condition. Flexibility in varying the length, shape, distance offshore, and orientation of each unit would help optimize the performance of a stabilization scheme.

For a bay beach suffering erosion, due to reduction in supply upcoast and new breakwater construction for a harbor within an embayment, adding artificial headlands at appropriate locations can create several smaller bay beaches in static condition. The planform of these stable bays in static equilibrium can be sketched after simple calculations using equation 1, or aided by a prepared table of R/R_0.

On an eroding convex coast, such as a barrier beach or spit, its narrow section may be overwashed or breached by storm waves accompanied by storm surge. Most barrier beaches have retreated in the past decades due to the supply from upcoast being drastically reduced. Attention should be given to a geomorphic solution rather than by hard structure alone. A series of stable bay beaches is preferred to a system of groins. The headland control concept allows the units to be installed with variable length, inclination to the coast, distance offshore, and spacing between units. The distance offshore and space between units should be gradually reduced in a cascade manner, thus allowing the local coastal processes to approach natural conditions. The reference wave obliquity angle β used in equation 1 for the successive bays formed between headlands should vary, being greater at the farther downcoast, where closer spacing is also desirable.

Salient planform behind single detached breakwater

Waves diffract and refract in the lee of a detached breakwater and produce a nearshore current circulation system extending beyond the region of the shadow zone covered by wave diffraction. Consequently, sand accumulation in the form of salient or tombolo may occur behind the breakwater, depending mainly on the material available, the length of breakwater and its distance offshore from the original high-tide shoreline. Hsu and Silvester (1990) have demonstrated that the distance of the salient apex from the breakwater is preferred to that from the original high-tide shoreline, since the former relates more to the wave energy input and the effect of wave diffraction. Previous studies using the latter have resulted in unacceptable scatter when plotting the salient dimension versus the ratio of breakwater length and distance offshore.

They also illustrated that the curved beaches so formed can be related to the bay shape in static equilibrium, for which equation 1 can be applied. The analysis is limited to the case of normal wave incidence. The accretion assumes a parabolic form to a theoretical apex whose position can be determined using the dimensionless ratio of offshore distance to the length of the breakwater, for which a table has been available (Silvester and Hsu, 1997). This accretion is at the expense of the material beyond the extremities of the detached breakwater, and hence erosion at beaches downcoast. The final waterline at some distance either side of the breakwater will recede landward, implying beach erosion in this vicinity. However, to the knowledge of this author, this distance between the original high-tide shoreline and the receding one has not been fully quantified whether in laboratory experiments or in field conditions.

In the case of a salient formed in a multiple detached breakwater system, an accurate prediction method is not available. An attempt was made to analyze the planform using the same dimensionless ratios, as for single detached breakwater, and the gap width between units. However, the correlation based on the field data of the salients and tombolos collected in Japan did not yield reliable results for practical applications (Hsu and Silvester, 1989).

Stabilizing beaches downcoast of harbors

Many new harbors of variable capacities were constructed on coasts where no natural shelter is available, such as coastal indentation, offshore islands, headlands, capes, barrier spits, or reefs. This requires an initial breakwater to be built and possibly some extension over time as further commercial expansion demanding a large port, hence longer and larger structures. Even until recent time, harbor engineers and planners have considered that the construction of a harbor has priority over coastline evolution downcoast. They have often ignored the principle of geomorphology that would cause beach accretion in the lee of offshore breakwaters and erosion downcoast. Coastal scientists in Greece, Italy, Japan, and the United Sates have reported adverse effects on downcoast beaches (Silvester and Hsu, 1997).

Similar to the sand accumulation behind single detached breakwaters, salient forms in the lee of a harbor breakwater, inevitably cause beach erosion downcoast and a possible silting of the refuge basin at the entrance. Compared with the case of single detached breakwater, a harbor breakwater does not have a clear middle point along its longitudinal axis. With the need for a breakwater to be extended over time due to commercial demand for a larger port, it will produce a larger salient in the lee, and hence erosion further downcoast (Hsu *et al.*, 1993).

Hsu and Silvester (1996, p. 3986) have examined the situation of stabilizing beaches downcoast of harbor extensions. They believe that

> In geomorphologic terms, the sandy shoreline of a bay downcoast of a harbor may be stable or in static equilibrium, or could be in dynamic condition, if sediment is still being supplied from upcoast or from downcoast to form a salient predicted by a static bay shape equation. Should commercial expansion demand a larger port, the general solution is to run a breakwater form the headland or existing structure. This has the potential to create a new static equilibrium beach, often with accretion in the lee at the expense of beach erosion downcoast. It is strongly recommended that geomorphic approach be incorporated to stabilize downcoast beaches early in the planning stage of a harbor, or as remedial measures. By creating bay beaches in static equilibrium, the potential beach erosion downcoast of a harbor will be kept to a minimum or may be prevented completely (Hsu and Silvester, 1996).

Thus, instead of employing the conventional coastal protection method, in which seawalls, groins, and detached breakwaters were used without much success after the erosion downcoast has become serious, bay beaches of variable sizes should be considered to stabilize the eroding beaches. The geomorphic approach, involving the construction of stable bay beaches aided with beach nourishment, can be regarded as a permanent solution for this occasion. With the macroscopic view of the whole coastline rather than looking at a limited section, a better picture of the erosive problem will become clear. As such, erosion can be predicted long before it is likely to take place, and hence means can be devised to prevent it. Case studies have been reported for harbors in Australia, Japan, Taiwan, and South Africa (Hsu and Silvester, 1996).

Designing recreational beaches

With the economic development around the world there is a community movement demanding a better coastal living environment and full utilization of the coastal zone. As a result, many artificial beaches have been constructed for recreation and tourism, as well as for the restoration of eroding beaches.

Silvester and Hsu (1997) have detailed the requirement for the creation of a stable bay shape for artificial and recreational beaches. One basic requirement is to construct suitable protective structures for preventing sediment loss from the artificial beaches and for providing a calm wave environment. Detached breakwaters with beach nourishment and/or groins have been used in all the cases so far reported in the literature.

Around the mid 1980s, two major Japanese ministries responsible for the preservation and protection of the coast against natural and man-made disasters launched two similar beach protection schemes incorporating hard structures and artificial beaches. These were the coastal community zone (CCZ) projects administered by the Ministry of Construction and the integrated shore protection system (ISPS) under the Ministry of Transport (Hsu *et al.*, 2000).

Most of the beaches so designed are either partially or fully enclosed cells behind offshore breakwaters or between them, hence the term *pocket beaches*. Berenguer and Enriquez (1988, p. 1412) have observed that: "In the recent years, the increase of pocket beaches has been remarkable. In Mediterranean countries where tidal effects are insignificant, especially in Spain, Italy, France, and Israel, this system of beach regeneration has been widely used because of its rapid results and its economical nature. Japan also has a lot of experience in this kind of work. In Spain, there are over 40 beaches of this kind. They are usually situated on stretches of coastline where there is a lot of tourism and in which there is also a fairly rapid erosional process ..."

Coastal management and planning

People have used public beaches for various purposes. But, many beaches around the world are still being subjected to erosion, due to improper coastal practices, human greediness and local political pressure for the convenient use on the narrow coastal strips. Coastal protection has also arisen from improper coastal zoning for residential and public development too close to the edge of the ocean, without knowing the stability of the beach involved. Many coastal communities have paid an expensive price through this experience of combating with the recurrent erosion problem. Another example is that landscape architects and consultants, who may not be familiar with the principle of coastal geomorphology, are engaged by an authority to beautify the coastal strips for tourism. Some of the additional facilities so created may fall prey to a storm, so the taxpayer's money is wasted.

It is essential in modern coastal management that beach managers in coastal communities be able to apply the principle of coastal geomorphology, and "to examine such questions as where the beach sediment came from, whether it is still coming, and whether the beach is stable, or shows long-term accretions or erosion, as distinct from short-term cut and fill or seasonal alternations." (Bird, 1996).

For any development proposal on a bay beach, the stability of the existing beach should be investigated, prior to giving permission for residential development or structures to be built close to the water. The concept of the stable bay shape of Hsu and Evans (1989) is readily applicable to such cases, without the need of engaging in an expensive and time-consuming exercise, such as physical modeling and computer simulations. If the bay beach is in a dynamic condition, residential development and infrastructures should not be allowed seaward of the predicted static bay limit, plus a little buffer for a storm. Even if the bay is already in static equilibrium, a buffer distance is also required for a storm to act in the future. This would save the need for costly protective measures to be built later. A full investigation may then be carried out only after the preliminary desktop study using the principle of coastal geomorphology, if an elaborate exercise is still warranted.

Geomorphic engineering: a future perspective

Coates (1980) has designated a specific term of *geomorphic engineering* to combine the fields of geology, engineering, and environmental science together, which used to work independently on coasts. It is also important that biological processes in coastal geomorphology be considered. The traditional single task of coastal protection should gradually be replaced by the preservation of the coastal environment, by creating stable and safe beaches to coexist with nature. It is in such a manner that a careful balance among the conflicting requirements of protection, development, and conservation can be met, with planning and management using coastal geomorphology fulfilling the central role.

John R.C. Hsu

Bibliography

Berenguer, J.M., and Enriquez, J., 1988. Design of pocket beaches: the Spanish cases. *Proceedings, 21st International Conference on Coastal Engineering*. American Society of Civil Engineers, 2, pp. 1411–1425.

Bird, E.C.F., 1996. *Beach Management*. Chichester: John Wiley & Sons.

Coates, D.R., 1980. Geomorphic engineering. In Coates, D.R. (ed.), *Geomorphology and Engineering*. London: George Allen & Unwin, pp. 3–21.

Davies, J.L., 1980. *Geographical Variation in Coastal Development*, 2nd edn. London: Longman.

Dean, R.G., 1991. Equilibrium beach profiles: characteristics and applications. *Journal of Coastal Research*, **7**(1): 53–84.

Hsu, J.R.C., and Evans, C., 1989. Parabolic bay shapes and applications. *Proceedings Institution of Civil Engineers*, Part 2. London: Thomas Telford, 87, pp. 557–570.

Hsu, J.R.C., and Silvester, R., 1989. Salients leeward of multiple offshore structures. *Proceeding, 9th Australasian Conference on Coastal and Ocean Engineering*, pp. 347–351.

Hsu, J.R.C., and Silvester, R., 1990. Accretion behind single offshore breakwaters. *Journal of Waterway, Port, Coastal and Ocean Engineering*. American Society of Civil Engineers, **116**(3): 362–380.

Hsu, J.R.C., and Silvester, R., 1996. Stabilizing beaches downcoast of harbor extension. *Proceedings, 25th International Conference on Coastal Engineering*. American Society of Civil Engineers, 4, pp. 3986–3999.

Hsu, J.R.C., Uda, T., and Silvester, R., 1993. Beaches downcoast of harbours in bays. *Coastal Engineering*, **19**(1,2): 163–181.

Hsu, J.R.C., Uda, T., and Silvester, R., 2000. Shoreline protection methods—Japanese experience. In Herbich, J.B. (ed.), *Handbook of Coastal Engineering*. New York: McGraw-Hill, pp. 9.1–9.77.

Krumbein, W.C., 1944. Shore processes and beach characteristics. Washington, D.C.: Beach Erosion Board, US Army Corps of Engineers, Technical Memorandum No. 3.

MOC (Ministry of Construction) 1997. *1997 Coastal Handbook*. Tokyo: National Coast Association (In Japanese).

Moreno, L.J., and Kraus, N.C., 1999. Equilibrium shape of headland-bay beaches for engineering design. *Proceedings, Coastal Sediments '99*. American Society of Civil Engineers, 1, pp. 860–875.

Pethick, J., 1986. *An Introduction to Coastal Geomorphology*. London: Edward Arnold.

Silvester, R., 1960. Stabilization of sedimentary coastlines. *Nature*, **188**: 467–469, Paper 4749.

Silvester, R., 1979. A new look at beach erosion control. *Annual*, Disaster Prevention Research Institute, Kyoto University, **22**(A), pp. 19–31.

Silvester, R., and Hsu, J.R.C., 1993. *Coastal Stabilization: Innovative Concepts*. Englewood Cliffs, NJ: Prentice-Hall.

Silvester, R., and Hsu, J.R.C., 1997. *Coastal Stabilization*. Singapore: World Scientific, (Reprint of Silvester and Hsu, 1993).

Uda, T., 1997. *Beach Erosion in Japan*. Tokyo: Sankai Do (In Japanese).

Yasso, W.E., 1965. Plan geometry of headland bay beaches. *Journal of Geology*, **73**: 702–714.

Cross-references

Artificial Islands
Beach Erosion
Beach Features
Beach Nourishment
Coastal Changes
Coastal Processes (see Beach Processes)
Dynamic Equilibrium of Beaches
Headland-Bay Beach
History, Coastal Geomorphology
Human Impact on Coasts
Sandy Coasts
Shore Protection Structures

ENVIRONMENTAL QUALITY

Introduction

Marine environmental quality has been defined by Harding (1992, p. 23) as "the condition of a particular marine environment (coastline, estuary, bay, harbor, nearshore and offshore waters, open ocean) measured in relation to each of its intended uses. It is usually assessed quantitatively and requires both indices of condition and change, and established guidelines and objectives set by environmental, health, and resource agencies." The assessment of coastal marine environmental quality typically considers the intrinsic value of the environment as well as whether the environment supports certain intended uses such as swimming, shellfish harvesting, or mariculture. As noted by Harding (1992), it is necessary to investigate ecosystem attributes (e.g., primary productivity, biomass production, nutrient cycling, species diversity, and disease incidence) to accurately measure ecosystem changes due to multiple stresses, particularly those associated with pollution and physical habitat disturbances, which can greatly impact environmental quality of coastal systems. It is important not only to determine that ecological change has occurred by evaluating structural characteristics but also to understand the processes forging change in the state of ecosystems (e.g., eutrophication).

The coastal zone is a transition area encompassing both terrestrial and marine components where the interaction between the land and sea is both intense and dynamic (Sorensen and McCreary, 1990). Investigations of environmental quality in this critically important zone, therefore, must consider estuarine and coastal marine waters together with neighboring watersheds, which, are closely coupled to them. Because more than half of the world's human population resides within ~60 km of the coast (Goldberg, 1994; Alongi, 1998), many valuable coastal habitats (e.g., wetlands, estuaries, beaches, and the nearshore ocean) are threatened by escalating land development, urbanization, industrial expansion, and maritime activities. More specifically, contaminant inputs from point and nonpoint pollution sources, as well as the physical alteration and destruction of habitat due to shipping, marine transportation, marine mining, energy production (oil and gas recovery, and electric power generation), fisheries exploitation, aquaculture, coastline development, and other types of human intervention, have degraded environmental quality of the coastal zone. Overuse or misuse of coastal resources, such as overfishing and excessive marine mining, can severely deplete resources beyond the limits of sustainability. Poorly conceived development control programs and inadequate resource management have contributed greatly to many of the acute and insidious environmental problems observed in this zone.

Deficient fisheries management provides an example. Strong evidence exists that the annual global fisheries catch of more than 80 million mt is approaching the maximum production obtainable; ~90% of this catch derives from coastal areas (Barcena, 1992). In addition, 35% of the 200 major fish stocks worldwide are currently classified as overfished. For US fish stocks where sufficient data have been compiled, more than 40% are overutilized (Sissenwine and Rosenberg, 1993). Overfishing reduces fisheries production as well as genetic diversity and ecological resilience (Costanza *et al.*, 1998).

Many coastal habitats worldwide are in various degrees of degradation because of heavy human exploitation (Kennish, 1997; Alongi, 1998). Estuaries, beaches, and nearshore oceanic environments support multibillion dollar commercial and recreational activities. Shore tourism ranks among the most profitable industries in the United States. Estuarine and coastal marine fisheries contribute more than $20 billion annually to the US economy, and they supply employment for more than a million people (Pirie, 1996; Kennish, 2000). Most US commerce (98% by volume) is transported by water, with much of the goods passing through estuarine and marine waters (Kennish, 2000). Marine oil and gas production amounts to ~$16 billion annually (Pirie, 1996). The mining of sand, gravel, salt, phosphorite, and tin from beaches and nearshore marine substrates also supports multibillion dollar industries (Cronan, 2000; Pinet, 2000).

Stemming directly from human dependence on and usage of coastal resources and an often cavalier concern for conservation and long-term sustainability, shallow water estuarine and marine ecosystems—particularly in urbanized regions—are susceptible to multiple stressors (Yap, 1992; Kennish, 1997). While many adverse effects on these ecosystems are linked to development along the coast, activities in watersheds far removed from the coastal zone (e.g., dam construction, stream channelization, and careless land-use practices) can also be detrimental (McIntyre, 1992). Some local impacts may even originate from global sources, as in the case of certain widely distributed chemical contaminants (e.g., PCBs, lead, and radionuclides) delivered via atmospheric deposition. However, a global approach to maintaining marine environmental quality has not been effectively formulated. Currently, a patchwork of international and regional conventions, protocols, agreements, national laws, regulations, and guidelines exists to protect marine environments, although a comprehensive global marine environmental management framework is critically needed to preserve the integrity of marine systems (Côté, 1992). Hence, local and regional environmental management programs must be supplemented by an adequate global strategy to ensure long-term protection of the coastal zone (Nollkaemper, 1992). Aside from dealing with issues of science, this strategy must involve socio-economic policy, improved technologies, new educational initiatives, and international cooperation (Boehlens, 1992).

This contribution examines the environmental quality of coastal environments, assessing conditions in estuarine and nearshore marine waters and in the watersheds and airsheds that affect them. Environmental problems commonly encountered in the coastal zone are discussed along with the management programs instituted to remediate them. Furthermore, various measures of marine environmental quality are investigated to provide an analysis of coastal ecosystem health assessment.

Coastal environmental problems

There are multiple stressors in the coastal zone associated with anthropogenic activities that potentially influence environmental quality. These stressors can be divided into two major categories: (1) pollution, and (2) habitat and resource degradation. A vast array of industrial, municipal, and domestic pollutants enters coastal systems from land-based, atmospheric, and aquatic sources. The greatest concentrations of contaminants occur in urbanized and industrialized systems, notably those exhibiting "hot-spot" conditions (e.g., Boston Harbor, New York Bight Apex, Baltimore Harbor, and Puget Sound) (Kennish, 1992, 1997). Pollution is a serious concern because it degrades coastal water quality, causes both acute and insidious impacts on estuarine and marine organisms (e.g., behavioral, genetic, neurological, and reproductive abnormalities; debilitating diseases; and death), exposes humans to increased health risks, and results in diminished human use of coastal resources. Biotic communities inhabiting polluted environments often experience dramatic shifts in species composition, abundance, diversity, and dominance (Pearson and Rosenberg, 1978; Warwick *et al.*, 1987; Simboura *et al.*, 1995).

Many potentially damaging human activities extend beyond the introduction of pollutants to estuarine and coastal marine waters. Among the most significant are the physical degradation of habitats and overexploitation of living and nonliving resources. The draining of wetlands for land reclamation; siting of resorts, hotels, and marinas for purposes of tourism and recreation; and building of coastline industrial installations and harbors for marine transportation and commercial operations destroy many hectares of natural habitat each year. The loss and alteration of coastal habitats typically reduce vital spawning, nursery, or forage grounds for numerous organisms, and diminish the utility of these areas as sources of food and shelter (Yap, 1992; Kennish, 1997, 2000). When these impacts persist for a considerable period of time over extensive areas, significant changes in biotic communities usually arise, such as reduced species diversity and population abundance.

Overexploitation of living resources, where the rate of harvesting of a target species exceeds its reproductive capacity, has been responsible for the local collapse of fisheries in many regions of the world (Pinet, 2000). Overfishing is resistant to traditional resource management approaches, some of which are based on considerable uncertainty and may be ineffective or inappropriate (Costanza et al., 1998). In regard to nonliving resources, sand and gravel mining and other coastal mineral mining operations adversely affect benthic habitats and communities. In addition to disturbing the benthos, the mining of sand in nearshore areas can impact fisheries, deplete sand needed for beach replenishment, or exacerbate shore erosion.

Pollution

The Joint Group of Experts on the Scientific Aspects of Marine Pollution (GESAMP, 1982), an advisory body to the heads of eight organizations of the United Nations (i.e., UN, UNEP, UNESCO, FAO, IMO, IAEA, WHO, and WMO), has defined marine pollution as "the introduction by man, directly or indirectly, of substances or energy into the environment, resulting in such deleterious effects as harm to living resources, hazards to human health, hindrance to marine activities including fisheries, impairment of quality for use of seawater, and reduction of amenities." Many investigators do not consider this definition to be sufficiently comprehensive because it only addresses those anthropogenic activities that impact the marine environment as a consequence of the introduction of waste substances and energy (Bewers and Wells, 1992). Pollution differs from contamination which occurs "when a man-made input increases the concentration of a substance in seawater, sediments, or organisms above the natural background level for that area and for the organisms" (Clark, 1992). However, the concentration of this substance does not exceed the threshold beyond which measurable damaging effects are manifested.

There are five principal pathways by which contaminants enter estuarine and coastal marine environments: (1) nonpoint source runoff from land; (2) direct pipeline discharges; (3) riverine inflow; (4) atmospheric deposition; and (5) waste dumping (Capuzzo and Kester, 1987; Kennish, 1992, 1997, 2000; McIntyre, 1992). Prior to 1970, point sources of pollution, notably the direct dumping or pipeline discharge of waste materials, accounted for the largest fraction of pollutants in these environments. However, tighter legislative and regulatory controls commencing with passage of the Federal Water Pollution Control Act Amendments in 1972 and subsequent amendments in 1977 (Clean Water Act) and 1987 (Clean Water Act Amendments) greatly reduced point source pollutant inputs to estuarine and coastal marine waters. Most pollutants (>75%) now enter these systems via nonpoint source runoff from land (i.e., dispersed,

Table E4 Point and nonpoint sources of pollution in estuarine and coastal marine environments

Sources	Common pollutant categories
Point	
Municipal sewage treatment plants	BOD, bacteria, nutrients, ammonia, toxic chemicals
Industrial sewer overflows	Toxic chemicals, BOD
Combined sewer overflows	BOD, bacteria, nutrients, turbidity, total dissolved solids, ammonia, toxic chemicals
Nonpoint	
Agriculture runoff	Nutrients, turbidity, total dissolved solids, toxic chemicals
Urban runoff	Turbidity, bacteria, nutrients, total dissolved solids, toxic chemicals
Construction runoff	Turbidity, nutrients, toxic chemicals
Mining runoff	Turbidity, acids, toxic chemicals, total dissolved solids
Septic systems	Bacteria, nutrients
Landfills/spills	Toxic chemicals, miscellaneous substrates
Silvicultural runoff	Nutrients, turbidity, toxic chemicals

Source: US Environmental Protection Agency, National Water Quality Inventory, Washington DC, 1986.

diffuse, and uncontrolled sources) and atmospheric deposition (GESAMP, 1990). Non-point source pollution is particularly problematic because it cannot be easily regulated or controlled by land-based technological measures. Moreover, biotic impacts of nonpoint source pollution are difficult to assess. Table E4 lists the types of point and nonpoint sources of pollution, which commonly affect estuarine and coastal marine environments.

Estuarine and marine organisms respond to pollutant inputs in several ways depending on the bioavailability of contaminants; their uptake, accumulation, and disposition in the body; and the interactive effects of multiple contaminants (McDowell, 1993). When exposed to elevated pollutant concentrations, some individuals exhibit acute physiological changes manifested by impaired growth, reproduction, and immunological function. Others display behavioral changes and greater susceptibility to disease. Chronic exposure to pollution stress usually results in marked changes in population structure, including reduced abundance and biomass, increased mortality, and altered age/size distribution. Recruitment of an affected population also may be significantly altered. At the community level, responses to insidious and cumulative pollution impacts frequently include: (1) decreased species diversity, (2) loss of rare or sensitive species, (3) opportunistic species dominance, and (4) shifts in the age structure of longer-lived species (Howells *et al.*, 1990). In addition, trophic interactions can be conspicuously modified.

Thousands of pollutants reach estuarine and coastal marine waters each year from land-based sources. Among the most serious pollutants in terms of potential impacts on environmental quality are halogenated hydrocarbon compounds, polycyclic aromatic hydrocarbon compounds, trace metals, and nutrient elements. The halogenated hydrocarbons (e.g., PCBs, DDTs, CCDs, and CDFs) consist of some of the most ubiquitous, persistent, and toxic substances found in coastal environments. They pose serious health risks to aquatic biota. Similarly, polycyclic aromatic hydrocarbons (e.g., anthracene, fluorene, naphthalene, and pyrene) are widely distributed organic pollutants that in sufficient concentrations threaten broad groups of organisms. A number of these compounds, particularly the high-molecular-weight varieties, are carcinogenic, mutagenic, and teratogenic to many estuarine and marine organisms. Trace metals (e.g., cadmium, copper, lead, mercury, and zinc), when present above a threshold availability, also are toxic to numerous biotic groups. Among the range of pathological, physiological, and behavioral responses observed in organisms exposed to toxic levels of metal contaminants are tissue inflammation and degeneration, neoplasm formation, and genetic derangement; feeding, digestive, and respiratory dysfunction; and aberrant physiological, reproductive, and developmental activity. Table E5 provides a list of many of these priority pollutants, which are now being monitored by the National Oceanic and Atmospheric Administration (NOAA) in US estuarine and coastal marine environments.

One of the most rapidly growing problems in estuaries and coastal embayments of the United States and abroad is eutrophication associated with nutrient loading (Livingston, 1996, 2000). Excessive nutrient inputs, notably nitrogen and phosphorus, can rapidly stimulate phytoplankton growth and lead to toxic or nuisance algal blooms that often degrade water quality by contributing to high oxygen demands in bottom waters and the buildup of toxins (e.g., sulfides). Aside from being linked to oxygen depletion events, algal bloom may directly impact benthic organisms by smothering them or creating deleterious shading conditions. The development of anoxia or hypoxia in shallow embayments ultimately culminates in large-scale mortality of benthic populations and fish and the reduction of useful habitat. In severe cases (i.e., hypereutrophication), imbalances in trophic systems may arise, manifested by acute changes in food web components and the elimination of valuable estuarine and coastal marine species.

Table E5 Chemical contaminants measured in the NOAA National Status and Trends Program

DDT and its metabolites	Anthracene
2,4′-DDD	Acenaphthene
4,4′-DDD	Acenaphthylene
2,4′-DDE	4-ring
4,4′-DDE	Fluoranthene
2,4′-DDT	Benz(*a*)anthracene
4,4′-DDT	Chrysene
Chlorinated pesticides other than DDT	5-ring
Aldrin	Benzo(*a*)pyrene
Cis-Chlordane	Benzo(*e*)pyrene
Trans-Nonachlor	Perylene
Dieldrin	Dibenz(*a,h*)anthracene
Heptachlor	Benzo(*b*)fluoranthene
Heptachlor epoxide	Benzo(*k*)fluoranthene
Hexachlorobenzene	6-ring
Lindane (γ-HCH)	Benzo(*ghi*)perylene
Mirex	Indeno(1,2,3-*cd*)pyrene
Polychlorinated biphenyls	**Major elements**
PCB congeners 8, 18, 28, 44, 56, 66, 101, 105, 118, 128, 138, 153, 179, 180, 187, 195, 206, 209	Al, Aluminum
	Fe, Iron
	Mn, Manganese
Polycyclic aromatic hydrocarbons	**Trace elements**
2-ring	As, Arsenic
Biphenyl	Cd, Cadmium
Naphthalene	Cr, Chromium
1-Methylnaphthalene	Cu, Copper
2-Methylnaphthalene	Pb, Lead
2,6-Dimethylnaphthalene	Hg, Mercury
1,6,7-Trimethylnaphthalene	Ni, Nickel
3-ring	Se, Selenium
Fluorene	Ag, Silver
Phenanthrene	Sn, Tin
1-Methylphenanthrene	Zn, Zinc
	Tri-, di- and mono Butyltin

Source: NOAA, National Status and Trends Program, Silver Spring, Maryland, 2000.

Habitat loss and alteration

Much deterioration of estuarine and nearshore oceanic environments is directly attributable to development along the immediate shore and in coastal watersheds. Here, physical destruction of habitat represents the most overt form of environmental degradation (Linden, 1990; Viles and Spencer, 1995; Alongi, 1998; Kennish, 2000). A multitude of human activities and structures modifies or destroys coastal habitats, chiefly:

- Domestic, agricultural, and industrial construction.
- Wetlands drainage, landfill, and reclamation.

- Grid ditching, dredging and filling, and dredge material disposal.
- Stream channelization (for flood control), canal excavation, and freshwater diversions.
- Mining, silviculture, and mariculture operations.
- Support infrastructure (roads and utilities).
- Dams, reservoirs, dikes, levees, and impoundments.
- Human-induced land subsidence (associated with water, oil, and gas withdrawal).
- Bank protection (bulkheads, revetments, and riprap).
- Shore protection structures (jetties, groins, seawalls, and breakwaters).
- Harbors, marinas, and shoreline installations.
- Moored structures (e.g., docks, piers, pilings).
- Shipping and marine transportation.

Physical habitat impacts are manifested most conspicuously by structural changes in watershed drainage patterns, coastal wetlands, beaches, coastline and bathymetric configurations, and other geomorphological features. These impacts can be exacerbated by nonstructural alterations, such as the introduction of aggressive, nonnative plant species or spills of chemical contaminants. Overall, habitat losses have been dramatic in the coastal zone during the past century. For example, more than 50% of the original tidal salt marsh habitat in the United States has been destroyed by anthropogenic activities (Watzin and Gosselink, 1992). In Long Island Sound, dredging and filling operations were largely responsible for the ~25–35% reduction in tidal wetlands area documented during the 20th century (Strieb *et al.*, 1993). Nearly 25% of the original wetlands habitat of the Delaware estuary was eliminated by human activities over the 20-year period from 1954 to 1974 (Tiner, 1985, 1990). Development in the Barnegat Bay–Little Egg Harbor water-shed in New Jersey has resulted in the following habitat losses: (1) 13,731 ha or 20% of the upland forested area between 1972 and 1995; (2) 1,875 ha or 6% of the freshwater wet-lands area during the same period; and (3) 4,200 ha or 28% of the tidal salt marsh area during the past 100 years (Lathrop *et al.*, 1999). Most of the tidal salt marsh losses occurred between 1940 and 1970. Since passage of the Coastal Wetlands Law of 1970, the loss of tidal salt marsh habitat in this watershed has decreased to <1.5%. Similarly, vegetated wetlands area in Galveston Bay, Texas, declined by 17–19% between the 1950s and 1989 mainly as a result of dredging and filling, drainage and conversion of land areas, subsidence coupled to groundwater withdrawal, and coast-line modifications (Pulich *et al.*, 1991; Shipley and Kiesling, 1994). Direct human impacts accounted for at least 26% of the Louisiana wetland losses observed between 1955 and 1978 (Boesch *et al.*, 1994). Of the original tidal marsh area in San Francisco Bay (i.e., ~2,200 km^2), only ~125 km^2 of undiked marsh remains (Conomos, 1979). Diking and filling of wetlands bordering the bay have had a devastating impact over the years. On the southeast coast of the United States, ~11% of the coastal marshes has been diked and flooded for insect control (i.e., mosquitoes and sandflies) and to attract wintering waterfowl (Alongi, 1998). Nearly 10,000 ha of coastal wetlands in the United States are lost each year primarily because of development impacts (Kennish, 2000).

The effects of physical habitat destruction are even greater in many other countries, particularly in developing nations. For instance, ~75% of the mangrove forests in Puerto Rico and more than 50% of the mangrove forests of Southeast Asia have been cleared. Other countries with significant mangrove habitat losses include India, Venezuela, and Ghana. The destruction of mangrove forests in these countries has been principally ascribed to reclamation, timbering, and mariculture (Hatcher *et al.*, 1989; Ong, 1994; Eisma, 1998). Reclamation has been responsible for nearly the complete disappearance of natural salt marshes along the mainland coast of the Dutch Wadden Sea (Eisma, 1998).

Physical habitat destruction is perhaps most pronounced at ocean coastline resort areas. Here, heavy development and recreational pursuits significantly alter natural habitats. Developed coasts are replete with coastline high rises, condominiums, and lagoon housing. Global sea-level rise, compounded in some areas by local subsidence (e.g., northern Gulf of Mexico), poses a serious long-term threat to many shore communities. Relative sea-level rise increases the risk of damage from storm surges, hurricanes, and other natural events.

Of more imminent concern along many coasts is reduced sediment delivery as a consequence of increased dam construction and river diversions (Aubrey, 1993). Beach nourishment projects have been instituted nationwide to counter the effects of sand deficits and severe erosion problems. Many marine engineering structures have been erected off beaches to trap sediment moving alongshore (Nordstrom, 1994a,b). These shore-perpendicular features (i.e., groins and jetties), constructed of quarrystone, timber, or a combination of stone and timber, are designed to capture sediment entering from updrift areas. However, they frequently increase erosion downdrift, thereby creating beach problems for neighboring communities. In many regions, groin fields exist to reduce the rate of beach retreat and the seasonal fluctuations in beach profiles over extensive areas. They are blatant reminders of the mounting pressures of shore communities to stave off natural shore retreat.

Shore-parallel structures (e.g., seawalls) also provide an engineering strategy to control the forces of nature. These structures protect beachfront property as well as infrastructure from invasive damage of storm waves, flooding, erosion, and natural disasters (e.g., tsunamis). However, they often increase scour and erosion seaward, and commonly reduce beach width (Kennish, 2000).

Millions of dollars are spent annually in coastal states throughout the United States on beach revitalization programs and for the protection of beachfront structures. Much of this financial support is driven by the multibillion dollar tourism industry. However, future costs associated with beach restoration and shore-protection efforts are likely to escalate with eustatic sea-level rise, which is projected to increase by another 20–115 cm during the 21st century (Woodroffe, 1993).

Marine environmental quality

Pollution and habitat degradation potentially threaten the quality of coastal environments. Because of the dependence of human populations on coastal and marine resources, it is imperative for marine and coastal policy and management programs to formulate strategies that mitigate pollution and habitat loss. This will require a concerted effort on the part of government agencies (federal, state, and local), academic institutions, industries, stakeholders, and the general public to address the myriad of problems occurring in the coastal zone and the measures that must be implemented to restore, protect, and maintain the integrity and health of coastal environments.

Monitoring is vital to assessing marine environmental quality. Harding (1992) advocates a more integrated approach to environmental quality monitoring, incorporating attributes of marine ecosystems that, if impaired, may lead to significant changes in ecosystem structure, rather than focusing on effects at lower levels of biological organization. By monitoring processes (e.g., bioaccumulation, primary productivity, and eutrophication), assessment of the status and trends in eco-system health will lead to the development of more effective long-term environmental management strategies.

Marine environmental quality monitoring

Marine environmental quality monitoring programs should incorporate several components in assessing ecosystem health: (1) analysis of the main processes and structural characteristics of the ecosystem; (2) identification of the known or potential stressors; (3) development of hypotheses about how these stressors may affect the ecosystem; and (4) identification of measures of environmental quality and ecosystem health to test these hypotheses (Harding, 1992). Estuarine and marine monitoring efforts in the United States, Canada, and other developed nations are concentrating on the identification of the sources of stress (e.g., municipal and industrial discharges, ocean dumped materials, and oil spills), biotic exposure to contaminants (e.g., in the water column and bottom sediments), and pollution effects (i.e., on individuals, populations, communities, and ecosystems). Regular monitoring yields the data necessary for status and trends assessment (O'Connor, 1990, 1994; O'Connor *et al.*, 1994; O'Connor and Beliaeff, 1995).

An integral component of coastal resource management entails the development of comprehensive environmental quality monitoring programs that supply sufficient data to define the extent and severity of pollution or other anthropogenic impacts and provide the information needed to devise effective measures to protect and rehabilitate the natural environment and its resources (National Research Council, 1990). Hameedi (1997) emphasized that there are two principal reasons for conducting environmental monitoring: (1) to define the status and trends in environmental quality nationwide or in a broad spatial context; and (2) to generate a database to address specific coastal research management needs in a regional or local context. As discussed by Hameedi (1997), the dual-purpose of environmental monitoring requires the implementation of a survey program with a tiered structure based on a sparsely distributed network of stations spread over a broad spatial extent, as well as designated regions embedded within this network of stations that are tailored to support specific coastal resource management needs. In addition, these environmental monitoring programs must include meaningful biological measures of contaminant exposure, contaminant-induced adverse biological effects, and susceptibility of biota to environmental degradation.

Because marine environmental quality monitoring programs are typically designed to describe the condition of the natural environment and the health of its resources, they generally focus on the following: (1) assessment of temporal trends in the levels of environmental degradation due to chemical contaminants and other anthropogenic sources; (2) determination of the magnitude and spatial extent of environmental degradation due to chemical contaminants and other anthropogenic sources; (3) incidence of pathological conditions, physiological dysfunction, and biological abnormalities in sentinel organisms resulting from environmental degradation; and (4) formulation of composite indices or sets of indicators that communicate information about the condition of the natural environment and are tractable toward public policy objectives (Hameedi, 1997). The trend in recent years has been to derive integrated measures of coastal ecosystem health by aggregating water quality and biological measurements in a quantifiable framework. The Index of Biotic Integrity, Ecosystem Distress Syndrome, Sediment Quality Triad, and Benthic Index provide examples (Fausch *et al.*, 1984; Chapman *et al.*, 1987; Weisberg *et al.*, 1992; Hameedi, 1997).

Biological measures are now used more extensively by major marine environmental quality monitoring programs to assess coastal environmental conditions on local, regional, and national scales. Included here are national programs such as the Environmental Monitoring and Assessment Program (EMAP) administered by the US Environmental Protection Agency and the National Status and Trends (NS&T) Program administered by the NOAA (Paul *et al.*, 1992; Wolfe *et al.*, 1993; O'Connor, 1994). Both of these large-scale, long-term monitoring programs have generated comprehensive databases on temporal and spatial trends in environmental quality of US coastal waters.

National monitoring programs

The National Research Council (1990) stressed a decade ago that regional monitoring of coastal environmental quality is necessary because of the lack of sufficient data to properly assess environmental conditions of coastal habitats nationwide. Two main problems exist with this type of assessment: (1) difficulty in distinguishing natural from anthropogenic-induced changes in condition; and (2) inability to evaluate whether indicators of adverse change and the ecological consequences of such change are consistent over regional-sized areas (Hanson *et al.*, 1993). Most of the more than $100 million spent on coastal environmental monitoring in the United States covers the cost of local (site-specific) compliance monitoring (e.g., wastewater discharges) (O'Connor, 1990). However, during the past decade, more effort has been expended on implementing regional and national coastal environmental quality monitoring programs, such as EMAP, the NS&T Program, and the National Estuary Program (NEP).

Environmental monitoring and assessment program

EMAP, initiated in 1990, was developed in response to the demand for information on the degree to which existing pollution control programs and policies protect the nation's ecological resources (Schimmel *et al.*, 1994). The principal goal of the coastal and estuarine components of EMAP was to assess the condition of the nation's coastal ecological resources. More specific EMAP objectives were:

1. To estimate the current status, extent of changes, and long-term (decadal) trends in indicators of the condition of the nation's ecological resources on a regional basis with known confidence.
2. To monitor indicators of pollutant exposure and habitat condition and to seek associations between anthropogenic-induced stresses and ecological condition.
3. To provide periodic statistical summaries and interpretive reports on ecological status and trends to resource managers and the public (Paul *et al.*, 1992).

The focus of EMAP-Estuaries, in turn, was to measure the status and change in selected indicators of ecological condition in estuaries and to provide a quantitative measure of the regional extent of estuarine environmental problems, most notably sediment contamination, habitat loss, eutrophication, and hypoxia (Kennish, 2000). The EMAP-Estuaries design consisted of seven well-defined regions or provinces in the continental United States (i.e., Acadian, Virginian, Carolinian, Louisianian, Californian, Columbian, and West Indian Provinces), four in Alaska (i.e., Alaskan, Aleutian, Arctic, and Bering), and one in the Great Lakes. Excluding the Great Lakes Province, estuaries in this regionalized scheme are defined by the landward boundary of the maximum inland extent of tide, and by the distal end of the continental shelf. Climatic zones and prevailing ocean currents represented the principal criteria used to establish the boundaries of the provinces (Terrell, 1979; Bailey, 1983; Holland, 1990; Paul *et al.*, 1992).

The US Environmental Protection Agency established an integrated sampling strategy for near coastal ecosystems. It coordinated the near coastal component of EMAP, whenever possible, with the NEP and NOAA's NS&T Program. These programs, taken together, have provided for more effective and organized monitoring of environmental conditions that will facilitate delineation of the factors contributing most significantly to coastal environmental problems.

Although EMAP was coordinated with efforts of the NEP and NS&T Program, there are some fundamental differences. For example, sample site selection in EMAP was a vigorously random procedure, whereas the NS&T Program annually sampled at fixed locations. In addition, EMAP did not analyze chemical contaminant concentrations in biotic tissue (i.e., mussels or oysters), as in the case of the NS&T Program. However, both programs surveyed chemical contaminants in bottom sediments of estuaries. The goal of EMAP was to eventually sample and analyze sediments from about 800 sites nationwide, which represents approximately three times the total number of sites sampled by the NS&T Program (NOAA, 1991; Kennish, 1997).

In the case of the NEP, management conferences convened by the Administrator of the US Environmental Protection Agency examine environmental conditions and trends in nationally significant estuaries, identify the most significant problems in these systems, and develop action-oriented comprehensive conservation and management plans (CCMPs) to mitigate human impacts. The CCMPs have two primary goals:

1. To restore and maintain the chemical, physical, and biological integrity of nationally significant estuaries, including restoration and maintenance of water quality, a balanced indigenous population of shellfish, fish, and wildlife, and the recreational activities in these systems.
2. To assure that the designated uses of these estuaries are protected.

There are 28 estuaries which have been identified as NEP sites since 1987 (Table E6) (USEPA, 1994; Kennish, 2000).

Table E6 List of national estuary programs

Albemarle/Pamlico Sound, NC
Barnegat Bay, NJ
Buzzards Bay, MA
Casco Bay, ME
Charlotte Harbor, FL
Corpus Christi Bay, TX
Delaware Bay, DE/NJ/PA
Delaware Inland Bays, DE
Galveston Bay, TX
Indian River Lagoon, FL
Long Island Sound, CT/MA/NY/RI
Lower Columbia River, WA
Maryland Coastal Bays, MD
Massachusetts Bays, MA
Mobile Bay, AL
Morro Bay, CA
Narragansett Bay, RI/MA
New Hampshire Estuaries, NH
New York/New Jersey Harbor, NY/NJ
Peconic Estuary, NY
Puget Sound, WA
San Francisco Bay/Sacramento-San Joaquin Delta, CA
San Juan Bay, PR
Santa Monica Bay, CA
Sarasota Bay, FL
Tampa Bay, FL
Terrebonne-Barataria, LA
Tillamook Bay, OR

Source: National Estuary Program, US Environmental Protection Agency, Washington, DC, 2000.

National status and trends program

NOAA created the NS&T Program in 1984 to monitor trends of chemical contamination and to assess the effects of human activities on coastal and estuarine systems nationwide. Sediment and biotic samples (bottom-feeding fish and shellfish) are collected from a network of sites located around the coastline of the United States and analyzed for a broad suite of trace metals and organic chemicals (Table E5) (O'Connor, 1992; O'Connor and Beliaeff, 1995). This program investigates the trends of chemical contamination in space and time and determines biological responses to that contamination. The immediate goal is a regional and national assessment of coastal environmental quality (O'Connor, 1990).

The NS&T Program consists of four main components: (1) National Benthic Surveillance Project; (2) Mussel Watch Project; (3) Biological Effects Surveys and Research; and (4) Historical Trends Assessment. NOAA initiated the National Benthic Surveillance Project in 1984 and the Biological Effects Surveys and Research, and the Mussel Watch Project in 1986 (Kennish, 1997). Several biological measures are used to assess the significance of chemical contamination in the coastal zone. NS&T surveys employ these measures (i.e., biological effects in sediments, bivalves, and fish) to ascertain the spatial patterns and extent of toxicity, the prevalence and severity of biological impacts, and the relationships between the measures of effects and the concentrations of chemicals in each survey area. The severity and magnitude of impacts among survey areas nationwide also can be compared using these data (Long, 1998).

Since 1986, the Mussel Watch Project has sampled seven bivalve species nationwide to assess the extent of chemical contamination in US coastal waters. Among these sentinel organisms are the blue mussel *Mytilus edulis* on the East Coast from Maine to (Cape May) New Jersey, the American oyster *Crassostrea virginica* on the East Coast from Delaware Bay southward and in the Gulf of Mexico, the mussel *Mytilus californianus* on the West Coast, the oyster *Ostrea sandivicensis* in Hawaii, the mangrove oyster *Crassostrea rhizophorae* in Puerto Rico, the smooth-edged jewel clam *Chama sinuos* at one site in the Florida Keys, and the zebra mussel *Dreissena polymorpha* in the Great Lakes (O'Connor, 1992; O'Connor and Beliaeff, 1995). Because of interspecies differences in contaminant bioaccumulation—even if the surrounding environments are identical—and the limited geographical distribution of the species, data comparisons among sites must be conducted with caution.

Results of NS&T monitoring surveys indicate that chemical contaminants are not uniformly distributed in US coastal and estuarine environments. The highest contaminant levels are recorded in sediments and biota of urbanized systems, such as those near the cities of Boston, Newark (NJ), San Diego, Los Angeles, and Seattle. Regionally, the highest contaminant levels are registered in the northeastern states, with elevated contaminant levels being relatively rare in the southeastern states and along the coastline of the Gulf of Mexico. Coastal and estuarine sediments of the northeast shelf large marine ecosystem region (i.e., Virginia to Maine) have a disproportionately large number of sampling sites displaying "high" concentrations of chemical contaminants. For example, more than two-thirds of all the sampling sites in the United States with "high" concentrations for most contaminants occur in this region, although less than one-third of all sampling sites nationwide are found here. The greater contaminant levels in sediments of the northeastern states are a function of the larger population numbers and associated development pressures observed in this region (O'Connor, 1996).

National estuary program

The NEP employs a holistic approach to manage estuarine environments and resources. This consists of a comprehensive evaluation of environmental conditions across a complex array of interconnected and interdependent systems from upland habitats of watersheds to the open waters of embayments. Subsequent to assessing these conditions, strategies are formulated and plans implemented to remediate pollution problems and other anthropogenic impacts, and to improve the long-term viability and ecological integrity of the estuary.

Some of the most serious problems encountered at NEP sites are those that may lead to ecosystem state changes. Examples are nonpoint source pollutant inputs, overenrichment of nutrients, introduction of pathogens and exotic species, and overexploitation of fisheries or other resources. They usually require a systematic, detailed, basinwide assessment. As a result, NEP investigations focus on comprehensive ecosystem health issues. Characterization reports prepared for NEP sites synthesize extensive databases, and they usually address the following points:

1. Status and trends of the water quality, natural resources, and uses of the estuary.
2. Linkages between pollutant loadings and changes in the water quality, natural resources, and uses of the estuary.
3. Description of human impacts on the water quality, natural resources, and uses of the estuary.
4. Identification of the priority problems in the estuary and the selection criteria used to determine them.
5. Hypotheses of cause–effect relationships for the priority problems and the research necessary to establish relationships.
6. Likely causes of the priority problems, examining data-bases on nutrients, chemical contaminants, and natural resources.
7. Final lists, historic descriptions, and background information on priority problems to be addressed in the CCMP.
8. Environmental quality goals and objectives established for the estuary, which form the basis for the monitoring program developed to evaluate the effectiveness of actions implemented under the CCMP.
9. Knowledge of uncertainties in the databases which can be used to direct further data gathering and research efforts and is important to the development of an effective sampling design in the post-CCMP monitoring program (USEPA, 1994).

Action plans in the CCMP specify the measures that must be taken to address priority problems in the watershed and estuary (USEPA, 1992). Arrays of possible management activities are considered, and those deemed to be most likely to result in positive change are implemented. Estuary restoration and maintenance at NEP sites depend most greatly on the implementation of carefully crafted action plans.

The strength of the NEP is that it forges partnerships between government agencies, business and industry, academic institutions, and the general public with a unified goal—to protect and restore the health of nationally significant estuaries and to enhance their living resources (USEPA, 1994). Critically important objectives of this program are to increase public awareness and understanding of pollution problems, as well as other anthropogenic impacts, and to ensure public participation in the decision-making process. A consensus-building process is necessary to establish effective pollution abatement and control programs, to revitalize damaged habitats, and to properly manage exploited resources.

Coastal zone management: conclusions

In recent years, it has become clear that the resolution of coastal environmental problems involves more than purely sectoral approaches to the management of land and marine habitats and resources. For example, atmospheric deposition alone delivers substantial amounts of contaminants to coastal systems often from sources at great distances. Hence, the governing council of the United Nations Environmental Programs (UNEP) favors a regional approach to marine pollution control. GESAMP, in turn, periodically publishes global reports on the state of the marine environment. The London Dumping Convention, in force since 1975, deals with regional or global pollution issues mainly through national legislation and regulations of countries contracting to the convention.

Future management programs must focus on more inte-grated and well-coordinated initiatives to counter the current narrow conservation and protection strategies, wasteful land-use practices, unsustainable coastal development, resource- use conflicts, and escalating habitat degradation in the coastal zone (Hildebrand and Norrena, 1992). Integrated Coastal Zone Management (ICZM) offers the framework for resolving many of the complex problems now facing coastal communities, such as resource conservation, environmental quality and amenities, and balanced human uses of aquatic systems (Hameedi, 1997). ICZM has been defined as a dynamic process in which a coordinated strategy is developed and implemented for the allocation of environmental, socioculture, and institutional resources to achieve the conservation and sustainable multiple use of the coastal zone (Coastal Area Management and Planning Network, 1989; Hildebrand and Norrena, 1992). It involves multi-sectoral, resource planning, and management approaches when dealing with human activities in coastal areas (Clark, 1996).

For ICZM to be effective, it is necessary not only to understand the ecological problems occurring in the coastal zone but also the economic, legal, and policy issues related to coastal resource use (Hameedi, 1997). Administrative demands of implementing ICZM require a high degree of coordination among various levels of government and between different governments. Hildebrand and Norrena (1992) advise that nations consider initiating ICZM programs on a regional basis, which will enable coastal planners and managers to concentrate on the most severe environmental problems. These investigators support the continued development of national and subnational coastal management strategies, in concert with appropriate applications in specific coastal regions.

An integrated management approach is the most viable strategy for improving environmental quality in coastal zones worldwide. An ICZM program has several advantages over traditional sectoral management schemes because it enables broad participation of parties for the resolution of conflicts between a variety of economic and development needs. It promotes sustainability of coastal resources and long-term protection of environmental health. A successful ICZM program, therefore, will effectively manage development as well as conserve natural resources in a complex dynamic environment (Bossi and Cintron, 1990; Hildebrand and Norrena, 1992; Clark, 1996).

Michael J. Kennish

Bibliography

Alongi, D.M., 1998. *Coastal Ecosystem Processes*. Boca Raton: CRC Press.

Aubrey, D.G., 1993. Coastal erosion's influencing factors include development, dams, wells, and climate change. *Oceanus*, **36**: 5–9.

Bailey, R.G., 1983. Delineation of ecosystem regions. *Environmental Management*, **7**: 365–373.

Barcena, A., 1992. An overview of the oceans in Agenda 21 of the 1992 United Nations conference on environment and development. *Marine Pollution Bulletin*, **25**: 107–111.

Bewers, J.M., and Wells, P.G., 1992. Challenges for improved marine environmental protection. *Marine Pollution Bulletin*, **25**: 112–117.

Boehlens, R.G.V., 1992. From policies to science: strategies for marine environmental protection. *Marine Pollution Bulletin*, **25**: 14–17.

Boesch, D.F., Josselyn, M.N., Mehta, A.J., Morris, J.T., Nuttle, W.K., Simenstad, C.A., and Swift, D.J.P., 1994. Scientific assessment of coastal wetland loss, restoration, and management in Louisiana. *Journal of Coastal Research*, **20**(Special Issue): 103 pp.

Bossi, R., and Cintron, G., 1990. *Mangroves of the Wider Caribbean*. Nairobi: U.N. Environment Program.

Capuzzo, J.M., and Kester, D.R. (eds.), 1987. *Oceanic Processes in Marine Pollution*, Vol. 1, *Biological Processes and Wastes in the Ocean*. Malabar: Robert E. Krieger Publishing.

Chapman, P.M., Dexter, R.N., and Long, E.R., 1987. Synoptic measures of sediment contamination, toxicity, and infaunal community composition (The Sediment Quality Triad) in San Francisco Bay. *Marire Ecology Progress Series*, **37**: 75–96.

Clark, J.R., 1996. *Coastal Zone Management Handbook*. Boca Raton: Lewes Publishers.

Clark, R.B., 1992. *Marine Pollution*, 3rd edn. Oxford: Clarendon Press.

Coastal Area Management and Planning Network, 1989. *The Status of Integrated Coastal Zone Management: A Global Assessment*. Summary Report of a Workshop Convened at Charleston, South Carolina, July 4–9, Rosenstiel School of Marine Sciences, University of Miami, Miami, Florida.

Conomos, T.J., 1979. Properties and circulation of San Francisco Bay waters. In Conomos, T.J. (ed.), *San Francisco Bay: The Urbanized Estuary*. San Francisco: Pacific Division of the American Association for the Advancement of Science, San Francisco, California, pp. 47–81.

Costanza, R., Andrade, F., Antunes, P., van den Belt, M., Boersma, D., Boesch, D.F., Catarino, F., Hanna, S., Limburg, K., Low, B., Molitor, M., Pereira, J.G., Rayner, S., Santos, R., Wilson, J., and Young, M., 1998. Principles for sustainable governance of the oceans. *Science*, **281**: 198–199.

Côté, R.P., 1992. Marine environmental management: status and prospects. *Marine Pollution Bulletin*, **25**: 18–22.

Cronan, D.S. (ed.), 2000. *Handbook of Marine Mineral Deposits*. Boca Raton: CRC Press.

Eisma, D., 1998. *Intertidal Deposits: River Mouths, Tidal Flats, and Coastal Lagoons*. Boca Raton: CRC Press.

Fausch, D.D., Karr, J.R., and Yant, P.R., 1984. Regional application of an index of biotic integrity based on stream fish communities. *Transaction Am. Fish. Society*, **113**: 39–55.

GESAMP, 1982. Scientific Criteria for the Selection of Waste Disposal Sites at Sea. London: Inter-Governmental Maritime Consultative Organization, Reports and Studies No. 16.

GESAMP, 1990. *The State of the Marine Environment*. Oxford: Blackwell Scientific.

Goldberg, E.D., 1994. Coastal Zone Space—Prelude to Conflict? Paris: UNESCO, UNESCO Technical Report.

Hameedi, M.J., 1997. Strategy for monitoring the environment in the coastal zone. In Haq, B.U. (ed.), *Coastal Zone Management Imperative for Maritime Developing Nations*. Amsterdam: Kluwer Academic Publishers, pp. 111–142.

Hanson, P.J., Evans, D.W., and Colby, D.R., 1993. Assessment of elemental contamination in estuarine and coastal environments based on geochemical and statistical modeling of sediments. *Marine Environmental Research*, **36**: 237–266.

Harding, L.E., 1992. Measures of marine environmental quality. *Marine Pollution Bulletin*, **25**: 23–27.

Hatcher, B.G., Johannes, R.E., and Robertson, A.I., 1989. Review of research relevant to the conservation of shallow tropical marine ecosystems. *Oceangraphy Marine and Biology Annual Review*, **27**: 337–414.

Hildebrand, L.P., and Norrena, E.J., 1992. Approaches and progress toward effective integrated coastal zone management. *Marine Pollution Bulletin*, **25**: 94–97.

Holland, A.F., 1990. *Near Coastal Program Plan for 1990: Estuaries*. EPA 600/4-90-003, U.S. Environmental Protection Agency, Office of Research and Development, Narragansett, Rhode Island.

Howells, G., Calamari, D., Gray, J., and Wells, P.G., 1990. An analytical approach to assessment of long-term effects of low levels of contaminants in the marine environment. *Marine Pollution Bulletin*, **21**: 371–377.

Kennish, M.J., 1992. *Ecology of Estuaries: Anthropogenic Effects*. Boca Raton: CRC Press.

Kennish, M.J. (ed.), 1997. *Practical Handbook of Estuarine and Marine Pollution*. Boca Raton: CRC Press.

Kennish, M.J. (ed.), 2000. *Estuary Restoration and Maintenance: The National Estuary Program*. Boca Raton: CRC Press.

Lathrop, R.G., Jr., Bognar, J.A., Hendrickson, A.C., and Bowers, P.D., 1999. *Data Synthesis Report for the Barnegat Bay Estuary Program: Habitat Loss and Alteration*. Technical Report, Barnegat Bay Estuary Program, Center for Remote Sensing and Spatial Analysis, Cook College, Rutgers University, New Brunswick, New Jersey.

Linden, O., 1990. Human impact on tropical coastal zones. *Natural Resources*, **26**: 3–17.

Livington, R.L., 1996. Eutrophication in estuaries and coastal systems: relationships of physical alterations, salinity stratification, and hypoxia. In Vernberg, F.J., Vernberg, W.B., and Siewicki, T. (eds.), *Sustainable Development in the Southeastern Coastal Zone*. Columbia: University of South Carolina Press, pp. 285–318.

Livington, R.L. (ed.), 2000. *Eutrophication in Estuarine Systems of the Northeastern Gulf of Mexico*. Boca Raton: CRC Press.

Long, E.R., 1998. The use of biological measures in assessments of toxicants in the coastal zone. In Vernberg, F.J., Vernberg, W.B., and Siewicki, T. (eds.), *Sustainable Development in the Southeastern Coastal Zone*. Columbia: University of South Carolina Press, pp. 187–219.

McDowell, J.E., 1993. How marine animals respond to toxic chemicals in coastal ecosystems. *Oceanus*, **36**: 56–61.

McIntyre, A.D., 1992. The current state of the oceans. *Marine Pollution Bulletin*, **25**: 28–31.

National Research Council, 1990. *Managing Troubled Waters: The Role of Marine Environmental Monitoring*. Washington, DC: National Academy Press.

Nollkaemper, A., 1992. Marine pollution from land-based sources: towards a global approach. *Marine Pollution Bulletin*, **24**: 8–12.

NOAA, 1991. *Second Summary of Data on Chemical Concentrations in Sediments from the National Status and Trends Program*. NOAA Technical. Mem. NOS OMA 59, National Oceanic and Atmospheric Administration, Rockville, Maryland.

Nordstrom, K.F., 1994a. Beaches and dunes of human-altered coasts. *Prog. Geogr.*, **18**: 497–516.

Nordstrom, K.F., 1994b. Developed coasts. In Carter, R.W.G., and Woodroffe, C.D. (eds.), *Coastal Evolution*. Cambridge: Cambridge University Press, pp. 477–495.

O'Connor, T.P., 1990. *Coastal Environmental Quality in the United States, Chemical Contamination in Sediments and Tissues. A Special NOAA 20th Anniversary Report*. National Oceanic and Atmospheric Administration, Office of Ocean Resources Conservation and Assessment, Rockville, Maryland.

O'Connor, T.P., 1992. *Recent Trends in Coastal Environmental Quality: Results from the First Five Years of NOAA Mussel Watch Project*. National Oceanic and Atmospheric Administration, Office of Ocean Resources Conservation and Assessment, Rockville, Maryland.

O'Connor, T.P., 1994. The National Oceanic and Atmospheric Administration (NOAA) National Status and Trends Mussel Watch

Program: National monitoring of chemical contamination in the coastal United States. In Clothern, C.R., and Ross, N.P. (eds.), *Environmental Statistics, Assessment, and Forcasting*. Boca Raton: Lewis Publishers, pp. 331–349.

O'Connor, T.P., 1996. Coastal sediment contamination in the Northeast Shelf Large Marine Ecosystem. In Sherman, K., Jaworsky, N.A., and Smayda, T.J. (eds.), *The Northeast Shelf Ecosystem: Assessment, Sustainability, and Management*. Cambridge: Blackwell Science, pp. 239–257.

O'Connor, T.P., and Beliaeff, B., 1995. *Recent Trends in Coastal Environmental Quality: Results from the Mussel Watch Project*. NOAA Technical Report, National Oceanic and Atmospheric Administration, Department of Commerce, Rockville, Maryland.

O'Connor, T.P., Cantillo, A.Y., and Lauenstein, G.G., 1994. Monitoring of temporal trends in chemical contamination by the NOAA National Status and Trends Mussel Watch Project. In Kramer, K.L.M. (ed.), *Biomonitoring of Coastal Waters and Estuaries*. Boca Raton: CRC Press, pp. 29–50.

Ong, J. E., 1994. The status of mangroves in ASEAN. In Wilkinson, C.R. (ed.), *Living Coastal Resources of Southeast Asia: Status and Management*. Townsville: Australian Institute of Marine Science, pp. 52–75.

Paul, J.F., Scott, K.J., Holland, A. F., Weisberg, S.B., Summers, J.K., and Robertson, A., 1992. The estuarine component of the E.P.A's Environmental Monitoring and Assessment Program. *Chemical Ecology*, **7**: 93–116.

Pearson, T.H., and Rosenberg, R., 1978. Macrobenthic succession in relation to organic enrichment and pollution of the marine environment. *Oceanography and Marine Biology Annual Review*, **16**: 229–311.

Pinet, P.R., 2000. *Invitation to Oceanography*, 2nd edn. Boston, MA: Jones and Bartlett Publishers.

Pirie, R.G., 1996. US ocean resources 2000: a national plan for growth. In Pirie, R.G. (ed.), *Oceanography: Contemporary Readings in Ocean Sciences*, 3rd edn. New York: Oxford University Press, pp. 283–292.

Pulich, W.M., White, W.A., Castiglione, M., and Zimmerman, R.J., 1991. Status of submerged vegetation in the Galveston Bay system. In Shipley, F.S., and Kiesling, R.W. (eds.), *Proceedings of the Galveston Bay Characterization Workshop*. Webster (Texas): Galveston Bay National Estuary Program Publication GBNEP-6, pp. 127–132.

Schimmel, S.C., Melzian, B.D., Campbell, D.E., Strobel, C.J., Benyi, S.J., Rosen, J.S., and Buffum, H.W., 1994. *Statistical Summary: EMAP-Estuaries, Virginian Province—1991*. EPA/620/ R-94/005, U.S. Environmental Protection Agency, Narragansett, Rhode Island.

Shipley, F.S., and Kiesling, R.W. (eds.), 1994. *The State of the Bay: A Characterization of the Galveston Bay Ecosystem*. Webster (Texas): Galveston Bay National Estuary Program Publication GBNEP-44.

Simboura, N., Zenetos, A., Panayotidis, P., and Makra, A., 1995. Changes in benthic community structure along an environmental pollution gradient. *Marine Pollution Bulletin*, **30**: 470–474.

Sissenwine, M.P., and Rosenberg, A.A., 1993. US fisheries: status, long-term potential yields, and stock management ideas. *Oceanus*, **36**: 48–54.

Sorensen, J.C., and McCreary, S.J., 1990. *Institutional Arrangements for Managing Coastal Resources and Environments*. NPS/US AID Series, National Park Service, Office of International Affairs, Washington, DC: Coastal Management Publication No. 1 [Rev].

Strieb, M. (and the Living Resources Work Group), 1993. *Assessment of Living Resources. Comprehensive Conservation and Management Plan Supporting Document*. New York: U.S. Environmental Protection Agency, Stony Brook.

Terrell, T.T., 1979. *Regionalization of Coastal Ecosystems of the United States and its Territories*. Washington, DC: US Fish and Wildlife Service, Office of Biological Service, FWS/Physical OBS-79/80.

Tiner, R.W., Jr., 1985. *Wetlands of Delaware*. Cooperative Publication, National Wetlands Inventory Project, US Fish and Wildlife Service Region 5, Newton Corner, Massachusetts, and Division of Environmental Control, Delaware Department of Natural Resources and Environmental Control, Dover, Delaware.

Tiner, R.W., Jr., 1990. Pennsylvania's Wetlands: Current Status and Recent Trends. US Fish and Wildlife Service, National Wetlands Inventory, *Technical Report*. Newton Corner, MA.

USEPA, 1992. *National Estuary Program Guidance: Comprehensive Conservation and Management Plans Content and Approval Requirements*. EPA 842-B-92-002. US Environmental Protection Agency, Washington, DC: EPA 842-B-92-002.

USEPA, 1994. *A National Estuary Program Guidance: Technical Characterization in the National Estuary Program*. Washington, DC: US Environmental Protection Agency. EPA 842-B-94-006.

Viles, H., and Spencer, T., 1995. *Coastal Problems: Geomorphology, Ecology, and Society at the Coast*. London: Edward Arnold.

Warwick, R.M., Pearson, T.H., and Ruswahyuni, M., 1987. Detection of pollution effects on marine macrobenthos: further evaluation of the species abundance/biomass method. *Marine Biology*, **95**: 193–200.

Watzin, M.C., and Gosselink, J.G., 1992. *The Fragile Fringe: Coastal Wetlands of the Continental United States*. Technical Report, Louisiana Sea Grant Program, Louisiana State University, Baton Rouge, Louisiana, US Fish and Wildlife Service, Washington, DC, and the National Oceanic and Atmospheric Administration, Rockville, Maryland.

Weisberg, S.B., Frithsen, J.B., Holland, A.F., Paul, J.F., Scott, K.J., Summers, J.K., Wilson, H.T., Valente, R., Heimbuch, D.G., Gerritsen, J., Schimmel, S.C., and Latimer, R.W., 1992. *EMAP-Estuaries Virginian Province 1990 Demonstration Project Report*. Narragansett, Rhode Island: US Environmental Protection Agency, Environmental Research Laboratory, EPA 600R-92/100.

Wolfe, D.A., Long, E.K., and Robertson, A., 1993. The NS&T bioeffects surveys: design strategies and preliminary results. In Magoon, O.T., Wilson, W.S., Converse, H., and Tobin, L.T. (eds.), *Coastal Zone '93: Proceedings of the 8th Symposium on Coastal and Ocean Management*, Vol. 1. New York: American Society of Civil Engineers, pp. 298–312.

Woodroffe, C., 1993. Sea level. *Progress in Physical Geography*, **17**: 359–368.

Yap, H.T., 1992. Marine environmental problems: experiences of developing regions. *Marine Pollution Bulletin*, **25**: 37–40.

Cross-references

Beach Erosion
Cleaning Beaches
Coastal Engineering (see Shore Protection Structures and Navigation Structures)
Coastline Changes
Dams, Effect on Coasts
Demography of Coastal Populations
Dredging of Coastal Environments
Human Impact on Coasts
Mangrove Coasts
Meteorological Effect on Coasts
Mining of Coastal Materials
Monitoring, Coastal Ecology
Reclamation
Salt Marsh
Sea-Level Rise, Effect
Sediment Transport (see Cross-Shore Sediment Transport and Longshore Sediment Transport)
Shore Protection Structures
Tourism and Coastal Development
Water Quality

EOLIANITE

Eolianite, also known as aeolianite, (a)eolian calcarenite, calcareous (a)eolianite, dune limestone, dune sandstone, kurkar (Middle East) or kunkar (India), is a generally consolidated coastal rock formation consisting of at least partially lithified wind-blown sand. Strictly eolianite could include any wind-blown sediment, such as silt (loess, brickearth), clay (parna), and volcanic ash, but conventionally it is restricted to dune sand, although finer sediment is sometimes present. Typically, the dune sand has been partially cemented by secondary internal precipitation of carbonates from percolating groundwater. There is often evidence of several stages of such cementation. The proportions of carbonate vary, but are typically at least 50% and often more than 90%, derived from shell debris, coralline material, bryozoans, and foraminifera, which lived on the seafloor. There are varying proportions of non-calcareous sediment, mainly quartz (quartzose sandstones with less than 50% carbonate are termed quartz-arenites).

Eolianite was originally described from Bermuda by Sayles (1931), although it had been noted by explorers on the Australian coast, where Vancouver in 1791 thought it coralline while Charles Darwin in 1844 reported that it consisted of wind-blown sand and associated limestone. It is usually of Pleistocene age, and may be overlain by unconsolidated Holocene dune topography. It often passes well below present sea level: for example, the eolianites of the Nepean Peninsula in Victoria, Australia, are Pleistocene dune formations, which extend more than 140 m below, and up to 60 m above, present sea level (Bird, 1993). It contains fossil shells, plant remains, and occasional evidence of land animals.

Eolianite dune formations have been emplaced on the coast during Quaternary oscillations of land and sea level. During falling and low sea-level phases, sand derived from calcareous seafloor organisms was winnowed and carried landward by onshore winds to be deposited as transgressive dunes in the vicinity of the present coast. Similar sand was also carried shoreward during marine transgressions to form beaches and backshore dunes, as well as multiple beach and dune ridges parallel to the coastline.

Many eolianite coasts have a series of partly overlapping transgressive dune formations, indicating alternations of surface stability, when a capping of soil and vegetation developed, and instability, when the next transgressive dune moved in (Figure E8). Each dune formation is capped by a soil horizon (paleosol) overlying a calcrete (limestone) layer formed by subsoil precipitation of carbonates. Contemporary vegetation produced calcareous fossil root structures (rhizoconcretions) in the subsoil. The interbedded soils are typically terra rossa, incorporating eolian silt and clay accessions to the dune surface. Occasionally, the footprints of animals or birds are found on outcrops of sandstone, which were temporary land surfaces, buried when dune accretion resumed. Locally, there are soil pipes, cylindrical hollows formed where root penetration was followed by the local subsidence of soil into depressions deepened by solution and encased by precipitated calcrete. Each phase of landscape stability came to an end with the arrival of the next transgressive dune formation, marked by dune sandstone, often with laminations indicating the intermittent advance of a dune front, and sometimes with calcareous fossil stem structures formed around buried vegetation. Subsequent erosion may expose root and stem concretions as "petrified forests," as at Cape Bridgewater, Australia. In places there are unconformities because the earlier dune landscape was eroded before the next dunes arrived.

Commonly eolianites are capped by Holocene dunes, but where these are absent (or have been removed by erosion) exposed calcrete layers can develop karst topography (limestone mounds or ridges, pedestals, clay-floored vases and ditches, solution hollows). Some dune sandstone formations remain relatively unconsolidated, and can be mobilized by wind action as active dunes if the overlying calcrete crust is removed.

Eolianite dune formations are generally sufficiently lithified to be cut back as cliffs up to 60 m high, the harder components (calcrete layers) dominating the profile while caves and coves are excavated in the less consolidated dune sandstones. These cliffs are fronted by subhorizontal low-tide shore platforms formed where shore rock outcrops are lowered by solution processes to a level that has become indurated by further carbonate precipitation (Hills, 1971) (Figure E8).

Eolianite dune topography is found on the southern and eastern coasts of the Mediterranean, in Morocco, on the Red Sea and southern Arabian coast, in western India, around the Caribbean (Kaye, 1959), on the Brazilian coast, in South Africa, on oceanic islands such as Mauritius and Hawaii, and extensively along the western and southern coasts of Australia (Fairbridge and Teichert 1953; Semeniuk and Johnson, 1985). On the eastern coast of Australia, beaches and dunes are quartzose, and eolianites have not formed: the dunes remain unconsolidated, either active and mobile or retained by a vegetation cover. On the eastern coast of the Mediterranean, there is a similar transition, the eolianite (kurkar) dune topography of the Israel coast passing southward into more irregular, mobile dunes on the coast of the Gaza Strip, where the sands have an increasing proportion of terrigenous (Nile-derived) quartzose sediment.

The submergence and dissection of eolianite dune topography formed during low sea-level phases is illustrated off the western and southern coasts of Australia. Some submerged eolianite ridges have been planed off by wave action on the sea floor, but there are also remnants of submerged eolianite dune topography. Similar submerged eolianite dune ridges occur off the Bahamas, northeast Brazil, and Sri Lanka, and some of these have become reefs with coral and algal crusts.

Eric Bird

Bibiliography

Bird, E.C.F., 1993. *The Coast of Victoria.* Melbourne: Melbourne University Press.

Fairbridge, R.W., and Teichert, C., 1953. Soil horizons and marine bands in the coastal limestones of Western Australia. *Journal of the Royal Society of New South Wales*, **86**: 68–87.

Hills, E.S., 1971. A study of cliffy coastal profiles based on examples in Victoria, Australia. *Zeitschrift für Geomorphologie*, **15**: 137–180.

Kaye, C.A., 1959. Shoreline features and Quaternary shoreline changes. Puerto Rico. *US Geological Survey Professional Paper*, 317-B, pp. 49–140.

Figure E8 Cliff and shore platform cut in eolianite at Jubilee Point on the coast of Victoria. The cliff section shows two Pleistocene dune calcarenite formations capped by calcretes and paleosols, surmounted by unconsolidated Holocene dunes. The subhorizontal shore platform is exposed at low tide (photo E.C.F. Bird, Copyright—Geostudies).

Sayles, R.W., 1931. Bermuda during the ice age. *Proceedings, American Academy, Arts and Science*, **66**: 381–467.
Semeniuk, V., and Johnson, D.P., 1985. Modern and Pleistocene rocky shore sequences along carbonate coastlines, southwestern Australia. *Sedimentary Geology*, **44**: 225–261.

Cross-references

Beachrock
Beach Ridges
Cheniers
Coastal Soils
Dune Ridges
Eolian Processes

EOLIAN PROCESSES

Eolian processes are those processes relating to, or caused by the wind. The term eolian comes from the Greek *Aeolus*, the God of Wind. Eolian (aeolian—non-US equivalent) processes include the transportation and movement of sediments (particularly sand, clay, and silt), erosion of sediments, rocks and landforms, and the creation of dunes, sheets, and landforms. Eolian processes are most common on coasts and in arid environments, particularly hot and cold deserts.

Eolian sand transport is initiated when the wind velocity exceeds a critical threshold velocity. The threshold velocity required to initiate clay and silt transport is far greater than that of sand due to their platy nature and cohesive forces, but once motion is initiated, fine grained material (particularly silt) may stay in suspension for a considerable time period. Three main forms of grain transport take place, creep, saltation, and suspension. Creep refers to the movement of grains by rolling over the surface. Saltation (from the Latin *saltare*, to jump) is the bounding movement of grains in which grains rise almost vertically and then describe a long low trajectory downwind. Once saltation occurs, some energy is transferred to the surface by grain collisions and this aids creep and the ejection of further grains into the air stream. This process occurs at lower threshold velocities than that required for initiation of sediment transport and has been termed reptation by some authors (e.g., Lancaster, 1995). Suspension occurs where grains are lifted sufficiently to overcome gravity for some period of time. The initiation of eolian sediment transport and eolian processes are affected by surface conditions (e.g., moisture, cements, and bonding agents, vegetation) surface roughness and topography.

Eolian transport causes deflation, which may merely lead to the slight lowering of a surface (e.g., beach, dune or sand plain), but may also be the dominant process leading to the development of deflation hollows (and slacks), basins, and plains. In coastal and desert environments, deflation leads to the formation of blowouts, parabolic dunes, and wind-eroded plains. Here, deflation erodes the surface down to the seasonally lowest water table, a calcrete or indurated horizon, a palaeosol, or rock and boulder strewn surface. Eolian transport also causes abrasion of landforms. Wind transported sand grains abrade surfaces forming ventifacts (a polished, facetted and/or abraded stone or pebble), yardangs (sharp crested linear ridge of wind-eroded rock), and zeugen (desert rock pillars and yardangs, which are considerably undercut; Whittow, 1984).

In the coastal environment, eolian processes may lead to the development of various dune types. Where vegetation is important or present, coppice dunes, shadow dunes, foredunes, blowouts, and parabolic dunes may be formed. Where vegetation is less important and sediment supply is moderate to high, transgressive dunefields may occur and display a variety of dune types (e.g., sheets, barchans, transverse and oblique dunes, dome dunes, and star dunes) and associated landforms such as precipitation ridges and deflation plains and basins may be present (Hesp, 1999). Deserts may display a wide variety of eolian dune forms including dome dunes, barchans, transverse dunes, longitudinal and linear dunes, shadow dunes, sheets, and star dunes. Each tends to be related to sediment volume, wind directional variability, and average annual velocity (Cooke *et al.*, 1993). Vegetation processes are important in modifying eolian processes commonly leading to the complete colonization of once active dunefields in both coastal and desert environments.

Patrick A. Hesp

Bibliography

Cooke, R.U., Warren, A., and Goudie, A.S., 1993. *Desert Geomorphology*. London: UCL Press.
Hesp, P.A., 1999. The backshore and beyond. In Short, A.D. (ed.), *Handbook of Beach and Shoreface Morphodynamics*, Chicester: John Wiley and Sons, pp. 145–170.
Lancaster, N., 1995. *Geomorphology of Desert Dunes*. London: Routledge.
Whittow, J.B., 1984. *The Penguin Dictionary of Physical Geography*. Hamondsworth: Penguin Books.

Cross-references

Coastal Hoodoos (see Beach Processes)
Coastal Processes
Desert Coasts
Dune Calcarenite (See Eolianite)
Dune Ridges
Ripple Marks

EROSION: HISTORICAL ANALYSIS AND FORECASTING

Introduction

The ability to forecast future coastal positions has taken on increased importance as development along the world's coasts has risen dramatically over the past few decades. This is readily apparent, for example, in the United States where development has transformed many small beach villages into moderately to densely populated cities. Currently, about 350,000 structures are located within 500 feet of the US open-ocean and Great Lakes coasts (Heinz Center, 2000).

Rates of coastal erosion are calculated by monitoring the location of a representative geomorphic indicator, usually the high-water line (HWL; i.e., wet–dry boundary) or bluff line, over a specified time frame. Rates are obtained by measuring the location of two or more shorelines (hereinafter the term "shoreline" generally refers to the "high-water line," or "bluff line," unless noted otherwise, with no distinction made between cliff and bluff) from historical shoreline change maps. These maps are produced by digitizing shoreline change reference features from various sources, such as historical maps, aerial photographs, and Global Positioning System (GPS) surveys, followed by combining and overlaying the shorelines onto a common coordinate system. Historical shoreline change maps for the United States often contain four to eight or more plotted shorelines and can span 150 years. Erosion rates are typically calculated from the maps by plotting or generating a line perpendicular to the multiple shorelines and measuring the amount of movement over a period of time defined (and constrained) by the dates of the plotted shorelines. In many cases, subsets of the historical shorelines are used, particularly in areas where prolonged and perhaps permanent physical changes to the beach system have occurred (e.g., inlet openings) or where the construction of man-made structures (e.g., groins, jetties, and seawalls) makes older data unrepresentative of the long-term trend.

Various statistical methods have been proposed to calculate long-term erosion rates, but because of the paucity of data and uneven sampling, higher-order statistical methods have thus far proven no better at predicting future shoreline positions than "simple" methods, such as linear regression. The selection of appropriate shorelines to use in predictions is critical in obtaining a reliable long-term erosion rate. Some have suggested using only the more recent shoreline position data, erroneously assuming that older, historical maps are not sufficiently accurate. However, Crowell *et al.* (1991) have shown that historical maps known as "T" or topographic sheets, produced as far back as the mid-1800s by the United States National Ocean Service (NOS) (and predecessor agencies), are of sufficient accuracy for determining long-term rates of beach erosion. In addition, Crowell *et al.* (1993) and Galgano *et al.* (1998) demonstrated the importance of using data sets spanning 60–80 years or more to determine the long-term trend of erosion for forecasting future shoreline positions. Such long data spans are needed because the erosion/recovery cycle associated with severe storms can obscure the underlying trend of erosion for decades.

The principal steps involved in developing and analyzing historical shoreline change maps are as follows:

- Data selection: select appropriate map and imagery source data for the area to be studied;
- Data preparation: determine shoreline change reference feature to be monitored (usually the HWL or top edge of the bluff line); establish geographic control on source data and annotate features (i.e., shoreline segments and control points) to be digitized;
- Data correction: correct air photo distortions and identify and correct map distortions and errors;
- Data digitization: digitize and combine shorelines and plot them onto a suitable medium;
- Data analysis: calculate erosion rates from shore-perpendicular transects.

Sources of historical data

There are three principal data sources used in the production of historical shoreline change maps: maps, aerial photography, and GPS surveys. In the United States, the primary historical map data source is NOS T-sheets. The first T-sheets date back to 1835, but maps of 1835–45 vintage sometimes show inaccuracies in longitude. In 1846 NOS surveyors began using the telegraph as part of field mapping. This significantly improved longitudinal accuracy to the point that maps of post-1846 vintage are usually accurate enough for use in modern historical shoreline mapping studies.

NOS has periodically remapped coastal areas every 20–60 years. As such, two or three pre-1940s and one to two post-1940s T-sheets are generally available for most of the US East and Gulf Coasts (Anders and Byrnes, 1991). NOS T-sheets are an invaluable data source that expands the temporal range of the historical shoreline data from about 60-years (using aerial photography and GPS ground surveys) to approximately 150 years. A major prerequisite, however, is that the accuracy of each map be verified by analyzing current or "known" geo-positions of recoverable triangulation stations mapped on the historical NOS T-sheets. As an example, analysis of the accuracy of NOS T-sheets for a historical shoreline mapping project for Massachusetts demonstrated that 120 of 141 (85%) T-sheets were of sufficient accuracy for use in a study of long-term shoreline change (Crowell *et al.*, 1991). Map scale is obviously a major consideration—1 : 10,000 or larger is preferred, although maps at scales as small as 1 : 20,000 are often usable (Anders and Byrnes, 1991; Moore, 2000).

Vertical aerial photographs date back to the 1920s; however, aerial photography produced prior to the late-1930s is typically overly distorted and of insufficient resolution for use in historical shoreline mapping studies. Moreover, it was not until after World War II that air photos on the US coasts were taken with any degree of regularity. Many of the air photos were taken to document storm impact and consequently are of questionable use for determining long-term rates of beach erosion (Douglas *et al.*, 1998). Fortunately, however, several sets of post-World War II aerial photographs suitable for historical shoreline mapping studies are often available for a given area (Crowell *et al.*, 1997).

Field-based surveys using GPS have been increasingly used within the past seven years (Morton *et al.*, 1993). Shoreline position data are collected by driving a GPS-equipped off-road vehicle directly along the HWL. Geo-coordinates are collected at specified intervals, enabling real-time digitization of the HWL (Byrnes and Hiland, 1993; French and Leatherman, 1994). Table E7 provides a list of milestones in historical shoreline mapping.

Shoreline change reference features

A number of potential geomorphic features can be used to monitor historical shoreline changes. However, in many situations the HWL has been demonstrated to be the best indicator of the land–water interface for historical shoreline comparison studies. It is easily recognizable in the field (it was the boundary line mapped by early NOS topographers), it can usually be approximated from air photos (Stafford, 1971), and it represents a shoreline feature that has been accurately mapped (albeit sporadically) since the mid-1800s. Morton *et al.* (1993), however, notes that the HWL is not always the most diagnostic of long-term shoreline movement, particularly in areas with gently sloping beaches or where rates of erosion are low. In these and other situations, Morton argues that other morphological features, such as erosion scarp, crest of washover terrace, or vegetation line are more diagnostic of long-term change than the HWL. Use of these other indicators, however, severely limits the temporal span of source data. While these shoreline indicators are sometimes depicted on the historical NOS T-sheets (the only good source of pre-aerial photo shoreline position information), only the HWL was delineated with enough accuracy to be useful for temporal position comparisons.

For cliffed situations, the top edge of the bluff (employing the terms cliff and bluff interchangeably) is commonly used as a shoreline change reference feature. Unfortunately, historical maps depicting accurate bluff positions are rare, thereby limiting historical source data to the temporal span covered by aerial photographs (e.g., 1930s to present).

Table E7 Significant milestones in the development of historical shoreline change mapping

Lucke (1934)	Cartographically traced historical and current T-sheets to produce shoreline change maps showing temporal changes for Barnegat Inlet, New Jersey.
Plusquellec (1966)	Produced shoreline change maps compiled from vertical air photos for an area between Brigantine and Beach Haven Heights, New Jersey. The air photos were not corrected for tilt distortion.
Stafford (1971)	Took point measurements from aerial photographs in a study that documented coastal erosion along the Outer Banks of North Carolina. In this study, Stafford developed procedures that minimized scale and tilt distortions inherent in air photos. In addition, Stafford and Langfelder (1971) demonstrated that the mean high water line was the best indicator of land–water contact, and most subsequent (non-bluff) mapping studies have used this interface as the shoreline marker for comparative purposes (Leatherman, 1983).
Morton (1974)	Produced historical shoreline change maps for the entire coast of Texas. These maps were produced by cartographically tracing T-sheets and aerial photograph mosaics.
Dolan *et al.* (1978)	Developed a computer storage and retrieval system, called orthogonal grid address system (OGAS), that incorporated manually produced, unrectified aerial photography in erosion studies.
Leatherman (1982, 1983)	Developed a computer program, called metric mapping, that corrected T-sheets for scale distortion and rectified air photos for scale, tilt, tip, and yaw distortions. Metric mapping computer routines automated much of the data compilation and analysis. The program was used to generate historical shoreline change maps for large sections of the US east coast.
Everts *et al.* (1983)	Used cartographic and computer mapping techniques to compile historical shoreline change maps for a number of US mid- and south-Atlantic states.
Morton et al. (1993) Byrnes and Hiland (1993) French and Leatherman (1994)	Used GPSs in shoreline analyses and historical shoreline change mapping.
Overton *et al.* (1996) Moore *et al.* (1999) Coyne *et al.* (1999)	Used softcopy photogrammetry to measure shoreline change. The technique significantly improves upon earlier computer methods for rectifying aerial photography. For example, fewer ground control points are required, and the shoreline is digitized on-screen.

Other features are sometimes used to monitor historical bluff change. The State of Wisconsin, for example, has experimented with the mid-bluff contour, with the rationale that the middle of the bluff is typically the most stable portion of the slope because it is located below areas subject to frequent failure and above the associated debris or talus pile. A computer program selects the mid-bluff contour from a digi-tal elevation map, in theory, removing the effects of operator error when delineating the shoreline change reference feature. However, the temporal limitations of using the mid-bluff contour are even more restrictive than for using top edge of the bluff, as the quality of 1930s and 1940s aerial photography may prevent their use in determining the mid-bluff contour.

In summary, the HWL is generally the preferred shoreline indicator for use in documenting historical shoreline changes. However, certain beach characteristics (e.g., gently sloping beaches) or situations where the HWL is unavailable or meaningless (e.g., bluff situations in the Great Lakes) may require use of other shoreline indicators. Use of the HWL allows shoreline change comparisons that includes maps, aerial photographs, and ground surveys, thereby expanding the historical limits to the mid-1800s. Use of other shoreline indicators limits shoreline comparisons to aerial photographs and ground surveys, thereby constraining the historical bounds to the last 60 years.

Accuracy of source data

Accuracy of source data has been discussed in a number of papers (Morton, 1974; Anders and Byrnes, 1991; Crowell *et al.*, 1991; Gorman *et al.*, 1998; Moore, 2000). In general, most researchers acknowledge that historical and current aerial photographs provide an important source from which to monitor shoreline change and calculate rates of change. Questions regarding accuracy of older source data, particularly historical NOS T-sheets, spurred Crowell *et al.* (1991) to devise a way to determine the maximum errors one would expect to find using historical and current aerial photography and historical NOS T-sheets. The authors estimated "combined" errors by estimating error for each individual step in shoreline data compilation, including survey, digitization, mapping, and correction of photographic distortion. The individual estimated errors were combined using the theory of propagation of errors to obtain estimates of expected error in the digitized location of the interpreted shoreline (in this case, the high-water line). Expected errors were calculated for the following data sources (all at 1 : 10,000 scale): 1846–1930 T-sheets prepared prior to the use of aerial photography: 8.4–8.9 m; post-1950s T-sheets compiled using aerial photography: 6.1 m; non-tidal coordinated aerial photography: 7.5–7.7 m. These figures were based on analysis of T-sheets from the northeast coast of the United States. Importantly, short-term process variability factors, such as daily and monthly tidal changes and storm effects, were not included in the error estimates.

Preparation, rectification, and compilation of source data

In order to compare mapped, aerial photographed, and field surveyed shorelines, a common coordinate system must be chosen so that all source data can be properly aligned to ensure the most accurate determination of shoreline movement. Modern computer techniques have automated much of the compilation process, but a certain amount of manual preprocessing must completed prior to data digitization. The preprocessing varies by the type of input data.

Historical maps

To align historical maps to a specific coordinate system, one must identify a number of permanent and stable points or features on the map for which accurate and current geographic coordinates are known. The features are known as primary control points. For computer applications, a minimum of four primary control points are required, although six to eight points are commonly used to improve accuracy and aid in error analysis. These points are used to calculate the transformation necessary to change from one coordinate system to another and to correct for error factors such as scale differences, rotation, and uneven map shrinkage. Selection of primary control points is trivial if the source map contains current geo-coordinates, such as latitude and longitude tick-marks. However, historical maps often have latitude longitude lines based on obsolete data. For example, latitude–longitude coordinates for the US maps produced prior to 1927 must be updated to better standards such as the North American Datum of 1927 (NAD27), or preferably the more recent and accurate NAD83. Several Geographic information system (GIS) programs are currently available that automatically transform the pre-1983 digitized map data to the 1983 standard. If latitude–longitude coordinates are not available for the map, or if the coordinates are present but with an unknown datum standard, then other techniques must be used to update the map. One very useful technique is to search the map for survey stations or hard permanent features for which modern coordinates are known. In the case of NOS T-sheets, several *triangulation stations* are usually mapped for which current coordinates can be obtained.

Aerial photography

Since 1927, aerial photography and photographic methods have been used to provide topographic information along the coast, creating an important dataset for use in historical shoreline change studies. Air photos, however, are not map projections; corrections for distortion due to tilt, tip, and yaw (and sometimes relief) must be made prior to and/or concurrent with the compilation process so that shoreline positions obtained from air photos can be compared with those compiled from planimetric maps and ground surveys. Several computer programs are now available that automate air photo rectification in a process called "soft-copy photogrammetry" (Moore, 2000). The process involves creating photomosaics prepared from scanned and rectified aerial photographs. A series of ground control points are used to spatially reference the imagery to geo-coordinates. Pass points, which are identifiable landmarks common to two or more photographs (but not ground controlled), are supplemental and serve to reference the photos to each other (Coyne *et al.*, 1999). In summary, the rectification process transforms the imagery to a map projection and corrects or minimizes map distortions, such as tilt, scale differences, relief displacement, and radial lens distortion.

GPS surveys

Field-based shoreline surveys using GPS has been increasingly used since the early 1990s (Byrnes and Hiland, 1993; Morton *et al.*, 1993; French and Leatherman, 1994). Shoreline position data are collected by driving an off-road vehicle equipped with a GPS unit directly along the HWL. Geo-coordinates are collected at specified intervals, allowing real-time digitization of the HWL. GPS measurements are so accurate (within a meter) that measurement error is usually insignificant compared with the ambiguity inherent to a shoreline change reference feature.

Shorelines digitized from maps and air photos along with GPS digital surveys are combined and referenced to a common geo-coordinate base. The resultant hardcopy product is referred to as a historical shoreline map or historical shoreline change map. Most such maps usually contain a minimum of four shorelines, although six to eight shorelines are commonly displayed.

Calculation of long-term erosion rates

The primary reason for calculating long-term erosion rates is for land-use planning, ranging in purpose from providing information to property owners to establishing regulatory setback lines for coastal construction. Usually building setbacks are based on the average annual erosion rate (AAER) for the area, multiplied by a specified number of years, commonly 30 and 60 years. The computed setback is then measured landward from an erosion reference feature, whose movement would have the most impact on the structural integrity of a building. As such, the erosion reference feature is typically the top edge of a bluff, dune escarpment, vegetation line, or beach scarp. If none of these are present, then the HWL is often used. If the erosion rate at a site is, for example, 5 feet (approximately ≈1.5 m)per year, then the 30-year setback is 150 feet (≈45 m) and the 60-year setback is 300 feet (≈90 m) both of which would be measured landward from, for example, the top edge of the bluff.

The high valuation of coastal property in many countries underscores the importance of developing coastal erosion data that are both accurate and appropriate. Accurate rates of erosion can be determined in many ways, using all or subsets of shoreline data, but not all of the rates would necessarily be appropriate for the situation. Assuming that the source shoreline data are accurate, there are three additional important issues to consider prior to calculating appropriate long-term rates of change: (1) what temporal span of shoreline data should be used in determining the rate (i.e., should short-term, medium-term, or long-term data be used); (2) should post-storm shorelines be included; and (3) what statistical method should be used to determine the rate? Insight into the answers to these questions can be gained from the following example of an actual history of shoreline positions.

Consider the data shown in Figure E9 (Douglas and Crowell, 2000). The graph shows 10 shoreline positions from 1845 to 1997. The shoreline location data are from the Atlantic coast of Delaware, just south of the Indian River Inlet near a location called Cotton Patch Hill. The 1929, 1962, and 1963 shoreline positions were photographed or mapped immediately following major storms and are shown as open circles on Figure E9. Note that these three post-storm shoreline positions are inconsistent with the long-term trend line determined from the other seven shoreline positions. During a large storm, more erosion can occur in a few days than in the previous 50 years, which is often followed by an extended period of beach recovery. Note that accretion continued after the 1962 Ash Wednesday nor'easter for at least a decade (Figure E9). Therefore, determination of the long-term rate of beach erosion should not be undertaken without consideration of the role of severe storms on shoreline position.

The example above also serves to illustrate the proper statistical approach to the shoreline position forecasting problem. A correct methodology requires that the forecast model be in reasonable agreement with the actual physical situation. Figure E9 shows that there was a long-term trend of beach erosion, superimposed with episodes of severe erosion followed by extended periods of recovery. Use of post-storm shoreline positions in determination of the long-term trend will yield aberrant results. Unfortunately, the most common method used by researchers to calculate rates of erosion is the "end-point-rate" method. In this method, the analyst selects two representative shorelines, often the earliest and the most recent, to calculate the amount of shoreline movement. This distance is then divided by the time elapsed between successive shoreline positions. The advantages of this technique include simplicity in application, and two shorelines suffice to obtain a rate. A major disadvantage is that the position of one or both of the endpoint shorelines may be aberrant, and so the rate of change can be misleading for use in determining future shoreline positions. Moreover, data between the end points, of value in determining the rate of change, are not used and hence potentially useful information is ignored.

Another method becoming increasingly popular among researchers (see Crowell and Leatherman, 1999) is linear regression. With this method, a best-fitting straight line is determined that minimizes the sum of the squares of the lengths of vertical line segments drawn from the individual data points to the fitted line. The advantages of this method are that: (1) all data points (except for post-storm shorelines) are used in the rate calculation, thereby reducing the influence of spurious data points; (2) linear regression is an easily understood statistical analysis tool that can be performed with most spreadsheet programs and scientific pocket calculators, and (3) summary and related statistical techniques are available to test and measure the quality of the straight-line fit and estimate the variance of the data (Kleinbaum and Kupper, 1978). However, careful attention must be paid to the influence of severe storms on shoreline position when selecting the data to be used in the regression analysis. Douglas and Crowell (2000) present a methodology for evaluating shoreline position data and computing predicted shoreline positions and their uncertainty. Furthermore, only summertime data should be used because of large seasonal fluctations in beach width; combining summer and winter shorelines is almost like comparing apples and oranges— one can obtain nearly any answer that you wish, ranging from accretionary to a highly erosional trend.

A method for determining rates of erosion known as the minimum description length (MDL) criterion (Rissanen, 1989) was proposed by Fenster *et al.* (1993). This technique uses a type of complexity-penalty model criteria to determine the straight or curved line (polynomial) to fit the data set. While conceptually interesting, the MDL method de-emphasizes historical data to the extent that short-term shoreline perturbations unduly influence the data set, thereby hindering useful long-term (30- to 60-year) forecasts. Crowell *et al.* (1997) demonstrated that simple linear regression consistently gives better long-term forecasting results than does the MDL or the end-point rate technique.

Conclusions

Many difficulties can be encountered in accurately determining the long-term behavior of shorelines based on sparse data sets. The complexity increases if engineering changes (such as construction of groins or jetties) have been made to the shoreline, or if tidal inlets are opened as a result of a large storm. In addition, capes and spits often display a complex episodic cycle of erosion and accretion that is unrelated to storms. Thus, the highest-quality forecasts of long-term shoreline position need to consider meteorological and geomorphic factors and the role they play in the evolution of the shoreline position.

Mark Crowell, Stephen P. Leatherman, and Bruce Douglas

Bibliography

Anders, F.J., and Byrnes, M.R., 1991. Accuracy of shoreline change rates as determined from maps and aerial photographs. *Shore and Beach*, **59**(1):17–26.

Byrnes, M.R., and Hiland, M.W., 1993. Shoreline position and nearshore bathymetric change. Technical report prepared for Coastal Engineering Research Center.

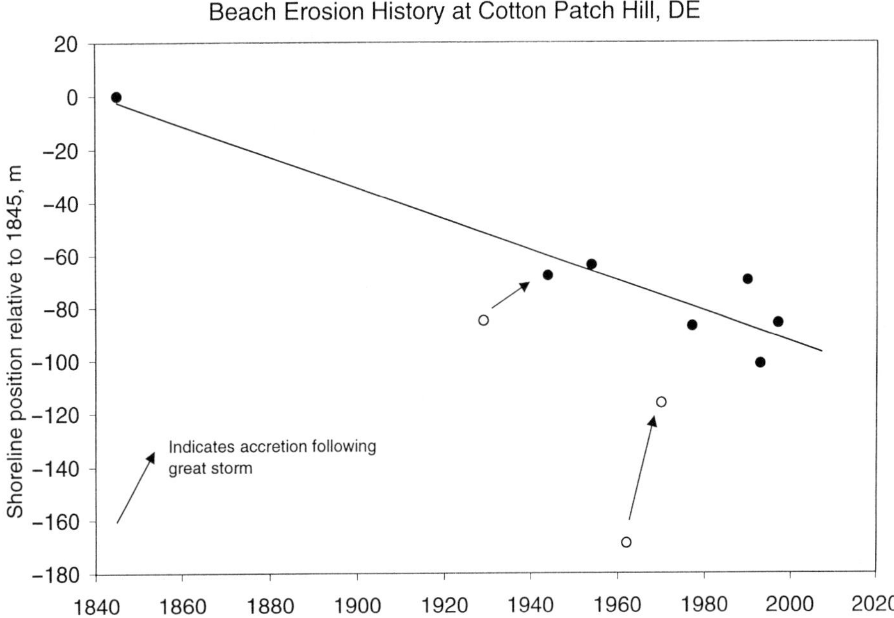

Figure E9 Graph of beach erosion history at Cotton Patch Hill, DE (after Douglas and Crowell, 2000).

Coyne, M.A., Fletcher, C.H., and Richmond, B.M. 1999. Mapping coastal erosion hazard areas in Hawaii: observations and errors. In Crowell, M., and Leatherman, S.P. (eds.), *Coastal Erosion Mapping and Management. Journal of Coastal Research*, **28**(Special Issue): 171–184.

Crowell, M., and Leatherman, S.P., eds., 1999. Coastal Erosion Mapping and Management. *Journal of Coastal Research*, **28**(Special Issue): 196 pp.

Crowell, M., Leatherman, S.P., and Buckley, M.K., 1991. Historical shoreline change: error analysis and mapping accuracy. *Journal of Coastal Research*, **7**(3): 839–852.

Crowell, M., Leatherman, S.P., and Buckley, M.K., 1993. Erosion rate analysis: long term versus short term. *Shore and Beach*, **61**(2): 13–20.

Crowell, M., Douglas, B.C., and Leatherman, S.P., 1997. On forecasting future U.S. shoreline positions: a test of algorithms. *Journal of Coastal Research*, **13**(4): 1245–1255.

Dolan, R., Hayden, B., and Heywood, J., 1978. A new photogrammetric method for determining shoreline erosion. *Coastal Engineering*, **2**: 21–39.

Douglas, B.C., and Crowell, M., 2000. Long term shoreline position prediction and error propagation. *Journal of Coastal Research*, **16**(1): 145–152.

Douglas, B.C., Crowell, M., and Leatherman, S.P., 1998. Considerations for shoreline position prediction. *Journal of Coastal Research*, **14**(3): 1025–1033.

Everts, C.H., Battley, J.P., and Gibson, P.N., 1983. Shoreline movements, report 1: Cape Henry, VA to Cape Hatteras, N.C., 1849–1980. Vicksburg, MS: *CERC Technical. Report 83–1*, 111 p.

Fenster, M.S., Dolan, R., and Elder, J.F., 1993. A new method for predicting shoreline positions from historical data. *Journal of Coastal Research*, **9**(1): 147–171.

French, G.T., and Leatherman, S.P., 1994. The use of global positioning systems in coastal studies. Washington, DC: FEMA, *Technical report*.

Galgano, F.A., Douglas, B.C., and Leatherman, S.P., 1998. Trends and variability of shoreline position. *Journal of Coastal Research*, **26** (Special Issue): 282–291.

Gorman, L., Morang, A., and Larson, R. Monitoring the coastal environment; part IV: mapping, shoreline changes, and bathymetric analysis. *Journal of Coastal Research*, **14**(1): 61–92.

Heinz Center, 2000. *Evaluation of erosion hazards*. Washington, DC: H. John Heinz III Center for Science, Economics and the Environment.

Kleinbaum, D.G., and Kupper, L.L., 1978. *Applied Regression Analysis and Other Multivariate Methods*. Pacific Grove: Duxbury Press.

Leatherman, S.P., 1982. New automated technique of historical shoreline mapping. Geological Society of America, Annual Meeting, New Orleans, LA.

Leatherman, S.P., 1983. Shoreline mapping: a comparison of techniques. *Shore and Beach*, **51**: 28–33.

Lucke, J.B., 1934. A study of Barnegat Inlet, New Jersey. *Shore and Beach*, **2**(2): 44–49.

Moore, L.J., 2000. Shoreline mapping techniques. *Journal of Coastal Research*, **16**(1): 111–124.

Moore, L.J., Benumof, B., and Griggs, G.B., 1999. Coastal erosion hazards in Santa Cruz and San Diego Counties, California. In: Crowell, M., and Leatherman, S.P., (eds.), *Coastal Erosion Mapping and Management. Journal of Coastal Research*, **28**(Special Issue): 121–139.

Morton, R.A., 1974. *Shoreline changes on Galveston Island: an analysis of historical changes of the Texas Gulf shoreline, Geological Circular 74–2*, Austin, TX: Texas Bureau of Economic Geology, 47 p.

Morton, R.A., Leach, M.P., Paine, J.G., and Cardoza, M.A., 1993. Monitoring beach changes using GPS surveying techniques. *Journal of Coastal Research*, **9**(3), 702–720.

Overton, M., Petrina, C., and Fisher, J., 1996. Determining shoreline position using historical photography and digital softcopy photogrammetry. ASPRS/ACSM Annual Convention, *Technical Paper*, Vol. 1, pp. 512–513.

Plusquellec, P.L., 1966. Coastal morphology and changes of an area between Brigantine and Beach Haven Heights, New Jersey. M.S. thesis, Department of Geology, University of Illinois, 115 p.

Rissanen, J., 1989. Stochastic Complexity in Statistical Inquiry. Singapore: World Scientific Press.

Stafford, D.B., 1971. An aerial photographic technique for beach erosion surveys in North Carolina, U.S. Army Corps of Engineers, Coastal Engineering Research Center, *Technical Memorandum No. 36*, 115 p.

Stafford, D.B., and Langfelder, J., 1971. Air photo survey of coastal erosion, *Photogrammetric Engineering*, **37**, 565–575.

Cross-references

Coastline Changes
Coasts, Coastlines, Shores and Shorelines
Cross-Shore Sediment Transport
Global Positioning Systems
Mapping Shores and Coastal Terrain
Photogrammetry

EROSION PROCESSES

Erosional processes along coastlines include: (1) the direct effects of hydraulic action, wedging, and cavitation by waves; (2) abrasion (corrasion), using sand, gravel, and larger rock fragments as tools; (3) attrition of the rock particles themselves during this abrasive action; (4) salt weathering or fretting; (5) erosion by organisms (bioerosion); and (6) chemical attack, or corrosion, which weakens the rocks and accelerates erosion. The rates of erosion by these processes are a function of the exposure of a coast to wave attack (wave energy and length of time of exposure), and the resistance of the materials to erosion and weathering.

Waves are the most dominant force causing coastal erosion. Ocean waves typically break at depths that range from about 1 to 1.5 times wave height. Because waves are seldom more than 6 m high, the depth of vigorous erosion by surf is usually limited to from 6 to 9 m below sea level. This theoretical limit is confirmed by observations of breakwaters and other coastal structures that are only rarely affected by surf to depths of more than 7 m. The change in water level caused by tides, however, allows waves to attack a greater range of elevations on the coast.

Wave impact directly exerts great pressure and can be a highly effective agent of erosion. The pressures produced by breaking waves have been measured by dynamometers, with reported values as high as 250 kg/m^2 (6,000 lb/ft^2). These measurements agree with theoretical estimates (Gaillard, 1904) based on wavelength and celerity. When the waves are long and their celerities are correspondingly high, the greatest pressures are generated.

When waves break against a cliff face, water is driven into cracks and joints and the air within the cracks is greatly compressed. The compressed air acts as a wedge, widening the fractures and loosening blocks of rock. As the wave recedes, the compressed air expands with explosive force (a process called cavitation), effectively quarrying out large blocks (Barnes, 1956). Such quarrying action has provided enough energy to move blocks weighing many tons. There are many reports of the movement of large blocks by wave action, including a 2,500-ton piece of concrete at Wick (Scotland), and a section of a breakwater weighing 1,700 tons that was completely overturned at Bilbao (Spain).

Another method of marine erosion is abrasion, the mechanical wearing and grinding away of rock surfaces by the friction and impact of rock particles. Also called corrasion, this process involves the action of sand, gravel, and large rocks picked up by the waves and driven back and forth against the exposed shore. Potholes can develop where large clasts are rotated by swirling water in the surf or breaker zone. The abrasion process includes dragging (friction), rolling, bouncing (saltation), and the artillery action of large fragments, which can shatter solid rock when hurled against a cliff. At Tillamook Rock on the Oregon coast, for example, rocks weighing as much as 297 kg have been picked up during severe storms and thrown through the windows of the lighthouse, 44 m above mean high water. Similarly, the windows of the Dunnet Head Lighthouse on the north coast of Scotland, more than 100 m above the high water mark, have been broken by rocks propelled by wave action.

Concentration of wave attack at the cliff base accelerates erosion and notches the base, permitting collapse or movement of the overlying material (Sunamura, 1992). Seepage of ground-water (especially along seams and fractures), frost action, and wind combine with the undercutting to produce cliff recession (Nott, 1990). Differences in the resistance of the rocks being eroded and episodic collapse may produce an irregular cliff face, subject to mass movements (Duperret *et al.*, 2002; Benumof *et al.*, 2000). The debris produced by cliff collapse is subsequently used in abrasion of the cliff base, and/or removed by wave action, allowing the wave attack to continue. Thus, cliff erosion is generated by two processes: notching at the base of the cliff by marine processes, and collapse and denudation of the entire cliff face by a combination of subaerial and marine processes (Duperret *et al.*, 2002).

The rate of cliff retreat, combined with rock type, climate (especially rainfall) and vegetation will determine the nature of mass movement down the cliff face. An extensive examination of Japanese cliff erosion, found that long-term cliff retreat is logarithmically related to rock strength and unrelated to cliff height (Sunamura, 1977, 1992). Benumof *et al.* (2000) also found that rock strength, rather than wave energy or slope, is the critical factor in coastal cliff retreat.

Mathematically, the major factors of basal erosion have been summarized in Sunamura's equation:

$$x = f(F_w, S_r, t),$$

where x is the basal cliff erosion rate, F_w is the wave induced force, S_r is the cliff material resistance, and t is the time. Belov *et al.* (1999) developed a mathematical erosion function (the so-called Belov–Davies–Williams (BDW) equation) which accounts for erosion forces and the strength of the rocks, and also for the exponential decrease in wave erosion intensity with height. Numerical models have been quite successful at simulating basal cliff erosion (Belov *et al.*, 1999).

The level of planation of the cliff base will usually lie immediately above the level of permanent rock saturation, where continued wetting and drying weakens the rock, allowing accelerated wave erosion. This level will, however, vary depending on wave energy, rock type and structure, platform width, and accompanying atmospheric processes (Stephenson and Kirk, 2000). The abrasion along the shore works horizontally to form a wave-cut cliff and a wave-cut platform. Exposure may be an important factor, as studies of shore platforms in Japan have shown that the mean platform width is about 40 m on the sheltered Inland Sea, 50 m on the more exposed Japanese Sea, and 60 m on the exposed Pacific coast (Trenhaile, 1997).

Attrition is the process of wearing down by friction, involving the abrasion, impact, and grinding together of clasts in motion, resulting in smaller and usually better-rounded particles (in Europe, abrasion is used as a synonym for attrition). The impact and rubbing together of the rock fragments during wave action, as well as the grinding of the blocks that fall as the cliff is undercut, cause the tools themselves to wear down to smaller sizes and become rounded. Studies of attrition of beach clasts include experiments and observations in natural settings. Kuenen (1964) experimentally correlated rate of attrition with relative resistance and median clast size. Early comparisons of natural sediment distributions were done by Landon (1930), but lack of control makes the conclusions questionable. The observed diminution of grain size along a coast could be also due to change in source material or to sorting by longshore transport.

Salt weathering or fretting takes place when salt crystals grow within pore spaces in rocks; the crystal growth expands the pore and may lead to fracturing of the rock (Wellman and Wilson, 1965; Mottershead, 1989). This may be the result of strong evaporation that draws salt solutions to the surface by capillary action. The salt-weathering process is most effective in areas exposed to salt spray and subject to alternative wetting and drying. Salt weathering can also produce a honeycomb pattern in the rocks called tafoni, which is most characterisitic of tropical and subtropical semiarid to moderately arid climates. In chalk cliffs, basal notching has been attributed to salt weathering and repeated wetting and drying cycles (Duperret *et al.*, 2002).

The chemical and solvent action of seawater may be especially important along limestone and chalk coasts. On some atolls, solution may be responsible for 10% of total erosion of the coast (Trenhaile, 1997). However, even non-calcareous rocks may weather more rapidly as a result of the chemical action of seawater. In some cases, chemical dissolution and salt fretting may act together.

Additional factors that may contribute to mechanical weathering in certain areas include large diurnal temperature variations and biological activity. Weathering due to temperature variations mainly affects the surface layers. The temperature distribution in rocks is non-uniform, and differences in rock composition yield thermal conductivity contrasts and large temperature gradients. This can cause thermal cracking of blocks, resulting in displacement along existing joints and fractures.

Biological erosion is especially important in tropical or subtropical areas having limestone coasts. Algae are probably the most important bioerosional agent on rocky coasts. The substrate under algal mats is exposed to the products of metabolism and organic waste, which can directly etch the rocks. Algae also support grazing organisms such as gastropods and echinoids, which abrade rock surfaces. Bioerosion, combined with the chemical solution of limestone by algae, produces the intricate scalloping known as phytokarst (Bromley, 1978). Boring and browsing organisms erode rocks most effectively in the inter-tidal zone, and bioerosion can eventually weaken the framework of reefs to the point where collapse takes place. Rates of bioerosion on horizontal and vertical limestone surfaces have been estimated at about 0.5–1 mm per year (Trenhaile, 1997).

In cold regions, where unconsolidated but frozen, ice-bonded sediments are exposed to wave erosion, convective heat transfer between seawater and the melting surface of the frozen cliff sediment (thermal abrasion) combines with erosive wave action and slope failure to undercut cliffed coasts (Kobayashi *et al.*, 1999). Wave erosion removes the unfrozen sediment, which would insulate the frozen sediment from further melting. The combined thermal and mechanical processes can lead to rapid erosional retreat (>1 m per year) during periods of open-water conditions. Tidally induced frost action occurs in the inter-tidal zone where rocks freeze when exposed to air and thaw when covered by the rising tide. This mechanism is particularly effective because of the frequency of freeze–thaw cycles. In coastal Maine, for example, there are an average of 133 freeze–thaw cycles each year in the inter-tidal zone versus only 30 or 40 cycles inland (Trenhaile, 1997).

Michael R. Rampino

Bibliography

Barnes, H.J., 1956. Cavitation as a geological agent. *American Journal of Science*, **254**: 493–505.

Belov, A.P., Davies, P., and Williams, A.T, 1999. Mathematical modeling of basal coastal cliff erosion in uniform strata: a theoretical approach. *Journal of Geology*, **107**: 99–109

Benumof, B.T., Storlazzi, C.D., Seymour, R.J., and Griggs, G.B., 2000. The relationship between incident wave energy and sea-cliff erosion rates. *Journal of Coastal Research*, **16**: 1162–1178.

Bromley, R.G., 1978. Bioerosion of Bermuda reefs. *Palaeogeography, Palaeoclimatology, Palaeoecology*, **23**:169–197.

Duperret, A., Genter, A., Mortimore, R.N., Delacourt, B., and De Pomerai, M.R., 2002. Coastal rock cliff erosion by collapse at Puys, France: The role of impervious marl seams within chalk of NW Europe. *Journal of Coastal Research*, **18**: 52–61.

Gaillard, D.B.W., 1904. *Wave Action in Relation to Engineering Structures*. Washington, DC: US Army Corps of Engineers.

Kobayashi, N., Vidrine, J.C., Nairn, R.B., and Saloman, S.M., 1999. Erosion of frozen cliffs due to storm surge on Beaufort Sea coast. *Journal of Coastal Research*, **15**: 332–344.

Kuenen, Ph. H., 1964. Experimental abrasion: surf action. *Sedimentology*, **3**: 29–43.

Landon, R.E., 1930. An analysis of beach pebble abrasion and transportation. *Journal of Geology*, **38**: 437–446.

Mottershead, D.N., 1989. Rates and patterns of bedrock denudation by coastal salt spray weathering: a seven-year record. *Earth Surface Processes and Landforms*, **14**: 383–398.

Nott, J.F., 1990. The role of sub-aerial processes in sea cliff retreat— a southeast Australian example. *Zeitschrift für Geomorphology, N. F.*, **34**: 75–85.

Stephenson, W.J., and Kirk, R.M., 2000. Development of shore platforms on Kaikoura Peninsula, South Island, New Zealand. II: The role of sub-aerial weathering: *Geomorphology*, **32**, 43–56.

Sunamura, T., 1977. A relationship between wave-induced cliff erosion and erosive force of waves. *Journal of Geology*, **85**: 613–618.

Sunamura, T., 1992. *Geomorphology of Rocky Coasts*. Chichester: John Wiley and Sons.

Trenhaile, A.S., 1997. *Coastal Dynamics and Landforms*. Oxford: Oxford University Press.

Wellman, H.W., and Wilson, A.T., 1965. Salt weathering: a neglected geological erosive agent in coastal arid environments. *Nature*, **205**: 1097–1098.

Cross-references

Bioerosion
Cliffs, Erosion Rates
Cliffs, Lithology versus Erosion Rates
Ice-Bordered Coasts
Notches
Rock Coast Processes
Shore Platforms
Wave-Dominated Coasts
Waves
Weathering in the Coastal zone

ESTUARIES, ANTHROPOGENIC IMPACTS

Introduction

Estuaries are highly stressed, physically controlled ecosystems characterized by wide variations in environmental conditions. A multitude of anthropogenic activities also affects these coastal ecotones and can significantly impact their water quality, habitats, and biotic communities. These activities are closely coupled to accelerated population growth and development in coastal watersheds. By the year 2025, the coastal population worldwide is expected to approach six billion people (Weber, 1994; Hameedi, 1997). Poorly planned human settlement in the coastal zone will lead to greater stresses on estuarine systems worldwide.

Much human activity is concentrated near estuaries because of their great commercial, recreational, and aesthetic value. Estuaries support multibillion dollar commercial and recreational pursuits, such as tourism, fisheries, mariculture, transportation, shipping, electric power generation, oil and gas recovery, and other endeavors (Kennish, 2000). Estuarine and coastal marine fisheries alone generate more than $20 billion annually for the US economy (Pirie, 1996). However, an array of environmental problems has arisen in response to greater human use of estuarine resources, most notably related to pollution inputs and habitat degradation.

Pollution and habitat impacts

Enrichment

Day *et al.* (1989) have identified five categories of estuarine human impacts: (1) enrichment; (2) toxins; (3) overfishing; (4) introduction of exotic species; and (5) physical alteration. Kennish (1992, 1997, 2000, 2001) has provided an overview of them. Enrichment refers to impacts primarily associated with the input of excessive amounts of nutrients, organic matter, and heat (calefaction or thermal loading). The overfertility of estuarine waters, particularly large inputs of nitrogen and phosphorus from point and nonpoint sources, promotes rapid algal growth (e.g., phytoplankton blooms) and biomass accumu-lation which can lead to hypoxic or anoxic conditions and reduced phytoplankton species richness. The eutrophication of estuarine waters often has devastating consequences, occasionally culminating in heavy mortality of benthic and finfish populations, including commercially and recreationally important forms, and causing significant changes in estuarine community structure and trophic dynamics. Phytoplankton blooms (including toxic and nuisance algal blooms) stimulated by nutrient enrichment usually decrease light attenuation, and the shading effect commonly reduces the production and areal distribution of submerged aquatic vegetation (Valiela, 1995; Valiela *et al.*, 1997; Livingston, 2001). This habitat change often results in the shift from a sea grass-dominated system to a phytoplankton-dominated system in shallow coastal bays (Kennish, 2000).

Estuarine eutrophication is an escalating problem worldwide because of increases in nutrient inputs from a burgeoning human population in the coastal zone. Influxes of nutrients from farmlands, stormwater runoff, wastewater discharges, groundwater seepage, atmospheric deposition, and other sources have contributed to nutrient loading concerns in numerous estuaries. Examples of such impacted US systems include Long Island Sound (New York), tributaries of Chesapeake Bay (e.g., Patuxent, Rappahannock, and York Rivers), Waquiot Bay (Massachusetts), Perdido Bay (Florida), and Barataria Basin (Louisiana). In the conterminous United States, Bricker *et al.* (1999) have documented 44 highly eutrophic estuaries and another 40 estuaries with moderately eutrophic conditions.

Organic carbon enrichment in the form of wastes from sewage and wildlife inputs, livestock and fish processing operations, and industrial effluents not only exacerbates eutrophication problems by raising inorganic nutrient concentrations but also compromises water and sediment quality by increasing pathogen concentrations. Overenrichment of sewage wastes is linked to elevated biochemical oxygen demand and reduced dissolved oxygen levels in estuaries. Sewage wastes can have significant impacts on estuarine and marine organisms (Costello and Read, 1994). For example, such chronic organic carbon enrichment degrades benthic communities, typically reducing species richness and favoring the proliferation of a few opportunistic forms, which frequently dominate the communities.

The calefaction of estuarine waters by the release of waste heat from electric generating stations influences the physiological and behavioral responses of organisms in near-field regions of outfall sites. Biotic impacts typically observed in these receiving waters are decreased primary production, failed reproduction, increased mortality (e.g., heat- and cold-shock mortality of fish), and altered community structure (Kennish, 1992, 1997, 2001). Behavioral adjustments may be dramatic, with many nektonic organisms exhibiting avoidance or attraction responses to the effluent. Significant changes in the species composition, abundance, and diversity of biotic communities commonly occur in outfall canals of electric generating stations.

Toxins

Most chemical contaminants enter estuarine environments from land-based sources via agricultural and urban runoff, municipal and industrial discharges, riverine inflow, and atmospheric deposition. Many have either acute or chronic adverse effects on organisms. Chief among these contaminants are halogenated hydrocarbon compounds, polycyclic aromatic hydrocarbons (PAHs), and heavy metals. All pose a potential threat to organisms, especially in hot spot areas of urbanized estuaries where more elevated concentrations of the contaminants are found (Daskalakis and O'Connor, 1995; O'Connor, 1996). Halogenated hydrocarbon compounds, PAHs, and heavy metals are particle reactive substances; they tend to rapidly sorb to fine-grained sediments and accumulate on the estuarine seafloor, which serves as a repository for the contaminants (Kennish, 1997). In bottom sediments of estuaries, the contaminants resist degradation and may persist essentially unaltered for many years. While it is important to monitor the absolute concentrations of the contaminants, their bioavailability must be considered when assessing biotic impacts (Hamelink *et al.*, 1994).

The halogenated hydrocarbons originate mainly from agricultural and industrial sources and include an array of biocides (e.g., insecticides, herbicides, and fungicides), low-molecularweight compounds (e.g., chlorofluorocarbons), and high-molecularweight chemicals (e.g., chlorinated aromatics and chlorinated paraffins). Some of the most notable synthetic organic compounds of this group (e.g., PCBs, DDT, and non-DDT pesticides) rank among the most toxic constituents in estuaries, and consequently are a high priority concern. They are suspected etiologic agents which have been linked to a number of disorders in estuarine animals such as reproductive abnormalities, altered endocrine physiology, aberrant developmental patterns, skin and liver lesions, and cancer (Kennish, 2001). DDT and PCBs are particularly hazardous substances because they biomagnify in food chains; therefore, they pose a potentially serious threat to humans who consume contaminated seafood products.

Polycyclic aromatic hydrocarbons are widely distributed contaminants that largely derive from fossil fuel combustion, land runoff, municipal and industrial wastewater discharges, and oil spills. Natural processes (e.g., volcanic activity) also generate PAHs, but their inputs to estuaries are low relative to human sources. PAHs are a problematic class of contaminants because the higher-molecular-weight forms are predominantly carcinogenic, mutagenic, and teratogenic, and the lower-molecularweight varieties can be acutely toxic. PAHs, when readily bioavailable, produce hepatic neoplasms (e.g., hepatocellular carcinoma, hepatocellular adenosoma, and cholangiocellular carcinoma) and other disorders in demersal fish and shellfish. They also are responsible for a range of sublethal effects manifested as biochemical, behavioral, physiological, and pathological aberrations in organisms that can contribute to changes in the structure of estuarine communities.

Heavy metals likewise are deleterious to estuarine organisms, causing reproductive and developmental abnormalities when present at elevated concentrations (Kennish, 1992, 1997). Orga-nometals (e.g., tributyl tin, alkylated lead, and methylmercury) are particularly toxic to these organisms. Riverine inflow, wastewater discharges, and atmospheric deposition deliver a large fraction of the total heavy metal burden found in estuaries. The combustion of fossil fuels and wastes from mining, smelting, electroplating, and refining operations are important sources. The concentrations of heavy metals in bottom sediments of estuaries are three to five orders of magnitude greater than those in overlying waters (Bryan and Langston, 1992; Kennish, 1998). Thus, benthic organisms may be most vulnerable to the harmful effects of these contaminants.

Overfishing

Fisheries overexploitation depletes estuarine fish and shellfish stocks and often cause their collapse, which simultaneously can generate dramatic ecosystem changes. For example, the collapse, of a major (piscivorous) fish population may lead to the rapid increase in its prey species, creating potential imbalances in community structure and the alteration of ecosystem function (Jennings and Kaiser, 1998; Pinnegar *et al.*, 2000). Many commercially important marine species are now overfished, and the exploitation patterns of others are unsustainable (Pauly *et al*, 1998). As a result, the trend has been to "fish down" marine food webs, rendering affected ecosystems impoverished and less valuable (Williams, 1998). Because the marine fisheries harvest worldwide has increased fourfold since 1950, the global catch has shifted from large piscivorous species to smaller planktivorous forms (Boehlert, 1996).

Examples of overexploited estuarine fisheries include the striped bass (*Morone saxatilis*), chinook salmon (*Oncorhynchus tshawytscha*), and delta smelt (*Hypomesus transpacificus*) in San Francisco Bay. The sea trout fishes (*Cynoscion arenarius and Cynoscion nebulosus*) in Sarasota Bay, Florida, also appear to be overfished, with landings having declined by 50% since 1960. Several fish and shellfish species likewise appear to be overfished in the Albemarle-Pamlico Sound system of North Carolina (Kennish, 2000).

Introduced species

The accidental or purposeful introduction of exotic species to an estuary increases the probability of change in the native species assemblages (Moyle, 1986). When the introduced species become overwhelmingly successful in their adapted environment, they can outcompete and exclude the native species, promoting shifts in the structure and trophic organization of communities. In extreme cases, the ecological balance of the system may be threatened, as evidenced by acute reductions in biodiversity. The adapted habitats of the introduced species commonly lack natural controls, enabling these forms to rapidly dominate the native plant and animal communities.

One of the most heavily "invaded" estuaries in the world is San Francisco Bay, where nearly 250 species of plants and animals have been introduced (Cohen and Carlton, 1998). More than half of the fish species in the delta region consist of exotic, non-native forms, and most macroinvertebrates along the inner shallow water habitats of the system are introduced species (Kennish, 2000). Among the prominent introduced species in this estuary are the striped bass (*M. saxatilis*), oyster drill (*Urosalpinx cinerea*), and smooth cordgrass (*Spartina alterniflora*). The Asian clam (*Potamocorbula amurensis*), an introduced bivalve that has outcompeted native clam species (e.g., *Macoma balthica* and *Mya arenaria*), has seriously decimated the phytoplankton community of Suisun Bay.

Other examples of the detrimental effects of introduced species include the invasion of a parasitic protozoan (*Haplosporidium nelsoni*) in Delaware Bay, which has nearly destroyed the American oyster (*Crassostrea virginica*) fishery. Major fouling problems have developed in Corpus Christi Bay, Texas, due to the introduction of the brown mussel, *Perna perna* (Kennish, 2000). Rapid expansion of the nutria, *Myocastor coypu*, a South American mammal introduced into the Gulf of Mexico, may have contributed to the loss of coastal wetlands habitat in Louisiana (Day *et al.*, 1989).

Physical alterations

Some of the most damaging impacts of human activities on estuaries are those that acutely modify or destroy habitats. They may occur within the estuarine basin (e.g., dredging and dredged material disposal, shipping operations, and mariculture) or along the estuarine shore (e.g., bulkheading, riprap, docks, piers, and marina facilities). However, many serious impacts also originate in coastal watersheds, particularly bordering wetlands. For instance, increased development and construction of impervious surfaces facilitate erosion and land runoff which can significantly alter water quality, hydrology, and salinity in the estuary. Sprawled development accelerates the fragmentation of natural habitats, thereby forming an impediment to the feeding and migration of wildlife.

Historically, habitat destruction and hydrological alteration have been pronounced in coastal wetlands. For example, the construction of canals, levees, and impoundments, as well as the diking and ditching of wetland surfaces, have destroyed marsh plant communities and have greatly modified drainage to estuaries. These effects have been conspicuous in Atlantic and Gulf Coast salt marshes. The impounding and drainage of wetlands habitat for agriculture and other purposes have eliminated ~50% of the original tidal salt marsh habitat in the United States (Kennish, 2001). In Puerto Rico, human activities have destroyed more than 70% of the mangroves fringing the coast (Alongi, 1998).

Activities in upland areas also can have a devastating effect on estuarine habitats. Mining and silviculture usually raise the concentrations of contaminants (e.g., metals), sediments, and other particulates transported to the coast. However, dam construction lowers these inputs as well as the downstream flow. The channelization of streams and rivers for flood control generally enhances freshwater discharges to estuaries. In contrast, the diversion of freshwater for agriculture, municipal, and industrial uses reduces freshwater inputs. This is exemplified by conditions surrounding San Francisco Bay, where more than 50% of the freshwater inflow is diverted elsewhere, mainly to satisfy agricultural demands (e.g., irrigation) which account for ~85% of the diverted water volume (Kennish, 2000). This freshwater diversion has significantly altered the salinity regime and the distribution of organisms in the upper estuary.

The withdrawal of groundwater, oil, and gas from subsurface formations has been responsible for the subsidence of coastal areas, the local retreat of estuarine high-tide shorelines and the loss and alteration of wetland and estuarine habitats. Such is the case at Galveston Bay and the lower Chesapeake Bay area. Acute subsidence also occurs along the northern Gulf of Mexico (i.e., Louisiana Coast) where sediment compaction, downwarping of the underlying crust, and eustatic sea-level rise are contributing to rapid coastline retreat.

Conclusions

Estuaries rank high on the list of the most heavily impacted aquatic systems on earth. This is largely attributed to accelerating population growth and development in the coastal zone and escalating use of estuarine resources. Most specifically, nutrient enrichment, chemical contaminant inputs, overfishing, introduction of exotic species, and physical alteration of habitats have adversely affected many estuarine systems. In some cases, the structure and trophic organization of biotic communities have been greatly altered. Ecological imbalances have raised concerns regarding the long-term health and viability of these critically important coastal environments.

New management strategies are being formulated and implemented in numerous estuaries and coastal watersheds nationwide to mitigate point and nonpoint sources of pollution and to mollify habitat impacts that threaten estuarine resources. Efforts are underway to restore damaged habitats in estuarine systems, including the revitalization of sea grass, salt marsh, and mangrove biotopes (Zedler, 2000). A major goal is to effectively manage estuaries to ensure sustainable exploitation of renewable resources while protecting the long-time health and viability of the systems. Such a program will help to maintain the ecological integrity of these invaluable coastal waterbodies.

Michael J. Kennish

Bibliography

Alongi, D.M., 1998. *Coastal Ecosystem Processes*. Boca Raton: CRC Press.

Boehlert, G.W., 1996. Biodiversity and the sustainability of marine fisheries. *Oceanography*, **9**: 28–35.

Bricker, S.B., Clement, C.G., Pirhalla, D.E., Orlando, S.P., and Farrow, D.R.G., 1999. *National Estuarine Eutrophication Assessment: Effects of Nutrient Enrichment in the Nation's Estuaries*. National Oceanic and Atmospheric Administration, National Ocean Service, Special Projects Office and the National Centers for Coastal Ocean Science, Silver Spring, Maryland.

Bryan, G.W., and Langston, W.J., 1992. Bioavailability, accumulation, and effects of heavy metals in sediments with special reference to United Kingdom estuaries: a review. *Environmental Pollution*, **76**: 89–131.

Cohen, A.N., and Carlton, J.T., 1998. Accelerating invasion rate in a highly invaded estuary. *Science*, **279**: 555–562.

Costello, M.J., and Read, P., 1994. Toxicity of sewage sludge to marine organisms: a review. *Marine Environmental Research*, **37**: 23–42.

Daskalakis, K.D., and O'Connor, T.P., 1995. Normalization and elemental sediment contamination in the coastal United States. *Environmental Science and Technology*, **29**: 470–477.

Day, J.W., Jr., Hall, C.A.S., Kemp, W.M., and Yàñez-Arancibia, A. (eds.), 1989. *Estuarine Ecology*. New York: John Wiley & Sons.

Hameedi, M. J., 1997. Strategy for monitoring the environment in the coastal zone. In Haq, B.U. (ed), *Coastal Zone Management Imperative for Maritime Developing Nations*. Amsterdam: Kluwer Academic Publisher, pp. 111–142.

Hamelink, J.L., Landrum, P.F., Bergman, H.L., and Benson, W.H. (eds.), 1994. *Bioavilability—Physical, Chemical, and Biological Interactions*. Boca Raton: Lewis Publishers.

Jennings, S., and Kaiser, M.J., 1998. The effects of fishing on marine ecosystems. *Advances in Marine Biology*, **34**: 201–352.

Kennish, M.J., 1992. *Ecology of Estuaries: Anthropogenic Effects*. Boca Raton: CRC Press.

Kennish, M.J. (ed.), *Practical Handbook of Estuarine and Marine Pollution*. Boca Raton: CRC Press.

Kennish, M.J., 1998. Trace metal-sediment dynamics in estuaries: pollution assessment. *Reviews of Environmental Contamination and Toxicology*, **155**: 69–110.

Kennish, M.J. (ed.), 2000. *Estuary Restoration and Maintenance: The National Estuary Program*. Boca Raton: CRC Press.

Kennish, M.J. (ed.), 2001. *Practical Handbook of Marine Science*, 3rd edn. Boca Raton: CRC Press.

Livingston, R.J., 2001. *Eutrophication Processes in Coastal Systems*. Boca Raton: CRC Press.

Moyle, P., 1986. Fish introductions into North America: patterns and ecological impact. In Mooney, H., and Drake, J. (eds.), *Ecology of Biological Invasions of North America and Hawaii*. New York: Springer-Verlag, pp. 27–43.

O'Connor, T.P., 1996. Coastal sediment contamination in the Northeast Shelf Large Marine Ecosystem. In Sherman, K., Jaworsky, N.A., and Smayda, T.J. (eds.), *The Northeast Shelf Ecosystem: Assessment, Sustainability, and Management*. Cambridge: Blackwell Science, pp. 239–257.

Pauly, D., Christensen, V., Dalsgaard, J., Froese, R., and Torres, F., Jr., 1998. Fishing down marine food webs. *Science*, **279**: 860–867.

Pinnegar, J.K., Polunin, N.V.C., Francour, P., Badalamenti, F., Chemello, R., Harmelin-Vivien, M.-L., Hereu, B., Milazzo, M., Zabala, M., D'Anna, G., and Pipitone, C., 2000. Trophic cascades in benthic marine ecosystems: lessons for fisheries and protected-area management. *Environmental Conservation*, **27**: 179–200.

Pirie, R.G. (ed.), 1996. US ocean resources 2000: a national plan for growth. In Pirie, R.G. (ed.), *Oceanography: Contemporary Readings in Ocean Sciences*, 3rd edn. New York: Oxford University Press, pp. 283–292.

Valiela, I., 1995. *Marine Ecological Processes*, 2nd edn. New York: Springer-Verlag.

Valiela, I., Kremer, J., Lajtha, K., Geist, M., Seely, B., Brawley, J., and Sham, C.H., 1997. Nitrogen loading from coastal watersheds to receiving estuaries: new method and application. *Ecological Applications*, **7**: 358–380.

Weber, P., 1994. It comes down to the coasts. *World Watch*, March/April, 20–29.

Williams, N., 1998. Overfishing disrupts entire ecosystems. *Science*, **279**: 809–815.

Zedler, J.B. (ed.), 2000. *Handbook for Restoring Tidal Wetlands*. Boca Raton: CRC Press.

Cross-references

Dams, Effects on Coasts
Demography of Coastal Populations
Dikes
Dredging of Coastal Environments
Environmental Quality
Estuaries
Human Impacts on Coasts
Marine Debris—Onshore, Offshore, Seafloor Litter
Oil Spills
Sea-Level Rise, Effect
Submerging Coasts
Tourism and Coastal Development
Water Quality

ESTUARIES

In the now voluminous coastal literature, the definition of "estuary" is highly variable, and frequently reflects the standpoint of the author. Pethick (1984) states that: "estuaries are among the largest and most complex of all landforms, so complex in fact that attempting to provide a basic definition has proved extremely difficult."

Indeed one writer describes the definition of an estuary as rather like pornography—difficult to define, but you know it when you see it!

Central to the concept is that *estuaries are the tidal mouths of rivers* (Pethick, 1984). The French geomorphologist Guilcher (1958) pointed out that the word estuary comes from the Latin *aestus*, meaning tide, but that in France it has a popular synonym, *etier*, and the two words denote that part of the river system where the tides (water level rise and fall) and tidal currents make themselves felt. This is essentially the definition adopted by Fairbridge (1980) who defined an estuary as: "an inlet of the sea reaching into a river valley as far as the upper limit of tidal rise." This concept is also accepted by Caspers (1967) who argues that "the upper limit of the estuary is determined not by salinity but by tidal forces. In other words it is determined hydrodynamically rather than hydrochemically."

Thus, in principle an estuary can be defined from several different perspectives, including from the point of view of the geomorphology, sedimentation processes, tidal influences, the physical oceanography/salinity structure, or the ecology. In simple terms the definition is perhaps best considered from the standpoint of either "form" or "function."

From the geomorphic perspective, an estuary is defined in the general sense as: "the mouths of rivers, widening as they enter the sea" (Bird, 1972). On a global scale they typically form in drowned river valleys. The Webster Dictionary (2nd edn) describes an estuary as "a passage, as the mouth of a river or lake, where the tide meets the river current; more commonly, a narrow arm of the sea at the lower end of a river." However, the authoritative Glossary of Geology, edited by Jackson (1997) defines an estuary variously as:

> (a) the seaward end or the widened funnel-shaped tidal mouth of a river valley where freshwater mixes with and measurably dilutes sea water and where tidal effects are evident; e.g. a tidal river, or a partially enclosed coastal body of water where the tide meets the current of a stream. (b) A portion of an ocean, as a firth or an arm of the sea affected by freshwater; e.g. the Baltic Sea. (c) a drowned river mouth formed by subsidence of land near the coast or by the drowning of the lower portion of a non-glaciated valley due to the rise of sea level.

In physical geography terms the *Webster Dictionary* (2nd edn) defines an estuary as: "a drowned river mouth, caused by the sinking of the land near the coast." Stamp, in his *Glossary of Geographical Terms* (1966), defines an estuary as a "tidal opening, an inlet or creek through which the tide enters; an arm of the sea indenting the land," but Stamp states that this definition is rare in modern use.

Cameron and Pritchard (1963) provide a definition of estuaries perhaps the most widely accepted by coastal scientists, as: "a semi-enclosed coastal body of water which has a free connection with the open sea and within which sea water is measurably diluted with fresh water from land drainage." This definition appears in most of the modern textbooks such as Pethick's (1984): "*An Introduction to Coastal Geomorphology*" and Black (1986): "*Oceans and Coasts.*" Such a definition leads to a classification of estuaries as a function of their salinity structure, focusing on the oceanographic properties determined by the physical dimensions in relation to river flows, salinity distribution, and tidal conditions. Another influencing factor is the Coriolis effect which becomes important for wide estuaries. Based upon these physical characteristics, estuaries may be classified as follows (Bowden, 1967): (1) *Salt wedge estuaries* occur where river flows dominate the salinity structure. In such cases, the river waters flow seawards as a layer of freshwater separated from the underlying saltwater by a sharp density interface. This reduces the turbulence and mixing between the two separate water bodies, and the saltwater extends as a down-sloping wedge along the bottom into the head of the estuary. (2) *Two layer flow with entrainment.* Again two distinctive water bodies of different salinity and density are apparent in the vertical, but when internal waves at the interface break, saltwater is entrained upwards into the surface freshwater, thereby increasing its salinity. To compensate for this entrainment there is an upstream directed residual flow along the bottom. It is this current that facilitates transport of fine suspended material upstream into the upper reaches of the estuary. (3) *Two layer flow with vertical mixing.* In relatively shallow estuaries the turbulence generated by the tidal flows generates mixing through a greater proportion of the water body. Fresher water is mixed downwards and saline water upwards, but there is no sharp density interface boundary. However, there are still two layers of flow and a bottom landwards residual current. (4) *Vertically homogeneous estuaries.* When tidal currents are strong in relation to the freshwater input, or in shallow estuaries where form drag induces turbulence throughout the flow, the vertical mixing becomes complete throughout the water column, and there is no measurable variation in salinity from the surface to the bottom. There is, of course, a horizontal gradient of salinity from the estuary head to the mouth.

Wind may have an important influence on the estuarine density driven circulations. Wind stress on the water surface produces a net transport of surface water, and the waves generated increase the intensity of vertical mixing. These processes on occasions can override the normal estuarine density circulation.

Physiographically estuaries are of four identifiable types (Healy and Kirk, 1982). These are: (1) *Fiords*, characterized as former valley glaciers in mountainous terrain, with steep sides, a deep basin, and in many cases a sill at its mouth. Such features occur in Norway, Scotland, southern Chile, and southwest New Zealand. But there are also more complex lowland fiords of the Baltic type, formed originally as meltwater depressions under the Pleistocene ice sheet, and flooded with the postglacial sea-level transgression. (2) *Drowned river valleys* or *rias* occur widely along mid-latitude coasts, and occur in their purest form as meandering river valleys along coasts with hard crystalline basement rocks with low sedimentation rates. In many cases, complete valley systems were drowned resulting in a dendritic pattern of arms. Along active sedimentation coasts the rias are often modified as they connect with the open sea, typically exhibiting sediment infill of the drowned river valley to form modern tidal flats (Figure E10) and low alluvial plains developed over the estuarine sediments. They may possess barriers across the mouth of the embayment, such as occurs on the Oregon and New Zealand coasts (Figure E11). Where major faulting occurs the rias may become fault-aligned as in New Zealand's Marlborough Sounds. (3) *Barrier enclosed estuarine lagoons* occur along the low sandy barrier coasts of the world, such as the Atlantic seaboard of the United States or the eastern coast of Australia (Figure E12). The older term "bar built" estuaries is genetically misleading, and should be avoided, as few barriers originate from "bars." The enclosing barriers typically comprise Holocene dune barrier-islands and spits as is typical along the

Figure E10 Drowned river valley with active infill sedimentation forming muddy tidal flats, Raglan Harbour, New Zealand (photo: T. Healy).

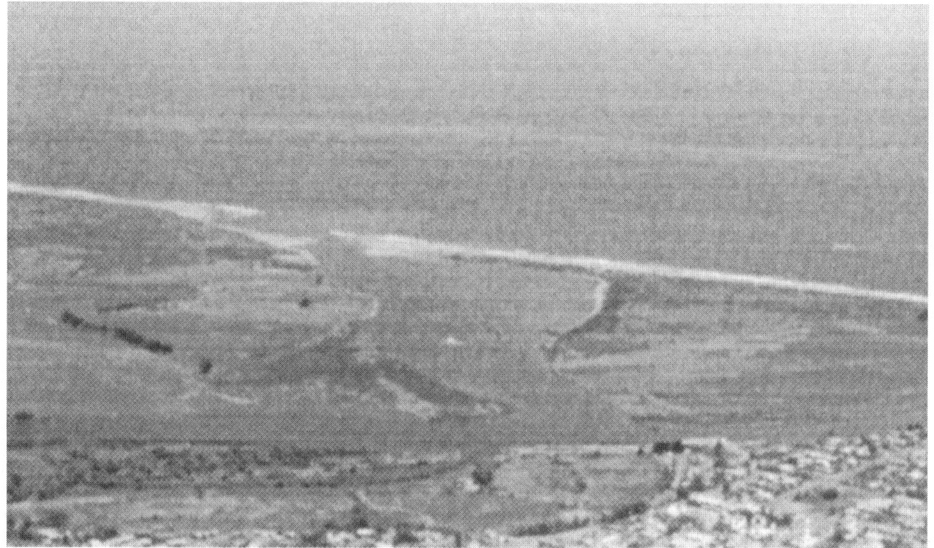

Figure E11 Estuary of the Waioeka River mouth, Bay of Plenty, New Zealand, illustrating a Holocene sandy barrier formation across the river mouth (photo. T. Healy).

Figure E12 The shallow estuarine lagoon of Maketu, Bay of Plenty, New Zealand, enclosed by a Holocene barrier spit. The entrance is a tidal inlet system (photo: T. Healy).

Figure E13 The down faulted structural graben of the Firth of Thames, New Zealand, into which flows the large Waihou River. The name "Firth" is a geomorphic misnomer, as this structural estuary did not originate from ice scour (photo: T. Healy).

Atlantic and Gulf coasts of the United States, but that is not always the case. Some enclosing barriers may comprise high (up to 200 m) Pleistocene dunes, as for the very large harbors on the west coast of the North Island of New Zealand, and sometimes the enclosing barriers may be of gravel, and even boulders, as for example at Nelson, New Zealand. These lagoonal-type estuaries are characterized by shallow depths and frequently broad expanses of intertidal flats become exposed at low tide. Where they occur along littoral drift coasts the entrances are characterized by narrow tidal inlets with ebb and flood tidal delta systems. Because of the overall shallow lagoon, the estuaries are well mixed in terms of their salinity structure, although there may be salinity variation from the mouth to the head of the lagoonal estuary. In arid coasts, the lagoonal waters may become hypersaline. (4) *Structural and tectonically induced estuaries* are typified by downthrown grabens (San Francisco Bay), tectonic depressions (Botany Bay, NSW), or flooded calderas (Banks Peninsula, New Zealand). These tend to be large-scale features (Figure E13).

Clearly, there is a geomorphic gradation both within and between large estuaries. For example, a large barrier-enclosed estuarine lagoon may possess arms in its upper reaches that exhibit the geomorphic characteristics primarily of a drowned river valley, and salinity characteristics of a salt wedge and two-layer flow. Conversely, a drowned river valley estuary with overall dendritic outline may possess barriers across its mouth (e.g. Chesapeake Bay, or Raglan Harbour in New Zealand).

However, all of these geomorphic categories of estuaries feature an element of "drowning" from rise of the postglacial sea level, which reached its approximate present level some ~6,000 years ago (Russsell, 1967). However, continued drowning of an estuary is rarely identified in modern times, despite the present global warming and its consequent sea-level minor rise.

On the contrary, active sedimentation within estuaries overall outweighs any modern sea-level rise effect. This is especially so for high sediment load major rivers such as the Amazon, the Yellow River of China, and the Mississippi, where rapid sedimentation and the growth of deltas has greatly modified the estuary over the past ~6,000 thousand years. Indeed, estuaries are well recognized as natural sediment traps, and are ephemeral features of the geological record. As Russell notes, most estuaries have changed considerably from the active sedimentation during the past few millenia, and many estuaries are headed for extinction. Rapid sediment input into estuaries from the inflowing river catchments, acts to fill them up from the land. But (especially for lee coast estuaries) diabathic onshore transport across the shelf provides sediment at the coasts that sweeps into the estuary mouths, and acts to infill estuaries from the sea. In addition, the saline density circulation within estuaries and mixing of mud-laden freshwaters from the catchment with the saline marine waters, leads to flocculation of fine particles, which residual bottom currents tend to transport toward the head of the estuary, where the muds deposit and accumulate.

Within most estuaries tides are the major energy source for mixing fresh and saltwater, and for resuspending sediment from the bed and transporting those sediments seawards or landwards. Frictional effects dissipate the tidal wave energy within the estuary, and deform the tidal wave. But this is offset by convergence of the tidal wave as it progresses up the estuary. Depending upon the balance between tidal wave convergence compared with frictional dissipation effects, the tidal range within an estuary can be larger than at the open sea before decreasing landwards toward the river (a hypersynchronous condition), or the frictional effects can dominate and the tidal amplitude decreases landward from the mouth, a hyposynchronous condition. Most ria type estuaries are a hypersynchronous and attain maximum current strength in middle or landward reaches, whereas for the barrier enclosed estuaries the maximum current strength occurs at the narrow tidal inlet.

In many estuaries, there exists an essentially closed circulation system for fine suspended particles, which causes them to become entrapped. This occurs due to the formation of "turbidity maxima" and from the "settling lag" effect. In the latter process, there is a time lag between the turn of the tide (when current velocity is zero) and when the concentration of fine suspended particle concentrations become minimum. This settling lag is due to the time it takes for the fine material to settle through the water column after the reduction in flow velocities (and reduction in turbulence), and thus the particle is carried some distance beyond the point where it starts to sink. Thus for a flood tidal current near high water, a certain amount of fine sediment settles toward the landward areas. The lag effect can explain why in many estuarine tidal flats there is a much greater proportion of mud particles than in the adjoining open sea, and why moving landwards up the estuary the particle size gradually fines (Postma, 1967).

A "turbidity maximum" of suspended sediment concentration often occurs in the upper reaches of many estuaries, particularly in the partially mixed types, where suspended sediment concentrations may reach 10 to 100 times greater than either seaward or further up the river, and may reach 1–10 g/L. This feature occurs in a wide variety of estuaries over the world, and occurs because fine particle settling is trapped in the null zone where river current converges with the landward estuarine flow near the bottom (Nichols and Biggs, 1985).

The formation of highly concentrated fluid mud layers at the bottom may also be part of the sediment dynamics. Fluid mud is a highly concentrated sediment suspension with a sediment concentration of between 10,000 and 300,000 mg/L, and can originate from the turbidity maximum at slack water. It may take the form of mobile or stationary suspensions (Kirby, 1988). But fluid mud can also form from wave action (Healy *et al.*, 1999).

A typical feature of shallow estuaries is the estuarine "turbid fringe." This is generated by the wave action as the tide rises over the intertidal flats and reagitates the fine particles into suspension. Thus even in days of very light wave action a zone of brown turbid waters is observed around the shore of the estuary. Fine sediment transport in the estuary is also affected by specific processes such as aggregation of the fine particles, deposition, erosion, and consolidation. Classically, the aggregation of fine sediment particles was explained by the so-called electric double-layer theory. In recent decades, additional mechanisms have been identified including organic aggregation, bioflocculation, and pelletization.

Terry R. Healy

Bibliography

Bird, E.C.F., 1972. *Coasts.* Canberra: Australian National University Press.

Black, J.A., 1986. *Oceans and Coasts.* Dubuque: W.C. Brown.

Bowden, K.F., 1967. Circulation and diffusion. In Lauff, G.H. (ed.), *Estuaries.* Washington DC: American Association for the Advancement of Science, pp. 15–36.

Caspers, H., 1967. Estuaries: analysis of definitions and biological considerations. In Lauff, G.H. (ed.), *Estuaries.* Washington DC: American Association for the Advancement of Science, pp. 6–8.

Cameron, W.M., and Pritchard, D.W., 1963. Estuaries. In Hill, M.N. (ed.), *The Sea,* Vol. 2, New York: John Wiley and Sons, pp. 306–24.

Fairbridge, R.W., 1980. The estuary: its definition and geodynamic cycle. In Olausson, E., and Cato, I. (eds.), *Chemistry and Biogeochemistry of Estuaries.* New York: Wiley.

Guilcher, A., 1958. *Coastal and Submarine Morphology.* London: Methuen & Co.

Healy, T.R., and Kirk, R.M., 1982. Coasts. In Soons, J., and Selby, M., (eds.), *Landforms of New Zealand.* Auckland: Longman-Paul, pp. 80–104.

Healy, T., Mehta, A., Rodriguez, H., and Tian, F., 1999. Bypassing of dredged muddy sediments using a thin layer dispersal technique. *Journal of Coastal Research,* **15**(4): 1119–1131.

Jackson, J.A.(ed.), 1997. *Glossary of Geology.* Alexandria: American Geological Institute.

Kirby, R., 1988. High concentration suspension (Fluid Mud) layers in Estuaries. In Dronkers, J., and van Leussen, W. (eds.), *Physical Processes in Estuaries.* Berlin: Springer Verlag, pp. 463–487.

Nichols, M.M., and Biggs, R., 1985. Estuaries. In Davis, R.A. (ed.), *Coastal Sedimentary Environments.* New York: Springer Verlag, pp. 77–186.

Pethick, J., 1984. *An Introduction to Coastal Geomorphology.* London: Edward Arnold.

Postma, H., 1967. Sediment transport and sedimentation in the estuarine environment. In Lauff, G.H. (eds.), *Estuaries.* Washington: American Association for the Advancement of Science, pp. 158–179.

Russell, R.J., 1967. Origins of estuaries. In Lauff, G.H. (ed.), *Estuaries.* Washington: American Association for the Advancement of Science, pp. 158–179.

Stamp, D., 1966. *Glossary of Geographical Terms.* London: Longmans.

Cross-references

Barrier
Barrier Islands
Coastal Lakes and Lagoons
Estuaries, Anthropogenic Impacts
Muddy Coasts
Ria
Salt Marsh
Tidal Creeks
Tidal Inlets
Tidal Prism
Tides
Vegetated Coasts
Wetlands
Volcanic Coasts

EUSTASY

Eustasy refers to a globally uniform change in sea level. Suess (1888) originally attributed eustasy to crustal subsidence and sediment deposition. Removal or addition of water to oceans during glacial/interglacial cycles was another proposed cause. Fairbridge (1961) summarized these theories as follows: (1) *tectono-eustasy* (tectonic deformation of the ocean basins), (2) *sedimento-eustasy* (loading of crust by sediments), and (3) *glacio-eustasy* (removal/addition of water to the oceans by expansion or contraction of polar ice sheets). Another process is *hydro-isostatic* loading of the seafloor by glacial meltwater (Walcott, 1972; Clark *et al.*, 1978). Worldwide compilations of Holocene sea-level curves, however, revealed major regional differences, thus challenging the concept of eustasy (Pirazzoli, 1991).

Seismic reflection profiles on continental shelves, collected during petroleum exploration in the 1970s, showed similar patterns of "sequence boundaries," or regional unconformities, in passive margin

Table E8 Summary of processes affecting global sea level

Process	Rate (m/ka)	Duration (yr)	Amplitude (m)	References
Change in ocean water quantity				
Glacio-eustasy	10	10^4–10^5	120–150	Bard *et al.*, 1996; Fairbanks (1989)
Hydro-eustasy	0.5–5	10^4–10^5	~50	Hay and Leslie (1990)
Steric changes	1	10–10^4	10	Rona (1995)
Change in ocean basin volume				
Tectono-eustasy				
Ocean ridges	0.01	10^8	500	Rona (1995)
Plate collisions	0.002	10^7	100	Rona (1995)
Hot spots	0.001	10^8	100	Rona (1995)
Sedimento-eustasy	0.003	10^8	80	Rona (1995)
Glacio-isostasy	1–10	10^4	100	Peltier (1999)
Dynamic changes				
Geostrophic currents	—	1–10	0.01–1	Gornitz (1995b)
Atmospheric forcing	—	1–10	0.01	Gornitz (1995b)
ENSO	—	3–8	0.1–0.5	Gornitz (1995b)

sedimentary deposits from many parts of the world (see *Sequence stratigraphy*). Sequence boundaries have been widely interpreted as marking major episodes of synchronous eustatic sea-level falls, which are summarized in global sea-level curves (e.g., Vail *et al.*, 1977; Haq *et al.*, 1987). Although widely used, these curves have been criticized on grounds that sequence boundaries are uncritically accepted as eustatic, that magnitudes of sea-level fluctuations are largely speculative, and that boundaries have not been accurately dated (for good overviews, see Christie-Blick *et al.*, 1990; Miall, 1997). However, curves from various sources show certain common features, suggesting that at least some of these proposed sea-level events could be worldwide in scope.

Causes of eustasy

Sea-level integrates the effects of diverse geological and meteorological phenomena, such as glacio-eustasy, glacio-isostasy, plate motions, etc., which operate on differing spatial and temporal scales (Table E8). The relative magnitudes of these processes vary from place to place, thus tending to obscure any globally coherent sea-level signal. As a result, the search for a universal "eustatic" sea level curve has turned toward obtaining improved regional and local ("relative") sea-level curves (Tooley, 1993). Yet, as shown below, improved modeling of glacio-isostatic adjustments and detailed, accurately dated sea-level data from areas of known uplift enable reconstruction of global sea-level rise over the last 18,000 years and the past century.

Table E8 outlines processes affecting global sea level over various geologic timescales. These processes involve: (1) changes in the volume or mass of ocean water, (2) changes in the volume of the ocean basins, and (3) dynamic changes of the ocean surface.

Changes in the volume or mass of ocean water

Glacio-eustasy. Melting or accumulation of continental ice sheets during the last 2.5 million years has produced large changes in ocean water volume and global sea level. These oscillations of sea level over timescales of tens to hundreds of thousands of years are caused primarily by changes in solar radiation arising from periodic variations in the earth's orbit around the Sun (the "Milankovitch" cycles, Imbrie *et al.*, 1984).

Changes in sea level during glacial cycles have been analyzed indirectly using fluctuations in the ratio of ^{18}O to ^{16}O in deep-sea microorganisms, such as foraminifera from ocean sediment cores, and also by dating emerged coral reefs on tectonically rising islands (see *Geochronology*, *Coral Reefs, Uplifted*). Variations in the oxygen isotope ratio serve as a proxy for alternations in global ocean volume, and hence in sea-level, linked to changing masses of continental ice. Sea level curves for the past several hundred thousand years have been constructed from well-dated raised coral terraces on Barbados and Papua New Guinea (Gallup *et al.*, 1994). These data show highstands of sea level at approximately 193,000, 125,000, 100,000, and 83,000 years ago.

The post-glacial marine transgression of the last 18,000 years has been plotted by careful radiometric dating (^{14}C, and ^{230}Th-^{234}U) of corals from three widely separated oceanic islands with known tectonic trends and minimal glacio-isostatic influence. These three areas show a remarkably consistent and uniform rise in sea level of around 120–125 m, which therefore represents a true eustatic curve (Figure E14, Fairbanks, 1989; Bard *et al.*, 1996). Two periods of accelerated sea level rise are detected: the older at around 14,000 years BP (labeled MWP-1A) and the younger at 11,500–11,000 years BP (MWP-1B). These two events correspond to two major meltwater pulses associated with rapid breakup and melting of the ice sheets.

The process of glacio-eustasy is now well established for the late Cenozoic, but less certain is the extent to which Milankovitch cycles have operated in the more distant geologic past. Antarctica has been glaciated for at least 14 million, and possibly as much as 35 million years. The late Paleozoic Gondwana glaciation led to a major marine regression. A late Ordovician glaciation was centered on the Sahara Desert in north Africa. Beyond these, other glacial episodes are less well established, and the presence of 10^4–10^5-year frequencies in the stratigraphic record are not necessarily conclusive proof of Milankovitch cycles (Miall, 1997).

Hydro-eustasy. Redistribution of water among the major terrestrial hydrological reservoirs, in response to major climate shifts, can also affect the ocean's level. A potentially significant terrestrial reservoir on timescales of tens to hundred of thousands of years resides in groundwater (Hay and Leslie, 1990). Sediments may contain enough pore space to hold the equivalent of 76 m of sea level (or 50 m with isostatic adjustment).

Steric changes. Over timescales of decades to centuries, changes in ocean water density become important contributors to sea-level change. Warming, cooling, or changes in salinity of ocean water affect its volume (but not its mass content). Thermal expansion of the upper layers of the oceans, as the earth's climate warmed approximately 0.6°C during the last hundred years, could account for 0.3–0.7 mm/yr of the observed sea-level rise of 1–2.0 mm/yr during this period (Houghton *et al.*, 2001).

Changes in volume of ocean basins

Tectono-eustasy: On geologic timescales of tens to hundreds of millions of years, major factors controlling sea level include changes in seafloor spreading rates, cooling of oceanic crust, and passive margin thermotectonic subsidence (Rona, 1995). At times of rapid seafloor spreading, a comparatively larger volume of younger, warmer, more buoyant basaltic crust is created at mid-oceanic ridges. This displaces water, raising sea level on the continental margins. Slow seafloor spreading rates produce the opposite effect. Plate collision, thrusting, and terrane accretion can also cause changes in sea level on timescales of 10^7 years. Continental collision reduces land area due to compression, producing a corresponding increase in ocean area which can lead to a

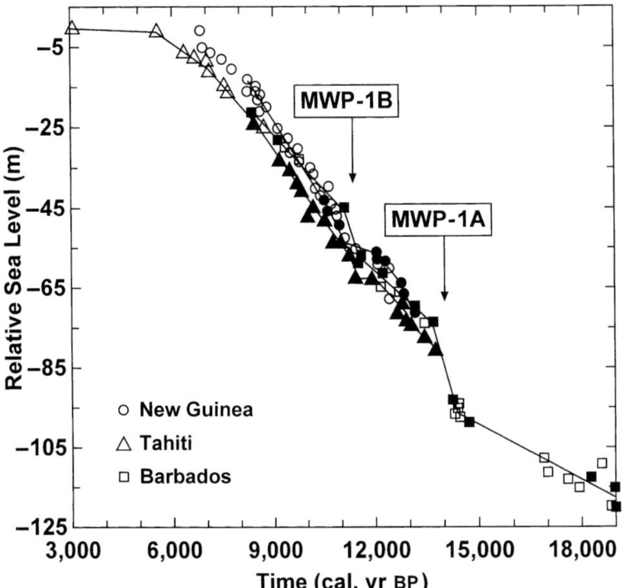

Figure E14 Sea-level curve of the post-glacial marine transgression. Based on radiocarbon and U-Th dating of corals from Tahiti, Barbados, and New-Guinea (modified from Bard *et al.*, 1996).

drop in sea level of up to 100 m (Table E8). Other tectono-eustatic mechanisms include hot-spot activity and deep mantle upwelling (Hallam, 1992).

Sedimento-eustasy: Displacement of water by sedimentation on continental margins and ocean basins displaces water, causing an apparent rise in sea level.

Glacio-isostasy Loading and unloading of ice masses during major glacial cycles induce non-uniform glacio-isostatic adjustments of the earth's crust (Table E8). Resulting changes in coastal elevation strongly affect the location of the earth's coastlines and local sea level. For example, much of Canada and Scandinavia, formerly covered by ice, are still rebounding isostatically and local sea level is falling. Conversely, eastern United States and the low countries of Europe, at the southern edge of the ice, are still subsiding, with locally rising sea level (Peltier, 1999). Subsidence accounts for close to half of the observed recent sea-level rise along the United States east coast (Gornitz, 1995a;b).

Geophysical models calculate the redistribution of mass accompanying the latest deglaciation (Peltier, 1999; Lambeck and Nakada, 1992). These models can be used to filter residual glacio-isostatic movements which contaminate the 20th century sea-level records based on tide-gauges. This has allowed an estimation of the climate warming-induced "absolute" or eustatic sea-level change (e.g., Peltier and Tushingham, 1989). An alternative approach employs late Holocene paleosea-level indicators to remove long-term geologic trends (Gornitz and Lebedeff, 1987; Gornitz, 1995a).

Dynamic changes

Atmospheric and oceanographic processes are responsible for fluctuations in sea level that last from 1 to 2 years to a decade or more, and are coherent over long distances. These fluctuations arise from ocean currents, and closely coupled oceanographic–atmospheric phenomena, such as the Northern Atlantic Oscillation (NAO), or the El Nino-Southern Oscillation (ENSO) in the Pacific Ocean (see *ENSO*; Table E8). During the El Nino phase of the ENSO cycle, anomalously warm water propagates eastward, splitting into north and south poleward-propagating Kelvin waves, causing sea level to rise (for several months) along the west coasts of the Americas (Komar and Enfield, 1987).

Conclusions

Sea-level change is a complex phenomenon, varying considerably over space and time. Nonetheless, considerable progress has been made in modeling of glacio-isostasy and in accurate dating of Holocene sea-level proxies, far removed from glacial influences. This information will enable separation of the present-day eustatic (global) sea-level signal from regional to local geological factors.

Vivian Gornitz

Bibliography

Bard, E., Hamelin, B. *et al.*, 1996. Deglacial sea-level record from Tahiti corals and the timing of global meltwater discharge. *Nature*, **382**: 241–244.

Christie-Blick, N., Mountain, G.S., and Miller, K.G., 1990. Seismic stratigraphic record of sea-level change. In *Sea-Level Change*, Washington, DC: NRC, pp. 116–140.

Clark, J.A., Farrell, W.E., and Peltier, W.R., 1978. Global changes in post-glacial sea level: a numerical calculation. *Quaternary Research*, **9**: 265–287.

Fairbanks, R.G., 1989. A 17,000-year glacio-eustatic sea level record: influence of glacial melting rates on the Younger Dryas event and deep-ocean circulation. *Nature*, **342**: 637–642.

Fairbridge, R.W., 1961. Eustatic changes in sea level. In Ahrens, L.H., Press, R., Rankama, K., and Runcorn, S.K. (eds.), *Physics and Chemistry of the Earth*, New York: Pergamon Press, pp. 99–185.

Gallup, C.D., Edwards, R.L., and Johnson, R.G., 1994. The timing of high sea levels over the past 200,000 years. *Science*, **263**: 796–800.

Gornitz, V., 1995a, A comparison of differences between recent and late Holocene sea-level trends from eastern North America and other selected regions. *Journal Coast Research* **17**(Special issue): 287–297.

Gornitz, V., 1995b. Monitoring sea level changes. *Climate Change*, **31**: 515–544.

Gornitz, V., and Lebedeff, S., 1987. Global sea level changes during the past century. In Nummedal, D., Pilkey, O.H., and Howard, J.D. (eds.), *Sea Level Fluctuation and Coastal Evolution*. SEPM Special Publication 41, 3–16.

Hallam, A., 1992. *Phanerozoic Sea-Level Changes*. New York: Columbia University Press.

Haq, B.U., Hardenbol, J., and Vail, P.R., 1987. Chronology of fluctuating sea levels since the Triassic. *Science*, **235**: 1156–1167.

Hay, W.W., and Leslie, M.A., 1990. Could possible changes in global groundwater reservoir cause eustatic sea-level fluctuations? *Sea-Level Change*, Washington, D.C., NRC pp. 161–170.

Houghton, J.T., Ding, Y., Griggs, D.J., Noguer, M., Vander Linden, P.J., Dai, X., Maskell, K. and Johnson, C.A., (eds.), 2001. *Climate Change 2001: The Scientific Basis*. Cambridge: Cambridge University Press.

Imbrie, J. *et al.*, 1984. The orbital theory of Pleistocene climate: support from a revised chronology of the marine O^{18} record. In Berger, A. *et al.* (eds.), *Milankovitch and Climate: Understanding the Response to Astromonical Forcing*, Dordrecht: Reidel, pp. 269–305.

Komar, P.D., and Enfield, D.B., 1987. Short-term sea level changes and shore-line erosion. In Nummedal, D., Pilkey, O.H., and Howard, J.D. (eds.), *Sea Level Fluctuation and Coastal Evolution*. SEPM Special Publication 41, 17–27.

Lambeck, K., and Nakada, M., 1992. Constraints on the age and duration of the last interglacial period and on sea-level variations. *Nature*, **357**: 125–128.

Miall, A.D., 1997. *The Geology of Stratigraphic Sequences*. Berlin: Springer-Verlag.

Peltier, W.R., 1999. Global sea level rise and glacial isostatic adjustment. *Glob. and Planet. Change*, **20**: 93–123.

Peltier, W.R., and Tushingham, A.M., 1989. Global sea level rise and the greenhouse effect: might they be connected? *Science*, **244**: 806–810.

Pirazzoli, P.A., 1991. World Atlas of Holocene Sea-Level Changes. Amsterdam: Elsevier Science Publication.

Rona, P.A., 1995. Tectonoeustasy and Phanerozoic sea levels. *Journal Coastal Research* **17**(Special issue): 269–277.

Suess, E., 1888. *Das Antlitz der Erde* vol. 2, Vienna: Tempske.

Tooley, M.J., 1993. Long term changes in eustatic sea level. In Warrick, R.A., Barrow, E.M., and Wigley, T.M. (eds.), *Climate and Sea Level Change, Observations, Projections, and Implications*, Cambridge: Cambridge University Press, pp. 81–107.

Vail, P.R., Mitchum, R.M., Jr., Todd, R.G. *et al.*, 1977. Seismic stratigraphy and global changes of sea level. In Payton, C.E. (ed.), Seismic Stratigraphy—Applications to Hydrocarbon Exploration. *American Association of Petroleum Geologists Memoriss*, **26**: 49–212.

Walcott, R.I., 1972. Past sea levels, eustasy, and deformation of the Earth. *Quaternary Research*, **2**: 1–14.

Cross-references

Changing Sea Levels
Coral Reefs, Uplifted
Geodesy
Glacio-Isostasy
Geochronology
Hydro-Isostasy
Sequence Stratigraphy

EUROPE, COASTAL ECOLOGY

Introduction

Europe occupies a little less than 7% of the total area of the world land surface. For its size its coastline is long, estimated at 143,000 km (Stanners and Bordeau, 1995). The European figure is derived from a map of 1:3 million scale and represents a conservative figure. The coastlines of Great Britain and Greece, for example, are 18,838 and 15,000 km, respectively, when the many islands are included. Though the figure for Great Britain is taken from measurements made at a scale of 1:50,000 and not directly comparable with the European figure they give an indication of the scale of the resource.

Europe's landmass is bounded by the Atlantic Ocean in the west and includes several enclosed seas (Figure E15). The basic fabric of the coast is formed from rocks which span the whole of the geological timescale from the Precambrian period (3,000–570 million years ago) to the rocks laid down in the Cretaceous–Tertiary period (140–2 million years ago). Superimposed on this structure, in the north, is the effect of the ice-ages of the Quaternary period (2 million to 10,000 years ago) and the glacial deposits, which were left behind as the ice melted. In the south, tectonic activity associated with volcanoes and earthquakes has also had an impact, though on a more limited geographical scale.

A combination of oceanographic factors (such as tides and sea-level change) as well as climatic effects, force the landward or seaward movement of sediments derived from erosion of the land or seabed. This provides opportunities for new coastal areas to develop as the material is deposited on or adjacent to the land. A variety of other variables, including the ameliorating effect of the warm Atlantic waters and westerly winds help to define a west to east climatic gradient, which is most clearly seen in the nature of the vegetation. Human activities have also been important, though more recent in their impact, causing modification or loss of habitat and "squeezing" the coastal zone (Doody, 2001). The description that follows provides an overview of the nature of Europe's coastline from Norway to the eastern Mediterranean and the flora and fauna which occur there.

The "nature" of Europe's coast

The coastline and marine waters of Europe are areas of great contrast with a high diversity of wildlife and landscape features. The underlying geology helps to define the coast and its landscape. Today, in many areas where the coastline is composed of "hard" rocks resistant to erosion, it may appear unchanging. However, during the 2 million years or so of the Quaternary Period the growth (and retreat) of the ice sheets eroded mountains and scoured river valleys. In the north and west from Norway to the Atlantic coasts of Ireland and western Britain, the combined effects helped to mold the landscape, exposing some underlying structures and covering others. This glaciated northern landscape exists today as a highly indented coast with exposed sea cliffs, coastal inlets (including fjords, fjards and firths, and a few, usually small estuaries), rocky shores, and pocket beaches. By contrast, the coasts of southern England, France, northern Spain, and Portugal include "softer" though resistant limestone molded by more recent events (such as erosion from the sea and land slip) to form cliffs and headlands. Hard igneous rocks of Precambrian age and resistant granites also abound in the northern Baltic. Here, the lie of the land is gentler and cliffs are much less evident. Isostatic rebound continues to create small flat offshore islands.

At the southern limits of the ice sheets, in the southern North Sea and the Baltic Sea particularly (Figure E15), major deposits of glacial sediment left behind as the ice retreated northwards have helped to create the present low-lying landscapes. These provided abundant soft sediments for reworking during the Holocene epoch of the last 10,000 years. Today eroding glacial till cliffs, sand dunes, salt marshes and tidal sand, and mud flats are abundant. These combine to form large estuaries in the meso- to macro-tidal southern North Sea and deltas on the micro-tidal shores of the southern Baltic. Within the matrix of sandy spits and bars and tidal flats, promontories of harder limestone rocks help form embayments in a few locations. Sub-tidal areas also tend to be dominated by softer sediments on the sea bed.

The Mediterranean Sea has both hard and soft rocks, the former predominating as karstic limestone cliffs, and more gently sloping rocky shores. Generally, the land rises uninterrupted from the sea. Where cliffs occur these can reach a height of over 150 m, as on the southern coast of Spain. Over the last 4,000 years or so erosion of the mountains has also helped to create large coastal plains as their river catchments have delivered sediment to the coast. In the micro-tidal areas of both the Mediterranean and Black Seas large deltaic systems such as the Ebro Delta in Spain, the Po Delta in Italy, and the Danube Delta in the Black Sea have grown partly in response to deforestation in the hinterland.

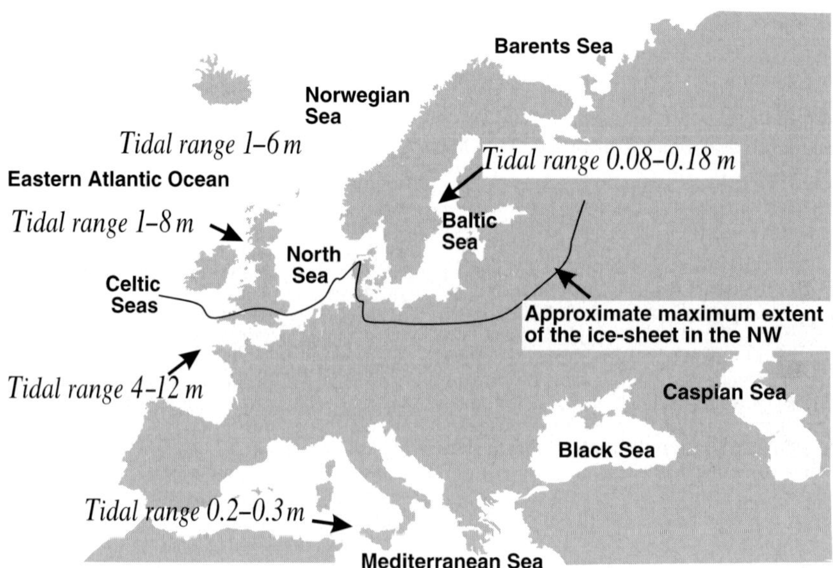

Figure E15 Europe's seas.

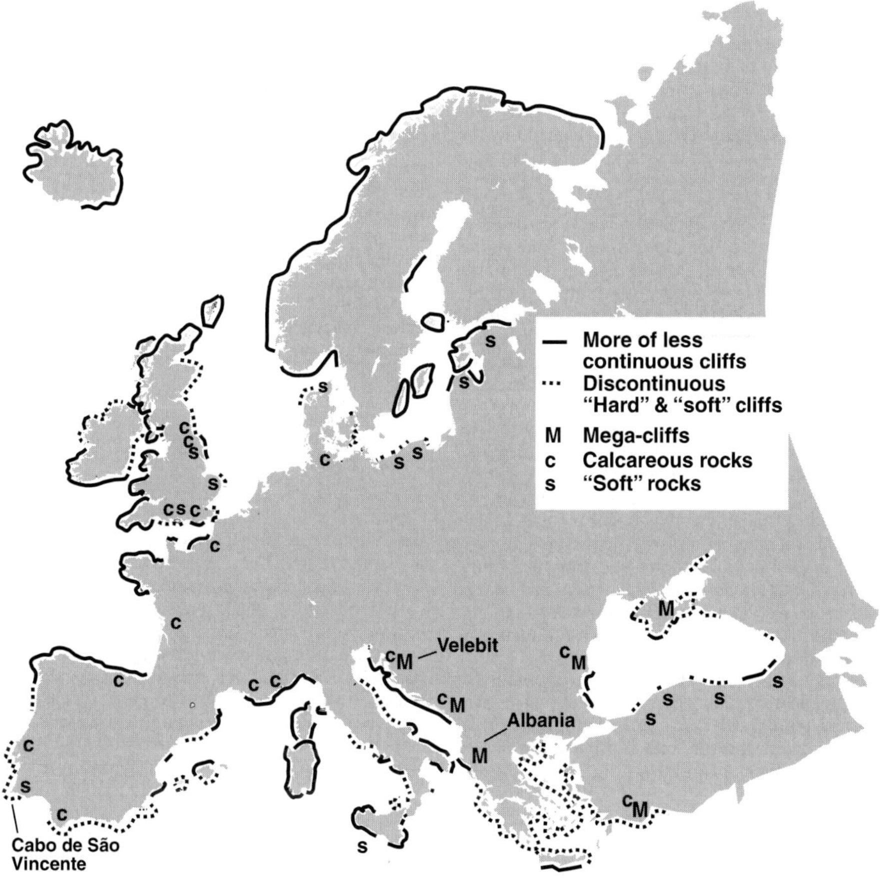

Figure E16 Distribution of the "cliffed" coastlines of Europe.

Figure E17 An elevated coastline (the Old Man of Hoy, Orkney Islands, Scotland).

Sea-level change, tidal range, and wave action

Long-term sea-level change is a major determining factor in the development of the coastline and coastal habitats. From around 7,000 years ago, when sea level began to stabilize at its present level, many of the coastal formations that are present today began to develop. Their evolution then and now depends on a combination of factors, including the availability of sediment and the influence of sea-level change, tidal range, and wave action.

With global sea levels rising today at about 1–2 mm per year (Warrick, 1993) there appears to be a general trend towards erosion and mobilization of sediments. Exceptions occur where isostatic rebound of glaciated areas is rising faster than sea level, causing a general rise in the land relative to the sea in areas such as Norway, the Gulf of Bothnia and around the margins of the Central Highlands of Scotland. Where the land continues to sink or is stable, relative sea-level rise tends to push coastal habitats toward the land. When this migration is restricted by resistant, rising landforms or artificial structures built to defend the land from erosion and/or flooding, the shore is squeezed into an ever narrower zone. This has important implications not only for the habitats themselves, but also the ability of the coastal landforms to provide a flexible response to storms and extreme tidal events. It may also impact on human use and infrastructure at the coastal margin.

Tides and tidal range provide a further mechanism for change. Where large amounts of material are available for transport extensive sedimentary habitats can be created. In the macro-tidal areas (spring tidal range >4 m) of the west, including the Celtic Seas, the Bristol Channel (where the tidal range, at 12 m is the second highest in the world) and the Atlantic coast of France, sediments are moved inland to fill inlets and bays, creating estuaries. In meso-tidal (spring tidal range 2–4 m) or micro-tidal areas (spring tidal range <2 m) the tides have progressively less influence. At the lower end of the tidal range the sediment movement is offshore, with rivers tending to exert a more dominant role. Waves and wave energy help determine the final configuration of the deltas as in the Baltic and the Mediterranean.

Climate

The coasts and seas of Europe encompass a wide range of climatic conditions from the Arctic Circle to the deserts of north Africa. Summer temperatures rarely rise above 10°C in most northern latitudes. This gradient is further influenced in the west by the marine waters, which are warmed by the Gulf Stream, creating an extreme maritime climate, which helps to define the Atlantic biogeographical zone encompassing much of northwest Europe (Polunin and Walters, 1985).

Unlike the areas further north the climate of the Mediterranean is very different. The winters are warmer (averaging 6°C) and the summers dry and hot. The range of temperature is much greater than on the Atlantic coast. The combination of thin limestone soils, dry climate and the long history of grazing and burning of the natural vegetation, has helped to create low-growing shrubby, drought and grazing resistant vegetation, which covers large areas of the Mediterranean. Since storms and waves are generated over a smaller sea area than the Atlantic, extreme exposure to gales is less evident than on the Atlantic coast.

Elevated landscapes—sea cliffs

Sea cliffs help to create some of the most spectacular and remote landscapes in Europe. They may have vertical or sloping faces, and in exposed locations are often drenched in salt spray, especially where they face the full extent of the Atlantic Ocean, and in the northern North Sea. Their orientation (and hence degree of exposure) and the nature of the underlying rock helps to determine the type of vegetation growing there. Although not considered in detail here stability is also an important factor helping to perpetuate shallow skeletal soils.

Sea cliff vegetation

Three main types of vegetation have been identified in the European Union Habitats and Species Directive, (European Com-mission, 1999). A brief description of the first two is given below.

Vegetated sea cliffs of the Atlantic and Baltic coasts. In exposed areas with hard rock cliffs, some of the best examples of maritime cliff vegetation in Europe are to be found. These include communities dominated by roseroot *Sedum rosea* and Scot's lovage *Ligusticum scoticum*, on the salt-spray drenched cliffs. Above this spray zone, the cliffs support communities that include arctic-alpine plants including purple saxifrage *Saxifraga oppositifolia*, moss campion *Silene acaulis*, mountain avens *Dryas octopetala* and in Scotland the endemic Scottish primrose *Primula scotica*. In some areas, the extent of the truly maritime zone can be small. In sheltered locations, such as the east coasts of Ireland and Scotland, the cliffs are only marginally exposed to the full power of the wind, breaking waves and salt spray. Here, the maritime influence is reduced and plants less resilient to exposure and salt spray grow in communities more typical of inland locations. Further south from south west Ireland, Wales and south west England to southern Portugal, spectacular west facing cliffs also occur. These are less resistant than the harder rocks further north and in some areas rapid rates of erosion (1 m or more per year) can be measured as, for example, in the chalk cliffs of the Channel coast. Although still exposed to the Atlantic Ocean and its gales, climatic amelioration also allows a richer flora to develop. In limestone areas of south west Britain, northern France, and Spain the cliffs provide important refuges for warm loving species, such as spider orchid *Ophrys sphegodes*, formerly typically found in inland grasslands. In some areas, cliffs provide the most spectacular natural rock gardens, as at Cape St. Vincent in Portugal, where the endemic *Cistus palinhae* is one of the many species present.

Vegetated sea cliffs of the Mediterranean coasts with endemic Limonium *spp.* Unlike the corresponding cliffs in the north and west, the incidence of exposure and hence the maritime nature of cliff vegetation in the Mediterranean is much reduced. Exceptions include the megacliffs of Croatia (Velebit) which represent a coastal habitat of exceptional biological importance. The habitats present range from emerging rocks with a calcifying algal community to sea water-drenched hyper-saline communities, including *Arthrocnemum glaucum* and *Atriplex portulacoides*, to exposed windswept cliff tops from 460 to 1,200 m high. Interspersed in more sheltered areas are woodlands and scrub. There is little elsewhere in the literature concerning these habitats though reference is made to similar cliffs in Albania and southwest Turkey (van der Maarel, 1993). These areas should probably be considered alongside the Atlantic woodlands of the northwest and the laurel forests of the Canary Islands for their conservation importance. The vegetation of all of these is to one degree or another dependent on its presence of moisture-laden air derived from strong onshore winds.

Sea cliffs and seabirds

Because of the close proximity to the rich waters of the Atlantic Ocean and the North Sea, coastal cliffs often support large colonies of nesting seabirds. The most common and conspicuous species are guillemot *Uria aalge*, kittiwake *Rissa tridactyla*, fulmar *Fulmarus glacialis* and razorbill *Alca torda*, which nest on ledges. Steep-sided granite cliffs may support large populations of gannet *Sula bassana*. On sloping cliffs where soil has accumulated, species which prefer to nest in burrows are found in high numbers such as puffin *Fratercula arctica* and manx shearwater *Puffinus puffinus*. Sea cliffs also provide nesting sites for a rare birds of prey such as the peregrine *Falco peregrinus* and the raven *Corvus corax*.

Seabird colonies do occur on cliffs and islands on more southerly coasts, though the populations, especially of cliff-nesting species are much smaller. Birds of prey are important and of these Eleonora's falcon *Falco eleonorae* is of special significance. It is a small falcon feeding mostly on small birds. It migrates northwards from Madagascar where it is widely dispersed feeding on a range of insects and in a wide variety of habitats. These include coastal cliffs and cultivated areas, rice fields, lightly wooded slopes or coastal plains. It breeds in colonies on steep rocky coastal cliffs and islands, mainly in the Mediterranean. The breeding season is later than other species and timed to coincide with passerine migrations, ensuring a supply of food for its young.

"Soft" rock landscapes—unconsolidated cliffs

Cliffs derived from unconsolidated deposits of clay (including glacial material), loam or other easily eroded sediments occur throughout the area. Landslip is common in sand and clays in the southern North Sea, southern Baltic Sea and on the south coast of England and in the sedimentary deposits of the Black Sea. Continuous erosion can result in the formation of high sloping unstable cliffs with little or no vegetation other than ephemeral communities. Where erosion is more episodic, the slopes can support scrub and even woodland on ledges remaining stable over periods of one hundred years or so. In these situations, some of the least "maritime" of the coastal cliff vegetation is found and may be indistinguishable from inland types.

The release of sediment during periods of erosion is an important factor in the development of coastal systems elsewhere. Preventing erosion in order to protect buildings or other human land uses can have consequences for habitats along the shore. As less sediment is available for movement by longshore drift opportunities for the creation of new beaches, themselves a form of coast protection, is restricted. In extreme cases adjacent habitats may be so deprived of sediment that they begin to erode.

Complex sedimentary systems

In contrast to cliffed landscapes, sedimentary landscapes are characteristically low-lying (Figure E18). They can include a variety of very different habitats and are often highly unstable. Tidal flats and salt marshes, sand dunes, and gravel are the main formations that help to define the ecosystem. Coastal wetlands may include some or all of these habitats though the tidal habitats, especially sand and mud flats, have a special significance and are described first. The short- to medium-term influence of tides and waves, freshwater inflows and sediment movement, or longer-term trends associated with sea-level change can result in the systems being very complex both spatially and temporally. Included within this section are the other components that make up the complex coastal sedimentary habitats, including the more terrestrial sand dunes and gravel structures.

Coastal wetlands

Coastal wetlands occur throughout Europe (Figure E19). Wolff (1989) estimates that intertidal areas cover 9,300 km^2, excluding small wetland areas in Iceland, Russia, Norway, Spain, Portugal, and Italy where little or no information is available. This implies that intertidal wetlands cover only 0.09% of Europe's land surface, a figure clearly highlighting the scarcity of the habitat type. They are most extensive in the major estuaries of the southern North Sea and the deltas of the southern Baltic and Mediterranean Seas. The Wadden Sea in Holland, for example, has approximately 123,900 ha and the Ebro Delta in Catalonia, one of the five biggest in the Mediterranean/Black Sea area has some 350,000 ha of tidal and former tidal land.

Estuaries. Estuaries, deltas, and lagoons lie within a range of coastal wetlands which have specific characteristics defined partly by the geographical location and degree of sediment input. In the north and west fjords and fjards, usually considered to be a special form of estuary, are specifically associated with glaciated areas and have limited inputs of sediment. The former are drowned glacial troughs, often associated with major lines of geological weakness. They have a close width–depth ratio, steep sides and an almost rectangular cross section. Sediment deposition is mostly restricted to the head of the fjord where the rivers usually discharge. The latter are glacial relicts on lowland coasts; ice-scoured rock basins, bars and small islands are characteristic features. Rias also have limited sediment and are formed when river valleys were inundated at the end of the last glacial period. They occur on the exposed western shores of Europe in macro- or meso-tidal areas. They are characteristically steep-sided with wooded slopes and occur in south west England and northern France, Spain, and Portugal.

Estuaries usually have major inputs of sediment. They are "partially enclosed bodies of water, open to saline water from the sea and receiving fresh water from rivers, land run-off or seepage" (Davidson *et al.*, 1991). They are complex systems, subject to a usually twice daily tidal rise and fall, they may have sand dunes or gravel structures enclosing sheltered shallow mud and sand shoals and salt marshes. They are particularly extensive in lowland areas with abundant sediment and a macro–meso-tidal range. Particularly large sites occur in the southern north sea, (in and around the Thames Basin, the delta region of Holland and in the Wadden Sea).

Estuarine fauna. A key component in the development of the ecological value of coastal wetlands, especially tidal estuaries, is their high productivity. This is derived from *in situ* primary production of salt marshes and in estuarine waters, as well as inputs from the land and the sea. This abundant source of food can result in very large numbers of a few dominant organisms. Typical densities on muddy shores of the mud snail *Hydrobia ulvae* can be up to 120,000 individuals present per square meter. The burrowing amphipod *Corophium volutator* is also numerous with up to 60,000 individuals per square meter. The segmented worms *Nereis diversicolor* (a ragworm) and on more sandy shores *Arenicola marina* (lugworm), are also often abundant and of particular importance as prey for wintering wading birds. In this context, estuaries in north west Europe are especially valuable and frequently support nationally and internationally important numbers of water birds (wildfowl and waders). Some indication of the significance of these estuaries can be gleaned from the location of the main wintering populations of the knot *Canutus canutus islandica* (Figure E20).

While these wintering bird populations are often the most obvious visible manifestation of the productivity of a wetland system a range of other species occur. These include sea fish using the shelter of the

Figure E18 Low-lying sedimentary coastal plain derived from sediments eroded from the hills of northern Albania.

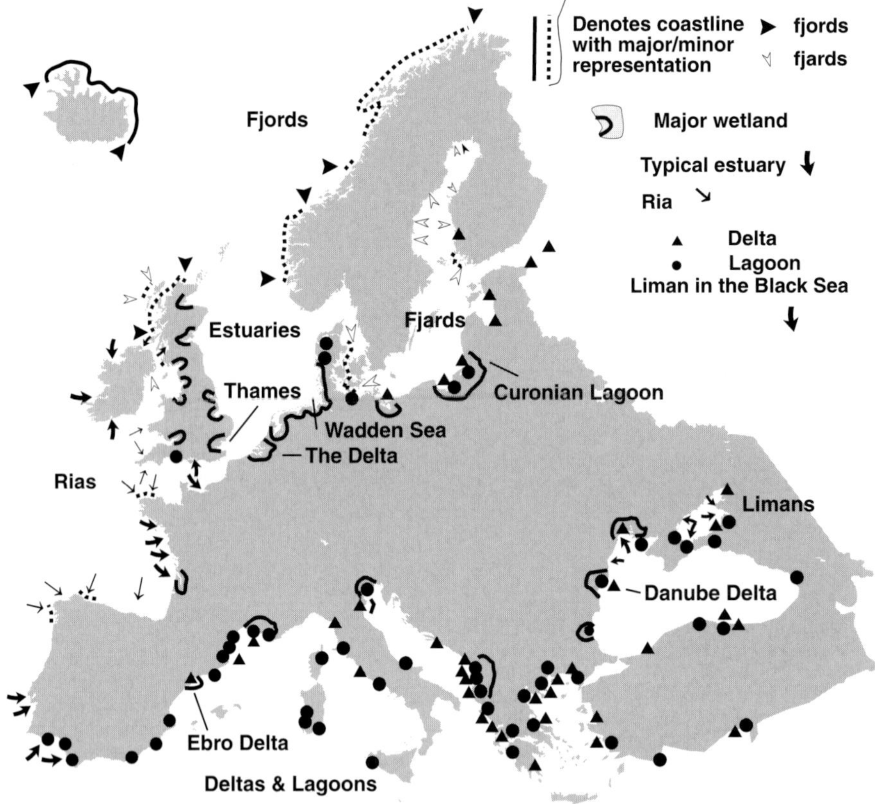

Figure E19 Distribution of coastal wetlands in Europe (adapted from Doody, 2001).

estuary to spawn, or as nursery areas. Diadromous fish are specialists of the estuarine environment. Catadromous species such as *Anguilla anguilla* (the common eel) move from freshwater to the sea where they spawn, passing on the way the salinity gradients in the estuaries and deltas. In contrast an andromous species such as *Salmo salar* (salmon) develops from eggs shed upstream in rivers. A major part of its life is spent outside the home rivers at sea. Important feeding areas are in the Atlantic along the west coast of Greenland, above the Faroe Islands, in the Norwegian Sea, and possibly in the Barents Sea. In the Baltic, the Curonian Lagoon has 46 freshwater, migratory, and marine fish including the Baltic salmon, reflecting its importance for these species.

Deltas and lagoons

Coastal deltas and lagoons are most characteristic of micro-tidal regions (with a tidal range <2 m), and hence in Europe are particularly abundant around the shores of the Baltic, Mediterranean, and Black Seas. Deltas are sedimentary coastal plains built up from material brought down by rivers to cover shallow offshore areas creating sedimentary features which protrude beyond the normal limit of the coastal margin. They are best developed in micro–meso-tidal areas where wave action is the dominant force in building the sand dunes, barrier islands bars, and salt marshes. Lagoons are shallow, virtually tideless, pond- or lake-like bodies of coastal salt or brackish water that are partially isolated from the adjacent sea by a sedimentary barrier, but which nevertheless receive an influx of salt water from that sea (Barnes, 1980; Guélorget and Perthuisot, 1992). Although they are mostly found in micro-tidal seas, they are also present on some macro-tidal coasts—such as the northern North Atlantic—where offshore deposits of gravel are to be found. Gravel then replaces the sand of micro-tidal seas as the barrier material. Europe possesses the smallest proportion of lagoonal coastline of any continent, with only 5% of its coast in this category (Cromwell, 1971). Several deltas, for example that of the Danube (and Nile), have developed within—and have now largely or completely obliterated—former lagoons enclosed by barrier-island chains. The limans of the Black Sea coast receive their salt input via seepage.

Fauna. Deltas and lagoons are also important for a variety of water birds. They support waders, wildfowl, grebes, and similar marshland birds many wintering or on passage, as well as a considerable number of nesting species. Among these black-winged stilt *Himantopus himantopus*, avocet *Recurvirostra avosetta* and herons, including purple heron *Ardea pururea* and large populations of a few species of breeding duck are significant. However, it is the spectacular flamingo *Phoenicopterus ruber roseus*, which is most readily appreciated by anyone visiting the Camargue or the Ebro Deltas in the western Mediterranean. Numbered in thousands in the Camargue it is a species dependent on wetlands which periodically dry out and become hypersaline. These conditions naturally occur in North Africa and the Mediterranean but it is the juxtaposition of "natural" lagoons with "artificial" salinas which provide especially suitable nesting conditions (Sadoul *et al*., 1998).

The lagoons also support invertebrate and fish faunas. These may include essentially freshwater or marine/estuarine species capable of withstanding a degree of brackishness as well as specialist lagoon species. Among the marine fishes are sea bass *Dicentrarchus labrax* and sole *Solea vulgaris* while carp *Cyprinus carpio* and pike *Esox lucius* live in freshwater. Eel *A. anguilla* and mullets (Mugillidae) bridge the salinity gap.

Salt marshes

Salt marshes include all halophytic plant communities and their associated animals, which are influenced by sea water. Typically they are regularly inundated by the tide, have sometimes rapid accumulations of fine sediment and in the absence of enclosure, include transitions to nontidal vegetation. They are at their most extensive on flat tidal plains in sheltered locations where abundant sediment is deposited and wave action restricted (estuarine marshes). These conditions are most favorable to the establishment of the pioneer salt marsh plants such as *Salicornia* spp., *Suaeda* spp. and *Spartina* spp. which are among the most frequent colonists in northern latitudes. Lagoonal marshes occur in association with lagoons and are most extensive in the micro-tidal areas. Salt marshes also occur on beach plains behind exposed or partially protected beaches and in the lee of barrier islands. In some areas where sediment is restricted or (as is the case on some rocky shores and cliff tops) absent, inundation is intermittent and dependent on storms or extreme high water levels. In these situations, "perched" salt marshes on exposed cliffs and narrow saltings of rocky shores "beach head" and

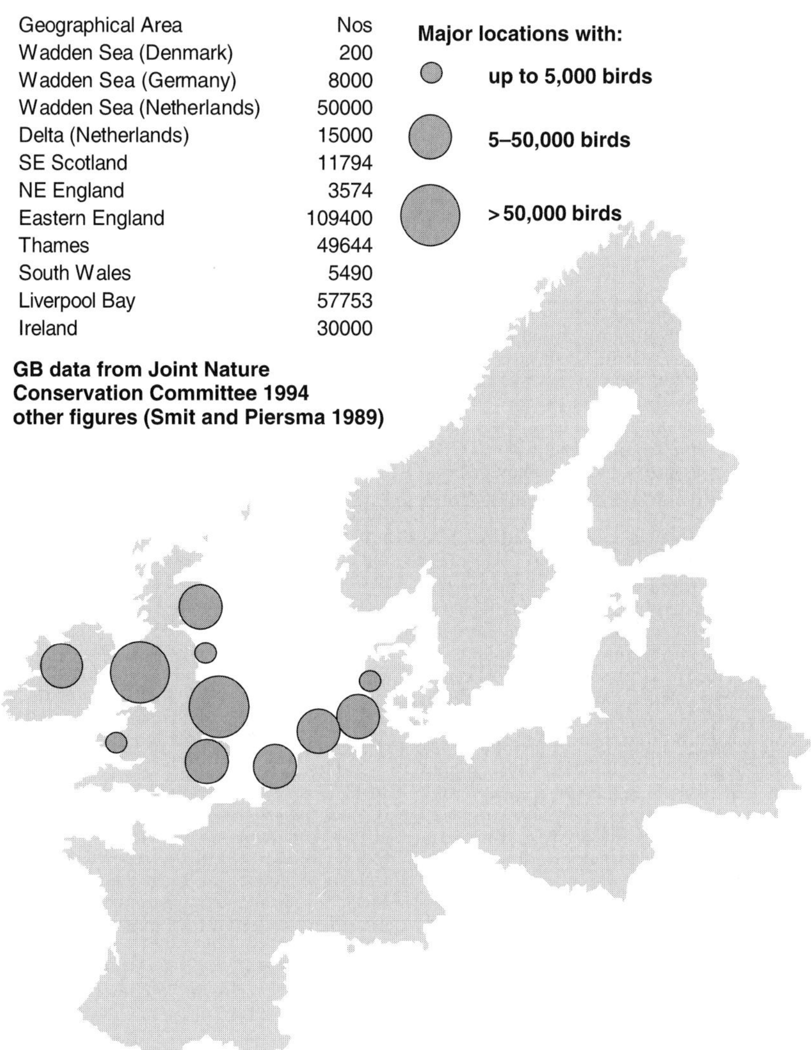

Figure E20 Distribution of the main wintering sites of the knot *Canutus canutus islandica* (adapted from Doody, 2001).

"loch head" salt marshes have vegetation similar to the typical salt marsh but without their rapid sedimentation (Doody, 2001a).

Distribution in Europe. Salt marshes occur throughout Europe and are particularly associated with the major macro-tidal estuaries of the southern North Sea and south and west England (Dijkema, 1984). Elsewhere in the northwest they occur as scattered small sites associated with rocky coasts. In the Baltic, they are more restricted, partly because of the very low salinity within this sea (Dijkema, 1990). In the Mediterranean, all the major deltas have extensive salt marsh communities. Elsewhere they again occur as scattered sites, though here less frequent than in the north. Their distribution is closely related to that of the coastal wetlands.

Vegetation succession. A key element in the development of salt marshes in sediment rich areas is the process of succession. A sequence of communities tolerant of different degrees of tidal submergence may develop, together with transitions to other, terrestrial vegetation. Transitions may be to reed swamp sand dune, gravel, freshwater, or woodland. This natural transition may be truncated by the construction of sea walls or other coastal defenses, with the loss of these upper salt marsh transitions and their associated wildlife. In some areas where a saline influence continues landwards of sea walls on enclosed unimproved salt marshes, coastal grazing marshes with brackish ditches may develop.

The early stages of colonization are represented by two main community types (European Commission, 1999), namely: ***Salicornia* and other annuals colonizing mud and sand** and ***Spartina* swards (*Spartinion maritimae*)**. These are typically referred to as pioneer marsh colonizing the lower levels of the tidal flats, most frequently inundated by the sea. Under these conditions, areas of mud, exposed for several days, are required for the early colonizers to become established. The communities which develop are composed of a few, mostly annual species occuring in mono specific stands, or in simple combinations. These typically include *Spartina* communities, most frequently *Spartina anglica* or stands of *Salicornia* spp., *Suaeda maritima* and *Aster tripolium* with *Puccinellia maritima* dominant. They are most frequently found in the macro–meso-tidal areas of the northwest.

The next stage in the successional process is typically refer-red to as mid-marsh. The distinction between low and mid-marsh is blurred but as the tidal height of the marsh increases additional species occur. These are dominated by the shrub *A. portulacoides* in the absence of grazing with increasingly *Festuca rubra* on grazed marshes. There is a wide variation in the composition of these communities and *Juncus gerardii* is another characteristic species. The presence of abundant *Armeria maritima* gives many marshes in the northwest a visually distinct appearance with a close-cropped, pink sward in summer.

The development of upper-marsh in the absence of enclosure provides some of the richest plant communities. Mid-upper marshes in the northwest are collectively classified within community type. **Atlantic saltmeadows (*Glauco-Puccinellietalia maritimae*)**. The presence of a number of species such as *Suaeda vera*, and *Frankenia laevis* in southeast England, more frequently found in Mediterranean regions suggest an overlap in distribution with. **Mediterranean salt meadows (*Juncetalia maritimi*)** and **Mediterranean and thermo-Atlantic halophilous scrubs (*Sarcocornetea fruticosi*)**. Transitions to swamp also occur in the absence of enclosure and are usually dominated by *Scirpus maritima* and *Phragmites australis*, which can be extensive in the upper reaches of estuaries. These communities form important transitions and may be particularly rich, especially where freshwater seepages or transitions to other habitats such as sand dunes occur. In the Baltic, the low salinity and micro-tidal range of the sea restricts the presence of the pioneer communities and extensive swamps dominated by *Phragmites* are more prevalent (Dijkema, 1990). Similarly in the Mediterranean the extent to which a true succession develops is restricted and narrow zones of a few mid-marsh species quickly merge into more extensive swamp. Some of the best examples of these occur in the still relatively undeveloped coastal deltas of northern Albania.

Fauna. Salt marshes present difficult environments for colonization by invertebrates due to changes in salinity and humidity caused by periodic tidal immersion. The fauna is a mixture of marine, freshwater and terrestrial species, which are adapted in various ways to the stressful environment. Marine species tend to occur lower down the marsh and often burrow to avoid desiccation. Terrestrial and freshwater species occur mostly in the upper marsh and transition zones, and have adapted to, or avoid immersion in saline water. The presence of individual invertebrate species is often closely associated with the type of vegetation which occurs.

Successful breeding by a variety of birds occurs on the upper levels of unenclosed salt marshes following the high spring tides in April and May. Although early nests may be destroyed by high tides, at least one brood is usually possible. The structural diversity of the marsh is also of crucial importance to some species such as redshank *Tringa totanus* and particularly high populations may be present in the best habitats. In winter, many of the most numerous species of wader feed almost exclusively on the invertebrates in the mud flats at low tide and are not dependent on the salt marshes as their main food supply. However, some also use the salt marsh for roosting at high tide, or as supplementary feeding around the pans and bare areas on the marsh (e.g., dunlin *Calidris alpina* and knot). Only the redshank and snipe, of the wading birds, use salt marsh in preference to the exposed tidal flat, probably because they like to be sheltered from view by the cover of vegetation. For these species the salt marsh represents the primary feeding location throughout most of the tidal cycle.

Other species, mainly grazing ducks and geese use the salt marsh as a source of food. The geese seek out open areas with close-cropped or low vegetation. Even here there are distinct preferences, and while the brent goose feeds on Eel-grass (*Zostera* spp.) in the low marsh, the stronger grey-lag goose feeds on the tougher grasses of the marsh itself. Of the ducks the wigeon feeds on both low and high marsh, normally at the waters edge. For most of these species the more palatable grasses favored under high stocking densities of cattle and sheep, are important to the continued use of the salt marsh by the grazing ducks and geese.

Sand dunes

Sand dunes develop wherever there is a suitable supply of sediment (within the size range 0.2–2.0 mm) which is moved onshore by the tide and then blown inland to form accumulations varying from a few centimeters to 40 m or more thick. The type of sand dune which develops depends on the underlying topography and climate. In most locations in the temperate regions of the world, vegetation plays an important role in the growth of the typical dune landscape, facilitating the accumulation of sediment (Ranwell and Boar, 1986). Bay dunes are the most common form and are associated with indented coastlines, often between headlands. Backshore dunes can be large and occur when both dominant and prevailing winds blow sand inland. On especially exposed sites, coastal cliffs may be covered with a veneer of sand blown up over the cliff to create "climbing dunes." Barrier islands tend to occur in exposed locations and may be built on gravel bars. Ness dunes, including cuspate forelands, occur where prevailing and dominant winds are in opposition. Sometimes progradation may be quite rapid and a series of low dune ridges form, interspersed with wet hollows. These are also found in areas where sea level is rising relative to the land as in southeast Norway, northeast Scotland, and Northern Ireland. Spit dunes occur when sediment is deposited at the mouth of an estuary. They take a variety of complex forms depending on the action of waves (longshore drift) and the force of the river flows to the sea. These also include the dune forms associated with deltas in micro-tidal seas.

Distribution in Europe. Sand dunes are distributed throughout Europe (Figure E21). In the areas where sediments are restricted there are a relatively large number of small sites, as in northern Norway and on the rocky, cliffed coasts of the exposed shores of the Atlantic from western Ireland to Spain, the karst coasts of the Mediterranean, and in the Black Sea. In the west where there are abundant coastal sand deposits, prevailing westerly winds blow the sand grains onshore to create large

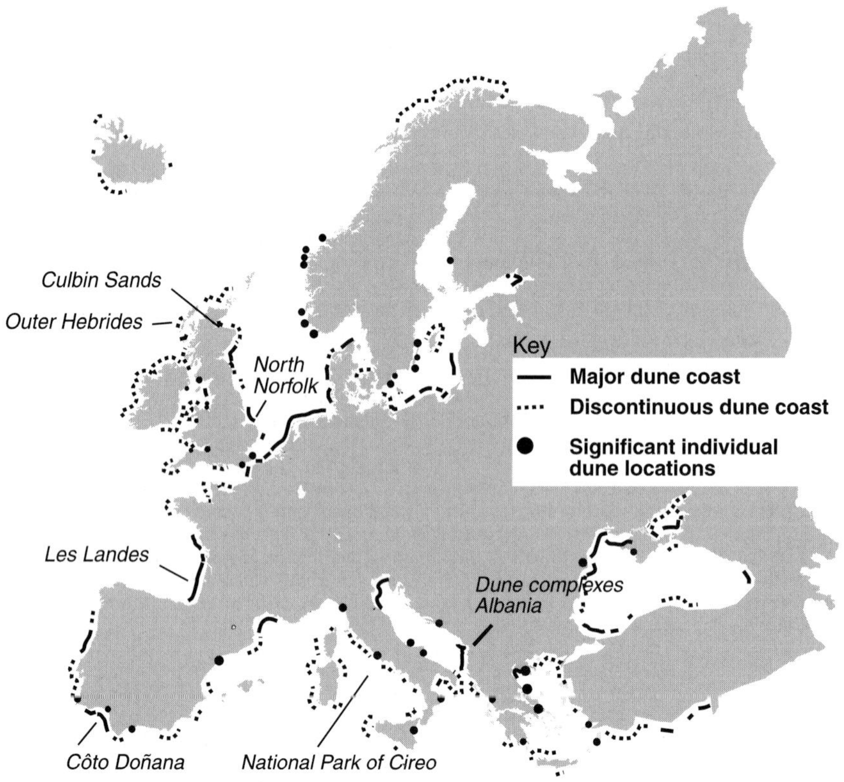

Figure E21 Distribution of sand dunes in Europe (adapted from Doody, 2001).

backshore systems such as those of the Outer Hebrides in Scotland. These include some of the best and largest examples of the extensive cultivated sandy plain or machair, also present in the west of Ireland. In northern Denmark massive accumulations of sand are forced onshore under the action of the prevailing winds from the North Sea. In France on similarly exposed coasts the dunes stretch many kilometers inland. Here, the area known as "Les Landes" has a special significance, and is one of the most extensive dunes in Europe.

On the North Sea coast and in the northern Baltic, dunes develop on a coastline, which is rising relative to sea level. Here, they are generally prograding with ridges becoming established as a sequence lying "parallel" to the coast. They are sometimes interspersed with damp hollows in which dune slack vegetation develops. In eastern Scotland some large examples occur, as at Culbin Sands a system with in excess of 4,000 ha, one of the largest areas of blown sand in Great Britain.

In the southern Baltic and southern North Sea where glacial sediments are particularly abundant, sand dunes are also extensive, for example, they make up 80% of the coastline of Poland. The predominant dune type is represented by sand bars and spits, which lie parallel to the coast (Poland and Wadden Sea). In southern Portugal and Spain, the dune systems also include barrier islands and spits. Rivers bring the sediment to the sea, which is then reworked by wave and tidal action and blown onshore. One the most important dune systems in Europe, lies within the Doñana National Park in Southern Spain. The park consists of beaches, foredunes, high mobile dunes (up to 30 m), and stabilized dunes which enclose a major wetland.

On the Mediterranean and Black Sea coasts dunes are often found in association with deltas. In Italy, dunes probably originated during the thermal optimum after the last glaciation (5,000 years ago). They have since been broken by rivers and their subsequent development has been a product of natural erosive forces and human activity. Today dunes are present around the whole coastline but only in a few protected areas, like the National Park of Circeo, can natural development be seen. Along the Greek coast there are many places where sand dunes cannot develop because the hills or mountains outcrop as rocky coastlines near to the sea. Sand dunes tend either to occupy a narrow fringe bordering flat areas of land or exceptionally form extensive dunes up to 10 m height, as in Western Peloponnesus.

Vegetation succession. The vegetation of sand dunes in temperate regions such as Europe can be grouped according to the successional sequence from the beach inland. These have been defined for the north west coastal areas as follows:

Strandline: This includes vegetation along the high tide line. It is usually ephemeral, salt tolerant and composed of a limited number of species;

Foredune: The first stage in sand deposition occurs here. The vegetation is limited in species diversity, dependent on its ability to withstand the influence of salt spray and trap moving sand;

Yellow dune: This represents the main and usually most rapid phase of dune growth. *Ammophila arenaria*, the main dune forming species, is able to withstand rapid burial by sand;

Dune grassland: Stabilized dune vegetation dominated by species of grass and herbs develops under the influence of grazing and in a moist climate. It is mostly found in northwest Europe;

Dune heath: Occurs on sand dunes where the calcium carbonate content of the soils is low, either because the original sand has a high proportion of silica or where leaching has taken place;

Scrub: Areas with low shrubs. These include woodland understorey species in the northwest and the "maquis" and other similar vegetation in the Mediterranean;

Dune slacks: Vegetation which develops under the influence of high water table and is often completely flooded in winter;

Woodland: Natural forest with various pine species or deciduous trees such as oaks. In most areas natural woodland is scarce.

The European Habitats Directive recognizes three major types of vegetation (European Commission, 1999). Of these the dunes of the "Atlantic, North Sea and Baltic" include at least 11 distinct types and those of the Mediterranean coast, 7 Most of the sand dunes described for northwest Europe appear to have a sequence of vegetation types which potentially includes all the more important successional communities from strandline (driftwalls) to yellow and grey dune, dune pasture and scrub, described above. (In areas where beach erosion is occurring some of the early stages of succession may be absent.)

The sequence of vegetation on the mid-Atlantic coast is similar to that occurring in the northwest. In the north of France, some of the botanically richer areas occur where dunes are composed of calcareous sand and lie against to chalk cliffs. The further south the dunes are formed, the more the southern elements of the flora appear.

On the Mediterranean coast the zonation described above is much less obvious. While *A. arenaria* still plays and important part in dune establishment, communities with dry grasslands and scrub are more prevalent. The following are recognized from the Manual of European Union Habitats (European Commission, 1999).

Crucianellion maritimae fixed beach dunes
Dunes with *Euphorbia terracina*
Malcolmietalia dune grasslands
Brachypodietalia dune grasslands with annuals
Coastal dunes with *Juniperus* spp.
Cisto-Lavanduletalia dune sclerophyllous scrubs
Wooded dunes with *Pinus pinea* and/or *Pinus pinaster*

Fauna. The dune fauna is not especially rich or abundant though the variety of slope, aspect and vegetation structure, coupled with the dynamic nature of many systems, make dunes one of the most suitable habitats for a range of invertebrates requiring open conditions. They are also frequented by warmth-loving species, which bask on the dry warm slopes even in the more northern latitudes. Hymenoptera (especially wild bees, digger wasps, spider wasps, and other solitary wasps) are able to build nests easily in the soft sand and are often found dwelling in sand dunes (Haeseler, 1989, 1992).

The dune avifauna includes breeding colonies of terns such as the sandwich tern *Sterna sandvicensis* and the little tern *Sterna albifrons* which nest on open sand. Other species such as the eider duck *Somateria mollissima* and the shelduck *Tadorna tadorna* nest in the ground vegetation in the north. Further south (including eastern Spain) tree nesting species of heron, such as the squacco heron *Ardeola ralloides* and the purple heron *Ardea purpurea* can be found.

Rabbits *Oryctolagus cuniculus* occur in abundance in all areas and are stock food for fox *Vulpes vulpes* populations of which inhabit mainland dunes. Polecat *Mustela putorius*, weasel *Mustela nivalis*, several species of bats, voles, mice, and shrews also occur in most of the areas, while the roe deer *Capreolus capreolus* and red squirrel *Sciurus vulgaris* are less frequent. Among reptiles and amphibians are sand lizard *Lacerta agilis*, natterjack toad *Bufo calamita* and the edible frog *Rana esculenta*.

Gravel

Gravel is the term applied to sediments larger in diameter than sand but smaller than boulders. A predominant particle size of over 2 mm separates gravel from sand (King, 1972) whereas at much over 200 mm diameter, boulders approximate to cliff habitat. Four environmental factors are responsible for the growth of a gravel beach. There must be an available supply of sediment in the size range mentioned above. At the same time waves, wind, and tidal currents should be favorable for its movement. Since this coincidence is unpredictable, considerable periods of time may exist when no movement takes place, these are interspersed by times of marked activity resulting in stable and mobile gravel habitats varying both in time and space.

Geomorphologists and ecologists have recognized five categories of gravel structures which vary in their oceanicity and therefore in their ecology. The simplest and commonest type is the fringing beach forming a strip in contact with the land along the top of the beach. Gravel spits form where there is an abrupt change in the direction of the coast. They often contain recurved hooks and a recurved distal end, a result of deflection of waves by refraction. Bars or barriers are effectively spits which have formed across estuary mouths or indentations in the coast. These three structures are basically foreshores, regularly washed by spray and storm waves and therefore possessing only limited or ephemeral vegetation over much of their area. Cuspate forelands and offshore barrier islands, on the other hand, are larger structures and are more terrestrial in nature. The former develop when gravel is available in large quantities and piles up in front of fringing beaches or spits and is then driven landwards by storm waves to form a series of roughly parallel ridges. Where wave approach can from two directions only, such apposition beaches will form into cuspate forelands (Figure E22). Large masses of gravel may also form offshore barrier islands under

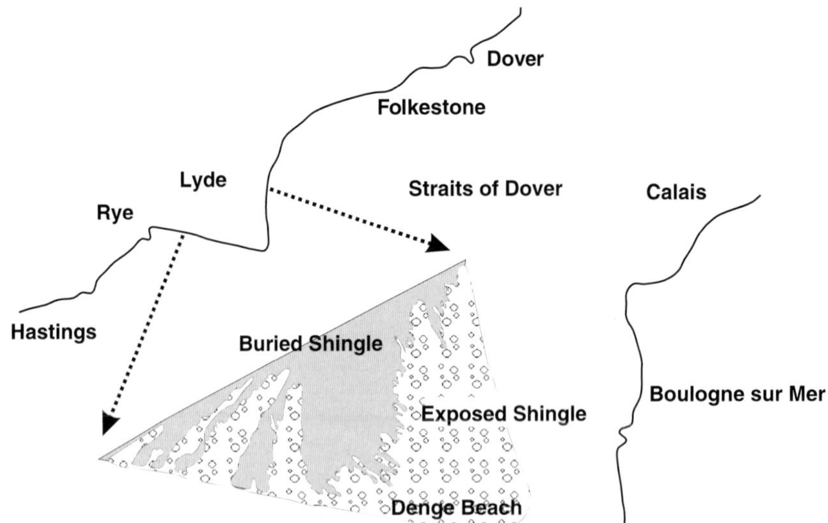

Figure E22 Dungeness southeast England—a cuspate foreland.

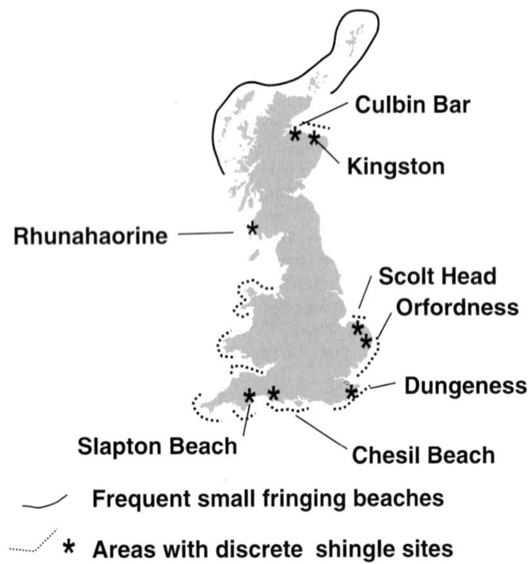

Figure E23 Distribution of gravel shores in the United Kingdom, some key sites are named.

conditions of shallow water and low-energy environments such as in the North Sea and the Baltic.

Distribution. Gravel deposits occur throughout Europe, especially in the north and west. Fringing beaches are widely distributed, though often small and narrow. Larger gravel structures occur as isolated and discrete entities. (Doody, 2001b). In the north, the fjords and sea lochs in both the meso-tidal and micro-tidal areas of the Atlantic, North Sea, and Baltic Sea support long stretches of gravel foreshores. Individual, more substantial structures are scattered throughout the area. The situation in the United Kingdom illustrates the nature of this occurrence (Figure E23). This is derived from extensive surveys undertaken in the 1990s and organized by the statutory conservation body, the Nature Conservancy Council.

Vegetation. Two plant communities identified in the European Union Habitats Directive (European Commission, 1999) help to define these formations: **Annual vegetation of drift lines** and **Perennial vegetation of stony banks**. Gravel shore vegetation includes *Mertensia maritima* prevalent in the north and *Glaucium flavum* and *Matthiola sinuata* in the south. Mobility is an over-riding consideration and the species which colonize are able to withstand periodic disturbance. These include *Lathyrus japonicus* and *Crambe maritima* colonizing the gravel just above the high water mark where the shore may remain stable for periods of 3–4 years. *Crambe maritima* is constant in the three sub-types of **Perennial vegetation of stony banks**, for the Baltic, Channel coast, and the Atlantic, respectively. Besides stability (or its converse mobility) beach composition and especially the amount of fine material present in the gravel matrix is another principal factor influencing the type of vegetation namely:

(1) gravel without matrix (relatively uncommon)—vegetation is limited to encrusting lichens and the most tolerant angiosperms such as *L. japonicus*;
(2) gravel with a sand matrix—dry yet stable conditions, contains species such as *Lotus corniculatus* and *Honckenya peploides*;
(3) gravel with a silt–clay matrix—ecologically related to salt-marshes but with freer drainage. The vegetation is similar to that of salt-marsh levées with *A. portulacoides* or *Glaux maritima*;
(4) gravel with an organic matrix—rotting seaweed is nutrient rich and supports populations of species that arrive as seed within tidal debris. Species include *Beta maritima* or *Atriplex* spp. where there is little drainage from the land, or with more humic soils reed *Phragmites communis* or *Iris pseudacorus* swamp may be present.

Fauna. The fauna associated with gravel structures largely consists of invertebrates (Shardlow, 2001) and avifauna. In the case of the former large numbers of species, such as bumble bees are found. Gravel beaches and structures provide suitable habitat for a small number of other species. These are generally ground nesting birds (Cadbury and Ausden, 2001) which make a simple scrape where they lay their camouflaged eggs. The importance of gravel as a habitat for breeding terns has led to the conservation of several sites and the need to create areas protected from human trampling affects. Few vertebrates other than domestic herbivores are found on gravel sites except hares and rabbits, though in some places foxes are important predators of ground nesting birds.

Human influences

This entry along with many text books describe coastal habitats in terms of their ecological or geomorphological development. Salt marshes and sand dunes, in particular are described as exhibiting primary succession. Classic studies in Great Britain covered salt marshes and sand dunes on the North Norfolk coast. All emphasized the natural status of the vegetation in areas where ecological processes appear to have been relatively free from human influence. The vegetation was thus described as a series of community types progressing from the early pioneer stages to more complex forms related to the physical factors influencing their development. While this is in part true, it is also the case that much of Europe's coastline has been influenced in one way or another by human activity. A few examples are discussed below.

Sea cliffs and slopes

Apparently inaccessible steep-sided sea cliffs and slopes are often seen as virtually unaffected by human activity. Certainly difficulties of access, low productivity, associated with a harsher climate mean that the harder rock of the north are less built on than their counterparts in some parts of the Mediterranean. However, grazing is and has been a traditional activity and even cliffs on more inaccessible islands have been and in a few cases continue to be plundered for sea birds and their eggs.

Sand dunes

Major tourist development, especially along the Mediterranean coast, has destroyed many sand dunes. Building directly onto the beach in some cases has completely obliterated some areas while sand extraction for use nearby has also taken its toll. Another activity is planting with non-native trees, especially conifers, to stabilize them. This has affected many systems throughout Europe and while the dune form is retained most of the dune plant and animal species are lost. One exception to this general picture is the large dune systems which have developed in association with major deltas on the coastline of Albania and in some parts of south east Turkey. Here, many different dune forms are present with a maximum height of about 50 m and width of up to 4.3 km, including huge beach plains with embryo dunes, parabolic dunes, blowouts, dune slacks, lakes, secondary barchans, and dune fields. They also include complete transitions to pine woodland with native *P. pinea* and *Pinus halipensis* as the forest tree. They are among the most important dune areas in Europe.

Secondary habitats coastal grazing marsh and salinas

The partial enclosure of salt marsh by the erection of a small earthen bank was probably one of the first human activities directly affecting coastal habitats. The aim was to extend the period of summer grazing by livestock (cattle and sheep), by preventing over-topping on spring tides. As the early techniques of enclosure improved, larger clay or earth banks were constructed enclosing consolidated salt marsh excluding the tide from larger areas. Historically the land was subsequently used as pasture, with little or no further "improvement". These areas of coastal grazing marsh are defined by the presence of permanent and semi-permanent grassland, drainage ditches, and enclosing earth dikes. The wildlife interest has developed alongside the traditional use of the area for agricultural.

The production of salt is one of the oldest industries known to man. Industrial salinas are large, artificial, and complex lagoon systems and a major source of salt produced today. The principal aim of any salt company is to produce a maximum tonage of salt per annum to meet the demands of commerce and industry. However, these artificial ecosystems are in fact, wetlands of international importance and areas with a rich and diverse flora and fauna, vertebrate and invertebrate communities. They host important breeding populations of aquatic birds, besides rare and endangered species. The special nature of Mediterranean salinas make them vitally important resting and refuelling sites for many thousands of shorebirds migrating between the Palearctic breeding grounds and their winter quarters in Africa (Sadoul *et al.*, 1998).

Grazing and other management

A key feature of many habitats throughout Europe is the long history of exploitation by human kind, including grazing by domesticated stock. This has had a profound influence on the type of vegetation which has developed. In the northwest grazing (by rabbits cultivated in warrens) helped to create species rich calcareous dune grassland and heathland, preventing the natural progression to scrub and woodland. In areas formerly used for grazing by domestic stock a reversion is taking place to secondary mixed scrub, broad-leaved woodland and pine forest. This is particularly prevalent in Finland today where invasion of open land by pine forest with *Pinus sylvestris* is common. Prior to this, open vegetation was maintained by the grazing of domestic animals.

Grazing and burning are intimately bound up with the development of the low, thorny scrub vegetation which predominates in the Mediterranean. It is also clear that many of the major deltas would not be as extensive without the influence of deforestation in the hinterland. Thus it can be seen that the rich diversity of habitats and associated species, which exist along the coastline of Europe are influenced to a greater or lesser extent by human activities. At the other extreme major developments including tourist growth in the Mediterranean has greatly reduced the ecological and landscape diversity of many areas to the detriment of wildlife and cultural heritage. The way in which this has altered the coastline of Europe and the approaches being adopted to monitoring the impact of development and policy initiatives designed to protect it is the subject of the entry on Monitoring, Coastal Ecology.

J. Pat Doody

Bibliography

Barnes, R.S.K., 1980. *Coastal Lagoons*. Cambridge: Cambridge University Press.

Cadbury, J., and Ausden, M., 2001. Bird communities of coastal shingle and lagoons. In Packham, J.R., Randall, R.E., Barnes R.S.K., and Neal, A. (eds.), *Ecology and Geomorphology of Coastal Shingle*. Westbury: Academic and Scientific Publishing, pp. 304–320.

Cromwell, J.E., 1971. Barrier coast distribution: a world wide survey. *Abstracts Volume of the 2nd National Coastal and Shallow Water Research Conference*, p. 50.

Davidson, N.C., Laffoley, D.d'A., Doody, J.P., Way, L.S., Gordon, J., Key, R., Drake, C.M., Pienkowski, M.W., Mitchell, R. and Duff, K.L., 1991. *Nature Conservation and Estuaries in Great Britain*. Peterborough: Nature Conservancy Council.

Dijkema, K.S. (ed.), 1984. *Salt Marshes in Europe*. Strasbourg: European Committee for the Conservation of Nature and Natural Resources, Council of Europe.

Dijkema, K.S., 1990. Salt and brackish marshes around the Baltic Sea and adjacent parts of the North Sea: their vegetation and management. *Biological Conservation*, **51**: 191–209.

Doody, J.P., 2001a. Perspective: the importance of conserving coastal shingle in Europe. In Packham, J.R., Randall, R.E., Barnes R.S.K., and Neal, A. (eds.), *Ecology and Geomorphology of Coastal Shingle*. Westbury: Academic and Scientific Publishing, pp. 421–440.

Doody, J.P., 2001b. *Coastal Conservation and Management: An Ecological Perspective*. Boston: Kluwer Academic Publishers.

European Commission, 1999. *Interpretative Manual of European Union Habitats. Eur.*, 15/2, European Commission, DG. Environment.

Guélorget, O., and Perthuisot, J.-P., 1992. Paralic ecosystems. Biological organisation and functioning. *Vie et Milieu*, **42**: 215–251.

Haeseler, V., 1989. The situation of the invertebrate fauna of coastal dunes and sandy coasts in the western Mediterranean (France, Spain). In van der Meulen, F., Jungerius, P.D., and Visser, J. (eds.), *Perspectives in Coastal Dune Management*. The Hague: SPB Academic Publishing, pp. 128–131.

Haeseler, V., 1992. Coastal dunes of the southern North Sea as habitats of digger wasps. In Carter, R.W.G., Curtis T.G.F., and Sheehy-Skeffington, M.J., Balkema, A.A. (eds.), *Coastal Dunes: Geomorphology, Ecology and Management: proceedings of the third European Dune Congress*, Galway, Ireland, 17–21 June 1992, Rotterdam, pp. 381–389.

King, C.A.M., 1972. *Beaches and Coasts*. London: Arnold.

Polunin, O., and Walters, M., 1985. *A guide to the vegetation of Britain and Europe*. New York: Oxford University Press.

Ranwell, D.S., and Boar, R., 1986. *Coast Dune Management Guide*. Abbots Ripton: Institute of Terrestrial Ecology, NERC.

Sadoul, N., Walmsley, J.G., and Charpentier, B., 1998. *Salinas and Nature Conservation*. Conservation of Mediterranean Wetlands, No. 9, Arles: Tour du Valat.

Shardlow, M., 2001. A review of the conservation importance of shingle habitats for invertebrates in the United Kingdom. In Packham, J.R., Randall, R.E., Barnes, R.S.K., and Neal, A. (eds.), *Ecology and Geomorphology of Coastal Shingle*. Westbury: Academic and Scientific Publishing, pp. 355–376.

Stanners, D., and Bourdeau, P. (eds.), 1995. *Europe's Environment: The Dobříš Assessment*. European Environment Agency (EEA). Luxembourg: Office for Official Publications of the European Communities.

van der Maarel E. (ed.), 1993. *Ecosystems of the World 2A. Dry Coastal Ecosystems—Polar Regions and Europe*. Amsterdam: Elsevier.

Warrick, R.A., 1993. Climate and sea level change: a synthesis. In Warrick, R.A., Barrow, E.M., and Wigley, T.M.L. (eds.), *Climate and Sea level Change—Observations, Projections and Implications*, Cambridge University Press, pp. 3–24.

Wolff, W.J., 1989. Getijdewateren. In Wolff, W.J. (ed.), *De internationale betekenis van de Nederlandse natuur. Een verkenning*. Gravenhage: SDU's, pp. 92–96.

Cross-references

Coastal Climate
Dune Ridges
Estuaries
Europe, Coastal Geomorphology
History, Coastal Ecology
Monitoring, Coastal Ecology
Rock Coast Processes
Salt Marsh
Vegetated Coasts

EUROPE, COASTAL GEOMORPHOLOGY

The coastlines of Europe cover a latitudinal range of 45° reaching from 80°N in Spitsbergen to 35°N in southern Crete, Greece. Thus, nearly all coastal types except for mangroves and coral reefs are present in Europe (Bird and Schwartz, 1985; Kelletat, 1995), including arctic ice coasts with calving glaciers, drifting sea ice contact and permafrost decay along cliffs on Spitsbergen or young volcanic coastlines on the Aeolian islands of Italy or Santorini/Greece. Processes of salt weathering associated with tafoni can be observed at the Costa Brava in Spain and the islands of Corsica and Sardinia or SW-Turkey, biogeneous coastal forms (Laborel, 1987) including vermetid boiler reefs and micro atolls are characteristics on Crete islands (Greece) as well as coastal cementation like aeolianites and beachrocks. Because of the rather long and complicated geologic and tectonic history of the European continent, the coastlines are greatly differentiated. Their total length in natural scales is several 100,000 km. Many of the scientific terms in coastal sciences derive from European examples, like ria, fjord, canale, tombolo, or Thyrrhenian, Sicilian, Calabrien for Pleistocene sea level highstands.

The framework of coastal forming in Europe

The geology of Europe varies from very old structures, the proterozoic shields of granite and gneiss in Scandinavia, to rather young like the vivid neotectonic areas—even in historical times—in the eastern Mediterranean (Pirazzoli *et al.*, 1996). The orogenic belts of the Caledonids are still visible in the coastal configuration of Scotland, the Varistids in southern England, and the imprints of the alpidic orogenesis can be observed along nearly all coastlines of the Mediterranean. Areas of longer subsidence are marked by large embayments like the North Sea basin, the Gulf of Biscaya or the intra- continental sea of the Baltic and the intercontinental sea of the Mediterranean. Extremely resistant silicate rocks survived thousands of years of wave attack, whereas Pleistocene moraines or Holocene dunes suffer from strong erosion and abrasion. Vertical movements influence coastal development in many regions, but by different causes: long lasting and ongoing glacio-isostatic uplift in Scandinavia (nearly 300 m, with a modern maximum of 9 mm/year, acc. to Åse, 1980), Iceland and Scotland (both with more than 120 m), salt intrusion in small places of the North Sea basin (e.g., Heligoland) active horst and graben structures in the eastern Mediterranean (Figure E24), sediment load in large deltas (Rhône, Ebro, Po, Figure E25) or local subsidence by gas and water drilling in the Venice area. All processes cause their respective characteristic coastal shape.

Furthermore, exogenic factors of coastal forming are manifold, and here in particular the climate and its influence on the ocean. The spectrum ranges from frost climates of average annual temperature below −2°C and sea ice cover of nine months to nearly 30°C of surface water temperature in summer. Extreme humid conditions with far over 1,000 mm of precipitation are typical for NW-Europe, but then again less than 200 mm/year of precipitation is measured in eastern Crete (Greece), where evaporation exceed precipitation by a factor of 10. The salt content of sea water ranges from nearly 0% in the inner Baltic Sea to at least 3.9% in the eastern Mediterranean. Moreover, wind and wave spectra are very different: in the west strong winter storms may cause severe destruction and flooding, while in small bays of the Baltic or the Mediterranean wave action is insignificant. Tidal range and the strength of tidal currents differ also along a west–east gradient: spring tide range is highest around southwestern England and the Britanny coast of Western France (Figure E26), reaching more than 10 m, but in the Baltic and the Mediterranean it usually is less than 0.5 m to zero. Nevertheless, low air pressure and extreme winds may cause storm surges up to 3 m above datum even in these enclosed basins. Even tsunami impacts can be detected such as along the Algarve coast of southern Portugal, southern Atlantic Spain, Mallorca island, Apulia, Cyprus (Kelletat and Schellmann, 2002), or in marshes in Scotland and in coastal lakes of western Norway (Bondevik *et al.*, 1997).

Figure E24 Satellite image of the Mugla-area in southwestern Turkey with ingressive coastal forms as horst and graben structures with straight fault coastlines as well as partial drowning of a terrestrial fluvial and denudational relief (© SI/ANTRIX/Euromap 1996, GAF 1997).

Specific coastal regions

European coastal regions with a specific environmental or geomorphic factor are illustrated in Figure E22: northern Europe including Scotland and Iceland present coastlines of emergence by glacio-isostatic uplift between near 0–285 m (Klemsdal, 1982; Lambeck, 1993; Kelletat, 1994; Steers, 1973). Neverthe-less, those coastal forms are predominantly marked by ingression, so that fjords (Figure E27), fjärds, bodden, förden (Figure E28) or partly submerged roches moutonnées as skerries (Figure E29) as well as partly drowned eskers, end moraines, or drumlins, the latter particularly in Denmark and southern Finland are characteristic (Granö and Roto, 1993; Duphorn *et al.*, 1995). Here, glacial erosion has carved bedrock far below sea level and therefore the late- and post-glacial transgression was not compensated by late- and post-glacial uplift. The area of glacio-isostatic uplift mostly corresponds with glacially sculptured coastal landscapes, in which the forms from the last glaciation are much clearer than those from the penultimate glaciation, which are smoothened by periglacial processes during the Weichselian. Another specificum of the northern sections in Europe is frost action along the coast including drifting sea ice contact, in particular in the inner and eastern Baltic as well as in the Barents Sea and in some inner parts of northern Norwegian fjords (Figure E26), where Gulf Stream influence is minimal and freshwater input is important.

Pleistocene glaciations have left behind specific materials along the coasts: moraines are washed out and now show boulder beaches as in the Aaland archipelago between Sweden and Finland or along the southern Baltic coastline, partly pushed into ridges by drifting sea ice (Philip, 1990). These boulders are an important natural protection against surf beat and erosion. Glacial outwash and fluvioglacial deposits are characterized by gravel and sand, which dominates the southern coasts of Iceland as well as the west coast of Denmark or the northwestern coast of Germany and the Netherlands. If sand supply from the ice ages is abundant, extended beaches are developed, sometimes accompanied by coastal dunes. These conditions may also be found along the southern Baltic coastlines (Baker *et al.*, 1990).

Figure E25 Satellite image of the delta of River Po, Italy. The progradation mostly is a young Holocene feature (© ESA, Eurimage 1989, GAF 1989).

In contrast to formerly glaciated landscapes, fluvial formed landscapes are dominant in the south of Europe, where rivers have carved into older bedrock during low sea levels. Postglacial transgression has transformed these valley mouths into rias as in Southern Ireland, England (Steers, 1976), Britanny in France and in NW-Spain, some of them widened by tides and waves into estuaries as in the mouth of the rivers Elbe (Germany), Thames (England), or Seine (France).

A third representative coastal area in Europe is formed due to faulting processes associated with horst and graben structures such as southern Greece including the Aegean Sea and western Turkey (Figure E24). A vivid neotectonic is still dislocating coastal sections vertically (Lambeck, 1995; Pirazzoli *et al.*, 1996, see also Figure E30). In more downward moved parts coastlines of ingression with drowned medium-to-steep slopes are common. Here, canale (Figure E31), short rias, transformed by karst and erosion processes into calas of different shapes are developed and as well drowned karst features like dolines or poljes dominate (Kelletat, 1995). In contrast, cliffs (Figure E32) and coastal terraces are typical for the uplifted parts of the landscape. The areas of (Pleistocene) uplift in southern Europe often correspond with young orogenic structures, representing excellent examples of Pleistocene marine terraces up to several 100 m above sea level today (Kelletat *et al.*, 1976).

As limestone is the predominant rock type in the Mediterranean area, coastal bioerosion with notches (mostly by grazing *Patella*) and rock pools (by grazing *Littorina neritoides*) as well as bioconstruction by calcareous algae (e.g., *Neogonio-lithon notarisii*) and vermetids (e.g., *Dendropoma petraeum*, Figure E33) are widely common in this part of the European coast (Kelletat, 1997). Another specification is the extended distribution of younger Pleistocene aeolianites along the Mediterranean coasts, whereas Holocene beach rocks are mostly restricted to the hot and dry sections of thek eastern Mediterranean (Briand and Maldonado, 1997).

There is only one area of medium to high tides in Europe, located between southern England and the Britanny coast of France where the spring tide range reaches around 10 m (Figure E26), diminishing to the inner German Bight and Scotland to medium values of 2–4 m. These areas may show extended tidal flats, with a sandy matrix where wave action and tidal currents are strong, and muddy in regions of weaker water movements (Figure E34; Verger, 1968; Ehlers, 1988; Hillen and Verhagen, 1993; Kelletat, 1998). The growth of green and brown macro algae (*Fucus* sp., *Laminaria* sp. and others) along the rocky shores is characteristic for the cooler European coastlines from northern France including the British Isles and Iceland to northernmost Norway (but excluding the Baltic). They partly give protection against strong surf beat, whereas in the Mediterranean sea grass may form a thick carpet on beaches after winter storms. Along parts of the Swedish, Finnish, Polish, or German coastline with a low salt content of seawater, coastal freshwater swamps with *Phragmites* sp. occur.

Coastal forms and sediments

The differentiation of coastal forms along the European coastlines is visualized in Figure E35 giving an impression of the general form of bays (rias and estuaries by fluvial action, fjords, fjärds, förden, and bodden by glacial sculpturing, canale by folding and faulting control, cala by karst and abrasion, or lagoons with barriers and spits up to 90 km in the eastern Baltic by longshore drift). The map also represents the different coastline types: rocky coastlines with cliffs (and sometimes intertidal platforms), sandy shores with beaches (Guilcher and King, 1961; King, 1972, including areas with beach rocks), muddy ones in tidal flats, deltaic by fluvial progradation (as in the areas of low tides and surf of the Mediterranean), or biogenic structures along limestone coasts with poor sediment supply.

Deeply incised fjords are the main coastal features of northern Iceland, northwestern Scotland and nearly all of Norway except of the easternmost parts where cliff sections around the North Cape and the Barents peninsula occur. In Scotland and the central north coast of Iceland wider rocky embayments in a low and hilly landscape smoothened by Pleistocene glaciation reveal more aspects of fjärds, which are typical for the Swedish coastline along the Baltic and some parts of Finland. Here, an old peneplain is slowly dipping seaward. Numerous skerries decorate all rocky Scandinavian coastlines, whereas to the periphery of former glaciation morainic sediments and fluvioglacial deposits dominate. Thus, the landscape is either characterized by partly drowned eskers, drumlins and end moraines, which are mostly well preserved in areas with low surf energy conditions, such as in the island archipelago of southeastern Denmark, or fast receding cliffs may have been developed like along the southern shores of the Baltic. As a rather unusual feature in the Scandinavian coastal landscape the so-called strandflat exists as an undulating rocky plain surrounding most of the coasts of Norway, but occurring likewise in northwestern Iceland. The form originates from a glacially smoothed very old peneplain now situated around sea level with a medium width of 16 km and a maximum one of 40 km. Roches moutonnées are the dominant features, and old inselbergs may rise from the peneplain to several 100 m of relative height. Outwash plains (sander), fed by meltwater and glacial-burst

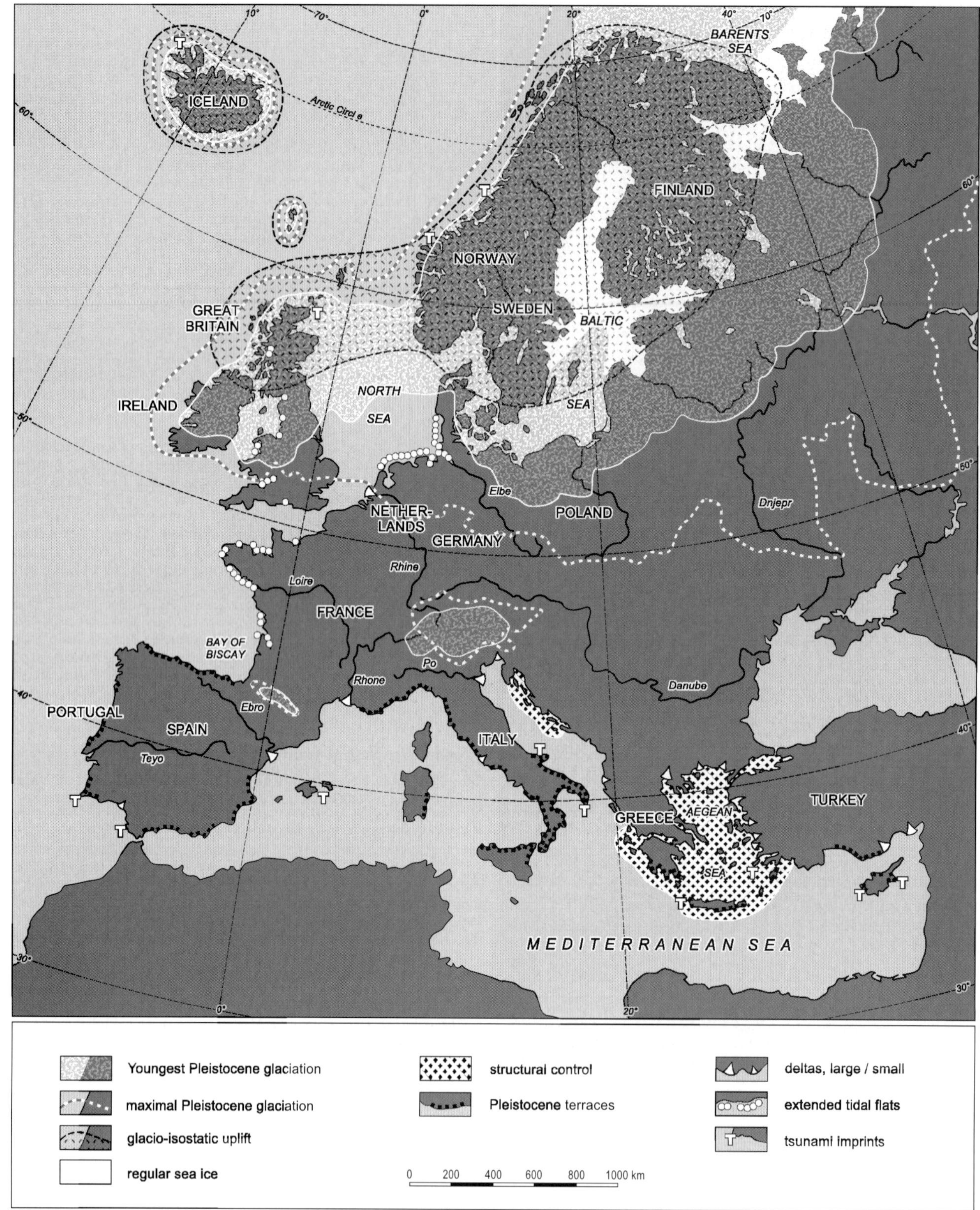

Figure E26 Map of the European coastlines showing special environmental parameters.

floods from the huge glaciers are typical for the southern part of Iceland. Therefore, the coasts show beaches and even barriers and lagoons. Along the southern shores of the Baltic very long spits fed by longshore drift from the west, large lagoons and intercalated cliffy sections occur. Along the Polish coast 52% are sandy beaches, 28% are fresh-water swamps, and only 20% show cliffs. At the German Baltic coastline this type gradually changes to ingressive forms from glacial and fluvioglacial origin like bodden and förden (Figure E34).

The longest section of muddy and sandy tidal flats in Europe with barrier islands and spits can be found from western Denmark all along

Figure E27 Satellite image of Sognefjord in western Norway, the deepest (more than 1,200 m) and longest fjord in northern Europe with about 200 km of length and a width of 1–8 km. Around the mouth of the fjord structural control of the coastline in very old crystalline rocks, worked out by several glaciations, can be identified (© EOSAT, GAF, AAF 1985).

the German North Sea coast and the Netherlands to Belgium and northernmost France. Further-more, inland marshes are typical (Dijkema, 1987), and along broader sandy beaches coastal dunes are well developed. Other areas of tidal flats are present in many ria embayments from Ireland to Great Britain and western France. They partly are rocky, exhibiting a rich variety of algal belts and hard encrustations of barnacles, mussels, etc. Northern England shows a great variety of spits, marshes, tidal bays, bars, and uplifted coastal terraces, whereas the southern part is dominated by the world famous high chalk cliffs along the English Channel and rias of different length and steep rocky headlands to the southwest. The southwestern coast of France along the Bay of Biscay shows the longest coastal dune belts of Europe with shifting sands of more than 100 m height.

The coastline of Northern Spain along the Bay of Biscay is rocky with cliffs and caves cut into the limestone. To the west narrow and deep rias can be observed, which change to the Portuguese coast into wider and shallower fluvial embayments. The coasts of Portugal, however, are mostly sandy with dune belts and lagoons, except of some rocky headlands and the high and steep cliffs developed in the hard limestone at the southwestern most point of Europe. They are constituted of soft rocks with numerous small landslides at the south-facing Algarve coast. Continuing to the Strait of Gibraltar the eastern Portuguese coastline and the Spanish one are mostly low and bordered by lagoons and dune belts along sandy beaches. Only the last 50 km exhibit rocky coastlines, headlands, and cliffs with smaller sandy beaches.

The volcanic archipelagos of Madeira, the Azores, and the Canary islands are mostly steep, some with mega cliffs up to more than 500 m height. They are not singularly formed due to abrasion, but in addition they represent head scars of huge submarine slides from the collapse of the high and growing volcanic edifices.

The coastlines of the Mediterranean are limited in their coastal forms because tides are nearly absent, and longshore drift is less efficient because of the smaller ocean areas. Thus, the limited tides and wave energy conditions are responsible for many deltas, among them the very large ones of Ebro, Rhône, or Po, and numerous others along the Italian and Turkish coasts. Headlands with cliffs in limestone including bioerosive notches alternate with longer beach sections at the Spanish Mediterranean coastline, whereas in southern France the western section consists of beaches, barriers, beach ridges plains, and the Rhône delta, and of rocky steep coastlines in the east with calanques and calas, which are dominant on the Balearic islands, as well. The most extended granitic coastal landscapes decorated by huge boulders from deep weathering during pre-Pleistocene times and Holocene tafoni can be found around the islands of Corsica and Sardinia. Here, also longer sections of coastal barriers and lagoons exist. Except for the young volcanic Aeolian islands peninsula Italy has beaches along nearly 70%

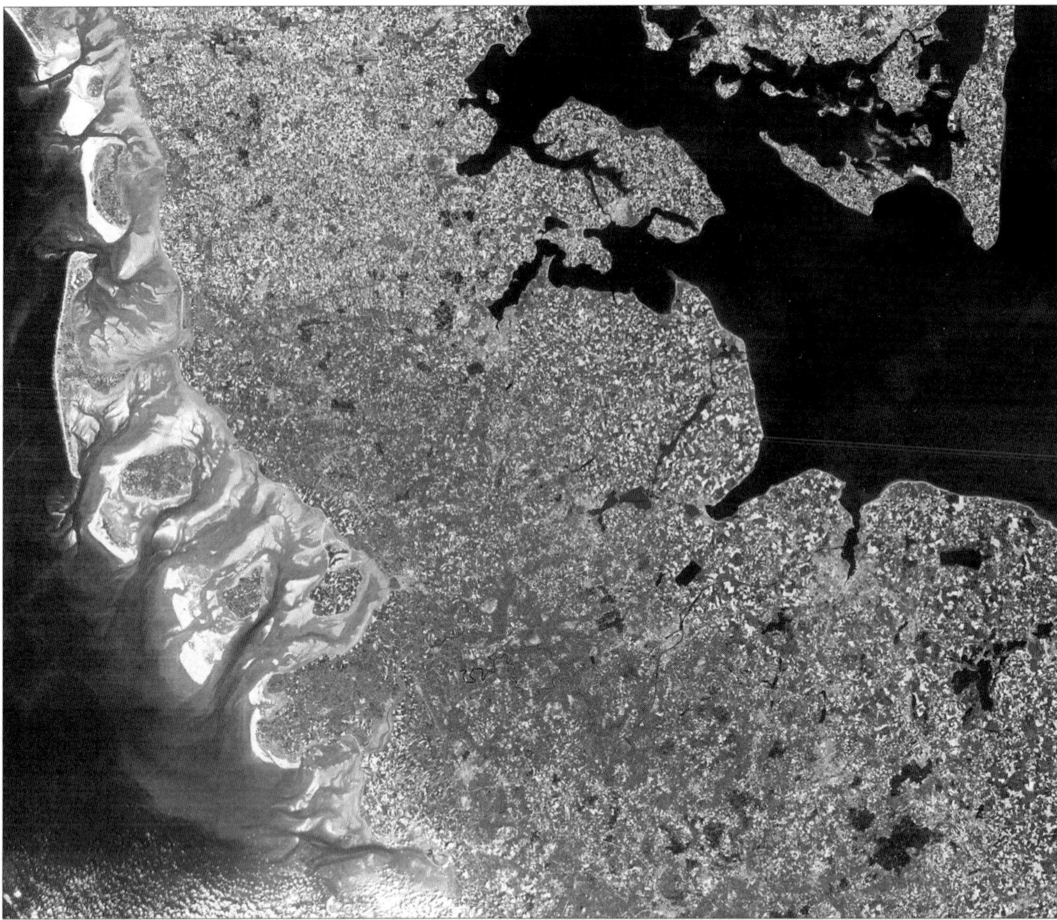

Figure E28 Satellite image of northern Germany and southern Denmark. To the west tidal areas with different types of islands (Pleistocene morainic hills, Younger Holocene marshes, very young spits and dunes), sandy bars and muddy flats with tidal rivers and ebb deltas can be seen. To the east at the Baltic coastline ingression into morainic and fluvioglacial relief on the Danish islands or into ria like förden is typical in an environment without tides (© ESA, Eurimage 1992, GAF 1995).

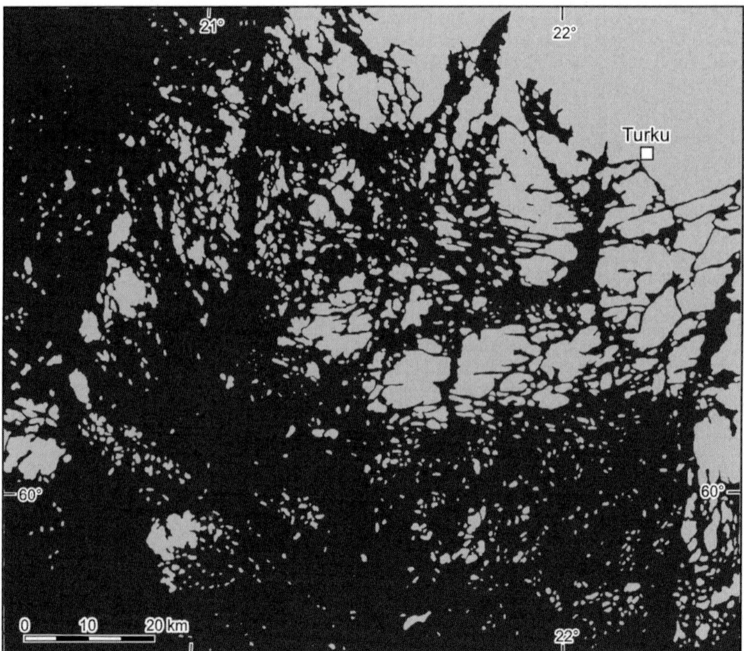

Figure E29 Numerous skerries are typical for the coastlines between Finland and Sweden in the northern Baltic Sea.

Figure E30 Bioerosive notches in limestone at the east coast of Rhodes (Greece) show step-like neotectonic uplift during the last 2,000 years with a maximum of about 2.5 m.

Figure E31 The Adriatic coast along Croatia is a typical area of parallel mountain ridges from Mesozoic folding, drowned by subsidence and post-glacial sea-level rise to form canale coastlines.

of its coastline, whereas in the southeast and on Sicily cliffs and Pleistocene coastal terraces are widespread. From Istria to northern Albania drowned landscapes with the typical parallel canale features occur (Figure E31), beaches are very rare. Another type of an ingressive coastal landscape can be found along the Greek and Turkish coasts. It is the various change of structural embayments as grabens and horst like long peninsulas (Figure E24), mostly showing drowning as the predominant process with medium to steeper slopes but a less amount of real cliffs. Limited sections of uplifted coastal terraces alternate with beaches along barriers and deltas. The islands of the Aegean Sea again are drowned remnants of an older land bridge. Due to the absence of longer rivers and the rather dry climate beach material is scarse. Neotectonics are an important process in coastal forming, as well as bioconstruction at the warmest easternmost parts of the Mediterranean, where many of the small beaches are cemented by Holocene beachrocks.

The maps do not show areas of artificial coastlines mostly due to the need for protection against storm floods in winter by groins, beach nourishment, sea walls, dikes, or land reclamation, most widespread around the North Sea (Figures E34 and E36) as well as along Mediterranean bathing beaches and harbors.

Many of the European coasts are endangered (Kelletat, 1988; Caputo *et al.*, 1991; Tooley and Jergersma, 1992; French, 1994; Bartolini and Carobene, 1996;). The primary cause hereby is the accelerated erosion along beaches—at least 75% of all European beaches suffer from this process—marshes and coastal dunes, which as a consequence costs billions of Euro/year for maintaining bathing beaches, coastal settlements, or farmland. The reason is a small general sea-level rise and a local accelerated one caused by deltaic sedimentation, down warping of a ring-like structure around the glacio-isostatic uplifted areas, or pumping of oil, gas, or groundwater from below the sea. Some of the reasons of the accelerated erosion may be located not directly along the coasts but far inland. The predominant cause is the increasing number of reservoirs holding back sediment, and the mining of sand, gravel, and cobble from river beds, in particular in the Mediterranean area. Consequently millions of tons of sediment are not transported to the coasts any longer. This is in striking contrast to a period of progradation and delta build up during late Holocene and historic times, in particular during the Greek and Roman antiquity and the early Middle ages, in some places even starting during the later Neolithic (Erol, 1983; Brückner, 1998). For several 1,000 years progradation was in particular responsible for the development of barriers, spits, or marshes in spite of the still slowly rising sea level. The reason was the ongoing extension of agriculture and deforestation, resulting in soil erosion and denudational processes apart from the coast. Coastal protection against the increased frequency of storms is another problem, which leads to the design of continuous new and

Figure E32 Chalk cliffs along the French coast of Normandy with stack and sea arch.

Figure E33 Living bioconstructional micro atoll by vermetids (*D. petraeum*) giving a very reliable sea-level indicator along the west coast of Crete, Greece.

higher dikes. It can be stated that overall the shape of some thousands of kilometers of coastline in Europe are the result of anthropogenic structures. However, beside that, heavy pollution due to tourist settlements or rivers bringing pulp mill sewage to the sea and oil spillage from the heavy road traffic additionally are causing severe problems in coastal areas. Another important negative effect is caused by the cleaning of ship tanks associated with tar pollution, in particular in the eastern Mediterranean.

Pleistocene and Holocene sea levels

In Northern Europe, glaciation has eliminated all traces of sea levels from former interglacial high stands, and those are mostly missing in areas without significant uplift like around the North Sea Basin and the English Channel. In contrast evidences of higher interglacial sea levels as sediments with marine fossils (e.g., the foram *Hyalinea balthica* for colder events in the older Pleistocene, or the tropical gastropod *Strombus bubonius* for the younger interglacials in the Mediterranean)

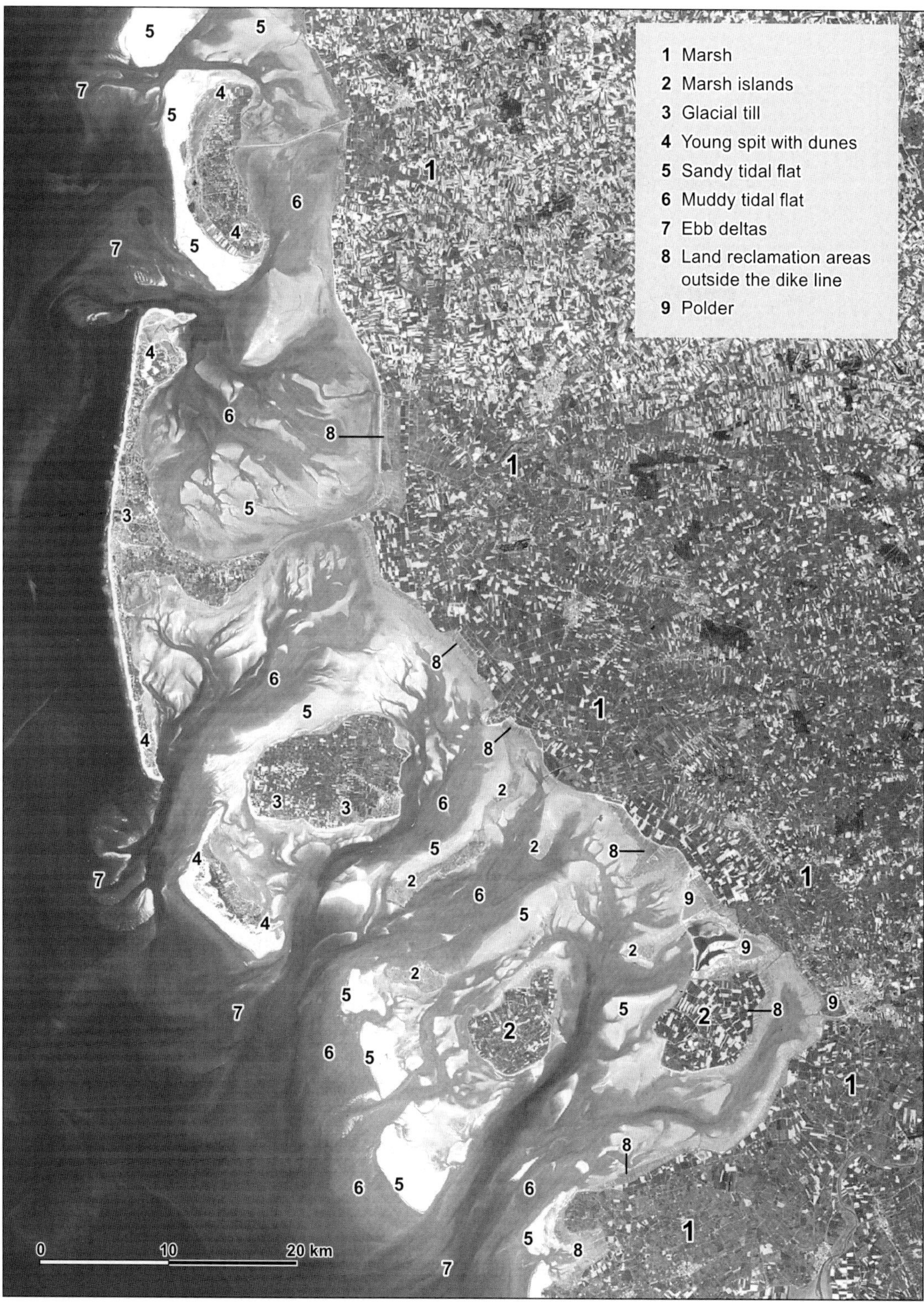

Figure E34 Satellite image of the tidal areas and islands along the west coast of northern Germany (© ESA, Eurimage 1992, GAF 1995).

or even as geomorphic features like coastal terraces are widespread in the geologic young environments of the alpidic orogenesis. Côte d'Azur between France and Italy, northern Sicily, or Apulia, and Calabria in southern Italy are world famous sites for the discussion of higher interglacial sea levels. At some places, such as in SW and NE Peloponnesos, Greece or southern Italy, terrace sequences reach several 100 m above sea level. Unfortunately, the absolute dating determination remains still rather difficult. The coasts and as well the islands of the eastern Mediterranean exhibit remnants of higher sea levels, but mostly in short sections, bordered by faults and areas of significant subsidence. The differentiation of substages within the same interglacial (like oxygen-isotopic stages 5e, 5c, 5a) is still under research. Nevertheless the study of high (or uplifted) interglacial sea level evidence has contributed greatly to the understanding of Mediterranean neotectonics.

Holocene sea level is still a topic of intensive discussion in the scientific community (Binns, 1972). In northern Europe, a very good

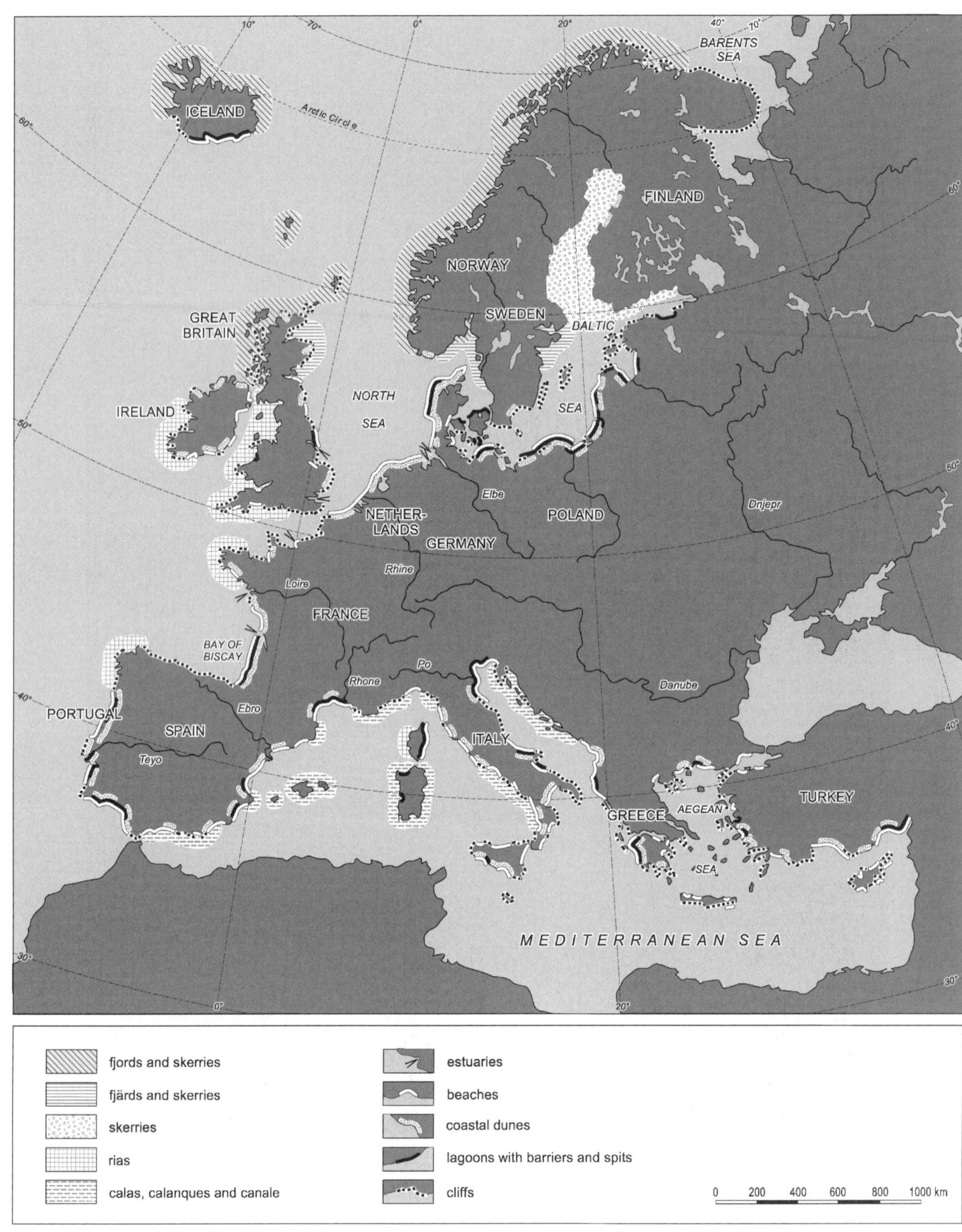

Figure E35 Forms and sediments along the coastlines of the European continent.

Figure E36 Land reclamation outside the marshes at the northern German North Sea coast, mostly as a tool for protection of dikes and farmland.

dating of glacio-isostatic rebound by the analysis of uplifted beach ridge sequences exists (Figure E33), whereas along the marshy German, Danish, Netherlands, and English coastlines of the North Sea geo-archaeology has much contributed to the identifying of former Holocene sea levels (Behre, 2001). Here, the sea-level curve of the last 6,000 years is characterized by at least two transgressions and regressions of 1–3 m. Some archaeological remains of the Bronze Age below high water level in the high tide range area of western France point to a sea-level rise of more than 3 m in this geologic stable region. Along the Mediterranean coastlines archaeology as well contributes to the identification of the Holocene sea-level curve: drowned ruins and quarries from Greek antique times, the position of Roman fish tanks, and even relics of early Christian times can be found below sea level. Beside these man-made structures, natural sea level indicators are very suitable for relative dating approaches, among them bioerosive notches (Figure E26) or bioconstructive elements like trottoirs or algal and vermetid rims (Figure E32 and Kelletat, 1997). They are much more reliable than beach ridge sequences, beachrock or terraces along deltas. In spite of literally hundreds of dated and measured sites, there is no general agreement about the eustatic sea-level movements in the Younger Holocene in the Mediterranean. But nevertheless these studies have led to a much better understanding of the ongoing and vivid neotectonics in this part of Europe, reaching a maximum of 9 m on western Crete (Greece) since the year AD 365 (Kelletat, 1996).

Dieter Kelletat and Anja Scheffers

Bibliography

Åse, L.A., 1980, Shore displacement at Stockholm during the last 1000 years *Geografiska Annaler*, **62A**(1–2): 83–91.

Baker, W.M., Jungerius, P.D., and Klijn, J.A. (eds.), 1990. Dunes of the European coasts. *Catena Suppl.* **18**: 227 p.

Bartolini, C., and Carobene, L., 1996, Italy's instability in the coastal domain. *Zeitschrft far Geomorphologie*, NF, **102**(Suppl.Bd.): 71–94.

Behre, K.-E., 2001. Holozäne Küstenentwicklung, Meeresspiegelbewegungen und Siedlungsgeschehen an der südlichen Nordsee. *Bamberger Geographische Schriften*, **20**: 1–18.

Binns, R.E., 1972. Flandrian strandline chronology for the British Isles and correlation of some European postglacial strandlines. *Nature*, **235**: 206–210.

Figure E37 Gravel beach ridge sequence in northernmost Norway up to +45 m due to glacio-isostatic uplift.

Bird, E.C.F., and Schwartz, M.L., 1985. *The World's Coastline*. New York, Van Nostrand Reinhold Company.

Bondevik, S., Svendesen, J.I., Johnson, G., Mangerud, J., and Kaland, P.E., 1997. The Storegga tsunami along the Norwegian coast, its age and run up. *Boreas*, **26**(1): 29–53.

Briand, F., and Maldonado, A. (eds.), 1997. Transformations and evolution of the Mediterranean coastline. CIESM Science Series No. **3**, Bulletin de l'Institut Océanographique, Monaco, No. Special 18, 243 p.

Brückner, H., 1998. Coastal research and geoarchaeology in the Mediterranean region. In Kelletat, D. (ed.), *German Geographical Coastal Research—The Last Decade*. Tübingen, 235–258.

Caputo, C., D'alessandro, L., La Monica, G.B., Landini, B., and Lupia Palieri, E., 1991. Present erosion and dynamics of Italian beaches. *Zeitschrift fur Geomorphologic*, NF, **81**(Suppl.Bd): 31–39.

Dijkema, K.S., 1987. Geography of salt marshes in Europe. *Zeitschrift fur Geomorphologic* NF, **31**(4): 489–499.

Duphorn, K., Kliewe, H., Niedermeyer, R.-O., Janke, W., and Werner, F., 1995. Die deutsche Ostseeküste. Sammlung Geologischer Führer, Bd. **88**, Berlin: Borntraeger.

Ehlers, J., 1988. *The Morphodynamics of the Wadden Sea*. Rotterdam/Brookfield, Balkema.

Erol, O., 1983. Historical changes on the coastline of Turkey. In Bird, E.C.F., and Abbri, P.F. (eds.), *Coastal Problems in the Mediteranean Sea*. Bologna: International Geographical Union Commission on the Coastal Environment, pp. 95–108.

French, J.R., 1994. Tide-dominated coastal wetlands and accelerated sea-level rise: a northwestern European perspective. In Finkl, Ch. W., Jr. (ed.), *Coastal Hazards—Perception, Susceptebility and Mitigation. Journal of Coastal Research*, **12**(Special Issue): 91–101.

Granö, O., and Roto, M., 1993. Geographical information system for the inventory of coastal characteristics in Finland. Frankfurt a.M.: Proceedings of International Coastal Congress ICC-Kiel '92, 519–527.

Guilcher, A., and King, C.A.M., 1961. Spits tombolos and tidal marshes in Connemara and West Ireland. *Proceedings Royal Irish Academy*, **61b**: 283–338.

Hillen, R., and Verhagen, H.J. (eds.), 1993. Coastlines of the Southern North Sea. New York, *American Civil Society of Engineers*, 363 p.

Kelletat, D., 1988. Das CORINE-Projekt "Coastal Erosion Risks" der Europäischen Gemeinschaft—eine Pilotstudie. *Hamburger Geogr. Studien*, **44**: 21–30.

Kelletat, D., 1994. Main trends of the Late- and Postglacial coastal evolution in northernmost Norway. *Essener Geographische Arbeiten*, **25**: 111–124.

Kelletat, D., 1995. Atlas of Coastal Geomorphology and Zonality. *Journal of Coastal Research*, Spec. Issue **13**.

Kelletat, D., 1996. Perspectives in Coastal Geomorphology of Western Crete, Greece. *Zeitschrift fur Geomorphologie*, NF, **102**(Suppl. Bd.): 1–19.

Kelletat, D., 1997. Mediterranean coastal biogeomorphology: processes, forms, and sea-level indicators. *Bull. de l'Institut Océanogr. Monaco*, numéro special, **18**: 209–226.

Kelletat, D., 1998. *German Geographical Coastal Research. The Last Decade*. Tübingen: Institute for Scientific Co-operation.

Kelletat, D., and Schellmann, G., 2002. Tsunamis on Cyprus: field evidences and 14C dating results. *Zeitschrift fur Geomorphologie*, NF, **46**(1): 19–34.

Kelletat, D., Schröder, B., Kowalczyk, W., and Winter, K., 1976. A synoptic view on the neotectonic development of the Peloponnesian coastal regions. *Zeitschrift der Deutsche Geologishen Gesellschaft*, **127**: 447–465.

King, C.A.M., 1972. *Beaches and Coasts*. London: St. Martin's.

Klemsdal, T., 1982. Coastal classification at the coast of Norway. *Norsk Geografisk Tidsskrift*, **36**: 129–152.

Laborel, J., 1987. Marine biogenic constructions in Mediterranean. *Scient. Rep. Port-Cros Nat. Park*, **13**: 97–126.

Lambeck, K., 1993. Glacial rebound of the British Isles—II. A high-resolution high-precision model. *Geophysical Journal of International* **115**: 960–990.

Lambeck, K., 1995. Late Pleistocene and Holocene sea-level changes in Greece and southwestern Turkey: a separation of eustatic, isostatic and tectonic contributions. *Geophysical Journal of International* **122**: 1022–1044.

Philip, A.L., 1990. Ice-pushed boulders on the shores of Gotland, Sweden. *Journal of Coastal Research*, **6**(3): 661–676.

Pirazzoli, P.A. *et al.*, 1996. Coastal indicators of rapid uplift and subsidence: examples from Crete and other eastern Mediterranean sites. *Zeitschrift fur Geomorphologie*, NF, **102**(Suppl Bd.): 21–35.

Steers, J.A., 1973. *The Coastline or Scotland*. Cambridge: Cambridge University Press.

Steers, J.A., 1976. *The Coastline of England and Wales*. 2nd ed. Cambridge: Cambridge University Press.

Tooley, M.J., and Jelgersma, S. (eds.), 1992. *The Impacts of Sea-Level Rise on European Coastal Lowlands*. Oxford: Basil Blackwell.

Verger, F., 1968. *Marais et wadden du littoral francais*. Bordeaux: Biscay.

Cross-references

Changing Sea Levels
Classification of Coasts (see Holocene Coastal Geomorphology)
Europe, Coastal Ecology
Geographical Coastal Zonality
Ice-Bordered Coasts
Machair
Notches
Paraglacial Coasts
Strandflat
Submerging Coasts
Tectonics and Neotectonics

F

FAULTED COASTS

Faulted coasts are chiefly associated with active continental margins where tectonic plates are colliding. They often correspond to bold coasts, characterized by continuously steep and straight scarps. However, terraced coastal areas dislocated by faults as well as fiord or estuary coasts controlled by faults have also to be included into the broad category of faulted coasts.

Sea cliffs and fault scarps

In some instances, coastal cliffs may be identified as fault scarps of direct tectonic origin which separate raised blocks of land from dropped ones that have been depressed below sea level. Plunging cliffs, cliffs which are not fronted by shore platforms or beaches and plunge abruptly into deep water, can be the product of recent faulting if down-thrown blocks have subsided to appreciable depths. In that case, and not only in sheltered waters and even if the resistance of rocks is not considerable, plunging conditions could persist long after faulting and the initial profile of the coast may long remain practically unaltered (Johnson, 1919). Cliff recession is very slow since erosional processes are inhibited by the reflection of incoming nonbreaking waves against vertical or nearly vertical walls rising out of deep water. Furthermore, deep waters at the cliff base prevent accumulation of sediment derived from alongshore transport or subaerial scarp weathering, resulting in the lack of abrasive materials which could increase the erosive power of the waves. However, in the course of time, hydrostatic pressure and hydraulic quarrying caused by the rise and fall of standing waves, and the compression of air in rock fractures eventually are able to produce notches and caves. Narrow erosional platforms may subsequently develop, allowing breaking waves to attack the cliff and destroy the plunging condition (Trenhaile, 1987).

Several coastal cliffs in New Zealand are thought to be actual fault scarps (Cotton, 1916). For instance, the cliffs along the Wellington fault, on the northwestern shore of Port Nicholson have been explained in this way. They slope at about 55°, show little evidence of marine modification at the shore level, and descend to a depth of about 20 m, close and parallel to the shoreline. In Russia, on the Murmansk coast, to the west of the White Sea, Archean granites and gneisses are faulted along a line which almost coincide with the present coastline. The foot of the cliffs extends to a depth of several tens of meters without alteration of slope. During storms, waves are reflected from the scarp wall and there are no wave-cut notches or fragments of benches (Zenkovich, 1967). In the United States, some steep cliffs along the coast of California are attributable to fault scarps. For example, a down-thrown block has dropped below sea level on the northeastern side of San Clemente Island, forming a straight cliff with no intervening shelving bottom (Shepard and Wanless, 1971). The Gulf of California, Mexico, gives nice examples of recent faulted coastlines, especially in the islands next to its western side where straight, continuous, and precipitous escarpments display facetted spurs and truncated structures, are not bordered by shelves, have deep water at their base, and show angular contact with the adjacent submarine basin (Shepard, 1950). All these features are evidence of no marine erosion after faulting that is probably recent. For instance, the sea cliff on the southwestern side of Angel de la Guarda Island corresponds to a fault scarp which steeply plunges to the bottom of a trench located at a depth of about 1,000 m. In volcanic areas, cliffs can be directly produced by faulting induced by vulcanicity. Sea cliffs devoid of a bordering shelf on the southern flank of Kilauea Volcano, in Hawaii, can be identified as direct normal fault scarps although, at least in some cases, they may correspond to gigantic slump scars.

Sea cliffs and fault-line scarps

Most tectonically controlled coastal cliffs correspond to fault-line scarps that are scarps produced by structurally induced erosion at ancient faults. As a matter of fact, on the one hand there has been little time for important faulting since the postglacial rise of sea level, on the other hand older original tectonic features are not retained for long because of the erosional effects of subaerial geomorphic processes during periods of low sea level. For example, along the steep Santa Lucia Mountain coast of central California, the continental shelf is in some parts less than 0.6 km wide, suggesting that this is a fault-line coast since a retreat of a relatively short distance, due to marine erosion, has followed faulting. In Japan, the straight montainous coastline which characterizes the northwestern part of Shikoku stretches along the so-called Median Tectonic Line. In the Malta archipelago, the coastline pattern of the main island is mainly controlled by faulting. Here, in the northern part, bays correspond to downthrown blocks that were partially submerged. In the southwestern part, spectacular cliffs which can be more than 200 m high are erosionally derived from a major northwest trending fault, known as the Maghlak fault that has been displaced vertically by at least 230 m and probably moved during the Quaternary (Paskoff and Sanlaville, 1978).

In northernmost Chile, where the oceanic Nazca plate is subducted under the continental South American plate, the extremely arid coast is characterized, over more than 800 km, by a striking, precipitous, northward trending cliff which is 700 m high on average. Because of its length and height, this scarp represents a major geomorphic feature at a worldwide scale. From Arica (18°S lat.) to Iquique (20°S lat.) the cliff is almost vertical and its top can reach 1,000 m above sea-level. At present, it is an active feature, dropping into the sea and receding under marine action. As a result, ephemeral streams of wadi type are cut by the scarp and hanging above sea level (Figure F1(A)). Landslides, triggered by earthquake shocks and affecting rock masses previously loosened by intense tectonic facturation, are also contributing to the cliff retreat. From Iquique down to 27°S lat., the cliff remains very high but not so steep. As a matter of fact, it is no longer an active one since

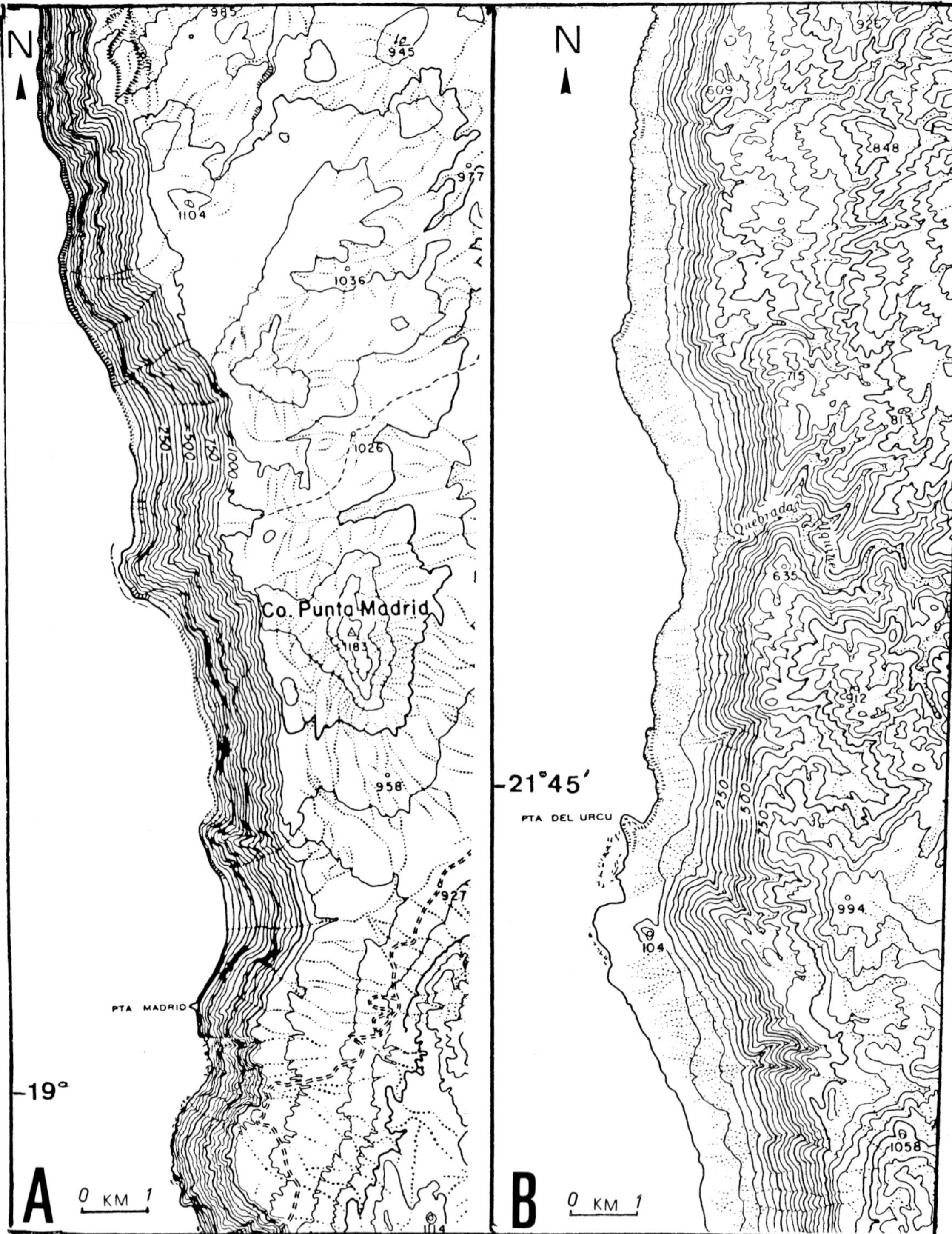

Figure F1 The very high cliff of northernmost Chile corresponding to a major fault-line scarp. (From topographic maps of the *Instituto Geográfico Militar* of Chile). (A) active cliff, (B) abandoned cliff with emerged wave-cut terraces at its base.

a belt of emerged wave-cut platforms, several hundred meters wide, extends at its base, indicating a relative fall in sea level (Figures F1(B) and F2). The cliff of northernmost Chile is a typical fault-line scarp (Paskoff, 1978). Detailed geological mapping failed to reveal faulting at the foot of the cliff whose lack of straightness is also inconsistent with a direct tectonic origin. Actually, the cliff results from an important faulting phase which generated major, north–south trending, *en échelons*, normal faults at the end of the Oligocene Epoch and uplifted the continental block which constitutes the present Coastal Cordillera. Subsequently, these faults retreated under marine erosion, over about 10 km which is here the width of the continental shelf, mainly during a major middle to upper Miocene transgression. Later differential crustal movements occurred. This explains why the cliff is an active one in its northern part where subsidence may be related to a collapse of the continental margin toward the Chile-Peru trench, and an abandoned one in its southern part where, on the contrary, an uplift motion has been taking place.

Coastal terraces and faulting

Former wave-cut platforms that have become emerged because of a fall of sea level due to some combination of eustatic and crustal processes are not unusual on coastal belts. When located on active margins, these elevated marine terraces generally show evidence of active faulting which has occurred in Quaternary time (Pirazzoli, 1994).

Figure F2 The very high cliff of northernmost Chile near 21°25′S lat. The toe of the scarp is covered by thick screes which are the result of salt-weathering and seismic shocks. Wave-cut platforms extent at its base. (Photo R. P. Paskoff).

In north-central Chile, the 110-km-long coast of the Talinay Heights area, between 30°S and 31°S lat., is characterized by a flight of several emerged Plio-Quaternary marine terraces which has been densely and strongly faulted, uplifted, and tilted (Ota *et al.*, 1995). Dip-slip normal faults, which are early Pleistocene in age, have a clear topographic expression, forming conspicuous scarps. The most significant one is the Quebrada del Teniente fault (Figure F3). Distinctively straight, NNW–SSE trending, it runs parallel to the present shoreline and crosses the entire area over a length of about 60 km, dislocating the terraces and generating a striking east-facing scarp. The maximum amount of vertical displacement reaches 52 m.

In eastern Tunisia, the Monastir surroundings (36°N lat.) are well-known for their Tyrrhenian (name used in the Mediterranean for the last interglacial marine episode, i.e., oxygen isotopic stage 5) deposits and for the term Monastirian which was incorporated to the standard Mediterranean terminology for Quaternary marine shorelines early in the 20th century. For a long time Monastir was the type locality for the 18–20 m terrace. This classic, widely quoted term was finally and definitely dropped when it was established that crustal deformation had occurred near Monastir (Paskoff and Sanlaville, 1981). The area is characterized by a triangular wave-cut platform (Figure F4) which dates back to the late Quaternary (oxygen isotopic substage 5e) and has been affected by two main faults. The Wadi Tefla fault which trends northeast is a normal fault which dislocates the platform in two levels and has a maximum vertical throw of 15 m. It was erroneously considered to be a former sea cliff separating two distinct marine terraces. The NNW–SSE Skanes–Khniss fault is a major fault of regional signficance which displays a vertical displacement that reaches 50 m. It also shows a sinistral strike-slip component characterized by an horizontal movement of at least 500 m. The fault has been active in historical times since it cuts a Roman mosaic on a floor of a house which was built in the 2nd century AD.

Fiord or estuary coasts and fault control

In plan view, coasts developed on intrusive and metamorphic terrains may show intersecting linear patterns of fiords or estuaries, depending on the latitude, marking former glacial or fluvial erosion exploiting the structural weakness of faults. For instance, in southernmost Chile, especially between 44°S and 55°S lat., fiords regionally named *canales*, penetrate inland. They correspond to glacial troughs which were eroded and overdeepened during the last Pleistocene glaciation by active Andean glaciers, mainly in Jurassic granitic rocks. The fiord pattern is strikingly controlled by fault lines which are mainly oriented N–S, NE–SW, and NW–SE. Fault lines were selectively exploited by eroding glaciers. Fiords form a dense network of long marine arms, characterized by straight stretches and right angles (Paskoff, 1996).

As already stated by Mclean (1982), the concept of faulted coast has now to be used in a broader sense than initially defined by the pioneers of coastal geomorphology. It encompasses not only cliffs controlled in some way or other by faulting, but also landforms found in the coastal

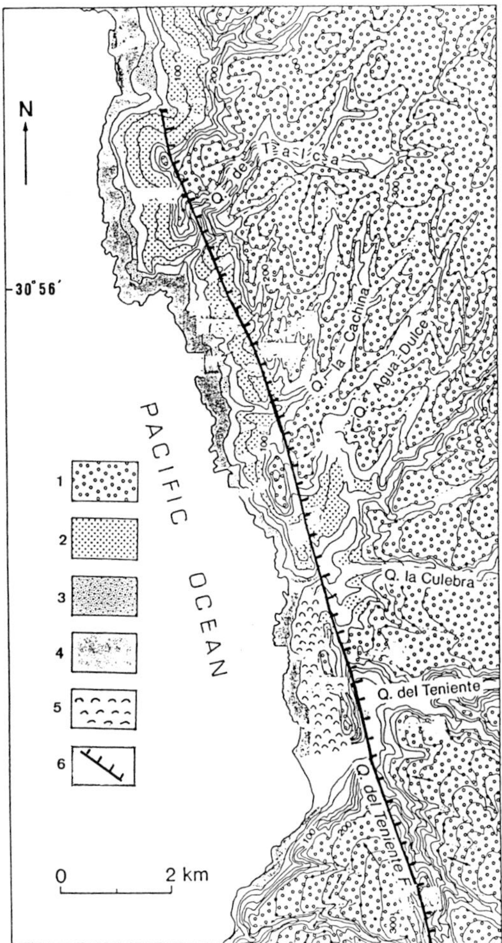

Figure F3 Geomorphological sketch of the southern part of the Talinay Heights coastal area in northcentral Chile (modified from Ota *et al.*, 1995). (1) Talinay I wave-cut terrace (Pliocene?), (2) Talinay II wave-cut terrace, (3) Talinay III wave-cut terrace (oxygen isotopic stage 9?). Talinay IV wave-cut terrace (oxygen isotopic stage 5), (5) dunes, (6) Quaternary fault.

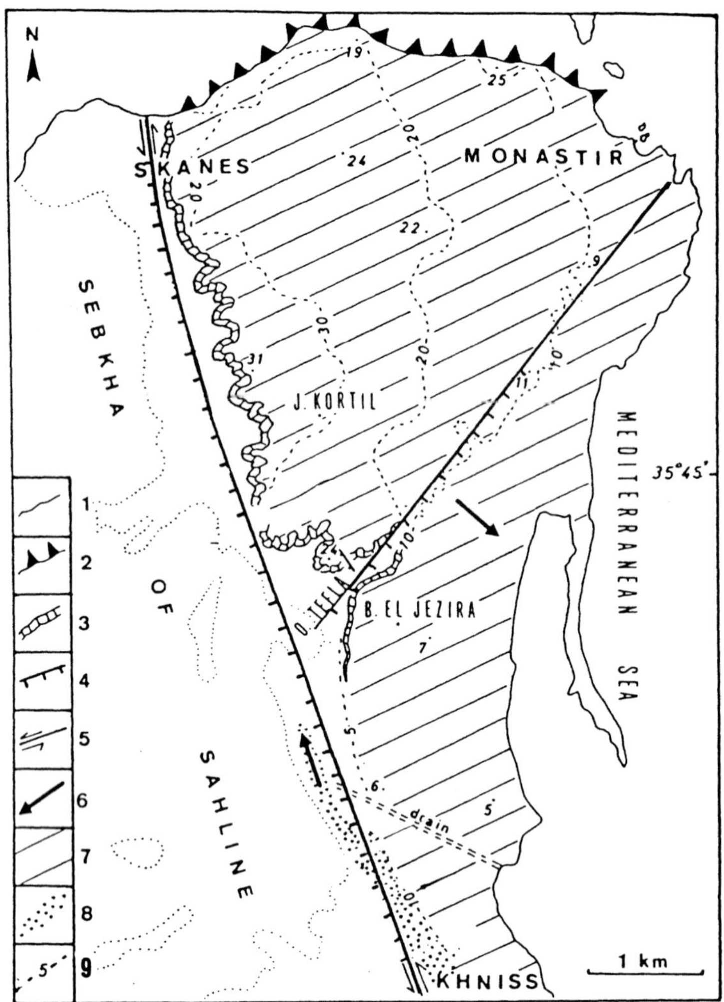

Figure F4 Geomorphological sketch of the Monastir area in eastern Tunisia (modified from Paskoff and Sanlaville, 1981). (1) low rocky coastline, (2) sea cliff, (3) erosional scarp of tectonic origin, (4) normal fault, (5) strike-slip fault, (6) tilting, (7) wave-cut platform dating back to the last interglacial age (oxygen isotopic substage 5e), (8) former coastal barrier dating back to the last interglacial age (oxygen isotopic substage 5e), (9) contour line in meters.

belt which are directly or indirectly related to lines along which vertical or horizontal movements of tectonic origin have taken place.

Roland P. Paskoff

Bibliography

Cotton, C.A., 1916. Fault coasts in New Zealand. *Geographical Review*, **1**: 20–47.

Johnson, D.W., 1919. *Shore Processes and Shoreline Development*. New York: John Wiley & Sons.

McLean, R.F., 1982. Fault coast. In Schwartz, M.L. (ed.), *The Encyclopedia of Beaches and Coastal Environments*. Stroudsburg: Hutchinson Ross Publishing Company, pp. 424.

Ota, Y., Miyauchi, T., Paskoff, R., and Koba, M., 1995. Plio-Quaternary marine terraces and their deformation along the Altos de Talinay, north-central Chile. *Revista Geológica de Chile*, **1**: 89–102.

Paskoff, R., 1978. Sur l'évolution géomorphologique du grand escarpement côtier du désert chilien. *Géogrophie Physiqueet Quaternaire*, **4**: 351–360.

Paskoff, R., 1996. *Atlas de las Formas de Relieve de Chile*. Santiago de Chile: Instituto Geográfico Militar.

Paskoff, R., and Sanlaville, P., 1978. Observations géomorphologiques sur les côtes de l'archipel maltais. *Zeitschrift für Geomorphologie NF*, **3**: 310–328.

Paskoff, R., and Sanlaville, P., 1981. Tyrrhenian deposits and neotectonic at Monastir, Tunisia. *Zeitschrift für Geomorphologie NF, Suppl.-Bd.* **40**: 183–192.

Pirazzoli, P.A., 1994. Tectonic shorelines. In Carter, R.W.G., and Woodroffe, C.D. (eds.), *Coastal Evolution*. Cambridge: University Press, pp. 451–476.

Shepard, F.P., 1950. Submarine topography of the Gulf of California, 1940, E.W. Scripps cruise to the Gulf of California, part III. *Memoirs of the Geological Society of America* **43**: 1–32.

Shepard, F.P., and Wanless, H.R., 1971. *Our Changing Coastline*. New York: McGraw-Hill.

Trenhaile, A.S., 1987. *The Geomorphology of Rock Coasts*. Oxford: Clarendon Press.

Zenkovich, V.P., 1967. *Processes of Coastal Development*. Edinburgh: Oliver and Boyd.

Cross-references

Changing Sea Levels
Cliffed Coasts
Cliffs, Erosion Rates
Geochronology
Shore Platforms

FORECASTING—See EROSION: HISTORICAL ANALYSIS AND FORECASTING

G

GEOCHRONOLOGY

In coastal research, the terms "geochronology" or "geochronostratigraphy" are used to determine timescales for coastal processes and coastal evolution utilizing relative stratigraphic techniques (morpho-, pedo-, bio-, lithostratigraphy) and absolute dating methods. Relative and absolute temporal scales and dating methods complement each other.

In the 1980s and 1990s, applications of the absolute dating of coastal forms, sediments, and processes were improved, new dating methods were established, and the precision and accuracy of existing age determination methods were considerably increased. Numerous texts describe dating techniques in Quaternary sciences in detail (e.g., Smart and Frances, 1991; Wagner, 1998).

Conventional and mass spectrometric radiocarbon dating methods, as well as ^{230}Thorium/^{234}Uranium isotope, and Electron Spin Resonance (ESR) age determination methods are the most commonly used absolute dating techniques in coastal research (Table G1). The applicability of Thermal and Optical Stimulated Luminescence dating to coastal geochronology is currently being tested. The commonly used Amino Acid Racemization (AAR) method is a relative chronostratigraphic method. The paleomagnetic dating method has proven useful for differentiating marine and littoral sequences between the Middle and Early Quaternary.

Radiocarbon (^{14}C) dating

The importance of this method for the dating of Holocene and late Pleistocene marine terraces, beaches, and other coastal features cannot be overestimated. It is the most common and widely applied dating method in geochronology.

Carbon-14 (^{14}C) is a radioactive carbon isotope with a half-life of 5,730 ± 40 years. The isotope originates from the upper atmosphere, where nitrogen atoms are bombarded by cosmic rays. The isotope ^{14}C is chemically converted to CO_2. This $^{14}CO_2$ mixes with atmospheric CO_2, which contains the stable isotopes ^{12}C and ^{13}C. The natural concentration of radioactive carbon in the atmosphere is approximately $1:10^{12}$. In living organisms, the ^{14}C content is in equilibrium with the CO_2 of their environments (atmosphere, freshwater, saltwater).

The CO_2 exchange between the organism and its environment discontinues with death, and ^{14}C decays to ^{14}N with a half-life of 5,730 years. The upper dating limit of this conventional technique lies at approximately 30,000 years ^{14}C BP (approximately five times the half-life of ^{14}C). Radiocarbon ages are expressed in years ^{14}C BP, with BP referring to the time before AD 1950. Human induced nuclear detonations in the 1950s and 1960s have changed the atmospheric ^{14}C composition.

There are two laboratory methods of radiocarbon dating: The conventional decay counting method which measures beta rays from the ^{14}C decay process; and the Accelerator Mass Spectrometry (AMS) technique which determines the ratio between the number of ^{14}C to ^{12}C atoms in a sample. The AMS technique requires only small samples (less than 1 mg). Dating is limited to 50,000 years due to sample or laboratory contamination.

Since the 1960s, the radiocarbon method has been used successfully for dating Holocene and Late Pleistocene samples. The ^{14}C method delivers reliable results for wood, charcoal, and plant matter (e.g., peat, seeds, leaves). However, the dating of humus, speleothems, caliche, bones, and teeth is frequently inaccurate. In coastal research, the ^{14}C dating method is widely applied to the dating of marine fauna (e.g., mollusk shells, coral) and microfauna (e.g., foraminifera, ostracods). It is an excellent tool for geoarchaeological research in coastal settings where it can be cross-checked with stratigraphies from artifacts (Brückner, 1997).

Variations in the production of atmospheric ^{14}C may be caused by sunspot activity, changes in the intensity of the earth's magnetic field, or variability in cosmic radiation. This atmospheric source of error can be reduced by calibrating ^{14}C dating results (at present the calibration table covers the last 20,000 years) with a dendrochronologic tree ring calendar (dendrochronology: an absolute dating method based on the counting of tree rings). In addition, marine samples may be influenced by the so-called "marine reservoir effect," which is caused by the variability in the ^{14}C concentration in seawater. Deep ocean water has a lower ^{14}C content than the atmosphere. The ^{14}C dating of marine fauna (mollusks, coral, microfauna) from shores influenced by upwelling deep ocean water results in ^{14}C ages that are too old. The global mean reservoir effect for marine carbonates is approximately 400 years and varies spatially and temporally.

Case study: ^{14}C dating of beach ridges

The littoral zone of northern Andréeland (northern Spitsbergen) is characterized by more than one hundred beach ridges ranging in elevation from present sea level to 70 m. They represent uplifted former coastlines (Brückner and Halfar, 1994; Brückner, 1996). Two conditions led to their formation: (1) The West Spitsbergen Current, the northernmost continuation of the Gulf Stream extending up to 80°N, created seasonally ice free conditions in fjords. This allowed wave action which caused the formation of beach ridges. (2) The current elevation of these beach ridges is a result of coastal uplift due to the glacio- isostatic rebound following the last deglaciation.

The radiocarbon dating results of mollusk shells (Figure G1) illustrate: (1) the onset of deglaciation and the beginning of beach ridge formation in the Allerød Interstadial (11,970 ± 60 years ^{14}C BP minus

Table G1 Some important dating methods in coastal research

Dating method	Dating material	Time scales to be dated	Major problems influencing dating results
Absolute dating methods			
Isotope dating methods			
Radiocarbon (^{14}C)	All carbon containing material, for example wood, charcoal, peat mollusk shells*, coral*, foraminifera*, ostracods*	Holocene up to Late Pleistocene conventional ^{14}C method: 20 to 30 ka ^{14}C BP; AMS ^{14}C method: approximately up to 50 ka ^{14}C BP	Variability in atmospheric ^{14}C (cosmic, Suess and nuclear bomb effects); contamination of samples by younger C bearing material; *marine reservoir effect
Thorium-230/ Uranium-234 ($^{230}Th/^{234}U$)	coral, mollusk shells$^+$, bones$^+$, teeth$^+$, speleothems$^+$, travertines$^+$	Holocene up to Late Middle Pleistocene Coral: approximately up to 200 ka BP moullusk shells; approx. up to 130 ka BP (less reliable)	Isotope mobilization, especially uranium migration (radiometric system may not be closed); variability of $^{234}U/^{238}U$ ratio in sea water
Radiation induced dosimetric dating methods			
Thermal and optical stimulated luminescence (TL, OSL)	dune sands, loess, littoral sands+	Holocene up to Late Pleistocene at least up to 70 ka bp	Incomplete bleaching; istope mobilization, especially uranium migration
Electron Spin Resonance (ESR)	coral, mollusk shells, foraminifera, some gastropods	Holocene up to Middle Pleistocene coral: up to 400–600 ka BP mollusk shells, foraminifera, gastropods: up to 200–300 ka BP (upper dating limit still unknown)	Istope mobilization, especially uranium migration; recrystallization of sample
Relative dating methods			
Paleomagnetic dating	all material with stable magnetic remanescence, for example fine grained coastal sediments	Middle and Early Quaternary, limited to paleomagnetic reversals and events	Destruction of primary magnetic signal and acquisition of secondary magnetization
Amino Acid Racemization (AAR)	mollusk shells	From several 100 to several 100 ka BP	Results require calibration through absolute dating techniques; racemization rates may vary between genera; diagenesis temperature histories may differ between geographic regions

+ frequently used, but little suited for dating method
* see remarks in the right column

reservoir effect of approximately 400 years) due to the arrival of the West Spitsbergen Current; (2) that fjords were blocked by sea-ice during the Younger Dryas cold phase. This period of no beach ridge formation ranged approximately from 11,100 to 10,600 ^{14}C BP (minus reservoir effect of approximately 400 years); (3) that the major glacio-isostatic uplift occurred between 12,000 and 9,200 years ^{14}C BP (minus reservoir effect of approximately 400 years) and amounted to more than 140 m. In approximately 12,000 ^{14}C BP, eustatic sea level was at least 70 m lower than today. Since then the oldest beach ridges have been uplifted from less than 70 m below up to 70 m above present sea level.

Thorium-230/Uranium-234 isotope ($^{230}Th/^{234}U$) dating

The Thorium-230/Uranium-234 ($^{230}Th/^{234}U$) dating method is the most frequently used uranium isotope age determination technique. In coastal research, this method is utilized for dating Middle and Late Quaternary coral and mollusk shells.

This method is based on the fact that coral and mollusk shells take up the mother nuclide Uranium-234 without its daughter nuclide Thorium-234. Uranium-234 decays to Thorium-230, which has a half-life of 75.4 ka. Therefore, the upper dating limit of this method is reached at approximately 350 ka BP (as a rule of thumb: dating limit equals five times half-life).

The uranium and thorium isotope content can be determined α-spectrometrically or mass spectrometrically (Thermal Ionization Mass Spectrometry = TIMS). The mass spectrometric method requires smaller sample sizes (approximately 0.5–0.2 g) and is more precise (analytical error is approximately 1%).

The precision of the Th/U dating method has significantly improved due to advances in analytical techniques, however, numerous sources of error still limit the accuracy of the Th/U dating results. One key assumption in Th/U dating is that neither the mother nor the daughter nuclides relocated after the uranium accumulation in the coral structure or mollusk shell (a closed system). However, uranium is easily soluble in an oxidation environment and may have been added or removed. The resulting error in age determination can amount to tens of thousands of years and more. Partially recrystallized coral samples are neither suited for Th/U dating, since uranium and thorium may be mobilized in the process of the formation of calcite through the recrystallization of the aragonitic coral structure.

Mollusk shells are more prone to uranium and thorium migration than coral. Therefore, they are less suited for Th/U dating (Kaufmann *et al.*, 1971). While coral absorb approximately 3 ppm uranium from seawater, mollusk shells take up most of their uranium (approximately 0.2–4 ppm) after death from ground and seepage water. The majority of this uranium is taken up within the first few centuries and millennia after their death (Schellmann and Radtke, 1997). However, later isotope exchanges with the surrounding environment are possible. The variability of the $^{234}U/^{238}U$ ratio in seawater can also cause errors in the Th/U dating of Pleistocene coral and mollusk shells (Bard *et al.*, 1992).

Despite all potential sources of error, mass spectrometric Th/U dating of coral was utilized for the extension of the ^{14}C calibration curve to up to 30 ka BP (ka = 1000 years; Bard *et al.*, 1990). The current knowledge of Last Interglacial sea-level changes (between 75 ka to 130 ka BP) is largely based on α- and mass spectrometric Th/U dating of fossil coral. Even the Th/U dating results of Penultimate Interglacial coral reefs (around 200 ka BP) can reach considerable accuracy. The Th/U dating of Middle Pleistocene and older coral samples is not reliable due

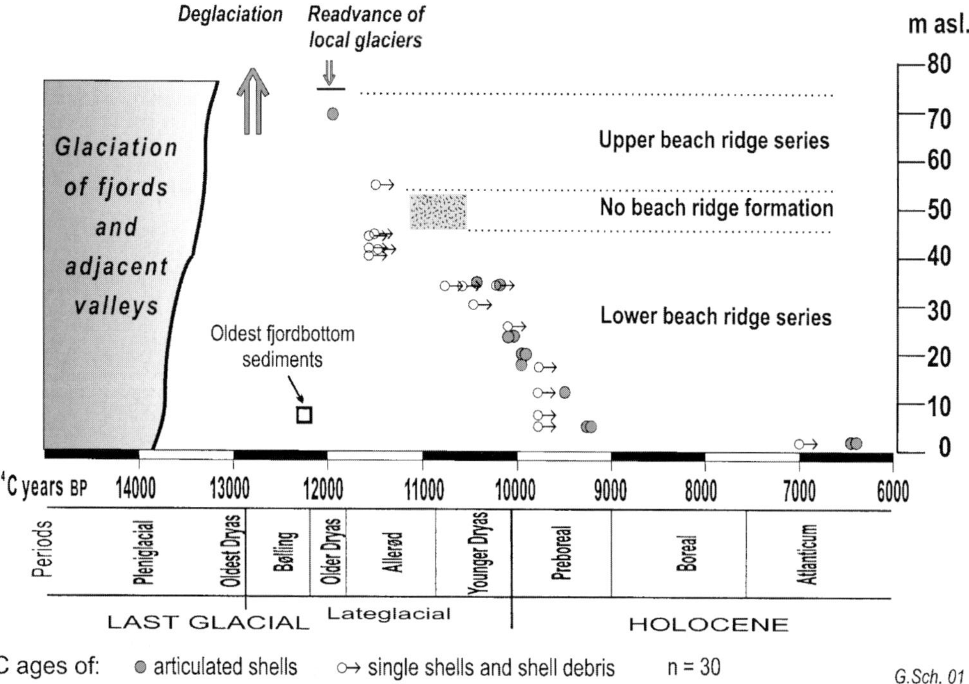

Figure G1 Deglaciation and evolution of beach ridges in Northern Spitsbergen (Woodfjord, Wijdefjord) based on ^{14}C datings. Beach ridges [in m asl] are represented by circles. All ages are uncalibrated ^{14}C years BP. The average reservoir effect for marine carbonates (approximately 400 years) needs to be considered for the correlation of the ^{14}C ages to geologic time periods (e.g., Younger Dryas). Articulated shells give reliable ^{14}C ages for beach ridge formation. The ^{14}C ages of coastal features based on single shells and shell debris are frequently overestimated since this material may have been relocated.

to their increased weathering and the still unknown $^{234}U/^{238}U$ ratio in seawater during those times.

To summarize, Th/U dating of fossil coral produces accurate results up to 200 ka BP, while the Th/U dating of mollusk shells is less reliable.

Thermal and optical stimulated luminescence (TL, OSL) dating

Luminescence dating techniques determine the time period that has elapsed since a charge population trapped within natural crystalline minerals was last reset. The fundamental concept of this method (Prescott and Robertson, 1997; Wintle, 1997; Aitken, 1998) is the use of naturally occurring semiconducting crystalline minerals, such as quartz and feldspars, as natural dosimeters. These minerals record their exposure to naturally occurring ionizing radiation generated by the decay of uranium, thorium, potassium, and to a minor extent, by rubidium and cosmic rays. Ionizing radiation energy moves electrons from their stable state in a crystal lattice to electron traps. Luminescence techniques measure the signals that are related to the relocation of these electrons.

The luminescence signal is the light emitted from crystals when they are stimulated by heat or light after having received a radiation dose (natural or artificial radiation dose). The number of electrons migrating to traps increases with time and dose of natural radiation. The traps are linked to a variety of defects within the crystal structure. These "trapped" electrons are attached to defect sites, where they are stable until a further amount of energy is thermally or optically introduced. For sedimentary material, such as dune or littoral sands, the resetting event (called "bleaching") is the exposure of the material to daylight prior to sediment deposition. After bleaching, the process starts all over again. Luminescence techniques date the time that has elapsed since the last exposure to daylight (during transport) or heating of the sediment.

Luminescence emissions are called thermoluminescence (TL) or optical stimulated luminescence (OSL) depending on the type of energy used to release the trapped charge during measurements. Infrared stimulated luminescence (IRSL) and green light stimulated luminescence (GLSL) are OSL techniques. The OSL signal is more sensitive to light than the TL signal. Exposure to daylight resets the OSL signal within a few minutes and the TL signal within several hours.

Luminescence techniques measure the ionizing radiation dose (equivalent dose, D_E) that has accumulated in the sample since the last reset of charge. One of the three methods is generally applied to determine D_E: (1) additive dose method, (2) regeneration method, (3) partial bleach method. Lumi-nescence techniques can be conducted on many aliquots of a sample (multiple aliquot approach), or on one aliquot of a sample with multiple measurements (single aliquot approach), or on one grain of a sample (single grain approach).

Measurements of the environmental ionizing radiation rate (dose rate or annual dose, D') are required for the age determination of a sample. The sample age is calculated by dividing D_E [in Gy] by the annual dose D' [in Gy/ka]. Luminescence methods have been successfully used for the generation of deposition chronologies up to at least 70 ka with precisions of 5–15%.

Luminescence methods are applied to a wide range of depositional environments. Aeolian sediments, including loess or coastal dune sands, are best suited for luminescence dating since the transport prior to their deposition guarantees their exposure to light. Littoral, lacustrine, or fluvial sediments were subject to transport mechanisms not necessarily ensuring exposure to light. However, these sediments may be suited for luminescence dating since light penetrates into the water column depending on water depth and the concentration of suspended sediment. The complete bleaching of sediments during their deposition is likely in shallow marine environments that are located in regions with intense insolation. Sediments cannot be dated using TL or OSL when a radioactive disequilibrium exists (e.g., uranium uptake or excess).

Electron spin resonance (ESR) dating

The first systematic studies and applications of Electron Spin Resonance (ESR) dating of coral and mollusk shells were conducted in the 1980s. Since then, a large number of Pleistocene coastal features and sediments have been dated using ESR.

ESR dating as well as luminescence techniques (TL, OSL) are radiation induced dosimetric dating methods. They are based on the radiation damage in minerals that increases with their age and is caused by natural radioactivity and cosmic rays. These damages generate paramagnetic centers and radicals that can be measured using ESR.

An ESR age derives from the ratio of the accumulated radiation dose D_E to the antecedent natural dose D'. The accumulated radiation dose D_E (equivalent dose) is determined by an "additive dose method." For this method, 20 or more aliquots of the same sample are stepwise irradiated with increasing γ-doses. Trapped electrons increase in number causing the amplitude of the ESR dating signal to increase. The D_E is determined by extrapolating the resulting dose response curve to zero ESR intensity. The former natural dose D′ is estimated based on the content of radioactive sources (e.g., uranium, thorium, potassium) in the sample and its environment, the intensity of cosmic rays, and present moisture conditions. D′ is determined through laboratory analysis or through field measurements using a portable γ-spectrometer. The correct determination of D′ and D_E is essential for any ESR age estimate. Rink (1997) describes ESR dating techniques in Quaternary sciences and discusses their accuracy and precision.

Recently, both precision and accuracy of ESR dating of fossil coral and mollusk shells have improved due to the higher resolution of ESR measurements and due to new techniques for the determination of D_E (Schellmann and Radtke, 1997, 1999, 2001a). However, ESR ages can deviate by several thousand years if the natural radiation dose (D′) the sample was subject to in the past was altered by uranium or thorium mobilization or by changing moisture conditions. Problems arise (1) from the heterogeneity of the surrounding sediments, and (2) since shells are systems open to uranium migration.

ESR dating results of mollusk shells that were embedded in heterogeneous beach sediments are most affected by these sources of error. Dating errors can reach more than 15%. Therefore, ESR dating results for mollusk shells from littoral deposits are frequently limited to the mere differentiation between Holocene, Last Interglacial, Penultimate Interglacial, and older sediments (Schellmann and Radtke, 2000).

The dating of coral originating from relatively homogeneous reef limestone can generally be accomplished with errors of approximately 10%. Substages of sea-level highstands (e.g., Last Interglacial Oxygen Isotope Substages 5a, 5c, and 5e) can be dated due to improvements in the determination of the accumulated radiation dose D_E.

ESR datings of coral from the Late Pleistocene are now as accurate as those dated using the Th/U method (Figure G2). Therefore, the application of both methods simultaneously is recommended for the dating of coral up to 200 ka BP.

The ESR method allows for the dating of coral and mollusk shells for a larger time span than any other Quaternary dating method. The lower dating limit lies at a few thousand years BP. The ESR method should theoretically allow for the dating of coral and mollusk shells that are several million years old. However, numerous ESR dating results for mollusk shells and coral illustrate that the upper dating limit lies at 200–300 ka and 400–600 ka, respectively.

The upper dating limit is largely dependent on changing conditions due to diagenesis and weathering. For example, the recrystallization of the aragonite coral structure or mollusk shell can cause considerable underestimates of ages, since it eliminates previously stored radiation.

ESR has been widely applied to dating sequences of raised coral reef terraces with ages of up to 500–600 ka, for example, terraces on Sumba island (Pirazzoli *et al.*, 1991), and on Barbados (Radtke and Grün, 1988). Mollusk shells with ages ranging from 200 to 300 ka were also dated using ESR, for example, mollusk shells at the Patagonian Atlantic coast (Schellmann, 1998; Schellmann and Radtke, 2000). Most recently, Late Pleistocene gastropods in Aeolianites on Cyprus were accurately dated using ESR (Schellmann and Kelletat, 2001).

Case study: ESR dating of Pleistocene coral reef terraces in southern Barbados

The ESR dating of the fossil coral reef terraces on Barbados delivers an impressive example for the application of this method (Figure G3). In southern Barbados, these terraces rise up to 121 m above present sea-level. The ESR ages increase with height up to 361 ka. The age–height relationships allow for the calculation of neotectonic uplift in Barbados and of paleo- sea-level changes (Radtke, 1989).

Amino Acid Racemization (AAR) dating

The Amino Acid Racemization (AAR) dating technique is a relative age determination method for marine mollusk shells (Miller and Brigham-Grette, 1989). AAR values are applied to timescales ranging from hundreds to hundreds of thousands of years. The method is based on the rate of racemization and epimerization of indigenous amino acids preserved in mollusk shells. The temperature history during diagenesis and the rate of hydrolysis of residual proteins influence the protein degradation processes.

Several uncertainties in the applicability and reliability of AAR remain, since (1) racemization rates may vary between genera, and (2) diagenesis temperature histories may differ between different geographic settings. The AAR method seems to be reliable for distinguished regional aminostratigraphic units only. Since AAR is a relative dating technique, its results need to be calibrated using absolute dating methods.

Paleomagnetic dating

Paleomagnetic dating determines the natural remanant magnetization of sediments and correlates it to the known paleomagnetic polarity stratigraphy. Permanent magnetism is carried by ferromagnetic minerals that are minor constituents of almost all rocks. Magnetic minerals align to the earth's magnetic field and keeps that orientation when they are deposited or precipitated in the sedimentary matrix during early diagenesis. Thompson (1991) lists criteria for the application of this method.

The paleomagnetic reversals at the base of the so-called Brunhes and Matuyama chrons at 0.78 Ma (normal to reverse) and 2.6 Ma (reverse to normal), respectively, are the most important polarity changes of the earth's magnetic field in recent geological history. The Jaramillo and Olduvai subchrons are two timespans within the normal polarized Matuyama chron. At least 18 short-lived polarity events occur within the Brunhes and Matuyama chrons. The most prominent short-lived event is the "Blake Event" (approximately 117 ka), which seems to be a worldwide reversed interval (Langereis *et al.*, 1997; Worm, 1997).

Sample applications of the paleomagnetic method include the dating of marine terraces in southern Italy (Brückner, 1980), and the dating of a sequence of beach ridges and coastal dunes at the Moroccan Atlantic coast (Brückner and Hambach, unpublished data). The paleomagnetic method requires a continuous sequence of coastal formations and a narrow sampling strategy. Paleomagnetic dating allows to differentiate within the Middle and Older Quaternary, where other dating methods fail.

Gerhard Schellmann and Helmut Brückner

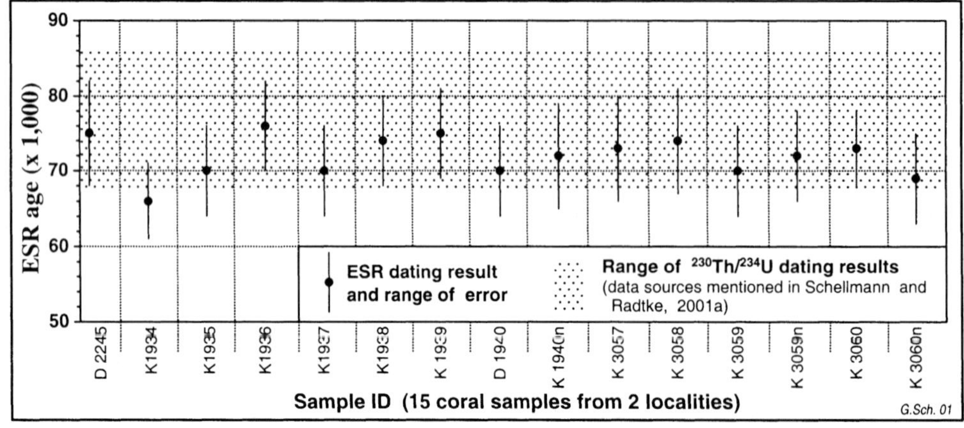

Figure G2 Comparison of ESR and Th/U datings of late Last Interglacial coral samples (OIS 5a) from the south coast of Barbados (modified after Schellmann and Radtke, 2001a).

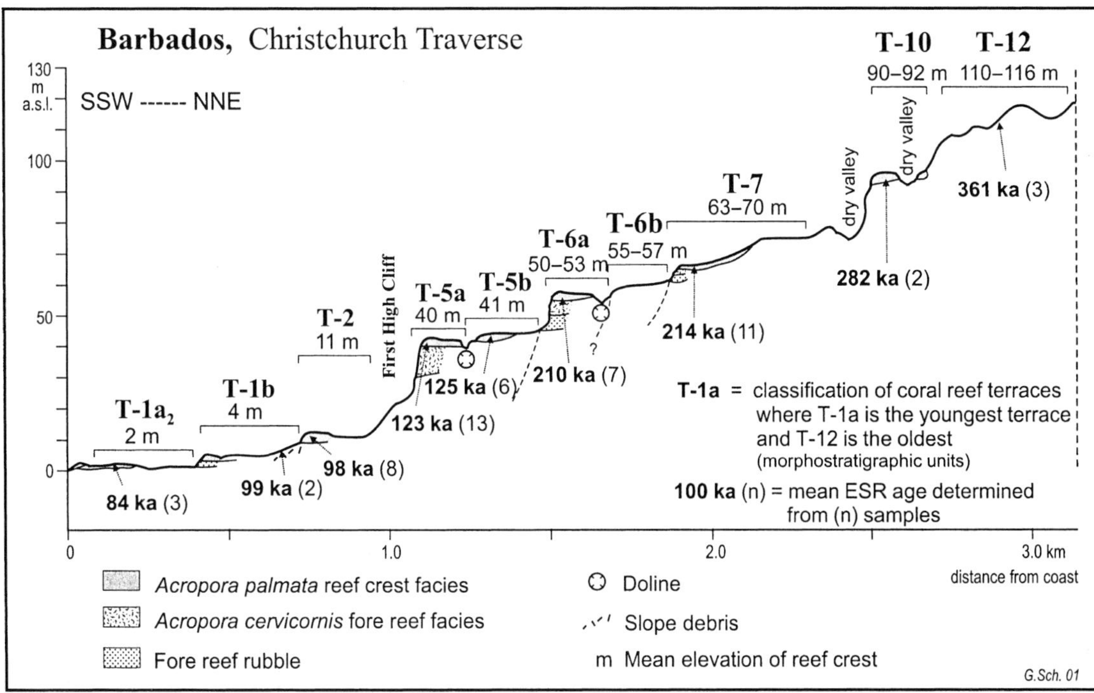

Figure G3 ESR dating of fossil coral reef terraces in southern Barbados (modified after Schellmann and Radtke, 2001b).

Bibliography

Aitken, M.J., 1998. *An Introduction to Optical Dating: The Dating of Quaternary Sediments by the Use of Photon-stimulated Luminescence*. Oxford: Oxford University Press.

Bard, E., Hamelin, B., Fairbanks, R.G., and Zindler, A., 1990. Calibration of the ^{14}C timescale over the past 30,000 years using mass spectrometric U-Th ages from Barbados corals. *Nature*, **345**: 405–410.

Bard, E., Fairbanks, R.G., and Hamelin, B., 1992. How accurate are the U-Th ages obtained by mass spectrometry on coral terraces. In Kukla, G.J. and Went, E. (eds.), *Start of a Glacial*. Berlin: Springer-Verlag, pp. 15–22.

Brückner, H., 1980. Marine Terrassen in Süditalien. Eine quartärmorphologische Studie über das Küstentiefland von Metapont. *Düsseldorfer Geographische Schriften*, 14, Düsseldorf, Germany: Düsseldorf University.

Brückner, H., 1996. Studies of beach deposits in northern Spitsbergen. *Heidelberger Geographische Arbeiten*, **104**: 375–389.

Brückner, H., 1997. Coastal changes in western Turkey—Rapid delta progradation in historical times. In Briand, F., and Maldonado, A. (eds.), *Transformations and evolution of the Mediterranean coastline*. CIESM Science Series, no. 3, 63–74 (Bulletin de l'Institut océanographique, numéro spécial 18. Musée océanographique, Monaco), Monaco.

Brückner, H., and Halfar, R.A., 1994. Evolution and age of shorelines along Woodfiord, northern Spitsbergen. *Zeitschrift für Geomorphologie N.F.*, **97**(Suppl.-Bd.): 75–91.

Kaufmann, A., Broecker, W.S., Ku, T.L., and Thurber, D.L., 1971. The status of U-series methods of mollusc dating. *Geochimica et Cosmochimica Acta*, **35**: 1155–1183.

Langereis, C.G., Dekkers, M.J., de Lange, G.J. Paterne, M., and van Stantvoort, P.J.M., 1997. Magnetostratigraphy and astronomical calibration of the last 1.1 Myr from an eastern Mediterranean piston core and dating of short events in the Brunhes. *Geophysical Journal International*, **129**, 75–94.

Miller, G.H., and Brigham-Grette, J., 1989. Amino acid geochronology: Resolution and precision in carbonate fossils. *Quaternary International*, **1**: 111–128.

Pirazzoli, P., Radtke, U., Hantoro, W.S., Jouannic, C., Hoang, C.T., Causse, C., and Borel-Best, M., 1991. Quaternary raised coral-reef terraces on Sumba island, Indonesia. *Science*, **252**: 1834–1836.

Prescott, J.R., and Robertson, G.B., 1997. Sediment dating by luminescence: a review. *Radiation Measurements*, **27**(5/6): 893–922.

Radtke, U., 1989. Marine Terrassen und Korallenriffe. Das Problem der quartären Meeresspiegelschwankungen erläutert an Fallstudien aus Chile, Argentinien und Barbados. *Düsseldorfer Geographische Schriften*, 27, Düsseldorf, Germany: Düsseldorf University.

Radtke, U., and Grün, R., 1988. ESR dating of corals. *Quaternary Science Reviews*, **7**: 465–470.

Rink, W.J., 1997. Electron Spin Resonance (ESR) dating and ESR applications in Quaternary science and archaeometry. *Radiation Measurements*, **27**: 975–1025.

Schellmann, G., 1998. Jungkänozoische Landschaftsgeschichte Patagoniens (Argentinien). Andine Vorlandvergletscherungen, Talent-wicklung und marine Terrassen. *Essener Geographische Arbeiten*, **29**: 216.

Schellmann, G., and Kelletat, D., 2001. Chronostratigraphische Untersuchungen litoraler und äolischer Formen und Ablagerungen an der Südküste von Zypern mittels ESR-Altersbestimmungen an Mollusken- und Landschneckenschalen. *Essener Geographische Arbeiten*, **32**: 75–98.

Schellmann, G., and Radtke, U., 1997. Electron Spin Resonance (ESR) techniques applied to mollusc shells from South America (Chile, Argentina) and implications for the palaeo sea-level curve. *Quaternary Science Reviews*, **16**: 465–475.

Schellmann, G., and Radtke, U., 1999. Problems encountered in the determination of dose and dose rate in ESR dating of mollusc shells. *Quaternary Science Reviews*, **18**: 1515–1527.

Schellmann, G., and Radtke, U., 2000. ESR dating stratigraphically well-constrained marine terraces along the Patagonian Atlantic coast (Argentina). *Quaternary International*, **68–71**: 261–273.

Schellmann, G., and Radtke, U., 2001a. Progress in ESR dating of Pleistocene corals—an approach for D_E determination. *Quaternary Science Reviews*, **20**: 1015–1020.

Schellmann G., and Radtke U., 2001b. Neue Ergebnisse zur Verbreitung und Altersstellung gehobener Korallenriffterrassen im Süden von Barbados. In Schellmann, Gerhard (ed.) Von der Nordessküste bis Neuseeland–Beiträge zur 19. Jahrestagung des Arbeitskreises "Geographie der Meere und Küsten" vom 24–27. Bamberg: *Bamberger Geographische Schriften*, **20**: 201–224.

Smart, P.L., and Frances P.D., (eds.), 1991. *Quaternary dating methods: a user's guide*. Cambridge; Quaternary Research Association, *Technical Guide No.4*, p. 233.

Thompson, R., 1991. Palaeomagnetic dating. In Smart, P.L., and Frances, P.D. (eds.), *Quaternary dating methods: a user's guide. Technical Guide No. 4*. Cambridge; Quaternary Research Association, pp. 177–198.

Wagner, G.A., 1998. *Age Determination of Young Rocks and Artifacts—Physical and Chemical Clocks in Quaternary Geology and Archaeology*. New York: Springer.
Wintle, A.G., 1997. Luminescence dating: laboratory procedures and protocols. *Radiation Measurements*, **27**(5/6): 769–817.
Worm, H.-U., 1997. A link between geomagnetic reversals and events and glaciations. *Earth Planet. Science Letters*, **147**: 55–67.

Cross-references

Coral Reefs, Emerged
Isostasy
Sea-Level Indicators, Geomorphic
Uplift Coasts

GEODESY

According to the traditional classical definition, attributed to F. R. Helmert in 1880, geodesy is *the science of the measurement and mapping of the earth's surface*. Nowadays, this definition is deemed to encompass determination of the topography of the earth's landmasses, the bathymetry of the seafloor, the shape of the surface of the ocean, and form and nature of the earth's internal and external gravity field.

History

The origins of modern geodesy can be traced to antiquity. The ancient Greeks speculated on the shape of the earth with Pythagoras, and later Aristotle, introducing the concept of a spherical earth. The first scholar to make a scientific estimate of the shape of the earth was Eratothenes of Alexandria (276–195 BC) who obtained a value for the circumference of earth to within 2% of modern estimates. A later estimate by Posidonius (135–51 BC) was used by Ptolemy (AD 90–168) to produce the monumental eight volume *Guide to Geography*, which contained maps of the world destined to be the definitive work for the next 15 centuries. Columbus, based on such maps, was to underestimate the distances from Europe to India and Asia and, in turn, discover the Americas. Advances in astronomy, instrumentation and physics in the 16th and 17th centuries led to Isaac Newton (1643–1727) and Christian Huygens (1629–95) proposing that the earth was not spherical, but flattened at the poles. This proposal led to some controversy and, it was not until the late 18th century, following increasingly accurate measurements on the earth's surface, that an ellipsoidal model for the shape of the earth was universally adopted.

Essential geodetic concepts

Definition of the size and shape of the earth does not usually involve surface topography or the bathymetry of the oceans, but the surface of mean sea level and its theoretical continuation under the continents. Topographic heights and bathymetry are measured relative to this hypothetical surface which is known as the geoid. By definition, the geoid represents an equipotential surface and is therefore a direct representation of the earth's gravity field. The geoid is a mathematically complex surface, its shape being strongly influenced by irregular mass distributions below the earth's surface. It is thus unsuitable as a reference surface for a geometric figure of the earth. For further details on the earth's gravity field and the geoid, the reader is referred to Heiskanen and Moritz (1967).

For geodetic calculations, such as computing geodetic latitude and longitude and defining map projections (e.g., Bugayevskiy and Snyder, 1995), an ellipsoid of revolution (a three-dimensional surface generated by rotating an ellipse 360° about its minor axis) is defined. This simple figure is known as a reference ellipsoid and is defined by a semimajor axis (equatorial radius) and semiminor axis (polar radius). The earth's semimajor axis is about 21 km greater than the semiminor axis, with the flattening being approximately 1/300. A commonly used reference ellipsoid is that associated with the World Geodetic System 1984 (WGS84) (NIMA, 1997).

For geodetic measurements, the concept of geodetic datum is introduced. A geodetic datum is a three-dimensional representation of the surface of the earth through a framework of discrete points whose coordinates and elevation (either above mean sea level or a reference ellipsoid) are accurately known. In individual regions, datum coordinates and computations are based on a local reference ellipsoid, which corresponds closely to the shape of the geoid in that area. Modern satellite techniques, such as the Global Positioning System (GPS) *(q.v.)*, rely on geocentric reference ellipsoids which have the center of the earth as the origin. To ensure compatibility between GPS coordinates and coordinates in the framework of the local geodetic datum, a mathematical function defining the transformation of the coordinates between the two systems must often be computed. A geodetic datum forms the basic structure for densification surveys which, in turn, can be incorporated into maps or Geographic Information System (GIS) *(q.v.)* databases. The concepts of geodetic datums and reference systems are covered in detail by Torge (1991).

Applications of Geodesy

Space geodetic measurement techniques (descriptions of very long baseline interferometry (VLBI), satellite laser ranging (SLR), and GPS are given in Seeber (1993)) are used to define international geodetic datums, such as the International Terrestrial Reference Frame (ITRF), which provide the basic infrastructure for positioning and navigation on the surface of the earth. These reference frames are used for global and regional plate tectonic studies, measurement of isostatic rebound, and connection of *tide gauges (q.v.)* for monitoring the levels of the world's oceans (see also *Sea-level datum (q.v.)*). They are also critical for satellite orbit tracking required for such applications as satellite *altimetry (q.v.)* and *SAR (q.v.)* studies. Global geodetic measurements are also used to derive parameters that can describe variations in the rotation of the earth (e.g., length of day, polar precession, and nutation), and the dynamics of the atmosphere, oceans, and solid earth (Lambeck, 1988).

Local geodetic measurements rely on survey instrumentation such as electronic distance meters (EDM), precise levels and *GPS (q.v.)* receivers. Such measurements are primarily used for mapping purposes, providing local control for *photogrammetry (q.v.)* and *remote sensing (q.v.)*, and for engineering applications such as monitoring physical changes in man-made structures and ground surfaces. For a complete appraisal of geodesy and terrestrial surveying, the classic text is Bomford (1980).

Mike P. Stewart

Bibliography

Bomford, G., 1980. *Geodesy*, 4th edn. Oxford, England: Clarendon Press.
Bugayevskiy, L.M., and Snyder, J.P., 1995. *Map Projections: A Reference Manual*. London; Bristol, PA: Taylor & Francis.
Heiskanen, W.A., and Moritz, H., 1967. *Physical Geodesy*. San Francisco: Freeman and Company.
Lambeck, K., 1988. *Geophysical Geodesy: The Slow Deformation of the Earth*. Oxford, England: Clarendon Press.
NIMA WGS84 Update Committee, 1997. *Department of Defense World Geodetic System 1984, Its Definition and Relationships with Local Geodetic Systems,* 3rd edn. National Imagery and Mapping Agency: TR 8350.2.
Seeber, G., 1993. *Satellite Geodesy*. Berlin; New York: de Gruyter.
Torge, W., 1991. *Geodesy*. Berlin; New York: de Gruyter.

Cross-references

Altimeter Surveys, Coastal Tides and Shelf Circulation
Geographic Information Systems
Global Positioning Systems
Photogrammetry
Remote Sensing: Wetlands Classification
Sea-Level Datums (See Tidal Datums)
Tide gauges

GEOGRAPHIC INFORMATION SYSTEMS

Introduction

Geographic Information Systems (GIS) are computerized spatial decision-making tools that coordinate place-based information and allow users to manage, analyze, and display geographic data. Since coastal features are typically spatial, coastal managers and scientists have incorporated this technology from its beginnings. However, coastal data often represent dynamic, multi-dimensional systems that present problems to this historically land-based technology.

History

The concept of a GIS originated from non-computerized decision-making techniques. One classic example is a decision-making technique developed by Ian McHarg, a landscape architect and author of *Design with Nature* (McHarg, 1969). McHarg's method for identifying suitable locations for development was based on defining the most favorable characteristics of various sets of information such as slope, soil type, geology, land use, and accessibility and then combining the layers into a composite map. Although McHarg's technique may be carried out in an analog fashion, it is a popular example of the power of applying digital computer technology.

The beginning of the modern, digital, GIS began in the late 1960s. Several efforts in developing digital spatial systems began concurrently and involved applications for urban planning, transportation, natural resources, and the US Census. The first use of the term GIS was the Canada Geographic Information System, implemented in 1964 to identify marginal agricultural lands for rehabilitation (Star and Estes, 1990; DeMers, 1997; Foresman, 1998; Wright and Bartlett, 2000).

Data models

A data model represents the spatial entity and its relationship to the real world (Goodchild, 1992). There are two primary data models in wide use today: raster models and vector models.

Raster models generally depict continuous data by representing data points in a regular pattern of cells. The size of the cells, for example, 10 m by 10 m, defines the geometric resolution and a series of row and column coordinates define the spatial extent. Continuous data commonly depicted with a raster model include surface elevation, precipitation, and satellite image data such as land cover (Bernhardsen, 1992).

Vector models are topological, representing discrete data with nodes and lines (Bernhardsen, 1992). The node may define a starting point, an intersection, an end point, or any point along a line. The topological model retains the relationship and connectivity between and among points, lines, and polygons. For example, a series of nodes and lines may define two adjacent counties; the shared border is captured and stored only once in a vector model and that relationship remains intact even when the geometry is changed.

A third type of data model may represent continuous or discrete data as a network and is most often applied in hydrology and transportation applications (Goodchild, 1992). A fourth model, the object model, follows a new era of object-oriented programming and overcomes some of the deficiencies of the common models (Neves *et al.*, 1999).

Functionality

A GIS typically uses hardware and software that allows the following functionality:

- Data input/capture
- Data storage/management
- Data manipulation
- Data analysis
- Data output and display

If data are not already in a digital form, data capture may be a time-consuming component of building a GIS. Data capture techniques include hand-tracing maps with a digitizing puck (mouse) and tablet, electronic scanning that converts points and lines, Global Positioning Systems (GPS) that digitally record surveyed locations, and manual entry of geographic coordinates. Acquiring digital spatial data from secondary sources typically requires a data conversion process, that is, converting from one proprietary format to another or converting from one coordinate system or map projection to another. Once geographic entities are digitally captured, assigning attribute identities is typically required. For example, a river (the entity) may be hand digitized and its name would be assigned separately as an attribute.

Data storage and management is conducted using the database management system (DBMS) component of a GIS. Users create, update, and store the geographic entity files and their associated attributes. Metadata, information about the history and extent of the spatial data, may similarly be stored and managed. Once collected and stored in a GIS database, spatial data may be manipulated (queried, classified, and edited) using tools inherent in GIS software packages.

The ability to analyze spatial data is the component that differentiates GIS from desktop mapping software. GIS software provides tools to analyze distance, direction, connectivity, and proximity through the use of spatial and attribute queries. The movement of humans or natural features through linear networks such as roads and streams may be modeled. Furthermore, data modeling tools allow the depiction of two-and three-dimensional characteristics of the earth's surface, subsurface, and atmosphere. Map overlay analysis involves combining two or more layers of information through logical operations. McHarg's suitability analysis is a perfect example of the use of map overlay analysis.

Finally, data output and display capabilities allow the GIS user to prepare graphic images of their products. The images may be output as paper maps or stored as electronic files for computer screen display or for use in web-based applications.

Coastal GIS

We use our coasts for food, minerals, waste disposal, industry, transportation, leisure, and tourism. With as much as 50% of the US population living on or near a coast, the economic and aesthetic importance of the coastal zone to humans is obvious. Coastal zones are also hazardous places with the threats of flooding, erosion, hurricanes, earthquakes, or tsunamis. The environmental movement of the 1960s and early 1970s highlighted the need for coastal stewardship and by 1972, the Coastal Zone Management Act was passed. With a focus on stewardship and multi-stakeholder management, coastal scientists sought new ways to implement their management directives. Since coastal data are spatial, GIS technology was recognized as a useful tool for coastal management decision-making.

In the early 1970s, coastal literature began to address the need for coastal zone management tools and GIS technology was implicated. The first GIS coastal applications were developed for particular tasks such as development permitting or tourism planning along coasts (Bartlett, 2000). Early efforts to apply GIS were thwarted by limited availability of software, hardware, and trained personnel. And still, with availability of proprietary GIS packages that are easier to use and to maintain, the built in functionality still lacks for coastal applications that require three-dimensional and temporal analysis.

Bartlett identified several primary issues in the "challenge of applying GIS to the coast" including data availability and data models (Bartlett, 2000). It is rare to find available and accessible data and collecting primary data is costly and time consuming. In acquiring secondary data one must ensure data quality issues such as completeness, accuracy, currency, and lineage. In developing a coastal GIS, as in any GIS development, primary and secondary data will most likely be required.

The structure of the data model has also been a limitation in coastal GIS applications. Raster-based models hold the typical advantages of simple programming and high volume storage, however, spatial resolution is compromised. Conversely, vector systems portray finer resolution but inappropriately portray boundaries that are "fuzzy." Dynamic segmentation (network models), object models, and others "hold considerable potential for advancing the utility of GIS within the coastal and marine environment" (Bartlett, 2000).

To date, a number of large coastal GIS-based projects have been implemented and the efforts were largely collaborative between federal governments, academic institutions, non-profit organizations, and local governments. In the United States, federal agencies such as the National Oceanic and Atmospheric Administration (NOAA), the US Geological Survey (USGS) and the Federal Emergency Management Agency (FEMA), have led national case studies. FEMA with a mandate from the National Flood Insurance Reform Act of 1994, has conducted case studies to determine the feasibility of mapping erosion hazards throughout the US coastal zone (Daniels *et al.*, 1998). FEMA contracted the Washington State Department of Ecology (with support from USGS, NOAA, and local agencies) to calculate a 60-year erosion hazard area for Pacific County, Washington. Pacific County is located north of the Columbia River and includes 60 km of dynamic shoreline, with both high erosion and high accretion rates. The data collection included digitizing historical coastlines from historical topographic sheets, determining contemporary high-tide shorelines from recent aerial photography, and digitizing the average high water line and vegetation line. They used GIS analysis methods to analyze the past, current, and future high-tide shorelines. The results show that from 1950 to 1995 the rates of change varied within the county from 28.3 m/year of accretion to 27.0 m/year of erosion; the mean rate was 0.9 m/year of accretion. The study determined that 626 structures were within the projected 60-year hazard area.

Future

Despite current limitations, GIS remains an important tool for coastal applications. As collaborative projects and research on the use of GIS in coastal applications continues, Bartlett offers that the key to a successful GIS is planning:

> It is essential to establish the user-base for whom the coastal zone GIS is intended, and the view of the coast and its component elements that is relevant to this constituency, since this operational context will have a direct bearing on the more technologically-oriented questions of hardware and software selection, and the overall architecture of the system. It will also define the information products expected from the system, and hence the types of data, and the processes performed on these, that are required in order to produce the desired output (Bartlett, 2000, p. 18).

One measure of the importance of GIS to coastal scientists is the increasing number of user-groups, web pages, and conferences organized around the theme. Examples include, the International Geographic Union's Commission on Coastal Systems and the Working Group on Marine Cartography of the International Cartographic Association collaborative international symposium known as CoastGIS and a web page devoted to maintaining a bibliography of coastal zone application of GIS (Fell *et al.*, 1997).

Shawn M. Boeser

Bibliography

Bartlett, D.J., 2000. Working on the Frontiers of Science: Applying GIS to the Coastal Zone. In Wright, D.J., and Bartlett, D.J., (eds.), *Marine and Coastal Geographic Information Systems*. London: Taylor and Francis, pp. 11–22.

Bernhardsen, T., 1992. *Geographic Information Systems*. Arendal: Viak IT.

Daniels, R.C. *et al.*, 1998. Coastline mapping and identification of erosion hazard areas in Pacific county, Washington. In *Proceedings of the Environmental Systems Research Institute* (ESRI) *User Conference '98*.

DeMers, M.N., 1997. *Fundamentals of Geographic Information Systems*. New York: John Wiley & Sons.

Fell, B. *et al.*, 1997. *Coastal Management: A bibliography of Geographic Information System Applications*. Charleston, WV: NOAA Coastal Services Center.

Foresman, T.W. (ed.), 1998. *The History of Geographic Information Systems: Perspectives from the Pioneers*. Upper Saddle River, NJ: Prentice-Hall.

Goodchild, M.F., 1992. Geographical Data Modeling. *Computers and Geosciences*, **18**(4): 401–408.

McHarg, I., 1969. *Design with Nature*. Garden City, NY: Doubleday & Company.

Neves, J.N. *et al.*, 1999. A virtual GIS room: interfacing spatial information in virtual environments. In Camara, A.S., and Raper, J. (eds.), *Spatial Multimedia and Virtual Reality*. London: Taylor and Francis.

Star, J. and Estes, J.E., 1990. *Geographic Information Systems: An Introduction*. Englewood Cliffs, NJ: Prentice Hall.

Wright, D.J., and Bartlett, D.J. (eds.), 2000. *Marine and Coastal Geographic Information Systems*. London: Taylor and Francis.

Cross-references

Airborne Laser Terrain Mapping and Light Detection and Ranging
Erosion: Historical Analysis and Forecasting
Global Positioning Systems
Instrumentation (See Beach and Nearshore Instrumentation)
Mapping Shores and Coastal Terrain
Monitoring, Coastal Geomorphology
Nearshore Geomorphological Mapping
Photogrammetry
RADARSAT-2
Remote Sensing of Coastal Environments
Synthetic Aperture Radar Systems

GEOGRAPHICAL COASTAL ZONALITY

The term "coastal zone" has been misleadingly used, even in scientific papers to describe the coastal environment or the region between the foreshore and the upper supratidal along a coastline. "Zones" in geoscience, however, should be named only for more or less latitude-parallel belts between the equator and the poles, which have been established in the first order by the different radiation of sunlight on our globe.

Nearly all macro-patterns on earth, which depend on climatic influences, like vegetation, soil formation, or forms of relief and inherent processes show a zonal distribution. For the latter, including both relic and contemporary systems, this has been accepted for decades (Murphy, 1968; Büdel, 1977); and the term "climatic geomorphology" has therefore been established. Regarding coastal features and processes, however, these have been looked upon for a long time mostly as azonal, because cliffs, beaches, lagoons and barriers, deltas, etc. can be found in all latitudes; and rock-depending forms are azonal, as well.

It was only in 1964 that Davies established a deductive classification of zonal coastal patterns based on prevailing wave and surf types, and in 1972 Davies introduced his pioneering book on "*Geographical Variation in Coastal Development*," giving an overview of the zonal coastal forms and processes, all inductively identified. Other authors then studied the global distribution of single features and their differentiation according to latitude, or the influence of humid or arid tropical climates on coastal landforms.

Almost two decades ago a "system of zonal coastal geomorphology" was developed by Valentin (1979). This system is limited to the definition of 51 types of non-coastal landforms which were characteristic of the Wisconsin (Würm) glacial climato-morphogenic zones. Many of these areas were in part drowned during the postglacial sea-level rise. The glacio-eustatic drowning is in fact the only "littoral" and global factor in this system, but it is not a zonal process since it can be observed worldwide. In his "Atlas of Coastal Geomorphology and Zonality" Kelletat (1995), however, applied the question of zonality in a strictly inductive sense, that is, interpreted from field evidence and areal distribution patterns, in contrast to previous deductive models. This is only one of the first steps and a complete system of geographical coastal zonality will remain an unsolved problem, so far as the general features of coastal evolution have been firmly established, and we are still far from this aim today. Some examples of zones and sub-zones on selected meridional transects may elucidate the problem (see Figure G4).

The zonality of the coasts of Europe and the Middle East is particularly influenced by two special conditions: the warming effect of the Gulf Stream in the north and the semi-enclosed oceanic basin of the Mediterranean Sea. Thus, it is only north of the Arctic circle in the fresh or brackish waters of fiords that reach far inland and into the Baltic Sea, where the coastal influence of sea ice becomes recognizable. Between coastal freezing of the Baltic and permanent sea ice around the North Pole another zone can be found without shore and sea ice contacts because of the warming of the Gulf Stream around northernmost Norway. In hard rocks, there are only fiords and skerries affected by contemporary coastal processes.

The next important morphodynamic boundary marks the equatorial limit of frost weathering which reaches 54°N and comprises parts of the German coasts and eastern coasts of the British Isles. Thick medium-sized algae are found in the rocky littoral (*Fucus*, *Laminariae*, and others) that extends to about 46°N in France. South of 44°N latitude in the Mediterranean, calcareous algae and encrusting vermetids create youthful organic rock formations, which sometimes grow together to form trottoirs and even miniature atolls, particularly south of 36°N latitude. South of 38°N latitude, the occurrence of beachrock increases as do coral reefs south of 30°N in the Red Sea; while south of 28°N, the northernmost outposts of mangroves can be found at the Sinai Peninsula.

The Mediterranean may be used to illustrate problems inherent in the field methods that are commonly applied. Davies (1972, p. 183) classified coasts within the zone of the "lower latitudes" and so did Valentin (1979), specifying them as "subtropical." Davies puts the Mediterranean as well as the whole of Australia into the same zone; whereas in Valentin's system, Australia is divided into a tropical and a subtropical zone. Both authors use the same major elements for classification, that is, the distribution of corals, mangroves, and beachrock. But it was only the existence of beachrock that caused the inclusion of the Mediterranean into the subtropical zone or into the zone of the lower latitudes, respectively. In the southern parts of the Australian coast beachrock does not exist, and the subtropical zone would be erroneously shown to extend beyond the beachrock distribution area. On the other hand, the boundary of the coral reef distribution area is practically the same as that of beachrock at the east coast; whereas, mangroves are found in some areas of the south coast far beyond the usual areas of the subtropics in the Northern Hemisphere. This is an important anomaly.

Along the east coast of North America, the zonality of presently active coastal processes is more closely spaced (Kelletat, 1989). The effects of sea ice on rock and beach coasts extend as far south as 46°N,

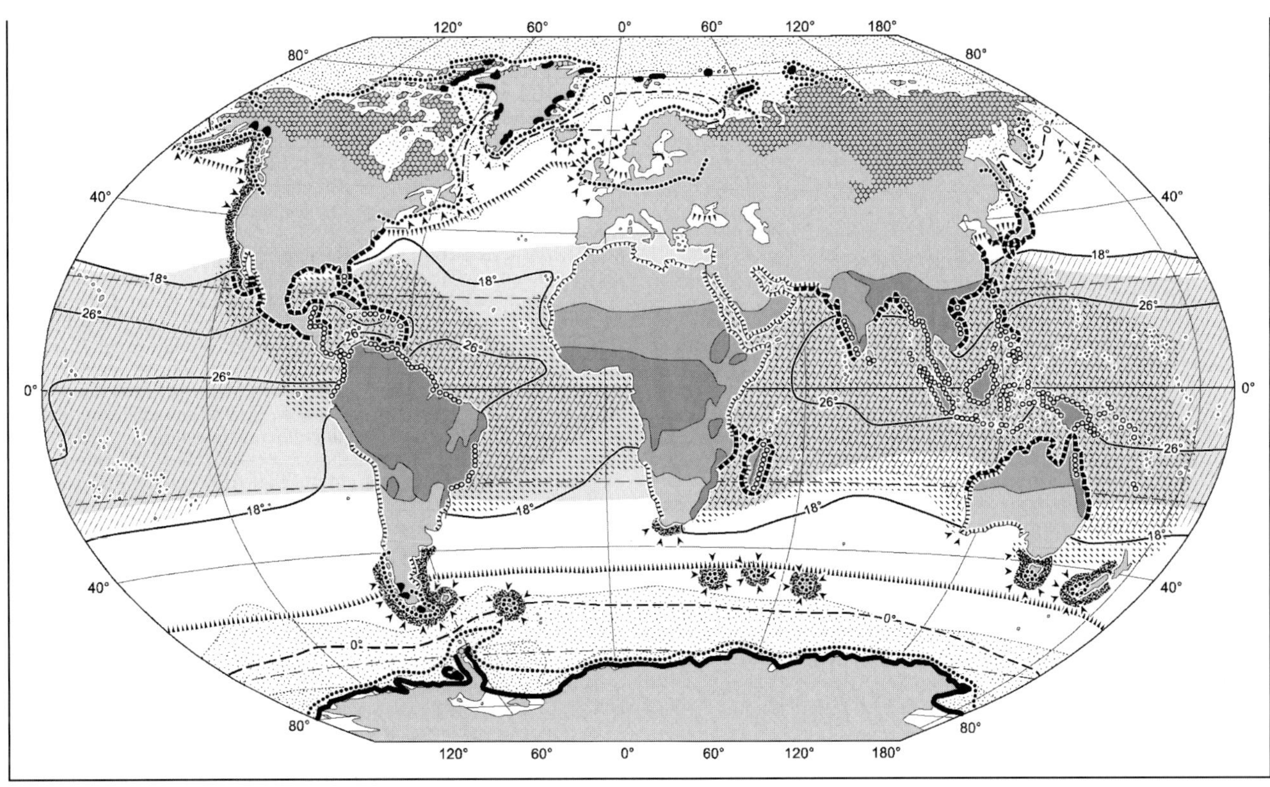

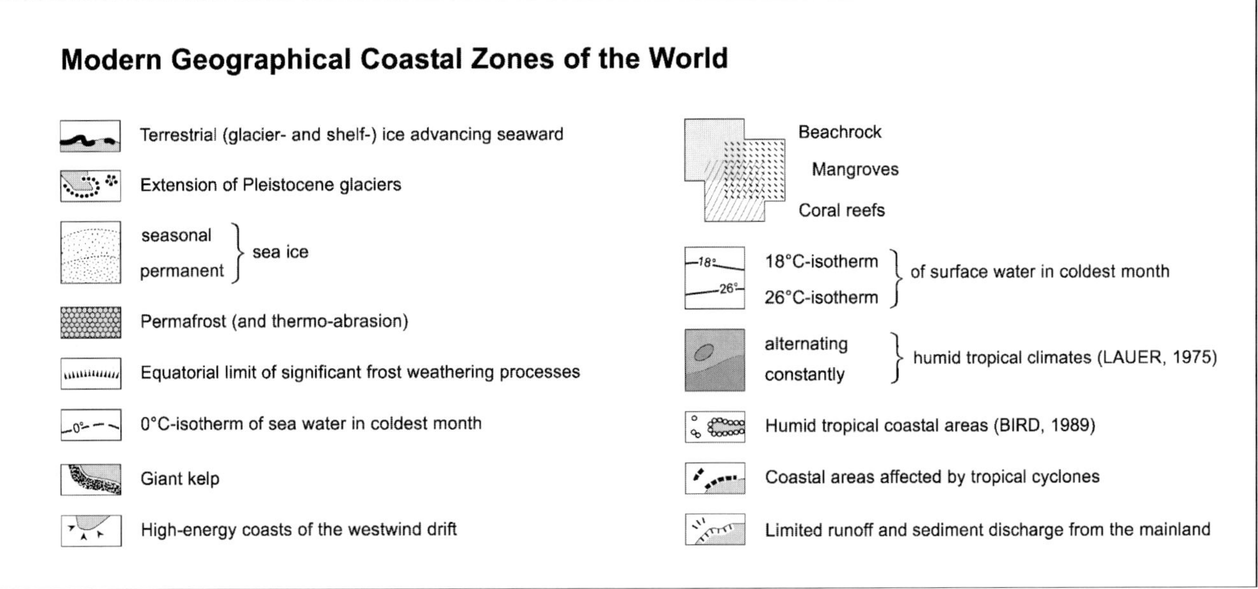

Figure G4 Modern geographical coastal zones of the world. (After Kelletat, 1995.)

and to around 40°N significant frost weathering processes can be observed. South of 35°N there are compound oyster reefs, that is organic rock formations, together with an increasing number of beaches with massive shell deposits. From 34°N to 30°N, hypersaline tidal flats occur in places almost completely free of vegetation (or only with microscopic plant material present), especially around the highest spring tide level. Latitude 28°N marks the poleward boundary of mangroves, 26°N that of the coral reefs. In total, there is only a separation of 20 latitudinal degrees between the coral reefs of the tropical zone and the sea ice effects of the subpolar zone, in comparison to at least 38° on the European side of the Atlantic Ocean. These findings may be contrasted with those on the coasts of eastern Australia and New Zealand; this region is the only one where the problems of zonality have been carefully studied (e.g., Kelletat and Seehof, 1986; Kelletat, 1988). Here, the morphological impact of both sea ice and frost on the coastal environment is missing. Giant kelp (*Durvillsea, Macrocystis*, and others) can be found as far north as 41°S, thus, effectively protecting the rocky coasts from heavy surf. Serpulids and sabellariae form solid organic substrates as far north as 38°S (eastern Australia) and 37°S (New Zealand) respectively, whereas oyster reefs reach as far south as 33°S (Australia) and 37°S (New Zealand).

Within tidal limits, hypersaline tidal flats almost free of vegetation are found as far south as 42°S in New Zealand and south of 28°S in eastern Australia. Latitude 38°S is the southern boundary of mangroves. Just south of the limit of massive coral reefs, that is 25°S, is the change from the tropical to the extratropical species in the rocky littoral.

The zones of giant subpolar algae and mangrove communities commonly reach within 3° latitude and sometimes even overlap. In contrast, the distance between them being 14° on the North American east coast and at least 18° in Europe. In Florida, the northern boundary of the mangrove zone is 2° north of that of the coral zone, whereas in Australia/New Zealand it is 13°, but in North Africa/Middle East (e.g., Sinai Peninsula) the coral reefs extend 2° farther north than the northernmost mangrove, *Avicennia marina* being the species in all areas. These differences cannot properly be explained yet, mainly because the extreme and mean values of both air and water climates differ significantly at these boundary lines.

Thus, both zonal and azonal features at the coasts are manifold, zonal aspects are mainly given by climatic and biogenic parameters (at least the latter partly depends on the first) and geodynamic, structural, petrographic, or sedimentological properties dominate the azonal aspects.

Most authors have tried to define a rather simple system of coastal zonality, either by separating tropical (including subtropical) and extratropical areas or by dividing the Earth's coasts into high-, mid-, and low-latitudes. In his synthesis of 1977 (pp. 181/182) Davies characterizes these zones as follows:

High-latitude coasts: They are frozen part of the year and seasonal ice is active on the beach. Characteristics are: low wave energy because of freezing and limited fetch, many pebbles at the beaches due to former glaciation and frost weathering, barriers developed well and widespread, dunes rather limited and tidal flats dominated by grassy vegetation. Cliffs are mainly formed by frost action, mass wasting is common, often due to permafrost decay. Further categories within high-latitude coasts may be determined by the duration of the sea-ice cover.

Mid-latitude coasts: They show high wave energy and frontal storm maxima, frequent cliffs in hard rock with quarrying and abrasion, sloping intertidal rock platforms, weaker developed barrier systems but with partly very extensive dune areas. Biogenic effects are believed to be insignificant, except for dunes and tidal flats. A differentiation of this broad zone may be made by the limit of Ice-Age glaciers deeply influencing the types of sedimentation along the coastline.

Low-latitude coasts: They are exposed to a relatively permanent approach of wind, waves, and swell and only reduced wave energy is active. Algae and corals contribute to coastal forming, and lithification of dunes and beaches may occur. The widespread coastal sediments are of fine-grain deposits. Dunes exist in limited areas only, and rocky coasts with horizontal rock platforms are restricted. Perpendicular cliffs are very rare (except in arid areas), salt weathering is typical as well as extensive tidal woodlands (mangroves). A differentiation should be made regarding the more humid and the more arid environments.

Although the spatial distribution of some important coastal phenomena is generally known, it is difficult or nearly impossible to define the zonal limits by means of climatic data (although radiation/climate are the main sources of the zonal patterns of the earth).

The 18°C-isotherm of surface water in the coldest month of the year is the thermal factor which sets limits for the coral reef zone. For other coastal features a definition is much more complicated; mangroves occur in very humid to extremely arid regions (e.g., precipitation nearly zero to thousands of millimeters per year) and within limits of low to high evaporation. Mean annual air temperatures at mangrove sites vary from about 28°C to nearly 15°C, those of the coldest month from more than 25°C to only 7°C, and coldest water temperatures from more than 25°C to nearly 6°C. As thermal limit for mangroves (*Avicennia* sp. or *Kandelia* sp. in Japan), several frost days with air temperatures of −4°/−5°C may be defined; whereas the number of months with mean temperatures below 20°C may reach from 12°C to zero (as well as the number of arid months).

The climatic data for beachrock are at least as heterogeneous as for the polar limits of mangroves (i.e., annual precipitation from nearly zero to several thousands of millimeters, low to high evaporation, zero to 12 arid months, mean annual air temperature between 15°C and more than 27°C, mean temperature of the coldest month down to 7°C, as well as water temperature, which may never reach 20°C during the year as in Namibia). When comparing the beachrock distribution on the islands of Oahu (Hawaii) and Madagascar, a clear relationship to the arid areas cannot be detected; microclimatic aridity may be more important than high water temperature. Both islands are situated in the core zone of beachrock distribution. In its marginal fringe like the Mediterranean Sea, beachrock in higher frequency and quality is clearly restricted to the warmer and more arid southern and eastern parts. It is only sporadic and poorly developed in its more humid and colder north and west.

In higher latitudes, the equatorial limits of giant kelp at the coasts show a similar and rather high differentiation (i.e., annual precipitation from around 300 to more than 1,300 mm, mean annual temperatures from 7.5°C to at least 17°C, and the coldest month even from 13°C down to −5°C). The limiting factor seems to be the air temperature, not exceeding a monthly average of 20°/21°C. The average water temperature may vary from below zero to more than 15°C. Twenty degrees centigrade seems to be lethal.

All features discussed here belong to the younger Holocene or even to contemporary decades, that is, an extremely short geologic period under a stable climatic regime. It is, however, difficult and till now partly impossible to commit to a satisfactory definition of limiting climatic conditions for most of the coastal features. In this respect, the related sciences are far from a consensus.

As a synthesis, a complex world map (Figure G4) may show the state-of-the-art in inductive coastal zonality.

For example, it is difficult to define zones and subzones by coastal forms because of their multi-temporal origin, that is from very different climates and, therefore, inherent with aspects of other zones. It may be prudent to restrict a consensus to the distribution of actual coastal processes (or those of the younger Holocene, when sea level was roughly at its present position). This will show a more homogeneous picture, comparable for all zones. A difficult, time-consuming problem that remains is: detecting and evaluating processes. Due to the present state of research, we usually attempt to infer the process itself from its result (i.e., we infer cementation from the occurrence of beachrock and horizontal rock platforms from the process of salt/frost weathering or water-layer leveling, etc.). A fundamental question is, whether we should define a zone or sub-zone by the dominant process (if one exists) or by a combination of more than one process. Does a zone or sub-zone need to show only adequate processes or their forms? Or should they be dominated spatially, or should a zone be characterized by a process even if its geomorphological effect and importance is minimal? Additional problems include insufficient or inadequate knowledge of coastal forms and processes worldwide as well as regionally and thematically.

In reference to coastal features, one is faced with problems that are dominated by azonal geodynamic, geological, structural, tectonic, petrographic, or sedimentological factors, and by seawater or surf conditions, by climatic parameters, or by biospheric processes, etc. Integration of all these categories into only one model of coastal zonality seems not to be acceptable.

Knowing that contemporary coasts were formed during the last 6,000 years, an extremely short timespan in geological and geomorphological times, the question arises as to whether the morphodynamic patterns now detected are representative of the last interglacial. Has the Holocene reached its climax conditions or is it still in process? An ancillary question regards zone shifts when climatic conditions changed. If it is true that coral reefs have their lethal winter temperature limit at 18°C and giant kelp at about 20°C in summer, a general change of 0.5 to 1.0°C (that may occur within decades) would transfer their zonal limits for hundreds of kilometers. Singular and azonal events like changing ocean currents or El Niño effects may re-inforce this tendency.

Being aware of the delicate balance of geomorphological zones and sub-zones during a given geological or climatic condition, it may be interesting to reconstruct those zones and sub-zones valid for former geological periods (which might have been longer lasting once they had reached their climax with certainty) and to learn from these results. On the other hand, by detecting coastal zones of former periods, it sometimes may be possible to deduce their environmental (i.e., paleoclimatological) parameters.

Dieter H. Kelletat

Bibliography

Büdel, J., 1977. *Klimageomorphologie*. Stuttgart: Springer.

Davies, J.L., 1964. A morphogenic approach to world's shorelines. *Zeitschrift für Geomorphologie, N.F.*, **8**: 127–142.

Davies, J.L., 1972. *Geographical Variation in Coastal Development*. Geomorphology Texts, 4; Edinburgh: Oliver and Boyd.

Davies, J.L., 1977. *Geographical Variation in Coastal Development*, 2nd edn, Geomorphology Texts, 4, Edinburgh, London: Longman.

Kelletat, D.H., 1988. Coastal zones of Australia and New Zealand, compared with those of the Northern Hemisphere. In 26th Congress International Geographical Union, Abstracts, I, p. 293, Sydney.

Kelletat, D.H., 1989. The question of "Zonality" in Coastal Geomorphology. With tentative application along the East Coast of the USA. *Journal of Coastal Research*, **5**(2): 329–344.

Kelletat, D.H., 1995. *Atlas of Coastal Geomorphology and Zonality*. Journal of Coastal Research, Special Issue, 13, Fort Lauderdale: Coastal Education Research Foundation.

Kelletat, D.H., and Seehof, G., 1986. Über die zonale Anordnung der gegenwärtigen Küstenformungsprozesse im Osten Australiens. *Berliner Geographische Studien*, **18**: 41–77.

Murphy, R.E., 1968. Landforms of the world. *Annals Association of American Geographers*, **58**, Map Suppl. 9.

Valentin, H., 1979: Ein System der zonalen Küstenmorphologie. *Zeitschrift für Geomorphologie, N.F.*, **23**: 113–131.

Cross-references

Beachrock
Changing Sea Levels
Classification of Coasts (see Holocene Coastal Geomorphology)
Coral Reefs
Glaciated Coasts
Holocene Coastal Geomorphology
Ice-Bordered Coasts
Mangroves, Geomorphology
Weathering in the Coastal Zone

GEOHYDRAULIC RESEARCH CENTERS

Introduction

Several geohydraulic research centers specialize in coastal science and engineering research. Among the largest are current or former government research facilities that have been partially or completely privatized. The centers have laboratory facilities used for both research and scale-model studies to aid in project design. In recent years, most centers have developed numerical models for project design. The following summarizes capabilities of the major geohydraulic research centers.

WL/Delft Hydraulics, Netherlands

WL/Delft Hydraulics located in Delft, the Netherlands, is one of the world's oldest and largest geohydraulic research centers. It was founded in 1927 and is one of the five large technological institutes in the Netherlands. These institutes were originally established by the Netherlands' government but are currently independent consulting and research organizations. WL/Delft Hydraulics provides advice and technical assistance on water-related projects that range from applied research and consultancy to multidisciplinary policy studies.

WL/Delft Hydraulics has a staff of about 400 with expertise in hydrology, hydraulics, morphology, water quality, and ecology. It addresses both construction and design issues related to the offshore, coasts, harbors, estuaries, rivers, and canals and also related environmental impact assessments. Research and development constitutes about 40% of the income of WL/Delft Hydraulics with 10% provided by the Netherlands' government for basic research, 5% for national research programs, 15% for joint research projects with the Rijkiswaterstaat of the Netherlands' government, and 5% related to research projects of the European Union. About 25% of income is for technical assistance in studies around the world.

WL/Delft Hydraulics has extensive coastal and hydraulic laboratory facilities with some designated since 1989 as large installations available for use by European Union members. Its Delta flume is 240 m long, 5 m wide, and 7 m deep and 1–5.5 m water depths are used in testing. The flume's wave generator can produce wave periods from 1 to 12.5 s and a 2.5 m regular wave height or a 1.9 m random wave height. Its Scheldegoot facility is a wave flume that is 0.55 m long, 1 m wide, and 1.2 m deep with a 1.2 m-high wave generator that can operate with water depths of 0.25–1.0 m and produce both periodic and random waves. The wave generator board is equipped for active wave absorption to avoid spurious long-period waves, and a second wave board is located at the flume's end for active wave absorption. The flume has a pump system allowing simulation of currents. The Vinje basin is a multidirectional wave basin with a variable length up to 60 m, a width of 26.4 m, and a maximum depth of 0.75 m. It has a directional spectral wave generator that can produce a maximum wave height of 0.3 m, a significant wave height of 0.15 m for a Pierson–Moskowitz spectrum, and wave periods between 0.4 and 3.0 s. The Schelde basin is a three-dimensional (3D) basin for studying wave attack on structures. The wave basin is 30 m long, 22.5 m wide, and 1.2 m deep with a maximum water depth of 1.0 m. The wave generator can produce waves with a maximum height of 0.4 m and a maximum significant height of 0.2 m.

WL/Delft Hydraulics has an oscillating water tunnel constructed to study sediment transport under controlled full-scale simulated wave conditions. The tunnel has a length of 14 m with an inner width of 0.3 m and an inner height of 1.1 m. Wave periods can be produced with periods between 5 and 15 s, maximum velocities of 2 m/s, and maximum acceleration of 2 m/s^2. WL/Delft Hydraulics also has a rotating annular flume with a mean diameter of 2.1 m, channel width of 0.2 m, and depth of 0.3 m. A circular lid on the surface drives flows, and the channel can be rotated in the opposite direction to minimize secondary currents at the bottom. Then too, there is a sand flume 98.7-m long including a 50 m flow section and 35 m-long testing section that is 1.5 m wide, and 1 m deep. The maximum depth with sand is 0.4–0.5 m. The maximum flow discharge is 0.8 m^3/s and maximum sediment flow is 0.3 m^3/h. In addition, WL/Delft Hydraulics has a dredging flume for testing of cutting and suction devices, trenching equipment, anchors, and jets. It has a variable length up to 50 m, width of 9 m (5.5 m research flume and 3.5 m settling basin), and a depth of 2.5 m.

The tidal flume at WL/Delft Hydraulics is 130 m long, 1 m wide, and has a depth of 0.1–0.9 m. Tides, water depth, salinity, and silt concentration can be controlled at the sea boundary, whereas freshwater discharge and silt concentration can be controlled at its river boundary.

Delft3D is a numerical modeling system developed by WL/Delft Hydraulics that can simulate 2D and 3D flows from tides, winds, density gradients, and waves. It has a central hydrodynamics module that drives ecological, sediment transport, wave, water quality, and morphological modules.

Coastal and Hydraulics Laboratory, United States

The Coastal and Hydraulics Laboratory (CHL) located in Vicksburg, Mississippi, USA, was formed in 1996 through a merger of the Hydraulics Laboratory (HL) founded in 1929 (HL was the original Waterways Experiment Station—WES) and the Coastal Engineering Research Center founded in 1930 (predecessor organization was the Beach Erosion Board). CHL is part of the seven-laboratory complex of the Corps of Engineers Research and Development Center (ERDC) that includes four laboratories at WES (Geotechnical and Structures, Environmental, and Information Technology Laboratories in addition to CHL) and the Construction Engineering Research Laboratory in Illinois, Cold Regions Research Laboratory in New Hampshire, and the Engineering Topographic Laboratory in Virginia. The ERDC has 2,100 employees with 1,200 at WES including 250 mostly engineers and scientists in CHL.

CHL has the largest coastal and hydraulic laboratory complex in the world with over 150,000 m^2 of laboratory facilities under roof. About a quarter of these facilities are used for coastal engineering research and scale modeling with the remainder used for hydraulic-engineering research and scale modeling. CHL has several wave flumes all with spectral-wave generators including the following: 105 m long, 2 m wide, 1.4 m water depth, 0.45 m wave height; 80 m long, 3.5 m wide, 1.4 m water depth, 0.45 m wave height; 65 m long, 3 m wide, 1 m water depth, 0.5 m wave height; 65 m long, 1.5 m wide, 1 m water depth, 0.5 m wave height; 60 m long, 0.6 m wide, 1 m water depth, 0.6 m wave height; 45 m long, 1 m wide, 0.6 m water depth, 0.2 m wave height; and 45 m long, 1 m wide, 0.6 m water depth, 0.2 m wave height. CHL's L-Shaped facility tests 3D stability of structures such as breakwaters and is 75 m long, 15 m wide at the wave generator, 25 m wide at the testing end; operates with water depths up to 1.5 m, and uses a wave generator with a wave board that is 4 m high and 15 m wide and produces waves 0.6 m in height.

CHL has about 30,000 m^2 of building space for 3D wave basins used for studying waves in harbors, movable-bed modeling, coastal processes, wave interaction with ships and other floating structures and other coastal-engineering applications. Large remote-controlled scale model ships (e.g., container ships, submarines, aircraft carriers) measuring about 7 m in length are employed to test ship maneuverability in directional-spectral seas with currents and wind. Long-wave harbor resonance is tested in a facility with a wetted area of over 4,000 m^2 and 14 spectral-wave generators each approximately 5 m long that can generate waves with periods from 1 to 10 s. CHL has two numerical ship and tow simulators for design of navigation channels that include a complete full-scale ship bridge and real-time visualization.

CHL's Long Shore Transport Facility is a unique 30-by-50 m (expandable to 80 m) and 1.4 m deep wave basin used to study longshore currents and sediment transport. Wave makers generate regular or irregular waves with heights up to 0.4 meters, periods of 1.0 to 3.5 seconds, and angles up to 20 degrees from shore normal. A longshore current re-circulation system has 20 independent upstream and

20 downstream flow channels each with a turbine pump that discharges up to 75 L/s. The system accurately controls cross-shore distribution of wave-driven longshore currents and re-circulates currents from downstream to upstream. Downstream flow channels have sediment traps that record sand accumulation and a dredging system that re-circulates sand collected in the traps to the upstream end.

CHL's new ESTEX Hyperflume facility used for conducting unsteady, nonuniform flow and sediment transport research has a 15 m by 18 m and 3 m deep basin connected to an 18 m by 107 m and 1.2 m deep basin and a parallel 3 m by 122 m and 1.2 m to 3 m deep basin.

CHL has software that it distributes at no cost, such as the Automated Coastal Engineering System (ACES). ACES includes well-known models such as the shoreline change model, GENESIS, and the cross-beach and dune model, SBEACH. CHL also has developed the Groundwater/Surface Water/ Watershed Modeling System (GMS, SMS, WMS), integrated modeling systems that are distributed by commercial vendors. These systems have about 9,000 users around the world. The CH3D model developed by CHL is a 3D modeling system for hydroenvironmental and sediment-transport problems.

DHI—Institute of Water and Environment, Denmark

The Danish Hydraulics Institute (DHI) located in Horsholm, Denmark, and VKI Institute for the Water Environment, merged January 1, 2000, to form DHI—Institute of Water and Environment. The VKI Institute for the Water Environment was an independent self-governing institute for water environment affiliated with the Danish Academy of Technical Service that specialized in environmental chemistry, environmental technology, and ecology. Annual funding by the Danish Agency for Development of Trade and Industry provided part of research and development at VKI. DHI specialized in management of the aquatic environment including hydrodynamic and environmental modeling, coastal hydraulics, and port and coastal structures. DHI—Institute of Water and Environment has 460 employees, 75 of whom are employed in international subsidiaries and branch offices, and 230 of whom were employed by the DHI prior to the merger.

DHI—Institute of Water and Environment operates shallow- and deep-water experimental basins for testing of wave agitation in harbors, stability of breakwaters, loads on offshore structures, response of floating structures, and wave hydrodynamics. One basin is 30 m by 20 m including a 3 m deep offshore basin (with 12 m deep pit) for 3D waves, currents, and wind. This basin is equipped with a 60-flap directional-spectral wave generator. Three other basins have dimensions of 32 by-30 m, 0.45 m deep; 30 by-30 m, 0.75 m deep; and 62 by-30 m, 0.45 m deep. It has three flumes, one 35 by-1.8 m; 1.8 m deep; 35 by-5.5 m, 0.8 m deep; and 28 by-0.74 m, 1.2 m deep. Other facilities also are available for use at the nearby Technical University of Denmark.

DHI—Institute of Water and Environment markets a wide variety of 1-D, 2-D, and 3-D numerical models including models of current and water levels to simulate tides, storms, cyclones, and tsunamis. Its 3D ocean circulation model (MIKE3) is used to simulate the coastal ocean when stratification is important. Other models are used for wave hindcasting, advection—dispersion, sediment transport, morphology, and water quality. In collaboration with the Technical University of Denmark, DHI developed the noncohesive sediment transport model STP that can simulate sand transport due to waves and currents including surf-zone transport. The model LITPACK is used to evaluate changes of a coast due to human activities. DHI's model, MIKE21, solves the Boussinesq equations and is used in port planning. DHI studies ship maneuvering using a numerical ship simulator that is driven by numerical hydrodynamic models.

HR Wallingford, United Kingdom

HR Wallingford located in Wallingford, United Kingdom, was founded in 1947 as the United Kingdom's Government's Hydraulics Research Organization, but was privatized in 1982. It has large-scale experimental facilities used for basic and applied research with 29,000 m^2 of covered laboratory space. Its UK Coastal Research Facility consists of a directional spectral wave generator in a basin with tidal and longshore current generating capabilities. It is 36-by-22 m with a depth range of 0.3–0.8 m. Its Flood Channel Facility is a large flume for investigating sediment transport in rivers and flow interactions with flood plains. This facility is 56-by-10 meters with a maximum discharge of 1.1 m^3/s. A Sloping Sediment Duct is used for investigating flow and sediment transport on flat to very steep slopes (up to 33 degrees). The working section is 6-by-0.6 m and is 0.25 m deep with a maximum flow of 0.15 cubic meters per second. The High Discharge Flume is for studying flow and sediment transport or equipment testing and has a working section of 25-by-2.4 m with depth up to 1.2 m and a maximum flow of 1.2 cubic meters per second. The Tilting Flume is for measurements of flow and sediment transport or surface runoff testing with a working section of 25-by-2.4 m, a depth of up to 0.6 m, and with a maximum tilt of 1 in 30 and maximum flow of 0.25 cubic meters per second. HR Wallingford is a partner with Electricité de France in the development and application of TELEMAC, a finite-element modeling system for flows, sediment transport, and water quality in two and three dimensions. Its model SeaWorks calculates wave and beach processes.

Canadian Hydraulics Centre

The Canadian Hydraulics Centre (CHC), located in Ottawa, Canada, is a government agency and a unit of the National Research Council, Canada's leading scientific research organization.

The CHC has extensive laboratory facilities in an 8,400 square-meter building. Its multidirectional wave basin is 50-m long, 30-m wide, and 3 m deep with a 15 m deep central pit with a 6-m diameter. The directional spectral wave generator can produce waves up to 0.7 m. CHC's shallow wave basin is 47 m long, 30 m wide, and 0.9 m deep. Wave generators produce long-crested waves up to 0.2 m high. Currents and tides also can be generated in the facility. Its coastal wave basin is 63-m long, 14 m wide, and 1.5 m deep and is equipped with a single wave generator that is 14 m long and can produce irregular long-crested seas. CHC's large wave flume is 97 m long, 2 m wide, and 2.7 m deep, and its wave generator can produce waves up to 1.1 m in height. A smaller wave flume is 63-m long, 1.22 m wide, and 1.22 m deep, its wave generator can produce waves with heights up to 0.25 m, and it has a pumping system that can produce flows up to 0.2 cubic meters per second.

CHC markets a variety of coastal-process software. Its wave-agitation software can be used for wave-climatology prediction in the open ocean, wave transformation into shallow water, waves in harbors, and waves produced by ships. It has 1D, 2D, and 3D hydrodynamic models for water levels and currents in lakes, rivers, estuaries, and coastal areas. Its sediment modeling includes modeling of beach evolution and erosion and deposition of cohesive and noncohesive sediments. CHC's environmental modeling can be used to predict the fate of oil or chemical spills.

Flanders Hydraulics, Belgium

Flanders Hydraulics, Antwerp, Belgium, is a research institute of the Belgium government that was established in 1933. Flanders Hydraulics advises the government on harbor and waterway design, coastal engineering, and storm surges and other floods. It also provides consultancy to private organizations.

Flanders Hydraulics has two wave flumes for breakwater stability and beach-profile- change research that are equipped with spectral wave generators with active wave absorption to eliminate wave reflections. A wave basin is used to study wave transformation and equilibrium studies. Its spectral wave generator can be moved to generate waves with angles of up to 30 degrees from normal, and longshore currents can be generated by pumping water in a closed circuit. Flanders Hydraulics has a circulating mud-erosion water tunnel used to study the erosion and deposition of cohesive sediments. The tunnel is closed with a measuring section 3-m long, 0.4 m wide, and 0.4 m high. The measuring section is followed by an introductory section 20 m long and 0.4-by-0.4 m square and an outlet section 3 m long.

Flanders Hydraulics operates a ship handling and maneuvering simulator used to both train pilots as well as to design hydraulic structures, harbor entrances, access channels, harbor turning basins, and other aspects of navigation channel design. Flanders Hydraulics also has facilities to study hydraulic structures that block upstream fish migration.

Kajima Research Institute, Japan

The Kajima Research Institute (KRI) was established in 1949 and is located in Tobitakyu, Chofu, Tokyo, Japan. It is a research institute of Kajima Corporation, a private-sector construction company. KRI has a laboratory facility built in 1975 that is in a 2,870 m^2 building. KRI has a wave basin that is 58 m long, 20 m wide, and 1.6 m deep, and its directional spectral wave generator can produce waves up to 0.6 m in height with wave periods from 0.5 to 5 s, and it can actively absorb reflected waves. The basin is equipped to generate tides. The KRI has a large wave flume 62 m long, 2 m wide, and 2 m deep, and its wave generator can produce waves up to 0.6 m in height and it can absorb reflected waves. It has a medium-size wave flume that is 60 m long, 0.7 m wide, and 1.5 m deep, and its wave generator can produce spectral waves with maximum heights of 0.5 m and absorb reflected waves. There is a 30 m section of glass windows. Currents with maximum discharge of 0.2 m^3/s can be generated in the flume.

Then too, KRI has developed a beach evolution predictive model, COASTLINE, that can evaluate waves and wave-induced currents, the effects of coastal development on sediment transport, wave-induced scour around a structure, and the efficacy of beach nourishment and shore protection.

Manly Hydraulics Laboratory, Australia

The Manly Hydraulics Laboratory is located in Manly Vale, a suburb of Sydney, Australia. It has 2,230 square meters of covered experimental space and 3,530 square meters of open-air experimental space. It has a 35-by-1 m and 1.4 m deep wave flume with adjustable floor and a random wave generator, an 11-by- 0.6 m and 0.8 m deep tilting flume with glass panels, and a 30-by-30 m and 1 m deep wave basin with a random wave generator.

Hydraulics and Maritime Research Center, Ireland

The Hydraulics and Maritime Research Center established in 1979 is a semi-autonomous unit within University College Cork, Ireland, and a part of the Coastal Zone Institute. Its Ocean Wave Basin is 25 m long, 18 m wide, and 1 m deep. There is a 9 m square pit in the center of the facility with a depth of 2.5 m. The basin has a directional spectral wave generator that consists of 40 flaps along the 18 m side of the basin. It has an active absorption system for reflected waves.

University of Delaware, United States

The University of Delaware's Center for Applied Coastal Research located in Newark, Delaware, USA, has an Ocean Engineering Laboratory that is among the largest university laboratories in the United States with a two-storey 30-by-38 m facility for laboratory experiments in coastal and ocean processes. Its directional wave basin is 20-by-20 m and 1.1 m deep and has a directional spectral wave generator with 34-flap wave paddles. Its precision wave tank is 33-by-0.6 m and 0.76 m deep with approximately 60% of its length with glass walls. Its wave generator can produce regular and irregular waves and water can be recirculated in the tank and produce a current of 30 cm/s. The facility has a spiral wave basin 8.5 m in diameter for small-scale coastal process studies where an 'infinite' beach is required. A recirculating Armstrong flume is 0.4 m wide and 0.6 m deep is used to generate hydraulic jumps. The Center also has a Sand Beach Wave Tank, a towing and wave tank with a cross section of 2-by-1 m and a piston wave generator.

Scripps Institution of Oceanography, United States

The Scripps Institution of Oceanography located in La Jolla, California, USA, has a hydraulics laboratory with facilities to address coastal and oceanographic research. The facility is in a building 31-by-24 m. Its wind-wave tank is 44.5-by-2.39 m and 2.44 m deep and has glass windows, the largest being 5.5 m long. The wave generator can produce a maximum wave height of 0.6 m. An open-circuit wind tunnel produces variable wind velocities ranging from 0 to 16 m per second. The laboratory has a wave basin with dimensions of 15.2-by-18.3 m and a water depth of 0.61 m. A unidirectional spectral wave generator can produce waves with heights as great as 0.22 m. A glass-walled wave tank is 33 m long, 0.5 m wide, and 0.5 m deep. The wave generator produces waves up to 0.25 m in height and there is a reversible flow system that can pump 0.041 cubic meters per second. A granular fluid mechanics test facility is 12.2 m long, 6.1 m wide, with a maximum depth of 3 m. An adjacent basin is 10 m long with a fluidizing channel that is 1 m^2 in cross section with a high-flow slurry pump used for sand transport. A stratified flow tank with a length of 30 m and a glass-walled test section 16 m in length, 1.1 m in width, and a depth of 1.1 m is used to study stratified flows with flow velocity uniform or in two separate layers flowing at rates from 0 to 1.3 m per second. The laboratory has an oscillatory flow tunnel tank with an overall length of 16 m and a test section 0.39-by-0.4 m. Maximum oscillatory peak-to-peak water particle displacement is 2 m. The maximum sinusoidal velocity is 1.8 m per second. A steady bi-directional flow up to 0.2 m per second can be superimposed on the oscillatory flow. There are sediment traps at both ends, and the tank can be tilted up to 9.1 degrees to simulate a sloping beach.

Queen's University, Canada

The Coastal Engineering Research Laboratory of Queen's University is located in Kingston, Ontario, Canada. The Laboratory is in an 1,860 square-meter building in the west campus of Queen's University. The Laboratory has a coastal model basin that is 21-by-21 m and equipped with a spectral wave generator. Tides and currents can be reproduced in the basin and a sand bed can be used in experiments. The Laboratory has an oscillating water tunnel for research on boundary layers, forces on structures, and other problems. It has three 61-m long wave flumes, two of which are equipped with spectral wave generators and one with a monochromatic wave generator. The Laboratory has a 1D model for simulating long-term beach morphology and is developing a quasi-3D model for study of shore processes.

Catalonia University of Technology

The Maritime Experimental and Research Flume was built in 1992 at the Maritime Engineering Laboratory of the Catalonia University of Technology for experimentation in maritime harbors and coastal engineering. The flume is 100 m long, 3 m wide, and 5 m deep. The facility has a spectral wave generator, a variety of instrumentation, and a bottom profiler for morphodynamic tests. The generator can produce maximum regular wave heights of 1.6 m. The facility is a part of the European Union large infrastructure program that makes facilities available for member-country use.

James R. Houston

Cross-references

Coastal Engineering (See Shore Protection Structures and Navigation Structures)
Computer Simulation Models (See Coastal Modeling and Simulation)
Engineering Applications of Coastal Geomorphology
Instrumentation (See Beach and Nearshore Instrumentation)
Numerical Modeling
Physical Models
Sediment Transport (See Cross-Shore Sediment Transport and Longshore Sediment Transport)
Surf Modeling
Wave Climate
Waves

GEOTEXTILE APPLICATIONS

The technological term "geotextile" has now come to stay in the context of coastal engineering. Perhaps, the original term "filter fabric" is more expressive, but it has fallen into comparative disuse. Geotextiles are really fabric materials used in a "geo," that is an earth-related, engineering context. These fabrics may be based on natural fibers or man-made synthetic fibers. The latter type of geotextiles are referred to as "geosynthetics." Geosynthetic materials are generally nonbiodegradable, whereas natural geotextiles are often biodegradable. Biodegradability would be an advantage or a disadvantage, depending upon the user needs and use contexts.

Most of the man-made geotextiles (i.e., geosynthetics) are made from polymers of various chemical types. Some common polymers used are polyethylene (PE), polypropylene (PP), polyvinylchloride (PVC), nylons, polystyrene (PS), and polyesters. Natural geotextiles are made from natural fibers, which are also polymers, largely based on cellulose and lignin.

Historical development

Pioneers of coastal engineering introduced the use of geosynthetics in the 1950s in the United States. According to Bruun (2000), some of the first coastal geotextile research in the United States was carried out at the University of Florida, Gainesville, where plastic filters were tested. Work on the use of nonwoven fabrics for soil reinforcement was done at the Rhone-Poulence textiles in France. Early development work on geotextiles was also conducted in the Netherlands, Britain, Germany, and the United States. Currently, the geotextile industry is quite active in research and development in the United States, through the establishment of the

Geosynthetic Research Institute at Drexel University. Additional details on the history of geotextiles can be obtained from Heerten and Kohlhasse (2000).

Material types and chemistry

Geotextiles are only a sub-family in the larger family of "Geosynthetics." Other sub-families include geogrids, geonets, geomembranes, geosynthetic clay liners (GCL), geopipes, and geocomposites. Geotextiles are woven or nonwoven fabrics. Geogrids are plastics formed into open grid-like configuration. Geonets are formed by continuous extrusion of sets of polymeric ribs which when opened give rise to a net-like configuration. Geomembranes are very thin sheets of polymeric materials. GCL's are thin layers of bentonite clay sandwiched between geotextiles or geonets. Geopipes are simply buried plastic pipes and represent one of the oldest members of the family of geosynthetics. On the other hand, geocomposites are the newest member of the family and are composed of either geotextiles and geonets, or geotextiles and geogrids, or geogrids and geomembranes, etc.

The material used for fabricating geotextiles is invariably some kind of polymer. A "polymer" is a giant molecule of high molecular weight, obtained by polymerizing one or more smaller units called "monomers." For example, the simple chemical "ethylene" is a monomer. When it is polymerized under suitable conditions, we get "polyethylene" (PE, or polythene). Similarly, the polymerization of simple monomers such as propylene, vinylchloride, or styrene produces PP, PVC, or PS, respectively. In some cases, two monomers are involved in polymerization. Thus, nylons are polyamide polymers formed by the polymerization of diamines and dicarboxylic acids (e.g., hexamethylene diamine and adipic acid). Polyesters such as polyethylene terephthalate (PET) are formed by the polymerization of ethylene glycol and terephthalic acid or derivatives or analogues of these monomers. Polymers commonly used in the geotextile industry include PE, PP, PVC, PET, PS, PA (nylons), and cellulose. Cellulose is a naturally occurring polymer based on glucose as the monomer.

Another recent development is the use of natural geotextiles, which include fabrics formed using coir, the natural fiber from the husk of coconuts. These geotextiles have exceptional strength due to the high content of a cellulose lignin polymer in them. Besides, they are naturally degradable, which may be attractive in certain applications such as natural stream bank restoration. Other natural fibers include jute, sisal, mixed coir-jute, and mixed sisal-jute.

Geotextile applications

Barrett (1966), in his well-known and classic paper, discusses the use of fabric materials as reinforcements behind precast concrete seawalls, and under precast erosion control blocks. At present there are several application domains for geotexiles, as discussed below (Heerten and Kohlhasse, 2000): (1) filters in erosion control structures (revetments, coastal dikes, etc); (2) separators in the foundation of groins and breakwaters, (3) fabric forms for sand filled bags or tubes as construction elements for groins, dikes, and dunes; (4) flexible scour protection mats at different offshore or coastal structures; (5) reinforcement in dredged material sites; (6) membranes for use in landfill caps; (7) bags for disposing dredged material from navigation channels; and (8) dikes for river training and coastal structures.

Koerner (1998) classifies the applications into six broad categories: (1) *Separation*—is the usage of geotextile to provide physical separation between two material types, while retaining their individual integrity and functions; (2) *Reinforcement*—involves using the geotextile to improve on the inherent strength of the system; (3) *Filtration*—is the use of the geotextile to retain the soil behind it while providing an outlet for dissipation of hydraulic forces; (4) *Drainage*—involves the concept of using the geotextile to provide sufficient fluid movement through it without much loss of the soil behind it; (5) *Containment*—is the concept of providing isolation between two surfaces with the aid of an impervious geotextile; and (6) *Combined uses*—which is the use of the geotextile for any combination of the above listed uses.

Design considerations

Important general considerations include the following: (1) *Function*—the type of use the geotextile should provide; (2) *Specifications*—the engineering properties that would be needed to satisfy that function; (3) *Availability*—the supply versus demand of that particular type of geotextile in the region; and (4) *Cost*—the price of obtaining and installing the geotextile in place. For small projects with financial constraints, design by cost and availability is used, which tries to obtain an optimal balance between funds available, area to be covered with geotextile, and unit price of the geotextile.

Key design properties of interest include the following: (1) *Physical properties*—includes specific gravity, mass per unit area of the geotextile, thickness, and stiffness; (2) *Mechanical properties*—includes compressibility, tensile strength, seam strength, burst strength, and elongation; (3) *Hydraulic properties*—includes porosity, percent openings in the geotextile, an equivalent opening size, permeability, and soil retention; (4) *Endurance properties*—includes potential for damage during installation, long-term strength loss, abrasion potential, and clogging potential; and (5) *Degradation properties*—includes potential for degradation from sunlight, temperature, oxidation, biological action, and hydrolysis.

Various international organizations provide standards for estimating the above properties, with the American Society for Testing and Materials (ASTM, 2000) providing the lead. Other leading organizations include the International Standards Organization (ISO), Permanent International Association of Navigation Congresses (PIANC), Geosynthetic Research Institute (GSI) and British Standards Institution (BSI).

Koerner (1998) discusses key aspects of design considerations for the many functional uses of geotextiles, a brief summary of which is presented here

- *Separation design criteria*: Separation function is primal to all geotextiles. Indeed many other functions depend, in their turn, on separation. The use of geotextiles as separators is illustrated by their placement between a soil subgrade (beneath) and a stone base (above). The important criteria here are burst resistance, tensile strength requirements, puncture resistance, and impact resistance.
- *Soil reinforcement design criteria*: Reinforcement of soil is another cardinal application of geotextiles. Seawall reinforcement, reinforcement of embankments, foundation reinforcement and *in situ* slope reinforcement are the important specific cases. Key properties include tensile strength, puncture resistance, and impact resistance.
- *Filtration design criteria*: Geotextile filters are used behind retention walls, around underdrains, beneath erosion control structures and as silt fences. Adequate fluid permeability and soil retention are the main requisites.
- *Drainage design criteria*: The two main design variations here are gravity drainage design and pressure drainage design. Various manufacturing textiles such as woven silt film, woven nonfilament, nonwoven heatbonded, nonwoven resin-bonded, nonwoven needle-punched, and hybrid systems, are used. Nonwoven needle-punched geotextiles are suited for drainage of fine-grained soil masses such as clays and silt. A high permeability and high soil retention are essential to this application.
- *Containment design criteria*: Application of geotextiles in containment include coastal landfills and other waste barriers. Key properties include low fluid permeability and high soil retention.

Summary

The use of geotexiles in coastal areas has been practiced for a very long time for shore protection needs. However, advances in manufacturing and marketing has led to innovative uses of geotextiles in the coastal zone in recent years. These include the use of sand filled geobags as the core of dunes, sand filled geobag breakwaters and reefs, geobag spur dikes, and deep ocean placement of contaminated sediments in geobags. The future of the geotextile seems limited only by human imagination.

Ram K. Mohan and C.G.R. Nair

Bibliography

ASTM, 2000. *Standard Method for Testing Soils and Rocks*. Philadelphia: American Society for Testing and Materials.

Barrett, R.J., 1966. Use of plastic filters in coastal structures. *Proceedings of the 16th International Conference On Coastal Engineering*. New York: American Society of Civil Engineers, pp. 1048–1067.

Brunn, P., 2000. *Port Engineering*. Houston: Gulf Publishing.

Heerten, G., and Kohlhase, S., 2000. Geotextiles in Coastal and Harbor Engineering. In Bruun, P. (ed). *Port Engineering*. Houston: Gulf Publishing, pp. 512–530.

Koerner, R., 1998. *Designing with Geosynthetics*. Upper Saddle River: Prentice Hall.

Cross-references

Bioengineered Shore Protection
Capping of Contaminated Coastal Areas
History, Coastal Geomorphology
Navigation Structures
Shore Protection Structures

GLACIATED COASTS

Introduction

There are a wide variety of glaciated coasts, usually with both erosional and depositional features. Coasts may be actively glaciated systems, such as Antarctica, southeastern Alaska, and Greenland, or formerly glaciated regions such as Scandinavia, Scotland, northeastern Canada and New England, the Pacific Northwest, and southern Chile. Formerly glaciated, or paraglacial coasts (FitzGerald and van Heteren, 1999) are modified by modern coastal processes, and by changes during deglaciation such as permafrost action, outwash, isostatic rebound, and proglacial lacustrine and marine deposition and erosion. Glaciated coasts characteristically have moderate to high relief, with abundant bedrock outcrop. They occur in cold temperate to polar regions where modern glaciers and ice sheets are stable, or where perennial ice was stable down to sea level in glacial climates. Formerly glaciated coasts are found as far as 40° south and north as a result of the Wisconsin (Würm) Last Glacial Maximum (LGM) 20–22 ka ^{14}C (Denton and Hughes, 1981; Andersen and Borns, 1994). Earlier Quaternary glaciations as well as Precambrian and Paleozoic glaciations may have extended even farther from the poles. At present, outlet glaciers from ice sheets meet the coast in Antarctica, Svalbard (Figure G5) and Greenland, extending to a maximum of 63°N (southern Greenland) and 63°S (Antarctic Peninsula). Valley and piedmont glaciers extend to the sea in Alaska (Brady Glacier, 58°N), Baffin Island (71°N), and Chile (47°S). This variability in latitudes demonstrates that relief, precipitation, ocean currents, and other climatic factors must be considered in evaluating changes in extent of ice sheets.

Study of the former extent of ice sheets and valley glaciers on coasts such as Antarctica, New Zealand, Chile, Alaska, and northeastern Canada provides an important proxy for climate change. Of particular interest are the relationships of events in the Northern and Southern Hemisphere (Moreno *et al.*, 2001), and to potential driving forces for climate change such as the Milankovitch theory of variations in insolation due to Earth orbital parameters. Other theories of climate forcing include dust, atmospheric CO_2, ocean currents, solar variability, internal ice-sheet dynamics, and deglacial freshwater flooding into the oceans (e.g., Broecker *et al.*, 1968; Berger, 1988; MacAyeal, 1993; Denton and Hendy, 1994; Broecker, 1997; Denton and Hall, 2000; Clark *et al.*, 2001).

Glaciated coast erosional features

Some characteristic erosional landforms include glacial troughs and fjords. Glacial troughs are immense channels cut by ice streams across the continental shelf in formerly glaciated areas, such as the Laurentian Channel off the Gulf of St. Lawrence and the Northeast Channel off the Gulf of Maine. Other troughs representing greater extent of present ice sheets are found around Antarctic, such as in the Ross Sea and Amundsen Sea. Besides the major ice-stream erosional features, ice sheets also channel flow around resistant rock outcrops and into less resistant valleys, creating characteristic streamlined hills and linear valleys. Erosional resistance is controlled by rock type (igneous and metamorphic versus sedimentary), grain size, weathering, spacing, and frequency of fractures and bedding planes, and structural orientation relative to ice flow. Small-scale streamlined outcrops, roches moutonnées, may form islands (Figure G6). These and less regular small islands are known as skerries. Large-scale streamlining of hills and valleys creates a rolling coastline of complex peninsulas and embayments, known as fjard coasts, as in Maine, Sweden, and other formerly glaciated regions (e.g., Andersen and Borns, 1994).

Figure G5 Coastal outlet glacier with terminal moraine at present sea level, Spitsbergen (June, 1976, courtesy of J.M. Demarest).

The classic coastal glacial erosion feature is the fjord. Fjords are steep-walled embayments hundreds of meters deep, tens of kilometers long, and kilometers wide. Their mouths are often restricted by a lip composed of bedrock and/or till. This lip may lead to a salinity density stratification in the fjord, and also acts as a dam that contains abundant proglacial and glaciomarine sediments within the deeper basins. Fjords are formed by valley glaciers, where streaming flow incises more efficiently vertically than laterally, and are subsequently drowned by glacial retreat and rising post-glacial sea level. Sedimentation in fjords is strongly controlled by geometry of the glacial erosion, degree of continuing glacial inflow, and by circulation effects of stratification. Commonly there are coarse deltaic sediments and morainal banks at the head, with fine muddy basins in the remainder of the system. Rapid sediment influx creates instability. Slumps, subaqueous fans, and channels created by sediment-gravity flows are common (Prior *et al.*, 1981; McCann and Kostaschuk, 1987).

Glacial depositional features

Glaciers deposit sediments at their bases, sides, and front. Landforms created by these deposits may form important coastlines on their own, such as Cape Cod, Martha's Vineyard, and Nantucket, or may anchor littoral drift systems, such as the southern coast of Long Island. Moraines and outwash plains core the outer and inner portions of the well known "flexed arm" of Cape Cod (Oldale, 1992). Reworking of till and stratified outwash from these sources provides abundant sand and gravel to the modern beach and barrier systems (Fisher, 1987; Leatherman, 1987). Drumlins are less continuous than moraines, often scattered in fields. Johnson (1919) describes a model for evolution of Boston Harbor and the Massachusetts Bay coast as erosion of drumlins supplied coarse sediment for barriers, until the drumlins were consumed. The coastline system then jumped back to the succeeding sets of drumlins and their sediment sources. This coastal evolution of drumlin fields is also found in Clew Bay, western Ireland. In settings where bedrock topography is more important for forming embayments, but glacial sediment sources such as drumlins, moraines, eskers, and kames

Figure G6 Roche moutonnée forming an island approximately 15 m long, in Pemaquid Pond, Maine, USA. Last Wisconsin ice flow was from right to left (175°T). Note the characteristic smooth, striated and polished stoss side and jagged, plucked lee side, formed on a resistant pegmatite dike intruding schists (D.F. Belknap photo, 7/08/2000).

provide the sediment for beaches and barriers, a similar saltating coastline mechanism may prevail. Boyd *et al.* (1987) and Forbes *et al.* (1991) show that the southern coast of Nova Scotia has retreated in this manner during Holocene transgression. Newfoundland has outstanding examples of drumlins and moraines sourcing baymouth barriers (Forbes *et al.*, 1995) (Figure G7). Pocket beaches and extended coastlines near these proximal glacial deposits are dominated by boulders, gravel, and sand (Kelley *et al.*, 2001). They may form pure gravel beaches, or mixed sand and gravel systems.

Sediment is also brought to the coast as glacial outwash, in valley-train deposits, in deltas and subaqueous outwash fans. These are generally sand and finer gravel, but may include outsized clasts carried by floods and ice rafts. The southern coast of Iceland contains numerous sandurs (outwash fans) that provide the basis for barrier and beach systems (Nummedal *et al.*, 1987). The southern coast of Alaska contains a rich variety of glaciated coastline types, including outwash plains, direct deposition from the Malaspina piedmont glacier into the littoral zone, and fjords with retreating valley glaciers (Molnia, 1980; Powell, 1983). The finest material is glacial rock flour: silt and clay particles carried primarily in suspension. Abundant rock flour causes rapid infilling of lagoons and embayments (up to several meters/yr) in proglacial and glaciomarine environments. The rapid accumulation of proglacial and glaciomarine fine sediments creates a concentric drape over underlying bedrock or previously deposited glacial sediments, resulting in a distinctive signature in seismic reflection profiles (Piper *et al.*, 1983; Belknap and Shipp, 1991). Later slower sedimentation and reworking produces ponded, flat-lying accumulations. Uplift of the glaciated coastline exposes glaciomarine mud, deltas and outwash fans, as well as moraines and drumlins to wave attack. Much of the coast of Maine, Puget Sound, the Baltic Sea, and similar locations consist of eroding bluffs of these materials. Clay bluffs in particular may be highly unstable, tending to fail as retrogressive slumps.

Glaciomarine deltas form where abundant coarse sediment causes progradation into the sea. On glaciated coasts, especially during deglacial phases, isostatic rebound, and relative sea-level change must be in balance for short periods in order to allow deltas, and other coastline indicators, to stabilize. Glaciomarine deltas are commonly of the classic Gilbert delta type (Figure G8). Topsets are braided stream deposits at the top of the section. These streams incised and reworked underlying units while the delta was prograding. Foresets are avalanche deposits at the terminus of the traction conveyor, the mouth of the stream. Sometimes it is possible to find the "roll-over" point where the process changed from traction to avalanching, allowing confident reconstruction of coastline position and relative sea level at that point (Figure G8). The third component is bottomsets, the suspension and sediment-gravity flow deposits found basinward of the foresets.

Glaciomarine environments

There are two major types of glaciomarine environments, tidewater glaciers and ice shelves (Powell, 1984). Tidewater glaciers have ice fronts in seawater in restricted embayments, fjords in particular, and produce smaller icebergs. Tidewater glaciers terminate in the ocean at an ice cliff, which is a calving front at or near grounding line (the position where the ice floats out of contact with the bottom) (Figure G9), either in the open ocean (Pfirman and Solheim, 1989) or in protected embayments (Powell, 1983). Many Alaskan coastal glaciers, such as in Glacier Bay, are presently retreating rapidly through this process of calving of icebergs. There are several lithofacies found in this environment seaward of the grounding line. Morainal banks are composed of diamicton, gravel, and marine sand and mud. Stratified sand and gravel subaqueous outwash fans interfinger with till and distal glaciomarine mud. These fans may be narrowly restricted in fjords, or broad composites under continental ice-sheet margins (Ashley *et al.*, 1991). Fans also are commonly channeled through traction under density underflows, and may be sites of active slumping. Active ice push can deform and override morainal banks and stratified deposits. Iceberg drop and dump deposits are a common admixture to this and more distal environments. Laminated sand and mud are characteristic of basins farther from the grounding line. These are produced from density overflows, interflows, and underflows. In the marine environment, however, the latter are rare, localized close to the source because of the lower density of fresh meltwater, unless charged with very high concentrations of sediment (2–3 gm/L). Laminated deposits are easily characterized as glaciomarine when they contain abundant ice-rafted detritus.

Ice shelves terminate in the ocean, with the ice edge distant from the grounding line. Ice shelves are broad and flat, and are restricted primarily to Antarctica today. Ice shelves calve immense tabular icebergs hundreds to thousands of square kilometers in area. Polar ice shelves are cold and do not produce abundant sediment. The Ross Sea and Amundsen Sea, for example, are blanketed by marine biogenic sediments draped over glacial deposits left by Holocene grounding-line retreat (Anderson *et al.*, 1983; Kellogg and Kellogg, 1988; Jacobs, 1989). Ice shelves that were abundantly productive of sediment were found during deglaciation after the Wisconsin (Würm) LGM, in particular where fed by ice streams such as the Scotian Shelf and Gulf of Maine (Bacchus *et al.*, 1997; Schnitker *et al.*, 2001).

Glacioisostatic rebound

The release of the immense weight of ice sheets causes rapid isostatic readjustment. As a consequence, raised beaches can be found preserved around many formerly glaciated regions such as Hudson Bay (Fairbridge and Hillaire-Marcel, 1977), Newfoundland (Liverman, 1994), Spitsbergen (Boulton *et al.*, 1982), and Scandinavia (Andersen and Borns, 1994). In Maine, raised beaches formed on the outer edges of glaciomarine deltas (Figure G10) and were raised to elevations 60–130 m above present sea level (Thompson, 1982). Understanding isostasy through study of deformed glacial coastlines has important implications for deep Earth structures such as viscosity of the mantle, thickness and elastic/plastic properties of the lithosphere. These reconstructions also help constrain poorly understood properties of former ice sheets, in particular their thickness.

Figure G7 Drumlins sourcing gravel barriers, Holyrood Bay, Newfoundland (D.F. Belknap photo 5/27/2000).

Figure G8 Gravel pit exposure into a glaciomarine highstand delta, Lamoine, Maine, USA. Top of exposure is approximately 70 m above present sea level (D.F. Belknap photo 10/12/1982). Topset–foreset contact approximates contemporaneous sea level. The "rollover" shows the preserved delta front in transition from near-horizontal topsets to foresets inclined near the angle of repose (30–33°).

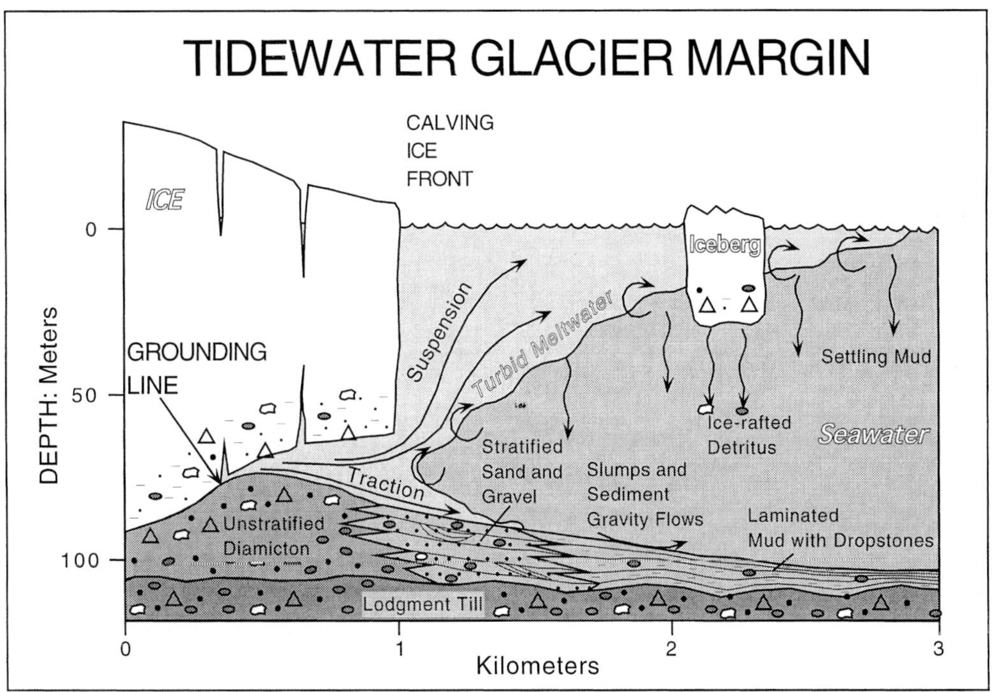

Figure G9 Model of a tidewater glacier calving margin, with morainal bank, stratified subaqueous fan, and distal mud and dropstone facies, after Thompson (1982, figure 3, p. 215; and Pfirman and Solheim, 1989).

Figure G10 Highstand shoreline formed ca. 13 ka in sand and gravel at the front of a glaciomarine delta on East Base, 75 m above present sea level (D.F. Belknap photo 10/12/1982).

Summary

Glaciated coasts are heterogeneous in morphology and sediment type. They occur in higher latitudes today, but the influence of former episodes of glaciation imposes a strong control on coastal evolution through preserved erosional and depositional landforms, and the variety of sediment types. One association that typifies the glaciated coastline is coarse gravel beaches adjacent to muddy embayments. Lack of long-term stability is another important characteristic, driven by isostatic changes, laterally variable sources of fine and coarse sediment, and a complex, deranged landscape upon which coastal environments form and migrate. Glaciated coasts are not only a widespread littoral environment in the modern and during glacial times, but study of changing distributions and types can supply critical proxy information for reconstruction of global climate change.

Daniel F. Belknap

Bibliography

Andersen, B.G., and Borns, H.W., Jr., 1994. *The Ice Age World*. Oslo: Scandinavian University Press.

Anderson, J.B., Brake, C., Domack, E., Myers, N., and Wright, R., 1983. Development of a polar glacial-marine sedimentation model from Antarctic Quaternary deposits and glaciological information. In Molnia, B.F. (ed.), *Glacial-Marine Sedimentation*. New York: Plenum Press, pp. 233–264.

Ashley, G.M., Boothroyd, J.C., and Borns, H.W., Jr., 1991. Sedimentology of late Pleistocene (Laurentide) deglacial-phase deposits, eastern Maine; an example of a temperate marine grounded ice-sheet margin. In Anderson, J.B., and Ashley, G.M. (eds.), *Glacial Marine Sedimentation: Paleoclimatic Significance*. Boulder, Colorado: Geological Society of America Special Paper 261, pp. 107–125.

Bacchus, T.S., and Belknap, D.F., 1997. Glacigenic features and shelf basin stratigraphy of the eastern Gulf of Maine. In Davis T.A., Bell, T., Cooper, A.K., Josenhans, H., Polyak, L., Solheim, A., Stoker, M.S., and Stravers, J.A. (eds.), *Glaciated Continental Margins: An Atlas of Acoustic Images*. New York: Chapman-Hall Pub. Co., pp. 213–216.

Belknap, D.F., and Shipp, R.C., 1991. Seismic stratigraphy of glacial-marine units, Maine inner shelf: In Anderson, J.B., and Ashley, G.M. (eds.), *Glacial-Marine Sedimentation; Paleoclimatic Significance*. Boulder, Colorado: Geological Society of America Special Paper 261, pp. 137–157.

Berger, A.L., 1988. Milankovitch theory and climate. *Reviews of Geophysics*, **26**: 624–657.

Boulton, G.S., Baldwin, C.T., Peacock, J.D., McCabe, A.M., Miller, G., Jarvis, J., Horsefield, B., Worsley, P., Eyles, N., Chroston, P.N., Day, T.E., Gibbard, P., Hare, P.E., and von Brunn, V., 1982. A glacio-isostatic facies model and amino acid stratigraphy for late Quaternary events in Spitsbergen and the Arctic. *Nature*, **298**: 437–441.

Boyd, R., Bowen, A.J., and Hall, R.K., 1987. An evolutionary model for transgressive sedimentation on the eastern shore of Nova Scotia. In FitzGerald, D.M., and Rosen, P.S. (eds.), *Glaciated Coasts*. San Diego, CA: Academic Press, pp. 87–114.

Broecker, W.S., 1997. Thermohaline circulation, the Achilles heel of our climate system: will man-made CO_2 upset the current balance? *Science*, **278**: 1582–1588.

Broecker, W.S., Thurber, D.L., Goddard, J., Ku, T.L., Matthews, R.K., and Mesolella, K.J., 1968. Milankovitch hypothesis supported by precise dating of coral reefs and deep-sea sediments. *Science*, **159**: 297–300.

Clark, P.U., Marshall, S.J., Clarke, G.H.C., Hostetler, S.W., Licciardi, J.M., and Teller, J.T., 2001. Freshwater forcing of abrupt climate change during the last glaciation. *Science*, **293**: 283–287.

Denton, G.H., and Hall, B.L. (eds.), 2000. Glacial and paleoclimatic history of the Ross ice drainage system of Antarctica. *Geografiska Annaler. Series A: Physical Geography*, **82**:(2–3), p. 293

Denton, G.H., and Hendy, C.H., 1994. Younger Dryas age advance of Franz Josef Glacier in the Southern Alps of New Zealand. *Science*, **264**: 1434–1437.

Denton, G.H., and Hughes, T.J., (eds.), 1981. *The Last Great Ice Sheets*. New York: John Wiley and Sons.

Fairbridge, R.W., and Hillaire-Marcel, C., 1977. An 8,000-year paleo-climatic record of the 'Double-Hale' 45-yr solar cycle. *Nature*, **268**: 413–416.

Fisher, J.J., 1987. Shoreline development of the glacial Cape Cod coastline. In FitzGerald, D.M., and Rosen, P.S. (eds.), *Glaciated Coasts*. San Diego, CA: Academic Press, pp. 279–305.

FitzGerald, D.M., and van Heteren, S., 1999. Classification of paraglacial barrier systems: coastal New England, USA. *Sedimentology*, **46**: 1083–1108.

Forbes, D.L., Taylor, R.B., Orford, J.D., Carter, R.W.G., and Shaw, J., 1991. Gravel-barrier migration and overstepping. *Marine Geology*, **97**: 305–313.

Forbes, D.L., Orford, J.D., Carter, R.W.G., Shaw, J., and Jennings, S.C., 1995. Morphodynamic evolution, self-organisation, and instability of coarse-clastic barriers on paraglacial coasts. *Marine Geology*, **126**: 63–85.

Jacobs, S.S., 1989. Marine controls on modern sedimentation on the Antarctic continental shelf. *Marine Geology*, **85**: 121–153.

Johnson, D.W., 1919. *Shore Processes and Shoreline Development*. Facsimile edn. 1972, New York, Hafner Pub. Co., p. 584.

Kelley, J.T., Belknap, D.F., FitzGerald, D.M., and Boothroyd, J.C., 2001. Quaternary sea-level change and coastal evolution in eastern Maine. In West, D.P., Jr., and Bailey, R.H. (eds.), *Guidebook for Geologic Field Trips in New England 2001*. Boston, MA: Annual Meeting of the Geological Society of America, pp. A1–A31.

Kellogg, T.B., and Kellogg, D.E., 1988. Antarctic cryogenic sediments: biotic and inorganic facies of ice shelf and marine-based ice sheet environments. *Palaeogeography, Palaeoclimatology, Palaeoecology*, **67**: 51–74.

Leatherman, S.P., 1987. Reworking of glacial outwash sediments along outer Cape Cod: development of Provincetown Spit. In FitzGerald, D.M., and Rosen, P.S. (eds.), *Glaciated Coasts*. San Diego, CA: Academic Press, pp. 307–325.

Liverman, D.G.E., 1994. Relative sea-level history and isostatic rebound in Newfoundland, Canada. *Boreas*, **23**: 217–230.

MacAyeal, D.R., 1993. Binge/purge oscillations of the Laurentide Ice Sheet as a cause of the North Atlantic's Heinrich Events. *Paleoceanography*, **8**: 775–784.

McCann, S.B., and Kostaschuk, R.A., 1987. Fjord sedimentation in northern British Columbia. In FitzGerald, D.M., and Rosen, P.S. (eds.), *Glaciated Coasts*. San Diego, CA: Academic Press, pp. 33–49.

Molnia, B.F., 1980. Twentieth-century history of the Gulf of Alaska coastline Cape Suckling to Cape Spencer. In Field, M.E., Bouma, A.H., Colburh, I.P., Douglas, R.G., and Ingle, J.C. (eds.), *Quaternary Depositional Environments of the Pacific Coast*. Pacific Coast Paleogeography Symposium 4, Pacific Section, Society of Economic Paleontologists and Mineralogists, Los Angeles, pp. 121–141.

Moreno, P.I., Jacobson, G.L., Jr., Lowell, T.V., and Denton, G.H., 2001. Interhemispheric climate links revealed by a late-glacial cooling episode in southern Chile. *Nature*, **409**: 804–808.

Nummedal, D., Hine, A.C., and Boothroyd, J.C., 1987. Holocene evolution of the south-central coast of Iceland. In FitzGerald, D.M., and Rosen, P.S. (eds.), *Glaciated Coasts*. San Diego, CA: Academic Press, pp. 115–150.

Oldale, R.N., 1992. *Cape Cod and the Islands: the Geologic Story*. East Orleans, MA: Parnassus Imprints.

Pfirman, S.L., and Solheim, A., 1989. Subglacial meltwater discharge in the open-marine tidewater glacier environment: observations from Nordaustlandet, Svalbard Archipelago. *Marine Geology*, **86**: 265–281.

Piper, D.J.W., Letson, J.R.J., Delure, A.M., and Barrie, C.Q., 1983. Sediment accumulation in low-sedimentation, wave-dominated, glaciated inlets. *Sedimentary Geology*, **36**: 195–215.

Prior, D.B., Wiseman, W.J., and Bryant, W.R., 1981. Submarine chutes on the slopes of fjord deltas. *Nature*, **290**: 326–328.

Powell, R.D., 1983. Glacial-marine sedimentation processes and lithofacies of temperate tidewater glaciers, Glacier Bay, Alaska. In Molnia, B.F. (ed.), *Glacial-Marine Sedimentation*. New York: Plenum Press, pp. 185–232.

Powell, R.D., 1984, Glacimarine processes and inductive lithofacies modelling of ice shelf and tidewater glacier sediments based on Quaternary examples. *Marine Geology*, **57**: 1–52.

Schnitker, D., Belknap, D.F., Bacchus, T.S., Friez, J.K., Lusardi, B.A., and Popek, D.M., 2001, Deglaciation of the Gulf of Maine. In Weddle, T.K., and Retelle, M.J. (eds.), *Deglacial History and Relative Sea-Level Changes, Northern New England and Adjacent Canada*. Boulder, Colorado: Geological Society of America Paper 351, p. 9–34.

Thompson, W.B., 1982. Recession of the late Wisconsinan ice sheet in coastal Maine. In Larson, G.J., and Stone, B.D. (eds.), *Late Wisconsinan Glaciation of New England*. Dubuque, Iowa: Kendall/Hunt Pub. Co., pp. 211–228.

Cross-references

Boulder Barricades
Changing Sea Levels
Climate Patterns in the Coastal Zone
Gravel Barriers
Ice-Bordered Coasts
Paraglacial Coasts

GLOBAL POSITIONING SYSTEMS

Developed and operated by the US Department of Defense (DOD), the Global Positioning System (GPS) is a satellite-based navigation system that allows the determination of precise location anywhere on earth. The major components of the GPS consist of the space segment, the control segment, and the user segment (Leick, 1994). Coastal scientists may use GPS techniques to collect field measurements for later implementation into a *Geographic Information Systems* (GIS) database.

The space segment includes a constellation of as many as 27 high altitude satellites complete with computers and atomic clocks. As of September 2000 there were 27 satellites (GPS World, September 2000). A full constellation consists of 24 satellites; the others serve as back-ups. The satellites orbit the earth in 12 h from an altitude of 20,286 km (12,600 miles).

The control center monitors the satellite system from five stations around the world: Hawaii, Ascension Island, Diego Garcia, Kwajalein, and Colorado Springs. A control center compiles an ephemeris that includes advance calculations of the satellite's orbit; forecasts of the ionospheric and atmospheric conditions are also uploaded to each satellite's computer. A control center additionally monitors the functioning of each satellite and can readjust the clock and locational offsets.

The user segment consists of portable receivers and the community of more than four million GPS users worldwide (Clinton, 2000). Although GPS receivers vary in their components and thus their capabilities, any receiver will determine location as X, Y, or Z coordinates, typically expressed as longitude, latitude, and altitude.

Satellite ranging

Both the receiver and the satellite generate an identical pseudo-random code that is transmitted simultaneously and allows the determination of signal travel time from the satellites. There are two types of pseudo-random code: C/A code and P code. Civilians use the lower frequency C/A code and the P code is reserved for the DOD (Hurn, 1989). When the receiver picks up a transmission from one satellite, it computes the distance by multiplying the velocity (speed of light) by travel time. The geometric principle of trilateration allows one to accurately measure a position with just three distance measurements (Figure. G11). Calculating the distance from the first satellite narrows the possible location to the intersection of a sphere (with a radius of the distance) and the sphere of the earth. A second distance measurement identifies a second intersecting sphere that further reduces the possible location. A third distance measurement produces a sphere that intersects the others at two possible locations: the points where the third sphere intersects the circle of the first two spheres. In three-dimensional space, trilateration requires a fourth measurement to positively identify one location and its altitude. One of the two locations identified with three measurements is often so obviously wrong, however, that a fourth measurement is not needed (Hurn, 1989).

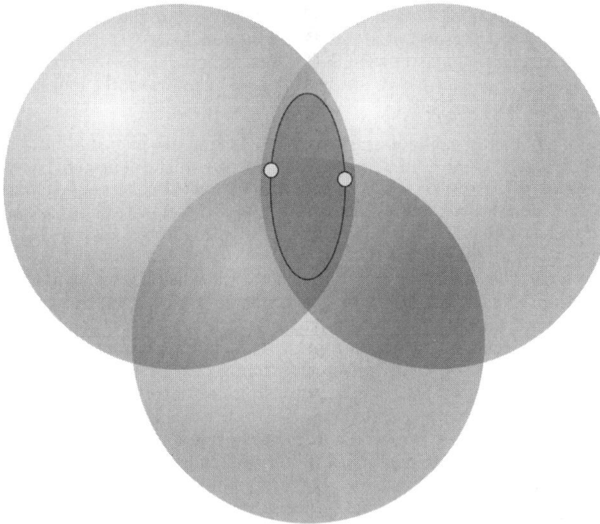

Figure G11 Trilateration with three distance measurements.

Sources of error

There are several types of error that can influence the accuracy of a typical GPS measurement by as much as 100 m (Steede-Terry, 2000). If the task requires accuracy better than that produced by standard GPS methods, differential GPS (DGPS) methods may provide accuracy to within a few centimeters.

Although the satellites carry atomic clocks that keep very accurate time, minimal clock error is still possible and will influence travel time calculations. Additionally, the less accurate receiver clocks may introduce error as well.

A second type of error occurs because radio signals only travel at the speed of light in a vacuum; the ionosphere and atmosphere can therefore slow down the GPS signals. A control center relies on the fact that the GPS satellites transmit signals on two frequencies thereby allowing for dual frequency error correction (Hurn, 1989). Modeling of ionospheric and atmospheric conditions provides another method of correction.

The assumption that the signals will travel in a straight line creates a third potential for error. Multi-path error is caused when a signal is reflected or bounced before reaching the receiver (Leick, 1994). GPS receivers are capable of processing the signals and determining that the earliest signal to arrive is the best choice.

A fourth type of error, geometric dilution of precision (GDOP), is related to the angle between the satellite and the receiver. Since a wider angle between the satellite and the receiver allows for more precise measurement, good quality receivers analyze the positions of the satellites and choose the best four satellites (Hurn, 1989).

Selective availability (SA), the DOD's reduction of the C/A code accuracy for military defense, is an intentional degradation of orbital and atomic clock data. In May of 2000, after determining that permitting an accurate C/A code had little effect on national security, President Clinton eliminated the practice of SA (Clinton, 2000). With SA turned off, GPS measurements, depending on the type of receiver, may provide a horizontal accuracy of 10–15 m (Steede-Terry, 2000). Vertical accuracy is more difficult because satellite signals are not detected below the visible horizon of the earth.

Differential GPS

Differential GPS provides a simple solution that corrects for many sources of error and can provide accuracy from 5 m to a few centimeters, depending on the type of receiver. DGPS relies on two receivers: one receiver is stationary (the base or reference station) and transmits from a known location; the second receiver is the roving device that collects locational data in the field. The receivers both transmit or receive pseudo-random code from the satellites but, since the base station location is already known, it uses that information to compute the travel time that each satellite signal should take. It then measures the actual travel time (if the satellite signal traveled in a straight line with no interference) and that difference becomes the error correction factor. Base stations are often established by government agencies, such as the US Coast Guard or the Federal Aeronautics Administration, and the receivers continually compute correction factors as the satellites orbit overhead. The correction factors are encoded in a computer file that users may download at a later time for post-processing. If a user requires "real time" correction because they are navigating, for example, some receivers are designed to receive radio signals or corrections directly. Real time correction is typically less accurate than post-processing because the correction occurs by averaging a number of readings in a few seconds, versus the potential for averaging several minutes of data collected by a stationary, roving, receiver (Steede-Terry, 2000).

Future

As GPS technology improves and becomes less expensive, the number of potential applications will continue to increase. The US government has estimated that by 2003 the GPS user community will double to more than eight million users and the market will increase to more than US $16 billion (Clinton, 2000). Despite its continuing importance for determining accurate location in navigation, surveying and recreation, the potential for GPS to supply very accurate time, that is, Universal Coordinated Time, will further it's usefulness.

Shawn M. Boeser

Bibliography

Clinton, W. President of the United States, 2000. *Improving the Civilian Global Positioning System (GPS)*. Washington DC: The White House.

GPS World., Sept. 2000. Satellite Update.

Hurn, J., 1989. *GPS: A Guide to the Next Utility*. Sunnydale, CA: Trimble Navigation Ltd.

Leick, A., 1994. *GPS Satellite Surveying*. New York: John Wiley & Sons Inc.

Steede-Terry, K., 2000. *Integrating GIS and the Global Positioning System*. Redlands, CA: ESRI Press.

Cross-references

Airborne Laser Terrain Mapping and Light Detection and Ranging
Erosion: Historical Analysis and Forecasting
Geographic Information Systems
Instrumentation (See Beach and Nearshore Instrumentation)
Mapping Shores and Coastal Terrain
Monitoring, Coastal Geomorphology
Nearshore Geomorphological Mapping
Photogrammetry
RADARSAT-2
Remote Sensing of Coastal Environments
Synthetic Aperture Radar Systems

GLOBAL VULNERABILITY ANALYSIS

Introduction

Climate can have great influence on our lives as shown by the great damage and loss of life in events such as Hurricane Mitch in Central America, the 1999 cyclone in Orissa, India, and the flooding in Mozambique in 2000. Such events could be intensified by climate change, making this issue a major challenge for the 21st century. This widespread concern has generated a global policy response including the Intergovernmental Panel on Climate Change (IPCC) and the United Nations Framework Convention on Climate Change (UNFCCC), whose signatories are committed, among other things, to "avoid dangerous climate change." The key policy issue is the relative merits of reducing greenhouse gas emissions (usually termed mitigation) and/or adapting to the impacts of climate change, with a mixed response being most realistic.

A major consequence of climate change is global sea-level rise that could cause serious impacts around the world's coast. In the context of coastal zones, the goal of vulnerability analysis for sea-level rise (and other coastal implications of climate change) is to assess the potential impacts on coastal populations and the related protection systems and coastal resources, including the ability to adapt to these changes. A range of methods for such analyses has been developed and these have been extensively applied at the national and sub-national level (e.g., IPCC CZMS, 1992; Klein and Nicholls, 1999). These varied studies are often based on different assumptions and scenarios, so they are difficult to synthesize to the larger scales most pertinent to the policy debate outlined above. Therefore, there have also been efforts at vulnerability analysis at the regional and global scale.

The first global vulnerability analysis was completed in 1992 and evaluated: (1) increased flood risk and potential response costs; (2) losses of coastal wetlands; and (3) changes in rice production, assuming a 1-m global rise in sea level (Hoozemans *et al.*, 1992). This was rapidly updated with a second edition (Hoozemans *et al.*, 1993). Here, only results for this second edition are discussed and henceforth this analysis is termed GVA1. The IPCC Common Methodology (IPCC CZMS, 1992) was followed throughout. These results influenced the United Nations Conference on Environment and Development (Rio de Janeiro, Brazil, 1992) (IPCC CZMS, 1992), and the World Coast Conference (Noordwijk, the Netherlands, 1993) (WCC '93, 1994), and are included in the IPCC Second Assessment Report (Bijlsma *et al.*, 1996). Subsequently, Nicholls *et al.* (1999) made a major improvement relative to GVA1 for the flood and wetland analysis. This was upgraded to a dynamic form, including improved impact algorithms, which can consider variable sea-level rise scenarios, the implications of growing coastal populations, and rising living standards (this analysis is henceforth termed GVA2). It is widely cited within the IPCC Third Assessment Report and has also contributed to a series of impact studies based on common climate and socioeconomic scenarios (e.g., Parry and Livermore, 1999).

This following description first outlines the key concepts of global vulnerability assessment. It then presents some selected methods of analysis, considering coastal population and flood risk, adaptation to increased flood risk and wetland loss. The results together with their validation and use are then considered, including the differences between the methods used. Lastly, possible developments in the near-future are considered.

Concepts, constraints, and approaches

In the present context, *vulnerability* is defined as the degree of capability to cope with the consequences of climate change and sea-level rise (Klein and Nicholls, 1999). As such, the concept of vulnerability comprises:

- the susceptibility of a coastal area to the physical and ecological changes imposed by sea-level rise;
- the potential impacts of these natural system changes on the socio-economic system;
- the capacity to cope with the impacts, including the possibilities to prevent or reduce impacts via adaptation measures. (This last factor is often termed "adaptive capacity").

However, there are four main barriers to the comprehensive vulnerability assessment, irrespective of the scale of assessment (Nicholls and Mimura, 1998):

- incomplete knowledge of the relevant processes affected by sea-level rise and their interactions;
- insufficient data on existing conditions;
- difficulty in developing the local and regional scenarios of future change, including climate change;
- the lack of appropriate analytical methodologies for some impacts.

For the global assessments, the availability of consistent and complete global databases on (1) the distribution, density and present status of the impacted resources and (2) the nature and probability of the impacting hazardous events was a major constraint. Coverage at a global scale was sometimes incomplete due to regional gaps, or only coarse resolution data was available.

All these problems necessitate careful consideration of what can realistically and usefully be assessed. After considering these limitations, global assessments have evaluated fairly simple parameters to date, with the main focus on impact assessment rather than adaptation assessment. GVA1 was limited to an assessment of the potential impacts on three distinct elements in the coastal zone and one possible adaptive response:

- *risk to population (and adaptation potential)*—population at risk of flooding and also potential protection upgrade costs;
- *ecosystem loss*—coastal wetlands of international importance at loss;
- *agricultural impacts*—rice production at change (in south, southeast, and east Asia only).

GVA2 refined the methods and results for the first two elements, but the underlying data is the same. Only these elements will be considered here.

In order to assess the vulnerability of a coastal zone to sea-level rise we need uniform procedures to compare and to integrate national and regional studies. To determine impacts in measurable and objective terms, the concept of *values at risk*, *values at loss*, and *values at change* were used. The concept of risk is defined as the consequence of natural hazardous events multiplied by the probability of the occurrence of these events, *excluding* the system response. The concepts of loss and of change are defined as the consequences of natural hazardous events multiplied by the probability of the occurrence of these events, *including* the system response. The "risk"-approach is considered appropriate to assess the consequences of sea-level rise on the probability of episodic hazards such as flood impacts on the coastal population and economy. System response is excluded because it is difficult to predict how flood events may change the behavior of the population in the long term. The concepts of "loss" and "change" are appropriate for impacts on ecosystems and agricultural production, respectively, because it is the long-term consequences of sea-level rise that are the most important factors influencing the magnitude of the impacts.

To examine the capacity to cope with flooding, a set of protection measures was developed with these impact studies to enable the comparison of the impacts of sea-level rise "with- and without increased protection measures." While one may be susceptible to increased flooding, if one can easily afford to upgrade defenses, there is little cause for concern and overall vulnerability is low.

Lastly, the scale of assessment needs to be considered. Much of the available data for these studies was only available at national resolution and was of uncertain quality. Therefore, some of the underlying data as well as the assumptions about physical processes, physical and socio-economic boundary conditions, limit the accuracy of the national-scale results. This is especially true since the last major revision of the underlying databases was in 1993. However, the errors appear to be unbiased so regional and global estimates are expected to be more robust (Nicholls, 1995). Therefore, all the results are aggregated to 20 regions and the global scale, and this is the output of the analysis that has been utilized elsewhere. The national data and results are available in Hoozemans *et al.* (1993) for reference purposes, although the limits of the accuracy and validity of these detailed results should be born in mind.

Some of the methods are now considered in more detail.

Coastal population and flood risk

Storm surges are temporary extreme sea levels caused by unusual meteorological conditions. The resulting coastal flooding is a major issue damaging livelihoods, causing great distress and in the extreme, loss of life. As many as two million people may have been killed by storm surges in the last 200 years, mainly in south Asia (Nicholls *et al.*, 1995). Sea-level rise will raise the mean water level, and hence allow a given surge to flood to greater depths and penetrate further inland. Changes in storm tracks, frequencies, and intensity would also change surge characteristics, but in the absence of credible scenarios, this factor is considered constant in time within this analysis.

The concept of risk is considered appropriate in the context of assessing the consequences of sea-level rise on flooding for the population in the coastal zone. As rising sea levels intensify flood hazards, some human response might be expected. However, this response is not considered as such a prediction was considered unrealistic given that it involves human choice (ranging from migration to increased protection). Therefore, a high Population at Risk (PaR) indicates the need for some kind of a response. Possible protection costs against flooding are considered in the next section.

Based on the definition of risk, PaR is defined as the product of the population density in a certain risk zone and the probability of a hazardous flooding event in this risk zone. The resulting number is interpreted as the average number of people expected to be subject to flooding events per time unit (/year). Hence, PaR has also been termed "average annual people flooded" (Nicholls *et al.*, 1999). The "risk"-value is able to reflect changes in:

- the population living in the risk zone (coastal flood plain);
- the flood frequency due to sea-level rise;
- the protection standard of defenses.

As a general approach, the following steps were undertaken in GVA1 to determine the PaR for the various scenarios:

- Assessment of the height of the maximum flood level theoretically threatening the low-lying coastal zone, taking into account present and possible future extreme *hydraulic and geophysical conditions*.
- Determination of the *flood-prone area* and calculation of the area contained between the coastline and the maximum flood level.
- Assessment of the *present state of protection* against flooding.
- Determination of the coastal *population densities* for the present and future state.
- Determination of the PaR with and without measures, with and without sea-level rise, and for conditions in the years, 1990 and 2020.

To estimate global PaR with a reasonable accuracy, the world's coast was divided into 192 coastal zones based on the 181 coastal countries (as existed in the early 1990s). For each of the 192 coastal zones, a database was developed and an identical stepwise calculation scheme was followed to arrive at a coastal zone-specific PaR-number for each scenario (e.g., Figure G12). The database contained the following elements:

(1) the maximum area of the coastal flood plain after sea-level rise;
(2) the flood exceedance curve for storm surges from a 1 in 1 year event to a 1 in 1,000 year event;
(3) the average coastal population density in 1990;
(4) the occurrence or absence of subsidence; and
(5) the standard of coastal protection.

Three fundamental assumptions are that (1) the coastal flood plain has a constant slope; (2) the population is distributed uniformly across the coastal zone; and (3) if a sea defense is exceeded by a surge, the entire area behind the sea defense is flooded.

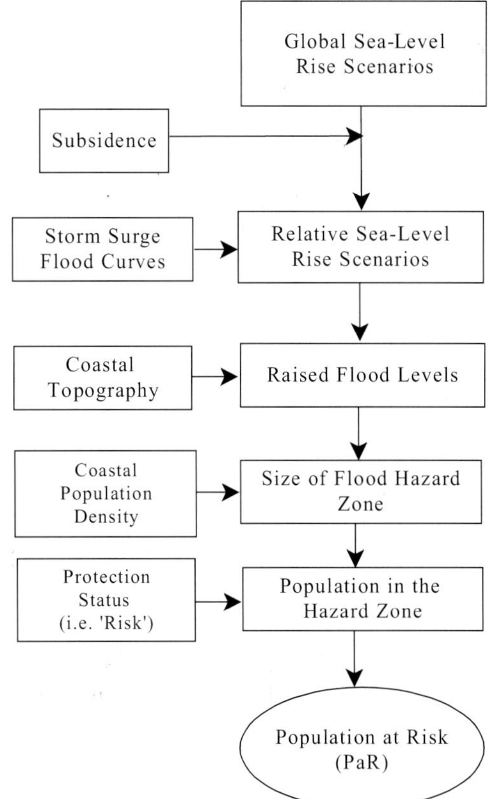

Figure G12 The flood model algorithm as used in GVA2 (modified from Nicholls, 2000).

Calculations proceed as shown in Figure G12. Estimates of four storm surge elevations (1 in 1 year, 1 in 10 years, 1 in 100 years, and 1 in 1,000 years) are raised by the relative sea-level rise scenario and converted to the corresponding land areas threatened by these different probability floods assuming a uniform coastal slope. These areas are then converted to people in the hazard zone using the average population density for the coastal area. Lastly, the standard of protection is used to calculate PaR. These national estimates are then aggregated to regional and global results.

In GVA1, a 1-m rise in sea level was imposed on the 1990 (and 2020) world. Coastal population density was estimated and used directly. In subsiding coastal areas, 15 cm of subsidence was assumed. Protection standards were estimated indirectly as discussed below. Lastly, only impacts of the expansion of the flood plain were considered, although this was amended by Baarse (1995).

In GVA2, a more dynamic approach was followed in which climate and socioeconomic scenarios both reflect realistic timescales. The 1990 coastal population density was increased (or decreased) at twice the rate of national growth. This is simply projecting present trends (Bijlsma *et al.*, 1996). In coastal areas subject to subsidence, a uniform subsidence of 15 cm/century was applied to the entire coastal area, although it is recognized that this is only a first approximation.

There are no global databases on the level of flood protection. Therefore, GVA1 adapted the World Bank classification of less developed, middle and high developed nations to estimate this parameter indirectly and used the GNP/capita in 1989 as an "ability-to-pay" parameter (Table G2). GVA2 used the same concept, but the algorithm was improved to reflect: (1) existing defense standards; (2) the greater costs of protecting deltaic areas against flooding; and (3) the increasing risk of flooding within the coastal flood plain as sea levels rise. The minimum standard of protection in 1990 was assumed to be 1 in 10 years, reflecting that people do not choose to live in highly flood-prone areas. Deltaic areas have a longer land–water interface than elsewhere, and a greater need for water management within the extensive low-lying areas that are protected, substantially raising protection costs. Based on expert judgment, the protection classes shown in Table G3 were

Table G2 Protection Classes used in GVA1

GNP/capita (US$) (or ability-to-pay)	Protection class (PC)	Protection status	Design frequency
<600	PC 1	low	1/1 to 1/10
600–2400	PC 2	medium	1/10 to 1/100
>2400	PC 3	high	1/100 to 1/1000

Table G3 Revised protection classes used in GVA2, allowing for deltaic and non-deltaic coasts

GNP/capita (US$)				
If deltaic coast	If non-deltaic coast	Protection class (PC)	Protection status	Design frequency
<2400	<600	PC 1	low	1/10
2400–5000	600–2400	PC 2	medium	1/100
>5000	2400–5000	PC 3	high	1/1000
—	>5000	PC 4	very high	1/1000

selected. Lastly, the increase in flood risk within the existing flood plain produced by sea-level rise is estimated by reducing the protection class as sea-level rises.

In GVA2, two protection scenarios are considered:

- constant protection (i.e., constant 1990 levels); and
- evolving protection in phase with increasing GNP/capita.

The evolving protection scenario is more realistic based on observed trends during the 20th century. It should be noted that evolving protection only included measures that would be implemented *without* sea-level rise—that is, there are no proactive adaptation measures to anticipate sea-level rise. These two protection scenarios allow us to examine how such evolving protection might reduce vulnerability to sea-level rise.

Adaptation to flooding

To estimate realistic first order national-scale protection costs within the constraints of the GVA1, a simple modular approach was adopted (Hoozemans and Hulsbergen, 1995). This assumed that the protection class is upgraded by one class (e.g., PC 1 increases to PC 2—Table G2). Then the revised PaR is calculated together with the protection costs as outlined below. The regional protection costs are compared with the regional GNP to quantify their relative cost, and hence the relative capacity to implement such measures.

The method aims to address the wide range of existing coastal defense types and their related costs. The following factors were used:

- the lengths of low-lying coastline (or coastal areas) to be protected,
- a set of six standardized coastal defense measures to be applied,
- standard unit costs for each type of defense measure,
- individual national cost factors, to take account of local cost factors.

For each country, the national costs (C_N) were found by applying the following summation at the national scale for all partial stretches of coastline in that country or coastal area which need protection:

$$C_N = \Sigma\, l_c \cdot c_m \cdot c_f \qquad \text{(Eq. 1)}$$

where l_c is the coastal length, c_m is the unit defense measure cost, and c_f is the national cost factor.

Regionally and globally, the aggregated cost is found by summing the respective national costs. This approach does *not* produce a basis for national coastal defense planning. However, the modular setup provides a realistic and practical framework for subsequent, more detailed analyses, to improve the accuracy and local relevance of the individual modules.

For each coastal country, an evaluation was made of the present types of sea defenses. It was assumed that new or upgraded defenses will be based on this experience, and hence the preferred type of defense options were selected by expert judgment. This selection should also account for matters like soil conditions, elevation and wave-exposure of the shore, the availability of construction materials, and the value of the direct hinterland.

The length of coastline vulnerable to flooding was determined from the earlier World Coast Estimate (WCE) study by adding the length of low coast, the length of city waterfronts, and the length of low coast of islands. Because there were no suitable global databases on geomorphology/defense status (e.g., dunes, dikes, salt marshes) within each country, a typical coastline was selected, based on the dominant coastline type per country.

Six types of defense measure were considered:

(1) *Stone-protected sea dike*;
(2) *Clay-covered sea dike*;
(3) *Sand dune*;
(4) *Tourist beach maintenance (beach nourishment)*;
(5) *Harbors and industrial areas (upgrade)*;
(6) *Elevation of low-lying small islands*.

Defense measures 1, 2, and 3 apply in most cases. However, the effective implementation of any measures requires a well-functioning technical and organizational infrastructure (as this is a key element of adaptive capacity). More demanding coastal defense works like those used to close off large estuaries were deliberately omitted from the standard list, although their application will be the most economical solution in some cases, as experience in England, the Netherlands, and Japan illustrates.

Cost estimates for the standard protection measures were established in 1990 US dollars. They are based on the following assumptions and conditions, which draw strongly on Dutch technical experience:

- Standard defense constructions are defined for each situations, including dimensions, construction material, and construction methods.
- The schematized designs are based on well-established procedures.
- Standard unit costs are derived from the Dutch situation, including provisions for all the costs, including design, execution, taxes, etc.
- Construction methods and cost estimates are based on the assumption of construction in one continuous operation per project.
- The hydraulic regime determined in the flood analysis is considered in design, increasing adaptation costs in areas with large surges.

Coastal wetlands and loss

Natural systems may also be impacted by sea-level rise. Coastal wetlands are defined as salt marshes, mangroves, and associated unvegetated intertidal areas (and here exclude features such as coral reefs and shallow-water sea grasses). Wetlands are not impacted by short-term fluctuations in sea level such as tides and surges, but they are susceptible to long-term sea-level rise and show a dynamic and nonlinear response (Nicholls *et al.*, 1999). Therefore, we are considering potential losses. In this case, it is important to consider the wetland response to sea-level rise to make credible impact estimates.

All the evidence shows that coastal areas with a small tidal range are more susceptible than similar areas with a large tidal range. Loss of coastal wetlands due to sea-level rise can be offset by inland wetland migration (upland conversion to wetland). However, in coastal areas without suitable low-lying areas, or in low-lying areas protected against flooding, wetland migration cannot occur, producing a coastal squeeze.

A database of the type, area, and location of most coastal wetlands of international importance was created as part of GVA1. It mainly comprised sites recognized by the Ramsar Treaty. It is missing data for certain regions such as Canada, the Gulf States, and the former Soviet Union.

GVA1 identified those wetlands that might be threatened by a 1-m rise in sea level, but did not project actual losses. These threatened wetlands were identified based on coastal geomorphology, tidal range, and local population density.

A nonlinear model of coastal wetland response to sea-level rise was developed in GVA2 (Figure G13). The modeling effort is split into two parts (1) vertical accretion and (2) wetland migration. To model vertical accretion, the availability of sediment/biomass for vertical accretion is parameterized using critical values of non-dimensional relative sea-level rise (*RSLR**):

$$\text{RSLR}^* - \text{RSLR/TR} \qquad \text{(Eq. 2)}$$

where RSLR is the relative sea-level rise scenario and TR is the tidal range on spring tides. Hence, wetlands in areas with a low tidal range are more susceptible to sea-level rise than wetlands in areas with a higher tidal range. (The rate of relative sea-level rise was implicit being defined by the 95-year period of interest). A critical value of RSLR*

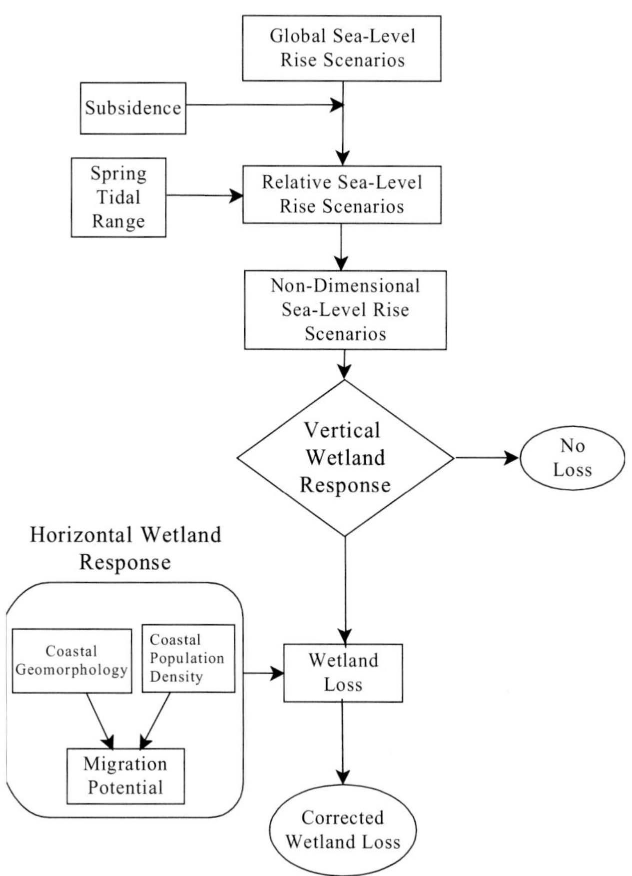

Figure G13 The wetland loss model algorithm used in GVA2 (modified from Nicholls *et al.*, 1999).

($RSLR^*_{crit}$) distinguishes two distinct wetland responses to sea-level rise in terms of vertical accretion:

(1) $RSLR^* \leq RSLR^*_{crit}$, no wetland loss as wetland accretion ≥ sea-level rise; and
(2) $RSLR^* > RSLR^*_{crit}$, partial or total wetland loss as wetland accretion < sea-level rise.

If wetland loss occurs, it is modeled linearly using the excess sea-level rise up to $RSLR^* = RSLR^*_{crit} + 1$. Above this rise, (near-) total loss is assumed and wetlands will only survive if there is inland wetland migration. This simple model captures the nonlinear response of wetland systems to sea-level rise and the association of increasing tidal range with lower susceptibility to loss. The literature stresses the large uncertainties concerning quantitative wetland response to sea-level rise, so a range of values for $RSLR^*_{crit}$ which encompasses the available information were selected (Nicholls *et al.*, 1999). The wetland sites are aggregated to the 192 coastal areas defined in the flood analysis, except for eight continuous national coasts that were subdivided because of a large variation in tidal range.

To model wetland migration, the same approach as GVA1 was used. The natural potential for the migration of the coastal wetlands under sea-level rise was evaluated for each wetland site using the global coastal geomorphic map of Valentin (1954) (showing the limited work on global-scale coastal typology in the last 40 years!). Three classes of migration behavior are recognized: (1) migration is possible; (2) migration is impossible; and (3) migration is uncertain. In the latter cases, losses were calculated assuming both migration and no migration and this contributes to the uncertainty between the low and high range of the results. In areas where migration is possible, the population density was estimated for the 2080s in a consistent manner to the flood analysis. If this population density exceeded 10 inhabitants/km^2, it was assumed that wetland migration would be prevented by flood protection structures. In areas where wetland migration is possible, wetland losses are assumed to be zero (i.e. wetland migration compensates for any losses due to inundation).

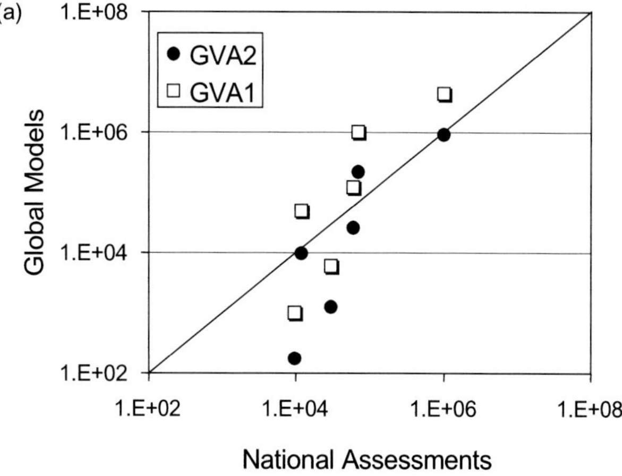

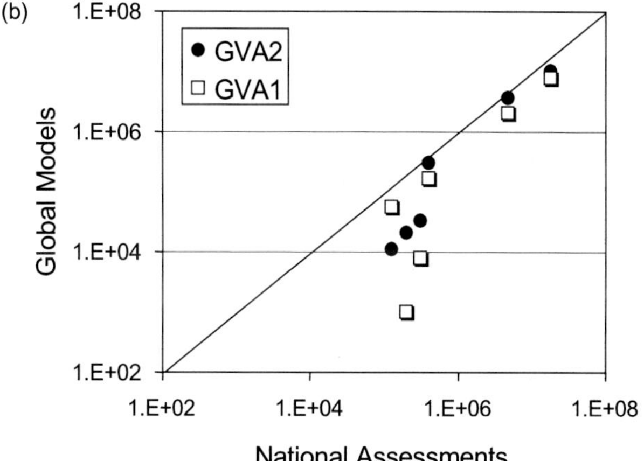

Figure G14 National PaR-estimates from GVA1 and GVA2 against independent national-scale vulnerability assessments for six countries (Egypt, Germany, Guyana, the Netherlands, Poland, and Vietnam): (a) no sea-level rise in 1990, and (b) 1 m sea-level rise in 1990 (modified from Nicholls *et al.*, 1999).

Validation/interpretation of the GVA

Validation of any model is a critical step, which increases confidence in the absolute quality and interpretation of the results. However, global assessments can also be interpreted as relative impact measures. Given that global assessments provide internally consistent results, the relative impacts may be more reliable than the absolute impacts. Therefore, it is suggested that users interpret the results from both an absolute and a relative perspective.

For the flood analysis, an independent data set of the impact parameters was developed via national-scale vulnerability assessments (Nicholls, 1995, 2000). While these national-scale results consider the impacts of sea-level rise on the 1990 world without any socioeconomic changes, the results can be used to validate the global flood model for these scenarios. In broad terms, the results for GVA2 are of the right order of magnitude for the three parameters assessed, and are also an improvement over the results in GVA1 (e.g. Figure G14). Therefore, the changes to the methods in GVA2 are justified.

Validation of the protection cost estimates suggested that they are broadly reasonable (Nicholls, 1995). The validation of the wetland loss models remains limited due to a lack of suitable calibration data.

Results

Globally, about 200 million people live in the coastal flood plain (below the 1 in 1,000 year flood elevation). GVA1 estimated that PaR is 50 million people/year in 1990. Most of these people live in deltaic areas in the developing world. The expansion of the flood plain due to a 1 m sea-level rise will increase PaR to 60 million people/year based on 1990 population. Allowing for the additional factor of increased flood

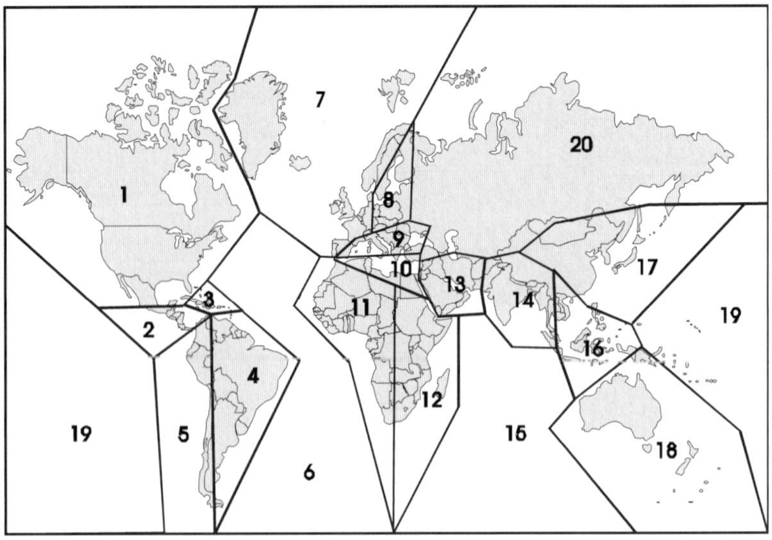

COASTAL REGIONS

Figure G15 The 20 regions used in the GVA (from Hoozemans *et al.*, 1993).

frequency within the existing coastal flood plain, PaR doubles to 120 million people/year (Baarse, 1995). Upgrading coastal protection infrastructure against a 1 m rise in sea level as outlined above could collectively cost US $1,000 billion (1990 dollars), or 5.6% of the 1990 Global World Income. In this case, PaR is reduced from 60 to 7 million people/year, so there would be substantial benefits to coastal inhabitants. As these cost estimates assume an instantaneous rather than a progressive response and do not consider erosion in non-tourist areas or the costs of water management and drainage, they are more useful as a relative cost rather than an absolute adaptation cost.

Coastal wetlands are already declining at 1%/year to indirect and direct human activities. They would decline further due to a 1 m rise in sea level: more than half of the world's coastal wetlands could be lost.

The coastal regions defined in Figure G15 have different problems in the GVA1. Six regions are vulnerable to the loss of coastal wetlands: North America; Central America; South America Atlantic Coast; north and west Europe; northern Mediterranean; and Pacific large islands (GVA regions 1, 2, 4, 7, 9, and 18). Nine regions were considered to be vulnerable with respect to both flood impacts/response costs and loss of coastal wetlands: the Caribbean islands; west Africa; the Indian Ocean small islands; east Asia; the Pacific small islands; the southern Mediterranean; south Asia; southeast Asia; and east Africa (GVA regions 3, 11, 15, 17, 19, 10, 14, 16, and 12). Relative flood impacts are significant for small island settings, but the absolute impacts are highest in south, southeast, and east Asia.

The improved and validated approach of GVA2 suggests that under the 1990 situation, PaR is 10 million people/year. Even without sea-level rise, PaR is likely to increase to the 2050s due to increasing coastal populations. A 1 m rise in sea level produces a 14-fold increase in PaR given the 1990 world, rather than a 3-fold increase as found by Baarse (1995). Therefore, in the absence of adaptation, the impacts of sea-level rise on flooding are much more dramatic than previously realized. Evolving protection reduces the magnitude of flood impacts, but the relative increase in people flooded given sea-level rise is still dramatic. Under a lower sea-level rise scenario of 38 cm by the 2080s, the global increase in flooding will be seven-fold compared with the situation without sea-level rise. Most of these people will be flooded so frequently that some response seems inevitable. The most vulnerable regions are similar to GVA1, comprising large relative increases in the small island regions of the Caribbean, Indian Ocean, and Pacific Ocean small islands (GVA regions 3, 15, and 19), and large absolute increases in the southern Mediterranean, west Africa, east Africa, south Asia, and southeast Asia (GVA regions 10, 11, 12, 14, and 16).

Wetland losses given a 1 m rise in sea level could approach 46% of the present stock. Taking a 38 cm global scenario by the 2080s, between 6% and 22% of the world's wetlands could be lost due to sea-level rise. When added to existing trends of indirect and direct human destruction, the net effect could be the loss of 36–70% of the world's coastal wetlands, or an area of up to 210,000 km^2. Therefore, sea-level rise is a significant additional stress which makes the prognosis for wetlands even more adverse than existing trends. Regional losses would be most severe on the Atlantic coast of North and Central America, the Caribbean, the Mediterranean, and the Baltic. While there is no data, by implication, all small island regions are also threatened due to their low tidal range.

The major change in flood impacts from GVA1 to GVA2 reflects the more realistic assumptions in GVA2 concerning the present protection status, and its degradation as sea-level rises. The most important effect on PaR is the increased risk of flooding in the existing flood plain, rather than the expansion in the size of the flood plain as sea levels rise. Wetland losses in GVA1 and GVA2 are difficult to compare, but results for the common 1 m scenario are similar. The main benefit of GVA2 is its more flexible form.

What have we learned/next steps?

The analyses described here have proven the concept and utility of global vulnerability assessment for policy analysis. The results confirm that global sea-level rise could have a range of serious impacts on the world's coasts if we fail to plan for these changes. Further, these impacts will be greater in some regions than others with parts of Asia, Africa, and small island regions most adversely impacted. This is an important result to be considered by the UNFCCC policy process. However, what to do is not evaluated by the existing analyses.

In scientific terms, the rigor of developing such models gives improved insights into the functioning of the integrated (i.e., coupled natural-human) coastal system from the local to the global scales. For instance, this work explicitly considers the relationship between local wetland accretion and elevation change rates and potential regional and global wetland losses. It also explores how increased protection of human activities in low-lying areas may exacerbate wetland vulnerability under a scenario of rising sea level. Continued research efforts will provide important insights that will be useful to research programs such as the International Geosphere–Biosphere Programme, Land–Ocean Interactions in the Coastal Zone Project (IGBP-LOICZ) (Holligan and de Boois, 1993) and more generally improve the scientific basis of long-term coastal management.

Policy analysis for climate change requires flexible tools, which can link different emission scenarios all the way to potential impacts and adaptation potential. This will allow exploration of a wide range of sea-level rise (and other climate change) and socioeconomic scenarios, including different mixtures of mitigation and adaptation options. The experience with developing GVA1, and its improvement to GVA2 provide the basis to develop such tools. Important needs for future models include operation at a finer resolution, a better description of impact processes, and the facility to include different response and adaptation pathways, among other improvements. A European Union research project called Dynamic and Interactive Assessment of National, Regional, and Global Vulnerability of

Coastal Zones to Climate Change and Sea-level rise (or DINAS-COAST) is exploring these issues and will report in 2004.

Robert J. Nicholls and Frank M.J. Hoozemans

Bibliography

Baarse, G., 1995. *Development of an Operational Tool for Global Vulnerability Assessment (GVA): Update of the Number of People at Risk due to Sea-level Rise and Increased Flood Probabilities.* The Hague, The Netherlands: Ministry of Transport, Public Works and Water Management, CZM-Centre Publication No. 3, p. 15

Bijlsma, L., Ehler, C.N., Klein, R.J.T., Kulshrestha, S.M., McLean, R.F., Mimura, N., Nicholls, R.J., Nurse, L.A., Perez Nieto, H., Stakhiv, E.Z., Turner, R.K., and Warrick, R.A., 1996. Coastal zones and small islands. In Watson, R.T., Zinyowera, M.C., and Moss, R.H. (eds.), *Impacts, Adaptations and Mitigation of Climate Change: Scientific-Technical Analyses.* Cambridge: Cambridge University Press, pp. 289–324.

Holligan, P., and de Boois, H. (eds.) 1993. *Land-Ocean Interactions in the Coastal Zone: Science Plan.* Stockholm: International Geosphere–Biosphere Programme, IGBP Report No. 25.

Hoozemans, F.M.J., and Hulsbergen, C.H., 1995. Sea-level rise: a worldwide assessment of risk and protection costs. In Eisma, D. (ed.), *Climate Change: Impact on Coastal Habitation,* London: Lewis Publishers, pp. 137–163.

Hoozemans, F.M.J., Marchand, M., Pennekamp, H.A., Stive, M., Misdorp, R., and Bijlsma, L., 1992. The impacts of sea-level rise on coastal areas: Some global results. In *Proceedings "The Rising Challenge of the Sea,"* Margarita Island, Venezuela, March 9–13 1992. NOAA, Silver Spring, Md. pp. 275–292.

Hoozemans, F.M.J., Marchand, M., and Pennekamp, H.A., 1993. *A Global Vulnerability Analysis: Vulnerability Assessment for Population, Coastal Wetlands and Rice Production on a Global Scale,* 2nd edn. the Netherlands: Delft Hydraulics.

IPCC CZMS, 1992. *Global Climate Change and the Rising Challenge of the Sea.* Report of the Coastal Zone Management Subgroup. IPCC Response Strategies Working Group, the Hague: Rijkswaterstaat.

Klein, R.J.T., and Nicholls, R.J., 1999. Assessment of coastal vulnerability to sea-level rise, *Ambio,* **28**: 182–187.

Nicholls, R.J., 1995. Synthesis of vulnerability analysis studies. *Proceedings of WORLD COAST 1993,* Ministry of Transport, Public Works and Water Management, the Netherlands pp. 181–216.

Nicholls, R.J., 2000. An analysis of the flood implications of the IPCC Second Assessment global sea-level rise scenarios. In Parker, D.J. (ed.), *Floods.* London: Routledge, pp. 148–162.

Nicholls, R.J., and Mimura, N., 1998. Regional issues raised by sea-level rise and their policy implications. *Climate Research,* **11**: 5–18.

Nicholls, R.J., Mimura, N., and Topping, J., 1995. Climate change in South and Southeast Asia: Some implications for coastal areas. *Journal of Global Environment Engineering,* **1**: 137–154.

Nicholls, R.J., Hoozemans, F.M.J., and Marchand, M., 1999. Increasing flood risk and wetland losses due to global sea-level rise: Regional and global analyses. *Global Environmental Change,* **9**: S69–S87.

Parry, M., and Livermore, M. (eds.), 1999. A new assessment of the global effects of climate change. *Global Environmental Change,* **9**: S1–S107.

Valentin, H., 1954. *Die Kusten der Erde,* Berlin: VEB Geographisch-Kartographische Anstalt Gotha.

WCC'93, 1994. *Preparing to Meet the Coastal Challenges of the 21st Century.* World Coast Conference Report, Noordwijk, Nov. 1993, The Hague: Rijkswaterstaat.

Cross references

Changing Sea Levels
Classification of Coasts (See Holocene Coastal Geomorphology)
Coastal Subsidence
Demography of Coastal Populations
Deltas
Dikes
Global Warming, Effect (See Greenhouse Effect and Global Warming)
Greenhouse Effect and Global Warming
Natural Hazards
Sea-Level Rise, Effect
Small Islands
Wetlands

GLOBAL WARMING—See GREENHOUSE EFFECT AND GLOBAL WARMING

GLOSSARY OF COASTAL GEOMORPHOLOGY—See APPENDIX 5

GRAVEL BARRIERS

Definition

Gravel barriers are morpho-sedimentary units comprising beach face and rear landward-sloping unit whose surface sediment size ≥3ø. Surface sediment may be swash sorted by particle size and shape to form zoned facies. Internal structure shows matrix-supported as well as open-worked clast-supported stratification. Thus fine sediment (≤0ø) is an element of most gravel barriers. The type concept of sand, mixed sand and gravel, and pure gravel barriers (Kirk, 1980) is simplistic. Gravel barriers comprise sediment mixtures, the surface is but one coarse-clastic dominated facies set, as a function of incident wave energy, swash dynamics, and tidal range. Carter and Orford (1993) consider "coarse-clastic" a more appropriate descriptive term than "gravel."

Location and type

Most gravel (including shingle) literature concerns mid- to high-latitude barriers (mainly northern latitudes) associated with paraglacial coasts. These barriers tend to be transgressive and sediment depleted (Orford *et al.*, 1991). There are also gravel barriers, predominantly regressive and sediment abundant, in southern latitudes formed from mega-river supply or from reworked fluvial gravel fans associated with mountains. Gravel accumulations constructed from storm-generated coral reef debris can occur on tropical atolls.

Morphology

Transgressive barriers show a classic barrier morphology: narrow (<100 m), elongated longshore, and comprising one ridge crest up to 14 m high above mean sea-level. Ridge height is a function of sediment shape and size and breaking wave height. Breaches in transgressive barriers are usually initiated by cross-barrier hydraulic gradients, forced either landward by storm surges or seaward by terrestrial flood impoundment. A reduction in the fine-sediment component is likely in transgressive barriers. Regressive barriers show multiple beach ridges often with seaward trends in ridge elevation. Internal sediment facies variation reflects the importance of cuspate and berm morphology superimposed on prograded ridge sedimentation. What controls multiple ridge development is still widely debated: extreme storms, storminess phases, extreme tidal elevation, sediment supply and textural mix, and stochastic processes.

Essential concepts

Relative sea-level (RSL) change sets paraglacial barrier development tempo. Forbes *et al.* (1995) argued that longshore supply is controlled by RSL rise. Jennings *et al.* (1998) suggest that RSL rise controls types of gravel barrier development: spatial stability exists between 2 and 8 mm a^{-1} RSL rise. Below 2 mm a^{-1}, RSL rise is too slow for sediment replenishment and allows longshore reworking of a barrier. Above 8 mm a^{-1}, the high-tide shoreline is moving too fast for barrier concentration. Orford *et al.* (1995) identified barrier retreat rate as 1 m a^{-1} for each millimeter per annum RSL rise.

Barriers are defined as drift or swash-aligned, based on sediment supply. Paraglacial barriers related to till-covered basements are often fed from supply-limited sources of sediment (e.g., drumlins). Early to middle stages of drumlin erosion (high longshore supply) lead to associated longshore drift-aligned barriers (forced regression). As sediment supply reduces, the barrier becomes swash-aligned with both longshore cannibalization of shoreline barrier sediment and onshore movement of remnant sediment by overtop and overwash. Spatial fluctuations in sediment supply are not as important for barriers fed from unlimited supply sources.

Gravel barrier development is related more to storm wave than fair-weather conditions. Beach crest height is proportional to past breaker heights recorded at high water, hence the storm connection. Extreme run-up may flow over the crest top, transporting gravel to the crest (=over-top). Overtopping builds up the crest elevation, reducing the rate of subsequent overtopping and preventing barrier retreat. Extreme volumes of overtopping water can wash sediment down the barrier back-slope (=overwash). This lowers the crest height and allows smaller breaking waves to overwash. Consistent overwashing leads to landward migration of the gravel barrier. Quiescence and storminess periods may regulate short-term barrier retreat rates, but RSL rise controls long-term retreat.

Basement structure defines the barrier's accommodation space. When sediment supply is scarce then a crenulate coastal configuration is essential to define both shoreline hinge points and offshore bathymetric variation that causes refraction of incident wave by which barriers can accumulate sediment. Crenulate coastlines help transform drift-aligned gravel spits into swash-aligned gravel barriers as supply dwindles.

Conclusions

Swash-aligned barriers cannot rollover indefinitely so barrier breakdown will occur (Orford *et al.*, 1996). Sediment may come from rollover incorporation but this is outweighed by sediment loss to the barrier's extending length. Where sea level is rising the reducing return period associated with extreme water levels means that the barrier is more liable to overwashing and multiple phases of barrier-breaching. Wave refraction around the breaches forces the barrier remnants to fall back on themselves, with sediment moving onshore along transverse shoal lines to act as a source for new landward barrier construction. It is possible that swash-aligned barriers in rollover mode may be "scalped" in extreme storms (=overstepped). The upper barrier is swept onshore to form a new barrier leaving the old basement as a relict feature on the foreshore (Forbes *et al.*, 1991). The 20th century has seen a trend for fixing gravel barriers by artificially maintaining the crests, allowing urban development in their lee. The coastal squeeze is very pronounced behind such modified barriers with no space for rollover. The high likelihood of catastrophic gravel barrier failure is a salient problem for 21st century coastal zone management.

Julian D. Orford

Bibliography

Carter, R.W.G., and Orford, J.D., 1993. The morphodynamics of coarse clastic beaches and barriers: a short and long-term perspective. *Journal of Coastal Research*, **15**(Special issue): 158–179.

Forbes, D.L., Orford, J.D., Carter, R.W.G., Shaw, J., and Jennings, S.C., 1995. Morphodynamic evolution, self-organisation, and instability of coarse-clastic barriers on paraglacial coasts. *Marine Geology*, **126**: 63–85.

Forbes, D.L., Taylor, R.B., Orford, J.D., Carter, R.W.G., and Shaw, J., 1991. Gravel barrier migration and overstepping. *Marine Geology*, **97**: 305–313.

Jennings, S.C., Orford, J.D., Canti, M., Devoy, R.J.N., and Straker, V. 1998. The role of relative sea-level rise and changing sediment supply on Holocene gravel barrier development; the example of Porlock, Somerset, UK. *The Holocene*, **8**: 165–181.

Kirk, R.M., 1980. Mixed sand and gravel beaches: morphology, processes and sediments. *Progress in Physical Geography*, **4**: 189–210.

Orford, J.D., Carter, R.W.G., and Jennings, S.C., 1991. Coarse clastic barrier environments: evolution and implications for Quaternary sea-level interpretation. *Quaternary International*, **9**: 87–104.

Orford, J.D., Carter, R.W.G., McKenna, J., and Jennings, S.C., 1995. The relationship between the rate of mesoscale sea-level rise and the retreat rate of swash-aligned gravel-dominated coastal barriers. *Marine Geology*, **124**: 177–186.

Orford, J.D., Carter, R.W.G., and Jennings, S.C., 1996. Control domains and morphological phases in gravel-dominated coastal barriers. *Journal of Coastal Research*, **12**: 589–605.

Cross-reference

Beach Sediment Characteristics
Changing Sea Levels
Drift and Swash Alignments
Gravel Beaches
Paraglacial Coasts
Reflective Beaches
Sediment Budget

GRAVEL BEACHES

Introduction

Gravel beaches are accumulations of shore material formed into distinctive shapes by waves and currents and containing lithic particles in the gravel size range. The beach form is a generally seaward-sloping boundary between a water body and mobile sediment, and a flat or landward-sloping surface at the upper limit of the beach. One or more gravel ridges may exist in the subaerial profile. A bar might be present on the submerged profile. Figure G16 describes beach features to be discussed. Natural beaches composed of coarse material are common in high latitudes where much coastal sediment has a glacial transport history. They also occur at other latitudes, especially near eroding cliffs, platforms, and reefs (Carter, 1988, p. 130).

Terminology

Gravel beach is a term that includes shingle beach and mixed beach. Various investigators refer to gravelly beach material as having a median diameter between 5 and 75 mm, and shingle larger than 2 mm. The composition of the gravelly surface results from winnowing of finer sediments from the original material, or from active sorting of material during transport both onshore and alongshore. A shingle beach contains no or negligible amounts of sand and finer material. Mason *et al.* (1997) identified two categories of mixed beach. One is a broadly homogeneous mixture of sand and gravel, with some grading across shore and alongshore. The distinctly sandy region, if present, is usually exposed only during low spring tides. The second category is a composite beach, characterized by a wide, sandy, intertidal terrace flanked by a gravel ridge. The division between the two sediment types in this second beach category is marked by an abrupt change in slope. As the water level changes, beach reflectivity changes rapidly. Both types of mixed beaches may have complex stratigraphy. The surface layer of pebbles and cobbles is often underlain by gravel and pebbles of mixed sizes supported in a sand matrix. Surface material of the berm crest might not be wave-sorted gravel where the back beach or berm crest is quite narrow and fronts an eroding cliff, but composed of colluvium derived from the cliff face with particle size unrelated to the washover process.

Significant studies

The interactive influences involving beach material size, sediment transport, profile characteristics, and evolution of beach profile in response to changing wave energetics mark gravel beaches as different from sand beaches. Literature of shore processes contain fewer studies of gravel beaches than studies of sand beaches, possibly because sand beaches occur in parts of the world where their economic value to upland property and demand as a recreational asset are relatively greater. Chesil Beach in Dorset, United Kingdom stands out as a shingle beach with a long history of study of processes relating wave energetics to particle sorting and profile characteristics, one as early as 1853 (King, 1972, p. 307). Bluck (1967) and Orford (1975) reported studies of the control by particle shape and size on zonation of beach surface gravels. Van der Meer and Pilarczyk (1986) developed mathematical relationships between profile features and incident wave characteristics from tests of variable wave conditions and beach material sizes in a wave flume. Approximately 200 km of the Black Sea coastline has been reconstructed

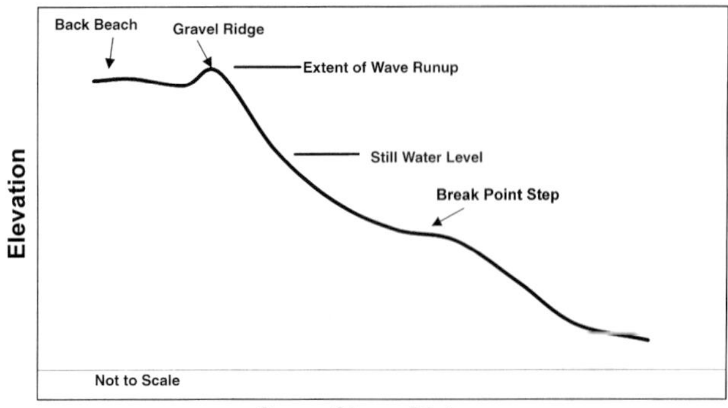

Figure G16 Idealized gravel beach profile features.

with designed gravel beaches (Zenkovich and Schwartz, 1987; Kiknadze, 1993). Kirk (1992) reported monitoring results of a mixed sand and gravel renourishment project, and developed a method for estimating the fill life.

Essential concepts

Gravel beach profile shape and particle size are hydraulically related. Beach morphology results from the near balance of forces acting on the beach material delivered to the nearshore zone. Hydraulic forces are directed onshore during wave advance. Opposing forces are weight and mechanical resistance of the particles being moved. Seaward directed forces are downslope component of weight and hydraulic forces associated with wave downrush, percolation flow in the beach face, and return flow in the surf zone. The mode of transport corresponding to predominant particle size and predominant form of wave energy reaching the nearshore zone influence beach morphology and wave energy dissipation rate across the profile. Gravel beaches are usually dominated by plunging breakers with no surf zone, with longshore transport occurring as bedload in the swash zone, and display higher reflectivity of the low-frequency wave energy (compared with that of a dissipative foreshore of a sandy beach).

Nearshore mechanics

The forms of energy dissipation and the modes of sediment transport make a gravel beach different from a sand beach. Wave energy is dissipated on a sand beach through turbulent dissipation in the wave breaking process through the surf zone, bottom friction, and work expended in suspending sediments. Swell wave reflection is lower for sand beaches and water infiltration into the beach face is relatively insignificant, compared with gravel beaches. Wave breaking at a gravel beach is closer to the high-tide shoreline, with the Irribarren Number well into the plunging breaker range. Gravelly beach material is moved as bedload or in close contact with the bed under most transport conditions. Bedload is a much less efficient mode of transport than suspended load. Plunging breakers at the shore confine longshore transport of gravel to the swash zone. Measurements of Mason *et al.* (1997) showed that swell wave reflection increased with beach gradient for gradients exceeding 0.06 (3.5°), but for lower gradients no relationship was observed. Reflection of wind waves showed no relationship with gradient, and never exceeded 40%. Powell (1988) determined from laboratory studies that wave energy reflected by a shingle beach was about 10% for all values of sea steepness greater than 0.02, but reflection increased rapidly as sea steepness decreased below 0.02. Greater energy reflection diminishes the energy available to create high runup.

Swash mechanics

Wave energy loss in runup and rundown is mainly through frictional loss on a sandy beach surface (Damgaard *et al.*, 1996, p. 11). With highly porous beach material, percolation of the swash into the surface accounts for most of the energy loss that would otherwise be available for producing higher runup. The greater porosity of the gravelly beach material creates a greater asymmetry of shear stress applied to swash zone surface material in wave uprush and downrush. As water from the advancing wave percolates into the beach face, the water table locally is raised and the volume of water available for transporting material seaward is diminished. Van Wellen *et al.* (1997) reported measurements on a shingle beach during a high-energy wave event of shore-normal runup, in which the average sediment transport rate in uprush was over three times that of the backwash.

Morphology

The resulting equilibrium slope of a gravel beach face is steeper, and the transport asymmetry is responsible for the creation of the berm that often is present at a gravel beach. Typical cross-sectional shape of a gravel beach includes a bar seaward of the break point, a step in the profile at the point of wave breaking, and a berm built near the limit of uprush. Laboratory tests of Powell (1988) showed that generally less than 2% of runups of random waves exceed the wave-formed beach crest, regardless of shingle size or wave conditions. Simpson (1995) determined that the elevation of the berm crest on a semi-protected gravel beach approximately matches the runup elevation having the 33% exceedance probability based on combined probability of tide levels and runup distribution of the two-year return period storm. Figure G17 describes the profile geometry and pertinent elevations. The inset shows the particle size distribution of that beach surface material. Figure G18 is a photograph of the upper beach at the transect described in Figure G17. The plan shape of gravel beaches often includes cusps. Cusps are associated with steep, reflective beaches and are more prevalent on shores composed of coarse material than fine material.

Application

The interest in developing quantitative prediction of geometry and transport rates for gravel and shingle beaches is increasing as their value is recognized for support of aquatic habitat and for shore protection, and as more development pressure has potential to disrupt balances that maintain gravel beaches. The features of a gravel beach that are of interest to the engineer, coastal planner, and beach management decision-maker are the elevation of the beach crest (or berm height), the beach slope (or beach gradient), and the location and dimension of the submerged bar. The designer is interested as well in particle size distribution and its effect on hydraulic conductivity. These elements determine the rate of wave energy dissipation, wave reflection, and wave runup height. Processes of interest are sorting, depth of disturbance, and cross-shore and longshore transport. These processes determine suitability of a constructed protective beach, the longevity of a beach nourishment project, and the potential impact of a littoral barrier in a gravel beach system. Artificial gravel beaches can create shore protection that is environmentally compliant and acceptable to the recreational user. More research is needed to fully specify the effect of sand content in mixed beaches during varying levels of wave attack. Quick and Dyksterhus (1994)

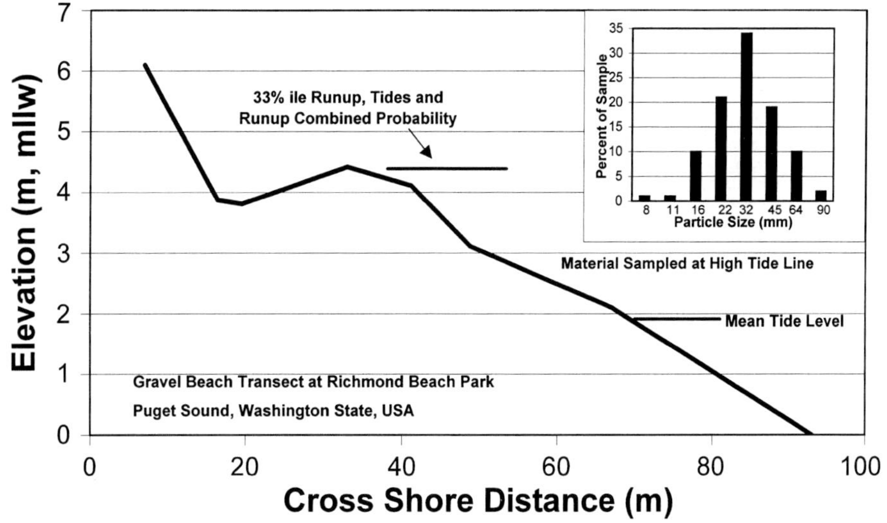

Figure G17 Gravel beach profile, particle sizes, and water level characteristics, example from moderate wave environment.

Figure G18 View of upper profile of gravel beach plotted in Figure G17 (Richmond Beach, Puget Sound, Washington, USA).

experimented with different heights of incident waves and beach material composed of sand and gravel. The same high level of wave attack that produced a steep, bermed profile with gravel caused strong offshore movement and flattening of the slope when the beach was a gravel mixture containing only 25% sand. Under medium wave attack, the mixed beach behaved generally as a gravel beach. This change in behavior with waves is important to designers who intend to design a beach to be dynamically stable in higher wave events. Generally, adding sand to a shingle beach can lower the beach gradient, and adding gravel to a sandy beach has no effect.

David P. Simpson

Bibliography

Bluck, Brian J. 1967. Sedimentation of beach gravels: examples from South Wales. *Journal of Sedimentary Petrology*, **37**: 128–156.

Carter, R.W. 1988. *Coastal Environments*. London: Academic Press Limited.

Damgaard, J.S., Stripling, S., and Soulsby, R.L., 1996. *Numerical Modelling of Coastal Shingle Transport*. H R Wallingford Report TR 4.

Kiknadze, A.G., 1993. Scientific basis of regulation of coastal processes. In Ruben Kos'yan (ed.), Coastlines of the Black Sea. *Proceedings of Coastal Zone '93*, American Society of Civil Engineers, pp. 201–213.

King, C.A.M., 1972. *Beaches and Coasts*, 2nd Ed. London: Arnold.

Kirk, R.M., 1992. Experimental beach reconstruction—Renourishment on mixed sand and gravel beaches, Washdyke Lagoon, South Canterbury, New Zealand. *Coastal Engineering*, **17**: 253–277.

Mason, T., Voulgaris, G., Simmonds, D.J., and Collins, M.B., 1997. Hydrodynamics and sediment transport on composite (Mixed Sand/Shingle) and sand beaches: a comparison. *Proceedings, Coastal Dynamics '97 Conference*, American Society of Civil Engineers, pp. 48–67.

Orford, J.D. 1975. Discrimination of particle size zonation on a pebble beach. *Sedimentology*, **22**: 441–463.

Powell, K.A., 1988. The dynamic response of shingle beaches to random waves. *Proceedings, International Conference on Coastal Engineering*, American Society of Civil Engineers, pp. 1763–1773.

Quick, Michael C., and Patricia Dyksterhuis, 1994. Cross-shore transport for beaches of mixed sand and gravel. *Proceedings, International Symposium: Waves—Physical and Numerical Modelling*, Canadian Society Civil Engineers, pp. 1443–1452.

Simpson, D.P., 1995. Determination of exceedance frequencies of wave height and runup in Puget Sound. *Proceedings, Puget Sound Research '95 Conference*, Seattle, Washington.

Van der Meer, J.W., and Pilarczyk, K.W., 1986. Dynamic stability of rock slopes and gravel beaches. *Proceedings, International Conference on Coastal Engineering*, American Society of Civil Engineers, pp. 1713–1726.

Van Wellen, E., Chadwick, A.J., Bird, P.A.D., Bray, M., Lee, M., and Morfett, J., 1997. Coastal Sediment Transport on Shingle Beaches. *Proceedings, Coastal Dynamics '97 Conference*, American Society of Civil Engineers, pp. 38–47.

Zenkovich, V.P., and Schwartz, M.L., 1987. Protecting the Black Sea—Georgian S.S.R. gravel coast. *Journal of Coastal Research*, **3**: 201–209.

Cross-references

Beach Sediment Characteristics
Cross-Shore Sediment Transport
Dynamic Equilibrium of Beaches
Gravel Barriers
Longshore Sediment Transport

GREENHOUSE EFFECT AND GLOBAL WARMING

Most of the coastal processes examined in this volume depend on climate, sea level, or both. Over long periods of time, climate has always fluctuated. The next century, however, may see unusually rapid changes because humanity is adding gases that absorb infrared radiation to our planet's atmosphere, and thereby enhancing the "greenhouse effect." This entry examines the greenhouse effect, the trends in climate, and expected future changes.

Greenhouse effect: the earth's changing energy balance

Any planet's temperature depends mainly on (1) the amount of sunlight received, (2) the amount of sunlight reflected into space, and (3) the extent to which the atmosphere retains heat. Figure G19 shows our

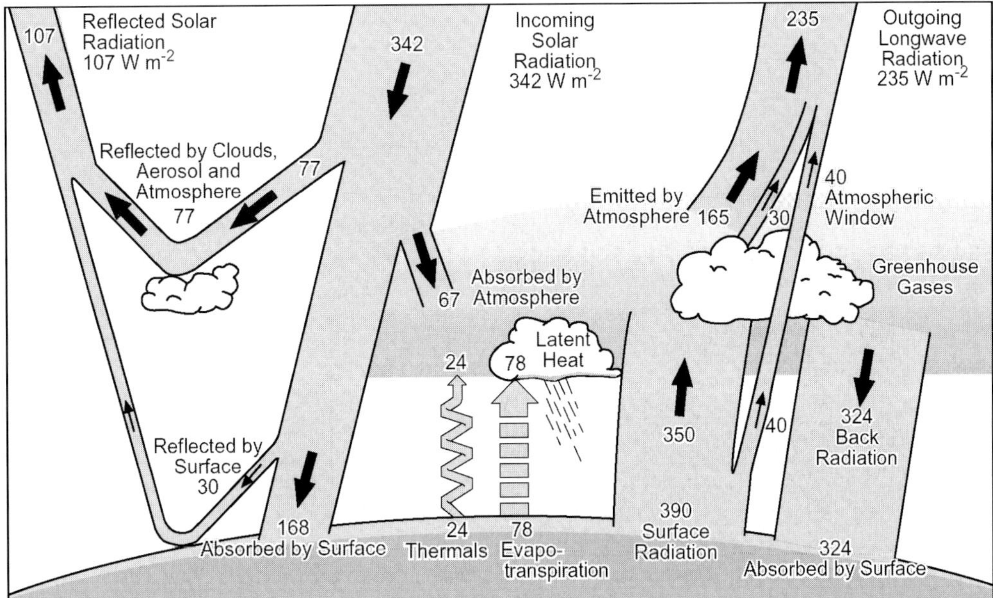

Figure G19 The earth's average annual energy balance.
Source: Kiehl, J.T., and Trenberth, K.E., 1997. Earth's annual global mean energy budget. *Bulletin of American Meterological Society* **78**: 197–208.

planet's annual energy balance, measured by watts of energy for each square meter of the earth's surface. Of the 342 W/m^2 reaching the atmosphere, 107 W/m^2 are reflected back into space by clouds, ice, and other reflective surfaces. Another 168 W/m^2 of sunlight is absorbed by the earth's surface, which re-radiates the energy toward space in the infrared part of the spectrum. But the water vapor and carbon dioxide found naturally in the atmosphere absorb all but 40 W/m^2 of this infrared radiation. This absorption warms the atmosphere, and the warmer atmosphere then re-radiates energy upward into space and downward back to the earth's surface. This feature of the earth's energy balance has long been known as the "greenhouse effect" (Arrhenius, 1896) because the atmosphere—like the glass panels of a greenhouse—allow sunlight to penetrate but retain heat. The importance of the greenhouse effect is apparent from Figure G19. The downward radiation from the atmosphere (334 W/m^2) is twice the energy received directly from sunlight. Without the greenhouse the effect, the earth's surface would be about 33°C (60°F) colder than it is currently.

A second important impact of the greenhouse effect, albeit less widely discussed, is the cooling of the stratosphere. The more a substance absorbs energy, the more rapidly it radiates away its energy, which means that greenhouse gases radiate energy more rapidly than other gases in the atmosphere. Therefore, higher concentrations of greenhouse gases create both a warming effect (increased ability to absorb infrared radiation) and a cooling effect (increased ability to radiate energy away). The warming effect is more important in the lower part of the atmosphere than the upper atmosphere. In the lower atmosphere there is more infrared radiation to absorb, because it is coming from the surface and gases below, and from the gases above, whereas in the upper atmosphere, infrared is only coming from below. Because the gases radiate energy away in all directions throughout the atmosphere, the increased greenhouse gases have a fairly uniform cooling effect. With uniform cooling, greater warming below and less warming aloft, the net effect is warming near the surface and cooling in the upper atmosphere.

Over the last two million years, changes in the timing and amount of sunlight striking the earth have induced ice ages and interglacial warm periods. During ice ages, global temperatures were about 5°C colder than today, and much of the Northern Hemisphere was covered with ice; during interglacial periods, temperatures have often been about 1°C warmer than today. Many scientists believe that cyclical fluctuations in the earth's orbit around the sun, and the tilt and wobble of the earth's axis slowly alter the amount and timing of sunlight received by the earth. The resulting long-term shift in climate is sometimes called the "Milankovich effect." (Milankovich, 1930; Hayes *et al.*, 1976).

The direct effects of changes in solar irradiation appear to have been substantially amplified by several climate feedbacks (IPCC, 1996).

- *The albedo feedback*: Warmer temperatures caused glaciers to retreat, exposing land and open seas, both of which reflect less light into space than snow and ice.
- *The water vapor feedback*: Warmer temperatures increas the absolute humidity of the atmosphere, which in turn increases the greenhouse effect from water vapor.
- *Carbon-cycle feedback*: Warmer temperatures reduce the solubility of CO_2 in water, leading to higher atmospheric concentrations of CO_2, enhancing the greenhouse effect.
- *Cloud feedbacks*: There is no data on how clouds behaved during previous interglacial cycles, and scientists are unsure whether clouds amplify or dampen the global warming from other factors.

Although changes in solar irradiation have been important in the past, changes in the greenhouse effect will probably be more important during the 21st century. Humanity is adding gases that absorb infrared radiation to the atmosphere, and thereby strengthening the greenhouse effect. Figure G20 provides estimates of the impact of these gases and other human activities on the amount of energy absorbed by the earth's surface. The first three bars in the chart are concerned with greenhouse gases, most of the remaining bars focus on activities that increase the extent to which sunlight is reflected or absorbed.

Greenhouse gases

The chief greenhouse gases are CO_2, methane, and nitrous oxide. Whenever oil, coal, gas, or wood are burned, carbon dioxide is released into the atmosphere. Approximately half of this CO_2 is soon absorbed by the oceans or by increased plant photosynthesis. The other half remains in the atmosphere for many decades. As a result, the average concentration of carbon dioxide has increased from around 275 parts per million (ppm) before the industrial revolution, to 315 ppm when precise monitoring stations were set up in 1958, to 368 ppm in 1999. This change has increased the amount of energy striking the earth's surface by about 1.5 W/m^2.

The remaining gases released by human activities appear to be adding about 1.0–1.5 W/m^2. Increased methane and halocarbons are both adding about 0.5 W/m^2. About two-thirds of the current emissions of methane into the atmosphere result from cattle farming, rice paddies, landfills, coal mining, oil and gas production, and several other human activities. Radiation from methane has increased by about 0.5 W/m^2 since the Industrial Revolution. Nitrous oxide is released by the use of nitrogen fertilizers, the burning of wood, and some industrial processes. Some of the most important halocarbons are the fully halogenated chlorofluorocarbons once used in aerosol spray cans, refrigerators,

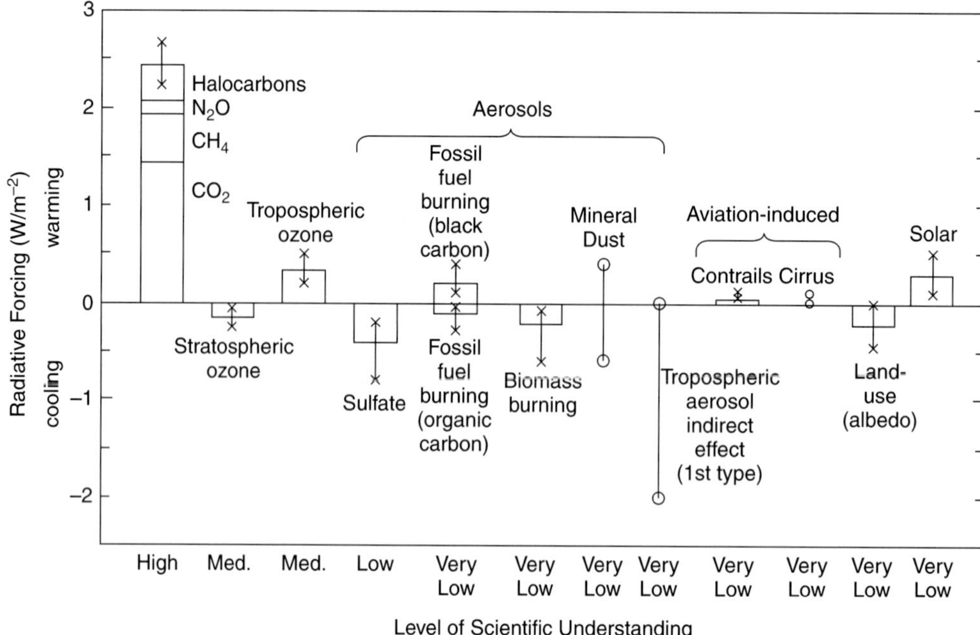

Figure G20 Estimates of changes in global annual average radiative forcing from various causes, 1750 to the Present. *Source*: IPCC (2001).

air-conditioners, electronics cleaning, and foam-blowing. The production of those gases is being phased out under the terms of the Montreal Protocol on Substances that deplete the Ozone Layer. Nevertheless, the gases will remain in the atmosphere for several decades.

Aerosols

Humanity is also increasing the extent to which the atmosphere and the surface of the earth reflect light back into space, which tends to cool the earth. Land uses can alter the reflectivity of the earth's surface; tropical deforestation appears to be the most important change, but it has only reduced the amount of sunlight absorbed by about 0.1 W/m^2. Changes in the reflectivity of the atmosphere, however, may be altering the earth's radiative balance as much as carbon dioxide, albeit in the opposite direction. Most importantly, society is adding very fine particles and droplets known as aerosols.

Aerosols have three different impacts on the global energy balance. First, they tend to reflect light back into space (the *direct cooling effect*). Second, large particles absorb sunlight and radiate infrared radiation in all directions, including toward the surface (the *direct warming effect*). Finally, aerosols can increase the density of clouds (the *indirect effect*).

The most important aerosols are sulfates and smoke particles. Power plants that burn coal, as well as copper, lead, and zinc smelters, release sulfur dioxide, which reacts with water vapor in the atmosphere to form sulfates. Sulfates currently reflect enough light back into space to reduce the amount of energy striking the earth's surface 0.2–0.8 W/m^2. Unlike most greenhouse gases, which remain in the atmosphere for decades or longer, precipitation removes most sulfates within a few weeks. As a result, sulfates tend to be concentrated in the areas immediately downwind of major industrial areas.

Smoke from oil and coal has an ambiguous impact: black carbon particles absorb sunlight and heat the atmosphere, while organic particles behave more like aerosols and reflect light back into space. IPCC (2001) estimates that black carbon provides a positive forcing of 0.1–0.4 W/m^2 with 0.2 the most likely value, and that organic carbon has a forcing of about −0.1 W/m^2. It is not clear whether mineral dust has a warming or cooling effect. A final confounding factor is the so-called indirect effect of aerosols on clouds. Aerosols appear to increase the concentration of droplets while decreasing the size of the typical droplet, which leads clouds to scatter more light back to space. Sulfates alone appear to have an indirect effect of −0.3 to −1.8 W/m^2, which may be greater than the direct cooling effect from sulfates.

Trends related to global climate change

Data on a wide variety of environmental indicators are consistent with the consequences that scientists generally expect to result from increasing concentrations of greenhouse gases. This section summarizes data reported in IPCC (2001).

Temperatures

Global temperatures are rising. Reasonably good data from thermometers are available for land surface temperatures from approximately 1860 to the present. These data suggest that the average land surface temperature has risen 0.3–0.8°C (0.6–1.5°F) in the last century (IPCC, 2001). Studies that combine land and sea measurements have generally estimated that global temperatures have warmed 0.4–0.8°C (0.7-1.4°F) in the last century. As Figure G21 shows, about two-thirds of this warming took place during 1900–1940. Global temperatures declined slightly from the 1940 to 1975, but rose more rapidly during the last 25 years than in the period before 1940.

Surface temperatures are not rising uniformly. The winters between 50°N and 70°N (the latitude of Canada and Alaska) are warming relatively fast, while summer temperatures show little trend. Urban areas are warming somewhat more rapidly than rural areas, because of both the changes in land cover and the consumption of energy occurring in densely developed areas (the latter is known as the "urban heat island" effect). All of the contiguous United States warmed during the last century, except for a portion of the southeast. No one has adequately explained why this region cooled while other areas warmed.

Nighttime low temperatures appear to be rising about twice as rapidly as daytime highs. This decreased diurnal temperature range is generally occurring in areas where cloud cover is increasing. Unfortunately, the data required to analyze the diurnal temperature range are available only for the period 1950–1993, and only about half the world's land areas.

The temperature trends for the atmosphere vary by altitude. For the period 1979–1999, satellite observations and weather balloons show very small warming trends of 0.04–0.05°C/decade in the lower troposphere (surface to 8 km), even though the earth's surface warmed about 0.15°C per decade over the same period. By contrast, during the period 1958–79, there was almost no change in the average surface temperature, while the lower troposphere warmed close to 0.2°C/decade. There seems to be no trend in the upper troposphere (8–18 km). Finally, the stratosphere is cooling: Observations from weather balloons suggest that the lower stratosphere (15–23 km above the surface) cooled about 0.1–0.3°C/decade from 1959 to 1979, and 0.5–1.2°C/decade from 1979 to 1999. The more comprehensive satellite observations available since 1979 suggest a cooling of about 0.5°C in the lower stratosphere.

Scientists are still debating the importance of the various atmospheric temperature trends (or lack thereof). Although some nonspecialists are confused by cooling in the upper atmosphere, climatologists have long expected that greenhouse gases would cool the stratosphere, as discussed above. Moreover, in addition to the expected greenhouse

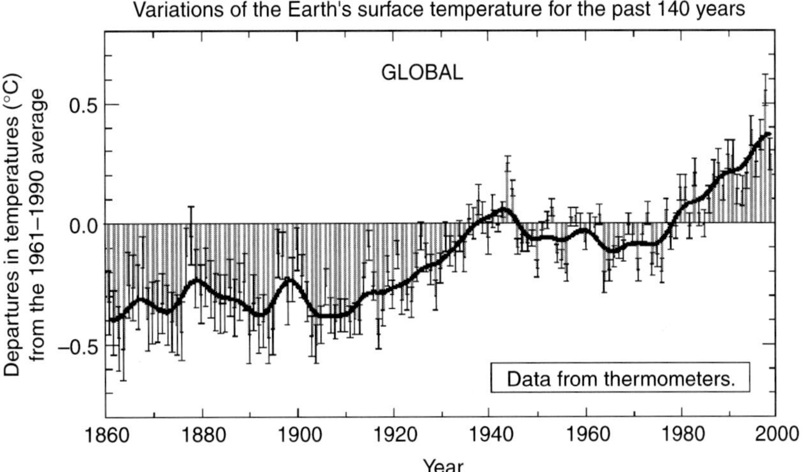

Figure G21 The temperature at the earth's surface based on thermometers, 1860–2000. The solid line is a 20-year moving average. Error bars for annual observations represent two standard deviations.
Source: IPCC (2001).

cooling, ozone depletion from chlorofluorocarbons probably cool the stratosphere: ozone absorbs incoming ultraviolet radiation and thereby warms the stratosphere, but less ozone means less warming from ultraviolet.

The lack of a recent warming in the troposphere is harder to explain. Climate models generally have projected that the lower atmosphere would warm by about the same magnitude as the earth's surface. The Intergovernmental Panel on Climate Change (2001) believes that about half of the discrepancy can be explained by "observation biases" caused by differences in coverage and the timing of volcanic eruptions and the El Nino cycle. The other half of the discrepancy may represent a real difference over the 20-year period. Because the surface and the troposphere are two different entities, it is not necessarily surprising that they are following different trajectories.

The world's oceans are also warming. The average sea surface temperature has risen approximately 0.6°C in the last century. One would expect the lower layers of the ocean to warm more slowly, because of the great heat-absorbing capacity of the oceans. The upper 300 m appear to have warmed at a rate of approximately 0.3°C/century. Around Bermuda, at the 1,500–2,500 dbar depth, temperatures warmed 0.05°C/decade from 1922 to 1995. Between 800 and 2,500 m below the surface, ocean temperatures have warmed about 0.5°C in the area bounded by 52–66° W and 20–32°N (IPCC, 2001).

Is the last century's warming the start of a new trend, or simply part of a natural cycle that has always caused temperatures to rise and fall? Because thermometer data is sparse before 1860 outside of a few countries, scientists must rely on tree rings, corals, ice cores, pollen, deposits from mountain glaciers, and—in some cases—historical documentation. These sources are much less precise than thermometers, and in some cases they are able to show direction of change but not the actual temperature. Nevertheless, their estimates have been validated by comparisons with the instrumental record. Figure G22 shows the best available reconstruction of temperatures over the last one thousand years. These data suggest that Northern Hemisphere temperatures fluctuated within a 0.4°C range from AD 1000 to 1900, before warming approximately 0.8°C in the last century. The final decade of the second millennium appears to have also been the warmest. Because this warming is more than twice the standard deviation of previous fluctuations, it is probably not simply a random fluctuation.

The hydrologic cycle

Precipitation has increased by about 2% over the world's continents since 1900. Based on over 20,000 data stations, this estimate is statistically significant, but geographically variable. Precipitation increased 9–16% on average over the land areas between 30°N and 85°N, but only about 3% in the areas between 30°S and 55°S. Overall, precipitation declined between 10°N and 30°N.

Throughout most of the United States, precipitation has increased 5–10% in the last century. Along the northern tier states and in much of Canada, rainfall has increased 10–15%. Much of this increase has taken place between September and November; in the southeast, rainfall has actually declined between June and August. Rainfall is also tending to be more concentrated in heavy downpours. At the beginning of the 20th century, only 9% of the nation experienced a storm each year in which more than 2 in. of precipitation fell in a 24-h period. In recent decades, such a severe storm has occurred each year over close to 11% of the nation (see Figure G23). The trend toward more concentrated rainfall is apparent in Canada and much of Europe as well.

Humidity is also increasing. In the United States, both relative humidity and the dew point have been rising, particularly at night. There is some evidence that absolute humidity is increasing worldwide at several different altitudes.

Clouds, by contrast, may offset this warming. Because clouds reflect sunlight back into space and absorb outgoing infrared radiation, and increase in clouds can tend to limit daytime warming but increase the warming at night. Observations indicate that cloud cover has increased over the United States and Soviet Union. These increases are highly correlated with the decline in the diurnal temperature range.

Snow and ice cover

The world's ice cover can be divided into five categories: polar ice sheets, small glaciers, snow cover, ice shelves, and sea ice. The polar ice sheets of Greenland and Antarctica contain enough ice to raise sea level about 70 m. Most of those ice sheets are in areas that are too cold for ice to melt even if temperatures warm a few degrees. No one has demonstrated that greenhouse gases are responsible for any trends in the mass of these ice sheets.

The world's small glaciers and ice caps retreated enough to raise sea level about 5 cm since 1900. In the tropics, where snow is only found at high altitudes, the freezing line is about 100 m higher than two decades ago and glaciers are retreating in all areas. The European Alps glaciers have retreated enough to expose areas that have been covered by ice for several thousand years, revealing, for example, the 5,000-year-old Oetzal "ice man" and various ancient artifacts.

Snow cover in the Northern Hemisphere has decreased about 10% since 1966. Satellite records show that snow cover has retreated earlier in the spring and summer, but it has not changed during fall and winter. Assessments based on weather station records suggest that the area covered by snow on average over the course of a year during the last decade was less than any decade during the last century. During winter and fall, the amount of snow cover may have increased slightly.

Between 1978 and 1996, the area of sea ice has decreased in the Northern Hemisphere, according to satellite data, especially during summer. Over the last century, sea ice has declined about 15% during summer, 8% during spring, and very little during autumn and winter. The seasonality of the decline contradicts the calculations of most climate models, which project greater reductions of sea ice during winter. Sea ice appears to have *increased* about 3% in Antarctica in the last two decades; but the data are insufficient for estimating a longer-term trend.

The large Ross and Filchner-Ronne ice shelves in western Antarctica, hundreds of meters thick, do not have a discernible trend. Several smaller

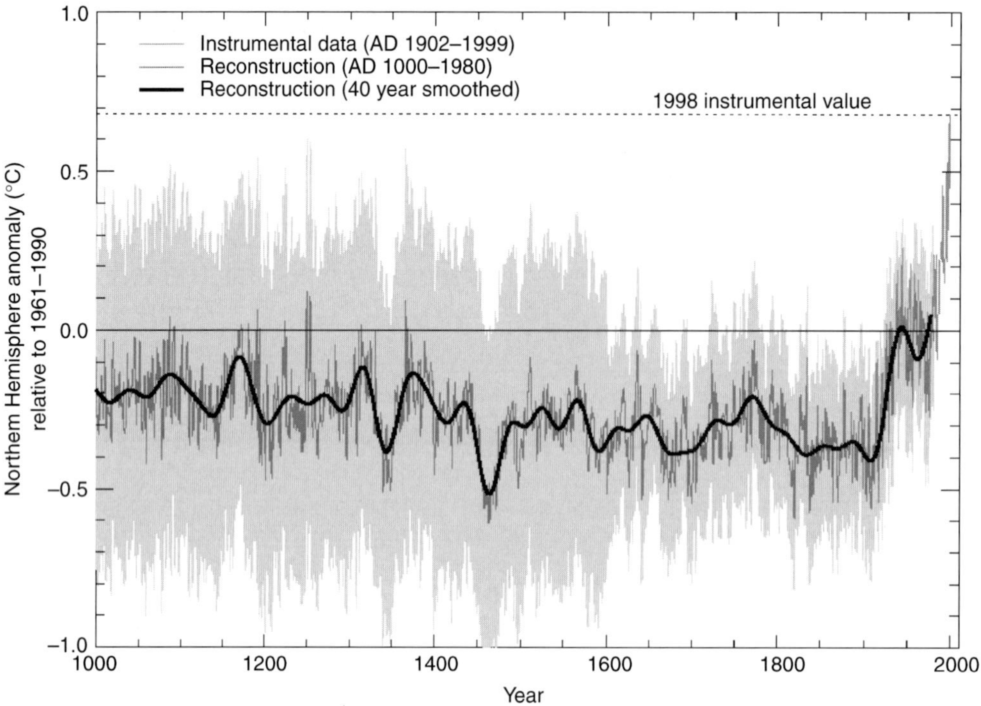

Figure G22 Average temperature of the Northern Hemisphere based on geological indicators, AD 1000–2000. Data based on tree rings, corals, ice cores, and historical records for 1000–1900, and thermometers from 1900–2000. The shaded area represents two standard deviations *Source*: Mann *et al.* (1999) as modified by IPCC (2000).

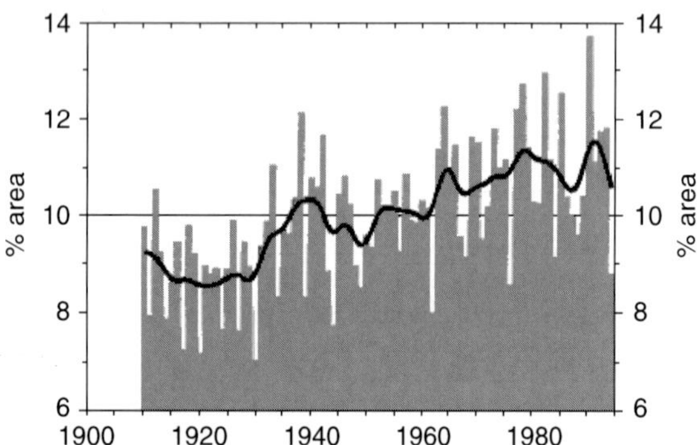

Figure G23 Portion of the United States experiencing at least 2 in. of rain in a single calendar day, at least once in a given year. *Source*: Tom Karl (1995, 1993), National Climatic Data Center.

ice shelves around the Antarctic Peninsula, however, have been retreating, possibly as a result of the 2°C warming in that region over the last 40 years. Climate models have not attributed this region's warming to greenhouse gases.

Sea level

Table G4 summarizes the various contributors to sea level rise. During the 20th century, sea level rose 10–20 cm, about half of which can be attributed to global warming. Approximately 2–4 cm resulted from the melting of small glaciers and ice caps, which most researchers attribute entirely to global warming; and models estimate that global warming should have caused ocean water to expand enough to raise the sea another 3–7 cm. Global warming appears to have had a negligible impact on the polar ice sheets. However, the long-term adjustment of the Antarctic ice sheet to the warming since the last Ice Age may have caused a contribution up to 5 cm. Changes in terrestrial water storage have both added to (e.g., groundwater depletion, runoff from urbanization, and wetland loss) and subtracted from sea level (e.g. dams, infiltration from irrigation), but the net effect may still be significant. The best estimate of the contributions to sea-level rise (7 cm) is about half the best estimate of observed sea-level rise (15 cm), implying either that one of the contributors is substantially underestimated, or the true value of most of the contributors is near the high end of the uncertainty range.

Projections of the future

Emissions, concentrations, and radiative forcing

The extent and speed at which humanity changes the climate will depend to a large extent on the rate at which society adds additional greenhouse gases to the atmosphere. If we continue to use today's technology, then as populations and the economy grow, emissions of CO_2 will continue to increase. Until a few decades ago, energy consumption grew at about the same rate as the Gross National Product. After oil supplies were disrupted in 1973 and 1979, however, people made substantial efforts to decrease energy consumption, including better insulation, smaller cars, and more energy efficient appliances. More recently, relatively low gasoline prices, economic growth, and demographic trends have led Americans to buy larger cars, minivans, and light trucks that consume more gasoline than the compact cars that were more in vogue a decade ago. Nevertheless, many businesses are continuing to institute energy conservation measures. As a result, US emissions have grown 1.2% per year since 1990, while the economy grew 3.1% per year.

Worldwide, emissions have grown approximately 1.7% per year since 1900, but only about 1.0% since 1990. CO_2 has increased several percent per year in some developing nations, but remained roughly stable in Western Europe, and declined due to the economic contraction in the nations comprising the former Soviet Union. Japan, Canada, and Australia experienced roughly the same emissions growth as the United States.

Will CO_2 emissions continue to grow in the next century? The answer depends both on the evolution of our way of life, and on whether governments take measures to reduce emissions. The IPCC (2000) has developed a large set of scenarios for analyzing emissions trajectories assuming that no policies are enacted to reduce CO_2 emissions. Table G5 illustrates some of the key aspects of four of the most commonly cited scenarios.

Higher economic growth does not necessarily mean higher CO_2 emissions. In IPCC's Scenario "A1T," world population increases from

Table G4 Comparison of observed sea-level rise with estimated rise in sea level attributable to various causes

	Presumed cause	Source of estimate			
Thermal expansion	Greenhouse gases	Model	0.3	0.5	0.7
Small glaciers and ice caps	Local warming	Measurements	0.2	0.3	0.4
Greenland	Regional warming	Model	0	0.05	0.1
Antarctica	Regional warming	Model	−0.2	−0.1	0.0
Greenland and Antarctic ice sheets	Long-term post-glacial adjustment	Measurements	0.0	0.25	0.5
Permafrost	Global warming	Model	0.00	0.025	0.05
Sediment deposition	Soil erosion	Calculations from measurements	0.00	0.025	0.05
Terrestrial storage	Dams and groundwater pumping	Calculations from observations	−1.1	−0.35	0.4
Total of component estimates	Various	Various	−0.8	0.7	2.2
Total measured sea-level rise		Observations	1.0	1.5	2.0

Source: Columns 1, 4, 5, 6 are from IPCC (2001), table 11.10. Columns 2 and 3 based on text of IPCC (2001), chapter 11.

Table G5 Characteristics of four key emissions scenarios

		Scenarios			
	Recent[a]	A1T	A1F1	A2	B1
Population (billions)	5.3				
2050		8.7	8.7	11.3	8.7
2100		7.0	7.1	15.1	7.0
Gross world product (trillions)	21				
2050		187	164	82	136
2100		550	525	243	328
Per-capita ($US) in 2050	3,900	21,500	18,850	7,256	15,632
Energy intensity (10^6 J/US$)	16.7				
2020		8.7	9.4	12.1	8.8
2050		4.8	6.3	9.5	4.5
2100		2.3	3.0	5.9	1.4
Nonfossil fuel energy (%)	18				
2020		21	15	8	21
2050		43	19	18	30
2100		85	31	28	52
Fossil CO_2 Emissions (GtC)	6.0				
2020		10.0	11.2	11.0	10.0
2050		12.3	23.1	16.5	11.7
2100		4.3	30.3	28.9	5.2
Cumulative CO_2 emissions 1990–2100 (GtC)		1,068	2,189	1,862	983
Atmospheric CO_2: 2100 (ppm)	367	570	970	870	540
CO_2 forcing (W/m^2) 1990–2050		1.62	2.24	1.90	1.46
CO_2 forcing (W/m^2) 1990–2100		2.39	5.15	4.42	2.06

[a] 1990 values except for atmospheric CO_2, which represents concentration in 1999.
Source: Information extracted from text and tables of IPCC (2000, 2001).

5.3 billion today to 8.7 billion by 2050, before declining to 7 billion in 2100. The gross world product grows by about 3.7% per year from 1990 to 2050, but only 2.2% from 2050 to 2100, leading to a 20-fold increase in per capita income. But in this scenario, increased wealth and stabilizing population comes along with increased reliance on non-carbon based energy, from 18% today to 85% in 2100. As a result, emissions of CO_2 increase from 6 Gt/yr in 1990 to 12.3 Gt/yr in 2050, but decline to 4.3 Gt/yr in 2100.

Emissions increase to 30 Gt/yr under scenario A1T, which assumes a similar amount of economic growth, but continued reliance on fossil fuels. Scenario A2 leads to similarly high CO_2 emissions, under very different assumptions. In this scenario, per capita income only doubles over the next century, because world population increases to 15.1 billion, while the gross world product only increases 2.3% per year. In spite of the seven-fold difference in year 2100, the major differences do not emerge until the middle of the 21st century. Therefore, the cumulative emissions from 1990 to 2100 are only about twice as great for the high-emissions scenarios for the low-emissions scenarios.

How will atmospheric concentrations change? Both the oceans and plants remove CO_2 from the atmosphere, through a number of processes that themselves depend on climate. Carbon dioxide dissolves into the upper layers of the ocean. Solubility, however, declines as temperatures increase. Moreover, the ability of the upper layers to take up CO_2 is limited, and is maintained by ocean circulation in which water with high CO_2 concentrations sinks while deep water with low concentrations of CO_2 is brought to the surface. If global warming decreases ocean circulation, then the oceanic removal of CO_2 would be slowed. The terrestrial biosphere can store some of the CO_2 released into the atmosphere, both because climate warming lengthens the growing season in cold regions and because CO_2 itself enhances photosynthesis. However, if insufficient water is available, warmer temperatures could also enable soils to release carbon dioxide into the atmosphere.

The models that have been developed to calculate how much CO_2 will remain in the atmosphere imply that CO_2 concentrations will increase throughout the next century for all of the IPCC emissions scenarios. IPCC models suggest that concentrations are likely to increase from the current concentration of 367 ppm to somewhere between 540 and 970 ppm, that is anywhere from 2 to 3.5 times the preindustrial concentration of CO_2. Today, the concentration of 367 ppm is about 90 ppm greater than the preindustrial concentration. Thus, IPCC's scenarios imply that in the next century humanity will increase atmospheric CO_2 by 2–6 times as much as we increased CO_2 in the last century.

The scenarios all imply somewhere between 1.5 and 2 W/m^2 additional radiative forcing through the year 2050, and anywhere from 2 to 5 W/m^2 through the year 2100. The IPCC undertook similar assessments for the other greenhouse gases, and the key aerosols. Considering all the greenhouse gases and aerosols, the next century seems likely to see an increased radiative forcing somewhere between 2 and 7 W/m^2,

Table G6 Change in radiative forcing due to greenhouse gases and aerosols (watts per square meter, global annual average)

Contributor	1750–1990	A1T	A1F1	A2	B1	B2
Forcing: 1990–2050						
Carbon Dioxide	1.46	1.62	2.24	1.9	1.46	1.37
Methane	0.48	0.25	0.3	0.27	0.04	0.02
Nitrous Oxide	0.15	0.08	0.18	0.17	0.12	0.08
Tropospheric Ozone	0.35	0.37	0.66	0.43	0.04	0.28
Stratospheric Ozone	−0.15					
Sulfates	−0.4	0.17	−0.07	−0.21	0	0.08
Black Carbon[a]	0.4	0.35	0.49	0.21	−0.16	0.17
Organic Carbon[a]	−0.3	−0.43	−0.61	−0.27	0.2	−0.21
Ozone Depleters[b]	0.328	−0.147	−0.147	−0.147	−0.147	−0.147
HFCs[c]	0.004	0.133	0.13	0.114	0.082	0.083
Fluorides[d]	0.006	0.019	0.019	0.02	0.012	0.019
	2.128	2.415	3.192	2.487	1.647	1.745
Forcing: 1990–2100						
Carbon Dioxide	1.46	2.39	5.15	4.42	2.06	2.73
Methane	0.48	0.14	0.51	0.59	−0.07	0.39
Nitrous Oxide	0.15	0.11	0.4	0.36	0.17	0.14
Tropospheric Ozone	0.35	0.11	0.89	0.87	−0.16	0.43
Stratospheric Ozone	−0.15					
Sulfates	−0.4	0.28	0.17	0.05	0.25	0.12
Black Carbon[a]	0.4	0.46	0.65	0.56	−0.2	0.44
Organic Carbon[a]	−0.5	−0.58	−0.82	−0.7	0.25	−0.54
Ozone Depleters[b]	0.328	−0.232	−0.232	−0.232	−0.232	−0.232
HFCs[c]	0.004	0.256	0.242	0.225	0.106	0.216
Fluorides[d]	0.006	0.052	0.052	0.061	0.028	0.049
	2.128	2.986	7.012	6.204	2.202	3.743

[a] IPCC uses slightly different categories for describing historic and future radiative forcing from black carbon and organic carbon. Table 6.11 reports values for historic fossil fuel black carbon, fossil fuel organic carbon, and the net effect from biomass, which includes both types of aerosol carbon. The projections in table 6.15 report total black carbon and total organic carbon, a convention that this table follows.

[b] Ozone depleters include CFC 11, 12, 113, 114, and 115; carbon tetrachloride, HCFC's 22, 141b,142b, and 123; halons 1211 and 1301, and Ch_3CCl_3.

[c] Includes HFC23, 125, 32, 134a, 143a, 152a, 227ea, 245ca, and 43-10mee.

[d] Includes carbon tetrafluoride, ethyl fluoride, butyl fluoride, and sulfur hexafluoride.

Source: Adapted from the text of chapter 6 and tables 6.1, 6.11, 6.14, and 6.15 of IPCC, 2001.

compared with a historic contribution of approximately 2 W/m^2 (Table G6). Carbon dioxide will continue to be by far the most important contributor, responsible for 70–80% of the anthropogenic greenhouse warming.

Future climate

Global temperatures. Since 1979, scientists have generally agreed that a doubling of atmospheric carbon dioxide increases the earth's average surface temperature by 1.5–4.5°C (3–8°F). More recent studies have suggested that the warming is likely to occur more rapidly over land than the open seas. Moreover, the warming in temperatures tends to lag behind the increase in greenhouse gases. At first, the cooler oceans will tend to absorb much of the additional heat and thereby decrease the warming of the atmosphere. Only when the ocean comes into equilibrium with the higher level of CO_2 will the full warming occur.

As a result of the delay induced by the oceans, climate scientists do not expect the earth to warm by the full 1.5–4.5°C (3–8°F) during the next century, even though the level of CO_2 is expected to more than double and other greenhouse gases would add to the warming. In 1995, the Intergovernmental Panel on Climate Change projected a warming of 1.0–3.5°C (1.8–6.3°F) by the year 2100; the more recent IPCC (2001) estimated a range of 1.4–5.8°C, largely due to the inclusion of a few emissions scenarios that seem very unlikely. Wigley and Raper (2001) estimated a 90% confidence interval that global temperatures will rise 1.7–4.9°C from 1990 to 2100.

Sea level. The warmer temperatures are expected to raise sea level by expanding ocean water, melting mountain glaciers, and melting parts of the Greenland Ice Sheet. Warmer temperatures also increase precipitation, as described below. Snowfall over Greenland and Antarctica is expected to increase by about 5% for every 1°F warming in temperatures. Increased snowfall tends to cause sea level to drop if the snow does not melt during the following summer, because the water will either be on land or in the ocean. (The amount of water in the atmosphere is less than the water it takes to raise the oceans by 1 mm.) Considering all of these factors, the IPCC estimates that sea level will rise 20–86 cm by the year 2100. Environmental Protection Agency (EPA, 1995) estimated that global sea level has a 50% chance of rising 45 cm (1.5 ft) by the year 2100, but a 1-in-100 chance of a rise of about 110 cm (over 3.5 ft).

Over the longer run, more substantial changes in sea level are possible. Some scientists believe that the West Antarctic Ice Sheet could slide into the oceans after a sustained warming, or if other factors raised sea level. The vulnerability of this ice sheet is poorly understood. It contains enough ice to raise sea level 6 m (20 ft), and coastal scientists generally agree that sea level was 20 ft higher than today during the last interglacial period, which was only slightly warmer than today. While some scientists have suggested that there is fossil evidence on the polar ocean floor that this ice sheet collapsed during the last interglacial period, there is no scientific consensus on this question.

Two risk assessment studies involving panels of experts (EPA, 1995; Vaughan and Spouge, 2002) have sought to quantify the potential magnitude of unlikely-but-possible scenarios under which west Antarctica makes a substantial positive contribution to sea level. Almost all of the experts have concluded that Antarctica is most likely to have a negligible contribution to sea level over the next century. Nevertheless, they all agreed that there is some risk that a catastrophic collapse of the ice sheet could occur over a couple of centuries if polar water temperatures warm by a few degrees, with a contribution of about 30 cm in the next century but more than a meter during each of the following centuries. Most of the scientists estimated that such a risk had a probability of between 1 and 5%, Because of this risk, as well as the possibility of a larger than expected melting of the Greenland Ice Sheet, the EPA study estimated that there is a 1% chance that global sea level could rise by more than 4 m (almost 14 ft) in the next two centuries. Although the IPCC (2001) did not include these extreme scenarios, they did acknowledge their scientific basis.

Glaciologists generally agree that a sustained warming of Greenland could cause an irreversible loss of the ice sheet there. Even in today's climate, the temperature at sea level is too warm for an ice sheet to persist; but most of the ice sheet is well above sea level and hence colder. IPCC (2001) reviews several modeling studies that suggest that 3°C and 5.5°C warming of Greenland temperatures would lead Greenland to contribute 1 and 3 mm/yr, respectively, over the next 1,000 years.

The rates of sea-level rise at specific locations will diverge from the global average rise, for two reasons. First, land movements caused by a variety of factors currently cause relative sea-level changes to vary from a drop of several millimeters per year in parts of Alaska and Scandinavia, to rises of more than 1 cm in a few areas where groundwater pumping has accelerated the rate of sea-level rise. Second, global warming may not have a uniform effect on sea level: Changing climate may alter local sea level through changes in winds, currents, and pressure. Within estuaries, changes in rainfall and evaporation may cause the mean tide level to diverge from the trends in the open ocean. Moreover, the earth's center of gravity would migrate away from the polar ice sheets if their ice was discharged into the oceans.

Regional climate change: climate models. Projecting the change in climate for particular regions is more difficult than estimating global warming. There is a general consensus that temperatures will warm throughout the United States. However, scientists are unable to say whether particular regions will receive more or less rainfall; and for many regions they are unable to even state whether a wetter or a drier climate is more likely.

Virtually all published estimates of how the climate could change in the United States are the results of computer models of the atmosphere known as "general circulation models." These complicated models are able to simulate many features of the climate, but they are still not accurate enough to provide reliable forecasts of how the climate may change; and the several models often yield contradictory results. For the time being, however, these models are about all we have to say how the climate may change in particular areas. This leaves many researchers with a rather unsatisfactory situation: the climate models have been reviewed as acceptable descriptions of global climate with explicit provisos that one should not use their results to protect regional climate change; but those who want to analyze site-specific impact of greenhouse gases invariably use them anyway.

Given the unreliability of these models, researchers trying to understand the future impacts of climate change generally analyze different scenarios from several different climate models. The hope is that, by using a wide variety of different climate models, one's analysis can include the entire range of scientific uncertainty. Even this assumption probably overstates the extent to which existing scientific understanding has been quantified. *The projections of climate change in specific areas are not forecasts but are merely reasonable hypothetical examples of how the climate might change.*

Regional temperatures

The historical temperature record shows that a rise in the global average temperature does not automatically imply that every part of the world warms. The cooling from sulfates may offset the warming in some areas. Moreover, natural fluctuations in the jet stream and other factors often can cause some areas to be unusually cool when others are warm. During the summer of 1988, for example, when the eastern United States suffered severely hot and dry weather, cold relatively deep ocean water began to flow to the sea surface off the mid-Atlantic Coast, keeping the coastal zone unusually cool. Scientists have not ruled out the possibility that global warming could induce such shifts, which could lead to little or no warming in some areas while other areas warm by much more than the 1.0–3.5°C (3–8°F) expected for the world as a whole.

One of the most seriously discussed regional shifts in climate concerns the Gulf Stream, which keeps northern Europe much warmer than the parts of Canada, United States, and Russia at similar latitudes. The current is largely propelled by the sinking water in the North Atlantic off the coast of Greenland. The warm Gulf Stream waters evaporate more, and have higher salinity levels, than the surrounding ocean. At the middle latitudes, the warm saline water remains at the surface because the higher temperature more than offsets the density increase caused by high salinity. At higher latitudes, however, the cold polar climate cools all the water to within a few degrees of freezing. With all the water at approximately the same temperature, the more saline Gulf Stream water sinks. Many climate models, however, predict that warmer temperatures will increase rainfall over the Gulf Stream enough to reduce its salinity, which may reduce the amount of sinking water, and weaken the current. The possibility of a sudden shutdown of this deepwater formation cannot be ruled out. Such a shutdown would cause Europe to cool even while the rest of the earth warmed!

The region of the United States that has been the most thoroughly examined is the area from 35°W to 50°N and 85°N to 105°W. Climate model results suggest that if global temperatures warm 2.6°C, the combined impact of aerosols and greenhouse gases is likely to warm this region approximately 1.5–3.5°C (3–6.5°F) during winter. The same models suggest, however, that summer temperatures will warm between 0°C and 0.5°C (less than 1°F). Our actual uncertainty for future temperature change is probably at least twice as great as these ranges suggest, because global warming is also uncertain.

Regional and hydrology

Water resources are sensitive both to rising temperatures and changes in precipitation. Although scientists expect global temperatures to rise approximately 0.5 to 1.5°C (1–3°F) by the year 2050, most climate models suggest that warming over land will be greater than the warming over the sea. Because higher temperatures increase evaporation and plant transpiration, rainfall would generally have to increase just to maintain current levels of water availability. Holding other factors constant, the potential for evaporation and transpiration increases about 5–10% per degree (C) throughout most of the United States (Waggoner and Revelle, 1990).

There is a general consensus that annual worldwide precipitation and evaporation will increase a few percent for every degree of warming; this prediction is consistent with current trends. But there is considerably less certainty about rainfall in particular locations, and whether the rainfall will increase enough to offset the increased evaporation. Many scientists, however, believe that middle latitudes such as that of the United States will see drier summers. Assuming that the land warms more than the sea, evaporation over the land will increase by more than the evaporation over the sea that produces rainfall. Thus, summer rainfall may not increase by as much as evaporation.

For specific locations, however, it is currently impossible to confidently project even the direction, let along the magnitude or timing, of the seasonal or even annual changes in precipitation. In the Central North American region, most models estimate that rainfall will increase approximately 10% during winter, but the models disagree about whether summer rainfall will increase or decrease. The scenarios that show an increase in precipitation also project warming of 4–5°C (7–9°F), which would generally cause evaporation to increase by 20–50%. Thus, the models generally suggest that summers in the United States will be drier.

Whether or not annual or seasonal rainfall increases, many climate models project that rainfall will occur in a smaller number of heavier storms. The intensity increase would result both because the warmer atmosphere holds more water vapor, and because greenhouse gases increase the radiative cooling in the upper atmosphere, which induces intense rainstorms. Several models project that even if summer rainfall in the mid-western US declines slightly, heavy rainstorms would occur about twice as often. The National Center for Atmospheric Research also expects fewer but heavier rainstorms. Data on existing trends are consistent with this general prediction.

Climate modelers have long suggested that drought may also become more severe in the mid-latitudes. Scientists at NASA have suggested that in the long run, a worldwide expansion of deserts is likely (Rind *et al.*, 1990). Today, deserts tend to be found at latitudes between 20° and 32°, where potential evapotranspiration is generally greater than the average rainfall. The NASA study showed that with a 4°C warming, the potential evapotranspiration would increase 30–40%, while precipitation would only increase 10–15%. Even in the nearer term, deserts may expand. Land areas are likely to warm more rapidly than the oceans; so evaporation on land is likely to increase more than evaporation from the oceans. Because oceanic evaporation ultimately drives precipitation, it follows that evaporation would be likely to increase by more than precipitation in most continental areas. Most of the recent model studies have confirmed this early NASA result (IPCC, 2001).

Storms

Hurricanes require water temperatures of at least 79°F (26°C) to form. They often gain energy when they cross warm currents such as the Gulf Stream, and lose energy when crossing cooler waters. Therefore, scientists and popular accounts of global warming have long suggested the potential for global warming to increase both the frequency and intensity of hurricanes. Nevertheless, hurricane experts seeking to verify this risk have generally questioned such a presumption. First, there is no evidence that hurricane frequency or intensity has been increasing as global temperatures warmed during the last century. Second, hurricanes also require that wind sheer be minimal. Wind sheer tends to be high in the Atlantic and low in the Pacific during El Nino years, which many models suggest will occur more frequently as global temperatures warm. Finally, hurricanes tend to occur more frequently when the Gulf

Stream and other major ocean currents are strongest; as mentioned above, many scientists expect these currents to weaken as the earth warms. Putting all of these factors together, IPCC (2001) concluded that we simply cannot say whether hurricanes will become more or less frequent; but that a modest intensification from those that do occur seems likely.

Extratropical storms such as northeasters seem likely to decline in frequency, because of the reduction in the temperature difference between the poles and the equator that is generally expected to occur. But the intense storms caused by deep low-pressure may become more severe, because these lows are exacerbated by the latent heating associated with increased evaporation at higher temperatures.

Implications for the coastal zone

The foregoing discussion suggests that global warming is likely to alter many of the environmental conditions in the world's coasts. Warmer temperatures will lead plant and animal species to move into what are now cooler areas. Such migrations will be relatively easy for migratory birds and fish that swim along the oceans; but it may be more problematic for corals which require a particular substrate, or fish that inhabit the northern Gulf of Mexico and thus are unable to swing any farther north. People whose vacation habits are motivated largely by climate may be among the most rapid species to adapt. Intense rainstorms may increase pollution runoff, while longer droughts may increase salinities in estuaries. The changing storm patterns may increase or decrease wave-driven beach erosion. The many impacts of sea-level rise are discussed in *Sea-Level Rise, Effect*.

Will these changes be rapid enough to matter to most people who inhabit the coastal zone, or do they merely constitute a changing background condition that must be considered by scientists and engineers but not the public at large? The available research suggests that sea level is already rising rapidly enough to matter to those who inhabit the most vulnerable areas, and so the accelerating rise is bound to be relevant to an increasing fraction of the world's coastal inhabitants. Rising temperatures are starting to have some effects, but it is far from clear what—if anything—people can do about them. The big unknown remains the possible changes in hydrology and storminess. No one knows whether those changes will be good or bad, or whether they will be gradual like sea-level rise or so sudden that by the time we know for sure what to expect, it will have already happened.

James G.Titus

Note: Project Manager for Sea Level Rise, US Environmental Protection Agency. Member Maryland and DC Bars. The opinions expressed do not necessarily reflect the views of the US Government, which reserves the right to duplicate this article for official use.

Bibliography

Arrhenius, S., 1896. On the influence of carbonic acid in the air upon the temperature of the ground. *Philosophical Magazine*, **41**: 237–276.

Environmental Protection Agency, 1995. *The Probability of Sea Level Rise*. Washington, DC: US Environmental Protection Agency.

Hayes, J.D., John Imbrie, and Shackleton, N.J., 1976. Variations in the earth's orbit: pacemaker of the Ice Ages. *Science*, **194**: 1121–1132.

IPCC (Intergovernmental Panel on Climate Change), 1996. *Climate Change 1995: The Science of Climate Change*. New York: Cambridge University Press.

IPCC (Intergovernmental Panel on Climate Change), 2000. *Emissions Scenarios*. New York: Cambridge University Press.

IPCC (Intergovernmental Panel on Climate Change). 2001. *Climate Change 2001: The Scientific Basis*. New York: Cambridge University Press.

Karl, T.R. *et al*., 1993. Recent variations of snow cover and snowfall in North America and their relation to precipitation and temperature variations. *Journal of Climate* **6**: 1327–1344.

Karl, T.R., Knight, R.W., and Plummer, N., 1995. Trends in high-frequency climate variability in the twentieth century. *Nature*, **377**: 217–220.

Kiehl, J.T., and Trenberth, K.E., 1997. Earth's annual global mean energy budget. *Bulletin of American Meteorological Society*, **78**: 197–208.

Mann, M.E., Bradley, R.S., and Hughes, M.K., 1999. Northern Hemisphere temperature during the past millenium. Inferences, uncertainties, and limitations. *Geophysical Research Letters*, **26**: 759–762.

Milankovich, M., 1930, Mathematical climatology and astronomical theory of climate change. In Koppen, W., Geiger, R. (eds.), *Handbuch der Klimatologie*, Vol. 1. Berlin Gebruder Borntrager, pp. 1–176.

Rind, D. *et al*., 1990. Potential evapotranspiration and the likelihood of future droughts. *Journal of Geophysical Research*, **95**: 9983–10005.

Vaughan, D., and Spouge, J., 2002. Risk estimation of collapse of the west antarctic ice sheet. *Climatic Change*, **52**: 65–91.

Waggoner, P.E., and Revelle, R.R., 1990. Summary. In Waggoner, P.E. (ed.), *Climate Change and U.S. Water Resources*. New York: John Wiley & Sons.

Wigley, T.M.L., and Raper, S.C.B., 2001. Interpretations of high projections of global mean warming. *Science*, **293**: 451–454.

Cross-references

Changing Sea Levels
Climate Patterns in the Coastal Zone
Coastal Climate
Coastal Temperature Trends
Demography of Coastal Populations
El Niño–Southern Oscillation
Eustacy
Meteorologic Effects on Coasts
Sea-Level Changes During the Last Millenium
Sea-Level Rise, Effect

GROSS TRANSPORT

Gross transport refers to the total movement of sand approximately parallel to the shoreline over a specified period, often one year (see entries on *Cross-shore Sediment Transport* and *Longshore Sediment Transport*). Waves approaching the shore at an oblique angle, such that the crests of the breakers are at an angle to the shoreline, generate longshore currents that convey the mobilized sand along the shore. In many locales, the wave approach direction varies over time such that this transport can be in either direction relative to the shoreline (see entry on *Waves*). Consequently, sand may move upcoast for an interval and then, as wave conditions change or the lesser effects of tidal or wind-driven currents intervene, reverse direction and move downcoast. Gross transport amounts or rates account for all transport regardless of direction. The units of gross transport are typically volumetric rates (cubic meters per year).

There are difficulties associated with estimating gross transport. First, an estimate requires comprehensive sets of data collected over long periods. These would typically include the height and the direction of breaking waves measured for long enough intervals to average out the randomness and often enough to include the real variation in waves brought about by changing atmospheric forcing. No autonomous method for measuring either the height or the angle of breakers has been developed. However, a technique for making autonomous measurements outside the surf zone and converting them to equivalent values for calculating longshore transport has been developed (Seymour and Higgins, 1978). Finally, the models for longshore transport are rudimentary and can result in estimates with very large errors, even when quality wave data are available. Only occasionally, special conditions exist that will allow direct measurement of gross transport (Seymour *et al*., 1981). One such condition is the construction of a jettied entrance that will function for a sufficiently long period as an effective trap for longshore transport in both directions.

The significance of gross transport to coastal engineers is related to understanding the fate of beach sand in the presence of some mechanism for sequestering or otherwise removing sand from the system. In other instances, the coastal engineer is usually concerned with the difference between the upcoast and downcoast transport (see entry on *Net Transport*). Entrances to tidally influenced bays or lagoons along sandy coasts will typically form shoals offshore and inshore of the entrance channel. These shoals can impound very large quantities of sand. If a new entrance is constructed (see entry on *Navigation Structures*) coastal engineers must be concerned with the impact on the adjacent beaches during the formation of these shoals. The availability of sand to form shoals is a function of the local gross transport so that this value is useful in determining strategies for remediation. Further, dredging of shoals in natural or constructed inlets may be required before they reach their equilibrium extents in order to maintain safe navigation channels. Again, the rates at which these channels will refill are a function of the gross transport. These data are useful in estimating the frequency and cost of dredging operations

Narrow continental shelves often contain submarine canyons (see entry on *Continental Shelf*). These underwater valleys often extend from close to the shore to great depths beyond the shelf. It is generally agreed that these canyons have been formed over geological time by erosion from episodic sand-laden gravity-driven flows (Heezen and Ewing, 1952). The presence of huge depositional fans at the canyon outlets indicates that these features can function as significant loss mechanisms for beach sand. The canyon head, near the shoreline, fills with sand delivered by longshore transport. Typically, the canyon will fill during strong longshore transport events regardless of their direction. Thus, gross transport rates provide an indication of the potential for losses at submarine canyons, which could have significant effects on the c ondition of associated beaches.

Richard J. Seymour

Bibliography

Heezen, B.C., and Ewing, M., 1952. Turbidity currents and submarine slumps, and the 1929 Grand Banks Earthquake. *American Journal of Science*, **250**: 849–878.

Seymour, R.J., and Higgins, A.L., 1978. Continuous estimation of longshore sand transport. In *Coastal Zone '78, Proceedings of the Symposium on Technical, Environmental, Socioeconomic and Regulatory Aspects of Coastal Zone Management*, American Society of Civil Engineers, 3, pp. 2308–2318.

Seymour, R.J., Domurat, G.W., and Pirie, D.M., 1981. A sediment trapping experiment at Santa Cruz, California. In *Proceedings of the 17th Coastal Engineering Conference*, American Society of Civil Engineers, 2, pp. 1416–1435.

Cross-references

Continental Shelves
Energy and Sediment Budgets of the Global Coastal Zone
Longshore Sediment Transport
Navigation Structures
Net Transport
Sediment Budget
Sediment Transport (See Cross-Shore Sediment Transport and Longshore Sediment Transport)
Waves

GROUND-PENETRATING RADAR

Originally designed for the purpose of environmental and geotechnical investigation of the shallow subsurface, the ground-penetrating radar (GPR) is a high-resolution geophysical technique which is being successfully used in stratigraphic research of sedimentary sequences. GPR was instrumental in imaging the internal architecture of a variety of recent geological settings: fluvial (Leclerc and Hickin, 1997; Roberts *et al.*, 1997), glaciofluvial (Beres *et al.*, 1995, 1999), periglacial (Jol *et al.*, 1996a; Busby and Merritt, 1999), aeolian (Schenk *et al.*, 1993; Harari, 1996; Jol *et al.*, 1998), and lake deltas (Jol and Smith, 1991; Smith and Jol, 1997). The application of GPR to coastal research has led to significant advances in our understanding of mesoscale (centimeters to 10's of meters) stratigraphy of marginal marine sequences and has already resulted in several radar facies models (Baker, 1991; FitzGerald *et al.*, 1992; Jol *et al.*, 1996b; van Heteren *et al.*, 1996, 1998; Smith *et al.*, 1999; Buynevich and FitzGerald, 2000; FitzGerald *et al.*, 2000). Knowledge of the principles and limitations of the GPR technique in coastal settings is essential if accurate and complete models of large-scale coastal development are attempted. A general introduction to the GPR method and discussion of its applications to coastal stratigraphic research are presented here.

Theoretical background and methodology

Ground-penetrating radar method is in many ways similar to seismic-reflection profiling, except that it uses electromagnetic (EM), rather than acoustic, waves with the frequency range of 10–2,000 MHz. The behavior of the EM waves, governed by Maxwell's equations, is controlled by such factors as the material's electric conductivity, magnetic permeability, and dielectric permittivity. The first two parameters control the attenuation and ultimately the penetration depth of the EM signal. Highly conductive (less dielectric) materials resulting in signal dissipation and loss. Sensitivity and frequency of the antenna, as well as the gain setting on the control unit also control the penetration and resolution of the GPR record (Figure G24; Davis and Annan, 1989; Conyers and Goodman, 1997; van Heteren *et al.*, 1998).

Dielectric permittivity is the ability of a material to become polarized and respond to incoming EM waves (von Hippel, 1954). Relative dielectric permittivity (ε_r) is a dimensionless constant which is obtained by comparing the permittivity of a particular material with that of a vacuum (which is one). This parameter controls the velocity of EM waves and is inversely related to it:

$$v = \frac{c}{\sqrt{\varepsilon_r}}, \qquad \text{(Eq. 1)}$$

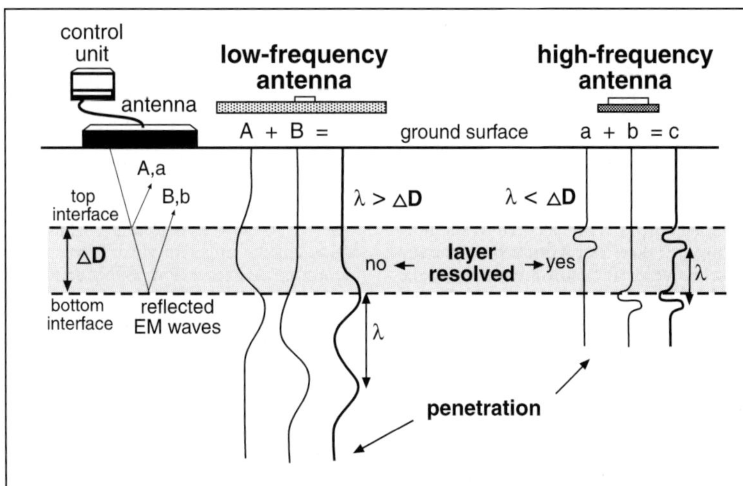

Figure G24 Generalized diagram depicting the differences in resolution and penetration depth for the low- and high-frequency antennae. Low-frequency wave (A) defines the top interface, but the distance between the interfaces (ΔD) is less than the wavelength (B). The relatively long wavelength (λ) of the resulting EM signal (C) does will not resolve both interfaces. A high-frequency EM wave defines both the top (a) and bottom (b) interfaces. The resulting wave (c) is shorter than the distance between the two interfaces. The penetration in this case, however, is less than with a low-frequency antenna (modified after Conyers and Goodman, 1997).

Table G7 Relative dielectric permittivity (ε_r) of common earth materials

Material	ε_r
Air	1
Ice	3–4
Freshwater	80
Seawater	81–87
Sand	
Unsaturated	3–7
Saturated	20–32
Unsaturated, with gravel	3.5–6.5
Saturated, with gravel	15.5–17.5
Silt	
Unsaturated	2.5–5.0
Saturated	20–30
Clay	
Unsaturated	2.5–5.0
Saturated	15–40
Bedrock	4–15

Source: Davis and Annan (1989); Conyers and Goodman (1997); van Heteren *et al.* (1998).

where v is the velocity measured in nanoseconds (1 ns = 10^{-9} s) and c is the speed of light (0.2998 m/ns). Typical values of relative dielectric permittivities for common geological materials are presented in Table G7. The magnitude of a particular subsurface reflection is directly related to the difference in dielectric properties of the two subsurface materials. From equation 1, the velocity of EM waves ranges from about 3 cm/ns in water to 30 cm/ns in air. Therefore, the groundwater table commonly appears as a strong sub-horizontal reflector on the radar trace. Above it, the unsaturated portion of the sedimentary sequence with lower permittivity values will be penetrated more readily by the radar signal. Consequently, a greater thickness of dry sediments can be represented on a particular record, compared with the part of the sequence below the water table, and will have a lower vertical exaggeration than the saturated portion. Once the velocities of EM waves or dielectric constants of specific lithologies (both saturated and unsaturated) are established and the time (t, ns) is obtained from the record, the approximate depth (D) to a particular target layer may be calculated using the two-way travel time between the ground surface and the layer in question, or between two layers as:

$$D = \frac{vt}{2} = \frac{ct}{2\sqrt{\varepsilon_r}}. \qquad \text{(Eq. 2)}$$

Sediment characteristics that affect the subsurface behavior of EM waves include pore-water and clay mineral content and their chemical composition, as well as magnetic properties of constituent minerals (Topp *et al.*, 1980; van Heteren *et al.*, 1998). High concentrations of ferromagnesian ("heavy") minerals (e.g., magnetite, ilmenite, garnet) that form as storm-lag deposits on beaches often produce prominent reflectors on GPR records. In some cases these sedimentological anomalies, as well as thick iron-stained horizons, may have high enough magnetic permeability values such that they attenuate the magnetic portion of the EM signal and preclude further penetration (Topp *et al.*, 1980).

Ground-penetrating radar profiles taken over areas with substantial relief need to be topographically corrected. This arises from the fact that the ground surface on the time-record output is represented as the horizontal surface. For example, on a profile taken over a dune, the dune surface will be depicted as a horizontal reflector and (sub-)horizontal layers of the under-lying sequence (e.g., dune/beach contact, washover deposit, peat horizon) will appear as concave-upward reflections (i.e., the mirror images of dune topography). For short segments of the trace, groundwater table, if present, may be assumed (or determined) to be nearly horizontal and thus aid in correcting the record.

The radar data are commonly collected in either step mode or continuous mode. In the step mode, the source of the signal (transmitter) and the receiver are placed on the ground, and after the reading is obtained, moved to the next location. Continuous mode involves the unidirectional movement of both transmitter and receiver (usually housed within one antenna box). The cart-mounted GPR system allows rapid collection of continuous traces over large areas and has proven to be an effective method of subsurface data collection. In recent years, the advances in radar technology and software have improved the resolution and processing of the field data. Using a series of parallel, closely spaced transects, a three-dimensional image of the subsurface, as well as plan-view time slices that correspond to various depths, can be obtained (Beres *et al.*, 1995, 1999; Conyers and Goodman, 1997). Most recently, a combination of time-domain reflectometry and sediment peels have been used by Van Dam and Schlager (2000) to identify the reflector-producing horizons and construct synthetic radar traces analogous to synthetic seismograms. Because the description and analysis of sedimentary facies on GPR records are similar in many ways to seismic-stratigraphic attributes (e.g. reflector configuration, frequency, continuity, etc.), the onshore records of the barrier lithosome can be integrated with the available nearshore and offshore seismic reflection profiles (van Heteren *et al.*, 1998).

Limitations of the techniques

The most obvious limitation of a specific GPR system setup are the penetration depth and resolution. Both of these attributes depend on the choice of the antenna, which is in turn governed by the research objectives (Jol, 1995; Smith and Jol, 1995). Figure G24 shows that the top and bottom interfaces of a subsurface layer can only be resolved if these are separated by at least one wavelength of the radar signal (cf. Davis and Annan, 1989). High-frequency antennae (500–2,000 MHz) are smaller in size and provide high resolution at the expense of relatively limited penetration (commonly less than 8–10 m). Antennae with frequency range of 12.5–50 MHz have poor resolution, but allow for a maximum probable penetration of 45–65 m (Smith and Jol, 1995). The optimal antenna frequencies for stratigraphic research are 100 MHz or 120 MHz, which allow penetration of 15–20 m (deeper in unsaturated sequences), while still providing high-resolution images.

One of the important considerations in GPR research is the attenuation of the EM signal by brackish to salty groundwater due to its high conductivity. As a result, sections of many profiles adjacent to a beach or backbarrier margin, or deeper portions of profiles affected by saltwater intrusion are often reflection-free (van Heteren *et al.*, 1998). In some locations, the barrier is wide enough such that most of the stratigraphy can be imaged by GPR without attenuation. Besides seawater, increase in clay content drastically increases conductivity thereby reducing or precluding signal penetration. The top of the clay unit itself, however, often appears as a strong reflector on the radar trace. As mentioned above, sedimentary deposits with high magnetic permeability may also limit signal penetration, while themselves providing strong reflectors. Aside from sedimentological characteristics, the distortion and amplitude change of the EM waves as they propagate into the earth may reduce the accuracy of the final depth/distance calculations.

Due to inherent differences in sediment properties from site to site, the velocity of EM waves may also vary (Davis and Annan, 1989; van Heteren *et al.*, 1996). The velocity values for a particular site may be calibrated through common mid-point analysis (Jol and Smith, 1991) or by measuring the depth to specific marker horizons in sediment cores and calculating velocity values from equation 2. In general, unless the geology of a study site is well known, sediment cores should be taken along the GPR profile to interpret the major reflectors. In turn, the radar images of dipping subsurface horizons (e.g., buried marsh layer, sloping bedrock surface, tidal inlet channel) may be used to maximize the coring effort by planning a core site in the area where the depth to a target reflector is minimal.

Examples of GPR images from coastal environments

Styles of barrier progradation

Many coastal areas have experienced progradational (seaward) growth. Geomorphic expression of this process, such as a series of beach ridges is often used to determine the origin, magnitude, orientation, and chronology of barrier growth (Tanner, 1995). However, in many cases dense vegetation, parabolic dune migration, or human development have modified or obscured the surface expression of barrier growth (Figure G25(A)). In such areas, subsurface records, complemented with sediment cores, may provide the only means of analyzing the erosional–depositional history of a barrier. In many instances the progradational of a barrier may be punctuated by erosion and shore retreat. At these sites, there may be no distinct morphological evidence of an erosional event, except occasional washovers or dune and berm scarps confined to the youngest portion of the barrier. These features, particularly in the earlier constructional history of the barrier, are often preserved as buried accumulations of coarse-grained sediments or heavy-mineral horizons that are rarely detected in the field. Such lithological anomalies may be observed in sediment cores, but their

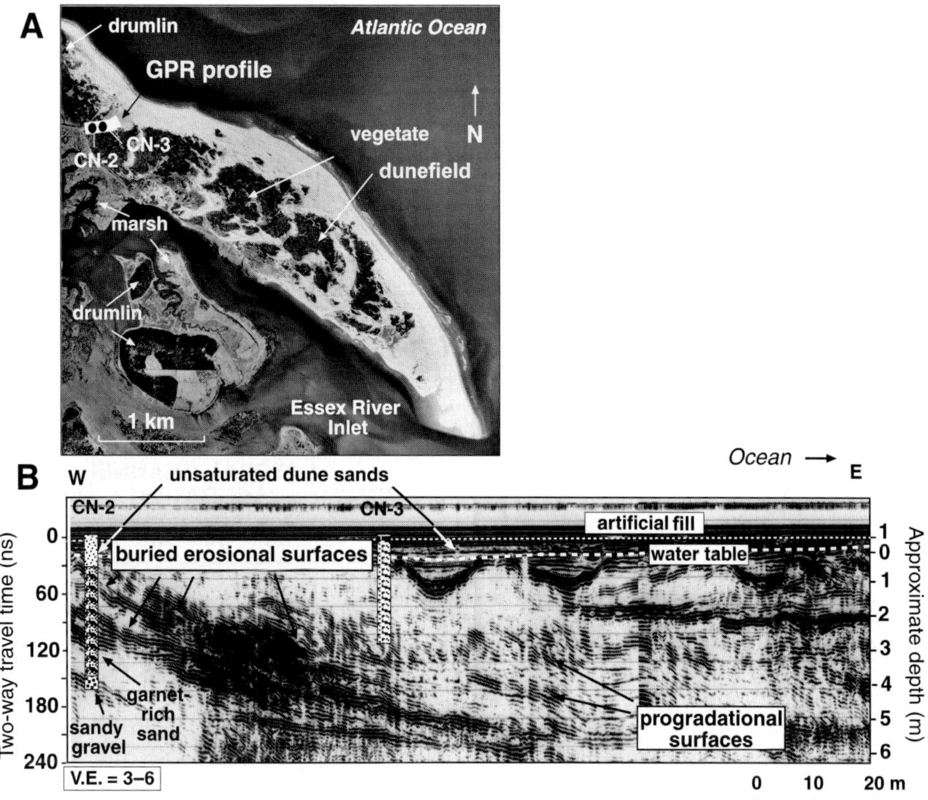

Figure G25 (A) Vertical aerial photograph of Castle Neck Barrier, Massachusetts, dominated by vegetated parabolic dunes. Note the absence of beach ridges. (B) Shore-normal GPR transect taken across a parking lot reveals a series of strong seaward-dipping reflectors in the landward segment of the GPR trace giving way to a sequence of less prominent, nearly uniformly spaced reflectors in a seaward direction. Sediment core CN-2 penetrated several layers of concentrated garnet-magnetite sands interbedded with quartz-rich units. All records are taken as a continuous trace with a 120 MHz antenna. See text for discussion.

geometry and continuity can only be confirmed in geophysical records (Figure G25(B)). For example, the GPR profile in Figure G25(B) illustrates a series of prominent tangential-oblique reflectors that represent buried erosional beach face and berm scarps. They grade into uniformly spaced sigmoidal-oblique reflectors in a seaward direction, which mark a period of increased sediment supply (Buynevich and FitzGerald, 2000).

Geometry of tidal inlet paleo-channels

Channel-fill sequences of tidal inlets may comprise a significant portion of the barrier lithosome and, in some instances, the locations and dimensions of former inlet channels can be detected with GPR. Mixed-sediment barriers are ideal for the recognition of inlet-fill structures. Due to large contrast between the coarse-grained channel lag and finer-grained channel-fill deposits, the outline of the channel often appears as a prominent concave-upward reflector. Figure G26 shows a paleo-inlet channel that has migrated along a retrograding, sand-and-gravel barrier as evidenced by a series of northward-dipping reflectors. Eventually, the inlet stabilized in one location and was infilled by sediment from a seaward source recorded as subhorizontal reflectors within the paleo-channel. Using GPR profiling, the locations of the former inlets can be mapped and compared with historical maps, where available. At least 18 historical inlets were mapped along Duxbury Beach, Massachusetts, where none exist today (FitzGerald *et al.*, 2000). In addition, such elements of inlet channel geometry as depth, width, and approximate length (using a series of records) can be determined. The elevation of the paleo-inlet channel relative to present sea level can also be estimated.

Stratigraphy of coastal lakes

Freshwater lakes and ponds of various origins (closed lagoons, glacial depressions, dune swales, deflation basins, etc.) are common features along many coasts. Their sedimentary fills serve as archives of depositional events that result from climatic, oceanographic, and geomorphic changes in coastal regions. Shifts from organic- to clastic-dominated deposition result in a sequence of interbedded layers with distinct lithological and dielectric properties, making these systems suitable for GPR research. Figure G27 shows a shore-normal transect taken across an ice-covered coastal lake. A steeply dipping reflector representing the bedrock surface can be traced to a depth of over 7 m below the lake surface. The seaward portion of the trace shows the margin of the barrier dunefield as a landward-dipping surface with several basinward-dipping internal reflections. Below the flat lake-bottom reflector, a sequence of wavy reflectors can be traced across the profile. These represent muddy, organic-rich lake-bottom sediments interbedded with aeolian sands. A prominent convex-up reflector with a transparent core is indicative of a buried dune. This interpretation is based on similar reflector configuration observed on the lake bottom adjacent to a recently migrating dune.

Summary

Ground-penetrating radar has proven to be a valuable tool for high-resolution imaging of antecedent geology, stratigraphy, and hydrogeology of coastal systems. Although saltwater attenuation presents a significant limitation in coastal lowlands, areas with moderate to high rainfall and relatively good sediment permeability often contain considerable freshwater lenses (5–20 m) which ensure good penetration of EM signal. Varying degrees of textural and compositional heterogeneity of sediments in many coastal sequences produce the lithological contrast necessary to generate subsurface reflections. These systems provide excellent natural laboratories for effective and detailed stratigraphic analysis using GPR profiling supplemented with sediment cores. Such studies have already significantly improved our knowledge of coastal development over a wide range of temporal (years to millennia) and spatial (centimeters to 10's of kilometers) scales and served to emphasize the complexity of coastal processes and resulting stratigraphic records.

Ilya Buynevich and Duncan M. FitzGerald

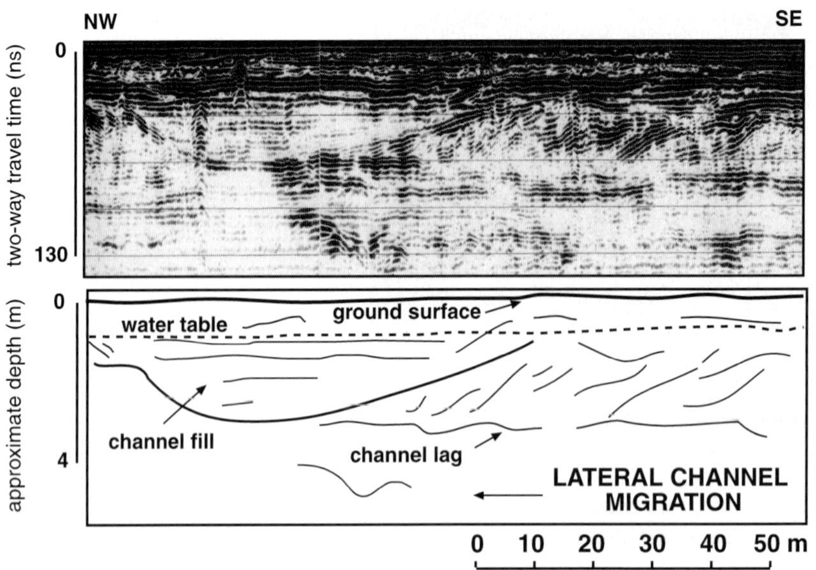

Figure G26 Shore-parallel profile and interpretation of a buried tidal inlet paleo-channel at Duxbury Beach barrier, Massachusetts. A series of dipping reflectors indicate the migration of the channel in a northward direction. A prominent concave reflector on the left represents the final position of the paleo-channel that was subsequently filled.

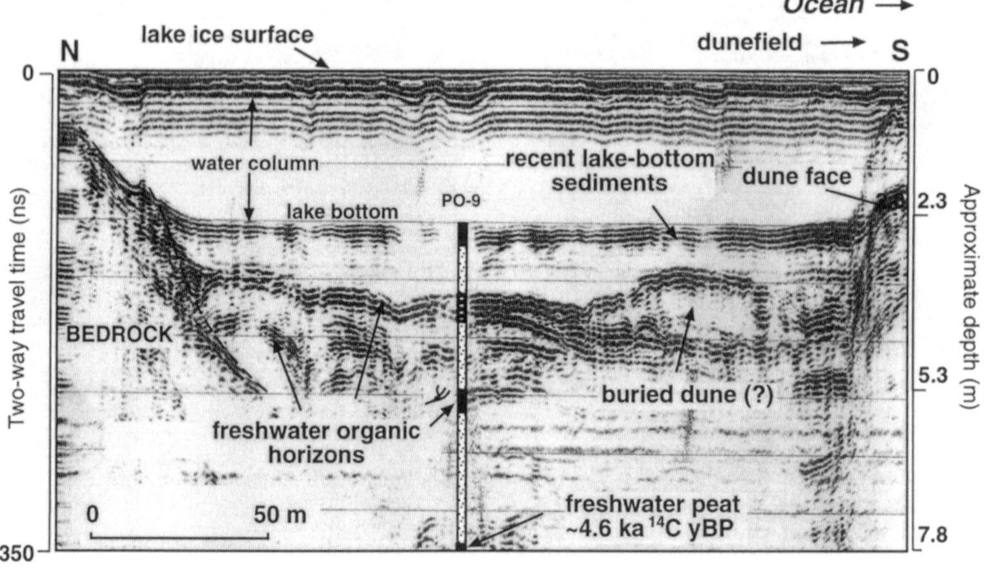

Figure G27 Radar transect taken over an ice-covered surface of Silver Lake, Maine. The undulating sub-horizontal reflectors are organic-rich lake-bottom deposits.
Note: The convex-upward reflection within the lake sequence interpreted as a buried dune.

Bibliography

Baker, P.L., 1991. Response of ground-penetrating radar to bounding surfaces and lithofacies variations in sand barrier sequences. *Exploration Geophysics*, **22**: 19–22.

Beres, M., Green, A., and Huggenberger, P., 1995. Mapping the architecture of glaciofluvial sediments with three-dimensional georadar. *Geology*, **23**: 1087–1090.

Beres, M., Huggenberger, P., Green, A., and Horstmeyer, H., 1999. Using two- and three-dimensional georadar methods to characterize glaciofluvial architecture. *Sedimentary Geology*, **129**: 1–24.

Busby, J.P., and Merritt, J.W., 1999. Quaternary deformation mapping with ground penetrating radar. *Journal of Applied Geophysics*, **41**(1): 75–91.

Buynevich, I.V., and FitzGerald, D.M., 2000. Styles of coastal progradation revealed in subsurface records of paraglacial barriers, New England, USA. *ICS-2000 Conference Programs and Abstracts*, Rotorua, New Zealand, p. 42.

Conyers, L.B., and Goodman, D., 1997. *Ground-penetrating Radar: An Introduction to Archaeologists*. Walnut Creek: AltaMira Press.

Davis, J.L., and Annan, A.P., 1989. Ground-penetrating radar for high-resolution mapping of soil and rock stratigraphy. *Geophysical Prospecting*, **37**: 531–551.

FitzGerald, D.M., Baldwin, C.T., Ibrahim, N.A., and Humphries, S.M., 1992. Sedimentologic and morphologic evolution of a beach-ridge barrier along an indented coast: Buzzards Bay, Massachusetts. In Fletcher, C., and Wehmiller, J. (eds.), *Quaternary Coasts of the United States: Marine and Lacustrine Systems*, SEPM Special Publication No. 48, pp. 64–75.

FitzGerald, D.M., Buynevich, I.V., and Rosen, P.S., 2000. Historical and geological evidence of former tidal inlets along a retrograding barrier: Duxbury Beach, Massachusetts, USA. *ICS-2000 Conference Programs and Abstracts*, Rotorua, New Zealand, pp. 52.

Harari, Z., 1996. Ground-penetrating radar (GPR) for imaging stratigraphic features and groundwater in sand dunes. *Journal of Applied Geophysics*, **36**(1): 43–52.

Jol, H.M., 1995. Ground penetrating radar antennae frequencies and transmitter powers compared for penetration depth, resolution and reflection continuity. *Geophysical Prospecting*, **43**: 693–709.

Jol, H.M., and Smith, D.G., 1991. Ground penetrating radar of northern lacustrine deltas. *Canadian Journal of Earth Sciences*, **28**: 1939–1947.

Jol, H.M., Young, R., Fisher, T.G., Smith, D.G., and Meyers, R.A., 1996a. Ground penetrating radar of eskers, kame terraces, and moraines: Alberta and Saskatchewan, Canada. *Proceedings of the 6th International Conference on Ground Penetrating Radar (GPR'96)*, Sendai, Japan, pp. 439–443.

Jol, H.M., Smith, D.G., and Meyers, R.A., 1996b. Digital ground penetrating radar (GPR): an improved and very effective geophysical tool for studying modern coastal barriers (examples for the Atlantic, Gulf and Pacific coasts, U.S.A.). *Journal of Coastal Research*, **12**: 960–968.

Jol, H.M., Vanderburgh, S., and Havholm, K.G, 1998. GPR studies of coastal aeolian (foredune and crescentic) environments: examples from Oregon and North Carolina, U.S.A. *Proceedings of the 7th International Conference on Ground Penetrating Radar (GPR'98)*, Lawrence, Kansas, pp. 681–686.

Leclerc, R.F., and Hickin, E.J., 1997. The internal structure of scrolled floodplain deposits based on ground-penetrating radar, North Thompson River, British Columbia. *Geomorphology*, **21**(1): 17–38.

Roberts, M.C., Bravard, J.P., and Jol, H.M. 1997. Radar signatures and structure of an avulsed channel: Rhone River, Aoste, France. *Journal of Quaternary Science*, **12**: 35–42.

Schenk, C.J., Gautier, D.L., Olhoeft, G.R., and Lucius, J.E., 1993. Internal structure of an aeolian dune using ground-penetrating radar. In Pye, K., and Lancaster, N. (eds.), *Aeolian Sediments: Ancient and Modern*, IAS Special Publication No. 16, pp. 61–69.

Smith, D.G., and Jol, H.M., 1995. Ground penetrating radar: antenna frequencies and maximum probable depths of penetration in quaternary sediments. *Journal of Applied Geophysics*, **33**: 93–100.

Smith, D.G., and Jol, H.M., 1997. Radar structure of a Gilbert-type delta, Peyto Lake, Banff National Park, Canada. *Sedimentary Geology*, **113**: 195–209.

Smith, D.G., Meyers, R.A., and Jol, H.M., 1999. Sedimentology of an upper-mesotidal (3.7 m) Holocene barrier, Willapa Bay, SW Washington, U.S.A. *Journal of Sedimentary Research*, **69**: 1290–1296.

Tanner, W.F., 1995. Origin of beach ridges and swales. *Marine Geology*, **129**: 149–161.

Topp, G.C., Davis, J.L., and Annan, A.P., 1980. Electromagnetic determination of soil water content: measurements in coaxial transmission lines. *Water Resourses Research*, **16**: 574–582.

Van Dam, R.L., and Schlager, W., 2000. Identifying causes of ground-penetrating radar reflections using time-domain reflectometry and sedimentological analyses. *Sedimentology*, **47**: 435–449.

van Heteren, S., FitzGerald, D.M., Barber, D.C., Kelley, J.T., and Belknap, D.F., 1996. Volumetric analysis of a New England barrier system using ground-penetrating radar and coring techniques. *Journal of Geology*, **104**: 471–483.

van Heteren, S., FitzGerald, D.M., McKinlay, P.A., and Buynevich, I.V., 1998. Radar facies of paraglacial barrier systems: coastal New England, USA. *Sedimentology*, **45**: 181–200.

von Hippel, A.R., 1954. *Dielectrics and Waves*. Cambridge: MIT Press.

Cross-references

Beach Stratigraphy
Coastal Sedimentary Facies
Hydrology of the Coastal Zone
Instrumentation (See Beach and Nearshore Instrumentation)
Monitoring Coastal Geomorphology
Paleocoastlines
Sequence Stratigraphy

GROUNDWATER—See HYDROLOGY OF THE COASTAL ZONE

H

HAZARDS—See NATURAL HAZARDS

HEADLAND-BAY BEACH

A headland is defined in common language as: (1) a point of usually high land jutting out into a body of water: promontory; (2) high point of land or rock projecting into a body of water. Therefore, a headland-bay beach is a beach whose shape is mainly conformed by the fact that it is located between such headlands, or at least adjacent to one. Some of the synonymous terms that can be found in literature to describe a headland-bay beach are: bay-shaped beach, pocket beach, zeta bay, bow-shaped bay, and half-heart bay. This type of feature shows a gradually changing curvature which Krumbein (1944) noted resembled that of a logarithmic spiral curve.

Johnson (1919) gave an incisive description of wave refraction caused by headlands along an embayed coastline, and Krumbein (1944) showed a simplified diagram of wave refraction into a bay lying to the lee of a headland. According to Yasso (1965), a headland was considered to be any natural or artificial obstruction that extended seaward from the coastline and caused a change in some element of the coastal wave pattern because of its presence.

Historical development

Observations of headland beach morphology in nature

Krumbein (1944) was the first author to describe a beach planshape—Halfmoon Bay, California (USA)—as being similar to the increasing radius of curvature found in the logarithmic spiral, although the restricted classification was removed some years later and the paper was reprinted in 1947. Yasso (1965) selected four US beaches for testing goodness of fit to the logarithmic spiral approximation. Poles for three of the best- fitting logarithmic spirals were located in close proximity to the seaward end of each headland. The spiral angle α was found to range between 41.26° and 85.64°. Berenguer and Enríquez (1988) reviewed data from 24 beaches around Spain and derived empirical correlations between geometrical characteristics of the layout of the offshore breakwaters and beach planshape. Moreno and Kraus (1999) performed fittings of analytical planshapes to a data set of 46 beaches in Spain and the United States and derived preliminary engineering design guidance including the proposal of a new functional shape.

Observations of headland beach morphology in the laboratory

Silvester (1960) tracked the time evolution of a beach in a physical model and observed that the beach between headlands tended to reach an equilibrium shape—logarithmic spiral shape—in response to persistent swell directions under a certain wave angle of attack. The model coastline was allowed to erode without replenishment of sand at the updrift end. Yasso (1965) pointed out that the conditions of Silvester's physical-model test (lack of continuous sediment supply to the updrift end of the model and close spacing of the headlands) suggested that the equilibrium form achieved in the model may not be identical to that achieved under natural conditions of sediment supply and wider separation of headlands.

Silvester (1970) performed additional model tests in which three different wave conditions were generated at three different angles of incidence to the alignment of a headland on an initially straight sandy beach. It was observed that the coastline developed three distinct curvature zones: first, a near circular section in the lee of the upcoast headland; second, a logarithmic spiral; and finally, a segment tangential to the downcoast headland. The time evolution of the spiral constant angle was plotted for the three incident wave angles tested, but only one wave condition was run long-enough as to see an asymptotic trend. A graphical linear relationship between the logarithmic spiral constant angle α and the wave angle β was provided based on the three angles tested.

Spătaru (1990) studied Romanian Black Sea beaches subjected to normal wave incidence by means of physical models. Equilibrium beach planshapes were considered to be described by arcs of circumferences, and provided design guidance based on geometry.

Numerical model approaches

Mashima (1961) constructed wave-energy diagrams from wind-rose diagrams and studied the configuration of stable coastlines based on energy considerations. When the wave energy is a semi-ellipse, the configuration of the stable coastline is approximately parabolic on which the tangent direction at the apex of the parabola coincides with the direction of the major axis of the wave-energy diagram.

LeBlond (1972) attempted to study how wave-induced longshore currents in the presence of a headland could erode a linear beach by developing a numerical model. LeBlond (1972) stated that if there existed a planimetric shape which the headland beach asymptotically approached, it must have the following properties: (1) it should be concave outwards, near the headland, and then convex outwards. (2) the sand transport should increase monotonically along it. (3) erosion, by causing the beach to be displaced normal to itself, should not qualitatively change the shape of the beach. LeBlond also pointed out that the logarithmic spiral did not satisfy the first condition because it is always concave outwards, and that there may be other curves satisfying all of the above three conditions, and one could not decide "*a priori*" which one will be the equilibrium one.

The main modification implemented by Rea and Komar (1975) with respect to previous numerical modeling efforts, was the combination of two orthogonal one-dimensional grids to simulate beach configurations, so that beach erosion could proceed in two directions without the necessity of a full two-dimensional array. Testing of the model in a hooked beach coastline configuration indicated that the coastline would always attempt to achieve an equilibrium configuration governed by the pattern of offshore wave refraction and diffraction and the distribution of wave-energy flux.

Walton (1977) presented an analytical model to describe the equilibrium shape of a coastline sheltered by a headland using a continuous wave-energy diagram consisting of representative offshore ship wave height and direction observations. It was found to produce coastline shapes similar to the logarithmic spiral shape for sheltered beaches in Florida. The model worked by establishing that the coastline orientation at a certain point was normal to the average direction of the so-called energy of normalized wave attack—that is, the energy which is allowed by the headland to reach the shore.

Yamashita and Tsuchiya (1992) constructed a numerical model for three-dimensional beach change prediction to simulate a pocket beach formation. The model consisted of three modules to calculate waves, currents, and sediment transport and beach change. The wave transformation module was based on the mild-slope equation of hyperbolic type; the current module was horizontally two-dimensional with direct interaction with sea-bottom change, which was evaluated by the sediment transport model formulated by Bailard (1982) in the third module.

In a theoretical work on the subject of headland-bay beaches, Wind (1994) presented an analytical model of beach development, where the shape of the headland-bay beach remained constant with time and expanded at a rate according to a time function. Wind's (1994) conceptual framework was based on knowledge of the existence of a headland-bay beach shape centered around a pole and that evolved in time in a more or less constant shape. If the position of the coastline is described by the radius r, the angle δ, and time t, the evolution of the coastline with a constant shape implies that the coastline might be described as

$$r(\delta, t) = r_0 f(\delta)\, e(t) \qquad \text{(Eq. 1)}$$

where r_0 is the constant, $f(\delta)$ is the shape function, and $e(t)$ the evolution function of the coastline in time. With respect to the time function, it was shown that in the diffraction zone it should follow a $t^{1/3}$ law, whereas for the refraction zone a $t^{1/2}$ law was found. This implied that the evolution of a headland-bay beach in the diffraction zone should initially be faster and on the long term, slower than the evolution in the refraction zones. With respect to the shape, the function shape $f(\delta)$ is expressed in terms of functions representing the diffracted wave field. The logarithmic spiral is obtained by taking the functions for the group velocity and the geometrical part for the driving force as constants.

Equilibrium planshapes of headland-bay beaches

Three functional shapes have been proposed to describe the equilibrium planshape of headland-bay beaches, namely the logarithmic spiral shape, the parabolic shape, and the hyperbolic tangent shape.

The *logarithmic spiral* (also named equiangular, or logistic spiral), first described by Descartes, was described as the curve that cuts radii vectors from a fixed point O under a constant angle α (Figure H1). The equation of the logarithmic spiral can be written in polar coordinates as

$$R = R_0\, e^{\theta \cot \alpha} \qquad \text{(Eq. 2)}$$

where R is the length of the radius vector for a point P measured from the pole O, θ is the angle from an arbitrary origin of angle measurement to the radius vector of the point P, R_0 is the length of radius to arbitrary origin of angle measurement, and α the characteristic constant angle between the tangent to the curve and radius at any point along the spiral. The pole of the spiral is identified as the diffraction point (Silvester, 1960; Yasso, 1965), and the characteristic angle of the spiral is a function of the incident wave angle with respect to a reference line. For headlands of irregular shape and for those with submerged sections, the diffraction point cannot be specified unambiguously, a problem entering specification of all equilibrium shapes. The reference line extends from the approximate location of the diffraction point to a downdrift headland. This shape is extremely sensitive to variations in the characteristic angle α because the angle enters the argument of an exponential function (Moreno and Kraus, 1999), being the practical consequence that α has to be accurately defined. For engineering application four unknowns must be found: location of pole (two coordinates), characteristic angle α, and scale parameter R_0. The shape of the log spiral is controlled only by α, with the parameter R_0 determining the scale of the shape. In fact, the functioning of R_0 is equivalent to setting a different origin of measurement of the angle θ. In other words, graphically the log spiral may be scaled up or down by turning the shape around its pole.

Values of α for headland-bay beaches reported in the literature range from about 45° to 75°. As α becomes smaller, the log spiral becomes wider or more open. There are two singular values for α: if $\alpha = 90°$, the log spiral becomes a circle, and if $\alpha = 0°$, the log spiral becomes a straight line.

Various authors have noted that fitting of the log-spiral shape is difficult in the downdrift section of the beach. It is a particular concern in attempting to fit to long beaches or to beaches with one headland. However, even in these situations, a good fit could be achieved for the stretch near the headland (Moreno and Kraus, 1999).

The *parabolic shape* of a headland-bay beach was proposed by Hsu *et al.* (1987) and is expressed mathematically in polar coordinates by equation 3 for the curved section of the beach and by equation 4 for the straight downdrift section of the beach (Moreno and Kraus, 1999),

$$\frac{R}{R_0} = C_0 + C_1\left(\frac{\beta}{\theta}\right) + C_2\left(\frac{\beta}{\theta}\right)^2 \quad \text{for } \theta \geqslant \beta \qquad \text{(Eq. 3)}$$

$$\frac{R}{R_0} = \frac{\sin\beta}{\sin\theta} \quad \text{for } \theta \leqslant \beta \qquad \text{(Eq. 4)}$$

where R is the radius to a point P along the curve at an angle θ, R_0 is the radius to the control point at angle β to the predominant wave-front

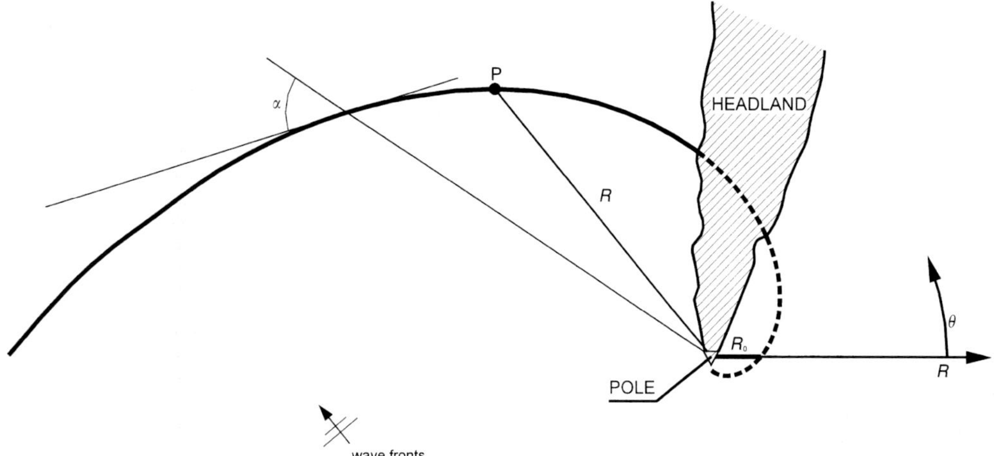

Figure H1 Definition sketch of the log spiral planshape.

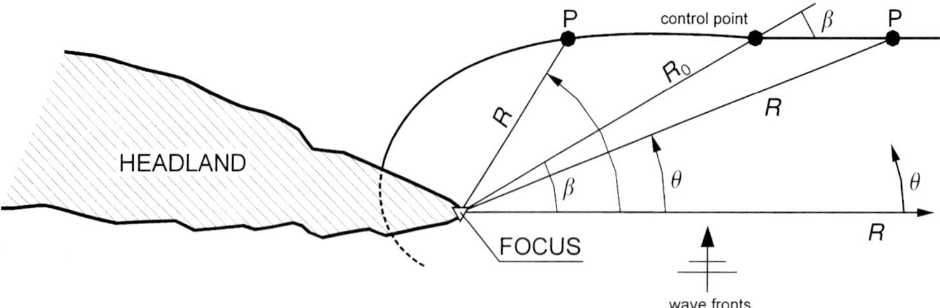

Figure H2 Definition sketch of the parabolic planshape.

direction, β is the angle defining the parabolic shape, θ is the angle between line from the focus to a point P along the curve and predominant wave-front direction; and C_0, C_1, and C_2 are the coefficients determined as functions of β—the coefficients C_0, C_1, and C_2 are listed in Silvester and Hsu (1993) form from β = 20° to 80° at a 2°-interval (Figure H2).

For this parabolic shape, the focus of the parabola is taken to be the diffraction point. The three coefficients needed to define the shape (see Silvester and Hsu, 1991, 1993) are functions of the predominant wave angle with respect to a control line. The control line is defined similarly to the case of the log-spiral shape as the line that extends from the diffraction point to a reference point, at an angle β between the control line and the predominant wave crest orientation. Downdrift of the reference point, the coastline is assumed to be aligned parallel to the incident wave crests. This shape pertains to that of a long straight beach with shape controlled by one headland. Values of β ranged in prototype beaches from 22.5° to 72.0°, whereas the variation in model beaches was from 30° to 72°.

Sensitivity tests were performed (Moreno and Kraus, 1999), where the response of the parabolic shape to a change in the value of the characteristic angle β and of R_0 was analyzed. The results proved that R_0 is a scaling parameter—the length of the control line, the alongshore extent of the shape decreases as β increases because the parabolic shape is defined only for θ ≥ β. In summary, the angle β controls the shape of the parabola, and R_0 controls its size. Because the control line intersects the beach at the point where the curved section meets the straight section of the beach, the sensitivity of the parabolic shape to errors in the estimation of the control point was examined. This was done by jointly changing R_0 and β while keeping the distance from the headland to the straight coastline constant. This observation means that the control point is not well-defined, that is uncertainty in selection of the control point and hence the corresponding joint combination the radius R_0 and β has little influence on the final result.

According to Moreno and Kraus (1999), the parabolic shape provides good fits for beaches with a single headland, because they consist of a curved section (well described by the portion of the beach protected by the headland) and a straight section (well describes the downdrift section).

González (1995) provided an improvement in the lack of definition of the location of the control point on the formulation of the parabolic shape by developing a relationship between β and the geometry of the beach and the incident wave climate according to the following equation:

$$\beta = \frac{\pi}{2} - \tan^{-1}\left[\frac{(0.1 + 0.63\, y/L)^{1/2}}{y/L}\right], \qquad \text{(Eq. 5)}$$

where y is the offshore distance from diffraction point to coastline, L the average wavelength in the lee area (between coastline and diffraction point).

For practical use of the parabolic shape, five unknowns need to be solved for: location of focus (two coordinates), characteristic angle β, scaling parameter R_0, and the orientation of the entire parabolic shape in plan view.

The *hyperbolic tangent shape* was developed by Moreno (1997) and proposed for engineering design of equilibrium shapes of headland-bay beaches by Moreno and Kraus (1999) to simplify the fitting procedure and to reduce ambiguity in arriving at an equilibrium coastline shape as controlled by a single headland (Figure H3). As mentioned above, it can be difficult to specify the location of the pole or focus, and the characteristic angle (angle between predominant wave crests and the control line) for developing a log-spiral shape or a parabolic shape. In addition, the log-spiral shape does not describe an exposed (straight) beach located far downdrift from the headland, so that another shape must be applied.

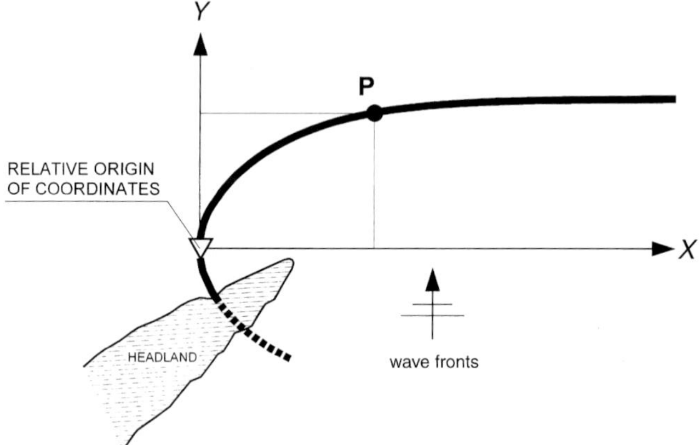

Figure H3 Definition sketch of the hyperbolic tangent planshape.

The hyperbolic tangent functional shape is defined in a relative Cartesian coordinate system as:

$$y = \pm\, a \tan h^{m}(bx) \qquad \text{(Eq. 6)}$$

where y is the distance across shore, x is the distance alongshore; and a (units of length), b (units of 1/length), and m (dimensionless) are empirically determined coefficients.

This shape has three useful engineering properties. First, the curve is symmetric with respect to the x-axis. Second, the values $y = \pm a$ define two asymptotes; in particular of interest here is the value $y = a$ giving the position of the downdrift coastline beyond the influence of the headland. Third, the slope dy/dx at $x = 0$ is determined by the parameter m, and the slope is infinite if $m < 1$. This restriction on slope indicates m to be in the range $m < 1$.

According to these three properties, the relative coordinate system should be established such that the x-axis is parallel to the general trend of the coastline with the y-axis pointing onshore. Also, the relative origin of coordinates should be placed at a point where the local tangent to the beach is perpendicular to the general trend of the coastline. These intuitive properties make fitting of the hyperbolic-tangent shape relatively straightforward as compared to the log-spiral and parabolic shapes, making it convenient in design applications.

Sensitivity testing of the hyperbolic tangent shape was performed (Moreno and Kraus, 1999) to characterize its functional behavior and assign physical significance to its three empirical coefficients: a controls the magnitude of the asymptote (distance between the relative origin of coordinates and the location of the straight coastline); b is a scaling factor controlling the approach to the asymptotic limit; and m controls the curvature of the shape, which can vary between a square and an S-curve. Larger values of m ($m \geq 1$) produce a more rectangular and somewhat unrealistic shape, whereas smaller values produce more rounded, natural shapes.

To fit the hyperbolic tangent shape to a given coastline, we must solve for six unknowns: the location of the relative origin of coordinates (two coordinates), the coefficients a, b, and m, and the rotation of the relative coordinate system with respect to the absolute coordinate system. Because of the clear physical meaning of the parameters, fitting of this shape can be readily done through trial and error.

Moreno and Kraus (1999) found the hyperbolic-tangent shape to be a relatively stable and easy to fit, especially for one-headland bay beaches. According to their work, the following simple relationships were obtained for reconnaissance-level guidance:

$$ab \cong 1.2 \qquad \text{(Eq. 7)}$$

$$m \cong 0.5 \qquad \text{(Eq. 8)}$$

The physical meaning of equation 6 is interpreted that the asymptotic location of the downdrift shoreline increases with the distance between the coastline and the diffracting headland. Equations 6 and 7 are equivalent to selecting one family of such hyperbolic tangent functions for describing headland-bay beaches, and these values are convenient for reconnaissance studies prior to detailed analysis. Equation 6 could be more precisely written as:

$$a^{0.9124}\, b = 0.6060 \qquad \text{(Eq. 9)}$$

Headland-control concept of shore protection

The headland-control concept of shore protection was first proposed by Silvester (1976) and further discussed by Silvester and Ho (1972), and was described as a combination of groins and offshore breakwaters at alongshore and seaward spacings such to create long lengths of equilibrium-bay beaches (Silvester and Hsu, 1993). The structural dimensions in proportion to beach length are much smaller, and headlands are spaced much farther apart than offshore breakwaters. Therefore, it is intended to be a "regional" means of shore protection. Because headlands form pocket beaches, they might best be applied in a sediment-deficient area or for stabilizing an entire littoral reach of coast.

Headland beaches compartmentalize the coastline and reorient it in the local compartments to be parallel to the wavecrests of the predominant wave direction. If a coast has a substantial change in wave direction annually, the headland-bay beach might not be as stable as beaches behind traditional detached breakwaters, or shorter headland-bay beach compartments would be required.

A headland-bay beach design requires that a tombolo forms or be created (as by beach fill) behind the anchoring headland. If this connection is lost, the pocket beach is destroyed, and sediment can move alongshore, between adjacent compartments. Headland-bay beaches function and have their main attribute in creating independent pocket beaches, for which there is little or no communication of sand alongshore. Therefore, a headland beach presents a total barrier to littoral drift and can only be considered as a shore-protection alternative if such a barrier would not pose a problem to adjacent beaches.

The assumption that there is a single predominant wave direction which controls the final coastline shape is questionable where long distances are involved. If the design goal is to stabilize a regional extent by multiple pocket beaches, the headland-control concept might be appropriate.

Luis J. Moreno

Bibliography

Bailard, J.A., 1982. Modelling on–offshore sediment transport in the surf zone. In *Proceedings of the 18th International Conference on Coastal Engineering*, American Society of Civil Engineers, 2, pp. 1419–1438.

Berenguer, J.M., and Enríquez, J., 1988. Design of pocket beaches. The Spanish case. In *Proceedings of the 21st International Conference on Coastal Engineering*, American Society of Civil Engineers, 2, pp. 1411–1425.

González, M., 1995. Morfología de playas en equilibrio. Planta y perfil. Ph.D. dissertation Santander, Spain: Universidad de Cantabria (in Spanish).

Hsu, J.R.C, Selvester, R., and Xia, Y.M., 1987. New charae teristics of equilibrium shaped bays. In *Proceedings of the 8th Australision Conference on Coastal and Ocean Engineering*, American Society of Cevil Engineers, pp. 140–144.

Johnson, D.W., 1919. *Shore Processes and Shoreline Development*. New York: John Willey & Sons.

Krumbein, W.C., 1944. Shore processes and beach characteristics. *Beach Erosion Board, Technical Memorandum*, No. 3. Washington, DC: U.S. Army Corps of Engineers.

LeBlond, P.H., 1972. On the formation of spiral beaches. In *Proceedings of the 13th International Conference on Coastal Engineering*, American Society of Civil Engineers, 2, pp. 1331–1345.

Mashima, Y., 1961. Stable configuration of coastline. *Coastal Engineering in Japan*, **4**: 47–59.

Moreno, L.J., 1997. *Critical review of the headland-concept of shore protection*. Master of Engineering Report. College Station, Texas, USA: A&M University.

Moreno, L.J., and Kraus, N.C., 1999. Equilibrium shape of headland-bay beaches for engineering design. In *Proceedings of the Coastal Sediments '99*, American Society of Civil Engineers, 1, pp. 860–875.

Rea, C.C., and Komar, P.D., 1975. Computer simulation models of a hooked beach shoreline configuration. *Journal of Sediment Petrology*, **45**: 866–872.

Silvester, R., 1960. Stabilization of sedimentary coastlines. *Nature*, **188**: 467–469.

Silvester, R., 1970. Growth of crenulate shaped bays to equilibrium. *Journal of Waterways and Harbors Division*, **96**(2): 275–287.

Silvester, R., 1976. Headland defense of coasts. In *Proceedings of the 15th International Conference on Coastal Engineering*, American Society of Civil Engineers, 2, pp. 1394–1406.

Silvester, R., and Ho, S.K., 1972. Use of crenulate shaped bays to stabilize coasts. In *Proceedings of the 13th International Conference on Coastal Engineering*, American Society of Civil Engineers, 2, pp. 1347–1365.

Silvester, R., and Hsu, J.R.C., 1991. New and old ideas in coastal sedimentation. *Review in Aquatic Sciences*, **4**(4): 375–410.

Silvester, R., and Hsu, J.R.C., 1993. *Coastal Stabilization: Innovative Concepts*. Englewood Cliffs: Prentice-Hall.

Spătaru, A., 1990. Breakwaters for the protection of Romanian beaches. *Coastal Engineering*, **14**: 129–146.

Walton, T.L., 1977. Equilibrium shores and coastal design. In *Proceedings of the Coastal Sediments '77*, American Society of Civil Engineers, 1, pp. 1–16.

Wind, H.G., 1994. An analytical model of crenulate shaped beaches. *Coastal Engineering*, **23**(3–4): 243–253.

Yamashita, T., and Tsuchiya, Y., 1992. A numerical simulation of pocket beach formation. In *Proceedings of the 23rd International Conference on Coastal Engineering*, American Society of Civil Engineers, 3, pp. 2556–2566.

Yasso, W.E., 1965. Plan geometry of headland-bay beaches. *Journal of Geology*, **73**: 702–714.

Cross-references

Bay Beaches
Dynamic Equilibrium of Beaches
Engineering Applications of Coastal Geomorphology
Longshore Sediment Transport
Shore Protection Structures
Wave Refraction Diagram

HEALTH BENEFITS

Introduction

Alternative, or parallel, medicine has been steadily gaining followers and its merits have been discussed in conferences held at traditional colleges of medicine. Osteopathy, chiropractic, naturopathy, naprapathy, acupuncture, Chinese traditional medicine, aromatherapy, thermalism, thalassotherapy, and others have entered the common vocabulary for some time, some celebrating their 100th anniversary on the scene. The latter two make use of ocean algae in packs, powders, and other forms.

Thalassotherapy and thermalism are medical approaches known for thousands of years and have gained access to hospitals, private practitioners' offices, and the hallowed halls of some universities' faculties of medicine. Sea products are also used in balneotherapy and aromatherapy. The first is a treatment technique based on bathing—and used in both thalassotherapy and thermalism—in which plant extracts, oils, marine products (mainly salts and algae) are added to the water. Aromatherapy is more controversial; it is based on the use of organic substances (essences) with massages or in bath and steam treatments;

centers are not numerous but it is seemingly *à la mode* in Great Britain where it is offered in 24 sites; a volume handling the topic in America has been published in 1999. Thonon-les-Bains is the only French center with a comprehensive program. (Valnet, 1995).

Thalassotherapeutics [θαλασσα, sea; θηραπεια, *care, treatment*] and thermalism [θερμοσ, warm, heat] got a boost in the 19th century when tourism lifted out of its infancy. The enthusiasm displayed by prestigious visitors to "health" centers—particularly of thermalism at the time—made it possible for "*cure*" stations to associate cultural, leisure, and even gambling activities with a treatment program. Napoleon III and his empress Eugénie contributed much, albeit mostly unintentionally, to the new fashion. A journey to a "*cure*" station took on a double aspect: a visit to a place to improve or restore one's health, the avowed aim of a trip, and of tourism and discovery, the hidden aim of the voyage, because a voyage it often was: roughly 7 h by railway from Paris to Plombières or to Deauville. Today the *curist* and *tourist are commonly* both one and the same person.

Tourism has become a full-fledged activity. Both the "treatment" and the desire to know a region and its cultural traits motivate the visitor. And for a modern person in search of his "better-being" simultaneously with his "well-being" marketing specialists play it from both sides: tourism uses thalasso/thermalism as a travel theme, while the cure-station uses tourism to diversify its offer. The aim is to "twin," to link, the medical quality of "cures" to the tourists' and curists' quality of life. An effort that has also to be directed toward children who may find relief, cure, or improvement of ailments related to breathing, skin, and developmental troubles, such as enuresis, growth, and fracture healing. *Mens sana in corpore sano*: the body will heal the better if the mind is kept away from the problem: tourism may thus in some cases become a component of the curative process.

It has been often said, lately, that the tourist is increasingly in search of a "return to the sources"; this is not a legend, but fact, and water, pure clean water that is, is commonly equated with health and well-being. A concern that has not escaped the attention of the European Union's Commission (Anonymous, 1996; Charlier, 1999). Hence an increased interest in "water centers" whether Tatabanya, Baden-Baden, Mangalia, Varna, Djerba, or Le Touquet-Paris-Plage. The blossoming of the "social tourism" as a consequence of paid leave, the explosion of mass tourism and a trend to spend holidays at considerable distance from one's home-base are powerful factors in making "exotic" destinations popular and sites, such as the Black Sea, both easily conceivable in vacation planning, and also very accessible. In Brussels for instance separate offices offer exclusively "health and tourism" products: *Thalasso* for one, *Thermalism* for another, are thriving on the fashionable Avenue Louise yet are patronized by a clientele from every social and income level.

The concept of health tourism

Efforts have been made for decades in Romania to develop tourism and seawater therapeutics along the Black Sea coast. Other coastal countries in the area—Bulgaria, Russia, Ukraine—have not been lagging either. The success, in France for instance, of thalassotherapy centers has been heralded by economists, tourism officials, and many health practitioners.

Thalassotherapy, and also thermalism, is thus certainly not new. What can be new is the updating of the facilities and the introduction of new technological advances. Health tourism is a powerful development tool and is undergoing an in-depth change. Where balneotherapy and exposure to coastal climate have been essentially a re-adaptive and convalescence treatment, a constantly growing segment of the clientele is made up of younger—and middle aged—people who seek an effective approach to reshaping, the *remise en forme*, encompassing not only physical reshaping but also a health restorative process.

This involves demands on computers, equipment, personnel qualification, research, thus also retooling and refurbishing, and a fresh outlook on approach and problem-solving. Marine muds have an economic potential in muds export to other countries, for example, the Italians provide muds to Bourbonne- les-Bains, France and the Israelis ship their Dead Sea muds to Plombières-les-Bains, France.

Competitive, concurrent or concomitant?

Thermalism, thalassotherapy, and lately aromatherapy, all claim scientifically proven curative effects. An abundant literature has been published recently, for instance in France, but often strongly publicity-slanted (François, 1999). Which does not mean that several treaties have not been written long before (Jacob, 1570, Russell, 1720). Boulangé (1995, 1997a and b; Boulangé *et al.*, 1989) of the University of Nancy, and others, like Collin (1995) and Constant *et al.* (1995), have made important contributions to the field [Larivière, 1958, Lance, 1988, Hérisson, 1989, Valnet, 1995]. Both thermalism and thalassotherapy use waters, muds, and thermal gases.

If thalassotherapy is sole in using algae (but not algal mud), seawater and its "aeration," techniques, however, appear to be quite similar in thermalism and thalassotherapy. They associated at the *Thermalies* of the 17th *Salon de la Santé* at Versailles in March 1999, a congress twinned with MEDEC *Salon de la Médecine*. As for aromatherapy it has not really wrought its place in the panoply of alternative medicines; it uses a variety of oils, extracted from plants; care with aromatherapy is not thus far eligible for benefits from the social security system in France, though included in "putting back into shape" treatments in some centers, including well-known ones such as for instance Thonon-les-Bains. With the notorious exception of Belgium, thermal and thalassotherapy cures are reimbursable medical expenses in most countries of the European Union and in some others as well.

There are more than a hundred thermal cure centers in France alone not counting those of overseas departments (La Réunion and Guadeloupe); thalassotherapy centers are a least a good fifty.

France with 50 centers, leads in numbers, with Germany in second place. Other stations surround the Black Sea, and are also located in Denmark, The Netherlands, and Belgium (Ostend, Knokke). Belgium (Spa, Limelette), and Luxembourg (Mondorf-les-Bains) have famed thermal centers and it is Spa that gave its name to the cure centers. The practice of "taking the waters" is far less common in the United States where it rather faded away during the in-between World Wars period, though Saratoga Spa is still "on the map."

Historical perspective

The map of Europe is literally strewn with places where the Romans tapped thermal waters and built their famous *thermae* from *the Baths of Caracalla* to those of Trier (Trêves), Lindesina (today Bourbonne-les-Bains), or Plombières dating from before the year 100. Roman Emperor Augustus was there, and so were scores of other leading figures of history such as writers Diderot, Chateaubriand, and rulers or their relatives like Laetitia Bonaparte (Napoleon I's mother), Napoléon III, Louis XV of France, German Emperor Wilhelm II to name but a very few.

Buchet (1985), describing medicine and surgery during the 1st century in Gaul, focuses on the role of thermal waters, but use of seawater for therapeutic aims was known in what are contemporary Egypt, France, Italy, and Greece as far back as 3000 BC. Nor was thalassotherapy a stranger in the medical arsenal of classical times. The knowledge and practice was spread by Celts, Gauls, and especially Romans (Grenier, 1960; Kretzschmer, 1966; Rameau, 1980; Buchet, 1985; Anonymous, 1991; Malissard, 1994). Bath, in Britain (the Romans' Aqua Sulis) got its name from the Romans' custom. Ancient Greeks placed considerable faith in the healing power of the sea. Greek poet Euripides (480–460 BC) wrote "The sea restores man's health," Greek philosopher Plato (428–437) "The sea washes all man's ailments," and 20 centuries later historian Jules Michelet (1798–1874) opined "*La terre vous supplie de vivre; elle vous offre ce qu'elle a de meilleur, la mer, pour vous relever . . . .*" But thalassotherapy faded away in the Western world imbued with Aristotelian logic, later nurtured by Gallileo (1568–1642) and Descartes (1596–1650), more recently by Pasteur (1822–95) and physiologist Claude Bernard (1813–78) (Larivière, 1958).

Springer traces it back, in modern times, to the Margate Royal Sea-Bathing Hospital, and famed Blackpool, England, has its spot in sea-water therapy history (Springer, 1935; Charlier, 1975). Russell (1720) of England, Barelli of Italy, Perochaud (and closer to us Rivière) of France, Benecke of Germany are credited as founders of the contemporary seawater therapy while Boulangé and his coauthors act as contem-porary spokesmen for French thermal-therapy (1995a,b). Thalassotherapy is, however, no longer limited to utilization of the maritime climate, but involves administration of seawater orally and by injection, use of the spray of water, of the pounding effects of waves, of heated seawater baths, and such even newer approaches as combining electroacupuncture and sea-water therapy. The 1935 four hundred seashore sanatoria and preventoria have multiplied during the second half of this century.

While centers of marine cures are numerous in France, Germany, Belgium, Russia, Ukraine, Romania (Eforie, Mangalia), Bulgaria (Varna), Israel, little or no interest has been shown for decades in the United States and is at best stagnant in Great Britain. In intense use before the 1920s, seawater injections though credited with healing nervous and blood diseases in children, fell in disuse (Larivière, 1958).

In the late 1930s near miraculous cures of nervitis, lumbagos, cellulitis, and obesity focused anew attention on the healing effects of sea-air and seawater (Gruber, 1968). Today injections and oral administration of seawater, even in minute quantities (in wise opposition with the Russell prescriptions of large quantities!) can claim serious therapeutic effects.

Rebirth

The resurgence of interest is coupled with new concepts. One hundred fifty years ago a report ventured that "therapeutics draw good results, every day, from the use of seawater and from salt springs; and although its use in baths and tubs is often not as advantageous as when taken in the surf, when the mechanical action of the fluid is added to its chemical action, one can still expect much of this (thalassic) therapeutic application . . ." (Translation, 1856).

In the 20th century, under the influence of Freudian writings and the psychosomatic philosophical movement, medical thinking split into a traditional scientific approach and "enthropological medicine" which includes acupuncture and thalassotherapy (Range, 1958). The coincidence of timing between the renewed interest in marine cures and the growing disenchantment with current ways of life in an ever increasingly technological society may be underscored. The return to the sources' desire is, of course, part of that trend.

Thalassotherapy is neither limited to nor solely based upon use of seawater: part of the treatment is the change of lifestyle, the new surroundings, the freeing of the individual from modern life's stresses, embodying aerosol- and helio-therapy. The tiny salt particles contained in sea-air (aerosols) work their way into the deepest parts of pulmonary alveoles and settle on their walls with a probably not negligible physiological effect (Woodcock and Blanchard, 1957). The high proportion of ultraviolet seaside sun rays influences favorably calcium metabolism. That natural oligo-element and others such as magnesium, manganese and cobalt, which buttress the organism's natural defenses, are also absorbed through warm seawater baths. Its biochemical properties make it a successful side-effect-free substitute for comfort medications.

Heated seawater causes a dilatation of cutaneous vessels and under water jet streams has the same beneficial effects as the pounding of the waves against the body and its spraying by sea foam. The initial shock of cold water in swimming pools has been looked at as a potential negative factor, particularly for older persons. However, most centers swimming pools are now adequately heated and wave machines provide the beneficial pounding.

Physiological effects of marine climates are reflected in a slowing down of the rates of breathing and heartbeat. The amplitude of the respiratory movement and pulmonary ventilation are increased, and so are the hematites in the blood and hemoglobin ratio, while heart contraction is reinforced; the body is thus better prepared for the beneficial impact of sea-water baths due to an increase of cutaneous exchanges. Many physicians recognize such additional symptoms as neuroendocrinic and growth stimulation, and an increase of diuresis and of basic metabolism.

Showers prior and after baths, overall or localized, exert a dual thermal and mechanical action on vessels and nerve endings; alternating of short cold- and warm-water sprays may well have the same tonifying effect as the Finnish sauna. Gynecological irrigations favor seawater's hypertonic action upon mucous tissues and penetrative ability. Nasal irrigations, aerosols, gargles help with sinus problems, ear-, nose-, and throat-ailments (Arehart, 1969).

The medical and pharmaceutical value of marine "products" has of course been proven. Didemnin-B, diazonomide-A, dolastin-10, and discodermolide are all potential cancer-fighting compounds derived from marine organisms, dwellers of the coral reefs ecosystems. Marine organisms produce chemical compounds—and over 6,000 unique compounds have been isolated with hundreds providing "drug leads"—with antiviral, antibacterial, and antifungal properties. The bryozoan *Bugula neritina* produces bryostatin-1, a potential anticancer agent, *Pseudopterogorgia elisabethae*, a soft-bodied coral known popularly as the Caribbean sea whip, produces anti-inflammatory pseudopterosins.

French and German pharmaceutical firms market vials of seawater tapped at 50 km offshore at depths of up to 20 m. It is claimed that such waters when purified provide cures for gastric troubles. With a reduced salinity the water remains nevertheless rich in magnesium and other oligo-elements, and free of chlorides, closely resembles blood plasma. Bread, crackers, and pasta made using seawater are marketed in France, Germany, and The Netherlands. Seaweeds play a significant role in cosmetology (DeRoeck, 1991).

Algae

Algae have been alternatively, and concurrently, praised and damned along coasts (Charlier, 1990; Charlier and Lonhienne, 1996) including the Black Sea (Bologa, 1985/86; Petrova-Karadjova, 1990; Bologa *et al.*, 1999). The European Union under its COST-48 program has encouraged research into their use and their eradication (Morand *et al.*, 1990; Guiry and Blunden, 1991; Schramm and Nienhuis, 1996). They have a role among others in food and feed, in cosmetology, methane- and fertilizer-production, and in therapy (Găstescu, 1963; Pricajan and Opran, 1970; Cotet, 1970).

The passage of algae components such as iron and cobalt through the skin is controversial for several physiologists. On the other hand, biomedical applications of *Lyngbya majuscula* are recognized by oncologists; this reef dwelling blue–green alga produces curacin-A which functions as an anti-proliferative that inhibits cell division, the mechanism by which cancer grows and spreads. Algae powder has been added to seawater baths and algae creams are sold in pharmacies and cosmetology stores.

They are present in some marine muds, and in some treatment centers their proportion is increased. German physicians had already in 1929 collected muds in a remote corner of Wilhelmshafen harbor. Romania has advertised them widely. French "marine cures" use an alga jelly mixed with wet sand heated in a double boiler. The mixture applied to the body slowly releases its heat. The ionic displacement of marine electrolytes and algal constituents through the skin is however, not universally agreed upon. Challenged 15 years ago, the practice is continued both in centers and on shores. German centers provide *pelotherapy* using silt packs rich in vegetal and mineral substances, rather similar to the moor-silt.

A mineral spring discovered in the royal residence of Ostend (Belgium) launched a thermalism and thalassotherapy center (1856) at one time catering to as many as 80,000 "curists." The vegetal marine mud, also in use here, is principally made up of compacted peat carried at strong high tides to the beach area. Dried, it is turned into a powder and mixed with a marine clay powder. For use in local applications the peloid is mixed with seawater and heated in a double boiler.

Some of the hyper-saline lakes of the Romanian Black Sea coast, according to the season have temperatures that may reach 27°C with an alkaline pH. The microfauna is abundant and at least 30 species of algae have been identified. The water level may fall to 14 cm of the adjoining Black Sea. The bottom muds, rich in amino acids and carcinoids, have a high rate of natural radioactivity. Some muds are sapropelic, with phyto-remains, particularly algae which are putrefied in an anaerobic environment. Lake Techirghiol, once a bay of 1,170 ha, now separated by a sand bar, is the source of the mud but irrigation of surrounding land caused a salinity decrease with resulting ecosystem changes.

Black Sea centers ring its shores; originally catering especially to their own nationals, they have increasingly drawn foreign visitors. Mangalia, southernmost resort, has attracted seamen since classical times; it forms with Eforie and Neptun, artificial creations, the Romanian cure complex. Blessed with a balmy climate, the center offers a therapy based on seawater and sapropelic mud use, sulfurous mesothermal springs; mud baths and application of mud poultices, it acquired some international reputation. The black, pasty, sapropelic mud comes from Lake Techirghiol, with a mineralization of 80 g/L. Concentrated mud extracts have been shipped to distant locations. At 150 m from the sea, the beach facilities follow the Egyptian method of open-air treatment. An air rich in iodine, magnesium, bromine, and sodium chloride creates an ambience particularly favorable for aero-ionization and insolation.

As in Germany, pelotherapy is also practiced with peat mud found in Lake Mangalia. Bicarbonated, hypotonic, mesothermal (26°C), radioactive, sulfurous water sprouts from springs on or near the Mangalia and Neptun beaches. The sapropelic mud rich in carbonaceous or bituminous matter has a plasticity value of 250 g, a thermal metric capacity index of 20.99. It is enhanced by bacteriostatic, bactericidal, and antiallergic qualities due to its high vitamin (C, E, B2 and B12), nicotinic acid, hormones, and organic content.

Economics

Setting aside the savings aspects in hospital days and pharmaceuticals consumption which benefit state and private insurance systems, and the individual, and considering the tourism aspects, it appears that thalassotherapy and thermalism are large earners and big employers. Taking, for example, the sole thermalism in France, in 1998 centers hosted 548,003 curists representing 9,864,054 "visitors" days for insured parties and 527,629 days for other curists, for a year's total of 10,391,683.

Many did not come by themselves but were accompanied by noncurists, representing an additional 300,000 persons.

A low-priced cure costs an average FRF 1,000 (€ 150) for a six-day stay, with a per person FRF 2,200 (€ 320) tab for food and accommodations (thus exclusive of such additional expenses as beverages, entertainment, sundry purchases). The income for the French centers, besides payment for medical services, exceeds thus by far 13,400,000 "days" × FRF 2200 = FRF 29,480,000,000 or approximately US$5,781,000,000, about EUR 4,494,000,000 [Exchange rates used in the calculations are FRF6.3 = US$1; Euro 1 FR = F 6.56]. It is furthermore estimated that the 100-plus thermal centers provide employment to at least 100,000 people.

Examining these numbers for a developing economy as for instance in Romania, one one-hundredth of this is 1,000 jobs and an income of US$46,740,000 for a single low-priced center. *Thermae* income amounts to, using the same formula, FRF 104,000,000 or about US$16,508,000 or EUR 15,857,000 [Exchange rates used in the calculations are FRF6.3 = US$1; EURO1 = FRF6.56]. An adjustment factor is naturally needed as wages and prices are clearly higher in France than in Romania. Nevertheless, thalassotherapy (and of course thermalism) are not to be looked at as an insignificant economic player (Guillemin *et al.*, 1994; Boulangé, 1995; Anonymous, 1997; Collin, 1997).

Revenues are generated by use of facilities, hotels, restaurants, but also by pharmaceuticals and cosmetics: the University of California has received in royalties for patented pseudopterosins over US$1,200,000, and the cosmetics firms have collected several millions more. Algae and muds can thus be "earners."

Treatment centers

Thalassotherapy stations have a long history with the largest number of stations in Germany and France. Most of the 22 German stations grew in importance during the last half century. They have attracted a large clientele and contributed substantially to the growth and expansion of touristic sites. France remains the leader with the largest number of stations, some catering to the well-to-do. Of over 40 hydro-linked health resorts found in Great Britain, the birthplace of sea-bathing hospitals, only Springs Hydro at Packington, Ashby de la Zouch in Leicestershire propose thalassotherapy; algae baths or seaweed body wraps are offered in Scotland (St. Andrews, Kingdom of Fife) and in London and balneotherapy at Newport Pagell (Buckinghamshire). Thalassotherapeutic facilities sprung up principally around the Mediterranean in Spain, Monaco, Greece, and Tunisia, and in Israel. Black Sea facilities acquired a solid reputation over the last decades. The development of a therapy *de pointe*, free from extravagant claims in up-to-date facilities can nurture a sustainable and rational growth of Romanian Black Sea resorts.

One may recall the medications developed some decades ago—and still in use—by Romanian Dr. Aslan (Gerovital, Aslanvital) against antioxidant (ao) ageing. Her work remained highly controversial to the point that some considered her therapy charlatanism. Is a similar risk conceivable with thalassotherapy and its sister cure thermalism? Indeed development must be considered on a long-range scale.

The introduction of new technologies such as the combination of electro-acupuncture with thalassotherapy is due to win over a new clientele. Thalasso-electro-acupuncture brought back from China relatively recently, has rapidly gained *droit de cité* and garnered enthusiasts in French centers.

Comparisons have been made among European centers and a critical analysis of the recent relevant bibliography published (Collin, 1995; Constant *et al.*, 1995; Graber-Duvernay, 1999). A study conducted by the French National Health Insurance System (*Caisse Nationale d'Assurance Maladie*) observed 3,000 persons who took a thermal cure during a span of three years and found a health improvement dealing with various ailments, for example, rheumatism, back pain, arterial problems (Morand *et al.*, 1990; Guiry and Blunden, 1991), in two-thirds of the group and a concomitant decrease in the length and frequency of hospital stays. Furthermore, the use of medicines dropped or was cut for 72.4% of the patients while the disbursements of the Health System for thermal care represent barely 0.22–0.43% and "*cures*," stays at thermal centers, represent 0,89–1% of the total medical "consumption." It is not preposterous to extrapolate these observations to the domain of thalassotherapy.

In fact, a thorough scientific project has been conducted by the department of hydrological and climatological medicine of the University of Nancy medical faculty and results made public in July 1999. The conclusions are positive and doubts as to the value of thermal cures seriously challenged. The congress (*assises*) of thermalism held in Toulouse, France in May 1999 (*Proceedings* were published in June 1999) confirms the medical dimension of thermalism. Positions held since 1996 seem thus appropriate. Furthermore, statistics show a substantial sustained drop in the frequency of hospital stays and the consumption of medicines.

The venture is foreseeably valid and sustainable. Boulangé in his 1989 and 1995 papers examining the scenario, both in the framework of the European unification and of the next century, for French thermalism—and there is no reason for not extrapolating these views to thalassotherapy—sees a bright future (Boulangé *et al.*, 1989).

With France and Germany still the leaders in centers, facilities and techniques, a noticeable increase in treatment centers has taken place in Tunisia and Morocco. One center in Austria, another in The Netherlands, import marine muds so it can claim thalassotherapy care. Marine muds of Israel are used in the Plombières (France) thermal center (Anonymous, 1999). A unique case is that of Ein Bokek, Israel on the Dead Sea. There is virtually no town but a cluster of hotels catering to patients seeking treatment mainly for (arthritic) psoriasis. The treatment commonly combines helio- and thalassotherapy; solar rays are filtered by an extra 300 m of atmosphere than elsewhere on earth and seawater here has a tenfold higher salt content. Results of treatments have been repeatedly commented in the *Journal of the American Academy of Dermatology* (1985) and the *International Journal of Dermatology* (1995, 1997).

Roger H. Charlier and Marie-Claire P. Chaineux

Bibliography

Anonymous, 1991. Les thermes romains. *Actes de la Table-Ronde organisée par l'Ecole Française de Rome*. Paris, Collect. Ecole Franç. de Rome No. **142**.

Anonymous, 1996. *European Commission Water Policy*. Brussels, Office for Official Publications of the Commission of the European Communities.

Anonymous, 1997. *Enquête; suivi d'une cohorte de 3000 curistes durant 3 ans*. Service National du Contrôle Médical-Caisse Nationale de l'Assurance Maladie des Travailleurs Salariés. *Etudes et Enquêtes. Echelon National du Service Médical*, **2**.

Anonymous, 1999. The Dead Sea experience. *Psoriasis Resource*, **1**(1/2): 1–4.

Anonymous, 1999. Discovering the Dead Sea. *National Psoriasis Foundation Bulletin*, **30**(3): 1–3.

Arehart, J.L., 1969. Oceanic drug chest. *Sea Frontiers*, **15**: 99–107.

Bologa, A.S., 1985/1986. Present state of seaweed production along the Romanian Black Sea shore. *Vie et Milieu*, **39**(2): 1–22.

Bologa, A.S. *et al.*, 1999. Distribution of planktonic primary production in the Black Sea. In Besiktepe, S. *et al.*, (eds.), *Environmental Degradation of the Black Sea: Challenges and Remedies*. Dordrecht, NL, Kluwer Academic Publishers, pp. 131–145.

Boulangé, L.M., 1997. Thermalisme ou thermalismes. In Boulangé, M.L. (ed.), *Les vertus des cures thermales*. Montpellier, Editions Espace 34.

Boulangé, L.M., 1997. *Les vertus des cures thermales*. Montpellier, Editions Espace 34.

Boulangé, M., 1995. La place du thermalisme à l'aube de l'an 2000. *Bulletin Social Science and Medicine [Luxembourg]*, 132, (**2**): 43–44.

Boulangé, M., 1995. Les effets généraux de la cure thermale. *Presse thermale et climatique*, **132**(4): 213–216.

Boulangé, M., Collin, J.F. and Irlinger, P., 1989. Perspectives du thermalisme français à l'aube de l'Europe. In Hérisson, C. (ed.), *Crénothérapie et réadaptation fonctionnelles*. Paris, Masson, pp. 259–264.

Buchet, L., 1985. *Médecine en Gaule. Médecine et chirurgie pendant les premiers siècles du moyen âge*. Paris: Picard.

Charlier, R.H., 1975, Thalassotherapy. In Fairbridge, R.W and Schwartz, M. (eds.), *Encyclopedia of Coasts and Beaches*. Englewood Cliffs: NJ, Hutchinson *et al.* (Volume in the *Encyclopedia of the Earth Sciences Series*, pp. 807–813).

Charlier, R.H., 1990. Algae—Resource or scourge? *International Journal of Environmental Studies*, **48**(1): 77–91 (2), 237–250.

Charlier, R.H., 1999. Aquacide—An urgent need to legislate to protect: Viewpoint and review. *International Journal of Environmental Studies*, **56**(2): 325–343.

Charlier, R.H. and Lonhienne, T., 1996. The management of eutrophicated waters. In Schramm, W. and Nienhuis, P.H. (eds.), *Marine Benthic Vegetation. Recent changes and the Effects of Eutrophication*, chapter II, pp. 46–78.

Collin, J.F., 1995. Situation comparée du thermalisme dans différents pays européens. *Bulletin of Social Science and Médicine [Luxembourg]*, **132**(2): 61–67.

Collin, J.F., 1997. Eléments d'évaluation médicale dans le domaine thermal. In Boulangé, L.M. (ed.), *Les vertus des cures thermales*: Montpellier, Editions Espace 34.
Constant, F., Collin, J.F., and Boulangé, M., 1995. Analyse bibliographique. Critique de la littérature médicale internationale dans le domaine des recherches thermales. *Bulletin of Social Science and Medicine)[Luxembourg]*, **132**(2): 49–59.
Cotet, P., 1970. Les lacs dobrogéens et leurs rapports génétiques avec les changements de niveau de la Mer Noire: *Travaux du Colloque National de Limnologie Physique [Romania] "Géographie des Lacs"*, 27–45.
De Roeck-Holtzhaver, Y., 1991 in Guizy and Blunden *op. cit*. pp. 83–94.
François, H., 1999. *Le thermalisme, source de santé*. Plombières- les-bains, Editions Ronald Hirle, collection *Calodae*, pp. 298.
Găstescu, P., 1963. *Lacurile din Republica Populara Romîna:* Bucuresti: Editura Academici Republicii Populare Romîne.
Graber-Duvernay, B., 1999. *Les acquis scientifiques du thermalisme en France*. Paris: Fédération thermale et climatique française, pp. 34
Grenier, A., 1960. *Les monuments romains: thermes, aqueducs. Archéologie Gallo-Romaine*. Paris: Presses Universitaires de France.
Gruber, M., 1968. The healing sea. *Sea Frontiers*, **14**: 74–86.
Guillemin, F., Constant, J.F. and Boulangé, M., 1994. Short and long term effects of spa therapy in chronic low back pain. *British Journal of Rheumatology*, **33**: 148–151.
Guiry, M.D. and Blunden, G. (eds.), 1991. *Seaweed Resources in Europe. Uses and Potential*. Chichester, UK: John Wiley & Sons.
Hérisson, C., 1989. *Crénothérapie et réadaptation fonctionnelles*. Paris: Masson.
Jacob, H., 1570. *Traité des admirables vertus des eaux de Bourbonne-les-Bains*: Bourbonne-les-Bains, L'Auteur.
Kretzschmer, F., 1966. *La technique romaine*. Bruxelles, Desclée-De Brouwer.
Lance, P., 1988. *Le guide "marine" de la thalassothérapie*. Paris: l'auteur, b.p. 102, F-75022, Cedex 01.
Larivière, A., 1958. *Les cures marines*. Paris: Aubier.
Malissard, A., 1994. *Les romains et l'eau*. Paris: Realia/Les Belles-Lettres.
Morand, P., Charlier, R.H., and Mazé, J., 1990. European bioconversion projects and realizations for macroalgal biomass: Saint-Cast-le-Guildo (France) experiment. *Hydrobiologia*, **204/205**: 301–308.
Petrova-Karadjova, V., 1990. Monitoring of the bloom along the Bulgarian Black Sea coast. *Rapports de la Commission internationale pour la mer Méditerranée*, **31**(1): 209.
Pricajan, A. and Opran, C., 1970. La protection et l'exploitation rationelle des lacs et des boues thérapeutiques. In *Géographie des Lacs, Travaux du Colloque National de Limnologie Physique*. Bucarest: Maison d'Editions de l'Etat.
Rameau, J.C., 1980. *Les thermes Gallo-Romains de Bourbonne-les-Bains*. Bourbonne-les-bains; Edit. de la Madeleine.
Range, cited in Larivière C.I., 1935. Thalassotherapy. *Archives Physical Therapy, X-Ray and Radium*, Springer, 662–666.
Russell, R., 1720. *A Dissertation Concerning the Use of Seawater in Diseases of the Glands*. Blackpool: The Author.
Schramm, W. and Nienhuis, P.H. (eds.), 1996. *Marine Benthic Vegetation Recent Changes and the Effects of Eutrophication*. New York/Heidelberg: Springer Verlag.
Translated from *Procès-Verbal de l'Académie Impériale Française de Médecine*, Séance du 5 février 1856.
Valnet, J., 1995. *Aromathérapie*. Paris: Maloine éditeurs.
Woodcock, D.C., and Blanchard, A.H., 1957. Bubble formation and modification in the sea and its meteorological significances. *Tellus*, **9**: 145–158.

Cross-references

Beach Use and Behaviors
Black and Caspian Seas, Coastal Ecology and Geomorphology
Europe, Coastal Ecology
Europe, Coastal Geomorphology
Human Impact on Coasts
Tourism and Coastal Development
Tourism, Criteria for Coastal Sites
Water Quality

HINDCASTING—See WAVE HINDCAST

HISTORY, COASTAL ECOLOGY

Introduction

The history of the investigation of the way in which coastal habitats and species interact with each other and respond to human intervention is relatively recent. Studies of the distribution of individual species populations, pattern and process in species and groups of species, and the relationship with the physical environment, has mostly taken place during the last century. An ecosystem approach which embodies all of the above and lays stress on the complexity of the interactions, is the most recent manifestation of the developing science. Nowhere is this more obvious than in the study of the ecology of coastal habitats, species, and coastal systems.

Some of the earliest work involved quantitative descriptions and successional studies of vegetation. These were undertaken around the end of the 19th century, and included work on the Dutch Wadden Sea islands and on sand dunes in Germany. Studies in Denmark also took place on dunes and were the forerunners of the study of plant ecology (van der Maarel, 1993).

This early work was carried out on habitats which were mostly, apparently free from human interference. Thus, coastal salt marshes and sand dunes with seemingly simple successional sequences were amongst the first to be studied in relation to the way they responded to natural forces. Understanding of the role of human activity was not considered to be a priority. This situation changed dramatically in last half of the 20th century as human populations put increasing pressure on marine and coastal ecosystems throughout the world. This entry attempts to describe some of the ecological principles which have been developed over the last 100 years or so, and discusses how they are applied to coastal systems today.

Early work, unravelling complex coastal systems

Studies of the autoecology and ecophysiology (of plants) dominated early coastal ecological work. Salt marshes and sand dunes were obvious candidates for studying ecological change because of their seemingly regular patterns of succession and their dynamic nature. Some of the best known British ecologists of the first half of the 20th century, for example, included detailed work on these coastal habitats (Tansley, 1949; Salisbury, 1952).The early descriptions were concerned both with the pattern of plant communities and also the process through which these patterns developed (e.g., chapman, 1976). Thus, coastal vegetation became recognized as a series of types progressing from early pioneer stages to more complex forms which could be related to the physical parameters affecting their development. The role of animals in this process was not considered and studies of animal populations continued along a largely separate path.

Plant zonation and succession

Zonations of seaweeds can be related to tidal influence and/or wave exposure, and on rocky shores these can be very pronounced (Stephenson and Stephenson, 1972). Similar patterns can be discerned for many coastal, terrestrial habitats, though the reasons for them may be less easy to interpret. Early studies looked at the role of individual plants in overcoming the rigors of what was considered to be a hostile environment, in an attempt to establish whether there were recognizable factors determining the sequence of development. *Suaeda fruticosa*, a plant of gravel shores, for example, was shown to be able to establish itself over time by slowing down the landward movement of the beach (Oliver and Salisbury, 1913). On sandy shores pioneer plants and other obstacles do the same by arresting the movement of sand grains (Figure H4). The growth habit of the plant itself, including its root system, soil water, and mineral nutrient relationships were the subject of classic studies on sand dunes (Salisbury, 1952) and helped to define the adaptations of individual species in overcoming environmental perturbations. Similar studies were carried out by others on sand dunes and salt marshes (e.g., Chapman, 1934).

The mechanism through which succession took place was perhaps most clearly elucidated for salt marshes. Primary colonizers such as *Salicornia europea* or *Suaeda maritima* (in north-west Europe) become established on accreting sedimentary tidal flats. The pioneer plants, tolerant of immersion in seawater, were shown to be dependant on periods in the early stages of plant establishment when they are free from tidal

Figure H4 Strandline and pioneer sand dunes, Magilligan Point, Northern Ireland.

Figure H5 Parallel zones on a salt marsh. *Spartina* expansion following erosion, Severn Estuary, England/Wales.

movement. As sediment height increases, the marsh is subject to progressively fewer tidal inundation, less sediment is deposited and a richer complement of plants and animals replaces the specialist salt-tolerant species (Ranwell, 1972). Thus, from an ecological point of view salt marshes provided evidence of primary succession, with the development of a parallel spatial zonation (Figure H5) which could be related to tidal inundation. The fact that this occurred apparently "largely without human interference" made them ideal for studying the processes associated with "natural" vegetation development.

By 1972 when Ranwell published his book on the ecology of salt marshes and sand dunes recognition of the importance of understanding the complex relationships, which determined the nature of salt marshes and sand dunes, was much more clearly understood (Ranwell, 1972). These and other early studies showed that there are zonations attributable to environmental gradients in coastal vegetation. Anyone looking at the early stages in salt marsh or sand dune growth will need little convincing that this is so. However, unravelling the precise relationships was much more difficult than it at first appeared.

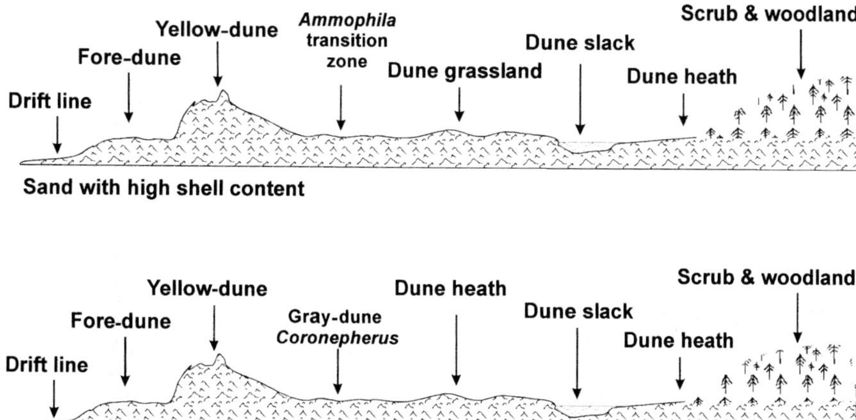

Figure H6 A sequence of sand dune succession on calcareous and acidic sand. Notice the dune building phase (high dune) followed by deflation and the development of more stable communities.

Salt marshes not only respond to tidal movement, but also to relative sea-level change. The fact that these operate over very different temporal scales is an important factor in understanding the mechanisms involved. The movement of estuary channels, which may cause erosion operates over yet a different timescale. The development of salt pans or the effects of rotting seaweed cause changes to the surface vegetation. Added to this is the fact that succession may be cyclical as sediments are exposed when erosion takes place.

Sand dunes also show forms of successional development in the early stages of growth by the accretion of sand, aided by specialist plants. However, once the main body of the dune is formed other processes come into play and the change from mobile foredunes and yellow dunes to grassland, heath, scrub, and woodland is rarely a straight forward progression, though it may be depicted as such (Figure H6). Blowouts occur with or without the intervention of man and can be the precursors of dune slacks. Similarly, the reprofiling of dune ridges under the influence of changing wind patterns bring an infinitely variable topography, the origins of which may be difficult or impossible to unravel.

Coastal networks

As the mechanisms through which the habitats developed were unravelled, so the sheer complexity of the relationship with the "natural" environment became more apparent. Individual species relationships, to some extent, provided only a first level in understanding of the way in which coastal systems and their biological components operate. The physical conditions in which plants and animals exist provide powerful controls on populations, especially where there are major perturbations in these conditions. Superimposed on this are the various natural cycles involving predator–prey relationships and intra- and inter-specific competition. Thus, it is not surprising that as the study of ecology developed, these complex relationships came under scrutiny. These interactions are the "stuff" of ecology and the studies have been many and various. Nowhere is this more apparent than in coastal wetlands.

Nutrient cycles and energy flows within wetland communities are key and often quoted examples. Attempts to describe the complex interactions between detritus, nutrient inputs, primary producers, detritus feeders, predatory fish, and birds are usually presented as "generalized" food webs. However, these pictures can only represent a very simplified view of the systems involved. A detrital food chain in a mangrove forest alone, for example, may include 11 different groups of animals with untold species involved.

The importance of geomorphology

Engineers were perhaps amongst the first to appreciate the interaction and complexity of whole coastal ecosystems. Writings early last century described the way the coast responded to natural conditions such as sea-level rise, tides, storms, wave action, wind, and rain and how these affected the shoreline (Wheeler, 1903; Carey and Oliver, 1918). Later studies were concerned with coastal erosion and land reclamation (Du-Plat-Taylor, 1931; Matthews, 1934). These helped lead the way to unravelling the relationships between the physical and biological components of the coastal environment, including an appreciation of the dynamic nature of the habitats involved.

The description of coastal landforms and the factors which have helped to shape them come under the general heading of geomorphology. Some of the most important texts, which have helped secure a better understanding of the coast and its management needs come from these studies. In Great Britain, during the 1940s and 1950s, a survey of the geology and history of change on the coast was undertaken and described in two classic works (Steers, 1946, 1972). These provided a much better understanding of the coastal landscape and its conservation needs and helped lay the foundation for much of the subsequent coastal policy. In America, comprehensive accounts of the development and the processes by which coastal systems came into being were led by geologists (e.g., Johnson, 1919). More recent academic publications look at examples from around the world and provide more detail, not only of the way in which the coast responds to the many natural driving forces, but also the influence of human use in shaping its condition (Bird, 1984; Carter, 1988; Carter and Woodroffe, 1994).

Human occupation and use

As has already been intimated above, from an ecological perspective, coastal habitats are often considered amongst the more natural ecosystems. This has led to the impression that the process of succession takes place in a sequence which is determined largely by natural forces. The classic studies of the salt marshes and sand dunes on the North Norfolk coast (Chapman, 1938, 1941, 1959) or the Dovey Estuary in west Wales (Yapp *et al.*, 1917) emphasized the natural status of the vegetation. However, this hides a long history of human use. On the sedimentary shorelines of the Wash, for example, artificial embankments to enable salt-making were present in some numbers in Roman times (Simmons, 1980). Hay making, oyster cultivation, turf and reed cutting, and samphire gathering all take place or have taken place on upper marshes throughout Europe (Dijkema, 1984). The deliberate planting of *Spartina anglica*, itself a hybrid fashioned from the interaction of a native and an introduced species in southern England (Marchant, 1967), has been a major influence on salt marshes throughout the world.

Occupation of sand dunes probably dates back several thousand years as archeological studies have shown. Many sites have examples of middens with the remains of shell fish in them. An analysis of flint tools suggests that settlers of Torrs Warren, a sand system in south-west Scotland, may have appeared between 5,500–7,000 years ago (Coles, 1964). It seems that cultivation has taken place since then and in 1572 three farm houses were recorded on the Warren. As long as 4,500 years ago a small settlement existed on the shores of Skaill Bay in Orkney. The Neolithic people which inhabited the site, known today as "Skara Brae," seem to have been agriculturists as well as hunters/gatherers (Ritchie and Ritchie, 1978) living on the coast until their village was overwhelmed by a sand storm. Since Medieval times, dunes have been used extensively as rabbit warrens and are grazed by domestic stock to the present day. Research on the origins of the deltas of the

Mediterranean show a link between population growth and decline, deforestation in the hinterland, and the growth and retreat of the deltas themselves.

This knowledge has altered our perception of coastal habitats. Far from being natural, many coastal systems are highly modified by human activity. Their ecology and conservation thus depends both on understanding the "natural" processes by which they develop and the impact of human action (Doody, 2001).

Conclusion

Unravelling the ecological interactions of species and their environment has been the main thrust of ecological studies for most of the last century. It has been suggested that this has been largely undertaken without any direct investigation of the human factor and its impact on the individual elements in the coastal environment, whether they are concerned with vegetation succession or change in animal populations. As our understanding of the relationship between the physical and biological components of the coastal environment has grown, so has concern for the impact of human use both on the coast itself and its ability to sustain human uses. This has brought into sharp focus the need for a more integrated approach to the study of coastal ecosystems. At the same time the role of ecology has moved substantially from the study of systems "apparently" free from human interference, to one where socio-economic forces are equally, if not more important, to understanding how coastal landscapes function.

Sustaining coastal habitats

The study of ecology has also formed the basis for the identification, protection, and management of areas of nature conservation importance. Indeed, it was the preeminent ecologists of the day who helped develop the present approach to the protection and management of nature reserves. In the United Kingdom, for example, in 1942 the British Ecological Society established a Nature Reserves Committee under the Chairmanship of Sir Arthur Tansley (Sheail, 1987). It was this committee which provided the foundation for the series of Nature Reserves and other protected areas that are so important to the conservation of the remaining areas of natural and seminatural habitats today.

In the early postwar period, the understanding of successional processes, the concept of climax vegetation, and the notion of zonation, helped to lead conservation managers toward a preoccupation with the protection of existing interests, identified when the sites were first assessed. To many the "naturalness" of the system was the prime reason for its conservation. However, as has been argued above, many so called "natural" systems are not natural at all. Most if not all, in temperate regions at least, have in some way been modified by human activity. The extensive unmodified lagoons and deltas with their apparently natural salt marshes and sand dunes in Albania and eastern Turkey, grow rapidly today because of deforestation in the hinterland. Even here drainage and land "improvement" for agriculture has destroyed large areas of transitional vegetation and pollution is a major consideration.

This understanding has important consequences both for the development of conservation policy and the ecological principles upon which it is based. Agenda 21, a program for action agreed at the United Nations, Earth Summit in Rio in 1992, aims to help achieve the twin goals of sustainable human development and the maintenance of biodiversity. The statements clearly point the way toward recognition of the interrelationship between the so called "natural environment" and human economic and cultural activity. If policies are to be developed which fulfill the aspirations both the politicians and the traditional conservationist, future ecological studies must include human use as a key component of the "natural" system.

The coastal squeeze. An important component of human use centers around the loss of coastal and tidal land. The process by which this affects the margin between the land and the sea can be described as "coastal squeeze." Here human action pushes the limits of intensive use of the land (for agriculture, industry, and the like, or land claim of tidal areas), toward the sea. At the same time, where sea level is rising relative to the land, the upper limits of tidal influence are pushed landward (Figure H7). Taken together, these effects cause a narrowing of the shore and the loss of coastal habitats. In its turn this can result in a reduction of the capacity of the shore to withstand, and recover from, episodic events including major storms. In low-lying coastal areas, for example, coastal properties are put at greater risk from flooding and/or erosion, as the protection afforded by a wide beach is reduced, as it becomes steeper and narrower.

The cumulative effects of habitat loss may also reduce the ability of the coast to recover from major environmental perturbations. This may include difficulties in replenishing living and nonliving resources. In their turn these result in an inability to sustain the economic fabric of some areas.

A question of sediments. Other factors, which have been increasingly recognized as being important to the sustainable use of the coast, include the nature and availability of sediments. In this context, there is also an increasing recognition that the reduction in availability of sediments, whether due to offshore exploitation, damming of rivers, or reducing the flow from longshore drift, has a predicable outcome. Many deltas have grown through the transport and deposition of sediments eroded from the hinterland following deforestation. Studies suggest that this situation has been reversed as the damming of rivers has reduced the available sediment supply. As a result, today the outer margins of many deltas are eroding as they become wave-dominated rather than ones where freshwater river flows exert the major force. This pushes the margins of the delta landward and with it increases saline water intrusion into the underground aquifer. The implications for the continued economic use of these areas for agriculture (e.g., rice cultivation), ground water abstraction, or problems associated with erosion are largely ignored by the political and economic forces, which continue to dictate land use policy and coastal management.

Ecological change as a healing force, a new paradigm?

Change is an important part of the development of coastal systems. This was most dramatically revealed by major events, such as the landslip between Axmouth and Lyme Regis (Devon, Dorset, southern England) on Christmas Day 1839, which created a chasm 1 km long and up to 122 m wide, taking with it a small village. The steady erosion of the cliffs at Dunwich (Suffolk, England) has now thrown all seven of the Medieval churches of this once-important port into the sea. At the same, time saltmarsh erosion and accretion are relatively natural phenomena, taking place in response to changes, for example, in the location of the tidal channels or, over a longer time period, in sea level. Recognizing these changes as factors to work with rather than against, may provide a more sustainable approach to coastal development.

In this context, the alliance between the ecologist, geomorphologist, and coastal engineer in re-creating new coastal habitats may be a first step toward accepting change as a means of securing more sustainable living on the coast. For example, the re-creation of salt marshes, or other tidal wetlands from land given back to the sea may not only secure new nature

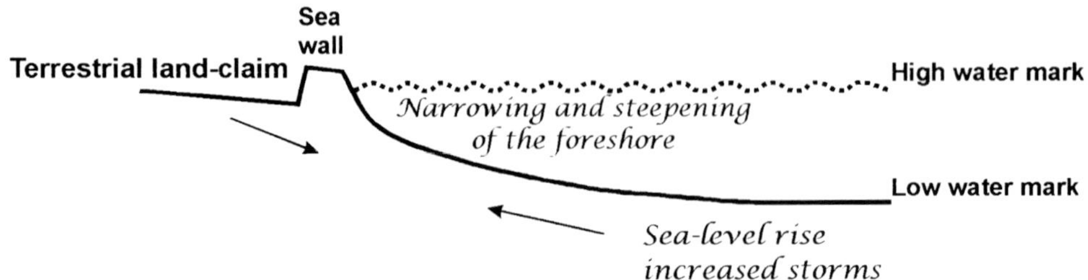

Figure H7 "Squeezing" the coast.

conservation opportunities, but also provide a more sustainable sea defense. Perhaps, ultimately we will emulate our ancestors, and in some areas initiate major change and then step back and let nature take its course. Under these circumstances the role of the ecologist may be paramount in predicting the outcome of a particular management option. This will require a marriage between the traditional approach which sees the ecological process as natural, and a more pragmatic one where human factors play a significant role. This will not reduce the importance of the "natural" succession, but set it in a wider human context. In the same way that our ancestors helped fashion the present coast and its biological and non-biological resources by accident, we can do it by design!

J. Pat Doody

Bibiliography

Bird, E.C.F., 1984. *Coasts—an Introduction to Coastal Geomorphology*. Canberra: Australian National University Press.

Carey, A.E. and Oliver, F.W., 1918. *Tidal Lands, a Study in Shore Problems*. London: Blackie and Son.

Carter, R.W.G., 1988. *Coastal Environments*. London: Academic Press.

Carter, R.W.G. and Woodroffe, C.D. (eds.), 1994. *Coastal Evolution, Late Quaternary Shoreline Morphodynamics*. Cambridge: Cambridge University Press.

Carter, R.W.G., and Woodroffe, C.D. (eds.), 1994. *Coastal Evolution, Late Quaternary Shoreline Morphodynamics*. Cambridge: Cambridge University Press.

Chapman, V.J., 1934. The ecology of Scolt Head Island. In Stress, J.A. (ed.), *Scolt Head Island*, Cambridge: W. Heffer & Sons, pp. 77–145.

Chapman, V.J., 1938. Studies in salt marsh ecology, I–III. *Journal of Ecology*, **26**: 144–179.

Chapman, V.J., 1941. Studies in salt marsh ecology, IV. *Journal of Ecology*, **29**: 69–82.

Chapman, V.J., 1959. Studies in salt marsh ecology, IX. Changes in salt marsh vegetation at Scolt Head Island. *Journal of Ecology*, **47**: 619–639.

Chapman, V.J., 1976. *Coastal Vegetation*. Oxford: Pergammon Press.

Coles, J.M., 1964. New aspects of Mesolithic settlement of south-west Scotland. *Transactions Dumfries and Galloway Natural History and Antiquarian Society*, **XII**: 67–98.

Dijkema, K.S. (ed.), 1984. Salt marshes in Europe. Strasbourg: Council of Europe, *Nature and Environment Series*, No. 30.

Doody, J.P., 2001. *Coastal Conservation and Management: an Ecological Perspective*. Boston: Kluwer Academic Publishers.

Du-Plat-Taylor, F.M., 1931. *The Reclamation of Land from the Sea*. London: Constable & Company.

Johnson, D.W., 1919. *Shore Processes and Shoreline Development*. New York: Prentice Hall.

Marchant. C.J., 1967. Evolution in *Spartina* (Gramineae). I, The history and morphology of the genus in Britain. *Journal of the Linnaean Society (Botany)*, **60**: 1–26.

Matthews, E.R., 1934. *Coast Erosion and Protection*. London: Charles Griffin & Company.

Oliver, F.W., and Salisbury, E.J., 1913. Vegetation and mobile ground as illustrated by *Suaeda fruticosa* on shingle. *Journal of Ecology*, **1**: 249–272.

Ranwell, D.S., 1972. *Ecology of Salt marshes and Sand Dunes*. London: Chapman and Hall.

Ritchie, A., and Ritchie, G., 1978. *The Ancient Monuments of Orkney*. Edinburgh: HMSO.

Salisbury, E., 1952. *Downs and Dunes, their Plant Life and its Environment*. London: G. Bell & Sons.

Sheail, J., 1987. *Seventy-Five Years in Ecology*. The British Ecological Society, Oxford: Blackwell Scientific Publications.

Simmons, I., 1980. Iron Age and Roman coast around the Wash. In *Archaeology and Coastal Change*, Thompson, F.H. (ed.), London: The Society of Antiquities, Occasional Paper No. 1.

Steers, J.A., 1946. *The Coastline of England and Wales*. Cambridge: Cambridge University Press.

Steers, J.A., 1972. *The Coastline of Scotland*. Cambridge: Cambridge University Press.

Stephenson, T.A., and Stephenson, A., 1972. *Life Between Tidemarks on Rocky Shores*. San Francisco: Freeman.

Tansley, A.G., 1949. *The British Islands and their Vegetation*. Cambridge: Cambridge University Press.

Van der Maarel, E., 1993. Dry coastal ecosystems: scope and historical significance. In E. van der Maarel (ed.), *Ecosystems of the World: Dry Coastal Ecosystems, Polar Regions and Europe*, Amsterdam, Elsevier, pp. 1–6.

Wheeler, W.H., 1903. *The Sea Coast*. London: Longmans, Green and Company.

Yapp, R.H., Johns, D., and Jones, O.T., 1917. The salt marshes of the Dovey Estuary. II, The salt marshes. *Journal of Ecology*, **5**: 65–103.

Cross-references

Deltas
Dune Ridges
Estuaries
Europe, Coastal Ecology
Monitoring, Coastal Ecology
Rock Coast Processes
Salt Marsh
Vegetated Coasts

HISTORY, COASTAL GEOMORPHOLOGY

> I am only too painfully aware how increasingly difficult it is to find time for a careful study of the work of our predecessors . . .
>
> (Geike, 1897)

The coast has been of primary concern to man since he first set foot upon it, and will undoubtedly continue to be so long into the future. It must have been "studied" by the earliest dwellers in their quest for food and also by later inhabitants for other reasons, such as a need of harbors or for defense. Men learned early about cliffs, rocky shores, and sandy beaches; they must also have discovered the operation and importance of tides, currents, and waves upon the shore. In light of man's long-continued interest in the coastal zone, it is surprising that its scientific study lagged so far behind that of many of the earth's other environments.

One explanation may well be that the shoreline, like so many of nature's other boundaries, failed to attract the attention of scientists. Scientifically, it was long a no man's land—neither "ocean nor land." Oceanographers were reluctant to tread where their ships would not go, and geologists were equally reluctant to tackle the sea near the shore.

Early observations and theories

Although the actual beginnings of coastal study are shrouded in the haze of history past, some of the speculations of the early Greeks and Romans have been preserved. They were aided by having a keen practical interest in things coastal; as early as the 4th century BC they were well acquainted with Mediterranean and Black Sea littorals. Their knowledge of coasts was advanced greatly by such men of action as Alexander the Great. The anonymous Greek *Periplus of the Erythraean Sea*, which described the coasts of the western Indian Ocean and even, albeit hazily, the coast east of India, was used by sailors during the days of Pliny. Although based on direct observation and unhindered by religious teachings, early theories about coasts were nonetheless tempered by the superstitions, legends, and myths that were in the heritage of all Mediterranean peoples at the time.

The presence of marine fossils far from the sea led Aristotle (384–322 BC) and others to conclude that the sea had previously occupied higher levels. Strabo (54 BC–AD 25) even went so far as to write that it frequently changed levels—rising at times, falling at others. The role of rivers in altering the landscape also attracted the attention of these natural philosophers, and, although they had several ideas about where and how river water originates, there seems to have been little doubt in their minds about what happens when it enters the sea. Aristotle, Strabo, and others recognized deltaic and alluvial plain deposits. They knew that river-carried sediment first shallows the sea, then converts it into marshland, and changes it eventually into dry, farmable land. Erosion as a coastal phenomenon was not ignored. Strabo, for example, noted that the ebb and flow of tides prevent deltas from advancing continuously outward into the sea.

The role of man as coastal agent was also studied. Strabo cited an example of the attempt to improve the harbor of Ephesus in the 2nd century BC:

> The mouth of the harbor was made narrow by engineers, but they, along with the king who ordered it, were deceived. . . . He thought the entrance would be deep enough for merchant vessels. . . . But the result was the opposite, for the silt, thus hemmed in, made the whole of the harbor . . . more shallow. Before this time the ebb and flow of the tides would carry away the silt and draw it to the sea outside (Russell, 1967, p. 299).

Such was the state of knowledge and theory during those Greek and Roman times at about the beginning of the Christian era. Although these notions persisted in Europe only until the fall of the Roman Empire, they were nonetheless preserved and nurtured in the Arab world during much of the time leading up to the European Renaissance.

The Renaissance

The translation of Ptolemy's *Geography* in the early 15th century was a major step toward the great discoveries of the 15th through the 18th centuries. Much of the actual knowledge held by the Greeks, Romans, and Arabs about coasts came from navigators, and so it was to continue for many centuries. Many of the voyages during this period (such as those of the Cabots, Verrazano, and Gómez along the northeast coast of North America) were strictly coastal; there was no attempt at colonization or at exploration of the interior. Penrose (1955, p. 147) wrote: "A fair coastal survey was the sole result—nothing more. . . ."

Such surveys added rapidly to the body of knowledge that was beginning to accumulate, and influenced those who would put their minds to coastal problems. Most of the material was pure description, and just what influence it actually had on the development of coastal morphology as a science is uncertain. Nonetheless, it was available for such thinkers as Leonardo da Vinci (1452–1519), who recognized terraces for what they are, and for Steno (1631–87), who established a depositional sequence in explanation of the strata he observed.

Bernhard Varenius (1622–50), because he successfully merged description with theory, has been credited with laying the foundations for geography as a science (Mather and Mason, 1939). These foundations are recorded in his *Geographia Generalis*, a book that was used as a text in British universities for a century by such notables as Isaac Newton. On the topic of coastal morphology, Varenius wrote, "The ocean in some places forsakes the shores, so that it becomes dry land where it was formerly sea" (Mather and Mason, 1939, p. 25). He reasoned that many factors may be involved in such a change, including those of erosion and deposition, ocean currents and tides, texture and structure, river flow, and wind. Process was very real in his science.

The influence of scripture

Nearly all "scientists" of the 17th, 18th, and much of the 19th centuries were influenced to greater or lesser degree by the Bible and the Church. In some cases, this acceptance was probably not unlike that of the Greek philosophers, influenced unknowingly as they were by mythology. The biblical quotations of greatest significance, as found in the King James version of the Bible are:

> Let the waters under the heaven be gathered together unto one place, and let the dry land appear . . . (Genesis, 1: 9). Let the waters bring forth abundantly the moving creature . . . (Genesis, 1: 20). And . . . the flood was forty days upon the earth; . . . and the waters prevailed exceedingly upon the earth; and all the high hills, that were under the whole heaven, were covered (Genesis, 7: 17, 19).

Because, according to the Bible, the separation of land from sea occurred two days before animal life in the sea was created, the utilization of *fossils* in explaining the history of the earth, as the Greeks had done, was heresy. However, the Bible offers the universal flood of Noah's time as a possible out, one that was used in many explanations. A major difficulty was time. Because of the strict application of scripture, time was unavailable as a basis for applying observable processes to earth history. The major challenge was how to present the facts as offered by landscape observations without running counter to theological precepts.

Theories of landscape development and coastal morphology

Within such a Church-dominated intellectual environment, it is not surprising that a number of theories were proposed to explain the earth and its surface forms. During the 18th and 19th centuries many were proposed, accepted, modified, merged, and abandoned. Contemporary and subsequent history has labeled many of these theories by their predominant theme, and some also by the name of their principal proponent. Examples include neptunism (Wernerism), plutonism, diluvialism, and fluvialism (Huttonism). These theories and their major advocates are treated at length in volume one of *The History of the Study of Landforms*, by Chorley *et al.* (1964). Basically, most of the earlier theories were catastrophic in approach, reflecting a high degree of biblical influence; later ones tended toward uniformitarianism, although there was much overlap. In any event, each theory in its own way has played a role in the development of coastal geomorphology.

The neptunists, of whom Werner (1749–1817) was the principal spokesman, invoked a primeval universal ocean in accounting for both stratigraphic sequence and landform types. The earth's rocks were formed, according to the neptunists, through the successive accumulation of chemical precipitates within the ocean. Although some surface forms were caused by submarine deposition and erosion during the presence of the universal ocean, most forms resulted from erosion during a rapid recession of oceanwater. From this viewpoint, every landform originally was coastal—at least, in the sense that it was formed by turbulent oceanic currents and waves.

The diluvialists, as exemplified by Buckland (1784–1856), held a theory that was similar in some ways to that of the neptunists. The catastrophe they invoked, however, was Noah's flood. In the pure form of diluvialism, the flood was more a destructive than constructive agent. Again, it was rapidly receding water that carved the landscape. Some of the many types of otherwise unexplained phenomena that were accounted for by the flood were erratics, underfit streams, and terraces both river and marine.

Most of the catastrophists recognized that coastal and riverine erosion and deposition did occur but at such a slow pace that they had no place in earthly or "Godly" schemes. Nonetheless, even before and especially during the dominance of catastrophism, uniformitarian ideas surfaced and by the mid-19th century were dominant.

The uniformitarians were united on the concept that currently operating processes were capable of creating present-day landscapes. They were not united, however, as to which processes were most important. In general, there were two camps: one, represented by Hutton (1727–97) and Playfair (1747–1819), believed that the river was the most important agent in landscape development; the other, represented by Lyell (1797–1875) and Ramsay (1814–91), advocated marine abrasion as dominant.

Form and process: 17th–19th centuries

Although it is basically correct to conclude that the dominance of the catastrophic schools during their heyday retarded the study of geomorphology, the study of *coastal forms and processes* actually progressed to some extent. Because such forms and processes were considered insignificant from the standpoint of the overall scheme of landscape development, they could be looked at without fear of countering religious dogma if one did not try to conclude too much. Coastal cliff erosion and deposition are so conspicuous, especially in parts of the British Isles, that they did not escape the consideration of layman, geologist, and engineer alike. Hutton and Playfair, although mainly fluvialists, recognized coastal processes. Hutton, for example, wrote, ". . . we never see a storm upon the coast, but that we are informed of the hostile attack of the sea upon our country" (1788; quoted in Chorley *et al.*, 1964, p. 39). Playfair, in the same vein, emphasized the obviousness of coastal erosion when he wrote, "If the coast is bold and rocky, it speaks a language easy to be interpreted." He also noted that once fragments of rock are detached they "become instruments of further destruction . . ." (1802; quoted in Chorley *et al.*, 1964, p. 60). Thus, throughout the period of, and subsequent to the Renaissance, statements appeared that indicated some relatively advanced thoughts about coasts, some of which did overstep the bounds of Church dogma.

For example, John Ray (1627–1705), a keen observer, went so far as to propose that the combination of subaerial erosion and coastal cliff retreat would eventually reduce all land to a level below the sea. Guettard (1715–86), famous for his geological maps, believed the sea to be the major agent in land erosion, and that cliff coasts were the remnants of former extensive hill systems. He observed that sediments brought to the sea by rivers mixed with material eroded from adjacent cliffs and submerged rocks. However, he tempered this view by noting that the action of the waves would have little effect beneath the surface of the sea. Lavoisier (1743–94) adopted the Guettard idea that littoral beds are composed of materials from varied sources, but went a step further and noted that the coarsest materials are highest on the shore, and are followed downslope by coarse sands, fine sands, and clay. The width of each band, he maintained, varies with the steepness of the slope.

One of the most perceptive of the natural historians of this period was the little-recognized John Walker (1731–1803). He was the first effective teacher of geology at the University of Edinburgh where he held the chair of natural history from 1779 to 1803. His students included the geologists Playfair, Hall, and Jameson. His lecture notes—not published until 1966—contained many advanced notions that must have guided much of the thinking of the geologists of the early part of the 19th century. Some of Walker's notions, of relevance to coastal geomorphology, dealt with: continental drift—"... why not America from Europe and Asia and indeed every one continent from another"; coral reefs—he was apparently the first geologist to actually describe the growth of coral reefs; subsidence—he not only described the processes of alluviation but wrote that sediments "... are found in great quantity and to considerable depth, being the sediment of rivers...." (Walker, 1966, pp. 178, 183) anticipating R.J. Russell's work of 150 years later. Unlike most geologists of the time, Walker believed that sea sand was formed by the weathering (his term) of rocks rather than by chemical precipitation from the sea (Walker, 1966).

Marine planation

With Ray's 17th-century and Guettard's 18th-century views of marine erosion, coupled with the fact that even the most dedicated of fluvialists placed great importance on marine processes, it is not surprising that some men in the 19th century considered the sea as being more powerful than the river. Lyell was one of them, although he had not always been so. During his career he gave increasing importance to marine erosion, finally considering it to be the major modifier of the landscape. His ideas were modified by Ramsay, and eventually evolved into the theory of marine planation. Ramsay's concepts included two ideas: one, that the sea is capable of planing surfaces over which it moves, regardless of rock composition; the other, that unequal hardness in cliffs will result in differential erosion and the creation of an irregular coastline. He also maintained that marine planation accompanies shifts in sea level, using as evidence the presence of plains at different elevations above sea level. The escarpments between these levels he explained as old sea cliffs.

The importance of planation theory is emphasized by Chorley *et al.* (1964, p. 313): "Even when the idea of universal marine erosion had been discredited, the planation part of the theory lived on in Davis's cycle of erosion and in the writings of mid-20th-century geomorphologists."

As far as the coastline is concerned, Ramsay and others emphasized increasing irregularities because of differential erosion, a view that was not difficult to accept in the British Isles. However, Dana (1813–95), a confirmed fluvialist, disagreed. He believed that: "... waves tend rather to fill up the bays and remove by degradation the prominent capes, thus rendering the coast more even, and at the same time, accumulating beaches that protect it from wear" (1849; quoted in Chorley *et al.*, 1964, p. 363). These conclusions were based on observations made during Dana's four-year voyage in the Pacific Ocean.

During this period, thought was being given to a number of agents previously little considered. Hutton, it is true, had recognized the importance of chemical action in soil formation, and others had discussed the transport of matter being carried to the sea in solution. Nonetheless, it was von Richthofen (1833–1905), a staunch follower of Ramsay, who applied such ideas to coasts. He wrote:

> The weathering and loosening of rock by sea salts, carbonic acid, the formation of ozone, and the gripping of plants and animals—to which must be added the action of frost in higher latitudes—aids the mechanical action of the striking billows (1882; quoted in Mather and Mason, 1939, pp. 515–516).

Gilbert and Lake Bonneville

It is somewhat curious that the western explorations in the United States, especially those during the last half of the 19th century, should be important from the standpoint of the history of the study of coastal morphology. This importance is even more surprising when one realizes that much of the western field work combined with the studies being made on the Ganges (Everest, 1793–1874) and the Mississippi (Humphreys, 1810–83) rivers and in the heavy rainfall areas of the tropics in helping to reestablish the notion that the river is the dominant agent in geomorphology.

Especially significant are the coastal concepts presented by Gilbert (1843–1918) in *Lake Bonneville* (1890) and *The Topographic Features of Lake Shores* (1885). He treated a variety of shore-related topics, including beaches, cliffs, terraces, barriers, lagoons, waves, currents, undertow, backwash, and sorting, among many others. His deductions, based on intensive fieldwork, were lucidly presented. The main limitation to their usefulness was that, having been derived from work on lakes, they are not always applicable to oceanic situations. A major case in point is the concept of the bottomset, foreset, and topset bed composition of delta terraces, a concept that has little value when dealing with major oceanic deltas. Nonetheless, by describing coastal landforms in terms of physical processes, Gilbert set the style for present-day research in coastal geomorphology.

Gilbert, like most of the other geologists involved in the western explorations, was not tradition-bound and thus was able to distinguish between the relative importance of subaerial and marine processes, as he did in his Lake Bonneville research.

Possibly his most important contribution in the development of geomorphology is Gilbert's concept of grade, a concept that he used in the development of his ideas on beach equilibrium. He wrote: "... in order that the local process be transportation only, and involve neither erosion nor deposition, a certain equilibrium must exist between the quantity of the shore drift on the one hand and the power of the waves and currents on the other" (1885, p. 101).

Gilbert utilized lake shorelines and lake deposits as indicators of past climates and tectonic history. For example, he was able to correlate lake-terrace width with rock type in a lake's discharge channel and the lack of horizontality and parallelism of shores with orogenic movement.

Davis, Gulliver, and Johnson

The role of William Morris Davis (1850–1934) in the development of geomorphology has been analyzed many times (most thoroughly by Chorley *et al.*, 1973). Davis influenced in some way nearly every aspect of geomorphology, including coastal geomorphology. This influence was realized in several closely linked forms, including his own research and publications on coastal topics, the wide adoption of his concepts of the cycle of erosion, and the work of his students (especially Gulliver and Johnson) who emphasized the study of coasts.

Much of Davis's research on coastal problems was related to reestablishing support for Darwin's subsidence theory of coral-reef growth. By writing some three quarters of a century after Darwin, Davis was able to incorporate in his writings data about sea-level changes during the glacial period. He believed that the only way to properly understand coral reefs was through an examination of the "... physiographic features of the coasts, either insular or continental, that are bordered by fringing reefs or fronted by barrier reefs ..." (1928; quoted in Chorley *et al.*, 1973, p. 592) and not from the reefs themselves.

Another example of Davis's contributions in coastal geomorphology is his *The Outline of Cape Cod* (1896). It illustrates an attempt at geomorphic reconstruction: "Let the activities of the sea be resolved into two components: one acting on and off shore, the other along shore; and let the effects of the first of these components be now examined alone, ..." (Davis, 1896, p. 700).

Davis incorporated the ideas of Gilbert's beach equilibrium within his cyclic concepts: "Here the sea is able to do more work than it has to do. Its action is like that of a young river.... When a graded profile is attained, the adolescent stage of shore development is reached" (Davis, 1896, p. 701). This paper also presents numerous diagrams, a Davisian hallmark, illustrating the development of graded profiles (both normal and longitudinal) and of bars and spits.

The cyclic concepts presented in the Cape Cod paper preceded by three years the publication of his most famous and influential paper, *The Geographical Cycle* (Davis, 1899). The Cape Cod paper is only one example of the fact that the cyclic concept had been in Davis's mind for many years before the turn of the century. The influence of this concept in coastal geomorphology is further evidenced in the dissertation by Gulliver (1865–1919) that was entitled simply *Shoreline Topography* (1899) and was published in the same year as *The Geographical Cycle*.

Davis was continuously working with the cycle, and considered that it is "... not arbitrary or rigid, but elastic and adaptable...." For example: "Like the processes of surface carving, the processes of shoreline development are subject to variation with climate, from the work of the ice foot in polar regions to the work of coral reefs and mangrove swamps in the torrid zone" (Davis, 1905, p. 290).

The most influential of Davis's disciples was Douglas Johnson (1878–1944). Despite Gulliver's early coastal work under Davis's direction, it remained for Johnson to publish the first inclusive book dealing with coastal morphology, a book that Zenkovich (1967) considered to be the most complete theoretical study of coasts available. Entitled *Shore Processes and Shoreline Development* (Johnson, 1919), it was aimed at presenting an analysis of the forces operating along the shore together with a systematic discussion of the cycles of shoreline development. This book had a major influence on the study of coasts for at least 40 years, an influence that must be considered to have been detrimental in some regards. Johnson's emphasis on the importance of submergence and especially emergence in shoreline development, as well as the incorporation of these aspects of shore profile development in his classification scheme, delayed more meaningful approaches to coastal understanding for several decades. Nonetheless, much of his material is still useful, and should be consulted by any serious student.

Coastal geomorphology: World War I to World War II

Although the Davisian and Johnsonian evolutionary and qualitative approaches to geomorphology dominated coastal research from World War I to World War II, there were nonetheless a number of important developments that occurred during this period of time. One of the most significant of these developments was that coastal research became a respectable research endeavor after Johnson's book was published. Earlier most coastal research was of a practical nature—harbor construction, coastline defense, coastal mining, and the like. Few university scholars studied coasts for their own sake. Geologists, for example, often resorted to coasts, but only because coastal cliffs provided them with good exposures of the rocks they were studying, not because of any interest in their existence as coastal forms.

A second and concurrent development was the rapid rate at which human utilization of the coast developed. The increase resulted directly from increases in coastal populations and indirectly from man's increased mobility, desire for coastal vacations, and use of coastal resources. Fortunately, this increased utilization of the coast was accompanied by an increase in its study, although not to the extent that might have been desirable.

The sponsorship by the United States National Research Council of separate studies on tides and sea-level changes in the 1920s, the creation of the Commission of Coastal Studies in the USSR and the Coastal Engineering Research Center in the United States, and the publication of such volumes as *Recent Marine Sediments* (Trask, 1939) are examples of the organized endeavors that began between the two World Wars.

Despite a number of technical advances, there were surprisingly few actual substantive developments. In a very real sense, this period of time was transitional and set the stage for the rapid rise in coastal research that began after World War II. The way in which this transition proceeded might be illustrated by considering the way in which depositional landforms came to be recognized as significant elements of the landscape. One of the pioneers of depositional geomorphology was R.J. Russell (1895–1971). Trained in the Davisian School with its emphasis on erosion, Russell had to develop a completely new perspective in order to understand the depositional landforms he found when he moved to Louisiana. He utilized the thousands of cores taken during the drilling for oil in the Mississippi River delta, sets of detailed topographic maps (the drafting of which was prompted by a severe flood in 1927), and aerial photography (Russell, 1936). This research of Russell and his co-workers over a 20-year period led to the acceptance of the importance of three-dimensional studies in geomorphology, a clarification of the different types of subsidence, a reevaluation of Johnson's concepts of emergence and submergence and of Davis's concept of old age, and to new notions about sea-level fluctuations.

World War II and the impetus it gave coastal research

By the end of the 1930s, research on coastal problems, as well as that on other scientific topics, was beginning to recover from the difficult times that accompanied a worldwide depression when World War II erupted. Many scientific and engineering activities such as those being planned by the revitalizing Beach Erosion Board (BEB) of the United States had to be redirected toward war efforts. An even more extreme response to wartime demands was the dissolution of the Commission of Coastal Studies in the USSR mentioned above. Before its demise; however, the Commission sponsored a number of coastal engineers and geomorphologists (such as Vesolod Zenkovich) many of whom became very productive after the war.

In the case of the BEB, wartime coastal research efforts were focused on gathering information on foreign beaches. W.C. Krumbein (1944) of the United States and W.W. Williams (1947) from England, along with their associates, concentrated on those coastline characteristics that could impact beach landings including beach slope, sediment texture, wave climate, and tidal parameters. During the War, the BEB produced more than 50 reports on beaches around the world including some in the little previously studied tropical zones.

More important from the standpoint of coastal research than the descriptive details of many of the world's beaches that came from these efforts, were the war-generated developments that impacted virtually all of science. The developments included (1) the assemblage of scientists into research groups including those that became known as Research and Development Laboratories, (2) many technological advances in equipment useful to coastal research, (3) specialized educational opportunities at universities, (4) the spread of both scientists and scientific research around the world, (5) the multiplication of regional, national, and international organizations devoted to geomorphology including coastal morphology, (6) an increasing emphasis on scientific conferences, workshops, and symposia, and (7) an increase in opportunities to publish on basic, as well as applied, topics.

Coastal research during the third quarter of the 20th century: a period of expansion

One of the major advances that blended many of these developments was the creation, in 1952, of the Commission on Coastal Sedimentation within the International Geographical Union (IGU)—a commission whose name was changed a few years later to the Commission of Coastal Geomorphology (Schou, 1964). The senior members of the Commission— A. Schou (Denmark), J.A. Steers (England), V. Zenkovich (USSR), R.J. Russell (USA), and A. Guilcher (France)— were among the most productive geomorphologists during and immediately following World War II. With the diverse topical and regional backgrounds they and numerous corresponding members represented, they recommended a very diverse set of topics that needed research (Schou, 1964).

A couple of years earlier (1950) coastal engineers began holding biennial conferences from each of which stemmed proceedings volumes on coastal topics. By 1996 more than 2,300 papers had been produced, most of which have relevance to coastal geomorphology.

Whereas the 1950s showed much progress, some coastal geomorphologists still considered the subject immature. For example, Williams (1960) observed that:

> . . . theories of coastal behaviour have been based on the most simple visual observations, and there is no doubt that some of these observations have been misleading . . . [however, he added further that] today new techniques are being developed which should lead to a more certain knowledge of the subject (p. xi).

Similarly, as late as 1968, Russell still considered ". . . the subscience of coastal morphology as one in relative infancy . . ." but, like Williams, one with great promise. During the 1950s and 1960s, a number of coastal scientists began to base their research on the concept of process and response. They in general followed the procedure advocated by Strahler (1952) for fluvial morphologists. Strahler stated that:

> . . . dynamic-quantitative studies require, first, a thorough morphological analysis in order that the form elements of landscapes may be separated, quantitatively described, and compared from region to region (p. 1118).

Such a procedure proved to be especially valuable for morphologic research in the coastal zone because of the great number of forms and processes present there and because of the complex nature of the interrelationships that exist between them. The process-response models that resulted from such studies proved of value in both the production of specific types of coastal behavior and the provision of a clearer understanding of the integrated nature of the coastal system.

During the period of the 1950s and 1960s, such varied techniques as radioactive tracers, aqualung diving, satellite imagery, electron microscopy, and high-speed computers began being used to provide data and analyses about coastal forms and processes at scales both larger and smaller than had been possible during prewar years (Walker, 1977).

Although the scientific study of coasts had traditionally been in the hands of western Europeans and Americans, during the years following World War II it truly became international. Evidence of this development is indicated by: the frequency with which international symposia were being held; by the increasing numbers of research papers being produced in non-Western countries (Walker, 1976); and by the increasing frequency of research along arctic, desert, and tropical coasts.

In the third quarter of the century the number of publications in coastal geomorphology increased dramatically. Some of the books, like that of Douglas Johnson 40 years earlier, were broadly based. Included were Guilcher's *Morphologie Littorale et Sous-Marine* (1954), Zenkovich's *Processes of Coastal Development* (1967), Bird's *Coasts* (1969), King's *Beaches and Coasts* (1972), and Davies' *Geographical Variation in Coastal Development* (1973). Others such as Ippen's *Estuary and Coastline Hydrodynamics* (1966), Shepard and Wanless' *Our*

Changing Coastlines (1971), and Komar's *Beach Processes and Sedimentation* (1976) deal with special aspects of coastal science.

Also, during this period of time, collections of papers began to become common. Many of them were special issues of standard periodicals, such as *Dynamics and Morphology of Sea Coasts* edited by Longinov (1969) as Volume 48 of the *Transactions of the Institute of Oceanology*; Tedrow and Deelman's *Soil Science* (1975), which is devoted to soil formation in sediments under water; and Fairbridge's *Contributions to Coastal Geomorphology* (1975), and Kaiser's *Küstengeo- morphologie* (1968), as special issues of *Zeitschrift für Geomorphologie*. Yet another category of volumes that resulted from symposia include: *Estuaries*, edited by Lauff (1967); *Waves on Beaches and Resulting Sediment Transport*, edited by Meyer (1972); *Coastal Geomorphology*, edited by Coates (1973); *Nearshore Sediment Dynamics and Sedimentation*, edited by Hails and Carr (1975); and *Research Techniques in Coastal Environments*, edited by Walker (1977). Still, a third type of compilation was developed toward the end of this period. Especially valuable from the standpoint of the development of coastal concepts, its volumes brought together the key papers representing the development of particular topics. A prime example is the *Benchmark Papers in Geology*, of which *Spits and Bars* (1972) and *Barrier Islands* (1973), both edited by Schwartz, and *Beach Processes and Coastal Hydrodynamics*, edited by Fisher and Dolan (1977) are especially appropriate.

Contemporaneous with the publication of volumes such as those listed above was the development of a number of concepts that modified the focus of many coastal researchers. In 1962, Per Bruun, in response to the emerging concern over sea-level rise, proposed what is now known as the *Bruun Rule* (Schwartz, 1967). In essence, it states that a beach will maintain equilibrium through concurrent erosion and deposition as sea level rises. Criticized and modified subsequent to its proposal, the *Bruun Rule* nevertheless continues to be the focus of numerous studies on into the 21st century.

Although coastal sediment transport had been the subject of discussion for decades, the coastal circulation cell concept in relation to the compartmentalization of erosion, transport, and deposition of sediment along coasts was proposed in the 1960s. Bowen and Inman (1966) applied the concept in California while Stapor (1971) used it in connection with a Florida study.

Probably the most innovative development to appear during the third quarter of the 20th century was the scheme proposed by Inman and Nordstrom (1971) in which plate tectonics was used as the basis for classifying coasts. With this scheme coastal morphologists were forced into giving "... more thought to long-term but continuing processes in their attempts at explaining present form and location" (Walker, 1975, p. 4).

The last quarter of the 20th century: a period of sophistication and diversification

Whereas the third quarter of the 20th century was one of rapid expansion in coastal research, the last quarter might be considered as one in which research became more sophisticated and diversified. Improved equipment, expanded multidisciplinary cooperation, modified methodologies, and enhanced funding contributed to a number of new research avenues in coastal geomorphology. Possibly one of the most important realizations, following from the cell concept, was the recognition of the interrelationships that exist between different parts of coastal systems.

Even though coastal research became more diversified during the 1980s and 1990s, the research dedicated to beach morphology and processes continued to receive the most attention. Much of this attention is devoted to small-scale hydrodynamics and sediment transport as it occurs in the surf/swash zone. During the 1980s, several large-scale field experiments were conducted on and in the surf zone. Included were the *Near-shore Sediment Transport Study* (Seymour, 1987), the *Duck Experiments* (Mason *et al.*, 1987), and the *Canadian Coastal Sediment Study* (Willis, 1987), among others (Horn, 1997). Not surprisingly, considering the complicated nature of such a dynamic zone, different views as to the nature of swash dynamics developed. One group, as exampled by Guza *et al.* (1984), holds that swash is dominated by low-frequency infragravity motions and can be attributed to standing long waves; whereas, another group, represented by Hibberd and Peregrine (1979), holds that swash motion is mainly driven by incident waves that collapse at the shoreline and propagate up the beach face.

For the geomorphologist the value of these hydrodynamic studies is in their help in predicting sediment transport and beach morphology. New instruments, such as the optical backscatter sensor (OBS) and the acoustic backscatter sensor (ABS) are enabling the measurement of suspended sediment concentrations in the surf zone and are being used in conjunction with acoustic doppler velocimeters (Downing *et al.*, 1981).

As noted above much of the research along the shoreline has been concerned with the contact zone of the beach and surf. Nevertheless, increasing attention, especially during the past two decades, has been given to mudflats and coastal dunes and especially to their marsh/mudflat and beach/dune systems (Viles and Spencer, 1995). Although mudflats, like the marshes and mangroves behind them, traditionally were considered wasteland and thus prime areas for reclamation, their importance ecologically, once recognized, has served to bring them into the research spotlight.

In China, where mudflats have been reclaimed for millennia, basic research into their characteristics was initiated primarily by the Coastal and Estuarine Research Laboratory (later one of China's Key Laboratories) in Shanghai under the leadership of J. Chen, and the School of Geoscience, Nanjing University. Much of the earlier mudflat research was on mudflat distribution but by the late 1980s it included examination of how tidal currents influence grain size distribution (Wang, 1989) and mudflat sediment exchange (Tang-Yinde, 1989). The stability/ erodibility of mudflats during tidal cycles and in response to mudflat biotic activity have also been the subject of recent research (Viles and Spencer, 1995). During the 1990s, several investigators began dealing with the effects of sea-level rise on muddy coasts (El Ray *et al.*, 1995; Han *et al.*, 1995).

The seminal research on sand dunes by Bagnold (1941) served as the rationale for dune research for more than 30 years following World War II. It was not until the late 1970s that some researchers began to emphasize the distinctions between interior and coastal dunes. An increased appreciation for the facts that the presence and stability of dunes are important in coastal protection, that human activity can impact heavily on dunes, and that coastal dunes play an important role in the beach/dune system led to an increase in the quantity and variety of research on coastal dunes during the last quarter of the 20th century.

Much of the new research on coastal dunes has been devoted to the determination of vertical velocity profiles across beaches and within dune fields and to the measurement of sediment transport by winds (Carter *et al.*, 1990). Included is the basic research by Hotta (1988) and by Hsu (1977) who examined the boundary-layer conditions in both sand dunes and coastal ice ridges along an arctic coastline. Hydrodynamic research, which has become so important in many aspects of coastal research, has been applied to coastal dunes by Hesp (1988) and his colleagues. For example, Hesp used morphodynamics as a basis for classifying foredunes by linking vegetation and morphology to nearshore processes. Other factors that have received attention recently are the importance of beach wetness to sediment transport (Jackson and Nordstrom, 1998) and large-scale budgeting in the beach/dune system (Illenberger and Rust, 1988).

Dune research, like that on beaches, has become very international in scope. Some of the most important morphologic research on coastal dunes has been conducted in Australia, South Africa, The Netherlands, Japan, the United Kingdom, and the United States.

Among the other coastal topics that have recently attracted increased attention are shore platforms, sea-level rise and coastal erosion, extraterrestrial coastal geomorphology, and humans as geomorphic agents.

The focus in shore platform research has been on morphology and processes (Stephenson, 2000). For example, Trenhaile and Byrne (1986) concentrated on tides and sea-level change, Sunamura (1992) emphasized wave dynamics, and Viles and Naylor (2002) investigated the role of biogeomorphic processes on platform development.

Along with the proliferation of research on global warming and sea-level change since the 1970s, has come a surge in the investigations devoted to the impact of sea-level change on the coastline. Because all types of coasts are affected when sea level changes, it is not surprising that research of such impacts has been done on beaches, reefs, estuaries, deltas, and even coastal cliffs (Viles, 1989). Morphological and ecological changes in coastal marshes have been intensively examined by Reed (1995), mangrove response by Woodroffe (1995), foredune erosion by Carter and Stone (1989), and, as mentioned above, mudflats by Han *et al.* (1995). This kind of research is destined to intensify even more so long as sea level continues to rise as it has been doing the past few decades.

Extraterrestrial geomorphology is one of geomorphology's newest subdisciplines (Baker, 1993). It, like most other themes in geomorphology, has a long history (see e.g., Gilbert, 1893). Not surprisingly, the field of extraterrestrial geomorphology also has a coastal component (Baker, 2001). Parker *et al.* (1993), after examining high-resolution images of Mars, concluded that that planet not only had liquid water in the past but also possessed lakes and even oceans. Recent studies have identified relic lake sediments, deltaic deposits, and eroded "massifs" (resembling the wave-eroded headlands in Lake Bonneville) (Parker and

Currey, 2001). It is believed that other extraterrestrial bodies such as Venus and Titan, the moon of Saturn, may also support such features and thus provide research opportunities for additional coastal geomorphologists well into the future.

The importance of humans as agents of geomorphologic change has long been recognized and documented (Thomas, 1956; Turner, 1990). The coastal zone, as one of the world's unique landscapes, has been impacted intensively along much of its extent. Indeed, along the coastline of some of the world's most densely populated coastal zones, the bulk of the shore is now artificial. This provides an opportunity and challenge to coastal geomorphologists because, as Viles (1990) noted:

> Studies on coastal landforms, processes and change are becoming more and more relevant as human intervention (at scales ranging from sand mining on a single beach to global warming) increases (p. 238).

The rapid increase in human intervention in the coastal zone is intimated by the fact that in the year 2000 the number of coastal dwellers was equal to the global population of 1950 (Haslett, 2000). It is not surprising then that many coastal geomorphologists as well as coastal engineers are heavily involved with coastal research.

Much of this research has been focused on coastline protection (Walker, 1988; Pilarczyk, 1990), beach nourishment (Finkl and Walker, 2002), beach stabilization (Silvester and Hsu, 1991), beach management (Bird, 1996), and wetland restoration (Mitch and Gosselink, 2000). Sherman and Bauer (1993), referring to this trend, predict that:

> . . . human altered coastal systems will be a major focus of research for coastal geomorphologists over the next twenty years . . . [and, equally as relevant] that coastal scientists will have less choice and less input to what their subject of study will be . . . (p. 240).

Organizations, conferences, proceedings, journals, and books: 1975–2000

The fourth quarter of the 20th century saw the number of organizations, conferences, proceedings, journals, and books dealing with coastal geomorphology continue to increase. In addition, the coastal component of a number of geomorphology organizations also increased in prominence. For example, at the First International Conference on Geomorphology (1985) more than one fifth of the papers presented dealt with fluvial and coastal topics. Coastal geomorphologic papers have continued as an important ingredient of subsequent conferences. Recently established journals (i.e., those founded since 1975) that publish coastal papers of interest to geomorphologists include: *Applied Ocean Research* (1979), *Coastal Society Bulletin* (1977), *Earth Surfaces Processes and Landforms* (1976), *Geomorphology* (1987), and the *Journal of Coastal Research* (1985). In addition, many long-established journals have recently published special numbers devoted to coastal topics. They include: *The Geographical Review* with a number called *Coastal Geomorphology* (1988); *Catena* with a number on *Morphology and Sedimentation in Fluvial–Coastal Environment* (1997); *Marine Geology* with numbers on *Beach Ridges* (1995) and *Large Scale Coastal Behavior* (1995); and *Physical Geography* with a number on *Coastal Dunes* (1994).

One of the most extensive series of special numbers devoted to coastal topics is that published as Special Issues by The Coastal Education and Research Foundation (CERF). Since 1986, more than 30 such special issues have been published under such titles as *The Effects of Seawalls on the Beach* (1988), *Impacts of Hurricane Hugo: September 10–22, 1989* (1991), *Coastal Hazards* (1994), *Sediment Transport and Buoyancy in Estuaries* (1997), and *Tidal Dynamics* (2001).

In addition to such special numbers are the numerous review articles that have recently appeared detailing the advances in coastal geomorphology. *Progress in Physical Geography*, for example, has had a number of status reports including: *Coastal Landforms* by S.B. McCann (1982), *Coastal Depositional Landforms; a Morphodynamic Approach* by L.D. Wright and B.G. Thom (1977), *Coastal Geomorphology* by H.A. Viles (1990), *Sea-level Rise as a Global Geomorphic Issue* by D.R. Stoddart and D.J. Reed (1990), *Beach Research in the 1990s* by D.P. Horn (1997), and *Mid-Holocene Sea-Level Change and Coastal Evolution* by A. Long (2001).

Possibly the most revealing characteristic in the development of coastal research subsequent to World War II is illustrated by the articles appearing in what must now be considered the bellwether journal of coastal science, the *Journal of Coastal Research* (JCR). Since its foundation in 1985 as an "International Forum for the Littoral Sciences" it has published some 1,400 articles in more than 18,000 pages. Although published in the United States, JCR reflects the international character of today's coastal research in that more than half of the articles stem from countries other than the United States. Of the total, the United Kingdom is represented by more than 80 articles; New Zealand, Australia, and South Pacific Islands by 116; Canada by 57; Latin America by 73; Scandinavia and The Netherlands by 63; the Orient by 51; and Russia by 24 (Finkl, 2002, personal communication).

Equally as revealing of the trend in research in coastal geomorphology is the specialized nature of the books that have appeared since 1975. Although prior to that date, as noted above, conference proceedings were often quite topical, books tended to remain inclusive. By 1980, however, coastal books began to reflect the maturation of coastal geomorphology in two significant ways. First, details about coastal environments had developed to a point where specialized volumes became justified and second, theories of coastal science had been enhanced sufficiently to lead to new approaches in considering the field especially in those portions of coastal geomorphology relevant to human activities.

The first category of books is well exampled by a series, edited by E.C.F. Bird for John Wiley & Sons, that includes *Coral Reef Geomorphology* (1988) by André Guilcher; *Coastal Dunes: Form and Process* (1980), edited by K. Nordstrom, N. Psuty, and B. Carter; and *Geomorphology of Rocky Coasts* (1992) by T. Sunamura. The second-type approach is represented by three books recently published by UK authors, namely: *Coastal Problems: Geomorphology, Ecology and Society at the Coast* by H. Viles and T. Spencer (1995); *Coastal Systems* by S.K. Haslett (2000); and *Coastal Defences* by P.W. French (2001).

Conclusion

Although, as noted in the introduction of this History of Coastal Morphology, the coast was late in attracting the attention of scientists, it eventually became one of the major subdisciplines in the field of geomorphology. The rapidity with which it has developed during the past few decades might have pleased R.J. Russell who, even as late as 1968, considered the field in its infancy. Since World War II, a number of conditions evolved to place coastal geomorphology on a sound footing. Advancements in monitoring systems, development of new theories, increased funding opportunities, broadening the scales of investigation, inputs from other disciplines, and debates of controversial conclusions have all played a part in the maturation of coastal geomorphology.

As has been true throughout the history of science each new discovery rewards investigators with intriguing questions. Coastal geomorphology is no exception. The variety of the earth's coastal forms coupled with the dynamics of the forces operating on them will tax the expertise of coastal geomorphologists far into the future.

H. Jesse Walker and Molly McGraw

Bibliography

Bagnold, R.A., 1941. *The Physics of Blown Sand and Desert Dunes*. London: Chapman and Hall.

Baker, V.R., 1993. Extraterrestrial geomorphology: science and philosophy of Earthlike planetary landscapes. *Geomorphology*, **7**(1–3): 9–35.

Baker, V.R., 2001. Editorial: extraterrestrial geomorphology: an introduction. *Geomorphology*, **37**(3–4): 175–178.

Bird, E.C.F., 1969. *Coasts*. Cambridge, MA: M.I.T. Press.

Bird, E.C.F. (ed.), 1996. *Beach Management*. Chichester: Wiley & Sons.

Bowen, A.J., and Inman, D.L., 1966. *Budget of Littoral Sands in the Vicinity of Point Arguello, California*. U.S. Army Coastal Engineering Research Center, 19p.

Carter, R.W.G., and Stone, G.W., 1989. Mechanisms associated with the erosion of sand dune cliffs, Magilligan, Northern Ireland. *Earth Surface Processes and Landforms*, **14**: 1–10.

Carter, R.W.G., Nordstrom, K. F., and Psuty, N.P., 1990. The study of coastal dunes. In Nordstrom, K., Psuty, N., and Carter, B. (eds.), *Coastal Dunes: Form and Process*. Chichester: John Wiley & Sons, pp. 1–14.

Chorley, R.J., Beckinsale, R.P., and Dunn, A.J., 1973. *The History of the Study of Landforms, Volume 2: The Life and Work of William Morris Davis*. London: Methuen and Co.

Chorley, R.J., Dunn, A.J., and Beckinsale, R.P., 1964. *The History of the Study of Landforms, Volume 1: Geomorphology Before Davis*. London: Methuen and Co.

Coates, D.R., 1973. *Coastal Geomorphology*. Binghamton: State University of New York.

Davies, J.L., 1973. *Geographical Variation in Coastal Development*. New York: Hafner Publishing Co.

Davis, W.M., 1896. The outline of Cape Cod. In *Geographical Essays*. New York: Dover Publications, 1954 reprint, pp. 690–724.

Davis, W.M., 1899. The geographical cycle. *Geographical Journal*, **14**: 481–504.

Davis, W.M., 1905. Complications of the geographical cycle. In *8th International Geographical Congress, Washington, 1904, Report*, pp. 150–163. (Reprinted in Davis, W.M., 1954, *Geographical Essays*. New York: Dover Publications, pp. 279–295.)

Downing, J.P., Sternberg, R.W., and Lister, C.R.B., 1981. New instrumentation for the investigation of sediment suspension processes in the shallow marine environment. *Marine Geology*, **42**: 19–34.

El Ray, M., Nasr, S., Frihy, O., Desouky, S., and Dewidar, K., 1995. Potential impacts of accelerated sea level rise on Alexandria Governorate, Egypt. *Journal of Coastal Research*, **14**: 229–307.

Fairbridge, R.W., (ed.), 1975. Contributions to Coastal Geomorphology. *Zeitschrift für Geomorphologie Supplement*, **22**.

Finkl, C.W., and Walker, H.J., 2002. Beach nourishment. In Chen, J.-Y., Eisma, D., Hotta, K., and Walker, H. (eds.), *Engineered Coasts*. Dordrecht: Kluwer Academic Publishers, pp. 1–22.

Fisher, J.S., and Dolan, R. (eds.), 1977. *Beach Processes and Coastal Hydrodynamics*. Stroudsburg, PA: Dowden, Hutchinson & Ross.

French, P.W., 2001. *Coastal Defences*. London: Routledge

Geike, A., 1897. *The Founders of Geology*. New York: Dover Publications, 1962 reprint.

Gilbert, G.K., 1885. The topographic features of lake shores. *U.S. Geological Survey Annual Report*, **5**: 65–123.

Gilbert, G.K., 1890. Lake Bonneville. Washington, DC: U.S. Geological Survey Monograph I, 438 p.

Gilbert, G.K., 1893. The moon's face: a study of the origin of its features. *Philosophical Society of Washington Bulletin*, **12**: 241–292.

Guilcher, A., 1954. *Morphologie Littorale et Sous-Marine*. Paris: Presses universitairies de France.

Guilcher, A., 1988. *Coral Reef Geomorphology*. Chichester: Wiley.

Gulliver, F.P., 1899. Shoreline topography. *Proceedings of the American Academy of Arts and Sciences*, **34**: 149–258.

Guza, R.T., Thornton, E.B., and Holman, R.A., 1984. Swash on steep and shallow beaches. *Proceedings of the 19th International Conference on Coastal Engineering*, pp. 708–723.

Hails, J.R., and Carr, A.P. (eds.), 1975. *Nearshore Sediment Dynamics and Sedimentation*. New York: John Wiley & Sons.

Han, M., Wu, L., Hou, J., Liu, C., Zhao, S., and Zhang, Z., 1995. Sea-level rise and the North China coastal plain: a preliminary analysis. *Journal of Coastal Research*, **14**: 132–150.

Haslett, S.K., 2000. *Coastal Systems*. London: Routledge.

Hesp, P.A., 1988. Surfzone, beach and foredune interactions on the Australian south east coast. *Journal of Coastal Research*, **3** (Special issue): 15–25.

Hibberd, S., and Peregrine, D.H., 1979. Surf and run-up on a beach: a uniform bore. *Journal of Fluid Mechanics*, **95**: 323–345.

Horn, D.P., 1997. Beach Research in the 1990s. *Progress in Physical Geography*, **21**(3): 454–470.

Hotta, S., 1988. Sand transport by wind. In Horikawa, K. (ed.), *Nearshore Dynamics and Coastal Processes*. Tokyo: University of Tokyo Press, pp. 218–238.

Hsu, S.A., 1977. Boundary-layer meteorological research in the coastal zone. In Walker, H.J. (ed.), *Geoscience and Man, Volume XVIII, Research Techniques in Coastal Environment*. Baton Rouge: Louisiana State University School of Geoscience, pp. 99–111.

Illenberger, W.K., and Rust, I.C., 1988. A sand budget for the Alexandria coastal dunefield, South Africa. *Sedimentology*, **35**: 513–521.

Inman, D.L., and Nordstrom, C.E., 1971. On the tectonic and morphologic classification of coasts. *Journal of Geology*, **79**: 1–21.

Ippen, A.T., 1966. *Estuary and Coastline Hydrodynamics*. New York: McGraw-Hill.

Jackson, N.L., and Nordstrom, K.F., 1998. Aeolian transport of sedimentation on a beach during and after rainfall, Wildwood, NJ, USA. *Geomorphology*, **22**: 151–157.

Johnson, D.W., 1919. *Shore Processes and Shoreline Development*. New York: John Wiley & Sons.

Kaiser, K. (ed.), 1968. Küstengeomorphologie, *Zeitschrift für Geomorphologie Supplement*, **7**.

King, C.A.M., 1972. *Beaches and Coasts*. New York: St. Martin's Press.

Komar, P.D., 1976. *Beach Processes and Sedimentation*. Englewood Cliffs, NJ: Prentice-Hall.

Krumbein, W.C., 1944. *Shore Processes and Beach Characteristics*. TM-8, U.S. Army Corps of Engineers, Beach Erosion Board, Washington, DC.

Lauff, G.H. (ed.), 1967. *Estuaries*. Washington, DC: American Association for the Advancement of Science.

Long, A., 2001. Mid-Holocene sea-level change and coastal evolution: *Process in Physical Geography*, **25**(3): 399–408.

Longinov, V.V. (ed.), 1969. *Dynamics and Morphology of Sea Coasts*. Institute of Oceanology (Moscow) Trans. 48 (translated from the Russian). Washington, DC: U.S. Department of Commerce.

Mason, C., Birkemeier, W.A., and Howd, P.A., 1987. Overview of DUCK85 nearshore process experiment. In *Proceedings of the Coastal Sediment '87*, ASCE, pp. 818–833.

Mather, K.F., and Mason, S.L., 1939. *A Source Book in Geology 1400–1900*. Cambridge, MA: Harvard University Press.

McCann, S.B., 1982. Coastal landforms. *Progress in Physical Geography*, **6**(3): 439–445.

Meyer, R.E. (ed.), 1972. *Waves on Beaches and Resulting Sediment Transport*. New York: Academic Press.

Mitch, W.J., and Gosselink, J.G., 2000. *Wetlands*, 3rd edn. New York: John Wiley & Sons.

Nordstrom, K.F., Psuty, N.P., and Carter, R.W.G., 1990. *Coastal Dunes*. Chichester: Wiley.

Parker, T.J., and Currey, D.R., 2001. Extraterrestrial coastal geomorphology. *Geomorphology*, **37**(3–4): 303–328.

Parker, T.J., Gorsline, D.S., Saunders, R.S., Pieri, D.C., and Schnelberger, D.M., 1993. Coastal geomorphology of the Martian Northern Plains. *Journal of Geophysical Research*, **98**: 11061–11078.

Penrose, B., 1955. *Travel and Discovery in the Renaissance*. Cambridge, MA: Harvard University Press.

Pilarczyk, K.W. (ed.), 1990. *Coastal Protection*. Rotterdam: Balkema.

Reed, D.J., 1995. The response of coastal marshes to sea-level rise: survival or submergence? *Earth Surface Processes and Landforms*, **20**: 39–48.

Russell, R.J., 1936. Physiography of lower Mississippi River delta. Baton Rouge: Louisiana Department of Conservation, *Geological Survey Bulletin*, **8**: 3–199.

Russell, R.J., 1967. Aspects of coastal morphology. *Geografiska Annaler, Series A*, **2–4**: 299–309.

Russell, R.J., 1968. Foreword. In Kaiser, pp. v–vii.

Schou, A., 1964. *Bibliography 1959–1963*. International Geographical Union, Commission in Coastal Geomorphology. *Folia Geographica Danica*, **X**(1).

Schwartz, M.L., 1967. The Bruun Theory of sea-level rise as a cause of shore erosion. *Journal of Geology*, **75**: 76–92.

Schwartz, M.L. (ed.), 1972. *Spits and Bars*. Stroudsburg, PA: Dowden, Hutchinson & Ross.

Schwartz, M.L. (ed.), 1973. *Barrier Islands*. Stroudsburg, PA: Dowden, Hutchinson & Ross.

Seymour, R.J., 1987. An assessment of NSTS. In *Proceedings of the Coastal Sediments '87*, ASCE, pp. 642–651.

Shepard, F.P., and Wanless, H.R., 1971. *Our Changing Coastlines*. New York: McGraw-Hill.

Sherman, D.J., and Bauer, B.O., 1993. Coastal geomorphology through a looking glass. *Geomorphology*, **7**: 225–250.

Silvester, R., and Hsu, J.R.C., 1991. New and old ideas in coastal sedimentation. *Reviews in Aquatic Science*, **4**(4): 375–410.

Stapor, F.W., 1971. Sediment budgets on a compartmented low-to-moderate energy coast in Northwest Florida. *Marine Geology*, **10**: 1–7.

Stephenson, W.L., 2000. Shore platforms: a neglected coastal feature?. *Progress in Physical Geography*, **24**(3): 311–327.

Stoddart, D.R., and Reed, D.J., 1990. Sea-level rise as a global geomorphic issue. *Progress in Physical Geography*, **14**(4): 441–445.

Strahler, A.N., 1952. Hyposemetric (area–altitude) analysis of erosional topography. *Geological Society of America Bulletin*, **63**: 1117–1142.

Sunamura, T., 1992. *Geomorphology of Rocky Coasts*. Chichester: John Wiley & Sons.

Tedrow, J.C.F., and Deelman, J.C. (eds.), 1975. *Soil Science*, **119**(1). Baltimore, MD: Williams and Wilkins.

Thomas, W.L., Jr. (ed.), 1956. *Man's Role in Changing the Face of the Earth*. Chicago: University of Chicago Press.

Trask, P.D. (ed.), 1939. *Recent Marine Sediments*. New York: Dover Publications, 1968 reprint.

Trenhaile, A.S., and Byrne, M., 1986. A theoretical investigation of Holocene development of rock coasts, with particular reference to shore platforms. *Geografiska Annaler*, **68A**: 1–14.

Turner, B.L., II (ed.), 1990. *The Earth as Transformed by Human Action*. Cambridge: Cambridge University Press.

Viles, H.A., 1989. The greenhouse effect, sea-level rise, and coastal morphology. *Progress in Physical Geography*, **13**(3): 452–461.

Viles, H.A., 1990. Coastal geomorphology in the 1990s. *Progress in Physical Geography*, **14**(2): 238–250.
Viles, H.A., and Naylor, L.A., 2002. The (W)hole story: broadening biogeomorphology research. Abstracts of Conference Papers. *Transactions, Japanese Geomorphological Union*, **22**(4): C-167.
Viles, H.A., and Spencer, T., 1995. *Coastal Problems*. London: Edward Arnold.
Walker, H.J., 1975. Coastal morphology. *Soil Science*, **119**: 3–19.
Walker, H.J. (ed.), 1976. *Coastal Research. Geoscience and Man*, Volume XIV. Baton Rouge: Louisiana State University.
Walker, H.J. (ed.), 1977. *Research Techniques in Coastal Environments Geoscience and Man*, Volume XVIII. Baton Rouge: Louisiana State University.
Walker, H.J. (ed.), 1988. *Artificial Structures and Shorelines*. Dordrecht: Kluwer Academic Publishers.
Walker, J., 1966. In Scott, H.W. (ed.), *Lectures on Geology*. Chicago, II : University of Chicago Press.
Wang, Y., 1989. The mudflat coast of China. *Canadian Journal of Fisheries and Aquatic Sciences*, **40**: 160–171.
Williams, W.W., 1947. The determination of gradients on enemy held beaches. *The Geographical Journal*, **109**: 76–92.
Williams, W.W., 1960. *Coastal Change*. London: Routledge and Kegan Paul.
Willis, D.H., 1987. The Canadian coastal sediment study: an overview. In *Proceedings of the Coastal Sediments '87*, ASCE, pp. 682–693.
Woodroffe, C.D., 1995. Response of tide-dominated mangrove shorelines in Northern Australia to anticipated sea-level rise. *Earth Surface Processes and Landforms*, **20**: 65–85.
Wright, D.L., and Thom, B.G., 1977. Coastal depositional landforms. *Progress in Physical Geography*, **1**: 412–459.
Yinde, T., 1989. A discussion on sediment exchange and transport formation in mudflats and its criterion. *Journal of Sediment Research*, **1**: 1–7.
Zenkovich, V.P., 1967. *Processes of Coastal Development*. New York: Wiley Interscience.

Cross-references

Beach Processes
Changing Sea Levels
Coastal Changes, Rapid
Coastal Climate
Coastal Zone Management
Geohydraulic Research Centers
History, Coastal Protection
Mapping Shores and Coastal Terrain
Monitoring, Coastal Geomorphology
Numerical Modeling
Physical Models
Sea-Level Rise, Effect

HISTORY, COASTAL PROTECTION

Though the claim of the Frisians to have been the first to devise an embryonic system of coastal defense is probably appropriate for the northern European area, others may equally assert their right to first place; the history of coastal engineering can be traced back in China to the East Han dynasty era. Indeed large coastal defense projects were initiated between about 25 and 220 before our era (Xu Qiwang, 1993). In the Mediterranean, likewise, coastal engineering had an "early start" particularly among Greeks, Etruscans, and Romans, and also Carthaginians, Minoans, Phoenicians, Sumerians, and Egyptians.

Classical times

Dikes were built under the reign of Apollonios Ptolaemeus II Philadelphis, Egyptian ruler, during 259–258 BC. He donated to his employee Stothoaetis, a large section of land in Ghoram, Fayoum (Philadelphia); where the owner decided to exploit the embankments by establishing a network of canals and dikes, well before "westerners" drained land in more northern areas. A cursory examination of a picture of the Alexandria Lighthouse, one of the seven wonders of the world of the classical times that crumbled to the bottom of the sea due to a 14th-century earth tremor, shows rather sophisticated dikes around Pharos Island. A French expedition funded by the Electricité de France brought back to the surface, in 1998, the colossal statue of Ptolaemeus II that stood in front of the lighthouse.

Coastal defense history is, naturally, closely linked to harbor creation and development. Primitive breakwaters were put in place occasionally with ramps allowing the top of the waves to pass over them. The Phoenicians had devised wave catchers by excavating holes and establishing trenches in the rocks lining the shores (Raban, 1988). Carved breakwaters were cut out of bedrock: a suitable wave-absorber profile was created which had a gentle grooved slope at the waterline. The classical example dating from the 2nd century BPE is still visible at Ventotene (Italy), it had an overspill. One may see in it an early version of today's Fontvieille breakwater (Principality of Monaco).

Greek and Etruscan breakwaters and seawalls consisted of rubble mounds topped by cut rocks; though no mortar was used, neighboring blocks were sometimes held together by clamps and joints made out of metal. With the discovery of pozzolanic ash hydraulic-cement, solid breakwaters could be built underwater, and several vertical composite concrete walls have been preserved from the 2nd century to the 5th century before our era. There are illustrations of a vertical breakwater made *in situ* within a wooden frame and tie-rods. Toe protection against scouring was provided occasionally by a bronze slab (Oleson, 1988).

Not only had these engineers mastered the art of erecting cofferdams for construction "in the dry," but they also thought of caissons, forerunners of contemporary building methods. Watertight wooden cellular caissons were used to cast large concrete breakwaters, for example, at Caesarea. "Permeable" breakwaters and "arched moles" were installed in various sites (Franco, 1996a). Apparently well before the Dutch dyke-builders thought of sinking old ships, the Romans sank old hulls, filled them with concrete and had a breakwater placed in no time, under Claudius' reign, Caligula's "monster" ship was sunk at Ostia (50) to provide a breakwater (Testaguzza, 1970). Remnants are still visible at 4 km from the Fiumicino airport.

Trajan built a rubble mound breakwater on an "island," reshaped by nature and workers leading to a subsequent mild slope profile (Franco, 1996b).

Medieval times and Renaissance

Grillo (1989) reports that the earliest written document dealing with shore protection dates back to 537 when fagines, wicker faggots, supplemented by timber piles and stones, held up earthen dikes adding their protection to that of the dunes.

"Timber and rock revetments and groynes have been used (in the Venice area) until 1700 to halt beach erosion and silting" notwithstanding a lengthy transport for rocks and short life span of wood (Franco, 1996b). Strict environmental regulation governing shore protection can be traced back to legal documents of 1282 and 1339: prohibition to cut or burn trees from coastal forests, to pick mussels from rock revetments, to let cattle walk the dikes, to remove sand, and vegetation, from beach or dune, and to export materials used in coastal defense (Grillo, 1989). The *Magistrato alle acque*, created in 1501, invited suggestions to reduce the high cost of the coast defense, so, in the 18th century, there appeared rip-rap revetments, gabions, staircase-placed limestone blocks, and the use of mortar and steel links and flexible steel strips became common.

Whereas beaches protect the littoral, they had to be maintained, and artificial nourishment with offshore dredged sand was initiated as early as the 17th or 18th century, hence long before California used the method (1919).

Massive *murazzi*, devised by Zentini and his team were constructed from 1741 on; with an average width of 12 m (39 ft), their crest peaked at 5 m (16 ft) above mean sea level. It took 40 years to construct a protective seawall 20 km (12.4 miles) long (Charlier and De Meyer, 1998). *Murazzi* have withstood the test of time, though storms took their toll and toe protection was eventually provided by a rubble mound structure and, more recently, jet-grouting diaphragms. Commenting on the Genoa breakwater, an actual fortification with a superstructure, Franco (1996b) underscores its importance as in 1245 it was proclaimed a "pious work" thereby compelling every citizen of the Republic of Genoa to provide in his will for the breakwater's maintenance.

Of course no Renaissance technology review can pass over Leonardo da Vinci whose talents included hydraulics and is the father of a proposed triangular-shaped island breakwater. He as well championed the credo of "working with Nature," rather than against it: *ne coneris contra ictum fluctus: fluctus obsequio blondiuntur (Nature should not be faced bluntly and challenged, but wisely circumvented)*.

Franco (1996b) has virtually provided a catalog of Italian designed breakwaters: use of irregular blocks with pozzolanic concrete crown and large rock armor porosity, (Crescentio in 1607), a monolithic superstructure over a leveled rubble mound foundation; (De Mari in 1638), armored with precast blocks (San Vincenzo mole at Naples, 1850), vertical composite structures (1896), and caisson construction (1915, 1931, 1936, 1938, 1995).

Contemporary approaches

If new approaches to coastal defenses were slow, and little new technology was introduced during the centuries—though improvements and refinements were often made—the pace of change accelerated considerably in the 20th century, even more so during the last decades. The large beach nourishment achievement in Belgium (1980s) has been surpassed, profile nourishment and berm feeding have been implemented—aimed simultaneously at restoration and protection—not less than 40 alternative methods have been proposed and tried out. But the problem has not been solved; for instance, all 30 of the US coastal states suffer from erosion and some see Hawaii's tourist industry in jeopardy.

At the turn of the 20th century, the response to coastal erosion remained the construction of hard defense structures: groins, jetties, breakwaters, revetments, gabions, placement of tetrapods, and the like. However, the groins placed at Miami Beach and Long Island N.Y. did not stall retreat of the shore. The same situation prevailed in Europe where beaches were shrinking in Denmark, Germany, The Netherlands, Belgium, and France. The Mediterranean beaches were not spared either. The problem is worldwide. Variants were tried out, sometimes meeting limited success, that is, floating, permeable, offshore breakwaters. If hard structures do protect or extend beaches on one face, the downdrift side is starved.

Beach planners, engineers and geologists proposed to artificially renourish beaches and the approach was tried on northern California beaches in the early 1920s. Since then the "re-charging" of beaches has been carried out around the globe, wherever there were sufficient funds to undertake an operation that is not inexpensive. Major schemes were undertaken in the United States, Belgium, and France. The beaches require regular additions of sand and in some cases a major storm may carry back to sea a major part of the artificial deposit.

Artificial beach nourishment entails many operations among which selection of the materials and of the source spot, method of material transportation, study of the waves and weather climate, of the physical and biological impacts. Improvements on the simple method of direct material dumping have been sought and thus appeared profile feeding, establishment of a feeder berm (e.g., De Haan, Belgium), and combinations of hard and soft defenses. Beach dewatering has been presented as a new alternative (e.g., Carolinas in the United States) but in fact it is more a complementary than an alternative method. At any rate it helps a beach retain the nourished material for a longer span of time. Sand backpassing and by-passing have proven valuable approaches (Hillsboro Inlet, Florida; Durban, Republic of South Africa); sand is transferred from one side of a structure or formation to the other which is starved. In compensation dredging, a somewhat similar method, the material is dredged and carried to another site that needs to be nourished.

There are over 40 methods that have been proposed and/or patented to halt coastal erosion. Though most have merits, most also have objectional side effects or hard-to-accept environmental impacts. Two of them, Berosin® and Beachbuilder®, apparently are free of them but are, at least temporarily, unaesthetic. The latter of the two attempts to promote beach accretion by using the power of the erosive waves to build up the beach.

Still other approaches have been tried: artificial reefs, artificial or restored dunes, fields of algae or synthetic fronds, creation of inlets for seawater. However, one should not lose sight of the fact that shore landward migration is inexorable, that people can only slow down the inland progress of the sea, and that only nature can definitely reverse the trend.

Conclusion

In the 19th and 20th centuries, coastal protection against an advancing sea has been centered on a variety of hard structures (groins, breakwaters, seawall, tetrapods, etc.) and artificial nourishment. Environmental concerns have steadily played a more important role. Some 40 types of alternative schemes have been proposed over the last decades, with a large number of them faulted for negative environmental impacts. As the economic consequences of the landward migration of the shore are often disastrous, the search of solutions remains.

Roger H. Charlier

Bibliography

Charlier, R.H. and De Meyer, C.P., 1998. *Coastal Erosion: Response and Management*: Heidelberg and New York: Springer Verlag.

Crescentio, 1607, cited in Franco 1996b. De Mari, 1638, cited in Franco 1996b.

Franco, L., 1996a. Ancient Mediterranean harbours: a heritage to preserve. *Journal of Ocean and Coastal Management*, **30**(2/3): 115–151.

Franco, L., 1996b. History of coastal engineering in Italy. In Kraus, N. (ed.), *History and Heritage of Coastal Engineering*. New York: American Society of Civil Engineers, pp. 275–335.

Grillo, S., 1989. *Venice Sea Defences*. Venice: Ed. Arsenale (in Italian).

Oleson, J.P., 1988. The technology of Roman harbors. *International Journal of Naval Architecture*, **17**(2): 147–157.

Raban, A., 1988. Coastal processes and ancient harbour engineering: *Proceedings of the 1st International Symposium Cities on the Sea—Past and Present. B.A.R. International Series,* **404**: 185–261.

Testaguzza, O., 1970. *Portus.* Rome: Julia (in Italian).

Xu, Q., 1993. Brief history of marine development in Ancient China. *Abstracts of the Fifth International Congress on the History of Oceanography*, p. 58.[full text in Proc. Vth Int. Conf. Hist. Dc. 2002].

Cross-references

Beach Drain
Beach Nourishment
Bioengineered Shore Protection
Bypassing at Littoral Drift Barriers
Coastal Zone Management
Dredging of Coastal Environments
Engineering Applications of Coastal Geomorphology
Human Impact on Coasts
Navigation Structures
Shore Protection Structures

HOLOCENE COASTAL GEOMORPHOLOGY

Normally forming processes on the earth need time measured in geological scales (i.e., many thousands to millions of years) to show adequate results with mature forms. In contrast to this general rule all coastal forms we can see along the world's coastline are not older than 6,000–6,500 years, because the oceans had not reached this level before this date; coming from a lowstand near −100 m during the last cold phase with extended glaciers (last Ice Age). The Holocene as a geological epoch, however, started about 10,000–11,000 years ago with the shifting to modern climate and temperatures, but it took several thousand years to melt the extended ice sheets and fill the oceans again. Therefore, the coastal forms we see today belong only to the last phase of the Holocene. There might be some inherited features incorporated in them, formed during former sea-level highstands in the warmer periods of the ice ages, the last time being about 100,000 years ago.

The world's coastlines are the most extended geomorphological features on earth, measuring several 100,000 km on small-scale maps, but more than 1 million km in nature. They are developed in all the different geotectonic and geodynamic situations, along all petrographic units of hard rocks and sediments and in all climatic and biogeographic regions of the world. Therefore, their geomorphological inventory and the amount of forming processes are numerous. Besides the tectonic situation (stable, subsiding, rising) and the type of rock (more or less resistant, massive or stratified, limestone or silicate, hard rock or loose sediments) the kind of wave impact along a certain coastline is the most important factor, itself depending on other facts like depth of nearshore water, strength of wind, tidal range, and others. These elements all have a wide range; in particular the tides, from microtidal (less than 1 m), to macrotidal (with spring-tide ranges exceeding 10 m as in southern England or along the French coast, or even more than 15 m in the Fundy Bay of eastern Canada). These tides distribute wave energy and currents on a wide horizontal area, resulting in tidal flats exposed during low water, marshes from high water and storm deposition, or mangrove fringes in warmer latitudes. Surf beat, with effects on coastal abrasion, depends—besides the water depth—on strong winds from one direction over a long time and on a wide ocean area (called fetch), and the availability of particles as abrasive agents in the surf. Nearly zero energy coasts in calm climates may change to those with gale forces and surf waves more than 10 m high around the southern parts of South America, South Africa, or Tasmania/New Zealand, but even in warmer latitudes hurricanes may occur with similar geomorphological effects. They usually are accompanied by storm surge, that is, rising water driven by strong wind against the coastline and rising additionally by very low barometric pressure. This has the same effect of

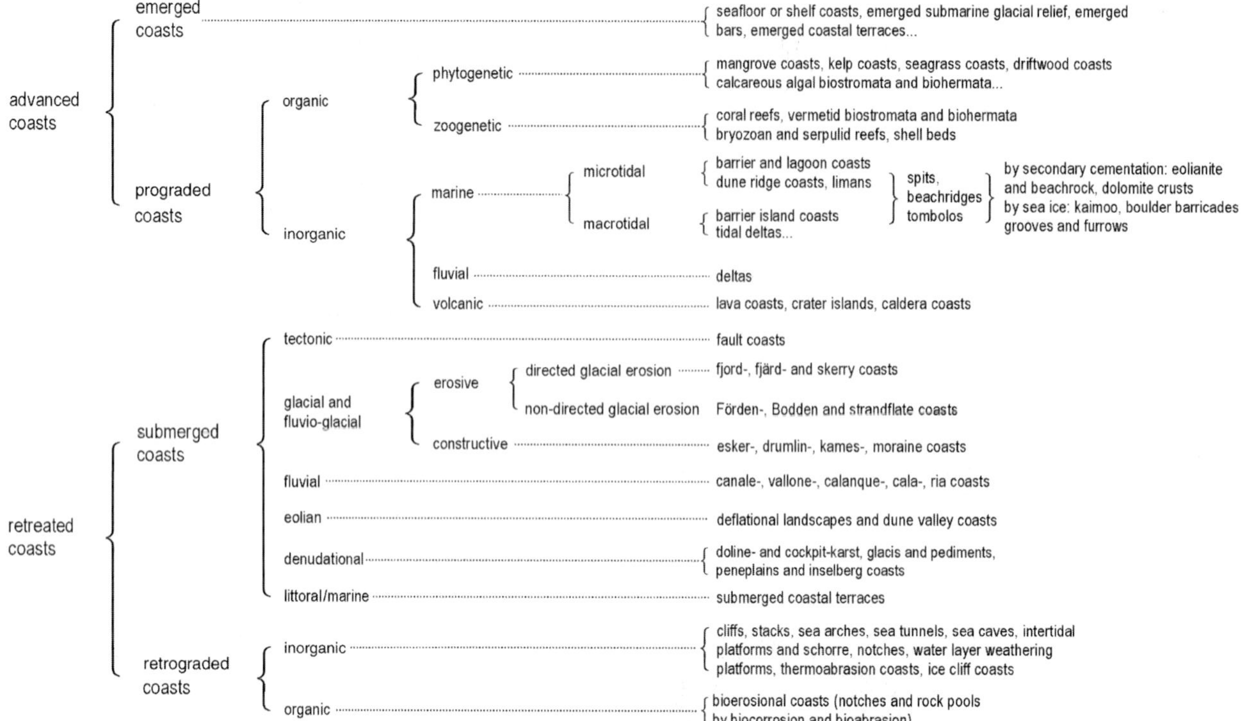

Figure H8 Genetic classification of the world's coastlines (based on Valentin,1952; and modified by Kelletat, 1999).

flooding on low-lying coastal lands as winter storms in higher latitudes. In the highest latitudes, however, sea-ice cover may suppress wave action during many months of the year, but this ice may have geomorphological effects when moving along a coastline during breakup in spring time. Beside the influences of the lithosphere, atmosphere, and hydrosphere the biosphere at the coastlines of the world will promote growth, as well. There are mangrove coasts, coral reef coasts, or those with dominant kelp, sea grass, or driftwood.

All in all, the world's coastlines show a latitudinal zonality (Kelletat, 1995) from sea-ice belts to coral reefs, depending on climate and water temperature in the first order, and on aridity or humidity, in the second order. Even small secular sea-level variations (by neotectonic or isostatic movement of the land, or eustatic changes of the water body itself) have a strong influence on coastal forms: relative rising sea levels lead to more erosion or drowning, falling sea levels to emergence and accumulation. Congruent with the ongoing sea-level rise during the last 1–2 centuries on the order of 1–3 decimeters (maybe partly caused by the anthropogenic greenhouse effect with global warming) a significant erosion of beaches and coastal dunes as well as wetlands has been observed. Holocene sea-level variations, however, are still under debate. It seems that there are a large number of different sea-level hypotheses depending on geotectonic factors as well as eustatic ones, sediment load, compaction, and more (Pirazzoli, 1991).

The prehistory of all modern coastal forms is a rising sea level (i.e., about 100 times faster than today's rising) from the last glacial maximum to the present warm phase. This steady transgression on formerly terrestrial environments could not leave marked forms but reworked the weathered mantle on the drowned landscapes, thus generating abundant coastal sediment. The transformation of the landscape by waves, however, was minimal, and the type of "primary coasts" with partly submerged terrestrial forms dominated nearly everywhere. After reaching the modern level about 6,000 years ago, the balance between marine and littoral processes on one side and terrestrial ones on the other side was concentrated along a fixed level, leading to a significant transformation and sculpturing the many types of "secondary coasts," dominated by littoral processes. The most striking evidences for abrasion and coastal destruction are cliffs, sea arches, sea caves, or stacks. Their amount of recession depends on surf energy and rock resistance but can reach several 100 m during Holocene times, leaving slightly inclining rock platforms in front of the cliffs. There is not only surf that destroys the coasts, but salt weathering or bioerosion may be important in certain environments, as well. Advancing coasts show beaches with accompanying coastal dunes (possibly cemented into beachrock or eolianite in more arid environments), sometimes formed into spits or barriers (or chains of barrier islands in regions with higher tidal range), or marshlands, whereas deltas can develop even along submerging coasts, if sediment discharge is strong enough. Beside the destructive and constructive coastal forms the category of ingressive features is very variable, representing many relief forms of the earth in a partly drowned status (fjords, rias, tropical karst, etc.).

A fully genetic classification of the world's coastlines is presented in Figure H8: All belong either to the class of advanced or retreated coasts, advancing may be caused by emergence or progradation, retreating by submergence or erosion. Next categories point to organic or inorganic processes or types of partly drowned terrestrial relief, tidal range, etc. (Figure H8). The coasts—in particular the accumulative ones—are important archives for even small environmental changes, because their ecosystems are very sensitive. They more and more fill the gap between terrestrial proxies (from inland ice, peat, lakes) and oceanic ones (from deep-sea sediments).

On a worldwide scale, the coasts are more and more often being transformed by human beings, either directly (land reclamation, coastal protection with seawalls and dikes, mining of beaches, aquaculture in mangrove areas, and others), or indirectly, by drilling oil or water from the ground, with the consequence of subsidence (Mississippi Delta, Venice), trapping river sediments in reservoirs far from the coast with the consequence of starving beaches, polluting nearshore waters by pesticides or sediment suspension from deforestation and agriculture, with the consequence of coral reef destruction, etc., or by global warming by emission of greenhouse gases. Textbooks on coastal geomorphology mostly give the impression of a consecutive, slow development in coastal environments. New research brings into discussion, whether sudden events of extreme energy (such as tsunami) may be as important or more important for the development of some coastal forms and deposits. All in all many questions remain to be solved in coastal geomorphology, and the coastlines of the world are by far not adequately investigated.

Further reading on this subject may be found in the following bibliography.

Dieter Kelletat

Bibliography

Bird, E.C.F., 1984. *Coasts. An Introduction to Coastal Geomorphology*. Oxford: Blackwell.

Bird, E.C.F., and Schwartz, M.L. (eds.), 1985. *The World's Coastline*. New York: Van Nostrand Reinhold.
Carter, R.W.G., 1988. *Coastal Environments*. London: Academic Press.
Carter, R.W.G., and Woodroffe, C.D., 1994. *Coastal Evolution. Late Quaternary Shoreline Morphodynamics*. Cambridge: Cambridge University Press.
Davies, J.L., 1964. A morphogenic approach to world shore-lines. *Zeitschrift für Geomorphologie, NF*, **8**: 127–142.
Davies, J.L., 1972. *Geographical Variation in Coastal Development*. London: Longman.
Fairbridge, R.W. (ed.), 1968. *The Encyclopedia of Geomorphology*. New York: Reinhold.
Kelletat, D., 1995. Atlas of coastal geomorphology and zonality. *Journal of Coastal Research*, Special issue **13**(1): 286.
Kelletat, D., 1999. *Physische Geographie der Meere und Küsten*. Stuttgart: Teubner.
Paskoff, R., 1994. *Les littoraux. Impacts des aménagements sur leur évolution*. Paris: Masson.
Pirazzoli, P.A., 1991. *World Atlas of Holocene Sea-Level Changes. Elsevier Oceanography Series* 58. Amsterdam
Valentin, H., 1952. Die Küsten der Erde. Beiträge zur allgemeinen und regionalen Küstenmorphologie. Petermanns Geographische Mitteilungen, Ergänzungsheft 246, Gotha.
Zenkovich, V.P., 1967. *Processes of Coastal Development*. London: Oliver & Boyd.

Cross-references

Classification of Coasts (see Holocene Coastal Geomorphology)
Coral Reef Coasts
Coastal Subsidence
Coastline Changes
Holocene Epoch
Mangrove, Coastal Geomorphology
Salt Marsh
Storm Surge
Tides
Vegetated Coasts

HOLOCENE EPOCH*

The Holocene, or "wholly recent," Epoch is the youngest phase of earth history. It began when the last glaciation ended, and for this reason is sometimes also known as the post-glacial period. In reality, however, the Holocene is one of many interglacials which have punctuated the late Cainozoic Ice Age. The term was introduced by Gervais in 1869 and was accepted as part of valid geological nomenclature by the International Geological Congress in 1885. The International Union for Quaternary Research (INQUA) has a commission devoted to the study of the Holocene, and several International Geological Correlation Programme (IGCP) projects have been based around environmental changes during the Holocene. A technical guide produced by IGCP Subproject 158B ("Palaeohydrological Changes in the Temperate Zone") represents a comprehensive account of Holocene research methods (Berglund, 1986). Since 1991, there has also existed a journal dedicated exclusively to Holocene research (*The Holocene*, published by Arnold).

During the Holocene, the earth's climates and environments took on their modern, natural form. Change was especially rapid during the first few millennia, with forests returning from their glacial refugia, the remaining ice sheets over Scandinavia and Canada melting away, and sea levels rising to within a few meters of their modern elevations in most parts of the world. By contrast, during the second half of the Holocene, human impact has become an increasingly important agency in the modification of natural environments. A critical point in this endeavor was when *Homo sapiens* began the domestication of plants and animals, a process which began in regions like the Near East and Mesoamerica very early in the Holocene, and which then spread progressively to almost all areas of the globe. For short histories of the Holocene, see Roberts (1998) and Bell and Walker (1992).

Although there are different schools of thought about how the Holocene should be formally defined (see Watson and Wright, 1980), the most common view, and one which is supported by INQUA, is that the Holocene began 10,000 radio-carbon (^{14}C) years ago. But ^{14}C chronologies count AD 1950 as being the "present day" and also underestimate true, or calendar, ages by several centuries for most of the Holocene. Nonetheless, there is evidence of a global climatic shift remarkably close to 10,000 ^{14}C yr BP (years before present), often involving a sharp rise in temperature (see Atkinson *et al.*, 1987).

Various attempts have been made to subdivide the Holocene, usually on the basis of inferred climatic changes. Blytt and Semander, for instance, proposed a scheme of alternating cool–wet and warm–dry phases based on shifts in peat stratigraphy in northern Europe. Some researchers believe there is evidence of a "thermal optimum" during the early-to-mid part of the Holocene. During the 1980s, the Cooperative Holocene Mapping Project (COHMAP) members established a comprehensive paleoclimatic database for the Holocene (Wright *et al.*, 1993), and showed that variations in the earth's orbit were the principal cause of differences in climate between the early Holocene and the present day. For this reason, the early Holocene is unlikely to provide a good direct analog for a future climate subject to greenhouse-gas warming (Street-Perrott and Roberts, 1993).

Neil Roberts

*Updated from the Encyclopedia of Environmental Science (1999) by the author and printed with the permission of Kluwer Academic Publishers.

Bibliography

Atkinson, T.C., Briffa, K.R., and Coope, G.R., 1987. Seasonal temperatures in Britain during the past 22,000 years reconstructed using beetle remains. *Nature*, **325**: 587–592.
Bell, M., and Walker, M.J.C., 1992. *Late Quaternary Environmental Change: Physical and Human Perspectives*. London: Longman; New York: Wiley.
Berglund, B. (ed.), 1986. *Handbook of Holocene Palaeoecology and Palaeohydrology*. New York: Wiley.
Roberts, N., 1998. *The Holocene. An Environmental History*. Oxford: Blackwell.
Street-Perrott, F.A., and Roberts, N., 1993. Past climates and future greenhouse warming. In Roberts, N. (ed.), *The Changing Global Environment*. Oxford: Blackwell.
Watson, R.A., and Wright, H.E., Jr., 1980. The end of the Pleistocene: a general critique of chronostratigraphic classification. *Boreas*, **9**: 153–163.
Wright, H.E., Kutzbach, J.E., Webb, T., Ruddiman, W.F., Street-Perrott, F.A., and Bartlein, P.J. (eds.), 1993. *Global Climates for 9000 and 6000 Years Ago*. Minneapolis, MN.: University of Minnesota Press.

Cross-references

Geochronology
History, Coastal Gemorphology
Holocene Coastal Gemorphology
Sea-Level Changes During the Last Millenium

HONEYCOMB WEATHERING

Honeycomb weathering produces extensive networks of small cavities that form on rock surfaces. These patterns initially develop as many shallow depressions, but continued development produces deep chambers that are separated by thin septa of unweathered rock (Figure H9). Individual cavities are typically several centimeters in width and depth, the shape often being controlled by bedding planes, foliation, or other structural features of the rock in which they occur. In many localities the holes occur in association with a hardened surface layer formed when dissolution of ferruginous minerals has been followed by precipitation of ferric hydroxides near the outcrop surface. The thickness of this hardened layer may range from a few millimeters to several centimeters.

Honeycomb weathering has worldwide distribution, typically found in coastal outcrops and inland deserts. The most-studied occurrences are those of Australia, the western United States (Figure H10), and South Victoria Land, Antarctica. This type of weathering is most commonly observed in sandstone, but it also occurs in granite, gneiss, schist, gabbro, and limestone. The mode of formation of these cavities is not well established, and several hypotheses have been advanced. Possibly more than one type of mechanism may be involved.

In formulating any explanation it is necessary to account for the extremely selective nature of the erosion. One possibility is that the rock possesses internal variations in hardness, composition, or porosity. This conflicts with the fact that honeycomb weathering commonly occurs in rocks that have a high degree of physical and chemical homogeneity. Thus,

Figure H9 Honeycomb weathering in sandstone near Baku, Azerbaijan on the west coast of the Caspian Sea. (Photo, M.L. Schwartz.)

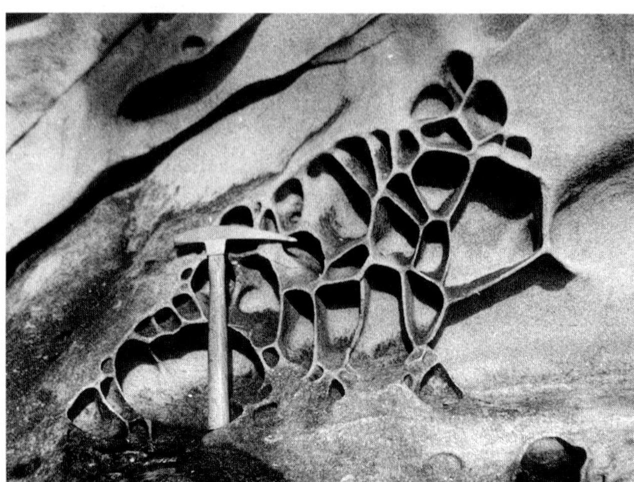

Figure H10 Honeycomb weathering in sandstone at Larrabee State Park, near Bellingham, Washington, USA. (Photo George Mustoe.)

it is more likely that attack by the weathering agents occurs in a differential fashion. Early investigators invoked a diverse variety of geomorphic processes to explain honeycomb weathering, but these cavities are now generally accepted to be caused by salt weathering, where evaporation of wave splash or saline pore water produces salt crystals that wedge apart mineral grains (Evans, 1970). Chemical dissolution of silicate minerals may also play an important role (Young, 1987). Just how these process works to produce a delicate honeycomb pattern remains an enigma. Mustoe (1981, 1982) believed that coatings of lichens and algae on the rock surface control the pattern of cavity development. Laboratory experiments by Rodriguez-Navarro *et al.* (1999) suggest that erosion may be related to variations in wind currents flowing over the rough outcrop surface.

George Mustoe

Bibliography

Evans, I.S., 1970. Salt crystallization and rock weathering, a review. *Revue de Geomorphologie Dynamique*, **19**: 153–177.

Mustoe, G.E., 1981. The origin of honeycomb weathering. *Geological Society of America Bulletin*, **93**: 108–115.

Mustoe, G.E., 1982. Cavernous weathering in the Capitol Reef desert, Utah. *Earth Surface Processes and Landforms*, **8**: 517–526.

Rodrigeuz-Navarro, C., Doehne, E., and Sebastian, E., 1999. Origins of honeycomb weathering: the role of salts and wind. *Geological Society of America Bulletin*, **111**: 1250–1255.

Young, A.R.M., 1987. Salt as an agent in the development of cavernous weathering. *Geology*, **15**: 962–966.

Cross-references

Bioerosion
Cliffs, Lithology versus Erosion Rates
Coastal Climate
Coastal Hoodoos
Coastal wina Effects
Desert Coasts
Notches
Shore Platforms
Tafone
Weathering in the Coastal Zone

HOODOOS—See COASTAL HOODOOS

HUMAN IMPACT ON COASTS

Introduction

Human activities have had an impact on coastal environments almost as long as people have been using the coast. It was not long before attempts to control erosion resulted in various types of structures such as jetties, groins, and seawalls. Access to the beaches of barrier islands resulted in fill-type causeways. Development of barrier islands accessed by these causeways resulted in various types of construction that have negatively impacted on the coastal zone. Harbors and the navigational channels leading to them have resulted in some problems for estuaries. Overall, there are many ways whereby human activities have caused problems for a broad spectrum of coastal environments. Some of these are a result of direct acts of development and others are indirectly the result of these activities. The following discussion will consider some of the more obvious and problematic human impacts. Many of the examples will come from Florida where such development-related activities have been underway for several decades and where we can learn from our experiences.

Direct impact

Hard protection

Among the first, the most widespread, and the most problematic human impacts on the coast is the erection of hard structures: those that are immobile and that are not modified by coastal processes. This category includes seawalls, groins, breakwaters, and jetties. Their primary purpose is to protect the coast from erosion and to stabilize tidal inlets. The variety of such structures is tremendous (CERC, 1984), and a comprehensive discussion of them would fill a book. The basic characteristics of each general type and some examples will be provided to acquaint the reader with such structures.

Figure H11 Rip-rap in front of concrete seawall. This is one of the most common types of erosion protection on the open coast environment.

Seawalls of various types have been around for more than a century. The basic situation is protection of the open coast landward of a beach from wave attack. Such structures may be poured concrete, metal sheet piling, pressure-treated wood, or various types of rip-rap; well-placed boulders. Sometimes longard tubes or other sand-filled plastic or cloth tubes are used as temporary protection. Such an approach is viewed as a semi-hard type of protection.

There are problems with these types of structures. Regardless of their type, they tend to be temporary. Eventually, time or severe storms will cause them to fail or become useless for various reasons. Some of the problems associated with seawalls are the scouring that occurs at the base. In many cases, these walls are vertical which takes the full impact of the waves. Rip-rap placed in front of the seawall dissipates wave energy but continued wave attack will eventually result in failure (Figure H11).

Groins are placed perpendicular to the shore in an attempt to keep sediment from being carried away by longshore currents. Most of these are overdesigned and act like dams along the beach causing downdrift erosion (Figure H12). The best performing groins, also among the largest, are along the North Sea coast of The Netherlands and Germany (Figure H13).

Jetties and stabilized inlets are also a problem for beach erosion. They act as large dams, prohibiting longshore transport from moving across the inlet. This is, of course, their purpose because they are constructed in order to stabilize the inlet and maintain its navigability. The downdrift erosion produced by such jetties is chronic (Figure H14) and can only be avoided by some type of sediment or bypass system. Another problem with jetties is that many leak sediment or are short enough to allow sediment to pass around the end so that regular dredging becomes a necessary procedure for maintenance of inlet navigation. Attached breakwaters also have similar problems.

Detached breakwaters (Figure H15) are probably the most benign of the hard open coast structures. These are typically parallel to the shore and are designed to provide some combination of protection from shore erosion and safe mooring for small boats. The negative effects are that they commonly cause a salient to form in the lee of the structure that in many cases forms a tombolo. This connection between the structure and the shore is an unwanted occurrence, which can sometimes be eliminated by lowering the breakwater to permit some wave action and therefore, longshore currents to keep sediment from accumulating. In some cases, the shore erodes even in the presence of these structures leaving them in place but without function. Removal is generally very expensive; commonly more than their emplacement.

Figure H12 Groin showing a significant amount accumulation on the updrift side.

Soft protection

More recently there has been a dramatic and nearly total shift to soft protection for erosion control along the open coast. This has come in two basic approaches: (1) beach nourishment, and (2) vegetation and protection of dunes. These approaches have received the endorsement of the engineering and the environmental community alike. The result is much more esthetic shore protection, however, these techniques are not permanent solutions to coastal erosion. The only permanent solution is abandonment of the barrier islands and open coast shores.

Figure H13 Large groin on the North Sea coast where there is little difference in accumulation of beach sand on either side of the structure.

Beach nourishment. Although not new as a method of erosion control, beach nourishment became a standard beginning in the early 1980s (NRC, 1995). The basic scheme is to artificially rebuild the beach that was removed by erosion, and to do it as close to the original, natural beach as possible. The most important factor for such a construction project is the location of a sediment source that is similar to that on the natural beach and that is of sufficient volume to do the job. Most of these projects require an average of about 1 million m^3 of beach quality sand being placed on the eroding shore area. Some are smaller and a few have been much larger.

For a variety of reasons, the borrow site for such beach sand is typically seaward of the beach rather than landward, although a few upland sources have been used. Most of the offshore sources have been the ebb-tidal deltas associated with tidal inlets, and large sand bars or old beach deposits from the present shoreface. Some projects have been nourished using beach quality material dredged from inlets but these are typically at the smaller end of the spectrum. In addition to the limitations posed by the availability of beach quality sand, the distance of transport is a major consideration because this is what comprises most of the cost of the project.

The design of the nourishment project is undertaken by coastal engineers and requires considerable input of coastal processes, storm surge levels at the site in question, desired protection, and historical characteristics of beaches at the site. Commonly, the elevation of the construction berm is associated with the surge level of a 10–20-year storm. In many parts of the Gulf Coast of the United States that would be at about 1.5–2 m above mean sea level. Some extreme elevations along the Atlantic Coast have been constructed at 3–3.5 m above mean sea level.

Figure H14 Jetty at a tidal inlet showing a significant amount of downdrift erosion as the result of the inability of sediment to cross the structure and the inlet.

Figure H16 Construction of a new beach using a conveyor belt to offload sand from a large barge.

Figure H15 Photograph of a detached breakwater designed to protect the shore landward of it. (Courtesy of D. FitzGerald.)

Figure H17 Pumping and spreading nourishment sand in the construction of a new beach.

Width is also variable but is almost always at least 15–20 m and sometimes will be up to 50 m.

Once the borrow material has been located, all of the permits are obtained, contracts let, and pre-nourishment surveys completed, then dredging and construction begins. The typical approach is to use a large suction dredge to remove the sand, pump it through pipes or onto barges and eventually be pumped onto the subject shore. Barges are used for transport when the borrow site is more than a couple of kilometers from the construction site. Large tractors are then used to spread the new beach into the specifications in the construction plans (Figures H16 and H17).

Such nourishment projects typically cost from about $3/m^3 up to $15/m^3 meaning that many of these projects are over US$10 million. Overall, this has been a very successful approach. Some areas such as in New Jersey and parts of North Carolina have not been uniformly successful but others such as the Florida Gulf Coast have been successful beyond predictions (Trembanis and Pilkey, 1998). We are still learning about the best way to conduct such protection activities but the cost/benefit ratio on most of them has been high on the benefit side. Probably, the most successful of these projects, and the largest, is the one at Miami Beach (Figure H18). It was completed in 1980 at a cost in excess of US$60 million but is performing very well and is protecting billions of dollars of upland properties.

Dune protection and stabilization. The other environment that is critical to protection of the upland environment and properties is the dunes that are immediately landward of the beach. Destruction of dunes for development purposes has been eliminated along most coasts and preservation of these dunes is a high priority. In many places dunes have been rebuilt, vegetation has been established, and other measures have been put in place to stimulate dune growth and to preserve those that are there.

Actual construction of dunes is not a widespread practice although small dunes are commonly built to initiate dune development. This approach is most effectively used after a severe storm has caused removal of all or portions of foredunes or if there have been no dunes along a backbeach area. Most commonly, the approach for stimulating dune growth is through planting of appropriate vegetation. This vegetation provides a very efficient mechanism for trapping wind-blown sand from the beach, and dunes form very quickly (Figure H19). These plantings are even irrigated in some areas. Other efforts to enhance dune growth are various types of fencing that will trap sand. Originally, the same type was used that has been used in northern climates for trapping snow along the highways but that has now given way to various types of biodegradable material that will deteriorate when buried for some time. In some countries such as The Netherlands, rows of twigs and shrubbery are planted to trap the wind-blown sand.

Once established, there are methods of preserving the dunes. The most common and most effective is construction of walkovers to prevent foot erosion from people. This is widespread along coasts where there is considerable development and traffic to the beach. In more remote areas, the paths to the beach are simply developed so as to be in opposition to the prevailing and predominant wind directions. The paths also may have multiple changes in direction to prevent wind from blowing along them and eroding the sand.

Soft shore protection is now the standard due to its compatibility with the natural coastal environment and its esthetically pleasing appearance. Although costs are high and maintenance is mandatory, this approach has prevailed for about two decades and will likely do so in the future.

Figure H18 Photograph of Miami Beach before and after construction of the huge beach nourishment project. (Photo courtesy of U.S. Army, Corps of Engineers.)

Indirect impact

There are numerous ways in which human activities along the coast can indirectly impact the behavior of coastal environments. These range from the obvious situations where jetties at an inlet may impact beach erosion kilometers down the beach to more subtle situations where activities in an estuary can influence open coast morphodynamics. The discussion here will focus on the activities that impact on tidal inlet stability. All of these activities are involved, one way or another, with coastal development.

Before proceeding farther in the discussion it is important to briefly consider the important factors in inlet morphodynamics. First, the volume of water that passes through an inlet during a given tidal cycle is the tidal prism, a water budget. The prism is the product of the area of the back-barrier area served by the inlet and the tidal range. It is the tidal prism that determines the size and stability of the inlet. Large prisms tend to maintain stability in the inlet position and cross-sectional area whereas small tidal prisms lead to instable inlets that migrate and often close. It is the latter condition that is a main reason for the construction of jetties as considered in an earlier part of the discussion.

There are various types of human activities in the back- barrier/estuarine environments that can impact on the tidal prism and therefore, on tidal inlets. In nearly every case the result of these activities is a decrease in tidal prism and a lack of stability in the tidal inlet(s) involved.

Figure H19 Artificially planted vegetation that is designed to trap sand and promote growth of dunes.

Figure H20 A fill-type causeway constructed between the mainland and a barrier island.

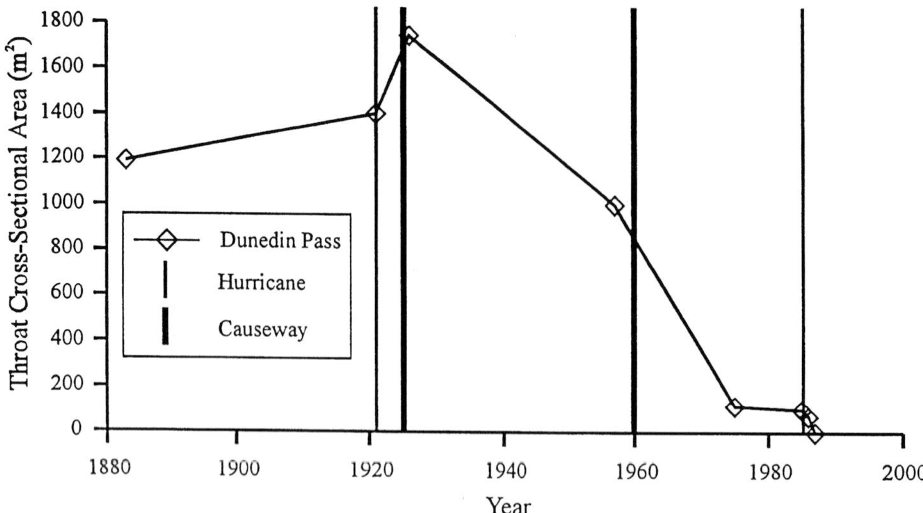

Figure H21 The effects on Dunedin Pass of the construction of two causeways along the central Gulf Coast of the Florida peninsula. (Modified from Lynch-Blosse and Davis, 1977.)

Figure H22 Finger canal development on the back-barrier area as the result of dredge and fill construction.

Construction of fill-type causeways

One of the first major activities of barrier island development is the construction of roadways to access these recreational and resort areas. This was typically accomplished by dredge and fill construction of causeways (Figure H20), many of which were built in the United States in the early 1920s. This was the time that many private vehicles were on the road and people wanted to "drive to the beach."

Such causeways acted like dams in the area between the mainland and the barrier islands. They became efficient tidal divides because, except for navigational passages, there was essentially no tidal flux across these ribbons of fill. As a consequence, there was significant change in the tidal prism of the tidal inlet(s) that served this back-barrier area. The result was change in the inlet stability, typically a reduction in the size and commonly the tendency for migration. As an example, Dunedin Pass in the Clearwater area of the Florida Gulf Coast (Lynch-Blosse and Davis, 1977) showed a major reduction in its cross-sectional area shortly after construction of the Clear-water Causeway in 1922 and then after construction of the Dunedin Causeway in 1964 it was reduced substantially again (Figure H21). This same phenomenon has also taken place on the coasts of Texas, New Jersey, and North Carolina.

Another type of human activity that has had a major influence on inlet stability is the dredge and fill activity that has taken place on the landward side of the barrier islands. The typical situation on this side of a barrier island is domination by some type of wetland environment; either salt marsh or in the low latitudes, mangrove communities. As pressures of development increased and space on barriers became limited these wetlands were included. Such lands brought only a small price because they were deemed worthless. The potential for development was, however, very good. These wetlands were converted to buildable uplands by dredge and fill techniques. Finger canals were dredged through the wetland and the spoil was cast to the side thus producing a supratidal finger of land which when stabilized by seawalls, was suitable for building houses. These finger canals and their associated small peninsulas of land are widespread along many back-barriers but are most common in Florida (Figure H22).

Obviously such development practices have terrible negative impacts on the wetland environments, essentially destroying all of them. Fortunately, they were stopped in the 1960s. The impact of such activities on the tidal prism is also significant. These practices of dredge and fill construction reduce the area of the back-barrier which in turn reduces the tidal prism (Davis and Barnard, 2000). A comparison of two examples from the west-central coast of Florida shows the contrast. Caladesi Island is a virtually pristine barrier island that is about the same size and shape as Long Key, a fully developed barrier. The outlines of these barriers display how the area of the back-barrier has been reduced (Figure H23). An extreme example of this has occurred in Boca Ciega Bay in the St. Petersburg area of Florida (Davis, 1989). This large back-barrier estuary has had its surface area reduced by more than 25% since the 1920s.

Construction of the intracoastal waterway

The Intracoastal Waterway (ICW) is a dredged and maintained navigational channel that extends from Brownsville, TX across the Gulf Coast and then continues along the east coast of the United States up to New England. It was originally constructed for commercial traffic to protect vessels, especially barges, from severe weather and energetic wave conditions. Through most of its extent, the ICW has a design width of 50 ft (~15 m) and a depth of 8 ft (~2.5 m). This channel cuts through some land areas but mostly follows along the open water, back-barrier areas. Along most of its extent the spoil from dredging was cast alongside in piles forming small islands. Aerial photos along many stretches show dozens of these islands that have become valued for fishing and as bird rookeries. The negative side of this dredge spoil is that in many places it was dumped on wetlands and destroyed the area covered. Now a few decades old, the ICW is in need of major maintenance dredging in many locations. Whether or not this happens depends on plans to dispose of the dredge spoil. It is unlikely that disposal of this spoil will be permitted along the channel for environmental reasons. This spoil is not suitable for beach nourishment because of its high mud content and its toxic content at some locations. The cost of disposing it in deep-water offshore is probably too high to be a viable solution. A solution is still not forthcoming.

Another important impact of ICW dredging is its impact on the tidal prism. In narrow back-barrier areas where the prism is comparatively small, such a channel can divert the tidal prism from a natural inlet, along its path up and down the channel. Such a circumstance took place

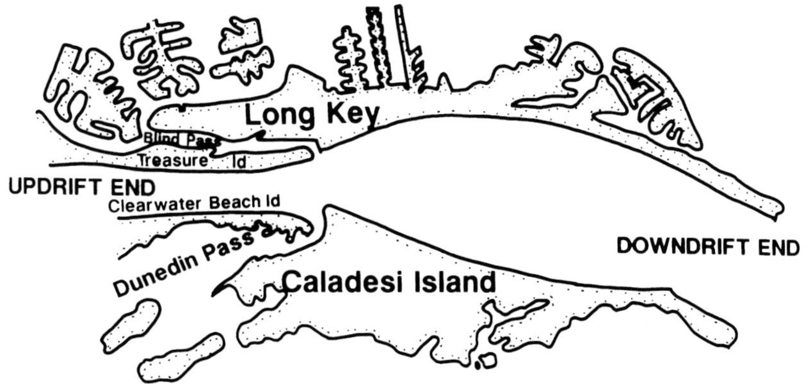

Figure H23 Comparison of the natural shore of Caladesi Island and the drastically modified shore of Long Key, two originally similar barrier islands on the Gulf Coast of Florida. (From Davis, 1989.) Reproduced by permission of the ASCE.

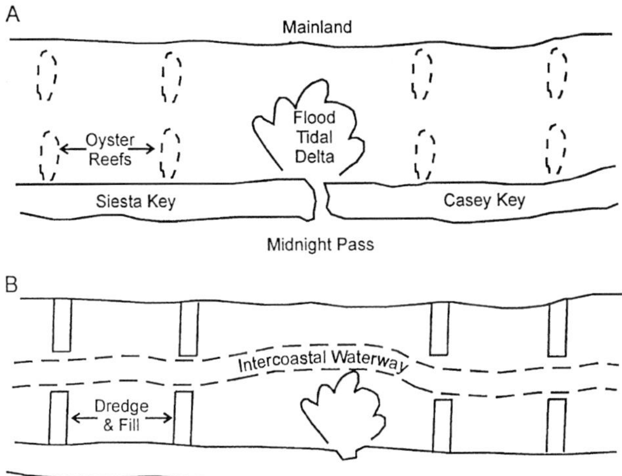

Figure H24 Schematic maps of Little Sarasota Bay showing (A) the natural conditions and (B) the results of dredging of the Intracoastal Waterway and dredge and fill of oyster reef areas.

in Little Sarasota Bay, FL in conjunction with the ICW dredging and some other construction practices. The result was closure of a natural tidal inlet that had been open throughout historical time (Davis *et al.*, 1987).

Coincidently with the dredging of the ICW was the dredge and fill construction over several elongate oyster bars that were covered with mangroves. Dredge material was placed on these features and sheet piling was used to contain it. These areas now have several houses on each one. The result is that all circulation within Little Sarasota Bay is now channeled through the ICW and the natural inlet that served this bay, Midnight Pass, has been closed due to its greatly reduced tidal prism (Figure H24).

Summary

Human impacts along the coastal zone are numerous, widespread in kind, and typically detrimental to the environments where they take place. This discussion has focused on the open coast impacts emphasizing the beach and inlet environments. Other types exist, especially in various coastal bay environments.

In general, the impacts on these environments are the result of development pressures for more space to be occupied by residential or commercial properties. Because the land along the coast is so expensive, the pressures are great; both economical and political. The consequences have been disastrous in nature. Fortunately, most governmental jurisdictions have taken action to prevent such activities in the future; at least in the developed countries. Many of the developing countries are still way behind in their planning and management of the coast. Unfortunately, there are indications that too many of them are not learning from the many mistakes that have already taken place and that are well-documented. The other unfortunate circumstance is that even though we have rules and ordinances in place, there seem to be too many exceptions that are granted to violate these regulations.

Richard A. Davis, Jr.

Bibliography

CERC, 1984. *Shore Protection Manual*, Volume 1. Coastal Engineering Research Center. Vicksburg, MS: U.S. Army Corps of Engineers.

Davis, R.A., 1989. Management of drumstick barrier islands. In Stauble, D.M. (ed.), *Barrier Islands: Process and Management*. New York: *ASCE, Coastal Zone '89*, pp. 1–16.

Davis, R.A., Jr., and Barnard, P.L., 2000. How anthropogenic factors in the back-barrier area influence tidal inlet stability: examples from the Gulf Coast of Florida, U.S.A. In Pye, K., and Allen, J.R.L. (eds.), *Coastal and Estuarine Environments: Sedimentology, Geomorphology, and Geoarchaeology*. Geological Society, London, Special Publications 175, pp. 293–303.

Davis, R.A., Hine, A.C., and Bland, M.J., 1987. Midnight Pass, Florida: inlet instability due to man-related activities in Little Sarasota Bay. New York: *ASCE, Coastal Sediments '87*, pp. 2062–2077.

Lynch-Blosse, M.A., and Davis, R.A., 1977. *Stability of Dunedin and Hurricane Passes, Pinellas County, Florida*. New York: *ASCE, Coastal Sediments '77*, pp. 774–789.

National Research Council, 1995. *Beach Nourishment and Protection*. Washington, DC: National Academy Press.

Trembanis, A.C., and Pilkey, O.H., 1998. Summary of beach nourishment along the U.S. Gulf of Mexico shoreline. *Journal of Coastal Research*, **14**: 407–417.

Cross-references

Beach Nourishment
Bioengineered Shore Protection
Dredging of Coastal Environments
Environmental Quality
Estuaries, Anthropogenic Impacts
Navigation Structures
Shore Protection Structures
Tidel Inlets
Tidal Prism

HYDROLOGY OF THE COASTAL ZONE

Concepts and definitions

In the broadest sense, the hydrology of the coastal zone could include the distribution and movement of any water in the coastal zone; however, in practice, this term generally refers to ground water. A few studies have looked at ground water movement in coastal barriers, water table

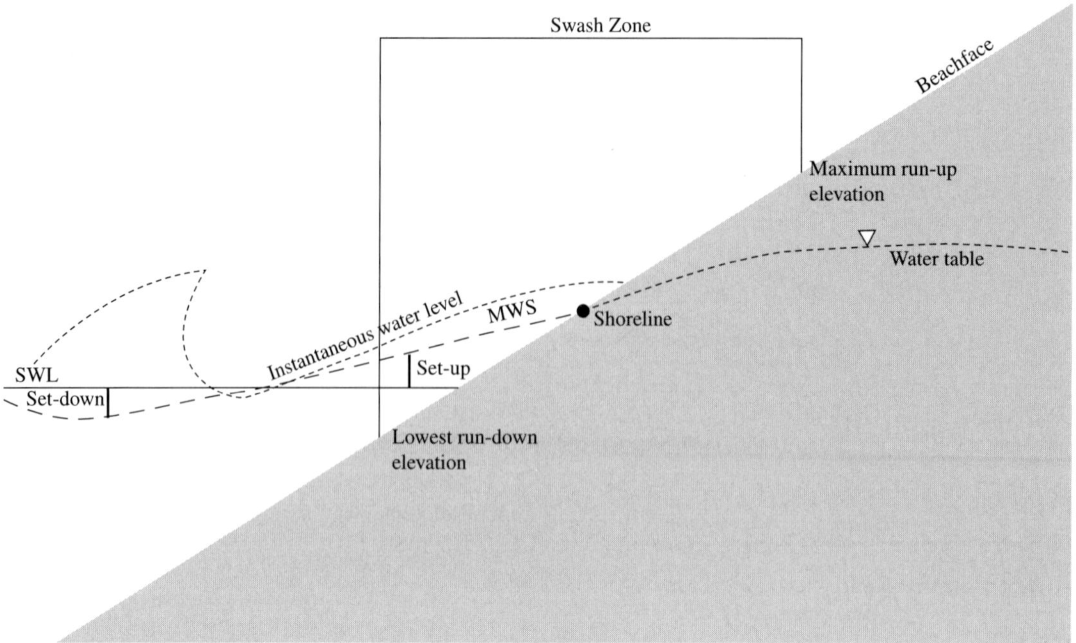

Figure H25 Definition sketch of surface and subsurface water levels in the swash zone.

fluctuations in estuarine environments and gravel beaches, or sand moisture content effects on aeolian sediment transport. However, most studies of coastal ground water dynamics have concentrated on ground water in beaches, and in particular, in the swash zone of sandy beaches. The *swash zone* is the section of the foreshore where final wave-energy dissipation occurs, which is alternately covered and exposed by water motions.

The complex interaction of surface and subsurface water in the swash zone are defined below and illustrated in Figure H25. The *still water level* (SWL) is the water surface in the hypothetical situation of no waves. When the local water-surface elevation is averaged over a time span much longer than incident and infragravity periods but shorter than the tidal period, the result is the local mean water level, which traces the *mean water surface* (MWS). The MWS in the surf and swash zones generally has a gradient which balances the change in the *radiation stress*, defined as the excess flow of momentum due to the presence of waves. Changes in radiation stress are balanced by changes in hydrostatic pressure; in other words, by changes in water level. This difference is known as set-up or set-down. *Set-up* is a wave-induced increase in the MWS, whereas *set-down* is a wave-induced decrease in the MWS. Set-down occurs seaward of the breakers, where radiation stress is at its maximum. The positive gradient due to radiation stress is balanced by a negative water-surface gradient, resulting in a lowering of the MWS to below SWL. Set-up occurs inside the surf zone, where the decrease in radiation stress due to energy dissipation is balanced by the raising of the MWS to above SWL. As long as energy dissipation continues, set-up continues to increase in the onshore direction and is greatest at the shoreline. The concept of set-up is discussed in more detail in the entry on *Surf Zone Processes*.

The *shoreline* is the position where the MWS (including the set-up) intersects the beachface; in other words, the line of zero water depth. The shoreline represents the land–water boundary, which moves across the intertidal beach at a range of frequencies from incident waves (3–15 s) to tides (daily or twice daily for high tide–low tide cycles, and approximately 2 weeks for spring—neap cycles). The limits of shoreline excursion define the boundaries of the swash zone, which migrates up and down the foreshore of the beach over a tidal cycle. The seaward and landward limits of the swash zone are, respectively, the point of collapse of the wave or bore, and the landward limit of wave action. There are two components to the water motions in the swash zone. The first is *swash* (also referred to as *uprush*), which is a landward-directed flow characterized by the upslope transport of water. The second component of the cycle is the *backwash*, which is the downslope movement of the water which follows maximum run-up. The uprush–backwash cycle is essentially an oscillation superimposed on the maximum MWS (including set-up) inside the surf zone. Total wave run-up represents the combined effect of set-up and swash at incident and infragravity frequencies.

The *maximum swash height*, or *maximum run-up elevation*, is the maximum vertical height above SWL reached by the uprush. Wave *run-down elevation* is the lowest vertical height reached by the backwash of a wave before the uprush of the next wave begins to run up the beachface. The run-down elevation may be below SWL.

The terminology used by ground water hydrologists to describe subsurface water may not be familiar to coastal researchers; therefore, these terms are defined in some detail here, and are illustrated in Figure H26. The best general reference on ground water hydrology is Freeze and Cherry (1979), from which many of the definitions in this entry are taken.

The beach water table is generally considered to be the continuation of the MWS inside the beach; however, a more physically correct definition of the *water table* is an equilibrium surface at which pore water pressure is equal to atmospheric pressure. The water table may also be referred to as the *phreatic surface*. *Pore water pressure* is the fluid pressure in the pores of a porous medium relative to atmospheric pressure. Below the water table, pore water pressure is greater than atmospheric pressure; above the water table, pore water pressure is less than atmospheric pressure. Hydrologists generally use the term *ground water* to refer to water below the water table, where pore water pressures are positive, and use *soil water* to describe water above the water table where pore water pressures are negative (subatmospheric). However, to equate beach sediment with a soil would be misleading, so in beach hydrology, the term *ground water* is commonly used to mean any water held in the sand below the beach surface. The *phreatic zone* is the permanently saturated zone beneath the water table. The *vadose zone*, which is sometimes called the *zone of aeration* or the *unsaturated zone*, is the unsaturated region of a beach sand body extending from the water table to the sand surface. In the saturated (phreatic) zone, pore spaces are filled with water and pore water pressures are equal to or greater than atmospheric pressure. In the zone of aeration, the pores are filled with both water and air and pore water pressures are less than atmospheric. For this reason, beach ground water zones are better defined by pore water pressure distribution than by saturation levels. A *capillary fringe* develops immediately above the water table as a result of the force of mutual attraction between water molecules and the molecular attraction between water and the surrounding sand matrix. The capillary fringe may also be referred to as the *tension-saturated zone*. (Ground water hydrologists often use the terms *tension* or *suction*—which can be used interchangeably—to describe a pressure which is negative relative to atmospheric pressure). In the capillary fringe, pore spaces are fully saturated, but the capillary fringe is distinguished from the water table by the fact that pore water pressures are negative. The thickness of the capillary fringe in sand beaches may vary between a few millimeters to nearly a meter, and it may extend to the sand surface. Some workers also refer to an intermediate zone which may occur above the capillary zone where the degree of saturation may vary, but remains less than 100%.

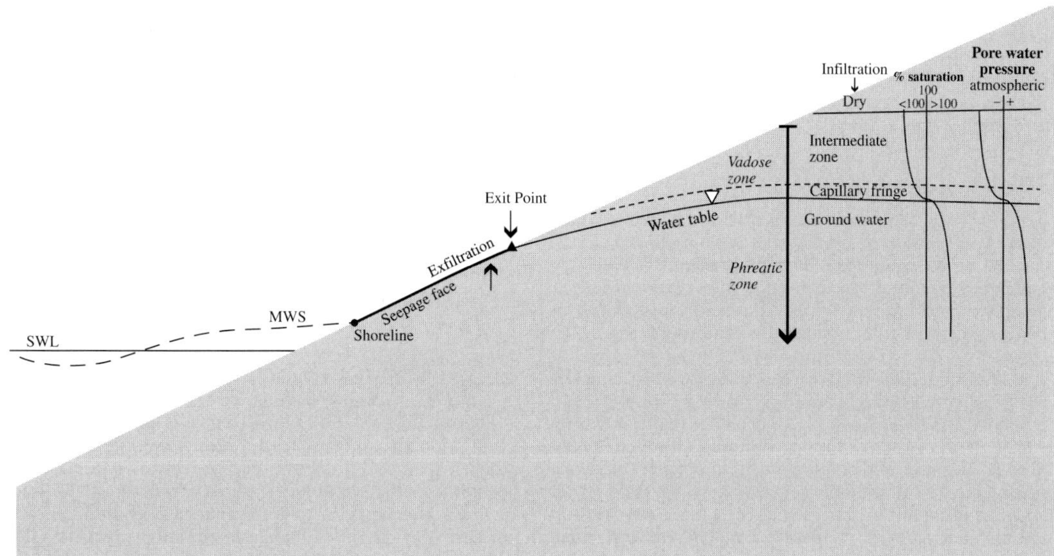

Figure H26 Definition sketch of beach ground water zones when the water table is decoupled from the tide.

Importance of beach ground water and swash zone processes

Swash zone and beach ground water processes are of interest to geomorphologists who wish to determine beach erosion or deposition, or aeolian sediment transport, to marine biologists who are interested in intertidal fauna, and to engineers who require data on run-up, particularly on coastal structures such as breakwaters. Over the past few decades, data on the position of the shoreline, which is directly dependent on swash zone processes, have emerged as one of the principal sources of information for monitoring coastal change. In some cases, the shoreline position (identified as the maximum extent of run-up) is used to establish legal boundaries, setback lines or flood hazard zones.

Marine biologists have an interest in the swash zone, as the distribution and type of macrofauna inhabiting the intertidal zone of sandy beaches appears to be related to the swash climate. Both the interstitial fauna and the macrofauna of sandy beaches are directly affected by swash and ground water processes: the former by infiltration, which is responsible for flushing oxygen and organic materials into the sand, and the latter by swash dynamics and the position of the seepage face, which influence tidal migrations and burrowing (McArdle and McLachlan, 1991). Differences in the spatial distribution and abundance of beach fauna have been explained in relation to sediment size and beach slope, but have not yet been related successfully to swash dynamics.

Coastal engineers have long recognized the need for a better understanding of swash zone processes, largely concentrating on the measurement and modeling of run-up on structures such as breakwaters. Such studies are needed to establish design criteria, particularly the elevation of the structure required to prevent overtopping by the run-up of extreme waves. Recently, the commercial possibility of modifying beach water table elevation to control beach erosion has been recognized, and several studies have investigated the use of beach dewatering as an alternative to hard engineering practices. Beach dewatering works by lowering the water table artificially through a system of buried drains and pumps (see the entry on *Beach Drain* for further details). Other engineering applications where knowledge of beach ground water dynamics is important include water quality management in closed coastal lakes and lagoons, and the operation of water supply and sewage waste disposal facilities in coastal dunes, contaminant cycling in estuaries and coastal water resource management issues such as salt water intrusion into coastal aquifers, wastewater disposal from coastal developments, and pollution control.

An understanding of swash and beach ground water dynamics is also important in the modeling of beach profile evolution. At present, most beach profile and shoreline change models either do not include, or vastly simplify, sediment transport processes in the swash zone. Cross-shore sediment transport models have demonstrated considerable success in predicting eroding beach profiles on relatively fine sand beaches. Predictions of accretionary events and the behavior of coarser sediment beaches are generally less good, particularly in the inner surf zone and swash zone (Schoones and Theron, 1995). Since an accretionary event is defined by the deposition of sediment above mean sea level, the lack of detailed knowledge of swash and beach ground water dynamics is probably an important factor in the inability of beach profile models to simulate accretionary events accurately.

Erosion and accretion of the beach profile, and the resulting movement of the position of the shoreline, are a direct result of sediment transport processes occurring in the swash zone and inner surf zone. Beach ground water—swash dynamics provide an important control on swash zone sediment transport, which affects the morphology of the intertidal beach by controlling the potential for offshore transport or onshore sediment transport and deposition above the SWL. Cyclic erosion and accretion of the beachface as a result of relative elevations of the beach water table and swash have been substantiated by researchers for many years. Most of these studies suggest that beaches with a low water table tend to accrete and beaches with a high water table tend to erode. Recent observations indicate that flows in the swash zone can also affect the beach profile seaward of the intertidal profile, influencing sediment transport in the bar region.

Swash and beach ground water interaction may play a particularly important role in profile evolution and sedimentation patterns on macrotidal beaches. Many macrotidal beaches have two, and sometimes three, distinct beach zones: a flat, dissipative low-tide beach and a steeper, more reflective high-tide beach. There is generally an abrupt decrease in beach slope on macrotidal beaches where the water table intersects the beachface, which may be also marked by a change in sediment size between coarse and fine material. The interaction between beach ground water and swash flows may provide a mechanism for the shore-normal sorting of coarse and fine material that is often observed on macrotidal beaches, with dissipative sand low-tide terraces at the base of steep reflective high-tide gravel ridges. For example, Turner (1993a) surveyed 15 macrotidal beaches on the Queensland coast in Australia, and found that a decrease in sediment size was strongly correlated to an increase in the relative extent of the lower gradient (saturated) lower region of the intertidal profile. Turner (1995) developed a simple numerical model that incorporated the interaction of the tide and the beach water table outcrop. This model predicted the development of a break in slope resulting from landward sediment transport and berm development across the alternately saturated and unsaturated upper beach, while the profile lowered and widened across the saturated lower beach. Hughes and Turner (1999) gave different empirical equations for equilibrium slope on unsaturated and saturated beachfaces.

Common to all these studies is the observation that when the water table outcrops above the tide, two zones are distinguished: a lower saturated zone that promotes downslope (offshore) sediment transport, and an upper region that alternates between saturated and unsaturated conditions, with upslope (onshore) sediment transport potentially enhanced by infiltration. However, the relative importance of infiltration is not yet known, and will be discussed further below in the section on *Mechanisms of surface–subsurface flow interaction*.

Behavior of beach ground water

Turner and Nielsen (1997) identified a number of mechanisms which have been associated with observed beach water table oscillations: seasonal variations, barometric pressure changes associated with the passage of weather systems and storm events, propagation of shelf waves, and infragravity and incident waves. However, the majority of research has concentrated on tide-induced fluctuations of the beach water table.

A number of studies since the 1940s have described the shape and elevation of the beach water table as a function of beach morphology and tidal state. The majority of these studies have been limited to measurements of water table elevations across the beach profile, although limited observations of both longshore and cross-shore ground water variations have been reported. The elevation of the beach water table depends on prevailing hydrodynamic conditions such as tidal elevation, wave run-up and rainfall, and characteristics of the beach sediment that determine hydraulic conductivity, such as sediment size, sediment shape, sediment size sorting, and porosity. Observations of beach water table behavior show that the water table surface is generally not flat. Several authors have showed that the slope of the water table changes with the tide, sloping seaward on a falling tide and landward on a rising tide. The slope of the water surface has been found to be steeper on a rising tide than on a falling tide. Other researchers have measured water table elevations with a humped shape, with the hump near the run-up limit. Water table oscillations have been shown to lag behind tidal oscillations. Observed water table elevations are asymmetrical, as the water table rises abruptly and drops off slowly compared to the near-sinusoidal tide which drives it. For a given geometry, the lag in water table response is due mainly to the hydraulic conductivity of the beach sediment. With increasing distance landward, the lag between the water table and the tide increases and the amplitude of the water table oscillations decreases. However, Raubenheimer *et al.* (1999) found that fluctuations at spring-neap frequencies are attenuated less than fluctuations at diurnal or semidiurnal frequencies. Wave run-up, tidal variation, and rainfall may produce a *superelevation*, or *overheight*, of the beach water table by raising the elevation of the beach water table above the elevation of the tide.

Many researchers have observed that the beach acts as a filter that only allows the larger or longer period swashes to pass. Both the amplitude and the frequency of the ground water spectrum decrease in the landward direction. The further landwards the given ground water spectrum, the narrower its band and the more it is shifted toward lower frequencies. Infiltration into the beach also acts to reduce the amplitude and frequency of the input swash energy. Based on observations such as these, the beach has often been described as a low-pass filter (meaning that only lower frequency oscillations are transmitted through the beach matrix). High-frequency, small waves are damped and their effect is limited to the immediate vicinity of the intertidal beachface slope, whereas low-frequency waves can propagate inland. Comparison of run-up and ground water spectra shows a considerable reduction in dominant energy and also a shift in dominant energy toward lower frequencies.

Decoupling between the tide and the beach water table occurs when the ground water exit point becomes separated from the shoreline (shown in Figure H26). This occurs because the rate at which the beach drains is less than the rate at which the tide falls, so the tidal elevation generally drops more rapidly than the water table elevation and decoupling occurs, with the water table elevation higher than the tidal elevation. The *exit point* is the position on the beach profile where the decoupled water table intersects the beachface. After decoupling occurs, the position of the exit point is independent of the MWS until it is overtopped by the rising tide. Below the exit point, a *seepage face* develops where the water table coincides with the beachface. The seepage face is distinguished by a glassy surface. The seepage face is different from the water table in that its shape is determined by beach topography. However, water on the seepage face is at atmospheric pressure, as is water on the water table. The extent of the seepage face depends on the tidal regime, the hydraulic properties of the beach sediment, and the geometry of the beachface; thus the degree of asymmetry in water table response will vary between beaches. The exit point is generally assumed to mark the boundary between a lower section of the beach which is saturated and an upper section which is unsaturated; however, this assumption is probably an oversimplification.

Modeling beach ground water dynamics

An *aquifer* is a saturated geologic unit that can transmit significant quantities of water under ordinary hydraulic gradients (Freeze and Cherry, 1979). An *unconfined aquifer*, or *water table aquifer*, is one in which the water table forms the upper boundary. Beach ground water systems are generally treated as unconfined aquifers because commonly the upper boundary to ground water flow is defined by the water table itself rather than by some surface layer of impermeable material (Masselink and Turner, 1999). The beach ground water system is underlain by an impermeable boundary at a depth which is often unknown. The rate of flow (or *specific discharge*) of water through unconfined aquifers, u, is given by Darcy's Law:

$$u = -K\frac{\partial h}{\partial z} \qquad \text{(Eq. 1)}$$

where h is hydraulic head (units of length, L), z is the vertical coordinate (L), and K is hydraulic conductivity (LT^{-1}).

Darcy's Law is valid as long as flow is laminar, which is a reasonable assumption for sand beaches; however, this may not be the case for gravel beaches. Darcy's Law shows that the rate of ground water flow is proportional to the hydraulic gradient, or slope of the water table. The *hydraulic gradient* (dh/dz) is the change in hydraulic head (h) over distance; in this case, the change in elevation (z). Water flows down the hydraulic gradient in the direction of decreasing head. The *hydraulic head* (h) is the sum of the elevation head (z) and the pressure head (ψ), and is measured in length units above a datum. There is no standard datum that is used in beach hydrology, but many researchers use the elevation of an impermeable layer beneath the beach sediment, so that the vertical coordinate z is measured from the impermeable base. Some workers have considered the hydraulic head in a beach ground water system to be the elevation of the free water surface, or water table elevation. However, this is only true when there is no vertical component to the flow; in other words, when Dupuit–Forcheimer conditions apply (see below). *Hydraulic conductivity*, K, may be defined as the specific discharge per unit hydraulic gradient. The hydraulic conductivity reflects the ease with which a liquid flows and the ease with which a porous medium permits the liquid to pass through it, and relates the mean discharge flowing through a porous substance per unit cross section to the total gravitational and potential force. Hydraulic conductivity has units of velocity, usually ms^{-1} in the case of beach ground water systems. Hydraulic conductivity should be distinguished from *permeability* (also referred to as *intrinsic* or *specific permeability*), denoted by k, which is the measure of the ability of a rock, soil, or porous substance to transmit fluids and refers only to the characteristics of the porous medium and not to the fluid which passes through it. Permeability has dimensions of L^2.

Ground water hydrologists generally model water flow using Darcy's Law in combination with an equation of continuity that describes the conservation of fluid mass during flow through a porous medium. A common approach to modeling beach ground water flow in response to tidal forcing in sandy beaches uses the one-dimensional (1-D) form of the Boussinesq equation:

$$\frac{\partial h}{\partial t} = \frac{K}{s}\frac{\partial}{\partial x}\left(h\frac{\partial h}{\partial x}\right) \qquad \text{(Eq. 2)}$$

where h is the elevation of the water table (L), t is time (T), K is hydraulic conductivity (LT^{-1}), s is the specific yield (dimensionless), and x is horizontal distance (L). The *specific yield*, which is also known as the *drainable porosity*, is defined as the volume of water that an unconfined aquifer releases from storage per unit surface area of aquifer per unit decline in water table (Freeze and Cherry, 1979).

It should be noted that specific yield, or drainable porosity, is not the same as porosity, and the two terms should not be used interchangeably. *Porosity* is the volume of the voids in a sediment or rock divided by the total volume of the sediment or rock. Porosity is denoted by n, and is usually reported as a decimal fraction or percent. The *volumetric water content*, θ, is defined as the volume of water in a sediment or rock sample divided by the total volume of the sediment or rock. In saturated conditions, where the pores are filled with water, the volumetric water content, θ, is equal to the porosity, n. In unsaturated conditions, where the pores are only partially filled with water, the volumetric water content, θ, is less than the porosity, n.

There is also a difference between specific yield and *specific storage*, which is defined as the volume of water that a confined aquifer releases from storage under a unit decline in hydraulic head (Freeze and Cherry, 1979). The term specific storage refers to a unit decline in hydraulic head below the water table, in an aquifer which is saturated. Releases from storage in unconfined aquifers (such as beach sediments) represents an actual dewatering of the pores, whereas releases from storage in confined aquifers represent only the secondary effects of water expansion and aquifer compaction caused by changes in the fluid pressure (Freeze and Cherry, 1979).

The main assumption in using equation 2 is that ground water flow in a shallow aquifer can be described using the Dupuit–Forchheimer approximation. Dupuit–Forchheimer theory states that in a system of shallow gravity flow to a sink when the flow is approximately horizontal, the lines of equal hydraulic head or potential are vertical and the gradient of hydraulic head is given by the slope of the water table. In effect, the theory neglects the vertical flow components. Using Dupuit–Forchheimer theory, two-dimensional (2-D) flow to a sink can be approximated as 1-D flow, and the resulting differential equation (equation 2) is relatively easily solved. In beaches which are underlain by relatively impermeable material it is likely that Dupuit–Forchheimer theory provides an adequate description of ground water flow, and field studies such as those of Baird *et al.* (1998) and Raubenheimer *et al.* (1999) support this assumption.

Where Dupuit–Forchheimer assumptions do not apply, for example, in artificially drained beaches, the beach aquifer should be considered as a 2-D flow system. One approach is to assume that the water table is a free surface or flow line so that

$$\frac{\partial h}{\partial t} = \frac{K}{s}\left(\frac{\partial H}{\partial z} - \frac{\partial h}{\partial x}\frac{\partial H}{\partial x}\right) \quad \text{(Eq. 3)}$$

where H is the total or hydraulic head (L) and z is vertical distance (L). As in equation 2, h is the elevation of the water table (L), t is time (T), K is hydraulic conductivity (LT^{-1}), s is the specific yield (dimensionless), and x is horizontal distance (L). Equation 2 is much easier to solve than equation 3 and should be used whenever the assumption of near-horizontal flow through the beach sand is generally met.

Several analytical and numerical models have been developed which are able to predict beach water table fluctuations in response to tides (Nielsen, 1990; Turner, 1993b,c; Li *et al.*, 1997a; Baird *et al.*, 1998; Raubenheimer *et al.*, 1999). These Boussinesq models, based on solutions to equation 2, have been successful in reproducing observed fluctuations of the beach water table at diurnal and higher tidal frequencies, and also reproduce observations such as the shape and slope of the beach water table, the lag and landward attenuation of beach water table oscillations, and seepage face development. However, these models generally underpredict the water table elevations under conditions when wave effects are important. Models of beach water table fluctuations that incorporate wave effects have been developed only very recently. Nielsen *et al.* (1988) proposed the use of a linearized version of the Boussinesq equation (equation 2) with an additional term to model water table fluctuations in the zone of run-up infiltration:

$$\frac{\partial h}{\partial t} = \frac{Kd_a}{s}\frac{\partial^2 h}{\partial x^2} + U_1(x, t) \quad \text{(Eq. 4)}$$

where d_a is the aquifer depth and $U_1(x, t)$ is the infiltration/ exfiltration velocity per unit area. As in equation 2, h is the elevation of the water table (L), t is time (T), K is hydraulic conductivity (LT^{-1}), s is the specific yield or drainable porosity (dimensionless), and x is horizontal distance (L). Li *et al.* (1997b) and Li and Barry (2000) have developed more complicated models to predict wave-induced water table fluctuations; however, none of the models which include wave effects have yet been tested against field or laboratory data.

Finally, beach ground water models have not yet been linked to swash hydrodynamic and sediment transport models, although Turner (1995) modeled beach profile response to ground water seepage using an equilibrium net transport parameter. In particular, models of swash–ground water interactions do not yet incorporate the physical processes such as infiltration and ground water outflow which are thought to influence sediment transport in the swash zone. The relative importance of these mechanisms is where the greatest areas of uncertainty arise.

Mechanisms of surface–subsurface flow interaction and implications for sediment transport

Several mechanisms have been suggested to explain why beaches with a low water table tend to accrete and beaches with a high water table tend to erode. The mechanisms which are proposed most frequently are infiltration and exfiltration. The terminology used to discuss these mechanisms requires some clarification, as different terms may be used by hydrologists, engineers, and other coastal scientists. The physical process of interest is that of vertical flow within a porous bed and/or through a permeable boundary. Vertical flow exerts a force within the bed called *seepage force*, which is defined as a force acting on an individual grain in a porous medium under flow, which is due to the difference in hydraulic head between the front and back faces of the grain (Freeze and Cherry, 1979). The seepage force, F, is exerted in the direction of flow and is directly proportional to the hydraulic gradient, and is given by

$$F = \rho g \frac{\partial h}{\partial z} \quad \text{(Eq. 5)}$$

where ρ is the density of the fluid (ML^{-3}), g is acceleration due to gravity (LT^{-2}), and $\partial h/\partial z$ is the hydraulic gradient (dimensionless). In the convention used here, a positive hydraulic gradient represents a downward-acting seepage force and a negative hydraulic gradient represents an upward-acting seepage force.

The vertical flows which produce this seepage force have been referred to in a number of different ways in the literature: bed ventilation, suction and blowing, piping, seepage erosion, ground water sapping, etc. In the case of beach hydrology, however, the terms infiltration and exfiltration are most commonly used. *Infiltration* is the process by which water enters into the surface horizon of a soil or porous medium, such as beach sediment, in a downward direction from the surface by means of pores or small openings. Infiltration is often used interchangeably with *percolation*, which more correctly refers to the flow of water through a soil or porous medium below the surface. Recently, the term *exfiltration* has been used to describe outflow from the bed. Infiltration/exfiltration velocity may also be referred to as seepage velocity.

Grant (1946, 1948) was among the first to suggest a link between beach ground water behavior and swash zone sediment transport, proposing a simple conceptual model which has been highly influential in beach hydrology research. Grant defined a dry foreshore as one with a low water table and an extensive infiltration zone. On a dry foreshore, most of the water infiltrates rapidly into the sand above the water table, which reduces the flow depth of the swash and thus the velocity, allowing sediment deposition. Grant's conceptual model also described conditions on a wet foreshore, one whose water table is high and contiguous with the surface of most of the foreshore. He reasoned that when the beach is in a saturated condition throughout all of the foreshore the backwash, instead of being reduced by infiltration, retains its depth and is augmented by the addition of water rising to the surface of what he called the effluent zone (the seepage face). Grant also noted that ground water outcropping at the beach surface can cause dilation or fluidization of the sand grains, allowing them to be entrained more easily by backwash flows.

The logic of Grant's conceptual model has led many researchers to concentrate on the effects of infiltration losses on beach accretion and erosion, suggesting that infiltration losses during swash provide the main mechanism by which beach accretion occurs above the SWL. Because the swash and backwash are relatively shallow, a small change in water volume due to infiltration could significantly decrease the energy available for sediment transport. Within the swash zone, rapid water table fluctuations due to swash infiltration into the capillary fringe may also influence sediment mobility. Ground water flow at deeper levels within the beach is also influenced by infiltration during swash uprush, although the hydraulic gradients developed tend to be small.

Although most researchers have concentrated on infiltration/ exfiltration and possible effects on swash/backwash asymmetry, Nielsen (1992) suggested that vertical flow within the beach alters the sediment transport characteristics due to a modification of the effective weight of the sediment, which will act to stabilize the bed under infiltration or destabilize under exfiltration. Turner and Nielsen (1997) identified several mechanisms by which vertical flow through a porous bed could affect swash zone sediment transport, including an alteration in the effective weight of the surface sediment due to vertical fluid drag and modified shear stresses exerted on the bed due to boundary layer thinning due to infiltration or thickening due to exfiltration. Turner and Masselink (1998) identified a number of effects of vertical flow through a porous bed: the angle of attack at which the main flow contacts the particles is altered; dead water is flushed out of the top bed layer, increasing the exposed surface area of a particle to the main flow; and the changed wake behind a particle not only affects that particle but others in its lee. Turner and Masselink (1998) summarized the effect of these processes on the boundary layer, with stream lines being drawn closer to the sediment–fluid interface under infiltration and moved away from the sediment–fluid interface under exfiltration. The result is a vertical shift of the boundary layer velocity profile, with an increase of shear stress at the bed under infiltration and a decrease under exfiltration.

Experimental work on the influence of seepage flows within sediment beds provides conflicting results concerning the effect on bed stability. These contradictory results may be because the effects of seepage force and boundary layer thinning tend to oppose each other. While infiltration results in a stabilizing seepage force, simultaneous boundary layer thinning has the opposing effect of enhancing sediment mobility and *vice versa* for exfiltration (Hughes and Turner, 1999). The relative importance of these opposing effects depends on the density of the sediment and the permeability of the bed (Nielsen *et al.*, 2001). Although recent work by Baldock and Holmes (1998) showed that sediment transport over a fluidized bed in the presence of a steady current may be little different from that over a normal sediment bed, they also suggested that a seepage flow might have a significant effect on sediment transport during sheet flow. In their experiments, the bulk motion of a top layer which was many grain diameters thick was sometimes observed during exfiltration or no seepage, but was suppressed by infiltration. Sheet flow conditions are likely to occur during backwash and probably also during the uprush.

Nielsen (1997) proposed a revised Shields parameter that includes the effects of infiltration/exfiltration:

$$\theta_m = \frac{u_{*0}^2(1 - \alpha(w/u_{*0}^2))}{gd_{50}(s - 1 - \beta(w/K))} \qquad \text{(Eq. 6)}$$

where w is the seepage velocity (LT^{-1}, with infiltration negative), u_{*0}^2 is the shear velocity without seepage (LT^{-1}), s is relative density (dimensionless: ρ_s/ρ, where ρ_s is the density of the sediment and ρ is the density of the fluid), K is hydraulic conductivity (LT^{-1}), g is acceleration due to gravity (LT^{-2}), d_{50} is median grain diameter, and α and β are constants, defined by Neilsen *et al.* (2001) as 16 and 0.4, respectively. The factor β is intended to quantify the increase of the particle's weight due to the vertical seepage velocity, and is 1 for particles in the bed but considerably smaller for particles on the surface (Nielsen *et al.*, 2001). The modified Shields parameter in equation 6 was designed to account for the opposing effects of infiltration, as the extra term in the numerator represents the increase in shear stress due to the thinning of the boundary layer and the extra term in the denominator represents the effect of the downward seepage drag on the effective weight of the grains (Nielsen *et al.*, 2001). Equation 6 suggests that for a fixed sediment density, as grain size (and therefore, hydraulic conductivity) decreases, the stabilizing effect will increase. Therefore, finer quartz sands ($d_{50} < 0.58$ mm) are likely to be stabilized by infiltration, whereas the net effect of infiltration on beaches of coarser sediment may be destabilizing (Nielsen, 1997). Nielsen *et al.* (2001) extended this analysis to show that infiltration is likely to enhance sediment mobility for dense, coarse sediment where $\alpha(s-1) > \beta(u_{*0}/K)$ and impede sediment motion for light, fine sediment where $\alpha(s-1) < \beta(u_{*0}/K)$.

Turner and Masselink (1998) also followed this approach, but included the effects of the seepage flow on the bed shear stress. They used their modified Shields parameter, which incorporated an additional through-bed term, to calculate the swash-zone transport rate in the presence of infiltration/exfiltration relative to the case of no vertical flow through the bed. Their modeling showed that altered bed stresses dominated during uprush, indicating enhanced sediment mobility relative to the case of an impermeable bed. They found that altered bed stress effects were also dominant during backwash; however, the net effect of combined seepage force and altered bed stress was less pronounced during backwash than during uprush. Turner and Masselink (1998) concluded that the effects of combined seepage force and altered bed stress enhanced net onshore sediment transport on a saturated beachface.

Nielsen *et al.* (2001) conducted laboratory measurements to investigate the effects of infiltration on sediment mobility of a horizontal sand bed under regular nonbreaking waves under conditions of steady downward seepage, and compared these to measurements without infiltration. Their experiments showed that infiltration had the effect of reducing the mobility of 0.2 mm sand, and they suggested that infiltration effects on sediment mobility in the swash zone would be minor if infiltration rates are in the range reported by other researchers, where $w < 0.15K$.

Although infiltration and exfiltration are the primary mechanisms by which ground water flow is thought to influence sediment transport in the swash zone, the potential of beach ground water fluctuations to cause bed failure due to instantaneous fluidization has also been considered. *Fluidization* of sediment occurs when the upward-acting seepage force exceeds the downward-acting immersed particle weight. In particular, it has been suggested by a number of workers that tidally induced ground water outflow from a beach during the ebb tide may enhance the potential for fluidization of sand, and thus the ease with which sand can be transported by swash flows. However, tidally induced ground water outflow alone is unlikely to be sufficient to induce fluidization, because hydraulic gradients under the sand surface will tend to be relatively small, generally of the order of the beach slope (1 : 100–1 : 10). In addition, Turner and Nielsen (1997) found that, rather than fast water table rise in the swash being the cause of upward flow (and hence potential fluidization), rapid water table rise within the swash zone resulted from a small amount of infiltration of the swash lens. However, upward-acting swash-induced hydraulic gradients which are capable of fluidizing the bed have been measured within the top few centimeters of the beach. Horn *et al.* (1998) presented field measurements of large upward-acting hydraulic gradients which considerably exceeded the fluidization criterion, which occurs when the upward-acting hydraulic gradient is greater than (i.e., more negative than) about −0.6 to −0.7 (in the convention used here). The mechanism responsible for these upward-acting hydraulic gradients is not clear.

Baird *et al.* (1996) argued that fluidization is only generally possible in the presence of swash on a seepage face. As a swash flow advances over the saturated beach surface there will be a rapid increase in pore water pressures below the beach surface. When under swash flow, the beach sediment behaves like a confined aquifer. The sediment is saturated and movement of water into the beach is extremely limited since changes in porosity due to expansion and contraction of the mineral "skeleton" will be minimal. However, water pressures will propagate rapidly through the sediment. As the swash retreats there will be a release of pressure on the beachface, potentially giving large hydraulic gradients acting vertically upwards immediately below the surface. The resultant seepage force associated with these upward-acting hydraulic gradients could be sufficient to induce fluidization of the sand grains at the surface. They showed theoretically how hydraulic gradients in the saturated sediment beneath swash can exceed, or at least come close to, the threshold for fluidization.

Baldock *et al.* (2001) compared field measurements of swash-induced hydraulic gradients in the surface layers of a sand beach to the predictions of a simple (1-D) diffusion model based on Darcy's Law and the continuity equation. The model allows for dynamic storage (within the sediment–fluid matrix) due to loading/unloading on the upper sediment boundary. The model predicted minimal hydraulic gradients for a rigid, near fully saturated sediment which were in accordance with measurements close to the seaward limit of the swash zone. The model also provided a good description of the measured hydraulic gradients, both very close to the surface and deeper in the bed, for the region of the beach where the beach surface is frequently exposed between swash events. These model-data comparisons suggest that the surface layers of a sand beach store and release water under the action of swash, leading to the generation of relatively large hydraulic gradients as suggested by Baird *et al.* (1996). However, the model was not able to predict the very large near-surface negative hydraulic gradients observed by Horn *et al.* (1998), although, for the same swash events, the agreement was good deeper in the bed. Baldock *et al.* (2001) concluded that the very large upward-acting hydraulic gradients observed in the upper part of the bed were not simply due to pressure propagation during swash loading/unloading or swash-generated 2-D subsurface flow cells. Instead, they suggested that these very large negative hydraulic gradients are probably generated by alternative mechanisms; possibly due to non-hydrostatic pressures developing within the sheet flow layer that occurs during backwash.

The implications of these hydraulic gradients for sediment transport are not clear. Vertical seepage forces are not themselves capable of transporting sediment; however, this process may act to provide readily entrainable material which is then available for transport, onshore under uprush or offshore under backwash. Nielsen *et al.* (2001) noted that their experiments indicated only the effect of infiltration/exfiltration on sediment mobility and did not necessary suggest anything about the direction of net sediment transport. This is likely to be affected by other factors such as the phase relationship between infiltration/ exfiltration-induced effects on sediment transport and swash flows. For example, Blewett *et al.* (1999) measured events where large upward-acting hydraulic gradients occurred when the head of water at the surface, and therefore, the uprush or backwash flow, was zero. Under these conditions, even if the sediment were to be fluidized, it would not be transported. However, in other data sets, Blewett *et al.* (1999) reported measurements with upward-acting hydraulic gradients of −1.7, which were more than sufficient to fluidize the bed. These hydraulic gradients lasted for approximately 4 s in waves with a period of 6.3 s under a falling head of water, initially as deep as 40 mm, and under offshore-directed flows of 0.7–1.4 ms^{-1}. This suggests a possible erosional mechanism under backwash. Clearly, the phasing between these potentially destabilizing hydraulic gradients and swash flows is critical to the potential for sediment transport.

Nielsen *et al.* (2001) argued that if the beachface tends to be fluidized during backwash as suggested by Horn *et al.* (1998), a mechanism must exist to enhance sediment transport during the uprush in order to balance this effect—otherwise the beach would rapidly disappear. They suggested that this balancing effect might be delivered by fluidization due to strong horizontal pressure gradients near bore fronts. However, Hughes *et al.* (1997) suggested an alternative mechanism, arguing that onshore transport in the uprush is likely to be significantly influenced by turbulence and sediment advection from bores arriving at the beachface. The lack of a clear mechanism for onshore transport highlights the complexity of sediment transport processes in the swash zone, as the exact nature of the relationship between swash flows, beach ground water, and cross-shore sediment transport is not yet known.

For another aspect of this discussion see Otvos (1999).

Diane Horn

Bibliography

Baird, A.J., Mason, T.E., and Horn, D.P., 1996. Mechanisms of beach ground water and swash interaction. In *Proceedings of the 25th International Conference on Coastal Engineering*, American Society of Civil Engineers, pp. 4120–4133.

Baird, A.J., Mason, T.E, and Horn, D.P., 1998. Validation of a Boussinesq model of beach ground water behaviour. *Marine Geology*, **148**: 55–69.

Baldock, T.E., and Holmes, P., 1998. Seepage effects on sediment transport by waves and currents. In *Proceedings of the 26th International Conference on Coastal Engineering*, American Society of Civil Engineers, pp. 3601–3614.

Baldock, T.E., Baird, A.J., Horn, D.P., and Mason, T., 2001. Measurements and modelling of swash-induced pressure gradients in the surface layers of a sand beach. *Journal of Geophysical Research*, **106**(C2): 2653–2666.

Blewett, J.C., Holmes, P., and Horn, D.P., 1999. Measurement and modelling of swash hydrodynamics. In *Proceedings of Coastal Sediments '99*, American Society of Civil Engineers, pp. 377–392.

Freeze, R.A., and, Cherry, J.A., 1979. *Groundwater*. Englewood Cliffs, NJ: Prentice Hall.

Grant, U.S., 1946. Effects of groundwater table on beach erosion. *Geological Society American Bulletin*, **57**: 1952 (Abstract).

Grant, U.S., 1948. Influence of the water table on beach aggradation and degradation. *Journal of Marine Research*, **7**: 655–660.

Horn, D.P., Baldock, T.E., Baird, A.J. and Mason, T., 1998. Field measurements of swash induced pressure gradients within a sandy beach. In *Proceedings of the 26th International Conference on Coastal Engineering*, American Society of Civil Engineers, pp. 2812–2825.

Hughes, M.G., and Turner, I.L., 1999. The beachface. In Short, A.D. (ed.), *Handbook of Beach and Shoreface Morphodynamics*. Chichester: John Wiley & Sons, Ltd, pp. 119–144.

Hughes, M.G., Masselink, G., Hanslow, D., and Mitchell, D., 1997. Towards a better understanding of swash zone sediment transport. In *Proceedings of Coastal Dynamics '97*, American Society of Civil Engineers, pp. 804–813.

Li, L., and Barry, D.A., 2000. Wave-induced beach groundwater flow. *Advances in Water Resources*, **23**: 325–337.

Li, L., Barry, D.A., and Pattiaratchi, C.B., 1997a. Numerical modelling of tide-induced beach water table fluctuations. *Coastal Engineering*, **30**(1/2): 105–123.

Li, L., Barry, D.A., Parlange, J.-Y., and Pattiaratchi, C.B., 1997b. Beach water table fluctuations due to wave run-up: capillarity effects. *Water Resources Research*, **33**: 935–945.

Masselink, G., and Turner, I.L., 1999. The effect of tides on beach morphodynamics. In Short, A.D. (ed.), *Handbook of Beach and Shoreface Morphodynamics*. Chichester: John Wiley & Sons, Ltd, pp. 204–229.

McArdle, S., and McLachlan, A., 1991. Dynamics of the swash zone and effluent line on sandy beaches. *Marine Ecology Progress Series*, **76**: 91–99.

Nielsen, P., 1990. Tidal dynamics of the water table in beaches. *Water Resources Research*, **26**: 2127–2134.

Nielsen, P., 1992. *Coastal Bottom Boundary Layers and Sediment Transport*. Singapore: World Scientific.

Nielsen, P., 1997. Coastal groundwater dynamics. In *Proceedings of Coastal Dynamics '97*, American Society of Civil Engineers, pp. 546–555.

Nielsen, P., Davis, G.A., Winterbourne, J.M., and Elias, G., 1988. Wave setup and the water table in sandy beaches. New South Wales Public Works Department, Coastal Branch, *Technical Memorandum* T.M. 88/1.

Nielsen, P., Robert, S., Møller-Christiansen, B., and Oliva, P., 2001. Infiltration effects on sediment mobility under waves. *Coastal Engineering*, **42**(2): 105–114.

Otvos, E.G., 1999. Rain-induced beach processes; landforms of ground water sapping and surface runoff. *Journal of Coastal Research*, **15**: 1040–1054.

Raubenheimer, B., Guza, R.T., and Elgar, S., 1999. Tidal watertable fluctuations in a sandy ocean beach. *Water Resources Research*, **35**: 2313–2320.

Schoones, J.S., and Theron, A.K., 1995. Evaluation of ten cross-shore sediment transport/morphological models. *Coastal Engineering*, **25**: 1–41.

Turner, I.L., 1993a. Beach face permeability, the groundwater effluent zone, and intertidal profiles of macro-tidal beaches: a conceptual model. In Thomas, M. (ed.), *Catchments and Coasts of Eastern Australia*. Department of Geography, University of Sydney Monograph Series 5, pp. 88–99.

Turner, I.L., 1993b. The total water content of sandy beaches. *Journal of Coastal Research*, **15**(Special issue): 11–26.

Turner, I.L., 1993c. Water table outcropping on macro-tidal beaches: a simulation model. *Marine Geology*, **115**: 227–238.

Turner, I.L., 1995. Simulating the influence of groundwater seepage on sediment transported by the sweep of the swash zone across the intertidal profile of macrotidal beaches. *Marine Geology*, **125**: 153–174.

Turner, I.L., and Nielsen, P., 1997. Rapid watertable fluctuations within the beachface: implications for swash zone sediment mobility? *Coastal Engineering*, **32**: 45–59.

Turner, I.L., and Masselink, G., 1998. Swash infiltration–exfiltration and sediment transport. *Journal of Geophysical Research*, **103**(C13): 30,813–30,824.

Cross-references

Beach Drain
Coastal Wells
Cross-Shore Sediment Transport
Depth of Disturbance
Submarine Groundwater Discharge
Surf Zone Processes
Tides
Waves

I

ICE-BORDERED COASTS

Introduction

Since the publication of *The Encyclopedia of Beaches and Coastal Environments* (Schwartz, 1982), the term "cold coasts" has come into common use even serving as a chapter title in the book *Coastal Problems*. The authors, Viles and Spencer (1995), use the 1961 definition by R.L. Nichols that cold coasts ". . . are those where there is or has been abundant sea ice, lake ice, water-terminating glaciers or deeply frozen ground" (p. 254). The advantage of such a definition is that it avoids the latitudinal restriction placed by such locational designators as Arctic and Antarctic and thus can accommodate lower latitudinal examples including the tidewater glaciers of Chile and southern Alaska and the presence of sea ice along the Labrador and Hokkaido coasts or even the coast of Spain during the Pleistocene.

This entry treats two of these types of cold coasts: namely, water-terminating glaciers and sea ice. Glacial ice is land-derived and tends to be perennial; sea ice, on the other hand, develops seaward of the coastline and is usually seasonal. They both can serve as erosional, transportational, and depositional agents along coastlines, although the rates and intensities of their action vary greatly between the two and with time and location.

Glacial ice and the coast

One of the conspicuous features of the landscape at present and during much of the past three million years is glacial ice. Although today it is dominant (as a coastal feature) only in Antarctica, Greenland, and a few smaller islands in high latitudes, it is still sufficiently abundant to affect more than 35,000 km of the world's coastline. Glacial ice impacts the coastline in several ways including burying and depressing the coast as it moves across it into the sea, flowing in preexisting valleys into the sea as tidewater glaciers, and replacing the traditional shore with an ice margin that then interfaces directly with the sea.

Ice sheets, such as those in Antarctica and Greenland, serve as the source of ice that moves under its own weight and in response to gravity from inland to sea. Glacial ice is so extensive in Antarctica that only about 5% of the coastline is ice-free (Figure I1). Greenland contrasts with most of Antarctica in that glacial ice does not dominate the coastal zone but flows through relatively few high passes in the coastal mountains (Figure I2). A major exception is where inland ice reaches Melville Bay along a 460 km front.

A further characteristic of ice sheets is that their great weight depresses the landmasses upon which they form. Near the coastline the depressed continental shelf slopes downward inland thereby reversing the general slope of the coastal zone. Large parts of Antarctica and Greenland have been depressed to such extents that their present elevation is below sea level. Although, with rebound, much of this depressed land would be above sea level, large areas beneath the West Antarctic Ice Sheet (which averages 440 m below sea level) would still be a series of islands with lengthy coastlines even after the rebound that would accompany deglaciation (Paterson, 1994).

Ice shelves, which are large ice masses floating on the sea, range in thickness up to several hundred meters. They are especially common in Antarctica (Figure I1), but also occur as small shelves in Greenland, Ellesmere Island, and Franz Josef Land. The ice shelves of Antarctica fill many of the continent's embayments giving the continent a nearly circular form excepting the northward trending Antarctic Peninsula.

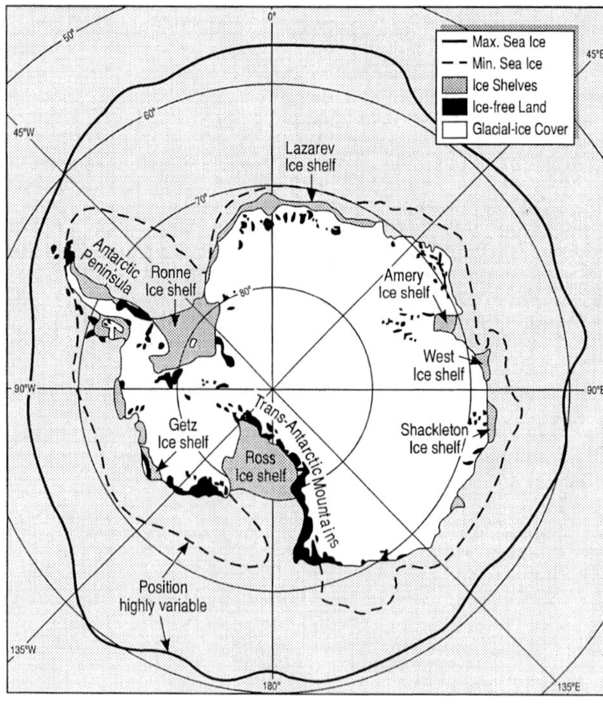

Figure I1 Ice and the Antarctic coastline (modified from Mellor, 1964; Paterson, 1994; and Crossley, 1995).

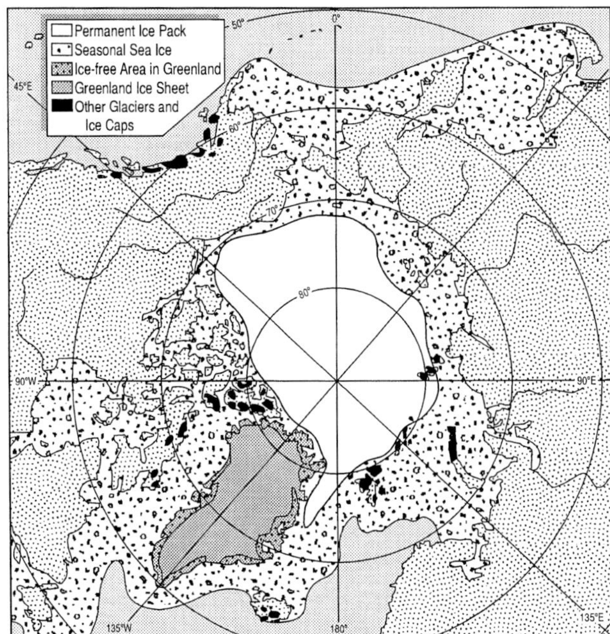

Figure I2 Ice and the Northern Hemisphere Coastline. Note that within the ice-free coastal area of Greenland there are numerous small ice caps (modified from Mellor, 1964; Field, 1975; and Williams and Ferrigno, 1995).

Figure I3 Ice cliff, Antarctica. (Williams and Ferrigno, 1995. Photograph by Charles Swithinbank, courtesy of Richard S. Williams, United States Geological Survey.)

Thus, the 11,000 km (more than one-third of the total) of Antarctica's coastline occupied by ice shelves, possesses ice cliffs (Figure I3) at the ocean interface. These ice cliffs are impressive coastal features as illustrated by the Ross Ice Shelf (Figure I1) where the front edge is more than 200 m thick with a 20–30 m cliff above water. The ice landward of the Ross ice cliff thickens to 700 m at land's edge (Robe, 1980). The exposed ends of these floating glaciers are impacted by the same agents and processes as the coast proper including waves, tides, currents, and sea ice.

Many of the coastal glaciers in Antarctica (~38%) and most in the Northern Hemisphere do not have floating shelves or tongues but are grounded at their termini (Powell and Domack, 1995). For example, part of the west coast of the Antarctic Peninsula has glaciers that have formed ice walls at the shore because of undercutting and calving at the ice front where they override a gravelly beach (Robin, 1979).

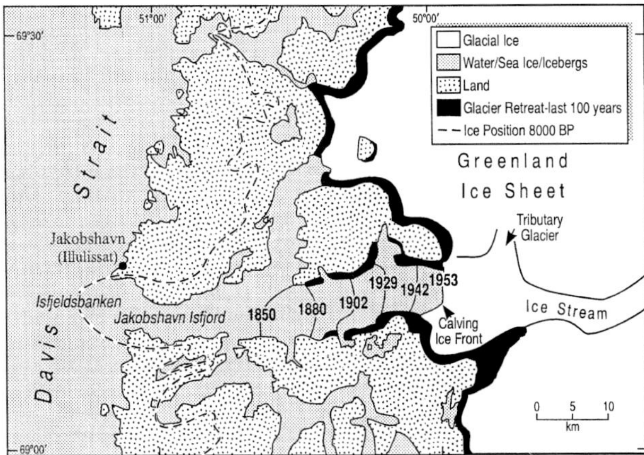

Figure I4 Tidewater glacial retreat (1850–1953) and ice-margin retreat in Jakobshavn Isfjord (modified from Williams and Ferrigno, 1995).

Tidewater glaciers are found in nonpolar areas such as southern Alaska and Chile as well as at higher latitudes. In the Arctic, tidewater glaciers are present in Jan Mayen, Svalbard, Novaya Zemlya, Severnaya Zemlya, Ellesmere, Baffin Island, Bylot Island, and Devon Island as well as Greenland and Alaska (Figure I2). The termini of temperate tidewater glaciers are grounded as are many in higher latitudes (Hambrey and Alean, 1992). Most tidewater glaciers are confined in fjords of variable length and terminate at some distance inland from their mouth. Many are very long, such as the combined 350 km long Nordvestfjord and Scoresby Sund in east Greenland. Such fjords not only have glaciers at their inner ends but also are icebound with icebergs and sea ice. Because of the nature of the coastal mountain rim around the Greenland Ice Sheet the coastal impact is highly varied at the front of outlet glaciers. Although many flow into the sea, others terminate on land (Figure I4) and in glacially created lakes (Warren, 1991).

A major characteristic of glaciers is their changeable rate of advance and retreat. Thus, their position and therefore their impact on the coastlines they border is continuously varying. Jakobshavn Isbrae (Figure I4) in west Greenland, for example, retreated 26 km between 1851 and 1953 (Williams and Ferrigno, 1995) exposing sizable sections of the fjord's shore that it formally bordered.

The most important mechanism for the loss of glacial ice, whether it be the large ice shelves of Antarctica or the smaller tidewater glaciers of Alaska, is calving at their termini. Calving produces icebergs that range in size from very small, as those produced in constricted fjords, to very large such as those tabular icebergs that calve from ice shelves. On March 17, 2000, iceberg B-15, with a length of 295 km and a width of 37 km calved from the Ross Ice Shelf. It broke into two parts a few months later (Lazzara *et al.*, 1999).

Icebergs, once formed, essentially become floating islands. The act of calving increases the number of sides in contact with the sea, hastening their disintegration. Many of them float for years as exampled by those ice islands that have been used as research bases in the Arctic Ocean. Others become trapped, even if temporarily, in sea ice and some become grounded in shallow water. They can also be erosional agents, often in association with the floating pack ice that forms around them, creating deep gouges in the nearshore bottom.

Glaciers, whether they terminate onshore, at the shore, along fjords or at some distance offshore, are major morphological agents. If terminating inland, meltwater drainage carries the sediment formed by the glacial scour that accompanies the advancing ice to the sea creating depositional facies along the shore. Those glaciers that terminate at the shore and those that have overridden the shore leave behind ice-scoured surfaces and depositional forms including a variety of morainal types. Such coasts, once released from their overburden of ice, rebound as is happening in northern Canada, Scandinavia, and many parts of Antarctica especially on the Antarctic Peninsula. Some of the raised beaches on the Peninsula are as much as 60 m above present sea level (Kirk, 1985).

Sea ice and the coast

Sea ice, one of the most variable elements in the oceanic system, varies seasonally in areal coverage by nearly 500% in the Southern Hemisphere but by less than 200% in the Northern Hemisphere (Figures I1 and I2).

These percentages show that sea ice is more strongly seasonal in the Antarctic than the Arctic. It is again a reflection of the nature of the two ocean areas involved. Because the Arctic Ocean is surrounded by land and the Antarctic Ocean surrounds a continent, ice formation and movement and therefore, impact on the coastlines is quite different. Most of the Arctic Ocean sea ice is of the multi-year type whereas most (~85%) of that of the Antarctic is first-year ice.

During the Arctic winter, sea ice not only affects the coasts surrounded by the Arctic Ocean but also extends south to locations such as Hokkaido, Japan in the North Pacific Ocean and New Foundland in the North Atlantic (Figure I2). In contrast, during summer virtually all Arctic coasts are free of sea ice for varying lengths of time. Exceptions include northern Greenland, Ellesmere, and part of the Canadian Arctic Archipelago where sea ice may last throughout the year. Of major interest and concern is the great variability in sea ice cover and thickness. An analysis of satellite passive microwave observations shows those areas exhibiting negative trends in the sea ice season are larger than those exhibiting positive trends (Parkinson, 2000). If such negative trends continue, ice-free periods along Arctic coasts will continue to lengthen.

During the austral winter, sea ice extends north several hundred kilometers from Antarctica even reaching 55°S in the Indian Ocean (Figure I1). During that period of the year, sea ice is present along the entire coastline although nearshore in many locations are polynyas—some maintained by katabatic winds. As summer approaches, sea ice drifts out and away from most of the coast except for a few locations where it remains attached to the shore throughout the summer (Wadhams, 2000).

Sea ice and its impact on coasts during summer

For a few months during summer most of the coasts in both the Antarctic and Arctic are ice-free. However, in the Arctic Ocean especially, the permanent (although highly mobile) pack is often close enough to shore to dampen waves and thus reduce their impact on the coast. During those periods of time when the pack retreats from the coast thereby increasing fetch over coastal waters, storms can cause severe erosion as happened at Barrow, Alaska in October 1986 (Walker, 1991). The shorter the ice-free period and the narrower the shore lead, the more limited the wave action alongshore.

Under certain wind and current conditions pack ice can move onto shore even at the height of the summer season causing some ice scour and sediment transport. This situation is especially true along the Beaufort Sea coast.

Freeze-up and the ice foot

The factors affecting the timing of freeze-up include temperature, wind, snow, waves, tides, and the nature of the shoreline. When the temperature is lowered to the value at which seawater freezes, ice forms on the foreshore and within the interstices of shore sediments. Ice buildup occurs through the addition of the spray, swash, slush ice, and ice floes brought by waves and tides plus the addition of snow. The accumulation becomes the ice foot, a major characteristic of ice-bordered shore (Figure I5). The form, structure, and extent of the ice foot varies with shore gradient as well as tidal range and wave conditions during formation (Taylor and McCann, 1983). A gently sloping shore face and a high tidal range favor the development of a wide ice foot whereas variable wave conditions often produce complex structures (Owens, 1982).

The ice foot, composed of a mixture of sediments, snow and ice, rests on a beach surface that is also frozen. As waves approach the ice foot, they continue to deliver sediment from offshore either as loose particles or as material already incorporated into the pancake and brash ices that are added to the ice foot mix (Evenson and Cohn, 1979).

The ice foot is bottomfast and immobile. At its seaward edge it abuts floating ice that moves vertically with tidal and wave action. They are separated by tidal cracks along a line which has been referred to as a "hinge zone" (Forbes and Taylor, 1994). The degree of roughness of the sea ice at the hinge zone tends to increase with increasing tidal range.

Once freeze-up is complete, wave action on the coast ceases and any sediment transport by longshore currents is confined to locations seaward of the bottomfast ice zone. However, exceptions do occur. In the case of the Beaufort Sea, for example, severe winter storms can produce override with ice being forced over the ice foot high up on the shore. These features are known among the Inuit as "ivu."

Although bottomfast ice usually extends out to water depths of two or more meters, shorefast ice extends out over deeper water to distances of as much as 20–30 km (Taylor and McCann, 1983) where it merges with drifting pack ice. Shorefast ice is relatively immobile especially when present between islands as in the Canadian Arctic Archipelago. However, along open coasts, as those facing the Beaufort and Chukchi Seas, shorefast ice may be subject to occasional drift. The area between the outer limit of shorefast ice and the drifting pack, known as the "stamukhi" zone, is characterized by large pressure ridges some large enough so that they last through the summer.

Ice melt and breakup

Although shore leads in the sea ice may open early during the breakup period, the ice foot is not directly affected. Its ice melts in place. The rate and timing depend mainly on temperature conditions and depending on the amount of ice and snow present, may last into summer (Figure I5). During winter, snow accumulates on the irregular surface and may be quite thick especially if there is a cliff behind the beach. As sea ice begins to move, especially with offshore winds, it breaks apart and floats the outer edges of the bottomfast ice. In the process, much of the sediment that has been incorporated into the sea ice is transported off and alongshore. The last ice to be removed from the coast are the large ice masses that become stranded and often buried onshore. Their melt rate is affected by the sediment that may cover them (Figure I6).

In contrast to the shore ice melt and breakup that occurs along the exposed shore is that occurring out from river mouths. Off river mouths the gradients of subaqueous deltas are usually more gentle than those along other coastlines so that the bottomfast sea ice zone is wider. During river breakup, floodwaters progress out over the sea ice and in the larger rivers beneath it in the subice distributary channels that do not freeze to the bottom. The water flowing over the sea ice deposits much of its sediment on top of the ice before it reaches pressure-ridge cracks or potholes out from the bottomfast/shorefast ice boundary where it drains to continue flowing seaward (Walker, 1974). The drainage through these holes (Figure I7) create in the bottom what Reimnitz has labeled strudel-scour holes (Reimnitz and Bruder, 1972). Most of the sediment deposited on top of the sea ice is later deposited in the delta as the ice melts. However, some of it is transported seaward and alongshore as the offshore ice begins to drift.

Figure I5 Stranded ice-foot at Barrow, AK in June.

Figure I6 Five-meter-high sea ice pileup with accumulated snow onshore during melt season. Note the gradual settling of incorporated beach materials as ablation occurs.

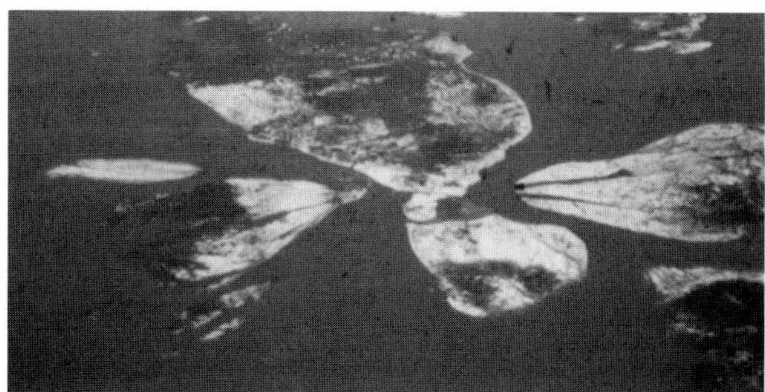

Figure I7 Drainage hole in sea ice at the Colville delta front. The floodwaters from the river drain from the ice creating strudel scour in deltaic sediments.

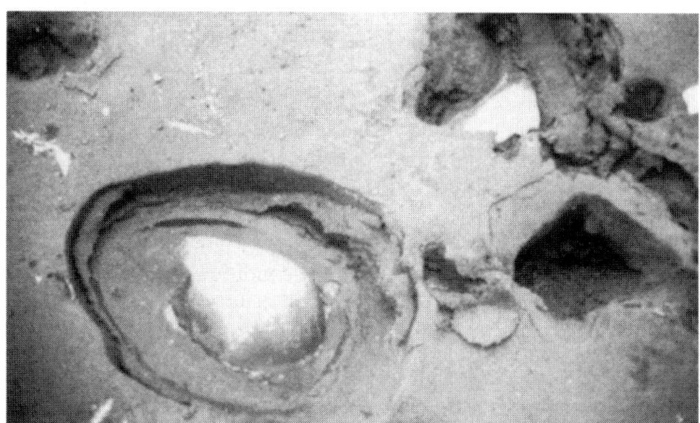

Figure I8 Small kettles forming in the beach at Wainwright, AK as stranded ice blocks melt.

Sea ice: its role in erosion, transportation, and deposition

In addition to the transport of sediment and the development of strudel mentioned above, sea ice is involved in other geologic processes along coastlines. The movement of sea ice on the beach and near the shore can produce scour marks and ridges. The uneven bottom of sea ice and especially when present as pressure-ridge keels gouge bottom materials as the ice drifts along the shore or up to the beach. As sea ice rides up onto the shore it both erodes and transports beach materials some distance above the high-tide shoreline. If the moving ice is impeded, ice pileups develop. They form not only at the high-tide shoreline but also at the hinge-line that joins bottom- fast and shorefast ice and around grounded floes. Such ice pileups impact heavily on both shore and nearshore sediments. Pileup heights of 20 m are common; some grow to double that (Forbes and Taylor, 1994).

The beach which can have a very irregular surface at the end of the melt season because of the presence of ice-push ridges, kettles left by the melting of stranded ice blocks (Figure I8), and a variety of scour forms is generally reworked by waves during the ice-free season. Only those forms that are especially large or that have been produced at the back of the beach may last through the summer.

Sea ice like glacial ice serves as both an agent of erosion and of protection. In the case of sea ice, the period of time it protects the coast from erosion varies from hours or days to year round in some very sheltered locations, whereas in the case of glacial protection the time period may be reckoned in millennia. With the rapid changes taking place in the icescapes of the world (e.g., the 15% reduction in sea-ice cover in the Arctic Ocean in the past 20 years (Krajick, 2000)), the impact on the coastlines of ice-bordered coasts is continually in flux.

H. Jesse Walker

Bibliography

Crossley, L., 1995. *Explore Antarctica*. Cambridge: Cambridge University Press, Australian Antarctic Foundation.

Evenson, E.B., and Cohn, B.P., 1979. The ice-foot complex: its morphology, formation and role in sediment transport and shoreline protection. *Zeitschrift für Geomorphologie, N.F.*, **23**(1): 58–75.

Field, W.O., 1975. *Mountain Glaciers of the Northern Hemisphere: Atlas*. Hanover, NH: U.S. Army, CRREL.

Forbes, D.L., and Taylor, R.B., 1994. Ice in the shore zone and the geomorphology of cold coasts. *Progress in Physical Geography*, **18**(1): 59–89.

Hambrey, M., and Alean, J., 1992. *Glaciers*. New York: Cambridge University Press.

Kirk, R.M., 1985. Antarctica. In Bird, E.C.F., and Schwartz, M.L. (eds.), *The World's Coastline*. New York: Van Nostrand Reinhold Company, pp. 1047–1053.

Krajick, K., 2000. Arctic life, on thin ice. *Science*, **291**: 424–425.

Lazzara, M.A., Jezek, K.C., Scambos, T.A., MacAyeal, O.R., and Van der Veen, C.J., 1999. On the recent calving of icebergs from the Ross Ice Shelf. *Polar Geography*, **23**(3): 201–212.

Mellor, M., 1964. *Snow and Ice on the Earth's Surface*. Hanover, NH: U.S. Army, CRREL.

Nichols, R.L., 1961. Characteristics of beaches formed in polar climates. *American Journal of Science*, **259**: 694–708.

Owens, E.H., 1982. Ice foot. In Schwartz, M.L. (ed.), *The Encyclopedia of Beaches and Coastal Environments*. Stroudsburg, PA: Hutchinson Ross Publishing Company, pp. 480–481.

Parkinson, C.L., 2000. Variability of Arctic sea ice: the view from space, an 18-year record. *Arctic*, **53**(4): 341–358.

Paterson, W.S.B., 1994. *The Physics of Glaciers*, 3rd edn. Trowbridge: Pergamon.

Powell, R., and Domack, E., 1995. Modern glaciomarine environments. In Menzies, J. (ed.), *Modern Glacial Environments*. Oxford: Butterworth-Heinemann, pp. 445–486.

Reimnitz, E., and Bruder, K.F., 1972. River discharge into an ice-covered ocean and related sediment dispersal, Beaufort Sea coast of Alaska. *Geological Society of America Bulletin*, **83**: 861–866.

Robe, R.Q., 1980. Iceberg drift and deterioration. In Colbeck, S.C. (ed.), *Dynamics of Snow and Ice Masses*. New York: Academic Press, pp. 211–259.

Robin, G. de Q., 1979. Formation, flow, and disintegration of ice shelves. *Journal of Glaciology*, **24**(90): 259–271.

Schwartz, M.L. (ed.), 1982. *The Encyclopedia of Beaches and Coastal Environments*. Stroudsburg, PA: Hutchinson Ross Publishing Company.

Taylor, R.B., and McCann, S.B., 1983. Coastal depositional landforms in northern Canada. In Smith, D.E., and Dawson, A.G. (eds.), *Shorelines and Isostacy*. London: Academic Press, pp. 53–75.

Viles, H., and Spencer, T., 1995. *Coastal Problems: Geomorphology, Ecology and Society at the Coast*. London: Edward Arnold.

Wadhams, P., 2000. *Ice in the Ocean*. Singapore: Gordon and Breach Science Publishers.

Walker, H.J., 1974. The Colville River and the Beaufort Sea: some interactions. In Reed, J.C., and Sater, J.E. (eds.), *The Coast and Shelf of the Beaufort Sea*. Arlington: The Arctic Institute of North America, pp. 513–540.

Walker, H.J., 1991. Bluff erosion at Barrow and Wainwright, Arctic Alaska. *Zeitschrift für Geomorphologie Supplement*, **81**: 53–61.

Warren, C.R., 1991. Terminal environment, topographic control and fluctuations of West Greenland glaciers. *Boreas*, **20**: 1–15.

Williams, R.S., Jr., and Ferrigno, J.G., 1988, 1993, 1995, 1998. *Satellite Image Atlas of Glaciers of the World*. Washington, DC: United States Geological Survey, Professional Papers 1386B, 1386C, 1386E, 1386I.

Cross-references

Antarctica, Coastal Ecology and Geomorphology
Arctic, Coastal Ecology
Arctic, Coastal Geomorphology
Glaciated Coasts
Paraglacial Coasts

INDIAN OCEAN COASTS, COASTAL ECOLOGY

Introduction

The coastal zone is the area covered by coastal waters and the adjacent shore lands, strongly influenced by each other. Coastal lands are some of the most productive and invaluable habitats of the biosphere, including estuaries, lagoons, and coastal wetlands. They are a place of high-priority interest to people, commerce, military, and to a variety of industries. Because it contains a dense population, the coast undergoes environmental modification and deterioration through reclamation, dredging, pollution, industry, and anthropogenic activities.

India has a vast coastline of approximately 7,000 km along the Arabian Sea in the west and the Bay of Bengal in the east. The western coastal plains lie between the Western Ghats and the Arabian Sea, further split into the Northern Konkan Coast and the Malabar Coast. The eastern coastal plains on the other hand, lie between the Eastern Ghats and the Bay of Bengal. Indian coasts have a large variety of sensitive ecosystems—sand dunes, coral reefs, sea grass beds, wetlands, mudflats, and rocky and sandy shores.

As shown in Figure I9, there are number of backwaters, estuaries and coastal lagoons that support the rich and diverse flora and fauna (Figure I10). These coastal habitats are considered to be highly productive areas in terms of biological productivity and one of the "*hotspots*" of marine biodiversity. Over 11,000 faunal (10,400 invertebrates and 625 vertebrates) and over 800 floral (624 algae, 50 mangrove, 32 angiosperms, 71 fungi, 14 lichens, 12 sea grass) species have been identified from Indian coastal areas (Untawale *et al.*, 2000; Anon, 2002). The majority of the shallow coastal and backwater areas form an important spawning and nursery ground for commercially important fishes, molluscs, crustaceans, and various other species that constitute the coastal fishery of India.

Type of marine habitats

Mangroves

Mangroves are woody plants that grow at the interface between land and sea in tropical and subtropical latitudes where they exist in conditions of high salinity, extreme tides, strong winds, high temperatures,

Figure I9 Some of the important coastal lagoons and marine biosphere reserves of India.

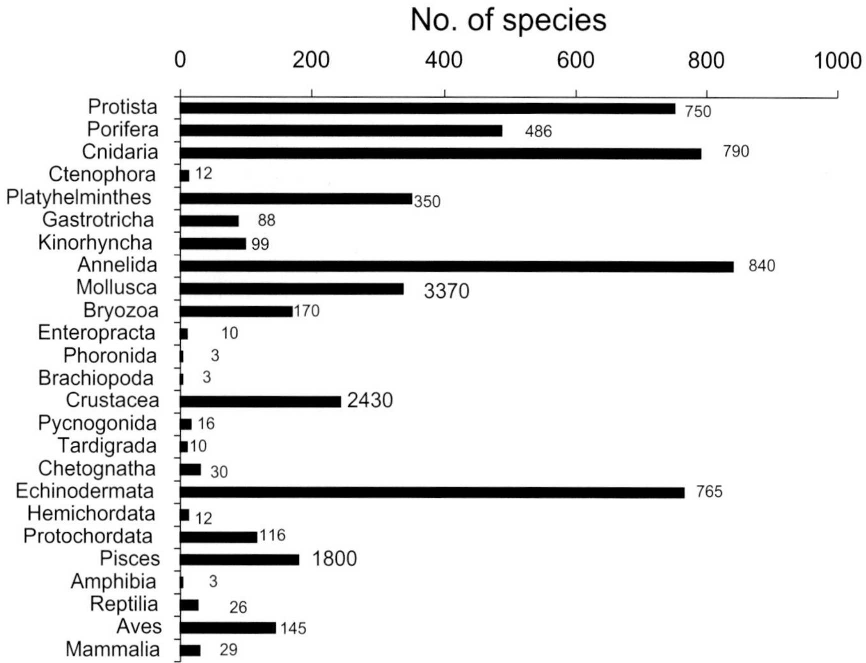

Figure I10 Marine faunal diversity of India. (After Anon, 1997; Untawale *et al.*, 2000; Anon, 2002)

Figure I11 Estuarine mangrove habitat along the Indian coast.

and muddy, anaerobic soils (Kathiresan and Bingham, 2001). Mangroves are usually present in estuarine and muddy shores (Figure I11), but can also be found on sand peat.

They are a complex and highly productive ecosystem that forms the interface between land and sea. The mangroves are widespread along the east and west coast of India. A detail of areas covered by mangrove habitat is given in Table I1.

On the Indian subcontinent, the mangrove ecosystems are distributed within the intertidal or tidal, supra-tidal or subaerial deltaic zones of both the east coast, facing the Bay of Bengal, and the west coast, facing the Arabian Sea. Mangrove flora of India comprises 50 exclusive species belonging to 20 genera, and 37 mangrove-associated floral species (Jagtap *et al.*, 2002; Upadhyay *et al.*, 2002). The maximum species diversity occurs in the Mahanadi delta along the Orissa coast, with 36 mangrove species present. Mangroves in India are estimated to cover about 4,871 km^2 (Upadhyay *et al.*, 2002). The mangrove ecosystem on the Indian subcontinent is of three types.

1. *Deltaic mangroves*: These are found along the mouth of different major estuaries on the east coast, and two gulfs (Gulf of Kachehh and Gulf of Khambhat) on the west coast. However, deltaic mangroves cover up to 53% of the total Indian mangals, which are estimated to be about 2,560 km^2, out of which the Gangetic delta popularly, known as *Sunderbans*, alone covers about 78%. About 48 species of mangroves have been recorded from the east coast (Upadhyay *et al.*, 2002). Mangrove distribution is scattered along the west coast with stunt growth and less species diversity.

Table I1 State-wise mangrove forest cover in India (in 1999)*

States along the east coast	Area (km^2)
Tamil Nadu	21
Andhra Pradesh	397
Orissa	215
West Bengal	2125
Andaman & Nicobar Islands	966
Sub-total	3724
States along the west coast	
Karnataka	3
Goa	5
Maharashtra	108
Gujarat	1031
Sub-total	1147
Grand Total	4871

* Source: Forest survey of India (1999)

2. *Coastal mangroves:* These are found along the intertidal coastlines, minor river mouths, sheltered bays, and backwater areas of the west coast. They extend from Gujarat to Kerala, and constitute to 12% of the mangals area of India. Due to less freshwater supply the mangals in the west coast are less, sparse, and show stunted growth. About 41 species have been reported from the west coast (Jagtap *et al.*, 2002).

3. *Island mangroves:* They are found along the shallow but protected intertidal zones of bay islands, Lakshadweep and Andamans. These are estimated to be about 16% (800 km^2) of the total mangrove area. About 30 true species of mangroves have been recorded from the island areas.

Species of *Avicennia* and *Aegicera* are dominant vegetation in the Godavari–Krishna and Cauvery deltaic system while *Ceriops decandra, Sonneratia apetala* are dominant on the Mahanadi delta. About 33 species of mangroves have been reported from the Gangetic Sunderbans, with species such as *Heritiera fomes, C. decandra, Xylocarpus* spp., *Lumnitzera* sp., *Sonneratia alba, Kandelia candel, Nypa fruticans*, and *Phoenix paludosa*. The mangroves of the West Bengal are dominated by *Excoecaria agallocha, C. decandra, S. alba, Avicennia* spp., *Bruguiera gymnorhiza, Xylocarpus granatum, Xylocarpus moluccensis, Aegiceras corniculatum*, and *Rhizophora mucronata*.

Species such as *R. mucronata, Rhizophora apiculata, Avicennia officinalis, Avicennia marina, Ceriops tagal, E. agallocha*, and *Acrostidum aureum* are most dominant along the west coast. Mangrove habitats harbor a variety of flora and fauna species (Figures I12 and I13). Until recently, mangroves were treated as unwanted plants and were used largely as a source of timber and charcoal. Therefore, mangrove ecosystems have been severely depleted during the last two decades. According to recent surveys, deforestation has destroyed about 44% and 26% of mangroves along the west and east coast, respectively (Upadhyay *et al.*, 2002). It is only in recent years that they have been recognized as ecologically vital. Mangroves play a very important role in protecting the shore from major erosion. The ecosystem forms an ideal nursery for the juvenile forms of many economically important fish and prawn species. A large percentage of the detrital food, which supports a variety of young fish and shrimps, is generated from mangroves.

Management

Traditionally, mangroves have been utilized for their wood, mainly for construction, fuel, and stakes. The bark of *Rhizophora* and *Bruguiera* spp., are used for tannin extraction. The leaves and fruits of *Avicennia* are used as fodder. *Avicennia* forests in the Gulf of Kachchh are constantly grazed by cattle. Mature fruits of *Sonneratia* as well as young fruits of *C. tagal* are consumed as vegetables in the human diet. Similarly, the young shoots of *A. aureum* and *Salicornia brachiata*, associated fern and an obligate halophyte, respectively, are also used as vegetables (Jagtap *et al.*, 2002). Mud from the mangrove regions is used as manure for paddy and coconut fields. The roots and leaves of *Derris heterophylla* are used for narcotizing and stupefying fish. Extracts of *Acanthus* leaves are used for rheumatic disorders, while that of *Bruguiera* species for high blood pressure and *Rhizophora* extracts as a cure for jaundice. The bark and the leaves of *E. agallocha*, though poisonous, are used to cure rheumatism.

Mangrove fisheries

Mangroves have a rich and diverse fish assemblage and the habitat is commercially exploited for capture as well as captive fisheries. The capture

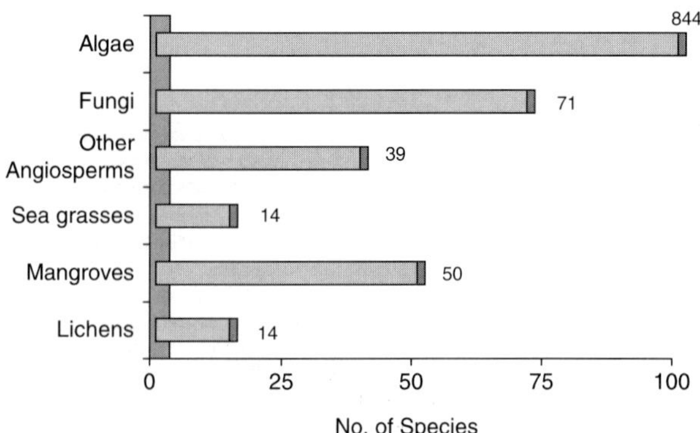

Figure I12 Floral diversity in mangrove habitat. (Compiled from Anon, 2002)

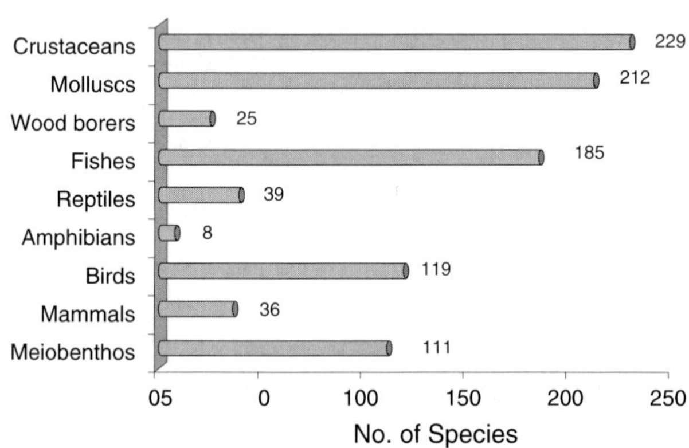

Figure I13 Faunal diversity in mangrove habitat. (Complied from Untawale *et al.*, 2000)

fisheries mainly consist of various species of bivalve and gastropod, crabs, prawns, and fishes from the proper mangrove regions and estuarine waterways. The captive fishery includes fish and prawn farming in the mangrove regions as well as mussels and oyster culture in the estuarine region. The salt-affected, water-logged, tidal regions in the vicinity of mangrove environments are commonly used for paddy-cum-prawn farming and salt production.

Lagoons

Coastal lagoons are shallow water bodies lying parallel to the coastline and separated from the open sea by a narrow strip of land or salt bank (Figure I14). They are a very rich and fragile natural ecosystem. As shown in Figure I9, lagoons are distributed all along the Indian coast.

There are eight important lagoons along the east coast; they are the Chilika, Pulicat, Pennar, Bendi, Nizampatnam, Muttukadu, Muthupet. Chilika is the largest brackish water lagoon and Pulicat is the largest saltwater lagoon on the eastern seaboard of India. Chilika lagoon (Figure I15) is spread over 1,100 km^2 while Pulicat lagoon is spread over 350 km^2.

There are nine important lagoons along the west coast of India (Figure I9). They are Vembanad, Ashtamudi, Paravur, Ettikulam, Veli, Murukumpuzha, Talapady lagoon of the Bombay coast, and the Lakshadweep lagoons (Kavaratti and Minicoy). Vembanad and Ashtamudi are the largest coastal and backwater lagoons found in Kerala.

Halophila spp., *Thalassia* spp., *Cymodocea* spp., *A. marina, Acanthus* spp., *Xanthium* spp., *Acacia* spp., *Gracilaria, Asterionella* spp., *Enteromorpha* spp., are some of the floral species found in the lagoons of India.

The benthic fauna of lagoons constitute various species of, foraminiferas, nematodes, gastrotrichs, oligochaetes, polychaetes, calanoids, amphipods, isopods, decapods, tanaids, and molluscs. Among all the lagoons, Chilika has the richest biodiversity. Ecologically, it is

Figure I14 Coastal lagoon at Vadhawan.

Figure I15 Chilika lagoon (Orissa, India).

endowed with a wealth of flora and fauna (Figure I16). A total of 788 faunal species has been reported form the Chilika (Ingole, 2002). The majority of these are aquatic and almost 29% (225) are fish species (Figure I16). About 61 are protozoan species, 37 nematodes, 29 platyhelmenthes, 31 polychaetes, 58 decapods and brachyuran crabs, 37 amphibians and reptile, 136 molluscs, 18 mammals, and 156 bird species.

Chilika Lake contributes to a major portion of the fish catch in the region. The lagoons provide an excellent opportunity for aquaculture and get a good foreign exchange. The rich demersal fishery (especially of prawn, crabs, and molluscs) supports over 80,000 fishermen from 122 villages. The lake supports one of the largest populations of waterfowl during winter season. The area is known as an ideal habitat for crocodile, dolphin, and a variety of birds. An area of 15.53 km^2 of this lake has been designated as wildlife sanctuary. However, Chilika has been facing some natural and manmade problems, particularly frequent shifting of the mouth region, reduced seawater inflow, siltation, and encroachment. With a rise in human population in the lagoon periphery, pollution from domestic sewage, pesticide, agriculture, chemical, and industrial effluent have become major threats to the lagoon ecosystem. The lagoon was rapidly shrinking at the rate of about 2 km^2 per year. According to Ingole (2002) fish catches which used to be about 8,500 tonnes per year in 1980s were dwindling during 1985–2000.

However, due to the timely action taken by the Chilika Development Authority (CDA, Government of Orissa), the lake environment is being restored under the "Chilika Development Plan." Substantial increases

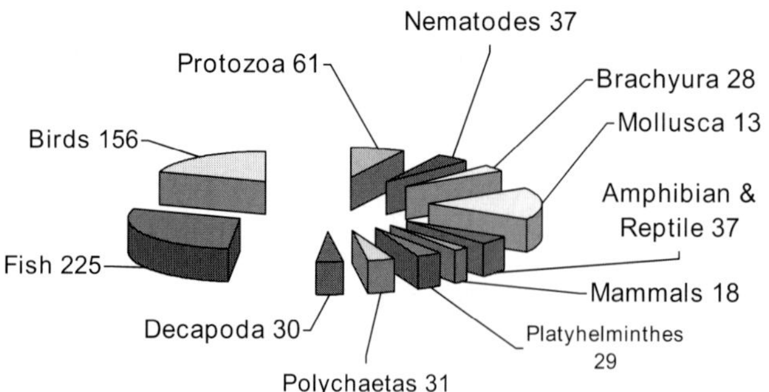

Figure I16 Biodiversity of Chilika Lake ((No. of species); Ingole, 2002).

in the fish production (11,989 tonnes) during 2001–02 (Ingole, 2002) clearly demonstrated the efforts of CDA toward the sustainable development and conservation of this important ecosystem. Recently, due to the rapid recovery of the lake ecosystem, the CDA Authority has been awarded the International *Ramsar Award* and Chilika is also included in the list of *Ramsar Sites*.

Estuaries

Estuaries are an integral part of the coastal environment. They are the outflow regions of rivers, making the transitional zone between the fluvial and marine environment. Estuaries are the focal point of studies, and activities. Fourteen important estuaries have been reported along the east coast and west coast of India (Anon, 1997; Qasim, 1999). Hooghly, Rushikulya, Godavari and Krishna, Edaiyur-Sadras, Araniar, Ennore, Cooum, Adyar, Uppanar, Vellar, Kollidam, Kaveri, Agniyar, Kallar along the east coast, and Ashtamudi, Korapuzha, Beypore, Periar, Kadinamkulam, Vembanad, Netravathi and Gururpur, Gangolli, Pavenje, Kali, Narmada Amba, Purna, Mandovi, and Zuari estuary on the west coast of India.

Biodiversity in the estuaries is very impressive. Some floral species are *Oscillatoria* spp., *Enteromorpha* spp., *Spirogyra* spp., *A. marina, Excoecaria* spp., and *Sonneratia* spp. Various species of polychaetes, crustaceans, molluscs, echinoderms, and fish are the faunal component of estuaries.

Estuaries are semi-enclosed water bodies and thus, they provide a natural harbor for trade and commerce. They are also effective nutrient traps and provide a vital source of natural resources for people and are used for commercial, industrial, and recreational purposes. They also act as nursery grounds for a variety of shrimps and finfish and are the best settling places for clams and oysters.

Indian estuarine ecosystems are deteriorating day by day through human activities, and the dumping of an enormous quantity of sewage into the estuary has drastically reduced the population of spawning fishes. It has also caused considerable ecological imbalance and resulted in the large-scale disappearance of the flora and fauna. Introduction of untreated municipal wastewater and industrial effluents into these water bodies has led to serious water pollution including heavy-water pollution, which becomes bio-magnified and reaches people through the food chain.

Mudflats

Mudflats develop in sheltered places of the intertidal area. Twice each day, water flows in and out with the tides, filling or draining the flat. The mudflat receives nutrients from the tidal flow and the nearby marsh, particularly as it decays. This means that mudflats have rich plant and animal communities.

They are important as sedimentation areas and provide a rich source of organic material for the endo- and epibenthic community. Because of this high availability of organic substances, oxygen content in the pore water is rather low and may limit chemical and biological degradation processes.

Phytoplankton and zooplankton are abundant, as are mud snails. Filter-feeding animals such as oysters and clams live in mudflats because of the availability of plankton. Fish and crabs move through the flats at high tide. Birds and predatory animals visit tidal flats at specific times for feeding.

(A)

(B)

Figure I17 Sandy beaches of Goa (Central west coast of India).

Sandy beaches

The sandy beaches seem to be barren. There is no lush growth of macroalgae and apart from sea grasses there is no obvious plant life. Few animals live on the surface and most of them live below the sand. The organisms found here have suitable burrowing mechanisms, which may take the form of proboscis, parapodia of a polychaeta, and foot of molluscs.

Vast stretches of sand are seen along the east and west coast of India forming the boundary between land and sea. The beaches are subjected to the forces of waves, tides, currents, and winds. Sandy beaches along the Gujarat coast are limited between muddy and rocky shores. Although sandy strips along rocky cliffs are also observed along the Maharashtra and Goa coast, beaches along the central west coast of India are of sandy nature. The sandy shores of Goa and Karnataka (Figure I17) are of limited width, while Kerala has extensive sandy beaches interspersed with coastal lagoons. Tamil Nadu has sand strips along the deltaic shores and rock-bound beaches. Beaches along the Andhra Pradesh coast are of limited width interspersed by the rivers Godavari, Krishna, and their tributaries. Orissa has extensive sand strips at Konark—Puri. The coast of Andaman and Nicobar Islands have sand strips interspersed with bluffs, rocks, or shingle along the coastline. The Lakshadweep atolls have long stretches of coralline sandy beaches with unique vegetation.

Macrofaunal species (benthic organisms having body size >0.5 mm) such as *Donax incarnates, Donax spiculum, Donax faba, Donax scortum, Suneta scriptta, Mactra* spp., *Paphia malabarica, Bullia melanoides, Umbonium* spp., Oliva *spp., Emerita holthuisi, Eurydice* spp., *Gastrosaccus* spp., *Ocypode ceratophthalma, Ocypode macrocera, Ocypode platyarsus, Dotilla intermedia, Glycera alba, Lumbriconereis latreilli, Onuphis eremita* are some of the common species found on the sandy beaches of India.

Sand dunes

The term "sand dune" reflects the image of vast amounts of shifting sand, barren of plants, and hostile to human habitation. Hot and dry winds shape and arrange the sand in geometric and artistic patterns.

Dunes are of two types. The first type is found in the extremely dry interior desert such as Rajasthan in India and the other type is the coastal sand dune, which occur along the coast of India.

Dunes are composed of wind-blown sand. Fore-dunes are built up at the back of the beaches on the crest of berms and dune ridges where vegetation or other obstacles trap wind-blown sands. During periods of coastline advancement, successful dunes may develop to form a series of parallel dunes (Figure I18).

The common vegetation found in the fore-dune are *Hydrophylax maritima, Ipomoea pes-caprae* (Saurashtra), *Canavalia maritima, Cyperus arenarius, I. pes-caprae, Launea sarmentosa, H. maritima* (Orissa), *H. maritime, I. pes-caprae, C. maritima, C. arenarius, I. pes-caprae, L. sarmentosa, H. maritime, Sporobolus virginicus* and *Zoysia matrella* (Andhra Pradesh), *I. pes-caprae, L. sarmentosa, Dactyloctenium aegypticum, C. arenarius* (West Bengal), *I. pes-caprae, Spinifex littoreus* (Goa). In the parabola dune the following vegetation have been recorded along the west coast of India: *Halopyrum mucronatum, Borreria articularis, Lotus garcinii, Asparagus dumosus, Enicostema hyssopiflorum, Peplidium maritimum, Cassytha filiformis*, and along the east coast of India *Euphorbia rosea, Geniospermum tenuiflorum, Phyllanthus rotundifolius, S. littoreus, Goniogyne hirta, Perotis indica, Brachiaria reptens, Elusine indica, Rothia indica, Trianthema pentandra*.

Sand dunes throughout the world have been recognized for their ecological significance. The coastal dune vegetation acts as shelter belts protecting the inner land. When the dune system and the vegetation are destroyed for short-time benefits in the name of development, disaster occurs. Floods and cyclones on the east coast of India are eye-opening examples. The cause and effect of destruction are not short-termed or locally limited.

Rocky shores

In contrast to sandy beaches where many individuals live unseen in the soil, many of the plants and animals of the rocky shores are conspicuously displayed. Another feature of the rocky shore is the distinctly noticeable zonation of plants and animal communities.

The rocky coast shows a richer fauna than that of the sandy beach and is varied in composition. The rock and crevices give shelter to numerous crabs, molluscs, and fish. *Grapsus grapsus, Tectarius trochoid, Littorina scabra, Littorina angulifera, Littorina undulat, Trohcus* sp., *Cellana radiata, Planaxis sulcatus, Thais bufo, Drupa* sp., *Hemifusus pugulinus, Perna viridis, Nereis* spp., amphipods, isopods, Holothuria, Sea Urchin, nudibranch, *Aplasia* spp., are some of the common fauna of the rocky shores. Common flora of the rocky shore are different species of *Porphyra, Gracilaria,* and *Enteromorpha, Padina, Ulva, Gelidium, Sargassum,* and *hypnea, Chaetomorpha* spp.

Sea grass

Sea grasses are submerged flowering plants (angiosperms) that have adapted to life in the sea. They differ from what we refer to as "seaweed," in that they are plants with vessels and well-defined root and shoot systems. Sea grasses have been able to successfully colonize the marine environment because of five properties:

- the ability to live in a salty environment;
- the ability to function normally when fully submerged;
- a well-developed anchoring system;
- the ability to complete their reproductive cycle while fully submerged;
- the ability to compete with other organisms under the more or less stable conditions of the marine environment.

Sea grass habitats are mainly limited to the mudflats and sandy regions from the lower intertidal zones to a depth of ca. 10–15 m along the open shores and in the lagoons around islands. The major sea grass meadows in India occur along the southeast coast (Gulf of Mannar and Palk Bay) and along a number of the islands of Lakshadweep in the Arabian Sea and of Andaman and Nicobar in the Bay of Bengal. The largest area (30 km^2) of sea grass occurs along the Gulf of Mannar and Palk Bay, while it is estimated that ca. 1.12 km^2 occurs in the lagoons of major islands of Lakshadweep. A total of 8.3 km^2 of sea grass cover have been reported from the Andaman and Nicobar Islands (Jagtap *et al.*, 2003).

The sea grasses of India consists of 14 species belonging to 7 genera. The Tamil Nadu coasts harbor all 14 species, while 8 and 9 species have been recorded from Lakshadweep, and Andaman and the Nicobar group of Islands, respectively. The east coast supports more species compared to the west coast of India. The main sea grasses are *Thalassia hemprichii, Cymodocea rotundata, Cymodocea serrulata, Halodule uninervis* and *Halophila ovata*. Species such as *Syringodium isoetifolium* and *Halophila* spp., occur in patches as mixed species. Gulf and bay estuaries mostly harbor low numbers of species, dominated by *Halophila beccarii* in the lower intertidal region and by *Halophila ovialis* in the lower littoral zones.

Sea grass beds are:

- Major primary producers (food manufacturers) in the coastal environment. Their high primary productivity rates are linked to the high production rates of associated fisheries;
- Stabilizers of bottom sediments; they provide protection against erosion along the coastline;
- Important nutrient sinks and sources, that is they help in the recycling of nutrients.

With root-like stems, which extend horizontally under the sea bottom, sea grasses act to stabilize the sediment. These sediments, that

Figure I18 Typical sand dune found along the east coast of India.

would otherwise settle on coral and prevent contact with sunlight, tend to accumulate and become trapped in the sea grass. Turtle grass, the most common type of sea grass thrives in areas that are protected from wind-driven current and surf. The broad leaves of turtle grass act as huge filters, removing particles from the water and depositing them as fine sediment. These sediments often contain organic matter, which contribute to the high productivity of this habitat. For this reason, the sea grass habitat attracts various species of fish, conch, lobster, turtles, and manatees for feeding and breeding. Numerous species of reef fish use sea grass as a protective nursery, hiding amid the grass from predators. Moreover, adult fish that hide in the coral reef during the day and venture out at night to feed, take advantage of the rich source of food that exists in the sea grass.

The natural causes of sea grass destruction in India are cyclones, waves, intensive grazing, and infestation of fungi and epiphytes as well as "die-back" disease. Exposure at the ebb tide may result in the desiccation of the bed. Strong waves and rapid currents generally destabilize the meadows causing fragmentation and loss of sea grass rhizomes. The decrease in salinity due to extensive freshwater run-off also causes disappearances, particularly of the estuarine sea grass bed.

Anthropogenic activities such as deforestation in the hinterland or mangrove destruction, construction of harbor or jetties, loading and unloading of construction materials as well as anchoring and moving of boats and ships, dredging and discharge of sediments, land filling and untreated sewage disposal are some of the major causes of sea grass destruction of India.

As of yet, there is no specific legislation protecting sea grasses, although generally the Fisheries Department is responsible for this habitat. Green Reef recommends that in addition to controlling the use of pesticides and decreasing run-off, dredging activities need to be strictly monitored and limited. Sea grass beds should be thought of as an indication of the health of the ecosystem; when they begin to disappear, we know there is trouble.

Coral reefs

Coral reefs are among the most biologically productive, taxonomically diverse, and aesthetically important living organisms among all the aquatic ecosystems.

The major coral formation in India is around the Lakshadweep (816 km^2) and Andaman Islands (960 km^2), as well as in the Gulf of Mannar (94 km^2) and the Gulf of Kachchh (406 km^2). Reefs in the Gulf of Kachchh, Gulf of Mannar, and Andaman and Nicobar Islands are mostly of the fringing type, with a few platform, patch and atoll reefs, and coral pinnacles. Lakshadweep Islands on the contrary are mostly atolls with a few coral heads, platform reefs, and sand cays. Coral, though rare in occurrence, are reported at many locations along the west coast.

About 44 species of scleractinian coral and 12 species of soft corals occur in the Gulf of Kachchh (Figure I19). The submerged reefs of these areas harbor 18 species of stony coral and have 45% coverage. *Acropora, Porites, Pseudosiderastrea*, and *Favia*, and one species of soft coral *Juncella juncea* are common. Two species of hard coral (*Pseudosiderastrea tayamai* and *Porites lichens*) have been recorded in patches along the central west coast. Large well-developed hard colonies have also been sighted at Colaba near Mumbai.

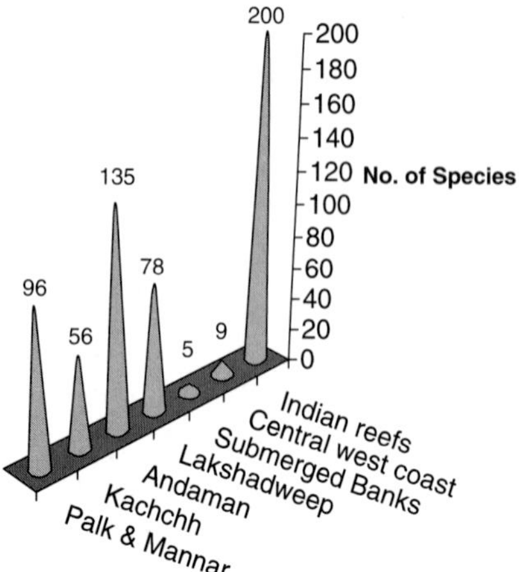

Figure I19 Coral distribution along the Indian coast.

Malvan is considered to have the richest marine biodiversity along the central west coast. *Porits*, a stony coral is most common in Malvan waters. Nine species including a rare coral species belonging to the genera *Coscinaraea, Favites, Goniastrea, Synaraea* and *Pseudosiderastrea, Cyphastrea, Turbinaria* have been recorded in this region. Angria bank—a submerged reef off Ratnagiri has stony corals. A few patches of scleractinian corals have been reported in subtidal waters off the Goa coast (Rodrigues *et al.*, 1998). Five coral species have been recorded from Gaveshani Bank off Malpe along the Karnataka coast (Qasim and Wafar, 1979; Wafar, 1990). *Pocillopora eydouxi* has recently been reported from Vishakhapatnam on the east coast.

About 96 species belonging to 36 genera of Scleractinian and 7 species of Ahermatypic corals are reported from Palk Bay and the Gulf of Mannar (Figure I19). *Montipora* and *Acropora* represent 40% of the species in the Palk Bay. Thirty species of stony corals are reported from Manauli reef, of which the dominant corals are *Acropora, Porites, Goniastrea, Favia, Pocillopora* and *Montipora*. Krusadai island has a well-developed coral reef and sustains 96 species belonging to 26 genera of hermatypic corals.

Massive occurrences of corals provide much needed protection from waves to the coastline; and coral productivity yields a multitude of flora and fauna dependant on the coral ecosystem. Coral reefs also provide opportunities for skin diving, under-water photography, sport fishing, and shell collection, thus, providing a vital stimulus to the tourist industry. Coral reefs are often exploited for calcium carbonate—a raw material for many lime-based industries. The fishery resources of the reefs are extremely rich and diversified. They are also exploited for their beautiful associated fauna such as molluscs and ornamental fish.

Due to population pressure, most of the coral reefs have become extremely vulnerable to industrial development and pollution along the coastline. Unless protection is offered to the coral reefs, most of them will diminish in size in the future and ultimately die. The reefs that will probably flourish are on the atolls of Lakshadweep and some of the islands of Andaman and Nicobar Islands.

Marine protected areas

There are 26 Marine Protected Areas (MPAs) in India comprised of national parks and wildlife sanctuaries declared as coastal wetlands; especially mangrove, coral reefs, and lagoons, under Wildlife (Protection) Act, 1972. These 26 MPAs are located in 7 coastal states and 2 union territories and cover 13% of the total coastal wetland area of the country. Some of these MPAs are shown in Figure I9 and Table I2. The Gulf of Kachchh Marine Sanctuary and Marine National Park, the Gulf of Mannar National Park, and the Wandoor Marine National Park (Andaman Islands) have been established primarily to protect marine habitats.

Table I2 Marine national parks and sanctuaries along the Indian coast (Anon, 2002).

States	Location	Name of the biosphere
Gujarat	Gulf of Kachchh	National Marine Park, Bird Sanctuary, and coral reef
Maharashtra	Mahaim (Mumbai)	Natural Mangrove Park
	Malvan	Marine Park (coral reef)
Goa	Chorao island	Bird Sanctuary
Lakshadweep islands	Kawaratti, Agatti, Minicoi islands	Marine Park (coral reef)
Tamil Nadu	Gulf of Mannar	National Marine Park, Sanctuary, and coral reef
	Muthupet	Reserve Mangrove Forest
Andhra Pradesh	Coringa	Wildlife Sanctuary
Orissa	Nalaban (Chilika)	National Marine Park and Bird Sanctuary
	Bhittarkanika	Wildlife Sanctuary (Turtle project)
West Bengal	Haribanga	National Marine Park and Bird Sanctuary (Tiger project)
Andaman and Nicobar Island	Wandoor	National Marine Park

Management of coastal ecology

As shown in Figure I20, there is great demand for industrial development along the Indian coast. This in turn amounts to the increase in pressure on the coastal zone due to concentration of population, development of industries and ports, discharges of waste effluents and municipal sewage, and spurt in recreational activities which have adversely affected the coastal environment. Coastal resources are affected by activities far distant from the coast (viz., deforestation, damming of rivers, bunding/barraging of the coastal water bodies, river sand mining, discharge of pesticides, heavy metals, domestic and factory wastes, garbage, and other substances that are harmful to the coastal areas. Lagoons, estuaries, wetlands, and nearshore shallow water areas are particularly vulnerable to these activities. Considering the urgent need for protecting the Indian coast from degradation, the Government of India enacted "Coast Regulation Zone (CRZ) Act, 1991." The area influenced by tidal action up to 500 m from High Tide Line (HTL) and the land between the Low Tide Line (LTL) and the HTL has been declared as Coastal Regulation Zone (CRZ).

Mangroves and coral reef ecosystem

The mangroves are constantly threatened by increasing anthropogenic pressures such as indiscriminate cutting, reclamation mainly for agriculture and urbanization, fuel and construction, and overgrazing by domestic cattle. Mangrove ecosystems along the west coast, particularly in the States of Gujarat, Maharashtra, and Kerala, have been degraded to a large extent. However, West Bengal (Sundarbans), Orissa, and the Andaman and Nicobar Islands still form the best mangrove ecosystems of India. Considering various estimates for the mangrove cover in the country, 30% of the mangrove area was reclaimed for different anthropogenic activities during the period 1975–90 (Jagtap *et al.*, 1993).

Realizing the importance of mangroves and coral reefs, the Government of India initiated efforts for their conservation and management. They were declared as ecologically sensitive areas under the Environment (Protection) Act, 1986, banning their exploitation, and followed by a CRZ Notification 1991 prohibiting development activities and disposal of wastes in the mangroves and coral reefs. The Ministry initiated a plan-scheme on conservation and management of mangroves and coral reefs in 1986 and constituted a National Committee to advise the Government on relevant policies and programs. On the recommendations of this committee, 15 mangrove areas in the country have been identified for intensive conservation (Anon, 1997 & 2002; Jagtap *et al.*, 2002).

Considering the importance of coral reefs and the factors responsible for their deterioration, Andaman and Nicobar Islands, Lakshadweep Islands, Gulf of Mannar, and Gulf of Kachchh have been identified for conservation and management. Efforts have been initiated to establish Indian Coral Reef Monitoring Network (ICRMN) to integrate various activities on coral reefs through national and international initiatives.

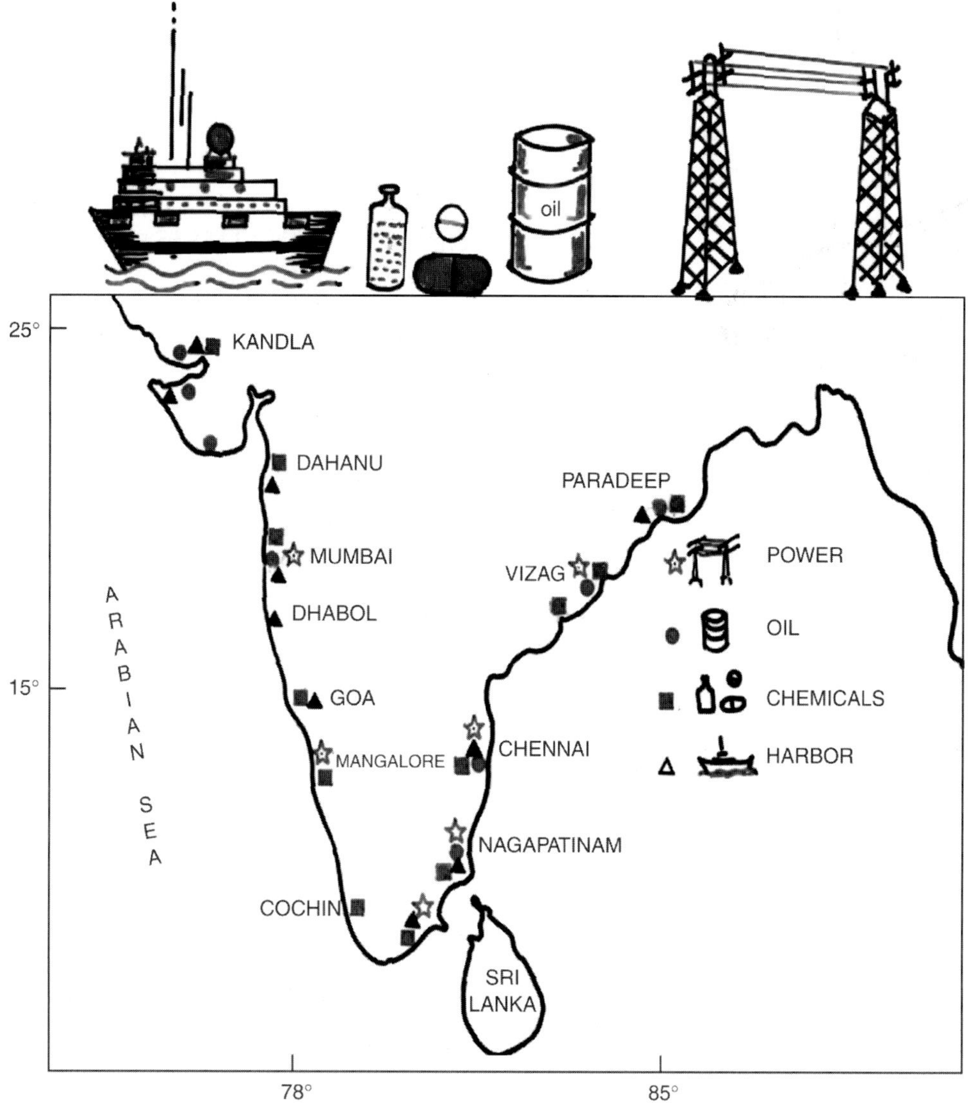

Figure I20 Patterns of industrial development along the Indian coast.

Institutions of database networking and capacity building and training on coral reefs have been identified. With the establishment of the National Mangrove Committee (NATMANCOM), attempts are being made to protect, conserve, and restore the mangrove habitats. Mangrove regions in the country have been categorized presently under the Ecologically Sensitive Zone; vide CRZ Act of the country. As per the CRZ Act, no development in the mangroves or in the vicinity is allowed prior to an environmental impact assessment (EIA) and clearance from the Ministry of Environment and Forests (MoEF), Government of India. Few of the mangrove regions in the country have been conserved as Biosphere reserves for germplasm and wildlife sanctuaries.

Shrimp culture activities

The aquaculture industry is growing at a faster rate than many of the sectors of the coastal zone in India. Socially, its product is seen as a currency. However, if aquaculture expansion is not regulated, its long-term consequences will be felt in the quality of water bodies. In fact some adverse effects of shrimp culture are already seen along the east coast of India, particularly along the coast of Andhra Pradesh.

Sandy shores

Mining of beach sand is a widespread activity in many of the coastal areas. Though sand is an important constituent of construction activities, it has to be borne in mind that it is these sand deposits that provide the natural protection to the coast from erosion. Storms, waves, currents, and wind temporarily displace vast quantities of beach sand that is then held in storage as sand bars. These sand bars then become the protectors of the coasts against those forces, which finally return the sand to the beach. Thus, removal of sand from any part of the beach can aggravate the erosion and recession of the beach front altogether.

For the above reasons and also for the sustainable management of the coastal resources, many countries are now developing Coastal Zone Management (CZM) strategies, and some have already begun to adopt such programs. The CZM has to manage, develop, and conserve natural resources and, while doing so, it has to integrate the concerns of all relevant sectors of the society, economy, and prosperity.

A major thrust in implementing the CRZ notification is in the conservation of coastal resources, to achieve their sustainability, and long-term protection of its natural assets. The criterion for sustainable use is that the resource shall not be harvested, extracted, or utilized in excess of the quantity that can be produced or regenerated over the same period. It is important to learn the acceptable limits of coastal environmental degra-dation and the limits of sustainability of coastal resources. Hence, in order to achieve the goals set forth in the notification, it is imperative to have first-hand-information on the present land use practices and availability of the resources in the coastal zone. Contribution No.3897 of NIO, Goa.

Baban Ingole

Bibliography

Anon, 1997. *Estuaries of India: State of the Art Report*. ENVIS, Publication Series, 2/97.

Anon, 2002. *National Biodiversity Strategy and Action Plan—India*. Ministry of Environment and Forests, Government. of India.

Forest Survey of India, 1998. State of Forest Report 1999. Dehradun, India: Forest Survey of India.

Ingole, B.S., 2002. Increased macrobenthic density and diversity: indicator of recovery of Chilika Lake from environmental degradation. *Proceedings of the. National Seminar on Creeks, Estuaries and Mangroves—Pollution & Conservation*, pp. 213–215.

Jagtap, T.G., Chavan, V., and Untawale, A.G., 1993. Mangrove ecosystem of India: a need for protection. *Ambio*, **22**(4): 252–254.

Jagtap, T.G., Murthy, P.S., and Komarpant D.S., 2002. Mangrove ecosystem of India: conservation and management. In Hosetti, B.B. (ed.), *Wetlands Conservation and Management*, Jaipur, India: Pointer Publishers, pp. 35–67.

Jagtap, T.G., Komarpant, D.S., Rodrigues, R., 2003. The seagrasses of India. In *World Atlas of Seagrasses*, 113–120.

Kathiresan, K., and Bingham, B.L., 2001. Biology of mangrove and mangrove ecosystems. *Advances in Marine Biology*, **40**: 84–254.

Qasim, S., Z., 1999. *The Indian Ocean: Images and Realities*. New Delhi: Oxford and IBP.

Qasim, S. Z., and Wafar, M.V.M., 1979. Occurrence of living corals at several places along the west coast of India. *Mahasagar-Bulletin of National Institute of Oceanography*, **12**: 53–58.

Rodrigues, C.L., Caeiro, S., and Raikar, S.V., 1998. Hermatypic corals of the Goa coast, west coast of India. *Indian Journal of Marine Science*, **27**: 480–481.

Untawale, A.G., Dhargalkar, V.K., and Deshmukher, G.V., 2000. Prioritization of Potential sites for marine Biodiversity conservation in India. in- Setting Biodiversity conservation priorities for India. WWF for Nature-India Publication, pp 104–131.

Upadhyay, V. P., Rajan, R., and Singh, J. S., 2002. Human-mangrove conflicts: The way out. *Current Science*, **83**(11): 1328–1335.

Wafar, M.V.M., 1990. Global warming and coral reefs. In Victor Rajamanickam, G. (ed.), *Sea Level Variation and its Impact on Coastal Environment*. Tamil University Publication No. 131, pp. 411–432.

World Atlas of Seagrasses. In Green, E.P. and Short, F.T. (eds.), Berkeley: University of California Press, pp. 113–120.

Cross-references

Aquaculture
Bioconstruction
Coastal Lakes and Lagoons
Coral Reef Coasts
Estuaries
Human Impact on Coasts
Indian Ocean Coasts, Coastal Geomorphology
Indian Ocean Islands, Coastal Ecology and Geomorphology
Mangroves, Ecology
Mangroves, Geomorphology
Muddy Coasts
Vegetated Coasts
Wetlands

INDIAN OCEAN COASTS, COASTAL GEOMORPHOLOGY

The coastal geomorphology of the Indian Ocean coast, with special reference to coasts of Pakistan, India, SriLanka, Bangladesh, and Myanmar, is mainly governed by the processes associated with monsoons.

Pakistan

The coastline of Pakistan, from the Iranian border on the west to the Indian border on the east is about 990 km long. This coastline is one of the active tectonic regions. The coast here is associated with a narrow continental shelf, except off Indus delta. The coast of Pakistan is divided into the Makran coast, Las Bela coast, Karachi coast, and Indus Delta coast.

The Makran coast, with approximately 473 km length, from the Iranian border to Ras Malan, consists of long sandy beaches associated with either wide coastal plains or valleys landward. These plains and valleys are interrupted by uplifted marine terraces at places. Also the Makran hill ranges, which lie about 32 km from the coast, become part of the coast at Ras Malan with massive headlands. Spits and bars are common seasonal morphologic units along sandy beaches. At places, well-developed beach ridges are seen. Dasht is the only river which brings a small quantity of sediment from land to the Arabian Sea along this coast.

The Las Bela coast, with about 260 km length, extends from Ras Malan to Ras Muari. The Ras Malan range is made up of sandstone and shelly limestone and presents gorges and cliffs as high as 600 m, which drop directly to the sea. Followed by this, on the east, is the Las Bela plain. The coast here consists of a series of beach ridges, sand dunes, bars, tidal flats, and lagoons with mangroves (Bird and Schwartz, 1985). Between the Ras Malan ranges and the Las Bela main valley notable mud volcanoes are present, the largest among them is called Chandragup (Snead, 1964). On the eastern side of the Las Bela valley, promontories of limestone are present. Here marine terraces at different elevations have wave cut sea caves and blow holes (Snead, 1966). Along this coast, the Hab River joins the Arabian Sea at its mouth compound bars, shallow lagoons, and sandy beaches are present.

The Karachi coast, about 48 km in length from Ras Muari to Clifton beach, consists of low rocky cliffs and sandy beaches of almost equal length. Marine terraces, sea caves, and arches are common in the sandstone and shale rocks. Sandy barrier beaches, spits (the longest one 15 km in length), shallow lagoons, tidal flats, and salt evaporation ponds are common along the sandy beaches.

The Indus delta coast is about 200 km long with uniform landforms namely large tidal channels with mudflats in between, barrier bars and spits with hooks, and beaches, and a few small mangrove shrubs. The sand bars and delta channels are dynamic in nature as they change their morphology due to tidal currents, waves and channel floods. The coastal area in this region is very flat and therefore up to 6.5 km from the coastline it is submerged during high tide. The river Indus brings a large quantity of sediment from land to sea and joins the Arabian Sea along this coast.

India

Most of the early literature on Indian coastal geomorphology was essentially of a descriptive nature based on the nature, location, and relationships of the landforms and sea level. Ahmad's (1972) was possibly the first and only book on coastal geomorphology of India, and contains data collected from large-scale maps and inferences drawn on the nature of the coasts. In addition, there are some isolated studies by Vaidyanathan (1987), Baba and Thomas (1999). The Space Application Centre (SAC, 1992) has carried out a comprehensive study on the coast using LANDSAT and IRS data. The information on Indian coastal geomorphology presented in this article is based on these and many more isolated published studies.

India has about a 7,500-km long coastline. The coastline of India has been undergoing morphological changes throughout the geological past. The sea level fluctuated during the period of last 6,000 years and recorded marked regression during the period between 5,000 and 3,000 years before present (Rajendran *et al.*, 1989). The present coastal geomorphology of India has evolved largely in the background of the post-glacial transgression over the preexisting topography of the coast and offshore (Baba and Thomas, 1999).

The major rivers that cut across the coast and bring large quantities of water and sediment to the coast from Indian continent are the Ganges, Brahmaputra, Krishna, Godavari, and Cauvery on the east coast, and the Narmada and Tapti on the northwest coast. In addition, there are about 100 smaller rivers, these also supply considerable quantities of water and sediment. While larger rivers have well-developed deltas and estuarine systems, almost all the small rivers have estuarine mouths with extensive mud flats and salt marshes and some of them with estuarine islands.

The continental shelf of India is very wide on the west coast with about 340 km in the north, tapering to less than 60 km in the south. The shelf is narrow along the east coast. The coastline on the west receives southerly winds that bring high waves during the monsoons (June–September). The east coast generally becomes active during the cyclones of the northeast monsoon period (October–November). The tidal range varies significantly from north to south. It is around 11 m at the northwest, 4.5 m at the northeast and around 1 m at the south.

Considering geomorphic characteristics, the Indian coast is divided into two categories, namely coasts on the west coast of India and coasts on the east coast of India. The coast on the west coast of India differs from the east in that there are practically no deltas on the west coast. The coastline here is modified by headlands, bays, and lagoons at irregular intervals. There is distinct evidence of the effect of neotectonics in some sections (Vaidyanathan, 1987). The east coast on the contrary is known for the number of deltas especially along the northern portion, West Bengal and Orissa coast. Deltas in the southern portion have helped in recognizing ancient channels, ancient beach ridges, former confluences, and strandlines.

West coast of India

Though there are a large numbers of small rivers bringing enormous quantity of sediment to the Arabian Sea along the west coast, deltas are not formed, possibly due to the high-energy condition of the coast. Beach morphological changes along the west coast are controlled by the southwest monsoon. The maximum morphological changes occur during early monsoons (June–August). During this period most of the material is transported to the offshore and some alongshore. Most of the material appears to be returned again during the fairweather season.

The west coast of India is further divided into the Gujarat coast, Maharashtra, Goa and northern Karnataka coast, and southern Karnataka and Kerala coast based on their geomorphological distinctions. The coastal area of Gujarat is the largest in the country with about 28,500 km^2. The coast in Gujarat, from west to east, varies from a deltaic coast, the irregular drowned prograded coast, the straightened coast, the spits and cuspate foreland complex, and the mudflat coast. The Gujarat coast is further divided, from west to east, into five regions, namely Rann of Kutch, the Gulf of Kutch, the Saurashtra coast, the Gulf of Khambat, and the South Gujarat coast, based on coastal geomorphic characteristics.

The Rann of Kutch remains saline desert for the larger part of the year and is further divided into the Great Rann and the Little Rann. On the west of the Great Rann of Kutch is the area of the lower Indus deltaic plain which is characterized by tidal creeks and mangroves. The coastline in the Gulf of Kutch has extensive mudflats and is highly indented with a number of cliffed rocky islands (Baba and Thomas, 1999). Migration, joining of different creeks, reorientation of tidal current ridges, and regression of the sea are seen and are related to tectonically activated lineaments. The coast here is fringed by coral reefs and mangroves. Algae, salt marsh, dunes, and salt pans are very common. The Saurashtra coast has numerous cliffs, islands, tidal flats, estuaries, embayments, sandy beaches, dunes, spits, bars, bays, marshes, and raised beaches at some places. The coast, in Gulf of Khambat is indented by estuaries and consists of mudflats, dunes, and beaches. Here mudflats are seen at different levels and paleo-mudflats have been related to regression. The south Gujarat coast is relatively uniform and is indented by a series of creeks, estuaries, marshes, and mudflats. The Gujarat coast, from Great Rann to the south Gujarat coast, presents evidence for both emergent and submergent coasts.

The Maharashtra, Goa, and northern Karnataka coasts are characterized by pocket beaches flanked by rocky cliffs, estuaries, bays, and at some places mangroves. Beaches in southern Goa and some places along northern Karnataka, however, are long and linear in nature with sand dunes. The Mandovi and Zuari estuarine system in Goa is the largest in this part of the coast. Mudflats are found mainly along estuaries and creeks. Rocky promontories on the Maharashtra coast are made up of Deccan basalts whereas in the south they are mainly of granite gneisses. A number of raised platforms can be seen all along the coast. There are a few islands along the southern parts of this coast near Karwar. This coastal stretch is typical of a cliffy coastline with raised platforms and strong evidence of a submergent coast. The beaches in Goa and northern Karnataka are well-studied and classified as stable beaches with seasonal morphological changes and annual cyclicity (Nayak, 1993).

The southern coast of Karnataka is characterized by long linear beaches, estuaries, spits, mudflats, shallow lagoons, islands, and a few patches of mangroves. Satellite image studies revealed northward shifting of the mouth of estuaries along this coast (SAC, 1992). Beach erosion is severe in some areas along this stretch. The Kerala coast is known for the presence of laterite cliffs, rocky promontories, offshore stacks, long beaches, dunes, estuaries, lagoons, spits, and bars. Using Landsat images, three sets of sand dunes have been identified. The mud banks are unique transient nearshore features appearing during monsoons (Mathew and Baba, 1995) at Kerala. They are unique phenomenon occurring at particular locations along the Kerala coast during the southwest monsoon season, which act as natural barriers to coastal erosion. Along the coast, sand ridges, extensive lagoons, and barrier islands (700 landlocked islands) are indicative of a dynamic coast. About 420 km of the 570 km coastline is protected by seawalls and about 30 km of the coast is undergoing severe erosion. Maximum loss of material has been reported along the southern sections. The predominant southwest wave approach during monsoons, result in northerly littoral drift with varying speed. Some parts of the Kerala coast are known for rich heavy-mineral deposits. The characteristic coastal geomorphology provides an ecosystem, which supports both agriculture and fisheries. Evidences of both emergent and submergent coasts are available for the southern Karnataka and Kerala coast.

East coast of India

The deltaic systems of the east coast experience the high sedimentation rate and periodic cyclones which result in extensive floods.

The east coast in the south, along Tamil Nadu and Pondicherry, is straight and narrow except for indentations at Vidyaranyam. The major landform along this coast is the presence of a large delta formed due to the Cauvery River and its tributary system. The other landforms are mudflats, beaches, spits, coastal dunes, rock outcrops, salt pans and strand features. At a few places mangrove systems, and at Gulf of Mannar and Rameshwaram fringing and patchy reefs, are seen. Deposition and erosion have been reported at different beaches along this stretch. Rich heavy-mineral deposits have been reported at Muttam–Manavalakuruchi.

The coastline of Andhra Pradesh, mainly the deltaic coast, is 640 km long and comprised of bays, creeks, extensive tidal mudflats, spits, bars, mangrove swamps, marshes, ridges, and coastal alluvial plains. Inundations are seen in the extreme south of the Andhra Pradesh coast, that is, in the saltwater lagoon of Pulicate lake and also between the Godavari and Krishna deltas. The Kolleru lake, situated in the interdelta, formed due to coalescence of the deltaic deposits of the rivers and later it cut off from the sea (SAC, 1992), is shrinking on the northern side. The deltaic and southern coasts are rich in agriculture and aquaculture production. The deltaic coast is well vegetated with mangroves. The Pulicate lake has extensive tidal flats and a 12 km long spit. In the north, residual hills and ridges are seen close to the sea. Rocky outcrops and bay beaches are seen here. Storm wave platforms, sea caves in rocks, and cliffs are common coastal features in the north. A critical examination of the relief chart off the region around the Krishna River confluence has indicated the presence of extensive banks in the shelf zone (Varadarajulu *et al.*, 1985). The islands of the Krishna delta front are intertidal and submerged to a large extent during spring tide. The Krishna delta front has been growing through spits and barrier bars (SAC, 1992).

The Orissa coast is a site of deposition formed and controlled by the Mahanadi and Brahmani–Baitarani deltas. Mudflats, spits, bars, beach ridges, creeks, estuaries, lagoons, flood plains, paleo-mudflats, coastal dunes, salt pans, and paleo-channels are observed all along the Orissa coast. The Chilka lagoon is the largest natural water body of the Indian coast. The inlet mouth of Chilika lake is exposed to high annual northward littoral drift and observed to migrate about 500 m northward per year (Chandramohan *et al.*, 1993). The width of beaches at Orissa vary. Littoral transport of sediments in the coastal region is a strong process. The coast is also exposed to severe cyclones. Turbidity in the nearshore as well as in the estuarine region is very high. Progradation of the coastal region in the north of the Devi estuary, and drifting of beaches has been observed. The Bhitarkanika and Hatmundia mangrove reserves are as extensive as 190 km^2. Gopalpur is rich in heavy minerals. Prominent and well-developed sand dune deposits containing monazite, zircon, rutile, ilmenite, and sillimanite occur along the southern coast of Orissa.

The West Bengal coast represents a typical deltaic strip with almost a flat terrain. The Hooghly and its distributaries form the conspicuous drainage system and forms an estuarine delta. The major geomorphic features are mudflats, bars, shoals, beach ridges, estuaries, a network of creeks, paleo-mudflats, coastal dunes, islands like sagar and salt pans. The Sundarbans, one of the largest single block of halophytic mangrove systems about 1,430 km^2, of the world need a special mention.

SriLanka

SriLanka has a coastline of about 1,700 km including that of the Jaffna lagoon. It is a tectonically stable tropical island consisting mainly of Precambrian rocks, and in the northwest, Miocene limestones, and Quaternary sediments. The central portion of the island is a highland surrounded by lowland coasts. Two-thirds of the island's coastline consists of sandy beaches bounded by Precambrian headlands (Swan, 1979). The remaining one-third of the coastline, in the northwest and north, consists of sedimentary rocks. Beach material is predominantly terrigenous. Coastal dunes occur along some sections depending on prevailing energy conditions.

The continental shelf between the Gulf of Mannar and Pak Strait in the northwest and north, respectively, is considerably wide across to India. Elsewhere it is narrow.

The coastline of the island is affected by northeast and southwest monsoons. Wave energy is relatively low in the north and northwest because of shallow seas and barriers. In general, however, beaches are open to seasonal strong wave action. In the north and northwest where energy is low, sheltered lagoons with mangroves, estuaries, barrier beaches, spits, and tidal flats are common. Corals forming fringing and small barrier reefs are also seen. Beaches here are narrow and composed of coarse calcareous material. In the Pak Bay and Gulf of Mannar many depositional morphological units are seen. They include, intertidal barriers, multiple sand bars, dunes, and in the southern part of the Gulf of Mannar a stable sand spit growing toward the northwest. Net sediment movement toward the north along the west coast causes this spit to maintain a stable sand body. From this spit to Colombo in the South, important morphological features that are seen are relict beaches, sand dunes, flood plains, deltas, lagoons, and swamps backed by raised beaches. Dune deposits here overlie limestone. From Colombo, further to the south, lateritized Precambrian rocks form promontories. A sandstone reef offshore, opposite Colombo, acts as a barrier to incoming large waves and the supply of sand material. Raised beaches up to 8 m above sea level are seen at Colombo. Further south the coastline is smooth and sandy, with bays and headlands, backed by raised beaches, flood plains, swamps, and laterite terraces. Along the southwest coast, wave energy is high and sand supply is poor and therefore the coast is undergoing severe coastal erosion. Yun-Caixing (1989) studied coastal erosion and protection using remotely sensed data between Colombo and the southernmost point of the island. The southernmost part of the island consists of low platforms of resistant granitic rocks. The coast here is indented and morphologic units seen are promontories, cliffs, barrier beaches, lagoons, and swamps.

The east coast adjacent to the southern tip, consists of wide coastal plains and low coastlands. Headlands are spaced far apart, and behind long barrier beaches are lagoons, and estuarine deltas. Sand-rich rivers traverse this sector. This change in coastal morphology is in response to a change in geological structure (Cooray, 1967). Further north along the east coast, there are two linear submarine structures, namely the Great and the Little Basses (reef) ridges. These ridges are composed of calcareous sandstone (Throckmorton, 1964). The landforms of bedrock and sand dunes are replaced by broad flood plains, river terraces, and lagoons, further north small barrier beaches are present. A series of large lagoons which are interconnected are called Batticaloa lagoon, a major feature along east coast. Further north, the coastline is made up of bays and headlands of coral, backed by beach ridges and lagoons. Estuaries, deltas, lagoons, and bay-head barrier beaches are common features along the coast. Along the northeast, bays and headlands backed by raised beaches, lagoons, and low residual rises are the common morphological features. Old beach deposits and dunes are seen, which are rich in ilmenite and rutile minerals.

Bangladesh

The coastline of Bangladesh is around 654 km long from the Indian border in the west to the Myanmar border in the east. This excludes tidal channels and delta estuaries. If estuaries, islands, and tidal channels are included it is more than 1,320 km long. The Bangladesh coast is divided into four parts from west to east; Sundarbans, cleared Sundarbans, Meghna, and Chittagong. Except for the last one Chittagong, the coastline is low, swampy, and rapidly changing and composed of sediments of the Quaternary period in large alluvial basins. The source is from two vast river systems, the Ganges and the Brahmaputra.

The Sundarbans are thick mangrove and nipa palms swamps, with a total distance of about 280 km (about 195 km in Bangladesh) from the Hooghly River in India to the Tetulia River in Bangladesh. About 68 km long, sundarban forests have been cut and destroyed. This area, is presently, being used for extensive farming. Tidal estuaries, flat marshy islands, creeks and channels, banks of soft muds and clays with thick mangrove and nipa palms are characteristic features of the Sundarban coast. It represents the older deltaic plain of the Ganges with the presence of old beach ridges in the western swamps. The Meghna is a single main channel which after collecting water and material from the Ganges, Brahmaputra, and Meghna rivers, joins the Bay of Bengal. The characteristic feature is the series of extensive shoals called the Meghna flats developed at the mouth of the River Meghna. These shoals are barren mud and sand bodies. This strongly supports a drowned coastal region.

The geomorphic history of the deltaic plain, which includes Sundarbans and Meghna, is continuous shifting of the river course. In recent times, the Ganges has shifted to the east, resulting in the Meghna as a major course. The shifting is explained as tectonic by Morgan and Mcintire (1956). Sediment supply and tectonic history at the delta with reference to the last glacial period is explained by Chowdhury (1996).

The Chittagong coast extends 274 km between two rivers, namely River Feni in the north and River Naf on the Myanmar border. Small beaches and broad sand flats between headlands along this coast are the common features. There are many islands and shoals found along this stretch of the coast.

Myanmar

The Myanmar coast is about 2,300 km long from the Bangladesh border to the border of Thailand. The coast is divided into three parts namely the Arakan, Irrawaddy, and the Tenasserim.

The Arakan coast runs parallel to a mountain chain of strongly folded Mesozoic and Tertiary rock. Near the Bangladesh border, the coast is elongated with steep-sided rocks and islands, but further south the coast consists of estuarine channels, mangrove forests, patchy coral reefs, and islands. The coast is an example of an emerged coast with many raised beaches and old sea cliffs. Another significant feature of this coast is the presence of mud volcanoes which form temporary islands. With wave action, coming in slowly, they transform to shoals.

The Irrawaddy delta coast runs west to east, and is a large delta with deposition of silt and sand. The delta features a number of shoals, estuarine distributaries, channels, and mangrove forests. From the delta region a large volume of sediment is shifted to the east to the Gulf of Martaban by southwest waves during monsoons.

The Tenasserim coast is composed of rocky promontories, valleys, estuaries, mangrove-fringed creeks, and sand spits. Estuarine lagoons and bays are silted up and transformed into mangrove swamps and saline marshy lands. Beaches are rich in heavy minerals, namely ilmenite and monazite. Some beaches are also backed by coastal sand dunes.

G.N. Nayak

Bibiliography

Ahmad, E., 1972. *Coastal Geomorphology of India*. New Delhi: Orient Longman.

Baba, M., and Thomas, K.V., 1999. *Geomorphology of the Indian Coast, Strategy for Sustainable Development in the Coastal Area*. New Delhi: Ministry of Environment and Forests.

Bird, E.C.F., and Schwartz, M.L. (eds.), 1985. *The World's Coastline*. New York: Van Nostrand Reinhold Company Inc.

Chandramohan, P., Sanilkumar, V., and Nayak, B.U., 1993. Coastal processes along the shorefront of Chilka lake, east coast of India. *Indian Journal of Marine Sciences*, **22**(4): 268–272.

Chowdhury, K.R., 1996. Coastal evolution in Bangladesh during the Holocene period. *30 International Geological Congress, Beijing*, China, **2**.

Cooray, P.G., 1967. *Introduction to the Geology of Ceylon*. Colombo: National Museums.

Mathew, J., and Baba, M., 1995. Mudbanks of Southwest Coast of India—Part II wave–mud interactions. *Journal of Coastal Research*, **II**(1): 179–187.

Morgan J.P., and Mcintire W.G., 1956. Quaternary geology of the Bengal Basin. *Coastal studies Institute Technical Report 9*. Baton Rouge: Louisiana State University.

Nayak, G.N., 1993. *Beaches of Karwar—Morphology, Texture and Mineralogy*. Panaji, India: Rajhauns Vitaran.

Rajendran, C.P., Rajagopalan, G., and Narayanaswamy, 1989. Quaternary geology of Kerala—evidence from radio-carbon dates. *Journal of Geological Society of India*, **33**: 218–222.

Snead, R.E., 1964. Active mud volcanoes of Baluchistan, West Pakistan. *Geographical Review*, **54**: 546–560.

Snead, R.E., 1966. *Physical Geography Reconnaissance Las Bela Coastal Plain, West Pakistan*. Baton Rouge: Louisiana State University Press.

Space Application Centre (SAC), 1992. *Coastal Environment, Scientific Note*. Ahmedabad: SAC.

Swan, B., 1979. Areal variations in textures of shore sands. *Srilanka Journal of Tropical Geography*, **49**: 72–85.

Throckmorton, P., 1964. Appendix A. In Clarke, A.C., *The Treasure of the Great Reef*, London: Barkar.

Vaidyanathan, R., 1987. Coastal geomorphology in India. *Journal Geological Society of India*, **29**: 373–378.

Varadarajulu, R., Harikrishna, M., Chitti-Babu, P., and Chakravarthy, P., 1985. Geomorphology of the coastal zone around Kristen River confluence, Mahasagar, **18**(2): 265–272.

Yun-Caixing, 1989. Analysis of coastal erosion and inundation using remotely sensed data, *China-Ocean-Engineering*, **3**(2): 229–242.

Cross-references

Barrier Islands
Coral Reef Coasts
Coastal Lakes and Lagoons
Desert Coasts
Indian Ocean Coasts, Coastal Ecology
Indian Ocean Islands, Coastal Ecology and Geomorphology
Mangroves, Geomorphology

INDIAN OCEAN ISLANDS, COASTAL ECOLOGY AND GEOMORPHOLOGY

Geographically, the Indian Ocean islands (Figure I21) range from oceanic to continental, geologically from volcanic, limestone, granite, metamorphic to mixed, and physiographically from low to high. Most of these types of islands, though, are not sharply separated.

Oceanic islands are those considered never to have been part of, or connected with, any continental landmass. Their biota is commonly poor in diversity, with unbalanced or uneven representation of taxa, compared with those of continents or continental landmasses. Chagos archipelago, Diego Garcia, and Cocos-Keeling are some such examples. Continental islands may vary in dimensions from subcontinental sizes down to small rocky outposts, the essential characteristics being their continental type rocks and their history showing a former land connection to an adjacent continent. Madagascar, the Malay archipelago, Seychelles, Sri Lanka, and Indonesian islands are examples of continental islands. Seychelles in the north-west Indian Ocean is an extreme case that is totally isolated today, but in Mesozoic and possibly early Tertiary time it was connected to Madagascar. Volcanic islands are rather small (1–100 km across) but often very high, ranging in elevation from 500 m to 3,000 m. They occur generally in irregular clusters, in sub-rectangular patterns or in long lines. Coral islands appear either as an accumulation of coral sand and gravel on the surface of coral reefs or as a slightly emerged limestone platform of formerly live coral not more than a few meters above mean low water. Barrier Islands are constructed entirely by the terrigenous or bioclastic sands from barrier beaches and are built up by longshore drift, probably first as offshore bars, and gradually gaining size later by eustatic oscillations, dune building, and colonization by vegetation.

Because of the very high number of the islands within some island groups in the Indian Ocean it would be difficult to describe them all. Instead, salient features of major groups are given below (see also Table I3).

Western Indian Ocean

Gulf of Kachchh islands

The 42 islands of the Gulf of Kachchh (22°15′N–23°40′N; 68°20′–70°40′E) are the northernmost coralline or sandstone based islands in India. Almost uninhabited, the vegetation inland consists only of shrubs. Several of the islands have dense mangrove patches on the coast, 34 islands have fringing reefs (often called as patch reefs) confined to intertidal sandstones or wave-cut, eroded, shallow banks. The region is tectonically unstable and evidence of uplift can be seen in the form of raised reefs near the mouth of the Gulf, not far from extant islands.

The coastal geomorphology and the fauna and flora of the islands are influenced considerably by the sediment depositional regime, high-velocity tidal currents (up to 5 knots), and a large range in environmental parameters (e.g., temperature 15°–30°C, salinity 25–40). The extreme conditions also limit coastal biodiversity to 37 species of corals and a smaller number of other invertebrates. However, algal growth along these coasts can be substantial at certain times of the year. The mangroves already constrained by high salinity and high tidal exposure also have been heavily impacted due to felling for fuel and fodder. Areas around some of the islands have earlier been good pearl oyster and chank fishing grounds, and one of the islands is even called Chank island. However, overexploitation has decimated both these fisheries.

Laccadive–Chagos ridge

Lakshadweep islands. These are the northern-most islands of the Laccadive–Chagos ridge (9°–12°N; 72°–74°E). Located about 200–400 km off the southwest coast of India, this part of the ridge comprises of 12 atolls, 3 reefs, and 5 submerged banks. Of the 36 islands on the atolls, only 10 (Minicoy, Kalpeni, Andrott, Agatti, Kavaratti, Amini, Kadamat, Chetlat, Kiltan, and Bitra) are inhabited. The northernmost Bitra Island is the smallest inhabited island in India. Among these, Minicoy is separated from the rest by the 9° channel. It is culturally and linguistically closer to the Maldivian islands.

Basically coralline, and no more than a few square kilometers in area, all these islands are low-lying with profuse coral growth all around. The only cultivated plant is coconut, besides a few vegetable and horticultural

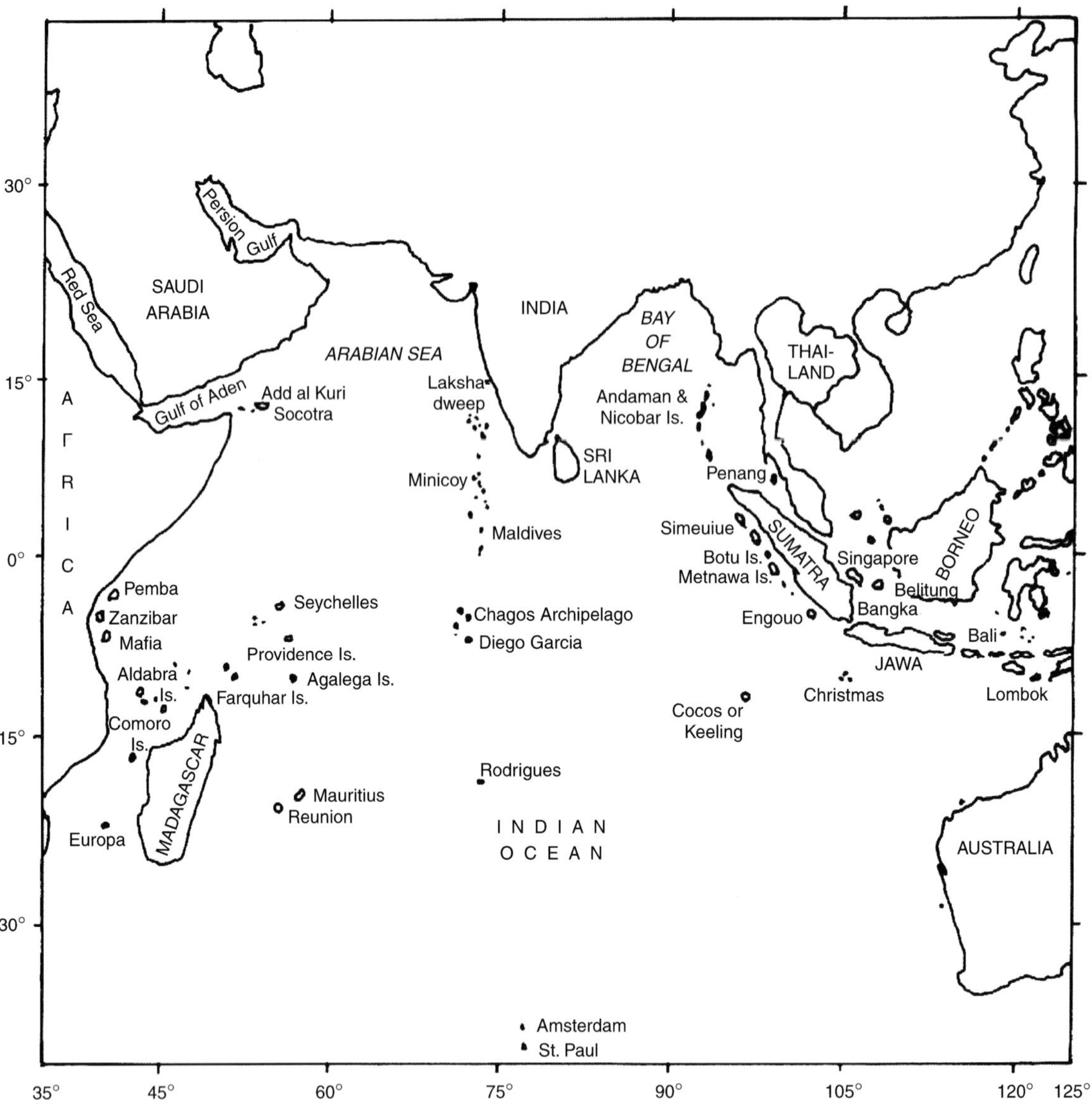

Figure I21 Major islands of the Indian Ocean.

plants introduced from the mainland. The coast toward the lagoon is sandy and habited by sand dune flora *Spinifix* and *Ipomea*. The seaward shore is rocky and typical of all oceanic atolls, with a steep drop in profile. Radiocarbon dating of the storm beach at Kavaratti island gave an age of about 6,000 BP indicating their recent origin. Some of the uninhabited islands are only sand cays; one of them is an important nesting ground for seabirds.

The littoral and sublittoral fauna and flora have been studied reasonably well. The known biodiversity status is as follows: hard corals—104 species, soft corals—37 species, fishes—163 species, invertebrates—about 2,000, and algae—119 species. The bleaching event of 1998 has, however, caused a serious reduction in coral biodiversity. Shore erosion and silting are additional causes for loss of coral cover and reduction in species abundance.

Maldivian islands. The double chain of Maldivian islands (7°N–0.5°S, 73°E) is the largest part of the Laccadive-Chagos ridge that extends southwards from India to the center of the Indian Ocean. The more-than 1,200 islands, clustered in 19 groups of atolls, are entirely low-lying. The geologic history of the island chain is a complicated picture of sea-level changes, reef and carbonate platform development, and erosional events.

As would be expected in the case of small islands on the atolls, the coastal ecology of the islands is reef-dominated. The reefs, though principally atolls, have the unusual features of broken rims that consist of numerous patches or faroes, many of them with islands, and the presence of lagoonal islands which are simply knolls with their emergent surfaces capped with vegetation.

Biological and ecological information on Maldivian islands is rather poor, with stress on only some groups. About 200 species of corals under 60 genera have been recorded so far. No comprehensive checklist of other groups of coastal marine fauna exists; however, the cowry shells and groupers (40 species) are important components of reef biodiversity. Similarly, descriptions of zonations of the reefs are known from some islands but detailed studies of the ecology, either at community or at species level, are scarce.

Islands of Chagos archipelago. The Chagos archipelago (5°–8°S 71°–73°E), the southern part of the Laccadive–Chagos ridge, consists of five coral atolls with islands, besides several reefs that are partially or wholly exposed at low tide. The five atolls are: Great Chagos Bank, Peros Baňhos, Salomon, Egmont, and Diego Garcia. The number of islands on these atolls varies from 4 in Diego Garcia, to 24 in Peros Baňhos. The total land occupied by these islands is about 40 km^2.

Most of these islands are located on the atoll rims with elevation of no more than 2–3 m. Raised reefs with small, uplifted, and vertical cliffs rising to over 6 m occur in two atolls—southern Peros Baňhos and northwestern part of the Great Chagos Bank. Isotopic dating of fossil corals in the emerged beach rock of the islands and some extant corals

Table 13 Indian Ocean Islands

Name of island	Position	Location	Geomorphology	Major disturbance
Barrow islands (Middle Boodie, Pasco & Double islands)	20°40′–20°58′S, 115°18′–115°30′E	West Australia (56 km off Pilbara coast)	Fossiliferous Miocene Limestone	Oil fields
Christmas island	10°35′S, 105°35′E	290 km south of Indonesian islands	Guano deposits	Anchorage and phosphate loading dock
Cocos–Keeling islands	12°00′S, 96°56′E	South east Indian Ocean	Atoll	Storms, earthquakes, red tides
Dampier archipelago (six main and many small islands)	20°20′S–20°50′S, 116°20′E–117°10′E	Eastern Indian Ocean (northwest Australia)	Igneous rock of Archean age	Dredging for shipping
Houtman Abrolhos islands (Total 35 islands)	28°15′–29°00′S, 113°30′E–114°05′E	Eastern Indian Ocean		Oilfield
Kimberley coast reefs		Northwestern coast of Australia	Steep rocky shores	Iron ore mining
Broome	17°58′S, 122°14′E			
Wyndham	15°28′S, 128°06′E			
Holothuria and Lacepede reefs	16°52′S, 122°08′E			
Adele	15°31′S, 123°09′E			
Monte Bello islands and Lowendal islands	20°20′S–20°33′S, 115°27′E–115°37′E	West northwest of Dampier	Vacant Crown islands	Tourism and oil exploration activities Reef damage by *Acanthaster plancii*
Muiron island and Ningaloo island	21°40′S, 114°2′E	Eastern Indian Ocean	Limestone islands. Part barrier and part fringing reefs	Overexploitation of marine resources Petroleum tenements
Rowley shoals		Northwest Australian shelf	Atoll	Fishing, anchor damage
Mermaid reefs	17°7′S, 119°36′E			
Clarke reef	17°10′S, 119°20′E			
Imperieuse reef	17°35′S, 119°56′E			
Scott reef and Seringapatam reef	14°5′S, 121°51′E 13°40′S, 122°00′E	–	Atolls	Exploitation of molluscs and holothurians
Bahrain Mubarraq, Sitra and other islands	26°N, 50°E	Persian Gulf, Saudi Arabia coast	Limestone and deserts	Oil and gas
Lampi island (30 islands)	10°50′N, 98°10′E	Near Burma	Coral reef	Undisturbed
Moscos island	13°50′N–14°20′N	Burma Sea	Rocky shoreline	Illegal logging, collection of turtle eggs
Chagos archipelago				
Blenheim reef	5°12′S, 72°29′E	East Chagos Archipelago	Atoll	Undisturbed
Diego Garcia	7°20′S, 72°25′E	South of Chagos	Atoll	Military presence, dredging and blasting
Egmont atoll (five islands)	6°40′S, 71°20′E	West of Chagos	Atoll	Undisturbed
Peros Bañhos atoll	5°20′S, 71°55′E	Northwest of Chagos	Atoll	Undisturbed
Salomon atoll	5°20′S, 72°15′E	North of Chagos	Atoll	Undisturbed
Speakers bank	5°32′S, 72°25′E	North of Chagos	Submerged atoll	No manmade changes
Victory bank	5°32′S–72°15′E	–	Submerged atoll	Undisturbed
Comoros island	12°S, 44°E	North of Mozambique	Four high islands, fringing reef	Dredging
Mayotte barrier reef	12°30′S, 45°10′E	Southernmost islands of Comoros group	Extinct volcano surrounded by reef	Siltation from erosion, mining of coral rocks
Andaman islands	10°30′N–14°N, 92°E–93°E	Bay of Bengal	Emerged part of mountain chain	Massive siltation due to deforestation
Nicobar islands	6°30′N–9°30′N, 93°E–94°E	Bay of Bengal	Emerged part of mountain chain	Overexploitation of corals and shells for ornamental use
Lakshadweep islands	9°N–12°N, 72°E–74°E	Arabian Sea	Atolls	Dredging, mining, siltation, and *A. plancii* infestation
Indonesian islands				
Kepulauan Aru	0°50′–1°25′N, 131°16′E	Irian Jaya	Coral island	
Pulau Mapia	1°1′N–134°10′E	Irian Jaya	Coral island	
Raja Ampat (proposed wildlife reserve)	0°25′N–130°08′E	Irian Jaya	Coral island	
Sabuda-Tataruga	2°30′N–130°50′E	Irian Jaya	Coral island	

Table 13 *(Continued)*

Name of island	Position	Location	Geomorphology	Major disturbance
Karimunjawa (proposed wildlife reserve)	5°50′N–110°20′E	Java	Coral island	
Pulau Seribu (strict nature reserve)	5°26′–5° 37′S, 106°24′–106°37′E	Java	Coral island	Fishing activities
Karimata	1°30′S, 109°00′E	Java Kalimantan	Coral island	
Pulau Maratua (Karang Muaras proposed strict nature reserve)	1°50′–2°10′N, 118°30′–118°55′E	Java Kalimantan	Coral island	
Pulau Sangalaki (Marine park)	2°05′N–118°15′E	Java Kalimantan	Coral island	
Pulau Semama (wildlife reserve)	2°05′N–118°15′E	Java Kalimantan	Coral island	
Aru Tenggare (proposed marine reserve/marine park)	6°35′S – 7°11′S, 134°12′– 135°E	Moluccas	Coral island	Fishing and exploitation of marine resources
Kepulauan Kai Barat Tayandu (proposed marine multiple use reserve)	5°30′S, 133°0′E	Moluccas	Coral island	
Pulau Angwarmase	7°55′S, 131°20′E	Moluccas	Coral island	
Pulau Banda (marine park)	4°30′S, 130°E	Moluccas	Volcanic origin	
Pulau Pombo (marine park)	3°31′S, 128°22′E	Moluccas	Coral island	Overexploitation
Pulau Renyu–Pulau Lucipara (proposed strict nature reserve)	5°40′S, 127°50′E	Moluccas	Coral island	Overexploitation
Komodo National Park	8°35′S, 119°30′E	Nusa Tenggara	Rocky island	Dynamite fishing
Pulau Rakit	8°35′S, 118°E	Nusa Tenggara	Coral island	
Pulau Satonda	8°10′S, 117°45′E	Nusa Tenggara	Volcanic	
Kepulauan Peleng	1°50′S, 123°14′E	Sulawesi	An archipelago	
Kepulauan Sangihe	3°45′S, 126°35′E	Sulawesi	Islands and reefs	
Pulau Kakabia	6°55′S, 122°30′E	Sulawesi	Islands and reefs	
Pulau Pasoso	0°10′S, 119°45′E	Sulawesi	Limestone island	
Pulau Smalona	5°10′S, 119°10′E	Sulawesi	Coral islands	
Selat Muna	5°05′S, 122°05′E	Sulawesi	Coral islands	
Spermonde islands	5°30′S, 119°10′E	Sulawesi	Coral islands submerged reef, patch reefs	Dynamite fishing, sedimentation
Tiga island	3°50′S, 123°15′E	Sulawesi	Islands and reefs	
Togian island	0°10′N–0°40′S, 121°32′E–122°12′E	Sulawesi	Extended mountains surrounded by limestone	Dynamite fishing
Tukang Besi	5°35′S, 123°50′E	Sulawesi	Archipelago	
Muara Siberut (five islands)	1°30′S, 99°25′E	Sumatra	Islands with coral reefs	Overfishing
Pulau Weh	6°0′S, 95°30′E	Sumatra	Coral reefs	
Sheedvar island	27°0′N, 53°E	Iran	Coral reefs	
Pulau Paya (group of islands)	6°02′N–6°05′N, 99°54′–100°04′E	Malaysia	Coral islands	Fishing activities
Pulau perak	4°41′N, 98°56′E	Malaysia	Conical rocky islands	
Pulau Tiga	5°44′N, 115°40′E	Malaysia	Volcanic island	Legal protection and management
Sembilan islands	2°03′N, 100°33′E	Malaysia	Volcanic island	
Pulau Bohey dulong Pulau Bodgaya Pulau Tetagan	4° 38′N, 118°46′E	Malaysia	Volcanic island	Domestic pollution, overexploitation of marine fauna
Pulau Sipadan	4°05′N, 118°40′E	Malaysia	Coral reef	
Pulau gaya Pulau Sapi Pulau Mamutik Pulau Manukan Pulau Sulung	6°04′N–5°55′N, 116°E	Malaysia	White sandy beaches with rocky interruption	Coral mining, overexploitation of reef fauna
Maldives (a group of atolls)	7°N–0°30′S, 73°E	Central section of Chagos archipelago	Atolls	Growing urbanization, domestic pollution
Mauritius	20°S, 58°E	180 km northeast of Réunion island	Remains of old emerged coral reefs	
Cargados Carajos shoals	16°23′S, 59°27′E	North northeast to Mauritius	Coral islands	Exploitation of marine products, Guano mining
Île Plate	19°53′S, 57°39′E	–	Volcanic rock	Dynamiting, fishing, anchoring, and boat grounding

Table 13 *(Continued)*

Name of island	Position	Location	Geomorphology	Major disturbance
Rodrigues	19°42′S, 63°25′E	North eastern Mascarene island	Volcanic island	Public interference
Mozambique				
Ilhas da Inhaca e dos Portugueses (Inhaca islands)	26°S, 33°E	Southern most island of Mozambique	Pleistocene dune rock –	Goat overgrazing denudation, public interference
Primeira and Segundo islands	16°S–17°S, 38°E–41°E		–	
Quirimba islands	10°45′S–12°42′S, 41°E		–	
Réunion islands				
Réunion	21°7′S, 53°32′E			
Europa and Basses de India	22°20′S, 40°20′E	In Mozambique channel	Atolls	Undisturbed
Iles Glorieuses	11°30′S, 47°20′E	100 km northwest of Europa	Atolls	Undisturbed
Tromelin	15°52′S, 54°5′E	390 km east of Madagascar	Atolls	Undisturbed
Saudi Arabia				
Al Wajhto Qalib	26°16′N, 36°28′E	Northern Red Sea	Rocky island	Undisturbed
Farasan archipelago	17°40′N, 42°10′E	Red Sea	Coral islands, swamps, and reefs	Undisturbed
Seychelles	5–10°S, 45–56°E	NE of Madgascar		
Platte island	5°50′S, 55°E		Granitic and coralline islands	– –
Doivre island	5°50′S, 55°E			
Aldabra island	9°25′S, 46°25′E	North of Mozambique channel	Atoll, limestone	–
Bird island	3°43′S, 55°13′E	Northern edge of Seychelles	Sandy clay and calcareous	–
Dennis island	3°48′S, 55°40′E	–		Phosphate mining
Cousin islands	4°20′S, 55°40′E	Southwest Seychelles	Granite island	Tourism
Curieuse island	4°05′S, 55°43′E	–	–	–
Singapore		–	–	Growing urbanization
Pulau Hantu	1°20′N, 103°50′E			
Pulau Suelong	–			
Pulau Salu	–			
SriLanka				
Trincomalee and Pigeon island	8°N, 82°E	16 km from Trincomalee	Extended part of Indian mainland (limestone-granite)	Mining for lime, dynamite fishing, tourist pressure, overexploitation of marine fauna
Sudan				
Mukkarwar island	20°50′N, 37°17′E	North of port Sudan	Well-formed reefs	
Sanganeb atoll	19°45′N, 37°25′E	Red Sea	Atoll	*A. plancii* infestation, overexploitation
Suakin archipelago	19°14′N, 37°51′E			Fishing, tourism pollution, and human interference
Sultanate of Oman				
Daymaniyat islands (10 islands)	23°45′N, 58°10′E	Gulf of Oman	Corals	Litter, *A. plancii* infestation
Tanzania				
Latham island	6°50′S, 39°50′E	64 km off Dar es Salaam	Atoll	Undisturbed
Mafia island	7°40′S, 40°40′E	South of Dar es Salaam		–
Maziwe island	5°30′S, 39°5′E	Northern Tanzania		–
Thailand				
Koh Larn	12°55′N, 100°47′E	Northern Gulf of Thailand	Coral reefs	Tourism and pollution
Koh Sak				
Koh Krok				
Koh Phuket	8°N, 98°20′E	Andaman Sea	Rocky extended mountains	Extensive sedimentation, dredging, mining

gave an age not more than 5,200 years BP, indicating that all these islands are relatively recent in origin.

Being coralline islands, the coastal morphology is typically characterized by vast and profuse reef growth. Mangroves and associated flora are absent. The coastal and inland flora consists primarily of native vegetation (*Tournifortia*, *Scaevola*, and *Casuarina*), disturbed by coconut plantations in inhabited islands. None of the 250 species of flora are endemic. Two faunal groups—birds and turtles—are important biological components. The islands provide nesting grounds for over 50 species of seabirds and several species of green and hawksbill turtles.

Seychelles

A total of about 42 granitic and 74 coralline islands, spread over 5°–10°S and 45°–56°E, with a total land area of 455 km^2 comprise the Seychelles. The inner Seychelles islands to the north are granitic, remnants of ancient Gondwanaland, rugged, mountainous, and rise up to 1000 m in the Morna Seychellois on Mahé island. Coralline islands and atolls of the Amirantes, Farquhar, and Aldabra groups, spreading westwards and southwards from the granitic group, constitute the second group. They are composed of numerous low islands or atolls in several clusters, each located on top of volcanic structures of various sizes. Principal granitic islands in the Seychelles archipelago are Mahé, Praslin, Silhouette, La Digue, Curieuse, Felicité, North Island, St. Anne, Providence, Frigate, Denis, Cerf, and Sea Cow island. Among the coralline islands and atolls, Alphonse, Bijoutier, St. François, St. Pierre, Astove, Assumption, Coetivy, and Aldabra are the major ones. Some of the coralline islands, though relatively low-lying by comparison with granitic islands, are often taller, reaching as much as 8 m above sea level. These include Aldabra, Assumption, Astove, Cosmoledo, and St. Pierre. These limestone atolls have formed on top of volcanic structures and rise from water depths of over 2,000 m. The high limestone islands are also characterized by terraces that reflect changes in sea levels during the last glacial cycle. The Amirantes island group is the second largest group after Inner Seychelles and comprises 10 islands and atolls and several shoals and submerged reefs. The major atolls in the Farquhar (or Providence) group are Farquhar and Providence, each rimmed with several islands. St. Pierre Island in this group is a circular, uplifted atoll, with coastal cliffs rising to 10 m. In the Aldabra group, Aldabra and Cosmoledo atolls have several islands on their rims, whereas Astova and Assumption are elevated atolls: in the Assumption island, reef rocks rise to 7 m above sea level, with dunes on the east and south rising to nearly 30 m.

Coastal geomorphology of all the islands is characterized by the presence of reefs. Three major types—fringing, platform, and atoll reefs—are recognized. The granitic islands have well-developed fringing reefs. Among the outer islands, several are raised platform reefs (e.g., Assumption and St. Pierre) and others are atolls (e.g., St. Joseph, St. François).

There are pronounced differences in the coastal ecology of these islands. The northern islands lie in the path of the east-flowing Equatorial Counter Current whereas the southern islands lie in the path of the west-flowing South Equatorial Current. Besides, the granitic islands receive more rainfall and are often forested. As a result, the nutrient regimes in the coastal waters are distinctly different between these two groups. The high nutrient levels around the granitic islands favor dominance of crustose coralline algae and frondose macroalgae whereas nutrient-poor waters around the coralline islands support a hermatypic coral dominance. The raised reefs were used as nesting sites by seabird colonies and have been extensively mined for guano.

Description of coastal ecology has primarily been with reference to coral reefs, though not all have been studied as extensively as Mahé or Aldabra islands. The coral diversity is more or less same between granitic and outer islands, the former with 51 genera and the latter with 47 genera. The total species count is 161, and this does not include those of *Acropora*. Though the coasts with well-developed reefs can be expected to sustain a high faunal and floral diversity, there is still poor documentation from most islands. The recorded forms include 128 species of marine caridean shrimps, 49 species of brachyuran decapods, 150 species of echinoderms, 450 species of molluscs, about 1,000 species of fish, 8 species of sea grasses and 4 species of turtles. Among these the gigantic land tortoise of Aldabra atoll and the double coconut, *coco de mer*, of Mahé Island are unique.

Mascarene islands

Mauritius. Mauritius (20°S, 58°E) represents the southern part of the Mascarene Plateau, which is an arcuate series of banks extending for 2,000 km from the Seychelles Bank. The Mauritius island is volcanic in origin and is composed of olivine basalt and doleritic basalt.

Mauritius, along with several small adjacent islands, spreads over 1,865 km^2. The northern part of the island is a plain while the center is a plateau rising to a peak height of 826 m at Piton de la Rivière Noire and bordered by low mountain remnants of a large volcano. The south of the island is largely mountainous.

The crenulated coastline, exposed to varying wave activity, extends to about 200 km. The southwest coast is made up of basaltic rocks while carbonate sands, the bulk of which is coral debris, largely cover the remaining parts. A large submarine platform with extensive fringing coral reefs that cover three-fourth of its coastline surrounds the island. There are also the remains of old, emerged coral reefs, recalcified to varying degrees, indicating recent uplift.

Sugarcane and tea cultivation are the major revenue sources for the islands. Conversion of forest lands to plantations has, however, reduced the original forest cover to less than 1%. The other important revenue sources for the island are coastal fishing and tourism. Reef fisheries yield about 200 tons per year and reef tourism caters to more than 300,000 visitors a year.

Considering that the entire coastline is reef-rimmed, the corals- and reef-associated fauna and flora are important components of coastal biodiversity. The reefs cover an area of about 300 km^2, with a maximum reef width of 4 km. Mauritius reefs are notable for the absence of reef flats; consequently, sediments accumulate in the lagoon providing a favorable environment for sea grass growth. A total of six sea grass species—*Thalassiodendron ciliatum, Syringodium isoetifolium, Halophila ovalis, Halophila stipulacea, Halodule universis*, and *Halodule* sp.—are known from Mauritius.

A total of about 186 species of corals belonging to more than 50 genera have been reported from the Mascarene archipelago, with 75% of these recorded in Mauritius reefs. The fish diversity, with 263 species, is also high. The molluscan fauna is another important biological constituent of Mauritius reefs. A detailed survey has revealed the presence of more than 3,500 species, with approximately 10% of them being endemic. These include the Imperial Harp shell *Harpa costata* and the cowry *Cypraea mauritiana*, besides species like *Clanculus mauritianus* and *Bursa bergeri*. Diversity of marine macro algae is also high: 127 species, mainly red and green algae, have been recorded from the littoral zone.

Higher freshwater flux, more siltation, and high humidity favor the growth of mangroves on the northeast and east coasts of Mauritius. Dense mangroves dominated by *Rhizophora mucronata* cover an estimated area of 20 km^2.

Rodrigues. Rodrigues island (area 110 km^2; 19°42′S, 63°25′E) and two cays at 10°24′S and 56°38′E (known as Agalega island) are part of the Republic of Mauritius. Rodrigues, the smallest of Mascarene islands, is of volcanic origin and consists of subhorizontal basaltic flows. The northeastern part of the island is mountainous but not very tall, the peak height not exceeding 400 m.

Like Mauritius, Rodrigues island is also reef-rimmed, with a wide expanse of reef platform extending without a break for 90 km around the island, providing a fringing reef cover of about 200 km^2. Presence of reef flats, with a width ranging from a low of 50 m in the east to as much as 10 km in the west distinguishes Rodrigues from Mauritius. The reef flats also provide habitat for large sea grass beds, though the species diversity is much less; only two species—*H. ovalis* and *Halophila balfouri* are known from Rodrigues. Muddy accumulations, hence development of mangroves, are rare.

Coral species diversity is high, with a similar number of species as in Mauritius. Information on other coastal marine fauna and flora are scarce; however, the small islands around Rodrigues are important nesting sites for brown noddy, lesser noddy and white tern.

Réunion. Réunion (21°7′S, 53°32′E) is the most southwesterly of the Mascarene islands. Covering an area of 2,512 km^2, Réunion is a large Hawaiian-type volcano that includes an older part, the massif of Piton des Neiges (peak height 3,069 m), incised by three large cirques, and an active volcano to the southeast. Several small islands—Tromelin to the north of Réunion, Europa and Bassas de India atolls in the Mozambique Channel, Juan de Nova off the west coast of Madagascar, and Iles Glorieuses to the north of the Mozambique Channel, are the other island dependencies of Réunion. These are relatively very small. Tromelin is no more than a cay of 1.1 km^2, Juan de Nova is a raised fossil reef of 9.6 km^2 and the Grande Glorieuse Island covers only 3.9 km^2.

The coast of Réunion is generally rocky, with low cliffs cut in lavas. Sectors of low coast correspond with the three large depositional cones built below the three cirques and show pebbly beaches while the rare sandy beaches are related to embryonic fringing reefs. Because of the relatively young age, reefs are less developed, discontinuous, narrow with their widest part no more than 550 m, and cover only an area of 7.3 km^2. In contrast with Mauritius and Rodrigues, reef platforms are abundant on Réunion but muddy accumulations and mangroves are totally absent. Lack of organic and terrigenous material also limits the diversity of sea grasses and their extent: only one species, *H. stipulacea*, that too not in abundance, has been known from Réunion.

Coastal marine fauna and flora have been better inventoried in Réunion than in the other two islands. This includes 40 species of dinoflagellates, 150 species of macro algae, 120 species of corals, 90 species of hydroids, 2,500 species of molluscs, and more than 650 species of fish. Coastal fisheries, however, are relatively less developed, with no more than 1,500 tons yield per year, of which only 100–150 tons are truly reef fishes, Tourism, likewise, is also less developed compared with Mauritius.

Madagascar

Madagascar is a fragment of a lost continent (Lemuria) and this severance from the ancient land mass is evident from the sheer drop of mountain into ocean depths of 3,000 m or so, especially on the eastern side. The island's rocks, volcanic structure, besides the subsoil formed of granites, gneiss, and crystalline schists, warm water springs and frequent earthquakes also provide evidence for this origin. Paleontological evidence for this comes from the remains of many large prehistoric birds and even the present day fauna of the island has a special individuality. The large number of endemic flora also confirms that Madagascar is an island that has been long since isolated from other regions. Madagascar is also one of the largest islands in the world (5,87,000 km^2) with a coastline of about 4,000 km. The east and west coasts are asymmetrical in physiography. The east coast presents an almost unbroken appearance, with few bays and indentations. The continental shelf here is narrow and coral reefs and mangroves are poorly developed. The west coast, on the other hand, has a broad continental shelf and has the majority of the island's reefs and mangroves.

Reefs and mangroves are important coastal ecosystems of Madagascar. The reefs cover an area of 200 km^2. Most of the west coast has large tracts of tidal marshes (4,250 km^2) of which 3,200 km^2 are populated by mangroves. Sea grass beds are also extensive on the west coast. Emergent fossil reefs up to 10 m above present sea level are found in the far northwest coast. A barrier reef, 10–16 km offshore, also exists at the edge of the continental shelf.

There is an extensive bibliography of the various coastal and marine fauna and flora, synthesized from numerous studies of many French scientists. Most of the information is from Toliera and Nosy-Bé and biodiversity of the whole island could be still higher than what is known—200 species of corals, 1,500 species of fishes, 28 species of sponges, 227 species of echinoderms, 1,158 species of mollusks, 779 species of crustaceans, 121 species of worms, 182 species of ascidians, 108 species of algae, besides 5 species of turtles and 32 species of mammals.

Comoros islands

The Comoros archipelago consists of four major islands—Grande Comore, Anjouan, Moheli and Mayotte—at the northern end of the Mozambique Channel (12°S, 44°E). All these islands are of volcanic origin and mountainous, and are surrounded by numerous coralline and granitic islets. The Grand Comore is the largest island among these, with an area of 1,131 km^2. While three of these islands have fringing reefs, Mayotte is surrounded by a 140 km long barrier reef lying 13–15 km offshore. The coastal features include the mangroves, which, in some islands, are expanding due to influx of terrestrial sediments from hillsides.

Among these islands, only Mayotte has been studied to some extent. These studies are essentially related mainly to the description of the barrier reef and its faunal and floral composition, since it is one of the few barrier reefs in the world, and the best developed in the Indian Ocean. The reef, which had a good live coral cover, was heavily impacted during a bleaching event in 1983 related to El Nino.

Islands off Tanzania

The islands off the coast of Tanzania are Mafia, Pemba, Unguja (Zanzibar), and those of Songo Songo archipelago. Mafia island (7°40′S, 40°40′E), along with the four small adjacent islands, are continental islands off Rufiji delta. The coastline consists of vast stretches of sheltered and exposed fringing reefs and mangroves besides beds of sea grasses, algae, and soft corals. The Unguja island, slightly larger than the Mafia island, as well as the Mnemba island lying to the north, also have coral development all around the coast. The Pemba island is 62 km long and 22 km wide, with a reef area of 1,100 km^2 along its coastline. All these islands show evidence of several terraces, along with indication of a relatively recent subsidence. Another feature common to these is the 3,300–3,400 m of marine sediments, ranging from Miocene to Cretaceous, underlying them.

The Songo Songo archipelago (8°30′S, 39°30′E) consists of a 7 km long island, with four smaller islands in the vicinity. As with other islands off Tanzania, these islands support some of the largest expanses of shallow water coral reefs, with the estimated reef cover of about 40–50 km^2. No other remarkable coastal features are known from these islands.

Islands off Mozambique coast

The islands off the Mozambique coast are grouped into Quirimbas archipelago, Primeiras and Segundas archipelago, Bazaruto archipelago, and Inhaca and Portuguese islands. Besides these, the Mozambique, Goa, and Cobras islands are located just 4 km off the mainland coast.

The Quirimbas archipelago (10°45′–12°42′S) comprises a 200 km chain of 32 islands along with numerous reef complexes. The Primeiras and Segundas archipelagos, located at 16°12′– 17°17′S consists of 10 islands and two reef complexes. The Bazaruto archipelago consists of five islands located between 21°30′ and 22°10′S. All these islands are small, with the largest no more than 25 km^2 in area and all lie close to the coasts.

Coastal morphology of these islands is composed of grasslands, scrubs, and mangroves, with varying degrees of development in the different islands. All of these, however, support good fringing reef growth. As with most islands, it is the coral fauna and fish that were widely studied. About 50 genera of reef building corals and 300 species of fish are known from Quirimbas archipelago. A total of about 155 molluscan forms, with 6 endemic species among them, have been reported from Bazaruto archipelago. The western coasts and the area between the islands and the mainland coast have good sea grass beds.

Socotra island

The Socotra (8°N, 53°E), off the mouth of Gulf of Aden, is an island that has survived the subsidence of the great primeval continent, which embraced present-day Africa, the Middle East, southern Asia, and the Northwestern part of the Indian Ocean. It is a fairly large island (3,582 km^2) with its mountainous interior rising to 1,520 m. The coastline is varied, consisting partly of low-lying plains and partly of steep limestone cliffs, edging an undulating plateau (500–600 m high) that covers much of the island.

Bahrain

This consists of a group of low-lying islands, largely of limestone outcrop and desert, off the Saudi Arabian coast. Ranging from a rocky out crop (Jidda Island) to the large Bahrain Island (660 km^2), these are low and sandy islands, except for clusters of barren rocky hills in the center.

Eastern Indian Ocean

Gulf of Mannar islands

A chain of 20 islands (8°45′–9°16′N, 79°4′–79°29′E) constitutes the coralline islands of the Gulf of Mannar between southeast India and Sri Lanka. None of these Islands is inhabited. Spreading over not more than 2 km^2 individually, most of these islands have only shrubs as vegetation and occasionally some patches of mangroves. The fringing reef growth in profuse all around the islands. The sea grass beds associated with the reefs have been important feeding grounds for the Dugong species.

Geologically, these islands are connected with those of northern Sri Lanka through a series of shallow banks called Adam's bridge between Rameswaram in India and Talaimannar in Sri Lanka. As a consequence, there is a good similarity in island geomorphology, coastal ecology, and fauna and flora among these islands.

Sri Lanka

Sri Lanka lies off the southeast tip of India between 6° and 10°N and 80° and 82°E (65,610 km^2). The island is basically a central mountain mass of Pre-Cambrian crystalline rocks ringed by a broad coastal plain. The highest point is the Pidurutalagala peak (2,700 m). The plains are fairly level in the north but the extensive soft limestone deposits are broken elsewhere by outcrops of the main rock core. The coastal region is characterized by fringing reefs, shallow lagoons, marshes, and many sandy bars, especially in the north. Though not a very large island, three distinct climatic features are evident: the low country Dry Zone that receives low rainfall, the East Coast Plains that experience one monsoon, and the Low Country Wet Zone which receives rainfall from both the monsoons.

Coastline features vary considerably, from sandstone to granite, but detailed studies are scarce. Along the west coast, coral growth is mainly on ancient sandstone and on the east coast, it is on gneiss or granite outcrops. Fringing reefs are found only along 2% of its 1,585 km coastline and not all of these are comprehensively mapped. Surveys have mainly been carried out for reef-building corals and fishes. A total of about 183 species of corals and over 350 species of fishes have been recorded. Though a number of faunal groups occur in these reefs, their systematic records are not known. Mangroves occur along the southwest, northwest, and northern coasts. The total area covered by mangroves is 36 km^2. *Rhizophora, Avicennia, Excoecaria, Lumnitzera, Aegiceras* and *Sonneratia* are the mangrove genera recorded.

Extensive damage to coastline habitats has been recorded where reefs and mangroves are abundant. The reefs are mined for lime manufacture and the mangrove wood is used to fuel the limekilns. Often coastal forests are also exploited for use as fuel in the limekilns. Though detailed information on the impacts are not available, damages to coastal habitats at local scales are considerable.

Andaman and Nicobar islands

These are the emerged parts of a mountain chain that stretches from the Arakan Yoma in Myanmar to the islands of Indonesia. Spread meridionally between 6° and 14°N, and between 91° and 94°E in the Bay of Bengal, these islands number more than 500, of which only 38 are inhabited. All the islands are mountainous, sedimentary in nature, and have fringing reefs towards the east. The total area covered by these islands is 8,293 km^2. The Andaman group of islands is separated from the Nicobar group by the 10° channel which has a heavy tidal flow and difficult to navigate with conventional crafts. As a result, the biogeography of the Andaman has more of Malay affinities whereas that of the Nicobar has Indonesian affinities.

Most parts of these islands are covered with thick forests and the low-lying areas are covered with mangrove swamps. Biodiversity of the islands is quite high, with an abundance of corals, fishes, algae, turtles, and the unique saltwater crocodile. Avifauna is more endemic in nature, with distinct local species of eagles, parakeets, and orioles. The islands have a few land mammals like deer and elephants, which were introduced from the mainland India.

Indonesian islands

Indonesia is an island nation of 13,700 islands having a coastline of about 60,000 km. The islands form a region of tectonic instability, marked by frequent earthquakes and volcanic eruptions. These tectonic movements have also shifted out of the sea some of the numerous coralline reef formations along the island coast.

Typically tropical in climate, the larger islands (e.g., Java, Sumatra, Borneo, and Sulewasi) have varied coastal geomorphology ranging from mangrove-bordered shores through estuarine deltas to coral reefs. Borneo is the third largest island in the world and lies between the Sulu Sea, Java Sea, and South China Sea. It is densely forested, with extensive swampy lowlands in the southern and southwestern coastal areas.

Smaller islands are more coralline in nature. The Indonesian region is known for the highest diversity of corals and mangrove species, in the latter case practically all known mangrove species from the new world are present here.

Cocos or Keeling islands

These islands (12°S, 96°56′E), numbering 27 and covering a total area of 30 km^2, lie about 960 km southwest of Sumatra (not to be confused with Cocos Island, a small uninhabited island of 26 km^2 area off Costa Rica). The largest among them is the West Island, with an area of 3.2 km^2. These are extremely coralline islands but also with a luxuriant growth of land vegetation that includes coconut palms, sugarcane, and banana. No quantitative accounts of other fauna are available but the coconut eating crab *Birgus latro* is an interesting species from these islands.

Christmas island

Christmas island, with an area of 128 km^2, and lying 650 km south of Java Head, is a strongly uplifted island surrounded by high cliffs of coral limestone. It is well-known for its extremely rich phosphate deposits and the coconut crab *B. latro* (another Christmas island is an atoll in the Line islands, central Pacific).

In summary, the major features of Indian Ocean islands are:

- Wide range in size, from sand cays to some of the largest islands in the world.
- Coralline origin in most of the oceanic regions and granitic or sedimentary origin in coastal regions.
- Coasts characterized by fringing reefs in almost all cases, and mangroves and other wet lands in most others.
- Endemism in some islands, with native flora and fauna.
- Most of the smaller islands are uninhabited. Where settlements have taken place, marked erosion in biological diversity and resources are noticeable.

Suggested further reading on this subject may be found in the bibliography listing.

M. Wafar, Sayeeda Wafar, and Prasanna Yennavar

Bibliography

Bird, E.C.F., and Schwartz, M.L., 1985. *The World's Coastline*. New York: Van Nostrand Reinhold.

Chandra, S., Arunachalan, B., and Suryanarayanan, V., 1993. *The Indian Ocean and its Islands: Strategic, Scientific and Historical Perspectives*. New Delhi/Newbury Park/London: Sage Publications.

Guilcher, A., 1988. *Coral Reef Geomorphology*. Chichester: John Wiley and Sons Ltd.

McClanahan, T.R., Sheppard, C.R.C., and Obura, D.O., 2000. *Coral Reefs of the Indian Ocean: Their Ecology and Conservation.* NewYork: Oxford University Press.

Qasim, S.Z., 2000. *Indian Ocean in the 21st Century: Linkages and Networking*. New Delhi: Sai Publishers.

Sengupta, R., and Desa Ehrlich, 2001. *The Indian Ocean. A Perspective*, Volume 1 and 2. New Delhi and Calcutta: Oxford & IBH Publishing Company.

Cross-references

Atolls
Cliffed Coasts
Coral Reef Coasts
Coral Reef Islands
Coral Reefs
Coral Reefs, Emerged
Indian Ocean Coasts, Coastal Ecology
Indian Ocean Coasts, Coastal Geomorphology
Mangroves, Geomorphology Ecology
Mining of Coastal Materials
Rock Coast Processes
Small Islands
Tors

INGRESSION, REGRESSION, AND TRANSGRESSION

A transgression is a landward shift of the coastline while regression is a seaward shift. The terms are applied generally to gradual changes in coast line position without regard to the mechanism causing the change. In addition, these terms usually are applied to changes over periods greater than 10^3 years as can be expected to be recorded by facies distributions in the geologic record or the stratigraphic interpretation of seismic reflection. "Transgressions" and "regressions" are commonly used, for example, to refer to coast line changes due to glaciations, which cause both eustatic sea-level changes and subsidence or rebound. Of particular significance is the "Holocene transgression" which corresponds to a eustatic rise in sea level of between 100 and 130 m and between 18,000 and 6,000 yr BP. The terms also have been applied, however, to changes occurring over shorter time scales in, for example, Lake Chad.

"Ingression" refers to the advance of marine conditions into more-or-less confined areas, like the drowning of a river valley (Schieferdecker, 1959, terms 1260 and 1840, as cited in Jackson, 1997) or to the infiltration of water to an interior low-lying area of land creating an inland body of water. In the former application, at least, neither mechanism or time period is implied although the tendency seems to be to use "ingression" to refer to more rapid, if not catastrophic, transitions rather than to more gradual changes. An example might be the marine invasion of a glacial lake by breeching of a morainal barrier during post-glacial, sea-level rise. A rise in relative sea level will also raise the water table in coastal aquifers causing the appearance of lakes and ponds but the same could be accomplished by changes in recharge (percipitation–evapotranspiration).

Facies relationships and implications

Regressions or transgressions can be due to any combination of (1) eustatic sea-level rise or fall; (2) subsidence or uplift; and (3) sedimentation or erosion. In sedimentology, transgressions are also referred to as "retrogradation" in which a depositional environment due to the rate of creation of accommodation space exceeds the sediment supply (Curray, 1964; Nichols, 1999). This results in a landward shift of coastal facies belts, although it is possible that the advance of the marine conditions had occurred so quickly that sediment deposition could not remain in equilibrium with changing conditions. In this case, relict, subaerial sediments may be found drowned in place or shallow water deposits covered discontinuously by deep water facies. Transgressions are usually associated with a rise in relative sea level due to eustatic sea-level rise and/or coastal subsidence. Even during periods of stable sea level or slowly rising sea level, however, erosion of coastal deposits can result in a transgression.

Regressions are also referred to as "progradations" in which the sediment is supplied at a higher rate than the potential space for sediments to be accommodated is created. In the geologic record, this can be preserved by up-column transitions to distinctly shallow water facies. A regression is usually associated with a falling sea level, but even in the face of a stable or slightly rising local sea level, the shoreline may still be displaced in a seaward direction by the rapid deposition of coastal sediments. Progradation is, for example, associated with delta formation. The relationship between the rate sea level changes and the rate of deposition or erosion was discussed by Curray (1964).

Onlaps and offlaps

The terms seem to have entered the literature associated with "onlaps" or "offlaps." Onlaps are overlapping relationships of shallower water sediments over deeper water sediments, which progressively pinch out toward the margins of a sedimentary basin. "Marine transgressive sequence" is used as a synonym for onlaps. Offlaps occur when progressively younger sediments have been deposited in layers offset seaward, often associated with an upward coarsening. Curray, accordingly, came to refer to the occurrence of a shift in the shoreline independent of the evidence found in the geologic record for these changes.

Further suggested readings are also included in the following bibliography.

Henry Bokuniewicz

Bibliography

Carter, R.W.G., 1988. *Coastal Environment*. New York: Academic Press.

Curray, J.R., 1964. Transgressions and regressions. In Mitter, R.L. (ed.), *Papers in Marine Geology*. New York: Macmillan.

Jackson, J., 1997. *Glossary of Geology*. 4th edn. Alexandria: American Geological Institute.

Kennet, J.P., 1982. *Marine Geology*. Englewood Cliffs: Prentice-Hall.

Leeder, M., 1999. *Sedimentology and Sedimentary Basins from Turbulence to Tectonics*. Oxford: Blackwell Science.

Nichols, G., 1999. *Sedimentology and Stratigraphy*. Oxford: Blackwell Science.

Schieferdecker, A.A.G. (ed.), 1959. *Geological Nomenclature*. Gorinchem: Royal Geological and Mining Society of the Netherlands.

Vail, P.R., Mitchum, R.M., and Thompson, S., 1977. Global cycles of relative changes in sea level. In Payton, C.E. (ed.), *Seismic Stratigraphy—Application to Hydrocarbon Exploration*. American Association of Petroleum Geologists, Memoir 26, pp. 83–97.

Cross-references

Changing Sea Levels
Coastal Sedimentary Facies
Sea Level Indicators, Geomorphic
Sequence Stratigraphy

INSTRUMENTATION—See BEACH AND NEARSHORE INSTRUMENTATION

ISOSTASY

Numerous observations point to a complex and changing relationship between land and sea surfaces throughout geological time. In some localities elevated coral reefs, wave-cut rock platforms, and molluscs embedded in their original marine sediments attest to past sea levels having been higher than present. At other sites, drowned forests and submerged sites of human occupation point to sea levels having been locally lower than present. These observations represent a measure of relative sea-level change which can involve a land-movement signal as well as an ocean-volume signal. The indicators of submerged or elevated coastlines therefore point to one of three occurrences: land has moved up or down, ocean volumes have changed, or both have occurred simultaneously.

Tectonic process operating within the earth have caused uplift and subsidence throughout the Earth's history, resulting in relative sea-level change on a wide range of spatial and temporal scales. They include uplift and subsidence at convergent plate margins where the relative sea-level change is usually episodic and abrupt but cumulative over long periods of time resulting in, for example, the marine mollusc beds high in the Andes of South America that were first described by Charles Darwin. The tectonic processes also include slower and longer-duration events such as the initiation of continental rifting and sea floor spreading with the concomitant changes in the displacement of water by the developing ocean ridge system. Long-term thermal contraction of the cooling outer layers of newly created ocean crust at the ocean ridge results in sea-floor subsidence, creating basins into which sediments accumulate, thereby magnifying the subsidence. Large volcanic edifices stress the earth and cause more local subsidence and deformation of the earth's surface in the vicinity of the load.

At the same time that the tectonics events shape the earth's surface and shift the relative positions of land and sea surfaces, ocean volumes also change, largely because of climate-driven changes in the extent of glaciation of the planet. During extended cold periods large ice sheets form, extracting water from the oceans and lowering sea levels. As the climate warms up sufficiently to melt the ice sheets sea levels again rise. Such glacial cycles have occurred at intervals throughout much of the earth's history but they have been most significant during recent times, the Quaternary period, for which the record has not yet been wholly overprinted by the subsequent tectonic and land-shaping events.

The combined result of the tectonics and glacial cycles is a sea-level signal that has varied significantly in time as well as being geographically variable. The record of this variability is, however, far from complete, and to be able to model and predict the migration of coastlines, an understanding and separation of the underlying causes of sea-level change is essential. Isostatic processes are key elements in this understanding and separation.

The isostatic process

Isostasy is the tendency of the earth's crust and lithosphere—the upper, effectively elastic layer of the earth—to adjust its vertical position when loaded at its surface by, for example, ice, water, volcanos, or sediments. For this purpose, the earth can be represented to a good approximation as a spherically symmetric body with a fluid core of about 3,400 km radius. The upper layer is called the lithosphere and includes the crust. Its thickness is typically between 50 and 150 km, varying with the tectonic history of the region. The lithosphere is characterized by being relatively cold and to behave elastically when subjected to load stresses below a critical failure limit. The mantle, between the lithosphere and core, is at a temperature that is relatively close to the melting point of terrestrial materials. As a result the mantle flows viscously, with characteristic relaxation times of 10^4–10^5 years, when subject to non-hydrostatic stress. It is this zonation of a "rigid" lithosphere over "viscous" mantle that gives validity to the isostatic models.

The simplest representation of isostasy is by "local" response models which are statement of Archimedes' principle: a load of heights h, density ρ, placed on the earth's surface results in a subsidence of the underlying surface of $\delta = h\rho/\rho_m$, where ρ_m, the density of the mantle, exceeds the density of the lithosphere (see Figure I22(A)). This model assumes that the crust or the lithosphere has no shear strength (or has failed under the load) and overlies a fluid mantle. This model, while unrealistic in many respects, is nevertheless useful for estimating magnitudes of crustal deflection beneath loads. For example, under a 3 km thick ice sheet the crust is predicted to deflect by about 1 km. A more reasonable model is one in which the load is supported by both the "elastic" strength of the crust-lithophere and by the buoyancy forces at the base of the layer (Figure I22(B)). In this model, the mantle also behaves as a fluid and it provides a reasonable description of the earth's response to loads with time constants that are longer than the relaxation times of the mantle. These models have been extensively used to represent the response of the earth to sediment loads or to volcanic loads. They are usually referred to as regional isostatic models.

When the load duration is of the order 10^3–10^5 years any load-generated stresses that have propagated into the mantle will not have relaxed and the viscosity of the mantle must be taken into account. In these cases the isostatic models are usually represented by an elastic layer over a viscous or viscoelastic halfspace or, in the case of global problems, by spherical shell models of an elastic lithosphere over a viscoelastic mantle and fluid core. Both lithosphere and mantle may be represented by some degree of layering in physical properties (elastic moduli, viscosity, and density). Formulation of these spherical response models are well developed and solutions for the surface deformation under complex

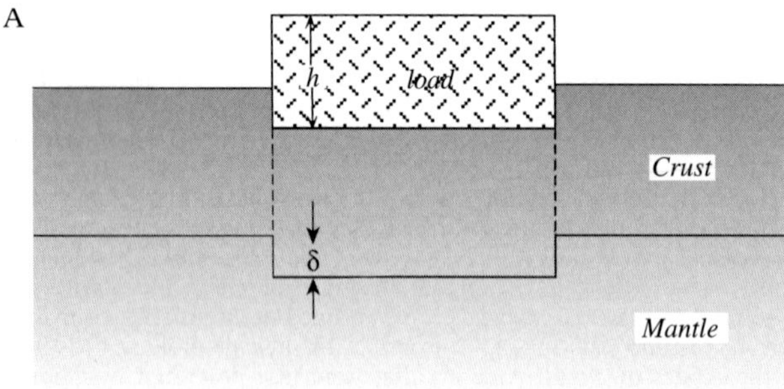

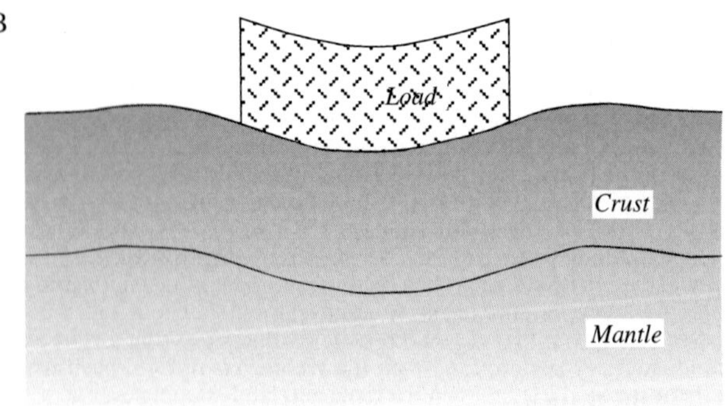

Figure I22 Models (A) local isostasy (B) regional isostasy. In (A) the load is supported by the buoyancy force at the base of the crust or lithosphere, whereas in (B) the load is also supported by the elastic stresses created in this layer. As the load diameter in (B) increases the isostatic response at the center of the load approaches that of local isostasy.

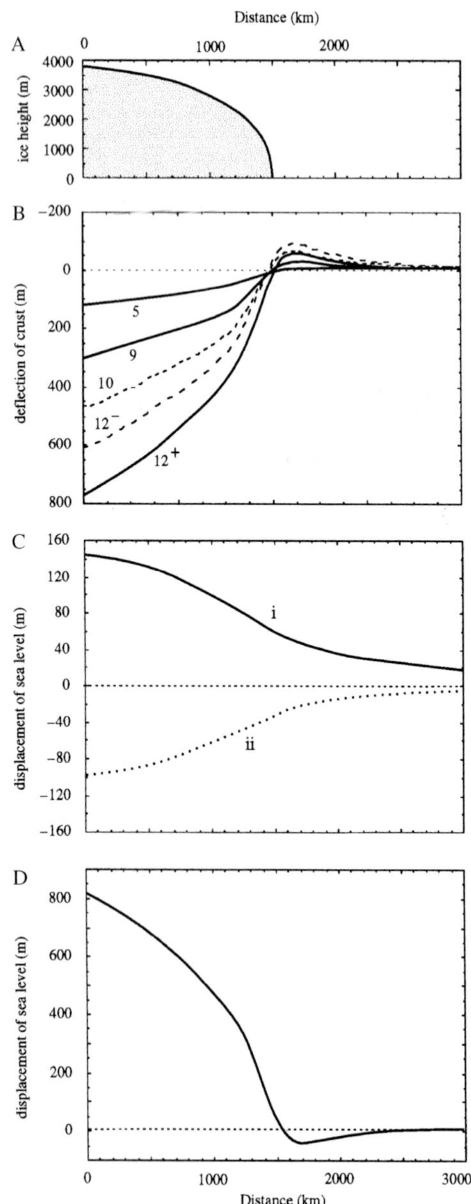

Figure I23 (A) Radial cross section of axisymmetric ice sheet. (B) Deformation of the earth's surface under the ice that has loaded the earth for 20,000 years (curve 12+). At 12,000 years ago the load is removed instantaneously. The initial response is elastic (curve 12−) and this is followed by viscoelastic creep, the surface being shown at 10,000 years (10), 9,000 years, and 5,000 years ago. (C) The gravitational attraction of the ice load, represented as the deflection of the geoid (i), and the change in geiod from the change in the planets gravity due to the deformation of earth under the load (ii). The results are shown for a period before unloading starts. (D) The relative sea-level change, due to the combination of crustal deformation, change in gravitational attraction, and ocean volume change long after the load has been removed. The sea level is expressed with respect to its present position. If the a coastline formed near the center of the load soon after the ice melted, it would now be at nearly 800 m elevation.

surface load geometries exist. Figure I23 illustrates an example of surface deformation where a large-diameter axi-symmetric ice sheet has been instantaneously removed. The rheology (viscosity structure) of the planet is realistic (see Figure I27, below) and the results indicate that the crustal readjustment continues for thousands of years after the unloading is completed.

In addition to the surface deformation, the gravity field of the planet also changes under the load: the shape of the envelope containing the mass is modified by the deformation and material is redistributed within this envelope. At the same time there is a redistribution of the material on the surface: sediments are transported from mountains into basins, or the meltwater from land-based ice sheets flows into the oceans. Surfaces of constant gravitational potential—surfaces on which the gravity vector is everywhere perpendicular—therefore, change with time as the load and planetary response evolve. One such equipotential is the geoid, the shape of the ocean. (If the ocean is not an equipotential surface then the gravity vector has a component along the surface and ocean currents result until an equilibrium state is reached; thus in the absence of winds and other perturbing forces, the ocean will be an equipotential surface. This is called the geoid.) Figure I23 illustrates the change in the equipotential surface resulting from the unloading. It includes a contribution from the surface load itself—the ice "attracts" the ocean water and pulls the ocean surface up around it (curve *ii*)—and a contribution from the earth's deformation (curve *i*). The illustration is for the period while the ice is intact and when melting starts both curves will evolve with time.

The example in Figure I23 illustrates that relative sea-level change resulting from the removal of the ice sheet contains several elements. The crust is displaced radially, the ocean surface is deformed by the redistribution of surface and internal mass, and water is added to the ocean. The rebound resulting from the melting (or growth) of the ice sheet is referred to as glacio-isostasy. The water added to (or withdrawn from) the oceans has its own isostatic effect and is referred to as hydro-isostasy.

The combined glacio–hydro-isostatic processes are of global extent. The melting of an ice sheet in one location modifies sea level globally, not just by changing the amount of water in the ocean but because of the planet's isostatic response to the changing surface load of ice and water. Other loading processes, such as by sediments or volcanic loads, are usually more local in their consequences. Also, these tectonic process generally occur on longer time scales so that the mantle response can usually be approximated as a fluid, and the local or regional isostatic models are mostly appropriate.

Glacio-isostasy

Ice sheets represent surface loads that reach radii in excess of 1,000 km and thickness approaching 3 km. These loads are large enough to deform the earth and to produce substantial changes in sea level as illustrated in Figure I23. Glacio-isostasy is the major cause of sea-level change in areas of former glaciation. When a large ice sheet melts the rebound of the crust is of larger amplitude than the rise in sea level resulting from the addition of the meltwater to the oceans (typically 120–130 m, see Figure I29 below) from all of the ice sheets. If ΔV_i is the change in volume of ice on land and A_o the area of the ocean, then this second signal is

$$-\frac{\rho_i}{\rho_w}\int \frac{1}{A_o(t)}\frac{d}{dt}\Delta V_i(t)dt$$

where ρ_i, ρ_w are the densities of ice and water, respectively, and both A_o and ΔV_i are functions of time. This contribution is referred to as the ice-equivalent sea-level change.

Because of the viscosity of the mantle, the crust continues to rise long after the ice has vanished and sea level appears to have fallen since deglaciation. This is seen in the Gulf of Bothnia and northern shores of the Baltic Sea, as well as in the Hudson Bay area of northern Canada. For these locations near former centers of glaciation the rebound signal dominates and the observed sea-level curves are characteristic relaxation curves (although only the post-glacial part of the change is recorded) (Figure I24 (Angermanälven)). Near the ice margins the rebound is reduced in magnitude and may become comparable to the rise resulting from the increase in ocean volume. Now the time dependence of the sea-level change becomes more complex, with its character depending on the relative importance of the two contributions. In Figure I24(Andøya), for a site just within the ice-sheet margin, the rebound initially dominates but later, because of the melting of other and distant ice sheets, the ocean volume increase becomes the dominant factor and sea level rises until a time when all ice sheets have melted. The remaining signal is a late stage of the relaxation process and sea levels continue to fall up to the present.

The rate and magnitude of the sea-level change is a function of the earth's viscosity and the ice history: of the duration of the ice load, of its areal extend, and of its thickness. The importance of the rebound phenomenon is that it provides a means of estimating the earth's rheology: if climate models and geomorphological observations constrain the ice geometry through time, then observations of sea-level change provide a constraint on the mantle viscosity. If the ice models are not sufficiently well-known then it becomes possible to learn something about the ice sheets as well. Figure I25 illustrates observational results for sea-level change across Scandinavia. Here, the ice sheet reached its maximum at about 20,000 years ago and most melting occured between about 16,000 and 10,000 years ago. As the ice retreated, coastlines formed on the emerging land providing a comprehensive description of the rebound across northern Europe. The rebound did not cease at the time melting ceased and coastlines have continued to retreat in formerly glaciated regions up to the present. This can be seen in tide gauge records across the Baltic, with present sea-level falling locally at rates approaching 1 cm/yr in the northern part of the Gulf of Bothina. Figure I26 illustrates the rate of crustal rebound and to obtain relative sea-level change these values must be increased by about 1–1.5 mm/yr. Coastlines here continue to retreat despite other factors that may contribute to an increase in global ocean volume.

Glacio-isostasy does not cease at the ice sheet margins. Because the mantle flow generated by the changing surface load is constrained within a deformable shell, when some areas are depressed under a growing load others are uplifted. The latter areas form a broad zone or swell around the area of glaciation of amplitude that may, depending on the size of the ice sheet, reach a few tens of meters. When the ice sheet melts this peripheral swell subsides and for island or coastlines on it sea level will be seen to be rising at a rate that is over and above the ice-volume equivalent contribution (Figure I24(Store Bælt)). Beyond the Scandinavian relic ice margins this occurs in areas of the North Sea and as far away as the western and central Mediterranean and here the sea level continues to rise even when all melting has ceased. Beyond the North American ice sheet this zone of recent crustal subsidence and marine flooding occurs as far away as the southern USA and Caribbean.

Observations of sea level within and beyond the former ice margins provide the principal source of information on mantle viscosity. A typical result for northwestern Europe is illustrated in Figure I27 where the rebound phenomenon provides a good constraint on the viscosity of the upper mantle. The main features of the viscosity profile include a lithosphere of thickness 65–75 km, a relatively low value for the viscosity of the mantle immediately below the lithosphere, and increasing viscosity with depth, particularly at a depth of about 700 km. Analyses for different regions produce comparable results although actual values for the viscosity and lithospheric thickness may differ because of the possibility that the rheology is laterally variable. The determination of such variability is one of the important research areas in glacio-isostasy.

While the glacio-isostatic models are well understood, one of the key limitations of their application is the inadequate knowledge of the former ice sheets. The ice margins at the time of the Last Glacial Maximum, some 20,000 years ago, are usually well-defined by geomorphological markers but the timing of their formation is not always known. This occurs particularly where the ice margins stood offshore and left few datable traces of both the time of their formation and of their retreat. Also, the ice thickness cannot usually be inferred from observational evidence alone and is inferred instead from glaciological and climate models. The sea-level observations can nevertheless help constrain the ice models in important ways. Thus, the total ice volumes in the models for all the major ice sheets must yield a global sea-level curve that is consistent with the changes observed far from the ice sheets (see hydro-isostasy). Also, details in the ice models can also be derived from the sea-level data from sites within and near the former ice margins. The shape of the sea-level curve from a near-margin site (Figure I24) changes quite rapidly with distance from the former ice margin, with the signal evolving from that for a central-load site to that for a site on the peripheral swell, and observations across the margin can constrain the former ice distribution within the ice-marginal region. One of the more recent research directions in glacio-isostasy is the use of this sea-level and crustal-rebound evidence to improve models of the ice sheets during the last deglaciation phase.

Hydro-isostasy

As ice sheets melt, the additional water entering the world's oceans loads the sea floor, load stresses are propagated through the elastic lithosphere into the mantle, the newly stressed mantle material flows toward unstressed regions and the sea floor subsides. The shape and holding-capacity of the ocean basin is thereby modified and the ocean water is redistributed, changing sea level. This adjustment of the earth

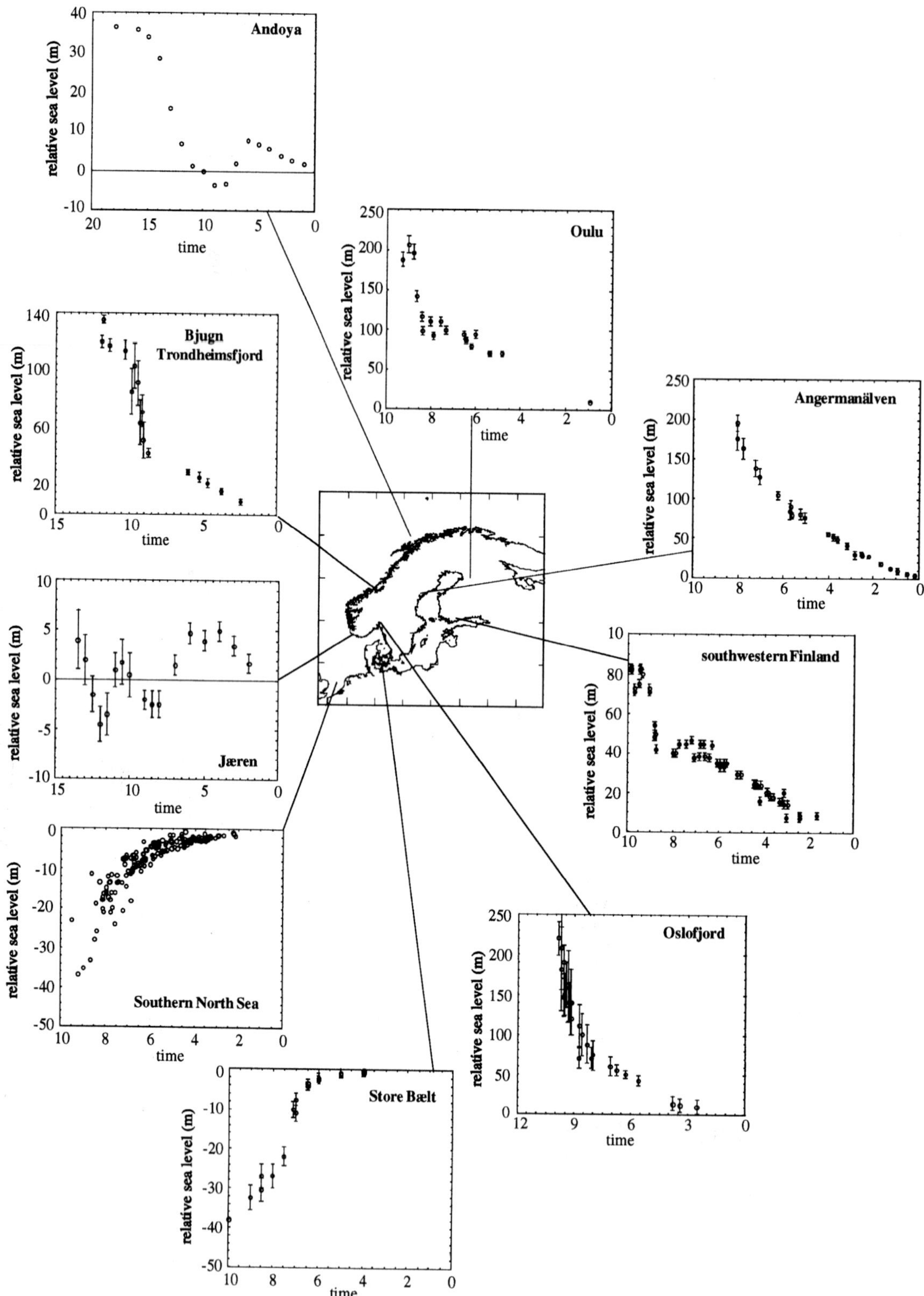

Figure I24 Observed relative sea-level change for sites in Scandinavia illustrating some of the spatial variability in the response. The ice sheet covered all of Scandinavia and spread onto the German, Polish, and Russian plains. Retreat started at about 18,000 years ago and the final disappearance of ice occurred at about 10,000 years ago. The time scale used in these plots corresponds to the radiocarbon time scale which differs from a calendar time scale by about 10–15% for this interval (1 C^{14}years ≈ 1.1–1.15 calendar year).

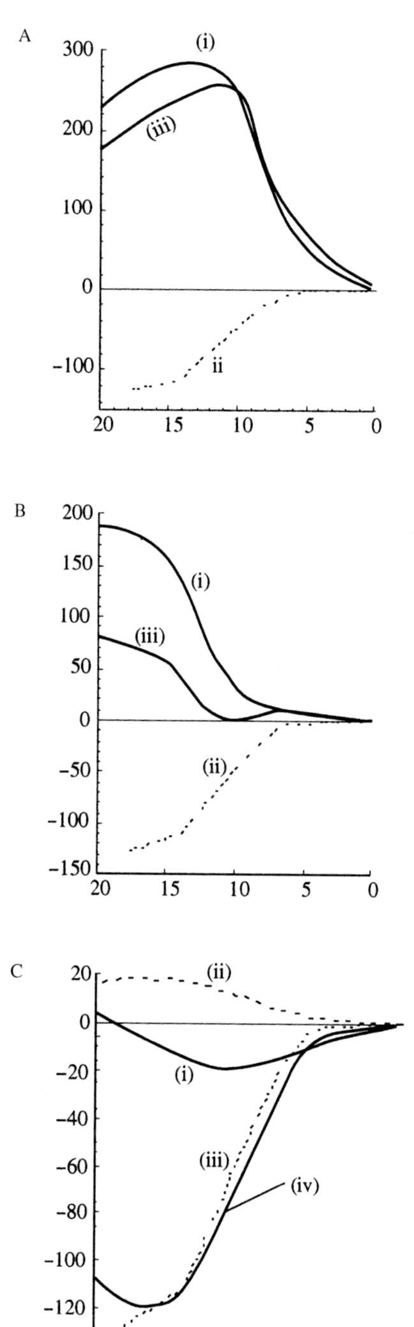

Figure I25 Schematic contributions to sea-level change from the glacio-isostatically driven crustal rebound and increase in ocean volume from meltwater. (A) For a location near a former center of glaciation where the rebound (i) exceeds the rise in sea level (ii) from the added melt-water, (iii) is the total change. (B) For a location near the former ice margin where the two contributions are of comparable magnitude but of opposite sign. (C) For a location beyond the ice margin where the crustal uplift is replaced by subsidence. The effect of the water load (ii) now is important as well as the ice-load effect (i). The meltwater contribution is given by (ii) and the total change by (iv).

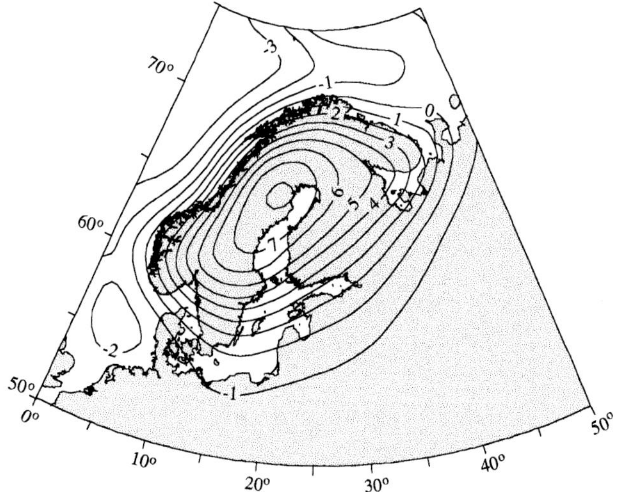

Figure I26 Present rates of crustal uplift (in mm/yr) of Scandinavia based on rebound models and on observed rates from tide gauges across the region (from Lambeck *et al.*, 1998; with permission of Blackwell Publishing).

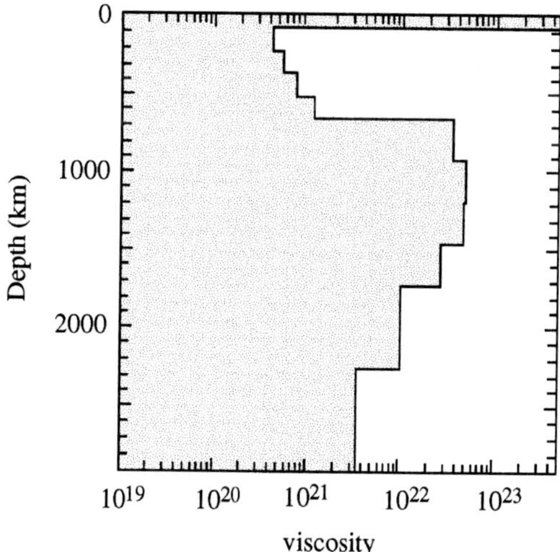

Figure I27 Profile of mantle viscosity (in units of Pa s) inferred from glacial rebound analysis of European sea-level data (G Kaufmann, with permission).

under the time-dependent water load and the concomitant sea-level change is referred to as hydro-isostasy. Since the onset of the last deglaciation, sea levels have risen on average by about 120–130 m and the additional load has been sufficient to modify the shape of the earth. This is a result of the long wavelength nature of the water load. Loads of dimensions less than the thickness of the lithosphere are supported mainly by the strength of the lithosphere and the resulting surface deformation is small. But large-dimension loads effectively see through the lithosphere and are supported by the much more ductile mantle which flows even under small changes in the stress field.

At continental margins the hydro-isostatic deformation of the earth's surface describes quite complex patterns because of the geometry of the load. The lithosphere acts as a continuous elastic layer or shell and the continental margin is dragged down by the subsiding ocean lithosphere but, because of the asymmetry of the load, not by the same amount as in mid-ocean. At the continental coastlines, therefore, the subsidence will be less than it would be in mid-ocean. At the same time, some of the mantle material flowing away from the stressed oceanic mantle flows beneath the continental lithosphere, causing minor uplift of the interior. The net effect of the ocean volume increase is a seaward tilting of the continental margin which will be seen as a variable sea-level signal across the shelf. This effect is clearly seen for tectonically stable continents that lie far from former ice sheets, as in the case of Australia.

While the ice sheets are still melting the dominant sea-level signal here is from the increase in ocean volume and the glacio- and hydro-isostatic effects are second order. But when melting ceases the on-going isostatic effects come into their own. Now sea level appears to be falling at the coastal site as the ocean waters recede to fill the still-deepening ocean. In consequence, small sea-level highstands are left behind with peak amplitudes of 1–3 m occurring at the time global melting ceased (Figure I28). Such highstands are common features along many continental margins and manifest themselves as relic shorelines or fossil corals above the present formation level or habitat. If the coast is deeply indented, sites at the heads of gulfs, being furthest away from the water load, experience greatest uplift while offshore islands experience least uplift. This differential movement provides a direct measure of the viscosity of the mantle across the continental margin.

Like the glacio-isostatic effect, the water load does not only deform the surface of the earth, it also results in a redistribution of mass and a change in gravity and in the shape of equipotential surfaces. The total sea-level changes associated with the hydro-isostasy include, therefore, both the crustal radial deflection and the associated geoid change. Also, the glacio- and hydro-isostatic effects are closely linked when their cause is the deglaciation of the last ice sheets. Near the edge of the ice sheets, for example, the water is pulled up (Figure I23) and the waterload is increased above what would result from a uniform distribution of the meltwater over the entire ocean. Here the hydro-isostatic signal is a function of the magnitude of the glacio-isostatic effects. Elsewhere, the broad zone of crustal rebound surrounding a large ice sheet may occur in an oceanic environment. Then, when the ice sheet melts this swell subsides, increasing the volume of the ocean basin, water is withdrawn from other parts of the ocean, and a further global adjustment of sea level occurs. Thus, the treatment of hydro–glacio isostasy requires a global and consistent formulation that ensures that these various interactions are included.

The hydro-isostatic signal is an on-going one even when major melting of the world's ice sheets ceased about 6,000–7,000 years ago. Thus sea-level change today will contain a small but not insignificant component of hydro-isostatic origin (cf. Figure I28 for the Australian region). This signal must, of course be superimposed upon any other changes, including possible global warming signals. The results indicate that sea levels around the Australian margin are slowly falling under the combined glacio–hydro-isostatic response to the past melting of the large ice sheets (with the possible exception of Tasmania where the glacio-isostatic effect of Antarctic ice volume changes becomes significant, canceling out the hydro-isostatic signal such that little overall

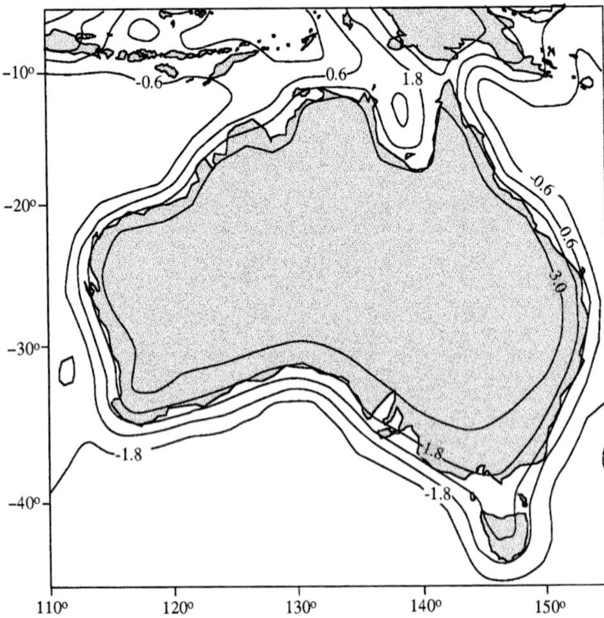

Figure I28 Sea level at 6,000 years ago around the Australian margin illustrating the effect of hydro-isostasy as a tilting of the margins of the continents. Sea levels are present relative to present mean sea level. Contour intervals are 1.2 m. The present-day rate of change in mm/yr given approximately by dividing the contour values by 6 and changing the sign, the resulting negative value for most locations indicating a fall in sea level from isostasy alone.

change now occurs). Similar isostatic effects will be present at all coastline, increasing in magnitude as the locality approaches the regions of former glaciation.

The importance of the sea-level observations far from the ice margins is that because the glacio–hydro isostatic effects are relatively small (10–15% of the total signal) they provide an estimate of the change in volume of the oceans when corrected for the isostatic effects: the observed sea level, less the isostatic correction yields the ice-equivalent sea level defined above and hence an estimate of the change in ocean volume ΔV_i. Several long records, extending back to the Last Glacial Maximum, of local sea-level change exist which provide evidence for the change in ice volume since this time. They indicate (Figure I29) that maximum ice volumes globally were $(50\text{–}55) \times 10^6$ km^3 greater than today but, they do not indicate necessarily where this extra ice was stored. To resolve that issue recourse to the study of the glacio-isostatic process from formerly glaciated regions is necessary.

Sediment and volcanic loading

Large accumulations of sediment occur along many of the continental margins reaching, in some instances, a thickness of 10 km or more. The rate of accumulation is usually slow and continuous, occurring over periods of tens of millions of years with the sources of sediments coming from continental interiors where tectonic processed have caused uplift and erosion processes have carried the sediments to the sea. Examples include the Bay of Bengal, the northwestern margins of Europe, the eastern margin of North America, and the Gulf of Mexico. Thick accumulations of sediments are possible because of the subsidence of the lithosphere under the growing sediment load. With the above model of local isostasy an ocean basin of depth d_o can, with adequate sediment supply, lead to a maximum subsidence of $d_o\rho_s(\rho_s-\rho_m)$ where ρ_s is the density of sediments. This assumes that the basin is ultimately filled to sea level. For $d_o = 4$ km, $\rho_s = 2.5$ g cm^3, $\rho_m = 3.5$ g cm^{3}the maximum thickness of sediment that can be attained is about 10 km. However, in this case the deeper sediments will have been deposited in water depths initially of $d_o = 4$ km, whereas the characteristics of the fauna preserved in the basin sediments usually indicate that deposition invariably occurred in relatively shallow waters. Isostasy alone, therefore, cannot produce thick sediment sequences but it does act as an amplifier of subsidence that is the result of other processes: in this case mostly the thermal contraction of ocean lithosphere as it cools from an initially hot layer formed at the ocean ridges and then moves away from the heat source.

On short time scales, sediment loading can lead to substantial coastal subsidence. This may occur in conjunction with deglaciation cycles where sediments are eroded from the continents during the deglaciation stage and delivered to coastal environments at some later stage. An example of such subsidence occurs along the US coast of the Gulf of Mexico, particularly for the Mississippi delta. Here, coastal subsidence occurs at rates approaching 10 mm/yr and are attributed in part to the isostatic response to recently delivered sediments, but also in part to the extraction of fluids from the sediments and the associated compaction. Here, as in most isostatic problems, several factors will contribute to the observed signal.

Volcanic loading of the crust provides another example of isostasy at work. Large volcanic complexes form on the sea floor, and elsewhere, because of upwelling convection currents in the mantle that lead to an injection of magma into the crust and ultimately onto the surface as volcanos. The mantle source regions for the magma appear to be long-lived and as the lithosphere moves over the earth's surface under the forces of plate tectonics, a trail of volcanos is left on the surface. The Hawaiian chain provides the type example. Other examples include the Society Island chain whose current center of volcanic activity lies to the east of Tahiti. The subsidence of the lithosphere beneath the volcano is adequately described by the regional isostatic model in which the load is supported by the elastic stresses within the lithosphere and by the buoyancy force at the base of the layer (Figure I22(b)). Because of the elastic properties of the lithosphere small peripheral bulges, concentric about the center of loading, develop and any islands located in this zone at the time of volcano development are uplifted by some tens of meters. An example of this is provided by the uplifted atoll that forms Henderson island, southeast of Pitcairn island. This small island about 200 km from the volcanic island of Pitcairn, appear to have been uplifted some 20–30 m at the time of Pitcairn's formation, perhaps 700,000 years ago. The location of the zone of maximum peripheral uplift provides a measure of the flexural wavelength of the lithosphere, a parameter that characterizes the physical response of the layer to loading. With time, some relaxation of the loading stresses can be expected to occur within this layer such that the isostatic state evolves

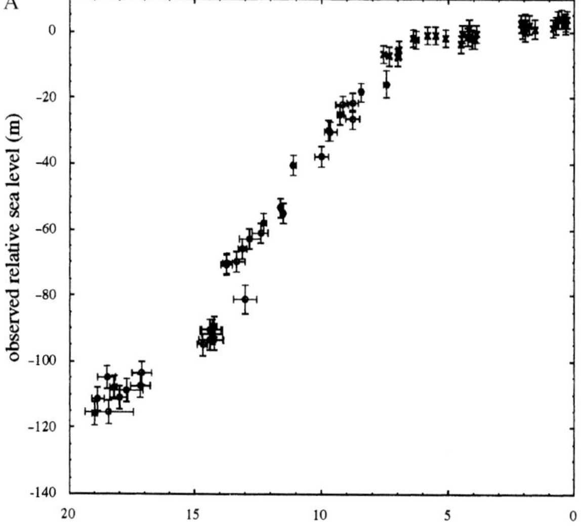

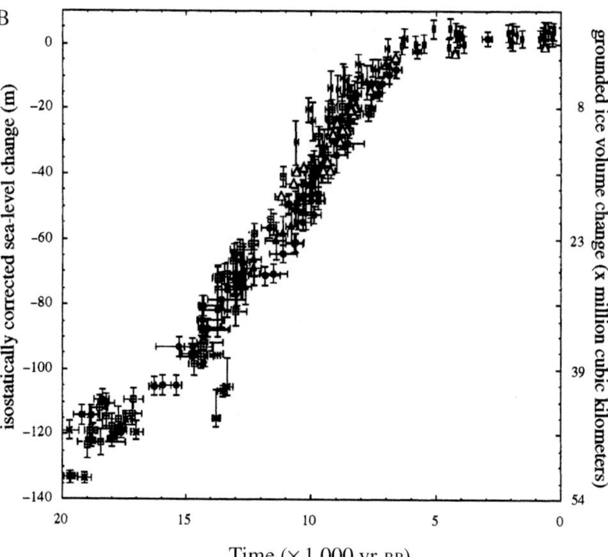

Figure I29 Sea-level change for the past 20,000 years. (A) A record of observed local relative sea-level change from Barbados and other Caribbean sites, and (B) isostatically corrected sea level from a number of sites distributed globally and combined into a single ice-equivalent sea-level curve. Scale on the right hand side gives the corresponding change in volume of ice on land and grounded on shallow sea floor.

slowly from regional to local isostasy and that the volcano slowly subsides. Thus Tahiti, a relative young volcanic load of about 1–2 million years, may be subsiding at a rate of about 0.2 mm/yr or less.

These examples of vertical movements driven by sediment or volcanic loading illustrate the interaction that occur between the various isostatic contributions to sea-level change. To estimate rates of tectonic uplift or subsidence, heights of identifiable coastlines are measured with respect to present sea level. Thus, the fluctuations in sea level of glacio-isostatic origin must be known, but these fluctuations are inferred from the same observational evidence. An important research area is to develop methods for separating out these effects, through observational improvements and through improved modeling of the physical processes.

Suggested further reading on this subject may be found in the following bibliography.

Kurt Lambeck

Bibliography

Lambeck, K., and Johnston, P., 2000. Response to "What about asthenosphere viscosity? By W. Fjeldskaar" *Geophysical Journal International*, **142**: 277–281.

Lambeck, K., 1988. *Geophysical Geodesy: The slow Deformations of the Earth*. New York: Oxford University Press.

Lambeck, K., Smither, C., and Ekman, M., 1998. These of glacial rebound models for Fenoscandinavia based on instrumented sea- and lake level records. *Geophysical Journal*, **135**: 375–387.

Peltier, W.R., 1998. Postglacial variations in the level of the sea: implications for climate dynamics and solid-earth geophysics. *Reviews in Geophysics*, **36**: 603–689.

Watts, A.B., 2001. Isostasy and Flexure of the Lithosphere Cambridge University Press.

Cross-references

Changing Sea Levels
Coastal Changes, Gradual
Coastal Changes, Rapid
Coastal Subsidence
Endogenic and Exogenic Factors
Eustasy
Geodesy
Glaciated Coasts
Ingression, Regression, and Transgression
Paleocoastlines
Sea-Level Change During the Last Millennium
Sea-Level Rise, Effect
Submerged Coasts
Tidal Datums
Uplift Coasts

J

JET PROBES

Introduction

Jet probes are used for a variety of purposes (e.g., underwater cable routing, marine archaeology, coastal engineering) and are usually deployed in conjuction with other data-collection techniques such as hydrographic surveys (to determine water depths and map existing bottom conditions), subbottom profile survey (to identify near subbottom stratigraphy, 3–7 m depth), side-scan sonar survey (to identify morphological variations, and natural and man-made obstructions on the seabed), and vibratory coring (to acquire direct physical information of nearsurface sediments). Jet probe surveys acquire indirect physical information on subsurface lithology by surveying the thickness and stratigraphic layering of sedimentary covers on land or underwater. A jet of either air or water is used to penetrate the sand cover; the latter, however, is only applicable underwater (USACOE, 2002).

Most jet probe surveys, in the service of coastal engineering for shore protection via beach renourishment, provide a rapid means for determining the nature of unconsolidated sedimentary deposits that occur underwater. Because jet probes have no cutter head and depend only on the power of a water jet to penetrate bottom sediments, they are restricted to use in shallow waters (i.e., the effective range of operating depths is usually from 1 to about 30 m) that overlie unconsolidated (loose) sandy deposits. Clear water is desirable, but not essential, because it facilitates site location, maneuverability of jet probe equipment over the bottom, and visual estimation of turbidity plumes that are created by water-jetted penetration of a pipe down through the sediment (CBNP, 1995). Jet probing finds application in marine archaeology, geotechnical studies that feature searches of seabed deposits for beach-compatible sands that can be placed on degraded beaches, and geological investigations that attempt to determine the thickness of sand covers on the seafloor or on lakebeds. Although widely deployed in many different kinds of environments and for various applications by scientists and engineers, jet probing probably finds most extensive application in coastal sand searches (e.g., Meisburger and Williams, 1981; Meisburger, 1990; Keehn and Campbell, 1997; Finkl *et al.*, 1997, 2000, 2003; Andrews, 2002) that require reconnaissance surveys of bottom types or verification of geophysical survey data (e.g., subbottom profiles, side scan sonar surveys).

Grab samples provide information about surficial seafloor sediments, whereas vibracore and jet-probe samples can penetrate down into the sediment layers. Vibracore samples are relatively inexpensive to obtain and can recover the long and relatively undisturbed cores that are required to assess the composition and grain sizes of the materials, as well as to establish the stratigraphy of the deposits (e.g., Meisburger and Williams, 1981). Water jets are less expensive than cores (CBNP, 1995; USACOE, 2002), involving the water-jetted penetration of a pipe down through the sediment in order to determine the layering, as opposed to (undisturbed) core retrieval for splitting and analysis.

Marine archaeology

This tool assists marine archaeologists in determining the nature or presence of materials or features that lie within or underneath bottom sediments (Anon, 1996). On archaeological sites, the jet probe is manually driven through various nonconsolidated sediments on the seabottom where the probing pipe goes through soft strata until it hits bedrock, a cemented stratum, compacted clay, or artifact. This tool provides information regarding the location and elevation of buried ancient waterline features (indicators of previous sea-level positions) and other geomorphological data. Ultimately, the information enables the archaeologist to reconstruct shallow coastal-marine sedimentary environments, local surficial stratigraphic sequences, and other geological features that can then be dated and calibrated with archaeological finds.

Stratigraphic studies

Coastal-marine stratigraphic studies often rely on a range of techniques that are used to compile various kinds of information, that is, related to layering of different kinds of materials on the seabed (e.g., Toscano and Kerhin, 1990; Wells, 1994). Data are commonly derived from several independent studies viz. surface sediment samples, vibracores, and seismic records to compile an assessment of Quaternary stratigraphy, as, for example, in the Paranaguá Bay Estuary in southern Brazil (Lessa *et al.*, 2000). Estuarine environments often provide ideal conditions for jet probing because there is a range of unconsolidated materials related to coarse- and fine-grained facies. Fluvial- continental deposits often occur with paleo-valleys as substratum for more recent sedimentation. These kinds of estuarine environments are often characterized by the intercalation of trans- gressive–regressive mud and sand facies that can be effectively studied using jet probes in conjunction with other techniques.

Underwater surveys of lakebeds often use jet probes to assist in reconnaissance verification of sedimentary bottom types, especially where sediment samples and grain-size analyses are eventually required. Lakebed studies often combine jet probing with underwater video investigation as independent lines of inquiry. Jet probe surveys to determine the thickness of sand cover are based on differentiation of the kinds of materials that are penetrated by the jet probing. On the American Great lakes, for example, the presence of diamictites (tills) that have been eroded from truncated drumlins to produce cobble–boulder lag deposits on the lakebed can limit the effeciveness of jet probes, as would any other substantial impediment to penetration of sedimentary layers (e.g., Stewart, 2000).

Assessment of sand resources and mining

Sandy shores occur along about 13% of the world's coastline (Coleman and Murray, 1976) and it is estimated that today about 75% of these shores are eroding (Bird, 1985). Beach erosion is thus a common problem along sandy coastlines and it is necessary to artificially renourish beaches because they provide natural protection from storms and have economic value (Finkl and Walker, 2002). The location of materials that are suitable for beach renourishment becomes an issue for best management practices that have to consider environmental concerns, methods of shore protection, storminess, and impact of exploration procedures to locate sand bodies on the seafloor. Even though sand sources differ from region to region around the world, there is a commonality to the need for good-quality sand and methods of looking for adequate long-term supply, as described, for example, by Anders *et al.* (1987), Conkright *et al.* (2000), and Walker and Finkl (2002).The salient problem then, is how to best locate sand sources that are appropriate for beach nourishment. Although inland sand sources are often suitable from a textural and compositional point of view for beach replenishment, their location away from the coast requires overland transport that can pose significant placement problems along the shore. Offshore sources of beach-quality sand are thus most often sought as geotechnical and economic reserves. Inner continental shelves host a range of coastal (e.g., beach ridges, dunes, nearshore bars, flood- and ebb-tidal deltas, estuarine sands) and marine sediments (e.g., shoals, banks, ridges, terraces, blanket deposits) as well as terrestrial deposits (e.g., glaciofluvial materials on valley floors, winnowed tills, coarse-grained alluvial terraces, and plains), all of which have been drowned and modified by rising sea levels during the Holocene (e.g., Toscano and York, 1992). Offshore sands that are suitable for beach renourishment are a sought-after and coveted commodity because in many regions they are in dwindling supply (e.g., Freedenberg *et al.*, 2000).

Advanced geophysical and geotechnical procedures are often backed up or verified by low tech efforts, such as jet probing, that are essential to the efficiency and economic success of offshore sand searches. General procedures for the exploration and development of borrows were summarized by James (1975) and Meisburger (1990), for example, who emphasized the value of collaborative approaches. Figure J1 shows the main sequential steps in modern data collection that are followed in sophisticated coastal sand searches that integrate diverse techniques. As shown in the figure, jet probe surveys are typically conducted at the reconnaissance level in conjunction with a suite of independent, but related, geophysical and geotechnical survey operations that provide specific kinds of information that collectively elucidate sediment thickness, lateral continuity, structural relationships, and composition. It is important to note that the first step in sand resource assessment is the review of historical data, an effort that is essential to proper appreciation of prior efforts and conclusions. Jet probe logs (Figure J2) are archived because they contain useful information that may be required later. A typical jet probe log, as shown in Figure J2, includes the usual kinds of locational information; date acquired, water depth, top and bottom divers, start- and end-times, etc. Notes in the log include important information that is related to the length of pipe, penetration depth, jet pump capacity, weather conditions, turbidity levels, and characteristics of the sand (grain size, percentage of silt content, color).

Sand searches for beach nourishment and protection commonly employ jet probe surveys (e.g., see Meisburger and Williams, 1981; CBNP, 1995; Walther, 1995; Freedenberg *et al.*, 2000; Finkl *et al.*, 1997, 2000). Usually conducted as a reconnaissance field survey (cf. Figure J1), the procedure is often misunderstood and the least utilized tool in sand

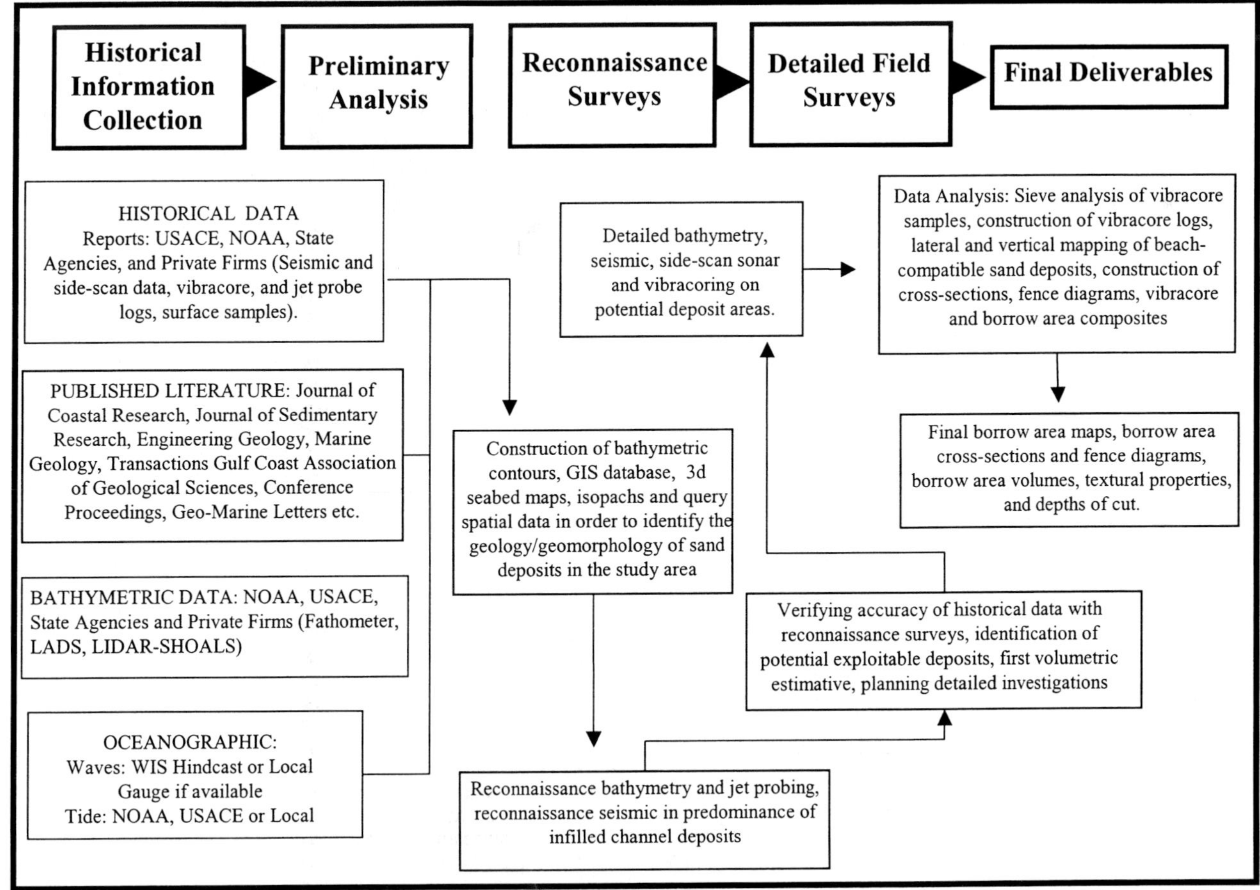

Figure J1 Flow diagram showing the organization, routing, and sequential application of coastal sand searches that are normally deployed on the inner continental shelf. Note that investigations begin with review of historical data, including proprietary reports and works in the public domain, and proceed to the construction of electronic databases that interface with GIS frameworks. Reconnaissance jet probing assists in the verification of historical data and provides focusing criteria for conducting detailed surveys. After review of historical, laboratory, and field data, sand resources with the greatest potential for use as beach sediments are identified as borrow sites. Jet probe surveys provide critical information in the evaluation of offshore sand resources and help identify which deposits are exploitable.

JET PROBE LOG

PROJECT: LONGBOAT KEY JET PROBES

JET PROBE: LBJP-02-14

COORDINATES:	DATE: 03/27/02	WATER DEPTH: 24.8
N = 1,132,829	START TIME: 1441	TOP DIVER: MTL
E = 408,682	END TIME: 1447	BOTTOM DIVER: JW

NOTES	ELEV.	DEPTH	SYMBOL	DESCRIPTION
LENGTH OF PVC PIPE: 20.0' PENETRATION DEPTH: 18.0' JET PUMP TYPE: BRIGGS & STRATON 3.0 HP GAL/HR: 8460 DIAMETER OF PIPE: 1.5"	-24.8	0		SEA FLOOR
SUPPORT VESSEL: CPE II POSITIONING: DIFFERENTIAL GPS USCG BEACON NAVIGATION SYSTEM: "HYPACK"				
WEATHER: WIND: DIR: NORTHWEST SPEED: 10–15 Kt. WAVES: DIR: NORTHWEST HEIGHT: 1–3 Ft. CURRENT: DIR: NORTH SPEED: SLIGHT	–37.8	0–13		SAND (SW) With Some Shell Hash
SAND SAMPLES TOP:–24.8 Ft., 0.17 mm, Silt: 1.59% Munsell Color: 5Y–7/1 Light Gray MID:–33.8 Ft., 0.14 mm, Silt: 2.07% Munsell Color: 5Y–7/1 Light Gray BOTTOM: –42.8 Ft., 0.15 mm, Silt: 2.46% Munsell Color: 5Y–7/1 Light Gray	–42.8	13–18		SAND (SW) Fine Grained
		18		REFUSAL
TURBIDITY: TOP (0'–9'): Moderate BOTTOM: (9'–18'): Moderate		19–24		
DRAWN BY: MDA CHECKED BY: JLA JOB NO: 8488.47				

Figure J2 Example of a jet probe log showing the kind of information that is logged in verbal or numerical formats along with graphic displays of sediment composition. These digital logs are part of a GIS framework and the information contained in them can be queried for special purposes. Note that some information is back loaded into the logs because it is obtained subsequent to field logging. Granulometric analyses, for example, report median grain sizes for clasts (e.g., sand grains) and particulate matter (e.g., percent silt content).

search investigations. Usually deployed after preliminary assessment of historical data (e.g., geophysical and geotechnical information), comprehension of the regional geology and geomorphology, and computer aided analysis (including GIS summaries), jet probing should verify previously indicated field conditions. Jet probe surveys thus perform a valuable function in sand searches and their relevance and importance should not be underestimated as a time- and cost-saving effort.

Reconnaissance bathymetric and jet probe surveys are also used to verify hydrographic features with widely spaced bathymetric surveys, historical surface sand samples, jet probes, core sites, and other potential sand features in the study area. Reconnaissance bathymetric surveys groundtruth and verify the National Oceanic and Atmospheric Administration hydrographic data in selected areas of potential sand deposits. The reconnaissance bathymetry should be compared with historical bathymetry to identify areas where sand has accumulated by natural coastal processes or offshore dredge disposal. An example of reconnaissance jet-probe survey is shown in Figure J3 for a portion of the southwestern coast of Florida. Here, on the wide continental shelf of the eastern Gulf of Mexico, a range of sedimentary deposits overlie a karstified limestone peneplain that extends seaward from the Florida peninsula (Evans *et al.*, 1985). Although the karst surface is somewhat irregular due to dissolution of the carbonate rocks, drowned valleys are infilled and planar areas are covered by blanket deposits and ridges.

Inlets along this coast, which produce deltaic deposits, show no strong regional trends and are stable in terms of channel width, length, geographic position, and orientation (Vincent *et al.*, 1991; Finkl, 1994). The low wave energy regime that influences sediment accumulation at inlets in this region enhances construction of large ebb-tidal deltas, which store enormous quantities of sand (Davis *et al.*, 1993). Flood-tidal deltas along the west-central Florida coast are relatively inactive due to small tidal ranges, sheltered lagoons, and ebb-dominated inlets (Davis and Klay, 1989; Finkl, 1994). The wide Continental shelf offshore southwest Florida, described by Davis (1997) and which gently slopes seaward toward the central basin of the Gulf of Mexico, maintains shallow depths to 9 km offshore to the 10 m isobath. Shelf morphologies and coastal (inlet) morphodynamics impact spatia distributions of mineral resources (Wright, 1995), large-scale coastal behavior (Short, 1999), and barrier island evolution (Oertel, 1979).

Various types of sand ridges (linear accumulations of sand bodies) are common on inner shelves along many shores the world over (viz. Duane *et al.*, 1972; Swift and Field, 1981; McBride and Moslow, 1991). These topographically positive sedimentary accumulations on the seafloor are recognized as relict sand bodies that formed in response to prior stillstands of mean sea level (MSL) when sea levels were lower than those of today. On the shelf off southwestern Florida, for example, prominent seabed morphologies include linear sand ridges, some of

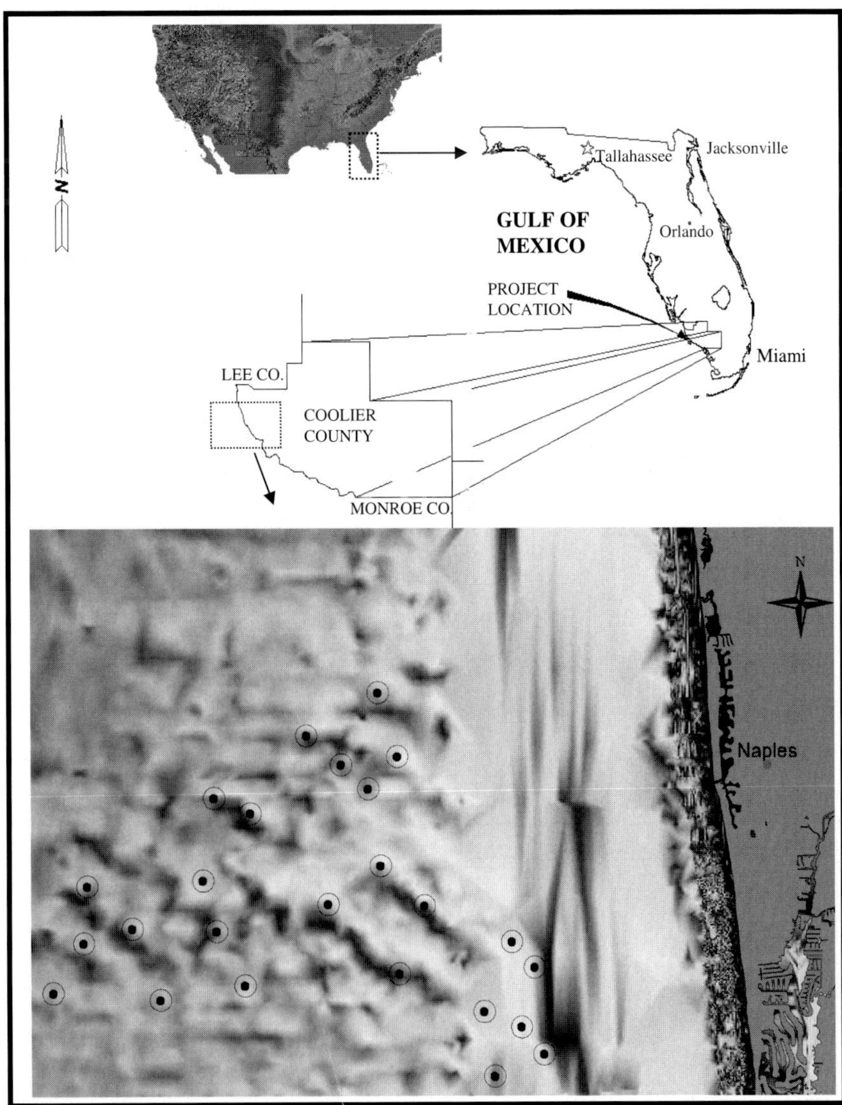

Figure J3 Jet probe location diagram showing bathymetry in terms of a graduated grayscale ramp so that sedimentary accumulations on the seafloor may be inferred from bathymetric highs. Jet probe locations, identified by the circled dots, are strategically placed to provide information related to the deposit thickness. Note the placement of jet probes on sand ridges.

which extend continuously for distances greater than 6 km. These deposits formed during the Flandrian Transgression (most recent Holocene trend in sea-level rise) (Davis, 1997). Depressional (negative topographic) features are incised into the karst surface and some surficial marls. When the continental shelf was exposed to subaerial geomorphic processes during low stands of sea level, streams cut into the karstified surface and persisted as valleys until sea level rose and they became infilled with recent marine and terrigenous sediments.

Figure J3 shows the distribution of reconnaissance jet probe locations of Naples, Florida. The jet probe locations are strategically placed on the basis of hydrographical, geophysical, geotechnical, and geological (including geomorphological) information (cf. Figure J1) that indicates the presence of sand deposits. As illustrated here by the shaded bottom relief on the lower left side of the Figure J3, caused by sedimentary accumulations on the seafloor, jet probes are sited on ridges, inter-ridge depressions, sand flats, and in other areas to verify thickness of sedimentary covers (or lack thereof). Emphasis must be placed on the fact that jet probes are not randomly sited on the seabed just to see what is there; rather, siting is intelligently coordinated with all collateral data that is related to the nature of bedrock surfaces and sedimentary accumulations on the seafloor. Reconnaissance jet probing is a strategy that is conducted as part of an overall coordinated methodology to define the presence of beach-quality sands on the seafloor.

Jet probes are thus taken in areas that show promise for sand deposits and to confirm historical vibracore logs. Jet probes and surface sand samples provide an indication of the thickness and characteristics of the unconsolidated sediment layers. With two dive teams, consisting of a geologist and a support diver, generally 8–15 jet probes can be obtained in a day depending on water depths, weather, and sea conditions (Andrews, 2002).

Geologists who are proficient in SCUBA diving, operate the jet probe by penetrating a graduated 7 m water pressure pipe into the ocean bottom and making observations as it passes through the sediment layers. The geologist is on the bottom and the support diver stays at the upper end of the probe to hold it upright against the current (Figure J4). The support diver also observes the turbidity level changes from above as silt is washed out of the probe hole (becoming suspended in the water column) during penetration of the seafloor. The geologist on the bottom observes the graduated scale on the probe and by the "feel" of the objects it encounters, makes mental notes of the depths of each change in texture, which are afterwards incorporated into the field log (cf. Figure J2). An experienced diver-geologist can distinguish layers such as shell, rubble, sand, peat, clay, and rock. The probe is jetted to the total length of the pipe (usually 7 m) or until it encounters a layer that it is unable to penetrate. On the Florida Gulf coast, karstified limestone formations on the inner continental shelf (e.g., Evans *et al.*, 1985; Hine *et al.*, 1998) usually limit jet probe penetration because sand deposits are less than 7 m thick. Nearly contiguous offshore sand ridges (described above), which are related to ebb-tidal deltas, and paleo barrier island, beach, and surf zone environments constitute the major source of beach renourishment sand on the central coast of west Florida. Jet probes, which are ideally suited to quickly and economically measure the thickness of thin sand deposits (i.e., <7 m in thickness), are therefore widely used in this geomorphic setting to determine the isopachs of shelf deposits that often occur in the form of sand sheets (shoals) or low ridges.

To obtain sediment samples from various depths, wash borings are obtained by the following methods. The geologist, who directs the jet probe into bottom sediments, takes two sample bags that are labeled "mid-depth" and "bottom of hole." The support diver, near the water surface, takes one sample bag labeled "surface sample." The probe is driven to its total depth of penetration, point of refusal (caused by hard layers, large floater, or bedrock) or maximum length of pipe. If that depth is 6 m, for example, the probe is pulled out and a second hole is

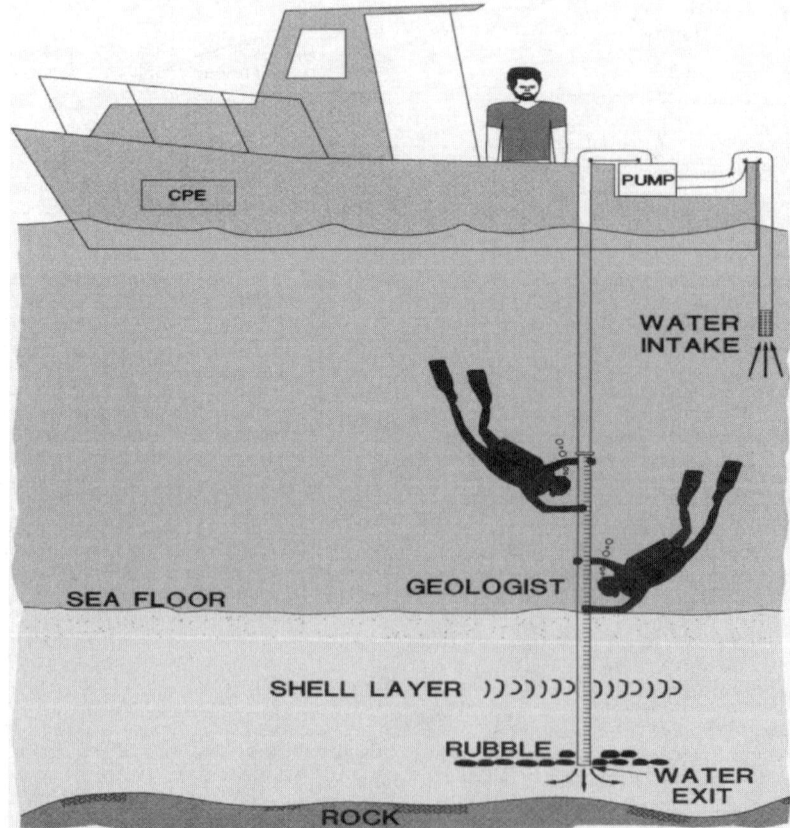

Figure J4 Schematic diagram showing the procedure for jet probing bottom sediments on the seafloor. Note that the portion of pipe that penetrates into the sediments contains graduated marks so that sediment thickness can be accurately determined. The geologist-diver works at the lower level near the seafloor and is proficient at estimating the nature of the materials probed by the "feel" of the pipe as it penetrates to refusal or reaches the end of the pipe.

probed to a depth of 3 m, 2–4 m up current from the first hole. The geologist pulls the first probe and the support diver signals the boat to haul the probe to the surface. The geologist takes a sample of the material that has formed a mound (spoil pile) around the probe hole and places it in the "bottom of hole" sample bag. A subsample of the material forming a mound around the second (shallower) hole is placed in the bag labeled "mid depth." The support diver, after the jet probe is hauled to the surface workboat, swims toward the bottom while moving against the current at about 2 m from the probed area (first two holes) and obtains an undisturbed "surface" sample from the bottom.

The subsamples removed from the washout mounds provide a representative bulk sample of the material that the probe passed through and which was jetted to the surface by the water pressure in the pipe. Materials comprising the washout mounds are deposited in the reverse order of the actual stratigraphic layers in the bottom sediments. Wash borings tend to have inherently low slit contents because the fine-grained particles, which have lower specific gravities than larger grains, tend to remain in the water column as suspensiods. The denser grains thus settle annularly in a mound around the jet probe. Suspension of fine-grained materials (typically silt plus clay and possibly organics) produces turbidity clouds in the water, which are quickly dispersed by currents. It is essential for the near-surface support diver to estimate changes in the turbidity plumes issuing from the jet probe so that the presence of fine-grained sediments is not under estimated from inspection of the heavier wash borings that quickly settle out of the water column. With experience, estimates of fines at different depths can be surprisingly accurate. Even though these samples (spoil from jet probing and estimates of fines) are extremely useful in the selection of areas for additional investigation, they are not meant to supplement or replace vibracores when defining borrow sites.

Upon returning to the surface workboat, both the diver- geologist and support diver immediately relay their underwater jet probing observations (i.e., depths of penetration, nature of the materials in different layers, and levels of turbidity that were associated with different depths) to the second onboard geologist who records this information in a permanent logbook. The descriptions relayed to the logbook should also include information that is relevant to characterization of the seafloor surface viz. sand ripples, algae, sea grass, surface rubble, or other observations. This information is often used to assist in the interpolation of sidescan sonar data. The sand samples are cataloged and notes on the texture (grain size) and color are recorded.

To prepare jet probe data for inclusion in reports, data that were recorded in the logbook are digitally entered into a jet probe log that is formatted in a manner similar to vibracore logs (see *Vibracores*). Sand samples are sieved to determine grain size and compared, in both wet and dry states, with a Munsell soil color chart. Representative samples are archived in small sample bags for presentation, reporting, and review. An example of data compilation for a jet probe survey is summarized in Table J1, which shows the classification of the jet probe, local relief of the surrounding seafloor, penetration of the probe, grain size, turbidity, and other relevant observations. Classification of the jet probe is important to interpretations of the survey because a single probe does not determine the viability of a deposit. The classification reported here is not universal, just an indication of what kind of system might be devised to show the resource potential of a probed area. Categorization of the "area of influence" for a single jet probe is comprehended by the application of "buffers," whereas multiple jet probe penetration defines a deposit. The buffer concept for jet probes represents an area that expands or contracts, depending on local sedimentary and geomorphological conditions. A sand sheet deposit will, for example, have a larger buffer zone around each jet probe because these kinds of deposits tend to be rather uniform over relatively large distances. The buffer around a probe on isolated sand ridges or in valley fills (i.e., drowned fluvial valleys, delta distributaries, tidal channels) will be a smaller zone because these kinds of deposits have limited lateral extents and conditions of sedimentation change in relatively short distances away from the probe. Local relief of the seabed in this area, increased by the presence of sedimentary bodies, is an indication of penetration depth for jet probes. Figure J5 demonstrates the observation with a fairly good correlation coefficient ($R^2 = 0.3935$). Once a survey is completed and the full range of parameters is appreciated and incorporated into an electronic database (see below), each jet probe is back classified so that it indicates the location of potential sand resources to be further investigated by refined geophysical (seismic and sidescan sonar) and geotechnical (vibracore) methods. Each jet probe is thus classified into one of five categories that range from unsuitable to a high potential for use. The categories are defined in Table J1 and it is important to note that application of the buffer concept in a spatial context on maps permits the recognition of sands (and the associated seafloor texture as seen in sidescan sonar images, three-dimensional bathymetric models, or isobathic expression of geomorphic units) that are potentially useful in beach replenishment projects. Grain size is determined by granulometric procedures from the subsamples collected by the geologist manning the jet probe. Turbidity is reported as estimated in the field and is a rough guide to the percent silt in the deposit (which is accurately determined later in the laboratory). Other observations included in Table J1 refer to the presence of rock fragments (e.g., limestone rubble, coral fragments), whether grain sizes fine or coarsen upwards or downwards, or any other property that should be noted.

Modern jet probe surveys are interfaced with advanced navigational software and differential GPS that make it possible to incorporate data into GIS database systems in such a way that reconnaissance-level surveys can be easily updated by new information and to facilitate efforts to groundtruth geophysical and geotechnical surveys. Table J1 is an example of the kind of jet probe-related information that can be extracted from GIS databases or queried for specific purposes. GIS analysis rings also facilitate querying procedures that can locate potential targets for sand mining activities and that can also point to areas where sediment texture and compositional information is insufficient to make reliable conclusions as to the presence of quality-sand sources. On the basis of jet probe data and other information, specific sand deposits are identified for detailed field surveys. Summary reports are usually prepared in a composite GIS framework, that is, in an electronic database and maps that help estimate sand volumes and approximate costs for detailed investigation of each potential borrow area, based on characteristics such as grain size and distance from the beach nourishment site on the shore and dredging suitability.

Conclusion

Jet probes are used to obtain information related to surficial sediment thickness on land and in shallow coastal waters. On land, jet probes may be powered by air or water pressure forced through a length of pipe. Jet Probes represent a good low-cost survey method for reconnaissance surveys on the sea and lakebeds. They are applicable to archaeological investigations and stratigraphic studies of thin sedimentary sequences, but it is in the search for beach-compatible sediments on the inner continental shelf that they find greatest use.

As a coastal resource tool, jet probes are often underutilized because researchers tend to use more sophisticated survey methods in the belief that greater value is received from greater expenditure. Jet probes are, however, an economical way of determining not only the thickness of sedimentary bodies but also their composition, grain size, compaction, and inclusions of rock fragments or other materials. When used collectively with a defined area as a specialized reconnaissance survey method, jet probes provide groundtruthing for geophysical, geotechnical, geological, and geomorphological interpretations of the seabed sediments. The main drawback for jet probing is that operators need to acquire sensitive skills for interpreting the "feel" of probe penetration. With some practice, however, geologist-driver operators can become proficient estimators of the various parameters that are normally associated with jet probe surveys. The most widespread application of jet probing is in coastal sand searches because increased knowledge of offshore sand resources is required for beach nourishment projects. Maximum water survey depths for jet probing are limited to about 30 m and the depth of penetration to the length of pipe that is easily handled underwater, usually about 7 m. For practical considerations, the minimum operating depth in water is about 1 m. As the search for sand resources intensifies, due to increasing erosion of beaches and coastal land loss on protective barrier islands and shoals, jet probes will increasingly serve as comparatively inexpensive procedures for evaluating seabed sediments on inner continental shelves.

Advanced geophysical and geotechnical procedures are essential for the accurate definition and location of sand resources on the inner shelf; however, these resources can be optimized if backed up or verified by low tech and less costly efforts, such as jet probing, that are essential to the efficiency and economic success of offshore sand searches. Advancements in positioning and navigation software and hardware that can be interfaced with GIS systems in the field permit analysis of spatial data associated with the jet probes in a timely fashion that increase survey efficiency and applicability. Although combinations of modern marine exploration techniques have contributed to the cost-effectiveness and success of sand search investigations, they are of reduced value if they are not accompanied by logical and rational planning of surveys in accordance with local geology and geomorphology.

Charles W. Finkl and Lindino A. Benedet

Table J1 Summary of field results for a jet probe survey off Charlotte County, southwestern Florida.

Jet Probe (#)	Category (rank)[a]	Relief (ft)	Penetration (ft)	Grain size (mm)	Turbidity (estimated)	Observations and notes
1	3	N/Av[b]	19	0.18	M to H[c]	Four feet of silty sand with clay balls on top/ no refusal
2	4	N/Av	12	0.24	H; H to M	Three feet of sand with silt/clay on top
3	5	N/Av	14	0.42	L to M	Slightly fining upwards
4	5	N/Av	20	0.31	L to M	Slightly fining upwards/no refusal
5	4	5	5	0.23	L to M	Fining upwards (1 ft layer), rock on bottom
6	4	5	7	0.35	H	Two feet of silty sands on top, 1 ft rubble on bottom
7	3	3	4	0.23	M to H	Fining upwards, rock on bottom, not well-defined ridge
8	2	4	3.5	0.19	M to H	Homogeneous, rock on bottom
9	1	4	3	0.16	H	One-foot thick layer of silty sand on top
10	2	5	7	0.17	L to M	About 4% silt and very fine sand
11	3	4	3	0.33	M to H	Fining upwards, relatively thin
12	1	4	3.5	0.17	H	Silty sands, 0.5 ft of rubble
13	2	5.5	4.5	0.17	M to H	One layer of silty sand on top
14	2	4.5	7.5	0.15	L to M	Fining upwards, very fine sediments
15	3	5	4	0.71	H	Fining upward, shell fragments
16	3	4.5	5	0.19	M	Finer than 0.2
17	2	5	8	0.14	H	Silty sand, too fine and H turbidity
18	4	5	9	0.16	M to H	Finer top, 0.5 mm visual estimate of bottom sand
19	3	6	16	0.14	M to H	Finer top, coarser on bottom (0.23 visual estimate)
20	2	1	3	0.22	H	Finer silty sand in top layer
21	4	6	8	0.23	M to H	Fining upwards, 1′ silty sand on top, 0.5′ rubble bottom
22	4	4	6	0.23	M to H	Fining upwards, at least 4 ft of fine–medium sand
23	3	4.5	4	0.95	M to H	Shell fragments, some silt in top 2 ft
24	4	7	7	0.52	M to H; L to M	At least 4 feet of clean sand, silty sand in top 2 ft
25	3	4.5	6	0.17	L to M	Clean sediments but too fine
26	5	3.5	5	0.23	L to M	Five feet of clean sand
27	5	4	8	0.37	L to M	Eight feet of clean sand, fining upward
28	4	6	6	0.19	L to M; M to H	Somewhat finer-grained than neighbors
29	5	3	5	0.29	M; M to H	Four and one-half feet of clean sand
30	2	2	2	0.41	M to H	Missed the top of the ridge
31	3	3.5	3	0.28	M	Missed the top of the ridge
32	3	3	3	0.22	H	Limited penetration, high turbidity
33	3	4	4	0.19	M to H	Fine sand, M to H turbidity levels
34	1	1	0	—	—	Trough before reef gave a "false-ridge" impression
35	1	3	3	0.16	H	Fine sediments, high turbidity, limited thickness
36	1	2	2	0.22	H	Limited thickness, high turbidity
37	2	3.5	3	0.34	H	Limited thickness, high turbidity
38	2	3	5	0.2	M to H	Silty sand on top, relatively high silt %
39	3	3	4	0.54	M	Shell fragments, 3 ft of clean sand
40	3	3	3	0.6	H	High turbidity, limited thickness
41	1	1	2	0.19	H	Trough after outcrop, limited thickness and silty sediments
42	3	3	4	0.6	M to H	Limited thickness, shell fragments, 1 ft layer of rubble 3 to 4′
43	4	2.5	5	0.57	L to M	One foot of rubble from 4 to 5
44	5	3.5	6	0.5	M	Five and one-half feet of coarse gray sand
45	2	3	2	0.22	L to M	Missed depositional area
46	5	3.5	11	0.25	M	Homogeneous sediment distribution
47	4	N/Av	14	0.17	L	Coarsening upward
48	2	N/Av	15	0.13	L	Sediments <0.15, but coarsening upwards and low turbidity
49	2	N/Av	15	0.13	L	Sediments <0.15, but coarsening upwards and low turbidity
50	2	N/Av	15	0.15	L	Sediments <0.15, but coarsening upwards and low turbidity

Table lists major criteria that are useful for the interpretation of sand deposits. Information that is summarized in tubular form assists in the identification of materials that are suitable for beach replenishment.

[a] Buffers divide jet-probed sedimentary deposit thickness into four categories based on sand quality and dredging capabilities, as follows: (1) Unsuitable: Deposit is less than 0.5 m thick, or mean grain size <0.17 mm, or there are high levels of turbidity during jet probing. (2) Marginally useful: Deposit is less than 1 m thick, or mean grain size is <0.2 mm, or the deposit is thicker with larger grain sizes but there is high turbidity, or the presence of silty sands or rubble layers. (3) Conditionally usable: Deposit thickness is between 0.5 and 1 m with relatively good quality sediments containing a mean grain size greater than 0.2 mm, but there are limiting factors such as limited thickness of sand bodies, or high turbidity levels, or good penetration but sediments analyzed <0.2 mm but visual description on other layers was >0.2 mm mean diameter of sand grains. (4) Potentially useful: Deposits thicker than 1 m but less than 1.5 m and sand grain sizes are between 0.2 and 0.25 mm; there is moderate to low turbidity. (5) High potential for use: Deposit is more than 1.5 m thick and sand grain size is more than 0.25 mm, there is moderate to low turbidity.

[b] N/Av = not available.

[c] The terms low, medium, and high are relative estimates of silt content based on visual interpolation of turbidity plumes.

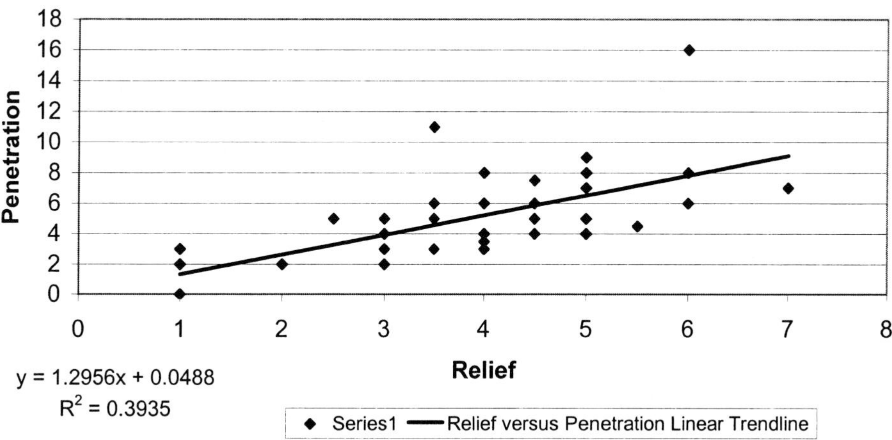

Figure J5 Linear regression analysis for a subset of jet probes collected offshore Naples, Florida, showing that jet-probe penetration increases with increasing local relief of sediments on the seafloor. The survey area included a series of sand ridges with intervening troughs. The troughs contained a thin veneer of sand (generally less than 0.5 m) over limestone bedrock whereas the sand ridges had a local relief up to at least 2.5 m (units on the graph are in feet).

Bibliography

Anders, F.J., Hansen, M., and McLellan, N., 1987. *Atlantic Coast beach Protection Project: Ocean City, Maryland (Draft Final Report)*. Vicksburg, MS: US Army Corps of Engineers, CERC-WES, 60p.

Andrews, J.A., 2002. Finding and developing cost effective beach compatible sand sources in the Gulf of Mexico, lessons learned in Florida, Louisiana and Texas. *Proceedings 2002 National Conference on Beach Preservation Technology*. Tallahassee, FL: Florida Shore and Beach Preservation Association, pp. 43–55.

Anon, 1996. The underwater excavations in Caesarea 1995–1996. Haifa, Israel: University of Haifa, The Leon Recanati Institute for Maritime Studies, *C.M.S. News, Report No. 23*, December 1996. [Now RIMS Newsletter]

Bird, E.C.F., 1985. *Coastline Changes—A Global Review*. Chichester, UK: Wiley.

Coleman, J.M., and Murray, S.P., 1976. Coastal sciences-recent advances and future outlook. *Science, Technology, and the Modern Navy*. Arlington, VA: Department of the Navy.

Committee on Beach Nourishment and Protection (CNBP), 1995. Appendix F: Project construction and sediment sources, transfer, and placement. In *Beach Nourishment and Protection*. Washington, DC: National Academy Press, pp. 267–293.

Conkright, R.D., Williams, C.P., and Christiansen, L.B., 2000. *Offshore Sand Resources in Northern Maryland Shoal Fields*. Baltimore, MD: Maryland Geological Survey File Report No. 00-2, 94p.

Davis, R.A., Jr., 1997. Geology of the Florida coast. In Randazzo, A.F., and Jones, D.S. (eds.), *Geology of Florida*. Gainesville, FL: University of Florida Press, pp. 155–168.

Davis, R.A., Jr., and Klay, J.M., 1989. Origin and development of Quaternary terrigenous inner shelf sequences, southwest Florida. *Transactions of the Gulf Coast Association of Geological Societies*, **39**: 341–347.

Davis, R.A., Jr., Klay, J.M., and Jewell, P., 1993. Sedimentology and stratigraphy of tidal sand ridges—southwest Florida inner shelf. *Journal of Sedimentary Petrology*, **63**(1): 91–104.

Duane, D.B., Field, M.E., Meisburger, E.P., Swift, D.J., and Williams, S.J., 1972. Linear shoals on the Atlantic inner continental shelf, Florida to Long Island. In: Swift, D.J., Duane, D.B., and Pilkey, O.H. (eds.), *Shelf Sediment Transport: Process and Pattern*. Stroudsburg, PA: Dowden, Hutchinson, and Ross, pp. 447–498.

Evans, M.W., Hine, A.C., Belknap, D.F., and Davis, R.A., 1985. Bedrock controls on barrier island development: West-central Florida coast. *Marine Geology*, **63**: 263–283.

Finkl, C.W., 1994. Tidal inlets in Florida: their morphodynamics and role in coastal sand management. In Viggosson, G. (ed.), *Proceedings of the Hornafjördur International Coastal Symposium*, Reykjavik, Iceland, Icelandic Harbour Authority, pp. 67–85.

Finkl, C.W., and Walker, H.J., 2002. Beach nourishment. In Chen, J., Eisma, D., Hotta, K., and Walker, H.J. (eds.), *Engineered Coasts*. Dordrecht, The Netherlands: Kluwer, pp. 1–22.

Finkl, C.W., Khalil, S.M., and Andrews, J.L., 1997. Offshore sand sources for beach replenishment: potential borrows on the continental shelf of the eastern Gulf of Mexico. *Marine Georesources and Geotechnology*, **15**: 155–173.

Finkl, C.W., Khalil, S.M., and Spadoni, R.H., 2000. A geomor- phological approach to carbonate sand beaches. In Magoon, O.T., Robbins, L.L., and Ewing, L. (eds.), *Carbonate Beaches 2000: First International Symposium on Carbonate Sand Beaches*. Reston, VA: American Society of Civil Engineers, pp. 53–66.

Finkl, C.W., Andrews, J.A., and Benedet, L.A., 2003. Shelf sand searches for beach renourishment along Florida Gulf and Atlantic coasts based on geological, geomorphological, and geotechnical principles and practices. *Coastal Sediments 2003*. Reston, VA: American Society of Civil Engineers.

Freedenberg, H., Hoenstine, and Dabous, A., 2000. Preliminary identification of sand resources in federal waters along the central Florida east coast. *Proceedings 2000 National Conference on Beach Preservation Technology*. Tallahassee, FL: Florida Shore and Beach Preservation Association, pp. 247–257.

Hine, A.C., Locker, S.D., Harrison, S.E., Edwards, J.H., Donohue, B., and Twichell, D.C., 1998. *Proceedings of the 1998 National Conference on Beach Preservation Technology*, Tallahassee, FL: Florida Shore and Beach Preservation Association, pp. 260–275.

James, W.R., 1975. Techniques in Evaluating Suitability of Borrow Material for Beach Nourishment. Ft. Belvoir, VA: US Army Corps of Engineers, CERC Technical Memorandum No. 60.

Keehn, S., and Campbell, T.J., 1997. The renourishment planning and design process for Captiva Island, Florida. *Proceedings of the 1997 National Conference on Beach Preservation Technology*. Tallahassee, FL: Florida Shore and Beach Preservation Association, pp. 75–89.

Lessa, G.C., Meyers, S.R., and Marone, E., 2000. Holocene stratigraphy in the Paranquá Bay Estuary, south Brazil. *Journal of Sedimentary Research*, **68**(6): 1060–1076.

McBride, R.A., and Moslow, T.F., 1991. Origin, evolution, and distribution of shoreface sand ridges, Atlantic inner shelf, USA. *Marine Geology*, **97**: 57–85.

Meisburger, E.P., 1990. Exploration and sampling methods for borrow areas. Vicksburg, VA: US Army Corps of Engineers. *CERC Coastal Engineering Technical Report CERC-90-18*.

Meisburger, E.P., and Williams, S.J., 1981. Use of Vibratory Coring Samplers for Sedimemt Surveys. Vicksburg, VA: US Army Corps of Engineers, *CERC Coastal Engineering Technical Aid No. 80-4*.

Oertel, G.F., 1979. Barrier island development during the Holocene recession, southeastern United States. In Lethearman, S.P. (ed.), *Barrier Islands: From the Gulf of St. Lawrence to the Gulf of Mexico*. NewYork: Academic, pp. 273–290.

Short, A.D., 1999. *Handbook of Beach and Shoreface Morphodynamics*. Chichester: Wiley, 329p.
Stewart, C.J., 2000. *Lower Great Lakes Erosion Study: Toussaint River, Ohio*. Buffalo, New York: US Army Corps of Engineers, Buffalo District.
Swift, D.J., and Field, M.E., 1981. Evolution of a classic sand ridge field: Maryland sector, North American inner shelf. *Sedimentology*, **28**: 461–482.
Toscano, M.A., and Kerhin, R.T., 1990. Subbottom structure and stratigraphy of the inner continental shelf of Maryland. In Hunt, M.C., Doenges, S.V., and Stubbs, G.S. (eds.), *Studies Related to Continental Margins, Years Three and Four Activities*. Austin, TX: Bureau of Economic Geology, University of Texas.
Toscano, M.A., and York, L.L., 1992. Quaternary stratigraphy and sea-level history of the U.S. middle Atlantic Coastal Plain. *Quaternary Science Reviews*, **11**: 301–328.
US Army Corps of Engineers staff (COE), 2002. Erosion, transport, and deposition of cohesive sediments. In *Coastal Engineering Manual—Part III*. Vicksburg, MS: EM 1110-1-1100. 42p.
Vincent, C.L., Corson, W.D., and Ginerich, K.J., 1991. *Stability of Selected United States Tidal Inlets*. Vicksburg, MS: US Army Corps of Engineers, GITI Report No. 21.
Walker, H.J., and Finkl, C.W., 2002. Beach nourishment: case studies. In Chen, J., Eisma, D., Hotta, K., and Walker, H.J. (eds.), *Engineered Coasts*. Dordrecht, The Netherlands: Kluwer, pp. 23–59.
Walther, M.P., 1995. Looking for sand sources further offshore, case study: Venice Phase II: Sand Wars and sand shortages and sand holding structures. *Proceedings of the 8th National Conference on Beach Preservation Technology*. Tallahassee, FL: Florida Shore and Beach Preservation Association, pp. 1–16.
Wells, D.V., 1994. *Non-Energy Resources and Shallow Geologic Framework of the Inner Continental Margin off Ocean City, Maryland*. Baltimore, MD: Maryland Geological Survey Open File Report No. 16.
Wright, L.D., 1995. *Morphodynamics of Inner Continental Shelves*. Boca Raton: CRC Press.

Cross-references

Archaeology
Coastal Sedimentary Facies
Mapping Shores and Coastal Terrain
Monitoring, Coastal Geomorphology
Nearshore Geomorphological Mapping
Offshore Sand Banks and Linear Sand Ridges
Offshore Sand Sheets
Shelf Processes
Surf Zone Processes
Vibracore

JOURNAL LISTING—See APPENDIX 2

K

KARST COASTS

Karst coasts are defined as coasts made of calcareous rocks and showing distinctive geomorphic characteristics related to their lithologic nature (Jennings, 1985). In some cases, these original features are directly due to seawater action and marine weathering. Halokarst is a word sometimes used to designate such erosion forms (Fairbridge, 1982). In other cases, karst coasts correspond to the exposure by coastal retreat or to the drowning by the postglacial transgression of subaerial or underground karst features, which are actually the result of limestone solution in continental environments (Trenhaile, 1987).

Coastal karstification

Seawater corrosion

The word corrosion, introduced by A. Guilcher, includes different chemical, physicochemical, and biological processes operating on carbonate-rich rocks in coastal environments and resulting in specific erosional features.

The ability of seawater to dissolve calcium carbonate has long been a contentious issue. Nowadays, the most widespread opinion is that no real evidence favoring solution has been produced. It appears that coastal seawater is saturated or oversaturated with calcium carbonate and data-showing solution at night, through emission of carbon dioxide by green algae living in pools, is not wholly convincing. However, recent work seems to indicate that chemical erosion has been underestimated (Miller and Mason, 1994). It is now proved that, at least in tropical environments, undersaturation of inshore waters may occur at night with respect to calcite and at any time with respect to aragonite and high magnesian calcite, accounting for some 10% of the erosion in coralline limestones (Trudgill, 1976).

Bioerosion (*q.v.*), a term for the removal of rock by the direct action of living organisms, is generally acknowledged to play the greatest role in the development of coastal corrosional features, not only in the tropics where an enormously varied marine biota live on calcareous substrates, but also in higher latitudes (Kelletat, 1988). Algae are probably the most important erosive organisms, both in the intertidal and the supralittoral zones. Endolithic cyanophyta are boring organisms that actively contribute to the rock destruction. Fungi and lichens also are effective rock borers. Grazers consuming the microflora, such as the gastropods *Littorina* and *Patella*, can cause mechanical rasping of rock surfaces, which have been weakened by the penetration of endolithic algae. The chiton *Acanthopleura* has hard teeth that enables it to erode resistant limestones. Borers are responsible for excavations into the substratum. Penetrating habits of *Lithophaga, Lithotrya, Cliona* are frequently mentioned. *Lithophaga* acts through mechanical boring facilitated by acid secretion, which causes a softening of the rock. The sponges pertaining to the genus *Cliona*, which are able to bore microscopic to macroscopic excavations in limestones, play a particularly important role in the disintegration of rock substrates. Also worms, such as *Polydora*, may be active borers in calcareous substrates. Biological erosion is of great significance on the limestone coasts, which may justify the term of biokarst which has been proposed for the resulting forms (Spencer, 1985).

In the supralittoral zone, physicochemical processes operate jointly with bioerosion. Spray action, implying wetting and subsequent drying, leads to salt crystallization and causes rock disintegration. Eolianites are especially prone to such kind of weathering.

The rate of sea corrosion in calcareous rocks has been measured in a great number of sites all over the world. A figure of about 1 mm per year may be considered as an average rate.

Shore platforms

Bare erosional platforms (see entry on *Shore Platforms*) may be found on low carbonate-rich rocky coasts where waves are not supplied with clastic tools of allocthonous origin, which enhance their mechanical action. Figure K1 represents a typical profile of a corrosional shore platform from the Mediterranean which can be used as an illustration of a littoral karst in subtropical, low wave energy, and microtidal conditions. There is a general zoning of forms between low water mark and the area reached only by spray. The main feature of the midlittoral (intertidal) zone is represented by a platform, several meters in width, which is called trottoir (*q.v.*), a French word for sidewalk (Figure K2). In its upper part, the platform shows pools with overhanging edges and, lower down, wide shallow pools with flat bottoms, the so-called "vasques," which are separated by low, narrow, continuous, sinuous rims made of residual rock or built by calcareous organisms. Seawards, before terminating abruptly in a small vertical infralittoral cliff, the platform is often characterized by an overhanging ledge made by vermetids (*Dendropoma*) and calcareous algae (*Neogoniolithon*). Sometimes, a fossil trottoir, a few tens of centimeters above the active one, characterized by crater-shaped pools, is found, pointing to a higher relative sea level during the Holocene. Further up, in the supralittoral zone where spray is acting, above an overhanging cliff pitted by alveoles and vermiculations, jagged and sharp lapiés are found (Figure K3). They are separated by deep pools with overhanging rims. In fact, wave energy, tidal range, and mainly seawater temperature are the most important parameters which explain a great variety of shore platforms in carbonate rocks. In cool temperate regions, the trottoir is unknown, whereas in warm seas the presence of deep notches and protruding visors is noticeable (Guilcher, 1953).

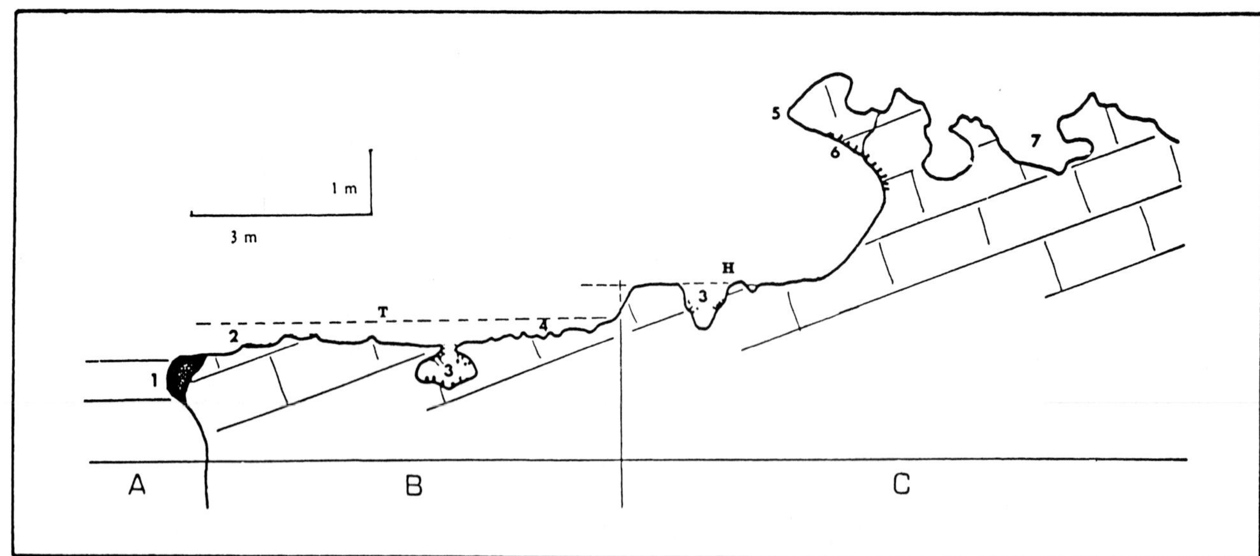

Figure K1 Schematic profile across a limestone shore platform in the Mediterranean, after R. Dalongeville (1977). A, infralittoral zone; B, midlittoral zone; C, supralittoral zone; H, elevated trottoir indicating a Holocene relative sea level higher than the present one; T, trottoir; 1, vermetid ledge; 2, shallow pool of vasque type; 3, pool with overhanging sides; 4, alveoles; 5, overhanging cliff; 6, vermiculations; 7, lapiés and pools.

Figure K2 Typical trottoir developed into an upper Pleistocene eolianite, northern Israel coast. (Photo R.P. Paskoff.)

Notches

Coastal *notches* (*q.v.*) are indentations due to lateral cutting by sea corrosion, which is particularly active on calcareous-rich rocks in tropical waters (Figure K4). Cliffs, low rocky shores, or wave-thrown large boulders on coral reefs may be affected by notches, which are good indicators of sea-level position, especially where the tidal range is low and the coastal environment sheltered. Site exposure is the most important factor and two main types of notches are to be distinguished (Pirazzoli, 1986): (1) tidal notches, in relatively protected sites and cut in the intertidal zone, which are relatively narrow; (2) surf notches in exposed sites and cut above high tide level. In the case of the first type, when the undercutting is well developed, being 2–3 m deep, the notch roof often forms an overhanging rock ledge, called the visor. Deep tidal undercut on low stacks and isolated blocks may result in mushroom-shaped rocks. Surf notches have a distinctive morphology due to water turbulence and spray action. Organic accretion around high tide level by calcareous algae and vermetids protects the substrate calcareous rock and inhibits sea action, meanwhile erosion proceeds above, forming an asymmetric notch with generally a short roof and a developed floor which eventually may form a bench protruding seaward. Simultaneously, bioerosion, is responsible for the development of another notch around low tide level. Consequently, double notches are not necessarily evidence of a relative sea-level change.

Figure K3 Upper Pleistocene eolianite affected by lapiés and pools in the supralittoral zone, south of Tangier, Morocco. (Photo R.P. Paskoff.)

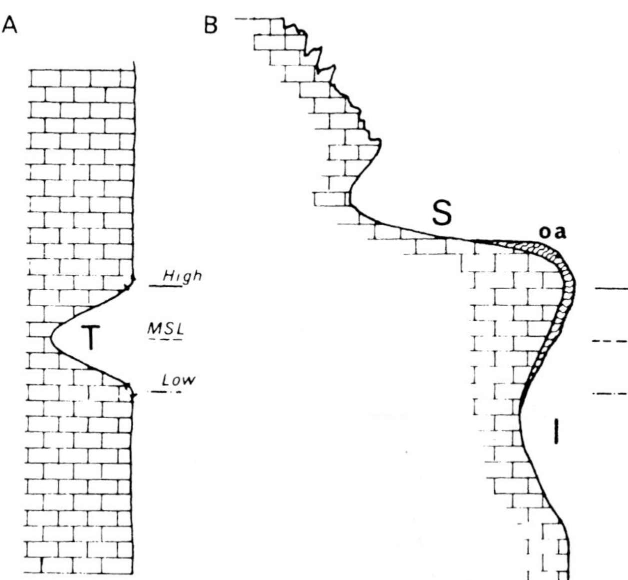

Figure K4 Sea corrosion notch profiles, after P.A. Pirazzoli (1986). A, sheltered environment; B, swell exposed environment; I, infralittoral notch; oa, organic accretion; S, surf notch; T, tidal notch.

Solution pipes in eolianites

Calcareous eolianites or eolian calcarenites are subaerially cemented paleodunes (see entry on *Dune Calcarenite*). They characterize many coastal regions which are semiarid at present and they may also be found in those which were semiarid at some stages during the Pleistocene, a period of important sea-level oscillations and climate changes. They have been described from the Mediterranean, the Canary Islands, South Africa, southern and western Australia, Bermuda. Diagenesis by continental waters of the carbonate-rich sand deposits, largely composed of fragments of marine organisms, took place in a vadose environment and generally includes dissolution of unstable aragonite and high-Mg calcite, and precipitation of relative low-Mg calcite.

Such eolianites may be affected by piping, which produces tubular underground conduits (Figure K5). For instance, in northern Tunisia, conspicuous pipes, 20–40 cm in diameter and a few meters deep have been reported (Paskoff, 1996). They are cylindrical in shape, taper vertically downward, and occur in aggregated clusters. They show a red coat which is a hardened calcitic crust and are filled by red silts and sands, sometimes strongly cemented. In the early literature, it was suggested that pipes represent the pseudomorphs of former tree stumps buried under advancing sand dunes. They were also simply explained as random solutional features corresponding to points where percolating water happened to converge and caused localized subsurface dissolution. A more appealing explanation was recently put forward for the Bermuda pipes by S.R. Hervitz (1993) who suggested that the cylindrical vertical conduits are the products of the stemflow of tree species capable of acidifying intercepted rain water and funneling large quantity of it down their trunks. The result is a subsurface dissolution forming pipes, which extend vertically downward through eolianites.

Exposed and submerged terrestrial karst on coasts

There are coasts where karst landforms can be followed from the land into the sea practically without any substantial modification. A conspicuous example of such a situation is given at Along Bay, near Haiphong, in northern Vietnam, where a typical tower karst, developed in a humid tropical environment, has been submerged and makes up an archipelago of islands and islets in a shallow sea (Figure K6).

Exposed karstic landforms

As a result of coastal erosion, continental karstic features may have been exposed and modified by marine processes. For instance, the nearly vertical chalk cliffs (see entry on *Chalk Cliffs*) of the Normandy coast, in France, show typical solution forms of terrestrial origin, which have been revealed by the shoreline retreat. There are examples of inherited karstic caves debouching in the face of the cliffs. Others have been captured by sea caves developed at the foot the cliffs under mechanical wave attack. Coastal erosion has also exposed deep cylindrical hollows, which originally formed as solution pipes. The famous arch and pillars at Etretat owe their existence to marine action in a highly karstified area. The crest of the Normandy chalk cliffs shows a crenulated appearance,

Figure K5 Vertical section of a solution pipe on an active sea cliff developed into an upper Pleistocene eolianite in northern Tunisia. (Photo R.P. Paskoff).

which is due to dry valleys, locally called *valleuses*, now hanging as a result of coastline recession. Near Cascais, in central Portugal, on a rock coast, bare and deeply developed lapiés are exposed in the upper mid-littoral zone because decalcification red silts which once covered them have been stripped off by swash denudation. In Australia, west of Melbourne, the Port Campbell coast illustrates the effects of marine erosion processes on a highly karstified area of Miocene limestones. Sinkholes and caves have been cut by wave action (Baker, 1943).

Drowned karstic landforms

The interaction between continental processes, sea-level variations, and wave erosion is well illustrated by subaerial or underground features which developed in limestone terrains, such as enclosed depressions or caves, and are now invaded by the sea. In this respect, Malta gives conspicuous examples of what could be called marine sinkholes, originally karstic features developed in coralline limestones (Paskoff and Sanlaville, 1978). On Gozo Island, Dewra Bay is a semi-circular cove, measuring 340 m in diameter, which results from marine erosion breaching a doline wall whose eastern half alone has been preserved and whose bottom is now largely occupied by the sea. An islet, Fungus Rock, is the last remnant of the destroyed western wall (Figure K7). Qwara, an immediately neighboring landform, is an identical circular sinkhole, 400 m in diameter and 70 m deep, bounded by vertical walls, which remained unbreached, but has been nevertheless partially inundated through a karstic gallery now connecting the open sea with the depression. Blue Grotto, in the southern part of the main island, corresponds to another kind of cove which is due to the sea eroding into a large cave system and causing subsequent roof collapse. In Asturias, Spain, different types of periodically or permanently flooded dolines, by salt water, have been described (Schülke, 1968). Marine charts of the subsident Dalmatian coast, in Croatia, clearly show broad enclosed basins of polje type, which are completely submerged (Baulig, 1930). The extended Novigrad Bay is regarded as a polje invaded by the sea. The freshwater lake of Vratna, on the island of Cres, appears to occupy the floor of an uvala, a large depression resulting from the coalescence of sinkholes, which was deepened when sea level was lower than now and became a permanent lake as a result of the postglacial transgression. In the vicinity of Marseille, in France, giant lapiés have been identified at a depth of 40 m in the Veyron Bank. In the same area, the partly submerged Cosquer cave, which was discovered in 1991 and became notorious for its exceptionally nice paintings and engravings dating back to the upper paleolithic period, has only one entrance located at 37 m

Figure K6 Islands and islets corresponding to a submerged tropical karst, Along Bay, northern Vietnam. (Photo M. Paskoff.)

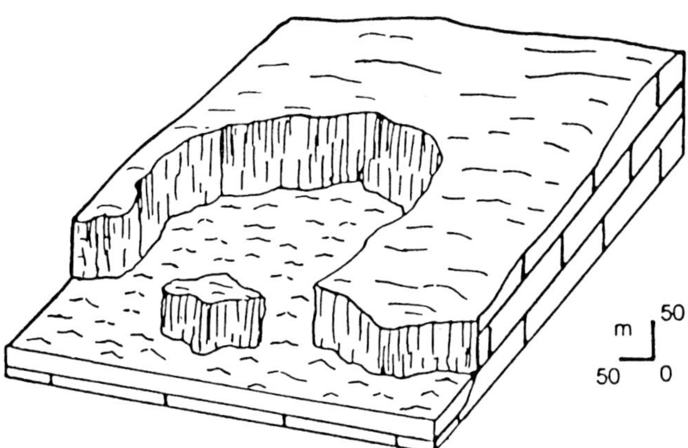

Figure K7 Dweyra Bay and Fungus Rock, Gozo Island, Malta: a semi-circular cove created through sea invasion of a doline (after Paskoff and Sanlaville, 1978).

under present sea level. The limestone littoral of Provence, as others calcareous coasts in the Mediterranean, shows many active submarine springs or resurgences, for instance the one at Port-Miou, near Cassis, east of Marseille, located at a depth of 12 m.

Calanques or calas

Calanque from the French coast of Provence or *cala* from the Balearic islands in Spain are words which designate a narrow, short, and drowned valley with steep sides, developed in limestone terrains and continued inland by a dry course. Several types have been identified, as in Malta, for example (Paskoff and Sanlaville, 1978). Some are of ria type. They were deepened during glacial periods of low sea level and pluvial climate, which facilitated the cutting of deep ravines by stream action, subsequently submerged in their lower portion by postglacial transgressions, as it was the case with the Holocene sea-level rise (Figure K8). The amazing pattern of digitate calanques, which makes the site of the capital city, Valletta, one of the finest anchorages in the world, corresponds to a branching valley system whose deeply and extended drowning resulted from a marked subsidence in this area. Other calanques are linked with continental karst processes. They appear, at least partly, related to the formation of subterranean caves and conduits by freshwater solution of limestones along faults. Later on, wave erosion opened these cavities and the rushing of marine water caused the roofs to collapse. Such a case raises an important question about the extent to which karst features can develop below the sea without any change in the relative level of land and sea. If the response is positive, calanques may form without a marine transgression being necessary.

Karstified coral reefs

According to some authors, barrier reef and atoll morphology (see entry on *Coral Reefs*) is fundamentally karst induced (Purdy, 1974). This theory, sometimes called "the karstic saucer theory" (Guilcher, 1988) and still in discussion, states that the shape of such reef forms derives from antecedent and horizontal coralline platforms, which were emerged and modified by subaerial solution processes. It is logical to assume that all reefs which are today at or near sea level were emergent during the glaciation periods. Being emerged, they became karstified. Rainfall and percolating water action is supposed to be more rapid toward the interior of the calcareous platforms than around their steep edges where runoff is rapid. The result is a saucer-shaped surface dissolution with a central depression and a peripheral raised rim interrupted by ravines through which a part of the water escaped, the other part percolating through the limestone. Subsequent submergence of karst-eroded platforms started a revival of coral growth and the ramparts resulting from previous subaerial karstification became barrier reefs or atolls.

The karstic saucer theory, which is not in contradiction with the Darwin's subsidence theory of coral reef formation and can be combined with it, is supported by geomorphic observations. Floors of many lagoons are characterized by numerous upstanding pinnacles or knolls which rise from various depths and have living coral on their surfaces. They are generally interpreted as remnants of karstification during low sea-level glacioeustatic phases and compared with the tower karst developed in rainy tropical environments. Lagoons may also show pits which are considered as submerged sinkholes or blue holes. At Mayotte, a Comoro island, the bottom of the lagoon shows several enclosed depressions with steep sides, 60–70 m deep, lying at approximately 20 m below the surrounding floor. These features are obviously the result of a subaerial karstification. Mataiva, an atoll in the Tuamotu archipelago, has a reticulated lagoon formed by a network of about 70 pools of varying sizes, with an average depth of some 10 m, separated by shallow ridges. This strange honeycombed pattern whose exact conditions of formation remain uncertain is thought to derive from a complicated evolution during which tropical phases of karstification are supposed to have occurred when the structure was emerged. So, it appears that the sea-level lowerings related to the Pleistocene glaciations have in many cases determined a karstification of preexisting coral structures.

Figure K8 Typical calanque in the island of Gozo, Malta. It corresponds to a partially drowned valley, with steep sides, cut into coralline limestones. (Photo R.P. Paskoff.)

Roland P. Paskoff

Bibliography

Baker, G., 1943. Features of a Victorian limestone coastline. *The Journal of Geology*, **6**: 359–386.

Baulig, H., 1930. Le littoral dalmate. *Annales de Géographie*, **39**: 305–310.

Dalongeville, R., 1977. Formes littorales de corrosion dans les roches carbonatées au Liban. *Méditerranée*, **3**: 21–33.

Fairbridge, R.W., 1982. Karst coast. In Schwartz, M.L. (ed.), *The Encyclopedia of Beaches and Coastal Environments*. Stroudsburg: Hutchinson Ross Publishing Company, pp. 500–502.

Guilcher, A., 1953. Essai sur la zonation et la distribution des formes littorales de dissolution du calcaire. *Annales de Géographie*, **62**: 161–179.

Guilcher, A., 1988. *Coral Reef Geomorphology*. Chichester: John Wiley & Sons.

Hervitz, S.R., 1993. Stemflow influences on the formation of solution pipes in Bermuda eolianite. *Geomorphology*, **6**: 253–271.

Jennings, J.N., 1985. *Karst Geomorphology*. Oxford: Blackwell.

Kelletat, D., 1988. Quantitative investigations on coastal bioerosion in higher latitudes: an example from northern Scotland. *Geoökodynamic*, **9**: 41–51.

Miller, W.R., and Mason, T.R., 1994. Erosional features of coastal beachrock and eolianite outcrops in Natal and Zululand, South Africa. *The Journal of Coastal Research*, **10**: 374–394.

Paskoff, R., 1996. Upper Quaternary eolianites in coastal Tunisia: a review. *Zeitschrift für Geomorphologie N.F.*, **102**(Suppl.-Bd.): 135–149.

Paskoff, R., and Sanlaville, P., 1978. Observations géomorphologiques sur les côtes de l'archipel maltais. *Zeitschrift für Geomorphologie N.F.*, **3**: 310–328.

Pirazzoli, P.A., 1986. Marine notches. In O. van de Plassche (ed.), *Sea-level Research: A Manual for the Collection and Evaluation of Data*. Norwich: Geobooks, pp. 361–400.

Purdy, E.G., 1974. Reef configurations: cause and effect. In L.F. Laporte (ed.), *Reefs in Time and Space*. Tulsa: Society of Economic, Paleontologists and Mineralogists, pp. 9–76.

Schülke, H., 1968. Quelques types de dépressions fermées littorales et supralittorales liées à l'action destructive de la mer (Bretagne, Corse, Asturies). *Norois*, **57**: 23–42.

Spencer, T., 1985. Limestone coastal geomorphology: the biological contribution. *Progress in Physical Geography*, **12**: 66–101.

Trenhaile, A.S., 1987. *The Geomorphology of Rock Coasts*. Oxford: Clarendon Press.

Trudgill, S.T., 1976. The marine erosion of limestones on Aldabra Atoll, Indian Ocean. *Zeitschrift für Geomorphologie N.F.*, **26** (Suppl.-Bd.): 164–200.

Cross-references

Atolls
Bioerosion
Chalk Coasts
Coral Reefs
Eolianite
Notches
Rocky Coasts
Shore Platforms
Trottoirs
Weathering in the Coastal Zone

KLINT

The term klint, widely used in countries around the Baltic Sea, was originally a Danish and Swedish word synonymous with *klev*, signifying an escarpment in sedimentary rocks. Usually it comprises a line of marine abrasion or ancient (pre-Quaternary) fluviatile erosion scarps. In Swedish, the word occurs synonymously with *grike*, signifying a type of hollow formed by karst weathering (Martinsson, 1958), and also means mountaintop or bioherm (coral reef) hillock. Also the term *fjällglint* is used for rock formations in a Scandinavian mountain ridge (*fjällkedjan*). In German literature the word *glint* is preferred and in such form was adopted also in Estonia, the classical klint area (Tammekann, 1940). The word in English is often recorded as *clint*, in Russian *glint* (more rarely *klint*), in Latvian *glints* (*klints* in Latvian means *rock*), and in Lithuanian *klintas* is used. The corresponding Estonian word is *paekallas*; for separate klint lobes and promontories the word *pank* is also used.

Distribution and structure

The most well-known klint in the Baltic Sea area is the Ordovician. It consists of an almost continuous, but indented and lobated arc, from the western coast of the Island of Öland (Västra Landborgen) in Sweden over the Baltic Sea via the north coast of Estonia to Lake Ladoga in Russia (Figure K9). This monumental escarpment, up to 56 m high (at Ontika in Estonia) and 1,200-km long, is called the Baltic Klint (Figure K10). Its basal part consists of Cambrian rocks, dominated by sand- and siltstones and soft "Blue clays." The hard crest layers of the klint, however, which primarily cause the steepness of the escarpment, consist of Ordovician limestones (Orviku, 1940). In the westernmost area, on the Island of Öland, the klint crest is developed exclusively in Middle Ordovician limestone beds. A submarine Ordovician klint, as a morphological feature, has been identified on sea charts and with

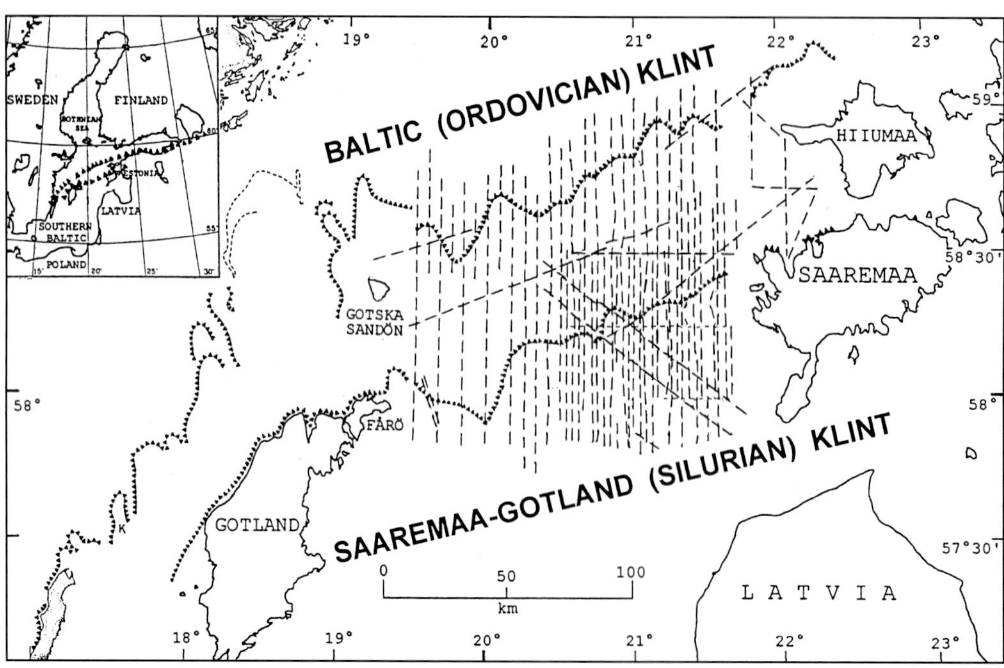

Figure K9 The distribution of klint (black triangles) in the Baltic Sea area. Compiled by R. Vaher, based on the data published by A. Martinsson (1958) and I. Tuuling (1998). Dashed lines mark seismic profiles.

Figure K10 The Pakerort Klint on Pakri Peninsula, northwestern Estonia. (photo by A. Miidel.)

geophysical sounding methods (Figure K9). The Baltic Klint, called North-Estonian Klint in the Estonian mainland, is dissected by river valleys forming over 20 picturesque waterfalls (Aaloe and Miidel, 1967). A submarine Vendian–Cambrian klint line in the scientific literature indicates not only the escarpment line, but also the border between the Fennoscandian Shield and the Russian Platform.

To the north of the Baltic (Ordovician) Klint on the seafloor, Vendian and Cambrian scarps and to the south step-by-step Silurian, Devonian, and Carboniferous klints crop out. The Saaremaa–Gotland (Silurian) Klint in Silurian strata forms an arc of more or less lobed scarps along the northwest coast of Gotland Island in Sweden, on the seabottom and on the north coast of Saaremaa Island in Estonia, following to Central Estonia. The westernmost part of this klint arc limits the shelf on which two Karslö islands, Store Karlsö and Lilla Karlsö, are situated. The larger part of the klint crest on Gotland reaches between 40 and 50 m above sea level. On Saaremaa, the height of cliffs is up to 22 m at Mustjala and on the Estonian mainland rarely more than 10 m. Devonian and Carboniferous klints in the contemporary topography differ from each other in scarps of different height and length.

Formation

Some klints in the Baltic Sea area began to develop during the Late Silurian and pre-Middle Devonian continental period (Puura *et al.*, 1999). Later the prolonged pre-Pleistocene erosional–denudational processes were of utmost importance, operating upon the main features of structural changes in the Precambrian crystalline basement and the sedimentary cover. The latter has a gentle southward inclination (6–18'). Due to the tectonic uplift, new areas were influenced by the lateral river flows forming questa-like topography with steep northern and gentle southern slopes. The influence of the old drainage systems upon the klint formation is evident. The Ordovician klint system developed throughout the entire length by the erosion of the soft, mainly Cambrian sandstone and clay strata, overlain by much more resistant limestones, which determined the retreat of the klint and caused the relative steepness of the feature. The bottommost part of the Saaremaa–Gotland Klint also consists of softer rocks in mainland Estonia, for example, from marls of the Jaani Regional Stage, overlain by the Jaagarahu limestones, a large part of which consists of reefs, which cause the dissected and lobated appearance of the klint. Zigzag contour lines in many places are dependant on tectonic joints.

During the ice ages the klints were influenced by glacial erosion and, after the retreat of the continental ice, the Ordovician and Silurian klints were strongly changed by wave action of the Baltic Sea. This influence was different, because different parts of the klints rose above the sea level at different times (Orviku and Orviku, 1969). If the Ordovician Klint at Ontika in northeastern Estonia was under the influence of waves during the Baltic Ice Lake more than 11,000 years ago, then the Island of Osmussaar in northwestern Estonia appeared above the sea level only some 2,000 years ago.

The term *klint* is used more and more not only in the Baltic Sea area, but also in North America and western Europe for the mentioning of steep escarpments with monoclinal bedding of sedimentary rocks. There is also a specific type of klint (glint)-line lakes in Norway and Scotland, which were formed between the ice and escarpments.

Anto Raukas

Bibliography

Aaloe, A., and Miidel, A., 1967. Eesti pangad ja joad. Tallinn, Eesti Raamat.

Martinsson, A., 1958. The Submarine Morphology of the Baltic Cambro-Silurian area. *Bulletin of the Geological Institutions of the University of Uppsala*, **38**(21): 11–35.

Orviku, K., 1940. Lithologie der Tallinn-Serie (Ordovizium Estland). *Acta et commentationes Universitatis Tartuensis*. A. **36**: 1.

Orviku, K., and Orviku, K. jun., 1969. Über die Beständigkeit der Entwicklung der Küsten im Estländischen Küstengebiet in der Spätglazialzeit, im Holozän und in Gegenwart. Eesti NSV Teaduste Akadeemia Toimetised. *Keemia*Geoloogia*, **18**(2): 128–139.

Puura, V., Vaher, R., and Tuuling, I., 1999. Pre-Devonian landscape of the Baltic Oil-Shale Basin, NW of the Russian Platform. In Smith, B.J., Whalley, W.B., and Warke, P.A. (eds.), *Uplift, Erosion and Stability: Perspectives on Long-term Landscape Development*. Geological Society, London, Special Publications, **162**, pp. 75–83.

Tammekann, A., 1940. The Baltic Glint. A Geomorphological Study. Part I. Morphology of the Glint. Publicationes Instituti Universitatis Tartuensis Geographici, 24.

Tuuling, I., 1998. Shipborne geophysical study of an Ordovician–Silurian carbonate platform, Farö–Hiiumaa area, northeastern Baltic Sea. Department of Geology and Geochemistry, Stockholm University, No. 301. Stockholm.

Cross-references

Changing Sea Levels
Cliffed Coasts
Cliffs, Erosion Rates
Europe, Coastal Geomorphology
Geographic Terminology
Rock Coast Processes

L

LAGOONS—See COASTAL LAKES AND LAGOONS

LANDSLIDES—See MASS WASTING

LATE QUATERNARY MARINE TRANSGRESSION

The total volume of water in the world's oceans exhibits a nearly perfect negative correlation with global ice volume; when one increases the other decreases. This is known as *glacial eustasy* (first proposed by Maclaren, 1842). The balance between global ice volume and ocean water volume is controlled by climate. At the last glaciation maximum some 20,000 radiocarbon years BP large quantities of water were withdrawn from the oceans and accumulated in the form of extensive continental ice caps. We may try to reconstruct past glacial volume changes by the following three means:

(1) the recording of corresponding sea-level positions, which are affected by numerous other variables;
(2) the recording of corresponding oxygen isotope variations, which are affected by other factors, too, not least temperature;
(3) volumetric estimates of corresponding ice caps, which is quite a rough method (presently stored ice in Antarctica, Greenland, and alpine glaciers are estimated in this way).

Though all three methods have their limitations and problems, there is a general agreement that the 20 ka glacial eustatic lowering was on the order of 120 m, as seen in sea level (e.g., Fairbanks, 1989), in oxygen isotope values (e.g., Shackleton, 1987), and in glacial volume (Flint, 1969). According to Chappell *et al.* (1996) there is an excellent agreement between the coral record of the Huon Peninsula in New Guinea and deep-sea oxygen isotope records.

The glacial eustatic rise in sea level as a function of the switch from ice age conditions at about 20 ka to interglacial climatic conditions is known as "the postglacial transgression." It commenced some 20,000 radiocarbon years ago and ended at around 5,000 radiocarbon years BP. This rise in sea level was neither smooth nor globally consistent.

Figure L1 gives the combined view of sea-level changes as established from coral reefs in Barbados, from sea-level data off west Africa, and from the eustatic component as calculated from multiple sea-level records in northwest Europe.

Oscillations were induced both by glacial eustatic variations and by the interaction with others factors acting on sea level. The largest glacial eustatic oscillations were those associated with the high-amplitude climatic changes at around 13–10 radiocarbon Ka ago, that is, the period including the classical climatic oscillations of the Bölling Interstadial, the Older Dryas Stadial, the Alleröd Interstadial, and the Younger Dryas Stadial (Fairbanks, 1989; Mörner, 1993).

Regional variations in amplitude and fine-structures of the transgression were induced by additional variables (Mörner, 2000). Those variables are deformation of the geoid relief (Mörner, 2000), internal adjustment to loading changes (e.g., Peltier, 1998), changes in earth's rotation, and the ocean circulation system (Mörner, 2000). Therefore, the actual sea-level rise after the last glaciation maximum differ significantly from place to place over the globe as illustrated by the atlas of Holocene sea-level curves (Pirazzoli and Pluet, 1991).

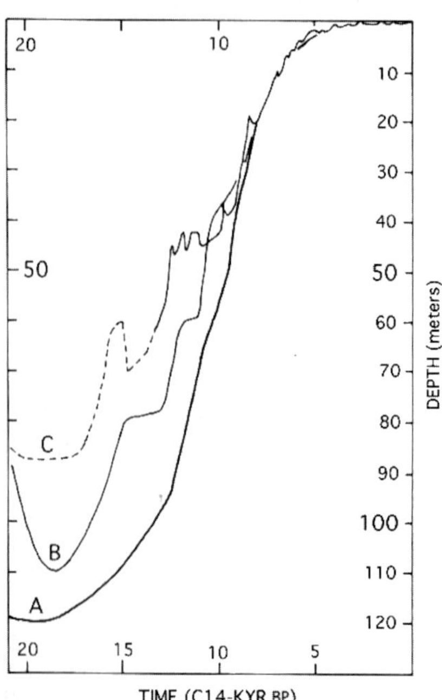

Figure L1 The postglacial transgression illustrated by the sea-level changes recorded in (A) Barbados, (B) west Africa, and (C) northwestern Europe. Depth in meters and age in C14-years BP.

Global sea level was dominated by the glacial eustatic rise in sea level up to about 5,000 radiocarbon years BP. After that, it was dominated by the redistribution of ocean masses.

Nils-Axel Mörner

Bibliography

Chappell, J., Omura, A., Esat, T., McCulloch, M., Pandelfi, J., Ota, Y., and Pillans, B., 1996. Reconsiliation of Late Quaternary sea level derived from coral terraces ot Huon Peninsula with deep sea oxygen isotope records. *Earth and Planetary Science Letters*, **141**: 227–236.

Fairbanks, R.G., 1989. A 17,000-year glacio-eustatic sea level record: influence of glacial melting rates on the Younger Dryas event and deep-ocean circulation. *Nature*, **342**: 637–641.

Flint, R.F., 1969. *Glacial and Quaternary Geology*. New York: John Wiley & Sons.

Maclaren, C., 1842. The glacial theory of Professor Agassiz of Neuchatel. *American Journal of Science*, **42**: 346–365.

Mörner, N.-A., 1993. Global change: the high-amplitude changes 13–10 ka ago—novel aspects. *Global Planetary Changes*, **7**: 243–250.

Mörner, N.-A., 2000. Sea level changes and coastal dynamics in the Indian Ocean. *Integrated Coastal Zone Management*, Spring-Ed., 17-20. ICG Publ. Ltd.

Peltier, W.R., 1998. Postglacial variations in the level of the sea: implications for climate dynamics and solid-earth geophysics. *Reviews of Geophysics*, **36**: 603–689.

Pirazzoli, P.A., and Pluet, J., 1991. *World Atlas of Holocene Sea-Level Changes*. Amsterdam: Elsevier Oceanographic Series, **58**, pp. 1–300.

Shackleton, N.J., 1987. Oxygen isotopes, ice volume and sea level. *Quaternary Science Review*, **6**: 183–190.

Cross-references

Changing Sea Levels
Coastal Changes, Gradual
Coastal Changes, Rapid
Coastline Changes
Geodesy
Holocene Epoch
Sea-Level Changes During the Last Millennium
Sea-Level Rise, Effect

LIFESAVING AND BEACH SAFETY

Lord, Lord! methought, what pain it was to drown!
What dreadful noise of waters in mine ears!
What ugly sights of death within mine eyes!
Methought I saw a thousand fearful wrecks,
Ten thousand men that fishes gnawed upon.

King Richard III
By William Shakespeare

Beaches are a major attraction for people throughout the world. In the United States, it is estimated that 85% of all tourism revenue stems from visits to coastal areas (Center for Marine Conservation, 2000). Like any natural area, however, hazards exist that can result in injury or death. Managing these hazards is key to ensuring that beaches can be safely enjoyed.

Drowning is the most serious problem related to beach and water use. Annual deaths by drowning, whether from floods, sinking ships, recreational swimming, home accidents, and other causes, easily outstrip deaths from war and terrorism. In 1997, for example, 4,051 Americans died by drowning (National Center for Health Statistics, 1997). According to the Centers for Disease Control and Prevention, drowning is the second leading cause of injury-related death for American children aged 1–14 years (National Center for Health Statistics, 1997). And death is not the only outcome of distress in the water. It has been found that for every child who drowns, 17 or more are treated at hospitals for complications related to near-drowning (Wintemute, *et al.*, 1988).

Beyond drowning, many other hazards exist in the aquatic environment. Spinal injury from shallow water diving occurs with unfortunate regularity, often caused by inadvertently diving into a submerged sandbar or bodysurfing over a wave and striking bottom. Scuba related illness has existed since the inception of scuba and, while the safety of equipment has improved, so has the volume of people scuba diving. Boating accidents, both recreational and commercial, are a source of injury and death. And of course, all of the common ailments related to any other physical activity occur in and around the water. All of these, and many more, have increased attention paid to accident prevention and, in particular, the provision of lifeguards at coastal beaches.

Ocean Beach in San Francisco provides a stark example. In 1998, three persons drowned on the first day of summer at this beach, which is beset by strong surf, consistent rip currents, and relatively cold water. No lifeguards were provided by the federal government, which owns the beach as part of a national park. Instead, strongly worded signs had been used in an effort to warn people away from use of the water. This was clearly not fully effective at preventing drowning.

As the summer of 1998 progressed, further drownings occurred, eventually totaling seven. There was heavy media attention and advocacy by groups like the International Life Saving Federation (ILS) and the United States Lifesaving Association (USLA) for the immediate provision of lifeguards.

Under public and political pressure, the National Park Service ultimately relented and provided preventive beach rescue services beginning at the start of the summer of 1999. During that summer, there were no drownings. Lifesaving services have continued since.

Lifesaving history

In historical terms lifesaving, as an organized response to persons in distress in the water, is young. The rescue of shipwrecked sailors appears to have spurred some of the earliest organized efforts. China's Chinkiang Association for the Saving of Life was established in 1708 as the first of its kind in the world (Shanks *et al.*, 1996). It eventually came to involve staffed lifesaving stations with specially designed and marked rescue vessels.

In the Netherlands, the Maatschappij tot Redding van Drenkelingen (Society to Rescue People from Drowning) was established in Amsterdam in 1767, primarily to address problems of drowning in the numerous open canals in Amsterdam. This society remains in existence today, now promoting a wide variety of drowning prevention initiatives. English lifesaving efforts began in 1774, though boating rescue operations were not initiated until 1824 (Shanks *et al.*, 1996).

In 1787, the Massachusetts Humane Society began what was to become a lifesaving movement in the United States that evolved into the US Life-Saving Service (USLSS). The USLSS was eventually composed of an extensive national network of coastal lifesaving stations staffed by government paid lifesavers, and credited with saving over 170,000 lives. In 1915, this organization joined with the Revenue Cutter Service to become the US Coast Guard.

It was only in the late 1800s that swimming, then known as bathing, began to emerge as a widely popular form of recreation. When ocean resorts were built, in places like Atlantic City and Cape May, New Jersey, drowning quickly emerged as a problem. Various drowning prevention methods were implemented, including the use of lifelines in the water—fixed ropes to which bathers could cling.

When these approaches proved inadequate, police were assigned to lifesaving duties in Atlantic City. Eventually though, police resources became strained by this responsibility. Instead, a corps of lifeguards was employed in 1892. In Cape May, efforts began with rescue rings hung on bathhouses and the provision of dories on the beach that could be used for rescue. By 1865, hotels began hiring persons to staff the surfboats. Later, a municipal lifeguard operation was begun that continues to the present day.

Both the American Red Cross and the YMCA initiated efforts in the early 1900s to teach Americans to swim and to rescue each other when in distress. This grew into nationwide networks of swimming instruction and lifeguard training that exist today, with a focus on pools and inland beaches.

While the need for prevention and rescue services was evinced by drownings, the elemental lifesaving techniques were just that—rudimentary steps that one swimmer could use to rescue another—person to person and often without equipment. The rescue equipment, what there was of it, was adapted from other disciplines, such as the devices that had been used by the USLSS to rescue sailors from the sea.

Surfboats, similar to those used by the USLSS, were adapted for use by lifeguards to row to swimmers in trouble. They remain in use in a few areas of the United States. The predominant method of rescue though, was by swimming to the victim.

One of the greatest difficulties for swimming lifesavers was the struggle sometimes required to overpower a panicked victim before the rescue could be completed. The line and reel (landline), was an early solution. A lifeguard would swim out to the victim while attached to the line, clutch the victim, and would be rapidly pulled back to shore by others.

This method had the advantage of quick retrieval, but there were some disadvantages too. The line produced drag, which could slow approach to the victim; it required two or more persons to operate; it was inadequate in cases of multiple rescues simultaneously occurring at different locations; and, it could become tangled. In Atlantic City, use of the line was discontinued after a lifeguard was strangled by the device. Nevertheless, it was widely used elsewhere for decades and is still in use in a few areas.

As an alternative, lifeguards fastened an eight-foot line and shoulder harness to a life ring. The lifeguard would swim out with the life ring, push it to the victim, and tow the victim to a dory or to shore. This avoided contact with the victim, but like the line and reel, the life ring created significant drag in the water.

Captain Henry Sheffield, an American with a variety of aquatic accomplishments to his credit, was touring Durban, South Africa in 1897 when he designed the first "rescue can," also called the "rescue cylinder," for a lifesaving club there (Brewster, 1995). It was made of sheet metal and pointed on both ends, with the same over-the-shoulder harness and line as had been used on life rings. The advantage was that it moved much more smoothly through the water, providing little drag. A disadvantage was that the heavy metal and pointed ends could cause injury.

In Australia, the first volunteer lifesaving "club" was founded at Bondi in 1906. Prior to that, ordinances had proscribed swimming, but civil disobedience eventually resulted in making swimming permissible and prompting the need for lifesaving services. Surf Life Saving Australia, one of the largest volunteer organizations in the world today, grew out of the Australian tradition that began at Bondi, of voluntarily guarding the beach. Today, some Australian lifesavers are paid, but most are still volunteers.

In 1907, George Freeth was brought from Hawaii to Redondo Beach, California to help promote a seaside resort. Billed as, "the man who walks on water," Freeth was the first person to surf on the American West Coast and is considered by many to be the first California beach lifeguard, as he made many rescues of persons in distress. According to the USLA though, it was not until the legendary Duke Paoa Kanhanamoku visited California in 1913 and introduced his redwood surfboard to Long Beach, California lifeguards, that the surfboard was adopted as a rescue tool. Later, the term "rescue board" would be coined (D'Arnall *et al.*, 1981).

The initiation of lifeguard services has often resulted from drownings and, even today, drownings tend to spur the provision or augmentation of such services. In 1918, 13 people drowned in a single day in San Diego, California, spurring the creation of a lifeguard service that now counts some 240 lifeguards providing response to coastal emergencies 24 h a day, throughout the year.

In 1935, Santa Monica, California lifeguard, Pete Peterson, seeing a need for a device that could be wrapped around the victim for greater security in the surf, produced the first rescue tube as an inflated device. Though it was vulnerable to weather, it became quite popular among lifeguards, and even more so when, in 1964, it was made of foam rubber, hot dipped with a rubber coating. Known to many veteran lifeguards as the "Peterson Tube," or just the "Peterson," it is in wide use today.

The rescue tube is an excellent tool for surf rescue because the victim is less likely to become separated from the rescuer in breaking waves. It is a particularly valuable option for semiconscious or unconscious victims, though it is less useful for multiple victim rescues, since its design is intended for a single victim. Interestingly, it has now become a common tool at pools and waterparks. So a device developed by a single lifeguard for a particular environment has come to be used in all aquatic environments throughout the world.

Captain Sheffield's sheet metal rescue "can" came to be constructed of aluminum in 1946, which lightened it substantially and allowed the ends to be rounded, but it was still heavy and itself presented a hazard. Then, Los Angeles County lifeguard Bob Burnside developed an improved rescue buoy, made of plastic, with handles on each side. It was on the beach in 1972 and greatly improved upon the ability of lifeguards to safely effect rescues, particularly in open water environs.

The "Burnside buoy" continues to be the rescue device of choice in situations where a highly buoyant and hydrodynamic float is needed to handle multiple victim rescues, since several victims can easily hold onto it at once. As a symbol of modern lifeguarding, it may have even come to eclipse the life ring.

The invention of the swim fin has changed lifesaving too. Lifeguards with fins are much faster in their swimming approach to victims and have the power to easily rescue several victims at a time, fighting the very currents that caused distress in the first place. Fins are particularly valuable in areas where the surf breaks gradually offshore and where distress may occur far from the beach. For some lifeguard agencies, swim fins are a required tool for swimming rescues and each lifeguard has a pair available at all times.

Rescue boards, a variation of surfboards and one of the original rescue devices, have been perfected to include handles for extra victims, specially designed decks for knee paddling, and lighter material. The lengths range from 3 to 4 m, with longer boards being more buoyant and faster, but heavier. Using a rescue board, a well-trained lifeguard can move quickly over the water and keep eight or more victims afloat. These devices can also be carried easily atop lifeguard emergency vehicles.

Statistics

The volume of beach use has expanded tremendously, as has the work of lifesavers. For 1998, the USLA reported the following from major reporting beach lifeguard agencies (Table L1).

Rip currents

USLA statistics show that over 80% of rescues at surf beaches are due to rip currents. This phenomenon is caused by a variety of factors. First, wave action pushes water up the slope of the beach. Then, gravity pulls it back to sea level. As it seeks to return, the water takes the path of least resistance, which sometimes causes it to be concentrated in

Table L1 Beach Lifeguard Statistics—1998

Beach attendance	
Total	256,721,418
Rescues	
Total	63,088
Primary cause	
Rip current	27,030
Surf	3,141
Swiftwater	142
SCUBA	112
Cliff rescues	
Total	75
Boat rescues	
Total	2,618
Passengers	3,207
Vessel value	$56,012,701
Boat assists	
Total	6,487
Passengers	15,865
Vessel value	$87,693,000
Preventive actions	
Total	2,735,889
Medical aids	
Total	209,317
Major	9,529
Minor	199,788
Drownings	
Total	111
Unguarded area	104
Guarded area	7
Fatalities	
Total	43
Enforcement actions	
Total	618,111
Warnings	594,899
Boat/PWC	16,127
Citations	6,219
Arrests	866
Lost and found persons	
Total	23,958
Public safety lectures	
Total	60,979
Number of students	405,561

Note: The addition of the "*Primary Causes*" of "Rescues" will not add up to the "Total" rescues because some agencies do not specify the cause.
Source: United States Lifesaving Association.

currents of water moving away from shore. These currents in the ocean are called rip currents.

Rip currents have three major components. The *feeder* is the main source of supply, composed of water that has been pushed up the beach by wave action. The *neck* is a relatively narrow river of water within the ocean moving back to sea. The *head* is typically a mushroom shaped area of water as the rip current disperses outside the surfline.

Wherever there is regular surf, there will be some form of rip currents. These currents vary in intensity according to wave energy, as well as bottom conditions. For example, a rocky bottom with channels can foment the formation of strong rip currents, as can reefs parallel to shore. They can also form due to channeling of water by undulations in the sand bottom, jetties, groins, and piers.

The USLA has identified several different types of rip currents (Brewster, 1995), including:

- *Fixed rip currents*: These are found on sand beaches, and remain in the same place so long as underlying sand conditions remain the same.
- *Permanent rip currents*: These remain in the same area year round and are usually seen on beaches with rocky bottoms, near groins, or piers, where the underlying structure never changes and rip current intensity varies only with swell size and direction.
- *Flash rip currents*: These currents occur suddenly and unexpectedly typically due to sets of waves that are higher in size than other waves and bring unusually high volumes of water ashore quickly. They may form regardless of an obvious differentiation in underlying beach structure.
- *Traveling rip currents*: These currents usually occur on sandy beaches and move along with the prevailing swell direction as longshore currents move water along the beach.

Worldwide, it would appear that the highest volume of rescues by lifeguards, by a significant margin, exists in the southern California counties of Los Angeles, Orange, and San Diego. There, regular and strong wave action and resulting rip currents, combined with high, year-round beach attendance, ensure that many persons will need rescue. In 1998, the USLA reports that lifeguards in these three counties effected 43,882 rescues, which represented 66% of all rescues reported to USLA by American beach lifeguard agencies that year. In contrast, Surf Life Saving Australia reports that a total of 12,948 rescues were effected by Australian surf lifesavers in the 1998/99 season.

Drowning prevention

Another important statistic is "preventive actions." Preventive actions are typically warnings to swimmers and others to avoid areas of hazard that might result in distress or drowning. Not all agencies report these actions to USLA, but USLA statistics show that for every rescue, there are at least 43 preventive actions by lifeguards. Clearly, this is an essential action, without which the number of rescues and drownings might be much higher. As such, the value of lifeguards extends well beyond the reactive service of rescuing someone in need, to active prevention.

According to the USLA

> USLA has calculated the chance that a person will drown while attending a beach protected by USLA affiliated lifeguards at 1 in 18 million (.0000055%). This is based on the last ten years of reports from USLA affiliated lifeguard agencies, comparing estimated beach attendance to the number of drownings in areas under lifeguard protection.

Historically, lifeguards have typically been placed on beaches and the level of lifeguard coverage increased only after drownings have occurred. This may be partially due to a view that swimming and related activities are considered discretionary, recreational activities, and to some, worthy of a lesser level of attention than more common or necessary activities. Nevertheless, the impact of death resulting from drowning, for whatever reason, is the same as that from other causes.

Another problem confronting lifesaving is the lack of standardized systems for rating the ambient hazards at swimming beaches that might dictate specific levels of lifeguard protection. The wide variety of factors that increase the likelihood of distress and drowning are quite complex. Such factors include attendance levels, weather, water temperature, surf, strength of rip currents, swimming skills of users, etc. Recent efforts by Professor Andrew Short on behalf of Surf Life Saving Australia have produced a system to consider these many factors and develop appropriate preventive strategies (Short, 1997). It has been effectively applied in several areas. Further work and testing of this system is underway.

Beach signs

Statistics indicate that the provision of lifeguards results in heightened safety and drowning prevention. In some areas, however, passive warning systems, such as flags and signs, are the only source of drowning prevention measures. This is particularly true at areas with low attendance or relatively benign ocean conditions. However, as was demonstrated at Ocean Beach in San Francisco, where extremely strongly worded signs were initially employed in place of lifeguards, signs alone are of limited value in drowning prevention.

At present, there is no internationally recognized standard for beach signage. Since many beachgoers are tourists, this lessens the likelihood that preventive signs will be read or understood. Regardless of the reason, signs seem of very limited value. One study of beach signs found that 85–90% of beachgoers did not recall having seen the signs posted at beaches where they were recreating.

Modern lifesaving

The major national lifesaving organizations of the world have organized themselves into a single worldwide confederation, known as the International Life Saving Federation. Through this organization, lifesavers exchange information, extend lifesaving aid to countries lacking preventive programs, and meet to vie in lifesaving competition.

Many changes have taken place in lifesaving over the years. Effective modern lifeguarding involves a carefully orchestrated system of drowning prevention in which lifeguards in elevated towers oversee designated areas of water and cross-check the edges of areas of responsibility of neighboring lifeguards. Some areas employ a so-called "Tower 0" system, which involves a permanent, elevated structure with an enclosed observation deck overseeing an entire beach area, sometimes more than a mile in length.

The lifeguard in Tower 0 acts as something of an overseeing traffic controller who coordinates activities of lifeguards in beach level towers, sending backup when they make rescues or need assistance for medical aid. This system depends on mobile backup services from emergency vehicles and boats.

The need for emergency vehicles initially arose as lifeguards were expected to patrol larger areas and were summoned away from their regularly assigned stations to emergencies elsewhere. Lifeguards in emergency vehicles can respond over long distances to deliver rescue personnel and equipment to remote areas or simply to better cover longer beaches. They are now an essential element of backup at major beaches that reduce the need for personnel and increase the rescue equipment that can be transported to assist at a rescue.

Vehicles can also provide emergency backup to other lifeguards stationed a significant distance away. The public address systems on lifeguard vehicles are invaluable for delivering preventive warnings to swimmers and communicating to lifeguards in the water.

Rowed dories continue to be used in a few areas of the world for rescue, but since the advent of motorized vessels with compact engines, motorboats have become the rescue vessel of choice in most areas. The 10 m Baywatch boats of Los Angeles County, with their inboard motors, were some of the first boats designed specifically for lifesaving. They allow lifeguards to rescue multiple victims, as well as to respond to offshore boating emergencies. The only limitation of these vessels is that they must remain outside the breaking surf, where victims are sometimes trapped.

When Australian lifesavers first modified commercially available inflatable vessels to operate effectively as rescue boats in the surf environment and dubbed them IRBs, they pioneered one of the most striking modern advancements in lifesaving. Unlike hard hull vessels, these 4 m boats with small outboards are able to navigate the largest and most powerful surf to rescue distressed swimmers and surfers. Even in the unfortunate event of a capsizing, these vessels are easily righted and their outboards rehabilitated for operation. They have proven their worth time and time again in the most inclement conditions.

It was not long ago that personal watercraft (PWCs; also known by the trade name Jet Ski) were introduced to the waterways. To some they are seen as a noisy irritant, or a water toy more suited to an amusement park; however, personal watercraft have transformed boating recreation, becoming one of the most popular types of recreational boat.

Initially, few saw the PWC as a viable rescue tool. Now however, thanks in particular to pioneering efforts by Hawaiian lifeguards and loaner programs provided by some manufacturers, PWCs are employed by many lifeguard agencies as rescue boats. These boats are extremely quick, powerful, usually unaffected by capsizing, and can be operated by a single rescuer; although they work best when two lifeguards work in concert using a towed rescue sled.

As the boats available to lifesavers have changed and adapted, so have the technologies that make boat rescues more effective. Global positioning systems, for example, can now be found aboard many of the larger and more advanced lifeguard rescue boats. And radio direction finders allow boat operators to hone in on an emergency radio signal from a boat in distress.

The most sophisticated rescue implement used by lifeguards today is aircraft, particularly helicopters. In some areas, lifeguards have made arrangements with a local police or rescue helicopter service to assist them in times of need; but in Rio de Janeiro, Brazil, Durban, South Africa, New Zealand, and Australia, helicopters have long been a basic tool of the lifeguard agencies themselves. They allow quick access to offshore or remote emergencies, unimpeded by traffic or surf conditions. They also allow for rapid evacuation of the injured to hospitals. When not used for emergencies, helicopters are excellent platforms for observation, patrols of remote areas, searches, and dissemination of public information.

Communication has been a perennial challenge for lifesaving. Particularly at beaches, where lifeguards may cover broad expanses of shifting sand, remote areas, and the open sea, there is a need to quickly summon backup or advise others of an emergency in progress. Whistles, megaphones, and flags were some of the first tools used for this purpose, but their value declines rapidly as distance increases. For this reason, two-way radios have become the communication tool of choice.

First aid is another area of tremendous change. Bottled oxygen, suction devices, and one-way masks are all highly recommended pieces of first aid equipment, which many lifesavers have available. So are spinal and cervical immobilization devices. A recent innovation in this area is the floating backboard. Implements to protect against communicable disease have also become a must.

All of these improvements in equipment, technology, and rescue techniques, along with the exploding attendance levels at aquatic areas, now demand training levels among lifesavers that go well beyond the early days. Effective lifesaving has always called for superior resuscitation skills, but with advancements in techniques this has come to require tens of hours of training in both resuscitation and professional level cardio-pulmonary resuscitation. In some areas, lifeguards are trained to the level of paramedics, and many full-time lifeguards are emergency medical technicians.

General rescue skills have been advanced also. In America, lifeguards assigned to coastal beaches typically receive 80 to well over 100 hours of basic training before they are given basic lifesaving duties. Full-time lifeguards, who work year-round in California, Florida, and Hawaii receive many additional hours of training. In the European Union, the minimum standard to become a lifeguard is now some 300 h. Other countries exceed even these levels.

In addition to the responsibility for water rescues, some lifeguards have been called upon to perform specialized emergency services in and around the aquatic environment. This has required special training and equipment appropriate to the task.

Expanding lifeguard services

From the rocky Irish coastline, to the soft sandstone of the California coast, people headed for the beach are sometimes stranded on the cliffs above. Some lifeguards are trained and equipped in high-angle rescue to pluck them from their plight. In San Diego, where over 25 of these rescues are performed each year, a special rescue vehicle with a crane aboard is used to lower rescuers and raise victims to safety.

All lifeguards have some degree of responsibility for controlling the swimmers they protect and regulating activities, but in some places that responsibility has advanced to regular law enforcement power. Lifeguards employed by the State of California and Volusia County, Florida are empowered as police officers, carrying firearms, and enforce even the most serious crimes that occur on their beaches.

Some of the other specialized services provided by lifeguards include marine firefighting and flood rescue. At least two lifeguard agencies, San Diego and Los Angeles County, have expanded upon the emergency callback system used by other lifeguard agencies to staff lifeguards 24 h a day to respond to the many calls that come in to their dispatch centers during nighttime hours.

Lifeguards are also participating in mutual aid networks within their communities. As many lifesaving groups align themselves with police and fire agencies, they are increasingly called upon to act in concert with other public safety providers when natural disasters strike.

Lifeguards around the world are taking their preventive responsibilities one step further through the development of youth programs. Called nipper or junior lifeguard programs, they train youngsters in ways to safely use the waters and encourage their later participation as lifesavers. Many thousands participate in these programs each year.

Lifesaving competition too, has become tremendously popular, involving some 600,000 people annually. Many of the early international exchanges of lifesaving information came through the solidarity brought about through competition and this tradition continues with events sponsored by the International Life Saving Federation and its member federations, as well as other groups. These competitions encourage not only information exchange and international goodwill, but also inspire lifesavers to maintain the high levels of fitness needed to effectively save lives in the water.

With the continually increasing responsibilities, technologies, and training, the role of the lifesaver has been transformed over the years. The lifesaver has acquired an internationally recognized and tremendously positive image of a well prepared, physically fit, and versatile person ready for any emergency that might develop in or near the water. This has benefited water safety, but it has also burnished the image of lifeguards, now more likely than ever to be seen as providing an integral layer of essential public safety protection, allowing safe use of a sometimes hazardous coastline.

B. Chris Brewster

Bibliography

Brewster, B. Chris (ed.), 1995. *The United States Lifesaving Association Manual of Open Water Lifesaving*. Englewood Cliffs, NJ: Brady/Prentice Hall.

Center for Marine Conservation, 2000. *Report on the Health of the Oceans*. Washington, DC: Center for Marine Conservation.

D'Arnall, Douglas *et al.*, 1981. *Lifesaving and Marine Safety*. Piscataway, NJ: New Century Publishers.

National Center for Health Statistics (NCHS), 1997. *National Mortality Data, 1997*. Hyattsville, MD: NCHS.

Shanks, Ralph *et al.*, 1996. *The US Life-Saving Service*. Petaluma, CA: Costano Books.

Short, Andrew D., 1997. *Australian Beach Safety and Management Program*, International Medical-Rescue Conference. Leuven, Belgium, International Life Saving Federation.

Wintemute, G.J. *et al.*, 1998. The epidemiology of drowning in adulthood: implications for prevention. *American Journal of Preventive Medicine*, **4**: 343–348.

Cross-references

Beach Use and Behaviors
Coastal Currents
Environmental Quality
Rating Beaches
Rip Currents
Sandy Coasts
Surf Zone Processes
Surfing
Water Quality

LIGHT DETECTION AND RANGING (LIDAR)—See AIRBORNE LASER TERRAIN MAPPING

LITTER—See MARINE DEBRIS

LITTORAL

In the vernacular, "littoral" refers to a shore or coastal region from the Latin *litus*, shore. In the technical usage, "littoral" and its associated, derivative nomenclature are variously defined primarily depending on the disciplinary context. Even within a discipline, the terms tend to be used with some elasticity and in a semi-quantitative sense due to the quantitative imprecision of boundaries defined by "high tide" or "low tide," "ordinary surf," etc., or by primary and secondary biotic transitions (Figure L2).

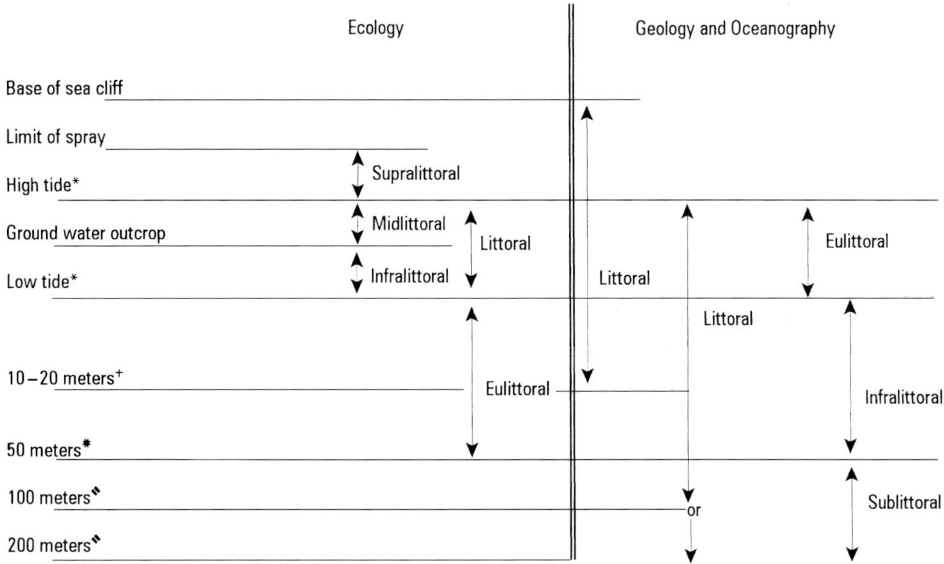

* This may be a level implied by a wrack line or other tide marks or a more rigorous elevation, i.e., mean high spring tide or mean low spring tide (or anything in between).

+ Estimated depth of vigorous wave generated turbulence.

* Estimated lower boundary of euphotic zone.

* Or the water depth at the shelf break.

Figure L2 Vertical boundaries for various usage of "littoral" and its derivatives.

The first technical use of "littoral" in English was in a biological context (Forbes and Hanley, 1853, as cited in Hedgpeth, 1957) to designate the space between tide marks, or the intertidal. "Supralittoral" and "sublittoral" were first employed in a similar context (Lorenz, 1863, as cited in Hedgpeth, 1957) with supralittoral being near the shore but above the level of high tide and sublittoral defined by water depth between two fathoms (3.65 m) and fourteen fathoms (25.60 m).

Ecological context

Marine ecologists have used the term to not only include the region between the tides but also the substrate wetted by ordinary surf above the level of high tides or "spray zone." In modern usage, the "supralittoral," or supratidal generally has referred to the spray zone immediately above either the highest high tide (Nybakken, 1993) or the mean high spring tide (Eisma, 1997) where the substrate is commonly, more-or-less, moistened by extreme tides, waves, and spray. This has also been called "extratidal." Ecologists divide the zone between the tide levels, into a "mid-littoral" from highest high tide to the level at which the water table on the beach outcrops at the beach surface; and an infratidal from the level of the water table outcrop to the elevation of the lowest low tide (Nybakken, 1993). "Eulittoral" refers to that part of the littoral zone less than 50 m in depth (American Geological Institute, 1960), 50 m, or between 40 and 60 m, being the lower limit at which "more abundant attached plants can grow" (Sverdrup *et al.*, 1942 citing Ekman, 1935). In practice, these terms are not so much defined by the physical processes controlling the landward distribution of sea salt but by the flora and fauna characteristic of such an influence.

In limnology, "littoral" refers to the zone below the highest reach of typical waves in a calm season (i.e., where the ground is dampened by waves) and above the maximum depth of substrate supporting rooted plants with floating or emergent leaves or flowers (e.g., Pillsbury, 1970; Jackson, 1997). Strictly speaking, it is possible for the littoral zone of a lake to cover the entire submerged area.

Geological and oceanographic context

When applied to the marine environment by marine geologists "littoral" generally refers to the zone between high tide and low tide. Classification of these zones was set forth by Johnson (1919) with reference to earlier investigations and to "legal authorities"; "shore" and "beach" both were synonymous for the intertidal and also referred to as "littoral." More strictly speaking the "littoral" is that region between mean high water spring tide, and mean low water spring tide (Eisma, 1997). Such a definition recognizes the complexity of the tidal signal and vicissitudes of a particular high tide level around some average position. "High tide" (or "low tide") is an imprecise term because these levels vary from one tidal cycle to the next. Harmonic analysis provides dozens of tidal constituents, acting at different periods, which conspire to provide any instantaneous high tide (or low tide). In other geological context, the "littoral" zone has a lower boundary at the depth of vigorous wave-generated turbulence, generally 10–20 m or, in one case, one-third the wave length of storm generated surface waves, and an upper limit at the seaward foot of sand dunes on the shore of the base of beach ridges or sea cliffs (Fairbridge, 1968). In a common usage, where the lower boundary of the littoral zone is at the level of low tide, "sublittoral" is the region below the low tide level to a depth of 100 m (American Geological Institute, 1960) or 200 m (Nybakken, 1993). This has also been referred to as the "shelf zone."

Oceanographers have tended to use "littoral" to designate the broad region from the high tide elevation down to the compensation point, generally between 200 and 400 m in the shelf. Alternatively, it is used as a synonym for "neritic" referring to the region below the low tide level to a depth of 200 m or to the shelf break (Kuenen, 1950, p. 313; Jackson, 1997). In this usage "eulittoral" can be synonymous with "intertidal." "Infralittoral" is the shallow (or inner) sublittoral generally the euphotic zone or between low tide and the depth compatible with the occurrence of phanerograms or photophilous algae (Visser, 1980, term 2839). Alternatively, the lower limit has been cited as about 50 m (Hedgpeth, 1957). "Sublittoral" is the region between water depths of 50 and 200 m.

The terms, without the "intertidal," "supratidal," "subtidal" synonyms of course, may also be applied to the shores of tideless seas where regular water level variations due to, perhaps, storm surges, seiches, etc. are observed. In light of these various usages, these terms are best specifically defined in articles and some care may need to be exercised when reading the literature.

Henry Bokuniewicz

Bibliography

American Geological Institute, 1960. *Dictionary of Geological Terms*. Garden City, NY: Anchor Press Doubleday.
Eisma, D., 1997. *Intertidal Deposits, River Mouth, Tidal Flats and Coastal Lagoons*. Boca Raton, FL: CRC Press.
Ekman, S., 1935. *Tiergeagraphie des Meeres*. Leipzip: Akad. Verlagsgesellsch.
Fairbridge, R.W. (ed.), 1968. *Encyclopedia of Geomorphology*. New York: Reinhold Book Corporation.
Forbes, E., and Hanley, S., 1853. *A History of British Mollusca and Their Shells*. London: John van Voorst.
Hedgpeth, J.W., 1957. Classification of marine environments. *Geological Society of American Memoir*, **67**(1): 17–28.
Jackson, J., 1997. *Glossary of Geology*, 4th edn. Alexandria, VA: American Geological Institute.
Johnson, D.W., 1919. *Shore Processes and Shoreline Development* (1965 facsimile). New York: Hefner Publishing Co.
Kuenen, Ph.H., 1950. *Marine Geology*. New York: John Wiley.
Lorenz, J.R., 1863. *Physikalische Verhaltnisse und Vertheilung der organismen in Quarnerischen Golfe*. Vienna: Kais. Kon. Hof. und Statdtsdr.
Nybakken, J.W., 1993. *Marine Biology: An Ecological Approach*. 3rd edn. New York: Harper Collins.
Pillsbury, R.W., 1970. *The Encyclopedia of Biological Sciences*, 2nd edn. In Peter Grey (ed.), New York: Van Nostrand Reinhold Co., pp. 507–508.
Sverdrup, H.V., Johnson, M.W., and Fleming, R.H., 1942. *The Ocean*. Englewood Cliffs, NT: Prentice-Hall.
Visser, W.A., 1980. *Geological Nomenclature*. Gorinchem: Royal Geological and Mining Society of the Netherlands.

Cross-references

Beach Features
Coastal Boundaries
Hydrology of Coastal Zone
Tidal Environments
Tides

LITTORAL CELLS

A littoral cell is a coastal compartment that contains a complete cycle of sedimentation including sources, transport paths, and sinks. The cell boundaries delineate the geographical area within which the budget of sediment is balanced, providing the framework for the quantitative analysis of coastal erosion and accretion. The sediment sources are commonly streams, sea cliff erosion, onshore migration of sand banks, and material of biological origin such as shells, coral fragments, and skeletons of small marine organisms. The usual transport path is along the coast by waves and currents (longshore transport, longshore drift, or littoral drift). Cross-shore (on/offshore) paths may include windblown sand, overwash, and ice-push. The sediment sinks are usually offshore losses at submarine canyons and shoals or onshore dune migration, rollover, and deposition in bays and estuaries (Figure L3).

The boundary between cells is delineated by a distinct change in the longshore transport rate of sediment. For example, along mountainous coasts with submarine canyons, cell boundaries usually occur at rocky headlands that intercept transport paths. For these coasts, streams and cliff erosion are the sediment sources, the transport path is along the coast and driven by waves and currents, and the sediment sink is generally a submarine canyon adjacent to the rocky headland. In places, waves and currents change locally in response to complex shelf and nearshore bathymetry, giving rise to subcells within littoral cells (e.g., Figures L4–L6).

The longshore dimension of a littoral cell may range from one to hundreds of kilometers, whereas the cross-shore dimensions are determined by the landward and seaward extent of the sediment sources and sinks. Littoral cells take a variety of forms depending on the type of coast. Cell forms are distinctive of the following coastal types: collision (mountainous, leading edge), trailing-edge, marginal sea, arctic, and coral reef. The first three types are determined by their position on the world's moving plates while the latter two are latitude dependent.

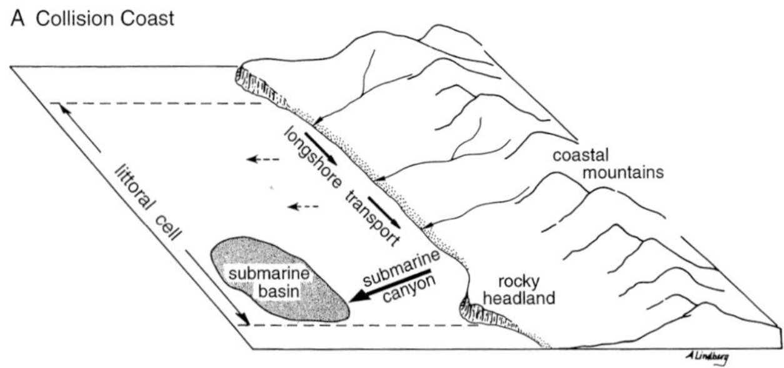

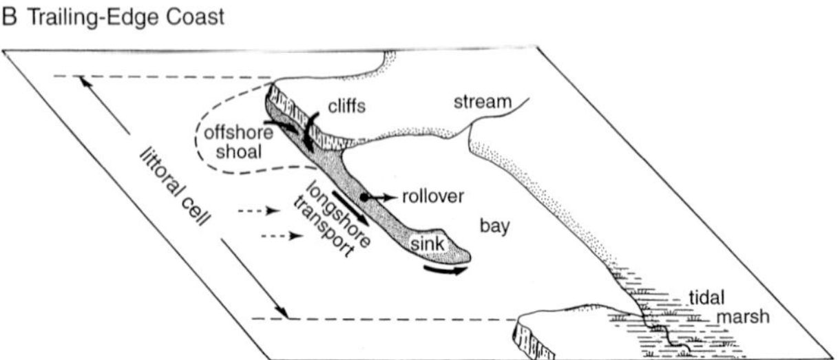

Figure L3 Typical (A) collision and (B) trailing-edge coasts and their littoral cells. Solid arrows show sediment transport paths; broken arrows indicate occasional onshore and offshore transport modes (after Inman, 1994).

Background

The concept of a littoral cell followed from the observation that the southern California coast was naturally divided into discrete sedimentation cells by the configuration of the coastal drainage basins, headlands, and shelf bathymetry. The principal sources of sediment were the rivers, that periodically supplied large quantities of sand to the coast. The sand is transported along the coast by wave action until the longshore drift of sand is intercepted by a submarine canyon that diverts and channels the flow of sand into offshore basins (Figure L3(A)). It was found that littoral cells, because they contain a complete cycle of sedimentation, provided the necessary framework for balancing the budget of sediment. These concepts were first presented at the International Geological Congress, Copenhagen (Inman and Chamberlain, 1960). The littoral cell now plays an important role in the US National Environmental Protection Act (1974) and the California Environmental Quality Act (1974), and it has become a necessary component of environmental impact studies. In the realm of public policy and jurisdictions, the littoral cell concept has led to joint-power legislation that enables municipalities within a littoral cell to act as a unit (Inman and Masters, 1994).

The configuration of littoral cells depends on the magnitude and spatial relations among the sediment sources, transport paths, and sinks. These in turn have been shown to vary systematically with coastal type. Because the large-scale features of a coast are associated with its position relative to the margins of the earth's moving plates, plate tectonics provides a convenient basis for the first-order classification of coasts (Inman and Nordstrom, 1971; Davis, 1996). This classification leads to the definition of three tectonic types of coast: (1) collision coasts that occur on the leading edge of active plate margins where two plates are in collision or impinging on each other, for example, the west coasts of the Americas; (2) trailing-edge coasts that occur on the passive margin of continents and move with the plate, for example, the east coasts of the Americas; and (3) marginal sea coasts that develop along the shores of seas enclosed by continents and island arcs, for example, coasts bordering the Mediterranean Sea and the South and East China Seas.

It is apparent that the morphologic counterparts of collision, trailing-edge, and marginal sea coasts become, respectively, narrow-shelf mountainous coasts, wide-shelf plains coasts, and wide-shelf hilly coasts. However, some marginal sea coasts such as those bordering the Red Sea, Gulf of California, Sea of Japan, and the Sea of Okhotsk are narrow-shelf hilly to mountainous coasts. A more complete coastal classification includes the latitudinal effects of climate and other coastal forming processes such as ice-push and scour and reef-building organisms. The examples of the latter two coastal types described here are (4) arctic form

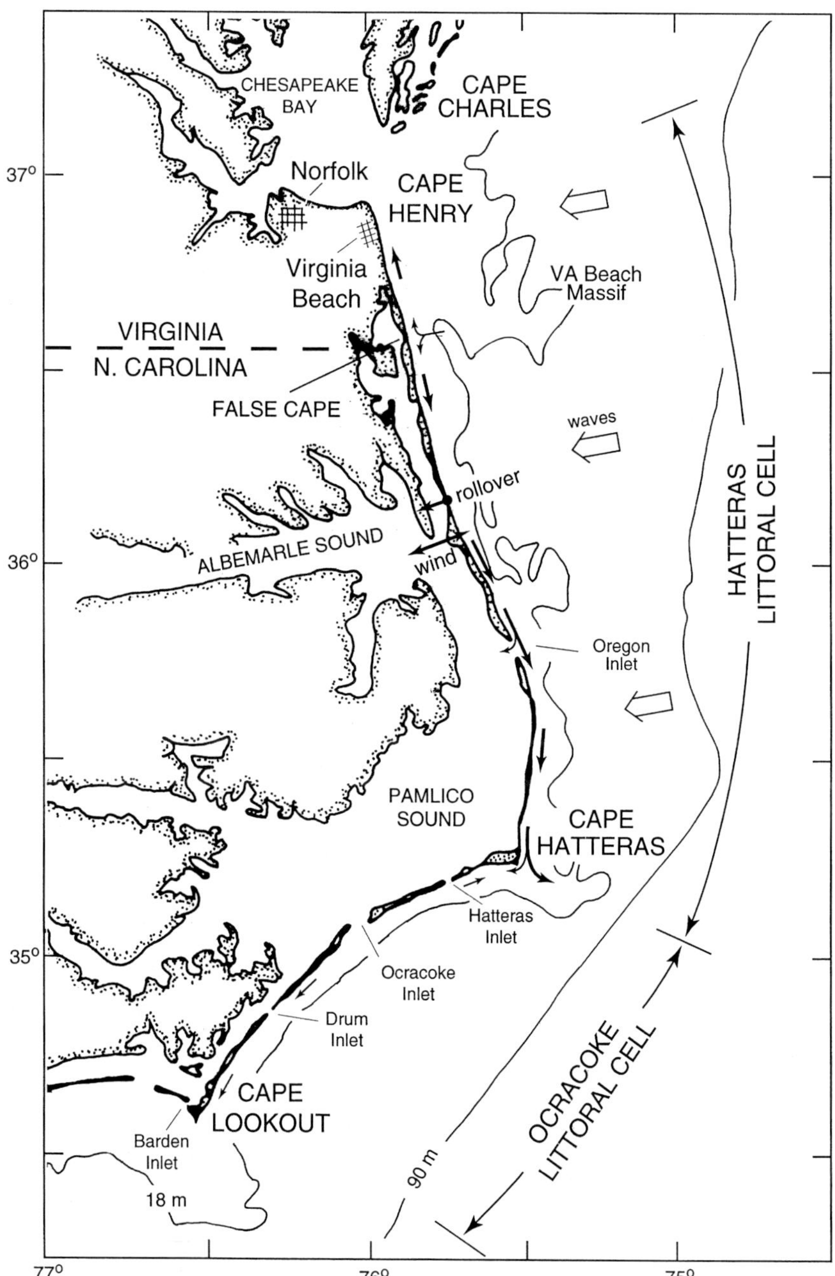

Figure L4 Hatteras and Ocracoke Littoral Cells along the Outer Banks of North Carolina (after Inman and Dolan, 1989).

of cryogenic coasts; and (5) coral reef form of biogenic coasts. The kinds of source, transport path, and sink commonly associated with littoral cells along various types of coast are summarized in Table L2.

Collision coasts

Collision coasts form at the active margins of the earth's moving plates and are best represented by the mountainous west coasts of the Americas. These coasts are erosional and characterized by narrow shelves and beaches backed by wave-cut sea cliffs. Along these coasts with their precipitous shelves and submarine canyons, as in California, the principal sources of sediment for each littoral cell are the rivers that periodically supplied large quantities of sandy material to the coast. The sand is transported along the coast by waves and currents primarily within the surf zone like a *river of sand*, until intercepted by a submarine canyon. The canyon diverts and channels the flow of sand into the adjacent submarine basins and depressions (Figure L3(A)).

However, in southern California most coastal rivers have dams that trap and retain their sand supply. Studies show that in this area the yield of sediment from small streams and coastal blufflands has become a significant replacement for river sediment. Normal wave action contains sand against the coast and, when sediment sources are available, results in accretion of the shorezone. However, cluster storms associated with El Niño–Southern Oscillation events as occurred in 1982/83 produced beach disequilibrium by downwelling currents that carried sand onto the shelf (Inman and Masters, 1991). The downwelled sediment is lost to the shorezone when deposited on a steep shelf such as that off Oceanside, California, or it may be returned gradually from a more gently sloping shelf to the shorezone by wave action. The critical value of slope for onshore transport of sand by wave action varies with sand size, depth, and wave climate, but for depths of about 15–20 m it is approximately 1.5% (1.0 degree).

Trailing-edge coasts

Trailing-edge coasts occur along the passive plate margins of continents and include the coasts of India and the east coasts of the Americas. The mid-Atlantic coast of the United States, with its wide shelf bordered by

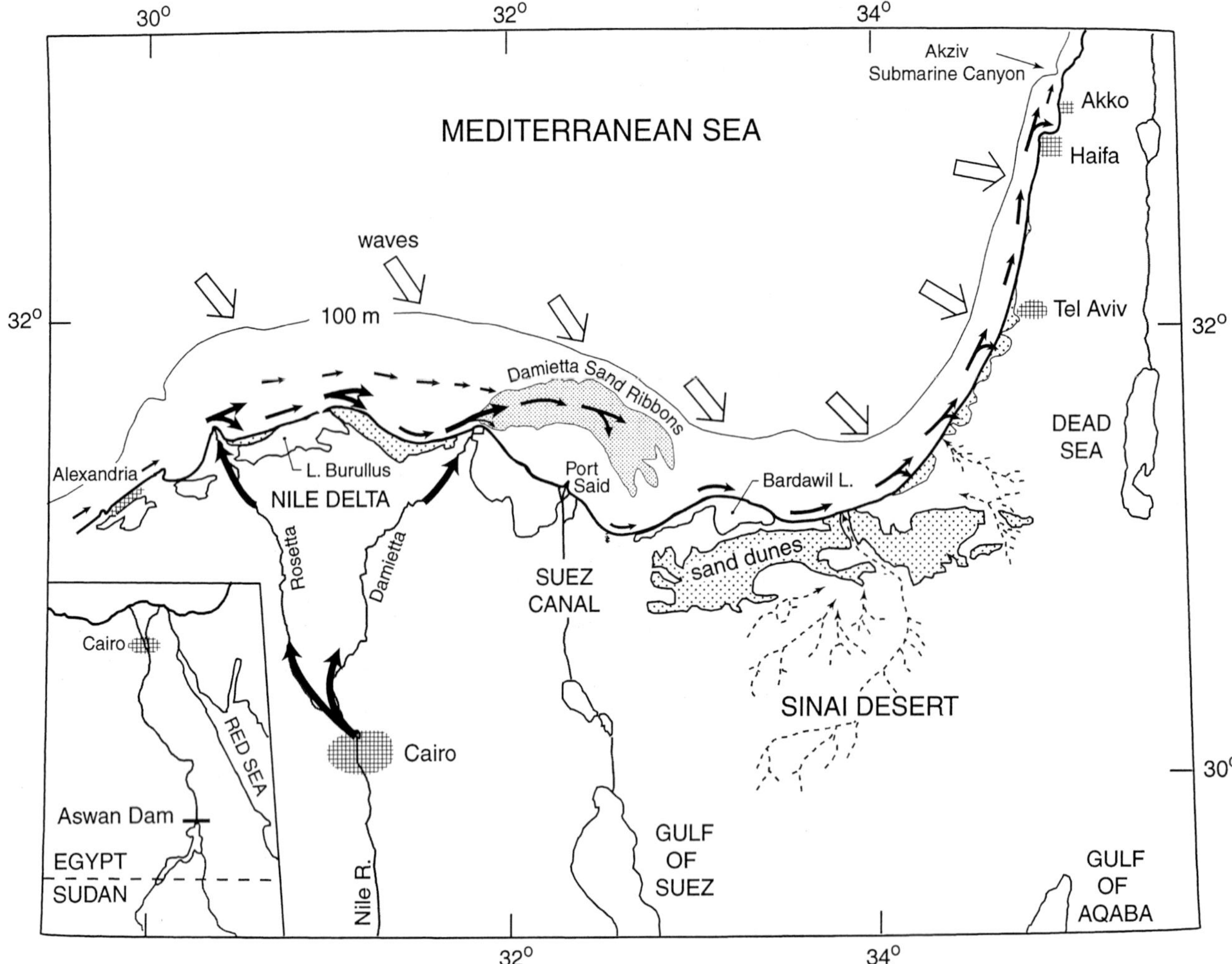

Figure L5 The Nile Littoral Cell extends along the southeastern Mediterranean coast from Alexandria, Egypt to Akziv Submarine Canyon off Akko, Israel. Sediment transport paths shown by solid arrows (after Inman and Jenkins, 1984).

coastal plains, is a typical trailing-edge coast where the littoral cells begin at headlands or inlets and terminate at embayments and capes (Figures L3(B) and L4). This low-lying barrier island coast has large estuaries occupying drowned river valleys. River sand is trapped in the estuaries and does not usually reach the open coast. For these coasts, the sediment source is from beach erosion and shelf sediments deposited at a lower stand of the sea, whereas the sinks are sand deposits that tend to close and fill estuaries and form shoals off headlands. Under the influence of a rise in relative sea level, the barriers are actively migrating landward by a rollover process in which the volume of beach face erosion is balanced by rates of overwash and fill from migrating inlets (e.g., Inman and Dolan, 1989). For these coasts, the combination of longshore transport and rollover processes leads to a distinctively "braided" form for the *river of sand* that moves along the coast.

The Outer Banks of North Carolina, made up of the Hatteras and Ocracoke Littoral Cells, extend for 320 km and are the largest barrier island chain in the world (Figure L4). The Outer Banks are barrier islands separating Pamlico, Albemarle, and Currituck Sounds from the Atlantic Ocean. These barriers are transgressing landward, with average rates of shoreline recession of 1.4 m/yr between False Cape and Cape Hatteras. Oregon Inlet, the only opening in the nearly 200 km between Cape Henry and Cape Hatteras, is migrating south at an average rate of 23 m/yr and landward at a rate of 5 m/yr. The net southerly longshore transport of sand in the vicinity of Oregon Inlet is between one-half million and one million cubic meters per year.

Averaged over the 160 km from False Cape to Cape Hatteras, sea-level rise accounts for 21% of the measured shoreline recession of 1.4 m/yr. Analysis of the budget of sediment indicates that the remaining erosion of 1.1 m/yr is apportioned among overwash processes (31%), longshore transport out of the cell (17%), windblown sand transport (14%), inlet deposits (8%), and removal by dredging at Oregon Inlet (9%). This analysis indicates that the barrier system moves as a whole so that the sediment balance is relative to the moving shoreline. Application of a continuity model to the budget suggests that, in places such as the linear shoals off False Cape, the barrier system is supplied with sand from the shelf (Inman and Dolan, 1989).

Marginal sea coasts

Marginal sea coasts front on smaller water bodies and are characterized by more limited fetch and reduced wave energy. Accordingly, river deltas are more prominent and are often important sources of sediment within the littoral cell. Elsewhere, barrier island rollover processes are similar to those for trailing-edge coasts. Examples of marginal sea coasts include the shores of the Gulf of Mexico with the prominent Mississippi River delta, the seas bordering southeast Asia and China with the Mekong, Huang (Yellow), and Luan river deltas, and the Mediterranean Sea coasts with the Ebro, Po, and Nile river deltas.

Although the Mediterranean area is associated with plate collision, the sea is marginal with restricted wave fetch and prominent river deltas. The Nile Littoral Cell extends 700 km from Alexandria on the Nile Delta to Akziv Submarine Canyon near Akko, Israel, one of the world's longest littoral cells (Figure L5). Before construction of the High Aswan Dam, the Nile Delta shore was in a fluctuating equilibrium between sediment supplied by the river and the transport along the coast. Now the sediment source is erosion from the delta, particularly the Rosetta promontory, in excess of 10 million m^3/yr. The material is carried eastward in part by wave action, but predominantly by currents of the east Mediterranean gyre that

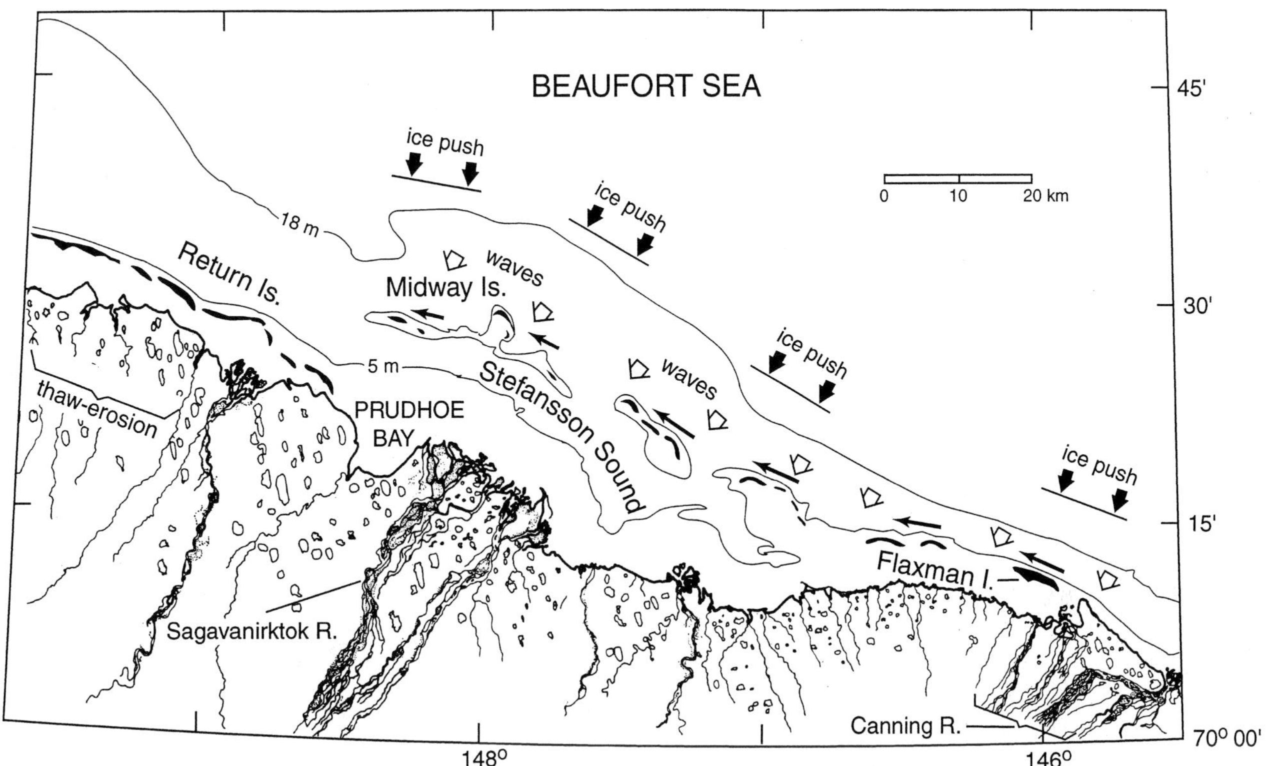

Figure L6 Flaxman Littoral Cell extending 100 km from the mouth of Canning River to the Midway Islands. The barrier chain of islands is enclosed by the 5 m depth counter. Major axis of oriented thaw lakes are normal to the direction of summer winds (after Inman, 1994).

Table L2 Typical source, transport path, and sink for littoral cells of various coastal types

Coastal features	Collision	Trailing-edge	Marginal sea		Arctic form of cryogenic	Coral reef form of biogenic
Morphology	Narrow-shelf mountainous	Wide-shelf plains	Narrow-shelf mountainous	Wide-shelf hilly	Wide-shelf [a] plains	Coral reef
Latitude/climate	Temperate and subtropical	Temperate and subtropical	Temperate and subtropical	Temperate and subtropical	Arctic	Tropical
Forcing[b]	Waves (1–10 kw/m)	Waves (1–5 kw/m)	Fetch-limited waves (1–2 kw/m) Tides[c]	Fetch-limited waves (1–2 kw/m) Tides[c]	Winter ice-push Summer waves	Waves (1–10 kw/m)
Littoral cell						
Sediment source	Rivers Cliffs Blufflands	Headlands Cliffs Shelves	Rivers Deltas	Rivers Deltas	Shelf Rivers Thaw-erosion	Reef material
Transport path	Longshore (river of sand)	Longshore and, rollover[d] (braided river of sand)	Longshore	Longshore and rollover[d]	Ice-push Rafting Longshore	Reef surge channels to beach, longshore to awa
Sink	Submarine canyons Embayments Dune migration	Estuaries Shoals Rollover Dune migration	Various including submarine canyons	Embayments Shoals Rollover Dune migration	Shoals Spit-extension	Awa channels to shelf

[a] All high latitude coasts appear to be trailing-edge coasts.
[b] Average incident wave energy-flux per meter of coastline (Inman and Brush, 1973).
[c] Tides may be important along any ocean coasts, but are sometimes amplified in marginal seas.
[d] Rollover processes include overwash and dune migration.

sweep across the shallow delta shelf with speeds up to 1 m/s. Divergence of the current downcoast from Rosetta and Burullus promontories forms accretionary blankets of sand that episodically impinge on the shoreline. The sand blankets move progressively downcoast at rates of 0.5–1 km/yr in the form of accretion/erosion waves. Along the delta front, coastal currents augmented by waves transport over 10 million m^3/yr, and the longshore sand transport by waves near the shore is about 1 million m^3/yr (Inman and Jenkins, 1984; Inman *et al.*, 1992).

The Damietta promontory causes the coastal current from the east Mediterranean gyre to separate from the coast and form a large stationary eddy that extends offshore of the promontory, locally interrupting the sediment transport path. The jet of separated flow drives a migrating field of sand ribbons northeasterly across the delta (Figure L5). The ribbons arc easterly than southeasterly towards the coast between Port Said and Bardawil Lagoon (Murray *et al.*, 1981). The Damietta sand ribbons form the eastern edge of a subcell within the Nile Littoral Cell.

Off Bardawil Lagoon, the longshore sand transport is about 500,000 m^3/yr and gradually decreases to the north with the northerly bend in coastline. This divergence in the littoral drift of sand results in the build up of extensive dune fields along the coasts of the delta, Sinai, and Israel. This sediment loss by wind blown sand constitutes a major "dry" sink for sand in the Nile Littoral Cell.

Arctic coasts

Arctic coasts are those near and above the Arctic Circle (66°34′N Latitude) that border the Arctic Ocean and whose littoral cells have drainage basins in North America, Europe, and Asia. Tectonically, Arctic coasts are of the stable, trailing-edge type, with wide shelves backed by broad coastal plains built from fluvial and cryogenic processes. The coastal plains are permafrost with tundra and thaw lakes. A series of barrier island chains extends along the Beaufort Sea coast of Alaska (Figure L6). For these coasts, cryogenic processes such as ice-push and permafrost thaw compete with river runoff, waves, and currents as important sources, transport paths, and sinks for sediment. Ice-push is a general term for the movement of sediment by the thrust of ice against it. Some common features include ice-push ridges and mounds, ice-gouge, ice pile-up, ride-up rubbling, and bulldozing.

During the nine months of winter, Arctic coasts are frozen solid and coastal processes are entirely cryogenic. Wind stress and ocean currents buckle and fracture the frozen pack ice into extensive, grounded, nearshore, pressure-ridge systems known as stamukhi zones. The stamukhi zone is a shear zone of ice grounded in 10–25 m depth that molds and moves shelf and barrier island sediment. The keels from the individual pressure ridges groove and rake the bottom, plowing sediment toward the outer barrier islands. Ice-gouge relief up to 2 m occurs across the shelf to depths of about 60 m (Barnes *et al.*, 1984).

Winter is terminated by a very active transitional period of a few days to a few weeks during spring breakup when a combination of factors associated with ice movement, waves, and currents, and extensive fluvial runoff all work in concert along the coast. The grounded ridges in the stamukhi zone break up and move, producing ice-push features and vortex scour by currents flowing around the grounded ice, creating an irregular bottom known as ice-wallow topography. Closer to shore, vertical drainage of river floodwater and sediment through cracks in the shorefast ice form large strudel-scour craters in the bottom (Reimnitz and Kempema, 1983).

Finally, a short summer period occurs in which the ice pack withdraws from the Beaufort Sea coast forming a 25–50 km wide coastal waterway. Although the summer season is short, storm waves generated in the band of ice-free water transport relatively large volumes of sand, extending barrier islands and eroding deltas and headlands. The summer processes are classical nearshore phenomena driven by waves and currents as shown by the beaches and barrier island chain beginning with Flaxman Island in the vicinity of Prudhoe Bay (Figure L6). The sediment sources include river deltas, onshore ice-push of sediment, and thaw-erosion of the low-lying permafrost sea cliffs. Thaw-erosion rates of the shoreline are typically 5–10 m/yr in arctic Russia and, over a 30-year period, averaged 7.5 m/yr for a 23-km coastal segment of Alaska's Beaufort Sea coast midway between Point Barrow and Flaxman Barrier Islands (Reimnitz and Kempema, 1987).

The Flaxman Barrier Island chain extends westward from the delta of the Canning River. It appears to be composed of sand and gravel from the river, supplemented by ice-push sediments from the shelf (Figure L6). The prevailing easterly waves move sediment westward from one barrier island to the next. The channels between islands are maintained by setdown and setup currents associated with the Coriolis effect on the wind-driven coastal currents. The lagoons behind the barrier islands appear to have evolved in part from collapse and thaw-erosion of tundra lakes (Wiseman *et al.*, 1973; Naidu *et al.*, 1984).

However, even the summer period is punctuated by occasional "Arctic events," including ice-push phenomena and unusually high and low water levels associated with storm surges and with Coriolis setup and setdown, a phenomenon whose intensity increases with latitude. The active summer season ends with the beginning of fall freeze-up.

Coral reef coasts

Coral reef coasts are a subset of the broader category of biogenous coasts where the source of sediment and/or the sediment retaining mechanism is of biogenous origin as in coral reef, algal reef, oyster reef, and mangrove coasts. Coral reefs occur as fringing reef, barrier reef, and atolls, and they are common features in tropical waters of all oceans at latitudes within the 20°C isotherm.

Although the concept of the littoral cell applies to all types of coral reef coast, the most characteristic are littoral cells along fringing reef coasts bordering high islands, where both terrigenous and biogenous processes become important. Reefs may be continuous along the coast or occur within embayments. In either case, the configuration of the fringing reef platforms themselves incorporates the nearshore circulation cell into a unique littoral cell (Figure L7). The circulation of water and sediment is onshore over the reef and through the surge channels, along the beach toward the awas (return channels), and offshore out the awas. An awa is equivalent to a rip channel on the sandy beaches of other coasts (Inman *et al.*, 1963).

Along coral reef coasts, the corals, foraminifera, and calcareous algae are the sources of sediment. The overall health of the reef community determines the supply of beach material. Critical growth factors are light, ambient temperature, salinity, and nutrients. Turbidity and excessive nutrients are deleterious to the primary producers of carbonate sediments. On a healthy reef, grazing reef fishes bioerode the coral and calcareous algae and contribute sand to the transport pathway onto the beach.

The beach behind the fringing reef acts as a capacitor, storing sediment transported onshore by the reef-moderated wave climate.

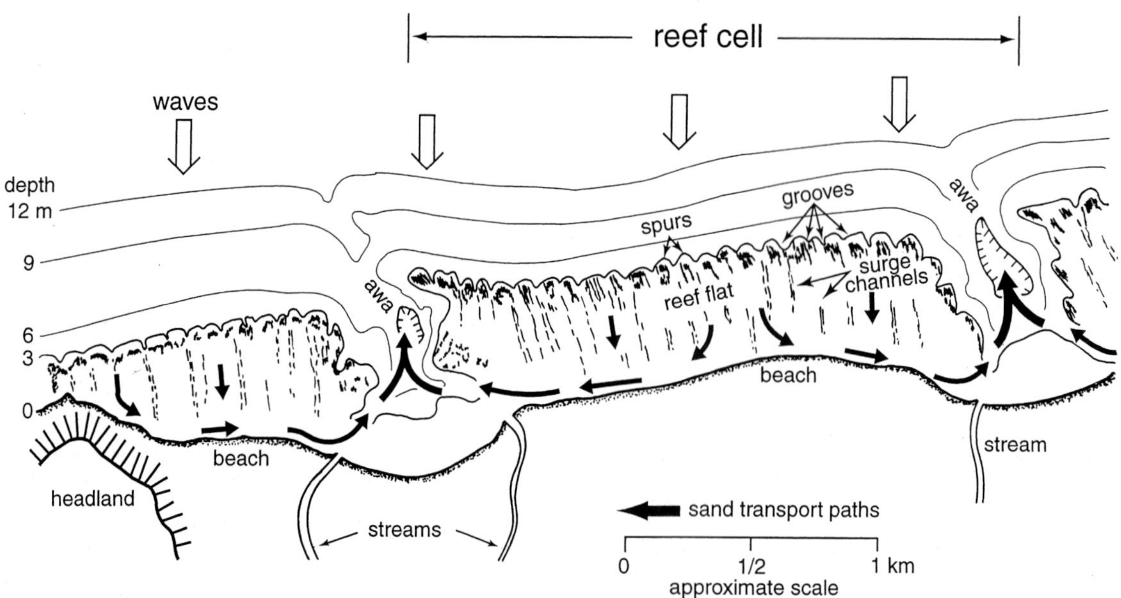

Figure L7 Schematic diagram of littoral cells along a fringing reef coast (after Inman, 1994).

It buffers the shoreline from storm waves, and releases sediment to the awas. In turn, the awas direct runoff and turbidity away from the reef flats and out into deepwater. Where the reef is damaged by excessive terrigenous runoff, waste disposal, or overfishing, the beaches are imperiled.

Douglas L. Inman

Bibliography

Barnes, P.W., Reimnitz, E., and Schell, D.M. (eds.), 1984. *The Alaskan Beaufort Sea—Ecosystems and Environments*. Orlando, FL: Academic Press.

Barnes, P.W., Asbury, J.L., Rearic, D.M., and Ross, C.R., 1987. Ice erosion of a sea-floor knickpoint at the inner edge of the stamukhi zone, Beaufort Sea, Alaska. *Marine Geology*, **76**: 207–222.

Bowen, A.J., and Inman, D.L., 1966. Budget of littoral sands in the vicinity of Point Arguello. *US Army Corps of Engineers*, Coastal Engineering Research Center, **19**, 41.

Davis, R.A., Jr., 1996. *Coasts*. Upper Saddle River, NJ: Prentice Hall.

Inman, D. L., 1994. Types of coastal zones: similarities and differences. In K.F. Turekian and D.F. Boesch, chairmen. *Environmental Science in the Coastal Zone*, National Research Council, pp. 67–84.

Inman, D.L., and Brush, B.M., 1973. The Coastal Challenge. *Science*, **181**, 20–32.

Inman, D. L., and Chamberlain, T.K., 1960. Littoral sand budget along the southern California coast. In Volume of Abstracts, *Report of the 21st International Geological Congress*, Copenhagen, Denmark, pp. 245–246.

Inman, D.L., and Dolan, R., 1989. The Outer Banks of North Carolina: budget of sediment and inlet dynamics along a migrating barrier system. *Journal of Coastal Research*, **5**: 193–237.

Inman, D.L., and Frautschy, J.D., 1965. Littoral processes and the development of shorelines. In *Coastal Engineering (Santa Barbara Specialty Conference)*. American Society of Civil Engineers, p. 1006.

Inman, D.L., and Jenkins, S.A., 1984. The Nile littoral cell and man's impact on the coastal zone of the southeastern Mediterranean. In *Proceedings 19th Coastal Engineering Conference*. American Society of Civil Engineers, Vol. 2, pp. 1600–1617.

Inman, D.L., and Masters, P.M., 1991. Budget of sediment and prediction of the future state of the coast. In *State of the Coast Report, San Diego Region, Coast of California Storm and Tidal Waves Study*. U S Army Corps of Engineers, **5**, p. 43.

Inman, D.L., and Masters, P.M., 1994. Status of research on the nearshore. *Shore & Beach*, **62**: 11–20.

Inman, D.L., and Nordstrom, C.E., 1971. On the tectonic and morphologic classification of coasts. *Journal of Geology*, **79**: 1–21.

Inman, D.L., Elwany, M.H.S., Khafagy, A.A., and Golik, A., 1992. Nile Delta profiles and migrating sand blankets. In *Proceedings 23rd Coastal Engineering Conference*. American Society of Civil Engineers, Vol. 3, pp. 3273–3284.

Inman, D.L., Gayman, W.R., and Cox, D.C., 1963. Littoral sedimentary processes on Kauai, a subtropical high island. *Pacific Science*, **17**: 106–130.

LaBelle, J.C., Wise, J.L., Voelker, R.P., Schulze, R.H., and Wohl, G.M., 1983. Alaska Marine Ice Atlas. *University of Alaska*, Arctic Environmental Information and Data Center, Anchorage, Alaska 99501, 302.

Murray, S., Coleman, J.M., Roberts, H.H., and Salama, M., 1981. Accelerated currents and sediment transport off the Damietta Nile promontory. *Nature*, **293**: 51–54.

Naidu, A.S., Mowatt, T.C., Rawlinson, S.E., and Weiss, H.V., 1984. Sediment characteristics of the lagoons of the Alaskan Beaufort Sea coast and evolution of Simpson Lagoon. In Barnes, P.W. *et al.* (eds.). *The Alaskan Beaufort Sea, Ecosystems and Environments*. Orlando, FL: Academic Press, pp. 275–292.

Reimnitz, E., and Kempema, E., 1983. High rates of bedload transport measured from infilling rate of large strudel-scour craters in the Beaufort Sea, Alaska. *Continental Shelf Research*, **1**: 237–251.

Reimnitz, E., and Kempema, E.W., 1987. Thirty-four-year shoreface evolution at a rapidly retreating Arctic coastal site. In Hamilton, T.D., and Galloway, J.P. (eds.), *Geologic Studies in Alaska by the U.S. Geological Survey during 1986*. U.S Geological Survey Circular, Vol. 998, pp. 161–164.

US Army Corps of Engineers *Engineering and Design—Coastal Littoral Transport* (www.usace.army.mil/inet/usace-docs/eng-manuals/em1110-2-1502/toc.htm).

Wiseman, W.J., Jr., Coleman, J.M., Gregory, A., Hsu, S.A., Short, A.D., Suhayda, J.N., Walters, C.D., and Wright, L.D., 1973. Alaskan arctic coastal processes and morphology. Louisiana State University, *Technical Report* 145, 171.

Cross-references

Arctic, Coastal Geomorphology
Barrier Islands
Classification of Coasts (see Holocene Coastal Geomorphology)
Climate Patterns in the Coastal Zone
Coasts, Coastlines, Shores, and Shorelines
Coral Reefs
Deltas
El Niño–Southern Oscillation
Energy and Sediment Budgets of the Global Coastal Zone
Littoral Drift Gradient
Sediment Budget
Tectonics and Neotectonics

LITTORAL DRIFT GRADIENT

It is well known both empirically, and theoretically from sediment transport modeling that rates of net littoral drift or longshore sediment transport rates (m^3/yr), vary progressively along a littoral drift coastline—see, for example, figure 9-9 in Komar (1998, p. 386; also pp. 434–435). Disparity in net littoral drift rates along the coastline may arise from several causes, but fundamentally important is variation of the angle of the breaking waves, which drive the alongshore sediment transport. This typically results from change in direction of the wave power resultant relative to the coastline alignment, or from irregularities in nearshore bathymetry influencing the wave shoaling and refraction processes, for example, wave focusing. Other causes include the loss of the littoral drift to a sedimentation "sink" such as a dredged tidal inlet, an offshore submarine canyon, or the action of headlands and cuspate forelands partially blocking and accumulating the littoral drift. The general effect is that progressively alongshore in the net drift direction, some segments of coastline have increasing rates of net littoral drift, and others decreasing rates. This alongshore variation in net littoral drift rates constitutes a *littoral drift gradient*. Van de Graaff and Bjiker (1988) state: "In many cases a gradient in the longshore sediment transport is the main reason of the erosion problems of sandy coasts." An alongshore increase in sediment transport rates in the net drift direction creates a *positive* littoral drift gradient; conversely a decrease in transport rates in the net drift direction results in a *negative* littoral drift gradient.

Littoral drift gradients have significant implications for the beach geomorphology. Consider a given point on a coastline with *positive littoral drift gradient* (increasing rates of net drift alongshore). Because more sediment is being transported away from the sector of beach at that given point than is being replenished from updrift, a net beach sediment budget deficit would be expected. Accordingly, the beach will likely exhibit long-term geomorphic manifestations of an erosive beach, such as a tendency for a dissipative beach state, faceted dune faces, frequent occurrences of lag deposits of surficial laminae of heavy mineral concentration, or cobble–pebble concentrations on the beach surface, slow beach recovery after an episodic erosive event, and a steepening of the beach-nearshore profile. Conversely, for a sector of beach exhibiting a *negative littoral drift gradient* (decreasing rates of net littoral drift alongshore), more material is being supplied from updrift than is being transported away in the net drift direction, and the beach will tend toward a positive sediment budget projecting an accretionary beach state with concomitant geomorphic features of well-developed beach berm, well-formed accreting frontal dune, and, in certain morphodynamic circumstances, a well-formed nearshore bar, and low gradient beach-nearshore profile. On a small scale, a littoral drift gradient is manifest as updrift accumulation against a jetty or groin with leeside down-drift erosion.

In situations of marked littoral drift gradient but little long-term beach morphological change, the effects of the littoral drift gradient may be compensated for by onshore or offshore (diabathic) sediment transport mechanisms. Thus on "lee coasts," onshore sediment drift from the shelf and shoreface, induced by longer period swell waves, assisted by quasi-permanent offshore winds and concomitant bottom return upwelling currents, may mask the erosive effects on the beach induced by a positive (increasing alongshore) net littoral drift gradient. Conversely, for "exposed coasts" facing predominantly onshore winds, the wind forced waves, and downwelling current effects induce an erosive morphodynamic beach state, which may mask the potential beach sediment accretion associated with a negative (decreasing alongshore) net littoral drift gradient.

Terry R. Healy

Bibliography

Komar, P.D., 1998. *Beach Processes and Sedimentation*, 2nd edn. Englewood Cliffs, NJ: Prentice Hall.

Van de Graaff, J., and Bjiker, E.W., 1988. Seawalls and shoreline protection. *Proceedings 21st International Conference on Coastal Engineering*. Reston, VA: American Society of Civil Engineers, pp. 2090–2101.

Cross-references

Beach Erosion
Littoral Cells
Longshore Sediment Transport
Wave Focusing

LOG-SPIRAL BEACH—See HEADLAND-BAY BEACH

LONGSHORE SEDIMENT TRANSPORT

Longshore transport refers to the cumulative movement of beach and nearshore sand parallel to the shore by the combined action of tides, wind, and waves and the shore-parallel currents produced by them. These forces usually result in an almost continuous movement of sand either in suspension or in bedload flows (see entry on *Cross-Shore Sediment Transport*). This occurs in a complex, three-dimensional pattern, varying rapidly with time. At any moment, some sand in the area of interest may have an upcoast component while other sand is moving generally downcoast. The separation of the total transport into components parallel and perpendicular to the shore is artificial and is done as a convenience leading to a simpler understanding of a very complex environment. To be meaningful, the rate of longshore transport must be averaged over intervals of at least many wave periods and is typically predicted or measured over much longer times, ranging up to a year (see entries on *Gross Transport* and *Net Transport*). It is also summed algebraically, at least conceptually, in the direction perpendicular to the shoreline. The result is that the transport rate is defined either as the volume of sand passing a point on the shore in a unit time (Watts, 1953), or the immersed weight of sand passing per unit time (Inman and Bagnold, 1963). The latter method accounts for the average density of the sand, which can vary significantly between beaches comprised principally of carbonate rather than quartz sand.

The integral of the transport rate over time results in an estimate of the longshore transport in units of volume or weight, as appropriate. If this integration is done without reference to the direction of movement, upcoast or downcoast, the result is termed gross transport (see entry on *Gross Transport*). On the other hand, if the sign of the direction is included, the result is termed net transport (see entry on *Net Transport*). Net transport can vary from a tiny fraction of the gross transport to a value nearly equal to it, depending on the characteristics of the forcing functions at the site. Estimates of both gross and net values have engineering significance. Jettied harbor entrances, for example, may shoal at a rate related to the gross longshore transport passing them. If the jetties themselves are partial barriers to longshore transport, the net transport can yield guidance to the estimation of accumulation or erosion of sand on either side of the entrance, providing aid in the design of remediation measures.

Longshore transport is the result of a longshore current that conveys sand put in suspension or mobilized on the seabed by waves (see entries on *Waves* and *Coastal Currents*). The longshore current is usually dominated by the flows induced by waves approaching the shoreline at an angle, although this current can be enhanced or reduced by wind-driven or tidal currents (see entry on *Tides*). Larger, more energetic waves mobilize more sand and produce stronger longshore currents so that the magnitude of longshore transport is directly related to the incident wave energy as well as to the angle of wave incidence.

One of the interesting features of longshore transport is that, for the most part, it is impossible to discern directly. The magnitude of the rapid shore-perpendicular motions of the individual sand particles are typically so much greater than the longshore velocity that they prevent the perception of the longshore motion. Even moving back from the particle dynamics and viewing the whole beach face does not provide any clues. If the longshore transport into a reach of shore equals the transport out; the result is no visible change in the configuration of the beach. Only when the transport rate changes along the shore, because of a change in the wave height or approach angle for example, or because of the construction of a barrier such as a groin, does the beach change in a manner than can be readily detected. Typically, these changes occur slowly such that images or measurements over intervals are required to sense them. The terms *littoral drift* or *littoral transport* are occasionally used interchangeably with *longshore transport*.

Prior to the 20th century, it was generally assumed that tidal currents provided the longshore motivation for beach sand. Komar (1976, pp. 183–190) provides an interesting history of the development of the various explanations for the generation of longshore currents by the oblique approach of waves. The relationship between longshore transport and the properties of the incident waves that formed the basis for most of the research in the second half of the 20th century was first suggested by Eaton (1951). The longshore transport volumetric rate was assumed to be proportional to the product of the longshore component of the wave energy flux, evaluated at the breaker zone and the sine of the angle that the breakers formed with the shoreline. This product was variously called the longshore component of either the wave power or the wave energy flux. Inman and Bagnold (1963) modified the theory to correct the volume to an immersed weight, which had the benefit of making the constant of proportionality nondimensional.

Local values of longshore transport are difficult to measure because it typically is a mixture of suspended and bedload transport (see entry on *Energy and Sediment Budgets of the Global Coastal Zone*). Instruments have been developed to measure suspended sediment concentrations at a point, but no satisfactory method of measuring bedload has yet been demonstrated. Largely based on inferences from the suspended sediment concentrations, it is generally believed that longshore transport is at a maximum in two zones. One is under the breaking waves and the other is in the swash zone on the beach face.

The simple analytical expressions for longshore transport, which attempt to predict the total transport across a shore-normal line, have been extended into two-dimensional numerical models. These two-dimensional models, in general, contain no additional physical insights into the sediment transport models, but they do allow the prediction of shore evolution caused by gradients in the longshore transport rate (Hanson and Kraus, 1989).

Richard J. Seymour

Bibliography

Eaton, R.O., 1951. Littoral processes on sandy coasts. *Proceedings. First Conference on Coastal Engineering*, American Society of Civil Engineers, Long Beach, CA, October 1950, pp. 140–154.

Hanson, H., and Kraus, N.C., 1989. GENESIS: generalized model for simulating shoreline change. US Army Corps of Engineers, Waterways Experiment Station, *Technical Report* CERC-89-19.

Inman, D.L., and Bagnold, R.A., 1963. Littoral Processes. In Hill, M.N. (ed.), *The Sea*, Vol. 3. New York: Interscience, pp. 529–533.

Komar, P.D., 1976. *Beach Processes and Sedimentation*. Englewood Cliffs, NJ: Prentice-Hall.

Watts, G.M., 1953. *A study of sand movement at South Lake Worth Inlet, Florida*. US Army Corps of Engineers, *Beach Erosion Board Technical Memo* 42.

Cross-references

Beach Processes
Coastal Currents
Energy and Sediment Budgets of the Global Coastal Zone
Gross Transport
Littoral Drift Gradient
Net Transport
Numerical Modeling
Tides
Waves

M

MACHAIR

Machair is a Gaelic word which applies to those areas of ancient sand dune systems mainly in the Hebrides of Scotland, but also in west Ireland, where long dune grasses have been superseded and the topography is essentially a low, plain surface (Ritchie, 1979). Its main characteristics are summarized below,

1. A level, low-lying, surface at a mature stage of geomorphological evolution, which is part of a very old fully vegetated coastal sand-dune system and is normally marshy in winter.
2. A base of blown sand which has a significant percentage of shell-derived materials and a narrow range of grain sizes.
3. Lime-rich soils with a pH value normally greater than 7.0.
4. A sandy grassland-type vegetation with long dune grasses and other early dune species having been eliminated.
5. Evidence of a history of anthropic interference including heavy grazing (especially rabbits, sheep, and cattle), rotational cultivation and, in places, artificial drainage.
6. A moist, cool location with characteristically strong onshore winds.

The origin of this dune system was more than 7,500 years ago, as a product of the rapid Flandrian marine transgression which began to level off at this time in this region. Powerful onshore Atlantic wave and wind energy transported preexisting glacial, fluvio-glacial, and shell-derived sediments across the extensive low gradient shallow continental shelf. Machair topography is influenced strongly by the underlying, preexisting landforms of the land extension of this shelf. As the sand-body transgressed landwards old organic and sand-rich deposits became exposed on the foreshore. More than 14 sites have been investigated for stratigraphy, pollen, and carbon 14 dating which provide evidence for the sequence of machair evolution. (Ritchie, 1985).

In the Outer Hebrides the total volume of sand in the beach, dune, and machair system has been more or less constant and, has therefore been subjected to extensive recycling. Nevertheless some ongoing supply from shells on the marine shelf is a possible addition to the original sediment bank. The most commonly used model for machair landforms is summarized in Figure M1, which shows the distinction between the actively retreating coastal edge and the mature inland surfaces. Typically, machair will occupy more than 90% of the area of the total dune system. Most low machair plains are deflation surfaces which are close to the winter water table. Machair can also be produced at the landward margin of the system by redepositional encroachment into marsh or loch basins.

Submergence has produced strands and sand flats between and inland of some machair systems. The drying-out time at low tide is insufficient for these areas to provide for secondary sand dune development. The open Atlantic beaches, however, do exhibit normal beach and dune exchanges associated with a retreating system. Wind-erosion with net landwards redeposition is also found in some inland features such as sand hills and ridges, normally initiated by severe overgrazing (Angus and Elliot, 1992). This erosion–redeposition process is the main mechanism for long-term machair evolution and landwards extension. Human interference including artificial drainage has also affected soils and landforms. There is an extensive literature of archaeological research from Bronze Age times, especially in the Western Isles, which has added greatly to the detailed evidence of machair evolution (Gilbertson *et al.*, 1999).

The study of machair has been dominated by work done in the Outer Hebrides, notably in the Uist Islands where the machair is almost continuous along the entire Atlantic coastline, and its typology has been dominated by the particular circumstances of these islands. In west Ireland, machair often occurs in bayhead locations (Carter, 1990). In the Inner Hebrides, islands such as Tiree are almost entirely covered by dunes and machair, and the general model which depends on coastal submergence and finite sand supply may not be wholly applicable. Even on the Uist islands machair has developed on different timescales, as determined by location, offshore bathymetry, and coastal configuration (Ritchie and Whittington, 1994).

Machair is not therefore a unique type of coastal sand dune landform. Its age, topography, calcareous soils, and general dependence on geomorphological recycling can be replicated elsewhere but the word "machair" has landscape and cultural connotations which are associated with a distinctive history of land use and tenure which is now preserved only in the Gaelic fringes of northwestern Europe.

William Ritchie

Bibliography

Angus, S., and Elliott, M.M., 1992. Erosion in Scottish machair, with particular reference to the Outer Hebrides. In Carter, R.W.G. *et al.* (eds.), *Coastal Dunes: Geomorphology, Ecology and Management, for conservation*. Rotterdam: Balkema, pp. 93–112.

Carter, R.W.G., 1990. Geomorphology of the Irish coastal dunes. Catena Supplementary **18**: 31–39.

Gilbertson, D.D., Schwenninger, J.L., Kemp, R.A., and Rhodes, E.J., 1999. Sand drift and soil formation along an exposed north Atlantic coastline: 14,000 years of diverse geomorphological, climatic and human impacts. *Journal of Archaeological Science*, **26**: 439–469.

Ritchie, W., 1979. Machair development and chronology in the Uists and adjacent islands. Proceedings of Royal Society of Edinburgh. **77B**: 107–122.

Ritchie, W., 1985. Inter-tidal and sub-tidal organic deposits and sea level changes in the Uists, Outer Hebrides. *Scottish Journal of Geology*, **21**: 161–76.

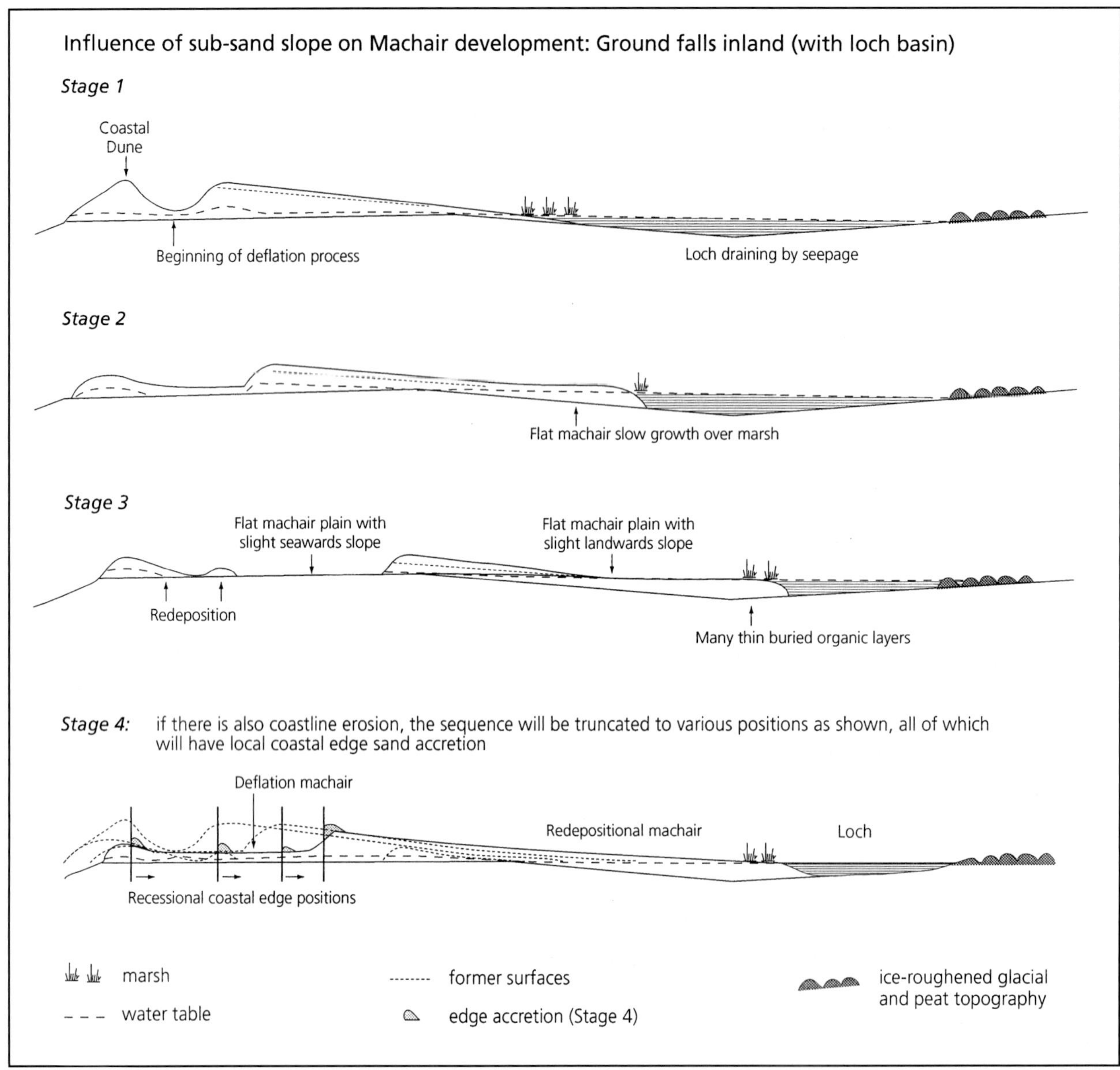

Figure M1 General model of machair evolution.

Ritchie, W., and Whittington, G., 1994. Non-synchronous aeolian sand movements in the Uists: the evidence of the intertidal organic and sand deposits at Cladach Mór, North Uist. *Scottish Geographical Magazine*, **110**: 40–46.

Cross-references

Archaeology
Changing Sea Levels
Dune Ridges
Eolian Processes
Human Impact on Coasts
Sandy Coasts
Sea-Level Rise, Effect

MANAGED RETREAT

Managed retreat is a collective term for the application of coastal zone management and mitigation tools designed to move existing and planned development out of the path of eroding coastlines and coastal hazards. This strategy is based on a philosophy of moving out of harm's way, and is proactive in recognizing that the dynamics of the coastal zone should dictate the type of management employed (e.g., identify and map the hazards as a basis for establishing regulations to move property and people away from migrating and/or storm-impacted coastlines).

The term "managed retreat" also is used in a more restrictive sense where shore-protection structures are removed selectively to allow natural coastal environments to be reestablished. For example, Viles and Spencer (1995) describe the creation of a small marsh on Northey Island, Blackwater estuary, Essex, England, by lowering a 200 m section of seawall and building a spillway to allow tidal inundation to be reestablished. This approach of letting parts of a coastline erode in a controlled way to create habitat and manage the coast in a way sympathetic to nature also is known as *managed realignment* (French, 1997). Managed realignment has the advantage that the sediment budget is reestablished.

Need for managed retreat

The Second Skidaway Conference on America's Eroding Shoreline concluded:

> ...the American shoreline is retreating. We face economic and environmental realities that leave us two choices: (1) plan a

strategic retreat now, or (2) undertake a vastly expensive program of armoring the coastline and, as required, retreating through a series of unpredictable disasters. (Howard *et al.*, 1985)

That conclusion applies to developed coasts globally. The recommendation for strategic retreat is synonymous with managed retreat.

The 15 years following the Skidaway Conference proved their predictions to be accurate with the exception that beach nourishment replaced armoring as the preferred engineering method of stabilizing coastlines. Armoring has increased globally (Nordstrom, 1994), and is still a common response to coastline erosion at the individual-property level in the United States. Beach nourishment is proving costly (US Army Corps of Engineers, 1994; Valverde *et al.*, 1999). In the Caribbean and along the Atlantic and Gulf Coasts, the damage from hurricanes is rising (e.g., Hugo, 1989; Andrew, 1992; Opal, 1995; Georges, 1998). Their impact has induced random retreat at the individual-property level, and forced communities to reexamine their coastal zone management strategies. And, although the greenhouse effect is a subject of debate, sea level is rising for most of the world's coastlines, and the rate of rise is increasing.

At the close of the 20th century, a report by the Heinz Center for Science, Economics and the Environment estimated that 10,000 coastal structures in the United States were within the estimated 10-year erosion zone (Leatherman, 2000). As of 1998, coastal counties in the United States exclusive of the Great Lakes, had a total flood insurance coverage of $466,874,000,000 (H. John Heinz III Center for Science, Economics and the Environment, 2000). The best option for many of these properties and their communities is managed retreat. Although "retreat" strikes a negative cord for some, elements of the strategic retreat option increasingly are being incorporated into coastal zone management.

The shift from engineering to "Soft" solutions

Historically, the method of choice to protect beachfront buildings and property was to hold coastlines in place through engineering by armoring (Table M1). By the 1950s and 1960s the realization that coastal buildings were subjected to higher winds and flooding (even those behind seawalls) led many states and communities to adopt more stringent building codes to strengthen buildings in the coastal zone. Coastal management was segmented both in locale and application (e.g., each community or agency focusing on a limited coastal reach or single problem). On barrier islands, the focus was often on the high-tide shoreline rather than a holistic management approach for an entire island or chain of islands. By the 1970s, the US national experience dictated that something be done to control the losses incurred from hurricanes and great storms like the 1962 Ash Wednesday Storm. The tremendous loss of habitat also was being recognized as salt marshes and shell fisheries were lost or closed. The results were two-fold: the National Flood Insurance Act of 1968 (also the result of persistent property loss on riverine floodplains), and the Coastal Zone Management Act of 1972. Building requirements were upgraded, and many coastal states began to define critical environments and control development through permit processes. Communities and states adopted approaches such as zoning and set back requirements. By the early 1980s, Integrated Coastal Zone Management (ICZM) (Clark, 1995) or simply Integrated Coastal Management (ICM) defined strategies in which a variety of management tools were being combined (Table M1). Cincin-Sain and Knecht (1998) define ICM "as a process by which rational decisions are made concerning the conservation and sustainable use of coastal and ocean resources and space" and "is designed to overcome the fragmentation inherent in single-sector management approaches ..."

Continued beach loss and the associated losses of storm protection, recreational use, aesthetics, and beach economy, led to greater interest in "soft" solutions such as beach nourishment to combat erosion. Beach nourishment, however, is a modern equivalent of the engineering "fix" to hold the line, and does not recognize the natural dynamics of coastline retreat in areas where sea level is rising. This approach has proven drawbacks including ongoing costs, diminishing sand supplies, shorter half-lives of nourished beaches, and environmental impacts from dredging and sand placement.

Common regulatory methods, such as building codes, requirements for structures to be elevated, and controls on development density and type through land-use planning and most zoning, may lessen the impact of storms, but these methods do not remove development from the hazard zone. In some cases, vulnerability to hazards is increased. Furthermore, these approaches do not recognize coastline retreat as coastal adjustment takes place in response to sea-level rise, changes in sediment supply, variable wave regime, and other controls of coastline equilibrium. High-density development along shores all over the world demonstrate that land-use planning either has not worked or coastal management has come after the fact.

Table M1 General property damage mitigation options on the beachfront (modified after Bush *et al.*, 1996)*

Hard stabilization
- Shore-parallel
  - Seawalls
  - Bulkheads
  - Revetments
  - Offshore breakwaters
- Shore-perpendicular
  - Groins
  - Jetties[a]

Soft stabilization
- Adding sand to beach
- Beach replenishment
- Beach bulldozing/scraping
- Increasing sand dune volume
  - Sand fencing
  - Raise frontal dune elevation
  - Plug dune gaps
- Vegetation
  - Stabilize dunes (oceanside)
  - Marsh (soundside)

Modification of development and infrastructure (control through zoning, building codes, insurance eligibility requirements)
- Retrofit homes
- Elevate homes choose elevated building sites
- Lower-density development
- Curve and elevate roads
- Block roads terminating in dune gaps
- Move utility and service lines into interior or bury below erosion level

Managed retreat
- Abandonment
  - Unplanned
  - Planned
- Relocation
  - Active (relocate before damaged)
  - Passive (rebuild destroyed structures elsewhere)
- Long-term relocation plans (zoning, land use planning)
- Setbacks
  - Fixed
  - Rolling
- Acquisition
- Avoidance: recognize hazard areas and avoid
  - Tidal inlets (past, present and future)
  - Swashes
  - Permanent overwash passes
  - Wave-velocity zones

* These management options are listed in increasing order of preference for Integrated Coastal Management (ICM). Historically, early management usually focused on shoreline stabilization, relying on a single mode of armoring. Various mitigation tools have been added to management plans, often in response to deficiencies in earlier plans that were revealed by the impact of the most recent storm.
[a] Jetties are built specifically to protect harbor entrances or maintain inlets, and are not constructed to protect coastlines. They are listed here because they impact adjacent shorelines, and that impact must be considered in management schemes.

Methods of managed retreat

"Retreat" is sometimes used for setbacks (e.g., Clark, 1995), or has been viewed simply as denying property owners the right to construct shore-hardening structures, forcing abandonment (Sturza, 1987). Managed retreat, however, implies applying an appropriate management strategy from a menu of tools, including stabilization techniques in some cases (e.g., particular urban coastlines). Specific retreat mitigation techniques include: abandonment, relocation, setbacks, land acquisition, and avoidance.

Abandonment

Abandonment may be unplanned, or part of a planned strategy of retreat. Historically, abandonment is often an unplanned, post-storm response to destruction of buildings and land loss (e.g., bluff retreat so that reconstruction is impossible). Fallen houses in the water or on the beach are a common sight along open-ocean coastlines after hurricanes and northeasters. Similar scenes are common in the Great Lakes and large embayments like Chesapeake Bay. Destruction may be so complete that the property is abandoned. Ruins of houses destroyed in Hurricane Gilbert in 1988 remained 12 years later along Mexico's Yucatán coast. Entire villages on barrier islands and along eroding bluffed shores have been abandoned. The village of Broadwater, Hog Island, VA, was abandoned in 1933 after losses to storms and shore erosion, although in part it was a short-term planned abandonment as houses were relocated off the island. Earlier, Cobb Island, VA, and Edingsville Beach, SC, had met similar fates. West coast abandoned towns include Bayocean, OR, and Cove Point, WA. Over the last century, 29 villages have been abandoned (lost) to the sea along England's Yorkshire coast.

Planned abandonment can be incorporated into managed retreat in several ways. Long-term planned abandonment can follow what is sometimes called the "do nothing" approach. Buildings are regarded as having a fixed life span, and when their time comes to fall into the sea, bay, or lake, no attempt is made to protect them. Buildings are razed either just before or after failing. Planned abandonment can be achieved by prohibiting post-storm reconstruction, or by requiring relocation landward of the revised post-storm setback control line. The original South Carolina Beachfront Management Act would only allow habitable structures damaged beyond repair (two-thirds or greater damaged) to be rebuilt landward of the no-construction zone (Beatley *et al.*, 1994). In part because of a poorly written law, post Hurricane Hugo enforcement led to the famous court case of *Lucas v. South Carolina Coastal Council* in which the plaintiff prevailed, resulting in the law being rewritten and softened. Rebuilding after storms can be discouraged by other methods as well, such as denial of flood insurance and other subsidy programs.

Relocation

For an existing building, the most obvious way to avoid a hazard is to move away from it! For developed coasts relocation is an essential component of managed retreat. So it is with an eroding or shifting high-tide shoreline.

Active relocation is undertaken by moving a building back either before it is threatened, or, if threatened, before it is damaged.

Passive relocation is achieved by rebuilding a destroyed structure in another area, away from the shore and out of the coastal hazard zone.

Long-term relocation usually implies a broader strategy through community zoning or land-use plans that identify a frontal zone of buildings likely to be impacted by known erosion rates or predicted flood levels from storm surge and coastal flooding. These building are then scheduled for relocation over an extensive period (assuming they will not be lost in coming storms). In effect, this is an engineered retreat, and on a barrier island the plan may include creation of new land on the soundside of the island for relocating the structures (Viles and Spencer, 1995). Artificial island migration is achieved, however, because barrier islands are usually backed by sensitive marshes and wetlands, the approach is questionable and raises complicated issues of property rights and changing ownership. This approach is more easily achieved in moving communities off of riverine floodplains and non-barrier coasts.

Even where setbacks are used, a retreating coastline will catch up with the property, and relocation will again become an option. Relocation is often the best economic option (Table M2) even though the up-front cost may be high. One can find examples along almost every coastline where armoring is used in which the cost of seawalls, groins and breakwaters, or nourishment, over the lifetime of the property, exceeds the value of the property, and greatly exceeds the cost of moving the structure.

The 1987 Upton–Jones Amendment to the National Flood Insurance Program (NFIP) allowed owners of threatened buildings to use up to 40% of the Federally insured value of their homes for building-relocation purposes (Wood *et al.*, 1990). The law recognized relocation as a more economical, more permanent, and more realistic way of dealing with long-term erosion and flood problems. The NFIP would pay a relatively small amount to assist relocating or razing a threatened house, rather than paying a larger amount to help rebuild it; only to see the rebuilt house destroyed in a subsequent storm, and paying to rebuild again. By March 1995, North Carolina had claims for over 70 relocations and 168 demolitions, and accounted for over 60% of all coastal claims under the program. The National Flood Insurance Reform Act of 1994 ended the Relocation Assistance Program as of September 23, 1995, replacing the Upton–Jones program with the National Flood Mitigation Fund.

Table M2 The advantages and disadvantages of relocating buildings back from a retreating shoreline (modified after Bush *et al.*, 1996)

Advantages
- building moved out of hazard zone, or is less vulnerable to hazards
- natural shoreline processes allowed to continue
- preserves the beach and associated value to community
- high probability of one-time-only cost (economical in the long term)
- cost savings because no public or private money spent on stabilization

Disadvantages
- high initial cost
- politically difficult
- building site must be deep enough to allow suitable moveback, or an alternative site must be purchased
- structure must be of a type and design/construction that allows it to be moved (e.g., a wood-frame house is easier to move than a cinder-block house on a poured concrete slab)
- coastal land is lost

Demolition and relocation activities are eligible for grant assistance under the program, but now compete with other mitigation approaches, including elevation and flood-proofing programs, acquisition of flood-zone properties for public use, beach nourishment, and technical assistance. Some states have encouraged relocation with similar programs (e.g., Michigan) (Platt *et al.*, 1992), or require houses to be moveable through the building permit process (e.g., New York).

The relocation alternate often is regarded as too expensive or technically impossible, but the move of the famous Cape Hatteras, NC lighthouse in 1999 again proved the feasibility and economic wisdom of this alternative (Pilkey *et al.*, 2000). Relocation is not a new mitigation strategy. Lighthouses have been relocated in North America since the 19th Century. Entire communities have relocated by choice or by necessity when they can no longer defend against the ravages of nature. Discouraged by continual hurricane damage, the citizens of Diamond City, NC relocated in 1899, disassembling their houses and barging them to their new locations. Rice Path, NC relocated because of encroaching sand dunes. FEMA's web site gives examples of recent success stories of relocation. Moving houses and communities off of riverine floodplains is not uncommon (e.g., English, IN; Rhineland, MO; Glasgow, VA).

Deep property lots are an important element in planning for future relocation Deep lots allow homeowners to relocate houses threatened by erosion to another location on their own property. In effect, lot depth determines possible future on-site relocation. While relatively deep lots are found in some coastal communities, new developments are often designed to maximize the number of dwelling units, resulting in small lots. Despite this trend, some communities, such as Nags Head, NC, are now requiring deep lots (oceanside to soundside on barrier islands) in order to provide for relocation.

Setbacks

Setbacks as the name implies are a management tool to keep structures out of extreme-to-high hazard zones, or at least at a distance from the hazardous processes (e.g., coastal erosion, v-zone flooding, storm surge). Klee (1999) reviews two types: "stringline" and "rolling" setbacks. A stringline setback simply requires that new construction be a fixed distance inland from a reference line (e.g., the back of the beach, the vegetation line, the crest of the dune line). The regulatory line is not adjusted for changes such as storm impact. A rolling setback is one in which the regulatory line shifts landward as the high-tide shoreline erodes (e.g., as the bluff edge, back beach, or dune toe retreats).

Although setbacks often are defined as creating zones in which no buildings or structures are allowed, in reality most setback regulations allow for variance application, and in some jurisdictions, liberal granting of variances circumvents management intent.

How far back is a "safe" building setback? The answer is difficult and will vary from place to place according to erosion rates and state and local regulations. No uniformity exists between coastal states' setback regulations in terms of how they are defined or applied (see Leatherman, 2000, table 4.4 for state-by-state summary).

While setbacks put some distance between buildings and the shore, that distance does not remain constant. When high-tide shoreline retreat catches up to the buildings, the original setback distance is of no consequence. Once again, the relocation or abandonment options must be considered.

Acquisition

Land acquisition can be an important component of a managed retreat plan. Land in the public trust through federal, state, and local ownership usually provides benefits in terms of conservation, providing public access to the shore, contributing to recreational and tourism needs, preserving aesthetics, and protecting habitat. Most coastal states have land acquisition programs and governments can purchase land through negotiated purchases where owners voluntarily sell land, or, less common, by eminent domain (condemnation proceedings). Other strategies include tax incentives, donations of conservation easements, trading of land, and transference of development rights. Condemnation usually results in a much higher cost for the land. Most land acquisition programs are hampered by a lack of funding. Florida and California are states with fairly successful programs. Just how well a publicly owned urban coastline can serve its citizens is demonstrated by Chicago's 18-miles of continuous public parkland.

Avoidance

The best way not to experience a hazard is to avoid it! Although the decision not to locate in a hazardous area may not seem like retreat, including areas where no development is allowed because of specific hazards, critical habitats, or sediment sources, is usually part of managed retreat. In this case, zoning can contribute to safe siting of structures away from coastal hazards. In part, setbacks reflect an avoidance approach, however, as noted, setbacks are temporary because coastline retreat will eventually reach buildings that met the original setback requirement. An eroding coastline may be more than just a hazard. Sacrificial coastline may be necessary to preserve the down-drift sediment budget, and Hooke (1998) gives an example in which the South Wight Borough Council (England) does not allow coastal defense works so a cliff-line will continue to erode and provide sediment to the beaches of Sandown Bay.

Again, disincentives may be used in an effort to encourage people not to build in high-hazard zones or in areas of critical habitat. Federal laws such as the 1982 Coastal Barriers Resources Act (COBRA) and the 1990 Coastal Barriers Improvement Act (CoBIA) have designated areas in which development is allowed, but buildings are not eligible for federal flood insurance or any post-storm federal assistance such as small business loans and funding to rebuild infrastructure. After Hurricane Fran in 1996, however, such assistance apparently did go to the community of North Topsail Beach, NC which is located in a COBRA unit.

Nags Head, North Carolina: managed retreat at work

The managed-retreat approach has been successfully implemented by the town of Nags Head, NC. This mitigation strategy stems from a desire to protect Nags Head's family beach atmosphere that attracted the residents in the first place (Bush *et al.*, 1996). Recognizing that hurricanes are inevitable, the Nags Head Repetitive Loss Plan and Floodplain Management Plan's implementation includes an extensive list of pre-storm mitigation measures, town response during a storm event, and post-storm mitigation and reconstruction measures (Nags Head, 1995).

The town adopted building standards more restrictive than required by either FEMA or the North Carolina Coastal Area Management Act (CAMA). Incentives are used to encourage development to be located as far back from the ocean as possible, including strict setbacks (minimum standard of 150 ft (45.7 m) setback from mean high water). Because small, single-family structures are much easier to move, the town has limited the development of oceanfront hotels and condominiums. Deep lots running perpendicular to the shore provide considerable room for relocation. Prior to rebuilding after a storm the Town may require adjoining lots in common ownership to be combined into a single lot. New construction of wood frame, multi-story, multi-family, buildings is not permitted. Strict limits are set on the amount of impervious surfaces within the oceanfront zoning districts that further reduces the amount of real property at risk. The post-storm measures include building moratoriums, policies on reconstruction, and a program for rapid acquisition of land.

The general theme of Nags Head's mitigation plan is based on the recognized history of coastline retreat, and that it is far better to adopt a policy of planned retreat than to wait for a disaster to force retreat. That philosophy is not new to residents of North Carolina's Outer Banks. A landmark property in Nags Head is the Outlaw House, named for the Outlaw family. This structure has been moved back 600 ft (183 m) from the retreating high-tide shoreline in five separate moves over 100 years.

The cost of moving buildings is the best economic strategy because the solution is long-term compared with relying on beach nourishment with an estimated cost of approximately $2 million per mile (1.6 km). The area's relatively high wave energy would require additional nourishment every three years resulting in an average annual cost of $3 million. This expenditure would continue as long as replenishment was the chosen mitigation technique. By comparison, the cost of removing structures from the threatened areas is much less. As of the early 1990s, Nags Head had accounted for 78 of the 379 (21%) Upton–Jones petitions submitted nationwide, 55 of which had been approved (Williams, 1993). Of these 55, 35 petitions requested funds for demolition at an average cost of $74,409, and 19 requested funds for relocation at an average cost of $30,211 (Williams, 1993). Similarly, an estimated cost for a beach nourishment program along a 4.5 (7.6 km) mile reach of South Nags Head shoreline was about $9 million every 3 years compared with the retreat option estimated at about $2 million every 20–25 years.

Relocation is a viable coastal management tool, and need not be considered only for single-family houses. When a structure is moved, the danger is reduced (Table M2).

The 10/100-year relocation concept

The difficulty of applying a managed retreat strategy is exemplified by areas such as the Myrtle Beach Grand Strand, SC, Miami Beach, FL, and other great oceanfront resort communities where a vast number of high-rise condominiums and hotels are right on the high-tide shoreline. At present, beach replenishment is economically feasible for these communities because of the large number of people that use the beaches and the significant revenue generated. The Miami Beach replenishment project, the most successful on the east coast in terms of replenished beach lifetime, has lasted for over 20 years. Along parts of the Grand Strand, SC, replenishment has to be repeated almost yearly. A time will come, however, when the economics of replenishment will no longer be acceptable. The increasing sand volumes needed, the declining sand supplies, and escalating project costs will make nourishment a less acceptable management tool. The time is approaching when serious consideration will have to be given to managed-retreat alternatives such as relocation and land acquisition.

Although the argument is that development along urban coasts is either not feasible or too entrenched to consider managed retreat, the alternative is both feasible and, perhaps, preferable for some communities. The International Association of Structural Movers says that moving large structures is technologically feasible, though expensive. Recall also that relocation can mean demolishing a building and rebuilding its replacement elsewhere. The unanswered question is economics.

Urban communities and owners of large buildings should not exclude managed retreat as a management tool, and need to begin researching the economics of this option. One possibility is a *10/100-year relocation plan* in which a relocation strategy is developed within 10 years and implemented as necessary over the following century (Bush *et al.*, 1996). Cost comparisons of traditional relocation or relocation by demolition and rebuilding should be evaluated against the long-term feasibility of continuing the replenishment option (e.g., the projected sea-level rise, financing requirements, identifying and acquiring distant sand resources, a timetable for obtaining necessary permits, etc.). Whether buildings can be relocated on the present property or off property, within the community or outside must be ascertained. What are the options and questions yet to be raised? A 10-year planning window should set the stage for implementation. Plans will vary by community and coastal type, and will take decades to implement. Virtually all coastal communities will need such programs of managed retreat over the next 100 years, or they will fulfill the prediction of retreating as the result of a series of coastal calamities.

The need for long-term managed retreat

In summary, to hold the line against the sea-level rise for all of the world's developed coasts is unrealistic. Managed retreat may provide the best set of tools for mitigating coastal hazards and reducing property losses. Avoidance remains the best solution for undeveloped and lightly developed areas, while various forms of relocation are the long-term solution for even urbanized shores. Coastal land acquisition is one method of meeting both of these goals, however, greater funding will be needed for future acquisition to succeed. Setbacks are a temporary solution, even when redefined periodically as rolling setbacks. In order for managed relocation to work, integrated land-use planning and zoning efforts will have to take a broader, holistic approach. For example, barrier-island management policies must consider the entire island, moving from a focus on site-specific and linear (island front) regulation to a whole-island perspective, and from shore hardening/hold-the-line programs to approaches which concentrate on preservation, augmentation, and repair of the natural systems.

William J. Neal, David M. Bush and Orrin H. Pilkey

Bibliography

Beatley, T., Brower, D.J., and Schwab, A.K., 1994. *An Introduction to Coastal Zone Management.* Washington DC: Island Press.

Bush, D.M., Pilkey, O.H., and Neal, W.J., 1996. *Living by the Rules of the Sea.* Durham, NC: Duke University Press.

Cicin-Sain, B., and Knecht, R.W., 1998. *Integrated Coastal and Ocean Management: Concepts and Practices.* Washington DC: Island Press.

Clark, J.R., 1995. *Coastal Zone Management Handbook.* Boca Raton, FL: Lewis Publishers.

French, P.W., 1997. *Coastal and Estuarine Management.* London, UK: Routledge.

H. John Heinz III Center for Science, Economics and the Environment, 2000. *The Hidden Costs of Coastal Hazards; Implications for Risk Assessment and Mitigation.* Washington DC: Island Press.

Hooke, Janet, 1998. *Coastal Defense and Earth Science Conservation.* Bath, UK: The Geological Society.

Howard, J.D., Kaufman, W., and Pilkey, O.H., 1985. *National Strategy for Beach Preservation,* Second Skidaway Institute of Oceanography Conference on America's Eroding Shoreline. Savannah, GA: Skidaway Institute of Oceanography.

Klee, G.A., 1999. *The Coastal Environment: Toward Integrated Coastal and Marine Sanctuary Management.* Upper Saddle River, NJ: Prentice Hall.

Leatherman, S.P., Chair, Erosion Hazards Steering Committee, H. John Heinz III Center for Science, Economics and the Environment, 2000. *Evaluation of Erosion Hazards.* Washington DC: The H. John Heinz III Center for Science, Economics and the Environment.

Nags Head, 1995. *Town of Nags Head Repetitive Loss Plan and Floodplain Management Plan.* Nags Head, NC: Town of Nags Head.

Nordstrom, K.F., 1994. Developed Coasts. In Carter, W., and Woodroffe, C.D. (eds.), *Coastal Evolution: Late Quaternary Shoreline Morphodynamics.* Cambridge, NY. Cambridge University Press, pp. 447–509.

Pilkey, O.H., Bush, D.M., and Neal, W.J., 2000. Lessons from lighthouses: shifting sands, coastal management strategies, and the Cape Hatteras lighthouse controversy. In Schneiderman, J.S. (ed.), *The Earth Around Us.* New York, NY: W.H. Freeman and Company, pp. 198–220.

Platt, R.H., Miller, H.C., Beatley, T., Melville, J., and Mathenia, B.G., 1992. *Coastal Erosion: Has Retreat Sounded?* Boulder, CO: Institute of Behavioral Science, University of Colorado, Program on Environment and Behavior, Monograph No. 53, 195p.

Sturza II, R.P., 1987. The Retreat Alternative in the Real World: The Kill Devil Hills Land-Use Plan of 1980. In Platt, R., Pelczarski, S., and Burbank, B. (eds.), *Cities on the Beach.* Chicago, IL: University of Chicago, Department of Geography, Research Paper No. 224, pp. 289–294.

US Army Corps of Engineers, 1994. *Shoreline Protection and Beach Erosion Study: Yearly Cost Data for Constructed Projects (1950–1993),* accompaniment to *Shoreline Protection and Beach Erosion Control Study Phase I: Cost Comparison of Shoreline Protection Projects of the US Army Corps of Engineers.* Institute of Water Resources Report, 94-PS-1, January 1994.

Valverde, H.R., Trembanis, A.C., and Pilkey, O.H., 1999. Summary of beach nourishment episodes on the U.S. East Coast Barrier Islands. *Journal of Coastal Research,* 15: 1100–1118.

Viles, H., and Spencer, T., 1995. *Coastal Problems: Geomorphology, Ecology and Society at the Coast.* London, UK: Arnold.

Williams, J.M., 1993. *An Examination of the Risks of Coastal Development and Some Possible Mitigation Methods: A Case Study of Nags Head, North Carolina.* Master's project, Durham, NC: Duke University, School of the Environment.

Wood, W.L., Chairman, Committee on Coastal Erosion Zone Management, 1990. *Managing Coastal Erosion.* Washington DC: National Academy Press.

Cross-references

Coastal Boundaries
Coastal Zone Management
Economic Value of Beaches
Global Vulnerability Analysis
Greenhouse Effect and Global Warming
Sea-Level Rise, Effect
Setbacks

MANAGEMENT—See COASTAL ZONE MANAGEMENT

MANGROVES, ECOLOGY

Introduction

Mangroves have always been considered as marginal ecosystems for at least three main reasons. First, the global mangrove area does not exceed 180,000 km^2 representing less than 2% of the world's tropical forest resources. Second, their discontinuous distribution, at the land and sea interface of tropical and subtropical coastlines, is primarily characterized by tidal regimes, which is a unique forest habitat. Third, the frequent wide fluctuations of environmental factors (dissolved oxygen, salinity, organic, and inorganic suspended matter) have induced in mangrove flora, a complex range of adaptations, lacking in other woody species, unable to compete or to survive in these highly variable and adverse environmental conditions (low oxygen content in soils, sulfate toxicity, high NaCl in water and soils, exposure to hurricanes and surges, muddy soils, instability, etc.). Yet, these ecosystems are highly productive with an average primary productivity often higher than that of neighboring continental forest types.

Many species of invertebrates and vertebrates of commercial value use mangrove habitat for food and shelter during their life cycle. Most mangrove ecosystems around the world have been depleted during the 20th century. Until the 1980s they have been extensively converted to other uses. For the last 20 years, many mangrove areas have come under full or partial protection, and restoration programs are being implemented in almost every one of the 70 countries possessing mangroves.

Most of the mechanisms and processes regulating mangrove ecosystems; primary productivity, food webs, nutrient fluxes, physiological adaptations of plants and animals, etc. are still poorly known, and this fragmentary knowledge is mainly restricted to species of commercial value.

Present distribution of mangroves

Six geographical zones have been recognized (Chapman, 1976; Snedaker, 1982; Rao, 1987; Saenger and Bellan, 1995; Duke *et al.*, 1998).

With rare exceptions, mangroves are restricted to coastal areas where mean monthly air temperatures, in winter, are higher than 20°C and where ground frost is unknown (Figure M2). The tallest (up to 35 m tall) and more dense mangroves are found in bioclimatic conditions with high annual rainfall (>2000 mm) and a short dry season (<3 dry months). They can survive in arid areas (Persian Gulf, Mauritania, Red Sea), in the form of low or dwarf, monospecific stands (Dodd *et al.*, 1999). The largest contiguous surface area of mangroves, covering more than 6,000 sq. km, is located in the upper Bay of Bengal, on the delta of the Ganges.

Recent estimates (Spalding *et al.*, 1997) indicate that the total mangrove area is about 180,000 sq km, most of it being located in South and Southeast Asia (Table M3). A few nations dominate these area statistics. For example, of the approximately 70 countries with this ecosystem, Indonesia, Australia, Brazil, Nigeria have about 43% of the world's mangroves. Indonesia alone has 23% and 12 countries have two-thirds (Table M3). Political and management decisions relating to mangrove stands of these countries will have significant effects on the global status in the future (Hamilton and Snedaker, 1984). It is assumed that at least 30% of these ecosystems are degraded or very degraded.

Productivity goods and services of mangrove forests

Our general knowledge of mangrove structural properties, above ground biomass and litter production (Saenger and Snedaker, 1993) is rather

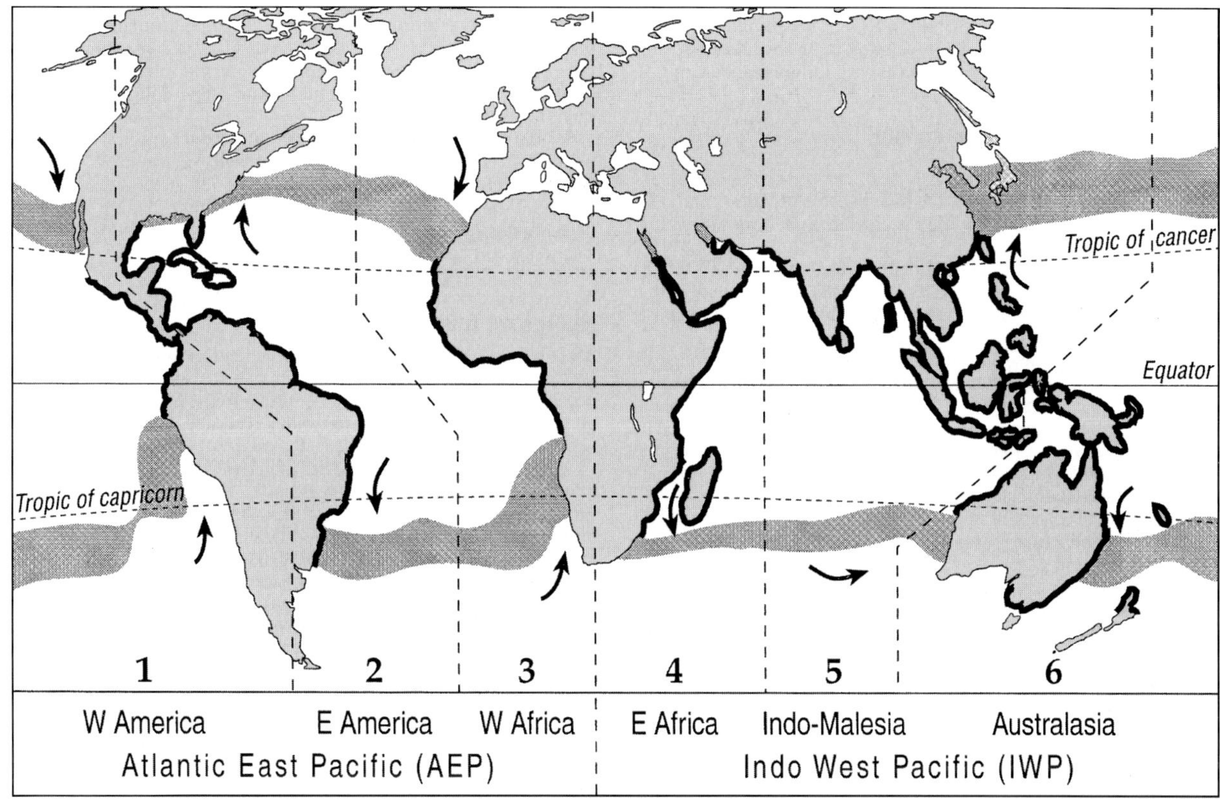

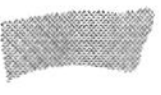

Figure M2 Main biogeographic mangrove area (redrawn from Duke, 1998).

Table M3 Estimates of mangroves areas (after Spalding *et al.*, 1997)

Region	Global (km^2)
South and Southeast Asia	75,173 (41.5%)
Australasia	18,789 (10.4%)
The Americas	49,096 (27.1%)
West Africa	27,995 (15,5%)
East Africa and the Middle East	10,024 (5,5%)
Total area	181,077
Major countries	**(km^2)**
Indonesia	42,500
Australia	11,500
Brazil	13,800
Nigeria	10,500
Malaysia	6,400
India	6,700
Bangladesh	6,300
Cuba	5,500
Mexico	5,300
Papua New Guinea	4,100
Colombia	3,600
Guinea	2,900
Total	119,100

well-advanced as extensive work has been carried out, especially in Malaysia (Sassekumar and Loi, 1983), Australia and New Zealand (Duke 1988; Woodroffe *et al.*, 1988), USA (Lugo *et al.*, 1980; Twilley, 1982; Lahmann, 1988), Brazil (Adaime, 1985), French Guyana (Fromard *et al.*, 1998). The following general conclusions can be drawn:

- in equatorial and sub-equatorial areas, mangrove height is about 12–20 m, rarely exceeding 25 m, and the litterfall varies from 12 to 16 t. ha^{-1} yr^{-1} (dry material).
- near the tropics, their average height is about 8–12 m and the known mean annual litterfall is of the order of 8 t. ha^{-1}.
- along warm temperate coastlines, where mangroves are exceptionally found (New Zealand), their height does not exceed 4 m and the litterfall declines to 4 t. ha^{-1} yr^{-1}.

High productivity and short retention times of organic matter in most mangrove ecosystems make them extremely important, not only for fisheries (Snedaker and Lugo, 1973). In the last 25 years, it has been shown that mangroves throughout the world serve a multitude of functions for the people inhabiting them (ISME, 1993), but more and more foresters and shrimp farmers are looking at the profits that can be derived from converting the mangroves into fish and shrimps ponds. As clearly shown by recent world satellite surveys (Spalding *et al.*, 1997; Blasco and Aizpuru, 2001) conversion of mangroves to aquaculture, to agriculture, and to urban development are the main causes of mangrove destruction.

Large-scale conversion to agriculture is conspicuous in the Gangetic delta (India and Bangladesh) where the population density exceeds 800 inhabitants km^{-2}, in Myanmar (Irrawady delta) and in most coastal areas of West Africa, from Senegal to Nigeria. Paddy fields, sugar cane, orchards, and oil palm are the main products. Concerning the conversion to aquaculture, fish ponds (*Chanos chanos*) have replaced most mangroves in the Philippines, whereas shrimp farming in Thailand, Indonesia (*Penaeus monodon, Penaeus merguiensis*) and Equador (*Penaeus vannamei*) has converted large areas of mangroves in recent years. In 1977, it was estimated that 1.2 million a of mangroves in the Indo-pacific region has been converted to aquaculture ponds (Saenger *et al.*, 1983). In many countries (Vietnam, Malaysia, Thailand, etc.) highly pyritic soils have lead to high acidity, causing major difficulties for the operators of aquaculture ponds with spectacular declines in yields (less than 400 kg ha^{-1} yr^{-1} instead of 1,400 initially). And the development of acid sulfate soils in drained mangrove areas of West Africa has had catastrophic consequences with failure of rice crops.

Mangroves versus chronic coastal environmental problems

Many investigators have described the building role of *Rhizo-phora*. The actual efficiency of sediment retention is very unequal from one place to another. The coastal afforestation program which is in progress in Bangladesh (Saenger and Siddiqi, 1993), with more than 200,000 ha, demonstrates that plantations with *Sonneratia apetala* contribute both to the acceleration of land accretion and to the stabilization of the exposed soft sediments.

In the last few years, new findings have been published concerning the role of mangroves as biochemical barriers to pollutants. Although metal mass balances in mangroves are still at an exploratory stage, with net import and net export occurring in different oceanographic cycles, the final balance, after a long period of monitoring, seems to be a net import of trace metals, as a result of complex physical and chemical coastal processes (Clark *et al.*, 1997; Lacerda, 1998).

Recent studies in coastal Brazil have shown that large heavy-metal concentrations (Hg and Zn) are retained in the rhizosphere sediments under very refractory chemical forms, making their uptake extremely difficult. They are preferably accumulated in perennial tissues, such as below ground biomass and trunks, whereas their concentrations in leaves remain extremely low. Although this is a new field of research there are strong and convergent indications that mangrove soils and plants minimize pollution by heavy metals in tropical coastlines (Lacerda *et al.*, 2000).

Concerning trace metals concentration and distribution in mangrove animals (Zn, Cd, Cu, Cr, and Pb), few studies have been carried out on mangrove oysters (*Crassostraea*) and mussels (*Mytella guyanensis*). No general conclusion can be drawn so far because of the scarcity of reliable data.

Flora, zonation, and role of salinity

The earliest mangrove fossil materials (pollen, fruits, hypo-cotyls) have been recorded from Brazil, Europe, Asia and Australia, etc. during the lower Cenozoic, about 55 million years before present (Eocene).

Throughout the world's tropics, growing preferably on muddy soils of deltaic coasts, in lagoons and estuarine shorelines, about 70 species of trees and shrubs, including putative hybrids, are considered as exclusive to the mangrove habitat (29 genera and 20 families) and about 24 are important but not exclusive.

The commonest species belong to three genera: *Rhizophora*, with nine taxa bearing conspicuous stilt roots, *Avicennia*, with eight taxa having dense, slender aerial roots known as pneumatophores and *Sonneratia* (nine taxa) with stout, conical pneumatophores (Table M4).

One of the most conspicuous and enigmatical features in mangrove ecosystems is the zonation of species. Such a common spatial patterning seems to be the result of complex processes involving dispersal strategies, plant succession mechanisms, physiological attributes of species, interspecific interactions, and seasonal fluctuations of physico-chemical parameters (Snedaker, 1982; Smith, 1992).

The importance of salinity has been stressed by most mangrove specialists (Saenger *et al.*, 1983; Tomlinson, 1986; Hutchings and Saenger, 1987; Duke *et al.*, 1998). Each species of tree and shrub has its own tolerance to salinity concentrations. This partly explains the upstream and downstream distribution of each species. *Avicennia integra, Heritiera fomes, Sonneratia lanceolata* occur upstream, or in freshwater-dominated habitats, where tidal penetration is limited and the average annual water salinity is lower than 20 ppt. The majority of remaining mangrove species (Table M4) are found in the intermediate part of the estuaries (all members of the Rhizophoraceae, *Avicennia alba, Avicennia officinalis, Excoecaria, Lumnitzera, Pelliciera, Xylocarpus*, etc.), indicating that their optimum salt requirements oscillate between 20 and 30 ppt. A few woody species are most generally found at the mouth of estuaries, in downstream and sea-front sectors, thriving year round in highly saline environmental conditions (30–45 ppt); this is the case for *Avicennia marina, Rhizophora mangle, Bruguiera gymnorrhiza, Laguncularia racemosa, and Sonneratia alba.*

Avicennia marina has the largest, almost continuous, biogeographical distribution of any mangrove species, extending from east Africa to the Red Sea and Pakistan, to the Indian Ocean and the Southern Pacific. Some of its ecotypes are found in the Arabian Gulf, occupying one of the driest mangrove habitats in the world, in which salt concentrations may reach levels that are beyond the physiological limit for other species (Dodd *et al.*, 1999). Under laboratory conditions, *A. marina* can survive a very wide range of salinities (Ball, 1996, 1998), from distilled water to salt concentrations as high as 175 ppt. The exact mechanisms of the physiological adaptation to saline environments, involving osmoregulation, ion compartmentation and selective ion uptake by roots, salt excretion, are still unclear.

Several attempts have been made to correlate species richness with important environmental factors, especially in Southeast Asia and Australia (Duke, 1992; Ball and Pidsley, 1998). Although temperature, rainfall, tidal and soil peculiarities, water salinity, etc. are known to be determining factors, the interplay between such factors and species richness is not properly understood.

Mangrove fauna

Most mangrove fauna can be subdivided in three main groups: resident species (occurring primarily in mangroves), seasonal migrants, and occasional mangrove species. In contrast to the list of mangrove plants given in Table M4, associated biota recorded from mangrove often occur in other habitats, adjacent to mangrove.

As a general rule, many species of terrestrial mammals are found in mangrove ecosystems but none seem to be strictly confined to mangroves. Even the estuarine crocodile (*Crocodylus porosus*) one of the largest species, and the endangered Bengal tiger (*Panthera tigris tigris*), common in the mangroves of the Ganges, are not restricted to these ecosystems. In the same way, fruit bats and nectar-feeding bats which often roost in mangroves, the crab-eating Macaque (*Macaca fascicularis*) and the Proboscis monkey (*Nasalis larvatus*) are more abundant in mangroves than in other habitats.

The use of mangrove ecosystems as nursery grounds for larval and juvenile fish was first demonstrated by Odum and Heald (1972) in Florida. The group of fish restricted to mangroves is mainly that of mudskippers, belonging to the gobiid subfamily Oxciderninae, with two genera (*Periophtalmus* and *Periophthalmodon*) and 5–10 species. The main peculiarity of mudskippers which are common practically throughout tropical mangroves, is their ability to withstand exposure to air especially during low tide. According to Hutchings and Saenger (1987) the main importance of mangroves to fish, is the availability of food, especially prawns, for juveniles, and protection, as piscivores are often under-represented in mangroves. Fish fauna is now rather well-known for most mangrove regions (Tholot, 1996).

Most mangrove birds are found in a variety of other habitats but mangroves provide secure roosting sites at high tide and the food for many species. The most spectacular population is probably that of scarlet ibis (*Eudocimus ruber*) in Trinida's Caroni mangrove, gathering more than 20,000 birds.

Mangroves are known to serve as nursery grounds for young crabs and juvenile prawns and as a source of seed for aquaculture. The mud crab (*Scylla serrata*) is probably the most popular Crustacea in coastal Southeast Asian countries. Among molluscs, several oysters (*Crassostrea commercialis, Crassostrea lugubris, Crassostrea iredalei, Crassostrea malabonensis*) and the green mussels (*Perna vividis*) are cultured commercially, whereas the cockle (*Anadara granosa*) has a great commercial value, particularly in Malaysia and Thailand.

Air-breathing Arthropods (Myriapods, Arachnids, and Insects), are probably extremely numerous in all mangroves (Murphy, 1985) in terms of specific biodiversity, but little work has been done so far on the insect fauna. They usually remain in air-filled cavities during high tide; a conspicuous "cave fauna" has been described in the mud-lobster mounds (*Thalassina*). Diptera are richly represented in these ecosystems where highly distinctive and unique genera have been identified; some families are also characteristic of mangroves (Tethinidae, Canaceidae, etc.). Likewise, among Lepidoptera, some butterflies and moths are found only in mangroves, the most diversified being perhaps among Nymphulinae. Finally, some wasp and bee taxa (Hymenoptera) are endemic to mangroves where they may play a noteworthy commercial role (high quantities of natural honey are collected from the mangroves, in Bangladesh).

The complex interactions between invertebrates and plants are very poorly known (Robertson *et al.*, 1990), unless they are conspicuous. Many propagules (*Rhizophora, Bruguiera, Avicennia*) are destroyed by borer beetles and a torticoid moth causes mass mortality among *Sonneratia* seedlings.

Mangrove microorganisms

Microbiota are known to play essential roles in mangrove food chain and biogeochemical cycles (carbon, sulfur, nitrogen, iron, phosphorus,

Table M4 Trees and shrubs of the world's mangroves (based on Saenger *et al.*, 1983; Duke *et al.*, 1998)

Flora									
Genus	Species	Form	Biogeographic regions (see Figure M2)						
Acanthus	*ebracteatus* Vahl.	S					5	6
	ilicifolius L.	S					5	6
Acrostichum	*aureum* L.	F	1	2	3	4	5	6
	danaeifolium Langsd.	F	1	2				
	speciosum Willd.	F					5	6
Aegialitis	*annulata* R. Br.	S					5	6
	rotundifolia Roxb.	S					5	
Aegiceras	*corniculatum* (L.) Bl.	S					5	6
Aglaia	*cucullata* Roxb.	T					5	
Avicennia	*alba* Blume	T					5	6
	bicolor Standl.	T	1					
	germinans L.	T	1	2	3			
	integra	T						6
	marina Forsk. Vierh	S/T	1			4	5	6
	officinalis L.	T					5	6
	rumphiana Hall. f.	T					5	6
	schaueriana Stapf.	T		2				
Bruguiera	*cylindrica* (L.) Blume	T					5	6
	exaristata Ding Hou	S/T						6
	gymnorrhiza (L.) Lam.	T				4	5	6
	hainesii C.G. Rogers	T					5	6
	parviflora Roxb.	T					5	6
	sexangula (Lour.) Poiret	T					5	6
Camptostemon	*philippensis* Becc.	T					5	
	schultzii Mast.	T						6
Ceriops	*australis*	S/T						6
	decandra Griff.	S/T					5	6
	tagal C.B. Rob.	S/T				4	5	6
Conocarpus	*erectus* L.	S/T	1	2	3			
Cynometra	*iripa* Kostel	S					5	6
Diospyros	*littoralis*	T						6
Dolichandrone	*spathacea* K.	T					5	6
Excoecaria	*agallocha* L.	T					5	6
	indica Muell. Arg.	T					5	
Heritiera	*fomes* Buch. Ham.	T					5	
	globosa Kost.	T					5	
	littoralis Aiton	T				4	5	6
Kandelia	*candel* L. Druce	S/T					5	
Laguncularia	*racemosa* Gaernt.f.	S/T	1	2	3			
Lumnitzera	*littorea* Voigt	S/T					5	6
	racemosa Willd.	S/T				4	5	6
	× *rosea*	S						6
Mora	*oleifera* Ducke	T	1					
Nypa	*fruticans* Van Wurmb.	P		2	3		5	6
Osbornia	*octodonia* F. Muell.	S/T					5	6
Pelliciera	*rhizophorae* Pl. Triana	T	1	2				
Pemphis	*acidula* Forster	S				4	5	6
Phoenix	*paludosa* Roxb.	P						6
Rhizophora	*apiculata* Blume	T					5	6
	mangle L.	S/T	1	2	3			
	mucronata Lam.	T				4	5	6
	racemosa G. Mey	T	1	2	3			
	samoensis (Horchr.) S.	T						6
	stylosa Griff.	S/T					5	6
	× *harrisonii*	T	1	2	3			
	× *lamarckii*	T					5	6
	× *selata*	T						6
Scyphiphora	*hydrophyllacea* Gaertn.	S					5	6
Sonneratia	*alba* J. Smith	T				4	5	6
	apetala Buch.-Ham.	T					5	
	caseolaris (L.) Engl.	T					5	6
	griffithii Kurz	T					5	
	lanceolata Bl.	T					5	6
	ovata Backer	T					5	6
	× *gulngai*	T					5	6
	× *urama*	T						6
	alba x *ovata*	T					5	
Tabebuia	*Palustris*	S	1					
Xylocarpus	*granatum* Koenig	T				4	5	6
	mekongensis	T					5	6

P, Palm; S, Shrub; F, Fern; T, Tree.

etc.), under very peculiar anaerobic and aerobic conditions. Fungi and bacteria may become associated, in mangrove sediments, providing intensive, rapid, and almost continuous recycling of organic matter, during floods, under anoxic conditions, as well as during exposure to air at low tides. Given the high temperature of water and sediments (25–35°C), the primary production is rapidly attacked and decayed.

The role of fungi in the processes of degradation of lignocellulose materials, is still obscure; very little is known, as few species have been tested for their ability to degrade the organic matter (Jones and Hyde, 1988). Even the taxonomic inventory is only now beginning. About 100 species are known today (73 Ascomycotina, 25 Deuteromycotina, and only 2 Basidiomycotina).

From a biogeographical point of view it appears that most mangrove fungi are distributed across the tropical world (Jones, 1996). This could be due either to their very efficient dispersal mechanisms or to an early evolution, before the separation of landmasses.

According to Boto (1988) much further research is needed to properly understand, the exact role of microbiota, especially of bacteria, in mangroves.

Conclusion

These are a few of the many areas that briefly illustrate the extreme complexity of mangrove ecosystems and the numerous issues to be solved. There are a number of poorly understood processes involved in the productivity and biogeography of mangrove species. Some preliminary studies (Duke, 1995; Maguire and Saenger, 2000), are indicating infraspecific variation having a genetic basis. The exact genetic variation within mangrove taxa is unknown.

But there is a danger that most of the natural processes and interactions will never be properly understood, because the mangroves of the world, except in Australia, are being transformed, fragmented, degraded, or converted at a very rapid pace, as a consequence of the population growth in the tropical coastal world (Field, 1996).

F. Blasco, M. Aizpuru and J. Besnehard

Bibliography

Adaime, R.R., 1985. Produçao do bosque de mangue de Gamboa Nobrega. Ph. D. thesis, Universidade de Sao Paulo.

Ball, M.C., 1996. Comparative ecophysiology of mangrove forest and tropical lowland moist rainforest. In Mulkey, S.S, Chazdon, R.L, and Smith, A.P. (eds.), *Tropical Forest Plant Ecophysiology*, New York: Chapman and Hall, pp. 461–496.

Ball, M.C., 1998. Mangrove species richness in relation to salinity and waterlogging: a case study along the Adelaide River floodplain, northern Australia. *Global Ecology and Biogeography Letters*, **7**: 73–82.

Ball, M.C., and Pidsley, S.M., 1998. Establishment of tropical mangrove seedings in relation to salinity. In *Proceedings of Workshop on Research and Management in Darwin Harbour*. North Australia. Research Unit, Darwin, pp. 123–124.

Blasco, F., and Aizpuru, M., 2001. Depletion of the mangroves of Asia. *Wetlands* (*Ecology and Management*, **9**(3): 245–256.)

Boto, K.G., 1988. *The carbon cycle mangrove microbiology*. New Delhi: UNESCO-UNDP, pp. 41–59.

Chapman, V.J., 1976. *Mangrove Vegetation*. Vaduz: Cramer.

Clark, M.W., Mcconchie, D., Saenger, P., and Pillsworth, M., 1997. Hydrological controls on copper, cadmium, lead and zinc concentrations in an anthropogenically polluted mangrove ecosystem, Wynnum, Brisbane, Australia. *Journal of Coastal Research*, **13**: 1150–1158.

Dodd, S.R., Blasco, F., Rafii, Z.A., and Torquebiau, E., 1999. Mangroves of the United Arab Emirates: ecotypic diversity in cuticular waxes at the bioclimatic extreme. *Aquatic Botany*, **63**: 291–304.

Duke, N.C., 1988. Phenologies and litter fall of two mangrove trees, *Sonneratia alba* Sm. and *S. caseolaris* (L.) Engl. and their putative hybrid. S. x gulngai N.C. Duke. *Australian Journal of Botany*, **36**: 473–482.

Duke, N.C., 1992. Mangrove floristics and biogeography. In Robertson, A.I., and Alongi, D.M. (eds.), *Tropical Mangrove Ecosystems*. Washington DC: American Geophysical Union, pp. 63–100.

Duke, N.C., 1995. Genetic diversity, distributional barriers and rafting continents—more thoughts on the evolution of mangroves. *Hydrobiologia*, **295**: 167–181.

Duke, N.C., Ball, M., and Ellison, J.C., 1998. Factors influencing biodiversity and distribution gradients in mangroves. *Global Ecology and Biogeography Letters*, **1**(7): 27–47.

Field, C.D., 1996. *Restoration of Mangrove Ecosystems*. Okinawa, Japan: ISME.

Fromard, F., Puig, H., Mougin, E., Marty, G., Betoulle, J.M., and Cadamuro, L., 1998. Structure, above ground biomass and dynamics of mangrove ecosystems: new data from French Guiana. *Oecologia*, **115**: 39–53.

Hamilton, L.S., and Snedaker, S.C., 1984. *Handbook for Mangrove Area Management*. Honolulu, HI: East-West Center.

Hutching, P.A., and Saenger, P., 1987. *Ecology of Mangroves*. University of Queensland Press, St Lucia, Brisbane.

ISME, 1993. *Conservation et utilisation rationnelle des mangroves de l'Amérique latine et de l'Afrique*. Okinawa.

Jones, E.B.G., 1996. Marine mycota of Malaysian mangroves. *Wallaceana*, **79**: 11–12.

Jones, E.B.G., and Hyde, K.D., 1988. Methods for the study of marine fungi from the mangroves. In Agate, A.F. (ed.), *Mangrove Microbiology*. New Delhi: UNESCO-UNDP, pp. 9–27.

Lacerda, L.D., 1998. *Biogeochemistry of Trace Metals and Diffuse Pollution in Mangrove Ecosystems*. International Society of Mangrove Ecosystems. Okinawa.

Lacerda, L.D., Machado, W., and Moscatelli, M., 2000. Use of mangroves in landfill management. *GLOMIS Newsletters: Electronic Journal*.

Lahmann, E.J., 1988. Effects of different hydrological regimes on the productivity of *Rhizophora mangle* L. A case study of mosquito control impoundments at Hutchinson Island. Saint Lucie County, Florida. Ph. D. thesis. Miami: University of Miami.

Lugo, A. E., Twilley, R.R., and Patterson-Zucca, C., 1980. The role of black mangrove forests in the productivity of coastal ecosystems in south Florida. Corvallis: Oregon. Report to E.P.A. Corvallis Environmental Research Laboratory.

Maguire, T.L., and Saenger, P., 2000. The taxonomic relationships within the genus *Excoecaria* L. based on leaf morphology and rDNA sequence data. *Wetlands Ecology and Management* **8**: 19–28.

Murphy, D.H., 1985. A survey of animal kingdom as represented in mangroves. Introductory training course on mangrove ecosystems. Department of Zoology National University, Singapore.

Odum ,W.E., and Heald., E.J., 1972. Trophic analysis of an estuarine mangrove community. *Bulletin of Marine Sciences*, **22**: 671–738.

Rao, A.N., 1987. Mangrove ecosystems of Asia and the Pacific. In UNDP/UNESCO, RAS/79/002 (ed.), *Mangroves of Asia and the Pacific, Status and Management*. New Delhi: UNESCO, pp. 1–48.

Robertson, A.I., Giddins, R., Smith, T.J., 1990. Seed predation by insects in tropical mangrove forests. *Oecologia*, **83**(2): 213–219.

Saenger, P., and Bellan, M.F., 1995. *The Mangrove Vegetation of the Atlantic Coast of Africa*. Toulouse: LET, University of Paul Sabatier.

Saenger, P., and Siddiqi, N.A., 1993. Land from the sea: the mangrove afforestation program of Bangladesh. *Ocean and Coastal Management*, **20**: 23–39.

Saenger, P., and Snedaker, S.C., 1993. Pantropical trends in mangrove above-ground biomass and annual litter fall. *Oecologia*, **96**: 293–299.

Saenger, P., Hegerl, E.J., and Davie, J.D.S. (eds.), 1983. *Global Status of Mangrove Ecosystems*. Gland: IUCN.

Sassekumar, A., and Loi, J.J., 1983. Litter production in three mangrove forest zones in the Malay Peninsula. *Aquatic Botany*, **17**: 283–290.

Smith, T.J., 1992. Forest structure. In Robertson, A.I. and Alongi, D.M. (eds.), *Tropical Mangrove Ecosystems*. Washington, DC: American Geophysical Union, pp. 101–136.

Snedaker, C.S., and Lugo, E.A., 1973. *The Role of Mangrove Ecosystems in the Maintenance of Environmental Quality and a High Productivity of Desirable Fisheries*. Gainesville: Center for Aquatic Sciences, University of Florida.

Snedaker, S.C., 1982. Mangrove species zonation: why? In Sen, D.N., and Rajpurohit K.S. (eds.), *Task for Vegetation Science*, Volume 2. Dr. Junk Publishers, pp. 111–125.

Spalding, M., Blasco, F., Filed, C.D. (eds.), 1997. *A World Mangrove Atlas*. Okinawa: International society for mangrove ecosystems.

Tholot, P., 1996. *Les poissons de mangrove de Nouvelle Calédonie*. Paris: ORSTOM.

Tomlinson, P.B., 1986. *The Botany of Mangroves*. Cambridge: Cambridge University Press.

Twilley, R.R., 1982. Litter dynamics and organic carbon exchange in black mangrove (*Avicennia germinans*) basin forests in a South-West Florida estuary. Ph. D. thesis, Gainesville: University of Florida.

Woodroffe, C.D., Bardsley, K.N., Ward, P.J., Hanley, J.R., 1988. Production of mangrove litter in a macrotidal embayment, Darwin Harbour, N.T., Australia. *Estuarine Coastal Shelf Sciences*, **26**: 581–598.

Cross-references

Coral Reef Coasts
Desert Coasts
Mangroves, Remote Sensing
Vegetated Coasts

MANGROVES, GEOMORPHOLOGY

Mangroves are halophytic shrubs and trees that grow in the upper part of the intertidal zone on the shores of estuaries and lagoons, and on coasts sheltered from strong wave action, as in inlets or embayments or in the lee of headlands, islands, or reefs. They grow sparsely on rocky shores and coral reefs, where their roots penetrate fractures in the rock, and on sandy substrates, but are more luxuriant, forming dense scrub and woodland communities, on muddy substrates and shoals exposed at low tide, particularly where there is a supply of muddy sediment. Where wave energy is low they spread forward to the mid-tide line, but as wave action increases along a coastline the mangrove fringe thins out and disappears. On the other hand, mangroves colonize areas that have become more sheltered as the result of the longshore growth of sand bars, spits, or barriers.

The width of a mangrove fringe generally increases with tide range, and on macrotidal coasts can attain several kilometers, as on the tide-dominated shores of gulfs and estuaries in northern Australia, where wide mangrove areas are backed by sparse salt marshes and saline flats flooded during exceptionally high tides and summer rains. Within the humid tropics mangroves grow to forests with trees 30–40 m high, as on the west coasts of Malaysia and Thailand, in Indonesia, Madagascar, and Ecuador. On drier and cooler coasts within, and outside, the tropics (extending to about 30°N and up to 39°S) mangroves generally form extensive scrub communities (Chapman, 1976).

Where mangroves are spreading seaward there is an abundance of seedlings and young shrubs on the adjacent mudflats and a smooth canopy rising landward as the trees increase in age and size. A receding mangrove coast is indicated by a microcliff and the truncation of the mangrove canopy, so that trunks are exposed with trunks and roots being undercut, or where the mangroves have died, and any seedlings fail to survive.

Mangroves are physically adapted to survive in a marine tidal environment. Some species (e.g., *Avicennia marina, Sonneratia alba*) have root systems with networks of pneumatophores, snorkel-like breathing tubes that project vertically from the muddy substrate (Figure M3), and allow the plant to respire when the tide falls. With the aid of these structures, mangroves can grow in areas submerged by the sea at each high tide, but the need for several hours subaerial exposure between each submergence sets a seaward limit, usually close to mid-tide level. Pneumatophores occupy a roughly circular zone around the stem of each plant, and can be very numerous and closely spaced, with densities of up to 300/m^2. Other mangroves, such as *Rhizophora, Bruguiera, Ceriops*, and *Lumnitzera* species, have subaerial prop or stilt roots that branch downward to the mud and support the stems.

Formation of mangrove terraces

Sediment carried into mangroves by a rising tide is retained by the filtering network of stems, pneumatophores, or prop roots as the tide ebbs. This leads to the gradual building up of a depositional terrace between high spring and high neap tides, with a transverse slope (usually $<1:50$), descending seaward to the outer edge of the mangroves and continuing across the lower intertidal zone, which is either unvegetated, or has patches of seagrass. Terrace formation is also aided by the presence of a subsurface root network, which binds the accreting sediment. The depositional terrace under mangroves is similar to that formed beneath salt marshes (*q.v.*).

Mangroves initially colonize substrates that are stable or slowly accreting, but once established they promote further accretion because they diminish current flow and wave action (Augustinus, 1995). A mangrove fringe also shelters nearshore waters from winds blowing off the land, and thus reduces seaward losses of mud from the shore, whereas shoreward drifting by waves produced by onshore winds and rising tides continues.

On some coasts, the pioneer mangroves are *Avicennia* spp., with pneumatophores that trap muddy sediment and raise the intertidal surface until it is colonized by *Rhizophora* and other mangrove species (Figure M4). On other coasts *Rhizophora* species occupy the seaward margin. To the rear, sedimentation proceeds more slowly, but eventually the terrace is built up to high spring tide level, where the mangroves may be backed by salt marsh or unvegetated dry mudflats. Most mangrove substrates are dominated by muddy sediment, but there is also peat accumulation where the mangrove community generates large quantities of organic matter from decaying leaves, stems, and roots, and from the various organisms that inhabit mangrove areas. In due course, these can form peat deposits, raising the substrate level: on the coast of Florida *Rhizophora* is growing on vertically accreting peat deposits.

Figure M3 Pneumatophores projecting from a muddy intertidal surface around *A. marina* shrubs in Westernport Bay, southeastern Australia. (Copyright—Geostudies.)

Figure M4 Mangroves (*A. marina*) advancing on to intertidal mudflats in Cairns Bay, north-eastern Australia. The darker zones to the rear are *Rhizophora* and other mangrove species. (Copyright—Geostudies.)

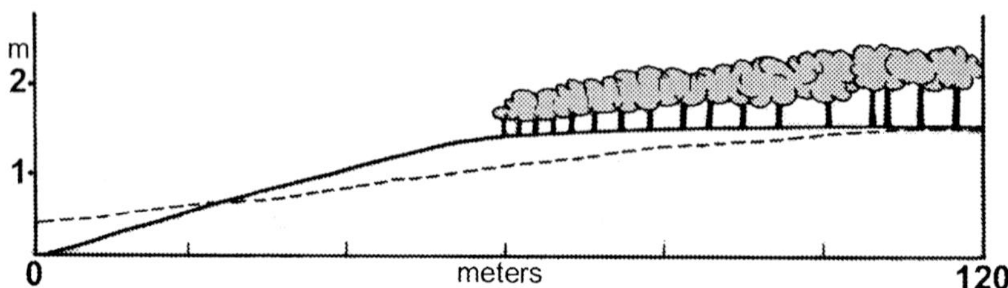

Figure M5 A mangrove terrace formed by vertical accretion of sediment as mangroves spread seaward. If the mangroves then die, or are removed, the intertidal profile is degraded (dashed line). If they revive, the mangrove terrace is restored. (Copyright—Geostudies.)

As the terrace attains high spring tide level it is submerged only by infrequent exceptionally high tides, occasional storm surges, or river flooding. Sedimentation is thus very slow, but accretion of peat and drift litter may raise the substrate to levels where it can be colonized by freshwater and land vegetation, a process that is aided by emergence on coasts where sea level is falling relative to the land.

Sections through mangrove terraces (exposed in the banks of tidal creeks or in microcliffs at the seaward edge) generally show stratified deposits, with layers of fine sand or organic material within the mud. These variations are related to wave conditions, storm waves washing fine sand into the mangroves, and mud accretion continuing as the tides rise and fall in calm weather. Although often characterized as tide-dominated morphology, most mangrove terraces are influenced by wave action as the tide rises and falls. When the depositional terrace is submerged, occasional storm surges may wash sand, gravel, and shelly material up into the mangroves to be deposited as cheniers (*q.v.*). Former cheniers may be indicated by lenticular deposits of coarser sediment within mud in mangrove terrace stratigraphy.

Rates of accretion in mangroves can be measured by laying down marker layers of colored sand, coal dust, or similar material on the surface, and returning to put down borings and measure the thickness of sediment added subsequently. There are often difficulties where burrowing crabs churn up the substrate. Changes can also be measured on implanted stakes. In Westernport Bay, Australia, such measurements showed sustained mud accretion of up to 4.5 cm/yr in the *Avicennia* fringe, with slower deposition at higher levels and continuing vertical fluctuations on adjacent mudflats. In the absence of mangrove vegetation the substrate remains a mobile intertidal slope. The pattern of mud accretion is strongly correlated with the density of the pneumatophore networks which trap and retain muddy sediment, and low accretion mounds form above the general level of the muddy shore within pneumatophore networks around isolated *Avicennia* trees. When pneumatophore networks were simulated artificially by planting a network of wooden stakes a similar accretion mound formed within the staked area, but when the stakes were removed the accreted sediment was quickly dispersed. Mangroves with pneumatophores thus trap sediment as effectively as salt marsh plants (Bird, 1986). Mangroves with prop roots, such as *Rhizophora* spp., may be less effective in trapping mud, but depositional terraces have been formed on coasts fringed by such mangroves.

The widening of a mangrove terrace depends on a continuing sediment supply, as on deltaic and estuarine shores close to the mouths of rivers or within embayments where intertidal muddy areas are extensive. Changes within river catchments, such as deforestation, increase soil erosion and augment the rate of sediment yield to the coast, thereby accelerating accretion in intertidal areas and promoting the spread of mangroves. This has occurred in the Segara Anakan lagoon in southern Java, where a greatly increased sediment yield resulted in siltation and the rapid advance of the mangrove fringe (Bird and Ongkosongo, 1980).

If mangroves are killed or cleared, their substrate is soon lowered by erosion. In Westernport Bay, Australia, mangroves were extensively cleared in the mid-19th century, and the depositional terraces that had formed beneath them were degraded to a steeper transverse slope and dissected, so that the roots of former mangroves were laid bare. Subsequent recolonization of these areas by mangroves was followed by mud accretion, which has rebuilt the depositional terrace (Figure M5) (Bird, 1986).

Nevertheless, the idea that mangroves are land-builders has not been universally accepted (see discussions by Vaughan, 1910; Davis, 1940; Carlton, 1974). Some have suggested that mangroves merely occupy intertidal areas that become ecologically suitable as the surface is raised by accretion, independently of any effects of vegetation, implying that the

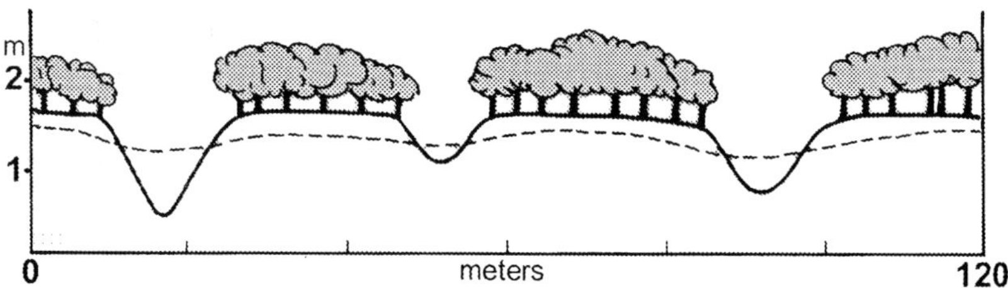

Figure M6 Transverse section across a mangrove terrace with intervening tidal creeks. If the mangroves die, or are removed, the creeks become wider and shallower (dashed line). If they revive, their cross-profile is restored. (Copyright—Geostudies.)

depositional terraces would have developed even if mangroves had not been present (Watson, 1928; Scholl, 1968). Others have deduced an interaction between colonizing mangroves and intertidal deposition (Thom, 1967). It is possible that *Avicennia* and other mangroves with pneumatophores promote accretion and coastline progradation as they spread forward on to the intertidal zone, whereas *Rhizophora* and other mangroves without pneumatophores are less effective in trapping sediment.

Tidal creeks in mangrove areas

As mangrove terraces build up in the form of a sedimentary wedge, there are alternations of tidal submergence as water invades the mangroves and emergence as it drains off. The ebb and flow of the tide forms and maintains a system of tidal creeks, the dimensions of which are related to the volume of water entering and leaving as the tide rises and falls. Typically dendritic and intricately meandering, these are channels within which the tide rises until the water floods the marsh surface; then they become drainage channels into which some of the ebbing water flows from the mangroves. They are thus like minor estuaries, particularly, where they receive freshwater from hinterland runoff. Where cheniers are present the mangrove creek network may be reticulate, as on the Niger delta (Allen, 1965), while straight parallel creeks are more often found where the tide range is large, the transverse gradient small, or the rate of seaward spread of mangroves rapid, as on Hinchinbrook Island, north-eastern Australia.

In the early stages of the formation of a mangrove terrace, tidal creeks are relatively wide and shallow in cross section (like the broad channels on intertidal mudflats), but as the mangrove terrace expands they become narrower and deeper, and their banks higher and steeper. Oversteepening results in local slumping, when blocks of compact mud, often with clumps of mangrove vegetation, collapse into the creek, especially where the banks have been burrowed by crabs.

If the mangrove vegetation dies, or is cleared away, the tidal creeks become wider and shallower, and if revegetation then occurs they again narrow and deepen (Figure M6). There are also changes as tidal creeks meander or migrate laterally, so that mangrove trees are undermined on one bank while mangrove seedlings colonize sediment deposited on the other. Channel banks may oscillate in response to variations in the volume of tidal ebb and flow; they advance during phases of local accretion when the channel diminishes, and recede during episodes of tidal scour, when the channel is enlarged.

Seaward margins of mangroves

Mangrove terraces are being eroded on coasts that are now receiving little or no sediment. Low cliffs have been cut into their seaward margins, particularly on deltaic coasts where the sediment supply has been reduced because of dam construction or the natural or artificial diversion of a river outlet. The seaward edge of mangrove terraces is often undercut by a muddy microcliff up to a meter high. Microcliff recession may be accompanied by continuing vertical accretion of muddy sediment in the mangroves, building up the terrace even though seaward advance has come to an end. In some places, the cliffing results from lateral movement of a tidal channel, undercutting the outer edge of the mangroves, but generally it is due to larger waves reaching the mangroves as the result of deepening of the lower intertidal zone, either because of progressive entrapment of nearshore sediment drifting into the upper vegetated area, or because of continuing submergence of the coast in response to a rising sea level (Guilcher, 1979).

It may be that, as on the sides of developing tidal creeks, seaward margins become oversteepened and cliffed, particularly during occasional storm wave episodes. Cliffing of this kind is repaired if there is an abundant supply of sediment to restore the profile, permitting mangroves to spread again, but if there is a sediment deficit a mangrove cliff will persist and recede until the mangrove terrace has been completely removed.

If global warming proceeds, mangroves are likely to spread to suitable habitats beyond their present poleward limits. A rising sea level will however, impede the seaward advance of mangroves, and increase erosion on their seaward margins, except where there is a compensating increase in sediment supply. It has been estimated that a sea-level rise of more than 1.2 mm/yr will lead to widespread destruction of mangroves and erosion of their substrates (Ellison and Stoddart, 1991).

Eric Bird

Bibliography

Allen, J.R.L., 1965. Coastal geomorphology of eastern Nigeria: beach ridge, barrier islands and vegetated tidal flats. *Geologie en Mijnbouw*, **44**: 1–21.

Augustinus, P.G.E.F., 1995. Geomorphology and sedimentology of mangroves. In Perillo, G.M.E. (ed.), *Geomorphology and Sedimentology of Estuaries*, Amsterdam: Elsevier, pp. 333–357.

Bird, E.C.F., 1986. Mangroves and intertidal morphology in Westernport Bay. *Marine Geology*, **77**: 327–331.

Bird, E.C.F., and Ongkosongo, O.S.R., 1980. *Environmental Changes on the Coasts of Indonesia*. Tokyo: United Nations University.

Carlton, J.M., 1974. Land-building and stabilization by mangroves. *Environmental Conservation*, **1**: 285–294.

Chapman, V.J., 1976. *Mangrove Vegetation*. Vaduz: Cramer, 447p.

Davis, J.H., 1940. The ecology and geological rôle of mangroves in Florida. *Publications of the Carnegie Institution*, **524**: 303–412.

Ellison, J.C., and Stoddart, D.R., 1991. Mangrove ecosystem collapse during predicted sea-level rise: Holocene analogues and implications. *Journal of Coastal Research*, **7**: 151–165.

Guilcher, A., 1979. Marshes and estuaries in different latitudes. *Interdisciplinary Science Reviews*, **4**: 158–168.

Scholl, D.W., 1968. Mangrove swamps: geology and sedimentology. In Fairbridge, R.W. (ed.), *Encyclopaedia of Geomorphology*. New York: Reinhold, pp. 683–688.

Thom, B.G., 1967. Mangrove ecology and deltaic geomorphology: Tabasco, Mexico. *Journal of Ecology*, **55**: 301–343.

Vaughan, T., 1910. The geologic work of mangroves in southern Florida. *Smithsonian Miscellaneous Collection*, **52**: 461–464.

Watson, J.D., 1928. Mangrove forests of the Malay Peninsula. *Malay Forest Records*, **6**: 1–275.

Cross-references

Cheniers
Mangroves, Ecology
Mangroves, Remote Sensing
Muddy Coasts
Peat
Salt Marsh
Tides
Vegetation Coasts

MANGROVES, REMOTE SENSING

Introduction

Most research activities and published documents (at least 95%) on the biology and ecology of mangroves concern the factors influencing the productivity, the biodiversity, and the biogeographical distribution. Naturally, the biodiversity itself and the adaptive mechanisms to salinity and waterlogging constitute major fields of research. In comparison, published documents on the use of remote sensing for mangrove studies are in a minority.

In recent years, "Remote Sensing" has become a term applied to all kinds of information acquired by satellites, although in its broad sense, it refers to the gathering and analysis of images acquired by sensors and cameras located at some distance from the target of study including aircrafts, balloons, etc. (Haines-Young, 1994). The main advantages of satellite over aerial photographs have often been advocated, especially by foresters, who rightly insisted that large-scale surveys are feasible through computer processing. In addition, monitoring events such as erosion, degradation processes, human impacts, flooding, phenology, etc. can be carried out as each satellite makes regular overhead passes on a fixed orbit.

A brief review of the most commonly used remote sensing technology for mangrove studies is given here. The capabilities and constraints of spaceborne instruments are outlined. This entry focuses only on conspicuous achievements from a few square kilometers to the whole tropical coastal world.

Sensors characteristics

The application of remote sensing tools to mangrove forests began tentatively in the early 1970s with some optimistic attempts to use Landsat MSS data, with a ground resolution of about 60–80 m (*"ground or spatial resolution"* is *"the minimum distance between two objects that a sensor can record distinctly"*— Simonett, 1983), in huge mangrove areas, especially in the Gangetic delta which has the largest contiguous mangroves in the world. The main hope was to increase the frequency of forest observations and inventories, especially in remote fast changing areas, where access and fieldwork are particularly difficult. At best, the repeated frequency was 16 days and 4 spectral bands were available in the visible (VIS) and near-infrared (NIR) wavelengths. Since 1982, Landsat satellites have an instrument (Thematic Mapper) with a ground resolution of 30 m and 7 spectral bands including middle-infrared (1.57–1.78 μm) and a thermal-infrared (2.10–2.23 μm). After the launching in 1986 of the first high-resolution satellite, SPOT 1, operating at 20 m in a multispectral mode and at 10 m in a panchromatic mode, and an improved repeat cycle, the possibility of acquiring accurate data, for almost all mangroves of the world, has been conspicuously increased. This is how the first World Mangrove Atlas was achieved and published (Spalding *et al.*, 1997), followed by the first worldwide mangrove inventory carried out by the European Community (Aizpuru *et al.*, 2000).

These instruments, including radar, generate a signal only in each of a small number of very broad bands (Table M5). In the case of radar, SAR (Synthetic Aperture Radar) data, the spatial resolution is rather good (6–25 m) and these products are widely available since the end of

Table M5 A summary of the main satellites used for vegetation mapping

Satellites, sensors, launch year, and distance to earth	Band	Spectral bands	Spatial resolution	Overhead passes and swath	Suitable mapping scale	Discrimination
AVHRR	1	0.58–0.68 μm			Global and regional	
NOAA	2	0.725–1.1 μm	HRPT, LAC: 1 km	Daily		Forest/nonforest
(since 1978)	3	3.55–3.95 μm	GAC: 4 km			
(860 km)	4	10.5–11.3 μm	GVI: 15 km	2,700 km	1/1,000,000	Biological rhythms
	5	11.5–12.5 μm			–1/10,000,000	
LANDSAT MSS	4	0.45–0.6 μm		16 days	National	
(1972)	5	0.56–0.7 μm	56 m × 79 m			Physiognomy
(915 km and	6	0.67–0.8 μm		180 km	1/200,000–	
705 km)	7	0.78–1.1 μm			1/1,000,000	
LANDSAT TM	1	0.45–0.52 μm				
(1982)	2	0.53–0.61 μm				
(705 km)	3	0.62–0.69 μm	30 m	16 days	Local	Phenology
	4	0.78–0.91 μm				
	5	1.57–1.78 μm		180 km		Physiognomy
	7	2.10–2.35 μm				
	6	10.4–12.6 μm				
SPOT 1, 2, 3	1	0.500–0.590 μm		26 days		Dominant floristic
HRV	2	0.615–0.680 μm	20 m		1/50,000–	groups
(1986)	3	0.790–0.890 μm		60 km	1/200,000	
(830 km)	P	0.510–0.730 μm	10 m			
ERS-1(1991)	SAR	5.3 GHz			Local	Physiography
and ERS-2 (1995)			30 m	100 km	1/50,000	Vegetation cover?
(785 km)	C	5–7 cm			1/100,000	Soil water content
Radarsat (1995)	SAR	5.3 GHz	Multi-resolutions	24 days	1/50,000	Same +
(798 km)	C	5–7 cm	25–100 m	100–500 km	1/250,000	crops
JERS-1 (1992)	SAR	1.2 GHz	20 m	44 days	1/50,000	Same
(568 km)	L	24 cm		75 km		
ENVISAT (2000)	SAR	5.3 GHz	multi-resolutions	35 days	1/50,000	Same +
(800 km)	C	5–7 cm	25 m	100–400 km	—	coastal zones
			150–1 km		1/1,000,000	topography
	B0	Blue 0.44–0.47 μm				
SPOT4	B2	Red 0.61–0.68 μm	1 km	daily	All scales	
VEGETATION	B3	PIR 0.79–0.89 μm				Forest types
	MIR	MIR 1.58–1.73 μm		2,250 km	since	Main crops
and						Phenology
					1/25,000	Physiognomy
SPOT4	B1	Green 0.50–0.59 μm				Dominant floristic
HRVIR	B2	Red 0.61–0.68 μm	20 m		—	groups
	B3	PIR 0.79–0.89 μm		60 km		Fire monitoring,
(1998)	MIR	MIR 1.58–1.73 μm			1/5,000,000	etc.
(830 km)	P	0.59–0.75 μm	10 m			

the 1990s from a number of recent spacecrafts (ERS-1, ERS-2, JERS, Radarsat ...). Mangroves have already been studied at various research levels with radar, which have the ability to penetrate clouds, permitting very frequent repetitive observations. Most of these instruments are primarily dedicated to physical oceanography and polar observation. Their use for environmental applications is yet very limited. When SAR data are used for mangrove studies, they are often processed in combination with optical data as one of the major issues is the suppression of the random noise associated with SAR data (the speckle), which induces strong limitations for the delineation of coastal units and land cover types (Mougin *et al.*, 1993; Kushwaha *et al.*, 2000; Phinn *et al.*, 2000; Proisy *et al.*, 2000).

More complex hyperspectral scanning systems are now providing new types of data sets. They measure the intensity of the radiations received from coastal ecosystems in each of a large number and very narrow bands (50–300 bands, about 2 or 3 nm each). MODIS (MODerate resolution Imaging Spectro-radiometer) is already in space on EOS (Earth Orbiting System) satellite launched in 1999. Several comparable instruments are airborne: VIFIS (Variable Interference Filter Imaging Spectrometer) with 64 spectral bands, AVIRIS (Airborne Visible InfraRed Imaging Spectrometer), CASI (Compact Airborne Spectrographic Imager), etc. Since 1981, tens of thousands of photographs, including color infrared, have been taken from the space shuttle orbit and stereoscopic coverages are available. To date, very high-ground resolution data (about 1 m) have not been made available to scientists mainly for security reasons. They are progressively appearing on the market (IKONOS data).

All these sophisticated instruments are generating an incredible amount of digital georeferenced data, covering all the ecosystems of the coastal world in a repetitive manner. Have we been able to develop the necessary technology to adapt the quality and the Timescale of observation to the magnitude and rapidity of human-induced degradations in mangrove areas?

Basic principles

Each satellite has its own technical properties designed for specific missions. Whatever satellites are used, the physical principle remains almost constant, based on the fact that different mangrove subtypes show different reflectance patterns. Spaceborne sensors measure the solar radiation that are reflected or radiated by the "targets." The "reflectance" of a given ecosystem (a mangrove with *Nypa fruticans* or a salt marsh with Chenopodiaceae) is the ratio between the reflected solar energy and the incident solar energy. In order to minimize the distortions induced on each signal by the atmosphere and to take into account the sensors–object–sun geometry, several corrections have to be applied to raw recorded data before they are processed (i.e., atmospheric and geometric corrections). Ultimately, it is quite obvious from Table M5 that spatial and temporal resolutions determine the type of information that can be derived from satellites (Holben and Fraser 1983; Graetz, 1990; Blasco and Aizpuru 2001). It is implicit that permanent global monitoring of mangroves, at high spatial resolutions (Landsat TM, SPOT HRV and HRVIR, IRS1C-LISS-III, ERS, etc.), is practically not feasible and that local mangrove monitoring at low spatial resolution (NOAA-AVHRR, SPOT-VEGETATION) is also impossible (Justice 1985; Townshend and Justice, 1986; Malingreau *et al.*, 1989).

The different coastal ground units reflect differently in the VIS parts of the spectrum (dense continental green trees, wet clayey soils, sandy beaches, mangroves ...) as well as across all the wavelengths from the ultraviolet to the thermal-infrared and microwaves (Figure M7). Some signals are not reflected but are radiated by ground units especially in the thermal-infrared. The different patterns of reflectance from different ecosystems are often termed "spectral signatures."

Usually the "signal mangrove" is distinct from others coastal ecosystems for two main reasons:

1. It is necessarily confined to the nearshore tropical intertidal zone.
2. It is the result of two main signatures often recognised in the VIS and NIR domains:
    - The evergreen character of mangrove trees leads to strong reflectance in the NIR channel (0.7–1.1 μm), generally coded in red on a color composite.
    - The permanently wet soils which have a noteworthy reflectance in the red channel (0.6–0.7 μm), generally coded in green on a color composite.

This is why dense mangroves always appear dominantly in a reddish violet color, turning to bluish violet in open mangrove stands. A given tall dense mangrove stand with *Rhizophora* sp. (Malaysia) may have the same "signature" as a low thicket with *Avicennia marina* (New Zealand). This is one of the major limitations of these technologies, especially in forestry in general, where one of the main goals is to discriminate the species of trees and to estimate the structure of the forest and the size of its components.

In order to transfer experimental results to field situations, several "vegetation indices" have been proposed. The most commonly used is the Normalized Difference Vegetation Index (NDVI). It relies upon differential reflectances of the mangrove canopy (which is always green) in the red wavelength (here the response is mostly determined by the absorption band by the chlorophyll) and in the NIR wavelength, where the response is the result of scattering determined by the cuticles of leaves and the density of the cover.

$$\mathrm{NDVI} = \frac{\mathrm{NIR} - \mathrm{VIS(red)}}{\mathrm{NIR} + \mathrm{VIS(red)}}$$

The thermal domain (4 to 12 μm or bigger wavelength range) has not been exploited so far in mangrove deforestation processes, although a local sharp increase in surface brightness temperature could most probably be related to the replacement of dense mangrove stands by the so called saltish "blanks" which are barren soils, almost denuded of any vegetation. One of the limitations of the thermal domain is due to the fact that the relationship between the size of barren mangrove area and the temperature of the concerned pixels is difficult to establish. Thermal-infrared radiation allows rapid measurement of variations of

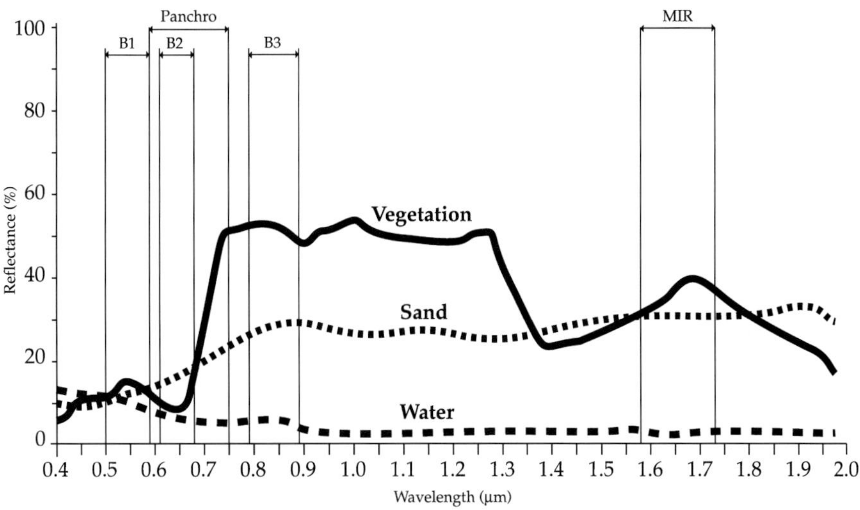

Figure M7 SPOT4-HRVIR: Spectral signatures from representative land cover types in mangrove areas.

water surface temperature. On the other hand, most features to be analyzed and mapped in coastal areas are narrow and linear. They are not always observed from space and may be lost at inadequate resolutions.

To date, the most common resolutions of satellites used for coastal surveys range between 10 and 30 m (SPOT, IRS, Landsat TM). When these resolutions are available, they lead to rather accurate maps at 1/50,000–1/100,000 scales. When these resolutions are not available, a number of statistical problems arise linked to difficult comparisons of maps drawn at different scales.

For each object and each pixel, the ecologist has at his disposal its reflectance, its longitude and latitude. This is why remote sensing data are usually exploited with the help of Geographic Information Systems (GIS). These databases and software combine time-independent parameters (geology, physiography, average climatic parameters, etc.) and a number of parameters whose values vary with time (phenology of trees, stand density, radiometry, etc.).

A lot of frameworks have been proposed to guide the existing classification systems; they combine various concepts from remote sensing, plant biogeography, landscape ecology, etc. (Phinn *et al.*, 2000).

Processing methods

In terms of forestry and vegetation classification, mangroves belong to the broad class of "closed evergreen tropical forest." Mangrove subtypes are identified according to the species composition, the structure of the stand and the ecological status (salinity of water and soils, climatic aridity, etc.), but neither the floristic composition nor structural properties nor environmental parameters are perceptible from space, which identifies only the spectral and textural parameters of the top canopy. This explains the low statistical separability of dense mangrove subclasses in the VIS and NIR wavelengths of high-resolution satellite data. This relates to the canopy architecture and the textured nature of reflectance patterns in which variance within forest classes can be greater than between forest classes (Hill, 1999; Kay *et al.*, 1991). Then, the easiest way to increase the spectral separability is to consider that the criteria of density of the cover is the starting point of any mangrove classification system.

Many approaches have been suggested for satellite image processing. None of them is yet fully convincing. They logically fall into four main groups:

1. *The visual interpretation* is probably the commonest but least explicit. In this case, a digital image has been converted into a photographic product which is studied by eye, with reference to field data. The delineations obtained with this empirical method are often interpreter-dependent.
2. *In the unsupervised classification*, the data are digitally processed with automatic enhancement techniques. One of major limitations of these approaches is caused by the fact that there are sometimes no significant difference between the spectral properties of mangroves and other neighboring ecosystems.
3. *Supervised classifications* are the most frequently described in the scientific literature. Field data, assisted or not with a Global Positioning Systems (GPS), are used for the selection of training samples to direct this type of classification, using several algorithms such as the maximum likelihood, the minimum distance, etc. (Figures M8 and M9). Practically, the scientists use existing maps as a reference in order to obtain a preliminary image segmentation, separating mangrove areas from other plant communities.
4. Finally, *temporal series*, principal component analysis, bands ratios, vegetation indices which convert multispectral information into a single index, etc. constitute another group of methods rather often employed.

According to the aims of the study and the concerned area, each method has its own advantages and insufficiencies. To evaluate each method, a rigorous quantification of its accuracy would be necessary. However, such a test is very rarely carried out (Green *et al.*, 1998).

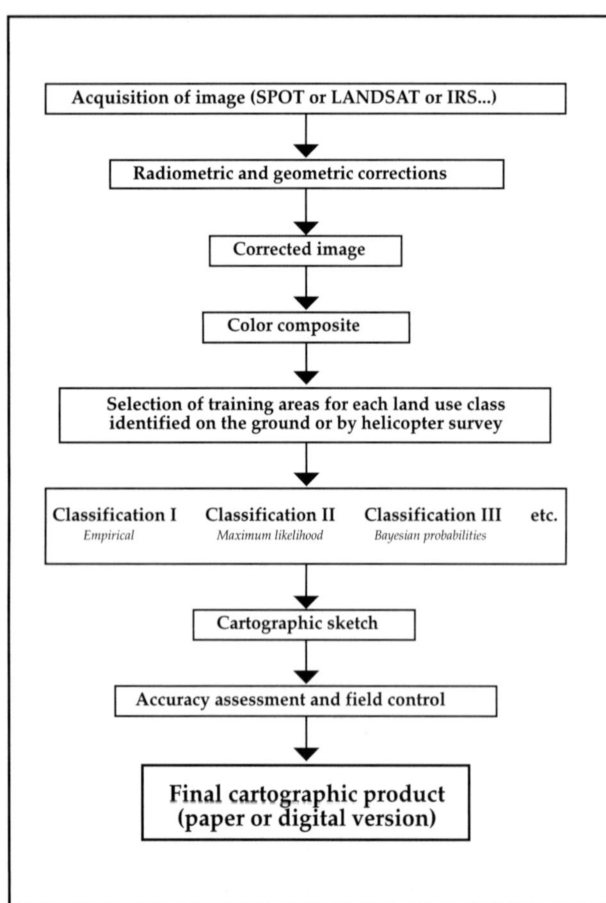

Figure M8 Classical approach for mangroves mapping using remote sensing data (supervised classification).

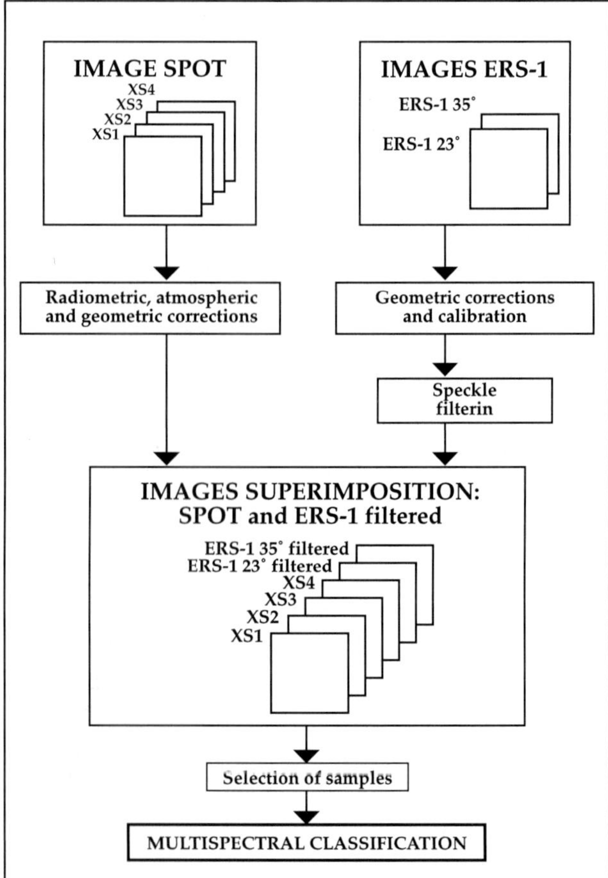

Figure M9 Synergy between SPOT and ERS-1, general methodology.

Mangrove features discriminated from space

The level of discrimination achieved with remote sensing products varies from one sensor to another and according to the processing method. As a general rule, when the amount of field data increases, the quality of the image processing improves. Almost all authors using aerial photos distinguish floristic classes labeled by genera and species. This level of discrimination which separates *Rhizophora apiculata* from *A. marina* stands is impossible at present from space. With spaceborne sensors, the easiest and primary objective is to separate mangroves from non mangrove vegetation (Biña *et al.*, 1980). Some authors confess their inability to discriminate satisfactorily these two classes with SPOT XS (Green *et al.*, 1998).

Several attempts have been carried out to distinguish stand density classes (Aschbacher *et al.*, 1995) and height classes (Gao, 1999). All these individual attempts are to be considered as preliminary research findings. From a practical point of view, it appears that seven most useful physiognomic classes can be currently detected and mapped, with or without GIS assistance, from space products:

Dense natural mangroves. This is the most important class, often located in protected areas. These stands are multi-specific and the ground coverage exceeds 80%.

Degraded mangroves where the ground coverage by trees and shrubs is about 50–80%. The spectral signal integrates the response of chlorophyllous elements and water-soaked soils.

Fragmented mangroves (or mosaics) where the remaining mangrove trees have a ground coverage about 25–50%. In such a case, the signal is primarily determined by the moist soils although the response of the remaining green vegetation is noticeable.

Leafless mangroves. As mangroves are almost always evergreen trees, a strong absorption in the NIR band (0.70–0.95 μm) has to be considered as totally abnormal, being induced either by a mass mortality of mangrove trees (Gambia, Côte d'Ivoire, etc.) or by an unexplained disease (cryptogamic, virus, insects, etc.).

Mangrove deforestation areas or clearfelled mangroves. Forest exploitation and clearfelling are dominant causes for the destruction of mangrove ecosystems (Saenger *et al.*, 1983). Any opening in a mangrove canopy can be detected from space because corresponding pixels have been replaced either by water at high tide or by crusts of sodium chloride deposits during the dry season, at low tide.

Mangrove converted to other uses. The most conspicuous human impacts on mangrove ecosystems is their conversion to shrimp ponds (Thailand, Ecuador, Viet Nam, Indonesia, etc.) or to agriculture (mainly paddy fields in Asia and West Africa). The signals of irrigated crops, mainly paddy fields and sugarcanes, being very distinct from that of mangroves (strong absorption in the NIR band), the delineation of these conversions has become routine work from space.

Restored mangroves and afforestation areas. Recently accreted intertidal zones bearing dense vegetation often correspond to mangrove restoration sites (Field, 1996) or to afforestation activities. The monitoring of such areas has become effective from space at the mouths of the Ganges (Bangladesh) where the rate of survival and growth of *Sonneratia apetala* Buch-Ham is distinctly different from one island to another (Saenger and Siddiqi 1993; Blasco *et al.*, 1997). Another interesting example is found in the Mekong delta (Viet Nam), near the capital city Ho-Chi-Minh, where *Rhizophora apetala* Bl. has been extensively replanted. In any case, dense monospecific planted stands have a high photosynthetic activity causing high absorption of photons and low response in the wavelength 0.6–0.7 μm, which make them rather easily discriminated from space.

Accuracy assessments carried out so far for each classification algorithm or for each discriminated class, lead to extremely variable results (Hudson and Ramm, 1987). In a vast mangrove area like the Ganges (India and Bangladesh), the "dense mangrove class" which is the largest class, covering an area exceeding 6,000 km^2, may present a classification performance of 90%, although some restored mangrove areas are often assigned either to fish ponds or to rice fields or to salt marshes or even to algal deposits, because the high physiognomic diversity of planted mangroves (age of the plantation, density, species diversity, substrate, etc.) induces very diverse spectral responses. For small areas, the existing possibilities to improve the accuracy of the maps and to increase the number of discriminated classes have been described by Ramsey *et al.* (1998).

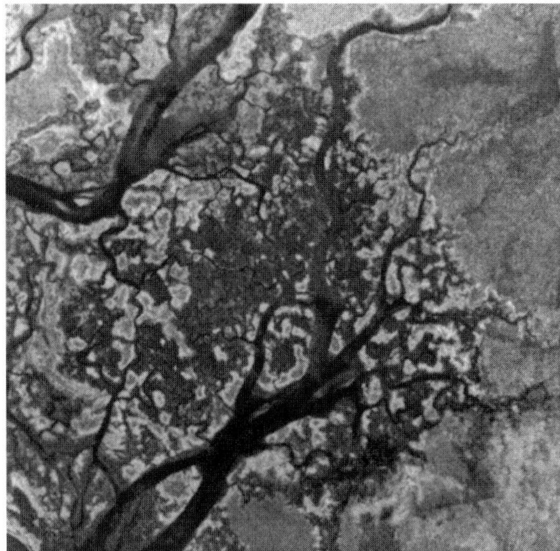

Figure M10 Mangrove studies from space (part of SPOT data KJ 022/333 dated 21/10/93). The fragmentation of the habitat appears clearly. Dark patches correspond to mangrove types. ((c) cnes 1993-distribution Spot Image.)

Conclusion and issues

According to the latest mangrove resource assessment (Aizpuru *et al.*, 2000) carried out with space products, the total extent is about 170,000 km^2 and the world's mangrove regression during the last decade has been about 1,030 km^2 y^{-1}. A critical analysis shows that remote sensing utilization for mangrove studies is still limited for several technical reasons. It appears from what has been said that the enormous commitments made to promote remote sensing technology during the last 30 years have not yet given access to the data actually needed by modern ecological researchers, that is, identification of trees, structure of stands, physical or biological stresses, sediment load and geochemistry of brackish waters, etc.

Regarding the inventory and monitoring of mangrove ecosystems using satellite data, there is no worldwide standard method which could be applied straight away. At local levels, a periodic survey of mangroves from space is operational (Figure M10). The permanent monitoring of these ecosystems, at local and regional levels, is premature.

Oil spills detection from space is by no means an easy task. Such pollution events are causing great damage every year in mangrove ecosystems of Nigeria, the Caribbean, the South China Sea, etc. either by oil tanker accidents, by illegal tanker cleaning, or by oil leaks from platforms or wells, etc. Recent advances have been achieved primarily because hydrocarbon compounds at the surface of the sea reduce water surface roughness which can be detected on SAR images.

The accuracy of each mangrove map and statistics is extremely difficult to assess rigorously. It is probably included between 50 and 90% in most studied cases. Combining data sets generated from field surveys, aerial photographs, especially color infrared, and high-resolution satellite products (TM or SPOT and SAR) lead to a more accurate appraisal of mangrove resource and local ecological conditions than does the analysis of high resolution satellite data alone. The merger of these sources of data produces classified maps, which include many features not separable solely by existing space data.

Finally, remote sensing specialists recognize that the technology has so far been less successful in coastal areas than in continental zones. The reasons lie in questions of spatial and temporal scales and on the physics of coastal signals often distorted by marine aerosols and warped by the proximity of the ocean (Cracknell, 1999). However, the synergy, between optical and SAR data and the new data provided by hyperspectral scanning systems and by very high-ground resolution tools, leave little doubt that a new breed of remote sensing is emerging.

F. Blasco, M. Aizpuru and D. Din Ndongo

Bibliography

Aizpuru, M., Achard, F., and Blasco, F., 2000. Global assessment of cover change of the mangrove forest using satellite imagery at medium to high resolution. In *EEC Research Project no 15017-1999-05 FIED ISP FR*, Joint Research Center, Ispra.

Aschbacher, J., Ofren, R.S., Delsol, J.P., Suselo, T.B., and Charrupat, T., 1995. An integrated comparative approach to mangrove vegetation

mapping using remote sensing and GIS technologies: preliminary results. *Hydrobiologia*, **295**: 285–294.
Biña, R.T., Jara, R.B., and Roque, C.R., 1980. Application of multilevel remote sensing survey to mangrove forest resource management in the Philippines. In *Proceedings of the Asian Symposium on Mangrove Development, Research and Management*, 28–29 August. University of Malaya, Kuala Lumpur.
Blasco, F., and Aizpuru, M., 2001. Depletion of the mangroves of Asia. *Wetlands Ecology and Management* **9**(3): 245–256.
Blasco, F., Gauquelin, T., Rasolofoharinoro, M., Denis, J., Aizpuru, M., and Calderou, V., 1997. Recent advances in mangrove studies using remote sensing data. *Australian Journal of Marine and Fresh Water Research*, **49**(4): 287–296.
Cracknell, A.P., 1999. Remote sensing techniques in estuaries and coastal zones. *International Journal of Remote Sensing*, **19**(3): 485–496.
Field, C.D., 1996. *Restoration of Mangrove Ecosystems*. Okinawa: ISME.
Gao, J., 1999. A comparative study on spatial and spectral resolution of satellite data in mapping mangrove forests. *International Journal of Remote Sensing*, **20**(14): 2823–2833.
Graetz, R.D., 1990. Remote sensing of terrestrial ecosystem structure: an ecologist's pragmatic view. In Hobbs, R.J., and Money, H.A., (eds.), *Remote Sensing of Biosphere Functioning*. *Ecological Studies*, Volume 79. New-York: Springer Verlag.
Green, E.P., Clark, C.D., Edwards, A.J., and Ellis, A., 1998. Remote sensing techniques for mangrove mapping. *International Journal of Remote Sensing*, **19**(5): 935–956.
Haines-Young, R., 1994. Remote sensing of environmental change. In Roberts, N. (ed.), *The Changing Global Environment*. Oxford: Blackwell Publishers, pp. 22–43.
Hill, R.A., 1999. Image segmentation for humid tropical forest classification in Landsat TM data. *International Journal of Remote Sensing*, **20**: 1039–1044.
Holben, B.N., and Fraser, R.S., 1983. Effects of atmosphere and view and illumination geometry on visible and near infrared radiance from AVHRR. *Remote Sensing of Environment*, **9**: 115–129.
Hudson, W.D., and Ramm, C.W., 1987. Correct formulation of the Kappa coefficient of agreement. *Photogrammetric Engineering and Remote Sensing*, **53**: 421–422.
Justice, C.O., Townshend, C.J., Holben, B., and Tucker, C.J., 1985. Analysis of the phenology of global vegetation using meteorological satellite data. *International Journal of Remote Sensing*, **6**: 1271–1318.
Kay, R.J., Hick P.T., and Houghton, H.J., 1991. Remote sensing of Kimberley rainforests. In Mckenzie, N.I., Johnston, R.B., and Kendrick, P.G. (eds.), *Kimberley Rainforests*. Chipping Norton: Surrey Beatty & Sons, pp. 41–51.
Kushwaha, S.P.S., Dwivedi, R.S., and Rao, B.R., 2000. Evaluation of various digital image processing techniques for detection of coastal wetlands using ERS-1 SAR data. *International Journal of Remote Sensing*, **21**: 565–579.
Malingreau, J.P., Tucker, C.J., and Laporte, N., 1989. AVHRR for monitoring tropical deforestation. *International Journal Remote Sensing*, **10**: 855–867.
Mougin, E., Lopès, A., Hery, P., Marty, G., Le Toan, T., Fromard, F., and Rudant, J.P., 1993. Multifrequency and multipolarisation on mangrove forests of French Guyana during SAREX-92 experiment. Preliminary results. In *Workshop Proceedings Sarex 92*, Paris.
Phinn, S.R., Menges, C., Hill, G.J.E., and Stanford, M., 2000. Optimizing remotely sensed solutions for monitoring, modeling and managing coastal environments. *Remote Sensing of Environment*, **73**: 117–132.
Proisy C., Mougin, E., Fromard, F., and Karam, M.A., 2000. Interpretation of Polarimetric Radar Signatures of mangrove forests. *Remote Sensing of Environment*, **71**: 56–66.
Ramsey, E.W., Nelson, G.A., and Sapkota, S.K., 1998. Classifying coastal resources by integrating optical and radar imagery and color infrared photography. *Mangroves and Salt Marshes*, **2**: 109–119.
Saenger, P., and Siddiqi, N.A., 1993. Land from the sea: the mangrove afforestation program of Bangladesh. *Ocean and Coastal Management*, **20**: 23–39.
Saenger, P., Hegerl, E.J., and Davie, J.D.S., 1983. *Global Status of Mangrove Ecosystems*. Gland: IUCN.
Simonett, D.S., 1983. The development and principles of remote sensing. In American Society of Photogrametry (ed.), *Manual of Remote Sensing*, 2nd edn. Fall Church, pp. 1–35.
Spalding, M., Blasco, F., and Field, C., 1997. *World Mangrove Atlas*. Okinawa: ISME.
Townshend, J.R.G., and Justice, C.O., 1986. Analysis of the dynamics of African vegetation using the NDVI. *International Journal of Remote Sensing*, **9**: 187–236.

Cross-references

Airborne Laser Terrain Mapping and Light Detection and Ranging
Altimeter Surveys, Coastal Tides and Shelf Circulation
Geographic Information Systems
Global Positioning Systems
Mangroves, Ecology
Mangroves, Geomorphology
Mapping Shores and Coastal Terrain
Monitoring, Coastal Ecology
Photogrammetry
RADARSAT-2
Remote Sensing of Coastal Environments
Remote Sensing: Wetlands Classification
Synthetic Aperture Radar Systems

MAPPING SHORES AND COASTAL TERRAIN

Maps of coastal features and their associated physical, chemical, and biological attributes depict spatial relationships at an instant in time. Despite the dynamic state of nearshore coastal terrain, large interior and upland areas away from the shore remain relatively stable. Maps of these relatively stable areas provide a basis for many activities ranging from interpreting the geologic history of a region to long-term planning and management of coastal resources. Maps of shores and coastal terrain continue to be the most useful and effective way of communicating spatial data to scientists, government officials, resource managers, planners, and the general public. Whether in hard copy or digital form, maps serve as the basic tool for synthesizing and presenting complex scientific information about past, present, or future environmental conditions. Maps also provide inventories and establish baseline conditions for documenting and monitoring coastal change.

Modern mapping of shores and coastal terrain evolved substantially during the 20th century. As recently as the early 1900s, classical stratigraphic nomenclature was used universally to subdivide coastal strata on the basis of named formations and their interpreted geologic ages. A consequence of this historical geologic method of mapping was that all late Pleistocene and Holocene sediments, regardless of their origin, were identified as undifferentiated Quaternary alluvium (Qal). Although classical stratigraphic coastal mapping persisted until the 1960s, earlier observations by physical geographers and geomorphologists laid the groundwork for recognizing that patterns of topography and soils could be used to interpret preserved depositional features such as ancient rivers, deltas, beaches, and dunes. By the 1930s, morphological mapping criteria were well established in the Gulf Coast region of the United States where the petroleum industry promoted detailed mapping of coastal depositional surface structures. This need-driven systematic regional mapping of coastal plain sediments in conjunction with an improved understanding of coastal processes eventually led to alternative coastal and subsurface mapping concepts including morphostratigraphy, genetic stratigraphy, and eventually sequence stratigraphy.

The sources of data used to map coastal terrain varies widely depending on the applications and anticipated needs to characterize the coastal lands and surrounding submerged areas. Continuous spatial coverage is provided by conventional aerial photographs and advanced satellite images (including radar, Advanced Very High Resolution Radiometer (AVHRR), and interferometric Synthetic Aperture Radar (SAR), whereas line or profile data are available from geophysical surveys (seismic reflection, airborne electromagentic, ground-penetrating radar), lidar missions, and Global Positioning Systems (GPS) surveys. Although point data are by definition discontinuous, most point data such as textural and geochemical attributes measured at sample sites or monitoring stations can be interpolated and mapped to provide continuous spatial coverage if the spatial distribution of discrete sample sites is sufficiently dense. Most of the recent advances in technology used as sources of data or for mapping shores and coastal terrain are discussed in other sections of this encyclopedia.

Maps of coastal terrain are presented in a variety of formats and at various scales depending on the intended use. Small- scale maps are commonly generalized and serve as executive summaries, whereas large-scale maps are normally intended as work maps because they provide the necessary detail for site-specific analyses. Before the common use of computers capable of storing and manipulating large electronic files, coastal maps were hand-prepared by skilled cartographers. The pre-electronic methods of coastal mapping and types of maps produced by state and federal agencies in the United States were summarized by Ellis (1978). Since coastal maps were printed for mass distribution, a popular format has been large atlases containing bound or unbound sheets of maps. Large format atlases are still widely used for convenience despite the high coasts of printing and problems associated with filing and storing oversize documents.

Standard techniques for comparing and displaying multiple layers and generating derivative map products now involve the use of Geographic Information Systems (GIS). Furthermore, compact disks and internet websites are used routinely for rapid dissemination and mass distribution of electronic map images.

Coastal mapping strategies

Maps of shores and coastal terrain are prepared for many different purposes. Early mapping strategies were developed primarily for scientific purposes and the newly acquired knowledge that would improve our understanding of earth history. Later, it was realized that maps of coastal terrain have exceptional societal value when applied to coastal zone management objectives. Examples of coastal mapping applications are public policies and regulations that depend on credible scientific data such as inventories of natural resources and identification of hazard prone areas.

Morphostratigraphy

Morphostratigraphy mapping of coastal terrain involves correlating surficial features by integrating physical characteristics such as landform preservation, slopes, elevations, soil composition, and degree of dissection. Generally, the purpose of morphostratigraphic mapping is to reconstruct the paleogeography and thus the geologic history of a region. An example of the morphostratigraphic approach is the map of extensive coastal plain terraces prepared by Cooke (1930). Although morphostratigraphic mapping is no longer widely used for generating coastal map products, it is a valid technique for some applications.

Genetic stratigraphy and depositional systems

Observant field geologists in Europe and North America have long recognized the genetic implications of many common sedimentary deposits (e.g., fluvial, marine, glacial). However, the concepts of genetic stratigraphy and mapping coastal depositional systems and depositional facies tracts were not formalized until the late 1960s. Genetic stratigraphy incorporates lithofacies and biofacies attributes, three-dimensional geometries, vertical sedimentary successions, and lateral facies relationships to interpret the origins of sedimentary deposits and to reconstruct the paleogeography at the time of deposition (shore position, depositional strike, and dip).

Fisher *et al.* (1972) used the advanced concepts of genetic stratigraphy to map the late Quaternary depositional systems of the Texas coast as a principal framework for environmental geologic applications. The regional environmental geologic maps of coastal Texas were published as an atlas series that included explanatory text and extensive tables of statistics for each coastal region. Within each region, maps of topography and bathymetry, active physical processes, biologic assemblages, man-made features, current land use, physical properties, and mineral resources complemented the maps of coastal depositional systems. The environmental geologic atlas focused on the upland costal terrain, whereas a companion atlas containing a series of maps of the same regions and at the same scale focused on the coastal wetlands as well as sediment textures, sediment geochemistry, and benthic organisms of the adjacent submerged lands (White *et al.*, 1983). Together, the environmental geology and submerged lands atlases represented a remarkable achievement considering that a large number of diverse color-separated maps and texts were systematically prepared and published for an entire coastal region before the advent of computer-assisted mapping and commercial development of a GIS.

Natural resources

Maps of coastal terrain are also used to delineate the natural biological and economic resources of a region. Maps of biological resources include the distribution and composition of diverse wetlands, reefs (coral and oyster), faunal assemblages, fishery habitats, and areas of environmental protection, or concern such as breeding or nesting grounds. In the United States, coastal wetlands are mapped periodically as part of the National Wetlands Inventory (Cowardin *et al.*, 1979).

Economic coastal resources can be specific mapping targets, or they can be derived from other map types, such as maps of depositional systems or coastal subenvironments. Typical mineral resources that are mapped include sand and gravel for construction aggregate and beach nourishment, shell deposits as a substitute for lime or road material, and beach placers and heavy mineral deposits that are mined for their metals and precious stones. *Mining of coastal materials* is discussed in another section of this encyclopedia by Osterkamp and Morton.

Coastal processes and natural hazards

During the latter half of the 20th century as coastal populations throughout the world rapidly increased, the need also dramatically increased for maps depicting areas threatened by natural coastal hazards such as beach erosion, storm flooding, storm-surge washover, earthquakes and liquefaction, tsunamis flooding, landslides, subsidence, and active faulting. Most developed countries have national or regional mapping projects designed to identify the areas of greatest vulnerability where most people, infrastructure, and economic development are at risk. Many state and provincial governments also have produced hazard maps designed to minimize the loss of lives and property damage associated with short-term high-energy events such as storms, tsunamis, and landslides or long-term permanent hazards such as submergence.

Baseline inventories and coastal monitoring

Maps of coastal terrain can serve either as catalogs of natural resources or as historical snapshots of dynamic conditions. As coastal maps grow older, one of their principal uses is to document changes in the status or trends of coastal features and attributes. Much of our current understanding of coastal dynamics originated as a result of comparing old coastal charts and maps with more modern depictions. Some common parameters used to document coastal changes are topography and bathymetry, *shoreline position* (defined here as the high-water line or high-tide shoreline), nearshore morphology (sea cliffs, barrier islands, tidal flats, and inlets), wetland distribution, sediment composition and texture, sediment geochemistry (trace elements, metals, pollutants), and land use. Most of these variables are subject to natural change and all of them are susceptible to changes induced by human activities.

Coastal mapping and monitoring have become routine activities of many government agencies because they promote acquisition of historical data that can be used to quantify environmental change. These activities also allow compilation of empirical data for developing and testing predictive models and forecasting future conditions. Physical and chemical parameters that are suitable for detecting and monitoring rapid environmental changes (geoindicators) have been selected and described by an international working group (Berger and Iams, 1996). The preferred coastal geoindicators of environmental change, which are shoreline position and wetlands distribution, are commonly presented on maps of coastal shores and terrain.

Coastal mapping applications

Global syntheses

During and after World War II, global mapping of shore types was a high priority because of important military applications. Of critical concern were the coastal conditions that would be encountered during amphibious operations as well as other factors such as vehicular trafficability and suitability of the coastal materials for road and airport construction. Global coastal classifications typically involve aerial photograph and map compilations that compare and contrast various attributes of the shore and adjacent land.

Most modern classifications use an integrated approach that emphasizes tectonic setting (passive or trailing margins, active or leading margins, marginal seas), morphology (coastal plains, mountains, sea cliffs), biogenic characteristics (coral reefs, marshes, mangrove forests, swamps), depositional origin (deltaic, barrier–lagoon, chenier plain, eolian, glaciated), and terrain composition (gravel, sand and mud, bedrock, ice). These classifications can be combined with climatic overprints (tropical, temperature, glacial, desert) and modifying processes that shape the landscape such as relative sea level (emergent, submergent), tide range, wave energy, wind regime, and volcanic activity to produce a wide range of map units or layers. Examples of these global coastal classifications were presented and discussed by Putnam *et al.* (1960), and Davies (1980), among others.

Hazards mitigation

A modern trend in computer-assisted mapping of coastal terrain integrates sophisticated computer applications and a GIS to generate indices or some other quantitative attribute that can be used to classify coastal terrain with respect to some anticipated hazard. Cooper and McLaughlin (1998) reviewed the purpose and techniques of 18 coastal mapping applications based on multivariate indexing. These techniques typically employ an additive or multiplicative algorithm of weighted variables or factors to generate a quantitative index. The indices can be analyzed using various multivariate statistical techniques to group the quantitative results into classes that represent levels of hazard vulnerability or risk. This type of computer-assisted quantitative analysis has the appearance of being objective and highly precise, but uncritical acceptance of the computed results can lead to erroneous scientific conclusions and policy implications. Expert levels of scientific knowledge and experience are critical to the outcome of the methods and are absolutely necessary to test the mapped results for both accuracy and reliability.

Resource protection and management

Some mapping applications are intended specifically for environmental protection and resource management. Examples are the Environmental Sensitivity Index (ESI) maps (Michel *et al.*, 1978). ESI mapping represents a conceptual advancement that recognizes different susceptibilities to environmental damage from oil spills depending on characteristics. For example, marshes and other wetlands are highly sensitive to environmental damage from oil spills, whereas concrete seawalls are not. The ESI method of classifying features has gained wide acceptance and is now a standard resource management tool used to develop contingency plans in the event of an oil spill or to minimize environmental damage during the cleanup operations of a spill.

Integrated maps and predictive models

Now that coastal change is recognized as an important societal issue, questions are asked about how much land will be lost in the future, where the shore will be at some particular time, and which communities will be flooded if sea level continues to rise. Several methods (models) have been developed to map projected shoreline positions or zones of flooding based on assumptions regarding past shoreline positions or estimated rates of future sea-level rise. Unfortunately, all the predictive models are limited because they cannot anticipate significant changes in the factors that cause or control coastal changes and therefore the forecasts may not be very accurate. Despite large uncertainties regarding the model results, some planners may want to examine maps based on model predictions because they provide at least some basis for deciding about future use and development of the coast.

Model-derived maps that forecast future coastal conditions can be either qualitative or quantitative. Qualitative predictions of coastal evolution are based on general understanding of how nearshore environments respond to changing oceanic conditions. Studies of modern coasts show that a rapid rise in sea level will cause narrowing of some barrier islands and accelerate the migration of other barriers while saltwater marshes will replace fresh and brackish water marshes. Also during a rapid rise in sea level uplands are converted to wetlands, flood plains are enlarged, and the area that would be inundated by storms of historical record are increased. These nonquantitative predictions of coastal change are useful for dramatizing what will happen in the future, but they are of little use when it comes to knowing where and when the changes will occur.

Maps displaying quantitative predictions of future coastal conditions rely on statistical models, geometric models, or numerical (deterministic) models. Even though all of these models can be used to predict future coastal conditions, they are based on completely different assumptions and analytical methods. For example, statistical models do not attempt to understand the causes of coastal change. Instead, they depend on actual observations that presumably include the important parameters that cause coastal change. Geometric models emphasize how coastal change is controlled by elevations, slopes, and shapes responding to increased water levels. Numerical models attempt to explain coastal change as a series of equations that are written to represent physical conditions and coastal processes.

Both geometric and numerical models of coastal change commonly rely on the concept of a nearshore profile that is in equilibrium with the coastal processes. Coastal engineers have suggested that offshore profiles are smooth and have a concave shape that is controlled only by the size of sand grains and the dissipation of wave energy (Dean, 1991; Bodge, 1992). Based on these and other assumptions, the generalized shape of the offshore profile is expressed as a mathematical equation (Bruun, 1962; Dean, 1991) that relates the profile shape to sediment characteristics. Investigations of offshore profiles, however, show that a single mathematical expression does not adequately represent all offshore profiles (Bodge, 1992). Pilkey *et al.* (1993) discussed the assumptions of the equilibrium profile and presented strong arguments that challenge the validity of the concept. Because an equilibrium profile does not exist at most coastal sites, they also questioned the validity of coastal change models that incorporate equilibrium profile conditions. An incomplete understanding of complex coastal processes and the lack of an equilibrium profile are the main reasons why geometric and numerical models are unable to give reliable predictions of coastal changes several decades into the future.

Maps derived from statistical models

Simple statistical models assume that coastal change in the future will be similar to that recorded in the past. Therefore, the historical record of observed changes is the best predictor of future changes. Simple statistical models also reduce the observed coastal changes to a single value, which represents the average rate of movement. Dolan *et al.* (1991) summarized the most common statistical analyses of shoreline movement and described the advantages and disadvantages of each technique. However, none of the linear time-averaging techniques used to calculate rates of change are appropriate if the historical record contains large reversals in the trend of coastal change. To accommodate irregular shoreline movement, Fenster *et al.* (1993) developed a statistical method of analyzing historical shoreline changes and determining which data should be used to predict future changes.

Maps dereived from statistical models have distinct advantages because the historical data are real and easy to obtain, and the analysis is easy to understand. These maps also present some disadvantages. The data are not broadly applicable, the analysis assumes uniform (linear) responses even though they may be irregular (nonlinear), statistical analyses can be strongly biased by data clusters and anomalous events, and physical processes summarized in coastal change records may not adequately represent future conditions. The most severe limitation of historical projections is that they are incapable of accurately predicting future responses if some condition is greatly altered. Predictions of climate change (Titus, 1988) indicate that the rate of sea-level rise will probably accelerate and other factors such as variable substrate composition, sediment influx, and storm activity could invalidate the extrapolation of even recent rates of coastal change.

Maps derived from geometric models

Maps generated from geometric models are based on the premise that coastal change is caused primarily by a relative rise in sea level. They also employ several simplistic assumptions such as a smooth, curved equilibrium profile (no offshore bars) that does not change shape as the beach retreats. Also these models allow only for onshore and offshore movement of sediment (no net alongshore movement) and they presume a water depth on the profile beyond which no sediment is eroded or deposited. These generalized assumptions must be valid in order for the models to make accurate predictions. However, none of the assumptions can be universally demonstrated with field data (Pilkey *et al.*, 1993).

Simple coastal submergence models, such as the one used by Daniels (1992), employ ground slopes, elevations, and projected sea levels to predict future shoreline positions and areas of permanent inundation. This static topographic technique, which does not account for coastal erosion or sediment transport, is used to estimate areas of inundation, potential losses of wetlands caused by flooding, or transformation of wetland types. Coastal flooding models that assume one-dimensional passive inundation may be adequate for predicting inundation and land loss around estuaries, but they are inappropriate for predicting inundation and land loss around estuaries, but they are inappropriate for predicting coastal change along ocean beaches. This is because simple submergence models may greatly underestimate the landward retreat of shores that erode as a result of sea-level rise.

Bruun (1962) presented the first and most frequently applied geometric model that graphically relates shoreline recession to a relative rise in sea level. Most numerical models employ the Bruun Rule or a similar relationship to estimate the horizontal movement of the shoreline for a particular sea-level rise scenario. The original mathematical expression of the Bruun Rule assumes (1) an equilibrium offshore profile, (2) material eroded onshore is directly deposited offshore with no gain or loss in sediment volume, (3) only cross-shore transport occurs, (4) the increase in offshore profile elevation is equal to the rise in water level so that water depth remains constant, (5) the profile shape remains unchanged as it is shifted landward and upward, and (6) there is a water depth on the profile beyond which there is no active sediment transfer. The stringent closed-system requirements of an equilibrium profile, fixed closure depth, negligible net alongshore transport, and conservation of sediment volume across the same profile cannot be met at most coastal sites.

The fundamental issue involving predictive geometric models is the shape of beach and nearshore profiles for it is this parameter that determines the horizontal displacement of the shoreline relative to an incremental rise in sea level (Bruun, 1962). According to the Bruun Rule, shoreline recession is 50–100 times the rise in relative sea level (Komar *et al.*, 1991); however, paleogeographic maps reconstructed from the late Wisconsin/ Holocene sea-level history show that shorelines actually retreated 1,500–2,500 times the vertical rise in sea level over broad continental shelves.

Some field tests have supported the general concept of the Bruun Rule (Hands, 1983) at least along coasts where profiles could rapidly equilibrate relative to the rise in sea level. However, the Bruun Rule commonly does a poor job of predicting changes at a specific site. If the Bruun Rule only approximates general erosion trends, then it may have little relevance to many map applications. Komar *et al.* (1991) recommended using large error bars with coastal predictions derived from the Bruun Rule as a reminder of the large uncertainty associated with the method.

Geometric models predict only maximum potential coastal change and therefore they are unable to accommodate such things as the time lag before equilibrium conditions are reached. Another major deficiency of most geometric models is that they fail to take into account sediment transport or its long-term equivalent, sediment budget. An exception to this general statement is the geometric model of Everts (1987), that does include gradients of alongshore sediment transport.

Maps derived from combined statistical and geometric models

Some methods of predicting coastal change combine long-term rates of change, determined by air photo methods (taken as background change), with shoreline retreat predicted by the Bruun Rule. Future sea-level scenarios, such as those forecast by Environmental Protection Agency (EPA) (Titus, 1988), provide the input for estimating probable magnitudes of sea- level rise for the period of interest. An example of the hybrid method of mapping predicted coastal response was presented by Kana *et al.* (1984) in their analysis of potential inundation at Charleston, South Carolina.

Although most coastal change models focus on coastal erosion and inundation, one model has been developed to map wetland changes as a result of predicted sea-level rise. Park *et al.* (1989) developed the SLAMM model (*S*ea *L*evel *A*ffecting *M*arshes *M*odel) to analyze what impact a long-term (>100 years) accelerated rise in sea level would have on the composition and distribution of coastal wetlands. The model starts with initial conditions (wetland classes and elevation at a particular site) then predicts future conditions in time steps by combining geometric inundation (sea-level rise scenario) with coastal erosion (Bruun Rule). Although the model does not explicitly simulate salinity changes, it does accommodate sediment accumulation as well as inland wetland migration and conversion of biotic assemblages. Results of the study (Park *et al.*, 1989) suggested that nearly half of the existing marshes and swamps in the contiguous United States would be lost to open water if sea-level rises 1 m during the next century. These models also predict a net loss of wetlands because old wetlands will be destroyed faster than new marshes can form.

Maps derived from numerical (deterministic) models

Most numerical models, also known as deterministic models, presume that coastal changes are mainly caused by wave energy. Like the geometric models, most of the deterministic models also assume a smooth, curved offshore profile that does not change shape as the beach retreats, only onshore and offshore movement of sediment, and a water depth on the profile beyond which no sediment is eroded or deposited. Most of these models also assume that sea level remains constant for the period of time that coastal change is being predicted. At many coastal sites these assumptions are either invalid or oversimplifications. Most numerical models are designed to predict changes of short coastal segments and for brief periods (less than a decade). They are intended to evaluate the effects of coastal structures on shoreline evolution or to simulate specific conditions such as storm-induced beach erosion or bathymetric changes (Kriebel and Dean, 1985). An extensive review and critique of numerical model assumptions and limitations was provided by Thieler *et al.* (2000).

Numerical models of coastal change require site-specific values for such parameters as wave climate, alonshore and crossshore sediment transport, and sediment budget. The common lack of local oceanographic and geological data coupled with the fact that nearshore hydrodynamics are nonlinear and therefore, nonadditive means that prediction confidence rapidly declines after the first few years of simulation. Subsequent simulations are further hampered by a poor understanding of nearshore physical relations, especially the relations of sediment transport to forcing events and profile recovery after storms that is necessary as a starting point for the next simulation. The result of this uncertainty is a probability distribution of shoreline positions with confidence bands that define an envelope of possible future shoreline positions. Verification of these models is also hampered by the need for detailed oceanographic data collected for the same time period as the observed shoreline changes. This generally means a short historical record when both shoreline movement and oceanographic data were available. Numerical models also rely heavily on intuition and extensive local experience of the user at the site being modeled. If the local data and engineering expertise are not available, then the model results may be erroneous.

Kriebel and Dean (1985) formulated a procedure for estimating cross-shore sediment transport resulting from the nearly instantaneous beach and dune erosion during a storm. Although this model is based on the equilibrium profile concept, it addresses the problem of maximum erosion potential not being achieved because of rapidly changing conditions during the storm. Instead, it emphasizes nearshore profile adjustment that depends on the storm surge. This model employs a generalized beach/dune profile where the dunes form the onshore limit of sediment motion. Thus, it is not applicable to overwash beaches where dunes are low or absent and surface elevations are below the storm surge elevations. The Kriebel and Dean model has some applications with regard to delineating erosion hazard zones and locating coastal structures, but it addresses only one phase of beach cyclicity and therefore, it is inappropriate for predicting long-term coastal changes.

GENESIS (Generalized Model for Simulating Shoreline Change) is a one-dimensional numerical model used to predict changes in shoreline position caused by coastal structures (Hanson, 1989). In addition to the basic assumptions of numerical models, GENESIS assumes that all sediment transport is alongshore and it does not recognize onshore and offshore sediment movement. The model can handle a shoreline up to 100 km long, but a prediction period of only 10 years. Basic input parameters are starting shoreline position, wave statistics, beach profiles and bathymetry, boundary conditions, and the configurations of engineering structures. Although GENESIS is capable of simulating longer shorelines and greater durations than most other models, it is not applicable to open-coast changes that are tidally dominated, storm-induced, or caused by water-level fluctuations. Its greatest utility is for predicting transitions from one beach stability state to another (Hanson, 1989).

Advanced mathematical models that can accurately predict coastal changes are still in the formative stages of development, because the coastal processes being simulated are complex and existing equations do not adequately describe sediment transport across the beach and offshore profile (Komar *et al.*, 1991). Furthermore, there is a general lack of field data (wave climate. wave-field transformation, nearshore currents, sediment budget, offshore bathymetry) for calibrating the models. Although some of the numerical models incorporate future magnitudes of sea-level rise, a fully three-dimensional model has not been developed that will distinguish among different pathways of coastal evolution depending on variable rates of sediment supply and sea-level rise. For example, slow rates of sea-level rise typically allow eroding barrier islands to maintain a dune ridge that retards erosion. In contrast, rapid rates of rise cause dune breaching, washover, and eventually barrier migration. During highest rates of sea-level rise the barrier is drowned in place, overstepped, and partially preserved on the inner shelf. Furthermore, the models do not adequately provide for variable sediment textures. The existing models have been developed, tested, and verified for sandy beaches but not for muddy shores despite the fact that many eroding coasts are composed of thin sand beaches overlying muddy estuarine and marsh deposits.

Hazards such as beach erosion and permanent flooding of many coastal regions were reasonably consistent and predictable before large-scale high-value economic development, because unaltered processes and the geologic framework primarily controlled them. However, post-development human modifications have caused large-magnitude imbalances in the natural forces. As a consequence of this induced disequilibrium, future predictions of coastal change will be more difficult to make and will require better quantification and incorporation of human alterations and interventions such as land reclamation, shore stabilization, and beach nourishment (Morton and McKenna, 1999).

Existing needs and future directions

In most developed or developing countries, recent increases in coastal populations and the increased levels of education and interest of coastal residents have led to a dramatic increase in the demand for coastal maps. Now, coastal regulators and property owners alike can gain easy access to the basic map products that are used to establish and enforce coastal zone management policies. While the demand for coastal maps has increased, the earth and biological sciences have evolved from being qualitative and descriptive to being much more quantitative and process-oriented. A result of these improved technical capabilities has been the generation of more quantitative maps that are used to convey rates of change or levels of risk, and to forecast future impacts of changing processes or coastal conditions.

Quantitative hazard mapping (Gornitz *et al.*, 1994; Shaw *et al.*, 1998) gives the appearance of being precise, but values of risk or vulnerability that are selected arbitrarily have no specific validity with regard to physical processes and probable future conditions. In fact, there is no scientific way of equating assigned scores or rankings to predictable coastal responses. Ordinal rankings are useful for relative comparisons of hazard vulnerability, but they are commonly subjective and depend entirely on the values assigned to the individual factors and then how the calculated scores are assigned to a particular level of risk. Clearly, there is a need for better representation of integrated hazard risk and vulnerability of coastal areas. The trend toward quantitative indices to assess coastal hazard vulnerability is commendable, but there needs to be more scientifically objective critical analyses of both the input and the results. For example, what does coastal vulnerability actually translate to: is it increased beach erosion, or increased flooding of the landform? The implied coastal responses to a particular hazard, for example, to a relative sea-level rise, are not well-defined and because of the ambiguity, the vulnerability indices are difficult to evaluate.

The ordinal rankings or scores for mapping coastal vulnerabilities are questionable for the following reasons. First, there is too much subjectivity in deciding which physical parameters to include. There are a number of parameters that seem obvious (land elevation, rates of relative sea-level change, land composition, vulnerability to storm impacts), but many other parameters are commonly included and there is no objective way of deciding how many parameters to include, and which set of parameters and algorithms provides the most accurate results. In addition, there is the problem of how to integrate each potential hazard into an aggregated value. Currently, the reported methods take parameters that are quantifiable (rates of shoreline movement or sea-level change, tide range, wave energy) and convert them to an index assuming that the cumulative interactions of the parameters are either additive or multiplicative. There is no physical theory involved in this approach, simply a forced numerical ranking. Another problem involves weighting individual parameters or groups of parameters to achieve a desired result. And finally, there is the problem of how to express the integrated value in terms of low, intermediate, or high risk. An example would be the use of statistics, such as the application of quartile statistics, to achieve a subdivision of low, moderate, high, and very high hazard risk. This approach is convenient and numerically defensible, but the results may not have anything to do with the actual responses of beaches and barriers to rising sea level and potential increased storminess in the future.

The hazard index algorithms commonly used typically yield comparable, but untenable hazard rankings for areas that are remarkably different from a geological and oceanographic perspective. For example, results of the hazard assessment by Gornitz *et al.* (1994) suggest that the Chandeleur Islands of Louisiana and the barrier island at Nags Head, North Carolina are equally vulnerable to future sea-level rise. Coastal geologists familiar with both of these areas agree that the rapidly retreating low-profile washover barriers of the Chandeleur chain are much more vulnerable to both short-term ephemeral flooding by storms as well as long-term permanent inundation by a eustatic rise in sea level than the barrier coast at Nags Head. This type of application error is a result of trying to integrate disparate data into a common coastal hazard index by using simple additive (linear) assumptions that do not accurately portray the physical processes and properties of the nearshore environments.

Rates of shoreline change are generally reported as single values without the benefit of error bars indicating the uncertainty of projected shoreline positions. Furthermore, errors associated with the predicated rates of change are magnified by as much as 60 times when they are used to map projected erosion zones (National Research Council, 1990; Morton and McKenna, 1999). Therefore, minimizing the uncertainty of these predictions should be a primary objective of coastal research.

Another limitation of many coastal maps is that they become obsolete quickly. In contrast to classical stratigraphic maps of coastal terrain that were not time-dependent, many maps prepared today for risk assessment (beach erosion, flood-prone areas), resource inventory (wetland distribution), or pollution characterization (contaminated sediments) are subjected to relatively rapid change. Because these maps can be outdated quickly, there is a need for planned periodic revision.

Maps of shores and coastal terrain are much more advanced technologically and more specific in their application now than when they were first prepared. A consequence of these improvements is the present requirement of increased scientific accuracy. In the past, mislocation of a geological contact was of academic concern, but it had essentially no impact on public policy and coastal regulations. Now maps of flood or washover hazard zones, setback lines, and rates of shoreline movement can dramatically influence potential land use and economic value of coastal property. As the severity of the consequences of coastal change in densely developed areas increases (coastal erosion, sea-level rise), so does the need for accurate scientific maps.

Robert A. Morton

Bibliography

Berger, A.R., and Iams, W.J. (eds.), 1996. *Geoindicators: Assessing Rapid Environmental Changes in Earth Systems*. Rotterdam: A. A. Balkema.

Bodge, K.R., 1992. Representing equilibrium beach profiles with an exponential expression, *Journal of Coastal Research*, **8**: 47–55.

Bruun, P., 1962. Sea level rise as a cause of erosion. *Journal of Waterways and Harbors Division, ASCE*, **WW1**: 117–55.

Cooke, C.W., 1930. Correlation of coastal terraces. *Journal of Geology*, **38**: 577–589.

Cooper, J.A.G., and McLaughlin, S., 1998. Contemporary multidisciplinary approaches to coastal classification and environmental risk analysis. *Journal of Coastal Research*, **14**: 512–524.

Cowardin, L.M., Carter, V., Golet, F.C., and LaRoe, E.T., 1979. *Classification of Wetlands and Deepwater Habitats of the United States*. U.S. Fish and Wildlife Service, FWS/OBS-79/31, 131p.

Daniels, R.C., 1992. Sea-level rise on the South Carolina coast: two case studies for 2100. *Journal of Coastal Research*, **8**: 56–70.

Davies, J.L., 1980. *Geographical Variation in Coastal Development*. London: Longman.

Dean, R.G., 1991. Equilibrium beach profiles: characteristics and applications. *Journal of Coastal Research*, **7**: 53–84.

Dolan, R., Fenster, M.S., and Holme, S.J., 1991. Temporal analysis of shoreline recession and accretion. *Journal of Coastal Research*, **7**: 723–744.

Ellis, M.Y., 1978. *Coastal Mapping Handbook*. Washington, DC: U.S. Government Printing Office.

Everts, C.H., 1987. Continental shelf evolution in response to a rise in sea level. In Nummedal, D., Pilkey, O.H., and Howard, J.D. (eds.), *Sea-level Fluctuation and Coastal Evolution*. Society of Economic Paleontologists and Mineralogists, Special Publication 41, pp. 49–57.

Fenster, M.S., Dolan, R., and Elder, J.F., 1993. A new method for predicting shoreline positions from historical data. *Journal of Coastal Research*, **9**: 147–171.

Fisher, W.L., Brown, L.F., McGowen, J.H., and Groat, C.G., 1972. *Environmental Geologic Atlas of the Texas Coastal Zone, Galveston-Houston Area*. Austin: University of Texas, Bureau of Economic Geology (one of seven atlases).

Gornitz, V.M., Daniels, R.C., White, T.W., and Birdwell, K.R., 1994. The development of a coastal risk assessment database: vulnerability to sea-level rise in the U.S. southeast. In Finkl, C. W. (ed.), *Coastal Hazards, Perception, Susceptibility, and Mitigation. Journal of Coastal Research*, **12** (Special issue): 327–338.

Hands, E.B., 1983. The Great Lakes as a test model for profile responses to sea level changes. In Komar, P.D. (ed.), *Handbook of Coastal Processes and Erosion*. Boca Raton, Fl: CRC Press, pp. 167–189.

Hanson, H., 1989. GENESIS—a generalized shoreline change numerical model. *Journal of Coastal Research*, **5**: 1–27.

Kana, T.W., Michel, J., Hayes, M.O., and Jensen, J.R., 1984. The physical impact of sea level rise in the area of Charleston, South Carolina. In Barth, M.C., and Titus, J.G., (eds.), *Greenhouse Effect and Sea Level Rise*. New York: Van Nostrand Reinhold Company, pp. 105–150.

Komar, P.D., Lanfredi, N., Baba, M., Dean, R.G., Dyer, K., Healy, T., Ibe, A.C., Terwindt, T.H.J., and Thom, B.G., 1991. The response of beaches to sea-level changes: a review of predictive models. *Journal of Coastal Research*, **7**: 895–921.

Kriebel, D.L., and Dean, R.G., 1985. Numerical simulation of time-dependent beach and dune erosion. *Coastal Engineering*, **9**: 221–245.

Michel, J., Hayes, M.O., and Brown, P.J., 1978. Application of an oil spill sensitivity index to the shoreline of lower Cook Inlet Alaska. *Environmental Geology*, **2**: 107–117.

Morton, R.A., and McKenna, K., 1999. Analysis and projection of erosion hazard areas, Galveston and Brazoria counties, Texas. In Crowell, M., and Leatherman, S.P. (eds.), 1999. *Coastal Erosion Mapping and Management. Journal of Coastal Research*, **28**(Special issue): pp. 106–120.

National Research Council, 1990. *Managing Coastal Erosion*. Committee on Coastal Erosion Zone Management, Marine Board, Washington, DC: National Academy Press.

Park, R.A., Trehan, M.S., Mausel, P.W., and Howe, R.C., 1989. Coastal wetlands in the twenty-first century: profound alterations due to rising sea level. In Davis, F.E. (ed.), *Proceedings of the American Water Resources Association Annual Conference, Wetlands: Concerns and Successes*, Tampa, FL, pp. 71–80.

Pilkey, O.H., Young, R.S., Riggs, S.R., Smith, A.W., Wu, H., and Pilkey, W.D., 1993. The concept of shoreface profile of equilibrium: a critical review. *Journal of Coastal Research*, **9**: 255–278.

Putnam, W.C., Axelrod, D.I., Bailey, J.P., and McGill, J.T., 1960. *National Coastal Environments of the World*. Los Angeles: University of California.

Shaw, J., Taylor, R.B., Forbes, D.L., Ruz, M.-H., and Solomon, S., 1998. Sensitivity of the Coasts of Canada to Sea-Level Rise. *Geological Survey of Canada Bulletin*, **505**: 79p.

Thieler, E.R., Pilkey, O.H., Young, R.S., Bush, D.M., and Chai, F., 2000. The use of mathematical models of predict beach behavior for U.S. coastal engineering: a critical review. *Journal of Coastal Research*, **16**: 48–70.

Titus, J.G. (ed.), 1988. *Greenhouse Effect, Sea Level Rise and Coastal Wetlands*. Washington, DC: Environmental Protection Agency, EPA-230-05-86-013.

White, W.A., Calnan, T.R., Morton, R.A., Kimble, R.S., Littleton, T.G., McGowen, J.H., Nance, H.S., and Schmedes, K.E., 1983. *Submerged Lands of Texas, Corpus Christi Area: Sediments, Geochemistry, Benthic Macroinvertebrates, and Associated Wetlands*. Austin: University of Texas, Bureau of Economic Geology (one of seven atlases).

Cross-references

Beach Erosion
Beach Features
Classification of Coasts (see Holocene Coastal Geomorphology)
Coastal Changes, Gradual
Coastal Changes, Rapid
Coastal Subsidence
Coastal Zone Management
Coasts, Coastlines, Shores, and Shorelines
Erosion: Historical Analysis and Forecasting
Geographic Information Systems
Global Positioning Systems
Monitoring, Coastal Geomorphology
Natural Hazards
Nearshore Geomorphological Mapping
Oil Spills
Remote Sensing of Coastal Environments
Sea-Level Rise, Effect
Wetlands

MARINE DEBRIS—ONSHORE, OFFSHORE, SEAFLOOR LITTER

Introduction

Marine debris is a problem that affects beaches/coastlines and the seafloor at all depths, and its impact is of global significance. It has been recognized as a serious pollutant for around 30 years, but has only gained widespread recognition in the past decade or so. Cleanup schemes, particularly those requiring public participation, have led to greater public awareness. This though does not appear to have led to any great reduction in the amounts of debris being found on beaches worldwide. Marine debris has been defined as "any manufactured or processed solid waste material (typically inert) that enters the marine environment from any source" (Coe and Rogers, 1997, p. xxxi). Marine debris is also often termed marine or beach litter. The sources of this form of pollution may be from the land (e.g., direct from beach users or from rivers) or from the ocean itself (e.g., from ships or offshore installations). Once in the marine environment debris may remain for many years, particularly if it is plastic and numerous worldwide beach-based debris studies have recorded plastic as the dominant material (e.g., Garrity and Levings, 1993, in Panama; Jones, 1995, in Australia). Indeed, plastics have been considered an environmental and pollution threat to the marine realm whose importance will incrementally increase through the 21st century (Goldberg, 1997). The problems created are chronic and potentially global, rather than acute and local or regional as many would contemplate.

Sources

Marine debris sources can be broadly classified into two groups: seaborne sources and land-based sources. Most concern has focused on debris discharged from vessels, but there is now extensive evidence that landborne discharges are a major source of marine debris. It has been stated that landborne sources are believed to be a much more significant contributor of pollutants to the marine environment than are vessels (Faris and Hart, 1995). Land-based debris can enter the sea through rivers, or can be blown, washed, or discharged directly from land. The absence of sewage treatment installations, the presence of combined sewer overflows, storm water discharges, run-off from landfills sited nearby rivers and in coastal areas, the absence of waste services or landfills in rural areas, recreational beach users, and fly tipping, all contribute to debris ending up on beaches or in the oceans.

Identifying from which source debris has originated is an altogether more difficult task. On occasions the source of the pollution is clear and local (Walker *et al.*, 1997), but all too often the sources are not so obvious and can be international either in terms of shipping, or land-based litter from other continents, for example, American litter on west coast European shores (Olin *et al.*, 1995). The movement patterns, sinks, and degradation rates of marine debris are still not completely understood, although there is significant recent research in this area (Williams and Simmons, 1997). In these circumstances, one cannot generalize or make assumptions about sources, site-specific measurements will almost always be required (Earll *et al.*, 2000). Possible sources, whether geographical or socio-demographic, can only be established by recording the maximum amount of detail concerning the debris item (Williams *et al.*, 1999). At present, there is no accepted methodology that enables researchers to link litter items to their source, the conceptual step taken to link litter to a source requires the following:

- the identity of the item is known or at least described systematically,
- the function and application of the item is understood, and
- that quantities of the item are measured.

Studies of why littering takes place in particular situations, for example, fly-tipping, or failure to use port reception facilities, are likely to be very important in the future as greater emphasis is placed on this aspect.

Case studies—extent of problem

The source of marine debris found on beaches worldwide varies widely. As previously stated, there are certain site-specific elements that influence the source of litter and these can differ from beach to beach. Comparisons of debris amounts are generally complicated by differences in methodology among studies, beach substrates, and environmental factors influencing the transport of debris items. Although comparisons are difficult, certain similarities can be noticed. In a survey of debris along the Caribbean coast of Panama, Garrity and Levings (1993) found that 56% of the items were made of plastic; 89% of this plastic debris being related to consumer or household goods; that the country of origin of the debris was related to distance from the survey site. Garrity and Levings (1993) concluded that (1) local household waste, (2) shipping, and (3) nearshore marine activities were the major sources of debris. They found no evidence of substantial input from industrial, recreational, or offshore commercial fishery sources.

Corbin and Singh (1993) in a study of Caribbean island coastlines showed that the amount and kind of items found were associated with types of coastal activities and variations in population density. Even though the study area was a busy lane for liners and other ships passing through the Panama Canal, little evidence was found of debris from distant sources or debris discarded from cruise ships washing up on the coast. A study of marine debris at Bird Island, South Georgia, by Walker *et al.* (1997) helps to illustrate the problem of generalizing the sources of such wastes. The findings were that the source of much of the marine debris was from local fisheries, with the majority of debris originating from jettisons by long-line fishing vessels.

Williams and Simmons (1997) conducted surveys on beaches fringing the Bristol Channel, UK, an estuarine area with relatively low levels of shipping. Very low amounts of foreign material were encountered during the study, suggesting low levels of ship discards. In contrast to this, studies carried out by the Tidy Britain Group in other parts of the United Kingdom have found that the primary sources of debris within their study area originated from shipping vessel sources. As stated earlier, comparisons between locations are difficult, any generalization about sources, persistence, and dynamics of marine litter would therefore be unwise.

Results from studies initiated in the Mediterranean showed that certain portions of this area do not have large amounts of litter emanating from seaborne or riverine sources, but rather from the high numbers of beach users. The importance of location is shown in that, "there are indications that most Mediterranean coastal litter is land-based, in contrast to the reported marine-based litter on the western European shores" (Gabrielides *et al.*, 1991, p. 437). Debris on beaches is a world-wide problem and there is no region that has escaped this form of pollution.

Problems—socioeconomic—tourism, fishing, health, aesthetic

Effects on humans

Numerous studies of beach litter have commented on the potential danger to visitors, mainly from foot lacerations caused by stepping on glass or discarded ring pull tabs (Olin *et al.*, 1995; Williams *et al.*, 2000). Other, more dangerous items have been encountered on beaches that are less expected. Munitions and containers of corrosives have been found washed ashore along with pyrotechnics, packaged hazardous goods. A further example occurred in 1993, off the coast of France, with an accident involving the ship "*Sherbo*" in which 60,000 bags of a pesticide similar to nerve gas were lost overboard (Olin *et al.*, 1995).

Attention has turned recently to the less obvious health risks that can feature on beaches. These items are medical waste and sewage-related debris. Although the risks are considered to be relatively low, any contact with infected sanitary products or fluids in syringes or other medical equipment, or ingestion of any of these could cause disease. Forty needlestick accidents on bathing beaches were reported between 1988 and 1991 to the UK Public Health Laboratory Service Communicable Disease Surveillance Centre. Medical wastes have appeared on holiday beaches and in some places sharp containers are now being issued to lifeguards, who are advised not to go barefoot on these beaches (Philipp, 1993). Studies carried out in Panama by Garrity and Levings (1993) also encountered significant levels of medical waste.

Debris on beaches is not just a health problem for those visitors who always remain on land, as bathers are also at risk. It provides information on ocean debris even though it is uncertain whether beach litter is representative of the ocean litter (Jones, 1995). It is however, the *only* realistic indicator of the amount and type of debris present in the ocean (Walker *et al.*, 1997). Sewage-related debris on a beach would seem to suggest that the adjacent waters are contaminated with sewage which means a health risk to sea users. Bathers exposed to sewage-contaminated water have a higher risk of skin and ear infections. In 1990, it was reported to the UK House of Commons that the aesthetic quality of recreational waters is becoming more important as the public becomes increasingly aware of the risks (House of Commons Environment Committee, 1990).

Public attention to problems relating to the coastal zone have been based more upon public perception than on any scientific knowledge or evaluation of sources, fates, and environmental effects. Associations have been made between the public perception of items affecting the aesthetic appearance of bathing water and bathing beaches and the gastrointestinal symptoms experienced after bathing in sewage polluted water (Nelson *et al.*, 1999). Another viewpoint could be that the public debate on sewage in bathing water has rarely made any distinction between the aesthetic impact and actual health risk.

Economic effects

The problem of litter in the environment leads not only to potential health risks, but also to economic losses. Stranded debris has direct and indirect social and economic costs to coastal communities, with the financial strains imposed by such debris not always easy to quantify, or to appreciate. Economic loss has been split into two areas, first loss to fisheries, and also loss to tourism.

Fisheries. The economic impact of debris on fishing has been studied over many years. Such losses have occurred due to the fouling of trawl nets by bottom debris, blocking of water intake pipes by plastic sheeting and, propeller foulings (Jones, 1995). Damage to ships following collisions with debris at sea have also been reported. Costs result mainly from repair of damage and lost time.

"Ghost fishing" affects commercial fishing interests. This hazard occurs as a result of lost or abandoned nets and traps, which leads to the capture of target and nontarget species. This will reduce reproductive potential, as immature fish that have not produced offspring are removed from the population (Pollard *et al.*, 1999). Large items of debris are capable of tearing nets and other fishing gear and the presence of certain debris can lead to entire catches being discarded. Data is limited as to the costs incurred due to these encounters with litter.

An extensive study carried out by Nash (1992), concerned the impacts of debris on a group of subsistence fishermen. The findings were similar to others relating to commercial fishing, including propeller entanglements, fouling of nets, damage to fishing gear. One exception was that during the gathering of shellfish and mollusks by hand, waste such as glass can lead to foot or hand injury.

An important distinction between commercial and subsistence fisherman is that even a minor decrease in yield can lead to a lack of provision for the latter with respect to basic needs, such as food. This can lead to abandonment of fishing completely (Nash, 1992). The knowledge that marine debris can cause livelihoods to be lost might be a greater spur for authorities to deal with the problem than knowing about the damage to wildlife.

Aesthetic quality, perception, and tourism. The loss of tourism and recreational potential are very real impacts of marine debris. A coastal community that relies heavily on tourism for its livelihood can have its income severely depleted by marine debris (Corbin and Singh, 1993). Perhaps the greatest impact associated with marine litter is not to organisms, but to the economic loss associated with the reduction of amenities. The money that can be made, or indeed lost, from tourism and related industries is enormous; the UK maritime leisure industry is worth £8 billion Sterling a year, with £6 billion relating to seaside holidays (Maritime Technology Foresight Panel, 1996).

The aesthetic value of beaches can be reduced by the appearance of plastics, sewage-related debris, and other items of litter. People prefer to visit clean beaches, with both land and water free of litter, rather than those containing various assortments of marine debris. The public may avoid certain beaches if they find their appearance unacceptable (Williams *et al.*, 2000). The effect of aesthetic issues on the amenity value of marine and riverine environments has been defined by the WHO as: loss of tourist days; resultant damage to leisure/tourism infrastructure; damage to commercial activities dependent on tourism; damage to fishery activities and fishery-dependent activities; damage to the local, national, and international image of a resort (Philipp, 1993). Many of these problems are manifest in developing regions such as the small island states of the South Pacific, where natural resources may be limited and economic development is largely dependent upon coastal tourism (e.g., Gregory, 1999). Particular problems lie with waste disposal and management whether it is generated on land or vessels at sea. On an atoll or small high island an ever-expanding mountain of waste is difficult, if not impossible to handle. There are sharply conflicting interests between the sophisticated demands of most tourists and the environmental degradation inflicted upon local inhabitants who also have aspirations for a better lifestyle.

When considering aesthetics, one must remember that it is usually a subjective and intangible concept. Aesthetics is a branch of philosophy concerned with the essence and perception of beauty and ugliness. Aesthetics also deals with the question of whether such qualities are objectively present in the things they appear to qualify or whether they exist only in the mind of the individual. In essence, whether objects are perceived by a particular mode—the aesthetic mode, or whether instead the objects have, in themselves, special qualities—aesthetic qualities.

With regard to marine debris and aesthetics it would seem that the perception is almost universal amongst the population and therefore, the consideration of litter is perhaps less subjective. Cause and effect relationships have been established regarding public perception and lost revenue. Beach closures along with public perception of contaminated bathing areas in 1987 and 1988 resulted in approximately US$2 billion of lost revenue for New Jersey and New York states, the losses were ascribed to debris (Rees and Pond, 1995).

As well as losses from tourism there are continual costs of beach cleanup efforts that take time and money. Cleaning the coast costs local authorities thousands of US dollars per year, additional costs are incurred when hazardous containers are found and have to be recovered from beaches. The cities of Santa Monica and Long Beach in California, USA, each spent more than US$1 million in 1988–89 to clean their beaches and costs continue to rise (Kauffman and Brown, 1991). A European example is the Swedish Skagerrack coast where more than 6,000 m^3 of litter was collected in 1993. Approximately 9,000 working days over 4–5 months with a total cost of around £1 million Sterling, gives the fiscal price of clearing marine litter at £156/m^3 (Olin *et al.*, 1995). Harbor authorities in the United Kingdom also have to pay for the costs of keeping navigational channels free from litter. At Studland, Dorset, UK, one million visitors per year along a 6 km stretch of beach results in 12/13 tonnes of litter collected weekly in the summer months at a cost of £36,000 Sterling per annum (Williams *et al.*, 2000). If any area is consistently polluted with debris then this can lead to falls in property values (Rees and Pond, 1995).

Biologic interactions

The impacts of marine debris on wildlife are generally divided into two groups: entanglement and ingestion although fouling organisms and blanketing effects also warrant serious consideration (Winston *et al.*, 1997). Entangled animals can drown, be fatally or seriously wounded, or have reduced ability to catch food, travel, or avoid predators. Ingested material can block and damage digestive tracts and reduce feeding (Jones, 1995). It is estimated that over one million birds and 100,000 marine animals and sea turtles die each year from entanglement in, or ingestion of, plastics (Faris and Hart, 1995). Of the 115 species of marine mammal, 47 have been known to become entangled in and/or ingest marine debris (Pollard *et al.*, 1999).

Entanglement

Entanglement of marine animals in debris can be broadly split into four areas:

- Large items of debris trap animals, which may result in the drowning of air-breathing species, asphyxiation of fish species that need constant movement to respire, or death by starvation or predation (Pollard *et al.*, 1999). Large or heavy pieces of debris are also liable to drag animals down.
- Smaller items of debris greatly increase drag factors. This will lead to an increased vulnerability to predators and a decreased ability to forage, which ultimately leads to starvation.
- Smaller debris items can become snagged on the seafloor trapping animals, or entangling birds and other animals on land.
- Entangled objects can tighten around the animal leading to restrictions in growth. This can lead to death or inhibit the ability to reproduce (Faris and Hart, 1995), and can also affect feeding.

The dangers to marine animals and birds caused by entanglement in man-made debris have been well documented. In areas of particularly heavy maritime traffic or where oceanic currents naturally accumulate surface material, these problems can be particularly acute (Walker *et al.*, 1997).

A study carried out by Lucas (1992) on Canadian beaches between May 1984 and September 1986 produced data on beach litter composition and entanglement of marine animals. Results found that Harbor and Gray seals were entangled on Sable Island beaches in strapping, net, rope, and other items. Of 241 Gray seal pups handled during research, 2.5% were entangled. Further findings included, seabirds tangled in trawl net, six-pack yokes, and balloon ribbons; a Sable Island horse was also found on the beach, with both hind legs entangled in a bundle of plastic strapping. The discovery of the entangled horse indicates the threat posed to terrestrial animals, as well as marine species, from marine debris.

Ingestion

The problem of ingestion appears to have attracted less attention and research than the entanglement of animal species. Plastic ingestion often leads to a less acute effect than entanglement; this could be due to the gradual accumulation of plastic debris in the guts of some animals. Some species may be able to regurgitate or excrete debris, but some plastics do not appear to pass through the intestines of certain seabirds as there is a marked absence of debris from droppings (Faris and Hart, 1995).

Epibionts, encrusters, fouling, and associated biota

Freely drifting plastic artifacts and other synthetic materials provide habitats for many opportunistic colonizers, and may act as attachment surrogates for natural floating substances such as logs, pumice, and some surface-dwelling, free-swimming larger marine animals. Studies of beach-cast plastic debris from shores of the western North Atlantic and the Southwest Pacific have revealed more than 100 epibiont and associated motile taxa (Winston *et al.*, 1997). The initial colonizers following biofilm development, are filamentous algae, hydroids, ascidians, and other soft fleshy organisms. These do not long survive desiccation and disintegration once exposed to the elements in harsh beach environments. As a consequence, the record is biased toward resistant, hard-shelled, and crustose organisms, that typically includes barnacles, bryzoans, tube worms, mollusks, foraminifera, and coralline algae, as well as some more resistant sponges and hydrozoans. Of these, the most common taxon is bryozoa with over 60 identified species represented. The extent of bryozoan cover and species diversity is latitudinally dependent. Species richness is greatest in low latitudes and decreases polewards in both hemispheres (Winston *et al.*, 1997).

The biologic communities of pelagic plastics may find side-by-side associations of related species inhabit quite different environmental niches. A single item recovered from a northern New Zealand beach hosted barnacles typical of sheltered shores (*Balanus modestus*), more exposed coasts (*Balanus trigonus*), and drifting objects (*Lepas anatifera*) and another carried a motile crab fauna represented by common algal dwellers, rocky shore taxa, and a pelagic species. There is also evidence that some other taxa may reproduce as they are buoyed along on their floating debris island (Winston *et al.*, 1997). Larger floating objects or aggregations of debris may also attract resident schools of fish, which in turn bring birds and marine predators.

There is evidence that passively drifting islands of plastic and other debris may be a vector for local, regional, and transoceanic dispersal of marine organisms and perhaps even some terrestrial ones (Gregory and Ryan, 1997). For example, the common Indo-Pacific oyster *Lopha cristagalli* has been found on a southernmost New Zealand beach attached to a tangled mass of rope, while Florida debris carried a previously unrecorded bryozoa (*Thalamoporella* sp.) similar to a Brazilian species (Winston *et al.*, 1997). It has also been suggested that some terrestrial flora and fauna elements could be picked up during a stranding episode, to be later floated off and carried away by offshore winds (Gregory, 1991). While pelagic plastics may have less potential than ballast waters for the introduction of aggressive, habitat-harming alien taxa, it is not a threat that should be ignored. Gregory (1991) suggested that alien species rafted on drifting plastic could pose threats to the biota of sensitive and/or protected nearshore environments and perhaps the delicately balanced terrestrial ecosystems of small oceanic islands. These are factors that need to be taken seriously by those having stewardship responsibilities for conservation or heritage estate. An example is Codfish Island lying a short distance offshore from Stewart Island, southern New Zealand. This is a managed refuge for a small population of a large flightless parrot, the kakapo (*Strigops habroptilus*) which is nearing extinction. The arrival of rats, mustellids, or cats on the island through rafting from the mainland some 4 km away could be disastrous for the survival of this species.

Public perception

The appearance of clear water does not necessarily mean that the water is uncontaminated, but the presence of certain items on a beach may, however, imply poor microbiological water quality. Likewise, a beach that is free from any trace of litter does not imply that the sanitary quality of the sand is good. Particular litter items attain a higher degree of emotional response within the general public than others. Sewage-related debris (SRD), and medical and hazardous items arouse greater levels of offense, or feelings of unpleasantness, than do more general items of litter such as beverage containers or confectionery wrappers. Sewage-derived debris has a greater social impact than any other aesthetic pollution environmental parameter (Williams *et al.*, 2000). The UK House of Commons Committee stated that "while the risk of infection by serious disease is small, the visible presence of fecal and other offensive materials carried by the sewerage system can mean serious loss of amenity and is therefore an unacceptable form of pollution" (House of Commons Environment Committee, 1990, p. xvii).

Methodologies

Surveys can be focused on beaches, seas, or rivers where debris is used as an indicator of oceanic, riverine, estuarine, or lake conditions (Williams *et al.*, 1999). Many studies monitoring marine debris have concentrated on specific items or categories: for example, Jones (1995) dealt with fishing debris. Other studies though have been less specific and these have assessed areas of land or water for amounts and composition of marine debris (e.g., Corbin and Singh, 1993). Beach surveys are often based on relatively small areas of study, with low numbers of surveyors involved in the collection of data. Larger-scale studies often require many more people to collect data if they are to be completed at low cost within an acceptable time frame, and not all of these surveyors can be expected to have had previous experience of carrying out litter surveys. However, the use of members of the public or local interest groups in such studies has the added value of raising public awareness and indirect education (Williams *et al.*, 1999). This can be witnessed in public participation schemes such as Beachwatch in the United Kingdom, the Campaign for Marine Conservation in the United States, and Pitch-In-Canada.

Surveys are used to determine the amount and type of debris in a specified area at a certain time and to determine how types and amounts of debris change with time. Studies may be simple enumeration studies, assessing types and litter quantities, or they can be more detailed, indicating age and origin of items. For example, Gabrielides *et al.* (1991), Corbin and Singh (1993), Jones (1995), and Walker *et al.* (1997).

Beach studies face problems in that the amount of debris is influenced by beach dynamics, oceanic circulation patterns, weather, debris characteristics, cleaning operations, and offshore recreation, and commercial practices (Faris and Hart, 1995). The many different methods employed in collecting data for beach debris surveys make result comparisons very difficult. There is as yet no single accepted methodology for assessing beach litter.

Campaigns and initiatives to combat marine debris

There are a number of campaigns and public participation schemes that aim to raise awareness and reduce the marine debris problem. Education and public awareness are key elements in the reduction of marine debris. Public involvement in beach litter management takes two forms: Direct action such as beach cleanups and monitoring; and indirect action, such as education, award schemes, and legislation. The involvement of the public in beach monitoring and cleanup programs has a dual advantage in that it allows a large sample size to be achieved, and raises awareness among society which will then translate into effective individual action to reduce litter at source. Some of the campaigns worldwide are: The Center for Marine Conservation in the United States; Coastwatch Europe; Beachwatch in the United Kingdom, run by the Marine Conservation Society; and Pitch-In-Canada. There has been some concern that where volunteers are involved in the collection of data that it can lead to spurious results. Trials by the Tidy Britain Group in the United Kingdom showed that volunteers frequently incorrectly identify litter items. An opposing view has been presented in other research (Williams *et al.*, 1999), although it has been found that particular *items* are consistently misidentified by the public, for example, cotton bud sticks (Q tips).

Beach cleaning

Beach cleanups provide a way of collecting data on the types and quantities of marine debris. Beach cleans cannot permanently solve the problem of marine debris as they do not reduce quantities at source, even though there is intense pressure to clean a beach, especially by authorities wishing to promote tourism. However, cleanups are really only applicable locally, are expensive if undertaken by mechanical means and labor-intensive. Conversely, if volunteers are employed the costs are minimal. However, cleanups *per se* do not resolve the problem if they do not address the *issues of prevention at source* and it is the links to sources that represents the future challenge (Earll *et al.*, 2000; Williams *et al.*, 2000).

There are, in essence, two methods of beach cleaning: Mechanical beach cleaning involves motorized equipment utilizing a sieve effect which scoops up sand and retains the litter, therefore, it is not selective. Most sieve machines are coarse-grained allowing items such as cigarette stubs and cotton bud sticks to pass. The use of mechanical beach cleaners may threaten the stability of some beaches, through the removal of organic matter which forms the "glue" holding sand grains together (Pollard *et al.*, 1999). The passage of such vehicles over the beach interferes with beach ecology and it is a costly method. This method though is limited in that it cannot be used on pebble beaches. The advantages of such mechanical cleanups are that it is fast and can provide an apparently pristine beach for visitors and can cover a large area. In areas with hazardous or sanitary waste it negates the need for picking up material so reducing potential health risks to individuals (Williams *et al.*, 2000). The alternative to mechanical methods is manual beach cleaning. These are often carried out where the expense of a mechanical device is prohibitive, or the substrate is not receptive to such machines. Manual cleans organized as community events on small areas may ensure that the beach is cleaned of small items missed by mechanical cleans (Pollard *et al.*, 1999).

The offshore: (pelagic marine debris)

While significant quantities of land-sourced debris and litter have been reported from harbor and inshore waters for some time, knowledge of amounts and distributions is limited. Heyerdahls's (1971) observations from the raft *Ra* on its slow drift across the equatorial Atlantic provided an initial demonstration of the extent to which surface waters were becoming contaminated by pelagic marine debris. Whether it is for coastline or high seas surveys it is convenient to separate plastic litter into four size categories (micro-litter < 1 mm; meso- 1–10 mm, mostly pellets or nibs of virgin resin; macro—mostly degradational flakes and smaller items to 10 cm; mega—larger items >10 cm). Systematic investigations to establish quantities and distribution of pelagic plastic litter have been sporadic and are based on either surface towed neuston (or pleuston) nets or have used sighting surveys from vessels on passage. The former have focused primarily on meso-litter, mostly plastic pellets or nibs, and the latter on macro- and mega-litter items identifiable with the naked eye from a vessel's deck or bridge.

There is little information available about the quantities and distribution patterns of plastic micro-litter. The source lies in some propriety hand cleaners and cosmetic preparations, and air-blast cleaning media as well as from degradation and disintegration of larger debris items. There can be little doubt that micro-litter is now globally dispersed.

Plastic meso-litter, mainly in the form of nibs or pellets of virgin polystyrene and polyethylene, has a universal presence in oceanic surface waters. The greatest densities have been noted in coastal and shelf waters off major urban and manufacturing centers—some quoted maximum pellet densities include >100,000/sq. km off the eastern seaboard of North America; >40,000/sq. km in waters of Cook Strait, New Zealand; 1,500/sq. km in the Sargasso sea; and 1,500–3,600/sq. km in the Cape Basin Region of the South Atlantic west of South Africa.

Mega-litter quantities have been reported from all marine waters since the casual and anecdotal comments made by Heyerdahl (1971) brought this problem to the fore. Distribution patterns for plastic litter in all size categories across the high seas are similar. The greatest densities, whether measured by weight or item count are to be consistently found in coastal and shelf waters adjacent to and down drift from major urban and manufacturing regions. On the open ocean, distant from land-based sources it tends to concentrate along oceanic fronts and in large eddy systems or gyres.

Concentrations of macro- and mega-litter are also present along many shipping routes particularly those of the North Atlantic and North Pacific. They are much less across the South Pacific where shipping traffic is sparser and industrial developments are fewer and distant.

The seafloor: (benthic marine debris)

The seafloor from intertidal and shallow sublittoral to outer shelf, slope, and abyssal depths has been identified as an important sink for marine debris (Goldberg, 1997). An early demonstration of this came with the recognition of plastic film accumulating on the floor of the Skaggerack by Hollström (1975). The problem is now appreciated to be a global one with many observations made by divers, through video footage from remotely operated vehicle (ROV's) as well as sampling by bottom trawls. Data has been obtained from varying depths and at many widely separated places. Latterly there have been several studies presenting substantial data on types, amounts, and distribution of marine debris on the seafloor, and although bottom trawl sampling is the preferred technique, methodologies vary, making comparisons difficult; for example, 6.5 m beam trawl pulled for 25–90 min (Kanehiro *et al.*, 1995); haul of 6 h at 3.5 knots (Stefatos *et al.*, 1999); benthic tows along a 1.85 km track (Hess *et al.*, 1999); Moore and Allen (2000), towed along isobath for 10 min at 0.8–1.2 m/s; trawl times of 5–30 min, and also estimates of densities from a submersible along tracks of 730–6,500 m (Galgani *et al.*, 2000); furthermore, in each of these studies, the categories of marine debris identified differ.

The quantities of sunken litter being reported are high. Litter densities on the seafloor of central Tokyo Bay, Japan, ranged from ca. 25,000–ca. 60,000 items/sq. km (Kanehiro *et al.*, 1995). Of this, plastics comprised 80–85% with fishing-related items between 2.7 and 9%. Quantities had not significantly changed over a four-year period (1989–93) and land-based sources were considered to be of most importance. Stefatos *et al.* (1999), recognized that marine debris concentrations on floors of the enclosed Patras and Echinadhes Gulfs, western Greece, reached 240 and 89 items/sq. km, respectively. They noted that these differences could be related to land-based sources for the former and shipping traffic in the latter. From studies of inshore waters around Kodiak Island, Alaska, Hess *et al.* (1999) showed that fisheries-related and other plastic debris quantities were greatest in inlets (20–25 items/sq. km) and least in open waters outside inlets (4.5–11 items/sq. km). These differences were considered to reflect variations in fishing effort and water circulation patterns. Moore and Allen's (2000) shelf survey of the southern California Bight, ranked quantities of anthropogenic and natural debris into four broad categories (trace, low, moderate, high) on the basis of number and weight of items determined from standardized trawl times along isobaths between depths of 20 and 200 m. Bathymetrically, the proportion of area with anthropogenic debris increased with increasing distance along a broad offshore front, from inner to outer shelf. This suggested a source that lies in disposal practices from boating activities. The most comprehensive and thorough reports are those coming from European and western Mediterranean waters (see Galgani *et al.*, 2000). Densities found were highly variable between and within separate sampling areas. Near metropolitan areas they could exceed 100,000 items/sq. km but elsewhere maximum values were lower (50,000 items/sq. km in the Bay of Biscay; 600 items/sq. km in the North Sea 200 km west of Denmark). It was also noted that concentrations of debris (to densities >50,000 items/sq. km) were encountered at depths of >2,000 m on floors of canyons along the Mediterranean coast of France. Variations in distribution patterns were attributed to geomorphologic factors, local anthropogenic activities, and land-based river inputs.

Mechanisms by which the mostly neutrally buoyant plastics in marine debris reaches the deep-seafloor are poorly understood. Significant quantities of land-sourced materials on submarine canyon floors to considerable distances offshore, suggest rapid transport through nearshore zones and entrainment in bottom hugging currents. Density increases following rapid and heavy fouling may be sufficient to permanently sink them. On the other hand, grazers may clean covered surfaces leading to "yo-yo like" episodes of submergence and resurfacing until permanent settlement to the seafloor is effected. As well as biofilm development, plastic sheeting may also attract nonliving detritus, which with photodegradation and progressive embrittlement leads to density increases taking it to the seafloor without the need for invoking downwelling and/or entrainment.

The epibionts of benthic plastic debris are not as well known as those of pelagic items. Accounts are limited (e.g., Hollström, 1975) but indicate a hard ground biota characterized by bryozoans, sponges, and foraminifera, with barnacles, mollusks, and polychaetes is typical. At shallow, photic zone depths, there is development of crustose (coralline) red algae as well as soft brown and green algae. Bryozoa are generally the dominant epibiont of both pelagic and benthic plastics.

Plastic sheeting together with larger, more solid items and discarded fishing gear is an undesirable addition to the deep-seafloor and potentially damaging to the environment (Goldberg, 1997). The blanketing effects of sheeting may damage biotas of both soft sediment and rocky hard ground substrates at all depths from intertidal to the abyss. They may lead to anoxia and hypoxia induced by inhibition of gas exchange between pore water and seawater (Goldberg, 1997). Ironic as it may seem, could benthic plastic debris standing above the seafloor enhance or enrich local biotic diversity in the short term, for in the long term it is doomed to permanent internment in a slowly accumulating sediment cover?

Degradation

Breakdown of plastics mainly takes place through photodegradation which leads to surficial cracking followed by embrittlement and ultimately complete disintegration into powder. Biodegradation is seldom important with most plastics that enter the marine realm. Physical abrasion is also a mechanism for the breakdown of plastics along coastlines—particularly high energy cliffed and rocky shores. Degradation performance is generally measured through changes in tensile strength and viscosity although UV and laser spectroscopy are other approaches. Several studies have shown that the rates of weathering of polyethylene and other plastics are substantially reduced when floating in seawater compared to those when exposed outdoors to normal atmospheric conditions. Enhanced photodegradable polyethylene also degrades more slowly under marine conditions. Alternatively, expanded polystyrene foam is known to deteriorate more rapidly in seawater than on atmospheric exposure. Material that has been buried for some time in beach/riverine sediments retains much of its tensile strength and may be exhumed during episodes of erosion (Gregory, 1999). Plastics sinking to the deep seafloor will not be subject to photodegradation and if resistant to biodegradational processes will be preserved there until burial is completed.

A.T. Williams, D.T. Tudor and M.R. Gregory

Bibliography

Coe, J.M., and Rogers, D.B. (eds.), 1997. *Marine Debris: Sources, Impacts and Solutions*. New York: Springer-Verlag.

Corbin, C.J., and Singh, J.G., 1993. Marine debris contamination of beaches in St. Lucia and Dominica. *Marine Pollution Bulletin*, **26**(6): 325–328.

Earll, R.C., Williams, A.T., Simmons, S.L., and Tudor, D.T., 2000. Aquatic litter, management and prevention—the role of measurement. *Journal of Coastal Conservation*, **6**(1): 67–78.

Faris, J., and Hart, K., 1995. *Seas of Debris: A Summary of the Third International Conference on Marine Debris*. North Carolina Sea Grant, UNC-SG-95-01.

Gabrielides, G.P., Golik, A., Loizides, L., Marino, M.G., Bingel, F., and Torregrossa, M.V., 1991. Man-made garbage pollution on the Mediterranean coastline. *Marine Pollution Bulletin*, **23**: 437–441.

Galgani, F., Leaute, J.P., Moguedet, P., Souplet, A., Verin, Y., Carpentier, A., Goraguer, H., Latrouite, D., Andral, B., Cadiou, Y., Mahe, J.C., Poulard, J.C., and Nerisson, P., 2000. Litter on the seafloor along European coasts. *Marine Pollution Bulletin*, **40**: 516–527.

Garrity, S.D., and Levings, S.C., 1993. Marine debris along the Caribbean coast of Panama. *Marine Pollution Bulletin*, **26**(6): 317–324.

Goldberg, E.D., 1997. Plasticizing the sea-floor: an overview. *Environmental Technology*, **18**: 195–202.

Gregory, M.R., 1991. The hazards of persistent marine pollution: drift plastics and conservation islands. *Journal of the Royal Society of New Zealand*, **21**: 83–100.

Gregory, M.R., 1999. Plastics and South Pacific Island shores: environmental implications. *Ocean and Coastal Management*, **42**: 603–615.

Gregory, M.R., and Ryan, P.G., 1997. Pelagic plastics and other persistent synthetic debris: a review of Southern Hemisphere perspectives. In Coe, J.M., and Rogers, D.B. (eds.), *Marine Debris: Sources, Impacts, and Solutions*. New York: Springer, pp. 49–66.

Hess, N.A., Ribic, C.A., and Vining, I., 1999. Benthic marine debris, with an emphasis on fishery-related items, surrounding Kodiak island, Alaska, 1994–1996. *Marine Pollution Bulletin*, **38**: 885–890.

Heyerdahl, T., 1971. *The 'Ra' Expeditions*. London: George Allen and Unwin Ltd.

Hollström, A., 1975. Plastic films on the bottom of the Skagerrak. *Nature*, **255**: 622–623.

House of Commons Environment Committee. 1990. *Fourth Report. Pollution of Beaches*, Volume 1. London: HMSO.

Jones, M.M., 1995. Fishing debris in the Australian marine environment. *Marine Pollution Bulletin*, **30**(1): 25–33.

Kanehiro, H., Tokai, T., and Matuda, K., 1995. Marine litter composition and distribution on the sea-bed of Tokyo Bay. *Fisheries Engineering*, **31**: 195–199.

Kauffman, J., and Brown, M., 1991. California marine debris action plan. In Magoon, O.T., Converse, H., Tippie, V., Tobin, L.T., and Clark, D. (eds.), *Coastal Zone '91. Proceedings of the 7th Symposium on Coastal and Ocean Management*. Long Beach, CA, July 8–12, 1991, New York: ASCE. USA, pp. 3390–3406.

Lucas, Z., 1992. Monitoring persistent litter in the marine environment on Sable island, Nova Scotia. *Marine Pollution Bulletin*, **24**(4): 192–199.

Maritime Technology Foresight Panel, 1996. *Report of the Working Group on Coastal Waters and Maritime Leisure*. pp. 3–4.

Moore, S.L., and Allen, M.J., 2000. Distribution of anthropogenic and natural debris on the mainland shelf of the southern California Bight. *Marine Pollution Bulletin*, **40**: 83–88.

Nash, A., 1992. Impacts of marine debris on subsistence fishermen—an exploratory study. *Marine Pollution Bulletin*, **24**(3): 150–156.

Nelson, C., Williams, A.T., Botterill, D., Rees, G., and Richards, C., 1999. Beach health risk assessment and pollution perception. In Stephen T. Trudgill, Des E. Walling, and Bruce W. *Water Quality: Processes and Policy* Webb (eds.), New York: John Wiley and Sons Ltd, pp. 65–72

Olin, R., Carlsson, B., and Stahre, B., 1995. The west coast of Sweden—the rubbish tip of the north sea. In Earll, R.C. (ed.), *Coastal and Riverine Litter: Problems and Effective Solutions. Coastal Management for Sustainability*. Candle Cottage, Kempley, Glos., UK, pp. 12–18.

Philipp, R., 1993. Community needlestick accident data and trends in environmental quality. *Public Health*, **107**: 363–369.
Pollard, S., Davis, C., and Harris, E., 1999. *Beachwatch '98. Nationwide Beach-Clean and Survey Report*. Ross-on-Wye, England: The Marine Conservation Society.
Rees, G., and Pond, K., 1995. Impacts: aesthetics, health and physical clearance. In Earll, R.C. (ed.), *Coastal and Riverine Litter: Problems and Effective Solutions. Coastal Management for Sustainability*. Candle Cottage, Kempley, Glos., UK. pp. 5–7.
Stefatos, A., Charalampakis, M., Papatheodorou, G., and Ferentinos, G., 1999. Marine debris on the seafloor of the Mediterranean Sea: examples from two enclosed gulfs in western Greece. *Marine Pollution Bulletin*, **36**: 389–393.
Walker, T.R., Reid, K., Arnould, J.P.Y., and Croxall, J.P., 1997. Marine debris surveys at Bird Island, South Georgia 1990–1995. *Marine Pollution Bulletin*, **34**(1): 61–65.
Williams, A.T., and Simmons, S.L., 1997. Estuarine litter at the river/beach interface in the Bristol Channel, UK. *Journal of Coastal Research*, **13**(4): 1159–1165.
Williams, A.T., Pond, K., Tudor, D.T., Jansen, H., and Liu, H.B., 1999. The robustness of litter transect data collection by different survey groups. In Özhan, E. (ed.), *Proceedings of the MEDCOAST 99-EMECS 99 Joint Conference: Land Ocean Interactions—Managing Coastal Ecosystems*. 9–13 November 1999, Antalya, Turkey: MEDCOAST, Middle East Technical University, Ankara, Turkey, pp. 715–725.
Williams, A.T., Pond, K., and Phillipp, R., 2000. Aesthetic aspects. In Bartrum, J., and Rees, G. (eds.), *Monitoring Bathing Waters*, London & New York. Chapter 12. E & F N Spon, pp. 283–311.
Winston, J.E., Gregory, M.R., and Stevens, L.M., 1997. Encrusters, epibionts, and other biota associated with pelagic plastics: a review of biogeographical, environmental and conservation issues. In Coe, J.M., and Rogers, D.B. (eds.), *Marine Debris: Sources, Impacts and Solutions*. New York, Springer. pp. 81–97.

Cross-references

Aquaculture
Cleaning Beaches
Coastal Zone Management
Economic Value of Beaches
Environmental Quality
Natural Hazards
Rating Beaches
Tourism and Coastal Development
Water Quality

MARINE PARKS

There is a wide and growing range of marine parks. They include designated natural marine areas, such as the coral reefs off the coast of Al Fujayrah in Abu Dhabi, which were declared the country's first marine parks in 1995, the Great Barrier Reef, and artificial tourist facilities, such as Seaworld California in San Diego and Marineland Canada near Niagara Falls. The latter often include captive marine life, including marine mammals, as well as recreational features and structures such as water-slides. They attract large numbers of visitors and for many people are their first introduction to the marine world. They are, however, controversial because of their use of captive mammals, such as dolphins, although concern for animal welfare means that the best operate with very high standards of care and carry out important research. Some natural marine parks use underwater observatories to provide dry access to the undersea world, for example, in the Marine Park and Underwater Observatory at Coral World on the northeast coast of St. Thomas in the US Virgin Islands. The majority, however, depend upon boat and diving access to the marine world, for example, in the Pulau Weh Marine Park in western Indonesia and the Malindi Marine Park on the coast of Kenya.

The terminology used to describe Marine Parks varies from country to country depending upon the specific legal designation used, and the term "Marine Park Area" is also commonly used. Many marine parks also form parts of a worldwide pattern of Marine Protected Areas (MPAs), which include a wide range of legislative and conservation practice. However, not all marine parks are marine protected areas and not all marine protected areas form parts of marine parks.

Historical background

Although the First World Conference on National Parks held in 1962 expected governments of coastal countries to address the creation of marine parks or reserves to protect underwater areas as urgent, the response was very mixed, and in many countries very little action occurred.

Ray (1976) suggested that marine areas could be reserved to achieve several objectives:

1. Conservation of habitat
2. Protection of species habitat
3. Conservation of important breeding areas
4. Conservation of aesthetic values
5. Protection of cultural and archeological areas
6. Provision of sites for interpretation, education, tourism, and recreation
7. Provision of areas for research
8. Provision of areas for monitoring the effects of human activities, and
9. Acting as areas in which to train personnel in protected area management.

In 1988, the International Union for the Conservation of Nature (IUCN) defined a MPA as "any area of intertidal or sub tidal terrain, together with its overlying water and associated flora, fauna, historical and cultural features, which has been reserved by law or other effective means to protect part or all of the enclosed environment." The IUCN Commission on National Parks and Protected Areas (1994) recognized that protected areas were managed for six main reasons:

1. Strict protection (i.e., strict nature reserve or wilderness area, e.g., Laut Banda, Indonesia);
2. Ecosystem conservation and recreation (i.e., National Park e.g., Ras Mohammed National Park, Egypt);
3. Conservation of natural features (i.e., Natural Monument);
4. Conservation through active management (e.g., habitat or species management area, e.g., Galapagos marine reserve, Ecuador);
5. Landscape or seascape conservation and recreation (protected landscape or seascape, e.g., Northern Sporades, Greece);
6. Sustainable use of natural resources (managed resource protected area, e.g., Kiunga Marine National Reserve, Kenya).

IUCN proposed that marine parks should be large (more than 1,000 ha), and have both high marine conservation value and high recreational potential. Educational and recreational use in these areas should be controlled at a level which ensured that the area's natural or near natural state is maintained. In contrast, marine reserves should be areas which have some outstanding ecosystem feature and/or species of flora or fauna of natural or scientific importance or they should be representative of a particular marine area. They may contain fragile ecosystems or life forms, be areas of biological or geological diversity, or be of particular importance for the conservation of genetic resources. These reserves should be large enough to ensure the integrity of the area, to protect communities and species, and to maintain natural processes. However, IUCN also expects that they can accommodate a wide range of human activities which are compatible with their primary goal in marine and estuarine settings (Box M1).

When MPAs are selected, their special importance is usually because (Norse, 1993) they are:

1. areas of high diversity;
2. areas of high endemism;
3. areas of high productivity;
4. spawning areas that serve as a source of recruits;
5. nursery grounds; and
6. migration stopover points and bottlenecks.

Thus, although there is strong support for the co-establishment of marine parks and MPAs, the reasons for the designation of MPAs can be in conflict with the expectation that marine parks are for recreation as well as conservation. Furthermore, as Gibson and Warren (1995) stress, legislative requirements for the creation of MPAs depend upon a combination of national and international law. For example, the Law of the Sea Convention (articles 192 and 194) imposes an "obligation on States to protect and preserve the marine environment and requires

Box M1 IUCN recommends that a national system of marine protected areas should have the following objectives

- To protect and manage substantial examples of marine and estuarine systems to ensure their long-term viability and to maintain genetic diversity
- To protect depleted, threatened, rare or endangered species and populations and, in particular, to preserve habitats considered critical for the survival of such species
- To protect and manage areas of significance to the life cycles of economically important species
- To prevent outside activities from detrimentally affecting the MPAs
- To provide for the continued welfare of people affected by the creation of MPAs
- To preserve, protect, and manage historical and cultural sites and natural aesthetic values of marine and estuarine areas, for present and future generations
- To facilitate the interpretation of marine and estuarine systems for the purposes of conservation, education, and tourism
- To accommodate within appropriate management regimes a broad spectrum of human activities compatible with the primary goal in marine and estuarine settings
- To provide for research and training, and for monitoring the environmental effects of human activities, including the direct and indirect effects of development and adjacent land-use practices.

Based on Kelleher and Kenchington (1992, p. 9)

them to take action to prevent, reduce or control pollution from any source." In addition, there are obligations under the International Convention for the Prevention of Pollution from Ships 1973 (MARPOL) and the Ramsar Convention (1971), and through the International Maritime Organisation and the UNEP Regional Seas Programme to identify areas which are sensitive and should be avoided by shipping. It is not surprising, therefore, that there is so much variety amongst the world's marine parks.

Marine parks in practice

Practice varies from country to country, but many of the fundamental principles for marine park conservation and management were developed in the Great Barrier Reef of Australia. Different approaches were adopted by Japan, but both countries provide good examples of approaches to marine park establishment.

Australia's Great Barrier Reef Marine Park Act 1975 is one of the earliest examples of applying the conservation philosophy of the World Conservation Strategy. The Great Barrier Reef is the world's largest system of corals and associated life forms. It extends along the north-east coast of Australia for almost 2,000 km from the Tropic of Capricorn to about 9°S. It is a complex maze of 2,500 individual reefs ranging in size from under 1 ha to over 100 sq. km. The Reef Region exceeds 350, 000 sq. km (135, 000 sq. miles)—an area slightly larger than Poland. In the south, the Reef is wider than 100 km and is characterized by patch reefs separated by narrow winding channels or open water. To the north, the Reef is much narrower, occasionally less than 30 km, with a series of linear ribbon reefs at its eastern edge.

The Great Barrier Reef has a long history of use for fishing, with Australian aboriginal fishermen using it 15,000 years ago. With the arrival of Europeans, it became widely used for commercial fishing, including pelagic and demersal fish, prawns, scallops, turtles, and beche de mer. In addition, it is a prime location for scientific research and for a wide range of recreational activities. Between 1960 and 1983 the number of trawlers fishing within the Reef region grew from 250 to 1,400 with a turnover of about $(A) 40 million in 1983 (Kelleher, 1985). Tourism is a very important economic activity throughout the region with more than two million visitor trips to the Reef, island, and adjacent mainland in the 12 months ending in March 1980 (Kelleher, 1985). By 1997, the annual value of tourism was estimated to exceed $(A)1 billion and the direct economic value of commercial fisheries at about $(A)200 million. There are an estimated 24,300 privately registered boats involved in recreational fishing as well. In 1997, the Great Barrier Reef Marine Park Authority (GBRMPA) recorded 1.6 million visitor-days. Accessibility has improved greatly as high-speed catamarans with speeds between 25 and 35 knots have brought the Reef within 1 hr travel time from mainland towns.

The Great Barrier Marine Park Act 1975 (Section 7(1)) established the GBRMPA with the following functions (Kelleher, 1985):

(a) To make recommendations to the Minister in relation to the care and development of the Marine Park including recommendations from time to time, as to
  (i) the areas that should be declared to be parts of the Marine Park and
  (ii) the regulations that should be made under this Act
(b) To carry out, by itself or in co-operation with other institutions or persons, and to arrange for any other institutions or persons to carry out, research and investigations relevant to the Marine Park;
(c) To prepare zoning plans for the Marine Park in accordance with Part V;
(d) Such functions relating to the Marine Park as are provided for by the regulations;
(e) To do anything incidental or conducive to the performance of any of the foregoing functions.

The Park is divided into six Sections, each with a zoning plan, which aims to allow any reasonable activity, but is designed to ensure that

1. the natural qualities of the Park are conserved for users, both today and in the future;
2. suitable areas are identified where reasonable uses are permitted;
3. incompatible activities are separated; and
4. these activities do not cause unacceptable damage (Kelleher, 1985).

GBRMPA followed five phases in establishing zoning plans for each Section of the park (Kelleher, 1985):

1. All the available information on the section in industry and government reports and the scientific literature and maps was collected, collated, and studied.
2. A public participation program, using both specially prepared information as well as extensive use of the media, advised of the intention to prepare a zoning plan and sought information from the public.
3. A draft zoning plan was prepared using both the information review and the information provided by the public.
4. A second public participation program using the same approaches was used to gather comments on and reactions to the draft zoning plan.
5. After taking all comments received into account the final zoning plan was submitted to the Minister to seek approval before laying the plan before Parliament. After 20 days without any motions against the plan in either House of the Australian Parliament, the Minister published the date upon which the zoning plan would come into force. The first zoning plan, for the Capricornia Section covering over 11, 800 sq. km (4,558 sq. miles), went into force in July 1981.

The GBRMPA thus came into being after extensive public consultation and was firmly established within the legal structures of Australia. Marine Parks need to have both the support of the wide variety of stakeholders in them and it is essential that appropriate legal arrangements are in place to maintain their protected status. However, the legal processes used may not be consistent with the customary practices of indigenous peoples who live and use them. Their interests must be safeguarded, not least because they are often custodians of information about the ecology and heritage of these sites.

Monitoring and research are very important for both the establishment of Marine Parks and their sustainable management. The GBRMPA identified three information and research needs (Kelleher, 1985):

1. Information management, to ensure that information is available for analysis and interpretation.
2. Analysis of use to understand how the area is used, the effects of these uses on physical, biological, and economic processes, and their intensity, patterns, and rates of change. Changes which may be due to external factors such as sea-level rise or climate change and effects which are related to the uses themselves can be identified.
3. Resource analysis building up an inventory of the physical, chemical, human, and biological processes.

This approach to resource analysis, use analysis and information management has provided a model for the approach to marine parks. It acknowledges that marine parks are valued economic and social resources where the use of the resources must be monitored, managed, and if necessary changed to ensure the longer-term survival of the park's habitats and economy. The process of marine park management is a continuous one and in July 1998 the GBRMPA was restructured in order to strengthen its focus upon four major critical issues:

1. fisheries;
2. tourism and recreation;
3. water quality and coastal development; and
4. conservation, biodiversity, and World Heritage.

These four themes have increasingly become concerns for many marine parks and protected areas, for sometimes they are victims of their own success. As they attract visitors, so the pressures both within the parks and along adjacent coasts have grown. At the same time, there has been a growing recognition that management of the water quality of these areas depends upon management of the catchments that drain toward them. The emphasis has shifted toward more holistic management of the coastal zone within which the marine parks are sited.

Whereas, the GBRMPA approach applies to a specific very large feature, the Japanese approach is designed as a frame-work for the development of marine parks, many of which are small in comparison, throughout Japan. In 1931, Japan passed a National Parks Law to allow the creation of parks for protection of scenery and for recreation (Marsh, 1985). Between 1934 and 1936, 12 areas were designated, including some coastal sites, the largest being the Inland Sea (Seto Naikai). The Natural Parks Law of 1957 defined the purpose of parks as

> the protection of places of scenic beauty, and also, through the promoted use thereof, as a contribution to the health, recreation and culture of the people.

The Law provided for three types of parks: National Park, Quasi-National Parks, and Prefectural Nature Parks. By the mid-1980s, Japan had 27 National Parks, 54 Quasi-National Parks, and 297 Prefecture Nature Parks, many of which were coastal in location. Following the First World National Parks Conference in 1962 which drew international attention to marine parks, the Nature Conservation Society of Japan (NCSJ) set up a marine parks investigation committee in 1964. In 1966, the Ministry of Health and Welfare financed the investigation and in 1967, the NCSJ committee became an independent foundation: the Marine Parks Center of Japan. In 1970, the Natural Parks Law was amended to allow the creation of marine parks within national parks. Ten were designated immediately and a further 12 in 1971. By 1985, there were 23 Marine Parks adjacent to 10 National and 13 Quasi-National Parks. They included 57 Marine Park Areas (special zones within marine parks) totaling 2,387 ha and ranging in size from 3.6 ha to over 233 ha. Marine Parks aim at "preserving beautiful underseascapes." Criteria for selection required (Marsh, 1985) that:

1. both land and sea areas surrounding the Marine Park Area are designated as National or Quasi-National Park and that nature conservation on land can be fully ensured;
2. the seabed topography is typical of the area and undersea fauna and flora abundant;
3. seawater should be transparent and unlikely to become turbid or polluted;
4. water depths are not more than 20 m;
5. there are slow currents and the park is sheltered from violent waves;
6 on land, there should be enough space to construct such facilities as landing piers, rest houses, visitor centers, and car parks;
7. there should be coordination with local fisheries concerning their use of the Marine Park, and;
8. the slightest risk of destruction of under-seascapes by all kinds of industrial exploitation should be prevented.

A marine park is defined as the total area comprising three zones:

1. Ordinary Area—including that part of the park within 1 km of the coastline.
2. MPA: the areas within the park having specific features warranting the most protection.
3. Ordinary or buffer zone including the area within 1 km of the MPA.

In practice, most Japanese marine parks are divided into two zones: the Ordinary Area zone including the sea within 1 km of the MPA or the coastline and the MPA core zone.

For some countries, the criteria are simpler, but explicit. For example, the 38 Malaysian marine parks were established in order to protect and conserve (1) the marine ecosystem, especially the coral reefs, for the management of the fisheries resources in the coastal waters in order to maintain/increase fish landings and (2) the coral reefs for research on biodiversity, and for the purposes of education and recreation/eco-tourism. There is thus a very strong emphasis on the economic importance of these areas for fisheries and tourism.

In contrast, in some regions where mass-tourism was established early and has ceased to grow rapidly, marine protected areas may be an effective way to restore damaged environments. For example, after tourism to Spain slowed down in the 1990s, it decided to establish a protected area every 30 km along the coast to ensure the maintenance of marine and terrestrial fauna and flora. Many sites that have been partly damaged could be restored, in particular around the main areas of interest for tourism (Gubbay, 1995). Spain has created some 25 protected areas on the Mediterranean coastline, 6 related to the coast or sea. MPAs come under the jurisdiction of the Ministry of Agriculture, Fisheries and Food and include a wide range of sites both in reasons for designation and in size (Table M6).

In contrast, Finland has designated MPAs within the context of the needs of a Regional Sea: the Baltic. Finland is a signatory to the Helsinki Convention that requires its members to protect the marine environment and biodiversity (HELCOM, 1996). Both marine and coastal biotopes are included within the framework of HELCOM initiatives for the protection of species, habitats, and ecological processes including the establishment of coastal and marine Baltic Sea Protected Areas (BSPAs). This is unusual in that it recognizes that MPAs are often natural units which cross national marine boundaries and that cooperation within Regional Seas is essential. Finland has eight proposed BSPAs, most of which are already protected as National Parks. For example, the Archipelago National Park, established in 1983, forms the core of a larger Archipelago Sea Biosphere Reserve established by UNESCO in 1994. It consists of around 1,000 islands and islets with important bird areas forming clusters separated by large expanses of open water. Although it is open to the public, there are restrictions, including prohibition of landing, especially during the nesting season.

Italy has designated more than 70 marine and coastal protected areas, including Marine Natural Reserve (MNR), Nature Reserve (NR), Special Protection Area (SPA), Fisheries Reserve, Biogenic Reserves, Recreation Parks, State Reserves, Marine Reserves and Parks, and Marine Sanctuaries. The designations take into account such treaties as the Mediterranean Action Plan and the Ramsar Convention (WCMC, 2001). As well as subscribing to internationally recognized initiatives, due to its position sharing borders and waters with seven states, Italy

Table M6 Spanish MPAs and other Spanish-protected areas of international importance

MPAs (Mediterranean coast)	Biosphere Reserves	World Heritage Site
Medas Islands Marine Reserve (40 ha)	Donaña National Park	Salinas de Ibiza y Formentera Nature Reserve
S' Arenal Regional Protected Landscape (400 ha)	Island of Minorca	
Tabarca Marine Reserve (1463 ha)	(including protection of the sea adjacent to protected core areas)	(protecting oceanic *Posidonia* sea grass beds)
Columbretes Natural Park and Marine Reserve (5,766 ha)		
Archipelago de Cabrera National Park (Balearic Islands) (10,000 ha)		
Cabo de Gata Nature Park and Marine Reserve (26,000 ha)		

Source: World Conservation Monitoring Centre (WCMC).

has agreements with them for protection of the marine and coastal environment. For example, an international marine reserve was established between Sardinia and Liguria (Italy), Corsica (France), and Provence (France and Monaco) in 1993. There is thus considerable variety in the approach taken within Europe as areas that have often been exploited for centuries are given protection and restoration is undertaken.

In North America, both the United States and Canada have identified a wide range of areas which should be designated as MPAs, including marine parks. MPAs are frequently protected primarily to conserve important biological or historical features, but many are also marine parks although they are not so designated because they allow access for recreational activities. In Michigan, for example, the Great Lakes Bottomland Preserves are designated in order to protect shipwrecks of historical importance. The Michigan Department of Natural Resources Underwater Salvage Committee has emphasized that, whereas underwater parks would provide physical facilities for recreational divers such as slipway access and interpretation programs, bottomland preserves are intended to be set-aside areas where little or no salvage will be allowed and they are recognized as areas of distinctive historic and/or recreational interest.

The United States has a very large number of MPAs, but only one area specifically designated as a Marine Park (Julia Pfeiffer Burns MP in California) and there are 13 State Marine Parks (2 in Washington and 11 in Alaska). There is a single Aquatic Park at Estero Bay in Florida. This is not uncommon. South Africa, for example, has only two Marine Reserves and the United Kingdom has no Marine Parks, although it has a very large number of MPAs.

Canada has a long-established system of national parks, the first being established in 1885 in part of what is now the Banff National Park. However, its marine national park system only came into effect in 1986 with the National Marine Parks Policy. In contrast to many countries, Canada has based its marine parks system on the fundamental principle that it should protect the country's major biogeographical provinces, specifically the Arctic, Atlantic, and Pacific marine environments and the Great Lakes, by sampling their representative, outstanding, and unique characteristics. This is based upon a broad-scale hierarchical system of biogeographic units or "marine regions" (Mondor, 1992). In establishing the marine parks, the emphasis has been upon ensuring that there is a steady movement toward each marine region being represented: for example, the St. Lawrence River Estuary marine region is represented by the Saguenay Marine Park. For countries that are working toward establishing marine parks, this well-tried approach is recommended by IUCN, but for many less-developed countries, there is neither the funding nor information to carry out such a comprehensive approach. Establishment of marine parks may be as a result much more ad hoc.

The variety of existing practice in selection of MPAs has led IUCN to draw upon best practice and experience to identify criteria for deciding upon areas within MPAs or for determining their boundaries (Box M2).

Box M2 IUCN criteria for inclusion within MPAs and delimiting boundaries.

- Naturalness
- Biogeographic importance
- Ecological importance
- Economic importance
- Social importance
- Scientific importance
- International or national significance
- Practicality/feasibility

Seven-stage sequence of decision-making in establishing and managing an MPA

(1) Legal establishment of boundaries
(2) Zoning
(3) Enactment of zoning regulations
(4) Specific-site planning
(5) Specific-site regulation
(6) Day-to-day management
(7) Review and revision of management

At each stage, the following should be taken into account

- Geographical habitat classification
- Physical and biological resources
- Climate
- Access
- History
- Current usage
- Management issues and policies
- Management resources

Economic value of marine parks

The economic value of marine parks has been the key driver for much of their development especially in tropical areas where coral reefs are often the main attractions and where, as a result, considerable tourism income is generated. Marine parks both attempt to provide appropriate levels of protection to the natural resources of the parks and to maximize tourist income. These objectives, however, are often incompatible. However, organizations such as the Marine Conservation Society have played a role in developing codes of conduct for divers, which point out the risks of permanent damage to coral. There may be conflicts between traditional users of newly designated parks, especially fishermen, and in some areas, for example, the Seychelles, steps were taken to allow traditional uses to continue. Dixon and Sherman (1990) showed that the Virgin Islands National Park produced considerable profits for the local economy.

Involvement of the community

Ultimately, the success and sustainability of marine parks depends upon the people who use them as visitors and the local community that often depends upon them for its livelihood. So, for example, in the Balicasag Island Municipal Marine Park in the Philippines key elements in its development included involving the project community worker into the community and baseline data collection, education, organization of a resource management core group, building responsibility around beneficial projects, and formalizing the community in management of the park. Continuity of management was essential for success—it means that the local community has a consistent point of liaison within the park (White and Dobias, 1990). If visitors enjoy the parks, but cause irretrievable damage to the ecosystems and disrupt the livelihoods of the human communities which depend upon food from the sea, they will destroy for future generations the same opportunity to enjoy and wonder at the beauty of the undersea world. Users are expected to respect and to help preserve the parks. This depends upon responsible use and it is now common to inform recreationists, especially divers, of good practice. Typically, this information is available both on websites as well as at the parks themselves. It usually focuses on

1. *Good diving practice*, such as obtaining local diving information from qualified local diving operator before diving and maintaining proper buoyancy control. This includes being aware of and obeying all the marine parks regulations.
2. *Prevention of direct damage* to coral reefs, by using rest floats provided by the park authorities, and using only the mooring buoys in the parks. Dropping anchors on reefs causes serious damage.
3. *Avoidance of impacts* on the underwater plants, animals, and fish. This means being aware of potential impacts on aquatic life through interactions with it; leaving litter, feeding fishes or provoking them, and collecting corals and shells for souvenirs should be avoided. Underwater photography is encouraged instead. There is also an expectation that any environmental disturbances or destruction observed during dives are reported to marine park centers so that park staff can prevent further damage or reduce its impact.
4. *Involvement in and support* for marine park activities, such as education and practical conservation work to help conserve the marine habitats and species.

Note. The International Union for the Conservation of Nature (IUCN) World Conservation Monitoring Centre (WCMC) in Cambridge, England, is the most important source of information regarding protected sites. Further reading on this topic may be found in Cognetti (1990), Forest and Park Service (2000), Halsey (1985), Kenchington (1991), Kenchington and Hudson (1984), and Salm and Clark (1984).

Conclusion

Marine parks, MPAs, and similar designations have a common goal: better understanding of the marine environment for the benefit of all who depend upon them. They depend upon sustained resources and incomes, the latter often from tourism, but this can be a source of conflict. Fortunately, there is a growing awareness, often because of the excellent public education programs that many parks run, that these are areas which must be nurtured and managed with the highest quality of understanding of their natural and human systems. Lien and Graham (1985) recognized that the establishment of marine parks would be difficult but rewarding. Marine ecosystems would present new challenges, partly because of the natural characteristics and partly because legal and constitutional problems would confront park managers and planners. They anticipated that research, management, and education would present technological problems which would need new and creative approaches. It would be untrue to say that all those challenges have been overcome, but the spread of marine parks, the common acceptance that these represent a window on the ocean commons and the increased use of new technologies to explore, understand, explain, and nurture the marine ecosystems mean that the challenges which Lien and Graham foresaw have been identified and confronted. As these areas come under greater pressure during the early 21st century, the challenges will remain.

Vincent May

Bibliography

Cognetti, G., 1990. Marine research and conservation of Mediterranean coastal habitats. Nature and Environment Series No. 50, Strasbourg: Council of Europe.

Dixon, J.A., and Sherman, P.B., 1990. *Economics of Protected Areas: A New Look at Benefits and Costs*. London: Earthscan Publications Ltd.

Forest and Park Service, 2000. The principles of protected area management in Finland: guidelines on the aims, function and management of state-owned protected areas. *Nature Protection Publications of the Finnish Forest and Park Service. Series B*, **54**.

Gibson, J., and Warren, L., 1995. Legislative requirements. In Gubbay, S. (ed.), *Marine Protected Areas: Principles and Techniques for Management*. London: Chapman and Hall, pp. 32–60.

Gubbay, S. (ed.), 1995. *Marine Protected Areas: Principles and Techniques for Management*, London: Chapman and Hall.

Halsey, J.R., 1985. Michigan's Great Lakes bottomland preserves. In Lien, J., and Graham, R., (eds.), *Marine Parks and Conservation: Challenge and Promise*, Toronto: National and Provincial Parks Association of Canada, Volume 2 pp. 65–76.

HELCOM, 1996. Third periodic assessment of the state of the marine environment of the Baltic Sea, 1989–1993. *Baltic Sea Environmental Proceedings*, No. 64A.

IUCN, 1994. *Guidelines for Protected Area Management Categories*. Gland, Switzerland: IUCN Commission on National Parks and Protected Areas with the assistance of the World Conservation Monitoring Centre.

Kelleher, G., 1985. The Great Barrier Reef marine park. In Lien, J., and Graham, R. (eds.), *Marine Parks and Conservation: Challenge and Promise*, Volume 2. Toronto: National and Provincial Parks Association of Canada, pp. 17–28.

Kelleher, G., and Kenchington, R., 1992. *Guidelines for Establishing Marine Protected Areas*. A Marine Conservation and Development Report. Gland, Switzerland: IUCN.

Kenchington, R., 1991. Tourism development in the Great Barrier Reef Marine Park. *Ocean and Shoreline Management*, **15**: 57–78.

Kenchington, R.A., and Hudson, B.E.T., 1984. *Coral Reef Management Handbook*. Jakarta: UNESCO.

Lien, J., and Graham, R. (eds.), 1985. *Marine Parks and Conservation: Challenge and Promise*. Toronto: National and Provincial Parks Association of Canada.

Marsh, J.S., 1985. Japan's marine parks in preserves. In Lien, J., and Graham, R. (eds.), *Marine Parks and Conservation: Challenge and Promise*, Volume 2. Toronto: National and Provincial Parks Association of Canada, pp. 29–44.

Mondor, C.A., 1992, Planning for Canada's system of national marine parks. In Kelleher, G., and Kenchington, R. (eds.), *Guidelines for Establishing Marine Protected Areas*. Gland: A Marine Conservation and Development Report IUCN, pp. 49–59.

Norse, E. (ed.), 1993. *Global Marine Biodiversity: A Strategy for Building Conservation into Decision Making*. Washington, DC: Center for Marine Conservation.

Ray, G.C., 1976. Survey of critical marine habitats and requirements for their conservation. *Collected Abstracts of the International Conference on Marine Parks and Reserves*, May 12–21, Tokyo, Japan.

Salm, R.V., and Clark, J.R., 1984. *Marine and Coastal Protected Areas: Guide for Planners and Managers*. Gland: IUCN.

White, A.T., and Dobias, R.J., 1990. Community marine tourism in the Philippines and Thailand: a boon or bane to conservation? In Miller, M.L., and Auyong, J. (eds.), *Proceedings of the 1990 Congress on Coastal and Marine Tourism, Honolulu*, Volume 2. Newport: National Coastal Resources Research and Development Institute, pp. 453–459.

Cross-references

Archaeology
Coastal Zone Management
Conservation of Coastal Sites
Coral Reefs
Economic Value of Beaches
Environmental Quality
Human Impact on Coasts
Organizations (see Appendix 3)
Tourism and Coastal Development

MARINE TERRACES

A marine terrace is any relatively flat, horizontal, or gently inclined surface of marine origin, bounded by a steeper ascending slope on one side and by a steeper descending slope on the opposite side. In temperate regions, marine terraces often result from marine erosion (abrasion or denudation) (*marine-cut* terraces, or shore platforms) or consist of shallow-water to slightly emerged accumulations of materials removed by shore erosion (*marine-built* terraces). In intertropical regions they may also result from bioconstruction by coral reefs and accumulation of reef materials (reef flats).

The development of a series of elevated, stepped marine terraces usually corresponds to the superimposition of eustatic changes in sea level and of a tectonic uplifting trend. Uplifted terraces act in this case as a continuous tape recorder, each step developing when the rising sea level overtakes the rising land. Each raised terrace, eventually capped by marine and/or alluvial materials, is a fossil counterpart of the present-day terrace, platform, or reef flat. For relatively rapid uplift rates (e.g., >1 mm/yr), each marine terrace corresponds to a different interglacial period or stage and ages usually increase with elevation, the uppermost levels being also the less well-preserved. For slower uplift rates, marine terraces may have a polycyclic origin, sea level returning again at the terrace level after a period of exposure to weathering.

When reconstructing past sea-level changes from the study of datable raised marine terraces, two assumptions are usually made: (1) that the eustatic sea-level position corresponding to at least one raised terrace is known and (2) that the uplift rate has remained essentially constant in each section. From these assumptions, the eustatic sea level can be calculated for each dated terrace (Pirazzoli, 1993). Bloom and Yonekura (1990) proposed a calculation procedure in which the assumption of a constant rate of dislocation becomes unnecessary when a sequence of dated emerged terraces is found at different heights on several transects.

Example of a sequence of stepped Quaternary marine terraces: Sumba Island, Indonesia

An exceptional sequence of raised coral reef terraces is preserved at Cape Laundi, on the north coast of the island, between the present sea level and an ancient patch reef now at 475 m above sea level. At least 11 terraces are wider than 100 m, 6 of them even being over half-a-kilometer-wide. Most of these terraces are polycyclic in origin and can be subdivided into a number of secondary levels by more or less continuous scarps or by alignments of fossil reef ridges. ESR, Th/U, and ^{14}C date estimations of almost unrecrystallized corals, most of them in growth position, have made possible the identification of the younger terraces (up to isotope Stage 9 (ca. 330 ka ago) and of those corresponding to isotope Stage 15 (ca. 600 ka ago). When the most likely uplift trend deduced from the present altitude of dated terraces (ca. 0.5 mm/yr) is extrapolated to the whole raised section, most geomorphological features appear to correspond to interglacial stages, up to Stage 27 (ca. 0.99 Ma) for the uppermost patch reef (Pirazzoli *et al.*, 1993) (Figure M11).

Paolo A. Pirazzoli

Figure M11 The uppermost sequence of marine terraces at Cape Laundi, Sumba Island, Indonesia. Behind the very wide Terrace II (in the foreground, at elevations between 135±10 m and 110 ± 5 m, developed during isotope Stages 9 and 7), six major steps of coral reef terraces, visible in the background up to 475 m, developed between about 330 ka and one million years ago (Photo B466, September 1988).

Bibliography

Bloom, A.L., and Yonekura, N., 1990. Graphic analysis of dislocated Quaternary shorelines. In *Sea-Level Change. Studies in Geophysics*. Natl. Washington, DC: Academic Press, pp. 104–115.

Pirazzoli, P.A., 1993. Global sea-level changes and their measurement. *Global and Planetary Change*, **8**: 135–148.

Pirazzoli, P.A., Radtke, U., Hantoro, W.S., Jouannic, C., Hoang, C.T., Causse, C., and Borel Best, M., 1993. A one million-year-long sequence of marine terraces on Sumba Island, Indonesia. *Marine Geology*, **109**: 221–236.

Cross-references

Changing Sea Levels
Coral Reefs Emerged
Eustasy
Paleocoastlines
Pleistocene Epoch
Sea-Level Indicators, Geomorphic
Tectonics and Neotectonics
Uplift Coasts

MASS WASTING

Mass wasting, the movement of material downslope by gravity, occurs as slopes evolve toward stable, equilibrium forms. Active coastal slopes are often in short rather than long-term stability, because of wave undercutting, oversteepening, and the removal of basal debris. For convenience, a distinction is made in this discussion between the types of mass wasting that occur on rock and cohesive clay coasts. Some slope movements occur on both types of coast, however, and many cliffs consist of variable combinations of rock and clay. Nevertheless, translational slides and the free fall of material from steep slopes tend to be typical of rock coasts, whereas deep rotational slumps and shallower slides and flows of wet material are more common on cohesive clay coasts (Trenhaile, 1987, 1997; Sunamura, 1992; Viles and Spencer, 1995).

Rock coasts

Fresh rock surfaces and the presence of debris at the foot of cliffs testify to the importance of rock falls on many coasts. Although they occur more often than deep-seated slides, most falls are much smaller, and they are therefore less frequently reported. Rock falls are particularly common in well-fractured rocks where there is undercutting by wave action, bioerosion, or chemical solution, or differential erosion of weaker rocks at the cliff base. In cool environments, rock falls tend to occur when frost is most active, although this may reflect high water pressures in rock clefts generated by snow melt and precipitation in spring and autumn, rather than the effects of frost action. It has also been suggested that tides can generate damaging fluctuations in water pressure in rock cliffs. This could explain why the size and frequency of rock and slab falls in the Bay of Fundy, for example, increase eastwards, as the tidal range increases. Many falls are caused by the reduction in confining pressures resulting from cliff erosion and retreat, and the formation of tension cracks parallel to the cliff face. Deep tension cracks in massive, cohesive rocks allow large slabs, as opposed to rock rubble, to be released. Slabs can also topple or overturn by forward tilting, particularly in rocks that consist of columns defined by joints, cleavage, or bedding planes which dip into the rock mass. Topples are able to occur on slopes whose gradients are lower than those associated with landslides.

Rock and slab falls, sags, and topples and other surficial failures are generally the result of weathering, basal erosion, and hydrostatic pressures exerted by water in rock clefts, whereas deep-seated slides occur in rocks that have been weakened by alternate wetting and drying, clay mineral swelling, or deep chemical weathering. Translational movements occur along straight slide surfaces. They are structurally controlled, and are usually associated with seaward dipping rocks, alternations of permeable and impermeable strata, massive rocks overlying incompetent materials, or argillaceous and other easily sheared rocks with low bearing strength. The failure surface is frequently a bedding, cleavage, or joint plane. Deep-seated rotational slumps are most common in thick and fairly homogeneous deposits of clay or shale, but more irregular slumps can occur in rock, where they tend to combine the characteristics of slumps with those of translational slides.

Coastal landslides have been particularly well-documented in the Chalk along the southern coast of England and along the Pacific Coast of the United States. In southern England, most slides have occurred where chalk and other arenaceous and calcareous sedimentary rocks are underlain by clay, shales, marls, or mudstones. Most slides are deep-seated rotational movements, although they also involve mudslides, slab failure, surficial sliding, toppling, sagging, and rock falls. Many slides have occurred in seaward dipping igneous and sedimentary rocks on the coast of northern Oregon, and along seaward-plunging synclines in southern California. A bentonite bed in shales and limestones provided an incompetent horizon for sliding at Portuguese Bend, near Los Angeles, whereas the large block glide at Point Firmin, about 8 km to the east, occurred along the bedding planes of shales that dip seawards at between 10° and 22°.

Cohesive masses fail when there is a critical combination of slope height and angle, in relation to the strength and bulk density of the material. The occurrence of joints, bedding planes, faults, and other discontinuities are more important, however, than the strength of the rock *per se*. Terzaghi (1962) considered the effect of the joint pattern on the critical slope angle of cliffs—the steepest gradient which can be maintained in long-term stability. His analysis showed that slope stability in jointed rocks largely depends upon the joint pattern and orientation, and to a lesser extent upon the effective cohesion of the rock mass. The principles of rock mechanics are increasingly being applied and incorporated into simulation models to determine failure mechanisms and to account for erosion rates and variations in the form of coastal cliffs, and to account for the formation and shape of marine caves (Davies and Williams, 1986; Allison and Kimber, 1998). In southern Wales, for example, it has been shown that translation failures commonly result from the formation of tension cracks near the cliff face, and toppling where there are tension cracks and basal undercutting. Translational failure is dominant where mudstones are prevalent at the cliff base, but toppling is more common where limestones provide a fulcrum at the foot of the cliff. The safety factor is inversely related to the ratio of the depth of undercutting at the cliff base to the distance of the tension crack from the cliff face. Numerical modeling, based on a range of geo-technical parameters, has also been used to predict the most likely modes of failure in the cliffs of southwestern Wales (Davies *et al.*, 1998).

Landsliding events are generally triggered by the build-up of groundwater, and they tend to take place during or shortly after snowmelt, or prolonged and/or intense precipitation. Groundwater reservoirs provide a water supply that is less dependent on seasonal rainfall, and further instability can result from the ponding of water in surface depressions that were created by older landslide events. Water washes out beds of sand in potential slide masses, causing swelling and the generation of pressures in argillaceous rocks, it generates high pore and cleft water pressures, and it softens colluvial materials allowing them to flow. The lubricating effect of water is not considered to be important, however, and the increase in weight of the potential slip mass as a result of water absorption is only of minor significance. Water can initiate slides in permeable rocks that overlie impermeable materials, and where massive, heavy rocks are on top of wet, impermeable clay or shale. Water from septic systems, irrigation, runoff disruption, and other human activities are playing increasingly important roles in some areas (Griggs and Trenhaile, 1994). For example, septic tanks and cesspools associated with residential development may have exacerbated slope instability at Portuguese Bend in southern California, and irrigation may have played a similar role at Point Firmin.

There is also a close cause-and-effect relationship between wave action, which undercuts and steepens coastal slopes, and the occurrence of landslides. Therefore, the degree of protection afforded to a cliff by beach material can be of enormous importance in determining its stability. The damming of rivers and the building of groins and other coastal structures has reduced longshore sediment transport in some areas, depleting beaches and exposing the cliffs to more vigorous wave action. For example, at Newhaven, in southern England, the build-up of coarse clastic material behind a jetty has provided protection to the cliff behind, whereas interruption of the littoral drift by the development of Folkestone Harbor may have been responsible for the increased activity of slides in the late 19th and early 20th centuries.

Changing sea level, whether on a geological or tidal time scale, causes the water table to fluctuate within potential slide masses. Pore and cleft water pressures vary with tidal ebb and flow, and cliffs may be most unstable when there are high water tables and low tides. Landslides can occur during high tidal periods, however, because of the increased ratio between pore pressures and the total overburden at the base of the slide (i.e., increased buoyancy at the toe of the slope) (Muir-Wood, 1971). In addition to the buoyant effect of the water, it has been suggested that the Portuguese Bend landslide in southern California is most unstable during high tidal periods because of increased pore water pressure, and saturation of the slide material and the basal shear plane at the base of the slide.

Cohesive clay coasts

The erosive resistance of a cohesive material is complex—it depends upon the water content and the properties of the pore water and eroding fluid, as well as the clay content and its plasticity, structure, consolidation pressure, and compressive or shear strength. Small pieces of clay can be eroded by pitting and flaking, but larger units can be removed by spalling along fractures and sandy planes.

Cliffs of sand and gravel tend to erode more rapidly than silt and clay, although this can be partly offset by the formation of protective, coarse-grained beaches at the cliff base. The accumulation of coarse sediment determines the accessibility of the cliff base to wave action. Rates and modes of cliff erosion can therefore vary according to the amount of beach material and the morphology of the beach profile (Jones and Williams, 1991; Shih and Komar, 1994). Eroded fine-grained cohesive sediment is generally carried offshore as suspended load, and as it provides no protection to the cliff, there is generally no relationship between the height of a clay cliff and its rate of erosion.

Because of variations in wave energy, temperature, precipitation, and groundwater pressures, the subaerial and marine erosion of cohesive coasts in temperate regions is often markedly seasonal in nature (Brunsden, 1984). The strength and stability of clay materials is affected by variations in the groundwater level. Mass wasting also occurs where grains are removed by seepage or piping of outflowing groundwater, which causes back sapping and collapse of the free face, often by toppling. Although seepage is generally associated with coarse silts to fine sands, it can also occur in clays, and it often develops at the base of water bearing fine, predominantly cohesionless, material underlain by an aquiclude.

There is a strong relationship between the frequency and mobility of slope movements in cohesive sediments and the presence of clays with a high proportion of swelling minerals. For example, mudslides in Denmark occur in clays that are dominated by swelling montmorillonite minerals, and the most mobile and furthest advanced portion of a mudflow in northeastern Ireland has the highest montmorillonite and illite to kaolinite ratios. Although slides occur in materials with the highest proportion of swelling clay minerals and the highest plasticity on St. Lucia and Barbados, soil plasticity is also sensitive to the types of exchangeable metal cations held in association with the clay minerals. The occurrence of large amounts of sodium in montmorillonite-rich clays, for example, increases their plasticity.

Shallow sliding occurs in cohesive sediments as mudslides or, if the water content is very high, as mudflows (Figure M12). High moisture content is not required for flow initiation in fine-grained, cohesionless materials, but it must be above the liquid limit if there is a high proportion of clay. Mudslides have a bimodal shape, consisting of fairly steep feeder flows and gently sloping accumulation flows composed of single or overlapping lobes of clay and other debris. Shallow rotational sliding produces miniature slump blocks along the frontal edge, which has a steeper slope than the accumulation area. The feeder zones begin to move with the seasonal rise in groundwater levels. Surface and subsurface movements in the accumulation zone are largely by shearing at its boundaries, and material can be incorporated in the mudslides from below. Other material is added to the flows by falls and slides from the bowl-shaped head zone and from the sides, and it may help to trigger rapid surges, usually following periods of heavy rainfall. Deep-seated rotational landslides also occur where basal erosion is rapid enough to remove the mudflow debris and then steepen the coastal slope.

Hutchinson's (1973) form/process classification of cliffs in the London Clay of southeastern England is based on the relative rates of basal marine erosion and subaerial weathering (see *Cliffed Coasts*):

1. Type 1 occurs where the rate of basal erosion is broadly in balance with weathering and the rate of sediment supply to the toe of the slope by shallow mudsliding. The slope undergoes parallel retreat and erosion only removes slide material, as opposed to *in situ* clay (Figure M12).
2. Type 2 cliffs occur where basal erosion is more rapid than weathering. The cliff experiences cyclical degradation, with slope steepness varying between upper and lower values corresponding to toe erosion and degradation and deep-seated, slump-dominant modes, respectively (Figure M13).
3. Type 3 cliff slopes develop when a cliff is abandoned by the sea and debris is carried to its foot by a series of shallow rotational slides. Abandoned cliffs therefore often have a steeper upper slope on which landsliding is initiated, and a flatter lower slope where colluvium accumulates.

Spatial variations in cliff type along the shore of Lake Erie reflect variations in wave intensity and the type of glacial sediment, whereas

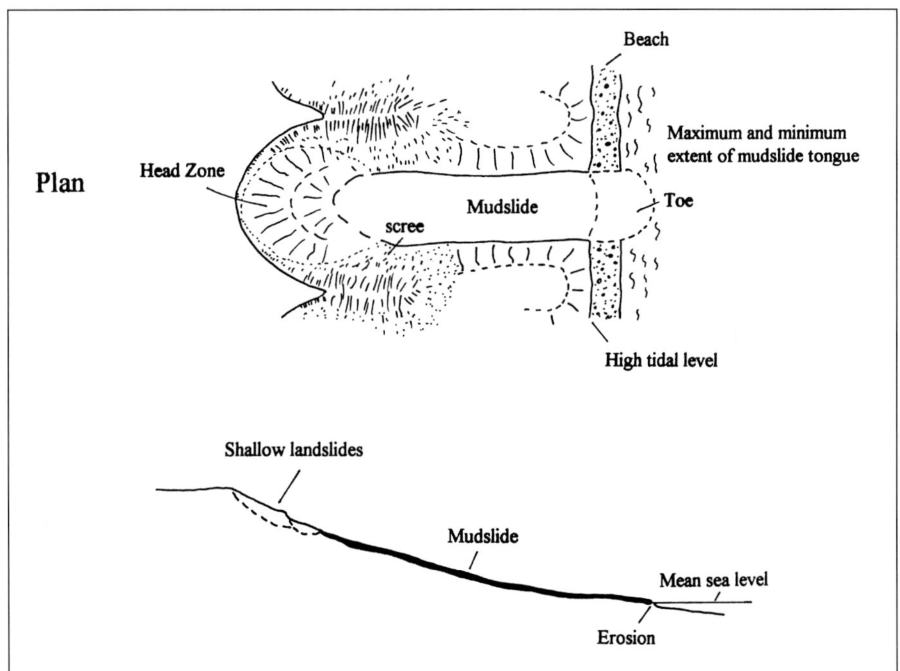

Figure M12 Mudslide in London Clay cliff with moderate wave erosion (from Hutchinson, 1973 by permission of Building Research Establishment Ltd, UK).

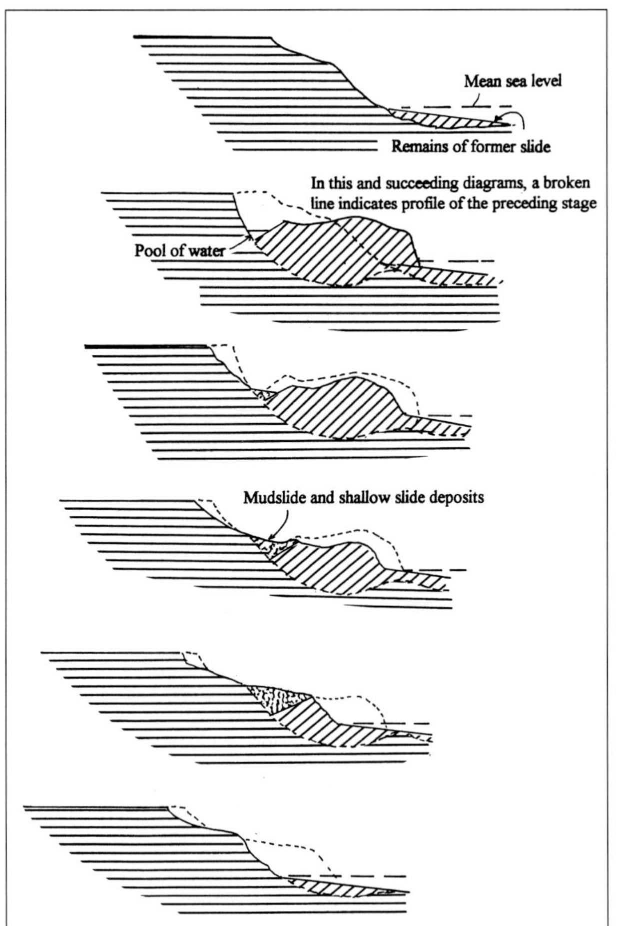

Figure M13 The cyclical behavior of London Clay cliffs with strong wave erosion (from Hutchinson, 1973 by permission of Building Research Establishment Ltd, UK).

temporal changes occur in response to changes in lake level. Type 1 cliffs develop during periods of normal lake levels, when there are moderate rates of cliff retreat; type 2 during periods of high lake levels and rapid cliff retreat; and type 3 during periods of low water levels and slow basal erosion. Quigley *et al.* (1977) distinguished four types of cyclical instability in this area, according to variations in cliff height, wave height, and rates of cliff retreat. Although Hutchinson's classification is also generally applicable to cohesive cliffs in Denmark, glacio-isostatic recovery and changes in relative sea level are responsible for some significant differences.

Hutchinson's model is generally applicable to a wide range of cohesive sediments, but alternate models of cliff retreat are relevant to specific areas. For example, shallow, planar landslides have been found to be more important than larger but less frequent rotational landslides in glacial sediments in northeastern Ireland. Bromhead (1979) considered that the transition from one type of cliff to another largely depends upon the nature of the material at the crest of the slope, and to a lesser degree on its groundwater hydrology, rather than on the relative rates of subaerial and marine erosion. He suggested that mudsliding dominates if the material at the crest of the slope is similar to that in the rest of the slope. Mudsliding is inhibited, however, if there is stronger or better drained material at the crest. The lack of protective debris at the cliff foot then allows wave action to erode and oversteepen the cliff, ultimately precipitating a deep-seated rotational landslide. These slides tend to be of the multiple rotational type when the cliffs are capped with a thick, hard, and jointed caprock.

Alan S. Trenhaile

Bibliography

Allison, R.J., and Kimber, O.G., 1998. Modelling failure mechanisms to explain rock slope change along the Isle of Purbeck Coast, UK. *Earth Surface Processes and Landforms*, **23**: 731–750.

Bromhead, E.N., 1979. Factors affecting the transition between the various types of mass movement in coastal cliffs consisting largely of over-consolidated clay with special reference to southern England. *Quarterly Journal of Engineering Geology*, **12**: 291–300.

Brunsden, D., 1984. Mudslides. In Brunsden, D., and Prior, D.B. (eds.), *Slope Instability*. Chichester: Wiley, pp. 363–418.

Davies, P., and Williams, A.T., 1986. Cave development in lower Lias coastal cliffs, the Glamorgan Heritage Coast, Wales, UK. In Sigbjarnarson, G. (ed.), *Iceland Coastal and River Symposium Proceedings*. Reykjavik: National Energy Authority, pp. 75–92.

Davies, P., Williams, A.T., and Bomboe, P., 1998. Numerical analysis of coastal cliff failure along the Pembrokeshire Coast National Park, UK. *Earth Surface Processes and Landforms*, **23**: 1123–1134.
Griggs, G.B., and Trenhaile, A.S., 1994. Coastal cliffs and platforms. In Carter, R.W.G., and Woodroffe, C.D. (eds.), *Coastal Evolution*. Cambridge: Cambridge University Press, pp. 425–450.
Hutchinson, J.N., 1973. The response of London clay cliffs to differing rates of toe erosion. *Geologia Applicata e Idrogeologia*, **8**: 211–239.
Jones, D.G., and Williams, A.T., 1991. statistical analysis of factors influencing cliff erosion along a section of the west Wales Coast, UK. *Earth Surface Processes and Landforms*, **16**: 95–111.
Muir-Wood, A.M., 1971. Engineering aspects of coastal landslides. *Proceedings of the Institute of Civil Engineers*, **50**: 257–276.
Quigley, R.M., Gelinas, P.J., Bou, W.T., and Packer, R.W., 1977. Cyclic erosion-instability relationships: Lake Erie north shore bluffs. *Canadian Geotechnical Journal*, **14**: 310–323.
Shih, S.-M., and Komar, P.D., 1994. Sediments, beach morphology and sea cliff erosion within an Oregon Coast littoral cell. *Journal of Coastal Research*, **10**: 144–157.
Sunamura, T., 1992. *Geomorphology of Rocky Coasts*. Chichester: Wiley.
Terzaghi, K., 1962. Stability of steep slopes on hard unweathered rock. *Géotechnique*, **12**: 251–270.
Trenhaile, A.S., 1987. *The Geomorphology of Rock Coasts*. Oxford: Oxford University Press.
Trenhaile, A.S., 1997. *Coastal Dynamics and Landforms*. Oxford: Oxford University Press.
Viles, H., and Spencer, T., 1995. *Coastal Problems*. London: Edward Arnold.

Cross-references

Cliffs, Erosion Rates
Cliffs, Lithology versus Erosion Rate
Hydrology of Coastal Zone
Klint
Rock Coast Processes

METEOROLOGICAL EFFECTS ON COASTS

The most dramatic and long-lasting meteorological impact on many coasts is in response to storms. Virtually, every continent on earth is variously impacted by storms, the degree to which being a function of many factors including storm intensity, duration and path, as well as antecedent geology of the inner shelf and coast. Cyclones that exert important controls on coasts are generally categorized as hurricanes, tropical, and extratropical storms. Land- and sea-breezes are observed along many coasts and are in response to differential temperatures during day and night; onshore winds during the day develop nearshore sea state, whereas offshore flow in the evening causes wave decay close to shore. Neither effect can equal the impacts of waves, currents, and winds generated during cyclones. The low latitudes are dominated by tropical storms and hurricanes, whereas the mid- and higher-latitudes experience extratropical storms and weather fronts. Frontal systems are associated with extratropical cyclone development.

Mid-latitude cyclones and their associated fronts significantly impact all of the coasts of the United States, although the frequencies of major storms decrease from north to south along both the Atlantic and Pacific coasts. Along the Pacific northwest and New England coasts, mid-latitude cyclone frequencies can range upwards to two or three storms per week during winter months, with average frequencies decreasing to one or less a month in southern Florida and southern California. Seasonally, mid-latitude cyclones occur year round along northern coasts, although frequencies and intensities are much lower in the summer. Local and regional coastal configuration and orientation strongly modify exposure to storm processes. For example, Cape Hatteras on the Atlantic and Point Reyes on the Pacific coast are particularly vulnerable given their orientation. Similarly, the northern Gulf of Mexico experiences significant coastal change during frontal events given its general east–west orientation. As an example of rapid and intensive cyclogenesis, the "Storm of the Century," an intense extratropical cyclone, comparable in strength to a category 1 hurricane, formed in the western Gulf of Mexico in March 1993 and inflicted major damage along the Gulf Coast's beaches and low-lying areas (Schumann *et al.*, 1995).

Cold fronts are usually initiated by outbreaks of cold air (Polar air masses) from Canada, which advances south, and southeast where they ultimately encounter warm air. Each year up to 30 fronts move south and significantly impact the Gulf of Mexico and southeast Atlantic beaches. As the frontal systems push south, strong southerly winds ahead of the fronts generate deepwater waves whose significant wave height may exceed 5 m. Considerable water level set up occurs along the ocean-facing beaches. A considerable degree of beach erosion and dune breaching accompanies this pre-frontal stage, particularly along low-lying coasts such as Louisiana. This pre-frontal period can last up to several days preceding the arrival of the front. During this period, a considerable volume of water is funneled into the bays, lagoons, and sounds behind the barrier coasts causing elevated water levels. As the front moves across the coast a dramatic wind shift from the south to the north occurs over a matter of hours. While nearshore waves are suppressed in the Gulf and Atlantic during this period of strong northerly winds, high frequency, steep waves are generated in the bays and lagoons where fetch lengths permit. During this post-frontal phase significant wave heights approaching 3 m have been measured in some bays fringing the northern Gulf. Research now suggests that these frontal passages cause dramatic changes to the soundside beaches of the southeast and Gulf bay shores and is the primary process behind chronic coastal erosion.

Each year several of the mid-latitude cyclones evolve into powerful coastal storms with typical storm tracks from southwest to northeast along the Atlantic coast. These storms have traditionally been called Nor'Easters because of the strong winds that blow from the northeast ahead of the warm fronts. The most severe occur in October and November and again in March and April. These storms usually develop and intensify as they move parallel to the coast toward the northeast. The Ash Wednesday storm of 1962 was one of the most powerful Nor'Easters to impact the East coast of the United States. Generating deepwater significant wave height in excess of 10 m, the Ash Wednesday storm caused severe erosion along a 1,000 km stretch of coast from North Carolina to Long Island, New York. Because of the danger to life and property and the importance of predicting the explosive dynamics of storm development, a concerted research effort has been directed toward enhancing our comprehension of these systems. After the Ash Wednesday storm, a detailed compilation of powerful coastal storms was compiled (Mather *et al.*, 1964) and more recently, the major coastal storms from New England to Florida have been classified by synoptic weather situations and evaluated in terms of changing frequencies of severe events through many decades (Dolan and Davis, 1992).

Tropical storms and hurricanes also impact the entire US mainland coast from New England to Texas, and to a much lesser extent, southern California. The season extends from June through November and thus there is overlap with mid-latitude cyclones in the fall. Unlike mid-latitude cyclones, tropical storms and hurricanes originate over the eastern tropical Atlantic, Caribbean Sea, and Gulf of Mexico. Tracks are generally westward over the tropical Atlantic then recurve northward to the North American coast (Figure M14).

Hurricanes are much smaller than mid-latitude cyclones, and unless the track is close to and parallel with the coast, the severe impacts are restricted to much shorter segments of the coastline (Muller and Stone, 2001).

A direct strike by a major hurricane is more devastating than mid-latitude cyclones because wind speeds, storm surges, and waves are generally much higher. Because of the counter-clockwise circulation of surface winds in a hurricane in the Northern Hemisphere, wind speeds, surge, and waves are generally higher to the right of the system than to the left. Severe morphological change and destruction to infrastructure along coasts is most often limited to a coastal segment no more than 25–50 km to the right of the location at which the storm makes landfall (Muller and Stone, 2001). At many locations in the southeastern United States, tropical storms impact the coast as frequently as once every three to four years on average, whereas the recurrence of major hurricanes is much more infrequent. As an example, in south Louisiana and south Florida the recurrence interval for these storms is every 10 to 15 years on average, and only once every 100 years along the more sheltered coast of Georgia (Elsner and Kara, 1999; Muller and Stone, 2001). In recent years, there has been considerable research conducted on decadal or longer storm-track patterns which result in long runs of years with very limited storm activity followed by shorter runs of years with frequent storms (Gray, 1999). Since the mid-1990s, the clustering of storms has resulted in a marked increase in beach erosion along significant portions of the southeastern United States (Stone *et al.*, 1996, 1999). In turn, this has resulted in a marked increase in new beach nourishment and renourishment projects costing tens of millions of US dollars.

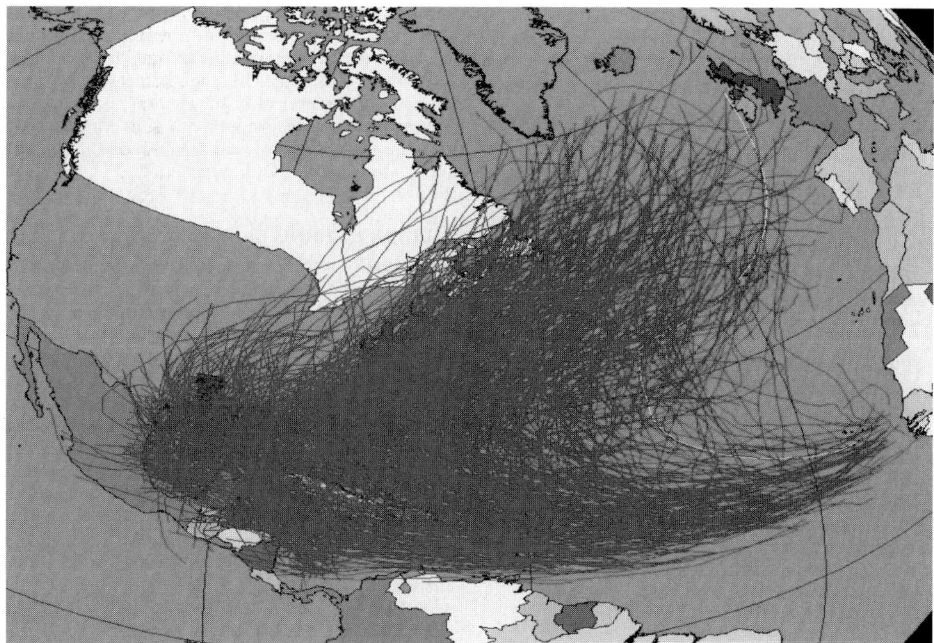

Figure M14 North Atlantic hurricane tracks from 1886–1996. (Data obtained from National Oceanic and Atmospheric Administration, National Hurricane Center.)

The coast of southern California and Baja California are also susceptible to tropical storm and hurricane impacts that move from southern Mexico northwestward parallel with the coast. Each hurricane season there are numerous tropical storms and hurricanes generated off the southwestern coast of Mexico, however, most storms eventually move to the west out to sea. The rare storm that affects southern California is feared more for heavy rains and subsequent flooding rather than high winds, surge, and waves. One of the most memorable of these events caused much unexpected flooding in the Los Angeles basin in September 1939 (Hurd, 1939).

The coasts of the Great Lakes are subject to frequent mid-latitude cyclones, normally moving from southwest to northeast across the lakes and eventually down the St. Lawrence River valley to the Atlantic. Frequencies of mid-latitude cyclones and associated fronts can be as high as six or more per month from October to May, with frequencies and especially intensities much lower in summer. Similar to the Atlantic coast, the truly destructive great storms occur most frequently during fall and spring. Cold fronts in the late fall and early winter, when lake waters are much warmer than cold polar air behind the fronts, are also associated with Great Lakes snow squalls occurring along the eastern and southeastern lake shores. This is pronounced where the orographic effects of hilly terrain amplify the lake effect snow squalls. These are most apparent over the Keweenaw Peninsula in Upper Michigan, and southeast of Lake Erie and Ontario in New York.

Along the Pacific coast from the Canadian to Mexican border, mid-latitude cyclones are again the primary storm events, with the season extending normally from October through April. There are more storms at northern rather than southern locations, with southern California experiencing approximately one to two storms at most per month. These storms tend to originate along the eastern coast of Asia or over the Gulf of Alaska, so that the storms arrive in an occluded stage, with the warm and cold fronts not particularly noticeable at the surface. Instead, the warm tropical air in these systems has typically been lifted above the surface, resulting in very large storms with gale-force winds along the coasts, and widespread precipitation occurring as snow in the mountains. Precipitation and melted snow can generate mudflows and floods that endanger and destroy homes in the narrow canyons near the coast. These storms tend to stall along the coast so that the duration of storm weather can often last for two to three days, with high waves generated by southwesterly winds causing significant erosion of beaches along the Pacific coast. This problem is exacerbated when the storms coincide with high astronomical tides.

Gregory W. Stone and Robert A. Muller

Bibliography

Dolan, R., and Davis, R.E., 1992. Rating northeasters. *Mariners Weather Log*, **36**(3): 4–11.

Elsner, J.B. and Kara, A.B., 1999. *Hurricanes of the North Atlantic*. New York: Oxford.

Gray, W.M., 1999. On the causes of multi-decadal climate change and prospects for increased Atlantic basin hurricane activity in the coming decades. *10th Symposium on Global Change Studies*, American Meteorological Society, pp. 183–186.

Hurd, W.E., 1939. Weather on the Atlantic and Pacific oceans: North Pacific ocean. *Monthly Weather Review*, **67**(9): 356–358.

Mather, J.R., Adams, H., III, and Yoshioka, G.A., 1964. Coastal storms of the eastern United States. *Journal of Applied Meteorology*, **3**: 693–706.

Muller, R.A. and Stone, G.W., 2001. A climatology of tropical storm and hurricane strikes to enhance vulnerability prediction for the southeast U.S. coast. *Journal of Coastal Research*, **17**: 949-956.

Schumann, S.A., Moser, J., Johnson, G.A., Walker, N.D., and Hsu, S.A., 1995. An overview of a strong winter low in the Gulf of Mexico, 12–13 March 1993. *National Weather Digest*, **20**(1): 11–25.

Stone, G.W., Grymes, J.M., Armbruster, C.A., and Huh, O.K., 1996. Overview and impacts of hurricane opal on the Florida coast. Eos, *Transactions of the American Geophysical Union*, **77**: 181, 184.

Stone, G.W., Wang, P., Pepper, D.A., Grymes, J.M., Roberts, H.H., Zhang, X., Hsu, S.A., and Huh, O.K., 1999. Studying the importance of hurricanes to the Northern Gulf of Mexico Coast. Eos, *Transactions of the American Geophysical Union*, **80**: 301, 305.

Cross-references

Beaufort Wind Scale
Climate Patterns in the Coastal Zone
Coastal Climate
Coastal Wind Effects
Natural Hazards
Storm Surge
Wave Climate
Wave Environments
Waves

MICROTIDAL COASTS

Coasts where the tidal range (difference between successive high and low tide levels) does not exceed 2 m are commonly referred to as microtidal (Hayes, 1979; Davies, 1980; Cooper, 1994). Such coasts may be composed of a variety of materials and occur in all latitudinal zones and in a variety of energy settings. Their common characteristics are derived from the fact that their small tidal range focuses marine action (via waves and tidal currents) into a relatively narrow vertical range. Hayes (1979) identified a generalized link between tidal range and coastal morphology. Microtidal coasts were characterized by long, narrow barriers with abundant washover features, well-developed flood-tidal deltas, and small ebb-tidal deltas. These characteristics have often led to application of the term "wave-dominated" to such coastlines, although Davis and Hayes (1984) demonstrated that this is not universally true. Some microtidal coasts do exhibit dominance of tidal currents over waves and some macrotidal coasts exhibit wave-dominance. The west peninsular coast of Florida is a good example of a microtidal coast on which wave energy (and sediment supply) is low and yet large tidal deltas develop that are more characteristic of tide-dominated conditions (Davis and Hayes, 1984). Microtidal coasts are widespread in oceanic areas (most Indo-Pacific islands), in some semi-enclosed seas (Mediterranean, Baltic, Gulf of Mexico), on parts of continental margins (S.W. Africa, S. Australia, central Brazil, Antarctica) and on lake shorelines (US Great lakes, African Rift lakes).

Open coast morphology

Wave processes on microtidal coasts are generally the principal agent of morphological change and sediment transport. Waves constantly expend their remaining energy at the same elevation and thus their effectiveness in coastal modification and sediment transport is greatly enhanced in comparison with areas of greater tidal range.

On coasts where little sediment is available, wave action that is concentrated in a narrow zone may give rise to a well-developed intertidal shore platform, an erosional feature produced by undercutting and removal of rock or semi-consolidated material. In tropical or warm temperate zones, the production of an erosional notch and shore platform on limestone coasts may be aided by chemical processes of weathering. Trenhaile (1980) has established that tidal range does exert a strong control on shore platform morphology.

Where sediment is available, wave processes are effective in reshaping sediment bodies (beach and beachface) by cross-shore and longshore transport. Depositional shores on microtidal coasts thus tend to be rapidly modified by wave processes to attain a form of equilibrium. Examples of such forms include static equilibrium (where wave energy is equalized along the shore), dynamic equilibrium (where sediment input is balanced by sediment losses), and graded equilibrium (where sediments are sorted along the shore by size such that at all points they just exceed the transport capacity of waves) (Carter, 1988). Concentration of energy in a narrow vertical band enhances the development of coastal cells that develop in response to wave and current conditions. Some of the best literature on microtidal coasts is based on studies in the US Great Lakes shores and the lack of tidal influence has enabled some basic concepts in coastal geomorphology to be investigated (Dubois, 1975).

Because of the limited intertidal area and relatively narrow beaches that limit the potential for entrainment of sand by wind, sand dunes tend to be poorly developed on microtidal coasts and those that are present often owe their existence to oblique onshore winds and/or a falling relative sea level tendency. In contrast, the lack of regular wetting of the beach surface by tidal action may enhance the potential for deflation by wind on microtidal beaches.

River influences

Rivers that enter the sea on microtidal coasts experience limited tidal influences in their lower reaches. Thus, estuaries tend to be shorter than those of areas with greater tidal range. In addition, if sediment is available in the vicinity of the river mouth, this may be reworked by waves to form a rivermouth barrier that encloses the estuary mouth temporarily or permanently. In a review of the effects of tidal range on tidal inlet morphology, Hayes (1979) noted that microtidal examples tended to have well-developed flood-tidal deltas and relatively poorly developed ebb-tidal deltas as a result of high wave energy. Deltas in microtidal coastal areas exhibit a range of morphologies in accordance with the level of wave energy. The Mississippi delta, for example, has a distinctive birds foot morphology indicating the relative ineffectiveness of waves in modifying it, whereas the Sao Francisco in Brazil comprises a series of beach ridges that represent the wave modification of deltaic sediment delivered to the coast. In few instances, however, does a single process operate to the exclusion of all others and it is worth noting in this regard that the Mississippi delta does contain chains of barrier islands that are the wave-reworked sections of the delta and which form after fluvial dominance has diminished.

Human modification

Microtidal coastlines do not contain appreciable intertidal areas and their relative constancy of water level has rendered them favorable for certain forms of human exploitation, particularly for shipping and more recently for recreation. Microtidal coastal areas have recently seen an increase in marina development, particularly in milder climatic regions such as the Mediterranean and Gulf of Mexico.

Development of recreational facilities on microtidal beaches has the advantage of a relatively constant beach area at all tidal stages and thus avoids the regular variability of available space that characterizes beaches of high tidal range. This apparent stability, is however, often deceptive in the longer term. Storms on microtidal coasts may produce elevated sea levels through surges that temporarily raise water levels and which may enhance erosion of backshore areas. Since such storms may be relatively infrequent, a number of such areas have been deemed stable and have been developed. Subsequent erosion of these areas by enhanced wave action has threatened developments and given rise to a succession of shore protection works.

It is interesting to note that human modification of tidal inlets (for example, by barrage construction or inlet closure) may diminish tidal range and produce a change in coastal type in the vicinity of the inlet to microtidal conditions.

J. Andrew G. Cooper

Bibliography

Carter, R.W.G., 1988. *Coastal Environments*. Cambridge: Cambridge University Press.

Cooper, J.A.G., 1994. Lagoons and microtidal coasts. In Carter, R.W.G., and Woodroffe, C.D. (eds.), *Coastal Evolution*. Cambridge: Cambridge University Press, pp. 219–265.

Davis, R.A. and Hayes, M.O., 1984. What is a wave-dominated coast? *Marine Geology*, **60**: 313–329.

Davies, J.L., 1980. *Geographical Variation in Coastal Development*. Harlow: Oliver & Boyd.

Dubois, R.N., 1975. Support and refinement of the Bruun Rule on beach erosion. *Journal of Geology*, **83**: 651–657.

Hayes, M.O., 1979. Barrier island morphology as a function of tidal and wave regime. In Leatherman, S.P. (ed.), *Barrier Islands*. London Academic Press, pp. 3–22.

Trenhaile, A.S., 1980. Shore platforms: a neglected coastal feature. *Progress in Physical Geography*, **4**: 1–23.

Cross-references

Atolls
Barrier
Beach Erosion
Beach Processes
Beach Ridges
Bioerosion
Cliffed Coasts
Deltas
Dune Ridges
Eolian Processes
Erosion Processes
Human Impact on Coasts
Longshore Sediment Transport
Rock Coast Processes
Sandy Coasts
Shore Platforms
Storms (see Meteorological Effects)
Storm Surge
Tides
Wave-Dominated Coasts

MIDDENS—See SHELL MIDDENS

MIDDLE AMERICA, COASTAL ECOLOGY AND GEOMORPHOLOGY

Introduction

There are several reasons why geomorphology, ecology, and integrated coastal management, have gained attention globally in Mesoamerica (Middle America, Central America). From the coasts of this region, energy, materials, and food are extracted, cities are developing, industries and port are growing, transport and tourism are in expansion, and general infrastructure and environmental technology marketing are increasing (Windevoxhel *et al.*, 1999; Yáñez-Arancibia, 1999, 2000). Under this approach, during the 1990s more than 75 million US dollars have been destined for initiatives and regional projects on integrated coastal zone management, only in Central American countries, by international agencies (Rodriguez and Windevoxhel, 1996).

In this framework the "coastal zone" is a broad geographic space of interactions between the sea, the land, freshwater drainage, and the atmosphere, in which the principal interchanges of materials and energy are produced between the marine and terrestrial ecosystems.

Traditionally, Central America includes the following countries: Belize, Guatemala, Honduras, El Salvador, Nicaragua, Costa Rica, and Panama. Nevertheless, from a "mesoamerican" point of view and from an ecological approach because of tropical latitudes, and littorals in two oceans—Atlantic and Pacific—southern Mexico and northern Colombia will be considered, in this paper, as part of the complex of Middle America, because all are integrated components of the functional structure of coastal zone in Central America. The central America coastal zone posseses extensive scenic and geographical wealth as well as great ecological biodiversity (ecosystems, habitats, forcing functions, biological species, functional groups).

Middle America is a "neotropic mosaic." A mosaic of authors' training and experiences; a mosaic of roots and cultural heritage resources; a mosaic of social development; a mosaic of ecosystems and ecological approaches; a mosaic of biogeographical regions and biodiversity; a mosaic of climatic zones; a mosaic of pristine areas as well as highly degraded zones. Because of this, to deal with common terms of reference, the purpose of this is to focus on the tectonic history, the regional geography and geomorphology, and to analyze the major ecological systems, as the key focus for understanding the geomorphology/ecology of coastal zones in Middle America (i.e., continental shelf, coastal lagoons and estuaries, mangroves, and coral reefs).

Central America—Antilles tectonic history

The tectonic history of Central America and the Antilles is extremely complex (Burke *et al.*, 1984; Buskirk, 1992). Geological studies are fragmentary, and the application of plate tectonic theories to detailed areas has been controversial (Donnelly, 1985; Coates *et al.*, 1992; Suárez *et al.*, 1995). Measurements of present-day plate motions can give a picture of recent movements but the older history of the area is more difficult to interpret; much geological evidence is missing.

The tectonic hypothesis of Duncan and Hargraves (1984) provides the best framework for the Buskirk (1992) zoogeographic analysis for understanding Mesoamerican ecological relationships. Based on plate tectonics determined from the geometry of hotspots on the earth's mantle, the Duncan and Hargraves model reconstructs the evolution of the Caribbean region as follows. In the Cretaceous, as the North and South American plates moved farther apart, subduction movements of the Pacific ocean floor under the west side of the Americas formed the proto-Antilles island arc. In the late Cretaceous, an oceanic plateau formed behind that arc and began to push it northeastward between the continents. This motion was halted at about the Eocene and Oligocene when the arc collided with the Bahamas platform. At this time, subduction of the Pacific ocean floor under the west side of the oceanic platform, in combination with uplift of older Cretaceous island arc features, formed the lower Central American arc. Since the Oligocene there has been eastward movement of the Caribbean plate. Subduction of the Atlantic ocean crust beneath the eastern margin of the Caribbean plate has formed the Lesser Antilles island arc. A subduction zone along the Pacific side of Mesoamerica forms the western boundary of the Caribbean plate.

To this framework, Buskirk (1992) added information about the Greater Antilles used in preparing the sketches in Figure M15. Solid

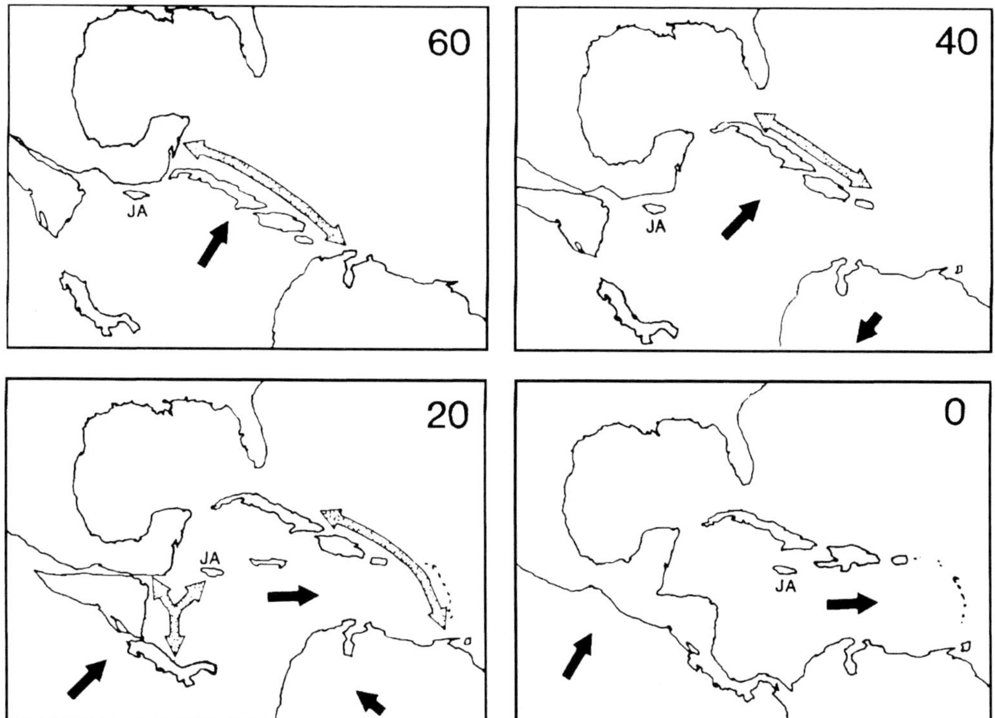

Figure M15 Hypothetical sketch of Mesoamerican and Caribbean history at 60 (Paleocene), 40 (late Eocene), 20 (early Miocene), and 0 (Present) millions years before present (after Buskirk, 1992). Land outlines do not indicate shoreline at those times but are merely for recognition of the present-day terrains. Solid arrows indicate the predominant direction of tectonic movement. Stippled arrows indicate suggested dispersal pathways.

arrows indicate the direction of major tectonic movements. Land outlines do not indicate coastlines at the time, but are merely for recognition of the relative location of present-day terrains. The position of the Honduras–Nicaragua and Costa Rica–Panama blocks is based on the model of Duncan and Hargraves (1984). Although each of the blocks might be considered a tectonic unit for the last 50 million years, both have been changing extensively during that time with much shearing and internal deformation.

It is suggested that the Greater Antilles island arc (Cuba, northern Hispaniola, and Puerto Rico) moved northeastward during the late Cretaceous after fragmenting its western end (to form Jamaica) in a collision against Mexico–Guatemala. Pollen data from the Eocene and Oligocene of Cuba and Puerto Rico indicated that there was frequent dispersal between the island of this northern Caribbean arc (Buskirk, 1992). Jamaica was largely inundated during this time. Since the mid-Tertiary it has moved as the eastern end of the Nicaraguan block.

In Figure M15 (after Buskirk, 1992), probable prevalent pathways of dispersal at 20, 40, and 60 million years ago are suggested by the stippled arrows. Probably, there was a period of dispersal between North and South America in the Cretaceous or early Tertiary, despite the lack of geological evidence for an emergent island arc at that time. In about the Miocene (beginning 23 million years ago), as the Lesser Antilles island chain became prominent, there was dispersal from South America into Puerto Rico, Hispaniola, and Cuba. As Jamaica become emergent, it initially was colonized mostly by the Central American terrapins.

Following the emergence of Jamaica, therefore, geological changes altered the potential for dispersal of terrestrial organisms between Jamaica and Mesoamerica. In the Miocene, the emerging island was situated closer to Central America than presently, and shallow reefs or emergent limestone banks could have provided stepping-stone conditions for dispersal between Jamaica and the Mesoamerican island chain. During the Pliocene, as Jamaica uplift increased, tectonic activity slowly moved the island eastward as the rise between it and the Nicaragua subsided.

Regional geography and geomorphology

It is not only the tectonic history or the relative areas of continental shelves and the open tropical oceans we need to quantify, if we want to understand the basis of Mesoamerican coastal ecosystem productivity. Rather, we must also know the geographical and geomorphological characteristics of continental shelves, shallow bays, and coastal lagoons and estuarine system features. Are they narrow or wide, flat or steep, what are their superficial deposits, what is the geology of the coastline that lies behind them, and do they lie on the east or the west of the Mesoamerica continental mass? All these characteristics—quite apart from the circulation of water masses over them and the seasonality of river discharges—determine the kind and quantity of ecosystem productivity and biomass production (Deegan *et al.*, 1986; Longhurst and Pauly, 1987).

Tectonic activity causes shelves to be narrow on active, subducting margins as along the Pacific coast of Mesoamerica, and allows them to be wider along passive margins, as on the eastern coast of the American continent (Kolarsky *et al.*, 1995; Parkinson *et al.*, 1998). The insular arc in the Caribbean protects shelves from wave action and allows the development of relative wide shelves, such as the Campeche Sound (Mexico), the Magdalena basin (Colombia), or the Orinoco delta (Venezuela) in the northern and southern end of the Atlantic Mesoamerican coastal zone. But earthquakes associated with coastal subduction has been clearly correlated with benthic organisms (Cortés *et al.*, 1992).

Major rivers can modify the morphology of continental shelves: the Grijalva/Usumacinta (Mexico), Dulce (Guatemala), Magdalena (Colombia), and Orinoco (Venezuela) rivers have all widened the adjacent continental shelves and have built fans of sediment down the continental slope on the Atlantic coast of Mesoamerica, even the influence of the Amazon from Brazil must be considered as important in the coastal ecology of Middle America. Figure M16 shows the astonishing manner in which a few major tropical river systems dominate the global transport of sediments from the continents to the ocean coast in Middle America. Recent references for Colombia are (Restrepo and Kjerfve, 2000a,b). This amount of sediment discharged to the coastal zone is reflected in the distribution of the sediments of tropical continental shelves and seas, and in the high turbidity of coastal seas in the monsoon-type regions of Mesoamerica (i.e., Tabasco-Campeche Mexico, Costa Rica, Panama–Colombia).

The superficial geology of continental shelves is determined by processes that also determine coastal morphology, working in concert with several important biogenic processes of sediment production. Shelf sediment in the Middle America coastal zone includes inorganic components such as sand derived from weathering of rock, and transported by wind, longshore currents, or rivers, and also silts and clays that are mostly river borne. The organic components of shelf sediments include terrestrial plant debris in all stages of decomposition which is either riverborne as particulates, or flocculated from dissolved organic material at the fresh/saltwater interface in the estuaries (i.e., Gulf of Fonseca shared by El Salvador, Honduras, and Nicaragua; Sierpes delta in Costa Rica, both areas in the Pacific coast). The remaining shelf deposits are oolite sand formed by precipitation from dissolved calcium; shelf-sand and large particles of calcareous material derived from molluscan benthos; reef corals and coraline algae, produced largely *in situ* and not transported great distances; and finally, small particulate organic material from planktonic communities of plants and animals.

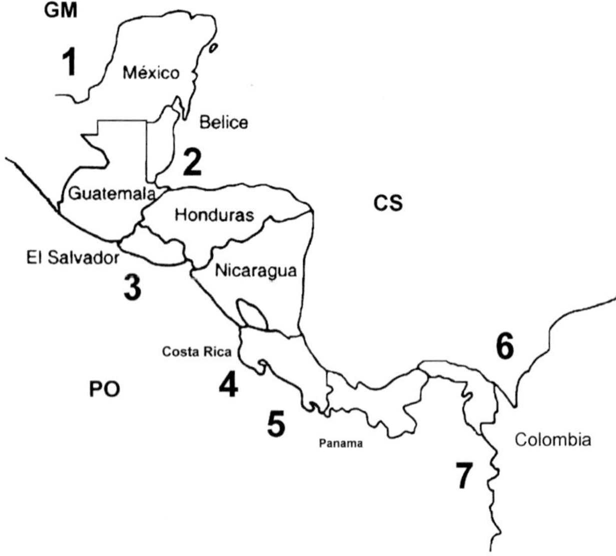

Figure M16 River transport of water and silt to the continental shelf from the major drainage basins for Middle America. Water discharge in $m^3\ s^{-1}$, and sediment yield in million tons per year. (1) Mexico, Grijalva-Usumacinta 4,400 $m^3\ s^{-1}$; 126×10^6 tons yr^{-1}. (2) Guatemala, Rio Dulce 1,100 $m^3\ s^{-1}$. (3) El Salvador/Honduras/Nicaragua, 834 $m^3\ s^{-1}$. (4) Costa Rica, Gulf of Nicoya Tempisque River 50 $m^3\ s^{-1}$. (5) Costa Rica, Terraba-Sierpes Deltaic River 338 $m^3\ s^{-1}$; 2×10^6 ton yr^{-1}. (6) Colombia, Magdalena Deltaic River 7,200 $m^3\ s^{-1}$; 144×10^6 tons yr^{-1}. (7) Colombian western discharge 8,020 $m^3\ s^{-1}$; 16×10^6 tons yr^{-1}. GM = Gulf of Mexico, CS = Caribbean Sea, PO = Pacific Ocean.

Contrary to general conceptions, tropical continental shelves dominated by the effects of mangroves or reef corals are more the exception than the rule (Longhurst and Pauly, 1987). Mangroves are important only where deltaic or a low-lying coastal plain occurs (Kjerfve, 1990; Wolanski and Boto, 1990; Jimenez, 1994; Kjerfve, 1998; Yáñez-Arancibia and Lara Domínguez, 1999); and coral reefs are important only where negligible amounts of terrigenous material reach their habitat (Wolanski and Choat, 1992; Kjerfve, 1998; Cortes, 2002).

Major ecological systems

Continental shelf

The eastern Mesoamerican shelf is much more complex, and dominated by terrigenous deposits principally near the mouths of the Grijalva/Usumacinta, Dulce, Magdalena, Orinoco rivers, where spectacular mud regions dominate. This is also true in the Terraba-Sierpes river delta on the western coast of Costa Rica. The continental coastline of Central America is much folded and fractured by tectonic activity, and has basins formed by calcareous deposits. The wide, very shallow Yucatan shelf has offshore banks (notably Campeche Bank), at least one being of atoll form, and the shelf sediments are predominantly calcareous. Along the northern South America coast (the southern end of Middle America), which itself is much altered by large tectonic faults, a shelf is lacking in the Pacific, but because of the outgrowth of the Magdalena estuary in the Colombian Atlantic a significant portion of shelf exists. Off Venezuela, the shelf deposits are terrigenous and the shelf itself is complex, with the deep Cariaco Trench intruding. Offshore, the Antilles and Caribbean islands are rocky, or bear coral reefs, and the geological environment resembles the western Pacific, an impression strengthened by the occurrence off Belize and Cancun (Mexico) of a Barrier reef complex similar in structure if not in extent to the Great Barrier Reef of Australia.

The western coast of Middle America is dominated by an active subduction zone almost the full length of the continent. Coral reefs and banks are largely absent, and mangroves dominate only along parts of Central America (Chiapas Mexico, Guatemala, Gulf of Fonseca El Salvador/Honduras/Nicaragua, Terraba-Sierpes Costa Rica, and Colombian coasts). Despite the absence of major rivers, terrigenous material is very important. The continental shelf is very narrow. In the Panama Bight, silt and terrestrial organic material contribute to muddy deposits, especially closer to the coast; here, the outer shelf tends to have sandier deposits. The shelf is very narrow or absent off Middle America, where depths fall away down the continental slope to 6,000 m in the Middle America Trench. Only off Panama and the Gulf of Tehuantepec in southwestern Mexico is there a significant width of shelf (Tapia García, 1998); elsewhere, only in isolated bays and gulf are there shallow areas resembling continental shelf habitats.

Coastal lagoon systems

Coastal lagoons here include lagoons, estuaries, and deltaic plains, in the sense of Day and Yáñez-Arancibia (1982), Kjerfve (1986), Yáñez-Arancibia (1987). These systems occur in all tropical oceans (Kjerfve, 1994). They are a specially prominent topographic feature on both the Atlantic and Pacific coasts of Middle America, including the northern (Mexico) and southern (Colombia) end (Yáñez-Arancibia, 1999). Many stretches of coastline are backed not by solid land, but by coastal lagoons behind barrier island, open to the sea to various degrees, with or without rivers, and of great significance in some coastal fisheries as nursery grounds (Lasserre and Postma, 1982; Kjerfve, 1994).

Lagoons not associated with coral reefs are most strongly developed on coasts with a history of submergence during the Holocene post-glacial sea-level rise during the last 10,000 years or so (Kjerfve, 1986, 1990). Lagoons may also form behind the cuspate spits that accrete at the mouth of open estuaries and delta mouths, as in Tabasco/Campeche Mexico (Yáñez-Arancibia and Day, 1988). Special evaporative mechanisms may also build lagoons in arid or semiarid regions, characterized by ephemeral inlets (Kjerfve, 1994). The formation and maintenance of a lagoon's coastal barrier depends on a balance between the supply of sedimentary material and its removal to deeper water by wave action. Sediment may be supplied to the outer side of the barrier by longshore currents and normal wave action, or may enter the lagoon itself by beach "washover" during storms as in Amatique Bay Guatemala; Terminos Lagoon, Mexico, Gulf of Fonseca El Salvador/Honduras/Nicaragua (Yáñez-Arancibia and Day, 1988; Yáñez-Arancibia *et al.*, 1999). It may also accumulate by the settlement of riverborne silt, by flocculation of organic material at the saltwater–freshwater interface, or by local estuarine plant production especially by a mangrove ecosystem and associated coastal wetlands (i.e., La Encrucijada Chiapas, Mexico; Tortugero, Costa Rica).

The primary determinant for the existence of lagoon–barrier coasts is a relative small tidal range (Kjerfve, 1986). In microtidal (<2 m) environments as in Middle America, wave action is especially important in maintaining linear barrier islands. There is a relationship between tidal range and the width and number of the mouths of tidal channels by which lagoons open to the sea. In the Middle America region, there are few, narrow entrances, and barrier islands are long and narrow. Also of great significance for lagoon geomorphology is the pattern and amount of river discharge that they receive and must carry to the sea; in many places, remains of terrestrial plant cover destroyed by deforestation are having important effects on the sedimentary regime within lagoons. Precipitation/evaporation rate is a further important factor in the evolution and environmental behavior of lagoons once formed. Lagoons in Middle America are in a state of continuing evolution, and many are already filled in by sea-sand transport and the accumulation of both terrigenous sediments and biogenic material. Nevertheless, the remaining complex dominates the coastal ecology in Central America and it is an important factor in the coastal productivity and subsequent fisheries in the region. This family of coastal lagoons differs from those that are formed in association with coral reefs as in Yucatan Peninsula, Mexico, or Belize.

Shelf–estuary–lagoon relationships

There is a general relationship between the amount of water and terrigenous sediments discharged onto continental shelves, and the production there of fish that has been studied particularly well in the Gulf of Mexico (Deegan *et al.*, 1986; Sánchez-Gil and Yáñez-Arancibia, 1997), confirming the linkages between shelves, estuaries, and coastal lagoons, coupled with river discharge budgets, estuary area, intertidal area, coastal vegetation area, and fishery harvest (Figure M17). Such relationships probably can also occur in the Gulf of Fonseca influence area, the Gulf of Nicoya and Gulf Dulce influence areas, and Terraba-Sierpes estuarine plume, in the Pacific coast of Middle America, assumed by analyzing the reference of Jesse (1996), von Wangelin and Wolff (1996), Wolff (1996), Vargas (1996), Leon-Morales and Vargas (1998), Dittmann and Vargas (2001). Seasonally, there is a natural alternation between dry-season conditions with the river discharge, and rainy season when the river discharge dominate the deltaic conditions, as well as the winds regime and tidal range throughout the year.

As we might expect from the greater relative area of reef, rock, and especially soft bottom habitats off deltaic systems in the Atlantic coast of Middle America, Longhurst and Pauly (1987), based on selected references, can distinguish four species assemblages of fish that comprise the coastal demersal resources in the tropical west central Atlantic coastal ocean:

1. *Lutjanid community*. Fauna of rock, coral, and coral sand from Florida to Brazil, which is dominant on the Bahamas, the Antilles, and the other Caribbean islands; and on the coast of Yucatan to Panama (Balistidae, Logocephalidae, Lutjanidae 14 spp., Pomadasyidae 3 spp., Serranidae 11 spp., Synodidae), see Longhurst and Pauly (1987), Lowe-McConnell (1987), Arreguín *et al.* (1996).
2. *Subtropical sciaenid community*. Fauna of soft deposits from the Gulf of Mexico to at least Cape Hatteras, and especially well developed near river mouths influence, including the Mississippi and Grijalva/Usumacinta deltas (Branchiostegidae, Clupeidae, Gerreidae, Polynemidae, Serranidae, Bothidae 19 spp., Sciaenidae 19 spp.), see Longhurst and Pauly (1987), Sánchez-Gil and Yáñez-Arancibia (1997).
3. *Tropical sciaenid community*. Fauna of the soft inshore muddy and soft sand deposits from the southern coast of the Caribbean to Cape Frio in Brazil (Dasyatidae, Ariidae 6 spp., Clupeidae, Gerreidae, Heterosomata 3 spp., Ephippidae, Pomadasyidae, Sciaenidae 19 spp.), see Longhurst and Pauly (1987), Lowe-McConnell (1987).
4. *Sparid community*. Fauna of the sandy and muddy sands of the subtropical regions north of Cape Hatteras and south to Cape Frio with very attenuated representation through the tropical region, that is, Yucatan to Venezuela (Ariidae, Carangidae 9 spp., Clupeidae, Mullidae, Sciaenidae, Sparidae 10 spp., Synodidae or Synodontidae),

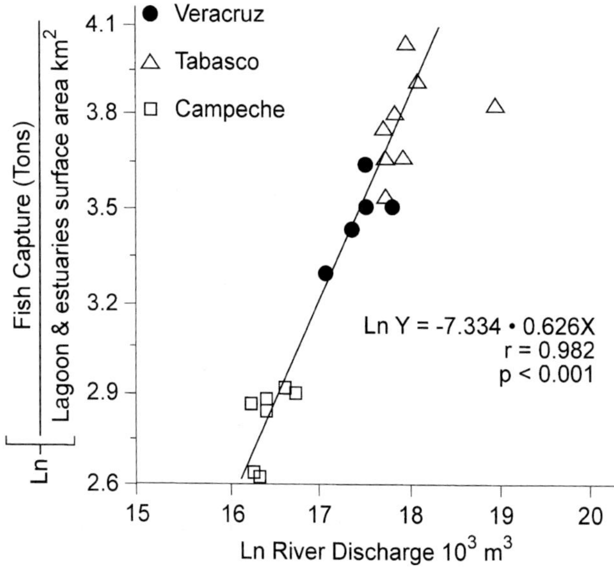

Figure M17 Linear logarithmic regression between fishery harvest per unit open waters (lagoons and estuaries) and average river discharge. The data are from the states of Veracruz, Tabasco, and Campeche in Mexico.

see Longhurst and Pauly (1987), Lowe-McConnell (1987), Sánchez-Gil and Yáñez-Arancibia (1997).

In the Pacific of Middle America important progress has been done for fish and benthos species assemblages, and after analyzing Jesse (1996), Wolff (1996), Vargas (2001), and Dittmann and Vargas (2001), some analogies can be found in the coastal demersal resources in the tropical coastal zone in the Mesoamerican Pacific in comparison with the Caribbean coast.

Because relatively few fish species are wholly adapted to a life cycle within a lagoon–estuarine system *sensu-stricto*, Longhurst and Pauly (1987) pointed out a controversial discussion on the topic. But it is clear—nevertheless—that in the tropic of Middle America, and especially in monsoon-type areas where estuarinization of the continental shelf occur, the distinction between estuaries and the neritic sea is slighter than in non-monsoon type areas (Yáñez-Arancibia, 1985; Yáñez-Arancibia *et al.*, 1994). This approach can be easily applied in Tabasco/Campeche, Mexico; Amatique Bay, Guatemala; Magdalena delta shelf Colombia; Gulf of Fonseca, El Salvador/Honduras/Nicaragua; Gulf of Nicoya, Costa Rica.

There are also sufficient descriptions of specialized estuarine fish fauna for it to be clear that this is often a reality in the tropics of Middle America and adjacent sub-tropical regions, where the size of the estuary and its tidal regime permit a relatively long flushing period. If the Guntherian hypothesis of estuarine-dependence of continental shelf fish stocks can not be valid for the tropical regions as a whole (Longhurst and Pauly, 1987), at least it can be applied without any doubt in tropical America especially in soft bottom communities on the shelf associated with extensive deltaic systems, river discharge, and broad areas of coastal vegetation (Deegan *et al.*, 1986; Pauly, 1986, 1998; Pauly and Yáñez-Arancibia, 1994; Sánchez-Gil and Yáñez-Arancibia, 1997; Baltz *et al.*, 1998; Chesney *et al.*, 2000; Lara-Dominguez, 2001, i.e., Figure M17).

As a consequence, some important points have emerged from those studies; (1) The utilization of the lagoon–estuarine environment is an integral part of the life cycle of numerous fishes, particularly in the Neotropics (Middle America). (2) The lagoon–estuarine environment is mainly utilized by juveniles and young adults. (3) There are a greater number of fish species in tropical and subtropical lagoon–estuarine ecosystems than in comparable temperate or boreal systems. (4) Second-order consumers are more abundant and diverse than first-order consumers or top carnivores. (5) Functional components are (a) freshwater spawners occurring in waters <10 ppt salinity, (b) brackish water groups limited to 10–34 ppt salinity, (c) marine spawners occurring in waters > or = 35 ppt salinity. (6) Three groups of fish occur in lagoon–estuarine systems (a) resident species, those which spend their entire life cycle within coastal lagoons, (b) seasonal migrants, those which enter the lagoon during a more or less well-defined season from either the marine or the freshwater side and leave it during another season, (c) occasional visitors, those which enter and leave the lagoon without a clear pattern within and among years. To these, two other groups may be added: (d) marine, estuarine-related species, those which spend their entire life cycle on the inner sea shelf under the estuarine plume influence, and (e) freshwater, estuarine-related species, those which spend their entire life cycle in the fluvial-deltaic river zone, associated with the upper zone of the estuarine system. Finally, if the "recruitment" both biological and fishery, utilize coastal lagoons and estuaries, those resources are "estuarine-dependent," but if that resources utilize regularly the option of the estuarine plume on the continental shelf, they are "estuarine-related" or estuarine opportunistic.

An advanced approach, following shelf–estuary–lagoon relationships, and the compressive models of trophic fluxes in key coastal Mesoamerican ecosystems, has recently been done. Construction of mass-balance trophic models of aquatic ecosystems, including complete food webs, from primary producers to top predators, allows numerous inferences on ecosystem status and function and can serve as a base for dynamic simulation models, as well as comprehension of shelf–estuary–lagoon relationships and the coastal ecological processes evolved. In Mesoamerica, this is a clear coastal ecological perspective, that is, in the southern Gulf of Mexico (Pauly *et al.*, 1999), in the Dulce Gulf (Wolff *et al.*, 1996), and in the Gulf of Nicoya (Wolff *et al.*, 1998).

For instance, the Dulce Gulf acts differently from most tropical coastal ecosystems, as it is dominated by biomass and energy flow within the pelagic domain and resembles rather an open ocean system than an estuarine one. Due to its low benthic biomass and low overall productivity there seems no potential for a further development of the demersal and semi-demersal fishery inside the Gulf; and an increase of the fishing pressure on pelagic fish would seriously threaten the large resident predators, such as dolphin, sharks, and large birds species (Wolff *et al.*, 1996). In the Gulf of Nicoya, the model shows that shrimps occupy a central role within the Gulf as converters of detritus and other food into prey biomass for many predators, that seem to be simultaneously affected by the over exploitation of this resource (Wolff *et al.*, 1998), as occurs in classical tropical coastal estuary–shelf relationships.

In the southern Gulf of Mexico, the trophic model shows the estimation of the primary production required to sustain fisheries offshore, coupled with tropical lagoon–estuarine systems (Terminos Lagoon), which can be determined given knowledge of fishery catches, from the trophic levels of the exploited groups represented in those catches, and the transfer efficiency between trophic levels (Pauly *et al.*, 1999).

These examples will be more useful the smaller the scale that is used for the spatial stratification of the overall models; also more local models will mean more consideration of local data and constraints and hence a more realistic overall model and more realistic aggregate statistics (i.e., primary production required to sustain the fisheries of those key coastal ecosystems in Mesoamerica).

Mangroves, coral reefs, islands, and other related ecosystems

Not only coral reefs, or river basins and associated wetlands, are a common coastal landscape in Middle America, but the most important critical habitats are the mangroves in the coastal zone (Jiménez, 1994; Kjerfve, 1998; Yáñez-Arancibia and Lara-Domínguez, 1999). The relative importance of mangroves for each Central America country is illustrated by Windevoxhel *et al.* (1999). This rough representation shows that mangroves are the most important forest formation in certain countries and they should be given priority in management and conservation. The entire coast is characterized by the presence of mangroves, with nine species present in five genera on the Pacific, and four genera on the Caribbean and the Atlantic coast of Mesoamerica. Central America possess 8% of the world's mangroves, and represents 7% of natural forest in the region (Jiménez, 1994; Suman, 1994). The greater extensions of mangrove on the Pacific are found off the coast of southwestern Mexico, Guatemala, Costa Rica, Panamá, and the Gulf of Fonseca (Table M7). Pacific mangroves maintain lesser biodiversity associated with root systems than those in the Caribbean, where the small changes of tides provide conditions of environmental stability for root-associated organisms, including important sea grass beds associated with mangrove systems, as in Campeche and Yucatan Peninsula Mexico, Honduras, Nicaragua, and Costa Rica. However, Pacific mangroves are more extensive due to the greater tides and coastal topography. These conditions offer greater areas that exclude competitors and favor facultative halophytes with saline intrusion. On the Caribbean and the Atlantic coast of Mesoamerica, the most extensive areas of mangrove are found in Honduras and Nicaragua, and also southeast México.

In general terms, Pacific coast coral reefs are less extensive and diverse than those on the Caribbean (Cortés, 2002). Live coral formations have been described in El Salvador, Costa Rica, and Panama. Coral communities in the Pacific are richer off the southwestern coast of Costa Rica, and particularly in Panama, where at least 21 species have been reported (Cortes, 1997a). On the Caribbean side, coral reefs can be found in all of the countries in the Atlantic coast of Mesoamérica (Cortés, 1997b). The Belize coral barrier or reef system (Mexico, Belize, Guatemala, Honduras) runs more than 220 linear kilometers and contains atolls and other formations practically unique in the Caribbean Sea. More than 60 species of coral have been reported to exist on the coast (Woodley, 1995; Gibson and Carter, 2002), but the total number of species associated with the coral reef system is still uncertain. Offshore the Antilles and Caribbean island are rocky, or bear coral reef, and the geological environment recalls that the western Pacific, an impression strengthened by the occurrence off Belize and Cancun (Mexico) of a barrier reef complex similar in structure, if not, in extent, to the Great Barrier Reef of Australia.

The Pacific coast has long sandy beaches with a broad range of texture and color, and predominant clear coastal waters. Beaches on the Caribbean and the Atlantic coast are less extensive as a result of current patterns and sea cycles, as well as the oceanographic and geomorphologic factors, and because the presence of important river basins, predominant turbid coastal waters, with the exception of Yucatan Mexico, Belize, and southeastern Costa Rica.

Islands and islets abound on the Caribbean coast. There are some 2,400. In Campeche Mexico, Isla del Carmen is a barrier sandy island on the Usumacinta delta; but mostly associated with coral formations, as in Yucatan Peninsula Mexico (Cayo Arcas, Alacranes, Chinchorro, Isla Mujeres, and Cozumel), Belize (the Keys), Honduras (Bay Islands

Table M7 Biophysical characteristics of the Central American coastal zone (from Windevoxhel *et al.*, 1999, with permission from Elsevier)

Biophysical aspects	BEL	GUA	HON	E.S.	NIC	C.R.	PAN	Total
National Territory (km^2)	22,965	108,889	112,088	20,935	118,358	50,900	77,082	511,217
Population (millions) 1994	0.209	10.322	5.497	5.641	4.275	3.334	2.611	31,889
Density (Pop/km^2) 1994	9.1	94.8	49.0	269.5	36.1	65.8	33.9	62.4
% Population in the CMZ	39	26	15	13	24	7	50	21.6
Length of coast (km)	250	403	844	307	923	1376	2500	6603
Coast/territory rate	0.01	0.003	0.007	0.001	0.008	0.03	0.03	0.01
Contin. Platform, 200 m (km^2)	8250	12,300	53,500	17,700	15,800	57,300	237,650	
EEZ area (thousand of km^2)	n.d.	99.1	200.9	91.9	159.8	258.9	306.5	1117.1
Mangrove areas (ha)	11,500	16,000	14,5800	26,800	15,5000	41,000	170,800	566,900
Coral reefs (km)	474	1	364	1	455	2.5	320	1617.5
Surface drainage Pacific (%)	0	21	18	100	10	53	69	39
Surface drainage Carib. (%)	100	79	82	0	90	47	31	61

and Cochinos Keys), Nicaragua (Miskitos, Cisne, and Maiz Keys), and Panama (Bocas del Toro and San Blas Archipelago). In comparison, there are few islands on the Pacific coast except for Panama, with some 200. There is a small group of islands in the Gulf of Fonseca (including Meanguera, Meanguerita, Amapala, and El Tigre). A group of eight islands are found in Costa Rica's Gulf of Nicoya, with the Chira island and Murcielago island located to the north, and Isla del Caño 11 km in front of Peninsula de Osa. Coco Island, 500 km to the southwest of the continent, marks the most distant territorial point of the Central America region.

Physiographic variability is also present on the sea bottoms. The Mesoamerican Trench extends all along the Central America Pacific and has a maximum depth of 6,662 m. The Caribbean Cayman Trench has a maximum depth of 7,680 m, with depths up to 2,000 m off Belize.

The largest regional outcrops of ocean waters are found off the Pacific gulfs of Fonseca, Panama, and Papagayo. These are caused by the Caribbean's seasonal winds, which push water out to the sea causing outcrops of colder water richer in nutrients. This effect is also true in the Gulf of Tehuantepec in the southwestern Mexico caused by the seasonal winds coming from the Gulf of Mexico throughout the Tehuantepec Itsmo. As it is typical of tropical seas, the Caribbean's surface water is mixed very little with colder deep waters which are richer in nutrients, and as a result the open seas are low in primary productivity. The greatest wealth, in terms of Caribbean productivity, is associated with the presence of coral reefs, mangroves, and other important ecosystems (i.e., deltas) on which regional fishing depends.

Global ecological scenario: synthesis

The physiographic, hydrological, climatic, physiochemical, bathymetric characteristic, and habitat heterogeneity, littoral currents, and extension of continental shelf, previously described determine productivity, as well as the quantity and distribution of Central America's coastal-marine resources. Likewise, this distribution has historically conditioned utilization of resources and their relation to socio-economic development in the region. Traditional numbers indicate that Central America possess 6,603 km of coast, representing approximately 12% of the Latin America and Caribbean coast. These contain some 567,000 ha of mangroves, 1,600 km of coral reef, and about 237,650 km of continental platform where multiple activities of economic and social importance are carried out (Windevoxhel *et al.*, 1999). The region has potential for utilizing more than 1.1 million km^2 of exclusive economic zone (Table M7). Nevertheless, and following the "Mesoamerican" approach focused on in this entry, these numbers should be much more impressive considering the states of Chiapas, Campeche, Yucatan, and Quintana Roo in southern Mexico, as well as Colombia, and this is true for the length of the marine coast and its natural resources, shelf to 200 m depth, exclusive economic zone, urban population in coastal cities, average annual volume—metric tons—of goods loaded and unloaded as crude oil/gas products and dry cargo, tourism, and fish products (Yáñez-Arancibia, 1999).

The Central America coast consists of numerous peninsulas, gulfs, and bays favoring a high degree of physiographic diversity. Extensive intertidal zones and well-developed coastal barriers encircle great coastal protected waters. On the Pacific, coastal cliffs are highly developed in Costa Rica, and partially developed in El Salvador, Nicaragua, The Gulf of Fonseca, Panama, and Colombia; while Guatemala and Mexico have no coastal cliffs. On the Caribbean side, the coast tends to be quite flat and cliffs are nonexistent due to less drastic geological and geomorphologic processes, with some local exception in the Dulce River low basin in the Atlantic coast of Guatemala. There are very important mega-lagoon/estuarine systems in Mesoamerica, that is, Terminos Lagoon Mexico (Yáñez-Arancibia and Day, 1988), Amatique Bay, Guatemala (Yáñez-Arancibia *et al.*, 1999), Gulf of Fonseca, El Salvador/Honduras/Nicaragua (Gierloff-Emden, 1976), Gulf of Nicoya and Gulf Dulce Costa Rica (Voorhis *et al.*, 1983; Vargas, 1995; Lizano, 1998), and the Panama Bay.

Climatic conditions vary latitudinal along the Pacific coast. A dry zone from northern Costa Rica to Guatemala experiences water shortage for at least five months; transition toward a system of greater moisture is presented in Guatemala; rainfall is extremely heavy in southwestern Costa Rica and Panama, where shortage occurs no more than one of two months (Jiménez, 1999). The particular characteristics of the Caribbean coast determine differences between this area and the Pacific. For example, while tides on the Pacific reach up to 6 m (extreme), those on the Caribbean and the Atlantic of Mesoamerica coast are around 30 cm (extreme). Dominant winds on the Caribbean coast produce waves up to 3 m, higher than those on the Pacific. Pacific coast rivers are short and highly dynamic, and discharge significant volumes of sedimentation from May to November during the rainy season. Rivers on the Caribbean tend to be longer and discharge is greater and more stable as a result of topographical conditions and almost year-round precipitation, and in the case of the northern end of Mesoamerica (Pacific and Atlantic of Mexico) with a significant seasonal pulse in September–October.

Alejandro Yáñez-Arancibia

Bibliography

Arreguín, F., Munro, J.L., Balgos, M.C., and Pauly, D. (eds.), 1996. Biology, fisheries and culture of tropical groupers and snappers. *ICLARM Conference Proceedings*, **48**.

Baltz, D.M., Rakocinski, C.F., and Fleeger, J.W., 1993. Microhabitat used by marsh-edge fishes in Louisiana estuary. *Environmental Biology of Fishes*, **36**: 109–126.

Baltz, D.M., Fleeger, J.W., Rakocinski, C.F., and McCall, J.N., 1998. Food, density, and microhabitat: factors affecting growth and recruitment potential of juvenile saltmarsh fishes. *Environmental Biology of Fishes*, **53**: 89–103.

Burke, K., Cooper, C., Dewey, J.F., Mann, P., and Pindell, J.J., 1984. Caribbean tectonics and relative plate motions. *Geological Society of America Memories*, **162**: 31–63.

Buskirk, R.E., 1992. Zoogeographic and plate tectonics relationships of Jamaica to Mesoamerica. In Darwin, S.P. and Welden, A.L. (eds.), *Biogeography of Mesoamerica: Proceedings of a Symposium*. New Orleans: Tulane University Press, pp. 9–16.

Chesney, E.J., Baltz, D.M., and Thomas, R.G., 2000. Louisiana estuarine and coastal fisheries and habitats: perspectives from a fish's eye view. *Ecological Applications*, **10**(2): 350–366.

Coates, A.G., Jackson, T.B., Collins, L.S., Cronin, T.M., Dowsewitt, H.J., Bybell, L.M., Junge P., and Ovando, J.A., 1992. Closure of the Isthmus of Panama: the marshare marine record of Costa Rica and western Panama. Geological Society America. Bullets, **104**: 814–828.

Cortés, J., 1997a. Biology and geology of coral reefs of the eastern Pacific. *Coral Reefs*, **16**(Suppl.): S39–S46.

Cortés, J., 1997b. Status of the coral reefs of the Caribbean of Central America. *Proceedings of the 8th International Coral Reef Symposium*, Panama, **1**: pp. 335–340.

Cortés, J. (ed.), 2002. *Latin American Coral Reefs*. Elsevier Science Amsterdam: Ltd.

Cortés, J., Soto, R., Jimenez, C., and Astorge, A., 1992. Earthquake associated mortality of intertidal and coral reefs organisms (Caribbean of Costa Rica). *Proceedings of the 7th International Coral Reef Symposium. Guam*, **1**: 235–240.

Day, J.W., and Yañez-Arancibia, A., 1982. Coastal lagoons and estuaries: ecosystem approach. *Ciencia Interamericana (Mar. Sci.)*, OEA Washington, DC, **22**(1–2): 11–26.

Deegan, L.A., Day, J.W., Gosselink, J.G., Yáñez-Arancibia, A., Soberón, G., and Sanchez-Gil, P., 1986. Relationships among physical characteristics, vegetation distribution and fisheries yield in Gulf of Mexico estuaries. In Wolfe, D.A. (ed.), *Estuarine Variability*. New York: Academic Press Inc., pp. 83–100.

Dittmann, S., and Vargas, J.A., 2001. Tropical tidal flat benthos compared between Australia and Central America. In Reise, K. (ed.), *Ecological Comparisons of Sedimentary Shores*. Berlin, Heidelberg: Springer-Verlag, pp. 275–293.

Donnelly, T.W., 1985. Mesozoic and Cenozoic plate evolution of the Caribbean region. In Stehli, F. G., and Webb, S. D. (eds.), *The Great American Biotic Interchange*, New York: Plenum Press, pp. 89–121.

Duncan, R.A., and Hargraves, R.B., 1984. Plate tectonic evolution of the Caribbean region in the mantle reference frame. *Geological Society of America Memoires*, **162**: 81–93.

Gibson, J., and Carter, J., 2002. The reefs of Belize. In Cortés, J. (ed.), *Latin American Coral Reefs*. Amsterdam: Elsevier Science Ltd.

Gierloff-Emden, H. G., 1976. *La Costa de El Salvador: Monografia Morfológica Oceanográfica*. Ministerio de Educación, Dirección de Publicaciones, San Salvador, pp. 1–285.

Jesse, S., 1996. Demersal crustacean assemblages along the Pacific coasts of Costa Rica: a quantitative and multivariate assessment on the RV Victor Hansen CR Expedition 1993/1994. *Revista de Biologia Tropical*, **44**(Suppl. 3): 115–134.

Jiménez, J. A., 1994. *Los Manglares del Pacífico de Centromérica*. Heredia Costa Rica: Editorial Fundación UNA.

Jiménez, J. A., 1999. Ambiente, distribución y características estructurales en los manglares del Pacífico de Centro América: Contrastes climáticos. In Yáñez-Arancibia, A., and Lara-Domínguez, A.L. (eds.), *Mangroves Ecosystems in Tropical America*. INECOL Mexico, UICN/ORMA Costa Rica, NOAA/NMFS Silver Spring MS, USA, pp. 51–70.

Kjerfve, B., 1986. Comparative oceanography of coastal lagoons. In Wolfe, D. A. (ed.), *Estuarine Variability*, New York: Academic Press Inc., pp. 63–81.

Kjerfve, B., 1990. *Manual for Investigating Hydrological Processes in Mangrove Ecosystems*. UNESCO/UNDP. New Delhi, India, pp. 1–79.

Kjerfve, B. (ed)., 1994. *Coastal Lagoon Processes*. Elsevier Oceanography Series 60. Amsterdam: Elsevier Science Publications B.V.

Kjerfve, B. (ed)., 1998. *CARICOMP-Caribbean Coral Reef, Seagrass and Mangrove Sites.* Coastal Region and Small Island Paper, 3. Paris: UNESCO, 347pp.

Kolarsky, R. A., Mann, P., and Montero, W., 1995. Island arc response to shallow subduction of the Cocos Ridge Costa Rica. In Mann, P. (ed.), *Geologic and Tectonic Development of the Caribbean Plate Boundary in Southern Central America*. Geological Society of America Special Publication, 295, pp. 235–262.

Lara-Domínguez, A.L., 2001. Ecological structure of estuarine fish communities: habitat linkages among dominant species groups in terminos lagoon, Mexico. Ph.D. dissertation. Department of Oceanography and Coastal Sciences, Baton Rouge: Louisiana State University.

Lassserre, P., and Postma, H. (eds.), 1982. Coastal lagoons. *Oceanolofica Acta*, **5**(4): 1–461.

Leon-Morales, R., and Vargas, J. A., 1998. Macroinfauna of a tropical fjord-like embayment: Golfo Dulce Costa Rica. *Revista de Biologia Tropical*, **46**(Suppl. 6): 81–90.

Lizano, O. G., 1998. Dinámica de las aguas en la parte interna del Golfo de Nicoya ante altas descargas del río Tempisque. *Revista de Biologia Tropical*, **46**(Suppl. 6): 11–20.

Longhurst, A.R., and Pauly, D., 1987. *Ecology of Tropical Oceans*. San Diego, CA: Academic Press Inc.

Lowe-McConnell, R. M., 1987. *Ecological Studies in Tropical Fish Communities*. Cambridge, UK: Cambridge University Press.

Parkinson, R.W., Cortés, J., and Denyer, R., 1998. Passive margin sedimentation in Costa Rica's North Caribbean coastal plain, Rio Colorado. *Revista de Biologia Tropical*, **46**(Suppl. 6): 221–236.

Pauly, D., 1986. Problems of tropical inshore fisheries: fishery research on tropical soft bottom communities and the evolution of its conceptual base. In Borgese, E.M., and Ginsburg, N. (eds.), *Ocean Year Book*, Volume 6. London: The University of Chicago Press, pp. 29–54.

Pauly, D., 1998. Tropical fishes: patterns and propensities. *Journal of Fish Biology*, **53**(Suppl. A): 1–17.

Pauly, D., and Yáñez-Arancibia, A., 1994. Fisheries in coastal lagoons. In Kjerfve, B. (ed.), *Coastal Lagoon Processes*. Elsevier Oceanography Series, 60. Amsterdam: Elsevier Science Publication B.V., pp. 379–399.

Pauly, D., Arreguin, F., Browder, J., Christensen, V., Manickchaud, S., Martinez, E., and Vidal, L., 1999. Towards a stratified mass-balance model of trophic fluxes in the Gulf of Mexico. In Kumpf, H., Steidinger, K., and Sherman, K. (eds.), *The Gulf of Mexico Large Marine Ecosystem: Assessment, Sustainability and Management*. Malden, MA: Blackwell Science Inc., pp. 278–293.

Restrepo, J.D., and Kjerfve, B., 2000a. Magdalena river: interannual variability (1975–1995) and revised water discharge and sediment load estimates. *Journal of Hydrology*, **235**: 137–149.

Restrepo, J.D., and Kjerfve, B., 2000b. Water discharge and sediment load from the Western slopes of the Colombian Andes with focus on Rio San Juan. *Journal of Geology*, **108**: 17–33.

Rodriguez, J. J., and Windevoxhel, N. J., 1996. Analisis regional de la situación de la zona marino costera de Centroamérica. *Technical Report* IDB. Washington DC, UICN/ORMA, pp. 1–131.

Sánchez-Gil, P., and Yáñez-Arancibia, A., 1997. Grupos ecológicos funcionales y recursos pesqueros tropicales. In Flores, D., Sánchez-Gil, P., Seijo, J. C., and Arreguín, F. (eds.), *Análisis y Diagnóstico de los Recursos Pesqueros Críticos del Golfo de México*. Universidad Autónoma de Campeche, EPOMEX Serie Científica 7, pp. 357–389.

Suárez, G., Pardo, M., Domínguez, J., Ponce, L., Montero, W., Boschini, I., and Rojas, W., 1995. The Limón, Costa Rica earthquake of April 22, 1991: back arc thrusting and collisional tectonic in a subduction environment. *Tectonics*, **14**: 518–530.

Suman, D. (ed.), 1994. *El Ecosistema de Manglar en América Latina y la Cuenca del Caribe: Su Manejo y Conservación*. New York: RSMAS University of Miami, The Tinker Foundation.

Tapia García, M. (ed.), 1998. *El Golfo de Tehuantepec: El Ecosistema y sus Recursos*. Iztapalapa, México D.F: Universidad Autónoma Metropolitana.

Vargas, J. A., 1995. The Gulf of Nicoya estuary, Costa Rica: past, present, and future cooperative research. *Helgolander Meeresunters*, **49**: 821–828.

Vargas, J. A., 2001. Ecological dynamics of a tropical intertidal mudflat community. In Nordstrom, K. F., and Roman, C. T. (eds.), *Estuarine Shores: Evolution, Environments and Human Alterations*. New York: John Wiley & Sons Ltd, pp. 355–371.

Von Wangelin, M., and Wolff, M., 1996. Comparative biomass spectra and species composition of the zooplankton communities in Dulce Gulf and Gulf of Nicoya, Pacific coast of Costa Rica. *Revista de Biologia Tropical*, **44**(Suppl. 3): 135–155.

Voorhis, A., Epifanio, C. E., Maurer, D., Dittel, A. I., and Vargas, J. A., 1983. The estuarine character of the Gulf of Nicoya, an embayment on the Pacific coast of Central America. *Hydrobiologia*, **99**: 225–237.

Windevoxhel, N. J., Rodríguez, J. J., and Lahmann, E. J., 1999. Situation of integrated coastal zone management in Central America: experiences of the UICN wetlands and coastal zone conservation program. *Ocean and Coastal Management*, **42**(2–4): 257–282.

Wolanski, E., and Boto, K. G. (eds.), 1990. Mangrove oceanography and links with coastal waters. *Estuarine Coastal and Shelf Science*, **31**(5): 503–743.

Wolanski, E., and Choat, M. (eds.), 1992. Coral reef oceanography. *Continental Shelf Research*, **12**(7/8): 779–1002.

Wolff, M., 1996. Demersal fish assemblages along the Pacific coast of Costa Rica: a quantitative and multivariate assessment based on the RV Victor Hensen CR Expedition 1993/1994. *Revista de Biologia Tropical*, **44**(Suppl. 3): 187–214.

Wolff, M., Hartmann, H. J., and Koch, V., 1996. A pilot trophic model for Golfo Dulce, a fjord-like tropical embayment, Costa Rica. *Revista de Biologia Tropical*, **44**(Suppl. 3): 215–231.

Wolff, M., Koch, V., Chavarria, J. B., and Vargas, J.A., 1998. A trophic flow model of the Gulf of Nicoya, Costa Rica. *Revista de Biologia Tropical*, **44**(Suppl. 3): 63–79.

Woodley, J. D., 1995. Tropical Americas regional report of the issues and activities associated with coral reef and associated ecosystems. *Report Prepared for the 1995 International Coral Reef Initiate Workshop*. Dumaguete City, Philippines.

Yáñez-Arancibia, A. (ed.), 1985. *Fish Community Ecology in Estuaries and Coastal Lagoons: Towards an Ecosystem Integration*. ICML-UNAM, PUAL, Editorial Universitaria, México D.F.

Yáñez-Arancibia, A., 1987. Lagunas costeras y estuarios: cronología, criterios y conceptos para una clasificación ecológica de sistemas costeros. *Revista Sociedad Mexicana Historia Natural*, **39**: 35–54.

Yáñez-Arancibia, A. (ed.), 1999. Integrated coastal management in Latin America. *Ocean and Coastal Management*, **42**(2–4), 77–368.

Yáñez-Arancibia, A., 2000. Coastal management in Latin America. In Sheppard, C. (ed.), *Seas at the Millennium: An Environmental Evaluation*. Amsterdam: Elsevier Science Ltd., pp. 457–466.

Yáñez-Arancibia, A., and Day, J. W. (eds.), 1988. *Ecology of Coastal Ecosystems in the Southern Gulf of Mexico*. Baton Rouge: Coastal Ecology Institute LSU, Organization of American States, Washington DC, Mexico D.F. UNAM Editorial Universitaria.

Yáñez-Arancibia, A., and Lara-Domínguez, A. L. (eds.), 1999. *Mangroves Ecosystems in Tropical America*. INECOL Mexico, UICN/ORMS Costa Rica, Silver Spring MS: NOAA/NMFS.

Yáñez-Arancibia, A., Lara-Domínguez, A. L., and Pauly, D., 1994. Coastal lagoons as fish habitats. In Kjerfve (ed.), *Coastal Lagoon Processes*. Elsevier Oceanography Series, 60, Ámsterdam: Elsevier Science. Publicatios. B.V., pp. 363–376.

Yáñez-Arancibia, A., Zárate, D., Gómez, M., Godines, R., and Santiago Fandino, V., 1999. The ecosystem framework for planning and management of the Atlantic coast of Guatemala. *Ocean and Coastal Management*, **42**(2–4): 283–317.

Cross-references

Barrier Islands
Coastal Climate
Coastal Lakes and Lagoons
Coastal Zone Management
Continental Shelves
Coral Reefs
Deltas
Estuaries
Holocene Coastal Geomorphology
Mangroves, Ecology
Mangroves, Geomorphology
Muddy Coasts
Salt Marsh
Sandy Coasts
Shelf Processes
Tectonics and Neotectonics
Vegetated Coasts
Wetlands

MINING OF COASTAL MATERIALS

Mining is the process of extracting rock and mineral material from near-surface sources of the earth. It may include small, local borrow pits with no supporting infrastructure or large-scale operations with permanent facilities for removing, loading, and transporting the resource. Considering only coastal areas, those adjacent to the shoreline and extending seaward through the breaker (surf) zone (U.S. Army Coastal Engineering Research Center, 1966) and landward through the highest surfaces subject to modern processes of wave alteration, the area represented (based on estimates by Kuenen, 1950) is a little less than 1% of the global surface. The coastal zone, as used here, includes all parts of barrier islands, lagoons, tidal flats, and mangrove swamps, but does not include embayments and fiords, permanently exposed parts of deltas, lands bordering estuaries (such as the Bay of Fundy), and other inland tide-influenced areas.

Energy conditions within the coastal zone, as evidenced by the landforms, are very high. The zone typically encompasses a transition from terrestrial fluvial and eolian to marine processes, but in high latitudes may be transitional from glacial or periglacial to marine ice-shelf conditions. Everywhere coastal areas are subject to retreats and advances of the shore owing to deposition or erosion by oceanic currents, isostasy, and especially, change in sea level related to storage and melting of glacial ice. Recently, a relative rise in sea level has been augmented or induced in some low-lying coastal areas as a result of large-volume subsurface fluid withdrawal. The geologic history of these processes defines the changing positions of coastal zones relative to the rock and mineral resources stored in and supplied to them, and determines the volumes and concentrations of materials that may have economic value.

Most of the economically recoverable coastal-zone resources worldwide are sand and gravel deposits and heavy-mineral placers locally embedded in them. Occurrences of these resources are typically results of combined rock-fragment transport by water and ice from terrestrial areas, reworking by both marine and eolian currents, and either sea-level stability or inundation by rising sea level. Conversely, where tectonics or marine-circulation patterns produce nondeposition of terrigenous clastics or erosion of rocky shores, limestone and building stone may be exposed and available for mining. Authigenic mineralization in nearshore areas, especially of phosphorite, may require specific conditions of upwelling of nutrient-rich seawater in relatively stable areas of continental margins.

Sand and gravel

Deposits of sand and gravel are the most widespread and are probably the most economically important non-energy resource of global coastal zones (Cruickshank and Hess, 1975; Williams, 1986). Estimated global reserves of sand and gravel on continental margins are 2 orders of magnitude greater than those of terrestrial supplies; the proportion of these deposits that occur within coastal zones is uncertain but is no doubt substantial (Cruickshank, 1974). Principal uses of the resource include aggregate in cement and asphalt for construction, road base, earth fill, beach restoration, and a variety of industrial products dependent on silica sand and other sand-sized minerals (Langer and Glanzman, 1993). Increasingly during the latter 20th century, deposits of coastal-zone sand and gravel near urban areas were mined as an alternative to the extraction and shipping of dwindling terrestrial supplies from greater distances. During this same period there was also a shift from onshore to offshore mining. Because sand with physical properties comparable to those of beaches is needed for beach replenishment, shallow offshore sand accumulations near the sand-deficient beaches, such as shoreface, tidal deltas, and shoals, have become preferred mining sites.

Movement of erosion products from land areas by late-Quaternary fluvial and glacial action, and the flooding of the deposits by elevated sea level related to deglaciation and eustatic change, are the main processes by which the enormous reserves of global sand and gravel have developed and been preserved (Williams, 1986). Where wide continental shelves are tectonically inactive, these sand and gravel resources mostly remain concentrated in the nearshore depositional environments, generally being moved only short distances by marine currents and wave action.

Mining of coastal-zone sand and gravel only recently has become important on a global scale as fluvial sand and gravel deposits near coastal cities have been depleted. Harbingers of this trend especially have been urban areas of small- to medium-sized islands, where limited onshore supplies of fluvial sand and gravel have been nearly exhausted, leading to the dredging and use of coastal and offshore supplies. Since 1987, for example, an average of nearly 25 million tons of sand and gravel has been dredged annually from deposits off coastlines of Great Britain (Hitchcock *et al.*, 1999), accounting for perhaps a fourth of the national demand. In Japan, about half of recent aggregate needs have been met by mining marine supplies, and almost all remaining sand and gravel reserves of Puerto Rico are offshore (Rodriguez, 1994). Extending the estimates of Williams (1986), 20% or more of the sand and gravel needs of the Netherlands and Denmark are met by nearshore sources; adjacent to the New York metropolitan area about 1.5 million m^3 of coastal-zone sand and gravel are extracted annually, in part as an aid to navigation.

Placer minerals

Placers, heavy minerals (specific gravity > 2.85) released by rock weathering and generally concentrated mechanically by fluvial or littoral processes, occur in coastal areas where the minerals are derived locally from bedrock or from rocks of inland drainage basins that supply detritus to the coastal zone. Following transport from the rock source to a coastal environment, placers may be concentrated by tides, waves, and nearshore currents wherever shore processes are neither strongly erosive nor depositional (Rajamanickam, 2000). Thus, depending on provenance, the ability of water, ice, and wind to transport mineral particles varying in size and specific gravity, and fluctuations of sea level, coastal-zone placers occur as flooded fluvial and glacial deposits, disseminated beach deposits, or eluvial deposits (Garnett, 2000a; Kudrass, 2000; Yim, 2000).

Fluvial placers include nodules of gold and cassiterite (SnO_2) that have been transported short distances from their sites of origin by high-energy streamflows and deposited among coarse alluvial deposits of stream channels. Although wave action leads to concentration of these high-density minerals, deposition typically occurs during periods of lowered sea level when fluvial processes moved the placers into what is now the coastal area. Coastal-zone deposits of gold have been mined in Alaska, the South Island of New Zealand, southern Argentina and Chile, and the Philippines (Garnett, 2000a), and deposits of cassiterite have been extracted from coastal Indonesia, Malaysia, Burma, Thailand, China, Australia, and Cornwall, Great Britain (Yim, 2000). Although of relatively low specific gravity (3.5), but having extreme hardness and thus resistance to abrasion, diamond also occurs as placers in flooded paleo-river courses of coastal (and deeper marine) zones, often having originated far inland (Garnett, 2000b). Coastal diamond deposits are mined from Namibia and western South Africa; mining of diamonds off northern Australia, southern Kalimantan (Indonesia), and elsewhere generally has not proven profitable (Garnett, 2000b).

Disseminated beach deposits, or beach placers (specific gravity 2.85 to about 5.5), are not sufficiently heavier than most sediment particles (specific gravity about 2.7) to permit concentration by processes of fluvial sorting. Instead, sand-sized fluvial deposits of these light heavy minerals, especially rutile (TiO_2), ilmenite ($FeTiO_2$), magnetite (Fe_3O_4), zircon ($ZrSiO_4$), garnet ($Fe_3Al_2(SiO_4)_3$, monazite ($[Ce,Y]PO_4$), sillimanite (Al_2SiO_5), and, to a lesser degree, diamond, become concentrated in surf zones through wave sorting during periods of stable or rising sea level (Kudrass, 2000; Rajamanickam, 2000). Coastal-zone deposits of these minerals have been mined in Australia (rutile, zircon, garnet, monazite), South Africa (diamond, ilmenite), New Zealand (magnetite), India (ilmenite, zircon, rutile, garnet, monazite, sillimanite), Brazil (zircon, monazite), and Florida (zircon).

Other than gold and cassiterite, which are very heavy and tend to remain near the host rocks from which they were derived, the principal eluvial mineral of economic importance is phosphorite, formed mainly of varieties of apatite ($Ca_5[F,OH](PO_4)_3$) and defined as having a P_2O_5 content of at least 5% (Riggs, 1979). Unlike other heavy minerals, exploitable deposits of coastal-zone phosphorite can be formed authigenically by deposition of nutrient phosphorus during periods of low sedimentation in zones of upwelling submarine currents. The phosphorite becomes concentrated through microbial processes and later, if affected by sea-level change, further concentration of primary phosphate grains may occur by coastal-zone wave action (Riggs, 1979; Burnett and Riggs, 1990). Examples of economically minable phosphorite deposits occur along the coasts of Chile and Peru and in the Atlantic Coastal Plain of Florida and North Carolina (Riggs *et al.*, 1991).

Limestone and shell

The mining of coastal-zone limestone largely is limited to areas where terrestrial resources are otherwise unavailable within economical depth or transport distance, or are preferred owing to features such as shell content or purity. Examples are shallow-marine limestones bordering volcanic islands, coral and oyster reefs, and chalks and coquinas of nearly pure $CaCO_3$. Along tectonically active coastlines where terrestrial sediment is poorly stored, carbonate rocks of any age may be exposed and available for extraction. In contrast, shallow-marine lime precipitates and organic carbonates, principally shells, accumulate mainly on stable continental platforms having virtually no detrital deposition. Exceptions are those lagoonal areas where shell concentrations are so large that the muddy matrix can be removed economically. Where shells of marine organisms are concentrated as unlithified calcareous clasts, the resource may be mined as a source of lime for cement in addition to its aggregate properties (Cruickshank and Hess, 1975). In the Bahamas, deposits of precipitated aragonite and aragonitic oolites are mined as a source of pure calcium carbonate. Although these sources of lime, measured in billions of tons along the Grand Bahamas Bank, for example, may be mined locally, the value of limestone and shell deposits extracted annually is small compared to those of sand and gravel and heavy minerals.

Building stone and crushed rock

Coastlines of tectonically active areas, volcanic islands, and areas subject to isostatic rebound following glaciation typically have a narrow coastal zone bordered by bedrock. Where the mining of building stone and crushed rock for aggregate from inland sources is restricted owing to land use or environmental concerns, rock may be quarried at the landward edge of the coastal zone. Historically, these rock quarries have been small and the amount of rock extracted for building stone and aggregate has been of minor economic importance. Beginning about 1990, however, restricted access to available inland rock reserves in areas of high population density, such as western Europe, resulted in the establishment of superquarries at coastal exposures of granite, quartzite, gneiss, limestone, and sandstone. Examples of active superquarries, those yielding at least 5 million tons of rock annually, are in Scotland, Norway, and Mexico. Other coastal superquarries are planned for the Shetland Islands, the Western Isles of Scotland, Canada, Ireland, Norway, and Spain (Pearce, 1994).

Evaporites

The mining of offshore evaporite deposits, particularly halite, gypsum, and anhydrite, is limited mostly to coasts of countries where inland supplies are not readily available and importation costs are prohibitive. Thick beds of evaporites are mined in several regions of the United States, for example, but the needs of the Philippines are satisfied partly by the extraction of about 0.6 million tons of marine salt annually (Lyday, 1995). Similar quantities of salt are mined in Indonesia but the portion taken from coastal or marine environments is not specified (Kuo, 1995). Particularly along coasts of arid lands, such as those of the Arabian Peninsula and the Caspian Sea, evaporation of runoff and near-surface ground water concentrates dissolved solids beneath sabkhas. Extensive resources of potassium and magnesium in these areas, both as brines and in bedded evaporites, are likely to be extracted from these areas in the early 21st century.

Potential environmental impacts

Mining is a human disturbance that can have adverse environmental consequences if steps are not taken to prevent alteration of physical processes and degradation of water quality. In some well-documented examples, coastal sand extraction clearly has lowered beach and dune elevations, which subsequently has caused increased storm flooding, washover, and beach erosion (Zack, 1986; Nichols *et al.*, 1987; Webb and Morton, 1996). A well-documented example of damage due to reduced beach elevation, washover, and sea-cliff erosion occurred in the early 1900s at the fishing village of Hallsands, on the south Devon coast of England. The village was destroyed during intense storms following extensive dredging of tidal-zone sand and gravel (Worth, 1923; Robinson, 1961).

Where fluvial sand and gravel have been mined from stream channels entering the ocean or in lagoons, excavation pits filled with anoxic water may alter nearshore circulation patterns and degrade water quality. In other examples, subtidal sand extraction for beach nourishment has altered nearshore currents and wave refraction patterns, causing or accelerating erosion of the adjacent beaches (Combe and Soileau, 1987; Rodriguez, 1994). Diamond mining along the west coast of South Africa has disrupted the intertidal fauna, destroyed kelp beds, adversely affected lobster habitat, and vast areas of dunes have been destroyed to get to the underlying placer deposits. Controversy surrounds the possible adverse biological effects of hydraulic dredging in lagoons and other nearshore waters. Clearly, dredging of living reefs is deleterious, but the conclusions are less certain about the permanent damage to benthic organisms caused by increased concentrations of suspended sediment associated with offshore dredging (Blake *et al.*, 1996).

In most countries, sand mining from beaches and dunes begins as small, local operations because the resource is abundant and extraction costs are low. These operations, now illegal in the United States, continue until government authorities recognize that the practices cause permanent environmental damage, and then policies are formulated to mitigate future damage from sand extraction. Depending on the government's ability to enforce the regulations, mining practices may continue illegally where the economy is depressed. Most countries with coastal rock resources suitable for superquarry development are imposing standards for licensing designed to protect scenery and recreation, marine biota, and water and beach resources.

W.R. Osterkamp and R.A. Morton

Bibliography

Blake, N.J., Doyle, L.J., and Cutler, J.J., 1996. Impacts and direct effects of sand dredging for beach renourishment on the benthic organisms and geology of the West Florida shelf. Final Report for the Minerals Management Service, *OCS Report* 95-0005, 109p.

Burnett, W.C., and Riggs, S.R. (eds.), 1990. *Phosphate Deposits of the World, Volume 3, Neogene to Modern Phosphates*. Cambridge: Cambridge University Press.

Combe, A.J., and Soileau, C.W., 1987. Behavior or man-made beach and dune, Grand Isle, Louisiana: Coastal Sediments '87. *American Society of Civil Engineers*, **2**: 1232–1242.

Cruickshank, M.J., 1974. Model for assessment of benefit/cost ratios and environmental impacts of marine mining operation. *Proceedings of the International Symposium on Minerals and the Environment*. London: Institute of Mining and Metallurgy.

Cruickshank, M.J., and Hess, H.D., 1975. Marine sand and gravel mining. *Oceanus*, (Fall issue): 32–44.

Garnett, R.H.T., 2000a. Marine placer gold, with particular reference to Nome, Alaska. In Cronan, D. S. (ed.), *Handbook of Marine Mineral Deposits*. Boca Raton: CRC Press, p. 67–101.

Garnett, R.H.T., 2000b. Marine placer diamonds, with particular reference to southern Africa. In Cronan, D. S. (ed.), *Handbook of Marine Mineral Deposits*. Boca Raton: CRC Press, pp. 103–141.

Hitchcock, D.R., Newell, R.C., and Seiderer, L.J., 1999. Investigation of benthic and surface plumes associated with marine aggregate mining in the United Kingdom. Final Contract Report for the U.S. Department of the Interior, Minerals Management Service: Coastline Surveys Limited, Gloucestershire, UK.

Kuenen, P.H., 1950. *Marine Geology*. New York: Wiley & Sons.

Kudrass, H.R., 2000. Marine placer deposits and sea-level changes. In Cronan, D.S. (ed.), *Handbook of Marine Mineral Deposits*. Boca Raton: CRC Press, pp. 3–12.

Kuo, C.S., 1995. The mineral industry of Indonesia. In *Minerals Yearbook*, Volume III. U.S. Geological Survey, Miscellaneous and Special Book, pp. 79-86.

Langer, W.H., and Glanzman, V.M., 1993. Natural aggregate—building America's future. U.S. Geological Survey Circular 1110, 39p.

Lyday, T.Q., 1995. The mineral industry of Philippines. In Minerals Yearbook, Volume III. U.S. Geological Survey, Miscellaneous and Special Book, pp. 147–152.

Nichols, M., Cerco, C., Clinton, G., and Martinez, R., 1987. Coastal dunes for protection and sand resources. *Proceedings of the 5th Symposium on Coastal and Ocean Management*, **2**, Washington, DC, pp. 1302–1311.

Pearce, Fred, 1994. Rush for rock in the highlands. *New Scientist*, **141**: 1907.

Rajamanickam, G.V., 2000. Light-heavy minerals on the Indian continental shelf, including beaches. In Cronan, D.S. (ed.), *Handbook of Marine Mineral Deposits*. Boca Raton: CRC Press, pp. 13–26.

Riggs, S.R., 1979. Petrology of the Tertiary phosphorite system of Florida. *Economic Geology*, **74**: 195–220.

Riggs, S.R., Ellison, S.P., Jr., Fisher, W.L., Galloway, W.E., Jackson, M.L.W., and Morton, R.A., 1991. Mineral resources and geopressured-geothermal energy. In Salvador, Amos (ed.), *The Gulf of Mexico Basin: The Geology of North America*. Boulder: *Journal of the Geological Society of America*, pp. 495–528.

Robinson, A.H.W., 1961. The hydrography of Start Bay and its relationship to beach changes at Hallsands. *Geography Journal*, **131**: 63–77.

Rodriguez, Rafael, 1994. Sand and gravel resources of Puerto Rico. In *Selected Issues in the USGS Marine and Coastal Geology Program*: San Juan, Puerto Rico: U. S. Geological Survey, 2p.

U.S. Army Coastal Engineering Research Center, 1966. Shore protection, planning, and design. Government Printing Office, Washington, DC: *Technical Report* No. 4, 3rd edn., 580p.

Webb, R.M.T., and Morton, R.A., 1996. Impacts of mining and urbanization along the north coast of Puerto Rico: a case study of the San Juan metropolitan area. *Abstracts with Programs 1996, 109th Annual Meeting, Geological Society of America*, p. A80.

Williams, S.J., 1986. Sand and gravel deposits within the United States exclusive economic zone: resource assessment and uses: *Proceedings of the 18th Annual Offshore Technology Conference*, Houston, TX, pp. 377–384.

Worth, R.H., 1923. Hallsands and Start Bay, Part 3. *Transactions, Devon Association*, **55**: 131–147.

Yim, W.W.-S., 2000. Tin placer deposits on continental shelves. In Cronan, D. S. (ed.), *Handbook of Marine Mineral Deposits*. Boca Raton, FL: CRC Press, pp. 27–66.

Zack, Allen, 1986. Effects of sand removal on the shallow aquifer in the vicinity of the Camuy mangrove forest, Puerto Rico. *Proceedings of the 3rd Caribbean Islands Water Resources Congress*, Puerto Rico Water Resources Association, pp. 41–48.

Cross-references

Beach Nourishment
Beach Processes
Beach Sediment Characteristics
Changing Sea Levels
Environmental Quality
Shelf Processes

MODELING PLATFORMS, TERRACES, AND COASTAL EVOLUTION

Terraces and their associated platforms and sea cliffs are the wave-cut and wave-built features associated with the land–water interface of seacoasts and lake shores. Along ocean coasts, they are the primary signature of the stillstands in water level during the transgressions and regressions of Pleistocene and Holocene epochs. The mechanics of terracing are fundamental to understanding the evolution of today's coastlines with their platforms, sea cliffs, barriers, spits, and capes. Coastal evolution models must incorporate processes that treat sediment transport and deposition as well as the abrasion and cutting of bedrock formations. This can be accomplished by coupled models, one treating the mobile sediment and the other bedrock cutting.

Background

Early studies of terraces, platforms, and sea cliffs include the work of de Beaumont (1845), Cialdi (1866), Fisher (1866), and Gilbert (1885). Gilbert's study of the active topographic features along the shores of the Great Lakes, supplemented by the visually distinct lake levels of the Pleistocene fossil shores of Lake Bonneville in Utah, provided the most detailed insight into the formation of platforms and terraces. He describes the wave-quarried hard rock platforms as *wave-cut terraces* backed by sea cliffs and the depositional features comprised of littoral drift as *wave-built terraces*. He also studied the terrace relations to changing lake level. Emery (1960), Shepard (1963), and others used this nomenclature with *marine terrace* as a more general term for wave-cut and wave-built terraces along ocean coasts.

However, as pointed out by Trenhaile (1987), platforms may not be wave-cut but formed by other processes such as solution. He prefers *shore platform* as a more general term for rock surfaces of low gradient within or close to the intertidal zone. He uses the term *wave-cut terrace* to refer to the specific category of platform formed by waves (Trenhaile, 2002). Sunamura (1992) classifies shore platforms developed during the present sea level as (1) sloping, (2) horizontal, and (3) plunging cliff. A more descriptive nomenclature for the latter two would be (2) step platform and (3) submerged platform. Generally, types (1) and (2) develop from cliff recession, with the step in platform (2) caused by differential erosion of rock strata. Plunging cliff platforms have sea cliffs that extend below the present water surface before joining a submerged platform. The submerged platform is a remnant feature from rapid sea-level rise and/or land subsidence. In what follows we will discuss numerical modeling of wave-cut terraces. These features consist of rock platforms backed by sea cliffs that were formed during the present stillstand in sea level as well as relic terraces now found on the continental shelf buried under Holocene sediment (Figure M18).

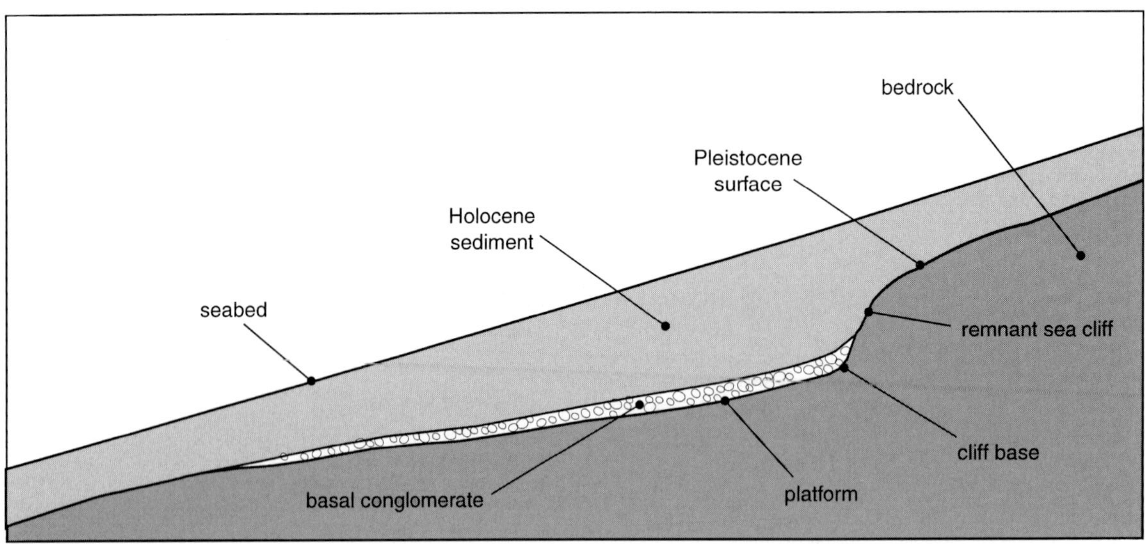

Figure M18 Illustration of wave-cut terrace notched into the bedrock and now found on the shelf by seismic profiling below a cover of Holocene surface sediment.

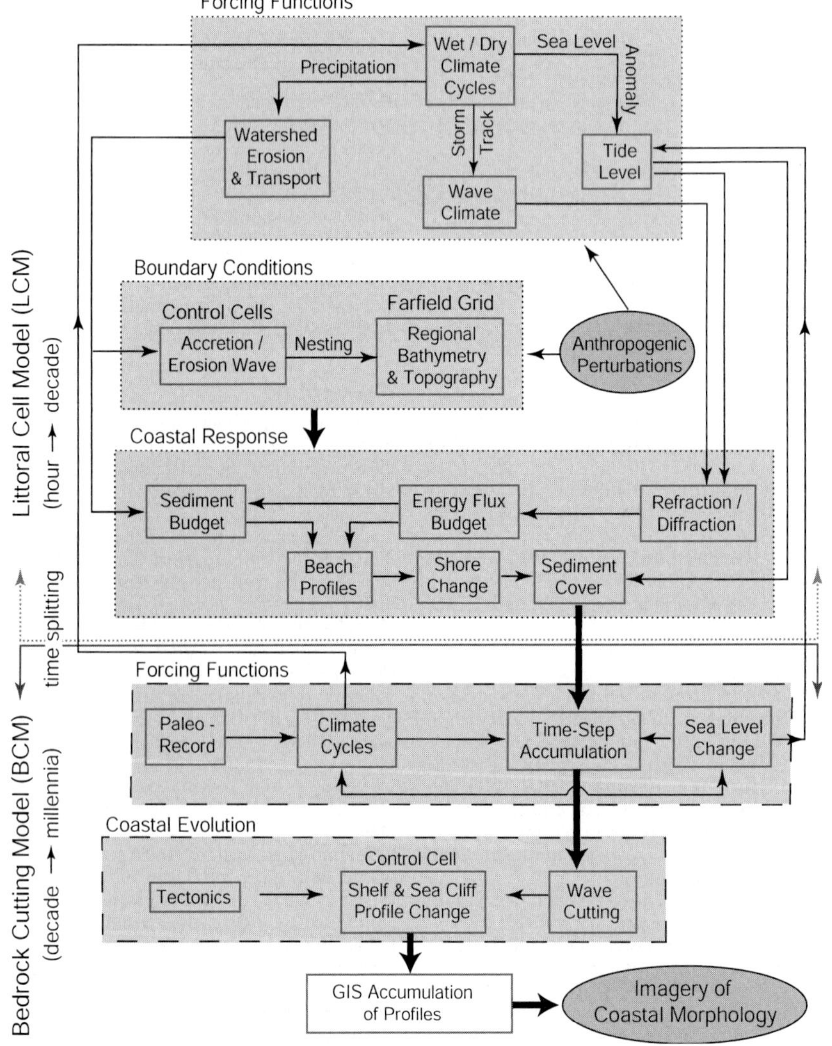

Figure M19 Architecture of the Coastal Evolution Model consisting of the LCM (above) and the BCM (below). Modules (shaded areas) are formed of coupled primitive process models (after Inman *et al.*, 2002).

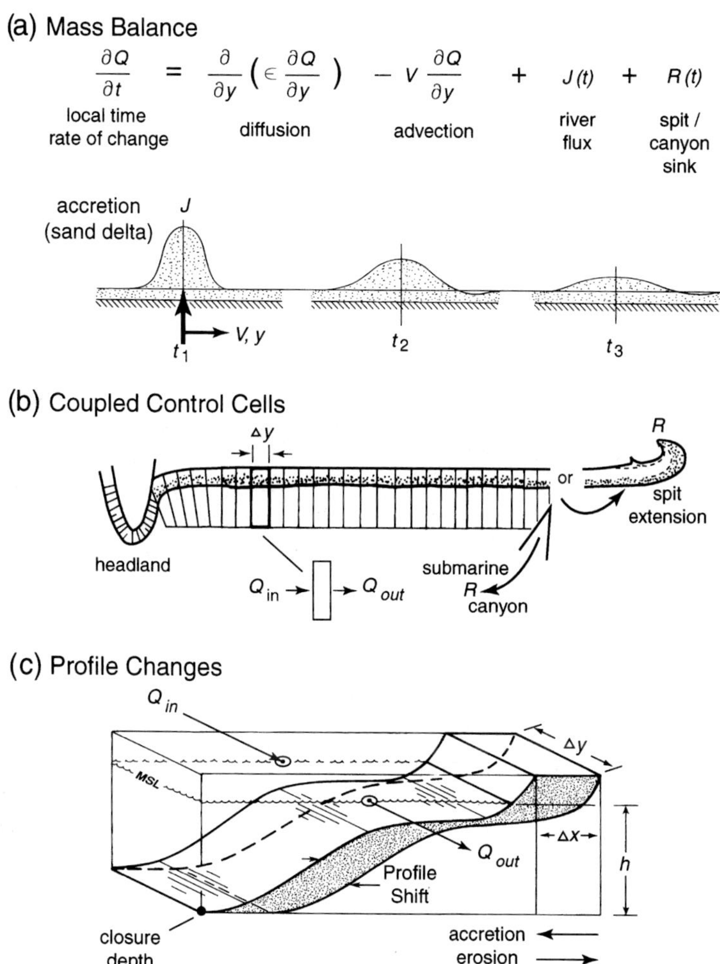

Figure M20 Computational approach for modeling shoreline change (after Inman *et al.*, 2002).

The present configuration of coastlines and their associated terraces and sea cliffs retain vestiges of the previous landforms from which they have evolved. Coastal evolution is a Markovian process where the present coastal features are dependent on the landforms and processes that preceded them (Inman and Nordstrom, 1971). This means that modeling coastal evolution must move forward in time from past known conditions and be evaluated by the present before proceeding to the future. Thus, paleocoastlines with their wave-cut terraces become time and space markers for modeling coastal evolution.

Numerical modeling of landforms

Physical models have provided insights and guidance to many of the processes leading to coastal evolution (e.g., Inman, 1983; Sunamura, 1992). Generally, these efforts are limited by the uncertainties between laboratory experiments and the time and space scales of the landforms they represent. These uncertainties are circumvented by numerical models where the temporal and spatial scales of the landforms are applied to the laws governing geomorphology.

Numerical modeling of landform evolution is a rapidly expanding field driven by the need to understand the environmental consequences of climate change and sea-level rise. Numerical modeling has been enabled by the revolution in computational power, graphical representation, and ever-expanding digital databases of streamflow, wave climate, sediment flux, and landform topography (Inman and Masters, 1994). Several two-dimensional models have been developed for the formation of wave-cut and wave-built terraces at various sea-levels. Storms *et al.* (2002) describe a process-response model for the development of barrier beaches during sea-level rise, and Trenhaile (2002) developed a model for the formation of rock platforms during changing sea level.

The Storms *et al.* (2002) model uses energy and mass flux balances to solve for incremental changes in the cross-shore profiles of mobile sediment in the Caspian Sea. On the other hand, the Trenhaile model uses a force-yield criterion to calculate the incremental erosion of steep, rocky submarine slopes. The latter model does not balance the budget of energy flux and, for certain selections of model parameters, requires a greater expenditure of energy in cutting rock than is available in the incident waves. This deficiency can be overcome by the energetics-based rock cutting model of Hancock and Anderson (2002). Developed for the formation of strath terraces in the Wind River valley during the Quaternary, the model includes sediment transport, vertical bedrock cutting that is limited by alluvial cover, and lateral valley-wall erosion. When reformulated for wave-forcing and sea-level change, their approach is applicable to wave-cut terraces.

Architecture of a coastal evolution model

Here, we describe the broad outlines of a three-dimensional coastal evolution model developed under funding from the Kavli Institute (Inman *et al.*, 2002). The model is functionally based on a geographic unit known as a littoral cell. A littoral cell is a coastal compartment that contains a complete cycle of sedimentation including sources, transport paths, and sinks. The universality of the littoral cell makes the model easily adaptable to other parts of the world by adjusting the boundary conditions of the model to cells characteristic of different coastal types (see entry on *Littoral Cells*).

The Coastal Evolution Model (Figure M19) consists of a Littoral Cell Model (LCM) and a Bedrock Cutting Model (BCM), both coupled and operating in varying time and space domains determined by sea level and the coastal boundaries of the littoral cell at that particular time. The LCM accounts for erosion of uplands by rainfall and the transport of mobile sediment along the coast by waves and currents, while the BCM accounts for the erosion of bedrock by wave action in the absence of a sedimentary cover. During stillstands in sea level along rock coasts, the combined effect of bottom erosion under breaking waves and cliffing by wave runup carves the distinctive notch in the shelf rock of the wave-cut terrace (Figure M18).

In both the LCM and BCM, the coastline of the littoral cell is divided into a series of coupled control cells (Figure M20). Each control cell is a small computational unit of uniform geometry where a balance is obtained between shoreline change and the inputs and outputs of mass and momentum. The model sequentially integrates over the control cells in a downdrift direction so that the shoreline response of each cell is dependent on the exchanges of mass and momentum between cells, giving continuity of coastal form in the downdrift direction. Although the overall computational domain of the littoral cell remains constant throughout time, there is a different coastline position at each time step in sea level with similar sets of coupled control cells.

Time and space scales used for wave forcing and shoreline response (applied at 6 h intervals) and sea-level change (applied annually) are very different. To accommodate these different scales, the model uses multiple nesting in space and time, providing small length scales inside large, and short timescales repeated inside of long time scales.

The LCM (Figure M19, upper) has been used to predict the change in shore width and beach profile resulting from the longshore transport of sand by wave action where sand source is from river runoff or from tidal exchange at inlets (e.g., Jenkins and Inman, 1999). It has also been used to compute the sand-level change (farfield effect) in the prediction of mine burial (Inman and Jenkins, 2002).

Bedrock erosion

The BCM (Figure M19, lower) models the erosion of bedrock by wave action during transgressions, regressions, and stillstands in sea level. Because bedrock cutting requires the near absence of a sediment cover, the boundary conditions for cutting are determined by the coupled mobile sediment model, LCM. When LCM indicates that the sediment cover is absent in a given area, then BCM kicks in and begins cutting. BCM cutting is powered by the wave climate input to LCM but applied only to areas where mobile sediment is absent. Time-splitting logic and feedback loops for climate cycles and sea-level change are imbedded in LCM together with long runtime capability to give a numerically stable couple with the BCM.

Bedrock cutting involves the action of wave-energy flux to perform the work required to notch the country rock, abrade the platform, and remove the excavated material. Both abrasion and notching mechanisms are computed by wave-cutting algorithms. These algorithms provide general solutions for the recession of the shelf and sea cliff. The recession is a function of the amount of time that the incident energy flux exceeds certain threshold conditions. These conditions require

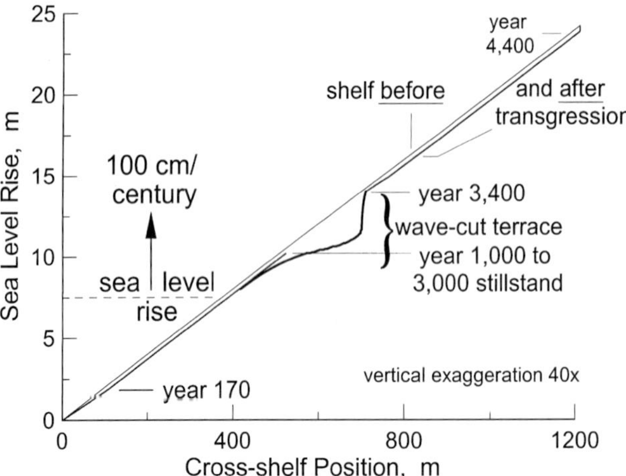

Figure M21 Example of BCM showing change in initial 2% shelf slope (thin line) due to wave cutting during a transgression/stillstand/transgression sequence (heavy solid and dashed line) (after Inman et al., 2002).

sufficient wave-energy flux to remove the sediment cover, and a residual energy flux that exceeds the erodibility of the underlying bedrock. The erodibility is given separate functional dependence on wave height for platform abrasion and wave notching of the sea cliff.

The erodibility for platform abrasion increases with the 1.6 power of the local shoaling wave and bore height, commensurate with the energy required to move the cobbles in the basal conglomerate that abrade the bedrock platform (Figure M18). As a consequence, recession by abrasion is a maximum at the wave breakpoint and decreases both seaward and shoreward of that point. In contrast, the erodibility of the notching mechanism is a force-yield relation associated with the shock pressure of the wave bore striking the sea cliff (Bagnold, 1939; Trenhaile, 2002). The shock pressure is proportional to the runup velocity squared, and its field of application is limited by wave runup elevation. Wave pressure solutions (Havelock, 1940) give a notching erodibility that increases with the square of the wave runup height above water level.

An example of terrace cutting by the BCM is shown in Figure M21 where a constant sea-level rise of 100 cm/century over a continental shelf sloping 2% was interrupted by a 2,000-year stillstand. The wave cutting was driven by a two-decade continuous wave record reconstructed for the southern California shelf by wave monitoring (Inman *et al.*, 2002). This data was looped 220 times to provide forcing over the 4,400-year long simulation. Inspection of the figure shows that the shelf slope receded about 15 m during the periods of rapid sea-level rise. During the 2,000-year stillstand, a wave-cut terrace was formed with about a 150 m wide wave-cut platform and a 3 m high remnant sea cliff. These dimensions are in approximate agreement with evidence of wave-cut terraces along the California coast (Inman *et al.*, 2002). However, models of terrace cutting at paleo-sea levels will always require input of proxy wave climate appropriate for the location being modeled as well as the proper erodibility coefficients for the bedrock at that location (see entries on *Climate Patterns in the Coastal Zone*, and *Energy and Sediment Budget of the Global Coastal Zone*).

Douglas L. Inman, Scott A. Jenkins and Patricia M. Masters

Bibliography

Bagnold, R.A., 1939. Committee on wave pressure: interim report on wave-pressure research. *Journal Institution of Civil Engineers*, **12**: 201–226.

Cialdi, A., 1866. Sul moto ondoso del mare e su le correnti di esso specialmente su quelle littorali. Roma: *Tipografia delle Belle Arti*, 693 p.

de Beaumont, L.E., 1845. Levees de sable et de galet, *Lecons de Geologie Pratique*, (Septieme lecon), Volume 1. In Bertrand, P., Paris [Excerpted in part In Schwartz, M.L. (ed.), 1973. *Barrier Islands*. Stroudsburg: Dowden, Hutchison & Ross].

Emery, K.O., 1960. *The Sea off Southern California*. New York: John Wiley & Sons.

Fisher, O., 1866. On the disintegration of a chalk cliff. *Geological Magazine*, **3**: 354–356.

Gilbert, G.K., 1885. The topographic features of lake shores. *U.S. Geological Survey. Fifth Annual Report*, 1883–84, Washington, DC.

Hancock, G.S., and Anderson, R.S., 2002. Numerical modeling of fluvial strath-terrace formation in response to oscillating climate. *Geological Society of America Bulletin*, **114**: 1131–1142.

Havelock, T.H., 1940. The pressure of water waves upon a fixed obstacle. *Proceedings of the Royal Society London*, A, **175**: 409-421.

Inman, D.L., 1983. Application of coastal dynamics to the reconstruction of paleocoastlines in the vicinity of La Jolla, California. In Masters, P. M., and Flemming, N. C. (eds.), *Quaternary Coastlines and Marine Archaeology*. London: Academic Press, pp. 1–49.

Inman, D.L., and Jenkins, S.A., 2002. Scour and burial of bottom mines, a primer for fleet use. University of California, San Diego, Scripps Institution of Oceanography, SIO Reference Series 02-8, text, fig. & appen.

Inman, D.L., and Masters, P.M., 1994. Status of research on the nearshore. *Shore and Beach*, **62**: 11–20.

Inman, D.L., and Nordstrom, C.E., 1971. On the tectonic and morphologic classification of coasts. *Journal of Geology*, **79**: 1–21.

Inman, D.L., Masters, P.M., and Jenkins, S.A., 2002. Facing the coastal challenge: modeling coastal erosion in southern California. In Magoon, O.T. *et al.* (eds.), *California and the World Ocean '02*. Reston, VA: American Society of Civil Engineers.

Jenkins, S.A., and Inman, D.L., 1999. Sand transport mechanics for equilibrium in tidal inlets. *Shore and Beach*, **67**: 53–58.

Shepard, F.P., 1963. *Submarine Geology*, 2nd edn. New York: Harper & Row.

Storms, J.E.A., Weltje, G.J., Van Dijke, J.J., Geel, C.R., and Kroonenberg, S.B., 2002. Process-response modeling of wave-dominated coastal systems: simulating evolution and stratigraphy on geological timescales. *Journal of Sedimentary Research*, **72**: 226–239.
Sunamura, T., 1992. *Geomorphology of Rocky Coasts*. New York: John Wiley.
Trenhaile, A.S., 1987. *The Geomorphology of Rock Coasts*. New York: Oxford University Press.
Trenhaile, A.S., 2002. Modeling the development of marine terraces on tectonically mobile rock coasts. *Marine Geology*, **185**: 341–361.

Cross-references

Climate Patterns in the Coastal Zone
Energy and Sediment Budgets of the Global Coastal Zone
Littoral Cells
Shore Platforms

MODES AND PATTERNS OF SHORELINE CHANGE

Introduction

It is estimated that worldwide 70% of all beaches are eroding (Bird, 1985); in the United States this percentage may approach 90% (Heinz Center, 2000). Nearly every developed shoreline in the United States is retreating, and the coast is on a collision course with seaside development. This problem is manifest in the United States where heavily developed barrier islands are experiencing sea-level rise and beach erosion. For example, Galgano (1998) demonstrated that more than 86% of US East Coast beaches are eroding. With property values estimated at over $3 trillion for the US East and Gulf Coast barriers, these policy decisions invariably have important economic consequences for coastal communities. Consequently, a precise understanding of shoreline (defined as the high water line or wet–dry boundary) change using accurate shoreline change models and geomorphic characteristics is paramount.

Managing beach erosion is a difficult challenge for government agencies. The costs of erosion are often burdensome. Also, there is a paucity of long-term beach erosion data, and we still do not thoroughly understand how shorelines change over time. Consequently, policy decisions are regularly based on statistical modeling without a comprehensive understanding of patterns of shoreline change (Leatherman, 1993). The methodology frequently employed is to determine shoreline change rates for an entire reach using simple statistical analyses of erosion.

Barrier islands are naturally dynamic landforms and inevitably change position and shape depending upon the changing relationship between coastal processes, the geologic framework, and human intervention. To date, a great deal of research has focused on understanding beach (profile) dynamics and defining shoreline (planform) change. Initially these studies focused on beach structure and beach-shoreface equilibrium models, generally over limited spatial and temporal scales-measured on the order of years and decades (e.g., Dean, 1977; Bruun, 1988). More recently, coastal geomorphologists have presented new theories for their long-term change and evolution over much longer time frames (e.g., Hayes, 1979; Leatherman *et al.*, 1982; Anders *et al.*, 1990; List and Terwindt, 1995; McBride *et al.*, 1995).

Modes of shoreline change

It is generally accepted that there is geographic diversity in coastal landforms. This diversity is a result of the differences and combined effects of coastal processes and antecedent geology; shoreline change and coastal configuration are the integrative result. A "mode" of shoreline change is defined as a discrete pattern of shoreline movement identified through recognition of a unique change in shape, a rate of movement, or a cycle of change.

Modes of shoreline change may occur as two-dimensional (linear) or three-dimensional (alongshore) transformations operating over variable temporal and spatial scales, caused by the interaction of coastal processes within the constraints of geologic controls and sometimes with anthropogenic influence (e.g., inlet jetties). A mode of shoreline change can likewise depict a specific stage in an ordered sequence of shoreline development, whereby successive modes are dependent on preceding environmental factors and conditions for their existence (e.g., mesotidal barrier inlets). Modes of shoreline change identified to date are given in Table M8.

What relevance can be associated with the delineation and analysis of modes of shoreline change? Understanding modes of shoreline change provides a standardized geomorphic framework upon which shoreline change data can be more accurately interpreted. Purely statistical methods of shoreline change analysis (e.g., the "black-box" approach) often yield misleading results. Defining modes of change based on geomorphic principles and hence understanding the reasons for shoreline change yields a practical management tool. For many years, coastal scientists have been mapping historical shoreline positions to calculate rates of change (Heinz Center, 2000). Nonetheless, traditional uses of these data (e.g., statistical prediction of future shoreline positions for building setback codes) has been rather confined and flawed (Douglas *et al.*, 1998). Determination of an average shoreline rate of change and an understanding of the temporal and spatial variability associated with this change can be useful only if the causes of the change are recognized and understood.

Wave versus tide energy

Processes that operate along the coast give rise to changes in barrier island size, shape, orientation, and position. Price (1953) first illustrated how wave and tidal energy were primary controls on shoreline processes and beach planform. In a later publication, Price (1956) suggested that the bottom profile and geologic structure of the nearshore have important influences on beach morphology based on his analysis of low-energy coastlines in the Gulf of Mexico. He concluded that the most important geomorphic control on depositional coasts is the type and magnitude of hydrologic energy in the region. Davies (1964) illustrated that the two most significant hydrologic factors are wave energy and tidal current energy, both influenced by tidal range.

Hayes (1975) was the first to interpret the combined influence of wave energy and tidal range on barrier island morphology on a global scale. As a result of Hayes' (1975) work and later research by Nummendal *et al.* (1977), morphogenetic nomenclature of tide-dominated, wave-dominated, and mixed energy became widely accepted classifications for barrier systems (McBride *et al.*, 1995). Hayes (1979) showed how tidal range influences barrier island morphology because the effectiveness of wave action is attenuated, and tidal current activity is increased as vertical tidal range increases. Hence, in microtidal ($T_R < 2$ m), wave-dominated conditions, barrier islands are long and narrow, and are characterized by only a few ephemeral inlets. In contrast, short, stubby "drumstick" barrier islands are manifest in mesotidal ($T_R = 2$–4 m) conditions. In this case, barrier morphology is linked to the presence of relatively stable inlets. Large ebb-tidal deltas are associated with these inlets and are common on mesotidal coastlines. The interaction of these large ebb-tidal deltas with wave energy, combine to play a significant role in shaping the coastal configuration and morphology of adjacent barrier islands by accumulating and storing large quantities of sand-sized sediment which becomes available to the island on occasion and by inducing littoral current reversals driven by wave refraction. The end result is the characteristic "drumstick" shape (Hayes, 1979).

Davis and Hayes (1984) concluded that the "standard" definitions of wave-dominated (microtidal) and tide-dominated (mesotidal) coasts are

Table M8 Proposed shoreline classification types based on modes of shoreline change (after Leatherman, 1993)

Type	Classification
1	Simple linear retreat
2	Alternating erosion and accretion
3	Progressive erosion alongshore
4	Storm punctuated erosion
5	Inlet-induced erosion
6	Cyclic shoreline change
7	Mesotidal barrier behavior
8	Apparent island rotation
9	Pre-Holocene controlled configuration
10	Progressive erosion temporally
11	Cape-like features
12	Spit elongation

based solely on the analysis of moderate wave-energy environments and are not adequate to explain the numerous exceptions to these broad classifications. Hayes' (1975, 1979) classification is based largely on coasts with moderate wave energy. In their analysis of Gulf of Mexico shoreline types, Davis and Hayes (1984) explained that tidal prism (the volume of water that must be exchanged through a tidal inlet) represents one of the more important, but frequently discounted factors in determining the morphology of barrier islands. In the Gulf of Mexico, which has a diurnal tide, there is no relationship between tidal range and tidal prism because of the large lagoons and bays. Nonetheless, large tidal prisms, especially in areas of low wave energy, can explain large, well-developed ebb-tidal deltas and the characteristic drumstick shape. They suggest that a continuum between processes exists, and it is therefore possible to identify three types of coasts: (1) wave dominated, (2) tide dominated, and (3) one balanced between tides and waves.

Cape development

The importance of wave energy and tidal currents to the morphology of coastal landforms has been used to explain the development and morphology of cape-like features (Davies, 1964). Over the past three decades, the genesis and morphology of cape-like features has led to a great deal of study and scientific speculation. Most early models attributed the formation of cape-like features to accretional processes through the action of waves and currents, eddies from large ocean currents, and geologic controls. Hoyt and Henry (1971) observed that that the popular models of cape formation based on modern processes (e.g., wave refraction, sediment transport and deposition) did not explain the association of capes with major rivers and their similarity to ancestral capes in the region. Hoyt and Henry (1971) rejected the premise that waves and currents transport sufficient sediment out of adjacent embayments because it is not consistent with wave refraction and the concentration of wave energy on promontories. Their analysis of the Carolina capes suggests that erosion is the predominant process shaping the modern coastal configuration. The Carolina capes are most likely the result of the retreat and reworking of ancestral Pleistocene river deltas during Holocene sea-level rise. Moslow and Heron (1981) supported the theory that the modern coastal configuration of these well-known capes can be attributed to the erosion and gradual landward retreat of formally larger antecedent capes.

Finkelstein (1983) proposed a model for the development of sedimentary capes on straight shorelines through development of a large recurved spit at a littoral barrier. Using Fishing Point, Virginia at the southern end of Assateague Island as an example, Finkelstein (1983) demonstrated that cape-like features might occur where an inlet traps a large volume of littoral sediment. In the case of "cape" Assateague, a large volume of longshore sediment (115,000 m^3/yr) is trapped updrift of Chincoteague Inlet. This cape has developed since 1859 and has a pronounced influence on the shoreline by sand-starving the barriers islands to the south (Rice and Leatherman, 1983). The impoundment of longshore sediment in the cape has starved the downdrift beaches to the extent that a 5-km offset is observed on the islands to the south, and erosion rates of 10+ m/yr are observed for tens of kilometers to the south. The net result is a highly concave shoreline extending 35 km south of the cape, termed an "arc of erosion" (Leatherman, 1993).

Dolan *et al.* (1979) hypothesized that capes will develop along the Virginia barrier island chain. In their analysis, the authors postulated that the pronounced growth of the bulbous updrift ends of drumstick barrier islands were an indication of cape development. Unfortunately, they did not recognize the cyclic change of accretion and erosion of the bulbous ends of drumstick barrier islands that operates on timescales of decades to centuries. Hence, their prediction of cape development was flawed.

Leatherman *et al.* (1982) conducted a geomorphic analysis of long-term (>100 yr) shoreline change maps and geomorphic data to explain the coastal configuration and varied patterns of retreat of the Virginia barrier islands. They showed how sediment supply, controlled by the net southerly littoral drift, and the influence of inlets played a critical role in individual island adjustments. Antecedent Pleistocene topography beneath the barrier islands influences inlet position and stability. Differential subsidence and littoral sediment starvation have triggered a more rapid landward migration in the northern group of islands. Finally, wave refraction over the inner continental shelf has induced wave focusing on the southern two groups of barrier islands, causing more changes in orientation than the northern group. The presumed capes are in reality drumstick barrier islands behaving as described by Hayes (1979), and the recorded changes in the coastal configuration of the Virginia barrier islands suggests that a smoother shoreline configuration (e.g., opposite of capes) can be anticipated in the future (Leatherman *et al.*, 1982).

Tidal inlets

Tidal inlets represent the most important feature along a sandy coastline in relationship to littoral processes. The preeminent role of inlets in the cyclic nature of barrier island morphodynamics was introduced by Hayes (1975, 1979). Dean and Walton (1975) indicated that tidal inlets represented the largest sediment sink of beach sand along the coast and are believed to be responsible for much of the beach erosion in Florida.

Hayes (1975) pointed out that microtidal inlets are more dynamic than their mesotidal counterparts. Mesotidal inlets are more stable and are strongly associated with Pleistocene drainage channels. Halsey's (1979) research along the Delmarva Peninsula strongly suggested that inlet channels are correlated with relict thalwegs (the deepest part of a stream channel) of ancient streams. Further, the higher interfluves become the loci or "nexus" for barrier island formation.

The critical role played by inlets, natural and jettied, in controlling the morphology and orientation of shorelines is well documented. There is growing research dedicated to determining the spatial extent of the downdrift effect of inlets (FitzGerald and Hayes, 1980; Leatherman *et al.*, 1987). The critical role played by inlets is their cumulative influence on the regional orientation and composition of coastal reaches.

Tidal inlets are dynamic features and can dramatically affect the morphology of adjacent barrier shorelines (Hayes, 1975). The processes of inlet migration can bring about erosion and deposition to flanking barrier islands. This arc of erosion is possibly the most dramatic influence of tidal inlets on adjacent barrier islands. Finkelstein (1983) illustrated how Chincoteague Inlet served as a littoral barrier and caused an arc of erosion extending 35 km downdrift. Wave refraction associated with large ebb-tidal deltas has been shown to cause reversals in the longshore transport system, resulting in onshore or alongshore sediment transport and ultimately unique changes in beach planform (Hayes, 1979).

The degree to which inlets modify the adjacent shorelines is a function of their tidal range. In most instances tidal range correlates well with the size of tidal inlets. South Carolina offers an excellent example of the role of tidal inlets on coastal orientation and morphology. FitzGerald *et al.* (1978) illustrated how the number of inlets along the South Carolina coast is correlated to tidal range. Along the northern segment of the South Carolina coastline, the mean tidal range is 1.5 m. In this area, tidal inlets comprise less than 2% of the total shoreline. In contrast, the southern segment of this coast has a tidal range of 2.2–2.5 m. Accordingly, the amount of tidal inlet shoreline increases to 25%. Since the processes related to ebb-tidal deltas can directly control the morphology of adjacent barrier islands, tidal inlets will have a greater influence on shoreline morphology along barrier island coastlines with increased tidal ranges (FitzGerald *et al.*, 1978). Nummendal *et al.* (1977) noted that the erosional and accretional nature of South Carolina's barrier islands is intimately connected with the change of tidal inlets. Increasing tidal ranges south of Cape Romain dictate larger and more numerous tidal inlets and therefore increased seaward transport of littoral sediments. FitzGerald *et al.* (1978) proposed that the three primary types of shoreline change in South Carolina were associated with three types of tidal inlets.

Brown (1977) recognized three distinct geomorphic zones along the South Carolina coast. The northern segment, or the arcuate strand, is relatively stable through time and is characterized by few inlets. A cuspate delta occupies the central portion of the coastline. In the south where there are numerous inlets, barrier islands with highly variable shoreline change rates predominate (Brown, 1977). Anders *et al.* (1990) showed in their analysis of shoreline movements in South Carolina that the greatest variability of shoreline change rates was in the vicinity of coastal inlets. These findings were supported by Hubbard *et al.* (1977) who pointed out that the arcuate strand beaches of northern South Carolina are stable everywhere except in the vicinity of tidal inlets. Only where there are 15–20 km between inlets does one escape their influence on the shoreline. These observations were supported on a more quantitative basis by Galgano (1998), who demonstrated that tidal inlets control shoreline change on 70% of the beaches on the mid-Atlantic coastline.

Leatherman (1984) demonstrated the dramatic influence of jettied inlets on the long-term evolution of a barrier island. In a natural system, it is generally accepted that the long-term evolution of a barrier island will follow a general model of landward migration through time. In this scenario, the entire system (e.g., the barrier, bay, and mainland shoreline) will migrate as a geomorphic unit and retain the same general configuration. Leatherman (1984) provided an example where the system has been altered to the extent that large-scale evolution of a coastal unit is fundamentally changed. Ocean City Inlet was stabilized by jetties in 1934/35. These jetties had a very rapid and pronounced influence on the morphology of northern Assateague Island. The jetties interrupted a longshore sediment flow of 140,000 m^3/yr. Further, the ebb-tidal delta captured an estimated 6,000,000 m^3 of sand, thus completely depriving the downdrift beaches of sand and creating an arc of erosion that extends some 10 km downdrift. Northern Assateague Island, which eroded at an historical rate of 0.6 m/yr prior to inlet stabilization, reached rates averaging 11.5 m/yr (Leatherman, 1984).

Leatherman (1984) indicated that the implications of this dramatic increase in erosion will have important consequences on the long-term evolution of northern Assateague Island, Maryland. The northern segment of the island will not migrate landward as an entity as some predict in accordance with accepted models of shoreline change. If sea level alone were the driving influence behind this migration, this assumption would be essentially correct, but sea-level rise represents only a small percentage of the erosion potential and the bay is shrinking in width over time. Leatherman (1984) theorized that there would be a loss of a section of the barrier island with the opening of the mainland to ocean waves in the foreseeable future, similar to what occurred at Nauset Beach, Chatham, Massachusetts (Leatherman, 1984).

Leatherman *et al.* (1987) modeled the erosion of northern Assateague Island to quantitatively determine its future movement. The authors proposed that the arc of erosion will continue to migrate southward through time, and this pattern of shoreline evolution can be mathematically modeled. The northernmost 10 km of the island currently represents the arc of erosion. There is no terminal end to this arc in a temporal sense. In other words, the arc of erosion will continue to migrate southward through time. Further, the area of maximum erosion will migrate downdrift so that the erosion rate will represent a non-steady-state condition. Leatherman *et al.* (1987) determined that because of the island's low elevation and narrow width, a storm will cause inlet breaching by the year 2020 with loss of the northern portion of the island as the sand is welded onto the mainland (unless artificial beach nourishment is undertaken).

Barrier migration

Tidal inlets have been ascribed an increasingly important role in barrier island migration. Most existing barrier islands are believed to have originated 7,000–8,000 years ago and have migrated to their modern positions with post-glacial sea-level rise. A number of theories describing the mechanisms for barrier island migration have emerged. Perhaps, the most widely accepted theory is the concept of continuous, albeit intermittent, barrier island migration by shoreface retreat (Leatherman, 1983). Research by Kraft (1971), Kraft *et al.* (1975), and Belknap and Kraft (1985) using extensive core samples and radiometric data demonstrated how barrier islands migrate landward and upward on the continental shelf during periods of sea-level rise. The landward migration or "roll-over" of barrier islands in a transgressive environment is a central issue in the study of modes of shoreline change.

Early researchers (e.g., Godfrey and Godfrey, 1973) held that overwash was the most important agent in landward migration. Field data supported what was intuitively obvious—that wave-driven overwash sediments were transported across the island, causing bayside accretion. Hence, landward migration was accomplished by a steady progression of beach erosion and bayside accretion. McGowan and Scott (1975) supported this hypothesis suggesting that overwash events transported considerable volumes of sand to the landward side of the island. Therefore, in a condition of dynamic equilibrium, the barrier island profile does not change its size or shape, but instead maintains its volume and migrates landward with equal magnitudes of seaside erosion and bayside accretion.

Later, work by Armon (1979) in Canada and Leatherman (1979) in Maryland countered that inlets played the predominant role in barrier island migration. Using extensive on-site surveys of overwash fans on Assateague Island, Leatherman (1979) showed that overwash processes were most important for building the vertical elevation of the barrier island. But overwash processes do not transport sufficient volumes of sediment to the bayside to maintain barrier island width. Barrier island width is maintained by inlet processes through the creation of large flood tidal deltas. Relict flood deltas serve as the platform onto which overwash and aeolian sediments accumulate.

Leatherman and Zaremba (1986) researched the long-term evolution of Nauset Spit, Cape Cod, Massachusetts to determine the relative influence of inlet and overwash processes on barrier island migration. The lateral growth of the spit is largely controlled by inlet dynamics as inlets migrate downdrift (south), whereas overwash during major storms events caused substantial widening of this very narrow barrier spit in local areas. Overwash is the primary means of transporting sediment across barriers backed by glacial headlands or extensive salt marshes because the establishment of an inlet channel is precluded. Overwash in these circumstances tends to be the most effective barrier migration process through interaction with aeolian transport and dune-building processes in the subaerial zone. This vertical accretion permits the barrier island to maintain its elevation during landward retreat.

Time frames for barrier migration

While the mechanisms and geometry for barrier island migration and modes of shoreline change are intuitively well understood, the time frames decidedly are not. Much speculation exists in the literature and some envision that barrier island migration occurs extremely rapidly—on the order of 50–100 years. The implied assumption in their argument is that the barrier will conserve its mass and move as a unit. Others imply that in some situations, barrier islands are overtaken by sea level and drown in place (e.g., Rampino and Sanders, 1980). This barrier island drowning model has never been observed or supported by data (Panageotou and Leatherman, 1986). Recent studies (Schwab *et al.*, 2000) substantiate the findings of Panageotou and Leatherman (1986) that the 18-m isobath off Fire Island, New York is not a drowned barrier island.

Time frames for shoreline movement are a subject of much debate, and the literature is replete with examples of shoreline movements at all timescales. Moody (1964) showed how a single, large storm event could displace the shoreline by a hundred meters in a matter of a few days, but a continuum exists between short-term storm events to long-term evolution (e.g., order of centuries). In addition, a very high degree of post-storm recovery has been observed, even after great storms (Morton *et al.*, 1994; Douglas *et al.*, 1998).

The prevailing theme in geomorphic literature is that barrier islands are continuously migrating landward in response to sea-level rise. In doing so, they maintain their mass through equal amounts of seaside erosion and bayside deposition (from overwash events). However, quantitative studies at Fire Island, New York and Hatteras Island, North Carolina indicated that both sides of the islands are eroding rather than migrating (Leatherman, 1987). Therefore, beach erosion does not necessarily signify barrier island migration. Migration denotes a change in the barrier island's centroid; this is not accomplished by erosion alone. Barrier island migration is a time-averaged phenomenon. This migration is not a continuous process, but instead is caused by episodic and site-specific, storm-generated events over longer-time frames.

Leatherman (1987) presented geomorphic and shoreline change map data that indicated that many barrier islands are thinning by bayside and oceanside erosion to some critical barrier width. Once the barrier island has achieved this critical width, migration by inlet and overwash processes will become effective in barrier island translocation. McBride *et al.* (1995) observed this slimming-down process in their study of long-term barrier island behavior in the Gulf of Mexico. They concluded that when shore position is monitored over time, shoreline change could be quantified and classified into a number of geomorphic response types as a function of scale. These studies indicate the need for the longest possible period of record of shoreline position data in order to make the correct geomorphic interpretation.

McBride *et al.* (1995) conducted a quantitative analysis of historical shoreline position over a spatial scale of 10–100 km. Their analysis provides a scientific basis for documenting process-response relationships that shape regional coastal morphodynamics. Even though "megascale" shoreline change studies are typically under-sampled spatially, this type of data is essential for formulating realistic research and management strategies regarding form/process relationships. The authors identified eight geomorphic response types to classify barrier coasts on a "mega-scale;" that is coastal reaches of 10–100 km at time frames of decades to centuries (Table M9). The authors inferred that sea-level rise is one of the major factors that controls the occurrence of these geomorphic response types along the barrier chains in Georgia and Louisiana. In regions where sea-level rise occurs at lower rates, sediment supply appears to be the dominant factor.

Antecedent geology

Contemporary literature offers a growing number of works that assign increasing importance to the role of antecedent geology controlling coastal configurations and modes of shoreline change. Many early researchers understood that there were important geologic controls

Table M9 Megascale geomorphic response types (after McBride *et al.*, 1995)

Classification	Response Type
Simple	Lateral movement
	Advance
	Dynamic equilibrium
	Retreat
Complex	In-place narrowing
	Landward rollover
	Breakup
	Rotational instability

influencing coastal configurations. Price (1953) was perhaps the first to establish the important influence of antecedent topography on coastal landforms, but this research was hindered by the absence of quantitative and stratigraphic data.

Kraft (1971) conducted studies of Holocene sediments in coastal Delaware, illustrating that Holocene sediments are deposited over a complex Pleistocene unconformity. Large, modern depositional features, such as spits, baymouth barriers, and tidal deltas in Delaware are in large measure related to the presence and relative relief of the subsurface Pleistocene topography. Kraft (1971) showed how relict drainage patterns influenced the location of modern tidal inlets and shoreface sediment types. Further, Pleistocene headlands served as focal points for barrier island formation. In similar research, Halsey (1979) suggested that inlet channels along the Delmarva Peninsula were strongly correlated to relict stream thalwegs, and interfluves have become the loci for barrier island formation.

Kraft *et al.* (1975) presented a detailed study of the geologic structure of the Delaware coast and the influence of Pleistocene topography on modern barrier island change. Relict drainage systems, Pleistocene headlands (subsurface and subaerial) and relict barrier formations combined to create a complex system of "coastal anomalies" in what is an otherwise straight coastal configuration. These "coastal anomalies," or bulges and re-entrants along the coast, were quantified by detailed shoreline change mapping of the Delaware coast by Galgano (1989). This mapping indicated the spatial diversity of modes of shoreline change within a coastal compartment.

Davis and Kuhn (1985) conducted research along the western coast of Florida and determined that antecedent geology played a vital, but very different role in controlling shoreline change. This segment of the Florida coast is atypical for a microtidal coast in that it is composed of numerous drumstick barrier islands with stable inlet channels and large ebb shoals. Core samples and seismic profiling revealed that pre-Pleistocene geologic formations were responsible for this somewhat uncharacteristic situation. The data revealed that a subtle, but well-defined limestone ridge of Miocene age with a relief of approximately 1 m underlies the modern barrier islands. Further, shoreward of the barriers, the irregular limestone surface slopes gradually toward the Gulf of Mexico. Davis and Kuhn (1985) indicate that the limestone platform is instrumental in controlling the location of the barrier islands and attenuating wave energy.

Oertel *et al.* (1992) interpreted the function of antecedent topography in barrier lagoon development. Microtidal lagoons are relatively wide and open, with few marshes. Conversely, mesotidal lagoons are narrow with numerous marshes. The authors theorize that pre-Holocene topography is the key determining influence for this phenomenon. Oertel *et al.* (1992) suggested that lagoon floors along tide-dominated coasts reflect the antecedent topography, which is dominated by well-defined fluvial channels and interfluves. The interfluves produce shallow areas in the lagoon, which become colonized by marshes. In the microtidal setting, the floors of barrier lagoons are initially smooth after formation.

The major theme in much of this research is that passive margin coastlines, such as the US Atlantic coast, are significantly influenced by the geologic framework. Some of these barriers have been shown to be perched barriers, resting on a pre-Holocene structure that controls modern beach dynamics and morphology (Kraft *et al.*, 1975). Along other segments of the Atlantic coast, bathymetric features on the inner shelf modify waves and currents, ultimately affecting patterns of erosion and deposition. Riggs *et al.* (1995) presented a detailed appraisal of the influence on geologic structure on the North Carolina coast. Their research suggested that some shoreline features in North Carolina are controlled by pre-Holocene stratigraphic framework and beaches are perched on pre-Holocene sediments.

Riggs *et al.* (1995) subdivided the coastal zone into two distinct segments. North of Cape Lookout, the geologic framework consists of a Quaternary sequence; while the segment to the south is dominated by Tertiary and Cretaceous units that crop out across the coastal plain and along the shoreface. Superimposed on this regional framework is a relict drainage system resulting in a series of fluvial channels infilled by modern sediments separated by larger interfluves; this results in shorefaces which are either non-headland or headland-dominated.

Statistical modeling

One of the fundamental objectives of coastal geomorphology is to determine shoreline trends. To this end, coastal scientists have applied different statistical techniques to predict future shoreline positions. In the past several years, this practice has assumed increased significance because many coastal states are using historical shoreline change data to project shoreline position for application in land use policies, primarily in establishing building setback lines and insurance zones (Heinz Center, 2000).

The selection of an appropriate statistical model for shoreline change is a matter of some contention because it is critical to the final result. This problem is amplified because shoreline position data are typically limited and spaced irregularly through time. The principal issue is which shoreline positions represent the actual trend (i.e., not post-storm or wintertime shoreline data), what is the minimum acceptable period of record, how far into the future can the trend be usefully predicted, and which technique will best extrapolate the actual trend? The fundamental consequence of statistical modeling, regardless of the relative complexity, is that it will lose its physical and practical meaning if the geomorphic setting is not considered (Leatherman, 1993).

Fenster *et al.* (1993) proposed a new method, the Minimum Descriptor Length (MDL), for predicting shoreline change rates and forecasting future shoreline positions. This method selects the model (e.g., linear regression or polynomial) on the basis of the trend of the most recent data. It is predicated on the appearance of trend reversals within the data record and depends on the persistence of the data. The authors, however, neglected to disclose the quality of the prediction beyond a few years, nor did they adequately test the model against a spectrum of real data. Crowell *et al.* (1997) used the MDL to predict trends using a temporally rich set of sea-level rise data, replete with interannual variations, to assess the ability of the MDL to predict future trends against the established record of sea-level rise. That research demonstrated that the MDL is not a useful predictor of long-term trends. Crowell *et al.* (1997) concluded that in spite of interannual variability in the data, the most reliable *long-term* forecast of shoreline position will be made from linear regression using the longest possible time series. Finally, Galgano and Douglas (2000) illustrated the necessity to use more than 80 years of data to determine a shoreline change trend. At time frames less than 80 years simple calculations of change rate can actually produce the wrong sign (i.e., accretion versus erosion), and will likely produce a two or three sigma error in the trend. Furthermore, Galgano and Douglas (2000) demonstrated that the incorporation of post-storm and winter shoreline position data violate the assumptions of the linear regression model when mixed with summer shoreline position data.

Conclusions

Beach erosion and changes in coastal configurations are complex physical processes encompassing a number of natural and human-induced factors. These alterations drive the evolution of coastal configurations that generally conform to observable patterns; these modes of shoreline change are defined as a discrete pattern of shoreline movement identified through recognition of a unique change in shape, a rate of movement, or a cycle of change. Natural conditions that drive these changes include such variables as sea-level rise, tidal variations, wave energy, sediment supply, antecedent geology, seasonal variations in wave energy, and the episodic influence of storms. Humans influence the spatial and temporal variability of shoreline change by building structures (e.g., groins and jetties), dredging, damming rivers, and beach nourishment projects.

Spontaneous variations in natural processes and the effects of human activity on the coast drive spatial and temporal variations in shoreline change and the evolution of coastal landforms. Therefore, we should not anticipate that shoreline change rates and coastal configurations will remain uniform through time, instead we should realize that accelerations and declarations in rates of erosion/accretion and trend reversals could occur. Reliance on statistical methods of shoreline change analysis can often yield misleading results, whereas defining modes of change based on geomorphic principles and hence understanding the reasons for shoreline change provides a practical management tool.

Francis A. Galgano and Stephen P. Leatherman

Bibliography

Anders, F.J., Reed, D.W., and Meisburger, E.P., 1990. Shoreline movements; Report 2, Tybee Island, Georgia to Cape Fear, North Carolina, 1851–1983. *Technical Report* CERC-83-1, Vicksburg: U.S. Army Corps of Engineers Waterways Experimental Station.

Armon, J.W., 1979. Landward sediment transfers in a transgressive barrier island system, Canada. In Leatherman, S.P. (ed.), *Barrier Islands*. New York: Academic Press, pp. 68–81.

Belknap, D.F., and Kraft, J.C., 1985. Influence of antecedent geology on stratigraphic preservation potential and evolution of Delaware's barrier system. *Marine Geology*, **63**: 235–262.

Bird, E.C.F., 1985. *Coastline Changes—A Global Review*. Chichester, UK: John Wiley Interscience.

Brown, P.J., 1977. Coastal morphology of South Carolina. *Southeastern Geology*, **18**: 259–264.

Bruun, P., 1988. The Bruun Rule of erosion by sea-level rise: a discussion on large scale two- and three-dimensional usage. *Journal of Coastal Research*, **4**: 627–648.

Crowell, M., Douglas, B.C., and Leatherman, S.P., 1997. On forecasting future U.S. shoreline positions. *Journal of Coastal Research*, **13**: 1245–1255.

Davies, J.L., 1964. A morphogenic approach to world shorelines. *Zeitschrift fur Geomorphologic Bd.* **8**: 27–42.

Davis, R.A., and Hayes, M.O., 1984. What is a wave dominated coast? *Marine Geology*, **60**: 313–329.

Davis, R.A., and Kuhn, B.J., 1985. Origin and development of Anclote Key, West Peninsular Florida. *Marine Geology*, **63**: 153–171.

Dean, R.G., 1977. Equilibrium beach profiles: U.S. Atlantic and Gulf Coasts. *Ocean Engineering Technical Report No. 12*, Newark: University Of Delaware.

Dean, R.G., and Walton, T.L., 1975. Sediment transport processes in the Vicinity of inlets with special reference to sand trapping. *Estuarine Research*, Volume 2, New York: Academic Press.

Dolan, R., Hayden, B., and Jones, C., 1979. Barrier island configuration. *Science*, **204**: 401–402.

Douglas, B.C., Crowell, M., and Leatherman, S.P., 1998. Considerations for shoreline position prediction. *Journal of Coastal Research*, **14**: 1025–1033.

Fenster, M., Dolan, R., and Elder, J.R., 1993. A new method for predicting shoreline positions from historical data. *Journal of Coastal Research*, **9**: 147–171.

Finkelstein, K., 1983. Cape formation as a cause of erosion on adjacent shorelines. In *Proceedings of the Coastal Zone '83*. American Society of Civil Engineers, pp. 620–640.

FitzGerald, D.M., and Hayes, M.O., 1980. Tidal inlet effects on barrier island management. In *Proceedings Coastal Zone '80*, American Society of Civil Engineers, p. 2355.

FitzGerald, D.M., Hubbard, D.K., and Nummendal, D., 1978. Shoreline changes associated with tidal inlets along the South Carolina coast. In *Proceedings of the Coastal Zone 1978*. American Society of Civil Engineers, pp. 1973–1994.

Galgano, F.A., 1989. Shoreline recession and nearshore response: the Atlantic coast of delaware, *1845–1987*. M.A. thesis, College Park: University of Maryland.

Galgano, F.A., 1998. Geomorphic analysis of modes of shoreline behavior and the influence of tidal inlets on coastal configuration, U.S. East coast. Ph.D. Dissertation, College Park: University of Maryland.

Galgano, F.A., and Douglas, B.C., 2000. Shoreline position prediction: methods and errors. *Environmental Geosciences*, **7**(1): 23–31.

Godfrey, P.J., and Godfrey, N.M., 1973. Comparison of ecological and geomorphic interactions between altered and unaltered barrier island systems in North Carolina. In Coates, D.R. (ed.), *Coastal Geomorphology*. Binghampton, NY: State University of New York, pp. 239–258.

Halsey, S.D., 1979. Nexus; new model of barrier island development. In Leatherman, S.P. (ed.), *Barrier Islands*. New York: Academic Press, pp. 185–210.

Hayes, M.O., 1975. Morphology of sand accumulation in estuaries. In *Proceedings of the 2nd International Estuarine Research Federation Conference*, Myrtle Beach, SC, pp. 3–22.

Hayes, M.O., 1979. Barrier island morphology as a function of tidal and wave regime. In Leatherman, S.P. (ed.), *Barrier Islands*. New York: Academic Press, pp. 1–27.

Heinz Center, 2000. *Evaluation of Erosion Hazards*. Washington, DC: The H. John Heinz III Center for Science, Economics, and the Environment.

Hoyt, J.H., and Henry, V.J., 1971. Origin of capes and shoals along the southeastern coast of the United States. *Geological Society of America Bulletin*, **82**: 59–66.

Hubbard, D.K., Barwis, J.H., and Nummendal, D., 1977. Sediment transport in four South Carolina inlets. In *Proceeding of the Coastal Sediments '97*. American Society of Civil Engineers, pp. 797–814.

Kraft, J.C., 1971. Sedimentary facies patterns and geologic history of a holocene marine transgression. *Geological Society of America Bulletin*, **82**: 2131–2158.

Kraft, J.C., Allen, E.A., Belknap, D.F., John, C.J., and Maurmeyer, E.M., 1975. Delaware's Changing Shoreline. Technical Report No. 1. Dover, DE: Delaware State Planning Office.

Leatherman, S.P., 1979. Migration of Assateague Island, Maryland by inlet and overwash processes. *Geology*, **7**: 104–107.

Leatherman, S.P., 1983. Barrier dynamics and landward migration with Holocene sea-level rise. *Nature*, **301**: 415–418.

Leatherman, S.P., 1984. Shoreline evolution of North Assateague Island, Maryland. *Shore and Beach*, **52**: 3–10.

Leatherman, S.P., 1987. Time frames for barrier island migration. *Shore and Beach*, **55**: 82–86.

Leatherman, S.P., 1993. Modes of shoreline behavior: erosion rate analysis using geomorphic principles. In *Proceedings of the International Coastal Symposium*, pp. 218–223.

Leatherman, S.P., and Zaremba, R.E., 1986. Dynamics of a northern barrier beach: Nauset Spit, Cape Cod, Massachusetts. *Geological Society of America Bulletin*, **97**: 116–127.

Leatherman, S.P., Rice, T.E., and Goldsmith, V., 1982. Virginia barrier island configuration: a reappraisal. *Science*, **215**: 285–287.

Leatherman, S.P., Dean, R.G., Everts, C.E., and Fulford, E., 1987. Shoreline and sediment budget analysis of north Assateague Island, Maryland. In *Proceedings of the Coastal Sediments '87*. American Society of Civil Engineers, pp. 1460–1471.

List, J.H., and Terwindt, J.H.J., 1995. Large scale coastal behavior. *Marine Geology*, **126**: 1–3.

McBride, R.A., Byrnes, M.R., and Hiland, M.W., 1995. Geomorphic response type model for barrier coastlines: a regional prospective. *Marine Geology*, **126**: 143–159.

McGowan, J.H., and Scott, A.J., 1975. Hurricanes as geologic agents on the Texas coast. In Cronin, L.E. (ed.), *Estuarine Research*, Volume 2, New York: Academic Press.

Moody, D., 1964. Coastal morphology and process in relation to the development of submarine sand ridges off Bethany beach, Delaware. Ph.D. dissertation, Baltimore: Johns Hopkins University.

Morton, R.A., Paine, J.G., and Gibeaut, J.C., 1994. Stages and durations of post-storm beach recovery, southeastern Texas coast, USA. *Journal of Coastal Research*, **10**: 884–908.

Moslow, T.F., and Heron, S.D., 1981. Holocene depositional history of a microtidal cuspate foreland cape: Cape Lookout, North Carolina. *Marine Geology*, **41**: 251–270.

Nummendal, D., Oertel, G.F., Hubbard, D.K., and Hine, A.C., 1977. Tidal inlet variability—Cape Hatteras to Cape Canaveral. In *Proceedings of the Coastal Sediments '77*. American Society of Civil Engineers, pp. 543–562.

Oertel, G.F., Kraft, J.C., Kearney, M.S., and Woo, H.J., 1992. A rational theory for barrier–lagoon development. Quaternary Coasts of the United States: Marine Lacustrine Systems, SEPM Special Publication No. 48, pp. 77–87.

Panageotou, W., and Leatherman, S.P., 1986. Holocene–Pleistocene stratigraphy of the inner shelf off Fire Island, New York: implications for barrier island migration. *Journal of Sedimentary Petrology*, **56**: 528–537.

Price, W.A., 1953. *Shorelines and Coasts of the Gulf of Mexico*, Volume 1. College Station, TX: Texas A&M University Contributions in Oceanography and Meteorology, pp. 5–16.

Price, W.A., 1956. The low energy coast and its new shoreline types on the Gulf of Mexico. *Inqua*, **1**: 159–166.

Rampino, M.R., and Sanders, J.E., 1980. Holocene transgression in south-central Long Island, New York. *Journal of Sedimentary Petrology*, **50**: 1063–1080.

Rice, T.E., and Leatherman, S.P., 1983. Barrier island dynamics: the eastern shore of Virginia. *Southeastern Geology*, **24**(3): 125–137.

Riggs, S.R., Cleary, W.J, and Synder, S.W., 1995. Influence of inherited geologic framework on barrier shoreface morphology and dynamics. *Marine Geology*, **126**: 120–142.

Schwab, W.C., Thieler, E.R., Allen, J.R., Foster, D.S., Swift, B.A., and Denny, J.F., 2000. Influence of inner-continental shelf geologic framework on the evolution and behavior of the barrier-island system between Fire Island inlet and Shinnecock Inlet, Long Island, New York. *Journal of Coastal Research*, **16**: 408–422.

Cross-references

Barrier Islands
Beach Erosion
Beach Processes
Coastline Changes
Coasts, Coastlines, Shores, and Shoreline
Erosion: Historical Analysis and Forecasting
Littoral Cells
Mapping Shores and Coastal Terrain
Microtidal Coasts
Tidal Inlets
Tidal Prism
Tide-Dominated Coasts
Wave- and Tide-Dominated Coasts
Wave-Dominated Coasts

MONITORING COASTAL ECOLOGY

Introduction—human action

Early human coastal development was probably in harmony with coastal processes, modifying the habitats rather than destroying them. As human populations have increased and land-use intensified, these habitats have come under greater pressure. Intensive agriculture, afforestation, and infrastructure development, including the mass tourism boom in many areas, have resulted in the destruction of natural areas especially along the shores of the Mediterranean (Doody, 1995). The impact on the environment increased the perception that coastal areas are vulnerable. This has led to the remaining areas, especially those dominated by "natural" landscapes, to be considered particularly precious. National and international legislation seeks to prevent further damage and destruction in these "fragile" environments.

Faced with mounting pressure on coastal and marine resources, there is increasing concern about the ability of the coast to sustain the many uses to which it is put, including those involving socio-economic development. This has been brought into sharp focus with the recognition that global warming is a reality and that one of its consequences is a rising sea level. If sustainability of use is to be achieved then understanding coastal processes and human influence on them is a prerequisite for management, policy, and planning. The way in which this understanding is achieved lies in the survey of landforms, habitats and species, and monitoring or surveillance of change in relation to natural process and the impact of human activities.

An historical legacy

Knowledge of the status and distribution of landscapes, habitats, and species provide essential baselines against which to measure change. Some of the earlier historical maps are important in this context. While they do not provide detailed information about habitats or species distribution, they can give great insights into long-term change. Early maps of the Suffolk Coast in eastern England around the time of Henry VIII, for example, show the configuration of the coast in the vicinity of Orford, then an important port (Anon, 1979). When compared to more recent maps a picture of change can be built up, which gives an indication of the scale and time scale over which it occurs (Figure M22).

This historical perspective is important to our understanding of the coast and coastal systems and the way in which they react to the impact of human actions over time scales longer than that of a single generation. In the United Kingdom, for example, there are a number of early texts which look at coastal issues from the point of view of the engineer and discuss these in terms of "Coast erosion and protection" (Wheeler, 1903; Matthews, 1934) and "Shoreline problems" (Carey and Oliver, 1918), or in relation to the creation of new land from the sea (Du-Plat-Taylor, 1931). Wheeler (1903) provides a series of geographical descriptions of the coast of England and short essays on northern France, Belgium, and Holland. A more systematic survey of the coastline of England and Wales was undertaken at the request of the Ministry of Town and Country Planning in 1943 (Steers, 1946) which was later extended to Scotland (Steers, 1973). These studies give a description of the geology, geomorphology, and geography of the coast and helped form the basis for planning policy on the coastline of Great Britain in later years.

Though useful, this historical approach can only give a broad indication of change. It does not provide the more detailed information required for the management of individual sites and resources or unravel the many and often complex changes, which characterize the coast. For this targeted surveys are needed, together with a detailed research, surveillance and monitoring. These aspects are dealt with next.

Definitions

Survey

Surveying involves the collection of both quantitative and qualitative information about a feature or features using a standardized methodology, but where there is no pre-formed view on the likely findings. Thus, other than identifying what will be surveyed (a species, plant community, habitat, or human activity) the primary aim is to obtain information on the nature, scale, and location of the chosen subject.

Surveillance

Repeat surveys provide a means of identifying change over time, for example, in the status of habitats and species or the rate of exploitation of a natural resource. As with the original survey there is no predetermined notion as to what change might be expected. The purpose of the new survey is to establish the nature of any change from the previous "norm."

Monitoring

This implies the need to assess and understand the outcome of a particular course of action, such as the management of a habitat or species or the extent to which resource depletion is sustainable. This may relate to compliance with a standard or deviation from an acceptable state.

Survey

In order to assess the significance of a particular feature it is essential to know where it is, how widespread and how much of it exists. Survey is the first basic tool in assessing the value of a particular coastal feature and as an aid to its conservation and management.

Habitat survey

Systematic surveys of landscapes, habitats, and species are a first stage in understanding the coast especially when considering its conservation. For example wide-scale habitat surveys are used to describe the nature conservation resource within a country as an aid to the selection of sites for statutory protection. In the mid-late 1980s in the United Kingdom, the nature conservation agencies (notably the Nature Conservancy Council) commissioned a series of coastal habitat surveys of Great Britain covering salt marshes (Burd, 1988), sand dunes (Dargie, 1993, 1995; Radley, 1994) and shingle structures (Sneddon and Randall, 1993). Attempts have been made to bring information together at a wider European scale but these are hampered by the variety of definitions and survey methodologies used and can only give a broad indication of the nature and scale of the resource (Dijkema, 1984; Doody, 1991). Site-specific surveys are often much more detailed and are used as a basis for developing management strategies such as those for the conservation of the flora and vegetation of the Wadden Sea islands, in the southern North Sea (Dijkema and Wolff, 1983).

Species recording

Establishing the location and population numbers of birds, (wintering and breeding) is also an essential prerequisite for the selection of sites to

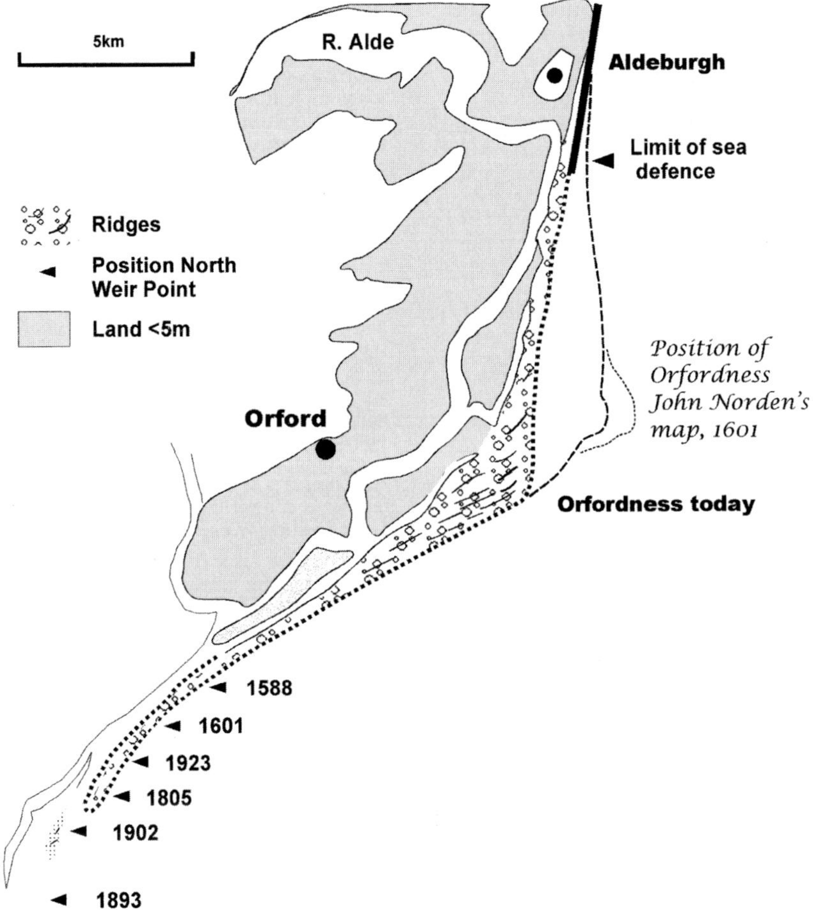

Figure M22 Change along the Suffolk coast between 1530 and 1963 derived from maps in Anon, 1977. (Adapted from Doody, 2001.)

protect rare and specialized species. It is important to know where the majority of a population of a restricted species resides during critical periods of its life cycle. For example, most of the world population of the gannet *Morus bassanus* nest on the coast of Great Britain and Ireland. In the case of this species it is even more important for the protection of the species, to know that 90,000 breeding pairs (approximately 5% of the total population) occupy one small site, the Bass Rock in south east Scotland. Similar considerations apply to other vulnerable and specialist species such as marine turtles which need to come ashore to nest on sandy beaches. Knowing where these are, and the number of breeding individuals is an essential prerequisite for any conservation program.

The results of habitats and species surveys can therefore provide:

- Summarized data from local to regional and national levels;
- Detailed maps showing distribution of vegetation, habitats, or breeding species;
- The basis for selection of important conservation areas;
- An assessment of the degree of conservation protection.

This information alone, tells us little about change, whether it is in terms of distribution, life-cycle movements, or population status. For this more frequent surveys are needed.

Identifying and understanding change—species studies

Managing the coast, whether for rare plants, specialist animals, important habitats, or human activities requires information on change and the causes of the change. If these show detrimental effects, either to the resources themselves, or knock-on effects to other interests, then it is essential to know what these are. By way of illustration, three elements can be discerned when monitoring individual species:

1. A species is rare or of special interest and managed to protect the population—are we being successful?
2. The species is invasive and we need to assess if our control measures are effective.
3. The species provides an indicator of change in the environment (Keddy, 1991).

Each of these is considered by reference to three examples:

1. Bird migration studies.
2. The expansion of *Spartina anglica* into many salt marshes throughout the temperate regions of the world.
3. Population studies of sea birds.

Bird migration

Determining whether a resident animal species is stable, declining, or increasing is relatively easy since recording breeding success or overall population numbers may be all that is required. However, unravelling the status of a migrant species is much more difficult. In this context, a systematic approach has been adopted for monitoring bird populations involving the use of bird-ringing techniques. The original purpose of this was to unravel the mysteries of bird migration. However, knowing the pattern of movement is only part of the story. These studies also provide a vital tool for helping to determine the life history and population dynamics of many bird species. The information needed includes the distances traveled, the stopping off (refuelling) points, and the routes over which migration takes place, as well as studies of the bird's breeding success and physical condition. Armed with this information it is possible to begin developing conservation strategies for species which may breed in the Arctic, winter in estuaries in the southern North Sea or even as far away as South Africa (Figure M23).

The importance of these studies is reflected in North America by the fact that the US Fish and Wildlife Service maintains a Division devoted to migratory bird management. Organizations undertaking such survey work, include the International Shorebird Survey which has been

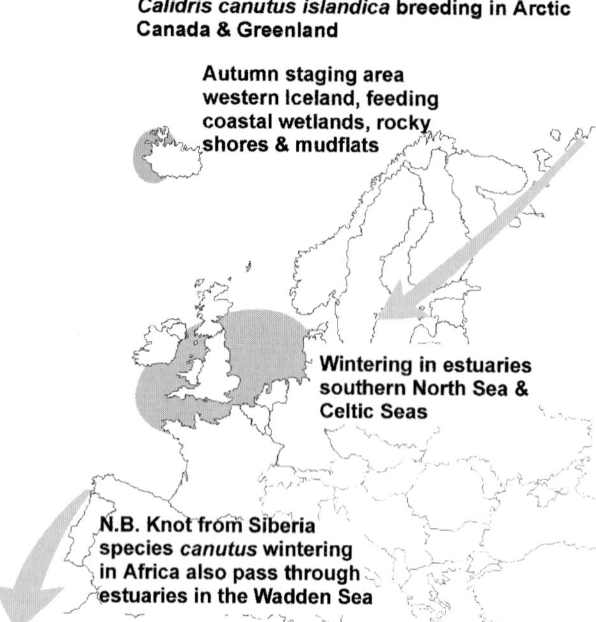

Figure M23 Migration patterns of the knot *Calidris canutus islandica* (see also British Trust for Ornithology, Migration Atlas Project). (Adapted from Doody, 2001.)

gathering standardized information on the numbers of shorebirds congregating at migratory stopover sites in the spring and fall since 1974. Standardizing and coordinating national studies using agreed practices for data collection and computerization are also recognized as part of the process. In North America, a "Migration Monitoring Network" was established in 1998 to coordinate activities including establishing standards and guidelines for the operation of monitoring programs. In Europe, in order to understand the significance of different parts of the species life cycle and the influence of human and other factors upon them, a European Union for Bird Ringing coordinates continent-wide research projects. This work is also used to monitor bird populations against international conventions and is a prerequisite for effective protective measurements for the many declining bird species, especially in their more vulnerable locations.

In the United Kingdom, the WeBS (Wetland Bird Survey) Low Tide Counts are part of a joint scheme coordinated by the British Trust for Ornithology. In this case, counts are made on about 20 selected estuaries each winter (November–February) to determine the distribution of birds during low tide and to identify important feeding areas that may not be recognized during Core Counts which, on estuaries, are mostly made at high tide. In this context, British estuaries are known to be important wintering areas for several waders (including the knot) that breed in Arctic Canada and Greenland, and staging areas for those that breed in Siberia and winter in Africa. The pattern of bird ringing recoveries, combined with the knowledge of breeding success and observed changes in wintering numbers provide the basis for assessing the status of the species. Evidence from bird counts in estuaries in the southern North Sea suggest that knot numbers have recovered from losses in the 1980s, though only to about 70% of its numbers in the 1970s. The reasons why the numbers have not returned to their previous levels are not clear. Loss of intertidal habitat through land-claim and deterioration in the quality of habitat at its wintering sites are implicated, in the absence of major change in breeding success. This has important implications for policy in relation to land claim around the estuaries of the southern North Sea, where large wintering populations of this species occur (Doody, 2001).

Spartina invasion

The invasion of non-native species and their effect on a wide variety of habitats and species is a major conservation (and in places economic) concern. The precise pattern of change is often impossible to predict. For example, no one could have anticipated the impact of a chance meeting between two species originating on opposite sides of the Atlantic on salt marshes throughout the world. An American plant, *Spartina alterniflora* was brought to Southampton, in England in ships ballast in the early 1800s. Hybridization between this species and the British *Spartina maritima* prior to 1870 produced a sterile hybrid *Spartina* × *townsendii*. Further change resulted in a natural fertile amphidiploid *S. anglica* being formed (Hubbard and Stebbings, 1967). At first the plant was heralded as an aid to land claim though latterly, as the new species spread in just a few decades to many estuaries in England and Wales, it began to cause concern to conservationists and others (Adam, 1990; Gray and Benham, 1990).

Monitoring this change took place through a variety of ad hoc approaches including field survey (Hubbard and Stebbings, 1967; Burd, 1988) and more localized research studies, Langstone Harbour (Haynes, 1986) and Poole Harbour (Gray and Pearson, 1986). The overall picture, which emerges is of a highly invasive species, which has had a profound effect on salt marshes both in England and throughout the world. These studies were not coordinated, yet together they provide a fascinating account of the scale and impact of change. This example serves to emphasize that there is often no one approach to monitoring and that especially over the medium to long-term a variety of methodologies, if properly interpreted, can provide an understanding of what is happening. From this an assessment of policy and management responses is possible, which in the case of *Spartina* has resulted in major efforts to control its spread both in England and Wales, and in the United States (Wecker, 1998).

Seabirds as indicators of change?

Once the status of a species has been determined, it is possible to relate any change, whether it is declining or expanding, to environmental or other human factors. These changes can be used not only as a means of assessing the population itself, but also as an indication of the state of other environmental conditions. For example, in the coastal environment the status of breeding seabirds might be used as an indication of the health of fish stocks. However, before such a causal link can be established, between a decline in a seabird population and a reduction in fish stocks other factors must be eliminated. These include:

- suitability of the nesting site (increases in grazing pressure may make burrows used by petrels and puffins more susceptible to collapse);
- incidence of predation (increases in predatory birds due to increased fishing effort and greater quantities of discarded offal causing death to young and/or reduced availability of food as parents are forced to disgorge before reaching their young);
- impact of offshore oil pollution and death of adult birds;
- changes in weather patterns.

Any or all of these could have an adverse effect on population size, which may hide the long-term implications of over-fishing. However, once these factors are recognized, and analyzed, change in animal populations can provide an important indicator of wider environmental damage. One of the special strengths of the use of bird population information, lies in the standardized methodology used for collecting data and the relatively long time frame over which the monitoring has been carried out.

Indicators for integrated coastal zone management

The above discussion shows how change in individual populations provides information, which is not only relevant to the conservation or control of the species being recorded, but can also be an indicator of the state of the wider environment. The examples given above involve quite simple interactions, though even these require detailed investigation to establish why change is occurring. When looking at the coast more widely, especially when socio-economic issues are included, assessing cause and effect, and hence policy implications can become highly complex and a real challenge. Making sense of the plethora of sometimes contradictory information, requires a systematic approach to its use. Although approaches differ in detail the basic sequence is the same and involves identifying the pressures which influence the state of the environment and the impact on human health and ecosystems. Society responds with various policy measures, such as regulations, information and taxes, or management action, which can be directed at any part of the system. The way in which this process develops and the factors important to the way in which it operates is illustrated next.

Taking stock

Taking stock of information on the environment is the first stage in any strategy to identify key issues for policy formulation and management action. In the United Kingdom in 1984 the need for an assessment of the status of the North Sea and its resources was recognized. As part of this assessment "a comprehensive description of the coastal margin of the North Sea and Celtic Seas, their habitats, species and human activities" was initiated. The resulting documents (Doody *et al.*, 1993; Barne *et al.*, 1995–98) and associated electronic publications provide a description of the whole of the coastal and marine environment of the United Kingdom and Ireland, the human use to which it is put and policies and organizations responsible for management. The subjects covered were determined by the wide range of stakeholders involved in the process, who also helped to identify the key information requirements and issues on the basis of their own knowledge and understanding. The results gave an insight into what has happened and a baseline to assess the need for action and monitor the effectiveness of policy decisions and management.

State of the coast reporting

The scale and range of information available to facilitate the understanding of the coast and its policy and management requirements at national level also is large and complex. When more than one nation state is involved the situation becomes even more complex. Different approaches, methodologies, and languages across national boundaries increase the barriers to the collection and collation of information. However, despite this State of the Environment reporting is used to address issues in geographical areas, including several, which cross national boundaries. A comprehensive review of the state of the Arctic has been prepared (Nilsson, 1997). In Europe, in addition to the European Union's general State of the Environment Report the "Dobris Assessments," there are other reports covering the regional seas, for example, the "Mediterranean Sea: Environmental State and Pressures" (Izzo, 1998). Few of these reports deal specifically with the coast most are either general State of the Coast reports or, as in the last case, are largely concerned with marine issues. However, as a methodology all have the characteristics of being comprehensive in their approach and including a variety of key stakeholders in their preparation.

Key issues

Stock taking and/or State of the Coast Reports provide a means of identifying key issues. These are important for the identification of indicators used in predicting change and the efficacy of policy implementation and management action. The key issues also help to identify key pressures on the system; another important element in reporting on the state of the system. A list of the issues most relevant to the coastal zone is given above, derived from various sources (Table M10).

Table M10 A list of key issues adapted from publications covering different levels of administration in the European Union

Key issues (not specifically coastal) derived at European level are:
- Socio-economy
- Transport
- Agriculture
- Energy
- Tourism
- Industry
- Households
- Air quality and ozone (stratospheric and tropospheric)
- Climate change
- Water stress
- Eutrophication
- Acidification
- Noise
- Biodiversity and landscapes
- Exposure to chemicals
- Health and environment
- Land and soil degradation
- Waste management

(after Stanners and Bourdeau, 1995)

In the North Sea key issues are
- Heavy metal contamination
- Artificial radionuclides
- Oil
- Shipping
- Nutrients
- Fisheries
- Organic contaminants
- Protection of habitats and species

(after Quality Status Report, 1993)

More specific key issues identified for the coastline of England and Wales
- Sea-level change and increased storminess
- Quality of bathing waters and beaches
- Loss of habitats and implications for biodiversity
- Pollution by hazardous substances
- Development pressures on the coast

(after Environment Agency, 1999)

Understanding the coast—research

Research is vital to understanding the way in which coastal systems develop, especially in helping to unravel cause and effect when considering the impact of policy and management action. This is particularly important in such a dynamic environment where coastal processes are often key in determining the nature of the coast and its ability to sustain human use.

Research

There are extensive research programs throughout the world designed to provide the necessary understanding of the impact on and limits to the use of the coast and its resources. Two approaches will be used to illustrate some of the techniques employed to identify the nature and scale of change and the underlying processes which "drive" the "natural" coast, namely *Lacoast* (*La*nd cover changes in *Coast*al zones) which uses satellite images to assess land-cover changes in coastal zones and *LOIS* (the *L*and *O*cean *I*nteraction *S*tudy) which is designed to gain an understanding of, and an ability to predict, the nature of environmental change in the coastal zone around the United Kingdom.

Lacoast. The aim of the Lacoast project is to assess quantitative changes of land cover/land use in European coastal zones during the last decades. It is being undertaken by the Agriculture and Regional Information Systems (ARIS) unit and the Space Applications Institute (SAI) at the European Union's Joint Research Centre. The project covers the entire coastal zones of 10 European Member States and includes a 10 km wide strip of land bounded by the coastline. This project provides quantitative estimates of land cover and land use change in the coastal zone, focusing in particular, on those caused by human activities. Where changes are observed, an exercise to identify and interpret the factors responsible for change is undertaken. The results so far have provided data relevant to environmental status and policy formulation at national and European scales helping to:

- identify major change at national levels;
- identify hot spots of change that when combined with other information can show trends in development pressures; and
- show the nature of human activity and hence the key issues requiring a change of policy and/or management action.

LOIS. The Land–Ocean Interaction Study, a 6-year multi million pound project (1992–98) funded by the United Kingdom's Natural Environment Research Council, involved more than 360 scientists from 11 institutes and 27 universities. It collected a vast amount of data and is a major research investment aimed at improving the understanding of the way in which coastal systems interact. As its title implies, it was designed primarily to help elucidate the relationships that exist between the exchange, movement, and storage of materials at the land–ocean boundary. Some 300 papers have been submitted to academic journals, possibly half the number that will finally be published, and a series of CD-ROMs produced (Natural Environment Research Council, 1998). Thus, as a research enterprise it is deemed to have been very successful.

A review of the main findings of the study suggests that there are some important broad conclusions to be drawn from the work—conclusions that have a more general value to the process of integrated management. For example, new insights to nutrient budgets show that 90% of the fluxes occur during only 5% of the time, that is, "pulses" are very important in delivering nutrient loads to the sea. This finding is important to any monitoring program designed to measure changes in nutrient levels since it would be easy to miss these important events. The work has also helped establish the validity of technologies, such as remote sensing, in identifying vegetation patterns, thus providing the possibility for a more cost-effective methods of survey and monitoring. A better understanding of sediment budgets and their relationship to erosion in an estuarine system has also been established. This leads the way to interpreting land– sea-level change and the implications for erosion and flooding.

Conclusion

Determining the status of the coastline its landscapes, habitats, and species concentrations and human interactions is a product of a wide variety of methodologies. These range from anecdotal observations to sophisticated, targeted research programs using modern technologies. Two phases can be discerned in the process of developing an understanding of the impact of human use on the ability of the environment to continue to sustain that use. The first involves a "stock take" of existing information which can be summarized as including the following:

1. *Geological, geomorphology, climate and sea level* variables, which provide the driving forces for the development of coastal systems.
2. *Coastal habitats and species* information, which set a baseline for the selection of important conservation areas (e.g., the European Union Habitats Directive).
3. *Human uses and activities* including the location and scale of infrastructure and intensity of use.

This can be described a "passive" stage in the context of monitoring as it is concerned initially with the location and scale of the resource or activities which take place there, rather than attempting to identify change and the causes of change. Undertaking an audit of coastal information as described above will also help to identify gaps in information and hence the need for updated or new surveys, and to assess their scale and frequency. Defining key issues for policy formulation and establishing the key human uses requiring surveillance or monitoring, is the next important step. Armed with this analysis it should be possible to develop a more "active" interaction between the provision of information and decision making. This may be summarized under four headings:

1. *Monitoring* to help identify and assess impacts, determine action and give feedback on its effectiveness.
2. *Surveillance* to identify unforeseen change and where appropriate ensure compliance with agreed legislative or other control mechanisms.
3. *Prediction* of the possible outcome of policy decisions whether in reaction to an unforeseen incident (e.g., oil pollution) or "natural" event (e.g., sea-level rise).
4. *Assessing* the effectiveness of action (management) is the final stage if we are to learn from good and bad practice.

One key aspect of the coast is its unpredictability. Identifying change without knowing what that change might be (surveillance) is therefore potentially very difficult. However, retrospective comparisons of maps, published works, the use of satellites and research (e.g., into sediment fluxes) as has been shown above, all help understanding. Twice daily tidal cycles are predictable, storms and tidal surges are not. It is often these events which cause substantial change. Similarly, cliff erosion rarely takes place on a small and regular basis and none of these unpredictable events are easily monitored. However, detailed surveillance of individual habitats can be effective and help elucidate wider scale change.

Monitoring—salt marshes and sea-level change

Salt marshes are important habitats and occur around the coast throughout the world. In addition to their nature conservation interest as habitats for specialist plants and animals, they also form part of the functioning system of coastal wetlands (in estuaries and tidal deltas). In some areas, salt marshes are eroding and a number of factors have been implicated including pollution and sea-level rise. However, a question arises as to whether this erosion is part of the natural dynamic or a longer term, and ultimately more significant change, resulting from global sea level rise or other more local effects caused by human intervention. Clearly determining the appropriate policy response will depend on understanding the extent to which human action is responsible for the observed change.

Visual and anecdotal evidence from south east England suggested that the salt marsh resource had eroded at a considerable rate during the 1980s and early 1990s. This in turn appeared to be the reason why a number of sea-walls were being undermined with an increased risk of flooding to adjacent low-lying land (Figure M24). In an area already experiencing a relative rise in sea level of 5.4 mm per year due to isostatic change and sea-level rise this suggested that threats to life and property were intensifying. Detailed field survey and comparison between sets of aerial photographs appeared to confirm anecdotal evidence and the salt marshes were assessed as eroding at an alarming rate (Burd, 1991).

On the face of it sea-level rise appears to be the main cause of loss. However, even if it is the main agent, many other factors come into play. Tides and tidal range, sediment availability, and the nature of the coast all influence the development of salt marsh. Local weather conditions including rainfall, discharge rates of rivers, and the state of the tide all contribute to the incidence and severity of flooding. Erosion rates depend on the strength of the feature being affected by the erosive force. Changes in sea level *per se* (eustacy) are the result of global forces associated with the atmospheric temperature whether due to human influences or not. In areas where the land level is stable or sinking (isostacy) there will be an inundation of the coast. Given all these factors it may be impossible to determine the precise reasons for the loss of salt marsh. Despite this, the fact remains that salt marshes are eroding and as a consequence sea-walls are being undermined and land threatened with flooding. Thus, the measurement of change in salt marsh provides a very powerful indication that an adverse change is taking place and one which needs to be addressed.

Figure M24 Land below the 5 m contour potentially at risk from coastal flooding and over-topping by the sea in Great Britain. (Adapted from Doody, 2001.)

Satellites as a means of monitoring coastal change

Using satellites to look at the environment of the earth is an important part of the space program. Remote Sensing Satellites provide information on a variety of features including physical oceanography, polar science, and climate research. However, satellites are currently under-utilized as a source of surveillance and monitoring data for coastal zone management (Doody and Pamplin, 1998). Part of the problem lies in the absence of any strong link between those concerned with the management and conservation of coastal areas and those interpreting satellite data or developing new missions. Examples of the former might be the "field ecologist" or conservation site manager, the latter the satellite "technocrat."

The key point in relation to the use of satellites lies in the fact that more frequent observations can be made and that these can, despite the initial costs of the data acquisition, may be more cost-effective than carrying out detailed and time consuming field survey. Although not all coastal habitats lend themselves to this approach and while cliffs (Figure M25) present special difficulties, flat expanses of tidal land (Figure M26), sand dunes and shingle beaches, and structures do not.

Figure M25 Cliffed landscapes require oblique photography to show features, the chalk cliffs of Beachy Head, southern England (original photograph JPD).

Figure M26 Tidal sandy flats are easily discerned from satellite images, Brancaster, North Norfolk, England (original photograph JPD).

Time scales for change are also an important factor. While satellites can cover large areas and record cumulative effects they are less good at measuring more rapid change. Until recently, a further impediment was the scale of resolution, which limited the level of detail that could be obtained. As the frequency of survey and resolution improves so will the ability of the imagery to provide more effective surveillance of the coastal zone. This will not negate the need for more traditional forms of survey and monitoring. Sloping landforms, detailed vegetation studies and measurements of river flows, sediment movement, tides, etc. will still require traditional approaches to research, survey, and monitoring. The point is that satellites can and should be used more extensively for monitoring and surveillance of the coastal zone.

Indicators for policy response and management action

The importance of the coast to the economic and social fabric of society in many parts of the world, increases the need for more widely based monitoring, encompassing both environmental and economic issues. Linking these into an assessment of the effectiveness of policy and

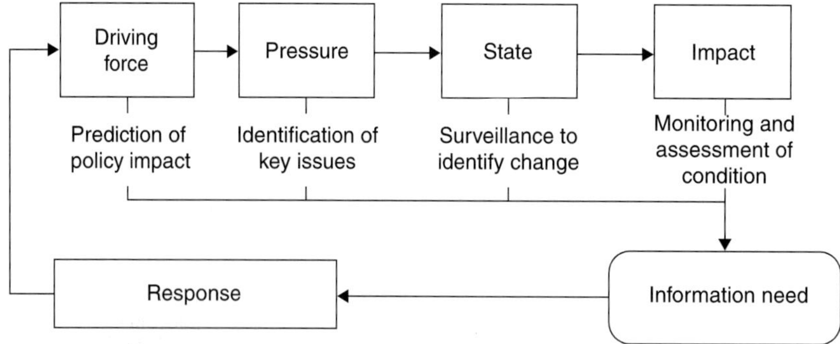

Figure M27 The European Environment Agencies modified United Nations' Pressure–State–Response model. This shows driving forces (e.g., industry and transport) producing pressures on the environment (e.g., polluting emissions), which then degrade the State of the environment and have impacts on human health and ecosystems. Society responds with various policy measures, such as regulations, information and taxes, or management action which can be directed at any other part of the system.

action, add another dimension to the complexity of the monitoring system. Information on future trends on the state of the environment and prospects for socio-economic and sectoral aspects are crucial for determining progress against policy targets and to ascertain among other things:

- whether current policy measures are to be expected to deliver the required improvements taking into account trends in external factors;
- whether additional polices might be considered as necessary to achieve the expected improvements;
- whether new policy needs are likely to emerge in uncovered areas.

At this stage, in the discussion monitoring has moved from being a concern of individual sectors or research academics, to being at the center of policy formulation and decision making. Developing indicators as an integral part of a strategy for policy formulation is a common approach (Figure M27). Using a variety of techniques involving, existing information and new data derived from remotely sensed sources and traditional survey, monitoring and research, it is possible to establish an integrated approach which matches the needs of the coastal zone. For example, we should be able to measure salt marsh change, relate this to sea-level rise, and provide a means of assessing different options for sea defense according to economic and social criteria.

The use of remote sensing techniques will not provide information on bird migration routes or the population density and success rate of breeding birds. Nor will it replace survey and surveillance of narrow beaches or vertical cliffs, or detailed vegetation maps needed for site management. These forms of data acquisition will still be required, as will the view of the informed "specialist," or "local" knowledge. When linked to the more traditional forms of survey, monitoring and research, and other programs designed to synthesize and summarize existing data, an interactive information system which provides support for coastal management decisions can be devised. This will require a multidisciplinary approach, if the demands for more integrated management of the coast and in the coastal zone are to be realized. The communication options afforded by the Internet also provide an opportunity to involve, and exchange information between the many relevant sectors operating on the coast. Achieving integration between the historical legacy of information, survey, surveillance, and monitoring as well as research, and across the environmental and the socio-economic divide is the challenge for monitoring environmental and human well-being and hence the future of the coast.

J. Pat Doody

Bibliography

Anon, 1977. *Orford Ness, A Selection of Maps Mainly by John Norden*, Presented to J.A. Steers, Heffer & Sons, Cambridge.

Barne, J.H., Robson, C.F., Kaznowska, S.S., and Doody, J.P., 1995–98. *Coasts and seas of the United Kingdom*. 17 Volumes, Peterborough: Joint Nature Conservation Committee.

Burd, F., 1988. *The saltmarsh survey of Great Britain—an Inventory of British saltmarshes*. Research and Survey in Nature Conservation, No. 17, Peterborough: Nature Conservancy Council.

Burd, F., 1991. *Erosion and Vegetation Change on the Saltmarshes of Essex and North Kent between 1972 and 1988*. Research and Survey in Nature Conservation, No. 42. Peterborough: Nature Conservancy Council.

Carey, A.E., and Oliver, F.W., 1918. *Tidal Lands: A Study in Shoreline Problems*. London: Blackie & Son.

Dargie, T.C.D., 1993. *Sand Dune Survey of Great Britain—a National Inventory. Part 2 Scotland*. Peterborough: Joint Nature Conservation Committee, Coastal Conservation Branch.

Dargie, T.C.D., 1995. *Sand Dune Survey of Great Britain—A National Inventory. Part 3 Wales*. Peterborough: Joint Nature Conservation Committee, Coastal Conservation Branch.

Dijkema, K.S. (ed.), 1984. Salt Marshes in Europe. *Nature and Environment Series,* No. 30, Strasbourg: Council of Europe.

Dijkema, K.S., and Wolf, W.J., 1983. *Flora and Vegetation of the Wadden Sea Islands and Coastal Areas*. Report 9 of the Wadden Sea Working Group, Leiden: Stitching Veth tot Steun aan Waddenonderzoek, pp. 305–308.

Doody, J.P. (ed.), 1991. *Sand Dune Inventory of Europe*. Peterborough/ Leiden: Joint Nature Conservation Committee/ European Union for Coastal Conservation.

Doody, J.P., 1995. Infrastructure development and other human influences on the coastline of Europe. In Salman, A.H.P.M., Berends, H., and Bonazountas, M. (eds.), *Coastal Management and Habitat Conservation, Proceedings of the 4th EUCC Congress, Marathon, Greece*, April, 1993. Leiden: EUCC, pp. 133–152.

Doody, J.P., 2001. *Coastal Conservation and Management: An Ecological Perspective*. Boston: Kluwer Academic Publishers..

Doody, J.P., and Pamplin, C., 1998. Information and ICZM—lessons from the European Union's Demonstration Programme on Integrated Management of Coastal Zones. *Proceedings of the 27th International Conference on Remote Sensing*, 8–12 June 1998, Tromso, Norway.

Doody, J.P., Johnson, C., and Smith, B., 1993. *Directory of the North Sea Coastal Margin*. Peterborough Joint Nature Conservation Committee.

Du-Plat-Taylor, F.M., 1931. *The Reclamation of Land from the Sea*. London: Constable & Co.

Environment Agency, 1999. *The State of the Environment of England and Wales: Coasts*. London: The Stationary Office.

Gray, A.J., and Benham, P.E.M., 1990. *Spartina anglica—A Research Review*. London: Institute of Terrestrial Ecology, HMSO.

Gray, A.J., and Pearson., J.M., 1986. Spartina marshes in Poole Harbour, Dorset with particular reference to Holes Bay. In Doody, J.P. (ed.), *Spartina anglica in Great Britain*, Focus on nature conservation, No. 5, Attingham Park: Nature Conservancy Council, pp. 11–16.

Haynes, F.N., 1986. *Spartina* in Langstone Harbour, Hampshire. In Doody, J.P. (ed.), *Spartina anglica in Great Britain*, Focus on nature conservation, No. 5, Attingham Park: Nature Conservancy Council, pp. 5–10.

Hubbard, J.C.E., and Stebbings, R.E., 1967. Distribution, dates of origin and acreage of *Spartina townsendii* (salt marshes in Great Britain. *Transactions of the Botanical Society of the British Isles*, **7**: 1–7.

Izzo, G. (Co-ordinator) 1998. *Mediterranean Sea: Environmental State and Pressures*. Copenhagen: European Topic Centre, Marine and

Coastal Environments, under contract to the European Environment Agency.
Keddy, P.A., 1991. Biological monitoring and ecological prediction: from nature reserve management to national state of the environment indicators. In Goldsmith, F.B. (ed.), *Monitoring for Conservation Ecology*, London: Chapman & Hall, pp. 249–267.
Matthews, E.R., 1934. *Coast Erosion and Protection*. London: Charles Griffin & Co.
Natural Environment Research Council, 1998. *Land–Ocean Interaction Study. LOIS Overview, Version 2*, CD-ROM, Plymouth: Centre for Coastal and Marine Studies.
Nilsson, A., 1997. *Arctic Pollution Issues: A State of the Arctic Environment Report*. Oslo: Arctic Assessment and Monitoring Programme.
North Sea Task Force, 1994. *North Sea Quality Status Report*. London: Oslo and Paris Commissions.
Radley, G.P., 1994. *Sand Dune Survey of Great Britain—A National Inventory. Part 1 England*. Peterborough: Joint Nature Conservation Committee, Coastal Conservation Branch, UK.
Ranwell, D.S., 1967. World resources of *Spartina townsendii* (sensu lato) and economic use of *Spartina* marshland. *Journal of Applied Ecology*, **6**: 239–256.
Sneddon, P., and Randall, R.E., 1993. *Coastal Vegetated Shingle Structures of Great Britain: Main Report*. Peterborough: Joint Nature Conservation Committee, Coastal Conservation Branch.
Stanners, D., and Bourdeau, P., 1995. *Europe's Environment—The Dobris Assessment*. Copenhagen: European Environment Agency.
Steers, J.A., 1946. *The Coastline of England and Wales*. Cambridge: Cambridege University Press.
Steers, J.A., 1973. *The Coastline of Scotland*. Cambridge: Cambridge University Press.
Wecker, M., 1998. *Integrating biological control in the integrated pest management program for* Spartina alterniflora *in Willapa Bay*. Unpublished report. Seattle: University of Washington.
Wheeler, W.H., 1903. *The Sea Coast. 1. Destruction, 2. Littoral Drift, 3. Protection*. London: Longmans, Green & Co.

Cross-references

Environmental Quality
Europe, Coastal Ecology
History, Coastal Ecology
Monitoring, Coastal Geomorphology
Remote Sensing of Coastal Environments
Remote Sensing: Wetlands Classification
Salt Marsh
Vegetation Coasts

MONITORING COASTAL GEOMORPHOLOGY

Coastal engineering and research, management of natural resources, beach and wetland restoration, navigation improvements, and military operations all share the need for copious amounts of data. These data typically are used to evaluate and monitor a specific reach of the coast. Ideally, a coastal monitoring program employs a multidiscipline approach in diagnosing the beach and nearshore zone. Many large programs such as the US Army Corps of Engineers' Shinnecock Inlet Study (Morang, 1999; Militelo and Kraus, 2001; Pratt and Stauble, 2001) or the Kings Bay Coastal and Estuarine Physical Monitoring and Evaluation Program (Kraus *et al.*, 1994) collect data for a range of physical conditions such as wave climate, the morphology of the beach and nearshore surface, and accretion and erosion trends across and along the shoreline (high-tide line or coastline) under investigation. These collections allow scientists and engineers to understand the coastal processes and their variability in response to waves, currents, winds, and tides.

The high-energy, often hazardous coastal environment is one of the most difficult places on earth in which to collect data. Consider the enormous forces concentrated in the narrow coastal zone: in hours, beaches disappear; in days, new inlets are cut; in a generation, rock cliffs crumble. Even the most massive coastal works have often been buried in sand, swept away, or pounded into rubble. How are instruments expected to survive in this harsh environment? Despite the engineering challenges, scientists and engineers, who never feel that they have enough data, continue to develop innovative techniques and instruments to help them answer some of the elusive questions about the sea and the land it touches.

This entry summarizes the various types of *field* data that are typically collected for coastal research and engineering and briefly introduces some of the most common instruments. We exclude data developed in physical or numerical models, although clearly these are important tools also used by coastal scientists. Biological monitoring is covered in a companion paper in this volume.

The quality of a field study depends on several factors:

1. Scientists must recognize the many problems, assumptions, and limitations in field data collection and make adjustments for them before attempting an interpretation. For example, a single cross-shore beach profile may not be representative of the beach topography during much of the year (e.g., possibly the survey was made after a storm when the beach was in an unusually eroded state).
2. The appropriate phenomena must be monitored to answer the engineering or scientific problem being investigated. For example, wave data are needed to choose appropriate stone size for a breakwater. But, an offshore ocean wave gauge may not provide useful wave statistics if the coast has complex topography with headlands and bays. A nearshore gauge is needed here.
3. Many of the techniques used to monitor processes or structures in the coastal zone are exceedingly complex. Inexperienced users should consult experts to prevent making important decisions based on poor data or data that do not truly answer their questions. Example 1: It is easy to purchase aerial photographs and interpret some sort of a shoreline (simply draw a line at the edge of the water). But what does this "shoreline" really mean, and can it be compared with shorelines identified from other dates that were based on different field techniques or criteria? Example 2: Offshore sand sources are being examined with geophysical techniques to determine if the sand is suitable for beach renourishment. Low-frequency acoustic systems may detect adequate quantities of material, but the resolution is too low to reveal that the sand contains interbedded layers of coral fragments. Here, a combination of low- and high-frequency systems should have been used, combined with core samples. Example 3: A series of cross-shore beach profiles show that a significant amount of sand was lost from the subaerial beach after a storm battered a recently placed beach fill, leading an analyst to conclude that this sand is lost. However, the profiles were wading-depth only. If sled profiles that spanned the entire active zone had been collected, the analyst might have detected that much of the sand lost from the beach had accumulated in the offshore sand bars and that the total sand volume in the active zone was almost unchanged.

Classes of monitoring techniques

Coastal monitoring can be divided into three general classes of techniques, each of which use specific instruments and analysis methods (Table M11). Note that a comprehensive study might employ instruments and data from all three classes.

1. *Remote sensing methods*: instruments provide information about the land and the sea from a distance without being in physical contact (e.g., aerial photography, laser imaging).
2. *In situ instruments*: instruments are placed *in* the media being studied, such as current meters moored in the ocean.
3. *Sampling methods*: devices retrieve a sample of the material being examined (i.e., water, ice, sediment, biological material) so that the scientist can conduct more detailed examination in a laboratory.

Remote sensing and geophysical methods

Remote sensing methods include familiar and well-proven technologies such as aerial photography (first used in World War I), satellite imaging systems in the 1970s, and laser bathymetry in the 1990s. See Philipson (1997), Henderson and Lewis (1998), Lillesand and Kiefer (1999), and Rencz and Ryerson (1999) for manuals on various sensing technologies. All remote monitoring methods require measuring and recording some form of acoustic or electromagnetic energy and then relating the resulting data to specific earth parameters (Table M12).

The term "geophysics" is defined as the "study of the earth by quantitative physical methods" (Bates and Jackson, 1984, p. 209). Geophysical methods are a form of remote sensing in that a researcher uses a tool to remotely image the seafloor or the strata below. The result is a depiction of the subsurface geology, a mathematical model based on varying acoustic impedances of air, water, sediment, and rock. The model, which must be interpreted, is based on numerous assumptions, and the user must always remember that the real earth may be very different

Table M11 Classes of monitoring systems for coastal studies

Type	Characteristics	Examples	Advantages	Disadvantages
Remote sensing	Distant from the area or media being studied (without physical contact)	• Air photographs • Video monitoring • Electromagnetic geophysics • Acoustic geophysics	• Broad coverage • Cover large area in short time • Suitable for hostile environments (military applications or harsh weather)	• Interpretation is a model—must be verified with some field data or *a priori* knowledge. • Resolution often too coarse for shoreline mapping or morphology studies • Occasionally challenging or conflicting interpretation
In situ	Measures parameter from within the media	• Current meters • Wave gauges • SBT (salinity bathythermographs) • Strain gauges • Tracking devices on turtles, whales	• Often better resolution than remote systems • Good temporal coverage • Can sometimes be used to monitor changing phenomena not accessible remotely	Point source sampled over time—must interpolate spatially
Sample recovery	Retrieves a portion of the material	• Plankton tows • Coring devices • Seafloor grab samplers • Sediment traps	• Allows detailed analysis in laboratory of the "real" material • Re-analysis possible for different properties	• Point data only—may not be representative spatially or temporally • Preservation and archiving of samples is expensive

Table M12 Remote sensing systems

Energy type	Examples	Common applications
Electromagnetic	Radar, Synthetic Aperture Radar (SAR)	Ocean wave measurement Ground-penetrating radar Object identification (military)
	Laser	Terrestrial topography (LIDAR) Bathymetry (SHOALS)
	Multi-spectral	Satellite (LANDSAT, SPOT, AVHRR)
	Visual wavelengths	Aerial photography (features, terrain)
	Infrared	Aerial photography (plant life)
Acoustic	Low frequency (<1 kHz)	Petroleum exploration, deep penetration Earth properties research, earthquake monitoring
	Medium frequency (0.5–7 kHz)	Seismic equipment (boomers, sparkers): near-surface strata, geohazards, sand resources, rock outcrops
	High frequency (12–450 kHz)	Bathymetry: sea bottom features, nautical charting Side-scan sonar: seafloor features, archaeology, shipwrecks, pipelines, breakwaters Multi-beam bathymetry: structures, breakwaters, detailed sea-bottom inspection

than the model printed on paper or displayed on his monitor. This warning notwithstanding, geophysical (particularly acoustic) methods, have proven to be extremely powerful tools in numerous coastal applications, including:

- Determining water depth (bathymetric or hydrographic surveys)
- Imaging the sea bottom to identify surficial sediments, measure bottom features such as ripples, and locate abandoned structures and hazardous debris
- Measuring the thickness of strata to locate suitable quantities of sand for beach renourishment
- Mapping gas pockets, rock outcrops, and other geological hazards
- Identifying coral, fish reefs, and other biologically sensitive areas
- Mapping cultural and archaeological resources such as sunken ships and pipelines.

Echo sounders or depth-sounders, side-scan sonar, and subbottom profilers are three classes of equipment commonly used to collect geophysical data in marine exploration. All three are acoustic systems that propagate sound pulses in the water and measure the lapsed time between the initiation of the pulse and the arrival of return signals reflected from target features on or beneath the seafloor. Other geophysical methods, such as magnetic, gravity, and electrical resistivity, laser line scan and electronic still cameras can be used in specialized engineering applications (Griffiths and King, 1981; Reed, 2001), but are not as common in reconnaissance coastal studies.

Until the 20th century, measuring water depth consisted of the slow and laborious use of sounding poles and lead lines (Shalowitz, 1962). The introduction of acoustic echo sounding after World War I revolutionized charting, and for the following 70 years, single-beam acoustic depth-sounders were used for most bathymetric surveys. In the United States, offshore waters are normally surveyed by the National Oceanic and Atmospheric Administration (NOAA) (Umbach, 1976), while navigation channels, canals, and rivers are surveyed by the US Army Corps of Engineers (USACE; USACE, 1994). Most nations have standards that specify procedures to achieve required horizontal (positioning) and vertical (water depth) accuracies. The International Hydrographic Organization (IHO) has published universal standards for bathymetric surveys for the organization's member states (IHO, 1998). An exhaustive bibliography of hydrographic charting technology is listed in IHO (2001). Multi-beam echo sounders, becoming increasingly common now, are improvements on the traditional single-beam systems because they allow very detailed imaging of underwater structures and topography (Figure M28) (Cowls, 2000).

Side-scan sonar provides an image of the aerial distribution of sediment, surface bed forms, and large features such as shoals and channels. It can thus be helpful in mapping directions of sediment transport and areas of deposition or erosion. Belderson *et al.* (1972), Leenhardt (1974), Flemming (1976), Mazel (1985), and Fish and Carr (1990) provide additional details on the use and theory behind side-scan sonar. Skilled interpreters have long been able to use side-scan sonographs to visually identify seabed sediments, but this practice normally required some *a priori* knowledge of the survey area. New acoustic approaches have recently been developed that mathematically distinguish seafloor characteristics based on scattering parameters and waveform shapes (Whitehead and Cooper, 2001). Side-scan sonar also is a powerful tool for offshore engineering and construction, where it is used to survey structures, inspect rock jetties and breakwater units, track pipelines and telephone cables, and identify hazards such as shipwrecks (Figure M29). Up through the 1980s, most side-scan systems recorded on analog paper charts, but during the 1990s, modern computers have revolutionized the display of side-scan data, making possible real-time three-dimensional (3D) display of side-scan and multi-beam data (McAndrew, 2001).

Subbottom profilers, as the name implies, are used to examine the stratigraphy below the seafloor. The principles of subbottom seismic profiling are fundamentally the same as those of acoustic depth-sounding, but subbottom acoustic transmitters and receivers employ lower frequency, higher power signals to penetrate the seafloor (Sheriff, 1977; Sieck and Self, 1977). "High-resolution" generally means that the surveys are intended for engineering purposes or for identifying strata and structures in the uppermost 50 or 60 m of the sediment column. Typical applications include reconnaissance geological surveys, foundation studies for offshore platforms, hazards surveys to locate buried debris and gas pockets, and surveys to identify mineral resources (sand for beach renourishment). Figure M30 is an example of subbottom profiling in a tropical area with coquina limestone reefs.

Figure M28 Multi-beam acoustic data from Shinnecock Inlet, Long Island, New York, December 1998. Sand waves are evident in the inlet, and the highly irregular bottom near the west jetty indicates scour holes. North is to the top of the image. (Data collected and processed by State University of New York, Stony Brook.)

Ground-penetrating radar (GPR) uses electromagnetic energy to image subbottom sediments. Radiowave energy is transmitted through the sediment and reflects from materials as a function of variations in dielectric constants and electrical resistivity (Daniels, 1989; Sellmann *et al.*, 1992). The main limitation of GPR is that it must be used in freshwater environments such as the Great Lakes or reservoirs, but FitzGerald *et al.* (1992) and van Heteren *et al.* (1994) have successfully used GPR to delineate structure and stratigraphy of beach ridges in New England.

A single geophysical method rarely provides enough information about subsurface conditions to be used without sediment samples or additional data from other geophysical methods. Each geophysical technique typically responds to different physical characteristics of earth materials, and correlation of data from several methods provides the most meaningful results. Morang *et al.* (1997a) provide more details on geophysical systems and provide tables that list frequencies and resolution. *All geophysical methods rely heavily on experienced operators and analysts*. Inexperienced users should seek help both in contracting for surveys and in interpreting records.

Aerial photography is a traditional technology that continues to be very useful in monitoring coastal morphology, evaluating changes in wetlands and deltas, and mapping urban areas. Historical photographs, often dating to the 1930s, are invaluable in evaluating 20th century changes to beaches and coasts. Figure M31 is an example of photographs taken three days after the Great New England Hurricane in 1938. These photographs document the opening of Shinnecock Inlet, Long Island, New York. Photogrammetry methods are covered in another entry in this volume.

Since the mid-1990s, Light Detection and Ranging (LIDAR) survey systems, based on lasers mounted in aircraft or helicopters, have been used around the world to obtain high-resolution topographic measurements. Most of these systems are for land use only, but one particular system, the Scanning Hydrographic Operational Airborne Lidar Survey (SHOALS), is now regularly used by the USACE to survey coastal waters and inlets. The system is based on the transmission and reflection of a pulsed coherent laser light from an aircraft or helicopter equipped with the SHOALS instrument pod and with data processing and navigation equipment (Lillycrop and Banic, 1992; Estep *et al.*, 1994; Irish *et al.*, 2000; West *et al.*, 2001). In operation, the SHOALS laser pulses 400 times per second and scans an arc across the aircraft flight path, producing a survey swath equal to about half of the aircraft altitude (usually 300–500 m). A strongly reflected light return is recorded from the water surface, followed closely by a weaker return from the seafloor. The difference in time of the returns corresponds to water depth. Data density can be adjusted by flying higher or lower, at different speeds, or by selecting different scan widths.

SHOALS has revolutionized hydrographic surveying in shallow water for several reasons:

1. The most important advantage is that the system can survey up to 35 km^2/h, thereby densely covering large stretches of the coast in a few days. This capability enables almost instantaneous data collection along shores subject to rapid changes.
2. The system can be mobilized quickly, allowing broad-area post-storm surveys or surveys of unexpected situations such as breaches across barriers.
3. SHOALS can survey directly from the nearshore zone across the beach, allowing efficient coverage of shoals, channels, or breaches that would normally be impossible or very difficult to survey using traditional methods, especially in winter. The system can now survey bluffs up to 10 m high, allowing a single platform to provide topographic and bathy-metric data.

The main limitation of SHOALS is its critical dependence on water clarity. Maximum survey depth is over 30 m in clear water, like the Caribbean Sea, but the laser signal is drastically attenuated if the water is turbid due to suspended sediment, algae, or surf. Sometimes, a survey has to be delayed until a particular time of the year to be successful in an area that experiences river runoff or algae blooms. For example, the SHOALS survey at Shinnecock Inlet, on Long Island, New York, had to be carefully scheduled to avoid algal blooms (Figure M32). Because of the immense amount of data that the system collects, data processing, archiving, and management are a challenge.

Airborne LIDAR Bathymetry (ALB) is a major advance in military operations because of its ability to rapidly survey unfamiliar littoral

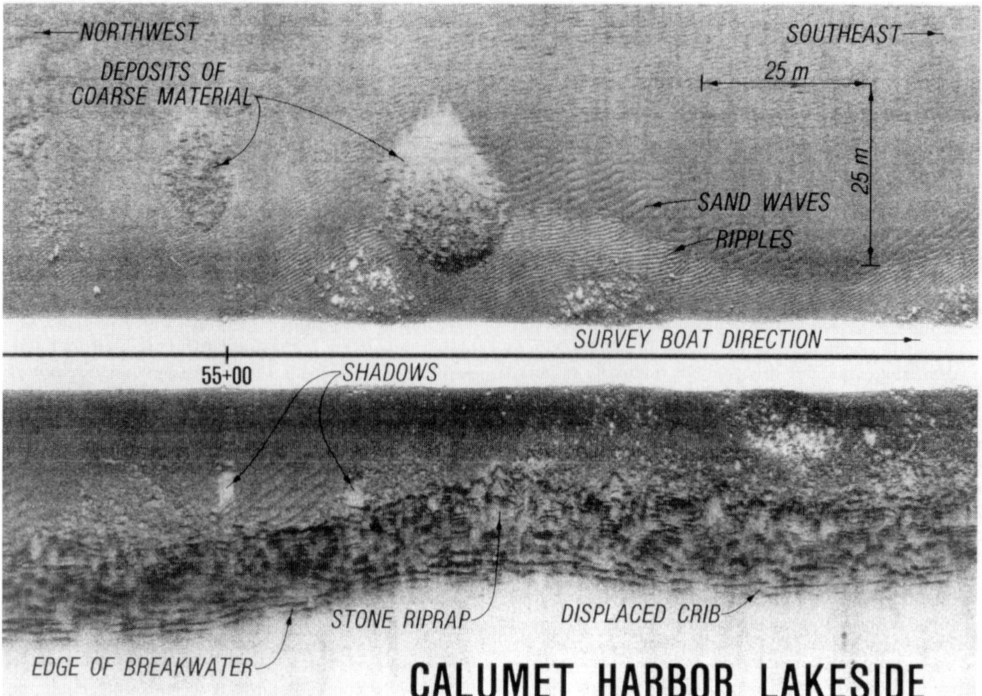

Figure M29 Annotated side-scan sonar record from Calumet harbor, southern Lake Michigan. The sonograph reveals a subtle displacement of some of the wood crib units and also shows how some construction material was deposited on the lake floor.

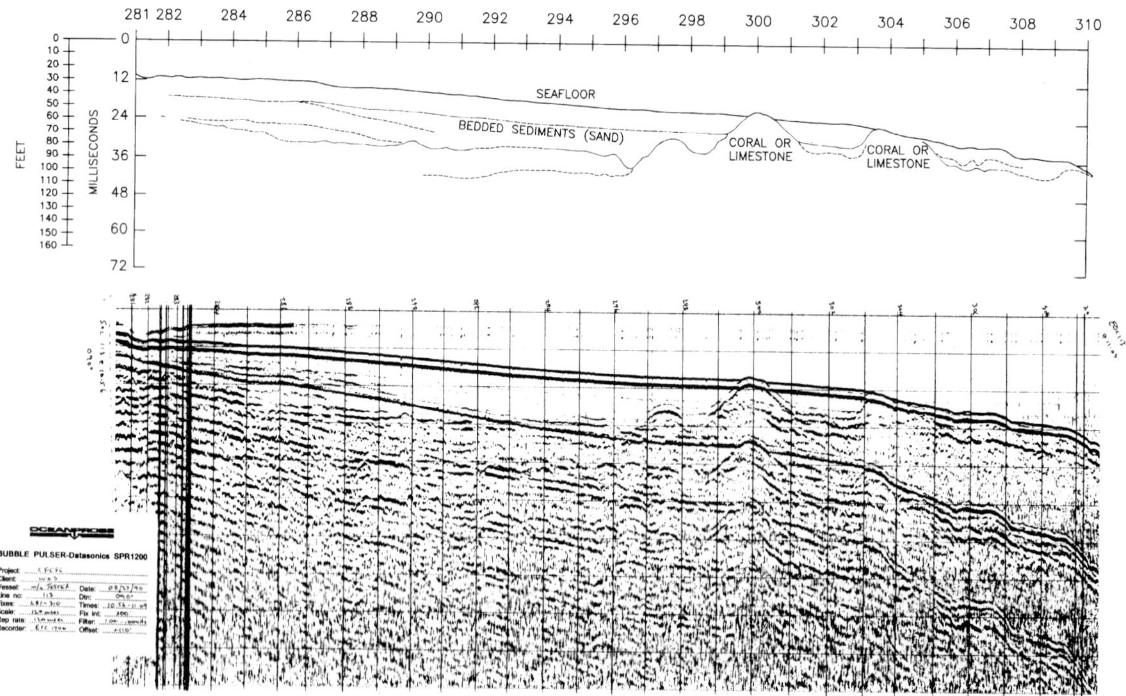

Figure M30 Analog subbottom profiler record from offshore Palm Beach County, Florida. Shore is to the left. The lump near Fix 300 is a coral or limestone outcrop. This is an example of data valuable for locating offshore sand resources to use for beach renourishment.

areas that may be subject to military operations (West *et al.*, 2001). Conventional survey methods require the deployment of small boats equipped with single- or multi-beam acoustic equipment along with terrestrial surveyors to perform beach profiles. These traditional methods are slow, are highly susceptible to weather, and expose the surveyors to hostile fire.

A similar technology is Airborne Topographic Mapper (ATM), an aircraft-based system that provides highly detailed topographic data of beaches. The National Aeronautics and Space Administration (NASA) and a group of partners are in the process of mapping the entire coast of the continental United States in unprecedented detail.

In situ technologies

The most common coastal applications of *in situ* systems are monitoring water flow and meteorological properties. Measuring waves and currents

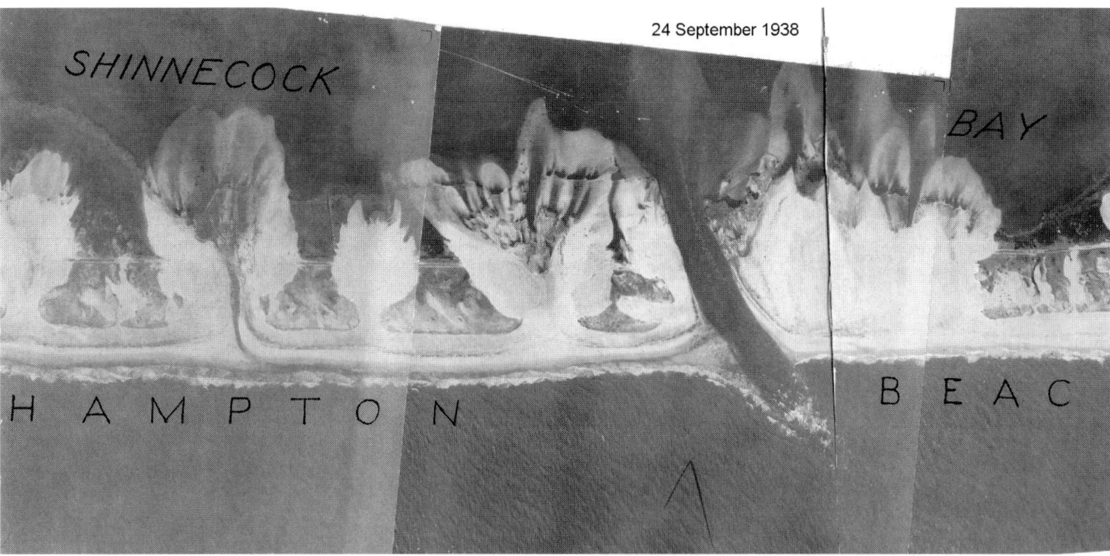

Figure M31 Vertical aerial photograph taken September 24, 1938 (three days after the Great New England Hurricane crossed the south shore of Long Island, New York). This image shows the new Shinnecock Inlet and numerous washover fans. Historical images like this are valuable in tracing the origin and behavior of coastal features (photographs from Beach Erosion Board Archives).

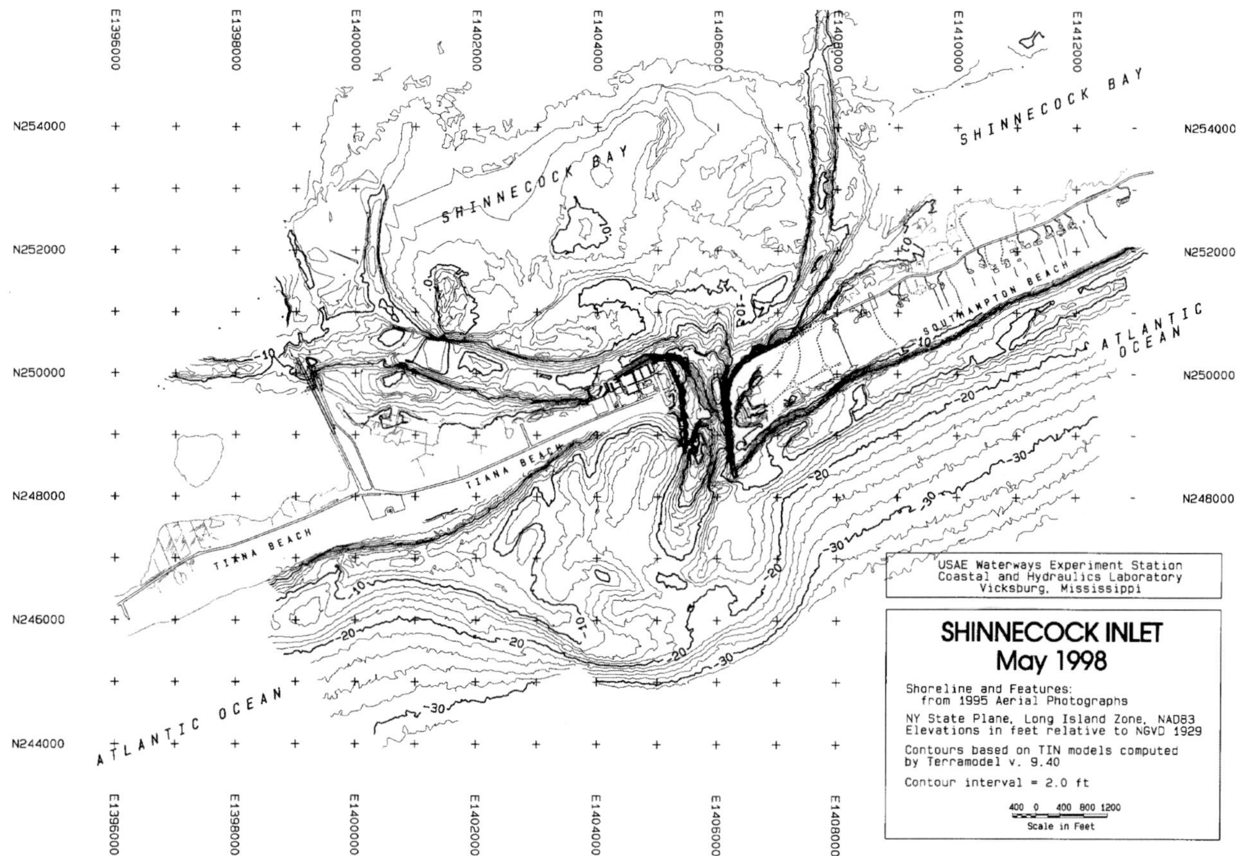

Figure M32 Example of contoured SHOALS hydrographic LIDAR data from Shinnecock Inlet, Long Island, New York, July 1989. Contouring based on 37,700 points, one-tenth of the original data density. This shows how an ebb shoal has evolved since 1938 when the photographs in the previous figure were taken (from Morang, 1999).

in the high-energy, hazardous nearshore zone is one of the more challenging endeavors of coastal engineering and research. Measurement programs must be thoroughly planned *before* any gauges are deployed to ensure that useful data are collected. A scientist must:

1. Determine what data units and analyzed products are needed to answer the critical engineering or scientific questions at the site.
2. Determine how long the gauges must be at the site (i.e., several years or just during the winter season).
3. Consider placing gauges in locations that are compatible with previous field programs.
4. Evaluate environmental constraints such as ice or trawler activity.
5. Be sure that enough funding is allocated for the analysis of the data.

Wave gauges

Two general types of wave gauges are available: nondirectional and directional. In general, directional gauges and gauge arrays are more expensive to build, deploy, and maintain than nondirectional gauges. Despite the greater cost and complexity, the former are preferred for most projects because the directional distribution of wave energy is an important parameter in many applications, such as sediment transport analysis and calculation of wave transformation. Wave gauges can be installed in buoys, placed directly on the sea or lake bottom, or mounted on existing structures, such as piers, jetties, or offshore platforms. Deployment of wave gauges is expensive and must be balanced against other project requirements. There are no hard-and-fast rules regarding how much wave or current data to collect—more is always better! Table M13 (from Morang *et al.*, 1997b) provides broad guidelines for wave gauging.

A wave-gauging program requires several steps:

- Planning (determine goals, allocate budget).
- Instrument deployment, servicing, and recovery.
- Data recovery (either real-time or periodic for internal-recording gauges).
- Data reduction, quality control, and generation of statistics.
- Display of statistics and summary results.

Wave data analysis is a complex topic in itself. To a casual observer on a boat, the sea surface usually appears as a chaotic jumble of waves of various heights and periods moving in many different directions. Most wave gauges measure and record a signal indicative of the changing elevation of the water surface. Unfortunately, when this signal is simply plotted against time, it provides little initial information about the characteristics of the individual waves that were present at the time the record was being made. Therefore, further processing is necessary to obtain wave statistics that can be used by coastal scientists or engineers to infer what wave forces have influenced their study area. Earle and Bishop (1984), Horikawa (1988), and Earle *et al.* (1995) provide background in the mathematics of wave analysis.

Another difficulty encountered in wave analysis is that different researchers are often inconsistent in their use of technical terms, and users are urged to be cautious of wave statistics from secondary sources and to be aware of how terms were defined and statistics calculated. For example, "significant wave height" (H_s) is defined as the average height of the highest one-third of the waves in a record. How long should this record be? Are the waves measured in the time domain by counting the wave upcrossings or downcrossings? Most automated processing programs *estimate* significant wave height by performing spectral analysis of a wave time series in the frequency domain and equating $H_s = H_{m0}$. The latter equivalency is usually considered valid in deep and intermediate water but may not be satisfactory in shallow water (Horikawa, 1988). To prevent collecting the wrong data when specifying or planning a wave measurement program, researchers are urged to consult a standard list of wave parameters such as the one prepared by the IAHR Working Group on Wave Generation and Analysis (1989).

Current meters

Currents can be measured by two general approaches. One of these, *Lagrangian*, follows the motion of an element of matter in its spatial and temporal evolution. The other, *Eulerian*, defines the motion of the water at a fixed point and determines its temporal evolution. Lagrangian current measuring devices are often used in sediment transport studies, in pollution monitoring, and for tracking ice drift. Eulerian, or fixed, current measurements are important for determining the variations in flow over time at a fixed location, such as in a tidal inlet or harbor entrance channel. Recently developed instruments combine aspects of both approaches.

Four general classes of current measuring technology have been tested or are in use (Appell and Curtin, 1990):

- Lagrangian methods: drifters and dye.
- Spatially integrating methods: experimental (not used often).
- Point source and related technology: moored current meters.
- Acoustic Doppler Current Profilers (ADCP) and related technology (fixed point and boat-mounted).

Table M13 Suggested wave gauge placement for coastal project monitoring

I. High-budget project (major harbor, highly populated area)

A. Recommended placement:

1. One (or more) wave gauge(s) close to shore near the most critical features being monitored (e.g., near an inlet). Although nearshore, gauges should be in intermediate or deepwater based on expected most common wave period. Depth can be calculated from formulas in the *Coastal Engineering Manual* (USACE, 2002).
2. In addition, one wave gauge in deepwater if needed for establishing boundary conditions of models.

B. Schedule:

1. Minimum: 1 year. Monitor winter/summer wave patterns (critical for Indian Ocean projects).
2. Optimum: 5 years or at least sufficient time to determine if there are noticeable changes in climatology over time. Try to include one El Niño season during coverage for North American projects.

C. Notes:

1. Concurrent physical or numerical modeling: Placement of gauges may need to take into account modelers' requirements for input or model calibration. Placement in shallow water may be needed to test wave transformation models or wave hindcasts.
2. Preexisting wave data may indicate that gauges should be placed in particular locations, or gauges may be placed in locations identical to the previous deployment to make the new data as compatible as possible with the older data. Long, continuous data sets are extremely valuable.
3. Hazardous conditions: If there is a danger of gauges being damaged by anchors or fishing boats, the gauges must be protected, mounted on structures (if available), or deployed in a location that appears to be the least hazardous.

II. Medium-budget project

A. Recommended placement:

1. One wave gauge close to shore near project site.
2. Obtain data from nearest deepwater buoy (for US waters, NOAA, National Data Buoy Center (NDBC) buoys).

B. Schedule: minimum 1-year deployment; longer if possible.

C. Notes: same as IC above. Compatibility with existing data sets is very valuable.

III. Low budget, short-term project

A. Recommended placement: gauge close to project site.

B. Schedule: if 1-year deployment is not possible, try to monitor the season when the highest waves are expected (usually winter, although this may not be true in areas where ice pack occurs or in tropical storm areas).

C. Notes: same as IC above. It is critical to use any and all data from the vicinity, anything to provide additional information on the wave climatology of the region.

The large number of instruments and methods that have been tested underscores that detection and analysis of fluid motion in the oceans is exceedingly complex. The difficulty arises from the large continuous scales of motion in the water. "There is no single velocity in the water, but many, which are characterized by their temporal and spatial spectra. Implicit then in the concept of a fluid 'velocity' is knowledge of the temporal and spatial averaging processes used in measuring it" (McCullough, 1980, p. 106). In shallow water, particularly in the surf zone, additional difficulties are created by turbulence and air entrainment caused by breaking waves, by suspension of large concentrations of sediment, and by the physical violence of the environment. Trustworthy current measurement under these conditions becomes a daunting task.

Lagrangian methods: Dye, drogues, ship drift, bottles, temperature structures, oil slicks, radioactive materials, paper, wood chips, ice floes, trees, flora, fauna, and seabed drifters have all been used to study the motion of the oceans (McCullough, 1980). A disadvantage of all drifters is that they are only quasi-Lagrangian sensors because, regardless of their design or mass, they cannot exactly follow the movement of the water. Nevertheless, they are particularly effective at revealing surface flow patterns if they are photographed or video recorded on a time-lapse basis. Simple drifter experiments can also be helpful in developing a sampling strategy for more sophisticated subsequent field investigations. Floats, bottom drifters, drogues, and dye are used especially in the littoral zone where fixed current meters are adversely affected by turbulence (e.g., see Ingle, 1966). Resio and Hands (1994) examined the use of seabed drifters and commented on their value in conjunction with other instruments.

Point source (Eulerian) and related technology: In channels, bays, and offshore, direct measurements of the velocity and direction of current flow can be made by instruments deployed on the bottom or at various levels in the water column. Two general classes of current meters are available: mechanical (impeller-type) and electronic. Mechanical meters, based on rotors or propellers that turn in the water, have been largely phased out in favor of electronic instruments. These have the advantages of rapid response and self-contained design with no external moving parts to get fouled or corroded. They can be used in real-time systems and can be used to measure at least two velocity components, but only at a single depth. Vertical or horizontal arrays of meters can be deployed to measure the distribution of velocities vertically through the water column or across a specific area such as a channel.

Acoustic Doppler Current Profilers: These profilers (known as ADCPs) operate on the principle of Doppler shift in the backscattered acoustic energy caused by moving particles suspended in the water. Assuming that the particles have the same velocity as the ambient water, the Doppler shift is proportional to the velocity components of the water within the path of the instrument's acoustic pulse. The backscattered acoustic signal is divided into parts corresponding to specific depth cells, often termed "bins." The bins can be various sizes, depending upon the depth of water in which the instrument has been deployed, the frequency of the signal pulse, the time that each bin is sampled, and the acceptable accuracy of the estimated current velocity (Bos, 1990). ADCPs have revolutionized current measurement in all water depths, but a great advantage of using ADCPs in shallow water is that they provide profiles of the velocities in the entire water column, allowing more comprehensive views of water motions than do strings of multiple point source meters. An ADCP can be moored at a fixed point to monitor the change in water velocities over time, or it can be mounted on a boat to measure flow patterns over an area, such as across an inlet (Figure M33). ADCP data are inherently noisy, and signal processing and averaging are critical to successful performance.

Physical sampling methods

Despite the explosive development of remote sensing and geophysical technology during the last half of the 20th century, there continues to be a need to collect actual samples of the sediment or rock underlying a study area. There are several reasons for this continued reliance on what is sometimes regarded as a primitive desire to touch the sediment. First, any remote sensing technique is still "remote." It is an interpretation of what is down there and is hopefully an accurate picture, but this interpretation (or model) still needs field confirmation to provide ground-truthing. Hence, a seismic profiling study is often accompanied by a coring program. Second, many sediment characteristics can still be obtained only by analysis of the actual samples. For example, sieving is still the best way to determine the distribution of grain sizes in a sample. For geotechnical investigations, samples are needed to determine strength, organic content, compressibility, and permeability (USACE, 1996). Surficial sediments provide information about the energy of the environment as well as the long-term processes and movement of materials, such as sediment transport pathways, sources, and sinks.

Three types of devices are available for retrieving sediments in the marine environment: grab samplers, dredges, and corers. For soft surface

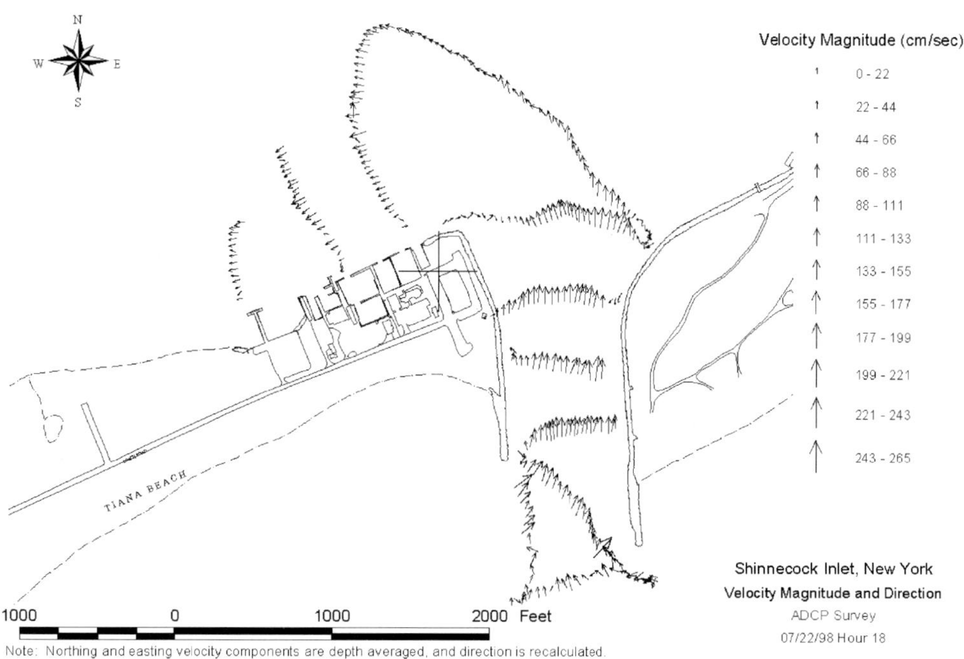

Figure M33 Analysis of ADCP current-meter surveys in Shinnecock Inlet, Long Island, New York. The instrument was mounted on a boat that made transects across the inlet. The arrows show the direction and magnitude of the flow during the flood tide. The tide enters the inlet from the Atlantic Ocean and diverges in the bay, with much of the flow moving to the west (from Pratt and Stauble, 2001).

materials, a variety of grab-type samplers of different sizes and design have been developed (Bouma, 1969; USACE, 1996). Most consist of a set of opposing, articulated scoop-shaped jaws that are lowered to the bottom in an open position and are then closed by various trip mechanisms to retrieve a sample. Some grab samplers are small enough to be deployed and retrieved by hand, but most require some type of lifting gear (Table M14). For hard seafloor areas, steel dredges can be dragged along the bottom from a boat, but dredges suffer from selective recovery (only loose boulders may be recovered), and positional information is poor.

For all sediment sampling, it is necessary to obtain a sufficient quantity of material to compute a statistically valid grain-size distribution, and if there is gravel in the sample, many liters of sample may be needed. For example, if the maximum particle size (<10% of the sample) is 2.4 mm, only 0.1 kg of sample is required, but if the sample includes cobble of 64 mm, 50 kg are needed. Required quantities are specified in manuals prepared by the standards organizations of various countries (e.g., British Standards Institution, 1990; American Society for Testing and Materials (ASTM), 2001).

Although surficial samples are helpful for assessing recent processes, they are typically of limited value in a stratigraphic study because grab devices usually recover less than 15 or 20 cm of material. To recover subbottom sediments, a type of coring system is needed. One of the simplest is the vibratory corer (vibracorer), commonly used by geologists to obtain samples in lacustrine, shallow marine, and coastal environments (Fuller and Meisberger, 1982a,b; USACE, 1996). Vibracoring is described in a companion entry in this volume. The core may be up to 4 or 5 m long, which is adequate for borrow site investigations, near-surface cross sections, and other coastal studies.

Cores can be invaluable because they allow a direct, detailed examination of the layering and sequences of the subsurface sediment. The

Table M14 Subaqueous sediment sampling without drill rigs and casing

Device	Application	Description	Penetration depth	Comments
Grab samplers				
Petersen grab[a]	Large, relatively intact "grab" samples of seafloor	Clam-shell type grab weighing about 450 kg with capacity about 0.4 ft^3 (0.1 m^3)	To about 10 cm	Effective in water depths to 600 m (or more with additional weight).
Corers (for subsurface sampling)				
Harpoon-type gravity corer	Cores 1.5–6-in., (3.8–15.2 cm) diameter in soft to firm sediments	Vaned weight connected to coring tube dropped directly from boat. Tube contains liners and core retainer	To about 10 m	Maximum water depth depends only on weight. Undisturbed (UD) sampling possible with short, large-diameter barrel
Free-fall gravity corer	(As above for harpoon type)	Device suspended on wire rope over vessel side at height above seafloor of about 5 m and then released	Soft sediment to about 5 m. Firm sediment to about 3 m	(As above for harpoon type)
Piston gravity corer (Ewing gravity corer)	2.5-in. (6.35 cm) sample in soft to firm sediments	Similar to free-fall corer except that coring tube contains a piston that remains stationary on the seafloor during sampling	Standard core barrel 10 ft (3.0 m); additional 10-ft sections can be added	Can obtain high-quality UD samples
Piggott explosive coring tube	Cores of soft to hard bottom sediments	Similar to gravity corer. Drive weight serves as gun barrel and coring tube as projectile. When tube meets resistance of seafloor, weighted gun barrel slides over trigger mechanism to fire a cartridge. The exploding gas drives tube into sediments	Cores to 1−7/8 in. (4.75 cm) and to 10-ft lengths have been recovered in stiff to hard materials	Has been used successfully in 6,000 m of water
Norwegian Geotechnical Institute gas-operated piston	Good-quality samples in soft clays	Similar to the Osterberg piston sampler except that the piston on the sampling tube is activated by gas pressure	About 10 m	
Vibracorer[b]	High-quality samples in soft to firm sediments. Dia. 3.0 in. (7.6 cm)	Apparatus is set on seafloor. Air pressure from the vessel activates an air-powered mechanical vibrator to cause penetration of the tube, which contains a plastic liner to retain the core	Length of 20 ft (6 m) Rate of penetration varies with material strength. Samples a 20-ft core in soft sediment in 2 min	Maximum water depth about 60 m
Box corer	Large, intact slice of seafloor	Weighted box with closure of bottom for benthic biological sampling or geological microstructure	To about 0.3 m	Central part of sample is undisturbed

Notes:
[a] Similar grab devices include the Ponar®, whose jaws close when the device strikes the seafloor, and the Ekman, whose jaws are activated by a messenger. A messenger is a metal plug that slides down the steel cable and strikes a spring-loaded jaw-control mechanism. Other types are the Van Veen and Smyth-Macintyre.
[b] Vibracorers vary greatly in size. Many are custom-made at universities or consulting companies. Some are electrically or mechanically powered, and many use 3.0-in. (7.6 cm) thin-walled aluminum irrigation pipe as a combination core barrel and liner.
Source: Adapted from HUNT (1984) and other sources.

sequences provide information regarding the history of the depositional environment and the physical processes during the time of sedimentation. Depending upon the information required, the types of analysis that can be performed on the core include grain size, sedimentary structures, identification of shells and minerals, organic content, microfaunal identification (pollen counts), X-ray radiographs, radiometric dating, and engineering tests. If only information regarding recent processes is necessary, then a box corer, which samples up to 0.6 m depths, can provide sufficient sediment. Because of its greater width, a box corer can recover undisturbed sediment from immediately below the seafloor, allowing the examination of bedding microstructure and lamination. These structures are usually destroyed by traditional vibratory or rotary coring.

If it is necessary to obtain deep cores, or if cemented or very hard sediments are present, the only alternative is rotary coring. Truck- or skid-mounted drilling rigs can be conveniently used on beaches or on barges in lagoons and shallow water. Offshore, rotary drilling becomes more complex and expensive, usually requiring jack-up drilling barges or four-point anchored drill ships. An experienced crew can drill and sample 100 m of the subsurface in about 24 h. Information on drilling and sampling practice is presented in Hunt (1984) and USACE (1996).

Mapping and shoreline change

Value of comparing historical with modern data

Collecting field data is just one phase of a coastal monitoring study. Analyzing, interpreting, and organizing the results are an equally important phase because this is how findings are displayed and communicated to engineers, coastal managers, and policy-makers. Several levels of analysis are possible. The simplest typically is the measurement of linear changes of coastal features over time. Historical charts, modern maps, aerial photographs, and LIDAR or topographic data can reveal details on:

- Long- and short-term advance or retreat of the shore. Shoreline change data are critical for coastal managers tasked with establishing setback lines and guiding growth in the coastal zone, especially in low areas subject to flooding.
- The impact of storms, including barrier island breaches, overwash, and changes in inlets, vegetation, and dunes.
- Human impacts caused by coastal construction, dune destruction, or dredging.
- Compliance with permits, illegal filling, and dumping.
- Biological condition of wetlands, estuaries, and barrier islands.
- Susceptibility of urban areas to storm flooding and catastrophic events (e.g., hurricanes) by means of storm surge models.

If historical 3D (bathymetric and topographic) data are available, volumetric comparisons between the old and modern surveys can provide quantitative information on:

- Longshore sediment movement.
- Shoaling or siltation associated with tidal inlets, river mouths, estuaries, and harbors.
- Sediment changes on ebb and flood shoals and in inlet channels.
- Nearshore bathymetry changes over time.
- Migrations of channel thalwegs.

Coastal engineers often make volumetric comparisons to compute amounts of sediment trapped by structures, examine the growth of shoals in navigation channels, determine dredging requirements, compute dredging contract payment, and evaluate post-dredging channel conditions. Volumetric analyses are also used to monitor the performance of beach renourishment projects.

Definitions and map datums

Many coastal zone features and subdivisions are difficult to define because temporal variability or gradational changes between features obscure precise boundaries. In addition, nomenclature is not standardized, and ambiguity is especially evident in the terminology and zonation of shore and littoral areas. If an ambiguous term is not precisely defined (for instance, the ever-controversial "shoreline"), the intended boundaries may differ greatly. Therefore, one of the most critical issues when combining historical and recent geographical data is to establish what datums have been used. A datum is "a fixed or assumed point, line, or surface, in relation to which others are determined; any quantity or value that serves as a base or reference for other quantities or values" (Bates and Jackson, 1984, p. 127). For coastal engineering and geologic studies, both horizontal (geographic location) and vertical (distance above sea level or other surface) datums must be established. Readers are referred to Umbach (1976), USACE (1994), Gorman *et al.* (1998), Moore (2000), and textbooks on geodetic surveying and photogrammetry for more background information on this complex topic.

Many possible datums can be used to monitor historical changes of the shoreline. The complexities around this term are discussed in a companion entry in this volume. Readers in the United States are also referred to the National Shoreline Data Standard, a draft standard prepared by NOAA, which defines shoreline as,

> The line of contact between the land and a body of water. On Coast and Geodetic Survey nautical charts and surveys, the shoreline approximates the mean high-water line. In Coast Survey usage, the term is considered synonymous with coastline (Shalowitz, 1962).

In many situations, the high water line (hwl) has been found to be the best indicator of the land–water interface, the coastline (Crowell *et al.*, 1991, 1993). The hwl is easily recognizable in the field and can sometimes be identified in aerial photographs by a line of seaweed and debris (wave run-up line). The wet-dry line, also visible on air photos, is more difficult to interpret because it is a function of water table, which in turn is a function of tide level, recent rainfall, and other factors. The datum printed on historical US National Ocean Service (NOS) T-sheets (topographic) is listed as mean high water. Fortunately, the early NOS topographers approximated hwl during their survey procedures (Shalowitz, 1962). Therefore, direct comparisons between historical T-sheets and modern aerial photograph interpretations are possible. However, even years ago, coastal scientists recognized that the hwl as mapped from aerial photographs was a dynamic feature and potentially a less reliable indicator of shoreline position than morphologic features on the shore that remain relatively constant from day to day. Some of these morphologic indicators are the berm crest, base of the dune, permanent vegetation line, crest of the overwash terrace, and shore protection structures, all of which are unaffected or only slightly altered by short-term changes in water levels (Morton, 1997). Therefore, many recent field surveys using Global Positioning System (GPS) instruments mounted on all-terrain vehicles are defining the shoreline based on these morphologic features. Whether these shorelines can be directly compared with older air photo interpretations is problematic.

A discussion of water level datums and shoreline definition is beyond the scope of this entry. An introduction to this complicated topic is presented in NOS (1988). The classic *Shore and Sea Boundaries* (Shalowitz, 1962, 1964; Reed, 2000) are exhaustive references on technical and legal issues relating to datums in the United States. USACE (1994) provides details on establishing tidal datums for coastal hydrographic surveys. In the Great Lakes of North America, water levels fluctuate on yearly and on longer-term cycles due to meteorologic and hydrologic conditions over the continent. In the lakes, charts and water depths are referenced to the International Great Lakes Datum of 1985 (Coordinating Committee, 1992).

Cross-shore profile surveys

Topographic and nearshore profile surveys provide one of the most direct and accurate means to assess geologic and geomorphic changes on the shoreface to water depths of 10 or 15 m. Beach profiles conducted over time can document erosion and accretion in the coastal zone, shoreline changes, sand bar movement, and dune volume changes. They provide the basic data for evaluating what happens to sand placed in beach nourishment projects (Weggel, 1995).

Permanent or semi-permanent benchmarks are required to reoccupy profile sites over successive months and years. These benchmarks should be located on a stable land feature at the landward end of the profile lines to minimize their likelihood of being damaged in storms. The locations of survey monuments must be carefully documented and referenced to other survey markers or to control points. Then, the profile data can be used to evaluate changes in sea level, changes in volume of sand on the beach, or other phenomena that require reference to established regional datums. The ability to accurately reestablish a survey monument is critical because it ensures that profile data collected over many years will be comparable. Locations that might experience dune erosion should be avoided, and care should also be taken to reduce the visibility of benchmarks to minimize vandalism. (Unfortunately, damage caused by vandals is an irritating and expensive problem at all coastal projects, even those far from urban areas.) With the increasing use of GPS receivers by surveyors, the rigorous need for duplicate survey monuments may be reduced. This is an evolving technology, and

Table M15 Example of Profile Survey Scheme for monitoring beach fill (after Stauble, 1994)

Time	Times/year	Number of profiles
Pre-fill	2	Collect within fill area and at control locations in summer and winter months to characterize seasonal profile envelope (beach and nearshore to closure depth).
Post-fill	1	Collect all profiles immediately after fill placement at each site (beach and offshore) to document fill volume. Collect control profiles immediately after project is completed.
Year 1	4	Four quarterly survey trips collecting all beach and offshore profiles out to depth of closure. Begin series during the quarter following the post-fill survey.
Continue Year 1 schedule to time of next renourishment (usually 4–6 years). If project is a single sand placement, taper surveys in subsequent years:		
Year 2	2	6- and 12-month survey of all beach and offshore profiles.
Year 3	2	6- and 12-month survey of all beach and offshore profiles.
Year 4	1	12-month survey of beach and offshore profiles.

Note: (1) If project is renourished, repeat survey schedule from post-fill immediately after each renourishment to document new fill quantity and behavior. (2) Monitoring fill after major storms is highly desirable to assess fill behavior and storm-protection ability. Include both profile and sediment sampling. Conduct less than one week after storm conditions abate to document the beach and offshore response.

for now we still recommend that two monuments per survey line be established. US Army Corps of Engineers standards for survey monuments are specified in USACE (1990).

When planning a beach profiling study, both the frequency of the sampling and the overall duration of a project must be considered. There are no definitive guidelines for the timing of profile lines, but for most sites, summer and winter surveys are recommended. Resurveying profiles over a period of more than one year can be of substantial help in understanding seasonal changes. In addition, supplemental surveys can be made after big storms to determine their effects and measure the rate of recovery of the local beach. Table M15 outlines a suggested survey schedule for monitoring beach fill projects.

The longshore spacing of profiles must be carefully planned. Profiles should be at close enough intervals to show any significant changes in lateral continuity. Reviewing locations of historical shorelines in the study area is one way to establish the gross limits of the area that should be examined in detail, particularly along rapidly changing coasts. Often, a spacing of 1,000 ft (300 m) is used in studies in the United States, but closer spacing may be needed near inlets, within project boundaries, or at the ends of spits or barriers that experience frequent changes, and fewer profiles may be needed in the more stable portions of barrierislands.

For proper coverage, profile lines should extend landward of the zone that can be inundated by storms, usually behind the frontal dunes. But, often it is not necessary to survey across an entire barrier island. For example, shore and dune deposits that are now inland from the modern shoreline may only be affected by marine or lacustrine processes during the most severe storms. Aerial photographs of these interior areas may be adequate to show morphologic changes. Seaward, the profiles should extend deep enough to include the portion of the shoreface where most sediment moves (i.e., to beyond closure—discussed in another entry in this volume). Areas subject to dune breaching and overwash require special efforts to establish a reusable benchmark system.

The land portion of a profile is typically surveyed with a stadia rod and a total station. The method is inexpensive and can extend offshore to about 1 m water depth (or deeper if the rodman is adventurous). The nearshore seafloor is often surveyed by a sled that is towed by boat out into the water from about +1.5 m to closure depth. This results in overlap between onshore rod surveys and sled surveys to assure that the two systems are recording the same elevations. With careful field technicians, vertical accuracy for sled surveys is in the range of ± 0.15 m (Clausner *et al.*, 1986). If offshore surveys are conducted by boat-mounted echo sounder, overlap with rod surveys is often not possible because most boats cannot survey in water shallower than about 2 m. Also, acoustic surveys are difficult in the surf zone because bubbles in the water attenuate the acoustic pulses. Jet skis have proven to be a useful way to obtain data in the difficult surf zone or in tidal inlets when a towed sled is not available or is not practical due to logistical conditions (Dugan *et al.*, 1999). With advances in remote sensing technology and data processing methods, remote sensing may replace field profile surveys in some circumstances (Judge and Overton, 2001).

Figure M34 shows an example of profile data collected at Westhampton Beach, Long Island, New York (facing southeast towards the Atlantic Ocean). In many areas around the world, beaches show a distinct pattern of summer growth and winter erosion. But along the south shore of Long Island, analysis of profiles has revealed that the seasonal effect is subtle (Morang *et al.*, 1999). Storms occur during both the winter and the summer, and although the berm and dune crest often retreat and bars move offshore during each event, recovery typically is rapid as littoral material moves back onshore. For this study, the ability to detect subtle sand volume changes along the Long Island shore has underscored the value of a time series of topographic data spanning several years.

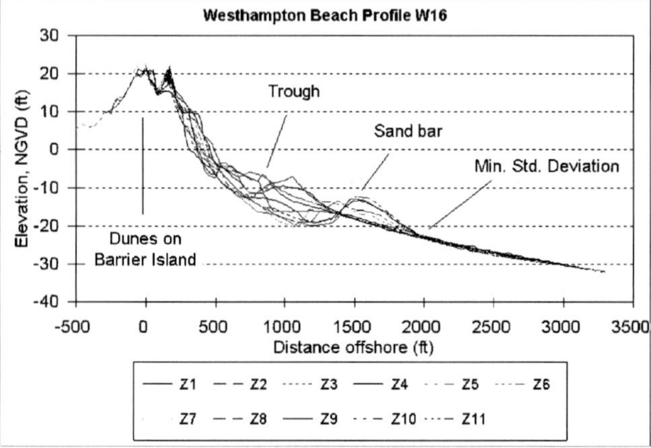

Figure M34 Example of cross-shore profile surveys from Westhampton Beach, Long Island. Surveys were made at six month intervals and demonstrate significant movement of the bar and trough. The location where the profiles converge, at the position of minimum standard deviation, is sometimes interpreted as the depth of closure.

Conclusions

Inexpensive computing and data storage technology and innovative sensors have brought about a revolution in marine data collection during the last two decades that shows no sign of ending. Advances have occurred in four areas: (1) sensors; (2) interactive or real-time data analysis of data; (3) data integration among different sensors and navigation systems; (4) data display and presentation, particularly with respect to Geographic Information System (GIS) organization and display. These new technologies have allowed marine scientists to examine phenomena in much greater detail than ever before and with much greater positional (navigation) accuracy. However, despite these advances, some areas of the coastal zone continue to be hazardous and difficult for any form of *in situ* data collection or monitoring. This is especially true of the breaker zone, where few instruments can survive during storms and where it is still difficult to obtain trustworthy bathymetry. As a result, some basic processes, such as the cross-shore distribution of longshore currents, are still little understood. Perversely, the more data we collect, the more we realize that we still know so little about marine processes in the coastal zone and the delicate interplay between hydrodynamics, geology, and biology.

Andrew Morang and Laurel T. Gorman

Bibliography

Appell, G.F., and Curtin, T.B. (eds.), 1990. *Proceedings of the IEEE Fourth Working Conference on Current Measurement, Current Measurement Technology Committee of the Oceanic Engineering Society*. Current Measurement Technology Committee of the Oceanic Engineering Society. New York: Institute of Electrical and Electronics Engineers.

American Society for Testing and Materials (ASTM), 2001. *2001 Annual Book of ASTM Standards, Section 4, Construction*, Vol 4.08, Soil and Rock: D 420–D 5779. Philadelphia, PA: American Society for Testing and Materials.

Bates, R.L., and Jackson, J.A. (eds.), 1984. *Dictionary of Geological Terms*, 3rd edn. Garden City, NY: Anchor Press/Doubleday.

Belderson, R.H., Kenyon, N.H., Stride, A.H., and Stubbs, A.R., 1972. *Sonographs on the Sea Floor*. Amsterdam: Elsevier Publishing Company.

Bos, W.G., 1990. A comparison of two doppler current profilers. In Appell, G.F., and Curtin, T.B. (eds.), *Proceedings of the IEEE Fourth Working Conference on Current Measurement*. Current Measurement Technology Committee of the Oceanic Engineering Society. New York: Institute of Electrical and Electronics Engineers, 207–214.

Bouma, A.H., 1969. *Methods for the Study of Sedimentary Structures*. New York: John Wiley-Interscience.

British Standards Institution, 1990. *Methods of Test for Soils for Civil Engineering Purposes*. BS 1377-1, 2, 3, and 4. London: British Standards Institution.

Clausner, J.E., Birkemeier, W.A., and Clark, G.R., 1986. Field comparison of four nearshore survey systems. Vicksburg, MS: US Army Engineer Waterways Experiment Station, *Miscellaneous Paper CERC-86-6*.

Coordinating Committee on Great Lakes Basin Hydraulic And Hydrologic Data, 1992. IGLD 1985 Brochure on the International Great Lakes Datum 1985. Washington, DC: US Government Printing Office.

Cowls, S., 2000. How to Choose a Multibeam. *International Ocean Systems*, **4**(4): 4–6.

Crowell, M., Leatherman, S.P., and Buckley, M.K., 1991. Historical shoreline change: error analysis and mapping accuracy. *Journal of Coastal Research*, **7**(3): etc 839–852.

Crowell, M., Leatherman, S.P., and Buckley, M.K., 1993. Shoreline change rate analysis: long term versus short term data. *Shore and Beach*, **61**(2): 13–20.

Daniels, J., 1989. Ground penetrating radar. In *SAGEEP '89–Proceedings of the Symposium on the Applications of Geophysics to Engineering and Environmental Problems*. Englewood, CO: Environmental and Engineering Geophysical Society, pp. 62–142.

Dugan, J.P., Vierra, K.C., Morris, W.D., Farruggia, G.J., Campion, D.C., and Miller, H.C., 1999. Unique vehicles used for bathymetry surveys in exposed coastal regions. In *U.S. Hydrographic Conference '99*. Rockville: MD: The Hydrographic Society of America.

Earle, M.D., and Bishop, J.M., 1984. *A Practical Guide to Ocean Wave Measurements and Analysis*. Marion, MA: Endeco Incorporated.

Earle, M.D., Mcgehee, D., and Tubman, M., 1995. Field wave gaging program, wave data analysis standard. Vicksburg, Mississippi: US Army Engineer Waterways Experiment Station, *Instruction Report CERC-95-1*.

Estep, L.L., Lillycrop, W.J., and Parson, L.E., 1994. Estimation of maximum depth of penetration of a bathymetric lidar system using a Secchi disk database. *Marine Technology Society Journal*, **28**(2): 31–36.

Fish, J.P., and Carr, H.A., 1990. *Sound Underwater Images, A Guide to the Generation and Interpretation if Side Scan Sonar Data*. Cataumet, MA: American Underwater Search and Survey, Ltd.

Fitzgerald, D.M., Baldwin, C.T., Ibrahim, N.A., and Humphries, S.M., 1992. Sedimentologic and morphologic evolution of a beach-ridge barrier along an indented coast: Buzzards Bay, Massachusetts. In Fletcher, C.H., III, and Wehmiller, J.F. (eds.), *Quaternary Coasts of the United States: Marine and Lacustrine Systems*. Tulsa, Oklahoma: Society for Sedimentary Geology, (SEPM) Special Publication No 48, pp. 65–75.

Flemming, B.W., 1976. Side-scan sonar: a practical guide. *International Hydrographic Review*, **53**(1): 65–92.

Fuller, J.A., and Meisberger, E.P., 1982a. A simple, ship-based vibratory corer. *Journal of Sedimentary Petrology*, **52**(2): 642–644.

Fuller, J.A., and Meisberger, E.P., 1982b. A lightweight pneumatic coring device: design and field test. Vicksburg, MS: US Army Engineer Waterways Experiment Station, Coastal Engineering Research Center, Miscellaneous Report No. 82–8.

Gorman, L.T., Morang, A., and Larson, R.L. 1998. Monitoring the coastal environment; Part IV: Mapping, shoreline change, and bathymetric analysis. *Journal of Coastal Research*, **14**(1): 61–92.

Griffiths, D.H., and King, R.F., 1981. *Applied Geophysics for Geologists and Engineers—the Elements of Geophysical Prospecting*. Oxford: Pergamon Press.

Henderson, F.M., and Lewis, A.J. (eds.), 1998. *Manual of Remote Sensing, Volume 2, Principles and Applications of Imaging Radar*, 3rd edn. New York: Wiley Interscience.

Horikawa, K. (ed.), 1988. *Nearshore Dynamics and Coastal Processes: Theory, Measurement and Predictive Models*. Tokyo, Japan: University of Tokyo Press.

Hunt, R.E., 1984. *Geotechnical Engineering Investigation Manual*. New York: McGraw-Hill.

IAHR Working Group on Wave Generation and Analysis, 1989. List of sea-state parameters. *Journal of the Waterway, Port, and Ocean Engineering, American Society of Civil Engineers*, **115**(6): 793–809.

IHO, 1998. *IHO Standards for Hydrography and Nautical Cartography*, 4th edn., Ref. S-44. Monaco: International Hydrographic Bureau (Available in English, French, Spanish).

IHO, 2001. *Reference Texts for Training in Hydrography*, 3rd edn. Monaco: International Hydrographic Bureau.

Ingle, J.C., Jr., 1966. *The Movement of Beach Sand*. New York: Elsevier.

Irish, J.L., McClung, J.K., and Lillycrop, W.J., 2000. Airborne lidar bathymetry: the SHOALS system. *PIANC Bulletin, No. 103-2000*, pp. 43–53.

Judge, E.K., and Overton, M.F., 2001. Remote sensing of barrier island morphology: Evaluation of photogrammetry-derived digital terrain models. *Journal of Coastal Research*, **17**(1): 207–220.

Kraus, N.C., Gorman, L.T., and Pope, J., 1994. Kings Bay coastal and estuarine physical monitoring and evaluation program: coastal studies. Vicksburg, MS: US Army Engineer Waterways Experiment Station (in 2 volumes), *Technical Report CERC-94-9*.

Lillesand, T.M., and Kiefer, R.W., 1999. *Remote Sensing and Image Interpretation*, 4th edn. New York: Wiley Interscience.

Lillycrop, W.J., and Banic, J.R., 1992. Advancements in the U.S. Army Corps of Engineers hydrographic survey capabilities: the SHOALS system. *Marine Geodesy*, **15**: 177–185.

Leenhardt, O., 1974. Side scanning sonar—a theoretical study. *International Hydrographic Review*, **51**(1): 61–80.

Mazel, C., 1985. Side-scan sonar training manual. Salem, NH: Klein Associates.

McAndrew, R., 2001. Data processing—real time and post. *International Ocean Systems*, **5**(5): 17–19.

McCullough, J.R., 1980. Survey of techniques for measuring currents near the ocean surface. In Dobson, F., Hasse, L., and Davis, R. (eds.), *Air–Sea Interaction—Instruments and Methods*. New York: Plenum Press, pp. 105–126.

Militelo, A., and Kraus, N.C., 2001. Shinnecock Inlet, New York, site investigation, Report 4, Evaluation of flood and ebb shoal sediment source alternatives for the west of Shinnecock interim project, New York. Vicksburg, MS: US Army Engineer Waterways Experiment Station, Technical Report CHL-98-32.

Moore, L.J., 2000. Shoreline mapping techniques. *Journal of Coastal Research*, **16**(1): 111–124.

Morang, A., 1999. Shinnecock Inlet, New York, site investigation, Report 1, Morphology and historical behavior. Vicksburg, MS: US Army Engineer Waterways Experiment Station, *Technical Report CHL-98-32*.

Morang, A., Larson, R.L., and Gorman, L.T., 1997a. Monitoring the coastal environment; Part III: Geophysical and research methods. *Journal of Coastal Research*, **13**(4): 1964–1985.

Morang, A., Larson, R.L., and Gorman, L.T., 1997b. Monitoring the coastal environment; Part I: Waves and currents. *Journal of Coastal Research*, **13**(1): 111–133.

Morang, A., Rahoy, D.S., and Grosskopf, W., 1999. Regional geologic characteristics along the South Shore of Long Island, New York. In *Proceedings of Coastal Sediments '99*. Reston, VA: American Society of Civil Engineers, pp. 1568–1583.

Morton, R.A., 1997. Gulf Shoreline movement between Sabine Pass and the Brazos River, Texas: 1974 to 1996. Geological Circular 97-3, Bureau of Economic Geology. Austin, TX: The University of Texas at Austin.

National Ocean Service, 1988. Fantastic Tidal Datums. Physical Oceanography Division, National Ocean Service, National Oceanic and Atmospheric Administration. Washington, DC: Department of Commerce.

Philipson, W. (ed.), 1997. *The Manual of Photographic Interpretation*, 2nd edn. Bethesda, MD: American Society for Photogrammetry & Remote Sensing.

Pratt, T., and Stauble, D., 2001. Shinnecock Inlet, New York, Site Investigation, Report 3, Selected Field Data Report for 1997, 1998, and 1998 Velocity and Sediment Surveys. Vicksburg, MS: US Army Corps of Engineers, Engineer Research and Development Center, *Technical Report CHL-98-32*.

Rencz, A.N., and Ryerson, R.A. (eds.), 1999. *Manual of Remote Sensing, Volume 3, Remote Sensing for the Earth Sciences*, 3rd edn. New York: Wiley Interscience.

Reed, M., 2000. *Shore and Sea Boundaries, Volume 3, The Development of International Maritime Boundary Principles through United States Practice*. US Department of Commerce, National Oceanic and Atmospheric Administration. Washington, DC: US Government Printing Office, 432 p.

Reed, T.B., 2001. Drinking from the fire hose—stop worrying and love the byte, a review of how to deal with seafloor mapping, data- overload, greater resolution, and bandwidth. *Sea Technology*, **42**(6): 16–25.

Resio, D.T., and Hands, E.B., 1994. Understanding and interpreting seabed drifter (SBD) data. Vicksburg, MS: US Army Engineer Waterways Experiment Station, *Technical Report DRP-94-1*.

Sellmann, P.V., Delaney, A.J., and Arcone, S.A., 1992. Subbottom surveying in lakes with ground-penetrating radar. Hanover, NH: US Army Corps of Engineers Cold Regions Research & Engineering Laboratory, *CRREL Report 92–8*.

Shalowitz, A.L., 1962. *Shore and Sea Boundaries, Volume 1, Boundary Problems Associated with the Submerged Land Cases and the Submerged Lands Acts*. Publication 10–1, US Department of Commerce, Coast and Geodetic Survey. Washington, DC: US Government Printing Office, 420 p.

Shalowitz, A.L. 1964. *Shore and Sea Boundaries, Volume 2, Interpretation and Use of Coast and Geodetic Survey Data*. US Department of Commerce, Coast and Geodetic Survey. Washington, DC: US Government Printing Office, Publication 10–1, 749 p.

Sheriff, R.E., 1977. Limitations on resolution of seismic reflections and geologic detail derivable from them. In Payton, C.E. (ed.), *Seismic Stratigraphy—Applications to Hydrocarbon Exploration*. Memoir 26. Tulsa, Oklahoma: American Association of Petroleum Geologists, 3–14.

Sieck, H.C., and Self, G.W., 1977. Analysis of high resolution seismic data. In Peyton, C.E. (ed.), *Seismic Stratigraphy—Applications to Hydrocarbon Exploration*. Tulsa, OK: American Association of Petroleum Geologists, Memoir 26, pp. 353–385.

Stauble, D.K., 1994. A physical monitoring plan for northern Assateague Island, Maryland. Philadelphia, PA: US Department of Interior, National Park Service.

Umbach, M.J. (Commander, NOAA), 1976. *Hydrographic Manual*, 4th ed. Washington, DC: National Ocean Survey, National Oceanic and Atmospheric Administration, US Department of Commerce. (updates 1979, 1980, and 1981).

US Army Corps of Engineers, 1990. *Survey Markers and Monumentation*. Engineer Manual EM 1110-1-1002. Washington, DC: Headquarters, US Army Corps of Engineers.

US Army Corps of Engineers, 1994. *Hydrographic Surveying*. Engineer Manual EM 1110-2-1003. Washington, DC: Headquarters, US Army Corps of Engineers.

US Army Corps of Engineers, 1996. *Soil Sampling*. Engineer Manual EM 1110-2-1907. Washington, DC: Headquarters, US Army Corps of Engineers.

US Army Corps of Engineers, 2002. *Coastal Engineering Manual*. Engineer Manual EM-1110-2-1100. Washington, DC: Headquarters, US Army Corps of Engineers (in six parts).

van Heteren, S., FitzGerald, D.M., and McKinlay, P.A., 1994. Application of ground-penetrating radar in coastal stratigraphic studies. In *GPR '94—Proceedings of the Fifth International Conference on Ground Penetrating Radar*. Waterloo, Ontario, Canada: Canadian Geotechnical Society and Waterloo Centre for Groundwater Research, pp. 869–881.

Weggel, J.R., 1995. A primer on monitoring beach nourishment projects. *Shore & Beach*, **63**(3): 20–24.

West, G.R., Lillycrop, W.J., and Pope, B., 2001. Utilizing airborne Lidar bathymetry technology for REA. *Sea Technology*, **42**(6): 10–15.

Whitehead, E.J., and Cooper, P.S., 2001. An acoustic approach to seabed discrimination and classification. *International Ocean Systems*, **5**(4): 16–21.

Cross-references

Airborne Laser Terrain Mapping and Light Detection Ranging
Beach and Nearshore Instrumentation
Beach Sediment Characteristics
Coastal Warfare
Global Positioning Systems
Jet Probes
Mapping Shores and Coastal Terrain
Monitoring Coastal Ecology
Photogrammetry
Remote Sensing of Coastal Environments
Remote Sensing: Wetlands Classification
Synthetic Aperture Radar Systems
Tracers
Vibracore

MUDDY COASTS

Muddy coast is defined as a coastal depositional environment which exhibits muddy sediments as a major component of the sedimentary morphodynamic system. These morphodynamic deposits possess textural characteristics containing a high proportion of silt and clay, and exhibit identifiable sub-tidal, intertidal, and supra-tidal stratigraphy. Such deposits tend to form extensive low-gradient morphological surfaces, and are often manifest as broad intertidal flats, colloquially termed "mudflats."

The visually obvious muddy coastal sedimentary deposits and geomorphic forms occur mainly within the intertidal zone of coastal fringe waters, and are clearly evident as surficial deposits forming tidal flat and drainage channel deposits. Muddy coastal depositional environments also occur in nearshore sub-tidal locations characterized by relatively low energy hydrodynamic conditions, such as in shallow estuaries and embayments. But direct marine processes influencing muddy deposition extend inland as far as storm surge tides are effective, and especially in low-gradient coasts, some muddy coast deposits occur on land areas above the normal high-tide level.

In principle, silt and clay particles comprising the suspended sediment load in coastal waters will deposit (1) if there is sufficient particle concentration such that enhanced deposition takes place, (2) if the suspended load particles are advected into quiet waters where the fine particles gradually sink below wave agitation and turbulence levels and deposit on the bottom, and (3) where salt wedge and "estuarine circulation" mixing processes at the head of estuaries induce flocculation of the fine particles.

Definition of "mud"

In sedimentological terms "mud" is easily defined. Texturally "mud" is a detrital deposit composed of particle sizes smaller than 63 μm (i.e., 0.063 mm), and includes the textural classes of *silt*, 0.063–0.004 mm or 6φ–10φ in the sedimentological (phi) classification, and *clay*, which is generally accepted as being composed of particle sizes smaller than 0.004 mm (4 μm) or smaller than 10φ (Folk, 1968; Leeder, 1982).

However, muddy coastal deposits almost always contain a proportion of organic matter—typically ~3–5%-originating from the importation of fine suspended sediments into the coastal zone of the muddy sediment deposition. The organic matter includes both terrigenous (from fine vegetation detritus) and marine biogenic matter, as well as originating from *in situ* biogenic processes, such as mucus and feces from worms and other benthic organisms inhabiting the surficial muds. In many locations organic matter contributions include fine plant detritus originating from such diverse sources as mangrove stands, or adjacent "pastures" of sea grass, marsh grasses, and wetlands. Biogenic contribution to muddy sedimentary deposits may also include small shell fragments, minute sea urchin spines, and fragments of living and dead diatoms, foraminifera, ostracods, and coccoliths (Augustinus, 2002; Wang *et al.*, 2002a).

Additional mineral matter occurs in the mud deposits, including fine sands, terrigenous matrix fragments, and shelly gravel fragments. Composition of the mineral matter is influenced by local supply and may consist of quartz, feldspar and mica, and small proportions of heavy minerals.

Typically, mud deposited in the intertidal zone is soft, pliable, of high plasticity, thixothrophic, and contains a large volume of water within its physical structure. Practically, a person may sink up to their knees in wading through intertidal muddy deposits. This thixotrophic property suggests that in geotechnical terms mud behaves either as a plastic or a fluid substance indicating that mud deposits exhibit properties rather different from noncohesive clasts in the sedimentary environment.

In mud-rich shallow waters of muddy coastal areas, the seafloor may be characterized by "fluid mud"—a state in which the particles occur partially as concentrated colloidal state, and the viscous substance possesses extremely high water content (Mehta and Hayter, 1989; Mathew and Baba, 1995).

Muddy coastal sediments also possess interesting geochemical characteristics, related to the chemical reactions associated with anoxic conditions and microorganism activity within the mud deposits. Below the oxic surface layers, usually only a few centimeters thick, the mud often looks blue-black and exhumes a pungent sulfurous smell.

Morphodynamic types of muddy coast

For expansive open ocean tidal flats, for example, of China, muddy deposits are an integral component. The muddy deposits may range from several hundred meters to more than 10 km in extent (Wang *et al.*, 2002b). But muddy coastal deposits may also exist in restricted lenses a few tens of meters wide. Wang *et al.* (2002a) identifies several morphodynamic categories of muddy coast:

Tidal flats: They are the typical geomorphic form of expansive muddy coasts and are characterized by intertidal surfaces with low slopes of order 0.5–1.0 : 1000. Distinct morphological and sedimentary zonation features can be identified (Wang *et al.*, 2002b). Generally, a low relief salt marsh zone is developed within the supra-tidal zone, which may be surmounted by shelly chenier beaches or ridges, or artificial stopbanks (dikes) on populated flat lowland coasts. Below high-tide level, true "mudflats" are located in the upper part of the intertidal flat. Clay and silt layers are often developed in the central parts of the tidal flats, and the lower-to-sub-tidal sectors become more sandy–silty or silty–sand, with wave ripple bedform features developed on the surface. Dendritic meandering tidal creeks are common on broad muddy tidal flats, formed by the scouring of ebb-tidal waters as they drain off the large areas of intertidal flat upon the falling tide. Erosion scour features, both tidal current and wave induced, are often observed on the central sectors (Wang *et al.*, 1990). For broad muddy tidal flats, spatially varying morpho-sedimentary zonation is evidently associated with different hydrodynamic forcing processes acting on the flat.

Inner deposits of barrier-enclosed lagoons: Within the landward side of barrier-enclosed lagoons, along the southeast coast of the United States, for example, a veneer of deposited mud is typically found. Such muddy deposits are typical of barrier lagoons, and may be associated variously with wetlands, oyster beds, and estuarine reentrant valley head deposits. Source of the muddy deposits is mainly from input from the surrounding land catchment, and the deposits are usually thin (0.5–2 m) but areally extensive in the available area, and overlie sands of the barrier lagoonal system.

Enclosed sheltered bay deposits: Muddy deposits of varying types may occur in sheltered embayments. Geomorphically, such embayments may be tectonically induced by block faulting, or form from drowned river valleys consequent upon the Holocene sea-level transgression. The deposits normally originate from the surrounding catchments and occur particularly in the tributary valley heads.

Estuarine drowned river valley deposits: Most funnel-shaped estuaries are drowned river valleys, which have significant freshwater input and within which there is strong mixing between the fresh and saline marine waters. The resulting "estuarine circulation" causes mud to be transported landwards on the bottom of the estuary, enhancing the tendency for deposits to become markedly more muddy toward the head of the estuary or bay. The freshwaters also carry suspended sediment loads of silt and clay, especially in river floods, and these tend to undergo flocculation and deposition upon mixing with the saline waters in the upper part of the estuary. The result is a concentration of muddy deposition at the heads of the estuaries.

Supra-tidal (storm surge) mud deposits: Lowland coasts with a broad shallow shelf, especially where they exist in the form of a large funnel, such as the German Friesland bight, the Bohai Bay in China, the Gulf of Thailand, or the Gulf of Bangladesh, may be subject to strong storm surges. Where there is abundant nearshore submarine muddy deposits, wave action during the storm surge event reagitates the silts and clays resulting in waters of high suspended load sediment concentration. These mud-laden waters can be swept onshore, mix with high suspended load river waters, and a mud layer deposited in the supra-tidal zone. Such supra-tidal deposits are found in China, and in Bangladesh as a result of storm surges and tropical cyclones. The material deposited tends to be mainly silt.

Chenier plains: Shelly chenier ridges can form plains as described by Schofield (1960) and Woodroffe *et al.* (1983) for the Firth of Thames in northern New Zealand. The shells comprise bivalves such as cockles, and during storm events become transported landwards, or otherwise move as wave-induced littoral drift along an active chenier beach face. The chenier ridges typically form shell spits overlying thick deposits of structureless mud, and the seaward-most chenier may comprise the active beach face surmounting a broad intertidal mudflat.

Swamp marsh and wetland deposits: This type of muddy coast deposit occurs in temperate coastal environments beyond the limit of mangroves. Good examples are found along the low terrain coasts which occur on a broadscale around the Dutch and German North Sea coasts, and in the Wash of England. Other well-known examples are found on the landward side of the extensive barrier enclosed lagoon system that is found around the southeast United States and the Gulf coast.

Mangrove forest and swamp deposits: Mangrove forests occur widely in the tropics and subtropics and are typically associated with muddy deposits. Their root mat assists in the trapping of fine suspended sediments. Mangrove stands also occur on the landward shores of coral platform reef and enclosed lagoons, and likewise may induce a veneer of muddy deposits to occur over the coral (Schaeffer-Novelli *et al.*, 2002; Wolanski *et al.*, 2002).

Mud veneer deposits on eroded shore platforms: In this type of muddy coast, an eroded shore platform has subsequently been layered with a veneer of muddy sediment. These types of muddy coast occur on a relatively small-scale and in sheltered harbor environments where there is a high source of muddy material either from adjacent catchments or from sublittoral deposits in the nearshore. Volumes of mud involved are relatively small. Examples occur in the harbors of northern New Zealand.

Ice deposited mud veneer: On Arctic shores, a muddy flat may be formed by ice rafts transporting mud and boulders, grounding and melting out in the thaw so that mud and boulders are jointly deposited on the tidal flats (Dionne, 2002).

Sub-littoral mud deposits: Along many coasts in sheltered and semi-enclosed bays where there is an abundant source of fine sediment supply and not particularly strong tidal currents, mud deposits occur in the sublittoral zone. These deposits may be of the fluid mud type, or occur as offshore mud banks, as off the coast of Kerala, India (Mathew and Baba, 1995). On occasions of strong wave agitation and shorewards wind-induced surface current, the agitated fine mud sediment particles may be eroded and suspended on the wave turbulence creating an estuarine "turbid fringe" and swept onshore and deposited as a mud layer either in the high tidal zone or the intertidal zone.

Beaches on muddy coasts

Although the dominant visual morphological feature of muddy coasts is often the broad intertidal flat, this is really just one component of a beach system. The extensive intertidal flat may also be part of a littoral drift system. In many estuarine beaches the following sequence can be observed: (1) a supra-tidal section with mainly storm surge silt and clay deposits active in storm surge events, and sometimes may have wind blown sand cappings; (2) a narrow, steep ($\sim 10°$), active beach face comprised of sand and shelly gravel sediment; (3) a broad intertidal flat containing muddy sediment and often *Zostera* beds; and (4) a subtidal zone containing mainly silts. Of course many variations exist, for example, the intertidal flat may be mainly sand and biogenic gravel with relatively small amounts of mud, and the subtidal sector can contain coarser sediments. But the important point is that the entire system is an active beach and sediment may be transported by wave and current processes over (diabathic transport) and along (parabathic transport) its components. Thus, shells and even terrigenous gravels originating in the subtidal zone may be transported onshore to accrete on the upper beach face. As tidal waters submerge and drain the intertidal flats, agitation of the surface sediments by waves creates a *turbid fringe* of high suspended load sediment—a typical feature of muddy coasts.

Stratigraphy and sedimentary structures of muddy coastal deposits

Wang *et al.* (2002a) identify that tidal flat muddy deposits are often stratified in fine laminae, the delicate bedding being caused by the interlamination of thin layers of very fine sand material in the more argillaceous basic substrate. The lamination of thin bedding is the result of varying hydrodynamic conditions, including the obvious flood and ebb oscillating tidal flows, but also reflecting neap and spring, storm, and seasonal changes. The bedding is seldom strictly parallel, but instead is generally streaky and lenticular. The alternating fine-sandy and clay layers wedge out rapidly in lateral extent and their thickness is not uniform. Such alternating bedding indicates frequent reworking of sediments under the influence of tidal currents of varying strength and direction, and the influence of meteorological factors, especially storm winds and associated waves.

Stratigraphically, modern coastal mud deposits tend to be thin relative to true deep-sea marine muds. This reflects the limited time—only some ~6,500 years—since the Holocene sea level reached its approximate present level, as well as the shallow depositional environment, which is typically energetic on a day-to-day Timescale. In some areas the deposits, especially clays, appear to be structureless and massive, but in other deposits there are often rare lenses of sand and shell—the result of episodic storm processes (Park and Choi, 2002).

In stratigraphic section, across broad tidal flats such as occur in China, shelly chenier deposits overlie the supratidal clay-rich muds. Typically, these muds laid down by storm surge processes, are clay-rich, massive, lensoid-shaped, and also contain relatively high organic matter, especially plant fragments. The supra-tidal clays overlie the silty muds of the upper tidal flats, which overlie the alternating fine sand and clay lenses of the mid-tidal area. These in turn overlie the very fine sand and coarse silt of the low tidal zone flat (Wang *et al.*, 2002b).

Sedimentary structures found in muddy coastal deposits include ripple bedforms on the sandier sediments. Within the larger tidal creeks with high current speeds and transporting sandy bedload, bedforms of megaripples, and current-cross-bedding is typical. Upon the mudflats, bioturbation structures produced by organisms inhabiting the tidal flats are also evident, while polygon cracks are to be found on the supra-tidal deposits. These sedimentary structures reflect a range of dynamic processes on the muddy coast tidal flats.

Geographical spread of muddy coast

Muddy coasts are developed in a wide range of coastal climatic and oceanographic environments in the world, from Arctic oceanic coast to the tropics (Wang *et al.*, 2002a). Their detailed geographic location and nature are presented in detail by Flemming (2002). Extensive well-formed, broad expansive intertidal muddy coasts are present around the Yellow River delta in the Bohai Sea of China, as well as the abandoned Yellow River delta in North Jiangsu Province along the Yellow Sea, and the southern embayments of the Changjiang River. Extensive sectors of muddy coast also occur along the east coast of the Americas, with the Amazon River mouth and the northern coasts of Guiana being of immediate note. Extensive muddy coast also occurs along the shores of Canada's Bay of Fundy, around the seasonal ice covered St. Laurence estuary, along the Atlantic east and southeast coasts of the United States, around the Mississippi delta and Vicinal coasts of the Gulf of Mexico, and inside several of the larger bays along the west coast of North America from California to Washington. In Europe, muddy coast deposits occur around the North Sea and Baltic coasts, such as Dollart, Lay and Jade Bays in Germany, the funnel-shaped mouths of large rivers (Elbe, Weser) in the east and north Friesian Bight, and in The Wash and other estuaries of east England. These exhibit many different examples and types of muddy coast. Muddy coast is widely evident in the tropical and monsoonal countries of Asia, including much of the coasts of India, Bangladesh, Thailand, Malaysia, Indonesia, and Vietnam. The northeast coasts of Australia, as well as the islands of the Pacific including New Zealand, New Guinea, and even the coral reef islands, contain areas of muddy coastal deposits. For the African continent, muddy coast occurs mainly within the tropical belt of both the east and west coasts.

It is evident that muddy coasts are characteristic of all continents (perhaps with the exception of Antarctica which typically exhibits a steep rocky profile for its minor coastal sectors of ice free coastline). They occur in a wide variety of the global coastal environments, but can vary widely in size and morphogenetic environment. However, muddy coasts tend to be particularly conspicuous in tropical zones, especially in the Asia region, and the muddy coastal deposits of largest extent are associated with major continental river discharges.

Factors leading to formation of muddy coast

A number of factors facilitate and enhance muddy coast formation (Wang *et al.*, 2002a).

Abundant fine-grained sediment supply

This is the most important factor in influencing the formation of muddy coastal deposits of various types. Fundamentally, if a major source of fine sediments continuously provides muddy sediments at a faster rate than the hydrodynamic conditions can remove them, then muddy coastal deposits will form. Conversely, if muddy source material is not available in sufficient quantities within a coastal environment then muddy deposits do not form. Sources of sediment for concentration to form muddy coast are many and varied, but include:

Adjacent rivers. Adjacent rivers with high concentrations of fine suspended sediment loads. The classic example is the Huangho River of China, which drains a huge catchment including highly erodible loess regolith, which provides the bulk of the fine sediments to the Bohai Bay. Other examples include the Amazon (draining a huge and partly steep catchment of highly weathered tropical soils), the Fly River of New Guinea, the Mekong River of Vietnam/Laos (likewise draining large and partly steep catchments of highly weathered tropical soils), and the Waipaoa River of New Zealand (draining a steepland catchment of highly erodible montmorillonitic clay-rich Tertiary rocks). In the latter cases, the mountainous steepland catchments are also subject to episodic, intense, orographically induced rainfalls, factors which enhance the high volume of sediment load delivered to the littoral zone.

Supply of sediment from offshore. Often sediment that forms or contributes to coastal muddy deposits may be of "secondary origin" in the sense, that it is reworked from offshore. In this sense, the reworked mud deposits are "palimpsest." Examples include the onshore transfer of fine sediments from the ancient river deposits along the northern Jiangsu coast (Wang *et al.*, 2002b), and the onshore transfer of eroded glacial till deposits from the North and Baltic seas.

Erosion of coastal sedimentary deposits and cliffs. In many enclosed and semienclosed seas, active erosion of the cliffs provides a limited source of fine material for muddy coastal deposits. In the Kiel Bay of the Western Baltic the surrounding coast is formed of glacial till containing a high proportion of fine sediment (Healy *et al.*, 1987). Rapid erosion of the boulder clay cliffs (~0.3–1.0 m/a) provides considerable material for local muddy coastal deposits. Where there has been rapid cliff erosion and continuing supply of muddy sediment the shore platform may subsequently become veneered with mud. Likewise in sheltered steepland embayments containing an estuarine lagoon fronted by barrier spits (Healy *et al.*, 1996) erosion of the cliffs produces mud deposits overlying the early Holocene marine sands.

Tidal range

A macro tidal range (>4 m) is often regarded a necessary factor in explaining extensive broad muddy-coast deposits. However, a general dearth of sediment supply or high wave energy may restrict tidal flat development (Davis, 1983). Certainly, the broad expansive intertidal flats and associated muddy deposits of the Jiangsu coast of China are very wide and have associated a macro tidal range. Other extensive muddy coastal deposits associated with large tidal ranges include the Bay of Fundy and The Wash of England (Amos, 1995). But in other locations of muddy coasts the tidal ranges may be meso (2–4 m) or micro (<2 m), and still extensive muddy deposits of various types occur, for example, around the coast of Indonesia, India, the southeast coast of the United States off the mouth of the Hwang-ho (Yellow River) of China, the Fly River of New Guinea, and around northern New Zealand.

A fundamental question is: why should a large tidal range be associated with broad intertidal muddy coastal deposits? Evidently, the answer lies in a different morpho-depositional mechanism from sandy and mixed sand gravel beaches, combined with the thixotrophic nature of the muddy deposits. While sandy beaches are characterized by noncohesive grains, and thus undergo regular morphodynamic change in response to varying wave conditions, they typically assume some modal morphodynamic beach state. However, beaches on muddy coasts tend not so obviously to change their morphology, although the beach is active over time. This is because: the individual muddy sediment clasts possess cohesive qualities and thus the surface layer of sediments is not so easily moved around by the waves and currents as are noncohesive sandy sediments. Rather, in energetic wave conditions the surface layers of mud particles are reagitated and taken into suspension within the water column. Moreover, the sediments in purely muddy beaches tend to be deposited in episodic storm surge events with a net onshore movement of suspended sediment providing the material to become deposited on top of the existing muddy flat sediments. Finally, the muddy bulk sediment possesses little strength, and is subject to fluid and plastic flow, and this results in a low angle of natural repose within the intertidal zone.

Thus, "pure" muddy coast (i.e., without a landward surmounting narrow beach face of sand or shelly gravel or chenier beach) typically exhibits a muddy upper intertidal and supra-tidal zone of subdued relief, which undergoes relatively little morphodynamic change compared to a sandy beach. The essential point seems to be that the mode of formation of the intertidal muddy deposits is by net onshore sediment movement, and that the morphodynamics of deposition are such that an essentially flat topographic surface results. Providing there is abundant and continuing fine sediment supply then the extent or width of a muddy intertidal flat becomes a function of the tidal range (Wang *et al.*, 2002b).

Conclusion

Muddy coast geomorphology and sedimentary depositional systems occur widely around the earths' coastlines, but compared to sandy coasts have been relatively little studied. Texturally, "mud" is defined based upon particle size, and comprises mineral material of silt and clay sizes. Muddy coasts are defined as possessing distinctive muddy deposits, typically but not exclusively manifest as broad intertidal muddy flats. Additional geomorphic features found on muddy coasts include chenier ridges and beaches, veneer muddy deposits, and shallow offshore muddy deposits, while associated biologic features may occur such as mangrove stands and salt marshes. Muddy coasts occur widely in geographic distribution from the tropics to the Arctic. Their fundamental *raison d'existence* is abundant mud-sized sediment particles in the coastal environment, but the existence of broad intertidal mudflats is often associated with a large tidal range.

Terry R. Healy

Bibliography

Amos, C., 1995. Siliclastic tidal flats. In Perillo, G. (ed.), *Geomorphology and Sedimentology of Estuaries. Advances in Sedimentology,* Volume 53. Amsterdam: Elsevier, pp. 273–306.

Augustinus, P., 2002. Biochemical factors influencing deposition and erosion of fine grained sediment. In Healy, T.R., Wang, Y., and Healy, J.A. (eds.), *Muddy Coasts of the World: Processes, Deposits and Function*. Amsterdam: Elsevier, pp. 203–228.

Davis, R.A., 1983. *Depositional Systems. A Genetic Approach to Sedimentary Geology*. Upper Saddle River: Prentice-Hall.

Dionne, J-C., 2002 Sediment content of the ice-cover in muddy tidal areas of the turbidity zone of the St. Lawrence estuary and the problem of the sediment budget. In Healy, T.R., Wang, Y., and Healy, J.A. (eds.), *Muddy Coasts of the World: Processes, Deposits and Function*. Amsterdam: Elsevier, pp. 463–478.

Flemming, B.W., 2002. Geographic distribution of muddy coasts. In Healy, T.R., Wang, Y., and Healy, J.A. (eds.), *Muddy Coasts of the World: Processes, Deposits and Function*. Amsterdam: Elsevier, pp. 99–202.

Folk, R.L., 1968. *Petrology of Sedimentary Rocks*. Austin: Hemphill's.

Healy, T.R., Sneyd, A., and Werner, F., 1987. First approximation mathematical model of sediment budget and shore profile development for a semi-enclosed sea; example of Kiel Bay, Western Baltic. *Mathematical Geology*, **19**: 41–56.

Healy, T.R., Cole, R., and de Lange, W., 1996. Geomorphology and ecology of New Zealand shallow estuaries and shorelines. In Nordstrom, K.F., and Roman, C.T. (eds.), *Estuarine Shores*. New York: John Wiley and Sons.

Leeder, M. R., 1982. *Sedimentology: Process and Product*, London: Unwin.

Mathew, J. M., and Baba, M., 1995. Mudbanks of the southwest coast of India. II: Wave-mud interactions. *Journal of Coastal Research* **11**(1): 179–187.

Mehta, A.J., and Hayter, E.J., 1989. High concentration cohesive sediment transport. *Journal of Coastal Research* (Special issue) **5**: 230p.

Park, Y.A., and Choi, K.S., 2002. Late Quaternary stratigraphy of the muddy coastal deposits, west coast of Korea. In Healy, T.R., Wang, Y., and Healy, J.A. (eds.), *Muddy Coasts of the World: Processes, Deposits and Function*. Amsterdam: Elsevier, pp. 391–345.

Schaeffer-Novelli, Y., Cinton-Molero, G., and Soares, M.L., 2002. Mangroves as indicators of sea level change in muddy coasts of the world. In Healy, T.R., Wang, Y., and Healy, J.A. (eds.), *Muddy Coasts of the World: Processes, Deposits and Function*. Amsterdam: Elsevier, pp. 245–262.

Schofield, J.C., 1960. Sea level fluctuations during the last 4000 years as recorded by a chenier plain, Firth of Thames, New Zealand. *New Zealand Journal of Geology and Geophysics*, **3**: 467–485.

Wang, Y., Collins, M.B., and Zhu, D., 1990. A comparative study of open tidal flats: the Wash (U.K.), Bohai Bay and west Yellow Sea (Mainland China). *Proceedings of International Symposium on the Coast Zone*. China Ocean Press. pp. 120–130.

Wang, Y., Healy, T., and Members of SCOR Working Group 106, 2002a. Definitions, properties and classification of muddy coasts. In Healy, T.R., Wang, Y., and Healy, J.A. (eds.), *Muddy Coasts of the World: Processes, Deposits and Function*. Amsterdam: Elsevier, pp. 9–18.

Wang, Y., Wu, X., and Zhu, D., 2002b. Tidal flats and associated muddy coast of China. In Healy, T.R., Wang, Y., and Healy, J.A. (eds.), *Muddy Coasts of the World: Processes, Deposits and Function*. Amsterdam, Elsevier, pp. 319–346.

Wolanski, E., Spagnol, S., and Lim, E.B., 2002. Fine sediment dynamics in the mangrove fringed muddy coastal zone. In Healy, T.R., Wang, Y., and Healy, J.A. (eds.), *Muddy Coasts of the World: Processes, Deposits and Function*. Amsterdam: Elsevier, pp. 279–292.

Woodroffe, C.D., Curtis, R.J., and McLean, R.F., 1983. Development of a chenier plain, Firth of Thames, New Zealand. *Marine Geology*, **53**: 1–22.

Cross-references

Beach Sediment Characteristics
Cheniers
Coastal Sedimentary Facies
Mangroves, Coastal Geomorphology
Salt Marsh
Storm Surge
Tidal Flats
Tidal Flats, Open Ocean Coasts

N

NATURAL HAZARDS

Natural hazards are physical phenomena that expose the coastal zone to risk of property damage, loss of life, or environmental degradation. *Rapid-onset* hazards last over periods of minutes to several days. Examples include major cyclones, accompanied by high winds, waves, surges, and tsunamis—giant sea waves set off by earthquakes, volcanic eruptions, or submarine landslides. *Slow-onset* hazards develop incrementally over longer time periods. Examples include coastal erosion and sea-level rise.

The vulnerability of the world's coastlines to natural hazards varies considerably, because climate, tectonism, and other physical variables, such as bathymetry, shelf width, landform, lithology, and coastal configuration change from place to place. For example, although tropical cyclones may form wherever sea surface temperatures exceed 26°C, landfall occurs most commonly in particular regions such as southeast Asia, the Caribbean Sea, or southeastern United States (Figure N1). Tsunamis are most prevalent around the Pacific Ocean, due to the seismicity and volcanicity associated with convergent-plate boundaries along the circum-Pacific "Ring of Fire."

Rapid coastal development and urbanization expose increasing numbers of people to natural hazards in a dynamic and unstable environment. Globally, approximately 400 million people live within 20 m of sea level and within 20 km of a coast (Small *et al.*, 2000). In the United States, over 139 million people (53% of the population) live in coastal counties (Culliton, 1998). Anticipated climate changes will greatly amplify risks to coastal populations. By the end of this century, rise in sea level by 2–5 times present rates could lead to more frequent flooding and inundation of low-lying regions, worsening beach erosion, as well as loss of ecologically

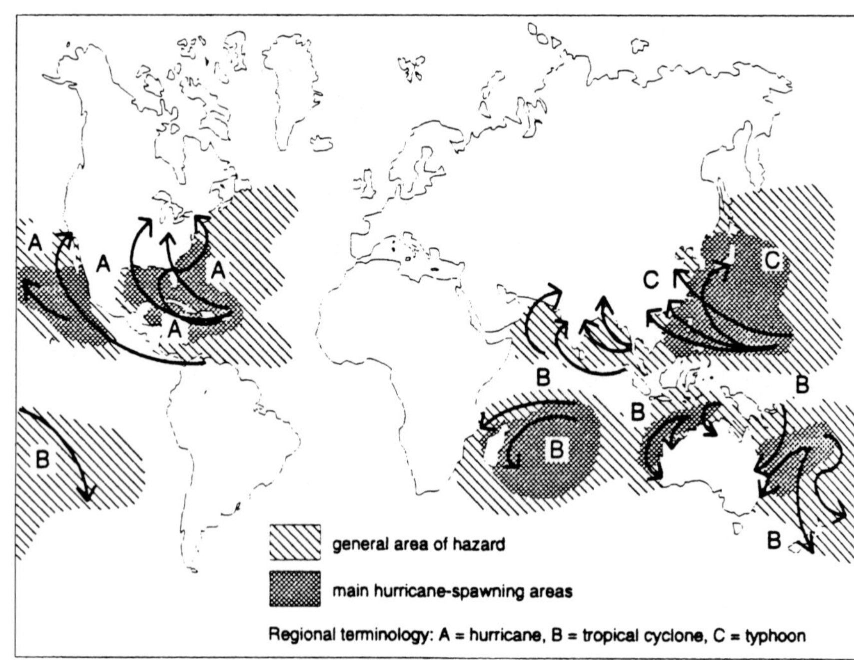

Figure N1 Location of tropical cyclone breeding areas (from Alexander, 1993).

productive coastal wetlands, and saltwater intrusion into coastal aquifers and estuaries (IPCC, 2001a,b).

Rapid-onset hazards

Tropical cyclones

Tropical cyclones (also called *hurricanes* in the Atlantic basin, *cyclones* in the Indian Ocean, and *typhoons* in the western Pacific) are major low-pressure systems that develop and intensify over the open ocean between ±5° and 25° latitude (Figure N1). Hurricanes form where sea surface temperatures rise above 26°C, under conditions of high relative humidity, strong evaporation rates, weak vertical wind shear, and atmospheric instability (Henderson-Sellers *et al.*, 1998). In the Atlantic basin, the hurricane season lasts from June to November, peaking between August and early October.

A tropical cyclone consists of a rotating atmospheric low-pressure system, ranging from less than one hundred to several hundred kilometers in diameter. At the center is the "eye"—a zone of calm atmosphere and nearly clear sky. It is surrounded by the eyewall, where the strongest winds of the hurricane are concentrated (Figure N2). Spiraling around the eyewall are rainbands of high winds and torrential rains. In the Northern Hemisphere, the cyclonic system rotates counterclockwise around the eye, while simultaneously moving forward along a northwesterly track. Thus, winds are stronger on the cyclone's right side, due to the additive effect of the mean forward velocity of the storm and the rotational speed around the eye (Coch, 1995).

A number of factors contribute to hurricane damage. One of these is the damage produced by very high wind speeds (minimum 119 km/yr). In addition, hurricanes cause severe flooding due to the combination of heavy rainfall, waves, and storm surge. In coastal areas, a major hazard is flooding by the storm surge, which is a dome of water produced by the low barometric pressure and strong wind shear (see *Storm Surge*). The area at greatest risk lies to the right of the eyewall at landfall; the risk falls off with increasing distance. The hazard also depends on the mean forward velocity of the storm: a slowly moving storm can cause flooding over several tidal cycles. The angle at which the hurricane track crosses the coast also determines the hazard. A hurricane track which parallels the eastern coastline of a continent keeps the stronger right side of the cyclone seaward, thereby reducing the damage potential to land. On the other hand, a coast-normal track crossing land may cause serious damage on the right side of the storm (Coch, 1995).

The intensity of hurricanes is classified from 1 to 5 on the Saffir–Simpson scale, with 5 representing the most severe category (Table N1). This scale rates hurricanes on the basis of central air pressure (the lower the pressure, the more intense the storm), sustained wind velocity, storm surge potential, and potential for property damage. The three most powerful US hurricanes since 1900 includes a category 5 storm that struck the Florida Keys in 1935, Camille in 1969 (category 5; Louisiana-Mississippi), and Andrew in 1992 (category 5; south Florida, Louisiana) (NOAA, 2001).

The deadliest US hurricane (category 4) struck Galveston, Texas in September, 1900, killing over 8,000 people. The next most lethal hurricanes include several that swept over the Florida Keys in 1919, 1928, and 1935 (categories 4 and 5), and New England in 1938 (category 3). In the United States, deaths caused by hurricanes have decreased substantially in recent years, because of the advance warning provided by weather satellite tracking. In other parts of the world, however, tropical cyclones continue to exert a high toll. Between 1960 and 1981, tropical cyclones have killed 386,200 people in Bangladesh, 24,930 in India, and 7,480 in Vietnam (Alexander, 1993, p. 156).

Hurricane behavior in the Atlantic basin has not changed detectably over the last 60 years, although intense hurricanes (categories 3–5) were more frequent in the 1940–1960s and late 1990s (Landsea *et al.*, 1999). Yet the costliest hurricanes have occurred in recent years (i.e., Andrew in 1992 [category 5, Florida, Louisiana, $26.5 billion in damages], Hugo in 1989 [category 4, South Carolina, $7 billion], and Floyd in 1999 [category 2, mid-Atlantic & NE U.S., $4.5 billion]; NOAA, 2001). Extensive coastal development over the last few decades has increased the amount of property at risk to devastating storms (Van der Vink *et al.*, 1998; Chagnon and Chagnon, 1999).

Extratropical cyclones

Extratropical cyclones are the dominant type of storm responsible for major coastal flooding and beach erosion at mid-latitudes. These storms occur most frequently between October and April in the Northern Hemisphere, peaking in winter. Although wind speeds and surge heights are much lower than in hurricanes, extratropical cyclones inflict considerable damage because of their greater spatial extent (typically over 1,000 km versus 100–150 km for hurricanes), high waves, and duration spanning several tidal cycles at a given location (Davis and Dolan, 1993; Dolan and Davis, 1992). Areas at risk to flooding from extratropical cyclones include the Atlantic Coast of the United States north of 36°N, northwestern Europe, and northeastern Asia.

US Atlantic coast winter storms (*northeasters* or *nor'easters*) have been classified into five categories (5 is the most severe) based on significant wave height, $H_{1/3}$ (average of highest one-third of waves), duration,

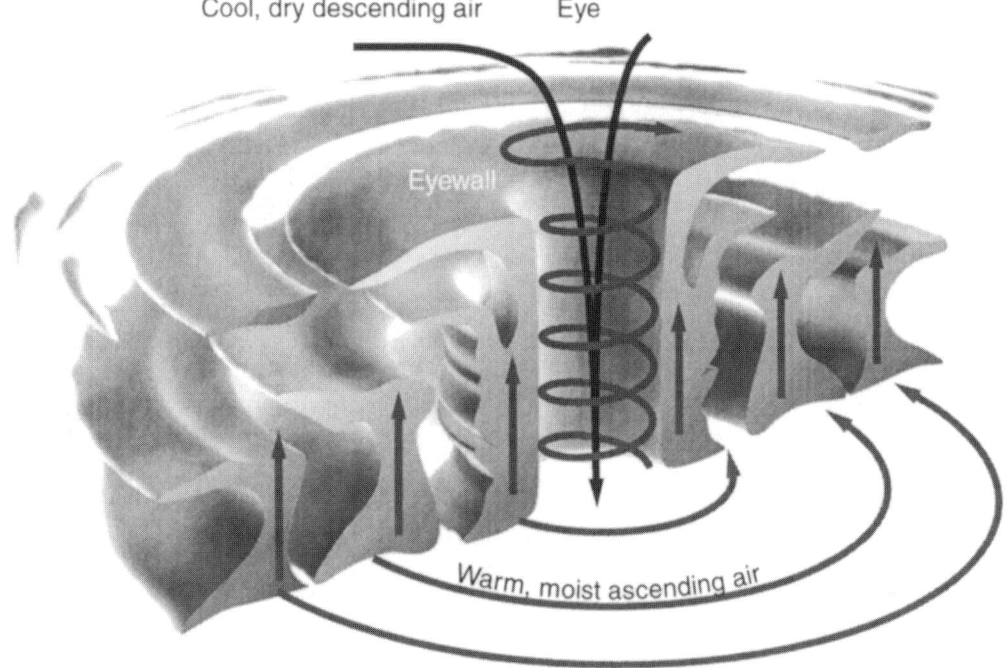

Figure N2 Structure of a hurricane.

Table N1 The Saffir–Simpson scale of hurricane intensity (after NOAA 1997)

Category	Central pressure (millibars)	Winds (km/h)	Surge (m)	Damage
The Saffir–Simpson scale				
1	⩾980	119–153	1.2–1.5	Minimal
2	965–979	154–177	1.8–2.4	Moderate
3	945–964	178–209	2.7–3.6	Extensive
4	920–944	210–249	3.9–5.5	Extreme
5	<920	>249	>5.5	Disaster

Effects
1 Minimal damage, mainly to unanchored mobile homes, trees, shrubs. Minor coastal flooding.
2 Some damage to roofs, windows, mobile homes, piers, vegetation. Coastal flooding; small boats break moorings.
3 Some structural damage to small buildings; mobile homes destroyed. Flooding destroys coastal small structures. Floodwaters may cover terrain below 1.5 m.
4 Major damage to lower floors of coastal buildings. Significant beach erosion. Potential flooding of terrain below 3 m inland as far as 9.6 km, requires mass evacuation.
5 Roofs blown off buildings. Many smaller buildings destroyed or blown away. Damage to flower floors of buildings below 4.5 m elevation, within 460 m of shoreline. Massive evacuation may be necessary.

Table N2 Dolan–Davis scale for the classification of Atlantic Coast nor'easters (after Dolan and Davis, 1994)

Storm class	Frequency (percent) (N = 1564)	Significant wave height (m)	Duration (h)	Power (m^2h)
The Dolan–Davis scale				
1 Weak	50.3	2.0	8	32
2 Moderate	25.1	2.5	19	107
3 Significant	21.6	3.2	35	384
4 Severe	2.5	5.0	62	1420
5 Extreme	0.5	6.8	97	4332

Effects
1 Minor beach erosion. No property damage.
2 Moderate beach erosion; minor dune erosion. No property damage.
3 Significant beach and dune erosion. Moderate property damage.
4 Severe beach and dune erosion. Overwash damage on low-profile beaches. Community-wide loss of structures.
5 Extreme beach erosion, dune destruction. Massive overwash in sheets and channels. Extensive regional-scale property losses in million of dollars.

D, and power, P, where $P = H^2_{1/3} \times D$ (Dolan and Davis, 1992, 1994; Table N2). (The energy of a wave, hence its destructive capability, is proportional to $H^2_{1/3}$.) A qualitative assessment of the relation of storm class to coastal damage is also indicated in Table N2.

Nor'easters affecting the eastern United States have varied in number and severity over time, but do not show any long-term trends (Zhang *et al.*, 2000). Severe storms peaked during the late 1970s and again in the early 1990s (Dolan and Davis, 1994). Significant nor'easters within the last 40 years include the "Ash Wednesday" storm (March 6–7, 1962), the Halloween storm (October 31, 1991), and two other powerful storms on December 11–12, 1992, and March 13–14, 1993. The "Ash Wednesday" storm generated 10-m-high waves in the open ocean and caused over US$300 million in property damage along 1,000 km of the Atlantic coast (Davis and Dolan, 1993). This storm was especially destructive because it lasted for five tidal cycles. The December 1992 nor'easter hit the mid-Atlantic and New England coasts with waves over 6 m in some locations (Davis and Dolan, 1993) and brought the entire transportation system of New York City to a virtual standstill (USACOE/ FEMA/NWS, 1995). Several months later, the March 1993 nor'easter (the so-called "storm of the century") created blizzard conditions over most of the East Coast and churned waves over 4.5 m along much of the East Coast. The Ash Wednesday storm of 1962 and the Halloween storm of 1991 rank as class 5 nor'easters (Davis and Dolan, 1993).

Northwestern Europe has also experienced severe winter storms. On February 1, 1953, a North Sea storm swept over the Netherlands, breaching dikes, inundating 150,000 ha, and killing over 1,800 people. That storm also caused major damage in eastern Great Britain (Alexander, 1993, p. 138). In 1990, a series of eight powerful gales, some with hurricane-force winds, hit western Europe between late January and early March (Munich Re, 1993).

Storm surges

A storm surge is the elevation of ocean water level above that expected from astronomical tides, caused by a passing storm. The surge results from the reduced atmospheric pressure, which causes sea water to rise by around 1 cm for each drop in pressure of 1 mb, and from the wind stress, which literally pushes water toward the coast. Factors contributing to the surge height include the storm intensity, width and slope of coastal shelf, and coastal configuration. The total flood height also depends on the tidal stage (see *Storm Surge*).

The flood surge is a major cause of coastal erosion (see below). On exposed barrier islands, sand washes from the beach and dunes, and redeposits bayward in overwash fans. The surge also cuts channels or tidal inlets across the barrier. As the surge ebbs, seaward-flowing water creates further erosion. The flood surge damages buildings and other structures by direct wave impact, hydraulic lift of waves and water beneath raised buildings, wave energy reflected from protruding structures such as groins and jetties, as well as battering from floating debris (Coch, 1995).

Regions of the world most vulnerable to surges from tropical cyclones include low-lying, deltaic coasts of southeast Asia, Bangladesh, China, the Philippines, the southeastern United States, the Caribbean, and northern Australia (Figures N1, N3). Surges from extratropical cyclones have also produced extensive flooding in the Netherlands, Germany, Italy, as well as the eastern United States (see on page 682).

Bangladesh is particularly vulnerable to storm surges because of its low topography, shallow continental shelf, high tidal range, tendency for coastal cyclones to converge near the apex of the Bay of Bengal, and high population density especially on low-lying islands (Murty and Flather, 1994). Storm surges caused over 220,000 deaths in 1970 and 150,000–200,000 more in 1991 (Alexander, 1993, p. 542). In the Netherlands, catastrophic storm surges killed 80,000 people in 1281, and 100,000 in 1421. The 1953 storm claimed around 1,900 lives (including England; Munich Re, 1993). The Netherlands has subsequently embarked on a massive program to strengthen its national sea defenses in the central Netherlands to be able to withstand the 1/10,000 year flood (De Ronde, 1991). In Venice, a storm flood of 1.94 m above the local datum inundated most of the historic city on November 3–4, 1966. A mean sea level rise of only 34 cm would reduce the return period of this flood from 165 to only 15 years (Pirazzoli, 1991).

Tsunamis

A tsunami is a series of waves caused by the vertical displacement of ocean water, triggered by an earthquake, volcanic eruption, landslide, or more rarely, the impact of a bolide. Most tsunamis occur on islands and shores around the Pacific Ocean. In contrast to ordinary water waves, tsunamis have long wavelengths (from 150 to 700 km), long periods (10–60 min), and low amplitudes (≤ 1 m). They propagate with enormous speeds (600–800 km/hr) over open ocean. The velocity of propagation is given by $v = \sqrt{(g \cdot h)}$, where v is velocity, g is the acceleration due to gravity, and h is water depth. The wavelength is $L = v \cdot T$, where L is wavelength and T is the period (Alexander, 1993; Bell, 1999).

The danger of tsunamis increases dramatically as the waves approach shallow water, where they slow down but increase tremendously in height. Water levels rise rapidly up to 20 m above normal sea level. Occasionally, a "bore" or wall of turbulent water moves up narrow bays or estuaries. The highest wave may occur hours after several lower ones have passed (Coch, 1995). The height of the wave as it arrives at the shore is related to the original amount of water displaced, distance from source, nearshore bathymetry, and coastal configuration (Bell, 1999). The destructiveness is linked to the run-up, or maximum vertical height (Table N3).

Tsunamis are usually caused by displacement of water due to vertical movement of the seafloor along faults, during major earthquakes associated with plate subduction or other tectonic events. Waves are formed as the displaced water mass attempts to return to an equilibrium state. However, major volcanic eruptions can also set off tsunamis. For example, the catastrophic tsunami produced by the Krakatau eruption in 1883 generated waves up to 30 m high in places that killed over 30,000 people in Java and Sumatra (Coch, 1995). Less commonly, submarine landslides can also cause tsunamis, as apparently happened in Papua New Guinea in 1998 (Tappin *et al.*, 1999).

Table N3 Tsunami intensity scale (after Bell, 1999; Soloviev, 1978)

Intensity	Wave height at run-up (m)	Frequency in Pacific Ocean
I	0.5	once per hour
II	1	once per month
III	1	once per eight months
IV	4	once per year
V	8	once in three years
≥VI	16	once in ten years

Effects

I *Very slight.* Weak waves detectable only on tide gauge records.
II *Slight.* Waves noticeable only on very flat shores.
III *Rather large.* Some flooding on low coasts; slight damage to small structures; light boats swept away; reversed flow up estuaries or rivers.
IV *Large.* Flooding of shore. Damage to coastal structures, embankments and dikes. Smaller ships and larger sailing vessels swept away. Floating debris.
V *Very large.* General flooding of shore to some depth. Considerable damage to quays, heavy structures near shore; extensive debris; vessels except for large ships swept away; large bores in estuaries; people drowned.
≥VI *Disastrous.* Partial to complete destruction of man-made structures along shore. Flooding of coasts to great depths. Large ships severely damaged. Trees uprooted or broken by waves; many casualties.

Around 90% of destructive tsunamis occur around the Pacific Basin at an average rate of 2 per year (Alexander, 1993). The Hawaiian Islands have suffered close to 160 such events in the past 200 years. Waves up to 8 m, generated by an earthquake in the Aleutian Islands in 1946 struck Hilo, Hawaii, killing 159 people and causing US$26 million ($1946) in damage. The tsunami caused by the 1960 Chilean earthquake produced waves up to 10.7 m at Hilo, killing 61 people. Crescent City, California, was struck by waves up to 6.3 m high from the 1964 Alaska earthquake, which caused 11 fatalities and US$7.5 million in property damage. A devastating tsunami in 1998 that killed around 2,200 people in Papua New Guinea may have been caused by an earthquake-induced underwater landslide (Tappin *et al.*, 1999). A tsunami that battered various sites along the Pacific coast of Japan in January, 1700, has been traced to a massive earthquake (estimated magnitude 9) on the Cascadian subduction zone (Satake *et al.*, 1996).

The Atlantic Coast of North America is also potentially at risk to destructive tsunamis, because of slope instabilities on the continental shelf. Discharge of natural gas along fractures along the outer continental shelf edge off southern Virginia-North Carolina could eventually destabilize the shelf edge, causing massive submarine landslides and tsunamis (Driscoll *et al.*, 2000). Underconsolidated sediments on the New Jersey continental slope are also potentially susceptible to slumping (Dugan and Flemings, 2000).

Slow-onset hazards

Coastal erosion

Around 70% of the world's sandy beaches are retreating at present (Bird, 1985). The prevalence of beach and cliff erosion (The NRC [1995] defines coastal erosion more narrowly as the volumetric loss of sand from a beach by waves, currents or other processes, whereas recession is the landward [linear] displacement of the coastline over time.) from many parts of the world, even where human impacts are minimal, implies an underlying global cause, such as the recent sea-level rise (IPCC, 2001; see below), which may have exacerbated other more localized processes. Among these are differences in rock resistance, coastal subsidence, diminution in sediment yield of rivers (because of reduced precipitation or upstream entrapment in artificial reservoirs), increase in wave attack due to greater number or severity of coastal storms, longshore drift, beach mining, or cliff quarrying, and presence of engineering structures that intercept sediment flow or enhance scour (Bird, 1996).

In the United States, the average erosion rate along the Gulf Coast is 1.8 m/yr, with 63% of the shoreline retreating (Dolan *et al.*, 1985). Mean erosion rates in Louisiana are 4.2 m/yr— a consequence of the high relative sea-level rise due to land subsidence (Penland *et al.*, 1989). Erosion rates along the Atlantic Coast average 0.8 m/yr; 79% of the coastline shows some degree of erosion (Dolan *et al.*, 1985). The Atlantic side of the southern Delmarva Peninsula is an erosional "hotspot," possibly related to anomalously high subsidence rates near the mouth of Chesapeake Bay and jetties at Ocean City, Maryland, which have curbed southward littoral drift to Assateague Island (NRC, 1995). Widespread coastal erosion has also occurred as a result of major nor'easters (see above).

Coastal erosion is less severe on the Pacific Coast (Dolan *et al.*, 1985). Nonetheless, 28.6% of the California coast is at "high risk" to natural hazards, with an additional 36.8% "requiring caution" (Griggs, 1994). California's most extensive coastal hazard comes from eroding cliffs and bluffs. Cliff failure arises from wave attack, subaerial erosion, mass movements, and earthquakes. Beaches and dunes are also prone to erosion, particularly following high waves and tides during El Niño events (Griggs, 1994; Storlazzi and Griggs, 2000; see also *El Niño–Southern Oscillation, ENSO*). Damming of many rivers has reduced the sediment supply to the sea, limiting beach recovery following episodic erosion events.

Beach nourishment and dune restoration are now a major means of shore protection against storm damage and long-term erosion (NRC, 1995). Beach nourishment or restoration consists of placing sand that has usually been dredged from offshore or other locations onto the upper part of the beach. The process must be repeated at intervals which depend on local factors such as beach profile, average sand grain size, and anticipated losses due to storms and other processes. An estimated US$2.4 billion (adjusted to $1996) has been spent renourishing Atlantic, Gulf Coasts, and Great Lakes beaches since the 1920s (Program for the Study of Developed Shorelines, 1999). Over US$480 million have been spent replenishing East Coast beaches between 1990 and 1996 alone (Valverde *et al.*, 1999).

Sea-level rise

Sea-level rise represents a slow-onset hazard whose main impacts will become apparent only decades from now. Mean global sea level has been increasing by around 1–2.0 mm/yr over the last 100–150 years. This is the most rapid rate within the last few thousand years and is probably linked to the 20th century warming trend of ~0.6°C (Gornitz, 1995; IPCC, 2001a). The recent sea-level rise comes mainly from melting of mountain glaciers and thermal expansion of the upper ocean layers, with less certain contributions from polar ice sheets. Anticipated global warming could increase rates of sea-level rise by factors of 2–5 by the end of this century, greatly amplifying risks to coastal populations (IPCC, 2001a,b). The major consequences of accelerated sea-level rise are permanent inundation of the shoreline, more frequent coastal flooding, increased erosion, and saltwater intrusion (IPCC, 2001b).

The impacts of sea-level rise will not be globally uniform, because of local variations in vertical crustal movements, topography, lithology, wave climatology, longshore currents, and storm frequencies. Low gradient coastal landforms most susceptible to inundation include deltas, estuaries, beaches and barrier islands, coral reefs and atolls. Regions at risk include the Low Countries of Europe, eastern England, the Nile Delta, Egypt, the Ganges–Brahmaputra, Irrawaddy, and Chao Phraya deltas of southeastern Asia, eastern Sumatra and Borneo (Figure N3). In the United States, the mid-Atlantic coastal plain, the Florida Everglades, and the Mississippi Delta will be especially vulnerable.

Coastal wetlands will be among the most severely affected ecosystems since they lie largely within the intertidal zone. The ability of a salt marsh (or mangrove) to keep pace with rising sea level depends on relative rates of submergence versus vertical accretion. In the United States, present rates of marsh accretion are generally keeping up with relative sea-level rise, except for parts of Louisiana and Chesapeake Bay (NRC, 1987). Louisiana, with a relative sea-level rise of ~10 mm/yr, is losing around 140 km^2 of wetlands annually (Boesch *et al.*, 1994).

Coral islands, with average elevations of only 1.5–2 m above present sea level are also at high risk. If projected rates of sea-level rise approach 10–12 mm/yr, near the upper limit of coral growth, some slow-growing species may drown, but a slight rise in sea surface temperatures may prove even more detrimental to coral reef survival (IPCC, 2001b). The additional stress of increasing sea-surface temperatures may inhibit the ability of corals to keep pace with sea-level rise. Increasingly frequent coral "bleaching" occurrences in recent years have been linked to warmer than normal sea surface temperatures, in part associated with El Niño events. Tropical ocean islands and reefs at risk include the Maldives, many Pacific and Caribbean islands, and Australia's Great Barrier Reef.

Coastal flooding due to storm surges will increase in frequency with sea-level rise, even if the number and strength of storms do not change (Zhang *et al.*, 1997). The flood return period is very sensitive to very minor increases in sea level. For example, by the end of this century, the return period of the 100-year flood in New York City would be reduced to around 33 years, even at present rates of sea-level rise (2.7 mm/yr; see *Storm Surge*).

Coastal erosion is likely to increase as sea-level rises. Waves will be able to cut into cliffs at a higher level, triggering mass movements and slumping in poorly consolidated sediments. Barrier islands respond to rising ocean levels by eroding on their ocean sides and depositing sand by overwash on the bay side, a process called "barrier rollover" (Figure N4).

The high-tide shoreline's response to sea-level rise is often estimated using the Bruun Rule, which states that a typical concave-upward beach profile erodes sand from the beachfront and deposit it offshore, so as to maintain constant water depth (see *Beach Erosion; Coastline Changes*). High-tide shoreline retreat depends on the average slope of the shore profile. A 1 m sea level rise could cause the beach to retreat by as much as 100–200 m, depending on the slope. The Bruun Rule has been generalized to account for landward migration and upward growth of the barrier. Other high-tide shoreline response models involve sediment budget analysis, dynamic approaches, or historical trend analysis (NRC, 1987; Douglas *et al.*, 1998).

As sea level rises, saltwater will generally penetrate further up estuaries and rivers, as well as infiltrate into coastal aquifers, which could contaminate urban water supplies. Excessive groundwater pumping has already induced an upward migration of the saltwater–freshwater interface in many coastal localities. Upstream migration of the salinity front due to sea-level rise is analogous to that occurring under present drought conditions.

Vivian Gornitz

Figure N3 Major low-lying deltaic areas of the world. Tracks of tropical cyclones are also shown.

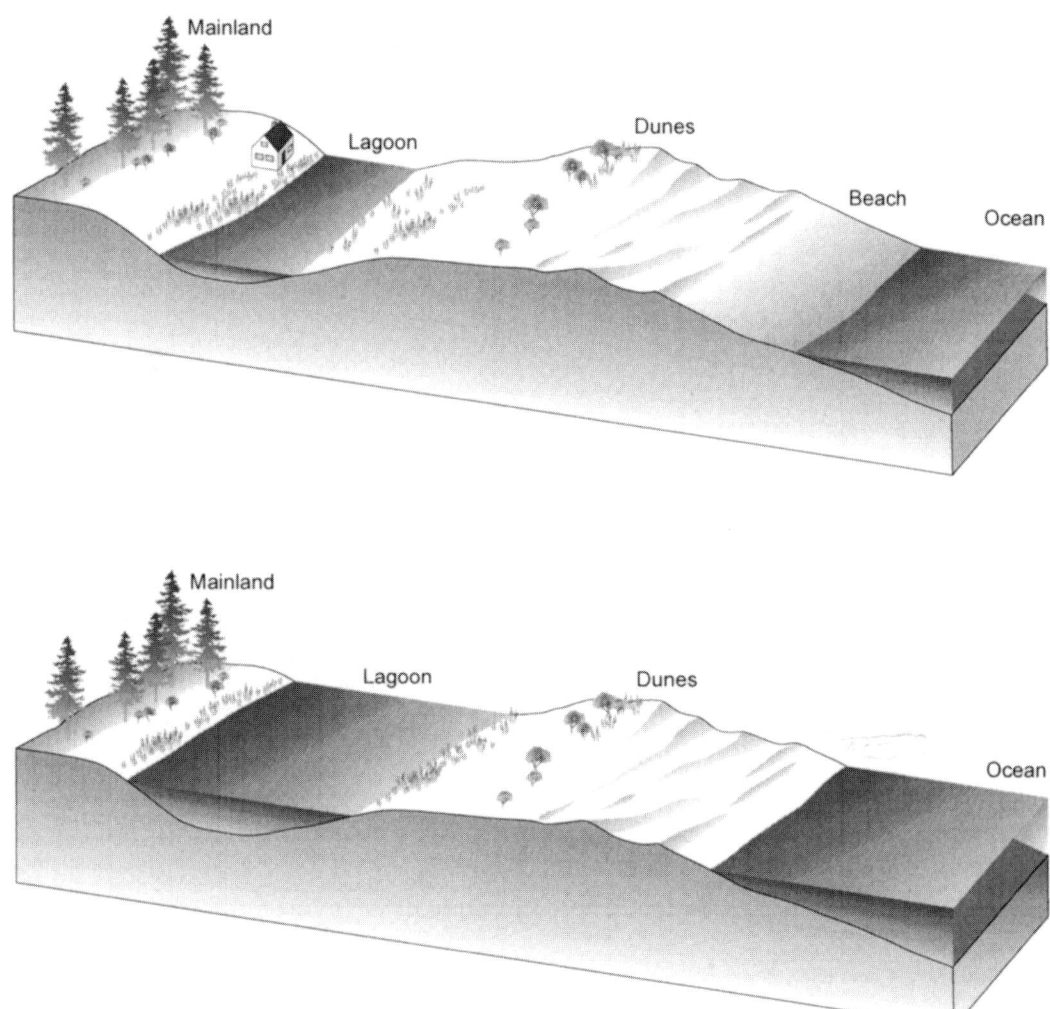

Figure N4 Effects of sea-level rise on barrier islands. Top, present sea level; bottom, future sea level.

Bibliography

Alexander, D., 1993. *Natural Disasters*. New York: Chapman & Hall.

Bell, 1999. *Geological Hazards: Their Assessment, Avoidance, and Mitigation*. London: E &FN Spon (Routledge).

Bird, E.C.F., 1985. *Coastline Changes: A Global Review*. Chichester: John Wiley & Sons.

Bird, E.C.F., 1996. *Beach Management*. Chichester: John Wiley & Sons,

Boesch, D.F., Josselyn, M.N., Mehta, A.J., Morris, J.T., Nuttle, W.K., Simenstad, C.A., and Swift, D.J.P., 1994. Scientific Assessment of Coastal Wetland Loss, Restoration and Management in Louisiana. *Journal of Coastal Research.* Special Issue 20.

Chagnon, S.A., and Chagnon, D., 1999. Record-high losses for weather disasters in the United States during the 1990s: how excessive and why? *Natural Hazards*, **18**: 287–300.

Coch, N.K., 1995. *Geohazards: Natural and Human*. Englewood Cliffs, NJ: Prentice Hall.

Culliton, T.J., 1998. *Population: Distribution, Density and Growth. Silver* Spring, MD: NOAA's State of the Coast Report.

Davis, R.E., and Dolan, R., 1993. Nor'easters. *American Science*, **81**: 428–439.

De Ronde, J.G., 1991. *Rising Waters: Impacts of the Greenhouse Effect for the Netherlands*. The Hague: Rijkswaterstaat.

Dolan, R., and Davis, R.E., 1992. An intensity scale for Atlantic coast northeast storms. *Journal of Coastal Research*, **8**: 840–853.

Dolan, R., and Davis, R.E., 1994. Coastal storm hazards. In Finkl, C.W., Jr. (ed.), *Coastal Hazards: Perception, Susceptibility and Mitigation. Journal of Coastal Research.* Special Issue No. 12, pp. 104–114.

Dolan, R., Anders, F., and Kimball, S., 1985. Coastal Erosion and Accretion. In *National Atlas*, Reston: US Geological Survey.

Douglas, B.C., Crowell, M., and Leatherman, S.P., 1998. Considerations for shoreline prediction. *Journal of Coastal Research*, **14**: 1025–1033.

Driscoll, N.W., Weissel, J.K., and Goff, J.A., 2000. Potential for large-scale submarine slope failure and tsunami generation along the US mid-Atlantic coast. *Geology*, **28**: 407–410.

Dugan, B., and Flemings, P.B., 2000. Overpressure and fluid flow in the New Jersey continental slope: implications for slope failure and cold seeps. *Science*, **288**: 289–291.

Gornitz, V., 1995. Monitoring sea level rise. *Climate Change*, **31**: 515–544

Griggs, G.B., 1994. California's coastal hazards. In Finkl, C.W., Jr. (ed.), *Coastal Hazards: Perception, Susceptibility and Mitigation. Journal of Coastal Research.* Special Issue No. 12, pp. 1–15.

Henderson-Sellers, A., Zhang, H., Berz, G., Emanuel, K., Gray, W., Landsea, C., Holland, G., Lighthill, J., Shieh, S-L., Webster, P., and McGuffie, K., 1998. Tropical cyclones and global climate change: a post-IPCC assessment. *Bulletin of American Meteorological Society*, **79**: 19–38.

IPCC, 2001a. Intergovernmental Panel on Climate Change. In Ding, Y., Grigg, D.J., Noguer, M., van der Linden P.J., Dai, X., Maskell, K., and Johnson, C.A. (eds). *Climate Change 2001: The Scientific Basis*. Cambridge: Cambridge University Press.

IPCC, 2001b. Intergovernmental Panel on Climate Change. In McCarthy, J.J., Canziani, O.F., Leary, N.A., Dokken, D.J., and White, K.S. (eds). *Climate Change 2001: Impacts, Adaptation, and Vulnerability*. Cambridge: Cambridge University Press.

Landsea, C.W., Pielke, R.A., Jr., Mestas-Nunez, A.M., and Knaff, J.A., 1999. Atlantic basin hurricanes: indices of climatic changes. *Climate Change*, **42**: 89–129.

Munich Re, 1993. *Winter Storms in Europe: Analysis of 1990 Losses and Future Loss Potential*. Munich: Munich Reinsurance Company.
Murty, T.S., and Flather, R.A., 1994. Impact of storm surges in the Bay of Bengal. In Finkl, C.W., Jr. (ed.), *Coastal Hazards: Perception, Susceptibility and Mitigation. Journal of Coastal Research*. Special Issue No. 12, pp. 149–161.
National Research Council, 1987. *Responding to Changes in Sea Level: Engineering Implications*. Washington: National Academy Press,
National Research Council, 1995. *Beach Nourishment and Protection*. Washington, DC: National Academy Press, 334p.
NOAA, 2001. *The Deadliest, Costliest, and Most Intense United States Hurricanes from 1900 to 2000 (and other frequently requested hurricane facts)*. NOAA Technical Memoris NWS TPC-1, J.D. Jarrell and C.W. Landsea. (http://www.aoml.noaa.gov/hrd/Landsea/deadly/index.html)
Penland, S., Ramsey K.E., McBride, R.A., Moslow, T.F., and Westphal, K.A., 1989. Relative sea level rise and subsidence in Louisiana and the Gulf of Mexico. *Louisiana Geological Survey Coastal Geology of Technical Report No. 3*, 65p.
Pirazzoli, P.A., 1991. Possible defenses against a sea-level rise in the Venice area, Italy. *Journal of Coastal Research*, **7**: 231–248.
Program for the Study of Developed Shorelines, 1999. *The U.S. Beach Nourishment Experience: New England, East Coast Barrier Islands, Gulf of Mexico, and Great Lakes Shorelines*. Durham: Duke University.
Satake, K., Shimazaki, K., Tsuji, Y., and Ueda, K., 1996. Time and size of a giant earthquake in Casadia inferred from Japanese tsunami records of January 1700. *Nature*, **379**: 246–249.
Soloviev, S.L., 1978. Tsunamis. In *The Assessment and Mitigation of Earthquake Risk*, Paris: UNESCO, 91–143.
Small, C., Gornitz, V., and Cohen, J.E., 2000. Coastal hazards and the global distribution of human population. *Environmental Geosciences*. **7**: 3–12.
Storlazzi, C.D., and Griggs, G.B., 2000. Influence of El Nino–Southern Oscillation (ENSO) on the evolution of central California's shoreline. *Geological Society of American Bulletin*, **112**: 236–249.
Tappin, D.R. *et al.*, 1999. Sediment slump likely caused 1998 Papua New Guinea tsunami. *EOS, Transactions of American Geophysical Union*, **80**(30): 329,334,340.
U.S. Army Corps of Engineers/FEMA/National Weather Service, 1995. Metro New York Hurricane Transportation Study. Interim Technical Data Report.
Valverde, H.R., Trembanis, A.C., and Pilkey, O.H., 1999. Summary of Beach Nourishment Episodes on the U.S. East Coast barrier islands. *Journal of Coastal Research*, **15**: 1100–1118.
Van der Vink, G. *et al.*, 1998. Why the United States is becoming more vulnerable to natural disasters. *EOS, Transactions of American Geophysical Union*, 79, 533, 537.
Zhang, K., Douglas, B.S., and Leatherman, S.P., 1997. East Coast strom surges provide unique climate record. *EOS, Transactions of American Geophysics Un.*, **78**: 389, 396–397.
Zhang, K., Douglas, B.S., and Leatherman, S.P., 2000. Twentieth-century storm activity along the U.S. East Coast. *Journal of Climate*, **13**: 1748–1761.

Cross-references

Barrier Islands
Beach Erosion
Changing Sea Levels
Coastal Changes, Gradual
Coastal Changes, Rapid
Coastline Changes
El Niño–Southern Oscillation (ENSO)
Global Warming (see Greenhouse Effect and Global Warming)
Sea-Level Rise, Effect
Storm Surge
Storms (see Meteorological Effects)
Tsunami

NAVIGATION STRUCTURES

Nations with access to oceans or to large seas, rivers, and lakes utilize the resource of water-borne transport for defense, commerce, and recreation. Structures that protect and promote boat and ship traffic are among the oldest and largest engineered works along the coast. Masters (1996) discusses coastal harbors constructed in the Mediterranean from about 4,000 years ago, including a Phoenician harbor with almost continuous use for the past 3,000 years. Franco (1996) describes the engineering and functioning of pre-Roman harbors starting from the first harbor of Alexandria, Egypt, built by the Minoans some 3,800 years ago, and he gives examples of Roman and Italian design of harbor structures, including contributions by Leonardo da Vinci.

Safe passage through harbors, inlets, and river mouths requires (1) sheltering of vessels from breaking waves that form the surf zone, (2) a relatively straight and permanent channel, and (3) a channel that does not unexpectedly fill with sediment (called sediment shoaling or simply "shoaling"). Engineering works that promote reliable navigation are called coastal navigation structures. The two major coastal navigation structures are jetties and breakwaters. This entry discusses the functioning of jetties and breakwaters, and their interaction with the coast.

Jetties

Jetties are typically placed at inlets and river mouths to maintain and promote navigability. A jetty is an engineered structure extending from the shore into a body of water and is usually constructed to serve three purposes. First, jetties partially or fully block waves, sheltering vessels as they pass through shallower water and the breaking waves of the surf zone to reach deepwater. In this way, jetties act as a breakwater, described in the next section. Second, jetties stabilize the location of the inlet, river, or harbor entrance, and in doing so they direct and confine the tidal and river current. By stabilizing the entrance location and directing the current, the location of the navigation channel tends to remain fixed in location and at least partially scoured by the current. Third, jetties block the movement of sediment transported alongshore (the littoral or shore drift), keeping the channel clear. Channel location and removal of shoals that impede navigation is accomplished by dredging.

If a single jetty stabilizes the navigation channel, it is built updrift of the entrance to block the predominant waves and littoral drift. A single jetty stabilizes the location of an inlet or river mouth by preventing its migration through spit extension from the updrift side. Typically, however, jetties are built in pairs that flank the harbor, inlet, or river mouth. As much as possible, jetties are aligned such that vessels exit directly into the incident waves. On a long and open coast, jetties are aligned perpendicular to the shore, because wave refraction causes wave crests to be parallel to shore. If the jetties protect a harbor or entrance near a headland or other area where the crests of the incident waves are not parallel to the coast, then one or more jetties may be constructed at an angle to the shore or have a seaward segment aligned at an angle to better block the waves.

Jetty length is usually such as to extend across the surf zone that can be present under navigable wave conditions. As a rule of thumb, the jetties are extended to a depth approximately that of the designed navigable depth. The elevation of the landward ends of jetties is typically above sea level to prevent sand moving along the coast from passing over the structure. The seaward end may gradually taper below the water and become submerged to allow currents and waves to pass around the structure without direct impact. However, the seaward ends of most jetties are above water and armored with larger stone to withstand direct impacts by waves. In the Unites States, jetties are typically constructed of stone that is cut according to certain specifications and fit in place. These structures are referred to as rubble mound jetties.

Because jetties intercept sand moving along the coast, provision is usually made to bypass the beach-quality sand or sediment to the downdrift side (Seabergh and Kraus, 2003). Sediment that falls into the channel is dredged and can be pumped to the downdrift nearshore or beach to prevent or reduce erosion. At longer jetties or where the wave climate is severe, beach-quality sediment may be dumped in the nearshore if it is infeasible or extremely expensive to place the material on the beach.

Another means of bypassing sand is by construction of a weir jetty. A weir is a purposeful low section in a jetty located near the shore. A typical weir is about 300 m long and with elevation about mean sea level or lower. A weir section is always designed together with a deposition basin that collects sediment passing over the weir before it reaches the navigation channel (Seabergh, 1983; Weggel, 1983). Sand can then be pumped from the relatively sheltered area of the weir to the downdrift beach.

Many large jetties have been in place for more than a century. As a result, they may induce a regional coastal response, similar to the response of a beach to a natural headland, producing updrift accretion and progradation of the beach, and downdrift erosion and shore recession. Construction or extension of jetties will also tend to move ebb-tidal deltas further offshore from their natural position (Pope, 1991), with portions of the original shoal that is located to the side of the narrow ebb current tending to return to shore.

Breakwaters

A breakwater is an engineered offshore structure that protects a harbor, anchorage, beach, or shore area by creating a sheltered region of reduced wave height. Breakwaters are always sufficiently high to afford protection against storm waves, and they are typically constructed parallel to the shore. A general principle is to align them to create the greatest shadow zone for either the direction of the predominant waves or the direction of the larger incident waves. Breakwaters must withstand direct impact of waves and are constructed of large stone or pre-cast armor units. Because breakwaters tend to be long and high, sediment moving along the beach can collect behind the structure.

In the United States, breakwaters are typically constructed as rubble stone mounds or with pre-cast concrete armor units that come in various shapes. In some countries, especially those where large stone is unavailable, caisson breakwaters are common. Caissons are pre-cast hollow concrete containers that are towed on site and then sunk in place by filling them with sand and small stone.

In coastal terminology, a mole is a massive solid-fill protective structure extending from the shore into deeper water, formed of masonry and earth or large stones, and serving as a breakwater or a pier. A mole is not designed to stabilize an inlet or river mouth and typically functions as a breakwater and part of harbor infrastructure.

Nicholas C. Kraus

Bibliography

Franco, L., 1996. History of coastal engineering in Italy. In Kraus, N.C. (ed.), *History and Heritage of Coastal Engineering*. New York: ASCE Press, pp. 275–335.

Masters, P.A., 1996. Paleocoastlines, ancient harbors, and marine archeology. *Shore & Beach*, **64**(3): 8–17.

Pope, J., 1991. Ebb delta and shoreline response to inlet stabilization, examples from the southeast Atlantic coast. *Proceedings Coastal Zone '91*, American Society of Civil Engineers, pp. 643–654.

Seabergh, W.C., 1983. Physical model study of weir jetty design. *Proceedings Coastal Structures '83*, American Society of Civil Engineers, pp. 876–893.

Seabergh, W.C., and Kraus, N.C., 2003. Progress in management of sediment bypassing at coastal inlets: Natural bypassing, weir jetties, jetty spurs, and engineering aids in design. *Coastal Engineering Journal*, **45**(4): 533–563.

Weggel, J.R., 1983. The design of weir sand by-passing systems. *Proceedings Coastal Structures '83*, American Society of Civil Engineers, pp. 860–875.

Cross-references

Bypassing at Littoral Drift Barriers
Dikes
Engineering Applications of Coastal Geomorphology
History, Coastal Protection
Longshore Sediment Transport
Shore Protection Structures

NEARSHORE GEOMORPHOLOGICAL MAPPING

Introduction

Widespread mapping of seabed topography was made possible through the introduction of seismic methods during World War II. Methods have improved dramatically since then and large surface areas of the deep-sea floor and continental shelf are now routinely mapped as, for example, in the Geological Long-Range Inclined Asdic (GLORIA) project based on long-range sidescan sonar surveys used by the US Geological Survey to map large segments of the Exclusive Economic Zone (EEZ) at depths generally greater than 400 m. The digital system, designed at the Institute of Oceanographic Sciences (now the Southampton Oceanography Center, Challenger Division, United Kingdom), was specifically designed to map the morphology and texture of seafloor features in the deep ocean. The GLORIA imagery, for example, displays a plethora of subsea geologic and geomorphological features such as volcanic edifices, fault scarps, channels, levees, slump scars, and crustal lineaments. Textural and tonal differences in the digital imagery show large sediment bedforms and varying sediment types. Ironically, more information has been collected from the outer continental shelf and deep-sea floor than from shallow water close to shore.

The nearshore zone is often neglected due to the difficulty of working in this technically hostile environment, which is characterized by spatially and temporally dynamic biophysical conditions of high-energy coastal ecosystems, irregular and shallow depths, variable and dangerous surf conditions, and strong currents. There are thus many dangers associated with marine survey in shallow waters, the surf zone being particularly notorious for both men and equipment. With the advent of modern high-resolution satellite imagery (e.g., SPOT, IKONOS), airborne sensor platforms such as LIDAR (Light Detection and Ranging), and development of digital aerial photography that can be georectified on the fly, mapping nearshore submarine topography becomes not only feasible but also practical for multipurpose applications. Passive satellite and photographic sensors that detect bottom reflected radiation restrict the method to clear, non-turbid Class II waters. Turbid nearshore waters along muddy coasts are unsuitable for classification of bottom types using satellite imagery and aerial photography. Muddy coasts, which are commonly associated with tropical mangrove systems and temperate salt marshes, are estimated to occupy about 75% of the world's coastline between 25°N and 25°S Latitudes (Fleming, 2002). These regions and other muddy coasts mainly associated with large deltaic environments (Wang and Healy, 2002) are generally excluded from satellite and airborne imagery. Coral reefs, sandy beach, and rock platform environments present more favorable settings where nearshore turbidity is minimal for at least part of the year. Under clear water conditions, maximum depth penetration is at least 30–40 m and so mapping can proceed in a shore-parallel swath of variable distance offshore, depending on the shoreface gradient.

Active remote sensing platforms such as airborne LIDAR are able to detect changes in bathymetry under a variety of conditions where the nearshore zone may be quite turbid. These developing methods, which allow rapid acquisition of topographic data, provide enormous amounts of information in the form of digital elevation points (laser repetition rates of 10,000 pulses per second can, for example, provide elevations of surface points with a nominal spacing of 2 m; RMS errors of less than 10 cm can be achieved for elevation points; overlapping lines can be flown to minimize non-scanned coastal segments and increase point density). These methods enable large-scale quantitative mapping of coastal features and are more cost effective than conventional surveying techniques. High-accuracy airborne LIDAR mapping, based on recent advances in Global Positioning System (GPS), Inertial Measurement Unit (IMU), laser ranging, and microcomputers has been applied to wave, beach, sea cliff, coastline position, and nearshore bathymetric surveys (e.g., Hwang *et al.*, 1998; Brock *et al.*, 1999; Krabill *et al.*, 2000; Stockdon *et al.*, 2002). The SHOALS system (a combined bathymetric/topographic LIDAR data source based on the acronym: Scanning Hydrographic Operational Airborne Lidar Survey) is another similar application of scanning laser mapping in the coastal zone that has found particular application in the vicinity of inlets and passes (e.g., Irish and Lillycrop, 1997, 1999). A 13-km^2-test area in Florida Bay, for example, was surveyed in just 12 h using SHOALS to demonstrate that airborne LIDAR bathymetric technology can be a valuable and cost effective tool for surveying large shallow water areas (Parson *et al.*, 1997). Complex morphologic features identified in Florida Bay included extensive shallow water networks of mud banks, cuts, and basins, among other features. Instead of using contours or isobaths to show terrain variability, modern computer displays associated with these laser-mapping systems often employ digital color ramps that gradationally distinguish one elevation from another by variation in colors. The result is a colorful map that shows transitional gradations in relief. Additionally, a technique of using false "sun illumination" can highlight artifacts, error sources, and features that are not always evident on color-coded bathymetry (e.g., Hogarth, 2002). These maps convey maximum information when they are combined with morphologic units that are interpreted and sea-truthed (i.e., based on information derived from visual observation, grab samples, jet probes, vibracores, etc.) from satellite or airborne imagery. Still other digital mapping procedures that are proving to be extremely useful in coastal areas are those that can now merge land-based topographic with marine-based bathymetric data into a seamless digital elevation model (DEM) (e.g., Milbert and Hess, 2001; Gesch and Wilson, 2002). The advantages of integrating USGS topographic data and NOAA hydrographic data in coastal DEM areas are

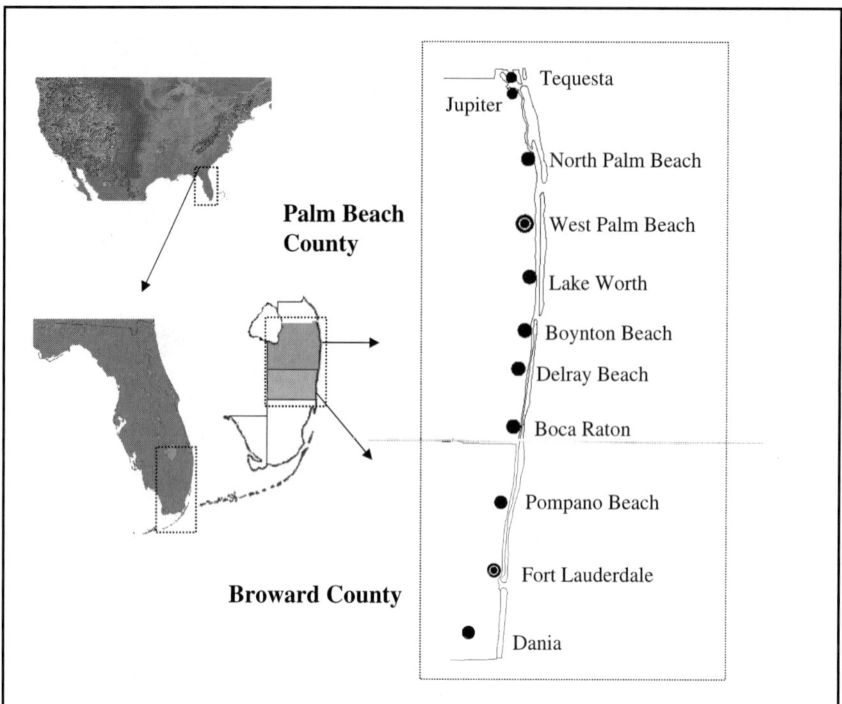

Figure N5 Location diagram showing the location of the Florida peninsula within the conterminous United States and the general Florida study area in Palm Beach and Broward counties along the southeast coast, as shown in the callout enlargements. The cities of West Palm Beach and Fort Lauderdale are near the center of their respective coastal counties. This 109-km stretch of subtropical coast is covered by 455 vertical, stereo-paired, digital, color, and georectified aerial photographs.

recognized as essential components of key baseline geospatial datasets for geologic, biologic, and hydrologic studies in the vicinities of tidal inlets and passes, lagoons, and estuaries. Highly accurate digital terrain models (DTMs), derived using high-accuracy photogrammetric techniques provide spatial richness that supports detailed study of coastal dune and beach processes (e.g., Judge and Overton, 2001).

Although high-resolution surveys in shallow waters are based on a wide range of sensor equipment and platforms (e.g., Gorman *et al.*, 1998; Byrne *et al.*, 2002), there is still a place for interpretation of aerial photography, especially for very detailed close order work in inshore areas. Aerial photography has a long history of application to document coastal features and of use in resolution of coastal problems such as flooding and erosion. Indeed, many coastal features are seen in some of the earliest commercially available aerial scenes as oblique images, as shown in this 1920s image of a long barrier spit in Broward County, Florida, near the present city of Fort Lauderdale (Figures N5 and N6). The early imagery in this case is extremely important to studies of coastal geomorphology because without this information, it is likely that the true coastal barriers along this coast would have been overlooked and ignored. Studies by Finkl (1993), for example, emphasized the crucial role of aerial photography for interpreting coastal morphology. Dietz (1947), for example, was among the first researchers to present a comprehensive analysis of the importance of aerial photographs in the study of shore features and processes. He pointed out the scientific value of aerial photographs in studying underwater features. A cogent and informative summary of aerial photography in the coastal zone is provided by El-Ashry (1977), who collected a range of examples covering multiple applications in different parts of the world. A more recent example of comprehensive use of aerial photographs may be found in Owens (1994) where aerial surveys covered over 40,000 km of Canadian coastline to assist in the compilation of oil spill response manuals. Oblique and vertical aerial photographs are used throughout the manual to illustrate different types of coastal environments, morphologic features, and shore processes. Other examples of new technologies, advanced techniques, and new applications (not limited to aerial photography) are found in a series of international conferences on remote sensing for marine and coastal environments sponsored by ERIM, a subsidiary of Veridian, Inc. (PO Box 134008, Ann Arbor, MI 48113). Vertical, stereo-paired aerial photographs find many applications, singly or in combination with other techniques, in coastal studies not only to relatively long-term shore variation in coastal planforms or coastline positions (e.g., Smith and Zarillo, 1990; Shoshany and Degani, 1992; Thieler and Danforth, 1994; Moore, 2000), but also in studies of short-term coastline changes that use fully automated techniques for determining the high-water line (e.g., Shoshany and Degani, 1992). Jiménez *et al.* (1997), for example, used temporally close order aerial photography (a series of seven aerial surveys over four months) to analyze highly dynamic coastal features in very flat, microtidal deltaic areas in Spain. Acquisition systems that collected temporal data to monitor the behavior of a dissipative multiple bar system in a nearshore zone near Terschelling, The Netherlands, were based on vertical aerial photogrammetry and echo sounding (Ruessink and Kroon, 1994) as were similar efforts that monitored nearshore submarine morphology at Duck, North Carolina (Guan-Hong Lee and Birkemeier, 1993), and along the Wanganui coast, New Zealand (Shand *et al.*, 1999). Alam *et al.* (1999), by way of another example, used high-resolution vertical black and white aerial photographs (1 : 20,000), using a zoom transferscope, to prepare a preliminary terrestrial geomorphological map of Cox's Bazar Coastal Plain, southeast Bangladesh.

Although the subject matter of nearshore geomorphological mapping is global in extent, this discussion necessarily focuses on the use of digital aerial photography for submarine geomorphological mapping along the nearshore zone of southeastern Florida. The example elucidated here is representative of many coastal areas that contain sandy beaches and rocky shores that front seabeds characterized by hardgrounds and unconsolidated sediments. The methodology is thus not restricted to tropical and subtropical carbonate environments, but applies to mid- and high-latitude coasts with low nearshore turbidity.

The nearshore environment of southeast Florida

Some of the more obvious topographic features of the inner continental shelf along the southeast coast of Florida were appreciated several decades ago when they were mapped by the US Army Corps of Engineers as hardgrounds and inter-reefal sand flats (Duane and Meisburger, 1969). The shore-parallel "reef" (comprised by rock and coral-algal components) system, composed of inshore exposure of the Pleistocene Anastasia Formation (a cemented, quartzitic, molluscan grainstone that formed in beach and shallow-water nearshore environments) (Stauble and McNeil, 1985; Davis, 1997) and coral-algal reef tracts (referred to as the Florida Reef Tract) (Lidz *et al.*, 1997),

Figure N6 Bay mouth bar and barrier spit near Fort Lauderdale, Florida, ca. 1925 (looking from east to west). The shallow open freshwater body on the right side of this oblique aerial photograph is Lake Mabel, now converted into the Port Everglades Turning Basin. A 9-km long barrier spit, situated approximately 100 m offshore, continued alongshore from Fort Lauderdale to Dania passing in front of the bay mouth bar. Inlet cutting in Fort Lauderdale beheaded the updrift part of the barrier spit turning it into a barrier island that migrated shoreward over a three to five year period. During the migratory phase or barrier island rollover, approximately 6–10 $\times$ 10^6 m^3 of sediment was moved shoreward (Finkl, 1993). Shown here is welding of the barrier spit to the bay mouth bar. The remaining southern (left side of photo) portion of New River Sound opens to a temporary natural inlet. By 1930, the entire barrier island (formerly a barrier spit) was welded to the mainland. When the Intracoastal Waterway (ICWW) was dredged through the tidal marsh (left center of photo), the ICWW—beach tract of land was mistaken for a barrier island and is still so designated today. (Partial scanned image from an original 1925 print by Fairchild Aerial Surveys, Inc., Long Island City, New York.)

increases in depth offshore as a giant staircase. These parabathic reef tracts extend southwards into the Florida Keys and represent approximate positions of paleocoastlines extensionally offshore and vertically within particular tracts as prior sea-level stands were revisited through time (Finkl, 1993, 1994). Sedimentary troughs that contain admixtures of clean, free-running sands, discontinuous lenses or stringers of silts and clays separate the reef tracts, or carbonate rubble accumulations deposited in association with paleo-inlets that cut through the reef tracts.

Most of the Holocene coast was characterized by extensive spits that extended downdrift and alongshore for tens of kilometers (cf. Figures N5, N6) (Finkl, 1993, 1994). The spits, which rose several meters above mean sea level, were stabilized by herbaceous and phanerophytic vegetation and protected shoreward sounds leading to coastal bays and estuaries. Inlet cutting and stabilization in the early 1900s initiated the demise of barrier spits (cf. Figures N5, N6) and islands by jetties and canalization of coastal wetlands, as described by Finkl (1993), to form the present shore. Bedrock of the Anastasia Formation is exposed onshore or buried at a shallow depth below present-day beaches (Figure N7). Most berms contain beach sands less than 2 m in thickness so that some beaches are stripped of sediment during storms to expose the underlying bedrock during part of the year (Finkl, 1994). Dunes fronting back beaches were commonly leveled for high-rise development so that today, incipient dunes only develop where buildings or infrastructure is setback from the shore. Seawalls that preclude dune formation back many beaches along this developed shore.

Morphologic features

The methodology employed in the identification of coastal geomorphic features was based on interpretation of digital, color aerial photographs at an acquisition scale of 1 : 3,900. In addition to on-screen digitizing of black and white images (derived from scanned color prints to a resolution of 300 dpi), the stereo-paired color prints were also utilized to enhance visual inspection of details and to resolve complicated photo patterns. Most photographs showed about 25% land and 75% ocean to provide dune-beach features and assist with location on the ground. Prior to mapping, two hundred and fifty-seven aerial photographs (23 $\times$ 23 cm color photographic prints) for the Palm Beach County shore were visually perused to determine the range of coastal features present in the study area. Compilation of a summary list of geomorphic features was organized as a general or organizational classification scheme with major categories for beach, bar and trough, dunes, rock outcrops, seabed morphosedimentary features, geologic structures, suspended sediments, and engineered structures. The morphologic forms and engineered structures that were noted are listed in Table N4.

Identification of the various morphologic features was based on direct recognition image interpretation strategy, but also employed field observations, information from grab samples and drill core, interpretation by inference, and probabilistic interpretation. Campbell (1996) provides a cogent summary of different image interpretation strategies and details disciplined procedures that enable the interpreter to relate geographic patterns on the ground (and under water) to their appearance in the image. Subaerial features such as dunes and infrastructure are identified by discretely different visual information. Because much of the native dune system has been destroyed by construction, many areas are interpreted through *a priori* knowledge. Visible dune sands and dune vegetation in other areas permitted application of a direct recognition strategy. Features in the backshore regions were identified by field observation and direct recognition. Beach berms and wrack lines were easily identified on the photographs due to distinct tonal variations. Interpretation of foreshore features was somewhat more difficult because this zone is partially submerged and acted upon by swash. The beachface is a predominant feature of the foreshore and can be conveniently divided into the upper (light-toned) and lower (dark-toned) beachface. Interpretation of geomorphic features of the inshore is sometimes hindered by complexity of spatial patterns. Sharp tonal contrasts facilitate interpretation of bars and troughs, which are the main sedimentary components of the inshore zones. Hardground features are identified by their dark tones, regular structural (lineament) patterns, and coarse texture of the rock surfaces. Many of the offshore features are identified using probabilistic interpretation as well as direct recognition. *A priori* knowledge of the existence of coral reefs (as mapped by Duane and Meisburger, 1969) in the offshore zone increases the probability of seeing reefs in the aerial photographs. Smooth tonal variations usually permitted differentiation of sandy bottom features from rock outcrops that are associated with disrupted blocky patterns and coarse textures.

Although a variety of interpretive techniques are available to assist in the identification of landforms in the coastal zone, it is necessary that the persons preparing the morphologic map have knowledge of coastal features prior to embarking on a project such as this. Preparation of a

Figure N7 Erosion of perched beach and dunes in Boca Raton, central Palm Beach County, Florida, showing exposure of bedrock. (A) Remains of a subaerial beach that has been stripped off the underlying rock platform, cut into the Anastasia Formation, by an Atlantic Northeaster winter storm. Wave attack from the northeast tends to erode beaches and deposit the sediments offshore. (B) Rocky platform that underlies the perched beach shown in (A), upper part of photo center. (C) Reentrant surge channel that has cut across the berm and is shown here nipping foredune sediments. The surge channel, which follows a cross-shore fracture in the bedrock, brings turbulent water to the backshore. Note the salient environments along this eroded shore where perched beaches (A) with berm thicknesses generally less than 2 m are stripped away by storm surge to expose rocky platforms (B) that lie at about mean sea level and how ingression is facilitated by structural weakness in the bedrock (C). (Photo: Lindino Benedet, Florida Atlantic University, Boca Raton, Florida; photo taken in Winter 2001.)

comprehensive legend before mapping (cf. Table N4) assists in the rapid completion of the project. Modification of some morphologic units can take place as mapping proceeds, however.

The preceding paragraphs briefly outline some of the salient techniques for interpreting coastal morphologic features from vertical, stereo-paired aerial photography. Interpretation assumes familiarity with landscapes and seascapes, including polygenetic landform sequences that are so common along the world's coastline. There is no substitute for careful fieldwork, which must at some stage precede interpretive efforts that attempt to deduce what is on the ground from aerial imagery. Of the numerous reference works that show in pictorial or graphic formats physical features of the coast, the following are noteworthy for their wide scope of topical coverage, geographic focus, clarity of presentation, or detailed orientation to specific classes of morphologic features: Bird (1976), Bradley (1958), Emery and Kuhn (1982), Forbes and Syvitski (1994), Guilcher (1988), Hopley (1988), King (1959), McGill (1958), Nordstrom (1990), Nordstrom *et al.* (1990), Pethick (1984), Russell (1967), Schwartz (1982), Short (1993), Trenhaile (1987), Zenkovich (1967). In contrast to these general reference works are those that focus specifically on the southeast Florida coastal zone viz. White (1970), Duane and Meisburger (1969), Finkl and DaPrato (1993), Finkl (1993, 1994), Finkl and Esteves (1997), Finkl and Bruun (1998), Brown (1998), Khalil (1999), Warner (1999), Finkl and Khalil (2000), Finkl and Benedet (2002), etc. Prager and Halley (1997), while conducting a survey of bottom types in Florida Bay using sequential aerial photography, studied the distribution of sea grass and its influence on wave energy impacting mud banks. Bottom types in this shallow, low-energy embayment that are identified by Prager and Halley (1997) include: open sand, hard bottom, open mud, mud bank suite, mixed bottom suite, sparse sea grass, intermediate sea grass, and dense sea grass. The subtropical, open ocean coast of southeast Florida provides a wider range of morphologic types than might at first be anticipated for a coast that is commonly regarded as a sediment-rich, sandy beach shore. The frequent occurrence of rock forms, both subaerial and submarine, lends diversity to an otherwise potentially monotonous stretch of shore. Examples of morphologic units occurring in this subtropical zone of the western North Atlantic Ocean are first described in terms of the hard rock units comprised by coral and algal reef materials and lithified coquina and sandy deposits. These hard rock morpho-units are emphasized because antecedent topography influences present coastal configuration in Florida (e.g., Evans *et al.*, 1985; Kelletat, 1989; Finkl, 1993, 1994; Finkl and DaPrato, 1993; Lidz *et al.*, 1997), and as described for other areas, for example, by Sanlaville *et al.* (1997), Pilkey (1998) and Dillenburg *et al.* (2000). The occurrence and distribution of sedimentary geomorphic features is then considered, often in relation to the geologic framework of these coastal environments in southeast Florida.

Reef tract and hardground morphologic features

The term *hard shore* refers to the occurrence of sand, gravel, cobbles, boulders, or bedrock as opposed to the antonym *soft shore*, which identifies shore composed of peat, muck, mud, soft marl, or marsh vegetation (Bates and Jackson, 1979). *Hardground*, on the other hand, denotes a zone on the seafloor, usually a few centimeters thick, where the sediment is lithified to form a hardened surface that is often encrusted, casehardened, or solution-ridden. Colloquial usage in the Florida environment restricts hard shore and hardground to subaerial and submarine outcrops of bedrock or lithified sediments. The term hardground is further unrestricted in definition as it is used in reference to coral reefs, coral-algal reefs, and exposure of bedrock such as the Anastasia Formation. When corals or coral-algal reefs are known, explicit identifying terminology is applied but often, especially in remote sensing analyses, differentiation of specific units is not possible without groundtruthing and hence the general term is applied. The term *reef tract* specifically refers to coral and coral-algal reefs whereas hardground refers to any hard surface, regardless of composition or origin.

Because coral reefs and coral-algal reefs, which comprise the Florida Reef Tract, and hardgrounds appear as many distinct forms on the aerial photographs, they are divided into different map units as follows: abrasion platforms, fluted inshore/offshore rock reefs, parabathic hardground stringer, coral-algal rock reef, and sandy bottom with rock outcrop. Linear disjunctive segments that parallel the coast make up what are locally referred to as the first, second, and third reefs that collectively characterize this part of the Florida Reef Tract. Sandy deposits cover parts of the coral reef and inshore Anastasia Formation.

Table N4 Classification of salient coastal morphological forms and engineering structures identified from large-scale aerial photographs for the southeast coast of Florida, centered on Palm Beach and Broward counties (cf. Figure 1)

Level 1[a]	Level 2[b]	Level 3[c]
Beach	Berm	
	Cusp	
	Beachface	Lower beachface
		Upper beachface
		Sub-beach step
Bar-and-trough topography	Channels in bar	
	Crescentic bar, trough	
	Longshore bar	Welded
		Transverse
	Rip channel	
	Low tide terrace	
	Trough	Beachface trough
		Shoreface trough
		Crenulated trough
		Infilled nearshore trough
Dunes	Undifferentiated dunes	
Rock outcrop (intertidal and submarine)	Abrasion platform	
	Coral-algal reef (reef tract)—Hardground (rock reef) [Complex unit]	Beachface outcrop
		Outer coral-algal rock reef
		Fluted rock reef
		Fluted offshore rock reef
		Parabathic hardground stringer
Sedimentary morphological features	Sandy bottom	Light-toned running sand
		Dark-toned running sand
		With rock outcrop
		Featureless sandy bottom
		Structurally controlled sandflat
	Tidal delta (ebb, flood shoal)	
Suspended sediments	Turbidity plume	
Engineering structures	Dredged inlet	
	Seawall	
	Rock revetment	
	Groin	
	Pier	
	Submerged breakwater	
	Jetty	

[a] Level 1 groups the most general morphologic forms and engineering works.
[b] Level 2 categorizes the broad forms of Level 1 into specific types.
[c] Level 3 incorporates specific types of Level 2 that have distinguishing characteristics that permit further differentiation.

Fluted inshore/offshore rock reefs

Several morphologically distinctive types of hardground occur in southeast Florida. Some coastal segments are characterized by *patterned hardgrounds* (rock reefs) with somewhat regularly spaced distinct cross-shore channels, referred to here as *flutes* (Figure N8). The flutes average about 1–10 m in width and their length extends the width of the outcrops that can range upwards of 30–50 m. The flutes are clearly evident in the photograph (left) but the unit is mapped (right) without detailing the position of flutes. Some of the more prominent examples of fluted rock reefs are contiguous for more than 400 m in a shore-parallel direction. The flutes occurs in outcrops of the Anastasia Formation that are often cracked or fractured into large blocks that make up a general reticulate pattern due to increased differential erosion of these zones of weakness. The small diabathic channels or flutes follow the initial or pre-existing structural depressions that were later subaerially etched into the carbonate rock surface prior to drowning by the Flandrian sea-level rise.

Abrasion platforms

Abrasion platforms (Figure N8), a type of shore platform (e.g., Pethick, 1984; Trenhaile, 1987; Sunamura, 1992) that occur in the intertidal and inshore zone, are developed on outcrops of the Anastasia Formation, as described by Duane and Meisburger (1969) for the coast of southeast Florida. The platforms, which show signs of weathering, solution, and wave-induced erosion, are generally quite narrow (less than 50–100 m) and may extend intermittently alongshore for distances of 1,500 m or more. These calcareous platforms also occur as blocky, jagged outcrops arranged in sets along the shore and are separated by sandy flats, as shown in Figure N8. Structural patterns (left photograph) in the Pleistocene bedrock show clear shore- parallel lineations with sand partially filling chemically weathered depressions that lie below more resistant facies that have been abraded by wave action to produce concordance of ridge levels.

Parabathic hardground stringer

Another common type of hardground that occurs in this region is referred to as a parabathic hardground stringer (Figure N9). These narrowly elongate morphological features are referred to as *stringers* because they tend to occur as long, ridged rock outcrops surround by sandy seafloor. These inshore rock outcrops are narrow parabathic forms that are usually less then 7–10 m wide but may be more than 200 m long. These stringers usually occur in groups and are separated by sandy deposits. These topographically positive seabed features represent more chemically resistant and structurally competent facies of the Anastasia Formation or coral-algal reef. The stringer shown here (dark lines in center of the left photograph) trends in a general north–south direction along the axis of strike of the now-lithified banks (including coquina beds) of the Anastasia Formation and is partly overlain by coral-algal rock reef (bottom center of photograph, left, and geomorphological map, right). Extensions of buried hardground that are emergent above surrounding sand formations, such as bars, often form hardground stringers.

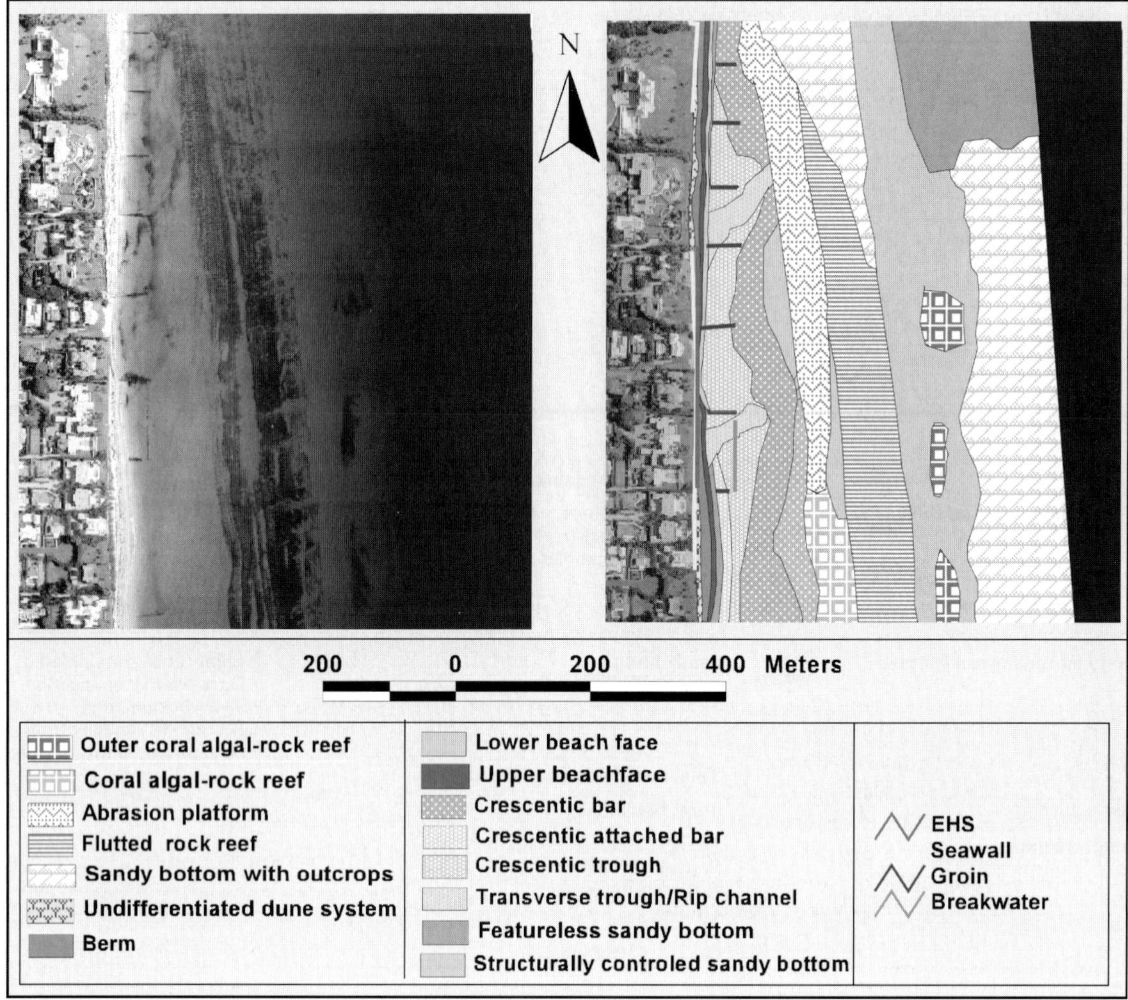

Figure N8 Rock outcrops and sandy bottoms along the southeast coast of Florida in the Town of Palm Beach, Palm Beach County. The complexity of geomorphic units is seen in parabathic and diabathic spatial distribution patterns, that is, differentiation of bottom types alongshore and across the shoreface. The occurrence of rock units on the seafloor is a most striking feature that is denoted by several categories of hardground: coral-algal reef, abrasion platforms, fluted rock reefs. The map of geomorphic units (right), interpreted from the aerial photograph (left), identifies complex features associated with the landward-most portions of the Florida Reef Tract, first and second reef systems. Isolated blocks of coral-algal rock associated with the 2nd reef system lie about 400 m offshore and poke through a thin veneer of sand that overlies slightly lower levels of the same rock unit (referred to as structurally controlled sandy bottom). The main features of the 1st reef are a beveled abrasion platform, rough carbonate rock reef of the Anastasia Formation, and a seaward portion identified as fluted rock reef that contains diabathic channels. Sedimentary features, composed of mixed siliciclastics and biogenic carbonate sands, include sandy bottoms and bar-and-trough topography. The rhythmic wave pattern of a large crescentic bar-and-trough system migrates alongshore, landward of the 1st reef. Rip current channels pass through the crescentic trough in a general NE direction. The beach is subdivided into berm, upper beachface, and lower beachface. A groin field and an erosional hot spot (zone of accelerated shore retreat) are additionally identified on the map.

Coral-algal rock reef

The northern extension of the Florida Reef Tract (see Lidz *et al.*, 1997), which extends northward from the Florida Keys, contains a seriously degraded coral reef environment that has been stressed by a variety of factors (Lidz and Hallock, 2000) to the point where many coral communities have died and their skeletons now serve as foundations for algal growth. The algal encrustations form a kind of biogenic carbonate rock that is mixed with surviving corals; the complex morphological structure (coral reef plus algal encrustations) is thus referred to as a *coral-algal rock reef* in an effort to identify its true status as a conditionally stable but degrading coral reef system. Bottom morphologies associated with this unit are thus complex intergrades that are undifferentiated due to interpretive difficulties to consistently separate true corals from coral-algal units. Outcrops of the Anastasia Formation sometimes occur in close proximity to these biogenic units (see Figures N8 and N9), further complicating the interpretive process. Varieties of these parabathic forms, which are mapped for simplicity as one unit, are visually contiguous as a distinct bottom type for distances ranging up to 4 or 5 km alongshore. Bottom sands separate successional coral reef systems (viz. the first, second, and third reefs) and also reefs from outcrops of the Anastasia Formation.

Sandy bottom with rock outcrop

This is a complex, intergraded bottom type (see Figures N8, N9, and N10) with spatial patterns that are too intricate or relationally detailed to warrant separation at the mapping scale. The unit is essentially a variety of hardground bottom type with a veneer of sand that ranges in thickness from a few millimeters to several tens of centimeters in thickness. The sandy veneer is so thin that structural features of the underlying rock are still visible, especially where slightly positive relief features emerge through the sand cover. Unconsolidated, free-running sands that overlie portions of the reef tract or intermittent rock outcrops characterize these bottom-type features. Because corals that have been buried by sediment are now dead, they are simply referred to as carbonate rock.

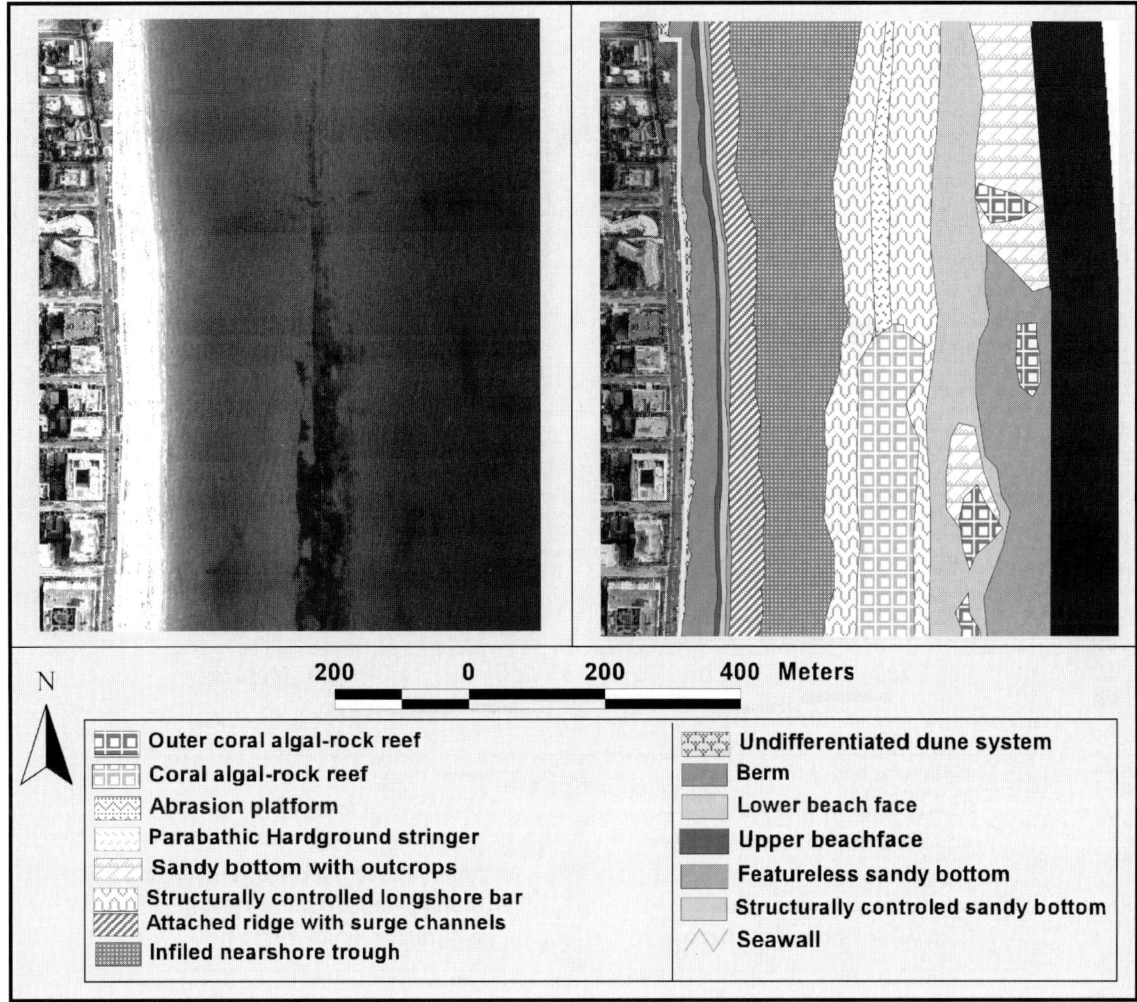

Figure N9 Sandy shore with submarine rock outcrops along the southeast coast of Florida a few kilometers south of the Town of Palm Beach, Palm Beach County. The aerial photograph (left) and interpreted geomorphological map (right) are companioned here to show salient characteristics of this shore. This sandy coastal segment is characterized by beach (berm, upper and lower beachface), bars and troughs, parabathic hardgrounds (coral-algal and rock reef), and sandy bottoms. Bedrock, which is close to the surface of the seabed here, is a component that is noted in composite units viz. sandy bottom with outcrops (which shallowly flanks coral-algal rock reefs) and structurally controlled sandy bottom (which shallowly overlies bedrock). The dominant parabathic geomorphic distribution patterns reflect shore-parallel rock outcrops on the seafloor. The parabathic attached ridge (welded to the lower beachface) with diabathic surge channels merges seaward with a partially infilled nearshore trough that together makes up the sandy nearshore environment. Structural control of sedimentary features and geomorphic patterns is evident in the vicinity of the coral-algal rock reef materials of the 1st Reef System in the Florida Reef Tract, the landward boundary of which lies about 300 m offshore. Note the northward extension of the hardground as a parabathic hardground stringer (about 500 m in length) in the upper central portion of the map (right). The black area on the seaward-most portion of the geomorphological map is uninterrupted due to depth constraints.

Rock outcrops occur primarily as parabathic features, but diabathic patterns with sediment between the outcrops can be found in some areas. Sedimentary infills immediately adjacent to the rock outcrops are relatively thin, usually less than 1 m thick. Sandy bottom with rock outcrop may occur in almost any geographic location along the shore as a reminder that bedrock or buried coral reef is never far away.

Sediment-composed morphologic features

A large part of the inner continental shelf, loosely analogous to what some researchers refer to as the shoreface (e.g., Wright, 1995), contains sedimentary covers that are surficially modified by currents to form distinct morphologic units. The various kinds of morphologic features, composed of silicilastic and carbonate sediments, occur in beach sands or as bars and troughs. Under conditions of relatively stable or slowly rising sea level, typical of the last 6,000 years, the shoreface may approach conditions of steady state equilibrium (as defined by Schumm, 1991). On shorter time frames, instantaneous or event timescales, the upper shoreface is subject to continual change with relaxation times on the order of days (Larson and Kraus, 1995). As Cowell *et al.* (1999) point out, the upper shoreface chases a dynamic equilibrium (Schumm, 1991) as boundary conditions vary stochastically, without ever quite making it. This section of the southeast Florida coast is no exception as beaches and inshore–offshore sedimentary features are seen to change over the long run, from winter to summer wave climates, to occurrences of winter northeasters and summer tropical storms. This morphologically dynamic zone features a range of sedimentary morphodynamic subaerial and submarine types that are identified from aerial photography in terms of coastal dunes, beaches, bar and trough systems, deltas, and sandy bottoms.

Coastal dunes

Dunes along this shore are generally compromised by construction for coastal infrastructure, recreational facilities, condominiums, or private estates. With the destruction of the primary dune system by development along this coast in the early part of the 20th century, there is little space left for contemporary dune building processes to operate.

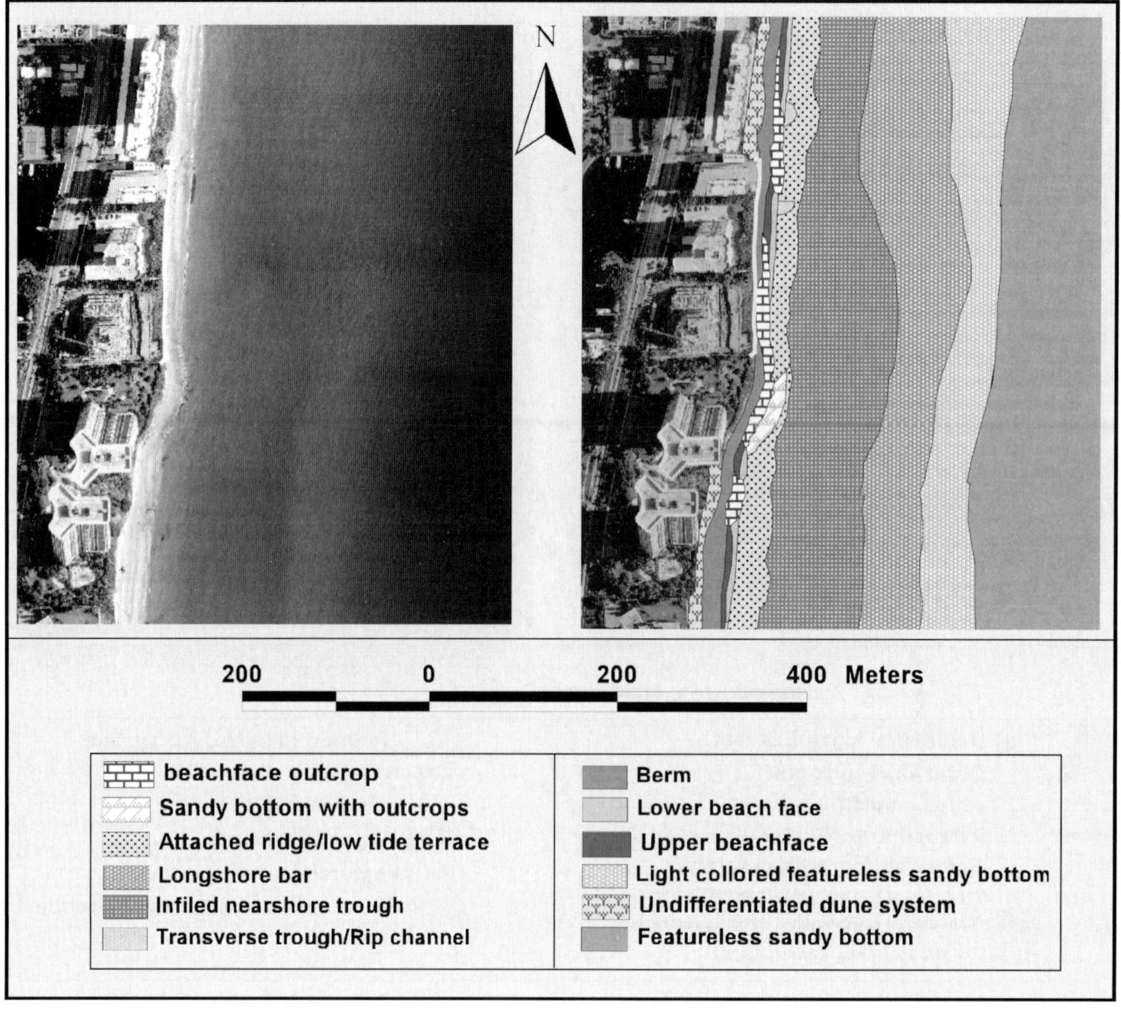

Figure N10 Example of beachface outcrops along the southeast coast near Boca Raton, Florida. This coastal segment shows beach sediments perched on top of bedrock (Anastasia Formation) that often helps to stabilize the shore and sometimes forms a natural salient, as seen in the lower left hand corner of the photograph (left) and in the interpreted map (right). Bedrock outcrops in parts of the low tide terrace where the sand cover is relatively thin and lineations in the rock structure are evident. Transverse troughs or rip channels transect the low tide terrace in the upper part of the photograph and are shown in the map. A bar and trough system up to 200 m in width grades seaward into flat lying featureless sandy bottom. Inshore rock outcrops are a prominent hard bottom feature along the shore and their presence is a clear influence on coastal configuration, when seen in plan view.

Nevertheless, some areas contain small foredunes that develop within a limited framework of environmental controls. Coastal dunes are found above the high water marks of sandy beaches and are formed by wind transport of predominantly loose, sand sized (2.0–0.05 mm) sediment of different origins, as described by Carter *et al.* (1990). Because of their limited developmental status and restricted geographic occurrence, no attempt is made here to differentiate types of coastal dunes. Dunes are mapped here, as shown in Figures N8–N12, as undifferentiated dune system. Severely restricted in geographic occurrence, the dunes may be somewhat wider (up to 40–50 m in width) (Figure N8, geomorphic map on right) to somewhat narrower (less than 20 m) (Figures N11 and N12; geomorphic maps on right).

Beach

As explained by Wright and Short (1984), there is considerable variability in concept, definition, and morphological properties of beaches. Although a common feature of many coasts, beaches can be deceptively complex systems that require careful attention to their biophysical characteristics and environments of formation. The term *beach* was originally used to designate the loose wave-worn shingle or pebbles found on English shores and is still used in this sense in some parts of England (Johnson, 1919, p. 163). In common parlance, the term is used locally for the seashore or lakeshore area. More closely defined usage refers to the temporary accumulation of loose water-borne material that is in active transport along, or deposited on, the shore between the limits of low and high water. From a geomorphological point of view, the beach is comprised by the unconsolidated material that covers a gently sloping zone, typically with a concave profile, and which extends landward from the low-water line to a place where there is a change in material or physiographic form (e.g., dune) or a line of permanent vegetation. Relying on these general concepts of beach as a morphodynamic unit, unconsolidated sandy material deposited along the shore zone between low and high water was mapped as beach (see units for berm and beachface in Figures N8–N12). Beaches in this area contain admixtures of siliciclastics and carbonates of biogenic origin from shells and calcareous algae. In the southern part of the southeast coast of Florida, beach sands are mostly carbonates and in the northern part they are mostly silicates. In the central part of this coastal segment, the beach sands are nearly equally differentiated into silica and carbonate components. All beach sands are shallowly underlain by the Anastasia Formation, which extends several kilometers inland (Finkl and Esteves, 1997) and continues as a shelf deposit along the southeast coast of the United States to at least the latitude of central North Carolina. Morphodynamically, these beaches key out as mostly intermediate types in the beach classification system developed by Short (1999) but the unit is not differentiated for mapping purposes. Based on beach width and planform configuration, it is evident that beaches are eroding, accreting, or stable. An example of an eroding beach, that is, one with a retreating coastline, is shown in Figure N11 where a condominium complex just out into the

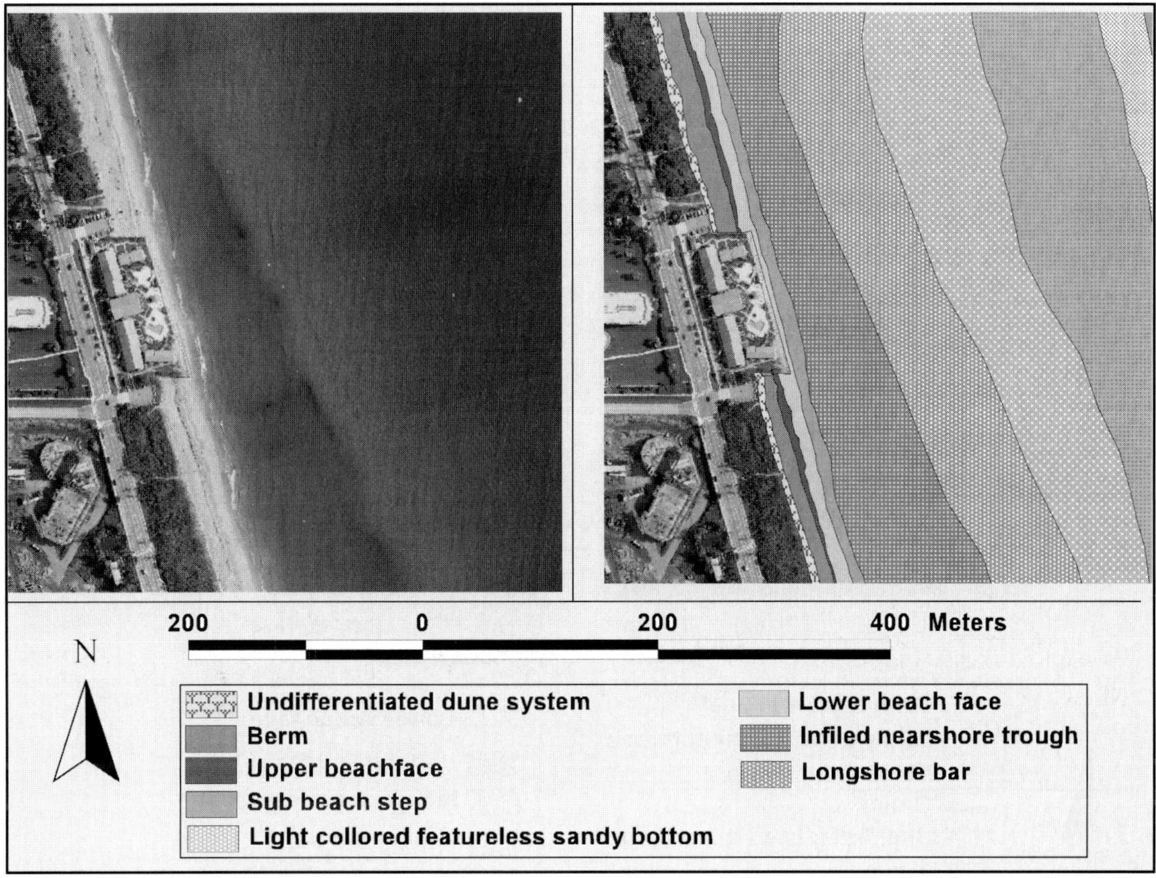

Figure N11 Example of a simple sandy shore with a series of shore-parallel sedimentary morphoforms viz. coastal dunes, beach, bar and trough, and featureless sandy bottom near Jupiter, Florida. The mapped polygons (right) correspond to major features interpreted from the aerial photograph (left) showing a truncated dune system restricted by the coastal highway (A1A), a wider berm on the northern updrift side of the recreational complex built on the beach, a sediment starved beach on the southern downdrift side of the structure, and the presence of a well-developed lower beach face and sub-beach step. The infilled nearshore trough flanks the seaward portion of the rectilinear beach face. Bars and troughs alternate in a seaward direction as depth increases. The Florida Reef Tract lies farther offshore and is not shown in this diagram. This coastal segment typifies a sediment-rich sequence of geomorphic forms and illustrates a gradation of bottom types that are commonly experienced along so-called soft coasts. The impression is, however, misplaced because the southeast coast of Florida is rock controlled (cf. Figures N8–N10).

sea. This facility sits astride the dune system, berm, and upper beachface; its seaward-most portion terminates on the lower beachface. The berm is widened on the northern updrift side of the complex and narrowed on the southern downdrift side that is now starved of sediment because this structure acts as a partial littoral drift blocker. Figure N9 shows a stable or accreting beach with a berm nearly 90 m in width.

Rhythmic features and berms were interpreted from the aerial photographs and mapped as separate units. The *beachface*, that section of the beach normally exposed to the action of wave uprush (i.e., the foreshore of the beach), was subdivided into two units, the *upper and lower beachface*. This distinction was interpreted from the aerials on the basis of geomorphic expression of the cross-shore profile, the *beach scarp* providing a consistently recognizable division, except along beaches where municipalities deliberately destroy the scarp by scraping. Beach scarps of 1–2 m can be hazardous barriers that block access to the water and they are thus destroyed, especially when they form in newly renourished beaches.

Cusps. These ephemeral features occur intermittently throughout the study area and belong to the class of "typical beach cusps," (versus storm cusps) as defined by Dolan and Ferm (1968). Cusp embayments average about 6 m wide (measured between horns) and have wavelengths of about 13–15 m. Beach cusps occur sporadically along the beach, usually in groups of rhythmic topography about 300 m long. These crescentic, concave-seaward features at the shoreline occur at different locations at different times of the year; in general, however, cuspate features seem to occur along coastal segments where reef tracts or hardgrounds outcrop in the inshore zone. Their morphometries (quasi-regular spacing alongshore) differ according to localized oceanographic processes that are influenced by bathymetry and bottom type. This rhythmic shoreline topography is associated with *intermediate beach systems*, as described by Hughes and Turner (1999). Beach cusps are not shown on any of the figures.

Berms. Berms are inter- to supratidal shore-parallel terraces or ridges that consist of a steeper seaward sloping beachface, topped by a shore-parallel crest. As explained by Hesp (1999), the term *berm* has come to be synonymous with the intertidal beach and backshore, as an asymmetric landform comprising a seaward swash slope (usually the steepest component; the "riser" in terrace morphological terminology), a crest, and a low, often linear, sometimes concave beach top slope (the "tread"). The term *berm* has often been taken to mean the nearly horizontal portion of the beach (i.e., the tread; see Davis, 1982). These low, impermanent, nearly horizontal or landward sloping benches, which occur as a backshore terrace on a beach, are formed by material deposited by storm waves. Some narrow native beaches (beaches that have not been renourished) have no berm while some of the wider, accreting beaches immediately updrift from jettied inlets may have several. Well-developed, wide berms are usually associated with renourished beaches that may extend more than 100 m seaward of frontal dunes. Eroded beaches may have narrow berms and along some critically eroded coastal segments there is no beach. Figure N10 (geomorphic map, right), for example, shows a rock outcrop immediately seaward of a seawall with no intervening beach. The seaward portion of the rock outcrop is partly overlain by a low-tide terrace. Berm crests define the seaward boundary of the dry beach. Some alongshore segments were mapped as a berm/upper beachface complex unit because it

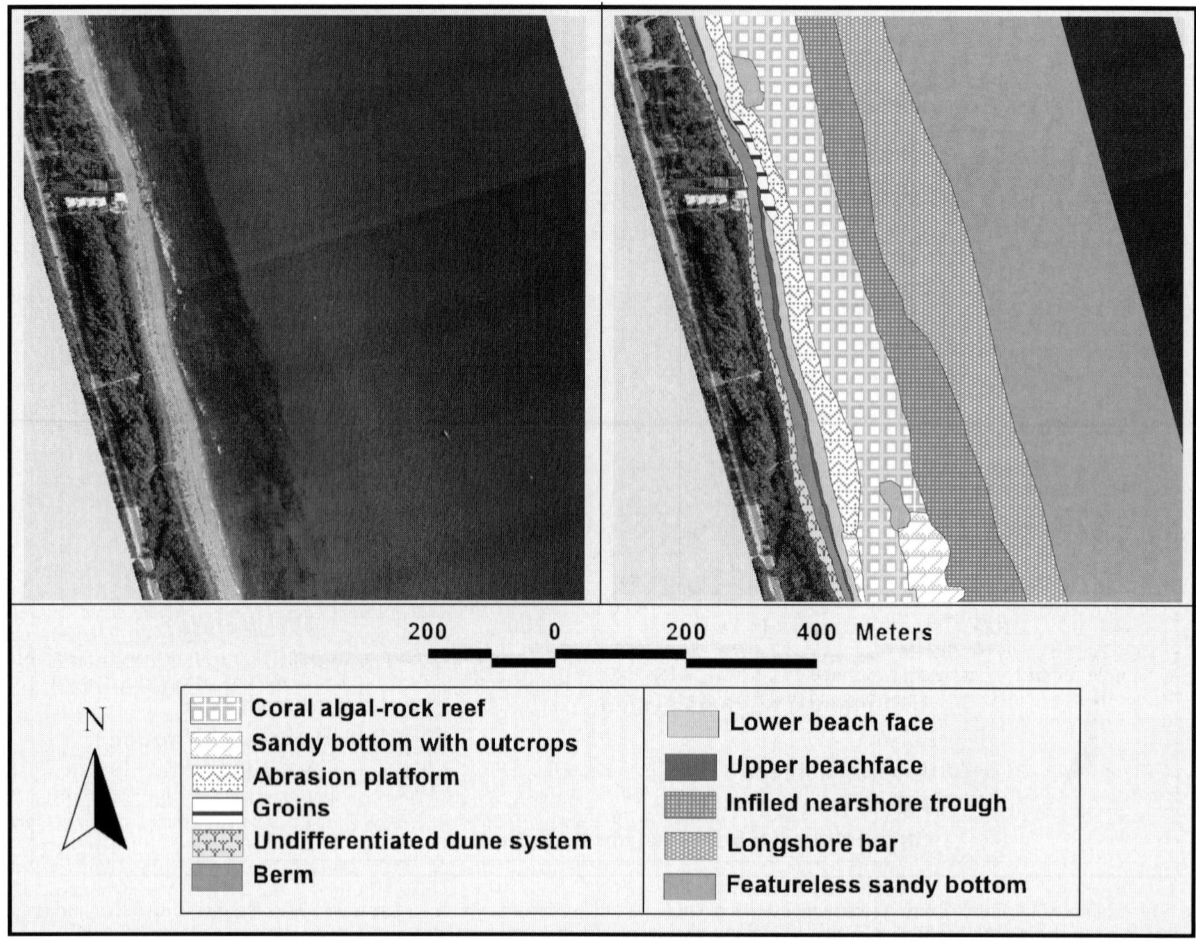

Figure N12 Example of alongshore rock outcrops on the southeast coast of Florida near Tequesta, Palm Beach County. Hardbottom types are clearly evident in the aerial photograph (left) by their dark tones, coarse texture, and linear patterns. The interpreted map (right), showing salient geomorphological units, delineates the presence of coral-algal rock reef seaward of the well developed abrasion platform that is about 100 m in width and 800 m in length. The beach is perched on top of the abrasion platform, which extends landward under the beach. A bar and trough system flanks the seaward side of the coral-algal reef, part of the 1st reef in the Florida Reef Tract. Undifferentiated sandy bottom extends at least 500 m offshore to about 30 m depth, below which photointerpretation is not possible. When hard bottoms occur this close inshore, sands from the perched beaches are often stripped away by storm waves and currents during northeasters leaving an extensive rock surface exposed to the base of coastal dunes.

was not possible to differentiate these two units on the aerial photographs. Small beach scarps separating the berm and upper beachface, which did not allow consistent identification on the aerials, compromised interpretation. On some beaches the scarp is mechanically swept, which combines the berm and upper beachface into a continuous unit, to make a user-friendlier beach. Beach morphology is extremely variable and the units mapped here (berm, upper/lower beachface) relate to the time of photo acquisition.

Upper/lower beachface. The *beachface* is that section of the beach that is normally exposed to the action of wave uprush, that is, the foreshore of the beach. The beachface is defined as being in a state of dynamic equilibrium when the net sediment transport, averaged over several swash cycles, is zero (Hughes and Turner, 1999). Thus, if offshore sediment transport is dominant, the beachface profile becomes flatter; and conversely, if onshore transport prevails, the profile becomes steeper. Beachface profiles rarely exhibit a uniform and planar gradient; the typical beachface is concave seaward but the beachface slope may vary alongshore. In this study, the *upper beachface* was distinguished form the *lower beachface* because they are morphologically distinct and are easily separated on aerial photographs due to different tone and pattern (cf. Figures N8–N12). The upper beachface extends from the landward limit of the lower beachface to the berm crest or foot of the beach scarp. The lower beachface is the seaward sloping portion of the foreshore where swash action begins to occur, that is, where waves break for the last time and swash moves up the beachface as a bore. The width and form of the beachface is significant because this is the feature that ultimately reflects or dissipates wave energy. In most cases, the lower beachface is less than 5 m wide.

Bar and trough systems

Study of cross-shore profiles has resulted in the notion of "winter" and "summer" profiles; winter storm waves removed sand from the berm forming a breakpoint bar, whereas calmer wave climates in summer induced landward migration of the bar with subsequent welding to the beachface to form a non-barred cross-shore profile (e.g., Fox and Davis, 1973; Aagaard and Masselink, 1999). Subsequent research suggested that beach response to wave climates was cyclic, rather than seasonal, and so the terms "barred" and "non-barred" appeared more appropriate (Greenwood and Davidson-Arnott, 1979). The separation between barred and non-barred profiles is now reported to relate to the direction of cross-shore sediment transport. Offshore sediment transport results in erosion of the beach and formation of barred profiles, whereas onshore sediment transport causes beach accretion and non-barred profiles (Aagaard and Masselink, 1999). Relationships between nearshore bar morphology and wave climates is complex and site specific, but periods of high or low wave energy impart distinctive bar types and sequences alongshore. Aagaard and Masselink (1999) report that in general, transverse bar morphology is found at the low energy end of the spectrum, whereas linear bar morphology characterizes high wave energy levels. Crescentic bars develop under intermediate wave energy conditions. All of these classical types of bar and trough systems are seen along the southeast Florida coast.

In addition to classical bar and trough forms, there are local or site-specific morphological variations and combinations that resulted in identification of the following main units: longshore bar, structurally controlled longshore bar, crescentic bar, crescentic attached bar, attached ridge with surge channels, crescentic trough, transverse trough/rip channel, and infilled nearshore trough. Longshore bars are low, elongate sand ridges that are built up mainly by wave action and which occur some distance from the shoreline. Longshore bars generally extend parallel to the shoreline and are submerged at least by high tides and are separated from the beach by an intervening trough (see Figures N8–N12). Several terms have been applied to these kinds of features (e.g., ball, offshore bar, submarine bar, barrier bar) as described here in their simplest situational occurrence. There is a wide range of morphological variation where bars and associated troughs occur as single linked units (e.g., Figure N12) or in multiple associations in rectilinear to strongly curved (wavy) planforms (e.g., Figures N8–N10). Bars and troughs may be continuous along the shore for many kilometers or they may occur as disjunctions continuously or sporadically along the shore (cf. Figures N8 and N12), parallel to shore or inclined at an angle (e.g., Figure N8). When longshore bars have migrated shoreward and become welded to the beachface as relatively high ridges of sand, they are termed welded bars or, as referred to here, as *attached ridges* (Figure N9).

Longshore bar. Along the southeast Florida coast, longshore bars occur in the inshore zone as single bars or as part of multiple bar and trough systems (Figures N10–N12). The bars are sediment ridges of running sand that generally parallel the shore. Some bars are rather narrow, for example, about 15 m wide, and continue without break for distances longer than 1.5 km. Occasionally, one end of transverse bars are welded to the shore either singly or in multiple sets (Figure N8).

Crescentic bars and crenulated troughs. Crescentic bars are rhythmic shore features that tend to have hyperbolic wavelengths on the order of 115–160 m and are usually located about 12–60 m from the shoreline (e.g., Figure N8). Crescentic bars are sometimes welded to the shore (Figure N8) but welded or not, they are associated with troughs that are either infilled nearshore or crenulated. The term *crenulated trough*, as applied here, describes certain inshore coastal segments where sedimentary troughs occur in the inshore zone adjacent to crescentic bars. The adjacent border of the crescentic bar forms their borders so that a convex seaward portion of a crescentic bar forms the concave seaward crenulated trough. Crenulated troughs may form on either the landward or seaward side of crescentic bars, and they may occur singularly (inside one cycle of the crescentic bar) or they may connect to form a crenulated trough that stretches 4 km or more alongshore.

Infilled nearshore trough. As defined here, infilled nearshore troughs are elongate sedimentary depressions that are associated with bars. Because the troughs tend to become partially infilled under subsequently calmer conditions than during their formation during storms, they are referred to as being infilled. These depressional features of the inshore zone range in width up to 150 m cross-shore and may extend for several kilometers alongshore (Figures N10–N12). As reported by Lippman and Holman (1990), longshore-bar-trough states are variable and the features may be highly mobile under certain wave conditions, whereas Wright *et al.* (1985) concluded that straight or rhythmic longshore bar and trough states could remain relatively stable under certain conditions. The bar and trough systems described here seem to remain positionally stable from year to year, based on observations of interannual photography. The presence of reefs and rock outcrops on the seafloor in the study area apparently complicates present comprehension of bar–trough positional stability.

Shoreface trough. These troughs, lying seaward of the lower beachface, contain variable water depths but all are several meters deeper compared with the longshore bars or hardground features to which they are adjacent. Shoreface troughs may extend uninterrupted for long distances, for example, up to several kilometers alongshore. Occupying areas of lower elevation compared with surrounding morphologic features, usually coral-algal rock reefs or structurally controlled sandflats, these troughs are commonly floored by hardbottom with thin sand veneers. Structural features of carbonate hardbottom are often evident in the aerial photographs, verifying the thin cover of overlying sand. The troughs more or less parallel landward variations in the Florida Reef Tract or follow inshore lineations in the Anastasia Formation.

Sandy bottom

Sandy sediments are a dominant feature of the seafloor on the inner continental shelf, occupying about 30–40% of the mapped areas (e.g., Brown, 1998; Warner, 1999), the remainder being coral-algal reef or exposure of bedrock as hardgrounds. The sandy deposits are mostly sheets overlying the karstified bedrock as basin infills. Corollary data from seismic survey shows that the sand lying between reefs in the Florida Reef Tract are shallow basinal-type deposits with stringers or lenses of fine grained (silt plus clay) sediments and occasional coarse rubble mounds in the vicinity of paleo-inlets (now submerged reef gaps) (e.g., see Finkl and Bruun, 1998; Khalil, 1999). Sedimentary infills here average several tens of meters in thickness, but present a generally monotonous flat topography at the surface between reefs. Because of the flat, rippled surface patterns seen on the aerial photographs, these features are referred to as sandflats or simply as sandy bottom. The sandy deposits are more or less of uniform grain size at the surface and thus appear similar and are easily recognized in the interpretive process. Some areas, however, are partially covered by anchored or bottom-drifting algal mats or less commonly with sea grass beds giving a dark-toned appearance in the imagery. Because these areas are so distinctive in appearance and thus mutually exclusive, they were separated as different mapping units and referred to simply as light- and dark-toned running sands. Areas of sand sheets appear similar to the inter-reefal sand flats but instead of occurring offshore they tend to occur more inshore and shallowly overlie karst limestone. The underlying limestone is evident in the imagery by its reticulate structural patterns that poke through the thin sedimentary cover. These sandy bottoms thus present complicated image patterns that are combinations of sedimentary cover over outcrops of rough bedrock surfaces. Although wide ranges of variable patterns are associated with this unit, they are grouped into one category of occurrence for mapping purposes.

Light- and dark-toned running sand. Unconsolidated running sands without vegetation or algal mats appear as smooth, fine textured, bright patterns on the aerial photographs. The "light-toned" photograph interpretation descriptor is used to differentiate this unit from the darker-toned seabed sands. These clean running sands occur mainly in the offshore zone (e.g., Figures N10 and N11), as rather wide (about 60–120 m in width) bars that parallel the shore or as sand sheets. Collateral data (e.g., side-scan sonar) indicates that these sandy bottoms may extend to several (6–7) meters in thickness and overlie carbonate substrates of the Florida Reef Tract. Dark-toned running sands are differentiated on the basis of the areal distribution and density of benthic algae or biological material that is integrated within the surface-most sediments. These dark-toned running sand sheets occupy large areas of seafloor and in contradistinction to the parabathic light colored bars and sheets, they display many different irregular shapes that are determined by the presence of organic matter at the sediment surface. These sands also infill depressions between reef tracts and occur as sheets of variable thinness over hardgrounds.

Structurally controlled sandflats. As defined here, these submarine sandy areas of flat lying sediments are bordered on both landward and seaward margins by hardground ledges. Figure N8, for example, shows marine sediments that have infilled solutional depressions in rock reefs. These sandy areas are laterally and vertically controlled by preexisting or antecedent topography that was previously subaerially exposed and then drowned by the Flandrian Transgression. The bounding ledges, which are clearly evident in aerial photographs, generally rise about 1–2 m above the seafloor to interiorly trap sediment. In some areas, the karst structures of the underling limestone show through the thin sedimentary veneer or are propagated upwards through the sedimentary cover to show the limestone's reticulate structural pattern. The structurally controlled sandflats tend to be irregular in overall shape, but their borders create zigzag or saw-tooth patterns. These sandy areas rarely contain algal mats and the sands are clean.

Tidal delta. Shoals associated with tidal inlets occur as flood-tidal deltas and ebb-tidal deltas along this microtidal (less than 2 m) coast with semi-diurnal tides. Tidal prism normally has a large influence on the size and shape of natural inlets, but inlets in the study area are cut through bedrock of the Anastasia Formation making them positionally stable (Finkl, 1993). Because the major ports have stabilized deep navigational entrances with cuts 15–18 m deep and jetties, which in turn results in practically no bypassing of sediments, only smaller inlets have well-developed deltas of the classical form in shallow water. These wave-dominated ebb-tidal deltas (Davis, 1997), always completely submerged along this coast, are sometimes deformed southwards due to structural impacts associated with the downdrift curvature of jetties and the net littoral drift from north to south. Bypass bars are often associated with the ebb-tidal deltas.

End products, applications and insight

Detailed coastal mapping, based on interpretation of aerial photographs, results in the preparation of special purpose morphologic maps along specified coastline segments. Mapping areas may be defined on the basis of coastal physiography or administrative units such as counties. Interpretation of georectified digital aerial photography has the advantage of being incorporated into spatial analysis programs such as ArcView®. Maps prepared in this manner can be compared with preexisting coverages for research and visual display. Mapping of coastal morphology provides new insight into the topologies of coasts that provides better understanding of process–form relationships. Classification of coastal morphologies elucidates associations of natural features and differentiates groups of related features. The delineation of coastal morphologies provides a basis to model coastal landform development and provides baseline data for subsequent change detection analysis.

Detailed coastal morphological maps (e.g., Brown, 1998; Khalil, 1999; Warner, 1999) are useful because they provide a record of coastal biophysical conditions that in turn may find application, for example, in: (1) searches for beach-quality sand that is required for beach renourishment, (2) studies of coastline stability where offshore morphology may determine onshore anchor points or erosional hot spots (localized zones of accelerated beach erosion), (3) in the development of computer models of coastal processes, (4) assessment of the structure and integrity of biological habitats, (5) background geological and geotechnical information for engineering works, (6) coastal hazards research related to surge flooding, overwash, sediment erosion and deposition, location of rip currents, and (7) investigations of coastal geomorphology, evolution of coastal configurations, and coastal/marine landform classification. The methodology is applicable to clear Class II waters where there is low turbidity in the water column during parts of the year when aerial photography can be acquired. Many rocky, sandy beach, and coral reef coasts are thus amenable to study using this technique. Muddy coasts and highly turbid waters are excluded from interpretation of bottom features using aerial photography.

These kinds of morphological studies emphasize the importance of underlying geomorphology to coastline configuration (in planform) and submarine beach profiles. With decreasing natural availability of littoral-drift sediment resulting from dredged inlet construction, the sand layer overlying the Anastasia Formation will continue to decrease in thickness. Consequently, an increased importance of underlying geology in describing profile variability and coastline configuration should be expected. It also follows that a greater portion of the beachface will be comprised of rock outcrops and a greater amount of littoral sediments may be permanently lost to the offshore. This positive feedback mechanism persists at the expense of artificial beach renourishment projects (Brown, 1998). As discussed by Finkl (1993), the limestone base will naturally inhibit shore erosion at points where exposed as submerged hardground in the nearshore zone (Figure N13) and thus control the alteration of beach profiles, shoreline configuration, as well as nearshore processes. Submarine headlands are thus seen as representing paleo-topographic highs in front of modern shores and modify the incoming wave climate (Pilkey *et al.*, 1993). Morphological mapping here clearly shows that rock-control of a sediment-starved beachface is a reality. As shown in Figures N3 and N13, perched beaches may be eroded away during storms and dunes threatened by waves but the rock platforms protect the shore. Sand eventually becomes reestablished on top of the bedrock and a new beach is again perched on top of the platform. These snapshots confirm on the ground what is interpreted from the aerial photographs and verifies the spatio-temporal importance of coastal materials that influence shoreline morphology.

It is the interpretation of aerial photography that makes this morphologic data crucial for coastal morphodynamic studies. Given more than one set of digital aerial photographs, multiple GIS coverages of the delineations of coastal geomorphology can be used in numerous overlay operations in order to gain insights into the planform coastal dynamics (e.g., Finkl and DaPrato, 1993; Brown, 1998). Essential to this topic is the development and implementation of methodologies by which to continually map and assess the coastal data acquired by present and future reconnaissance efforts.

Figure N13 Exposure of the Pleistocene Anastasia Formation in Boca Raton, Florida. This marine abrasion platform has truncated the coquinoid facies of the limestone bedrock base that occur along the Atlantic shore. The seaward margin of the platform along this rocky shore, which occurs as a pronounced ledge up to a meter or so in height, serves as a natural seawall. Note waves breaking offshore in the surf zone, tripped by hardgrounds, and the wave in the photo center that has crashed against the vertical wall of the ledge. There is a duality to this rock platform in that it helps to protect the shore from wave attack but at the same time inhibits onshore sediment, except under high-energy conditions. (Photo: Lindino Benedet, Florida Atlantic University, Boca Raton, Florida; photo taken in Winter 2001.)

Conclusions

In this example of detailed coastal mapping and classification, several important points emerge to illustrate the advantages of making morphological inventories based on interpretation of aerial photographs. That mapping based on georectified digital imagery is faster and more cost effective than conventional survey methods is hardly surprising. There are, however, many surprising elements to remotely sensed coastal surveys such as this one. Facts derived from visual and digital records are more reliable and informative than general impressions or suppositions that are often used in coastal planning, management, and engineering schemes. The southeast Florida shore is commonly referred to as a sandy beach coast with abundant sedimentary deposits offshore. Results of mapping here showed that beaches are shallowly perched over a limestone base, beach deposits are relatively thin at < 2 m in thickness (visible after large storms when beach sands are washed from buried shore platforms), there are many indicators of cross-shore processes (e.g., fluted rock reefs), there is a large expanse of hard bottom types in inshore and offshore zones, and that extremely narrow subaerial beaches occur inshore of reef gaps. Overall observations, based on analysis of the resulting morphological maps, indicate that this coastline morphology is lithologically controlled, the seabed is composed of drowned karst topography overlain locally by unconsolidated sedimentary bodies, and that coastal barriers such as the Florida Reef Tract modify coastline configuration.

Additional advantages that accrue from this kind of detailed mapping of coastal morphological features include the ability to define morphodynamic zones. Recognition of coastal process zones provides a rational basis for coastal management because certain topographic forms are morphodynamically significant, allowing interpretation of dominant processes. Although the technique is limited to clear Class II waters, significant stretches of the world's coastline warrant application of aerial photographic interpretation in conjunction with other methods of investigation.

Charles W. Finkl

Bibliography

Aagaard, T., and Masselink, G., 1999. The surf zone. In Short, A.D. (ed.), *Handbook of Beach and Shoreface Morphodynamics*. Chichester: Wiley, pp. 72–118.

Alam, M.S., Huq, N.E., and Rashid, M.S., 1999. Morphology and sediments of the Cox's Bazar Coastal Plain, south-east Bangladesh. *Journal of Coastal Research*, **15**(4): 902–908.

Bates, R.L., and Jackson, J.A., 1979. *Glossary of Geology*. Alexandria, VA: American Geological Institute.

Bird, E.C.F., 1976. *Coasts*. Canberra: Australian National University Press.

Brock, J.C., Sallenger, A., Krabill, W.B., Swift, R.N., Manizade, S.S., Meredith, A., Jansen, M., and Eslinger, D., 1999. Aircraft laser altimetry for coastal process studies. *Coastal Sediments '99*, pp. 2414–2428.

Brown, K.E., 1998. *Morphological Analysis of Beach Profiles in Relation to Shoreline Change*. Master's thesis, Florida Atlantic University, Boca Raton, Florida, 144p.

Byrne, S., Clifford, B., Simmons, W., Depner, J., Reed, B., Moestikiwati, J., and Smith, G., 2002. Processing data for seafloor mapping: Integration and metrics. *MTS Journal*, **35**(4): 20–32.

Bradley, W.C., 1958. Submarine abrasion and wave-cut platforms. *Geological Society America Bulletin*, **69**: 967–974.

Campbell, J.B., 1996. *Introduction to Remote Sensing*. New York: Guildford.

Carter, R.W.G., Nordstrom, K.F., and Psuty, N.P., 1990. The study of coastal dunes. In Nordstrom, K.F., Psuty, N.P., and Carter, R.W.G. (eds.), *Coastal Dunes: Form and Process*. Chichester: Wiley, pp. 1–16.

Cowell, P.J., Hanslow, D.J., and Meleo, J.F., 1999. The shoreface. In Short, A.D., (ed.), *Handbook of Beach and Shoreface Morphodynamics*. Chichester: Wiley, pp. 9–71.

Davis, R.A., 1982. Beach. In Schwartz, M.L. (ed.), *Encyclopedia of Beaches and Coastal Environments*. Stroudsburg, PA: Hutchinson and Ross, pp. 140–141.

Davis, R.A., 1997. Geology of the, Florida coast. In Randazzo, A.F., and Jones, D.S. (eds.), *The Geology of Florida*. Gainesville FL: University of Florida Press, pp. 155–168.

Dietz, R.S., 1947. Aerial photographs in the geological study of shore features and processes. *Photogrammetric Engineering*, **13**(4): 537–545.

Dillenburg, S.R., Roy, P.S., Cowell, P.J., and Tomazelli, L.J., 2000. Influence of antecedent topography on coastal evolution as tested by the Shoreface Translation-Barrier Model (STM). *Journal of Coastal Research*, **16**(1): 71–81.

Dolan, R., and Ferm, J.C., 1968. Concentric landforms along the Atlantic coast of the United States. *Science*, **159**: 627–629.

Duane, D.B., and Meisburger, E.P., 1969. *Geomorphology and Sediments of the Nearshore Continental Shelf: Miami to Palm Beach, Florida*. US Army Corps of Engineers, CERC *Technical Memorandum No. 29*, 47p.

El-Ashry, M.T. (ed.), 1977. *Air Photography in Coastal Problems*. Stroudsburg, PA: Dowden, Hutchinson & Ross.

Emery, K.O., and Kuhn, G.G., 1982. Seacliffs: their processes, profiles, and classification. *Geological Society America Bulletin*, **93**: 644–654.

Evans, M.W., Hine, A.C., Belknap, D.F., and Davis, R.A., 1985. Bedrock controls in barrier island development: West-Central Florida coast. *Marine Geology*, **63**: 263–283.

Finkl, C.W., 1993. Pre-emptive strategies for enhanced sand bypassing and beach replenishment activities: a geological perspective. *Journal of Coastal Research*, **18**(Special issue): 59–89.

Finkl, C.W., 1994. Management strategies for enhanced sand bypassing and beach replenishment in the southeast Florida coastal zone: potential for application of new technologies. *MTS '94 Conference Proceedings* (Marine Technology Society Annual Meeting, Washington, DC), pp. 107–114.

Finkl, C.W., and Benedet, L., 2002. Aerial Remote Sensing of Beach Systems along a Barrier Reef Coast in Southeast Florida: Implications for Management of Common Beach Types and Chronic Downdrift Erosion. *46th Annual Meeting Florida Shore and Beach Preservation Association* (Captiva Island, Florida, 25–27 September 2002) [Tallahassee: Florida Shore & Beach Preservation Association].

Finkl, C.W., and Bruun, P., 1998. Potential for manipulating wave action in port and coastal engineering. *Journal of Coastal Research*, **26**(Special issue): 1–10.

Finkl, C.W., Jr., and DaPrato, G.W., 1993. Delineation and distribution of nearshore reefs in subtropical southeast Florida coastal environments using Thematic Mapper imagery. *MTS '93 Conference Proceedings* (Marine Technology Society Annual Meeting, Long Beach, California), pp. 90–96.

Finkl, C.W., Jr., and Esteves, L.S., 1997. A preliminary analysis of Neogene depositional environments in Broward County, Florida. *Florida Scientist*, **60**(2): 94–103.

Finkl, C.W., and Khalil, S., 2000. Coastal mapping and classification: a new "old" tool for coastal managers. *Proceedings, 13th Annual National Conference on Beach Preservation Technology*. Tallahassee FL: Florida Shore & Beach Preservation Association, pp. 297–313.

Fleming, B.W., 2002. Geographic distribution of muddy coasts. In Healy, T., Wang, Y., and Healy, J.-A. (eds.), *Muddy Coasts of the World: Processes, Deposits and Function*. Amsterdam: Elsevier, pp. 99–201.

Forbes, D.L., and Syvitski, J.P.M., 1994. Paraglacial coasts. In Carter, R.W.G., and Woodruffe, C.D. (eds.), *Coastal Evolution*. Cambridge, England: Cambridge University Press, pp. 373–424.

Fox, W.T., and Davis, R.A., 1973. Simulation model for storm cycles and beach erosion on Lake Michigan. *Geological Society of America Bulletin*, **89**: 1769–1790.

Gesch, D., and Wilson, R., 2002. Development of a seamless multi-source topographic/bathymetric elevation model of Tampa Bay. *MTS Journal*, **35**(4): 58–64.

Gorman, L., Morang, A., and Larson, R., 1998. Monitoring the coastal environment; Part IV: Mapping, shoreline changes, and bathymetric analysis. *Journal of Coastal Research*, **14**(1): 61–92.

Greenwood, B.G., and Davidson-Arnott, R.G.D., 1979. Sedimentation and equilibrium in wave-formed bars: a review and case study. *Canadian Journal of Earth Sciences*, **16**: 312–332.

Guan-Hong, L., and Birkemeier, W.A., 1993. Beach and Nearshore Data: 1985–1991 CERC Field Research Facility. *Technical Report CERC-93-3*, 13p.

Guilcher, A., 1988. *Coral Reef Geomorphology*. New York: Wiley.

Hesp, P., 1999. The beach backshore and beyond. In Short, A.D. (ed.), *Handbook of Beach and Shoreface Morphodynamics*. Chichester: Wiley, pp. 145–169.

Hogarth, P., 2002. Cost effective swath bathymetry. *MTS Journal*, **35**(4): 33–45.

Hopley, D., 1988. *The Geomorphology of the Great Barrier Reef*. New York: Wiley.

Hughes, M., and Turner, I., 1999. The beachface. In Short, A.D. (ed.), *Handbook of Beach and Shoreface Morphodynamics*. Chichester: Wiley, pp. 119–144.

Hwang, P.A., Walsh, E.J., Krabill, W.B., Swift, R.N., Manizade, S.S., Scott, J.F., and Earle, M.D., 1998. Airborne remote sensing applications to coastal wave research. *Journal of Geophysical Research*, **103**(C9): 18791–18800.
Irish, J.L., and Lillycrop, W.J., 1997. Monitoring New Pass, Florida, with high density Lidar bathymetry. *Journal of Coastal Research*, **13**(4): 1130–1140.
Irish, J.L., and Lillycrop, W.J., 1999. Scanning laser mapping of the coastal zone: the SHOALS system. *ISPRS Journal of Photogrammetry and Remote Sensing*, **54**: 123–129.
Jiménez, J.A., Sanchez-Arcilla, A., Dou, J., and Ortiz, M.A., 1997. Analyzing short-term shoreline changes along the Ebro Delta (Spain) using serial photographs. *Journal of Coastal Research*, **13**(4): 1256–1266.
Johnson, D.W., 1919. *Shore Processes and Shoreline Development*. New York: Wiley.
Judge, E.K., and Overton, M.F., 2001. Remote sensing of barrier island morphology: Evaluation of photogrammetry-derived digital terrain models. *Journal of Coastal Research*, **17**(1): 207–220.
Kelletat, D., 1989. The question of "zonality" in coastal geomorphology—with tentative application along the East Coast of the USA. *Journal of Coastal Research*, **5**(2): 329–344.
Khalil, S.D., 1999. *Geomorphology of the Southeast Florida Inner Continental Shelf: Interpretations Based on Remote Sensing*. Master's thesis, Florida Atlantic University, Boca Raton, Florida, 136p.
King, C.A.M., 1959. *Beaches and Coasts*. London: Arnold.
Krabill, W.B., Wright, C.W., Swift, R.N., Frederick, E.B., Manizade, S.S., Yungel, J.K., Martin, C.F., Sonntag, J.G., Duffy, M., Huslander, W., and Brock, J.C., 2000. Airborne laser mapping of Assateague Island National Seashore Beach. *Photogrammatric Engineering & Remote Sensing*, **66**: 65–71.
Larson, M., and Kraus, N.C., 1995. Prediction of cross-shore sediment transport at different spatial and temporal scale. *Marine Geology*, **63**: 495–500.
Lidz, B.H., Shinn, E.A., Hine, A.C., and Locker, S.D., 1997. Contrasts within an outlier-reef system: Evidence for differential Quaternary evolution, south Florida windward margin, U.S.A. *Journal of Coastal Research*, **13**(3): 711–731.
Lidz, B.H., and Hallock, P., 2000. Sedimentary petrology of a declining reef ecosystem, Florida Reef Tract (U.S.A.). *Journal of Coastal Research*, **16**(3): 675–697.
Lippman, T.C., and Holman, R.A., 1990. The spatial and temporal variability of sand bar morphology. *Journal of Geophysical Research*, **95**: 1575–1590.
McGill, J.T., 1958. Map of coastal landforms. *Geographical Review*, **48**: 402–405.
Milbert, D.G., and Hess, K.W., 2001. Combination of topography and bathymetry through application of calibrated vertical datum transformations in the Tampa Bay region. *Proceedings Coastal Tools '01* (Charleston, South Carolina, 8–11 January 2001). Charleston: NOAA, Coastal Services Center, CD-ROM NOAA/CSC/20114-CD.
Moore, L.J., 2000. Shoreline mapping techniques. *Journal of Coastal Research*, **16**(1): 111–124.
Nordstrom, K., 1990. The intrinsic characteristics of depositional coastal landforms. *Geographical Review*, **80**: 68–81.
Nordstrom, K., Psuty, N., and Carter, W.R.G., 1990. *Coastal Dunes: Form and Process*. Chichester: Wiley.
Owens, E.H., 1994. Canadian coastal environments, shoreline processes, and oil spill cleanup. Ottawa, Canada: Environment Canada, *Report EPS 3/SP/5*, 328p.
Parson, L.E., Lillycrop, W.J., Klein, C.J., Ives, R.C.P., and Orlando, S.P., 1997. Use of Lidar technology for collecting shallow water bathymetry of Florida Bay. *Journal of Coastal Research*, **13**(4): 1173–1180.
Pethick, J., 1984. *An Introduction to Coastal Geomorphology*. London: Arnold.
Pilkey, O.H., 1998. A new view of the shoreface. *Journal of Coastal Research*, Special Issue No. **26**(Suppl.): 108–109.
Pilkey, O.H., Young, R.S., Riggs, S.R., Smith, S., Wu, H., and Pilkey, W.D., 1993. The concept of shoreface profile of equilibrium: A critical review. *Journal of Coastal Research*, **9**(1): 255–278.
Prager, E., and Halley, R., 1997. Open File Report-U.S. Geological Survey. Reston, VA: US Geological Survey, Report: OF 97-0526
Russell, R.J., 1967. *River Plains and Sea Coasts*. Berkeley, CA: University of California.
Ruessink, B.G., and Kroon, A., 1994. The behaviour of a multiple bar system in the nearshore zone of Terschelling, The Netherlands: 1965–1993. *Marine Geology*, **121**: 187–197.
Sanlaville, P., Dalongeville, R., Bernier, P., and Evin, J., 1997. The Syrian coast: A model of Holocene coastal evolution. *Journal of Coastal Research*, **13**(2): 385–396.
Schumm, A.A., 1991. *To Interpret the Earth: Ten Ways to be Wrong*. Cambridge: Cambridge University Press.
Schwartz, M.L., 1982. *The Encyclopedia of Beaches and Coastal Environments*. Stroudsburg, PA: Hutchinson Ross.
Shand, R.D., Bailey, D.G., and Shepherd, M.J., 1999. An inter-site comparison of net offshore bar migration characteristics and environmental conditions. *Journal of Coastal Research*, **15**(3): 750–765.
Short, A.D., 1993. *Beaches of the New South Wales Coast*. Sydney: Australian Beach Safety and Management Program.
Short, A.D., 1999. *Handbook of Beach and Shoreface Morphodynamics*. New York: Wiley.
Shoshany, M., and Degani, A., 1992. Shoreline detection by digital image processing of aerial photography. *Journal of Coastal Research*, **8**(1): 29–34.
Smith, G.L., and Zarillo, G.A., 1990. Calculating long-term shoreline recession rates using aerial photographic and beach profiling techniques. *Journal of Coastal Research*, **6**(1): 111–120.
Stauble, D.K., and McNeil, D.V., 1985. Coastal geology and the occurrence of beachrock: central Florida Atlantic coast. *Field Guide for the Annual Meeting of the Geological Society of America*, Vol. 1, 27p.
Stockdon, H.F., Sallenger, A.H., List, J.H., and Holman, R.A., 2002. Estimation of shoreline position and change using airborne topographic lidar data. *Journal of Coastal Research*, **18**(3): 502–513.
Sunamura, T., 1992. *Geomorphology of Rocky Coasts*. Chichester: Wiley.
Thieler, E.R., and Danforth, W.W., 1994. Historical shoreline mapping (I): Improving techniques and reducing positioning errors. *Journal of Coastal Research*, **10**(3): 549–563.
Trenhaile, A.S., 1987. *The Geomorphology of Rock Coasts*. Oxford: Clarendon Press.
Wang, Y., and Healy, T., 2002. Definition, properties, and classification of muddy coasts. In Healy, T., Wang, Y., and Healy, J.-A. (eds.), *Muddy Coasts of the World: Processes, Deposits and Function*. Amsterdam: Elsevier, pp. 9–18.
Warner, M.T., 1999. *Analysis of Coastal Morphodynamic Zones Based on Detailed Mapping in Palm Beach County, Florida*. Master's thesis, Florida Atlantic University, Boca Raton, Florida, 110p.
White, W.A., 1970. *The Geomorphology of the Florida Peninsula*. Tallahassee, FL: Florida Bureau of Geology, Geological Bulletin No. 51, 164p.
Wright, L.D., and Short, A.D., 1984. Morphodynamic variability of surf zones and beaches: A synthesis. *Marine Geology*, **56**: 93–118.
Wright, L.D., Short, A.D., and Green, M.O., 1985. Short-term changes in the morphologic states of beaches and surf zones: an empirical model. *Marine Geology*, **62**: 339–364.
Wright, L.D., 1995. *Morphodynamics of Inner Continental Shelves*. Boca Raton, FL: CRC Press.
Zenkovich, V.P., 1967. *Processes of Coastal Development*. Edinburgh: Oliver and Boyd.

Cross-references

Bars
Beach Features
Beach Processes
Coasts, Coastlines, Shores and Shorelines
Coral Reef Coasts
Coral Reefs
Mapping Shores and Coastal Terrain
Photogrammetry
Remote Sensing of Coastal Environments
Rhythmic Patterns
Shoreface
Surf Zone Processes

NEARSHORE SEDIMENT TRANSPORT MEASUREMENT

Definitions and key parameters

Nearshore sediment transport is driven by the combined motions of water waves and currents. Due to the complications caused by orbital wave motion, methods developed for measuring sediment transport in largely unidirectional environments may not be applicable in measuring coastal sediment transport. To make matters even more complicated, the relatively orderly wave motion becomes disrupted by the extremely turbulent wave breaking at the transition between land and ocean, often referred to as the surf zone or breaker zone. In the following, the term nearshore is used in a general sense, representing a variety of coastal environments.

Generally, sediment transport rate is calculated as

$$Q = \frac{1}{t}\int_0^t \int_0^{x_n} \int_0^h C_s(x, z, t)\, U_s(x, z, t)\, dz\, dx\, dt, \quad \text{(Eq. 1)}$$

where x_n is the width of a certain nearshore zone over which the transport is computed; h is the water depth; U_s is the particle velocity; C_s is the sediment concentration; x and z are the horizontal and vertical coordinates; and t is the time. The resultant Q is the total transport rate integrated over a water depth h across a horizontal scale x_n, and averaged over a time period of t. Sediment concentration C_s is a scalar quantity described only by its magnitude. Under most circumstances, C_s is greater near the bottom and decreases logarithmically upward. Therefore, it is important for sediment-concentration measurement to extend as close to the bottom as possible. In order to use equation 1 to calculate sediment transport, simultaneous current measurement is also important. It is worth emphasizing that sediment transport rate, the product of a scalar C_s and a vector U_s, is a vector in the same direction as U_s.

In coastal zones, especially in environments controlled strongly by wave motions, U_s and C_s vary rapidly with time and space. In a high-energy environment, such as the surf zone, sediment transport is extremely active. Measuring U_s and C_s simultaneously with satisfactory temporal and spatial resolutions is difficult. The product of U_s and C_s yields sediment flux

$$\overrightarrow{F_s(x, z)} = \overrightarrow{U_s(x, z)} \times C_s(x, z). \quad \text{(Eq. 2)}$$

Under many circumstances, directly measuring the sediment flux is technically easier than measuring the U_s and C_s separately. However, precise measurements of U_s and C_s will not only yield transport rate, but also shed light on transport processes. In complicated coastal waters, the sediment concentration C_s and particle velocity U_s are influenced by many hydrodynamic and sedimentary factors. It is beyond the scope of this entry to examine the many factors that determine U_s and C_s. A comprehensive summary was provided by Van Rijn (1993).

Sediment transport is often described in three modes: bed load, suspended load, and wash load. As often as these terms are used, they are not as clearly defined, especially from a measurement perspective. The wash load describes the portion of very fine particles that are transported by the water but are not represented in the bed. Wash load is typically neglected or incorporated into suspended load during transport measurements. Conceptually, bed load is usually defined as the part of the total load that is in frequent contact with the bed during transport. It primarily includes grains that roll, slide, or jump along the bed. The suspended load is the part of the total load that moves without frequent contact with the bed controlled by the agitation of fluid turbulence.

The concepts of bed load and suspended load provide a visual representation of the modes of particle movement. But in practice, bed load and suspended load, as they are defined above, cannot be distinguished during field or laboratory measurements. From a measurement perspective, bed load is often simply defined as the part of the total load that travels below a certain level, which may be defined differently by different researchers, often on the basis of the measurement device. Therefore, bed load and suspended load measured during different studies may not be directly comparable due to different definitions. Caution should therefore be exercised in using these terms in nearshore sediment transport measurement and clear definition should be provided.

Fundamentals of nearshore sediment transport measurement

One of the first and rather critical decisions in planning a nearshore sediment transport measurement project is to determine the appropriate temporal and spatial scales. "Coastal zone" is a general term describing many different types of coastal environments, including tidal inlets, surf zones, estuaries, tidal deltas, etc. Sediment transport is dominated by different processes in these different environments. The spatial scale should be adequate to describe the characteristics of the specific coastal environment. Coastal sediment transport often demonstrates apparent periodicity controlled by the cycles of water motion and sediment supply, such as wave cycles, tidal cycles, seasonal wave conditions, seasonal cycles of riverine sediment supplies, etc. The temporal scale should be able to resolve the periodicity that is of interest to the specific study.

The most appropriate temporal and spatial scales are also influenced by the objectives of the sediment transport measurement, and different measurement methods are suitable for different scales. Generally speaking, finer temporal and spatial scales may provide more accurate measurement at the target location during the study period. However, fine scale studies tend to be short-term with limited regional coverage and may not entirely capture regional characteristics and yield a representative regional long-term transport rate. Therefore, fine scales tend to be used in studies where the main objective is to quantify transport physics. On the contrary, coarser temporal and spatial scales focus on providing regional and long-term measurement, although local details and short-term variations tend to be neglected. For studies with the objective of obtaining long-term regional transport rate, coarse scales may, therefore, be more suitable than fine scales. Determining proper temporal and spatial scales comprises an important portion of the nearshore sediment transport measurement. As mentioned earlier, coastal sediment transport demonstrates various periodicities, and although it is unlikely that a single study will be able to resolve all of them, it is important to incorporate certain periodicities, especially when average values are to be measured and calculated.

It is worth emphasizing that both current and sediment-concentration profiles typically demonstrate logarithmic trends with rapid changes in the vertical direction. The slight errors in determining elevations may be exaggerated during vertical interpolation and calculation of sediment flux. The logarithmic shape of the profile should be taken into consideration during the vertical interpolation and sum, and calculation of depth-averaged values.

Since sediment transport rate is a vector, determining its direction is as important as determining its magnitude. The direction of the sediment flux is equal to the direction of particle velocity, or current velocity in most cases. The majority of current sensors measure the velocity along two perpendicular axes, and the vector sum yields the magnitude and direction of the total current. A proper coordinate system provides many conveniences in describing the transport direction. One of the most commonly used coordinate systems aligns one axis parallel to the shore and one perpendicular. However, this longshore and cross-shore system may not be the most efficient for coasts with complicated features such as inlets and river mouths, where careful consideration of the dominant regional features may be necessary.

Methods for measuring nearshore sediment transport

From its definition, sediment transport rate can be obtained by simultaneously measuring sediment concentration and particle velocity, or by measuring sediment flux directly. However, in dynamic and complicated coastal zones, direct measurement of sediment flux is sometimes not possible due to technical difficulties, and as a result, alternative indirect methods are often used. In the following, nearshore sediment transport measurements are discussed in two categories: direct and indirect measurement. Direct measurements obtain the sediment flux or via the product of sediment concentration and particle velocity. Indirect measurements involve the measurements of other quantities, such as morphological changes, from which sediment flux can be deduced.

Direct measurement of sediment transport rate

Numerous methods are available for measuring sediment concentration and particle (or flow) velocity. Usually, the principle difficulty is measuring concentration and velocity simultaneously at the same location. The most straightforward measurement of sediment concentration is to determine the weight of sediment particles in a unit volume of water, obtained from the study area. A commonly used method consists of an array of sampling bottles with a trigger device to open and close the bottles (Kana, 1979). A considerable area can be covered by each sampling event. Temporal coverage of this kind of measurement is usually limited because of the manual-driven sampling. Also, sampling interval is generally short limited by the typically small amount (no more than several liters) of water sample. In a wave-dominated environment, this method may not be able to provide a representative average over an entire wave cycle due to its short sampling. The short sampling interval also makes the simultaneous velocity measurement difficult. However, in less dynamic environments with limited wave influence, such as semi-enclosed estuaries, where sediment concentration may not change quickly, the instantaneous sampling may provide reasonable representation of average value. Direct deployment of sampling bottles is often replaced by a pump-suction device. By regulating the rate of intake, the sampling duration can be increased to provide measurement of average value over several wave cycles (Nielsen, 1984).

A variety of turbidity sensors are available for measuring sediment concentration. Among them, the optical backscatter (OBS) is commercially available and is probably the most commonly used. The OBS can be deployed and synchronized with current sensors to provide measurement of sediment flux. The automated and self-recording concentration sensors are capable of sampling at high frequency to resolve sediment suspension events, such as those driven by wave motions (Beach and Sternberg, 1992; Osborne and Greenwood, 1993). A limitation of sensors like OBS is that the output is significantly influenced by sediment properties, and therefore, it requires calibration using *in situ* sediment. Bottom sediment from the study site is often used for the calibration. However, it is not uncommon that bottom sediment is substantially different from sediment in suspension (Wang *et al.*, 1998a). Therefore, calibration using bottom sediment may introduce uncertainties in computing the OBS concentration. Comparison with direct concentration measurement as described in the previous paragraph is always desirable. The self-recording sampling allows long-term measurements, however, biological fouling, especially during the summertime necessitates periodic cleaning of the sensors. The spatial coverage of the turbidity sensors is often limited by their relatively high cost, since at least three are required for the measurement of a concentration profile.

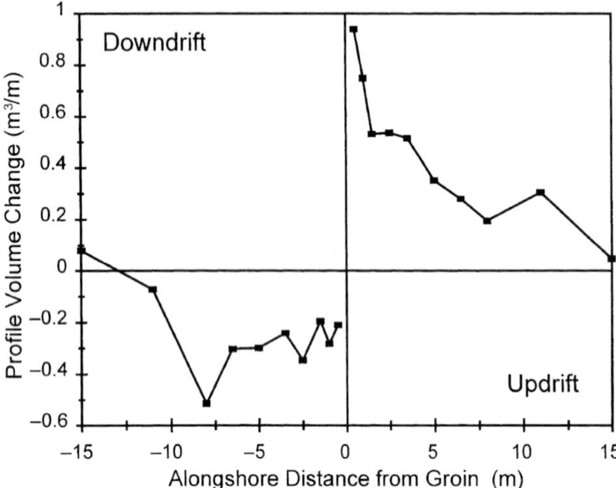

Figure N14 Sand accumulation at the updrift of a short-term impoundment, and erosion at the downdrift.

In order to obtain sediment flux, currents must be measured simultaneously with the concentration. Measuring current velocity is more complicated than measuring sediment concentration since velocity is a vector. Current measurement has been conducted for decades and the technology is fairly mature. The most commonly used current sensor is the electromagnetic current meter (EMCM), which measures currents along two perpendicular axes. Recently, the three-dimensional Acoustic Doppler Velocimeter (ADV), which measures particle velocity directly, has been used more and more frequently. The Acoustic Doppler Current Profiler (ADCP), which is capable of measuring three-dimensional current profiles throughout the water column, is becoming a promising technique in measuring coastal currents.

Sediment transport rate can be obtained by integrating sediment flux, which is commonly measured using sediment traps. Generally, there are two types of sediment traps, categorized by their functionality. Type 1 functions by reducing the intensity of the hydrodynamics to such an extent that the sediments that are moving with the fluid settle into the trap. Type 2 functions by providing a selective obstruction to sediment movement, for example, via the use of sieve cloth. The selective obstruction blocks the movement of the sediments and retains them in the trap, while allowing the majority of the fluid to pass through. Because it allows the passage of fluid, type 2 traps generally impose fewer disturbances to hydrodynamic conditions than type 1. Disturbance to hydrodynamic conditions may have a significant influence on the efficiency of sediment traps, as well as on the transport processes.

Due to the complexity of coastal hydrodynamics, no universal trap design functions well in all coastal environments. In the following paragraphs, sediment traps used to measure surf zone longshore sediment transport rate are discussed as examples. An example of the type 1 trap is described in Wang and Kraus (1999). A temporary shore-perpendicular structure was deployed as an obstruction to block longshore sediment transport, causing the sediment carried by the longshore current to be impounded updrift of the structure. The sediment flux was obtained by quantifying the volume of sediment accumulated at the updrift location during the period of experiment (Figure N14), which ranged from 2 to 6 h for the measurements conducted by Wang and Kraus (1999). Since the structure blocked longshore sediment supply, downdrift erosion occurred. According to mass conservation, the updrift accumulation should be equal to the down-drift erosion. The achievement of mass balance during the measurement provides a comprehensive check for data quality, which is valuable for non-repeatable and non-controllable field measurement. The disturbance the structure causes to hydrodynamics may introduce uncertainty in the measurement. Some permanent shore protection structures may function as total traps. Quantification of beach changes in the vicinity of these structures may provide reasonable estimates of regional sediment transport rate averaged over several months to a year (Dean, 1989).

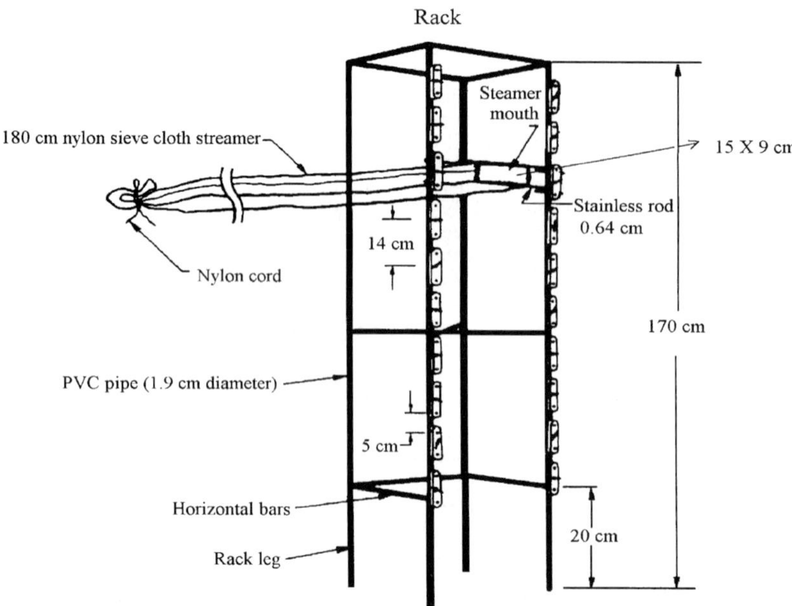

Figure N15 Design of the streamer sediment traps (modified from Wang *et al.*, 1998b).

The streamer sediment trap developed by Kraus (1987) and Wang *et al.* (1998b) provides an example of the type 2 trap (Figure N15). The streamer traps are made from sieve cloth of a mesh size dependent upon the size of the sediment in the study area. The traps not only allow water to flow through the sieve cloth but also move with the cross-shore wave motion, and therefore, impose minimal disturbance to hydrodynamics. The magnitude of the sediment flux is measured by the weight of sediment trapped in the streamer bag over the period of the experiment. The traps were operated manually, and the sampling interval ranged from 3 to 15 min, representing the value averaged over 60–200 waves. The longshore component of the sediment flux was measured by orienting the trap opening in the longshore direction. The streamer traps were mounted on a vertical rack, providing measurement of the vertical distribution of sediment flux. By deploying the trap array at different locations in the surf zone, cross-shore distribution of longshore sediment transport was also measured (Wang, 1998). Both the short-term impoundment and streamer sediment traps are fine-scale methods yielding nearly instantaneous measurements of longshore sediment transport, and are therefore more suitable for studying transport physics than obtaining regional long-term rates.

Indirect measurement of sediment transport

Generally, indirect methods can be classified into 2 categories. Method 1 provides an estimate of sediment transport rate via tracing the movement of a small amount of specially marked sediment particles (sediment tracers), from which the movement of all the grains is deduced. Method 2 estimates the rate of sediment transport by quantifying morphological changes resulting from net sediment transport.

Although an indirect method, sediment tracers were the primary method for measuring surf zone longshore sediment transport for over four decades (Ingle, 1966). The principal of the tracer method can be simplified as that a relatively thin layer of sand is assumed to move along shore at some average advection speed. The total longshore transport rate in the surf zone is then calculated as

$$Q = x_b Z \frac{P_2 - P_1}{t}, \qquad \text{(Eq. 3)}$$

where Z is the thickness of the moving layer, x_b is the surf zone width, P_1 and P_2 are locations of the tracer mass at the beginning and end of the experiment, and t is duration of the measurement. Accurately estimating Z through the vertical distribution of the tracer may be rather difficult (Kraus, 1985). Determination of location P_1 is straightforward, controlled by the initial tracer injection. Location P_2 can only be determined statistically and is influenced by many factors including the sampling scheme (White and Inman, 1989). The theoretical foundation and practical difficulties of the tracer method have been critically examined by Galvin (1987) and Madsen (1987).

Transport rate can also be estimated from morphological changes. The fundamental assumption is that morphological changes are caused by sediment accumulation or erosion, which are closely related to the rate of net sediment transport. For example, longshore transport rate can be estimated from the alongshore migration of barrier spits. From the volume change of ebb-tidal deltas, sediment transport in the vicinity of tidal inlets can be estimated (Davis and Gibeaut, 1990). These indirect methods are typically of coarse scales and tend to provide estimates of regional, long-term transport rate. Difficulties inherent in these indirect methods may include: (1) uncertainties in determining transport direction; (2) uncertainties involved in distinguishing the morphological changes caused by other processes such as tectonic subsidence and uplifting; and (3) uncertainties in estimating sediment bypassing. Also, morphological changes may be dominated by random extreme events.

Summary

Measuring nearshore sediment transport rate is a difficult task and no single method is universally applicable. A sound understanding of regional hydrodynamics and sediment-transport processes is important in adopting and executing the most suitable method for sediment-transport measurement. The first step in planning a transport measurement project is to determine the optimal temporal and spatial scales, which are partially controlled by the objective of the study. Some methods, such as sediment tracing or manual sediment trapping, are suitable for short-term measurement in a small region, while automated monitoring with self-recording instrumentation is more suitable for long-term study in a small, focused area. The few methods that are suitable for long-term, regional study are predominantly indirect techniques involving the examination of regional morphological changes. It is worth emphasizing that sediment transport rate is a vector and is described by not only a magnitude but also a direction. Sometimes, the directionality is simplified by studying the transport component in a pre-determined direction, such as transport in the longshore or cross-shore direction, respectively.

Ping Wang

Bibliography

Beach, R.A., and Sternberg, R.W., 1992. Suspended sediment transport in the surf zone; response to incident wave and longshore current interaction. *Marine Geology*, **108**: 275–294.

Davis, R.A., Jr., and Gibeaut, J.C., 1990. Historical morphodynamics of inlets in Florida: Models for coastal zone planning. Florida Sea Grant College Program, *Technical Paper 55*.

Dean, R.G., 1989. Measuring longshore sediment transport with traps. In Seymour, R.J. (ed.), *Nearshore Sediment Transport*. New York: Plenum Press, pp. 313–337.

Galvin, C., 1987. Vertical profile of littoral sand tracers from a distribution of waiting time. In *Proceedings of Coastal Sediments '87*, American Society of Civil Engineers, pp. 436–451.

Ingle, J.C., 1966. *The Movement of Beach Sand*. New York, Amsterdam: Elsevier Science Publishers.

Kana, T.W., 1979. Suspended sediment in breaking waves. University of South Carolina, Department of Geology, Columbia, South Carolina, *Technical Report No. 18-CRD*.

Kraus, N.C., 1985. Field experiments on vertical mixing of sand in the surf zone. *Journal of Sedimentary Petrology*, **55**: 3–14.

Kraus, N.C., 1987. Application of portable traps for obtaining point measurement of sediment transport rates in the surf zone. *Journal of Coastal Research*, **2**: 139–152.

Madsen, O.S., 1987. Use of tracers in sediment transport studies. In *Proceedings of Coastal Sediments '87*, American Society of Civil Engineers, pp. 424–435.

Nielsen, P., 1984. Field measurement of time-averaged suspended sediment concentration under waves. *Coastal Engineering*, **8**: 51–72.

Osborne, P.O., and Greenwood, B., 1993. Sediment suspension under waves and currents: time scales and vertical structures. *Sedimentology*, **40**: 599–622.

Van Rijn, L.C., 1993. *Principles of Sediment Transport in Rivers, Estuaries and Coastal Seas*. The Netherlands: Aqua Publications.

Wang, P., 1998. Longshore sediment flux in the water column and across the surf zone. *Journal of Waterway, Port, Coastal and Ocean Engineering*, American Society of Civil Engineers, **124**: 108–117.

Wang, P., and Kraus, N.C., 1999. Longshore sediment transport rate measured by short-term impoundment. *Journal of Waterway, Port, Coastal and Ocean Engineering*, American Society of Civil Engineers, **125**: 118–126.

Wang, P., Davis, R.A., Jr., and Kraus, N.C., 1998a. Cross-shore distribution of sediment texture under breaking waves along low-wave energy coasts. *Journal of Sedimentary Research*, **68**: 497–506.

Wang, P., Kraus, N.C., and Davis, R.A., Jr., 1998b. Total rate of longshore sediment transport in the surf zone: field measurements and empirical predictions. *Journal of Coastal Research*, **14**: 269–283.

White, T.E. and Inman, D.L., 1989. Transport determination by tracers—B: application of tracer theory to NSTS experiments. In Seymour, R.J. (ed.), *Nearshore Sediment Transport*. New York: Plenum Press, pp. 115–128.

Cross-references

Beach Erosion
Bypassing at Littoral Drift Barriers
Coastal Processes (see Beach Processes)
Cross-Shore Sediment Transport
Erosion Processes
Gross Transport
Longshore Sediment Transport
Monitoring, Coastal Geomorphology
Nearshore Wave Measurement
Sediment Budget
Sediment Suspension by Waves
Tracers

NEARSHORE WAVE MEASUREMENT

Definitions and key parameters

Generally, water waves are described by two length parameters, wave height and wavelength, and one temporal parameter, wave period. Wave height is the vertical distance between the wave crest and trough. Wavelength is the horizontal distance between two successive wave crests or alternatively, wave troughs. Wave period is the time needed for two successive crests or troughs to pass a spatial reference point. Direction of wave propagation is also an important parameter and critical when computing wave-induced sediment transport vectors. Scientific convention describes the wave direction as "direction to which it propagates" measured clockwise from the x-axis. In practice, however, wave direction is often reported as "direction from which the waves propagate," similar to the description of wind direction. It is necessary to specify the wave direction convention to avoid confusion. Depth over which the waves propagate is also important and is necessary in linking wave height and length to other parameters such as wave-induced water particle velocities and accelerations. Based on linear wave theory, wavelength and wave period are related through the dispersion relation as

$$L = \frac{g}{2\pi}T^2 \tanh\frac{2\pi h}{L}, \qquad \text{(Eq. 1)}$$

where L is the wavelength, g is the gravitational acceleration, T is the wave period, and h is the water depth. Although non-linear properties become increasingly more significant in shallow water than in deepwater, nonlinearities have rarely been considered during general wave measurements. Small amplitude (also referred to as Airy or Linear) wave theory has been used widely in coastal science and engineering research (Dean and Dalrymple, 1991).

A commonly used two-dimensional description of a small amplitude progressive wave is

$$\Phi(x, z, t) = -\frac{H}{2}\frac{gT}{2\pi}\frac{\cosh k(h + z)}{\cosh kh}\sin\left(kx - \frac{2\pi}{T}t\right), \qquad \text{(Eq. 2)}$$

where Φ is the velocity potential, k is the wave number defined as $2\pi/L$, x and z are the horizontal and vertical coordinates, respectively, and t is the time. The horizontal (u) and vertical (w) water particle velocities can be determined from the velocity potential as

$$u = -\frac{\partial\Phi}{\partial x}, \qquad \text{(Eq. 3)}$$

$$w = -\frac{\partial\Phi}{\partial z}. \qquad \text{(Eq. 4)}$$

Describing waves in the real world using one sinusoid, such as equation 2, is a significant simplification. A more realistic description would be a superposition of a large number of sinusoids, variable in direction. From a simplistic perspective, the objective of wave measurement can be generalized as an attempt to obtain the parameters of equation 2 to provide an optimal sinusoidal representation of real-world waves and their dominant direction of propagation. The purpose of this entry is to provide a summary of field methods used to measure these parameters, especially in nearshore regions. This review is prefaced, however, by an acknowledgment that the measurement of waves, and in particular their direction, has been one of the more difficult tasks in observational coastal engineering and oceanography.

Fundamentals of wave measurement

For the convenience of discussion, wave measurement is described in the following two categories: (1) measuring the sinusoidal parameters typically including wave height, period, and orbital velocities; and (2) measuring the direction of wave propagation. Measuring wave direction is more costly and instrumentation-intensive, and is more complicated with regard to data analysis, than measuring nondirectional waves.

Non-directional wave measurement

Generally, three approaches have been used to measure nondirectional waves and include measuring (1) the vertical acceleration of water particles; (2) orbital velocities of water particles; and (3) fluctuations in water surface. Approach (1) involves the use of wave-rider buoys, which are most commonly used in deep and intermediate depth wave measurement. Buoy use in coastal waters is limited by its relatively low sampling frequency and inability to measure water depth. Approach (2) requires complicated data processing procedures to obtain wave height. Approach (3) is most commonly used in coastal wave measurements and provides the most direct measurement of wave height and period. Most applications use wave height and period to describe waves, while orbital velocities are usually calculated based on a certain wave theory, for example, using equations 3 and 4 if linear wave theory is applied.

Directional wave measurements

In order to obtain wave direction, it is necessary to measure the nondirectional parameters simultaneously with one of the following: (1) the direction of water-surface slope; or (2) the direction of the water particle orbital velocity vector. Measuring the water-surface slope is accomplished using wave-rider buoys by recording the pitch and roll of the buoy. The water-surface slope can also be measured by deploying a specially arranged array of at least three water-surface variation sensors. Measuring the velocity vector is more frequently used to obtain directional nearshore waves by combining a velocity sensor with the water-surface variation sensors.

Wave direction can also be obtained through remote sensing techniques, such as radar and satellite imagery and has made considerable strides in large-scale ocean observing systems. A major advantage of this approach is that it provides a two-dimensional view, instead of a point measurement. Remote sensing methods remain to be innovative and promising for future wave measurements.

Wave sensors

For the convenience of discussion, wave sensors are divided into two categories, including (1) surface sensors and (2) subsurface sensors. There are two types of surface sensors, wave buoys, moving with the wave surface and recording the vertical acceleration of the wave motion, and surface piercing wave staffs which are generally attached to a fixed structure and measures the water-level fluctuations. The most commonly used subsurface sensors are the pressure transducers, measuring the pressure variations induced by water-level fluctuations. Water surface movement can also be measured with upward looking acoustic sensors mounted on the bottom, which measure the elevation of the air–sea interface.

Surface sensors

As discussed earlier, wave buoys are typically used in deepwater, where structures are not available to attach wave staffs and the water is too deep for bottom-mounted sensors. The application of buoys in coastal waters is seriously limited by its inability to measure water depth. Nevertheless, most ocean observing systems rely heavily on buoys for sea state measurement, and by and large, have provided unique data sets particularly during storms.

There are three main types of wave staff: resistance wire, capacitance, and electromagnetic transmission line staffs. The operating principle of this instrument is that the staff is a component of an electronic circuit which produces voltage changes proportional to the length of the staff that is either submerged or is not submerged. In turn this provides a measure of water-level fluctuation. Potential locations for wave-staff deployment are confined by mounting requirements. The mounting confinement also limits the options of incorporating directional measurements using wave-staff arrays. Biological fouling and electronic drifting limit the long-term applications of wave staffs.

Subsurface sensors

The most commonly used wave sensor in coastal waters is the bottom-mounted pressure transducer, which measures the pressure variation induced by water-level fluctuations. Directional waves can be measured by combining a pressure sensor and a velocity sensor. This combination comprises probably the most commonly used directional system for nearshore wave measurement, and is often called a PUV gage (Morang *et al.*, 1997).

A primary limitation of bottom-mounted pressure and current sensors is the increased attenuation of signal strength with increased water depth, referred to as depth attenuation. This results in a decreasing signal-to-noise ratio with increasing water depth. The magnitude of depth attenuation is a function of frequency, with high-frequency signals attenuating more rapidly than low-frequency signals. Depth attenuation of signal strength can be quantified by comparing measurements taken near the water surface and near the bottom (Figure N16).

Similar energy spectral density was measured both near the surface and the bottom for low-frequency components, while for high-frequency components, greater energy spectral density was measured near the surface than that measured near the bottom.

Depth attenuation can be corrected during data analysis. A commonly used procedure, derived from linear wave theory, applies a frequency-dependent depth attenuation factor, $K(f)$

$$K(f) = \frac{\cosh[k(z_s + h)]}{\cosh(kh)}, \qquad \text{(Eq. 5)}$$

where z_s is the depth of the sensor below the water surface (negative downward). Wave number and wavelength cannot be directly measured by the pressure sensor, or other point sensors, but are calculated based on the dispersion relation equation 1. The measured power spectral density, $E_m(f)$, is corrected as

$$E_c(f) = \frac{E_m(f)}{K^2(f)}, \qquad \text{(Eq. 6)}$$

where $E_c(f)$ is the corrected power spectral density. The attenuation-correction factor $K(f)$ increases exponentially as the frequency increases (Figure N17).

When pressure spectra are corrected by the above methods (equations 5 and 6), increasing energy density is sometimes seen for frequencies significantly greater than the frequency of maximum wave energy. This is shown by the sharp increase toward the tail of the corrected power spectrum for the sensor that was 2.9 m below the surface (Figure N16). This is a result of the fact that the attenuation correction amplifies both the signal and noise. The high-frequency noise can be amplified significantly when divided by a very small $K^2(f)$, causing considerable error in computing wave height and sometimes, peak wave period. Therefore, the spectra should be truncated at a high-frequency cutoff, above which incorrect noise amplification has occurred. Noise amplification and the determination of a high-frequency cutoff depend on a variety of factors including the characteristics of the individual sensor and data collection system, the depth of the sensor, and the height and period of the waves.

Proper determination of the cutoff frequency has significant influence on the calculation of significant wave height and other statistical wave parameters in coastal waters, such as in estuaries, where local wind-generated short-period waves often dominate. It is common for these waves to have peak periods of 2–4 s (0.50–0.25 Hz). As shown in Figure N17, if the sensor is mounted 4 m below the surface, a correction of over 10^3 will be applied to signals that are higher than 0.5 Hz. In other words, the noise beyond 0.5 Hz will be amplified 1,000 times. For open-ocean waves, components shorter than 2-s period are negligible. High-frequency cutoffs ranging from 3 to 6 s are often applied for open ocean wave measurements. However, in coastal waters, especially in semi-enclosed coastal water bodies, wave components with a period less than 2 s may not be neglected. Determination of the proper high-frequency cutoff may be dictated by local conditions, and careful study is necessary. The use of a fraction of the peak wave period may not be appropriate for bi-modal spectra, which are common in coastal waters.

In recent years, Doppler technology (e.g., the Acoustic Doppler Current Profiler—ADCP) has been used broadly to measure current–velocity profiles throughout the water column. There has been considerable interest in exploring their efficacy as a wave sensor, and the use of upward-looking ADCPs to measure both waves and currents was pursued and

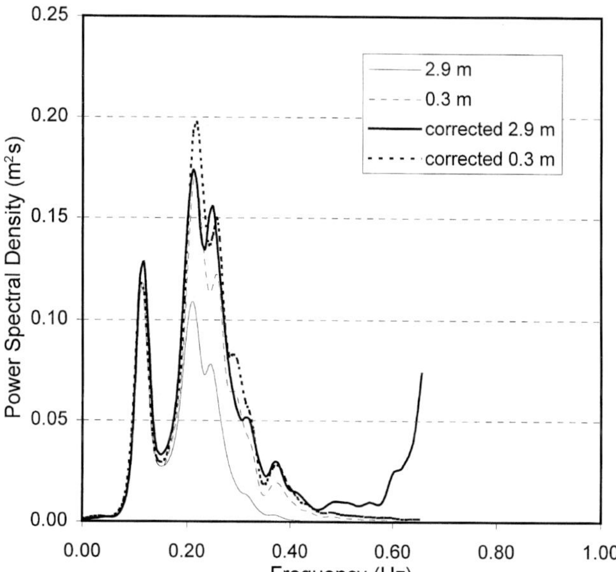

Figure N16 Wave spectral measured at different water depths, 2.9 and 0.3 m below surface, and depth-attenuation correction.

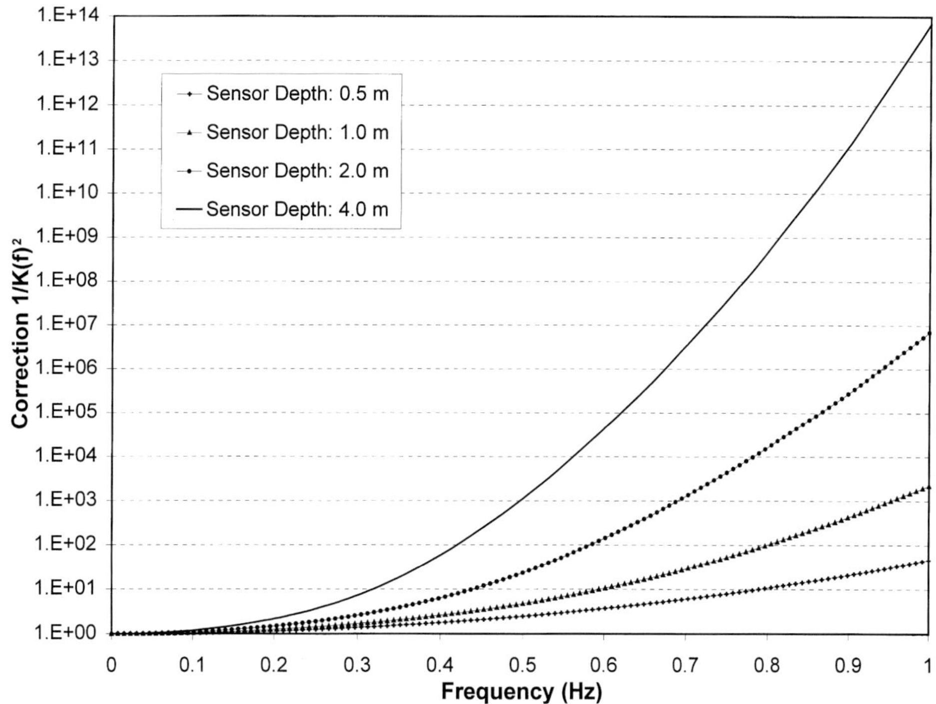

Figure N17 Increase of depth-attenuation correction with increase of frequency.

proved to be reasonably successful (Terray *et al.*, 1999). The spectrum of wave energy can be estimated from both the ADCP velocity measurements, and the direct echo-location of the surface along the four slant beams. It is worth noting that the ADCP is substantially more costly than the above wave sensors. The tremendous advantage is that a properly designed ADCP package is capable of yielding probably the most comprehensive point measurement of marine hydrodynamics including directional wave and the three-dimensional current profile.

Sampling schemes

Generally, a certain wave sampling scheme is composed of three aspects: (1) sampling rate of the sensor, often expressed in term of samples per second, or Hertz; (2) length of the sample; and (3) interval between each sampling events. The sampling rate and length are determined largely by the characteristics of the waves to be measured. The interval between samples is determined by several factors including the objectives of the research and capabilities of power supply and data storage.

The sampling rate should be rapid enough to capture the wave-induced water-level fluctuation. The minimum frequency that can be detected by a certain sampling rate, referred to as the Nyquist (f_N) or folding frequency, is

$$f_N = \frac{1}{2dt}, \qquad \text{(Eq. 7)}$$

where dt is the sampling rate. For example, if the sampling rate is 2 Hz, the minimum frequency that can be resolved is 1 Hz. However, it is worth mentioning that 2 Hz sampling may not capture the full magnitude of the 1 Hz signal, although the frequency is resolved. It is always desirable to sample at a faster rate than twice of the Nyquist frequency, and the latter should be determined with adequate knowledge of the shortest frequency that needs to be resolved, not the peak wave frequency. Each sampling event is typically called a wave burst. The duration, or length, of a wave burst is based on the peak wave period. As a rule of thumb, in order to obtain reliable statistical values of wave parameters, the length of a wave burst should be at least 150 times the peak period.

A general knowledge of the study area is important in determining the optimal sampling rate and wave-burst length. It is desirable to have a rapid sampling rate and long wave burst. However, these are often limited by the capacities of the power supply and data storage, especially for the bottom-mounted systems, which are typically powered by batteries and have a fixed amount of data storage. Optimal sampling balances the storage-consuming fast sampling and the power-consuming long wave burst with the most effective frequency coverage and reliable statistical properties.

The interval between wave bursts is an additional important parameter in planning a wave measurement program. Wave conditions in coastal waters can change quickly, thus it is preferable to have shorter intervals between wave bursts. However, more frequent wave bursts require more power and data storage. If the main goal of a wave measurement program is to obtain long-term averages of wave conditions, longer intervals between wave bursts, such as 3 or even 6 h, may be acceptable. If the main goal is to capture wave growth or development and dissipation of storm waves, shorter intervals, such as 1 h, are preferred. More than one sampling mode can be used, for example, a storm mode with 1 h interval and regular mode with a 3 h interval.

Wave data analysis

Nondirectional wave analysis

As discussed earlier, one sinusoidal form is not adequate to describe waves in the ocean. Waves are best presented in a statistical manner by means of spectral energy density and probability distribution. Numerous procedures exist for wave data analysis, largely based on concepts from time-series analysis and statistics. Details of wave analysis, especially that of directional waves, is complicated and beyond the scope of this entry. In the following, a procedure recommended by the Coastal and Hydraulic Laboratory of the US Army Corps of Engineers (Earle *et al.*, 1995), is discussed as an example. This procedure has been widely used in computing wave parameters, such as wave height, period, and direction. It is important to be consistent and to adopt a commonly used data processing procedure to ensure compatibility.

A key component of wave data processing is spectral analysis. Spectral analysis determines the distribution of wave variance as a function of frequency. Wave variance is proportional to the wave height squared, which in turn is proportional to wave energy. Thus, wave variance spectra are often referred to as wave-energy spectra. Spectral analysis is conducted using the Fourier Transformation. Wave spectra are typically calculated with units of variance/frequency so that integration over a frequency range provides the total variance of the wave components within the frequency range. Significant wave height (H_{mo}), which was originally defined as the average of the one-third highest waves in a wave record, can be calculated as

$$H_{mo} = 4 \times \sqrt{\sum_{1}^{m} E_c(f)\,df}, \qquad \text{(Eq. 8)}$$

where m is the number of discrete Fourier frequencies in the frequency band, and df is the small frequency interval over which the $E_c(f)$ is calculated. Another commonly used wave parameter is the peak wave period, which is defined as the reciprocal of the frequency of maximum wave variance. Other wave height parameters, such as the root-mean-square wave height (H_{rms}) and 1/10 maximum ($H_{1/10}$) wave height can be calculated assuming a Rayleigh probability distribution of wave heights:

$$H_{rms} = \frac{\sqrt{2}}{2} H_{mo}. \qquad \text{(Eq. 9)}$$

$$H_{1/10} = 1.80\,H_{rms} = 1.27 H_{mo}. \qquad \text{(Eq. 10)}$$

Directional wave analysis

Wave direction is important for applications such as harbor and structure design, wave refraction studies, and calculation of sediment transport rates (Massel, 1996). A directional spectrum, $E(f, \theta)$, is defined so that $E(f, \theta)df d\theta$ is the wave variance in a small frequency range df, and direction range dθ. Integration over all directions provides the frequency spectrum $E(f)$:

$$E(f) = \int_0^{2\pi} E(f, \theta) d\theta. \qquad \text{(Eq. 11)}$$

It is beyond the scope of this entry to examine in detail the measurement and determination of directional spectra (Earle and Bishop, 1984). Based on derivations of Longuet-Higgins *et al.* (1963), the directional spectrum can be simplified and represented by the first five Fourier coefficients

$$E(f, \theta) = \frac{a_0}{2} + \sum_{n=1}^{2} [a_n \cos(n\theta) + b_n \sin(n\theta)]. \qquad \text{(Eq. 12)}$$

The first five Fourier coefficients are calculated based on the cross spectrum analysis of two time series, such as water elevation and horizontal velocity vectors. A cross spectrum consists of a co-spectrum and a quadrature spectrum. The co-spectrum represents the variance of the in phase components of the two time series. The quadrature spectrum represents the variance of the out of phase components.

Since PUV gages are most commonly used in coastal water to obtain directional spectra, a simplified procedure of PUV data analysis is discussed in the following as an example. The five Fourier coefficients are calculated as

$$a_0(f) = \frac{C_{pp}(f)}{K_p^2(f)\pi}, \qquad \text{(Eq. 13)}$$

$$a_1(f) = \frac{C_{pu}(f)}{K_p(f)\,K_u(f)\,\pi(2\pi f)}, \qquad \text{(Eq. 14)}$$

$$b_1(f) = \frac{C_{pv}(f)}{K_p(f)\,K_u(f)\,\pi(2\pi f)}, \qquad \text{(Eq. 15)}$$

$$a_2(f) = \frac{C_{uu}(f) - C_{vv}(f)}{K_u^2(f)\,\pi(2f\pi)}, \qquad \text{(Eq. 16)}$$

$$b_2(f) = \frac{2C_{uv}(f)}{K_u^2(f)\,\pi(2f\pi)}, \qquad \text{(Eq. 17)}$$

where $C_{pp}(f)$ is the power spectral density of pressure at frequency f, $C_{pu}(f)$ and $C_{pv}(f)$ are the co-spectrum densities between pressure and velocities u and v, respectively, $C_{uv}(f)$ is the co-spectrum density between u and v, $C_{vv}(f)$ and $C_{uu}(f)$ are the power spectral densities of v and u, respectively, and $K_p(f)$ and $K_u(f)$ are the depth attenuation correction coefficients for pressure and velocity, respectively. There are two methods to calculate wave angle ϕ; using a_1 and b_1, or a_2 and b_2

$$\phi_m = \tan^{-1}\frac{b_1}{a_1} \quad \text{(Eq. 18)}$$

or

$$\phi_p = \frac{1}{2}\tan^{-1}\frac{b_2}{a_2}. \quad \text{(Eq. 19)}$$

Equation 18 is more frequently used, probably due to the relative ease in determining the quadrant. The result from using equation 19 is often referred to as the principal wave angle to differentiate it from the mean angle from equation 18. The quadrature spectrum is not used in the PUV data processing because the pressure and velocity sensors are typically at the same location and the signals are in phase. When an array of water-level sensors is used, the quadrature spectrum is necessary to examine the variance of the out of phase components.

Summary

Measuring waves in coastal zones is a complicated procedure and requires careful planning. Wave conditions in coastal zones can vary significantly from region to region. Coastal waves may also be substantially different from typical deep ocean waves in terms of spectral composition. Hardware and software developed for deep ocean swell measurement, such as large diameter disk buoys, may not be directly applicable to coastal wave measurements. Planning a coastal wave measurement project should include several considerations including objectives of the program, a general knowledge of the waves in the study area, types of instrumentation, sampling scheme, and data retrieval and analysis procedures. For wave measurements in semi-enclosed, high-frequency environments, it is important to have sensors close to the water surface to minimize the loss of important, high-frequency signals. A sound knowledge of the technical characteristics of different wave sensors and wave analysis procedures are also important in assembling an optimal system for a coastal wave measurement program.

Ping Wang and Gregory W. Stone

Bibliography

Dean, R.G., and Dalrymple, R.A., 1991. *Water Wave Mechanics for Engineers and Scientists*. Singapore: World Scientific.

Earle, M.D., and Bishop, J.M., 1984. *A Practical Guide to Ocean Wave Measurement and Analysis*. Marion, MA: Endeco, Inc.

Earle, M.D., McGehee, D., and Tubman, M., 1995. Field Wave Gaging Program, Wave Data Analysis Standard. Vicksburg, MS: US Army Engineer Waterways Experiment Station, Instruction Report, CERC-95-1.

Longuet-Higgins, M.S., Cartwright, D.E., and Smith, N.D., 1963. Observations of the directional spectrum of sea waves using the motions of a floating buoy. In U.S. Naval Oceanographic Office, *Ocean Wave Spectra*. Englewood cliffs, NJ: Prentice-Hall, pp. 111–136.

Massel, S.R., 1996. *Ocean Surface Waves: Their Physics and Prediction*. Singapore: World Scientific.

Morang, A., Larson, R., and Gorman, L., 1997. Monitoring the coastal environment; Part 1: waves and currents. *Journal of Coastal Research*, **13**: 111–133.

Terray, E.A., Brumley, B.H., and Strong, B., 1999. Measuring waves and currents with an upward-looking ADCP. *Proceedings of IEEE 6th Working Conference on Current Measurement*, IEEE Press, pp. 66–71.

Cross-references

Coastal Climate
Monitoring, Coastal Geomorphology
Wave Climate
Wave Environments
Wave Hindcasting
Wave Power
Wave Refraction Diagram
Waves

NET TRANSPORT

Net transport refers to the difference between the total upcoast and the total downcoast movement of sand approximately parallel to the shore over a specified period, often one year (see entries on *Cross-Shore Sediment Transport* and *Longshore Sediment Transport*). Waves approaching the shore at an oblique angle, such that the crests of the breakers are at an angle to the shoreline, generate longshore currents that convey the mobilized sand along the shore (see entry on *Waves*). In many locales, the wave approach direction varies over time such that this transport can be in either direction relative to the shore. Consequently, sand may move downcoast (positive direction) for an interval and then, as wave conditions change or the lesser effects of tidal or wind-driven currents intervene, reverse and move upcoast (negative direction). Summing the volumes of sand transport over the period of interest and taking into account their signs yields net transport. The customary units of net transport are volumetric rates (cubic meters per year). The terms littoral drift and longshore drift are occasionally used to mean either net transport or gross transport (see entry on *Gross Transport*) so caution must be taken in their interpretation.

In some special locations, because of physical barriers or climatic anomalies, waves approach from only one direction. In this case, the net transport can be identical to the gross transport. In most instances, however, this is not the case. Studies of the potential for longshore transport, based upon directional measurements of nearshore waves at a large number of coastal locations, have shown that the net transport is often a very small difference between large values of positive and negative transport. For example, at Oceanside Harbor in California, during the entire year of 1980, the net transport potential observed was less than 1% of the gross transport (Castel and Seymour, 1986)

Richard Seymour

Bibliography

Castel, D., and Seymour, R.J., 1986. Coastal data information program longshore sand transport report, February 1978–December 1981. Institute of Marine Resources, University of California, *IMR Reference No. 86-2*. 216 pp.

Cross-references

Cross-Shore Sediment Transport
Energy and Sediment Budgets of the Global Coastal Zone
Gross Transport
Longshore Sediment Transport
Sediment Budget
Waves

NEW ZEALAND, COASTAL ECOLOGY

Introduction

New Zealand has the fourth largest exclusive economic zone (EEZ) in the world (over 4 million km^2), and a long indented coast. The EEZ ranges from subtropical areas to subantarctic areas, though we confine our review to mainland New Zealand (i.e., North, South, and Stewart Islands). Our discussion considers areas from the upper limit of saltwater penetration to the edge of the continental shelf (mean depth 200 m), though we focus on coastal regions. Coastal regions supply New Zealanders with food, recreation, and a livelihood, with high-value near-shore capture fisheries and aquaculture being important contributors to the nation's economy.

The important features of the complex current patterns around New Zealand are becoming better known. The west coast is generally characterized by turbid, cold, productive seas, served in some areas by upwelling of nutrient-rich waters. Subtropical water impinges on northern New Zealand, and may reach down both coasts of the North Island in summer, influencing catches of pelagic gamefish. Interannual variation in the extent of those currents is large, and in recent years many previously unrecorded fishes and other taxa have been observed. The Subtropical Convergence (STC), a global oceanic front separating warmer subtropical waters to the north from cooler subantarctic waters to the south, intersects New Zealand's South Island. The STC passes around the southern end of the South Island, then follows the east coast of the South Island before moving offshore near Christchurch.

Intertidal habitats of New Zealand's coast include mangrove forests, sand and mudflats, salt marshes, surf beaches, and rocky platforms. Subtidally there are rocky reefs, harbor channels, sandy beaches, sounds, and fiords. Kelp forests occupy reefs, and solitary corals occur, and although there are no true coral reefs, reef-building corals do occur at the Kermadec Islands, 800 km northeast of Auckland. Over the 25 years after 1975, knowledge of coastal ecosystems has increased greatly, though Schiel (1991) was still able to suggest that very few processes structuring nearshore communities were understood.

Review of major ecosystems

Estuaries

The major biological habitats occupying intertidal areas in estuaries are mangroves, *Spartina* grass, and mud- and sand-flats. Mangroves occupy estuaries in the northern half of the North Island, and are thought to have important sediment-stabilizing and ecological roles there. Studies of mangrove populations have quantitatively and experimentally examined demography, litter production, and the inter-relationships between sediments and mangroves, but there are no comprehensive analyses of their effects on other fauna. Marsh grasses have been most intensively investigated in more southern areas, where attempts to control the introduced *Spartina* have had variable success. Sea grass (*Zostera* sp.) beds also form important habitat on intertidal areas throughout New Zealand. In recent years, some areas of sea grass are thought to have declined, perhaps due to various human activities, such as trampling on rocky intertidal platforms, or more subtle effects, perhaps related to water quality at the level of entire catchments. The other important plant of harbors is the green seaweed *Ulva lactuca*. Blooms of this species have occurred in several New Zealand estuaries, and are thought to indicate eutrophication. The seaweed also accumulates on shores of harbors after storms and rots, causing problems for local homes.

Throughout New Zealand the cockle *Austrovenus stutchburyi* is prominent in intertidal faunas of more sheltered shores, being harvested commercially throughout its latitudinal range. Several estuaries (e.g., Manukau, Firth of Thames; Figure N18) are important foraging areas for populations of birds that migrate to the Northern Hemisphere. The best-known faunal components of those areas are clams, and intensive studies of interactions between sedimentology and biology have been undertaken in the Manukau Harbour (e.g., Thrush *et al.*, 1997).

The dominant subtidal features of northern harbor channels are dense populations of pipi *Paphies australis*, often in association with abundant 11-armed starfish *Coscinasterias muricata*. The dense covers of shell associated with such areas probably modify sediment mobility, and the abundances of shellfish are possibly high enough to influence water quality. Harbors are thought to be important to stocks of commercially important fishes, and the fish fauna of harbors and estuaries in northern New Zealand is being investigated in a comprehensive program at present (2000). The fauna of Otago Harbour, near Dunedin and areas of Stewart Island have also been described, and feature broadly similar organisms to those in more northern areas. One important aspect of population biology that is not well understood is how the population biology of even common species varies along latitudinal gradients. Casual observations suggest large differences in the densities

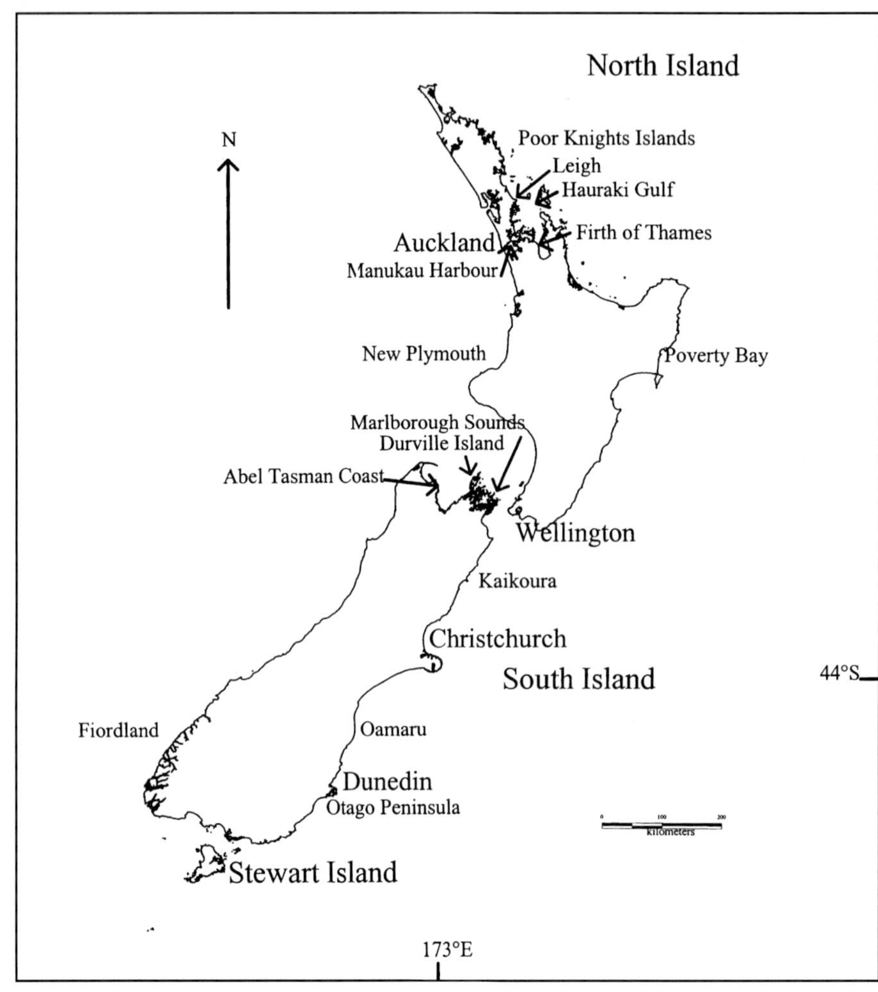

Figure N18 Map of New Zealand, showing localities mentioned in the text.

and sizes of seaweeds, intertidal bivalves, echinoderms, and fishes between northern and southern populations, but as yet there have been few focused studies on latitudinal patterns. One study has compared sites on the east and west coast of the South Island, but it is not clear whether such differences relate to broader-scale oceanographic processes or some other environmental difference between coasts.

Intertidal rocky shores

Considerable research investigating the rocky reef platforms in eastern South Island has been undertaken by University of Canterbury, with some additional comparative information available for the west coast. On the east coast, sea grass *Zostera* sp. and Neptune's necklace seaweed *Hormosira banksii* may occupy large areas of reef. Experimental investigations indicate complex patterns of demography of the sea grass that relate to position, and physical variables such as standing water and temperature. The seaweed is affected by trampling, but recovery is complex and influenced by season, location, and presence of corallines. Similar effects have been shown for trampling of coralline turfs in northeastern New Zealand.

Lower reaches of wave-exposed rocky shores of the southern half of the South Island are occupied by very large laminarians (Schiel, 1994). Earlier studies indicated geographic differences in the influence of grazers on the extent occupied by the bull kelp *Durvillaea antarctica*, and other patterns in morphology that were related to wave action. Research into the seaweed karengo at Durville Island is investigating the possibility of enhancing populations for iwi (local Maori).

Subtidal reefs

Most knowledge of subtidal reef ecology stems from work done in northeastern New Zealand, though additional information exists for New Plymouth, the Wellington coast, the Marlborough Sounds, the Abel Tasman coast, and Kaikoura (Figure N18). The northern studies have emphasized the role of seaweeds in structuring habitats, the importance of small mobile invertebrates such as amphipods in contributing to the productivity of those reefs, and the role of humans in modifying the populations of organisms on those reefs (see below).

Seaweed morphology has been shown to be important in several disparate ways in northeastern New Zealand. Surface area of seaweeds may be important to photosynthetic processes, as ammonium metabolism appears confined to surface layers of cells. A series of studies have shown that the secondary productivity of reefs in northeastern New Zealand is highest in stands of finely structured seaweeds, and that under certain circumstances the excretory products of the small animals might be of value as a nitrogen source to the seaweeds. Studies near Leigh documented large differences in life expectancy, morphology, and palatability to grazers for wave-exposed and wave-sheltered populations of the large fucalean *Carpophyllum flexuosum*. Investigations currently being undertaken near Dunedin examine the influence of wave action on large laminarian seaweeds directly by measuring wave forces.

The sea urchin *Evechinus chloroticus* forms prominent "barrens"—areas devoid of seaweeds—in northeastern New Zealand. Further south the extent of those barrens may be diminished, and in some areas, such as near New Plymouth, extensive barrens can occur in the presence of small abundances of *Evechinus*. In northern New Zealand, the large diadematid sea urchin *Centrostephanus rodgersii* may be abundant, and in deeper water the endemic diadematid *Diadema palmeri* may also be common. The other prominent grazers of New Zealand reefs are gastropods, which may be highly abundant both on the rocky seabed and on the fronds of seaweeds. Experimental investigations suggest that it is the grazing activities of *Evechinus* that structure seaweeds on reefs, and that gastropods mainly respond to the grazing of echinoids.

Seaweeds have also been shown to be important in structuring populations of fishes. Some fishes recruit into seaweed stands on reefs, others use detached seaweeds floating offshore as shelter (and may be moved onshore as a result of that association), still others feed mainly in areas where those seaweeds do not occur, and there is a growing base of knowledge regarding the biochemical pathways whereby herbivorous fishes process seaweeds. Small mobile crustaceans such as amphipods that contribute most of the secondary productivity on reefs, form an important an component of diet for nearly all reef fishes.

Open coast beaches and inner shelf

The dynamics of fauna occupying sediments of the inner shelf are relatively unknown, though much research attention has been focused on the effects of dredging of scallops and trawling. Again, dense populations of bivalves occur, some of which are implicated in structuring populations of other organisms. The best known of these is the horse mussel *Atrina zelandica*, dense beds of which occur throughout New Zealand. *Atrina* has a fragile shell, appears to have strong and weak year classes, is known to influence sedimentation and meiofauna, and to be influenced by physical disturbances such as dredging and anchoring. Other bivalves are important in other habitats; dense populations of the venerid *Tawera spissa* occur in northeastern New Zealand, and in Marlborough Sounds large areas are occupied by live and dead individuals of the dog-cockle *Glycymeris laticostata*. Beds of both *Tawera* and *Glycymeris* are sufficiently dense that they potentially structure the fauna of areas. Although large amounts of drift seaweed may occur on the seabed offshore, most investigations have examined its decay onshore, or its role in concentrating plankton at the sea surface.

Studies of sewage outfalls and dredge spoil disposal in offshore areas such as Tauranga and Gisborne generally suggest that the influence of those disturbances are localized and/or difficult to separate from natural variation. Biological investigations of coastal sediments have lagged behind physical oceanographic studies; in recent years the physical oceanography of the East Auckland Current, Manukau Harbour, Poverty Bay have all been described in some detail. The biological consequences of those processes are only well understood for the East Auckland Current, however, and there are few, if any, studies of the influences of such physical processes on organisms occupying sediments.

Fiords

The fiords of southwest South Island have an estuarine circulation system, which is maintained by the input of freshwater from the prodigious rainfall (>6 m per annum). Fiordland has a long complex coastline of inlets and islands, providing shelter from the prevailing westerly winds. Rainwater runoff from forests leach tannins from leaves and stains the surface fresh-water layer (up to 6 m thick) brown, limiting light penetration. Divers must pass through a murky freshwater layer in which visibility is poor, before bursting through into the clear saltwater below. In the low-light environment antipatharian corals may commonly be found as shallow as 10 m, and other deepwater emergent species also occur there. There are strong gradients in the distribution of several species along the length of the fiord (e.g., large brown seaweeds), and the freshwater layer influences grazing on mussels by starfishes. Starfishes are sensitive to salinity, whereas the mussels are able to resist freshwater incursions to some extent, so that mussels have a refuge in shallow water.

The Fiordland area is a renowned tourist destination, fishing area, source of hydro-electric power, and has been mooted for freshwater export. There is a readily accessible underwater observatory and a proposal for submarine tours. Visitors are concentrated in one readily accessible fiord, and other areas are subject to much lower visitor densities. However, some renowned dive destinations are under pressure from the large numbers of divers visiting. That effect is magnified as several of the species that are drawcards (such as red and black corals) are fragile and may be broken by careless divers. Such problems are likely to emerge with continuing diver pressure on areas such as Poor Knights Islands and Cape Rodney—Okakari Point Marine Reserve, near Leigh.

Marine birds and mammals

The best known New Zealand higher marine vertebrates are Hector's dolphin, seals, and penguins in southern New Zealand, and the huge flocks of wading birds in northern harbors. Hector's dolphin is endemic, permanently coastal, and thought to be endangered due partly to its susceptibility to set nets. Marine mammals and penguins are important tourist drawcards in southern New Zealand, particularly on the Otago Peninsula. Sperm whales are also abundant off the Kaikoura coast feeding in the productive waters overlying the Kaikoura Canyon. Conflicts between fisheries and Hooker's sea lion occur in some offshore fisheries, and modifications to fishing practices are used to minimize seal mortalities. In some areas of northeastern New Zealand feeding by birds in intertidal areas may be sufficiently intense to locally reduce abundances of clams.

Fisheries and aquaculture

New Zealand promotes a broad range of capture fisheries and aquaculture round its coast. Important coastal capture fisheries include spiny lobster *Jasus edwardsii* and snapper *Pagrus auratus*. Value of exports is increasing faster than the volume caught, reflecting at least partly improved handling techniques. Studies in several areas suggest that fisheries (both recreational and commercial) are a major influence on the marine environment and populations of marine animals. Environmental concerns are emerging regarding the environmental effects of some fisheries, and the role of commercial fishers is increasingly one of stewardship. The indigenous Maori people also have an important and growing stewardship role. Recreational fisheries can be important to populations of both the fish and the fishers, with snapper in the Hauraki Gulf, blue cod in the Marlborough Sounds, and kahawai throughout the country being among the most targeted species.

Studies of the benthic impact of trawl and dredge fisheries have revealed reductions in faunal richness and damage to other species. Some of those studies have revealed patterns at broad spatial scales that are consistent with fisheries being a dominant influence on benthic ecology. Concern regarding benthic impacts has led to closure of areas in northern New Zealand to scallop dredging. There are clear advantages for highly targeted, "clean" fisheries, which minimize bycatch. The benefits of such fisheries may extend into the overseas markets, where increasingly greater returns can be obtained for animals or plants that are captured in environmentally benign ways. Further, capture in such ways may have advantages in product quality and ultimately price, as greater shelf life follows from selective, highly controlled harvest.

The aquaculture of GreenShell™ mussels *Perna canaliculus* is a major industry, particularly in the Marlborough Sounds, but farms are found from the north of the North Island to Stewart Island. Mussels are grown on longlines, from which droppers bearing the mussels hang down (usually 10–12 m). Growers hope to harvest mussels within 14–17 months of seeding out, and aquaculture leases in productive areas are highly sought after. The environmental effects of such activities have been little investigated, and concerns regarding sustainability and benthic impacts are mounting. Localized depletion of phytoplankton within farms has been documented, but increasingly there are concerns of depletion at the level of entire embayments. Measurements of currents, phytoplankton abundances, and computer models are being used to assess the viability of proposed farms. Environmental effects of salmon farming have been studied, but that activity occurs in only a few locations, mainly in southern New Zealand. Other cultured species include Pacific oysters, cockles *A. stutchburyi,* and developed paua (abalone) *Haliotis iris* ventures. As yet few pharmaceutical products are being harvested from aquaculture ventures. As yet few pharmaceutical products are being harvested from aquaculture ventures, though these have considerable potential for high returns.

Introduced species

Some of New Zealand's important coastal fisheries are for introduced species. The Pacific oyster was introduced in the late 1960s or early 1970s, and is now an important aquaculture industry. Sea-run trout and salmon provide valuable recreational fisheries in the South Island, with a trawl bycatch of salmon having been important in the past. However, not all such introductions have positive outcomes, and international experiences (some as close as Tasmania) indicate that introduced species can have major negative effects on many aspects of coastal ecology. Other introduced species will colonize New Zealand shores, and it appears that managing the inevitable arrivals is the focus of current efforts. Pamphlets detailing candidate invaders are circulated by NZ Ministry of Fisheries, with the aim of detecting exotic arrivals early.

A pest management plan has devised for the Asian kelp *Undaria pinnatifida*. However, that species is atypical of introduced species in many ways, in that it is commercially valuable, and that it has an extremely dispersive phase, linked to human activites. *Undaria* has spread throughout southern New Zealand, as far north as Gisborne in Poverty Bay, and in a wide range of environments from harbors to wave-exposed shores. Intense efforts are being made to prevent its spread at Stewart Island, although the success of the seaweed elsewhere has been variable. In Marlborough Sounds, *Undaria* has flourished in sheltered areas, persisted on shallow sills that are impacted by ferry wash, survived well on marine farms that it has recruited to, but failed to persist in subtidal areas adjacent to farms in several areas. It appears that human activities such as boating and aquaculture are the major vector, that grazers on natural shores are able to limit the abundance of *Undaria* in sheltered area but perhaps not in wave-exposed areas (e.g., Oamaru), and that the abundance and species composition of animals that occupy *Undaria* are distinct from natural seaweed stands on the same shores. There are conflicting opinions regarding whether the species should be exploited as a valuable resource or whether it should be extirpated.

Another invader, the Asian date mussel *Musculista senhousia*, is confined to northern harbors and appears to have limited capacity for long-term occupation of areas. It forms matlike localized beds, but the beds appear not to persist. In the Marlborough Sounds some marine farms have had serious problems with recruitment of a solitary ascidian *Ciona intestinalis*. It is thought to compete with mussels for food, and heavy recruitments are sufficient to diminish growth. Concerns regarding biosecurity are mounting, and an awareness of the threat posed by organisms in ballast water has prompted studies of the fauna of ballast tanks, and investigations of ways of treating it to ensure that foreign organisms would not survive. However, other species are able to survive externally on ship hulls (e.g., *Undaria*), and to protect the value of New Zealand's marine resources, great emphasis will have to be directed into surveillance of our borders, and studies of the biodiversity and susceptibility of natural populations. The experience with *Undaria* indicates that studies elsewhere are not able to be directly transferred to the New Zealand situation.

Marine reserves

New Zealand has been at the forefront of marine reserve creation since 1975, and as of October 2000 has 16 marine reserves in which all marine life is protected under the Marine Reserves Act 1971. One of the limitations of that legislation is that it permits marine reserves to be established solely for the purpose of scientific study, and public enthusiasm for marine reserves as a conservation method seems to have outstripped the need for scientific investigation. Several other areas are protected under different legislation, and designated as marine parks. There are also several taiapure and mataitai reserves, which are under the control of local Maori to varying degrees. Marine reserve legislation is currently under review to determine whether the current Marine Reserves Act 1971 is appropriate. It seems likely that it will be reshaped so as to give a broader range of reasons for protecting particular areas.

These protected areas have provided considerable information regarding ecological interactions in coastal habitats. The best-known marine reserve is the Cape Rodney to Okakari Point Marine Reserve, near Leigh, north of Auckland, which has been fully protected for almost 25 years (as of 2000). Partly because of its ready access, but also perhaps because of the duration of protection, it has provided considerable information regarding interactions among organisms on coastal reefs. Fish and spiny lobster populations have been investigated in great detail, and there is some indication that neighboring areas benefit from the existence of the reserve. The most interesting development has been the loss of sea urchin-dominated "barren grounds" (areas with no seaweed) at Leigh, and the reduction of their area in several other northern reserves. It has been suggested that the replacement of barren grounds by kelp forests in reserves reflects the high abundance of predators that consume sea urchins, though that conclusion is contentious. Those studies are of considerable importance internationally, since only in New Zealand have completely protected marine reserves been established for sufficient duration to determine long-term effects.

Scientific investigations targeting blue cod and spiny lobsters have recently been undertaken in more southern marine reserves, and have generally found evidence of increased abundances and/or size similar to those done in North Island marine reserves. As more studies are undertaken in reserves the generality of effects will emerge. It is not surprising that humans have important effects on populations of marine organisms, given the intensity of fishing effort, improved positioning capacity via GPS, and improved quality of access due to changes in both roading and vessel capabilities. Strong gradients of fish abundance and size in relation to access and/or protection status are obvious in many fisheries.

Summary

New Zealand's coast contains most physical environments and biological habitats common in temperate regions worldwide. Knowledge of physical and biological environments is increasing rapidly. Direct and indirect human activities are important influences on marine systems. Coastal developments, introduced species, aquaculture and marine reserves will be important influences on nearshore marine systems in the future.

Russell Cole and Conrad Pilditch

Bibliography

Schiel, D.R., 1991. Coastal biology and the New Zealand Journal of Marine and Freshwater Research, 1967–91. *NZ Journal Marine and Freshwater Research*, **25**: 415–427.

Schiel, D.R., 1994. Kelp communities. In Hammond, L.S., and Synnot, R.N. (eds.), *Marine Biology.* Melbourne: Longman Cheshire.

Thrush, S.F., Pridmore, R.D., Bell, R.G., Cummings, V.J., Dayton, P.K., Ford, R., Grant, J., Green, M.O., Hewitt, J.E., Hines, A.H., Hume, T.M., Lawrie, S.M., Legendre, P., McArdle, B.H., Morrisey, D., Schneider, D.C., Turner, S.J., Walters, R.A., and Whitlatch, M.R., 1997. The sandflat habitat: scaling from experiments to conclusions. *Journal of Experimental Marine Biology Ecology*, **216**: 1–9.

Cross-references

Aquaculture
Climate Patterns in the Coastal Zones
Estuaries
Mangroves, Coastal Ecology
Marine Parks
New Zealand, Coastal Geomorphology and Oceanography
Rock Coast Processes
Sandy Coasts
Vegetated Coasts

NEW ZEALAND, COASTAL GEOMORPHOLOGY AND OCEANOGRAPHY

Historical background

New Zealand's coastal geomorphology exhibits a myriad of coastal landforms and sedimentary deposits, varying from estuarine types of fiords, rias, barrier enclosed estuarine lagoons, eroded calderas, and flooded grabens; to rocky cliffs and shore platforms, raised coastal terraces; to modern sandy Holocene barrier islands and spits, modern long sandy beaches and bayhead pocket beaches; and various types of nearshore sedimentary facies, to Pleistocene paleo-barriers; to various types of mixed sand-gravel beaches, boulder beaches and barriers; to muddy tidal flats and shelly gravel chenier ridges and beaches.

From the early 1900s, the classical geomorphology of New Zealand was widely expounded in the publications of Sir Charles Cotton, and especially in his book "*Geomorphology of New Zealand Part 1: Systematic*" first published in 1922 and revised as "*Geomorphology: An Introduction to the Study of Landforms*" (Cotton, 1942). About one-fifth of this classical text was devoted to coastal processes and landforms, including especially submerged and emerged coasts. Cotton (1974) later published the book "*Bold Coasts*" comprising a series of his papers on rias, cliffs, and rocky coasts and drawing widely on New Zealand examples. Cotton's pioneer work was focused primarily on deductive interpretation of erosional geomorphic features such as cliffs, headlands, and "old hat" platforms, from the Davisian geomorphic cycle perspective. Little emphasis was accorded coastal sedimentary deposits or investigation of coastal processes, although naturally at that time there was little technological capability to measure coastal geomorphic processes. Likewise some of the early New Zealand geologists and geographers, viz. P. Marshall, J. Bartrum, and G. Jobberns, wrote deductive papers on aspects of coastal geomorphology, for example, the origin of gravel in the Hawkes Bay beaches, and the origin of shore platform high level benches around the Auckland area. Jobberns was the first specialist "coastal geomorphologist," and researched the raised strandlines of North Canterbury (McLean, 1977). But at that time there was no fundamental research involving quantitative measurement of coastal sedimentary deposits or landforming processes. Healy and Kirk (1982) addressed this issue, and presented a more modern synthesis of New Zealand coastal geomorphology.

Figure N19 Professor Roger McLean, who may be regarded as the "father" of modern coastal geomorphology in New Zealand.

J.C. Schofield (1960) is credited with undertaking the first substantive surveys combined with shallow cores and dating to obtain quantitative data for New Zealand coastal evolution. His pioneering paper on the origin of a sequence of 13 progradational chenier ridges at the muddy head of the Firth of Thames attempted to relate progradational phases to oscillations of late Holocene sea levels. Later Schofield (1970) also published the first substantive regional beach sedimentological investigations based upon textural and mineralogical analysis of multiple samples from representative sand beaches of the northern half of the North Island.

The first modern systematic beach studies—involving quantitative measurement of form and process of Canterbury mixed sand-gravel beaches (McLean and Kirk, 1969)—were established at University of Canterbury in the 1960s by R.F. McLean, who may be regarded as the "father" of modern coastal geomorphology in New Zealand (Figure N19). McLean's student, R.M. Kirk, has continued coastal geomorphic studies at Canterbury since the 1970s. In 1973, sandy beach and estuarine sedimentary and process research was initiated at University of Waikato by T.R. Healy, where since 1980, and in collaboration with the pioneering developments in numerical simulation of two- and three-dimensional flows, waves, and sediment transport modeling of K.P. Black, the largest undergraduate and graduate school for coastal geomorphic and oceanographic studies in the country developed in the 1990s. Coastal geomorphic studies were also established at University of Auckland from the late 1970s, initially under the leadership of R.F. McLean and subsequently by K. Parnell. In the 1980s, a coastal unit with emphasis on numerical modeling was established within the then Ministry of Works and Development, but with restructuring of scientific research in the early 1990s, this group became part of the National Institute of Water and Atmosphere (NIWA), and under the leadership of T.M. Hume, has since grown into the largest group of coastal geomorphic and oceanographic researchers with strong national standing in New Zealand.

Environmental factors influencing New Zealand coastal geomorphology and processes

The distinctive diversity of coastal geomorphic types and sedimentary deposits characteristic of the extensive New Zealand coastline evolves from a combination of geological structure, tectonic and seismic history, lithology, a mid-latitude oceanic setting for wave and tidal processes, Pleistocene events, and climatic influences.

Geological structure and lithology. A series of lineal axial ranges comprise the essential structural backbone of New Zealand. In the North Island, these consist of lightly metamorphosed and intensely jointed Mesozoic greywacke, and in the South Island, of greywackes and metamorphosed Paleozoic schists of Otago and the alpine fault zone through to gneiss and granodiorites of south Westland. On-lapping the axial ranges are younger Tertiary rocks, predominantly soft clay-rich siltstones and some limestones, while ancient volcanic mounds punctuate the landscape. In the central North Island the active Taupo Volcanic Zone (TVZ), stretches from Lake Taupo to beyond the Bay of Plenty coast. The northern North Island exhibits evidence of back arc andesitic and acid volcanism, with basaltic volcanism around Auckland and Northland. This diversity of lithology provides a wide variety of topographic forms and sediment for distinctive beach types and coastal landscapes.

Influence of structure and Pleistocene events. New Zealand, like Japan and northern California, sits astride a plate margin where the coastal geomorphology closely reflects the adjacent lithology as well as historical geological events. Mesozoic and Tertiary orogenies created the basic axial ranges, the backbone of the country. Ongoing plate margin evolution saw the development of the transcurrent Alpine Fault stretching from the south of the South Island along the southern Alps and across Cook Strait to Wellington and through to Hawkes Bay. Along this transverse fault system the active earthquakes and tectonic dislocations have influenced the coastal geomorphology by creation of drowned transverse fault aligned valleys (rias) of the Marlborough Sounds and uplifted coastal terraces around Wellington.

An important element of structure is the alignment of the main islands comprising New Zealand along the plate margin and active subduction zone. Subduction of the Pacific plate under the central North Island results in a number of geologic processes including the active

TVZ, rapid tectonic uplift, folding, faulting, and stratal slumping with associated high seismic activity. These seismic and tectonic (uplift and downsinking) processes in the geological evolution, which are continuing at present, have occurred on both a regional and local scale. They enhance denudation of the hilly and highland catchments, providing high sediment loads to the coasts. Regional downsinking has helped create the rias of the Marlborough Sounds, while regional uplift in the central North Island has resulted in incised rivers without significant estuaries (e.g., Waitara River) cut into coastal terraces.

Events in the Pleistocene also played a major role in molding the modern coastal landscapes. Severe valley glaciation in the south Westland gneissic province produced glaciated U-shaped valleys, which upon drowning in the Holocene transgression have become classical fiords. The intense physical weathering of the fractured greywackes and schists, associated with the glaciations in the Southern Alps, provided outwash sands and gravels to build up alluvial fan deposits, the most striking example of which is the Canterbury Plains. These subsequently became blanketed in loess. Periglacial weathering action in the ranges of the North Island assisted in delivering a large volume of gravel sediment to the lowland plains and coast, notably in Hawkes Bay. Of course at that time the lowland plains extended 100–120 m below present sea level, so that many sand deposits were reworked over the inner continental shelf as sea level rose with the post-glacial transgression.

Influence of volcanism. Both modern and ancient volcanic influences are evident in the coastal geomorphic landscape. Eroded caldera in the ancient volcanic mound of Banks Peninsula today contain the harbors of Lyttelton and Akaroa. Volcanic deposits have also had an important influence in the central North Island west coast where erosion of lahar flows provides both cobbles and boulders for veneer beaches over eroding shore platforms, as well as the distinctive black titano-magnetite heavy mineral sand for the extensive sand littoral drift systems of the North Island west coast. Propylitized Miocene andesites and Pliocene rhyolites and ignimbrites comprising the Coromandel Ranges are back arc volcanics and have provided distinctive mineralogy for the beach sands of the embayed east Coromandel Peninsula.

Around the Auckland isthmus some of the numerous Pleistocene basalt volcanic cones are now eroded headlands, with the outstanding landscape feature being the recently formed island of Rangitoto, which erupted as Hawaiian-type fluid basalt flows, and is surmounted by a scoria cone crater active as recently as 600 years ago. Pleistocene and modern active volcanism from the TVZ has demonstrably influenced coastal evolution in the Bay of Plenty, where denudation of the acid volcanic tephras has provided large volumes of sand to the littoral zone. This has accumulated in the broad embayments to create extensive Holocene dune ridge progradations. The eruption of Mt. Tarawera (1886), which blanketed the surrounding landscape with pumicious deposits, provides detailed evidence of the impact of a large influx of sandy sediment into the coastal littoral system, and the air-fall ash deposit on the coastal dunes allows reliable dating of progradation rates (Pullar and Selby, 1971; Richmond *et al.*, 1984). The ability to use the technique of tephrachronology has allowed reliable dating of Pleistocene and Holocene dune and barrier deposits and interpretation of rates of geomorphic evolution along both the west and east coasts of the North Island.

Wave and climate influences. The mid-latitude oceanic islands of New Zealand are dominated by southwesterly waves originating from the Southern Oceans, and driven by the prevailing southwesterly winds. These waves, typically of period ~8–11 s and height ~1–3 m, drive littoral drift systems within regional and local coastal compartments along the west coasts. On the east coast of the South Island and the southeast coast of the North Island the wave climate is likewise predominantly swell from the southern oceans. North of East Cape (North Island), the embayed coast—a "lee coast" from the prevailing westerly winds—is sheltered from the southerly swells and is subject to a mixture of distant Pacific swells and local storm and wind- generated waves, typically with periods of 7–9 s and heights of 0.5–1.5 m from variable onshore directions. Episodic storms, especially decaying tropical cyclones from the north, generate an important component of the wave climate which cause dune erosion problems.

The El Niño–Southern Oscillation (ENSO) plays a major role in the beach geomorphology and coastal oceanography. In El Niño conditions New Zealand experiences continual 15–20 knot winds from the southwest. This causes a surfeit of orographically induced rainfall and flooding particularly in Westland of the South Island, and enhancing sediment supply to the coast, while the constant onshore wind stress induces downwelling on the coast and inner shelf. But on the east coasts, particularly of the North Island, the continual offshore wind stress induces drought conditions, onshore transport of nearshore sands, and inner shelf oceanographic up-welling. As a result the east coast beaches become generally well accreted in El Niño conditions and the coastal upwelling brings nutrient-rich ocean waters to the inner shelf, which enhances conditions for algal and phytoplankton blooms in the nearshore coastal waters and harbors—a condition which severely impacts on fisheries and aquaculture industries. During the La Nina phase the opposite conditions occur, with more frequent, stronger northeasterly winds causing erosional tendencies for the northeast coast beaches, but upwelling induced algal blooms on the west coast. To date it is not clear whether the west coast beaches significantly accrete during La Nina conditions.

Tidal effects. Tides are of the semi-diurnal type. Walters *et al.* (2002) calculated the tidal amplitudes around the oceanic islands of New Zealand based upon the predominant M2 constituent, to produce an approximate mean tide range. Based upon the tidal gages at the "standard ports," the average tidal range is 1.8 m with spring means of 2.1 m and neap means of 1.4 m. The tidal ranges are predominantly micro (<2 m mean tide). Largest tidal ranges (>3.5 m mean tide) occur in Tasman and Golden bays in the northern South Island, and along the central west coast (e.g., Onehunga, Port Taranaki) and exhibit a mesotidal (>2 m to <4 m) mean range. However, spring tidal ranges in some areas approach 4 m. Lowest tidal ranges occur along the East Coast region. In most ports and harbors (e.g., Bluff, Lyttelton, Dunedin and Timaru ports and Manukau, Kaipara and Whangarei harbors) the tidal range is amplified relative to the open coast tidal range, indicating a hypersynchronous tidal condition, while two (Wellington, and Picton) are suppressed, indicating a hyposynchronous tidal condition. The difference in tidal phase between the west and east coast results in strong tidal currents through Cook Strait separating the main islands.

Regional coastal geomorphology

The New Zealand coastline can be classified into several distinctive geomorphic sectors (Figure N20).

High-energy North Island west coast. This sector is subjected to high wave energies with prevailing onshore winds, and a divergent littoral drift system, flowing northwards north of Cape Egmont and north of Wellington, and southwards south of Cape Egmont. The coastline is essentially curvilinear in alignment. Around Taranaki, the coastline exhibits low cliffs and rugged terrain, with gravel boulders littering the shore platforms cut into the lahar deposits which flowed from Mount Taranaki. The high wave energies and swell wave refraction around small promontories create numerous locations for good surfing waves, with renown "breaks" at Raglan and other locations. The distinctive fine "black ironsands" comprise titanomagnetite heavy mineral mainly eroded from the Taranaki lahars, which in some Pleistocene dune deposits is of sufficient concentration to mine (Waverly, Taharoa). There is a general reduction of heavy mineral concentrations away from Taranaki, both north and south. South of Cape Egmont, wave refraction induces a southeastward moving drift, but north of Cape Egmont there is a consistent northward drift to North Cape. Beaches along this sector typically exhibit a dissipative morphodynamic regime. Some of the littoral system is cliffed but the high-energy waves also move sand on the submarine cut platforms of the inner shelf. The coastal sector centered on Taranaki is also undergoing rapid tectonic uplift forming coastal terraces, so that rivers, such as the Waitara, are actively downcutting and no large estuaries occur. North of Kawhia Harbor a number of large shallow barrier enclosed estuarine lagoon harbors occur (Aotea, Raglan, Manukau, and Kaipara). Each is a tidal inlet type, but unusual in that the barriers enclosing the harbors comprise Pleistocene dunes up to 300 m high, and depending upon tidal discharge relative to wave exposure, these tidal inlets tend to have well formed fan shaped ebb-tidal deltas. Because of the high tidal ranges the currents in the inlet gorges tend to be strong—as much as 4 m/s.

Northeast embayed "lee" coast. Between North Cape and East Cape the coast is characteristically indented, including several large embayments with many islands (e.g., Bay of Islands, Hauraki Gulf). The embayments possess compartmentalized sandy littoral drift systems, some of which may be quite extensive (e.g., Bay of Plenty, Bream Bay, Doubtless Bay, Pakiri-Mangawhai), and contain extensive Holocene dune ridge sandy barrier progradation systems. The typically hilly catchments and active erosion of the soft Tertiary lithologies means that

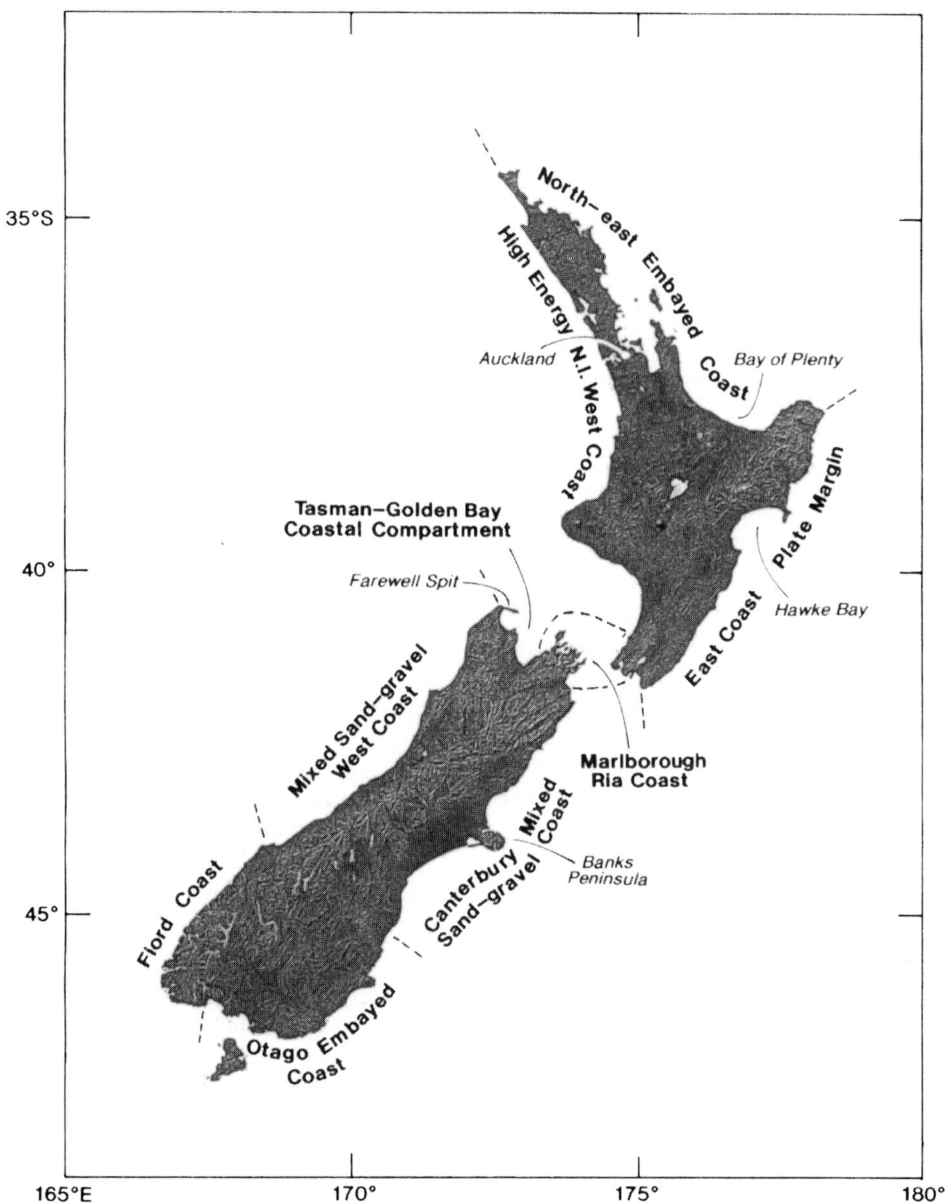

Figure N20 New Zealand coastal geomorphic provinces.

the drowned river valleys have to a large extent become infilled with sediment, so that the harbors are typically shallow and display extensive intertidal flats. At the mouth of many former drowned river valleys are Holocene sandy dune ridge barrier spits (e.g., Pauanui, Matarangi), which enclose shallow infilling estuarine lagoons. Within the Hauraki Gulf the soft Miocence alternating sandstone–siltstone flysch rocks have undergone rapid erosion in the Holocene and today display active cliffs and extensive shore platforms up to 400 m wide.

The Bay of Plenty contains the longest littoral system of the northeast region, and is dominated by low refracted swell waves of $T = 7–9$ s and $H = 0.5–1.0$ m. The bay is notable for containing the largest Holocene progradational extent of dune sands, 9 km in the Rangitaiki Plains, and also the largest barrier island (Matakana Island) in the country. The extensive Tauranga Harbor, geomorphically a barrier enclosed estuarine lagoon, is ~70% exposed intertidal flats at low tide, and possesses a unique enclosing barrier system comprising Holocene barrier tombolos and an intervening barrier island. The surfeit of sand in the Bay of Plenty system is derived from the adjacent TVZ, having been brought to the coast during the Pleistocene, and reworked across the shelf during the Holocene post-glacial transgression, with continuous additions from some of the larger rivers.

The Firth of Thames graben is not typical of this "lee" coast, but is notable for containing a major area of muddy coast, with deep Holocene mud deposits, broad intertidal mud deposits, and a well-developed shelly chenier ridge plain (Healy, 2002).

East Coast Plate Margin coast. Extending from East Cape to the Wairarapa and Wellington is a rocky, eroding high-energy, mainly cliffed coast with extensive shore platforms cut into the Tertiary rocks. The coast is surmounted by veneer sandy pocket beaches in shallow embayments, and often backed by raised marine cut terraces. This coast is seismically active and undergoing rapid uplift due to subduction processes. The general NE–SW alignment is broken by two large embayments, Poverty Bay and Hawke Bay. Poverty Bay lowlands comprise intercollated dune ridges and alluvial deposits. The high discharges from the Waipaoa River eject considerable muddy suspended sediment from the soft Tertiary mudstone catchments into the nearshore and continental shelf. Thus paradoxically, this coast is high energy but possesses very muddy sediment at shallow depths due to the high mud supply. Hawke Bay is likewise subjected to high muddy river discharges to produce muddy sediment at shallow depths, but the rivers here also deliver greywacke from the central axial ranges, so that the beaches are predominantly composed of mixed sand

and angular greywacke gravel. Resulting from the Napier earthquake of 1931, a large area of shallow seafloor and harbor was uplifted ~2 m to create new coastal lowland backed by degraded cliffs where marine processes are no longer active.

Marlborough Sounds ria coast. Fault aligned valleys, transverse to the coastal alignment, result in classical rias (*sensu stricto*) known as the Marlborough Sounds. The structural alignment and transcurrent faulting have created long linear rias. The steep hillslopes dip below the waterline with only minimal interruption to the profile and there is relatively little infilling sedimentation except at the valley heads. Thus the rias remain deep. Wellington Harbor on the southern North Island is likewise part of the ria coast.

Canterbury mixed sand–gravel coast. This is a distinctive and unusual littoral drift coastal sector formed from high-energy storm and swell wave processes eroding the alluvial outwash gravel deposits comprising the Canterbury plains (Kirk, 1980). The result is a well-formed steep berm beach comprised of mixed sand–gravel sediments which are fed partly by direct erosion of the outwash gravel deposits and partly from episodic inputs from the several large braided gravel river systems. Along sectors of the coast the active synoptic beach berm is surmounted by higher storm berms which reach almost 20 m above sea level adjacent to Banks Peninsula. Erosion of the outwash gravel deposits has resulted in low coastal cliffs behind the beach along the southern sector. South of Banks Peninsula backing dunes are rare, but they do occur on the sector north of the Peninsula. Banks Peninsula itself comprises a Miocene andesitic volcanic mound with eroded calderas creating present-day harbors. Eroded loess deposits blanketing the hills have provided muddy infill sediments within the caldera harbors.

Southeast Otago cliffed and mildly embayed coast. Low hills of southeast Otago result in a mainly cliffed coast with pocket sand beaches and where rivers reach the sea. The region receives large swells from the southern oceans, and the beaches typically exhibit high-energy morphodynamic conditions. Otago Harbor is a ria that has become largely infilled.

Fiord Coast of South Westland. The southwest tip of New Zealand comprises a gneiss and granodiorite massif standing up to 3,000 m above sea level. At the coast precipitous slopes drop down to, and below, sea level. Some boulder beaches occur but the coastline is predominantly steep. Pleistocene valley glaciations have carved a spectacular landscape of classical U-shaped valleys, forming classical fiords where they have become flooded at the coast (e.g., Milford Sound, Doubtful Sound). The location in the zone of "roaring forties" and high topography means that the region receives very high precipitation, which provides fluxes of freshwater to drive a fiord-type estuarine salinity structure and circulation. As the area comprises national park the natural environment of this spectacular forested coastline remains in an essentially pristine condition.

Mixed sand–gravel sector of the South Island west coast. Waves from the southern oceans drive a northward littoral drift with sediment being fed to the littoral system by the numerous braided gravel rivers draining from the Southern Alps. The beaches in the southern sector are mixed sand–gravel type, and the wave climate is high energy. As in the North Island west coast littoral system, sand evidently bypasses rocky cliffs, and ultimately feeds the large sand deposits of the 30 km Farewell Spit at the northern tip of the South Island.

Golden and Tasman bays sandy compartment. Located between the Marlborough Sounds and Farewell Spit are sandy compartments. Two features are remarkable here, namely the existence of a low dune ridge Holocene barrier island (one of only two in New Zealand) and the 13 km long Nelson boulder barrier spit, which encloses a shallow muddy lagoon of Nelson Haven. About half of the enclosed area has been reclaimed, and the region possesses one of the highest tidal ranges in the country.

Major coastal geomorphic and oceanographic research achievements in the modern era (post-1970)

In the 1960s, the modern era of quantitative and process-based coastal research in the universities commenced. Research was initially centered on shore platforms, and beach morphology and sediments based upon the contemporary concepts of the sediment budget and process response model (McLean, 1977). Hume *et al.* (1992) outline the major achievements of coastal geomorphic and oceanographic research until about 1990.

Estuarine sedimentation. In the 1970s, there were rising concerns about the impact of changing land use practices—especially urbanization—on the water quality, biology, and sedimentation in estuaries. Early integrated interdisciplinary studies were initiated on the Avon–Heathcote estuary, the Pauatahanui Inlet, and the Waitemata Harbor. Numerous systematic sedimentological studies followed, often undertaken as thesis projects from the universities, especially from the Waikato University group (Healy *et al.*, 1996). Most studies tended to concentrate on either the ecology or surficial sediments and only limited interdisciplinary investigations occurred.

Tidal inlets. New Zealand tidal inlets tend to have formed where Holocene barrier spits form across the embayment at the entrance to re-entrant valley estuaries. They are found predominantly in the north-eastern North Island where about 30 such features exist. They possess the typical morphological features of a narrow inlet gorge, and ebb and flood delta sand bodies, and are dynamic equilibrium systems. R.A. Heath investigated the tidal hydraulic equilibrium of the major inlets. The first quantitative hydrodynamics and sediment transport investigations were carried out by R.J. Davies-Colley around the entrance to Tauranga Harbor in 1976, where, prompted by port developments, the most comprehensive and detailed studies in a New Zealand inlet have since been carried out. Hume and Herdendorf (1990) provide a comprehensive compilation of morphological and empirical factors controlling New Zealand tidal inlet stability.

Numerical modeling of coastal hydrodynamics and sediment transport. Hydrodynamic numerical modeling of estuarine and inlet tidal flows was introduced to New Zealand relatively early by the pioneering work of K.P. Black when he developed his prototype one- and two-dimensional models for current flows and sediment transport in the tidal inlet at Whangarei Harbor (Black, 1983). For calibration and verification of the model, as well as wider understanding of the inlet physical system, an extensive field research program included hydrographic surveys, water level recorders, side-scan sonar mapping of the bottom sediments, a wide range of bottom photographs and sediment sampling taken by SCUBA diving, continuous recording current meters, drogue tracking of current flows, numerous tidal cycle vertical current profiles, and sediment traps, and underwater video records and sediment threshold experiments, and wave refraction studies. Subsequently modeling of estuarine and inlet hydrodynamics has been widely applied (Black *et al.*, 1999), initially in Tauranga Harbor, but also in most of the ports, and across a wide range of applications. In more recent years, the models have become three-dimensional (Black *et al.*, 2000) and applied to a range of EIA problems such as pollution dispersion associated with outfall effluents.

Tsunami modeling and research. New Zealand has experienced at least 32 identifiable tsunami events during its recorded history, but to date without the disastrous consequences experienced by Japan. The largest reported tsunami wave elevations were of order 10 m, originating primarily from local sources. The largest pan-Pacific tsunami recorded was the May 1960 Chilean tsunami which produced a wave elevation of 7 m at Whitianga in Mercury Bay. The most comprehensive tsunami research has been carried out by de Lange and Healy (1986) and their students. As part of disaster emergency planning and response, several tsunami scenarios have now been modeled for coastal sectors and harbors identified as potentially susceptible to significant tsunami impacts, for example, the Waitemata Harbor of Auckland city, the Bay of Plenty, Gisborne, Wellington Harbor, Port Lyttelton, and the Canterbury coast. Modern research is focusing on the ability of submarine slumping mechanisms to generate localized tsunami waves as occurred in March and May, 1947, centred around Tatapouri north of Gisborne.

Beach erosion and morphodynamics. As in many countries, severe beach erosion, especially when coastal property is placed at risk, spawns research programs on beach morphodynamics and beach budgets. The first integrated coastal erosion survey was along the Bay of Plenty coast in 1976–77, where episodes of erosion resulted in rapid duneline retreat, loss of houses, and construction of sea walls. The most spectacular example of erosion was along the embayed Holocene barrier spit of Omaha Beach, where in 1978 the seawall fronting a new subdivision was demolished at the height of a storm. This led to several studies seeking the cause and appropriate remediation, as well as litigation. Although the subject of much "expert" debate at the time, most would agree today that the cause was related to a long-term negative sand budget due to historical sand mining from the spit, consequent duneline retreat, and severe wave energy focusing in the storm. Other cases of erosion occurred locally as a result of interruptions to the littoral drift such as at inlets (e.g., Ohiwa Harbor) or downdrift of port developments (e.g., Port Napier and Port Taranaki), or due to sand mining from the beach (Papamoa in the Bay of Plenty). Concern for the sustainability and impact of long-term sand extraction from the nearshore zone along the 30 km Pakiri–Mangawhai coastal compartment resulted in a large and comprehensive study from 1996 to 1999. The study, lead by NIWA in collaboration with the universities of Waikato and Auckland, involved assessment of sand volumes onshore in the Holocene dune fields, investigation of the surficial sediment patterns from side-scan sonar mapping, bottom photographs and sedimentology, and shallow subsurface seismic profiling with calibrating vibrocore data. Historical beach profile data were analyzed for beach sediment budget. Oceanographic data were measured with recording current meters and wave gauges, and the wider Hauraki Gulf subjected to hydrodynamic and sediment transport modeling.

Beach–shore face–inner shelf sedimentation dynamics. It became evident from a number of studies that strict wave driven alongshore littoral drift is often difficult to ascertain for the compartmentalized New Zealand coast. The earliest studies linking the beach with the inner shelf sediments arose from investigations into severe beach erosion episodes at Omaha and Mangawhai beaches north of Auckland and the east Coromandel beaches, which had suffered during the "decade of erosion" from 1968 to 1978. The most detailed studies during the 1980s were on the east Coromandel "lee" shelf by B. Bradshaw and collaborators. These studies linked closely with the detailed research monitoring the dispersal of sandy dredged material on the inner shelf off Tauranga Harbor by the Waikato school. During the decade of the 1990s, NIWA undertook detailed research on inner shelf and beach sedimentary morphodynamic linkages offshore from the Katikati inlet to Tauranga Harbor.

Port developments and dredging issues. Requirements for port development EIAs have driven considerable research into the impact of dredging and spoil disposal. The most detailed work has been undertaken around the Port of Tauranga, a port which has expanded to become the largest export–import in the country during the last decade. The large Tauranga Harbor Study (1983–85), the most intensive harbor study to date, arose from the need to investigate whether dredging through the tidal delta inlet system had substantially affected the channel hydraulics and morphodynamic changes of the flood tidal delta and channels. Subsequent studies were undertaken on the dredge spoil dispersion on the inner shelf and shelf-beach interaction, as well as further studies on the impact of deepening the channels to take post-panamax sized vessels. Those physical and sedimentological impact studies were matched with ecological impact studies. Disposal of dredged material became a major environmental and litigation issue for the Port of Auckland, and resulted in an inquiry spawning numerous reports from the Dredging Options Advisory Group (DOAG).

Coastal hazard analysis and development setback planning. Assessment of coastal hazard and development setback has been a requirement for coastal local authorities since the 1970s. The first New Zealand application of quantifying the independent components of coastal hazard along a sandy duned coast was evolved by T.R. Healy for the setback of new subdivisions along the Bay of Plenty coast in 1976, and included quantitative assessment of the four independent parameters of long-term erosion (or accretion) trends, short (decadal) term "cut and fill" duneline fluctuations, assessment of expected sea-level rise effects, and the dune topographic stability factor. Summation of these four independent factors allows determination of an initial setback estimate, measured from the toe of the frontal dune, and based upon a planning time frame of 100 years (Healy and Dean, 2000). The methodology has been widely applied and is similar to that later proposed by J.G. Gibb (1981). Over recent years the methodology has been refined and the initial setback determination is now subjected to three tests, viz. (1) Is there sufficient reservoir of sand in the frontal dune to allow for the 1 : 100 year storm erosion episode? (2) Is the setback sufficient to allow for the 100 year storm surge? and (3) Is the setback sufficient to allow for tsunami washover? If the initial setback estimate is still subject to the above hazards, the setback is extended landwards beyond the zone of hazardous impact. The methodology may also be applied to existing subdivisions, in which the setback zone becomes a zone of non-further development.

Methodology for hazard analysis of development setback for cliffed coasts is given by Moon and Healy (1994) based upon identification of the mechanism of slope failure, and addition of a safety factor. For the cliffs cut into Miocene flysch deposits of the Waitemata Formation around the city of Auckland, the typical 100 year hazard setback is taken as ~16 m plus a safety factor of 7 m.

Innovative coastal environmental solutions and engineering applications. Early coastal engineering solutions to coastal erosion problems followed standard international practice of the early 20th century, and groins and seawalls were erected in many locations of localized coastal erosion. Most have subsequently failed or require expensive continuous maintenance. Since the 1970s, however, the environmental philosophy of allowing "nature to function in nature's way" has tended to hold sway, so that much scientific effort has been expended on understanding the coastal dynamic systems rather than resorting to "hard" engineering solutions. Many innovative applications have been demonstrated. Among the more notable are: (1) The application of *sediment transport modeling* at the design stage by K.P. Black in 1983 to determine the optimum length of bridge span relative to causeway length for the Tauranga Harbor bridge, which would minimize scour and shallow estuary morphodynamic impact. This resulted in a reduction of the steel bridge span and a saving of about $800,000 to the cost of the bridge construction. This application firmly demonstrated the value of numerical simulation as a major coastal processes tool. (2) The development of the concept of *artificial surfing reefs* by K.P. Black. This initiative originated during the late 1990s (Black *et al.*, 1997) with the dual purposes of coastal protection against beach erosion, and for enhancing breaking waves for surf board riders. The first arti-ficial reef so designed was constructed at Narrowneck on the Queensland Gold Coast in 1999. The reef units themselves are comprised of geotextile bags filled with dredged sand, a low cost construction system. Several such artificial surfing reefs are planned for New Zealand, including at Mt. Maunganui in the Bay of Plenty, New Plymouth and Opunake on the Taranaki coast, and at Summer Beach near Christchurch. (3) In the late 1990s, considerable work has been undertaken on the *re-design of Port Gisborne* to accommodate the large expansion of timber product for export coming on stream from 2005. An intensive field and modeling program has been undertaken and initial designs subjected to wave and sediment transport modeling (Healy *et al.*, 1998). This innovative application is to include substantial public amenity in the re-designed port-enclosing breakwater so that it acts as an artificial surfing reef to enhance a left hand break for the Pacific swells refracting into Poverty Bay. (4) *Artificial headlands for low-energy estuarine beach geomorphology* have been designed and implemented as a coastal erosion and coastal management option in the Waitemata Harbor, Auckland (2000). The artificial headlands use the principle that wave refraction into embayments at high tide will reduce sediment loss from the upper estuarine beach face by wave driven littoral drift, and induce retention of the sand and shelly gravel within the embayments. (5) The *use of dredge spoil for beach replenishment* has been applied at a few open coastal locations, including near Port Napier, Port of Tauranga, and Port Taranaki. These cases have used the concept of depositing the sandy dredge spoil as a berm in the shallow nearshore, with wave action taking the sediment on to the beach. The most detailed investigation has recently been for Port Taranaki (1997–99) where an extensive field program, including a large-scale tracer experiment, and supplemented by wave and current driven numerical simulation, has identified a suitable disposal site, which in conjunction with a planned artificial surfing reef, aims to re-establish the beaches of New Plymouth which had disappeared as a result of updrift port breakwater construction (McComb *et al.*, 2000).

Issues for the future

Research in the early years of the 2000 millennium is likely to concentrate on ongoing port developments and environmental impacts of dredging and disposal. Under development is investigation of the nature of mud deposition in ports and marinas, and attempts to induce its removal by "mud re-agitation" techniques, which requires ongoing field studies and numerical simulation of the results. Issues of the impact of sediment extraction from the littoral zone and the importance of closure depth and the links between diabathic and parabathic exchange will come under increasing scrutiny, and continuing developments can be expected in the frontier of morphodynamic numerical simulation of sedimentary bodies. Considerable effort is being undertaken to research beach and frontal dune behavior as part of "Dune-Care" programs in which local communities become stakeholders in the management and protection of the dunes. With the advent of modern data collection the application of marine GIS linked with dynamic modeling, presently unknown in the New Zealand context, will develop for wide application. Overall New Zealand has a sound basis of research achievements to tackle the coastal morphodynamic issues of the new millennium.

Terry R. Healy

Bibliography

Black, K.P., 1983. Sediment transport and tidal inlet hydraulics. Ph.D. thesis, University of Waikato, 331 pp.

Black, K.P., Andrews, C., Green, M., Gorman, R., Healy, T., Hume, T., Hutt, J., Mead, S., and Sayce, A., 1997. Wave dynamics and shoreline response on and around surfing reefs. 1st International Surfing Reef Symposium, Sydney, pp. 1–11.

Black, K.P., Green, M., Healy, T., Bell, R., Oldman, J., and Hume, T., 1999. Lagrangian modelling techniques simulating wave and sediment dynamics determining sand bodies. In Harrf, J., Lemke, W., and Stattegger, K. (eds.), *Computerized Modeling of Sedimentary Systems*, Berlin: Springer, pp. 3–21.

Black, K.P., Bell, R.G., Oldman, J.W., Carter, G.S., and Hume, T.M., 2000. Features of 3-dimensional barotropic and baroclinic circulation in the Hauraki Gulf, New Zealand. *New Zealand Journal of Marine and Freshwater Research*, **34**: 1–28.

Cotton, C.A., 1942. *Geomorphology. An Introduction to the Study of Landforms*. Christchurch: Whitcombe and Tombs Ltd, 505 pp.

Cotton, C.A., 1974. *Bold Coasts*. Wellington: A.H. & A.W. Reed.

de Lange, W.P., and Healy, T.R., 1986. New Zealand Tsunamis, 1840–1982. *New Zealand Journal of Geology and Geophysics*, **29**: 115–134.

Gibb, J.G., 1981. Coastal hazard mapping as a planning technique for Waiapu County, East Coast, North Island, New Zealand. Wellington: National Water and Soil Conservation Organisation, 63 pp.

Healy, T., 2002. Muddy coasts of mid-latitude oceanic islands on an active plate margin—New Zealand. In Healy, T.R., Wang, Y., and Healy J-A. (eds.), *Muddy Coasts of the World: Processes, Deposits and Function*, Amsterdam: Elsevier, pp. 263–278.

Healy, T.R., and Dean, R.G., 2000. Methodology for delineation of coastal hazard zone and development setback for open duned coasts. In Herbich, J.B. (ed.) *Handbook of Coastal and Ocean Engineering*, New York: McGraw-Hill, pp. 19.1–19.30.

Healy, T., and Kirk, R.M., 1982. Coasts. In Soons, J.M., and Selby, M.J. (eds.), *Landforms of New Zealand*, Auckland: Longman Paul, pp. 81–102.

Healy, T.R., Cole, R., and de Lange, W., 1996. Geomorphology and ecology of New Zealand shallow estuaries and shorelines. In Nordstrom, K.F., and Roman, C.T. (eds.), *Estuarine Shores: Evolution, Environments and Human Alterations*, New York: John Wiley and Sons, pp. 115–154.

Healy, T., Stephens, S., Black, K.P., Gorman, R.M., and Beamsley, B.J., 1998. Numerical and physical process studies for Port of Gisborne redesign for the 21st century. *Journal of Coastal Research Special Issue*, **26**: 304–311.

Hume, T.M., and Herdendorf, C.E., 1990. Morphologic and hydrologic characteristics of tidal inlets on a headland dominated, low drift coast, northeastern New Zealand. *Journal of Coastal Research Special Issue*, **9**: 527–563.

Hume, T.M., Bell, R.G., de Lange, W.P., Healy, T.R., Hicks, D.M., and Kirk R.M., 1992. Coastal oceanography and sedimentology in New Zealand, 1967–91. *New Zealand Journal of Marine and Freshwater Research*, **26**: 1–36.

Kirk, R.M., 1980. Mixed sand and gravel beaches: morphology, processes and sediments. *Progress in Physical Geography*, **4**: 189–210.

McComb, P.J., Black, K.P., Healy, T.R., and Atkinson, P.N., 2000. Coastal and sediment dynamics at Port Taranaki, New Zealand: a large, multi-faceted, field experiment. *Coastal Structures '99*, Rotterdam: Balkema, pp. 823–832.

McLean, R.F., 1977. Coastal research in New Zealand. *Geoscience and Man*, **14**: 135–144.

McLean, R.F., and Kirk, R.M., 1969. Relationships between grain size, size-sorting and foreshore slope on mixed sand-gravel beaches. *New Zealand Journal of Geology and Geophysics*, **12**: 138–155.

Moon, V., and Healy, T., 1994. Mechanisms of coastal cliff retreat and hazard zone delineation in soft flysch deposits. *Journal of Coastal Research*, **10**: 663–680.

Pullar, A.S., and Selby, M.J., 1971. Coastal progradation of Rangitaiki Plains, N.Z. *New Zealand Journal of Science*, **14**: 419–434.

Richmond, B., Nelson, C.S., and Healy, T.R., 1984. Sedimentology and evolution of Ohiwa Harbour, a tidally-dominated estuary in the Bay of Plenty. *New Zealand Journal of Marine and Freshwater Research*, **18**: 461–478.

Schofield, J.C., 1960. Sea level fluctuations during the last 4000 years as recorded by a chenier plain, Firth of Thames, New Zealand. *New Zealand Journal of Geology and Geophysics*, **3**: 467–485.

Schofield, J.C., 1970. Coastal sands of Northland and Auckland. *New Zealand Journal of Geology and Geophysics*, **13**: 767–824.

Walters, R.A., Goring, D.G., and Bell, R.G., 2002. Ocean tides around New Zealand. *New Zealand Journal of Marine and Freshwater Research*, **35**: 567–579.

Cross-references

Barrier
El Niño–Southern Oscillation
Gravel Barriers
Longshore Sediment Transport
Surfing
Submerging Coasts
Tidal Inlets
Tsunami

NORTH AMERICA, COASTAL ECOLOGY

The North American coastal ecosystems range in latitude from tropical Mexico to arctic Canada and have long been identified as among the most productive in the world (Odum, 1963). In reality, there is a broad range of annual productivities (ranging from several hundred to several thousand grams carbon) as well as species diversity. In the Arctic few species dominate the coastal ecosystems, while in the south along the Mexican coasts species diversity is especially rich (Lot *et al.*, 1993). Widely varying in geomorphology and chemistry, they comprise: hypersaline lagoons, cheniers, coralline islands, temperate estuaries and barrier islands of the US Atlantic coast; as well as vast stretches of beach and permafrost coastlines of northern Canada and a myriad of rocky intertidal communities and embayments along the Pacific Coast to Mexico (Odum *et al.*, 1974). As in most ecosystems, solar radiation exerts primary control on productivity and ultimately impacts distributions of the numerous species. Also critical in influencing the structure and function of these coastal systems are organic matter and nutrient exports from upland sources, as well as tidal energy. In contrast to terrestrial systems where precipitation is considered second in importance to input solar radiation, the force of astronomically driven tides is extremely important in structuring coastal systems which range along this coastline from below a centimeter in some marshes of Chesapeake Bay (Stevenson *et al.*, 2001) to over 10 m in the Bay of Fundy (Chmura *et al.*, 2001).

Water, critical in determining the productivity of terrestrial systems, is the key medium for exchange of energy and nutrients in coastal systems. Generally, inflows of freshwater from rivers are responsible for supplying organic materials and nutrients to coastal estuaries and lagoons as well as control of seasonal changes in salinity. Freshwater inflow plus tidal activity (along with bathymetric configuration of coastal water bodies) largely determine rates of mixing and flushing. Thus, the hydrological inputs at varying local and regional scales are critical in modulating responses to increasing nutrient loadings, which cause numerous changes in coastal ecosystems (Malone *et al.*, 1993; Turner and Rabalais, 1994; Bricker and Stevenson, 1996; Vorosmarty and Petersen, 2000).

Drainage basin and river discharge

Although the magnitude of freshwater runoff is responsible for structuring estuaries (Pritchard, 1967), water quality is critical in determining their productivity. Increasing the flux of relatively few critical elements (e.g., nitrogen, phosphorus, and silica) in freshwater discharges to coastal waters can cause eutrophication (i.e., abundant nutrients promote high phytoplankton biomass which consumes large amounts of oxygen as it decomposes) negatively impacting submersed grasses and traditional fisheries (Twilley *et al.*, 1985; Stevenson *et al.*, 1993). Water balance/runoff models, geographic information systems (GIS), and remote sensing are tools increasingly used to relate changes in land use to fluxes of nitrogen, phosphorus, and organic carbon into coastal systems. The generalized watershed loading function (GWLF) is an example of a widely used model to estimate nitrogen and phosphorus loading from runoff and groundwater in northeastern watersheds of the United States (Haith and Shoemaker, 1987). In this model hydrologic fluxes are driven by daily weather data, including rainfall, evapotranspiration, and snowpack melting (the last two inferred from temperatures). Groundwater fluxes are estimated from considering both vadose (unsaturated) and phreatic (saturated) contributions in watershed soils. Suspended sediment inputs, important in coastal turbidity and as vectors for trace metals (and other particle-reactive pollutants), have traditionally been estimated from the Universal Soil Loss Equation (Gottschalk, 1964).

Once nitrogen, phosphorus, and other materials enter receiving waters they are subject to bioprocessing by various biochemical processes. Remineralization of organic material at the edges of coastal watersheds occurs when bacteria interact with dissolved and particulate phases, in wetlands and rivers. First-order streams, are perhaps the most important elements of coastal watersheds in bioprocessing as the time and volume of water in contact with wetland and the benthos is greater than in higher order streams (Peterson *et al.*, 2001). Nitrogen, the most common element in fertilizer, transported as nitrate in streams or volatized to ammonium (NH_4^+), often ends up in coastal embayments (Pearl, 1985; Staver and Brinsfield, 1996). However, since biogeochemical buffering often occurs, actual nutrient delivery to coastal waters is not as great as watershed yield coefficients of particular land uses might suggest.

(A)

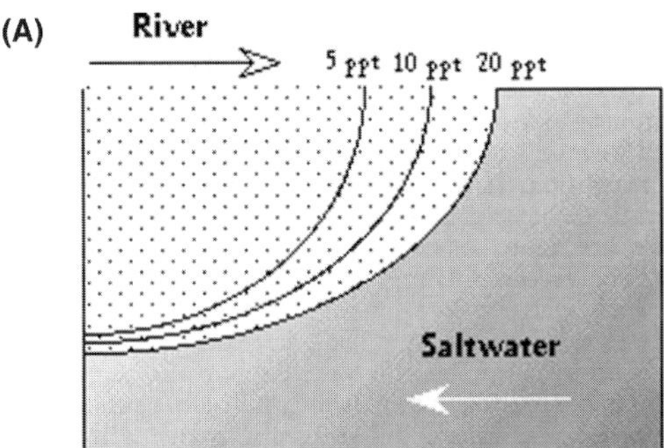

(B)

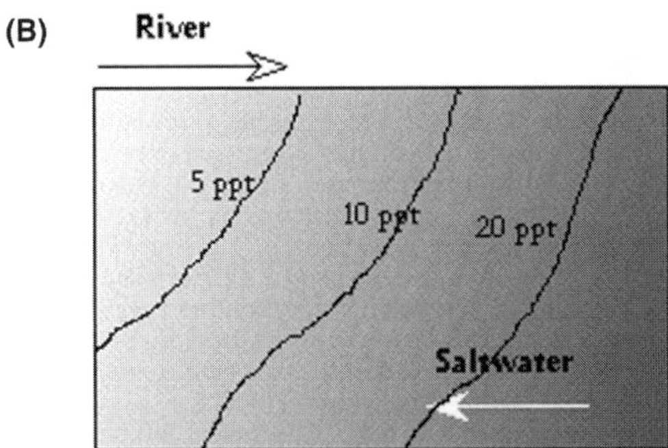

(C)

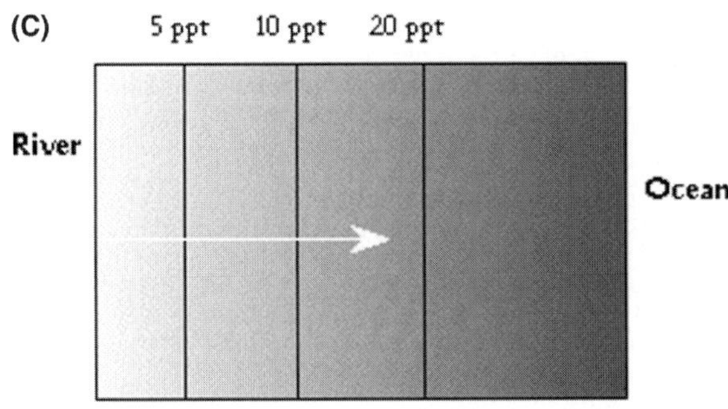

Figure N21 Types of general estuarine circulation and mixing. (A) Salt wedge or highly stratified system, (B) Partially mixed system, (C) Fully mixed system.

Sources of mixing

Waves, tides, and rivers control mixing in marginal environments. In estuaries, where major rivers meet the ocean, mixing is predominantly the result of tidal and river forces (Pritchard, 1967). Although there is a expansion of water surface area in estuaries with the incoming tide, particularly those of the low relief Atlantic coastal plain, the total net volume of water remains approximately the same throughout the tidal cycle; total estuarine volume rarely changes much (Berner and Berner, 1996). Essentially, incoming salty oceanic waters are transported landwards along the bottom balancing freshwater loss above the pycnocline in estuaries (Pritchard, 1954). Mixing occurs along the salinity gradient resulting in complex circulation patterns which are often estuary specific and impact the dispersion of sediment particles and other pollutants (Nichols and Biggs, 1985). These gradients generally depend on whether river discharge dominates the system, or tides (i.e., tidal currents).

A salt-wedge circulation occurs where inflowing river discharge greatly exceeds the volume of the incoming tide (expressed as the tidal prism) (Figure N21(A)). In situations where the river velocities are high, freshwater can over-ride the saltwater, yielding a sharp interface where mixing occurs by internal waves at the interface lifting parcels of saltwater into the inflowing river water (Nichols and Biggs, 1985). This distinct boundary is revealed by the tight clustering of isohalines (lines connecting water depths of equal salinity) where salinities can change from 5 to 30 ppt over vertical distance as little as 50 cm (Postma, 1980).

A partially mixed system occurs where stronger tidal currents diminish domination of the circulation by river flow, and the resulting greater tidal mixing produces a less sharply defined boundary between fresh and saltwater (Figure N21(B)). The stronger tidal currents produce both upward mixing of the river water and downward mixing of the saltwater. Partially mixed estuaries can be identified by a balance between river discharge and tidal prism. Fully or well-mixed circulation in an estuary occurs where the tidal prism is significantly larger than river discharge, and the tidal currents retard any tendency toward stratification of fresh and saltwater (Figure N21(C)). The result is that the estuary becomes well mixed with depth, and salinities vary only laterally, becoming more saline down estuary.

As critical to the fate of nutrients and pollutants in coastal waters is the flushing time, that is, the time it takes for the existing volume of freshwater to be replaced by river discharge (Aston, 1980), or

$$\tau = V_f/R,$$

where τ is the flushing time, V_f is the volume of freshwater in the estuary, and R is the river discharge into the estuary. Generally, the flushing times for less stratified and well-mixed estuaries tend to be longer than for more stratified, especially salt-wedge systems, where the fresh river water can flow out the estuary. Longer flushing times translate into longer periods for nutrients to be removed from estuary waters, with the visible evidence of such removal being an overall increase in phytoplankton biomass. In contrast, the faster flushing times for stratified estuaries, phytoplankton nutrient uptake is less, and phytoplankton debris is more likely to be exported from the estuary, slowing nutrient recycling (Berner and Berner, 1996). Moreover, stratified estuaries are more likely to become nutrient traps for remineralized nutrients in bottom waters (Redfield *et al.*, 1963).

Waves are important mechanisms of mixing in coastal waters. In salt wedge and stratified estuaries, internal waves produce mixing along the boundary of river discharge and incoming flood waters. In shallow water areas, waves, especially during storms, can fully mix the entire water column. At the same time, such waves may resuspend fine-grained particulates from the bottom. Wave mixing and resuspension also predominate as agents of mixing in shallow coastal lagoons, although considerable mixing from tidal exchange occurs within the vicinity of inlets. A measure of the amount of exchange that occurs in coastal lagoons can be estimated by comparing the volume of the lagoon to the *tidal prism* of the inlet. This is in essence an index of the amount of inlet "water capture"—that is, what volume of the lagoon is exchanged by a particular inlet. In large coastal lagoons, serviced by several inlets, significant variations in inlet tidal prisms probably influence circulation within the lagoon. Net drift in the lagoon is controlled by the tidal prism of the largest inlet.

Biogeochemistry of estuaries

Spatial and temporal changes in the concentration of nutrients and other chemical constituents of the waters of Atlantic Coast estuaries have had considerable attention, especially as indicators of eutrophication. Studies have focused on sources of chemical constituents, vectors (if associated with particulates), cycling and transformations, and sinks.

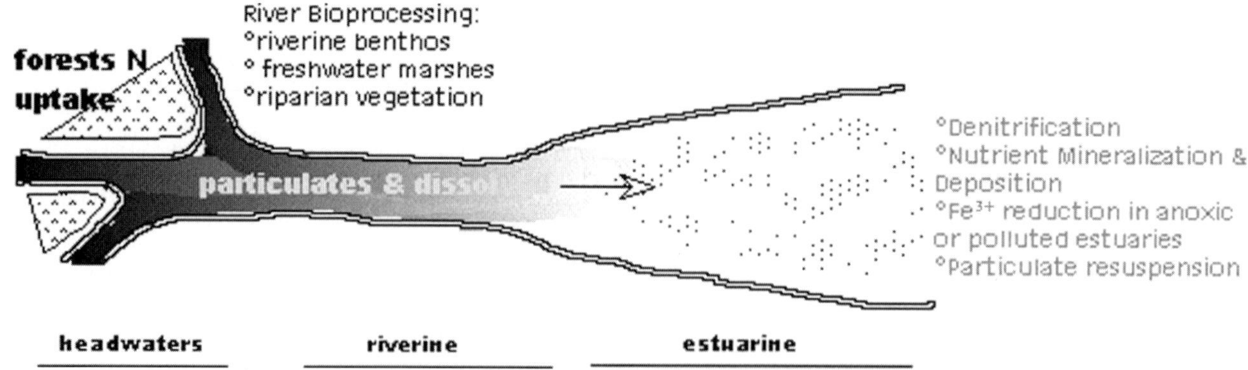

Figure N22 General biochemical interactions for a North American coastal plain estuary (based on Berner and Berner, 1996).

In simplest terms, the challenge of estuarine biogeochemistry is distinguished between terrestrial, marine and *autochonous* materials, and their fates (Figure N22 shows a general model of estuarine biochemical interactions).

Riverine constituents other than nutrients

Among the common chemical nonorganic constituents of coastal plain estuaries are Fe, Al, and Si. These constituents are commonly regarded as being of fluvial origin and indicative, when in high concentrations, of high rates of weathering within the drainage basin. Si, though naturally abundant in river water as dissolved silica, can also originate from groundwater sources. Various planktonic radiolarians and diatoms also contribute significant percentages. Dissolution of biogenic silica, however, is slowed by high concentrations of Fe^{3+} and Al^{3+} (Aston, 1980). Because much of the Al and Fe entering estuaries is often adsorbed on to fine clay particles, rapid flocculation of river clays between 5 and 15% salinity probably removes most of these metals (Aston, 1980). Moreover, this suggests that dissolution rates of biogenic silica are probably greater in the higher salinity (lower) parts of estuaries. Cations, such as Mg^{2+}, Ca^{2+}, K^{+}, and Na^{+}, estuarine waters are also largely derived riverine sources. In the rivers that flow through the highly weathered saprolitic rocks that compose much of the Piedmont of the Atlantic Seaboard's watersheds, concentrations of these ions tend to be high.

Dissolved organic matter (DOM) is also believed to be largely a function of riverine inputs. There is considerable seasonal as well as year-to-year variability in DOM, depending on river discharge and regional temperatures.

Nutrient constituents: phosphorus. The origin of phosphorus in estuarine waters largely reflects the influx of phosphate minerals in organic particulates and dissolved phosphate. Increasingly, the anthropogenic origin of phosphorus in sewage and other effluents has become a major concern in many North American estuaries. In Chesapeake Bay, Boynton *et al.* (1995) suggested that the total phosphorus load since European settlement had increased 16.5-fold, presently averaging about $0.065\,t\,km^{-2}\,yr^{-1}$. In contrast to nitrogen, annual phosphorus loads largely reflect basin-wide runoff inputs, not groundwater inputs. Thus, the anthropogenic effects tied to land cover and land use changes (especially in regard to runoff efficiency as well as fertilizer application) are critical to understanding pathways for phosphorus (and nitrogen) export to coastal waters. The decline in forest cover, in particular, has been demonstrated to be a major factor in rising nutrient loads.

Once in coastal waters, particularly estuaries, phosphorus (as phosphate) is subjected to inorganic and biological controls. Phosphate "buffering" by being adsorbed on to mineral surfaces (Pomeroy *et al.*, 1965) is perhaps the major inorganic control. Depending on the phosphate concentration, adsorption surface available (i.e., mineral species and abundance) pH (lower salinities and more acidic pHs than seawater) and, possibly redox (higher oxidation states in estuarine sediments possibly releasing phosphates), phosphate buffering can be a principal means of moderating sewage outfalls in coastal plain estuaries (Aston, 1980).

The biological cycling of phosphorus in US Atlantic Coastal Plain estuaries has been shown to exhibit a seasonal cycle. Concentrations of dissolved phosphates in ambient waters are highest in summer (Taft and Taylor, 1976). Bacteria and, to a lesser extent, phytoplankton (depending on turbidity and its effects on light penetration and intensity) play the principal roles in uptake of phosphorus in the phosphorus cycle (Berner and Berner, 1996). However, the uptake of phosphorus by such organisms may be temporary since consumption by protozoans and other filter feeders with excretion and death can release back to the water column dissolved and particulate organic phosphate compounds (Aston, 1980). Ultimate removal may occur upon to adjacent ocean waters.

Nutrient constituents: nitrogen. The impact of increased loadings of nitrogen upon the ecology of coastal waters of North America has been perhaps even greater than that of phosphorus. The input of nitrogen (as nitrate) to Atlantic Coast rivers increased by 30% during the late 1970s (Smith *et al.*, 1987). Coastal waters are generally nitrogen limited due to denitrification by bacteria of nitrate (NO_3^-) under anaerobic conditions, low nitrogen fixing due to low light conditions (Howarth, 1988), and the toxicity of sulfides to nitrogen-fixing bacteria (Mitsch and Gosselink, 1995). Fluxes of nitrogen into coastal waters from rivers occur both as dissolved nitrate and particulate forms, though the latter probably only predominate during high precipitation events, particularly if these events coincide with the application of fertilizers within local drainage basins (Figure N23). Considerable research in Chesapeake Bay (Staver and Brinsfield, 1996) has underscored the importance of groundwater discharge as the major source of nitrogen loading of coastal estuaries either directly into coastal waters, or by discharges into tributaries as stream baseflow. Bachman and Phillips (1996) estimated that 40% of the total nitrogen load of Chesapeake Bay reflects groundwater discharge. The data on groundwater loadings of nitrogen from other areas in North America are limited, but this source is clearly potentially significant in areas with extensive marshes that are cut by numerous tidal creeks (Seagle *et al.*, 1999).

Direct input of nitrogen from the atmosphere has been demonstrated over the last two decades to be a major source of nitrogen to North American coastal waters (Paerl, 1985), as it is globally averaging 10–25% of the global atmospheric nitrogen input (Duce, 1991). Studies of atmospheric nitrogen loadings in the middle Atlantic coast show that wet deposition accounts for the atmospheric nitrogen, with nitrate being the most common form, though organic nitrogen can be important seasonally (Seagle *et al.*, 1999). Dry deposition, though accounting for less than half the atmospheric loading of nitrogen, nevertheless may yield as much 1.5–4.1 $kgN\,ha^{-1}\,yr^{-1}$ in the Chesapeake Bay region (Gardner *et al.*, 1996).

Overall, as Seagle *et al.* (1999) point out, though figures for atmospheric deposition of nitrogen like those for groundwater loadings are limited, the phenomenon reinforces the importance of land use in not only controlling runoff and ground-water nutrient inputs into North American coastal waters, but the impact of land use in controlling the fate of atmospheric loadings. In particular, knowing the spatial distribution of upland and riparian forests/shrublands as nutrient buffers clearly holds the key to managing the effects of nutrients on North American Coasts. In this respect, databases like US National Oceanic and Atmospheric Administration's (NOAA), Coastal Change Analysis Program (CCAP; Klemas *et al.*, 1993), which portray changes in land cover and land use, comprise essential tools for wise coastal management.

Biological interactions

The populations of marine and brackish water organisms that characterize the coastal waters of North America constitute the fundamental

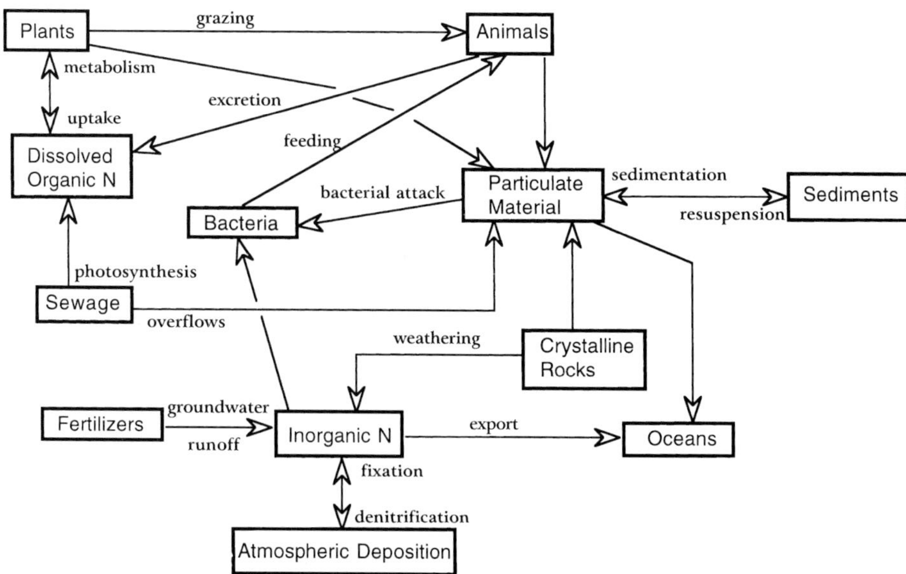

Figure N23 General nitrogen cycle for North America coasts (modified after Aston, 1980, with information from Boynton *et al.*, 1995).

components of a variety of ecosystems that range from the mangroves of south Florida to the high-latitude wetlands of Hudson Bay (Figure N23). The sheer variety of organisms, which vary broadly by latitude, and locally by salinity and season, is clearly beyond the scope of this entry. It is perhaps best to focus on examples of how these organisms comprise basic elements of North American coastal ecosystems, interact at various trophic levels, and influence overall ecosystem functioning.

Phytoplankton

Phytoplankton (diatoms, coccoliths, and dinoflagellates) populations in North American coastal waters, as along all coasts, depend on salinity, nutrient availability, and illumination. In estuaries, salinity plays an especially important role in the presence of species ranging from freshwater to brackish and fully marine species. Seasonal variations in salinity, particularly those associated with large spring runoff, may allow freshwater and lower salinity brackish species to migrate considerable distances down estuary. This down-estuary displacement, in highly stratified systems (e.g., salt-wedge systems), may ultimately prove disadvantageous, as internal wave mixing along salinity boundaries, may expose freshwater species in particular to osmotic stress. Euryhaline species are less likely to suffer such problems.

Nutrient influences on phytoplankton abundance reflect the similar riverine influxes that control nutrient dynamics. In Chesapeake Bay, low discharges of the Susquehanna River—the defining river of the estuary—and the other major tributaries, like the Potomac and James Rivers, in May through December probably underlie the nutrient limitations on phytoplankton productivity (Carpenter *et al.*, 1969). However, there is likely a greater year-round nutrient demand in most Atlantic Coast estuaries than inputs from external sources (absent significant sewage outfalls), and *in situ* processes in the estuaries must be supplying most of the remaining nutrient demand. These processes have been identified as: (1) remineralized nutrients excreted from herbivorous zooplankton and benthic organisms; (2) releases from sediment resuspension and interstitial sediment water; and (3) exchanges from nutrients in particulate forms and dissolved phases (Smayda, 1983).

In coastal lagoons and bays, which lack significant riverine inputs that produce changes in salinity and nutrient concentrations, light intensity in the euphotic zone above the compensation depth (where a balance occurs between rates of phytoplankton photosynthesis and respiration) is probably the major controlling factor. Resuspension of subtidal sediments is the principal mechanism by which turbidity and light attenuation occurs in lagoons, particularly in the open water lagoons that characterize the large, long thin barrier islands like Assateague Island and Atlantic City, New Jersey. Large-scale resuspension in these lagoons—many of which are quite shallow-occurs largely during storms. However, depending on tidal energies (and inlet tidal prisms) appreciable concentrations of subtidal materials may be entrained from the flanks of inlets and in the vicinity of shoals forming flood tidal deltas. Most of these materials probably settle out of suspension within a short distance of the inlets as tidal velocities rapidly slow. But in a lagoon with a large number of inlets, there could be significant diurnal variations in turbidity and illumination affecting phytoplankton. In the marshy lagoons behind drumstick barriers, outwelling of particulates from marshes during ebb is probably an additional source of suspended organic and mineral materials.

Lastly, though salinities in many coastal lagoons and bays do not vary as strongly seasonally as in estuaries, in areas away from inlets that may be blocked from ready tidal exchange by marshes, hypersaline conditions may prevail in late summer as evaporation rates increase, with concomitant effects on phytoplankton populations.

In estuaries of North America, plankton metabolism serves an important regulator of estuarine chemistry. Plankton photosynthesis is correlated with decreases in the concentration of carbon dioxide, ammonium, phosphate, nitrate, and silicate (Wolff, 1980). Moreover, oxygenation of local waters increases. In Long Island Sound, annual primary phtyoplankton productivity has been determined to be as much as 380 $gC\,m^{-2}\,yr^{-1}$, which is similar to the value of 308 $gC\,m^{-2}\,yr^{-1}$ determined for Narragansett Bay (Wolff, 1980; Smayda, 1983).

Benthic algae and macrophytes

The decline of sea grasses and parallel spread of benthic algae estuaries and coastal lagoons of North America have focused concern on the widespread phenomenon of coastal eutrophication. In coastal systems where high anthropogenic inputs of nutrients, coverage of benthic algae can be extensive, ranging form single cell species (like diatoms) to complex, multicellular species, forming mats. Studies of primary production of benthic in estuaries where nutrient concentrations are not yet excessive show comparatively low values, especially compared with phytoplankton (Wolff, 1980). Substrate characteristics have been suggested to be perhaps the most important limitation on benthic algae productivity. Single cell species prefer soft substrates like muds, whereas the complex multicellular species prefer hard substrates like shell debris, rocks, or shore construction, such as pilings or bulkheads.

The importance of benthic algae to the ecology of North America coasts has been underestimated, because often this component is measured as part of another system as in salt marshes where its productivity (but not biomass) sometimes rivals macrophytes. No such ambiguity relates to the role played by sea grasses, whose productivities exceed even phytoplankton. Sea grass biomass can reach 500 $gC\,m^{-2}\,yr^{-1}$ (Wolff, 1980) and Barsdate *et al.* (1974) indicate that 90% of production in a lagoon in Alaska is due to *Zostera*. The high productivity of sea grasses ensures that they serve as a major food sources for organisms ranging from gastropods to fish, though direct consumption is probably limited and most of their importance to food chains appears to be as detritus (Barsdate *et al.*, 1974; Stevenson, 1988).

As plants rooted in the sea bottom, sea grasses cycle a number of compounds from subtidal sediments into ambient waters through their leaves and by decomposition after death. Phosphate, in particular, may be transported up roots and through leaves of *Zostera maritima* at a rate of $0.066\,g\,m^{-2}\,day^{-1}$ (McRoy *et al.*, 1972).

Salinity, in lieu of anthropogenic influences, is the major control of the species distribution of sea grasses along North American coasts. *Zostera* and *Thalassia* can occur where salinities range through 35 ppt, whereas *Ruppia* and *Zanichellia* characterize brackish waters (Wolff, 1980).

Coastal wetlands: marshes and mangroves

The marshes—and the mangroves in south Florida—that characterize the North America coast occur in a variety of physiographic/depositional settings, but largely in areas protected from strong wave energies like lagoons and estuaries. In the United States southeast Atlantic Coast and Gulf Coast, coastal wetlands comprise a substantial part of the coastline, and individual marsh systems can easily range in size up to several thousand hectares or more. The contribution of large coastal wetlands to the ecology of coasts is clearly inescapable, but even the much smaller and more isolated marshes of New England up through the Canadian Maritimes are important to local ecosystems. Woodwell *et al.* (1979) demonstrated that significant interactions occurred between a salt marsh in Long Island Sound and local coastal waters.

Essentially, coastal wetlands serve several roles in North America coastal ecosystems: (1) as important components of several littoral biogeochemical cycles, especially nutrients and carbon; (2) as critical habitats for species ranging from invertebrates to birds; and (3) as sinks for mineral sediments. To these, may be their additional functions in coastline stabilization and erosion control, though rapidly eroding coastlines are seldom areas where marshes persist.

Coastal marshes are generally considered to be net sinks for inorganic nutrients and net exporters of organic materials (including associated nutrients) (Mitsch and Gosselink, 1995). Studies of nutrient exchanges in US Atlantic Coast marshes (e.g., Stevenson *et al.*, 1977) leave the general question of whether marshes are net exporters or importer of nutrients unsettled. Nixon (1980) concluded, in his review of available evidence at the time, that marshes were, overall, exporters of dissolved organic nitrogen (DON) and dissolved phosphorus, but importers of nitrite and nitrate. Nonetheless, it is clear that high levels of denitrification can occur in marshes. Gross denitrification at Great Sippewissett Marsh in Massachusetts, particularly from June through October, was almost twice the import of nitrate into the marsh from groundwater (Kaplan *et al.*, 1979).

The high net primary productivity has been cited as underscoring the importance of coastal wetlands as a source of organic detritus and dissolved organic carbon (DOC) in North American estuaries is clear. In eroding marshes the amount of organic detritus per hectare of marsh can be very large, particularly in brackish or estuarine marshes where the percentage of organic carbon by dry weight in marsh sediments is generally 50% (Stevenson *et al.*, 1985). However, whether all of these materials eventually serve as food sources in estuaries is a matter of debate. The concept that "outwelling" of organic materials from marshes buffers nutrient demand in estuaries has been suggested to be one of the major functions of tidal marshes in estuarine productivity (Odum, 1961). Actual support for this hypothesis has proven inconclusive. Haines (1979), using carbon isotope data, showed that carbon from strictly terrestrial plants dominated the tissues of fish and other herbivorous species, suggesting that whatever the quantity of carbon exported from marshes, it was not preferred as a primary food source.

Similar problems surround the functions of marsh as sediment sinks. While it is clear from sedimentological studies (Stevenson *et al.*, 1986) that marshes—and by extension, mangroves—comprise great stores of mineral sediment, the actual functions of coastal wetlands in littoral sediment budgets of North American coasts is not resolved. In Chesapeake Bay, Stevenson *et al.* (1988) suggested that suspended sediment deposition in marshes accounts for 11% of the total estuary sediment budget. Much of the effectiveness of coastal marshes lies in how tidal velocities vary between flood and ebb tides, with strong ebb domination of the tidal cycle promoting greater export of mineral than import. The few data on sediment fluxes in North American marshes implies that their functions as major sediment sinks are probably limited, with many marshes probably exporting more sediment than importing sediment during the tidal cycle (Stevenson *et al.*, 1988).

The case for providing critical habitat, especially for water birds and commercially valuable species of finfish and shellfish, is much stronger. It has been estimated that 90% of commercially valuable finfish and shellfish spawn and spend their juvenile stages in coastal marshes (Mitsch and Gosselink, 1995). More directly important to coastal ecology is the contribution of coastal marshes to sustaining a host of organisms in the marsh food web, particularly larval stages of filter feeding organisms like the oyster (*Crassotera virginica*), which as adults, contribute significantly to maintenance of estuarine water quality.

Animals

The variety of animals that characterize the coastal waters of North America include organisms that spend their lives ranging more or less freely in the water column like zooplankton and complex vertebrates, organisms that largely inhabit the benthos (mainly invertebrates among the metazoans and many unicellular species like various bacteria), and those that move from the benthos to water column, often depending on life stage. These animals, whether herbivorous or predatory, are often part of complex trophic interactions in littoral food webs, above all in estuaries. The occurrence of both pelagic and benthic species is related to turbidity, salinity, dissolved oxygen, and availability of food sources. The last, not surprisingly, is exemplified by the fact that zooplankton are often found in greatest abundance in North American coastal waters (as elsewhere) where numbers of phytoplankton, their principal food source, are highest (Miller, 1983). However, in the general ecological functioning of North American coastal systems, the role of animals can be seen to relate to aspects of coastal biogeochemical cycles: (1) remineralization of nutrients from grazing on phytoplankton; (2) mixing and resuspension of subtidal sediments, affecting turbidity and reintroduction of biogenic debris, often fecal material; and (3) the filtering of largely organic particulates by filter feeders like oysters and clams.

Grazing on phytoplankton and organic detritus, with the ultimate release of remineralized nutrients in the excreta from death of animals, is a major element of nitrogen and phosphorus cycles in estuaries. Though filter-feeding benthic invertebrates clearly play a part in this, the major regenerators of nutrients are zooplankton. Ingestion of detritus as a source of remineralized nutrients by copepods appears to be seasonally important along the US Atlantic Coast, with the months between March and May being the time of year when algal food sources are insufficient (Heinle and Flemer, 1976).

Mixing and resuspension of subtidal sediments, particularly organic particles, mainly reflects the activities of burrowing invertebrates. Bioturbation of subtidal sediments in estuaries not only homogenizes sediments down to depths of often a meter (*Mytilus*), but increases the potential for resuspension of sediments, especially fecal pellets. In Atlantic Coast estuaries like Chesapeake Bay, the annual amount of fecal material excreted by oysters can be as high as 1–2 metric tons per hectare (Nichols and Biggs, 1985). The potential resuspension of these materials, as well as other organic particulates, for later heterotrophy or bacterial attack by bacteria is a significant source of inorganic nitrogen (e.g., NO_3^-) and orthophosphates (Aston, 1980) in ambient waters.

The decline in the populations of filter-feeding organisms like the oyster has been cited as one of the major reasons for the crises in water quality along North America coasts. In Chesapeake Bay, the population of oysters at the time of initial European contact in 1608 was probably sufficient to filter all the Bay's waters in two to three weeks, whereas by the late 20th century it needed several decades (Newell, 1988). This considerable contribution to the littoral waters of North American coasts should not diminish the importance of the removal of remineralized nitrogen passing through the guts of oysters as fecal material which binds the sediment, and makes it less vulnerable to resuspension.

Threats to the ecology of North American coasts

Eutrophication

Coastal eutrophication in North America during the latter half of the 20th century has yielded poorer water quality, diminishing catches of commercially valuable species of finfish and shellfish, loss of habitat for other creatures, and lower recreational values. A recent report by the National Oceans Service of the US NOAA found that 65% of the total estuarine surface area could be considered as exhibiting moderate to high levels of eutrophic conditions, marked by depleted dissolved oxygen levels, loss of submerged aquatic vegetation, growth of macroalgae, increasing frequency of toxic algal blooms, and increasing levels of chlorophyll *a* (Bricker *et al.*, 1999). The middle Atlantic and Gulf Coasts, in particular, represent the most threatened coastal waters in North America.

Not the least of the many impacts of coastal eutrophication, are the mounting costs for saving what remains of rapidly vanishing coastal resources. The importance of the cost issue can be indicated by estimates of expenditures for controlling nutrients in the largest estuary in North America, Chesapeake Bay. Between 1985 and 1996, over US $3.5 billion were spent on nutrient control in the Bay's watershed (Butt and Brown, 2000). The cost of removing each kilogram of total nitrogen per year in the Chesapeake Bay alone probably ranges upwards of US $35 (Camacho, 1992). This cost, of course, does not reflect the large expenditures for the restoration of resources damaged by coastal eutrophication, encompassing funds for the monitoring and replanting of sea grasses, grants for improving the aquaculture of threatened species like the American oyster (*C. virginica*) in several Atlantic Coast estuaries, subsidies for the funding of alternative farming methods, and the like. But the dimensions of the threat posed by the eutrophication of North American coastal waters, epitomized by the rapidly enlarging "dead zone" in the Gulf of Mexico at the mouth of the Mississippi River (Turner and Rabalais, 1994), are such that delaying action may prove to incur costs well beyond the money needed for remediation.

Burgeoning human activities around coasts and in coastal watersheds underlie the phenomenon of coastal eutrophication. A major impediment to effective mitigation of the threat has been readily accessible, up-to-date information on the nature and extent of land cover changes and land use practices that contribute to loading of nutrients in coastal waters. The pace of change is such that growth of populations (regardless of the types of activities these people will be engaged in) could increase by 100% in many already heavily settled areas of the US Atlantic Coast (see Stevenson and Kearney, 1996).

Monitoring changes in land cover and land use is crucial to any assessment of the effects of human activities on coastal ecosystems. Satellite remote sensing has been particularly useful in this regard, and the C-CAP program in the United States (Klemas *et al.*, 1993) is a good example. This program produced the first, uniform information on land use and land cover changes in the US coastal zone using Landsat Thematic Mapper imagery. Application of remote sensing technologies in combination with Geographic Information Systems (GIS) and various spatial environmental models (e.g., Costanza *et al.*, 1990) are providing the broad synoptic tools for regional to coast-wide management of ecological impacts that attend population growth and development along North American coasts.

Providing information is not necessarily mitigation of ecological threats and damage, which require political action. Here, the progress has been more piecemeal, particularly in the United States with its combination of federal, state, and local jurisdictions, and overlapping and differing mandates between state and federal agencies. General water quality mandates, regulated by the US Environmental Protection Agency (EPA), are related to issues of direct or indirect risk to human health and seldom effectively address issues like coastal eutrophication that are usually loosely related to any specific human activity and tend to be regional to extra-regional in scope. Regional consortia of local and regional governments are perhaps best positioned to assume the various roles necessary for advocating (especially with local municipalities), enacting, and coordinating the measures for coastal mitigation. The Tri-State Commission—Maryland, Virginia, and Pennsylvania—for cleanup of the Chesapeake Bay is a good example of jurisdictional cooperation in a major effort to limit the influx of nutrients in a major coastal system of North America already burdened by extensive coastal eutrophication.

Land use policies aimed at controlling nutrient inputs span those advocating changes in regional agricultural practices to legislated controls on the types of land use permitted in critical areas of large watersheds feeding coastal systems. Controls on land use or density of development are not always popular, but are effective even where employed in only a limited fashion. In the 1980s, the State of Maryland enacted the Critical Areas Act mandating the creation of 1,000-foot (330 m) buffer zones around the Chesapeake Bay and its major tributaries. This buffer was largely intended to protect from encroaching development, upland forests and riparian zones that can attenuate nutrient runoff into coastal waters. In addition, the density of development in certain areas of the Bay coastline was further controlled, again to limit nutrient influx. To be sure, the Act met with mixed responses from various quarters, but subsequent studies did show that nutrient inputs were significantly diminished to the Bay (Marcus *et al.*, 1993).

Sea-level rise

The possibility of rapidly rising sea levels from global warming first became widely recognized almost 20 years ago, and was followed over the next two decades by numerous studies outlining what an acceleration in sea-level rise could mean for coastal systems. Though an accelerating sea-level rise could have a multitude of effects (both direct and indirect) on North American coastal ecosystems, at least two are likely: the potential extensive loss of coastal wetlands and increasing turbidity in Atlantic Coastal Plain estuaries from higher rates of shore erosion.

Estimates derived from historical surveys suggest that almost half of the coastal wetlands in the United States extant in 1900 had disappeared by the last quarter of the 20th century (Gosselink and Baumann, 1980). Many of these losses reflect development or other human activities, but overall it is safe to assume that a more generic cause, sea-level rise, was responsible. In recent years, regional studies ranging from the Mississippi Delta to the middle Atlantic Coast (cf. Stevenson *et al.*, 1986) indicate that marsh vertical accretion rates in many areas lag local rates of submergence. In areas where there are significant deficits in vertical accretion compared to sea-level rise, inventories using aerial photography and other techniques show high wetland losses. In the Mississippi Delta, annual losses of coastal marshes amount to almost 60 km^2 (Britsch and Kemp, 1990).

Projections for rates of global sea-level rise approaching 1 m per century would be catastrophic for the survival of coastal wetlands generally, in particular coastal marshes (Nicholls *et al.*, 1999). Admittedly, most research on the relations of coastal wetlands to sea-level rise has focused on middle latitude coastal marshes, but it is probably safe to assume high-latitude marshes in Hudson's Bay would fare the same. In addition, though the imminent widespread collapse of mangroves due to sea-level rise heralded almost a decade ago (cf. Ellison, 1993) was perhaps premature, the survival of mangroves in south Florida in such a scenario would be doubtful as well.

Despite some continuing disagreement concerning the overall ecological role of wetlands in estuaries and lagoons along North America's coasts, reviews by Nixon (1980), among others, make clear that loss of coastal wetlands would affect aspects of coastal ecology as various as nutrient cycling and fisheries. The economic costs of large-scale wetland decline and loss would be equally significant. Costanza *et al.* (1989) demonstrated that the size of shrimp harvests in Louisiana correlated well with marsh area in the Mississippi Delta.

The role of sea-level rise as the driver of long-term coastal erosion is well known. Nevertheless, the impact that an accelerated rate of sea-level rise would have on water turbidity in North American estuaries is just becoming evident. Detailed sediment budgets for many North America estuaries are generally not available, but it is clear that sediment inputs from shore erosion will increasingly loom larger as rates of sea-level rise increase. At present, shore erosion adds a volume of sediment into Chesapeake Bay every century equivalent to the District of Columbia (US Army Corps of Engineers, 1990). Much of the finer-grained eroded sediments will inevitably end up increasing the already high, suspended particulate load of estuaries, thereby further increasing water turbidity. With effects of high turbidity on estuarine biota that span damage to sea grasses to even the foraging success of predatory fish species like striped bass, mounting inputs of suspended sediments may offset any hard-won gains in water clarity from controls on upland erosion that have been emplaced in many areas in recent decades.

Baja California: a unique North American coast

In 1940, a year after the publication of *The Grapes of Wrath*, John Steinbeck with his friend, Edward Ricketts, spent five weeks exploring the Sea of Cortez. Exhausted and troubled by the controversy generated by his great novel, Steinbeck, connecting with his early undergraduate training as a marine biologist, surely could not have chosen a more unique coastal system in North America to visit. Surrounded by deserts, the Sea of Cortez contains some of the deepest basins (over 2,000 m) of any coastal region of the continent. About 4.5 million years old (Atwater, 1970), and formed by separation of the Baja Peninsula during the late Pliocene and Pleistocene, the area is seismically active, and characterized by occasional tsunamis, with volcanism occurring as recently as the Holocene (Gastil *et al.*, 1983). One of the most salient features of the Gulf, is its many islands (30–40) that fall into two general groups, northern and southern islands. The islands differ in origin, some being volcanic, others formed by uplift or submergence, and those next to the mainland created by narrow headlands separated by erosion (Gastil *et al.*, 1983).

The upper Gulf is an area characterized by some of the highest tides in the world, with spring tidal ranges in the vicinity of the Colorado River Delta reaching almost 10 m (Matthews, 1968). Tidal phase also changes from the delta to the middle Gulf, and though the tidal wave is modified by islands and, in particular, by shoals in the vicinity of the delta (Maluf, 1983). However, tidal phases for most of the Gulf appear to be largely semidiurnal in character, although they are regularly diurnal around Bahia Concepción and Guaymas during the month (Maluf, 1983).

The ecology of the Sea of Cortez is dominated by the seasonal climate variations and influence of the islands. Because the sea is too small to modify regional climates, water temperatures mirror seasonal changes in regional temperature, especially in the upper Gulf (Maluf, 1983). Together with the sometimes extreme range in salinities near the Colorado Delta, produced by changes in discharge of the Colorado, shallow-water biota show a greater interannual variations in composition (Maluf, 1983). The influence of the islands is reflected in bird, reptile, and mammal populations that change by location of the islands and their proximity to mainland coasts. Slight terrestrial faunal differences between the Baja Peninsula and the mainland Mexican Coast tend to characterize islands nearest either shore. Phytoplankton densities decline around islands clustered closest to mainland shores, where water turbidity is higher due to shore erosion and wave resuspension during storms (Thomson and Gilligan, 1983). Because water masses in the Sea of Cortez can be broadly partitioned into either being mainland-influenced or Pacific-influenced, there is said to be a pelagic species gradient within the upper and lower Gulf, but the actual gradient may run between the mainland shore and off-shore islands, reflecting the greater seasonal variation in sea surface temperature along mainland coasts (Thomson and Gilligan, 1983). Changes in benthic species also show mainland versus island differences, in this case largely due to the reefs that ring many of the islands, especially in the lower Gulf.

In the lower Gulf, a large, stable low oxygen zone, produced by decomposing phytoplankton, limits benthic species where it impinges upon shallower water areas (Maluf, 1983). At the mouth of the Gulf, ocean circulation disrupts, reinforcing the essentially confined-sea character of the Gulf of California.

Michael S. Kearney and J. Court Stevenson

Bibliography

Aston, S.R., 1980. Nutrients, dissolved gases, and general biogeochemistry in estuaries. In Olausson, E., and Cato, I. (eds.), *Chemistry and Biogeochemistry of Estuaries*. Chicester: John Wiley, pp. 233–295.

Atwater, T., 1970. Implications of plate tectonics for the Cenozoic tectonic evolution of western North America. *Geological Society of America Bulletin*, **81**: 3513–3536.

Bachman, L.J., and Phillips, P.J., 1996. Hydrological landscapes on the Delmarva Peninsula Part 2: estimates of base-flow nitrogen load to Chesapeake Bay. *Water Resources Bulletin*, **32**: 779–791.

Barsdate, R.J., Nebert, M., and McRoy, C.P., 1974. Lagoon contributions to sediments and water of the Bering Sea. In Hood, D.W., and Kelley, E.J. (eds.), *Oceanography of the Bering Sea*. Fairbanks: University of Alaska, pp. 553–576.

Berner, E.K., and Berner, R.A., 1996. *Global Environment: Water, Air, and Geochemical Cycles*. Upper Saddle River, NJ: Prentice Hall.

Boynton, W.R., Garber, J.H., Summers, R., and Kemp, W.M., 1995. Inputs, transformations and transport of nitrogen and phosphorus in Chesapeake Bay and Selected Tributaries. *Estuaries*, **18**(1B): 285–314.

Bricker, S.B., and Stevenson, J.C., 1996. Nutrients in coastal waters : a chronology and synopsis of research. *Estuaries*, **19**: 337–341.

Bricker, S.B., Clement, C.G., Pirhalla, D.E., Orlando, S.P., and Farrow, D.R.G., 1999. *National Estuarine Eutrophication Assessment: Effects of Nutrient Enrichment in the Nation's Estuaries*. Silver Spring, MD: Special Projects Office and the National Centers for Coastal Ocean Science, National Ocean Service, National Oceanic and Atmospheric Administration.

Britsch, L.D., and Kemp, E.R., 1990. *Land loss rates: Mississippi Deltaic Plain*. US Army Corps of Engineers, *Technical Report GL/90/02*.

Butt, A.J., and Brown, B.L., 2000. The cost of nutrient reduction: a case study of Chesapeake Bay. *Coastal Management*, **28**: 175–185.

Camacho, R., 1992. *Financial Cost Effectiveness of Point and Nonpoint Source Nutrient Reduction in the Chesapeake Bay Basin*. Rockville, MD: Cheapeake Bay Program Reduction Strategy Reevaluation. Interstate Commission for the Potomac River Basin, ICPRB Report 92-4.

Carpenter, J.H., Pritchard, D.W., and Whaley, R.C., 1969. Observations of eutrophication and nutrient cycles in some coastal plain estuaries. In *Eutrophication: Causes, Consequences, Correctives*. Washington, DC: National Academy of Sciences, pp. 210–221.

Chmura, G.L., Coffey, A., and Crago, R., 2001. Variation in surface sediment deposition on salt marshes in the Bay of Fundy. *Journal of Coastal Research*, **17**: 221–227.

Costanza, R., Farber, S.C., and Maxwell, J., 1989. Valuation and management of wetland ecosystems. *Ecological Economics*, **1**: 335–361.

Costanza, R., Sklar, F.H., and White, M.L., 1990. Modeling coastal landscape dynamics. *Bioscience*, **40**: 91–107.

Duce, E.P., 1991. Chemical exchange at the air-coastal interface. In Mantoura, R.F.C., Martin, J.-M., and Wollast, R. (eds.), *Ocean Margin Processes in Global Change*. Chicester: John Wiley, pp. 91–109.

Ellison, J.G., 1993. Mangrove retreat with rising sea level, Bermuda. *Estuarine, Coastal and Shelf Science*, **37**: 75–87.

Gardner, R.H., Castro, M.S., Morgan II, R.P., and Seagle, S.W., 1996. Nitrogen dynamics in forested lands of the Chesapeake Bay. In Nelson, S. (ed.), *Perspective on the Chesapeake Bay, 1995: Advances in Estuarine Science*, Chesapeake Bay Program, Scientific and Technical Advisory Committee (STAC), Chesapeake Research Consortium, Inc., *Publication No. 151*, pp. 1–36.

Gastil, G., Minch, J., and Phillips, R.P., 1983. The geology and ages of the islands. In Case, T.J., and Cody, M.L. (eds.), *Island Biogeography in the Sea of Cortez*. Berkeley, CA: University of California Press, pp. 13–25.

Gosselink, J.G., and Baumann, R.H., 1980. Wetland Inventories. *Zeitschrift fur Geomorphologie*, **34**: 173–187.

Gottschalk, L.C., 1964. Reservoir Sedimentation. In Chow, V.T. (ed.), *Handbook of Applied Hydrology*. New York: McGraw-Hill, 17-1–17-34.

Haines, E.B., 1979. Interaction between Georgia salt Marshes and coastal waters: a changing paradigm. In Livingston, R.J. (ed.), *Ecological Processes in Coastal and Marine Systems*. New York: Plenum Press, pp. 35–46.

Haith, D.A., and Shoemaker, L.L., 1987. Generalized watershed functions for stream flow nutrients. *Water Resources Bulletin*, **23**: 471–478.

Heinle, D.A., and Flemer, R.J., 1976. Flows of materials between poorly flooded tidal marshes and an estuary. *Marine Biology*, **35**: 357–375.

Howarth, R.W., 1988. Nutrient Limitation of Net Primary Production in Marine Ecosystems. *Annual Review of Ecological Systems*, **19**: 89–110.

Kaplan, W.L., Valiela, I., and Teal., J.M., 1979. Denitrification in a salt marsh ecosystem. *Limnology and Oceanography*, **24**: 726–734.

Klemas, V.V., Dobson, J.E., Ferguson, R.L., and Haddad, K.D., 1993. A coastal land cover classification system for the NOAA coast watch change analysis project. *Journal of Coastal Research*, **9**: 862–872.

Lot, A., Novelo, A., and Ramirez-Garcia, P., 1993. Diversity of Mexican aquatic plant flora. In Ramamoorthy, T.P. *et al.* (eds.), *Biological Diversity of Mexico: Origins and Distribution*. Oxford: Oxford University Press, pp. 577–591.

Malone, T., Boynton, W., Horton, T., and Stevenson, J.C., 1993. Nutrient loadings to surface waters: Chesapeake Bay case study. In Uman, M. (ed.), *Keeping Pace With Science And Engineering: Case Studies in Environmental Regulation*. Washington DC: National Academy of Press, pp. 8–38.

Maluf, L.Y., 1983. The physical oceanography. In Case, T.J., and Cody, M.L. (eds.), *Island Biogeography in the Sea of Cortez*. Berkeley, CA: University of California Press, pp. 26–48.

Marcus, W.A., Neilsen, C.C., and Cornwell, J., 1993. Sediment budget analysis of heavy metal inputs to a Chesapeake Bay estuary. *Environmental Geology and Water Sciences*, **22**: 1–9.

Matthews, J.B., 1968. The tides of Puerto Peñasco. *Journal of the Arizona Academy of Sciences*, **5**: 131–134.

McRoy, C.P., Barsdate, R.J., and Nebert, M., 1972. Phosphorus cycling in an eelgrass (*Zostera maritime* L.) ecosystem. *Journal of Fisheries Research Bd. Canada*, **27**: 1811–1821.

Miller, C.B., 1983. The zooplankton of estuaries. In Ketchum, B.H. (ed.), *Estuaries and Enclosed Seas. Ecosystems of the World 26*. Amsterdam: Elsevier Scientific Publishing Company, pp. 103–150.

Mitsch, W., and Gosselink, J.G., 1995. *Wetlands*. New York: Van Nostrand Reinhold.

Newell, R.I.E., 1988. Ecological changes in Chesapeake Bay, are they the result of overharvesting the American oyster, *Crassostera virginica*? In Lynch, M.P., and Krome, R. (eds.), *Conference Proceedings, Chesapeak Research Consortium Publication No. 129 (CBP/TRS 24/88)*, Solomons, MD: pp. 536–546.

Nichols, M.M., and Biggs, R.B., 1985. *Estuaries*. In Davis, R.A., Jr. (ed.), *Coastal Sedimentary Environments*. New York: Springer-Verlag, pp. 77–186.

Nicholls, R.J., Hoozemans, F.M.J., and Marchand, M., 1999. Increasing flood risk and wetland losses due to global sea level rise: regional and global analyses. *Global Environmental Change*, **9**: S69–S80.

Nixon, S.W., 1980. Between coastal marshes and coastal waters- a review of twenty years of speculation and research on the role of salt marshes in estuarine productivity and water chemistry. In

Hamilton, P., and MacDonald, K.B. (eds.), *Estuarine and Wetland Processes*. New York: Plenum Press, pp. 457–525.
Odum, E.P., 1961. The role of tidal marshes in estuarine production. *New York State Conservation*, **15**: 12–15.
Odum, E.P., 1963. *Ecology. Modern Biology Series*. New York: Holt Rinehart and Winston.
Odum, H.T., Copeland, B.J., and McMahan, E.A., 1974. *Coastal Ecological Systems of the United States*, Vol. I. Washington DC: The Conservation Foundation.
Paerl, H.W., 1985. Enhancement of marine primary production by nitrogen enriched acid rain. *Nature*, **315**: 747–749.
Peterson, B.J., Wolheim, W.M., Mulholland, P.J., Webster, J.R., Meyer, J.L., Tank, J.L., Marti, E., Bowden, W.B., Valett, H.M., Hershey, A.E., McDowell, W.H., Dodds, W.K., Hamilton, S.K., Gregory, S., and Morrall, D.D., 2001. Control of nitrogen by headwater streams. *Science*, **292**: 86–90.
Pomeroy, L.R., Smith, E.E., and Grant, C.M., 1965. The exchange of phosphate between estuarine water and sediments. *Limnology and Oceanography*, **10**: 167–172.
Postma, H., 1980. Sediment transport and sedimentation. In Olausson, E., and Cato, I. (eds.), *Chemistry and Biogeochemistry of Estuaries*. Chicester: JohnWiley, pp. 153–186.
Pritchard, D.W., 1954. A study of salt balance in coastal plain estuary. *Journal of Marine Research*, **13**: 133–144.
Pritchard, D.W., 1967. Observations of estuarine circulation in coastal plain estuaries. In Lauff, G.H. (ed.), *Estuaries*. Washington, DC: American Association for the Advancement of Science Special Publication 3, pp. 3–5.
Redfield, A.C., Ketchum, B.H., and Richards, R.A., 1963. The influence of organisms on the composition of sea-water. In Hill, M.N. (ed.), *The Sea*. New York:Wiley-Interscience, pp. 26–77.
Seagle, S.W., Pagnotta, R., and Cross, F.A., 1999. The Chesapeake Bay and Northern Adriatic Sea drainage basins: land-use and outputs of nutrients and pollutants. In Malone, T., Malej, A., Harding, L., Smodlaka, N., and Turner, E. (eds.), *Ecosystems at the Land–Sea Margin: Drainage Basin to Coastal Sea*. Washington, DC: American Geophysical Union Press, pp. 7–25.
Smayda, T.J., 1983. The phytoplankton of estuaries. In Ketchum, B.H. (ed.), *Estuaries and Enclosed Seas. Ecosystems of the World 26*. Amsterdam: Elsevier Scientific Publishing Company, pp. 65–102.
Smith, R.A., Alexander, R.B., and Wolman, M.G., 1987. Water-quality trends in the nation's rivers. *Science*, **23**: 1606–1615.
Staver, K.W., and Brinsfield, R.B., 1996. Seepage of groundwater nitrate from a Riparian agroecosystem into the Wye River estuary. *Estuaries*, **19**: 359–370.
Stevenson, J.C., 1988. Comparative ecology of submersed grassbeds in fresh, brackish and marine environments. *Limnology and Oceanography*, **33**: 867–893.
Stevenson, J.C., and Kearney, M.S., 1996. Shoreline dynamics on the windward and leeward shores of a large temperate estuary. In Nordstrom, K.F., and Roman, C.T. (eds.), *Estuarine Shores: Hydrological, Geomorphological and Ecological Interactions*. New York: John Wiley & Sons, pp. 233–259.
Stevenson, J.C., Heinle, D.R., Flemer, D.A., Rowland, R.J., and Ustach, J.F., 1977. Nutrient exchanges between brackish water marshes and the estuary. In Wiley, M. (ed.), *Estuarine Processes*, Vol. II, New York: Academic Press, pp. 219–240.
Stevenson, J.C., Kearney, M.S., and Pendleton, E.C., 1985. Sedimentation and erosion in a Chesapeake Bay brackish marsh system. *Marine Geology*, **6**: 213–235.
Stevenson, J.C., Ward, L.G., and Kearney, M.S., 1986. Vertical accretion rates in marshes with varying rates of sea-level rise. In Wolfe, D.A. (ed.), *Estuarine Variability*. New York: Academic Press, pp. 241–259.
Stevenson, J.C., Ward, L.G., and Kearney, M.S., 1988. Sediment transport and trapping in marsh systems: implications for tidal flux studies. *Marine Geology*, **80**: 37–59.
Stevenson, J.C., Staver, L.W., and Staver, K., 1993. Water quality associated with survival of submersed aquatic vegetation along an estuarine gradient. *Estuaries*, **16**: 346–361.
Stevenson, J.C., Rooth, J., Kearney, M.S., and Sundberg, K., 2001. The health and long term stability of natural and restored marshes in Chesapeake Bay. In Weinstein, M.P., and Kreeger, D.A. (eds.), *Concepts and Controversies in Tidal Marsh Ecology*. Dordecht, Netherlands: Kluwer Academic Press, pp. 709–735.
Taft, J.L., and Taylor, W.R., 1976. *Estuarine Processes*, Vol. 1. New York: Academic Press.
Thomson, D., and Gilligan, M.R., 1983. The rocky-shore Fishes. In Case, T.J., and Cody, M.L.(eds.), *Island Biogeography in the Sea of Cortez*. Berkeley, CA: University of California Press, pp. 98–129.
Twilley, R.R., Kemp, W.M., Staver, K.W., and Stevenson, J.C., 1985. Nutrient enrichment of estuarine submersed vascular plant communities. I. Algal growth and effects on production of plants and associated communities. *Marine Ecology Progress Series*, **23**: 179–191.
Turner, R.E., and Rabalais, N.N., 1994. Coastal eutrophication in the Mississippi River Delta. *Nature*, **368**: 619–621.
US Army Corps of Engineers, 1990. Chesapeake Bay Shoreline Erosion Study, Vol. 2. Baltimore, MD: Baltimore Corps of Engineers.
Vorosmarty, C.J., and Petersen, B.J., 2000. Macro-scale models of water and nutrient flux to the coastal zone. In Hobbie, J.E. (ed.), *Estuarine Science*. Washington, DC: Island Press, pp. 43–80.
Wolff, W.J., 1980. Biotic aspects of estuaries. In Olausson, E., and Cato, I. (eds.), *Chemistry and Biogeochemistry of Estuaries*. Chicester: John Wiley, pp. 264–295.
Woodwell, G.M., Hall, A.S., Whitney, D.E., and Houghton, R.A., 1979. The Flax Pond ecosystem study: exchanges of inorganic nitrogen between an estuarine marsh and Long Island Sound. *Ecology*, **60**: 695–702.

Cross-references

Estuaries
Global Vulnerability Analysis
Groundwater (see Hydrology of Coastal Zone)
History, Coastal Ecology
Monitoring, Coastal Ecology
North America, Coastal Geomorphology
Sampling Methods (see Monitoring, Coastal Ecology)
Salt Marsh
Sea-Level Rise, Effect
Vegetated Coasts
Water Quality
Wetlands

NORTH AMERICA, COASTAL GEOMORPHOLOGY

The shores of the North American continent span approximately 55° of latitude and 100° of longitude. The continent is bounded by four major coastal regimes associated with the Atlantic, Arctic, and Pacific Oceans, and the Gulf of Mexico. The length of open ocean coast, excluding non-barrier islands, exceeds 110,000 km, and is much longer if the Great Lakes, embayments, *lagoons*, and islands are included. The coastal morphology includes all of the major coastal types described in Shepard's (1937) classification. There are *mangrove* and coral systems in the southern reaches of the continent, and *ice*-dominated systems in the north. The common occurrence of fjords and rias attests to the delayed response of most of the North American coast to glacial processes and *eustatic* and *isostatic sea-level* changes.

Geomorphic provinces

The coast of North America (Figure N24) comprises parts of at least seven of the geomorphic provinces described in Graf (1987). The designations for these provinces, and their characteristics, are taken from Graf (1987), unless otherwise noted, and they are used to organize and introduce the principal features of coastal regimes discussed herein. The North American distribution of *muddy coasts* and coarse clastic coasts are presented separately.

Pacific Rim Province

The west coast of the continent, from the Mexico-Guatemala border to the Alaska Peninsula is included in the Pacific Rim Province. It is a region that is tectonically active, lying along the convergent boundaries of the Pacific, Juan de Fuca, and Cocos Plates with the North American Plate, although the Golfo de California is a spreading center. Most of this coast, other than Baja California and southern California, is a collision coast using the terminology of Inman and Nordstrom (1971). The general morphology of the Province is characterized by a narrow continental shelf and narrow or absent coastal plains backed immediately by steep and high mountain ranges. Coastal *cliffs* are common and estuarine or *barrier* systems are limited to the vicinity of drowned river mouths. Notable exceptions are the barrier systems along the eastern shore of the Golfo de California. Spring *tide* ranges vary from about

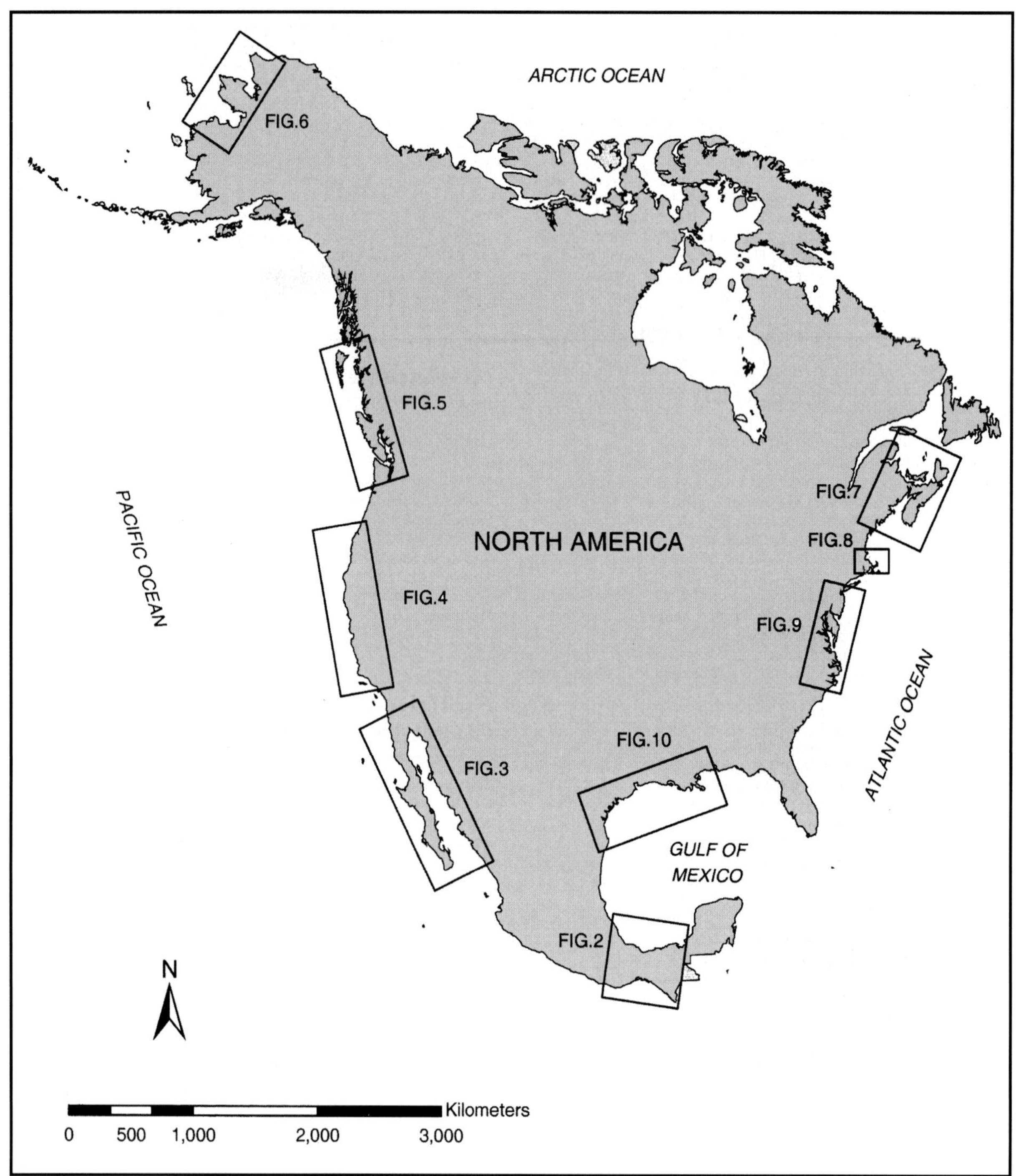

Figure N24 An outline of the North American continent. The insets mark regions represented in Figures N25–N33.

0.5 m along the Mexican coast near Acapulco, to 1.0–3.0 m along most of the rest of the Pacific coast to the US–Canada border. Ranges increase northward to more than 10 m in the vicinity of Anchorage, Alaska. Spring tide ranges reach 7 m in the northern reaches of the Golfo de California (Kelletat, 1995). The dominant source of wave energy along most of this coast is from ocean swell, although tropical-*storm*-induced *waves* are important in Mexico and (occasionally) southern California.

The Pacific coast of Mexico in the vicinity of Guatemala (Figure N25) is characterized by barrier/lagoon systems. *Dunes* are common on the barriers, and mangroves are present in many of the lagoons. *Coral reefs* occur sporadically along this coastline. Much of the coast north of the Golfo de Tehuantepec to Nayarit is cliffed where the Sierra Madre del Sur lies adjacent the coast. Along other reaches of this region, barrier/lagoon systems are extensive, especially east of Acapulco (Figure N25), in the vicinity of Zihuatenejo, and between Puerto Vallarta and Mazatlan.

There is another barrier/lagoon complex east of Los Mochis (Figure N26). The *sandy* barriers include coastal dune and *beach ridge* complexes. Many of the lagoons provide mangrove habitat (Kelletat, 1995). There are similar systems north from Los Mochis, but they become smaller and occur less frequently along the increasingly arid coastline of the Gulfo de California. Wave energy decreases northward into the Gulf, but the tide range increases to a maximum at the mouth of the Colorado River (Figure N26). The Colorado River *delta* comprises a

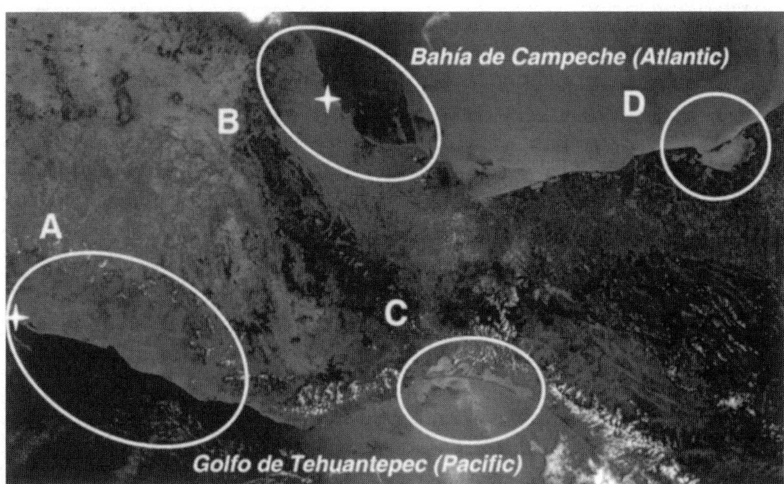

Figure N25 The Isthmus of Tehuantepec in tropical Mexico. This is a MODIS image (courtesy of NASA) from 29/03/02. (A) Sandy coastline extending east from Acapulco (✧) with barrier and lagoon systems. (B) Lowland coast in the vicinity of Veracruz (✧) includes sandy barrier systems and extensive marshes. A coral reef system is located just offshore from Veracruz. (C) The extensive barrier/lagoon/marsh systems of the Gulf of Tehuantepec. (D) The Laguna de Términos is about 50 km wide, fronted by a sandy barrier and backed by extensive marshes.

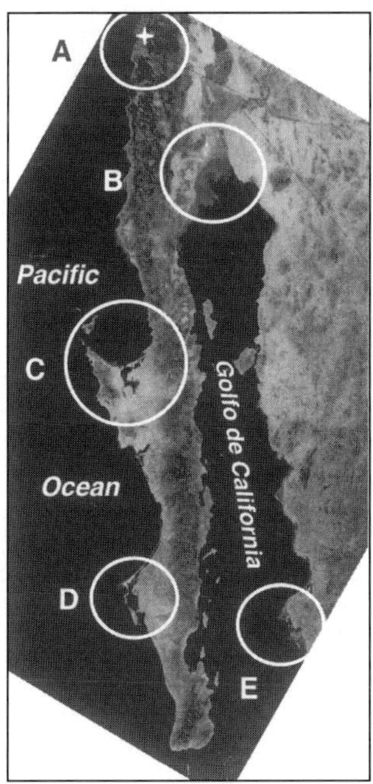

Figure N26 Baja California Peninsula (Mexico) and the Gulf of California. This is a MODIS image (courtesy of NASA) from 11/11/02. (A) San Diego, California, USA (✧). Note the small size of the bay relative to those at locations C and D in Mexico. (B) The delta of the Colorado River and the river's suspended sediment plume. (C) Laguna Ojo de Liebre and barrier systems in Viscaino Bay. (D) Bahia Magdalena and its extensive barrier systems. (E) Los Mochis and the northern edges of the barrier/lagoon/marsh systems of Sinaloa States.

large network of distributaries, estuaries, marshes, and saline lagunas. Many of these coastal environments are threatened by reduced fluvial discharge and sediment supply caused by large *dams* on the river (Carriquiry and Sánchez, 1999). The location of the delta is controlled by the San Andreas *Fault* that separates the North American plate on the east, from the Pacific plate, with Baja California, on the west.

The east coast of Baja California (Figure N26) is shaped largely by recent *tectonic* activity, with the Sierra de San Pedro Martir and the Sierra de La Giganta (both part of the Peninsular Range that runs north to Los Angeles) abutting the coast in the northern and southern Gulf, respectively. The coast is irregular, and frequently cliffed with pocket beaches between headlands. The arid climate produces few perennial streams to deliver sediment to the shore and the entire coast is low energy, so beaches tend to be relatively narrow. There are only a few, small barrier systems, at Santa Rosalia, for example, or La Paz. Baja California's west coast is also quite rugged. However, drainage basins are larger than those found along the Gulf coast, and sediment supplies are larger. The most prominent coastal features are Bahia Magdalena and Vizcaino Bay. The former is the largest of a series of bays created by the presence of a chain of *barrier islands* and *spits*. Coastal dunes are common on the barriers and the lagoons are backed by marshes or sand and salt flats. In Vizcaino Bay, Laguna Ojo de Liebre (Scammon's Lagoon), results from structural control on coastal development. The Sierra Vizcaino parallel the Sierra de La Giganta to create a large trough that has been filled with sediments from the mountain ranges. The barrier and lagoon systems are formed in these sediments along the bay. Extensive dune fields and salt flats occur across this complex. Raised marine terraces are found at elevations up to 150 m (Orme, 1998).

Near the US–Mexico border, the Tijuana River created a substantial *estuary* marsh system, and has provided the sediments that were transported northward to form the barrier spit that protects San Diego Bay from the open ocean (Figure N26). Most of the coast in this region has cliffs fronted by narrow sand or cobble beaches. This configuration, often backed by sets of marine terraces, is common to the rest of the Pacific coast, except where structural controls have formed basins or where rivers have formed estuaries. The city of Los Angeles, for example, is built across the structurally controlled Los Angeles Basin that has been filled with thick deposits (locally in excess of 6,000 m) of marine sediments. It is the largest coastal plain in California, not coincidentally occupied by the largest city. Los Angeles beaches are backed by *Holocene* (and older) dunes, although most dune surfaces have been urbanized. The southern California coastline, with the exception of the military reservation at Camp Pendleton in San Diego County, is also extensively human altered through beach nourishment, cliff stabilization efforts, and seawalls and groin fields (Griggs, 1998), and the extent of alteration rivals that of the New Jersey shore.

The southern California coastline is also characterized by well-defined sets of raised marine terraces. The Palos Verde Peninsula divides the coast of the Los Angeles Basin. There are 13 terraces on the peninsula, at elevations up to 411 m, the marine limit (Orme, 1998). The marine limit rises to about 600 m, with 15 terraces, in the vicinity of Santa Barbara, and falls to about 200 m near the Big Sur coast of central California. The Big Sur coastline (Figure N27) is formed of high marine cliffs carved into the Santa Lucia Mountains. Pocket beaches, of sand or coarser materials, occur in sheltered locations along this coast.

Monterey Bay (Figure N27) marks the separation of the Santa Lucia and Santa Cruz Ranges. There is a coastal, deltaic plain created by the Salinas River flowing between the mountain ranges. Wave energies along this coast are high, and most of the coastline is backed by dune

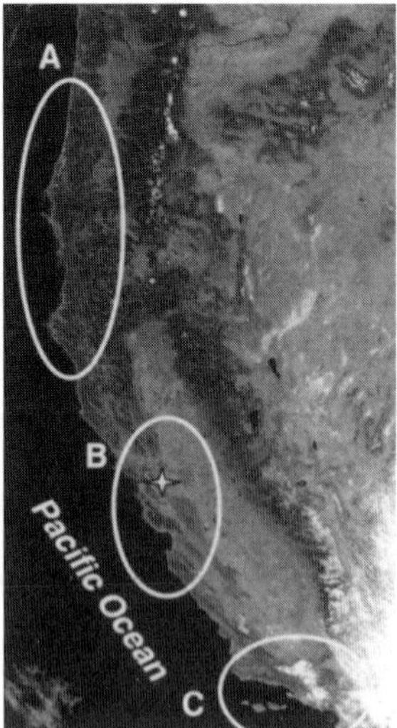

Figure N27 The coasts of Oregon and Cailfornia. This is a SeaWIFS image (courtesy of NASA) from 23/10/00. (A) The relatively straight coast of southern Oregon includes large barrier spit/estuary systems, and is backed by extensive coastal dunes. The major promontories are Cape Blanco in the north, and Cape Mendocino in California. (B) The central California coast from San Francisco (✧) to the Big Sur coast. San Francisco and the Bay are immediately west of the ✧. The large embayment to the south is Monterey Bay. (C) The west–east trending coast of southern California, including the Channel Islands. This coast is characterized by a series of small barrier/estuary systems separated by headlands and cliffs.

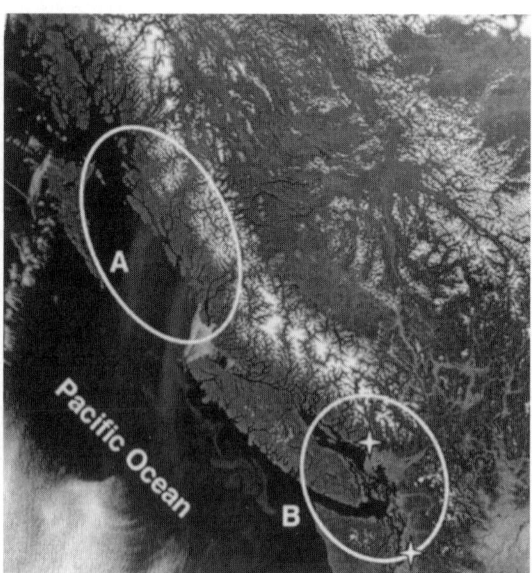

Figure N28 The west coast of Canada. This is a SeaWIFS image (courtesy of NASA) from 09/08/01. (A) The glaciated coast between Alaska and Vancouver Island. Rocky islands, fjords, and a paucity of sandy beaches are characteristic. (B) Vancouver, British Columbia (northern ✧) is on the northern edge of the Fraser River delta, on the Strait of Georgia. Seattle (✧) is on Puget Sound, connected to the Pacific through the Strait of Juan de Fuca. Fjords, rocky islands, and deltas are common.

complexes that extend inland, in locations, almost 10 km (Cooper, 1967). Coastal cliffs and low marine terraces back the coast from Santa Cruz, at the northern edge of Monterey Bay, north to San Francisco Bay. Noteworthy is the Devil's Slide coastline south of San Francisco, where steep, 250 m high cliffs have been eroded into the unstable Franciscan formation, and where, as the name implies, slope failures are common.

San Francisco Bay (Figure N27) is formed by the submergence of a Pliocene basin. The Sacramento River system flows into the bay, forming an inland delta and marsh complex. The distributaries are confined between levees and much of the delta surface is currently below sea level. During Pleistocene lowstands of sea level, the river debouched to the Pacific through the channel it had eroded between the Golden Gates. North of San Francisco is the Point Reyes peninsula, an outcrop of the Pacific Plate separated from the mainland by Tomales Bay, formed in a trough along the San Andreas fault line.

The California coast north to Cape Mendocino is extremely rugged, comprising high cliffs cut into the Coastal Range Mountains, and localized sand or gravel beaches. Numerous submarine canyons extend close to the shore. There are no large drainage systems reaching this coast and most of the reach remains inaccessible by land. North of the Cape, several large rivers, including the Eel, Mad, Klamath, and Smith, do reach the coast to form estuary and baymouth barrier systems such as the one that forms Humboldt Bay. Large coastal dune complexes have formed from the sediments blown off the beaches near these rivers.

The relatively straight coastlines of Oregon and Washington (Figure N27) lie adjacent to the subduction zones associated with the Juan de Fuca and Gorda Plates. As is common with collision coasts, rugged terrain and cliffs are characteristic along these shores. There are, however, numerous fluvial/estuarine systems with baymouth barrier spits. Good examples of these occur at Coos Bay, Newport, and Tillamook Bay in Oregon, and Willapa Bay, and Grays Harbor, Washington. There are also extensive dune systems along this coast (Cooper, 1958). Many of the dunes are confined to the barriers, but at several locations, such as near Florence, or Sand Lake, Oregon, they extend inland for kilometers.

The Pacific coasts of Canada and Alaska (Figure N28) show the marked influences of alpine glaciation. The coast of British Columbia shows a strong structural control on morphology, with the Georgia and Hecate depressions separating Vancouver Island and the Queen Charlotte Islands from the coastal mountain ranges on the mainland. The coast is deeply incised as a result of glacial erosion and the creation of numerous fjords. There are very few beaches along this entire coast, and they tend to be confined to the headwaters of the fjords where rivers deposit sands and gravels. There is a large delta system at the mouth of the Fraser River (Figure N28), creating a large coastal plain that is currently occupied by the city of Vancouver and its suburbs. Most of this coast, continuing into Alaska, is inaccessible by land.

The Pacific coast of Alaska (Figure N28) is similar to that of British Columbia, with three important distinctions. First, there is greater tectonic activity, especially along the Aleutian Megathrust, and this has spawned a chain of active *volcanoes* that have created the Aleutian Islands. Second, many of the Alaskan fjords retain tide-water glaciers that preclude any beach formation. Third, tidal ranges increase dramatically toward Anchorage, and this produces swift tidal *currents* that transport much of the fine sediment load from the coastal streams into deepwater. Most of this coastline is accessible only by water or air. About midway between Sitka and Cordova, Alaska, sits Lituya Bay, site of the largest "recorded" wave uprush or swash (just over 500 m) caused by an earthquake and resulting landslide, in 1958.

The Arctic coastline

The Arctic coasts of North America include portions of four geomorphic provinces: the Interior Mountains and Plateaus; the Rocky Mountains, the Arctic Lowlands; and the Canadian Shield. Over-arching characteristics of these coastal regimes are the effects of extensive sea ice, the effects of continental glaciation and isostatic adjustments, and the effects of permafrost. In addition to generating a suite of ice-related coastal landforms, such as ice foot, push ridges, thermokarst lagoons (Figure N29) or scour features, the presence of sea ice reduces or eliminates fetches, thereby reducing the wave energy available to reshape the coast (e.g., Trenhaile, 1990). Tide ranges are generally less than 1 m along the exposed arctic coasts, but increase to more than 4 m in parts of Hudson Bay and more than 6 m along parts of the Ungava Peninsula (Davies, 1980).

Interior Mountain Province

The coast between the Alaskan Peninsula and northern Kotzebue Sound (Figure N29) is part of the Interior Mountains and Plateaus Province. The regional morphology is dominated by a series of expansive, Quaternary coastal plains associated with a series of deltas. The

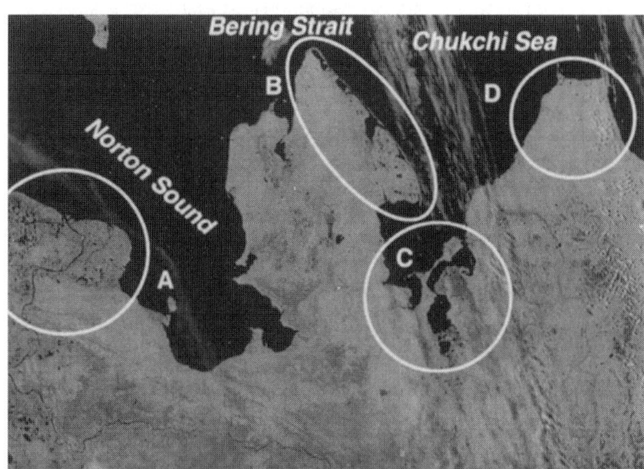

Figure N29 Alaska's Seward Peninsula (figure has west at the top). This is a MODIS image (courtesy of NASA) from 04/08/02. (A) The Yukon River delta, Characterized by fine sediments and thermokarst erosion. (B) North coast of the Seward Peninsula, with sandy barrier systems backed by lagoons and extensive marshes. Thermokarst erosion is common in the fine sediment deposits. (C) Kotzebue Sound and the Baldwin Peninsula. The southern shore has extensive coastal dunes. Most of the rest of the coast comprises deltaic mudflats with some thermokarst erosion. (D) Point Hope, at the western end of the Brooks Range.

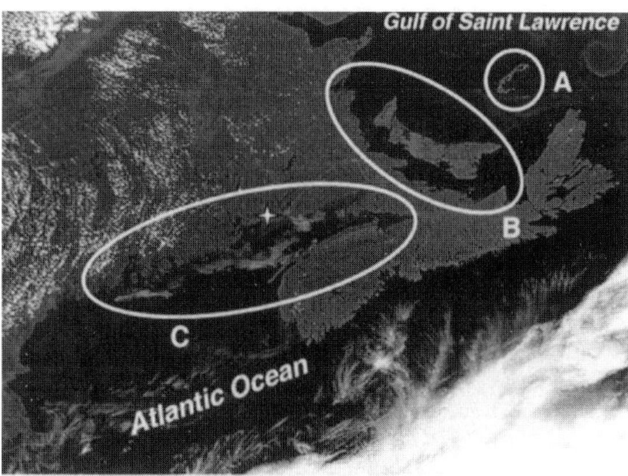

Figure N30 South eastern Canada. This is a SeaWIFS image (courtesy of NASA) from 20/07/01. (A) Îlesla Madeleine, Quebec. The islands are characterized by extensive sandy beaches and coastal dunes, with occasional rocky outcrops. (B) Prince Edward Island, Northumberland Strait, and the coastal plains of New Brunswick and Nova Scotia. The exposed coasts are characterized by barrier/estuary systems typically separated by cliffed headlands. (C) The Bay of Fundy and the glaciated coasts of New Brunswick and Maine. Muddy tidal flats are common in the Bay of Fundy east of Saint John (✧).

largest of these, the 3,000 km^2 Yukon delta, receives about 90% of the fluvial sediment yield in this Province (Walker, 1998). Coastal sediments are relatively fine-grained and mudflats (Flemming, 2002) and coastal dune fields (Walker, 1998) are common in the vicinity of the deltas.

Rocky Mountain Province

The Rocky Mountain Province is represented along a short distance of the Alaskan coast by the northern terminus of the Brooks Range. This coastline is characterized by permafrost cliffs with elevations between about 10 and 300 m (Walker, 1998), with the latter elevations occurring where the Brooks Range directly abuts the coast. Many of the cliffs are fronted by narrow, sandy beaches.

Arctic Lowlands Province

The Arctic Lowlands stretch from a point west of Point Barrow to the northern tip of the Boothia Peninsula, and includes a number of large, low relief islands. Banks, Victoria, Melville, and Prince of Wales Islands are the largest of these. Permafrost is ubiquitous in the Province, and many of the characteristic landscape elements of the coastal zone reflect this. Thermokarst depressions and pingos are common (Carter *et al.*, 1987). Coastal morphology reflects extensive control by glaciation and sea ice. This coast is ice-bound for much of the year, limiting wave-induced erosion. The eastern reaches of the lowlands are sheltered further by islands. Coastal sediments are typically derived from *bluff* erosion, glacial tills, or deltas, such as those of the Coleville and the MacKenzie Rivers. Much of the drainage of northern Canada debouches across the lowlands, creating extensive delta networks and delivering large sediment loads.

Canadian Shield Province

The Canadian Shield coast stretches from the relatively low relief of the Boothia Peninsula to the high relief coast of mainland Newfoundland (Labrador), and the northern coast of the Gulf of Saint Lawrence. It includes many islands, most notably Baffin Island. The landscape is dominated by the effects of Pleistocene glaciation (continental and alpine) and subsequent isostatic rebound. There is continuous permafrost along most of the low-lying coast, with sporadic permafrost found as far east (and south) as Labrador (Shilts *et al.*, 1987). Marine processes are minimized by short fetch distances caused by the presence of islands, and by the frequent presence of nearshore or shore-fast sea ice. The coasts of Baffin Island and much of Labrador are characterized by coastal cliff and fjord systems typical of high-relief *glaciated* coasts. The most distinctive large-scale morphological feature of the Shield coast is Hudson Bay, a flooded depression in the shield that has an east–west extent of about 800 km, and a north–south extent exceeding 1,000 km (including James Bay). The coast of southern and western Hudson Bay has emergent beach ridge systems that may extend inland up to 75 km (Martini *et al.*, 1980). Many of the beaches along eastern Hudson Bay through Ungava Bay are characterized by boulder strewn *tidal flats*. Rias are common along many reaches of the north coast of the Gulf of Saint Lawrence (Dubois, 1980).

Appalachian Mountains and Plateaus Province

The maritime provinces of Canada, and the coast of Maine are in the Appalachian Mountains and Plateaus Province. This includes the island of Newfoundland, Prince Edward Island, and Cape Breton Island. Most of the shores of this region display relict glacial landforms from the Pleistocene. The exposed northern coast of Newfoundland and the eastern coast of Nova Scotia are high wave-energy environments, characterized by *rocky* shores with till-derived, sand and gravel beaches in protected coves and estuaries. The coasts surrounding the southern reaches of the Gulf of Saint Lawrence are characterized by barrier bay systems. Sediments are derived mainly from the erosion of glacial tills or from offshore sources as a result of *transgression*.

The Bay of Fundy (Figure N30) is a low wave-energy environment that is dominated by tidal processes. Maximum spring tide ranges exceed 15 m in the southern arm of the bay, the Minas Basin, where large inter-tidal sand bodies are common. Wide, inter-tidal mudflats are common in the upper reaches of Chignecto Bay, Fundy's northern arm (e.g., Davidson-Arnott *et al.*, 2002).

Atlantic Coastal Plain Province

The Atlantic Coastal Plain Province reaches from Maine southward to the US–Mexico Border. Its morphological characteristics have been described by Walker and Coleman (1987). Tectonically, this is a passive, or trailing-edge coastline (Inman and Nordstrom, 1971) of low relief except where glacial or isostatic processes have created coastal cliffs in the northern parts of the Province. Most of the coastline in this Province is characterized by barrier island and estuary/bay systems, including many of the longest barrier islands on earth. Spring tide ranges on the open Atlantic coasts are typically less than 2 m, and less than 1 m along the coast of the Gulf of Mexico. Average annual wave energy decreases southward in this Province.

The coast from Maine south to New York (Figures N30–N32) displays the effects of continental glaciation during the Pleistocene, extensive submergence during the Holocene transgression, and a strong geological control on coastline features (Fitzgerald *et al.*, 2002). There are few sandy barriers north of Bigelow Bight (Figure N31). Drowned river mouths, moraines, and glacial outwash features are common. Examples of the former include Penobscot Bay, Maine, Boston Harbor

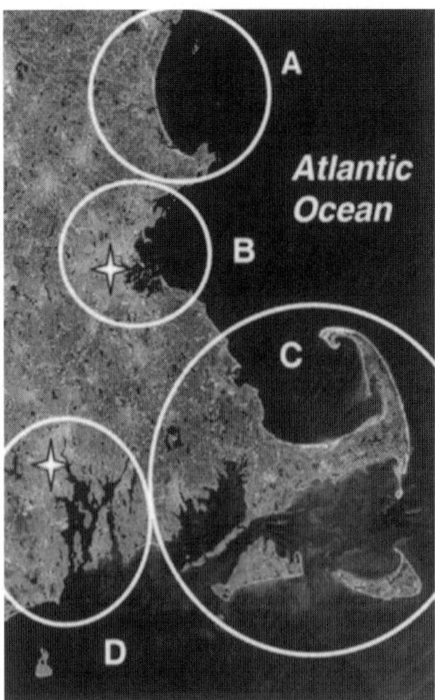

Figure N31 The southern New England coastline. This is a MISR image (courtesy of NASA) from 13/04/00. (A) Bigelow Bight, from Portsmouth, New Hampshire, past the Plum Island barrier, south to Cape Ann, Massachusetts. (B) Massachusetts Bay, including Boston Harbor and the City of Boston (✧). (C) The classic recurved spit of Cape Cod with sandy barrier systems common along the eastern shore, and the islands of Martha's Vineyard and Nantucket in the southwest and southeast. (D) The glaciated coastline in the vicinity of Narragansett Bay and Providence, Rhode Island (✧).

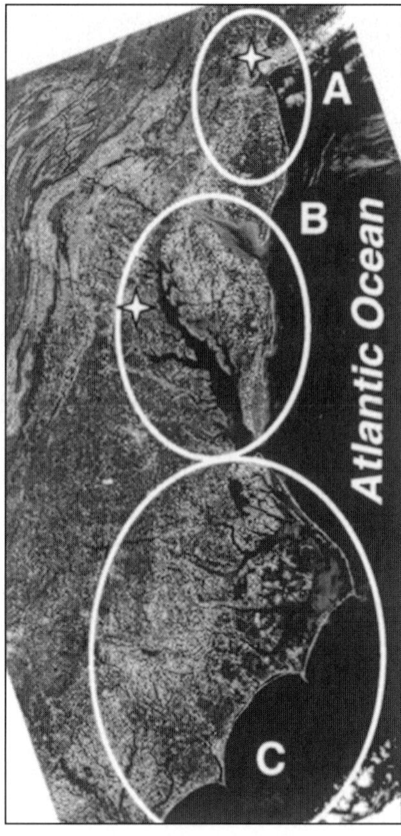

Figure N32 The mid-Atlantic seaboard of the United States. This is a MODIS image (courtesy of NASA) from 11/10/00. (A) New York Bight, western Long Island, New York City (✧), and the barrier systems of northern and central New Jersey. (B) Chesapeake Bay, a classic ria, and the Delmarva Peninsula. The ✧ indicates the approximate location of the Baltimore/Washington urban region. (C) The Carolina Capes—Hatteras, Lookout, and Fear (from north to south), with their long, sandy barriers, and Pamlico Sound.

(that includes drumlins, Figure N31), Narragansett Bay, Rhode Island (Figure N31), and the New York Bight (Figure N32). The dominant glacial sedimentation features include Cape Cod and the islands of Martha's Vineyard and Nantucket (Figure N31), and Long Island (Figure N32). Western and northern Cape Cod has a spine of moraine deposits, but the most dramatic coastal features are the large, recurved spits formed from reworked outwash. Barrier systems are common along the Cape Cod shore, and extensive sand dunes occur in the northern reaches. Similarly, Long Island, New York, comprises a morainal spine along its northern coast and an outwash plain along its southern coast. Glacial sands, eroded from the eastern end of the island, have been transported westward by longshore currents to form several barrier islands, including Fire Island. The New York Bight (Figure N32) is the drowned valley of the Hudson River, bounded to the east by Breezy Point, a barrier spit, and to the south by Sandy Hook, another barrier spit. Staten Island is formed by a remnant moraine and is near the southern limit of glacially affected coastline in North America.

The rest of the coastline in this geomorphic Province is dominated by series of barrier and lagoon systems. Fisher (1982) identified four barrier island coastlines in the United States, and all are in this Province: the Middle Atlantic States from New York (south shore of Long Island) through North Carolina; South Carolina and Georgia Sea Islands; Florida's east coast; and the Texas and Mexico coasts. The groups are distinguished not only by geography, but also by characteristic form assemblages. All of the barrier islands include coastal dunes (although perhaps substantially human-altered), and many of the larger islands have extensive dune systems.

In the Middle Atlantic States, the shores of New Jersey and the Delmarva peninsula (Figure N32) are morphologically similar. Both are characterized by barrier island chains often separated from the mainland by wide bays. Both coasts are terminated by large estuaries, Delaware Bay to the north, and Chesapeake Bay to the south formed by the submergence of fluvial systems. Along the northern New Jersey coast the direction of net *alongshore transport* is toward the north, terminating at Sandy Hook. From central New Jersey to Chesapeake Bay, transport along the Atlantic coast is to the south. The configurations of Cape May and Cape Charles result in part from this process. Kraft (1985) notes that this reach typifies the response of coastal plains to marine transgressions with elongated and often overlapping islands. The extensive use of shore protection structures and beach nourishment projects make the barrier coast of New Jersey one of the most extensively, human-altered shorelines in North America (e.g., Nordstrom, 2000).

The middle Atlantic coast of North Carolina also features nearly continuous barrier island chains (Figure N32). There are pronounced changes in orientation along this coast, with structural control across the boundary of the Chesapeake–Delaware Basin and the Cape Fear Arch (Walker and Coleman, 1987). The Holocene transgression submerged several large, low-relief drainage basins, leaving the elongated barrier chains configured as a set of capes, Capes Hatteras, Lookout, and Fear. The first two are well separated from the mainland by Albermarle and Pamlico Sounds. The larger barrier islands (e.g., Hatteras) have large, well-developed dune complexes.

The South Carolina and Georgia Sea Islands comprise a set of barriers with a distinctive morphology. Walker and Coleman (1987) describe three island types: Pleistocene remnants, Holocene barriers, and marsh islands (protected by the barriers). These barrier islands are much less elongated than their Middle Atlantic counterparts, with many more *inlets*.

The east coast of Florida between Jacksonville and Palm Beach is straight, other than the foreland caused by Cape Canaveral. There is no fluvial sediment delivery to the coast along this reach. Wave energies along this coast are moderate to low, but the region is subject to high-energy events caused by tropical storms. The barrier islands are elongated, and the bays behind them are also elongated with relatively few inlets. There are small dune systems on the barriers, sabellariid reefs in the inter-tidal zone (Kirtley and Tanner, 1968), and relict beach ridge plains on the mainland. Along the southern reaches of the Florida Peninsula, coral reefs replace the sabellariid colonies, there are low relief limestone cliffs, and calcium carbonate replaces silica as the dominant constituent of the beach sand. Mangroves occur in sheltered waters of the Florida *Keys*.

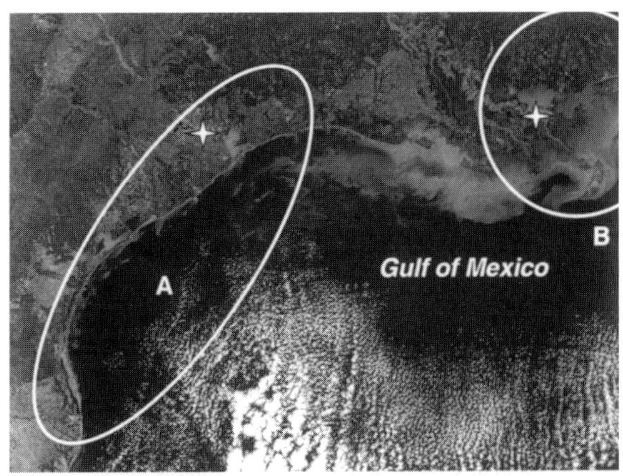

Figure N33 The Gulf of Mexico coasts of Texas and western Louisiana. This is a MODIS image (courtesy of NASA) from 13/11/02. (A) The coast of Texas from the Rio Grande River to the border with Louisiana. Long sandy barrier islands and extensive lagoon/marsh systems are characteristic of this low-lying coast. Houston (✧) is near the head of Galveston Bay. (B) The Mississippi River delta, characterized by extensive marshes and fine sediment deposition (note sediment plume from river mouth). New Orleans (✧) is located on the south shore of Lake Ponchartrain, with the latter created by the progradation of the delta.

The west coast of Florida is *micro-tidal* with very low wave energy. There are scattered barrier systems in the central sections of this reach, with mainly silica sands. Mangroves become locally important along the northern coasts, as wave energy north of Tampa Bay becomes trivial. In the Florida panhandle, wave energy increases westward, although tide ranges remain small. There are well-developed barrier, beach ridge, and coastal dune systems along this coast, especially from the Apalachicola River delta west to Pensacola.

The major morphological features of the US coast of the Gulf of Mexico are the Mississippi River delta and the barrier islands of Texas (Figure N33). There is minimal structural control along this coast, and the coastal plain is quite wide, 100 km or more in Texas and Louisiana (Walker and Coleman, 1987). Wave energy and tide ranges are low, although this region is also vulnerable to tropical storms. Beach sediments along the northwestern Gulf show strong influences of voluminous, fine sediment delivery by the Mississippi River and subsequent redistribution to the east by Gulf currents. The river drains more than 3×10^6 km^2, and discharges 580×10^9 m^3/yr of water and 330×10^9 kg/yr of suspended sediment (Meade, 1981). Delivery of this sediment load into the low-energy environment of the northern Gulf of Mexico has caused the progradation of the Mississippi River delta.

The Mississippi River delta is one of the largest, and most studied, deltas on earth (e.g., Walker and Coleman, 1987; Wells, 1996). Its classic birdfoot shape has grown more pronounced through historic times as the loss of *wetlands* has emphasized the configuration of the natural levees. Wetland loss, at rates averaging in excess of 50 km^2/yr for more than a half a century (Reed, 2002), represent the greatest rates of coastal land loss in North America. These losses are caused by oxidation of organic materials, subsidence of the delta deposits, and sea-level rise—the local, relative rate of sea-level rise exceeds 10 mm/yr (Reed, 2002). Coastal erosion contributes only slightly to the wetland loss rate, although it has contributed to the formation of several barrier island chains, such as the Timbaliers and Chandeleurs.

The coastline of Texas (Figure N33) is low energy and micro-tidal. There is a wide coastal plain of low relief, and the mainland coast is fronted by a series of long barrier islands and spits, including the longest undeveloped barrier in the world, Padre Island. The barriers are transgressive, being driven landward by sea-level rise. Beach sediments are also derived from the Mississippi River, and the four Texas streams that discharge through the barriers: the Brazos, San Bernard, and Colorado Rivers, and the Rio Grande (Aronow and Kaczorowski, 1985). The bays are also elongated, the Laguna Madre, for example, is about 200 km in length. Numerous small estuaries, from submerged drainage systems, form the landward perimeter of the bays. The entire coastal system is vulnerable to erosion, inundation, or washover, caused by tropical storms. Dune systems are common, especially along the southern coasts. Inlets are widely spaced, tend to have small ebb tidal deltas, and sediment *bypassing* is sporadic and event driven (Morton *et al.*, 1995). Galveston Island, once inundated by the most catastrophic hurricane in US history, is now the most human altered barrier in Texas, comprising groin fields, landfill, beach nourishment, and a seawall.

The Gulf coast of Mexico has several well-developed barrier/marsh/lagoon systems. South from the Rio Grande delta is the Laguna Madre and a long barrier spit and island system backed by a series of estuaries. This system extends from the Rio Grande south to the vicinity of La Pesca. The barriers and lagoons then become smaller and discontinuous until the large estuarine delta system of the Rio Panuco, near Tampico. Cabo Rojo is part of the Isla del Idolo barrier complex, including dunes and *beach ridges*, that protects the Laguna Tamiahua, and it is the only pronounced coastal feature in this reach. This coast is also in the range of tropical coral habitat.

Coral is also found off the Veracruz coast, and this reach is also characterized by large expanses of marshes, lagoons, and sandy barriers. Wave energy in this region is low, as is the tidal range. At about this location the coast begins a pronounced bend toward the northeast, and the large Laguna de Términos. Mangroves occur in sheltered locations within the lagoon systems, and dunes are common on the barriers. This coast changes largely in response to tropical storms and changes in water level. The western coast of the Yucatan Peninsula is dominated by sedimentation from large drainage systems, leading to the development of large beach ridge complexes associated with the deltas (Psuty, 1966). The eastern shore of the Yucatan is also a low-energy environment, with coral reef forming a barrier, and mangroves dominate in sheltered waters.

Distribution of muddy coasts

Few of the open ocean coasts of North America have extensive systems of muddy beaches. Their distribution is described by Flemming (2002), and the locations of the most extensive systems are summarized here. Muddy coasts are usually low-gradient, and often provide substrate for deltaic, lagoonal, or estuarine marsh systems. Typically, muddy coastal environments are found adjacent to large deltas, such as the Mississippi River delta (Figure N33) or where wave energy is low—in estuaries or embayments, such as the Bay of Fundy (Figure N30). Large, mud-dominated coasts are found in the lagoon systems of tropical Mexico (Figure N25), the Gulf of California, and the Baja California Peninsula (Figure N26) and are usually associated with marsh systems. There are few muddy-systems along the west coast of the United States, and those are confined to relatively small estuaries, except for the marshes around San Francisco Bay (Figure N27) and the Columbia River estuary. The muddy environments of the western Canadian coast are associated with the deltas of the Fraser and Skeena Rivers (Figure N28), and there are few substantial deposits along the Pacific coastline of Alaska. The rest of the Alaskan coast has muddy coastlines around Bristol Bay and in the vicinity of Norton Sound and the Yukon River delta (Figure N29).

Flemming (2002) notes that the major reaches of muddy coast in Arctic and Atlantic Canada are few, but include extensive mudflats in Hudson, James, and Ungava Bays, muddy marshes in the St. Lawrence estuary, and fringing mud deposits in the Bay of Fundy (Figure N30). The Atlantic coast of the United States is characterized by lengthy barrier systems that provide shelter for back barrier bays and marshes such as those in New Jersey and landward of the Carolina Capes (Figure N32). In all of these cases, the open Atlantic coasts are mainly sand. The estuaries of this coast, including Delaware and Chesapeake Bays (Figure N32), and Long Island Sound, provide sufficient protection from wave action to accumulate large muddy deposits, usually manifested in *salt marshes*. Similar systems are found around the Gulf of Mexico. Sandy barriers protecting muddy lagoons and marshes, with some estuaries, characterize the coasts of Mississippi, Alabama, and Texas (Figure N33), and much of the Mexican coast south to Belize. Notable for the extent of the deposits are Laguna Términos (Figure N25) and Laguna de Tamiahua. The Mississippi River delta (Figure N33) stands alone as the only significant reach of open coast to be characterized by muddy deposits. This reflects the river's extremely large, fine-sediment load, and the relatively low wave energy in the northern Gulf.

Distribution of coarse clastic coasts

Beaches composed of coarse clastic sediments, also referred to as shingle, gravel, pebble, cobble, or boulder beaches, are most common in *periglacial* or paraglacial environments, along cliffed coasts, or near the mouths of steep gradient coastal streams (Orford *et al.*, 2002). These beaches have steep profiles, and are relatively immobile compared to sand-dominated systems. Davies (1980) indicates that such deposits are of some degree of relative importance around most of the North American continent. Specifically, he categorizes these beaches as occasionally important along a reach of the Pacific coast roughly bounded by Acapulco (Figure N25) and San Diego (Figure N26), in Canada from northern Newfoundland to northern Nova Scotia (Figure N30), and along the reach from Maine (Figure N30) to New York (Figure N32). Davies classifies coarse clastic systems as important along the coasts of Canada and the United States north of Puget Sound on the west coast and north of Newfoundland on the east coast. They are also important along the coasts of Newfoundland and northern Maine.

Douglas J. Sherman

Bibliography

Aronow, S., and Kaczorowski, R.T., 1985. Texas. In Bird, E.C.F., and Schwartz, M.L. (eds.), *The World's Coastline*. New York: Van Nostrand Reinhold, pp. 129–145.

Carriquiry, J.D., and Sánchez, A., 1999. Sedimentation in the Colorado River delta and upper Gulf of California after nearly a century of discharge loss. *Marine Geology*, **158**: 125–145.

Carter, L.D., Higenbottom, J.A., and Woo, M., 1987. Arctic Lowlands. In Graf, W.L. (ed.), *Geomorphic Systems of North America*. Boulder, CO: Geological Society of America, pp. 583–628.

Cooper, W.S., 1958. *Coastal Sand Dunes of Oregon and Washington*. Boulder, CO: Geological Society of America, Memoir 72.

Cooper, W.S., 1967. *Coastal Dunes of California*. Boulder, CO: Geological Society of America, Memoir 104.

Davidson-Arnott, R.G.D., van Proosdij, D., Ollerhead, J., and Schostak, L., 2002. Hydrodynamics and sedimentation in salt marshes: examples from a macrotidal marsh, Bay of Fundy. *Geomorphology*, **48**: 209–231.

Davies, J.L., 1980. *Geographical Variation in Coastal Development*. London: Longman Group, LTD.

Dubois, J.M.M., 1980. Géomorphologie du littoral de la côte nord du Saint-Laurent: Analyse Sommaire. In McCann, S.B. (ed.), *The Coastline of Canada: Littoral Processes and Shore Morphology*. Ottawa: Geological Survey of Canada Paper 80-10, pp. 215–238.

Fisher, J.J., 1982. Barrier Islands. In Schwartz, M.L. (ed.), *The Encyclopedia of Beaches and Coastal Environments*. Stroudsburg: Hutchinson Ross, pp. 124–133.

Fitzgerald, D.M., Buynevich, I.V., Davis, Jr., R.A., and Fenster M.S., 2002. New England tidal inlets with special reference to riverine-associated inlet system. *Geomorphology*, **48**: 179–208.

Flemming, B.W., 2002. Geographic distribution of muddy coasts. In Healy, T., Wang, Y., and Healy, J.-A. (eds.), *Muddy Coasts of the World: Processes, Deposits, and Function*. Amsterdam: Elsevier Science, pp. 99–201.

Graf, W.L. (ed.), 1987. *Geomorphic Systems of North America*. Boulder, CO: Geological Society of America, pp. 643.

Griggs, G.B., 1998. California's coastline: El Niño, erosion, and protection. In Ewing, L., and Sherman, D.J. (eds.), *California's Coastal Natural Hazards*. Los Angeles, CA: University of Southern California Sea Grant Program, pp. 38–55.

Inman, D.L., and Nordstrom, C.E., 1971. On the tectonic and morphologic classification of coasts. *Journal of Geology*, **79**: 1–21.

Kelletat, D.H., 1995. Atlas of coastal geomorphology and zonality. *Journal of Coastal Research*, Special Issue 13.

Kirtley, D., and Tanner, W.F., 1968. Sabellariid worms; builders of a major reef type. *Journal of Sedimentary Petrology*, **38**: 73–78.

Kraft, J.C., 1985. Atlantic USA–Central. In Bird, E.C.F., and Schwartz, M.L. (eds.), *The World's Coastline*. New York: Van Nostrand Reinhold, pp. 213–219.

Martini, I.P., Cowell, D.W., and Wickware, G.M., 1980. Geomorphology of southwestern James Bay: a low energy, emergent coast. In McCann, S.B. (ed.), *The Coastline of Canada: Littoral Processes and Shore Morphology*. Ottawa: Geological Survey of Canada Paper 80-10, pp. 293–300.

Meade, R.H., 1981. Man's influence on the discharge of freshwater, dissolved material, and sediments by rivers to the Atlantic coastal zone of the United States. In *River Inputs to Ocean Systems*. Switzerland: UNEP, pp. 13–17.

Morton, R.A., Gibeaut, J.C., and Paine, J.G., 1995. Meso-scale transfer of sand during and after storms: implications for predictions of shoreline movement. *Marine Geology*, **126**: 161–179.

Nordstrom, K.F., 2000. *Beaches and Dunes of Developed Coasts*. Cambridge: University Press.

Orford, J.D., Forbes, D.L., and Jennings, S.C., 2002. Organisational controls, typologies and time scales of paraglacial gravel-dominated coastal systems. *Geomorphology*, **48**: 51–85.

Orme, A.R., 1998. Late Quaternary tectonism along the Pacific coast of the Californias: a contrast in styles. In Stewart, I.S., and Vita-Finzi, C. (eds.), *Coastal Tectonics*. London: The Geological Society, Special Publication No. 146, pp. 179–197.

Psuty, N.P., 1966. Beach-ridge development in Tabasco, Mexico. *Annals of the Association of American Geographers*, **55**: 112–124.

Reed, D.J., 2002. Sea-level rise and coastal marsh sustainability: geological and ecological factors in the Mississippi delta plain. *Geomorphology*, **48**: 233–243.

Shepard, F.P., 1937. Revised classification of marine shorelines. *Journal of Geology*, **45**: 602–624.

Shilts, W.W., Aylsworth, J.M., Kaszycki, C.A., and Klassen, R.A., 1987. Canadian Shield. In Graf, W.L. (ed.), *Geomorphic Systems of North America*. Boulder, CO: Geological Society of America, pp. 119–161.

Trenhaile, A., 1990. *The Geomorphology of Canada*. Toronto: Oxford University Press.

Walker, H.J., 1998. Arctic deltas. *Journal of Coastal Research*, **14**: 718–738.

Walker, H.J., and Coleman, J.M., 1987. Atlantic and Gulf Coast Province. In Graf, W.L. (ed.), *Geomorphic Systems of North America*. Boulder, CO: Geological Society of America, pp. 51–110.

Wells, J.T., 1996. Subsidence, sea-level rise, and wetland loss in the lower Mississippi River Delta. In Milliman, J.D., and Haq, B.U. (eds.), *Sea Level Rise and Coastal Subsidence*. Dordrecht: Kluwer Academic, pp. 281–311.

Cross-references

Barrier Islands
Bay Beaches
Beach Ridges
Cheniers
Cliffed Coasts
Coastal Lakes and Lagoons
Coral Reef Coasts
Coral Reefs
Deltas
Developed Coasts
Geographical Coastal Zonality
Human Impact on Coasts
Mangroves, Ecology
Mangroves, Geomorphology
Middle America, Coastal Ecology and Geomorphology
Muddy Coasts
Paraglacial Coasts
Sandy Coasts
Vegetated Coasts
Wetlands

NOTCHES

Littoral notches are more or less horizontal erosion features close to sea level into steeper coastal slopes. They prove that special processes of destruction around the surf level are much more intensive than in subtidal or supratidal positions. Their relation to tidal levels and their shape may differ from one type of notch to the other, because there are several genetic types of notches around the coastlines of the world (Kelletat, 1982; Pirazzoli, 1986).

Notches caused by melting

At ice cliffs of shelf ice, or even in driftng ice bergs, sharply incised notches can be detected at sea level, caused by melting processes of relatively warmer surficial waters, even in arctic or antarctic environments during summer. These forms may develop in a very short time (several days), but they are usually rapidly dislocated (mostly uplifted above sea level or tilted) by the moving of ice margins or tilting of ice bergs during the melting process, when the point of gravity will shift. Notches caused by melting processes may also occur in sediments of permafrost environments, if warmer seawater in the surf zone undermines steeper slopes. These forms are seldom stable for longer times; because by melting of the cliff, soft material will creep or slide into the surf zone and cover the notches.

Notches caused by chemical solution

In the scientific literature this kind of notch formation is very often described in carbonate rocks, because the sharp microforms resemble karstic features from terrestrial environments, and the seawater is believed to act as an aggressive agent upon limestones. This conclusion is incorrect, because seawater is always oversaturated (up to several 100%) by dissolved carbonates and is not able to destroy carbonates by solution. Solution may be responsible for notch formation in halites or gypsum rocks, but no descriptions from these environments exist.

Notches caused by salt weathering

In sedimentary or volcanic rocks with clasts of sandy or coarser grains salt weathering may act as a notch-forming process. It is often restricted to those levels above mean sea level, where salt water may rise through capillary action in the rock, transporting salts. By wetting and drying the grains in the rock may be loosened, producing honeycombs and tafoni of different sizes. Galleries of these smaller forms may decorate the rocky shores and, coalesced, form an irregular shaped kind of notch. The process dominates other littoral processes only in warm and arid regions with very limited surf action and tides, as in the eastern Mediterranean.

Notches caused by mechanical abrasion

This is a widespread (although only locally developed) notch type in hard rocks, where sand and pebbles or cobbles are present to be moved constantly in the surf zone. They polish the rock and abrade it. These notches show different profiles according to lithological characteristics, but are always very smooth, showing the rock's own color, because no algae or other organisms will survive the surf attack with abrading "weapons." The relation of this notch type to sea level or tidal levels may differ according to the availability of loose sediments: if only a limited amount is present, at a steep coast the polishing may only occur in the subtidal; and if a large amount is available, only the uppermost strata of the sediment can be moved against coastal rocks, so that the abrading and notching process is restricted to cliff parts in the upper surf zone. Therefore, a difference of the notch relative to mean sea level does not always mean relative sea-level variation. Abrasion notches are very often accompanied by smaller or larger coastal caves in parts of lesser resistant rocks. If rocky coasts dip into deeper water and no loose material is present in the surf zone, the extremely differing pressure of air and water under wave attack may destroy rocks with textural or structural weakness and break out larger portions of the cliff, but they will never form notches at plunging cliffs. In regions with drifting sea ice this effect (as well as frost action) may form notch-like features in less resistant rocks.

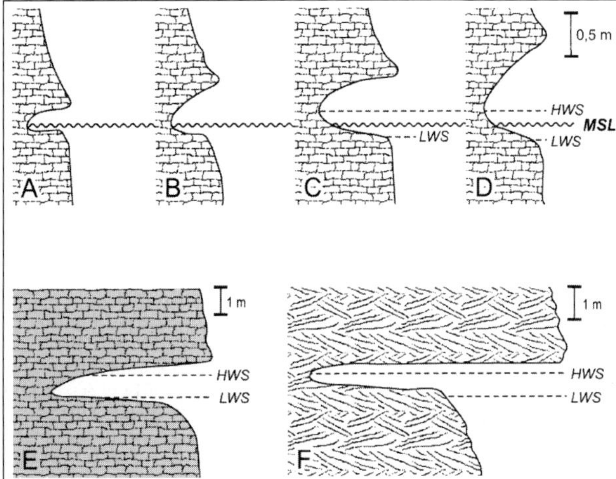

Figure N34 Shape of bioerosive notches and their relation to tidal levels (A–D from the Mediterranean): (A) tideless, sheltered; (B) tideless, exposed; (C) microtidal, sheltered; (D) microtidal, exposed; (E) microtidal, sheltered, in Neogene coral rocks of Palau, Micronesia; (F) microtidal, sheltered, in less resistant Pleistocene eolianites of the Bahamas.

Notches caused by bioerosion

This is by far the most widespread type of notch; but is restricted to carbonate rocks including carbonate sandstone (calcarenites) and sometimes dolomites, and to low latitudes (Kelletat, 1997). Its aspect is a horizontal back-carving of rocky shore faces, very often along extensive and continuous stretches of the coastline, with the smaller part under water and a wider above mean sea level. An important exception to this rule may be found at windward coasts in regions of strong and constant tradewinds, when bioerosion notches may be totally developed significantly above the high water mark. The shape of bioerosive notches differ according to tidal range and exposure (Figure N34). They are more open and wide with a tidal range larger than 1 m and more intensive surf, and more narrow with a nearly flat roof in very sheltered and almost tideless environments. These points are important if the relation of living and fossil notches to varying sea levels is discussed. In areas of large tidal ranges and extreme surf conditions bioerosional notches may be less developed or absent. Although bioerosive notches will undermine the cliff face rather rapidly (i.e., about 1–2 mm/year), collapse of cliffs by this process can only be detected in very less resistant rocks like eolianites. Looking closer to the shape of bioerosional notches, they always show a sharp microrelief in every part. The normal profile is asymmetrical and shows a rather flat bottom, an innermost knickpoint close to the bottom and a more (exposed) or less (sheltered) inclined notch-"roof" (Figure N34). The color of the rock in a notch is usually darker than in fresh cuts, ranging from greenish to dark grey and even black, the latter in levels above mean sea level. As a notch always indicates destructive processes near the surf horizon being more active than below and above, a special group of organisms forming notches should be living in these well-defined belts. In fact, we find here macroorganisms such as chitons or limpets (e.g., *Patella* sp.), used to living and grazing in permanently wetted but well-illuminated zones. Their grazing instrument is the so-called radula, their food being microscopic organisms like cyanophytes and chlorophytes (and sometimes fungi) (Schneider, 1976; Torunski, 1979). These are responsible for the rock colors: where wetting is guaranteed, these organisms live more epilithic, showing their green color; but in the tidal zone they escape from drying out and too much insolation by boring themselves into the rock, thus living as endoliths, their dried out parts showing darker colors. The specific penetration depth depends on the chance for photosynthesis and, therefore, is restricted to parts of a millimeter, the so-called light-compensation depth (which itself depends on the rock type with grain size and crystal forms). Under an scanning-electron microscope the boring patterns of the endoliths (bio-corrosion) can be detected as perforations of up to 800,000 per cm^2, and a three- to tenfold enlargement of the actively corroded surface, but no specific macroscopic geomorphological pattern is formed by this pro-cess. The latter is due totally to the limpets and other grazers, which wear down the surface in one grazing event by about 0.01–0.1 mm (bio-abrasion). Hence light can penetrate deeper into the rock, and the endoliths will bore deeper, as well, exhibiting a new sheet of food for the grazers. The restriction of the grazers to well-defined tidal (and wetting and drying) levels leads to the sharp horizontal bioerosive notches, visible after only some decades of grazing in the same level. All in all bioerosive notches, because of their good and rapid development and normally strong restriction to tidal levels, are among the best sea-level indicators at the coastlines of the world.

Dieter H. Kelletat

Bibliography

Kelletat, D.H., 1982. Hohlkehlen sowie rezente organische Gesteinsbildungen an den Küsten und ihre Beziehungen zum Meeresniveau. *Essener Geographische Arbeiten*, **1**: 1–27.

Kelletat, D.H., 1997. Mediterranean coastal biogeomorphology: processes, forms and sea-level indicators. In Briand, F., and Maldonado, A. (eds.), *Transformations and Evolution of the Mediterranean Coastline*. Monaco: Bulletin Institute of Océanography, No. Spécial, 18, pp. 209–226.

Pirazzoli, P.A., 1986. Marine Notches. In Van de Plassche, O. (ed.), *Sea-Level Research: a Manual for the Collection and Evaluation of Data*. Norwich: Geo Books, pp. 361–400.

Schneider, J., 1976. Biological and inorganic factors in the destruction of limestone coasts. *Contributions to Sedimentology*, **6**: 1–112.

Torunski, H., 1979. Biological erosion and its significance for the morphogenesis of limestone coasts and for nearshore sedimentation (Northern Adria). *Senckenbergiana maritima*, **11**: 193–265.

Cross-references

Bioerosion
Cliffed Coasts
Cliffs, Erosion Rates
Erosion Processes
Microtidal Coasts
Sea-Level Indicators, Biological in Depositional Sequences
Sea-Level Indicators, Geomorphic
Shore Platforms

NUMERICAL MODELING

Originally, numerical modeling mainly referred to the techniques employed for numerically solving a set of equations (algebraic or differential equations). However, the meaning has evolved and is now often used to also denote the selection of the equations to be solved. In turn, this involves describing the physical processes of interest in mathematical terms, which is also referred to as mathematical modeling. Formulating a set of governing equations might be straightforward on a theoretical, fundamental level, but in practice such equations could be impossible to solve (*i.e.,* use), not only because of insufficient computational power, but also because initial and boundary conditions as well as various types of forcing are not known or understood at the required level of detail. This is normally the case when dealing with different phenomena in nature, including the physical processes which control the response and development of coastal areas. These areas are shaped by forcing factors which act and produce responses at a wide range of spatial and temporal scales, often in a complex, nonlinear manner with strong feedback between forcing and response. For example, the scale of fluid motion in coastal areas extends from small-scale turbulence at a fraction of a wave period to sea-level rise occurring over centuries, causing responses in sediment transport and coastal evolution at scales from *ripples* to *barrier islands* (De Vriend, 1991; Larson and Kraus, 1995).

The above-described features of the coastal zone imply that numerical models are developed to describe a specific set of processes (at a certain time and space scale) and that the equations employed represent aggregated formulations in time and space, typically including coefficients and parameters with values which ultimately have to be determined through comparison with data. These properties are especially characteristic for *sediment transport* and coastal evolution numerical models, which are the main types of models discussed here. Numerical models of coastal hydrodynamics, such as wave transformation and nearshore circulation models, are more readily derived from a set of basic equations than sediment transport related models. In this respect, it is less complicated to select the equations in coastal hydrodynamics, although there are still many phenomena which are difficult to model, for example, wave breaking and flows associated with rip currents and bottom boundary layers.

Basic concepts

Numerical modeling involves selecting an appropriate set of equations to simulate the behavior of the coastal system under study and then solving the equations using suitable numerical techniques. Other elements of numerical modeling are verification, validation, and calibration (see Figure N35 for a description of the main elements in numerical modeling and how they are connected). In its classical meaning verification of the model involves "solving the equations right," whereas validation implies "solving the right equations" (Roache, 1997, p. 124). Thus, in the verification it is demonstrated that the equations are solved correctly yielding accurate results, which may be done by simulating an ideal case for which an analytical solution is available. The process of validation is typically more complicated since it involves establishing that the selected equations satisfactorily describe the system under consideration. In practice, this is done by employing measured data and showing that the numerical model can reproduce the measurements with some predefined accuracy. In general, numerical models include parameters and coefficients whose values have to be specified, which is done through calibration. In the calibration procedure, the model is used to simulate the coastal system using different sets of parameter values and the calculation result that is closest to the measurements (according to some criterion) defines the optimum parameter values, yielding a calibrated model. The available data set is normally split into two parts, where the first part is used for calibration and the second part for evaluation of how well the calibrated model can simulate the system. This evaluation of the calibrated model is often referred to as validation, although formally model validation should involve a more extensive confirmation that the model equations represent the coastal system.

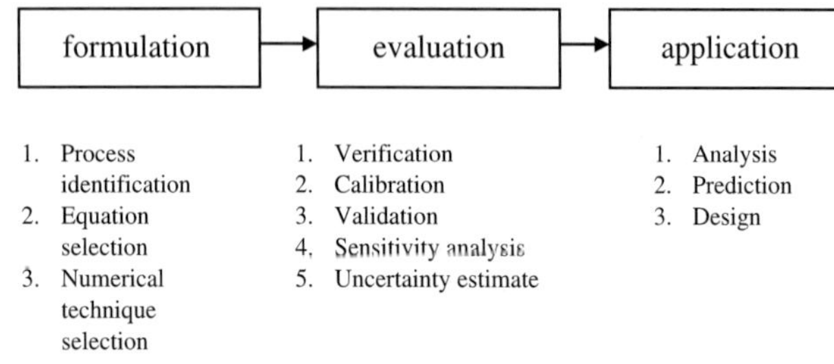

Figure N35 Schematized description of the main elements in numerical model development.

A unique solution cannot be found to a set of equations unless proper initial and boundary conditions are provided. Specifying boundary conditions implies giving the value of the quantity to be solved for (or some derivative of the quantity) at the boundary of the solution domain at all times. Similarly, the initial conditions yield information on the state of the system at the start of the simulation (*i.e.,* a value is assigned to all points under study in the solution domain). The values provided for initial and boundary conditions as well as for parameters and coefficients are collectively denoted as model input. In contrast, the calculated results are known as model output.

An important aspect of numerical modeling that is often not given enough attention is the uncertainty in the calculation results. Techniques to assess the uncertainties related to numerical errors, for example, discretization of the solution domain and the governing equations, are available (Roache, 1997). However, these measures are associated with model verification more than model validation. In order to quantify the overall uncertainty in a calculation, one typically has to resort to comparison with measurements, where some statistical measure is computed to characterize the deviation between measurements and calculations. The uncertainty of a model forecast (alternatively denoted as predictive skill) poses additional problems since there are no data available for such quantifications. Thus, one typically has to assume that the uncertainty of the forecast is approximately the same as that in reproducing the measurements used in the calibration/validation process.

The stability of a numerical model normally refers to how errors grow or decay in the solution. In the case of linear equations, it is possible to mathematically derive criteria for the stability, whereas this is not possible for nonlinear equations. The related term robustness is typically given a wider interpretation, not only focusing on the numerical scheme employed but on the model as such. Evaluating model robustness includes assessing the effects of changes in, for example, model structure, input parameter values, and numerical solution schemes, on the calculation result. If small modifications in model formulation or parameter values produce large changes in the result, model robustness is low, and vice versa. The robustness may be examined using sensitivity analysis where calculations are performed for a selected range of input values (or different model configurations) and the resulting output is analyzed in statistical terms to quantify model behavior. Monte-Carlo methods, where the input values are randomly selected from specified probability distributions, may be employed in this type of sensitivity analysis. For a validated model, Monte-Carlo methods are also highly useful to quantify the variability in the response of the coastal system being modeled. In practical applications with numerical models it is desirable to produce forecasts which include some statistical measures of the variability.

Solving the model equations, which normally constitute a set of ordinary or partial differential equations, requires numerical methods. A wide range of such methods have been developed and each one has its strengths and weaknesses (Abbott, 1979). The most common numerical methods in coastal applications are based on finite differences or finite elements. Traditionally, finite difference methods have been utilized more frequently in coastal numerical models because they are fairly straightforward in the formulation of the equations to be numerically solved and they handle time-dependent problems in a direct way. In these methods derivatives are written in terms of differences resulting in a set of algebraic equations to be solved. If the approximations of the derivatives are formulated in such a way that the variable of interest can be directly calculated at the next time level (or adjacent space level), the solution scheme is explicit. In cases where the difference scheme involves several time (space) levels simultaneously, making it necessary to employ iterative techniques in solving for the desired variable, the scheme is implicit. Finite element methods use spatial discretization of the solution domain into elements of a selected shape, and the variable of interest is interpolated over the element through a given approximating function. This function is obtained based on values of the variable defined at the nodes located in the element corners. Finite element methods are convenient for solving steady-state problems over solution domains of complex geometry where the variable of interest exhibits large spatial variability. An example from coastal applications is diffraction around *structures* in the nearshore.

Classification of numerical models

The most common classification of numerical models is with respect to the spatial dimension of the model, that is, a model can be one-(1D), two-(2D), or three-dimensional (3D) depending on whether the leading variables vary in one, two, or three dimensions, respectively. Additionally, time often enters as another dimension giving rise to the distinction between time-dependent and steady state (*i.e.*, time-independent) models. Another type of classification with regard to the temporal variation, typically employed in wave modeling, is into phase-resolving and phase-averaged models, depending on whether the variation during a wave cycle is modeled or not, respectively. The labeling of numerical models for coastal systems is often not stringent and schematized approaches to describe variations in some spatial dimensions have given rise to, for example, multi-layer (contour), 1.5D, and 2.5D models.

Other common ways of classifying numerical models are related to the following properties:

Probabilistic versus deterministic behavior: In a probabilistic model a random element has been added so that repeated simulations with the same input yields different output. A deterministic model always produces the same output for a specific input. Hybrid modeling approaches are often employed where a deterministic set of equations are used in combination with input generated from some probability distributions in order to simulate the variability of the system response (*i.e.*, a Monte-Carlo simulation approach).

Linear versus nonlinear behavior: A linear model implies that the principles of proportionality and superposition are applicable. Thus, the sum of two different input data sets produces an output which is the sum of the individual outputs for respective inputs. A nonlinear model does not follow these principles and in general includes equations that are much more difficult to solve numerically. Furthermore, some particular nonlinear equations give rise to solutions which display chaotic behavior characterized by aperiodic bounded dynamics with marked sensitivity towards the initial conditions. Such solutions have a behavior similar to a random system, although the equations generating the solutions are purely deterministic.

Distributed versus lumped description: A model where the leading variables are continuous functions of some spatial coordinates are known as a distributed model. In contrast, employing a lumped description implies that specific spatial domains are assigned single values to characterize the variable of interest (and other physical properties) and, typically, the time evolution of this variable is modeled. In practice, lumped descriptions are often employed to parameterize the variation at scales below the one modeled. For example, in estimating the bottom friction in the presence of ripples a single friction coefficient is normally used instead of calculating the detailed flow around the individual ripples.

Physically based versus black-box description: In physically based modeling the objective is to describe the system under study using the laws of physics, whereas in a black-box model the only concern is to develop a set of equations that reproduces the output from the input in the best manner. Statistical models from the field of time series analysis are typical examples of black-box models.

Numerical models for coastal evolution

An informative way of classifying numerical models of coastal systems is with regard to the characteristic time and space scales of the process or phenomenon under study. Nowadays there is a general agreement that it is not feasible to develop one model which would describe the entire coastal system, but a wide range of models is needed that simulate processes at different scales. A certain process is typically a result of factors acting at the representative time and space scales for the process, where factors at other scales have little influence and are neglected, parameterized, or handled through a statistical description. For example, in modeling beach erosion due to severe storms wind-induced waves and water level change are the primary forcing factors to consider, whereas sea-level rise or intra-wave turbulence are factors that are not directly taken into account. Similarly, barrier island migration may be a function of sea-level rise and there is no need to include wind-generated waves in a model of this process. Figure N36 provides an overview of different types of numerical models for the coastal areas based on the characteristic time and space scales for the models. In the following, a

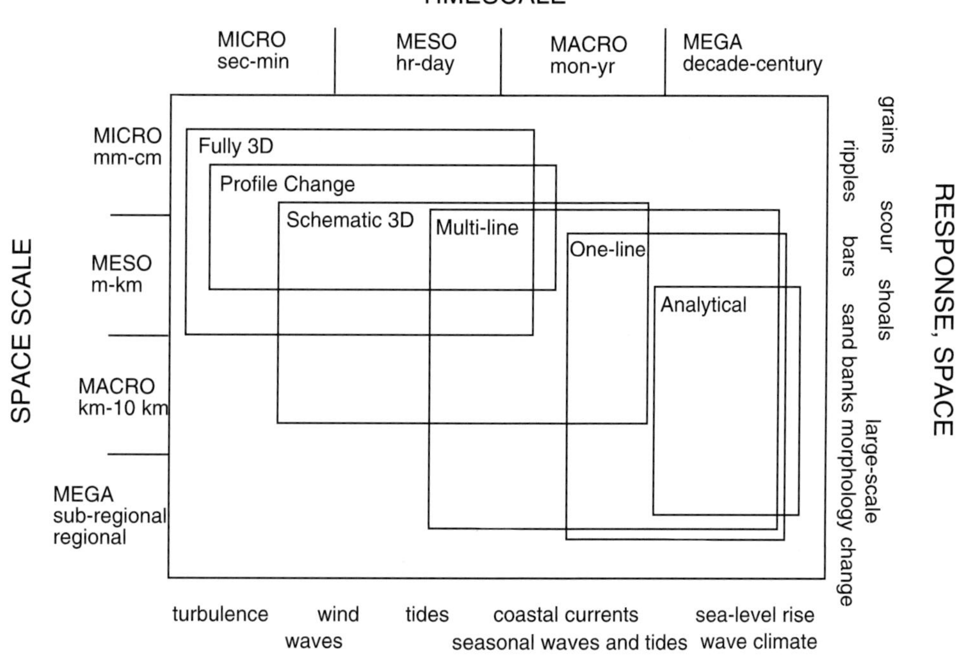

Figure N36 Classification of coastal evolution models in terms of characteristic spatial and temporal scales (after Hanson and Kraus, 1989; Larson and Kraus, 1995).

brief discussion is provided of the different types of models typically employed in simulating coastal evolution.

Historical development

Pelnard-Considère (1956) was the first to formulate a mathematical model to calculate the response of a beach to waves. His model only described the movement of the shoreline, assuming the beach profile to maintain a constant shape at all times. The theoretical work of Pelnard-Considère is the basis for many numerical models that have been successfully applied in calculating shoreline response to waves (Hanson, 1989). An extension of the one-line theory to include two contour lines was made by Bakker (1968), and Perlin and Dean (1978) further generalized this type of modeling to encompass an arbitrary number of contour lines. Watanabe (1982) made the first serious attempt to develop a complete three-dimensional numerical beach change model, which involved submodels to calculate nearshore waves, currents, sediment transport, and the resulting beach evolution. Initially, numerical modeling efforts aimed at simulating the main shape of the beach, whereas various types of morphological features overlaying this shape were considered secondary and not included in the modeling. However, later attempts were made to model such features, for example, longshore *bars* (Larson and Kraus, 1989) and different *rhythmic patterns* (Hino, 1974) including *offshore banks and linear ridges* (Falqués *et al.*, 1998).

Model overview

Analytical models of coastal evolution, as opposed to numerical models, are closed-form mathematical solutions of a simplified governing differential equation with the proper boundary and initial conditions. Analytical solutions have mainly been employed to determine shoreline change (e.g., Pelnard-Considère, 1956), although attempts have also been made to estimate *dune* erosion. Shoreline evolution is calculated under the assumption of steady wave conditions, idealized initial shoreline and structure positions, and simplified boundary conditions. A main assumption in the mathematical (either closed-form or numerical) shoreline change models is constancy of profile shape along the shore. Because of the many idealizations needed to obtain a closed-form solution, analytical models are typically too crude to use in planning and design, except possibly in the preliminary stage of project planning. Analytical solutions serve mainly as a means to make apparent trends in shoreline change through time and to investigate basic dependencies of shoreline change on waves and initial and boundary conditions. Larson *et al.* (1987) gave a comprehensive survey of new and previously derived analytical solutions of the shoreline change equation.

The *shoreline change numerical model* is a generalization of the analytical shoreline change model (Hanson, 1989; Price *et al.*, 1978). It enables calculation of shoreline evolution under a wide range of beach, wave, initial, and boundary conditions, and these conditions can vary in space and time. Because the profile shape is assumed to remain constant, the shoreline can be used in the model to represent beach position change. Thus, this type of model is sometimes referred to as a one-line model. The length of the time interval for which the model should be used is in the range of one year to several decades (see Figure N36), depending on the wave and sand transport conditions, accuracy of the boundary conditions, characteristics of the project, and whether the beach is near or far from equilibrium. The spatial extent of the region to be simulated ranges from 1 to 100s of kilometers depending on the same factors.

Principal uses of *profile response models* are prediction of beach and dune erosion produced by severe storms or hurricanes, and initial adjustment of beach fills to wave action and fill loss during a storm (Kriebel and Dean, 1985; Larson and Kraus, 1989; Schooness and Theron, 1995). This type of model only considers *cross-shore sediment transport* and neglects any differentials in the *longshore sediment transport* processes, that is, one profile line is sufficient for describing the beach response. During a storm event such a simplification is normally of adequate accuracy for engineering applications. The typical timescale of profile response models is hours to days for a storm event, whereas if long-time beach recovery or fill adjustment is investigated a timescale of months is of interest (Figure N36).

Three-dimensional beach change models describe changes in bottom elevation which can vary in both horizontal (cross-shore and longshore) directions. Therefore, the fundamental assumptions of constant profile shape used in shoreline change models and constant longshore transport in profile erosion models are removed. Although 3D-beach change models represent the ultimate goal of calculating sediment transport and coastal evolution, achievement of this goal is limited by our capability to predict *wave climates* and sediment transport rates. Thus, simplifying assumptions are made in schematic 3D-models, for example, to restrict the shape of the profile or to calculate global rather than point transport rates. Including both longshore and cross-shore transport rates typically results in numerical models which cover a wider range of scales than shoreline evolution and profile response models only (see Figure N36). However, the development of 3D models (schematic or full) still requires careful consideration of the processes to be simulated and their characteristic scales.

A natural extension of the shoreline change numerical model is to introduce some kind of schematized profile so that cross-shore transport and associated beach change may be described. In an *n-line model* (or *multi-layer model*) the movement of additional contour lines besides the shoreline is simulated (Bakker, 1968; Perlin and Dean, 1978). In these models the cross-shore transport rate is determined from a specified cross-shore distribution (e.g., in terms of wave, sediment, and profile shape parameters) or simply from the average slope between the contour lines. Although n-line models provide information on the cross-shore beach response, the profile shape is still fairly restricted and such models cannot simulate the response of complex morphological features where the profile depths do not increase monotonically with distance offshore (e.g., longshore bars).

Another method for including both cross-shore and longshore transport in a simplified way is to decouple the cross-shore and longshore calculations in some of the modules in the numerical model. For example, instead of calculating on a complete 2D horizontal grid, waves, currents, and sediment transport are computed separately at individual profile lines with the main coupling occurring through the sediment conservation equation solved on a 2D grid. *Decoupled models* (*i.e.*, *schematic 3D models*) can describe bottom topographies that are more complex than n-line models, although alongshore gradients in the modeled quantities should be sufficiently small to allow for a decoupling.

Fully 3D coastal evolution models represent the state-of-art of research and are not widely available for application. Waves, currents, sediment transport, and changes in bottom elevation are calculated point by point in small areas defined by a horizontal grid placed over the region of interest (solution domain). Such models can employ averaged quantities over the water column, in part or throughout the calculations, or resolve the variation through the vertical. Use of these models requires special expertise, and only limited application have been made on large and well-funded projects (Watanabe, 1982). Because fully 3D coastal evolution models are used in attempts to simulate fine details of waves, currents, and sediment transport, they require extensive validation and sensitivity analysis, which in turn has implications for the data requirements.

Future developments

The trend towards developing numerical models describing coastal processes over selected space and timescales will probably continue, yielding a cascade of models for simulating the coastal system at different scales. It is expected that some types of models at adjacent scales will merge, although models at a higher or lower level in the model hierarchy will mainly provide boundary conditions, aggregated forcing, or general constraints for a numerical model at a certain scale. A general development of different numerical models towards increased robustness and reliability will facilitate their use as tools in engineering projects as well as in interpretation and sensitivity analysis of the coastal system. At present, shoreline change and profile response numerical models are the most commonly used types of models for forecasting coastal evolution, but in the near future it is expected that various types of schematized 3D models will reach a state where they can be applied with confidence.

The time (and space) scale of interest in *coastal engineering* and *coastal zone management* is increasing, making additional demands on being able to model the behavior of the coastal system at these scales. Most numerical modeling of coastal evolution has been carried out at timescales from hours to decades, and the experience of modeling at longer scales is limited. However, considerable research is underway concerning coastal evolution at longer timescales, including both decadal, century, and millennium timescales. Thus, it is expected that in the near future reliable predictions of coastal evolution at timescales from decades to centuries can be made.

Hybrid models with both deterministic and probabilistic elements have not been exploited enough for simulating coastal systems. All too often, forecasts of coastal evolution are made without any attempt to assess the uncertainty in the predictions. Monte-Carlo simulation techniques are highly useful in obtaining quantitative estimates of the uncertainty and such techniques should be employed more frequently in numerical modeling, especially in engineering projects. This technique yields a range of possible solutions/predictions rather than a unique, single answer.

Magnus Larson

Bibliography

Abbott, M.B., 1979. *Computational hydraulics*. London: Pitman Publishing Limited.

Bakker, W.T., 1968. The dynamics of a coast with a groin system. *Proceedings of the 11th Coastal Engineering Conference*, ASCE, pp. 492–517.

De Vriend, H.J., 1991. Mathematical modeling and large-scale coastal behavior. Part 1: Physical processes. *Journal of Hydraulic Research*, **29**(6): 727–740.

Falqués, A., Calvete, D., De Swart, H.E., and Dodd, N., 1998. Morphodynamics of shoreface-connected ridges. *Proceedings of the 26th Coastal Engineering Conference*, ASCE, pp. 2851–2864.

Hanson, H., 1989. Genesis—a generalized shoreline change numerical model. *Journal of Coastal Research*, **5**(1): 1–27.

Hanson, H., and Kraus, N.C., 1989. Genesis: generalized model for simulating shoreline change. Report 1: Technical Reference. Vicksburg, MS: Coastal Engineering Research Center, US Army Engineer Waterways Experiment Station, *Technical Report CERC-89-19*.

Hino, M., 1974. Theory on formation of rip-current and cuspoidal coast. *Coastal Engineering in Japan*, **17**: 23–37.

Kriebel, D.L., and Dean, R.G., 1985. Numerical simulation of time dependent beach and dune erosion. *Coastal Engineering*, **9**: 221–245.

Larson, M., and Kraus, N.C., 1989. SBEACH: Numerical model for simulating storm-induced beach change. Report 1: Theory and model foundation. Vicksburg, MS: Coastal Engineering Research Center, US Army Engineer Waterways Experiment Station, *Technical Report CERC-89-9*.

Larson, M., and Kraus, N.C., 1995. Prediction of cross-shore sediment transport at different spatial and temporal scales. *Marine Geology*, **126**: 111–127.

Larson, M., Hanson, H., and Kraus, N.C., 1987. Analytical solutions of the one-line model of shoreline change. Vicksburg, MS: Coastal Engineering Research Center, US Army Engineer Waterways Experiment Station, *Technical Report CERC-87-15*.

Pelnard-Considère, R., 1956. Essai de théori de l'evolution des forms de rivages en plage de sable et de galets. *4th Journées de l'Hydralique*, Les Energies de la Mer, Question III, Rapport No. 1, 289–298.

Perlin, M., and Dean, R.G., 1978. Prediction of beach planforms with littoral controls. *Proceedings of the 16th Coastal Engineering Conference*, ASCE, pp. 1818–1838.

Price, W.A., Tomlinson, D.W., and Willis, D.H., 1978. Predicting changes in the plan shape of beaches. *Proceedings of the 13th Coastal Engineering Conference*, ASCE, pp. 1321–1329.

Roache, P.J., 1997. Quantification of uncertainty in computational fluid dynamics. *Annual Review of Fluid Mechanics*, **29**: 123–160.

Schooness, J.S., and Theron, A.K., 1995. Evaluation of 10 cross-shore sediment transport/morphological models. *Coastal Engineering*, **25**: 1–41.

Watanabe, A., 1982. Numerical models of nearshore currents and beach deformation. *Coastal Engineering in Japan*, **25**: 147–161.

Cross-references

Beach Erosion
Coastal Processes (see Beach Processes)
Coastal Modeling and Simulation
Coastline Changes
Cross-Shore Sediment Transport
History, Coastal Geomorphology
Human Impact on Coasts
Longshore Sediment Transport
Physical Models
Shore Protection Structures
Time Series Modeling
Wave Climate

O

OIL SPILLS

The term "oil spill" refers to the accidental release of liquid petroleum hydrocarbons to the environment. The visual nature of black oil coming ashore from a spill commonly attracts public interest on a national and even international scale, oftentimes falsely projecting damages far greater than actually occur. Of all areas, however, long-term damage from oil spillage is likely to be greatest within the ecologically rich coastal zone. Additionally, coastal tides and currents commonly increase damages as spilled oil is transported far from the site of the incident. The *Exxon Valdez* case provides a worst case example where over 2,000 km of Alaskan coastline was oiled from a single discharge location in Prince William Sound.

Principal reference material on oil spills can be found in the bi-annual proceedings of the International Oil Spill Conference available from the American Petroleum Institute (Washington, DC) and the annual proceedings of the Arctic Marine Oil Program sponsored by Environment Canada, Ottawa. Marine and coastal response guidance documents are available from the National Oceanic and Atmospheric Administration. National Research Council (1985) provides extensive reference material as well.

Quantity of oil entering the marine environment

In 1990, the level of petroleum hydrocarbons entering the marine environment was estimated by GESAMP (1993) as 2.35 million tons. (The common conversion factor is 1 ton = 7 barrels, 1 barrel = 42 US gallons.) Of this total, transportation-related activities accounted for 24% (564,000 tons) and included discharges from vessels as well as from terminals. In comparison, the estimated discharge from municipal and industrial sources is more than two times greater, 1.175 million tons or 50% of all oil entering the marine environment. The estimated quantities from other sources include atmospheric (305,500 tons, 13%), natural seeps (258,500 tons, 11%) and offshore exploration/production (47,000 tons, 2%). Note: 1 US gallon = 0.832 UK gallon.

Historically, there has been significant improvement in the prevention of accidental hydrocarbon discharges, particularly those related to marine transportation as well as those from offshore production. Figure O1 presents historic estimates for tons of oil spilled by tankers worldwide (Etkin, 2000). While a general decline in spillage is indicated, a single event during any year can highly influence the total spilled. For example, the loss of 223,000 tons from the *Amoco Cadiz* is larger than the total amounts spilled from 1992 to 1998. The number of tanker spills has also significantly declined, from a high of 3,153 spills in 1979 to an average of 1,200 from 1990 to 1998 (Etkin, 2000). Both spill size and frequency have declined in spite of the increase in the amount of oil shipped via tanker vessels.

Improvements to spill prevention are often associated with the aftermath of major spill events. After the *Exxon Valdez* incident in 1989, public scrutiny and international coverage of this event caused passage of US legislation and international agreements that strengthened requirements for spill-response equipment, training, and safety. In the United States, the passage of the Oil Pollution Act of 1990 will have national and international repercussions well into the future based on its requirement that all tankers be double-hulled by the year 2015 in order to enter US waters.

Hydrocarbon chemistry

Oil is composed of a complex mixture of hydrocarbons that varies highly by its source and subsequent refining. Once spilled, it is exposed to additional chemical and biological processes which further alters its composition.

The hydrocarbon component of oil is comprised of hundreds of organic compounds, principally divided into aliphatics and aromatics. In addition to hydrocarbons, oil may also contain small amounts of N, S, and O (termed NSO compounds), as well as widely varying concentrations of trace metals (e.g., V, Ni, Fe, Al, NA, Ca, Cu, and U). These metals and other unique components of the spilled oil are often used as a tracer to follow the distribution and long-term impacts of the spill.

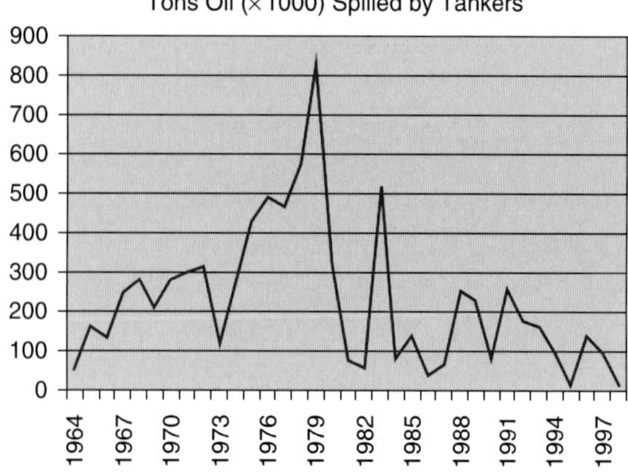

Figure O1 Tons of oil spilled worldwide from tanker vessels, 1964–98 (data from Etkin, 2000).

As crude oil is refined, the resulting product markedly changes in composition and physical properties (e.g., specific gravity, viscosity, pour point, and flash point), all of which influence the potential impact to the environment should it be released. The distillation process separates crude oil into refined products used for domestic and industrial purposes. These may be referred to as: light fractions (gasoline), middle distillates (kerosene, heating oil, jet fuel, and light gas oil), wide-cut gas oil (lube oil and heavy gas oil), and residuum or bunker oil.

For oil spill response planning, oils are commonly categorized into four groups (ITOPF, 1987). Group I has a specific gravity < 0.8, low viscosity (0.5–2.0 cSt at 15°C) and includes gasoline, naphtha, and kerosene. Group II has a specific gravity 0.80–0.85, average viscosity = 8 cSt at 15°C and contains light crude oils and gas oil. Group III has a specific gravity of 0.85–0.95, a higher viscosity averaging 275 cSt at 15°C and includes intermediate crude oils and medium fuel oils. Group IV are heavy crudes and fuel oils (e.g., bunker oils) that have a specific gravity of > 0.95 and high viscosities > 1500 cSt at 15°C. Group IV oils may include products that sink or have a buoyancy near that of water, greatly increasing the difficulty of recovery if spilled.

Spill weathering and fate

Once released to the environment, the spilled oil is exposed to a series of "weathering" processes that may alter its original chemistry and physical properties. These include spreading, advection, evaporation, emulsification, dissolution, dispersion, photo-oxidation, sedimentation, and biodegradation.

Spreading causes the formation of a relatively thin layer of oil on the water's surface after it is released. Affecting processes include gravity, friction, viscosity, and surface tension. The color of the oil on the water is related to thickness and has been used as a simple measure of the quantity of oil present. For example, a silver sheen has a thickness of approximately 0.0007 mm and represents approximately 100 L spread over 1 km^2. A brown sheen on the water's surface indicates a thicker layer (0.1–1 mm). When slicks are forced to the coastline by onshore winds, the slick may increase to a centimeter or more in thickness.

Advection is the movement of oil on the water's surface, influenced by currents and winds. Most predictions assume that oil will move in the direction and speed of the vector addition of current speed and direction plus 3% of wind velocity in the direction of the wind forcing. Shifting tides, eddies, and variable winds all add to the difficulty in predicting the real-time movement of the spill.

Evaporation is a major process influencing all aspects of oil chemistry and its physical properties. During evaporation, the light products (gasolines, kerosenes, etc.) escape to the atmosphere leaving a lesser amount of oil on the water, but increasing its specific gravity and viscosity. Evaporation can commonly remove 30–50% of light crude oils and refined products within a relatively short time period (hours to days).

While evaporation can substantially reduce the quantity of oil, the process of emulsification (a water-in-oil mixture) may significantly increase (double) the quantity of total oil mass by incorporating water within the oil structure. Emulsification commonly occurs by wave action and mixing. Oil weathering decreases as the oil emulsifies.

Dissolution is the transference of oil from the water's surface into solution within the water column. It is very minor accounting for 1% or less of the total oil spilled.

Dispersion is the process of forming oil particles or droplets in the water column as occurs during breaking wave conditions. The process of forming particles of ever smaller size enables the material to be biodegraded.

Photo-oxidation is the result of sunlight altering the composition of the oil. It accounts for only a small change (<1%) in the overall content of the spill. Direct effects of photo-oxidation on an actual spill have not been measured.

Sedimentation is the interaction of sedimentary particles and the spilled oil which may increase the density of the oil and carry it to the bottom. However, as most oils float, transport of substantial quantities to bottom sediments is relatively uncommon.

Biodegradation is based on the ability of bacteria and other organisms to degrade and convert the oil from a hydrocarbon to its fundamental elements. Biodegradation is able to act on small oil particles as the material is dispersed into the aquatic environment or when resident on shorelines. Oxygen and an adequate supply of nutrients are needed to sustain biodegradation.

Impact and reaction in coastal environments

Spilled oil can have a significant impact on the coastal environment, the level of which varies with physical processes (e.g., winds, tides, and currents) and with the geomorphology and ecology of the coast.

Coasts have been categorized and ranked in terms of sensitivity to oil spills by Gundlach and Hayes (1978). This system, called the Environmental Sensitivity Index (ESI) is widely used for spill contingency planning and to guide the response effort during spills. Common coastal types of the ESI are presented in Table O1.

The ESI shoreline ranking indicates that exposed rocky shores are the least sensitive to spilled oil, followed by beaches of varying sediment sizes (sands to gravels), and ending with the most spill-sensitive sheltered tidal flats, marshes, and mangroves.

Exposed rocky shores have the lowest sensitivity to oil spills. Where vertical, wave reflection tends to keep the oil offshore. In cases where the shore is terraced or wave cut, the sweep of the waves will remove most of the oil. The moderate to high wave activity present makes cleanup

Table O1 Characteristics of coastlines categorized in the Environmental Sensitivity Index (ESI). Coastlines are ranked 1–10, with 10 being the most sensitive (modified from NOAA, 1997)

ESI	Shoretype	Dominant sediment type and slope
1a	Exposed rocky shore or banks	Rocky = bedrock or boulders (>256 mm) of moderate-to-high slope Banks = marked by scarping, clays, and muds (<0.625 mm) are common
1b	Exposed seawalls and solid man-made structures	Vary from boulders and cobbles (>64 mm) to sand bags, solid concrete, sheet pile, or wood
2a	Eroding mud scarp on exposed beach	Silt and clay (< 0.0625 mm), very low-slope
2b	Exposed wave-cut platforms	Bedrock or boulders (>256 mm), low-slope backed by bluff or cliff
2c	Rocky shoals, bedrock ledges	Bedrock or boulders (>256 mm), low-slope
3a	Fine sand beach	Fine sand (0.0625–0.25 mm), low-slope (<5°)
3b	Scarps or steep slopes in sand	Sand = 0.0625–2.0 mm, marked by scarp or steep slope
4a	Medium-to-coarse sand beach	Grain size = 0.25–2.0mm, low-to-moderate slope (5–15°)
5	Mixed sand and gravel beach, bar or bank	Grain size = 1–64 mm, moderate slope (8–15°)
6a	Gravel beach or bar	Grain size > 2 mm, moderate–steep slope (10–20°)
6b	Riprap	Boulders (>256 mm), moderate–steep slope (>20°)
7	Exposed tidal flat	Coarse sand–mud (<2 mm), low slope (<3°)
8a	Sheltered rocky shore or scarp	Bedrock or boulders (>256 mm), moderate-to-steep slope (>15°)
8b	Sheltered riprap	Boulders (>256 mm), moderate–steep slope (>20°)
8c	Vegetated steeply sloping bluff	Soils (sand–mud) (<1 mm), moderate–steep slope (>15°)
9a	Sheltered tidal flat of sand or mud	Medium sand–mud (<0.5 mm), low slope (<3°)
9b	Vegetated low bank	Soils (sand–mud) (<1 mm), low–moderate slope (≥ 20°)
10a	Mangrove swamp	Mud (<0.625 mm), low slope (<3°)
10b	Salt marsh	(The type of vegetation type indicates the shoretype)
10c	Brackish/freshwater swamp	

difficult and in most cases unnecessary. However, these same areas may have high biological sensitivity, although the coast does not show long persistence of oil. For instance, the presence of bird colonies or seal haul-out areas on rocky shores will raise the sensitivity of the site. The ESI system uses symbols to designate these occurrences (NOAA, 1997).

On beaches, shore sensitivity increases with grain size. Coarser sediments commonly enable the deep penetration of oil and, because of greater changes in the form of the beach, burial is often deeper as well. In terms of cleanup, fine-grained sand beaches tend to be more compacted and flatter than coarse-sand and gravel beaches. This enables cleanup activities to be undertaken more easily, especially if the beach can be driven upon. As grain size increases and oil becomes more deeply buried, cleanup may become extremely difficult. At many spills, (e.g., *Exxon Valdez* (Alaska, USA) and *Amoco Cadiz* (Brittany, France)), pressure flushing was needed to remove oil from the gravelly sediments. Where oil was deeply buried (>1 m) in the upper berm, oiled sediments were physically relocated lower into the swash zone so that natural wave action would assist oil removal. As oil weathers, it becomes firmly adhered to the sediments, particularly gravels and cobbles. To remove oil, water may have to be heated which may cause additional damage to the intertidal plants and animals that survived the initial oiling. During spill events, cleanup operations are always reviewed in terms of potentially causing more harm than good. When more harmful than good, operations are stopped.

Tidal flats in the ESI ranking fall into two broad classifications depending on exposure. On exposed tidal flats, water saturates the sediments thereby inhibiting penetration into the flat. Oil tends to skim across the surface of the flat as the tide rises, eventually being deposited on the adjacent beach. However, clams, worms, and other organisms within the flat are likely to be exposed to high concentrations of hydrocarbons. During the *Amoco Cadiz*, hundreds of thousands of intertidal and shallow subtidal razor clams were killed by oil in the water column. In contrast to exposed flats, sheltered tidal flats have the potential for the long-term persistence of oil due to the lack of physical activity (waves and currents) to remove the oil and to the oil's affinity to bind with fine-grained materials.

Marshes are among the most sensitive of all temperate coastal habitats. Marshes are usually left to recover naturally unless oil quantities are thick and/or heavily mixed into the sediments. Heavy coatings (>3 cm) of oil at the *Metula* (1974) site in Patagonia, Chile, showed only minor recovery 25 years after the incident. Restoration of a damaged marsh can be greatly assisted by a replanting (as at *Amoco Cadiz*, Baca *et al.*, 1987). In cases where vegetation is coated but the sediments remain unoiled or only lightly oiled, recovery can be quite rapid depending on the season. The moderately oiled marsh at the *Julie N.* site in Maine (USA) during Autumn 1997 showed excellent recovery by Spring of the following year. The type of the oil spilled greatly influences the extent of damage and recovery. Lighter oils are generally more acutely toxic than heavier oils and crude.

Mangroves show sensitivity equal or greater than that of marshes. The *Peck Slip* (1979) site in Puerto Rico illustrated that tall mangrove trees and all associated life could be killed when the base of the tree and its sediments become heavily oiled. Recovery may be longer than that for marshes as years are necessary for the trees to regain their full height after oil concentrations have been sufficiently reduced to enable regrowth.

Response methods

Response activities to contain and recover spilled oil have not changed dramatically over the past 25 years. The principal methods of response in the United States is primarily a combination of mechanical and manual. Mechanical refers to the process of recovering oil by equipment such as skimmers or vacuum pumps. Manual recovery includes all aspects of physical labor, including shoveling, raking, hand-wiping, and collection using sorbent materials.

Alternative response methods are also available. In this category, dispersants are widely used outside the United States as a primary response tool. Most modern dispersants are essentially nontoxic and under appropriate conditions can enhance the dispersion of the slick into the water column, thereby making it available for biodegradation. Dispersant application is usually limited to offshore waters. Other agents are available for changing the oil's properties or "lifting" it from solid surfaces. Although rarely used, *in situ* burning, which burns oil at sea using igniters and a special fireproof boom, is gaining acceptance as a response tool. After the gross contamination has been removed by traditional means, the application of fertilizers to "biostimulate" existing bacteria may be used to augment naturally occurring bacterial populations which then degrade the remaining oil. Each alternative method has its place, and when combined with mechanical and manual recovery, provides a range of options to the spill responder.

Erich R. Gundlach

Bibliography

Baca, B.J., Lankford, T.E., and Gundlach, E.R., 1987. Recovery of Brittany coastal marshes in the eight years following the *Amoco Cadiz* incident. In *Proceedings 1987 Oil Spill Conference*. American Petroleum Institute, pp. 459–464.

Etkin, D., 2000. Worldwide Tanker Spill Statistics, Winchester: Environmental Research Consulting.

GESAMP (IMO/FAO/UNESCO/WMO/WHO/IAEA/UN/UNEP Joint Group of Experts on the Scientific Aspects of Marine Pollution), 1993. *Impact Of Oil And Related Chemicals And Wastes On The Marine Environment*. GESAMP Reports and Studies 50: 180pp.

Gundlach, E.R., and Hayes, M.O., 1978. Classification of coastal environments in terms of potential vulnerability to oil spill damage. *Marine Technology Society Journal*, **12**(4): 18–27.

ITOPF, 1987. *Response to Marine Oil Spills*. London: International Tanker Owners Pollution Federation.

NOAA, 1997. *Environmental Sensitivity Index Guidelines*, Version 2. Seattle: National Oceanic and Atmospheric Administration.

NRC, 1985. *Oil in the sea: Inputs, Fates and Effects*. Washington, D.C.: National Research Council, National Academy Press.

Cross-references

Beach Sediment Characteristics
Beaufort Wind Scale
Cleaning Beaches
Debris (see Marine Debris—Onshore, Offshore, Seafloor Litter)
Environmental Quality
Human Impact on Coasts
Rating beaches
Water Quality

OIL SPILLS, HIGH-ENERGY COASTS

Most oil spills at sea occur in relatively confined and sheltered localities, often from tankers in passage to or from oil terminals and ports. Natural processes, especially evaporation and other weathering effects quickly reduce the volume of oil at the surface but the more persistent residue requires expensive protection and clean-up procedures. Although normally less than is predicted at the time of the incident, substantial environmental and economic damage can occur along impacted coastlines and in nearshore zones. The Exxon Valdez oil spill in Alaska is the type-example of a large oil spill in a confined, low-energy location.

Some major oil spills occur in deep water more than 100 nautical miles from the nearest coastline. Normally weathering processes and, depending on wind and currents, the substantial time interval before the oil reaches a coastline minimize pollution damage. Chemical dispersants which are designed to breakdown the oil into tiny droplets for easier dilution, transport, and removal by microorganisms can be used but some nations prohibit their use nearshore or in confined sea and estuarine areas.

The *Braer* oil spill off Sumburgh Head in Shetland Islands, in 1993 epitomized a different type of oil spill when a large tanker ran aground on a high energy, rocky coastline (Figure O2). The *Sea Empress* oil spill off the coast of Wales, in 1996 also occurred in a moderately exposed coastline where strong tidal currents were also of considerable importance in the initial stage. In spite of the volume of oil involved and the initial "disaster" reaction to the grounding of the *Braer*, the true ecological and economic impacts were not substantial. The combination of gale-force winds, the incident energy of massive storm waves on the open Atlantic coastline and the strongly dissipative nature of the irregular cliff coastline led to rapid dispersal into a highly turbulent water column. Tidal and wave currents carried this oil, at depth, considerable distances offshore and alongshore. There was virtually no stranding of oil and the characteristic surface oil slick could not form due to the prevailing storm conditions. An exposed sand beach lies less than 1 km east of the position of grounding but, due to the power of backwash from the steep beach face, the oil and water mixture did not contaminate the intertidal zone. Although less than 1% of oil was involved, there were also local spray and aerosol effects due to the severity of wave action on the coastline.

Figure O2 Wreck of the oil tanker Braer on the south coast of the Shetland Islands in January 1993 (photograph kindly supplied by Mr. P. Fisher).

The *Braer* can be used to indicate that an oil spill on an exposed, high-energy coastline, especially a rocky and cliff coastline is unlikely to cause much environmental damage. Under such conditions, oil particles are likely to be transferred to the water column and, in time, lodged on the sea bed, perhaps at considerable distances offshore as a consequence of local and regional currents and water movements. Thus an oil spill on a high-energy coastline is more likely to create offshore, marine problems rather than to produce negative effects on the adjacent shore. In short, the critical difference is the presence or absence of a surface oil slick which is a consequence of local shelter and energy conditions at the time of the accident.

Further reading may be found in the following bibliography.

William Ritchie

Bibliography

Edwards, R., and Sime, H., 1998. *The Sea Empress Oil Spill*. London: Terence Dalton. 493pp.

Davies, J.M., and Topping, G., 1997. *The Impact of an Oil Spill in Turbulent Waters: The Braer*. Edinburg: H.M.S.O. 263pp.

RITCHIE, W., 1995. Maritime oil spills—environmental lessons and experiences with special reference to low-risk coastlines. *Journal of Coastal Conservation*, **1**: 63–76.

SELL, D. *et al.*, 1995. Scientific criteria to optimize oil spill clean up. *Proceedings of International Oil Spill Conference*, Long Beach, pp. 595–610.

Cross-references

Environmental Quality
Human Impact on Coasts
Marine Debris—Onshore, Offshore, Seafloor Litter
Oil Spills
Rock Coast Processes
Water Quality

OFFSHORE SAND BANKS AND LINEAR SAND RIDGES

Shelf sand banks and linear sand ridges are found on numerous modern and ancient continental shelves where sufficient sand exists and currents are strong enough to transport sand-sized sediment (Off, 1963; Snedden and Dalrymple, 1999; Dyer and Huntley, 1999). Sand banks and linear sand ridges are defined as all elongate coastal to shelf sand bodies that form bathymetric highs on the seafloor and are characterized by a closed bathymetric contour (Figure O3). Other terms used to refer to these specific bathymetric features include *linear shoals*, *shoreface ridges*, *shoreface-attached* or *detached ridges*, *shoreface-connected* or *disconnected ridges*, *tidal current ridges*, and *banner banks*.

Typically, these linear sand bodies have heights that are more than 20% of the water depth, lengths that range from 5 to 120 km, relief up to 40 m, and side slopes that average <1°. They are 0.5–8 km wide, asymmetrical in profile, and consist of unconsolidated fine-to-coarse sand or even gravel. Axes of these sand bodies are generally oriented shore-parallel, shore-oblique (10–50°), or shore-normal when compared to the adjacent coastline, but any orientation is possible. Sand banks and linear sand ridges are present in a wide range of water depths and found in estuaries, at tidal entrances, adjacent to coastlines (offshore headlands, spits, and barrier islands), on the exposed shelf, and along the continental shelf edge. Moreover, these sand bodies usually occur in groups with spacing of individuals on the order of 250 times the water depth, but solitary ridges do occur (Snedden and Dalrymple, 1999). Two or more linear ridges grouped together are referred to as a *sand ridge field*, but other terms like *ridge and swale topography* are also used.

Initial investigations regarding bathymetric irregularities on continental shelves date as early as the 1930s (Van Veen, 1935; Veatch and Smith, 1939). Since that time, investigations into the origin, morphology, and geology of sand ridges may be organized into three groups: (1) morphology and surficial sediment studies (e.g., Off, 1963; Uchupi, 1968; Duane *et al.*, 1972); (2) genesis and hydrodynamic regime (e.g., Swift and Field, 1981; Huthnance, 1982; Swift, 1985; McBride and Moslow, 1991; Snedden and Dalrymple, 1999), and (3) internal geology and stratigraphy (e.g., Penland *et al.*, 1989; Snedden *et al.*, 1994; Dalrymple and Hoogendoorn, 1997).

Classification

Many different classifications have been proposed for sand banks and linear sand ridges but the three most widely accepted are highlighted here. By the 1980s, leading investigators had classified linear sand bodies on continental shelves into two types—storm or tide built (Amos and King, 1984; Swift, 1985). In other words, either storm-generated or tide-generated currents were primarily responsible for ridge development and maintenance. Through time, however, researchers also recognized the importance of ridge precursors (i.e., sand body nuclei or initial irregularities, such as ebb-tidal deltas) and the subsequent hydrodynamic reworking of the nuclei (e.g., McBride and Moslow, 1991), which eventually led to two generic classifications (Dyer and Huntley, 1999; Snedden and Dalrymple, 1999).

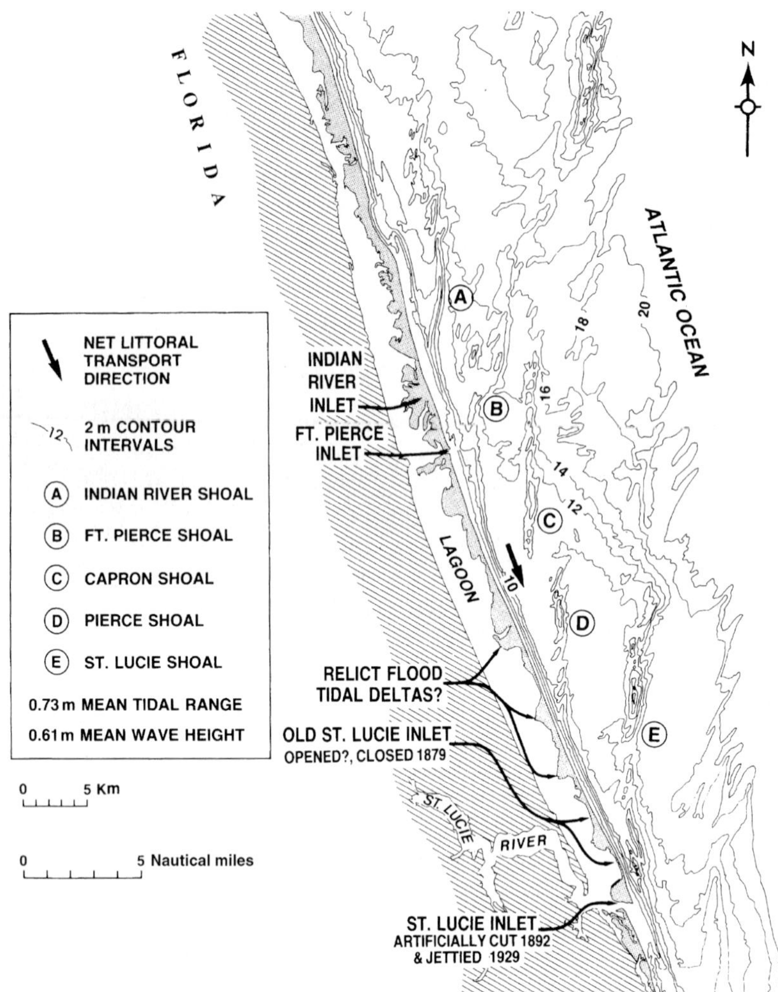

Figure O3 Example of shore-oblique shelf sand ridges, each characterized by a closed bathymetric contour. The Fort Pierce sand ridge field is located south of Cape Canaveral along the east coast of Florida, USA (from McBride and Moslow, 1991 with permission from Coastal Science). Orientations of other linear sand bodies may be shore-parallel or shore-normal.

Dyer and Huntley (1999) proposed a qualitative classification based on a generic relationship between different shelf sand bodies in light of their origin and development. They concluded that the morphodynamic stability models proposed by Huthnance (1982) were the most suitable at explaining the interaction between flow and sand body orientation. Their classification is composed of three primary ridge types and several subtypes as follows:

Type 1 Open shelf ridges
Type 2 Estuary mouth
- (a) Ridges (wide mouth)
- (b) Tidal delta (narrow mouth)
  - (i) Without recession (ebb-tidal deltas)
  - (ii) With recession (shoreface-connected ridges)

Type 3 Headland-associated banks
- (a) Banner banks (nonrecessional headland)
- (b) Alternating ridges (recessional headland)

Recognizing that the Huthnance stability model was applicable to both storm- and tide-built ridges, Snedden and Dalrymple (1999) proposed a unified model for ridge genesis and maintenance. They classified shelf sand ridges according to the amount of reworking of the original sand body (i.e., precursor irregularity) based on sedimentologic and stratigraphic evidence. As such, ridges were subdivided into three classes representing an evolutionary progression of increased ridge reworking and migration (Table O2). Class I ridges retain all of their original nucleus (i.e., precursor irregularity). Class II ridges are partially evolved sand bodies that have migrated less than their width and thus contain recognizable evidence of their original nucleus, whereas Class III ridges have migrated a distance equal to or more than their original width, thereby eroding virtually all evidence of their original nucleus.

Table O2 Classification and primary characteristics of shelf sand ridges and banks (modified from Snedden and Dalrymple, 1999)

	Class I	Class II	Class III
Type	Juvenile/static ridge	Partially evolved	Fully evolved
Precursor	Largely preserved	Partially preserved	Not preserved
Dynamics	Ridge stationary or rapidly buried	Ridge migrates < original ridge width	Ridge migrates ≥ original ridge width

Depositional setting and global occurrence

Sand ridge genesis is most favorable during transgression on continental shelves that are characterized by the following four conditions: (1) initial irregularities most commonly developed in the nearshore zone, (2) sufficient supply of loose sand, (3) sand-transporting currents (tidal or storm-driven), and (4) sufficient time for the sand to be molded into a ridge or ridge field (Snedden and Dalrymple, 1999).

Some of the best developed, *storm-maintained* linear sand ridges and fields are located on continental shelves along North America, South America, and northern Europe. Linear sand ridges on the Atlantic shelf of the United States reside offshore Long Island, New York, New Jersey, Maryland, North Carolina, and the east coast of central Florida (Duane *et al.*, 1972; McBride and Moslow, 1991). Numerous linear shoals and banks are also found along the northern Gulf of Mexico of the United States offshore the Florida Panhandle (McBride *et al.*, 1999), Louisiana (Penland *et al.*, 1989), and east Texas (Rodriguez *et al.*, 1999). Other classic localities include eastern Canada offshore Sable Island, Nova Scotia (Dalrymple and Hoogendoorn, 1997); northern Europe along the German Frisian Islands (Swift *et al.*, 1978); and South America offshore northern Argentina, Uruguay, and Brazil (Swift *et al.*, 1978). In contrast, well-developed, *tide-maintained* linear sand ridges or fields are located in the North Sea around England (Kenyon *et al.*, 1981; Belderson *et al.*, 1982) and other embayments characterized by >3 m tidal range (e.g., Gulf of Korea; Gulf of Cambay on Indian west coast; northern delta of Amazon River; northern end of Persian Gulf; Prince of Wales Strait just off Cape York, northernmost tip of Australia) as discussed by Off (1963).

Randolph A. McBride

Bibliography

Amos, C.L., and King, E.L., 1984. Bedforms of the Canadian Eastern seaboard: a comparison with global occurrences. *Marine Geology*, **57**: 167–208.

Belderson, R.H., Johnson, M.A., and Kenyon, N.H., 1982. Bedforms. In Stride, A.H., (ed.), *Offshore Tidal Sands, Processes and Deposits*. London: Chapman and Hall, pp. 27–55.

Dalrymple, R.W., and Hoogendoorn, E.L., 1997. Erosion and deposition on migrating shoreface-attached ridges, Sable Island, Eastern Canada. *Geoscience Canada*, **24**: 25–36.

Duane, D.B., Field, M.E., Meisburger, E.P., Swift, D.J.P., and Williams, S.J., 1972. Linear shoals on the Atlantic inner continental shelf, Florida to Long Island. In Swift, D.J.P., Duane, D.B., and Pilkey, O.H. (eds.), *Shelf Sediment Transport: Process and Pattern*. Stroudsburg: Dowden, Hutchinson, and Ross, pp. 447–498.

Dyer, K.R., and Huntley, D.A., 1999. The origin, classification, and modeling of sand banks and ridges. *Continental Shelf Research*, **19**: 1285–1330.

Huthnance, J.M., 1982. On one mechanism forming linear sand banks. *Estuarine and Marine Coastal Science*, 14: 79–99.

Kenyon, N.H., Belderson, R.H., Stride, A.H., and Johnson, M.A., 1981. Offshore tidal sand banks as indicators of net sand transport and as potential deposits. In Nio, S.D., Shuttenheim, R.T.C., and van Weering, T.C.E., (eds.), *Holocene Marine Sedimentation in the North Sea Basin*. Oxford: International Association of Sedimentologists Special Publication 5, pp. 257–268.

McBride, R.A., and Moslow, T.F., 1991. Origin, evolution, and distribution of shoreface sand ridges, Atlantic inner shelf, USA. *Marine Geology*, **97**: 57–85.

McBride, R.A., Anderson, L.C., Tudoran, A., and Roberts, H.H., 1999. Holocene stratigraphic architecture of a sand-rich shelf and the origin of linear shoals: northeastern Gulf of Mexico. In Bergman, K.M., and Snedden, J.W. (eds.), *Isolated Shallow Marine Sandbodies: Sequence Stratigraphic Analysis and Sedimentologic Interpretation*. Tulsa: Society of Sedimentary Geology (SEPM) Special Publication No. 64, pp. 95–126.

Off, T., 1963. Rhythmic linear sand bodies caused by tidal currents. *American Association of Petroleum Geologists Bulletin*, **47**: 324–341.

Penland, S.P., Suter, J.R., McBride, R.A., Williams, S.J., Kindinger, J.L., and Boyd, R., 1989. Holocene sand shoals offshore of the Mississippi River Delta plain. *Gulf Coast Association of Geological Societies Transactions*, **39**: 471–480.

Rodriguez, A.B, Anderson, J.B., Siringan, F.P., and Taviani, M., 1999. Sedimentary facies and genesis of Holocene sand banks on the East Texas inner continental shelf. In Bergman, K.M., and Snedden, J.W. (eds.), *Isolated Shallow Marine Sandbodies: Sequence Stratigraphic Analysis and Sedimentologic Interpretation*. Tulsa: Society of Sedimentary Geology (SEPM) Special Publication No. **64**, pp. 165–178.

Snedden, J.W., and Dalrymple, R.W., 1999. Modern shelf sand ridges: from historical perspective to a unified hydrodynamic and evolutionary model. In Bergman, K.M., and Snedden, J.W. (eds.), *Isolated Shallow Marine Sandbodies: Sequence Stratigraphic Analysis and Sedimentologic Interpretation*. Tulsa: Society of Sedimentary Geology (SEPM) Special Publication No. 64, pp. 13–28.

Snedden, J.W., Kreisa, R.D., Tillman, R.W., Schweller, W.J., Culver, S.J., and Winn, R.D., 1994. Stratigraphy and genesis of a modern shoreface-attached sand ridge, Peahala Ridge, New Jersey. *Journal of Sedimentary Research*, B64: 560–581.

Swift, D.J.P., 1985. Response of the shelf floor to flow. In Tillman, R.W., Swift, D.J.P., and Walker, R.G. (eds.), *Shelf Sands and Sandstone Reservoirs*. Tulsa: Society for Sedimentary Geology (SEPM) Short Course Notes 13, pp. 135–241.

Swift, D.J.P., and Field, M.E., 1981. Evolution of a classic sand ridge field: Maryland sector, North American inner shelf. *Sedimentology*, **28**: 461–482.

Swift, D.J.P., Parker, G., Lanfredi, N.W., Perillo, G., and Figge, K., 1978. Shoreface-connected sand ridges on American and European shelves: a comparison. *Estuarine and Coastal Marine Science*, **7**: 257–273.

Uchupi, E., 1968. The Atlantic Continental Shelf and Slope of the United States (Physiography). Washington, D.C.: U.S. Geological Survey Professional Paper 529-I, 30 pp.

Van Veen, J., 1935. Sand waves in the North Sea. *Hydrographic Review*, **12**: 21–28.

Veatch, A.C., and Smith, P.A., 1939. Atlantic submarine valleys of the United States and the Congo submarine Canyon. Boulder: *Geological Society of America Special Paper* 7, 101 pp.

Cross-references

Coastal Currents
Continental Shelves
Estuaries
Offshore Sand Sheets
Sediment Transport (see Cross-Shore Sediment Transport and Longshore Sediment Transport)
Shelf Processes
Storm Surge
Tidal Inlets
Tides

OFFSHORE SAND SHEETS

Sands sheets are common features of modern-day, shallow seafloors. In addition, the ancient rock record contains numerous examples of sheet sandstones in both coastline and shallow marine open shelf associations (e.g., Goldring and Bridges, 1973). The Cretaceous epicontinental interior seaway of North America, for example, contains extensive sand sheet deposits and has been the focus of intense research due to their petroleum-bearing nature (Brenner, 1978; Walker, 1983; Shurr, 1984; Kreisa *et al.*, 1986; Nummedal *et al.*, 1989; Walker and Eyles, 1990; Winn, 1990). Modern sand sheets can also provide societal and economic benefits such as good quality beach nourishment sand, strategic minerals and ores, and productive marine habitat. On the other hand, the mobile sediments and bedforms associated with sand sheets can create navigation hazards and uncover buried cables and pipes. Despite their prevalence and importance, little is known about their geometry, bounding surfaces, dimensions, internal structure, origin, evolution, composition, and texture. In part, the diversity of sedimentary processes that operate in shallow marine environments accounts for the paucity of sand sheet classification schemes and models (e.g., process, facies, etc.). The disparate definitions and descriptions found in the coastal and marine literature attest to this reality.

What is a sand sheet?

Sand sheet definitions range from mass accumulations of sand to a veneer or sheet-like body of surficial sediment (Twichell *et al.*, 1981; Stride *et al.*, 1982; Belderson, 1986). They can range in thickness from a few centimeters to tens of meters and have a lateral persistence of a few meters to tens of kilometers. Some genetic definitions include the migration and stacking of large bedforms (i.e., sand waves), especially during transgressions (Nio, 1976; Stride *et al.*, 1982; Walker, 1984; Belderson, 1986; Saito *et al.*, 1989). Sand sheets have also been given:

- lithofacies descriptions such as glauconite facies and calcareous facies
- informal designations such as "facies A" and "A sand"
- descriptive designations such as coarse sand, cross-bedded facies and surficial sand sheet
- environmental interpretations such as tidally dominated sand sheet facies, storm-dominated sand sheet facies, or transgressive sand sheet, and
- combinations of the above such as shallow marine siliciclastic facies.

(Swift *et al.*, 1971; Swift, 1976; Middleton, 1978; Knebel, 1981; Stride *et al.*, 1982; Walker, 1983, 1984; Belderson, 1986; Johnson and Baldwin, 1986).

Some studies of sand sheets on the eastern North American continental shelf have used the term surficial sand sheet to include a variety of (hierarchical) bedform types and sizes. For example, Swift *et al.* (1973) and Field (1980) delineated three scales of superimposed sand "bodies" on a storm-dominated, epicontinental shelf surficial sand sheet. From largest to smallest by size, these sand bodies include shoal retreat massifs (cape and estuarine; first order features of Johnson and Baldwin, 1986); linear sand ridges (second order features of Johnson and Baldwin, 1986); and ripple, megaripple, and sand wave bedforms. Stubblefield *et al.* (1975) showed that, in some locations on the Atlantic shelf, the base of the sand sheet occurs above the topographic base of the ridges due to storm–current deepening of the swales. Since the various components of the surficial sand sheet form as a result of differences in flow regimes, dissimilarities in lithologic and paleontologic characteristics can be expected. Thus, although the massifs and linear ridges are components of the surficial sand sheet, they are discrete sedimentary facies from the sand sheet facies.

Sand sheet settings

Surficial "sheets" of sand veneer large portions of marginal (pericontinental) and epeiric (epicontinental) seafloors. In some regions, sand sheets are found in association with sand ridges, while in others the sheets constitute individual entities. In particular, the sheets have been shown to be interspersed between sand ridges, to underlie the ridges, or to exist some distance from a ridge field (e.g., Swift *et al.*, 1972; Stride *et al.*, 1982).

Four primary physical processes control sediment transport, deposition and erosion on epi- and peri-continental shelves: ocean currents, meteorological (storm) currents, density currents, and tidal currents (Swift *et al.*, 1971; Johnson and Baldwin, 1986). Globally, 3% of the shelves are ocean current-dominated, 80% are storm-dominated, and 17% are tide-dominated (Swift *et al.*, 1984). The most extensively studied sand sheets within the storm-, ocean current-, and tide- dominated settings are located on the middle-Atlantic Bight of the North America shelf, the southeast African shelf, and the epicontinental seaways rimming the British Isles, respectively.

Oceanic current-dominated sand sheets

The primary process controlling the surficial sand sheet characteristics on oceanic current-dominated shelves is the inherently unsteady geostrophic boundary current. The best studied example of an oceanic current-dominated sand sheet is located along the southeast African shelf (Flemming, 1978, 1980, 1981). On the narrow central region of the southeast African shelf, a wide variety of bedforms is molded onto a sand sheet that is maintained by the powerful, geostrophic Agulhas Current (surface velocities >2.5 m s^{-1}), a component of the major Indian Ocean circulation (Pearce *et al.*, 1978). Sediment dispersal along this shelf is also influenced by the morphology of the upper continental margin, the wave regime, wind-driven circulation, and sediment supply. Other oceanic current-dominated sand regimes include the outer Saharan shelf, the South American shelf, the Newfoundland shelf, the Osumi Strait, and the trough region of the Korean Strait (Newton *et al.*, 1973; Johnson and Baldwin, 1986; Park and Yoo, 1988; Barrie and Collins, 1989; Ikehara, 1989).

Storm-dominated sand sheets

Storm-dominated sand sheets are typified by their intra- and inter-sheet bathymetric, hydraulic, sedimentologic, and stratigraphic variability. The most comprehensive studies of storm-dominated sand sheets are of those situated along the North American Atlantic shelf between Cape Cod and South Carolina (Swift *et al.*, 1972; Knebel, 1981; Pilkey *et al.*, 1981; Swift *et al.*, 1981); the Pacific shelf off northwest United States (Clifton *et al.*, 1971; Sternberg and Larson, 1976); the Bering Sea (Cacchione and Drake, 1979; Field *et al.*, 1981; Nelson *et al.*, 1982); the Gulf of Mexico (Uchupi and Emery, 1968; Nelson and Bray, 1970; Hobday and Morton, 1984); the Bahama Bank (Mullins *et al.*, 1980; Hine *et al.*, 1981); the southern Brazilian inner continental shelf (Figueiredo *et al.*, 1982); and the Argentina inner shelf (Parker *et al.*, 1982).

The sand sheet located on the United States' east coast continental shelf is one of the largest storm-dominated sheets in existence: Over 85% of the sea floor from Georgia to New England is veneered by sand (Knebel, 1981; Swift *et al.*, 1981). In general, the spatial and temporal distribution of processes collectively have had a diverse and widespread effect on the attributes of the sand sheet. The central and southern Atlantic shelf sand sheet is the product of an autochthonous regime formed by erosional shoreface retreat during the Holocene transgression (Swift *et al.*, 1972; Swift, 1976). The primary erosional agents have been surf-zone wave action, wave-driven currents, and wind-driven currents seaward of the surf zone. This erosional transgression has resulted in the overprinting of the relict nearshore topography by a palimpsest sand sheet which, in turn, is molded with ridges and swales (Swift *et al.*, 1972; Knebel, 1981). The sand sheet is discontinuous, ranging in thickness from 1 m to 10 m. The importance of antecedent topography is indicated by the sand sheet thickness which increases in the vicinity of Pleistocene shelf river valleys, that is, Hudson River, Great Egg River, and Delaware River, and shoal retreat massifs (Swift *et al.*, 1972, 1981, Knebel, 1981) and decreases on the paleo-interfluves.

Tide-dominated sand sheets

One of the earliest studied tidally dominated modern sand sheets is located on the epicontinental seaway rimming the British Isles and on the northwest European shelf. Since the 1930s, a plethora of information has been compiled regarding the sand sheet and sand bank facies in the North Sea and Irish Sea. This information includes surface morphological, textural, and process data (e.g., Johnson and Belderson, 1969; Stride, 1970; McCave, 1971; Terwindt, 1971). From these data, theoretical (conceptual) models have been constructed regarding large- and small-scale bedform development, maintenance and internal structure (e.g., Houbolt, 1968; Allen, 1980; Johnson *et al.*, 1981) and sediment transport paths (e.g., Stride, 1963; Johnson *et al.*, 1982).

The thin sand sheet (<12 m) flanking the British Isles has formed from the unmixing of Pleistocene-age deposits during the Holocene transgression (Nio, 1976; Stride *et al.*, 1982). Materials move along transport paths associated with longitudinal and transverse tidal velocity gradients which produce bedload parting and convergence zones. The net result of tidally dominated sediment transport and deposition is to produce a well-sorted sandy deposit where fine-grained material accumulates at the ends of the bedload transport paths (i.e., grain size is in quasi-equilibrium with peak tidal current velocity; McCave, 1970, 1971; Stride *et al.*, 1982; Belderson, 1986). Superimposed on this sheet is a wide variety of bedform types ranging from small-scale ripples to large-scale sand waves (e.g., Johnson *et al.*, 1981; Belderson *et al.*, 1982). Johnson *et al.* (1981) hypothesized that the form, grain size distribution, and internal structure of the sand sheet are related directly to bedform dynamics.

Within the tide-dominated setting, sand sheets can be located in a variety of environments including marine influenced estuary- and baymouths, peri- and epi-continental shelf seas, and straits. A review of the sand sheet attributes of each environment reveals that the majority of information known about the tide-dominated sand sheet comes from studies of sand sheets within epicontinental seaways and upstream converging funnel estuaries. More recently, Fenster (1995) conducted a comprehensive study of the tide-dominated sand sheet in Eastern Long Island Sound. This study showed that the evolution of the sand sheet was linked closely to erosion of a marine delta, changes in the geometry and accumulation space of the basin, sea-level changes, attendant nonlinear tidal wave distortion, and common estuarine processes such as the development and migration of tidal channels.

Belderson (1986) postulated that a continuum of sand sheet facies exists between the tide- and storm-dominated sand sheet end members. The surficial and internal characteristics of the sand sheet depend on the degree of storm- versus tidal-influence. In general, tide-dominated sand sheets contain abundant sand waves which face in the direction of the stronger of the peak ebb or flood currents. Thus, the internal configuration of tide-dominated sand sheets primarily consists of pervasive cross-bedding, evidence of rhythmic periodicities, large dip angles, and reactivation surfaces. In contrast, storm-dominated sand sheets are devoid of sand waves as oscillatory currents act as a destructive force on sand waves (Langhorne, 1982; Belderson, 1986). In these regions, hummocky megaripples develop in response to along-coast geostrophic flows combined with wave orbitals of similar, or weaker intensity and result in hummocky cross-stratification (Swift *et al.*, 1983). Thus, on the tide-dominated sand sheet, storm-wave and storm-current activity sporadically increases transport rates while on the storm-dominated sand sheets, sediment transport is controlled primarily by wave- and wind-induced currents (Belderson, 1986).

Gaps in knowledge

The majority of work conducted to date on the sand sheet facies has focused on the depositional environment and geographical context (e.g., Park and Yoo, 1988; Saito *et al.*, 1989), modern processes (e.g., Knebel, 1981), surface sediment textural characteristics (e.g., Hollister, 1973; Schlee, 1973), thickness (e.g., Knebel and Spiker, 1977), surficial morphology (e.g., Swift *et al.*, 1972; Twichell, 1983), formation (e.g., Stahl *et al.*, 1974; Johnson *et al.*, 1981), and expected development (e.g., Johnson *et al.*, 1981). Much less work has focused on the geometry, boundary characteristics, and internal configuration of the sand sheet facies, especially in tide-dominated regimes. Although surficial information can be useful for delineating sediment transport paths, a paucity of information regarding the internal structure of the sand sheet facies partially has been responsible for limited (or incorrect) interpretations of the rock record.

Table O3 Classification and attribute list of the tide-dominated sand sheet facies (from Fenster, 1995)

	Sand sheet environments				
	Estuaries		Shelf seas		
Attributes	Coastal plain type (upstream converging funnel)	Semi-enclosed (constricted mouth) (drowned river*)	Epicontinental	Pericontinental	Straits
Examples	Oosterschelde Estuary Fraser River Estuary Gironde Estuary Ord River Estuary Bristol Channel Outer Thames Estuary Severn Estuary	Long Island Sound Chesapeake Bay* Delaware Bay*	North Sea Northeastern Bering Sea	Georges Bank Nantucket Shoals Irish Sea	Taiwan Strait Malaca Strait Bungo Strait Korean Strait Bass Strait
Tidal range (m)	3–15	0.4–1.1 1–1.3*	1–4	1–8	1.5–4
Tidal system	Progressive, rectilinear	Standing (resonant co-oscillating) with progressive component; progressive, rectilinear can extend 5–10 km onto shelf*	Complex due to interference of more than one amphidromic system; rectilinear to rotary	Rotary	Progressive, rectilinear
Currents (m s^{-1})					
peak near-surface	0.6–1.3	0.4–2.2	0.5–1.0	0.3–0.8	0.3–1.0
peak near-bottom	0.5–0.9	0.3–0.6	0.3–0.6	0.55 times surface velocity	0.25 (ave.)
Dimensions (maximum):					
Length (km)	25	30	400	400	100
Width (km)	1–8	15	50	10–50	40
Area (km^2)	200	450	20,000	4,000–20,000	4,000
Thickness (m)		Variable, 1–24 m	12 (thickness has unknown relationship to original sand waves)	Variable, 1–20 m, up to 80 m	
Depth (m)	6–15 (can reach 35)	33–49	18–44	27–45	70–120
3D Geometry		Complex in three-dimensions due to lithofacies zonal variability	Complex in three-dimensions due to bed form sinuosity		
Grain size	Medium-to-coarse sands		Fine-to-coarse sand	Fine-to-very fine sand	Fine-to-coarse sand
Mean (mm)	0.32–0.65	0.56–1.07	0.15–0.40		0.13–0.50
Notes	Distribution influenced by mutual evasive ebb and flood channels, grain size gradient expected	Progressive decrease in mean grain size along velocity gradients (grain size in quasi-equilibrium with peak velocities)	Progressive decrease in mean grain size along velocity gradients (grain size in quasi-equilibrium with peak velocities)		

Table O3 *(Continued)*

Attributes	Sand sheet environments				
	Estuaries		Shelf seas		
	Coastal plain type (upstream converging funnel)	Semi-enclosed (constricted mouth) (drowned river*)	Epicontinental	Pericontinental	Straits
Sorting	Moderate-to-well sorted	Moderate-to-well sorted	well sorted	well sorted	Volcaniclastic,
Composition		Quartzose minor amounts of rock fragments, heavy minerals	Quartzose minor amounts of calcareous material, heavy minerals	Quartzose to subarkosic minor amounts of heavy minerals, rock, and shell fragments	consolidated mud clasts, quartz
Biogenic		Within bedform spatial and temporal density changes	Increase in faunal density and diversity away from zone of large sand waves; within bed form changes		
Surficial elements	Ubiquitous presence of large and/or small bed forms; asymmetry governed by stronger peak ebb or flood current; sand wave Ht $\leq$ 8 m 88–300 m	Ubiquitous presence of large and/or small bedforms-asymmetry governed by stronger peak ebb or flood current; sand wave Ht = 1–17 m, l = (most 5–15 m;	Ubiquitous presence of large and/or small bed forms; asymmetry governed by stronger peak ebb or flood current; sand wave Ht = 1–24 m, l < 400 m l = 150–750 m	Ubiquitous presence of large and/or small bed forms; asymmetry governed by stronger peak ebb or flood current; sand wave Ht = 1–10 m, most <3.4 m	Ubiquitous presence of large and/or small bed forms; asymmetry governed by stronger peak ebb or flood current; sand wave Ht = 0.5–15.3 m l = 100–300 m
Lower bounding surface	Erosional truncation due to migrating tidal channels	Erosional trunctation due to migrating tidal channels; erosional truncation due to migrating bed forms	Erosional truncation due to migrating bedforms	Erosional truncation due to ravinement surface	Erosional truncation due to ravinement surface
Lateral bounding surface	Controlled by: (1) sand availability (2) lower estuary physiography	Controlled by: (1) sand availability (2) lower estuary/bay geometry and physiography	Controlled by: (1) sand availability (2) current velocity (3) wave surge; sand/gravel boundary lies at near-surface mean spring peak tidal current = 1 ms^{-1}; outer limit of sand (sand/mud) lies at near-surface mean spring peak tidal current = 0.4 ms^{-1}; decrease in bedform size due to lack of sand	Controlled by: (1) sand availability (2) current velocity (3) wave surge (4) depth	Controlled by: (1) sand availability (2) current velocity (3) depth (4) antecedent geology
Internal structure	Erosional features and superposition of tidal channel sequences, megaripple structures,	Erosional features and superposition of tidal channel sequences, channel fill complex	Internal structure rel ated to location, 0°, basal surface of sheet; 1°, low angle		

	largescale cross bedding with reacti vation surfaces, mud couplets, tidal bundles and other rhythmic tidalites of various temporal and spatial scales	es ranging from complete (mature) to in complete (immature), erosional features associated with bed form trough migration including rising toepoint	basal surfaces of sand wave master bedding; 2°, generally pervasive low to moderately dipping sets of cross-strata with high-angle cross-bed dips; 3°, reactivation surfaces		
Vertical sequence	Pleistocene and Holocene sands unconformably overlie channel fill	Glaciolacustrine→ fluvial→low energy estuary→high energy estuary with varbility due to Holocene sediment supply, cut and fill processes	Successively formed sand wave complexes unconformably overlie Pleistocene gravel lags	Sand wave complexes unconformably overlie Pleistocene glacial diamict; separated by wave cut unconformity	
Secondary processes	Estuarine mixing and density gradients	Short-term enhancements due to estuarine circulation, storm forcing, flow divergence due to basin geometry and bathymetry	Geostrophic flows, storms, density gradients (water temperature)	Storms	Oceanic circulation
Partial reference list	Langhorne (1973); Nichols and Biggs (1985); Berné *et al.* (1988, 1993); Harris and Collins (1991); Stride and Belderson (1991)	Ludwick (1972); Bokuniewicz *et al.* (1977); Bokuniewicz (1980); Bokuniewicz and Gordon (1980); Knebel (1989); Fenster *et al.* (1990); Fenster (1995)	Harvey (1966); Belderson and Stride (1966, 1969); Johnson and Belderson (1969); Stride (1970); McCave (1971); Caston and Stride (1973); Belderson *et al.* (1982); Stride *et al.* (1982); Belderson (1986)	Twichell *et al.* (1981); Mann *et al.* (1981); Twichell (1983)	Keller and Richards (1967); Boggs (1974); Park and Yoo (1988); Arita *et al.* (1988); Malikides *et al.* (1989)

The lack of knowledge with respect to the surficial, subsurface, and facies characteristics of the sand sheets primarily has resulted from the difficulty involved in sampling the laterally extensive sand sheet. Consequently, most sand sheet analyses have relied on morphological and surficial sedimentological characteristics. Fenster (1995), however, did construct an evolutionary model and classification of tide-dominated sand sheets based, in part, on a regional assessment of modern-day surficial, internal, and boundary sand sheet characteristics (Table O3). More work is needed to define:

- the degree to which sand sheets incorporate relict units
- the three-dimensional distribution of subfacies within a sheet
- the controls on sheet shape (thickness and lateral continuity)
- lateral and bottom boundary characteristics, and
- within sheet characteristics (e.g., extent and continuity of unconformities).

Michael S. Fenster

Bibliography

Allen, J.R.L., 1980. Sand waves: a model of origin and internal structure. *Sedimentary Geology*, **26**: 281–328.
Arita, M., Joshima, M., Okamura, Y., and Ishibahi, K., 1988. Sea bottom (geological) sonar mosaic; Southern Bungo Strait. *Miscellaneous Map Series 28-1*, Geological Survey of Japan, one page map (1 : 10,000), 7 figures, 1 table, and text.
Barrie, J.V., and Collins, M.B., 1989. Sediment transport on the shelf margin of the Grand Banks of Newfoundland. *Atlantic Geology*, **25**: 173–179.
Belderson, R.H., 1986. Offshore tidal and nontidal sand ridges and sheets: differences in morphology and hydrodynamic setting. In Knight, R.J., and McLean, J.R. (eds.), *Shelf Sands and Sandstones*. Canadian Society of Petroleum Geologists Memoir 11, pp. 293–301.
Belderson, R.H., Johnson, M.A., and Kenyon, N.H., 1982. Bedforms. In Stride, A.H. (ed.), *Offshore Tidal Sands*. London: Chapman and Hall, pp. 27–57.
Belderson, R.H., and Stride, A.H., 1966. Tidal current fashioning of a basal bed. *Marine Geology*, **4**: 237–257.
Belderson, R.H., and Stride, A.H., 1969. Tidal currents and sand wave profiles in the north-eastern Irish Sea. *Nature (London)*, **222**: 74–75.
Berné, S., Auffret, J.-P., and Walker, P., 1988. Internal structure of subtidal sandwaves revealed by high-resolution seismic reflection. *Sedimentology*, **35**: 5–20.
Berné, S., Castaing, P., LeDrezen, E., and Lerocolais, G., 1993. Morphology, internal structure, and reversal of large subtidal dunes in the entrance to the Gironde Estuary (France). *Journal of Sedimentary Petrology*, **63**: 780–793.
Boggs, A., 1974. Sand wave fields in Taiwan Strait. *Geology*, **2**: 251–253.
Bokuniewicz, H.J., 1980. Sand transport at the floor of Long Island. In Saltzman, B. (ed.), *Advances in Geophysics*, Vol. 42. New York: Academic Press, 107–128.
Bokuniewicz, H.J., and Gordon, R.B., 1980. Sediment transport and deposition in Long Island Sound. In Saltzman, B. (ed.), *Advances in Geophysics*, Vol. 42. New York: Academic Press, pp. 69–106.
Bokuniewicz, H.J., Gordon, R.B., and Kastens, K.A., 1977. Form and migration of sand waves in a large estuary, Long Island Sound. *Marine Geology*, **24**: 185–199.
Brenner, R.L., 1978. Sussex Sandstone of Wyoming-Example of Cretaceous offshore sedimentation. American Association of Petroleum Geologists Bulletin, **62**: 181–200.
Cacchione, D.A., and Drake, D.E., 1979. Sediment transport in Norton Sound, Alaska: regional patterns and GEOPROBE system measurements. *US Geological Survey Open File Report*, pp. 79–1555.
Caston, V.N.D., and Stride, A.H., 1973. Influence of older relief on the location of sandwaves in a part of the southern North Sea. *Estuarine Coastal Marine Science*, **1**: 379–386.
Clifton, H.E., Hunter, R.E., and Phillips, R.L., 1971. Depositional structures and processes in the non-barred, high-energy nearshore. *Journal of Sedimentary Petrology*, **41**: 651–670.
Fenster, M.S., 1995. *The Origin and Evolution of the Sand Sheet and Ridge Facies: Eastern Long Island Sound*. Ph.D. dissertation, MA: Boston University.
Fenster, M.S., FitzGerald, D.M., Bohlen, W.F., Lewis, R.S., and Baldwin, C.T., 1990. Stability of giant sand waves in eastern Long Island Sound, U.S.A. *Marine Geology*, **91**: 207–225.
Field, M.E., 1980. Sand bodies on Coastal Plain shelves: Holocene record of the U.S. Atlantic inner shelf off Maryland. *Journal of Sedimentary Petrology*, **50**(2): 505–528.
Field, M.E., Nelson C.H., Cacchione, D.A., and Drake, D.E. 1981. Sand waves on an epicontinental shelf: Northern Bering Sea. *Marine Geology*, **42**: 233–258.
Figueiredo, A.G., Jr., Sanders, J.E., and Swift, D.J.P., 1982. Storm-graded layers on inner continental shelves: examples from Southern Brazil and the Atlantic coast of the central United States. *Sedimentary Geology*, **31**: 171–190.
Flemming, B.W., 1978. Sand transport patterns in the Agulhas current (south-east African continental margin). *Marine Geoscience Group*, University of Cape Town, Technical Report No. 10, pp. 57–60.
Flemming, B.W., 1980. Sand transport and bedform patterns on the continental shelf between Durban and Port Elizabeth (south-east African continental margin). *Sedimentary Geology*, **26**: 179–205.
Flemming, B.W., 1981. Factors controlling shelf sediment dispersal along the South-east African continental margin. In Nittrouer, C.A. (ed.), *Sedimentary Dynamics of Continental Shelves*; *Marine Geology*, **42**: 259–277.
Goldring, R., and Bridges, P., 1973. Sublittoral sheet sandstones. *Journal of Sedimentary Petrology*, **43**: 736–747.
Harris, P.T., and Collins, M.B., 1991. Sand transport in the Bristol Channel: bedload parting zone or mutually evasive transport pathways? *Marine Geology*, **101**: 209–216.
Harvey, J.G., 1966. Large sand waves in the Irish Sea. *Marine Geology*, **4**: 49–55.
Hine, A.C., Wilber, R.J., Bane, J.M., Neumann, A.C., and Lorenson, K.R., 1981. Offshore transport of carbonate sands along open, leeward bank margins: northern Bahamas. *Marine Geology*, **42**: 327–348.
Hobday, D.K., and Morton, R.A., 1984. Lower cretaceous shelf storm deposits, northeast Texas. In Tillman, R.W., and Siemers, C.T. (eds.), *Siliciclastic Shelf Sediments*. Society of Economic Paleontologists and Mineralogists Special Publication No. 34, pp. 205–213.
Hollister, C.D., 1973. Atlantic continental shelf and slope of the United States—texture of surface sediments from New Jersey to southern Florida. *US Geological Survey Professional Paper 529-M*.
Houbolt, J.J.H.C., 1968. Recent sediments in the southern bight of the North Sea. *Geologie Mijnbouw*, **47**: 254–273.
Ikehara, K., 1989. The Kuroshio-generated bedform system in the Osumi Strait, Southern Kyushu, Japan. In Taira, A., and Masuda, F. (eds.), *Sedimentary Facies in the Active Plate Margin*. Tokyo: Terra Scientific Pub. Co., pp. 261–273.
Johnson, H.D., and Baldwin, C.T., 1986. Shallow siliciclastic seas. In Reading, H.G., (ed.), *Sedimentary Environments and Facies*, 2nd edn., Oxford: Blackwell Scientific Publications, pp. 229–282.
Johnson, M.A., and Belderson, R.H., 1969. The tidal origin of some vertical sedimentary changes in epicontinental seas. *Journal of Geology*, **77**: 353–357.
Johnson, M.A., Stride, A.H., Belderson, R.H., and Kenyon, N.H., 1981. Predicted sand-wave formation and decay on a large offshore tidal-current sand-sheet. *Special Publications International Association of Sedimentologists*, **5**: 247–256.
Johnson, M.A., Kenyon, N.H., Belderson, R.H., and Stride, A.H., 1982. Sand transport. In Stride, A.H. (ed.), *Offshore Tidal Sands*. London: Chapman and Hall, pp. 58–94.
Keller, G.H., and Richards, A.F., 1967. Sediments of the Malacca Strait, Southeast Asia. *Journal of Sedimentary Petrology*, **37**: 102–127.
Knebel, H.J., 1981. Processes controlling the characteristics of the surficial sand sheet, U.S. Atlantic outer continental shelf. *Marine Geology*, **42**: 349–368.
Knebel, H.J., 1989. Modern sedimentary environments in a large tidal estuary, Delaware Bay. *Marine Geology*, **86**: 119–136.
Knebel, H.J., and Spiker, E., 1977. Thickness and age of surficial sand sheet, Baltimore Canton Trough area: *American Association Petroleum Geologists Bulletin*, **61**: 861–871.
Kreisa, R.D., Moiola, R.J., and Nøttvedt, A., 1986. Tidal sand wave facies, Ranco Rojo Sandstone (Permian), Arizona. In Knight, R.J., and McLean, J.R. (eds.), *Shelf Sands and Sandstones*. Canadian Society of Petroleum Geologists, Memoir II, pp. 277–291.
Langhorne, D.N., 1973. A sandwave field in the Outer Thames Estuzry. Marine Geology, **14**: 129–143.
Langhorne, D.N., 1982. A study of the dynamics of a marine sand wave. *Sedimentology*, **29**: 571–594.
Ludwick, J.C., 1972. Migration of tidal sand waves in Chesapeake Bay entrance. In Swift, D.J.P., Duane, D.B., and Pilkey, O.H. (eds.), *Shelf Sediment Transport: Process and Pattern*. Dowden, Stroudsburg, PA: Hutchinson, and Ross, Inc., pp. 377–410.
Malikides, M., Harris, P.T., and Tate, P.M., 1989, Sediment transport and flow over sandwaves in non-rectilinear tidal environment: Bass Strait, Australia. *Continental Shelf Research*, **9**: 203–221.

Mann, R.G., Swift, D.J.P., and Perry, R., 1981. Size classes of flow-transverse bedforms in a subtidal environment, Nantucket Shoals, North American Atlantic Shelf. *GeoMarine Letters*, **1**: 39–43.

McCave, I.N., 1970. Deposition of fine-grained suspended sediment from tidal currents. *Journal of Geophysical Research*, **75**: 4151–4159.

McCave, I.N., 1971. Sand waves in the North Sea off the coast of Holland. *Marine Geology*, **10**: 199–225.

Middleton, G.V., 1978. Facies. In Fairbridge, R.W., and Bourgeois, J. (eds.), *Encyclopedia of Sedimentology*. Stroudsburg: Dowden, Hutchison, and Ross, Inc., pp. 323–325.

Mullins, H.T., Neumann, A.C. Wilber, R.J., Hine, A.C., and Chinburg, S.J., 1980. Carbonate sediment drifts in the northern Straits of Florida. *American Association of Petroleum Geologists Bulletin*, **64**: 1701–1717.

Nelson, C.H., Dupré, W.R., Field, M.E., and Howard, J.D., 1982. Variation in sand body types on the eastern Bering Sea epicontinental shelf. *Geologie Mijnbouw*, **61**: 37–48.

Nelson, H.F., and Bray, E.E., 1970. Stratigraphy and history of the Holocene sediments in the Sabine-High island area, Gulf of Mexico. In Morgan, J.P. (ed.), *Deltaic Sedimentation: Modern and Ancient*. Society of Economic Paleontologists and Mineralogists, Special Publication No. 15, pp. 48–77.

Newton, R.S., Seibold, E., and Werner, F., 1973. Facies distribution patterns on the Spanish Sahara continental shelf mapped with side-scan sonar. *Meteor Forschungsergebnisse (English)*, C15, 55–77.

Nichols, M.M., and Biggs, R.B., 1985. Estuaries. In Davis, R.A. (ed.), *Coastal Sedimentary Environments*. New York: Springer-Verlag, pp. 77–186.

Nio, S.-D., 1976. Marine transgressions as a factor in the formation of sandwave complexes. *Geologie Mijnbouw*, **55**(1–2): 18–40.

Nummedal, D., Wright, R., Swift, D.J.P., Tillman, R.W., and Wolter, N.R., 1989. Depositional systems architecture of shallow marine sequences. In Nummedal, D., and Remy, R.R. (eds.), *Cretaceous Shelf Sandstones amd Shelf Depositional Sequences, Western Interior Basin, Utah, Colorado, and New Mexico*, 28th International Geological Congress Field Trip Guide T119, pp. 35–81.

Park, S.C., and Yoo, D.G., 1988. Depositional history of Quaternary sediments on the continental shelf off the southeastern coast of Korea (Korean Strait). *Marine Geology*, **79**: 65–75.

Parker, G., Lanfredi, N.W., and Swift, D.J.P., 1982. Seafloor response to flow in a southern hemisphere sandridge field: Argentine inner shelf. *Sedimentary Geology*, **33**: 195–216.

Pearce, A.F., Schumann, E.H., and Lundie, G.S.H., 1978. Features of the shelf circulation off the Natal coast. *South African Journal of Science*, **74**: 328–331.

Pilkey, O.H., Blackwelder, B.W., Knebel, H.J., and Ayers, M.W., 1981. The Georgia Embayment continental shelf: stratigraphy of a submergence. *Geological Society of America Bulletin*, **92**: 52–63.

Saito, Y., Nishimura, A., and Matsumoto, E., 1989. Transgressive sand sheet covering the shelf and upper slope off Sendai, Northeast Japan. *Marine Geology*, **89**: 245–258.

Schlee, J.S., 1973. Atlantic continental shelf and slope of the United States–Sediment texture of the northeastern part. *US Geological Survey Professional Paper 529-L*.

Shurr, G.W., 1984. Geometry of shelf-sand stone bodies in the Shannon Sand stone of Southezstem Montana. In Tillman, R.W. and Siemers, C.T. (eds.), Silicicastic Shelf Sediments. Society of Economic Poleontologists and Minerslogists, Special Publication No. 34, pp. 63–83.

Stahl, L., Koczan, J., and Swift, D., 1974. Anatomy of a shoreface-connected sand ridge on the New Jersey shelf: implications for the genesis of the shelf surficial sand sheet. *Geology*, **2**: 117–120.

Sternberg, R.W., and Larsen,, L.H., 1976. Frequency of sediment movement on the Washington shelf: a note. *Marine Geology*, **21**: M37–M47.

Stride, A.H., 1963. North-east trending ridges of the Celtic Sea. *Proceedings Ussher Society*, **1**: p 62–63.

Stride, A.H., 1970. Size and shape trends for sand waves in a depositional zone of the North Sea. *Geological Magazine*, **107**: 469–477.

Stride, A.H., Belderson, R.H., Kenyon, N.H., and Johnson M.A., 1982. Offshore tidal deposits: sand sheet and sand bank facies. In Stride, A.H. (ed.), *Offshore Tidal Sands*. London: Chapman and Holf, and Belderson, pp. 95–125.

Stride, A.H., and Belderson, R.H., 1991. Sand transport in the Bristol Channel east of Bull Point and Worms Head: a bed-load parting model with some implications of mutually evasive sand transport paths. *Marine Geology*, **101**: 203–207.

Stubblefield, W.L., Lavelle, J.W., and Swift, D.J.P., 1975. Sediment response to the present hydraulic regime on the central New Jersey shelf. *Journal of Sedimentray Petrology*, **45**: 337–358.

Swift, D.J.P., 1976. Continental shelf sedimentation. In Stanley, D.J., and Swift, DJ.P. (eds.), *Marine Sediment Transport and Environmental Management*. New York: Wiley, New York, pp. 311–350.

Swift, D.J.P., Sanford, R.B., Dill, C.E., Jr., and Avignone, N.F., 1971. Textural differentiation on the shore face during erosional retreat of an unconsolidated coast, Cape Henry to Cape Hatteras, western North American shelf. *Sedimentology*, **16**: 221–250.

Swift, D.J.P., Kofoed, J.W., Saulsbury, F.P., and Sears, P., 1972. Holocene evolution of the shelf surface, central and southern Atlantic shelf of North America. In Swift, D.J.P., Duane, D.B., and Pilkey, O.H. (eds.), *Shelf Sediment Transport: Process and Pattern*, Stroudsburg, PA: Dowden, Hutchinson, and Ross, Inc., pp. 499–574.

Swift, D.J.P., Duane, D.B., and T.F. McKinney, 1973. Ridge and swale topography of the Middle Atlantic Bight, North America: secular response to the Holocene hydraulic regime. *Marine Geology*, **15**: 227–247.

Swift, D.J.P., Young, R.A., Clarke, T.L., Vincent, C.E., Niedoroda, A., and Lesht, B., 1981. Sediment transport in the Middle Atlantic Bight of North America: synopsis of recent observations. *Special Publications of International Association of Sedimentalogists*, **5**: 361–383.

Swift, D.J.P., Figueiredo, A.G., Freeland, F.L., and Oertel, G.F., 1983. Hummocky cross-stratification and megaripples: a geological double standard? *Journal of Sedimentary Petrology*, **53**: 1295–1318.

Swift, D.J.P., McKinney, T.F., and Stahl, L., 1984. Recognition of transgressive and post-transgressive sand ridges on the New Jersey continental shelf: discussion. In Tillman, R.W., and Siemers, C.T. (eds.), *Siliciclastic Shelf Sediments*. Society of Economic Paleontologists and Mineralogists Special Publication No. 34, pp. 25–36.

Terwindt, J.H.J., 1971. Sand waves in the Southern Bight of the North Sea. *Marine Geology*, **10**: 51–67.

Twichell, D.C., 1983. Bedform distribution and inferred sand transport on Georges Bank, United States Atlantic Continental Shelf. *Sedimentology*, **30**: 695–710.

Twichell, D.C., McClennen, C.E., and Butman, B., 1981. Morphology and processes associated with the accumulation of fine-grained sediment deposited on the southern New England shelf. *Journal of Sedimentary Petrology*, **51**: 269–280.

Uchupi, E., and Emery, K.O., 1968. Structure of continental margin off Gulf Coast of the United States. *American Association of Petroleum Geologists Bulletin*, **52**: 1162–1193.

Walker, R.G., 1983. Cardium formation 2, Sand-body geometry and stratigraphy in the Garrington-Caroline-Ricinus Area, Alberta—The "ragged blanket" model. *Bulletin of Canadian Petroleum Geology*, **31**(1): 14–26.

Walker, R.G., 1984. Shelf and Shallow Marine Sands. In Walker, R.G. (ed.), *Facies Models*. Geosciences Canada, Reprint Series 1, pp. 141–170.

Walker, R.G., and Eyles, C.H., 1990. Topography and significance of a basinwide sequence-bounding erosion surface in the Cretaceous Cardium formation, Alberta, Canada. *Journal of Sedimentary Geology*, **61**: 473–496.

Winn R.D., Jr., 1990. Storm deposition in marine sand sheets: Wall Creek Member, Frontier Formation, Powder River Basin, Wyoming. *Journal of Sedimentary Petrology*, **61**(1): 86–101.

Cross-references

Coastal Sedimentary Facies
Continental Shelves
Ingression, Regression, and Transgression
Nearshore Sediment Transport Measurement
Offshore Sand Banks and Linear Sand Ridges
Ripple Marks
Sandy Coasts
Scour and Burial of Objects in Shallow Water
Sequence Stratigraphy
Shelf Processes
Shoreface

ORGANIZATIONS—See APPENDIX 3

P

PACIFIC OCEAN ISLANDS, COASTAL ECOLOGY

> The ecology of the tropics is an incredibly complex subject, probably far too complex to be grasped in its entirety by the human mind. Yet it is of the greatest urgency that it should be sufficiently understood so as to help man utilize tropical lands without utterly destroying the region eventually as a human habitat.
>
> Marie-Helen Sachet, 1967, Botanist, Smithsonian Institute (cf. National Biodiversity Team of the Republic of the Marshall Islands, 2000, inside front cover).

Adequate comprehension of the complexity of the vast number of relationships between organisms and their environment should be extended to both land and sea, especially where they are in relatively close proximity near the coastline.

Coastal ecology of isolated tropical Pacific islands offers many microcosmic examples of the forms and functions of species. It also focuses attention on the adaptation to a series of biogeochemical conditions presented by natural phenomena, and more recently by human activities. The ecosystems addressed in this selective review include coral reefs, marine, and freshwater wetlands, and coastal strand and forest. All of these can be found at, or relatively near tropical island coastlines within the world's largest single feature, the Pacific Ocean.

Ecological relationships along the coasts of Pacific Islands have been undergoing dynamic adjustment to tectonic and climatic changes at least throughout the last 1.8 million years (the Quaternary period). This synchronous and time-transgressive environmental change has continued into the Holocene Epoch, or recent postglacial warming period that began approximately 12,000 years BP. It is during this period that humans have had an increasing impact on the abiotic and biotic components of coastal ecosystems.

Here we focus on coastal ecology of Remote Oceania, which covers a vast area of the Pacific Ocean, and includes all those tropical Pacific Islands so isolated that humans were only able to find them within the past 4,000 years (see Figure P1). Within this huge region, the cultural impact of people locally, regionally, and perhaps globally must be, more or less, added to those natural factors which affect the changing nature of Pacific coastal ecology, such as volcanic eruptions, earthquakes, tsunami, and hurricanes (typhoons).

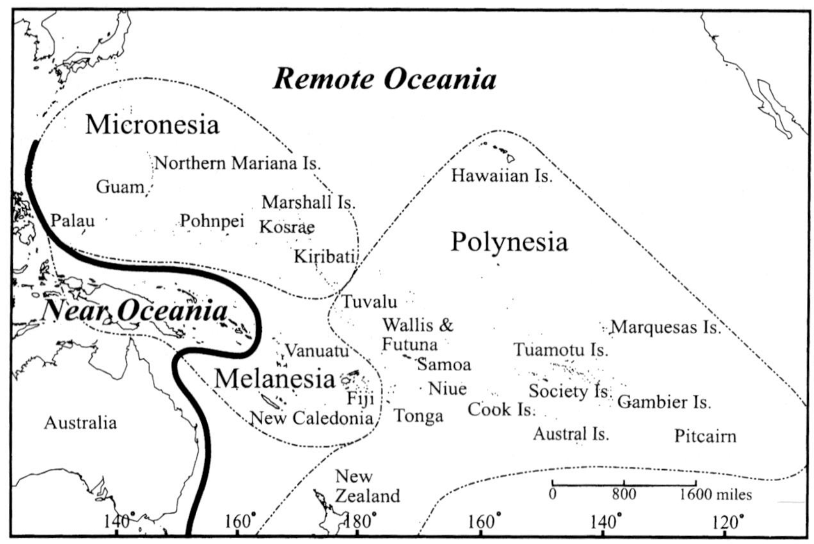

Figure P1 Map of Near and Remote Oceania (after Merlin, 2000).

The tropical Pacific Islands of Remote Oceania are basically the subaerial portions of hot-spot volcanoes (produced by melting anomalies), or limestone caps of reef development that are near or above the surface of the sea. The range of Remote Oceanic Islands include:

(1) High islands in various stages of volcanic development, geomorphological erosion, and subsidence or tectonic uplift (e.g., O'ahu, Nuku Hiva, Mo'orea, and Mangaia Islands);
(2) Atolls, which normally consist of an annular-shaped series of coral reefs with few to many low-lying islets comprised of wave-washed, unconsolidated reef, and other fauna or calcareous algal debris, more or less, surrounding a lagoon with all of the volcanic formation sunk below the sea level (e.g., Ulithi, Tarawa, and Takapoto Atolls);
(3) Raised Atolls or isolated reefs that have been uplifted or formed at higher stands of the sea, which are commonly referred to in the Pacific region as *makatea*, or raised reef islands (e.g., Fais, Nauru, Makatea, and Henderson Islands).

There are variants that, more or less, bridge the gaps between these islands categories (e.g., almost atolls such as Chuuk, Aitutaki, and Bora Bora Islands).

Thus the range of coastal ecosystems located near coastlines vary from those situated on the slopes of active volcanoes (e.g., some Hawaiian and Galapagos Islands), to those on older dormant or extinct volcanic high islands with well developed fringing and/or barrier reefs and few to many sandy beaches (e.g., Pohnpei, Kaua'i and Tahiti Islands), and those close to or along raised reefs (Atiu and Mitiaro Islands) and atoll shores (e.g., Ailinglaplap and Oeno Atolls).

Above and below the sea, a series of environments, or ecosystems, are distributed along tropical coastlines where inundation by sea water is continual, periodic, seasonal, or episodic in nature. These general environments include coral reefs, mangroves, freshwater wetlands, and strand communities. Their myriad inorganic and organic components interact with one another in wide variety of environmental relationships.

As volcanoes have subsided or been uplifted in concordance with climatically and/or tectonically controlled rising and lowering of the ocean level, those coastal species populations that have adapted to these environmental changes survived by effectively migrating in relationship to changing sea levels. In addition, environmental stresses produced by human use on land and in ocean have, more recently (during the past 1,000–4,000 years BP), also affected the distribution of species. It is generally assumed that these human impacts have accelerated in modern times; anthropogenic changes will be discussed below. Origins of the native species in the coastal ecosystems of the tropical islands in Remote Oceania are discussed first, followed by a description of the basic features of the ecosystems in which these species inhabit.

Origins of the species: biogeography and the filter effect

The ecosystem concept is a useful scientific form of reductionism designed to bring order to the diversity of life, including that found in the environments of the tropical coastal Pacific region. Whether or not ecosystems form and function in discrete, interrelated units is still an unresolved question for biologists. It has been argued that plants and animals in terrestrial ecosystems have independent ranges of distribution that often overlap in their adaptations to environmental parameters. This may also apply to marine species. Nevertheless, if only for systematic convenience, the ecosystem concept remains useful for scientific description and analysis.

It is clear that variable dispersal ability helps explain the distance-decay factor ("filter effect") determining which organisms can colonize remote geographic locations. Although many species adapted to tropical marine and nearshore land environments have long-distance dispersal mechanisms and are characteristically found over a vast region, there is consistent decline in the number of many species that become established in marine, and especially inland terrestrial environments over time in Remote Oceania (Carlquist, 1974; see Figure P2). For example, the total number of land birds on islands compared to the totals on neighboring islands and chains of islands in the South Pacific Region declines with distance east from the major source of immigrants in the New Guinea region, where there are 520 species, to 127 in the Solomons, 54 in Fiji, 33 in Samoa, and 17 in the Society Islands.

Indeed, a significant dropoff in the number of successful colonizing taxa can be verified for many groups of organisms found in tropical Pacific ecosystems, including those located at or near coastlines. For example, in the western tropical Pacific Islands of the Republic of Palau there are over 300 reef building coral species, while in remote Hawaii there are approximately 40 species (Grigg, 1983). And the distribution of damselfish (Pomacentridae) in the tropical Pacific drops off from 140 in the New Guinea region to 104 species in Fiji, 98 in Samoa, 81 in Tahiti, and 50 in Easter Island (Stoddart, 1992). Some of this oceanic barrier effect has been disrupted by humans who have brought many species (purposefully and unwittingly) with them over time, especially in the terrestrial island environments within the last two centuries (Mueller-Dombois and Fosberg, 1998; Kirch, 2000).

When distance from continents and other islands is combined with latitudinal proximity to the equator and rainfall distribution, the relative

Figure P2 Buoyant fruits of *Barringtonia asiatica* float long distances and eventually wash ashore viable in coastal areas such as shown here in the Solomon Islands (photograph by Mark Merlin).

biodiversity of coastal ecosystems in Remote Oceania can be explained. The further a coastal habitat is from the biological diverse continents and other islands, along with its relative distance from the equator in combination with declining annual rainfall, generally the fewer the number of naturally occurring species will be found. And this is much more pronounced for inland, high island environments (Stoddart, 1992).

Disturbance and succession

Environmental conditions in the ocean, atmosphere and on land, in themselves and in combination create periodic, sometimes catastrophic changes. Volcanic activity, including eruptions of both pyroclastics and lava, play a role in the periodic change of many ecosystems including those in the lowland coastal and nearshore environments. For example, basaltic lava flows erupting out of the active Kilauea volcano on the large island of Hawaii, often destroy coral reef formations as they enter the sea. However, the newly formed igneous surfaces provide the base for pioneer species to recolonize and initiate the successional processes. Over time and with limited disturbance, the ecosystem typically returns to a fully mature community of interacting organisms (Grigg and Maragos, 1974).

The erosional forces of mass wasting, along with stream, wind, and wave action, shape and reshape the coastal environments, both gradually and episodically. The relatively slow pace of disturbance and ecological recovery through biological succession, is occasionally punctuated by very intense environmental perturbations. Great waves of water generated in the Pacific Ocean can greatly alter the coral reefs and coastal lands over relatively short periods of time. These huge swells are often induced by high-energy land movements. Large earthquakes and/or giant landslides normally occur along subduction zones, and sometimes near or within archipelagoes of hot-spot volcanoes, particularly those in the youthful subaerial shield building stage (Moore *et al.*, 1994). Given these determinants of biodiversity according to geographical and climatic distribution, as well as ecological disturbance and succession, we can now review the coastal ecosystems or environments of Remote Oceania. These include coral reefs, seagrass beds, mangrove forests, coastal freshwater wetlands, swamps and marshes, strand communities, and coastal lowland forest.

Coral reefs

Coral reefs are biologically diverse oases of high-energy flow and nutrient cycling. Major organism groups found on coral reefs include algae, seagrasses, corals, and other animals. Coralline algae are often equal or greater contributors to the calcium carbonate structure of the reefs than coral organisms. The zonation and profusion of life in coral reef communities is determined by environmental parameters such as distance and depth from the coastline. Wave exposure, gradients in sedimentation, degree of salinity, and temperature are the primary factors causing this variation in abundance and composition in coral reef communities in Remote Oceania. Corals can be viewed as organisms with specific behavior, physiology, and structure, as well as colonies that serve as habitat for a broad diversity of other organisms. Furthermore, coral reefs themselves may be seen as ecosystems with a great variety of interactions among the organic and inorganic components that function in these environments (see Figure P3).

Over much of the tropical Pacific Ocean, the composition of coral reefs is frequently similar, with the genus *Acropora* often the major component of the bottom cover (Gulko, 1998). Isolated and younger coral reefs in Remote Oceania such as those in the Hawaiian Islands, however, differ significantly from those in other tropical areas. In the case of the Hawaiian reefs, they are not as well developed because of their relatively youthful geological age. Therefore, most reefs in the younger islands are small fringing reefs. Wave exposure is also important. The leeward coasts of the Hawaiian islands, for example, which are generally more sheltered, have reefs with more coral cover than the wave-pounded windward coasts (Grigg, 1983; Jokiel, 1987).

The younger coral reefs, which lack barrier reefs (e.g., those in the windward Hawaiian Islands and some of the southern Cooks Islands) are normally less productive than other reefs associated with many other older high islands or atolls in Remote Oceania. Less reef area and the lack of lagoons that collect coastal and terrestrial runoff in the very remote, and younger, windward Hawaiian reefs results in relatively poor supplies of nutrients; this in turn helps explain the comparatively low numbers of soft corals, sponges, tunicates, bivalves, and other filter-feeding animals. Unlike many other reefs in Remote Oceania, especially those associated with older high islands or atolls, the reefs of Hawaii are noticeably dominated by corals, even though there are relatively few species of corals on these reefs.

As noted above, because of the remote location of the Hawaiian Islands the coral reefs have considerably less biodiversity than other reefs, particularly those located much closer to the Indo-Pacific region to which they belong. One effect of this lower diversity is reduced specialization of reef-building corals in the Hawaiian Islands, which as a result have more extensive distributions of individual species than in less-isolated regions. Although the biodiversity of marine species is much less in the more isolated coral reefs of Remote Oceania, these reefs, such as those in Hawaii have particularly high levels of endemic marine species.

Figure P3 Coral reef in Chuuk lagoon with reef islets in the background (photograph by Mark Merlin).

Seagrass beds

Seagrasses are flowering plants, and are dissimilar to algae in that they produce true roots, stems, leaves, flowers (often inconspicuous), and seeds. Seagrass beds or meadows are found in marine or estuarine waters continuously flooded by saltwater. Most seagrass species are rooted in silty or sandy sediments located in shallow water up to about 7 m deep. Below this depth there is not enough sunlight for the seagrasses to survive.

Worldwide there are approximately 49 species of seagrasses in 12 genera. They are classified as belonging to at least two monocotyledonous families, Hydrocharitaceae and Pota-mogetonaceae, which should not to be confused with members of Poaceae, the grass family. Sixteen species of seagrasses are found in the Pacific Island region. However, little research has been done on seagrasses in this region, and some scientists suspect that new species remain to be described in the tropical Pacific (den Hartog, 1970; Mathieson and Nienhuis, 1991). Many species of marine algae, as well as numerous animals, are also found in seagrass areas.

Seagrass beds are generally located in a mixed species zone running parallel to shore, but normally separated from the mangrove vegetation by a narrow band of nonvegetated sand or silt. In some areas, the seagrass beds extend out over large areas of the fringing reef flat and often cover interior areas such as embayments.

Like the mangrove forests (described below), the seagrasses, with their dense rhizome and root systems, serve as traps for silts and sediments washed into the sea from land areas. Seagrass beds are extremely productive, providing food as well as shelter for a number of animal species, including sea cucumbers, clams, some fish such as siganids, dugongs, and some reptiles such as the green sea turtle (*Chelonia mydas*). Throughout its huge geographical range in the tropical oceans of the world, the green sea turtle commonly consumes species of seagrasses in the genera of *Thalassia*, *Cymodocea*, and *Halophila*, all of which are found along many coastal areas of the tropical Pacific Ocean.

Seagrass beds are very significant in coastal food webs; in fact they typically sustain more invertebrates and fish than nearby areas lacking seagrasses. A multitude of epiphytic algae and associated small animals live on seagrasses, and several kinds of mollusks, and some fish obtain most of their nutrition by feeding on these epiphytes. Substantial amounts of detritus form as microorganisms colonize dying seagrass tissue; and because of this most animals found in seagrass beds are detritivores (Bortone, 2000).

Mangrove forests

Mangroves forests are unique and successful adaptations to harsh environmental restraints. They are found in saltwater-influenced swamps, and are composed of woody plants (trees and shrubs) adapted to areas affected by ocean tides (Chapman, 1976). These distinctive forest communities are located in many subtropical and tropical areas around the world, covering approximately 60–70% of all tropical coasts (Por and Dor, 1984). They are typically found on muddy reef flats of coastal areas.

Mangrove ecosystems are found between the latitudes of 32°N and 38°S, along the tropical coasts of Africa, Australia, Asia, the Americas, and many tropical oceanic islands. There are varying scientific classifications of what constitutes a mangrove plant depending upon how strictly the term "mangrove" is defined. The worldwide number of species classified as mangrove plants varies considerably (e.g., 54 according to Tomlinson, 1986, and 75 according to Field, 1995). The greatest diversity of mangrove species exists in Southeast Asia.

The biogeographic filter effect discussed above also applies to mangroves. For example, in the coastal saltwater environments of the Republic of Palau there are nine known species; further east in similar coastal environments of Yap there are only seven known species; in the Marshall Islands, much further to the east in the Pacific Ocean, there may be only three species, all of which are not very common; and much further east in the Hawaiian Islands there are no native species of mangroves.

The plants of the mangrove forests are very important for the environment (or ecosystem). They stabilize coastal areas by trapping and holding sediment washed down from inland areas. In some areas, where rainfall washes the thick mud from the interior down to the mangrove swamps, the mass of tangled roots of the woody plants prevent much of this sediment from washing out and smothering the coral reefs and seagrass beds. In those areas where the mangroves are located on the muddy reef flats, the plants of these forests also offer some coastal protection from normal wave action and strong storm surges (Lugo and Snedaker, 1974, see Figure P4).

As "producers," the mangrove plants convert energy from the sun, through the process of photosynthesis, into useful proteins and carbohydrates in their leaves and other parts of the plants. Eventually these plant parts drop off into the water and are carried out with the tides into nearshore waters forming a rich "nutrient soup." This supplies food for nearshore fisheries and the reef animals. Mangroves also provide an

Figure P4 Mangrove swamp bordered by fringing coral reef on the ocean side and coastal forest on the land side of Pohnpei Island (photograph by Mark Merlin).

important source of nourishment for many animal species that live within these marine swamps, including various types of shellfish, finfish, skinks, geckos, insects, and birds. In addition, several of these species use the mangroves as nurseries for their young.

All trees that live in the mangrove must have a means of obtaining oxygen since the tide comes in and water regularly covers the roots. Some woody mangrove plants have roots above the ground. These aerial prop roots have special air canals that allow the trees and shrubs to absorb oxygen, which is hard to get from the thick mud where the mangroves grow.

Rhizophora apiculata and *R. mucronata* are common mangrove trees in many tropical Pacific coastal areas; they can be identified quite easily by their hanging, aerial, prop roots. Other relatively common mangroves, such as *Sonneratia alba* and *Lumnitzera littorea* can be recognized by the exposed, cone (or pencil) shaped roots that stick up out of the muddy soil from extensive, cable-like roots. These exposed cone shaped roots, known as pneumatophores, serve as breathing organs in the swampy areas. Still other mangrove trees, such as *Xylocarpus granatum* can be identified by their very distinctive buttressed (or arched) roots that spread out like cables (or ribbons). The extraordinary wavy roots stretch out, allowing this large mangrove tree to grow in the loose soil or thick mud. Another plant often associated with the mangrove forest is *Nypa fruticans* (Nipa Palm), which has many uses (Merlin and Juvik, 1996).

The mangrove vegetation can be divided into zones from the edge of the land out into the lagoon, or from the shore out onto the muddy fringing reef flat. First, there is the intertidal, border zone where lowland forest and mangrove plants meet. It is a narrow zone along the high tide line with little or no vegetation. This zone usually has shallow mud deposits with patches of shrubs and trees. Some muddy areas in this zone emerge at low tide. This inner margin of the mangrove forest is usually quiet and protected (Teas, 1983). The mangrove forest vegetation at the edge of the intertidal border often has more species of mangrove plants than other zones in the mangrove forest.

On the outer seaward fringe or channels in the mangrove forest where mangrove plants meet the sea, there is often an overhang of branches above the water. The overhanging branches have two basic forms, depending upon the dominant mangrove tree species fronting the reef flat or lagoon waters. The more common form of overhanging branches is dominated by the *Rhizophora* species. These trees (5–10 m, 15–30 ft), with a green bushy canopy and reddish brown, interlocking, aerial prop roots, often form a dense wall of vegetation. Less common is the form of overhanging branches dominated by groves of *S. alba*, a mangrove tree with the single thick trunk, open understory, and conical-shaped, breathing roots (pneumatophores) sticking out of the mud. Some species of clams and finfish are also taken from the mangrove forest areas.

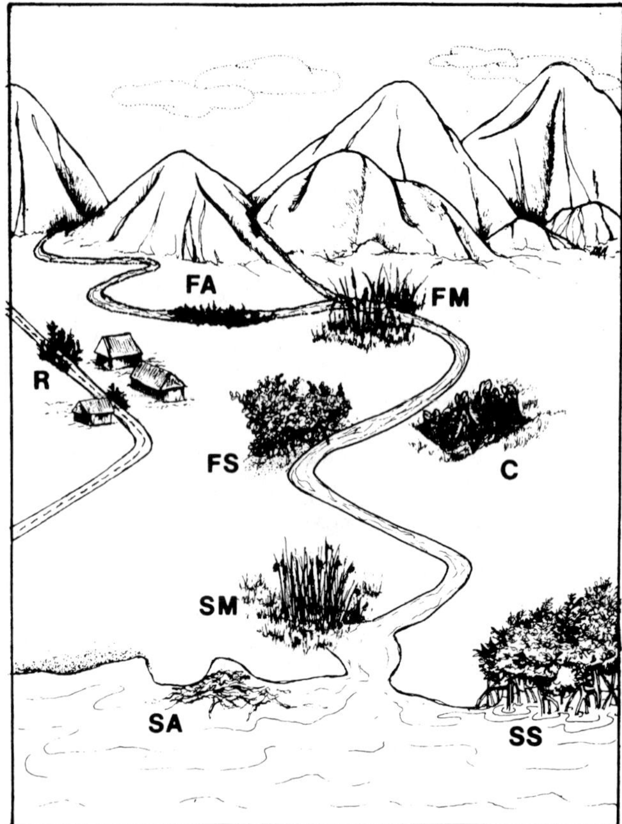

Figure P5 Schematic of wetland types: SA = seagrass beds; SM = coastal saltwater marsh; SS = mangrove swamp; FA = freshwater aquatic; FM = freshwater marsh; FS = freshwater swamp; C = cultivated wetlands; and R = ruderal wetlands (US Army Corps of Engineers).

Coastal freshwater wetlands, swamps, and marshes

Coastal freshwater wetlands, swamps, and marshes consist of those environments often flooded or saturated by ground or surface freshwater for periods of time long enough to support only plants that can grow in water-logged soils. Tropical freshwater wetland communities in Remote Oceania can be divided into five types: freshwater aquatic vegetation, freshwater marsh vegetation, freshwater swamp forest, cultivated wetland, and ruderal wetland (Stemmerman, 1981; Merlin *et al.*, 1992, 1993, see Figure P5).

Freshwater aquatic vegetation develops in areas permanently flooded with freshwater. This vegetation type is found in slowmoving streams, reservoirs, irrigation ditches, natural and artificial ponds, and open bodies of water surrounded by freshwater marshes. The plants in this aquatic vegetation grow submerged, partially submerged, or floating in the freshwater. They lack well-developed structural support provided by the more or less rigid trunks, stems and branches of plants growing in soil on land. Most of the species in this type of vegetation are annuals. Some of the species found in the freshwater aquatic vegetation can become troublesome weeds that multiply rapidly and choke waterways. Two examples of aquatic plants that may produce this kind of problem in Remote Oceania are "water lilies" (*Nymphaea* spp.) and "water hyacinth" (*Eichhornia crassipes*), both alien species introduced into Remote Oceania during the 20th century.

Freshwater marsh vegetation contains mostly nonwoody species, especially grasses and sedges that have roots in the water-logged soil, and are typically difficult to move through. Crops cannot be cultivated in the marshy areas without drainage. The vegetation of marshy areas is dominated by large sedges (Cyperaceae), grasses (Poaceae), and other herbs.

Freshwater swamp forests that have not been heavily exploited by people are mainly dominated by woody plants, such as *Terminalia* spp. and *Barringtonia racemosa*, which grow on soils that are often, if not always, water-logged. The swamp forests are found in depressions or other poorly drained places, usually behind the beach strand. In the past many freshwater swamp and marsh areas were cleared or severely altered in Remote Oceania for farming purposes. This often took place in areas with rich muddy soil where taro plants, such as *Colocasia esculenta* and *Cyrtosperma chamissonis* could be cultivated. Many of these former freshwater swamps are presently covered with secondary vegetation containing plants that recolonize this kind of wetland after the original vegetation has been heavily disturbed. In those wetlands where considerable disturbance has taken place, and no root crops are being cultivated, successional plants such as *Hibiscus tiliaceus* and *Phragmites karka* are commonly found. The environmental protection of those areas with remaining stands of swamp forests, such as those with *Terminalia carolinensis* found on Pohnpei and Kosrae Islands in the Eastern Caroline Islands of Micronesia, should be seriously considered because of their unique and interesting vegetation, as well as other aspects of resource conservation.

Cultivated wetland exists in areas where traditional "root crops" are grown. In many wetland areas, cultivated plants have replaced natural freshwater marsh and swamp plants. As noted above, traditional, edible taro species are planted in this largely artificial or man-made vegetation (see Figure P6).

Ruderal wetland vegetation is comprised mostly of weedy or successional species that are often aliens now common throughout the tropical Pacific. It occurs frequently in "waste" areas or terrain subject to flooding and interference by humans. This includes places such as rubbish dumps, roadside verges, and bombed sites.

Strand and lowland coastal forest

The flora and fauna of the coastal land ecosystems of Remote Oceania represent, more or less, an attenuated Indo-Malayan biota, containing relatively few plant and animal species, and even fewer endemics. The

Figure P6 Cultivated wetland on Peleliu Island with *C. esculenta* taro (photograph by Mark Merlin).

size and composition of the biota in these ecosystems is largely a function of the distance from the Indo-Malayan region, size of the environment, hurricane (typhoon) frequency, droughts, salinity, native and introduced plants and animals, wars, and other human-induced disturbances (Manner, 1987). Strand and lowland coastal forest includes ecosystems comprised of terrestrial organisms occurring near the coastline. The coastal or littoral vegetation in the tropical region of Remote Oceania is found in a narrow zone near the edge of the sea on open, sandy, and rocky shores. It also typically includes all of the native vegetation on islets comprising atolls.

Plants with different life forms such as small trees, shrubs, vines, and herbs, are found in the coastal or littoral vegetation. Although they can be very different, in some ways, these plants share several things in common. Many produce seeds that float and are able to survive for long periods of time immersed in saltwater. The seeds of these plants are often moved long distances by ocean currents. The plants of the coastal vegetation also have to be able to survive under very sunny, often windy and salty conditions, near the seashore. At the edge of the sandy and rocky shores, the nonwoody plants commonly have fleshy leaves and sap that is salty. Most of these specially adapted coastal plants, as well as animals such as seabirds, are found along many, if not most, tropical Pacific shores (Merlin and Juvik, 1996; Merlin *et al.*, 1997).

The plants that are found in and make up the coastal vegetation can be divided into more or less four separate types of littoral plant communities, including the herbaceous strand, littoral shrub land, pandanus scrub, and littoral forest (Whistler, 1994). Herbaceous strand or nonwoody vegetation is located above the high-tide mark of the ocean on both sandy and rocky or coral rubble shores (see Figure P7). Most strand species are perennials with deep taproots. Typical plants in the strand include creeping vines such as *Ipomoea pes-caprae* and *Vigna marina*. Littoral shrub land is typically located on windy coastal ridges and slopes, or on the seaward edges of coastal forests. Normally the dominant plant of this community is *Scaevola taccada*. Pandanus scrub is found usually on rocky, often exposed, windswept shores. As indicated in its name, this community is dominated by *Pandanus tectorius*, also known as "screwpine." *Pemphis acidula* shrubs or small trees are common on rocky, limestone coasts. Littoral forest is the most common kind of vegetation found along the tropical shores of Remote Oceania, typically just behind the strand vegetation. This forest is usually dense and is sometimes dominated by a single tree species. Common trees in this community include *Barringtonia asiatica*, *Hernandia sonora*, and *Callophyllum inophyllum*. The typical closed lowland wet forest is dominated by the widespread *Pisonia grandis*.

A variety of native animals reside in the coastal vegetation, especially where human disturbance has been low or non-existent and the less-remote islands. Among these animals are large numbers of breeding seabirds, some landbirds such as Pacific pigeons (*Ducula pacifica*) and fruit-doves (*Ptilinopus* spp.), some reptiles, and coconut crabs (*Birgus latro*).

Human impact

Human encroachment on marine coastal environments and adjacent areas, has produced some dramatic effects. The discussion that follows describes some of the more important ecological impacts that people have had on coral reefs, seagrass beds, mangrove forests, coastal wetlands, and coastal land communities.

Natural disturbances are certainly major, if not the most important factors affecting coral reefs (Grigg and Dollar, 1990). Nevertheless, the possible threats of human-induced global warming, including changes in sea level and increased frequency of intense low pressure system storms and El Niño Southern Oscillation events, are among many serious problems associated with human impact on coral reef ecosystems, including those in Remote Oceania. Among the more important problems are anchor damage, coral bleaching, coastal development, coral diseases, fish feeding, effects of fishing, tropical fish collecting, introduced species, marine tourism, sewage and eutrophication, effects of oil spills and heavy metals, effects of sedimentation, turtle tumors, effects of ultraviolet radiation on coral species, and effects of human divers (Richmond, 1993; Gulko, 1998).

The development and implementation of effective management plans to maintain the health and productivity of coral reefs requires a solid basis of scientific information and interpretation (Birkeland, 1997). Unfortunately, much of this information is still inadequate. For example, in the Hawaiian Islands, invasive alien algae and excessive fishing for food and the aquarium trade are present and dangerous threats to the nearshore coral reefs. However, a basic ecological understanding of how these pestiferous algae have become successfully established is still undetermined. In this case, more data is needed to understand the inherent factors such as growth parameters and reproductive strategies, as well as the anthropogenic impact on factors such as nutrient regimes and herbivore grazing pressure. These intrinsic and human-exacerbated factors interact ecologically to control the abundance of the alien algae (Hunter and Evans, 1995).

As noted above, alien invasions are further complicated if additional human-induced stresses are imposed on ecosystems. Vigorous coral reef ecosystems are normally dominated by reef-building corals and coralline algae, with macroalgae and algal turfs typically limited to places on the reefs that are comparatively more difficult for herbivores to reach. In the case of Hawaii, algal growth (native and non-native species) has been favored in some areas because of excessive fishing and

Figure P7 Coastal forest on Islet of Majuro Atoll with coral reef at low tide in the foreground (photograph by Mark Merlin).

eutrophication, and this has resulted in severe and perhaps permanent changes in reef community structure (Borowitzka, 1981; Steven and Larkum, 1993; Maragos *et al.*, 1996; McClanahan, 1997).

Seagrass meadows form very productive ecosystems and provide nutrition and protection for many kinds of fish and invertebrates. They also help fasten the seafloor and maintain or improve water quality. Some natural disturbances have major effects on seagrasses. These include storms, floods, and droughts. For example, heavy rainfalls or extended droughts can severely affect the normal inflows of freshwater in the marine seagrass beds. Acute changes in salinity may result, and consequently have very adverse effects on the seagrasses. Intense storms may also redistribute coastal sediments through increased water current flows and larger influxes of fresh water. Increased turbidity may result, reducing the light available for seagrass photosynthesis. In addition, in extreme cases, huge quantities of sediment rearrangement can bury some seagrasses.

Anthropogenic disturbances on seagrasses can also be quite severe. Seagrass communities, especially those located in estuaries, are adapted to tolerate cyclic natural disturbances. Continuous heavy disturbances induced by human activities, however, may have severe impact on seagrasses. Water pollution, foreshore development, and recreational use of waterways are some of the major problems affecting the health of seagrass beds.

Intense terrestrial soil erosion can lead to excessive runoff that may limit the sunlight available to seagrasses. This tends to stunt their growth, or even kill the plants. Reductions in the area of seagrass growth can reduce number and extent of animals dependent upon these plant species. Nutrient inputs from agricultural runoff and urban areas can increase to the point that abnormal algal blooms develop. Such blooms decrease light penetration and limit the range and productivity of seagrasses. Reductions in available light may also result from dredging and fish trawling activities that increase turbidity through sediment resuspension. In several areas, inappropriate ship movement in shallow waters where seagrasses are found is another critical problem.

Extensive damage to marine and estuarine habitats due to the unintentional destruction of the seagrasses has been documented in several areas. Among the few animals that feed, more or less, exclusively on seagrasses, are green turtles and dugongs, and since these species are listed as threatened animals, their continued existence is dependent upon preservation of seagrass beds.

Mangroves, the "rainforests of the sea," dominate a large portion of the world's subtropical and tropical coastlines, covering an estimated area of 22 million hectares. People in tropical areas throughout the world, including Remote Oceania, utilize the raw materials of trees of the mangrove forests FOR numerous purposes (Baines, 1981). Unfortunately, during last several decades, the global area in mangroves has diminished at an increasing rate due to a variety of human activities. These disturbances include harvesting for firewood, charcoal, resins, and building materials, as well as freshwater diversion, dredging for land reclamation projects, and drainage to facilitate road construction. To some extent these impacts are also occurring in Remote Oceania (Hamilton and Snedaker 1991).

Mangrove habitat has a special role in maintaining the health of the marine environment and human industries that depend on its abundant life. The increasing removal of mangrove forests for a variety of reasons is having significant impact on fishing industries and tourism. The natural hatchery function of mangrove forest helps restock many marine populations, and mangrove forests have become popular attractions for ecotourism. Like other plants, mangrove trees and shrubs require oxygen. However, a coat of oil, even a thin one, can block the breathing pores on the aerial roots of mangroves and may kill them. Therefore, oil spills can be very dangerous for mangroves.

The introduction of alien species in some cases is also causing profound changes in the ecology of coastal environments in Remote Oceania. For example, as noted above, in the Hawaiian Islands, the introduction of invasive algal species (e.g., *Gracilaria salicornia*) has had significant impact in some coral reef regions. In these same islands, the introduction of mangroves trees (particularly *Rhizophora mangle*, see Figure P8) to the coastal zone of Hawaii, which previously had no native mangrove species, has created a series of alien mangrove forests, especially on O'ahu Island (Juvik and Juvik, 1998). Today, these mangroves intrude upon historical sites and critical habitat for a number of endangered shorebirds.

Low lying, freshwater swamp forests, just inland of mangroves and above tidal influence, are greatly disturbed in many areas of Remote Oceania. Most native vegetation of these freshwater swamp forests has been cleared or modified by human activity. Traditional and modern expansion of taro cultivation, as well as the commercial plantation development and the negative impacts of road development, have greatly affected the freshwater swamp forest habitat in many tropical Pacific Islands. There is an urgent need to protect the widely threatened, native freshwater swamp ecosystems in many areas of the tropical world, including those in Remote Oceania.

The majority of people in Remote Oceania have traditionally lived relatively close to the coastlines of the isolated islands of this region, and this settlement plan has continued into modern times. People have thus utilized the coastal zone for many purposes including the construction of housing. Therefore, the coastal flora and fauna in many areas has been modified significantly by human activity. Over the years, a great many native trees, including the often dominant trees of *P. grandis* and others,

Figure P8 Alien mangrove, *R. mangle* in coastal habitat on O'ahu, Hawai'i (photograph by Mark Merlin).

have been replaced with useful introduced, subsistence-oriented food species such trees of coconut (*Cocos nucifera*), breadfruit (*Artocarpus altilis*), and Tahitian chestnut (*Inocarpus fagifer*).

During the last two centuries or so, much of the area of the traditional cultivated coastal forests of Remote Oceania has been converted to, more or less, single stands of commercial coconut and other plantations. Traditionally, native peoples have left much of the native strand vegetation and forest immediately adjacent to the coastline for protection from strong winds and storm surge. However, modern urban, industrial, military, and tourist development has resulted in the removal of all or most of the natural components of the coastal land ecosystems in many areas of the islands of Remote Oceania. Even on remote atolls, because of traditional farming practices, little remains of the native forests, except on those that remain uninhabited.

Mark D. Merlin

Bibliography

Baines, G.B.K., 1981. *Mangrove Resources and their Management in the South Pacific*. Noumea, New Caledonia: South Pacific Regional Environment Programme.

Borowitzka, M.A., 1981. Algae and grazing in coral reef ecosystems. *Endeavour*, **5**: 99–106.

Bortone, S.A. (ed.), 2000. *Seagrasses: Monitoring, Ecology, Physiology, and Management*. Boca Raton, Fla.: CRC Press.

Birkeland, C., 1997. Implications for resource management. In Birkeland, C. (ed.), *Life and Death of Coral Reefs*. New York: Chapman & Hall, pp. 411–435.

Carlquist, S., 1974. *Island Biology*. New York: Columbia University Press.

Chapman, V.J., 1976. *Mangrove Vegetation*. Vaduz: J. Cramer.

den Hartog, C., 1970. *The Sea-grasses of the World*. Amsterdam: North-Holland Publishing Co.

Field, C.D., 1995. *Journey amongst Mangroves*. Okinawa, Japan: International Society of Mangrove Ecosystems.

Grigg, R.W., 1983. Community structure, succession and development of coral reefs in Hawaii. *Marine Ecological Progress Series*, **11**: 1–14.

Grigg, R.W., and Dollar, S.J., 1990. Natural and anthropogenic disturbance on coral reefs. In Z.E. Dubinsky (ed.), *Coral Reefs: Ecosystems of the World*. Vol. 25. New York and Amsterdam: Elsevier Scientific Publishing Co. Inc., pp. 439–452.

Grigg, R.W., and Maragos, J.E., 1974. Recolonization of hermatypic corals on submergedlava flows in Hawaii. *Ecology*, **55**: 387–395.

Gulko, D., 1998. *Hawaiian Coral Reef Ecology*. Honolulu: Mutual Publishing.

Hamilton, Lawrence, S., and Snedaker, Samuel, C. (eds.), 1991. *Handbook for Mangrove Area Management*. New York and Honolulu: United Nations Environment Programme and East-West Center, Environment and Policy Institute.

Hunter, C.L., and Evans, C.W., 1995. Coral Reefs in Kaneohe Bay, Hawaii: two centuries of western influence and two decades of data. *Bulletin of Marine Science*, **57**: 501–515.

Jokiel, P.L., 1987. Ecology, Biogeography and evolution of corals in Hawaii. *Trends in Ecology and Evolution*, **2**: 179–182.

Juvik, S., and Juvik, J.O., 1998. *Atlas of Hawaii*. Honolulu: University of Hawaii Press.

Kirch, P.V., 2000. *On the Road of the Winds: An Archaeological History of the Pacific Islands before European Contact*. Berkeley: University of California Press.

Lugo, Ariel E., and Snedaker, S.C., 1974. The ecology of mangroves. *Annual Review of Ecology and Systematics*, **5**: 39–64.

Manner, H.I., 1987. Atoll flora and vegetation. *Alafua Agricultural Bulletin*, **12**: 67–80.

Maragos, J.E., Crosby, M.E., and McManus, J.W., 1996. Coral reefs and biodiversity: a critical and threatened relationship. *Oceanography*, **9**: 83–99.

Mathieson, C., and Nienhuis, P.H. (eds.), 1991. Seagrass ecosystems in the Tropical West Pacific. *Ecosystems of the World 24. Intertidal and Littoral Ecosystems*. New York: Elsevier.

McClanahan, T.R., 1997. Primary succession of coral-reef algae: differing patterns on fished versus unfished reefs. *Journal of Experimental Marine Biology and Ecology*, **218**: 77–102.

Merlin, M., 2000. A history of ethnobotany in Remote Oceania. *Pacific Science*, **54**: 275–287.

Merlin, M., and Juvik, J., 1996. *Irá me Neeniier nón Chuuk: Plants and their Environments in Chuuk*. Honolulu: Program on Environment, East-West Center.

Merlin, M., Capelle, A., Keene, T., Juvik, J., and Maragos, J., 1997. *Keinikkan Im Melan Aelon Kein: Plants and Environments of The Marshall Islands*. Honolulu: Program on Environment, East-West Center.

Merlin, M., Jano, D., Raynor, W., Keene, T., Juvik, J., and Sebastian, B., 1992. *Tuhke en Pohnpei: Plants of Pohnpei*. Honolulu: Environment and Policy Institute, East-West Center.

Merlin, M., Taulung, R., and Juvik, J., 1993. *Sahk Kap Ac Kain In Acn Kosrae: Plants and Environments of Kosrae*. Honolulu: Program on Environment, East-West Center.

Moore, J.G., Normark, W.R., and Holcomb, R.T., 1994. Giant Hawaiian underwater landslides. *Science*, **264**: 46–47.

Mueller-Dombois, D., and Fosberg, F.R., 1998. *Vegetation of the Tropical Pacific Islands*. New York: Springer Press.

National Biodiversity Team of the Republic of the Marshall Islands, 2000. *The Marshall Islands—Living Atolls Amidst the Living Sea, The National Biodiversity Report of the Republic of the Marshall Islands*. Santa Clarita, CA: St. Hildegard Publishing Company.
Por, F.D., and Dor, I. (eds.), 1984. *Hydrobiology of the Mangal: the Ecosystem of the Mangrove*. The Hague and Boston: W. Junk.
Richmond, R.H., 1993. Coral reefs: present problems and future concerns resulting from anthropogenic disturbance. *American Zoologist*, **33**: 524–536.
Stemmerman, L., 1981. *A Guide to Pacific Wetland Plants*. Honolulu: US Army Corps of Engineers.
Steven, A., and Larkum, A.W.D., 1993. Encore: the effects of nutrient enrichment on coral reefs. *Search*, **24**(8): 216–220.
Stoddart, D.R., 1992. Biogeography of the Tropical Pacific. *Pacific Science*, **46**(2): 276–293.
Teas, H.J. (ed.), 1983. *Biology and Ecology of Mangroves*. The Hague and Boston: Dr. W. Junk.
Tomlinson, P.B., 1986. *The Botany of Mangroves*. Cambridge, UK: Cambridge University Press.
Whistler, W.A., 1994. *Flowers of the Pacific Islands Seashore*. Honolulu: Isle Botanica.

Cross-references

Atolls
Changing Sea Levels
Climate Patterns in the Coastal Zone
Coastal Subsidence
Coral Reef Islands
Coral Reefs
Coral Reefs, Emerged
Estuaries
Greenhouse Effect and Global Warming
Holocene Coastal Geomorphology
Human Impact on Coasts
Mangroves, Ecology
Natural Hazards
Pacific Ocean Islands, Coastal Geomorphology
Salt Marsh
Sea-Level Rise, Effect
Small Islands
Storm Surge
Uplift Coasts
Volcanic Coasts
Wetlands

PACIFIC OCEAN ISLANDS, COASTAL GEOMORPHOLOGY

Scattered across one-third of the earth's surface, Pacific Islands have some of the world's most fascinating landscapes (Menard, 1986; Nunn, 1994, 1998a). Yet notwithstanding their common attributes (comparatively small size, remoteness, young geology, topographic and structural simplicity, simple climate, and soil patterns) it would be wrong to suppose that Pacific Island coastal landforms are uniform within this vast area.

Much variation stems from latitude. Tropical Pacific Islands within the belt of tropical cyclones (hurricanes, typhoons) generally have coastal landforms which manifest the occasional impact of high-energy waves and winds while those outside this belt may not.

The primary cause of coastal variations in geomorphology lies with island type, of which three major kinds can be recognized in the Pacific;

(1) volcanic islands, often slowly subsiding, generally high, well vegetated with broad coastal plains around river mouths;
(2) high limestone islands, often rising, commonly cliffed, dense vegetation often only locally, with little lowland close to the shore; and
(3) atoll islands, usually rising no more than 3 m above mean sea level, made largely from unconsolidated materials accumulated on reef flats.

Most Pacific Islands are concentrated in the ocean's southwest quadrant because this is where the processes largely responsible for island formation dominate. Most Pacific Islands also occur within the tropics where coral reefs grow. Within this broad picture, volcanic islands generally occur either in arcs parallel to convergent plate boundaries or in chains in intraplate (midplate) locations. Most high limestone islands are either associated with (former) lines of plate convergence or places where flexure of the intraplate lithosphere has taken place. Most atolls occur in tropical intraplate locations marking the places where volcanic islands have sunk.

Coastal geomorphology

Owing to the approximately 120-m (postglacial) rise in sea level between the Last Glacial Maximum some 17,000–21,000 years ago and 5,000–4,000 years ago, all Pacific Island coasts exhibit signs of drowning. For many years, it was considered by some that the "Micronesia Curve" of sea-level rise, which supposed sea level to have been rising continuously since the Last Glacial Maximum (Bloom, 1970), was applicable to the Pacific Islands. More recent work has shown that instead, throughout this region, sea level reached a maximum around 5,000–4,000 years ago and has since fallen 1–2 m (Figure P9; Nunn, 1995). Thus, superimposed on

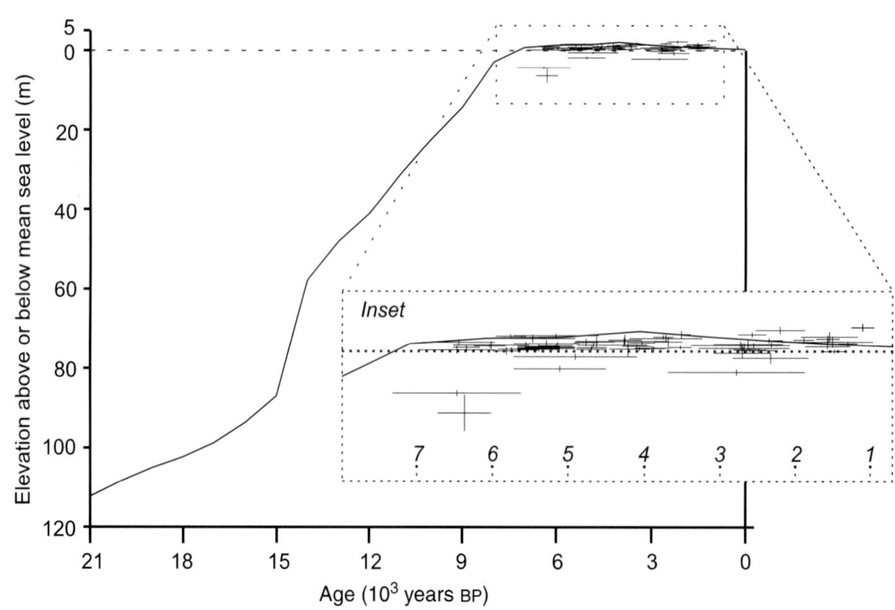

Figure P9 The course of postglacial sea-level changes in the Fiji Islands (after Nunn and Peltier, 2001). The solid line is the predicted sea-level change from the ICE-4G model. The data points are paleosea-level indicators from Fiji, enlarged in the inset to show the general concordance between empirical and predicted data.

the drowned nature of many coasts are the effects of a slight recent emergence. Superimposed on this in turn are the effects of tectonics (land-level changes) which affect many islands, particularly those along convergent plate boundaries.

Postglacial warming of ocean surface waters led to the recolonization of many nearshore and shallow areas by corals and associated reef organisms. As sea level continued to rise, corals grew upwards so that the reef surface remained within the photic zone. In some places, such as Tarawa in Kiribati and parts of French Polynesia, coral reefs apparently managed to "keep up" with rising sea level (Nunn, 1994). Elsewhere, as in most parts of the tropical southwest Pacific, for example, such reefs grew upwards at slower rates yet managed to "catch up" to sea level when its postglacial rise slowed around 6,000–5,000 years ago. In a few places, notably to the west of the islands of Samoa, reefs believed to have been established at sea level during the early postglacial were unable to grow upwards fast enough and are not visible at the surface today. These are characterized as "give up" reefs in the nomenclature of Neumann and McIntyre (1985).

The importance of coral reefs in the dynamics of modern coastlines in the tropical Pacific Islands cannot be overstated. Many beaches and sand islands are made solely from sediment created on reefs and, where those reefs become degraded and cease to be productive, beaches and sand islands can become severely eroded.

There has been a regrettable tendency to explain all environmental changes within the postsettlement history of Pacific Island coasts by human actions. While this may be true in large part of the last 100 years or so, there is clear evidence that at earlier times natural changes overwhelmed human endeavors. One of the most significant such occasions was the "AD 1300 event" in which a slight cooling, sea-level fall and increased El Niño frequency, caused widespread reef-surface death and infilling of coastal embayments, impeded lagoon-water circulation, and invoked major societal responses in Pacific Island societies (Nunn, 2000a; Nunn and Britton, 2001).

The following sections look at Pacific Island coastal geomorphology for each of the three major island types identified above.

Coastal geomorphology of volcanic islands

All oceanic islands in the Pacific began life as ocean-floor volcanoes but not all retain a recognizable volcanic form. Those which do are commonly young (<5 million years old) and/or in intraplate locations. Except around the mouths of large rivers, such islands typically have narrow coastal plains and, beneath their fringing reef platforms (where these exist), plunge steeply offshore. Most such coastal plains were formed only after the sea reached its present level 5,000–4,000 years ago and stabilized long enough (after its long postglacial rise) to erode platforms along island coasts. When sea level fell subsequently, these platforms emerged and became draped with sediment to form the coastal plains which have been the favored sites for human habitation in the Pacific Islands throughout the late Holocene (Figure P10).

Late Holocene sea-level fall also led to the outgrowth of both coral reefs and those coastlines composed of unconsolidated sediments, the infilling of coastal lagoons with sediment (both terrigenous and marine), and the subsequent establishment of characteristic ecosystems such as mangrove swamps (see below).

There is considerable controversy about the precise impact of early humans on Pacific Island landscapes, some regarding it as considerable (Kirch and Hunt, 1997), others as generally slight (Nunn, 1999a, 2001). Typical of the former is the scenario constructed for Aneityum Island in Vanuatu, where the earliest settlers are believed to have settled in upland areas (the valley floors then being too swampy), burned the forest which existed there, thus releasing large quantities of soil and sediment into the valleys which, after about 1,000 years of occupation, filled them up sufficiently for them to be settled by the people. The latter explanation sees humans as the passive agents in a landscape which was being changed solely by emergence of the land, in this case as a result of sea-level fall and tectonic uplift.

Coastal geomorphology of high limestone islands

High limestone islands are commonest in the Pacific along convergent plate boundaries where tectonic processes are generally most active and most variable in time and space. Typically such islands were formerly subsiding volcanic islands which developed a reefal capping before being uplifted. A good example are the Lau Islands of eastern Fiji which rise from a volcanic arc which became inactive and began subsiding some 5 million years ago. Subsequent uplift in the Plio–Pleistocene, perhaps due to the heating-up of a detached, partly subducted slab of lithosphere, elevated these islands as much as 315 m above sea level

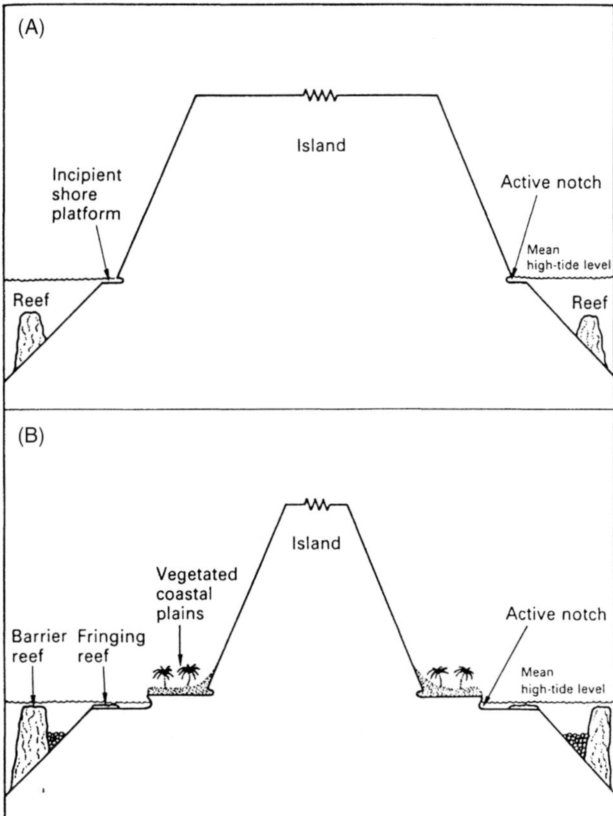

Figure P10 Formation of coastal plains around volcanic islands during the Holocene (from Nunn, 1994). (A) The time of the Holocene sea-level maximum. Sea level had been rising for the previous 11,000–13,000 years so little lateral erosion of the coastline had been accomplished. Reefs struggled to catch up with rising sea level; until they did so, a high-energy window was open. Potential settlers would have found this island type unattractive. (B) Following a fall of sea level in the late Holocene, the shore platforms cut at the higher sea level would have emerged and become covered with terrigenous and marine detritus. Barrier reefs would have caught up with sea level, the high-energy window would have been closed so that coastline erosion would have become relatively subdued. Fringing reefs would have developed and the lagoon become shallower. Potential settlers would have found this island type comparatively attractive.

(Nunn, 1998b). Another example is the island Niue, which has been gradually rising up a flexure in the lithosphere east of the Tonga Trench for around 500,000 years, and is a fine example of an emerged atoll with the former ring reef and atoll lagoon well preserved 70 m above sea level (Hill, 1996).

Owing to a lack of running water and thus only comparatively little subaerial fluvial development, most high limestone island coasts are cliffed, often riddled with long sub-horizontal epiphreatic caves marking former stands of the water table (controlled by sea level). Coastal embayments on such islands tend to be few and formed from the drowning of karstic hollows or collpased caves.

A variant on high islands made solely from limestone is the *makatea* islands in which a volcanic island is fringed by high limestone, the products of fringing-reef uplift (Nunn, 1994). Examples abound in the southern Cook Islands. Their coasts are similar to those of other high islands, one difference being that because of the fertility of their inland areas, they tend to have been comparatively densely populated so that human modifications to their coasts are sometimes considerable.

Coastal geomorphology of atoll islands

Atoll (and barrier-reef) islands—narrow strips of land formed on broad atoll (or barrier) reef flats from the accumulation of associated sediments—are entirely coastal. Rarely rising more than 3 m above sea level, atoll islands generally develop on the windward sides of atoll reefs. In the tropical Pacific, most atolls lie in the belt of easterlies and thus have islands exclusively along their eastern sides; good examples are Kapingamarangi Atoll in Micronesia and Tarawa Atoll in Kiribati. Some atolls are occasionally affected by storm waves coming from the opposite direction to windward, and thus display islands of coarser sediment along their leeward sides than those to windward; examples include Nanumaga Atoll in Tuvalu (McLean and Hosking, 1991). Most atoll reefs enclose a shallow subcircular lagoon and in those cases where the lagoon is small relative to sediment supply, it may become infilled; such is the case with Nui and Vaitupu Atolls in Tuvalu (McLean and Hosking, 1991).

Atoll coasts are among the most vulnerable in the Pacific because of their often/largely unconsolidated character. Ephemeral atoll islands named cays can be distinguished from those which are more permanent named *motu* (Nunn, 1994). *Motu* are atoll islands which owe their durability to the formation of various rock armors such as beachrock, conglomerate platforms (*pakakota*), or phosphate rock (Figure P11). *Pakakota* occur widely in the Tuamotus. Even more enduring among such islands are those which have experienced a degree of emergence (uplift and/or sea-level fall) and thus have a core of fossil reef within the sediment pile. Many of the larger atoll islands, such as several in Kiribati and the Marshall Islands, may have such emerged-reef cores.

Tectonic controls on island coastal geomorphology

Islands which are sinking or rising comparatively rapidly exhibit distinctive coastal landforms. Subsiding islands generally show an unusual degree of coastal embayment, as recognized first by Dana (1872). Where such islands are composed of a single volcano, drained radially, a characteristic stellate island may form; the island Ono in the Kadavu group of southern Fiji is a classic example. Faulted coasts which are subsided may well be quite straight; they are particularly common along the sides of young high subaerial volcanic islands, as in parts of Samoa. Although large deltas are understandably rare on Pacific Islands, where they occur, they are usually associated with local subsidence. In such instances, sediment supply to the delta front usually exceeds the effects of subsidence, and characteristic delta forms result. Examples include the Rewa Delta on Vitilevu Island in Fiji.

Uplifting islands in the tropical Pacific are commonly marked by staircases of emerged coral reefs, comprising fossil reef surfaces separated by terrace risers. Dating of reef staircases on the Huon Peninsula in Papua New Guinea (Chappell, 1974) and elsewhere in the southwest Pacific have given rise to precise chronologies of sea-level and tectonic change during the Quaternary. Most coasts of uplifting islands are cliffed or have only very slightly developed coastal plains. Many of these appear to have been formed only during the period of comparative (land–sea) stability between bursts of uplift, commonly associated with large earthquakes and are thus termed coseismic. Coseismic uplift is known to affect many islands along convergent plate boundaries in the southwest and western Pacific. Coseismic uplifts with average magnitudes of 0.74 m affect Tongatapu Island in Tonga approximately every 870 years (Nunn and Finau, 1995). The Fiji island Vatulele is thought to experience coseismic uplift of some 1.8 m approximately every 1,400 years (Nunn, 1998a).

Figure P11 Beachrock exposed along the northern (leeward) coast of Eluvuka (Treasure) Island in the Mamanuca Group of western Fiji. Exposure of this beachrock is thought to be due to erosion of the beach in successive tropical cyclones during the 1990s followed by persistent northwesterly winds which have prevented the return of sand cover [digital photo 26—P. Nunn].

Challenges for the 21st century

Pacific Islands and the submarine ridges from which many rise are among the steepest structures on earth and, while the ocean contributes to their stability, it cannot prevent occasional structural failures. Owing to their infrequency, the nature and (potential) effects of such failures have become clear only recently. A groundbreaking study of the structural failure of Johnston Atoll (Keating, 1987) was followed by others, summarized by Keating and McGuire (2000). It has been estimated that a major oceanic island flank failure occurs once every 25 years (Siebert, 1992). Flank collapses along the Hawaiian Ridge have been studied and dated using deeply submerged fossil reefs (Moore and Moore, 1984; Moore *et al.*, 1994). Such failures/ collapses are clearly capable of causing the disappearance of parts of islands, perhaps even whole islands, and producing giant waves which could wrought major changes to Pacific coasts. Their study is of considerable applied interest.

Pacific Island coasts composed of unconsolidated sediments are particularly vulnerable to change as sea level changes. While short-term changes, particularly during El Niño events, can have severe consequences for reef health, it is longer-term sea-level rise, particularly within the past 200 years or so, that has brought about significant changes to many Pacific Island coasts. Based on largely anecdotal evidence (more formal sources generally being absent from this region), it is clear that some coasts have been inundated and/or eroded laterally by as much as 200 m within the last 100 years. Recent warnings have been issued about the disappearing beaches of the Pacific Islands (Coyne *et al.*, 1996; Nunn, 1999b).

Such problems have been compounded by the human inhabitants of Pacific Island coasts. Since humans came to occupy most Pacific Islands in larger numbers and to make more demands on their coasts, so their landscapes have been modified in consequence (Nunn *et al.*, 1999). Much coastal vegetation has been removed and/or replaced by less suitable species. Mangroves in particular have been a major casualty of 20th century "development" in the Pacific Islands; the case studies of Ovalau and Moturiki Islands in Fiji show how those settlements which have removed their fringing mangrove forests now have severe problems of coastal erosion and inundation compared to those which preserved them (Nunn, 2000b).

Patrick D. Nunn

Bibliography

Bloom, A.L., 1970. Holocene submergence in Micronesia as the standard for eustatic sea-level changes. *Quaternaria*, **12**: 145–154.

Chappell, J., 1974. Geology of coral terraces, Huon Peninsula, New Guinea: a study of Quaternary tectonic movements and sea level changes. *Geological Society of America, Bulletin*, **85**: 553–570.

Coyne, M., Mullane, R., Fletcher, C., and Richmond, B., 1996. Losing Oahu: erosion on the Hawaiian coast. *Geotimes*, December 1996: 23–26.

Dana, J.D., 1872. *Corals and Coral Islands*. New York: Dodd, Mead.

Hill, R.J., 1996. Niue and adjacent seamounts. In Suva, Fiji: South Pacific Applied Geoscience Commission. Meylan, M.A., and Glasby, G.P. (eds.), *Manihiki Plateau, Machias and Capricorn Seamounts, Niue, and Tonga Trough: Results of Tui Cruises*, Vol. 10. SOPAC Technical Bulletin, pp. 31–44.

Keating, B., 1987. Structural failure and drowning of Johnston Atoll, Central Pacific Basin. In Keating, B.H., Fryer, P., Batiza, R., and Boehlert, G.W. (eds.), *Seamounts, Islands and Atolls*. Washington: American Geophysical Union (Monograph 45): 49–59.

Keating, B.H., and McGuire, W.J., 2000. Island edifice failures and associated tsunami hazards. *Pure Applied Geophysics*, **157**: 899–955.

Kirch, P.V., and Hunt, T.L. (eds.), (1997). *Historical Ecology in the Pacific Islands: Prehistoric Environmental and Landscape Change*. New Haven, CT: Yale University Press.

McLean, R.F., and Hosking, P.L., 1991. Geomorphology of reef islands and atoll motu in Tuvalu. *South Pacific Journal of Natural Science*, **11**: 167–189.

Menard, H.W., 1986. *Islands*. New York: Scientific American Books.

Moore, J.G., and Moore, G.W., 1984. Deposit from a giant wave on the island of Lanai, Hawaii. *Science*, **226**: 1312–1315.

Moore, J.G., Bryan, W.B., and Ludwig, K.R., 1994. Chaotic deposition by a giant wave, Molokai, Hawaii. *Geological. Society of America, Bulletin*, **106**: 962–967.

Neumann, A.C., and MacIntyre, I., 1985. Reef response to sea-level rise: keep-up, catch-up or give-up. In *Proceedings of the 5th International Coral Reef Congress*, **3**: 105–110.
Nunn, P.D., 1994. *Oceanic Islands*. Oxford: Blackwell.
Nunn, P.D., 1995. Holocene sea-level changes in the South and West Pacific. *Journal of Coastal Research*, Special Issue **17**: 311–319.
Nunn, P.D., 1998a. *Pacific Island Landscapes*. Suva: Institute of Pacific Studies.
Nunn, P.D., 1998b. Late Quaternary tectonic change on the islands of the northern Lau-Colville Ridge, southwest Pacific. In Stewart, I.S., and Vita-Finzi, C. (eds.), *Coastal Tectonics*, Vol. 146. Geological Society, London, Special Publications, pp. 269–278.
Nunn, P.D., 1999a. *Environmental Change in the Pacific Basin: Chronologies, Causes, Consequences*. London: Wiley.
Nunn, P.D., 1999b. Pacific Island beaches—a diminishing resource? *Asia-Pacific Network for Global Change Research (APN) Newsletter*, **5**(2): 1–3.
Nunn, P.D., 2000a. Environmental catastrophe in the Pacific Islands about AD 1300. *Geoarchaeology*, **15**: 715–740.
Nunn, P.D., 2000b. Coastal changes over the past two hundred years around Ovalau and Moturiki Islands, Fiji: implications for coastal-zone management. *Australian Geographer*, **31**: 21–39.
Nunn, P.D., 2001. Ecological crises or marginal disruptions: the effects of the first humans on Pacific Islands. *New Zealand Geographer*, **57**: 11–20.
Nunn, P.D., and Britton, J.M.R., 2001. Human–environment relationships in the Pacific Islands around AD 1300. *Environment and History*, **7**: 3–22.
Nunn, P.D., and Finau, F.T., 1995. Late Holocene emergence history of Tongatapu island, South Pacific. *Zeitschrift für Geomorphologie*, **39**: 69–95.
Nunn, P.D., and Peltier, W.R., 2001. Far-field test of the ICE-4G (VM2) model of global isostatic response to deglaciation: empirical and theoretical Holocene sea-level reconstructions for the Fiji Islands, Southwest Pacific. *Quaternary Research*, **55**: 203–214.
Nunn, P.D., Veitayaki, J., Ram-Bidesi, V., and Vunisea, A., 1999. Coastal issues for oceanic islands: implications for human futures. *Natural Resources Forum*, **23**: 195–207.
Siebert, L., 1992. Threats from debris avalanches. *Nature*, **356**: 658–659.

Cross-references

Atolls
Beachrock
Cays
Changing Sea Levels
Cliffed Coasts
Cliffs, Erosion Rates
Coral Reef Coasts
Coral Reefs Islands
Coral Reefs
Coral Reefs, Emerged
Faulted Coasts
Mangroves, Ecology
Mangroves, Geomorphology
Pacific Ocean Islands, Coastal Ecology
Submerged Coasts
Submerging Coasts
Uplift Coasts
Volcanic Coasts

PALEOCOASTLINES

As R.W. Fairbridge (1992) noted, "Paleogeography is an exercise in imaginative insight. One needs first to observe and understand the dynamic physical processes and landforms and then transfer and reapply these concepts to past landscapes." In essence, the study of paleocoastlines requires the application of fundamental concepts of geology. *Uniformitarianism* and its modern version *actualism* provide the conceptual basis for modern sedimentary environmental patterns to be used in analogous comparisons with older facies in the construction of paleogeographies. Equally Walther's dictum on *the correlation of sedimentary facies* provides the conceptual basis for differentiating *possible* conformable lateral and vertical facies patterns from the *improbable* in the delineation of ancient coastal landscapes. Climate, including glacial and other cycles, tectonics and eustasy are long-term forcing factors. Nevertheless, knowledge of local relative sea-level fluctuations are requisite to precise paleogeographic reconstruction.

Major advances in paleoenvironmental reconstructions in the mid to late 20th century were driven by the needs of the petroleum and coal industries for precision in sedimentary facies studies. For instance, Harlan Fisk, and later his students Hugh Bernard and Rufus LeBlanc, pioneered in intensive three-dimensional analysis of fluvial, deltaic, and coastal/nearshore marine facies. Emphasis was placed on sedimentary processes and depositional geometries of the various sedimentary environmental lithosomes, including lateral and vertical conformable and disconformable relationships. Studies of sedimentary environmental lineaments, geometries, internal sedimentary structures, and grain sizes and composition coupled with floral and faunal (fossil) analyses and ^{14}C dates therefrom allow precise definition of ancient landforms, physical processes, water salinities, and their lateral and vertical distributions. These factors are a requisite to the accurate reconstruction of ancient coastal landscapes. Kraft and Chrzastowski (1985); Belknap *et al.* (1994), and Kraft *et al.* (1987) as well as many other papers in the three referenced volumes of the Society for Sedimentary Geology (SEPM) provide many modern examples of sedimentary facies studies relevant to paleocoastline delineation.

Interdisciplinary study of paleocoastlines

Perhaps nowhere are paleocoastline morphologies of more interest than in the environs of historical and archaeological sites. Here, archaeology, history, and the Classics both complement and supplement each other in a synergistic manner clearly more effective than the sums of the disciplines applied separately. In the Aegean Sea, on the western coastline of Anatolia, sea level rose to its present position about 6,000 BP. Since then, sea level dropped 1–2 m relative to land, rising again to its present level. From the Würm glacial low stand to 6,000 BP fluvial and coastal environments were transgressed by nearshore marine environments, infilling valleys incised during Würm and earlier times. Since then, the Küçük Menderes (ancient Cayster River) has prograded 15 km seaward burying the marine sediments of the ancestral Gulf of Ephesus with a thin veneer of fluvial deltaic and barrier accretion plain sediments. The impact of this major change in coastal configuration had a profound effect on the peoples occupying the region from Neolithic time to present. Since the arrival of Ionian Greek colonists 3,000 years BP, the residents of the ancient city of Ephesus, its many harbors over 3 millennia of time, the Artemision (Temple of Diana of the Ephesians, one of the seven wonders of the ancient world), and the later Turkish peoples of the region have been continually adapting to major changes in coastal environmental morphologies (Kraft *et al.*, 2000).

The floodplain and delta of the Küçük Menderes have been much altered by 20th century drainage and irrigation works. The coastal barriers form a damming effect that leads to widespread seasonal swamps and short-term floods in the wet season. Many palimsests of morphic features may be observed from air photo, satellite imagery, and correlated with a carefully documented Austrian survey in the late 19th century (Figure P12). Coastal barrier accretion ridges only occur in the lower 2.5 km of the delta-floodplain. Thus, in much of the earlier, relatively sheltered marine embayment from 6,000 BP to Late Byzantine time (ca. 1,300 AD) fluvial-deltaic progradation was into a shallow marine prodelta shelf, probably with birdfoot like distributaries and flanking swamps, ponds, and marshes.

Details of sedimentary environmental lithosome geometries, sediment characteristics, micro- and macro-faunal and ^{14}C dates were attained by an intensive coring program (over 120 drill cores, Kraft *et al.*, 2000, 2001). Figure P13 is a schematic cross section based on 26 drill cores in the vicinity of the Artemision (Temple of Diana) constructed in the 7th century BC and finally destroyed by earthquake in the 3rd century AD. The Artemision was built on an alluvial fan of a flanking river and eventually, after destruction, buried therein. However, we can also show that the site of the Artemision was at one time, pre-3,000 BP, a coastline of the marine embayment. Therein lies an interplay with legend. *Callimachus: Hymn III to Artemis* tells us that "Amazons, lovers of battle set up a wooden image under an oak, in seaside Ephesos (II) and hippo offered a holy sacrifice to you; around the oak they danced you a war dance . . . Afterward around that wooden image, wide foundations were built . . . ." Our paleogeographic evidence confirms the coastline under the Artemision site in Bronze Age times. Further, we can position this "wonder of the ancient world" in a coastal setting overlooking the ancient Gulf of Ephesus, as late as Hellenistic time, ca. 200 BC. Eventually, two sets of alluvium inundated and buried the ruins of the Temple of Artemis (as shown in Figure P13).

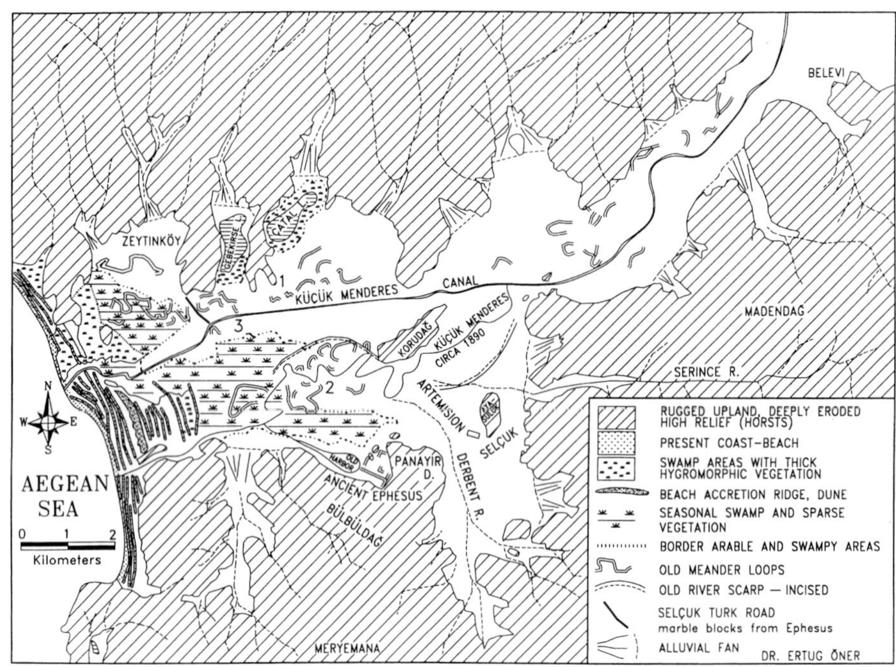

Figure P12 Holocene Epoch geomorphic elements of the Küçük Menderes (ancient Cayster River) in western Anatolia (from Kraft *et al.*, 2000. With permission, Österreichischen Archäologisches Institut.).

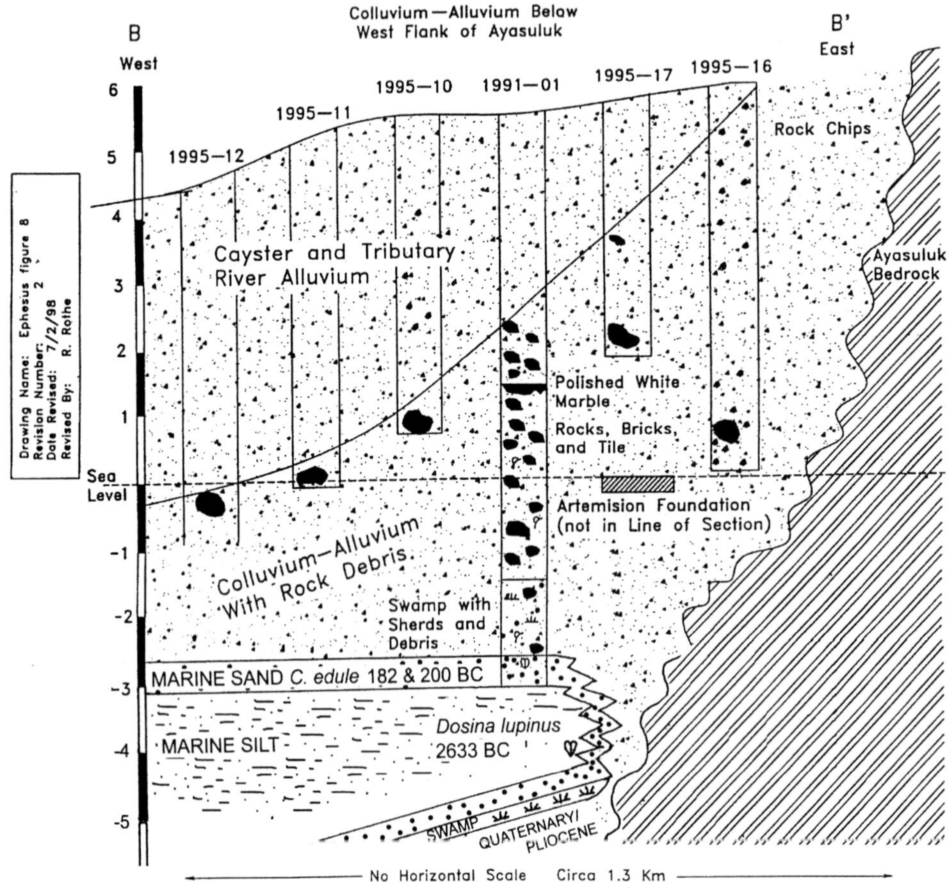

Figure P13 A schematic cross-section through the environs of the Artemision (Temple of Diana of the Ephesians, one of the Seven Wonders of the Ancient World). Constructed mid-7th Century BC, destroyed by earthquake 3rd Century AD, based on 26 drill cores (from Kraft *et al.*, 2000. With permission, Österreichischen Archäologisches Institut.).

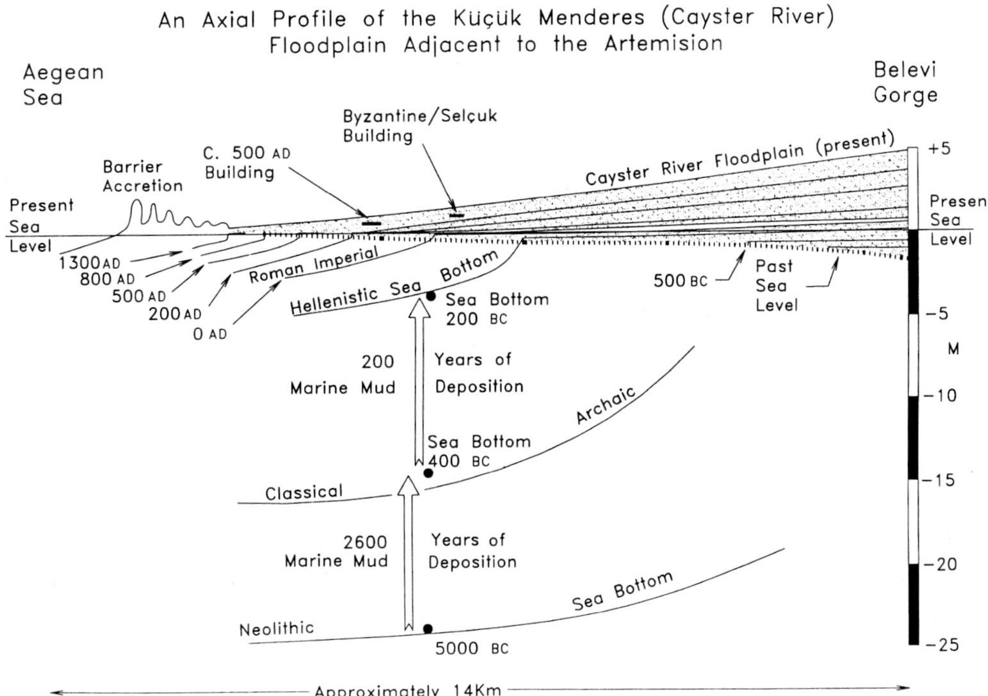

Figure P14 Cross-section of the lower Küçük Menderes (Cayster River) delta-floodplain. Seven thousand years of depositional valley infill, including 25 m of marine/estuarine sediments covered by a thin veneer of prograding floodplain alluvium and barrier accretion plain sands (from Kraft et al., 2000. With permission, Österreichischen Archäologisches Institut.).

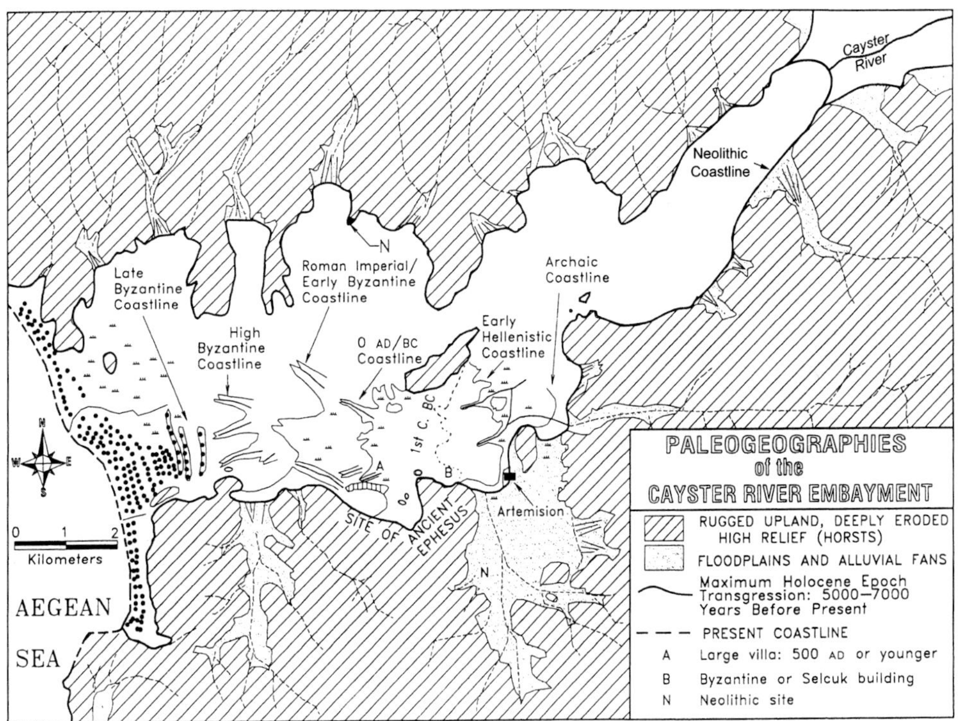

Figure P15 Paleocoastline progradation in the Küçük Menderes (Cayster River) embayment of the Aegean Sea in the vicinity of the ancient city of Ephesus and the Artemision (after Kraft *et al.*, 2001).

Using drill core evidence of sedimentary facies in their lateral and vertical distribution, we can show that the major portion of the Holocene strata that infilled the ancestral valley of the Küçük Menderes River was deposited in an open marine embayment, initially clear water, followed by distal and nearshore prodelta silts (Figure P14). Deposition was more rapid in the past 2,400 years, as the delta prograded and alluvium aggraded. A few archaeological sites in the floodplain provide minimal dates for the delta advances.

Figure P15 is a summary delineation of delta coastline progradation for the southern portion of the former marine embayment. The Neolithic coastline shows the maximum extent of the marine embayment of mid-Holocene time, ca. 6,000 BP, while, the Archaic coastline of 600 BC continues in an irregular configuration of distributaries, marsh, swamps, etc. until significant wave generated littoral transport of sands in barrier accretion started ca. AD 1,300. The city of Ephesus varied in location over 4 km along the southern flank of the embayment, eventually being built on portions of the delta including massive fill. Many documents attest to the "conflict or fight" with the river progradation over more than a millennium of time.

One of the clearest examples of negative environmental impact is a quote from Strabo (XIV 1, 24), "The city has both an arsenal and a harbor. The mouth of the harbor was made narrower by the engineers, but they, along with the King who ordered it, were deceived as to the result. I mean Attalus II Philadelphus (159–138 BC); for he thought that the entrance would be deep enough for large merchant vessels—as also the harbor itself, which formerly had shallow places because of silt deposited by the Cayster River—if a mole were thrown up at the mouth, which was very wide, and therefore ordered that the mole should be built. But the result was the opposite, for as the silt, thus hemmed in, made the whole of the harbor, as far as the mouth, more shallow. Before this time the ebb and flow of the tide would carry away the silt and draw it to the sea outside."

The literature abounds with writings of the problems that the ancient Ephesian peoples had with maintenance of their harbors over the millennia of existence of Ephesus, capital of the ancient Kingdom of Asia (Kraft *et al.*, 2000, 2001). Clearly interdisciplinary approaches to the study of paleocoastlines increases our abilities to precisely delineate ancient coastal configurations.

John C. Kraft

Bibliography

Belknap, D.F., Kraft, J.C., and Dunn, R.K., 1994. Transgressive Valley-Fill Lithosomes: Delaware and Maine. In Dalyrymple, R.W., and Zaitlin, B.A. (eds.), *Incised Valley Systems: Origin and Sedimentary Sequences*. Tulsa: SEPM (Society for Sedimentary Geology), SEPM Special Publication No. 51, pp. 304–320.

Fairbridge, R.W., 1992. Holocene marine Coastal Evolution in the United States. In Fletcher, C.H., III, and Wehmiller, J.F. (eds.), *Quaternary Coasts of the United States: Marine and Lacustrine Systems*. Tulsa: SEPM (Society for Sedimentary Geology), SEPM Special Publication No. 48, pp. 9–20.

Kraft, J.C., and Chrzastowski, M.J., 1985. Coastal Stratigraphic Sequences. In Davis, R. A. (ed.), *Coastal Sedimentary Environments*, 2nd edn. New York: Springer-Verlag, pp. 626–663.

Kraft, J.C., Chrzastowski, M.J., Belknap, D.F., Toscano, M.A., and Fletcher, C.H., III, 1987. The Transgressive barrier–lagoon coast of Delaware: morphostratigraphy, sedimentary sequences and responses to relative rise in sea level. In Nummedal, D., Pilkey, O.H., and Howard, J.D. (eds.), *Sea-Level Fluctuation and Coastal Evolution*. Tulsa: Society of Economic Paleontologists and Mineralogists, Special Publication No. 41, pp. 129–143.

Kraft, J.C., Kayan, I, Brückner, H., and Rapp, G., Jr., 2000. Geologic Analysis of ancient landscapes and the harbors of ancient Ephesus and the Artemision in Anatolia. Ergänzungshefte, Jahresheften des Österreichischen Archäologisches Institut, **69**: 175–233.

Kraft, J.C., Kayan, İ, and Brückner, H., 2001. The geological and paleogeographical environs of the Artemision. In Muss, U. (ed.), Der Kosmos der Artemis von Ephesos, Wien: Österreichischen Archäologisches Institut, Sonderschriften Band 37, 123–133, Tafel 5–6.

Cross-references

Archaeological Site Location, Effect of Sea-Level Changes
Changing Sea Levels
Coastal Changes, Gradual
Coastal Sedimentary Facies
Coastal Warfare
Coastline Changes
Holocene Coastal Geomorphology
Mapping Shores and Coastal Terrain

PARAGLACIAL COASTS

Definition

Paraglacial coasts are defined as those occurring on or close to glaciated terrain, where glacial erosion features or deposits have a recognizable effect on the nature and evolution of coastal morphology and sediments (Forbes and Syvitski, 1994). They thus encompass a range of coastal environments and landforms from proglacial fjord and outwash coastlines to sand- and gravel-dominated coasts deriving their sediment supply from glacigenic deposits, but where the glacial imprint is older and subdued. Such coastlines also occur in settings where other phenomena such as *periglacial* and related processes (including permafrost, ground ice, thaw subsidence, sea ice) or *human impact* (*q.v.*) may appear to dominate the coastal landscape. Essential features of paraglacial coastlines include a sediment source or physiographic setting in which the glacial origin exerts a continuing influence on the form and evolution of the coast.

Geographic extent

The geographic range of paraglacial coasts is broadly coincident with the maximum extent of the last major continental glaciation in the *Pleistocene Epoch* (*q.v.*), approximately 20,000 years ago (Figure P16). The most extensive paraglacial coasts (broadly defined by ellipses in Figure P16) are those of the Arctic circumpolar region, including parts of the North American Pacific and Atlantic coasts, Greenland, Iceland, northwest Europe including the Baltic, and many parts of the Arctic coasts of Canada, Norway, and Russia. Most of the Antarctic coast, where not presently glaciated, forms an equivalent southern circumpolar

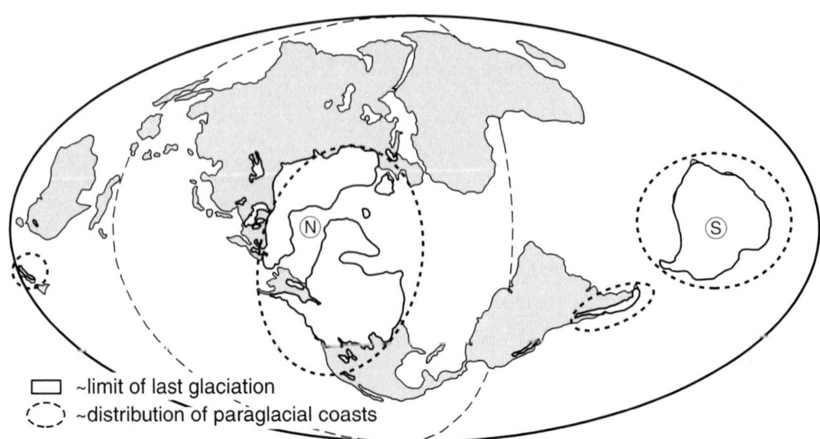

Figure P16 Approximate extent of last major glaciation and the geographic distribution of paraglacial coasts.

paraglacial region. Two other areas in the Southern Hemisphere showing extensive glacial influence on coastal development are the South Island of New Zealand and the west coast of South America in southern Chile, extending to Tierra del Fuego, where the southernmost coast of Argentina is also included.

Recognition and terminology

The term *paraglacial* was originally applied to river systems (Church and Ryder, 1972). In this context, it is used to describe the effects of disequilibrium excess sediment supply following deglaciation and the subsequent relaxation or decrease of sediment supply through surface reworking and downstream transport with time (cf. broken line in Figure P17B). While he did not use the term *paraglacial*, Johnson (1925) recognized the distinctive nature of coastal morphology directly arising from glacial erosion or indirectly sourced from glacial deposits such as moraines, drumlins, and outwash sediments. The term was adapted in the early 1980s for application to coasts where glacigenic sediment supply or morphology remain dominant controls (Forbes and Syvitski, 1994). Since that time, it has been most frequently employed in relation to coarse-clastic beaches and barriers sourced from glacial deposits (e.g., Forbes *et al.*, 1995; Orford *et al.*, 1996) but has also been applied to sediment sinks in coastal embayments (e.g., Carter *et al.*, 1992; Shaw and Forbes, 1992) as well as to glaciated shelves.

Classification

Paraglacial coasts can be classified into nine broad categories:

- sediment-starved coasts dominated by glacial erosion—fjords, overdeepened basins, and coastal barrens;
- depositional systems in confined fjord-head and proximal basin settings;
- open outwash coasts (not embayed or confined);
- basin-margin or open coasts with abundant glacigenic sediment in coastal cliffs;
- drumlin coasts characterized by sediment source switching and time-varying supply;
- embayed coasts with sediment supply controlled by sea-level change;
- coasts with patchy glacial cover or terraced outwash sediments—supply limitation or restricted access;
- forced-regressive coasts with limited sediment supply; and
- boulder-rich coasts in otherwise sediment-starved settings.

It is obvious from this list that the geomorphic expression may be highly variable and that sediment supply can range from negligible to abundant. However, apart from cases of significant fluvial or glaciofluvial sediment input (fjord-head deltas and outwash coasts), previous work on paraglacial systems has emphasized the rationing of sediment supply from glacigenic sources as a major control (and often the dominant influence) on coastal evolution in paraglacial settings. In this entry, we consider two examples of time-varying sediment supply and its implications for morphosedimentary outcomes at the coast.

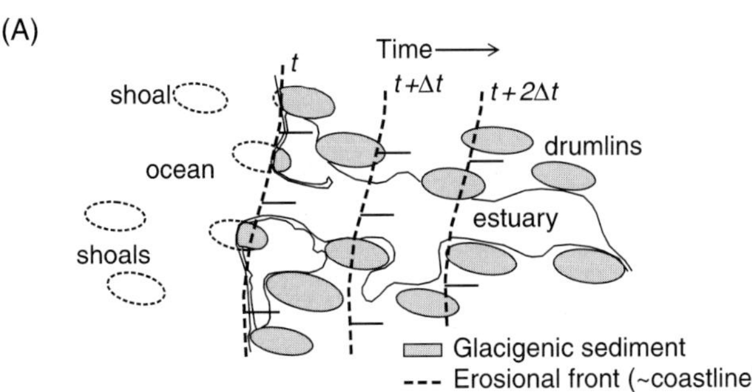

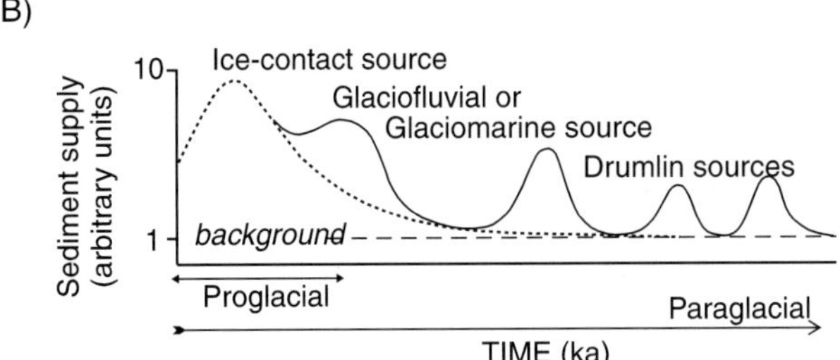

Figure P17 (A) Schematic illustration of marine transgression and passage of a coastal erosional front through a drumlin field. (B) Schematic time variation of sediment supply in paraglacial coastal systems (modified after Forbes and Taylor, 1987).

Sediment supply control

Coastal morphology and deposits are strongly dependent on the geological setting, climate, oceanographic environment, coastline orientation and exposure, and the rate and sign of relative sea-level change. Sediment supply is another critical variable, often in paraglacial settings the dominant factor defining the geomorphological response (Orford *et al.*, 1996). If sufficiently large, sediment supply may counteract the transgressive effects of relative sea-level rise and produce a regressive sequence on a prograding coast (Forbes *et al.*, 1995).

Persistent changes in relative sea level, related to postglacial *eustasy* (*q.v.*), *hydro-isostasy*, and *glacio-isostasy*, are near-ubiquitous on paraglacial coasts, where they exert a profound influence on coastal morphology and development, including control of delayed shore-zone access to glacigenic sediment sources. However sea-level change is neither fundamental nor unique to paraglacial environments and may occur on all types of coast throughout the world. The implications of these relative sea-level changes for coastal evolution in paraglacial systems include highly varying rates of landward retreat or seaward advance of the coastline, and significant variation in the rate of access to glacigenic sediment supplies.

On paraglacial coasts, sediment is predominantly derived from erosion of glacial deposits. In many cases these represent the only significant source. This has profound implications for the nature of coastal sedimentary facies and shore-zone morphology. Grain-size distributions in glacial deposits run the full gamut from clay to large blocks the size of houses. Glaciolacustrine and glaciomarine facies may contain a large proportion of clay and silt (glacial flour) with small proportions of sand and minor fractions of ice-rafted gravel. In other cases, the sediment source facies may include glaciofluvial outwash, till, or other ice-contact diamicts (Figure P17). The grain size in ice-contact sediments can range over more than six orders of magnitude (Forbes and Syvitski, 1994) and coarse clastic components (pebble and coarser) are often abundant. This leads to a preferential tendency for the development of gravel (including sand-pebble, cobble, and boulder) beaches on paraglacial coasts. Where the underlying bedrock is a soft sandstone or similarly friable lithology with a large proportion of sand, the ice-contact facies may be relatively sandy and large volumes of coastal sand derived from both bedrock and the overlying glacial units can feed extensive sandy beaches, barriers, nearshore bars, coastal dunes, and tidal inlet deposits, as in the southern Gulf of St. Lawrence.

The spatial disposition of source sediments on paraglacial coasts also exerts a major influence on patterns of coastal evolution. Landward movement of a coastal *erosional front* across a glaciated landscape (Figure P17A), as occurs with marine *transgression* under rising sea level, can lead to rapid mobilization of glacial deposits. The volume and spatial distribution of sediment may be discontinuous and vary widely, from a thin till veneer to larger volumes concentrated in distinctive glacial landforms such as eskers, kames, deltas, moraines, and drumlins. In such cases, the start of erosional incision (such as when barrier retreat exposes a back-barrier hillslope to wave action for the first time) can mark the beginning of significant sediment release. In the case of marine trangression through a drumlin field or multiple moraine complex, this process may involve repetitive injections of finite sediment volume into a coastal region (Figure P17B), spawning cyclic patterns of littoral-zone deposition with barrier growth, supply depletion, barrier breakdown, and landward transfer to begin the cycle over. This drumlin coast model has formed one of the dominant paradigms for understanding paraglacial coasts over the past 15 years (e.g., Boyd *et al.*, 1987; Carter *et al.*, 1990; Forbes *et al.*, 1995; Orford *et al.*, 1996).

Prograded bayhead barriers provide another example of time-varying sediment supply as a dominant control on coastal geomorphology. Experiments with numerical simulation of barrier growth in confined embayments with a thin till veneer provide useful insights on the role of relative sea-level change and sediment supply. Published examples (e.g., Forbes *et al.*, 1995) are based on coastal embayments in southeast Newfoundland. The plan view (Figure P18A) shows a narrowing embayment with ocean wave approach up the axis of the bay to a barrier complex at the head. Breaking of refracted waves along the shore is assumed to mobilize all unlithified glacial cover within reach of the waves and transport it alongshore to the head of the bay. Assuming

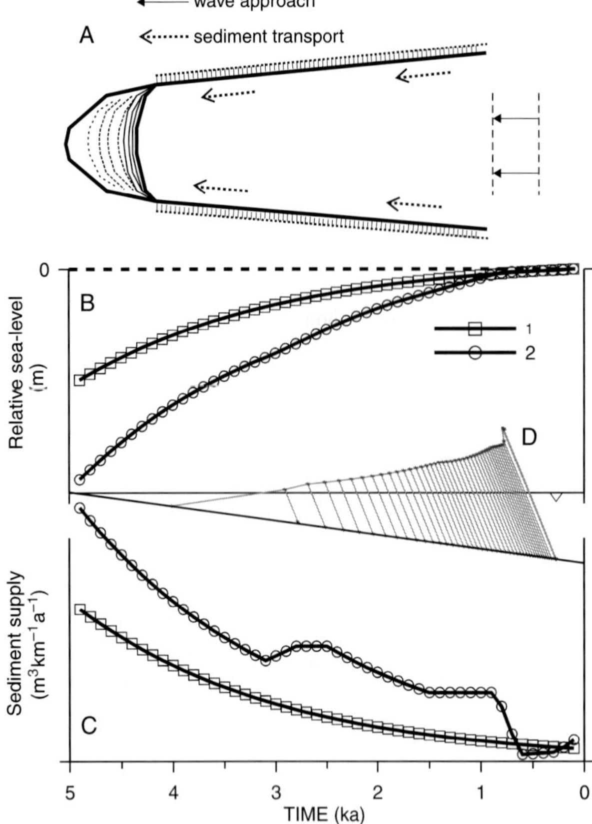

Figure P18 Numerical simulation of beach-ridge barrier growth in a bayhead setting. (A) Schematic plan view of embayment. (B) Arbitrarily selected relative sea-level histories. (C) Simulated sediment supply to bayhead barrier corresponding to the two sea-level histories shown above. (D) Model output showing schematic section through simulated beach ridges with high transgressive storm ridge at seaward margin.

simple geometry, the supply of sediment to the bayhead barrier is a function of access to erodible material, controlled by the rate of rise in relative sea level through time. The more rapid the rate of relative sea-level rise (Figure P18B), the greater the sediment supply; subtle variation in the rate of sea-level change can produce significant variation in the supply (Figure P18C). Progradation of the bayhead barrier (Figure P18D) is a function of sediment supply and accommodation space (increasing depth and bay width). Therefore, with diminishing relative sea-level rise, typical of the last few millennia in Newfoundland, the rate of simulated barrier progradation decelerates in response to reduced sediment input. As the progradation rate decreases, the potential for severe storm impact leading to overtopping and growth of the active beach ridge may increase. The morphological outcome is a seaward-rising sequence of barrier ridges, flooded at the back and reworked into a high transgressive storm ridge at the front.

These examples demonstrate the critical importance of sediment supply and the ways in which glacial inheritance may direct and ration the delivery of sediment to the coast. They also show the propensity for high spatial and temporal variation of sediment supply in some paraglacial settings. Some systems suffer from slow supply reduction, with distinctive morphological outcomes (Figure P18). Others experience highly fluctuating sediment budgets (Figure P17), resulting in abrupt coastal changes when supply from a given source is exhausted or other stability thresholds are exceeded (Forbes *et al.*, 1995). Delayed paraglacial sediment input to coastal systems, ultimately controlled by access through changing sea level, can prolong the paraglacial signal, extending the relaxation time through many thousands of years. Coastal morphology in glaciated regions of the mid- to high latitudes over a large part of the globe is thus affected to this day by the physiographic and sedimentary legacy of the last major glaciation.

Donald L. Forbes

Bibliography

Boyd, R., Bowen, A.J., and Hall, R.K., 1987. An evolutionary model for transgressive sedimentation on the Eastern Shore of Nova Scotia. In FitzGerald, D.M., and Rosen, P.S. (eds.), *Glaciated Coasts*. San Diego: Academic Press, pp. 87–114.

Carter, R.W.G., Orford, J.D., Forbes, D.L., and Taylor, R.B., 1990. Morphosedimentary development of drumlin-flank barriers in a zone of rapidly rising sea level, Story Head, Nova Scotia. *Sedimentary Geology*, **69**: 117–138.

Carter, R.W.G., Orford, J.D., Jennings, S.C., Shaw, J., and Smith, J.P., 1992. Recent evolution of a paraglacial estuary under conditions of rapid sea-level rise: Chezzetcook Inlet, Nova Scotia. *Proceedings of Geological Association*, **103**: 167–185.

Church, M., and Ryder, J.M., 1972. Paraglacial sedimentation: a consideration of fluvial processes conditioned by glaciation. *Geological Society of America Bulletin*, **83**: 3059–3072.

Forbes, D.L., and Syvitski, J.P.M., 1994. Paraglacial coasts. In Carter, R.W.G., and Woodroffe, C.D. (eds.), *Coastal Evolution: Late Quaternary Shoreline Morphodynamics*. Cambridge: Cambridge University Press, pp. 373–424.

Forbes, D.L., and Taylor, R.B., 1987. Coarse-grained beach sedimentation under paraglacial conditions, Canadian Atlantic coast. In FitzGerald, D.M., and Rosen, P.S. (eds.), *Glaciated Coasts*. San Diego: Academic Press, pp. 51–86.

Forbes, D.L., Orford, J.D., Carter, R.W.G., Shaw, J., and Jennings, S.C., 1995. Morphodynamic evolution, self-organisation, and instability of coarse-clastic barriers on paraglacial coasts. *Marine Geology*, **126**: 63–85.

Johnson, D., 1925. *The New England—Acadian Shoreline*. New York: Wiley [facsimile edn. 1967, New York and London: Hafner].

Orford, J.D., Carter, R.W.G., and Jennings, S.C., 1996. Control domains and morphological phases in gravel-dominated coastal barriers of Nova Scotia. *Journal of Coastal Research*, **12**: 589–604.

Shaw, J., and Forbes, D.L., 1992. Barriers, barrier platforms, and spillover deposits in St. George's Bay, Newfoundland: paraglacial sedimentation on the flanks of a deep coastal basin. *Marine Geology*, **105**: 119–140.

Cross-references

Eustasy
Glaciated Coasts
Pleistocene Epoch

PEAT

Peat forms from the accumulation of plant remains in waterlogged, anaerobic conditions that inhibit decomposition. In coastal situations peat forming environments occur within the intertidal zone, typically as *salt marsh* or *mangrove* peat, to the landward transition to freshwater environments, *wetlands* and *bog*. Accumulation requires a relatively low-energy environment, such as an *estuary* or behind *barriers* or *dunes*. Geographically the range is from the high latitudes to tropics, limited by water balance and the local conditions that constrain plant growth. If it is too dry, the plant material will decompose rapidly and peat will not accumulate.

Peat comprises authochthonous material, decomposed plant material deposited *in situ*, and allochthonous material, that which is transported to the site of deposition. The latter includes organic material, plants and fauna, and minerogenic material, usually clay, silt, and sand size fractions but occasionally coarser. The organic component includes the visible, or macrofossil, remains and a fine organic matrix. Separate constituents of this matrix that are indistinguishable with the naked eye include totally decomposed organic material and microfossils, such as pollen, spores, diatoms, thecamoebians, and foraminifera (see Troels-Smith, 1955, a classic work on the description of peat and other unconsolidated sediments, and recently summarized in chapter 11 of Jones *et al.*, 1999). The balance between organic and minerogenic sediment covers the full transition between the end members, with the intermediates described by terms such as sandy peat (predominance of organics) and organic sand (predominance of minerogenics). Peat-forming plant communities exhibit a spatial zonation in response to a series of environmental gradients; the most important of which in the coastal zone is usually ground surface elevation with respect to the tidal regime (Waller, 1994, chapters 3, 4, and 5). The spatial zonation changes through time, described as plant succession, as a result of autogenic and allogenic processes. Autogenic processes are those directional changes that operate within an ecosystem, trending toward equilibrium with the environment, the classical concept of climax vegetation. Allogenic processes are those operating externally to the vegetation. Examples include changes in sea level or sediment budget.

Where the balance of processes result in net accumulation, peat provides an excellent archive of environmental change. Successional changes that occur through time are recorded in the stratigraphic column. The physical, chemical, and biological properties of the sediment help to reconstruct the continuum from tidal flat, through marsh to wetland or bog. Although the classification into discrete units is difficult and somewhat arbitrary, it is relatively easy to reconstruct the trends or direction of change. This involves both the horizontal component of change (e.g., vegetation succession) and the vertical component (e.g., *sea-level change*). Early scientific studies, such as the geological surveys of coastal wetland areas carried out in the 19th century (or earlier) in northwest Europe, described the visible plant remains (macrofossils) within peat layers and from these inferred environmental change. With scientific advances during the 20th century, an increasing range of techniques has been used to analyze peat (van de Plassche, 1986). While macro-fossils, especially the roots and stems of the plants that were growing in the sediment, provide extremely good evidence of site conditions at the time of formation, their preservation is very variable and because only a small volume of sediment may be available (e.g., from a borehole) they may not give a reconstruction of the whole community or wider environment (Waller, 1994). From the microscopic sediment matrix the most widely studied element has been the microfossil content (especially pollen, spores, diatoms, and foraminifera).

Analysis of peat layers within a sediment sequence and the microfossils contained in the different layers can give reconstructions of a wide range of different environments, ranging from the coastal environments of previous interglacials, through the whole of the current interglacial to the present day. Peat can form quickly on the geological timescale. For example, the marshes around the Cook Inlet, Alaska subsided by up to 1.7 m during the great earthquake (M_w 9.2) of 1964 and were rapidly buried by tidal flat silt deposits (Figure P19). The 1964 marsh peat is now exposed along the banks of tidal creeks beneath minerogenic sediments, more than 1 m thick in some places, that grade upwards with an increasing organic content to the present marsh surface where there is already a centimeter or more of peat accumulating. While this example of recent marsh burial was the result of rapid coseismic change in land level, the gradual submergence of marsh peat beneath tidal flat minerogenic sediment records relative sea-level change due to nonseismic processes, especially the interaction of *eustasy* and *isostasy* (e.g., Allen, 2000).

Buried peats contain critical information on the long-term dynamics of coasts prior to the period of scientific observation and measurement. The threat of large earthquakes, comparable to the 1964 Alaska event, occurring in Washington, Oregon and British Columbia is clearly indicated by the multiple peat layers buried beneath many of the coastal marshes in these states (e.g., Shennan *et al.*, 1996). Figure P20 shows a peat layer from the Johns River, Washington. The sequence from 105 to 69 cm shows the colonization of a marsh onto tidal flat. As the marsh peat accumulates, low-marsh plants, shown by the Chenopodiaceae and *Aster*-type pollen, are replaced by high- marsh plants (e.g., indicated by *Potentilla* pollen). At the same time, marine (polyhalobian) and brackish (mesohalobian) diatoms give way to increasingly freshwater (oligohalobian) and salt-intolerent (halophobes) diatom species. The top of the peat marks the rapid return to tidal flat sedimentation, interpreted as the result of an earthquake that caused about 1 m of ground subsidence. The radiocarbon age of the peat indicates this occurred about 300 years ago. Other peat layers from the estuary indicate a total of eight such events during the last 5,000 years. The earthquakes result from the collision of the Juan de Fuca and American plates.

Ian Shennan

Figure P19 Salt marsh and tidal flats at Girdwood, in the upper Turnagain Arm of the Cook Inlet, Alaska. The dead spruce trees are rooted in a peat layer that had developed over the tidal flat sediments visible in the exposure. Coseismic subsidence during the earthquake of 1964 dropped the forest to a level below high tides. Tidal inundation killed the trees and gray tidal silt accumulated over the peat and around the tree trunks. At this location the silt is approximately 50 cm thick. The present salt marsh, dominated by sedges, represents the first stage in the renewed growth of peat as gradual land uplift since 1964 allows saltmarsh, and eventually, freshwater environments to develop once more.

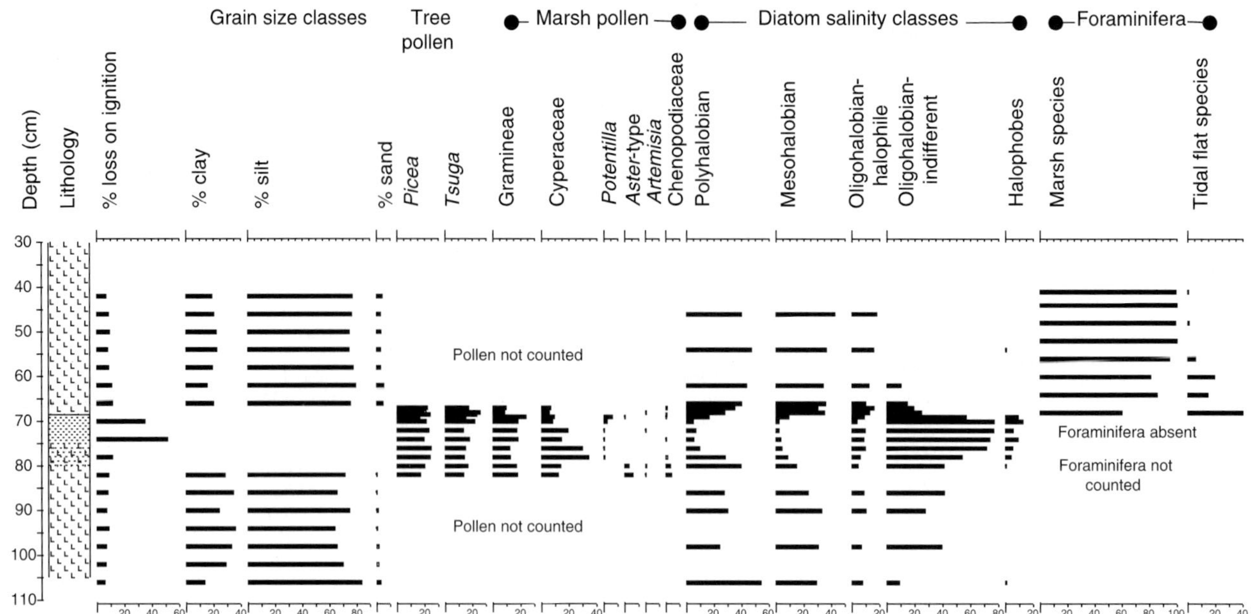

Figure P20 A buried peat layer from Johns River, Washington (after Shennan *et al.*, 1996). The peat, from 68 to 80 cm, lies between silt that represents tidal flat enviroments. The pollen, diatom, and foraminifera form the peat and the two silt layers record different stages of marsh development, then rapid burial caused by an earthquake ~300 years ago. The horizontal axes show percentage abundance for selected species for three microfossil types: pollen, diatoms, and foraminifera. The full data set comprises microfossils from over 100 different species.

Bibliography

Allen, J.R.L., 2000. Morphodynamics of Holocene salt marshes: a review sketch from the Atlantic and southern North Sea coasts of Europe. *Quaternary Science Reviews*, **19**: 1155–1232.

Jones, A.P., Tucker, M.E., and Hart, J.K. (eds.) 1999. *The description and analysis of Quaternary stratigraphic field sections*. Technical Guide 7, London: Quaternary Research Association.

Shennan, I., Long, A.J., Rutherford, M.M., Green, F.M., Innes, J.B., Lloyd, J.M., Zong, Y., and Walker, K.J., 1996. Tidal marsh stratigraphy, sea-level change and large earthquakes, 1: a 5000 year record in Washington, USA. *Quaternary Science Reviews*, **15**: 1023–1059.

Troels-Smith, J., 1955. Characterization of unconsolidated sediments. *Danmarks Geologiske Undersøgelse*, Series IV, **3**: 38–73.

van de Plassche, O., (ed.), 1986. *Sea-Level Research: A Manual for the Collection and Evaluation of Data*. Norwich: Geo Books.

Waller, M. (ed.), 1994. *The Fenland Project 9: Flandrian Environmental Change in Fenland*. East Anglian Archaeology Report No. 70, Cambridge: Cambridgeshire Archaeological Committee.

Cross-references

Barrier
Bogs
Coastal Soils
Dunes and Dune Ridges
Eustasy
Isostasy
Mangrove Coasts
Salt Marsh
Vegetated Coasts
Wetlands

PERIGLACIAL—See ICE-BORDERED COASTS

PHOTOGRAMMETRY

Photogrammetry can be defined simply as the science of making reliable measurements from photographs. Unlike a map, however, a photograph contains a number of distortions that require correction before accurate measurements can be made. A number of photogrammetric techniques can be employed to remove these distortions and obtain useful measurements. In coastal studies, photogrammetric techniques are commonly employed to establish the positions of historical and modern features-of-interest (e.g., shorelines (defined as the high-water line or wet-dry boundary), cliff edges, dune positions, etc.). Historically, the focus of study has been overwhelmingly on the use of vertical aerial photography to derive accurate shoreline positions, although photogrammetric applications using ground-based photography, videography and integration with other types of remotely sensed data (e.g., lidar) are becoming widespread. Most often, a time series of feature positions is compiled for the purpose of studying coastal dynamics, such as the evolution of geomorphic features, or determining rates of coastal change (e.g., shore erosion and accretion). These time series have also been used to guide the delineation of coastal erosion or flood hazard areas and building setback lines.

Historical development

Early studies of coastal erosion (e.g., Stafford and Langfelder, 1971) used point measurements made on individual air photos to determine shoreline rates of change. The procedure is straightforward: stable reference points such as buildings and road intersections are selected on air photos taken in different years, and the distance between these points and the shoreline reference feature (typically the boundary between wet and dry beach sand, which approximates high water) is measured. These measurements are then multiplied by the nominal scale of the aerial photograph to obtain ground distances. The change in distance over time yields a rate of shoreline change at that location. Subsequent shoreline delineation methods expanded upon this basic approach, using techniques such as photo-enlargement or a Zoom Transfer Scope to match a base map (e.g., Dolan *et al.*, 1978, 1980; Smith and Zarillo, 1990). These methods permitted a continuous representation of the shoreline to be mapped from the photos, rather than discrete points, and were a significant improvement over point measurements in terms of the amount of information that could be extracted from aerial photography.

One of the primary limitations in early historical shoreline mapping studies was the inability to quickly and accurately correct aerial photographs for their inherent distortions. The first integrated coastal photogrammetry software package to address this need was the Metric Mapping System (MMS) (Leatherman, 1983; Clow and Leatherman, 1984). For aerial photographs, this package furnished the ability to correct individual photographs for image displacements and produce shoreline positions on a plotted map. The Digital Shoreline Mapping System (DSMS) (Thieler and Danforth, 1994a,b) expanded upon this approach to include a full photogrammetric adjustment of large groups of photographs and produce data suitable for plotting shoreline change maps in a Geographic Information System (GIS). Both the MMS and DSMS software packages utilized x–y tablet digitizers to obtain vector data from aerial photographs. While this approach kept the amount of data acquired from the photographs to a manageable size, much useful visual information was not used. More recently, however, advances in computer storage and processing capability allow photogrammetric techniques to be applied to digitally scanned photographs. This process, known as softcopy photogrammetry, has recently seen increased use in coastal studies. A comprehensive review of available shoreline mapping techniques is furnished by Moore (2000).

Characteristics of aerial photography

The basic principles involved in the extraction of geographic data from aerial photographs are derived from the geometric relationships between image space and object space. Image space refers to the world inside the camera (i.e., the photographic image and measurements obtained from it). Object space refers to real-world geographic coordinates outside the camera.

In distortion-free space, points in image space can be projected to points in object space. This relationship is based on the principle of collinearity: the perspective center of the camera lens (which is considered a point), an image point and its corresponding ground point all lie on the same straight line (Figure P21). Aerial photographs, however, are subject to a number of distortions introduced at various stages in the photographic process that perturb the collinearity condition. These perturbations affect both image space and object space.

The image space coordinate system is defined by the locations of the fiducial reference marks on a photograph, the calibrated focal length, and the geometric distortion characteristics of the lens system in the camera. Image space is described by a three-dimensional, rectangular Cartesian coordinate system with the origin located at the principal point (the center of an aerial photograph as defined by fiducial reference marks around the edge of the photograph). For aerial photographs, the x-axis is typically positive in the direction of flight. The z-axis corresponds to the optical axis of the camera.

The distortions affecting image space result from lens distortion and film deformation. All camera lenses have measurable distortions and optical defects that affect the representation of image points on film. Lens distortions can be radial or tangential. Radial distortion is symmetric around the principal point and is caused by optical defects in the lens.

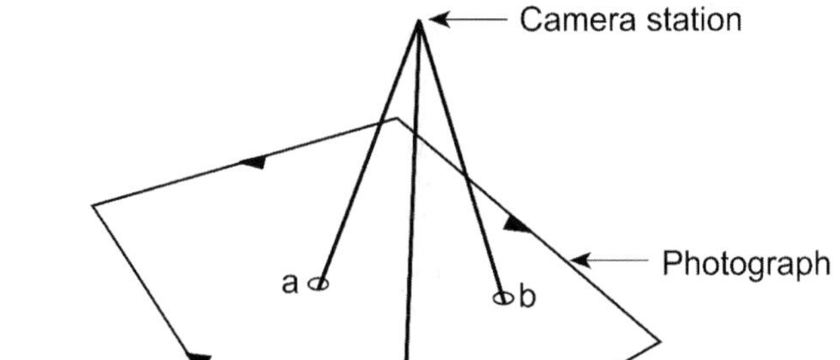

Figure P21 In distortion-free space, a projective relationship exists between image space (points on an aerial photograph) and object space (points on the ground). The camera station, an image point (a, b, c), and its corresponding ground point (A, B, C) all lie on the same straight line (after Thieler and Danforth, 1994b).

Tangential distortion is symmetric along a line through the principal point and results from the lens being slightly off-center in the camera. In a well-adjusted camera, however, only radial distortions are present.

The magnitude of lens distortion is highly variable. Some lenses used today have up to 0.110 mm radial distortion (American Society of Photogrammetry, 1980). Similar and sometimes greater amounts of lens distortion commonly are present in photographs taken prior to World War II, which brought an increased demand for accurate photography and improved lens manufacturing techniques. This is a particularly important consideration in using historical coastal photography. In most modern camera systems, however, lens distortion is negligible.

Two types of film deformation exist. Deformation can be introduced in the camera during the aerial survey or in subsequent processing. Film buckling, for example, may occur during the photographic survey due to irregularities in temperature, humidity or film spool tension in the camera. Further deformation is introduced not only in the development of the original negatives, but also in each generation of prints and transparencies (typically used by coastal researchers) made from the original negatives. The end result of these deformations is a photograph that no longer represents accurately the true geometric relationships between the fiducial reference marks and image points in the photo.

In addition to deformation occurring in the camera, the amount of film deformation present in a given photograph depends upon the age and type of material (glass, film, or paper), processing techniques used, and the temperature and humidity at the time measurements are made. Standard diapositive (transparency) film is generally stable within 0.005 mm. Photographic paper, however, is far less stable and may change in size up to 1% during processing alone (American Society of Photogrammetry, 1980).

The characteristics of object space cause image points on film to be displaced (as opposed to distorted) from their true position as a result of three factors: relief displacement, tilt displacement, and atmospheric refraction. Relief displacement is caused by changes in ground elevation within a photo that cause objects closer to the camera to be larger (i.e., at a larger scale) than those farther away. Relief displacement takes place radially from the nadir (Figure P22). Objects higher than the ground elevation at the point where the nadir intersects the ground (the ground nadir; Figure P22) are displaced outward; objects lower than the ground nadir point are displaced inward.

The determination and magnitude of relief and tilt displacements have been widely discussed in the context of coastal photogrammetry (e.g., Anders and Byrnes, 1991; Crowell *et al.*, 1991). Tilt displacement occurs due to the inability to keep the aerial camera perfectly leveled during photography. Some degree of tilt is always present in an aerial photograph. On a tilted photograph, the sense of displacement depends on whether the image point is on the low or high side of the isometric parallel (Figure P22). Points on the low side of the isometric parallel are displaced outward from the isocenter; on the high side they are displaced inward. Points on the isometric parallel are not displaced.

Atmospheric refraction (the bending of light rays through the atmosphere) also causes photograph image points to be displaced. The displacement occurs radially outward from the nadir. The magnitude of the displacement depends on the aircraft flight height, direction of the optical axis relative to the ground, and the focal length of the camera. For most coastal applications, atmospheric refraction is negligible because low-altitude photography is used.

Extracting geographic data from aerial photographs

The traditional approach to analytical photogrammetry (American Society of Photogrammetry, 1980) is composed of three steps that remove the perturbations described above, and exploit geometric relationships between overlapping aerial photographs to extract geographic data: (1) preprocessing; (2) aerotriangulation; and (3) postprocessing. Preprocessing reduces measured image coordinates to the image space coordinate system described above, as well as removes systematic errors such as lens distortion and film deformation effects. The goal of preprocessing is to transform measured image data from a photograph to an idealized image space in which distortions do not exist. In other words, the measured coordinates are refined to remove image space distortions before the data are used in subsequent processing.

Aerotriangulation is used to solve simultaneously for the camera position of each photograph in a large group of overlapping photographs (a strip or a block), as well as the coordinates of unknown ground points. The absolute orientation performed in an aerotriangulation adjustment is basically an extension of the technique of space resection, which determines the six elements of exterior orientation for a photograph, including the position (latitude, longitude, and elevation), and attitude (roll, pitch, and yaw; designated ω, ϕ, and κ, respectively)

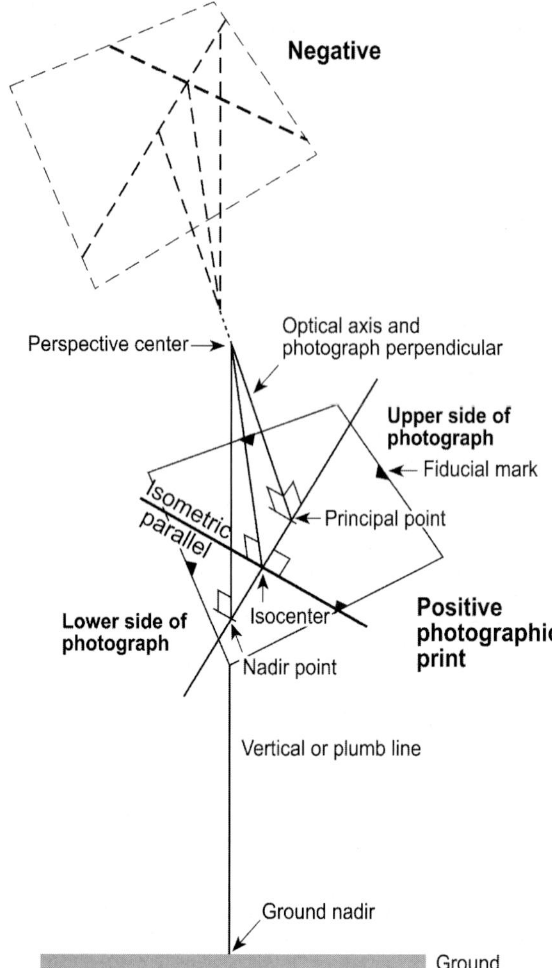

Figure P22 Definition sketch of terms used to describe the various elements of a tilted aerial photograph. Some degree of tilt is always present in an aerial photograph, which causes image points to be displaced from their true position (modified after American Society of Photogrammetry, 1980).

of the aerial camera. Atmospheric refraction effects can also be removed during space resection or aerotriangulation by applying a correction function each time the orientation of the camera is updated in the solution process.

Postprocessing typically involves transforming the camera position information into instrument settings used in stereoplotters or other photogrammetric equipment in order to compile basemaps or generate rectified photographs (also known as orthophotographs). In digital photogrammetry, camera information is input into computer software that performs orthorectification on scanned photographs. The result is a geographically corrected photograph that may be used directly in a GIS.

Ground control

The common requirement of all forms of photogrammetric manipulation is adequate ground control information so that quantitative, geographically accurate information can be extracted from the photographic data. Thus, establishing an adequate ground control network is critical. This is particularly true in studies using historical aerial photography, since points on the ground are the only source of position information.

A ground control network is a set of points that appears in one or more photographs, and provides the basic means of establishing a correspondence between the photographs and the ground. There are two types of points used in photogrammetry: ground control points and tie points. A point appearing in one or more photos for which information about its location is known (e.g., one or more of latitude, longitude, and elevation) is called a control point or ground control point. The image space coordinates of ground control points and their corresponding geographic coordinates are used to establish the collinear relationship between image space and object space (Figure P21).

There are a number of ground control data sources, including maps, field surveys and geodetic control tables. The locations of well-defined points shown on maps can be digitized, converted to geographic coordinates and used as ground control points. These points are termed "supplemental control points" because they are obtained from a map rather than by direct field survey. Supplemental control points, consisting primarily of buildings and road intersections, are the primary source of ground control for many historical shoreline mapping projects because other data are unavailable.

A tie point is defined as a point appearing in two or more photos, for which a corresponding ground position is not known. Tie points are used to "pass" or extend control between overlapping photos. These points are used in addition to ground control points to establish the relative orientation of photos to each other, such as is done when viewing a pair of overlapping photos through a stereoscope. Tie points commonly include features such as trees, buildings, and road intersections.

For most applications, at least four and preferably six to nine tie points are needed to provide adequate control for a given stereo-pair, and should be distributed throughout the photograph. Only a few control points are needed to establish the geographic orientation of a group of photographs in object space; most points used to control a photograph may simply be tie points.

Ideally, the exact planimetry (latitude, longitude, elevation) would be known for a large number of spatially and temporally well-distributed ground control points throughout all photographs. This is never the case, however, and ground control points of varying quantity and quality must be used when constructing the control network for a given mapping project.

In historical shoreline change studies, when several sets of photographs spanning several years of the same geographic area are used, the common ground control and tie points form a model that establishes correlations between images, in space and through time, that are readily exploited in aerotriangulation and error analysis.

It is often problematic, however, to furnish an adequate quantity and distribution of ground control and tie points due to the nature of photography along the shoreline and changes in coastal environments over time. Most photographs that include the shoreline, for example, are devoid of usable control points seaward of the shoreline. Coastal areas may also change rapidly over time, due to natural processes or human development, which reduces the number of stable points that can be used as ground control or tie points. In these situations, it is often necessary to use additional overlapping photos taken of more landward areas in order to augment the control network for the shoreline photographs.

Modern techniques and current investigations

Several different methods of digital rectification (Table P1) are available through the various softcopy photogrammetry software packages currently available. Not all rectification techniques, however, are equally accurate. The two most robust rectification methods are full orthorectification based on viewing and manipulating digital stereo-pairs, and pseudorectification, in which stereo-pairs are used but there is no ability to manipulate stereo-imagery. These two techniques both apply photogrammetric principles to produce orthophotographs, and involve the creation of a Digital Terrain Model (DTM) to remove relief displacement from the imagery. Generation of DTMs involves taking the user-input measurements (ground control and tie points) and deriving the DTM from a stereo model through interpolation (Ackerman, 1996). In many softcopy photogrammetry systems, DTM collection is automated but the user can edit the model. Editing a DTM (which is only possible with stereo-viewing photogrammetry) increases the accuracy of the ground surface model, allowing more complete removal of relief displacement in the orthophotograph. Using an unedited DTM may affect the spatial accuracy of the orthophotograph, and may not approximate rapid changes in topography. Another rectification technique, single frame resection, can be used to remove distortions associated with the camera system, but does not remove aircraft attitude or relief displacements. The final method available is warping (also known as "rubber sheeting"). In this case, no parameters are input to remove displacements; instead, ground control points are used to "stretch" the pixels of an individual image to a best georeferenced fit.

In comparing the various methods described above in coastal change applications, recent studies (Hapke and Richmond, 2000) have found that a shoreline position digitized on a stereo-based orthophotograph is

Table P1 Comparison of softcopy photogrammetry processing techniques

Technique	Process (result)	Products
Full orthorectification (stereo-viewing)	Input camera system parameters (remove camera distortions) Input tie points (remove aircraft attitude distortions) Add ground control points (georeference imagery) Create DTM (remove relief distortion) Edit DTM (correct for detailed topography) Aerotriangulate (tie all photos together, orthorectify)	Orthophotographs Stereo imagery High-accuracy DTMs
Pseudorectification	Input camera system parameters (remove camera distortions) Input tie points (remove aircraft attitude distortions) Add ground control points (georeference imagery) Create DTM (remove relief distortion) Aerotriangulate (tie all photos together, orthorectify)	Orthophotographs Uneditable DTMs
Single frame resection	Input camera system parameters (remove camera distortions) Add ground control points (georeference imagery) Least squares resample (statistical best-fit)	Pseudo-orthophotograph (relief distortion not removed)
Warping (rubber sheeting)	Add ground control points (georeference imagery) Stretch pixels to best fit approximation	Georeferenced image

the most reliable in terms of reproducing shoreline position. The technique of warping is the least reliable method and can be very inconsistent. In areas of high relief, or where high accuracy (submeter) is required, use of a stereo-edited DTM is recommended.

The recent proliferation of digital photogrammetry and GIS software has greatly increased the accessibility of aerial photograph analysis. Many software packages can be run on a standard desktop computer and do not require formal training in photogrammetry or GIS. Using such software requires scanning the aerial photographs to convert to digital format. There are several options as to the media of the aerial photography and the choice of scanners, all of which may have noticeable impact on the spatial accuracy of the final orthophotograph.

The first step in digital aerial photograph analysis is to convert the photograph into a digital format using a scanner. The most accurate choice, a geometrically correct photogrammetric scanner, will produce minimal distortion. This ensures that the spatial distribution of objects in the original photography is reproduced accurately in the digital version. Photogrammetric scanners provide the most accurate digital imagery, but these scanners and/or the contracting of photogrammetric scanning may be too costly for smaller-scale coastal mapping projects. As a less-expensive (and less-accurate) substitute, high-quality graphic design scanners or desktop scanners are available. Graphic and desktop scanners are not designed to assure geometrically accurate data conversions, and thus may introduce distortion to the digital imagery, which results in nonsystematic positional errors in the resulting orthophotograph (Hapke *et al.*, 2000). A further limitation of desktop scanners is that the scanning resolutions rarely exceed 800 dpi (dots per inch), whereas most photogrammetric workflows recommend 1,200 dpi or higher.

Standard aerial photography is typically available from the original negatives in two different media: contact (paper) prints and diapositives (positive film transparencies). Although contact prints cost less and are readily available, as noted above photographic paper may undergo stretching, shrinkage, and warping, resulting in positional errors. Diapositives are a much more stable media, since film is much less likely to undergo distortion.

Most digital softcopy photogrammetry software packages provide an overall RMS (root mean square) value as an indication of error associated with rectification. US National Map Accuracy Standards (NMAS) (Falkner, 1995) state that in order to conform to the accuracy standards, 90% of identifiable stationary objects should be accurate to within a specified RMS error. In a study comparing media types and scanners, Hapke *et al.* (2000) found that at a scale of 1 : 12,000, the only combination of scanner and media type that conforms to NMAS is a diapositive scanned on a photogrammetric scanner. The RMS error increases with the lower precision scanner and with the use of contact prints versus diapositives. The NMAS standards also require that the maximum error at any given point, either horizontal or vertical, does not exceed three times the magnitude of the RMS. For images scanned on a nonphotogrammetric scanner (either graphic design or desktop), this value is almost always exceeded in the X and Y, and in every case in the Z direction. If a coastal mapping application involves delineation of absolute position (e.g., establishing setback lines, determining hazard zones at the scale of individual property boundaries, etc.), or if the resulting data could be used for such applications, then it is especially important to use a technique that adheres to the NMAS.

Table P2 Suggested scanner and media types for coastal research

Project scope and examples	Recommended scanner and media
Absolute position: setbacks, hazard zones, elevation changes, short-term changes	Photogrammetric scanner, diapositives
Long-term erosion rates: bluff or shoreline change	Graphic arts scanner, diapositives
Thematic mapping: vegetation, land use, watershed mapping	Desktop scanner, diapositives

For other types of coastal applications, larger errors may be quite acceptable (e.g., rate calculations, especially over long periods of time), such that using contact prints and/or nonphotogrammetric scanners is appropriate to the study design. These methods, however, may introduce large nonsystematic errors into the final orthophotograph. These nonsystematic errors are most likely due to stretching or shrinking of the contact prints. Table P2 provides an overview of the recommended scanner and media choices for a variety of coastal research activities utilizing aerial photography. The image quality of scanned contact prints is significantly degraded as compared to scanned diapositives, regardless of the quality of the scanner. This may result in the inability to measure ground control and tie accurately.

The full stereo-orthorectification process requires the generation of a DTM that defines topography within the stereo-overlap region of images using a series of points and is required if the distortion from terrain relief is to be removed from the orthorectified images. Two formats of DTMs may be generated from stereo-images: grid or TIN (triangulated irregular network). The grid is a regularly spaced network of points where the elevation of individual points can be edited but the network spacing must remain constant, and thus the horizontal position of each point cannot change. In a TIN model, however, not only the elevation of points may be edited, but they may also be added in areas where a greater density is desired, or deleted in problematic areas. Another advantage of TIN models is the ability to add breaklines, which allows for more accurate definition of subtle topographic changes. Breaklines are crucial to accurately define the topographic signal of narrow or sharp features such as beach scarps, cliff edges, or dune ridges in the surface model. A breakline is a manually entered line composed of a series of points that are incorporated into the DTM. Breaklines can only be added to the model while viewing in stereo, as the operator must be able to identify the elevation change in order to correctly place the line.

DTMs generated from stereo-models can be used to derive volume, and hence volumetric changes. However, the automated process of DTM generation has distinct, nonsystematic errors that must be corrected if DTMs are to be used for accurate surface modeling and volumetric change analyses. Automated collection with no editing is problematic in images of coastal regions due to typical low-contrast beach conditions and high visual interference areas like the swash zone, where the water has moved and thus varies in position from one image

(A)
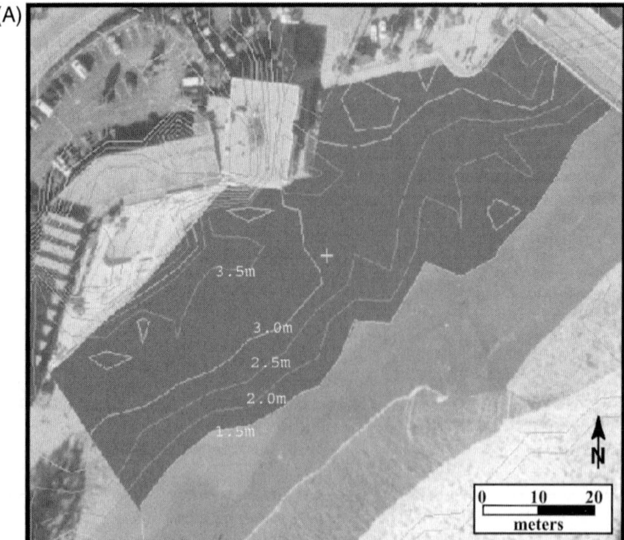

(B)

Figure P23 Topographic contours generated from an edited grid DTM of Cowell Beach, Santa Cruz, California, for two time periods: (A) January 27, 1998 and (B) March 6, 1998. The seaward extension of the area under which the volume is calculated is the 1.5-m contour. The volume of sand above a zero datum on January 27 is 49,740 m^3, as calculated from the topographic model. Almost one half the volume of sand on this beach was lost during the 1998 El Niño winter. By March 6, the volume of sand above a zero datum is 24,380 m^3. Note how the position of the 1.5-m contour has changed between January 27 and March 6.

to the next. In order to utilize the DTMs for time-series volumetric studies, careful editing is crucial.

Beach volume (Hapke and Richmond, 2000) and/or dune volume (Brown and Arbogast, 1999) can be calculated using an edited TIN format DTM derived from each date of photography (Figure P23). For the volume calculations, a base elevation must be defined, and 0 m (MSL) is typically used. However, surface elevations below the 0.5 to1.5-m zone contour on a subaerial beach are commonly in, or very near, the swash zone that creates severe interference when viewing in stereo and thus greatly reduces the accuracy of the contours. Merging other topographic data in high interference areas or areas prone to shadow (e.g., base of seacliff, scarp, or dune) is highly desirable if such data is available. In addition, once ground control points are measured within a block or strip of air photos, other sources of DTM data can be used for final orthorectification. These sources include lidar as well as existing contour maps.

Technological advances in surveying techniques and digital data collection are rapidly influencing the field of photogrammetry, and will eventually result in reducing the currently required processing time by several orders of magnitude. As described above, orthorectification and DTM generation both require a network of accurate ground control points that are visually identifiable on imagery. In order to remove displacements associated with aircraft attitude and topographic relief, a network of tie points are used to connect the images and then the ground control points must be digitally located in order to successfully aerotriangulate and rectify imagery. Airborne kinematic GPS used in conjunction with the collection of the photographic data eliminates the need to collect ground control points as well as measure tie points on the imagery. Changes in aircraft position from frame to frame can be recorded using an inertial measurement system as the photography is being collected; the attitude values (ω, ϕ, κ) can be entered into processing software. Airborne kinematic GPS can be used to record simultaneously the exact location of the center of the camera lens as the instant exposure occurs, thus eliminating the requirement of locating photo-identifiable ground control points in a strip or block of images (Lucas and Lapine, 1996).

For use in traditional softcopy photogrammetry, photographs are presently converted from film to digital format using a photogrammetric scanner. With continued advances in the resolution of sensor arrays in digital cameras, it is inevitable that photographic collection will eventually be dominated by digital systems.

Summary

The use of photogrammetric techniques in coastal studies has progressed in concert with technological advances over the past several decades. Early studies of coastal change used simple point measurements at control point locations on individual photographs. This was a time-consuming process and did not yield a great amount of information relative to the effort involved. In most cases, a simple rate of shoreline change at discrete locations was produced. Modern applications of photogrammetry, although computerized and somewhat automated, are similarly time-consuming. The data produced from this process, however, enables the delineation and measurement of a variety of parameters, such as morphology, position and volume changes in coastal environments that are of use to a wide audience including coastal managers and scientists.

E. Robert Thieler and Cheryl J. Hapke

Bibliography

Ackerman, F., 1996. Techniques and strategies for DEM generation. In Greve, C. (ed.), *Digital Photogrammetry: An Addendum to the Manual of Photogrammetry*. Bethesda: American Society for Photogrammetry and Remote Sensing, pp. 135–141.

American Society of Photogrammetry, 1980. *Manual of Photogrammetry*, 4th edn. Falls Church: American Society of Photogrammetry.

Anders, F.J., and Byrnes, M.R., 1991. Accuracy of shoreline change rates as determined from maps and aerial photographs. *Shore and Beach*, **59**(1): 17–26.

Brown, D.G., and Arbogast, A.F., 1999. Digital photogrammetric change analysis as applied to active coastal dunes in Michigan. *Photogrammetric Engineering and Remote Sensing*, **65**: 467–474.

Clow, J.B., and Leatherman, S.P., 1984. Metric mapping: an automated technique of shoreline mapping. In Falls Chorch, Virginia *Proceedings, 44th American Congress on Surveying and Mapping*. American Society of Photogrammetry, pp. 309–318.

Crowell, M., Leatherman, S.P., and Buckley, M.K., 1991. Historical shoreline change: error analysis and mapping accuracy. *Journal of Coastal Research*, **7**: 839–852.

Dolan, R., Hayden, B., and Heywood, J., 1978. A new photogrammetric method for determining shoreline erosion. *Coastal Engineering*, **2**: 21–39.

Dolan, R., Hayden, B.P., May, P., and May, S., 1980. The reliability of shoreline change measurements from aerial photographs. *Shore and Beach*, **48**(4): 22–29.

Falkner, E., 1995. *Aerial Mapping Methods and Applications*. Boca Raton: Lewis Publishers.

Hapke, C., and Richmond, B., 2000. Monitoring beach morphology changes using small-format aerial photography and digital softcopy photogrammetry. *Environmental Geosciences*, **7**: 32–37.

Hapke, C., Willis, C., and D'Iorio, M., 2000. Going digital: error evaluation of media and scanner types for aerial photographic analyses in

coastal change studies. In *Proceedings of the American Shore and Beach Preservation Association Annual Meeting*, Aug. 6–10, Kaanapali, Maui.
Leatherman, S.P., 1983. Shoreline mapping: A comparison of techniques. *Shore and Beach*, **51**: 28–33.
Lucas, J.R., and Lapine, L.A., 1996. Airborne GPS. In Greve, C. (ed.), *Digital Photogrammetry: An Addendum to the Manual of Photogrammetry*. Bethesda: American Society for Photogrammetry and Remote Sensing, pp. 39–41.
Moore, L.J., 2000. Shoreline mapping techniques. *Journal of Coastal Research*, **16**: 111–124.
Smith, G.L., and Zarillo, G.A., 1990. Calculating long-term shoreline recession rates using aerial photographic and beach profiling techniques. *Journal of Coastal Research*, **6**(1): 111–120.
Stafford, D.B., and Langfelder, J., 1971. Air photo survey of coastal erosion. *Photogrammetric Engineering*, **37**: 565–575.
Thieler, E.R., and Danforth, W.W., 1994a. Historical shoreline mapping (I): improving techniques and reducing positioning errors. *Journal of Coastal Research*, **10**: 549–563.
Thieler, E.R., and Danforth, W.W., 1994b. Historical shoreline mapping (II): application of the Digital Shoreline Mapping and Analysis Systems (DSMS/DSAS) to shoreline change mapping in Puerto Rico. *Journal of Coastal Research*, **10**: 600–620.

Cross-references

Beach Features
Coastal Boundaries
Geographic Information Systems
Global Positioning Systems
Mapping Shores and Coastal Terrain
Nearshore Geomorphological Mapping
Profiling, Beach
Remote Sensing of Coastal Environments

PHYSICAL MODELS

In the context of coastal science and engineering, the term physical model can be defined as the physical reproduction of the environment of interest in the laboratory, normally at reduced scale. To many readers, the use of physical models in an age of increasingly sophisticated numerical, or computer models must appear antiquated. Indeed, Leonardo da Vinci employed the technique to examine a variety of hydraulic problems in the 1500s (American Museum of Natural History, 1996)! Physical models retain their usefulness in the study of coastal processes and design for a variety of reasons, including:

1. When scaled correctly, physical models can simulate complex turbulent phenomena to a degree of accuracy that is not possible with analytical and numerical models.
2. The inherent control that laboratory experiments afford allows for the careful investigation of dominant physics, and the evaluation of cause-and-effect relationships among the various forcings of interest. These advantages are not attainable via field experiments because of the prohibitive expense of long-term coastal measurements having the requisite spatial coverage and resolution.
3. Physical models facilitate the development of visual-based, "hands on" knowledge of the environment and problem being investigated. This advantage can be particularly important when dealing with complex dynamics, and/or when examining the impact of coastal engineering design alternatives.

Physical models are however, not without disadvantages. These include:

1. Scale effects, which are present in virtually any reduced-scale physical model. These are associated with the improper scaling of one or more relevant forces or flow features. The degree of influence of scale effects depends on the relative importance of the poorly scaled variable(s) to the problem of interest. The optimization of model accuracy in the presence of scale effects remains part of the "art" of physical modeling.
2. The possibility that the natural environment cannot be fully reproduced because of (e.g.) incomplete representation of all natural forcings, or inadequate model dimensions leading to boundary effects.
3. The high expense associated with the construction and implementation of physical models. When comparing the costs of physical and numerical models, however, one should consider the full range of costs associated with each, including, for example, the developmental costs associated with a numerical model, and the field data acquisition necessary to calibrate and validate both types of models.

Similitude

Similitude between a reduced-scale physical model and the full-scale prototype is simply defined as the condition when the reduced-scale model is an exact reproduction of the full-scale system, or prototype, being examined. There are three basic forms of similitude, including:

- Geometric—wherein the ratios of corresponding lengths in the model and prototype are identical, for example, $(L_{model}/L_{prototype}) = (d_{model}/d_{prototype})$. In other words, the model and prototype have the same shape.
- Kinematic—wherein the ratios of corresponding flow velocities and accelerations in the model and prototype are identical, for example, $(V1_{model}/V1_{prototype}) = (V2_{model}/V2_{prototype})$, where *V1* and *V2* represent velocities at corresponding locations in the model and prototype.
- Dynamic—wherein the ratios of corresponding forces in the model and prototype are identical, for example, $F1_{model}/\ F1_{prototype}) = (F2_{model}/F2_{prototype})$, where *F1* and *F2* represent forces acting on corresponding locations in the model and prototype.

Assurance of geometric similarity is relatively straightforward, and requires knowledge only of the linear dimensions of the physical system to be modeled. These dimensions must then be reduced in equal proportion in all directions—vertical, lateral, and longitudinal. This assures that the reduced-scale model accurately reproduces both the scale and the shape of the prototype. In the case of reduced-scale models of coastal systems, the large spatial extent of the domain of interest and/or the restricted size of the laboratory facility, will at times require the use of a "distorted" model, in which the vertical and horizontal scales are reduced by different ratios. The use of distorted models is normally avoided because of the problems assuring kinematic and dynamic similarity in such models.

Assurance of kinematic and dynamic similarity requires knowledge, if not of the detailed physics of the full-scale system being modeled, then at a minimum knowledge of the important variables that describe the system. These variables include forces such as gravity and pressure, fluid characteristics such as density and viscosity, and flow characteristics such as velocity. One must first determine which among the various variables are important to the problem at hand. Proper representation of these variables can then assure that both kinematic and dynamic similarity are achieved. If, as is often the case in coastal systems, the number of variables is large, the proper scaling of each individual variable can become problematic. As an alternative, one can group the various variables into dimensionless parameters, which, if properly reproduced in the physical model, can assure similarity. Of course, the challenge lies in the selection of the dimensionless parameters. This process is known as *dimensional analysis*, and is accomplished either by inspection, using experience and/or knowledge of the physical system, or through a more formal process such as the Buckingham-π theorem (e.g., see Fischer *et al.*, 1979). For an excellent treatment of scaling considerations for a variety of practical coastal problems, the reader is referred to Hughes (1993).

The construction of physical models for coastal systems is usually accomplished through the use of one or both of the following dimensionless parameters: the Froude number: $\mathrm{Fr} = V/(gL)^{1/2}$, where V is the velocity of interest, L is the length of interest, and g is the acceleration due to gravity; and the Reynolds number: $\mathrm{Re} = VL/\nu$, where ν is the kinematic viscosity of the fluid. Both of these dimensionless parameters are constructed via the ratio of the force of interest—the force due to gravity in the case of the Froude number and the force due to viscosity in the case of the Reynolds number—to the inertial force. We discuss each parameter below.

Froude number similarity

The Froude number is employed in virtually every physical model study dealing with coastal processes, because of the importance of free surface (gravity) effects. Using our definition, $\mathrm{Fr} = V/(gL)^{1/2}$, Froude number similarity is achieved by assuring that:

$$\frac{V_m}{(g\,L_m)^{1/2}} = \frac{V_p}{(g\,L_p)^{1/2}}$$

where the subscripts "m" and "p" denote model and prototype values, respectively. Since gravity remains constant at model scale, this relation can be rewritten as:

$$\frac{V_m}{V_p} = \frac{L_m^{1/2}}{L_p^{1/2}}$$

which indicates that the scale reduction for the velocity is equal to the square root of the length scale reduction. Alternatively, since velocity is (length/time), we can restate this by saying that the scale reduction for time is equal to the square root of the length scale reduction. In other words, using geometric similarity and Froude number similarity, and employing a model that has (undistorted) length scale reduction of 9, the corresponding timescale reduction is equal to 3. If one was to use this model to simulate a full-scale 12-h flow event, for example, one would employ a 4-h event in the model.

Reynolds number similarity

The Reynolds number was originally employed to predict the onset of turbulence in flow through pipes. It is commonly used in physical model studies that address problems where viscous forces are expected to play an important role. Examples of such problems in the fields of coastal science and engineering include fluid flow within porous media such as sand or a stone structure, and boundary layer flows. Using our definition, $Re = VL/\nu$, Reynolds number similarity is achieved by assuring that:

$$\frac{V_m L_m}{\nu_m} = \frac{V_p L_p}{\nu_p}$$

This relation can be rewritten as:

$$\frac{V_m}{V_p} = \frac{L_p \nu_m}{L_m \nu_p}$$

which indicates that the scale reduction for the velocity is in this case equal to the *inverse* of the length scale reduction, multiplied by the viscosity scale reduction. If we assume that the same fluid is employed in the model as exists in the prototype, we can restate this by saying that the scale reduction for time is equal to the *square* of the length scale reduction. In other words, using geometric similarity and Reynolds number similarity, and employing a model that has (undistorted) length scale reduction of 9, the corresponding timescale reduction is equal to 81.

Clearly from this discussion, both Froude number similarity and Reynolds number similarity cannot be achieved for a given flow regime without much difficulty (e.g., using a model fluid having a different viscosity than that of fresh or salt water). It is therefore left to the coastal physical modeler to ascertain whether viscous or gravity forces dominate the system being examined. Once this decision is made, the physical model is developed using geometric similarity and either Froude or Reynolds number similarity. In most cases involving coastal systems, Froude number similarity is chosen, because of the dominance of free surface (gravity) effects. However, care must be taken to assure that viscous forces—which in such cases have been assumed to be relatively small in the prototype—are not large (relative to gravity forces) in the model. This most often requires avoiding large length scale reductions, which give rise to small Reynolds number flows at model scale, leading to potentially large viscosity effects. One example of such a dilemma is the small interstitial spaces in a model stone breakwater structure that has been reduced in scale to the point of creating very low Reynolds number flows within the spaces, and therefore unrealistic behavior of both the water motion and the stone response. An example of overcoming this problem can be found in the use of small models for slow-moving surface vessels such as sailing yachts. When such models are scaled according to geometric and Froude number similarity, the Reynolds number is too low to assure that a natural, turbulent boundary layer exists along the hull of the vessel. "Turbulence stimulators," in the form of a roughened bow of the vessel or small roughness elements placed along the leading edge of the vessel, serve to artificially stimulate the generation of turbulence at model scale.

Challenges in coastal physical modeling

Scale effects

The inherent difficulties associated with assuring both Froude number and Reynolds number similarity represent a formidable challenge to modelers seeking to simulate the behavior of the coastal environment over large prototype scales. Typically, such models are employed in the simulation of coastal wave processes, such as wave diffraction around a proposed breakwater, or wave refraction and shoaling in the nearshore region. In these cases, as in virtually every coastal physical model, the proper representation of surface wave dynamics is critical to success, and so Froude number scaling is employed along with geometric scaling.

The largest three-dimensional physical model facilities, or wave basins, have horizontal dimensions of the order 100 m. Even so, if one desires to model a coastal system over horizontal scales of 10 km or greater (e.g., to examine the high-tide shoreline impact of a coastal structure, geometric similitude requires a length scale reduction of the order 100). This presents immediate difficulties related to scale effects, the most obvious being the requirement that the water depth must likewise be reduced at model scale by the scale reduction factor of order 100. If, for example, the offshore depths at full scale vary over a range of 1–10 m, the model depths must range from 0.01 to 0.1 m, or 1–10 cm. These shallow water depths give rise to viscous dissipation in the model which is not present at full scale.

Although the modeler may be tempted to employ larger water depths in the physical model so as to avoid the unwanted wave attenuation at model scale, this is strongly discouraged. The use of vertical scale distortion alters the wave kinematics by altering the vertical distribution of velocity and pressure. This in turn alters the wave transformation processes, including refraction, diffraction, and shoaling. Rather than employ scale distortion, the modeler should first investigate whether the viscous effects are appreciable—particularly in terms of the problem being examined. This can be accomplished, for example, by comparing the reduced-scale wave behavior with that expected from either field data or analytical or model results. Such comparisons are in fact advisable regardless of whether scale effects are suspected. If the modeler concludes that scale effects may impact the study results, a decision must be made between proceeding—perhaps accepting that the model results are only qualitative representations of the full-scale system—or reconstructing the model at closer to full scale so as to reduce the relative importance of the model-scale viscous effects.

An additional concern regarding scale effects arises in cases when wave–structure interactions are being examined (e.g., in the model analysis of the influence of a breakwater on wave propagation). The examination of these interactions requires accurate representation of wave transmission and reflection at the structure face. This requires accurate representation of the wave–structure interaction and wave-induced flows. Since this localized behavior is not likely to be well reproduced at model scale because of the lack of Reynolds number similarity and the importance of viscous effects in such behavior (particularly for porous structures such as stone breakwaters), it is probable that the large-scale, three-dimensional model discussed here will not be adequate to address the problem. This situation is one example of a scenario in which a two-dimensional (cross-shore and vertical) physical model can be effective. Two-dimensional physical model facilities, that is "traditional" wave tanks, allow for modeling cross-shore processes and wave–structure interaction at scales larger than those normally utilized in three-dimensional facilities. The use of closer-to-full-scale models assures that even if Froude number scaling is used in the construction of the model, the model-scale Reynolds number is sufficiently high to assure that realistic turbulent flow exists (e.g., in the gaps within a model stone structure). It is in fact not uncommon for a coastal engineering physical model study to be composed of two separate components: (1) a two-dimensional study to examine wave–structure interaction (wave run-up, transmission, reflection, etc.), and (2) a three-dimensional study of wave transformation and other processes, which uses the results of the two-dimensional study in model design and interpretation.

Moveable bed models

The proper simulation of sediment transport dynamics remains one of the most challenging problems facing physical modelers. And yet, the simulation of sediment transport is essential if one is to examine the coastal environment in a complete manner, since we know that bottom topography is dynamic, changing over timescales of many wave periods in response to bottom shear stresses and currents associated with surface waves. In fact, knowledge of the "equilibrium" beach geometry and bottom topography (or nearshore profile) is often a desired output of coastal physical model studies. Since the scope and cost of field observations required to achieve the same results are prohibitive, and since numerical model algorithms to address the problem must still be considered an area of active research, physical model studies remain an important tool in addressing the problem.

For an excellent discussion regarding moveable bed modeling, the reader is referred to Kamphuis (1985) and Kamphuis (2000). A fundamental problem with moveable bed models is that the model sediment grain size obtained via the same geometric scale ratio used for the hydraulic model development will in all likelihood be so small as to introduce scale effects (e.g., cohesive forces, that are not present in the prototype).

For many problems of interest, the selection of the sediment grain size scale reduction factor begins with the assessment of whether bedload or suspended load sediment transport dominates the coastal system being examined. For example, we might expect that suspended sediment transport dominates the environment in the breaking wave region, or surf zone while bedload transport dominates moveable bed dynamics in deeper, less energetic regions of the coast.

In the case of bedload-dominated systems, a reasonable approach appears to be to select the length scale reduction factor based on the ratio of the prototype and model sand grain diameter, with the model sand grain diameter being chosen as to avoid cohesive forces, or roughly speaking, greater than 0.06 mm. Additionally, the model sand must have the same density as that of the prototype sand. In practice, this requires either a large model facility in order to accommodate scale reduction factors of order 10, or the good fortune of dealing with prototype sand having grain diameters measuring several millimeters.

In the case of suspended load-dominated systems, one must correctly scale the sandfall velocity, as this becomes a critically important parameter. Dean (1973) proposed the use of a dimensionless parameter that includes a measure of the erosion capability of the wave and the bias toward deposition represented by the fall velocity of the sand. This parameter, now commonly referred to as the Dean number, can be written:

$$\frac{H}{w\,T}$$

where H is the wave height, T is the wave period, and w is the sediment fall velocity.

The fall velocity of the model sediment can be selected by forcing similarity of the Dean number between model and prototype, using geometric similarity for the wave height and Froude number similarity for the wave period. Since sediment fall velocity is a strong function of grain diameter, this provides the scale reduction factor for the model sediment grain size, which is expected to differ from the length scale reduction factor.

Concluding remarks

Numerical models continue to become more sophisticated in their handling of the complex, turbulent coastal environment. Improved instrumentation allows for the collection of high-resolution data that are being used to both improve our understanding of coastal dynamics, and to provide essential data for numerical model calibration and verification. However, our knowledge of the complex interactions of waves, currents, and (moveable) bottom bathymetry across scales ranging from wave period to several years, has not progressed to the point where accurate simulations of coastal processes can be made solely with the use of numerical models. It would appear therefore, that physical models will remain an important tool for coastal scientists and engineers for the foreseeable future. It is likely in fact that for many problems (e.g., the influence of structures on the coastal environment), the future holds promise of stronger-than-ever collaboration between the physical and numerical modeling community, with physical models being used to provide valuable insights and verification data to the developers of numerical model algorithms.

Michael S. Bruno

Bibliography

American Museum of Natural History, 1996. *Codex Leicester, a Masterpiece of Science*. New York: American Museum of Natural History.

Dean, R.G., 1973. Heuristic models of sand transport in the surf zone. *Proceedings of the Conference on Engineering Dynamics in the Surf Zone*, Sydney, pp. 208–214.

Fischer, H.B., List, E.J., Koh, R.C.Y., Imberger, J., and Brooks, N.H., 1979. *Mixing in Inland and Coastal Waters*. New York: Academic Press.

Hughes, S.A., 1993. *Physical Models and Laboratory Techniques in Coastal Engineering*. New Jersey: World Scientific.

Kamphuis, J.W., 1985. On understanding scale effects in coastal mobile bed models. In Dalrymple, R.A. (ed.), *Physical Modeling in Coastal Engineering*. Rotterdam: A.A. Balkema, pp. 141–162.

Kamphuis, J.W., 2000. Designing with models. *Proceedings of, International Conference Coastal Engineering*, Sydney: American Society of Civil Engineers.

Cross-references

Beach Processes
Erosion: Historical Analysis and Forecasting
Geohydraulic Research Centers
Numerical Modeling
Shore Protection Structures
Wave Climate

PLACER DEPOSITS

A placer is any waterborne deposit of sand or gravel that contains concentrated grains of valuable minerals such as gold or magnetite, grains that had originally been eroded from bedrock but were then transported and concentrated by the flowing water. Placers can be found in rivers (alluvial placers) and on the coast, particularly in beaches (beach placers). In some locations, the ore minerals that initially were concentrated into a beach placer have been blown inland by coastal winds to form mineral-rich dune sands. Beach placers also have been submerged by the rising level of the sea, and are now found as relict placers within sediments on the continental shelf. However, some placer deposits found in continental shelf sands may have originated in that environment, formed by waves and currents that transported the shelf sands and concentrated the ore minerals.

Coastal placers, including those in beaches, dunes, and on the continental shelf, have been mined for gold, platinum, diamonds, the titanium-bearing minerals rutile, monazite and ilmenite, cassiterite (containing tin), magnetite (iron), zircon (zirconium), and garnet (used for making sand paper). Coastal placers are found throughout the world, with the economically most significant deposits being in Alaska (gold), India (magnetite), New Zealand (magnetite), Australia (rutile, monazite, and zircon), Indonesia (cassiterite), and South Africa (diamonds).

Placers are formed under the action of flowing water, by a current or the to-and-fro movement of water beneath waves, which transport the sand and gravel, leading to the concentration of the placer minerals due to their high densities. The range of mineral densities is shown in Figure P24, illustrating that platinum and gold have extremely high densities, while cassiterite, magnetite, etc. (generally referred to as heavy minerals) range in densities from about 3.5 to 7.0 g/cm^3. Of importance, these densities are greater than those of quartz and feldspar (the light minerals), which have a density on the order of 2.65 g/cm^3. Included in Figure P24 is the density of water, 1.00 g/cm^3 for fresh water and slightly greater for sea water.

Most deposits of sand on the continents are composed mainly of quartz and feldspar, with only small percentages of heavy minerals. When transported by currents and waves, the processes of selective grain sorting can occur where the quartz and feldspar are transported more easily and at higher rates, leaving behind a concentrated deposit of heavy minerals (Komar, 1989). In that the heavy minerals are generally dark in color, the result is the formation of a "black sand" deposit; if it contains a valuable ore mineral, the deposit then also qualifies as a placer.

The process of formation of black sand and placer deposits can be readily observed on beaches, especially during a storm that erodes back the beachface and berm, transporting the eroded sand to offshore bars. The tan-colored grains of quartz and feldspar are easily picked up by the swash of the waves at the shore, and are quickly transported offshore, leaving behind the heavy minerals as a concentrated black-sand layer on the eroded beach. In that the different heavy minerals have a range of densities and colors (e.g., pink garnet versus black magnetite), some selective grain sorting between the heavy minerals may also be seen (Komar, 1989). With the return of smaller waves following the storm, the quartz and feldspar sand moves back onshore, burying the black-sand layer. The search for beach placers, therefore, often necessitates digging trenches through the beach sand in order to reach the black-sand deposit at depth.

Having initially formed within the active swash zone of the beach, the placer minerals may be washed inland by still stronger storms, or blown there by the wind to form dunes. If the area of coast is accreting due to an abundant supply of sand, with the beach environment slowly shifting seaward, beach placer deposits dating hundreds to thousands of years old may be found well inland. This is the case on the southeast coast of Australia, where research has focused on the formation of such deposits

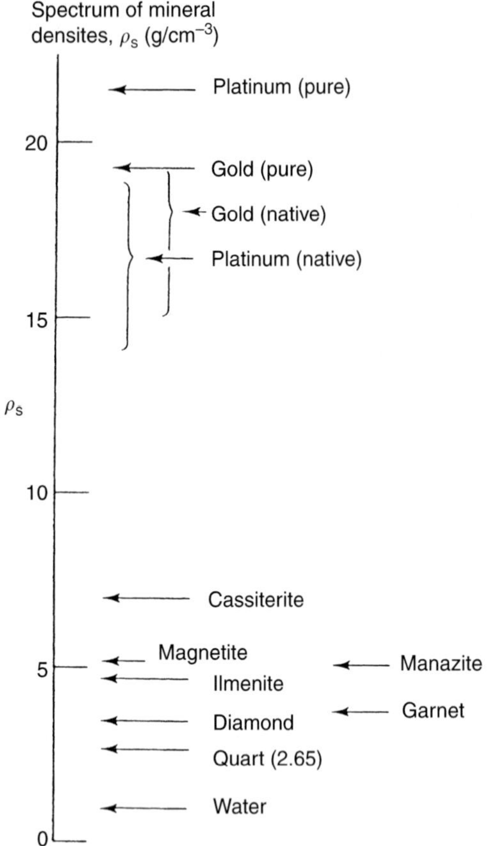

Figure P24 The range of densities of quartz grains and representative minerals found in placers (from Komar, 1989). With permission of CRC Press.

(Roy, 1999), which are also the deposits most actively mined at present, the mineral resource on the modern beach having been exhausted by mining in the past. The search for placer deposits also has extended out onto the continental shelf off the southeast coast of Australia, but generally with disappointing results. Research suggests that ocean waves may preferentially transport the heavy minerals and placer minerals onshore from the shelf sands, contributing to their concentration in the beaches, but depleting the shelf sands (Cronan, 2000). The principal minerals that are being mined from continental shelf placers are gold (Alaska), cassiterite (Indonesia), and diamonds (South Africa).

Paul D. Komar

Bibliography

Cronan, D.S., 2000. *Handbook of Marine Mineral Deposits*. Boca Raton: CRC Press.

Komar, P.D., 1989. Physical processes of waves and currents and the formation of marine placers. *Reviews in Aquatic Sciences*, **1**: 393–423.

Roy, P.S., 1999. Heavy mineral beach placers in southeastern Australia. Their nature and genesis. *Economic Geology*, **94**: 567–588.

Cross-references

Beach Features
Beach Sediment Characteristics
Beach Stratigraphy
Cross-Shore Sediment Transport
Eolian Processes
Mining of Coastal Materials
Shoreface
Surf Zone Processes

PLEISTOCENE EPOCH

The Pleistocene comprises most of the geological epoch of the Quaternary. It started at the end of the Tertiary 1.8–2.3 million years ago with a remarkable cooling of the earth's climate. The beginning of the Pleistocene is set when the first "cold guests," such as the mollusk *Cyprina islandica* or the foraminifer *Hyalinea balthica*, appeared in the Mediterranean (1.6 million years ago), or at the boundary of the Gauss-Normal paleomagnetic period to the Matuyama-Reverse paleomagnetic epoch 2.3–2.4 million years ago (Figure P25). It ended with the establishment of a warmer climate similar to modern conditions about 10,000–11,000 years ago, called the Holocene. Most characteristic of the Pleistocene was not only cooling, but a multitemporal change between cold and warm periods, established by oxygen isotope measurements (ratio of $^{16}O/^{18}O$ in foraminifera of the tropical Pacific). There were more than 20 of these cool and warm periods, the cold ones becoming real ice ages during the younger part of the Old Pleistocene less than 1 million years ago, but inland ice was present for a much longer time on the Antarctic continent. The terrestrial records, however, are much more incomplete than those from deep sea sediments, and the number and exact dating of cold and warm phases is under debate for all periods older than about 200,000 years (see also *Geochronology*).

Mostly the Pleistocene is set synonymous with the ice ages. During the cold periods, glaciers formed in arctic and antarctic latitudes moving equatorwards to about 50°N in Europe and even 40°N in North America. High mountain areas have been glaciated even along the equator. The cooling was much more intensive in higher latitudes, reaching 10–15° compared to equatorial regions, where about 4° lower temperatures than today occurred, so that smaller belts of coral reefs as well as tropical rain forest could survive. The ice ages also were epochs of dryer climate and a significant shifting of all vegetation belts toward lower latitudes, while forests were extinct in higher and most of the middle latitudes. The tundra-like grasslands were characterized by large mammals like the mammoth and mastodon, long-horned bison, cave bear or sable-toothed tiger as well as camels and horses, most of them extinct at the end of the Pleistocene epoch, perhaps accelerated by hunting man. The many and intensive climatic changes during the Pleistocene epoch certainly have had impacts on the evolution of plants and animals including man, and the Pleistocene is the period of human development from early hominids to *Homo sapiens sapiens* and its distribution over all continents.

With the rhythm of cold and warm periods (the first lasting many thousand to tens of thousands of years, the latter several to many thousand of years) the balance of liquid and frozen water on the globe changed significantly, leading to a lowering of the oceans in the amount of 100 m to more than 150 m during the ice ages and sea levels not far from today's during the peak of the warmer periods. This evidently has had a remarkable influence on coastal-forming processes, but we have little information on coastal phenomena of the sea-level lowstands. Much more evidence could be gained, including dating of the sea-level oscillations, in areas of ongoing uplift, where marine or littoral terraces have been developed by abrasion (such as in southern Italy) or by reef construction with high staircases of coral reef terraces found on the island of Barbados or on the Huon Peninsula of Papua, New Guinea. From this evidence, and from paleotemperature curves gained from foraminifera of tropical deep sea cores, we have learned that sea level was highest during the last interglacial period around 100,000–125,000 years ago, reaching about +6 m. Its remnants are widespread around the coastlines of the world. The last lowstand during the peak of the last glaciation (called Würm, Weichsel, Vistulian, Wisconsin or just LGM = Last Glacial Maximum) was around −100 to −120 m about 23,000–18,000 years ago. This was the last phase with wide exposed shelf areas and landbridges, enabling mankind to spread and bridge distances which now are covered by the ocean again.

Further reading on this topic may be found in the following bibliography.

Dieter H. Kelletat

Bibliography

Dawson, A.G., 1992. *Ice Age Earth. Late Quaternary Geology and Climate*. London, New York: Routledge.

Ehlers, J., 1996. *Quaternary and Glacial Geology*. Chichester, UK: Wiley.

Riser, J.A.M., 1999. *Quaternary Geology and the Environment*. Chichester, UK: Springer.

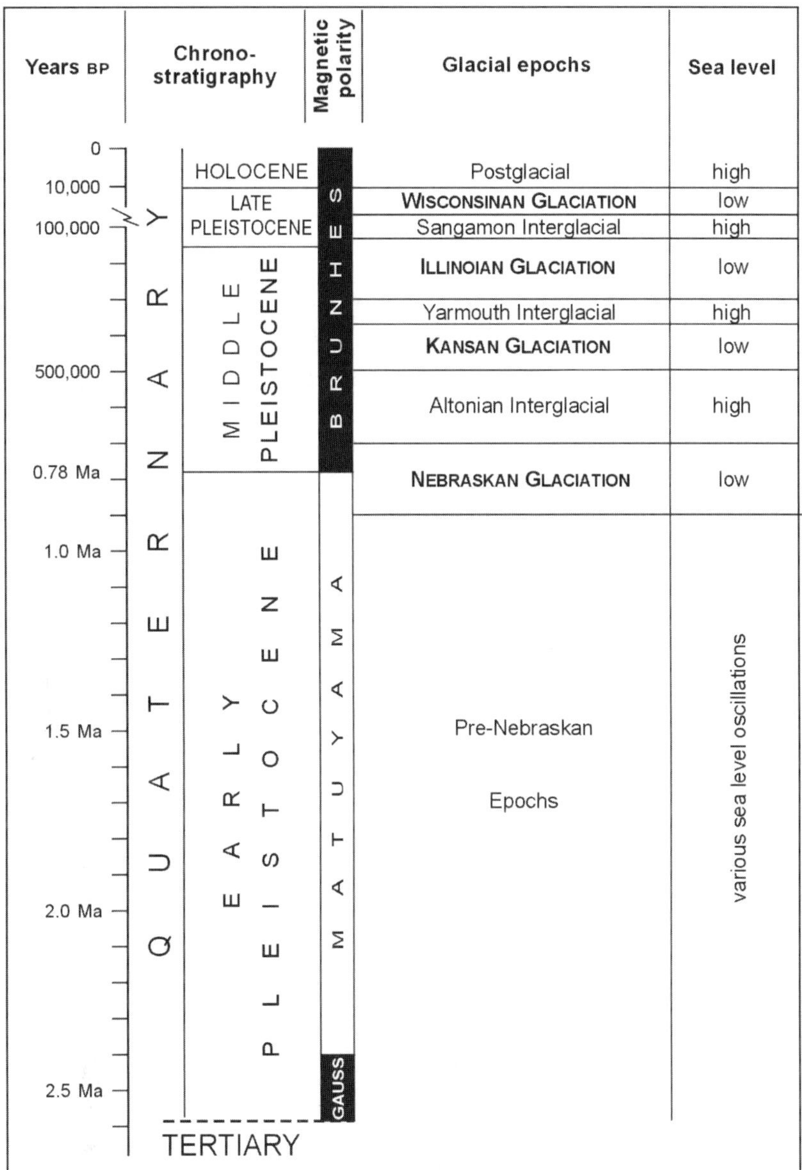

Figure P25 Main units of the Pleistocene epoch.

Shackleton, N.J., and Opdyke, N.D., 1976. Oxygen Isotope and Paleomagnetic Stratigraphy of Pacific Core V 28–239: Late Pliocene to Latest Pleistocene. *Geological Society of America, Memoir*, **145**: 449–464.

Cross-references

Geochronology
Holocene Epoch
Holocene Coastal Geomorphology

PLUVIAL LAKE SHORE DEPOSITS

Pluvial lakes are lakes that existed in enclosed basins in the interior of continents during times of enhanced rainfall, reduced evaporation, or a combination of both. These lakes in many cases left a valuable record of paleoclimates in the form of lake sediments, of erosional shore features, and of gravelly to sandy shore deposits. Pluvial lake shore features have been recognized in many parts of the world, but are particularly well known from North America. Here, they were first described, identified, and interpreted in classical studies by geological pioneers I.C Russel (1885), and G.K. Gilbert (1890) who worked on the shore deposits of Lake Bonneville ("the proto Great Salt Lake"), a pluvial lake that existed during the Wisconsinan/ Weichselian Ice Age.

In the last few years, Lake Bonneville shore deposits have received renewed attention. Many of the internal structures of these deposits remained unknown because of slumping or other types of outcrop deterioration. Only where gravel pits, road building, or fluvial dissection afforded access, could internal sedimentary structures be defined and interpreted (Oviatt and Miller, 1997). Now, however, the advent of Ground Penetrating Radar (GPR) permits identification of previously unknown sedimentary structures. For instance, Smith *et al.* (1997) described gravelly shore deposits from Lake Bonneville, and identified a sequence of sedimentary processes that led to the deposition of gravelly shore features.

Other gravelly shores exist further west, in California and Nevada. One of the pluvial lakes that existed here was Lake Lahontan (Figure P26) which left a variety of shore features. This is illustrated for the Jessup embayment, Nevada, by Adams and Wesnousky (1998, and references therein). The reader is referred to this excellent description of shore features and shore processes. The authors were successful in separating shore features formed during both transgressive and regressive as well as highstand phases of lake-level changes. Examples of other pluvial lakes that existed are Lake Russel (the "proto-Mono Lake") and Lake Manly (Figure P27) that filled the central part of Death Valley.

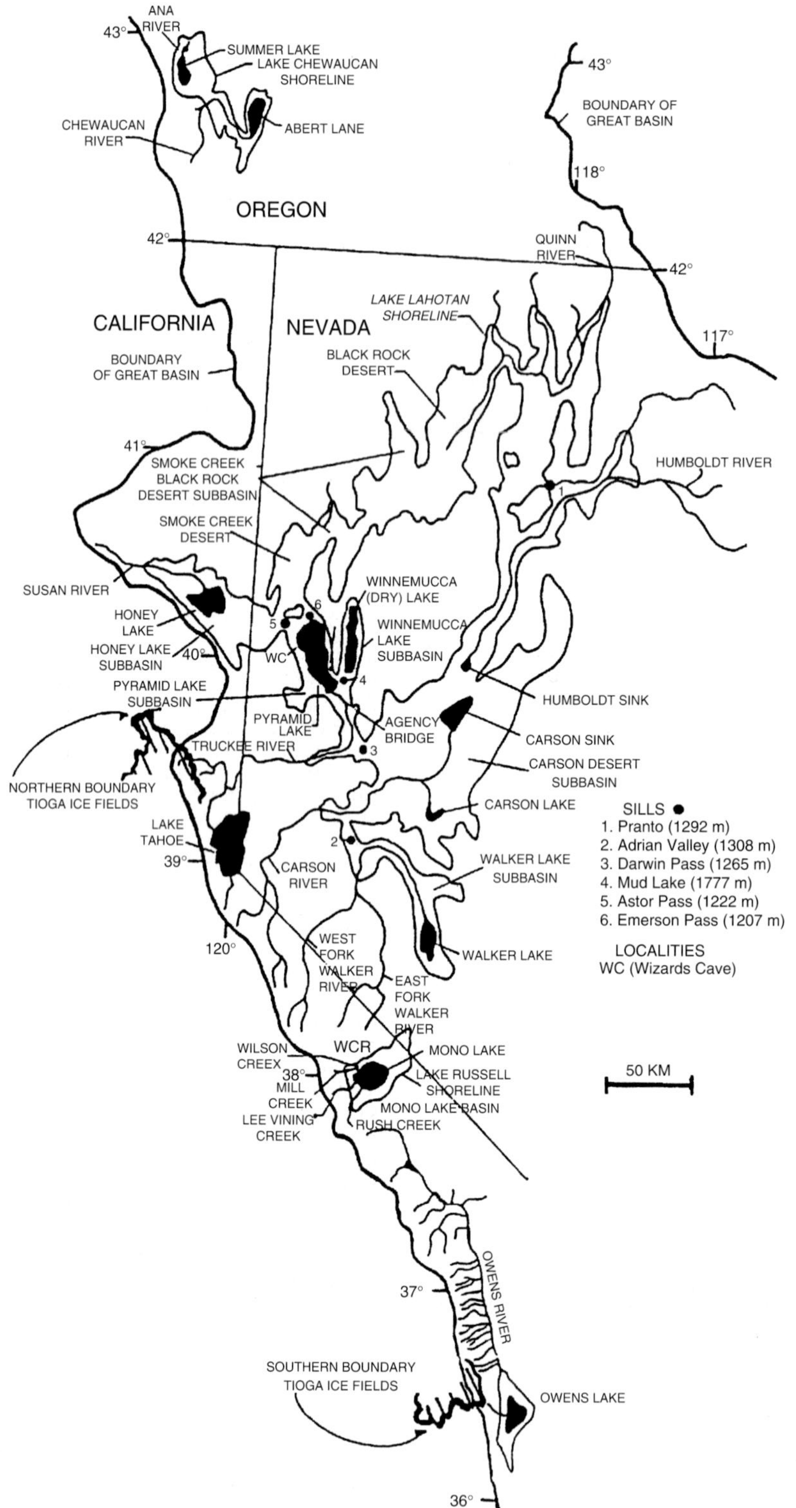

Figure P26 Map of Lake Lahontan, from Benson (1999). Reproduced with permission of American Geophysical Union.

Although these pluvial lakes left a legacy of both erosional and depositional features, this entry is primarily concerned with depositional features, consisting mainly of beach gravel. After a general description of beach gravel, we will illustrate these deposits, based on a study of a deposit from Lake Manly by Ibbeken and Warnke (2000).

Beach gravel consists of clasts which "tend to be well rounded, well stratified, and well sorted within strata" (Adams and Wesnousky, 1998, p. 1319). Clast sizes range from pebbles to cobbles or larger. In most cases beach-gravel deposits are clast supported, but may have a matrix of coarse sand to granules (Adams and Wesnousky, 1998). Depositional features are various types of barriers, spits, etc., similar to equivalent features on marine coastlines, and produced by washover processes during storm events, longshore transport, and simple aggradation on the beachface. However, the rapid lake level changes can produce lakeshore deposits with a history and internal architecture "notably more complex than their marine counterparts" (Blair, 1999, p. 217). Blair (1999) studied the Churchill Butte shore deposit in western Nevada, about 52 km east-southeast of Reno. The age of the deposit is about 14.5 ka. The shore remnant is easily accessible because it is bisected by US Hwy Alt. 95, leading from Fallon, Nevada to Carson City, Nevada. The locality affords an excellent introduction to gravelly lakeshore deposits, based upon the analysis of the deposit provided by Blair (1999). The following description is extracted from Blair (1999) to which the reader is deferred for more detailed information. "The beach deposit consists of a lakeshore barrier spit and a lake lower-shoreface spit platform. The lakeward side of the barrier consist of beachface deposits (1–10-cm-thick beds of granules and pebbles) sloping 10–15°. These interfinger downslope with thicker and less steep lakeward-dipping beds of pebble gravel of the lake upper shoreface. Interstratified with these beds are high-angle cross-beds that dip southward, alongshore. Landward-dipping (15–20°) sets form the proximal backshore of the barrier, deposited by overwash processes during storms. Fossiliferous sand and mud exist landward of the barrier in what was a lagoon, separated from the lake by the barrier. The lake lower shoreface consists of a southward prograding spit platform constructed by longshore drift, and is characterized by 'Gilbert' foresets of pebble gravel dipping southward 16°." The various facies identified by Blair (1999) are described in Table P3.

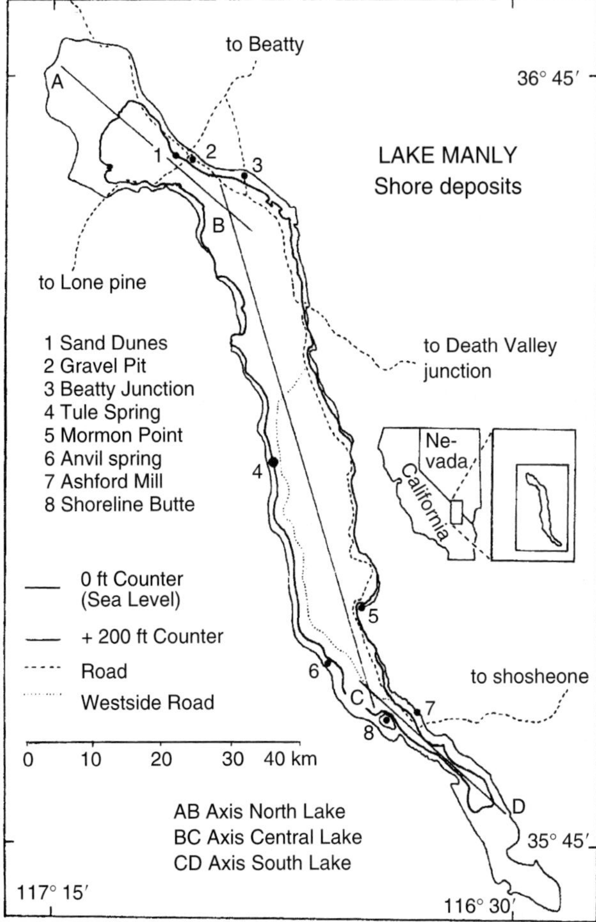

Figure P27 Sketch map of several, major known localities of shore features in Death Valley National Park. Zero feet (0 m) and 200 ft (61 m) contour lines indicated. From Ibbeken and Warnke, 2000. Reprinted with permission from Kluwer Academic Publishers.

The Hanaupah-Fan shore deposit at Tule Spring, Death Valley National Park, California

During the Pleistocene, a series of lakes existed in the enclosed basin of Death Valley, collectively referred to as Lake Manly. Higher lake levels existed during pre-Wisconsinan/Weichselian ice ages. For instance, the higher Beatty Junction bar complex (Figure P28) was deposited during the Illinoian (?) according to Orme and Orme (1991) although Klinger (2001) favors a younger age, based on soil development on the main spit. The reason(s) for higher lake levels in pre-Wisconsinan times are still in debate. A comprehensive description of shoreline elevations is provided by Meek (1997). During the Wisconsinan, the lake level oscillated about the 0 m contour line in the valley, and left a series of shore features that have been and continue to be, the subject of investigations. (Note added in proof: Machette *et al.*, 2003, how consider the deposit described below, to be 128–145 ka, correletive with oxygen-isotope stage 6).

Here, we describe a remarkable, gravelly shore deposit that exists west of Tule Spring (Figure P27). It can be seen from the Westside Road in Death Valley National Park, and can be reached by a short hike across the lower fan surface. The deposit consists of a gently sloping, WSW-ENE elongated ridge, about 600 m long, 165 m wide, and 8 m high (Figure P29). Its surface extends from −12 to +28 m in elevation. The sedimentary inventory consists of cross-stratified gravel beds (Figure P30) of various size ranges dipping in all directions (beachface and overwash sets), of horizontal berm gravel beds, and horizontal silt layers, probably lagoonal overwash deposits landward of a barrier. The deposit is the erosional remnant of a once much larger deposit that extended from the lakeshore east into the lake. Waves from both the north and the south eroded fan materials, and produced a sediment body with a complex architecture. Although at first we favored the idea that the deposit was formed during lake-level rise, albeit with major oscillations (Ibbeken and Warnke, 2000), subsequent ground-penetrating radar GPR profiles (unpublished data) show that the gravel beds have a predominant offlap architecture. As a working hypothesis we suggest that the deposit started to form during a highstand of Lake Manly, but at an unknown elevation since the uppermost part has been lost by erosion. Falling lake level caused the deposit to be extended eastward, and produced the sloping surface that exists today, albeit modified by faulting. In a final phase, a discordant gravel layer was deposited over the entire surface of the deposit. This uniform gravel layer is quite distinct from the surrounding fan surfaces. It is relatively fine grained, better sorted, and densely packed. Rock varnish is very well developed (Figure P31), resulting in a dark surface color that makes the deposit recognizable on aerial photographs. The exact depositional history and facies architecture still remains to be worked out, but the deposit has the potential of yielding valuable information on wave approaches and wave strength during the existence of Late Wisconsinan Lake Manly.

Detlef A. Warnke and Hillert Ibbeken

Table P3 A list of sedimentary facies identified by Blair (1999) from a beach deposit of Lake Lahontan. Note the attitude of the beds with respect to the paleoshore (modified after Blair, 1999)

Facies	Sedimentary features	Depositional environments
A	Unsorted, unstratified, angular, muddy bouldery pebble cobble gravel	Subaerial bedrock-fringing colluvium
B	Thickly bedded, matrix-supported, muddy cobble pebble gravel	Subaerial alluvial-fan debris flows
C	Lakeward-dipping low-angle (5–15°) beds of granular fine to medium pebble gravel	Lake beachface and upper shoreface
D	Landward-dipping (10–20°) foresets of sandy granule fine to medium pebble gravel	Proximal washover in a back-barrier pond
E	Irregular pebbly granule gravel, fossiliferous sand, mud, and diatom beds	Distal back-barrier pond
F	Poorly sorted, sandy granule pebble gravel in south-dipping (10–16°) foresets	Spit-front foresets, lower lake shoreface
G	Burrowed and oscillation ripples, granular sand in south-dipping (1–6°) toesets	Spit-front toesets, lower lake shoreface
H	Horizontally laminated silt and thickly bedded mud and clay	Lake bottom, below storm wave base

Figure P28 Beatty Junction bar complex. This is the main spit (Spit B) of Klinger (2001).

Figure P29 The Hanaupah-Fan shore deposit at Tule Spring. The view is from the Hanaupah-Fan toward the east. Geologists for scale.

Figure P30 Sedimentary features of the Hanaupah-Fan shore deposit. The number plate is 25 × 25 cm for scale. Modified from Ibbeken and Warnke (2000). (A) Debris flow near the base of the deposit. Southeastern limit of deposit. (B) NE dipping beachface sets, capped by berm crest sets. Northeastern limit of deposit. (C) Shore gravel encrusted by tufa. Northeastern limit of deposit. (D) SE dipping beachface deposits. Southern slope of deposit. (E) Beachface (?) deposits. Southwestern limit of deposit. (F) North-dipping beachface deposits, capped by berm crest sets. Northwestern limit of deposit.

Figure P31 Smooth, varnished surface of deposit.

Bibliography

Adams, K.D., and Wesnousky, S.G., 1998. Shoreline processes and the age of the Lake Lahontan highstand in the Jessup embayment, Nevada. *Geological Society of America Bulletin*, **110**: 1318–1332.

Benson, L., 1999. Records of millenial-scale climate change from the Great Basin of the western United States. *American Geophysical Union Monograph*, **112**: 203–225.

Blair, T.C., 1999. Sedimentology of gravelly highstand shoreline deposits, Churchill Butte, Nevada, USA. *Sedimentary Geology*, **123**: 199–218.

Gilbert, G.K., 1890. Lake Bonneville. *US Geological Survey Monograph*, **1**: 438 pp.

Ibbeken, H., and Warnke, D.A., 2000. The Hanoupah-Fan shoreline deposit at Tule Spring, a gravelly shoreline deposit of Pleistocene Lake Manly, Death Valley, California, USA. *Journal of Paleolimnology*, **23**: 439–447.

Klinger, R.E., 2001. Beatty Junction bar complex. In Machette, M.N., Johnson, M.L., and Slate, J.L. (eds.), *Quaternary and Late Pliocene Geology of the Death Valley Region: Recent Observations on Tectonics, Stratigraphy, and Lake Cycles. US Geological Survey Open-File Report* 01–51, pp. A40–A49.

Machette, M., Phillips, F., and Slate, J., 2003. New cosmogenic ^{36}Ce ages for velict soils on alluvial tans in Death Valley, California. ***Geological Society of America Abstracts with Programs***, **35**: 251–258.

Meek, N., 1997. The elevations of shorelines in Death Valley. *San Bernardino County Museum Association*, **44**: 75–84.

Orme, A.J, and Orme, A.R., 1991. Relict barrier beaches as paleoenvironmental indicators in the California desert. *Physical Geography*, **12**: 334–346.

Oviatt, C.G., and Miller, D.M., 1997. New explorations along the northern shores of Lake Bonneville. *BYU Geology Studies*, **42**: 345–371.

Russel, I.C., 1885. Geological history of Lake Lahontan, a Quaternary lake of northwest Nevada. *US Geological Survey Monograph*, **11**: 288 pp.

Smith, D., Currey, D., and Oviatt, C., 1997. Lake Bonneville classic depositional shore features: geochronology, geomorphology, stratigraphy and sedimentology. *Geological Society of America 1997 Annual Meeting Field Trip* No. 17, 26 pp.

Cross-references

Bars
Beach Features
Gravel Barriers
Paleocoastlines

POLDERS

Poldering is practiced for reclaiming arable land in many countries of the world in lacustrine, riverine, and coastal lowlands in areas with impeded drainage.

The art of poldering will be exemplified for the Flemish-Dutch-Northern German lowlands around the eastern shores of the North Sea. Poldering essentially consists of isolating a certain area by diking and improving the drainage of this area by expelling the surplus water. So poldering requires three major abilities: the art of diking, of drainage and of the discharge of surplus water.

The art of poldering in medieval times

The earliest diking and reclamation of peat areas was reported from around 800 AD. Monasteries and cloisters, being centers of knowledge, played an important role in the development of techniques and abilities for diking, poldering, and river training. The dikes were preferentially located on the natural levees of the creeks and rivers. Land reclamation required detailed local knowledge of the topography and hydrography of the area.

In early times dewatering was accomplished by digging of ditches according to the (small) slopes of the terrain. A major technical invention was the so-called "ebb-valve" which opened when the outside water level was lower and closed by the water pressure while the outer level was higher. It was only since 1200 AD that larger channels were dug to connect the major drainage creeks and small channels in order to facilitate the discharge of surplus water. But it was still a drainage under natural gradients, thus on the supratidal areas. Similar practices were employed in the riverine areas by digging of ditches, almost parallel to the main river courses, thus using the natural downslope gradient of the terrain.

It was in the 14th century that the sluice was developed, which made it possible to have different levels in and outside the polder. The sluice allows ships to pass the obstruction.

Poldering and land reclamation not only had beneficial effects by increasing the extension of arable lands. On the negative side, poldering decreases the area of the supratidal and higher intertidal lands. This reduced the tidal storage capacity. The drainage was more concentrated in the channels, which eroded. As a result the tidal wave could penetrate further inland and because of the funnel shape of estuaries and tidal basins the tidal range increased too. Not only the HW levels were heightened but also the storm surge levels. This is especially problematic in coastal lowlands experiencing a sea-level rise.

Dewatering resulted in bio-oxidation and compaction of the peat and clay layers, which obstructed the drainage. In addition, the lowering of the land surface by compaction was increased by the excavation of peat for domestic use and for salt extraction by an ever-increasing population. Many disastrous floods have been reported, and considerable acreage of land was lost.

The art of poldering in the 15th to 18th century

The problem of the drainage of the ever-sinking polders was solved effectively in the beginning of the 15th century by the introduction of the windmill as a pumping device. Much more water from lower levels could be pumped up and discharged and greater areas (combined polders) could be handled. From then on, the way of poldering showed a new setup (Figure P32) consisting of diked areas with ditches draining into internal drainage channels, which in turn drain into natural waters. Also in the riverine areas, the development of the windmills offered tremendous opportunities for improvement of the drainage. A large number of windmills were built here too. These developments resulted in new, larger reclamations, a well-balanced and organized hydrological drainage system and better water management.

However, windmills were not applied everywhere in the coastal and riverin areas. Windmills were costly (equivalent of US $500,000 at present) and have a low capacity. In addition, they could only function effectively at a wind speed above some 4 Beaufort, effectively only some 25% of the time (Van der Ven, 1993). Other areas were too low to be handled by windmills or required too high a discharge capacity. This was especially the case in Flanders and the northern part of the Netherlands and Germany. These areas could only be poldered after better techniques became available.

Poldering in the 19th and 20th century

Steam-powered pumping stations took over the function of the numerous windmills, because they were more powerful and cheaper. Again because of their greater capacity they made the reclamation of larger lakes possible (Figure P33). In addition, this resulted in an increase in scale of the polder units and a rationalization of the system of drainage channels. In the 20th century motor and electric power increased the capacity of the drainage devices and the size of the polders further. After completion in 1932 of the Closure Dike, which separated the former Zuiderzee from the Wadden Sea (Figure P33) three new modern polders were constructed in Lake IJssel between 1937 and 1968, totally covering some 145,000 ha.

Recent developments

In the last decade a discussion was started about depoldering some polders by opening dikes or substantially reduce the drainage. Two arguments are at stake. The first is the expected consequences of an increased sea-level rise viz. a heightening of the HW levels along the coast and the increase in precipitation in the drainage areas of the major rivers, which will negatively affect the safety of the dikes and drainage of the polders areas.

The second argument is that because of the developments of agriculture in the European Community (EC) the need for arable land has reduced considerably and more and more land becomes available for other purposes, including housing, recreation, and landscape restoration and nature conservation. So there is a shift in priorities in rural planning. Several plans are under consideration viz. the use, as in former times, of overflow dikes, to temporarily store water in the polders during peak river floods to reduce the top water levels. Another plan is to reduce the drainage of some higher polders, creating lakes in which housing, floating, or on heightened lands, recreation and nature conservation are foreseen. A third type of plan is to depolder areas in the upper parts of estuaries in order to increase the tidal volume and the tidal currents to reduce the necessary dredging in the shipping routes.

Conclusion

Poldering requires special techniques and abilities: precision leveling and positioning, adequate information of water levels, current velocities, drainage systems and facilities, diking, discharge techniques, and water management policies and organization. In past time these challenges

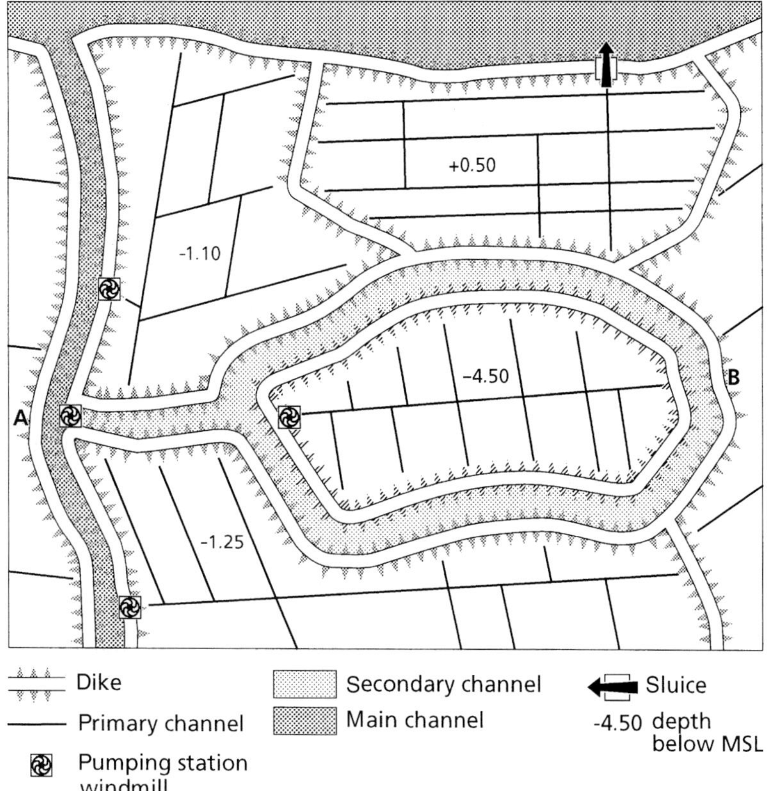

Figure P32 Drainage system in larger polders (from: A compact Geography of the Netherlands. With Permission of the Royal Dutch Geographical Society).

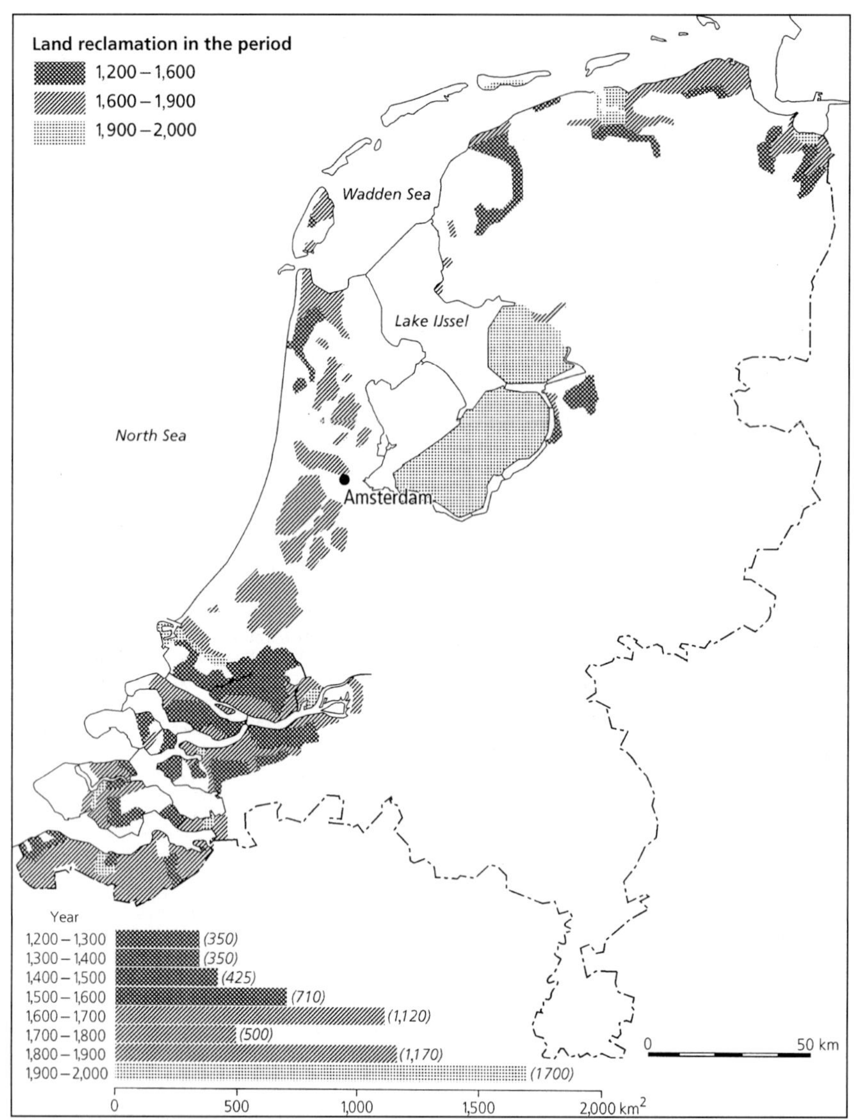

Figure P33 Land reclamation in the Netherlands (from: A compact Geography of the Netherlands. With Permission of the Royal Dutch Geographical Society)

forced the ever-increasing population to solve these problems. In recent times, facing problems of climatic change, there is a need for more space for water management and depoldering of some areas is a serious option.

Joost H.J. Terwindt

Bibliography

Van der Ven, G.P. (ed.), 1993. *Man-Made Lowlands. History of Water Management and Land Reclamation in The Netherlands*. Utrecht: Matrijs.

Cross-references

Alluvial-Plain Coasts
Changing Sea Levels
Coastal Subsidence
Coastal Zone Management
Dikes
Human Impact on Coasts
Reclamation
Salt Marsh
Tidal Environments
Wetlands
Wetlands Restoration

PRESSURE GRADIENT FORCE

Horizontal and vertical forces called pressure gradient forces drive fluids. In vector equations this is symbolically represented by $\vec{\nabla}p$, which is shorthand for "grad p." A pressure gradient force is a spatial expression of Newton's Second Law, $F = ma$, as applied to the ocean or the atmosphere.

Figure P34 shows a Cartesian coordinate system with the x-axis pointing eastward, the y-axis pointing northward, the z-axis pointing upward in the opposite direction of the gravity vector $\vec{g}$. The x–y plane is parallel to a level surface and may be defined as coincident with the equipotential surface, known as the marine geoid, situated approximately a meter or so below sea level. An equipotential surface is everywhere orthogonal to the local gravity vector where the geopotential $\phi = gz$ is a constant, and pressure (p) is a function of x, y, z.

The largest oceanic pressure difference, $\partial p = p_{z_2} - p_{z_1}$, occurs in the vertical over a distance $\partial z = z_2 - z_1$. The ratio $\partial p/\partial z$ is defined as the vertical pressure gradient and is given by the hydrostatic equation $\partial p = -\rho g \partial z$ where ρ is the density of seawater ($\rho_{\text{seawater}} \approx 1025\,\text{kg}\,\text{m}^{-3}$). If written in the form $(1/\rho)(\partial p/\partial z) = -g$, the hydrostatic equation defines a balance between the vertical Pressure gradient force per unit mass and the acceleration due to gravity. In m.k.s. units $(1/\rho)(\partial p/\partial z)$ is dimensionally $[m^3/kg]\,[N m^{-2}/m] = N/kg$, which is force (newtons) per unit mass (kilograms), when pressure difference is expressed in newtons per square meter ($\text{N}\,\text{m}^{-2}$) and vertical distance is expressed in meters (m).

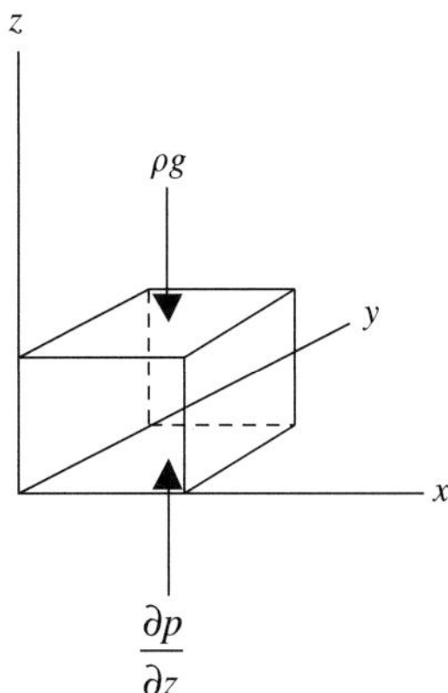

Figure P34 A cubic meter of sea water in hydrostatic balance where the upward pressure gradient force per unit volume $\partial p/\partial z$ is equal to the downward gravity force per unit volume ρg.

In terms of ocean currents, the most important pressure gradient force per unit mass is in the horizontal, $(1/\rho)(\partial p/\partial x)$ or $(1/\rho)(\partial p/\partial y)$ in the east–west direction and the north–south direction, respectively. Recalling that the partial derivative ∂p is taken with respect to a level surface ∂x or ∂y, the horizontal pressure gradient force may be expressed as height departures of a constant pressure surface. From the hydrostatic equation $p_2 = -\int_{Z_2}^{D=0} \rho g \partial z = \rho \partial g h_2$ and similarly for p_1, the pressure difference on a level surface is just $\partial p = p_2 - p_1$. Thus the east–west pressure gradient force per unit mass $(1/\rho)(\partial p/\partial x) = g(\partial h/\partial x)$, and likewise for the north–south direction $(1/\rho)(\partial p/\partial y) = g(\partial h/\partial y)$, where h is the height of the sea surface above the level surface.

In a similar fashion, the pressure gradient force may be expressed in terms of the gradient of geopotential ϕ. Since $\phi = gz$, and $(\partial \phi/\partial x) = g(\partial h/\partial x)$, the horizontal pressure gradient force per unit mass is just $\partial \phi/\partial x$ in the east–west direction, and $\partial \phi/\partial y$ in the north–south. Oceanographers commonly calculate the geopotential anomaly (also called the dynamic height anomaly) from measurements of temperature and salinity versus depth, and compute horizontal currents with respect to a deep reference surface (which is assumed to be a level surface), using the geostrophic equation.

Geostrophy further illustrates the role of horizontal pressure gradients in the sea. Consider the geostrophic equation $fv_g = (1/\rho)(\partial p/\partial x)$, where f is the Coriolis parameter $(1.459 \times 10^{-4} \sin \phi)$, $\partial p/\partial x$ is the east–west direction pressure gradient force per unit volume, and v_g is the north–south geostrophic current. Recall that pressure is an integral $p = \int_h^D \rho g dz$ from the sea surface ($z = h$) to some deep layer where $z = D$. Derivatives of an integral require application of Leibnitz's Rule, which yields:

$$\rho f v_g = \rho g \frac{\partial h}{\partial x} + \int_h^D g \frac{\partial \rho}{\partial x} dz$$

The term on the left-hand side is the force per unit volume due to earth's rotation creating a geostrophic current, v_g. On the right-hand side, the first term is proportional to the slope of the sea surface with respect to a level surface, and the second term is that due to the internal horizontal gradient of the density field. The term $\rho g(\partial h/\partial x)$ is the barotropic pressure gradient force per unit volume, from which the surface geostrophic current $v_g(z = 0)$ is calculated, and $\int_h^D g(\partial \rho/\partial x)\, dz$ is the baroclinic pressure gradient force per unit volume.

To illustrate these terms, consider a crossing of the Gulf Stream from Charleston, South Carolina to Bermuda. The sea surface slopes upward to the east a total of about 1 m and thus the barotropic term is positive. The subsurface horizontal density gradient $\partial \rho/\partial x$ however is negative, and at some depth $z = D_0$ the integral of the baroclinic term will be equal in magnitude but opposite in sign to the barotropic term. Thus at this horizon $v_g(z = D_0) = 0$, a depth called the level of no motion. In the offing of Charleston, D_0 typically is assumed to occur at about $z = 2000$ m.

Further reading on this subject may be found in the following bibliography.

George A. Maul

Bibliography

Knauss, J.A., 1997. *Introduction to Physical Oceanography*. Second Edition. Upper Saddle River, NJ: Prentice Hall.

Ocean Circulation, Second Edition, 2001. The Open University. Oxford: Butterworth-Heinemann.

Sverdrup, H.U., Johnson M.W., and Flemming, R.H., 1942. *The Oceans, their Physics, Chemistry, and General Biology*. Englewood Cliffs, NJ: Prentice-Hall.

Cross-references

Coastal Currents
Coastal Upwelling and Downwelling
Geodesy
Remote Sensing of Coastal Environments
Submarine Groundwater Discharge
Vorticity

PROFILING, BEACH

Surveys of the beach profile are conducted to locate the high-tide shoreline or land and sea boundary, to determine the shape or morphology of the beach, to evaluate the performance of shore-protection projects, and to monitor the volume of sediment in the beach. Legal or regulatory marine boundaries are usually determined as the location of a certain elevation related to a tidal datum such as mean high water, requiring a survey of the upper beach profile to the foreshore (see *Beach Profile* entry for terminology). Marine boundary issues and surveys are described by Shalowitz (1962, 1964) and Flushman (2001) and are not discussed here. This section describes surveys made to determine beach morphology and measure beach volume.

Beach profile surveys include considerations that are not part of traditional land surveying procedures or of bathymetric or hydrographic surveys. Knowledge of coastal processes is necessary to understand operations in surf zone waves, longshore current, and rip currents, as well as in the beach morphology itself. For example, the surveyor can "go off line" in strong currents, and survey points should be closely spaced in areas of the profile having steep gradients such as at the dune toe, scarp, and longshore bars. Surveyors who are familiar with the morphology of the beach profile will pay attention to its detail.

The length and spacing of survey lines (often called transects) alongshore depends on the purpose of the survey. A beach profile survey extends seaward from an established control point on land along a predetermined heading, typically along a line normal to the local trend in the high-tide shoreline. The starting point might be placed at or landward of the cliff, dune, or seawall at the coast. The survey proceeds seaward to the depth of closure (Kraus *et al.*, 1999), where the profile does not change for "long lines" and to wading depth for "short lines." If long lines are surveyed, the area of active sediment movement by waves on the nearshore beach profile is encompassed, and change in beach volume can be measured by taking the difference of two surveys performed at different times.

Methods of beach profile surveying

Land (inshore) survey

An early method for making an inshore survey that requires only a stadia or survey rod was devised by Emery (1961). An observer records levels on the survey rod by sighting parallel across to the horizon as the rod holder moves along the beach profile. This method is still employed to make inexpensive and adhoc inshore surveys. The standard rod-and-transit method of surveying has been replaced by readily available "total survey stations," which register data electronically and provide distance and elevation through setting of an angle to the baseline and measurement of the distance of the instrument to a survey monument. A total station measures time of flight of a beam transmitted from a light-emitting diode to and reflected from the survey prism. Such surveys provide the highest accuracy (subcentimeter) in horizontal and vertical position. Differential Global Positioning Systems (DGPS) can also be used. These typically give 1–2-cm accuracy in the horizontal and 2–4-cm accuracy in the vertical, and are most suited for reconnaissance or wide-area surveys or if measurement of beach volume is not the survey purpose.

The reproducibility or precision of the elevation measurement obtained from a total survey station was investigated in a beach environment with heat waves reflecting from the water and high humidity, which is typical (Kraus and Heilman, 1998). The standard deviation in the fixed height of a survey rod is plotted in Figure P35. For up to a kilometer from the survey station, precision is less than 1 cm. Depending humidity and glare, a total station has a range of 1–3 km, involving atmospheric conditions and presence of reflections of sunlight from the sea surface. Sled surveys are more labor intensive and require more time than boat surveys.

Marine (offshore) survey

Traditionally, offshore surveys have been done by boat serving as a platform for an acoustic echo sounder. Such a survey brings great potential inaccuracy because of motion of a vessel in waves and requirement for a tidal correction to be made for the water level throughout the survey period. Tide corrections are difficult to define for great longshore and offshore distances from a tide gauge. In addition to being inaccurate, echo sounder surveys by boat have the difficulty of joining to the inshore survey. Gibeaut *et al.* (1998) discuss procedures and improvements that can be made for echo sounder surveys.

For greatest accuracy of beach profile measurement, a sea survey sled is recommended (Grosskopf and Kraus, 1994). A total station (or a transit) can sight on survey prisms mounted on a mast of a sled that is towed to sea by boat. Figure P36 shows a sea sled with a mast containing a halo of survey prism targets that can be reached from a wide angle to minimize relocation of the survey station alongshore. A survey sled can be winched to shore or towed from shore to measure the entire subaqueous beach profile, eliminating a wading-depth inshore survey. Sled accuracy in the vertical is on order of ±1 cm. Recently, kinematic differential GPS receivers have been placed on personal watercraft (jet skis) that enables near-decimeter vertical accuracy (Wamsley and Edge, 2001).

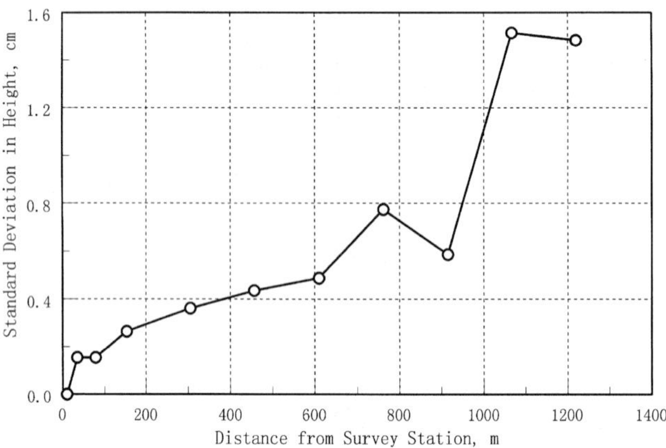

Figure P35 Standard deviation in measured height of prism at fixed elevation for 10 measurements taken at each specified distance from total survey station.

Overlapping of inshore and offshore surveys

A necessary and nonomittable aspect of a beach profile survey is to assure that there is adequate overlap between the inshore and offshore surveys. Assuming that the inshore survey is most reliable, overlap of the two surveys on each transect serves as quality control for the offshore survey and provides an adjustment for it, if necessary. An overlap distance of about 30 m and at least five survey points is recommended.

Sources of error

Highest accuracy is necessary to measure the beach profile if beach fill volume is to be tracked through time. An apparent small error in elevation (say, 10 cm) of the profile over several kilometers of beach can introduce uncertainty larger than the volume of sand originally placed. High accuracy, taken to be at centimeter level in elevation, is not necessary if profile morphology alone is being investigated.

A beach profile survey produces a set of distance-elevation datapoint pairs on specified shore-normal transects established along the shore, usually at fixed intervals, but sometimes becoming more dense in areas of interest and telescoping to a greater interval if located far from an area of interest. These transects are typically reoccupied to perform surveys through time. Such transects may not be reoccupied exactly in successive surveys, and slightly different location of the survey point and angle of the survey over a profile morphologic feature may produce slight variations between surveys that do not represent actual change.

The greatest source of inaccuracy in a beach profile survey is use of an inaccurate measurement method for the offshore survey, such as a boat-mounted echo sounder. Unless the water is calm and the water level relatively unchanging, such as in the Great Lakes, accuracy may be ±30 cm and not systematic between the transects of the same survey and between surveys at different times for the same transect. Overlapping of the inshore and offshore surveys will give a measure of this inaccuracy.

Consistency in operation of the survey station is desirable. Operators will sight somewhat differently on a target. To avoid possible operator

Figure P36 Sea sled on foreshore, North Padre Island, Texas.

bias between surveys, the same person should sight the survey station. In this way, any bias by a given operator will tend to cancel if taking a difference between surveys to document volume change and change in profile morphology. Errors due to beach and sea bottom properties fall partially in the category of operator error, in that different people holding a survey rod will place the rod on the beach or bottom with slightly different pressure. Affixing a small horizontal plate (typically about 5 cm in diameter) to the bottom end of the survey rod will minimize this error. Sea sled runners are made wide so as not to penetrate the seafloor.

Monumentation and documentation errors must be avoided if different organizations are involved in separate surveys at the same site. Different organizations tend to place transects in different locations and assign different names to the same transects; as much as possible, this confusion should be avoided through planning and communication. Monuments should be well controlled and tied to a permanent baseline with recovery feasible if a few monuments are lost. As much as possible, permanent (deep-driven) monuments should be placed on fixed land or landward of the dunes to reduce possibility of erosion.

Survey planning

Well-developed survey plans improve data quality. Data should be examined in the field by, for example, comparing previous surveys done at the same transect. Specialized profile survey analysis software such as the Beach Morphology Analysis Package (BMAP) (Sommerfeld *et al.*, 1993; US Army Corps of Engineers, 1995) makes data checking, editing, and archiving convenient. Field examination can identify datum errors, errors in spacing across shore such as not containing enough detail, and overlap mismatches.

Only experience at a given coast and comparison of profile surveys on adjacent transects can assure proper alongshore spacing. On an open coast with little curvature and no coastal structures, a spacing of 300 m between transects has proven satisfactory for measurement of beach volume. Long-term monitoring surveys might use 1-km spacing for documenting long-term trends. Near inlets and structures, the spacing may be reduced in step fashion to 500, 250, 100, and even 50 m. Local design, permitting, and construction requirements may also dictate the appropriate spacing.

Beach profile surveys are normally considered as a synoptic representation of the beach at a given time. Therefore, the full survey or logical subsets of the survey should be completed in the shortest possible time. Attention to the weather forecast will reduce interruptions by adverse seas and precipitation. Prior to a survey, the local authorities should be notified and permissions obtained, as necessary, both to access the beach and to avoid delays in obtaining permissions. Coordination with the city manager, city engineer, park ranger, police, lifeguards, and others may be necessary. It is recommended that a brochure be prepared to distribute to bystanders so that the attention of the surveyors is not interrupted.

All necessary equipment and backups, such as batteries, should be prepared to avoid delays in the survey. Pairs of highly visible temporary markers placed on each transect aid in aligning the inshore and land survey operators. Two-way radios, flags, gloves, tools, and first aid kit are part of the standard equipment needed for any kind of survey. A camera for taking pictures at each profile is also recommended. Further information on performing beach profile surveys can be found in Grosskopf and Kraus (1994).

Nicholas C. Kraus

Bibliography

Emery, K.O., 1961. A simple method of measuring beach profiles. *Limnology and Oceanography*, **6**: 90–93.

Flushman, B.S., 2001. *Water Boundaries: Demystifying Land Boundaries Adjacent to Tidal or Navigable Waters*. Wiley Series in Surveying and Boundary Control. New York: Wiley.

Gibeaut, J.C., Guitierrez, R., and Kyser, J.A., 1998. Increasing the accuracy and resolution of coastal bathymetric surveys. *Journal of Coastal Research*, **14**(3): 1082–1098.

Grosskopf, W.G., and Kraus, N.C., 1994. Guidelines for surveying beach nourishment projects. *Shore & Beach*, **62**(2): 9–16.

Kraus, N.C., and Heilman, D.J., 1998. Comparison of beach profiles at a seawall and groins, Corpus Christi, Texas. *Shore & Beach*, **66**(2): 4–13.

Kraus, N.C., Larson, M., and Wise, R.A., 1999. Depth of Closure in Beach-Fill Design. *Proceedings 12th National Conference on Beach Preservation Technology*, Tallahassee, Florida: Florida Shore & Beach Preservation Association, pp. 271–286.

Shalowitz, A.L., 1962. *Shore and Sea Boundaries. Volume 1: Boundary Problems Associated with the Submerged Lands Cases and the Submerged Lands Act*. Publication 10-1, US Department of Commerce, Coast and Geodetic Survey.

Shalowitg, A.L., 1964. *Shore and Sea Boundaries. Volume 2: Interpretation and Use of Coast and Geodetic Survey Data*. Publication 10-1, US. Department of Commerce, Coast and Geodetic Survey.

Sommerfeld, B.G., Mason, J.M., Larson, M., and Kraus, N.C., 1993. Beach morphology analysis package (BMAP). In Stauble, D.K., and Kraus, N.C. (eds.), *Beach Nourishment Engineering and Management Considerations. Proceedings Coastal Zone '93*, American Society of Civil Engineers, pp. 162–175.

US Army Corps of Engineers (USACE), 1995. Beach Morphology Analysis Package (BMAP), Version 1. CETN II-34, U.S. Army Engineer Research and Development Center, Coastal and Hydraulics Laboratory, Vicksburg, MS.

Wamsley, T., and Edge, B.L., 2001. Shoreline monitoring program on the Texas Coast by real-time kinematic differential global positioning system. *Shore & Beach*, **69**(1): 25–31.

Cross-references

Beach and Nearshore Instrumentation
Beach Profile
Coastal Boundaries
Coasts, Coastlines, Shores, and shorelines
Global Positioning Systems
Mapping Shores and Coastal Terrain
Monitoring, Coastal Geomorphology
Nearshore Geomorphology
Tidal Datums

R

RADARSAT-2

RADARSAT-2, the second in a series of Canadian spaceborne Synthetic Aperture Radar (SAR) satellites, was built by MacDonald Dettwiler, Richmond, Canada. RADARSAT-2, jointly funded by the Canadian Space Agency and MacDonald Dettwiler, represents a good example of public–private partnerships. RADARSAT-2 builds on the heritage of the RADARSAT-1 SAR satellite, which was launched in 1995. RADARSAT-2 will be a single-sensor polarimetric C-band SAR (5.405 GHz).

RADARSAT-2 retains the same capability as RADARSAT-1. Morena et al., 2004 For example, the RADARSAT-2 has the same imaging modes as RADARSAT-1, and as well, the orbit parameters will be the same thus allowing co-registration of RADARSAT-1 and RADARSAT-2 images. Furthermore, radiometric and geometric calibration is maintained thus permitting correlation of time series data for applications such as long-term change detection (Luscombe and Thomson, 2001).

The following features of the RADARSAT-2 system are thought to be the most significant in terms of their impact on existing and new applications.

Polarization modes. Three polarization modes: Selective, Polarimetry, Selective Single.

Resolution. 3 m ultra-fine mode and a 10-m Multi-Look Fine mode.

Programming lead time. Programming is defined as the minimum time between receiving a request to program the satellite and the actual image acquisition. Routine image acquisition planning is base-lined at 12–24 h, and emergency acquisition planning is base-lined at 4–12 h.

Processing. Routine processing is base-lined at 4 h; emergency processing is base-lined at 3 h; and 20 min for processing a single scene.

Re-visit. Re-visit is defined as the capability of the satellite to image the same geographic region. Re-visit is improved through the use of left- and right-looking capability.

Georeference. Image location knowledge of <300 m at down-link and <100 m post-processing.

RADARSAT-2 polarimetry modes

The RADARSAT-2 polarimetric capability is considered to be the most significant in terms of increasing the information content of the SAR imagery, and is subsequently discussed in more detail. To date, SAR data have been widely available from single channel (single frequency and polarization) spaceborne radars including ERS-1 and 2, JERS-1, and RADARSAT-1. RADARSAT-2 provides polarized data, and is the first spaceborne commercial SAR to offer polarimetry data.

The intent here is not to outline polarimetry theory, but to present the concepts in an intuitive manner so that those not familiar with polarimetry can understand the benefits of polarimetry and the information available in polarimetry data. Many articles are available that discuss polarimetry theory, applications, and provide excellent background information (Ulaby and Elachi, 1990; Touzi *et al.* 2004). Notwithstanding the inherent complexity of polarimetry, polarimetry in its simplest terms refers to the orientation of the radar wave relative to the earth's surface and the phase information between polarization configurations.

RADARSAT-1 is horizontally polarized meaning the radar wave (the electric component of the radar wave) is horizontal to the earth's surface (Figure R1). In contrast, the ERS SAR sensor was vertically polarized, implying the radar wave was vertical to the earth's surface. Spaceborne SAR sensors such as RADARSAT-2, ENVISAT, and the Shuttle Imaging Radar have the capability to send and receive data in both horizontal (HH) and vertical (VV) polarizations. Both the HH and VV polarization configurations are referred to as co-polarized modes. A second mode, the cross-polarized mode, combines horizontal send with vertical receive (HV) or vice versa (VH). As a rule, the law of reciprocity applies and HV ≅ VH (Ulaby and Elachi, 1990).

A unique feature of RADARSAT-2 is the availability of polarimetry data, meaning that both the amplitude and the phase information are available. The amplitude information is familiar to SAR users, but the phase information is likely new and rather nonintuitive. In its simplest term, phase can be thought of as the travel time for the SAR signal: the travel time is the two-way time between the sensor and the earth, and includes any propagation delays as a result of surface or volume scattering. It is the propagation delays and the scattering properties of the HH and VV polarization configurations that make polarimetry data so powerful.

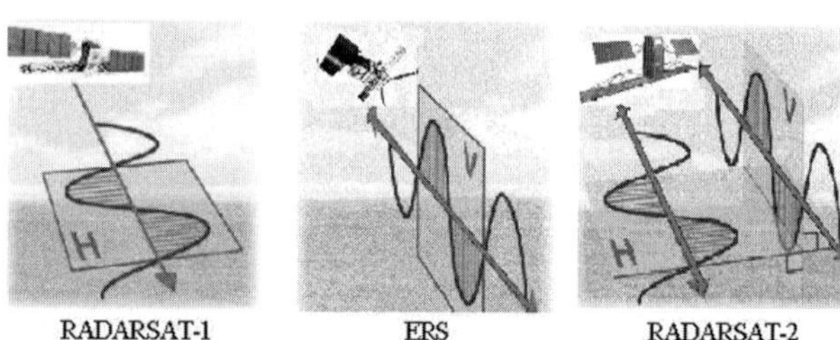

Figure R1 Orientation of horizontal (H) and vertical (V) polarization. Typical transmit and receive polarizations are HH, VV, and HV (adapted from the CCRS website).

Table R1 RADARSAT-2 modes. Beam mode name, swath width, swath coverage, and nominal resolution

	Beam mode	Nominal swath width (km)	Swath coverage to left or right of ground track (km)	Approximate resolution Rng × Az (m^2)
Selective Polarization	Standard	100	250–750	25 × 28
	Wide	150	250–650	25 × 28
Transmit H or V	Low incidence	170	125–300	40 × 28
Receive H or V or (H and V)	High incidence	70	750–1000	20 × 28
	Fine	50	525–750	10 × 9
	ScanSAR wide	500	250–750	100 × 100
	ScanSAR narrow	300	300–720	50 × 50
Polarimetry				
Transmit H and V on alternate pulses	Standard QP	25	250–600	25 × 28
Receive H and V on every pulse	Fine QP	25	400–600	11 × 9
Selective Single Polarization				
Transmit H or V	Multiple fine	50	400–750	11 × 9
Receive H or V	Ultra-fine wide	20	400–550	3 × 3

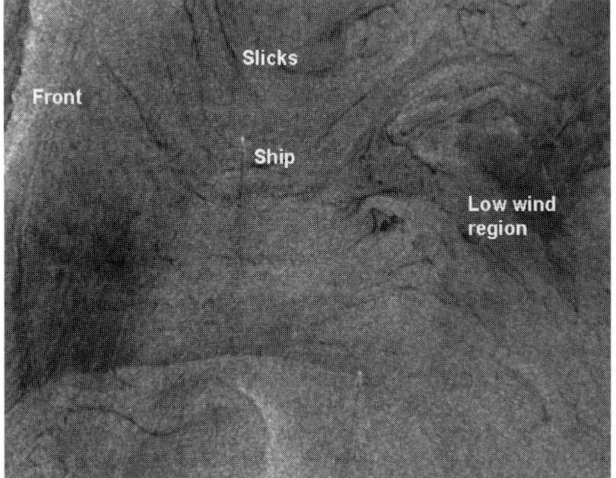

Figure R2 RADARSAT-1 SAR image acquired September, 1998 off the coast of Alaska. Typical ocean features and targets are shown. (Canadian Space Agency, 1998).

The RADARSAT-2 program has adopted the following terms to define the polarization modes (Table R1): Selective Polarization, Polarimetry, and Selective Single Polarization. Selective Polarization and Selective Single Polarization modes imply the availability of amplitude data, but no interchannel phase data. For example, amplitude data may be HH, VV, or HV imagery. In contrast, the polarimetry mode (also called quad-polarized) implies the availability of both amplitude and interchannel phase information. The amplitude information is the same as the Selective Polarization and Selective Single Polarization modes, but adds phase information, such as the co-polarized phase difference.

Marine applications

Marine applications of SAR data can be divided into three main categories: atmospheric phenomena, ship detection, and ocean features (Figure R2).

Atmospheric phenomena include the effect of large-scale atmospheric features such as hurricanes on the ocean surface. Although SAR images through the hurricane cloud-structure, the variability of the hurricane wind speed produces changes in the ocean surface-roughness that the radar detects. For example, the low-wind regime at the eye of the hurricane looks very different than the outer high-wind edges. The radar sensitivity to wind-induced roughness can also be used to map ocean-surface wind speed and direction. The use of VV polarization will be the preferred polarization configuration, largely due to a better radar response relative to HH or HV configurations.

Ship detection is optimal under low wind conditions, HH polarization, and large incidence angles. When co-polarized data and small incidence angles are used, there is increased radar return from the ocean surface, thus reducing the contrast between the ocean and the ship. The use of cross-polarized data (e.g., HV), however, enhances ship detection at small incidence angles due to the weaker return from the ocean surface, but similar return from the ship. Through the application of target decomposition algorithms, quad-polarized data can be used for ship detection and classification (Yeremy *et al.*, 2001).

Ocean features include the detection of eddies, fronts, slicks, currents, surface waves, and internal waves. Radar return from the ocean surface is due to Bragg scattering. Bragg scattering is stronger for VV polarization, thus VV polarization is predominantly used versus HH or HV. The use of quad-polarized data will add significantly to the information content of the SAR imagery.

Gordon C. Staples

Bibliography

CASI, 2004. *Canadian J. of Remote Sensing, RADARSAT-2* Special Issue, **30**(3): 365 pp.

Luscombe, A., 2001. *RADARSAT-2 Product Specification*, Richmond, Canada: MacDonald Dettwiler RN-SP-50–9786.

Morena, L., James, K., and Beck, K., 2004. An Introduction to the RADARSAT-2 mission, *Canadian Journal of Remote Sensing*, **30**(3): 221–234.

Ulaby, F., and Elachi, C. (eds.), 1990. *Radar Polarimetry for Geoscience Applications*. Norwood, MA: Artech House.

Yeremy, M., Campbell, J., Mattar, K., and Potter, T., 2001. Ocean Surveillance with Polarimetric SAR. *Canadian Journal of Remote Sensing*, **27**(4): 328–344.

Cross-references

Remote Sensing of Coastal Environments
Synthetic Aperture Radar Systems

RATING BEACHES

Introduction

Beaches are the number one recreational destination for Americans and Europeans, and a beach culture has developed worldwide. Nothing restores the body and soul like a stay at the beach. We are naturally drawn to the rhythmic pounding of the waves as if returning to our primordial beginnings. Recreational opportunities abound, and everyone, but perhaps children most of all, loves sand.

People are flocking to the shore in ever-increasing numbers for sun and fun. But most want much more from a beach experience—people are searching for real getaway places where they can escape from urban confinement and everyday pressures. The shore offers freedom from the "hemmed-in" feeling as we gaze out from the beach at the seemingly endless sea. The fresh, salty air invigorates the body as the sheer beauty and dynamic interplay between the waves and beach captures our imagination and refreshes our psyche.

So what do people look for in a beach? Water quality is probably the first concern. Polluted water can ruin any beach; the washing ashore of medical wastes along northern New Jersey beaches a decade ago nearly wrecked the local coastal economy as tourism plummeted. Coastal waters are the recipient of much of the nation's wastewater. Already stressed by pollution from upland sources, coastal environments are also adversely affected by the onslaught of contaminants from coastal development and wastes dumped at sea. Sewage-associated wastes are an eyesore, particularly along parts of the US urbanized northeast coast, where inadequate treatment systems still exist. The Gulf coast states, especially Louisiana, have the greatest accumulation of ship galley wastes on their beaches. All of these materials degrade nearshore coastal waters and foul beaches (Center for Marine Conservation, 1990). While concern about pollution is mounting as coastal development continues, some areas are addressing long-standing problems (e.g., construction of the massive Boston Harbor sewerage system). Actually, US beaches are generally quite clean compared to those in many countries, where pollution concerns receive far less attention and funding.

There is a worldwide coastward migration of the population, which itself is burgeoning at over 6 billion at present. What some oceanographers refer to as the ring around the bathtub is some of the most expensive real estate in the world. These areas are rapidly urbanizing, and there is much public concern about the quality of coastal areas. For example, it is estimated that by the year 2010, the coastal population will increase almost 60% in the United States. (NOAA, 1990). As the population trend continues, many of the qualities that attracted people initially are diminishing.

Crowded beaches such as Jones Beach, New York, where the sunbathing towels run together like a gigantic patchwork quilt, accommodate a huge number of people, but few would rate this as the best beach. If coastal communities are too successful in encouraging development, as Ocean City, Maryland has been in the last few decades, then overall quality also drops.

Beach weather and water temperature can greatly limit planned activities. The rugged beauty and sereneness of Oregon's beaches appeal to the wilderness enthusiasts, but the water is much too cold for swimming, and frequent rainy days can spoil any plans for sunbathing. While the picturesque Northwest Pacific coast may be described as the most beautiful, others prefer the palm-treed beachscapes of Hawaii. Which is the most beautiful beach? Shakespeare correctly noted that "Beauty lies in the eyes of the beholder." But all of us can probably agree on a few basic characteristics, which include the physical condition, biological quality, and human use and development of beaches.

Beach awards and ratings

Beach awards and evaluation systems in Europe have been identified as a valuable tool for promotion of beach tourism (Williams and Morgan, 1995). The proliferation of awards and awarding bodies in the United Kingdom, however, has led to low public awareness and distrust of their validity. Those currently in use in the United Kingdom include the European Blue Flag Award (administered by the Foundation for Environmental Education in Europe; FEEE, 1997), the Seaside Award given by the Tidy Britain Group, and the Good Beach Guide, a book published annually by the Marine Conservation Society (MCS, 1997). All of these awards are based on a limited number of factors and do not approach coverage of all measurable aspects of the beach environment (Williams and Morgan, 1995).

Over 2,000 beaches in 18 countries presently participate in the European Blue Flag Award with the numbers growing through time. Only one-third of beach users have a reasonable understanding of the award criteria, which are largely based on water quality; 11% recognized the Blue Flag itself and 7% actually thought the symbol represented danger (Williams and Morgan, 1995).

The Seaside Award was introduced in the United Kingdom in 1992 and is administered by the Tidy Britain Group (a NGO with some government funding). The award criteria are similar to the Blue Flag (e.g., water quality, beach cleanliness, and high standards of facilities and management). While the Blue Flag Award requires strict water quality standards, the Seaside Award system is less rigid so that more UK beaches qualify for this recognition (Williams and Morgan, 1995).

The Good Beach Award is published annually by the MCS and available for sale to the general public at bookstores. Beaches must exhibit a high standard of water quality and a low probability of sewage contamination (which is a major problem for many European beaches). Over 1,000 UK beaches were assessed, and 136 were recommended in 1997. There must be no sewage outfalls adjacent to the beach, bathing must be safe, and there must be no excessive marine litter or sewage—related debris (Williams and Morgan, 1995).

During the past decade, researchers in the United States (Leatherman, 1991, 1997) and Williams and Morgan (1995) in the United Kingdom have devised beach rating systems that attempt to take into account all measurable beach aspects. The two rating surveys are quite similar; beaches are scored for 50 parameters on a scale from one to five (Table R2). In the United States, 650 public recreational beaches were evaluated by Leatherman (1991) and 182 UK beaches were rated by Williams and Morgan (1995). Kapalua on Maui, Hawaii was the first National Winner in 1991 with a score of 92%, and Porthmeir along the English Cornwell coast was listed as the top UK beach in 1995 (at 86%).

For a number of the criteria, a beach user's preference was assumed (e.g., wide beaches are preferable to narrow ones). Quantitative values were attributed to all categories to the extent possible, but some were judged on a purely subjective basis (e.g., vistas far and near). No weighting was attached to the 50 parameters relative to each other in either scheme (Leatherman, 1991; Williams and Morgan, 1995).

US beaches ratings

A beach rating survey was designed to provide an objective appraisal of the major public recreational beaches along the US Atlantic, Gulf, and Pacific coasts. About 650 beaches were evaluated nationwide on the basis of 50 criteria with a sliding scale to quantify the elusive quality factor. In-state coastal experts provided information and assisted Leatherman (1991) in this evaluation. Only the open ocean and Gulf coast beaches were rated. Therefore, seashores along Long Island Sound in New York and Connecticut were not considered nor were the many small beaches in major bays such as the Chesapeake and San Francisco. Puget Sound is perhaps the most desirable coastal property in the State of Washington, but here again it is an inland marine water body. In addition, Alaska's long coastline with many sandy to cobbly beaches was not evaluated, although the Valdez oil spill spotlighted this area's scenic beauty and the small pocket beaches where the oil tended to accumulate and pool.

A battery of factors were arrayed in order to allow for a quantitative comparison of the various beaches. The relevant criteria are those which influence beach quality as broadly defined. The factors considered in this analysis are of three types: physical, biological, and human use and impacts (Table R2). These ranged from 1 (poor) to 5 (excellent). This approach follows that developed by Leopold (1969) in her quantitative comparison of aesthetic factors for rivers.

The survey was designed to reflect general beach usage with swimming water (see especially factors 5–14) being of primary importance. A water temperature scale for optimal and tolerable conditions for swimming and bathing (Figure R3) was also designed to aid in quantifying factor 5 of the questionnaire (see Table R2). In general, pristine beaches with limited development scored much better than the overdeveloped and overcrowded urban resort areas. A profile can be developed for each beach based on the 50 factors evaluated; Figure R4 illustrates this graphical representation for Kapalua on Maui, Hawaii—the top-rated beach nationally in 1991. For visual comparison, one of the nation's worst beaches (Pike's Beach in the New York City area) is presented (Figure R5). This rating of America's best beaches has been conducted for the past 10 years, and the results have appeared in the popular media. Specialty categories were determined by using a subset of the data. Figure R6 illustrates the ranking for the US northeast region as presented in America's Best Beaches (Leatherman, 1998).

Some coastal specialists have questioned the objectiveness of the beach rating scale. For example, white and pink sand are the most highly rated, while gray sand is assigned the lowest rating. While pure white sand does cause much glare, the sugar-white beaches of the Florida panhandle are considered the most beautiful by sunglass-wearing tourists. Also, the pink sand beaches of Bermuda are something to behold. Others have pointed out that the 50 parameters are all equally weighted and that some factors are more important than others. This consideration was partly dealt with by using a suite of factors that are all related to one variable such as pollution (Table R2, factors 21, 23, 27, 29–31, 43, 44, and 48) or beach safety (Table R2, factors 10–14, 21, 23, 28, 40, 41, and 49). Therefore, multiple factors are used to delineate important criteria in rating beaches.

Table R2 Beach rating questionnaire

	1	2	3	4	5
Physical factors					
1. Beach width at low tide	narrow <10 m	10–30 m	30–60 m	60–100 m	100+ m wide
2. Beach material	cobbles	sand/cobbles	coarse sand		fine sand
3. Beach condition or variation	erosional		stable		depositional
4. Sand softness	hard				soft
5. Water temperature	cold/hot				warm (70–85°F)
6. Air temprature (midday)	<60°F >100°F				80–90°F
7. Number of sunny days	few				many
8. Amount of rain	large			little	
9. Wind speeds	high				low
10. Size of breaking waves	high/dangerous				low/safe
11. Number of waves/ width of breaker zone	none	1	2	3	4+
12. Beach slope (underwater)	steeply sloping bottom				gently sloping bottom
13. Longshore current	strong				weak
14. Rip currents present	often				never
15. Color of sand	gray	black	brown	light tan	white/pink
16. Tidal range	large (>4 m)	3–4 m	2–3 m	1–2 m	small (<1 m)
17. Beach shape	straight				pocket
18. Bathing area bottom conditions	rocky, cobbles, mud				fine sand
Biological factors					
19. Turbidity	turbid				clear
20. Water color	gray				aquablue/turquoise
21. Floating/suspended human material (sewerage, scum)	plentiful				none
22. Algae in water	infested				absent
23. Red tide	common				none
24. Smell (seaweed, rotting fish)	bad odors				fresh salty air
25. Wildlife (e.g., shore birds)	none				plentiful
26. Pests (biting files, ticks, mosquitos)	common				no problem
27. Presence of sewerage/runoff outfall lines on/across the beach	several				none
28. Seaweed/jellyfish on the beach	many				none
Human use and impacts					
29. Trash and litter (paper, plastics, nets, ropes, planks)	common				rare
30. Oil and tar balls	common				none
31. Glass and rubble	common				rare
32. Views and vistas local scene	obstructed				unobstructed
33. View and vistas far vista	confined				unconfined
34. Buildings/urbanism	overdeveloped				pristine/wild
35. Access	limited				good

Table R2 *(Continued)*

	1	2	3	4	5
36. Misfits (nuclear power station; offshore dumping)	present				none
37. Vegetation (nearby) trees, dunes	none				many
38. Well-kept grounds/ promenades or natural environment	no				yes
39. Amenities (showers, chairs, bars, etc.)	none				some
40. Lifeguards	none				present
41. Safety record (deaths)	some				none
42. Domestic animals (e.g., dogs)	many				none
43. Noise (cars, nearby highways, trains)	much				little
44. Noise (e.g., crowds, radios)	much				little
45. Presence of seawalls, riprap, concrete/ rubble	large amount				none
46. Intensity of beach use	overcrowded				ample open space
47. Off-road vehicles	common				none
48. Floatables in water (garbage, toilet paper)	common				none
49. Public safety (e.g., pickpockets, crime)	common				rare
50. Competition for free use of beach (e.g., fishermen, boaters, waterskiers)	many				few

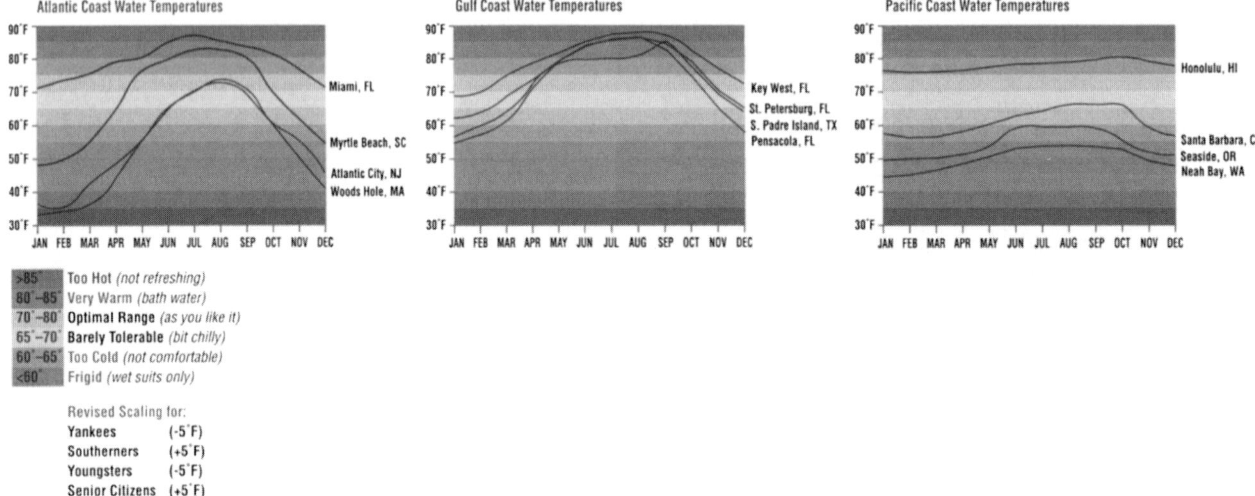

Figure R3 Water temperature scale for bathing and swimming.

America's most famous beaches were rarely rated the best ones for overall quality. For instance, the beaches at world famous Waikiki Beach in Honolulu, Hawaii have experienced progressive erosion to the point that the beach simply does not exist in some sections. Coney Island in New York city was a very popular beach in the 1960s, but crime and other problems have taken their toll.

Beach users survey

Morgan (1999) developed a questionnaire to query recreational beach users in Wales, UK. Overall, scenic quality was rated as the most important factor in the beach environment, but it must be remembered that Welsh beaches are "cold water" and not that conducive to swimming. Sand and water quality were also highly rated in this innovative study.

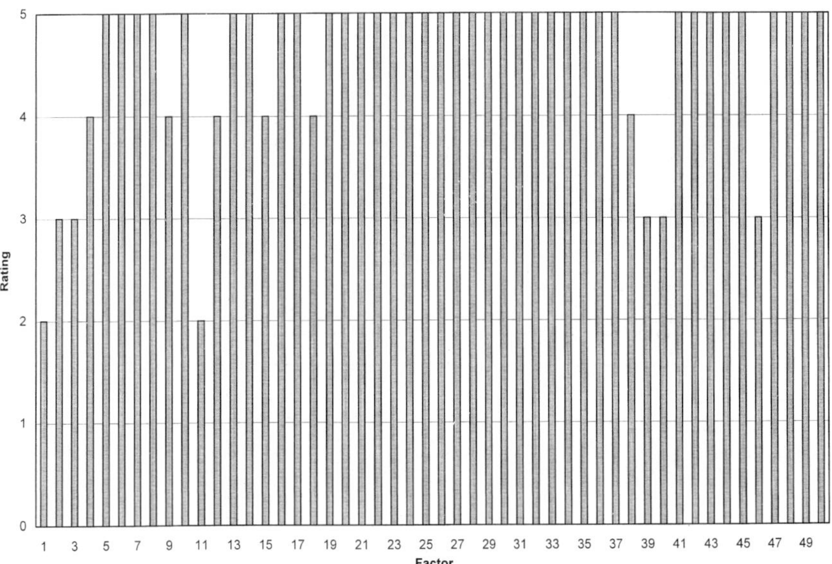

Figure R4 Beach profile for top-rated Kapalua Bay Beach, Maui, Hawaii.

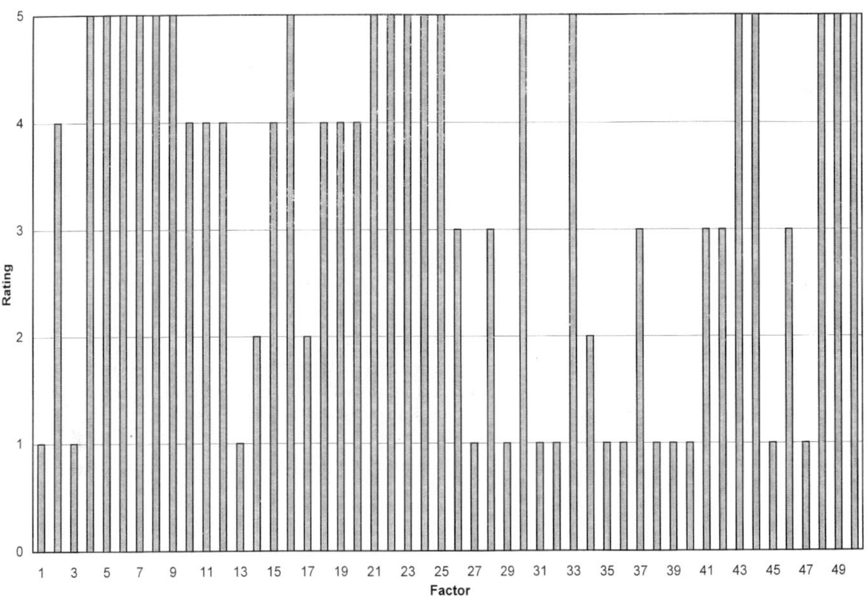

Figure R5 Beach rating profile for Pikes Beach, New York. Note that the "sawtooth" graph indicates a less aesthetically pleasing and overall lower quality beach.

There were many observed differences in beach user preferences, depending upon the type of beach (e.g., urban to rural) that the user preferred to visit. Some vacationers wanted all the amenities and "creature comforts" of a beach resort (e.g., good hotels and restaurants, many water-based activities, and nightlife) compared to others who preferred the natural characteristics of a beach (e.g., fauna and flora, scenery, and camping). Morgan (1999) found that high environmental quality was a prerequisite for all beach users, emphasizing the high level of public concern for this aspect.

Summary

Various beach awards have been ongoing in Europe for decades, but the use of scientific criteria and ratings have evolved subsequently to the efforts begun in the US beaches are always changing because of storm impact, beach nourishment, pollution problems, etc. so the rankings vary year to year.

Stephen P. Leatherman

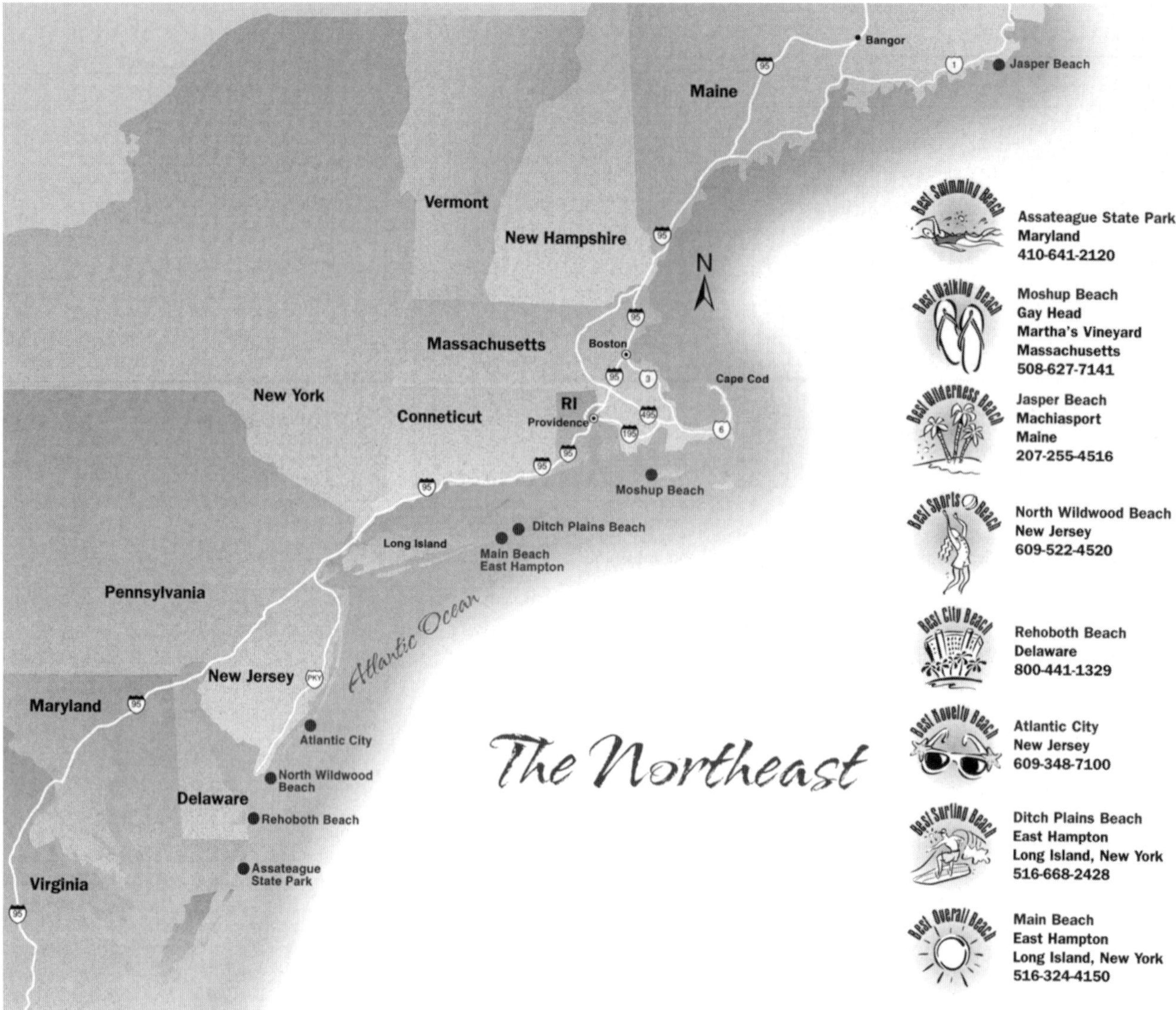

Figure R6 US beaches have been evaluated by region and category (Leatherman, 1998). Each region offers a different venue and recreational advantage. While swimming is the favorite activity of most beachgoers, many people just enjoy walking along beaches for the exercise or to enjoy the scenery. Purists prefer wilderness beaches, but many vacationers are looking for creature comforts; the top city beaches offer all the amenities.

Bibliography

Center for Marine Conservation, 1990. *Cleaning North America's Beaches*. Washington, DC.

FEEE, 1997. *The Blue Flag Awards of 1997*. Foundation for Environmental Education in Europe.

Leatherman, S.P., 1991. *Top 20 Beaches in Continental U.S. and Hawaii*. College Park, MD: University of Maryland.

Leatherman, S.P., 1997. Beach rating: a methodological approach. *Journal of Coastal Research*, **13**(1): 253–258.

Leatherman, S.P., 1998. *America's Best Beaches*. Laboratory for Coastal Research. Miami: Florida International University.

Leopold, L.B., 1969. Quantitative comparison of some aesthetic factors among rivers. Denver: U.S. Geological Survey Circular 620.

Marine Conservation Society, 1997. *Readers Digest Good Beach Guide*, London: David & Charles.

Morgan, R., 1999. Preferences and priorities of recreational beach users in Wales, U.K. *Journal of Coastal Research*, **15**(3): 653–667.

National Oceanic and Atmospheric Administration, 1990. *50 Years of Population Change Along the Nation's Coasts, 1960–2010*. Rockville, MD: National Ocean Service.

Williams, A.T., and Morgan, R., 1995. Beach awards and rating systems. *Shore and Beach*, **63**(4): 29–33.

Cross-references

Beach Sediment Characteristics
Cleaning Beaches
Environmental Quality
Human Impact on Coasts
Lifesaving and Beach Safety
Marine Debris, Onshore, Offshore, and Seafloor Litter
Reclamation
Surfing
Water Quality

RECLAMATION

Although normally associated with the rehabilitation of land, reclamation in the coastal context refers to the exclusion of marine or estuarine water from formerly submerged land. The basic idea of reclamation is to win land from the sea, to displace water and to create new land (Plant *et al.*, 1998, p. 563). The resulting land surface normally extends from the existing coastline and should be well above the level reached by the sea. Reclamation differs from the building up of shallow offshore grounds to form *artificial islands* (*q.v.*) (Kondo, 1995). It also differs from *polders* (*q.v.*) (CUR, 1993, p. 230, 244) in which the level of land subject to seasonal or permanent high water level is protected by dikes, and flood control and water management are important aspects.

Historical and geographical brief

The origins of reclamation probably date back to humankind's efforts in reclaiming land from estuaries and along coasts for their homes, livestock, and crops. For example, in the first few centuries, artificial clay mounds ("terps") were constructed by people in the lower parts of the Netherlands (Pilarczyk, 2000). It is likely that reclamation initially involved the draining of low-lying land and marshes. A survey of reclamation in temperate countries shows reclamation was primarily for agricultural purposes. Later, expanding population and increasing needs for industrial expansion and dock development led to a more focused objective of gaining land from the sea. It started with small-scale reclamation in almost all of the larger cities and ports situated on estuaries and coast throughout the world (Cole and Knights, 1979) (Figure R7). In fact, the reclamation of estuaries was better addressed, especially in the second-half of the 19th century to provide dock and harbor facilities to meet world trade expansion during the period of rapid industrialization in Europe and the United States (Kendrick, 1994).

Any mention of gaining land from the sea cannot avoid the Netherlands because of the massive scale of reclamation in a country where the battle against the sea has become an integral part of the way of life. Reclamation was carried out in conjunction with poldering and the construction of sea defense works. Many lessons were learnt not only in engineering aspects but also in the physical and economic planning of reclaimed land. Another country in which reclamation of its estuaries and low-lying coastal areas has been carried extensively is Japan, which has at least 400 years of reclamation history. It has considerable experience in the consolidation and settlement of fill material, particularly in the treatment of soft foundations, such as the wide application of the sand drain method and the sand compaction method (Watari *et al.*, 1994).

The need for reclaimed land for various developments varies worldwide and becomes particularly crucial around existing ports and cities and in countries where land scarcity is a factor, for example, in Singapore and Hong Kong (Walker, 1988) (Figure R7). Faced with land scarcity, Singapore has resorted to large-scale reclamation from the 1960s although reclamation dated back to the 19th century. Through various projects, it has reclaimed about 10% of its area, which totaled 647 km^2 in 1998. Under the Concept Plan to accommodate a population of 4 million from the present 3.2 million, another 100 km^2 have to be reclaimed. This would amount to 25% in land area since the 1960s (Figure R8). With current technology, it is not cost-effective to reclaim in waters deeper than 15 m. Another limitation is that reclamation encroaches into the sea space for shipping lanes and anchorage.

Reclamation methods

Land won from the sea is normally carried out by raising the level of previously submerged land above sea level using materials dredged from the sea or excavated from the land. The reclaimed site is where reclamation is to take place and eventually forms the platform for required land use. The borrow area is where such material is obtained for reclamation and the disposal area is where material, unsuitable for fill, is to be disposed.

Depending on the type of landfill, there are two major types of reclamation and each has its advantages and disadvantages (Plant *et al.*, 1998, p. 197, 561; CUR, 1993, p. 244).

1. In the "drained" reclamation, fill is placed on the reclaimed site. The advantage is that existing soft marine material can be left in place, reducing the amount of fill required. The disadvantage is the longer time required for settlement and a greater uncertainty in the duration and rate of consolidation. This could increase the costs in follow-up structures to accommodate a larger settlement. Artificial drainage is required to reduce drainage paths and accelerate settlement (Plant *et al.*, 1998, p. 561).
2. In the "dredged" reclamation, landfill by is by replacement, in which soft marine material is removed and replaced by imported fill. This means a larger volume of fill but the advantage of faster consolidation. The Hong Kong International Airport at Chek Lap Kok is an example of "dredged" reclamation in which three-quarters of the airport platform was reclaimed from the sea using a combination of dredging mud from the seabed, borrowing marine sand, excavation of islands, and construction of seawalls.

Site conditions have a strong influence on reclamation and factors such as climate, site access and site geology are important (Plant *et al.*, 1998, p. 15). Inaccurate information on these factors has led to delays, disruptions, disputes, and operational failures. For example, heavy

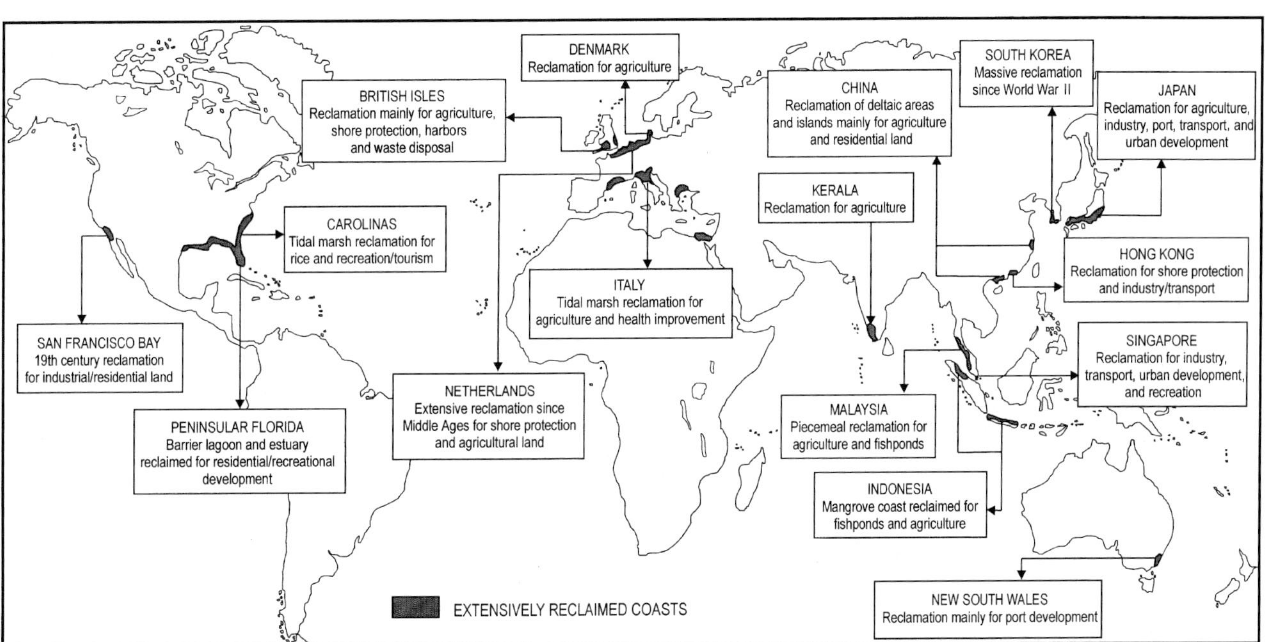

Figure R7 Major reclaimed coasts in the world.

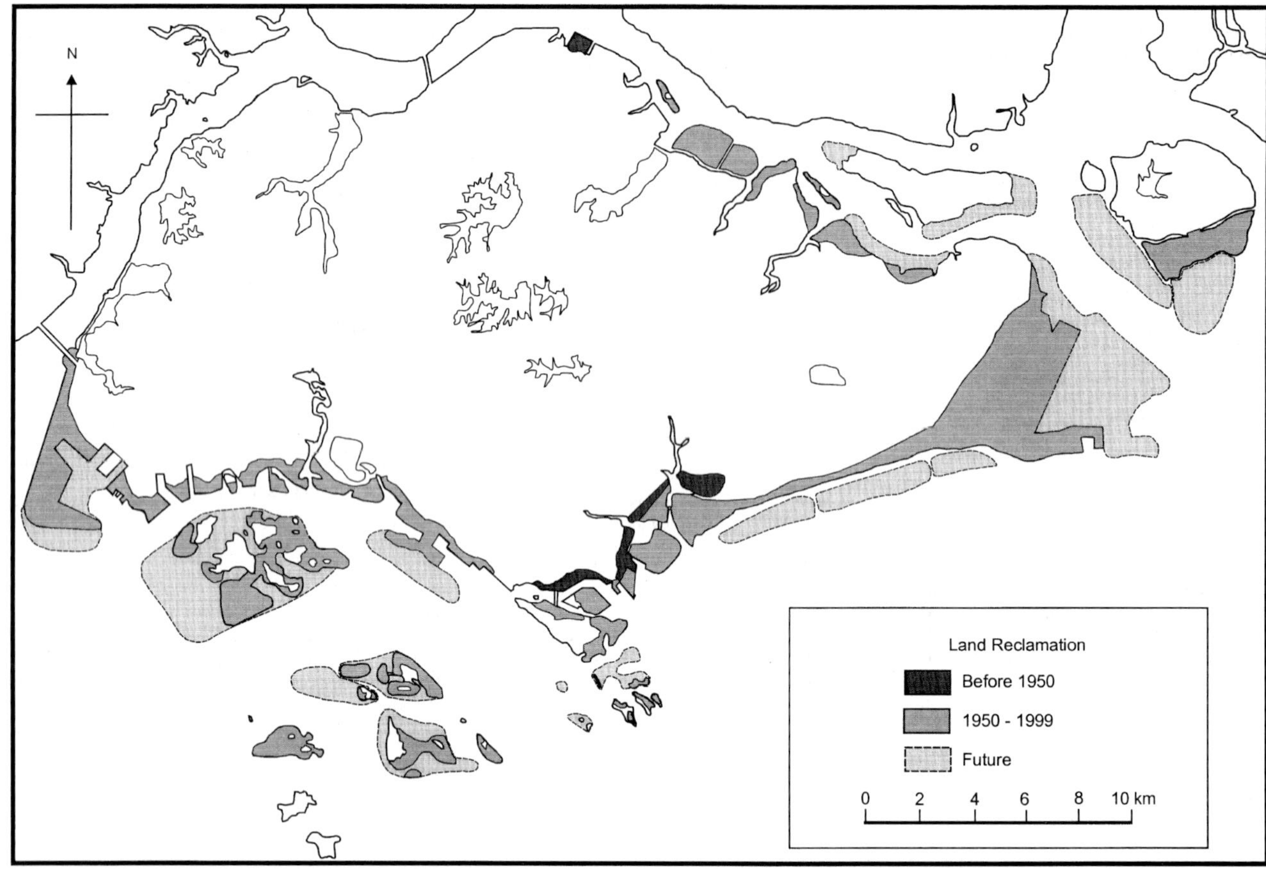

Figure R8 Reclaimed land in Singapore.

rainstorms can halt reclamation operation. Marine conditions as influenced by climate can be equally significant, for example, water depth and wave conditions influence the operation of dredgers. Seabed conditions can vary widely and water depth and type of materials influence both settlement and stability. A low-energy environment is advantageous as fill material can be placed on the seabed and not easily scoured by wave action. The availability of power, water, and communication services at the site can influence the design of reclamation.

Much of the early reclamation efforts often capitalized on the availability of fills such as quarry waste, urban refuse, excavation spoil, building demolition spoil, dredged material from ports and navigation channels, or the byproduct of some dredging operation. It was also carried out in connection with shallow tidal, marshy, or mangrove areas. Usually, for specific reclamation projects, materials cut from hills or seabed sand have been the most common fill. But with increasing scarcity of such fill materials, alternative sources, such as marine clays were dredged from the seabed. Marine aggregates, primarily sand and gravel from offshore-submerged sources, are also used in reclamation projects. In 1989, 7.7 million m^3 and more than 9 million m^3 were used for reclamation works in Denmark and Netherlands, respectively (Bokuniewicz, 2000).

Various techniques have been in use to deal with the reclamation of coastal areas with soft foundations. For example, in Japan, the removal of the soft material and replacement by sand was the common method. The lack of sand and environmental problems, such as increased turbidity of water and the problem of disposal of dredged material, saw the end of this method for large-scale reclamation in 1975. This was replaced by the sand drain, which first came into use in 1952. From 1970–83 some 230,000 sand drains, each about 20 m long, were used for seabed foundation improvement in Japan. Driving-type sand compaction was used in 1966 in offshore construction and supplemented by the deep mixing method (DMM) from the mid-1970s (Watari *et al.*, 1994).

For its earlier phases of reclamation, Singapore obtained landfill from the cutting of hills in the construction of public housing where extensive platforms were required. In the east coast reclamation, the fill was excavated by bucket wheel excavators and transported by belt conveyors to a loading jetty for loading to barges and dumped directly into the reclaimed site. Where water depth limited the movement of barges, the fill was unloaded by a reclaimer and conveyor system. Bulldozers and dump trucks then spread, graded, and compacted the reclaimed land (Yong *et al.*, 1991). The reclaimed land required little settlement after compaction.

As materials from land become less readily available, Singapore has to import from neighboring countries or obtain them from marine resources. The typical procedure in large-scale coastal reclamation is as follows (Yong *et al.*, 1991; Chuah and Tan, 1995).

1. Seabed stabilization is first carried out in several ways by the excavation of soft material from the seabed, dredging to form a sandkey trench or the installation of sand compaction piles or vertical drains. Sand compaction piles were first used in 1989. A hollow pipe casing is driven into soft material, filled with sand, then compacted, and the casing is withdrawn. Sand piles were placed at close intervals. This method strengthens the clay seabed, obviated dredging, and thus avoids the costs of disposal.
2. A sandkey is formed from transported sand towed in by hopper barges. In some cases, where conditions are suitable, a sand wall is also constructed along the coast to be reclaimed. Sand barges build up a stockpile outside the sand wall.
3. Reclamation is carried out by direct dumping or hydraulic filling using cutter suction dredgers and pumps or trailer hopper suction dredgers. Where sand is sucked from the stockpile, it is spread into the fill area by a floating spreader. If the depth is too shallow, sand is pumped through overland pipes.
4. After filling, compaction is carried out by rollers. Shore trimming and shore protection works (geotextile placing, stone placing, and hand-pitching) complete the reclamation. The reclaimed land is usually ready for development after 1–5 years.

The above procedure has been applied to the reclamation of a group of islands located about 1 km south of Jurong Industrial Estate to form a single island zoned for petrochemical and chemical industries. The shape of the reclaimed island is to give a maximum land area that could be possibly reclaimed and a coastline with harbor basins that have no adverse conditions on water currents, sedimentation, and navigation (Chuah and Tan, 1995).

Ground treatment

The magnitude and rate of settlement of *in situ* soils and fills is a major concern in reclamation. The settlement depends on the type of fill, method of placement, and use of reclaimed area (Plant *et al.*, 1998, p. 131). Ground treatment is any process in which the properties of the ground are improved or changed. Various methods are used to speedup settlement, each has its own merits.

Surcharging or preloading is the process in which the ground surface is loaded with additional mass. A stockpile of material is required and the material is moved from location to location until the process is completed (rolling surcharge). This procedure avoids importing vast quantities of additional material. Where rockfill is used, surcharging is an effective means of reducing future settlement in rockfilled areas (Plant *et al.*, 1998, p. 417) and less disruptive to the geotextile over rockfill. Fine-grained dredged materials take up to 2–3 years or more to consolidate under self-weight before surcharging is used. Although surcharging avoids the costs of handling large quantities of surcharge fill material and its eventual disposal, the whole process can take up to a decade or more for a 200-acre site (Thevanayagam *et al.*, 1994).

Vibrocompaction refers to ground treatment in which heavy vibrators are inserted into loose granular soils and then withdrawn leaving a column of compacted soil in the ground. This reduces creep and vibration-induced settlement of the ground surface during follow-up construction activities. It is a fast method of achieving a high degree of compaction for a specific material. In dynamic compaction, the ground is compacted by high-energy impacts using a tampering weight. It can be used for all fill types but its effective depth is limited to about 10 m.

To facilitate rapid settlement associated with very soft clayey material, water conduits have to be installed for easy dissipation of water. In the past, this was achieved by vertical columns of sand drains installed at a grid of about 1–3 m. In recent years, sand drains are being replaced by wick drains or band-shaped drains, which make installing faster and easier. In the reclamation of a seabed underlain by thick deposits of soft clay, soil improvement is required. One of the most popular procedures is to combine prefabricated band-shaped drains with preloading.

In the reclamation for Hong Kong International Airport at Chek Lap Kok, seabed conditions varied widely in terms of water depth, thickness of marine mud to be dredged, and underlying compressible strata. These had a strong influence on both settlement and stability of the reclaimed land. The time taken for settlement was difficult to determine as the settlement of fill after construction depends on the type of fill, method of placement, and use of filled area. Instruments monitored settlement and allowed analytical settlement models to be calibrated and updated as reclamation and follow-up works progressed. An extra 0.5 m was provided for future settlement. For ground treatment, surcharge, vibrocompaction, and dynamic compaction were used (Plant *et al.*, 1998).

Coastal protection

Land gained from sea has to be protected from erosion by waves and currents. Depending on coastal conditions, various types of halophytes can be used to protect reclaimed land. For example, marram grass (*Ammophila*) is used in Europe (Cole and Knights, 1979). In Singapore, idle reclaimed land along Changi Coast Road has been effectively colonized by beach vegetation, such as *Ipomoea pes-caprae* and trees such as *Casuarina equisetifolia*.

More often, reclaimed land is protected by shore protection works. In accordance with current international practice, seawall construction allows for some risk of damage under extreme conditions. A high factor of safety will result in a disproportionate effect on costs. As the philosophy on coastal protection changes, this is also reflected in the protection of reclaimed land. For example, in Japan, coastal protection began with seawalls, followed by groins and then detached breakwaters (Hsu *et al.*, 2000).

Although Singapore is not in a high wave-energy environment, various types of coastal protection works were required for the reclaimed land depending on several variables, such as the usage of reclaimed land, ease and speed of construction, site constraints, cost of construction, and predominant wave approach. Three major groups of shore protection works for reclaimed land in Singapore can be identified.

1. Reclaimed land with beaches. Initially, a seawall was used but this was superceded by a series of breakwaters acting as headlands between which beaches can be developed. These were used in the east coast because of the low wave energy, predominant wave approach from the southeast, and net littoral drift to the west. Beaches were formed in J-shaped bays with their upcoast curves in the east and downcoast straight sectors in the west.
2. Reclaimed land for port purposes. A sand key provides the broad base for the seawall to be constructed and withstand loads, for example, in Jurong. Sheet piles are used for deep-water frontage where wharf construction is required.
3. Reclaimed land retained behind a marine retaining wall. Various types of structures are used. Generally, stone bunds are constructed with stones large enough to withstand dynamic lifting and to absorb forces of waves on their outer face. Stone bunds were constructed around some islands with gaps for beaches to form (Wong, 1985). Various prefabricated structures were also used. Caissons or huge reinforced concrete boxes were towed to sea by tugboats, then filled with sand and positioned on the seabed. These are costly and used where a rocky seabed is available. Depending on their size to be used, L-blocks of 30–70 tons require a compacted rubble foundation and are positioned by cranes. These were used in the reclamation of Marina Bay and Tanjung Rhu.

Depending on the size of area to be reclaimed, the condition of the seabed which may require stabilization, and the necessity of shore protection to protect reclaimed land from erosion, the following approaches can be identified for reclamation in Singapore. For small areas, such as offshore islands, site preparation is followed by seabed stabilization for the construction of breakwaters to prevent erosion from currents and waves and followed by the filling of sand. For large areas extending from the mainland, site preparation is followed by seabed stabilization, filling of sand, and shore protection works. For large swampy areas, settlement is a problem and additional time and costs are required for shore protection measures. Site preparation is thus followed by sand filling and shore protection works.

New developments

Reclamation often involves work on land and sea, which is collectively referred to as land operations and marine operations, respectively. The land operation is basically civil engineering and quarrying operations and the scale of earthworks is typical of large mining operations. The marine operation usually involves dredging and coastal protection. Dredging is defined as "underwater excavation of seabed material, transportation of the materials to a discharge area, and subsequent discharging of dredged material" (Plant *et al.*, 1998, p. 266). In recent years, reclamation has seen new developments in both land and marine operations.

Modern reclamation relies on some heavy specialist equipment to recover (dredge) from the burrow area, transport, and place the material over the reclaimed site. Cutter-suction dredgers and trailing-suction hopper dredgers (also called trailer dredgers) have been developed to dredge a greater volume of materials in a short time and at lower costs, often in deeper waters. A trailing-suction hopper dredger is a self-proposed ocean-going vessel fitted with special dredging equipment. The hopper capacity of these dredgers has doubled from 10,000 m^3 in the 1980s to 23,000 m^3 and set to treble to 33,000 m^3 in 2000 (Riddell, 2000). A cutter-suction dredger differs from a trailing-suction hopper dredger in that the former is effectively stationary during dredging, has more control over the dredging process and is also capable of dredging harder material. Depending on water depth, a cutter-suction dredger trails as dredging progresses. The dredged material is projected through a nozzle at its bow in a process called "rainbowing" as land is reclaimed. Hopper barges are for direct dumping of material and are confined to sheltered waters.

The increasing use of geosynthetics in coastal and harbor engineering has also found its place in reclamation. Geosynthetics is a generic name given to various materials that are synthesized for use with geological materials to improve or modify their behavior. One major success was the construction of storm-resistant structures over soft soils at the coast in the Netherlands (Rao and Sarkar, 1998). In Japan, geosynthetics have been commonly used for more than three decades in land reclamation involving a soft clay foundation. They are mainly used for surface stabilization to increase bearing capacity and to reinforce the base of fill and for foundation improvement to replace sand drains in facilitating drainage (Akagi, 1998). Probably the world's largest sewn single sheet of geotextile used for reclamation took place in Singapore in one phase of the reclamation for Changi Airport. A 180-ha pond, 2,000 m in length and 750 m and 1,050 m in width at both ends, created earlier by the borrowing of sand contained slurry-like material 3 to 20 m thick. The initial spreading of sand over the pond failed as mud burst through the sand cap. Remedial measures started with the removal of some of the exposed slurry. Then, the geotextile sheet of 1,060,000 m^2 sewn from 5×90 m^2 rolls was laid across the silt pond to strengthen the foundation soil. This was supplemented by prefabricated band-shaped vertical drains to accelerate the consolidation time by shortening the drainage path (Na *et al.*, 1998). This example shows that high-strength geotextile can strengthen foundation soil that is extremely soft.

The technique of vacuum consolidation is being used in reclamation. This is a process in which a vacuum is applied to a soil mass to produce a negative pore pressure, leading to increased effective stress that leads to consolidation. Although the technique was known in the early 1950s, it was not widely used until the 1980s because of high costs and implementation difficulties. Its application was made possible by recent technological advances in geotextiles and efficient and cost-effective prefabricated vertical drains (wick drains) (Shang *et al.*, 1998). The technique has been field-tested for on-land vacuum consolidation in countries such as the Netherlands, France, Malaysia, Sweden, Japan, and China. It shows considerable promise as an economically viable method to replace or supplement surcharge fill and the potential to strengthen weak sediments on the seabed adjacent to or beneath water, or consolidate fine-grained hydraulic fills during construction. The technique can also be used with prefabricated horizontal drains and selective placement of dredged materials for new land reclamation (Thevanayagam *et al.*, 1994). It was applied to 480,000 m^2 of reclaimed land in Tianjin Harbor, China, where individual sites treated with vacuum ranged from 5,000 to 30,000 m^2, illustrating that it was especially attractive for hydraulic fills and in reclamation sites with a shortage of surcharge fills but that have an easy access to a power supply. Currently, research on the method is on numerical modeling of the consolidation process, the application of geotextiles over larger areas, and the development of high-efficiency vacuum equipment (Shang *et al.*, 1998).

During earthquakes, especially in Japan, some reclaimed land can undergo a complex phenomenon called liquefaction in which loose sandy deposits change into a liquid state. The liquefaction sites are related to the age of reclamation, the methods used, and type of material. The remediation methods depend on large-scale or localized remediation and are implemented during or after completion of land reclamation. With wide experience in this area, Japan has produced a handbook on remediation measures (Port and Harbor Research Institute, 1997) that can be used in other seismically active regions where reclamation has been carried out.

Compared with the past, environmental considerations are given serious attention in modern reclamation and associated works (Bates, 1994). The main areas of impact are at the dredging site, the transportation route, and the reclamation site. The potential adverse effects of dredging include the release of contaminants into the water, increased turbidity, disturbance to the seabed, erosion, noise, oil spillage, blanket cover, and consequent loss of habitat. To some extent, the impacts resulting from dredging and dumping may be overcome by changing the type of dredger and method of dumping. With the concern on the impacts on water quality and ecology, there is a need to develop methods to minimize environmental impact, such as modeling to examine environmental impacts of reclamation projects.

Conclusion

Reclamation continues to be a significant means in providing land to meet the needs of expanding population and economic development, especially in small countries and coastal cities where land is scarce. It brings about a complete change in the coastal environment. Its implementation involves the input of various disciplines, for example, physical geography, geology, soil mechanics, loose boundary hydraulics, land drainage, coastal engineering in the planning and design of reclamation, and ecology, with the increasing concern on loss of natural habitats. New developments have to try to overcome the technical constraints, make reclamation low cost, take advantage of available soft material and other fills, such as incinerator ash, and to reclaim further into increasing depths.

P.P. Wong

Bibliography

Akagi, T., 1998. Reclamation with geosynthetics. In Sarkar, S.S. (ed.), *Geohorizon: State of Art in Geosynthetic Technology*. Rotterdam: Balkema, pp. 89–96.

Bates, A.D., 1994. Dredging and dredgers. In Abbott, M.B., and Price, W.A. (eds.), *Coastal, Estuarial and Harbour Engineer's Reference Book*. London: Chapman and Hall, pp. 655–670.

Bokuniewicz, H., 2000. Marine aggregate dredging. In Herbich, J.B. (ed.), *Handbook of Coastal Engineering*. New York: McGraw-Hill, pp. 18.1–18.9.

Center for Civil Engineering Research and Codes (CUR), 1993. *Hydrology and Water Management of Deltaic Areas*. Rotterdam: Balkema.

Chuah, S.G., and Tan, D.T.L., 1995. Reclamation of Jurong Island. In *1995 Seminar on Engineering for Coastal Development (ECD 1995 Proceedings, Volume 10)*. Tokyo: The Kozai Club, pp. 111–119.

Cole, G., and Knights, B., 1979. An introduction to estuarine and coastal land reclamation and water storage. In Knights, B., and Phillips, A.J. (eds.), *Estuarine and Coastal Land Reclamation and Water Storage*. Farnborough: Saxon House, pp. 3–20.

Hsu, J.R.C., Uda, T., and Silvester, R., 2000. Shoreline protection methods—Japanese experience. In Herbich, J.B. (ed.), *Handbook of Coastal Engineering*. New York: McGraw-Hill, pp. 9.1–9.77.

Kendrick, M.P., 1994. Estuarial problems. In Abbott, M.B., and Price, W.A. (eds.), *Coastal, Estuarial and Harbor Engineer's Reference Book*. London: Chapman and Hall, pp. 615–635.

Kondo, T., 1995. Technological advances in Japan's coastal developments: land reclamation and artificial islands. *Marine Technology Society Journal*, **29**: 42–49.

Na, Y.M., Choa, V., Win, B.M., and Arulrajah, A., 1998. Use of geosynthetics for reclamation on slurry like soil foundation. In Yanagisawa, E., Moroto, N., and Mitachi, T. (eds.), *Problematic Soils*. Rotterdam: Balkema, pp. 767–771.

Pilarczyk, K.W., 2000. Design of dikes and revetments—Dutch practice. In Herbich, J.B. (ed.), *Handbook of Coastal Engineering*. New York: McGraw-Hill, pp. 3.1–3.104.

Plant, G.W., Covil, C.S., and Hughes, R.A. (eds.), 1998. *Site Preparation for the New Hong Kong International Airport*. London: Thomas Telford. p. 576

Port and Harbor Research Institute, 1997. *Handbook on Liquefaction Remediation of Reclaimed Land*. Rotterdam: Balkema.

Rao, G.V., and Sarkar, S.S., 1998. Geohorizon—overview of geosynthetics foe [sic] the environment. In Sarkar, S.S. (ed.), *Geohorizon: State of Art in Geosynthetic Technology*. Rotterdam: Balkema, pp. 1–11.

Riddell, J., 2000. Dredging: opportunities and challenges for 2000 and beyond. Terra et Aqua, No. 78, March 2000.

Shang, J.Q., Tang, M., and Miao, Z., 1998. Vacuum preloading consolidation of reclaimed land: a case study. *Canadian Geotechnical Journal*, **35**: 740–749.

Thevanayagam, S., Kavazanjian, E., Jr., Jacob, A., and Juran, I., 1994. Prospects of vacuum-assisted consolidation for ground improvement of coastal and offshore fills. In Rollins, K.M. (ed.), *In-Situ Deep Soil Improvement*. New York: American Society of Civil Engineers, Geotechnical Special Publication No. 45, pp. 90–105.

Watari, Y., Fukuda, N., Aung, S., and Yamanouchi, T., 1994. Japanese reclamation techniques for coastal and offshore areas with soft foundation. In Balasubramaniam, A.S. (ed.), *Development in Geotechnical Engineering: From Harvard to New Delhi, 1936–1994*. Rotterdam: Balkema, pp. 543–551.

Walker, H.J. (ed.), 1988. *Artificial Structures and Shorelines*. Dordrecht, The Netherlands: Kluwer Academic Publishers.

Wong, P.P., 1985. Artificial coastlines: the example of Singapore. *Zeitschrift für Geomorphologie*, **57**: 175–192.

Yong, K.Y., Lee, S.L., and Karunaratne, G.P., 1991. Coastal reclamation in Singapore: a review. In Chia, L.S., and Chou, L.M. (eds.), *Urban Coastal Area Management: The Experience of Singapore*. ICLARM Conference Proceedings 15. Manila, Philippines: International Center for Living Aquatic Resources Management, pp. 59–67.

Cross-references

Artificial Islands
Beach Drain
Bioengineered Shore Protection
Dredging of Coastal Environments
Geotextile Applications
Polders
Shore Protection Structures
Wetlands Restoration

REEFS, NON-CORAL

Although most modern reefs are communities of coral and coralline algae that live in clear, well-lit tropical and subtropical waters, there are many different groups of reef-forming organisms that are found on living and ancient reefs. The modern non-coral reefs thrive in a wide range of environments extending from sponge reefs in the arctic to non-photosymbiotic algae and Halimeda reefs found near methane seeping faults at depths of 600 m (Wood, 1999).

Definitions of reefs

The word, reef, is derived from the Norwegian word, rif, which means rib. In nautical terms, reef refers to a narrow chain of rocks, shingle or sand lying at or near to the surface of the water. When early sailing ships explored the tropical waters of the South Pacific, they encountered ring-like reefs of coral that enclosed a lagoon which they called "atoll" after "atolu," the Malayalam name for the Maldives Islands. In the more restrictive modern use of the word, reef denotes a rigid, wave-resistant framework constructed by large skeletal organisms (Ladd, 1944). While living coral reefs on atolls are wave resistant and contain a framework of corals and algae, boreholes drilled beneath the reefs consist of rubble, sediment, and voids (Hubbard *et al.*, 1990). Many ancient carbonate buildups, that are referred to as reefs, show that the original coral framework is almost completely destroyed by deep burial and diagenesis. A broader definition of reef, which would encompass both modern and ancient non-coral reefs, has been proposed by Rachel Wood: "a reef is a discrete carbonate structure formed by in-situ organic components that develops topographic relief upon the Seafloor" (Wood, 1999, p. 5).

Reef-forming organism on non-coral reefs

The earliest recognized reefs are composed of stromatolites which were found on Phanerozoic carbonate platforms dating back to 2.5 Ga. Stromatolites are finely laminated microbialites produced by photosynthetic blue-green algae (cyanobacteria) that form a range of morphologies including domes, columns, and mounds (Reitner, 1993). Ancient stromatolite reefs were constructed on preexisting carbonate ramps and rimmed shelves (Grotzinger, 1989). Living stromatolites are found in intertidal zones on Lizard Island in the Great Barrier Reef, in Shark Bay, Australia, and in submerged tidal channels on Lee Stocking Island in the Bahamas. Thrombolites, which also form reefs, are non-laminated microbial structures which often have a mottled or bioturbated appearance.

Archeocyathids were the first metazoan reef-forming organisms and are found in Lower Cambrian limestones. The archeocyathids are large sponges with double-walled inverted conical calcareous skeletons (Debrenne and Zhuravleb, 1994). The first bryozoan reefs appeared in deep cold waters perhaps related to the presence of microbial mounds in the Lower Ordovician (Pratt, 1989). Stenolaemate bryozoa colonies of clonally calcified chambers, which formed reefs in the Lower Ordovician and died out during the Permian, reappeared as gymnolaemate bryozoans in the Jurassic and expanded during the Cretaceous and Eocene. Phylloid algae are calcified algae of platy, cup, and encrusting leaf-like forms that inhabited many late Paleozoic reefs.

Rudistid reefs are common in the Jurassic to Cretaceous limestones. Rudists are heavily calcified, heterodont bivalves in which the hinge and ligament have been modified forming a complete uncoiling of both valves (Skelton, 1991). The large lower (right) valve is conical, cylindrical, or coiled and the upper (left) valve is flattened. Most rudists were semi-infauna, soft sediment dwellers, but they often colonized storm-generated debris forming large rudistid reefs.

Large colonies of the common oyster, *Crassostera virginica*, are found in intertidal to subtidal environments including sounds and estuaries where the salinity is between 5 and 30 ppt. The oyster spat becomes cemented to old oyster shells and forms mounds of oysters which are commonly known as oyster reefs. When the buried oyster beds are exposed as fossils, they take on a reef-like form, but do not resemble the modern coral reefs.

Conclusions

Most living reefs are composed of coral and coralline algae and consist of a wave-resistant framework constructed by large skeletal organisms. A broader definition of a reef is "a discrete carbonate structure formed by in-situ organic components that develops topographic relief upon the sea floor" (Wood, 1999, p. 5). A wide variety of non-coral organisms including coralline algae, stromatolites, archeocyathids, bryozoans, rudists, and oysters formed reefs in the geologic past and many are still forming reefs today.

William T. Fox

Bibliography

Debrenne, F., and Zhuravleb, A.Yu, 1994. Archaeocyathan affinity: how deep can we go into the systematic affiliation of an extinct group? In Balkema, E. (ed.), *Sponges in Space and Time. Proceedings of the 4th International Poriferan Congress*, Amsterdam, pp. 3–10.

Grotzinger, P.J., 1989. Facies and evolution of Precambrian depositional systems: emergence of the modern platform archetype. In Crevello, P.D., Wilson, J.L., Sarg, J.F., and Read F.F. (eds.), *Controls on Carbonate Platform and Basin Development*. Society of Economic Paleontologists and Mineralogists Special Publication 44, pp. 79–106.

Hubbard, D.K., Miller, A.I., and Scaturo, D., 1990. Production and cycling of calcium carbonate in a shelf-edge reef system (St. Croix, USVI): applications to the nature of reef systems in the fossil record. *Journal of Sedimentary Petrology*, **56**: 335–360.

Ladd, H.S., 1944. Reefs and other bioherms. National Research Council, Division of Geology and Geography, *Annual Report 4*, Appendix K, pp. 26–9.

Pratt, B.R., 1989. Small Ordovician patch reefs, laval Formation (Chazy Group), Caughnawaga, Monteral area, Quebec, In Geldsetzer, H.H., James N.P., and Tebbutt (eds.), *Reefs. Canada and Adjacent Area*, Volume 13. Canadian Society of Petroleum Geologists, pp. 506–509.

Reitner, J., 1993. Modern cryptic microbialite/metazoan facies from Lizard Island (Great Barrier Reef, Australia), formation and concepts. *Facies*, **29**: 3–40.

Skelton, P.W., 1991. Morphogenic versus environmental cues for adaptive radiations. In Schmidt-Kittler N., and Voegel K. (eds.), *Constructional Morphology and Evolution*. Berlin: Springer-Verlag pp. 375–88.

Wood, R.A., 1999. *Reef Evolution*. Oxford: Oxford University Press.

Cross-references

Algal Rims
Australia, Coastal Ecology
Bioherms and Biostromes
Caribbean Islands, Coastal Ecology and Geomorphology
Coral Reef Coasts
Coral Reefs

REFLECTIVE BEACHES

Definition and classification

Reflective beaches are systems where there is minimal wave-energy dissipation by breaking and therefore most energy is reflected by the nearshore morphology. In cases of strong reflection, individual reflected waves can be seen propagating away from the foreshore. Guza (1974), in his study of beach cusp formation, was apparently the first to use the term reflective beach. He distinguished between reflective and dissipative beaches (*q.v.*) using the surf-scaling parameter, ε:

$$\varepsilon = \frac{\alpha\omega^2}{g\tan^2\beta},$$

where α is the wave amplitude at breaking, ω is the wave radian frequency ($\omega = 2\pi/L$, where L is wave length), g is the gravity constant, and β is the beach slope in degrees. The proportion of incident wave energy that is reflected from the beach increases as ε decreases. For beaches where ε is larger than 20, most energy is dissipated by the turbulence associated with wave breaking. Where ε is less than about 2.5, most wave energy is reflected off the foreshore, and such beaches are designated as reflective. Thus, Guza (1974) used the relative degree of reflection or dissipation of incident waves as a rationale for the classification

Figure R9 A reflective gravel beach near Malin Head, Co. Donegal, Ireland. Note steep foreshore and cusps. Maximum height of the collapsing breakers is less than 0.5 m, with a period of about 6 s.

of beaches. This approach was subsequently subsumed under the rubric of nearshore morphodynamics.

Nearshore morphodynamics

The concept of nearshore morphodynamics was developed to characterize systems where form and process are closely coupled through feedback mechanisms. On beaches, waves (*q.v.*) interact strongly with sediments and morphology, and the form of wave breaking is one manifestation of these interactions. For a given wave steepness, H/L (where H is wave height), the breaker type will change as the nearshore slope changes. On a very low gradient slope, spilling breakers should occur. As the gradient increases, there should be a progression through plunging and collapsing breakers. Finally, on very steep beaches, surging breakers should occur (Galvin, 1968). For a constant nearshore slope, the same sequence of breaker types will occur as wave steepness decreases. Breaker type is closely associated with the expenditure of wave energy in the nearshore (e.g., reflection or dissipation) and the development of nearshore morphology. The morphology, in turn, controls breaker type. These relationships are the underlying bases for the concept of nearshore morphodynamics (see summary by Wright and Short, 1984). The recognition of characteristic sets of dynamic relationships provides the basis for using morphodynamic regimes (or states) as a means for classifying beach types. For example, collapsing or surging breakers occur on reflective beaches. This contrasts with dissipative beaches (*q.v.*), where spilling breakers are common. Plunging waves tend to occur on the intermediate beach states of the morphodynamic model (i.e., systems where $20 \geq \varepsilon \geq 2.5$), where neither reflection nor dissipation dominates the nearshore energy response.

Characteristics of reflective beaches

In cross section, morphodynamically reflective beaches display the classic form of "swell" or "summer" beach profiles (e.g., Sonu and Van Beek, 1971). According to Wright and Short (1984), other distinguishing characteristics include steep nearshore and beach slopes ($\tan \beta$ between about 0.10 and 0.20), and, typically, relatively coarse sediment sizes. Coarse clastic beaches, therefore, tend to be reflective. The subaerial beach tends to be narrow with a pronounced step at the foot of the foreshore. Low-energy reflective beaches are approximately two-dimensional alongshore. Higher-energy systems frequently include well-developed beach cusps on the foreshore. These cusp systems are presumed to be caused by low-mode, subharmonic edge waves. Incident wave energy is a maximum at or near the beach face. The classic reflective system displays only one coincident set of breakers (Figure R9), and substantial energy remains at the landward extremity of uprush. On meso- and macrotidal beaches, the nearshore system may be reflective only at higher tidal stages and dissipative at low tide (Short, 1991; Masselink and Hegge, 1995). Short and Hesp (1982) have linked the reflective beach state to the formation of small foredunes that are eroded frequently. This linkage is a key concept in the development of beach-dune interaction models (e.g., Sherman and Bauer, 1993).

Douglas J. Sherman

Bibliography

Galvin, C.J., 1968. Breaker type classification on three laboratory beaches. *Journal of Geophysical Research*, **73**: 3651–3659.

Guza, R.T., 1974. Excitation of edge waves and their role in the formation of beach cusps. Unpublished Ph.D. dissertation, University of California, San Diego, p. 102.

Masselink, G., and Hegge, B., 1995. Morphodynamics of meso- and macrotidal beaches: examples from central Queensland, Australia. *Marine Geology*, **129**: 1–23.

Sherman, D.J., and Bauer, B.O., 1993. Dynamics of beach-dune systems. *Progress in Physical Geography*, **17**: 413–447.

Short, A.D., 1991. Macro–meso tidal beach morphodynamics—an overview. *Journal of Coastal Research*, **7**: 417–436.

Short, A.D., and Hesp, P.A., 1982. Wave, beach and dune interactions in southeastern Australia. *Marine Geology*, **48**: 259–284.

Sonu, C.J., and Van Beek, J.L., 1971. Systematic beach changes on the outer banks, North Carolina. *Journal of Geology*, **79**: 416–425.

Wright, L.D., and Short, A.D., 1984. Morphodynamic variability of surf zones and beaches: a synthesis. *Marine Geology*, **56**: 93–118.

Cross-references

Bars
Beach Features
Beach Processes
Dissipative Beaches
Rhythmic Patterns
Sandy Coasts
Surf Zone Processes
Waves

REMOTE SENSING OF COASTAL ENVIRONMENTS

Coastal ecosystems are transitional environments that are sensitively balanced between open water and upland landscapes. Worldwide, they exhibit extreme variations in areal extent, spatial complexity, and temporal variability. Sustaining these ecosystems requires the ability to monitor their biophysical features and controlling processes at high spatial and temporal resolutions but within a holistic context. Remote sensing is the only tool that can economically measure these features and processes over large areas at appropriate resolutions. Consequently, it offers the only holistic approach to understanding the variable forces shaping the dynamic coastal landscape. Remote sensing must be able to adjust to these spatially and temporally changing conditions and also be able to discriminate subtle differences in these systems. As a result, remote sensing of coastal ecosystems is a complex undertaking that needs to incorporate not only the ability to define the observable hydrologic and vegetation features, but also the scale of measurement.

Historical development

Since the 1960s, remote sensing has been used to describe a new field of information collection that includes aircraft and satellite platforms carrying cameras to electro-optical and antenna sensor systems (Jensen, 2000). Up to that time, camera systems dominated image collection and photographic media dominated the storage of the spatially varying visible (VIS) and near-infrared (NIR) radiation spectral intensities reflected from the earth to aircraft platforms. Beginning in the 1960s, electronic sensor systems were increasingly used for collection and storage of earth's reflected radiation, and satellites were posed as an alternative to aircraft platforms. Advances in electronic sensors and satellite platforms were accompanied by an increased interest and use of radiant energy not only from the VIS and NIR wavelength regions but also from the thermal and microwave regions. In 1983, the American Society of Photogrammetry and Remote Sensing adopted a formal definition of remote sensing as "the measurement or acquisition of information of some property of an object or phenomenon, by a recording device that is not in physical or intimate contact with the object or phenomenon under study" (Jensen, 2000, p. 3). Although, others extended this definition to encompass the new technologies established for data collection, all definitions implicitly suggest that the property measured should describe a feature occupying a finite volume at a certain spatial and temporal position.

Remote sensing can include mapping of the earth's magnetic and gravitational fields and monitoring activities based on mechanical vibrations such as marine profiling by sonar and seismic exploration (Slater, 1980; Horler and Barber, 1981), or it can be applied to fields as diverse as cosmology to medical imaging. Historically, however, remote sensing has been used primarily to describe the collection of information transmitted in the form of electromagnetic radiant energy from an aircraft or satellite that is relevant to the earth's natural resources. Following this description, most remote sensing applications can be generally described by considering, (1) the nature of the probing signal, (2) the characteristics of the sensor and sensor platform, and (3) the interaction of the signal with the target.

Electromagnetic spectrum

The electromagnetic spectrum describes the distribution of energy per wavelength (λ) or within an interval of consecutive wavelengths referred to as spectral bands ($\Delta\lambda$). Standardized to the speed of light in a vacuum, the electromagnetic spectrum is used to categorize general similarities of electromagnetic radiation in terms of changes in wavelength. The spectrum begins at the short wavelength cosmic rays and extends to the long wavelength radio waves (Figure R10). Ultraviolet radiation below 0.3 μm is removed by atmospheric ozone absorption, and between 0.3 and 0.4 μm atmospheric scattering reduces image contrast to levels generally unacceptable for satellite remote sensing applications (Slater, 1980). From about 0.4 μm to about 0.7 μm (VIS) little absorption occurs in clear and unpolluted atmosphere, although scattering is higher in this region than at longer wavelengths. In practice, the VIS is subdivided into the blue, green, and red regions. Above 0.7 μm, remote sensing applications are concentrated in atmospheric transmitting regions or windows bridging strong absorption bands primarily related to water vapor, ozone, and carbon dioxide. The region from 0.7 to 1.3 μm defines the NIR region, and the combined VIS and NIR regions are commonly referred to as the VNIR. The middle infrared (MIR) region from 1.3 to 8.0 μm is sometimes partitioned into a shortwave infrared (SWIR) region from 1.3 to 2.5 μm dominated by reflectance and a region from 2.5 to 8.0 μm dominated by emission. The MIR contains high atmospheric water vapor and carbon dioxide absorption bands and includes an atmospheric window from about 3 to 4 μm and a region of strong absorption between 5 and 8 μm. Ozone absorption between 9 and 10 μm interrupts the thermal region extending from 8 to 14 μm. Poor atmospheric transmission and the lack of sensitive detectors and instrumentation prohibit remote sensing applications between around 14 μm and 1 mm. Starting at 1 mm and extending up to 1 m, the microwave region is subdivided into regions from the shortest (*K*) to the longest (*P*) wavelength. Beyond the microwave region begins the region of television and radio frequencies. The electromagnetic spectrum does not end abruptly, but frequencies less than 3 kHz corresponding to

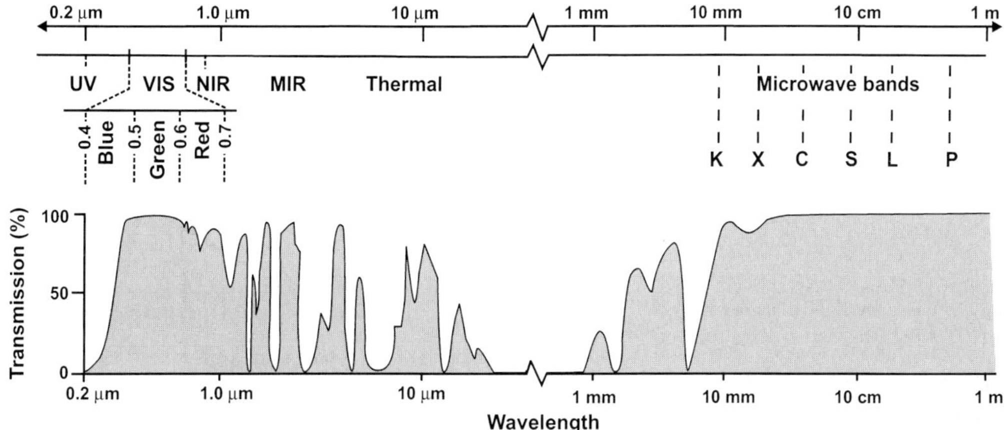

Figure R10 The upper figure shows the general locations and ranges of various wavelength regions. Note the microwave region is used by passive microwave (passive measurement of emitted energy) and radar (active) remote sensing systems. The lower figure depicts atmospheric transmission. Satellite remote sensing applications are normally carried out in spectral regions of high transmission (shaded) and avoid regions of high blockage.

wavelengths longer than 10^5 m are not used. Mechanical vibrations including sound and seismic waves begin at frequencies below 20 kHz. Marine profiling by sonar, the audio analog to radar, is in the 200-Hz range (Slater, 1980).

Platforms and sensors

Aircraft color infrared photography is useful in providing detailed biophysical, high-quality information about coastal ecosystems; unfortunately, turnaround for new map production is relatively slow. Satellite remote sensing provides holistic but detailed information on a regional as well as repetitive basis, and it is the only feasible approach to successfully overcome many intractable problems related to mapping and monitoring of coastal ecosystems. Satellite remote sensors are increasing in number, in type, and in operational usefulness, and will provide the basis for integrated remote sensing applications (Lillesand and Kiefer, 1994; Jensen, 2000). Visible to thermal sensors have the longest history and have shown promise in mapping and monitoring coastal wetland type, health, biomass, and water quality. Since the late 1970s, microwave has gained importance in wetland mapping (Lewis *et al.*, 1998). Microwave sensors extend the past capabilities of visible to thermal sensors in mapping coastal ecosystems by adding the potential for higher canopy penetration, more detailed canopy orientation and density information, and 24-hr-a-day collections nearly independent of weather conditions. Aircraft data will continue to provide calibration surveys, algorithm verification, testing of new sensor systems, specialized sensor collections, and in some cases local disaster response.

Remote sensing sensors measure radiant flux over bands defined by a range of wavelengths (e.g., VIS). A radiometer is a remote sensing instrument that measures radiant flux at any distance over a single band and a spectroradiometer measures over multiple bands. The specification of the instrument determines the spectral coverage, spectral, angular (or spatial or ground resolution), and radiometric resolutions. These factors clearly control the type and value of the data obtained.

Imaging and nonimaging sensors used in remote sensing use either electro-optical (VIS to thermal) or antenna detectors (microwave). Normally, nonimaging sensors are radiometers that collect accurate data over wide spectral regions where the spatial and spectral aspects are less important (Elachi, 1987). The signals returned to the sensor represent the scene spatial characteristics as rows and columns of pixels (discrete picture elements) within an image. Resampling is sometimes necessary to construct a continuous image representation of the scene (orbital and sensor characteristics), and almost always necessary to georeference the image to an earth coordinate system (e.g., latitude/longitude). Resampling tends to reduce confidence in the pixel location and blur the information contained in spatially adjacent pixels by adding a component of within-image spatial covariation. The pixel dimensions and location on the image represent an estimate of the spatial dimensions (spatially averaged ground area) and scene location that contributed to the pixel value.

Sensors commonly used can be separated into two types, active and passive. Active sensor systems both transmit and receive reflected or scattered radiant flux. The light detection and ranging (LIDAR) and radio detection and ranging (RADAR) are common active sensor systems. These systems track the time difference between the transmission of the emitted pulses of energy and the arrival of the scattered return at the sensor. The distance to or range of the target is then directly obtained from the time difference. Active systems can also control the nature of the radiation used to probe the target. For instance, the wavelength placement, bandwidth ($\Delta\lambda$), polarization, and the angle of incidence can be controlled by the sensor and sensor platform. And because the energy source is part of the system, active systems can operate day and night and in the case of radar during most weather conditions. These features allow a greater control over the application of remote sensing techniques, and their use in monitoring earth's resources is increasing.

Passive remote sensing primarily uses the sun as the source of electromagnetic energy. Planck's blackbody law describes the sun's energy distribution with respect to frequency and wavelength. A perfect blackbody absorbs and re-emits all electromagnetic radiation impingement upon it, while a partial emitter (gray body, e.g., water) is spectrally similar but of lower amplitude, and a selective emitter (e.g., quartz) is spectrally selective compared to the general shape of a blackbody. The sun's emitted radiation closely approximates a blackbody at 6,000 K where energy emissions are mostly contained between 0.3 and 2.5 μm, peaking near 0.55 μm. Solar radiation transferred through the earth's atmosphere and scattered and reflected by its surface is generally referred to as *solar reflected radiant flux* (Slater, 1980). In addition, solar heating of the earth's atmosphere and surface produces a secondary source of energy that has a distribution similar to radiation emitted from a blackbody at approximately 300 K. Energy emitted by the earth starts around 2.5 μm and peaks near 10 μm. Radiation emitted by the earth's atmosphere and surface is referred to as *self-emitted thermal radiant flux* (Slater, 1980). Radiant flux below about 2.5 μm represents solar-reflected radiant flux while above 6.0 μm it represents self-emitted thermal radiant flux. Between 2.5 and 6.0 μm, the relative amounts of each flux depend on the target reflectance, emissivity, temperature, and atmospheric transmittance. Lowered atmospheric transmittance between 2.5 and 5.5 μm normally results in the self-emitted flux dominating this region even when the surface has a high reflectance. Even though the self-emitted thermal radiant flux is very low at microwave wavelengths, the atmosphere is nearly transparent permitting successful application of passive microwave remote sensing.

Target interactions

Surface irradiance ($I_{T\lambda}$) (e.g., solar in passive, instrument in active) interactions with earth's features can be partitioned following Kirchhoff's law (a restatement of the conservation of energy) as the proportion of $I_{T\lambda}$ reflected ($\rho(\lambda)$), transmitted ($\tau(\lambda)$) and absorbed ($\alpha(\lambda)$), that is, $\rho(\lambda) + \tau(\lambda) + \alpha(\lambda) = 1$. These interactions are commonly described as finite volume or surface averages of discrete elements, such as algal cells in water, leaves in a canopy, pebbles on the soil surface, and aerosol particles in the atmosphere. If the aggregate properties associated with these elements are independent of changes in $I_{T\lambda}$ and the method of measurement, they are referred to as inherent, the desired quantity to extract from the target radiance ($L_{T\lambda}$) (Bukata *et al.*, 1995). Other definitions include the specific conditions, particularly the view (θ_v) and local incident (θ_i) angles (Figure R11), within the definition of inherent optical properties. Under such measurement conditions, inherent optical properties whether describing discrete scatters or averages, reflect, transmit, and absorb the same fraction of $I_{T\lambda}$ unless the material's inherent properties change. Unless the sensor is within the target volume (transmittance), a remote sensing sensor only measures the net result of the reflectance, transmittance, and absorption summed over the target (pixel) (including depth) and generalized to a single source and view (normally θ_v and θ_i) as r_λ. The resultant target reflectance (r_λ) represents the measured fraction of $I_{T\lambda}$ reflected ($I_{R\lambda}$) ($r_\lambda = I_{R\lambda}/I_{T\lambda}$, range = 0–1). The source of r_λ (and ultimately $L_{T\lambda}$) is surface and volume scattering, partitioned based on the transmittance depth (Whitt *et al.*, 1990).

VNIR and MIR

If the conditions of the measurement and $I_{T\lambda}$ are clearly specified, r_λ is related to the averaged target properties through the averaged inherent optical properties ($\rho(\lambda)$, $\tau(\lambda)$, and $\alpha(\lambda)$) throughout the surface or volume. However, the $L_{T\lambda}$ is commonly recorded but not irradiance ($I_{T\lambda}$), thus, a factor similar to r_λ and tied to $I_{T\lambda}$ by a geometric distribution related to $L_{T\lambda}$ is needed. In VNIR and MIR, the bidirectional (θ_v and θ_i, bistatic) reflectance distribution function ($BRDF_\lambda$) describes the fraction of $I_{T\lambda}$ reflected over a solid angle (Ω) at θ_v, or the surface distribution of $L_{T\lambda}$ ($BRDF_\lambda = L_{T\lambda}/I_{T\lambda}$). $BRDF_\lambda$ is a function of the incident

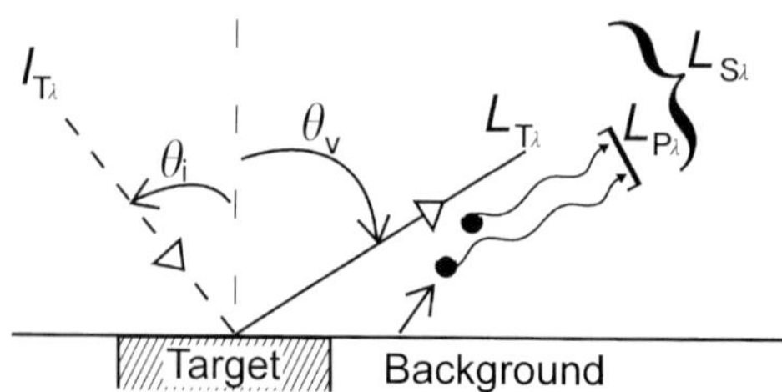

Figure R11 $I_{T\lambda}$ depicts solar irradiance at a sun zenith angle of θ_i on a horizontal surface target, but the same depiction can refer to radiance from an active source at an incidence angle of θ_i (in this case θ_i and θ_v may be equal although the direction of the incident and reflected (backscattered) fluxes would be opposite). $L_{T\lambda}$ depicts the reflected or scattered radiance from the target at θ_v, the sensor view zenith angle. $L_{P\lambda}$ depicts nontarget radiance added to $L_{T\lambda}$ from atmospheric scattering (● being the scatter center) and from areas surrounding the target (background). $L_{S\lambda}$ is the radiance at the sensor.

(ϕ_i) and view (ϕ_v) azimuths, θ_i, θ_v, and Ω_s subtended by the source at a point on the surface, and Ω_v subtended by the entrance pupil of the sensor at the surface, that is, $BRDF_\lambda(\theta_I, \phi_i, \theta_v, \phi_v, \Omega_s, \Omega_v)$ (Jensen, 2000).

Emitted energy

In thermal radiometry, the target or source emissions depend on the target contact kinetic temperature (KT) and the emissivity ($\varepsilon(\lambda)$). The $\varepsilon(\lambda)$ of a target equals $1 - \rho(\lambda)$ (when $\varepsilon(\lambda) = \alpha(\lambda)$, all transmitted incident Φ_λ is absorbed) and is the ratio of the emission spectral characteristics to a blackbody at the same temperature. The $\varepsilon(\lambda)$ relates KT to the self-emitted Φ_λ ($T_R(\lambda)$) from the target (i.e., $T_R(\lambda) = \varepsilon(\lambda)^{1/4} \cdot KT$). Because of this relationship, materials with equal KTs but different $\varepsilon(\lambda)$s will have different $T_R(\lambda)$s or in terms of the sensor, $L_{T\lambda}$s. Passive microwave intensity is the product of $\varepsilon(\lambda)$ and KT and is usually reported as brightness temperature. The $\varepsilon(\lambda)$ determines the energy from the effective emitting layer that is transferred across the soil surface, and it is dominated by surface soil moisture, and soil moisture dampens thermal and microwave emissions. In nonvegetated areas (e.g., deserts, oceans), atmospheric influences (including clouds) pose a problem to retrieval of the $T_R(\lambda)$ and therefore KT based on thermal radiometry. In vegetated areas, the vegetation adds to and attenuates the soil emissions; thus, $T_R(\lambda)$ detected at the sensor as $L_{T\lambda}$ contains emitted information proportional to both the vegetation and the soil layer. In passive microwave, the single scattering albedo and the optical depth describe microwave interactions and emissions from the overlying vegetation layer. In most cases, the effects of the single scattering albedo appear small and can be incorporated into the optical depth or set to a constant value.

Radar

The scattering properties of discrete targets (in isolation) are described by the radar cross-section (RCS, m^{-2}). As in optical cross-sections, the RCS symbolizes the interaction cross-section or the target backscatter reflectivity, not the actual target area (Raney, 1998). Flux reflected or scattered to the sensor is the product of the incident radiant flux and RCS normalized by propagation losses and the area of the receiving antenna (Zebker *et al.*, 1990). In typical resource applications, the radar signal recorded per pixel is the coherent summation of reflected (backscattered) energy from all scatterers (relative to the wavelength) in the distributed or diffuse target back to the sensor (Massonnet and Feigl, 1998). As such, the recorded flux depends on the target or pixel area. The backscatter coefficient ($\sigma°(R, A)$ at a specific range (R) and azimuth (A) location) generated from the calibrated return and normalized by the target area (corrected for the local incidence angle) (where, $\sigma°(R, A) \cdot \Delta R \cdot \Delta A / \sin \theta_i(R)$ is equivalent to RCS) represents the measured fraction of incident Φ_λ backscattered ($\Phi_{b\lambda}$) from the target, $\sigma° = \Phi_{b\lambda}/\Phi_\lambda$ (Raney, 1998). As in reporting BRDF (albeit at one angle, monostatic), $\sigma°$, at a specific wavelength, incident direction and polarization, is most closely related to the size, shape, orientation, and composition (primarily water content) properties of the diffuse target. Changes in $\sigma°$ reflect the variability of these diffuse targets to send the incident energy back to the sensor (Massonnet and Feigl, 1998). A positive $\log(\sigma°)$ in decibels implies focusing energy toward the sensor, while a negative number implies focusing energy away from the sensor (Elachi, 1987).

Synthetic aperture processing (focusing) creating synthetic aperture radar (SAR) (NASA, 1989) is used to improve the radar's spatial resolution. Within the processing, signal return variability is related to successive observations of the same area but from slightly different positions and somewhat different fine details in neighboring pixels with the same RCS (grainy appearance of image) (Elachi, 1987). Increasing the number of looks (statistical averages of the radar returns) reduces these effects but decreases the effective spatial resolution. Contrary to the total solar irradiance, polarimetric radar is capable of synthesizing well-defined polarization states represented in the linear case by horizontal and vertical orientations. After standard processing and image construction, phase information related to the distance between the sensor and the target can be linked to the polarimetric return (Elachi, 1987; Zebker *et al.*, 1990). An interferogram is the resulting phase difference between two SAR images collected either from two antennae (bistatic) from a single platform or one antenna (monostatic) at two different times (Massonnet and Feigl, 1998). In the latter case, the direction of observation and wavelengths must be identical, and in practice, the input images are collected from the same satellite in the same orbital configuration and focusing or synthetic aperture processing (SAR) of the original image data are identical.

Both imaging and non-imaging radar sensors are commonly used in today's remote sensing applications. Altimeters use radar's ranging capabilities to measure the surface topography profile to centimeter-level precision at relatively high pixel spatial resolutions. This level of precision requires precise measurement of time and information extracted from the shape and slope of the returned pulse. Scatterometers assess the average scattering properties over large areas and within narrow spectral bands and provide directional capability by including more than one antenna (Elachi, 1987). They provide high precision backscatter measurements that cover large areas but at low pixel spatial resolutions (Cavanie and Gohin, 1995). Imaging SAR sensors are most often used for resource mapping because of their ability to provide high pixel spatial resolution data at multiple wavelengths, polarizations, and incident angles.

In all cases, objects of comparable size to the microwave wavelengths (mainly 2–30 cm in satellite imaging of earth's resources) most strongly influence the microwave scatter. At a constant incidence angle, canopy components such as leaves and stems normally interact with microwave wavelengths from about 2 to 6 cm, and trunks and limbs at longer wavelengths (10–30 cm). Longer wavelengths (>30 cm) provide more information about the surface properties but little about the canopy volume. Use of cross polarizations (e.g., horizontal send and vertical return, HV) and higher incidence angles, however, enhance volume scattering relative to lower (more vertical) incidence angles and like polarizations (e.g., VV, HH).

Applications of remote sensing in coastal ecosystems

Polarization, angles of incidence and view, proportions of direct and scattered irradiance, and surface roughness all work to modify and build a directional reflectance character that becomes the target radiance ($L_{T\lambda}$). $L_{T\lambda}$ can be further altered from the surface to the satellite ($L_{S\lambda}$) by attenuation and addition of path radiance ($L_{P\lambda}$) (Figure R11), especially in the VIS but also in the NIR, MIR, and thermal regions and further modulated from the sensor to its representation on the image. In short, uncovering the relationship between $L_{S\lambda}$ output as a pixel in a grid-based image representation and the inherent properties of the target is often highly complex, and many times may be impossible to fully determine. Accountability can be built into the analysis by linking the image data to site-specific measurements through physical-based models. Greater accountability can diminish the reliance on gathering site-specific data and ultimately advance the operational and accurate representation of the temporal and spatial distributions of biophysical features in the scene.

The generation of the bidirectional reflectance distribution function ($BRDF_\lambda$) or radar backscatter coefficient ($\sigma°$) is required when target inherent properties are sought or when the spectral contrast between the target ($L_{T\lambda}$) and its surroundings at the sensor prevents successful classification of the target. Classifications can be improved by generation of $BRDF_\lambda$ or $\sigma°$; for example, biomass estimates are improved when generated from $L_{T\lambda}$ (top of canopy after atmospheric correction) versus $L_{S\lambda}$ (top of the atmosphere). Inferential relationships have more promise of extension over space and over time and the detection and mapping of subtle scene features when based on $BRDF_\lambda$ or $\sigma°$. The consistent success of these detection and monitoring methods based on remote sensing data requires a rudimentary estimation of the relative extent the target material reflects, transmits, and absorbs surface irradiance ($I_{T\lambda}$) under various geometric and $I_{T\lambda}$ conditions.

Classification

Most often, classification has relied on the differential interactions and responses of VNIR and MIR (e.g., spectral signature or $L_{T\lambda}$) to changes within the coastal scene to provide spectral classes uniquely linked to the type and state of coastal features. These classifications generally use simple image-based parametric statistical models (e.g., clustering techniques, principal component analysis, canonical correlation, discriminant analysis) that do not require detailed information about the plant, canopy, or water optical properties. The developed relationships are commonly limited to conditions existent only during the data collection, and are not necessarily extendable temporally or spatially. Integrating data from multiple remote sensors (VIS to microwave, hybrid models) has successfully improved the spatial detail of coastal vegetation classifications (Ramsey *et al.*, 1998), as have newer classifications based on nonparametric statistical models (e.g., neural network analysis) that allow greater control in linking the classification to the biophysical characteristics.

Classification of the radar images can follow the commonly used procedures of VNIR and MIR classifications (Ramsey *et al.*, 1998), while also enhancing the use of neighborhood information or image texture in the classification. Radar image texture is a combination of the system and scene features. System texture (speckle) can be diminished by increasing the number of looks or can be estimated and removed by averaging the radar return over undisturbed water bodies. The improved or corrected measure is a more accurate representation of scene texture and a more pertinent input into the classification process. A more direct method is developed on rule-based logic (van Zyl, 1989; Dobson *et al.*, 1995). In one method, returns associated with single or combinations of SAR sensors of different wavelengths and polarizations from sites with known structural characteristics are used to generate a progressive classification hierarchy. In another, rules based on predicted polarimetric return signatures from different features are used to classify targets within the scene.

Biophysical features

Vegetation type classifications are not always the primary objective of the remote sensing application. Often the objective is to directly link the remote sensing data to biophysical variables that describe the coastal ecosystem. Of the biophysical variables, leaf area index (LAI) is probably the most sought, and it can be related to wetted or total biomass and in certain instances primary productivity and even CO_2 exchange.

In the VNIR, the ability to map canopy LAI and productivity is based on two facts (Horler and Barber, 1981; Smith and Morgan, 1981). First, although biomass and yield depend ultimately on light absorption, the usually close correlation between leaf reflectance and transmittance provides a basis for remote sensing reflectance measurements to be successfully applied in agronomic applications. Second, there is a striking attenuation difference between the red and NIR. NIR is not significantly absorbed and is nearly equivalent to above-canopy flux while the red is strongly absorbed and is therefore nearly equivalent to flux transmitted between the canopy leaves. Thus, the degree of canopy shading is related to both the LAI and the differential interactions of red and NIR as described numerically by a vegetation index (VI). In the VNIR, two types of VI are commonly used to transform remote sensing data into estimates of LAI: those based on ratio transforms and those based on orthogonal transforms. These VIs can be modified and possibly improved by altering the required input bands, or by precorrecting the image data to account for atmospheric influence, but variability remains related to the canopy $BRDF_\lambda$, or primarily to canopy structure.

Because the three-dimensional (3-D) distribution of water within the canopy has the greatest influence on microwave interactions, and because wetted biomass is related to vegetation water content, passive microwave and radar remote sensing are sensitive to LAI, and in turn, biomass variations. Similar to VNIR to MIR ratios and differences, biophysical variables such as LAI can be related to radar copolarized (e.g., HH, VV) and cross-polarized (e.g., VH) ratios and differences (Wegmuller and Werner, 1997). The vegetation biomass distribution (canopy structure) and quantity (water content) are also related to canopy optical depth that in turn is related to emission variability at microwave wavelengths (Wigneron *et al.*, 1995).

Vertical canopy profiling

Vertical canopy profiling is an indirect result of using multispectral remote sensing systems. Canopy profiling can be used to estimate canopy architecture, an indicator of species variety, phenological stage, and present and past vigor (Malet, 1996). In general, the longer wavelengths transmit further into fully formed vegetation canopies relative to shorter wavelengths. Canopy structure, or the spatial distribution and orientation of the canopy elements, however, also influences canopy penetration and must be removed or accounted for before the variable return can be used to describe the canopy architecture. Typically, NIR to MIR wavelengths transmit from 8 to 10 leaf layers (equivalent LAIs) into the canopy and VIS wavelengths from 2 to 3 LAIs. Although there are notable exceptions, active sensors offer a greater ability to profile the response from various depths within the vegetation canopy as compared to passive sensors. The use of shorter to longer radar wavelengths, multiple incident angles probing with a single band, and multiple polarizations can offer variable depths of transmittance although the analogy becomes less straightforward in canopies with convoluted branching (Ramsey, 1998). LIDAR offers the most direct canopy profiling. By using the allometric relationship between tree height and diameter-at-breast-height, LIDAR can provide volumetric representation of the canopy structure.

Vegetation stress

One of the greatest challenges to coastal remote sensing is detecting plant stress before irreversible losses occur due to changes in inundation, flushing water salinity, and other external forces as a result of sea-level rise and shoreline alteration and protection. Although broadband VNIR to MIR remote sensing applications have detected broad indicators of vegetation stress, hyperspectral systems have identified specific spectral features related to stress from metal contamination to deficient foliar water content (Card *et al.*, 1988). Radar is sensitive to vegetation stress through changes in water content, and because optical depth is linked to water content, passive microwave and thermal radiometry are closely related to plant stress. Chlorophyll_a fluorescence can also be used to assess vegetation stress by providing estimates of photosynthetic capacity (Carter *et al.*, 1996). A passive technique using Fraunhofer line radiometers (FLR) detects the absorbed photosynthetic radiant flux (about 3%) re-emitted as fluorescence by taking advantage of the relatively strong Fraunhofer absorption lines in the solar irradiance (Horler and Barber, 1981; Carter *et al.*, 1996). One of the strongest Fraunhofer lines is located in the chlorophyll fluorescence peak providing a convenient method for identifying plants suffering from metal toxicity and water stress. Active laser-induced fluorescence (LIF) sensors also provide the ability to assess the fluorescent properties of the leaf pigments (e.g., chlorophyll). Both passive FLR and active LIF offer new capabilities to isolate the alteration or change in dominance of specific pigments as an early indicator of vegetation stress.

Thermal radiometry

In dry environments (e.g., nearly constant emissivity (ε_λ)), the rate of temperature change in response to variable surface solar irradiance (I_T) can be used to uniquely identify the target material (e.g., soil composition, mineral) (Lillesand and Kiefer, 1994). To characterize the rate materials respond to temperature changes (thermal inertia), radiant temperature ($T_R(\lambda)$) measurements are collected in the early morning and the afternoon. This temperature difference indicates the variable heat capacity of the different materials and can be used to map landcover variation. Heat capacity mapping has been applied mostly in nonvegetated regions for identifying geologic materials. Intense heat sources, however, can often be directly observed. Fires can exhibit self-emitted thermal fluxes down to about 3.0 μm, and some volcanic lava flows as low as the NIR (0.7–1.3 μm).

Area mixtures

In any remote sensing application, all pixels are weighted mixtures of different scene features, even when the pixel nearly matches the target feature's mean spatial extent (e.g., adjacency effects, boundary pixel landcover mixtures). Mixture models are used to extract the occurrence of specific scene features from composite mixtures (e.g., trees, water, bare ground). In a linear mixture model, weighted combinations of specific scene features (e.g., spectral endmembers) are combined to completely reconstruct every spectral signature ($L_{S\lambda}$ or $L_{T\lambda}$) as represented on the image. This reconstruction allows target features to be detected and the percent occurrence in each image pixel to be determined. In nonlinear endmember analysis, the interaction between target features is included. Mixture models can be based on broadband sensor data; but most successful applications are based on hyperspectral sensor data (Adams *et al.*, 1986).

Soil moisture content

The NIR to MIR regions are used to estimate soil moisture where the increase in soil moisture (as in the presence of standing water) dampens the return to the sensor. Successful studies have relied more on direct determinations with thermal radiometry and passive microwave and radar (Idso *et al.*, 1978; Ulaby *et al.*, 1983; Shutko, 1992; Kostov and Jackson, 1993; Chanzy *et al.*, 1995; van de Griend *et al.*, 1996). As noted, decreases in moisture content are associated normally with decreases in the radiant temperature ($T_R(\lambda)$). In the thermal region, emissions are constrained to within 50 μm of the surface and this shallow depth can exhibit rapid temperature variations. Even though monitoring diurnal temperature variations in the thermal region may improve the moisture content estimation, microwave is the only remote sensing platform and technique that can provide soil moisture with reasonable precision and consistency.

Both radar and passive microwave sensors are sensitive to non-bound water, and in general, sandy soils hold less bound water than clays. Normally, microwave is sensitive to soil moisture content within the top 2–5 cm of the soil depth (Chanzy *et al.*, 1995). In exceptional conditions, radar can detect changes in moisture content at depths greater than 1 m, and passive microwave returns have been related to moisture content and groundwater at depths exceeding 1 m (Reutov and Shutko, 1992). Up to saturation, soil moisture acts to enhance the radar return at any given soil surface roughness height; in some cases, surface roughness variability can severely hamper the ability of radar to estimate changes in soil moisture (Ramsey, 1998). Radar is sensitive to soil moisture under short vegetation canopies; however, where moderately dense, the detection of soil moisture depends on the relative strength of the vegetation canopy and the incident flux interaction (Dobson *et al.*, 1995). Surface roughness increases also enhance passive microwave emissions and can enhance soil emissivity differences between wavelengths. For most natural surfaces, however, roughness is not a serious limitation, and in wet soils, emissivity differences may be minor (Wang *et al.*, 1987; Engman and Chauhan, 1995). In most soils, passive microwave emissions are practically independent of soil type, salinity, bulk density, and temperature variability (Shutko, 1992; Engman and Chauhan, 1995), but overlying vegetation attenuation increases as water content increases (van de Griend *et al.*, 1996). In both passive microwave and radar, in general, the transmission through the vegetation canopy increases with increasing wavelength. Thus, longer wavelengths, HH polarization, and steeper incident angles are preferred for sensing soil moisture through a vegetation canopy.

Shoreline placement

The delineation of land and water is important in coastal classifications, land loss, and shoreline displacement. Accurate mapping of coastal shoreline (defined as the high water line or wet–dry boundary) placement requires not only high spatial resolution sensors but also the spectral ability to provide contrast between open water and the regional nearshore material. Historically, optical sensors (especially photographic) have provided image data used for shoreline mapping, but more recently, radar image data have been used to construct shorelines (Lee, 1990; Ramsey, 1995). Coastline detection and automated tracing algorithms are being developed to provide dynamic shoreline construction. In the case of radar, scatter from roughened open water at times can limit the ability to differentiate land and water areas. In addition, a shoreline position is dynamic, especially in coastal regions experiencing high tidal ranges. Consequently, to truly represent the shoreline position, the variation in the tidal height relative to mean high tide and the occurrence of influencing forces such as wind set-up or set-down or abnormal river runoff must be accounted for at the time of the measurement.

Flood monitoring

Remote sensing can detect flooding under vegetation. VNIR to MIR have been used, but microwave remote sensing offers the greatest potential for the instantaneous and consistent determination of flood extent. Within the microwave region, the most extensive history of flood detection under vegetation has been associated with radar (Ramsey, 1998). The radar return from flooded forest is usually enhanced compared to returns from nonflooded forests. The enhancement is related to the double bounce mechanism where the signal penetrating the canopy is reflected off the water surface and subsequently reflected back toward the sensor by a second reflection off a tree trunk (Hess *et al.*, 1990). In contrast, diminished returns from flooded relative to nonflooded coastal marshes have been observed (e.g., Ramsey, 1995). The marsh grasses may calm the water surface accentuating specular reflection but without the grasses providing the double bounce (Ormsby *et al.*, 1985). As in soil moisture mapping, flood detection can occur only if transmitted through the canopy; thus, longer wavelengths, HH polarization, and steeper incident angles are preferred.

Topography

Coastal topography controls the hydrology of the coastal wetlands, and thereby the distribution and health of the coastal vegetation. Offshore coastal bathymetry is a result of the dynamic forces of local erosion and sedimentation and littoral drift. Mapping and monitoring the onshore topography and offshore bathymetry is of vital importance to the coastal engineer and resource manager. Historically and currently, indirect methods based on the simultaneous viewing of overlapping images (parallax) are commonly used to generate topographic information. Methods based on passive optical remote sensing also have been used to map coastal bathymetry, but these methods are limited by severe and variable attenuation by the water column materials (Ji *et al.*, 1992; Lyon *et al.*, 1992). Besides audio-mechanical systems (sonar), radar and LIDAR systems offer a direct and more consistent approach to surveying coastal topography and bathymetry.

After processing and most orbital contributions have been eliminated from the interferogram (radar phase difference image), slight remaining differences in the point of view of the radar sensor yield fringes that follow the topography (Massonnet and Feigl, 1998). These topographic contours can be used to generate a digital elevation model (DEM). Additionally corrected for elevation and local elevation gradient (slope) spatial variances, the interferogram can be related to finer resolution topographic changes from deformation (e.g., coastal volcanoes, surface subsidence, erosion, rebound, or deposition). In monostatic systems, success of this technique requires all scatters comprising the target (e.g., overlying vegetation, moisture content, inundation) remain unchanged between the time of the two radar image collections (Massonnet and Feigl, 1998). While successful application is problematic in vegetated environments, absolute stability, and thereby success, in dynamic and highly vegetated coastal environments is less probable. Frequent and variable flooding and the associated changes in soil and vegetation moisture contents add complexities that appear as random speckles in the interferogram. In coastal areas, these complexities may limit the absolute elevation and elevation change resolutions attainable with interferograms generated from monostatic systems.

Airborne laser altimetry (ALS) is the simplest application of LIDAR remote sensing. ALS surveys are primarily performed at 700–1,000 m above ground level in order to eliminate most atmospheric attenuation of the signal. As in radar systems, when properly calibrated to a stable platform, the time between the emitted and detected pulses is directly related to the range, and thereby to changes in the surface elevation. Vegetation interferes with the laser pulse and complicates conversion of the ALS image into a topographic surface. Reflectance of solar illumination into the sensor field of view within the laser operational bandwidth also corrupts the ALS signal. A correction for vegetation interference and contamination uses data collected near in time along multiple transects to develop a topographic precision estimate.

Bathymetry

ALS is also used to develop coastal bathymetry maps. Most current ALS systems use two wavelengths: a green band for high water penetration to the bottom and a NIR band with little to no water penetration and almost total reflectance from the surface. Use of the two-band system helps diminish errors resulting from platform altitude variation so that the time delay difference between the two return pulses is a direct bathymetric measure. Increased turbidity, however, results in lower spatial resolutions and increased water volume backscattering of the emitted pulse creates false echoes in the record. The low altitude and fairly narrow coverage of current ALS systems hamper the operational feasibility of these systems in regional assessments. Even with these limitations, of all the electromagnetic systems, the ALS systems may provide the only feasible mechanism for rapid and consistent detailed mapping of coastal bathymetry and wetland topography.

Water quality and submerged aquatic vegetation

Remote sensing of estuarine and coastal waters is primarily concerned with mapping the type, concentration, distribution, and dispersion of materials suspended and dissolved in the water (water quality) (Morel and Gordon, 1980; Ramsey and Jensen, 1990), and the type and distribution of submerged aquatic vegetation and bottom cover (e.g., rock, mud, sand, shell). To accurately map the water quality and bottom type, the radiant flux depth-intensity and spectral distribution must be estimated (Kirk, 1980). In a well-mixed water column, the underwater radiation environment is determined by reflection and refraction at the water–air interface (surface), absorption and scattering within the water body, and reflection from the bottom. Surface reflections of $I_{T\lambda}$, upwelling restrictions (due to refraction at the surface), added atmospheric path and background fluxes, and atmospheric attenuation result in the upwelling water volume flux (below the water surface) transferred through the surface ($L_{T\lambda}$) normally comprising only 3–5% of the sensor signal ($L_{S\lambda}$). Corrected for water surface reflection, transmission, and atmospheric influences, $L_{T\lambda}$ can be related to the inherent bulk absorption and scattering properties of the water and water materials and bottom reflectance in the shallow waters (Morel and Gordon, 1980; Carder and Steward, 1985).

Excluding bottom reflections, the inherent bulk absorption ($\alpha(\lambda)$) and backscatter ($b_b(\lambda)$) coefficients can be related to $L_{T\lambda}$ as $L_{T\lambda}/I_{T\lambda} = \mathrm{BRDF}_\lambda = C_F \cdot b_b(\lambda)/\alpha(\lambda)$, where backscattering is scattering into the hemisphere trailing the incident flux, and C_F incorporates the ratio of two subsurface upwelling fluxes and the water refractive index (Carder and Steward, 1985; Bukata *et al.*, 1995). In optically complex coastal waters, bulk $b_b(\lambda)$ and $\alpha(\lambda)$ are commonly related to bulk descriptors of biomass (e.g., chlorophyll-a [Chl]), suspended materials (SMs) (e.g., detritus, suspended inorganic particles), and dissolved organic carbon (DOC) (Bukata *et al.*, 1995). Use of these bulk descriptors normally provides a good and stable estimation of water quality with respect to location and time in coastal environments. The inherent bulk properties are related to the specific optical coefficients (i) as $b_b(\lambda) = \Sigma C_i(b_b)_i(\lambda)$ and $\alpha(\lambda) = \Sigma C_i\alpha_i(\lambda)$, where the subscript ($i$) represents one water component and C_i refers to the components concentration (e.g., $\alpha(\lambda) = a_w + C_{Chl}\alpha_{Chl}(\lambda) + C_{SM}\alpha_{SM}(\lambda) + C_{DOC}\alpha_{DOC}(\lambda)$, where a_w is absorption due to the water).

LIDAR systems can be used to stimulate fluorescence in chlorophyll pigments associated with phytoplankton, plants and corals, and fluorescent DOC (i.e., Gelbstoff) (Measures, 1984). As in passive optical remote sensing of water quality, the signal returned to the sensor can be corrupted by the atmosphere and by addition of reflectance from the bottom. The selection of the excitation (send) and emission (return) wavelengths is either fixed by the LIDAR system or optimized by laboratory measurements. Chlorophyll is normally determined by excitation in the low red and measuring the emission in the high red wavelength regions. Gelbstoff is linearly related to natural fluorescence (Otto, 1967) that has been used as a conservative tracer of riverine and ocean waters mixing. Natural fluorescence is normally determined by excitation in the ultraviolet and by measuring the fluorescence in the blue. As in canopy profiling, the fluorescent return can be scattered or self-absorbed before reaching the water surface, causing ambiguity in mapping the concentration of the fluorescent material. Raman intensity variability can be measured simultaneously with the fluorescent return and used to remove the effect of self-absorption, thereby creating spatially comparable fluorescent images (Bristow *et al.*, 1981).

Bottom reflectance

Seagrasses and bottom type (mud, sand, shell) mapping is an important aspect of coastal monitoring. In this case, the overlying water column attenuates the signal to and from the bottom. In both passive VNIR and LIDAR mapping, the same problems apply as in water quality monitoring; however, in this case, the overlying water column signal must be removed from or de-emphasized in the return signal (Ji *et al.*, 1992; Lyon *et al.*, 1992). Increasing water turbidity and absorption can severely restrict the ability of either method (passive VNIR and LIDAR) to accurately map bottom reflectance variations as does low contrast between the different bottom materials.

Surface films and salinity

Observation of surface water features, especially surface films, has long been recognized as an indirect method of mapping convergence zones, mixing zones, and internal wave fields in optical oceanography (Klemas, 1980). The natural and extracted oils and similar substances also can be observed by SAR systems because they tend to dampen the creation of surface waves, smoothing the water surface, and attenuating the SAR returns. This differential dampening enables SAR sensors to map and monitor surface spills, and because of its nearly all weather capabilities, SAR provides capabilities many times superior to VNIR and MIR. Laser Induced Flouresence (ultraviolet excitation, visible emission) has also been used to detect and classify oil slicks and oil film thickness (Measures, 1984).

Of the possible conservative tracers of water mass mixing, salinity is the most notable. Salinity is not directly measurable with VIS to thermal systems (ignoring extremely slight dependencies), although salinity changes have been inferred from changes in other water properties, such as fluorescence and suspended particle concentrations. A more direct measure is based on the definition of salinity as the concentration of dissolved cations and anions. Changes in these concentrations change the water's ionic strength, leading to changes in the dielectric properties of the water that are most apparent at microwave wavelengths. These changes are best observed as changes in the emissivity, although to accurately observed changes, microwave emissions must be corrected for water temperature and surface roughness variations. To accomplish this correction, microwave measurements are collected at two wavelengths, obtaining a direct method to map changes in water salinity (Shutko, 1985).

Sea surface temperature

One of the earliest uses of radiometry was to map the sea surface temperature (SST) and thereby map different water masses and physical dynamics (e.g., frontal convergence, upwelling). Along the same line, water temperature mapping of heated effluents into rivers, estuaries, and coastal oceans is used to monitor compliance of discharges. In each of these radiometric applications, the radiant temperature ($T_{R\lambda}$) is related to the 3–5 μm thick surface skin by converting via emissivity ($\varepsilon(\lambda$, water)) to the skin kinetic temperature (KT). On average, surface skin temperatures are about 0.3 K cooler than bulk temperatures, but differences can range from about +1 to −1 K (Emery *et al.*, 1995). Depending on the water stability or the amount of mixing, the bulk temperature can represent the well-mixed surface layer or the temperature gradient depth. Often the skin $T_{R\lambda}$ (radiometer measurements) is related to the bulk KT or SST by breaking the surface skin with buckets of water (i.e., bucket temperature). Ship intakes and moored buoys offer bulk temperature measurements at variable depths. These measurements are used to correct atmospheric influences (especially water vapor) and consequently directly relate the sensor signal ($L_{S\lambda}$) to the SST, aggregating skin effects into atmospheric correction (Minnett, 1995). A separate type of atmospheric correction relates $L_{S\lambda}$ to atmospheric attenuation and thereby to the skin KT. Atmospheric effects are inferred from spectral relationships based on multiple band measurements (e.g., $\mathrm{SST} = a_0 + a_\lambda \mathrm{KT}_\lambda$, where λ refers to one or a combination of bands) (Emery *et al.*, 1995; Minnet, 1995). Alternatively, measurements of the same target but at different sensor view angles provide a direct measurement of atmospheric influences ($\mathrm{SST} = b_0 + b_{\lambda,N}\mathrm{KT}_N + b_{\lambda,S}\mathrm{KT}_S$, where N is nadir, S is oblique views, and λ refers to one band or a band combination). Based on current correction techniques, skin and bulk surface water temperatures can be estimated to less than 0.5°C. Even though less influenced by atmospheric conditions, the coarse spatial resolution associated with passive microwave systems makes them less preferred than thermal radiometry. Further, even at longer wavelengths, passive microwave measurements have shown dependence on surface roughness as a function of wind speed (Shutko, 1985; Trokhimovski *et al.*, 1995).

Sea ice

Detection and monitoring the distribution and type (first year and multiyear ice) of coastal (Arctic) sea ice is important in marine mammal ecology, climate processes, and early detection of global warming (Piwowar and LeDrew, 1996). Operational methods until recently have been applicable only to broadscale spatial inventories. More recently, optical, passive microwave, and radar sensors have provided higher spatial resolutions; however, increasing spatial resolution beyond 1 km constrains the ability to operationally monitor global or hemispheric regions. The primary factor controlling the remote sensing of sea ice is emissivity ($\varepsilon(\lambda)$), and the major factor controlling emissivity is salinity (Comiso, 1995). In thermal radiometry and passive microwave, the effective emissivity decreases from the cold saline first-year to the cold desalinated multiyear ice. Emissivity also decreases from first-year to multiyear ice due to the decreased density and increased surface roughness resulting in increased surface scatter. Added to this overall change, emissivity varies with wavelength and polarization. First-year ice is nearly independent of wavelength and polarization while emissivities of both VV and HH polarizations associated with multiyear ice increase with wavelength (Comiso, 1995). SAR returns tend to increase from first-year to multiyear ice (Drinkwater, 1995). In radar imaging, shorter wavelengths scattered from the ice surface primarily respond to dielectric differences (salinity) and roughness, while longer wavelengths (>5 cm) penetrate the ice (multiyear >> first-year) and are returned via volume scattering. Use of longer wavelengths normally results in a higher return from the relatively lower salinity and density multiyear ice than other ice types (Drinkwater, 1995). Of the three sensors, optical sensors are constrained by persistent clouds and darkness in Arctic regions, and radar returns from roughened multiyear ice surfaces are more prone to confusion with other types of sea ice than are emissions sensed by passive microwave sensors (Hall, 1998). Integrated approaches seem to provide the best results and the added benefit of comparison and validation.

Wind speed and surface waves

Short gravity (>1.7 cm) and small capillary (<1.7 cm) waves are the surface water features observed with operational radar systems (2–30 cm wavelengths) (Elachi, 1987). Increasing near-surface wind speed increases the amplitude of these waves, intensifying surface roughness that in turn promotes increasing slope (specular or facet) reflections and point scatter (or Bragg). As wind speed is related to surface roughness, altimeter and scatterometer sensors can provide estimates of wind speed (U). Scatterometers provide regional coverage, but at coarse pixel resolutions, while altimeter measurements cover narrow swaths. In the case of altimeters, σ^0 is a result of near nadir specular reflections (facet) that decrease as surface roughness increases ($\sigma^0 \propto U^{-x}$) (Elachi, 1987; Dobson, 1995). Although designed for measuring open ocean winds, scatterometers have also been found useful in measuring winds in coastal and enclosed seas. σ^0 derived from scatterometer measurements increases with wind speed increases (at >25° incident angle) as $\sigma^0 \propto U^x$ (x is dependent on wavelength) and are principally a result of Bragg scattering (Topliss and Guymer, 1995). Scatterometers also include multiple azimuths, providing the ability to estimate the wind direction. Wind speed accuracies derived from scatterometers are about 2 m/s with a directional tolerance of 20° (Topliss and Guymer, 1995). Compared to open ocean measurements, altimeter and scatterometer coastal measurements are more difficult to explain based on dynamic processes and are generally plagued by three types of problems: (1) contamination from land–water mixing, (2) varying wind–radar relationships, and (3) substantial influences of temporal and spatial variations in SST resulting in incorrect estimates of σ^0.

Currents and waves

Indirect observation of convergence and divergence zones associated with currents and possibly internal waves through varying surface features is used in optical remote sensing of ocean features (Klemas, 1980). Radar systems, however, offer a nearly unimpeded source of mapping surface features related to ocean dynamics. Short gravity and capillary waves created by wind stress and mechanically (independent of wind stress) are modified (local slope and growth) by long-period gravity or internal waves and variable currents (fronts, eddies, upwelling, tidal circulation) and bottom topography (Topliss and Guymer, 1995). These spatially and periodic modulated small wave fields (bands of roughness) are detectable with altimeters, scatterometers, and SAR imaging. Dependent on the angle of incidence and wavelength, SAR returns can be dominated by either a mixture of Bragg scattering or specular reflections. SAR images are used to define wave direction and length and to map the location of convergence zones and currents.

Water surface topography

Instantaneous sea surface height (S_0) observed by an altimeter is the sum of the geoid (N, a level surface of the earth's gravity field associated with a motionless ocean surface regarded as time invariant), the permanent dynamic topography (ξ_0, related to ocean circulation, $N + \xi_0$ = mean sea level), the variable topography (ξ_v, e.g., ocean tides and waves and swells), orbital and propagation errors (e.g., sensor attitude corrections, barometric correction), and sensor noise ($S_0 = N + \xi_0 + \xi_v +$ error + noise) (Le Traon, 1995). Wave heights from altimeter measurements are related to the shape or rise time of the returned pulse (Dobson and Monaldo, 1995). Increasing wave heights increase the slope or rise time. Large surface-wave heights or swell heights are estimated by differencing wave energy ($\propto$ (wave height)^2) and wave energy associated with wave heights estimated from altimeter wind speeds. In principle, skewness of the generated wave height probability distribution can also be used to estimate the dominant wavelength in unimodal seas (Dobson and Monaldo, 1995).

The low-frequency harmonic rise and fall of the coastal tides can be observed with satellite altimetry with a precision of about 3–5 cm (Han, 1995). Limited sampling frequency associated with altimeter data, however, leads to aliasing shorter tidal periods into longer tidal periods causing ambiguities in tidal period evaluations. Conversely, extraction of tidal fluctuations from the surface height variability is necessary to recover long-term height variability (e.g., annual cycles). As in other coastal applications, increased problems are created by the high temporal and spatial variability in the surface height due to basin morphology (shape and shallow depths ($\approx$100 m)), variable river runoff (buoyancy), solar heating (e.g., SST), wind stress (e.g., Eckman drift), and the surface expression of subsurface features (e.g., sea mounds or submarines).

Elijah W. Ramsey III

Bibliography

Adams, J., Smith, M., and Johnson, P., 1986. Spectral mixture modeling: a new analysis of rock and soil types at the Viking Lander 1 site. *Journal of Geophysical Research*, **91**: 8098–8112.

Bristow, M., Nielsen, D., Bundy, D., and Furtek, R., 1981. Use of water Raman emission to correct airborne laser fluorosensor data for effects of water optical attenuation. *Applied Optics*, **20**: 2889–2906.

Bukata, R., Jerome, J., Kondratyev, K., and Pozdnyokov, D., 1995. *Optical Properties and Remote Sensing of Inland and Coastal Waters*. New York: CRC Press.

Card, D., Peterson, D., Matson, P., and Aber, J., 1988. Prediction of leaf chemistry by the use of visible and near infrared reflectance spectroscopy. *Remote Sensing of Environment*, **26**: 123–147.

Carder, K., and Steward, R., 1985. A remote-sensing reflectance model of a red-tide dinoflagellate off west Florida. *Limnology and Oceanography*, **30**: 286–298.

Carter, G., Jones, J., Mitchell, R., and Brewer, C., 1996. Detection of solar-excited chlorophyll_a fluorescence and leaf photosynthetic capacity using a Fraunhofer line radiometer. *Remote Sensing of Environment*, **55**: 89–92.

Cavanie, A., and Gohin, F., 1995. Sea-ice studies with scatterometer. In Ikeda, M., and Dobson, F. (eds.), *Oceanographic Applications of Remote Sensing*. Boca Raton, FL: CRC Press, Inc., pp. 9–366.

Chanzy, A., Bruckler, L., and Perrier, A., 1995. Soil evaporation monitoring: a possible synergism of microwave and infrared remote sensing. *Journal of Hydrology*, **165**: 235–259.

Comiso, J., 1995. Sea-ice geophysical parameters from SSM/I data. In Ikeda, M., and Dobson, F. (eds.), *Oceanographic Applications of Remote Sensing*. Boca Raton, FL: CRC Press, Inc., pp. 321–338.

Dobson, E., 1995. Wind speed from altimeters. Satellite measurement and corrections. In Ikeda, M., and Dobson, F. (eds.), *Oceanographic Applications of Remote Sensing*. Boca Raton, FL: CRC Press, Inc., pp. 223–236.

Dobson, E., and Monaldo, F., 1995. Significant wave height from altimeter. In Ikeda, M., and Dobson, F. (eds.), *Oceanographic Applications of Remote Sensing*. Boca Raton, FL: CRC Press, Inc., pp. 223–236.

Dobson, M., Ulaby, F., and Pierce, L., 1995. Land-cover classification and estimation of terrain attributes using synthetic aperture radar. *Remote Sensing of Environment*, **51**: 199–214.

Drinkwater, M., 1995. Application of SAR measurements in ocean-ice-atmosphere interaction studies. In Ikeda, M., and Dobson, F. (eds.), *Oceanographic Applications of Remote Sensing*. Boca Raton, FL: CRC Press, Inc., 381–396.

Elachi, C., 1987. *Introduction to the Physics and Techniques of Remote Sensing*. New York: John Wiley & Sons.

Emery, W., Wick, G., and Schluessel, P., 1995. Skin and bulk sea surface temperatures: satellite measurement and corrections. In Ikeda, M., and Dobson, F. (eds.), *Oceanographic Applications of Remote Sensing*. Boca Raton, FL: CRC Press Inc., pp. 145–165.

Engman, E., and Chauhan, N., 1995. Status of microwave soil moisture measurements with remote sensing. *Remote Sensing of Environment*, **51**: 189–198.

Hall, K., 1998. Remote sensing of snow and ice using imaging radar. In Henderson, F., and Lewis, A. (eds.), *Principals and Applications of Imaging Radar*. New York: John Wiley & Sons, Inc., Manual of Remote Sensing, pp. 677–703.

Han, G., 1995. Coastal tides and shelf circulation by altimeter. In Ikeda, M., and Dobson, F. (eds.), *Oceanographic Applications of Remote Sensing*. Boca Raton, FL: CRC Press Inc., pp. 45–56.

Hess, L., Melack, J., and Simonett, D., 1990. Radar detection of flooding beneath the forest canopy: a review. *International Journal of Remote Sensing*, **11**: 1313–1325.

Horler, D., and Barber, J., 1981. Principles of remote sensing of plants. In Smith, H. (ed.), *Plants and the Daylight Spectrum*. New York: Academic Press. pp. 43–64.

Idso, S., Hatfield, J., Reginato, R., and Jackson, R., 1978. Wheat yield estimation by albedo measurements. *Remote Sensing of Environment*, **7**: 273–276.

Jensen, J., 2000. *Remote Sensing of the Environment, An Earth Resource Perspective*. Upper Saddle River: Prentice Hall.

Ji, W., Civco, D., and Kennard, W., 1992. Satellite remote bathymetry: a new mechanism for modeling. *Photogrammetric Engineering and Remote Sensing*, **58**: 545–549.

Kirk, J., 1980. Spectral absorption properties of natural waters: contribution of the soluble and particulate fractions to light absorption in some inland waters of south-eastern Australia. *Australian Journal of Marine and Freshwater Research*, **31**: 287–296.

Klemas, V., 1980. Remote sensing of coastal fronts and their effects on oil dispersion. *International Journal of Remote Sensing*, **1**: 11–28.

Kostov, K., and Jackson, T., 1993. Estimating profile soil moisture from surface layer measurements—a review. *Ground Sensing SPIE*, **1941**: 125–136.

Lee, J., 1990. Coastline detection and tracing in SAR images. *IEEE Transactions on Geoscience and Remote Sensing*, **28**: 662–668.

Le Traon, P., 1995. Basin-scale oceanic circulation from satellite altimetry. In Ikeda, M., and Dobson, F. (eds.), *Oceanographic Applications of Remote Sensing*. Boca Raton, FL: CRC Press, Inc., pp. 79–96.

Lewis, A., Henderson, F., and Holcomb, D., 1998. Radar fundamentals: the geoscience perspective. In Henderson, F., and Lewis, A. (eds.), *Principals and Applications of Imaging Radar*. New York: John Wiley & Sons, Inc., Manual of Remote Sensing, **3**: 131–181.

Lillesand, T., and Kiefer, R., 1994. *Remote Sensing and Image Interpretation*. New York: John Wiley & Sons, Inc.

Lyon, J., Lunetta, R., and Williams, D., 1992. Airborne multispectral scanner data for evaluating bottom sediment types and water depths of the St. Marys River, Michigan. *Photogrammetric Engineering and Remote Sensing*, **58**: 951–956.

Malet, P., 1996. Classifying the geometry of canopies from time variations of red and near-infrared reflectance. *Remote Sensing of Environment*, **56**: 64–171.

Massonnet, D., and Feigl, K., 1998. Radar interferometry and its application to changes in earth's surface. *Reviews of Geophysics*, **36**: 441–500.

Measures, R., 1984. *Laser Remote Sensing, Fundamentals and Applications*. New York: John Wiley & Sons.

Minnett, P., 1995. Sea surface temperatures from the along-track scanning radiometer. In Ikeda, M., and Dobson, F. (eds.), *Oceanographic Applications of Remote Sensing*. Boca Raton, FL: CRC Press pp. 461–472.

Morel, A., and Gordon, H., 1980. Report on the working group on water color. *Boundary-Layer Meteorology*, **18**: 343–355.

NASA, 1989. Instrument panel report. *SAR Synthetic Aperture Radar* (Earth Observation System). Washington, DC: Earth Science and Application Division, NASA Headquarters, IIF, p. 233.

Ormsby, J., Blanchard, B., and Blanchard, A., 1985. Detection of lowland flooding using active microwave systems. *Photogrammetric Engineering and Remote Sensing*, **51**: 317–328.

Otto, I., 1967. Investigations on optical properties and water-masses of the southern North Sea. *Netherlands Journal of Sea Research*, **4**: 532–551.

Piwowar, J., and LeDrew, E., 1996. Principal components analysis of arctic ice conditions between 1978 and 1987 as observed from the SMMR data record. *Canadian Journal of Remote Sensing*, **22**: 390–403.

Ramsey, E., III, 1995. Monitoring flooding in coastal wetlands by using radar imagery and ground-based measurements. *International Journal of Remote Sensing*, **16**: 2495–2502.

Ramsey, E., III, 1998. Radar remote sensing of wetlands. In Lunetta, R., and Elvidge, C. (eds.), *Remote Sensing Change Detection: Environmental Monitoring and Applications*. Chelsea, Michigan: Ann Arbor Press, pp. 211–243.

Ramsey, E., III, and Jensen, J., 1990. The derivation of water volume reflectances from airborne MSS data using *in situ* water volume reflectances and a combined optimization technique and radiative transfer model. *International Journal of Remote Sensing*, **11**: 979–998.

Ramsey, E., III, Nelson, G., and Sapkota, S., 1998. Classifying coastal resources by integrating optical and radar imagery and color infrared photography. *Mangroves and Salt Marshes*, **2**: 109–119.

Raney, K., 1998. Radar fundamentals; technical perspective. In Henderson, F., and Lewis, A. (eds.), *Principles and Applications of Imaging Radar*. New York: John Wiley & Sons, Inc., *Manual of Remote Sensing*, **3**: 9–130.

Reutov, E., and Shutko, A., 1992. Estimation of the depth to a shallow water-table using microwave radiometry. *International Journal of Remote Sensing*, **13**: 2223–2232.

Shutko, A., 1985. The status of the passive microwave sensing of the water—lakes, seas, and oceans—under the variation of their state, temperature, and mineralization (salinity): models, experiments, examples of applications. *IEEE Journal of Oceanic Engineering*, **OE-10**: 418–437.

Shutko, A., 1992. Soil/vegetation characteristics at microwave wavelength. In Mather, P. (ed.), *Understanding the Terrestrial Environment: The Role of the Earth Observations from Space*. Washington, DC: Taylor and Francis.

Slater, P., 1980. *Remote Sensing, Optics and Optical Systems*. London: Addison-Wesley Publishing Company.

Smith, H., and Morgan, D., 1981. The spectral characteristics of the visible radiation incident upon the surface of the earth. In Smith, H. (ed.), *Plants and the Daylight Spectrum*. New York: Academic Press, pp. 3–20.

Topliss, B., and Guymer, T., 1995. Marine winds from scatterometers. In Ikeda, M., and Dobson, F. (eds.), *Oceanographic Applications of Remote Sensing*. Boca Raton, FL: CRC Press Inc., pp. 205–221.

Trokhimovski, Y., Bolotnikova, G., Etkin, V., Grechko, S., and Kuzmin, A., 1995. The dependence of S-band sea surface brightness and temperature on wind vector at normal incidence. *IEEE Transactions on Geoscience and Remote Sensing*, **33**: 1085–1088.

Ulaby, F., Razani, M., and Dobson, M., 1983. Effects of vegetation cover on the microwave radiometric sensitivity to soil moisture. *IEEE Transactions on Geoscience and Remote Sensing*, **21**: 51–61.

van de Griend, A., Owe, M., de Ruiter, J., and Gouweleeuw, B., 1996. Measurement and behavior of dual-polarization vegetation optical depth and single scattering albedo at 1.4- and 5-GHz microwave frequencies. *IEEE Transactions on Geoscience and Remote Sensing*, **34**: 957–965.

van Zyl, J., 1989. Unsupervised classification of scattering behavior using polarimetric data. *IEEE Transactions on Geoscience and Remote Sensing*, **27**: 36–45.

Wang, J., Engman, E., Mo, T., Schmugge, T., and Shiue, J., 1987. The effects of soil moisture, surface roughness, and vegetation on L-band emission and backscatter. *IEEE Transactions on Geoscience and Remote Sensing*, **25**: 825–833.

Wegmuller, U., and Werner, C., 1997. SAR interferometric signatures of forest. *IEEE Transactions on Geoscience and Remote Sensing*, **35**: 18–24.

Whitt, M., Ulaby, F., and Sarabandi, K., 1990. Polarimetric scatterometer systems and measurements. In Ulaby, F., and Elachi, C. (eds.), *Radar Polarimetry for Geoscience Applications*. Norwood, Maine: Artech House, Inc., pp. 191–272.

Wigneron, J., Chanzy, A., Calvet, J., and Bruguier, N., 1995. A simple algorithm to retrieve soil moisture and vegetation biomass using passive microwave measurements over crop fields. *Remote Sensing of Environment*, **51**: 331–341.

Zebker, H., van Zyl, J., and Elachi, C., 1990. Polarimetric radar system design. In Ulaby, F., and Elachi, C. (eds.), *Radar Polarimety for Geoscience Applications*. Norwood, Maine: Artech House, Inc., pp. 273–312.

Cross-references

Airborne Laser Terrain Mapping and Light Detection and Ranging
Altimeter Surveys, Coastal Tides and Shelf Circulation
Coasts, Coastlines, Shores, and Shorelines
Mangroves, Remote Sensing
Mapping Shores and Coastal Terrain
Nearshore Geomorphological Mapping
Photogrammetry
Remote Sensing: Wetlands Classification
Synthetic Aperture Radar Systems

REMOTE SENSING: WETLANDS CLASSIFICATION

Coastal wetlands are a highly productive and critical habitat for a number of plants, fish, shellfish, waterfowl, and other wildlife. Wetlands also provide flood damage protection, protection from storm and wave damage, water quality improvement through filtering of agricultural and industrial waste, and recharge of aquifers. After years of degradation due to dredge and fill operations, impoundments, urban development subsidence/erosion, toxic pollutants, entrophication, and sea-level rise, wetlands have finally begun to receive public attention and protection (Daiber, 1986). Heightened awareness of the value of wetlands has resulted in the need to better understand their function and importance and find ways to manage them more effectively. To accomplish this, at least two types of data are required: (1) information on the present distribution and abundance of wetlands; and (2) information on the trends of wetland losses and gains.

Coastal wetlands can be conveniently divided into four major types: salt marshes, coastal fresh marshes, coastal forested and scrub–shrub wetlands, and tidal flats (Field *et al.*, 1991). Each of these wetland types has different hydrologic requirements and is dominated by a different type of vegetative cover. As a result, their spectral signatures and detectability by remote sensors differ significantly. For instance, salt marshes are widely distributed along the Atlantic and Gulf coasts and are dominated by smooth cordgrass (*Spartina alterniflora*) and frequently include other grasses, such as salt hay (*Spartina patens*) and big cordgrass (*Spartina cynosuroides*). The relative purity and size of salt marshes makes it possible to map them from aircraft and satellites.

Freshwater marshes are relatively diverse and have a more mixed vegetative cover producing a more complex, composite spectral signature. Forested and scrub–shrub wetlands, characterized as woody communities, are regularly inundated and saturated during the growing season. Wooded wetlands resemble spectrally wooded uplands and are therefore difficult to distinguish from wooded upland areas. Combining RADAR data with Landsat TM helps to distinguish wooded uplands from wetlands by providing soil moisture conditions beneath the tree canopy.

Wetlands mapping

Most of the major wetlands mapping programs, conducted by the United States Geological Survey (USGS), the National Oceanic and Atmospheric Administration (NOAA), and the Environmental Protection Agency (EPA), and other agencies, are described in Kiraly *et al.* (1990). Traditionally the US Fish and Wildlife Service (FWS) has played a key role, conducting its first nationwide wetlands inventory in 1954, which focused on waterfowl wetlands. In 1974, the FWS established the National Wetlands Inventory Project (NWI) to generate scientific information on the characteristics of US wetlands, including detailed maps and status/trend reports. The maps are available as 7.5 min quads at a scale of 1:24,000. Most have been digitized, converting them from paper maps to a GIS (Geographic Information System)-compatible digital line graph (DLG) format (Tiner, 1985).

The NWI program produced a new classification system and a more rigorous definition of wetlands: wetlands are transitional areas between terrestrial and aquatic systems where the water table is usually at or near the surface or the land is covered by shallow water. Wetlands must also have one or more of the following attributes: (1) at least periodically, the land supports predominantly hydrophytes; (2) the substrate is predominately hydric soil; and (3) the substrate is non-soil and is saturated with water or covered by shallow water at some time during the growing season each year (Cowardin *et al.* 1979). The classification system developed for the NWI by Cowardin *et al.* (1979) is hierarchical, progressing from systems (Marine, Estuarine, Riverine, Lacustrine, and Palustrine) and subsystems (Tidal, Subtidal, Intertidal, etc.), at the most general levels, to classes, subclasses, and dominance types (Figure R12). While suitable for use with field data and aerial photography, this classification system proved too complex for satellite remote sensors, which lacked the spatial, spectral, and temporal resolution required by such detailed mapping efforts. More recently, satellite techniques are being tested for updating the NWI maps and publishing status and trend reports every 10 years. Most coastal states have used aerial photography to map their wetlands in great detail (e.g., scales of 1 : 2,400) in order to satisfy legal or planning requirements.

Remote sensing

Satellite remote sensing of wetlands was attempted, with limited success, as soon as the first Landsat MSS was launched. SPOT with its 20/10 m resolution and Landsat Thematic Mapper (TM) with its six reflected bands and 30 m spatial resolution significantly improved our ability to map large coastal marshes. Using Landsat TM data, NOAA has initiated the Coastal Change Analysis Program (C-CAP) in order to develop a nationally standardized database on land-cover and habitat change in the coastal regions of the United States C-CAP inventories coastal submersed habitats, wetland habitats, and adjacent uplands and monitors changes in these habitats on a one-to five-year cycle with a minimum mapping unit of several hectares. This type of information and frequency of detection are required to improve scientific understanding of the linkages of coastal and submersed wetland habitats with adjacent uplands and with the distribution, abundance, and health of living marine resources. Using a rigorous protocol, satellite imagery, aerial photography, and field data are interpreted, classified, analyzed, and integrated with other digital data in a GIS. The resulting land-cover change databases are disseminated in digital form to users (Dobson *et al.*, 1995).

C-CAP developed a classification system, shown in Table R3 to facilitate the use of satellite imagery (Klemas *et al.*, 1993). Two study areas in South Carolina were used to evaluate a modified C-CAP classification scheme, image classification procedures, change detection algorithm alternatives, and the impact of tidal stage on coastal change detection. The modified C-CAP Classification Scheme worked well and can be adapted for other coastal regions (Jensen *et al.*, 1993). Unsupervised "cluster-busting" techniques coupled with "threshold 3 majority filtering" yielded the most accurate individual date classification maps (86.7–92.3% overall accuracy; Kappa coefficients of 0.85–0.90). The best change detection accuracy was obtained when individual classification maps were majority filtered and subjected to "post-classification comparison" change detection (85.2% overall accuracy; Kappa coefficient of 0.82). The multiple date images selected for coastal change detection had to meet stringent tidal stage and seasonal guidelines (Jensen *et al.*, 1993; Jensen, 1996).

Henderson *et al.* (1999) performed a detailed accuracy assessment of coastal land-cover mapping results obtained for Long Island using Landsat TM data and the C-CAP protocol. Table R4 displays two columns of user accuracies for C-CAP classification categories obtained by Henderson *et al.* (1999). The first column shows the user accuracies for the classification based on the raw spectral data, while the second column shows the accuracies after the data was recoded and filtered using ancillary verification data sets. Table R4 indicates that originally there were considerable errors in some categories, such as the Palustrine Wooded and Cultivated. However, as shown in Table R4, incorporation of ancillary data layers (e.g., aerial photographs, NWI wetland maps, etc.) increased the user accuracies of most categories into the upper 90% range, with the lowest, "Cultivated," attaining 86% (Henderson *et al.*, 1999).

Table R3 C-CAP coastal land-cover classification system[a]

Upland	Wetland	Water and submerged land
Developed land	Marine/estuarine rocky shore	Water
Cultivated land	Marine/estuarine unconsolidated shore	Marine/estuarine reef
Grassland	Estuarine emergent wetland	Marine/estuarine aquatic bed
Woody land	Estuarine woody wetland	Riverine aquatic bed
Bare land	Riverine unconsolidated shore	Lacustrine aquatic bed (basin ≥20 acres)
Tundra	Lacustrine unconsolidated shore	Palustrine aquatic bed basin ≤20 acres)
Snow/ice	Palustrine unconsolidated shore	
	Palustrine emergent wetland	
	Palustrine woody wetland	

[a] Only the upper two levels are shown in this table. The third, more detailed level has been omitted.

Table R4 Comparison of user's accuracy by classification category for combined raw spectral images and composite (including ancillary data) imagery

Category	Raw spectral (%)	Composite imagery (%)
Bare	92.67	93.00
Cultivated	70.00	86.00
Developed	96.67	98.00
Grassland	84.47	95.00
Water	100.00	99.00
Palustrine wooded	46.67	97.00
Palustrine emergent	85.33	97.00
Estuarine emergent	81.13	100.00
Wooded	87.33	89.00

Source: From Henderson *et al.*, 1999. Reproduced by permission of Taylor & Francis.

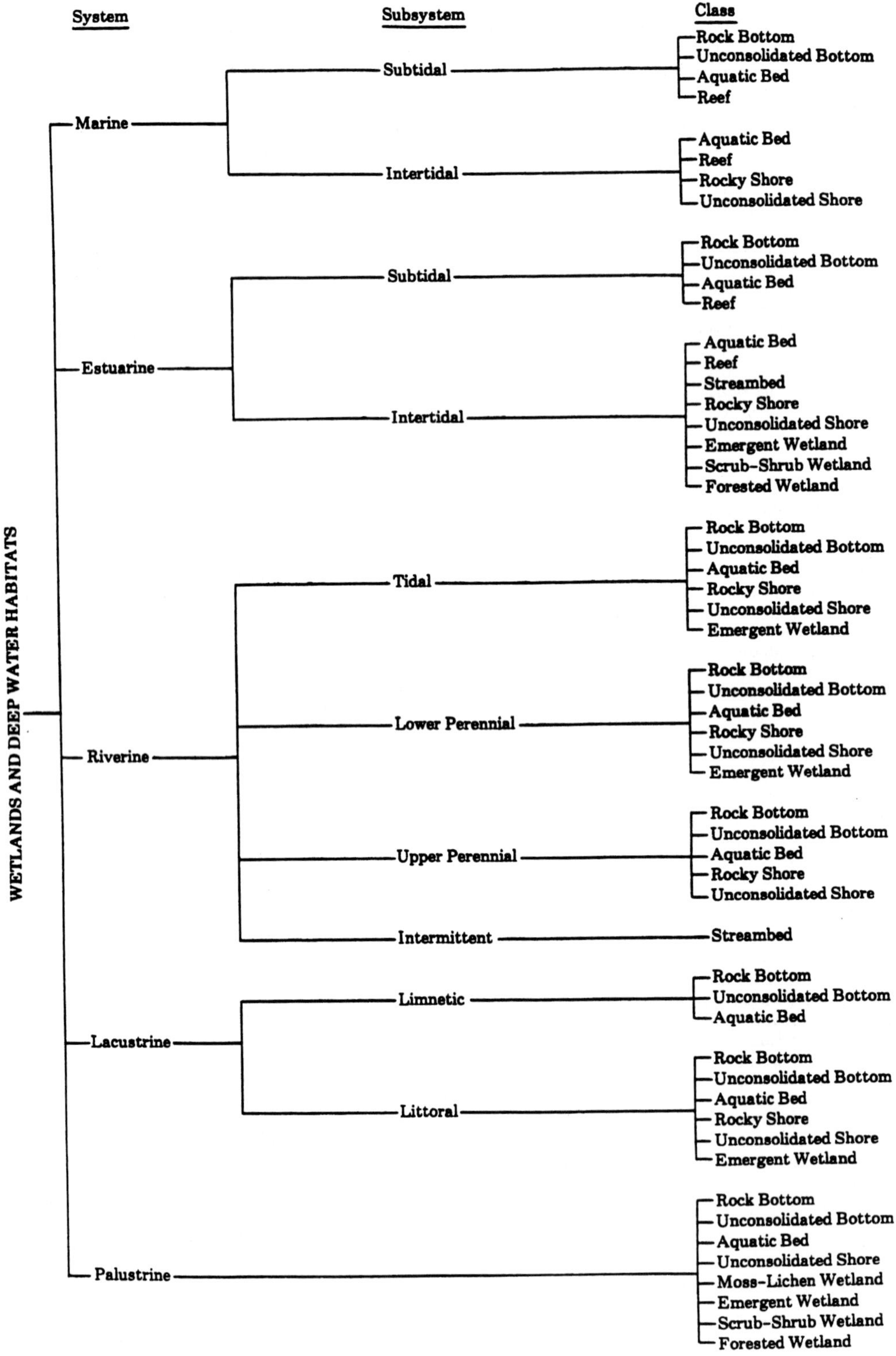

Figure R12 Classification hierarchy of wetlands and deepwater habitats showing systems, subsystems, and classes. The Palustrine System does not include deepwater habitats (Cowardin *et al.*, 1979, Fish and Wildlife Service).

Another way to improve the accuracy of wetland classifications derived from satellite imagery is to use multiple-date (multiple season) imagery. Multi-temporal Landsat TM imagery was evaluated for the identification and monitoring of potential jurisdictional wetlands located in the states of Maryland and Delaware (Lunetta and Balogh, 1999). A wetland map prepared from single-date TM imagery was compared to a hybrid map developed using two dates of imagery. The basic approach was to identify land-cover vegetation types using spring leaf-on imagery, and identify the location and extent of the seasonally saturated soil conditions and areas exhibiting wetland hydrology using spring leaf-off imagery. The accuracy of the wetland maps produced from both single- and multiple-date TM imagery were assessed using reference data derived from aerial photographic interpretations and field observations. Subsequent to the merging of wetland forest and shrub categories, the overall accuracy of the wetland map produced from two dates of imagery was 88% compared to the 69% result from single-date imagery. A Kappa test Z statistic of 5.8 indicated a significant increase in accuracy was achieved using multiple-date TM images. Wetland maps developed from multi-temporal Landsat TM imagery may potentially provide a valuable tool to supplement existing NWI maps for identifying the location and extent of wetlands in northern temperate regions.

Summary and conclusions

Coastal wetlands are valuable natural assets and must be protected and managed more effectively. To accomplish this, timely information on wetlands distribution, abundance, and trends are required. This information can be provided efficiently by remote sensors on aircraft and satellites. Since many wetlands occur in narrow, elongated patches and have complex spectral signatures, satellite sensors on Landsat and SPOT can provide accurate wetland maps only if multi-temporal images are used or significant amounts of ancillary data employed. Fortunately, technology, cost, and need are converging in ways that are making remote sensing and GIS techniques practical and attractive for wetlands mapping and coastal resource management (Lyon and McCarthy, 1995). With the launch of Landsat 7, the cost of TM imagery has dropped by nearly a factor of 10, decreasing the cost of mapping large coastal areas. New satellites, carrying sensors with much finer spatial (1–5 m) and spectral (200 bands) resolutions are being launched and may more accurately map and detect changes in coastal habitat. Advances in the application of GIS are helping to incorporate ancillary data layers to further improve the accuracy of satellite classification of coastal wetlands and land-cover.

Victor Klemas

Bibliography

Cowardin, L., Carter, V., Golet, F., and LaRoe, E., 1979. *Classification of Wetlands and Deepwater Habitats of the United States*. Washington, DC: U.S. Department of the Interior, Fish and Wildlife Service, Office of Biological Services.

Daiber, F., 1986. *Conservation of Tidal Marshes*. New York: Van Nostrand Reinhold Co.

Dobson, J., Bright, E., Ferguson, R., Field, D., Wood, L., Haddad, K., Iredale, H., III, Jensen, J., Klemas, V., Orth, R., and Thomas, J., 1995. NOAA coastal change analysis program (C-CAP): guidance for regional implementation. *NOAA Technical Report* NMFS 123. *A Technical Report of the Fishery Bulletin*, pp. 92.

Field, D., Reyer, A., Genovese, P., and Shearer, B., 1991. Coastal wetlands of the United States: an accounting of a valuable national resource. A Special NOAA 20th Anniversary Report. U.S. Department of Commerce, National Oceanic and Atmospheric Administration, and the National Ocean Service in cooperation with the U.S. Department of the Interior, Fish and Wildlife Service, pp. 59.

Henderson, F.M., Hart, T.F., Jr. Hearon, B.P., and Portolese, J.E., 1999. Mapping coastal ecosystems over a steep development gradient using C-CAP protocols. *International Journal of Remote Sensing*, **20** (4): 727–744.

Jensen, J., 1996. *Introductory Digital Image Processing: A Remote Sensing Perspective*, 2nd edn. Upper Saddle River: Prentice Hall.

Jensen, J., Cowen, D., Althausen, J., Narumalani, S., and Weatherbee, O., 1993. An evaluation of the coastwatch change detection protocol in South Carolina. *Photogrammetric Engineering and Remote Sensing*, **59**(6): 1039–1046.

Kiraly, S., Cross, F., and Buffington, J., 1990. Federal coastal wetland mapping programs. A Report by the National Ocean Pollutions Policy Board's Habitat Loss and Modification Working Group, Washington, DC: U.S. Department of the Interior, Fish and Wildlife Service, *Biological Report* **90**(18): pp. 174.

Klemas, V., Dobson, J., Ferguson, R., and Haddad, K., 1993. A coastal land cover classification system for the NOAA coastwatch change analysis project. *Journal of Coastal Research*, **9**(3): 862–872.

Lunetta, R., and Balogh, M., 1999. Application of multi-temporal Landsat 5 TM imagery for wetland identification. *Photogrammetric Engineering and Remote Sensing*, **65**(11): 1303–1310.

Lyon, J., and McCarthy, J., 1995. *Wetland and Environmental Applications of GIS*. Boca Raton: Lewis Publishers.

Tiner, R.W., Jr., 1985. *Wetlands of Delaware*. U.S. Fish and Wildlife Service, National Wetlands Inventory, Newton Corner, MA and Delaware Department of Natural Resources and Environmental Control, Wetlands Section, Dover, DE. Cooperative Publication.

Cross-references

Estuaries
History, Coastal Ecology
Monitoring, Coastal Ecology
Photogrammetry
Remote Sensing of Coastal Environments
Vegetated Coasts
Wetlands

RHYTHMIC PATTERNS

Beaches are seldom straight or smoothly curved in the longshore direction. Instead, they commonly include seaward projections of sediment, termed cusps, or embayments locally cut into the shore. Such features may be isolated, but more often occur in groups of alternating cusps and embayments that have a fairly regular spacing; they are then referred to as rhythmic patterns. There are several recognized types of rhythmic patterns, including beach cusps, sand waves, and giant cusps (Komar, 1998).

A wide range of spacings of rhythmic patterns can be found on beaches. Along the shores of ponds and small lakes the spacings between adjacent cusps may vary from less than 10 cm to 1 m. On ocean beaches with small waves, the spacing may be on the order of 2 m, while those built by large storm waves may be 50 m or more. Other rhythmic patterns, sand waves and giant cusps, have still larger spacings, typically ranging from 150 to 1,500 m, but with most being between 500 and 750 m, with the cusps projecting on average some 15–25 m seaward from the embayments.

The classification of rhythmic patterns

In the past, the classification of different types of rhythmic patterns has been based on the lengths of their spacings. Beach cusps were considered to have the smallest spacings, less than 25 m, while sand waves and giant cusps have larger spacings. These latter terms can be considered to be nearly synonymous, different names for the same or very similar features. Research in recent years has led to a better understanding of the formation of the various types of rhythmic patterns, and this now makes it possible to develop a genetic classification that depends on the processes of waves and currents that are responsible for their formation, rather than depending simply on their spacings (Komar, 1998). Furthermore, it is clear that rhythmic patterns having a wide range of spacings can be generated by a single mechanism, and more than one mechanism may be capable of producing rhythmic patterns having the same spacing. It is clear therefore that a genetic classification is needed, one that reflects the processes of formation. Such a genetic classification is depicted in Figure R13 (Komar, 1983), one that distinguishes between beach cusps, systems of rip-current embayments and cusps, series of transverse bars that produce cusps along the shore, and crescentic bars that are chiefly an underwater feature but can produce a rhythmic pattern that extends onto the dry part of the beach. In general, this order represents an increase in cusp spacings, with the latter three mechanisms yielding what had formerly been referred to as sand waves or giant cusps.

Beach cusps as a type of rhythmic pattern

The most easily recognized rhythmic pattern seen on beaches are the cuspate deposits of sand and gravel built by waves and known as beach cusps. Because of their marked regularity with nearly uniform spacings, beach cusps have attracted many observers and much speculation as to

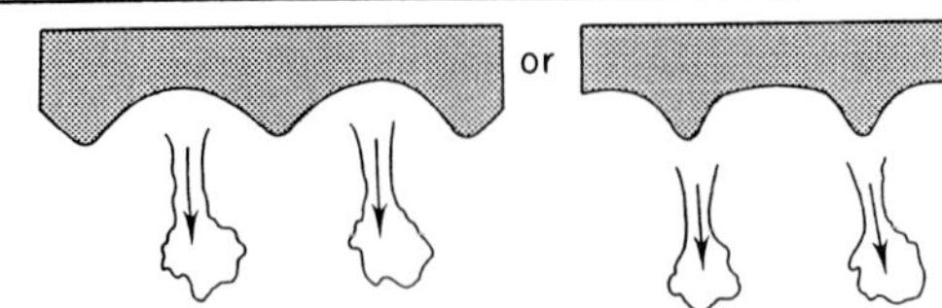

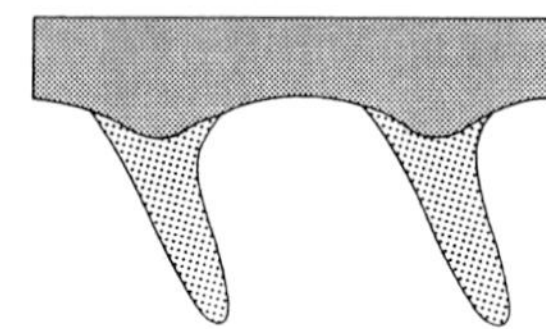

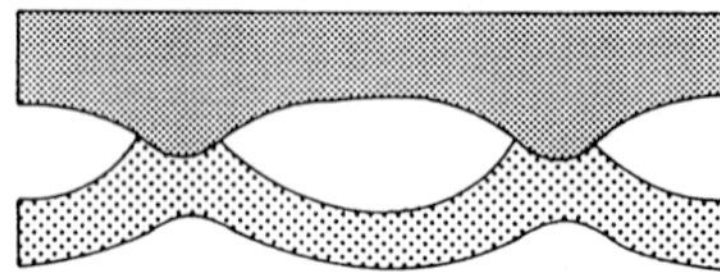

Figure R13 A genetic classification of rhythmic patterns where the origin of the series of cusps and embayments can be attributed to different processes of waves and currents (after Komar, 1983).

their origin. Arguments still persist with regard to the processes of wave motions and sediment transport that control their formation and determine the lengths of their spacings.

Beach-cusp formation is most favorable when the waves approach normal to the beach, that is, with their crests parallel to the longshore trend of the beach. This may explain why pocket beaches are particularly favorable sites for beach cusp formation. Furthermore, regular waves having long crest lengths are particularly conducive to cusp formation; beach cusps generally are not formed by irregular, confused seas.

Various investigators have described contrasting patterns of water circulation induced by the swash of waves around the cusps and within their embayments. In some cases there is an alternating surge inward and out of the embayments. Water flows from one embayment where the wave-swash runup has been a maximum, around the nearest cusp, and into a neighboring bay where it rushes up the beach face with the next wave. Thus, the maximum runup alternates in its timing between adjacent embayments. In other cases, the arriving waves break evenly along the beach, but the wave surge then piles up against the steep cusps and is divided into divergent streams that flow into the adjoining embayments. These streams head off the wave surge that had flowed directly up into the embayment. The two side streams from the cusps on either side meet at the center of the bay, and together form a seaward flow of considerable strength. This return flow can resemble a rip current, but unlike true rip currents, the flow is discontinuous and the mechanisms of formation are quite different. These contrasting patterns of water circulation observed around beach cusps suggests that more than one mechanism may give rise to their formation.

A number of hypotheses have been proposed to account for the formation of beach cusps. For a theory to be acceptable, it must account for the uniformity of spacing within an observed series of cusps, and the way in which this spacing is related to the wave parameters. The hypothesis that has been most successful in explaining the formation of beach cusps and accounting for their spacings is one based on the presence of edge waves in the surf zone (Guza and Inman, 1975). An edge wave is a type of wave that is trapped in the surf by refraction across the slope of the beach, moving in the longshore direction as it alternately refracts while moving offshore, then bends entirely around to return to the shore, reflects from the beach, and repeats the pattern of movement. The important result is that the presence of the edge wave affects the intensity and distance of swash runup on the beach, producing a regular runup spacing along the length of shore. This hypothesis explains the formation of beach cusps as the rearrangement of the sediment into a regular pattern of alternating cusps and embayments, corresponding to the longshore wave length of the edge waves and spacing of their maximum runup on the beach. The validity of this hypothesis has been demonstrated in the controlled conditions of laboratory wave basins, and by a few studies on ocean beaches that happened to be measuring edge waves at the same time beach cusps formed (see review in Komar, 1998). In that the longshore length of edge waves is determined by the wave period and slope of the beach, this hypothesis yields a mathematical equation that predicts the beach-cusp spacing (Guza and Inman, 1975). Measured beach-cusp spacings ranging from 0.1 to nearly 100 m have been shown to agree with this mathematical relationship (Komar, 1998). Thus, there is strong supporting evidence for the edge wave hypothesis of beach cusp formation.

There is, however, an alternative hypothesis that has been proposed to account for the formation of beach cusps, the so-called "self organization" hypothesis of Werner and Fink (1993). It envisions an initially smooth, straight beach, lacking cusps, but with waves arriving and swashing up the beach face with some degree of irregularity. According to the hypothesis, this irregularity in the wave swash produces variable amounts of beach sand movement along the shore, with a tendency for the sand to preferentially accumulate in a few isolated areas. The critical aspect of this hypothesis is that the zones of accumulated sand then affect the subsequent patterns of wave runup, thereby increasing the sizes of the accumulated sand and causing them to evolve into a pattern of beach cusps having regular spacings. It is this trend toward increasing regularity to which the name "self organization" refers. The main supporting evidence for this hypothesis comes in the form of computer simulation models that demonstrate the possibility of such an evolution. While not actually having mathematical relationships that predict the eventual beach cusp spacing, the computer models demonstrate that the spacing depends on the wave period and height, and on the beach slope, and in fact the model yields cusp spacings that are similar to those predicted by the edge-wave hypothesis. This similarity in prediction has made it difficult to distinguish which mechanism is responsible for the formation of beach cusps on ocean shores. It is possible that both hypotheses can, in different situations, account for the formation of beach cusps, and in some cases may actually work together.

As noted above, the early definition of "beach cusps" restricted them to spacings of 25 m or less, but provided no explanation for their formation. Although we are still uncertain as to the specific generation mechanism, the proposed hypotheses appear to offer satisfactory explanations, so the term "beach cusps" is now used in genetic classifications like that in Figure R13. For the most part the old and new uses of the term refer to the same rhythmic pattern, but we now recognize that edge waves and perhaps self-organization can generate beach cusps that have spacings up to 100 m.

Rip current embayments and cuspate shores

Within the series of rhythmic patterns diagrammed in Figure R13, generally the next larger form beyond beach cusps is the system of erosional embayments and intervening cusps formed by nearshore current systems that include seaward-flowing rip currents. In most instances the rip currents erode sand from the beach and transport it offshore, forming embayments at the rip-current positions, with cusps midway between. In rarer instances, particularly on steep beaches, coarse sand and gravel may accumulate at the shoreward ends of the rips, developing cusps at those positions. Rip currents, and hence the embayments and cusps, typically have spacings that range from tens to hundreds of meters. As such, the resulting rhythmic pattern corresponds to what has been referred to variously as sand waves or giant cusps.

The effect of the nearshore currents on the shore, forming series of embayments and cusps, is the surface expression of the underwater topography that is molded by the currents acting together with the waves. The seaward-flowing rip currents tend to erode channels across the full width of the beach within the embayments, segmenting the offshore bar. This leaves a system of cusps midway between rip currents, but each cusp seen on the shore is part of a shoal that extends out to the remaining segment of offshore bar.

This form of rhythmic pattern with embayments cut by rip currents is often important to property erosion in that the embayments narrow the beach width and remove most of the buffer protection offered to properties backing the beach. Although the rip embayments themselves do not usually produce much erosion of dunes and sea cliffs, they provide an area of deeper water where storm waves can approach close to shore before breaking against the coastal properties.

Rhythmic patterns produced by welded and transverse bars

A variety of sand bars have been observed in the nearshore that run obliquely to the longshore trend of the beach. These have been termed "welded" or "transverse" bars (Figure R13). An example is shown in Figure R14 on the ocean shore of Cape Cod, Massachusetts, consisting of a series of bars and a cuspate shore that has a distinctive longshore rhythmicity of several hundred meters. Therefore, the presence of welded or transverse bars can also give rise to a rhythmic pattern.

A number of suggestions have been made for the origin of bars that trend obliquely to the shore, and for the corresponding rhythmic pattern. It has been observed that when waves break at pronounced angles to the beach, the offshore bars that were originally parallel to the shore and segmented by evenly spaced rip currents, rotate to align themselves with the incoming wave crests. This may be the origin of the welded bars and rhythmic pattern seen in Figure R14.

Another type of oblique bar is found on coasts of low wave energy, for example in lakes or along the shore of the Gulf of Mexico. Referred to as transverse bars, they tend to occur in families that run parallel to one another, directed toward the offshore. At each point where a transverse bar joins the shore, a large cusp develops on the dry beach. Transverse bars can be fairly permanent features—examples on the shores of the Gulf of Mexico have been observed to persist in aerial photographs that span 25 years, showing little or no tendency to migrate alongshore during that time. Investigations have demonstrated that this type of transverse bar affects the paths of nearshore currents, with the current being concentrated over the bar and flowing offshore along its length, thereby perpetuating the bar's existence and extending it's length.

Figure R14 A series of welded bars and associated rhythmic pattern on Cape Cod, Massachusetts (photo courtesy of David S. Aubrey, Woods Hole Oceanographic Institution).

Crescentic bars and large-scale rhythmic patterns

Crescentic bars are one form of submerged offshore bars where rather than being linear, they have a regular lunate or crescentic shape (Figure R13), together with a uniform repetition along the length of beach. This regularity generally cannot be appreciated by observers on the dry beach, since most of the feature is underwater. However, there may be an associated series of cusps on the beach if the landward ends of the crescentic bars attach to the shore. In this instance, the presence of offshore crescentic bars leads to the development of another form of rhythmic pattern.

Crescentic bars are much larger features than beach cusps, and generally are somewhat larger than the rhythmic patterns due to rip currents or welded bars. In some instances, large crescentic bars form the outer-bar system of a beach, while the inner bar is linear and segmented by the more closely spaced rip currents. The range in lengths of crescentic bars is difficult to establish, since for many reported occurrences it is not possible to determine conclusively whether crescentic bars or some other form of rhythmic pattern is being described. Crescentic bars appear to range up to 2,000 m in length (Komar, 1998). At times there can be multiple crescentic bars on a beach, the further offshore the bars the larger their spacings. The corresponding rhythmic pattern on the beach would similarly have very large spacings between successive cusps.

Like beach cusps, the regularity in shapes and the even spacings of crescentic bars have inspired a number of suggestions as to their formation. The mechanism proposed by Bowen and Inman (1971), again by the movement of edge waves, provides the most reasonable explanation. In this case, however, important is the velocity of water movements associated with the edge waves, not their swash runup on the sloping beach which may be responsible for beach cusps. According to this mode of formation, beach sediment in the outer surf zone drifts about under the currents of the edge waves, until the sand reaches zones where the water velocity is low and the sand can accumulate. According to computer models of edge wave motions, this rearrangement of the sand would yield lunate-shaped bars that are remarkably similar to those observed on ocean beaches, a result that argues in favor of this hypothesis. Bowen and Inman (1971) conducted a series of laboratory wave-basin experiments that further confirmed this predicted sand accretion pattern, leading to the formation of crescentic bars. At this time, there is no reasonable alternative hypothesis for crescent-bar formation that satisfactorily accounts for their regularity in shapes and spacings, and for the formation of the associated rhythmic pattern.

Rhythmic patterns, irregular shores, and coastal erosion

Depending on the mechanism, rhythmic patterns may consist of alternating cusps and embayments whose spacings range from a few meters (beach cusps), to on the order of 100 m (rip-current embayments or welded bars), and on up to 500–1,500 m (crescentic bars). If one includes series of cuspate forelands or capes like those that exist along the southeast coast of the United States, the series can be extended up to tens of kilometers. When only one type of rhythmic pattern is present on the beach, it gives rise to a fairly regular spacing of alternating cusps and embayments. However, it is common for more than one type of pattern to occur simultaneously on a beach, and the summation of what are otherwise regularly spaced patterns can lead to an irregular beach and shore. An example is shown in Figure R15, the Cape Hatteras coast of North Carolina, photographed in 1970 before the lighthouse was moved (Dolan, 1971). Apparent are the series of large cusps and embayments, with a fair degree of regularity along the length of coast covered by the photograph. However, there is a level of irregularity, with some embayments being larger than others. Although the cause in this example is uncertain, it is probable that the irregularity of the rhythmicity is produced by the summation of two rhythmic patterns, that due to a rip-current cusps/embayment system, together with a larger-scale variation due to offshore crescentic bars. Of interest in this example, the respective embayments produced by rip currents and the crescentic bars appear to have combined in the area of the Cape

Figure R15 A rhythmic pattern of alternating cusps and embayments on the Cape Hatteras coast of North Carolina, photographed in 1970, with the largest embayment producing beach erosion and threatening the lighthouse (from Dolan, 1971).

Hatteras Lighthouse, resulting in the total loss of the beach and erosion of the dunes to the extent that the lighthouse was in danger. Therefore, an understanding of the origin and types of rhythmic patterns can be important to interpretations of the causes of coastal erosion problems.

Paul D. Komar

Bibliography

Bowen, A.J., and Inman, D.L., 1971. Edge waves and crescentic bars. *Journal of Geophysical Research*, **76**: 8662–8671.

Dolan, R., 1971. Coastal landforms: crescentic and rhythmic. *Geological Society of America Bulletin*, **82**: 177–180.

Guza, R.T., and Inman, D.L., 1975. Edge waves and beach cusps. *Journal of Geophysical Research*, **80**: 2997–3012.

Komar, P.D., 1983. Rhythmic shoreline features and their origins, In R. Gardner and H. Scoging (eds.), *Mega-Geomorphology*, pp. 92–112, Claredon Press, Oxford.

Komar, P.D., 1998. *Beach Processes and Sedimentation*, 2nd edn. Prentice-Hall.

Werner, B.T., and Fink, T.M. 1993. Beach cusps as self-organized patterns. *Science*, **260**, 968–971.

Cross-references

Accretion and Erosion Waves on Beaches
Bars
Beach Features
Beach Processes
Cuspate Forelands
Rip Currents
Surf Zone Processes

RIA

A ria is a long, narrow, often branching inlet formed by marine submergence of the lower parts of a river valley that had previously been incised below present sea level. Rias are the drowned mouths of unglaciated valleys, usually bordered by steep slopes rising to mountains, hills, or plateaux (Figure R16). The term is of Spanish origin, derived from large inlets on the coasts of Galicia such as the Ria de Arosa and the Ria de Muros y Noya, fingering far inland. They are known as abers in Brittany and Wales.

Von Richthofen (1886) defined a ria as a drowned valley cut transverse to the geological strike, but the Rias of Galicia do not meet this strict definition (Cotton, 1956). Perusal of coastal textbooks indicates that the term has come to be used as a synonym for a drowned valley mouth without any structural constraint. Rias generally have a dendritic (tree-like) outline, remaining open to the sea, as in Carrick Roads (Figure R17) in southwest England, Chesapeake Bay in the United States and Port Jackson (Sydney Harbor) in Australia. The long, straight valley-mouth gulfs on the southwest coast of Ireland, such as Bantry Bay, are examples of rias, even though they follow a geological strike that runs transverse to the general coastline. Where rivers have cut valleys across geological structures there may be tributaries that follow the geological strike, submerged to form a trellised pattern, as in Cork Harbor in southern Ireland. On the Dalmatian coast of the Adriatic Sea rias are elongated straits along valleys that follow the geological strike, linked to the sea by transverse channels.

Existing rias were formed by marine submergence during the Late Quaternary (Flandrian) marine transgression, but in southwest England there is evidence of several phases of valley incision during Pleistocene low sea-level phases, alternating with earlier ria formation

Figure R16 The ria at Aber Benoît, Brittany (photo: E.C.F. Bird: Copyright, Geostudies).

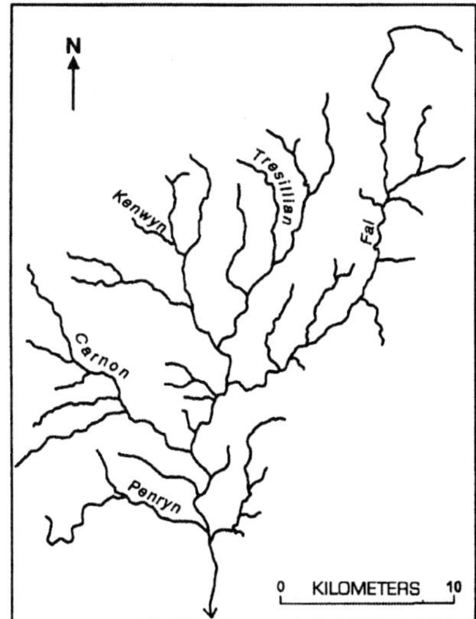

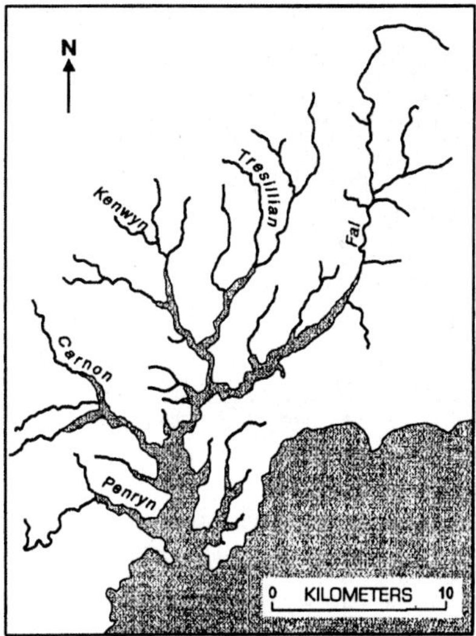

Figure R17 The Carrick Roads ria on the south coast of Cornwall, England. (Left) the river systems as they were 20,000 years ago, during the Late Pleistocene low sealevel phase. (Right) the present outlines, after partial submergence by the Late Quaternary marine transgression. (©Geostudies).

during interglacial marine transgressions and at least one Late Pleistocene higher sea-level phase indicated by an emerged beach (Kidson, 1977).

The Galician rias are generally wide and deep (up to 30 m) marine inlets in valleys which may have been shaped partly by tectonic subsidence and the recession of bordering scarps. Subsidence has probably contributed to the persistence of the ria at the mouth of Johore River in southeastern Malaysia, which remains a wide and deep inlet, whereas other Malaysian valley mouths have been infilled as alluvial plains, some with protruding deltas. Other rias persist because they were initially deep and sedimentary filling has been slow, as on the New South Wales coast (Roy, 1984).

Rias show varying degrees of sedimentary filling, partly from inflowing rivers and partly inwashed from the sea. There are marshy deltas at the heads of the several branches of Carrick Roads in Cornwall, while other rias in southwest England have banks of inwashed marine sand, particularly on the Atlantic coast, as in the Padstow ria, where at low tide the Camel River is narrow, flowing between broad exposed sandbanks that are submerged when the tide rises. Muddy sediment has been derived from periglacial deposits on bordering slopes.

Slopes bordering rias generally show only limited cliffing on sectors exposed to strong wave action, yielding sand and gravel to local beaches. Typically' there are sandy, gravelly, and rocky embayed shores, some spits, muddy sediment and salt marshes at their heads, and sandy shoals at their entrances (Castaing and Guilcher, 1995). The Rade de Brest is noteworthy for its several bordering sand and gravel spits (Guilcher *et al.*, 1957). Finer sediment is deposited in fringing tidal marshes and mudflats, and on the adjacent sea floor.

There is no clear distinction between a ria and an estuary, most rias being estuarine in the sense that inflowing rivers provide freshwater that meets and mixes with seawater moved in and out by the tide. Rias on high limestone coasts in the Mediterranean are known as calas or calanques, while those on arid coasts are termed sharms or sherms. In Chile' there is a transition southward from rias to fiords with increasing influence of glaciation on valleys.

Eric Bird

Bibliography

Castaing, P., and Guilcher, A., 1995. Geomorphology and sedimentology of rias. In Perillo G.M.E. (ed.), *Geomorphology and Sedimentology of Estuaries*. Amsterdam, Elsevier: pp. 69–111.

Cotton, C.A., 1956. Rias sensu stricto and sensu lato. *Geographical Journal*, **122**: 360–364.

Guilcher, A., Vallantin, P., Angrand, J.P., and Galloy, P., 1957, Les cordons littoraux de la rade de Brest. *Bulletin Comité Océanographique et Etude des Côtes*, **9**: 21–54.

Kidson, C., 1977. The coast of south-west England. In Kidson, C., and Tooley, M.J., (eds.), *The Quaternary History of the Irish Sea*. Liverpool: Seal House, pp. 257–298.

Roy, P.S., 1984. New South Wales estuaries: their origin and evolution. In: Thom, B.G., (ed.), *Coastal Geomorphology in Australia*. Sydney: Academic Press, pp. 99–121.

Von Richthofen, F., 1886. *Führer für Forschungsreisende*. Hanover: Janecke.

Cross-references

Dalmatian Coasts
Estuaries
Karst Coasts
Salt Marsh
Sharm Coasts

RIP CURRENTS

Definition, types, and early studies

Many of the world's beaches are characterized by the presence of strong, concentrated seaward flows called *rip currents*. Rips are an integral component of nearshore cell circulation and ideally consist of two converging longshore *feeder* currents which meet and turn seawards into a narrow, fast-flowing *rip-neck* that extends through the surf zone, decelerating and expanding into a *rip-head* past the line of breakers. The circulation cell is completed by net onshore flow due to *mass transport* between adjacent rip systems. Rip flows are often, but not always, contained within distinct topographic channels (Figure R18) and can be visually identified by darker streaks through the surf zone due to greater water depths, offshore moving foam/sediment patches, and surface turbulence created by the wave–current interactions. Rips are of great significance to coastal nearshore studies since they provide a major mechanism for the seaward transport of water and sediments, have a pronounced effect on nearshore morphology, aid in the dispersal of pollutants, and represent a major hazard to recreational beach users. It is therefore of some concern that many aspects of rip behavior, generation, and occurrence remain poorly understood. In fact, the term was first used by Shepard (1936) in order to distinguish rip currents from the misnomers *rip tide* and *undertow*, which are unfortunately still commonly used to describe rips today.

Rips are generally absent on pure dissipative and reflective beaches, but are a key component of sandy intermediate beach states as described by various microtidal beach models (e.g., Wright and Short,

Figure R18 Enhanced time-exposure image of topographically arrested accretion rips at Palm Beach, NSW, Australia. Rip channels appear as dark areas between bars (white regions) and are approximately 150–200 m apart. Note the absence of pronounced longshore feeder channels (image courtesy of G. Symonds, R. Holman, and R. Ranasinghe).

1984) and also occur, but are not as predominant, on macrotidal beaches. Short (1985) identified three types of rip currents: (1) *accretion rips* occur during decreasing or stable wave-energy conditions and are often topographically arrested in position (Figure R18) having mean velocities typically on the order of 0.5 m/s, but exceeding 1 m/s in high-energy surf zones (Brander and Short, 2000); (2) hydrodynamically controlled *erosion rips*, which occur under rising wave-energy conditions and are transient in location, having mean flows in excess of 1 m/s; and (3) *mega-rips*, which occur in embayments under high waves (>3 m) and can extend offshore for distances of more than 1 km, attaining velocities greater than 2 m/s. All are associated with localized erosion of the shoreline and often create rhythmic *rip embayments* termed *mega-cusps*. Relatively, permanent rips located adjacent to headlands, reefs, and coastal structures such as groins have been referred to as *topographically controlled rips*.

The primary limitation to our understanding of rips has been the logistical difficulty in obtaining quantitative field measurements. The first serious scientific attempts at describing rips (Shepard *et al.*, 1941; McKenzie, 1958) were largely qualitative and suggested that rips: (1) exist as a response to an excess of water built up on shore by breaking waves; (2) often display a periodic longshore spacing; (3) increase in intensity and decrease in number as wave height increases; (4) vary in location and intensity over time; and (5) flow fastest at low tide. Subsequent theoretical, laboratory, and field studies have attempted to explain these characteristics with varying degrees of success.

Rip generation and spacing

It is generally accepted that the primary mechanism behind the formation of rip currents is the presence of longshore variations in wave height which act to produce *wave set-up* gradients that drive water alongshore from regions of high water level to regions of lower water level. Bowen (1969) showed that these gradients are intrinsically related to variations in the longshore component of *radiation stress* (Longuet-Higgins and Stewart, 1964)

$$S_{yy} = \frac{E}{2} = \frac{1}{16}\rho g H^2,$$

where E is the energy, ρ is the water density, g is the gravitational constant, and H the wave height. Bowen (1969) demonstrated that within the surf zone, the longshore gradients in set-up and radiation stress act in the same direction to produce longshore feeder currents, whereas outside the surf zone, the S_{yy} gradient is balanced by a longshore variation in wave set-down and no longshore flow is produced.

Existing models for the generation of rip cell circulation have thus incorporated various mechanisms to account for the existence of longshore gradients in wave height/set-up and can be grouped into three main categories. The wave–boundary interaction model involves the modification of the wave field by non-uniform topography and/or coastal structures. The resulting convergence and divergence of wave rays due to wave refraction can produce regions of high and low waves such that rips can occur in the lee of offshore submarine canyons (Shepard and Inman, 1950), but more commonly adjacent to headlands and groins.

Wave–wave interaction models have been used to explain both longshore variations in wave height and regular longshore rip spacing. Bowen (1969) and Bowen and Inman (1969) showed both theoretically and in the laboratory that incident waves can generate synchronous edge waves which produce alternating patterns of high and low wave heights along the shore. Rips are produced at every other antinode with a rip spacing (L_r) equal to the edge wavelength given by

$$L_r = L_o \sin(2n + 1)\beta,$$

where L_o is the deep-water incident wave length, n is the edge wave mode, and β the beach slope. Dalrymple (1975) provided a model for long, straight beaches showing that the intersection of synchronous wave trains from different incident angles can also produce longshore set-up gradients and a regular rip spacing. The third type of mechanism, an instability model, was proposed by Hino (1974) who suggested that a longshore uniformity in set-up on plane beaches is unstable to any small disturbance caused by hydrodynamic or topographic factors and that predicted rip spacing was equal to four times the surf zone width. Subsequent studies based on direct field observations have shown that this ratio can range from 1.5 to 8.

It should be emphasized that validation of the above models has primarily been restricted to laboratory experiments. Using a long-term field dataset of rip spacing obtained by remote video images, Ranasinghe *et al.* (1999) showed that the models of Bowen (1969) and Dalrymple (1975) under-predicted observed rip spacing and that there was no evidence to support instability mechanisms as being responsible for rip spacing. Furthermore, the common acceptance of the edge wave model as an explanation for rip generation and spacing should be treated with caution since synchronous edge waves have not been measured in the field and the required interaction between edge waves and incident waves of the same period is believed to be restricted to steep, reflective beaches, an environment where rips are usually absent.

Based on field observations from Narrabeen Beach, NSW, Australia, Huntley and Short (1992) found that rip spacing increases with increasing wave height and surf zone width and with decreasing sediment size and beach gradient. Short and Brander (1999) used a global dataset to show that rip spacing is related to regional wave environments. Patterns of rip spacing were consistent within west coast swell ($L_r \cong 500$ m), east coast swell ($L_r \cong 200$ m), and Fetch-limited environments ($L_r \cong$ 50–100 m). Distinct scaling factors between the environments also applied to planimetric dimensions of the rip systems and were directly correlated to wave energy. Prediction of rip spacing and location remains problematic however, and it should perhaps be acknowledged that rip spacing is often irregular.

Topographic control and flow characteristics

The theoretical models for rip generation described previously are based on longshore variations in wave height, but on a beach consisting of alternating bars and offshore channels, Sonu (1972) found that under conditions of uniform longshore wave height, constant and extensive wave-energy dissipation across the bars, and local and intense wave breaking over the channels created a set-up gradient toward the channels. Set-up gradients generated in this manner support field data confirmation (Aagaard *et al.*, 1997; Brander, 1999) that rip flows are tidally modulated (Figure R19), since stronger flows at low tide would be expected with increased wave dissipation associated with shallower water depths over the bars.

Aagaard *et al.* (1997) used field measurements to show that rip velocity (u_r) can be predicted by

$$u_r = \frac{QL_r}{A_r},$$

where Q is onshore mass transport and A_r is the cross-sectional area of the rip channel. Based on computations for Q, rip velocity will increase with H^2 and decrease with longer wave periods, and will also increase with greater distances between rips and smaller channel areas. The latter is supported by field data by Brander (1999) who found a strong relationship between increasing rip velocity and decreasing channel area. However, the degree to which rip circulation is either controlled by antecedent topography or creates this topographic feedback effect through sediment transport processes remains unclear and it should be remembered that rips do occur on beaches without irregular topography.

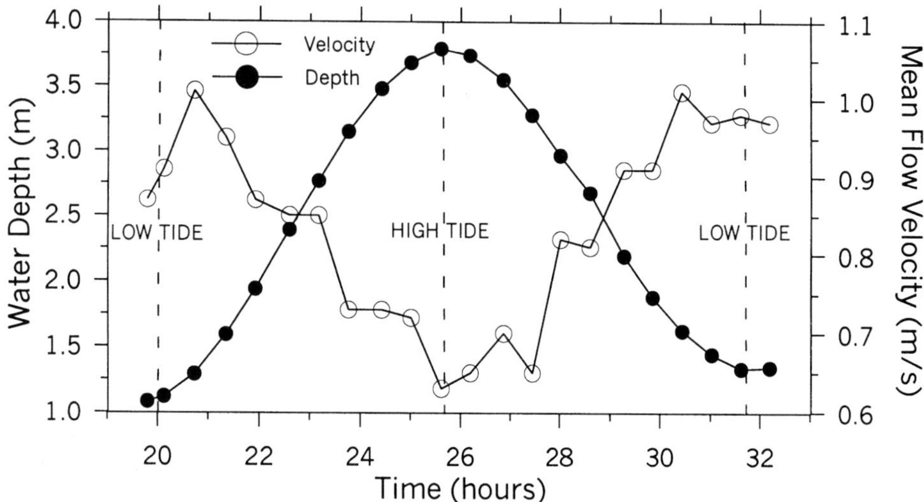

Figure R19 Tidal modulation of rip velocity at Muriwai Beach, New Zealand showing maximum and minimum flow strengths around low and high tide, respectively. Data points represent 34-min time averages recorded 0.9 m above the bed.

Field studies have also shown that rip velocities increase steadily from the feeders to the middle of the rip-neck, with strongest flows toward the middle of the water column, decreasing away from the bed and toward the surface (Sonu, 1972). Rip flows also experience low-frequency velocity pulses on the order of several minutes, but the forcing of this behavior in response to *infragravity waves, wave groups,* or *shear waves* has yet to be determined.

Robert W. Brander

Bibliography

Aagaard, T., Greenwood, B., and Nielsen, J., 1997. Mean currents and sediment transport in a rip channel. *Marine Geology*, **140**: 25–45.

Bowen, A.J., 1969. Rip currents. 1. Theoretical investigations. *Journal of Geophysical Research*, **74**: 5467–5478.

Bowen, A.J., and Inman, D.L., 1969. Rip currents. 2. Laboratory and field observations. *Journal of Geophysical Research*, **74**: 5479–5490.

Brander, R.W., 1999. Field observations on the morphodynamic evolution of a low-energy rip current system. *Marine Geology*, **157**: 199–217.

Brander, R.W., and Short, A.D., 2000. Morphodynamics of a large-scale rip current system at Muriwai Beach, New Zealand. *Marine Geology*, **165**: 27–39.

Dalrymple, R.A., 1975. A mechanism for rip current generation on an open coast. *Journal of Geophysical Research*, **80**: 3485–3487.

Hino, M., 1974. Theory on formation of rip current and cuspidal coast. In *Proceedings of the 14th International Conference on Coastal Engineering*. American Society of Civil Engineers, pp. 901–919.

Huntley, D.A., and Short, A.D., 1992. On the spacing between observed rip currents. *Coastal Engineering*, **17**: 211–225.

Longuet-Higgins, M.S., and Stewart, R.W., 1964. Radiation stresses in water waves—a physical discussion with applications. *Deep-Sea Research*, **11**: 529–562.

McKenzie, P., 1958. Rip-current systems. *Journal of Geology*, **66**: 103–111.

Ranasinghe, R., Symonds, G., and Holman, R., 1999. Quantitative characterisation of rip dynamics via video imaging. In *Proceedings of Coastal Sediments '99*. American Society of Civil Engineers, pp. 987–1002.

Shepard, F.P., 1936. Undertow, rip tide, or rip current. *Science*, **84**: 181–182.

Shepard, F.P., and Inman, D.L., 1950. Nearshore water circulation related to bottom topography and wave refraction. *Transactions of the American Geophysical Union*, **31**: 196–212.

Shepard, F.P., Emery, K.O., and LaFond, E.C., 1941. Rip currents: a process of geological importance. *Journal of Geology*, **49**: 337–369.

Short, A.D., 1985. Rip current type, spacing and persistence, Narrabeen Beach, Australia. *Marine Geology*, **65**: 47–61.

Short, A.D., and Brander, R.W., 1999. Regional variations in rip density. *Journal of Coastal Research*, **15**(3): 813–822.

Sonu, C.J., 1972. Field observation of nearshore circulation and meandering currents. *Journal of Geophysical Research*, **77**: 3232–3247.

Wright, L.D., and Short, A.D., 1984. Morphodynamic variability of beaches and surf zones, a synthesis. *Marine Geology*, **56**: 92–118.

Cross-references

Bars
Coastal Processes (see Beach Processes)
Coastal Currents
Lifesaving and Beach Safety
Sandy Coasts
Surf Zone Processes
Wave–Current Interaction
Wave-Dominated Coasts
Wave Environments

RIPPLE MARKS

General definition and description

Allen (1978) defined ripple marks as "... regular, ridge-like structures, transverse to current, which arise and are maintained at the interface between a moving, viscous fluid (water, air) and a moveable, non-cohesive sediment (usually sand) by interaction between fluid and transported sediment." Ripple marks fall principally into two classes: aeolian ripples and water-formed ripples. Fundamental work on aeolian ripples was undertaken by Bagnold (1941) as an army officer in the Lybian desert; he considered *ripples* to be constant in size with time once formed, whereas larger types grew with time, almost without limit. Aeolian ripples are influenced by saltation bombardment of sand creating ballistic ripples which lack internal structure, and have wavelengths related to saltation length. Water-formed ripples are created by lee eddy avalanches in the direction of sediment transport, possess well-defined internal structure, and have wavelengths controlled by grain size. Thus, despite outward similarities in form, there is little overlap in the mechanisms of genesis between subaqueous and aeolian forms.

Essential concepts and applications

Definitions based solely on morphology do not discriminate between what we "understand" to be ripple marks (small-scale bedforms) and genetically similar bedforms such as dunes, giant ripples, sand waves, or megaripples (large-scale bedforms). There appears to be a continuum in bedform morphology and sizes in both aeolian and subaqueous ripples, from the smallest forms found in silt to "giant" forms kilometers in length (Wilson, 1972; Ellwood *et al.*, 1975; Amos and King, 1984;

Ashley, 1990). So what are ripple marks, and how do they form? Darwin (1883) linked subaqueous ripples to vortices in the near-bed flow and subsequent sand transport. Exner (1925, from Allen, 1982) showed that they initiated from bottom irregularities, and were self-maintaining and self-organizing due to perturbations in the horizontal pressure gradient and sediment transport rate. They are considered to be the physical manifestations of bedload transport and the grain-to-grain interaction of the material in transport (Bagnold, 1963; Harms *et al.*, 1982; Middleton and Southard, 1984) and, as such, demonstrate the emergence of order out of the chaotic movement of individual sand grains within a viscous sub-layer at the bed. This order, according to Leopold *et al.* (1964), results from the creation of a *kinetic wave* in sediment flux not unlike traffic movement on a congested highway. Early classifications were based on the shape and size of bedforms (height, wavelength, asymmetry, planform, cord coherence) which resulted in the discrimination of: small-scale ripples, large-scale ripples (superimposed by smaller forms; Allen, 1968), short-crested ripples, intermediate-crested ripples, and long-crested ripples (Inman, 1957). Each ripple type was further classified on the basis of sinuosity, bifurcation, and continuity of the crestline into: straight, sinuous, linguoid, caternary, or lunate types (Allen, 1968). Classifications of ripples based purely on metrics do not consider genesis, and hence were considered deficient in two fundamental ways: (1) they could not be predicted, and (2) they could not be used to hindcast the conditions that formed them. Fundamental observations by H.C. Sorbey, and later Gilbert (1914) showed that current-formed ripples varied with flow type and flow intensity; the product of either oscillatory near-bed currents produced by waves, by unidirectional, turbulent currents; or by a combination of oscillatory and steady currents. The majority of these forms were found to be "*flow-transverse*"; that is, the crestline oriented normal to the direction of flow. Thus, from a genetic standpoint, ripple marks were primary classified as: wave (oscillation) ripples; current ripples; or combined-form ripples (Harms, 1975). In general, wave ripples are symmetrical, sharp-crested, and two-dimensional (2-D) in planform (or brick-pattern); current ripples, by contrast are generally asymmetrical showing a continuum on planform geometry from straight, through sinuous, to linguoid (Tanner, 1967); combined-flow ripples show a complex super-imposition of forms forming three-dimensional (3-D) ripples (Amos *et al.*, 1988; Arnott and Southard, 1990; Southard *et al.*, 1990). Bagnold (1956) and others suggested that small-scale and large-scale ripples were genetically different; referring to the latter as "dunes." The subsequent classification of large-scale bedforms and their distinction from ripples was presented by Allen (1985).

The classic work of Allen (1968), published in a book titled simply "*Current Ripples*," links clearly and elegantly the morphology, dynamics, and internal structure of current ripples with the near-bed unidirectional flows that created them. Current ripples occur in turbulent flows between the threshold for the traction and saltation/suspension of the rippled material at Froude numbers between 0.2 and 0.6 and flow Reynolds numbers between 5×10^{-3} and 10^{-5} (Tanner, 1978) and at mean grain sizes less than 600 μm. Allen (1985) reviewed the fundamental hydrodynamic research into the near-bed physical processes responsible for unidirectional bedform generation. A variety of phase-diagrams of bedform stability have resulted from this work largely expressed in 2-D: flow strength (power, pressure, or speed); and grain size (diameter, dimensionless diameter, or grain Reynolds number). Each scheme, according to Allen (1985) is "*restricted in applicability by the limitations of the database*," and are poorly understood for silt- and gravel-sized materials. Furthermore, most proposed 2-D phase relationships ignore the solid-transmitted (ballistic) part of the shear stress caused by the sand in motion (Bagnold, 1941). This ballistic contribution, important to the evolution of aeolian ripples, also appears important in the evolution of subaqueous ripples. As well, the possible feedback of ripple bed morphology into flow turbulence, and hence bed shear stress, has primary and secondary effects over that created by grain (skin) friction of an initial flat bed (Bagnold, 1963).

The classical work on the hydrodynamics of wave ripple formation, migration, and evolution was undertaken in the laboratory by Bagnold (1946) and Manohar (1955) and in the field by Inman (1957). Bagnold (1946) observed a complex interaction between the movement of sand grains, the structure of the benthic boundary layer (and in particular, the evolution of attached vortices), and ripple morphology. From these observations, rolling grain ripples, vortex ripples, and post-vortex ripples were defined; each form stable within a range of grain sizes and near-bed conditions of oscillatory flow. Later, Komar (1974) and Clifton and Dingler (1984) showed that the cord length of the wave ripples was linearly correlated to (and predictable by) either wave orbital diameter (orbital ripples), or grain size (anorbital ripples), or were transitional between the two (suborbital ripples). More importantly, the type of ripple and its orientation and cord length could be predicted from knowledge of (1) grain size, (2) water depth, (3) wave orbital diameter, and (4) wave orbital velocity (Sleath, 1984). The importance of this work lies in the power to discriminate wave conditions, and thus paleoenvironment (water depth, wave heights, etc.) in the geological past (Allen, 1981). However, one vital piece of information still remains to be introduced to understand the significance and behavior of ripple marks: the internal structure.

The link between ripple marks and internal structure came from early work on current ripples undertaken by van Straaten (1954) and later Reineck and Singh (1966) on tidal flats of the Waaden Sea. They showed conclusively that the internal structure of ripples, and the relationship to ripple form, provides a record of the evolution and migration of the ripple, and equally important, the direction and magnitude of net sediment transport. This attribute of bedforms has been explored and exploited by Harms *et al.* (1982) and later by Rubin (1987) who showed a complex suite of internal structures resulting from invariable or variable ripples. These ripples may be either transverse to the mean sand transport direction, longitudinal (parallel) to it, or oblique and may produce either 2-D or 3-D cross-bedding. Internal structures reveal that bedform superimposition is common. The cause of superimposition has been assigned to either a fluctuating flow in time (Allen, 1978), or to a multiple-boundary layer (Rubin and Hunter, 1987). However, field observation shows complex superimposed ripple patterns in nature resulting from wave–steady current interactions under storms (Amos *et al.*, 1988; Arnott and Southard, 1990). These patterns are further complicated by rotation of the flows leading to complex polygonal transitional patterns in the wave-formed ripple field (Allen, 1982), which merge or diverge dependent on the angle between the two flow types. At high angles of incidence, wave ripples and current ripples coexist in a steady state defined by the partitioned (wave or current) component of the bed shear stress; each ripple type responding to the flow as if oblivious of the superimposing stress (Young and Sleath, 1990). At low angles of incidence however, the bedforms coalesce into asymmetrical wave ripples, or multifrequency ripples with parallel crests (Allen, 1982).

Future research

Future studies on the interaction of combined flows on ripple genesis is required, particularly in relation to the net sediment transport direction and at the saltation/suspension threshold: the ripple "break-off region" (Grant and Madsen, 1982). The role of ballistic impacts to bedform evolution in subaqueous flows, also needs exploring, particularly for poorly sorted sand.

The feedback of ripple form on bed shear stress deserves attention: The late J. Ludwick once said that the world of sediment dynamics is divided into "*lumpers*" those who consider that bedforms influence the movement of sand, and "*splitters*" those who do not (assigning the frictional drag entirely to skin friction of the composite sediment grains). The literature on the relationship between form drag, turbulence, and ripple shape and size is inconsistent (Soulsby, 1997). Future work is needed on the feedback mechanism between bedform genesis and turbulence generation/dissipation within the benthic boundary layer. Only then will we know if the "*lumpers*" or "*splitters*" were right.

Carl L. Amos and Patrick L. Friend

Bibliography

Allen, J.R.L., 1968. *Current Ripples. Their Relation to Patterns of Water and Sediment Motion*. Amsterdam: North-Holland Publishing Company.

Allen, J.R.L., 1978. Ripple marks. In Fairbridge, R.W. and Bourgeois, J. (eds.), *The Encyclopedia of Sedimentology*. Stroudsburg: Dowden, Hutchinson & Ross.

Allen, J.R.L., 1982. *Sedimentary Structures, their Character and Physical Basis*, Volume I. Amsterdam: Elsevier.

Allen, J.R.L., 1985. *Principles of Physical Sedimentology*. London: Chapman & Hall.

Allen, P.A., 1981. Some guidelines in reconstructing ancient sea conditions from wave ripplemarks. *Marine Geology*, **43**: 59–67.

Amos, C.L. and King, E.L., 1984. Bedforms of the Canadian eastern seaboard: a comparison with global occurrences. *Marine Geology*, **57**: 167–208.

Amos, C.L., Bowen, A.J., Huntley, D.A., and Lewis, C.F.M., 1988. Ripple generation under the combined influences of waves and

currents on the Canadian continental shelf. *Continental Shelf Research*, **8** (10): 1129–1153.
Arnott, R.W., and Southard, J.B., 1990. Exploratory flow-duct experiments on combined-flow bed configurations and some implications for interpreting storm-event stratification. *Sedimentology*, **60**(2): 211–219.
Ashley, G.M., 1990. Classification of large-scale subaqueous bedforms: a new look at an old problem. *Journal of Sedimentary Petrology*, **60**(1): 160–172.
Bagnold, R.A., 1941. *The Physics of Blown Sand and Desert Dunes*. New York: John Wiley and Sons.
Bagnold, R.A., 1946. Motions of waves in shallow water: interactions between waves and sandy bottoms. *Proceedings of Royal Society of London, Series A*, **187**: 1–15.
Bagnold, R.A., 1956. The flow of cohesionless grains in fluids. *Philosophical Transactions of Royal Society of London, Series A*, **249**: 235–297.
Bagnold, R.A., 1963. Beach and nearshore processes. Part I, mechanics of marine sedimentation. In Thorne, C.R., MacArthur, R.C., and Bradley, J.B. (eds.), *The Physics of Sediment Transport by Wind and Water*. New York: American Society of Civil Engineers.
Clifton, H.E., and Dingler, J.R., 1984. Wave-formed structures and paleoenvironmental reconstruction. *Marine Geology*, **60**: 165–198.
Darwin, G.H., 1883. On the formation of ripple-mark in sand. *Proceedings of Royal Society of London*, **36**: 18–43.
Ellwood, J.M., Evans, P.D., and Wilson, I.G., 1975. Small scale aeolian bedforms. *Journal of Sedimentary Petrology*, **45**: 554–561.
Gilbert, G.K., 1914. The transport of debris by running water. *Professional Papers US Geological Survey*, 86.
Grant, W.D., and Madsen, O.S., 1982. Moveable bed roughness in unsteady flow. *Journal of Geophysical Research*, **87**(C1): 469–481.
Harms, J.C., 1975. Stratification produced by migrating bedforms. *Society of Economic Mineralogists and Paleontologists Short Course*, **2**: 45–61.
Harms, J.C., Southard, J.B., and Walker, R.G., 1982. Structures and sequences in clastic rocks. *Society of Economic Mineralogists and Paleontologists Short Course*, **9**: 8–51.
Inman, D.L., 1957. Wave generated ripples in nearshore sands. *US Army Corps of Engineers, Beach Erosion Board Technical Memorandum*, 100.
Komar, P.D., 1974. Oscillatory ripple marks and their evaluation of ancient wave conditions and environments. *Journal of Sedimentary Petrology*, **44**: 159–173.
Leopold, L.B., Wolman, M.G., and Miller, J.P., 1964. *Fluvial Processes in Geomorphology*. San Francisco: W.H. Freedman.
Manohar, M., 1955. Mechanics of bottom sediment movement due to wave action. *US Army Corps of Engineers, Beach Erosion Board Technical Memorandum*, 75.
Middleton, G.V., and Southard, J.B., 1984. Mechanics of sediment movement. *Publications SEPM Short Course Notes*, No. 3.
Reineck, H.E., and Singh, I.B., 1966. Primary sedimentary structures in the Recent sediments of the Jade, North Sea. *Marine Geology*, **5**(3): 227–235.
Rubin, D.M., 1987. *Cross-bedding, Bedforms and Paleocurrents*. Tulsa: Society of Economic Paleontologists and Mineralogists.
Rubin, D.M., and Hunter, R.E., 1987. Bedform alignment in directionally varying flows. *Science*, **237**: 276–278.
Sleath, J.F.A., 1984. *Sea Bed Mechanics*. New York: John Wiley & Sons.
Soulsby, R.L., 1997. *Dynamics of Marine Sands*. HR Wallingford Report SR, 466.
Southard, J.B., Lambie, J.M., Federico, D.C., Pile, H.T., and Weidman, C.R., 1990. Experiments on bed configurations in fine sands under bidirectional purely oscillatory flow, and the origin of hummocky cross-stratification. *Journal of Sedimentary Petrology*, **60**(1): 1–17.
Tanner, W.F., 1967. Ripple mark indices and their uses. *Sedimentology*, **9**: 89–104.
Tanner, W.F., 1978. Reynolds and Froude numbers. In Fairbridge, R.W., and Bourgeois, J. (eds.), *The Encyclopedia of Sedimentology*, Stroudsburg: Dowden, Hutchinson & Ross.
van Straaten, L.M.J.U., 1954. Composition and structure of Recent marine sediments in the Netherlands. *Leidsche geologische mededelingen*, **19**: 1–110.
Wilson, I.G., 1972. Aeolian bedforms—their development and origins. *Sedimentology*, **19**: 173–210.
Young, J.S.L., and Sleath, J.F.A., 1990. Ripple formation in combined transdirectional steady and oscillatory flow. *Sedimentology*, **37**(3): 509–516.

Cross-references

Bars
Beach Features
Beach Processes
Beach Sediment Characteristics
Beach Stratigraphy
Coastal Sedimentary Facies
Eolian Processes
Rhythmic Patterns
Scour and Burial of Objects in Shallow Water

ROCK COAST PROCESSES

Our ability to identify and measure the effect of rock coast processes has improved with the application of modern analytical techniques, geochronometric dating, and physical and mathematical modeling, but we are still largely ignorant of their precise nature (Trenhaile, 1987; Sunamura, 1992). The relative importance of rock erosional processes is often determined on the basis of ambiguous morphological evidence. Although the processes responsible for the slow lowering of rock surfaces have been inferred from micro-erosion meter data, the technique is unable to measure the dislodgement of large rock fragments by waves and frost.

It is difficult to obtain quantitative process data because of the imperceptible changes that generally occur on rock coasts within human lifetimes, the importance of storms and other high intensity—low frequency events, the lack of access to high and frequently precipitous cliffs, and the occurrence of exposed and often dangerous environments for wave measurement and subaqueous exploration. Changes in relative sea level and climate have also caused the nature and intensity of marine and subaerial processes to fluctuate through time, and because of slow rates of erosion, rock coasts often retain vestiges of environmental conditions that were quite different from today.

Mechanical wave erosion

Wave quarrying appears to be the dominant erosional mechanism in the vigorous storm wave environments of the middle latitudes, based on the frequent occurrence of fresh rock scars and coarse, angular debris consisting of joint blocks and other rock fragments on shore platforms and at the foot of cliffs. Weaker waves in polar and tropical regions also play important roles, however, in eroding weathered rocks and removing loose debris.

The forces exerted by waves on coastal structures depend upon their deep water characteristics, tidal elevation, and submarine topography. A broken wave may be a less effective erosional agent than a wave that breaks directly against a cliff or other steep, natural structure, but broken waves occur much more frequently. Therefore, the compression of air in joints and other structural discontinuities by broken waves is probably of much greater importance than the direct impact of waves on rocks (water hammer) and the generation of high shock pressures against near-vertical structures by breaking waves. Rock fragments and sand can be effective abrasional agents in the intertidal and shallow subtidal zones, although even large waves may be unable to agitate material sufficiently at the base of thick accumulations (Robinson, 1977). Gently sloping abrasional surfaces are generally much smoother than wave quarried surfaces, but deep grooves can develop where abrasion is concentrated along joint planes and other structural weaknesses. Potholes are approximately cylindrical depressions that form where sand or large clasts are rotated by swirling water in the surf or breaker zones. They are particularly common in the upper intertidal zone where abrasives are trapped at the foot of scarps, and in structural or erosional depressions, and they also inherit, and subsequently modify, corrosional hollows in calcareous rocks.

Air compression, water hammer, and other processes responsible for wave quarrying require the alternate presence of air and water, and they therefore operate most effectively in a narrow zone extending from the wave crest to just below the still-water level. Most mathematical models also suggest that standing, broken, and breaking waves exert the greatest pressures on vertical structures at, or slightly above, the water surface. Wave erosion on a rock coast must therefore be greatest at the elevation that is most frequently occupied by the water surface. Over long periods of time, the tidally controlled water surface is most frequently at, or close to, the neap high and low tidal levels, and wave action is increasingly concentrated within the neap tidal levels as the tidal range decreases (Trenhaile, 1987, 1997) (Figure R20).Therefore, in

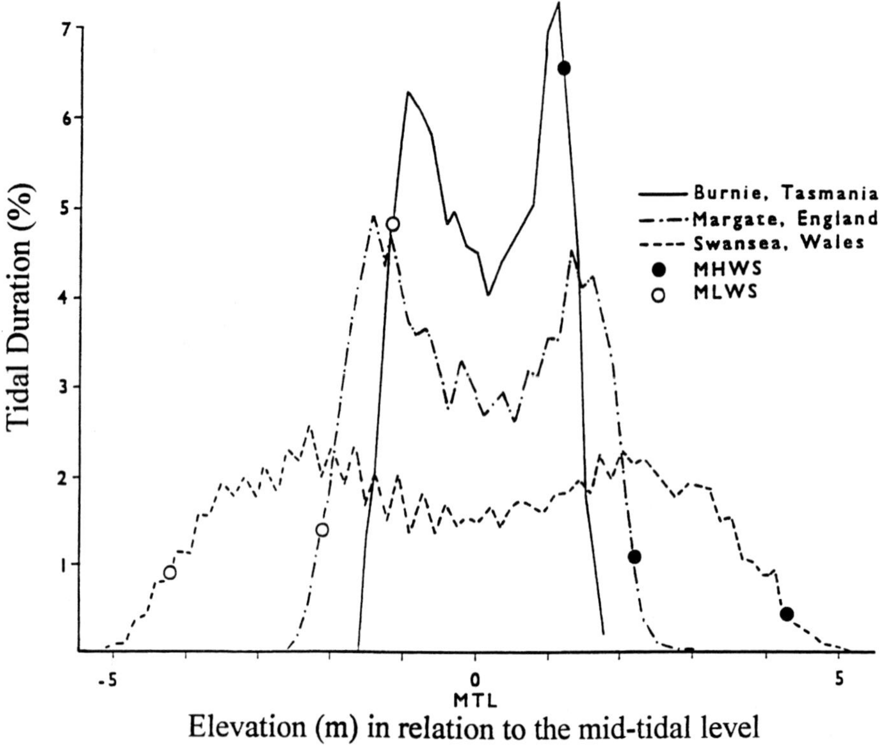

Figure R20 Tidal duration distributions—the amount of time that the water surface is at each intertidal elevation. MHWS and MLWS are the means of the high and low water spring tidal levels, and MTL is the mid-tidal level. (After Carr and Graff, 1982.)

controlling the elevation of the water surface, tides allocate the expenditure of wave energy and direct the work of mechanical wave erosional processes within the intertidal zone.

Although mechanical wave erosion is vertically distributed according to the tidal distribution of the water surface, it may be skewed toward the upper portions of the tidal range because of the occurrence of deeper water, and therefore higher waves, at high tide, and because large storm waves operate when sea level is meteorologically raised above the tidal level. In areas with low tidal ranges, this latter effect could elevate the zone of maximum erosion above the level of the high spring tides. Furthermore, as only vigorous storm waves, which operate at higher elevations than weak waves, are able to erode resistant rocks, whereas weaker waves, operating at lower elevations, are able to erode less resistant rocks, the difference between the level of greatest erosion and the most frequent tidal level may increase with the resistance of the rock (Trenhaile, 1987).

Weathering

Coasts are particularly suitable environments for physical and chemical weathering owing to alternate wetting and drying in the spray and intertidal zones, and the presence of salts. It is normally assumed that physical weathering is most important in high latitudes and chemical and biological weathering in low latitudes, in part because of the occurrence of suitable climates, but also because of fairly weak waves in these areas. Weathering also weakens rocks in the vigorous storm–wave environments of the middle latitudes, however, and it may be an essential precursor for the dislodgement and removal of rocks by weak waves in sheltered areas.

Physical weathering mechanisms include the alternate expansion and contraction of clay minerals under cycles of wetting and drying, temperature-dependent adsorption of water, thermally induced changes in volume, and frost action. The alternate expansion and contraction of clay minerals experiencing tidal- and weather-induced cycles of wetting and drying causes discontinuities to develop in shales and other argillaceous rocks, and also in the rocks that are adjacent to them. Clay minerals within small rock capillaries have negative charges that attract the positively charged ends of water molecules. This can breakup fine-grained, clay-rich rocks, which adsorb water and expand as temperatures rise, and desorb and contract as temperatures fall (Hudec, 1973). Temperature-dependent wetting and drying may be responsible for much of the field evidence that has traditionally been attributed to frost action, although several theories suggest that the two mechanisms could act together to generate more deleterious pressures within rocks than are generated by either mechanism acting alone. Thermally induced changes in volume also reduce the strength of some rocks. In southern Wales, for example, the expansion of limestones and the contraction of mudrocks under dry, hot conditions are responsible for diurnal variations in joint widths of up to 0.5 mm (Williams and Davies, 1987).

Although much remains to be determined about the mechanisms involved in frost action, we do have a general sense of the conditions that are most suitable for their operation. Coastal regions may be almost optimum environments. High levels of saturation can be attained in the supratidal and intertidal zones, and because intertidal rocks can freeze in air during low tide, and thaw in water during high tide, they experience many more frost cycles than in areas further inland. Intertidal frost action may also be particularly effective because of rapid changes in temperature caused by the sudden emergence and submergence of the rocks. Whereas rock temperatures rapidly increase when they are inundated in seawater, however, at least 5–6 h are needed for rocks to dissipate released latent heat and to cool to the freezing point of air. It is questionable whether critical levels of saturation can be maintained in the rocks over this period in the upper portion of the intertidal zone, and effective frost action in the lower portion of the intertidal zone may be inhibited by limited exposure to low air temperatures (Robinson and Jerwood, 1987). Although the presence of salts in solution can inhibit frost action, several studies have suggested that the greatest rock deterioration occurs in solutions that contain between 2 and 6% of their weight in salt; this suggests that frost action may be particularly effective in rocks that are saturated with seawater.

Frost- and temperature-dependent wetting and drying can only be effective erosional agents where there are suitable rocks, and waves that are strong enough to remove the coarse debris and prevent progressive burial of the cliff. Tidally induced frost action is inhibited at high latitudes by low water temperatures, but water and air temperatures suggest that it may occur at various times of the year in cool temperate regions (Trenhaile, 1987). Normal frost action, resulting from changes in air temperature, may also be more effective in the midlatitudes than in higher latitudes, where there are less frequent fluctuations about the freezing point. Atmospheric and tidally induced frost cycles are therefore probably most effective in cool, storm wave environments, and waves and frost also tend to be most effective on the same types of rock; strong wave action may therefore obscure or inhibit the effects of frost action in exposed areas.

Chemical and salt weathering are most important in warm temperate and tropical regions. Chemical weathering requires a good supply of water to promote chemical reactions, and more crucially, to remove the soluble products. Chemical reactions are accelerated by high temperatures in the tropics, but the lack of liquid water rather than low temperatures is probably primarily responsible for the fairly low rates of chemical weathering in high latitudes. Mechanical salt weathering occurs through the growth of crystals from solutions in rock capillaries, and crystal hydration and temperature-induced expansion. Chemical and salt weathering contribute to the formation of tafoni and honeycombs, the smoothing and lowering of shore platforms by the suite of processes collectively referred to as water layer levelling, case hardening, and the impregnation of joint planes by dissolved ions to form frame- or box-like structures, and the formation of various types of weathering pits (Trenhaile, 1987). The existence of a permanent level of saturation in the intertidal zone, separating a weak, weathered oxidation zone from a strong and largely unweathered saturated zone below, has been a basic tenet of Australasian workers for almost a century. Present evidence suggests that rocks can only be permanently saturated below the low tidal level, however, where they are constantly submerged.

There is continuing debate over the processes responsible for the sharp pinnacles, ridges, grooves, and circular basins that are characteristic of coastal limestones in the spray and splash zones. Although these features are similar to karren formed by freshwater on land, surface seawater is usually saturated or supersaturated with calcium carbonate. It has been suggested that solution could occur in rock pools at night, when the carbon dioxide produced by faunal respiration is not removed by algae. Lower pH then causes calcium carbonate to be transformed into more soluble bicarbonate. Solution can be inhibited or prevented by other biochemical processes, however, including dissolved organic substances coating rock surfaces and building complexes with calcium ions. Although chemical solution does appear to be possible in seawater, recent studies have provided support for the contention that marine karren and other characteristic features of limestone coasts are primarily bioerosional in origin.

Bioerosion

Bioerosion is the removal of the substrate by direct organic activity. It is probably most important in tropical regions, where there are fairly weak waves and an enormously varied marine biota living on coral, aeolianite, and other calcareous substrates. A wide range of techniques are used to breakdown rocks. Microflora and fauna that lack hard parts may use only chemical mechanisms, but other fauna secrete fluids that chemically weaken the rock, before mechanically abrading them with teeth, valvular edges, and other hard parts.

Microflora bore into rock and they change the chemistry of the water that is in contact with it—indeed cyanophyta (blue-green) and other algae may be the most important biological agents on rock coasts. Algae, lichen, and fungi are pioneer colonizers in the intertidal and supratidal zones and they allow subsequent occupation by gastropods, echinoids, chitons, and other grazing organisms that effectively abrade rock surfaces as they feed on epilithic and the ends of endolithic microflora. Grazing organisms are of enormous importance in some environments. For example, it has been estimated that they are responsible for about one-third of the surface erosion in the mid-tidal zone on Aldabra Atoll where sand is available for abrasion, and as much as two-thirds where sand is absent (Trudgill, 1976). At least 12 faunal phyla contain members that bore into rocks, especially in the lower parts of the intertidal zone. They include *Lithotrya* and other boring barnacles, sipunculoid and polychaete worms, gastropods, echinoids, *Lithophaga* and other bivalve molluscs, and Clionid sponges. Borers directly remove rock material, and they also weaken the remaining rock, making it more vulnerable to mechanical wave erosion and weathering. In the tropics, carbonate rocks favor chemical borers, but mechanical borers are active on a variety of substrates in temperate and cool seawater environments. There is a great deal of published data on bioerosional rates of erosion (Trenhaile, 1987), but they are of variable reliability and relevance. Nevertheless, most reported rates of erosion on vertical and horizontal limestone surfaces are between about 0.5 and 1 $mm\,yr^{-1}$, which may reflect the maximum boring rate of endolithic microflora.

Ice

Until recently, it was generally believed that coastal ice is an ineffective erosional agent in the coastal zone, but its potential contribution to the development of rock coasts in cold environments is now being reassessed. The formation of subhorizontal shore platforms in the South Shetland Islands has been attributed to fast ice freezing to the underlying bedrock, quarrying by grounded ice and stranded ice rocked by the tides, and abrasion by rocks frozen into the ice base (Hansom and Kirk, 1989). Ice-push, by wind-driven floating ice loaded with rock fragments, also assists gelifraction and frost wedging in quarrying gneissic joint blocks in macrotidal Ungava Bay. Much of the ice-foot melts in place and does not contribute to debris removal, but it may facilitate deep frost penetration by providing a thermal barrier to sporadic warming by tidal water. Frost weathering and the effects of ice abrasion, dislodgement, and quarrying are also considered to be the main processes responsible for the formation of wide, subhorizontal platforms in the upper St. Lawrence Estuary (Dionne and Brodeur, 1988). Although there is clear evidence of the erosive efficacy of shore ice in this area, weak slates and shales, high tidal range, strong currents, and large erratic blocks provide particularly suitable conditions for frost and ice action, and it remains to be determined whether these cold region mechanisms assume similar roles in less favorable places.

Mass movement

Active marine cliffs possess short rather than long-term stability because of undercutting, oversteepening, and the removal of basal debris by wave action. Mass movements therefore play an important role in the development of cliffed coasts, and there is a close relationship between the morphology of cliffs and the type of mass movement that takes place on them. Mass movements range, according to local circumstances, from the quasi-continuous fall of small debris to infrequent but extensive landsliding. Although rock falls are more frequent than deep-seated slides, they are generally much smaller. Falls occur in well-fractured rocks, especially where notches are cut into the cliff foot by waves, or, as in the tropics, by solution or bioerosion. Rock columns defined by joints or bedding planes also topple or overturn by forward tilting. Rock and slab falls, sags, and topples are essentially surficial failures induced by frost and other types of weathering, basal erosion, hydrostatic pressures exerted by water in rock clefts, and the reduction in confining pressures resulting from cliff erosion and retreat.

Deep-seated mass movements are triggered by groundwater build-up and basal undercutting. Translational slides usually occur where there are seaward dipping rocks, alternations of permeable and impermeable strata, massive rocks overlying incompetent materials, or argillaceous and other easily sheared rocks with low bearing strength. Slumps or rotational slides are common in thick, fairly homogeneous deposits of clay, shale, or marl. Sliding takes place in rocks that have been weakened by alternate wetting and drying, clay mineral swelling, or deep chemical weathering. Slides tend to occur during or shortly after snowmelt, or prolonged and/or intense precipitation. Water from septic systems, irrigation, runoff disruption, beach depletion through the building of coastal structures, and other human activities are playing increasing roles in some areas (Griggs and Trenhaile, 1994). The damming of rivers has also reduced the bed load reaching the coast, depleting beaches and exposing cliffs to more vigorous wave action.

Alan S. Trenhaile

Bibliography

Carr, A.P., and Graff, J., 1982. The tidal immersion factor and shore platform development. *Transactions of the Institute of British Geographers*, 7: 240–245.

Dionne, J.-C., and Brodeur, D., 1988. Frost weathering and ice action in shore platform development with particular reference to Québec, Canada. *Zeitschrift fur Geomorphologie, Supplement Band*, **71**: 117–130.

Griggs, G.B., and Trenhaile, A.S., 1994. Coastal cliffs and platforms. In Carter, R.W.G., and Woodroffe, C.D. (eds.), *Coastal Evolution*. Cambridge: Cambridge University Press, pp. 425–450.

Hansom, J.D., and Kirk, R.M., 1989. Ice in the intertidal zone: examples from Antarctica. In Bird, E.C.F., and Kelletat, D. (eds.), *Zonality of Coastal Geomorphology and Ecology*. Essener Geographische Arbeiten, **18**: pp. 211–236.

Hudec, P.P., 1973. Weathering of rocks in Arctic and Sub-arctic environment. In Aitken, J.D., and Glass, D.J. (eds.), *Canadian Arctic Geology*. Saskatoon: Geological Society Association of Canada. *Canadian Society for Petroleum Geologists Symposium*, pp. 313–335.

Robinson, D.A., and Jerwood, L.C., 1987. Frost and salt weathering of chalk shore platforms near Brighton, Sussex, UK. *Transactions of the Institute of British Geographers*, **12**: 217–226.
Robinson, L.A., 1977. Erosive processes on the shore platform of northeast Yorkshire, England. *Marine Geology*, **23**: 339–361.
Sunamura, T., 1992. *Geomorphology of Rocky Coasts*. Chichester: John Wiley.
Trenhaile, A.S., 1987. *The Geomorphology of Rock Coasts*. Oxford: Oxford University Press.
Trenhaile, A.S., 1997. *Coastal Dynamics and Landforms*. Oxford: Oxford University Press.
Trudgill, S.T., 1976. The marine erosion of limestone on Aldabra Atoll, Indian Ocean. *Zeitschrift fur Geomorphologie, Supplement Band*, **26**: 164–200.
Williams, A.T., and Davies, P., 1987. Rates and mechanisms of coastal cliff erosion in Lower Lias rocks. In Kraus, N. C. (ed.), *Coastal Sediments '87*. American Society for Civil Engineering, pp. 1855–1870.

Cross-references

Cliffed Coasts
Cliffs, Erosion Rates
Cliffs, Lithology versus Erosion Rates
Karst Coasts
Mass Wasting
Notches
Shore Platforms
Weathering Processes in the Coastal Zone

S

SALT MARSH

Salt marshes or saline wetlands are vegetated intertidal flats dominated by low-growing halophytic (salt-tolerant) shrubs and herbaceous plants, particularly grasses. Typically, salt marsh borders freshwater or brackish environments. Largely confined to temperate coastlines, they occupy a similar niche to tropical mangrove forests; that is, the upper intertidal zone of inlets, estuaries, lagoons, and embayments, or fronting the open sea where low-energy conditions persist (Frey and Basan, 1985). In warm temperate, subtropical, and some tropical regions, salt marsh and mangrove communities sometimes intermingle, but can be separated by definition on the basis of floristics or intertidal position.

Salt marsh originates with the spread of vegetation onto an accreting intertidal mudflat. Fine suspended sediments (silts and clays) and organic material washed in by tides, and subsequently trapped by roots of salt marsh vegetation, generate a gently sloping depositional terrace or platform between the high spring tide level and the mid-tide line (Bird, 2000). At a smaller scale, characteristic features of salt marsh include tidal creeks, levees, cliffs, and salt pans. The patterns and morphology of these features are collectively dictated by a number of factors: the extent of vegetative cover; the climatic, hydrographic, and edaphic influences on this vegetation; the availability, composition, deposition, and compaction of sediments; organism–substrate relationships; topography; tidal range; wave and current energy and stability of the coastal area (Frey and Basan, 1985).

The salt marsh platform is typically dissected by a system of meandering creeks and levees that channel the ebb and flow of tidal waters. Wide shallow creeks are common where the salt marsh is in an early developmental phase or where sediments have high sand content, but as the marsh platform rises or expands with continued accretion, well-defined steep-edged channels become more commonplace. The seaward edge of a marsh may grade smoothly into adjacent tidal flats, or it may end abruptly in a small vertical cliff up to several meters high, or the transition may be marked by an irregular series of channels separated by ridges (Haslett, 2000; Ke and Collins, 2002).

The wide variety of sediments underlying coastal marsh systems reflect the range of adjacent terrestrial and marine habitats from whence these originate. The shallow waters and low-velocity currents characteristic of marsh surfaces promote deposition of fine-grained inorganic sediments such as mud, clay, silt, and fine sands, and rarely gravels. Most are reworked and modified by roots of plants and the action of animals, leading to the deposition of organic components, including peats which are formed from the degradation of roots, stems, and leaves of marsh plants and animal skeletons including shells and carbonate tests (Chapman, 1977; Dawes, 1998).

Salt marshes support an abundance of organisms of both terrestrial and marine origin. They are highly productive ecosystems, but the extreme physiological and ecological stresses of the intertidal environment maintain characteristically low species diversities. Clear zonation patterns of vegetation and associated fauna are generally evident, influenced by factors such as salinity (which ranges from near ocean strength to near fresh in most systems), frequency and duration of tidal exposure, and climate (Haslett, 2000). Low-marsh communities toward the seaward edge normally comprise pioneering halophytes such as sea grasses (e.g., *Zostera*) and glassworts (e.g., *Salicornia*) through to those tolerant of brackish conditions (e.g., *Spartina*). High-marsh communities, being relatively more stable and terrestrially influenced, include more diverse assemblages of less salt-tolerant species such as rushes (e.g., *Juncus* spp.), grasses (e.g., *Puccinellia, Sporobolus*), and herbs (e.g., *Aster, Plantago*).

Vegetation cover of a salt marsh surface does not usually follow a continuous sequence from sea-edge to land. In addition to creeks and channels, highly saline, dry, or water-filled shallow depressions often feature in the surface of the marsh. They form where vegetation has either failed to establish, or where it has died back, or where drainage channels have slumped and blocked. Strong evaporation from the resulting pools of water or bare substrate surfaces advances the accumulation of salts, inhibiting seed germination and further establishment of plants, and giving rise to the conspicuous vegetation-free areas known as salt pans (Bird, 2000).

Terry R. Healy

Bibliography

Bird, E.C.F., 2000. *Coastal Geomorpholgy: An Introduction*. Chichester: Wiley.

Chapman, V.J., 1977. *Wet Coastal Ecosystems*. Amsterdam: Elsevier.

Dawes, C.J., 1998. *Marine Botany*, 2nd edn. New York: John Wiley and Sons.

Frey, R.W., and Basan, P.B., 1985. Coastal salt marshes. In Davis, R.A. (ed.), *Coastal Sedimentary Environments*. New York: Springer-Verlag, pp. 101–169.

Haslett, S.K., 2000. *Coastal Systems*. New York: Routledge.

Ke, X., and Collins, M., 2002. Saltmarshes in the West Solent (southern England): their morphodynamics and evolution. In Healy, T., Wang, Y., and Healy, J-A. (eds.), *Muddy Coasts of the World: Processes, Deposits and Function*. Amsterdam: Elsevier Science, pp. 411–440.

Cross-references

Coastal Soils
Deltas
Dikes

SAMPLING METHODS—See MONITORING, COASTAL ECOLOGY; MONITORING, COASTAL GEOMORPHOLOGY

SAND RIGHTS

Introduction

Sand Rights is a concept that merges the physical laws of sediment transport with societal laws of public trust. The basis doctrine is that human actions will not interfere, diminish, modify, or impede sand and other sediments from being transported to and along rivers, beaches, shores, or any flowing or windblown paths or bodies without proper restitution. Under this doctrine, projects should be designed or reevaluated to mitigate any interference that the project may have with sand transport.

As early as Justinian, cultures have understood that certain natural resources are incapable of private ownership. "By the law of nature these things are common to mankind—the air, running water, the sea and consequently the shores of the sea." (Institutes of Justinian 2.1.1, quoted in *National Audubon Society v. Superior Court* (1983) 33 Cal.3d 419, 433–34.) Traditionally water, navigation, fisheries, and tidelands had been covered by the public trust doctrine. In certain cases, states have extended the public trust doctrine to marine resources. The basis doctrine of Sand Rights would establish the application of public trust to inland and coastal sand resources.

Physical principles

The coast is a dynamic area—the junction of the land, the air, and the sea. The coastline can be divided into a series of littoral cells or segments to contain one or more sources of sand, mechanisms for moving sand along the coast, and a sink from which the sand cannot be carried back onshore by wave action. On the western US coast, headlands, rivers, and submarine canyons can often delineate the cell boundaries. On the eastern US coast, the littoral cells are defined differently, often as reaches between inlets, but the same processes apply. Similarly, littoral cells can be defined on the Gulf of Mexico and the Great Lakes. Sand is supplied to the coast from coastal streams, erosion of coastal cliffs, erosion of offshore reefs, or transport of offshore sediments onshore. Transport of sand up and down the coast is usually by waves and currents. Sediment sinks include submarine canyons, harbors, lagoons, sand dunes, and deep-water bars and shoals.

Coastal erosion is a major concern for many areas and sea-level rise will compound this problem. While coastlines have receded and advanced over the past centuries, human activities can interrupt or modify the supplies and delivery of this material to and along the coast. Dams and other flood control structures can block sediment and trap it in reservoirs or prevent channel erosion and reduce sediment supply. It has been estimated that hundreds of millions of cubic meters of sand are trapped behind dams in the Los Angeles, CA region and similar amounts are trapped by all the flood control structures on the Mississippi. The worldwide estimates for the amount of material trapped in reservoirs would be in trillions of cubic meters. In addition to trapping sediment, flood control structures reduce the carrying capacity of streams and thus the amount of sediment delivered to the coast. Finally, in-stream sand and gravel mining removes material that was being transported to the coast and reduces the overall amount of beach sediment that reaches the coast.

Coastal cliffs are a second source of beach material. Many coastal cliffs are composed of sandstone or contain a large percentage of sand-sized sediment. As these cliffs erode, they can supply significant amounts of beach quality material to the coast. When roads or railroads are built at the base of a coastal cliff, these structures can interfere with cliff erosion and reduce sediment delivery to the coast through road-clearing efforts. Supplies of sediment from coastal cliffs also can be reduced by the construction of seawalls or other armoring that prevents cliff erosion.

Waves can carry sand from offshore sources to the nearshore and onto the dry beach. These offshore sources, such as deep-water bars or ebb-tidal shoals, are often used for beach nourishment. Human activities of dredging from these offshore sources and depositing it on beaches has augmented the natural transport of offshore sediment to the coast, or made more sources of offshore sediments available as littoral material. However, if areas within the zone of on- and offshore transport are deepened, these areas will refill with littoral material that will be lost temporarily as beach material. Also, offshore structures can block the onshore transport of sediment.

Once sand is on the beach and in the nearshore area, waves and currents move it up and down the coast regularly. Structures, such as groins and jetties interrupt longshore transport of sediment and relocate areas of erosion and deposition. If there is no effort to place sediment downcoast of these structures or to by-pass sediment around the structures there will be accretion on the upcoast side and erosion on the downcoast side. The effect may be compounded by the construction of an entire series of structures. Offshore structures like breakwaters and reefs block or reduce wave energy shoreward of the structure. This modification of wave energy will alter the sediment carrying capacity in the nearshore and also the erosion and accretion areas.

Submarine canyons, deep-water bars, and sand dunes are natural sinks for coastal sediment. In general, once sediment is carried into one of these areas, it is no longer readily available as beach material. Harbors, inlets, and deepened navigation channels can also be temporary sinks. These features can function just like natural sediment sinks unless they are regularly maintained and the trapped material is reintroduced to the littoral system.

Finally, sand mining accounts for significant losses of beach material. Large-scale commercial sand mining has taken place on various beaches (e.g., Monterey, CA) and in various streams (e.g., San Juan Creek, Orange County, CA). This sand mostly goes for construction purposes, but it is also used in less familiar ways such as for making glass or pottery. Noncommercial sand mining also occurs. Every summer truckloads of sand are removed daily when seaweed, flotsam, and trash are raked from the beach. All these activities result in a net deficit of sand moving to and along the coast.

Societal significance of coastal sand

Coastlines have had importance throughout history. Fishermen built their homes and shops by the coast and gained access to the great food supply of the ocean. Oceans provided some of the earliest and most lasting modes of transportation and commerce routes. The coast has always had an important recreational value. The coast has also played an important role as a place of solace and spiritual renewal. It has become clear that everyone benefits from the coast.

Sand serves valuable purposes. It is a commercial resource, as is attested by the extensive sand-mining operations worldwide. Beaches are major recreational resources and tourist attractions for most coastal states and nations. Sandy beaches are also a major buffer and zone of protection for inland areas from coastal storms and hurricanes.

Beach erosion is a major problem in many areas. The coastline has never been static, yet, the problem of erosion has been exacerbated by those human activities that either have reduced supplies of sediment to the coast, or have changed its transport along the coast. As civilization demanded more from the coast, activities such as construction of cities and harbors interrupted natural supplies of coastal sediments. However, until recently, the connections between development far from the coast, changes to coastal sediment supplies and coastal erosion were not well-recognized.

Doctrine of public trust and sand rights

The doctrine of public trust traces it lineage to ancient Roman law that established certain types of property to be common to all people and incapable of private ownership. These common properties, *res communes*, include running water in the sea and the land beneath them. The sovereign or government is under obligation to manage these *res communes* lands for the public interest, in perpetuity. A private owner can have a bare legal title to the lands, the *jus publicum*, but this interest is subject to and restricted by the superior public interest.

The public trust doctrine was lost during the Middle Ages and was reinstated in England in the middle of the 16th century. Under this interpretation, the restraint was placed only upon the sovereign and only for tidelands, not for navigable waters. However, Parliament retained the power to enlarge or diminish the rights of the public over the tidelands, provided some legitimate public purpose was asserted.

The Roman and English versions of public trust have been carried over into American law where it is now firmly established. Since most land use law has been assigned to state and local governments, the development of public trust varies from state to state. Traditionally, public trust was reserved for navigation, commerce, and fisheries. Many states have expanded the use of public trust to water resources, protection of the environment, and recreational values.

In Florida, and states such as Oregon and Hawaii, the customary rights doctrine has been invoked to protect the public use of beaches. The Florida Supreme Court has observed:

> The beaches of Florida are of such a character as to use and potential development as to require separate consideration from other lands with respect to the elements and consequences of title. The sandy portion of the beaches are of no use for farming, grazing, timber production, or residency—the traditional uses of land—but has served as a thoroughfare and haven for fishermen and bathers, as well as a place of recreation for the public. The interest and rights of the public to the full use of the beaches should be protected. (*City of Daytona Beach v. Tona-Rama, Inc.* (1974) 294 So. 2d 73,77)

In California, the California Coastal Commission has used existing statutory authority to require mitigation of quantifiable sand losses. Both the losses from the construction of seawalls along coastal bluffs and the reductions to inland sediment supplies have been addressed through financial compensation into regional nourishment programs. Public agencies can use existing authorities to fund many nourishment projects.

Conclusion

The coast is one of the nations' most valuable areas and sand beaches are critical parts of the coast. The Sand Rights doctrine extends public trust to coastal beaches and those sand resources that are essential to perpetuation of coastal beaches. This doctrine requires that all decision-makers give careful consideration to proposed or existing projects that interfere with the delivery of sand to, or transport of sand along the beach. The basis doctrine is that human actions will not interfere, diminish, modify, or impede sand and other sediments from being transported to and along beaches, shores, or any flowing or windblown paths or bodies without proper restitution. Under this doctrine, projects should be designed or evaluated to mitigate any interference that the project may have with sand transport.

For further related reading see the following bibliography.

Katherine Stone, Orville Magoon,
Billy Edge, and Lesley Ewing

Bibliography

Bascom, W., 1980. *Waves and Beaches: The Dynamics of the Ocean Surface*, 2nd edn. New York: Anchor Books Press.

Ewing, L., and Sarb, S., 1998. California's program for mitigating sand supply impacts from seawalls. *Proceedings from Emerging Trends in Beach Erosion and Sand Rights Law*, June 3–5, Captiva Island, FL, pp. 232–248.

Inman, D.L., 1985. Damming of rivers in california leads to beach erosion. *Oceans '85: Ocean Engineering and the Environment*. Volume 1. Marine Technological Society and IEEE, **674**, pp. 22–26.

Magoon, O., and Edge, B.L., 1998. Sand rights—the fragile coastal balance. *Proceedings from Emerging Trends in Beach Erosion and Sand Rights Law*. June 3–5, Captiva Island, FL, pp. 1–19.

Magoon, O., and Edge, B.L., 1999. Bring back the beaches. *Civil Engineering*, December, 96.

Stone, K.E., 1999. Sand rights: a legal system to protect the shores of the sea. *Stetson Law Review*. Winter.

Stone, K.E., and Kaufman, B., 1988. Sand rights: a legal system to protect the shores of the sea. *Shore and Beach*, **56**(3): 8–14.

US Army Corps of Engineers, 1984. *Shore Protection Manual*. Washington, DC: Government Printing Office.

Cross-references

Beach Erosion
Beach Nourishment
Coastal Boundaries
Dams, Effect on Coasts
Erosion, Historical Analysis and Forecasting
Erosion Processes
Navigation Structures
Sediment Budget
Shore Protection Structures

SANDY COASTS

Introduction

Sandy coasts are those coasts dominated by an abundance of sand-size sediments (0.063–2 mm). The location of these coasts is a function of both sand sources and coastal processes. As a consequence, they are more abundant in certain climate and geological or plate settings. They are most prevalent in humid climates supplying abundant terrigenous sand to passive margin coasts and where exposed to more energetic wave and tidal environments that can both winnow out the finer mud and concentrate and deposit the abundant sands in a range of wave-tide features. In addition, sandy coasts are also influenced by secondary regional features such as geology of the hinterland and shore, geological inheritance, sources of shelf siliclastic and carbonate sands, and littoral drift.

Terrigenous sources

Sand is the most abundant sediment of the world's open coasts. It occurs from the poles to the tropics, but with latitudinal and regional maxima. Sand is globally abundant for two reasons. First, the ultimate source rock of most terrigenous sand is granite, which makes up the cores of all continents. As granite erodes the less resilient weaker minerals (feldspars) chemically weather to fines, leaving behind the harder sand-size quartz (or silica) grains together with minor percentages of sand-size heavy minerals, all of which are resilient to abrasion. Secondary sources are sedimentary and metasedimentary rocks, all of which contain variable percentages of sand-size material. The erosion and weathering of all these rocks potentially supplies boulders through mud. However, erosion and transport processes selectively erode and transport the fines most readily in suspension, then the sands as traction and bedload, while the coarser gravel, cobbles, and boulders become increasingly intransigent. Table S1 highlights the transportability of different size fractions.

Transport

Fines are readily transported in suspension by rivers and streams to the coast. At the coast their continued suspension and transport by waves and tidal currents result in their deposition being restricted to quieter estuarine, deltaic, or shelf locations. Sand on the other hand once it reaches the coast is deposited rapidly as bedload, building river mouth bars and deltas. It can then be reworked on- and longshore by waves into bars and beaches, and by tides into tidal sand waves and ridges, all shallow water and shore features. The size of gravel and boulders

Table S1 Sediment size and settling rates (from Short, 1999, © John Wiley & Sons Limited, reprinted with permission)

Size	Grain diameter	Time to settle 1 m	Distance traveled per hour in 1 m/s current
Clay	0.001–0.008 mm	Hours to days	3.6 km
Silt	0.008–0.063 mm	5 min–2 h	3.6 km
Sand	0.063–2.000 mm	5 s–5 min	10's m
Cobble	2 mm–6.4 cm	1–5 s	<<1 m
Boulder	>6.4 cm	<1 s	0

restricts their erosion, transport, and availability. They will only reach the coast at the mouth of short, steep rivers, along eroding rocky coasts and when delivered by glaciers.

Accumulation and recycling

For the above reasons, even though fines dominate many of the world's river sediment discharges their deposition is precluded from energetic shores. Sands however, are not only deposited right at the coast, but once deposited are ideally suited to reside there. Gravel and boulders will also remain at the shore once deposited, however, they are more restricted in extent to usually high relief and high latitude coasts. Quartz sand and particularly heavy minerals are extremely resistant to physical abrasions and can be reworked time and time again by rivers, waves, tides, and winds, and are termed polycyclic. Consequently, many sedimentary shores are composed of quartz sand because it is globally abundant in source rocks (20–40% of crust); it is the major bedload product of denudation; it is relatively easily transported during high river discharge events; it can reside in the most energetic coastal environments; it is resilient and over time and can accumulate along coasts, and can be reworked during rising sea levels, leading in places to massive polycyclic accumulations of Quaternary (shelf) sand deposits.

Carbonate sands

The second major source of coastal sand is marine carbonates that live from shallow intertidal waters to the inner shelf. While carbonate detritus is more easily broken and abraded by physical processes, their *in situ* and shore linear sources, can act as a continual supply, with waves and tidal currents eroding, abrading, and transporting the sand size and coarser material shoreward. Carbonate sand coasts in fact dominate large areas of the world's tropical and arid temperate coasts (Short, 2002) (Figure S1). The source of these sands are both the shallow coral algae and reefs in the tropics, while in more temperate latitudes shelf carbonates (molluscs, red algae, encrusting bryozoans, and echinoids) are delivered both during and subsequent to sea-level transgressions from depth as great as 100 m (Boreen and James, 1993).

Climate

Climate plays a major role in the occurrence of sand coasts. Most terrigenous sand is derived from denudation of the hinterland and transported by rivers to the coast. Regions of greatest chemical and physical weathering will potentially supply the greatest volumes of sediment including sand to the coast. Figure S2 shows the global distribution of mechanical erosion, the world's major rivers and their solid discharge. Using this as an indicator of bedload sand discharge, the greatest supply of sediment to the coast is associated with low to lower mid-latitude river systems (40°N–40°S). All the rivers and their headwaters are associated with humid tropical and mid-latitude climates that supply the higher rainfall and warmer temperatures to physically and chemically erode and transport the sediments. While there are some major rivers in the higher latitudes (>40°N), the smaller discharges and dominance of physical weathering supplies generally low quantities of predominantly coarser sediments (sand through boulders). The impact of both climate and latitude on shelf sediment supply is highlighted by Figure S3, which shows the latitudinal distribution of shell, coral, rock and gravel, sand and mud. Sand dominates the subtropics and lower mid-latitudes (20–40°), and while sand volumes are still large in the tropics, mud still dominates owing to the intense chemical weathering and lower-energy coasts.

Hinterland geology also plays an important secondary role, as sand-rich source rocks (granites, sandstone, metasedimentary) will supply more sand and less suspended sediment, than finer grained rocks (e.g., basalt, shale, limestone).

Plate setting

Location of a coast relative to its tectonic plate setting is also a major contributing factor in the location of sand coasts. Inman and Nordstrom (1971) classified the world's coast according to their tectonic setting, and noted the dominance of coastal plains and deltaic coasts, usually associated with sand coast, on passive margin coasts (Table S2). The reasons for this are highlighted in Table S3. The passive margin coasts are supplied with abundant sand and finer sediments by larger river systems, draining distant mountains. At the coast they supply abundant, mature, stable sand, and fines, which build extensive sand-rich coastal strand plains composed of deltas and barriers, as along the east coast of the Americas and India. In contrast, convergent coasts have high relief and short rivers which tend to supply, along with mass movement, a limited amount of coarse, poorly sorted unstable sediments, as along the west coast of the Americas.

Landforms

Deltas

Sand supplied by river to the coast is initially deposited in deltaic systems. The proportion of sand in a delta and the overall morphology of the delta is a function of both the fluvial sediment supply, the geological inheritance that influences the shape and size of depositional basin or accommodation space, and the contribution of the river flow, waves, tides, littoral currents, and winds to the reworking, redistribution, and

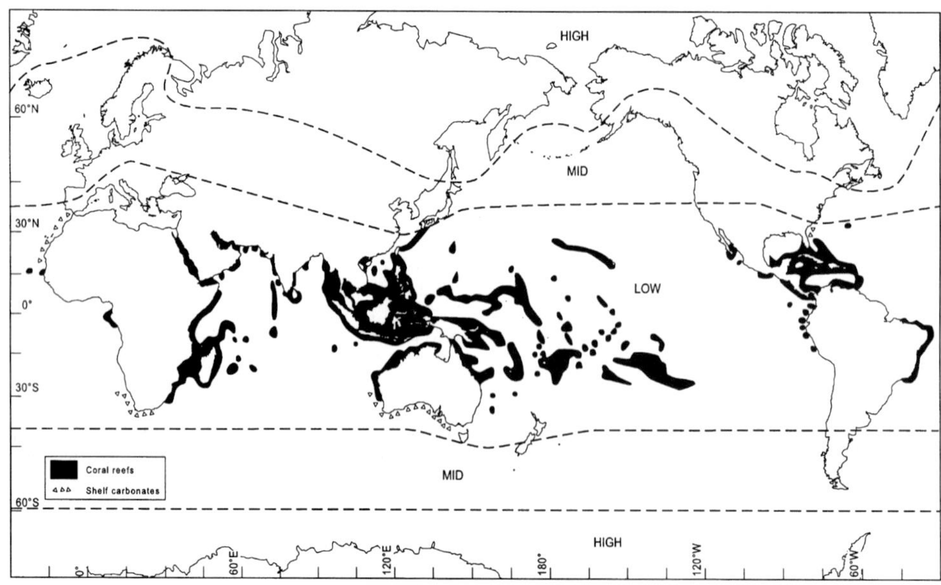

Figure S1 Extent of the low-, mid-, and high-latitude coasts, extensive coastal reefs and known areas of major supply of shelf carbonate sands to the coast (modified from Davies, 1980, and reprinted by permission of Pearson Education Limited).

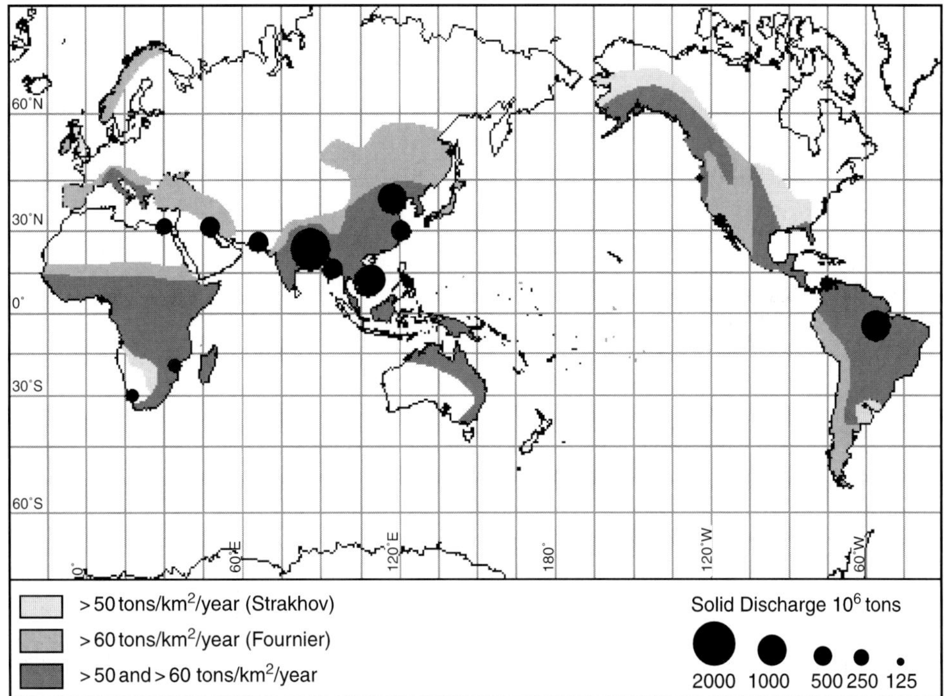

Figure S2 Global distribution of mechanical erosion. Circles are proportionate to the amount of solid discharge per annum by the world's major rivers (modified from Davies, 1980, and reprinted by permission of Pearson Education Limited).

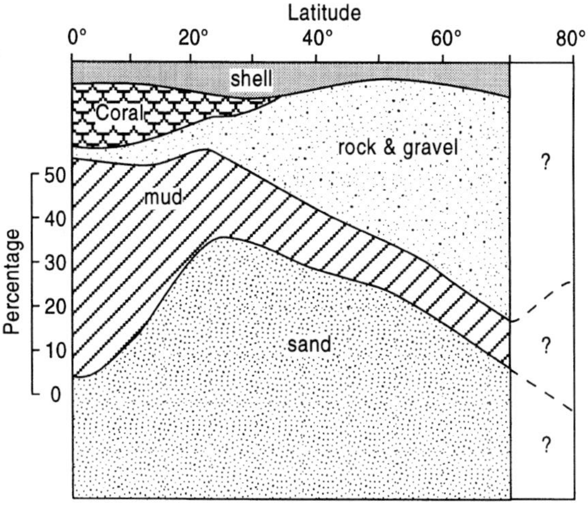

Figure S3 Relative frequency of occurrence of inner continental shelf sediment types by latitude (from Hayes, 1967, and with permission from Elsevier Science).

construction of the deltaic and adjacent depositional systems (see e.g., Davis, 1983). River-dominated deltas tend to deposit the sand as digitate river channel, levee, and river mouth bar deposits, as in the Mississippi, with limited shoreline sand deposits. Tide-dominated deltas accumulate the sand in subaqueous shore perpendicular to linear tidal sand waves and ridges, again with little sand at the shore. Only along wave-dominated deltas do extensive shore parallel beaches, barriers, and strand plains occur, which if exposed to strong winds may be capped by coastal sand dunes.

Beaches

Beach systems are the most common and visual expression of sand coasts. They form the shore of all sand coasts whether backed by deltas, barriers, and sand dunes or even fronted by sand flats. Beaches can however, be formed from fine sand through boulders, and like sand supply itself the sediment type is a function of tectonic setting (Table S3) and latitude/climate (Table S4). As Table S4 indicates sand beaches are more likely to occur in the low- to mid-latitudes, with beach type also a function of sediment size, and through climate, wave environment. Some of the world's longest beach systems occur along the east and Gulf coast of the United States and the central Brazilian coast of South America, with beaches up to several hundred kilometers in length in south Brazil. While the visible beach usually forms a narrow strip along the shore, most beaches also extend offshore to the limit of modal wave base. On high-energy coasts this may be as deep as 30 m and as far as 2–3 km offshore.

Dunes

All beaches composed of fine to medium sand and exposed to onshore winds, particularly strong winds, will be backed by coastal sand dunes, composed of fine well-sorted sand. The location and occurrence of dunes is also a function of latitude/climate (Table S4) with the world's largest coastal sand dunes occurring on passive margin coasts exposed to moderate to strong onshore winds, as along the Brazilian trade wind coast and the south to west-facing coasts of South Africa, southern Australia, South America, west coast United States, and parts of western Europe. Most coastal sand dunes extend up to a few tens to hundreds of meters inland reaching elevations of a few tens of meters. They can in very exposed locations migrate up to tens of kilometers inland and to elevations of over 100 m and up to 200–300 m in extreme locations. In these locations the coastal sand dunes become a very dominant and prominent feature of the coastal landscape.

Tidal sand deposits

Coastal sand deposits can also include tidal sand deposits, which are by definition located in the inter to subtidal. Tidal inlets and their ebb and flood deltas are a component of all barrier islands, and all barriers and beaches bounded by an inlet. They can also accumulate extensive sand deposits in estuaries, in the tidal deltas and some tidal flats. Tidal sand deposits occur at occasional inlets along wave-dominated barrier systems. They become more dominant with decreasing wave height and increasing tide range as inlets increase in number and size, while along tide-dominated shores intertidal sand flats and subtidal sand ridges dominate the coast and nearshore zones.

Table S2 Tectonic coastal types and shoreline types (from Short, 1999, © John Wiley & Sons Limited, reprinted with permission; modified from Inman and Nordstrom, 1971)

Tectonic setting and shoreline type	Mountainous	Hilly, narrow shelf	Plains, narrow shelf	Plains, wide shelf	Hilly, wide shelf	Deltaic	Reef	Glaciated	World's coast
Convergent	96.2							23.9	39.1
Passive-neo	1.0	69.5	15.5			3.8			4.3
Passive-Afro		21.3	73.3	1.0		20.7	16.7	47.8	6.8
Passive-Amero			11.2	98.0	11.3	37.2	49.8	28.2	35.4
Marginal	2.8	19.2		1.0	88.8	38.3	33.6		8.8
Total (%)	100	100	100	100	100	100	100		
World's coast (%)	39.4	5.0	4.7	31.9	7.1	1.2	1.3	4.6	100

Table S3 Coastal characteristics of convergent and passive margin coasts

	Convergent margin	Passive margin
Age	Young (1–10s of millions of years)	Old (100s of million of years)
Relief	Steep, mountainous	Low-gradient plains
Landforms	High mountains and volcanoes Narrow continental shelf Deep-sea trough	Coastal aggradation plains Wide, low continental shelf Continental slope
Tectonics	Active, earthquakes	Quiescent, stable
Weathering	Physical, mass movement	Chemical, fluvial, rivers
Drainage	Short steep streams	Long, meandering rivers
Sediments		
Quantity	Low	High
Size	Fine–coarse (mud-boulder)	Fine (mud-sand)
Sorting	Poor	Well
Color	Dark	Light
Composition	Unstable minerals	Stable minerals (quartz)
Coastal landforms	Rocky, few beaches	Extensive barriers and deltas
Examples	West coast Americas New Zealand Iceland Japan	East coast Americas Southern Africa Southern Australia North Alaska India

Table S4 Sediment and beach-dune characteristics on high-, mid-, and low-latitude coast (modified from Short, 1999, © John Wiley & Sons Limited, reprinted with permission)

Latitude	High	Mid	Low
Climate type	Polar	Temperate	Tropical
Latitudinal range	50°/60°–90°	30°/40°–50°/60°	0°–30°/40°
Sediments	Coarse (cobbles, shingle) dark, unstable minerals	Terrigenous and shelf quartz sand, temperate shelf carbonates	Fine terrigenous quartz sand, coral and algae reefs
Beach type	Reflective beach face Possible multi bar	Reflective to dissipative	Reflective
Other climatic impacts	Permafrost in barriers	Dune calcarenite in arid regions lithifies beaches and dunes	Beachrock lithifies intertidal beach
Dunes	Low and poorly developed owing to: coarse sand, light winds, short season, frozen winter surface, poor vegetation cover, prone to overwash	Largest on west-facing coasts in 40°S latitude, Well-developed: full range of dune types (foredunes through massive transgressive dunes)	Low to minimal owing to low waves and sediment supply and low winds Predominantly foredunes, some dune transgression on trade wind coasts

Barriers

All of the above, beaches, dunes, and tidal inlets are usually components of a larger sand barrier system. The barriers represent longer-term accumulations of wave-, wind-, and tide-deposited sand and other sediments. The barrier may be narrow transgressive barrier islands, wide regressive strand plains capped by beach and/or foredune ridges, or backed by larger coastal dune systems. Some of the world's most extensive sand barrier coasts include the barrier islands along the east and Gulf coast of the United States, the prograding attached barriers or embayed barriers as in much of southern Australia, and large barriers capped by dune systems as in parts of Brazil, South Africa, and Australia. Barriers can also contain tidal sand deposits and where not capped by dunes they are commonly backed by washover fans and aprons.

Large-scale sand coasts and landforms

As indicated in Tables S2 and S3, the largest accumulations of sand at the coast are associated with humid climates on passive margin coasts, in the lower mid to low latitudes. The most extensive sand coasts on each continent are:

North America: the southeast and Gulf coasts, both low-gradient passive margin coasts supplied by numerous rivers including the Mississippi, with the sand reworked onshore by a high-energy wave environment. Coastal dunes are however, poorly developed. Also parts of the west coast exposed to high shelf sediment supply and winds to build dunes.

South America: massive long-term sand supply to entire east coast, leading to an essentially sand barrier–dune coast from the Amazon south to Argentina. Long beach–barrier–dune systems for much of the coast.

Africa: beach, barrier and dune systems ring most of the continent, with sand supplies by local rivers, including the Nile, Niger, and Orange, and on the most exposed coasts by shelf supply of quartz, and in the south also carbonate sands.

Asia: sand dominates most exposed western (Europe—Mediterranean), southern (India), and eastern shores (Southeast Asia—China), with some substantial river systems and deltas and extensive sand barriers in south India and Sri Lanka.

Australia: 50% of passive margin coast consists of sand deposits, with low-energy beaches in north through high-energy beaches and dune systems across south. Supply from rivers and shelf in north, and from shelf quartz in southeast and carbonate in the south and west.

The largest single accumulation of coastal sand in the world is *Fraser Island* on Australia's east coast. The massive sand island is 125 km long and up to 25 km wide. It averages 100 m in elevation reaching 244 m and has an area of 1,840 km^2 and a conservative volume (above sea level) of 185 km^3. It is the largest and last of five near continuous sand islands which extend for 320 km along on a passive margin coast backed by a humid hinterland. The long-term (Pleistocene) accumulation of sand on the islands has been favored by numerous rivers supplying quartz sand to the updrift coast, predominately northern littoral drift toward the islands, coastal orientation to receive the southerly waves, and coastal inflection north of the island which places the island at the terminus of the littoral transport, bedrock headlands to tie the island, moderate to occasionally high-wave energy to move sand long and onshore, and moderate south-east trades to build successive layers of massive dunes up to 244 m high and 20 km wide.

Andrew D. Short

Bibliography

Boreen, T., and James, N., 1993. Holocene sediment dynamics on a cool-water carbonate shelf: Otway, southeastern Australia. *Journal of Sedimentary Petrology*, **63**: 574–588.

Davies, J.L., 1980. *Geographical Variation in Coastal Development*. 2nd edn. London: Longman.

Davis, R.A., Jr., 1983. *Depositional Systems*. Englewood Cliffs: Prentice Hall.

Hayes, M.O., 1967. Relationship between coastal climate and bottom sediment type on the inner continental shelf. *Marine Geology*, **5**: 111–132.

Inman, D.L., and Nordstrom, C.E., 1971. On the tectonic and morphologic classification of coasts. *Journal of Geology*, **79**: 1–21.

Short, A.D., 1999. *Handbook of Beach and Shoreface Morphodynamics*. Chichester: John Wiley & Sons.

Short, A.D., 2002. In Magoon, O.T., Robbins, L.L., and Ewing, L. (eds). Impact of carbonate sands on southern Australian beach systems. *Carbonate Beaches*. USGS & ASCE, Key Largo, 236–250.

Cross-references

Beach Sediment Characteristics
Barrier
Barrier Islands
Carbonate Sandy Beaches
Deltas
Dune Ridges
Muddy Coasts
Tidal Environments
Tidal Inlets
Wave-Dominated Coasts

SCOUR AND BURIAL OF OBJECTS IN SHALLOW WATER

Scour around objects on or near a sediment bed are caused by flow modification due to the object. The presence of the object generates vortices that locally change the bottom stress inducing changes in the sediment transport rate. The presence of an obstacle on a sediment bed where the stress is below the threshold (onset) of motion may induce local intensification and sediment transport, while an obstacle on a bed surface where the stress exceeds the threshold will alter the stress field, producing local depressions and ridges in the bed. The scour phenomenon occurs in unidirectional and oscillatory flow and in types of fluid ranging from air to water to sediment-laden turbidity currents and pyroclastic flows. Obstacles producing scour range from millimeter size grains to topographic features many meters high and kilometers in length. The resulting bedforms range from sand streaks and ripples to large desert dunes and scour moats and sediment drifts around seamounts in the deep ocean.

Here we are primarily concerned with scour in nearshore waters and on beaches where the flow is both unidirectional and oscillatory. Scour naturally occurs wherever a larger object occurs on an otherwise smaller grained bed. For example, a seashell or a kelp-rafted rock on the seabed will form scour features ranging in size from rhomboid marks around small objects to crater-like crescentic depressions twice the size of the shell or rock. On larger scales, rip currents scour large channels over and through longshore bars. Above the waterline, wind-blown scour features form around kelp clumps and rocks on the beach, while accretionary dunes and sand shadows form around hardrock outcrops on the coastal desert floor. In recent decades, the importance of scour in the burial of mines has led to increased interest in scour phenomena.

Introduction

Scour and scour marks are the erosional and accretionary bedform patterns that occur in the vicinity of obstacles that are on or near a sediment bed. Scour always involves some degree of perturbation in the flow system that changes the local pattern of erosion and deposition relative to that of the general flow. The primary scour pattern may be erosion as in the formation of crescentic scour around a rock on the beach (Figure S4) or depositional as in the dune deposits in the stagnation area of flow around an outcrop (Figure S5).

Any form that locally concentrates vorticity near the bed can elevate bed shear stress and initiate onset of grain motion, leading to local bed scour including bumps and depressions on the bed itself (Figure S6). Once initiated, a pattern of scour may spread down current in the form of a growing field of current ripples (Figures S6(B)–(D)), while vortex ripples under wave action may spread both against and with wave propagation from a single initiating irregularity in the bed (Inman and Bowen, 1962; Tunstall and Inman, 1975).

The most commonly studied scour patterns are those associated with single bluff (blunt) bodies placed on or protruding from the bed. There is an extensive engineering literature of the scour around the piles of bridges and piers (e.g., Collins, 1980; Chiew and Melville, 1987). Engineers refer to scour as *local* when it results from the effects of a structure on the flow pattern and *general* when it would occur irrespective of the presence of a structure; it is termed *clear water scour* when the bed upstream of a structure is at rest (e.g., Raudkivi, 1990).

In sedimentology, the interest has usually been in the scour pattern around individual objects, referred to as *scour marks* (e.g., Pettijohn and Potter, 1964; Reineck and Singh, 1975) or as *obstacle marks* (Allen, 1984, 1985). Allen further subdivides obstacle marks into *current crescents, current shadows*, and *scour-remnant ridges*. It appears, from the extensive literature on the subject, that current crescent and *crescentic scour mark* are the most general terms for the crescentic feature formed around an object on the bed, and that the feature may be either erosional as in Figure S4 or accretional as in Figure S5. The appearance of other associated features such as current shadow, scour-remnant ridges, and ripples are wake phenomena that depend upon the height to width aspects of the object, the nature of the flow system, and the type of sediment.

Bagnold (1941) introduced *sand shadow* for the various shapes of sand accumulation in the shelter of an obstacle. Allen's (1984) current shadow is based on Bagnold's sand shadow. Allen defines scour-remnant ridges as the small ridges of sand, snow, or mud preserved in place on the lee side of resistant objects after the surrounding material has been eroded away. Thus scour-remnant ridges are residual leeward

Figure S4 Wind-formed crescentic scour pattern around a rock (~30 cm diameter) on the beach berm, Coronado, CA. Wind blows from left to right (D.L. Inman photograph).

Figure S5 Wind-deposited crescentic pattern consisting of two barchan dunes formed around an outcropping ridge (~8 m high, 40 m long) on the desert floor, northwestern Gulf of California. Wind blows from upper left to lower right as indicated by sand shadows in the lee of bushes and by the double shadow from the sand drift passing along the sides of the ridge (D.L. Inman photograph).

phenomena while current (or sand) shadows are depositional and/or erosional leeward bedforms.

Characteristics of the flow around a vertical cylinder, such as a bridge pile in steady currents, have been investigated extensively (e.g., Shen *et al.*, 1969; Breusers *et al.*, 1977). It has been found that a horseshoe vortex above the scoured bed is a dominant factor in the scour process, and that the vortex has a close relationship with the bed profile near the cylinder. The vortex behavior caused by the object is thus a very important factor to consider in the estimation of bed scour as described under Scour Mechanics.

Relatively few studies have been conducted on scour induced by waves and currents. Nishizawa and Sawamoto (1988) and Sumer *et al.* (1992) have studied the flow around a slender vertical cylinder under waves using flow visualization techniques. They have reported relationships

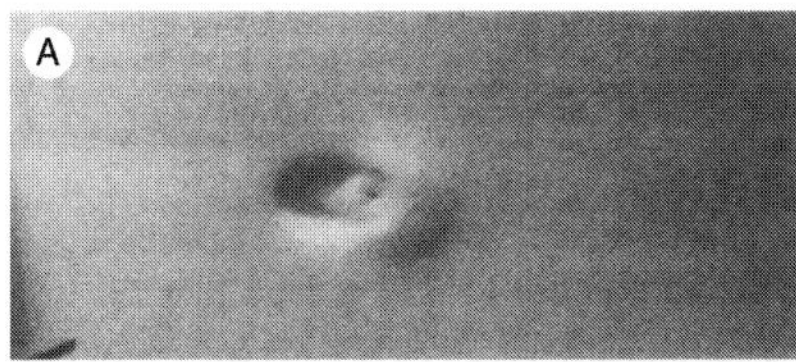

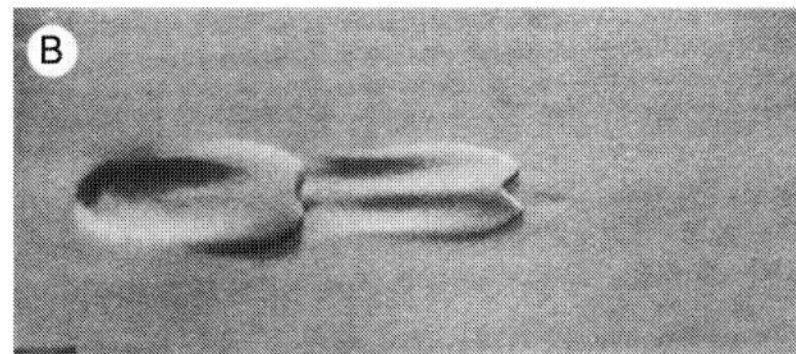

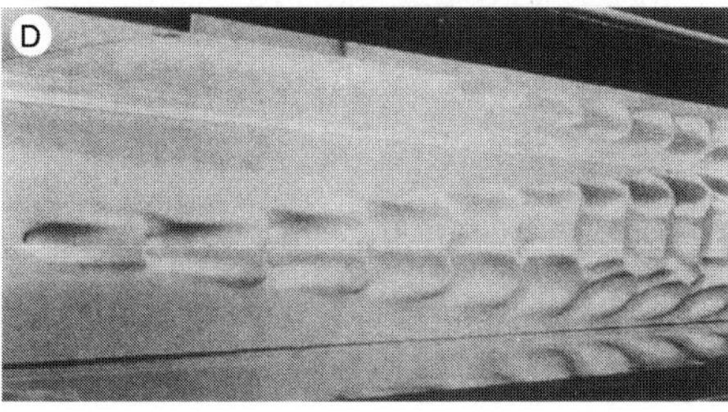

Figure S6 Progressive downstream propagation and lateral widening of a single perturbation (A) on a fine sand bed. Flume is 17 cm wide (sidewalls visible in panel as in (D) and bottom stress is near the threshold of grain motion (photographs by permission of John Dingler).

between the flow characteristics and nondimensional parameters of flow similitude such as the Strouhal number (acceleration forces/inertial forces) based on experimental results for a flat bottom. There is less literature for a scoured seabed, probably because these flows often show a highly complex three-dimensional rippled pattern owing to the complicated shape of natural obstacles and the unsteadiness of the main flow. Even for simple shapes such as a cylinder, unsteady three-dimensional flow over the bed generates vortices that govern the local scour process.

Continued scour around objects on a sand bed usually leads to complete burial of the object. Shells and rafted objects dropped to the seafloor eventually bury and disappear if the sand bed has sufficient thickness (e.g., Inman, 1957, Figures 20, 21). A study of mine scour and burial was conducted for the Office of Naval Research during 1953–56 on the sandy shelf off the Scripps Institution of Oceanography in La Jolla, CA, in water depths extending from the surf zone to 23 m (Inman and Jenkins, 1996). The mines were 1.6 m long cylinders, 60 cm in diameter. It was found that the burial process began with scour depressions around the mine and continued with rollover into the depressions (Figure S7). This process repeated itself until the mine was completely buried. Burial time ranged from hours and days in the surf zone to almost one year at 17 m depth.

Scour mechanics

The scour phenomenon around objects differs from other types of sediment flux in that the presence of an object on or near the bed induces local changes in an otherwise uniform pattern of bed stress, thereby causing local patterns of erosion and/or accretion that may differ from the general bedform pattern. The object may be either blunt or streamlined and the resulting scour pattern may be erosional, depositional, or both. Scour develops from a variety of mechanisms whose relative importance depends upon the scale and intensity of the flow and the relative size and shape of the obstacle. The most common and largest bedforms result from scour around bluff bodies where the formation of a horseshoe vortex generates a scour hole that begins on the upstream side and wraps around the object (Figure S4) as described below.

The mechanics of the scour around a body are inherent in the vorticity field generated when a fluid moves over a bed or solid surface. For example, consider the velocity profile above the bed and up current from an object on the bed (Figure S8). The shear near the bed in the bottom boundary layer generates vorticity between the layers of differing flow velocity creating a vorticity sheet. *Vorticity* is the angular momentum of a fluid element, while a *vortex* is the arrangement of many of these fluid elements into a pattern of angular motion.

The presence of an obstacle on the bed rearranges the sheet of vorticity in the boundary layer, and creates new vorticity by the shear flow around the obstacle. The flow disturbance of the obstacle creates a stagnation point (s') at the bed interface upstream of the obstacle. The bed vorticity in the approaching flow collects at the stagnation point forming a local excess of vorticity that organizes into a *forward bound vortex* (Figure S8) that moves the stagnation point (s) upstream of the vortex. The forward bound vortex initiates the scour process by causing intense velocity shear stress at the base of the obstacle. The incoming vorticity from the flow builds up in the bound vortex and the excess leaks around the base of the cylinder forming a pair of *trailing vortex filaments* on either side of the obstacle. The bound vortex with its pair of trailing filaments form a vortex system known as a *horseshoe vortex*. The trailing vortex filaments extend the region of scour from the upstream base of the obstacle, around the sides, and downstream. As the trailing filaments extend downstream, the vorticity of the filaments diffuse into the interior of the fluid, thereby slowing the filament rotation and weakening the shear stress on the bed. Consequently, the scour diminishes downstream of the obstacle forming a scour pattern around the obstacle known as *current crescent* or *crescentic scour mark*.

The horseshoe vortex and its associated crescent scour are nearfield bedform responses that occur over distances of about two obstacle diameters. Further downstream in the farfield the trailing filaments of the horseshoe vortex begin to entwine into a helical vortex system. At each crossover of the helical pairs, the induced velocities of the vortex system approach a null on the bed, allowing for complimentary depositional features such as ripple marks in the current shadow downstream of the crescentic scour. Initially, both the near and farfield scour and its shadow system are referred to as *forced* forms because these erosional and depositional responses are controlled by the length scales of the flow field around the obstacle. However, once the scour process erodes deeply into the bed, the crescentic scour features will modify the flow field around the obstacle. In turn, the modified flow field further modifies the form of the scour and the associated bedforms in the current shadow. The fully developed horseshoe vortex is a consequence of the scour depression. Therefore, once this feedback takes place, the scour depression becomes an *interactive* part of a fluid-bedform system where the bedform interacts with and extensively modifies the flow field above it.

Other scour mechanisms

Other scour mechanisms become important in very shallow water typical of the swash and backwash motion of wave runup on the beach face. These mechanisms are associated with thin flows where water velocity often exceeds the critical limit for wave propagation and where capillary waves become important. Also, in thin flows, common "V"-shaped ship waves are formed by small objects and induce stress perturbations on the sediment bed. As a consequence, the beach face often shows rhomboid marks caused by one or more of the mechanisms associated with thin flow (Figure S9). Large rhomboid marks 3–10 m across are known to occur on beaches following the backwash from tsunami waves (i.e., Shepard, 1963, figure 100(B)).

The flow regime over the beach face may be either subcritical ($u < \sqrt{gh}$) or supercritical ($u > \sqrt{gh}$) depending on the speed of the water u relative to the shallow water wave speed $\sqrt{gh}$, where g is the acceleration of gravity and h is the thickness of the flow. In either case, the height of small obstacles such as shells, pebbles, and the feathery antennae of filter feeding organisms that protrude above the bed are of the order of the flow thickness. Supercritical flow is readily perturbed by an obstacle on the bed and locally slowed to subcritical flow by small oblique hydraulic jumps (Henderson, 1966) upstream of and extending downstream from the obstacle in a V-shaped pattern. The turbulence of the hydraulic jump scours a corresponding V-shaped erosion pattern around the obstacle, often made strikingly visible by exposure of dark minerals in the laminated beach sand similar to that shown in Figure S9. Intersections of adjacent V-shaped jumps form the characteristic diamond pattern of the rhomboid ripple. These marks are distinguished by their long scour trail and because the vertex of the V-shape is always upstream of the obstacle, much like the crescentic scour of larger obstacles under subcritical flow conditions. However, the large supercritical

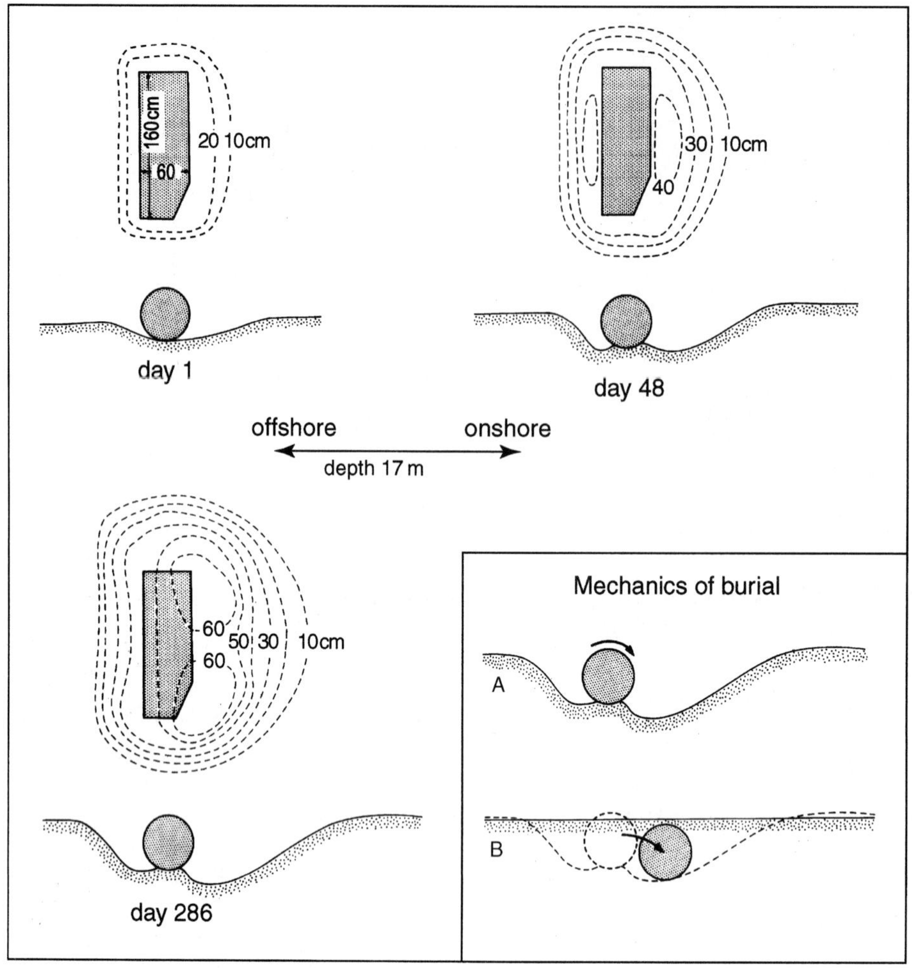

Figure S7 Scour and burial of cylindrical mine by wave action over a fine sand bottom off Scripps Beach, La Jolla, CA (after Inman and Jenkins, 1996; SIO Reference Series No. 96-13). Note vertical exaggeration in profile view.

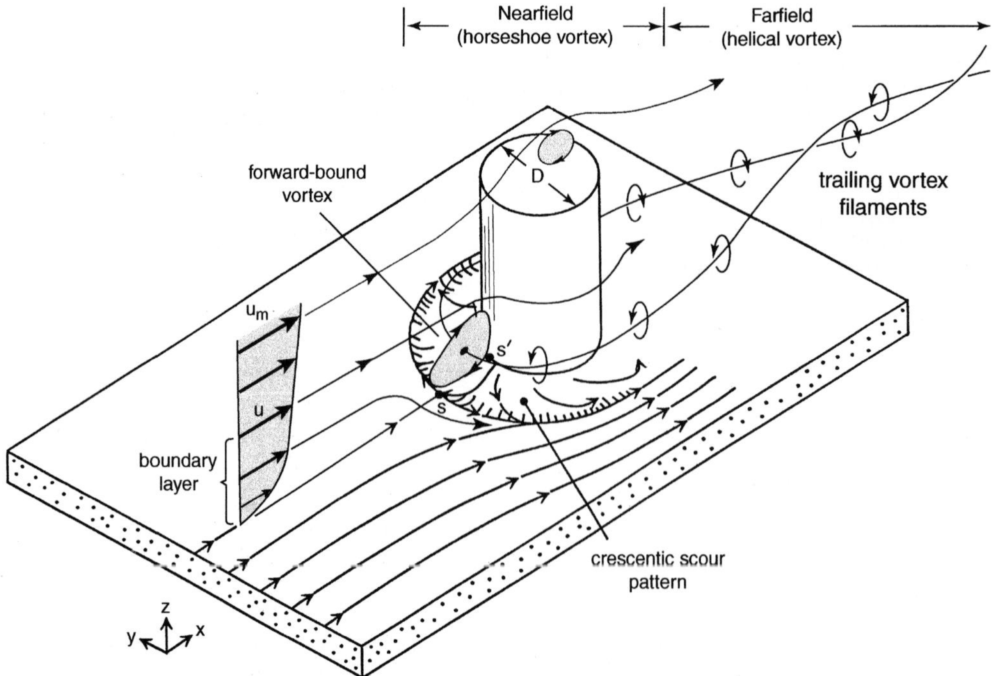

Figure S8 Definition sketch of fluid motion and scour features around an upright cylinder extending through the surface of a sediment bed under unidirectional flow (after Schlichting, 1979; Allen, 1984); compare with Figure S4.

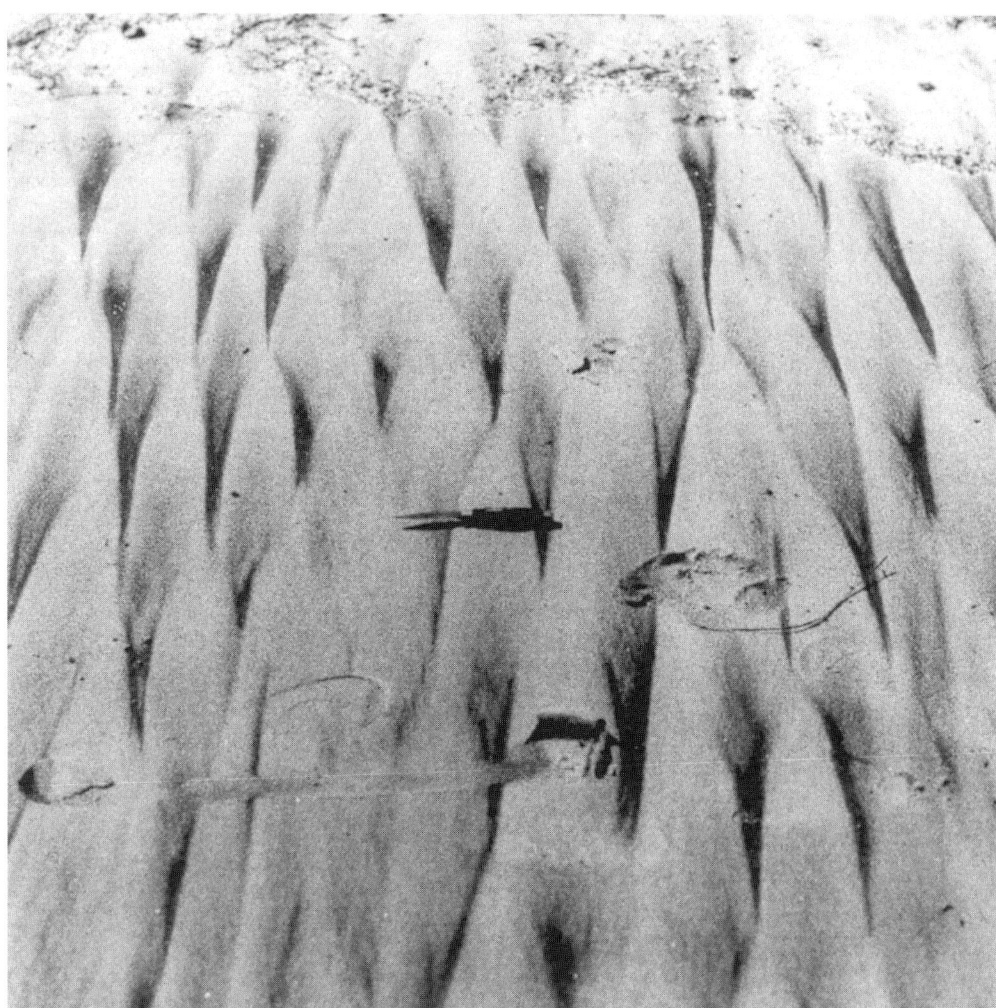

Figure S9 Rhomboid ripple marks on beach face at La Jolla, CA. Diamond pattern associated with flow divergence around antennae of a field of sand crabs (*Emerita analoga*). Photo looking seaward, knife (including blade) 12 cm, swash mark at top (D.L. Inman photograph).

rhomboid marks are less common than would be expected because supercritical flow over sand beaches rapidly develop *backwash ripples*, small-scale sand waves that parallel the beach contours and obliterate the large, extensive rhomboid marks that are found on the otherwise flat beach face.

There are at least three thin-flow mechanisms that may lead to the formation of rhomboidal patterns in subcritical backwash flow. These include the bow and stern wave mechanisms of ordinary ship waves (e.g., Stoker, 1957; Whitham, 1974), and the formation of capillary waves. Bow waves may form at the stagnation point upstream of an obstacle on the beach, and travel outward with the traditional Kelvin angle of the classical ship wave pattern. In contrast, thinning flow over the obstacle may become supercritical, then revert to subcritical flow with the accompanying small hydraulic jump downstream of the object. The wakes from both bow and stern waves form V-shapes that trail downstream of the object, with the vertex of the V on the upstream side of the obstacle for the bow-wave case and just downstream for the stern wave. The rhomboid ripple marks shown in Figure S9 were observed to have formed from the bow waves of the filter feeding beach crab *Emerita*. Sand crabs together with other beach-dwelling organisms, such as the bean clam *Donax gouldii* (Ricketts *et al.*, 1985), play a far more active part in shaping bedforms on sandy beaches than is generally recognized.

Often, the system of ship waves formed by small obstacles on the beach have short-length scales where surface tension forces become important. In this case, the gravity–capillary waves (Whitham, 1974) can propagate upstream as well as downstream of the object and produce a more parabolic-shaped wave and scour pattern near the object than for typical rhomboid marks.

Dougals L. Inman and Scott A. Jenkins

Bibliography

Allen, J.R.L., 1984. *Sedimentary Structures, Their Character and Physical Basis*. Amsterdam: Elsevier.

Allen, J.R.L., 1985. *Principles of Physical Sedimentology*. London, Boston: Geoge Allen and Unwin.

Bagnold, R.A., 1941. *The Physics of Blown Sand and Desert Dunes*. New York: William Morrow & Company.

Breusers, H.N.C., Nicollet, G., and Shen, H.W., 1977. Local scour around cylindrical piers. *Journal of Hydraulic Reasearch*, **15**: 211–252.

Chiew, Y.M., and Melville, B.W., 1987. Local scour around cylindrical piers. *Journal; of Hydraulic Research*, **25**: 15–26.

Collins, T.J., 1980. Investigating bridge scour. *Railway Track and Structures*, **7**: 6.

Henderson, F.M., 1966. *Open Channel Flow*. New York: The Macmillan Company.

Inman, D.L., 1957. Wave-generated ripples in nearshore sands. US Army Crops of Engineers. *Beach Erosion Board, Technical Memorandum*, 100.

Inman, D.L., and Bowen, A.J., 1962. Flume experiments on sand transport by waves and currents. In *Proceedings of the 8th Conference on Coastal Engineering*. Council on Wave Research, pp. 137–150.

Inman, D.L., and Jenkins, S.A., 1996. A chronology of ground mine studies and scour modeling in the vicinity of La Jolla. University of California, San Diego, Scripps Institution of Oceanography, *SIO Reference Series*, 96–13.

Nishizawa, M., and Sawamoto, M., 1988. Local scour around a vertical cylinder under the wave action. In *Proceedings of the 6th Congress of Advenced Ports Design*. International Association of Harbor Research, pp. 121–128.

Pettijohn, F.J., and Potter, P.E., 1964. *Atlas and Glossary of primary Sedimentary Structures*. New York: Springer-Verlag.
Raudkivi, A.J., 1990. *Loose Boundary Hydraulics*. Oxford: Pergamon Press.
Reinenck, H.-E., and Singh, I.B., 1975. *Depositional sedimentary Environments*. New York: Springer-Verlag.
Ricketts, E.G., Calvin, J., and Hedgpeth, J.W. (revised by Phillips, D.W.), 1985. *Between Pacific Tides*, 5th edn. Stanford, CA: Stanford University Press.
Schlichting, H., 1979. *Boundary-Layer Theory*, 7th edn. New York: McGraw-Hill Book Co.
Shen, H.W., Schneider, V.R., and Karaki, S., 1969. Local scour around bridge piers. American Society of Civil Engineers, *Journal of the Hydraulics Division*, **95**: 1919–1940.
Shepard, F.P., 1963. *Submarine Geology*, 2nd edn. New York: Harper & Row Publishers.
Southard, J.B., and Dingler, J.R., 1971. Flume study of ripple propagation behind mounds on flat sand beds. *Sedimentology*, **16**: 251–263.
Stoker, J.J., 1957. *Water Waves, the Mathematical Theory with Applications*. New York: Interscience Publishers, Inc.
Sumer, B.M., Fredsoe, J., and Christiansen, N., 1992. Scour around vertical pile in waves. American Society of Civil Engineers, *Journal of Waterways, Port, Coastal and Ocean Engineering*, **118**: 15–31.
Tunstall, E.B., and Inman, D.L., 1975. Vortex generation by oscillatory flow over rippled surfaces. *Journal of Geophysical Research*, **80**: 3475–3484.
Whitham, G.B., 1974. *Linear and Nonlinear Waves*. New York: John Wiley & Sons.

Cross-references

Accretion and Erosion Waves on Beaches
Beach Features
Coastal Warfare
Erosion Processes
Ripple Marks

SEA-LEVEL CHANGE—See CHANGING SEA LEVELS

SEA-LEVEL CHANGES DURING THE LAST MILLENNIUM

Introduction

The last millennium includes well-known periods of marked, possibly global, climatic change, such as the Medieval Warm Period (MWP) and the Little Ice Age (LIA), as well as the period of global industrialization, often referred to in climatic terms as the period of Modern Warming (MW). Some important questions, relating to these periods, are key to understanding the relationship between climate change and *changing sea levels* (*q.v.*). What was the pattern and amplitude of sea-level change during the LIA and the MWP? Can the MWP serve as an analog for present and future sea-level conditions? When did the rapid rise of sea level, as registered by tide-gauge measurements, commence? Has sea-level rise accelerated during the 20th century? Because sea-level records spanning the last millennium are recent from a geological point of view, they have particular relevance to issues of *global warming* (*q.v.*) and future climate. Given the historical timescale, the answers to the four questions rely on the interpretation of instrumental as well as geologic records.

In view of the importance of a good understanding of the most recent sea-level history, it is perhaps somewhat surprising that detailed records are rare. This can be ascribed to three causes. First, the length of observational records is relatively short. The oldest direct measurements of sea level started in Amsterdam in 1682. The Permanent Service for Mean Sea Level (PSMSL) database contains only eight records that start before 1850, the oldest being Brest in western France (1807). Second, conventional geological methods of *Holocene* (*q.v.*) sea-level reconstruction have inherent limitations. Diagrams of radiocarbon-dated sea-level indicators plotted in a time-depth diagram reveal underlying trends but usually do not yield sufficient precision to resolve oscillations of sea-level over the past 1,000 years, which are on the order of decimeters. Geomorphologic evidence, in the form of beach ridges, marine platforms, and rock notches, also has its limits in terms of resolving power. Accurate dating of sea-level positions is limited by statistical uncertainties of the radiocarbon method. Finally, human development has in many locations around the world impacted on sedimentary environments. In northwest Europe, for example, many salt marshes have been reclaimed or embanked and the once extensive coastal fen peatlands have become a rare feature.

This entry will review published records of sea-level change during the past millennium, with a primary focus on the salt marsh records from the northeast coast of the United States. It will highlight some of the common features in these records. The main aim of this review, however, is to discuss the pitfalls and problems that need to be resolved before any wider climatic interpretations can be made from the sea-level data.

Northeastern United States

In recent years, detailed analyses of the litho- and biostratigraphy of *salt marshes* (*q.v.*) along the northeast coast of the United States have yielded relatively complete and continuous records of sea-level changes spanning the last millennium. The vegetated surfaces of the salt marshes have remained close to the high tide mark throughout the middle and late Holocene while sea level has been rising, forming accumulations of highly organic sediment from which sea-level information can be extracted. Some of the sea-level records from the northeastern United States are depicted in Figure S10.

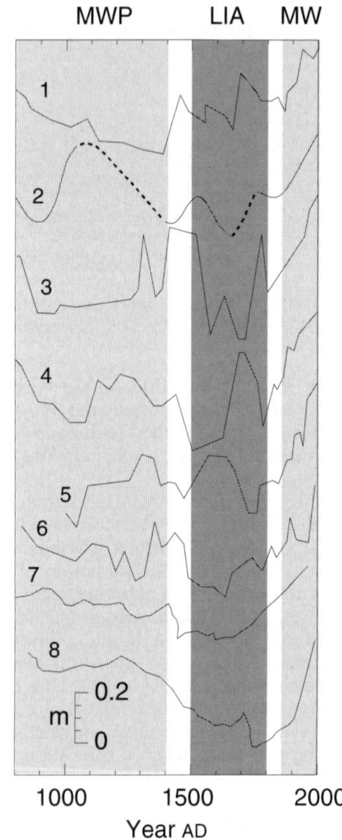

Figure S10 Sea-level records from Connecticut (1–6; compiled by van de Plassche, 2000) and Maine (7, 8), USA. (1, 2) Hammock River marsh. (3) Farm River marsh. (4, 5) East River marsh. (6) West River marsh. (7) Machiasport, eastern Maine (Gehrels, 1999). (8) Wells, southwestern Maine (Gehrels *et al.*, 2000). Millennial scale trends reflecting isostatic subsidence (1.0 m per 1,000 yr for Connecticut, 0.43 m per 1,000 yr for Machiasport, 0.30 m per 1,000 yr for Wells) have been removed from the relative sea-level curves. MWP, Medieval Warm Period; LIA, Little Ice Age; MW, Modern Warming.

Salt marsh sediments as recorders of sea-level change

The ability of salt marsh sediments to record sea-level rise is controlled by the balance between sediment accretion and sea-level rise. When sedimentation outpaces sea-level rise, a regression is recorded. Alternatively, when sea-level rise is faster than accretion, a transgressive overlap occurs. These landward and seaward environmental shifts are manifested in the lithostratigraphy of the marshes. For example, a regressive shift can be recognized by an organic-rich high marsh facies overlying a clay-rich low marsh facies, both containing identifiable remains of plant roots and rhizomes. Biostratigraphy is often more useful in that it can detect changes within a distinct lithostratigraphical unit, offering greater resolution. In the marshes of northeastern North America, foraminifera have proven to be the most useful biological sea-level indicators (Scott and Medioli, 1978). Different species of these single-celled organisms possess different tolerances to air and therefore live in distinct zones on the salt marshes, much like the vascular plants. Their high abundance and good preservation enable a quantitative assessment of former marsh heights from fossil foraminifera in cores. The most accurate way to calculate sea-level changes from reconstructed paleo-marsh surfaces is through the use of transfer functions based on modern foraminiferal distributions (Gehrels, 1999, 2000).

Besides the upward motion of sea level, the process of autocompaction provides accommodation space for sediments to accumulate (Gehrels, 1999; Allen, 2000). Autocompaction is defined as compression of a sedimentary package under its own weight (Kaye and Barghoorn, 1964; Allen, 1999). While the degree of autocompaction is sometimes difficult to assess, the process ensures that sea-level stillstands and even small sea-level falls are registered when sediment fills the accommodation space created by compression of the sediment package. In theory, a relative sea-level signal can be isolated when rates of sediment accretion and rates of autocompaction are determined. Sediment accretion is measured through detailed dating, while autocompaction can be determined from dating basal peat (Gehrels, 1999) or by a geotechnically based modeling approach.

Medieval Warm Period

The period from AD 800 to 1400 is often referred to as the MWP. Although warming may not have occurred synchronously on a global scale, the MWP offers the nearest reasonable historical analog to future global warming conditions. In Connecticut, van de Plassche *et al.* (1998) and van de Plassche (2000) recorded a sea-level rise of 0.25 m (corrected for isostatic subsidence) between AD 950 and 1000 and estimated that, during the MWP, sea level stood between 0 and 0.5 m higher than during the LIA. In an earlier study of the same salt marsh, however, Varekamp *et al.* (1992) could not find evidence for high rates of sea-level rise during the MWP. Nydick *et al.* (1995) documented slightly higher sea levels in one of two study sites a few kilometers further west. In eastern Maine, sea levels during the MWP were about 0.2 m lower than present-day levels, and 0.2 m higher than during the LIA (Gehrels, 1999). The error margins on these sea-level positions were determined statistically to be about ± 0.25 m. The record from southwestern Maine (Gehrels *et al.*, 2000) is better dated and shows a sea level close to the present height during the MWP.

Little Ice Age

Like the MWP, the duration and intensity of the LIA varies from place to place, but may be taken as the period from AD 1500 to 1800 in the northern hemisphere, although in Scandinavia AD 1900 is often considered as the end of the cooling. In Connecticut, Varekamp and Thomas (1998) recorded a slow rate of relative sea-level rise of 0.3 mm per year from AD 1300 to 1650, compared to a rise of 1.6 mm/yr averaged over the past 1,000 years. This finding was confirmed by van de Plassche *et al.* (1998) in a nearby marsh. In eastern Maine, sea level fell slightly between AD 1500 and 1700 (Gehrels, 1999), while in southwestern Maine, sea level fell after AD 1200 to reach a lowstand of 0.5 ± 0.25 m below present sea level between AD 1700 and 1800 (Gehrels *et al.*, 2000).

Past 300 years

Sea-level records from Connecticut and Maine are consistent in that they all show a rapid sea-level rise during the past several centuries. In southwestern Maine, a relative sea-level rise of 0.5 m has occurred since AD 1800 (Gehrels *et al.*, 2000). A similar rise occurred in eastern Maine (Gehrels, 1999) but the chronology is not well constrained. Records from Connecticut show a sea-level rise of 0.3–0.6 m since ~1700 AD superimposed on a long-term trend of 1 mm/yr (van de Plassche, 2000). In Dennis Creek, New Jersey, Varekamp, and Thomas (1998) documented a relative sea-level rise of 7 mm/yr since AD 1650, a rate which appears anomalously high. The rates of sea-level rise in the salt marshes of Connecticut and Maine for the past 300 years are comparable with *tide-gauge* (*q.v.*) observations during the past 100 years. The salt marsh records do not clearly resolve any acceleration of sea level in the past 150 years.

Other regions

Detailed high-resolution records from coastlines outside eastern North America do not exist, but several studies have shown fluctuations of sea level during the last 1,000 years. The studies discussed below are reviewed by Long (2000) and corresponding references can be found therein. In the South Pacific, sea levels were possibly up to 1 m higher than present levels during the MWP and up to 0.9 m lower than present levels during the LIA. Data from this region must be treated with caution as the sea-level index points come from many different islands thousands of miles apart and error margins were not considered. In a salt marsh on the south coast of Britain, increased sedimentation rates during the past 200 years might be related to accelerated sea-level rise. In the coastal plain of Belgium and northern France sea level did not reach a clear highstand during the MWP. Sea-level index points from the Frisian Islands in the Netherlands are imprecise, but a rapid sea-level rise may have occurred at some time during the past 800 years while sea-level rise possibly slowed down during the LIA.

Long instrumental records of sea-level change

Tide-gauge records long enough to span the climatic transition between the LIA and MW are sparse. Of the eight sea-level records commencing before 1850 that are held in the PSMSL database, seven are from Europe. The other one is from Bermuda and started in 1833. Woodworth (1990) analyzed the oldest European mean sea-level records, some of which are not officially part of the PSMSL database. Combining observations from Brest (starting in 1807), Sheerness (1834), Amsterdam (1700), and Stockholm (1774), he found an acceleration of 0.4 mm/yr per century. The Liverpool record, starting in 1768, shows an acceleration of 0.33 ± 0.10 mm per year per century, with a mean rise of 0.39 ± 0.17 mm/yr between 1768 and 1880 and 1.22 ± 0.25 mm/yr for the 20th century (Woodworth, 1999). Ekman (1999) estimated that sea-level rise, corrected for isostasy, increased at Stockholm from 0.0 ± 0.4 mm/yr between 1774 and 1884 to 1.05 ± 0.25 mm/yr between 1892 and 1991. In an earlier study, Mörner (1973) reconstructed from tide-gauge data at Amsterdam, Stockholm, and Warnemünde a stable "eustatic" sea level from 1682 to 1740, a fall of 0.25 mm/yr between 1740 and 1820 and a stable sea level from 1820 to 1840, followed by a rapid sea-level rise of 1.1 mm/yr around 1840 which lasted until 1950. The evidence from Florida for an acceleration since 1846 is weak (Maul and Martin, 1993), while Douglas (1992) found no statistical evidence to support a global sea-level acceleration after 1850. The feasibility of assessing a global rate and potential acceleration of sea-level rise from tide-gauge data may be questioned on the grounds that large parts of ocean basins are not represented by long records. For that reason, it may be argued that temporal variability of sea level should only be assessed on a regional scale.

Problems

Precision

The main problem in accurately reconstructing sea-level oscillations during the past 1,000 years is that the amplitude of fluctuations is on the same order as the precision of the sea-level indicators. The signal-to-noise ratio is often very low. Statistical analyses of the errors associated with the indicative meaning of microfauna in salt marshes, for example, show that they are on the order of ± 0.15 m at best (e.g., Gehrels, 2000). Surveying errors and other uncertainties add to the imprecision. Scott and Medioli's (1978) claim that the foraminifer *Trochammina macrescens* can indicate sea levels with a precision of ± 0.05 m is not statistically supported.

Salt marsh records of sea-level change are reliable only if sedimentation has been continuous. Variations in sedimentation rate should be adequately resolved by the dating of the sequence. Given the complex stratigraphy of many salt marshes, it is crucial that the site from which sea-level change is reconstructed is carefully selected.

Suitable sites for sea-level reconstruction may be found very near the highest tide level in broad, undisturbed areas of high marsh, away from tidal creeks and channels. In lower marsh areas and tidal flats, the precision of sea-level indicators is usually too low to resolve changes in height caused by accretion while sedimentation is often episodic. Clayey tidal flat and low-marsh facies are therefore best avoided, also because suitable material for dating is sparse. Some contradictions between sea-level records from the same salt marshes in Connecticut may be due to core site selection.

Near the highest tide level, accommodation space is minimal and its creation is controlled by a rising sea level and autocompaction (Allen, 2000). The stratigraphic record is hence (mostly) a reflection of sea-level change, rather than the static infilling of available accommodation space. Even during stillstands and sea-level reversals sedimentation and peat growth may continue, as the compression of the salt marsh sediments under their accumulating weight provides the accommodation space.

Dating

Radiocarbon measurement is the method commonly used for dating sea-level changes in the last 1,000 years. However, this method is limited in its precision due to the associated statistical counting errors. The radiocarbon ages, including their errors, need to be expressed into calendar years when rates of sea-level rise are calculated. The wiggles in the radiocarbon calibration curve provide particular problems for dating sediments that were deposited in the past 300 years. In Table S5, calibration results are given for two ^{14}C ages to illustrate radiocarbon age uncertainties. The ages represent, with 95% certainty (2σ), time spans of 74 and 131 years, respectively.

Pollen markers are often useful for dating sediments younger than 300 years, provided they are accurately matched with historical events. Lead and cesium radioisotopes provide means to date the last 100 years and enable comparisons with observational records.

Conclusions

Sea-level records for the past millennium come primarily from the northeast coast of the United States. In this region, there is sparse evidence for a significant highstand during the MWP. It appears that there has been a widespread sea-level low- or stillstand during the LIA, which has, to some extent, also been documented in other regions. Another common element in the US records is the rapid rise of sea level since about AD 1800. The longest European instrumental sea-level records have recorded an acceleration of sea-level rise into the 20th century. Records that start after AD 1850 are too short to detect any, possibly anthropogenic, acceleration in the rise of sea level.

The finding that sea-level rise was well underway before ca. 1850 would suggest that contemporary sea-level rise is, at least, partly the result of natural warming following the LIA. An acceleration in industrial times points to anthropogenic sea-level rise as an additional factor. Clearly, these results are not conclusive until they can be replicated in other localities and chronological control is improved.

An intriguing question is why the MWP did not produce any clear sea-level signal, even though temperatures were as high as they are today. Apparently, warmth does not necessarily produce higher sea levels through increased melting and thermal expansion of the ocean surface layer. The relationship between temperature and sea level is complicated by additional oceanographic factors. A hypothesis that needs to be tested is whether the velocity of the Gulf Stream can lower sea levels during warm spells. It has been shown from tide-gauge records that, along the east coast of North America, increase in Gulf Stream velocity produces low coastal sea levels (Ezer *et al.*, 1995) while transporting heat northwards at a higher rate thereby warming up the northern Atlantic (Kushnir, 1994). Bianchi and McCave (1999) produced the first record of deep-current strength south of Iceland showing high rates of flow around AD 1000, similar to today. More such records are needed to investigate the link between ocean current velocity and sea-level change along the east coast of North America.

Table S5 Two examples of calibration of radiocarbon ages into calendar years produced by the program Calib4

^{14}C age	1σ Calendar age (AD)	2σ Calendar age (AD)
500 ± 40	1408–39	1331–41, 1396–1450
200 ± 20	1661–70, 1674–80, 1741–42, 1766–70, 1779–87, 1788–98, 1940–54	1650–84, 1731–33, 1735–46, 1748–57, 1761–1809, 1927–54

While replication of sea-level records is important to substantiate any proposed link between rates of sea-level change and climate, it must be kept in mind that the response of sea level to a climatic change cannot be expected to be uniform across the globe. This is due to geoidal, steric, and oceanographic effects, which all act on regional scales. The search for global correlations of decadal-scale sea-level changes is therefore a futile exercise. Similarly, future sea-level rise will not be uniform, a point that is not taken into account by current Intergovernmental Panel on Climate Change (IPCC) predictions (Long, 2000).

The precision of sea-level reconstruction is limited by the vertical precision of sea-level indicators as well as by suitable chronological control. Reducing counting errors on radiocarbon measurements, ^{14}C wiggle-matching, the application of ^{210}Pb-dating and pollen markers, and development of new dating techniques, for example, aspartic acid racemization, will all contribute to decreasing dating uncertainties. It is possible that testate amoebae, which are found in narrow zones in the upper parts of salt marshes, can help increase the precision of sea-level reconstructions (Charman *et al.*, 1998). Statistical techniques, such as the use of transfer functions, offer a way to obtain objective quantitative assessments of the precision of sea-level reconstructions. Indeed, statistical techniques have been widely applied in many studies of paleo-environmental reconstruction, while sea-level research arguably lags behind in embracing a more quantitative approach.

Developments in sea-level research during the past decades have clearly shown that, in addition to the growth and decay of land-based ice, a multitude of factors contribute to sea-level change. Oceanographic factors, such as sea-surface dynamical and steric changes, become increasingly important when sea-level changes are investigated on centennial and decadal timescales. The role of sea-level change in decadal climatic variability, for example, its link with the North Atlantic Oscillation, is poorly understood. A major task for future sea-level research is to couple the changes of the surface of the ocean to other elements of the global climate system described by ocean-atmosphere models, in particular the thermohaline circulation.

Roland Gehrels

Bibliography

Allen, J.R.L., 1999. Geological impacts on coastal wetland landscapes: some general effects of sediment autocompaction in the Holocene of northwest Europe. The Holocene. **9**: 1–12.

Allen, J.R.L., 2000. Morphodynamics of Holocene salt marshes: a review sketch from the Atlantic and southern North Sea coasts of Europe. *Quaternary Science Reviews*, **19**: 1155–1231.

Bianchi, G.G., and McCave, N., 1999. Holocene periodicity in North Atlantic climate and deep-ocean flow south of Iceland. *Nature*, **397**: 515–517.

Charman, D.J., Roe, H.M., and Gehrels, W.R., 1998. The use of testate amoebae in studies of sea-level change: a case study from the Taf Estuary, South Wales, UK. *The Holocene*, **8**: 209–218.

Douglas, B.C., 1992. Global sea level acceleration. *Journal of Geophysical Research*, **97**: 12699–12706.

Ekman, M., 1999. Climate changes detected through the world's longest sea level series. *Global and Planetary Change*, **21**: 215–224.

Ezer, T., Mellor, G.L., and Greatbatch, R.J., 1995. On the interpentadal variability of the North Atlantic Ocean: model simulated changes in transport, meridional heat flux and coastal sea level between 1955–1959 and 1970–1974. *Journal of Geophysical Research*, **100**: 10559–10566.

Gehrels, W.R., 1999. Middle and late Holocene sea-level changes in eastern Maine reconstructed from foraminiferal salt marsh stratigraphy and AMS 14C dates on basal peat. *Quaternary Research*, **52**: 350–359.

Gehrels, W.R., 2000. Using foraminiferal transfer functions to produce high-resolution sea-level records from salt marsh deposits. *The Holocene*, **10**: 367–376.

Gehrels, R., Belknap, D., Black, S., and Newnham, R., 2000. Rapid recent sea-level rise in the Gulf of Maine unprecedented in the past millennium. *EOS Supplement, American Geophysical Union Transactions*, **81**: S275.

Kaye, C.A., and Barghoorn, E., 1964. Late Quaternary sea-level change and crystal rise at Boston, Massachusetts, with notes on the auto-compaction of peat. *Geological Society of America Bulletin*, **75**: 63–80.

Kushnir, Y., 1994. Interdecadal variations in North Atlantic sea surface temperature and associated atmospheric conditions. *Journal of Climate*, **7**: 141–157.

Long, A., 2000. Late Holocene sea-level change and climate. *Progress in Physical Geography*, **24**: 415–423.

Maul, G.A., and Martin, D.M., 1993. Sea level rise at Key West, Florida, 1846–1992: America's longest instrument record? *Geophysical Research Letters*, **20**: 1955–1958.
Mörner, N.-A., 1973. Eustatic changes during the last 300 years. *Palaeogeography, Palaeoclimatology, Palaeoecology*, **13**: 1–14.
Nydick, K.R., Bidwell, A.B., Thomas, E., and Varekamp, J.C., 1995. A sea-level rise curve from Guilford, Connecticut, USA. *Marine Geology*, **124**: 137–159.
Scott, D.B., and Medioli, F.S., 1978. Vertical zonations of marsh foraminifera as accurate indicators of former sea-levels. *Nature*, **272**: 528–531.
van de Plassche, O., 2000. North Atlantic climate-ocean variations and sea level in Long Island Sound, Connecticut, since 500 cal yr AD. *Quaternary Research*, **53**: 89–97.
van de Plassche, O., van der Borg, K., and de Jong, A.F.M., 1998. Sea level-climate correlation during the past 1400 yr. *Geology*, **26**: 319–322.
Varekamp, J.C., Thomas, E., and van de Plassche, O., 1992. Relative sea-level rise and climate change over the last 1500 years. *Terra Nova*, **4**: 293–304.
Varekamp, J.C., and Thomas, E., 1998. Climate change and the rise and fall of sea level over the millennium. *Eos, Transactions of the American Geophysical Union*, **79**(69): 74–75.
Woodworth, P.L., 1990. A search for accelerations in records of European mean sea level. *International Journal of Climatology*, **10**: 129–143.
Woodworth, P.L., 1999. High waters at Liverpool since 1768: the UK's longest sea level record. *Geophysical Research Letters*, **26**: 1589–1592.

Cross-references

Changing Sea Levels
Geochronology
Global Warming (see Greenhouse Effect and Global Warming)
Holocene Epoch
Peat
Salt Marsh
Sea-Level Indicators, Biological in Depositional Sequences
Tide Gauges

SEA-LEVEL DATUMS—See TIDAL DATUMS

SEA-LEVEL INDICATORS, BIOLOGIC

Use of subfossil fixed biological remains as Biological Mean Sea-level Indicators (BMSIs), (Laborel and Laborel-Deguen, 1994) or Fixed Biological Indicators (FBI) (Baker and Haworth, 1999) were initiated about 40 years ago (Donner, 1959; Van Andel and Laborel, 1964; Fevret and Sanlaville, 1966) and have gained recent impetus as the study of sea-level variations developed and took into account pluridisciplinary criteria. Use of FBI allowed reliable monitoring of recent sea-level variations along rocky coasts, stable or seismically active (Brazil, West Africa, Mediterranean, Japan, Australia).

Principle

On rocky shores, littoral fauna and vegetation currently develop in horizontal belts parallel to the water surface (Stephenson and Stephenson, 1949; Peres and Picard, 1964) which define several zones, where various eroding and building biological factors are at work (Figure S11). These zones are:

A *littoral fringe* (Stephenson and Stephenson, 1949) or *supralittoral* zone (Peres and Picard, 1964), wetted by surf where endolithic Cyanobacteria are dominating.

A *midlittoral zone* (Peres and Picard, 1964), submersed at regular intervals by waves or tides, where parallel vegetational belts are more developed. Eroding Cyanobacteria, patellaceous gastropods (limpets), and chitons (lower zone) are abundant. Rock-building agents (coralline rhodophytes) also occur.

An *infralittoral (sublittoral) zone*, from mean sea level (MSL) down to 25–35 m whose upper part bears brown algae, coralline rhodophytes, vermetid gastropods, oysters, annelids, cirripeds, and eroding agents such as clionid sponges, sea-urchins and rock-boring pelecypods (*Lithophaga*). A few species are restricted to the upper margin of that zone, but many display a clear-cut population limit at that level. Some

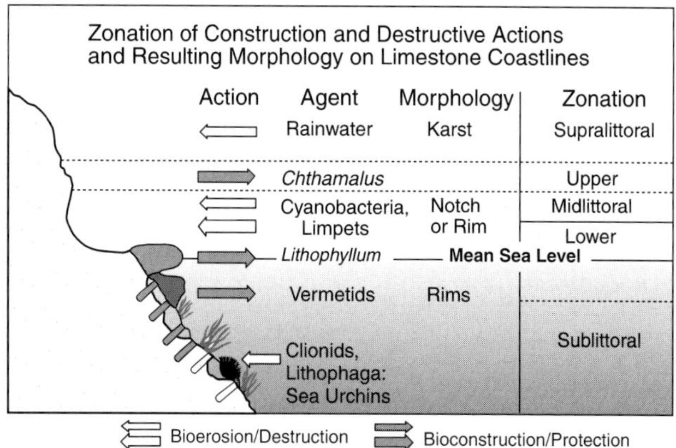

Figure S11 Division of erosion and construction on a vertical profile on limestones in temperate seas (modified from Laborel and Laborel-Deguen, 1994).

organisms build various reef or reef-like structures (bioherms, biostromes), or develop as an erosion-protecting cover.

On limestone coasts, the balance of bioerosion versus bioconstruction leads to various types of vertical profiles (Guilcher, 1953) such as "tidal" notches and horizontal "benches" or "tidal platforms" on soft rocks .

Definition of a Biological Mean Sea Level (BMSL)

The limit between the midlittoral and infralittoral zones defines a (BMSL) marked by a large and sudden increase in species diversity (Boudouresque, 1971). It also corresponds to such morphological features as the vertex of "tidal" notches or the inner edge of erosion "tidal" platforms (Focke, 1978; Pirazzoli, 1986).

Principal groups of plants and animals used as FBI

A small group of fixed plant and animal species such as the corallines *Lithophyllum lynoides* and *Lithophyllum onkodes*, vermetid gastropods of genus *Dendropoma* and *Spiroglyphus* (Laborel, 1986) and annelids such as *Idanthyrsus* and *Galeolaria* (Baker and Haworth, 1999) enjoy very narrow depth ranges located at that limit (or a little above or below) so they are currently used as FBI and considered to be among the most reliable indicators. They are all the more interesting since they often develop into algal rims, cornices, and other reef-like bioherms whose erosion-resistant remains are easy to spot and sample. Most other species have a wider vertical range but may be successfully used as FBI taking into account the upper limit of their populations (for infralittoral cirrhipeds, like *Balanus*, largely used in the archaeological study of ancient harbors (Morhange *et al.*, 1996), and *Lithophaga* holes). Scleractinian corals which develop in the infralittoral (sublittoral) zone belong to this category and the upper limit of coral construction is widely used as FBI in tropical waters (Hopley, 1986). The use of coral indicators, generally sampled by coring or drilling methods, has generated a large specific bibliography of its own that we shall not develop here.

For midlittoral species such as the small barnacles *Chthamalus* or *Elminius*, or some coralline rhodophytes, it is the lower limit of the population which is taken into account.

Population limits inside both midlittoral and sublittoral zones are considered as stable when MSL is constant. Aperiodic or seasonal short variations of sea level have little or no influence since most species have a long life and integrate sea-level variations on a yearly scale.

Field use of FBI

The FBI must provide information upon the direction and rapidity of past displacements of MSL and allow easy and accurate radiometric dating.

The elevation or submersion of a displaced coastline is defined as the altitudinal difference between the upper limit of the displaced remains and that of their present homologs, measured with the local BMSL as datum (Figure S12). No special reference (observed or calculated) to the actual water level is therefore necessary.

For species with a wide vertical range, best results are obtained when the uppermost limits of both fossil and living populations are well delineated. It must also be noted that biological sea-level marks are not

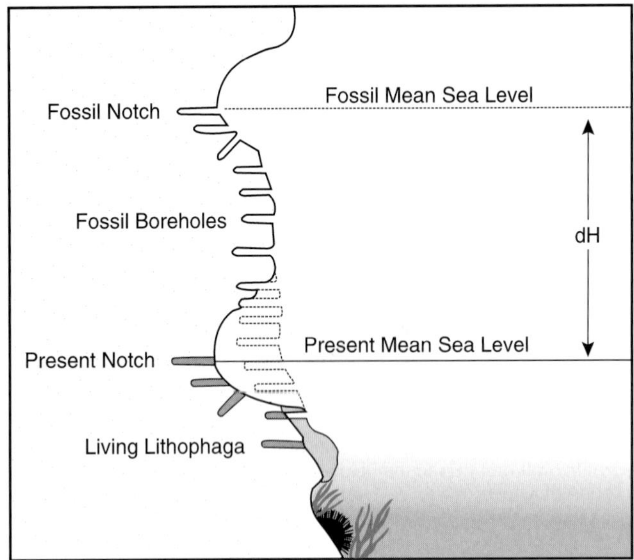

Figure S12 Measurement for the elevation of an ancient sea level (modified from Laborel and Laborel-Deguen, 1994).

perfect horizontal lines but may be warped, even on short distances, by local variations of hydrodynamism (Laborel and Laborel-Deguen, 1994). Measurements must then be done on a single vertical profile including both the fossil specimen and its present equivalent. Selecting a unique temporary bench mark for several measurements (Jardine, 1986), even at a horizontal distance of a few meters is therefore not recommended.

Discussion

An accuracy of ±5 cm was obtained in Crete (Thommeret *et al.*, 1981) for vermetid rims. In the western Mediterranean, a vertical accuracy of about ±10 to 20 cm is common on submerged lines of *L. lichenoides* (Laborel *et al.*, 1994). Lower accuracies (about ±50 cm or less) were obtained for Brazilian vermetids (Delibrias and Laborel, 1971) or for *Chthamalus* in surf-beaten crevices.

When sea level is falling slowly (about a few millimeters per year), the frailer sublittoral FBI species are killed, and their skeletons are eroded after a few years in the midlittoral zone. Reeflike structures, being stronger, may still be used as sea-level markers after long periods of midlittoral erosion. Preservation of frail details indicates rapid (even very rapid!) uplift (Thommeret *et al.*, 1981).

Determination of the velocity of a submergence movement is sometimes difficult but submersed *Lithophyllum* or vermetid rims may be preserved underwater for long periods of time.

The limits of the FBI method have been tested by comparison of field observation and direct sea-level measurement in an area of high volcanic activity (Morhange *et al.*, 1998). They proved to be reliable and sensitive in most cases, with the exception of short-lived oscillations of sea level (less than a few years) which are too rapid to be registered by biological growth.

Radiocarbon dating of aragonitic shells living in agitated surface sea water, is generally accurate. Calcareous algal bioherms can also be dated notwithstanding inner matrix and micritic cements. Direct isotopic counting methods make dating possible for small limestone fragments provided samples have suffered no contamination by alien carbon and the regional reservoir effect correction for sea water is known. Careful selection and cleaning of samples is always necessary.

Jacques Laborel and Françoise Laborel-Deguen

Bibliography

Van Andel, T., and Laborel, J., 1964. Recent high relative sea-level stand near Recife, Brazil. *Science*, **145**(3632): 580–581.

Baker, R.G.V., and Haworth, R.J., 1999. Evidence for the nature of late Holocene sea level fall on the New South Wales coast from fixed biological indicators: was the fall smooth or fluctuating. In Kesby, J.A., Stanely, J.M., McLean, R.F., and Olive, L.J. (eds.), *Geodiversity Readings in Australian Geography at the Close of the 20th Century*. Special Publications, 6, Canberra: Institute of Australian Geographers, pp. 351–360.

Boudouresque, C.F., 1971. Contribution à l'étude phytosociologique des peuplements algaux des côtes varoises. *Vegetatio*, **22**(1–3): 83–184.

Delibrias, G., and Laborel, J., 1971. Recent variations of the sea-level along the Brazilian coast. *Quaternaria*, **14**: 45–49.

Donner, J.J., 1959. The Late- and Post-glacial raised beaches in Scotland. *Annales Academiae Scientiarum Fennicae*, A, III, 53: 25pp.

Fevret, M., and Sanlaville, P., 1966. L'utilisation des Vermets dans la détermination des anciens niveaux marins. *Méditerranée*, **4**: 357–364.

Focke, J.W., 1978. Limestone cliff morphology on Curaçao (Netherlands Antilles), with special attention to the origin of notches and vermetid coralline algal surf benches. *Zeitschrift für Geomorphologie*, **22**: 329–349.

Guilcher, A., 1953. Essai sur la zonation et la distribution des formes littorales de dissolution du calcaire. *Annales de Géographie*, **331**: 161–179.

Hopley, D., 1986. Corals and reefs as indicators of paleo-sea-levels with special reference to the Great barrier Reef. In Van de Plaasche, O. (ed.), *Sea-Level Research: A Manual for the Collection and Evaluation of Data*. Volume 8. Norwich: Geo Books, pp. 195–228.

Jardine, W.G., 1986. Determination of altitude. In Van de Plaasche, O., (ed.), *Sea-Level Research: A Manual for the Collection and Evaluation of Data*. Volume 21. Norwich: Geo Books, pp. 569–590.

Laborel, J., 1986. Vermetids. In Orson van de Plassche (ed.), *Sea-Level Research: A Manual for the Collection and Evaluation of Data*, Volume 12. Norwich: Geo Books, pp. 281–310.

Laborel, J., and Laborel-Deguen, F., 1994. Biological indicators of relative sea-level variation and of co-seismic displacements in the Mediterranean area. *Journal of Coastal Research*, **10**(2): 395–415.

Laborel, J., Morhange, C., Lafont, R., Le Campion, J., Laborel-Deguen, F., and Sartoretto, S., 1994. Biological evidence of sea-level rise during the last 4500 years on the rocky coasts of continental southwestern France and Corsica. *Marine Geology*, **120**: 203–223.

Morhange, C., Laborel, J., Hesnard, A., and Prone, A., 1996. Variation of relative mean sea level during 2500 years on the northern shores of the Lacydon, the ancient harbour of Marseilles. *Journal of Coastal Research*, **12**: 841–849.

Morhange, C., Laborel, J., and Laborel-Deguen, F., 1998. Précision des mesures de variation relative verticale du niveau marin à partir d'indicateurs biologiques. Le cas des soulèvements bradysismiques de Pouzzoles, Italie du sud (1969–1972; 1982–1984). *Zeitschrift für Geomorphologie*, **42**(2): 143–157.

Peres, J.M., and Picard, J., 1964. Nouveau manuel de bionomie benthique en Mediterranée. *Recueil des travaux de la Station Marine d' Endoume*, **31**(47): 1–131.

Pirazzoli, P.A., 1986. Marine notches. In van de Plaasche, O. (ed.), *Sea-Level Research: A Manual for the Collection and Evaluation of Data*. Volume 12, Norwich: Geo Books, pp. 361–400.

Stephenson, T.A., and Stephenson, A., 1949. The universal features of zonation between tide-marks on rocky coasts. *Journal of Ecology*, **37**: 289–305.

Thommeret, Y., Thommeret, J., Laborel, J., Montaggioni, L.F., and Pirazzoli, P.A., 1981. Late Holocene shoreline changes and seismotectonic displacements in western Crete (Greece). *Zeitschrift für Geomorphologie, N.F., Bd.*, **40**(Suppl.): 127–149.

Cross-references

Algal Rims
Bioconstruction
Bioerosion
Bioherms and Biostromes
Coral Reefs
Littoral
Notches
Sea-Level Indicators, Biological in Depositional Sequences
Sea-Level Indicators, Geomorphic
Shore Platforms

SEA-LEVEL INDICATORS—BIOLOGICAL IN DEPOSITIONAL SEQUENCES

Introduction

Biological sea-level indicators have been discussed for *rocky intertidal coasts* but this section deals with biologic remains in a depositional sequence such as a marsh or *estuarine* (*q.v.*) deposit. Often the lithology will not provide sufficient information to delineate an accurate sea level from within a depositional sequence but biological remains in the form of fossils will greatly enhance the accuracy. Any good sea-level indicator must provide three essential elements: (1) accuracy, (2) preservability, and (3) be datable. Fossils within depositional sequences have the highest probability to provide all three of these elements. The type of fossil is very important because not all fossils will provide an accurate relocation of a former *sea level* (*q.v.*). The accuracy of a given fossil type depends mostly on the range of water depths that the organism occupies when living and it must be *in situ* when found in a sequence. However, even fossils with a large range may be useful to determine marine/freshwater transitions which may in turn sometimes be used as a sea-level indicator. There is a large range of organisms that serve as indicators; probably the most commonly used historically are "miscellaneous shells" which are not very accurate because they often are not in place and contain a large array of species with a broad band of water depth ranges. Some specific macro invertebrates such as mussels or oysters provide a range within ±2 m but this is not adequate for determining the smaller movements of the late *Holocene* (*q.v.*). By far, the most useful are microfossils because they occur in large numbers in small diameter cores which are often where the sea-level records come from in Holocene *submerged coastlines* (*q.v.*). In the following sections the various groups will be detailed in terms of where and how to use the various groups as well as a brief assessment of the accuracy possible with each group.

Plant groups

Marsh vegetation generally

There are many cases in the literature where sea levels have been reported based on either "undifferentiated *peat*" (*q.v.*), meaning the authors did not know if it was freshwater or marine (Emery and Garrison, 1967); in this case the only information that can be derived is that sea level was at about the depth (±5 m) of the deposit at the time of formation or lower if it was freshwater. This determination is useful in the absence of any other data but not useful for the problems of measuring late Holocene movements which are well within the error bars mentioned above. Sometimes it is known that the peat is a salt marsh deposit in which case it narrows the range to the upper half of the *tidal cycle* (Chapman, 1960, 1976). When one steps on a salt marsh it is visually obvious that there is a distinct plant zonation but the plants respond to more than just tidal exposure and often will have differing ranges even in adjacent marshes (Scott *et al.*, 1981, 1988) so the macro-vegetation alone is not a reliable indicator for subdividing salt marsh deposits in terms of sea level, especially in fossil peats.

On a broader scale macro-plant remains have been used to suggest trends of salinity change and hence *transgression* or *regression* (*q.v.*) but there are many factors that have to be considered since plants respond to many variables (Behre, 1986).

Many have also used buried forests as indicators of submergence (Heyworth, 1986) and again this gives only the indication of the sea level being below the forest at the time it was living. This is a very powerful tool, however, when looking at rapid changes in sea level such as earthquakes (Atwater, 1987) where the repeated submergence of a series of forests provides a sense of the periodicity of these events.

Microfossil plant remains

Apart from the variability of vegetation assemblages across the marsh, there is the problem of identifying macro-plant remains in a peat deposit since they tend to breakdown rapidly in the subsurface. However, some attempts have been made to use pollen which of course is the microscopic part of the sexual reproductive organs of all angiosperm plants. This meets with some success (Shennan *et al.*, 1998) but pollen is inherently reworked. Probably, more successful is the use of diatoms which are one-celled algae which leave a siliceous shell in the sediments (Shennan *et al.*, 1998, 1999). These fossils appear to leave a record that can partially subdivide the marsh into zones with varying accuracies (±20–50 cm at best).

The most useful technique involving microfossil plants is determining sea levels in an indirect way where *coastal ponds* either become submerged or emerged (Palmer and Abbott, 1986; Shennan *et al.*, 1996). If they are emerged they go from marine to freshwater and *vice versa* for submergence; the key is to be able to determine the sill depth of the basin and then relate that to the radiocarbon-dated transition. This method is usually accurate to within ±1 m depending on the depth and size of the basin (Laidler and Scott, 1996). Either diatoms (Palmer and Abbott, 1986) or dinoflagellates (Miller *et al.*, 1982) can be used in this manner. Dinoflagellates are also microscopic algae but they are organic-walled as opposed to the siliceous shells of diatoms. Both of these groups occur in large numbers such that 1 cm^3 of wet sediment is often sufficient to obtain a valid result (Haq and Boersma, 1978).

Animal groups

Macro-animal fossils

As discussed briefly above invertebrate macrofossils are limited in their value as sea-level indicators by their vertical range in the water column. Some groups such as *corals* (*q.v.*) or attached biological indicators have extremely narrow ranges and have been used to produce some of the best and longest sea-level curves (Fairbanks, 1989). However, these are limited in occurrence and rocky intertidal forms are often not well preserved, especially in a submergent regime. Other macro-invertebrate groups have been used in depositional sequences but their vertical ranges are usually quite high and hence accuracy is not good, especially the above-mentioned "miscellaneous shells." Peterson (1986) details the use of marine molluscs which are by far the most widely used macrofossil in Quaternary studies of raised marine deposits. He emphasizes the use of communities rather than single species to relocate sea level which is generally true for all organisms. Of all the species of molluscs perhaps *Mytilus* (the blue mussel) provides the highest accuracy (5 m) but even that range is high. However, in the raised deposits of northern Europe it is still very useful. However, as is the case with macro-vegetation, there is often insufficient material in small diameter cores that are often the basis for building sea-level curves, especially Late Holocene curves on submergent coastlines.

Micro-animal fossils

There are two principal groups of animal microfossils that have been used extensively in sea-level studies, ostracodes (van Harten, 1986) and foraminifera (Scott and Medioli, 1986). Ostracodes leave a calcareous shell as a fossil and that in itself presents a problem because many of the best deposits for sea-level studies are not conducive to the preservation of $CaCO_3$ hence the fossils are not present. van Harten (1986) suggests the resolution with ostracodes to be within 100 m which is not useful at all for modern sea-level studies but they can be used as accurate salinity tracers and hence suggest transgression and regression (Haq and Boersma, 1978).

On the other hand, foraminifera have both calcareous and agglutinated shells, agglutinated shells are resistant to dissolution in low pH, highly organic sediments, and often are very abundant in some highly organic deposits such as marshes (>5,000/10 cm^3, Scott and Medioli, 1980). Unlike plants foraminifera have been shown to be consistent in their vertical range in relation to tidal levels within a marsh sequence (Scott and Medioli, 1978, 1980, 1986) such that the same 8–10 species inhabit the world's marshes and specific assemblages can always locate the upper one-fourth of the tidal range. In some cases the higher high level can be accurately located to provide an incredible accuracy of ±5 cm (Scott and Medioli, 1978, 1980, 1986; Hayward *et al.*, 1999); this is by far the most accurate indicator now available but it is limited to coastal marsh areas and isolated peat deposits that are sometimes found offshore in marine surveys (e.g., Scott and Medioli, 1982). At the very least "undifferentiated peats" can now at least be determined to be either marine or freshwater which significantly increases their value as a sea-level indicator.

Archaeological remains

Although many assumptions must be made to use paleo-human occupation sites as sea-level indicators they have been used by many workers, especially on emergent coasts (Colquhoun and Brookes, 1986; Martin *et al.*, 1986). The main assumption is that paleo-humans did not carry their food that they gathered from coastal *estuaries* and *lagoons* (*q.v.*) far from the coastline. The most commonly used remains are shell middens or dumps of usually molluscs that are exposed in many areas but are probably the most spectacular on the South American east coasts of Brazil and Argentina (Martin *et al.*, 1986). Colquhoun and Brookes (1986) used a combination of shell middens and peats to reconstruct sea levels on the South Carolina (USA) coast.

Summary

The above is a small sampling of the most commonly used biological remains in depositonal sequences for determining sea levels. It is clear that some areas will provide a higher probability of determining an accurate and extended sea-level record than others and in some areas it may be impossible to obtain a sea-level record of any kind. Biological remains provide the best means of determining former sea levels because they are usually in a depositional sequence and they usually can supply the carbon required to obtain a ^{14}C date, without which you have a level but no way of knowing when the sea level was at that point. The key to determining accurate sea levels is being versatile in the approach taken to take advantage of what the record provides.

David B. Scott and Franco S. Medioli

Bibliography

Atwater, B.T., 1987. Evidence for great Holocene earthquakes along the outer coast of Washington State. *Science*, **236**: 942–944.

Behre, K.-E., 1986. Analysis of botanical macro-remains. In van de Plassche, O. (ed.), *Sea-level Research: A Manual for the Collection and Evaluation of data*. Norwich, UK: GEO Books, pp. 413–434.

Chapman, V.J., 1960. *Salt Marshes and Salt Deserts of the World*. London: Leonard Hill LTP.

Chapman, V.J., 1976. *Coastal Vegetation*, 2nd edn. Toronto: Pergamon Press.

Colquhoun, D.J., and Brookes, M.J., 1986. New evidence from the southeastern United States for eustatic components of late Holocene sea levels. *Geoarcheology*, **1**(3): 275–291.

Emery, K.O., and Garrison, L.E., 1967. Sea levels 7,000 to 20,000 years ago. *Science*, **157**: 684–687.

Fairbanks, R.G., 1989. A 17,000-year glacio-eustatic sea-level record: influence of glacial melting rates in the Younger Dryas event and deep ocean circulation. *Nature*, **342**: 637–642.

Haq, B.U., and Boersma, A., 1978. *Introduction to Marine Micropaleontology*. New York, Elsevier.

Hayward, B.W., Grenfell, H.R., and Scott, D.B., 1999. Tidal range of marsh foraminifera for determining former sea-level heights in New Zealand. *New Zealand Journal of Geology and Geophysics*, **42**: 395–413.

Heyworth, A., 1986. Submerged forests as sea-level indicators. In van de Plassche, O. (ed.), *Sea-level Research: A Manual for the Collection and Evaluation of data*. Norwich, UK: GEO Books, pp. 401–412.

Laidler, R.B., and Scott, D.B., 1996. Foraminifera and Arcellacea from Porters Lake, Nova Scotia: modern distribution and paleodistribution. *Canadian Journal of Earth Sciences*, **33**: 1410–1427.

Martin, L., Suguio, K., and Flexor, J.-M., 1986. Shell middens as a source for additional information in Holocene shoreline and sea-level reconstruction: examples from the coast of Brazil. In van de Plassche, O. (ed.), *Sea-level Research: A Manual for the Collection and Evaluation of Data*, Norwich, UK: GEO Books, pp. 503–524.

Miller, A.A.L., Mudie, P.J., and Scott, D.B., 1982. Holocene history of Bedford Basin, Nova Scotia: foraminifera, dinoflagellate and pollen records. *Canadian Journal of Earth Sciences*, **19**: 2342–2367.

Ortlieb, L., Barreintos, S., and Guzman, N., 1996. Coseismic coastal uplift and coralline algae record in northern Chile: the 1995 Antofagasta earthquake case. *Quaternary Science Reviews*, **15**(8–9): 949–960.

Palmer, A.J.M., and Abbott, W.H., 1986. Diatoms as indicators of sea-level change. In van de Plassche, O. (ed.), *Sea-level Research: A Manual for the Collection and Evaluation of Data*. Norwich, UK: GEO Books, pp. 457–489.

Peterson, K.S., 1986. Marine molluscs as indicators of former sea-level stands. In van de Plassche, O. (ed.), *Sea-level Research: A Manual for the Collection and Evaluation of Data*, Norwich, UK: GEO Books, pp. 129–156.

Scott, D.B., and Collins, E.S., 1996. Late mid-Holocene sea-level oscillation: a possible cause. *Quaternary Science Reviews*, **15**: 851–856.

Scott, D.B. and Medioli, F.S., 1978. Vertical zonations of marsh foraminifera as accurate indicators of former sea-levels. *Nature*, **272**: 528–531.

Scott, D.B., and Medioli, F.S., 1980. Quantitative studies of marsh foraminiferal distributions in Nova Scotia: their implications for the study of sea-level changes. *Cushman Foundation for Foraminiferal Research*, Special Publication 17, 58p.

Scott, D.B., and Medioli, F.S., 1982. Micropaleontological documentation for early Holocene relative sea level fall on the Atlantic coast of Nova Scotia. *Geology*, **10**: 278–281.

Scott, D.B., and Medioli, F.S., 1986. Foraminifera as sea-level indicators. In van de Plassche, O. (ed.), *Sea-level Research: A Manual for the Collection and Evaluation of Data*. Norwich, UK: GEO Books, pp. 435–456.

Scott, D.B., Williamson, M.A., and Duffett, T.E., 1981. Marsh foraminifera of Prince Edward Island: their recent distribution and application for former sea-level studies. *Maritime Sediments and Atlantic Geology*, **17**: 98–124.

Scott, D.B., MacKinnon, K.D., and Medioli, F.S., 1988. Salt marshes in Atlantic Canada. In *Wetlands of Canada*. Ottawa, National Wetlands Working Group and Canada Committee on Ecological Land Classification, pp. 265–282.

Shennan, I.A., Green, F., Innes, J., Lloyd, J., Rutherford, M., and Walker, K., 1996. Evaluation of rapid relative sea-level changes in North West Scotland during the last glacial–interglacial transition: evidence from Ardtoe and other isolation basins. *Journal of Coastal Research*, **12**(4): 862–874.

Shennan, I., Long, A.J., Rutherford, M.M., Innes, J.B., Green, F.M., and Walker, K.J., 1998. Tidal marsh stratigraphy, sea-level change and large earthquakes II: submergence events during the last 3500 years at Netarts Bay, Oregon, USA. *Quaternary Science Reviews*, **17**: 365–393.

Shennan, I.A., Scott, D.B., Rutherford, M., and Zong, Y., 1999. Microfossil analysis of sediments representing the 1964 earthquake, exposed at Girdwood Flats, Alaska, USA. *Quaternary International*, **60**: 55–74.

Van Harten, D., 1986. Ostracode options in sea-level studies. In van de Plassche, O. (ed.), *Sea-level Research: A Manual for the Collection and Evaluation of Data*. Norwich, UK: GEO Books, pp. 489–502.

Cross-references

Bogs
Coastal Lakes and Lagoons
Coral Raefs
Estuaries
Holocene Epoch
Ingression, Regression, and Transgression
Peat
Rock Coast Processes
Sea Level Indicators—Biological in Depositional Sequences
Submerged Coasts
Tides

SEA-LEVEL INDICATORS, GEOMORPHIC

Introduction

Several geomorphic features, erosional or depositional, develop near sea level. Some of them may be preserved after a change in sea level and can be used, therefore, as indicators of former sea-level positions. Erosional indicators can be preserved only in hard rock, and occur in a vertical range which depends on site exposure. They include notches, benches, trottoirs, platforms, abrasional marine terraces, strandflats, pools, potholes, sea caves, honeycomb features, and tafoni. For accurate sea-level reconstructions (Pirazzoli, 1996), it is essential therefore to refer the elevation of a former indicator to that of the active counterpart in the same place rather than to that of the present sea level. Depositional indicators include tidal flats, marine-built shore platforms and terraces, beaches, beachrocks, reef flats, and submerged speleothems. Erosional features are generally inadequate to date former sea levels, whereas marine deposits may include guide fossils or organic material liable to be dated radiometrically.

Erosional indicators

Several types of notches can be distinguished (Van de Plassche, 1986). Abrasion notches are often found near high-tide level at the boundary between a cliff and a shore platform, but they may develop at any level reached by wave action and give therefore only approximate indications of sea-level position. Much more precise are tidal notches, which are typical midlittoral erosional features, especially on limestone coasts. In sheltered conditions tidal notches occupy all the intertidal range, showing a vertex at about mean sea level (MSL) (Figure S13).

The formation of erosional benches and shore platforms is usually ascribed to the removal by waves of the weathered parts of cliffs. The lowest level of possible weathering corresponds to that of constant soakage by sea water, probably in the intertidal zone. Wide benches of abrasional origin are called platforms or terraces; they usually develop in the intertidal range, sloping gently seawards.

Not to be confused with tidal notches are surf notches, which may occur at higher elevations on limestone coasts exposed to persistent winds and strong surf and spray action; surf notches delimit the boundary between a surf bench and a cliff. On exposed coasts, organic accretions often develop near the outer edge of a bench, protecting the substrate rock, while erosion can proceed higher than the accretion level, widening a surf bench and undercutting a surf notch on the cliff. The elevation of surf benches depends upon site exposure; elevation as high as 2 m above high tide-level has been reported locally, but it may decline rapidly with decreasing wave exposure. On the Levant coasts, in the eastern Mediterranean, surf benches (called trottoirs) are quite common at elevations of 0.2–0.4 m above MSL.

In high latitude, strandflats are low shore platforms where the rock material, disintegrated by freeze–thaw alternation, has been washed away by wave action. The reliability of erosional coastal benches, platforms and strandflats as sea-level indicators depends on the identification and understanding of the processes by which these features were produced.

Coastal pools are flat-bottomed depressions resulting from local lowering of limestone benches. On sheltered shores they correspond to intertidal levels, whereas on exposed coasts they may occur in any supralittoral area in the reach of waves. Potholes are rounded depressions, usually deeper than wide, worn into solid rock by sand, gravel, pebbles, or boulders spun round by the force of waves. As they may be formed at various elevations, above or below sea level, they are not accurate sea-level indicators.

Sea caves are cavities excavated by erosion into a cliff in the range of wave action. They generally develop into weaker parts of the rock formation or are often of karstic origin. A sea cave open on both sides of a promontory is called a sea arch. Sea caves and arches are generally inaccurate sea-level indicators; however, in limestone formations, their floor, if regular and flat, may be related to a former low-tide position.

Honeycombs are small cavernous features, showing a cell-like structure, which may form on the surface of several granular rocks in the spray zone above high tide. Their development may be ascribed to temperature variations, chemical weathering, and wind corrasion. Honeycomb structures indicate the proximity of sea level below them, but with poor accuracy. Tafoni are cavernous weathering features of greater size (from a decimeter to several meters). Because they may occur also far away from the shore, tafoni should be avoided as sea-level indicators.

In uplifted areas of former ice sheets, low-tide positions corresponding to the marine limit (maximum recorded sea-level elevation) can be identified at the elevation where channels disappear from the surface of former glacio-marine deltas.

Depositional indicators

The grain size of marine deposits depends mainly on the energy of the water environment, which varies with exposure to waves and currents. Marine muds and clays can be deposited only in very calm water, either offshore at depth, or in very sheltered coastal basins. Tidal flats are marshy or muddy land areas which are covered and uncovered by the rise and fall of the tide. They are usually studied by analyzing samples collected using boring techniques. The upper part of tidal flats, which may reach but not exceed the elevation of extreme high tides, is usually very flat and colonized by halophytic vegetation. Flora and fauna still in growing position, and the local tidal range and topographic configuration, will often help to indicate the kind of environment in which they formed and the depth of their deposition. Well-chosen foraminiferal assemblages from coastal paleomarsh deposits may permit very precise (up to ± 10 cm) determinations of former sea level even in macrotidal areas. Botanical remains can also contribute to sea-level reconstructions, by determining the degree of salinity and therefore the position relative to mean high-water level. Assemblages of freshwater, brackish, and marine microflora (diatoms) can be used to infer marine transgressions and regressions (Van de Plassche, 1986).

The main difficulties in reconstructing former sea-level histories from tidal-flat deposits are given by the estimation of sediment compaction and tidal changes. Wet sediments have a low density. After their deposition, compaction is rapid during the first centuries, then decreases gradually. In the Mississippi Delta, for example, contemporary silty muds contain water to 70% of their volume. At a burial depth of 24 m there is about 15% loss by compaction and at about 1,000 m apparent subsidence may approach 70%. The highest rates of compaction correspond

Figure S13 Emerged tidal notches cut into a mushroom rock and, in the background, an emerged beachrock capped by green living algae, are consistent with the present-day low-tide situation and indicate recent stability in the relative sea level. Miyako Island (the Ryukyus, Japan). Local mean spring tidal range is 1.5 m; water level is 0.4 m below MSL (photo 6149, Feb. 1981).

to peat layers, which can reduce in volume as much as 90%. To survive, a tidal flat needs therefore rates of sedimentation high enough to compensate compaction phenomena. If sedimentation is inadequate, the tidal flat will be submerged; if it is excessive, its marginal, sheltered basin will be rapidly infilled. Tide characteristics depend narrowly upon the basin morphology, which is continually modified by sedimentation and compaction and eventually by sea-level changes. For example, in the Bay of Fundy, during the last 4,000 years, the high-tide level has increased 1.5 mm/yr faster than MSL.

Sand deposits are common where wave energy is moderate and are found on beaches and intertidal shores near lagoon entrances. Pebbles are generally found only in high-energy shores, usually on exposed beaches. Raised or submerged beaches are however, seldom used as sea-level indicators because of the wide uncertainties in determining a clear relationship between a beach sample and the MSL. Depositional shore platforms produced by the supply of marine sediments (sand, shingle, or shells) to the shore have been described by Zenkovich (1967) and called improperly "shore terraces"; broad coastal plains characterized by many low shelly ridges that run roughly parallel to the shore, like the "cheniers" in coastal Louisiana (Shepard and Wanless, 1971), also belong to the depositional shore platforms category. Their sea-level indications are similar to those provided by beaches.

Marine-built terraces develop near sea level and are made of accumulations of marine materials removed by shore erosion. Their landward part may be capped by alluvial deposits.

The term beachrock applies to beach sediments cemented by calcium carbonate in the intertidal zone (often in the upper part of it). Beachrocks are generally limited to areas of warm or temperate waters. Their upper surface usually shows a superimposition of beachrock slabs, each showing the same seaward slope as the nearby beach, delimited landward by basset edges. A beachrock is generally a good indicator of sea level, with a vertical uncertainty depending on the local tidal range. However, radiometric ages of beachrock samples (shells, coral debris) would provide only maximum dates for the sea level at which lithification occurred, because beach sediments may be reworked.

Reef flats are stony expanses of reef rock with a flat surface. They are generally situated in the lower part of the tidal range and capped by calcareous algae, patches of sand and coral debris, and occasionally a few coral colonies living in shallow pools or depressions. In the absence of moating phenomena, the uppermost corals in growth position, often appearing as microatolls, correspond to the spring low-tide level. When the reef flat surface is made of coral conglomerate, its elevation may be at any level in the intertidal range or in the reach of regular waves. The low-tide level at the time of the conglomerate lithification can be determined by petrological analysis as the vertical boundary between the former marine phreatic and marine vadose environments for the first generation of intergranular cements (Montaggioni and Pirazzoli, 1984).

Speleothems, such as stalactites or stalagmites, are usually formed in caves above sea level by the evaporation of mineral-rich water. When submerged in sea-flooded karst systems, the depth of speleothems indicates minimum sea-level rise. In the fortunate case of a marine biogenic cover capping the speleothem, both the marine bioconstruction and the speleothem deposits can be dated radiometrically.

In glacio-isostatically uplifted areas, the highest marks of marine action may be identified from the lower limit of perched glacial boulders, till and continuous terrestrial deposition, or from the upper limit of shore deposits, beach ridges, and *in situ* marine fossils.

Lastly on the continental shelves, seismic stratigraphy gives a picture of the way the rock strata have been piled on one another. Each transgressive–regressive cycle removes some sediment and leaves piles of new sediment, producing recognizable patterns such as erosional unconformities. These features can be used to identify the best places to collect borings to interpret sea-level changes. In the laboratory, diagenetic products and intergranular cementation identified by analysis of depositional material may also provide evidence of former marine zonation related to sea-level positions.

Paolo A. Pirazzoli

Bibliography

Montaggioni, L.F., and Pirazzoli, P.A., 1984. The significance of exposed coral conglomerates from French Polynesia (Pacific Ocean) as indicators of recent relative sea-level changes. *Coral Reefs*, **3**: 29–42.

Pirazzoli, P.A., 1996. *Sea-Level Changes: The Last 20000 Years*. Chichester: Wiley.

Shepard, F.P., and Wanless, H.R., 1971. *Our Changing Coastlines*. New York: McGraw-Hill.

Van de Plassche, O. (ed.), 1986. *Sea-Level Research: A Manual for the Collection and Evaluation of Data*. Geo Books: Norwich.

Zenkovich, V.P., 1967. *Processes of Coastal Development*. Edinburgh: Oliver & Boyd.

Cross- references

Beach Features
Beachrock
Cheniers
Marine Terraces
Muddy Coasts
Notches
Peat
Salt Marsh
Sequence Stratigraphy
Shore Platforms
Strandflat
Tidal Flats
Trottoirs

SEA-LEVEL RISE, EFFECT

Rising sea level is gradually inundating wetlands and lowlands, eroding beaches, exacerbating coastal flooding, raising water tables, and increasing the salinity of rivers, bays, and aquifers. In coastal areas where the human influence is slight, these effects are usually unimportant, as ecosystems and geological systems simply shift upward and inland with the rising water levels. Human systems, however, do not move inland so readily. As a result, the advancing sea and human activities are on a collision course in many developed areas. Caught in the middle, are the intertidal marshes, mangroves, beaches, and mudflats, as well as the plants and animals, that depend on them for their existence.

All of these effects may be accelerated in the coming decades if global warming accelerates the rate of sea-level rise. Global sea level rose approximately 18 cm in the last century, but warmer temperatures could increase that rise to 30–100 cm (see *Changing Sea Levels*).

Physical effects of sea-level rise

Coastline retreat: inundation and erosion

The most obvious implication of a rise in sea level is that land that is barely above sea level would be below sea level, and hence would be directly inundated if no measures were taken to hold back the sea. The easiest way to get a rough sense of the vulnerability to a large rise in sea level is to examine a topographic map to determine which areas are close to sea level. The first published assessment of the impacts of global warming on the US coast simply obtained estimates of the land below the 4.5 m (15-foot) contour, and assumed that those areas would be completely lost if the sea were to rise 4.5 m (Schneider and Chen, 1980).

Land elevations do not tell the whole story. Erosion can cause relatively high ground to be lost even if the sea does not rise enough to inundate it; and wetland accretion can allow marshes and swamps to persist even when the sea rises more than enough to inundate them. Moreover, depending upon the tide range in a given area, land whose elevation is greater than the projected sea-level rise might still be flooded by the tides. Finally, human activities can hold back the sea.

Inundation of low areas. Although the concept of inundation is simple enough, the quality of elevation information is insufficient to reliably estimate the amount of land, for example, within 1 m of mean high water. As Table S6 shows, topographic maps in most nations lack the vertical resolution necessary to estimate the land that could be inundated by a 1-m rise in sea level, and in many nations the elevation data is so poor that interpolations are probably meaningless. In the United States, the contour interval for the widely available topographic maps varies, generally from 1.5 to 6 m (5–20 ft). For much of the coast, better maps exist, but they are tucked away in government offices that created the maps for specific uses, such as flood insurance rates and the design of municipal drainage systems. A second problem is that available digital elevation data is usually much less precise than the printed topographic maps. In the United States, for example, the digital line graphs that contain the topographic information of the printed map, are available for a few states, but in most cases, the available digital elevation

Table S6 Vertical resolution of topographic maps in various nations

Nation/region	Units	Typical contour interval	Contour in areas with good coverage
Antigua	Meters	3–6	—
Argentina		None	
Bangladesh	Feet	50	1
Egypt	Meters	1	
Marshall islands	Feet	None	1
Mauritius	Meters	2	
India	Meters	20	10
Nigeria	Feet	100	
Senegal	Meters	40	5
United Kingdom	Meters	5	
Vietnam	Meters	1	
United States			
Northeast	Feet	20	10
Mid-Atlantic	Feet	10	3
Southeast/Gulf	Feet	5	2
Pacific	Feet	20–40	—

Source: Titus and Richman (2001).

model data provides modeled elevations based on interpolations between various points along a grid (Titus and Richman, 2001).

A recent study by the US Environmental Protection Agency developed maps of lands below the 1.5- and 3.5-m contours along the US Atlantic and Gulf coasts (Titus and Richman, 2001). The areas with the greatest amounts of land at risk include Florida, Louisiana, North Carolina, and the shores of Chesapeake and Delaware Bays (see Figure S14) Approximately 56,000 km^2 lie below the 1.5-m contour, while another 32,000 km^2 are found between the 1.5- and 3.5-m contours. Table S7 illustrates the current uses of the land below 1.5 m, which is about 75% wetlands, 20% farms and forests, and only 5% urban and residential. The land between 1.5 and 3.5 m, by contrast, is approximately 31% wetlands, 33% forest, 23% agricultural, and 13% urban and residential. Approximately 2,000 km^2 of low developed lands are on the bay sides of barrier islands, where people often filled wetlands with just enough sand to elevate the land above mean spring high water (Titus *et al.*, 1991). Approximately two million people reside below the 1.5 m contour.

Other nations have begun to assess vulnerability to sea-level rise. Table S8 summarizes those studies. China, United States, Bangladesh, the Netherlands, and Nigeria have the greatest amounts of very low land, in part because those nations all have major river deltas. On a percentage basis, however, coral atoll nations such as Marshall Islands, Kiribati, and Tuvalu (not shown) are the most vulnerable.

Wetlands. The majority of very low land is already inundated by high tide. Coastal marshes and swamps are generally found between the highest tide of the year and mean sea level. Because wetlands collect sediment and produce peat on which they can build, they have largely been able to keep pace with the historic rate of sea-level rise. Wetlands expanded inland as new lands were inundated, but because of sedimentation and peat formation, their seaward boundaries did not always retreat. Thus, the area of dry land just above the wetlands is less than the area of wetlands. The results in Table S7 corroborate this hypothesis: The dry land appears to be uniformly distributed at the low elevations. But the 42,000 km^2 of wetlands found below the 1.5-m contour is 30% more than the *total* area of land between 1.5 and 3.5 m . Therefore, *if* sea level rises more rapidly than wetlands can accrete, there will be a substantial net loss of wetlands (Figure S15). This loss could be further aggravated by increased wave erosion from deeper waters, and saltwater intrusion into freshwater swamps, which would tend to convert the swamps to open water.

Will wetlands be able to keep pace with sea-level rise? Published reports provide conflicting answers. As Figure S16 shows, the current rate of sea-level rise already appears to be too great for the marshes of Blackwater National Wildlife Refuge on the Eastern Shore of Chesapeake Bay in the United States, where one-third of the wetlands have eroded in the last few decades. Studies by Ellison and Stoddart (1991), Ellison (1993), Ellison and Farnsworth (1997) during the early 1990s concluded that Carribean and Bermudan mangroves could not accrete with a rise in sea level faster than about 1.5 mm/yr. Callaway *et al.* (1997) by contrast, concluded that mangroves are currently more than keeping up with higher rates of sea-level rise along the Gulf of Mexico. Parkinson *et al.* (1994) suggest that Carribean mangroves have recently shown accretion rates of 3.7 mm/yr, but that compaction could eventually reduce the net accretion to less than 2 mm/yr.

While scientific uncertainty about vertical accretion impairs our ability to forecast the landward retreat of the *seaward* boundary of coastal wetlands, economic and political uncertainty prevent us from projecting the inland advance of their *landward* boundary. If coastal areas are developed and protected from the rising sea, wetlands will not be able to advance inland, and in some cases, they will be squeezed between a rising sea and the protective structures.

Deltas are a special case. Large amounts of sediment can allow for unusually high accretion rates. At the same time, human activities that divert the sediment may prevent the sediment from reaching the wetlands. In Louisiana, flood control levees, the navigation infrastructure, and other human activities have disabled the natural processes with which the Mississippi Delta could otherwise keep pace with rising relative sea level; as a result, Louisiana is currently losing about 60 km^2 of wetlands per year (Louisiana Wetland Protection Panel, 1987). Dams along the Nile and Niger river have caused drastic erosion along the deltas in Egypt and Nigeria (Awosika *et al.*, 1992; El-Raey *et al.*, 1995). Future water management structures may have similar effects on Bangladesh, Iraq, Gambia, and other deltaic nations.

Erosion. A rise in sea level can cause an ocean beach to retreat considerably more than the retreat attributable to inundation alone (see Figure S17). The Danish coastal engineer Per Bruun (1962) demonstrated why: Beaches follow a characteristic profile determined primarily by the sediment size and the wave climate. The waves and winds regularly transport sand throughout the entire beach system, which extends from the dunes out to a depth of 10–30 m or so. Storms may erode the dunes and upper part of the beach; calm swell waves return sand back to the beach. If the sea rises 1 m, the profile will maintain its same shape with respect to the water level, which elevates the shore bottom by one 1 cm. This occurs because the swell that pushes sand from the bottom back onto the visible part of the beach can only reach so far below the surface. If the surface is 1 cm higher, the elevation down to which that swell can reach is also 1 cm higher. Hence, less sand is carried back onto the beach than would have been the case if the sea was not rising.

Bruun (1962) showed that beach profile adjustment requires the shore to retreat by an amount equal to sea-level rise divided by the average slope of the beach profile. Pure inundation, by contrast, equals sea-level rise divided by the slope of the land just above sea level. Because the beach just above sea level tends to be much steeper than the submerged portion, total coastline retreat is more than what one would expect considering only inundation. Studies applying the Bruun Rule have estimated that a 1 cm rise in sea level will generally cause beaches to erode 50–100 cm from New England to Maryland (Everts, 1985; Kyper and Sorenson, 1985), 200 cm along the Carolinas (Kana *et al.*, 1984), 100–1,000 cm along the Florida coast (Bruun, 1962), 200–400 cm along the California (Wilcoxen, 1986), 40–100 cm along the beaches of Alexandria, Egypt (El-Raey *et al.*, 1995), 25–600 cm along the beaches of Senegal (Niang-Diop, 1995), and 15 cm in the area around Recife, Brazil Erosion was less in the Brazilian site because of the natural protective features of coral reefs. In Senegal, erosion seems likely to account for only 1% of the potential land loss from a 1-m rise in sea level (Niang-Diop *et al.*, 1995), whereas it could account for 75% of the loss in Uruguay (Volonte and Nichols, 1995).

The importance of erosion does not result from its share of projected land loss so much as its universality. Much of the world's coasts are currently eroding, and rates are often fast enough to be noticed by coastal residents. The 30-cm rise in sea level likely along most coasts in the next 50–60 years would erode most sandy, muddy, and pebble shores 15–50 m. Because shorefront development tends to be less—and sometimes much less—than 30 m from the high water mark, projected erosion is likely to require a human response along the vast majority of developed coasts in the next few decades.

Barrier islands. Some of the most economically important vulnerable areas are the recreational resorts along the coastal barrier islands and peninsulas of the US Atlantic and Gulf coasts. Typically, the oceanfront block is 2–5 m above high tide; but portions of the bay side may be less than 1.5 m above high water. Many of these low bay sides are filled wetlands.

If human activities do not interfere, a barrier island can respond to sea-level rise by either (1) washing over landward and remaining intact, or (2) breaking up and drowning in place. Figure S18 provides a cross section of a barrier island washing over. This landward migration of the barrier island is analogous to erosion from the Bruun Rule, except that

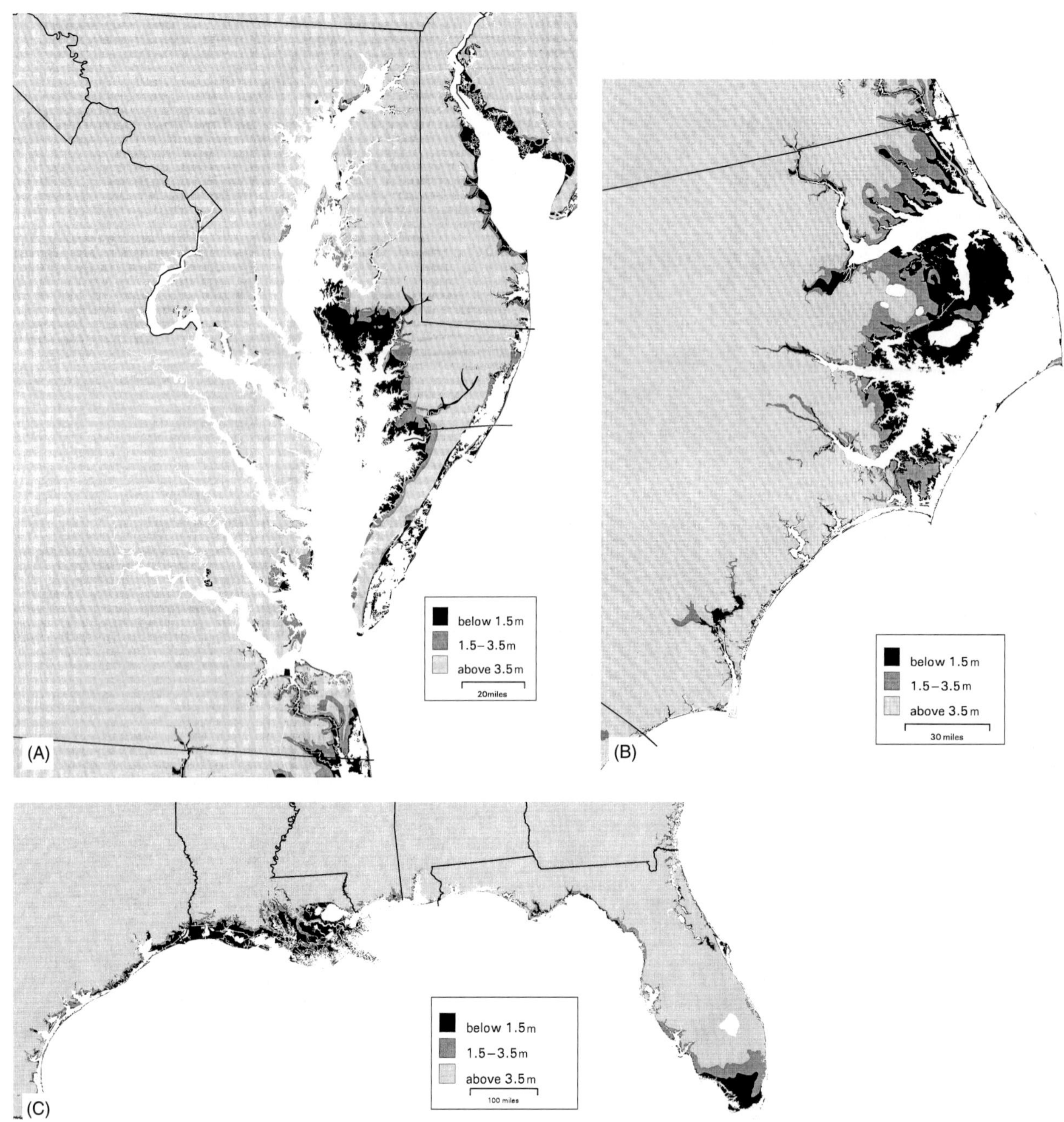

Figure S14 Maps of US lands close to sea level. These maps show the land below the 1.5- and 3.5-m contours along the coasts of (A) Chesapeake Bay, (B) North Carolina, and (C) the Gulf of Mexico. These maps are based on modeled elevations, not actual surveys or the precise data necessary to estimate elevations at specific locations. The map is a fair graphical representation of the total amount of land below the 1.5- and 3.5-m contours; but the elevations indicated at particular locations may be wrong. Although the map illustrates elevations, it does not necessarily show the location of future coastlines. Coastal protection efforts may prevent some low-lying areas from being flooded as sea level rises; and coastline erosion and the accretion of sediment may cause the actual coastline to differ from what one would expect based solely on the inundation of low land. These maps illustrates the land within 1.5 and 3.5 m of the National Geodetic Vertical Datum of 1929, a benchmark that was roughly mean sea level in the year 1929 (Titus and Richman, 2001).

the beach profile in this case goes all the way over the dune crest, across the island, to the bay, and maintenance of this more complicated profile requires sand to be washed landward. Generally, a barrier island erodes from the ocean side until it reaches a critical width, generally about 400–700 ft (Leatherman, 1979), after which the erosive forces of storms tend more to push sand landward onto the bay side of the island. The net effect of the washover process is similar to rolling up a rug; as the island rolls landward, it builds upward and remains above sea level. By contrast, Figure S19 illustrates the fate of Isle Derniere in Louisiana, which was unable to keep up with the (subsidence-induced) relative sea-level rise of 1 m per century.

Whether a particular island will break up or wash over depends on sediment supplies, the depths of the bays behind the island, and the rate of relative sea-level rise. For developed islands, however, the distinction between washover and breakup is largely academic: Most are much wider than the critical width necessary to permit island migration, and thus the islands would both erode from the ocean side and bay side. Moreover, development tends to impede landward migration: structures

Table S7 Current uses of land below the 1.5-m contour (km^2)

State	Total	Residental	Urban/industrial	Agriculture	Forest	Wetlands	Missing
AL	194.6	18.1	11.0	3.5	30.7	125.2	6.0
CT	62.9	13.1	9.6	1.1	12.9	26.0	0.2
DC	1.5	0.0	1.3	0.0	0.0	0.2	0.0
DE	387.3	8.8	1.6	72.3	27.3	277.2	0.0
FL	11,670.8	479.1	261.4	332.3	1,632.2	8,929.6	36.2
GA	1,732.5	34.6	17.9	15.6	177.8	1,484.0	2.8
LA	24,637.0	406.3	333.2	2,987.3	880.2	19,957.8	72.2
MA	363.5	60.4	43.5	8.7	61.1	178.6	11.2
MD	1,431.0	44.3	6.3	243.4	345.6	791.4	0.1
ME	382.1	49.7	19.2	23.2	175.0	112.1	3.1
MS	173.1	10.9	6.3	0.1	8.5	143.0	4.3
NC	5,512.2	132.8	70.4	610.3	1,204.0	3,486.6	8.1
NH	42.1	7.5	4.6	4.6	9.7	15.1	0.6
NJ	1,080.5	112.4	45.7	87.5	53.1	776.5	5.3
NY	239.1	69.2	34.5	7.0	17.3	110.1	1.0
RI	121.9	35.0	25.5	13.9	23.2	23.8	0.6
SC	2,328.8	46.6	29.6	187.9	252.5	1,809.1	3.2
TX	5,155.8	125.0	151.8	731.5	989.2	3,145.5	12.8
VA	330.8	40.4	25.9	32.9	65.8	165.8	0.0
Total (<1.5 m)	55,847.7	1,694.1	1,099.3	5,363.2	5,966.0	41,557.5	167.6
Total 1.5–3.5 m	31,929.1	2,603.8	1,411.3	7,236.2	10,654.0	9,995.3	28.5

Source: See Titus and Richman (2001).

Table S8 Estimated impacts of a 1-m rise in sea level

	Population affected (thousands)	Total land at risk (km^2)	Wetlands at risk	Capital at risk	Adaption/protection cost
Antigua	38	5	3	–	71
Argentina	–	3,400	1,100	5,000	1,800
Bangladesh	71,000	25,000	5,800	–	1,000
Belize	70	1,900	–	–	
Benin	1,350	230	85	118	400
China	72,000	35,000	–	–	
Egypt	4,700	5,800	–	59,000	13,100
Guyana	600	2,400	500	4,000	200
India	7,100	5,800	–	–	
Japan	15,400	2,300	–	85,000	156,000
Klribati	9	4	–	2	3
Malaysia	–	7,000	6,000	–	
Marshall islands	20	9	–	160	360
Mauritius	3	5	–	–	
Netherlands[a]	10,000	2,165	642	186,000	12,300
Nigeria	3,200	18,600	16,000	17,000	1,400
Poland	240	1,700	36	22,000	1,400
Senegal	110	6,100	6,000	500	1,000
St. Kitts-Nevis	NE	1	1	–	50
Tonga	30	7	–	–	–
United States	2,000	37,000	17,000	300,000	225,000
Uruguay	13,000	96	23	1,700	1,000
Venezuela	56,000	5,700	5,600	330	1,600

[a] In addition to the 6% of the land at risk, half the nation is below sea level but protected by dikes.
Source: Intergovernmental Panel on Climate Change (1996) for nations other than the United States. For United States, Yohe (1990), Titus *et al.* (1991).

block the landward transport of sand, and after storms deposit sand onto the streets, local public works departments generally bulldoze it back onto the beach, rather than allowing it to blow or wash to the bay side.

Flooding

Rising sea level increases the vulnerability of coastal areas to flooding for several reasons. By providing a higher base for storm surges, a 1-m rise in sea level (for example) would enable a 15-year storm to flood many areas that today are only flooded by a 100-year storm (Kana *et al.*, 1984; Leatherman, 1984). Moreover, beach erosion and wetland loss leave some properties more vulnerable to storm waves. Finally, higher surface- and groundwater levels reduce drainage and thereby increase flooding from rainstorms (Titus *et al.*, 1987).

All of these problems are already being experienced to some extent. The increased storm vulnerability resulting from erosion has led many states to fortify dunes; Louisiana has fortified entire undeveloped barrier islands (Louisiana Wetland Protection Panel, 1987). The more subtle problems of decreased drainage and higher water tables are also evident in the low parts of many low islands, where some of the streets and private lots are flooded each time it rains.

Many coastal areas are protected with levees and seawalls and would not necessarily experience inundation, erosion, or flooding. However, these structures have been designed for current sea levels; higher water levels would threaten the integrity of these coastal structures. For example, higher storm surges might overtop seawalls, and erosion could undermine them from below. In areas that are drained artificially, such as New Orleans, the increased need for pumping could exceed current pumping capacity (Titus *et al.*, 1987).

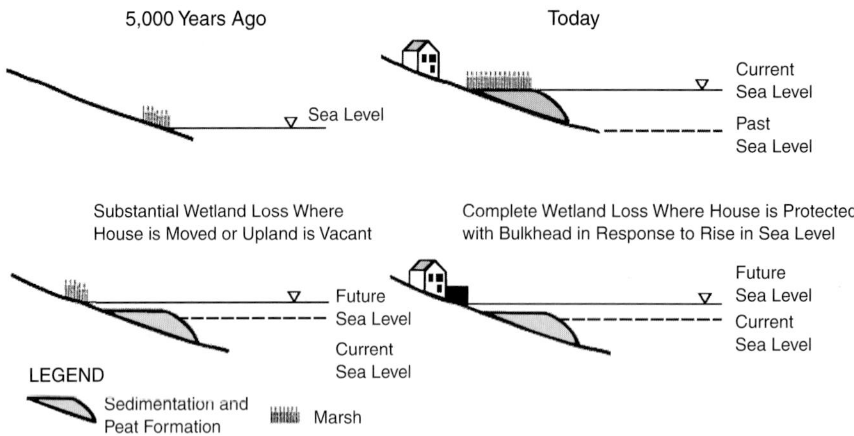

Figure S15 Evolution of a marsh as sea level rises (Titus *et al.*, 1991).

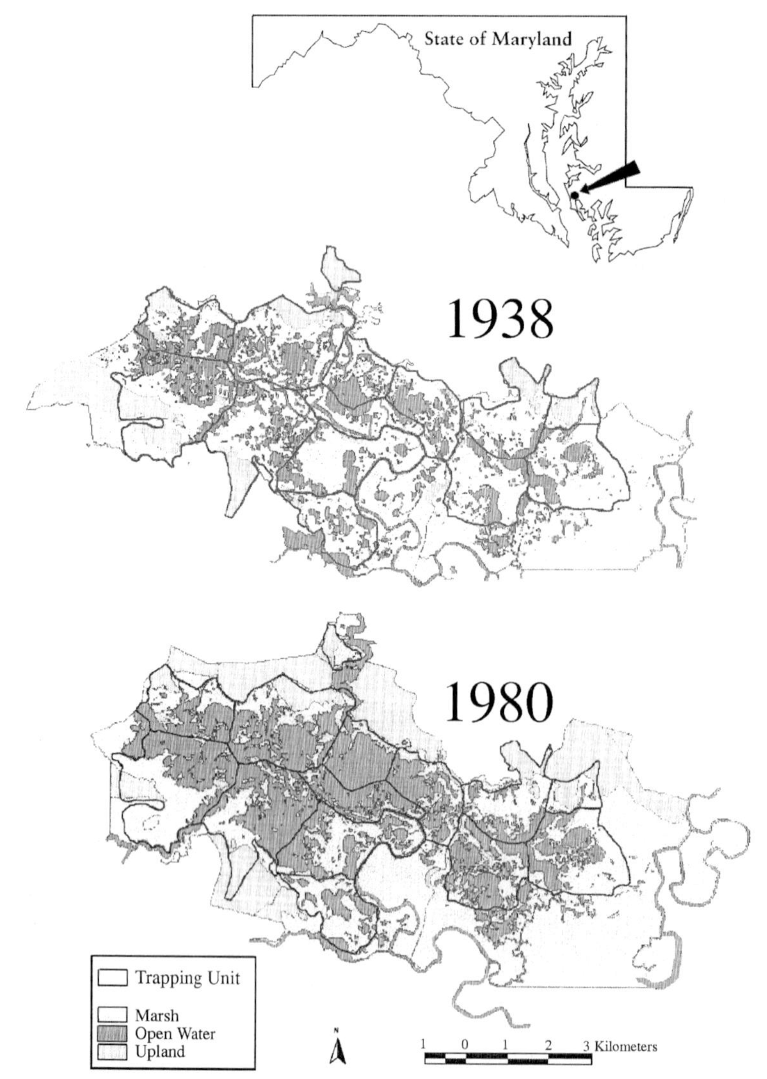

Figure S16 Blackwater Wildlife Refuge. Loss of wetlands at the Blackwater National Wildlife Refuge. This refuge, along the eastern shore of Chesapeake Bay, has lost approximately 50% of its coastal wetlands in the last 60 years, due to rising sea level, erosion, and saltwater intrusion. These maps show a 30% loss between 1938 and 1980. The extent of the loss is most evident when one focuses on particular trapping units within the refuge (trapping units are outlined in black and were delineated under land grants before the refuge was created). The increase in land area outside the trapping units resulted from land acquisition by the refuge (with permission of court J. Stevenson).

Saltwater intrusion

Rising sea level sends saltwater inland and upstream in rivers, bays, wetlands, and aquifers. New York, Philadelphia, and much of California's Central Valley rely on freshwater intakes that are slightly upstream from the point at which the water is salty during droughts. Residents and farmers in central New Jersey rely on the Potomac–Raritan–Magothy aquifer, which could become salty if sea-level rises (Hull and Titus, 1986). Miami's Biscayne aquifer is recharged by the freshwater Everglades, which are just inland of the salt-tolerant mangroves of South Florida. Although parts of the Everglades are a few meters above sea level, portions of the Everglades are less than 1 m above mean high water. If those areas become salty, a substantial part of the Biscayne aquifer would follow suit. The South Florida Water Management District already spends millions of dollars each year to prevent saltwater intrusion into the Everglades (Miller *et al.*, 1989).

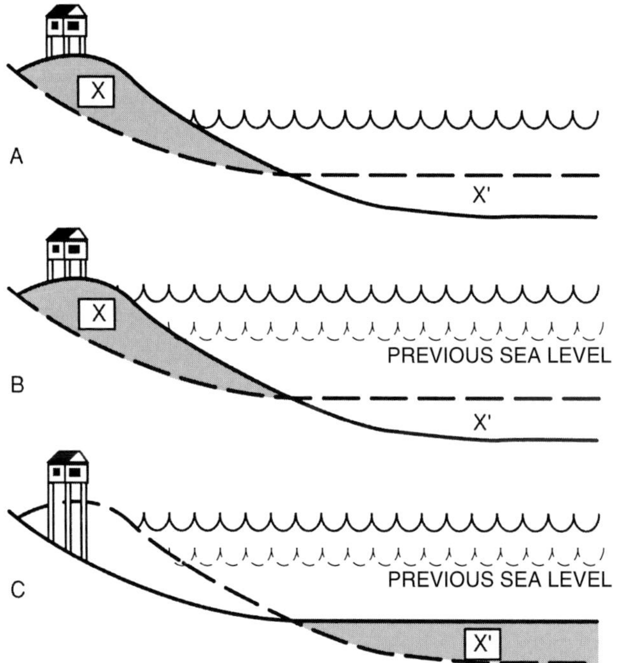

Figure S17 The Bruun Rule, (A) Initial condition; (B) immediate inundation when sea level rises; (C) subsequent erosion due to sea level rise. A rise in sea level immediately results in shoreline treat due to inundation; shown in a and b. However, a 3-foot rise in sea level implies that the offshore bottom must also rise 3 feet. The sand required to raise the bottom (X′) can be supplied for beach nourishment. Otherwise, waves will erode the necessary sand (X) from upper part of the beach as shown in (C).

Responses to sea-level rise

Structures to hold back the sea

For over five centuries the Dutch have used dikes to hold back the North Sea. These impermeable earthen walls protect the coast from storm surges, and lands below sea level from permanent inundation as well. Lands below sea level also require means to remove rainwater: An early solution was to dig ditches, and use windmills to pump water from the ditch up to a nearby canal. In many cases, tide gates keep water in the canal slightly below sea level by opening during low tide and closing the rest of the time. The Dutch approach has been applied in New Orleans (USA), Shanghai (China), and the Fens (UK). Storm-protection dikes protect land above sea level in many other areas. London, Providence (USA), and several other coastal cities are protected by large tide gates known as "storm surge barriers," which close only during storms.

To protect a shore from erosion, a structure need not be impermeable, but it must either block waves from hitting the shore or retain sediment in place. Vertical seawalls and bulkheads, and piles of rock placed at an angle—known as "revetments"—do both. Breakwaters limit the size of the waves that hit the shore, while groins can trap sand moving along the shore, albeit at the expense of erosion elsewhere.

Employing structures to hold back the sea can have adverse effects on the environment and public access to the shore. Any time a bulkhead or revetment is built between a developed area and a retreating shore,

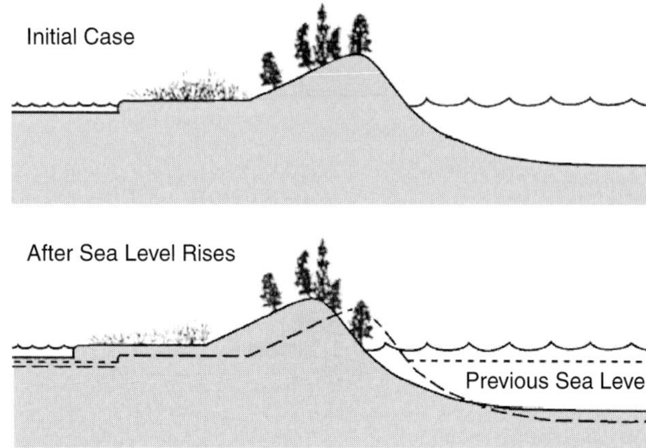

Figure S18 Overwash: the natural response of narrow barrier islands to rising sea level. As sea level rises, sand is washed to the bay side of the island, which increases the elevation of the island and allows it to migrate in land (Titus, 1990).

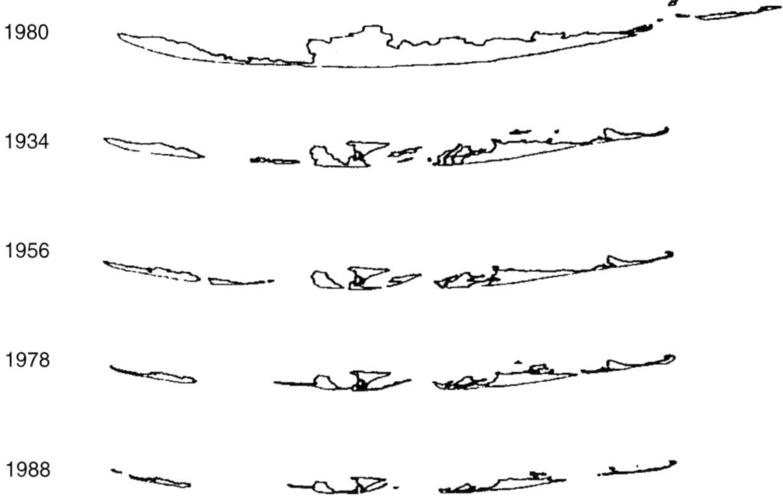

Figure S19 The breakup of Isle Derniere, Louisiana (now Isles Dernieres) (Louisiana Geological Survey).

eventually the wetlands, beaches, and mudflats between the dry land and the sea will be eliminated. Marshes and swamps are important habitat for many species or birds and fish, and also provide organic material to an estuary. Beaches and mudflats provide important habitat for horseshoe crabs and shorebirds. Estuarine beaches are particularly vulnerable because they are often only a few meters wide.

The public has a legal right to access along the intertidal wetlands and beaches under the "public trust doctrine," which is recognized by nations that follow common law and the civil law (Slade, 1990). Therefore, eliminating these lands can have the effect of preventing public access, effectively transferring ownership of the coast from the public to private landowners. Even where wetlands and beaches are not eliminated, dikes block the view of the waterfront and impair pedestrian access to the water. In the United States, people value access to ocean beaches more than estuarine shores; hence, most states prohibit coastline armoring along the ocean (and Gulf of Mexico) while allowing it along estuarine shores (Titus, 1998).

Elevating beaches and low dry land

Communities can retain natural shores and protect developed lands by placing additional soil, sand, or gravel onto the beaches and low land. After the 1900 hurricane killed 6,000 people, most of Galveston, Texas was elevated a few meters. Parts of Miami may have to be elevated because the soils are too permeable for effective pumping (Miller *et al.*, 1989) Land reclamation projects in San Francisco and Hong Kong (Nicholls and Leatherman, 1995) now include a safety margin for accelerated sea-level rise. Placing sand onto beaches is the most common response to erosion along developed ocean beaches in the United States.

A case study of Long Beach Island, New Jersey concluded that raising the entire island in place would be more feasible than encircling the island with a dike or allowing the island to retreat landward (Titus, 1990) (see Figure S20). A number of communities are gradually being elevated, although not necessarily as a conscious response to sea-level rise: Flood-prevention programs provide subsidies to elevate old homes on pilings. Communities undertake beach nourishment to elevate their

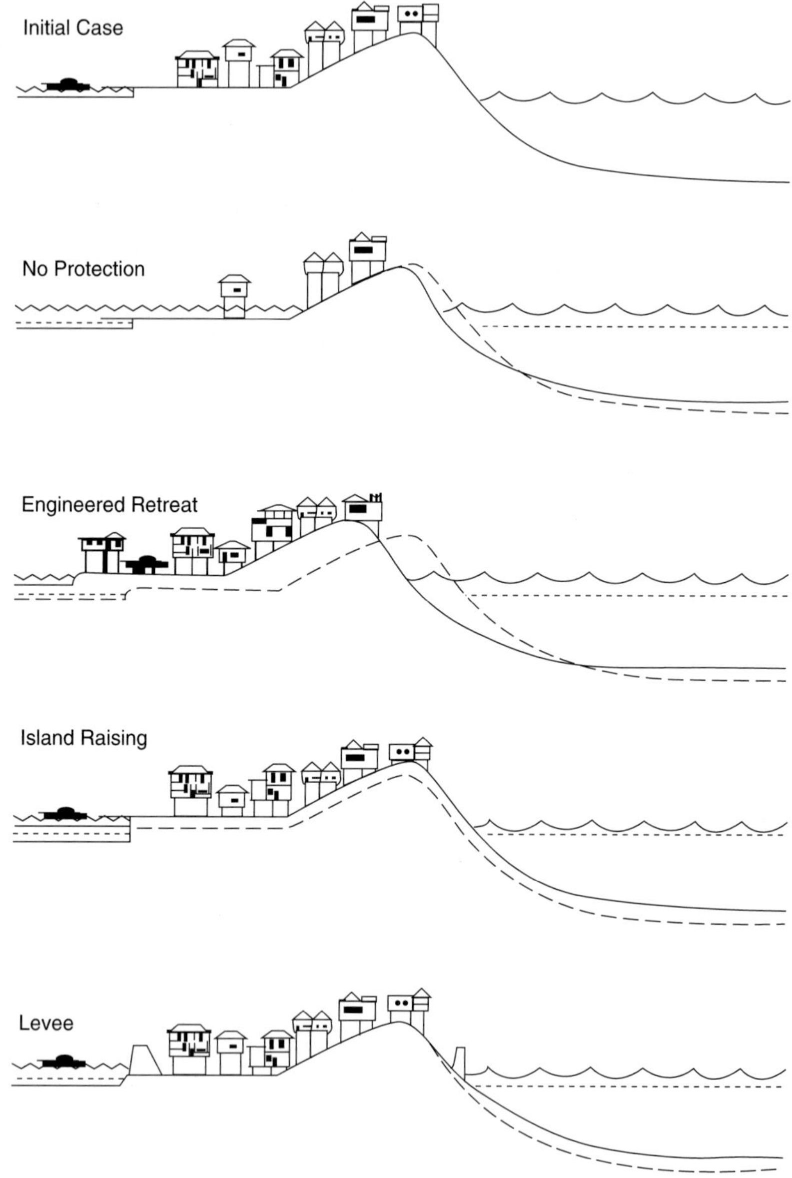

Figure S20 Responses to sea-level rise on developed barrier islands. Lightly developed islands may have no practical choice other than the "no protection" option, which would result in ocean side erosion and in some cases bayside inundation. Under the "engineered retreat" option, a community might tolerate ocean side erosion but move threatened structures to newly created bayside lands, imitating the natural overwash process that occurs with narrow undeveloped islands. A more common response is likely to be to raise entire islands as well as their beaches; although the sand costs are much higher than with an engineered retreat, existing land uses can be preserved. Finally, wide urbanized islands may choose to erect seawalls and levees (dikes); the loss of beach access and waterfront views, however, make this option less feasible for recreational barrier island resorts (Titus, 1990).

beach profiles. Finally, repaving projects elevate streets by tens of centimeters. The higher street often leaves surrounding home lots lower than the street, so that rainwater no longer drains from their yards into the street. As a result, homeowners bring in soil or gravel to elevate their lots enough so that they will drain into the street.

Allow shores to retreat

In lightly developed areas where land values are low relative to the cost of holding back the sea, nature simply takes its course. In most cases, the landowner bears the cost. However, the US flood insurance program sometimes pays for the cost of the lost structure (but not the land), and Australia directly compensates some landowners. The more difficult question arises if private erosion-control costs are low relative to property values, but the community wants to ensure that structures do not block the inland migration of coastal wetlands and beaches. The two primary approaches are:

(1) *Setbacks*, which prevents development of areas likely to be eroded; and
(2) *Rolling* easements, which allow development subject to the condition that the property will not be protected from rising water levels (Titus, 1998).

Setbacks are currently used to mitigate pollution runoff and to ensure that homes are safe from current flood risks. Several US states currently require an additional erosion-based setback, in which new houses are setback by an extra 20–60 times the annual erosion rate. The long-run success of this approach may be limited, however, because eventually the shore will erode to any setback line. The larger setback required to prevent development in any area threatened by rising sea level would require governmental compensation (Titus, 1998).

Rolling easements avoid these problems, by allowing development while informing property owners today that they will not be allowed to build bulkheads or fill the currently dry land if the effect in the future is to destroy wetlands or the public's right to the intertidal shore (Titus, 1998). This option requires neither a specific estimate of future sea-level rise nor large public land purchases, and it is economically efficient because it does not prevent owners from using their land unless the sea actually rises enough to inundate it. Moreover, incorporating this type of foresight into land-use planning can substantially reduce the economic costs of sea-level rise (Yohe *et al*., 1996).

Texas common law has recognized rolling easements for decades along its open coast. Maine has had regulations since 1987 explicitly informing homeowners that their houses will not be allowed to block the landward migration of wetlands or dunes even if the sea rises 3 ft. South Carolina's Beachfront Management Act, passed in response to the risks of a 1-foot rise in sea level, has been was modified to require rolling easements in some locations.

James G. Titus*

*Project Manager for sea-level rise, U.S. Environmental Protection Agency. Member Maryland and DC Bars. The opinions expressed do not necessarily reflect the views of the U.S. Government, which reserves the right to duplicate this article for official use.

References

Awosika, L.A., French, G.T., Nichols, R.J., and Ibe, C.E., 1992. The impacts of sea level rise on the coastline of Nigeria. In Coastal Zone Management Subgroup, Intergovernmental Panel on Climate Change. *Global Climate Change and the Rising Challenge of the Sea.*

Bruun, P., 1962. Sea level rise as a cause of shore erosion. *Journal of Waterways and Harbors Division* (ASCE), **88**: 116–130.

Callaway, J.C., DeLaune, R.D. *et al*., 1997. Sediment accretion rates from four coastal wetlands along the Gulf of Mexico. *Journal of Coastal Research*, **13**(1): 181–191.

Ellison, A.M., and Farnsworth, E.J., 1997. Simulated sea level change alters anatomy, physiology, growth, and reproduction of red mangrove (*Rhizophora mangle* L.). *Oecologia*, **112**(4): 435–446.

Ellison, J.C., 1993. Mangrove retreat with rising sea-level, Bermuda. *Estuarine Coastal and Shelf Science*, **37**(1): 75–87.

Ellison, J.C., and Stoddart, D.R., 1991. Mangrove ecosystem collapse during predicted sea-level rise—Holocene analogs and implications. *Journal of Coastal Research*, **7**(1): 151–165.

El-Raey, M., Nasr, S., Frihy, O., Desouki, S., and Dewidar, Kh., 1995. Potential impacts of accelerated sea-level rise on Alexandria Governorate, Egypt. *Journal of Coastal Research*, **14**(Special issue): 190–204.

Everts, C.H., 1985. Effect of sea level rise and net sand volume change on shoreline position at ocean city, Maryland. In *Potential Impacts of Sea Level Rise on the Beach at Ocean City, Maryland.* Washington, DC: Environmental Protection Agency.

Hull, C.H.J., and Titus J.G. (eds.), 1986. *Greenhouse Effect, Sea Level Rise, and Salinity in the Delaware Estuary*. Washington, DC: Environmental Protection Agency and Delaware River Basin Commission.

Kana, T.W., Michel, J., Hayes, M.O., and Jensen, J.R., 1984. The physical impact of sea level rise in the area of Charleston, South Carolina. In Barth, M.C., and Titus, J.G. (eds.), *Greenhouse Effect and Sea Level Rise: A Challenge for this Generation*. New York: Van Nostrand Reinhold.

Kyper, T., and Sorensen, R., 1985. Potential impacts of selected sea level rise scenarios on the beach and coastal works at Sea Bright, New Jersey. In Magoon, O.T. *et al*. (eds.), *Coastal Zone '85*. New York: American Society of Civil Engineers.

Leatherman, S.P., 1979. Migration of Assateague Island by inlet and overwash processes. *Geology*, **7**: 104–107.

Leatherman, S.P., 1984. Coastal geomorphic responses to sea level rise: Galveston Bay, Texas. In Barth, M.C., and Titus, J.G. (eds.), *Greenhouse Effect and Sea Level Rise: A Challenge for this Generation*. New York: Van Nostrand Reinhold.

Louisiana Wetland Protection Panel, 1987. *Saving Louisiana's Wetlands: The Need for a Long-Term Plan of Action*. Washington, DC: Environmental Protection Agency.

Miller, T., Walker, J.C., Kingsley, G.T., and Hyman, W.A., 1989. Impact of global climate change on urban infrastructure. In US EPA, *Potential Impacts of Global Climate Change on the United States. Appendix H. Infrastructure*. Washington, DC: U.S. Environmental Protection Agency.

Niang-Diop, I., Dennis, K.C., and Nichols., R.J., 1995. The impacts of sea level rise on the coastline of Senegal. In Coastal Zone Management Subgroup, Intergovernmental Panel on Climate Change, *Global Climate Change and the Rising Challenge of the Sea.*

Nicholls, R.J., and Leatherman, S.P., 1995. In McGregor, D., and Thompson, D. (eds.), *Geomorphology and Land Management in a Changing Environment*. Chichester: John Wiley and Sons.

Parkinson, R.W., Delaune, R.D. *et al*., 1994. Holocene sea-level rise and the fate of mangrove forests within the wider Caribbean region. *Journal of Coastal Research*, **10**(4): 1077–1086.

Schneider, S.H., and Chen, R.S., 1980. Carbon dioxide flooding: physical factors and climatic impact. *Annual Review of Energy*, **5**: 107–140.

Slade D.C. *et al*., 1990. *Putting the Public Trust Doctrine to Work*. Washington, DC: Coastal States Organization.

Titus, J.G., 1990. Greenhouse effect, sea level rise, and barrier islands. *Coastal Management*, **18**(1): 65–90.

Titus, J.G., 1998. Rising seas, coastal erosion, and the takings clause: how to save wetlands and beaches without hurting property owners. *Maryland Law Review*, **57**: 1281–1398.

Titus, J.G., and Richman, C., 2001. Maps of lands vulnerable to sea level rise: modeled elevations along the U.S. Atlantic and Gulf coasts. *Climate Research*, **18**: 205–228.

Titus, J.G., Kuo, C.Y., Gibbs, M.J., LaRoche, T.B., Webb, M.K., and Waddell, J.O., 1987. Greenhouse effect, sea level rise, and coastal drainage systems. *Journal of Water Resources Planning and Management* **113**: 2.

Titus, J.G., Park, R.A., Leatherman, S., Weggel, R., Greene, M.S., Treehan, M., Brown, S., and Gaunt, C., Yohe, G., 1991. Greenhouse effect and sea level rise: the cost of holding back the sea. *Coastal Management*, **19**(3): 171–204.

Volonte, C.R., and Nichols, R.J., 1995. Sea level rise and Uruguay: potential impacts and responses. *Journal of Coastal Research*, **14** (Special issue): 285–302.

Wilcoxen, P.J., 1986. Coastal erosion and sea level rise: implications for ocean beach and san Francisco's Westside transport project. *Coastal Zone Management Journal*, **14**: 3.

Yohe, G., 1990. The cost of not holding back the sea. *Coastal Management*, **18**: 403–432.

Cross-references

Barrier Islands
Beach Erosion

Changing Sea Levels
Coastal Zone Management
Estuaries
Greenhouse Effect and Global Warming
Sea-Level Changes During the Last Millennium
Setbacks
Shore Protection Structures
Small Islands
Washover Effects
Wetlands

SEDIMENT ANALYSIS AND CLASSIFICATION— See BEACH SEDIMENT CHARACTERISTICS

SEDIMENT BUDGET

Definition

As it pertains to coastal sedimentary systems, the sediment budget can be defined as the balance between changes in the volume of sediment stored in the system and the sum of the volumes of sediment entering or leaving the system. Examples of coastal sedimentary systems include estuarine areas composed of fine (mud-size) sediments and open-coast littoral systems most typically composed of sand-sized sediments. The focus here is on sedimentary systems of the littoral zone, the region of coast for which sediment transport is dominated by incident wave processes. This zone ranges from the intermittently dry beach (where wave swash dominates) to water depths of roughly 10–20 m (where the seabed first feels the impact of waves). A focus on the littoral zone reflects the objective of most sediment budget studies: to understand and/or predict long-term changes in the position of the coastline, the interface between the subaerial and subaqueous portions of the littoral zone.

Bowen and Inman (1966) provided one of the earliest sediment budget definitions for sandy coasts: "The procedure, sometimes referred to as the budget of sediments, consists of assessing the sedimentary contributions (credits) and losses (debits) and equating these to the net gain or loss (balance of sediments) in a given sedimentary compartment." In essence, the sediment budget is a sediment volume continuity equation with three terms that sum to zero: sediment flux in, sediment flux out, and volume rate of change within the system. Solving the sediment budget is, in principle, simply a matter of determining two of the terms and solving for the third. Komar (1996) provides a recent review of modern sediment budget techniques.

Spatial and temporal scales of the sediment budget

Although there are no absolute limitations on the spatial and temporal scales considered for sediment budget investigations, in practice spatial scales are on the order of tens of kilometers of coast or greater and temporal scales are on the order of tens of years or longer. In general, these are the scales relevant to the long-term management of coastal erosion. The sediment budget is thus concerned with the large-scale and long-term net result of all processes that transport sediment in the littoral zone and affect the position of the coastline. Individual sediment transporting events, such as storms, may not be relevant unless they contribute to long-term change in coastline position.

The littoral cell

Fundamental to the determination of the sediment budget is the identification of the sedimentary system, and in particular the boundaries of the system. In terms of the continuity equation, this is analogous to the definition of a control volume; for sediment budget studies the control volume is usually referred to as a littoral cell. No universally accepted standards exist for defining the littoral cell for all coasts—the coastal characteristics, the study objectives, and the method of sediment budget solution are all important factors in making this determination. One pragmatic, but broadly applicable, definition simply refers to littoral cells as "semi-contained entities where one can better develop a budget of sediments" (Komar, 1996, p. 18).

The placement of littoral cell boundaries is typically guided by the focus of most sediment budget studies: to understand and/or predict coastline change. In principle, the littoral cell encompasses the region of littoral sediments for which sediment volume changes are closely linked to coastline changes, while regions external to the littoral cell act as sediment source or sink areas for littoral cell gains or losses, respectively. In practice, it can be very difficult to objectively place littoral cell boundaries, with problems unique to defining the longshore, landward, and seaward boundaries.

Ideally, the longshore boundaries of the littoral cell are defined to minimize sediment exchange with other littoral cells along the coast. A classic example is a pocket beach bounded by rocky headlands that are presumed to act as barriers to littoral transport (Figure S21). Along many coasts, the longshore boundaries of the littoral cell are far less clear, especially on long stretches of sandy coast interrupted only by tidal inlets. If the littoral cell is chosen to encompass a large enough area, an assumption of no sediment exchange with adjacent littoral cells may be realistic. However, if a littoral cell defined in this way includes sections of coastline with widely differing rates of erosion or accretion, the value of the sediment budget as a tool for understanding and/or predicting coastline change will be limited. In practice, these larger cells are usually broken into smaller subcells to isolate sections of coast with quasi-uniform rates of change (e.g., Jarrett, 1991), with the underlying assumption that the processes acting within these smaller sections of coast are also quasi-uniform. Cell boundaries may thus lie within an uninterrupted section of coast, with the solution of the sediment budget requiring a simultaneous solution for all subcells with shared boundaries.

The landward boundary of the littoral cell is typically chosen as the base of an upland feature such as the toe of a dune or cliff, thus isolating littoral zone sediments from upland reservoirs of sediment that may serve as sources or sinks relative to the littoral cell. Although this is generally one of the most straightforward cell boundaries to define, significant erosion or accretion of the coastline over the time period of interest may require a moving cell boundary, with the associated complication that sediments may be repeatedly recycled between the littoral cell and upland features such as dunes.

The seaward boundary of the littoral cell can be especially difficult to define, in part because the processes of sediment transport in this environment are poorly understood. Typically, the seaward limit of the littoral cell is chosen on the basis of several simplifying assumptions. First, it is assumed that the littoral cell encompasses a coastal profile that maintains a constant form, generally known as an equilibrium profile (Dean, 1991), as the coast advances or retreats. With this assumption, littoral cell volume changes can be equated to coastline changes through a simple geometric relationship (e.g., Jarrett, 1991). It is further assumed that there is a seaward limit of significant sediment transport (Hallermeier, 1981), at a depth roughly corresponding to the seaward limit of the equilibrium profile. This depth, typically judged to be 8–15 m on the ocean coast, is often referred to as the "closure depth," a term

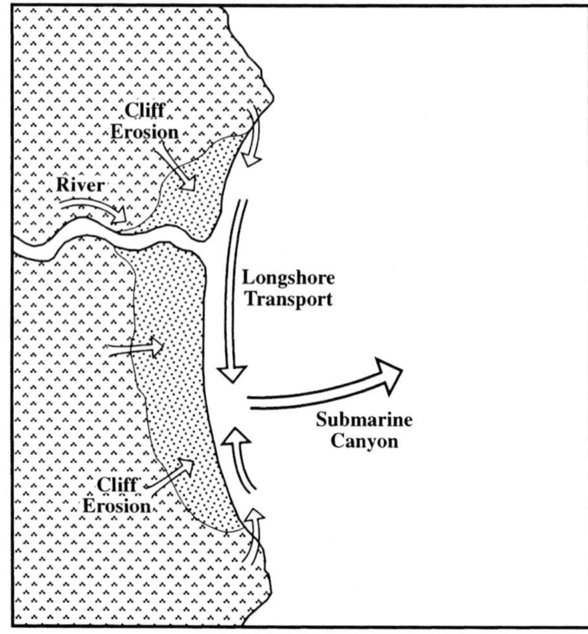

Figure S21 Schematic of a simple littoral cell consisting of a sandy pocket beach bounded by rocky headlands. Arrows represent pathways for sediment transport (after Komar, 1996).

derived from the depth at which sequential coastal profiles seem to "close," or exhibit no significant change over time. Although some profiles have been shown to maintain a reasonably constant form over many decades and simple equations exist for determining the closure depth as a function of the wave climate and sediment grain size (Hallermeier, 1981), profile data are typically inadequate for testing these assumptions rigorously in most study areas. Nevertheless, these simplifications provide an operational method for defining the seaward boundary of the littoral cell in the absence of more rigorous alternatives.

Quantifying sediment gains and losses within littoral cells

Regardless of the way in which the littoral cell is defined, a key challenge for all sediment budget studies is the determination of the sediment flux across cell boundaries. At each of the cell boundaries—longshore, landward, and seaward—there are many physical processes with the potential to add or remove sediment from the littoral cell. Although few of these processes are understood well enough to permit the direct calculation of the cross-boundary sediment flux integrated over the sediment budget timescale, a critical first step for all sediment budget studies is to identify the important sediment transport processes in the region examined. In most cases it is also useful to identify the sources and sinks of sediment associated with these processes, that is, the areas outside the littoral cell that either supply sediment to the littoral cell or store the sediment leaving the littoral cell. Quantifying sediment volume changes within these source or sink areas is often the most reliable method of quantifying gains or losses to the littoral cell. Table S9 summarizes the principal natural and anthropogenic processes responsible for sediment flux across littoral cell boundaries, and the typical sources and sinks of sediment associated with these processes.

Longshore transport

Waves breaking at an angle to the coast result in the longshore transport of sediment, also known as the littoral drift (see Komar, 1990 for review). This transport can occur both on the subaerial beach under the influence of wave swash, as well as in deeper water where steady longshore flows are generated by breaking waves. Perhaps one of the most important components of the sediment budget, the net longshore

Table S9 Principal processes of sediment transport and associated sources and sinks of sediment relevant to the sediment budget

Sediment transport processes resulting in littoral cell gains or losses	Sediment sources (for littoral cell gains)	Sediment sinks (for littoral cell losses)
Longshore transport	Adjacent littoral cells; inlet sand bodies	Adjacent littoral cells; inlet sand bodies
Cross-shore transport	Offshore areas; cliffs, dune scarps; inlet sand bodies	Offshore areas; washover fans; inlet sand bodies
Riverine transport	Upland areas	
Aeolian transport	Coastal dunes; washover fans	Coastal dunes
Gravity flows	Cliffs, scarped dunes	Submarine canyons
Biogenic processes	Calcareous sand of biogenic origin	
Human interventions	Offshore and upland areas used as a source for beach nourishment; updrift sand delivered by inlet bypassing	Dredge disposal sites; industrial uses of sand from beach mining; sand bodies trapped at structures including groins, jetties, and dams; reduction in source from naturally eroding upland features (e.g., cliffs and dunes); artificial dunes constructed by beach scraping

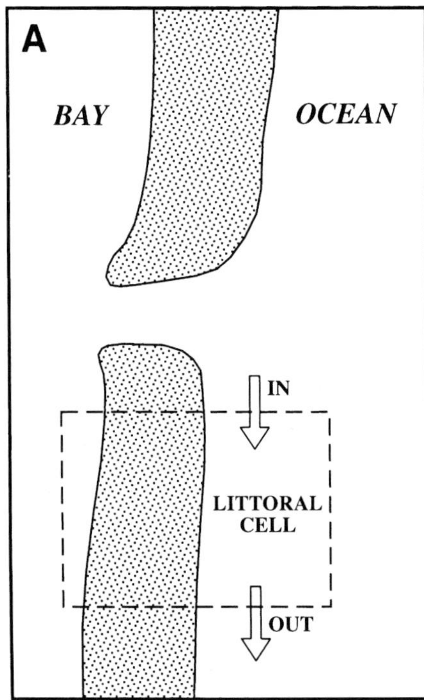

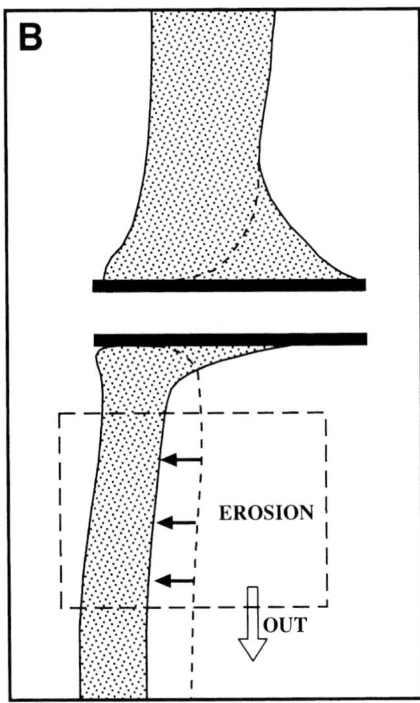

Figure S22 Application of the sediment budget on a barrier island coast. In (A) the rate of sediment transport out of the littoral cell is balanced by the rate of sediment bypassing of the inlet and the coastline is stable. In (B) the construction of inlet jetties has eliminated the transport into the littoral cell, causing a net loss to the littoral cell and coastal erosion.

transport is often readily apparent at coastal structures (e.g., jetties, groins) where sediment may be impounded (Figure S22(B)).

Two main approaches have been employed for estimating the littoral cell gains or losses due to longshore transport. A modeling approach employs empirically derived equations that relate the incident wave characteristics, in particular the wave height and direction, to the rate and direction of longshore transport (Komar, 1990). To provide estimates relevant to the sediment budget, the longshore transport must be integrated over the timescale of the sediment budget. Both the gross sediment transport (total, irrespective of direction along the coast) and the net transport may be of importance to the sediment budget depending on the nature of the littoral cell boundary. For example, in Figure S21, the gross longshore transport near the location of the submarine canyon is important if it is assumed that sand transported from either direction is trapped by the canyon. Alternatively, in Figure S22 only the net transport at the longshore boundaries of the littoral cell may be relevant to volume changes within the cell.

Unfortunately, these transport estimates are generally highly uncertain due to uncertainties in the equations themselves or simply because quality wave information is lacking. Frequently, this approach is used only to estimate the difference in the longshore transport between the longshore boundaries of the littoral cell, with the absolute magnitudes of transport adjusted so that the observed volume changes within the cell are explained (e.g., Jarrett, 1991).

As an alternative approach, littoral cell gains or losses due to longshore transport can be estimated, in some cases, by measuring changes in the volume of sediment contained in source or sink areas external to the littoral cell. For example, the growth of a terminal spit on a barrier island or the impounding of sediment updrift of a recently constructed barrier to longshore transport (e.g., Figure S22(B)) may provide quantitative information on the volume of sediment transported, or prevented from being transported, across a littoral cell boundary.

Cross-shore transport

Waves and current also transport sediment in the cross-shore direction, with the potential for net transport across the landward and seaward boundaries of the littoral cell. Although cross-shore transport can significantly influence the sediment budget, quantification of the gains and losses due to this process can be extremely difficult. Of the many important processes of cross-shore transport, few, if any, are understood to the degree that modeling or empirical approaches can provide quantitative estimates for the sediment budget. A further complication is that, for many of the relevant processes, the net transport across the littoral cell boundaries may be an extremely small fraction of the gross transport, yet this small net transport may still represent a major component of the sediment budget when integrated over the relevant timescale.

At the landward boundary of the littoral cell, wave swash—the up- and down-wash of waves breaking on the subaerial beach—represents one of the dominant cross-shore transporting processes. Wave swash can transport large quantities of sand onshore or offshore depending on many factors, including the sand texture and the continually changing characteristics of incident waves. Because much of this transport is cyclic (with sediment typically carried offshore during storms and onshore during fair weather) and does not result in net transport across littoral cell boundaries, the sediment budget contribution from wave swash may be minimal under most conditions. With increasing storm intensity, however, wave and water levels may become high enough to intersect upland features such as dunes and cliffs, resulting in erosion of these features (e.g., Ruggiero *et al.*, 1997). With the aid of gravity-induced slumping or debris flows, this results in a direct transfer of sediment from the upland feature to the adjacent littoral cell, a process that is not readily reversible (with the exception of aeolian transport, see below).

Quantifying the sediment budget contribution from the erosion of upland features is usually accomplished through a source and sink approach. With data on the topographic change of an upland feature, the volume of sediment delivered to the littoral cell can be estimated (e.g., Komar, 1983). However, as noted by Komar (1996) the volume of sediment delivered does not necessarily equal the volume of sand retained by the littoral cell if a component of the eroding sediment has a grain size incompatible with the littoral cell sediments (e.g., a silt or clay fraction will not remain in the littoral cell). In this case, only a fraction of the eroding upland feature volume can be assumed to contribute to the sediment budget.

When storm water levels and waves are high enough to overtop coastal dunes on barrier islands, offshore transport by wave swash may be replaced by onshore transport driven by both waves and steady currents associated with storm surge. The most common expression of this transport is a washover deposit or fan, consisting of sand removed from the littoral cell and deposited either on land or within the bay landward of a barrier island (e.g., Dolan and Hayden, 1981). As the storm intensity increases, the waves and currents associated with washover fans may become focused at a particular point along the coast, opening a new tidal inlet through the barrier island. Although new inlets are typically reclosed rapidly by natural processes or human intervention, large quantities of sand may be transported onshore though these temporary openings, forming deposits in the back-barrier bay similar to washover fans. Washover fans and deposits associated with new inlets represent a loss of sand from the littoral cell, which can be quantified by measuring the thickness and areal extent of such deposits (e.g., Kochel and Dolan, 1986). Estimating the long-term contribution from all storms over the timescale relevant to the sediment budget can be difficult, as washover and inlet deposits from past storms may be difficult to identify, both in terms of their volumes and their ages.

New inlets that remain open will subsequently develop a characteristic morphology consisting of ebb- and flood-tidal deltas on the seaward and landward side of the inlet channel, respectively. The processes of sediment redistribution associated with changes at tidal inlets are complex, involving both longshore and cross-shore transport, and may have a significant impact on the sediment budget of adjacent littoral cells (Rosati and Kraus, 1999). Quantifying the sediment budget terms associated with tidal inlets can be exceedingly difficult. Perhaps the most straightforward approach is the use of bathymetric change to quantify the volume of sand trapped (or released) by the inlet over the time period relevant to the sediment budget being formulated. Unfortunately, adequate bathymetric data are seldom available, and are of limited use for predicting the impact of major inlet changes (e.g., inlet opening, inlet closing, construction of inlet jetties, etc.). Frequently, assumptions are made of the efficiency of an inlet at trapping the gross longshore transport in order to arrive at an estimate of the inlet's sediment budget impact (Rosati and Kraus, 1999).

At the seaward boundary of the littoral cell, determining the sediment gains or losses due to cross-shore transport is generally extremely difficult. At present, sediment transport knowledge is inadequate for quantitative predictions, although studies suggest, at least qualitatively, that offshore areas can act as both sources (e.g., Williams and Meisburger, 1987) and sinks (e.g., Niedoroda *et al.*, 1985) of sediment relative to the littoral cell. Because quantification is difficult, it is often assumed that sediment transport is negligible seaward of the predicted closure depth, making the determination of sediment budget gains or losses at this littoral cell boundary unnecessary (in principle). As an alternate approach, some studies have assumed that major imbalances in the sediment budget—after accounting for all the sediment budget components that can be quantified—are attributable to sediment flux at the seaward boundary (e.g., Inman and Dolan, 1989). In almost all cases, however, this important component of the sediment budget remains a source of great uncertainty and warrants much further research.

Riverine sediment transport

The delivery of sand to the coast by rivers can represent a major component of the sediment budget. This is especially true on steep gradient coasts, such as typically found on the leading edge of continental plates, where rivers deliver their transport load directly to the coast. In contrast, the river valleys of low gradient, trailing edge coasts may be drowned, trapping sediment from rivers in estuaries before it can reach the coast.

A variety of techniques have been developed to estimate the sediment budget contribution from rivers, although all are associated with a high degree of uncertainty due to the difficulty of directly measuring the sediment load of a river integrated over the timescale relevant to the sediment budget. Komar (1996) reviews many of the methods that have been used, and makes the further point that, as in the case of sediment eroded from cliffs, only the fraction of delivered sand with a similar size as found in the littoral cell will contribute to the sediment budget.

Aeolian sediment transport

In many coastal areas, wind plays an important role in the flux of sediment across the landward boundary of the littoral cell. Upland features that are not well stabilized by vegetation, such as washover fans, may provide a source of sediment in areas dominated by offshore directed wind. In many areas, however, aeolian transport results in the formation of coastal dunes, which by their existence indicate that the net effect of wind transport is landward—a loss to the littoral cell. As for most sediment transport processes affecting the sediment budget, the direct calculation of aeolian transport's contribution to the sediment budget is not feasible, and a source and sink approach, whereby changes in dune volume are inferred to represent a littoral cell loss, may be the best available method (e.g., Inman and Dolan, 1989). Because coastal dunes are typically vulnerable to erosion by wave processes (see above), any sediment budget for which coastal dunes are important must take into account the possibility that dunes can represent a net source to the littoral cell, despite their earlier formation as a sink.

Gravity flows

At both the landward and seaward boundaries of the littoral cell, sediment transport driven primarily by the direct influence of gravity may represent another significant contribution to the sediment budget. At the landward boundary, slumping and other forms of mass wasting of cliffs or scarped dunes can deliver sediment almost instantaneously to the littoral cell. The processes initiating this transport are varied and complex, but in many cases the cross-shore transport of sediment by waves plays a key role (e.g., Everts, 1991). As described above, the contribution to the sediment budget can be estimated by quantifying the volumetric loss to the upland feature and accounting for the percent of eroded material expected to be size compatible with littoral cell sediments (e.g., Komar, 1983).

At the seaward boundary of the littoral cell, submarine gravity flows can cause significant losses to the sediment budget, especially along narrow-shelf coasts where submarine canyons extend landward into the littoral cell (e.g., Lewis and Barnes, 1999). These gravity flows are initiated when longshore transport delivers sand to the head of the canyon; periodically the deposit becomes over steepened to the point where it slumps and flows down the canyon as a turbidity current. The most straightforward means of quantifying the volume of sand lost from the littoral cell is by determining the volume change in the sediment sink area at the bottom of the canyon. Often, however, the necessary bathymetric change information is lacking and inferences about the volume of sediment lost through this process must be made by considering the balance of other sediment budget components.

Biogenic processes

In tropical and subtropical regions, a significant component of beach sand may be of biogenic origin, composed of calcium carbonate grains, rather than derived from terrestrial erosion (largely quartz or other silicate minerals). In these areas, the sediment budget may be strongly influenced by the biogenic production of carbonate material, such as corals, shells, and foraminiferal tests, which is reworked by physical processes into sand of the littoral cell. As described by Komar (1996), estimating the volumetric contribution to the sediment budget is difficult, because the mere presence of carbonate sand does not give the rate at which it is being produced and lost due to abrasion. Although some techniques have been developed (see Komar, 1996, for a description), much further research is needed to develop a reliable means of estimating this sediment budget component.

Human interventions

Along many coasts, the influence of human activities is becoming increasingly significant and must be accounted for in balancing the sediment budget. Human activities take many forms, but can be broadly categorized into activities that interfere with natural processes, and activities that move sediment directly. Interferences in natural processes typically result in the reduction or elimination of a sediment source, representing a loss to the littoral cell (tabulated as the creation of new sediment sinks in Table S9). The construction of dams can trap sediment in upland reservoirs, greatly reducing the riverine transport of sediment to the littoral cell. Armoring of eroding upland features, such as cliffs and dunes, prevents transporting processes such as gravity flows and cross-shore transport from delivering sediment that otherwise would have nourished the adjacent littoral cell. Groins and inlet jetties are designed to interrupt longshore sediment transport, creating new updrift sediment sinks for longshore transport (Figure S22). Estimating the impact on the sediment budget as a result of these engineering structures is, in many cases, more straightforward than for purely natural processes, as the volume of sediment impounded (or prevented from being eroded) can often be directly measured.

Human activities that move sediment directly may also result in sediment losses or gains to the littoral cell. Sediment gains include beach nourishment, whereby sand is placed within the littoral cell using a source area external to the littoral cell; and inlet bypassing, whereby sand on the updrift side of a tidal inlet is pumped to the downdrift side (representing a loss to the updrift littoral cell). Sediment losses include inlet channel dredging when the disposal site is outside the littoral cell, the mining of beach sand for industrial uses (becoming rare), and beach scraping following storms, whereby sand is bulldozed off the beach foreshore to form artificial dunes. The best available estimate of the sediment budget contribution of these engineering activities is, in most cases, the contractor's estimate of sediment volume delivered.

Effect of sea-level rise

Because sediment budget timescales are generally long, on the order of decades or longer, a rise in sea level relative to land has the potential to significantly influence coastline position. Though neglected in most sediment budget studies, sea-level rise occasionally has been treated as an agent of coastal change operating independently of other processes affecting the sediment budget (e.g., Inman and Dolan, 1989). The most common method of relating sea-level rise to coastline change is through a simple geometry-based model known as the Bruun Rule (Bruun, 1962), in which a rise in sea level results in an upward and landward translation of the equilibrium profile with a redistribution of sediment from onshore to offshore. If the redistribution of sediment is assumed to occur within the bounds of the littoral cell, the amount of coastline change predicted by the Bruun Rule can be removed, in principle, from consideration when balancing the sediment budget (Inman and Dolan, 1989).

Unfortunately, the approach of attributing a distinct component of coastline change to sea-level rise is not well founded by either field observations of the long-term profile response or through a processes-based modeling approach. In most cases, the uncertainty level in the overall sediment budget is enough to account for the component of coastline change that may be attributable to sea-level rise. Also, there is little justification for considering sea-level rise as an independent process, as sea-level rise by itself has no capacity to transport sediment. Rather, sea-level rise undoubtedly acts as a modifier of many transport processes influencing the sediment budget, although our understanding of these processes is currently too limited to quantify the effect of sea-level rise on coastline position. With the potential to be a significant and pervasive agent of long-term coastal change, sea-level rise clearly warrants much further research.

The balance: coastline change

In principle, the sediment budget can be balanced by summing the volumetric contributions of all relevant sediment transporting processes, giving the net volumetric change within the littoral cell. Coastline change can then be predicted from volume change by invoking the equilibrium profile assumption, whereby the profile maintains a constant form out to the closure depth while translating landward (erosion) or seaward (accretion). In many cases, however, great uncertainties remain in most sediment budget components, as described above, and only the end result—littoral cell volume change or the coastline change itself—is known with any degree of certainty. For this reason, the observed volume or coastline change is frequently used to adjust or calibrate the most uncertain terms in the sediment budget. The sediment budget is then used to make predictions of coastline change given expected changes in specific sediment budget components, most often those induced by human activities. Several simple examples are given below.

In the case of Figure S21, the most uncertain sediment budget component might be the down-canyon loss of sediment through gravity flows. By measuring the volume change within the littoral cell directly through bathymetric comparisons, or by estimating volume change from coastline change using the equilibrium profile assumption, the down-canyon loss can be estimated if the other important components of the sediment budget—here river input and cliff erosion—are reasonably well known. The utility of the sediment budget is then in making predictions of coastline change given a modification of a sediment budget component. For example, if cliff erosion is arrested by armoring a modification to the rate of coastline change can be predicted if it is assumed that the other sediment budget components remain unchanged after cliff armoring.

In another example, the construction of inlet jetties might interrupt the sediment bypassing a tidal inlet as in Figure S22. If a preconstruction sediment budget has been established, including estimates of longshore transport and any other significant components, then the impact of jetty construction on coastline change can be estimated by assuming a total or partial interruption in the longshore transport reaching the littoral cell, and again assuming that other components of the sediment budget remain unchanged.

Despite their uncertain nature, the sediment budget often represents the best available tool for gauging the impact of human interventions, as well as changes due to natural processes, on the rate of coastline change. Also, as stated by Komar (1996, p. 25), "the formulation of a sediment budget can also serve to organize what is known about a coastal area or littoral cell, and where the chief gaps still exist in our understanding, thereby providing a guide for future research."

Jeffrey H. List

Bibliography

Bowen, A.J., and Inman, D.L., 1966. Budget of littoral sands in the vicinity of Point Arguello, California. U.S. Army Corps of Engineers, Coastal Engineering Research Center, *Technical Memorandum* No. 19, 56pp.

Bruun, P., 1962. Sea level rise as a cause of shore erosion. *Journal of the Waterways and Harbors Division*, American Society of Civil Engineers, **88**: 117–130.

Dean, R.G., 1991. Equilibrium beach profiles: characteristics and applications. *Journal of Coastal Research*, **7**: 53–84.

Dolan, R., and Hayden, B., 1981. Storms and shoreline configuration. *Journal of Sedimentary Petrology*, **51**: 737–744.

Everts, C.H., 1991. Seacliff retreat and coarse sediment yields in Southern California. *Proceedings of the Coastal Sediments '91*. New York: American Society of Civil Engineers, pp. 1586–1598.

Hallermeier, R.J., 1981. A profile zonation for seasonal sand beaches from wave climate. *Coastal Engineering*, **4**: 253–277.

Inman, D.L., and Dolan, R., 1989. The outer banks of North Carolina: budget of sediment and inlet dynamics along a migrating barrier system. *Journal of Coastal Research*, **5**: 193–237.

Jarrett, J.T., 1991. Coastal sediment budget analysis techniques. *Proceedings of the Coastal Sediments '91*. New York: American Society of Civil Engineers, pp. 2223–2233.

Kochel, R.C., and Dolan, R., 1986. Role of overwash on a mid-Atlantic coast barrier island. *Journal of Geology*, **94**: 902–906.

Komar, P.D., 1983. The erosion of Siletz Spit, Oregon. In Komar, P.D. (ed.), *Handbook of Coastal Processes and Erosion*. Boca Raton, FL: CRC Press, pp. 65–76.

Komar, P.D., 1990. Littoral sediment transport. In Herbich, J.B. (ed.), *Handbook on Coastal and Ocean Engineering*. Houston: Gulf Publishing Co., pp. 681–714.

Komar, P.D., 1996. The budget of littoral sediments, concepts and applications. *Shore and Beach*, **64**: 18–26.

Lewis, K.B., and Barnes, P.M., 1999. Kaikoura Canyon, New Zealand; active conduit from near-shore sediment zones to trench-axis channel. *Marine Geology*, **162**: 39–69.

Niedoroda, A.W., Swift, D.J.P., Figueirdeo, A.G., and Freeland, G.L., 1985. Barrier island evolution, middle Atlantic shelf U.S.A. Part II: evidence from the shelf floor. *Marine Geology*, **63**: 363–396.

Rosati, J.D., and Kraus, N.C., 1999. Advances in coastal sediment budget methodology—with emphasis on inlets. *Shore and Beach*, **67**: 56–65.

Ruggiero, P., Komar, P.D., McDougal, W.G., and Beach, R.A., 1997. Extreme water levels, wave runup and coastal erosion. *Proceeding of the 25th International Conference on Coastal Engineering*, New York: American Society of Civil Engineers, pp. 2793–2805.

Williams, S.J., and Meisburger, E.P., 1987. Sand sources for the transgressive barrier coast of Long Island, New York: evidence for landward transport of shelf sediments. *Proceedings of the Coastal Sediments '87*. New York: American Society of Civil Engineers, pp. 1517–1532.

Cross-references

Barrier Islands
Barrier
Beach Erosion
Beach Nourishment
Beach Processes
Bypassing at Littoral Drift Barriers
Cliffs, Erosion Rates
Coastal Changes, Gradual
Coastline Changes
Cross-Shore Sediment Transport
Dams, Effect on Coasts
Depth of Closure on Sandy Coasts
Dune Ridges
Dynamic Equilibrium of Beaches
Energy and Sediment Budgets of the Global Coastal Zone
Eolian Processes
Erosion Processes
Gross Transport
Littoral Cells
Longshore Sediment Transport
Net Transport
Numerical Modeling
Sandy Coasts
Sea-Level Rise, Effect
Shore Protection Structures
Storm Surge
Wave-Dominated Coasts
Waves

SEDIMENT SUSPENSION BY WAVES

Suspended sediment under waves is defined as sediment that is picked up (entrained) from the seabed by the water and is kept entrained by the motion of the water. The suspension occurs as a result of the moving water entraining loosely consolidated sediments such as sand and carrying them up into the water column away from the seabed. When the water stops moving, the sediments eventually settle back down to the seabed under the influence of gravity. However, moving water can counteract the effects of gravity by exerting stress on the sediments and effectively dragging them up into the water column against gravity's pull. On the coast, waves arriving from the deeper ocean provide the necessary movement of water to lift sediments up off the seabed and transport them around.

Sediment transport processes

When waves interact with the seabed, they mobilize sediment and progressively change the shape of the coastline over time. The mobilization of sediment by waves may be thought of as taking place in two principal ways (Bagnold, 1963): (1) as a result of collisions between individual sediment grains, and (2) as a result of fluid stresses on individual sediment grains, both of which may maintain the motion of the sediment. The first process is termed bedload and commonly occurs with large sediments such as sands, gravels, and pebbles. These large-sized sediments tend to bounce off one another because the water would have to be moving quickly for sustained periods of time to keep them suspended. The second process is termed suspended load and commonly occurs with sand and smaller-sized sediments. These sediments are small enough for the water to drag them into suspension. Sediments like silt and clay that are smaller than sand size may be suspended for long periods of time and carried by the water over long distances before settling back to the seabed. These sediments form a subset of suspended material referred to as washload because they are suspended for long enough periods that they are washed away. Waves reaching coastlines are typically energetic enough to suspend sediments smaller than sand size and wash them away into deeper water.

Historically, bedload was considered to be the principal process in sediment transport and to account for the volumetric majority of sediment moved since it included the movement of large sediments. Recent work (e.g., Neilsen *et al.*, 1979; Sternberg *et al.*, 1989) has indicated that suspended load is the major contributor to sediment transport. Sediment carried in suspension can account for upwards of 80–90% of the volume of wave-transported sediment. So, even though bedload moves large sediments, the accumulated volume of smaller sediments moving in suspension comprises a significantly greater volume of sediment moved than bedload. Thus, the suspension of sediments under waves is an important factor in the redistribution of sediment and coastal evolution.

Sediment suspension mechanisms

In order to better understand the suspension of sediments under waves, several mechanisms have been proposed to account for the observed characteristics of suspension. Initially, diffusion, analogous to the mixing of concentrations at the molecular level in chemistry, was proposed as the mechanism for the suspension of sediment by Rouse (1937). By

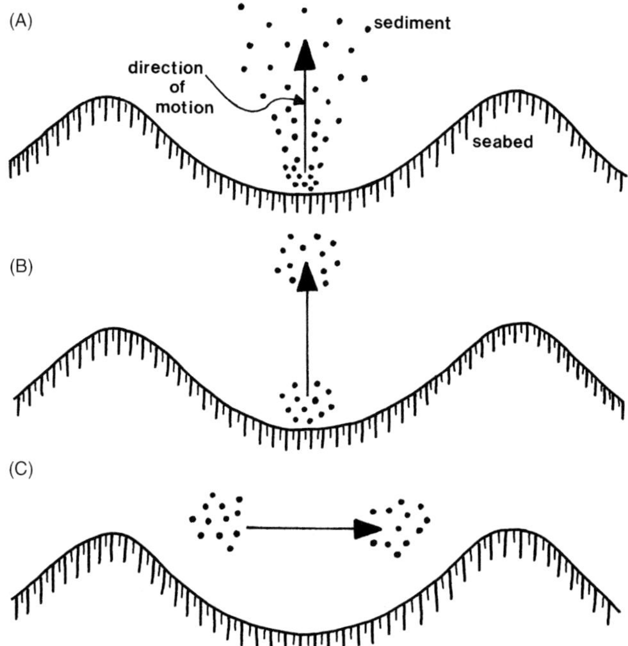

Figure S23 Mechanisms of sediment suspension: (A) diffusion, (B) convection, and (C) advection.

this mechanism, sediment grains behaved like molecules and moved down concentration gradients from areas of high concentration to areas of low concentration. Sediment suspension thus resulted from the movement of sediment away from the seabed (high concentration of sediment) into the water column (low concentration of sediment). Figure S23 represents the concept of diffusion showing the general pattern of motion of the water-sediment mixture. Currently, a molecular-type diffusion mechanism is commonly accepted for horizontal transport of sediments but is no longer accepted for vertical transport since the rate of diffusion in still water is not strong enough to overcome the force of gravity alone. However, diffusion can occur in moving water where the motion of the water counteracts gravity.

Additional mechanisms associated with the turbulent motion of the water were then proposed to supplant the molecular diffusion mechanism. These mechanisms, referred to as convection and advection, are represented in Figures S23(B) and (C), respectively. Convection and advection occur as a result of the drag of moving water on stationary particles towing the particles along with the water. This drag is referred to as a Reynolds stress and is a velocity-dependent force imparted on the sediment grains by the moving water. The faster the water, the greater the stress on the sediment. Convection occurs as a result of vertically directed Reynolds stresses that carry the water and sediment mixture upwards away from the seabed. The sediment is maintained in suspension by these fluid stresses that counteract the force of gravity. Advection occurs as a result of horizontally directed Reynolds stresses that carry the water and sediment back and forth across the seabed. Again the sediment is maintained in suspension by fluid stresses that counteract the force of gravity. Over rippled seabeds, convective and advective suspension of sediment are an extremely important factor due to near-seabed turbulence as water passes over the bedforms.

Flow separation and vortex propagation

The near-seabed turbulence that causes convective and advective suspension of sediment is influenced by the presence of bedform-induced flow separation. Flow separation occurs as a direct result of the conservation of energy within the water. Water energy is commonly described using the Bernoulli equation, which equates the total energy of the water as equivalent to the sum of the pressure of the water, the kinetic energy (related to water velocity) and the potential energy (related to the elevation of the water and gravity). Figure S24 illustrates the dynamics of flow separation over a bedform under unidirectional flow (after Honji *et al.*, 1980). In the figure, water is assumed to be flowing from left to right.

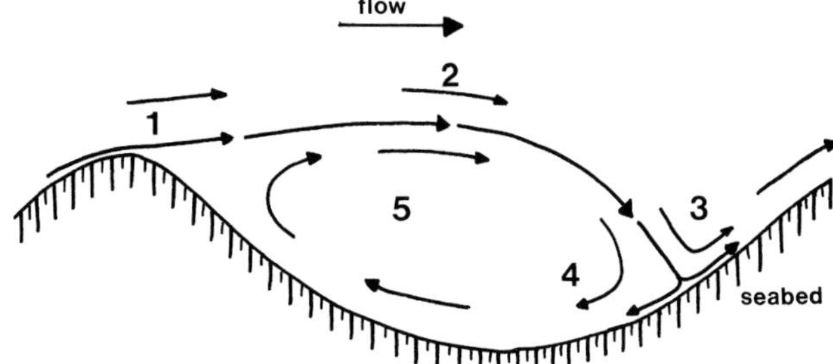

Figure S24 Schematic of flow separation and generation of a sediment suspending vortex (after Honji *et al.*, 1980).

Immediately over the ripple crest (1) the flow of water just above the seabed expands as the water column deepens into the trough of the ripple. Flow velocity decreases into the trough since there is a greater depth through which the same volume of water must travel than over the crest. In order to conserve energy in the direction of flow, the pressure of the water at the seabed increases to compensate for the decrease in velocity. If the water is moving fast enough, the increase in pressure at the seabed lifts the water upwards as a result of a pressure gradient away from the seabed and the flow separates. If the water is not moving fast enough, the change in pressure is not sufficient to lift the water away from the seabed. The flow of water (streamline) lifted off the seabed by the pressure difference (1) begins to accelerate as it is no longer slowed by the friction of the seabed (2). As the flow accelerates, velocity increases and, to compensate and conserve energy, pressure decreases. The decrease in pressure reduces the pressure gradient away from the seabed and the streamline reattaches (3). The streamline from 1 to 3 creates a free shear layer beneath which flow reverses (4) relative to the direction of the flow above the layer and forms a vortex of rotating water (5) in the lee of the ripple crest. The free shear layer has increased momentum relative to the surrounding water due to its larger velocity thereby creating a momentum gradient toward the seabed. The momentum gradient traps the rotating flow in the lee of the ripple where it entrains sediment.

Under a wave-dominated environment, the free shear layer that traps the suspended sediment forms and decays twice each wave cycle since the horizontal component of the oscillatory flow becomes zero twice each wave cycle. When the free shear layer decays, the vortex at the seabed is no longer trapped. The vortex generally has larger velocities than the water above and a momentum gradient is established away from the seabed. The upwards directed momentum gradient causes the vortex to rise away from the seabed, suspending sediment up into the water column (Tunstall and Inman, 1975). The vortex is both convected and advected away from the seabed and it diffuses over time into clouds of suspended sediment that are transported with the flow. The diffusion occurs down the concentration gradient from the vortex to the surrounding water. As the vortex decays, the Reynolds stresses suspending the sediment decay as well and the suspended material begins to settle out. The formation and subsequent ejection of vortices from the bed is an important mechanism for suspending sediment under waves and results in intermittent injection of suspended sediment into the water throughout a complete wave cycle.

Wave cycle model of sediment suspension

The intermittent nature of sediment suspension through vortex creation and ejection, or shedding, is significant because it leads to strong temporal variations in concentration and hence strong temporal dependencies in sediment transport. The temporal variations occur over the timescale of a complete wave cycle since the mechanisms controlling vortex shedding vary with the passage of each wave. The wave cycle model of sediment suspension is based on a model of periodic flow separation and vortex propagation over a rippled seabed that incorporates this intermittent suspension of sediment (Sleath, 1982). Figure S25 represents this conceptual model over various stages, or phase angles, of a wave cycle. The motion of water under a wave at the water surface is generally orbital, molecules of water traveling through a 360° orbit between wave crests. The wave orbit becomes flatter toward the seabed as the vertical motion of the water is muted and the water generally flows horizontally first in one direction as a wave crest passes and then horizontally in the other direction as the wave trough passes. This change in flow direction is referred to as oscillatory flow. Given this oscillation, the phase angle of the wave cycle may be referenced relative to the 360° of one complete

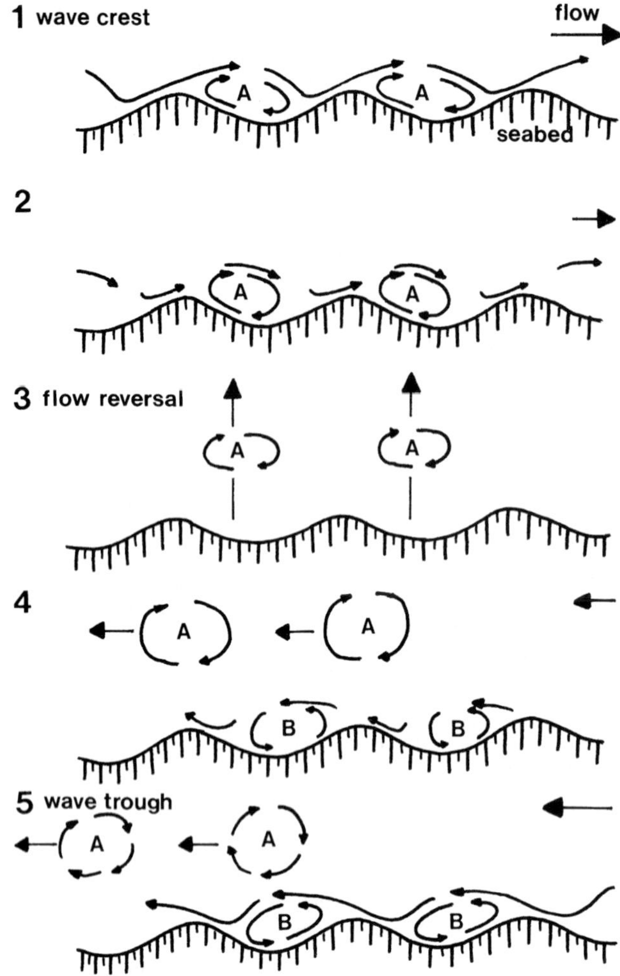

Figure S25 The wave cycle model of sediment suspension (after Sleath, 1982).

wave cycle. The start of the wave cycle may be set anywhere in the cycle but is commonly assigned to a maximum velocity in one direction under the wave cycle model. In Figure S25, the start of the wave cycle has been arbitrarily set to occur with the passage of a wave crest (1).

At the beginning of the wave cycle (1), phase angle $\theta = 0°$, the horizontal velocity of the flow is at a maximum to the right. Flow has separated over the lee slope of the ripple and has generated a vortex downstream of the crest (A). The vortex is trapped at the bed by the large momentum gradient directed toward the seabed from the free shear layer. The vortex entrains sediment from the seabed while trapped and reaches its maximum size and velocity. At this point, velocity is large but the suspended sediment concentration in the water column is small since the vortex is trapped at the seabed. As the horizontal flow to the right decreases in velocity (2) with the passage of the wave crest, phase angle $\theta = 45°$, the free shear layer decays resulting in a momentum gradient directed upwards away from the seabed. The vortex carries the entrained sediment as it begins to rise. When the horizontal velocity reaches zero (3) at the midpoint in the wave between the crest and the trough, phase angle $\theta = 90°$, the flow is on the point of reversal and the free shear layer has vanished. The vortex is ejected from the seabed by the upwards transfer of momentum and carries sediment into suspension. At this point, velocity is small and suspended sediment concentration in the water column is large since the vortex has been released upwards. With the collapse of the free shear layer the vortex is no longer constrained and the suspended material begins to diffuse under viscous and gravitational forces.

Once through the point of reversal, horizontal flow increases to the left (4) with the approach of the wave trough, phase angle $\theta = 135°$. The vortex ejected during the previous phase (A) is transported in the opposite direction to the flow which generated it. This vortex is deflected higher into the water column by the generation of a new free shear layer on the ripple slope opposite to the slope above which it formed. As the vortex (A) rises it diffuses into the water column and is advected across the bed by the water that is now flowing to the left. At the seabed a second vortex (B) is generated under the newly developing free shear layer. When the horizontal velocity reaches a maximum to the left (5) with the arrival of the wave trough, phase angle $\theta = 180°$, the flow of water is in the opposite direction to the flow at a phase angle $\theta = 0°$ (wave crest) and the second vortex (B) reaches its maximum size and velocity while trapped at the seabed. The velocity is now large again but suspended sediment concentration in the water is now dependent on the diffusing and advected vortices generated earlier in the wave cycle.

The second half of the wave cycle covering the passage of the wave trough and the arrival of the next wave crest, phase angles $\theta = 180°$ to $\theta = 360°$, go through these same five stages as the flow reverses from the left to the right. The second vortex (B) is ejected into the water column in the same manner as the first vortex (A) at the next flow reversal. The first vortex continues to diffuse and to be transported back and forth above the seabed so long as the fluid turbulence can maintain the sediment in suspension against the pull of gravity.

Not all waves will be large enough to generate flows at the seabed that will result in separation. The size of the wave needed to suspend sediment at the seabed will depend on the depth of water. If a series of waves pass by in succession, and each generates flows strong enough to develop vortices at the seabed, sediment suspended by the first wave may be kept in suspension until all the waves have passed. Such a series of waves is known as a wave group. Successive waves in a wave group inherit some turbulence from the preceding waves. This inheritance of turbulence leads to the suspension of sediment for periods of time longer than a single wave cycle. The extended suspension events result in the individual vortices coalescing into clouds of suspended material above the bed. Video observations (e.g., Dyer, 1980) of sediment suspension under wave groups show the formation of these suspension clouds and their subsequent decay once the wave group has passed.

Modeling sediment suspension

Given the conceptual wave cycle model of sediment suspension, it is clear that the marked spatial and temporal variations in sediment suspension are critical in determining overall sediment transport. Bailard (1984) developed a model for sediment transport based on energetics principles proposed by Bagnold (1963) but recognized that the use of time-averaged terms were a significant weakness to the model's predictive ability. The weakness is due to the fact that time-averaged terms do not represent the physical processes well (e.g., Osborne and Greenwood, 1993). The process of vortex shedding acts as a strong control on the suspension of sediment into the water column and provides a significant influence on the resulting concentration profile. Additionally, the time at which concentration reaches a maximum is out of phase with the time at which velocity reaches a maximum. The energetics approach relies on the velocity and concentration maximums occurring together.

Ideally, any model that is developed to predict sediment suspension under waves should account for the suspension characteristics outlined in the wave cycle concept of suspension. Neilsen (1993) investigated several different models for predicting sediment suspension under waves and concluded that simple models worked better than complex models. Simple models work better than complex models because the contribution of many of the variables in the process of sediment suspension is poorly understood as are the interrelationships between the variables. However, the simplest model, where sediment suspension is predicted from the value of the flow velocity, also fails to provide reasonable predictions because suspension maximums occur at a different phase to velocity maximums. Neilsen (1993) concluded that the shift in phase between velocity and concentration had to be included in the derivation of a model.

Atkins (1993) used the wave cycle model of sediment suspension to suggest that a simple model for sediment suspension could be derived as a function of the flow velocity shifted with respect to the suspended sediment concentration by a time lag equivalent to the phase shift between the two variables. The model predicted total suspended sediment concentration as the sum of convective, advective, and diffuse components. The components were predicted from the flow velocity using simple sinusoidal relationships. In experimental tests, the model was able to explain nearly 80% of the variation in sediment suspension measured under a synthesized irregular wave spectrum. Both peaks and lows in suspended sediment concentration were reasonably well predicted both in magnitude and through time. The success of the model suggests that the wave cycle model of sediment suspension under waves is an adequate starting point for further research into suspension mechanisms.

Conclusions

Sediment suspension under waves is a complex interplay between the currently identified mechanisms of convection, advection, and diffusion.

This interplay results in strong temporal relationships between water velocity and the concentration of suspended sediment. A group of successive waves energetic enough to suspend material can keep sediment suspended for extended periods of time. The predictive successes of simple models based on the physical mechanisms provides strong evidence that simple functional relationships between velocity and concentration may be used to predict sediment suspension under waves. A better understanding of the processes controlling the mechanisms of sediment suspension, their time dependencies, and their inter-relationships will lead to better predictive models of suspension and, ultimately, better models of sediment transport. Better models of sediment transport would then allow for better predictions of coastal evolution and increase our understanding of how and why coastlines change.

Rowland J. Atkins

Bibliography

Atkins, R.J., 1993. Sediment suspension under irregular "groupy" waves: a laboratory experiment. Unpublished Master of Science thesis, Toronto: University of Toronto.

Bagnold, R.A., 1963. Mechanics of marine sedimentation. In Hill, M.N. (ed.), *The Sea, Volume 3*. New York: Wiley-Interscience, pp. 507–528.

Bailard, J.A., 1984. A simplified model for longshore transport. *Proceedings of the 19th Coastal Engineering Conference*. New York: American Society of Civil Engineers, pp. 1454–1470.

Dyer, K.R., 1980. Velocity profiles over a rippled bed and the threshold of movement of sand. *Estuarine and Coastal Marine Science*, **10**: 181–199.

Honji, H., Kaneko, A., and Matsunaga, N., 1980. Flow above oscillatory ripples. *Sedimentology*, **27**(2): 225–229.

Neilsen, P., Svendsen, A., and Staub, C., 1979. Onshore–offshore sediment movement on a beach. *Proceedings of the 16th Coastal Engineering Conference*. New York: American Society of Civil Engineers, pp. 1475–1492.

Neilsen. P., 1993. Combined convection-diffusion modelling of sediment entrainment. *Proceedings. In Edge, B.L. (ed.), 23rd Coastal Engineering Conference*. New York: American Society of Civil Engineers, pp. 3202–3215.

Osborne, P., and Greenwood, B., 1993. Sediment suspension under waves and currents: time scales and vertical structure. *Sedimentology*, **40**: 599–622.

Rouse, H., 1937. Modern conceptions of the mechanics of turbulence. *Transactions of the American Society of Civil Engineers*, **102**: 436–505.

Sleath, J.F.A., 1982. The suspension of sand by waves. *Journal of Hydraulic Research*, **20**(5): 439–452.

Sternberg, R.W., Shi, N.C., and Downing, J.P., 1989. Continuous measurement of suspended sediment. In Seymour, R.J. (ed.), *Nearshore Sediment Transport*. New York: Plenum Press, pp. 231–259.

Tunstall, E.B., and Inman, D.L., 1975. Vortex generation by oscillatory flow over rippled surface. *Journal of Geophysical Research*, **80**(24): 3475–3484.

Cross-references

Beach Processes
Coastal Changes, Rapid
Cross-Shore Sediment Transport
Longshore Sediment Transport
Nearshore Sediment Transport Measurement
Ripple Marks
Surf Zone Processes
Waves

SEDIMENT TRANSPORT—See CROSS-SHORE SEDIMENT TRANSPORT; LONGSHORE SEDIMENT TRANSPORT

SEDIMENTARY BASINS

Regional subsidence in coastal areas is mainly due to the occurrence of sedimentary basins. Figure S26 indicates the major deltas of the world (Fairbridge and Jelgersma, 1990); many situated on top of sedimentary basins. Before giving some examples of basins situated in the coastal zone some information should be given about the origin of those basins.

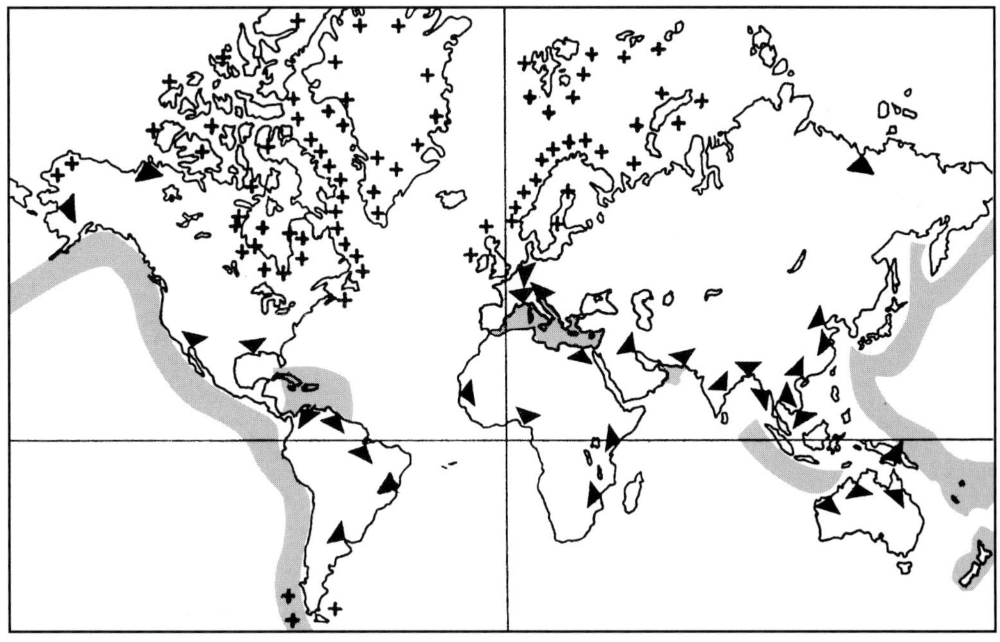

Figure S26 Location map of coastal instability (from Fairbridge and Jelgersma, 1990; with the kind permission of Kluwer Academic Publishers).

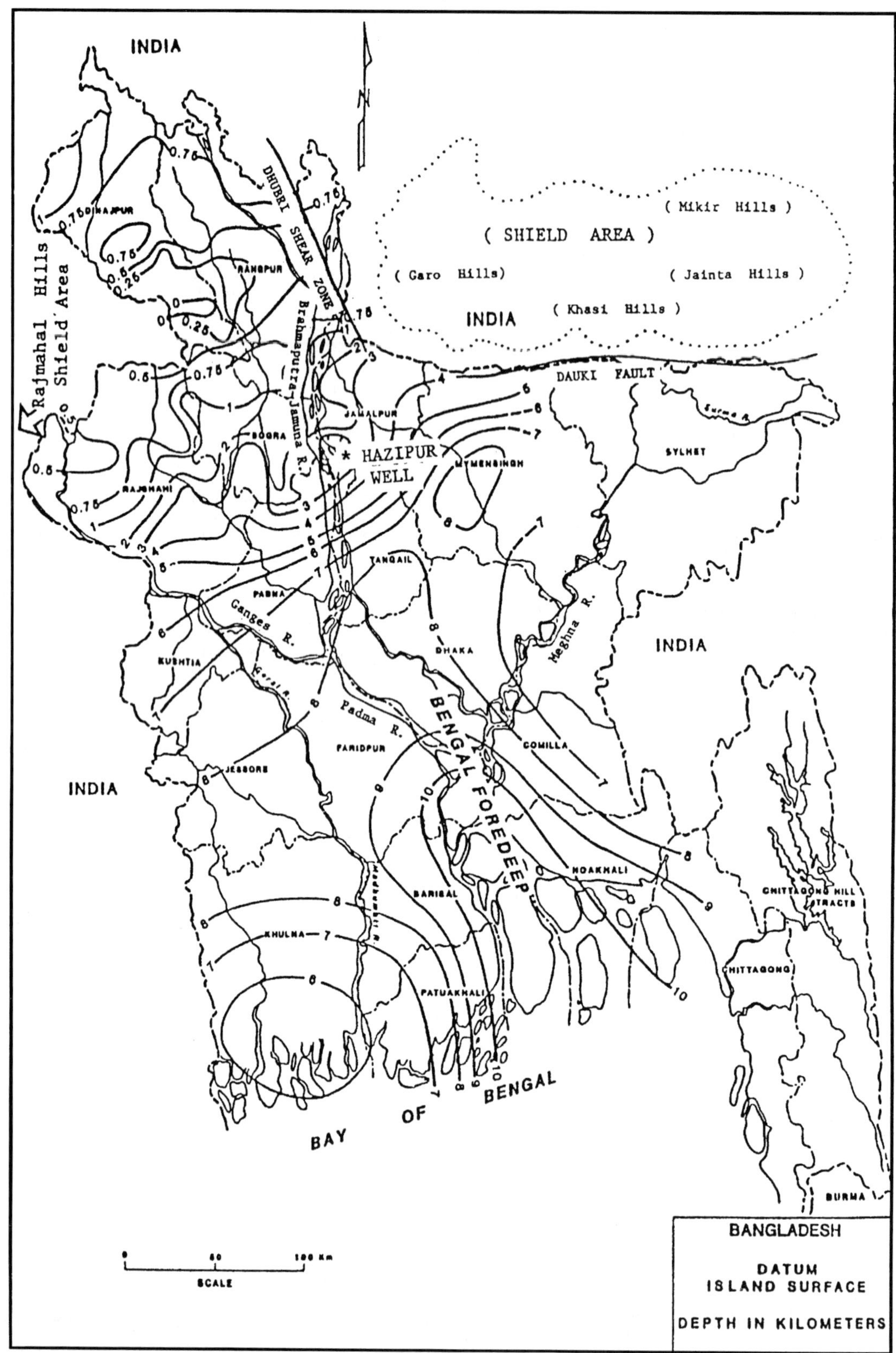

Figure S27 Depth to basement in the Bengal basin of Bangladesh (from Alam 1996; with the kind permission of Kluwer Academic Publishers).

Sedimentary basins are places of prolonged slow subsidence in general, compensated by sediment input of rivers and the sea. In recent time, however, this sediment input in several deltas was strongly reduced due to the construction of dams and reservoirs in the upstream area of the river and to other human interference. This so called "sand starvation" results in surface lowering and consequently in shore erosion. The subsurface layers of sedimentary basins have been subject to intensive investigations, many studies have been made of their stratigraphic sequences and their relation to sea-level changes.

During the last decades the concept of plate tectonics has given a more detailed idea as to how these basins are formed. The lithosphere consists of a number of plates which are in motion. Accordingly, sedimentary basins exist in an environment of the motion of these plates. Basins can either be formed by collision of plates or by lithospheric stretching of the plates. The first mentioned activity gives rise to basins due to flexures, they are present as foreland basins at the foot of mountain belts. The Bengal basin of Bangladesh and the Po basin of Italy are examples of these flexures and will be discussed below. Lithospheric stretching of plates causes fault zones which become later, due to thermal cooling, important areas of subsidence. Examples of these rift basins are the North Sea basin and the Niger basin. Another genetic class of rift basins are associated with strike-slip deformation. The Tertiary rift basins onshore and offshore of Thailand and Malaysia are associated with this phenomenon. A few samples of sedimentary basins present underneath important deltas are presented below.

The Po delta

The river plain overlying the Po–Veneto sedimentary basin is enclosed by the Alpine mountain chain in the north and by the Apennines in the south. Collision of the African plate with the European plate in the Late Cretaceous was responsible for this landscape. The Po–Veneto basin came into being and due to the slow but continuous subsidence, thick layers of sediment were deposited. Mapping the base of the Pliocene–Quaternary sequence has indicated several fault zones in the southest part of the plain. The Quaternary deposits reach a thickness of more than 1,000 m.

In geological terms, the lobate Po delta is a very young phenomena which came into being in historical time by an increase in sediment discharge caused by deforestation in the uplands and an increase in precipitation (Little Ice Age). In recent time reforestation, construction of dams and reservoirs, and human activities in the delta itself have caused sediment starvation, subsidence, and consequently shore erosion.

Ganges–Brahmaputra delta

The Bengal basin is a product of the collision of the Indian plate with the Burmese and Tibetan plate. From the Cretaceous onwards thick layers of sediments have been laid down in this basin due to tectonic and isostatic subsidence. The uplift and erosion of the Himalayas mountain range are the main sources of the sediments filling the basin. The thickness of the sedimentary sequences is demonstrated in Figure S27 by a contour line on the depth of the basin (Jones, 1985; Alam, 1996). Subsidence rates are thought to be 2–5 cm/century.

The Ganges–Brahmaputra delta and plain of India and Bangladesh have many fault zones which may have contributed to the shifting of river courses and the differences in surface lowering. Other tectonic phenomena are the occurrence of many earthquake epicenters. The Meghna fault zone, NE–SW direction, continues offshore in the Bay of Bengal and ends in the canyon of the Swatch of no Ground.

The alluvial and deltaic plain are subject to seasonal flooding by rivers and cyclonic surges. Due to the high tidal range the delta can be classified as tide-dominated.

Niger delta

The Niger delta has been developed since the Cretaceous when tectonic events culminated in seafloor spreading between the African and South American lithospheric plates. The post-drift period was characterized by instability and subsidence, due to thermal cooling, in the rift valleys and the rift margins. Due to this subsidence more than 8,000 m of Tertiary and Quaternary sediments are present (Figure S28) in the Niger sedimentary basin (Whiteman, 1982). Like all deltas in the world the recent delta sediments consist of a wedge of fine-grained sediments formed as a result of the post-glacial rise in sea level.

The coast of the delta is formed by small barrier islands separated by inlets and backed by an extensive area of mangrove swamps. The delta can be classified as wave-dominated with important input of river sediment. The latter has compensated for the subsidence but in recent time due to the construction of several dams and reservoirs the sediment supply is strongly reduced. This sediment starvation in the delta

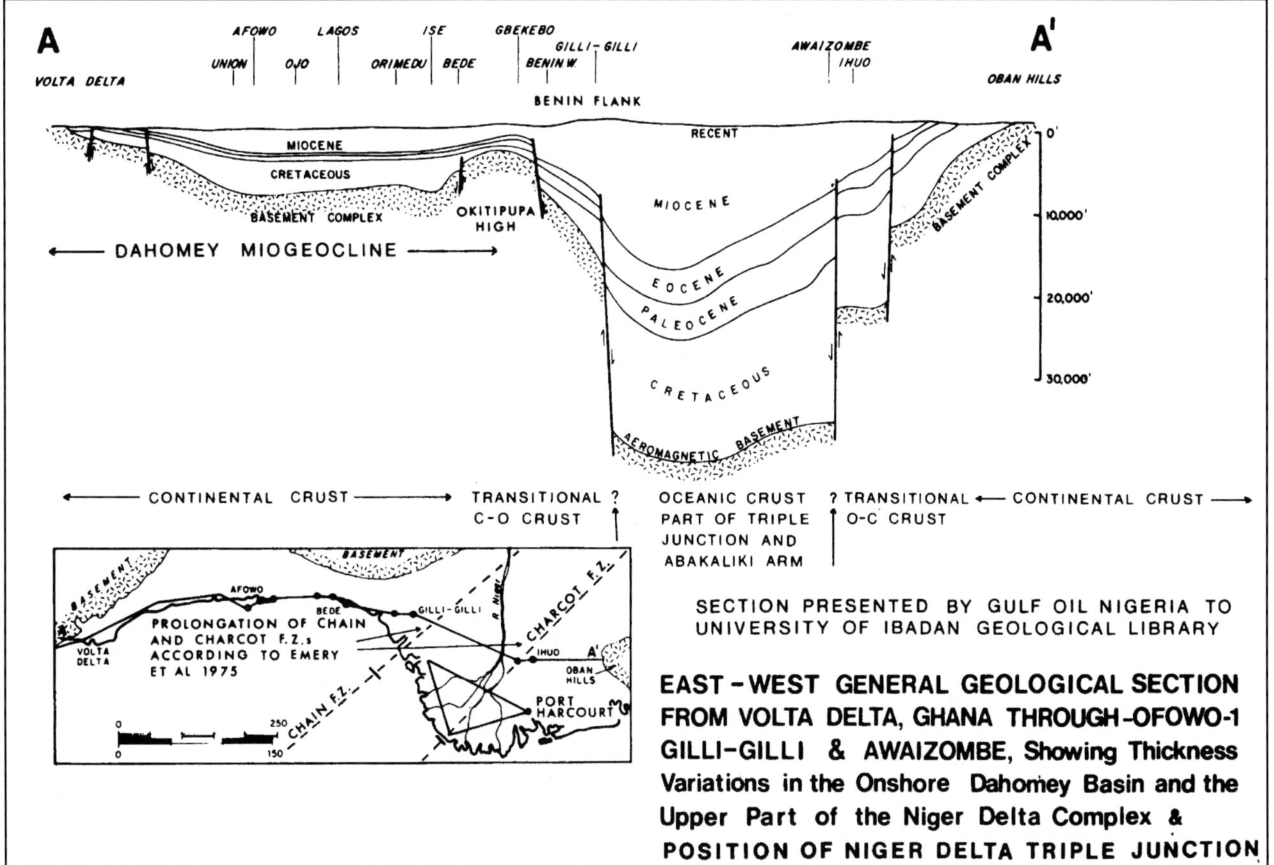

Figure S28 Cross-section of the Niger sedimentary basin (from Whiteman, 1982; with the kind permission of Kluwer Academic Publishers).

and the shore has resulted in important coastline retreat. Human activity in the delta, mining of gas, oil and water, has also contributed to surface lowering.

North Sea basin

The North Sea sedimentary basin developed during Late Cretaceous and Early Tertiary rifting phases marking the onset of seafloor spreading in the Arctic North Atlantic between the European and the North American–Greenland plate. At the same time the early phases of the Alpine orogeny took place.

Thermal cooling in the rift systems of the Early Tertiary caused important subsidence. Consequently, thick layers of Cenozoic sediments could be deposited. As demonstrated in Figure S29, the Cenozoic series reaches a maximum thickness of 3,500 m of which the Quaternary reaches a maximum of 1,000 m (Ziegler and Louwerens, 1977; Jelgersma, 1980). The Netherlands and NW Germany are situated on

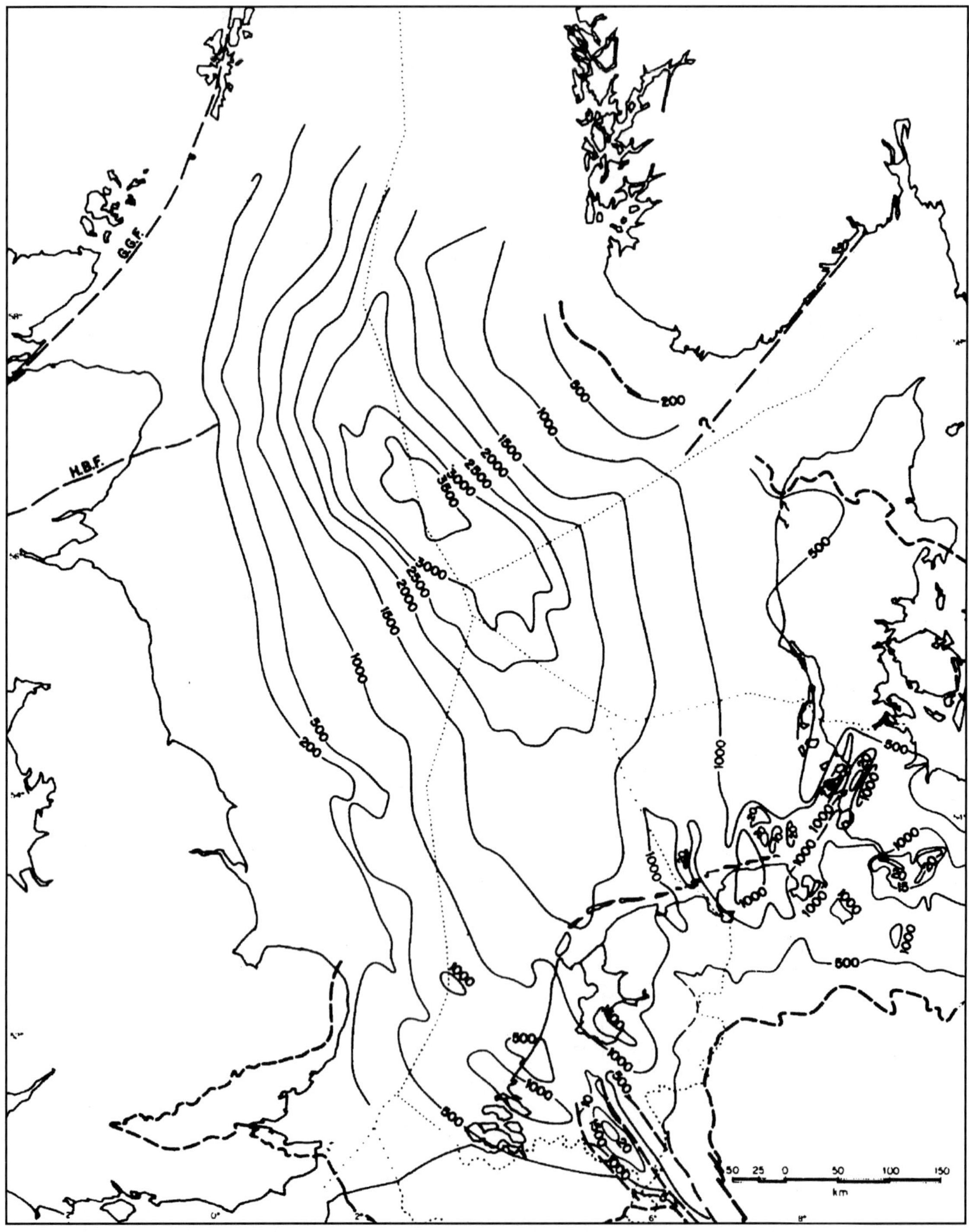

Figure S29 Depth contour of the base of the Tertiary (from Jelgersma, 1980, reproduced with permission of John Wiley & Sons Limited).

the edge of the subsiding North Sea basin. Tectonic subsidence occurs in these areas, contributing to a relative sea-level rise. The combined effect is measured by tide gauges, its amount being between 15 and 20 cm/century.

Mississippi delta

Sedimentation in the northern Gulf of Mexico has taken place since the end of the Cretaceous in a series of depocenters on the edge of the continental shelf. Sediment accumulation in the depocenters during the Neogene and the Quaternary reached high values as demonstrated in Figure S30 by the depth contour line of the base of the Pleistocene and a cross section of the depocenter (Woodbury *et al.*, 1973; Jelgersma, 1996).

The present Mississippi delta consists of a wedge of soft sediment deposited during the post-glacial rise in sea level. The river delta can be classified as river-dominated; the tidal range is nearly absent and serious wave attack occurs only during hurricanes.

Subsidence rates in the delta are locally very great; this is mainly due to human interference. Accordingly, much of the marshland surrounding the river delta itself has changed into lakes.

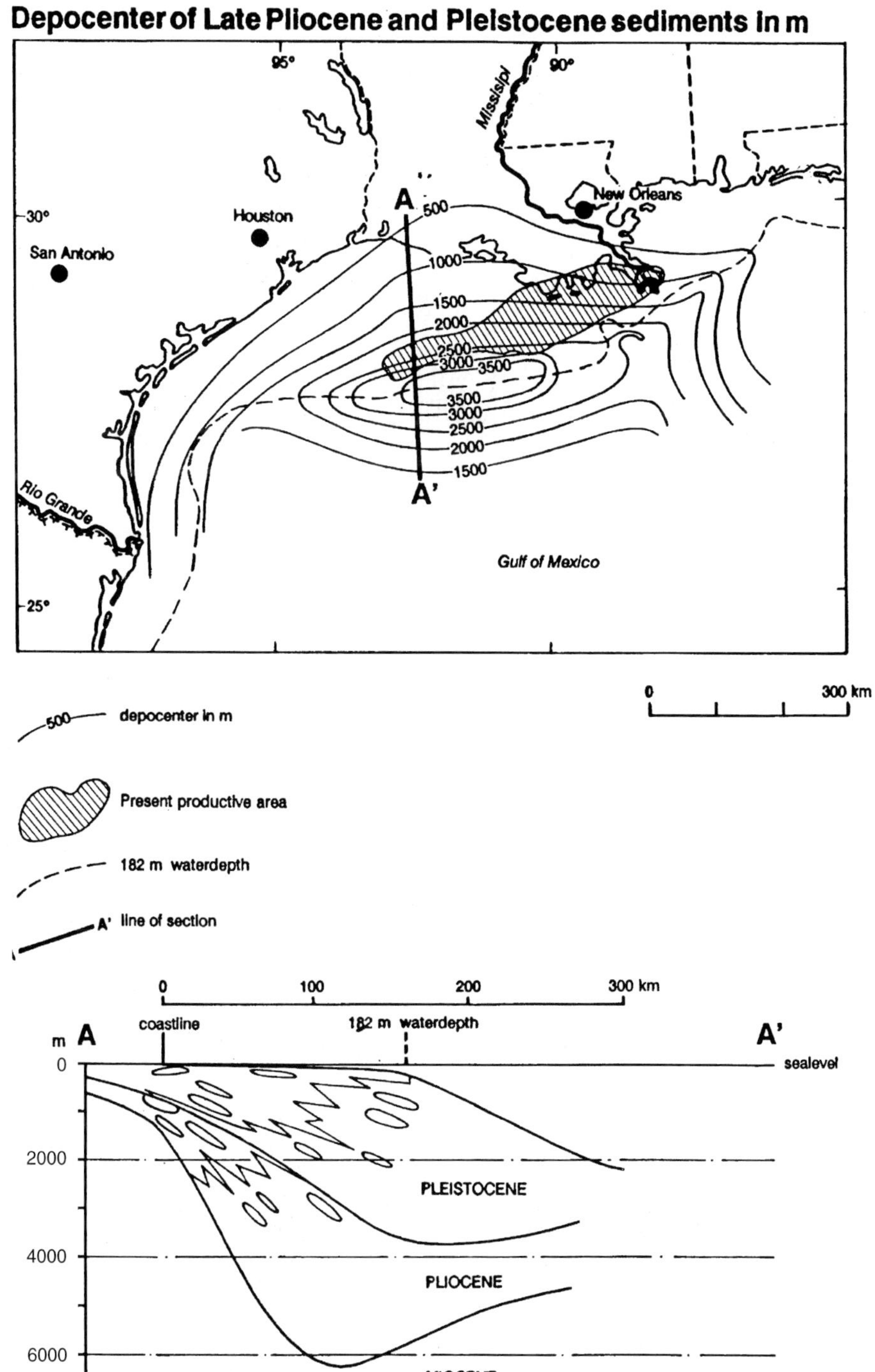

Figure S30 Depth contour of the base of the Pleistocene and a cross-section of the depocenter of the Mississippi (from Jelgersma, 1996; with the kind permission of Kluwer Academic Publishers).

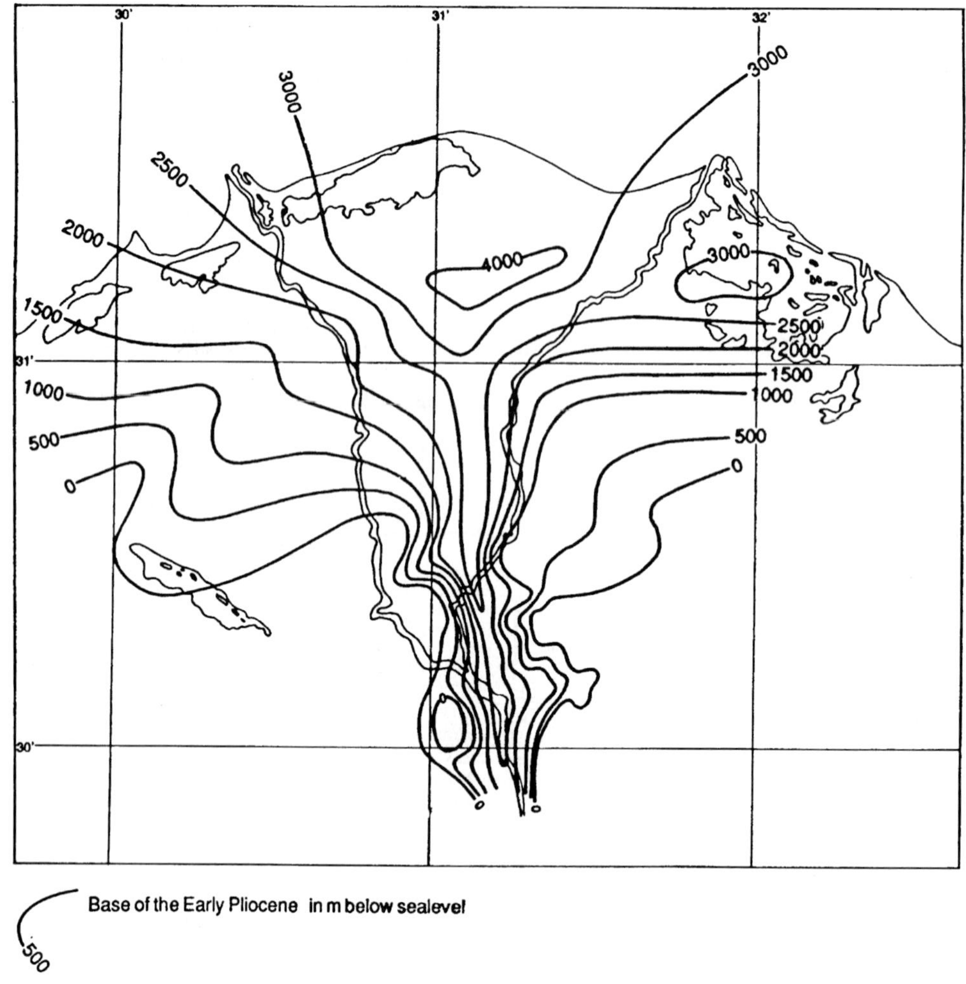

Figure S31 Depth contour of the base of the Early Pliocene in the Nile delta (from Jelgersma, 1996; with the kind permission of Kluwer Academic Publishers).

Nile delta

Crustal movements in the western Mediterranean during the Late Miocene obstructed the connection between the Atlantic Ocean and Mediterranean in the Straits of Gibraltar. The interrupted inflow from the Atlantic led to a gradual desiccation of the Mediterranean. Due to this lowering of the base level a deep canyon was cut down in the area of the recent Nile delta. This Late Miocene channel reaches in the northern embayment of the delta to a depth of 4,000 m. In the beginning of the Pliocene, a renewed opening between the Atlantic and the Mediterranean took place. The inflow of ocean water restored the level of the Mediterranean, consequently the Nile canyon was flooded and a thick layer of Pliocene, Pleistocene, and Recent deposits could be formed and is present in the subsurface of the delta.

In Figure S31, a contour map of the base of the Pliocene is presented; the Pliocene and the Quaternary deposits dip northwards reaching a thickness of 4,000 m (Said, 1981; Jelgersma, 1996). It is evident that in the northern embayment the subsidence reaches high values, probably caused by fault zones. The Nile delta has been a river-dominated delta; the input of river sediment compensating for the basin subsidence. During the last decades, due to the construction of the High Aswan dam, sediment input in the delta has been strongly reduced. This resulted in serious shore erosion and salt-water intrusion. The delta then changed from river- to wave-dominated.

Saskia Jelgersma

Bibliography

Alam, Mahmood, 1996. Subsidence of the Ganges–Brahmaputra delta of Bangladesh and associated drainage, sedimentological and salinity programs. In Milliman, John, D., and Haq, Bilal, U. (eds.), *Sea-Level Rise and Coastal Subsidence*. Dordrecht: Kluwer, pp. 169–192.

Fairbridge, R.W., and Jelgersma, S., 1990. Sea-level. In Paepe, Roland, Fairbridge, Rhodes, W., and Jelgersma, Saskia, (eds.), *Greenhouse Effect, Sea-Level and Drought. Nato ASI Series, Series C: Mathematical and Physical Sciences*, Volume 325. Dordrecht: Kluwer, pp. 117–145.

Jelgersma, Saskia, 1980. Late Cenozoic sea-level changes in the Netherlands and adjacent North Sea basin. In Morner, N-A. (ed.), *Earth Rheology, Isostasy and Eustasy*. New York: Wiley-Interscience, pp. 435–447.

Jelgersma, Saskia, 1996. Land subsidence in coastal lowlands. In Milliman, John, D., and Haq, Bilal, U. (eds.), *Sea-Level Rise and Coastal Subsidence*. Dordrecht: Kluwer, pp. 47–62.

Jones, P.H., 1985. *Geology and Groundwater Resources of Bangladesh*. Baton Rouge: P.H. Jones Hydro-Geology Inc.

Said, R., 1981. *The Geological Evolution of the River Nile*. New York: Springer-Verlag.

Whiteman, A.J., 1982. *Nigeria: its Petroleum Geology, Resources and Potential*, Volume 2. London: Graham and Trotman.

Woodbury, H.O., Murry, I.B., Pickford, P.J., and Akita, W.H., 1973. Pliocene and Pleistocene depocenters, outer continental shelf Louisiana and Texas. *American Association of Petroleum Geologists Bulletin*, **49**: 1893–1907.

Ziegler, P.A., and Louwerens, C.J., 1977. Tectonics of the North Sea. In Oele, E., Schuttenhelm, R.T.E., and Wiggers, A.J. (eds.), *The Quaternary History of the North Sea*. Acta University Symposium, Uppsala.

Cross-references

Coastal Changes, Gradual
Coastal Subsidence
Dams, Effect on Coasts
Deltaic Ecology
Deltas
Sequence Stratigraphy
Submerged Coasts
Submerging Coasts

SEISMIC DISPLACEMENT

When topographical changes are produced by an earthquake or a succession of earthquakes, a distinction is usually made between precursory displacements that occurred before the event (preseismic), the coseimic displacements during the event, and postseismic changes shortly later, often accompanying after-shocks.

Seismic displacements have a horizontal and a vertical component. The horizontal component is often clearly visible along fault lines activated by the earthquake, though more or less regular horizontal deformation may extend along the surface of nearby crustal blocks. In coastal areas, the vertical component is easily measurable and most important from a geodetic point of view, because it changes the relation to sea level.

As noted by Vita-Finzi (1986) earthquakes were being documented long before progressive uplift or depression attracted the attention of geologists. Several past events, important for their casualties and destructions, have been reported by ancient writers. Most past observers tend to say little or nothing about any ground movements accompanying an earthquake. Nevertheless, some exceptions exist that mention sudden seismic changes in the past, though seldom separating the effects of an earthquake from those of its after-shocks. According to several historical sources (Guidoboni *et al.*, 1994) in the summer of BC 426, when an earthquake struck the Gulf of Malia, in Greece, the central part of Atalante, near Euboea, was split open to the extent that ships could pass through, and some of the plains were flooded as far as 20 stades (ca. 4 km). It is also reported that in a winter night in BC 373, after a violent earthquake, the city of Helice, in the Gulf of Corinth, was drowned by the sea with all of its inhabitants. (Schwartz and Tziavos, 1979).

From ancient Japan, several examples of seismic displacements in coastal areas are reported by Imamura (1937). In AD 684, at the time of the Tosa earthquake in Shikoku Island, cultivated ground to the extent of 8.25 km^2 is said to have sunk beneath the sea. On September 4, 1596, at the time of a series of earthquakes, the island of Uryû-jima, slightly off the present city of Oita, with an area of 4 km N-S by 2.3 km E-W and a population of 5,000, sank beneath the waves, where it now lies at a depth of 30–40 fathoms.

Seismic changes in land level are often mere effects of the tilting of crustal blocks. In Shikoku Island, during the 1707 earthquake, subsidence at Kochi reached nearly 2 m, and upheaval 2.1–2.4 m near Yoshiwara, while two belts of subsidence occurred in the western half of the province, indicating that two contiguous crustal blocks had tilted, both of which dipping north.

Important seismic vertical displacements have been reported also from coastal regions in other parts of the world. In the delta area of the river Indus, at the time of the Kutch earthquake, on June 19, 1819, a fault scarp 3 m high appeared, extending for a distance of 80 km parallel to the coast. The region SE of this locality, about 5,200 km^2 in area began to sink, and within 3 h after the earthquake had changed into a sea.

In Chile, after the Valparaiso earthquake on November 19, 1822, the coast had risen 1 m, while at Quintero, some 24 km northwards, it was 1.2 m. According to Lyell (1875), the whole country from the foot of the Andes was raised on this occasion, the maximum rise being at a distance of about 3.2 km offshore.

Rigorous measurements of topographical changes produced by earthquakes started to be carried out toward the end of the 19th century. The geodetic investigation of ground deformation immediately after an earthquake, though often logistically problematic, offers the advantage that the area to be surveyed can be identified without much difficulty. The first post-event triangulation was done in Sumatra in 1892. The second survey was executed after the 1897 Indian earthquake and found a maximum vertical displacement of 3.6 m. Repeated leveling surveys carried out in seismic areas before and after a major earthquake, have helped to clarify the local pattern of crustal deformation. Tide gauge records, when located in areas affected by earthquakes, can also provide very precise, continuous information on all the preseismic–coseismic–postseismic sequence. Recently, complete control of land displacements can be furnished all over the world by satellite geodesy.

For the past, coastal coseismic displacements deduced from stepped elevated coastlines have been reported from several seismically active parts of the world. In some cases, detailed analysis of closely dated fossil coastlines (Figure S32) promises to reveal regional forms of the geoid

Figure S32 A crustal block approximately 200 km long, including all the western part of Crete and the nearby Antikythira Island (Greece), was uplifted and tilted coseismically by a great earthquake, probably on July 21, 365 AD. The upheaval reached 9 m in the southwestern part of Crete island (Pirazzoli *et al.*, 1996). In this photograph (No. 3980, Oct. 1977), taken near Piper Eliá, on the east side of Gramvousa Peninsula, Crete, continuous erosional marks left by the sea level before the uplift are well visible on the limestone cliff at about 5 m in elevation.

at successive time intervals and thus extend the brief period spanned by satellite data into the Holocene and possibly beyond.

Preseismic displacements are difficult to demonstrate for past events, as well as in historical times. This is because, in most cases, they are not reported. Preseismic movements are unexpected, may last for decades, and be sufficiently slow to escape notice, also in coastal areas. Such was the case, for example, for Cefalonia Island (Greece), before the 1953 earthquake that caused coseismic uplift of 30–70 cm along the south coast of the island. In the 1990s, however, a survey carried out along the uplifted coast (Pirazzoli *et al.*, 1994) was able to demonstrate that, near Karavomilos, a slow 15 ± 5 cm preseismic submergence must have occurred within a few years before the earthquake, therefore preceding the coseismic uplift that reached about 50 cm at this site.

Paolo A. Pirazzoli

Bibliography

Guidoboni, E., Comastri, A., and Traina, G., 1994. *Catalogue of Ancient Earthquakes in the Mediterranean Area*. Rome: Istituto Nazionale di Geofisica.

Imamura, A., 1937. *Theoretical and Applied Seismology*. Tokyo: Maruzen.

Lyell, C., 1875. *Principles of Geology*. London: J. Murray.

Pirazzoli, P.A., Stiros, S.C., Laborel, J., Laborel-Deguen, F., Arnold, M., Papageorgiou, S., and Morhange, C., 1994. Late-Holocene shoreline changes related to palaeoseimic events in the Ionian islands, Greece. *The Holocene*, **4**: 397–405.

Pirazzoli, P.A., Laborel, J., and Siros, S.C., 1996. Earthquake clustering in the eastern Mediterranean during historical times. *Journal of Geophysical Research*, **101**(B3): 6083–6097.

Schwartz, M.L., and Tziavos, C., 1979. Geology in the search for ancient Helice. *Journal of Field Archaeology*, **6**: 243–252.

Vita-Finzi, C., 1986. *Recent Earth Movements*. London: Academic Press.

Cross-references

Changing Sea Levels
Faulted Coasts
Submerging Coasts
Tectonics and Neotectonics
Uplift Coasts

SEQUENCE STRATIGRAPHY

Sequence stratigraphy is an informal chronostratigraphic methodology that uses stratal surfaces to subdivide sedimentary successions. Unlike most traditional lithostratigraphic units (NACSN, 1983), which are defined as regionally mappable packages (members, formations, groups) of similar lithologies (rock types), sequence stratigraphic units trend across traditional lithostratigraphic boundaries (Figure S33). This methodology owes its origins to the pioneering work of Caster (1934), Sloss (1963), Campbell (1967), and Asquith (1970). All of these workers documented that stratal surfaces trend across traditional lithostratigraphic boundaries, and concluded that stratal surfaces represent time-significant boundaries that can be used to define coeval packages of strata contained within different lithostratigraphic units (Figure S33).

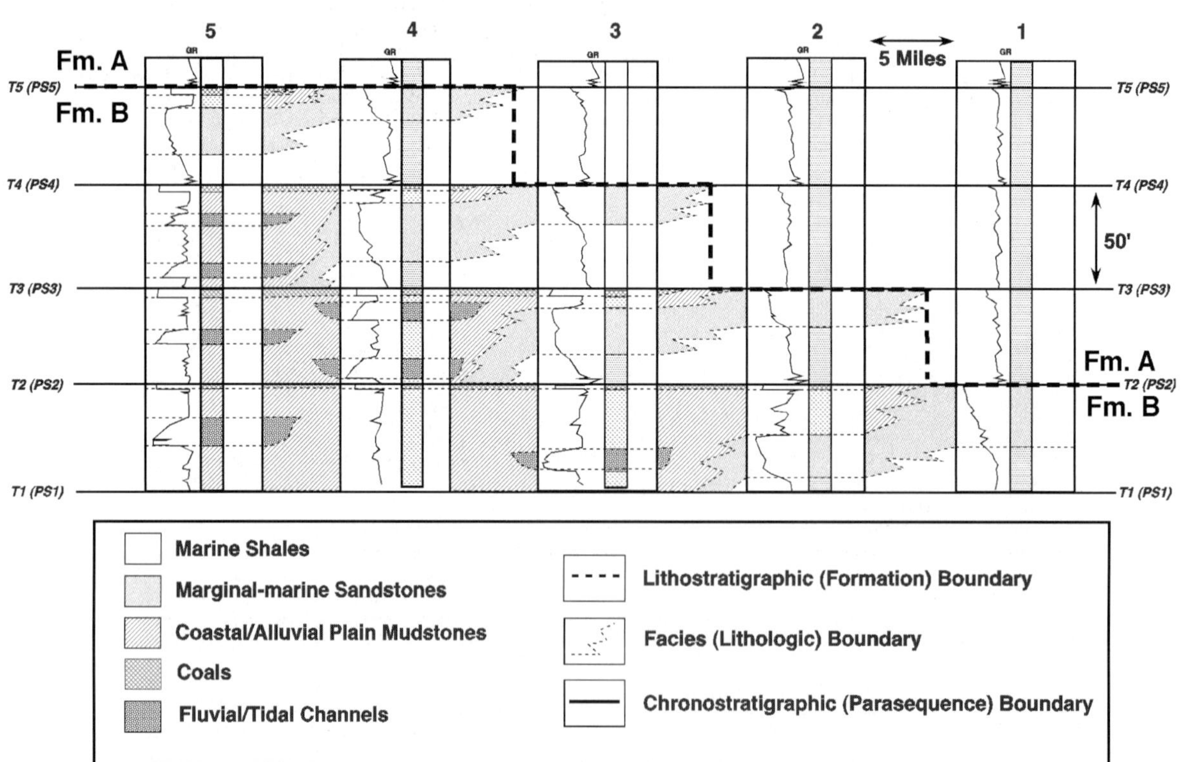

Figure S33 Idealized well-log cross section that illustrates the differences among facies, lithostratigraphic, and sequence stratigraphic (chronostratigraphic) correlations. Based on detailed correlation of well-log markers, five (5) regional flooding surfaces (parasequence boundaries) were identified. The parasequence boundaries represent chronostratigraphically significant surfaces that separate older strata below from younger strata above. Each parasequence contains a coeval facies succession that is dominated by marine shales downdip and coastal/alluvial plain deposits updip. The chronostratigraphic framework provided by the stratal (parasequence) correlations document that the traditional lithostratigraphic (formation) boundaries are time-transgressive. In the example provided, the lithostratigraphic boundary between Formation A (marine shale) and Formation B (marine shales, marginal marine sandstones, coastal plain mudstones, coals, and fluvial/tidal channels) occurs at time T2 (PS2) in Well 1 and time T5 (PS5) in Well 5.

The original concept of a sequence, a stratigraphic unit bounded by unconformities and their correlative conformities, was outlined by Sloss (Sloss *et al.*, 1949; Sloss, 1963). Using the Phanerozoic sedimentary succession of the North American craton as an example, Sloss defined six interregional unconformities and six unconformity-bounded units which he termed sequences. While this methodology gained little interest among most geoscientists in the 1950s, 1960s, and early 1970s, it did find a niche following among petroleum geologists who were using seismic data to determine subsurface stratigraphy. With the publication of the landmark "Memoir 26" on seismic stratigraphy by the American Association of Petroleum Geologists (AAPG) in 1977 (Payton, 1977), however, sequence stratigraphic methodology and concepts became an active topic of research and debate among geoscientists. Contained within this volume is the now classic 11-part paper written by researchers at Exxon Production Research Company (Vail *et al.*, 1977a). This entry outlined the fundamentals of seismic stratigraphy and offered the Depositional Sequence as the basic unit for stratigraphic analysis. Please note that while the unconformity-bounded sequence of Sloss (1963) covered thousands of feet and were typically a hundred million years in duration, the depositional sequences of Vail and others (1976) were commonly hundreds of feet thick and a million years in duration. As defined by Mitchum (1977), the *Depositional Sequence* is "… a stratigraphic unit composed of a relatively conformable succession of genetically related strata and bounded at its top and base by unconformities or their correlative conformities." The basic premise behind this methodology was the observation that "… primary seismic reflections are generated by physical surfaces in the rocks, consisting mainly of stratal (bedding) surfaces and unconformities with velocity-density contrasts." Furthermore, Vail and Mitchum (1977) concluded that since "… all the rocks above a stratal or unconformity surface are younger than those below it, the resulting seismic section is a record of the chronostratigraphic (time-stratigraphic) depositional and structural patterns and not a record of the time-transgress lithostratigraphy (rock stratigraphy)." Using datasets from the Cretaceous of North America and the Tertiary of South America, Vail *et al.* (1977d) illustrated that seismic stratigraphic surfaces trend oblique to traditional lithostratigraphic boundaries and define coeval facies successions among adjacent formations. Sequence stratigraphic analysis, as defined by Vail and Mitchum (1977), was based on the identification of stratigraphic units composed of a relatively conformable succession of genetically related strata termed depositional sequences. The upper and lower boundaries of depositional sequences are surfaces defined by unconformities and their correlative conformities. These surfaces are termed *Sequence Boundaries*. Vail and others (1976) believed that sequence boundaries (Figure S34) could be objectively identified on seismic as through-going surfaces defined by stratal (reflection) terminations both below and above. Stratal (reflection) terminations beneath sequence boundaries included truncation and toplap, while stratal (reflection) terminations above sequence boundaries included onlap and downlap (Mitchum, 1977). It should be noted, however, that in subsequent sequence stratigraphic publications (Van Wagoner *et al.*, 1990) downlap was no longer utilized as criteria for sequence boundary identification. It is now believed that downlap more commonly occurs within sequences, and should be used to define compartments (systems tracts) within sequences.

Because the characteristics of sequence boundaries and sequences appeared similar in a wide variety of basins through geologic time, Vail *et al.* (1977b) concluded that cycles of relative sea-level change were the primary controlling factor in sequence development. In the seismic stratigraphic methodologies outlined in Vail *et al.* (1977b), relative sea-level rises were interpreted by (coastal) onlap, relative sea-level still-stands by (coastal) toplap, and a relative sea-level falls by a downward (basinward) shift in (coastal) onlap (Figure S34). Within this paradigm, the physiographic break "*shelf edge*" along the basal sequence boundary serves as a major reference point for a given sequence. This shelf edge (Figure S34) permits delineation of lowstands, the interval of time when sea level (coastal onlap) is interpreted below the shelf edge, and highstands, the interval of time when sea level (coastal onlap) is interpreted above the shelf edge (Vail *et al.*, 1977b).

One final premise contained within the paper by Vail *et al.* (1977) is that global cycles of sea-level change are evident throughout the Phanerozoic, and that these cycles are globally synchronous. This contention was based on the belief of the Exxon researchers that the depositional sequences they defined on different continental margins were coeval and of similar magnitude (Vail *et al.*, 1977c). They concluded that these interpreted global cycles were records of geotectonic, glacial, and other large-scale processes and reflected major events of the Phanerozoic history. Based on this paradigm, Vail *et al.* (1977c) offered preliminary cycle charts to document the timing of the global (eustatic) events that controlled sequence development. Haq *et al.* (1987, 1988), as well as deGraciansky *et al.* (1998), have subsequently published updated versions of "The Cycle Charts."

With the publication of SEPM Special Publication Number 42 in 1988 (Wilgus *et al.*), seismic stratigraphy evolved into sequence stratigraphy. This volume also contained numerous articles by Exxon researchers who presented updated sequence models based on more modern seismic, as

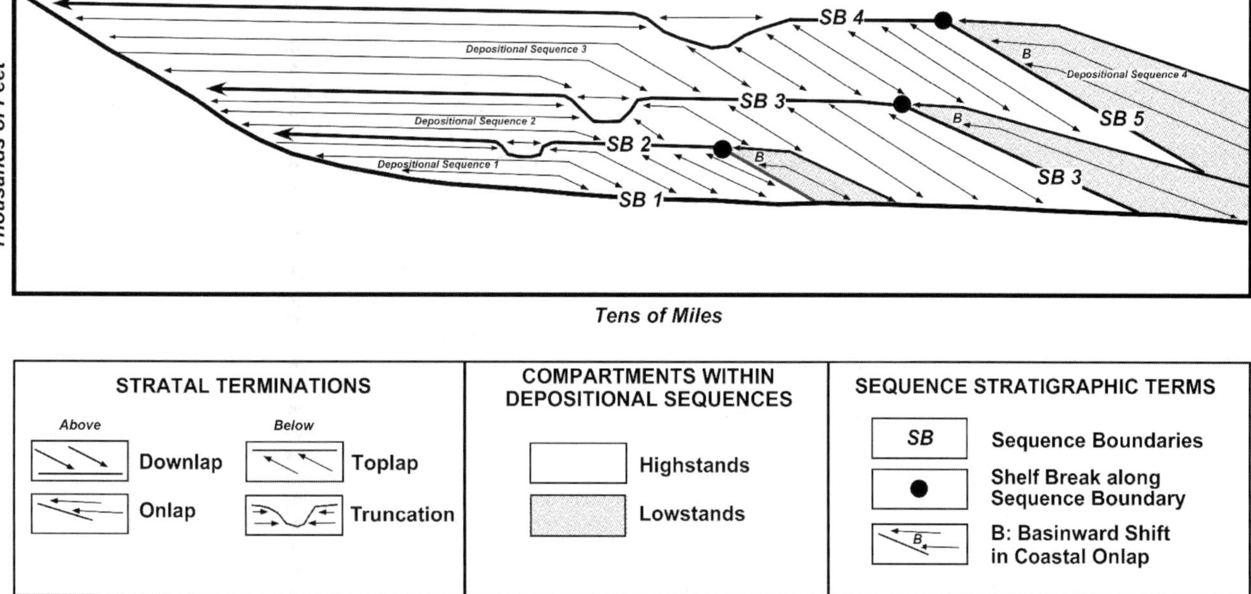

Figure S34 Seismic stratigraphic analysis, as outlined by Vail *et al.* (1977), was based on the identification of depositional sequences: a relatively conformable succession of genetically related strata. The upper and lower boundaries of depositional sequences are surfaces defined by unconformities and their correlative conformities (sequence boundaries). Sequence boundaries can be objectively identified on seismic as through-going surfaces defined by stratal terminations both above and below.

well as the integration of outcrop, core, and well-log data. A third publication by Exxon researchers (Van Wagoner *et al.*, 1990) continued to refine sequence stratigraphic methodologies based on outcrop, core, and well-log data. In this volume, Van Wagoner *et al.* (1990) proposed a hierarchy of stratal surfaces (beds, bedsets, parasequence, parasequence sets, sequences) all of which had chronostratigraphic significance. While the depositional sequence continued to be the fundamental stratal unit in sequence stratigraphy, parasequences were identified as the building blocks of sequences. *Parasequences* were defined as relatively conformable succession of genetically related strata bounded by marine-flooding or correlative surfaces (Van Wagoner *et al.*, 1990). Parasequence sets were defined as a succession of genetically related parasequences that form a distinctive stacking pattern (Van Wagoner *et al.*, 1990). Based on outcrop and subsurface observations Van Wagoner *et al.* (1988, 1990) proposed three basic compartments or *systems tracts* (lowstand, transgressive, and highstand) within sequences (*Figure* S35). These systems tracts were defined by major stratal boundaries, parasequence set stacking, as well as temporal and spatial position within sequences. Surprisingly, they were not defined by interpreted sea-level variations as their names might suggest. The lowermost compartment the *lowstand systems tract* is bounded by the sequence boundary at its base and the transgressive surface at its top (*Figure S35). The transgressive surface* is defined as first significant marine-flooding surface (parasequence boundary) across the shelf within a depositional sequence. In higher-relief basins (Figure S34), where the basal sequence boundary displays a well-defined "shelf break" (slope of 3–6 degrees), the lowstand systems tract consists of a basin-floor lowstand fan, a basinally restricted lowstand wedge, and an incised valley system that extends updip across the shelf, landward of the lowstand wedge (Van Wagoner et *al., 1990). In lower-relief "ramp-type" basins, where the basal sequence boundary lacks a well-defined "shelf break" (slope,* 1 degree), lowstands lack basin-floor fans (Van Wagoner et al., 1990). The lowstand wedge consists of aggrading parasequences that onlap at or below the depositional shelf break of the underlying sequence boundary. The *transgressive systems tract* is the middle compartment within sequences. It is bounded below by the transgressive surface, and above by the maximum flooding surface. The *maximum flooding surface* is defined as the maximum marine incursion (transgression) within a sequence. Parasequences within the transgressive systems tract form a distinctive retrogradational pattern. The *highstand systems tract* is the uppermost compartment within a sequence. It is bounded below by the maximum flooding surface and above by the overlying sequence boundary. Parasequences within the highstand systems tract form an aggradational and then progradational parasequence set.

As outlined by Van Wagoner *et al.* (1990) sequence boundaries, sequences, and systems tracts can be interpreted in terms of cycles of relative sea-level change. Sequence boundaries are interpreted to be the result of relative sea-level falls (Van Wagoner *et al.*, 1990). If slope failure, canyon development, and fluvial capture occur during this fall, mass transport complexes followed by submarine fans complexes can be locally deposited along the basin floor (Van Wagoner *et al.*, 1990). As the rate of relative sea-level fall decreases, reaches a stillstand, and slowly rises, fluvial incision and submarine fan deposition ceases. Coarse-grained braided fluvial and/or estuarine deposits aggrade within the incised valleys, and a lowstand wedge is deposited below the depositional shelf break of the underlying sequence boundary (Van Wagoner *et al.*, 1990). When the relative rate of sea-level rise increases toward its maximum, retrogradational parasequences of the transgressive systems tract are deposited (Van Wagoner *et al.*, 1990). As the rate of relative sea-level rise begins to slow and reach a stillstand, aggradational followed by progradational parasequences of the highstand systems tract are deposited (Van Wagoner *et al.*, 1990). As relative sea level falls once again, the next sequence boundary forms by fluvial incision, slope failure, and canyon development.

Since the publication of AAPG Memoir 26 in 1977, sequence stratigraphy has become a major research topic in sedimentary geology. Inclusion of industry, government, and academic researchers has produced healthy debate in a number of areas. From day one, a major issue of debate (Miall, 1986, 1992) was the contention by Vail *et al.* (1977c) that depositional sequences are the product of eustacy, and therefore form globally synchronous cycles that can be used as a basis for global chronostratigraphy (The

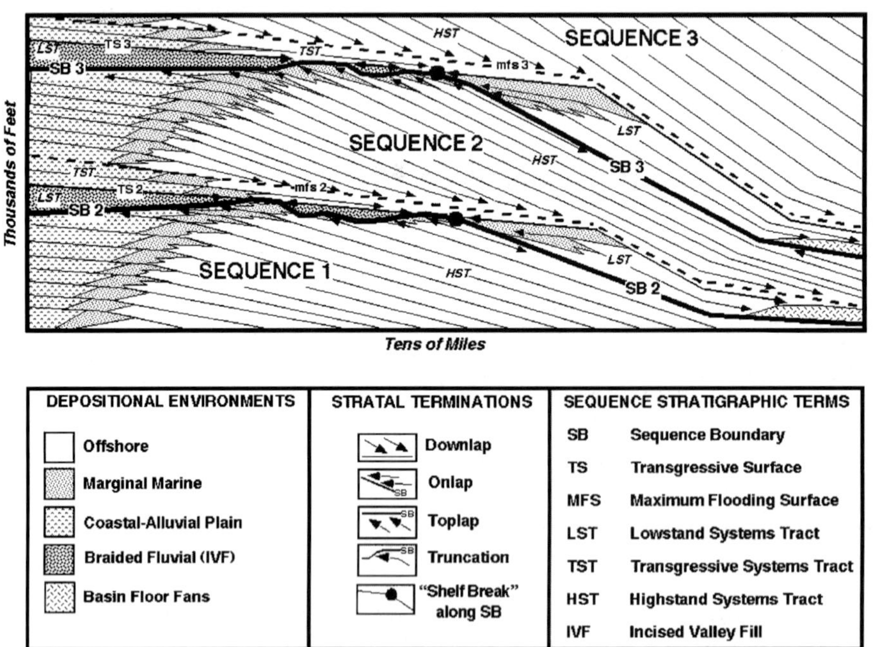

Figure S35 Idealized Model of a Depositional Sequence for a basin with distinct physiographic relief ("shelf break model"). In this updated sequence stratigraphic model, first proposed by Van Wagoner *et al.* (1988), three basic stratal compartments (systems tracts) were proposed: lowstand, transgressive, and highstand. The *lowstand systems tract* is the lowermost compartment within a sequence. It is bounded by a sequence boundary at its base and a transgressive surface at its top. The *transgressive surface* is defined as first significant marine-flooding surface (parasequence boundary) across the shelf within a depositional sequence. The lowstand systems tract typically consists of a basin-floor fan, a basinally restricted lowstand wedge (aggrading parasequence set) that onlaps the "shelf break" of the underlying sequence boundary, and an incised valley that extends updip across the "shelf-profile" of this sequence boundary. The *transgressive systems tract* is the middle compartment within a sequence. It is bounded below by the transgressive surface, and above by the maximum flooding surface. The *maximum flooding surface* is defined as the maximum marine incursion (transgression) within a sequence. Parasequences within the transgressive systems tract form a distinctive retrogradational parasequence set. The *highstand systems tract* is the uppermost compartment within a sequence. It is bounded below by the maximum flooding surface and above by the overlying sequence boundary. Parasequences within the highstand systems tract form an aggradational and then progradational parasequence set.

Cycle Charts). Discussions on the utility of defining and mapping sedimentary cycles (sequences) based on subaerial unconformities (*Depositional Sequences*: Van Wagoner *et al.*, 1990), maximum flooding surfaces (*Genetic Sequences*: Galloway, 1989), or transgressive surfaces (*T–R Sequences*: Embry, 1993) has also occurred. Even among practitioners of "depositional sequences," differences in sequence stratigraphic methodologies has arisen. The classic Exxon papers in AAPG Memoir 26 and Methods in Exploration Number 7 (Mitchum *et al.*, 1977; Van Wagoner *et al.*, 1990) defined sequences and systems tracts purely by stratal geometries. Within this context, interpreted sea-level variations, facies, or increments of geologic time were not considered in sequence boundary or system tract definition. Posamentier *et al.* (1992), however, based sequence boundary placement and system tracts definition on interpreted sea-level variations. Plint and Nummedal (2000) defined sequence boundaries by stratal termination, but systems tracts by interpreted sea level variations. Mellere (1994) based system systems tracts on facies change, not stratal compartments. Erskine and Vail (1988) based sequence boundaries and system tracts on interpreted cycles of geologic time. Finally, while Van Wagoner *et al.* (1990) contended that transgressive erosion was not a significant process in the rock record, others such as Plint and Nummedal (2000), strongly argue that transgressive erosion can significantly erode and modify highstand, as well as lowstand, depositional patterns.

Arthur D. Donovan

Bibliography

Asquith, D.O., 1970. Depositional topography and major marine environments, Late Cretaceous, Wyoming. *American Association of Petroleum Geologists Bulletin*, **54**(7): 1184–1224.

Campbell, C.V., 1967. Lamina, Laminaset, Bed, and Bedset. *Sedimentology*, **8**: 7–26.

Caster, K.E., 1934. The Stratigraphy and paleontology of northwestern Pennsylvania. *Bulletin of American Paleontology*, **21**(71): 1–185.

De Graciansky, P., Hardenbol, J., Jacquin, T., and Vail, P.R. (eds.), 1998. *Mesozoic and Cenozoic Sequence Stratigraphy of European Basins*. SEPM Special Publication No. 60.

Embry, A.F., 1993. Transgressive–Regressive (T–R) sequence analysis of the Jurassic succession of the Sverdrup Basin, Canadian Arctic Archipelago. *Canadian Journal of Earth Science*, **30**: 301–320.

Erskine, R.D., and Vail, P.R., 1988. Seismic stratigraphy of the Exmouth Plateau. In Bally, A.W. (ed.), *Atlas of Seismic Stratigraphy*, Tulsa, OK: AAPG Studies in Geology, pp. 163–173.

Galloway, W.E., 1989. Genetic stratigraphic sequences in basin analysis I: architecture and genesis of flooding-surface bounded depositional units. *AAPG Bulletin*, **73**: 125–142.

Haq, B.U., Hardenbol, J., and Vail, P.R., 1987. Chronology of fluctuating sea-levels since the Triassic. *Science*, **235**: 1156–1167.

Haq, B.U., Hardenbol, J., and Vail, P.R., 1988. Mesozoic and cenozoic chronostratigraphy and eustatic cycles. In Wilgus, C.K. *et al.* (eds.), *Sea-Level Changes. An Integrated Approach*. SEPM Special Publication No. 42, pp. 71–108.

Mellere, D., 1994. Sequential development of an estuarine valley fill: the Two Wells Tongue of the Dakota Sandstone, Acoma Basin, New Mexico. *Journal of Sedimentary Research*, **B64**(4): 500–515.

Miall, A.D., 1986. Eustatic sea level changes interpreted from seismic stratigraphy: a critique of the methodology with particular reference to the North Sea Jurassic record. *AAPG Bulletin*, **70**(2): 131–137.

Miall, A.D., 1992. Exxon global cycle chart: an event for every occasion. *Geology*, **20**: 787–790.

Mitchum, R.M., 1977. Seismic stratigraphy and global changes of sea level, Part 11: glossary of terms used in seismic stratigraphy. In Payton, C.E. (ed.), *Seismic Stratigraphy—Applications to Hydrocarbon Exploration*. AAPG Memoir 26, pp. 205–212.

Mitchum, R.M., Vail, P.R., and Thompson, S., 1977. Seismic stratigraphy and global changes of sea level, Part 2: the depositional sequence as the basic unit for stratigraphic analysis. In Payton, C.E. (ed.), *Seismic Stratigraphy—Applications to Hydrocarbon Exploration*. AAPG Memoir 26, pp. 53–62.

North American Commission on Stratigraphic Nomenclature, 1983. North American Stratigraphic Code. *The American Association of Petroleum Geologists Bulletin*, **67**(5): 841–875.

Payton, C.E., 1977. *Seismic Stratigraphy—applications to Hydrocarbon Exploration*. AAPG Memoir 26.

Plint, A.G., and Nummedal, D., 2000. The falling stage systems tract: recognition and importance in sequence stratigraphic analysis. In Hunt, D., and Gawthorpe, R.L. (eds.), *Sedimentary Responses to Forced Regressions*. London Geological Society Special Publication 172, pp.1–17.

Posamentier, H.W., Allen, G.P., James, D.P., and Tesson, M., 1992. Forced regressions in a sequence stratigraphic framework: concepts, examples, and exploration significance. *AAPG Bulletin*, **76**: 1687–1709.

Sloss, L.L., Krumbein, W.C., and Dapples, E.C., 1949. Integrated facies analysis. In Chester R. Logwell (ed.), *Sedimentary Facies in Geologic History*. The Geologic Society of America Memoir 39, pp. 91–124.

Sloss, L.L., 1963. Sequences in the cratonic interior of North America. *Geological Society of America Bulletin*, **74**: 93–114.

Vail, P.R., Mitchum, R.M., Todd, R.G., Widmier, J.M., Thompsom, S., Sangree, J.B., Bubb, J.N., and Hailelid, W.G., 1977a. Seismic stratigraphy and global changes in sea level. In Payton, C.E. (ed.), *Seismic Stratigraphy—applications to Hydrocarbon Exploration*. AAPG Memoir 26, pp. 49–212.

Vail, P.R., and Mitchum, R.M., 1977. Seismic stratigraphy and global changes of sea level, Part 1: Overview. In Payton, C.E. (ed.), *Seismic Stratigraphy—applications to Hydrocarbon Exploration*. AAPG Memoir 26, pp. 51–52.

Vail, P.R., Mitchum, R.M., and Thompson, S., 1977b. Seismic stratigraphy and global changes of sea level, Part 3: relative changes in sea level from coastal onlap. In Payton, C.E. (ed.), *Seismic Stratigraphy—Applications to Hydrocarbon Exploration*. AAPG Memoir 26, pp. 63–81.

Vail, P.R., Mitchum, R.M., and Thompson, S., 1977c. Seismic stratigraphy and global changes of sea level, Part 4: global cycles of relative changes in sea level. In Payton, C.E. (ed.), *Seismic Stratigraphy—Applications to Hydrocarbon Exploration*. AAPG Memoir 26, pp. 83–97.

Vail, P.R., Todd, R.G., and Sangree, J.B., 1977d. Seismic stratigraphy and global changes of sea level, Part 5: chronostratigraphic significance of seismic reflections. In Payton, C.E. (ed.), *Seismic Stratigraphy—Applications to Hydrocarbon Exploration*. AAPG Memoir 26, pp. 99–116.

Van Wagoner, J.C., Posamentier, H.W., Mitchum, R.M., Vail, P.R., Sarg, J.F., Loutit, T.S., and Hardenbol, J., 1988, An overview of the fundamentals of sequence stratigraphy and key definitions. In Wilgus, C.K. *et al.* (eds.), *Sea-Level Changes: An Integrated Approach*. SEPM Special Publication 42, pp. 39–48.

Van Wagoner, J.C., Mitchum, R.M., Campion, K.M., and Rahmanian, V.D., 1990. *Siliciclastic Sequence Stratigraphy in Well Logs Cores, and Outcrops*. AAPG Methods in Exploration Series, No.7.

Wilgus, C.K., Hastings, B.S., Kendall, C.G., Posamentier, H.W., Ross, C.A., and Van Wagoner, J.C. (eds.), 1988. *Sea-Level Changes: An Integrated Approach*. SEPM Special Publication #42.

Cross-references

Beach Stratigraphy
Changing Sea Levels
Coastal Sedimentary Facies
Geochronology
Ground-Penetrating Radar
Ingression, Regression, and Transgression
Late Quaternary Marine Transgression
Monitoring, Coastal Geomorphology
Seismic Survey

SETBACKS

Coastal management strategies reduce the risks associated with coastal hazards and protect coastal infrastructure, habitat, water quality, recreation values, and aesthetic properties. A *setback* is one type of regulatory method used by all levels of government to mitigate risks to coastal structures and to protect coastal resources. For example, following Hurricane Luis in 1995, three island territories in the eastern Caribbean (Antigua and Barbuda, Nevis, and St. Lucia) developed a shoreline (defined here as the high water line) management strategy involving the establishment of setbacks (Cambers, 1997).

What is a setback?

Taken literally, setbacks are a type of regulatory restriction that require coastal construction projects to "set back" a landward distance from a predetermined reference feature on the beach. This arrangement provides a buffer between a hazard area or natural area and coastal development. Setback lines often parallel (but are situated some distance inland from) a reference feature such as the high water line, vegetation line, dune crest, contour elevation, or cliff edge. Reference features are usually determined by the geomorphic and ecologic properties of a particular reach. The setback is the measured horizontal distance from the reference feature to the setback line. The method used to establish the horizontal distance is a key component in the process of establishing setbacks because coastal structures are permitted landward of the setback line, but not seaward without a variance. Therefore, setbacks can have substantial implications for state and local governments, managers, planners, developers, and citizens.

The overall purposes of a setback are to:

- provide a buffer between coastal development and coastal hazards, for example, flooding and erosion;
- protect or establish natural resources such as conservation areas;
- facilitate sustainable coastal development;
- ensure public access to the beach;
- maintain the aesthetic character of the coast.

Components and types of coastal setbacks

Coastal setbacks require one or more of the following components:

- a reference feature
- a feature deemed worthy of protection (for example, a natural resource)
- a method for establishing a setback line.

Reference features

Reference features provide a starting point from which to establish a setback distance. Reference features can be stationary (relatively permanent) or dynamic. Stationary reference features can include a surveyed elevation, dune toe or crest, vegetation line, road or other cultural feature (for example, lighthouse). Dynamic reference features typically include the shoreline (i.e., the intersection of the land and sea), a proxy for the shoreline (for example, high water line, HWL; mean high water line, MHWL; ordinary high water line, OHWL; normal water line, NHWL; etc.) or a cliff (also called bluff) edge. For example, Florida employs a setback from the seasonal high water line (SHWL) for coastal permitting. Wisconsin uses a statewide setback of 75 ft (22.9 m) from the ordinary high water mark (OHWM) and 75 ft (22.9 m) from the cliff edge to protect structures from cliff erosion.

Natural resources

Setbacks provide one method of managing coastal resources including their features, processes, and characteristics (Clark, 1974). Natural resources in the coastal zone can include coastal habitats and ecosystems (wetlands, estuaries, beaches, dunes, uplands), sand and gravel, small-town characteristics, visual openness and surf (Houlahan, 1989; Nelson and Howd, 1996; Klee, 1999). According to the US Fish and Wildlife Service (1995), more than 45% of all species listed as endangered or threatened reside within coastal habitats.

Setback line methods: fixed setbacks

"Fixed" setback lines are rigid in the sense that they do not depend on *a priori* measurements, but rather on a uniform, standard, or fixed distance landward of a reference feature. In this case, setback distances are determined prior to a permit application (Houlahan, 1989). For example, Delaware places setback lines 100 ft (30.5 m) landward from the seawardmost 7 ft (2.1 m) elevation above the national geodetic vertical datum (NGVD). In the Technical Belt of Poland, the Maritime Administration employs a setback of 200 m landward of the dune ridge or 100 m inland of the upper edge of a cliff. However, a case study conducted in Essex, England claimed that, along reaches of varied topography, a rigid setback line would have no topographical or planning relevance since the artificial boundary can bisect natural ecosystem and hydrologic boundaries (Bridge and Salman, 2000). Additionally, Klee (1999) considers fixed setbacks a "weak" approach because coastal construction can occur at existing setbacks regardless of current potential hazards.

Setback line methods: floating setbacks

"Floating" or variable-rate setbacks use a dynamic (nonrigid) natural phenomenon to determine setback lines and can change according to the topography of an area or by measurements of shoreline movement over time. Thus, floating setbacks reflect site- or reach-specific coastal processes and, in theory, allow for natural erosion to occur. Floating setback distances usually are established after a permit request (Houlahan, 1989). In most applications of variable-rate setbacks, setback lines typically are positioned in a landward direction of the reference feature by multiplying an "average annual erosion rate" (AAER in feet or meters per year) by a constant (time in years of setback protection). For example, the setbacks for single-family dwellings (defined as buildings with a floor area $<5{,}000$ ft^2 or 464.5 m^2) in North Carolina are determined by multiplying the AAER by 30 with a minimum 60 ft (18.3 m) setback. The constants and minimum setbacks can vary with building size, anticipated life-expectancy of a structure, or duration of a mortgage.

The setback distance, obtained by the variable-rate AAER method, is based on a prediction of a future shoreline position. This method for producing a setback requires accurate assessments of historical shoreline changes, usually in the form of a shoreline rate-of-change statistic. In some cases, erosion rate data are averaged or "smoothed" along the length of a coast and/or grouped into blocks of similar rate values to produce a representative rate for a segment of coast. When using a dynamic setback approach, shoreline change data must be updated periodically to reflect current and historical conditions. However, this approach may not explicitly account for extreme events such as storm surge.

Setback line methods: combined fixed and floating setbacks

Some US states require a combination of the two setback types to increase the setback a fixed distance landward of the distance given by the erosion rate. This strategy is used along the U.S.' Great Lakes (e.g., Michigan) to protect a cliff face and along some US Atlantic coastal states such as New Jersey and North Carolina to safeguard areas prone to severe erosion.

Methods used to establish "erosion rate-based" setbacks

Floating setback distances (in meters or feet) are obtained from extrapolating historical shoreline or cliff-line migration erosion rate or trend lines (in meters per year or feet per year) to a predetermined target date (in years). The goal of extrapolation is to provide the best estimate of a future shoreline or cliff-line position. In theory, the calculated setback distance reflects the actual distance the shoreline will migrate landward, year by year, over a number of years. The precision with which estimates of shoreline rates-of-change reflect actual changes and predict future changes depends on:

- the accuracy of the techniques used to identify and record historical shoreline positions (e.g., shoreline identification and rectification on historical maps, vertical aerial photographs, or Geographic Positioning Systems, GPS),
- the quality of the data base (e.g., map and photograph quality, number of maps and photos of shoreline positions available and used in the computation, the time period between measurements, the total time span of the record),
- the mathematical and statistical method used to calculate an historical shoreline rate-of-change—especially when faced with nonlinear, cyclic, or chaotic shoreline migration trends,
- the temporal variability of shoreline movement,
- the proximity of the observations to actual changes in the trend (sampling bias),
- the quantitative techniques used to predict a future shoreline position based on historical trends, and
- an analysis and understanding of the coastal processes and human-induced shoreline modifications that govern shoreline migration trends.

The data used to determine erosion rate-based setbacks consist of a number of shoreline positions obtained from historical maps, maps compiled from aerial photography, aerial photographs and GPS. Analyses of historical shoreline movement trends begin by overlaying a set of georeferenced maps and/or rectified aerial orthophotographs for a coastal area. Next, transects or monuments of known geographic coordinates are established perpendicular to a baseline and, preferentially, to the shoreline (i.e., estimate error is minimized as the baseline and shoreline become more parallel). At each transect or monument, a data set is produced of shoreline positions at specific dates from which trends or shoreline rates of change can be delineated.

Shoreline changes occur over a wide range of nested time scales. At one extreme, shorelines move in response to eustatic sea-level fluctuations and tectonism; at the other, changes result from the constant fluctuations of wind, wave, and tidal action. The shoreline position at any given time (and the ensuing direction and magnitude of shoreline movement) is a cumulative summary of all long- and short-term processes. However, aerial orthophotographs record instantaneous positions of the shoreline at a specific time, and maps often record shorelines surveyed from instantaneous to longer, but unknown time periods. Consequently, each recorded shoreline time/position data point possesses a degree of gross, systematic, or random error arising from attempts to locate precisely the shoreline datum (typically the HWL) from photographs and maps. This situation leads to erosion rates that contain various degrees of uncertainty. Historical shoreline rate-of-change statistics are, therefore, relatively long-term, linear summaries of several instantaneous, error-prone geographical shoreline positions. Dolan *et al.* (1991) discussed four methods used to calculate shoreline rates of change and showed how the potential sources of error can bias the final statistics.

Erosion-rate based setbacks are obtained by extrapolating the slope of a trend line (a rate-of-change statistic) which passes through a few (usually <10) historical shoreline position/time data points. In most cases, when setback distances are predicted from a line calculated by a linear regression, the y-intercept is omitted, the slope (rate-of-change) is retained, and the line is adjusted to pass through the most recent data point (not needed for predictions based on an end-point rate).

The accuracy of a shoreline prediction depends on the accuracy of both the data and the historical trend estimates based on those data (Fenster *et al.*, 1993; Douglas and Crowell, 2000; Galgano and Douglas, 2000). The challenge of establishing accurate setback lines comes from (1) using a limited number of error-prone shoreline position data to delineate actual historical trends and (2) knowing when to rely on a 30 year or more projection of a shoreline positions based on an historical *linear* erosion rate estimate. A rate-of-change statistic computed from shoreline position/time data sets implicitly assumes that shoreline movement is constant and uniform (i.e., linear) through time—often not the case in reality. In fact, shoreline movement can be linear, nonlinear, cyclic, or (perhaps) chaotic (Eliot and Clarke, 1989; Fenster *et al.*, 1993). The popularity of linear extrapolation for setback determination is due chiefly to its simplicity. As with any empirical technique, no knowledge of or theory regarding the sand transport system is required. Instead, the cumulative effect of all the underlying processes is assumed to be captured in the position history. Thus, an assumption implicit in the procedure for delineating erosion-rate setback lines is that the observed historical rate-of-change is the best estimate available for predicting the future. Regardless of the potential limitations of this assumption, the use of linear models to predict the future avoids problems (such as overfit, underfit, and/or accelerations) that arise when using nonlinear models to extrapolate. Thus, it is best to constrain projections of shoreline position to a line or series of lines.

The NRC (1990), Morton (1991), and many others have discussed the highly nonlinear nature of processes that transport sediment in the coastal zone. For shorelines that experience short-term fluctuations in shoreline migration or longer-term migration nonlinearity, linear estimation methods will vary in their ability to approximate the long-term historical trend. Fenster *et al.* (1993) developed and demonstrated a method to employ nonlinear models, when called for by statistical significance tests of historical shoreline trend data, as an intermediate step to find the best line, or range of lines, with which to estimate future shoreline positions. Moreover, they investigated techniques for extrapolating shoreline behavior and discuss what constitutes a significant change in shoreline trend (especially in the absence of structures or processes known to have altered the underlying system). The goal of this shoreline prediction technique was to allow assessment of the stability of a long-term trend relative to intermediate (the 50 years of aerial photography) and short-term (decennial) trends, thereby best relating past shorelines to those which can be expected in the future. In this way, *linear* predictions using a pair of lines representing the range of probable future shoreline positions (and the degree of prediction uncertainty) can be based on linear *or* curvilinear historical shoreline migration trends. This approach, used in concert with knowledge of the process-response system (if known), was designed to avoid problems associated with using nonlinear models for predictions and to provide a probabilistic basis for predicting future shoreline locations. Crowell *et al.* (1997) conducted a thorough review of this technique and found, using average annual sea-level data as a surrogate for a few instantaneous shoreline positions, that a simple linear regression surpasses other forecasting methods unless known physical changes have occurred to the system. Douglas and Crowell (2000) and Galgano and Douglas (2000) have examined errors associated with common shoreline forecasting methods.

Although process information for a coastal reach is critical to understanding long-term geomorphological changes, often these data are not available. In addition, the relative contribution of all the process variables that contribute to shoreline erosion or accretion may be difficult to determine. For these reasons, coastal management strategies have relied mostly on extrapolating shoreline migration data that use time as a surrogate for the processes that produce shoreline change. Other deterministic shoreline modeling routines lack universal reliability (e.g., Theiler *et al.*, 2000). Research efforts should strive to produce reliable process-based models that can predict shoreline erosion from the synergistic processes that produce changes to the sediment budget.

Several attempts have been made to identify and isolate the processes responsible for producing shoreline changes (the signal) and deviations from the signal (noise). For example, Leatherman *et al.* (2000) asserted that global eustatic sea-level rise drives coastal erosion and that a relationship exists between long-term average rates of shoreline erosion and the rate of sea-level rise (i.e., the shoreline retreat is 150 times the rate of sea-level rise). However, arguments posed by Pilkey *et al.* (2000), Sallenger *et al.* (2000), and Galvin (2000) that question this relationship underscore the difficulty involved in isolating specific oceanographic processes with their coastal (i.e., shoreline) responses.

The fundamental difficulties involved in relating hydrodynamics to sediment transport are further illustrated by recent research dealing with the relationship between tropical and extratropical storms and the long-term shoreline migration history of a coastal reach. Douglas and Crowell (2000) and Honeycutt *et al.* (2001) contend that, despite the fact that large storms can cause enough beach erosion to require a decade or more for recovery, it is appropriate to eliminate storms from shoreline change data bases. On the other hand, based on analyses of a reach along the wave-dominated Outer Banks of North Carolina, Fenster *et al.* (2001) concluded that the exclusion of storm-influenced data points is neither warranted nor prudent because such values do not constitute outliers, and they do not increase substantially the range of uncertainty surrounding predicted future shoreline positions. In addition, the added value of reducing uncertainty with the inclusion of more data points outweighs the potential advantages of excluding storm-influenced or storm-dominated data points.

Spatial considerations

The spatial considerations involved in establishing erosion-rate based setbacks stem from the need to project the setback line some landward distance from a continuous reference line. Therefore, extrapolations must occur at some predetermined spacing (transects) along a shore-parallel baseline that contains an infinite number of points. This situation raises a question, namely, where along a shore-parallel baseline, that is, at what spacings, should extrapolations occur? Although Geographic Information Systems (GIS) and high-speed digitizing techniques allow investigators to assemble almost unlimited samples for assessments of along-the-coast patterns of variation, there is value in knowing the relationship between sample size (e.g., number of digitized points) and the accuracy of an average rate-of-change estimate for a particular reach or an unspecified location along-the-shore. Rate-of-change values will be more useful for erosion-rate based setback delineation and are less likely to be contested when obtained by a scientifically and statistically sound approach. Although many coastal management agencies use arbitrary transect spacings (e.g., 50 m, 100 m), Dolan *et al.* (1992) used geostatistics (i.e., the theory of regionalized variables) to provide nomograms by which optimal sample sizes can be determined. In this case, optimal sample size is defined as the amount of data required to support the inferences of a particular study. Too few data produce inconclusive results; oversampling is inefficient. This study showed, in part, that the confidence in estimation of rate-of-change values in the spatial domain (due to spatial continuity) far exceeds the confidence in rate values calculated at a sample location in the temporal domain because along-the-shore rate-of-change values exhibit a high degree of spatial autocorrelation (nearest neighbor effect). In fact, the 50 m spacing of transects along Hatteras Island, North Carolina provides an excellent estimate of rates between transects (accurate to ±0.5 m/yr).

Administration and setback implementation

In addition to differences in the type of setback used by a particular agency, variations exist among nations and states with respect to how setbacks are administered and who administers them. Additionally, the technical standards and methods for establishing those standards vary widely. For example, some US states apply setbacks only to new construction, while others apply to both new and old development. Also, some states differentiate among classes of buildings (e.g., single-family dwelling, multi-family structures, and industrial/commercial buildings). For example, North Carolina uses data averaging of similar shoreline recession rates at neighboring transects to produce a single shoreline recession rate for a "block" of shoreline. On the other hand, Massachusetts calculates recession rates every 100 m along the shore.

Twenty-three of the twenty-nine US coastal states and territories (including American Samoa, Guam, Marianas Islands, Puerto Rico, and Virgin Islands) utilize regulatory setbacks as a coastal zone management tool (Bernd-Cohen and Gordon, 1998). According to the Heinz Center Report (2000), ten US states use fixed methods (43.5%), five use floating methods (21.8%), and four use a combination of the two (17.4%). All five US territories used fixed setback methods. Most of the US coastal states use state-controlled setbacks, although two states exclusively use local setbacks (California and Washington) and five states have no setbacks (Heinz Center, 2000). In most cases, states with setback programs use local government agencies to administer the program. These setback programs can be administered on either a mandatory or a voluntary basis.

In Europe, the use of setback policies is increasing in erosion-prone coastal nations. In a study of national policy and legislation in selected European Union and Baltic States, Bridge and Salman (2000) discuss the use of setback lines in the management of coastal zones for nine selected nations and list three types of setback methods in practice within Europe: (1) shore-parallel linear (fixed); (2) contour (a distance from a shore reference feature based on elevation); and (3) Exclusive Economic Zones (offshore protection). Of the nine listed nations, eight have established fixed setbacks: Denmark, Finland, Germany, Norway, Poland, Spain, Sweden, and Turkey. The buffer zone varies widely from 5 m to 3 km. England stands as the exception with no formal coastal setback policy in place. As in the United States, the types and implementation of setbacks within the European Union and Baltic nations vary. Among the cities and regions located along the Russian Baltic Sea coast, a lone local statutory regulation establishes a 1 km wide setback for the economically and militarily strategic Kaliningrad region. No other setback policy is in place within Russia's coastal zone.

Advantages and disadvantages

The advantages of using setbacks as a regulatory method include:

- prevent structural loss and damage from erosion
- protect coastal habitat and water quality
- provide open space for natural shoreline environment
- provide recreation and beach access
- allow natural erosion/accretion cycles to occur
- can contribute to sustainable management of coastal systems.

The disadvantages are:

- may not provide adequate protection, for example, fixed methods may not account for topography or variations in coastal erosion processes, buffer zone may not mitigate impacts
- may limit tax base (economically restrictive)
- erosion-rate based data have accuracy limitations
- may not address existing structures
- a strategy is needed to deal with structures located near "old" long-term predictions
- shore-parallel, linear setback lines do not include the marine zone
- enforcement may depend on cultural attitudes and administrative context.

Michael S. Fenster

Bibliography

Bernd-Cohen, T., and Gordon, M., 1998. *State Coastal Program Effectiveness in Protecting Beaches, Dunes, Bluffs, and Rocky Shores: A National Overview*. Final Report, National Ocean and Atmospheric Administration, National Ocean Service.

Bridge, L., and Salman, A., 2000. *Policy Instruments for ICZM in Nine Selected European Countries*. Final study report prepared for the Dutch National Institute for Coastal and Marine Management.

Cambers, G., 1997. *Planning for Coastline Change: Guidelines for Construction Setbacks in the Eastern Caribbean Islands*. CSI info 4, Paris: UNESCO.

Clark, J., 1974. *Coastal Ecosystems: Ecological Considerations for Management of the Coastal Zone*, DC: The Conservation Foundation, Washington.

Crowell, M., Douglas, B.C., and Leatherman, S.P., 1997. On forecasting future U.S. shoreline positions: a test of algorithms. *Journal of Coastal Research*, **13**: 1245–1255.

Dolan, R., Fenster, M.S., and Holme, S.J., 1991. Temporal analysis of shoreline recession and accretion. *Journal of Coastal Research*, **7**: 723–744.

Dolan, R., Fenster, M.S., and Holme, S.J., 1992. Spatial analysis of shoreline recession and accretion. *Journal of Coastal Research*, **8**: 263–285.

Douglas, B.C., and Crowell, M., 2000. Long-term shoreline position prediction and error propagation. *Journal of Coastal Research*, **16**: 145–152.

Eliot, I., and Clarke, D., 1989. Temporal and spatial bias in the estimation of shoreline rate-of-change statistics from beach survey information. *Coastal Management*, **17**: 129–156.

Fenster, M.S., Dolan, R., and Elder, J.F., 1993. A new method for predicting shoreline positions from historical data. *Journal of Coastal Research*, **9**: 147–171.

Fenster, M.S., Dolan, R., and Morton, R.A., 2001. Coastal storms and shoreline change: Signal or noise? *Journal of Coastal Research*, **17**: 714–720.

Galgano, F.A., and Douglas, B.C., 2000. Shoreline position prediction: methods and errors. *Environmental Geosciences*, **7**: 23–34.

Galvin, C., 2000. Comment to "Sea level rise shown to drive coastal erosion" by Leatherman *et al.* (2000), *Eos Transactions*, AGU, **81**: 437–440.

Heinz Center, 2000. *Evaluation of Erosion Hazards Summary*. A collaborative project of The H. John Heinz III Center for Science, Economics, and the Environment.

Houlahan, 1989. Comparison of construction setbacks to manage development in coastal hazard areas. *Coastal Management*, **17**: 219–228.

Honeycutt, M.G., Crowell, M., and Douglas, B.C., 2001. Shoreline position forecasting: impact of storms, rate-calculation methodologies, and temporal scales. *Journal of Coastal Research*, **17**: 721–730.

Klee, G.A., 1999. *The Coastal Environment: Toward Integrated Coastal and Marine Sanctuary Management*. Englewood Cliffs, NJ: Prentice-Hall.

Leatherman, S.P., Zhang, K., and Douglass, B.C., 2000. Sea level rise shown to drive coastal erosion. *Eos Transactions*, AGU, **81**: 55–58.

Morton, R.A., 1991. Accurate shoreline mapping: past, present, and future. *Coastal Sediments '91*, pp. 997–1010.

National Research Council, 1990. *Managing Coastal Erosion*. Washington, DC: National Academy Press.

Nelson, C., and Howd, P., 1996. Surf as a coastal resource: the wave of the future. *Geotimes*, 19–22.

Pilkey, O.H., Young, R.S., and Bush, D.M., 2000. Comment to "Sea level rise shown to drive coastal erosion" by Leatherman *et al.* (2000), *Eos Transactions*, AGU, **81**: 436.

Sallenger, A.H., Morton, R.A., Fletcher, C., Theiler, E.R., and Howd, P., 2000. Comment to "Sea level rise shown to drive coastal erosion" by Leatherman *et al.* (2000), *Eos Transactions*, AGU, **81**: 436.

Theiler, E.R., Pilkey, O.H., Young, R.S., Bush, D.M., and Chai, Fei, 2000. The use of mathematical models to predict beach behavior for U.S. coastal engineering: a critical review. *Journal of Coastal Research*, **16**: 48–70.

US Fish and Wildlife Service (FWS), 1995. Protecting coastal ecosystems. *Endangered Species Bulletin*, September/October.

Cross-references

Caribbean Islands, Coastal Ecology and Geomorphology
Coasts, Coastlines, Shores, and Shorelines
Coastal Zone Management
Erosion Processes
Managed Retreat
Mapping Shores and Coastal Terrain
Meteorological Effects on Coasts
Natural Hazards
Sea-Level Rise, Effect
Storm Surge

SHARM COASTS

A sharm coast is a type of irregular but mostly broad embayment with a limited extension inland, which can be found along the coast of the Red Sea; best developed in the southern part of the Sinai peninsula (Sharm el Sheikh) or in the middle part of the Saudi Arabian coastline (Schmidt, 1923; Bird, 2000). The sharms do interrupt broader coral-reef belts, but may be decorated with small reef benches. Mostly the inner parts of the sharms show beaches (often with beachrock) at the end of dry valleys (wadis), but at some places sharms exist without a connection to valley-like forms inland. There are several interpretations for the formation of the sharm embayments, and most probably several of them together are responsible: the lack or interruption of coral reefs maybe explained by sediment discharge and suspension at the mouth of the wadis, or by too much freshwater impact (possibly in the form of groundwater seepage). Other explanations include cutting of a former reef body by fluvial erosion during sea-level lowstands in glacial times, or overdeepening of the seabed in front of valley mouths where coral reef growth could not keep up with the rising sea-level in postglacial times. Some of the sharms are simply ria-like forms, that is, drowned lower parts of former valleys; others may be formed by slumping or sliding of oversteepened fringing reef bodies or submarine slopes, but no evidence for this process has ever been checked out in the Red Sea.

Dieter H. Kelletat

Bibliography

Bird, E., 2000. *Coastal Geomorphology, An Introduction*. Chichester: John Wiley and Sons.

Schmidt, W., 1923. Die Scherms an der Rotmeerküste von el-Hedschas. *Petermanns Geographische Mitteilungen*, **69**: 118–121.

Cross-references

Asia, Middle East, Coastal Ecology and Geomorphology
Coral Reef Coasts
Desert Coasts
Ria

SHELF PROCESSES

Introduction

The low slopes and gentle relief of most continental shelves immediately suggest that they were formed through some combination of marine erosion and sedimentation. Although this is largely true, the specific modes of origin are often complex. The processes of marine sedimentation and erosion are episodic and strongly related to the water depth. Seaward of the zone of breaking waves, these processes are generally related to storm activity. As a consequence, the timescale of change is long compared to the timescale at which significant changes in sea level are known to occur. In many areas, the timescale of tectonic change is also similar. This means that the continental shelves of the world have been alternately submerged and exposed throughout the numerous sea level excursions of the Pleistocene and earlier geologic epochs. The overall topographic characteristics and local relief features of most continental shelf areas result from a combination of subaerial and submarine processes.

The large-scale geologic setting and the long-term history determine the overall configuration of the shelf. The processes of the erosion and sedimentation can be effective in modifying features caused by tectonic warping or faulting. Depending upon the intensity of the wave-current climate, range of water depths, and degree of consolidation, the shelf processes can be effective in forming both broad scale and local relief.

The hydrodynamic regime

A discussion of shelf processes logically begins with a description of the behavior of waves and currents, with special attention to how they bring about marine sediment transport. The actual processes of marine sediment transport are complex, and a detailed discussion is beyond the scope of this entry. Briefly, we paramaterize the entrainment of bottom sediment by the fluid shear stress acting on a unit of bed area. This shear stress is proportional to the vertical gradient of fluid velocity acting immediately above the seabed. Thus, both the magnitude of the flow and its vertical structure contribute to its ability to entrain bottom sediment. Shelf currents arise from various combinations of tide, surface wind stress, and internal pressure gradients. Normally, the current bottom boundary layer is meters to tens of meters in thickness. Waves, on the other hand, have boundary layers that are generally less than 10 cm in thickness because the flow reverses during each wave cycle. The wave boundar layer is thin and it readily produces large shear stress acting on the seabed.

There are many factors that distinguish the marine environment from other environments with active sediment transport. Unlike rivers, the magnitude and direction of currents on the shelf are constantly changing. At the same time, the size and period of the waves are also varying. The orbital motion of the waves decreases exponentially with water depth and becomes insignificant at depths equal to half the deepwater wavelength. Inside of this depth, both waves and currents contribute to the shear stress acting on the bottom sediments. In many cases, especially during storms, the major stress results from the waves. A common condition is that the waves are primarily responsible for entraining the bottom sediment, and the currents determine the magnitude and direction of the sediment transport. There is no distinct depth limit for shelf sediment transport due to waves and currents. Instead, the frequency at which storm events are powerful enough to cause sediment entrainment simply diminishes with increasing water depth across the whole width of most shelves.

When waves and currents occur together there is an interaction because turbulence is generated by the intense shear in the wave boundary layer which interacts with the turbulence in the bottom boundary layer of the current. The effect is that the current experiences greater drag than would occur without the waves. Entrainment and transport of bottom sediment are also enhanced.

Marine sediment dynamics

Several factors related to the condition of the seabed influence how sediment is entrained. Local bottom roughness, which may result from the presence of sand ripples, worm burrows, or shell fragments, produce more drag on the current. Part of this enhanced shear is applied to the sediment surface and part results from drag that does not act directly to entrain particles. Thus, identical currents passing over adjacent seafloor segments with different bottom roughness characteristics will apply different magnitudes of bottom shear stress.

The bottom sediments may be granular or cohesive. Granular sediments are stabilized in their "at-rest" positions only by gravity. As a general rule, sediments with less than 10–15% clay particles exhibit granular behavior (Metha *et al.*, 1989). Sediments with higher concentrations of clay particles, or significant amounts of organic material, tend to resist entrainment because of binding effects.

Empirical relationships have been developed to relate the entrainment of granular bottom sediment to the fluid shear stress acting on the bed. At present there is no method to predict the onset of entrainment of cohesive sediments without direct measurements, either in the field or in specially equipped laboratory flumes.

Once the waves, currents, or a combination of both entrains the bottom sediment, it is subject to transport and deposition. As in rivers, the sediment is transported as suspended load, bedload, and dissolved load. But because waves play an important role in marine sediment dynamics, and because current conditions change rapidly, the sediment transport processes are noticeably different in shelf environments. Pure bedload transport is restricted to medium and coarser sand. These particles settle rapidly and are rarely a significant fraction of the suspended load outside of the nearshore zone.

There is no physical limit to the water depth at which bedload transport occurs. There are some places where bedload sand transport happens in mid-shelf and outer-shelf environments. However, the necessary high fluid stress rarely occur in the ordinary conditions in shelf environments. Only in certain places are tidal or wind-driven currents fast enough to cause this form of sediment transport.

Suspended sediment transport is the dominant mode in most continental shelf environments. Under most conditions, active deposits of medium and coarser sand are restricted to the beach, surf zone, and shoreface environments (Dean, 2001). Mud, silt, and fine sand sediment, entrained from the seabed, are dispersed upward into the shelf water column by turbulent eddies in the boundary layer of the currents and waves. In a stable transport condition, the turbulence and the vertical profile of the concentration of suspended sediment particles act to cause an upward flux that balances the settling speed of the suspended sediment. If the flow intensifies, more sediment will be entrained and both the concentration and height of the suspended layer will increase. The reverse of these processes cause deposition.

The presence of suspended particles alters the structure of the bottom boundary layer. The vertical concentration gradient acts to suppress the turbulence which, in turn, reduces the stress applied to the seafloor. The presence of a vertical density gradient, brought about by the vertical gradient in the concentration of suspended particles, can severely limits the thickness of the suspended layer. Both numerical model results and scale model observations have shown that the boundary layer of large storm waves can become saturated with suspended sediment. A sharp density gradient develops at the top of the boundary layer and prevents the sediment from escaping upward into the mean flow.

In most shelf environments suspended sediments are repeatedly entrained, transported, and redeposited until they are finally buried and protected (Niedoroda *et al.*, 1989). Cohesive deposits generally undergo an aging process after their burial that strongly affects their susceptibility to reentrainment. New deposits have high water contents and are weak. Over periods of days to months these sediments dewater and are bioturbated. Both of these effects make these sediments more resistive to reentrainment.

The intensity of boundary layer turbulence is also significantly reduced where the seafloor is covered by a smooth mud layer. The absence of boundary roughness allows a laminar boundary layer to develop to a significant thickness. This tends to suppress the shear stress transmitted to the bottom sediment. The entrainment of the sediment is thus suppressed.

Shelf sediment transport processes

The combined effects of waves and currents dominate shelf sediment transport. In most places, currents alone are inadequate to entrain the bottom sediment. Thus, under calm conditions there is little to no sediment transport. Shelf sediment transport is mainly accomplished during storm events. The magnitude of the transport is a nonlinear function of the intensity of the bottom currents and size of the waves. The magnitude of sediment transport rises sharply as these two forcing parameters increase. A major storm can be responsible for more sediment transport than occurs in many smaller storms. Because patterns of waves and currents do not repeat themselves exactly in a succession, the net transport of sediment over long time intervals results from the sum of individual displacements during each individual storm event. This makes the dispersion of sediment particles in most shelf environments quite complex.

There are cases where sediment is entrained and transported primarily by shelf currents. The shear stresses that are imparted to the bottom sediments on the open shelf by hurricane-driven currents can be large enough to affect significantly entrained sediment (Murray, 1970). In some locations around the British Islands strong tidal currents routinely cause sediment transport (McCave, 1972). In other places, such as the shelf off southeast Florida, a strong offshore current provides sufficient stress to continuously transport sediment.

There continues to be some dispute about whether density-driven bottom currents, similar to turbidity currents, occur on continental shelves. It has been proposed by Hayes (1963), that as a storm-surge return-flow floods back over barrier islands, it becomes sufficiently sediment charged to persist as a density-driven flow as far as the mid-shelf region. The Texas shelf, where these storm deposits have been observed, has been re-examined by Snedden and Nummedal (1989). They found that the beds associated with Hurricane Carla are better explained as resulting from a more normal downwelling flow in the latter portion of the storm.

A form of gravity-driven sediment-laddened bottom flow has recently been observed. Using a combination of direct measurements from an instrumented tripod and acoustic images, it was possible to show that fine sediment was entrained within the wave boundary layer during major storms on the Eel River shelf. The sediment was suspended by the intense shear in the wave boundary layer at a concentration that inhibited vertical mixing across the top of the boundary layer. Gravity, acting down the bottom slope, propelled the sediment suspension across the shelf to near the shelf edge. This is a wave-entrained bottom flow rather than a classic turbidity current because the source of intense turbulence is the waves rather than the mean flow.

Mass gravity flows in the form of mudflows and related turbidity currents are known to develop in shelf environments off major river deltas (Coleman and Prior, 1982). These can be associated with the venting of natural gas, or as a result of liquefaction due to excitation by surface storm waves. In either case large pockets of mud become unstable and flow downslope (Bea *et al.*, 1983). Mudflows of this type are capable of triggering turbidity currents (Niedoroda *et al.*, 2003b).

Hyperpycnal flows are density currents caused by sediment-ladden flows from a river mouth entering still water bodies. These have been observed in reservoirs (Syvitski *et al.*, 1998). These authors show that hyperpycnal flows are rare in continental shelf situations because the density contrast between saltwater and river water causes the river plumes to be buoyant.

Coastal plumes are an important sediment transport and dispersal mechanism on shelves. Depending on the size of the river watershed, the flood discharges may or may not be associated with local storm conditions. In small watersheds the correlation of storm conditions and a river flood plume is stronger. Under these conditions the plume is often driven along the shore by the storm winds, and temporary deposits form beneath it. Subsequent storm events that are not associated with river floods then shift the temporary deposits from the inner-shelf to the outer-shelf. This process has been well documented on the Eel River shelf (Wheatcroft *et al.*, 1999).

Sediment-rich flood plumes from large river systems often arrive at the river mouth during ordinary weather conditions. Because of the effects of the earth's rotation these discharges tend to turn cum sole (to the right in the Northern Hemisphere) and become dynamically coupled to the coastline. These plumes of brackish water with high concentrations of suspended sediments commonly persist for hundreds of kilometers along the Louisiana-Texas coast, the coast of the central United States, the south coast of the northern and many other places.

Temporary mud patches have been observed in the mid-Atlantic Bight off New Jersey and Long Island. These form during the calmer summer months on the mid- and inner-shelf. It is thought that under prolonged calm conditions an initial mud deposit begins to collect in a sheltered deeper area within the ridge and swale bottom topography. The presence of even a thin layer of mud makes the bottom hydrodynamically smooth. This inhibits the production of turbulence in the bottom boundary layer and promotes additional settling of suspended particles. The initial mud patch can thus grow. These persist until the stormy winter season.

Thicker, more persistent and extensive deposits of soft mud have been observed on the inner- and mid-shelf off Louisiana and east Texas (Suhayda, 1977). The water content of the mud is high so that it develops wave motions when forced by storm waves in the overlying water column. The wave energy is dissipated in the moving mud to the point that shelf areas behind these mud deposits are sheltered from the storm waves. Local fishermen are known to utilize these refuges in storms.

At high-latitudes ice is an agent of shelf sediment transport. The Canadian Atlantic shelf, the shelves of some of the large islands in the High Canadian Arctic and the shelf around Antarctica (Anderson, 1989) are places where a significant amount of ice-rafted sediment is deposited. This sediment comes from the melting of icebergs that have calved from glaciers. Stranded ice produces one of the oddest modes of shelf sediment transport that is important only very locally. Ice keels, either from icebergs or from the undersides of sea ice ridges, rake through the shelf sediments and produce distinctive gouges (scours) that can be several meters deep, tens of meters wide, and kilometers long (Niedoroda, 1991; Weeks *et al.*, 1991).

Shelf sedimentary sources

Sediment that is transported on the shelf originates either from land sources or from the reworking of previous deposits. Shelves in tectonically active areas tend to derive most of their sediment from the land. Shelves in stable tectonic areas are generally sediment-starved because rising sea level has caused the river sediments to be trapped within large coastal plain estuaries (Swift *et al.*, 1991).

On most coasts the coastline provides a line source of sediments to the adjoining continental shelf. The sediment is moved alongshore in the surf zone, in coastal plumes and in the accelerated flows of the upper shoreface. Coastal engineers have adopted the concept of a depth-of-closure (Hallermeier, 1978) for use in planning beach nourishment programs. This concept is based on the observation that the changes of beach and nearshore profiles are constrained within an envelope. The upper and lower envelope limits converge towards each other at the seaward extent suggesting that all of the sand simply exchanges between the beach and offshore bars without net loss. Actually, there generally is a net loss, but it is small compared with the large profile changes brought about by the cycle of storm flattening and recovery during quiet periods. Along coastlines undergoing net erosion this small loss to the upper shoreface provides a line source of sediment to the shelf.

Many of the discrete coastal sediment sources (rivers, tidal inlets, estuaries) have been discussed above. There is often a perturbation in the shelf sedimentary regime adjacent to these sources in the form of shoals, tidal sand waves or shifting deposits of sediment. In most cases, shelf processes spread the sediment delivered from these sources. Because shelf flows forced by the tide and wind are dominantly shore-parallel within a few kilometers of the coastline, this dispersal of sediment is also dominantly in the coast-parallel direction.

Much of the sediment on broad, sandy continental shelves results from the reworking of preexisting deposits. Where coastal systems are retreating in the face of rising sea level the mid-shoreface exposes older strata (Swift *et al.*, 1991). As the coastal system migrates landward these exposed strata become buried by lower shoreface deposits. This is the origin of a transgressive sand sheet that extends across the entire shelf. It is a common feature of broad sandy shelves and is formed by a combination of shoreface retreat and storm the reworking over the period since the last glacial maximum.

Shelf morphodynamics

Most continental shelves and their major features are composed of unconsolidated sediments. As these are exposed to considerable fluid power, in the form of waves and currents, it is predictable that the morphology of shelf features, and the shelves themselves, represent a time-averaged morphological adjustment of the climate of waves and currents. Niedoroda and Swift (1981) have shown that, although all unconsolidated shelves have a concave-upwards profile, the steepness of the inner shelf profile and the distribution of depths in the mid-shelf correlate with the general intensity of the wave-current climatology. The profile of shelves in stormy areas have a steeper inner-shelf shelf and overall greater depth.

The shoreface is defined according to the concept that it represents the time-averaged response of the shelf surface to the controlling hydrodynamics acting on a sloped surface (Niedoroda and Swift, 1981; Niedoroda *et al.*, 1985a,b). It has been recognized that the shoreface profile translates landward in the face of sea-level rise with little change to its overall shape (Bruun, 1962; Cowell *et al.*, 1995). Moody (1964) measured shoreface adjustments to severe storms off the Delaware coast and showed that the profile alternately steepened and flattened about a mean shape. Niedoroda and Swift (1981) argued that the shape of the shoreface was maintained by a time-averaged balance of sediment transport processes that involved the action of downwell and upwelling storm flows, the corresponding timing of maximum wave conditions during the storm events, the offshore bias caused by the bottom slope and diminished sediment suspension with depth and an onshore bias to sand transport during the period between storms. Measurement programs by Niedoroda *et al.* (1984), Wright *et al.* (1994), and have demonstrated the action of these components, but a complete balance of processes has yet to be successfully demonstrated.

The mode of origin and maintenance of some of the major shelf features has recently been studied. Sand waves are very common in the mid-shelf ridge and swale morphology in the North Sea, Canadian and American Atlantic shelf, and the northeast Gulf of Mexico, among other areas. Recent work by Nemeth *et al.* (2001) has examined these features with stability analyses. A system of equations covering both the shelf hydrodynamics and sediment dynamics has been developed. These are used to show that, under a range of current and wave conditions, a flat shelf surface is not stable. Differential sediment transport tends to produce a "bumpy" surface due to an interaction between the vertical velocity and turbulence profiles and the shape of shelf surface. Over time, only the bottom shape perturbations (bumps) within a narrow range of spacings (wavelengths) will grow. These become the sand waves. The characteristic timescale for producing these features in an energetic mid-shelf location is on the order of centuries to millennia. Once produced, the sand waves persist because they are the stable configuration of the shelf surface to the time-averaged wave-current climate of the area. Nemeth *et al.* (2001) have shown that, in areas where there is a net bias to the shelf flows the sand waves slowly migrate while maintaining their forms. Without this bias, sand waves can remain fixed in location and form.

Calvete *et al.* (2001) have used a similar mathematical method to explain the existence of shoreface connected sand waves. Their analyses show that these features are the stable response mode of the seafloor in the lower shoreface region to the wave–current climatology weighted according to the scale of the induce sediment transport events. The timescale for developing these features is in the range of centuries to millennia. They can be stable or they can migrate longshore (usually in the direction of the apex of the angle their axes make with the shoreline) and shoreward as the shoreface retreats.

The time-averaged morphodynamics of the shelf and shoreface profiles have been represented in numerical models (Niedoroda *et al.*, 1995, 2001, 2003a; Stive and deVriend, 1995; Zhang *et al.*, 1997, 1999; Carey *et al.*, 1999). In these studies the physics of shelf sediment transport are represented in time- and space-averaged forms so that individual events, such as storms, are not resolved. Both depth-averaged cross-shelf (1D-V) and plan-view (2D-H) representations have been developed. These models are in part based on the concept proposed by Clarke *et al.* (1983). This states that provided the characteristic sediment particle transport excursion distance in individual events (typically storms) is small relative to the scale of the shelf, then the time-averaged sediment transport can be represented as a Fickian process similar to turbulent diffusion, but at very large time- and length-scales. Thus the effects of individual storm flows, tidal currents, currents due to shelf waves and internal waves, and other fluctuating currents are represented by time-averaged diffusion coefficients, and all results are limited to terms that are long compared with the timescale of characteristic shelf sediment transport events.

These models have been able to reproduce the dependence of the shelf profiles on the major parameters that include the storm climate, sediment supply, and rate of sea-level rise (or fall). The characteristic responses to changes of these parameters are shown to vary from hours in the surf zone to millennia in the mid- to outer-shelf zones. The controlling parameters are the same as the "Sloss parameters" identified by Swift *et al.* (1991) based on more empirical observations.

This work that resulted in this entry has been assisted by the ongoing research in the ONR EuroStrataform Project (Contract Number N00014-03-C-0134).

Alan W. Niedoroda

Bibliography

Anderson, J.B., 1989. Antarctica's glacial setting. In Anderson, J.B., and Molnia, B.F. (eds.), *Glacial Marine Sedimentation: Short Course in Geology 9*. Washington, DC: American Geophysical Union, pp. 11–57.

Bea, R.G., Wright, S.G., Sircar, P., and Niedoroda, A.W., 1983. Wave-induced slides in South Pass Block 70, Mississippi Delta. *Journal Geotechnical. Engineering*, **109**: 619–644.

Bruun, P., 1962. Sea-level rise as a clause of sure erosion. *Journal Waterways & Harbors Division*, **88**: 117–130.

Calvete, D., Falques, A., deSwart, H.E., and Walgreen, M., 2001. Modeling the formation of shoreface-connected sand rages on the storm-dominated inner shells. *Fluid Mechanics*, **441**: 169–193.

Carey, J.S., Swift, D.J.P., Steckler, M., Reed, C.W., and Niedoroda, A.W., 1999. High-resolution sequence stratigraphic modeling 2: effects of sedimentation processes. In *Numerical experiments in stratigraphy: the recent advances in the stratigraphic and sedimentologic computer simulations*. SEPM Special Publication, 62, 151–164.

Clarke, T.L., Swift, D.J.P., and Young, R.A., 1983. A stochastic modeling approach to the fine sediment budgets all of the New York Bight. *Journal Geophysical Research*, **88**: 9653–9660.

Coleman, J.M., and Prior, D.B., 1982. Deltaic environments of deposition. In Scholle, P.A., and Spearing, D. (eds.), *Sandstone Depositional Environments. American Association Petroleum Geologist*, **31**, 139–178.

Cowell, P.J., Roy, P.S., and Jones, R.A., 1995. Simulation of large-scale coastal change using a morphological behavior model. *Marine Geologist*, **126**: 45–63.

Dean, R.G., 2001. Thirty years of beach nourishment in Florida, and where is all the beach nourishment sand going? In *Proceedings Annual Conference 45*, Florida Shore and Beach Preservation Association.

Hallermeier, R.J., 1978. Uses for a calculated limit depth to beach erosion. In *Proceedings 16, ICCE, Hamburg*, American Society of Civil Engineers, pp. 1493–1512.

Hayes, M.O., 1963. Hurricanes as geologic agents: case studies of Hurricanes Carla, 1961 and Cindy, 1963. *University of Texas Bureau of Economic Geology, Report of Investigations*, 61.

McCave, I., 1972. Sediment transport and escape of fine-grained sediment from shelf areas. In Swift, D.J.P., Duane, D.B., and Pilkey, O.H. (eds.), *Shelf Sediment Transport: Process and Pattern*. Stroudsburg, PA: Dowden Hutchinson & Ross, pp. 215–248.

Metha, A.J., Hayter, E.J., Parker, W.R., Krone.R.B., and Tetter, A.M., 1989. Cohesive sediment transport I: process description. *Journal Hydrological Engineering*, **115**: 1076–1093.

Moody, D.A., 1964. Coastal morphology and processes in relation to the development of submarine sand ridges off the Bethany Beach Delaware. Ph.D. dissertation, John Hopkins University.

Murray, S.P., 1970. Bottom currents near the coast during hurricane Camille, *Journal Geophysical Research*, **75**: 4579–4582.
Nemeth, A., Hulscher, S.J.M.H., and van Damme, R.M.J., 2001. Numerical simulation of sand wave evolution in shallow shelf seas. In *Proceedings Coastal Dynamics '01*, American Society of Civil Engineers, pp. 1048–1057.
Niedoroda, A.W., 1991. Sea ice gouging. In Herbich, J. (ed.), *Handbook of Coastal and Ocean Engineering, V.2*, Houston, 7X: Gulf Publishing Company, 997–1048.
Niedoroda, A.W., and Swift, D.J.P., 1981. Maintenance of the shoreface by wave orbital currents and mean flow: observations from the Long Island coast, *Geophysical Research Letters*, **8**: 337–348.
Niedoroda, A.W., Swift, D.J.P., Hopkins, T.S., and Ma, C.M., 1984. Shoreface morphodynamics on wave-dominated coasts, *Marine Geology*, **60**: 331–354.
Niedoroda, A.W., Swift, D.J.P., Figueiredo, G.L., and Freeland, G.L., 1985a. Barrier island evolution, Middle Atlantic Shelf, Part II: evidence from the seafloor, *Marine Geology*, **63**: 363–396.
Niedoroda, A.W., Swift, D.J.P., and Hopkins, T.S., 1985b. The shoreface. In Davis, R.A. (ed.), *Coastal Sedimentary Environments*. New York: Springer-Verlag, pp. 534–624.
Niedoroda, A.W., Swift, D.J.P., and Thorne, J.A., 1989, Modeling shelf storm beds: controls all the bed thickness and bedding sequence. In Morton, R.A., and Nummedal, D. (eds.) *Proceedings GCSSEPM Foundation 7th Annual Research Conference*, pp. 15–39.
Niedoroda, A.W., Reed, C.W., Swift, D.J.P., Arato, H., and Hoyanagi, K., 1995. Modeling shore-normal large-scale coastal evolution, Marine Geology, **126**: 181–199.
Niedoroda, A.W., Reed, C.W., Stive, M.J.F., and Cowell, P., 2001. Numerical simulations of coastal-tract morphodynamics. In *Proceedings Coastal Dynamics*, American Society of Civil Engineers, pp. 403–412.
Niedoroda, A.W., Reed, C.W., Das, H., Donaghue, J., Koch, J., Wang, Z.B., and Stive, M.J.F., 2003a. Modeling large-scale morphodynamics of complex coastal systems. In *Proceedings of Coastal Sediments '03*, American Society Civil, Engineers.
Niedoroda, A.W., Reed, C.W., Hatchett, L., Young, A., Lanier, D., Kasch,V., Jeanjean, P., Orange, D., and Bryant, W., 2003b. Analysis of past and future debris flows and turbidity currents generated by slope failures along the Sigsbee Escarpment in the deep Gulf of Mexico. In *Proceedings Offshore Technology Conference*, OTC 15162.
Snedden, J.W., and Nummedal, D., 1989. Sand transport kinematics on the Texas continental shelf during Hurricane Carla, September 1961. In Morton, R.A., and Nummedal, D. (eds.), *Proceedings GCSSEPM Foundation 7th Annual Research Conference*, pp. 63–76.
Stive, M.J.F., and deVriend, H.J., 1995. Modeling shoreface profile evolution, *Marine Geology*, **126**: 235–248.
Suhayda, J.N., 1977. Surface waves and bottom sediment response. *Marine Geotechnical*, **2**: 135–146.
Swift, D.J.P., Phillips, S., and Thorne, J.A., 1991. Sedimentation on continental margins: lithofacies and depositional systems *Special Publications of Internal Association of Sedimentologists*, **14**: 89–152.
Syvitski, J.P., Morehead, M.D., and Nicholson, M., 1998. HYDROTREND: a climate-driven hydrologic-transport model for predicting discharge and sediment load to lakes or oceans, *Computer Geoscience*, **24**(1): 51–68.
Weeks, W.F., Barnes, P.W., Rearic, P.M., and Reimnitz, E., 1991. Statistical aspects of ice gouging on the Alaskan shelf of the Beaufort Sea, U.S. Army Cold Regions Research and Engineering Lab. *Report*, 83–21.
Wheatcroft , R.A., Borgeld, J.C., Born, R.S., Drake, D.E., Leithold, E.L., Nittrouer, C.A., and Sommerfield, C.K., 1999. The anatomy of an oceanic flood deposits, *Oceanography*, **9**: 158–162.
Wright, L.D., Xu, J.P., and Madsen, O.S., 1994. Across-shelf benthic transport on the inner shelf of the Middle Atlantic Bight during the Halloween storm of 1991. *Marine Geology*, **118**: 61–77.
Zhang, Y., Swift, D.J.P., Niedoroda, A.W., Reed, C.W., and Thorne, J.A., 1997. Simulation of sedimentary facies on the Northern California Shelf. *Geology*, **25**: 635–638.
Zhang, Y., Swift, D.J.P., Fan, S., Niedoroda, A.W., and Reed, C.W., 1999. Two-dimensional numerical modeling of storm deposition on the Northern California shelf, *Marine Geology*, **154**: 155–168.

Cross-references

Changing Sea Levels
Continental Shelves
Coastal Currents
Cross-Shore Sediment Transport
Deltas
Depth of Closure on Sandy Coasts
Erosion Processes
Longshore Sediment Transport
Offshore Sand Banks and Linear Sand Ridges
Offshore Sand Sheets
Paleocoastlines
Physical Models
Sequence Stratigraphy
Shoreface
Wave–Current Interaction

SHELL MIDDENS

Shell midden refers to anthropogenic deposits containing noticeable amounts of shell, that is, calcareous invertebrate tests. Such deposits are common in marine coastal areas from subarctic to tropical latitudes throughout the world. They also occur along rivers and lakes, where freshwater molluscs comprise the shell consitutents.

Shell midden is not an analytically rigorous term in the sense of human activity, but a descriptive label that identifies the most superficially recognizable constituent of the deposit. *Shell-bearing site* is a more accurate label, although cumbersome and not likely to be adopted (Claasen, 1991). *Midden* originally meant domestic refuse deposited around a house, but shell-bearing sites in the label further obscures the importance of other constituents of the deposit. These sites do not solely represent shell-gathering activities by people. Other food remains such as fish or plants may be very abundant and may, in fact, represent a more significant economic activity at that location, but such remains generally require screening or microscopic analysis of samples to identify.

One sense in which shell middens or shell-bearing sites are a meaningful category is in terms of the depositional and preservational environment represented. As deposits, shell middens share the attributes of having increased porosity, permeability, and alkalinity (Stein, 1992). The alkaline environment created by the leaching of calcium carbonate from the shell and the neutralizing of groundwater, which is generally weakly acidic, creates excellent preservation for organic materials such as shell, bone, and antler, Also, because shell middens are located near aquatic habitats, they frequently are affected by saturation by water (Stein, 1992).

History of research

Investigations of shell middens in Europe, North America, and other continents began in the mid to late 1800s by naturalists, but once they were recognized as anthropogenic, became largely the domain of archaeologists in the 20th century (Classen, 1991; Stein, 1992). In the early part of the 20th century, it was not uncommon for shell middens to be excavated simply to recover artifacts for determining culture historical sequences, essentially disregarding the shell constituents. Most research, however, has focused on the unique aspects of shell middens, their rich invertebrate content, their relationship to coastal resources, and the unique depositional and postdepositional processes that affect them. By far the greatest amount of work has been expended on reconstructions of subsistence and diet, and relating this to the development of marine adaptations among human groups. Closely related is methodological research concerning appropriate sampling and quantification strategies for characterizing midden contents and methods for determing seasonality of deposits. Interest in how shell midden deposits were formed, in terms of the human behavioral components, began with the earliest observations, but studying formation and diagenesis from a geoarchaeological point of view is a more recent development. Middens have always been viewed as informing on ancient environments and habitats, from the position of the coast to the nature of the neritic environment, but the integration with studies of coastal ecology, geomorphology, and neotechtonism has increased dramatically in recent years. Studies of middens have been applied in legal cases concerning the traditional subsistence practices and territories of native peoples.

Behavioral component of shell deposition

A century of research at shell-bearing sites, combined with ethnohistoric and ethnoarchaeological studies around the world (Wessen, 1982; Trigger, 1986; Waselkov, 1987; Claasen, 1991) has resulted in the recognition of a great deal of variability in the cultural activity that created shell-bearing sites. Layers with dense shell may be deposited in association with major village settlements, annually re-used seasonal encampments, temporary wayfaring encampments, or specialized nonhabitation resource procurement sites. Shell middens commonly have more than 20 different taxa of marine invertebrates, each of which is deposited as the result of a distinct chain of activities. Multiple harvesting methods for multiple taxa might be employed at a specific location, and several different marine habitats might be exploited from a single base camp on foot or by watercraft. Shellfish might have been collected regularly, only seasonally, or incidentally to the procurement of other resources, and again, this is different for each taxa represented. They may have been consumed immediately or preserved for storage or trade, or used as bait rather than as food (Claasen, 1991). Processing ranged from eating shellfish raw to a variety of methods of roasting, steaming, boiling, drying, and smoking. These activities leave a wide variety of thermal features from pits to surface pavements of stone used as heating elements.

Shell refuse was disposed of in various ways. It was sometimes burned, and might be discarded adjacent to the habitation area, into the intertidal zone, or over a riverbank. In a short-term occupation, living areas and refuse discard areas might be discreet, but when a location was re-used over hundreds or thousands of years, these areas might shift, so that houses and other structures were built into former refuse areas or refuse disposed of in former house areas. In a shore setting the midden accumulation itself might make a significant contribution to accretion and become the most desirable place for habitation as it built up above the water level. Midden areas were sometimes later used as cemeteries, which adds to the impetus for protecting these resources against erosion and development.

Industrial, in contrast to domestic, uses of shell also create or contributed to shell accumulations, Marine shells have been widely used for making ornaments, such as beads and pendants, or as an inlay materials. They have also been used as musical instruments, for example, Aztec conch shell horns, and as a trading standards (e.g., wampum). Use of shell to manufacture tools is less common, but includes the manufactures of mussel shell knives and whaling harpoon heads in northwest North America, fishhooks in the Polynesian areas, and heavy pounders and adzes made from huge *Tridacna* shells in the Pacific. Tools and ornaments of marine shell were made not just for local consumption, but were traded far inland on all continents. Ground shell was used as temper in some preshistoric pottery wares. Shell also was used as a raw materials in construction (Blukis Onat, 1985). Crushed, it makes a permeable substrate, and in Mesoamerica was ground to make lime for plaster and cement. In the Pacific Islands, it was common for people to extend their land by building bulkheads, and then fill behind them with a mixture of refuse and materials dredged up from the reef flats (thus aiding in keeping canoe channels clear as well) which included coral and shells. Shell industries that involved mining shell for construction, or processing shells for non-subsistence trade items, likely created distinctive types of shell middens.

Anthropogenic versus natural shell accumulations

Recognition of the anthropogenic origin of some shell beds began with observations by naturalists in the 19th century. Criteria for distinguishing cultural shell accumulations from other fossils shell beds has continued to be an area of interest as archaeologists have continued to encounter regionally specific nonanthropogenic shell deposits and develop additional criteria to aid in field recognition. Locally dense accumulations of shell can result from activities of animals such as birds or otters that repeatedly break and leave shells in localized areas, or through beach processes that sort sediments by size, shaped, and density. Natural death assemblages in subtidal marine deposits are less similar, but can be confusing when these are elevated relative to contemporary sea levels. A further problem is the recognition of cases where cultural middens are redeposited by shore processes or by later cultures (Ceci, 1984).

A typical shell midden is likely to have associated artifacts such as tools, ornaments, and manufacturing waste, although these are often quite sparse. Thermally altered rocks, a ubiquitous byproduct of many food processing activities, are often abundant. Other evidence of thermal processes can include ash lenses, charcoal, lenses of oxidized sediment, and calcined shell and bone. The fine matrix in a midden is typically very dark, from charcoal and decomposed organic materials, although there may also be lenses of wood ash, or oxidized soil. The microstratification present in most middens reflects different dumping events as well as constructed features such as surface hearths, roasting or steaming pits, pavements of rock used as heating elements under drying racks, or postholes from large structures such as houses or small structures such as drying racks.

The faunal assemblages in most prehistoric shell middens are diverse, and at the same time, highly selected. The diverse array of marine invertebrates may well represent multiple habitats. Yet not all the locally available invertebrates were collected, and generally a limited size range of individuals. Selectivity is also indicated in the varying compositions of the strata and lenses of the deposit, which are dominated by different associations of taxa representing specific harvesting events. Numerous species of birds, fishes, and mammals can also be present. Flotation to recover carbonized plant remains has rarely been applied, but can reveal a diverse array of wood charcoal and edible plant tissues.

Shells in middens may be whole or fragmented, depending on how they were collected and processed, and depending also on postdepositional alterations such as trampling, or deliberate use for engineering purposes. Fragmentation often follows specific patterns that indicated particular processing methods, such as splitting chiton plates to remove them. Also, the fragmented edges are sharp and angular, not rounded. Attachments of other animals are not found on the interiors of shells, indicating the organisms were alive shortly before deposition. Pieces of weathered shell may occur, but as minor constituents.

Cultural middens can be found in a wide range of geomophological conditions because their deposition is not causally related to shore processes in the way that beach deposits are. Humans may transport shells to a location elevated above the shore, or toss refuse shells into the subtidal zone. Size varies, from small lenses that resulted from a single episode of harvesting and processing, to sites hundreds of meters in length, and thicknesses of over 10 m.

Of the traits described above, associated artifacts and constructed features are the most consclusive for recognizing *in situ* cultural midden deposits, but also the least likely to be observable in a field survery situation without extensive exposure. When observations are limited to shell scattered over a surface or a limited vertical exposure in a cut bank, the criteria of matrix color presence of thermally altered rock and charcoal, and diversity of faunal remains are often more useful. The fragmentation, weathering, rounding, and other alternation of the shells is particularly useful in distinguishing different transportation mechanisms, and the length of time elapsed between death of the organism and deposition. Natural death assemblages in marine deposits exhibit the least degree of movement, often including articulated shells with little fragmentation or mechanical damage. They also typically represent a single habitat, and include abundant juveniles. Beach deposits may incorporate some very fresh shells, but generally contain a significant amount of shells that died some time ago and are essentially part of the sediment load. These will be whole or fragmentary shells that are surface abraded and rounded on the exterior and exposed interior anatomical surfaces, as well as on fractured edges. Older shell may also have been used as substrate for other animals, indicated by attachment of animals such as barnacles, sponges, or bryozoa, on both interior and exterior surfaces. Animals with indications of death by predators, such as drill holes, are more common than in a cultural deposit. The diversity of taxa in a shell midden will be underestimated in field observations (compared with screened samples), but this criteria is usually sufficient to distinguish middens from accumulations of shells created by animals, which tend to focus on a more limited range of species.

Recent shell middens, that is, those formed in the historic period, are potentially quite different from prehistoric ones. Colonization of many areas by people of European descent and changes in traditional diets have led to very different uses of shellfish. Disposal practices are different, too, including incinerating and burying materials in back yards, and use of municipal garbage dumps. The practice of aquaculture, mass production for distant markets, and the use of mechanical equipment results in a different signature characterized by high selectivity, perhaps even a single species. Historic-era middens are most definitively identified by the presence of introduced shellfish species or associated artifacts of recent vintage. In some areas, prehistoric shell middens were mined for shell for use in road construction, for making lime, or as soil amendment, creating extensive secondary deposits. Microstratification and the pressure of constructed features are the best criteria for showing that the deposit is *in situ* and has been redeposited.

Research issues

Shell middens have played a majore role in terms of attempting to understand the evolution of maritime adaptations around the globe, yet the degree to which humans exploited or relied on marine intertidal resources at different points in human development is less well known than for terrestrial resources because changing sea levels have biased the archaeological record, drowning much of the Pleistocence and early Holocene record (Bailey and Parkington, 1988). The extensive record at the Klasies River Mouth site in South Africa shows use of marine resources, including molluscs, beginning 60,000 years ago (Thackeray, 1988), and a number of shell-bearing sites older than 20,000 years have been documented in Europe, Asia, and Africa (Claasen, 1991). Yet most recorded shell middens are Holocene in age. This is interpreted as evidence of increasing resource intensification in the last 10,000 years, correlated with increased population growth and decreasing mobility and territory size during which human groups turned from high-ranked resources such as large herbivores and incorporated more resources that are low-ranked in terms of energy return such as molluscs and other invertebrates into their subsistence (Bailey, 1978, 1983).

Reconstruction of prehistoric subsistence through the quantitative analysis of midden contents has been a major theme since the mid-20th century. Much of this research suffers from methodological problems such as inadequate sampling and assuming all shellfish represent food, and estimates of the human population derived from shellfish quantities are particularly flawed (Claasen, 1991). Related efforts have focused on determining the seasonality of harvesting shellfish, and by extension, the season of occupation. A number of studies have estimated season of death for species with annual growth rings by determining the proportion of the annual growth that had been achieved, using thin sections or surface measurements of ring width. The validity of the estimates of season of death suffers from reliance on modern samples, and the extension of these seasonality estimates to estimating the seasonality of habitation is weakened by inadequate attention to sampling and assuming that shellfish were harvested for immediate consumption as food, ignoring the possibility of storage and of nonfood uses (Claasen, 1991). A focus on the food value of the animal remains in middens continues (Erlandon, 1988), but archaelogists are more cautious about generalizing from small samples and make more use of comparative regional information.

Sampling problems are considerable, and not easily resolved. A relatively small excavation may remove, from primary context, hundreds of thousand of pieces of shell, all essentially artifacts. The information context of the shell itself becomes redundant long before the information content of other aspects of the site, such as features, bird bones, tools, is adequately sampled. Because of the high cost of collecting and analyzing shell and other small midden constituents (Koloseike, 1970), archaeologist commonly retain limited samples of the excavated materials. Results are frequently noncomparable between projects because of variation in sampling intensities and size fractions. Strategies for selecting the samples spatially are frequently non-probabilistic and not based on the growing understanding of the internal spatial structure of shell middens. For example, the use of a few column samples to characterize the midden, and especially to discuss, changes over time, is questionable given research by Campbell (1981) which shows that the shell composition of middens may vary as much horizonatally as vertically. In the analysis process, archaeologists variously choose weight, number of identified specimens, and minimum number of individuals to quantify the relative abundance of taxa, adding to the noncomparability of samples. Each measure provides different estimates of abundance, even changing rank order, because of differential fragmentation and recognizability of parts of different taxa.

Geoarchaeological approaches to shell midden formation

The use of geoarchaeological methods to understand the physical processes involved in the deposition and diagenesis of shell deposits is a much more recent endeavor. Examination of physical traits of midden sediments, such as sediment texture, clast orientation, soil chemistry, and surface weathering on particles is increasingly being used to interpret the physical process of transportation, bioturbation, and leaching (Stein, 1992). Diagenetic changes, especially those related to saturation can create apparent stratification that has been misinterpreted by archaeologists interpreting culture history.

Environmental reconstruction

The role of shell middens in reconstruction of coastal paleoenvironments ranges from simply using them as indicators of paleo-coastlines to inferring nearshore habitals from the molluscan species present, to using them as a means of dating geological deposits that reflect coastal accretion or neotectonic events (Grabert and Larsen, 1973). The need for caution and independent evidence to sort out changes due to habitat versus those due to cultural effect is widely recognized. Molluscan assemblages in coastal sites are indicators of coastal ecology and habitat as determined by substrate, exposure, salinity, and water temperature (Rollins *et al.*, 1990) but they also reflect human selection. The most effective work is interdisciplinary and uses multiple lines of independent evidence (e.g., Sanger and Kellogg, 1989; Shackelton, 1988).

Shell middens are playing a growing role in coastal geomorphological and neotectonic research because they are a specific type of deposit that can be assumed to represent a coastal position, although not causally related to a specific short position as a beach deposit might be. Shell middens contain abundant datable material, including historically diagnostic artifacts (McIntire, 1971), radiocarbon datable organic material, and shells datable by isotope analysis (Thackeray, 1988) and have been used to provide bracketing dates for subsidence events and tsunami deposits.

Summary

Despite the methodological caveats that plague research on shell middens, they have been shown to reveal a wealth of information on prehistoric adaptations and on coastal paleoenvironments. Interdisciplianary research efforts are increasing and new analytic techniques are being developed. Unfortunately the rate of destruction of these unique coastal resources is likewise increasing due to development. Although shell middens, like other archaeological sites, are protected by cultural resource protection and management laws in most countries, enforcement is inconsistent or nonexistent. With the increased rate of development added to the natural erosion processes that have always affected coastal sites, we are in danger of losing much of this nonrenewable resource. It is not only a scientific data base for interpreting prehistory and paleonenvironments, but also a historic record and direct physical connection to traditional places and practices for native peoples in many coastal areas.

Sarah K. Campbell

Bibliography

Bailey, G.N., 1978. Shell middens as indicators of postglacial economies: a territorial perspective. In Mellars, P. (ed.), *The Early Postglacial Settlement of Northern Europe*. London: Duckworth, pp. 38–63.

Bailey, G.N., 1983. Problems of site formation and the interpretation of spatial and temporal discontinuities in the distribution of coastal middens. In Flemming, N.C., and Master, P.M. (eds.), *Quaternary Coastlines and Marine Archaeology*. London: Academic Press, pp. 559–582.

Bailey, G.N., and Parkington, J., 1988. Introduction. In Bailey, G.N., and Parkington, J. (eds.), *The Archaeology of Prehistoric Coastlines*. Cambridge: Cambridge University Press, pp. 1–10.

Blukis Onat, A.R., 1985. The multifunctional use of shellfish remains: from garbage to community engineering. *Northwest Anthropological Research Notes*, **19**: 201–207.

Campbell, S.K., 1981. *The Duwamish No. 1 Site: A Lower Pudger Sound Shell Midden*. Office of Public Archaeology, Institude for Environment Studies Research Report 1. Seattle, WA: University of Washington.

Ceci, L., 1984. Shell midden deposits as coastal resources. *World Archaeology*, **16**: 62–74.

Claasen, C., 1991. Normative thinking and shell-bearing sites. In Schiffer, M.B. (ed.), *Archaeological Method and Theory, Vol. 3*. Tucson, AZ: University of Arizona Press, pp. 249–298.

Erlandson, J., 1988. The role of shellfish in prehistoric economies: a protein perspective. *American Antiquity*, **53**: 102–109.

Grabert, G.F., and Larsen, C.E., 1973. Marine transgressions and cultural adaptation: preliminary tests of an environmental model. In Fitzhugh, W. (ed.), *Prehistoric Maritime Adaptations of the Circumpolar Zone*. The Hague: Mouton, pp. 229–251.

Koloseike, A., 1970. Coast of shell analysis. *American Antiquity*, **35**: 475–480.

McIntire, W.G., 1971. Methods of correlating cultural remains with stages of coastal development. In Sears, J.A. (ed.), *Introduction to Coastal Development*. Cambridge, MA: Massachusetts Institute of Technology Press, pp. 188–203.

Rollins, H.B., Sandweiss, D.H., and Rollins, J.C., 1990. Mollusks and coastal archaeology: a review. In Lasca, N., and Donahue, J. (ed.), *Archaeological Geology of North America*, Centennial Special Vol. 4, Boulder, CO: Geological Society of America, pp. 467–478.

Sanger, D., and Kellogg, D.C., 1989. Prehistoric archeology and evidence of coastal subsidence on the coast of Maine. In Anderson, W.A., and Borns, J.H.W. (eds.), *Neotectonics in Maine*. Augusta, ME: Maine Geological Survery Bulletin No. 40, pp. 107–126.

Shackelton, J.C., 1998. Reconstructing past shorelines as an approach to determining factors affecting shellfish collecting in the prehistoric past. In Bailey, G., and Parkington, J. (eds.), *The Archaeology of Prehistoric Coastlines*. Cambridge: Cambridge University Press, pp. 11–21.

Stein, J.K., 1992. The analysis of shell middens. In Stein, J.K. (ed.), *Deciphering a Shell Middens*. New York: Academic Press, pp. 1–24.

Thackeray, J.F., 1988. Molluscan fauna from Klasies River, South Africa. *South African Archaeological Bulletin*, **43**: 27–32.

Trigger, B.G., 1986. *Native shell mounds of North America*. New York: Garland.

Waselkov, G.A., and Schiffer, M.B. (ed.), 1987. Shellfish gathering and shell midden archaeology. In *advances in Archaeological Method and Theory, Vol. 10*. Orlando, FL: Academic Press, pp. 93–210.

Wessen, G., 1982 Shell middens as cultural deposits: a case study from Ozete. Ph.D. dissertation, Department of Anthropology, Pullman, WA: Washington State University.

Cross-references

Archaeological Site Location, Effect of Sea-Level Changes
Archaeology
Geochronology
Human Impact on Coasts

SHINGLE BEACH—See GRAVEL BEACH

SHORE PLATFORMS

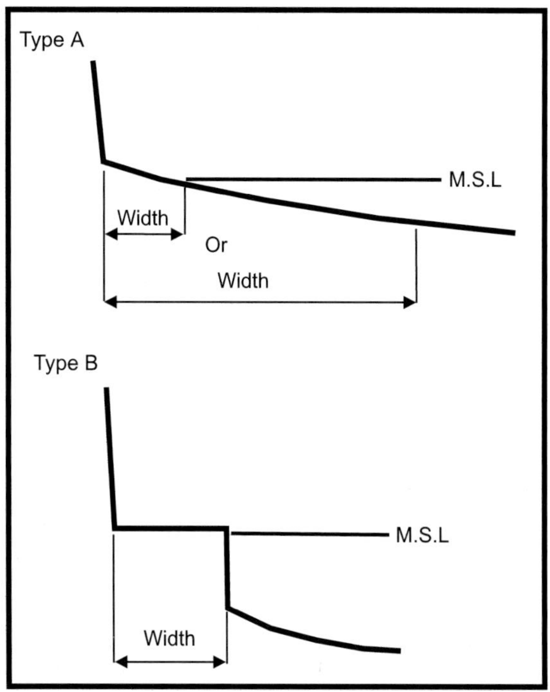

Figure S36 Two morphologies of shore platforms, Type A and Type B (adapted from Sunamura, 1992).

Shore platforms are horizontal or gently sloping surfaces backed by a cliff, eroded in bedrock at the shore. The erosional origin of these surfaces is evident because they cut across and expose geological structures. Shore platforms have long been classified using a tripartite scheme with respect to elevation in relation to the tide. Thus high-tide, intertidal, and sub-tidal platforms have been identified. Another classification uses the two most common profile forms, either the sloping platform (commonly 1–5°) or the horizontal platform. In recent times these have been referred to as Type A and Type B platforms, respectively (Figure S36). Shore platforms are common along much of the world's coastline occurring in all but the very highest coastal latitudes, as well as lakes. However, the total percentage of coastline composed of shore platforms is unknown. Associated with shore platforms are a variety of coastal features such as sea caves, ramps, notches, potholes, ramparts, and low-tide cliffs.

Many different terms synonymous with shore platform have been used, often having quite different genetic and morphological meanings. Such terms include: rock bench, high-water rock platform, Old Hat Type platform, abrasion platform, shore bench, storm-wave platform, marine bench, sloping wave-bench, intertidal platform, sea-level shore platform, wave-cut terrace, surf-cut terrace, coastal platform, bench, abrasion bench, denuded bench, tidal bench, wave-cut bench, rock platforms, high-water rock ledges, wave ramp, wave-cut platform, wave-cut terrace, wave-cut shore platform and bedrock platform. Clearly, the wide variety of terms used reflects different interpretations of morphology and processes by individual workers, and of the relationship between a surface and sea level. The term "shore platform" has been the most widely used, because it has no genetic connotations and is the most appropriate term since the development of shore platforms and the processes involved are still not fully understood.

The origin of shore platforms has been the subject of debate for 150 years. This debate is concerned with the relative roles of marine and subaerial processes in the development of shore platforms. Advocates of a marine origin have often referred to shore platforms as "wave-cut" platforms. The marine origin for shore platforms has dominated the literature but there is little quantitative evidence to support such a mode of development. The same is also true for subaerial weathering. Little quantitative research of either mechanism has occurred. Investigators have too often relied on interpretative approaches based on morphological features, despite numerous warnings of the ambiguity of morphology as an indicator of process. Subaerial weathering has often been ascribed a secondary role in platform genesis. It has been proposed that the remarkably horizontal nature of some shore platforms results from a secondary plantation of the surface by subaerial processes following initial cutting by marine forces. A small group of researchers have suggested that weathering is a primary process. In this scenario, waves serve only to remove the debris formed by weathering, but without the weathering there would be no platform. A third view is that both marine and subaerial processes play strong roles, but very few workers have suggested both processes are equally important.

Mechanical wave erosion includes impact forces, cavitation, hydrostatic pressure, and compression of air in joints spaces. Air compression results from the sudden inrush of water into a crack or space and causes an explosive shock upon the sudden release of the compressed air. Cavitation occurs when high water velocities at the bed cause a drop in pressure. If vapor pressure is attained then the rock surface can be damaged by the sudden formation and destruction of vapor pockets. Hydrostatic pressures increase with the depth of water. Since depth changes frequently under waves, variations in pressure occur at the bed, but for these variations to cause erosion the force exerted must exceed the strength of the rock. It remains to be determined whether or not these variations cause erosion. Impact forces include shock pressure and water hammer but these are restricted in extent because specific circumstances are required for them to operate effectively. Water hammer is the impact between a body of water and a solid and only operates if no air is trapped between the water and solid. The generation of shock pressures require the trapping of an air pocket by the front of a breaking wave as it impacts against a vertical structure. Shock pressure and water hammer are also dependent on the occurrence of breaking waves impacting directly on a platform or against the cliff. Since water depth determines the position of wave-breaking relative to the shore, true impact forces from breaking waves seldom operate on shore platforms. Larger waves break in deeper water further from the shore so that it is even less likely very large storm waves generate these forces. Only in situations where the depth of water in front of a platform is exactly suited to the breaking, water depth relationship, is there any likelihood of shock pressure and water hammer occurring and then only at the seaward edge of a platform. Water depths on platforms at high tide and even during storm and wave setup do not permit large breaking waves to impact directly against the cliff. More common is for the bore of a broken wave to shoal across a platform surface. While forces exerted are less than those generated by shock pressure and water hammer the greater frequency of air compression in joint spaces is likely to mean that it is the most important of marine processes. Associated with marine processes is abrasion, which is the wearing away of bedrock by the to and fro movement of sediments under wave action. Under more violent conditions sediments can be hurled against cliffs, acting like missiles. The effectiveness of abrasion relies on a suitable sediment and thickness of deposit. Harder sediment lying over softer bedrock will be more effective than sediments similar in nature to the platform rock. When sediments are too thickly deposited on a platform a protective role occurs. Wave tank experiments have provided some evidence that breaking and broken waves are capable of initiating shore platforms in blocks made of sand and plaster. Ice grounding by wave action and wind has be identified as an important development process on shore platforms in high latitudes.

Studies utilizing the Schmidt Hammer test have confirmed weathering occurs on platforms. Subaerial weathering includes both mechanical and chemical processes, such as solution, salt weathering, and wetting and drying. Some evidence exists to link higher erosion rates with greater numbers of wetting and drying cycles. The number of cycles is determined largely by the tide and varies with change in elevation within the tidal range. Salt spray, wave splash and rainfall contribute to and prevent wetting and drying depending on the timing in relation to the tidal cycle. The number of wetting and drying cycles also show seasonal variations with more in summer months than winter since drying is facilitated by higher temperatures. The growth of algae on platform surfaces reduces the number of cycles by preventing drying, especially in winter months. A subaerial process often identified is water-layer weathering. This involves a combination of weathering processes but seems to be most reliant on wetting and drying and to a lesser degree salt weathering and solution. Water layering morphology occurs where pools of water accumulate and are afforded time to evaporate and the surface dry. Pools up to several meters in diameter are surrounded by raised rims 2–5 cm high that occur along fracture lines the rock. These ridges develop along fissures because they hold water and undergo fewer cycles of wetting and drying. Water-layer weathering operates wherever seawater can accumulate and evaporate. Thus, it can occur from the low-tide level and has been reported as high as 24 m above sea level where spray accumulates. In high latitudes frost shattering has been proposed as significant in platform development. Although frost shattering is not limited to higher latitudes, it has been reported as a significant process on chalk platforms on the southeast coast of England.

Biological activity on shore platforms causes erosion but also inhibits other erosion processes. Bio-erosion can be subdivided into mechanical and chemical. Solution processes operate where biota release CO_2 into seawater, lowering the pH and in turn causing calcium carbonate to convert to calcium bicarbonate which has the effect of weakening the rock. Algae probably contribute significantly to erosion relative to other organisms. Endolithic algae have a wide vertical distribution at the shoreline and weaken rock by boring. Both endolithic and epilithic algae cause variations in the chemistry of water through metabolic functions and probably cause chemical erosion. Algae also serve as food for grazing organisms such as chitons and gastropods. Molluscs that graze on endolithic algae can achieve considerable erosion on coastal bedrock. Some chitons and gastropods carve out home scars that appear as impressions similar in size to their own shell in rock surfaces. The protective role of biota on shore platforms has been recognized but it is poorly understood. Dense and continuous mats of algae often cover significant areas of shore platforms during winter months in temperate environments. During warmer months desiccation removes these algae exposing the platform. Algae coverage protects the substrate from mechanical wave erosion and reduces some weathering processes by preventing drying. Thick growths of kelp such as *Durvillea antarctica* on the seaward edge of platforms dissipate wave energy through motion but also contribute to erosion when they are removed. Holdfasts thrown upon platforms often contain sizable pieces of rock.

Shore platform morphology can be characterized as being either near horizontal or gently sloping. Sloping platforms (Type A) grade into the sea whereas horizontal platforms (Type B) terminate at the seaward edge with a marked cliff that drops into the sea. This drop is sometimes referred to as the low-tide cliff. Platform morphology is largely controlled by geological factors. Evidence suggests that platform type is a function of the compressive strength of rock, with Type A platforms developed in softer rock than Type B. However, wave energy also appears to play a role with Type B platforms subjected to higher wave energy than Type A platforms. A demarcation between the two morphologies can be made based on the relative differences in the magnitude of the wave energy and rock strength. Variation in the gradient of platforms has been attributed to tidal range with steeper gradients generally occurring where the tide range is the greatest. This relationship holds best for platforms in macro-tidal environments and weakens as tidal range becomes smaller. So that platforms in micro-tidal areas have the lowest gradients although sloping Type A platforms are common. However, explanations for platform gradient employing tidal range ignore the role of geology.

Platform elevation is an important, if poorly understood, component of morphology. This is surprising given the wide use made of raised platforms for reconstructing sea level and tectonic histories in many coastal settings. Traditionally the view has been that platforms develop down to the low-tide level. To date it is unknown to what elevation platform surfaces develop in relation to the sea. This problem is more complicated because of the tendency of some platforms to be located at or above the high-tide level while others are intertidal. It becomes difficult to determine whether higher platforms have developed at that elevation in relation to the present sea level or if they have emerged or relate to a higher stand of sea level. Attempts have been made to link elevation with wave energy and geology. Interestingly the field evidence does not show a consistent relationship between elevation and wave energy. Some examples occur where higher platforms are found on headlands and lower ones in embayments, examples showing no such relationship have also been reported. Again this reflects the role of geology in determining platform morphology. Platform elevation has been found to increase as the compressive strength of the rock increases. The depth of water in front of platforms has also been shown to be an important control on platform elevation. This is because water depth controls the type of wave arriving and therefore the amount of energy delivered to the platform. If both geology and wave energy are considered together better explanations of platform elevation can be made.

Width is taken as the horizontal distance from the edge of a platform exposed at low tide to the cliff platform junction, and has often been associated with exposure to wave energy. Wider platforms might be expected where exposure to wave energy is greatest. A number of examples exist to support this proposition, but examples can also be found that show an opposite relationship with wider platform in sheltered embayments. The lack of a clear relationship reflects the strong influence geology has in determining morphological characteristics. Narrower platforms have been found to occur in rock with higher compressive strengths. The expectation that there should be a relationship between width and wave energy is reliant on a marine origin for shore platforms. The absence of a clear association between width and the wave environment may also reflect the role of subaerial processes. The distance to the seaward edge of a platform is largely irrelevant to weathering processes. The wave energy needed to removed weathered debris is significantly less than that required to erode rock from *in situ*, so that platforms could be much wider than under a wave energy control model of platform width. Understanding platform width is also made difficult because there is uncertainty that the seaward edge of a shore platform retreats landward. This is important because if the edge does not retreat then it marks the original position of the coastline. It might be reasonable to expect that platform width is limited by wave attenuation across the widening platform. Widening ceases when waves are no longer capable of removing weathered debris or erosion of the cliff, the platform can be said to be in static equilibrium. If, however, the seaward edge does retreat then the width is determined by the relative rates of erosion of the cliff and the seaward edge. Initially the rate of retreat of the cliff should be greater than the seaward edge to allow a platform to develop. Increasing width would attenuate wave energy and cause cliff recession to slow and could not increase until the seaward edge retreat allowed wave energy to do more work by reducing the width of the platform. It has been suggested that once a platform has developed both the cliff and edge of a platform erode at the same rate so that the entire profile retreats landward in a type of dynamic equilibrium.

The rates at which shore platforms develop are of considerable interest since these may provide an indication of the ages of platforms. Determining the age is necessary since it is possible that some platforms are polygenetic, having developed previously when sea level was at a similar position as today, or a platform that initially developed at a lower sea level has emerged to be coincident with the present sea level and reactivated. There are two components to be considered with respect to the rate of development. The retreat of the cliff backing the platform and rate of vertical lowering of the platform. It remains to be determined whether or not the seaward edge of a platform retreats landward. Cliff retreat rates vary considerably both spatially and temporally, but commonly occur at less than 1 m a^{-1}. Vertical lowering rates have only been measured at a few locations worldwide and with only a few exception over short time periods of about two years. Generally rates are in the range of 0.5–1.5 mm a^{-1} with a mean of 0.95 mm a^{-1}. Although there is considerable variation in these rates depending on location on the platform, season, and rock type. On tropical limestone, mean rates of lowering of 1.97 and 1.25 mm a^{-1} have been reported. Mean erosion rates on temperate limestone and mudstone of 1.13, 1.48, and 1.53 mm a^{-1} were recorded. Erosion rates on schist and greywacke 0.625 and 0.37 mm a^{-1}, respectively, have been measured. The publishing of mean erosion rates belies the fact that there is a considerable range in the measured rates, in some instances rates as high as

10 mm a^{-1} have been reported. In settings where seasonally driven weathering processes such as wetting and drying are important higher erosion rates occur in summer compared with winter. In those setting where winter storm activity is dominant higher erosion rates have been measured in winter. Two studies have extend the length of vertical lowering records beyond 10 years. Both presented erosion rates that were the same order of magnitude as short-term studies. One based on an 11-year period was cautious about extrapolating short-term erosion rates while the other based on a 20-year period suggested that short-term data can be extrapolated at least to the decadal scale. Another difficulty is that all vertical lowering rates are only measured at a scale of millimeters. There are no reports of large block disintegration thus the contribution of erosion at larger scales is unknown.

Numerous attempts have been made to numerically model platform development. These fall into two categories. Those based on the Gilbertian notion that platform development only begins when the erosive force of waves (F_W) exceeds the resisting force of rock (F_R) so that development begins when $F_W \geq F_R$. The second approach is based on empirical field evidence. Models from both approaches have attempted to elucidate the types of equilibria, elevation, width, and gradient that platforms attain. Attempts have also been made to model the effect of sea-level fluctuations and wave forces on platform development. A fundamental problem is that all modeling attempts are based on the view that platforms are wave cut. Not surprisingly these models are then used to support the view that platforms have a wave-cut origin. As yet no model has been developed that is based on a subaerial origin. A number of models have also produced contradictory results. In some instances the equilibrium form predicted is dynamic while other models suggest a static form of equilibrium will be reached. This difference in equilibrium type arises because different assumptions about erosion of the seaward edge of the shore platform were used. Platform gradient has been modeled as a function of tidal range in some examples while others have used wave height, water depth in front of a platform, and rock strength.

Suggested further readings may be found in the following bibliography.

Wayne Stephenson and Robert Kirk

Bibliography

Stephenson, W.J., 2000. Shore platforms remain a neglected coastal feature. *Progress in Physical Geography*, **24**(3): 311–327.

Sunamura, T., 1992. *Geomorphology of Rocky Coasts*. New York: Wiley.

Trenhaile, A.S., 1987. *The Geomorphology of Rock Coasts*. Oxford: Oxford University Press.

Cross-references

Cliffed Coasts
Cliffs, Erosion Rates
Cliffs, Lithology versus Erosion Rates
Erosion Processes
Notches
Thalassostatic Terraces
Weathering in the Coastal Zone

SHORE PROTECTION STRUCTURES

Shore protection in its widest usage refers to the reduction or elimination of damage to the shore and backland as might be caused by flooding, wave attack, and erosion. The shore may consist of cliffs, reefs, beaches, and artificial or engineered structures that form part of the water and land interface. Shore protection structures can be classified as hard, soft, or a combination. Soft structures or soft methods of shore protection usually involve placement of beach-quality sediment, typically sand, directly on the beach, a process called beach nourishment or beach fill. The beach fill may be placed across the upper beach profile and as a dune system. In such designs, the beach berm protects the dune against erosion, and the dune protects the backland from flooding and wave attack. Another type of soft shore protection structure is a "nearshore berm," referring to placement of material in an approximate linear form along the shore to break storm waves and to feed material to the beach during times of accretionary wave conditions (Hands and Allison, 1991; McLellan and Kraus, 1991). Information about beach fill design for shore protection can be found in Dean (2003) and in Coastal Engineering Manual (2003).

This entry principally concerns hard shore protection structures designed to reduce erosion of sandy beaches. Hard structures should usually be placed together with beach fill, as emphasized below with respect to preservation of the neighboring beaches. The generic forms of hard shore protection structures are groins, detached breakwaters, and seawalls, and these have several variations.

Elements of planning and coastal sediment processes

In considering local shore protection, the neighboring shore is also a concern and must be included in considerations because waves and currents transport sediment along the coast. Shore protection actions taken at one section of coast will usually have consequences for neighboring sections, and the timescales of the change induced may be short (months) or long (decades). Therefore, a regional approach is best considered that includes understanding of the sediment budget for the coast. Information about shore protection in the wider sense is available in Marine Board (1995), Komar (1998), Coastal Engineering Manual (2003), and Dean and Dalrymple (2002).

Shore protection is part of coastal zone management in which the uses and functions of the shore, as well as the benefits and costs of maintaining it, are evaluated. Such an evaluation is usually done by an associated group of stakeholders, including property owners, regulatory agencies, and other organizations with interest in preserving the property, environment, and functionality of the particular shore in question.

In approaching a shore protection design to mitigate beach erosion, the cause of the erosion should first be identified to determine the general approach and optimal design. For example, a rule of thumb is that if high-tide shoreline recession is greater than 1–2 m/year for a long stretch of shore, beach fill alone is probably not cost effective. Local areas of the shore experiencing significant beach erosion as compared with the adjacent shore are called erosional hot spots (Kraus and Galgano, 2001). Structures may be an effective hot-spot countermeasure under certain situations. Both initial costs of construction and the maintenance costs should be considered. Shore protection is a form of infrastructure; therefore, maintenance of the structures and beach fill is part of the project life cycle.

It is convenient to classify sediment transport as being directed either alongshore or across shore. Longshore sediment transport occurs throughout the year, with storms often creating the most transport. For an observer standing on the coast, it is convenient to define left- and right-directed longshore sediment transport. The net transport is then defined as the difference of the right- and the left-directed transport, and the gross transport rate is defined as the sum of the left- and right-directed transport. Changes in the high-tide shoreline position, or erosion and accretion, are caused by differences in net transport from one location to another on the coast.

Cross-shore transport is directed either onshore or offshore. Offshore transport occurs during storms, when the beach is eroded and sediment transport is transported seaward to create a storm bar. Sediment may also be transported onto the land (overwash), and this material is called washover. After a storm and during summer or mild wave conditions, sand that is stored in storm bars is moved onshore to widen the beach. In the following, the predominant direction of transport enters in consideration of the action of a shore protection structure.

Groins

Groins are shore-normal structures (Figure S37) emplaced for the purpose of either (1) maintaining the beach behind them, or (2) controlling the amount of sediment moving alongshore. Because modern coastal engineering practice includes a regional perspective that incorporates the stability of downdrift beaches, groin construction is normally done together with beach nourishment so that sediment impounded by the groin or groins is replaced in the total system. If beach fill is not included with groin construction and maintenance, downdrift beaches may suffer (Nersesian *et al.*, 1992). Groins have varied shapes, such as the simple groin, spur or L-shaped groin, and T-head (Figure S38). These shapes were developed to better retain sediment by reducing offshore transport by the rip current that tends to form at the stems of groins, and by providing a sheltered region from waves.

Groins are designed to operate where there is appreciable longshore sediment transport. Groins will not function well if cross-shore transport is strong, such as in the US Great Lakes, where steep waves tend to move sand offshore that can then bypass the groins by transport along longshore bars. Groins are often placed as a system or a field. Figure S39

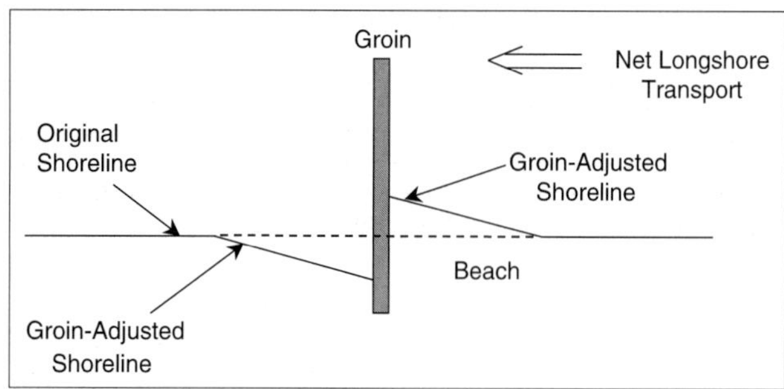

Figure S37 Definition sketch for beach response to a groin.

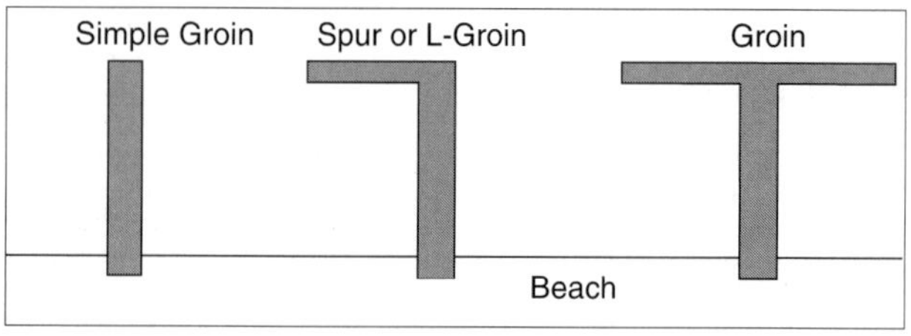

Figure S38 Examples of types of groins.

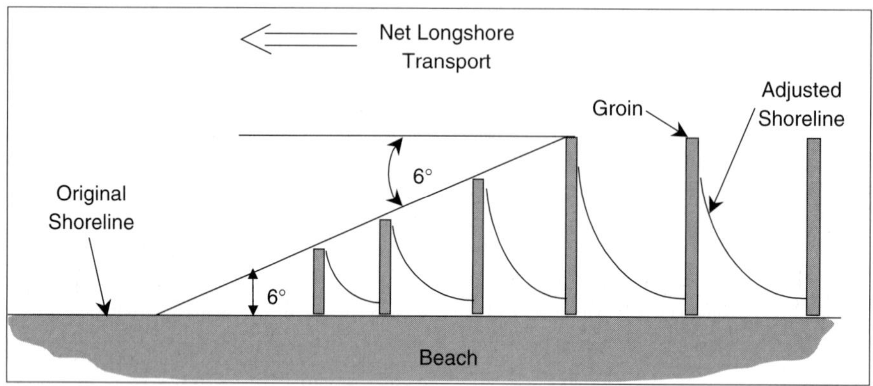

Figure S39 Tapered groins.

shows a schematic of a tapered groin field, with shortening of the groins in the direction of net transport to transition into the beach without structures.

Groin functioning and design can be parameterized in a numerical simulation model through three key processes (Kraus *et al.*, 1994) as (1) sediment bypassing around the seaward end of the structure, (2) structure permeability in allowing sediment to go through or over the groin, and (3) ratio of net to gross longshore transport rate, which controls in great part high-tide shoreline recession and advance at a groin, as schematically shown in Figure S39. Groin length determines the amount of bypassing. Typically, a groin will be shorter than the average width of the surf zone. A long groin will intercept too much sand moving alongshore and cause erosion of the downdrift beach by impounding or blocking sediment on the updrift side.

Groins may be a possible component of a shore-protection project and sand-management program under the following situations:

1. Where there is a divergent region of longshore transport, such as at the center of a crenulate-shaped pocket beach, or where the curvature of the coast changes greatly.
2. Where there is no source of sand, such as the downdrift side of a large harbor or inlet with jetties.
3. Where intruding sand is to be managed, such as to retain sand on the updrift side of a harbor to prevent shoaling of the channel and to stockpile the sediment for bypassing by land transport.
4. Where the sand transport rate is to be controlled or gated, such as to prevent undue loss of beach fill to unwanted areas.
5. Where an entire littoral reach is to be stabilized, such as on a spit, near a submarine canyon.
6. Where stabilization of the high-tide shoreline is required under extreme conditions.

Groins cannot prevent sand from moving offshore, such as during storms. For such situations, detached breakwaters serve better for shore protection.

Groins may be made porous or low to allow some amount of sediment to bypass them. Also, groins may be notched at the shoreward end to allow sand to pass through them that is being transported near the high-tide shoreline (Kraus, 2000). Such designs are intended to balance the amount of material bypassed to the amount retained under certain wave conditions and water level. Also, permeable and notched groins tend to minimize abrupt change in high-tide shoreline position on the sides of the groins, promoting a straight shoreline.

Detached breakwaters

Detached breakwaters are sometimes referred to as offshore breakwaters. They function to reduce wave energy on their landwards sides. Sediment will accumulate in the wave-sheltered region because the water is calmer and the longshore current behind the detached breakwater weaker than on adjacent shore that is open to full wave energy. Detached breakwaters are often constructed to be partially submerged, for example, at higher tide, to allow wave transmission over and, sometimes, through them. In this way, a certain amount of wave energy reaches the sheltered region and promotes sediment transport alongshore. Detached breakwaters are often built in groups, and this configuration is referred to as a segmented detached breakwater system. Detached breakwaters may be an appropriate shore protection measure in areas where cross-shore transport is large, or where wave energy must be reduced, such as at a change in orientation of the coast.

The response of the beach to detached breakwaters is controlled by at least 14 variables (Hanson and Kraus, 1990), making these structures more difficult to design than groins. The response of the beach to the presence of a detached breakwater (Figure S40) may be as a tombolo, for which the high-tide shoreline reaches the structure, or as a salient that describes a cuspate morphologic form that grows toward, but does not reach the structure. The type of beach response is determined by design requirements. Typically, it is desired to allow some amount of sediment to pass alongshore behind the structure. A potential design deficiency for detached breakwaters is to place them too far offshore. In such a case, the high-tide shoreline will show no response.

Main design parameters for detached breakwaters are distance offshore, length of structure, and transmission (depends on elevation of structure and its composition). The design takes into account the prevailing wave climate at the site and the desired beach response. For segmented detached breakwaters, the gap width between structures is also a key design parameter. Empirical design procedures have been developed that incorporate the geometric parameters of detached breakwaters (Pope and Dean, 1986; Rosati, 1990), whereas numerical simulation

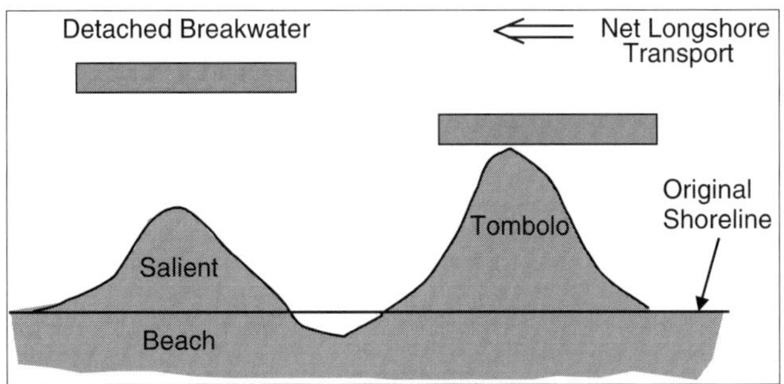

Figure S40 Definition sketch for beach response to detached breakwaters.

models can be applied to account for a wide range of conditions, including wave transmission (Hanson and Kraus, 1990). As depicted in Figure S40, a detached breakwater built closer to shore will tend to create a tombolo. Erosion or high-tide shoreline recession between structures can occur and must be accounted for in design.

Submerged detached breakwaters are sometimes called reefs, particularly if they have a broad crest. Reef breakwaters allow transmission, the wide crest reducing wave height behind them. Reef breakwaters often serve as an offshore retaining structure for sediment near the beach, acting to perch or elevate the beach.

Headland control is a variation of detached breakwaters, in which hard points or shore-connected detached breakwaters are constructed, often with orientation parallel to the crests of the predominant incident waves (Silvester and Hsu, 1993; Hardaway and Gunn, 1999). In this way, a curved headland beach, or a pocket beach if bounded laterally by two such structures, will form. The intent is to developed isolated beach compartments that will lose a minimum amount at their longshore ends, letting the high-tide shoreline achieve a form in equilibrium with the predominant direction of the incident waves.

Seawalls

Seawalls are constructed to prevent inland flooding from major storms accompanied by high water levels (storm surge) and large waves. Seawalls also fix the position of the land sea boundary if the sea reaches the structure. The main functional element of a seawall is the elevation to minimize overtopping from storm surge and wave runup. A seawall is typically a massive, stone and concrete structure with its weight providing stability against sliding forces and overturning moments. Whereas groins and detached breakwaters are constructed with preservation of the beach as a design goal, the purpose of seawalls is to protect the backland. Knowledge on seawalls and their interaction with beaches is compiled in a collection of papers found in Kraus and Pilkey (1988), and the literature on the subject has been brought up to date by Kraus and McDougal (1996).

Seawalls have many counterparts in nature, as do groins and detached breakwaters. Wiegel (2002a,b,c) gives an account of many types of seawalls and their performance.

Nicholas C. Kraus

Bibliography

Coastal Engineering Manual, 2003. Part V, Chapter 3, Shore Protection Projects, US Army Corps of Engineers, EM 1110-2-1100.

Dean, R.G., 2003. *Beach Nourishment: Theory and Practice*. Advanced Series on Ocean Engineering, Vol. 18, World Scientific Press, 420 pp.

Dean, R.G., and Dalrymple, R.A., 2002. *Coastal Processes with Engineering Applications*, Cambridge: Cambridge University Press.

Hands, E.B., and Allison, M.C., 1991. Mound migration in deeper water and methods of categorizing active and stable depths. *Proceedings Coastal Sediments 91*, ASCE, pp. 1985–1990.

Hanson, H., and Kraus, N.C., 1990. Shoreline response to a single transmissive detached breakwater. *Proceedings 22nd Coastal Engineering Conference*, ASCE, pp. 2034–2046.

Hardaway, C.S., and Gunn, J.R., 1999. Chesapeake Bay: design and early performance of three headland breakwater systems. *Proceedings Coastal Sediments 03*, ASCE, pp. 828–843.

Komar, P.D., 1998. *Beach Processes and Sedimentation*, 2nd ed. Englewood Cliffs, NJ: Prentice Hall.

Kraus, N.C., 2000. Groin notching—an innovative and effective concept. *Shore and Beach*, **68**(2): 18.

Kraus, N.C., and Galgano, F.A., 2001. Beach erosional hot spots: types, causes, and solutions. *Proceedings 14th Annual National Conference on Beach Preservation Technology*. Tallahassee, FL: Florida Shore & Beach Preservation Association, pp. 62–80.

Kraus, N.C., and McDougal, W.G., 1996. The effects of seawalls on the beach: Part I, an updated literature review. *Journal of Coastal Research*, **12**(3): 691–701.

Kraus, N.C., and Pilkey, O.H. (eds.), 1988. *The Effects of Seawalls on the Beach*. Special Issue 4, *Journal of Coastal Research*.

Kraus, N.C., Hanson, H., and Blomgren, S., 1994. Modern functional design of groins. *Proceedings 24th Coastal Engineering Conference*, ASCE, pp. 1327–1342.

Marine Board, 1995. *Beach Nourishment and Protection*, Committee on Beach Nourishment and Protection, National Research Council, The National Academies Press.

McLellan, T.N., and Kraus, N.C., 1991. Design guidance for nearshore berm construction. *Proceedings Coastal Sediments 91*, ASCE, pp. 2000–2011.

Nersesian, G.K., Kraus, N.C., and Carson, F.C., 1992. Functioning of Groins at Westhampton Beach, Long Island, New York. *Proceedings 23rd Coastal Engineering Conference*, ASCE, 3357–3370.

Pope, J., and Dean, J.L., 1986. Development of design criteria for segmented breakwaters. *Proceedings 20th Coastal Engineering Conference*, ASCE, pp. 2144–2158.

Rosati, J.D., 1990. Functional design of breakwaters for shore protection: empirical methods. Vicksburg, MS: US Army Engineer Waterways Experiment Station, *Technical Report, CERC-90-15*.

Silvester, R., and Hsu, J.R.C., 1993. *Coastal Stabilization: Innovative Concepts*. Englewood Cliffs, NJ: Prentice Hall.

Wiegel, R.L., 2002a. Seawalls, seacliffs, beachrock: what beach effects? Part 1. *Shore and Beach*, **70**(1): 17–27.

Wiegel, R.L., 2002b. Seawalls, seacliffs, beachrock: what beach effects? Part 2. *Shore and Beach*, **70**(2): 13–22.

Wiegel, R.L., 2002c. Seawalls, seacliffs, beachrock: what beach effects? Part 3. *Shore and Beach*, **70**(3): 2–14.

Cross-references

Beach Erosion
Beach Processes
Cross-Shore Sediment Transport
Gross Transport
History, Coastal Protection
Longshore Sediment Transport
Net Transport
Washover Effects

SHOREFACE

Introduction and origins of the term

The shoreface, a relatively steep surface that slopes away from the low-tide shoreline and imperceptibly merges with the flatter inner shelf or basin plain ramp, is an integral feature of nearly all clastic coasts (Figure S41). The term has been applied in a facies context in many geological interpretations of ancient coastal deposits in the rock record. This usage commonly invokes additional attributes that are unrelated to the morphologic feature and commonly unfounded.

Barrell (1912) coined the term "shore face" to describe the transition between the subaerial and subaqueous plains of a deltaic topset system: "The shore face is the relatively steep slope developed by breaking waves, a slope which separates the subaerial plain above form the subaqueous below." Johnson (1919) consolidated the term ("shoreface") and redefined it as the zone between the low-tide shoreline and the more nearly horizontal surface of the offshore. He expanded its usage to other types of coasts, as "the steeper, landward portion of the shore profile of equilibrium, of which the profile of the gently sloping subaqueous plain is the seaward continuation."

Development and maintenance of the shoreface

Early in the history of the concept, it was recognized that the shoreface is part of an equilibrium profile generated by shoaling waves. Many

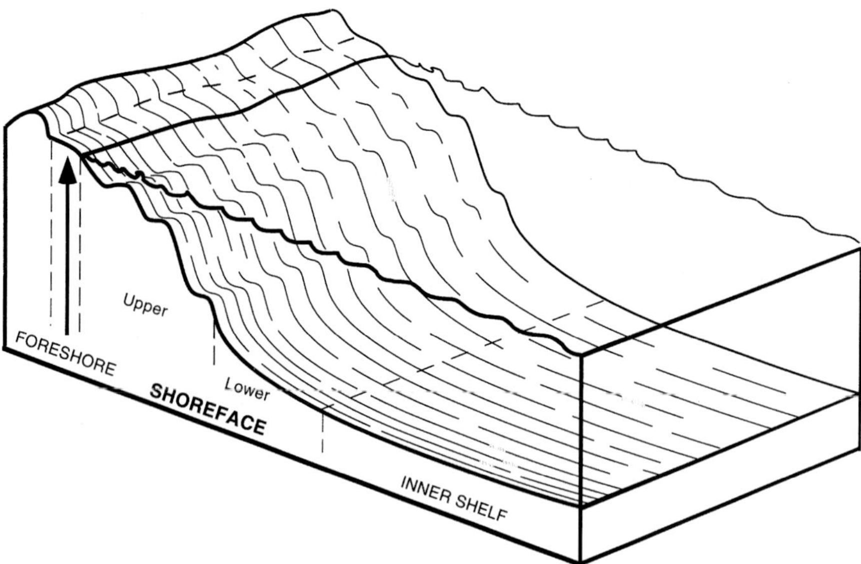

Figure S41 Sketch showing the character of a simple (progradational) shoreface. Shore-parallel breaker-bars, which may not occur on all shorefaces, are here used to separate it into "upper" and "lower" parts, which may be the only components recognizable in ancient shoreface deposits.

workers have assumed that the base of the shoreface is defined by fair weather wave base, but as Johnson noted in 1919, the two are unrelated. Wave base is commonly taken as a water depth equal to half of the length of the passing waves. As such it is primarily dependent on wave period. On the Pacific coast of the United States, long period (≥10 s) waves are common under fair-weather conditions, and their wave base lies in water depths of 78 m or more, well beyond the shoreface. On coasts receiving shorter period waves, the base of fair-weather waves lies in shallower water (e.g., about 20 m for 5 s waves), but coincidence of fair-weather wave base and the bottom of the shoreface is just that: a coincidence.

Although unrelated to wave base as such, the development and maintenance of a shoreface is at least partly controlled by aspects of shoaling waves. Niedoroda and Swift (1981, see also Niedoroda *et al.*, 1984) provide a plausible explanation for the maintenance of the shoreface, that balances the onshore transport of sand under (mostly) fair-weather conditions with the offshore transport during storms. In shallow water, the oscillatory flow of water induced by passing waves becomes asymmetric, with short relatively strong landward flow under the crest of a wave and a more prolonged weaker, seaward flow under the trough. This orbital velocity asymmetry drives sediment in a landward direction, building the shoreface. Storms sporadically interrupt this process and seaward flowing rip and geostrophic currents erode the shoreface. Much of the eroded sediment is subsequently returned to the shoreface during fair weather, maintaining its concave-up profile. Evidence for this combination of processes will be presented in a following section "Sedimentologic aspects of the shoreface."

Shoreface profiles

The shape of the shoreface profile on constructive (progradational) coasts differs from that of erosive (or transgressive) coasts. Progradation occurs where high rates of sediment supply (relative to base level change) cause a coastline to build laterally into the adjacent basin. Sediment is sufficient to provide a bottom profile in equilibrium with the shoaling waves. A simple concave-up profile typically results (Figure S42) that resembles the theoretical curve for landward increase in bottom orbital velocity asymmetry generated by shoaling waves (Figure S43). Because of the large volume of sediment available, the relief of progradational shorefaces reflects the energy of the coastal waves. On high-energy coasts like the southern coast of Washington State, the shoreface extends to a depth of 10 m (Figure S42(C)). On coasts with lower wave energy (e.g., the Gulf coast at Galveston, Texas), the shoreface bottoms out in shallower depths (Figure S42(A)).

Most coasts today are still in a transgressive or erosional phase, where the volume of sediment is insufficient to create a simple equilibrium profile. Commonly on such coasts, older deposits crop out on the shoreface, where they resist erosion by waves and coastal currents. Differential erosion creates irregularity in the shoreface profile (Figure S44), although many such profiles contain a simple "wave-tuned" component in their upper part, either because that is where unconsolidated sediment is concentrated or simply owing to intensified wave erosion in the shallowest water (Figure S45). Because transgressive coasts typically lack an abundance of sediment, the depth to which they extend reflects in part the topography crossed by the transgression. The relief of transgressive shorefaces is highly variable (Figures S44 and S45), but can exceed 20 m (Figure S44(C)).

Sedimentologic aspects of the shoreface

The textural character and sedimentary structures of shorefaces differ with energy and provenance. Typically, wave energy reaches a maximum in the shallower part of a shoreface, and coarse sand and gravel, if available, are concentrated in the active breaker zone adjacent to the shoreline (the coarsest typically being at the very base of the foreshore). The grain size typically decreases to seaward across a shoreface to fine or very fine sand at it base.

In some areas, the base of the shoreface marks the transition between sand and mud (Figure S42(A)), but in many places sand continues well out onto the inner shelf. On the coast of southwestern Washington State, for example, fine to very fine sand extends to water depths of 60 m or more. Although some geologists refer to the entire sandy part of a shoaling-upward coastal succession as shoreface deposits, there is no general basis for this interpretation. The proportion of sand, silt, and mud, depend on the sediment available and the wave-energy regime (Galloway and Hobday, 1996) and are not specifically related to the shoreface morphology.

An exception to the generally seaward-fining on most shorefaces occurs where gravel is present. Studies of modern gravel-bearing shorefaces (Shipp, 1984; Howard and Reineck, 1981) show a scattering of gravel at the base of the shoreface and the immediately adjacent inner shelf. Vertical successions of ancient gravelly shoreface deposits typically have a distinctive facies composed of scattered small pebbles in a matrix of fine- to very fine-grained sand in the interval 2–3 m below cross-bedded nearshore gravelly sand. The gravel at the base of the shoreface probably results from a combination of erosion of a shoreface during storms and its subsequent rebuilding in the storms aftermath as suggested by Niedoroda and Swift (1981). Gravelly sand eroded from the upper shoreface is carried seaward by rip currents or geostrophic flow and deposited at or near the base of the shoreface. Waves following the storm rework and winnow this material and transport the finer component landward. The coarsest part (typically small pebbles) is left behind as a "post storm lag."

The sedimentary structures of a shoreface depend on both wave energy and sediment texture. The shallower part consists of the

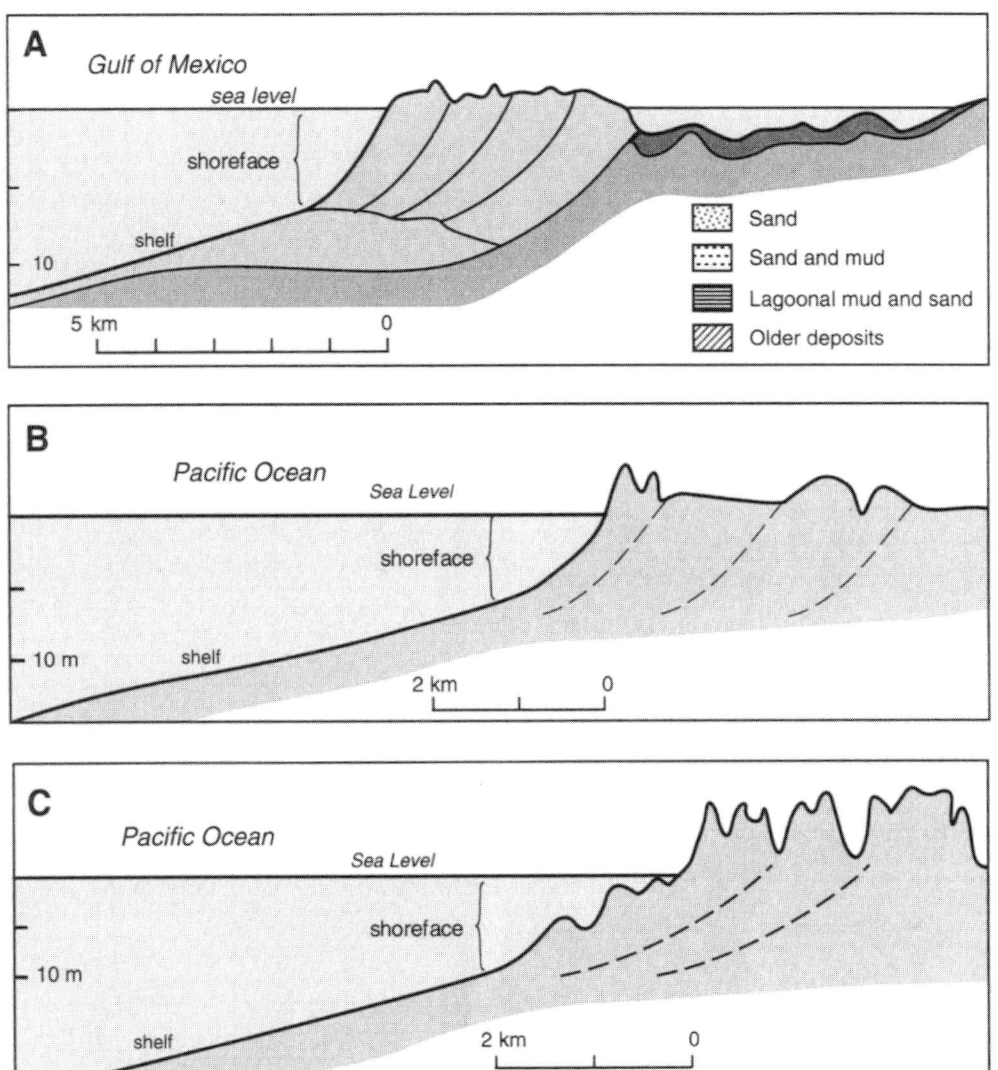

Figure S42 Shoreface profiles on progradational coasts, showing simple concave-up character. (A) Galveston, Texas, Gulf of Mexico (after Morton, 1994). (B) Nayarit coast of Mexico, Pacific Ocean (after Curray *et al.*, 1969). (C) Long Beach, Washington, Pacific Ocean (after Dingler and Clifton, 1994). Note differences in scale.

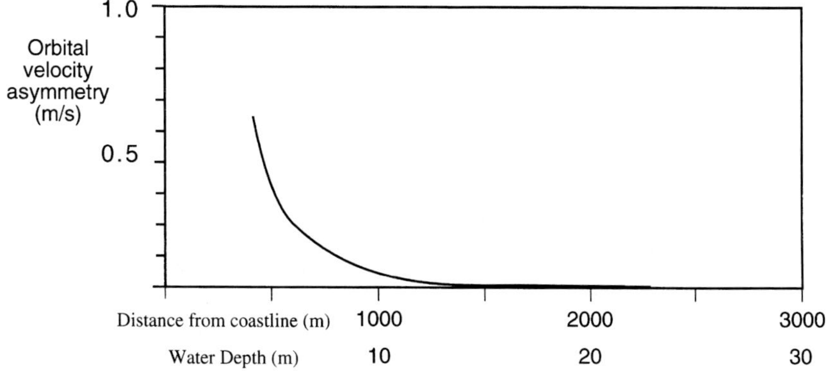

Figure S43 Predicted orbital velocity asymmetry (difference between the bottom velocities under the crest and trough of a shoaling wave, using Stokes second order wave theory) under 10-s waves with a deepwater wave height of 1.0 m over a bottom with a uniform slope of 1 : 100. Water depth at which velocity asymmetry becomes noticeable approaches the depth of the base of the shoreface off the Long Beach, Washington (Figure S41(C)) where these wave conditions obtain during fair weather.

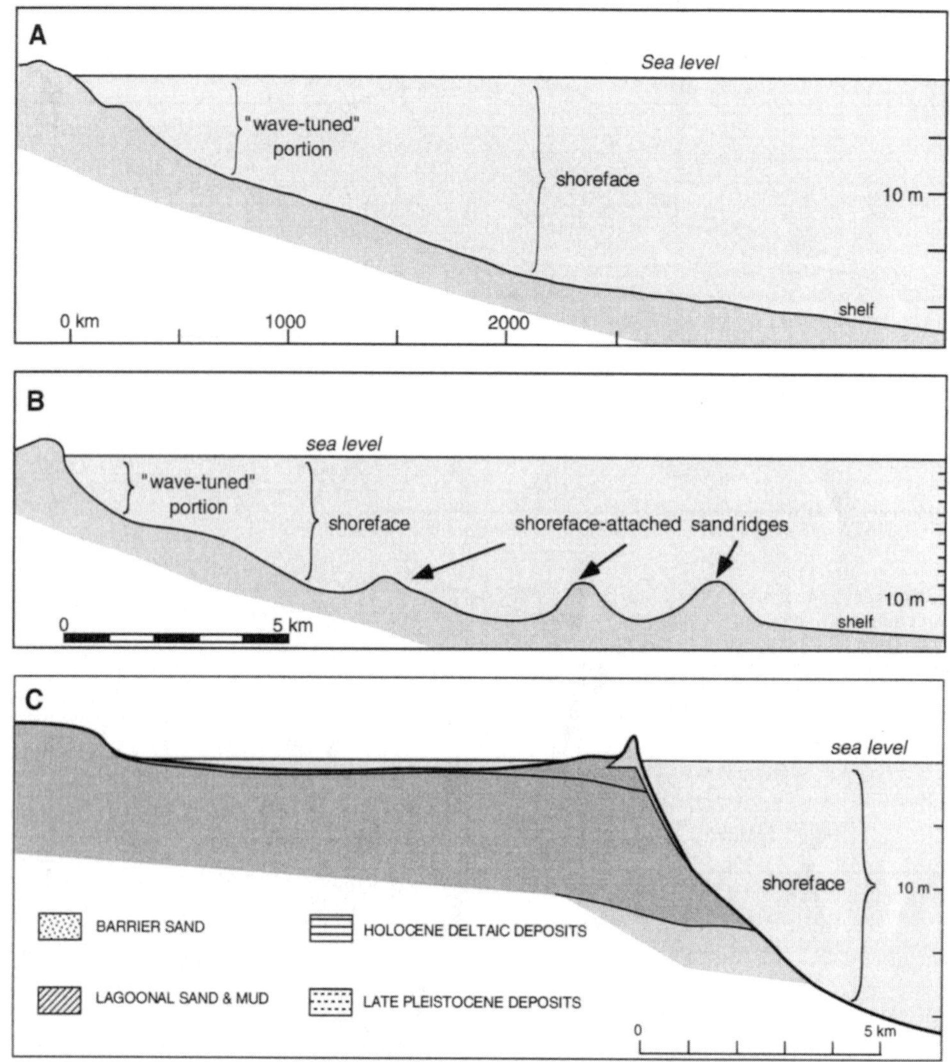

Figure S44 Shoreface profiles on transgressive coasts, showing variability of profile shape. (A) Long Island, New York (after Niedoroda *et al.*, 1984). (B) North of the Edisto River, North Carolina. (C) South Padre Island (after Morton, 1994). Topographic irregularity below "wave-tuned" portion in A and B presumably due to differential erosion of older deposits exposed on the shoreface during transgression. Note differences in scale.

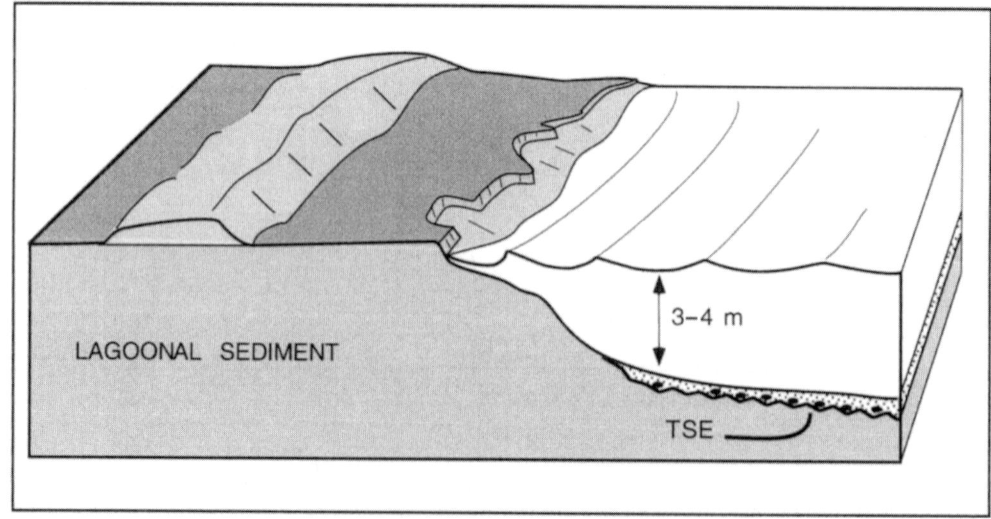

Figure S45 Erosional shoreface on a transgressing coast, Cape Romaine, South Carolina (after Hayes, 1994).

nearshore, in which bedforms are generated by passing waves or by longshore and rip currents of the nearshore circulation cell. Bedform size, shape, and orientation differ as a function of wave size and direction of approach and the grain-size of the sea bed material (Clifton and Dingler, 1984). Storms may generate hummocky or swaley cross-stratification. Guttercasts are a characteristic shoreface feature. Shore-parallel or other breaker bars are common features on the upper part of many shorefaces (Figure S41) and can serve to focus longshore and rip currents. Depending on infauna and sediment texture, bioturbation can obliterate much of the physical structures on the outer part of a shoreface.

Shoreface subdivisions

Many workers divide the shoreface into upper, middle, and lower components. Galloway and Hobday (1996) define the upper shoreface as the inner surf zone, the middle shoreface as the portion occupied by breaker bars systems and the lower shoreface as the area to seaward where the shoreface merges with the inner shelf. Howard and Reineck (1981) recognize three principal zones without relating them to shoreface morphology: a *nearshore* made up of clean sand in which physical sedimentary structures predominate, a *transition* in which fine and silty sand contain a mixture of biogenic and physical structures, and an *offshore* composed of sandy silt with only remnant stratification.

Shoreface zonation is less evenly applied to the stratigraphic record. Although geologists commonly subdivide ancient shoreface deposits into upper, middle, and lower facies (and some subdivide these further into proximal and distal components), they generally agree only on the character of the upper shoreface as an interval with abundant cross-bedding and other physical structures. This would probably include the middle shoreface of Galloway and Hobday (1996) where currents focused by bar-trough systems produce structures that resemble those of the surf zone. Deposits on the lower part of a shoreface may be indistinguishable from those on the adjacent inner shelf, unless gravel is present.

Otherwise the most realistic facies subdivision is that of an upper shoreface overlying a lower shoreface/inner shelf transition.

H. Edward Clifton

Bibliography

Barrell, J., 1912. Criteria for the recognition of ancient delta deposits. *Geological Society of America Bulletin*, **23**: 377–446.

Dingler, J.R., and Clifton, H.E., 1994. Barrier systems of California, Oregon, and Washington, In Davis, R.A., (Jr.) (ed.), *Geology of Holocene Barrier Island systems*, New York: Springer-Verlag, pp. 115–165.

Clifton, H.E., and Dingler, J.R., 1984. Wave-formed structures and paleoenvironmental reconstruction. *Marine Geology*, **60**: 165–198.

Curray, J.R., Emmel, F.J., and Crampton, P.J.S., 1969. Holocene history of a strandplain, lagoonal coast, Nayarit, Mexico. In Castanares, A.A., and Phleger, F.B. (eds.), *Coastal Lagoons, a Symposium*. Mexico: Universidad Nacional Autonma, pp. 63–100.

Galloway, W.E., and Hobday, D.K., 1996. *Terrigenous Clastic Depositional Systems: Applications to Fossil Fuel and Groundwater Resources*, 2nd edn. New York: Springer.

Hayes, M.O., 1994. The Georgia Bight barrier system. In Davis, R.A., Jr. (ed.), *Geology of Holocene Barrier Island Systems*, New York: Springer-Verlag, pp. 233–304.

Howard, J.D., and Reineck, H.-E., 1981. Depositional facies of high-energy beach-to-offshore sequence: comparison with the low-energy sequence. *American Association of Petroleum Geologists Bulletin*, **65**: 807–830.

Johnson, D.W., 1919. *Shore Processes and Shoreline Development*. New York: John Wiley.

Morton, R.A., 1994. Texas barriers. In Davis, R.A., Jr. (ed.), *Geology of Holocene Barrier Island Systems*, New York: Springer-Verlag, pp. 75–114.

Niedoroda, A.W., and Swift, D.J.P., 1981. Maintenance of the shoreface by wave orbital currents and mean flow: observations from the Long Island coast. *Geophysical Research Letters*, **8**: 337–340.

Niedoroda, A.W., Swift, D.J.P., Hopkins, T.S., and Ma, C.-M., 1984. Shoreface morphodynamics on wave-dominated coasts. *Marine Geology*, **60**, 331–354.

Shipp, R.C., 1984. Bedforms and depositional structures of a barred nearshore system, eastern Long Island, New York. *Marine Geology* **60**: 235–259.

Cross-references

Bars
Beach Sediment Characteristics
Beach Stratigraphy
Coastal Sedimentary Facies
Cross-shore Sediment Transport
Cross-shore Variation in Sediment Size
Dynamic Equilibrium of Beaches
Ripple Marks
Shelf Processes
Surf Zone Processes

SIMPLE BEACH AND SURF ZONE MODELS

Definition

Beach and surf zone models refers to conceptual models which provide some basic insight on the dynamic interaction between waves and beaches. In this entry, only simple aspects of these models will be discussed.

Background

The physical processes that mold beaches and that are at work in the surf zone are anything but simple. The best scientific minds that study these processes often disagree on what are the important cause and effect relationships at work in this complex and dynamic area. However, there are some relatively simple models which are useful for understanding important aspects of how beaches respond to waves, such as erosion and accretion. Beach erosion falls into two general categories: (1) When the littoral transport of sand, or possibly other sediment, to the beach is less than the littoral transport of sand from the beach, that is, a net loss in the longshore transport of sediment. (2) When sand moves from the subaerial (above water) beach to the submerged beach or the movement from the submerged beach to the subaerial beach. It is this latter category of offshore and onshore sand movement, referred to as cross-shore sediment movement, which will be regarded as erosion or accretion, respectively in this entry.

Summer beach–winter beach model

The simplest model for onshore and offshore sediment movement is commonly referred to as the summer beach–winter beach concept. By the 1950s, coastal scientists became aware of a strong seasonality of beaches in California. Generally, during winter storms waves would remove sand from the subaerial beach and deposit it offshore typically in bar formations. Milder wave conditions, during summer, would move the bars onshore and ultimately they would attach to the shore and rebuild the wider berm associated with the "summer beach." Figure S46 shows the profiles of a reflective and a dissipative beach. Using the terminology of Wright and Short (1984) reflective and dissipative beaches correspond roughly to summer and winter beaches, respectively. In interpreting Figure S46, note that there is very high vertical exaggeration.

Weather patterns in many parts of the world undermine the simple concept of a summer/winter beach pattern, and even in California, the concept is not too reliable. The important insight of the concept, that wider beaches are associated with a prolonged period of mild wave conditions and narrower beaches are associated with storm waves, is true.

U_t Beach index model

The U_t parameter was developed by Ahrens and Hands (2000) and it can be used like an index to classify wave events on beaches as either erosion or accretion. U_t can be calculated in terms of a sediment mobility number, N_{so}, and the deepwater wave steepness, H_{so}/L_o. The mobility number is defined: $N_{so}= H_o/\Delta d_{50}$, where H_o is the deepwater wave height offshore of the beach in question, d_{50} is the median grain diameter of sediment in the surf zone, and Δ is the relative density of the sediment given by, $\Delta = (\rho_s - \rho)/\rho$, where ρ_s and ρ are the density of the sediment and water, respectively. Deepwater wave steepness is the ratio of the deepwater wave height to deepwater wave length, L_o, where $L_o = gT^2/2\pi$, T is the wave period and g is the acceleration of gravity. U_t can be defined:

$$U_t \equiv [(N_{so}/33)(H_{so}/L_o)^{0.57}/\exp(8.3(H_{so}/L_o))]^{1/2.2} \qquad \text{(Eq. 1)}$$

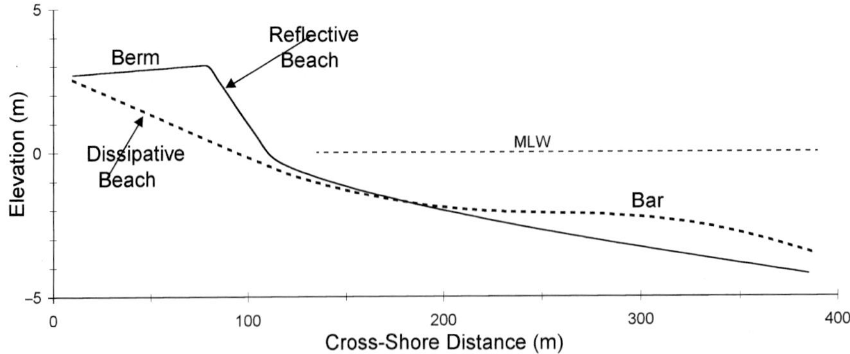

Figure S46 Typical profiles of the highest beach state, dissipative, and the lowest beach state, reflective.

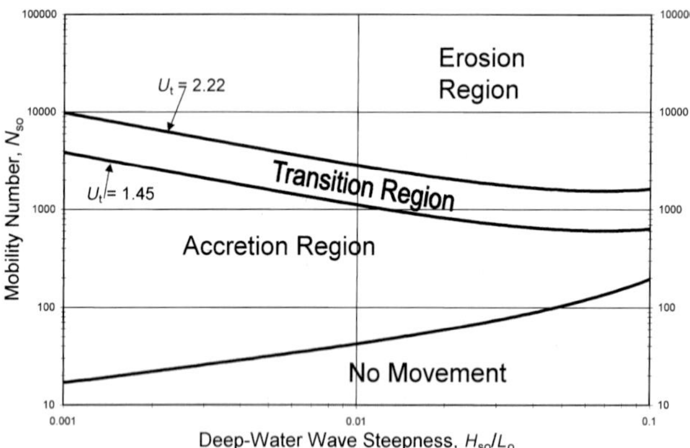

Figure S47 The U_t plane, showing erosion and accretion regions as defined by the U_t parameter.

The U_t parameter was calibrated using an extensive field data set of primarily qualitative observations of beach erosion or accretion, Kraus and Mason (1991). Data are from beaches all over the world, collected by many researchers, and include observations of 99 distinct wave events. The range of the parameter for this data set was, $0.42 \leq U_t \leq 6.56$, values of $U_t > 2.22$ were always observed to be erosion events and values of $U_t < 1.45$ were always observed to be accretion events. There was a transition region, $1.45 < U_t < 2.22$, where a mixture of erosion and accretion events occurred, but generally U_t did a good job of classifying events in the Kraus and Mason data set. Figure S47 shows the N_{so} versus H_{so}/L_o plane using the U_t parameter to determine various regions.

Certain features of Figure S47 should be noted because they provide insight into the erosion or accretion processes. Highly mobile sediment, that is, high mobility numbers, is associated with erosion and low sediment mobility is associated with accretion. The accretion region shrinks and the erosion region expands in going from low- to high-wave steepness. These are trends which are consistent with current understanding of erosion and accretion processes.

Antecedent beach conditions (ABC) model

There have been a number of attempts to develop a single parameter to quantify the tendency of sand to move onshore or offshore. The most useful simple parameter is the fall-velocity parameter given by $\Omega = H/wT$, where H is the wave height and w is the fall velocity of sediment in the surf zone. Some scientists have used the breaker height, H_b, on the beach to calculate Ω others have used the deepwater wave height, H_o, offshore of the beach to calculate Ω. Regardless of whether H_b or H_o is used, the principle is, if Ω is small, it indicates that sediment movement is predominately by bedload and there would be a net onshore movement of sediment. If Ω is large, it would favor offshore movement of sediment because there would be large quantities of suspended sediment which would be carried offshore by the bed return flow or, possibly, by rip-currents.

An interesting perspective on beach erosion and accretion can be gained by comparing the current value of the fall-velocity parameter, Ω(current), with the antecedent value, Ω(antecedent). Ω(current) is the average value of Ω for the day in question and Ω(antecedent) is a weighed average of Ω(current) starting yesterday and including the previous 30 days, Wright and Short (1984). The function is heavily weighted toward the recent past with the previous five days contributing 90% of the total weight. Wright and Short found that Ω(antecedent) correlated very well with the six fundamental beach states that they had identified. The six beach states can be ranked hierarchically from the highest state, dissipative, to the lowest state, reflective. Wave conditions which move a beach toward a higher state cause erosion and wave conditions which move a beach downstate cause accretion. These beach states have some very interesting quantum aspects which suggest analogies to quantum physics. Typical profiles for the end beach states are shown in Figure S46. A dissipative beach is the product of storm waves and has a flat beach face and a rather subtle bar or possibly subtle bars. A reflective beach is the product of a period of mild wave conditions and has a berm, steep beach face and no bars. Both dissipative and reflective beaches are relatively two-dimensional, although reflective beaches usually have pronounced cusps. The four intermediate beach states all have complex three-dimensional, dynamic topography, so, it is difficult to associate a single profile that is typical of intermediate states.

Figure S48 shows the Ω(current) versus Ω(antecedent) plane adapted from the model of Wright *et al.* (1985). The figure identifies the highest beach state, dissipative, and the lowest beach state, reflective, connected by a stable region. The stable region trends along the diagonal where Ω(current) = Ω(antecedent): that is, when the current conditions are approximately equal to the antecedent conditions, the beach state is stable. The stable connecting region contains the four intermediate beach states defined by Wright and Short (1984). Boundaries for the stable region shown in Figure S48 are approximate based on research by Wright *et al.* (1985). Also shown in Figure S48, are an erosion region and an accretion region. The erosion region is characterized by, Ω(current) > Ω(antecedent), and the accretion region by Ω(current) < Ω(antecedent).

To help illustrate the ABC model, a simple calm-storm-calm sequence is shown in Figure S48. Simple indicates that an extended calm period is interrupted as breaker heights increase monotonically from small values to large values during the height of the storm and decrease monotonically to small values as the storm wanes and another extended calm period begins. The sequence shown is not specific, but is consistent with breaker heights over an inshore bar, wave periods, and sand size that occur on the east coast of the United States or the coast of New South Wales, for example, $0.25 \leq H_b \leq 1.75$ m. Circles are shown along the path of the sequence to bracket one day intervals and arrows are used to indicate the time order of the sequence. The sequence starts at the lower left when a calm period is interrupted by the initial arrival of waves from a storm. With increasing breaker heights, the sequence quickly leaves the stable region and moves into the erosion region since Ω(current) quickly exceeds Ω(antecedent). As long as the storm intensity increases, Ω(current) increases causing Ω(antecedent) to increase too, this is the main period of erosion. However, at some point the storm intensity will start to wane and Ω(current) will start to decrease. During this initial waning period, Ω (antecedent) will continue to increase but the storm sequence will quickly leave the erosion region and drop into the stable region, and then into the accretion region. It is this initial waning period of the ABC model, when both Ω(current) and Ω(antecedent) are large, that is anomalous from the perspective of the

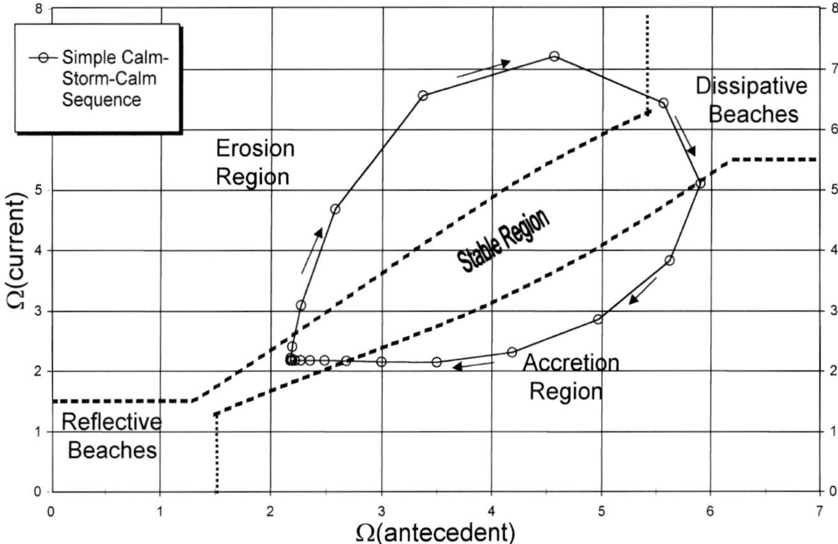

Figure S48 Current versus antecedent beach condition plane showing a "simple" calm-storm-calm sequence.

U_t Beach Index model and also other beach index models. The U_t model predicts that when wave conditions are energetic, $U_t > 2.22$, erosion will occur regardless of the antecedent beach conditions; the ABC model predicts that accretion will occur when Ω(current) is roughly 20% less than Ω(antecedent) regardless of how large Ω(current) is. These criteria can easily be in conflict during the early portion of a waning storm and current research indicates that the ABC model is the more realistic. As the storm continues to wane, the beach continues to rebuild, albeit slowly, until it returns to a stable condition similar to the condition at the beginning of the cycle.

Summary

Three models of beach and surf zone behavior are discussed. In order of increasing complexity they are: (1) summer–winter beach profiles, (2) the U_t Beach Index model, and (3) the antecedent beach condition (ABC) model. From a historical perspective, the summer–winter beach profile concept is interesting, but is certainly not a reliable model. The U_t model is a simple model to determine if beaches will erode or accrete for a given sediment size and wave conditions, but gives no information about beach profiles. The ABC is the most complex of the three models and is difficult to apply considering the level of information required. However, the ABC model is conceptually the most realistic and provides a great deal of information about beach and surf zone processes beyond erosion, accretion, and beach profiles (Wright and Short, 1984). The ABC model has also been widely accepted and generalized by other coastal scientists.

One of the most difficult aspects of the ABC model is to classify beaches and surf zones in the proper state. To help classify ABC beach states correctly, Lippmann and Holman (1990) developed computer enhanced video images of the surf zone to show bar and trough locations and shapes, and proposed some additional beach states. All of the models discussed are for microtidal beaches, that is, tide ranges less than 2 m. The ABC model has been generalized by Masselink and Short (1993) to include the influence of tides on beaches and surf zones and, therefore, to extend the usefulness of the ABC model to beaches with tide ranges greater than 2 m.

John P. Ahrens

Bibliography

Ahrens, J.P., and Hands, E.B., 2000. A simple perspective on cross-shore sediment movement. *Journal American Shore & Beach Preservation Association*, **68**(4): 3–14.

Kraus, N.C., and Mason, J.M., 1991. Field Data Set for Testing Beach Erosion and Accretion Predictive Criteria. Memorandum for Record, Coastal Engineering Research Center, Vicksburg, MS: US Army Eng. Waterways Experiment Station, 28 May 1991.

Lippmann, T.C., and Holman, R.A., 1990. The spatial and temporal variability of sand bar morphology. *Journal of Geophysical Research*, **95**(C7): 11575–11590.

Masselink, G., and Short, A.D., 1993. The effect of tide range on beach morphodynamics and morphology: a conceptual beach model. *Journal of Coastal Research*, **9**(3): 785–800.

Wright, L.D., and Short, A.D., 1984. Morphodynamic variability of surf zones and beaches: a synthesis. *Marine Geology*. **56**: 93–118.

Wright, L.D., Short, A.D., and Green, M.O., 1985. Short-term changes in the morphodynamic states of beaches and surf zones: an empirical predictive model. *Marine Geology*, **62**: 339–364

Cross-references

Beach Erosion
Beach Features
Cross-Shore Sediment Transport
Dissipative Beaches
Erosion Processes
Longshore Sediment Transport
Reflective Beaches
Rhythmic Patterns
Sandy Coasts
Surf Modeling
Surf Zone Processes

SLOUGHS—See ESTUARIES; TIDAL CREEKS

SMALL ISLANDS

Introduction

What is a "small island"?

The definition of a *small island* depends upon one's perspective. Any country in the family of nations can declare themselves a member of the Alliance of Small Island States (AOSIS). AOSIS nations include Antigua and Barbuda, Bahamas, Barbados, Belize, Cape Verde, Comoros, Cook Islands, Cuba, Cyprus, Dominica, Fiji, Federated States of Micronesia, Grenada, Guinea-Bissau, Guyana, Haiti, Jamaica, Kiribati, Maldives, Malta, Marshall Islands, Mauritius, Nauru, Niue,

Palau, Papua New Guinea, Samoa, Singapore, Seychelles, Sao Tome and Principe, Solomon Islands, St. Kitts and Nevis, St. Lucia, St. Vincent and the Grenadines, Suriname, Tonga, Trinidad and Tobago, Tuvalu, Vanuatu; observer nations include American Samoa, Guam, Netherlands Antilles, and the US Virgin Islands. AOSIS is a political organization of the United Nations, stemming from the 1990 Second World Climate Conference, which recognized Small Island Developing States (SIDS) as having a unique need in a changing environment.

From a geographic perspective, it is clear that AOSIS member states, and SIDS in general, are found in each major ocean, semi-enclosed, or marginal sea (see Figure S49). Some "SIDS" are low-lying non-island coastal countries with a decidedly marine climate. They all share a number of common development issues including energy, coastal and marine resources, waste management, biodiversity, sustainable tourism, population growth, and climate change, among others. Their socioeconomic circumstance commonly includes a rapidly growing population, and small gross domestic products with which to mitigate the effects of natural disasters and anthropogenic hazards. Most islands in AOSIS, and SIDS in general, are positioned in low latitudes, and thus are susceptible to common meteorological hazards such as tropical storms and flooding.

Geologically islands can be separated into seven island types: arc system islands; barrier system islands; coral reef islands; diapiric islands; estuarine and deltaic islands; intraplate islands; precontinent islands; and rift islands. Many island types are associated with plate boundaries and fault zones (e.g., rift and diapiric islands), others are located at passive margins or in the coastal zone, while yet others are well within the boundaries of plates and may or may not have associated coral reefs. Their petrographic characteristics range from basalt for rift and intraplate islands, to metamorphic and clastic sand and clay sedimentation for precontinent and estuarine islands, respectively, to the classic carbonates of coral reefs and the andesite of island arc system islands. Thus the cause of each island type varies considerably too, including subduction, volcanism at boundaries and within plates, coastal hydrodynamics, riverine terminus, and biological colonization upon a substrate. Other classification schemes exist but the above serves to illustrate the diversity of geophysical settings.

Small islands also tend to have distinctly different meteorological environments. While tropical islands all tend to have sea level air temperatures in the range of 20–30°C, precipitation patterns differ widely. The lush vegetation and high annual rainfall (>200 cm yr^{-1}) of the islands of the western Caribbean Sea are in stark contrast to the near desert conditions of Aruba, Bonaire, and Curaçao, barely 1,000 km due east. Much of the interannual variability in low-latitude rainfall is related to the annual cycle of the intertropical convergence zone (the ITCZ), and this in many places is punctuated by ocean-basin scale events such as the El Niño–Southern Oscillation (ENSO). Extratropical islands have equally complex climatological variability, including extreme winter precipitation at high latitudes and the formation of sea-ice. Again, no commonality exists in the biogeography of complex island ecosystems, because their environmental settings vary widely.

The physical notion of a small island then is well beyond the geopolitical perspective of AOSIS or SIDS. Humankind tends to categorize natural systems, but in the case of small islands this is quite problematical. Setting limits on the area of an island as a definition of "small" is as arbitrary as specifying latitude or population density or biodiversity. Yet as a working definition, it might be most productive to consider their shared range of natural hazards and the common socioeconomic consequences of limited infrastructure and response strategies to events both anthropogenic and geophysical. In addition, the notion of sustainable development needs to be included in the definition as a focus for well being. These concepts are addressed in the next sections.

Natural hazards

Tropical and extratropical storms tend to be one of the most severe natural hazards to small islands, and yet such events are a major source of freshwater, especially in low latitudes. The West Indian Hurricane, the western Pacific Typhoon, and the Indian Ocean Cyclone are similar tropical storms with varying degrees of severity. All coastal areas are at risk from the high surface winds, which can exceed 100 m s^{-1} with attendant rainfall in excess of 30 cm in 24 h. Extreme wind damage and flooding are common aftermaths of these storms, and it is not uncommon for years to pass before the socioeconomic infrastructure is restored. Tropical ecosystems have evolved to include the tropical storm as part of the natural cycle, and indeed it can be argued that storm events are needed from time to time to exercise the coupled land–air–sea system that defines the biological health of a small island.

Most of the toll in human terms from tropical storms is associated with storm surge and flooding. Many oceanic small islands are essentially mountains protruding above the sea surface and have very steep offshore submarine topography and narrow shelves. For storm surge height to build to the 5 m level of Hurricane Andrew in south Florida in 1992, a long and continuous coastline is necessary. Thus AOSIS countries such as Belize and Cuba are highly prone to extensive coastal flooding and loss of human life through drowning, which is the primary cause of death from tropical storms. The catastrophic loss of life, probably exceeding 250,000 persons, in the estuarine and deltaic islands of

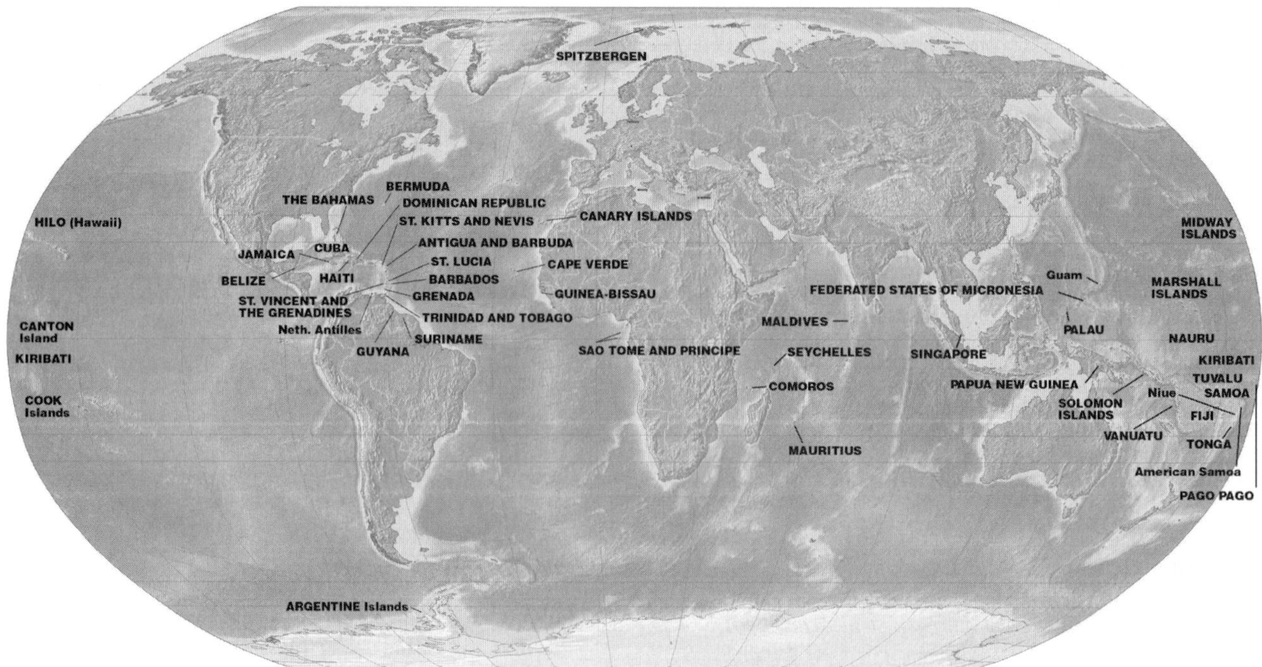

Figure S49 Location of islands or island systems mentioned in the text. Member nations of the AOSIS include several non-island countries. Small Island Developing States (SIDS) and AOSIS do not define "small" in terms of area or population (see IPCC, 1997).

Bangladesh in 1971, is not experienced on intraplate and island arc small islands due to the hydrodynamics of wind-stress induced storm surges. Modeling and observations suggest storm surges of 1–2 m, even in extremely powerful tropical storms, for small islands that are categorized as other than barrier islands, deltaic islands, or precontinent islands.

Many small islands are located in tectonically active regions, and thus are subject to numerous earthquakes. Many are volcanically active or were in their early history. The Hawaiian Islands are a classic example of intraplate islands that owe their existence to volcanism, and it is the basaltic base of these islands that offers corals the necessary substrate to colonize and grow. The low-lying atolls of the Midway Group represent the ultimate fate of Hawaii as it drifts to the northwest over the geothermal hotspot in the central Pacific that gives it genesis, and slowly subsides back into the deep. From a socioeconomic perspective, these and other tectonically active small islands require the necessary infrastructure to recover from a seismic event. It is an often-overlooked problem, especially in SIDS, to be able to recover from the aftermath of a major event, especially the control of fires, loss of electrical power, roads and bridges, and access to public health facilities for the rapidly growing human populations.

Seismic sea waves (tsunami) are another class of coastal hazard from a marine perspective. Although commonly thought of as a natural hazard of the Pacific Ocean, tsunamis are known to occur in all oceans, with historical references in the Mediterranean Sea as early as the 16th century BC. Most tsunami events are associated with strong, shallow focus, dip-slip submarine earthquakes, submarine or subaerial landslides or caldera collapse, submarine volcanoes, or asteroid impact. Not all of these geophysical events create a tsunami, it being especially difficult to forecast a tsunami from an earthquake alone. Small islands are particularly vulnerable to damage from tsunami events, and this has led to establishing warning systems facilitated by the Intergovernmental Oceanographic Commission of UNESCO and national agencies. In the Pacific, where major tsunami events seem to occur about every decade, an extensive array of seismometers, tide gauges, buoys, and communications systems can provide up to several hours of warning; in the Atlantic where the frequency of occurrence is perhaps twice a century, no warning system exists.

In the deep sea, a tsunami travels as a small-amplitude shallow-water wave with celerity $C = \sqrt{gH}$ where g is gravity and H is water depth. A typical celerity in the deep sea is 200 $\mathrm{ms^{-1}}$ and it can thus take 24 h for a tsunami to cross the Pacific Ocean, as did the wave from the 1960 Chilean earthquake. In smaller basins such as the Caribbean Sea, the total travel time will be an hour, or much less if the wave is locally generated, as was the case with the Virgin Islands tsunami of 1867. As the tsunami enters shallow water, conservation of energy requires that the amplitude of the wave must increase, and this has led to the great waves, some exceeding 30 m in height, rushing in to destroy coastal communities, ecosystems, and infrastructure. While wave height is a factor in the devastation, wave run-up is often the larger issue. Much like breaking surf on a beach rushes up the beach face, tsunami run-up has been recorded to reach 490 m. Immediate evacuation of the coastal zone is the only viable course of action if lives are to be spared.

Sea-level change on small islands varies as widely as the underlying geology. During the last 100 years or so, tide gauges have been operating to measure water level changes as is needed in tsunami warning systems, ENSO studies, and for determining the tides and chart datums. When the tide gauges are operated using accepted engineering practices including annual differential leveling surveys to juxtaposed benchmarks and regular visits by a trained observer, the long-term trend of relative sea level (RSL) can be discerned. It is the change of sea level relative to the land that is of socioeconomic consequence, and the local effects of vertical land motion, steric seawater changes, oceanic currents, atmospheric barometric pressure, wind-stress, and anthropogenic activities such as dredging and coastal construction complicate it. Many SIDS and member states of AOSIS have expressed great concern about global sea-level rise. For more information, see Pugh (2004).

Unfortunately, few small islands have information from quality-controlled water-leveling measuring systems to quantify relative sea level. Typically, a record-length exceeding three nodal tidal periods (18.61 years each) is needed to statistically discern change in relative sea level above the natural interannual variability associated with long-period tides and coupled atmosphere–ocean events such as ENSO, the Pacific Decadal Oscillation, or the North Atlantic Oscillation. A survey of RSL at selected small islands shows a wide variety of trends (see Figure S50), some showing sea-level rise, some sea-level fall, and some with no statistically significant change at all (see Table S10). Complicating the RSL issue is that even close-by islands can have very different trends, either upward or downward, especially intraplate small islands, deltaic islands, and those within island arc systems. It is simply not possible with current measurements to state unequivocally that small islands, or any coastal system, has an issue with relative sea level that can be attributed to global change, whether anthropogenic or geophysical.

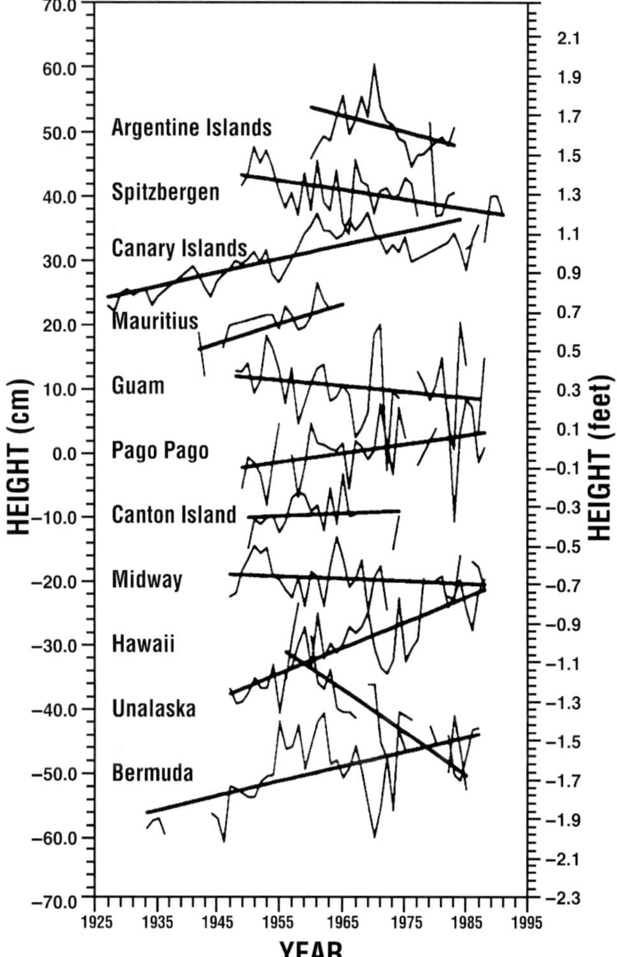

Figure S50 Relative sea level trends at selected small islands. The mean of each curve is offset by 10 cm for clarity. The heavy straight line is the linear least squares trend best fitting the annual mean values. See Table S11 for statistical summary and geographic coordinates. Redrawn from Maul (1996).

Anthropogenic hazards

The discussion of anthropogenic hazards on small islands, and especially to SIDS is meant to be no more exhaustive than the above material on natural hazards. For example the socioeconomic issue of RSL to a small island such as Key West, Florida, which has experienced a well documented 30 cm rise in sea level in the last 150 years (0.19 ± 0.01 cm $\mathrm{yr^{-1}}$), is quite different to that of an equally sized island in say Micronesia. Key West is a highly developed community with significant financial resources to mitigate RSL, and in fact due to extensive coastal engineering, the area of Key West island *per se* has *grown* by some 25% in the century and a half since records have been maintained. Most all developing AOSIS nations would not have the same experience.

Of the anthropogenic hazards identified in many studies, growth of the human population, deforestation, waste management, and marine exploitation are often mentioned. Again, no single small island will have the identical human-caused "hazards" as neighboring islands, but for many SIDS especially in tropical climes, similar patterns seem to arise. While there is cause for concern regarding global change as an instrument of stress to the small island community, it appears that it is local populations causing local problems that will exacerbate environmental degradation for the foreseeable future.

Table S10 Summary of least squares linear trend sea-level change statistics from selected small islands shown in Figure S50

PSMSL RLR Station	Latitude	Longitude	Years	n	Trend (mm yr^{-1})	Error ($\pm$mm yr^{-1})	r
Argentine Islands	65°15′S	64°16′W	1958–88	20	−2.6	1.3	−0.42
Bermuda	32°22′N	64°42′W	1932–88	45	+2.2	0.5	0.56
Canary Islands	28°29′N	16°14′W	1927–87	46	+2.0	0.2	0.80
Canton Island	02°48′S	171°43′W	1949–74	20	+0.3	1.0	0.07
Guam	13°26′N	144°39′E	1948–88	37	−1.0	0.8	−0.22
Hilo, Hawaii	19°44′N	155°04′W	1946–88	41	+3.9	0.5	0.80
Mauritius	20°09′S	57°30′E	1942–65	14	+3.0	0.9	0.67
Midway	28°13′N	177°22′W	1947–88	33	−0.5	0.5	−0.17
Pago Pago	14°17′S	170°41′W	1948–88	35	+1.3	0.6	0.35
Spitzbergen	78°04′N	14°15′E	1948–91	40	−1.6	0.5	−0.45
Unalaska	53°53′N	166°32′W	1955–88	21	−6.7	1.1	−0.82

Data base is the revised local reference (RLR) file of the Permanent Service for Mean Sea Level (PSMSL), Bidston Observatory, Birkenhead L43 7RA, United Kingdom.

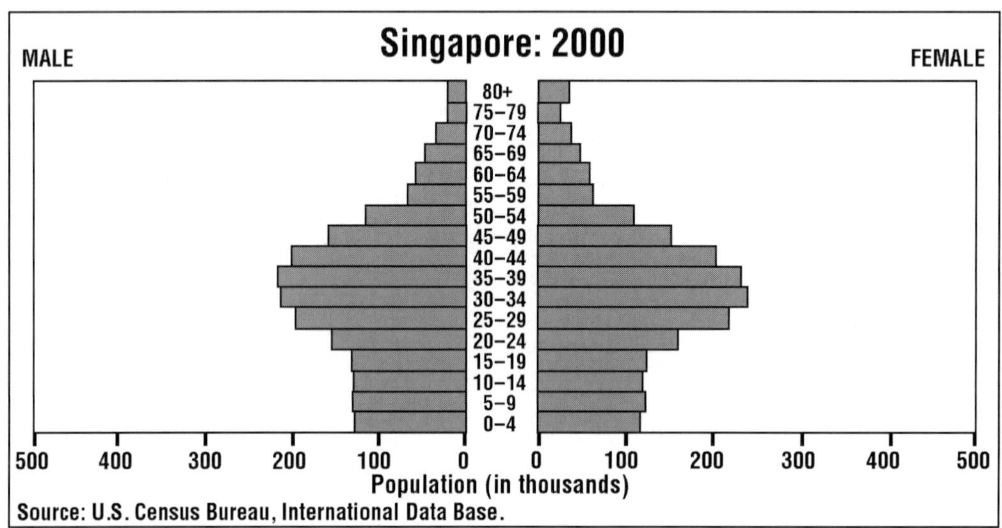

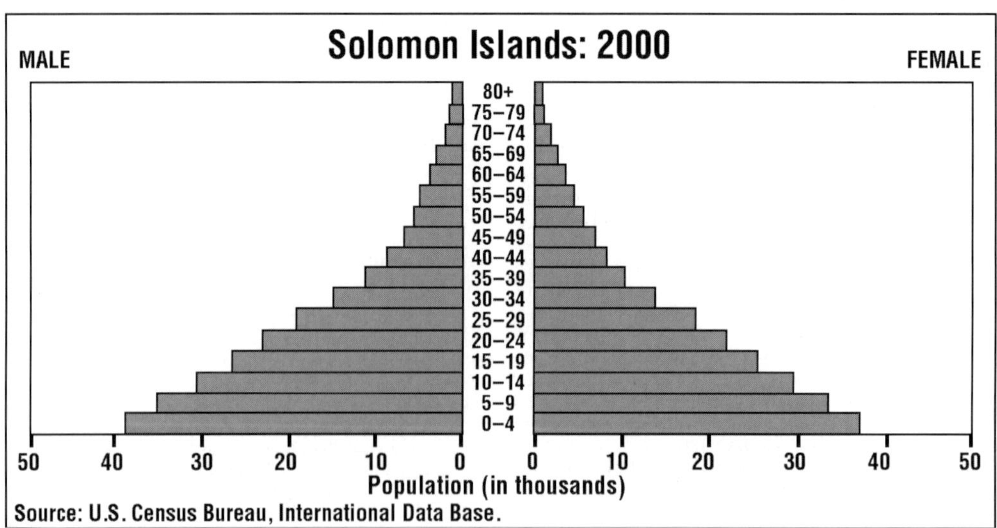

Figure S51 Population pyramid from two small island developing states. Upper panel shows a stabilizing population and the lower panel shows a rapidly growing population. Data from IDP (2000).

Demographics of many SIDS are characterized by a population pyramid that is wide at the base (i.e., many younger persons) and narrow at the apex (i.e., few older persons). In comparison, a society characterized by a population pyramid that is essentially cylindrical (i.e., the number of births, deaths, and distribution by age class are quasi-equal), is one of numerical stability (see Figure S51). Rapidly growing populations require matched growth in services such as education, medicine, food, freshwater, and infrastructure. Small islands are by definition limited in areal extent, and the rapid conversion of agricultural lands to housing, particularly in mountainous islands where cultivatable land is limited to the coastal zone, leads to marginalizing the ability of internal nutritional sustainability.

Deforestation is another consequence of anthropogenic pressure on small islands and other developing areas. Much of the deforestation is past history, such as the conversion of island forestlands to agriculture, as typified by the sugarcane industries of former European colonies in the Intra-Americas Sea. In tropical small islands, it has been the mangroves that have been affected by coastal forestry practices more than any other species. While in many cases mangrove trees have been used for construction, often they are simply removed for ascetic reasons associated with tourism. Not only has this severely impacted the role of these trees as the basis of the tropical marine food web, it has weakened the ability of the coast to resist erosion from tropical storms, tsunami, and sea-level rise.

Waste management in small islands is a critical issue. While in most cases the atmosphere will remove airborne pollutants, the oceanic circulation can cause waste to concentrate on the leeward side. Ship-generated waste, particularly that associated with the lucrative cruise ship trade, is a common problem in many small islands regardless of their latitude. Governmental agencies such as the International Maritime Organization and the United Nations Environment Programme have invested heavily in creating the facilities to manage such wastes, and yet with the growing pressure of other population-related activities on the environment, the coastal zones of small islands are increasingly stressing the marine ecosystem. Waste management of course is of a multiple nature, and while the focus herein is on the coast, the cause is both point-source and distributed.

Finally, considering anthropogenic hazards, the actions of human populations toward marine exploitation must be included. In the Caribbean Sea, for example, the actions of one nation upon another leads to political as well as environmental stress. Overfishing by one small island state upon another as a concerted practice has caused entire reef systems to become incapable of sustainable production. SCUBA divers have been known to sweep past an unprotected coral reef, and essentially remove the entire breeding population of numerous marine species. But the practice of using modern technology to efficiently scour a fishery is not limited to SIDS, as is quite evident in the regions of George's Bank and the Grand Banks. While overfishing has been used to illustrate the issue, souvenir hunting and jewelry harvesting of corals and other marine gems has equally led to exploitation of the marine environment well beyond a sustainable level.

Towards a sustainable environment

Small islands, in many cases, are tuned to the wealth associated with tourism. All the anthropogenic and natural hazards summarized above seem to converge on this central issue. The rugged beauty of isolated small islands, which were created by natural forces that would be considered catastrophic in a contemporary socioeconomic context, are just the scenery that lures visitors. James Mitchner's *Tales of the South Pacific* excites the imagination of fellow wanderers among the family of man. And it is that very attraction that stresses the ability of developing nations to sustain the economic growth necessary to support burgeoning human populations. Not achieving a stable population will lead to the failure of all services and sources of income. Ecotourism, while perhaps less lucrative than currently practiced tourism, is inherently sustainable and should be considered in the mix of solutions leading to sustainable economic development.

Two other critical issues seem to have arisen that cause and limit sustainability: sand mining and water availability.

Tourism and other growth industries require raw materials for construction and infrastructure development. All too often the most cost effective—at least at first—is inexpensive concrete products based on abundant beach sands. Healthy coral reefs in tropical small islands produce beach sands that balance the natural sediment transport to deepwaters as integrated over time. Parallel processes provide materials in other climes for development. Indigenous peoples used these supplies of nature without regard to over-exploitation because of low population densities in the coastal communities. This supply- and-demand balance is no longer viable within the context of exponential human population growth. Thus a paradigm shift is necessary for sustainability, not only regarding beach sand mining, but also for perhaps the most central human issue: freshwater supply.

Peoples of small islands, whether off the desert coast of Bahrain or the rain-forest coast of Singapore, or the frigid Argentine Islands, are focused on the supply of potable water. Sustainability requires that supply exceed demand. No human need except for air, at least on the diurnal-timescale, exceeds the requirement for safe and abundant drinking water. Earth does not have a water shortage problem; it has a cheap water shortage problem! Small islands are perhaps the quintessential expression of this issue. OTEC (ocean thermal energy conversion) and other ocean and coastal engineering solutions to the general issue of supply and demand of abundant freshwater through technology, have as yet realized their theoretical potential. While freshwater is the immediate issue, it is energy—inexpensive energy—that is the central issue, and it is not limited to SIDS.

Conclusions

Small islands are an enigma in many respects within the context of coastal science. They are subjected to storms of tropical and extratropical origin that seem to exacerbate the impact by anthropogenic coastal development. Sea-level change is seen to be extremely site specific, voiding the common perception that all such geographical entities are subject to inundation by external global forces beyond local control. Tsunami, the devastating seismic sea waves, are recognized as not just a Pacific Ocean hazard; highly populated coastal zones of the Atlantic Ocean being perhaps most vulnerable from a socioeconomic context. France and Portugal have recognized the danger of the next repeat of the 1755 Lisbon earthquake and tsunami by establishing a system to prevent a European catastrophe on the scale of the 1998 Papua-New Guinea tsunami.

Population density and pressure upon the coastal marine ecosystem by humankind is clearly a major issue of this quandary. Deforestation, mangrove forests in particular in tropical climates, has repercussions well beyond the erosion issues of sediment systems that provide the nourishment for tourism and other economic stimuli of developing island states. Development brings the inevitable issues associated with waste—what to do with all that stuff? Exploitation of the marine environment extends beyond coral reefs in the tropics, to include fisheries at all latitudes, tourism in pristine locales such as Antarctica, and turtles in the mid-latitude hatcheries of the beaches of east central Florida.

Next, there is the quandary of intellectual sustainability. Human population is, and has been for millennia, characterized by migration. The drain of educated young people not returning to their island homes all too often plagues SIDS. How is the international community of scholars going to create the social environment that encourages young professionals to share their knowledge in a developing nation? Educators in developed nations must encourage the return of the best and brightest scholars to homelands desperately in need of their skills, insight, and worldview perspective, and developing states must accept responsibility by creating policies for students sent abroad to want to return.

Finally there is the issue of "Who do you say is my neighbor?" How is the issue of life, liberty, and the pursuit of happiness, however, it is expressed in the social context of a particular small island society, executed? Clearly the "business as usual scenario" won't create a sustainable society. And in that context, small islands are not separable from the larger global continuum of humankind. Certainly AOSIS and SIDS provide a venue for discussing the special problems of societies with radically limited geographic boundaries. Geophysicists in the widest context can isolate the science peculiar to small islands. Solution of their peculiar problems truly requires a paradigm shift from the confrontational litigious nature of geopolitics to the notion of a shared destiny that does not extort funds and resources from one segment of humankind at the expense of another.

George Maul

Bibliography

IDP (International Data Base), 2000. *U.S. Bureau of the Census*. Washington DC: US Government Printing Office.

IPCC (Intergovernmental Panel on Climate Change), 1997. *The Regional Impacts of Climate Change, Chapter 9: Small Island States*. New York: Cambridge University Press.

Maul, G.A. 1996. *Small Islands: Marine Science and Sustainable Development*. Washington, DC: American Geophysical Union, Coastal and Estuarine Studies No. 51.

Pugh, D. 2004, *Changing Sea Levels*. Cambridge: University Press.

Cross-references

Barrier Islands
Changing Sea Levels
Coral Reef Islands

Demography of Coastal Populations
Mangroves, Ecology
Mining of Coastal Materials
Natural Hazards
Sea-Level Rise, Effect
Storm Surge
Tsunami

SOCIOLOGY, BEACHES—See BEACH USE AND BEHAVIORS

SOUTH AMERICA, COASTAL ECOLOGY

Geomorphologic and oceanographic characteristics of South America

South America (Figure S52) extends from tropical climatic zones (12°30′N) to cold polar zones (about 55°S), encompassing a great diversity of coastal and marine ecosystems. The tectonic history and geological factors, such as the present-day geomorphology and vertical motions of the coastline, influence the coastal and marine ecosystems of South America. Tectonically, South America is divided into two parts, the Andean chains to the west and a vast stable platform to the east, consisting of exposed Precambrian rocks and shallow sedimentary cover rocks (Kellogg and Mohriak, 2001). The Pacific Andean coastline is characterized by high relief, a relatively narrow shelf bordering a deep trench, small drainage basins, and rapid vertical motions of the coast. Low relief, broad shelf, and extremely large drainage basins and alluvial fans characterize the Atlantic coastline. Approximately 93% of South America's drainage is to the Caribbean and the Atlantic, away from the Andes, and provides the world's best example of continent-scale drainage control by plate tectonics (Kellogg and Mohriak, 2001).

The upper water circulation of Atlantic Ocean as a whole consists in its gross features of two great anticyclonic circulations or "gyres" separated over part of equatorial zone by the eastward flowing Equatorial Counter Current. In the South Atlantic, the upper-water gyre extends from the surface to a depth of about 200m near the Equator and to about 800 m at southern limits of the gyre at the Subtropical Convergence. The different portions of this gyre have different water properties. The South Equatorial Current flows west toward the American side of the South Atlantic. Part of the current crosses the equator into the North Atlantic, and is named the Guyana Current. The remainder turns south along the South American continent as the Brazil Current (Figure S52), which then turns east at about 30°S, at the Subtropical Convergence, and continues across the Atlantic as part of the Antarctic Circumpolar Current (Pickard and Emery, 1990). A contribution to the water in the South Atlantic comes from the Malvinas Current (or Falkland Current) flowing north from Drake Passage, up the coast of South America and reaches the Brazil Current at the Subtropical Convergence. The Brazil Current and Guyana Current are warm and saline, with a low concentration of nutrients, having come from the tropic region. The Malvinas Current is cold, with lower salinity and a higher concentration of nutrients, transporting subantarctic waters.

The circulation of the upper water of the Pacific is very similar in its main features to that of the Atlantic. In the South Pacific the superficial currents flow northward, from the Antarctic Circumpolar Current, forming the Peru (or Humboldt) Current (Figure S52). The Peru Coastal Current is closest to the coast and confined to the uppermost 200 m depth, transporting cold water (14–16°C) in summer (Pickard and Emery, 1990). The Peru Oceanic Current (down to 700 m depth) reaches higher velocities than Peru Coastal Current. Between these currents, the weak and irregular southward flow of the Peru (Humbouldt) Subsurface Countercurrent is usually sub-superficial but occasionally reaches the surface (Pickard and Emery, 1990). Along the Pacific coast of South America, an upwelling area extends from 4°S to 42°S. Peruvian coastal upwelling is peculiar because winds sustain the upwelling process throughout the year. The coastal upwelling system, which hardly comprises 0.02% of the total ocean surface, is of great significance because it determines the enormous productivity of Peruvian and Chilean coastal waters, representing almost 20% of the world's landings of industrial fish (Tarazona and Arntz, 2001). Wind intensity and persistence are highest in winter and lowest in summer. Wind stress, and consequently upwelling strength, has intensified over the last decades (Pickard and Emery, 1990). The oceanographic conditions along the Pacific coast are influenced by an interannual variability known as El Niño–Southern Oscillation (ENSO), which consists of a cold period (La Niña) and a warm period (El Niño), with anomalous warming of the eastern Pacific during about three months. El Niño occurs every 2–7 years, causing a deepening of the pycnocline and thus rendering the upwelling process inefficient (Tarazona and Arntz, 2001).

These complex assortments of geomorphologic and oceanographic characteristics result in very diversified environmental settings, allowing the occurrence of many different coastal ecosystems in South America. The occupation of coastal environments is expanding very fast, and in most cases, without adequate planning. The rising rates of coastal resources exploitation, the threat of global changes and sea-level rise and the economic losses due to environmental degradation, on the one hand making the coastal management a hard task, and on the other hand, these events evidence the urgency of planning the uses and conservation of these environments.

Simone Rabelo da Cunha

Sandy beaches ecosystems in South America

Sandy beaches constitute a dynamic interface between the land and the sea that stretch from the base of the dunes to the low tide mark or lower limit of the swash zone. Dunes, beaches, and surf zones are closely linked by the interchange of sand. The morphology of these dynamic and harsh environments is defined by the interaction of waves, tides, sediment characteristics, and topographic features, which combined characterize the morphodynamics of a beach (Short, 1996) (see *Beach Processes* and *Sandy Coasts*). Sandy beaches dominate ocean and estuarine shores of the South American coast. Pacific sandy beaches display a variety of morphodynamic environments. Venezuela, Colombia, and Ecuador present a narrow discontinuous stretch of pocket sand beaches, alternating with rocky shores and mangrove swamps. In Peru, which constitutes the southern extreme of mangrove swamps in the beginning of the warm-temperate Pacific coast, alternating exposed and sheltered beaches, together with active sand dunes, become prevalent. Chilean beaches are mainly found from 19°S to 42°S. Exposed sandy beaches with different morphodynamics alternate with rocky shores in the north (19–30°S) and intertidal sand flats at the mouth of rivers in south-central regions (38–42°S) (Jaramillo, 2001). Atlantic sandy beaches also display wide variations in morphodynamics. Sheltered, pocket beaches are found between Venezuela and northern Brazil, up to Cabo Frio (23°S). Pebble beaches, with a variable amount of rocky fragments, are also found. Southwards, within the warmtemperate biogeographic province, sandy beaches become more exposed and reflective, with coarse sands and steep slopes. A continuous (640 km) exposed, dissipative and microtidal (astronomic tides 0.5 m) sandy beach is found from Rio Grande do Sul (southern Brazil: ca. 29°S) to northeastern Uruguay (Barra del Chuy: 33°45′S), with fine to very fine, well-sorted sands, gentle slopes, well-developed frontal dunes, heavy wave action, a wide surf zone and large barometric tide ranges (Calliari *et al.*, 1996). The physiognomy of this coastline is mainly determined by the prevailing winds and wind-driven surf off the South Atlantic. Southwards, the exposure of beaches is ameliorated by the estuarine effect of the Río de la Plata, between Punta del Este and Montevideo (Uruguay) and also in the northeast of Buenos Aires Province (Argentina). Beyond the estuarine influence of the Río de la Plata River and up to the gulfs of northern Patagonia (around San Matías Gulf/Valdés Península: ca. 41°S), beaches are mainly characterized by fine sands and gentle slopes. These gulfs define the southern limit of the warm-temperate southwestern Atlantic province, which also demarcates a biotic transition in faunal composition. The cold-temperate coastlines south to 43°S, Patagonia, are mainly sand flats, pebble beaches, and long, high cliffs of Cenozoic marine sediments.

Pacific beaches of northern Chile and Peru present surf zones characterized by upwelling, high detrital inputs from kelp beds and the absence of surf diatoms (Arntz *et al.*, 1987). Contrastingly, exposed dissipative beaches of the Atlantic coast of southern Brazil, Uruguay, and Buenos Aires Province are characterized by non-periodic events of high concentrations (7×10^8 cells L^{-1}) of brown colored patches produced by surf zone diatoms such as *Asterionellopsis glacialis* (Odebretch *et al.*, 1995). Favorable surf circulation patterns promote greater retention of particulate primary production, thus allowing the beach/surf zone ecosystem of dissipative coastlines to function as semi-closed ecosystems (McLachlan, 1980). Long dissipative beaches of moderate to high

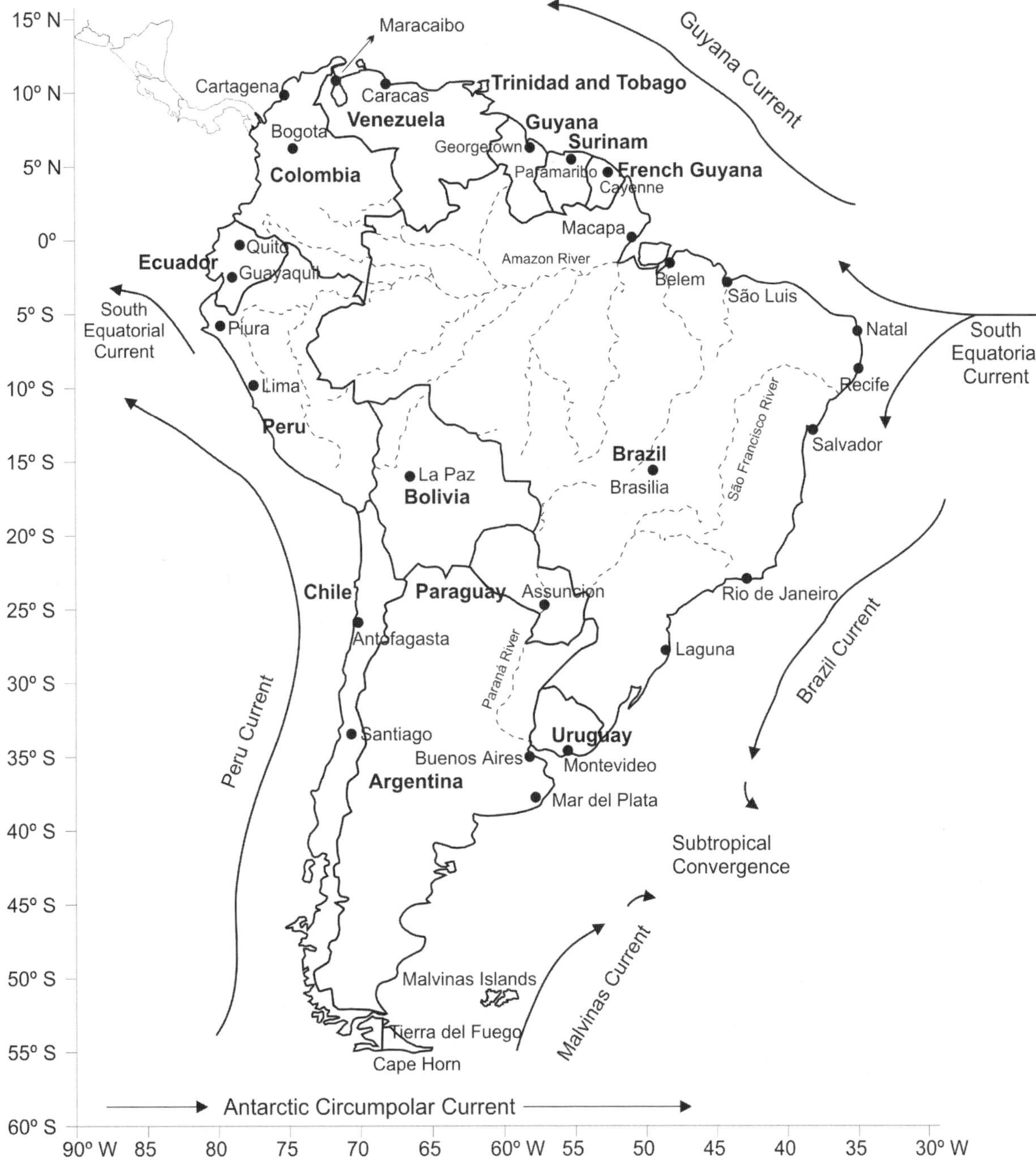

Figure S52 South America countries, showing larger rivers and main oceanographic currents.

energy, rip current activity and an associated dune system providing nutrients by groundwater flow, constitute suitable conditions primarily met in southern Brazil and Uruguay (Defeo and Scarabino, 1990). Abundant suspension feeders play an important role as efficient mineralizers, providing ammonium sources for surf diatoms. Phytoplankton at exposed reflective and sheltered shores is much less abundant, and food for macrofauna mainly comes from the sea as living and nonliving particulate organic matter, while debris and carrion are transported shoreward through the surf zones (McLachlan, 1980). Warm-temperate Atlantic beaches are also subject to seasonal blooms of toxic dinoflagellates. The strong dynamics of oceanic waters and the influence of the Río de la Plata also produce several toxic outbreaks (Méndez *et al.*, 1996).

Research on sandy beaches of South America has been mainly focused on macrofauna (community and population levels), particularly on exposed coastlines. Concerning communities, investigations deal with structure, seasonal dynamics and distribution patterns at spatial scales ranging from macro (km) to meso (individual beaches). Crustaceans, mollusks, and polychaetes are the most diverse groups. A review of existing information does not provide support to the well-known worldwide pattern of increasing number of species from temperate to tropical latitudes. The concurrent effects of morphodynamics, variations in tide amplitude and beach length, as well as the presence of localized areas with extremely high productivity due to upwelling events, mask the recognition of latitudinal trends. Indeed, the reverse pattern was found by Jaramillo (2001) in 3,000 km of Chilean coastlines (20–42°S): species richness increases southwards, from 15 (20–23°S) to 28 (40–42°S). Similar trends were reported by Dexter (1992) for 284 beaches around the world, who found an increase in species richness from tropical to cold-temperate beaches. Moreover, the number of species is not directly related to beach length or tidal amplitude: the greatest species richness in the Atlantic coast was recorded in pocket (short) sheltered and very sheltered beaches with sand sediments mixed with rock fragments (pebbles and cobbles), along São Sebastião

Channel, southeastern Brazil (23°43′S), where 31 species of crustaceans, 35 of mollusks, and 24 of polychaetes were documented for only one beach (Engenho d'Água; e.g., see Nucci *et al.*, 2001). Species richness, abundance, and biomass vary with beach morphodynamics: flat dissipative beaches have more diversity, abundance and biomass than reflective ones (Defeo *et al.*, 1992; Jaramillo, 1994). Intertidal suspension feeders are typical inhabitants in these high-energy dissipative beaches. On the contrary, only supralittoral species (insects, ocypodid crabs, some talitrid amphipods, and cirolanids) remain in truly reflective beaches.

Sandy beaches of South America display across-shore zonation of their macrofauna. In general, the upper shore or supralittoral of tropical and subtropical beaches is characterized by crabs (Ocypodidae), whereas on temperate beaches talitrid amphipods and cirolanid isopods prevail. The upper littoral or mid-shore fringe is dominated by cirolanid isopods and polychaetes, whereas the lower littoral and shallow sublittoral are mainly occupied by cirolanids, amphipods, bivalves, and hippid crabs of the Genus *Emerita*. Concerning tropical coastlines of Venezuela (ca. 10°N), the supralittoral tends to be dominated by the brachyuran *Ocypode quadrata*, talitrid amphipods and the cirolanid isopod *Excirolana braziliensis* (also on the middle shore), whereas the mid-littoral is mainly occupied by anomurans of the genus *Emerita* and *Lepidopa*, the mollusks *Donax denticulatus*, *Tivela* sp., *Terebra cinerea*, and *Olivella verreauxi*, and the polychaetes *Glycera* sp. and *Sthenelais limicola*. The shallow sublittoral is dominated by the gastropod *Mazatlania aciculata*, the echinoid *Mellita quinquiesforata* and the starfish *Astropecten marginatus* (e.g., see Penchaszadeh *et al.*, 1983). In Colombia, Pacific beaches with fine sands have higher species richness and abundance than the Atlantic counterparts with coarser grains (from 3°N to 11°N). Both coasts are dominated by the isopods *E. braziliensis* and *Exosphaeremona diminutum*, as well as the polychaetes *Scolelepis agilis* and *Hemipodus armatus* (Dexter, 1974). Sandy shores of Peru (about 8–12°S) are dominated by a low number of species with high biomass, notably the suspension feeders *Mesodesma donacium*, *Donax peruvianus* and *Emerita analoga* in the shallow intertidal (Arntz *et al.*, 1987). Other members are (Penchaszadeh, 1971): the sand crab *Ocypode gaudichaudii* and the cirolanid *E. braziliensis* (supralittoral and upper/mid-littoral), the polychaetes *Hemipodus triannulatus*, *Nephtys monilibranchiata*, and *Nephtys multicirrata* (mid-littoral) and the anomurans *Blepharipoda occidentalis* and *Lepidopa chilensis* (lower littoral and shallow sublittoral). In Chile, the most common species are the insect *Phalerisida maculata* (supralittoral) the cirolanid isopod *E. braziliensis* (supralittoral and upper littoral), the anomuran crab *E. analoga*, the bivalve *M. donacium*, and the polychaete *Nephtys impressa* (swash zone and shallow sublittoral), all of which occur between 20°S and 42°S (Jaramillo, 2001). The supralittoral ocypodid *O. gaudichaudii* found in northern Chile (20–23°S) is replaced by the talitrid amphipod *Orchestoidea tuberculata* in south-central Chile (32–42°S), as a result of latitudinal gradients in rainfall and sediment temperature.

On Atlantic coastlines, coarse sandy beaches in Surinam (ca. 6°N) show an impoverished fauna characterized by the presence of *O. quadrata* in the supralittoral (Swennen and Duiven, 1982). Sheltered sandy beaches of northern Brazil are dominated by crustaceans, mollusks, and polychaetes, commonly found on channels from Bahia to Rio Grande do Norte. The supralittoral is dominated by *O. quadrata* and the peracarids *Orchestia* spp. and *E. braziliensis*. Lower fringes present a low number of crustaceans and mollusks, depending on tidal exposure and the existence of reefs or freshwater runoff. Southwards (Rio de Janeiro State: ca. 23°S), exposed reflective/intermediate beaches harbor an average of 7 species, notably the talitrid amphipod *Pseudorchestoidea brasiliensis* (supralittoral), the isopod *E. braziliensis* (supralittoral and upper littoral), and the anomuran *Emerita brasiliensis* (lower littoral and shallow sublittoral), which together represent 95% of the macrofauna (Veloso and Cardoso, 2001). For roughly the same latitude, some 90 species of mollusks, crustaceans, and polychaetes appear in Engenho d'Água beach, along São Sebastião Channel (Brazil: ca. 23°42′S). The high species richness in these sheltered unconsolidated marine beaches, with rocky shores inundated by sand, is associated with higher habitat heterogeneity: rocky fragments form small tide pools and moist and shady microhabitats, which may widen the presence of some intertidal species (Denadai *et al.*, 2001). In one of the few comprehensive studies that included sublittoral fringes, Borzone *et al.* (1996) showed a marked increase in the number of species from the intertidal to sublittoral of sandy beaches of Parana, Brazil (25°30′S), particularly in reflective beaches.

Sandy beaches between Southern Brazil (29°S) and Argentina down to Golfo Nuevo (ca. 43°S) show a gradient from warm-temperate to cold-temperate faunistic composition, where the Río de la Plata acts as an effective ecological barrier (Escofet *et al.*, 1979). The wide chain of wave-exposed beaches along southern Brazil and Uruguay show maximum species richness (at least 23), density and biomass at the long chain (640 km) of dissipative beaches of fine sands and flat slopes, while the lowest values (5 species) occur in reflective beaches of Uruguay, with coarse sands and steep slopes (Defeo *et al.*, 1992). Crustaceans are the most represented (nine species), followed by mollusks (six), insects (five) and polychaetes (three). The cirolanid isopod *Excirolana armata* dominates in numbers, whereas the suspension feeders *Mesodesma mactroides*, *Donax hanleyanus*, and *E. brasiliensis* dominate in terms of biomass. The highest beach levels are occupied by *O. quadrata* (only juveniles are sporadically found in Uruguay), the amphipod *P. brasiliensis*, insects and the cirolanid isopod *E. braziliensis*. The polychaetes *Euzonus furciferus* and *Scolelepis gaucha* are found at mid-beach levels, together with the isopod *E. armata*. The higher intertidal levels are dominated by the amphipods *Bathyporeiapus ruffoi*, *Phoxocephalopsis zimmeri*, and *Metarpinia* sp., and the suspension feeders *M. mactroides*, *D. hanleyanus*, and *E. brasiliensis*, whereas the lower ones are occupied by the isopods *Macrochiridotea giambiagiae*, *Macrochiridotea lilianae*, and *Macrochiridotea robusta*, the polychaetes *Hemipodus olivieri* and *Sigalion cirriferum*, and the gastropods *Olivella formicacorsii*, *Buccinanops duartei*, *Olivancillaria vesica auricularia* and *Olivancillaria teaguei*. On Brazilian coastlines, *Callianassa mirim* (Callianassidae) is also found in this fringe (Gianuca, 1983).

Species richness and abundance decrease towards Río de la Plata (ca. 35–36°S), as a result of high salinity variability and decreasing exposure. Low abundances of *P. brasiliensis* and *E. armata* co-occur with some brackish water species. In Buenos Aires Province (ca. 37°S), the supralittoral species *O. quadrata* and *E. braziliensis* are not found, and *Orchestia platensis* and *P. brasiliensis* are rarely present. The intertidal is inhabited by the isopod *Cirolana argentina*, the clams *M. mactroides* and *D. hanleyanus*, and gastropods of the genus *Buccinanops* and *Olivancillaria* (Escofet *et al.*, 1979). The supralittoral of beaches of north Patagonian gulfs is inhabited by *O. platensis*; the upper littoral by *C. argentina*, the bivalve *Darina solenoides*, polychaetes *Scoloplos* sp., *Travisia* sp., and *Onuphis dorsalis*, and amphipods Haustoriidae and Phoxocephaliidae. The lower littoral and shallow sublittoral are represented by the bivalves *Tellina petitiana* and *Bushia rushi*, the gastropods *Olivella* sp., *Buccinanops globulosum*, *Olivancillaria arcellesis*, *Olivancillaria uretai* and *Olivancillaria urceus*, the isopod *Serolis gaudichaudii*, several amphipods (*Monoculopsis valentini*, *Stephensia haematopus*, Haustoriidae and Phoxocephaliidae) and polychaetes. *Callianasa* sp. and the polychaete *Arenicola brasiliensis* appear in sheltered beaches (Escofet *et al.*, 1979).

Faunistical zones across the beach also showed important variability, with aperiodic (even daily) and seasonal components (Brazeiro and Defeo, 1996). The across-shore position of patches varies according to the different susceptibility of each species to environmental variations. Generally, species richness increases from upper to lower beach levels in dissipative beaches, whereas in very reflective beaches the lower fringes tend to disappear. One or two faunistical belts are found in reflective beaches, and three or four in dissipative ones (Defeo *et al.*, 1992).

Life history traits, dynamics, and structure of sandy beach populations show clear geographical patterns. In one of the few large-scale studies conducted in South America, Defeo and Cardoso (2002) analyzed macroecological issues of the intertidal mole crab *E. brasiliensis* along some 2,700 km between Urca (Rio de Janeiro, Brazil: 22°57′S) and Arachania (Uruguay: 34°36′S). Most (11 from 14) life history traits of the crustacean *E. brasiliensis* were significantly correlated with mean water temperature of the surf zone, which could be summarized as follows: (1) a shift from continuous to seasonal reproductive and recruitment events from subtropical to temperate beaches; (2) an increase in individual sizes of the smallest ovigerous female, fecundity at size, predominance of females and individual weights from subtropical to temperate beaches; and (3) a decrease in male sizes, growth, and mortality rates towards temperate beaches. The concurrent effect of morphodynamics at a regional scale was also detected and in some cases masked clear latitudinal trends.

Lifespans appear to be also related to temperature, ranging from 1 to 3 years for the small, warm water species to >10 years in some cold-temperate clams (e.g., *M. donacium* in Chile), but most species live for 1–4 years and have relatively rapid growth to maturity. Wide seasonal growth oscillations occur in southern populations, with lowest rates during autumn and winter (Gómez and Defeo, 1999 and references therein); alternatively, tropical populations show continuous growth throughout the year.

Sandy beach populations are labile to climatic variability, which generates resurgences and mass mortalities. Arntz *et al.* (1987) showed dramatic fluctuations of the suspension feeders *D. peruvianus*, *M. donacium*, and *E. analoga* in Peru, as a response of the strong (ENSO) event of

1982–83. Following the mass mortalities of the dominant *M. donacium* as a response of a strong increase in sea surface temperatures during ENSO, *D. peruvianus* increased from 5% to 60–100%, and *E. analoga* increased from <1% to 29%. This suggests differential responses to climatic events and also potential interspecific interactions because of competitive release of resources by dominant macrofauna members. Variations in species composition along the Chilean coast have also been related to latitudinal gradients in rainfall and sediment temperature (Jaramillo, 2001).

Beach morphodynamics affect biodiversity on sandy coastlines. An increase in species richness, abundance, and biomass from reflective to dissipative beaches has been reported (Defeo *et al.*, 1992). However, a large-scale study conducted in Chile (Jaramillo, 2001 and references therein) showed that species richness, abundance, and biomass are higher at intermediate beaches, some of them located near areas of persistent upwelling. Populations that co-occur in contrasting environments are less sensitive to variations in beach morphodynamics, as revealed by comparisons of abundance, reproduction, recruitment, fecundity, growth, mortality, and burrowing time between reflective and dissipative beaches (Gómez and Defeo, 1999; Defeo *et al.*, 2001).

Most sandy beach populations have strong and persistent distribution patterns in response to an environment that is spatially and temporally structured by sharp, small-scale gradients (Defeo and Rueda, 2002). Specific habitat preferences across the beach generally determine significant correlations between abundance and mean grain size, beach face slope, sediment moisture and penetrability. Aggregations persist in time, but, in contrast to sessile species, the position of the patches varies according to the different susceptibility of each species to variations in physical (e.g., sediment moisture and temperature), and biological (swimming ability and burying) factors (Giménez and Yannicelli, 2000).

Unpredictable and strong short-term increase in tide ranges (i.e., up to sand dunes) generated by storm surges, wind-driven surf off the ocean and barometric tides could be a source of mortality. This mortality in many cases is size-dependent, as smaller intertidal organisms are more susceptible to being stranded in the upper littoral. Mass mortalities at dissipative beaches probably occur at a higher frequency than on reflective ones, where wave intrusion is mitigated by the steep slope on the lower shore.

Intra- and interspecific interactions are important in structuring sandy beach populations and communities. Results from long-term and large-scale field experiments, together with laboratory observations and field monitoring, suggest that population fluctuations in dissipative sandy shores are produced by the intertwined forces of environmental, density-dependent, and human-induced factors operating together at different spatial and temporal scales (Defeo, 1996a). Concurrent field sampling and laboratory experiments with cirolanid isopods showed that intra- and interspecific interactions would be of importance in population regulation (Defeo *et al.*, 1997). Local populations tend to present density-dependent growth and mortality. Density-dependent and density-independent forces acting together can jointly explain population fluctuations over time, as shown for the yellow clam *M. mactroides* in Uruguay (Lima *et al.*, 2000) and the guild of suspension feeders *M. donacium, D. peruvianus*, and *E. analoga* in Peru. The fact that density-dependent mechanisms are often manifested in dissipative beaches, with highest values of species richness and abundance, suggests that biotic interactions should be of utmost importance in these systems, whereas reflective beach populations should be mainly regulated by individual responses to the environment. Considering that these populations are highly spatially structured, potential compensatory and overcompensatory mechanisms have been shown to occur at small spatial scales in dissipative beaches with highest macrofauna abundance (Defeo, 1996a). Other studies, however, suggest that environmental harshness leaves limited scope for competitive interactions, as suggested for Chilean beaches (Jaramillo, 2001).

Predation by birds, gastropods, crabs, fishes, and insects usually generates high rates of size-dependent mortality, particularly at dissipative beaches with high macrofauna biomass. Research efforts were mainly focused on food habits of juvenile fishes in the inner surf zones of dissipative sandy beaches (e.g., Monteiro-Neto and Cunha, 1990). Predation could act as a selection pressure determining body size of macrofauna, and thus higher growth rates should be directed to diminish predation risks. An active selection of the site could also be invoked to decrease predation and desiccation risks.

Sandy beach ecosystems receive a variety of increasing anthropogenic impacts such as forestation, exploitation of coastal species, coastal development, pollution (disposal of liquid and solid wastes, freshwater discharges from wide plain basins used for agriculture and cattle rearing, agro-chemicals), and unplanned recreational use (Lercari and Defeo, 1999).

Sandy beach fisheries in South America rely on species with different life histories. Harvested stocks are the intertidal clams *Mesodesma* and *Donax*, mole crabs *Emerita* and supralittoral ghost crabs *Ocypode*. The main artisanal/recreational fisheries are based on the extraction *M. mactroides* (handpicking: Brazil, Uruguay and Argentina) and *M. donacium* (handpicking and diving: Peru and Chile). These fisheries have shown to be notoriously difficult to manage, because ocean beaches are open and extended systems readily accessible to commercial and recreational users, but also to unauthorized harvesters. Management measures are difficult to enforce and appear to be beyond the finances of most management agencies. High uncertainty in stock estimates, lack of basic biological knowledge, improvements in fishing power, low operating costs and a risk-prone management attitude, determined a trend towards overexploitation (Castilla and Defeo, 2001). Nevertheless, a successful management experiment was documented for *M. mactroides* of Uruguay, which included the experimental manipulation of the system based on a fishery closure for 32 months (Defeo, 1996a, 1998). The long-term study (8 years) showed that abundance, age composition, age-specific survival, fertility, and population growth rate of the yellow clam were significantly affected by adult density, recruitment variability, and fishing intensity (Brazeiro and Defeo, 1999). Population structure of the sympatric clam *D. hanleyanus* showed inter and intra-annual fluctuations of recruits and adults, with uneven periods of abundance related to fluctuations in the fishing effort targeted on *M. mactroides* (Defeo and de Alava, 1995). Operational management tools based on area-specific management plans (legal sizes and catch per fisher and fishing ground) were implemented in Chile through "Management and Exploitation Areas," defined as concessions allocated to fisher communities (Castilla and Defeo, 2001).

Disposal of liquid and solid wastes, accidental oil spills and deposition of mine tailings (Castilla, 1983) modify community composition, abundance, population structure, fecundity, and zonation of macrofauna in South America. Freshwater discharges from man-made canals also affected the habitat, community structure and abundance, fecundity and growth rates of *D. hanleyanus* (Defeo and de Alava, 1995) and *E. brasiliensis* (Lercari and Defeo, 1999). Mechanical disturbance of sands occasioned by recreational users during the summer seems to have little effect on macrofauna (Jaramillo, 2001).

Information is comparatively scarce or absent for plankton, meiofauna, vagile megabenthos, and nekton of the surf zone, as well as for birds and sub-terrestrial fauna inhabiting sand dunes. An ecosystem approach is needed for modeling networks of interactions between different components of food webs. Other processes and mechanisms affecting structure and functioning of sandy beach macrofauna (e.g., predation, commensalism, parasitism, and mutualism) are still little documented. There is considerable scope to elucidate the role of competition in structuring sandy beach communities, and its relative contribution according to morphodynamic states, exposure, and tidal regimes.

Biogeographic patterns in life history traits have not been adequately assessed in sandy beach ecology. Further research should be directed to clarify large-scale patterns, the relative contribution of factors influencing beach fauna and to decipher cause–effect relationships. At present, our knowledge of how dissimilar responses of populations could also result from locally adjusted genotypes or a combination of plastic and genetic responses is limited, and should be addressed by genetic studies throughout biogeographic ranges. Little is also known about dispersive abilities of meroplanktonic larval phases, and the mechanisms influencing larval distribution and connectivity between populations are still poorly understood (Defeo, 1996b). Research should also focus on planktonic stages and physical-oceanographic information (e.g., nearshore hydrodynamics) to determine the spatial scales at which the population dynamics is to be considered an open process, that is, if it is more related to the arrival rates of larvae than to post-settlement processes.

Long-term studies are scarce at both the community and population levels, and the consequences of natural or human-induced disturbances on the structure and dynamics of macrofauna are poorly known. Recent mass mortalities that occurred on a geographical range have been poorly documented and understood. Different approaches are needed to perform well-designed experimental and field studies directed to critically assess environmental impacts in these fragile ecosystems. This should be complemented by laboratory (microcosm) experiments to understand ecophysiological effects of pollutants and responses to abiotic factors coming from different human sources.

Sandy beach populations may be partially regulated by density-dependent processes of unknown extent. Future work should emphasize scale-dependent experimental manipulations of abundance, both through field and laboratory experiments. The spatial analysis of populations and the environment should be useful for monitoring changes in abundance, structure, and dynamics of populations.

Accumulation of toxins, such as those associated with algal blooms of toxic algae, can cause mass mortalities of suspension feeders or render them unsafe for human consumption. This limits the potential utilization of many species and creates the need for careful monitoring and management. As human impacts on coastal waters continue, blooms of toxin-producing phytoplankton could affect beach suspension feeders more. This should be a focus for future studies.

Omar Defeo and Anita de Alava

Rocky shore ecosystems in South America

Rocky shores are important ecosystems along the South American coast because of the great diversity of species and economic importance of some of these species, such as oysters, mussels, crabs, and fish. As a transitional ecosystem between terrestrial and marine environments, the rocky shore can be divided into three parts. The first, upper part of the rocky substrate is permanently exposed to the air (the supralittoral zone), the second part is only exposed during low tides (the intertidal or mid-littoral zone), and the third part is always submerged (the sublittoral zone). While the rocky substrate does not permit organisms to burrow, crevices, pits, and the accumulation of boulders create a three-dimensional matrix that provides different microhabitats for many kinds of organisms. Most of the organisms attach to the rock surfaces, thus providing additional microhabitats on which other organisms may settle. The sessile fauna on the intertidal zone is usually distributed in horizontal belts of dominant species, giving the zone a striped appearance called zonation. In the sublittoral zone, this distribution is much less marked and only some species show a depth zonation (see *Rock Coast Processes*).

On the Atlantic side of South America, traveling along the coast from north to south, rocky-shore communities first appear in the Santa Marta area of Colombia (11°13′N 74°14′W–11°20′N 74°05′W). For the most part, the coastal mountains, part of the Sierra Nevada de Santa Marta, plunge abruptly into the sea. However, a low and narrow rock (phyllite or quartz-diorite) platform exists, at least in the bays, which are often split up into small promontories and islands. The spatial distribution of this community was studied by Battaström (1980), who described the zonation as follows: the supralittoral is inhabited only by the gastropods *Littorina ziczac, Littorina angustior, Nodilittorina tuberculata*, and *Tectarius muricatus*; the upper intertidal zone has a bare appearance but is covered by blue-green algae and represents an impoverished upper part of the lower barnacle–vermetid zone; the lower intertidal zone is a narrow zone formed by barnacles (*Tetraclita* sp., *Chthamalus angustitergum, Megabalanus stultus*) and/or vermetids (*Petaloconchus varians, Spiroglyphus annulatus*); the sublittoral fringe is covered by a mixture of algae (*Ralfsia expansa, Sargassum* sp., *Laurencia papillosa*, and *Ectocarpus breviarticulatus*) and invertebrates (barnacles, chitons, limpets, boring organisms, and a multitude of crabs and snails living inside the algae); the sublittoral zone is characterized by the presence of encrusting coralline algae overgrown by a mat of macroalgae (most important are the rhodophyceans *Amphiroa* sp., *Jania* sp., *Hypnea* spp. *Ceramium nitens, Centroceras* sp., and *Laurencia* spp.) and many sessile and semi-sessile invertebrates (sponges, horny and scleractinian corals, the zoanthids *Palythoa* sp. and *Zoanthus* sp., sea anemones, scattered barnacles and chitons, the gastropods *Fissurella angusta* and *Acmaea* spp., the bivalves *Isognomon radiatus* and *Isognomon bicolor* and the ascidians *Styela canopus, Herdmania momus*, and *Pyura vittata*).

In Venezuela, rocky shores are present in the regions of Trinidad and the Paria Peninsula (11°N), but they are apparently poorly studied and no literature was found describing this area. Rocky shores are absent from the delta of the Orinoco River to the northern coast of Brazil, where the littoral is covered by a mangrove formation.

A few rocky shore ecosystems exist along the northeastern Brazilian coast, but hard substrates are commonly available as sedimentary rocky fringes along the shallow coast. Algae, coral communities, and other sessile invertebrates such as sponges, tunicates, bryozoans, and cnidarians usually inhabit these fringes. One of the few granitic rocky formations already studied is an intertidal and shallow sublittoral platform of boulders and gravel at Ponta Cabo Branco, Paraíba (7°S), which joins the fringing sedimentary reefs along the coast. These rocks form a unique ecosystem in the area composed of a rich sessile community of sponges (*Chondrilla nucula, Haliclona* sp., *Tedania ignis, Halichondria* sp.), tunicates (*Didemnum duplicatum, Didemnum psammatodes, Polysyncraton amethysteum, Eudistoma* spp.), cnidarians, macroalgas (*Ulva* sp., *Gelidium* sp.), oysters (*Crassostrea rhizophorae*) and associated fauna. In the sublittoral, corals (*Siderastrea stellata, Mussismilia hartti, Mussismilia hispida, Montastrea cavernosa, Agaricia agaricites*, and *Porites astreoides*) and the zoanthids *Palythoa* sp. and *Zoanthus* sp. are common invertebrates.

From central Brazil (Espírito Santo, 20°S) to Laguna in the south (Santa Catarina, 28°S), rocky shore environments are formed by granitic or basaltic rock, resulting from the erosion of the border of the Serra do Mar mountain chain, which lies parallel to the coastline. It is not a continuous ecosystem, but forms more or less extended outcroppings between sandy beaches. This is the principal rocky shore ecosystem along the coast of Brazil and also the most known. The first published descriptions of the community appeared between late 1940s and early 1950s (Oliveira, 1947, 1950; Joly, 1951, 1957) and since then, information on species distributions and communities has been found through many species surveys and ecological studies.

The following synthesis provides an overall picture of the community, and is based on work by Joly (1951, 1957), Oliveira (1947, 1950), Nonato and Pérès (1961), and Oliveira-Filho and Paula (1983). The supralittoral mostly comprises bare space used by the periwinkle gastropods *L. ziczac* and *Nodilittorina lineolata*, which are the most common and characteristic organisms at the lower part of this zone. Isopod crustaceans of the genus *Lygia* are also very common at the supralittoral.

The upper intertidal contains a dense belt of the barnacle *Chthamalus bisinuatus*, while *Tetraclita* and *Megabalanus* are found lower in the intertidal zone, but are not dominant space occupiers. Below the *Chthamalus* belt, *Brachidontes solisianus* is the dominant in terms of space used. However, in sites exposed to waves, *Perna perna* mussels can form dense beds in the mid and lower intertidal. More recently, *I. bicolor* has invaded the southeastern and southern coasts of Brazil and is replacing *B. solisianus* in some areas. The mid-intertidal zone is also colonized by many algae, such as *R. expansa, E. breviarticulatus, Centroceras clavulatum, Jania adhaerens, Acantophora spicifera*, and *Ulva* spp. On wave-exposed sites one can add *Porphyra* spp. (in winter), *Chaetomorpha antennina*, and many species of fleshy macroalgae, and coralline algae both articulated and encrusting. On the low intertidal and sublittoral the oyster *C. rhizophorae* can be abundant and common algal species are *L. papillosa, A. spicifera, Jania capillacea, Amphiroa fragillisima, Hypnea cervicornis, Rhodymenia pseudopalmata*, and *Corallina officinalis*. In many sites along the coasts of São Paulo, Paraná and Santa Catarina, the sabelariid polychaete *Phragmatopoma caudata* forms extended sandy reefs along the low intertidal zone. In the latter two states, the colonial ascidian *Eudistoma carolinense* forms a narrow belt below *Phragmatopoma* and comprises an important intertidal microhabitat for more than 117 species (Moreno and Rocha, 2001). Among the vagile invertebrates characteristic of the intertidal zone are some herbivorous mollusks, such as the limpets *Collisella subrugosa* and *Fissurella clenchi*, besides the periwinkles, which can migrate down from the supralittoral zone; the predator whelks *Stramonita, Pisania*, and *Leucozonia*; and the crabs *Pachygrapsus gracilis* and *Pachygrapsus transversus*.

A dense *Sargassum* spp. bed usually marks the upper sublittoral fringe and is probably the most abundant macroalgae in both tropical and subtropical sublittoral zones. It forms dense beds usually covering a thin layer of encrusting coralline algae, which are dominant space occupiers in many sublittoral sites, especially at places where grazing pressure is high. The most important herbivores at the sublittoral are the urchins *Arbacia lixula, Echinometra lucunter, Lytechinus variegatus*, and *Paracentrotus gaimardi*; chitons; gastropod mollusks of the genera *Aplysia, Astraea*, and *Tegula*; and fish, such as damselfish (*Stegastes* sp.), surgeons (Acanthuridae), and parrotfish (Scaridae). Sessile invertebrates common to the sublittoral zone are cnidarians of the genera *Palythoa* and *Zoanthus*, which form large encrusting colonies, and a variety of small arborescent hydrozoan colonies. Sponges are also very common and diverse—more than 120 species are known within the merely 25 km long São Sebastião Channel on the north coast of the State of São Paulo (Hajdu *et al.*, 1996). The most conspicuous bryozoan is the encrusting *Schizoporella*, but many small arborescent colonies are frequent, especially from the genus *Bugula*. Common ascidians are the encrusting colonial didemnids and the solitary pyurids and styelids (Rodrigues *et al.*, 1998).

The southern coast of South American, between Rio Grande do Sul (Brazil) and Rio de la Plata estuary is formed basically by a long sandy beach of 750 km in length. Very little rocky substrate is available along this coast that could represent an important geographical barrier for sessile organisms. Close to the borderline between Santa Catarina and Rio Grande do Sul (29°S), there is a small rocky outcropping formed by very tall volcanic rocky structures, called Torres ("towers" in English) because of the height of the rocks. Artificial rocky substrates are also present, such as the rocky jetties at the entrance of Lagoa do Patos, which also supports a diverse encrusting community.

Very few experimental studies have been undertaken to understand the dynamics of both intertidal and sublittoral rocky shore communities along the Brazilian coast. Caging experiments showed that littorinid grazing activity controls the abundance of microalgal populations in the supralittoral zone (Apolinário *et al.*, 1999). Experimental analysis of succession on cleared substrates in the intertidal zone revealed that *C. bisinuatus* recruits more on the *Brachidontes* zone and prefers granitic to basaltic substrates (Tanaka and Duarte, 1998). On the belt formed by *Sargassum cymosum* var. *nanum* in the low intertidal zone, succession is maintained in its earlier stages due to desiccation stress causing widespread algal mortality during the early summer (Paula and Eston, 1989). Experimentally cleared areas of the sublittoral zone revealed that *Sargassum stenophyllum* was both the competitive dominant with slow growth and an opportunist colonist (Eston and Bussab, 1990).

South to the Rio de la Plata River estuary in Argentina, there are loess platforms separated from a coastal cliff of 7–8 m by sand strips. The upper intertidal is inhabited by crusts of blue-green algae, while the mytilid bivalve *Bachidontes rodriguezi* is the dominant space occupier in the mid-intertidal zone with densities up to 33,000 ind m^{-2} and an associated community of around 40 species. The cause of this dominance is not only competitive abilities but also the absence of important predators at the intertidal zone (Lopez Gappa *et al.*, 1990). The barnacles *Balanus amphritite* and *Balanus glandula* are present in areas with fewer *Balanus rodriguezi*. The first species appears in Quenquén Harbor (38°34′S; 58°42′W) and the second has become abundant in Mar del Plata Harbor (38°0′S; 57°32′W) in recent years. There are no periwinkles in the Mar del Plata region, where they are replaced by the gastropod *Siphonaria lessoni*, which is abundant in both the intertidal and supralittoral zones.

Community distribution in the southern region of Argentina, the Chubut Province, is as follows: the supralittoral covered by various blue-green algae with the chlorophyte *Enteromorpha intestinalis* and the mollusks *S. lessoni* and *Brachidontes purpuratus* in tide pools; in the upper intertidal there is a striking absence of barnacles and the most important inhabitant is the gastropod *S. lessoni*; in the mid-intertidal there are three well-marked belts: *B. purpuratus*, the coralline alga *Corallinetum officinalis* and the phanerogams *Spartina montevidensis* and *Salicornietum ambigua*; the low intertidal is occupied by a mytilid belt (*Mytilus edulis platensis, Aulacomya ater*, and *B. purpuratus*); the sublittoral does not include the usual phaeophytes but instead has a wide belt of the chlorophytes *Codium fragile* and *Codium vermilara* establishing its upper limit (Olivier *et al.*, 1966).

Chile has an extended coast, about 2,600 miles long, which can be topographically divided into two very different regions: south of Chiloé (41°29′s), the coast is discontinuous, with mountains along the shore rising up to 3,000 m, comprising an eroded tectonic pattern of glaciated and non-glaciated fjords. North of Chiloé the coastline is very regular and fully exposed to the prevailing winds and waves, but there are geological differences: from Chiloé up to Navidad (33°57′S) the coastal range is made up of metamorphic shale of low elevation, from Navidad to Antofagasta (23°38′60S) it is mainly granitic rock, and north of Antofagasta it consists of volcanic rocks with sedimentary intrusions (Stephenson and Stephenson, 1972; Santelices, 1991).

Along the southern coast of Chile, south to Chiloé Island, the bad weather conditions and the limited access result in a lack of knowledge about the intertidal and sublittoral communities in these wave-exposed open coasts. Nevertheless, community descriptions are available from sheltered islands in the southernmost tip of South America. For instance, in the Beagle Channel the supralittoral zone contains several bands of lichens and the upper intertidal is covered by a mixture of algae (*Bostrychia mixta, Hildenbrandia lecannellieri, Pilayella littoralis, Adenocystis utricularis, Enteromorpha* spp., *Porphyra* spp., *Spongomorpha* spp.). Next, the mid-intertidal has mussels and barnacles together with high densities of the gastropods *Acmaea, Collisella, Nacella*, and *Siphonaria*, while the low intertidal is covered by pink encrusting coralline algae and high densities of *Nacella magelanica* and *Nacella mytilina*, with the brown alga *Lessonia vadosa* marking the lowest limit of the intertidal zone. The sublittoral communities of these habitats consist of belts of *Macrocystis pyrifera* in sheltered and semi-sheltered sites (Santelices, 1991).

In central Chile the supralittoral inhabitants are barnacles (*Jehlius cirratus*) in reduced densities, aggregations of *Littorina araucana* and *Littorina peruviana*, the algae *Porphyra columbina* and dark-red crusts of *H. lecanellieri*. The upper intertidal is covered by pure or mixed stands of chthamaloid barnacles (*Chthamalus scabrosus, J. cirratus*), while the mid-intertidal usually contains a belt of the mussel *Perumytilus purpuratus* mixed with the algae *Centroceras clavulatum, Enteromorpha compressa, Iridaea laminarioides, Ulva rigida*, and *Polysiphonia* spp. (Santelices, 1991).

Towards the northern coast of Chile, this same pattern of species distribution is observed for wave-exposed rocky habitats, but there is a northward reduction in the number of belt-forming algae species. *Macrocystis pyrifera* disappears north of Concepción (36°49′60S) where the dominant algae are *Lessonia trabeculata* in the sublittoral and *Lessonia nigrescens* along the sublittoral–intertidal fringe (Santelices, 1991). Rocky sheltered communities have not been well studied and the only description available is from Guiler (1959) in which the intertidal biota in Antofagasta Bay was studied. There the rocks are shale that forms a wide erosion platform covered by an association of *Pyura praeputialis* and *Corallina chilensis* immediately below the barnacle belt. The lower limit of the *Pyura* belt might be controlled by predation by the starfish *Heliaster helianthus* and *Stichaster striatus*. At the lower intertidal *Ulva* replaces *Corallina*; the large barnacle *Austramegabalanus psittacus* occurs in the lowest part of the *Pyura* belt and both organisms are covered by *Ectocarpus confervoides* and *Halopteris hordacea*. *Pyura chilensis* is widespread along the coast of Chile and do not form belts as *P. praeputialis* in Antofagasta region (Santelices, 1991).

Much experimental work has been done to examine species distributions in both the intertidal and sublittoral zones of the Chilean coast. *P. praeputialis*, for instance, maintains its dense intertidal beds by intraspecific self-facilitating mechanisms that enhance recruitment to the border of previously settled individuals (Alvarado *et al.*, 2001). The intertidal rocky community in Mehuin (39°25′60S), in the southern coast, is regulated by herbivory and competition; there are apparently no intertidal carnivores capable of controlling herbivore densities and high herbivore densities can destroy the red alga *Iridaea boryana* cover; depending on the season of the disturbance, the community may be dominated by the alga or by barnacles plus crustose algae (Jara and Moreno, 1984). In the sublittoral zone, experimental kelp canopy removal revealed that *M. pyrifera* control the species composition of understory algal community (Santelices and Ojeda, 1984), and, in central Chile, the removal of *L. nigrescens* results in a community of calcareous crustose alga *Mesophyllum* sp. in the presence of herbivory and large patches of *Gelidium chilense* in the absence of herbivory (Ojeda and Santelices, 1984). In northern Chile, the abundance of herbivores, algae morphology, plant density, water movement, and the egg case of elasmobranches which tie plant stipes together are the most important ecological factors for the persistence and stability of *L. trabeculata* beds (Vasquez, 1992).

A high degree of eco-geographic isolation seems to be a general characteristic of the intertidal and shallow sublittoral rocky communities of Chile. The result is a high degree of endemism and several of the endemic species occupy unique ecological niches with unknown parallels in comparable habitats elsewhere (Santelices, 1991). The causes of this isolation, and factors limiting the distribution of these species and communities are unknown and offer an important area of future research.

Oceanographic conditions and upwelling are strong influences on the coastal communities of Peru. This upwelling area associated with highly productive waters stretches from 4°S southward to 40°S, in central Chile, but the most intensive effects are seen in the Peruvian coast, especially during the winter (Tarazona and Arntz, 2001). Because of the upwelling process, intertidal and shallow sublittoral zones of northern Chile and Peru have high species diversity. In Peru, the forest-forming kelps (*Lessonia* spp. and *M. pyrifera*), the mussel *Argopecten purpuratus*, the gastropod *Thais chocolata*, the crabs *Cancer setosus, Cancer porteri*, and *Platyxanthus orbignyi*, and the sea urchin *Loxechinus albus* all occur in great numbers with a large biomass. Kelp forests form a sublittoral belt about 15 m wide and harbor numerous associated species. Mussels (*Perumytilus purpuratus, Semimytilus algosus*) also form beds in the intertidal zone, with more then 70 associated species. The structure of these communities tends to be controlled by grazers and predators, while the population dynamics of the algae *Macrocystis* is influenced by the upwelling and El Niño events (Tarazona and Arntz, 2001). As an example of this influence, the intense El Niño event of 1982–83 caused mass mortality of key species like the mussel *Semimytilus* and brown algae (*Macrocystis* and *Lessonia*) along the Peruvian rocky shores (Tarazona *et al.*, 1988).

In Colombia, the tectonic processes along the Pacific coast have given origin to abundant steep cliffs (>45° slope) and rocky shores with more gently sloping platforms, composed of boulders, pebbles, and gravel from cliff erosion. Along the northern coast and in the interior of Buenaventura Bay (3°52′60N), species have the following distributional patterns: desiccation-tolerant blue-green and green algae, a lichen species, periwinkles (*Austrolittorina aspera, Littoraria zebra*), the crab *Grapsus grapsus*, and the isopod *Lygia baudiana* inhabit the supralittoral.

About 20 species occupy the upper intertidal zone, among them the periwinkle *L. zebra*, the crab *Pachygrapsus transversus* on less wave-exposed cliffs, while barnacles (*Tetraclita, Chthamalus*), limpets (Fissurellidae, Acmaeidae, Siphonariidae), crabs (*G. grapsus*) and some green algae occur on exposed cliffs. The mid-intertidal zone is dominated by bivalves (*Brachidontes* sp., Isognomidae, and oysters) and the associated fauna is comprised by crabs of the families Xanthidae (*Eriphia squamata*) and Grapsidae (*P. transversus*), and the red coralline alga *Lithothamnion*. In the lower intertidal zone it is possible to find barnacles, anemones, sponges, the gastropods *Acanthina brevidentata* and *Thais kiosquiformis*, and some crabs. At this level an increasing number of boring organisms (bivalves of the families Pholadidae, Petricolidae, and Mytilidae and the ghost shrimp *Upogebia tenuipollex* contribute to the bioerosion of the rocky cliffs. Erosion rates can be as much as 300 $cm^3\ m^{-2} month^{-1}$ for igneous rocky cliffs and 450 $cm^3\ m^{-2}\ month^{-1}$ for sedimentary rocky cliffs (Cantera and Blanco, 2001).

The most important stresses on rocky shores are the trampling effects of tourists and fisherman, oil and sewage pollution, and selective collecting for food and obtaining mussel seeds for cultures. All these processes have been studied in different rocky shore ecosystems and showed marked influences in the intertidal community structure. Although many rocky shore ecosystems are located in protected areas inside ecological reserves, usually the goal of the reserve is to protect either the terrestrial or the underwater ecosystem while neglecting the intertidal zones. One of the few well-studied reserves is the Estación Costera de Investigaciones Marinas (ECIM) in Las Cruces (33°30′S), Chile, with a 5 ha human-exclusion exposed shore established in 1982. The intertidal community changes and food-web cascading effects, which occurred inside the reserve, showed that humans act as an efficient and selective keystone predator (Castilla, 1999).

The rocky shore communities throughout the South America coasts are reasonably well described. Yet, almost no information exists on the Patagonian coast of Argentina and the Ecuadorian coast. On the other hand, mechanisms of community structure are best studied in Chile and, very recently, Brazil. Apparently, the conservation of rocky shores has not been of primary importance for most countries environmental conservation policies. The few well-monitored known marine reserves are located in Chile.

Rosana Moreira da Rocha

Coral reefs ecosystems in South America

Coral reefs are highly diverse ecosystems that have been compared with the tropical rain forest. They are formed by coelenterates that secrete a calcium skeleton as well as by other calcium secreting organisms such as mollusks, corallinae algae, and sponges (see *Coral Reefs*). The structure formed by these organisms offer shelter and support for an incredible diversity of life, making coral reefs crucial to the culture and livelihood of millions of people in tropical coastal environments. Corals have on their living tissue symbiotic algae called zooxantelae that contribute with oxygen and organic compounds and receive protection from the coral colony. The phenomena known as bleaching is caused when the zooxantelae leaves the coral, usually due to stress related causes. In the last decades, mass bleaching events that have caused the death of many coral reefs have been reported from several parts of the world. These events, related to climate change, have added to other anthropogenic stresses, which are causing an alarming rate of degradation of coral reefs. The concerns with the health and conservation of coral reefs have led government and organizations to establish several programs for monitoring and protecting these ecosystems. The Global Coral Reef Monitoring Network (GCRMN) was established in 1995 with the objective of encouraging and coordinating monitoring of coral reefs at the government, community, and research levels around the world. Brazil, Colombia, and Venezuela are part of the Node of GCRMN for southern tropical America, together with Panama and Costa Rica (Wilkinson, 2000).

Coral reef formations in tropical South America are more developed in the Atlantic coast, as several important cold upwelling areas inhibit reef development in the Pacific coast. Coral reef formations on the Pacific side are found in a few reef patches on the coast of Colombia and on offshore islands. The Atlantic coast of South America is under strong continental influences, which introduces large amounts of sediment and inhibits the development of extensive coral reef formations in some areas, especially around large rivers such as the Amazon, Orinoco, and Magdalena. Coral reefs occur along the Atlantic coast of Colombia, Venezuela, and Brazil.

Colombia is the only South American country with both Pacific and Atlantic coasts and coral reefs. There are about 2,700 km^2 of coral reefs in the Caribbean waters, of which 75% are located in oceanic reef complexes. Along the mainland coast, there are fringing reefs on rocky shores such as the Santa Marta and Urabá areas (Garzon-Ferreira *et al.*, 2000).

In Venezuela, reefs occur along three Caribbean areas out of total 2,875 km of Caribbean and Atlantic coastline. Along the continental coast of Venezuela, the more developed coral formations occur in the Morrocony National Park and adjacent reefs, where more than 30 coral species can be found. The best reef formations are found 100 km offshore at Los Roques Archipelago, with 57 coral species and reefs growing up to 50 m depth (Garzon-Ferreira *et al.*, 2000).

Brazil is located in the central-oriental portion of South America with approximately 7,408 km of coastline running from 4°25′N to 33°45′S. Coral reefs are sparsely distributed along almost 3,000 km of coast (Maida e Ferreira, 1997), and their distribution and location is still poorly known (Castro and Pires, 2001). Laborel (1969) provided the most thorough qualitative description of Brazilian reefs (see also Maida and Ferreira, 1997 and Castro and Pires, 2001 for reviews). The Brazilian coast can be divided into three biogeographical realms. Intertropical (northern coast): comprises the northern coast from the French Guyana border to Cabo de São Roque. Tropical: the largest portion of the Brazilian coast, from Cabo de São Roque to Cabo Frio, Rio de Janeiro State. Subtropical: from Cabo Frio to the border with Uruguay. Reef formations, including some that are not true coral reefs, are present mostly along the tropical Brazilian coast (northeastern coast), although some coral growth also occurs at the northern region and in the southeastern coast up to São Paulo State. Coral diversity is low, with only 18 hard coral species, but 10 of these are endemic to Brazil. Of those, three species have an even more restricted distribution, only occurring on the reefs of Bahia State. Brazilian reefs present only one species of reef-dwelling soft coral, *Neopongodes atlantica*, which is an endemic form (Maida e Ferreira, 1997).

Reef formations such as those that are typical of the northeastern Brazilian coast are rare elsewhere, not displaying the distinctive zones generally observed in reefs around the world (Leão *et al.*, 1988). One of the main characteristics of the Brazilian reefs, are the constructions made by calcareous algae, from the group of the Melobesiae, and vermetid gastropods of the genus *Petaloconchus* and *Dendropoma*. These formations can be found on crystalline and eruptive rock, but are especially common on the seaward side of sandstone banks and coral reefs. They grow in the upper part of the reef front forming structures that are similar to the algal ridges of the Indo-Pacific reefs (Laborel, 1969). Endemic species like *Favia leptophylla, Mussismilia braziliensis, Mussismilia hartii*, and *Mussismilia hispida*, that among the principal reef builders, are archaic forms, the remnants of a tertiary fauna that was preserved in a refugium provided by the seamounts of Abrolhos bank during the last glaciation (Leão *et al.*, 1988).

On the northern coast the main geographic feature is the immense Amazonian estuary with a width of more than 350 km. The water of the Amazonian estuary, loaded with vast amounts of sediments, is transported by littoral currents to the north up to Guyana, forming an extensive barrier for coral reef development. According to Laborel (1969), the region has no reefs, only scattered coral growth, but further down the coast, about 80 km off São Luís, the capital of Maranhão State, lays a large coral bank called Parcel de Manuel Luis, whose existence has been known to navigators since the 17th century, mostly because of the danger it represented to navigation.

Reef formations are present along the northeastern coast from Cabo do São Roque (Rio Grande do Norte state) to the south of Bahia State. In Cabo de São Roque there is a group of oval shaped reefs located a few miles from the coast. These reefs are simple structures, usually formed by numerous pinnacles in a shallow sandy base. On the reef flat only two species of coral occur, *Siderastrea stellata* and *Favia gravida*. Calcareous algae Melobesiaea and the vermetid gastropods form an algal ridge. The seaward crest is dominated by *Millepora alcicornis*, followed by *M. hartii* on the slope and *Montrastea cavernosa* at greater depths. According to Laborel (1969) the main reef builder in this region is *S. stellata*.

In this region, from Natal to São Francisco River, the principal characteristics are the coastal sandstone banks and superficial coral reefs, disposed in various lines running parallel to the coastline along more than 600 km. The sandstone banks and superficial coral reefs form lines that are not continuous, and in the places where these formations are interrupted, the coast takes the form of small bays, normally with mangrove swamps whenever creeks and rivers are present (Maida e Ferreira, 1997). The sandstone banks are structures that can reach up to 10 km in length and 20–60 m wide. They can appear directly adjacent to the beach, or as submerged formations on the high tides (Dominguez *et al.*, 1990). In some areas up to three lines of reefs can be seen at one coastline, forming an effective protection to the shore. The length of each reef varies from 1 to 4 km, for the reefs that are exposed at low tide, and up to 10 km for the submerged reefs, such as the reefs located off Itamaracá Island, north of Recife (Dominguez *et al.*, 1990). The depth of surrounding waters is seldom greater than 10 m. In this area, the region of more extensive coral development is located between Recife and Maceió Cities. The Tamandare reef complex is the most studied area due to the presence of the Oceanographic Institute in Recife, established in 1958 by the Federal University of Pernambuco, and due to the work of Laborel (1969), who presented a detailed description of the area (Maida e Ferreira, 1997). Coral reefs in the area present a distinctive feature, given by their growth as isolated columns of 5–6 m high and expanded laterally on the top. Where the growths of these reef columns are dense, the reefs coalesce at their tops creating large structures with open spaces below the surface, forming a system of interconnected caves (Maida e Ferreira, 1997). The coral fauna of the reefs in this region is richer than up north. From the 18 species of stony coral described for the Brazilian coast, 9 species were described for this coast. The main reef builders in this region are the species *M. hartii* and *M. cavernosa*. The fish fauna is similar to the Caribbean fauna, but less diverse, with basically the same families but less species.

Further south, reefs disappear in the area around the mouth of the São Francisco River. The São Francisco River, with an average run-off of 3,300 $m^3 s^{-1}$, discharges large quantities of fine material and exerts considerable variation in salinity in the seawater for several kilometers off and along the coast, forming a large barrier for coral reef development over 100 km wide.

Coral formations are observed again south of the river mouth, along the coast of Bahia State. This coast has the higher diversity of reef formations, presenting superficial reefs; fringing reefs; large isolated mushroom shaped pinnacles, called "chapeirões," and large platform bank reefs (Leão *et al.*, 1988; Dominguez *et al.*, 1990). *M. braziliensis*, an endemic species to the Bahia coast, is the most important coral species, and the main reef builder. In the coast of Bahia State the continental shelf widens up, reaching over 200 km in the southern part of the coast. In this region, "Abrolhos," the largest and more diverse reef complex on the Brazilian coast, is located. Coral reef formations in the Abrolhos area are spread over an area of 6,000 km^2, up to 15 km long and 5 km wide. The whole region has been a National Marine Park since 1983, being the first Marine Park established in Brazil. A special permit is required from IBAMA (Brazilian Environmental Institute) and the Brazilian Navy to land on these islands. In Abrolhos there is observed the highest diversity of corals in Brazil. All hermatypic scleractinian corals and hydrocorals found in the Brazilian Coast are present in Abrolhos, from which seven species, including the principal constructors, are endemic forms. Two of the Brazilian species of scleractinian corals, *M. braziliensis* and *Favia leptophylla*, and the hydrocoral *Millepora nitida* only occur in the coast of Bahia.

South of Abrolhos, variations in water temperature are greater throughout the year, and there is a greater vertical temperature gradient. Reefs gradually disappear on the northern coast of Espirito Santo State in Rio de Janeiro State, few scattered points of coral growth are found in Cabo Frio, Angra dos Reis, and Ilha Grande, between the barriers represented by the cold upwelling waters around Cabo Frio, and the mouths of the rivers São Mateus, Mucuri, and Doce. The southern limit of coral growth is São Paulo State.

As in most regions of the world, anthropogenic impacts have been the main cause of degradation on coral reefs in South America. The human related activities that affect reefs in the region are the same that threaten most coral reefs around the world, such as land use practices that increase sedimentation, industrial, domestic and agricultural pollution, mining, over-exploitation of reef resources, and uncontrolled tourism. Coral reefs in the region support important biodiversity reservoirs and an expanding tourism industry (Garzon-Ferreira *et al.*, 2000). According to the report of the Status of Coral Reefs of the World (Wilkinson, 2000), bleaching events appear to have increased in frequency, but decreased in severity, throughout the 1990s, due to the global warming phenomena.

Beatrice Padovani Ferreira and Mauro Maida

Mangrove ecosystems in South America

Mangroves are transitional ecosystems between land and sea. Mangrove plants are adapted to salinity variation, waterlogged and hypoxic or reduced mud sediments (see *Mangroves*). Mangrove forests show best development in tropical protected shores, where low wave energy and abundant freshwater supply allow deposition and accumulation of fine organic mud and salinity range between 5 and 35 (Lugo and Snedaker, 1974). The extension of forests inland depends on tidal range and topography, and large forest belts can extend several kilometers landward from the sea. Under these optimal environmental conditions and humid areas, mangroves can attain their maximum growth and productivity. Examples of these are reported for Ecuador and Colombia, where red mangrove (*Rhizophora*) trees reach 40–50 m in height and more than 1.0 m in diameter (Lacerda *et al.*, 1993), and for coasts of Surinam, French Guyana, and northern Brazil, where black mangrove (*Avicennia*) trees reach 30–45, m in height and 0.7–1.0 m in diameter (Lacerda and Schaeffer-Novelli, 1992). The productivity of these systems, in terms of litterfall, varies from 3 to 10 metric ton $ha^{-1} year^{-1}$ (Lacerda *et al.*, 1993; Twilley *et al.*, 1997). Seaweeds can play an important role to mangrove productivity, representing 10–45% of total primary production of mangroves, not only in South America (Peña, 1998; Cunha, 2001), but also elsewhere (Steinke and Naidoo, 1990; Rodriguez and Stoner, 1990). Microalgae are certainly important, though not quantified. The high primary production, together with habitat heterogeneity, can sustain a large and diversified fauna, including transient and resident animals. Many of these are economically important, especially for artisanal users of mangrove. Around the world, many people are supported by mangroves' goods and services, as economic or food source (Hamilton and Snedaker, 1984; Saenger *et al.*, 1983).

South America mangroves, as much as New World mangroves, have a reduced number of tree species, contrary to Southeast Asia, which has nearly one hundred taxa of true mangrove trees (Tomlinson, 1986). There are only four genera: *Rhizophora* (Rhizophoraceae), *Avicennia* (Avicenniaceae), *Laguncularia*, and *Conocarpus* (Combretaceae). The species *Rhizophora mangle, Avicennia schaueriana, Avicennia germinans, Laguncularia racemosa*, and *Conocarpus erecta* have wide distribution. The species *Rhizophora harrisonii* and *Rhizophora racemosa* are more restricted to northern places of South America (Cintron and Schaeffer-Novelli, 1992). *Avicennia* is the most tolerant genus to environmental stress, sometimes dominating highly saline substrates or low-temperature areas, being able to attaining greater structural development in low salinity, disturbance-free environments (Cintron and Schaeffer-Novelli, 1992). Many plant species occur associated with mangrove forests, and the diversity can vary due to climatic conditions and proximity of other ecosystems, like rain forests. Despite this, some species frequently appear to be associated with mangroves around the world, including Atlantic and Pacific coasts of South America. The Malvaceae *Hibiscus tiliaceus* and the fern *Acrostichum aureum* are the most widespread (Cintron and Schaeffer-Novelli, 1983). Macroalgae occur colonizing roots and trunks of mangrove trees, rocks, stones, or shell fragments into the mangrove. Red algae of the genera *Bostrychia, Caloglossa, Catenella*, and *Polysiphonia* and the green algae of the genera *Cladophoropsis, Rhizoclonium*, and *Bloodleopsis*, dominate the macroalgal communities. However, other genera can also occur, especially in northern areas, where water transparency and salinity are high (Cordeiro-Marino, 1992).

South American mangroves occur on protected shores of all of the maritime countries except Chile, Argentina, and Uruguay. Along the Atlantic coast mangroves form a nearly continuous belt from northern countries to Laguna, Santa Catarina State, in southern Brazil (28°30′S), and the latitudinal limits are determined by the frequency, duration, and intensity of cold winter temperatures, rainfall and/or frost (Lacerda and Schaeffer-Novelli, 1992). Along the Pacific coast, the southern mangrove limit is the Tumbes River estuary, in Piura, northern Peru (5°32′S). The restricted distribution at the Pacific coast occurs due to climatic constraints generated by oceanographic conditions along the Peruvian and Chilean coasts, where the Andes Cordilleras and the tropical air currents diminishes rainfall, and the upwelling of cold waters of Peru (Humboldt) Current suppresses convective activity. These result in extremely arid climates, high soil salinity and almost absent freshwater input, limiting mangrove occurrence (Lacerda and Schaeffer-Novelli, 1992). The western limit of South American mangroves is the Galapagos Islands, off the Ecuadorian coast (Latitude 0° and longitude 91°00′W) (Lacerda *et al.*, 1993), and the eastern limit is Fernando de Noronha Islands, off the Brazilian coast (State of Pernambuco, 3°34′S; 32°24W) (Hertz, 1991). At the north of the South Atlantic coast, in Colombia, Venezuela, Guyana, Surinam, and French Guyana, mangrove occurrence is closely associated with protected areas and large rivers.

Peru has two mangrove forests. The larger one, Piura River has 58.5 km^2. The other one is located at Tumbes River estuary (5°32′S), and has only 3 km^2, and is the southern distribution limit of mangroves along Pacific coast. These forests are under a subtropical climate, with annual rainfall ranging from 66 to 300 mm, and temperature ranging from 18°C to 32°C. Peruvian mangroves are under high natural environmental pressure due to low rainfall that results in high salinity

(Echevarria, 1993). Occasionally the "El Nino" phenomenon, which occurs during summer, causes rapid increases in rainfall. During these events, intense rains induce large geomorphological changes in the coast, which strongly affect the mangrove forests. Human pressure, mainly deforestation for shrimp culture, also occurs, although mangroves are legally protected (Echevarria, 1993).

Mangrove forests in Ecuador occupy 161,770 ha, and more than 70% of these are located in the Gulf of Guayaquil (3°S, 80°W), in Guayas Province. This is the largest estuarine ecosystem on the Pacific coast of South America (Twilley *et al.*, 2001), and its fauna is described in South American mudflats. The dominant species is *Rhizophora harrisonii* followed by *R. mangle, A. germinans, L. racemosa*, and *Conocarpus erectus* (Twilley *et al.*, 2001). The structure of mangrove forest in this river-dominated estuary indicates optimum growth conditions with tree heights from 25 to 40 m, although some forests have trees up to 7 m (Twilley *et al.*, 1997), probably due to differences in soil phosphorus concentration and other variations on edaphic conditions. Rainfall is seasonal, with more than 95% of the precipitation occurring from December to May, causing seasonal river discharge ranging from $200\,m^3\,s^{-1}$ during dry season to $1{,}400\,m^3\,s^{-1}$ during wet season. Seasonality does not occur to soil salinity, neither to total litterfall of the forests, but occurs to the leaf component of litter fall (Twilley *et al.*, 1997). Mangrove forests with structural differences (mainly tree height) show different rates of litterfall, ranging from 6.5 to 10.6 ton ha^{-1} $year^{-1}$ (Twilley *et al.*, 1997). Owing to the construction of shrimp ponds and urban expansion along the shore of Estero Salado, mangrove areas of the Gulf of Guayaquil and the Guayas River estuary decreased from 159,247 ha in 1969 to 122,566 ha in 1995. Furthermore, defoliation of *Rhizophora, Avicennia*, and *Laguncularia* by larvae of *Oiketicus kirbyi* has contributed to loss of mangroves, though impacts vary interannually (Twilley *et al.*, 2001).

Colombia has 358,000 ha of mangrove, 90% of these along the Pacific coast and 10% along the Caribbean coast. The most important estuaries are Buenaventura bay (3°54′N; 77°W), in the Pacific coast of Colombia, and the delta of the Magdalena River (Colombia's largest river), that includes the large lagoon complex of Ciénaga Grande de Santa Marta (11°N; 75°45′W), in the central Caribbean coast of Colombia. Climatic differences between the Pacific and Atlantic coasts of Colombia, due to the effects of the Inter-Tropical Convergence Zone, have resulted in Pacific humid tropical forests and Caribbean dry tropical forests. These differences strongly influence mangrove forests structure and productivity, and Pacific mangroves show higher complexity and biomass, whereas Caribbean mangroves are smaller, with reduced crown and smaller biomass (Alvarez-Leon, 1993). Climate and fauna for these areas are described in South American mudflats. Mangroves are the most conspicuous vegetation of Ciénaga Grande de Santa Marta (central Caribbean coast), comprising 50,000 ha. The dominant species is the black mangrove *A. germinans*, followed by the red mangrove *R. mangle*, the white mangrove *L. racemosa*, and buttonwood mangrove *C. erecta*. The lagoon is dotted with small islets of *R. mangle*, and on the alluvial plain a 700 m wide belt of *Avicennia* and *Laguncularia* fringes the shores of the lagoon (Polania *et al.*, 2001). The average height of well-developed stands is 15 m, but 25 m tall trees and also dwarf stands are common (Alvarez-Leon, 1993). The maximum diameter at breast height (dbh) is 30 cm in *R. mangle*, 40 cm in *A. germinans* and 13 cm in *L. racemosa*. About eight species of algae are associated with mangrove prop roots, but they have not been quantified yet.

Mangrove habitats in Buenaventura Bay (Pacific coast) are comprised of *R. mangle, R. racemosa, A. germinans, L. racemosa, C. erecta*, and the associates *Pellicera rhizophorae* and *Mora oleifera*, these two occupying consolidated substrate (Cantera and Blanco, 2001). The mangrove habitats can be classified into three different physiographic types, according to Cintron and Schaeffer-Novelli (1983). Riverine mangroves are well developed (<35 m height, <20 cm dbh) along tidal creeks and estuarine zones of the bay. Bar mangroves are intermediate (<15 m height, <6 cm dbh), growing behind sand ridges near the mouth of the bay. Fringe mangroves growing in platforms resulting from bioerosion of sedimentary cliffs usually develop slowly, or are dwarf, but exceptionally they can exhibit development similar to riverine mangroves (6–9 m height, 4–11 cm dbh) (Cantera and Blanco, 2001). Several red and green algae (*Bostrychia, Catenella, Caloglossa, Bloodleopsis, Cladophora*, etc), as well as benthic diatoms and blue-green algae grow on prop roots, pneumatophores or form beds (Peña, 1998). The litter production ranges from 9.6 ton ha^{-1} $year^{-1}$ in bar mangroves to 11.4 ton ha^{-1} $year^{-1}$ in riverine mangroves, but is much lower in disturbed riverine mangroves, with 7.5 ton ha^{-1} $year^{-1}$ (Cantera and Blanco, 2001), and macroalgae contribute about 26% to total annual mangrove production (Peña, 1998).

Mangrove forests of Venezuela cover 250,000 ha, and major mangrove areas are the Orinoco River delta, the estuaries of the rivers San Juan, Limon, and San Carlos, and the lagoons la Restinga, Tacarigua, Cocinetas, and Sinamaica, and the Gulf of Cuare-Morrocoy Bay. Except for Orinoco River mangrove forests, which are typically fluvial and can attain over 40 m in tree height, all other mangrove forests of Venezuela are located in arid and semiarid regions (Conde, 2001). Annual rainfall in the Orinoco River watershed can be lower than 1,000 mm in northern lowlands and as high as 8,000 mm in southern high relief areas. In the delta, rainfall is markedly seasonal with two dry and two wet periods. The dominant vegetation in the delta is mangrove, which comprises 73% of Venezuela's mangrove forests. The dominant species are *R. mangle, A. germinans*, and *L. racemosa*, though *R. harrisonii* and *R. racemosa* have also been reported. Mangrove forests tend to form a 100 m wide belt along the margins of channels, with some individuals growing up to 20 m height and dbh about 25 cm. Mangroves are replaced landward by tall stands of herbaceous *Montrichardia*, followed by halophobic species (Conde, 2001).

The most part of mangrove forests in Trinidad and Tobago occur in Trinidad, which has about 7,000 ha of well-developed forests, used as timber resource and for tourism. These forests are important places for coastal birds. *R. mangle* is the most abundant tree species, and *R. harrisonii, R. racemosa* are apparently restricted to Trinidad. *A. germinans, A. schaueriana*, and *C. erectus* are also common (Bacon, 1993).

Brazil has 7,400 km of coastline and mangroves occur in a patchy fashion on 92% of this entire coastline. They reach from Oiapoque, Amapá (4°30′N) to Laguna, Santa Catarina (28°30′S) (Schaeffer-Novelli *et al.*, 1990), and mangrove area estimates range from 2,500,000 ha (Saenger *et al.*, 1983) to a more realistic 1,376,255 ha (Kjerfve and Lacerda, 1993). Brazilian tidal ranges decrease southward, from strongly macrotidal, with ranges greater than 4 m (8 m in some places) in the north, mesotidal (2 m) in most parts of the coast, and microtidal (0.2 m) in the south (Schaeffer-Novelli, 1993). Because of the large latitudinal gradient, mangrove forests structure are much more variable in Brazil than in other South America countries. Brazilian mangrove species are *R. harrisonii, R. racemosa, R. mangle, A. germinans, A. schaueriana, L. racemosa*, and *C. erectus*. *Hibiscus* and *Acrostichum* are very common plants associated with mangrove forests. The smooth grass *Spartina alterniflora* is usual along the entire coast, fringing mangrove and zones of accretion (Schaeffer-Novelli, 1993).

At the northern coast (4°30′N to 3°S) the climate is wet, with rainfall ranging from 2,000 to 3,250 mm $year^{-1}$. Mangroves develop very well north and south of the Amazon delta, reaching 20 m in height. Black mangrove *Avicennia* dominates mangrove forests, and *Rhizophora* stands are more frequent on estuaries under a more direct marine influence. *Laguncularia* is also common, especially in low salinity or backing *Rhizophora* fringes. In the Amazon delta region (1°40′N to 0°36′S) mangroves development and coverage are poor because of the overwhelming influence of the freshwater Amazon discharge, and are mixed with freshwater swamps (Schaeffer-Novelli *et al.*, 1990). Eighty-five percent of Brazilian mangroves occur along the 1,800 km length of this part of the coast, extending more than 40 km inland following the course of estuaries and rivers in the states of Pará and Maranhão. The Maranhão State has the most extensive (500,000 ha) and structurally complex mangrove forests (Kjerfve and Lacerda, 1993). Mangrove trees may reach 45 m height, some with dbh exceeding 0.8 m and above ground forest biomass about 280 ton ha^{-1}. The extension and complexity of these mangrove systems reflect hydrological and topographical characteristics of the coast.

The northeastern coast (3–13°S) has a dry climate, with a long and pronounced dry season. Annual rainfall (1,100–1,500 mm $year^{-1}$) is usually lower than potential evapotranspiration, and seasonal droughts and hypersalinity are usual. This coast is exposed to high-energy waves, characterized by sandy beaches and dunes, with reefs offshore. Mangroves develop poorly due to lack of freshwater runoff and prolonged droughts (Schaeffer-Novelli *et al.*, 1990), and have been restricted to protected areas in association with estuaries and coastal lagoons.

At eastern and southeastern coasts (13–23°S) a mountain chain (Serra do Mar) approaches the coast, restricting the width of the coastal plain. Shallow coastal lagoons are common behind narrow sandy spits. Rainfall (1,200 mm $year^{-1}$) is similar or higher than evapotranspiration, and there is no marked dry season. Where the Serra do Mar is very close to the coast, higher rainfall may occur. In areas protected from high energy, mangrove forests are large and well developed (Schaeffer-Novelli *et al.*, 1990), and trees can reach 6–15 m in height, mean dbh from 0.08 to 0.12 m and above ground forest biomass about 65 ton ha^{-1} (Kjerfve and Lacerda, 1993).

In the southern distribution of mangroves (24–28°30′S), rainfall largely varies in small spatial scale, ranging from 1,100 to 2,000 mm $year^{-1}$, being usually higher than evapotranspiration. Despite the water surplus, mangroves are not as well developed as in northern places, especially due to lower temperatures. *Avicennia* trees are taller than *Laguncularia* or *Rhizophora* in this region, reaching 10–15 m in height (Schaeffer-Novelli *et al.*, 1990). The mean of mangrove forests height vary from 2 to 7 m, with dbh from 4 to 12 cm and above ground forest biomass from 9 to 100 ton ha^{-1}, depending on position of the forest and flooding frequency (Tognella-De-Rosa, 2000; Cunha, 2001). South from 28°30′S cold winter temperatures inhibit mangrove occurrence, and salt marsh vegetation, as *Spartina, Scirpus, Juncus*, and other herbaceous plants colonize protected areas and tidal flats.

In all places in Brazil where detailed studies were made (as in Rio de Janeiro, São Paulo, Paraná, and Santa Catarina), there were not always clear zonation patterns for species. However, strong gradients for forests structure (mainly height and dbh) always occur from fringe to inner mangrove forests, reflecting the environmental gradients of flooding frequencies and certainly the geochemistry of the substrate (Schaeffer-Novelli *et al.*, 1990). Latitudinal and local gradients of mangrove forests structure strongly influence the patterns of forests productivity. Well-developed forests usually present higher biomass accumulation and higher litter fall than forests with shorter trees. In local scale, however, there is not always correlation between forest structure and litter fall, and it seems to be more related to differences in allocation patterns of production, and accumulation of biomass, which can vary due to edaphic differences (Cunha, 2001 and references therein). Productive processes are generally related to temperature, rainfall, and nutrients availability, but seasonal patterns of litter fall largely vary, and are not always directly related to productivity of the trees. Litter fall measurements are available mainly for southeastern and southern mangrove forests and vary from 3 to 10 ton ha^{-1} year^{-1} (Kjerfve and Lacerda, 1993 and references therein). Seaweed production was quantified only in Santa Catarina and varies from 0.3 to 1.8 ton ha^{-1} year^{-1}, where litter fall varies from 2.3 to 3.8 ton ha^{-1} year^{-1} (Cunha, 2001). In northern places, where temperature and water transparency are higher, seaweeds production certainly presents an important contribution to mangrove systems. Trophic interactions in mangrove systems are explained on South American mudflats.

Mangroves play an important role in South American economy, providing many goods and services for the human population (Tognella-De-Rosa, 2000). These include: coastline protection and stabilization, nursery for many economically important shellfish and crabs, and source of timber, firewood, charcoal, chemicals, medicine, and waterways for transport. In the north of South America, a great part of the shrimp fisheries is based on species, which depend on mangroves for completing their development. Timber, firewood, and charcoal seem to be the major uses of mangroves in South America, and mangrove bark is still a source of tannin in most countries (Kjervfe and Lacerda, 1993; Tognella-De-Rosa, 2000).

Despite their ecological and economical importance, South American mangroves are threatened by diverse natural and anthropogenic disturbances. Dredging and filling of channels, industrial and urban pollution have also resulted in large losses of mangrove areas. Low sheltered embayments often containing extensive mangroves are used for establishment of large industrial complexes, resorts, and extensive mariculture projects (Kjerfve and Lacerda, 1993; Schaeffer-Novelli, 1993).

Deforestation in mangrove areas has increased in the last years for all countries, mostly for conversion into salt ponds and mariculture, mainly for shrimp ponds. Land reclamation for building condominiums and marinas and illegal occupation are responsible for the most part of the deforestation, mainly in Brazil (Kjerfve and Lacerda, 1993). These contribute to the degradation of mangrove areas through physical fragmentation of landscape and the impoverishment or irreversible loss of genetic resources. Despite this alarming situation, mangrove forests are protected in almost all South American countries, and there is an increase in the preoccupation of public opinion in conservation matters. Simultaneously there is an increase in the quantity and quality of research in all South American countries in the recent years, especially for sustainable use and conservation of coastal ecosystems, including mangroves. Earlier studies had focused mainly on structural aspects of mangrove forests, but in the last 10 years many research groups around South America have been studying productive and litter dynamics, herbivory and competition, population and community dynamics, valuation of mangroves (economic ecology) biogeochemical and hydrological processes, and restoration and recreation of mangroves. Most of these works are just beginning, but the results already found have helped to plan the use and conservation of these amazing ecosystems.

Simone Rabelo da Cunha and Mônica Maria Pereira Tognella-De-Rosa

Mudflat habitats in South America

When shores become more protected from wave action, they become finer grained and accumulate more organic matter, thus, they become muddier. Mud particles accumulate where currents are low and most of the tidal flats of the world are associated with estuaries and similar embayments. Despite mudflats being characteristic of sheltered habitats, they are far from being static entities, since they represent areas of changing balance between erosion and deposition (Little, 2000). The coastal slope and the tidal range determine the areal extension of the tidal flats (see *Muddy Coasts*). In regularly flooded coasts, the tidal cycles determine the frequency and length of low tide exposure. In irregularly flooded areas, this exposition reflects the environmental unpredictability, mainly, due to the wind, rainfall, and evaporation effects. Waves, tides, and currents transport and sort the sediment particles, and also, determine their distribution, stability, and composition. Together with local climate and geomorphology, these factors constitute the environmental matrix of tidal flats, affecting the composition, abundance, and the distribution patterns of the organisms (Reise, 1985).

The intertidal flat is a rigorous environment for plants and animals, as they are intermittently exposed to the heat of the sun and to the drying action of air and wind. The variety and numbers of organisms gradually increases toward the lowest intertidal zone, which is exposed only during the lowest tides. Here, you will find not only typical intertidal species but also some essentially subtidal ones, able to survive out of water for short periods of time (Little, 2000). For mangrove flora, see "Mangrove Ecosystems of South America," in this entry.

The Gulf of Guayaquil (3°S, 80°W) of the coastal province of Guayas, Ecuador, is the largest (12,000 km^{-2}) estuarine ecosystem on the Pacific coast of South America. Surface water temperatures vary between 21.5°C and 25°C during the dry season and increase by 3°C during the wet season. Salinity decreases from 34 in outer regions to 30 (20 during the wet season) inside the gulf. Tides are 1.8 m near the upper boundary of the Gulf and increase to 3–5 m in the Guayas River estuary near the city of Guayaquil (Twilley *et al.*, 2001). The oysters *Crassostrea columbiensis* and *Crassostrea iridiscens*, the mangrove crab *Ucides cordatus*, the pelecypods *Mytella guayanenis, Mytella strigata* live in the intertidal of mangrove habitats, which are dominated by *R. harrisonii*. Mangrove fall rates vary from 6.47 to 10.64 ton ha^{-1} year^{-1}. A model of leaf litter dynamics suggests that geophysical energy (tides, river, discharge) controls the fate of mangrove leaf litter, though highest litter turnover rates are associated with the activity of the mangrove crab *Ucides occidentalis* (Twilley *et al.*, 1997). High tide predators like the blue crab *Callinectes sapidus* and the shrimps *Penaeus stylirostris, Penaeus vanamei*, that occur near mangrove areas (Twilley *et al.*, 2001), develop a severe predation on infauna in mudflats.

The Buenaventura Bay at the central Pacific coast of Colombia, due to the effects of the Inter-Tropical Convergence Zone and intense precipitation, is one of the most humid places in the world, with a mean annual air temperature of 25.9°C, and 228–298 days of precipitation per year. At high slack water, salinity ranges from 18 to 27 at the mouth of Buenaventura Bay to 4.8 at the Dagua River in the inner bay. Complex estuarine conditions and diverse habitats (sandy beaches, rocky shores, mangrove swamps, and mudflats) that maintain rich biological communities, characterize Buenaventura Bay as a diverse and productive ecosystem (Cantera and Blanco, 2001). Extensive mudflats occur around creeks in the inner bay with a rich macrofauna (157 spp.), dominated by deposit-feeders, most of which occupy the aerobic upper subsurface layers of sediments, while some bivalves and polychaetes burrow into deeper layers. The high temperatures and desiccation determine a poor diversity on upper littoral, with few dominants that remain buried most of time. Species richness (32) increases in the mid-intertidal and the lower intertidal zone (up to 67 species) displays high diversity and evenness and is inhabited by gastropods (*Natica, Nassarius, Anachis, Cerithium*), bivalves (*Tagelus, Anadara, Chione*) polychaetes (Amphinomidae, Capitellidae, Glyceridae, Nereidae), crabs (*Panoeus, Callinectes*), and gobiid fishes (Cantera and Blanco, 2001).

The lagoon complex of Ciénaga Grande de Santa Marta, located on the central Caribbean coast of Colombia, is part of the eastern delta of the Magdalena River (Colombia's largest river). The delta and the lagoon complex (1,321 km^{-2}) comprise the Ciénaga Grande (450 km^{-2}), the Ciénaga de Pajarales (120 km^{-2}), several smaller lagoons, creeks, and channels (150 km^{-2}), and mangrove swamps. The lagoon complex can be considered a euhaline–mixohaline system, with mean annual temperature of about 30 °C. Temporal and spatial salinity gradients are common, resulting from variable runoff, seawater intrusion, rainfall, and evaporation. During the dry season (249 mm rainfall), salinity varies between 30 and 40. During wet season periods (1,268 mm), salinity varies between 15 and 20 (short wet season) and close to zero on the long wet season. The Ciénaga Grande complex is a high productive system, where sequential pulses of phytoplankton (990 g cm^{-2} year^{-1}) and seasonal export of mangrove detritus (mangrove litter of 15.7 tons ha^{-1} year^{-1}), contributes to the carbon budget and sustains high secondary production in the lagoon (Polania *et al.*, 2001). Among the invertebrate fauna, mollusks are represented by approximately 98 species (66 genera and 48 families), of which 61 are of marine origin. Six species occur in mangrove sediments, *Melampus coffeus* being the most abundant among the three gastropods. The fiddler crabs *Uca rapax* and *Uca vocator*, the most abundant crabs in areas of mangrove sediments, can construct their burrows on tidal flat area and ingest detritus and macroinvertebrates there. The crabs *Eurypanopeus dissimilis, Pachygrapsus gracilis*, and *Petrolisthes armatus*, species of amphipods and the polychaete *Nereis virens* are abundant on muddy areas (Polania *et al.*, 2001). This polychaete is recognized as an important infaunal predator, which can control the nonpredatory infauna densities inside mudflat sediments (Reise, 1985). Fiddler crabs (*Uca* spp.), land crabs (*Cardissoma guanhumi, Ucides cordatus*), and grapsid crabs (mainly the genus *Sesarma*) are often fairly general feeders, depending mainly on scavenging the deposits and a certain amount of predation. The family grapsidae shows a distinct propensity to herbivore, that may include the scraping of epiphytic algae from the surface of mangrove roots, trunks and branches, and also eating the leaves and the reproductive products of mangrove trees. Others important predators in the Ciénaga Grande are the swimming crabs *Callinectes* spp., which invade the intertidal mangroves or mudflats from the subtidal areas, during flooding periods.

Another Caribbean habitat is the Maracaibo system, which acts as an assemblage of interactive brackish water bodies (>220 km^{-2}), comprised of the Gulf of Venezuela, Tablazo Bay, and the Maracaibo Strait, which connect Lake Maracaibo in the interior of the basin to the Caribbean Sea. Water temperature follows a seasonal pattern, with a minimum mean temperature of 29°C at 1-m depth in February and a maximum of 32.5°C in September. The resuspension of bottom sediments is the principal source of nutrients in the gulf Tablazo Bay, and the strait. Elevated phosphorus levels in the lake are largely due to terrestrial runoff and the nitrogen is a limiting factor under most conditions in the lake (Rodríguez, 2001). Dense mangroves (*R. mangle*) occur on muddy intertidal, and in subtidal waters the widgeon grass *Ruppia maritima* dominates. In muddy intertidal bottoms there are the pulmonate gastropod *Melampus coffeus* and high densities of fiddler crabs (*Uca cumulanta* with 330 burrows m^{-2} and *U. rapax* with 113 burrows m^{-2}) and the swamp ghost crab *U. cordatus* (10 burrows m^{-2}). The ubiquitous clam *Polymesoda solida* and the blue crab *C. sapidus* are abundant components in the submersed meadows. The sublittoral community of the Maracaibo system is not well known, however, the fact that 165 species of mollusks have been recorded suggests a diverse subtidal fauna (Rodríguez, 2001). Dominant components are the mytilid pelecypod *Mytella maracaiboensis* (up to 158 g dry weight^{-1} m^{-2}), the tubiculous amphipod *Corophium rioplatense*, and polychaetes, like *Heteromastus filiformis* (980 ind. m^{-2}), *Sigambra* sp. (1,380 ind. m^{-2}), *Streblospio* sp. (4,000 ind. m^{-2}), *Capitella capitata* (160 ind. m^{-2}), and *Nereis succinea* (360 ind. m^{-2}). The composition and distribution of species clearly follows a salinity gradient (Rodríguez, 2001).

In northeastern Brazil, the Itamaracá estuary (824 km^{-2}) shows characteristics of a tropical hot and humid ecosystem. Both salinity (27) and water temperature (26.8°C) are lower in the rainy season (February to August) than the salinity (34.1) and temperature (30.1°C) in dry season (September to January). Mangrove forests (28 km^{-2}), dominated by *R. mangle* and *L. racemosa*, occupy the lowlands along the Santa Cruz Channel and the lower part of tributaries. On mudflat habitats the detritus is the main diet item of macroconsumers, such as fiddler crabs *Uca* spp., the shrimps *Penaeus schimiti* and *Penaeus subtilis*, and the mollusks *Neritina virginea, Heleobia australis*, and *Tagellus plebeius* (Medeiros *et al.*, 2001).

In the southern part of the coastal plain of São Paulo State, Brazil (25°S, 48°W), the lagoon region and estuary of Cananéia extends for approximately 110 km. Tropical air masses prevail from the end of spring (September) and the end of summer (February) and cold fronts occur in April and May. The salinity is highly variable (0–34), and the input of freshwater and the introduction of sediments into the system, during the last 150 years, have modified physiographic and hydrologic characteristics, which influence biological structure and ecological functions of the lagoon region (Tundisi and Tundisi, 2001).

Among the invertebrate fauna (73 spp.), polychaete, crustacea, and mollusca dominate in number of species, while deposit feeding polychaetes characteristic of mudflat areas like *Loandalia americana, Laonice japonica, Clymene* sp., and *Clymenella* sp., are the most common species. The seasonal changes in macrobenthic abundance and diversity are attributed to changes in salinity, redox potential, sediment granulometry, and organic matter concentration (Tundisi and Tundisi, 2001).

Paranaguá Bay (612 km^{-2}) a subtropical estuarine system on the coast of Paraná State in southeastern Brazil (25°30′S, 48°25′W), is comprised of two main water bodies, the Paranaguá and Antonina bays (260 km^{-2}) and the Laranjeiras and Pinheiros Bays (200 km^{-2}). Mean salinity and water temperature in summer and in winter are 12–29°C and 23–30°C and 20–34°C and 18–25°C, respectively (Lana *et al.*, 2001). Mangroves colonize most intertidal areas around the bay (*R. mangle, A. schaueriana, L. racemosa, C. erectus*). *S. alterniflora* marshes colonize tidal flats or creeks as monospecific, discontinuous narrow (up to 50 m wide) belts in front of the mangroves (Lana *et al.*, 1991).

While salinity and environmental energy gradients appear to control large-scale distribution patterns in the bay (Lana, 1986), plant architecture and food availability seem to be the main source of small-scale macrofaunal variability (Lana *et al.*, 1991). The opportunistic gastropod *Heleobia australis* is dominant on the mudflats in the inner part of Paranaguá Bay. Two assemblages are characteristics of the low-energy areas in the central part of the bay: (1) one dominated by the polychaete *Clymenella brasiliensis* and the gastropod *Turbonilla* sp., in low-energy environments; (2) the other, dominated by the polychaetes *Owenia fusifirme* and *Magelona* spp., in moderate energy environments. Another sector of the bay, with silty-clay sediments on the lower intertidal flats, showed low diversity macrobenthic invertebrates, probably due to drastic fluctuations of salinity in 4–5 h periods. In this kind of environment, the dominants are the polychaetes *Laeonereis acuta* and *Heteromastus similis* (Lana, 1986).

The Patos Lagoon is a huge choked lagoon with a surface of 10,227 km^{-2}. It stretches in a NE–SW direction from 30°S to 32°12′S, where in the south part there are 971 km^{-2} of estuarine area (approximately 10% of the lagoon). The estuarine region exchanges water with the Atlantic Ocean through a 20 km long and 0.5–3 km wide inlet (Asmus, 1997).

As a consequence of reduced tidal influence (mean of 0.47 m) in the inlet and in the estuarine area, the salinity distribution lacks tidal variability but does correlate with wind forcing and variations in freshwater input on scales of hours to weeks (Garcia, 1997). The high frequency and low predictability of the salinity variations characterize the estuarine region of Patos Lagoon as an area chemically highly unstable (Niencheski and Baumgarten, 1997).

The macrobenthic community in the estuary is composed of approximately 40 spp., most of which are r-strategists with pronounced seasonal and interannual variations in abundance. The long and narrow entrance channel with unstable bottoms, the reduced tidal oscillations, the unpredictable wind and precipitation patterns, that cause a general absence of conservative gradients of salinity, may account for the low diversity of the macrobenthic fauna in the estuarine area (Bemvenuti, 1997a; Bemvenuti and Netto, 1998).

The larger part of the estuarine shallow shoals is dominated by intertidal and shallow mudflats (<1.5 m), either with or without the occurrence of the widgeon grass *Ruppia maritima* beds, but with epibenthic microalgal growth and occasional aggregations of macroalgae (mainly *Enteromorpha* spp.). The motile epibenthic organisms of mudflats are decapods like the blue crab *C. sapidus*, the shrimp *Farfantepenaeus paulensis*, the grapsid crab *Cyrtograpsus angulatus*, and the mud crab *Rhithropanopeus harrissii*. During summer juveniles of decapods and fishes exert a severe predation pressure on the macrobenthic community on estuarine mudflats (Bemvenuti, 1997a).

The epifauna of mudflats is mainly represented by the opportunist gastropod *H. australis* (Hidrobiidae), which densities may exceed 40,000 ind m^{-2} and achieves highest biomass of 246 g m^{-2}, though pronounced spatial and temporal changes in density are common. The densities of amphipods, isopods, and tanaidaceans, which are typical of lower reaches of intertidal flats, increase as a result of macroalgal aggregations, which supply habitat, food, and shelter against predators (Bemvenuti, 1997a).

The infaunal deposit feeder polychaete *L. acuta* reaches densities of 5,127 ind m^{-2} and a biomass of 28.26 g m^{-2} on intertidal and shallow mud habitats of Patos Lagoon estuary. The polychaetes *Nephtys fluviatilis* and *H. similis* have higher densities in mudflats but lower densities in the subtidal bottoms. Adults of both, *H. similis* and *L. acuta* attain refuge against predators by burrowing deeply (approximately 20 cm) in mud bottom habitats (Bemvenuti, 1997b). The deeply burrowing pelecypod *T. plebeius* form patches in mudflats and the deeper tube dwelling tanaid *Kalliapseudes schubartii*, in spite of a severe predation by fishes and decapods, attains densities up to 10,000 ind m^{-2}. This tanaid is a typical r-strategist, with intense reproductive activity in summer months, embryos marsupial protection, and intense recruitment, which maintains elevated densities in mudflats. In shallow habitats the pelecypod *Erodona mactroides* suffer high mortality during the first year and despite high densities (3,722 ind m^{-2}), the mean biomass rarely exceeds 105 g m^{-2} (Bemvenuti, 1997a,b).

Macrobenthic invertebrates display diverse feeding habits but detritus appears to be an obligate food item for most species in mudflats. Infaunal deposit feeders like *L. acuta* and the epifaunal *H. australis* occupy the first level of consumers. In general, the decapods are typically omnivorous and opportunistic feeders, which exploit different trophic levels as food items become available. The predation of infaunal polychaete *N. fluviatilis* on *H. similis* represents an important intermediate link between the non-predatory infauna and epifaunal predators (Bemvenuti, 1997c). The environmental stress of oligohaline estuarine systems are especially enhanced in the Patos Lagoon estuary and may have contributed a soft bottom community with wide trophic niches and abbreviated food chains (Bemvenuti, 1997c).

The Quequén Grande River (approximately 38°S and 59°W) that drains a basin of around 7,800 km^{-2}, has its estuarine area located between the cities of Necochea and Quequén, Buenos Aires Province, Argentina. Salinity shows remarkable fluctuations within the estuary (6–26), mainly to tidal cycle but also due to freshwater inflow (López Gapaetor *et al.*, 2001).

The infaunal macrobenthic community of intertidal flats of Quequén Grande River showed a very low species number, being mainly composed of four annelid species: the nereid *L. acuta*, the spionid *Boccardiella ligerica*, the tubificid *Ilyodrilus* cf. *frantzi*, and a species of *Capitella* (Lopéz Gapa *et al.*, 2001). The low biodiversity seems to be a characteristic feature of the brackish water environments of the Buenos Aires Province, since there are just three infaunal polychaetes in Quequén Grande estuary (Lopéz Gapa *et al.*, 2001), four in Samborombon Bay–Rio de la Plata (Ieno and Bastida, 1998), and five in Mar Chiquita coastal lagoon (Olivier *et al.*, 1972).

Bahía Blanca is a mesotidal coastal plain estuary in southwest of the Buenos Aires Province, Argentina. The estuarine area extends over about 2,300 km^{-2}, with extensive tidal flats (1,150 km^{-2}). The mean water temperature is 13°C, while salinity shows drastically differences (17) between the mouth and the head of the estuary (Perillo *et al.*, 2001). In the mudflats the polychaetes *L. acuta* and *Eteone* sp. occupy the lower and middle mesolittoral. The mollusks *H. australis* and *T. plebeius* inhabit the middle and upper mesolittoral. Dense populations of *Chasmagnathus granulata* represent the third association in the upper intertidal of salt marshes and mudflats (Elias, 1985). The grapsid crab *C. granulata* diet in marshes is dominated by pieces of marsh grass, while in the mudflats polychaetes, diatoms, ostracods, and nematodes predominate (Iribarne *et al.*, 1997).

The diverse and rich mudflats in the coastal habitats of South America are fascinating areas for ecological works, but in general, these areas are still not very well known. An evident lack of knowledge exists on the spatial–temporal patterns of the populations and communities, and of estimates of the secondary production of the macrofauna. Besides, it is also necessary to accomplish studies on the main processes that govern the mudflats. The biological interactions and the related physical variables must also be tested through field experiments. It is also strongly advisable that more attention be paid to the processes of low predictability, such as those related to the success of the establishment and the recruitment of invertebrates with pelagic larvae. It is also urgent that the development of long-term studies, which are powerful tools that allow distinction among the antropic effects of the natural effects, reach the variables and the resources of the ecosystems.

Carlos Emilio Bemvenuti

Seagrass beds in South America

Submerged macrophytes in general, and sea grass beds in particular, are recognized as ecologically important features of the coastal zone (Larkum *et al.*, 1989). Seagrasses actively construct and maintain extensive tidal flat structures in South America. Seagrasses provide food for herbivores, a habitat for other organisms, stabilize sediments, reduce or modify water movement and erosion, and present themselves as a substratum for colonization. In all these ways seagrasses add to the biodiversity and productivity of soft sediment environments and protect coastal areas from erosion.

The seagrasses of South America are not well known. Although South America's seagrasses continue to be the subject of some taxonomic debate, at least 10 seagrass species have been reported for the continent. Remarkably, seagrasses are almost absent from the Pacific coast of South America, the only seagrasses on the western continental coast being a couple of small populations of *Heterozostera tasmanica* in northern Chile at Coquimbo (Phillips, 1992). Intriguingly, this species is otherwise known only from Australia and it has been suggested that these are remnants of formerly widely distributed Chilean populations (Phillips, 1992). No seagrasses are known for Peru or Ecuador.

In the Caribbean, on the coasts and islands of Venezuela and Colombia, seagrasses can form very extensive meadows. *Thalassia testudinum*, turtle grass, is probably the most abundant seagrass, although it has not been reported further south. Plants are erect, with shoots up to 1 m, and grow in intertwined turf, forming extensive meadows on shallow sand or mud substrates from the lower intertidal to 20 m (Littler and Littler, 2000). *Syringodium filiforme* (manatee grass) has a similar distribution, the southernmost known populations occurring in Venezuela. This species differs from *Thalassia* in having cylindrical, narrow leaves, which form canopies up to 45 cm high. It grows in sand and mud down to 20 m (Littler and Littler, 2000).

Shoal grass, *Halodule wrightii*, which is common throughout the Caribbean and Brazil, has small supple, grass-like leaves. It is found growing on sand and mud from the intertidal down to 5 m (Littler and Littler, 2000). It has a tropical–subtropical distribution and is found in Colombia and Venezuela and in Brazil from Ceará to Paraná States (Phillips, 1992). The closely related *Halodule emarginata* is endemic to Brazil, occurring from Bahia to São Paulo States (Oliveira *et al.*, 1983). Two other species of *Halodule*, *Halodule lilianeae* and *Halodule brasiliensis*, have been reported as endemic to Brazil but the separation of these three endemic species, based on leaf tip characteristics, has been questioned and some authors consider them forms of *H. wrightii* (Creed, 2000).

Three *Halophila* (sea vine) species, *Halophila baillonii, Halophila engelmannii*, and *Halophila decipiens* grow in finer sands and sediments. *H. decipiens* is found in deepwater (to 30 m) in the southern Caribbean and has a tropical–subtropical distribution in the southwest Atlantic, stretching from the Brazilian State of Ceará to Rio de Janeiro (Oliveira *et al.*, 1983). *H. decipiens* can be found very shallow where turbid conditions exist because of run-off or pollution such as at Guanabara Bay, Rio de Janeiro, Brazil. *H. engelmannii* is a species restricted to the Caribbean, which has been reported at two locations in Venezuela and is only found down to 5 m depth (Littler and Littler, 2000). *H. baillonii*, which grows deeper, is found throughout the Caribbean but has also been reported twice (in 1888 and in the 1980s, though not since) at Itamaracá Island in the northeast of Brazil (Oliveira *et al.*, 1983).

Ruppia maritima, widgeon grass, is found sporadically from Venezuela down to Argentina, where it forms the southernmost populations of seagrasses in the world, at the Magellan Straits (Short *et al.*, 2001). Such records reflect the species' wide latitudinal distribution and tolerance to variable environmental conditions, as it can be found growing in coastal lagoons and estuaries with salinities from 0 to 39. At the Patos Estuarine Lagoon in southern Brazil, a large (about 120 km^2) area of *R. maritima* dominates the benthos and local primary productivity (Seeliger *et al.*, 1997).

An unattached leaf of what was reported as *Zostera* (Setchell and Gardner, 1935) has been found at Montevideo, Uruguay. Phillips (1992) commented that the leaf tip resembled that of *H. tasmanica* but that it was unlikely that it came from so far away as Chile. Seagrasses have not been reported for Guyana or Surinam, where information is very limited, although there is indirect evidence of seagrass beds in French Guyana. It is possible that *T. testudinum, H. wrightii, S. filiforme*, and *Halophila* spp. will be reported in the future, when surveys are carried out.

In South America, seagrasses are most abundant in the Venezuelan Caribbean region, in Colombia and sporadically at specific locations along the coast of Brazil. *Thalassia*, and to a lesser extent *Syringodium*, can be dominant, but in the Caribbean it is common to find *Thalassia* in mixed species stands with *Syringodium* or *Halodule*. *Halophila* species can also be found in monospecific or mixed species stands (Short *et al.*, 2001). In Venezuela, *T. testudinum* is widely distributed on the western, central, and eastern Venezuelan coast as well as around the islands (Vera, 1992). In Sucre State, northeastern Venezuela, *Thalassia* occupies 70% of the Cariaco Gulf, an area of about 290 km^2.

In classic models of Caribbean seagrass succession, *Halodule* (and sometimes *Syringodium*) are considered to be pioneers species (Gallegos *et al.*, 1994). *Halodule* is better able to occupy mobile sediments and facilitates the subsequent growth of *Thalassia* and *Syringodium*. In Brazil, where the competitors *Syringodium* and *Thalassia* are absent, *H. wrightii* is the seagrass that most frequently occurs in shallow waters where it can form extensive monospecific meadows, such as at Itamaracá Island, Pernambuco State (Magalhães and Eskinazi-Leça, 2000).

Of the deeper seagrasses, *H. decipiens* may be of great ecological importance because it may form extensive meadows most of which remain to be discovered. For example, at the Abrolhos Bank, southern Bahia State, Brazil, the suspicion that *Halophila* may be very abundant was recently confirmed in a Rapid Assessment Program (RAP) carried out in the region. Of 45 sites visited, *Halophila* was found at 18 at 5–22 m depth (Figueiredo, personal communication). As these sites were distributed over an area of about 6,000 km^2, the potential importance of *H. decipiens* in the region, especially in terms of primary productivity, could be enormous.

South American seagrasses are often found spatially close by or closely trophically linked to other marine and coastal ecosystems and habitats and this juxtaposition results in higher diversity (Creed, 2000). *Halodule, Syringodium*, and *Thalassia* are associated with shallow habitats without much freshwater input, such as reefs, algal beds, coastal lagoons, rocky shores, sand beaches, and unvegetated soft-bottom areas and nearby mangroves without too much salinity fluctuation. *Halophila* is associated with deeper reefs, algal and marl beds, and deeper soft-bottom vegetated areas. *R. maritima* can be found in lower (coastal lagoon, estuary, fish pond, mangrove, salt marsh, and soft-bottom unvegetated) and higher salinity (coastal lagoon, salt pond, soft-bottom unvegetated) habitats. Numerous physical, chemical, and biological interactions take place between these habitats.

Seagrass beds in South America are known to be important habitat for a wide variety of plants and animals and an enormous diversity of organisms is associated with the South American seagrasses. Groups that contribute most to the richness of seagrass systems in the Caribbean are polychaetes, fishes, amphipods, decapods, foraminifers, gastropods, and bivalve mollusks, macroalgae, and diatoms. Oligochaetes, nematodes, coelenterates, echinoderms, bryozoans, and sponges are other ecologically important groups. About 540 taxa (to genus or species level) of organisms are known to be associated with the Brazilian seagrasses (Creed, 2000). Over 100 macrofaunal and 46 epiphyte floral taxa have been identified in or on *Halodule* at Itamaracá Island, Brazil (Alves, 2000). In Venezuela, at the Caiaco Gulf, 112 mollusk species have been identified in *Thalassia* beds and 127 species of macroinvertebrates are associated with *Thalassia* beds at Mochimba Bay (Vera, 1992). 372 fish species have been reported associated with seagrasses at Bahia de Chengue, Colombia (UNESCO, 1998). Corals are also found in seagrass meadows throughout the Caribbean and Southwest Atlantic. For example, the corals *Meandrina brasiliensis* and *Siderastrea stellata*, the latter being endemic to Brazil, grow unattached in *H. wrightii* beds and are sold as souvenirs locally (Creed, 2000). At Bahia de Chengue, on the Caribbean coast of Colombia, corals of the genera *Manicina, Siderastrea, Millepora, Diaporia, Porites*, and *Cladocora* grow within the *Thalassia* beds (UNESCO, 1998).

Leaves of seagrasses are a substratum for epiphytic algae and invertebrates. Macroalgae in seagrass beds typically consist of members of the orders Dasycladales and Caulerpales, calcified chlorophytes (*Acetabularia, Halimeda, Penicillus, Udotea*) and rhodophytes such as *Acanthophora* and *Hypnea*. Calcareous chlorophytes which grow in the sediment in seagrass beds are major contributors to the sedimentary cycles of tropical South American shallow-water environments. Other macroalgae grow unattached within the bed as "drift" algae.

Seagrasses are extremely palatable to herbivorous fish, such as parrotfish and surgeonfish, and urchins. Seagrasses are not thought to be chemically defended, and their palatability is such that pieces of *Thalassia* have often been used in *in situ* bioassays to determine relative herbivore pressure on reefs. Such studies have elucidated the strong trophic link between coral reefs and seagrass beds in South America, which is best exemplified by grazing "halos." Herbivorous fish that graze seagrass beds abutting them create such halos. Because of predators, the herbivores do not distance themselves greatly from the reef; hence the halo. Sea urchins such as *Lytechinus variegatus* and *Tripneustes ventricosus* are also major herbivores of seagrass blades, consuming *Thalassia* and *Halodule*. Echinoids have a double role in the trophodynamics of the seagrasses. Not only are they major grazers on sea grass but also their feces are a food source for detritivores.

Because of the diversity of species assemblages, seagrass beds have been recognized as among the most productive fisheries areas in the Caribbean and southwestern Atlantic. Most of the living resources are associated directly (as food, habitat, or foraging areas) or indirectly (through export of primary production, larvae and juveniles to other habitats) with seagrass areas. Soft-bottom demersal fisheries exploit scianeids, mullets, sharks, penaeid shrimp, conch, loliginid squid, and octopods over seagrass beds and rough-bottom demersal and coral reef fisheries exploit snappers, groupers, grunts, and lobsters which are also found in seagrass habitat. Inshore pelagic fish may also hunt over the seagrass meadows. Other resources which are seagrass associates are turtles, crabs, oysters, sea urchins, sponges, stony corals, and seaweeds.

While the seagrasses of South America contribute to coastal protection, local productivity, and thus fisheries, there is hardly any information available about the value of seagrasses to the local economy. Economically important fish species such as the bluewing searobin (*Prionotus punctatus*), whitemouth croaker (*Micropogonias furnieri*), and mullet (*Mugil platanus*) are found and fished in Brazilian sea grass beds (Creed, 2000). Local fisheries exploit commercially important crustaceans such as blue crabs (*C. sapidus*), stone crab (*Menippe nodifrons*), lobster (*Panulirus argus* and *Panulirus laevicauda*), and shrimp (*Penaeus brasiliensis* and *Penaeus paulensis*), all of which are associated with seagrass beds. Other shellfish which are commercially collected from seagrass beds are clams (*Anomalocardia brasiliana, Tagelus plebeius, Tivela mactroides*), volutes *Voluta ebraea*, rockshells (*Thais haemastoma*), oysters (*Ostrea puelchana*), and cockles (*Trachycardium muricatum*) (Creed, 2000). In Chile, the Chilean scallop (*Argopecten purpuratus*) preferentially settles in *H. tasmanica* beds (Aguilar and Stotz, 2000).

Two threatened species, which feed directly on seagrasses from the Caribbean to Brazil, are the green turtle *Chelonia mydas* (Creed, 2000) and the West Indian manatee *Trichechus manatus* (Magalhães and Eskinazi-Leça, 2000). Both have benefited from specific conservation action sponsored privately and by the Brazilian Environmental Agency IBAMA (green turtles by the Projeto TAMAR and manatees by the Projeto Peixe-Boi Marinho). The black-necked swan *Cygnus melancoryphus* and the red-gartered coot *Fucila armillata* also feed directly on *R. maritima* in southern Brazil and Argentina but are not endangered (Seeliger *et al.*, 1997). Recently, the semi-aquatic capybara (*Hydrochaerus hydrochaeris*), which is the world's largest rodent, was observed feeding on *R. maritima* near Rio de Janeiro.

As seagrass beds occupy shallow, nearshore depositional environments, they are highly susceptible to damage by human activity. Pollution from land-based sources varies from country to country. Activities related to human settlements, agriculture and industry have been identified as major contributors to the pollutant loads reaching coastal and marine waters. The greatest threats are sewage, hydrocarbons, sediments, nutrients, pesticides, litter and marine debris, and toxic wastes. River loads are enhanced by erosion of watersheds caused by deforestation, urbanization, and agricultural activities. On the continent, the impact of moderate cultural eutrophication on seagrass ecosystem results in greater epiphyte levels and lower shoot density, leaf area and biomass of the seagrass. Sewage pollution has been reported throughout the continent.

Direct reports of impacts on seagrasses on the continent are few. However, pollution by heavy metals from sporadic mining and metal-working activities, by polychlorinated biphenyl congeners and organochlorine compounds and by nutrients from agricultural runoff and sewage discharge have all been reported. Effects of physical damage by anchors and trampling on seagrass and associated macroalgae have also been identified. Loss of water area, because of sediments produced after erosion due to deforestation, infilling for construction and dredging activities have also reduced the area occupied by South America's seagrasses. *R. maritima* has suffered from reduced freshwater inputs because of rice irrigation, population growth and lock construction (Seeliger *et al.*, 1997). It has been estimated that 100% of Chilean and Argentine, and 40% of Brazilian seagrasses are "highly threatened" (Creed, 2002). Thirty-six percent of Brazil's seagrasses are "moderately threatened" and 24% are in "low threat" areas.

There are five research groups currently studying seagrasses in South America (Colombia: Instituto de Investigaciones Marinas y Costeras (INVEMAR), Santa Marta; Venezuela: Instituto de Tecnología y Ciencias Marinas, Universidad Simón Bolívar, Caracas; Brazil: Laboratório de Ecologia Marinha, Universidade do Estado do Rio de Janeiro; Universidade Federal Rural de Pernambuco, Recife; Fundação Universidade Federal do Rio Grande—FURG, Rio Grande). Researchers are studying basic biology and ecology of the sea grasses in their region and participate in important regional or global programs. For example, in Venezuela, seagrasses are being monitored by the Coastal Ecosystem Productivity Network in the Caribbean (CARCOMP) Program at the Parque Nacional Morrocoy, which has been the subject of numerous studies, and at Punta de Mangle on the Isla de Margarida. There is also a CARICOMP site in Colombia at Bahía de Chengue within the Parque Natural Tayrona. These sites are used for comparisons of productivity throughout the Caribbean region (UNESCO, 1998). In Brazil three seagrass beds were recently included in the Global Sea grass Monitoring Network—SeagrassNet. At the Patos Lagoon, Rio Grande do Sul State, Brazil, a site where *Ruppia* beds are widespread and which has been studied for the past 25 years, is included in the Brazilian Long-term Ecological Research Program (PELD). In one way or another, all these programs aim to provide a perspective on long-term change in seagrass growth and productivity at regional or global scales.

Joel C. Creed

Marine mammals of South America

The coastal environments of South America are inhabited by a great diversity of marine mammals, from the orders Sirenia (manatees), Carnivora (otters and pinnipeds), and Cetacea (whales, dolphins, and porpoises). Sirenians and cetaceans have a fully aquatic existence, hardly ever coming ashore intentionally, while otters and pinnipeds leave the water from time to time, specially during the reproductive season. In order to better describe the species that occur along the coast of South America, each order will be dealt separately.

Order Sirenia—manatees

This order is divided in two families, Dugongidae and Trichechidae, and only the latter is represented in South America, by two species: *Trichechus inunguis*, the Amazonian manatee, and *Trichechus manatus*, the West Indian manatee. The Amazonian manatee is restricted to freshwater on the Amazon basin and will not be discussed here. The West Indian manatee inhabit coasts, estuaries, and major rivers from Rhode Island, United States, to Alagoas, Brazil (Reeves *et al.*, 2002). However, the distribution is not continuous, with gaps along this range. Despite the manatees' ability to move thousands of kilometers along continental margins, strong population separations between most locations were observed on the phylogenetic structure of *T. manatus* (Garcia-Rodriguez *et al.*, 1998). This is very important for its conservation, since that although the species has a relatively wide distribution, there is low genetic exchange between locations. Therefore, they should be managed separately. In Colombia this species is considered endangered, with hunting apparently increasing but incidental capture with nets still representing the species' major direct threat (Montoya-Ospina *et al.*, 2001). On the Venezuelan coast, a remnant manatee population exists in Lake Maracaibo, but none was found to occur along the more than 1,500 km of Caribbean coastline (O'Shea *et al.*, 1988). It is considered as critically endangered on the coast of Brazil, being subject to intentional and accidental mortalities and degradation of its habitat. Total populational size on the Brazilian coast is estimated to be approximately 500 animals (IBAMA, 2001).

Sirenians are the only herbivorous aquatic mammals, feeding on a large variety of coastal and freshwater vegetation, including several species of seagrasses, floating freshwater plants (*Hydrilla* and water hyacinths), and even the leaves and shoots of emergent mangroves (Berta and Sumich, 1999). In Brazil, it was found that shoalgrass, *H. wrightii*, is an important item of the manatee diet, being consumed together with many species of algae from the divisions Chlorophyte (*Caulerpa cupressoides, Caulerpa prolifera, Caulerpa racemosa, Caulerpa sertularioides, Halimeda opuntia*, and *Penicillus capitatus*) and Phaeophyte (*Dictiopteris delicata, Dictyota* sp., *Lobophora variegata*, and *Sargassum* sp.) (Cardoso and Picanço, 1998). Since each manatee may consume from 29.5 to 50 kg of plants per day (Würsig *et al.*, 2000), they are probably important in the ecology of seagrass beds.

Order Carnivora

Different from their terrestrial counterparts, aquatic carnivores are well adapted for foraging in water, being excellent swimmers and divers. Along the coasts of South America they are represented by one species from the Mustelidae (marine otter, *Lutra felina*), two of the Phocidae (southern elephant seal, *Mirounga leonina*, and leopard seal, *Hydrurga leptonyx*) and five of the Otariidae family (Juan Fernandez fur seal, *Arctocephalus philippii*; South American fur seal, *Arctocephalus australis*; Galápagos fur seal, *Arctocephalus galapagoensis*; Galápagos sea lion, *Zalophus californianus*; and South American sea lion, *Otaria flavescens*).

The Marine Otter is the only species of the genus *Lutra* that lives exclusively in marine environments. It ranges from the coasts of central Peru south to Cape Horn and along the Atlantic coast of Tierra del Fuego (Reeves *et al.*, 2002). Very little is known about this small otter, due to its secretive nature. It is believed that they feed mainly on shellfish, nearshore marine fish, and freshwater prawns.

The seven species of pinnipeds mentioned above are the ones most commonly found on the South American coasts. Vagrants of other species have been reported, but are not regularly observed. The Juan Fernandez and Galápagos fur seals, the Galápagos Sea Lion and the Leopard Seal have breeding colonies in islands around South America, while the other three species have breeding sites on the coast and are more commonly found in the nearshore environment. The breeding systems of these species are polygynous, with males competing among them for territories in the otariids or for dominance in a hierarchical system in the elephant seal. All species feed on fish (demersal or pelagic), cephalopods (octopuses and squid), and more occasionally crustaceans.

Order Cetacea

This order is divided in two suborders, Mysticeti (baleen whales) and Odontoceti (dolphins, porpoises, and sperm whales). Cetaceans occupy marine and freshwater environments in tropical, temperate, and polar regions. Of the 78 accepted species of cetaceans 50 are found on waters around South America (Jefferson *et al.*, 1993). They range from coastal specimens that do not go into waters deeper than 30 m, such as the Franciscana, *Pontoporia blainvillei*, to truly oceanic, deepwater species that only occasionally come closer to shore as the Sperm Whale, *Physeter macrocephalus*. Only seven species of cetaceans are endemic to South America: the Amazon River Dolphin, *Inia geoffrensis*; Commerson's Dolphin, *Cephalorhynchus commersonii*; the Chilean Dolphin, *Cephalorhynchus eutropia*; Tucuxi, *Sotalia fluviatilis*; the Estuarine Dolphin, *Sotalia guianensis*; Peale's Dolphin, *Lagenorhynchus australis*; and Burmeister's Porpoise, *Phocoena spinipinnis*.

They occupy different trophic levels, with whales being trophically more basal, consuming zooplankton, and most odontocetes being top predators consuming fish, cephalopods and even other marine mammals, as in the case of Killer Whales, *Orcinus orca*. Considering their large size and abundance, they probably play an important role in the structuring of marine communities. Even allowing for the unpredictability of strandings, cetacean carcasses are an important food source for terrestrial and benthic scavengers (Katona and Whitehead, 1988).

The occurrence of baleen whales on the coast of South America is restricted to the second semester of the year, during the austral winter. Humpback and Southern Right Whales (*Megaptera novaeangliae* and *Eubalaena australis*, respectively) use coastal areas and islands as breeding grounds, but do not feed while there. Thus their impact on local ecosystems is relatively low, apart from being a food source for other top predators (e.g., sharks) or scavengers when dead. However, seagulls have been observed feeding on skin from the back of Right Whales when they are on the breeding grounds (Rowntree *et al.*, 1998; Groch, 2001). On the other hand, odontocetes are present all year round and have important roles in the structuring of aquatic food chains, consuming a great variety of organisms. As a rule they are opportunistic, consuming the resource that is most available, including pelagic and demersal fish, cephalopods (octopuses and squids). However, some species are highly specialized, as the Sperm and Beaked Whales (families Physeteridae and Ziphiidae), which feed almost exclusively on squids.

André S. Barreto

Bibliography

Aguilar, M., and Stotz, W.B., 2000. Settlement of juvenile scallops *Argopecten purpuratus* (Lamarck, 1819) in the subtidal zone at Puerto Aldea, Tongoy Bay, Chile. *Journal Shellfish Research*, **19**: 749–755.

Alvarado, J.L., Pinto, R., Marquet, P., Pacheco, C., Guiñez, R., and Castilla, J.C., 2001. Patch recolonization by the tunicate *Pyura praeputilis* in the rocky intertidal of the Bay of Antofagasta, Chile: evidence for self-facilitation mechanisms. *Marine Ecology Progress Series*, **224**: 93–101.

Alvarez-Leon, R., 1993. Mangrove ecosystems of Colombia: utilization, impacts conservation and recuperation. In Lacerda, L.D., and Field, C.D. (eds.), *Proceedings of a Workshop on Conservation and Sustainable Utilization of Mangrove Forests in Latin America and Africa Regions*. ITTO/ISME Project PD114/90(F), **1**: 9–10.

Alves, M.S., 2000. Fauna associada aos prados de Halodule wrightii Aschers. In Barros, H.M., Eskinazi-Leça, E., Macedo, S.J., and Lima, T. (eds.), *Gerenciamento Participativo de Estuários e Manguezais*. Recife, Brazil: Universitária da UFPE, pp. 75–87.

Apolinário, M., Coutinho, R., and Baeta-Neves, M.H., 1999. Periwinkle (Gastropoda: Littorinidae) habitat selection and its impact upon microalgal populations. *Revista Brasileira Biologia*, **59**(2): 211–218.

Arntz, W.E., Brey, T., Tarazona, J., and Robles, A., 1987. Changes in the structure of a shallow sandy-beach community in Peru during an El Niño event. In Payne, A.I., Gulland, J.A., and Bink, K.H. (eds.), *The Benguela and Comparable Ecosystems. South African Journal of Marine Science*, **5**: 645–658.

Asmus, M.L., 1997. Coastal plain and Patos lagoon. In Seeliger, U.C., Odebrecht, C., and Castello, J. (eds.), Subtropical convergence marine ecosystem. The coast and the sea in the warm temperate southwestern Atlantic. Berlin Heidelberg: Springer Verlag, pp. 9–12.

Bacon, P., 1993. Conservation and utilization of mangrove forests in Trinidad and Tobago and Lesser Antilhes. In Lacerda, L.D., and Field, C.D. (eds.), *Proceedings of a Workshop on Conservation and sustainable utilization of mangrove forests in Latin America and Africa Regions*. ITTO/ISME Project PD114/90(F), **1**: 5–7.

Battaström, H., 1980. Rocky-shore zonation in the Santa Marta area, Colombia. *Sarsia*, **65**: 163–226.

Bemvenuti, C.E., 1997a. Benthic invertebrates. In Seeliger, U.C., Odebrecht, C., and Castello, J. (eds.), *Subtropical Convergence Marine Ecosystem. The Coast and the Sea in the Warm Temperate Southwestern Atlantic*. Berlin Heidelberg: Springer Verlag, pp. 43–46.

Bemvenuti, C.E., 1997b. Unvegetated intertidal flats and subtidal bottoms. In Seeliger, U.C., Odebrecht, C., and Castello, J. (eds.), *Subtropical Convergence Marine Ecosystem. The Coast and the Sea in the Warm Temperate Southwestern Atlantic*. Berlin Heidelberg: Springer Verlag, pp. 78–82.

Bemvenuti, C.E., 1997c. Trophic structure. In Seeliger, U.C., Odebrecht, C., and Castello, J., (eds.), *Subtropical Convergence Marine Ecosystem. The Coast and the Sea in the Warm Temperate Southwestern Atlantic*. Berlin Heidelberg: Springer Verlag, pp. 70–73.

Bemvenuti, C.E., and Netto, S.A., 1998. Distribution and seasonal patterns of the sublittoral benthic macrofauna of Patos Lagoon (South Brazil). *Revista Brasileria de Biologia*, **58**(2): 211–221.

Berta, A., and Sumich, J.L., 1999. *Marine Mammals—Evolutionary Biology*. San Diego, CA: Academic Press.

Borzone, C.A., Souza, J.R.B., and Soares, A.G., 1996. Morphodynamic influence on the structure of inter and subtidal macrofaunal communities of subtropical sandy beaches. *Revista Chilenia de Historia Natural*, **69**: 565–577.

Brazeiro, A., and Defeo, O., 1996. Macroinfauna zonation in microtidal sandy beaches: is it possible to identify patterns in such variable environments? *Estuarine Coastal and Shelf Science*, **42**: 523–536.

Brazeiro, A., and Defeo, O., 1999. Effects of harvesting and density-dependence on the demography of sandy beach populations: the yellow clam Mesodesma mactroides of Uruguay. *Marine Ecology Progress Series*, **182**: 127–135.

Calliari, L.J., Klein, A.H.F., and Barros, F.C.R., 1996. Beach differentiation along the Rio Grande do Sul coastline (Southern Brazil). *Revista Chilenia de Historia Natural*, **69**: 485–493.

Cantera, J.R., and Blanco, J.F., 2001. The estuary ecosystem of Buenaventura Bay, Colombia. In Seeliger, U.C., and Kjerfve, B. (eds.), *Coastal Marine Ecosystems of Latin America, Ecological Studies 144*. Berlin Heidelberg: Spriger Verlag, pp. 265–280.

Cardoso, E., and Picanço, M., 1998. Descricão da alimentacão natural fornecida aos peixes-boi (Trichechus manatus, Linnaeus,1758) em cativeiro do centro peixe-boi, Itamaracá/PE, in 8a Reunião de Trabalho de Especialistas em Mamíferos Aquáticos da América do Sul, Abstracts.

Castilla, J.C., 1983. Environmental impact in sandy beaches of copper mine tailings at Chañaral, Chile. *Marine Pollution Bulletin*, **14**: 459–464.

Castilla, J.C., 1999. Coastal marine communities: trends and perspectives from human-exclusion experiments. *Trends in Ecology and Evolution*, **14**(7): 280–283.

Castilla, J.C., and Defeo, O., 2001. Latin-American benthic shellfisheries: emphasis on co-management and experimental practices. *Reviews in Fish Biology and Fisheries*, **11**: 11–30.

Castro, C.B., and Pires, D.O., 2001. Brazilian coral reefs: what we already know and what is still missing. *Bulletin of Marine Science*, **69**(2): 357–371.

Cintron, G., and Schaeffer-Novelli, Y., 1983. Introduccion a la Ecologia del Manglar. Montevideo, Uruguay: ROSTLAC—UNESCO.

Cintron, G., and Schaeffer-Novelli, Y., 1992. Ecology and management of New World mangroves. In Seeleger, U. (ed.), *Coastal Plant Communities of Latin America*. New York: Academic Press.

Conde, J.E., 2001. The Orinoco River Delta, Venezuela. In Seeliger, U.C., and Kjerfve, B. (eds.), *Coastal Marine Ecosystems of Latin America Ecological Studies 144*. Berlin Heidelberg: Spriger Verlag, pp. 61–70.

Cordeiro-Marino, M., Braga, A., Eston, V.R., Fujii, M.T., and Yokoya, N.S., 1992. Mangrove macroalgal communities of Latin America: the state of art and perspectives. In Seeliger, U. (ed.), *Coastal Plant Communities of Latin America*. New York: Academic Press, pp. 51–64.

Creed, J.C., 2000. The biodiversity of Brazil's sea grass and sea grass habitats: a first analysis. *Biol. Mar. Medit*, **7**: 207–210.

Creed, J.C., 2002. Biodiversidade de gramas marinhas brasileiras. In Araújo, E.L., Moura, A.N., Sampaio, E.S.B., Gestinari, L.M.S., Carneiro, J.M.T. (eds.), Biodiversidade, Conservação e Uso Sustentável da Flora do Brasil. Recife, Brazil: UFRPE/Imprensa Universitária, pp. 127–131.

Cunha, S.R., 2001. Estrutura e produção das comunidades de Macroprodutores dos manguezais da Baía da Babitonga, SC. Ph.D. thesis. Universidade do Rio Grande.

Defeo, O., 1996a. Experimental management of an exploited sandy beach bivalve population. *Revista Chilenia de Historia Natural*, **69**: 605–614.

Defeo, O., 1996b. Recruitment variability in sandy beach macroinfauna: much to learn yet. *Revista Chilenia de Historia Natural*, **69**: 615–630.

Defeo, O., 1998. Testing hypotheses on recruitment, growth and mortality in exploited bivalves: an experimental perspective. *Canadian Special Publication of Fisheries and Aquatic Science*, **125**: 257–264.

Defeo, O., and Cardoso, R.S., 2002. Macroecology of population dynamics and life history traits of the mole crab *Emerita brasiliensis* in Atlantic sandy beaches of South America. *Marine Ecology Progress Series*.

Defeo, O., and de Alava, A., 1995. Effects of human activities on long-term trends in sandy beach populations: the wedge clam *Donax hanleyanus* in Uruguay. *Marine Ecology Progress Series*, **123**: 73–82.

Defeo, O., and Rueda, M., 2002. Spatial structure, sampling design and abundance estimates in sandy beach macroinfauna: some warnings and new perspectives. *Marine Biology*, **140**: 1215–1225.

Defeo, O., and Scarabino, V., 1990. Ecological significance of a possible deposit-feeding strategy in *Mesodesma mactroides* (Deshayes, 1854) (Mollusca: Pelecypoda). *Atlântica* (Brasil), **12**: 55–66.

Defeo, O., Brazeiro, A., de Alava, A., and Riestra, G., 1997. Is sandy beach macroinfauna only physically controlled? Role of substrate and competition in isopods. *Estuarine, Coastal and Shelf Science*, **45**: 453–462.

Defeo, O., Gómez, J., and Lercari, D., 2001. Testing the swash exclusion hypothesis in sandy beach populations: the mole crab *Emerita brasiliensis* in Uruguay. *Marine Ecology Progress Series*, **212**: 159–170.

Defeo, O., Jaramillo, E., and Lyonnet, A., 1992. Community structure and intertidal zonation of the macroinfauna in the Atlantic coast of Uruguay. *Journal of Coastal Research*, **8**: 830–839.

Denadai, M.R., Amaral, A.C.Z., and Turra, A., 2001. Spatial distribution of molluscs on sandy intertidal substrates with rock fragments in south-eastern Brazil. *Estuarine, Coastal and Shelf Science*, **53**: 733–743.

Dexter, D.M., 1974. Sandy-beach fauna of the Pacific and Atlantic coasts of Costa Rica and Colombia. *Revista de Biologia Tropical*, **22**: 51–66.

Dexter, D.M., 1992. Sandy beach community structure: the role of exposure and latitude. *Journal of Biogeography*, **19**: 59–66.

Dominguez, J.M.L., Bittencourt, A.C.S., Leão, Z.M.N., and Azevedo, A.E.G., 1990. Geologia do Quaternário costal de Pernambuco. *Revista Brasileira de Geociencias*, **20**: 208–215.

Echeverria, J., 1993. Mangrove ecosystems of Zarumilla-Tumbes, Peru Northern. In Lacerda, L.D., and Field, C.D. (eds.), *Proceedings of a Workshop on Conservation and Sustainable Utilization of Mangrove Forests in Latin America and Africa Regions*. ITTO/ISME Project PD114/90(F), **1**: 11–12.

Elias, R., 1985. Macrobentos del estuario de la Bahia Blanca (Argentina). I. Mesolitoral. *Spheniscus*, **1**: 1–33.

Escofet, A., Gianuca, N., Maytía, S., and Scarabino, V., 1979. Playas arenosas del Atlántico Sudoccidental entre los 29° y 43° LS.: consideraciones generales y esquema biocenológico. Memorias Seminario sobre Ecología Bentónica y Sedimentación de la Plataforma Continental del Atlántico Sur. Montevideo: UNESCO, pp. 245–258.

Eston, V.R., and Bussab, W.O., 1990. An experimental analysis of ecological dominance in a rocky subtidal macroalgal community. *Journal of Experimental Marine Biology and Ecology*, **136**: 179–195.

Gallegos, M.E., Merino, M., Rodriguez, A., Marbá, N., and Duarte, C.M., 1994. Growth patterns and demography of pioneer Caribbean sea grasses *Halodule wrightii* and *Syringodium filiforme. Marine Ecology Progress Series*, **109**: 99–104.

Garcia, C.A.E. 1997. Hydrographic characteristics. In Seeliger, U.C., Odebrecht, C., and Castello, J. (eds.), *Subtropical Convergence Marine Ecosystem. The Coast and the Sea in the Warm Temperate Southwestern Atlantic*. Berlin Heidelberg: Springer Verlag, pp. 18–20.

Garcia-Rodriguez, A.I., Bowen, B.W., Domning, D., Mignucci-Giannoni, A.A., Marmontel, M., Montoya-Ospina, R.A., Morales-Vela, B., Rudin, M., Bonde, R.K., and McGuire, P.M., 1998. Phylogeography of the West Indian manatee (Trichechus manatus): how many populations and how many taxa? *Molecular Ecology*, **7**: 1137–1149.

Garzon-Ferreira, J., Cortes, J., Croquer, A., Guzman, H., Leão, Z., and Rodriguez-Ramirez, A., 2000. Status of coral reefs in southern tropical America: Brazil, Colombia, Costa-Rica, Panama and Venezuela. In Wilkinson, C. (ed.), *Status of Coral Reefs of the World: 2000*. Cape Ferguson: Australian Institute of Marine Science.

Gianuca, N., 1983. A preliminary account of the ecology of sandy beaches in southern Brazil. In McLachlan, A., and Erasmus, T. (eds.), *Sandy Beaches as Ecosystems*. The Hague: W. Junk.

Giménez, L., and Yannicelli, B., 2000. Longshore patterns of distribution of macroinfauna on a Uruguayan sandy beach: an analysis at different spatial scales and of their potential causes. *Marine Ecology Progress Seriers*, **199**: 111–125.

Gómez, J., and Defeo, O., 1999. Life history of the sandhopper *Pseudorchestoidea brasiliensis* (Amphipoda) in sandy beaches with contrasting morphodynamics. *Marine Ecology Progress Series*, **182**: 209–220.

Groch, K.R., 2001. Cases of harassment by kelp gulls (*Larus dominicanus*) on right whales (*Eubalaena australis*) of southern Brazil. *Biotemas*, **14**: 147–156.

Guiler, E.R., 1959. Intertidal belt-forming species on rocky coasts of northern Chile. *Procedings of Royal Society of Tasmania*, **93**: 33–58.

Hajdu, E.M., Muricy, G., Berlinck, R.G.S., and Freitas, J.C., 1996. Marine poriferan diversity in Brazil: knowledge and management. In Bicudo, C.E.M., and Menezes, N.A. (eds.), *Biodiversity in Brazil: A First Approach*. São Paulo: CNPq.

Hamilton, L.S., and Snedaker, S.C., 1984. *Handbook for Mangrove Area Management*. Honolulu: United Nations Environment Programme and East West Center, Environmental and Policy Institute.

Herz, R., 1991. *Manguezais do Brasil*. São Paulo: Univervidade de São Paulo.

IBAMA, 2001. Mamíferos aquáticos do Brasil—Plano de ação, Versão II. Brasília: Instituto Brasileiro do Meio Ambiente e Recursos Naturais Renováveis.

Ieno, E.N., and Bastida, R.O., 1998. Spatial and temporal patterns in coastal macrobenthos of Samborombón Bay, Argentina: a case study of very low diversity. *Estuaries*, **21**: 690–699.

Iribarne, O., Bortolus, A., and Botto, F., 1997. Between-habitat differences in burrow characteristics and trophic modes in the southwestern Atlantic burrowing crab *Chasmagnathus granulata*. *Marine Ecology Progress Series*, **155**: 137–145.

Jara, H.F., and Moreno, C.A., 1984. Herbivory and structure in a midlittoral rocky community: a case in southern Chile. *Ecology*, **65**(1): 28–38.

Jaramillo, E., 1994. Patterns of species richness in sandy beaches of South America. *South African Journal of Zoology*, **29**: 227–234.

Jaramillo, E., 2001. The sandy beach ecosystem of Chile. In Seeliger, U., and Kjerfve, B. (eds.), *Coastal Marine Ecosystems of Latin America*. Berlin: Springer-Verlag, pp. 219–227.

Jefferson, T.A., Leatherwood, S., and Webber, M.A., 1993. FAO species identification guide. Marine mammals of the world. Rome: FAO.

Joly, A.B., 1951. Contribuição para o conhecimento da flora algológica marinha do Estado do Paranã. *Boletim do Instituto. Paulista de Oceanografia*, **2**(1): 125–138.

Joly, A.B., 1957. Contribuição para o conhecimento da flora ficológica marinha da baía de Santos e arredores. *Boletins da Faculdade de Filosofia, Ciências e Letras*. **14**: 1–169.

Katona, S., and Whitehead, H., 1988. Are Cetacea ecologically important? *Oceanography and Marine Biology: An Annual Review*, **26**: 553–568.

Kellogg, J.N., and Mohriak, W.U., 2001. The tectonic and geological environment of coastal South America. In Seeliger, U., and Kjerfve, B. (eds.), *Coastal Marine Ecosystems of Latin America*. Berlin: Springer-Verlag, pp. 1–16.

Kjerfve, B., and Lacerda, L.D., 1993. Mangroves of Brazil. In Lacerda, L.D. (ed.), *Conservation and Sustainable Utilization of Mangrove Forests in Latin America and Africa regions*. ITTO/ISME PD 114/90(F), pp. 245–272.

Laborel, J., 1969. Madréporaries et hydrocoralliaires recifaux des bresiliennes: systematique, ecologie, repartition verticale et geogrphie. *Annales de Institut Oceanographique*, Paris, **47**: 171–229 (in French).

Lacerda, L.D., and Schaeffer-Novelli, Y., 1992. Mangroves of Latin America: the need for conservation and sustainable utilization. *Mangroves Newsletter*, **5**: 4–6.

Lacerda, L.D., Conde, J.E., Alarcon, C., Alvarez-Leon, R., Bacon, P.R., D' Croz, L., Kjerfve, B., Polaina, J., and Vannuci, M., 1993. Mangrove ecosystems of Latin America and the Caribbean: a summary. In Lacerda, L.D. (ed.), *Conservation and Sustainable Utilization of Mangrove Forests in Latin America and Africa regions*. ITTO/ISME PD 114/90(F), pp. 1–42.

Lana, P.C., 1986. Macrofauna bêntica de fundos sublitorais não consolidados da Baía de Paranaguá (Paraná). *Nerítica*, **1**(3): 79–89.

Lana, P.C., Guiss, C., Disaró, S.T., 1991. Seasonal variation of biomass and production dynamics for above- and belowground components of a Spartina alterniflora marsh in a euhaline sector of Paranaguá Bay (SE Brazil). *Estuarine Coast Shelf Science*, **32**: 231–241.

Lana, P.C., Marone, E., Lopes, R.M., and Machado, E.C., 2001. The subtropical estuarine complex of Paranaguá Bay, Brazil. In Seeliger, U.C., and Kjerfve, B. (eds.), *Coastal Marine Ecosystems of Latin America, Ecological Studies 144*, Berlin Heidelberg: Spriger Verlag, pp. 131–145.

Larkum, A.W.D., McComb, A.J., and Shepherd, S.A. (eds.), 1989. *Biology of Seagrasses*. New York: Elsevier.

Leão, Z.M.N., Araujo, T.M.F., and Nolasco, M.C., 1988. Os recifes de coral da costa setentrional do Brasil. *Proceedings of Simpósio Internacional de Recifes de Coral*, **3**: 339–348.

Lercari, D., and Defeo, O., 1999. Effects of freshwater discharge in sandy beach populations: the mole crab *Emerita brasiliensis* in Uruguay. *Estuarine Coastal Shelf Science*, **49**: 457–468.

Lima, M., Brazeiro, A., and Defeo, O., 2000. Dynamics of a yellow clam (*Mesodesma mactroides*) population: recruitment variability, density-dependence and stochastic processes. *Marine Ecology Progress Series*, **207**: 97–108.

Little, C., 2000. *The Biology of Soft Shores and Estuaries*. New York: Oxford University Press.

Littler, D.S., and Littler, M.S., 2000. *Caribbean Reef Plants*. Washington, DC: Offshore Graphics.

Lopez Gappa, J.J., Tablado, A., and Magaldi, N.H., 1990. Influence of sewage pollution on a rocky intertidal community dominated by the mytilid *Brachidontes rodriguezi*. *Marine Ecology Progress Series*, **63**: 163–175.

López Gappa, J., Tablado, A., Fonalleras, M.C., and Adami, M.L., 2001. Temporal and spatial patterns of annelid populations in intertidal sediments of the Quequén Grande estuary (Argentina). *Hydrobiologia*, **455**: 61–69.

Lugo, A.E., and Snedaker, S.C., 1974. The ecology of mangroves. *Annual Review of Ecology and Systematics*, **5**: 39–64.

Magalhães, K.M., and Eskinazi-Leça, E., 2000. Os prados de fanerógamas marinhas. In Barros, H.M., Eskinazi-Leça, E., Macedo, S.J., and Lima, T. (eds.), *Gerenciamento participativo de estuários e manguezais*. Recife, Brazil: Universitária da UFPE, pp. 39–47.

Maida, M., and Ferreira, B.P., 1997. Coral reefs of Brazil: Overview e field guide. *Proceedings 8th International Coral Reef Symposium*, **1**: 263–274.

McLachlan, A., 1980. Exposed sandy beaches as semi-closed ecosystems. *Marine Environmental Research*, **4**: 59–63.

Medeiros, C., Kjerfve, B., Araujo, M., and Neumann-Leitão, S., 2001. The Itamaracá Estuarine Ecosystem, Brazil. In Seeliger, U.C., and Kjerfve, B. (eds.), *Coastal Marine Ecosystems of Latin America*. Ecological Studies 144, pp. 71–81.

Méndez, S., Severov, D., Ferrari, G., and Mesones, C., 1996. Early spring Alexandrium tamarense toxic blooms in uruguayan waters. *Proceedings 7th International Conference on Toxic Phytoplankton*. Paris: IOC, UNESCO, pp. 113–116.

Monteiro-Neto, C., and Cunha, L.P.R., 1990. Seasonal and ontogenetic variation in food habits of juvenile *Trachinotus marginatus* Cuvier, 1832 (Teleostei, Carangidae) in the surf zone of Cssino Beach, RS, Brazil: *Atlântica* (Brasil), **12**: 45–54.

Montoya-Ospina, R.A., Caicedo-Herrera, D., Millan-Sanchez, S.L., Mignucci-Giannoni, A.A., and Lefebvre, L.W., 2001. Status and distribution of the West Indian manatee, *Trichechus manatus manatus*, in Colombia. *Biological Conservation*, **102**: 117–129.

Moreno, T.R., and Rocha, R.M., 2001. The Biological substratum Eudistoma carolinense Van Name, 1945 in the Beach Itapema do Norte, Santa Catarina, Brazil. In Sawada, H., Yokosawa, H., and Lambert, C.C. (eds.), *The Biology of Ascidians*. Tokyo: Springer Verlag, pp. 271–277.

Niencheski, L.F., and Baumgarten, M.G., 1997. Environmental chemistry. In Seeliger, U.C., Odebrecht, C., and Castello, J. (eds.), *Subtropical Convergence Marine Ecosystem. The Coast and the Sea in the Warm Temperate Southwestern Atlantic*. Berlin Heidelberg: Springer Verlag, pp. 20–23.

Nonato, E., and Pérès, J.M., 1961. Observation sur quelques peuplements intertidaux de substract dur dans la région d'Ubatuba (État S. Paulo). *Cahiers de Biologie Marine*, **2**: 263–270.

Nucci, P.R., Turra, A., and Morgado, E.H., 2001. Diversity and distribution of crustaceans from 13 sheltered sandy beaches along São Sebastião Channel, south-eastern Brazil. *Journal of Marine Biological Association UK*, **81**: 475–484.

Odebrecht, C., Segatto, A.Z., and Freitas, C.A., 1995. Surf-zone chlorophyll a variability at Cassino Beach, southern Brazil. *Estuarine Coastal Shelf Science*, **41**: 81–90.

Ojeda, F.P., and Santelices, B., 1984. Ecological dominance of *Lessonia nigrescens* (Phaeophyta) in central Chile. *Marine Ecology Progress Series*, **19**: 83–91.

Oliveira, E.C., Pirani, J.R., and Giulietti, A.M., 1983. The Brazilian seagrasses. *Aquatic Botany*, **16**: 251–267.

Oliveira, L.P.H., 1947. Distribuição geográfica da fauna e da flora da Baía de Guanabara. *Memorias do Instituto Oswaldo Cruz*, **45**(3): 709–735.

Oliveira, L.P.H. 1950. Levantamento biogeográfico da Baía de Guanabara. *Memorias Instituto Oswaldo Cruz*, **48**: 363–391.

Oliveira-Filho, E.C., and Paula, E.J., 1983. Aspectos da distribuição vertical e variação sazonal de comunidades da zona das marés em costões rochosos do litoral norte do estado de São Paulo. I Encontro de macrófitas marinhas. *Instituto de Pesquisas da Marinha Publicacao*. **147**: 44–71.

Olivier, S.R., Escofet, A., Penchaszadeh, P., and Orensanz, J.M., 1972. Estudios ecológicos de la región estuarial de Mar Chiquita. Na. Com. Inv. Cient. (Buenos Aires, Argentina), I-Las comunidades bentónicas. *Anales de la Sociedad Cientifica Argentina*, **93**: 237–262.

Olivier, S.R., Paternoster, I.K., and Bastida, R., 1966. Estudios biocenóticos en las costas de Chubut (Argentina). I. Zonación biocenológica de Puerto Pardelas (Golfo Nuevo). *Bol. Inst. Biol. Mar.*, **10**: 3–74.

O'-Shea, T.J., Correa-Viana, M., Ludlow, M.E., and Robinson, J.G., 1988. Distribution, status, and traditional significance of the West Indian manatee *Trichechus manatus* in Venezuela. *Biological Conservation*, **46**: 281–301.

Paula, E.J., and Eston, V.R., 1989. Secondary succession on an exposed rocky intertidal algal community of the State of São Paulo, Brazil. Bolm Botânica, Univ. S. Paulo, **11**: 1–9.

Peña, E.J., 1998. Physiological ecology of mangrove-associated macroalgae in a tropical estuary. Ph.D. thesis. Columbia: University of South Carolina.

Penchaszadeh, P., 1971. Observaciones cuantitativas preliminaries en playas arenosas de la costa central del Perú, con especial referencia a las poblaciones de muy-muy (Emerita analoga)—(Crustacea, Anomura, Hippidae). UNESCO América Latina, DOCIN 6.

Penchaszadeh, P., De Mahieu, G., Farache, V., and Lera, M.E., 1983. Ecology of the sandy beach gastropod *Mazatlania aciculate* in Quizandal (Carabobo, Venezuela). In McLachlan, A., and Erasmus, T. (eds.), *Sandy Beaches as Ecosystems*. The Hague: W. Junk, pp. 655–660.

Perillo, G.M.E., Piccolo, M.C., Parodi, E., and Freije, R.H., 2001. The Bahia Blanca Estuary, Argentina. In Seeliger, U.C., and Kjerfve, B. (eds.), *Coastal Marine Ecosystems of Latin America, Ecological Studies 144*. Berlin Heidelberg: Spriger Verlag, pp. 205–217.

Phillips, R.C., 1992. The seagrass ecosystem and resources in Latin America. In Seeliger, U. (ed.), *Coastal Plant Communities of Latin America*. San Diego, CA: Academic Press, pp. 107–121.

Pickard, G.L., and Emery, W.J., 1990. *Descriptive Physical Oceanography, An Introduction*, Oxford, U.K.: 5th edn. Pergamon Press.

Polania, J., Santos-Martínez, A., Mancera-Pineda, J.E., and Botero Arboleda, L., 2001. The coastal lagoon Ciénaga Grande de Santa Marta, Colombia. In Seeliger, U.C. and Kjerfve, B. (eds.), *Coastal Marine Ecosystems of Latin America, Ecological Studies 144*. Berlin Heidelberg: Spriger Verlag, pp. 33–45.

Reeves, R.R., Stewart, B.S., Clapham, P.J., and Powell, J.A., 2002. *National Audubon Society Guide to Marine Mammals of the World*. New York: Alfred A. Knopff.

Reise, K., 1985. *Tidal Flat Ecology. An Experimental Approach to Species Interactions*. Berlin Heidelberg: Springer-Verlag.

Rodrigues, S. de A., Rocha, R.M., and Lotufo, T.M.C., 1998. Guia Ilustrado Para Identificação das Ascídias do Estado de São Paulo, São Paulo, Fapesp, Instituto de Biociências—USP.

Rodriguez, C., and Stoner, A.W., 1990. The epiphyte community of mangrove roots in a tropical estuary: distribution and biomass. *Aquatic. Botany*, **36**: 117–126.

Rodríguez, G., 2001. The Maracaibo system, Venezuela. In Seeliger, U.C., and Kjerfve, B. (eds.), *Coastal Marine Ecosystems of Latin America, Ecological Studies 144*. Berlin Heidelberg: Springer Verlag, pp. 47–60.

Rowntree, V., MacGuiness, P., Marshall, K., Payne, R., Sironi, M., and Seger, J., 1998. Increased harassment of right whales (*Eubalaena australis*) by kelp gulls (*Larus dominicanus*) at Península Valdés, Argentina. *Marine Mammal Science*, **14**: 99–115.

Saenger, P., Heagerl, E.J., and Davie, J.D.S., 1983. Global status of mangrove ecosystems. *Environmentalist*, **3**(Suppl. 3): 1–88.

Santelices, B., 1991. Littoral and Sublittoral communities of continental Chile, Chapter 14. In Mathieson, A.C., and Nienhuis, P.H. (eds.), *Intertidal and Littoral Ecosystems*, Ser. Ecosystems of the World 24. Amsterdam: Elsevier, pp. 347–369.

Santelices, B., and Ojeda, F.P., 1984. Effects of canopy removal on the understory algal community structure of coastal forests of *Macrocystis pyrifera* from southern South America. *Marine Ecology Progress Series*, **14**: 165–173.

Schaeffer-Novelli, Y., 1993. Brazilian mangroves. In Lacerda, L.D., and Field, C.D. (eds.), *Proceedings of a Workshop on Conservation and Sustainable Utilization of Mangrove Forests in Latin America and Africa Regions*. ITTO/ISME Project PD114/90(F), **1**: 12–13.

Schaeffer-Novelli, Y., Cintron, G., Adaime, R.R., and Camargo, T.M., 1990. Variability of mangrove ecosystems along the Brazilian coast. *Estuaries*, **13**: 204–218.

Seeliger, U., Odebrecht, C., and Castello, J.P. (eds.), 1997. *Subtropical Convergence Environments: The Coast and Sea of the Southwestern Atlantic*. Berlin: Springer Verlag.

Setchell, W.A., and Gardner, N.L., 1935. An occurence of Zostera on the east coast of South America. *Revista sudamericana de botanica*, **2**: 15–17.

Short, A., 1996. The role of wave height, period, slope, tide range and embaymentisation in beach classifications: a review. *Revista Chilenia de Historia Natural*, **69**: 589–604.

Short, F.T., Coles, R.G., and Pergent-Martini, C., 2001. Global seagrass distribution. In Short, F.T., and Coles, R.G. (eds.), *Global Seagrass Research Methods*. Amsterdam: Elsevier Science, pp. 5–30.

Steinke, T.D., and Naidoo, Y., 1990. Biomass of algae epiphytic on pneumatophores of the mangrove, Avicennia marina, in the St. Lucia estuary. *South African Journal Botany*, **56**(2): 226–232.

Stephenson, T.A., and Stephenson, A., 1972. *Life between Tidemarks on Rocky Shores*. San Francisco, CA: W.H. Freeman and Co.

Swennen, C., and Duiven, P., 1982. Numerical density and biomass of macrobenthic animals living in the intertidal zone of Surinam, South America. *Netherlands Journal of Sea Research*, **15**: 406–418.

Tanaka, M.O., and Duarte, L.F.L., 1998. Recruitment variation of the barnacle *Chthamalus bisinuatus* Pilsbry, 1916 in an exposed rocky shore in Southeast Brazil. *Bulletin of Marine Science*, **62**(1): 285–292.

Tarazona, J., and Arntz, W., 2001. The Peruvian Coastal Upwelling System. Chapter 16. In Seeliger, U., and Kjerfve, B. (eds.), *Coastal Marine Ecosystems of Latin America. Ecologial Studies 144*. Berlin, Heidelberg: Springer-Verlag, pp. 229–244.

Tarazona, J., Salzwedel, H., and Arntz, W., 1988. Oscillations of macrobenthos in shallow water of the Peruvian central coast induced by El Niño 1982–83. *Journal of Marine Research*, **46**: 593–611.

Tognella-De-Rosa, M.M.P., 2000. Manguezais catarinenses, Baía da Babitonga e Rio Tavares: Uma abordagem parcial dos atributos ecológicos e econômicos. Ph.D. thesis, Instituto de Oceanografia, Universidade de São Paulo.

Tomlinson, P.B., 1986. *The Botany of Mangroves*. Cambridge: Cambridge University Press.

Tundisi, J.G., and Matsumura-Tundisi, T., 2001. The lagoon region and estuary ecosystem of Cananéia, Brazil. In Seeliger, U.C., and Kjerfve, B. (eds.), *Coastal Marine Ecosystems of Latin America. Ecological Studies 144*, Berlin Heidelberg: Spriger Verlag, pp. 119–130.

Twilley, R.R., Pozo, M., Garcia, V.H., Rivera-Monroy, V.H., Zambrano, R., and Bodero, A., 1997. Litter dynamics in riverine mangrove forests in the Guayas River estuary, Ecuador. *Oecologia*, **111**: 109–122.

Twilley, R.R., Cárdenas, W., Rivera-Monroy, V.H., Espinoza, J., Suescum, R., Armijos, M.M., and Solórzano, L., 2001. The Gulf of Guayaquil and the Guayas River estuary, Ecuador. In Seeliger, U.C., and Kjerfve, B. (eds.), *Coastal Marine Ecosystems of Latin America. Ecological Studies 144*. Berlin Heidelberg: Spriger Verlag, pp 245–263.

UNESCO, 1998. *CARICOMP—Caribbean Coral Reef, Sea grass and Mangrove sites*. Paris: UNESCO.

Vasquez, J., 1992. *Lessonia trabeculata*, a subtidal bottom kelp in northern Chile: a case study for a structural and geographical comparison. In: Seeliger, U. (ed.), *Coastal Plant Communities of Latin America*. San Diego, CA: Academic Press, pp. 77–89.

Veloso, V.G., and Cardoso, R.S., 2001. The effects of morphodynamics on the spatial and temporal variation of the macrofauna of three sandy beaches on the Rio de Janeiro State, Brazil. *Journal of the Marine Biological Association of the UK*, **81**: 369–375.

Vera, B., 1992. Sea grasses of the Venezuelan coast: distribution and community components. In Seeliger, U. (ed)., *Coastal Plant Communities of Latin America*. San Diego, CA: Academic Press, pp. 135–140.

Wilkinson, C., 2000. *Status of Coral Reefs of the World 2000*. Cape Ferguson: Australian Institute of Marine Science.

Würsig, B.G., Jefferson, T.A., and Schmidly, D.J., 2000. *The Marine Mammals of the Gulf of Mexico*. College Station: Texas A and M University Press, The W.L. Moody, Jr., Natural History Series 26.

Cross-references

Beach Processes
Coastal Lakes and Lagoons
Coral Reefs
El Niño–Southern Oscillation
Estuaries
Human Impact on Coasts
Mangroves, Ecology
Mangroves, Geomorphology
Muddy Coasts
Rock Coast Processes
Sandy Coasts
Salt Marsh
South America, Coastal Geomorphology
Vegetated Coasts

SOUTH AMERICA, COASTAL GEOMORPHOLOGY

Investigation into the coastal geomorphology of Latin America does not have a lengthy history, but there are some indications that Latins and others are turning their attention to this important area (Psuty, 1970; Tavares Corrêa, 1996; Bittencourt *et al.*, 1999; Klein *et al.*, 2002). Large portions of the coastal zone remain unstudied in detail, and frequently only general descriptions exist (Putnam *et al.*, 1960; Dolan *et al.*, 1972; Bird and Schwartz, 1985). Questions about regional correlations of depositional and erosional features must await basic research into topical problems. However, the following interpretation of the coastal geomorphology of South America is presented as a base upon which to build many future layers of information.

Colombia

From Venezuela to the east to the border with Panama to the west, the Caribbean coast of Colombia is approximately 1,030 km in length. The Guajira Peninsula is the northernmost arm of the Andean mountain system, and the extreme northeastern tip is fronted by an active coral reef. Some evidence has been presented (Anderson, 1927) that there are raised coral platforms and fossil mollusks representing tectonic displacement of the land terminus. Sandy beaches without accompanying coral line the northwestern margin of the peninsula southward to Cape San Juan de Guia, where crystalline cliffs occur with pocket beaches.

South of Santa Clara the coastal zone is dominated by the delta of the Magdalena River. There is a broad arcuate delta in this area with several active distributaries. Great quantities of sediment are transported to the ocean to be reworked. Mangrove-covered mudflats are characteristic of this area, as is beach ridge and chenier topography. The fluctuating activity of the distributaries is responsible for episodes of accretion and erosion of the shore. Beach ridges and cheniers (*q.v.*) on the delta show the old coastal alignments, and their truncated forms provide evidence of reorientation of the coastline in past times. Near the Colombian–Panamanian border the Atrato delta provides the source of the sediments that form the shore features, but there is little change. Mangrove forest, distributary channels, and sand ridges along the distal margins continue to characterize the coast.

The 1,300km-long Pacific coast of Colombia begins as a high-cliffed coast at the Panamanian border. A portion of the flank of the Andes comes to the sea to create steep, vegetation-cloaked precipices plunging directly into the water. West (1956) described the region from south of Cape Corrientes to the border with Ecuador as having short drainage systems leading from the steep mountainous ridges, a shore consisting of fluvially derived sediments, and, with the exception of three places where the coast is cliffed, narrow sandy beaches lying between the ocean and dense mangrove forests. Large mudflats occurred in front of the beaches and in front of the fluvial plain at the foot of the mountains. In 1995, Martinez *et al.* reported for the first time in the periodical literature that most of the Cape Corrientes to Tumaco shore consisted of a series of barrier islands. Subsequently, Martinez *et al.* (2000) and Morton *et al.* (2000) described barrier washover events resulting from a relative sea-level rise. They attributed the relative rise in sea level variously to long-term subsidence, short-term seismic subsidence, and El Niño events.

Ecuador

The coast of Ecuador, including the shores of the Puna, Jambeli, and Galapagos islands, is about 2,500 km in length (Ayón and Jara, 1985). The northern coast is drenched with rainfall, and supports a dense vegetation down to the shore. However, at the southern margin of the country, the coast is stark and almost devoid of all vegetation. At both extremes, the coast consists of high cliffs fronted by a sandy fringing beach or pocket beaches. The high cliff extends consistently from near the border with Colombia southward to near Santa Elena. Near the latter location the coast takes a right-angle bend and trends southeastward. Along this portion of the coast, Sheppard (1930) described a series of marine terraces rising step-like to an elevation of 60 m. He interpreted these features as evidence of tectonic displacement. Ayón and Jara (1985) reported continental uplifting in this region as high as 200 m, evidenced by Pleistocene beachrock benches (locally called *tablazos*).

The Gulf of Guayaquil is an estuarine system whose shore is fringed by dense stands of mangrove on mudflats (Twilley *et al.*, 2000). Near the distal margins of the estuary, beach-ridge systems make up several of the exposed points. However, the southern margin of the estuary remains a cliffed coast. Southwestward toward Tumbes in Peru, a high-cliffed coast dominates with either a sandy fringe or pocket beaches.

Cliffs, shore platforms, and pocket beaches border the volcanic Galapagos Islands (Ayón and Jara, 1985). Eroded craters have resulted in steeply plunging cliffs, and shore platforms are cut into former lava flows. Uplift of some of the islands by about 100 m is shown by exposed sedimentary rocks and submarine lavas.

Peru

The hyperaridity of coastal Peru is responsible in large part for the specific geomorphologic features found there. The lack of precipitation presently limits the contribution of fluvially transported sediments to the littoral zone, and thus limits the development of beaches. During the Holocene, as the sea level was rising and melt-waters were coursing through the river valleys leading to the Pacific Ocean, there were great quantities of sediment discharged to be reworked by the waves and currents. Large sediment volumes were sent into the small estuaries at the river mouths and onto the narrow continental shelf. However, only a few rivers continuously reach the sea at present, and the situation has changed from that of several thousand years ago. The rise in sea level has continuously encroached upon the small estuarine locations, and scattered tectonic displacement has produced further alterations of the coastal features.

Peru is characterized by the presence of a 2,300km-long cliffed coast interspersed with pocket beaches and beaches fronting river mouths. The pocket beaches tend to be located where more easily eroded rock formations of sandstone, shale, or marl are exposed at the shoreline. The rocky promontories are usually composed of crystalline rock units. In a few places the cliffed coasts are cut into colluvial deposits that have moved down spacious interfluves during the Quaternary. At locations where the cliffs are in colluvial gravels and cobbles, a narrow fringing beach usually exists.

South of Pisco there is little or no coastal plain, and the coast usually consists of towering cliffs or small crescent-shaped embayments with narrow beaches. From Pisco north to Chiclayo there are a number of rivers that reach the sea and contribute to the development of localized coastal plains. According to Parsons and Psuty (1975), the mouths of the major river valleys have had a similar geomorphic history and a similar assemblage of landforms. Archaeologic evidence suggests that the fluvial plains found in these river mouths obtained their present characteristics about 1,000–1,500 yr BP. At that time the embayments were filled, and the leading margin of the fluvial plain included a classic beach/dune profile. The profile has been migrating inland since that period over a narrow backmarsh that has formed between the beach and the fluvial plain. Coastal sand dunes reach impressive size: 6 m high at the shore fringing the Sechura Desert and 20 m high at Negritos (Bird and Ramos, 1985). At the southern margins of the valleys, there is frequently a dynamic beach area where sand is being moved inland by the prevailing winds out of the southwest. The coastal dune ridge is frequently enlarged in this area and transformed into longitudinal dunes extending inland. In some instances, barcan dunes break off the sand sheet and migrate independently across the terrace surfaces.

Coastal displacement is evidenced in many locations of coastal Peru by raised shore platforms and raised beaches. Bird and Ramos (1985) reported Quaternary marine terraces (tablazos), incised by transverse valleys (quebradas), south of Zorritos. Other investigators have identified as many as 10 coastal terraces at elevations of up to 250 m at the mouth of the Ica River (Broggi, 1946) and up to 75 m in the Sechura Desert area. However, there is some question whether the displacement is part of regional uplift or local movement. Several detailed inquiries (Craig and Psuty, 1968; Parsons and Psuty, 1975) have shown that the movement is local because the platforms do not extend great distances. Further, flanking terraces in the valleys are Pleistocene depositional features. Most of the major river valleys contain no evidence of post-stillstand uplift (Psuty, 1978). In northern Peru, Richards and Broecker (1963) collected marine shells from several terraces that implied emergence of the platforms. A low terrace at 4.5 m had shells collected on its surface dated at 3,000 yr BP, whereas a high terrace of 75 m had shells assayed beyond 30,000 yr BP.

Chile

The coast of Chile trends parallel to the uplifted longitudinal coast ranges for about 4,400 km, or a total length of 35,000 km when all of the inlets and islands are considered (Araya-Vergara, 1985). The northern two-thirds of coastal Chile has characteristics similar to those of Peru. The Andean coastal range comes down to the sea, and the cliffed coast is interrupted by small pocket beaches or alluvial embayments where infrequent streams lead down to the shore. Numerous terraces appear along the cliffed headlands fronting the foothills. According to Börgel (1967), there are a series of steps reaching to 200 m, thought to be terraces of abrasion, north and south of Valparaiso. Paskoff (Fuenzalida *et al.*, 1965) believes that the highest terraces found in northern Chile at 250–400 m are probably Pliocene, whereas those below 250 m are likely to be Pleistocene. Numerous investigators (Fuenzalida *et al.*, 1965) have identified the considerable number and variety of terraces found in coastal Chile. It is unlikely that the terraces can be considered wholly as products of eustatic sea-level changes, as was suggested in the early investigations. Rather, the lack of uniformity in number, elevation, and kind points to localized tectonic events. However, one terrace at the 85/100 m level does appear to persist throughout much of northern and central coastal Chile. It is considered to have extensive deposits of marine gravel on its surface and also marine mollusks. It is possible that this surface represents an episode of regional displacement.

Rainfall increases southward in Chile, but there are no major coastal alluvial plains developing. The coast range comes to the sea and provides for only modest embayments. Pocket beaches prevail south from Valparaiso to Chiloé. However, the shores near the mouth of the Bio-Bio River and south of the Mataquito River are well developed, with sizeable active dunes penetrating inland. In this region of the Aruaco Province, Tavares Correa (1996) has investigated the potential of coastal dune management and "According to the aeolic potential, the dune environment of the three sectors were classified as for Conservation (South Sector), Stabilization (North Sector) or a combination of them (Central Sector). At the same time, these sectors were divided in small units which were assigned for activities like afforestation, stabilization, conservation, agriculture and grazing." There is also estuarine development at Valdivia, where a small fluvial plain has accumulated at the mouth of a river and a narrow belt of coastal features bounds its shoreward margin. Principal Component Analysis of sand from 16 beaches in the vicinity of Valdivia by Pino and Jaramillo (1992) identified two main groups, fully reflective sites with coarser particles and steeper profiles as compared with dissipative sites with intermediate characteristics. Jaramillo *et al.* (2002) subsequently studied the reflective–dissipative effect upon sandy beach microfauna on both sides of a concrete seawall in the same region.

South of the latitude of Puerto Montt the coastal configuration is dramatically changed. The coastal range becomes broken. The island of Chiloé retains many of the characteristics of the northern coast, but to the south the coast is altered by the processes of glaciation. A fjord coast is present south of 43°S latitude, with many channels extending entirely across the crestal portion of the coastal range into the interior passage. Many of the pocket beaches are cobble-strewn and exist only in sheltered areas. In this region of tidal marshes associated with environments of glacial retreat, earthquake-induced subsidence, and braided river channel migration, Reed (1989) investigated variations in marsh topography and morphology, and found "distinct morphological differences between the marshes developed on glacial outwash and river deposits, and those formed after a major subsidence event."

Weischet (1959) reports that southern Chile has several terrace levels, but they are not continuous through the region. They may be dissected remnants of a larger surface or products of local tectonic activity. However, the degree of glaciation is so thorough in this area that it is the characteristics of glaciation that must be considered rather than those of coastal processes. Patagonia is a region of straits, fjords, drumlins and roches moutonées, with glaciers that reach the sea (Araya-Vergara, 1985). In an 18th century readvance the bay at Lagun de San Rafaél was overrun by glaciers.

Argentina

Argentina is characterized primarily by a 5,700km-long cliffed coast with a narrow beach zone before it. The cliffs vary from only a few meters to the spectacular elevations of greater than 500 m south of Comodoro Rivadavia. Occasional areas of sedimentary accumulations do exist either as beaches fronting the cliffs or as substantial areas of beach ridges and coastal dunes.

The Río de la Plata estuary, 15,000 square km in extent, dominates the northern portion of the Argentinian coast (Schnack, 1985). From the mouth of the river to Cabo San Antonio, the shore is a tidal mudflat. Wave energies are low, and the fine-grained sediments derived from the fluvial system are not reworked into a beach form. However, Urien (1972) has interpreted beach ridge and chenier forms that were created in the period of 7,000–3,000 yr BP along the southern shore of the estuary. These features were part of the sand wedge that was being pushed up the continental shelf as the sea level was rising, and they developed prior to the silting of the estuary. Urien suggests that tectonic displacement of these forms has raised them to elevations of 9 m, where they and a marine terrace form the margin of Samborombón Bay. From Mar del Plata to Bahía Blanca, the coast consists of a low-cliffed shore with a narrow beach before it. Occasionally there are large dune fields leading from the beach. This portion of Argentina is an extension of the Pampas coming to the coast. South of Bahía Blanca the Negro and Colorado rivers transport considerable quantities of sand to the shore, and the beaches are extremely broad. The alluviation is not complete, however, for there are rocky islands and points located between these two river mouths. The Colorado delta is extensive, and there is evidence of coastal aggradation in the form of beach ridges and distributary elongation. Though the Negro delta is not nearly so extensive, it manages to fill its estuary. For further studies of sediment transport along the coast of Argentina see Isla (1997), Mar Chiquita; Kokot (1997), Punta Médanos; and Cuadrado and Perillo (1997), Bahía Blanca.

With the exception of well-developed beaches and associ-ated landforms at the Gulf of San Marcos and the Gulf of San Jorge, the southern half of Argentina is a cliffed coast. Investigators have identified a number of terraces that have been used to describe tectonic or eustatic displacement. Terraces ranging to 140 m have been noted in Patagonia. A 9 m terrace at Comodoro Rivadavia has been dated at 3,000 yr BP (Richard and Broecker, 1963), and has been used to suggest local tectonic movement. Dating beach gravels and beach ridges along the Patagonia coast, Rutter *et al.* (1990) found the penultimate or older glaciation shore at 24 and 41 m above mean sea level, the "intermediate" age shore at between 20 and 28 m, and "young" Holocene beach ridges at 8–12 m. Urien (1970) has indicated that a 9 m terrace also exists in Tierra del Fuego along parts of the Río de La Plata estuary and bordering the Paraná delta. However, he cautioned against attempting any broad correlations. More recent radiocarbon dating with corres- ponding altimetry over present sea level, at 15 locations along 3,500 m of the Argentina coast, by Codignotto *et al.* (1992) indicated relative uplift rates between 0.12 and 1.63 m/1,000 yr during the Holocene. A 20th century sea-level rise of from 1.6 to 3.5 mm/yr along much of the same sector has been reported by Lanfredi *et al.* (1998).

From Santa Cruz to the eastern tip of Tierra del Fuego the cliffs are cut into glacial morainic material. Occasional outcrops of bedrock are noted, as are pocket beaches (Etchichury and Remiro, 1967). At Punta Dungeness there is an excellent series of beach ridges created where currents converge at the point. The more protected western margin shows considerable accretionary history, whereas the exposed southeastern flank gives evidence of truncation of the ridges and extensive dune fields extending inland. Schnack (1985) has described the coast of Islas Malvinas as being indented and cliffy, with some raised Quaternary shore terraces.

Uruguay

The 600-km-long coast of Uruguay is extremely diverse for such a relatively short sector. The northern area consists of an extension of the barrier island system of southern Brazil. The sandy beach continues into Uruguay but narrows and becomes discontinuous, so that it becomes a series of sandy embayments set into Uruguay's crystalline Eastern Highlands Shield (Jackson, 1985). These embayments are not products of marine processes but, rather, the prior irregular topography of the shield encroached upon by a rising sea. In several places, the embayments contain small lagoons behind a sand barrier. This feature is the product of incomplete filling of small estuaries along the margin of the shield (Chebataroff, 1960).

From near Maldonado westward, the coast is the margin of the Río de la Plata estuary. For nearly this entire length there is a cliffed coast with a sand beach lying at its base. Delaney (1963, 1966) suggested that there are terraces cut into this cliff. And it was proposed by Urien (1970) that the beach sand fronting the cliffs is the product of local erosion of the bluffs rather than fluvial transport. The exposure of this portion of Uruguay to the southeast would tend to allow greater wave energies to reach this shore and favor the necessary erosion and sorting responsible for sandy beaches, while much of the rest of the estuary is a shallow muddy tidal flat. Occasionally mudflats dominate along the cliffed coast. The Río de la Plata is formed by the confluence of the Paraná and Uruguay rivers, which flow between Uruguay and Argentina, and owing to the shallow depth across this region, navigation can only be accomplished by dredging channels for large ships (Vieira and Lanfredi, 1996).

Brazil

The great size of Brazil allows for considerable diversity of coastal exposure and geomorphologic development. There are three principal portions of the 9,200km-long coast. The first is the area influenced by the Amazon River and its sediments; the second is the narrow coastal margin fringing the huge Brazilian Shield, creating an escarpment nearly adjacent to the ocean; the third is the southern area, where considerable quantities of sediments have accumulated to provide a barrier island formation.

The mouth of the Amazon River is a great estuary stretching for nearly 1,600 km (1,000 miles). Large quantities of sand and especially silt and clay are discharged by the river and accumulate along the shore margins. From the border with Surinam eastward to the Gulf of San Marcos, the fine-grained sediments blanket the shore and are cloaked with mangrove. The Gulf of San Marcos is another, much smaller, estuary that similarly contributes large quantities of fine-grained sediment.

East of the Gulf of San Marcos the shore begins to be characterized by sandy beaches lying before low hills. The sand beaches are interspersed with mangrove stands that dominate where local deltaic buildout occurs in association with short drainage systems leading off the eastern margin of the Brazilian Shield. Beginning in Rio Grande do Norte and continuing southward to the coastal margin of Alagoas state, the beach zone is severely attenuated. The dry climate and the short drainage systems limit the transport of sediment to the ocean margin. Further, this portion of Brazil is bordered by a fairly extensive coral reef. Where beach sediments have accumulated, there are also likely to be well-developed dune forms migrating inland over terrace surfaces.

South of Recife the coast is cliffed. The combination of cliffed coast and the presence of coral reef extends for 480 km (about 300 miles). Near Recife a small promontory has been investigated for evidence of a high sea level. Van Andel and Laborel (1964) have dated a fossil biogenous limestone that is encrusting granite a few meters above modern sea level as having been active 1,200–3,600 yr BP. The authors interpreted these dates to hypothesize that sea level was higher in that time period.

The sandy beach backed by an escarpment begins near the Alagoas–Sergipe border and continues south to Rio Grande do Sul. The beach often broadens in large curvilinear embayments, and there may be local mangrove stands, beach ridges, and deltaic buildout. However, the escarpment dominates the horizon, and the coastal geomorphic features occupy a small niche on the continental margin. Bittencourt *et al.* (1999) have proposed large-scale tectonic flexure as controlling geomorphic characteristics in this region. In the state of Paraná there is an extensive area of beach ridge development associated with the Maciel river. These ridges appear to resemble cheniers in that they are bounded by clayey deposits rather than forming a broad sandy surface. The beach ridges attain elevations of 10 m in their interior location, and gradually decrease to elevations of 2–3 m near the shore. Bigarella (1965) believed this was further evidence for a fluctuating and generally falling sea level. Where the state of São Paulo borders that of Paraná, Tessler and de Mahiques (1993) have shown that coastal sedimentary features, such as spits and sandbanks, are clearly indicative of longshore sediment transport directions.

Along the coast of Santa Catarina state there are a number of headland bay beach systems bounded by headlands or rock outcrops, where the shore assumes a curvature form. Klein *et al.* (2002) have investigated short-term beach rotation processes at three such sites. They found that erosion–deposition events varied with incident wave direction. And while there was no loss of sediment from a sector, original realignment was attained with a return to the prior wave direction.

The coastal margin of the state of Rio Grande do Sul is distinct from the rest of Brazil—it consists of a classic barrier island–lagoon sequence. Delaney (1963, 1966) described the geologic origin of the barrier island sequence as occupying a unique position in a geologic depositional basin with sediments being transported into it from several directions. Toldo Jr. *et al.* (2000) have suggested an average depositional rate of 0.75 mm/yr during the Holocene in the Lagoa dos Patos estuary. Certainly the positive sediment budget that existed in this area with the changing sea level caused the particular assemblage of broad sandy beach extending along the coast for 640 km and incorporating 120km-wide (Cruz *et al.*, 1985) beach ridge systems and large coastal dunes reaching 25 m in elevation. The northern margin of this Holocene coastal plain comes against a terrace surface with elevations of 15 m. It is probably of Pleistocene age, but whether its origin is wave-cut is unknown.

French Guiana, Surinam, Guyana

The coast of what was once referred to as the European Guianas is somewhat similar for its entire length of over 1,100 km. Basically, the shore is the product of massive quantities of fine-grained sediments discharged by the Amazon River that proceed to drift westward. Some small quantities of sand are contributed by the Amazon and by the smaller streams leading from the Guiana Shield to the shore.

The beach and inland coastal geomorphology of the Guianas is characterized by the active development of cheniers. These ridged features are coarse-grained deposits of sand and shell accumulating on a mud or clay foundation. Intermittent development of these coarser accumulations creates a series of sandy ridges bordered on either side by fine-grained sediments. At times the shore is a broad mudflat rather than a sandy beach. Several investigators (Zonnenfeld, 1954; Vann, 1959; Wells and Coleman, 1978) have described the massive clay waves that migrate along this coast and extend far out into the ocean. Turenne (1985) has described westerly migration of these waves or mudbanks as being at a rate of 1.3 km per year.

There is a type of pattern to the chenier or sand ridge development along this coast. A kind of apex forms near the west bank of the river mouth from which a series of ridges extends. The number and extent of the ridges decrease westward to the point where the coarser sediments are no longer found. Some of the river mouths, such as those of the Marowijne and the Suriname, show evidence of west bank progradation for several kilometers by means of mudflat development interspersed with sandy ridges. The ridge trends also provide evidence that the progradation sequence has not been continuous, because numerous ridges are truncated and new fulcrums have developed from which a fan-shaped series of ridges has spread.

Much of the mudflat area is colonized by mangrove. These trees line the shores in places as well as occupy the troughs between the ridges. Wave action appears to be attenuated by the extremely turbid waters, and thus the coastal clays are only infrequently disturbed. However, during the infrequent higher wave energies, there is considerable movement of the clay waves (see above). Large units of clay are displaced, and the sand and shell are sorted to accumulate as a beach on top of the clay base. These beaches continue to develop to form the cheniers.

Coastal lengths of these sectors are, respectively, French Guiana, 370 km (Turenne, 1985); Surinam, 350 km (Psuty, 1985); Guyana, 434 km (Schwartz, 1985).

Venezuela

Venezuela derives its name from late 15th century explorers finding lake dwellings on piles in mangrove swamps, and thus calling the area "Small Venice" (Ellenberg, 1985). Coastal Venezuela, 3,000 km in length, tends to be dominated by the northern terminus of the Andean mountain system. The distal portion of the Andes splits into two north-trending prongs and between them creates the depression occupied by Lake Maracaibo and the Gulf of Venezuela region. The eastern prong makes a sharp bend due east and establishes the northern margin of the country for most of its length. Modifications of the mountainous coastal topography are caused by deltaic development and by breaks in the Andean ridges.

The eastern coast of Venezuela is completely the product of the Orinoco River and its deltaic forms. A fairly large arcuate delta occupies the position from the Gulf of Paria south to the border of Guyana (van Andel, 1967). The southern third of the delta tends to be a series of coalescing distributaries that retain their fluvial forms as they discharge into the Atlantic. However, the northern two-thirds of the delta is lined with cheniers forming a well-developed shore. Mangrove forests occur at the frontal lobes of the southern margin of the delta and at the northern margin. However, in the area of chenier development, the mangrove is in the protected troughs between the ridges rather than at the exposed coast. Warne *et al.* (2002) have traced the evolution of the delta from a late Pleistocene sea-level lowstand to its present stand in the mid-Holocene.

They have also reported that the Volcán Dam, built on the Cãno Manamo to prevent flooding, expand agriculture, and enhance navigation, has had a negative effect on farming due to the presence of pyrite in the soil.

The Paria Peninsula is bounded on its southern side by sand beaches and fairly broad mudflats. The sand beaches are found in association with small streams that drain the high peninsula and contribute their coarser sediments. The eastern half of the south side of the peninsula is a steep rocky cliffed coast, as is the entire northern portion of this Andean extension. There are some pocket beaches, but not until the Gulf of Barcelona is a well-developed beach found. The shore consists of a broad curvilinear beach that has characteristics of barrier island development because several lagoons are formed behind it.

Westward beyond the Barcelona embayment, a cliffed coast appears once again with a number of pocket beaches. A well-developed sandy beach is located at the Triste Gulf; and there is a low, narrow isthmus connecting the Paraguaná Peninsula with the mainland, the result of recent uplift (Ellenberg, 1985). Along the Gulf of Venezuela most of the shore is sheltered from marine processes, and thus the features are of fluvial origin and the shores are marshy or lined with mangrove. However, that portion of the gulf exposed to swell waves from the northeast does have prominent coastal forms. The exposure to the northeast is also the dominant fetch direction for the tombolo to Paraguaná and the Triste Gulf beaches as well.

Tanner (1970) has shown that the western shore of the Gulf of Venezuela has a barrier island formation with a series of beach ridges prograding seaward over a distance of 7 km. The ridges are (low, (only about 1 m local relief) and there have been several interruptions in the accumulation sequence. These erosional breaks in beach ridge development are marked by longitudinal and parabolic dunes at the erosional shoreline whose form extends inland over older beach ridges. A number of beach ridge sectors along the Venezuelan coast are backed by sabkha salt lagooons (Ellenberg, 1985).

Norbert P. Psuty and Chizuko Mizobe

Bibliography

Anderson, F.M., 1927. Nonmarine tertiary deposits of Colombia. *Geological Society of America Bulletin*, **38**: 591–644.

Araya-Vergara, J.F., 1985. Chile. In Bird, E.C.F., and Schwartz, M.L. (eds.), *The World's Coastline*. New York: Van Nostrand Reinhold pp. 57–67.

Ayón, H., and Jara, W., 1985. Ecuador. In Bird, E.C.F., and Schwartz, M.L. (eds.), *The World's Coastline*. New York: Van Nostrand Reinhold, pp. 49–52.

Bigarella, J.J., 1965. Sand-ridge structures from Paraná coastal plain. *Marine Geology*, **3**: 269–278.

Bird, E.C.F., and Schwartz, M.L., 1985. *The World's Coastline*. New York: Van Nostrand Reinhold.

Bird, E.C.F., and Ramos, V.T., 1985. Chile. In Bird, E.C.F., and Schwartz, M.L. (eds.), *The World's Coastline*. New York: Van Nostrand Reinhold, pp. 53–56.

Bittencourt, A.C.S.P., Dominguez, J.M.L., and Ussami, N., 1999. Flexure as a tectonic control on the large scale characteristics of the Eastern Brazil Coastal Zone. *Journal of Coastal Research*, **15**(2): 505–519.

Börgel, R., 1967. Correlaciones fluviomarinas en la desembocadura del Rio Choapa. *Informaciones Geográficas*, **13–14**: 55–68.

Broggi, J.A., 1946. Las terrazas marinas de la Bahía de San Juan en Ica. *Sociedad Geológica del Peru, Boletin*, **19**: 21–33.

Chebataroff, J., 1960. Sedimentacion Platense. *Inst. Estudios Superiores, Rev.*, Año **4**(7): 544–566.

Codignotto, J.O., Kokot, R.R., and Marcomini, S.C., 1992. Neotectonism and sea-level changes in the coastal zone of Argentina. *Journal of Coastal Research*, **8**(1): 125–133.

Craig, A.K. and Psuty, N.P., 1968. *The Paracas Papers: Studies in Marine Desert Ecology*. Boca Raton, FL: Florida Atlantic University.

Cruz, O., Coutinho, P.N., Duarte, G.M., Gomes, A., and Muehe, D., 1985. Brazil. In Bird, E.C.F., and Schwartz, M. L. (eds.), *The World's Coastline*. New York: Van Nostrand Reinhold, pp. 85–89.

Cuadrado, D.G., and Perillo, G.M.E., 1997. Migration of intertidal sandbanks , Bahia Blanca Estuary, Argentina. *Journal of Coastal Research*, **13**(1): 155–163.

Delaney, P.J.V., 1963. Quaternary geologic history of the coastal plain of Rio Grande do Sul, Brazil. Baton Rouge, LA: Louisiana State University Press, *Coastal Studies Series No. 7*.

Delaney, P.J.V., 1966. Geology and Geomorphology of the Coastal Plain of Rio Grande do Sul, Brazil and Northern Uruguay. Baton Rouge, LA: Louisiana State University Press, Coastal Studies Series No. 15.

Dolan, R., Hayden, B., Hornberger, G., Zieman, J., and Vincent, M., 1972. *Classification of the Coastal Environments of the World, Part 1: The Americas*. Charlottesville, VA: University of Virginia, Department of Environmental Sciences.

Ellenberg, L., 1985. Venezuela. In Bird, E.C.F., and Schwartz, M.L. (eds.), *The World's Coastline*. New York: Van Nostrand Reinhold, pp. 105–113.

Etchichury, M.C., and Remiro, J.R., 1967. Los sedimentos litorales de la provincia de Santa Cruz entre Dungeness y Punta Desengaño. *Revista del Museo Argentino de Ciencias Naturales – Geología*, **6**: 323–376.

Fuenzalida, H., Cooke, R., Paskoff, R., Segerstrom, K., and Weischet, W., 1965. High stands of Quaternary sea level along the Chilean coast. In Wright, H.E., Jr., and Frey, D.G. (eds.), *International Studies on the Quaternary*. Boulder, CO: Geological Society of America, pp. 473–496.

Isla, F.I., 1997. Seasonal behaviour of Mar Chiquita tidal inlet in relation to adjacent beaches, Argentina. *Journal of Coastal Research*, **13**(4): 1221–1232.

Jackson, J.M., 1985. Uruguay. In Bird, E.C.F., and Schwartz, M.L. (eds.), *The World's Coastline*. New York: Van Nostrand Reinhold, pp. 79–84.

Jaramillo, E., Contreras, H., and Bollinger, A., 2002. Beach and faunal response to the construction of a seawall in a sandy beach of south central Chile. *Journal of Coastal Research*, **18**(3): 523–529.

Klein, A.H.F., Filho, L.B., and Schumacher, D.H., 2002. Short-term beach rotation processes in distinct headland bay beach systems. *Journal of Coastal Research*, **18**(3): 442–458.

Kokot, R.R., 1997. Littoral drift, evolution and management in Punta Medanos, Argentina. *Journal of Coastal Research*, **13**(1): 192–197.

Lanfredi, N.W., Pousa, J.L., and D'Onofrio, E.E., 1998. Sea-level rise and related potential hazards on the Argentine coast. *Journal of Coastal Research*, **14**(1): 47–60.

Martinez, J.O., Gonzalez, J.L., Pilkey, O.H., and Neal, W.J., 1995. Tropical barrier islands of Colombia's Pacific Coast. *Journal of Coastal Research*, **11**(2): 432–453.

Martinez, J.O., Gonzales, J.L., Pilkey, O.H., and Neal, W.J., 2000, Barrier island evolution on the subsiding central Pacific Coast, Colombia, SA: *Journal of Coastal Research*, **16**(3): 663–674.

Morton, R.A., Gonzalez, J.L., Lopez, G.I., and Corrêa, I.D., 2000. Frequent non-storm washover of barrier islands, Pacific Coast of Colombia. *Journal of Coastal Research*, **16**(1): 82–87.

Parsons, J.R., and Psuty, N.P., 1975. Sunken fields and prehispanic subsistence on the Peruvian coast. *America Antiquity*, **40**: 259–282.

Pino, M., and Jaramillo, E., 1992. Morphology, texture and mineralogical composition of sandy beaches in the south of Chile. *Journal of Coastal Research*, **8**(3): 593–602.

Psuty, N.P., 1970. Contributions to the coastal geomorphology of Latin America. In Lentneck, B., Carmin, R.L., and Martinson, T.L. (eds.), *Geographic Research on Latin America: Benchmark*. Muncie, IN: Ball State University, pp. 250–264.

Psuty, N.P., 1978. *Peruvian Shoreline Stability/Instability during Stillstand of Sea Level*. Lagos: International Geographical Union Regional Conference, Commission on Coastal Environments.

Psuty. N.P., 1985. Surinam. In Bird, E.C.F., and Schwartz, M.L. (eds.), *The World's Coastline*. New York: Van Nostrand Reinhold, pp. 99–101.

Putnam, W.C., Axelrod, D.I., Bailey, H.P., and McGill, J.T., 1960. *Natural Coastal Environments of the World*. Los Angeles, CA: University of California, Department of Geography.

Reed, D.J., 1989. Environments of tidal marsh deposition in Laguna San Rafael area, Southern Chile. *Journal of Coastal Research*, **5**(4): 845–856.

Richards, H.G., and Broecker, W., 1963. Emerged Holocene South American shorelines. *Science*, **141**: 1044–1045.

Rutter, N., Radtke, U., and Schnack, E.J., 1990. Comparison of ESR and amino acid data in correlating and dating Quaternary shorelines along the Patagonian coast, Argentina. *Journal of Coastal Research*, **6**(2): 391–411.

Schnack, E.J., 1985. Argentina. In Bird, E.C.F., and Schwartz, M.L. (eds.), *The World's Coastline*. New York: Van Nostrand Reinhold, pp. 69–78.

Schwartz, M.L., 1985. Guyana. In Bird, E.C.F., and Schwartz, M.L. (eds.), *The World's Coastline*. New York: Van Nostrand Reinhold, pp. 103–104.

Sheppard, G., 1930. Notes on the climate and physiography of Southwestern Ecuador. *Geological Review*, **20**: 445–453.

Tanner, W.F., 1970. Growth rates of Venezuelan beach ridges. *Sedimentary Geology*, **6**: 215–220.

Tavares Corrêa, C.H., 1996. Propuesta de uso suela en las dunas litorales de la provincia de Arauco, VIII Region, Chile. Unpublished Doctoral thesis in Environmental Sciences, Centro EULA-Chile. Concepcion: Universidad de Concepcion.

Tessler, M.G., and de Mahiques, M.M., 1993. Utilization of coastal geomorphic features as indicators of longshore transport: examples from the southern coastal region of the state of São Paulo, Brasil. *Journal of Coastal Research*, **9**(3): 823–830.

Toldo, Jr., E.E., Dillenburg, S.R., Correa, I.C.S., and Almeida, L.E.S.B., 2000. Holocene sedimentation in Lagoa dos Patos Lagoon , Rio Grande do Sul, Brazil. *Journal of Coastal Research*, **16**(3): 816–822.

Turenne, J.F., 1985. French Guiana. In Bird, E.C.F., and Schwartz, M.L. (eds.), *The World's Coastline*. New York: Van Nostrand Reinhold, pp. 93–97.

Twilley, R.R., Cardenas, W., Rivera Monroy, V.H., Espinoza, J., Suescum, R., Armijos, M.M., and Solorzano, L., 2000. The Gulf of Guayaquil and the Guayas River estuary, Ecuador. *Ecological Studies*, **144**: 245–263.

Urien, C.M., 1970. Les rivages et le plateau continental du Sud du Brésil, de l'Uruguay ey de l'Argentine. *Quaternaria*, **12**: 57–69.

Urien, C.M., 1972. Rio de la Plata estuary environments. *Geological Society of America Bulletin*, **133**: 213–234.

Van Andel, Tj.H., 1967. The Orinoco delta. *Journal of Sedimentary Petrology*, **37**: 297–310.

Van Andel, Tj.H., and Laborel, J., 1964. Recent high relative sea level stand near Recife, Brazil. *Science*, **145**: 580–581.

Vann, J.H., 1959. *The Physical Geography of the Lower Coastal Plain of the Guiana Coast*. New Orleans, LA: Louisiana State University, Department of Geography and Anthropology.

Vieira, J.R., and Lanfredi, N.W., 1996. A hydrodynamic model for the Rio de la Plata, Argentina. *Journal of Coastal Research*, **12**(2): 430–446.

Warne, A.G., Guevara, E.H., and Aslan, A., 2002. Late Quaternary evolition of the Orinoco Delta. *Journal of Coastal Research*, **18**(2): 225–253.

Weischet, W., 1959. Geographisches beobachtungen auf einer forschungsreise in Chile. *Erdkunde*, **13**: 6–21.

Wells, J.T., and Coleman, J.M., 1978. Longshore transport of mud by waves: northeastern coast of South America. *Geologie en Mijnbouw*, **57**: 353–359.

West, R.C., 1956. Mangrove swamps of the Pacific coast of Colombia. *Association of American Geographers Annals*, **45**: 98–121.

Zonnenfeld, J.I.S., 1954. Waarnemingen langs de kust van Surinam, Koninkl. *Tijdschrift van het Nederlandsch Aardrijkskundig Gerootschap*, **71**: 18–31.

Cross-references

Antarctica, Coastal Ecology and Geomorphology
Cheniers
Cliffed Coasts
Coral Reef Coasts
Deltas
Estuaries
Mangroves, Ecology
Mangroves, Geomorphology
Middle America, Coastal Ecology and Geomorphology
Muddy Coasts
Sandy Coasts
South America, Coastal Ecology

Reproduced with minor updates from Psuty, N.P., and Mizobe, C., 1982. South America, Coastal Morphology. In Schwartz, M.L. (ed.), *Encyclopedia of Beaches and Coastal Environments*. Stroudsburg, PA: Hutchinson Ross, pp. 765–770.

SPITS

General features of sand spits

Longshore currents lose their sediment transporting capacity at the down-current side of the break in coastline orientation, where the wave energy abruptly drops off, causing the currents to weaken (Horikawa *et al.*, 1988). Sediment accumulates at this location, forming elongate spits, which grow in the direction of the predominant longshore drift. The spits are classified according to their plane shape configuration as simple spits (with relatively linear features), recurved spits (with a distal end hooked landwards), and complex spits (with plural hooks). Figure S53 shows an example of the actual plane shape of a complex spit. Material supplied from the upcoast is transported alongshore to the distal portion, resulting in the extension of the spit. The configuration of the spit terminus depends mainly on (1) the longshore sediment transport rate; (2) wave diffraction and refraction processes; and (3) the interaction between dominant waves and waves arriving from different directions (Zenkovich, 1967). The episodic growth of hooks characterizes the development of a complex spit, as illustrated in Figure S54. A double spit, which is a set of spits extending toward each other from two adjacent locations, frequently develops on a coast where the sediment is

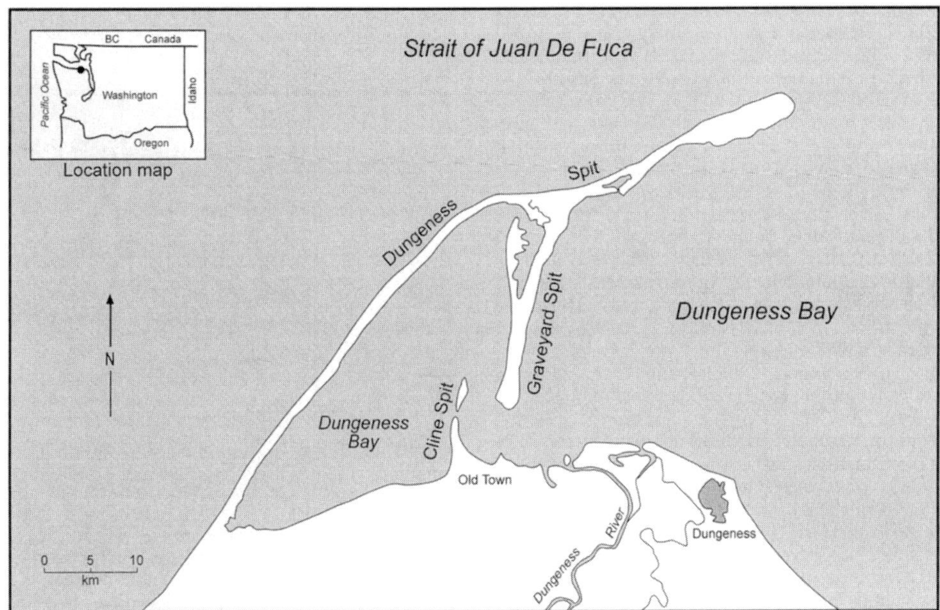

Figure S53 Dungeness Spit, a complex spit on the west coast of the USA (from Schwartz *et al.*, 1984, with permission of the *Journal of Coastal Research*).

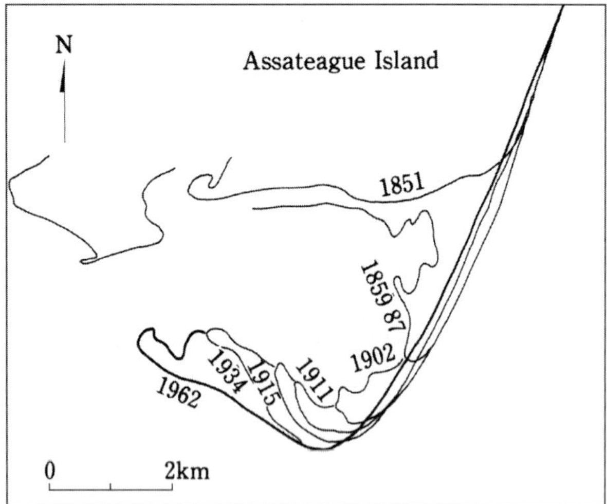

Figure S54 Growth of complex spit, located at southern Assateague Island, Virginia, USA (after Finkelstein, 1983).

transported alongshore in one direction at one time interval of long duration, and in the opposite direction at another time interval.

Longshore growth of spits results in the formation of barriers, with lagoons at their landward side. In addition to longshore spit elongation, barriers can be formed in close association with the postglacial sea-level rise: (1) a straight offshore bar emerged with sea-level rise and migrated landward to form a barrier; or (2) a preexisting subaerial sand mound in the vicinity of the shore moved landward keeping pace with sea-level rise to develop a barrier. Separation of a barrier by estuaries or tidal inlets leads to the formation of a barrier island. The origin of an actual barrier island is complex reflecting local wave climate, sediment supply, and relative rise of sea level (Schwartz, 1973).

Experiment of formation of sand spit and modeling

Sand spits can be reproduced by the movable bed experiment (Uda and Yamamoto, 1991). The initial contours of semicircular shape are made by using sand of median diameter of 0.28 mm in a wave tank, and the bed slope is one-fifth as shown in Figure S55. The initial shoreline is given by a semicircle of $r = 1.5$ m. Given incident wave height of 3 cm and wave period of 0.8 s, subsequent beach changes due to waves were measured. Figure S56 illustrates the shoreline changes under the 2.5-h wave action. The shoreline upcoast of No. 10 retreats, while it progrades downcoast of No. 10. The retreat rate of the shoreline in the eroded zone is faster in earlier stages, and its rate decreases with time. Sand transported by longshore drift accumulates near No. 13 and No. 14 located in the wave shadow zone to form a sand spit. The growth rate of the spit is also faster in earlier stages, and it gradually decreases with time.

Figure S57 shows the bottom contours measured after the wave action of 3 h The interval of the contour lines between 10 and 0 cm on land is very narrow at No. 1 through No. 7, whereas the interval between −2 and −8 cm is wide. This means that the scarp was formed on the foreshore, whereas a gentle slope is formed on the sea bottom due to the erosion. At No. 11 through No.15, the interval of the contour above 2 cm is similar to those at the initial state, but the contour intervals between 2 and 0 cm are widened and those between −2 cm and −16 cm are narrowed, indicating the formation of a flat plane on the foreshore and a steep slope below the sea surface.

The typical beach profiles representing the erosion and accretion zones, and the one in a neutral zone without erosion and accumulation of sand, are compared as shown in Figure S58. Along No. 2, located in the eroded zone, the shoreline significantly retreats with the formation of a scarp. The foreshore slope is constant, and the scarp is formed above the foreshore. During the erosion process, the slope of this scarp and the fore-shore slope remain constant, implying that the steep slope on land deforms as if it moves in parallel. Furthermore, the closure depth is 8 cm and a gentle slope is formed between the closure depth and the depth of 2.5 cm. Along No.13 in the accretion zone, most topographic change is observed below the sea surface. Sand accumulates within the wave run-up zone forming a flat beach surface. During the progradation of the shoreline, the foreshore slope remains constant. The bottom slope remains almost constant in the zone deeper than −2 cm, and the steep slope deforms as if it moves in parallel. In the eroded zone, no topographic deformation takes place below the critical depth for sand movement.

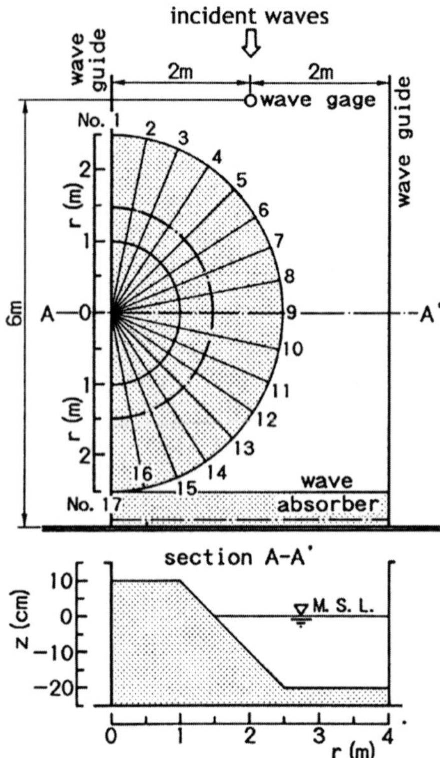

Figure S55 Model beach and alignment of measuring lines (from Uda and Yamamoto, 1991, with permission of the American Society of Civil Engineers).

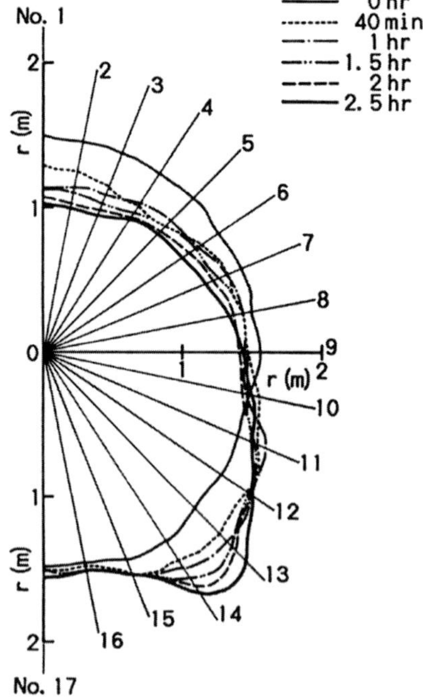

Figure S56 Change in shoreline configuration (from Uda and Yamamoto, 1991, with permission of the American Society of Civil Engineers).

In summary, the initial topography does not change in the upper part of the scarp. In the accretion zone, the initial topography does not change at levels higher than the maximum wave run-up height. Similarly there is no topographic change in the zone deeper than the critical depth where sediment is deposited. These profile changes can be modeled, as schematically shown in Figure S59.

Numerical simulation of formation of sand spit

Ashuton *et al.* (2001) show that coastline instability takes place when the incident angle of the deepwater waves relative to the shoreline exceeds 45°, and a sand spit can be automatically formed from an infinitesimal disturbance of the shoreline. In their method, the shoreline is divided into the infinitesimal meshes in the x- and y-axes, and unidirectional sediment movement forming a sand spit is reproduced by using this coordinate system. However, longshore sand transport is not provided by the deterministic method from the wave field. Watanabe *et al.* (2002) show that successive development of a sand spit can be numerically predicted using a one-line model with orthogonal curvilinear coordinate system fixed on the ever-changing shoreline. In their model, the breaker height and breaker angle are determined first on this coordinate system and longshore sand transport rate is obtained from these values, resulting in the shoreline calculation based on the continuity of sand. A new sea bottom shape is inversely assessed from the new shoreline configuration at some steps, assuming the profile changes in eroded and accreted areas, as shown in Figure S59, and a new wave field is calculated using this sea bottom data. Furthermore, smoothing of the shoreline configuration and redistribution of grid points are carried out to ensure the stability of the calculation. Figure S60(a) is the sand spit formation around an artificial sandy island measured in Alaska (Gadd, 1979a,b). A rectangular island of 137 m length and 99 m width was created in the very shallow sea of the depth of 0.91 m. Comparison of the shoreline is made between December 1976 and September 1978. Sand was transported from the northeast corner of the island due to the predominant waves from the northeast, and sand deposited to form two sand spits in the wave shadow zone at the southeast and northwest corners.

Figure S60(b) shows the calculation results assuming the incident wave height of 0.3 m and the wave period of 3 s. The wave field is predicted based on the wave orthogonal theory assuming irregular waves. In the calculation of the shoreline change using a finite difference method, time step of $\Delta t = 5.0 * 10^{-2}$ h is selected and the shoreline change was calculated up to 30,000 steps. Renewal of the depth data and resulting wave field was carried out at 300 steps (15 h). Furthermore, the closure depth, wave run-up height, $\tan \beta_E$ and $\tan \beta_A$ as shown in Figure S59 are set to be 40 cm, 20 cm, 0.1 and 0.3, respectively. The initial shoreline of a rectangular shape deformed with the successive recession of the shoreline at the northeast corner of the island. Eroded sand was transported by longshore drift forming two sand spits at the northwest and southeast corner of the island. Observed results are well reproduced by the numerical simulation. Figure S60(c) shows the seabed topography after change. A mild seabed was formed in the eroded zone, whereas in

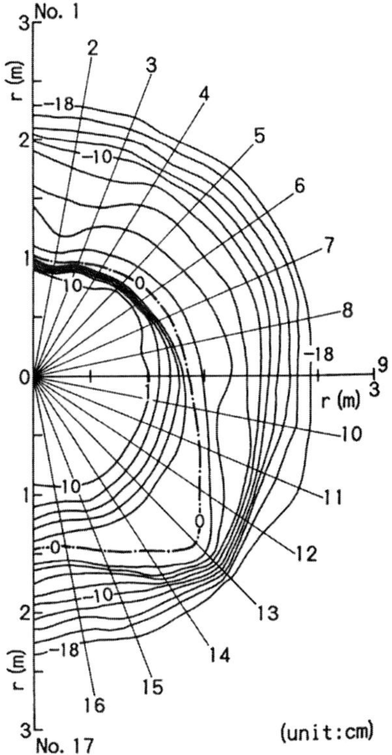

Figure S57 Beach topography after the formation of sand spit (from Uda and Yamamoto, 1991, with permission of the American Society of Civil Engineers).

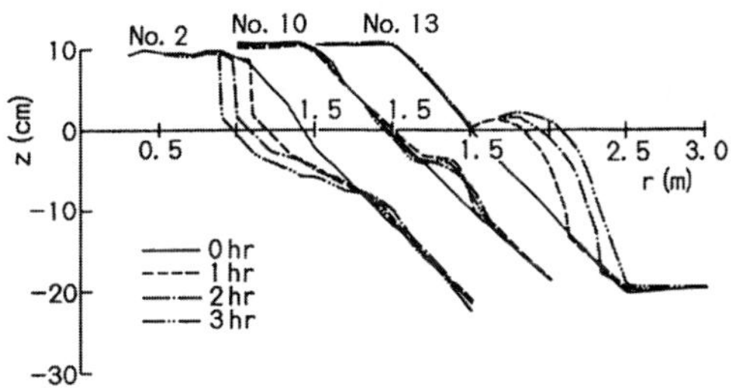

Figure S58 Beach profiles in eroded, accreted and neutral zones (from Uda and Yamamoto, 1991, with permission of the American Society of Civil Engineers).

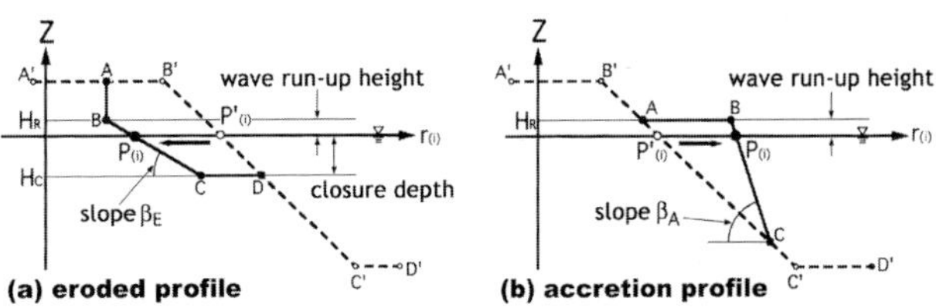

Figure S59 Model of beach profile change (Wantanabe *et al.*, 2002, *Journal of Coastal Engineering*).

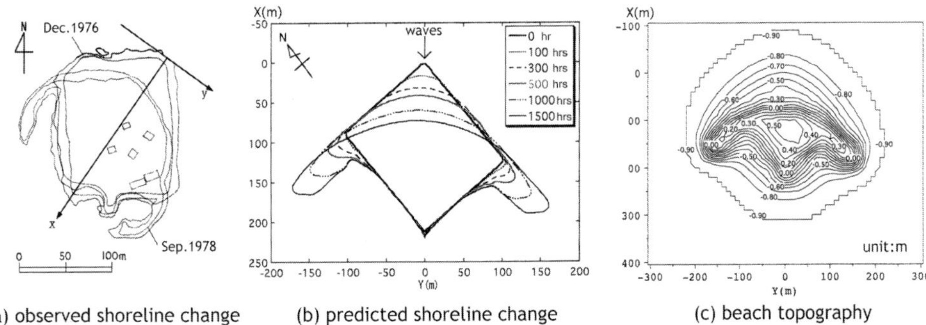

Figure S60 Change in shoreline configuration of an artificial island (from Gadd *et al.*, 1979 a,b, with permission of Tetra Tech Inc.).

the accretion area, a flat foreshore was formed due to the berm formation as well as the steep slope due to the falling of sand into deep sea. Thus, in the prediction of the development of the sand spit, coupling of changes in waves and sea bottom is important.

Takaaki Uda

References

Ashuton, A., Murray, A.B., and Arnault, O., 2001. Formation of coastline features by large-scale instabilities induced by high angle waves. *Nature*, **414**: 296–300.

Finklestein, K., 1983. Cape formation as a cause of eosion on adjacent shorelines. *Coastal Zone '83*, **3**: 620–640.

Gadd, P.E., Sonu, C.J., and Czerniak, M.T., 1979a. Gravel island slope protection, Beaufort Sea, Alaska. Pasadena, CA: *Tetra Technical Report*, TC-3286.

Gadd, P.E., and Czerniak, M.T., 1979b. Production island slope protection, Beaufort Sea, Alaska, Pasadena, CA: *Tetra Technical Report*, TC-3310.

Horikawa, K. (ed.), (1988). *Nearshore Dynamics and Coastal Processes*. Tokyo: University of Tokyo Press.

Schwartz, M.L. (ed.), 1973. *Barrier Islands*. Stroudsburgh, PA: Dowden, Hutchinson and Ross.

Schwartz, M.L., Fabbri, P., and Wallace, R.S., 1987. Geomorphology of dungeness spit, Washington, U.S.A., *Journal of Coastal Research*, **3**: 451–455.

Uda, T., and Yamamoto, K., 1991. Spit formation in lake and bay *Coastal Sediments '91*, **2**: 1651–1665.

Watanabe, S., Serizawa, M., Uda, T., and Ogawa, M., 2002. Development of predictive model of beach changes on a coastline with large curvature, *Journal of Coastal Engineering*, **49**: 501–506. (in Japanese.)

Zenkovich, V.P., 1967. *Processes of Coastal Development*. Oliver & Boyd.

Cross-references

Accretion and Erosion Waves on Beaches
Beach Features
Beach Processes
Coastal Modeling and Simulation
Modeling Platforms, Terraces and Coastal Evolution
Physical Models
Wave Refraction Diagram

STORM SURGE

A storm surge is the increase in ocean water level near the coast generated by a passing storm, above that resulting from astronomical tides. The atmosphere acts on the sea in two distinctly different ways. A reduction in the atmospheric pressure reduces the vertical force acting on a column of water beneath the sea surface, causing the sea water to rise, and *vice versa*. A decrease in atmospheric pressure of 1 mb will produce an increase in sea level of around 1 cm. This change is called the *inverse barometer* effect. The inverse barometer effect is seldom exactly observed in nature, because of the complex ways in which shallower waters of the continental shelves interact with passing atmospheric pressure systems.

Another major meteorological process contributing to the surge is the drag or stress on the sea surface due to the wind, measured as the horizontal force per unit area. Wind stress depends upon the wind speed and air density. The wind strength as well as direction relative to the coastline are important factors in elevating the sea surface. Wind blowing toward the coast causes a much greater rise in sea surface than wind blowing parallel to the shore. The effect of the wind on sea level increases inversely with water depth. Thus, the surge is amplified when wind blows over wide regions of shallow water, such as the North Sea or the northern end of the Bay of Bengal, off the coast of Bangladesh. The superposition of wave run-up on a high surge increases the storm hazard.

Two distinct types of meteorological disturbances produce major surges leading to coastal flooding and beach erosion (Dolan and Davis, 1994). These include tropical and extratropical cyclones. Tropical cyclones (known as *hurricanes* in the Atlantic basin, *typhoons* in the western Pacific, and *cyclones* in the Indian sub-continent) are intense low-pressure systems that strengthen over the ocean at low latitudes, where sea surface temperatures are at least 27°C (Landsea *et al.*, 1999). Tropical cyclones are usually small in extent but very powerful. The storm surge produced by a hurricane is a dome of water raised by the low barometric pressure, coupled with the strong wind shear (maximum sustained wind speeds of at least 119 km/h; surges 1–2 m), particularly on the right side of the low-pressure system. The hurricane's strength is greatest in the upper right-hand quadrant, as the high-velocity winds flowing counterclockwise around the eye (in the Northern Hemisphere) reinforce the lower-velocity steering winds which direct the forward motion of the cyclone (Coch, 1994).

Extratropical cyclones typically have lower wind speeds than hurricanes. Yet they often inflict considerable damage along the coasts because they cover a much wider area (> 1,000 km for extratropical cyclones versus 100–150 km for hurricanes) and are longer in duration, often persisting over several tidal cycles at a particular location (Dolan and Davis, 1994). South of Cape Hatteras, along the East Coast of the United States, tropical storms occur more frequently than extratropical storms, whereas further north, extratropical cyclones become dominant (Zhang *et al.*, 2000).

In addition to weather systems, surges are influenced by the configuration of the coastline and bathymetry. Surges are amplified by a wide continental shelf and where the coastline makes a right-angle bend, as, for example, at the apex of the New York Bight (Coch, 1994) or along the coast of Bangladesh (Murty and Flather, 1994). Interaction between surges and tides due to local topography may considerably change surge amplitude and phase.

The surge is amplified significantly when it coincides with astronomical tides (Wood, 1986). Above-average tides occur at new or full moon (spring tides), and at the solstices and equinoxes. These fortnightly or seasonal high tides are reinforced when they coincide with perigee (the closest approach of the moon to the earth in its orbit), or perihelion (the closest approach of the earth to sun). Tides, especially at high latitudes, are further enhanced during the 18.6-yr lunar nodal cycle, when the angle between the plane of the moon's orbit and the ecliptic is minimum (see *Tides*).

Many major extratropical storms have struck the eastern North American coastline at times close to perigean spring tides (Wood, 1986). One of the worst in almost 100 years was the "Ash Wednesday Storm" of March 6–7, 1962 which affected the entire Atlantic coast from South Carolina to Portland, Maine. This storm owed much of its destructiveness to its duration over five tidal cycles. Another major perigean spring storm was the "Saxby Tide" or "Saxby Gale" of October 4–5, 1869, which caused enormous damage along the Bay of Fundy, Canada (Desplanque and Mossman, 1999). Tides were nearly 2 m above the previous record. More recently, the entire New York metropolitan transportation system was paralyzed by flooding caused by a powerful nor'easter on December 11–12, 1992. With tides already above normal due to full moon, the water level at the Battery tide gauge peaked at 8.5 ft above National Geodetic Vertical Datum (7.8 ft above mean sea level; US ACOE/FEMA/ NWS, 1995).

Historical observations

Historical storm activity has been monitored using meteorological data and tidal records. Dolan and Davis (1994) have calculated wave heights from wind field data for each important storm along the East Coast, United States between 1942–92. They developed an extratropical storm classification scheme, similar to the Saffir–Simpson scale used for hurricanes, based on significant wave heights, duration, and storm "power," which they define as the square of significant wave height times duration. While this approach provides an index of storm damage potential, it does not specifically measure surge levels associated with these storms.

Another approach is based on measurement of hourly water levels recorded by a tide gauge. Since the water height includes surge, astronomical tide, and long-term sea level trends, the latter two components must be subtracted. Twentieth-century patterns of storm activity along the East Coast, United States have been reconstructed from hourly tide gauge data with records longer than 50 years, after removal of astronomical tide and sea-level factors (Zhang *et al.*, 2000). Storm indices based on surge number, duration, and "integrated intensity" (area under surge curve greater than 2 standard deviations) showed considerable interdecadal variations, but no consistent long-term trend during the 20th century.

The likelihood of future storm surge events can be calculated from daily or monthly surge maxima measured at individual tide-gauge stations, after appropriate correction for tides and secular sea-level trends (Ebersole, 1982). Statistical analyses of such data can then be used to prepare return period curves which show the probability of occurrence of a given surge (or flood) height (Figure S61). The total flood level (surge + tide + sea-level rise) is a more meaningful elevation for assessing impacts to coastal settlements and installations. The 100-year flood level is a commonly employed reference level (FEMA, 1997). It represents the flood height that has a 1% probability of occurrence in any given year.

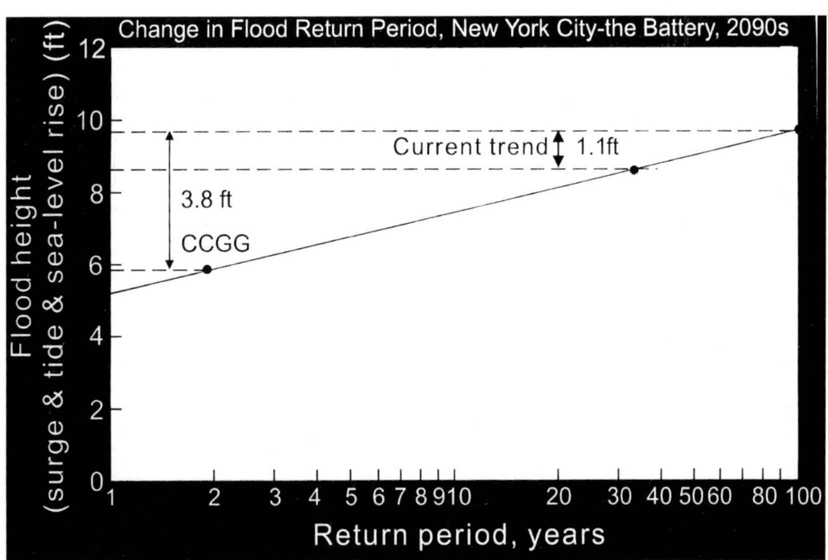

Figure S61 Flood return curve for New York City (based on data from S. Couch, US Army Corps of Engineers, New York District). The present 100-year flood level is 9.7 ft (upper dashed line). Extrapolating current rates of sea-level rise, the 100-year flood height is reduced to 8.6 ft (2.6 m) by the 2090s, with a return period of 33 years. A climate model (Canadian Climate Centre Model, greenhouse gases only, CCGG) projects a sea-level rise of 3.8 ft (1.15 m), with a return period of ~3 years (based on data in Gornitz *et al.*, 2002).

Modeling surges

The empirical data needed to derive return period curves are not available at all locations. Therefore, the storm-induced surge is often estimated by mathematical models, such as the WES Implicit Flooding Model (WIFM), tidal hydrodynamic model, frequently used by the US Army Corps of Engineers (Butler, 1978; Butler and Sheng, 1982). WIFM solves vertically integrated dynamic, shallow-water wave equations of fluid motion, incorporating information on bathymetry, topography, wave, and meteorological data in order to simulate coastal flooding. An important feature of this model is its ability to stretch the numerical grid, which allows a denser grid resolution in areas of particular interest.

The US National Weather Service has developed a numerical- dynamic surge model, SLOSH, for real-time forecasting of hurricane surges on continental shelves, open coastlines, and estuaries (Jelesnianski *et al.*, 1992). The model is activated by inputs of simple meteorological parameters (minimum central pressure, radius, storm track, and speed along track). The storm model balances surface forces including surface friction. Short period wave "run-up" is omitted; other longer-term wind–wave effects are generalized. SLOSH surge predictions are regularly applied to hurricanes affecting the US East and Gulf Coasts.

Another type of surge model treats the land–water interface as a moving boundary with inland grid cells becoming flooded as the water-level rises due to the storm surge or tide, and draining as the water recedes (Hubbert and McInnes, 1997). This model can readily be applied to projections of future sea-level rise.

Effect of sea-level rise on coastal flooding

Anticipated climate change is expected to accelerate rates of global mean sea-level rise by factors of 2–5, by the end of this century (IPCC, 2001). The rise in sea level will come from thermal expansion of the upper ocean layers, melting of mountain glaciers, with a larger uncertainty surrounding possible contributions from melting of the polar ice sheets.

The rise in mean sea level will be superimposed on storm surges from storms and astronomical tides. Locally, sea-level rise could be higher than the global mean change, due to land subsidence. Although the increase in flood height due to sea-level rise may only comprise a relatively small percent of the total elevation (assuming that the storm strikes at high tide), yet even a minor increase in sea level could significantly shorten the flood return period (Figure S61). For example, in New York City, the present-day 100-year flood height is slightly under 3 m. If sea level were to continue rising at current rates (~2.7 mm/yr), by the end of the century a flood of only 2.6 m would be equivalent to today's 100-year event. This flood has a return period of 33 years. In the most extreme scenario of sea-level rise (over a meter above present levels), the return period could be reduced to as little as 3–4 years (Gornitz, 2002). This could lead to much more frequent episodes of coastal flooding in low-lying areas (see also *Natural Hazards*).

Coastal surge hazards

Areas of the world most vulnerable to surges from tropical cyclones include the low-lying coasts of southeast Asia adjacent to major river deltas, the southeastern United States, and the northeastern coast of Australia. The convergence of several conditions places Bangladesh especially at risk to storm surges. The coast is nearly at sea level, the continental shelf is shallow especially in the eastern part of Bangladesh, the tidal range is high, and storms tracks have a tendency to recurve near the apex of the Bay of Bengal (Murty and Flather, 1994). A severe cyclone in November, 1970 produced surges of over 9 m, killing several hundred thousand people (Pugh, 1987). Another cyclone killed 11,000 people in May, 1985.

In the United States, the low-lying Gulf states, Florida, and the Carolinas are at high risk to hurricanes, although the Northeast is not immune. A hurricane that struck Galveston, Texas, on September 8–9, 1900 raised water nearly 5 m and killed almost 6,000 people (NOAA, National Weather Service). In September, 1938, another hurricane swept across Long Island and southern New England, pushing water levels 4.5–6 m above normal, and killed around 700 people (Ludlum, 1988). Other major damaging hurricanes include Hugo (South Carolina, 1989, category 4 on the Saffir–Simpson scale), Andrew (Florida, Louisiana, 1992, cat. 4), Camille (Mississippi, Louisiana, 1969, category 5), and an unnamed hurricane that struck the Florida Keys in 1935 (category 5).

Extratropical cyclones can also result in severe coastal flooding. Regions prone to surges from extratropical cyclones include the east coast of North America north of Cape Hatteras, the coasts surrounding the North Sea, and the Adriatic Sea. The Ash Wednesday Storm of March 1962, affecting much of the US east coast was mentioned above. Other recent major "nor'easters" include the December 11–12, 1992 storm, the March 13–14, 1993 storm, and the "Halloween Storm" of October 31 to November 1, 1991 (also known as "The Perfect Storm"—subject of a best-selling novel and film).

In northern Europe, the North Sea is exposed to Atlantic extratropical storms which move across shallow shelf waters. Furthermore, the gradual subsidence of the southern North Sea area makes low-lying regions of southeast England and the Netherlands increasingly vulnerable to flooding from surges. In January 31 to February 1, 1953, a major storm broke through the dike defenses of the Netherlands, flooding a large portion of the country (Pugh, 1987).

Vivian Gornitz

Bibliography

Butler, H.L., 1978. *Numerical simulation of tidal hydrodynamics: Great Egg Harbor and Corson Inlets*, New Jersey. Technical Report H-78-11, Vicksburg, MS: US Army Engineer Waterways Experi-ment Station, 117 p.

Butler, H.L., amd Sheng, Y.P., 1982. *ADI procedures for solving the shallow-water equations in transformed coordinates.* Proceedings 1982 Army Numerical Analysis and Computers Conference, ARO Report 82-3, pp. 365–380.

Coch, N.K., 1994. Hurricane hazards along the northeastern Atlantic coast of the United States. In Finkl, C.W., Jr. (ed.), *Coastal Hazards: Perception, Susceptibility and Mitigation. Journal of Coastal Research Special Issue No. 12*, pp. 115–147.

Desplanque, and Mossman, D.J., 1999. Storm tides of the Fundy. *Geological Review*, **89**: 23–33.

Dolan, R., and Davis, R.E., 1994. Coastal storm hazards. In Finkl, C.W., Jr. (ed.), *Coastal Hazards: Perception, Susceptibility and Mitigation. Journal of Coastal Research Special Issue No. 12*, pp. 103–114.

Ebersole, B.A., 1982. *Atlantic Coast Water-Level Climate.* WIS Report 7, Vicksburg, MS: US Army Engineer Waterways Experiment Station.

Federal Energy Management Agency (FEMA), 1997. *Answers to Questions About The National Flood Insurance Program*, Washington, DC: US National Government Printing Office.

Gornitz, V., Couch, S., and Hartig, E.K., 2002. Impacts of sea level rise in the New York city metropolitan area. *Global and Planetary Change*, **32**: 61–88.

Hubbert, G.S., and McInnes, K.L., 1997. A storm surge inundation model for coastal planning and impact studies. *Journal of Coastal Research*, **15**: 168–185.

Intergovernmental Panel on Climate Change (IPCC), 2001. Houghton, J.C., Ding, Y., Grigg, D.J., Noguer, M., van der Linden, P.J., Dai, X., Maskell, K., and Johnson, C.A. (eds). *Climate Change 2001: The Scientific Basis.* Cambridge: Cambridge University Press.

Jelesnianski, C.P., Chen, P., and Shaffer, W.A., 1992. *SLOSH: Sea, Lake, and Overland Surges from Hurricanes.* NOAA, National Weather Service, Silver Spring: NOAA Technical Report, NWS 48.

Landsea, C.W., Pielke, R.A., Jr., Mestas-Nunez, A.M., and Knaff, J.A., 1999. Atlantic basin hurricanes: indices of climatic changes. *Climatic Changes*, **42**: 89–129.

Ludlum, D.M., 1988. The great hurricane of 1938. *Weatherwise*, **41**: 214–216.

Murty, T.S., and Flather, R.A., 1994. Impact of storm surges in the Bay of Bengal. In Finkl, C.W., Jr. (ed.), *Coastal Hazards: Perception, Susceptibility and Mitigation. Journal of Coastal Research Special Issue No. 12*, pp.149–161.

Pugh, D.T., 1987. *Tides, Surges, and Mean Sea-Level.* Chichester: J. Wiley & Sons Ltd.

US Army Corps of Engineers/FEMA/National Weather Service, 1995. *Metro New York Hurricane Transportation Study. Interim Technical Data Report.*

Wood, F., 1986. *Tidal Dynamics: Coastal Flooding, and Cycles of Gravitational Force.* Dordrecht: D. Reidel Publishing Co.

Zhang, K., Douglas, B.C., and Leatherman, S.P., 2000. Twentieth-century storm activity along the US east coast. *Journal of Climate*, **13**: 1748–1761.

Cross-references

Changing Sea Levels
Meteorologic Effects on Coasts
Natural Hazards
Sea-Level Rise, Effect
Tide Gauges
Tides
Waves

STORMS—See METEOROLOGICAL EFFECTS

STRANDFLAT

The strandflat is a rim of gently sloping bedrock plain in front of higher land or coastal mountains (Klemsdal, 1982, 1985). The plain has a very irregular terrain with small differences of height. Most of the bedrock plain is covered with a thin mantle of loose material; only locally does the loose material have forms of its own. The gently sloping, undulating, flat produces, when meeting the sea, an uneven coastline with numerous bays, coves, inlets, islands, islets and skerries, all characteristics features of the skerry zone. The gradient of the supramarine zone varies between 5 and 25 m per km. The topography of the supramarine and skerry zones continues into the sea. At a depth between 30 and 60 m a break in the slope occurs and steeper slopes lead down to the paleic, old forms of the bankflats constituting part of the continental shelf. Inland, the strandflat terminates at heights between 30 and 80 m, where in most places a steep slope leads up to higher land, though in some places the transition is more gentle. The coastline of the mainland and the islands of the strandflat is rather long. The shore-zone is mainly either a gently sloping ice-smoothed rocky shore, without any post-glacial alteration, or a stony beach composed of reworked till. Only locally have post-glacial littoral processes produced pocket beaches of sand or clay (Klemsdal, 1982, 1985).

The term strandflat became a geomorphological term in 1894 when Reusch introduced it into the literature. Reusch put forth, from observation of strandflat localities along the Norwegian coast, a description of the strandflat and a discussion of its origin. Marine abrasion and frost weathering were said to be the main processes responsible for the development of the strandflat. Regarding the development of the strandflat, Nansen (1922) proposed frost weathering and sea ice frozen onto the shore, combined with waves breaking up the sea ice. Local glaciation, especially cirque glaciers, and also currents of inland ice, moving and spreading out in the coastal areas, were added to the picture by Holtedahl in 1929.

Today, the origin of the strandflat is closely associated with the general development of the landforms of the Scandinavian landmass. The land was exposed to denudation through the Mesozoic and the first part of the Tertiary. In a warm climate with dry and wet periods, denudation produced a land surface, the paleic surface (Gjessing, 1967), with well-rounded, mature landforms, close to a plain in the peripheral parts of the landmass, a propitious starting point for the development of the strandflat.

In the Tertiary the Scandinavian landmass was elevated and tilted, giving a steeper slope towards the west and northwest, bringing the paleic surface to different heights above sea level. Inland along the coast it varies from sea level to 500–700 m above sea level. Along the coast the elevation of the paleic surface may have been negligible.

Probably coincident with the uplift, the climate became temperate and more humid. Fluvial processes produced initial forms along a fluvial pattern in the paleic surface, most distinct in the west and northwest. Along the coast marine abrasion and denudation, favored by the old paleic surface, may have started the development of a peneplain, which can be easily fitted into the development of the strandflat.

Cirque glaciers, descending from higher coastal mountains onto the level surface along the coast, were important in the widening and splitting up of the strandflat. Valley glaciers and ice flows of the inland ice spreading out in the coastal areas (Holtedahl, 1929; Dahl, 1947), may also have taken part in the development. In interglacial times and at the beginning of the glacial periods frost weathering and marine abrasion were momentous, and together with mass movement and littoral transport, most active in the evolution of the strandflat.

Although its origin is still not settled, it is said that it is polygenetic (Holtedahl, 1959, 1960; Klemsdal, 1982, 1985); starting with the denudational processes of the paleic surface and stressing the frost weathering and marine abrasion, even though glacial and fluvial activity also have played a part in the development of the strandflat.

The distribution of the strandflat is closely connected with areas where Quaternary glaciations took place and where a harsh climate in Holocene favored frost weathering.

In Norway, the strandflat is found along the coast from the Stavanger area in the southwest to the western Finnmark in the northeast (Klemsdal, 1982, 1985). In eastern Finnmark, the strandflat is only a narrow strip of land, some broader but still narrow submarine parts, and no skerry zone. The maximum width is approximately 40 km, but as a mean the strandflat is 16 km wide. The supramarine part of the strandflat is normally between 5 and 10 km in width, but can reach 15 km. In other places, the strandflat is a few hundreds meters wide.

Furthermore, the strandflat is found in the archipelago of Svalbard, on Novaya Zemlya; on the peninsula of Taimyr; the archipelago on the west coast of Scotland; on Iceland; along the southwestern coast of Greenland; and in the arctic areas of Canada and Alaska. In the Southern Hemisphere, the strandflat is observed along parts of the Antarctic Peninsula and neighboring islands. The mentioned localities of the strandflat are all in a harsh climate where frost weathering takes place and where Quaternary glaciations occurred or influenced the processes.

Tormod Klemsdal

Bibliography

Dahl, E., 1947. On the origin of the strand flat. *Norsk Geografisk Tidsskrift*, **11**: 159–172.
Gjessing, J., 1967. Norway's paleic surface. *Norsk Geografisk Tidsskrift*, **21**: 69–132.
Holtedahl, H., 1959. Den norske strndflate med såerlif henblikk på den utvikling i kystområdene på Møre. *Norsk Geografisk Tidsskrift*, **16**: 285–303.
Holtedahl, H., 1960. The strandflat of the Møre-Romsdal Coast, west-Norway. Skrifter fra Norges Handelshøyskole. *Geogr. Avhandlinger* **7**: 35–43.
Holtedahl, O., 1929. On the geology and physiography of some Antarctic and Sub-Antarctic Islands. Scientific Results Norwegian Antarctic Expedition 1927–1928. *Det Norske Videnskaps Akademi Oslo*. **3**.
Klemsdal, T., 1982. Coastal classification and the coast of Norway. *Norsk Geografisk Tidsskrift*, **36**: 129–152.
Klemsdal, T., 1985. Norway. In Bird, E.C.F., and Schwartz, M.L. (eds.), *The World's Coastline*. New York, Van Nostrand Reinhold, 273–283.
Nansen, F., 1922. The strandflat and isostasy. *Vidensk. Skrifter I. Mat. Naturv. Klasse nr*. **11**.
Reush, H., 1894. Strandflaten, et nytt traek i Norges geografi. *Norges geologiske* Undersøkelser. **14**: 1–14.

Cross-references

Cliffs, Erosion Rates
Glaciated Coasts
Ice-Bordered Coasts
Paraglacial Coasts
Rock Coast Processes
Shore Platforms
Weathering Processes

STRUCTURES—See SHORE PROTECTION STRUCTURES; NAVIGATION STRUCTURES

SUBMARINE GROUNDWATER DISCHARGE

Background and definitions of terms

Groundwater comprises about 95% of all the freshwater on earth (Winter *et al.*, 1998). Of this groundwater, an unknown amount discharges to lakes, rivers, estuaries, bays, and oceans on an annual basis. Groundwater discharged to the marine environment (estuaries, bays, and the ocean) is known as submarine groundwater discharge (SGD), a term originally proposed by Johannes (1980). SGD occurs offshore as seepage (groundwater discharges from both unconfined and confined aquifers) (Simmons, 1989; Simmons and Reay, 1992), as recirculated seawater mixed with freshwater (Buddemeier, 1996), and in springs (Kohout, 1966). Additionally, SGD can occur in geo-thermal heat flows or vents (Kohout *et al.*, 1979; Schwerdtfeger, 1981; Nossin *et al.*, 1987), upwelling in reefs (Simmons and Love, 1987; Tribble *et al.*, 1992) and oceanic islands (Rougerie and Wauthy, 1993).

The chemical makeup of SGD varies spatially and temporally depending upon the composition of the aquifer material, residence time of the freshwater within the aquifer, onshore hydraulic heads, the amount of recirculation of seawater and groundwater in the nearshore, and the magnitude of the tidal fluctuations. The primary chemical characteristics (or *finger print*) of the water are functions of the original geochemical composition of the groundwater and the residence time in the aquifer and minerals contacted along the flowpath.

The geologic principle of *uniformitarianism* states that geologic processes today (including SGD) are similar to the processes of millions of years ago (Wicander and Monroe, 1993). Therefore, SGD processes occurring today in beaches, estuaries, and inlets are thought to be the same as past processes, since the formation of the continental plates, changed only by the conditions that drive them. This is generally true for all types of seepage, except for seepage that is influenced by anthropogenic factors such as drainage canals and aquifer pumpage. SGD is not a recently discovered phenomenon. The earliest account of springs in the ocean was presented by Kohout (1966) in which he indicated that Pausanius, who lived in the second century AD, noticed that off the coast of western Italy ... "there is water boiling in the sea and it has created an artificial island."

Importance of SGD in the coastal environments

The significance of SGD has become more widely appreciated in the last two decades, especially in efforts to improve understanding of marine geologic and geochemical processes, nearshore and offshore water quality anomalies, nutrient loading, and ecological impacts (Kohout, 1966; Johannes, 1980; Simmons, 1992; Zekster and Loaiciga, 1993; Buddemeier, 1996). Recent research, even by lay scientific journals, has been stressing the importance of SGD (Svitil, 1996). Anthropogenic nutrient input from stormwater runoff to the oceans and estuaries was previously believed to have been a major cause of eutrophication of coastal waters (Zekster and Laiciga, 1993; Lapointe and Matzie, 1996; US Environmental Protection Agency, 1998). Although groundwater was traditionally overlooked or underestimated in previous investigations, it has been demonstrated that groundwater input can be a major source of nutrient input to the oceans. Recent research has identified nutrient enrichment as one of the major processes responsible for the degradation of coastal waters. This enrichment, which causes problems ranging from anoxia to harmful algae blooms, is directly related to the transport of nutrient-enriched groundwater into coastal waters through SGD as demonstrated by Lapointe *et al.* (1990); Shinn *et al.* (1994); Harbor Branch Oceanographic Institute (1995). Groundwater and recharge areas are often contaminated by agricultural, urban fertilizers, animal waste, shallow injection wells, septic tanks, cesspools, and other nutrient sources. The result leads to increased concentrations of nitrogen (N) and phosphorus (P) in SGD (Simmons and Love, 1987; Lapointe and O'Connell, 1989; Shinn *et al.*, 1994, 1997; Finkl *et al.*, 1995; Finkl and Krupa, 2000; Finkl and Krupa, 2003) with disastrous long-term effects.

Governmental agencies and private organizations need SGD information to deal with water degradation issues, develop accurate numerical models, create tidal mixing models, make circulation and temperature predictions, and to establish correct boundary conditions for numerical groundwater and oceanic model simulations (Buddemeier, 1996). Accurate determination of SGD is essential because flow volumes are used to calculate mass balances and in the determination of nutrient fluxes to the coastal waters. However, experience has shown that quantifying SGD input is difficult because of uncertainties in groundwater flux measurements.

Occurrence of SGD

The rate and direction of seepage is influenced by hydraulic heads within the aquifer and by the underlying lithology, watershed topography, and seasonal weather patterns (Freeze and Cherry, 1979). Seepage is known to occur from the nearshore (Cable *et al.*, 1997a,b) up to about 100 km offshore (Manheim, 1967). The process of SGD is more complex than seepage to terrestrial water bodies because of the influence of bathymetry, micro-topology, tidal cycles, winds, currents, recirculation at or near the freshwater/saltwater interface, and the chemical, biochemical, geological, and ecological processes occurring at the freshwater/saltwater interface (Huettel and Gust, 1992; Buddemeier, 1996; Huettel *et al.*, 1996, 1998; Robinson, 1996; Shinn *et al.*, 1997; Uchiyama *et al.*, 2000).

Seepage occurs as baseflow (flow from saturated groundwater storage) and interflow (flow from subsurface storm flow) to the surface water (Davis and DeWiest, 1966; Freeze and Cherry, 1979). The baseflow component of seepage can be further divided into two primary velocity components. The first baseflow velocity component provides the normal daily discharge to the estuary/ocean and the second velocity component is controlled by daily tidal cycles. Interflow is the temporary saturated flow within the unsaturated portion of the aquifer. This flow generally begins some time after a rainstorm or seasonal climatic change (winter to spring) and occurs until hydraulic gradients within the aquifer have leveled or have dropped to background levels.

Indentifying and measuring SGD

Nearshore submarine groundwater discharge can be seen by visual observations and measured by pore water devices, seepage meters and results of isotope analysis (Simmons and Love, 1987; Harbor Branch Oceanographic Institute, 1995; Nuttle and Harvey, 1995; Moore, 1996; Cable *et al.*, 1997a,b). Offshore SGD can also be identified by side-scan sonar, near bottom echo sounders, seismic profiling, geophysical logging, conductivity, depth, and temperature (CDT) profiling, and deep-sea drilling (Manheim, 1967; Zektser and Meskheteli, 1988; Land *et al.*, 1995; Merchant *et al.*, 1996; Guglielmi and Prieur, 1997).

Measurement of seepage was initially developed for onshore applications, but adaptation of equipment and practices to the marine environment has recently made offshore applications commonplace. Using this relatively new and sophisticated technology, SGD rates and water quality parameters can be determined in the field. A comprehensive SGD study along coastal regions generally requires a variety of methods to best determine seepage rates. For example, monitor wells and piezometers can be used to determine horizontal and vertical gradients while water quality, radiochemistry, and seepage meter data are used to estimate the quantity and direction of seepage.

Coastal piezometers and groundwater wells

To determine the tidal effects from the horizontal and vertical movement of the groundwater/seawater boundary, Urish and Ozbilgin (1989) recommend the placement of one control well onshore, near the surf area. Onshore monitor wells and piezometers in shallow water can be installed using standard well technology, including casing advancement methods and jetting. Methods for installing monitor wells or piezometers underwater are adapted from land-based techniques, but are far more complicated and time-consuming (Lee and Cherry, 1978; Shinn *et al.*, 1994). Wells used to measure vertical gradients require short-screen intervals or open-ended pipes (Freeze and Cherry, 1979; US Army Corp of Engineers, 1993). Screen lengths can be as short as 0.15 m, but are generally 0.6 m. The screens should be sufficiently separated both horizontally and vertically, to calculate differences in hydraulic heads between the monitored groundwater zones and surface water (Harvey *et al.*, 2000, 2002).

Continuous logging of water level data from the piezometers, monitor wells, and surface water should be corrected to equivalent freshwater heads, if they are completed on the same flow path, to allow comparison of data (Senger and Fogg, 1990; Reich, 1996; Rasmussen, 1998). Tidal filtering of water level data should be done to obtain average vertical and horizontal gradients. To calculate the equivalent freshwater head, it is necessary to measure the transient salinities and/or total dissolved solids (TDS) concurrently with the water level measurements. All wells, piezometers, and the ground surface at each site should be surveyed vertically (elevation) and horizontally (longitude and latitude or state planar coordinates) to ensure accurate gradient calculations and site maps preparation.

Seepage meters

Since the 1930s, scientists have used a variety of chambers and seepage meters to study the benthic boundary layers associated with canal linings, rivers, and lakes (Gale and Thompson, 1975; Sonzogni *et al.*, 1977; Carr and Winter, 1980). One commonly used seepage meter is the modified drum apparatus devised by Lee (1977). This meter, commonly called a "Lee meter" or a conventional seepage meter, has been used in numerous marine studies (Lewis, 1987; Cable *et al.*, 1997a,b; Shinn *et al.*, 2002). The Lee meter is constructed from a standard metal drum (208 L) cut into thirds. The middle section is discarded and the two end sections are each prepared by using the existing hole on one end and drilling a similar hole on the other end. A connector is fitted to each hole; these are used to attach seepage bags after the meters are placed. The meters are installed into the ground with the open end downward, seepage bags are attached to the connectors and monitored.

Conventional seepage meters operate very simply: a change in water volume in the bag over time represents a seepage flux across the sediment interface. The quantity of water in the bag is compared with the initial volume in order to determine gain or loss of water. When the bag gains water, groundwater is discharging to surface water. Conversely, when the bag loses water, the surface water is recharging the groundwater. The change in quantity of water is averaged, per unit area, and usually converted to units such as milliliters per square meter per hour or centimeters per day.

Seepage can also be measured with automated seepage meters. These devices utilize electronic flowmeters located on the tops of seepage meters or ultrasonic flow meter sensors located on the top of a funnel apparatus (Reay and Walthall, 1992; National Aeronautics and Space Administration, 1992; Taniguchi and Fukvo, 1993; Paulsen *et al.*, 1997; Krupa *et al.*, 1998; O'Rouke *et al.*, 1999). Automated seepage meters can best deal with tidal cycles when continuous recorders are used. This allows the investigator to capture seepage rates over longer periods of time which includes storm events and normal tidal fluctuations, investigation duration (months versus days) and finally increases the statistical significance of the flux readings. Manual seepage meters capture data only over the time the bags have been installed on the meters.

Seepage meters are typically positioned within the first 100 m of the coast, according to Bokuniewicz and Zeitia (1980). Recent studies in Panacea, Florida and Perth, Australia, show that seepage typically decreases with distance from shore. However, multiple seepage meters are required at various locations in order to improve statistical significance and to evaluate spatial variability within a designated seepage location. The typical configuration of a Lee meter in the offshore environment is shown in Figure S62. Sediment changes, onshore topography, bathymetry, distance from shore, and ecological indicators such as sea grasses, corals, or algal mats should be considered when placing seepage meters. Figure S63 shows a typical nearshore seepage meter scenario with shallow and deep piezometer and conventional seepage meter.

Water budget—mass balance approach

The mass balance approach to estimation of SGD summarizes all inflows and outflows in a study area. Cherkauer (1998) states "the principle of conservation of the mass requires that the total quantity of groundwater entering the surface water must equal the net recharge within the watershed contributing flow." This approach assumes the system is in steady-state and there is no net change in the groundwater levels. Thus, the net infiltration (rainfall minus transpiration, evaporation, and anthropogenic uses) to the aquifer is equal to the SGD. Several investigators have completed large-scale regional water budgets (Buddemeier, 1996; Church, 1996), while others have attempted continental water budgets for Australia (Zektser *et al.*, 1983).

As with all models, the mass balance approach makes some assumptions about the system being modeled. These approaches and approximations overlook coastal system dynamics and may provide inaccurate results.

Chemical mass balance method

Natural tracers are also used to estimate SGD. Some common natural tracers include radioactive isotopes, silica, and chloride. Geochemical data can provide a long-term integrated picture of hydraulic processes in the aquifer and the groundwater/surface water exchange (Harvey *et al.*, 2000, 2002). If a particular solute occurs in groundwater and is not in other sources (or is in reduced concentrations) and the solute concentration in each source is known, the information can be used to infer the magnitude of SGD (Church, 1996; Moore, 1996, 1997; Burnett *et al.*, 2003). The magnitude of groundwater contributions has largely been ignored until recent isotope (^{236}Ra) work conducted in the South Atlantic Bight by Moore (1996). Moore indicates that groundwater contribution to the nearshore was approximately 40% of the riverine inputs during his study period.

Using seepage meters to measure SGD

When using seepage meters to make direct SGD flux measurements, the seepage meters, piezometers, and monitor wells should be installed two to three days prior to sampling to allow the sediments to rebound and the system to return to equilibrium. SGD measurements made before equilibration may result in erroneous seepage rates (Belanger and Walker, 1990). The installation of traditional and automated seepage meters on the seafloor or lake/river bottom is a simple task. The installers usually stand on the seepage meter, gently rocking back and forth until the meter is pushed about 15–20 cm into the sediment surface. This can be difficult when the seepage meter is more than 3 ft (1 m) below the surface of the water. Placement of the meter is important; the threaded hole is usually placed on the upslope side of the sediment to allow the movement of water, air bubbles, and gases from the seepage meter to the bag.

Gas collection in the seepage meter can become a problem because gasses are produced by the decomposition of organic material within the seepage meters. Gas production creates additional challenges with automated seepage measuring devices. Several investigators using electronic flowmeters to measure SGD found that air bubbles sometimes render the device useless until a change in the tidal direction or wave action forces the bubbles to move (Krupa *et al.*, 1998).

The sampling frequency of the traditional meters is variable and best determined by trial and error at each site. Broken seepage collection bags indicate higher seepage rates and require more frequent sampling. The sampling frequency may be as short as 0.5 h to as long as 12 h.

The seepage data must be adjusted to obtain accurate SGD calculations. Studies show that seepage meters require the use of a frictional coefficient to correct for the frictional losses in the meters and collection bags. (Shaw and Prepas, 1989; Belanger and Montgomery, 1992).

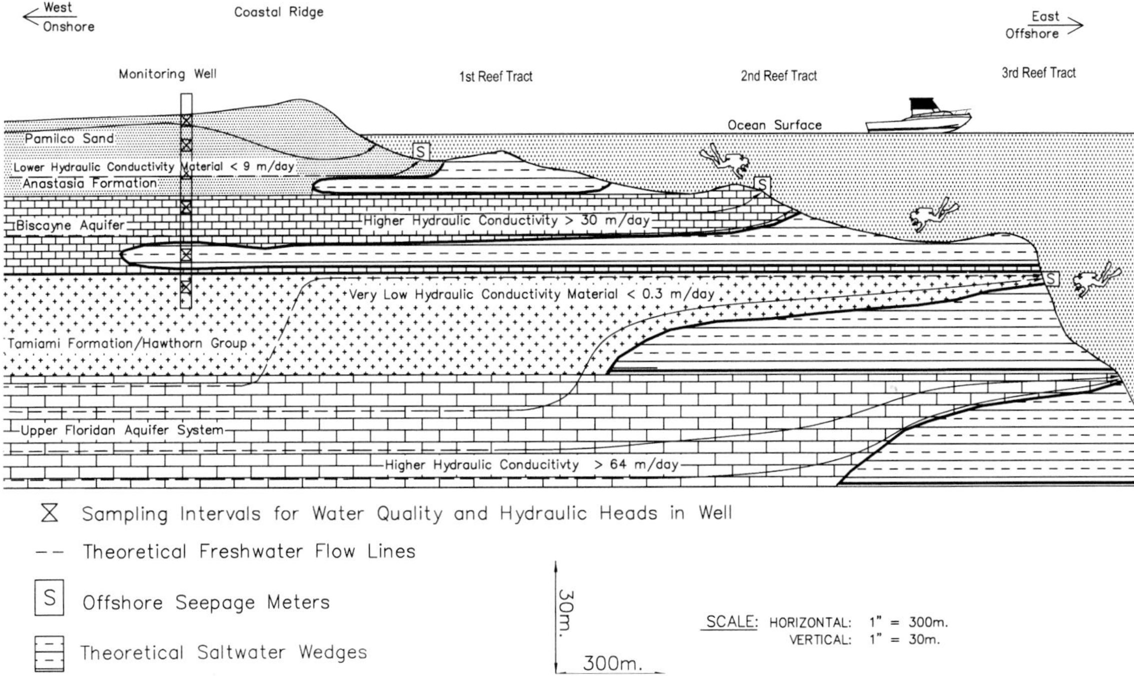

Figure S62 Generalized submarine groundwater discharge locations, lithostratigraphic, and theoretical flow line cross-section of the aquifer for Palm Beach County, Florida.

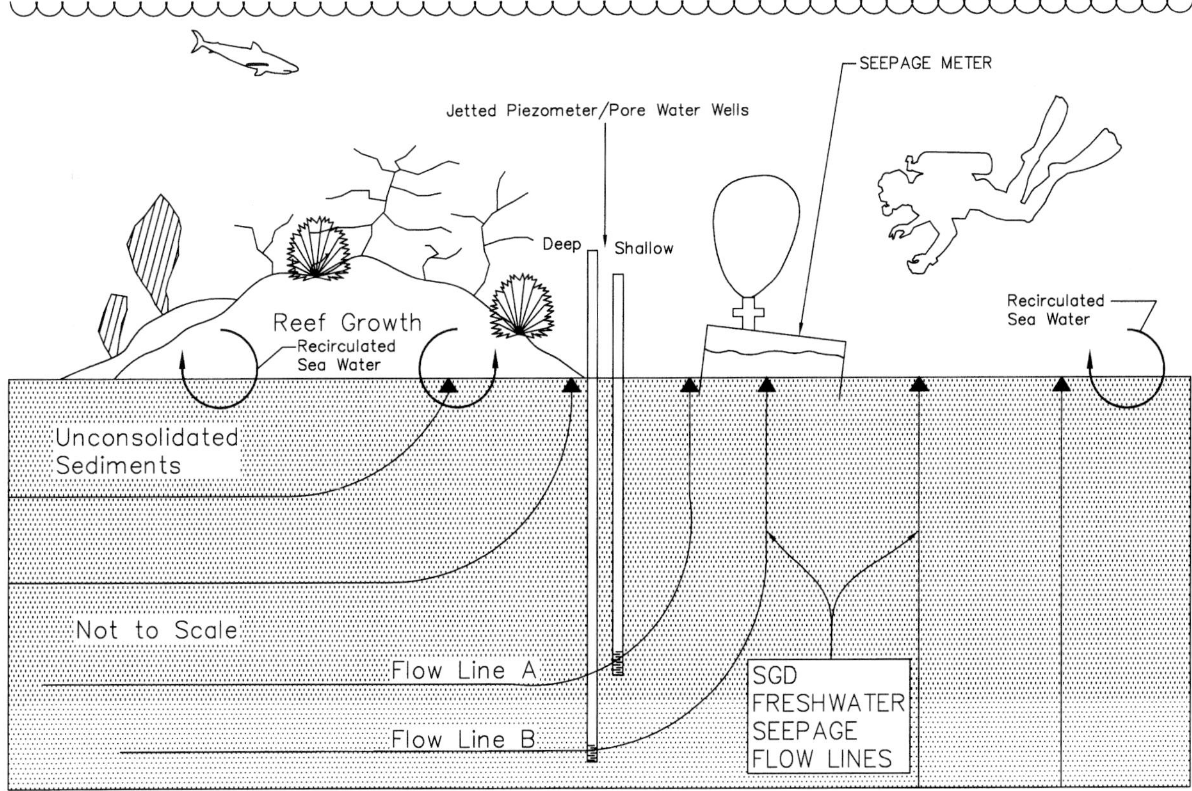

Figure S63 Typical nearshore seepage investigation set-up with theoretical freshwater flow lines, recirculated flow lines and deep and shallow monitoring piezometers.

Recent tank test results by Belanger indicate that different meter designs require different correction factors, and each meter requires different correction factors for discharge and recharge situations. The seepage rate is multiplied by the coefficient to calculate the actual SGD rate.

Factors affecting SGD results

Several factors besides the actual groundwater discharge rates affect SGD measurements and calculations. Gaps in data collection may cause errors. Many of the SGD measurement methods collect data for limited time periods and may miss the enhanced groundwater discharge during storm events. The size of the study area is also a concern because studies have found that SGD varies spatially and temporally. These spatial and temporal differences can be challenging when measurements are required over large areas.

Care must be taken in estimating the groundwater flux. It is important to avoid including oceanic and fluvial fluxes more than once. Riverine groundwater discharge must be calculated between the lowest gauging station and the mouth of the river and biogeochemical processes must also be considered.

Sediment lithology of coastal aquifers will vary based on depositional environments. In the field, any combination of confined and unconfined aquifer scenarios exists. Davis and DeWiest (1966), Strack (1975), Bear (1979), and Todd (1980) include solutions for analytical solution of groundwater flow from the aquifer to the coast for various aquifer combinations including unconfined, confined, stressed (pumping and drainage), and natural flow conditions.

SGD is highly complex and is rarely treated in terms of fully three-dimensional (3D) density dependent miscible fluid flow in a porous medium (Reilly, 1993). Instead, the system conceptualization is usually highly simplified to facilitate physical limitations and project budgets. The accuracy of analysis is dependent on the assumptions in the following three categories: (1) the mixing process, (2) the characteristics of the aquifer under study, and (3) the desired scale and detail of the resulting analysis (Reilly, 1993). These issues are discussed in depth, along with a brief discussion on the mathematical basis of different solution methods in Reilly (1993).

Relating local measurements to regional fluxes can be an arduous task and can result in misleading conclusions. The calculations used to transform water-level measurements in wells, seepage meter rates, and radioisotope data into a flux for an area can be complicated and time-consuming. Seasonal and tidal factors need to be understood and applied to *in situ* seepage results. Often, values obtained through analytical or numerical simulation were not verified in the field. Experience has shown that values determined by models may vary significantly from the estimates based on field measurements (Belanger and Walker, 1990).

Traditional mathematical models do not consider the cyclic or transient discharges from the groundwater component. Many analyses, such as the Darcy solution, assume steady-state conditions (Davis and DeWiest, 1966; Freeze and Cherry, 1979) and overlook the transient vertical position of the land/ocean interface (Urish and Ozbilgin, 1989).

Methodology can also affect seepage results. In traditional seepage meters, empty seepage bags appear to exert a slight suction that pulls water from the meter as the bags slowly unfold. Prefilling the bags reduces the anomalous short-term inflow and allows both discharge and recharge to be measured. Open water wave action and currents are reported to cause induced seepage rates (Libelo and MacIntyre, 1994; Shinn *et al.*, 2002). Harvey *et al.* (2002) collectively terms these methodology effects as apparent seepage rates.

A recent septic tank study near a tidally controlled estuary in Tequesta, Florida (US), demonstrates the tidal effect. *In situ* heat-pulse flowmeters in onshore monitor wells were used to measure horizontal velocities during high, low, and slack portions of the tidal cycles. The results showed groundwater velocities varying at low tide from 0.39 m/day to as high as 1.33 m/day in different layers of an unconfined aquifer (Harbor Branch Oceanographic Institute, 1995). Robinson (1996) and others report seepage rates as high as six to eight times higher on an ebb tide. A wide range of *in situ* velocities can occur under varying tidal conditions and with differing aquifer heterogeneities. Therefore, groundwater velocity ranges should be evaluated at each study location.

Additional factors that need to be considered in the collection of SGD from seepage meters are summarized below:

- Mechanical and physical factors affecting seepage meter performance (Libelo and MacIntyre, 1994; Harvey *et al.*, 2000; Shinn *et al.*, 2002)
- Meter-to-bag connectors and types of bags used (Lee, 1977; Shaw and Prepas, 1990; Belanger and Montgomery, 1992)
- Pre-filled volume in the bag (Lee, 1977; Shaw and Prepas, 1989; Belanger and Montgomery, 1992)
- Method of bag deployment and retrieval (Lee, 1977; Belanger and Montgomery, 1992; Harvey *et al.*, 2000)
- Gas produced from decomposition within the sediment (Harvey *et al.*, 2000).

Determination of seepage rates

Traditional methods of calculating SGD use one groundwater velocity value derived from hydraulic gradients and hydraulic conductivity estimates. Applying one velocity component to a particular solution can be misleading, as can the approach of using a simplified model to describe a complex system. If measured groundwater velocities and/or gradients are available, this information should be used. Total discharge for the aquifer (equation 1) summarizes the products of the two velocities (tidal, baseflow/interflow) and their respective cross-sectional areas (Figure S64). V_{tide} is defined as the discharge from the seepage face both during low tide and high tide. V_{baseflow} is the groundwater that is discharging regardless of the tide.

Empirical solution methods

Traditionally, seepage has primarily been estimated by applying Darcy's equation. This approach estimates discharge based on the hydraulic gradient between two points. The discharge is reported in units of length^3/time (L^3/T). Darcy (1856) defined the specific discharge shown below:

$$q = \left(\frac{Q}{A}\right), \qquad \text{(Eq. 1)}$$

where q is the specific discharge (length/time); Q is the is the total discharge or recharge per unit width (length^3/time); A is the cross-sectional area of the horizontal flow (length^2).

By knowing the head loss across the medium and the distance over which it occurred the Darcy equation can be expressed in a differential form:

$$q = -K\frac{dh}{dl}. \qquad \text{(Eq. 2)}$$

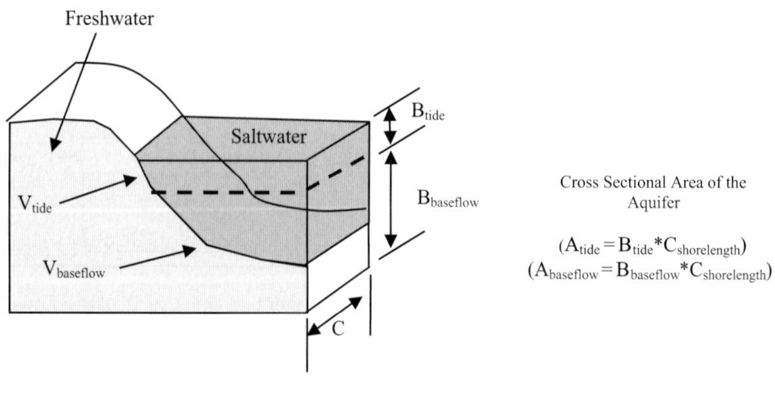

$$Q_{\text{total}} = (V_{\text{tide}} * A_{\text{tide}}) + (V_{\text{baseflow}} * A_{\text{baseflow}}) \qquad \text{[Eq. 1]}$$

Figure S64 Generalized cross-sectional groundwater velocities model for tidally-controlled aquifer.

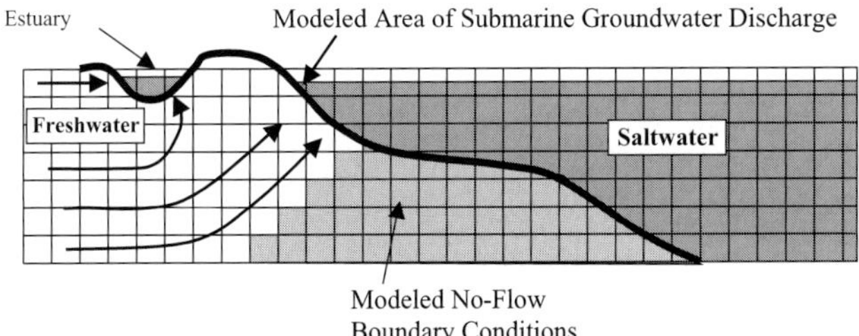

Figure S65 Typical finite difference grid used in modeling coastal boundary conditions in non-density dependent model.

Setting equations 1 and 2 equal to each other, and solving for Q (equation 3) yields the total discharge from the aquifer to the surface water body:

$$Q = -KA\frac{dh}{dl}, \qquad \text{(Eq. 3)}$$

where, Q is the total discharge (length^3/time); A is the cross- sectional area of the horizontal flow (length^2); K is the horizontal hydraulic conductivity (length/time); dh/dl is the hori-zontal hydraulic gradient (unitless).

Seepage calculations using Darcy's Law are limited because the solution assumes steady-state conditions and does not consider transient conditions, such as tidal action, rainfall, or onshore surface water levels.

To estimate the SGD for a portion of a coastline, the total discharge of freshwater per length of coast would multiplied by the length of coastline of interest.

$$Q_{\text{shore}} = -KA\frac{(h_1 - h_2)}{(l_1 - l_2)}L, \qquad \text{(Eq. 4)}$$

where, K is the horizontal hydraulic conductivity (length/time); L is the length of coastline (length); h_1 is the hydraulic head at point 1 (canal or aquifer water level); h_2 is the head at point 2 (generally the ocean); l_1 is the horizontal spatial location of point 1; l_2 is the horizontal spatial location of point 2.

If the SGD is occurring within a region of the aquifer that is sufficiently close to the shore and if the density of the groundwater varies, the density must be considered. The above equation can be modified to consider the fluid characteristics (Freeze and Cherry, 1979).

$$Q_{\text{shore}} = -\left(\frac{k\rho g}{\mu}\right)A\frac{(\text{h}_1 - \text{h}_2)}{(\text{l}_1 - \text{l}_2)}L, \qquad \text{(Eq. 5)}$$

where, k is the permeability (L/T^2); ρ is the density of the fluid (W/L^3); g is the gravitational constant (L/T^2); ρ is the dynamic viscosity (M/LT).

Estimation of actual flow velocities in the aquifer requires that the porosity be considered. Equation 3 is divided by porosity to yield the solute velocity of the flow system (Freeze and Cherry, 1979) as represented in equation 6:

$$V = \frac{q}{n} = \left(\frac{K}{n}\right)\left(\frac{dh}{dl}\right) \qquad \text{(Eq. 6)}$$

where V is the average linear groundwater velocity (length/ time); Q should be small "of is the specific discharge (length/time)"; n is the porosity.

This approach, which assumes one seepage rate for a length of coast, is valid when the flow is laminar and the groundwater is not experiencing significant diffusion. When seepage rates differ significantly along the shore it is appropriate to determine the discharge rates for each coastal segment (Cherkauer, 1998).

Numerical solutions

The problems discussed earlier in applying Darcy's equation to SGD also exist in the application of numerical calculations for both non-density and density dependent models. The primary considerations with numerical simulations are as follows:

1. Accuracy of the representation of the aquifer sediment layering, density of the fluids, and the vertical and horizontal hydraulic conductivity.
2. There is limited availability of coastal discharge data around the world. Many numerical modelers/simulators do not check the coastal discharge rates during the simulation process; it is considered an afterthought for the model simulations. The true discharge and spatial locations may or may not be correct. Recent seismic work in support of a modeling effort (Merchant *et al.*, 1996) off the North Carolina coast showed that paleofluvial channels and large-scale collapses constrain the possible discharge locations offshore discharge points.
3. Many numerical models are not designed to simulate density dependent flows, so modelers substitute simulated no-flow boundary conditions for the density difference in a stairstep fashion within the model domain (Figure S65) or use equivalent freshwater heads. Other models account for density but require data that is frequently unavailable and so may still predict an incorrect flux across the boundary condition. As with other numerical simulations, the validity of the application is a function of the modeler's experience.

Additional modeling considerations

Locating the groundwater/oceanic water interface is critical because the morphology, flow patterns (seepage), and solute transport vary significantly with location site (i.e., barrier islands, atolls). Urish and Ozbilgin (1989) state that vertical position of the interface can vary as much as 1.5 m vertically and 61 m horizontally, and is a function of sediment permeability, effective diffusivity, wave amplitude, and wave period (Rasmussen, 1998). Background data collection is essential to understanding the existing porewater salinity profiles at each location.

Additional long-term considerations are rising sea levels and resulting increases in groundwater levels. Nuttle and Portnoy (1992) indicate that relative sea level is rising at 30 cm per 100 years along the east coast of the United States. As a result, groundwater levels are being raised on a regional scale in low-lying areas. With the groundwater level closer to ground surface, decreased aquifer storage will increase runoff and reduce the amount of available recharge to the aquifer, subsequently decreasing SGD to coastal areas.

Empirical solutions versus numerical solutions

A comparison of solution methods for SGD. The South Florida Water Management District (SFWMD) has collected detailed measurements of surface water discharges and associated chemical surface water concentrations over a 9-year period for Palm Beach County, Florida (Finkl and Krupa, 2001). Detailed groundwater levels and groundwater chemical concentrations in this area have been collected by the United States Geological Survey and SFWMD over 10 years. The SFWMD used a regional model, known as the South Florida Water Management Model (SFWMM), to simulate the entire southern peninsula of Florida (SFWMD, 1993a,b). Shine *et al.* (1989) utilized a countywide numerical groundwater model (based on MODFLOW) to also estimate groundwater discharges to the Palm Beach County coast. Groundwater levels and surface water data were used in the models for calibration and verification.

Using these models, SGD values were calculated for the groundwater discharged from Palm Beach County to the offshore environment. For comparison, calculations based on Darcy's law were completed for the same length of coastline.

As seen in the above table, large variations in SGD estimation can occur in calculated and numerical model solutions (Table S11). The SFWMM had the lowest groundwater discharge value; this is likely because the model grid size (2 miles × 2 miles) does not allow the SGD process to be seen and measured. Two of the three solutions yielded SGD that equaled or exceeded the surface water inflows to the ocean. Corresponding groundwater and surface water concentrations for phosphorous and nitrogen were then calculated using the SGD fluxes to yield nutrient loading to the nearshore. It has been estimated that, on a worldwide basis, the total contributions of freshwater and nutrients to the ocean by SGD are roughly equivalent to the total contributions from riverine (in this case canal) discharges (Johannes and Hearn, 1985). Based on the estimated results and statements mentioned above, SGD is nearly equal to the surface water discharges in Palm Beach County, Florida.

Summary

Governmental agencies and private organizations need SGD information to better address water degradation issues and to create numerical models. Accurate determination of SGD is essential in calculating mass balances and in the determination of nutrient fluxes to the coastal waters. Experience has shown that quantifying this input is difficult because of uncertainties in the direct measurement of groundwater flux and in the simulation of groundwater fluxes.

Accurate knowledge of SGD is important because it can be an unseen hazard and can be used to assess environmental problems in coastal environments. The contribution of SGD to the coastal hydrologic regime is occasionally recognized in association with crescendo events and

Table S11 Comparison of three solutions of submarine groundwater discharge compared to actual surface water discharges and nutrient fluxes to offshore Palm Beach County, Florida

Discharge type	Water (million-meters^3/year)	Phosphorous flux (P) (metric tons)	Nitrogen Flux (N) (metric tons)
Surface Water Discharge			
Surface water (actual flow)	1,661	196[a]	2,471[a]
Groundwater Discharge			
Darcian flow solution[b]	2,211	414	5,727
USGS MODFLOW Application[c]	1,659	196	2,469
SFWMM SW/GW model[d]	50	6	75

[a] SFWMD, 1993b, average annual discharge for 9 years * average annual load for 9 years.
[b] Darcy's flow assumptions; $i^1 = 0.001$; $i^2 = 0.006$; $k^1 = 3.04$ m/day; $k^2 = 30.48$ m/day; coast length = 72.46 km; $b^1 = 18.2$ m; $b^2 = 45.7$ m; Load numbers total phosphorous (P) = 0.188 mg/L, Total Nitrogen (N) = 1.67 mg/L.
[c] Shine *et al.*, 1989, Average Annual.
[d] SFWMD, 1993a, Average Annual Discharge for 9 years.

concurrent marine algal blooms that degrade water quality, bottom habitats, and coral reef ecology (Finkl and Krupa, 2003). However, the more common situation is that SGD laden with nutrients from agro-urban activities on adjacent coastal plains causes environmental degradation so gradually that the cause and effect can escape public attention.

Steven L. Krupa and Cynthia J. Gefvert

Bibliography

Belanger, T.V., and Montgomery, M.T., 1992. Seepage meter error. *Limnology and Oceanography*, **37**: 1787–1795.

Belanger, T.V., and Walker, R.B., 1990. Ground water seepage in the Indian River Lagoon, Florida. In Tropical Hydrology and Caribbean Water Resources. *Proceedings International Symposium*. American Water Resources Association, pp. 367–375.

Bear, J., 1979. *Hydraulics of Groundwater*. New York: McGraw Hill, p. 569.

Bokuniewicz, H.J., and Zeitlin, M.J., 1980. Characteristics of the ground-water Seepage into Great South Bay. State University of New York at Stony Brook Marine Sciences Research Center Special Report 35, 80–2.

Buddemeier, R.W., 1996. Groundwater flux to the ocean: definitions, data, applications, uncertainties. In Buddemeier, R.W. (ed.), *Groundwater Discharge in Coastal Zone: Proceedings of an International Symposium. LOICZ Reports and Studies No. 8*, LOICZ, Texel, The Netherlands, pp. 16–21.

Burnett, W.C., Bokuniewicz, H., Huettel, M., Moore, W.S., Taniguchi, M., 2003. Groundwater and pore water inputs to the coastal zone. Biogeochemistry, **66**: 3–33.

Cable, J.E., Burnett, W.C., and Chanton, J.P., 1997a. Magnitude and variations of groundwater seepage along a Florida marine shoreline, *Biogeochemisty*, **38**: 189–205.

Cable, J.E., Burnett, W.C., Chanton, D.R., and Cable, P.H., 1997b. Field evaluation of seepage meters in the coastal marine environment, *Estuarine, Coastal and Shelf Science*. **45**: 367–375.

Carr, M.R., and Winter, T.C., 1980. An Annotated bibliography of devices developed for direct measurement of Seepage. Washington, DC: Unites States Geological Survey, *Open File Report 80–344*.

Cherkauer, D.S., 1998. The role of ground water in the pollution of urban coastal waters. *Workshop on Research Needs for Coastal Pollution in Urban Areas*, October 16–17, 1997; Milwaukee, WI, pp.22–31.

Church, T.M., 1996. An underground route to the water cycle. *Nature*, **380**: 579–580.

Darcy, H., 1856. *Les Fontaines Publiques de la Ville de Dijon*. Paris: Victor Dalmont.

Davis, S.N., and DeWiest, R., 1966. *Hydrogeology*. New York: John Wiley & Son.

Finkl, C.W., and Krupa, S., 2000. Environmental impacts of coastal-plain activities on sandy beach systems: Hazards, perception and mitigation. *Proceedings of the Brazilian Sandy Beaches Symposium: Morphodynamic, Ecology, Uses, Hazards and Management*. Itajaí, Santa Catarina, Brazil: Universidade do Vale do Itajaí, Centro de Ciencias Technológicas da Terra e do Mar, p. 29.

Finkl, C.W., and Krupa, S.L., 2003. Environmental impacts of coastal-plain activities on sandy beach system: hazards, perception and mitigation. *Simposio Brasileiro sobre Praias Arenosas*, Itajai, SC, Brazil.

Finkl, C.W., Jr., Krupa, S.L., and Giddings, J.B., 1995. Regional surface flows to tide and submarine groundwater discharges along the inner continental shelf of SE Florida. In *The 1st SEPM Congress on Sedimentary Geology*, St. Petersburg, FL: August 13–16, 1995 Program Abstracts.

Freeze, A., and Cherry, J., 1979. *Groundwater*. Englewood Cliffs, NJ: Prentice Hall.

Gale, W.F., and Thompson, J.D., 1975. A Suction Sampler for Quantitatively Sampling Benthos on Rocky Substrates in Rivers *Transactions of American Fisheries*, **2**: 398–405.

Guglielmi, Y., and Prieur, L., 1997. Location and quantification of submarine discharge of a confined aquifer with interstitial porosity; example of the alluvium aquifer of the lower var valley, mediterranean France. *Journal of Hydrology*, **190**(1–2): 111–122.

Harbor Branch Oceanographic Institute Inc., 1995. Tequesta Peninsula septic tank—water quality investigation. Jupiter. FL: *Loxahatchee River Environmental Control District*.

Harvey, J.W., Krupa, S.L., Gefvert, C.J., Choi, J., Mooney, R.H., and Giddings, J.B., 2000. Appendix for interaction between ground water and surface water in the northern Everglades and the relation to water budgets and mercury cycling. Reston, VA: United States Geological Survey, *Open File Report 00–168*.

Harvey, J.W., Krupa, S.L., Gefvert, C.J., Choi, J., Mooney, R.H., and Giddings, J.B., 2002. Interaction between ground water and surface water in the northern Everglades and the relation to water budgets and mercury cycling. Reston, VA: United States Geological Survey, *Water Resources Investigation* (02–4050).

Huettel, M., and Gust, G., 1992. Impact of roughness on interfacial solute exchange in permeable sediments. *Marine Ecology Progress Series*, **89**: 253–267.

Huettel, M., Ziebis, W., and Forster, S., 1996. Flow-induced uptake of particulate matter in permeable sediments. *Limnology and Oceanography*, **41**(2): 309–322.

Huettel, M., Ziebis, W., Forster, S., and Luther, G.W., III, 1998. Advective transport affecting metal and nutrient distribution and interfacial fluxes in permeable sediments. Elsevier: Netherlands. *Geochimica et Cosmochimica Acta*, **62**(4): 613–631.

Johannes, R.E., 1980. The ecological significance of the submarine discharge of groundwater. *Marine Ecology Progress Series*, **3**: 365–373.

Johannes, R.E., and Hearn, C.J., 1985. The Effects of submarine groundwater discharge on nutrient and salinity regimes in a coastal lagoon off Perth, Western, Australia. *Estuarine, Coastal and Shelf Science*, **121**: 789–800.

Kohout, F.A., 1966 Submarine springs: a neglected phenomenon of coastal hydrology. *Reprint from Central Treaty Organization's Symposium on Hydrology and Water Resources Development*, February 5–12, 1966. United States Geological Survey.

Kohout, F.A., Munson, R.C., Turner, R.M., and Royal, W.R., 1979. Satellite observations of a geothermal spring off Florida's West Coast. *Satellite Hydrology*, American Water Resources Association.

Krupa, S.L., Belanger, T.V., Heck, H.H., Brock, J.T., Jones, B.L., 1998. Krupaseep—The next generation seepage meter. *Journal of Coastal Research*, **25**: 210–213.

Land, L.A., Paul, C.K., and Hobson, B., 1995. Genesis of a submarine sinkhole without subaerial exposure: straits of Florida. *Geology*, **23**(10): 949–951.

Lapointe, B.E., and Matzie, W.R., 1996. Effects of stormwater nutrient discharges on eutrophication processes in nearshore waters on the Florida keys. *Estuarine, Coastal and Shelf Science*, **19**(2B): 422–435.

Lapointe, B.E., and O'Connell, J., 1989. Nutrient-enhanced growth of *Cladophora prolifera* in Harrington Sound, Bermuda: eutrophication of a confined, phosphorus-limited marine ecosystem. *Estuarine, Coastal and Shelf Science*, **28**: 347–360.

Lapointe, B.E., O'Connell, J.D., and Garrett, G.S., 1990. Nutrient couplings between on-site sewage disposal systems, groundwater's, and nearshore surface waters of the Florida Keys. *Biogeochemistry*, **10**: 289–307.

Lee, D.R., 1977. A device for measuring seepage flux in lakes and estuaries. *Limnology and Oceanography*, **22**: 140–147.

Lee, D.R.. and Cherry, J.A., 1978. A field exercise on groundwater flow using seepage meters and mini-piezometers, *Journal of Geological Education*, **27**: 6–10.

Lewis, John, B., 1987. Measurements of groundwater seepage flux onto coral reef: spatial and temporal variations. *Limnology and Oceanography,* **32**(5): 1165–1169.

Libelo, L.E., and MacIntyre, W.G., 1994. Effects of surface-water movement on seepage-meter measurements of flow through the sediment-water interface. *In Applied Hydrology, International Association of Hydrogeologists*: Hanover, **2**: 49–55.

Manheim, F.T., 1967. Evidence for submarine discharge of water on the Atlantic continental slope of the southern United States, and suggestions for further search. *Transactions of the New York Academy of Sciences*, **29**(7), 839–853.

Merchant, Sarah P., Snyder, Stephen, W., and Evans, David G., 1996. Modeling constraints on the origin of submarine ground-water discharge on the North Carolina Continental Shelf Boulder, CO: *Geological Society of America*, **28**(2): 38.

Moore, W.S., 1996. Large groundwater inputs to coastal waters revealed by ^{226}Ra Enrichments. *Nature*, **380**: 612–614.

Moore, S., 1997. High fluxes of radium and barium from the mouth of the Ganges-Brahmaputra river during low river discharge suggest a large groundwater source. *Earth and Planetary Science Letters*, **150**(1–2): 141–150.

National Aeronautics and Space Administration, 1992. *Flow Rate Logging Seepage Meter Patent Application N-93-19328*. Hampton, VA: NASA Patent Counsel, Langley Research Center.

Nossin, J.J., Sesoren, Atila, Kuran, Ibrahim Hakim, 1987. Karst and submarine freshwater discharge in the Antalaya Area, Southern Turkey. *International Conference on SPOT Early Assessment Programme. Centre National d'Etudes Spatiales*, Toulouse, France, pp. 719–729.

Nuttle, W.K., and Harvey, J.W., 1995. Fluxes of Water and solute in a coastal wetland sediment. 1: the contribution of regional ground water discharge. *Journal of Hydrology*, **164**: 89–107.

Nuttle, W.K., and Portnoy, J.W., 1992. Effects of rising sea level on runoff and ground water discharge to coastal ecosystems. *Estuarine, Coastal and Shelf Science*, **34**: 203–212.

O'Rouke, D., Paulsen, R.J., and Wong, T.W., 1999. Measuring submarine groundwater seepage using an ultrasonic flow meter and the drum method— a comparative study. *Conference on the "Geology of Long Island and Metropolitan New York"* April 24, 1999. Stony Brook, NY.

Paulsen, R.J., Smith, C.F., and Wong, T.-f., 1997. Development and evaluation of an ultrasonic groundwater seepage meter *Conference on the Geology of Long Island and Metropolitan New York Program with Abstracts*, April 19, 1997, pp. 88–97.

Rasmussen, L., 1998. Groundwater flow, tidal mixing and haline convection in coastal sediments. Master thesis, Tallahassee: Florida State University.

Reay, W.G., and Walthall, H.G., 1991. Remote semi-continuous flow rate logging seepage meter. *In Technology 2001: The Second National Technology Transfer Conference and Exposition*. San Jose, CA: NASA Conference Publication 3136, 2, 445–451.

Reay, W., and Walthall, H., 1992. Flow rate logging seepage meter. *National Aeronautics and Space Administration Patent Application.* NASA Case Number LAR 14835-1.

Reich, C.D., 1996. Diver-operated manometer: a simple device for measuring hydraulic heads in underwater wells. *Journal of Sedimentary Research*, **66**(5): 1032–1034.

Reilly, T.E., 1993. Analysis of ground-water systems in freshwater– saltwater environments. In William M. Alley (ed.) *Regional Ground-Water Quality*. New York: Van Nostrand Reinhold.

Robinson, M.A., 1996. A finite element model of submarine ground water discharge to tidal estuarine waters. Ph.D. Dissertation, Blacksburg, VA: Virginia Polytechnic Institute.

Rougerie, F., and Wauthy, B., 1993. The Endo-upwelling concept: from geothermal convection to reef construction. *Coral Reefs*, **12**:19–30.

Schwerdtfeger, B.C., 1981. On the occurrence of fresh water discharges. *Geologisches Jahrbuch Reihe, Hydrogeologie, Ingenieurgeologie*, **29**: 231–240.

Senger, R.K., and Fogg, G., 1990. Stream function and equivalent freshwater heads for modeling regional flow of variable density groundwater. 1. Review of theory and verification. *Water Resources Research*, **26**(9): 2089–2096.

Shaw, R.D., and Prepas, E.E., 1989. Anomalous, short-term influx of water into seepage meters. *Limnology and Oceanography*, **34**(7): 1343–1351.

Shaw, R.D., and Prepas, E.E., 1990. Groundwater–Lake interactions 1. Accuracy of seepage meter estimates of lake seepage. *Journal of Hydrology*, **119**: 105–120.

Shine, M.J., Padgett, D.G., Barfknecht, W.M., 1989. Ground water resources assessment of Eastern Palm Beach County, Florida. West Palm Beach, Fl: South Florida Water-Management District, *Technical Publication 89-4*; Part 1.

Shinn, E.A., Reese, R.S., and Reich, C.D., 1994. Fate and pathways of injection-well effluent in the Florida keys. *United States Geological Survey Open File Report 94-276*, 116.

Shinn, E.A., Reich, C.D., Hickey, T.D., and Tihansky, A.B., 1997. Geology and tidal pumping in the Florida Keys. *Abstracts. AAPG annual meeting Dallas TX.* pp. 106–107.

Simmons, G.M., Jr., 1989. The Chesapeake Bay's Hidden Tributary; Submarine Groundwater Discharge. In *Proceedings of Ground Water Issues And Solutions in The Potomac River Basin/Chesapeake Bay Region*. Dublin, OH: National Well Water Association, 9–29.

Simmons, G.M. Jr., 1992. Importance of submarine groundwater discharge (SGWD) and seawater cycling to material flux across sediment/water interfaces in marine environments. *Marine Ecology Progress Series*, **84**: 173–184.

Shinn, E.A., Reich, C.D., and Hickey, T.D., 2002. Seepage Meters and Bernoulli's Revenge. *Estuarine, Coastal and Shelf Science*, **25**(1): 126–132.

Simmons, G.M., Jr., and Love, F.G., 1987. Water quality of newly discovered submarine ground water discharge into a deep coral reef habitat. *Symposium Series for Undersea Research*, National Oceanic and Atmospheric Administration, Washington, DC (2), 155–163.

Simmons, G.M., Jr., and Reay, W.G., 1992. The phenomenon called submarine groundwater discharge. *United States Environmental Protection Agency, United States and American Water Resources Association Technical Publication Series*, 92-2, pp. 385–394.

Sonzogni, W.C., Larsen, D.P., Malueg, K.W., and Schult, M.D., 1977. Use of large submerged chambers to measure sediment–water interactions. *Water Research*, **11**: 461–464.

South Florida Water Management District, 1993a. Data retrieval for nutrient parameters; Total nitrogen and total phosphorous on all oceanic structures in Palm Beach County—June 15, 1993. West Palm Beach, Fl: *South Florida Water Management District*.

South Florida Water Management District, 1993b. Appendices and technical Information in Support of the lower east coast regional water supply plan. West Palm Beach, FL: *South Florida Water Management District*.

Strack, O., 1975. *Ground Water Mechanics*. Englewood Cliffs, NJ: Prentice Hall.

Svitil, K.A., 1996. Groundwater Secrets. *Discover*, September 1996, 28.

Taniguchi, M., and Fukuo, Y., 1993. Continuous measurements of ground-water seepage using and automated seepage meter. *Ground Water*, **31**(4): 675–679.

Todd, D.K., 1980. *Groundwater Hydrology*. New York: John Wiley & Sons, Inc.

Tribble, G.W., Sansone, F.J., Buddemeier, R.W., and Li,Y., 1992. Hydraulic exchange between a coral reef and surface water *Geological Society of America Bulletin*, **104**: 1280–1291.

US Environmental Protection Agency, 1998. Maui searches for answers to problem algae, *Nonpoint Source News Notes*. July/August 1998, **52**: 7–8.

US Army Corp of Engineers, 1993. *Installing Monitoring Wells/ Piezometers in Wetlands*. Vicksburg, MS: Water Resources Publication Technical Note HY-1A-3.1.

Uchiyama, Y., Nadaoka, K., Rolke, P., Adachi, K., and Yagi, H., 2000. Submarine groundwater discharge into the sea and associated nutrient transport in a sandy beach. *Water Resources Research*, **36**(6): 1467–1479.

Urish, D.W., and Ozbilgin, M.M., 1989. The Coastal ground-water boundary. *Ground Water*, **27**(3): 310–315.

Winter, T.C., Harvey, J.W., Franke, O.L., and Alley, W.M., 1998. ground water and surface water a single resource. *United States Geological Survey Circular 1139*. Washington, DC: United States Geological Survey.

Wicander, R. and Monroe, J.S., 1993. *Historical Geology*, 2nd edn. New York: West Publishing Company.

Zektser, I.S., and Meskheteli, A.V., 1988. Methods for studying submarine discharge of karst water. *International Association of Hydrological Sciences*, **176**(2): 677–681.

Zektser, I.S., and Loaiciga, Hugo A., 1993. Ground water fluxes in the global hydrologic cycle; past, present and future. *Journal of Hydrology*, **144** (1–4): 405–427.

Zektser, I.S., Dzhamalov, R.G., and Safronova, T.I., 1983. The role of submarine groundwater discharge in the water balance of australia *International Association of Hydrological Sciences*, (Louvain), International, IAHS-AISH Publication 142(1), 209–219.

Cross-references

Hydrology of Coastal Zone
Numerical Modeling
Shoreface
Water Quality

SUBMERGED COASTS

A submerged coast is defined as a coast resulting from the relative submergence of a landmass either through eustatic sea-level rise and/or crustal subsidence against subaerially produced forms and structures. The term carries no implication as to whether it is the land or the sea that has moved. In siliciclastics-dominated shelves, economic mineral deposits associated with submerged coasts include sand and gravel deposits, and, placer deposits such as diamonds, gold, cassiterite, and heavy mineral sands.

Recognition and dating

Submerged coasts are indicated by the occurrence of drowned subaerial features (Table S12). Because of the subaerial exposure of continental shelves during the last low sea-level stand, a paleosol with a lower moisture content and a higher density may exist beneath the Holocene marine deposits. Among the earliest drowned subaerial features recognized are the submerged forests off the English coast (Lyell, 1850). On continental shelves of the world, about 70% of the total area was identified by Emery (1968) to be covered by relic sediments including previously deposited subaerial, lacustrine, and paludal sediments. For sea-level changes during the past 20,000 years, marine zonation, erosional indicators, depositional indicators, archaeological remains, and historical data provide evidence for submergence (Pirazzoli, 1996).

For dating submerged coasts, the radiocarbon method and the U/Th method are widely used. The former was first applied to stable crustal regions for the reconstruction of late Quaternary sea-level history (Godwin *et al.*, 1958). In Barbados, the drilling and dating of coral reefs containing *Acropora palmata* indicated that sea level was at −121 ± 5 m around 17,000 yr BP (^{14}C) (Fairbanks, 1989). Subsequent dating of the deepest *A. palmata* using the U/Th method gave an age of ca. 19,000 yr BP corresponding to a sea level of −118 m (Bard *et al.*, 1990a). Far away from plate boundaries in Tahiti, a large sea-level jump was identified shortly before 13,800 yr BP with the U/Th method while radiocarbon ages on the same samples were significantly younger (Bard *et al.*, 1996). Dating of inner shelf sequences off Hong Kong (Yim *et al.*, 1990; Yim, 1999) also indicated a young age bias of pre-Holocene radiocarbon dates exceeding about 8,200 yr BP (^{14}C) while some of the shells with finite radiocarbon dates may instead be of last interglacial age. Pre-Holocene materials on the shelf are likely to yield unreliable radiocarbon dates compared with Holocene counterparts because of possible pedogenic alteration during low sea-level stand(s). For a comparison of radiocarbon and U/Th dates see Bard *et al.* (1990b). Because there is a maximum difference of about 3,500 years for a date of ca. 20,000 yr BP (^{14}C), the last glacial maximum (LGM) dated to 18,000 yr BP (^{14}C) may have occurred 21,500 sidereal years ago.

Table S12 Selected depositional and erosional features associated with submerged coasts

Depositional features	Erosional features
Terrestrial deposits (alluvium, colluvium, aeolianite, paleosols, peat, and other plant remains, etc.)	Notches Honey-combed rock surfaces Shore platforms
Karst, e.g., speleothems	Karst, e.g., caverns
Coral reef	
Beachrock	
Archaeological remains	

Sea-level minimum during the last glacial maximum

The global average depth of the shelf break at −130 m below present mean sea level (Shepard, 1973) provides a median value of sea-level lowering resulting from continental ice growth. Since this value is based on the compilation of hydrographic charts, it is probably free from bias (Bloom, 1983). Additional support for the sea-level minimum at −130 m is the change of about 10 m in global sea level through a 0.1‰ change in the oxygen-isotopic record (Shackleton and Opdyke, 1973) and the closeness to the median value sea-level position during the LGM identified on many shelves.

The radiocarbon dating of oolitic and biogenic carbonates associated with drowned beach barriers is a means of identifying the sea-level minimum during the LGM. On the Bengal Shelf, five samples of these materials from depths ranging from 125 to 133 m below the present sea level yielded dates ranging from 16,500 to 24,900 yr BP (^{14}C) (Wiedicke *et al.*, 1999).

Examples of submerged coasts

In North America, subaerial features of LGM to late glacial maximum age drowned by the Holocene transgression were reviewed by Bloom (1983). Terrestrial materials and landforms including mastadon and mammoth fossils, Indian artifacts, moraines, river channels, etc. were reported.

Clear-cut cases of coastal subsidence resulting from earthquakes have been documented. One such example is from near Haikou on Hainan Island in southern China where ruins of ancient villages, including a cemetery with engraved tombstones of 1604, were identified up to 10 m below present sea level (Y. Chen in Anonymous, 1983). A coastal area exceeding 10 km by 1 km was affected by subsidence during a Ming Dynasty earthquake recorded in historical documents. The earthquake was estimated to be of magnitude scale 8 and occurred on July 13, 1605.

A long history of submerged coasts has been found on "stable" shelves. Off Hong Kong, the study of offshore borehole sequences revealed a succession of paleo-desiccated crusts formed by the subaerial exposure of marine deposits during Quaternary low sea-level stands dating back to oxygen-isotope stages 2, 6, 8, and 10 (Yim and Tovey, 1995). This sequence is in agreement with the five interglacial–glacial cycles identified in the Vostok ice core in Antarctica (Petit *et al.*, 1999) confirming that the sea-level changes found were eustatically controlled. Figure S66 shows an example of an early Holocene submerged coast dated at ca. 8,000 yr BP (^{14}C) when sea level was at ca. −18 m below present mean sea level.

Micropaleontological studies of continental-shelf cores are useful in confirming submerged coasts when they penetrate the formerly exposed soil surface. A pollen sequence identified on the east Queensland shelf shows a transition from terrestrial vegetation (grasses and woodland), to salt marsh and mangroves (Grindrod *et al.*, 1999). This sequence is the reverse of mangrove successions recorded for regressive coasts and is compatible with a drowning coast. Similarly, a diatom sequence in a shelf core off Hong Kong with evidence for five marine transgressions showed a record of diatom preservation consistent with aging and pedogenesis during glacial period(s) (Yim and Li, 2000). The tests of diatoms were progressively destroyed through groundwater dissolution during the low sea-level stands with the lowest abundance found in deposits of oxygen-isotope stage 11, followed by stage 9 and stage 7. Deposits of stage 5 were similar in diatom abundance to stage 1 probably because of the "young" age.

Underwater mapping of rocky coasts using SCUBA is a possible way of identifying submerged coasts of late Quaternary age. Off Marseille submerged coasts were shown by the Grotte Cosquer archaeological

Figure S66 An early Holocene submerged coast when sea level was at ca. −18 m below present mean sea level. The exposure was available through seabed excavation for the foundation of the West Dam, High Island Reservoir, Hong Kong SAR. The Holocene marine deposits (darker) rest unconformably on a fining-upward sequence of Pleistocene alluvial deposits (paler). Wood and shells from near the base of the Holocene marine deposits yielded dates of ca. 8,000 yr BP (^{14}C) (Photograph courtesy of R.J. Frost).

site at a depth of about 60 m below present sea level and a paleo-coast at a depth of about 90 m below present sea level dated at 13,800 yr BP (^{14}C) (Collina-Girard, 1996). Underwater mapping off Corsica and the West Indies have revealed discontinuities at common depths of −11, −17, −25, −35, −45, −55 and −100 m below present mean sea level attributed to late Quaternary submerged coasts. However, it is possible that some of the submerged coasts are polycyclic in origin and dating is needed for their age verification.

Future work

Future work should address why late Quaternary sea-level curves do not record a sea-level fluctuation around 11,000 yr BP (^{14}C) attributable to the Younger Dryas. Is this caused by inadequate radiocarbon calibration? In order to obtain answers, samples from appropriate depths on the shelf are needed for dating.

The International Geological Correlation Programme project no. 396 "Continental shelves in the Quaternary" a five-year project initiated in 1996, has filled some of the gaps in knowledge on submerged coasts. Drill rigs have been developed to obtain core samples from shelves. Cable-route surveys undertaken on shelves are providing a wealth of information including cores for the study of submerged coasts since the LGM. An international effort similar to the Ocean Drilling Program operating on continental shelves would help to promote the study of Quaternary submerged coasts.

Wyss W.-S. Yim

Bibliography

Anonymous, 1983. Hainan earthquake 1605. *China Daily*, 25th May 1983.

Bard, E., Hamelin, B., Arnold, M., Montaggioni, L., Cabioch, G., Faure, G., and Rougerie, F., 1996. Deglacial sea level record from Tahiti corals and the timing of global meltwater discharge. *Nature*, **382**: 241–244.

Bard, E., Hamelin, B., and Fairbanks, R.G., 1990a. U-Th ages obtained by mass spectrometry in corals from Barbados: sea-level during the past 130,000 years. *Nature*, **346**: 456–458.

Bard, E., Hamelin, B., Fairbanks, R.G., and Zindler, A., 1990b. Calibration of the ^{14}C timescale over the past 30,000 years using mass spectrometric U-Th ages from Barbados corals. *Nature*, **345**: 405–410.

Bloom, A.L., 1983. Sea level and coastal morphology of the United States through the Late Wisconsin glacial maximum. In Wright, H.E.Jr. (ed.) *Late Quaternary Environments of the United States*. Mineapolis: University of Minnesota, pp. 215–229.

Collina-Girard, J., 1996. *Topography of the Continental Shelf off Marseille Scale 1:50 000*. Institut Geographique National, Aixen-Provence (in French with English legend).

Emery, K.O., 1968. Relict sediments on continental shelves of world. *Bulletin of the American Association of Petroleum Geologists*, **52**: 445–464.

Fairbanks, R.G., 1989. A 17,000-year glacio-eustatic sea level record: influence of glacial melting rates on the Younger Dryas event and deep-ocean circulation. *Nature*, **342**: 637–642.

Godwin, H., Suggate, R.P., and Willis, E.H., 1958. Radiocarbon dating of the eustatic rise in ocean level. *Nature*, **181**: 1518–1519.

Grindrod, J., Moss, P., and Van der Kaars, S., 1999. Late Quaternary cycles of mangrove development and decline on the north Australian continental shelf. *Journal of Quaternary Science*, **14**: 465–470.

Lyell, C., 1850. *Principles of Geology*, 8th edn. London: John Murray.

Petit, J.R., Jouzel, J., Raynaud, D., Barkov, N.I., Barnola, J.-M., Basile, I., Bender, M., Chappellaz, J., Davis, M., Delaygue, G., Delmotte, M., Kotlyakov, V.M., Legrand, M., Lipenkov, V.Y., Lorius, C., Pepin, L., Ritz, C., Saltzman, E., and Stievenard, M., 1999. Climate and atmospheric history of the past 420,000 years from the Vostok ice core, Antarctica. *Nature*, **399**: 429–436.

Pirazzoli, P.A., 1996. *Sea-Level Changes*. Chichester: Wiley.

Shackleton, N.J., and Opdyke, N.D., 1973. Oxygen isotope and paleomagnetic stratigraphy of equatorial Pacific core V28–238: oxygen isotope temperature and ice volumes on a 10^5 year and 10^6 year scale. *Quaternary Research*, **3**: 39–55.

Shepard, F.P., 1973. *Submarine Geology*, 3rd edn. New York: Harper and Row.

Wiedicke, M., Kudrass, H.-R., and Hubscher, Ch., 1999. Oolitic beach barriers of the last Glacial sea-level lowstand at the outer Bengal shelf. *Marine Geology*, **157**: 7–18.

Yim, W.W.-S., Ivanovich, M., and Yu, K.-F., 1990. Young age bias of radiocarbon dates in pre-Holocene marine deposits of Hong Kong and implications for Pleistocene stratigraphy. *Geo-Marine Letters*, **10**: 165–172.

Yim, W.W.-S., 1999. Radiocarbon dating and the reconstruction of late Quaternary sea-level changes in Hong Kong. *Quaternary International*, **55**: 77–91.

Yim, W.W.-S., and Li, J., 2000. Diatom preservation in an inner continental shelf borehole from the South China Sea. *Journal of Asian Earth Sciences*, **18**: 471–488.

Yim, W.W.-S., and Tovey, N.K. 1995. Desiccation of inner continental shelf sediments during Quaternary low sea-level stands. *Geoscientists*, **5/4**: 34–35.

Cross-references

Changing Sea Levels
Coastal Subsidence
Coastline Changes
Continental Shelves
Eustasy
Geochronology
Ingression, Regression and Transgression
Late Quaternary Marine Transgression
Paleocoastlines
Placer Deposits
Sea-Level Indicators, Geomorphic
Sedimentary Basins
Sequence Stratigraphy

SUBMERGING COASTS

A coast is submerging when the relative sea level rises above it. Submergence may be caused by sea-level rise, by land subsidence, or by the two.

Between about 20 and 6 kyr ago, when the melting of the Northern Hemisphere continental ice caps was completed, land-ice melting caused the global sea level to rise some 120 m, at an average rate of 8 mm/yr, but with peaks reaching 40–50 mm/yr during certain periods (Bard *et al.*, 1996). This rise caused the rapid submergence of huge continental shelf areas. Since about 6 kyr ago, the global sea level has remained almost stable in a high, interglacial position. Global sea-level rise for the last century is estimated to be of decimetric order. For the next century, climatic models that take into account increasing greenhouse effects, have predicted scenarios of sea-level rise between 0.09 and 0.88 m, with a central value of 0.48 m (IPCC, 2001).

Land subsidence may result from several (neo)tectonic processes. Its rate can be much variable in space and time. Since the peak of the last glacial times (i.e., during the last 20 kyr), in deep-sea areas hydro-isostatic displacements have been of the order of 40 m (about one-third of the relative sea-level rise, that is, an average rate of 2 mm/yr). This made the ocean to floor subside. When approaching coastal areas, subsidence rates lessened, depending on the water shallowness and the coastal topography. During the last 6 kyr such isostatic subsidence decreased exponentially. Near delta formations, however, sediment compaction and sedimento-isostasy have often been more significant, with rates of subsidence on the order of one to a few millimeters per year reaching the maximum values near the delta depocenters. During the last centuries, and especially in the 20th century, human-induced land subsidence was caused in many coastal areas by acceleration of compaction due to drainage, oil and gas extraction, or groundwater exploitation. Though the total amount of human-induced subsidence rarely exceeds a few meters, it was often obtained within a limited time range, generally a few decades, thus reaching during these periods dangerously high sinking rates.

As a result of the above processes, most of the world coastal regions have been submerging rapidly between about 20 and 6 kyr ago, due to the rapid postglacial eustatic rise. During the Holocene, most deltaic sequences began to accumulate between 8.5 and 5.5 kyr ago, with a modal age of about 7.5–7.0 kyr (Stanley and Warne, 1994). It is, therefore, only after that time, that is, after the end of the postglacial transgression, that land sinking may have started to be active in delta areas. Similar sinking, for recent sediment compaction, must have occurred, though at slower rates, also in most estuarine and lagoonal areas, and even in coastal plains. However, as long as the fluvial sediment input or the longshore drift could compensate for the land sinking, submergence phenomena could not start. It is especially during the last century that human action has been most active, not only in accelerating land sinking rates, but also in dredging from river beds or constructing breakwaters that cut off a longshore sediment supply.

Shallow coastal areas may also be submerged by marine erosion, in particular at the time of strong storm surges. In seismically active areas, coastal submergence may occur at the time of major earthquakes (Figure S67). One of the most impressive last century events of this kind was probably the great earthquake of March 27, 1964 in Alaska (magnitude $\geq$ 8.4, with an epicenter in the Prince William Sound area) when vertical crustal movements affected a region of at least 200,000 km^2, with a wide coastal zone of subsidence reaching a maximum of 2.2 m (Plafker, 1965). In the Rann of Kutch, on the border between

Figure S67 What is visible is not a fringing coral reef at sea level, but the result of a seismic subsidence that occurred in 1981 during an earthquake at Mavrolimni, in the Gulf of Corinth, Greece. Before the earthquake, Mavrolimni (literally Black Lake) was a lagoon, up to 10 m deep, connected to the sea by an opening to the north. It was delimited seawards by a sand-and-gravel barrier, a few meters wide, which was used as a mole for fishing boats. In 1981, the area subsided about 0.8 m and the top of the barrier was brought down just to sea level (Stiros and Pirazzoli, 1998) (Photo D554, August 1992).

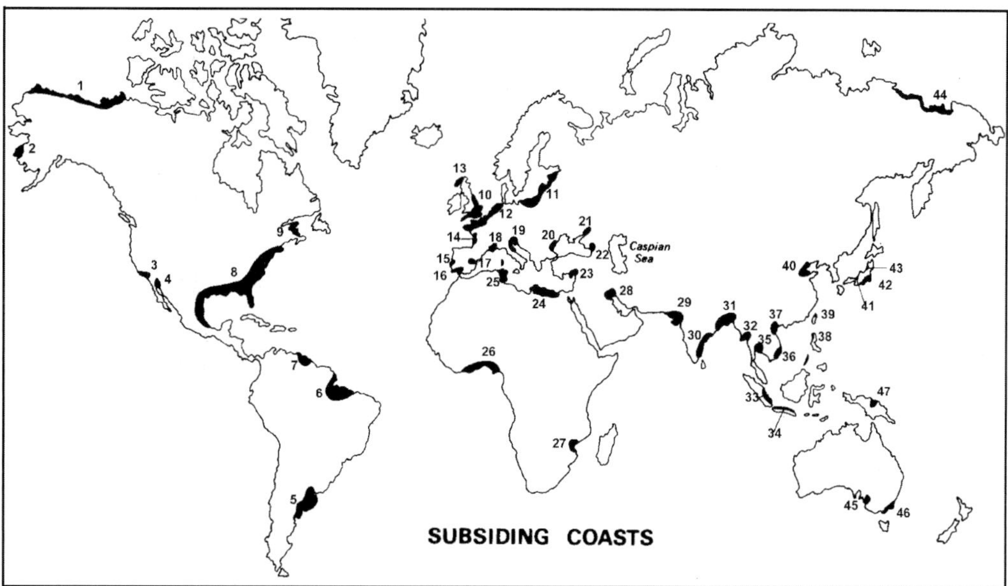

Figure S68 Sectors of the world's coastline that have been subsiding in recent decades, as shown by evidence of tectonic depression, increasing marine flooding, geomorphological and ecological indicators, geodetic surveys, and groups of tide gauges recording a rise of mean sea level greater than 2 mm/yr (assumed to be the present rate of global sea-level rise) over the past three decades. A distinction is made between submerging coasts (where sea level has risen relative to the land) and subsiding coasts (where there has been land subsidence). On the coasts of the Caspian sea level has risen by more than 2 m since 1977 as a result of an increase in water volume, possibly accompanied by some associated subsidence of the bordering land (from Bird, 1993, © Geostudies, reproduced with permission of John Wiley & Sons Limited). *Key to the map*: 1, Mackenzie delta and northern Alaska; 2, Yukon delta, Alaska; 3, Long Beach area, southern California; 4, Colorado River delta, head of Gulf of California; 5, Gulf of La Plata, Argentina; 6, Amazon delta; 7, Orinoco delta; 8, Gulf and Atlantic coast, Mexico and United States; 9, Maritime Provinces, Canada; 10, southern and eastern England; 11, the southern Baltic from Estonia to Poland; 12, North Germany, the Netherlands, Belgium and northern France; 13, Hebrides, Scotland; 14, Loire estuary and the Vendée, western France; 15, Lisbon region, Portugal; 16, Guadalquavir delta, Spain; 17, Ebro delta, Spain; 18, Rhône delta, France; 19, northern Adriatic from Rimini to Venice and Grado; 20, Danube delta, Rumania; 21, Eastern Sea of Azov; 22, Poti Swamp, southeastern Black Sea coast; 23, Southeast Turkey; 24, Nile delta to Libya; 25, Northeast Tunisia; 26, Nigerian coast, especially the Niger delta; 27, Zambezi delta; 28, Tigris-Euphrates delta; 29, Rann of Kutch; 30, Southeastern India; 31, Ganges–Brahamputra delta; 32, Irrawaddy delta; 33, Eastern Sumatra; 34, Northern Java deltaic coast; 35, Bangkok coastal region; 36, Mekong delta; 37, Red River delta, northern Vietnam; 38, Manila, Phillippines; 39, northern Taiwan; 40, Hwang-ho delta; 41, Maizuru, Japan; 42, Head of Tokyo Bay; 43, Niigata, Japan; 44, East Siberian coastal lowlands; 45, Port Adelaide region; 46, Corner Inlet region; 47, Sepik delta.

Pakistan and India, the 1819 earthquake resulted in a very wide area subsiding beneath the sea (Bird, 1993).

In volcanic areas, vertical movements can be very fast and lead to local submergence phenomena. This is the case for many calderas, produced by explosion, collapse, or even erosion. Relatively slower movements, probably produced by thermo-isostatic processes, may continue for long periods, that is, for centuries or even millennia in the case of dormant volcanoes, of even for million years for extinct volcanoes capped by oceanic atolls.

If the greenhouse-induced warming predicted by climatic models will be confirmed, a significant relative sea-level rise can be expected during the next centuries. According to IPCC (2001), if greenhouse gas concentrations were stabilized (even at present levels) sea level would nonetheless continue to rise for hundreds of years. It can therefore be expected that submerging coasts, presently confined mainly to sectors where the land has been subsiding (Figure S68), will become more extensive.

Paolo A. Pirazzoli

Bibliography

Bard, E., Hamelin, B., Arnold, M., Montaggioni, L., Cabioch, G., Faure, G., and Rougerie, F., 1996. Deglacial sea level record from Tahiti corals and the timing of global meltwater discharge. *Nature*, **383**: 241–244.

Bird, E.C.F., 1993. *Submerging Coasts*. Chichester: Wiley.

IPCC, 2001. *Climate Change 2001: The Scientific Basis*. Cambridge and New York: Cambridge University Press.

Plafker, G., 1965. Tectonic deformation associated with the 1964 Alaska Earthquake. *Science*, **148**(3678): 1675–1687.

Stanley, D.J., and Warne, A.G., 1994. Worldwide initiation of Holocene marine deltas by deceleration of sea-level rise. *Science*, **265**: p. 228–231.

Stiros, S.C., and Pirazzoli, P.A., 1998. Late Quaternary coastal changes in the Gulf of Corinth, Greece. Geodesy Laboratory, Department of Civil Engineering, Patras University, 49 p.

Cross-references

Changing Sea Levels
Coastal Subsidence
Sedimentary Basins
Isostasy
Tectonics and Neotectonics

SURF MODELING

The purpose of surf modeling is to realistically predict and quantify surf characteristics such as breaker heights, breaker types, breaker periods, and surf zone widths. Many surf models also provide other useful surf zone information such as wave setup and wave-generated longshore currents. Increasing computer capabilities and their availability, as well as research advances, have enabled determination of surf characteristics to evolve from use of relatively simple calculations, nomographs, or tabular methods, that are largely based on empirical relationships, to use of numerical computer models. These models solve mathematical equations that are based on the physics of surf zone hydrodynamics (see *Surf Zone Processes*). But, only relatively recently have computer surf models been developed and applied. In the surf zone, wave breaking, turbulence, nonlinear processes, and other complex phenomena combine to make modeling challenging. Here, the most important aspects of surf modeling for research and practical applications are described.

Modeled parameters

Owing to their potential adverse effects, breaker heights and breaker types are of most interest. A breaker's height is the distance between its trough (lowest point) and crest (highest point). Higher waves break further offshore where depths are deeper so that breaker heights vary across a surf zone. Because incident waves (see *Waves*) have varying heights with time, breaker heights also vary with time. The intervals between individual breakers, breaker periods, are the same as the periods of the incident waves and, thus, are more easily predicted than breaker heights and types.

Breakers are categorized as spilling, plunging, or surging (e.g., Wiegel, 1964; Dean and Dalrymple, 1984; Fredsøe and Deigaard, 1992; Smith, 2000). A breaker type, collapsing, is sometimes defined as an intermediate category between plunging and surging. A spilling breaker is characterized by an unstable crest that cascades down the shoreward face of the wave as a turbulent layer. For typical waves and bottom slopes of most beaches, spilling breakers occur most often. A plunging breaker is characterized by an unstable crest that curls over the shoreward face of the wave and plunges into the water below. Plunging breakers are of most concern for surf zone activities due to the forces of their crashing motion. A collapsing breaker is one for which the crest remains unbroken while the lower part of the shoreward face of the wave steepens and falls to produce a turbulent region. A surging breaker has a crest that remains essentially unbroken as it moves up a beach. For typical waves and bottom slopes of most beaches, surging breakers seldom occur. Most surf models lump collapsing breakers with surging breakers.

Because it represents the distance across possibly hazardous surf, the width of the surf zone is a key parameter. At any time, there is a distribution of wave breaking locations associated with the distribution of incident wave heights (see *Waves*). Higher waves break further offshore than lower waves. A practical definition is that surf zone width is the distance from the shoreline to where a trained conscientious observer sees the limit of depth-induced breaking over a reasonable time period, such as 10 min. This definition is preferable to defining the surf zone as the distance between the beach and the outermost breaker since the location of the outermost breaker varies with time. An analysis of over 600 video images of the surf zone shows that this practical definition results in surf zone widths that are most highly correlated to a location where 10% of the waves are breaking (Earle, 1999). Surf zone widths generally vary inversely with water levels that are caused, at most locations, mainly by astronomical tides (see *Tides*). For beaches without offshore bars, higher water levels cause waves to break on the steeper nearshore part of the beach resulting in more narrow surf zone widths. For beaches with offshore bars, higher water levels may allow waves to pass over the bar, or bars, without breaking while lower water levels may contribute to depth-induced wave breaking near the bar, or bars. An example of a 10 min average video image for Duck, North Carolina, shows wave breaking over an offshore bar and closer to the beach (Figure S69). For this example, the surf zone width along the indicated traverse was 44 m.

Longshore currents that move essentially parallel to the beach, also called littoral currents, are generated by waves breaking at angles relative to the beach. These currents are important for sediment transport and beach erosion applications as well as operations in the surf zone.

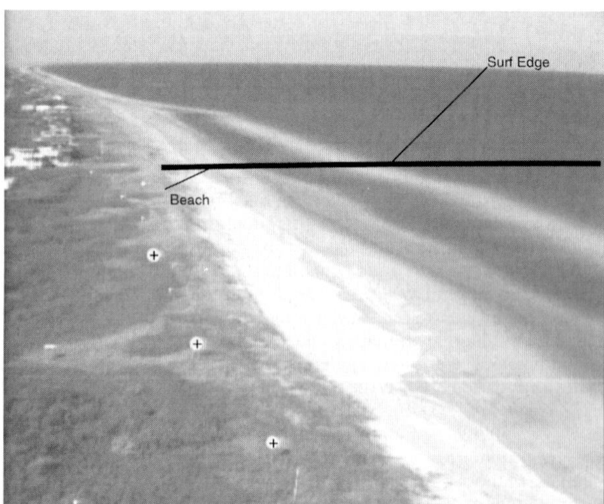

Figure S69 Ten minute average video image showing a surf zone with wave breaking over an offshore bar and at the shoreline. Surf zone width, 44 m, from the shoreline (labeled Beach) at the time of this image past the main outer region of wave breaking (labeled Surf Edge) was determined from this image along the indicated traverse.

Rip currents are hazardous intermittent currents that cannot be predicted reliably by present surf models. Rip currents are strong currents that are confined to a narrow jet that flows seaward from the surf zone in a direction perpendicular to the beach. Several methods by which rip currents are generated have been proposed, but successful modeling requires further research advances.

Other parameters may also be modeled, in some cases as part of the calculations to predict the parameters of main interest. However, most other parameters are generally considered more important for research than for practical applications. A possible exception is wave setup which is relatively simple to estimate and which somewhat affects the total water depth that largely determines breaker heights. Wave setup is a wave-induced increase in mean water level near a beach. Other parameters include wave runup, swash, water motions at periods longer than incident wave periods (e.g., infragravity waves), and cross-shore currents including undertow.

Estimation methods

Without using numerical models, estimates of where waves break and breaker heights can be obtained using simple procedures. The relationship that a wave breaks when its height (crest to trough) exceeds 0.78 multiplied by water depth has long been used in engineering practice as a good first estimate. In the ocean, many component waves, each with different wave heights, periods, and directions, combine to produce a wave spectrum. The root-mean-square wave height that is computed from a spectrum of waves is approximately given by 0.42 multiplied by water depth in a surf zone, where essentially all waves are breaking (e.g., Smith, 2000). These types of breaking criteria have often been combined with equations for wave shoaling and wave refraction (see *Waves* and *Wave Refraction Diagram*), assuming locally parallel depth contours, to develop engineering methods that yield useful estimates of breaker heights (e.g., Wiegel, 1964; Dean and Dalrymple, 1984; Smith, 2000).

Several surf "similarity" parameters that typically depend on breaker height, local bottom slope, and wave period are used to categorize breakers as spilling, plunging, or surging (e.g., Guza and Inman, 1975; Wright and Short, 1983; Smith, 2000). The various formulations provide reasonably similar results. Use of these parameters shows that spilling breakers are most common and that surging breakers are relatively rare.

Mainly for coastal engineering applications, estimation methods have also been developed for other types of surf zone information such as wave setup, runup, and longshore currents (e.g., Wiegel, 1964; Dean and Dalrymple, 1984; Fredsøe and Deigaard, 1992; Coastal and Hydraulics Laboratory, 1996, 1980–2000; Smith, 2000).

Numerical modeling

To characterize surf realistically, however, numerical models are needed. A recent research approach is numerically solving the hydrodynamic equations of fluid flow that govern shallow water waves in the time domain throughout a nearshore region. Boussinesq type models are probably the most common models of this type. Fredsøe and Deigaard (1992) and Svendsen and Putrevu (1996) provide overviews. Equations for variable nearshore bathymetry can be solved for water elevations and currents. Some models solve the fully nonlinear equations of motion that are usually expressed using a velocity potential. Time domain models are mathematically complex and computationally intensive. But, they can consider both cross-shore and longshore depth variations. These models have the potential to better predict wave-induced currents, including rip currents, than simpler one-dimensional models. Because there are several unresolved research issues, such as considering wave breaking, time domain models are not used yet for applied surf modeling.

Where a wave breaks is controlled largely by local depths in the immediate vicinity of breaking. Thus, one-dimensional surf models that consider local depth variations from outside of the surf zone to the shoreline are quite successful in determining breaker characteristics even though longshore depth variations are not considered. One-dimensional models are based typically on the energy transport equation given by

$$\frac{\partial(E_w C_g \cos\theta)}{\partial x} = -\langle \varepsilon_b \rangle$$

where E_w is total wave energy, C_g is wave group velocity, θ is wave direction relative to a normal to the beach, x is a distance from outside the surf zone to the land–water interface, and $\langle \varepsilon_b \rangle$ is the average rate of dissipation of wave energy per unit area of sea surface due to wave breaking. Dissipation due to bottom friction is small compared to that caused by wave breaking. E_w is given by

$$E_w = \frac{1}{8}\rho g H_{rms}^2,$$

where ρ is water density, g is the acceleration due to gravity, and H_{rms} is root-mean-square wave height.

Using various methods for calculating the dissipation rate, the energy transport equation is solved by numerical methods beginning outside the surf zone and progressing to the shoreline. This approach models energy and associated wave statistics rather than time varying characteristics of individual waves. A model that has been modified for several applications (e.g., Earle, 1999) was developed by Thornton and Guza (1983). Because most breakers are spilling breakers that resemble bores, equations for the energy dissipation from bores (e.g., Le Mehaute, 1962) are often employed. Other equations for wave height decreases caused by wave breaking can be used and provide generally similar results (e.g., Dally, 1990). Either wave-by-wave calculations (e.g., Dally, 1990, 1992) or probability calculations (e.g., Thornton and Guza, 1983) can be used to estimate the average dissipation. For wave-by-wave methods, wave heights are usually selected so that they represent a Rayleigh distribution (Longuet-Higgins, 1952) outside of the surf zone, but joint wave height and period probability distributions (e.g., Longuet-Higgins, 1983) may also be used. For probability methods that follow Thornton and Guza (1983), the average dissipation is obtained by integrating, over all wave heights, a bore dissipation function multiplied by the probability that a wave of a given height is or has broken.

Because it involves relatively few calculations and wave breaking probabilities can be used to estimate probabilities of different breaker types, a probability approach is attractive. Surf "similarity" parameters allow categorizing breakers as spilling, plunging, or surging. Given beach slope and wave period, values of breaker heights that delimit surging from plunging waves and plunging from spilling waves can be obtained. Using these limits, a wave breaking probability distribution (e.g., Thornton and Guza, 1983) can be integrated to provide breaker type probabilities that may vary across the surf zone.

This surf modeling description and the provided equations are reasonably general. Various formulations for breaker dissipation, wave breaking probability distributions, and breaker types have been used. Given an incident directional wave spectrum, $E(f, \theta)$, where f is wave frequency and θ is wave direction, the numerical approach involves calculating E_w, root- mean-square wave height, appropriate wave frequencies such as the frequency of maximum energy and the average frequency, and a suitable single wave direction that are contained in the complete equations. Wave direction is usually calculated so that concentration of wave energy in this direction provides the correct longshore momentum flux for later longshore current calculations. The energy transport equation is then solved numerically from outside the surf zone to the shoreline. Numerical integrations are also used for some calculations such as the average wave energy dissipation rate and wave breaking percentages. Fredsøe and Deigaard (1992) and Svendsen and Putrevu (1996) further describe several modeling approaches.

While modeling of only breaker characteristics may be performed in this manner, information about wave-generated longshore currents is often desired. Depth-averaged longshore currents are usually obtained using radiation stress theory (e.g., Longuet-Higgins, 1970a,b) or related equations. Doing this involves solving the longshore momentum equation given by

$$\tau_y^h + \frac{\partial}{\partial x}\left(\mu h \frac{\partial v}{\partial x}\right) - <\tau_y^b> + <\tau_y^w> = 0,$$

where V is the depth averaged longshore current, the first term is the longshore driving stress exerted by the waves, the second term represents horizontal mixing across the surf zone, the third term is the average longshore stress due to bottom friction (also a function of v), and the last term is the average wind- generated longshore stress. Including wind effects in this simple manner adds little to the computations. The parameter, μ, is a horizontal eddy viscosity and h is local water depth. The driving related to wave breaking is given by

$$\tau_y^h = <\varepsilon_b> \frac{\sin\theta}{c},$$

where c is wave phase speed.

Dissipation due to wave breaking is a key input for longshore current calculations. Thus, most surf models that calculate breaker information also calculate longshore currents. Breaker and dissipation calculations are first completed across the surf zone. The longshore momentum equation is then solved numerically for depth-averaged longshore currents. Formulation of the bottom stress involves the longshore current, the wave orbital velocities, and a bottom drag coefficient. Formulation of the wind stress involves the wind speed, wind direction relative to the beach, and a sea surface drag coefficient.

Water depths (bathymetry) along a traverse extending from seaward of the surf zone to the shoreline are critical to accurate surf modeling. Solving both the energy transport equation and the longshore momentum equation for one-dimensional models involves the local water depth, h, at each computational step along the traverse. In the energy transport equation, the wave group velocity and the average rate of dissipation of wave energy depend on water depth. In the longshore momentum equation, the longshore driving stress exerted by the waves and the horizontal mixing term depend on water depth. The bottom friction term also includes depth effects because it is a function of depth dependent wave orbital velocities. Because of the small spatial scales of cross-shore depth variations in surf zones, computational steps (calculation intervals) are typically on the order of 1 m. For two-dimensional or three-dimensional time domain models, there are similar depth requirements.

Lack of recent and accurate depth information at a beach of interest is often a limitation. Recent depth data are desirable because nearshore depths may change rapidly and substantially as a result of wave, current, and tide action. Accurate depth data are needed because wave breaking is strongly related to the water depth.

Equilibrium beach profiles (Dean, 1977) that have been used widely in coastal processes and engineering studies provide a method to estimate nearshore depths when actual data are not available. An equilibrium beach profile is given by

$$h(x) = A\, x^{2/3},$$

where h is water depth, x is distance from the mean sea level position of the shoreline, and A is a scale parameter that depends on sediment size. Equilibrium beach profiles were derived from an analysis of over 500 measured beach profiles and consider that beaches with more coarse sediments are steeper than beaches with finer sediments. An infinite slope theoretically exists at the shoreline, but surf models halt their calculations in extremely shallow water before this behavior causes problems. Also, equilibrium beach profiles do not consider the presence of offshore bars indicating the value of actual depth data when it is available.

Breaker heights and types, as well as surf zone widths, can be modeled fairly well by one-dimensional surf models, but longshore currents are often somewhat in error mainly for beaches with shallow offshore bars. At such beaches, modeled longshore current maxima are generally near the bar, but measured maxima are usually further shoreward. Paradoxically, approximately correct energy dissipation, that provides reasonable breaker characteristics and that drives longshore currents, provides relatively poorer longshore currents. Various improvements have been investigated including linear and nonlinear bottom stress equations, turbulent energy production by breakers, longshore pressure gradients caused by wave setup, and wave rollers representing spilling breakers. Turbulent energy and roller approaches delay driving of currents shoreward of breaker locations. Incorporating variable bottom friction to consider varying sediment sizes across the surf zone and to compensate for missing vertical turbulent effects shows promise. An example output of the US Navy's operational one-dimensional surf model with this modification depicts both breaker height and longshore current variations across the surf zone (Figure S70).

Because wave breaking is locally depth controlled and one-dimensional models provide suitable wave energy dissipation rates, research is investigating the use of one-dimensional surf models at multiple longshore locations to calculate the dissipation that subsequently drives two-dimensional current models. These more mathematically complicated models then consider longshore variations in both forcing, such as wave setup, and bathymetry (e.g., Putrevu *et al.*, 1995; Svendsen *et al.*, 1998). As earlier noted, time domain solutions of the hydrodynamic equations for wave motion are also being developed.

Modeling applications

Unlike earlier estimation methods, surf models can provide many types of information across the surf zone. One-dimensional models also can be run at different nearby locations to examine longshore surf variability caused by either incident wave variability or bathymetry changes. Thus, new applications are developing in addition to uses for basic information such as breaker heights. Coastal engineering appli-cations can employ more detailed information such as consideration of different

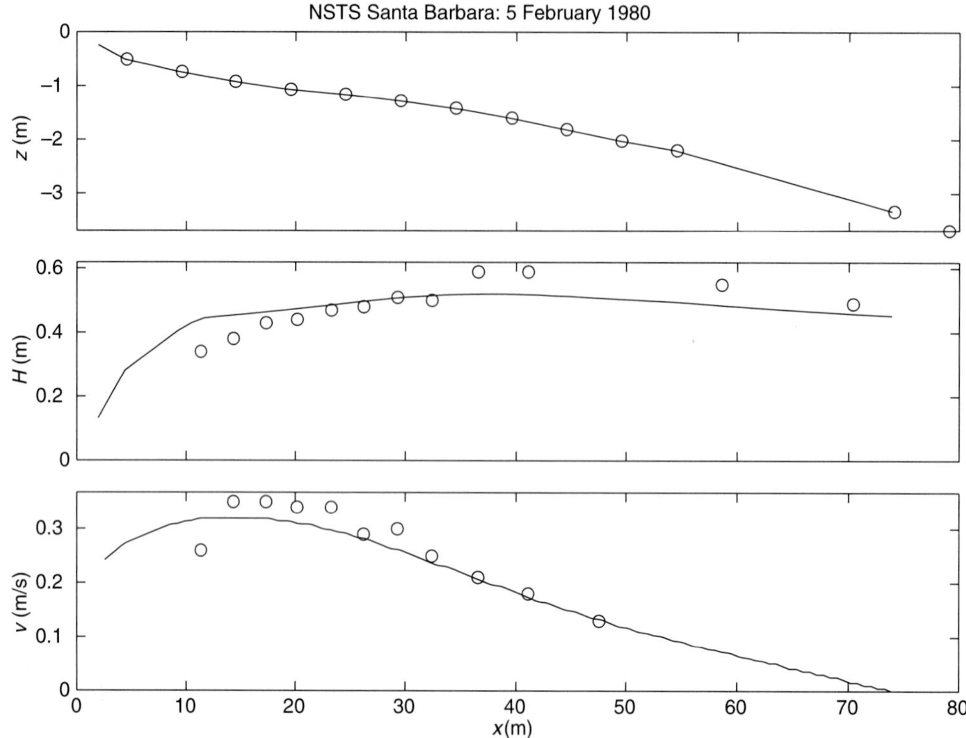

Figure S70 Example of a one-dimensional surf model output showing depth (top panel), root-mean-square breaker height (middle panel), and depth-averaged longshore current (bottom panel) across the surf zone. Measured data are marked by circles.

breaker types. Design of systems that operate in the surf zone, and planning of many types of activities, can be better accomplished with quantitative modeled information.

Information about surf is needed for coastal processes studies, such as beach erosion and sediment transport, and engineering analyses, such as design of coastal structures. The potentially adverse effects of surf on activities in the surf zone and the availability of surf models has resulted in increased use of these models to forecast surf conditions. The US Army Corps of Engineers developed a one-dimensional surf model for engineering applications (Larson and Kraus, 1991). The US Navy routinely operates a one-dimensional surf model to make surf forecasts for amphibious operations (Earle, 1999). Because various military systems require surf information for their designs, this model has also been used to develop surf statistics based on statistics of incident waves (see *Wave Climate* and *Wave Environments*) at locations of interest (Earle, 1999). Surf models have also been driven with real-time measured wave data to provide surf estimates without placing instruments in the hazardous surf zone (Nichols and Tungett, 1998) and have been linked with other models that provide incident wave conditions (Allard *et al.*, 1999).

Summary

A variety of surf models, primarily one-dimensional models, have been developed and applied. Accuracies of particular models are not reviewed here, but such information can be obtained from the references. Models that have been applied or operated routinely are sufficiently accurate for their intended applications. Research using time domain models shows that these models have significant promise particularly for improved calculations of currents. Most importantly, surf modeling provides quantitative surf information that considers the highly location specific nature of surf.

Marshall D. Earle

Bibliography

Allard, R.A., Hsu, Y.L., Miles, K., Mettlach, T., and Earle, M.D., 1999. Toward an operational coupled wave/surf forecasting system. *Proceedings of the Third Conference on Coastal Atmospheric and Oceanic Prediction and Processes*, Boston, MA: American Meteorological Society, pp. 350–355.

Coastal and Hydraulics Laboratory, 1980–2000. *Coastal Engineering Technical Notes*, numerous issues, Vicksburg: Waterways Experiment Station, US Army Corps of Engineers.

Coastal and Hydraulics Laboratory, 1996. *Automated Coastal Engineering System (ACES)*, version 2.0, numerous issues. Vicksburg: Waterways Experiment Station, US Army Corps of Engineers.

Dally, W.R., 1990. Random breaking waves: a closed-form solution for planar beaches. *Coastal Engineering*, **14**: 233–263.

Dally, W.R., 1992. Random breaking waves: field verification of a wave-by-wave algorithm for engineering application. *Coastal Engineering*, **16**: 369–397.

Dean, R.G., 1977. Equilibrium beach profiles: U.S. Atlantic and Gulf coasts, Newark: University of Delaware, *Ocean Engineering Technical Report 12*.

Dean, R.G., and Dalrymple, R.A., 1984. *Water Wave Mechanics for Engineers and Scientists*, Englewood Cliffs: Prentice-Hall.

Earle, M.D., 1999. Applied and operational surf modeling. *Shore and Beach*, **67**: 70–75.

Fredsøe, J., and Deigaard R., 1992. *Mechanics of Coastal Sediment Transport*, Singapore: World Scientific.

Guza, R.T., and Inman, D.L., 1975. Edge waves and beach cusps. *Journal of Geophysical Research*, **80**: 2997–3012.

Larson, M., and Kraus, N.C., 1991. Numerical model of longshore current for bar and trough beaches. *Journal of Waterway, Port, Coastal, and Ocean Engineering*, **117**: 326–347.

Le Mehaute, B., 1962. On non-saturated breakers and the wave run-up. *Proceedings of the Eighth International Conference on Coastal Engineering*, American Society of Civil Engineers, pp. 77–92.

Longuet-Higgins, M.S., 1952. On the statistical distribution of the heights of sea waves. *Journal of Marine Research*, **11**: 245–266.

Longuet-Higgins, M.S., 1970a. Longshore currents generated by obliquely incident sea waves, 1. *Journal of Geophysical Research*, **75**: 6678–6789.

Longuet-Higgins, M.S., 1970b. Longshore currents generated by obliquely incident sea waves, 2. *Journal of Geophysical Research*, **75**: 6790–6801.

Longuet-Higgins, M.S., 1983. On the joint distribution of wave periods and amplitudes in a random wave field. *Proceedings of the Royal Society of London*, **A, 389**: pp. 241–258.

Nichols, C.R., and Tungett, D.W., 1998. Buoys provide real-time surf data. *US Naval Institute Proceedings*, **124**: 71–72.

Putrevu, U., Oltman-Shay, J., and Svendsen, I.A., 1995. Effect of alongshore nonuniformities on longshore current predictions. *Journal of Geophysical Research*, **100**: 16119–16130.
Smith, J.M., 2000. Surf zone hydrodynamics. In *Coastal and Hydraulics Laboratory. Coastal Engineering Manual*. Vicksburg: Waterways Experiment Station, US Army Corps of Engineers.
Svendsen, I.A., and Putrevu, U., 1996. Surf-zone modeling. In Dally, W.R., and Zeidler, R.B. (eds.), *Coastal Dynamics '95*. Reston: American Society of Civil Engineers, pp. 13–32.
Svendsen, I.A., Sancho, F.E., Oltman-Shay, J., and Thornton, E.B., 1998. Modeling nearshore circulation under field conditions. In Edge, B.L., and Hemsley, J.M. (eds.), *Ocean Wave Measurement and Analysis*. Reston: American Society of Civil Engineers, pp. 765–776.
Thornton, E.B., and Guza, R.T., 1983. Transformation of wave height distribution. *Journal of Geophysical Research*, **88**: 5925–5938.
Wiegel, R.L., 1964. *Oceanographical Engineering*. Englewood Cliffs, Prentice-Hall.
Wright, L.D., and Short, A.D., 1983. Morphodynamics of beaches and surf zones in Australia. In Komar, P.D. (ed.), *CRC Handbook of Coastal Processes and Erosion*. New York: CRC Press, pp. 35–64.

Cross-references

Surf Zone Processes
Tides
Wave Climate
Wave Environments
Wave Refraction Diagrum
Waves

SURF ZONE PROCESSES

The surf zone can be defined as that relatively narrow strip of a body of water that borders the land, and which contains waves that are breaking due to the shallow water depth. However, because the tide level, incident waves, and local wind speed, and direction continually change, the width and character of the surf zone vary incessantly. Therefore, in a discussion of surf zone processes, the region of interest is actually the "nearshore" zone, herein defined as that region that is directly or indirectly affected by depth-induced wave breaking. Finally, a subregion called the "swash" zone is commonly delineated at the boundary between land and water, as that area which is alternately wetted and dried by wave uprush and backrush. These zones are indicated in Figure S71.

The depth-induced breaking of waves drives a progression of intertwined processes that ultimately leads to changes in the morphology of the beach itself, caused by erosion and accretion of sediment. These processes include the creation of the breaking wave roller, maintenance of residual turbulence in the water column, the creation of setup in the mean water level, generation of currents (cross-shore, longshore, and rip currents) and low-frequency motions, entrainment and suspension of sediment, and finally the transport of sediment.

Wave transformation in the nearshore

The phenomena that are most important to wave transformation in the nearshore are (1) refraction, (2) shoaling, and (3) decay due to depth-limited breaking. Refraction (analogous to the refraction of light) is the bending of the waves by variations in bathymetry, and generally tends to align the crests of the waves with the bathymetric contours. Therefore along a straight, smooth coastline, waves that approach from an oblique angle in deepwater may be almost shore-parallel by the onset of breaking in shallow water, as depicted in Figure S71. In this situation, refraction also serves to suppress the wave height. Along an irregular coastline, however, refraction will tend to focus wave energy on headlands and more subtle protruding bathymetric features, thereby locally increasing the wave height. In order to conserve the total amount of wave energy in the system, the wave height is correspondingly reduced in the embayments adjacent to the protruding feature.

As waves refract from deep to shallow water, the process of shoaling is also at work. In rudimentary explanation, as the waves move into shallow water they slow down, but consequently in order to maintain the total amount of energy flux, the wave energy (proportional to the square of the height) must increase. In the situation of a straight, smooth coastline, it is clear that the amplification of wave height due to shoaling typically exceeds the suppression due to refraction, as an increase in wave height prior to breaking is obvious to an observer on the beach.

When the water depth becomes too shallow to sustain the height of the growing wave, the wave becomes unstable and breaking ensues, as characterized in Figure S72. This point of incipient breaking is commonly estimated to be when the wave height reaches roughly 80% of the water depth. Incipient breaking is also dependent upon the wave period and the local bottom slope, for which many empirically derived formulas have been offered (e.g., see Weggel, 1972). In addition, the wind affects incipient breaking with a following wind causing waves to spill sooner, thereby widening the surf zone. An offshore wind delays breaking, thereby compressing the surf zone, and causes waves to plunge (Douglass, 1990).

As the water depth continues to decrease, breaking becomes fully developed and the wave height continues to decrease. The rate of decay of wave height depends predominantly on the bed slope. For slopes around 1/30, the incipient condition of 80% of the water depth continues

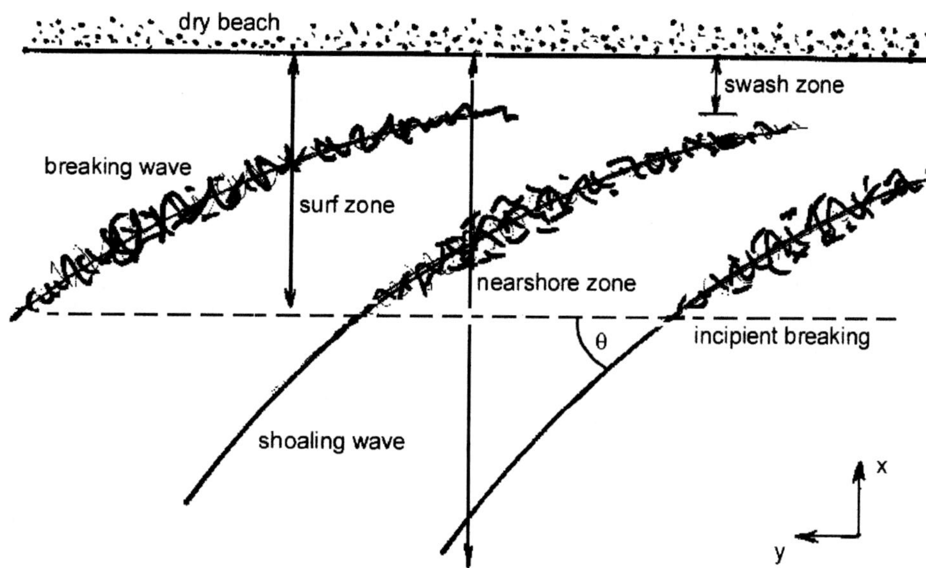

Figure S71 Definition sketch of the nearshore zone, showing overhead view of wave refraction, shoaling, and breaking.

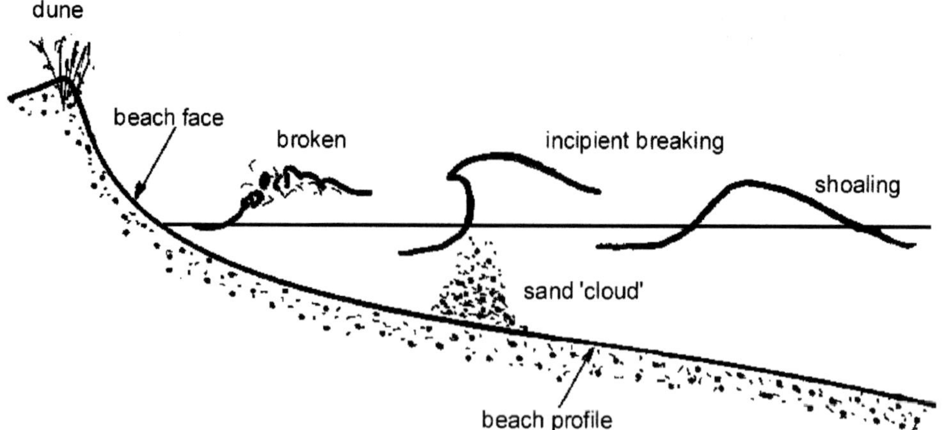

Figure S72 Cross-section of the nearshore zone, depicting wave transformation.

to hold true. For milder slopes, however, this value gradually decreases to as little as 35%, whereas for steep beaches the wave height might maintain a value equal to the local depth. More sophisticated models that describe breaking decay are available (e.g., Battjes and Stive, 1985; Dally *et al.*, 1985).

Because the beach face is typically steep (Figure S72), as the waves approach the mean water-level shoreline they collapse to form the characteristic strong up rush and downrush of the swash zone. This is the case especially during high tide. At low tide, however, the water level may intersect the beach at a much flatter segment of the profile, causing the waves to continue their decay as gently spilling breakers. In this situation, the swash zone may be extremely small, or may not exist at all.

The breaking wave roller

In watching the surf zone the most obvious feature is the "rollers," which characterize wave breaking. At breaking's onset, a portion of the energy that was previously transported in the organized wave motion is converted into a region of highly turbulent, aerated water that moves at roughly the wave celerity, as shown in Figure S73. Energy is dissipated in the form of heat in a shear layer that exists between the roller and the underlying organized flow, as well as inside the roller itself. In addition to its role in the energy budget of the surf zone, the roller is also a significant contributor to the mass and momentum balances (Svendsen, 1984).

Due to its highly turbulent, aerated nature, very few direct measurements of fluid motion in the roller have been obtained (see Mocke *et al.*, 2000), and so detailed knowledge is lacking. However, quantitative models for macro-properties of the roller exist, basically providing the evolution of the size of the roller as it crosses the surf zone (Dally and Brown, 1995).

Residual turbulence

As the roller passes by a particular spot in the surf zone, it leaves behind a layer of residual turbulence that expands downwards through the underlying water column as shown in Figure S73. This turbulence has important links to other surf zone processes, namely the suspension of sediment and the mixing of surf zone currents. The additional "stirring" by the residual turbulence establishes the vertical structure of surf zone currents, with stronger turbulence resulting in currents that are more uniform over depth. Residual turbulence contributes both directly and indirectly to lateral mixing of the currents as well, and is also responsible for maintaining concentrations of suspended sediment in the upper water column that far exceed those under non-breaking conditions.

The behavior of the residual turbulence left in the water column by the passing rollers has been found to mimic that of a free wake created by unidirectional flow past an obstacle. By applying principles from the basic theory for turbulent wakes, the temporal and spatial structure of the residual turbulence has been modeled to some degree (e.g., Sakai *et al.*, 1982; George *et al.*, 1994). The wave-period-averaged turbulence intensity and the vertical mixing in the wake have been quantified in terms of macro-properties of the roller itself (Dally, 2000).

Radiation stress

Although the water particle motion associated with ocean waves is oscillatory, if the momentum flux associated with each wave is mathematically integrated over the water column and then time-averaged over a wave period, one finds that the waves exert a mean residual stress on the water column. This stress, called the "Radiation Stress" (Longuet-Higgins and Stewart, 1964), is actually a stress tensor with three components. In a Cartesian coordinate system with x and y directions in the horizontal plane and z directed vertically upwards (as in Figures S71 and S73), two components, S_{xx} and S_{yy}, act like normal pressure in the x and y directions, respectively, whereas the third, S_{xy}, acts like a tangential shear stress. In the shallow water of the nearshore, they are each functions of the wave height (H) and local wave direction (θ), and are approximated by

$$S_{xx} \cong \frac{\gamma H^2}{8}\left(\frac{1}{2} + \cos^2\theta\right), \qquad \text{(Eq. 1a)}$$

$$S_{yy} \cong \frac{\gamma H^2}{8}\left(\frac{1}{2} + \sin^2\theta\right), \qquad \text{(Eq. 1b)}$$

$$S_{xy} \cong \frac{\gamma H^2}{8}\cos\theta\sin\theta, \qquad \text{(Eq. 1c)}$$

where γ is the density of water. Spatial gradients in these stress components slightly deform the mean water level in the nearshore, as well as drive surf zone currents.

Set-down and set-up

With a coordinate system oriented such that the x-axis is directed onshore, cross-shore gradients in S_{xx} due to shoaling of waves (i.e., an increasing wave height) actually depress the mean water level seaward of the surf zone. This "set-down" begins as waves enter transitional water and begin to feel the bottom, and gradually draws down the mean water level until reaching the point of incipient breaking. At this point the maximum set-down is observed to be roughly 2–4% of the height of the breaking wave (see Bowen *et al.*, 1968).

As breaking ensues and the roller develops, the mean water level first flattens for a short distance because there is a momentary balance between the decrease in S_{xx} and the increase in momentum flux in the roller as it grows in size. However, farther into the surf zone as both the wave and roller shrink, a "set-up" in the mean water level is created (see Figure S73), which reaches a maximum value at the shoreline. Although dependent on the bottom slope and the wave period, this maximum setup is roughly 12–18% of the breaker height. With a large swash zone, however, the maximum setup is difficult to define and measure, due to the intermittent presence of water.

The local wave-induced set-up is superimposed upon any large-scale deviation in mean water level forced by the winds. For example, offshore-directed winds apply a surface shear stress that tends to suppress the water level in the nearshore, whereas onshore-directed winds will push water up against the shore and hold it there, effectively elevating the mean water level as a "storm surge."

Cross-shore currents

Althought the water motion associated with a non-breaking wave is nearly balanced between forward and backward displacement, there exists a residual flux of water in the direction of wave propagation, which occurs mostly in the wave crest. In breaking waves, this flux is augmented by additional onshore flux of water in the aerated roller. Due to the presence of the shoreline, and with longshore-uniform conditions, this mean landward discharge in the upper part of the water column, must be locally compensated by a mean, offshore-directed "return-flow" in the lower water column, depicted in Figure S73. The return-flow current is often called "undertow," but is not to be confused with the rip current discussed below. Although the strength of the return-flow is only roughly 10% of the wave celerity, it is this weak, yet persistent current that is primarily responsible for the net offshore transport of sediment that occurs during storms (Dally and Dean, 1984; Dally and Brown, 1995).

The shear stress applied to the water surface by the wind also contributes forcing that affects the strength and structure of cross-shore current in the surf zone. An onshore-directed stress pushes additional surface water toward the beach which, again because of the presence of the shoreline boundary, must be balanced by an increased return-flow current in the lower water column. Consequently the onshore wind that often accompanies a storm further augments the wave-driven offshore current. With an offshore-directed wind the opposite occurs. That is (neglecting the wave-induced current for the moment), surface water pushed offshore by the wind stress is supplied by water drawn onshore in the lower water column.

Longshore currents

The mean flux of water along the beach is called the "longshore" current, and is also forced by both waves and local winds. For obliquely incident waves, the cross-shore gradient in S_{xy} (equation 1c) due to breaking creates a residual shearing thrust that is directed down the beach as depicted in Figure S74. In the outer surf zone this thrust is locally offset by an opposite-directed shear associated with the creation of the roller. However, as the roller subsequently starts to shrink, its thrust switches direction, and both thrusts now act in harmony to drive water down the beach (Osiecki and Dally, 1996). Consequently, the location where the longshore current is greatest is usually somewhat landward of the break point. The mean longshore thrust of the waves is balanced by the shear stress that the sand bed exerts on the water.

Seaward of breaking, there is no wave-induced forcing of the longshore current because the increase in S_{xy} associated with wave shoaling is exactly compensated by the decrease due to wave refraction. However, lateral mixing processes driven by both turbulence and residual convective acceleration enable the current inside the surf zone to drag the outside water along with it, creating a tail on the longshore current that diminishes in the offshore direction (Svendsen and Putrevu, 1994). Outside the breaker line, this lateral mixing stress is balanced by the mean bed stress of the current.

Depending upon its speed and direction, the wind can also be a significant contributor to forcing of the longshore current. In fact, it is often the primary driver during the initial stage of development of a storm. The alongshore component of the wind stress vector, acting on the water surface, can either act in harmony with the wave-driven forcing, or oppose it. Although a rare event, strong opposing winds can even arrest the wave-driven longshore current.

At the base of the beach face, the behavior of the longshore current is controlled by the character of the swash zone. For very flat profiles with little swash, the current speed approaches zero as depicted in Figure S74. For large swash zones though, the mean longshore current observed at the toe of the swash can be significant.

Rip currents and nearshore circulation

In situations where either the incident wave field or the under-lying bathymetry is not uniform along the beach, complicated circulation patterns can be established. These include the infamous "rip current" that is so dangerous to bathers. Not to be confused with the mild cross-shore current (undertow) described above, rip currents are very strong, yet locally confined, offshore-directed flows that can reach speeds more than 1.0 m/s, making them nearly impossible to swim against. One likely generation mechanism for a rip current is the presence of a low spot in a longshore bar, through which the majority of water carried landward over the bar by the waves along the adjacent beaches is returned. As shown in Figure S75, the rip current is fed by the mild, mean onshore transport of water that occurs in the wave crest and roller as discussed above, which accumulates to form longshore-directed feeder currents that search out the narrow, low section in the bathymetry to form the seaward-directed rip. Set-up is reduced in the rip channel but elevated on either side, which serves to drive the circulation cell. If the bathymetry is rhythmic in the longshore direction, rip currents can form at regularly spaced intervals, and can be present as long as the special bathymetry persists—sometimes even for several days (see Noda, 1972).

Another possible origin of regularly spaced rip currents is the interaction of waves approaching a beach from different directions (Dalrymple, 1975). The waves tend to enhance each other at regular intervals along the beach (antinodal zones), with zones of suppression in between (nodal zones). Where the waves reinforce, the setup is greater than where they negate. This results in structure in the mean water surface that drives water from the antinodes towards the nodes, which is again turned out to sea in the form of a rip.

Additional complex nearshore circulation patterns are often generated adjacent to coastal structures such as groins, jetties, piers, and breakwaters. Basically, the structure perturbs what might otherwise be a longshore-uniform wave climate, creating a region of reduced setup in its wave shadow, to which water is driven from remote regions of higher setup. The shadow can be the obvious result of wave blocking by the structure, as well as more subtle energy losses incurred due to "rubbing" of the waves against the structure.

Low-frequency motions

Often the incident waves arrive at the beach in distinct groups, especially if they were generated by a distant storm and local wind conditions are calm. This groupiness manifests itself as a low-frequency modulation of the wave height, where the wave height grows over several successive waves and then decreases over several more. Each group can last up to several minutes, sometimes with a notable period of calm in-between groups. The setup phenomenon described above responds directly to this rhythmic unsteadiness in wave height, and the gradually varying mean

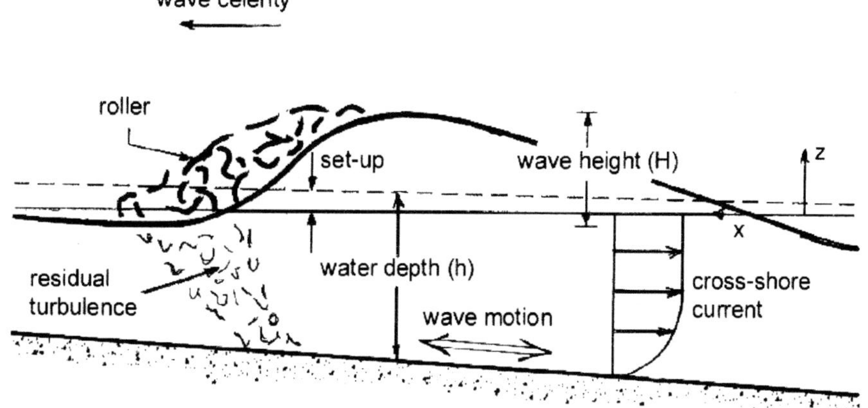

Figure S73 Hydrodynamic features of the surf zone (after Dally, 2000).

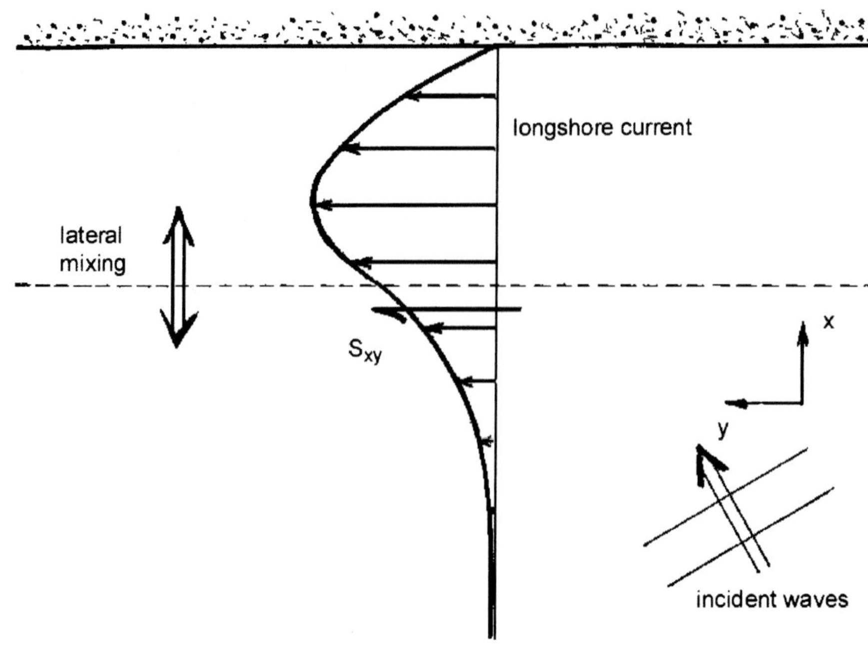

Figure S74 The longshore current, driven by the S_{xy} component of Radiation Stress for obliquely incident waves. Lateral mixing creates the "tail" of the current.

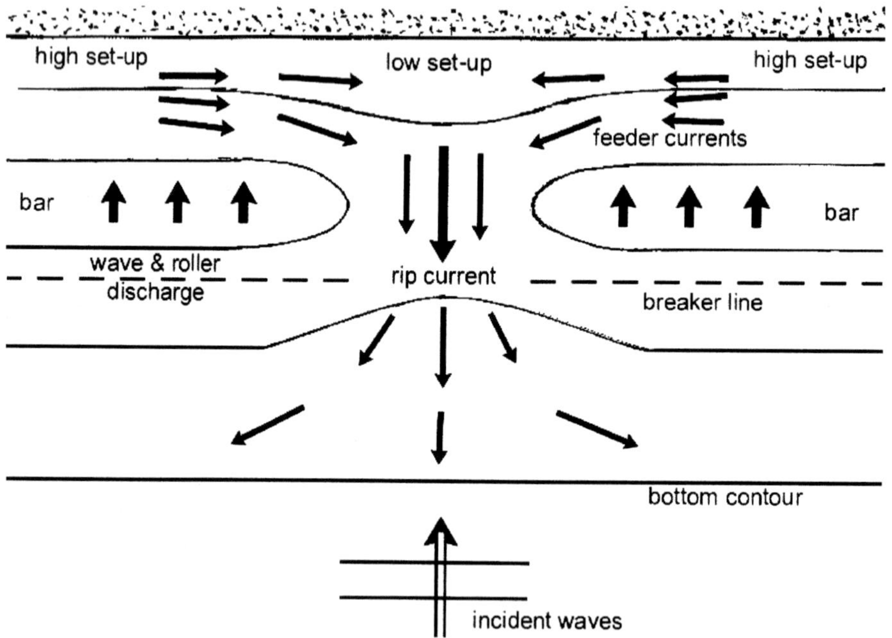

Figure S75 Rip currents generated at a low spot in the crest of a longshore bar.

water level that results is referred to as "surf beat." Surf beat can become particularly pronounced if the bottom slope is slight and the waves are large (i.e., a wide surf zone).

Because surf beat is a low-frequency motion ($f < 0.01$ Hz), some of this energy can be trapped against the coastline by refraction processes or can be amplified, particularly by reflection from headlands and shore-perpendicular structures. These long-period, "edge" waves actually propagate along the beach and, although measurable, are usually undetected by the naked eye (e.g., see Guza and Davis, 1974).

Groupiness in the waves also causes the mean forcing in the nearhore to vary in time, and consequently the cross-shore and longshore currents are unsteady. One additional low-frequency motion sometimes found in the surf zone is the "shear wave," which is actually a periodic instability generated by strong longshore currents. This phenomenon is detectable only with a current meter, as it does not directly influence the mean water-level elevation (Oltman-Shay *et al.*, 1989).

Sand entrainment and suspension

Although researchers continue to struggle with the extremely complicated problem of quantitatively modeling sediment entrainment and suspension in the surf zone, much is known qualitatively. In general terms, sediment is first entrained by the near-bed fluid motion, and then carried up into the water column by small-scale turbulence or larger-scale vortices. The most germane parameters are the size and gradation of the sediment (typically sand), its settling velocity in still water, and the strength and character of the water motion, which is usually quantified in terms of wave height and period, and current strength. Small sand grains with slow settling velocities are easily entrained and suspended in large quantities, whereas large, heavy particles understandably hug the bed. Larger waves suspend more sediment due to their stronger oscillatory velocities, whereas short-period waves support suspension by maintaining higher mean turbulence levels.

Outside the surf zone, the entraining water motion is comprised of oscillatory wave motion, mean currents, and low-frequency motions. Once entrained, sediment is mixed/ suspended higher into the water column by the turbulence generated at the bed as the waves and currents interact with bedforms such as ripples and mega-ripples.

Inside the surf zone, entrainment is enhanced by stronger wave motion, particularly the more pronounced flow that occurs under the crest of nonlinear, breaking waves. Suspension into the upper water column is greatly augmented by the residual turbulence left by the roller. Finally, the extreme flow conditions and large vortices generated by plunging breakers result in dramatic entrainment and suspension events.

At any particular location in the nearshore, the vast majority of entrainment and suspension occurs as the wave crest passes, creating a "cloud" of sediment (depicted in Figure S72). Outside the surf zone this cloud may be quite small, both in size and in concentration, especially if the entrainment is associated with sand ripples that are present on the bed. Near the break point, large, long-period waves can create what is known as "sheet-flow" conditions, in which the bed is scraped flat and, although there is strong entrainment, there is limited suspension due to a lack of turbulence. In both of these situations, the cloud of sand settles back to the bed before the next wave crest arrives. However, in the surf zone, and particularly for plunging breakers, the sediment cloud may take several wave periods to settle back to the bed, and may in fact be resuspended by a subsequent wave. It is for this reason that sediment concentrations inside the surf zone are generally several times greater than those outside (e.g., see Kana, 1979; Nielsen, 1992).

Farther into the surf zone, because the wave energy decreases due to breaking and the oscillatory motion abates, ripples will once again appear if the surf zone currents are weak. If the currents are strong, "mega-ripples" may be created, resembling those found in rivers. Finally, in the swash zone where wave collapse and runup occurs, sheet flow conditions return and the bed is smooth. In this region, sand entrainment and suspension is characterized as a slurry, rather than as a cloud.

Due to the randomness of the incoming waves, the creation of the sediment clouds is an intermittent process. In fact even in laboratory wave channels where so-called "regular" waves can be created, the details of the entrainment and suspension events vary from wave to wave because the underlying hydrodynamic processes are highly nonlinear. Consequently, time-series concentration measurements (typically made using either sonic or infrared instruments) are often averaged over many wave periods so that the basic structure can be identified.

To illuminate the concepts introduced above, Figure S76 presents idealizations of time-averaged measurements of suspended sand concentration observed in the outer surf zone in a large wave channel for regular, plunging breakers (see Barkaszi and Dally, 1992). Note that both the shape of the vertical profile and the overall magnitude of concentration change considerably as one moves from outside the break point (location "a"), to the plunge point (location "c") and into the surf zone (location "d"). However, as is typically the case the mean concentration decreases with elevation above the bed, as intuition would dictate.

Sediment transport and beach evolution

In the discussion of sand transport in the surf zone, and ultimately the changes in the beach that result from spatial gradients in transport, the water column can be split into two layers: (1) a "near-bed" layer, in which both the oscillatory wave motion and the mean current contribute to the displacement of sediment particles, and (2) a "fully suspended" layer where, effectively, only the mean current contributes to the net displacement (Dally and Dean, 1984). The thickness of the near-bed layer, d_f, is simply the distance that a sand grain falls in one wave period, given by $d_f = T \cdot \omega$, where T is the wave period and ω is the fall velocity in still-water. Of course on a natural beach, the incident waves have differing periods and the sand grains are not uniform in fall velocity, so the boundary between the near-bed and fully suspended regimes is not distinct. Also, when the water depth is less than the value of d_f, the nearbed regime extends over the entire water column, as is the case in the swash zone, for example.

The utility in defining the two regimes is that in the fully suspended layer, the average sediment transport rate is well approximated simply by multiplying the suspended sand concentration by the local current velocity. However, in the nearbed layer, because the sediment is in suspension for less than one wave period, the wave-induced oscillatory velocity does contribute to the net displacement, on average (Dean, 1973). Neglecting the mean current for the moment, and with suspension triggered as the wave crest passes, the sand in the lower half of the near-bed layer will experience a net shoreward displacement, whereas that in the upper half experiences a net seaward displacement, as shown in Figure S77. Because generally the concentration in a sand cloud decreases with elevation above the bed, it is apparent that in the nearbed layer the total net transport of sand is directed onshore. Reintroducing the mean current modifies this scenario of course, depending on its strength and direction.

Cross-shore transport and beach profile evolution

With this two-layer framework, the details of the sand transport processes that cause beach erosion and accretion can be described. At

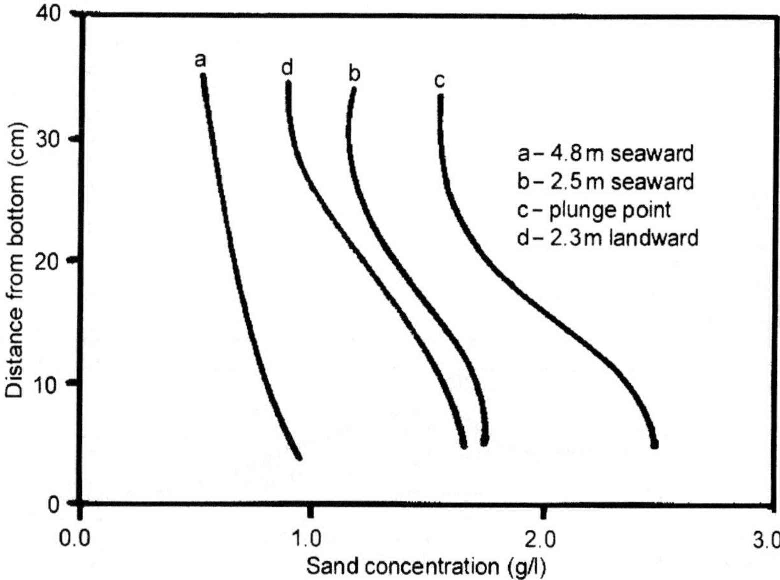

Figure S76 Idealized profiles of mean concentration of sand suspended by plunging breakers. Concentrations at the plunge point can be several times greater than elsewhere.

times when waves directly approach the beach ($\theta = 0$), the mean current in the surf zone and nearshore is comprised of only the offshore-directed undertow. Consequently, mean sediment transport in the fully suspended layer is directed solely offshore. If relatively little sand is being suspended above the nearbed layer, however, the total net mean transport will be onshore, and the beach face will accrete as it receives sand moved from the nearshore. When a storm develops, however, (1) the wave period shortens, thereby reducing the thickness of the nearbed layer, and (2) the residual turbulence levels increase, thereby carrying greater quantities of sediment into the fully suspended layer. Both of these characteristics enhance the offshore-directed transport to the point where it dominates the diminishing onshore transport, and the beach and inner surf zone consequently erode. The material is deposited in a longshore bar that develops near the point of incipient breaking where the entrainment processes weaken. When calmer conditions return and the mild onshore/nearbed transport once again dominates, this bar will gradually migrate onshore and supply sand to the swash zone, thereby rebuilding the beach face. This well-known sequence of beach erosion and recovery is illustrated in Figure S78.

From the discussion above it follows that if wave and water level conditions could be held constant, the beach would evolve until a state of "dynamic equilibrium" was attained, in which the onshore transport in the lower portion of the nearbed layer was balanced by the offshore transport in the upper portion and the fully suspended layer—a balance that must be maintained everywhere across the beach profile. Such a state is easily attained in a laboratory wave channel, but seldom approached in nature.

Longshore transport and coastline change

If waves approach the surf zone obliquely, both the longshore component of the oscillatory wave motion and the longshore current serve to carry sand along the beach. The flux of sand in the longshore direction in the fully suspended layer is again well approximated by the product of the mean concentration and the longshore current speed. In the nearbed layer, although sand initially suspended in the top of the layer may experience net upcoast displacement due to the oscillatory motion, the mean flux for the entire layer is downcoast, and is further enhanced by the longshore current. In the swash zone the uprush can occur at a strongly oblique angle, whereas the downrush typically does not. Consequently, the net longshore transport in the swash can be significant, and in fact is often the predominant zone of longshore transport in the nearshore (see Bodge and Dean, 1987).

If there is no longshore variation in the flux of sediment along the beach, then the beach profile and coastline remain stable. However, if the longshore transport is blocked by a natural immobile feature (e.g., a headland or reef/rock outcrop) or by a man-made structure (e.g., a groin or jetty), sand will accumulate on the updrift side and erode from

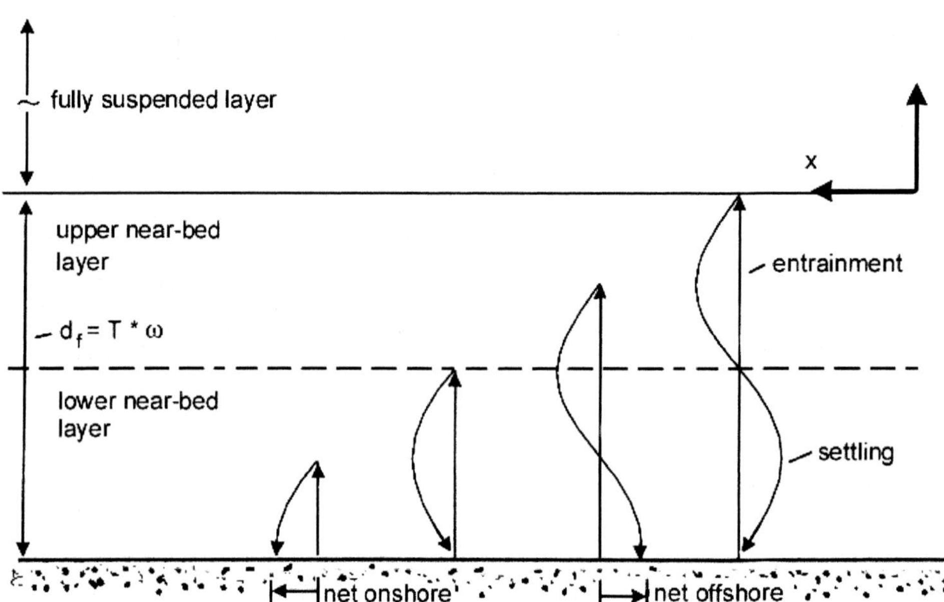

Figure S77 Sediment transport regions in the nearshore. Sand suspended in the near-bed layer returns to the bottom within one wave period, whereas that in the fully suspended layer does not (after Dean, 1973; Dally and Dean, 1984).

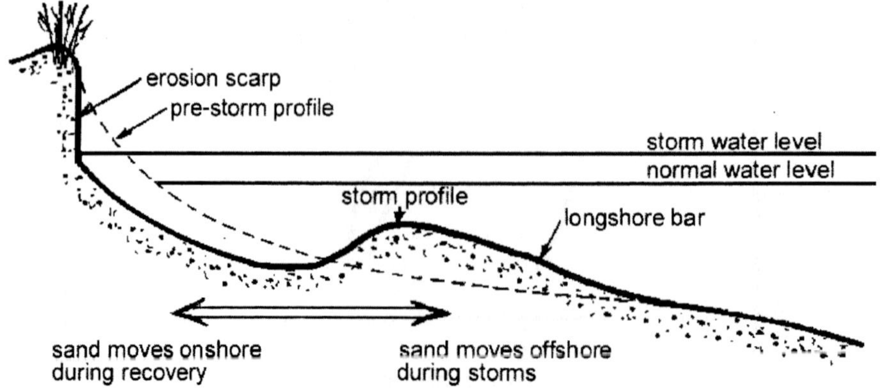

Figure S78 The cycle of storm-induced beach erosion (which creates a longshore bar), and subsequent recovery (which pushes the bar material back onshore).

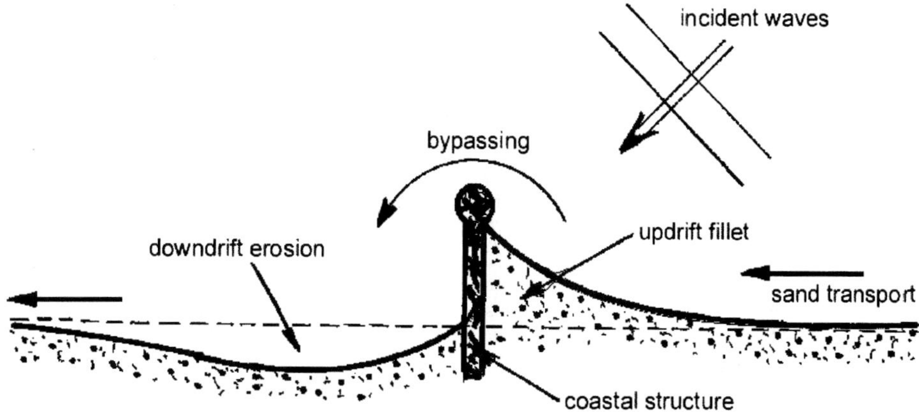

Figure S79 Shoreline change due to interruption of longshore transport of sand by a structure. The updrift fillet grows and the downdrift pocket erodes until the fillet is saturated and bypassing begins.

the downdrift, as shown in Figure S79. Also, on a seemingly unperturbed beach, longshore variations in offshore bathymetry cause variations in the nearshore wave climate, and consequently nonuniformity in the transport along the beach. This can result in erosion "hot spots," as well as otherwise mysterious zones of accretion. Finally, longshore variations in sand characteristics can lead to longshore gradients in transport and ultimately to the creation of a meandering coastline.

Additional suggested readings on this subject may be found in Battjes (1975), Dean and Dalrymple (1984), Fredsoe and Deigaard (1992), Komar (1998), Nadaoka and Kondoh (1982), and Whitford and Thornton (1996).

William R. Dally

Bibliography

Barkaszi, S.F., and Dally, W.R., 1992. Fine-scale measurement of sediment suspension by breaking waves at SUPERTANK. In *Proceedings of the 23rd Conference on Coastal Engineering*. American Society of Civil Engineers.

Battjes, J.A., 1975. Modeling of turbulence in the surfzone. In *Proceedings of the Symposium on Modeling Techniques*. American Society of Civil Engineers, pp. 1050–1061.

Battjes, J.A., and Stive, M.J.F., 1985. Calibration and verification of a dissipation model for random breaking waves. *Journal of Geophysical Research*, **90**: 9159–9167.

Bodge, K.R., and Dean, R.G., 1987. Short-term impoundment of longshore transport. In *Proceedings of Coastal Sediments '87*. American Society of Civil Engineers, pp. 469–483.

Bowen, A.J., Inman, D.L., and Simmons, 1968. Wave set-down and set-up. *Journal of Geophysical Research*, **73**(8): 2569–2577.

Dally, W.R., 2000. Roller momentum thickness and residual turbulence. In *Proceedings of the 27th Conference on Coastal Engineering*. American Society of Civil Engineers, pp. 59–69.

Dally, W.R., and Brown, C.A., 1995. A modeling investigation of the breaking wave roller with application to cross-shore currents. *Journal of Geophysical Research*, **100**(C12): 24,873–24,883.

Dally, W.R., and Dean, R.G., 1984. Suspended sediment transport and beach profile evolution. *Journal of the Waterway, Port, Coastal and Ocean Engineering Division*, **110**(1): 15–33, American Society of Civil Engineers.

Dally, W.R., Dean, R.G., and Dalrymple, R.A., 1985. Wave height variation across beaches of arbitrary profile. *Journal of Geophysical Research*, **90**(C6): 11917–11927.

Dalrymple, R.A., 1975. A mechanism for rip current generation on an open coast. *Journal of Geophysical Research*, **80**: 3485–3487.

Dean, R.G., 1973. Heuristic models of sand transport in the surf zone. In *Proceedings of the Conference on Engineering Dynamics in the Coastal Zone*. Sidney: Australia, pp. 208–214.

Dean, R.G., and Dalrymple, R.A., 1984. *Water Wave Mechanics for Engineers and Scientists*. Englewood Cliffs: Prentice-Hall.

Douglass, S.L., 1990. Influence of wind on breaking waves. *Journal of the Waterway, Port, Coastal and Ocean Engineering Division*, **116**(6): 651–663.

Fredsoe, J., and Deigaard, R., 1992. *Mechanics of Coastal Sediment Transport*. River Edge: World Scientific.

George, R., Flick, R.E., and Guza, R.T., 1994. Observations of turbulence in the surf zone. *Journal of Geophysical Research*, **99**(C1): 801–810.

Guza, R.T., and Davis, R.E., 1974. Excitation of edge waves by waves incident on a beach. *Journal of Geophysical Research*, **79**: 1285–1291

Kana, T.W., 1979. Suspended sediment in breaking waves. University of South Carolina, *Technical report No.18-CRD*.

Komar, P.D., 1998. *Beach Processes and Sedimentation*, Upper Saddle River: Prentice-Hall.

Longuet-Higgins, M.S., and Stewart, R.W. 1964. Radiation stress in water waves, a physical discussion with application. *Deep Sea Research*, **11**: 529–563.

Mocke, G., Govender, K., Alport, M., Smit, F., and Diedericks, G., 2000. Modelling of digitally imaged water levels and flow fields in the surf zone. In *Proceedings of the 27th Conference on Coastal Engineering*. American Society of Civil Engineers, pp. 1036–1049.

Nadaoka, K., and Kondoh, 1982. Laboratory measurements of velocity structure in the surf zone by LDV. *Coastal Engineering in Japan*, **25**: 125–145.

Nielsen, P., 1992. *Coastal Bottom Boundary Layers and Sediment Transport*. River Edge: World Scientific.

Noda, E.K., 1972. Rip currents. In *Proceedings of the 13th Conference on Coastal Engineering*. American Society of Civil Engineers, pp. 653–668.

Oltman-Shay, J., Howd, P.A., and Birkemeier, W.A., 1989. Shear instabilities of the mean longshore current, 2: field observations. *Journal of Geophysical Research*, **94**: 18,031–18,042.

Osiecki, D.A., and Dally, W.R., 1996. The influence of rollers on longshore currents. In *Proceedings of the 25th Conference on Coastal Engineering*. American Society of Civil Engineers, pp. 3419–3430.

Sakai, T., Inada, Y., and Sandanbata, I., 1982. Turbulence generated by wave breaking on beach. In *Proceedings of the 18th Conference on Coastal Engineering*. American Society of Civil Engineers, pp. 3–21.

Svendsen, I.A., 1984. Mass flux and undertow in a surf zone. *Coastal Engineering*, **8**: 347–365.

Svendsen, I.A., and Putrevu, U., 1994. Nearshore mixing and dispersion. *Proceedings of the Royal Society of London*, **445**: 561–576.

Weggel, J.R., 1972. Maximum breaker height. *Journal of Waterways, Harbors, and Coastal Engineering Division*, **98**(WW4): 529–548, American Society of Civil Engineers.

Whitford, D.J., and Thornton, E.B., 1996. Bed shear stress coefficients for longshore currents over a barred profile. *Coastal Engineering*, **27**: 243–262.

Cross-references

Bars
Beach Erosion
Beach Processes
Cross-Shore Sediment Transport
Coastal Currents
Erosion Processes
Longshore Sediment Transport
Surf Modeling
Wave Refraction Diagrams
Waves

SURFING

The origins of surfing are believed to date back to the ancient Polynesians, well before Captain James Cook became the first westerner to document surfboard riding, ca. 1777 (see Lueras, 1984). Simply hand-carved planks of wood, the first surfboards were large (up to 5 m in length), heavy, and difficult to control. Subsequent evolution during the 1920s to 1950s included hollowing the board, as well as shaping from balsa. In the early 1960s, the advent of synthetic foam and fiber-glass ushered in the modern era of surfing, and forever changed the nature of the sport by making surfboards shorter, lighter, faster, and more maneuverable. The modern surfboard is typically made of high-tech, composite materials with high strength-to-weight ratios. A custom-made board starts as a "blank," consisting of a foam core with one or more wooden stringers running its length to provide strength. After cutting to rough dimensions, the blank is shaped by hand, and the finished core is then covered with fiber-glass, using either polyester or epoxy resin. One or more fins, known as "skegs," are incorporated near the tail of the board to provide control and to enhance stability. Although a limited amount of laboratory and computer-simulation modeling of surfboards has been attempted (e.g., see Hornung and Killen, 1976), the evolution of board shape and design has been mostly by trial-and-error. Generally, long boards are used in gently spilling breakers, and shorter boards used in steep-faced, plunging waves.

"Stand-up" surfing is the most prevalent type of surfing, although body-boarding, where the surfer lies prone on a much shorter board and uses flippers to aid in wave catching, has gained significant popularity. Knee-boarding is a more rare form. Finally, in body-surfing no board is needed at all, with the surfer's own body used as a planing surface. This is arguably the "purest" form of surfing, practiced even by sea lions and bottlenose dolphins in the wild.

Catching a wave

In catching a wave, the surfer first takes up a position near the seaward edge of the surf zone, that is, at the break point of the largest incoming waves. As a suitable wave approaches, the surfer pivots to face towards shore and begins to paddle in order to gain forward momentum. With the face of the wave steepening, the surfer is lifted and the board tips forward, consequently accelerating in the direction of wave motion. The surfer then jumps to a semi-crouched position and slides down the wave face, thereby momentarily matching the speed of the wave. After "dropping in," the surfer executes a "bottom turn" in order to establish a path more tangential to the wave. From that moment on the surfer tries to maintain a position somewhere in the area immediately in front of the critical breaking region (the "pocket"), and so the ride is essentially a delicate reconciliation between the speed of the point of incipient breaking (the "break rate"), and the speed of the surfer (the "board speed"). If the break rate becomes greater than the maximum board speed that can be sustained by the surfer, the break point overtakes the surfer and the wave "closes out." If the break rate is less than the maximum sustainable board speed, the surfer has time to "carve," that is, maneuver up-and-down the face of the wave, as well as attempt other acrobatics. Maximum sustainable board speeds have been documented in excess of 18 m/s (40 mph) (Dally, 2001a).

Analysis of natural surfing breaks

What makes a particular site good for surfing? One of the first comprehensive studies of basic surfing mechanics, as well as the analysis of specific surfing breaks, was conducted by Walker (1974) for several famous breaks in Hawaii. Bathymetric surveys were performed, and estimates of surfer speed and "peel angle" (viewed from overhead, the angle between the wave crest and the path of the break point) were collected. The reason for these enduring, high-quality breaks was found to be the interplay of certain wave conditions—for example swell waves of sufficient height arriving from a specific direction—with the underlying reef bathymetry via the processes of wave refraction and shoaling. Recent surveys of other well-known surfing breaks around the Pacific Ocean confirms this, and detailed wave modeling demonstrates the importance of the interaction of bathymetric features of different scales in creating exceptional surfing breaks (Mead and Black, 2001a,b). In addition to bathymetric features, the interaction of waves and coastal navigation structures such as jetties has also created some well-known surfing breaks (Dally, 1990) such as at Sebastian Inlet, Florida (Figure S80). Finally, stochastic modeling has been utilized in attempts to quantify the "surfability" of a site, that is, the proportion of time in which good surfing waves are available (Dally, 1990, 2001b).

Engineering of man-made surfing breaks

For beaches that do not have the benefit of naturally induced, high-quality surfing breaks, the long-contemplated idea of enhancing waves for surfing using man-made structures has only recently been undertaken both in Australia (Pattiaratchi, 1997; Black *et al.*, 2001) and the

Figure S80 Surfer enjoying "First Peak" at Sebastian Inlet, Florida, which is a perennial break created by wave reflection from the northern jetty (photo by Gibber, courtesy of Eastern Surf Magazine).

United States. These artificial surfing breaks are submerged, mound-like structures, designed to both focus wave energy, thereby increasing the breaker height, as well as reduce the break rate by adopting an orientation that is strongly oblique to the crests of the incoming waves. Such reefs have been constructed out of concrete rubble units, as well as sand-filled geosynthetic bags.

William R. Dally

Bibliography

Black, K.P., Hutt, J.A., Mead, S., Jackson, A., and McGrath, J., 2001. Design of the gold coast artificial reef: surfing aspects. *Journal of Coastal Research*, **29**(Special issue): 115–130.

Dally, W.R., 1990. Stochastic modeling of surfing climate. In *Proceedings 22nd Conference on Coastal Engineering*, American Society of Civil Engineers, pp. 516–529.

Dally, W.R., 2001a. The maximum speed of surfers. *Journal of Coastal Research*, **29**(Special Issue): 33–40.

Dally, W.R., 2001b. Improved stochastic models for surfing climate. *Journal of Coastal Research*, **29**(Special issue): 41–50.

Hornung, H.G., and Killen, P., 1976. A stationary oblique breaking wave for laboratory testing of surfboards. *Journal of Fluid Mechanics*, **78**:.459–480.

Lueras, L., 1984. *Surfing: The Ultimate Pleasure*. New York: Workman Publishing.

Mead, S., and Black, K., 2001a. Field studies leading to the bathymetric classification of world-class surfing breaks. *Journal of Coastal Research*, **29**(Special Issue): 5–20.

Mead, S., and Black, K., 2001b. Functional component combinations controlling surfing wave quality at world-class surfing breaks. *Journal of Coastal Research*, **29**(Special issue): 21–32.

Pattiaratchi, C., 1997. Design studies for an artificial surfing reef at Cable Station, Western Australia. In *Proceedings of 1st International Surfing Reef Symposium*, Sydney, Australia.

Walker, J.R., 1974. Recreational surf parameters. University of Hawaii, Look Laboratory Report No.30, 311 p.

Cross-references

Geotextile Applications
Lifesaving and Beach Safety
Rating Beaches
Surf Zone Processes
Wave Refraction Diagram
Waves

SYNTHETIC APERTURE RADAR SYSTEMS

A synthetic aperture radar (SAR) is a remote sensing imaging system whose primary output product is a two-dimensional mapping of the radar brightness of a scene. Radar brightness is an expression of the scene's reflectivity in response to oblique illumination by microwave electromagnetic emissions. By definition, a SAR must be mounted on a moving platform, such as an aircraft or a satellite, and its illumination is directed to the side (and downward) to the surface. Its images are formed by scanning the area in two dimensions: range and azimuth. Range scanning is essentially at the speed of light, as radar pulses are transmitted and their reflections (backscatter) are received and recorded. Azimuth scanning is accomplished by the forward motion of the radar. SAR imaging "works" because the speed of light is very much greater than the along-track velocity of the radar. As the radar progresses along its flight path, the reflected signals from each transmission are collected and stored in memory. The heart of the system is the processor, which derives output image products from the stored data. The name "synthetic aperture" reflects the fact that signal processing algorithms replicate the signal collecting functions of a real aperture antenna array, thus "synthesizing" the effects of a physical antenna from a virtual data array.

One can visualize the key stages in the system through an optical comparison—a simple camera. A SAR's data memory is analogous to a record of the light field from a scene that may impinge on the front surface of a camera's lens, and the processor imitates (at least mathematically) the actions of the lens as it focuses the field and directs the refracted rays to the camera's film. As in a camera exposing black-and-white film, larger reflectivity (greater radar brightness) is expressed as lighter tones of gray tending toward white in a radar image. Lower reflectivity (less radar brightness) is expressed as tones of darker gray tending toward black. Colors may be artificially imposed on radar images to portray added dimensions, such as differences in reflectivity between two or more images, or differences in reflectivity as a function of wavelength or polarization.

A brief history

The original SAR idea is due to Carl Wiley in 1951. It was first reduced to practice by the University of Illinois, and by the Willow Run Laboratories of the University of Michigan. Several airborne systems were developed in the 1960s, although the technique was largely classified at the time. The first satellite SAR was embarked on Seasat (1978). Important earth-observing satellite SARs since then include Japan's J-ERS, the European Space Agency's ERS-1 and ERS-2, and Canada's RADARSAT (1995), which was still operational in June 2004.

Signal processing has been the pacing development in the history of SAR. Early SAR processors were analog optical contraptions, excited by lasers, and depended on relatively conventional photographic film for both the memory medium and the output image format. The first digital processors were developed for Seasat. In the fall of 1978, typically it would take more than 40 h to process one 50 km by 50 km scene, approximately one quarter of the nominal 100 km square Seasat SAR frame. Today, primary SAR image processing has progressed to the point that it is nearly transparent to most users of the data.

Basic concepts and parameters

Resolution is the minimum separation between two small and similar side-by-side reflectors such that they appear discretely in the image to be two, rather than merged into one. The distinguishing SAR performance attribute is the *azimuth resolution* realized through the synthetic aperture technique. The primary advantage of the SAR approach is that the radar-processor combination using data collected by a relatively small antenna can replicate the resolution that could otherwise be delivered only by an impossibly large antenna. For example, the Seasat SAR in its finest (or highest) resolution mode had about 6 m azimuth resolution. If a real aperture imaging radar were to have flown in place of Seasat, its antenna would had to have been about 16 km long (!) to achieve the same results. Azimuth resolution is proportional to range and inversely proportional to antenna length for a real aperture radar, but it is approximately equal to one half of the antenna length for a SAR. Seasat's SAR antenna was only 10 m long, not 16 km. Quite a difference, but not atypical of satellite systems that operate at 1,000 km range or so.

As microwave systems, most SARs employ relatively long wavelengths in contrast to optical or infrared systems. The wavelengths used by earth-observing satellite SARs include L-band (23-cm), S-band (10-cm), C-band (6-cm), and X-band (3-cm). RADARSAT is at C-band. At these wavelengths, roughness is the primary geophysical characteristic of the illuminated surface that determines the power of the backscattered signal, the radar brightness portrayed in SAR imagery. Surface roughness must be judged relative to the radar's wavelength, conditioned by the incident angle of the illumination at the surface, and other less important attributes. A surface appears rough to the radar as the topographic relief becomes comparable to the illuminating wavelength. A given surface always appears to become smoother as the angle of incidence is increased.

SAR backscatter may be enhanced considerably by favorable multi-reflection geometries. For example, stands of mangroves usually appear as bright areas, because the radar illumination bounces off both the water's surface and tree trunks, sending a relatively strong reflection back to the radar. The resulting image contrast is strongest for mangroves in calm (smooth) water.

Synthetic aperture formation requires that the data over the effective azimuth processing field must be phase-coherent. This leads to significant advantages for interferometric measurements, for example, which have demonstrated the ability to measure systematic surface displacements on the order of centimeters.

There is an inherent and not always pleasant consequence to phase coherence. For reflections from many distributed reflectors such as the ocean's surface, the random phases of the individual scattering facets cause mutual interference, leading to brightness modulation of the imaged field. This is known as speckle, whose variance is proportional to the mean radar brightness of the neighboring image area. Speckle may be reduced by averaging, which implies a signal-processing tradeoff with image resolution. “Looks” describes the number of statistically independent sub-images used to form the averaged image. The number of looks, divided by the product of the resolution parameters (in range and azimuth), is a fundamental SAR parameter, which is proportional to the Nyquist information capacity of the system. Usually, SAR range resolution is equal to the inverse range bandwidth (scaled to ground range), and azimuth resolution is approximated by half of the azimuth aperture length multiplied by the number of statistically independent looks. For example, the Seasat SAR used four looks in the processor, thus compromising its azimuth resolution to about 25 m, while reducing the speckle noise variance to a reasonable level (for many applications). The standard mode of RADARSAT, and the conventional modes of ERS image products all have comparable image quality, at least as rated by this norm. A significant increase in this parameter would imply a significant increase in SAR system cost.

All radar observations of scene reflectivity are in competition with additive radar noise. The additive noise is described by an equivalent reflectivity number, known as the noise equivalent reflectivity. Unlike speckle, the effect of additive noise is suppressed for higher relative levels of radar brightness, which can be due to either increased reflectivity or larger effective radar power.

Swath width, the span of the imaged area measured on the surface orthogonally to the satellite path, is the last of the top-level SAR image parameters. Swath widths typically are about 100 km for nominal 25-m 4-look image products. Higher resolution modes induce smaller swaths (e.g., 40 km). The extra-wide (~500-km) swaths available from RADARSAT’s ScanSAR mode are of particular interest to the oceanographic community, and can be interpreted to depict the detailed structure of nearshore wind fields. The trade-off implied by a ScanSAR image product is coarser resolution, which for a well-designed system can be offset somewhat by more looks, thus preserving image quality in spite of increased swath.

The SAR seen sea

For a distributed scene such as the sea, radar brightness is an expression of radar reflectivity, which in general is a function of instrumentation and of geophysical parameters. The instrumentation factors include polarization and wavelength of the radar illumination, and the incident angle at which the scene is observed. Geophysical parameters include surface roughness and conductivity, and environmental factors such as water temperature, air–sea temperature difference, wind speed, and surfactants, all of which have an impact on the ocean’s roughness at microwave scales. The radar reflectivity of the ocean tends to decrease with increasing incident angle, as shown in Figure S81.

Polarization of an emitted electromagnetic field is determined by the geometry of the antenna. The observed reflectivity depends on the polarization state of the receive antenna as well as of the transmitted field, where vertical (V) or horizontal (H) linear polarizations are the most commonly used for SAR. Most systems use the same antenna for both transmission and reception, leading to like-polarized reflectivity estimates. Ocean reflectivity is usually larger for VV polarization than for HH polarization.

For a wide range of illumination incidence, Bragg scattering is the dominant source of radar backscatter from the ocean. Bragg scattering is due to a systematic phase fit between the illuminating half-wavelength projected onto the sea surface, and a matching periodic ocean roughness component. For the wavelengths and incident angles of typical ocean remote sensing radars, Bragg waves range from long capillaries (~2 cm) to short gravity waves (~50 cm). “Roughness” in the microwave ocean remote sensing context implies a sensible population of these special Bragg surface wave components. These waves, which have relatively short lifetimes, are locally generated, and are modulated and advected by longer and more energetic seas. It follows that the larger wave structure of the ocean is observable by microwave systems primarily through patterns expressed in the shorter Bragg waves. This is known as the two-scale model of microwave ocean scattering.

Roughness at Bragg wavelengths is set up primarily by local wind stress. For a given incident angle and radar wavelength, there is a first order power law dependence between radar reflectivity and wind speed, as suggested in Figure S82. The slope of the quasi-linear part of the response is a function of wavelength, among other parameters. Within this quasi-linear region, a SAR’s radar brightness can be inverted to estimate wind speed, realizing in effect a high-resolution scatterometer.

Within limits, the power law is robust. The upper limit is approached as the radar reflectivity tends to saturate with increasing wind speed. The saturation wind speed is above 25 m/s for the radar parameters of RADARSAT at its steeper incident angles, for example.

The lower limit is more abrupt, and may be of considerable interest in coastal oceanographic applications. Bragg wave formation, and hence sensible radar brightness, occurs only for wind speeds above the lower cutoff threshold, which typically is about 2.5 m/s. Both the onset of Bragg waves and the level of reflectivity are functions of environmental conditions. If the system additive noise is sufficiently low such that the reflectivity threshold is visible, then relatively small changes in local environmental conditions may lead to large contrasts in radar brightness. For example, a change in relative air–sea temperature of only one degree or so can invert the boundary layer stability, which can be manifested as a relatively large change in radar reflectivity. These contrasts are often observed in oceanic SAR imagery, and have proven to be invaluable in tracking oil spills, for example. In general, the challenge is to deduce the appropriate geophysical cause from radar brightness patterns seen in the image data.

In oceanographic applications, the scene of interest is in motion, which can have noticeable consequences on oceanic SAR image characteristics. Gravity waves are characterized by the orbital velocities of their elemental surface constituents. These motions are transformed by a SAR into azimuth shifts of the corresponding image features. As a result, an image of azimuthal waves is distorted, an effect known as “velocity bunching.” Directional wave spectra can be deduced from

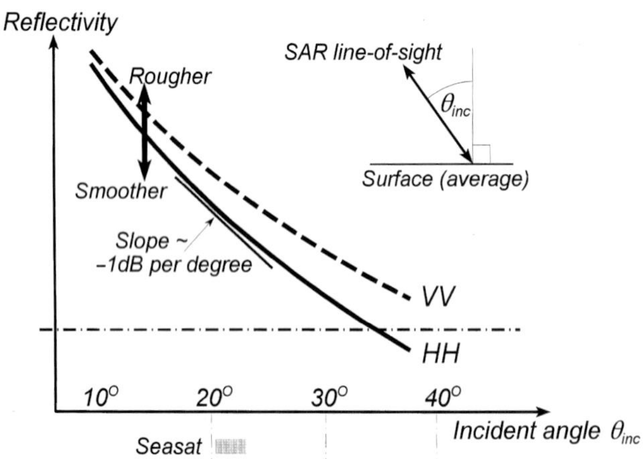

Figure S81 Radar reflectivity (backscattered power) tends to decrease with increasing incident angle, and is larger for rougher surfaces. The span of incident angles of several satellite SAR systems is indicated below the horizontal axis.

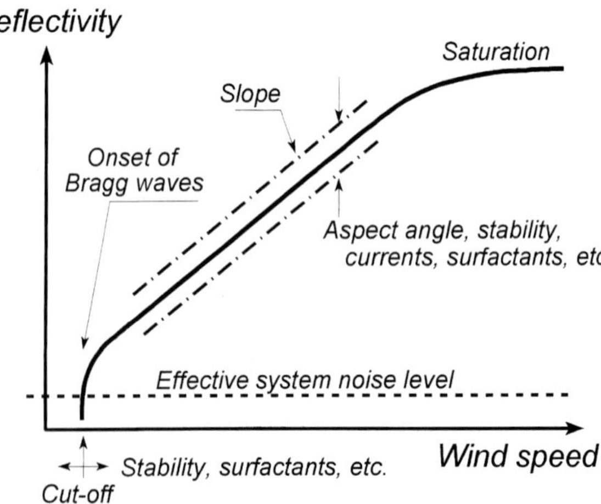

Figure S82 Between the lower and upper limits, the ocean’s reflectivity (relative power) increases in proportion to local wind speed. The slope and level of the proportionality depend on radar and environmental parameters. The lower limit is relatively abrupt; backscatter arises only above the minimum wind speed sufficient to excite Bragg waves matched to the radar. The upper saturation limit is more gradual.

oceanic SAR imagery, although extra effort is required to compensate for the degraded azimuthal wave components. The "wave modes" of ERS-2 and Envisat SAR have been designed with these characteristics in mind.

Summary

Image data from satellite-based SAR systems are in increasing demand. Applications span directional wave spectral estimation, ship detection and tracking, oil spill monitoring, fisheries monitoring, coastal wind analysis, coastal erosion, and storm tracking. The major limitation at this time is lack of timely data, due primarily to a shortage of operational systems. This limitation should be at least partially rectified in coming years.

R. Keith Raney

Bibliography

Mouchot, M.-C., and Garello, R., 1998. SAR for Oceanography. In Henderson, F.M., and Lewis, A.J. (eds.), *Principles and Applications of Radar Imaging*. New York: John Wiley and Sons.

Raney, R.K., 1998. Radar fundamentals: technical perspective. In Henderson, F.M., and Lewis, A.J. (eds.), *Principles and Applications of Radar Imaging*. New York: John Wiley and Sons.

Cross-references

Airborne Laser Terrain Mapping and Light Detection Ranging
Mangroves
Mapping Shores and Coastal Terrain
Photogrammetry
RADARSAT-2
Remote Sensing of Coastal Environments
Waves

T

TAFONE

Originally used to describe unusual weathering features found in Corsica, tafone (or tafoni, plural) has become established as the generic name for a type of cavernous weathering characterized by the existence of hollows or cavities that range in size from a few centimeters to a meter or more in diameter. Depth is variable, the cavities commonly being nearly hemispherical. The shape and orientation are strongly influenced by the rock fabric, causing the hollows to be elongated in the direction of foliation of bedding planes. Coalescence of the cavities produces mushroomlike shapes, natural arches, and other unusual sculptured forms. Eventually the outcrop surface may be destroyed by this process of expansion. When numerous small cavities occur, the resulting spongelike texture is termed honeycomb weathering. Both honeycomb weathering and tafone may occur independently, but often the two forms coexist and seem to originate from a similar process. Tafone occurs commonly in granitic rocks and in sandstone, but has also been observed in many other rocks. The rate of development may be rapid, cavities having formed in stone seawalls built less than a century ago.

Tafoni are abundant in the polar regions of Victoria Land, Antarctica, but they also occur in deserts of Australia, Africa, central Asia, and South America as well as in such humid regions as Hong Kong and the island of Aruba, West Indies. In North America, tafoni occurs in such diverse areas as the deserts of New Mexico, Utah, Arizona, and the coast of Washington state (Figure T1), and at mid-continent locations in Wisconsin and Illinois (Bryan, 1928; Blackwelder, 1929; Mustoe, 1982).

Tafoni are commonly found in outcrops having a hardened surface layer caused by precipitation of iron hydroxides or other compounds derived during weathering of the interior rock. The cementing action of these oxides produces a resistant rind. Any form of erosion that attacks the outcrop in a localized fashion will produce cavities, since penetration of the protective layer leads to rapid destruction of the weaker interior rock. Differential attack of the surface might result from variations in lithology, the presence of fissures or zones of high porosity that allow water to penetrate, or localized attack by organisms such as lichens. Because the formation of the protective ring requires moisture, this explanation is consistent with the observation that tafone seldom occurs in extremely dry environments, being more common along the margins of deserts than in the arid central regions. In many locations, tafoni occur mainly as coastal features presumably owing to the existence of a favorable microclimate, since even in extremely arid regions sea winds may cause the coastline to be relatively humid. The existence of a protective rind is most favored by an environment where some moisture is present but where periods of dryness allow evaporation to occur at the rock surface.

Tafoni typically form as a result of salt weathering, where evaporation of saline water triggers physical and chemical attack of the rock surface (Evans, 1970; Young, 1987; Rodriguez-Navarro and Doehne, 1999). In general, the distribution of tafone throughout the world correlates well with environments where salt crystallization occurs. In arid zones, salt weathering is concentrated where moisture is retained along the base of cliffs and undersides of boulders, where shadow weathering (tafone) may often be found. Along the coast, cavities are presumed to form by salt crystallization as wave-splash evaporates. Because salt weathering may attack in a relatively selective fashion, being controlled by variations in moisture or exposure to salt spray, tafoni might be formed even on outcrops where a hardened surface layer is absent, though the development of cavities would be enhanced when such a rind is present. Early explanations of tafoni invoking erosion by wind

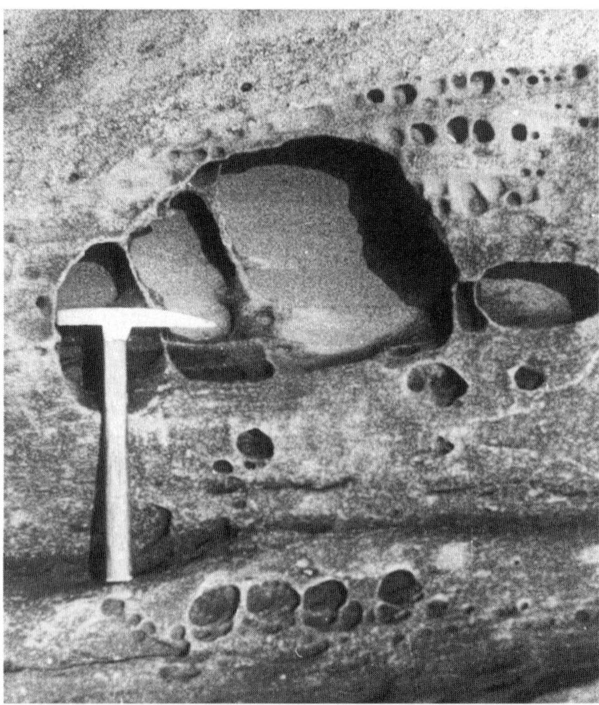

Figure T1 Tafone in arkosic sandstone, Larrabee State Park, near Bellingham, Washington USA (Photo, George Mustoe).

action, temperature fluctuation, or frost wedging have largely been abandoned owing to lack of substantiating evidence.

George Mustoe

Bibliography

Blackwelder, E., 1929. Cavernous rock surfaces of the desert. *American Journal of Science*, **17**: 393–399.

Bryan, K., 1928. Niches and other cavities in sandstone at Chaco Canyon, New Mexico. *Zeitschrift fur Geomorphologie*, **3**: 125–140.

Evans, I.S., 1970. Salt crystallization and rock weathering: a review. *Revue De Géomorphologie Dynamique*, **19**: 153–177.

Mustoe, G.E., 1982. Cavernous weathering in the Capitol Reef desert. Utah. *Earth Surface Processes and Landforms*, **8**: 517–526.

Rodriguez-Navarro, C., and Doehne, E., 1999. Salt weathering: influence of evaporation rate, supersaturation and crystallization. *Earth Surface Processes and Landforms*, **24**: 191–209.

Young, A.R.M., 1987. Salt as an agent in the development of cavernous weathering. *Geology*, **15**: 962–966.

Cross-references

Bioerosion
Cliffs, Lithology versus Erosion Rates
Coastal Climate
Coastal Hoodoos
Coastal Wind Effect
Desert Coasts
Honeycomb Weathering
Notches
Shore Platforms
Weathering in the Coastal Zone

TECTONICS AND NEOTECTONICS

Introduction

Rock deformation caused by the structure of the earth (e.g., folds, faults, joints, cleavage) is often the only kind of tectonic deformation considered by geological manuals, giving the impression that areas devoid of such type of deformation are "tectonically stable." However, according to the American Geological Institute (1960), "tectonic" is defined as "designating the rock structure and external forms resulting from the deformation of the earth crust." This definition implies that all processes which modify the external form of the crust, also when they result from forces external to the earth, have to be considered tectonic. This is the case, for example, for unidirectional vertical movements produced by earth surface processes of weathering and erosion (sedimento-isostasy), and also for rise and fall of the solid earth surface, especially in coastal areas, caused by external factors such as climate change (glacio-isostasy, hydro-isostasy) or eventually, at a smaller timescale, by the attraction of the sun and the moon (earth tides).

The term "neotectonics," ignored by most geological dictionaries or glossaries, was introduced by geodesists, geophysicists, and quaternarists during the last decades and defined in 1978 by the Board of the Neotectonic Commission of the International Association for Quaternary Research (INQUA) as "any active earth movement or deformation of the geodetic reference level, their mechanisms, their geological background (how old it ever may be), their implication for various practical purposes and their future extrapolation." This definition shows care not to isolate crustal movements from their geodynamic inheritance. Therefore, neotectonics has a wider scope than "active tectonics" (Wallace, 1986), which is defined as "tectonic movements that are expected to occur within a future time span of concern to society" and no real boundary back in time. It includes all timescales of movements, from instantaneous (seismic), 10–10^2 yr (geodetic), 10^2–10^4 yr (Holocene studies), 10^4–10^6 yr (Pleistocene studies), up to about 10^7 yr, if it is necessary to enable us to understand the origins of recorded movements (Mörner, 1980). The difference between former tectonics and neotectonics is therefore not far from that existing between an extinct fossil and a still living organism. Among the distinctive attributes of neotectonic studies listed by Stewart and Hancock (1994) are: (1) A wide range of methodologies and a variety of experts are commonly involved in a comprehensive study of the neotectonic history of a region, whereas paleotectonic structures are often investigated by structural geologists working on their own. (2) The neotectonician has the ability to compare inferences drawn from field observations with geophysical and geodetic data about rates and mechanisms of present-day processes. (3) Neotectonic displacement histories can be established with greater precision than paleotectonic ones, because Quaternary dating techniques allow relatively short time intervals to be detected. However, the approach by Stewart and Hancock (1994) mainly refers to active fault ruptures which are indeed privileged indicators of neotectonic displacements, but remains limited to boundaries of deformed crustal blocks, which may also remain obscured, without reaching the earth surface, or become obliterated by recent erosion or sedimentation processes.

In the following, various possible processes of tectonic/ neotectonic deformation in coastal areas are briefly reviewed, with special attention to vertical displacements which may be related to sea level. Significant examples of tectonic displacements are provided and rates of vertical movements are assessed. Some developments are partly inspired from ideas already discussed by Pirazzoli (1995).

Present-day and fossil coasts

The coastal outline has undergone major changes during the earth history. Some 300 Myr ago, when all continents are believed to have formed a single, huge landmass, Pangea, marine coasts could only exist along the perimeter of such a landmass. This supercontinent started to split apart about 200 Myr ago, first into two parts, Laurasia at the north and Gondwana near the South Pole, separated by a sea, Tethys. Subsequently Laurasia split out into North America, Europe, and North Asia, whereas Gondwana fragmentated into South America, Africa, Arabia, Madagascar, India, Australia, New Zealand, and Antarctica. Lastly, a slow drift has brought the various continents or continental fragments to their present state.

The reconstruction of the positions of the continents during the splitting or at the time of their coming together, during their migration, as well as the mechanisms of plate tectonics, make possible a comprehensive view of present-day and fossil coastal areas, with some marks of the latter now perched also near the top of mountain chains.

When the separation of two continents had been caused by the opening of a rift, continental margins are usually characterized by a low seismicity and a weak orogeny; in these structurally passive continental margins, which are devoid of an oceanic trench, the continental and the oceanic crust, which belong to the same plate, are joined. Such margins are frequent around the Atlantic Ocean and characterized by a well-developed continental shelf and a relative structural-tectonic stability (though they may be affected by important vertical deformation of isostatic origin).

On the other hand, where subduction processes occur, tectonics of continental margins are structurally active and show high seismicity and relief. Such margins, typical of Pacific coasts, correspond to areas where an oceanic plate is underthrusting beneath a continental plate. The continental shelf if present is very narrow, or even absent. Coastal California corresponds to a variant, with a sliding motion between the Pacific oceanic plate and the continental North-American plate along the San Andreas Fault.

In accretion zones, two continental plates are colliding. These active continental margins with strong seismicity and orogeny are generally located inside continents, far away from present-day coasts. However, at the first stages of collision, wide marine and coastal areas can be affected. This is the case in the present time of the Mediterranean region. Lastly, the coasts of epicontinental seas (bordering oceanic depths) form a special case, because the greater part of these shallow seas extending above continental shelves usually disappear during low sea-level periods and are therefore affected by significant hydro-isostatic vertical deformation.

Processes

Thermo-isostatic and volcano-isostatic tectonics

It is now generally accepted that plate tectonics modify, slowly but continuously, the shape of the oceanic basins. The rock of oceanic plates is formed by lava extrusions along oceanic ridges. After its formation, the oceanic crust moves slowly away and crosses the ocean, to be later destroyed in a subduction trench. When the crust spreads away from the ridges, it cools and thickens, thereby increasing in density. As a result the seafloor subsides isostatically, gradually submerging oceanic islands as they are carried laterally. Over several million years, thermo-isostatic subsidence will be on the order of kilometers. This can be verified easily in any world atlas, where the ocean depth is shown to increase gradually from 2,000–3,000 m above submarine ridges to 5,000–6,000 m or more above oceanic trenches. In tropical waters, the rate of thermo-isostatic subsidence is increased by the load of coral reefs, which have to grow vertically to maintain their sea-level position. The normal vertical evolution of an oceanic island in intraplate areas, when eustatic fluctuations of sea level are disregarded, would therefore be a gradual submergence, accompanied by relatively rapid erosion of exposed rocks by subaerial weathering and wave action. The existence of such general oceanic subsidence was first observed by Darwin, who used it to explain the different types of coral reefs by evolution from fringing reefs, to barrier reefs, and to atolls. Darwin did not know what the cause of subsidence was, but his theory was correct. We know today that oceanic intraplate subsidence is due to thermo-isostasy.

If an intraplate oceanic island is not subsiding, it is anomalous and requires explanation. Two main kinds of anomalies have been recognized: thermal rejuvenation and volcano-isostasy.

When the oceanic crust approaches a hot spot, which normally corresponds to an asthenospheric bump above a more or less fixed mantle melting anomaly, the normal cooling process is reversed, so that the crust is heated, becomes less dense and thinner, and thus rises (thermal rejuvenation process). Moving away from the hot spot area, cooling and subsidence predominate again.

Near hot spots, lava eruptions often occur. Their load produces on the lithosphere an isostatic depression under the volcanic mass and a peripheral raised rim at some distance (volcano-isostasy). According to empirical observations made around oceanic islands, the radius of the depressed area is generally less than 150 km from the load barycenter, whereas the peripheral rim develops at a distance of between some 150 and 300 or 330 km. When the translation movement of an oceanic plate over a hot spot has produced a line of islands, and two or more thermal rejuvenation spots form an alignment in the direction of plate movement, interaction between isostatically depressed areas and uplifted rims of nearby islands is possible and may produce complicated sequences of vertical deformation, with repeated phases of uplift and sinking.

Example of Quaternary thermo-isostatic uplift: Rurutu Island (South Pacific). Rurutu (22°30′S–151°20′W) is a basaltic island of the Austral–Cook island chain, approximately halfway between the hot spot near the active undersea Mac Donald volcano and the northernmost atoll of the Cook Islands (Palmerston). The volcanic basement of Rurutu has been dated 10.5 ± 2 Ma, whereas the age of the surrounding seafloor lies in the range of 40–90 Ma. The age of the volcanic basement of Rurutu is therefore consistent with the migration rate (10–11 cm/yr) of the Austral–Cook lithosphere with respect to the hot spot reference frame and the distance of about 1,350 km from Mac Donald. There is, however, a second hot spot along the Austral–Cook island chain, which generated some Cook Islands (Mauke, Atiu, Aitutaki) during the last 10 Myr. This second hot spot or plume is located near the present position of Rurutu (Dickinson, 1998). During the Quaternary, Rurutu entered the uplifting side of the hot spot swell. No new lava extrusion seems to have occurred, but gradual uplift took place, so that today the volcanic basement of Rurutu is surrounded by raised limestone tabular blocks, dated less than 1.85 Ma, reaching about 100 m in altitude. The limestone blocks finish in high vertical cliffs facing the ocean. Variations in the water table associated with Quaternary sea-level changes superimposed on gradual uplift have provoked karstic dissolution in the limestone at several levels (Figure T2). Dated remnants of the Last Interglacial coastline are now at +8–10 m in elevation; marks of Holocene sea stands have been identified at +1.7, +1.2, and +0.6 m (Pirazzoli and Salvat, 1992). The available evidence indicates that thermo-isostatic deformation has uplifted Rurutu Island at the average rate of 0.05–0.10 mm/yr since the early Pleistocene and at about 0.17 mm/yr in the Holocene.

Example of historical rapid thermo-isostatic deformation: "Temple of Serapis," Pozzuoli (Italy). This site demonstrates repeated rapid up and down displacements in the Phlegraean Fields caldera, near Naples. Burrows of *Lithophaga* mollusca in the three columns still left standing of the "Temple" (a Roman market probably built in the 2nd century BC near the Pozzuoli Harbor), clearly show that the ruins had subsided under sea level and then been gradually uplifted (Figure T3). Geomorphological, archaeological, historical, and radiometric data suggest a complex relative sea-level history. After the "Temple" was constructed, a subsidence of about 12 m took place. Elevated marine fossils indicate two submersion peaks, between the 5th and the 7th century AD, then again in the 13th to 14th century AD (Morhange *et al.*, 1999). An uplift of about 7 m followed, culminating in the Monte Nuovo volcanic eruption in AD 1538, and further subsidence. More recently, tide-gauge data from the Pozzuoli Harbor show two brief periods (1970–73 and 1982–84) of rapid uplift reaching a total of 3.2 m

Figure T2 The outer limestone cliff of Rurutu (Austral Islands, South Pacific) appears cut by networks of caves at several levels, remnants of karstic drainage corresponding to higher relative sea-level positions. The lowest line of caves, with the floor at +8–10 m, contains marine material which has been dated from the Last Interglacial. The cliff is also undercut by a slightly elevated deep notch dating from the Late Holocene (Photo 5743, April 1980).

Figure T3 The "Temple of Serapis" in Pozzuoli (Italy). Note the dark band on the columns produced by molluscan borings (the upper limit of which is slightly less than 6 m above the paving of the "Temple," which is partly flooded by a few centimeters of water) (Photo D388, October 1991). The same floor appears more submerged in photos taken in 1982–83 (Vita-Finzi, 1986, p. 10).

between 1968 and 1984, followed by slight renewed subsidence, still going on. Such vertical movements are understood as being of thermo-isostatic origin, with uplift corresponding to rock expansion under the action of heat beneath the caldera, and subsidence to contraction on cooling.

Vertical tectonic deformation near plate boundaries

When oceanic crust approaches the end of its travel, near a trench or a continental margin, the arrival of its mass in a subduction or a collision zone produces high seismicity and vertical tectonic movements become more complex.

In a subduction zone, the underthrusting side, which will be destroyed by plunging into the earth mantle, has a relatively simple tectonic behavior, differing notably from that of the overthrusting side. Before being submerged, the oceanic plate is generally subjected to arching phenomena, in order to make possible the change from a horizontal translation to a plunging beneath the overriding plate. This arching implies a wave-like flexuring of the lithosphere, with first a gradual uplift, as in the case of an oceanic island approaching a hot spot, than a gradual subsidence at accelerating rates. Case studies of such arching phenomena have been reported from Christmas Island, 200 km southwest of the Sunda Trench, from the Daito (Borodino) Islands, 150 km east of the Ryukyu Trench, and from the Loyalty Islands, at varying distances west of the New Hebrides Trench.

On the overthrusting side, various local or regional structural geodynamic factors may be superimposed. Here uplift is frequent and may be caused by: (1) piling up above the oceanic plate, on the inner side of the trench, of sediments too light to be subducted, which will raise the overthrusting edge isostatically; (2) elastic rebound phenomena linked to subduction faulting; (3) tilting of lithosphere blocks or other structural tectonic processes; and (4) volcanic activities. Subsidence in structural basins may also be significant.

Where subduction processes are prevented by the occurrence of a transform fault or by continental lithosphere on both sides of a plate boundary, collision or transform processes will occur with vertical movements which can become highly irregular. The best-known example of a transform fault crossing a coastal area is probably that of the San Andreas Fault, which extends south to north over a distance of about 1,300 km between the Gulf of California and the Mendocino Fracture Zone. South of the Mendocino triple junction, strike slip predominates along a broad and braided system of faults with a roughly northwesterly strike; blocks between faults are warped, folded, uplifted, depressed, and rotated. Farther northwestwards, vertical movements predominate (Crowell, 1986). In several areas along the San Andreas Fault evidence of uplift is missing, however, and aseismic uplift has also been reported from southern California.

The superimposed effects of long-term gradual uplift and of eustatic sea-level fluctuations often produce sequences of stepped marine terraces.

Example of stepped marine terraces in the Santa Cruz area, California. Six marine terraces, in which raised shell beds were first described by Darwin, indent a 60 km stretch of coastal topography. These terraces seem to have been raised by repeated slip earthquakes on the San Andreas Fault, with a return time of three to six centuries, uplifting the terraces at average rates between 0.13 and 0.35 mm/yr (Lajoie, 1986; Valensise and Ward, 1991).

Seismic displacements

In many seismic areas, vertical displacements of land may appear gradual in the long term, but in the short term they can consist of sudden vertical movements, separated by more or less long periods of quiescence or of gradual movement (Figure T4). The sudden movements usually take place at the time of great-magnitude earthquakes, which are often accompanied by surface faulting or folding and ground deformation. Land displacements occurring spasmodically at the time of an earthquake are called *coseismic*. Gradual displacements, often in opposition to the coseismic ones, may occur during the few years or decades preceding (*preseismic*) or following (*postseismic*) the coseismic event. The duration of the time interval between two coseismic events (*interseismic* period) is not perfectly regular and depends on the variability of the local tectonic stress accumulation. It may tend, however, to be repetitive statistically, with a return time which can vary, according to the seismotectonic area considered, from a few centuries to over ten thousand years. In coastal seismic areas, investigation and dating of former shorelines different from the present ones may be used to determine the age, distribution, and succession of vertical displacements. Of special interest for seismic prediction is the identification of the preseismic movements which may indicate the imminence of a great-magnitude earthquake.

Certain coseismic movements can be impressive. In Crete (Greece), a jerk of coseismic uplift raised the southwestern most part of the island as much as 9 m, probably on July 21, AD 365 (Pirazzoli *et al.*, 1996); in

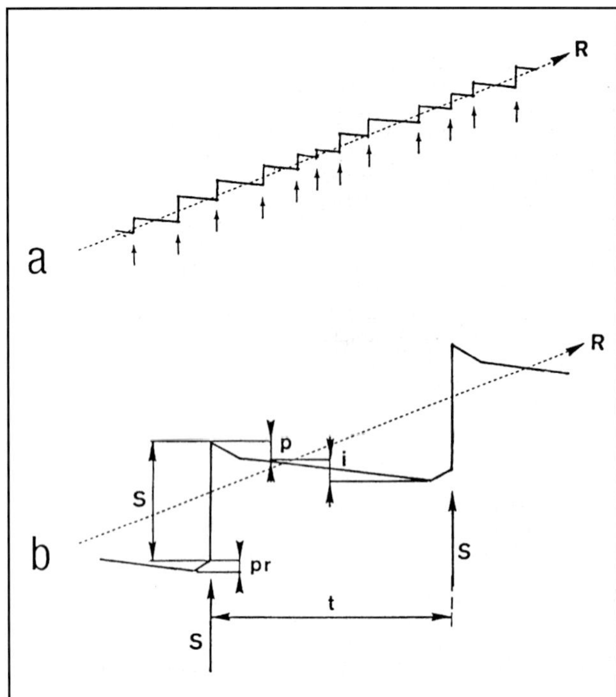

Figure T4 In uplifting seismic regions, the raising trend, though apparently gradual over the long term, may consist in the short term of sudden uplifting movements accompanying great earthquakes, separated by more or less long periods of seismic quiescence and gradual subsidence. pr, preseismic; S, coseismic; p, postseismic; i, interseismic; t, return period.

Alaska, the great earthquake of March 27, 1964 (magnitude ≥8.4) was accompanied by uplift reaching 10 m on land in Montague Island and more than 15 m offshore (Plafker, 1965).

Example of repeated Holocene coseismic deformation: Hatay (Turkey). On the Hatay coasts, clear evidence of uplifted Quaternary coastlines near the Orontes Delta have been recognized up to at least 140 m in elevation. The lower evidence, between 35 and 8 m above present sea level, consists of stepped, elevated, deltaic sedimentary surfaces, suggesting that the Orontes River has built here intermittent delta formations since at least the late mid-Pleistocene. Detailed study of late Holocene marine notches (Figure T5) and bioconstructed rims made of vermetids, oysters, and calcareous algae, have demonstrated that uplift has been episodic rather than gradual (Pirazzoli *et al.*, 1992). Two rapid land movements, probably of seismic origin, have been identified. The first movement, which occurred about 2,500 ± 100 years BP, was the strongest one and caused a local vertical displacement of about 1.7 m. The second movement (an uplift of 0.7–0.8 m) occurred around 1,400 years BP, probably at the time of the great earthquake followed by tsunami waves of May AD 526, which caused devastating damage in Antioch and prevented further use of its harbor Seleucia Pieria. The long-term trend of uplift in this coastal area can be ascribed to episodic reactivation of local fault lines, probably in connection with movements on the East Anatolian Fault system.

Glacio-isostatic and hydro-isostatic neotectonics

Transfers of ice or water masses between ice sheets and the ocean produce load displacements and therefore isostatic phenomena. The load of an ice sheet deforms the earth's crust. The resulting subsidence beneath the ice makes deeper material flow away, and raises an uplifted rim at a certain distance. When the ice sheet melts, an unloading occurs, resulting in uplift beneath the melted ice; the marginal rim will consequently tend to subside and move towards the center of the vanishing load. Part of these glacio- isostatic movements are elastic, and thus contemporaneous with loading and unloading. Due to the viscosity of the earth material, however, part of the movement will continue for several thousand years after the loading or unloading has stopped. Although most of the ice in Canada and Fennoscandia had disappeared between about 8,000 and 6,000 years BP, large vertical movements of uplift and subsidence continue in these areas today.

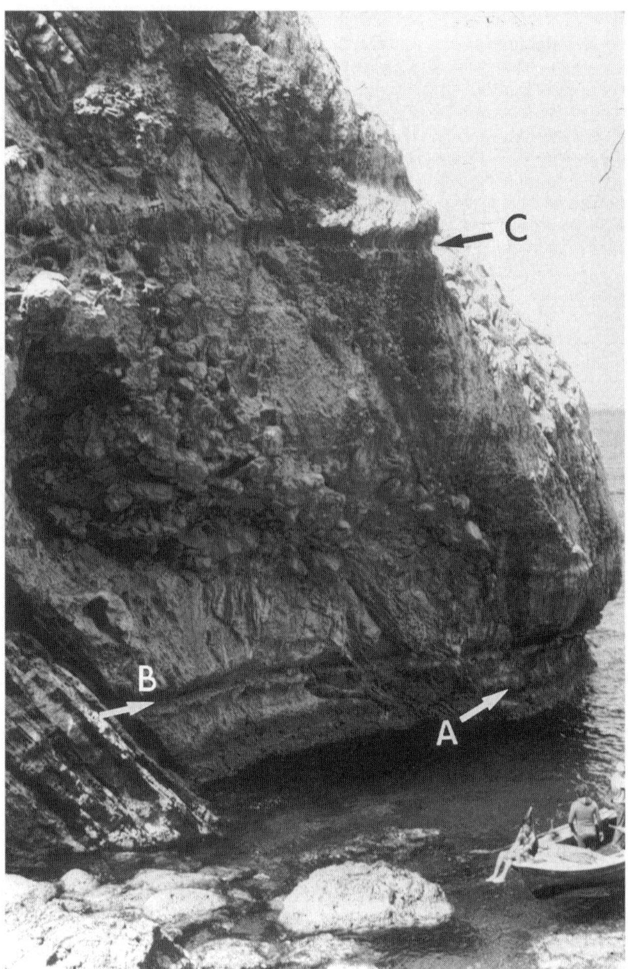

Figure T5 Two continuous notches cut into the limestone cliff at about (A) +0.7 m and (B) +2.0 m, mark the sea-level positions before two coseismic uplift movements which occurred some 1,500 and 2,500 years ago. (C) A similar undated notch at +12.3 m, may correspond to a Late-Pleistocene shoreline. South of the Orontes Delta, near the Turkish–Syrian boundary (Photo A739, May 1988).

It was during the second half of the 19th century, when the postglacial coastal record in Fennoscandia became accurate enough to yield a persuasive description of dome-like uplift, that the concept of glacio-isostasy could be asserted by De Geer, as had been done at about the same time for Lake Bonneville, in the United States, by Gilbert. During deglaciation, meltwater from ice sheets produces a considerable load on the ocean floor (on the order of 100 t/m^2 for a sea-level rise of 100 m), so that the sea bottom subsides (hydro-isostasy). In the upper part of gently sloping continental shelves, or in shallow seas where the postglacial water depth is less than the global change in sea level, the meltwater load will vary according to the local topography and bathymetry, generally increasing gradually towards the open sea. In this case, the hydro-isostatic constraints will produce a lithospheric flexure with a typical seaward tilting.

Small islands which rise steeply from the deep ocean floor are often considered as potentially good place to measure eustatic changes because, as discussed by Bloom (1971, p. 371) they "will be warped downward by the water load around them, but because the entire deep ocean floor is depressed, the volume of the ocean basin increases and sea level with reference to an island, or to a hypothetical buoy moored in deep water, should not change because of the isostatic deformation." In equatorial and tropical regions remote from the ice sheets, hydro-isostatic subsidence may be partly compensated by "equatorial ocean siphoning" (Mitrovica and Peltier, 1991). This mechanism, driven by the subsidence of those portions of the glacial forebulges which exist over oceanic regions, acts to draw water away from most equatorial and tropical regions, in order that the oceans maintain hydrostatic equilibrium, making some emergence possible. For this reason, slight Holocene emergence is frequent in tropical oceanic islands. This is the case, for example, for most Pacific atolls (Figure T6) in spite of their long-term trend towards thermo-isostatic subsidence.

Some geophysical models, based on the mathematical analysis of the deformation of a viscoelastic earth produced by surface loads, have been developed during the last two decades. They have been able to mimic, with an accuracy on the order of a few meters, sea-level changes reported from the field in various areas. These models, which assume a simplified earth structure and a melting history for all the continental ice loads which existed at the time of the last glacial maximum, have demonstrated that the rate, direction, and magnitude of crustal movements must have varied from place to place and, therefore, that no region can be considered as tectonically stable. Though they cannot replace field observation, these models have also been useful in providing a first-order approximation of the deglacial and postglacial sea-level history in areas where no field data are available.

Sediment-isostatic neotectonics

Weathering and gravitational forces cause continuous transfers of sediments in a one-way direction, from the continents to the oceans. This implies isostatic adjustments, with predominant uplift on the continents and subsidence in nearshore basins affected by rapid sedimentation. At the mouth of great rivers, when the rate of fluvial input overtakes the rate of sea-level change, deltaic sequences may tend to accumulate, especially on microtidal coasts. Recent sediment accumulation is affected by

Figure T6 Late-Holocene *scleractinian* corals have been left emerged in growth position, at the top of a pinnacle in the lagoon of Takapoto Atoll (Tuamotus), by a slight fall in the relative sealevel of probable hydro-isostatic origin (Pirazzoli and Montaggioni, 1988) (Photo 5478, March 1980).

compaction, which produces subsidence, but is also expected to form on the lithosphere surface an isostatic depression under the load and a marginal rim slightly rising at some distance. However, remaining in an area where general subsidence predominates, such rims may only appear as an area of lesser sinking. Lateral displacement of major delta branches have often occurred during the late Holocene; they imply, after some time, also a migration if their isostatic peripheral bulge.

The purely isostatic component is generally difficult to estimate, because observed rates of lowering include also sediment compaction, possible faulting, and for the past millennium also anthropogenic influences, such as conversion of wetlands to agricultural fields, river channelization, pumping, and withdrawal of water and diversion of water flow for irrigation. What can be measured is only the total amount of vertical displacement that has occurred during the Holocene delta formation, that is, since the postglacial sea-level rise began to decelerate, generally from about 8,000 to 6,500 years BP (Stanley and Warne, 1994). This vertical displacement varies locally, usually reaching a maximum on the outer delta plain close to the present coast.

Structural faulting and folding

Faults (and folds, which can develop ahead of blind faults propagating from depth towards the surface) are generally the main object of study for most structural tectonicians. In coastal areas, however, at least for vertical displacements, the most important neotectonic indicator is sea level. With high-tide shoreline marks, sea level leaves evidence of an altimetric datum all along the coast. An active fault may cross a rocky coast, leaving recognizable displaced high-tide shorelines on both sides (Figure T7). Such an occurrence is precious, because it may enable one to measure very precisely the vertical displacement between the footwall and the hanging wall of the fault, that is, the maximum amount of displacement when we are dealing with a tilted crustal block. In even more favorable cases former high-tide shorelines may be preserved, at least over a certain distance, at various levels on the fault plane; it will then be sufficient to map them along the coast to reconstruct tilting patterns and, if the high-tide shorelines can be dated, vertical displacement rates. Some caution is necessary, however; high coastal cliffs which are abnormally straight or gently curved are commonly suspected of being fault scarps. When uplifted evidence of former high-tide shorelines is missing from the cliff, the footwall of the fault may be underwater, concealed by sediments, or eroded, and this makes estimations of vertical movements hypothetical. But straight high cliffs may also result from differential erosion having removed a weaker seaward rock formation, without any tectonic implication. Many mappable faults may also have been formed by geologically ancient deformations, under tectonic regimes and stress situations long abandoned.

In most cases, active faults do not cut the coastline, but remain at a certain distance from it, on land or offshore; marks of uplift or subsidence indicated by former high-tide shorelines are therefore likely to indicate a lesser vertical displacement than at the fault scarp and a more complete survey is necessary before interpreting the vertical displacement observed.

Example of a paleotectonic coastal scarp: northernmost Chile. Between Arica and Iquique, except locally, an exceptionally high cliff directly drops into the sea, and is still retreating under wave action. In the surroundings of the Cerro Punta Madrid (18°57′S, 70°18′W) it reaches a height of about 1,000 m. This cliff derives from a large system of north–south oriented, *en échelons*, normal faults which formed at the end of the Oligocene epoch and whose scarps retreated under marine erosion during a Middle to Upper Miocene transgression. South of Iquique, the scarp is an abandoned cliff no longer receding under wave action; at its foot lies a shore platform, sometimes wider than 1 km, which has been uplifted to 50 m above present sea level (Paskoff, 1996).

Example of a neotectonic coastal scarp: southwestern Calabria (Italy). Calabria has been affected by strong uplift movements during the Quaternary. At least 12 Pleistocene levels of marine terraces have been identified, reaching as much as 1,350 m in altitude. Near Reggio Calabria, Last Interglacial marine deposits with the guide fossil *Strombus bubonius* at 157 m suggest that the average uplift rate exceeded 1 mm/yr during the last 125 ka. Near Palmi, the steepness of the coastal relief (Figure T8) and the narrowness (only a few hundred meters) of the continental shelf can be ascribed to active faulting, which is leaving abrupt scarps on the gneissic rock formations.

Example of neotectonic coastal folding. Along the southernmost coast of Taiwan (Hengchung Peninsula), the inner edge of the mid-Holocene

Figure T7 The surface of Late Pleistocene raised marine terraces has been displaced by a recent fault near Kupang, West Timor, Indonesia (photo B266, August 1988).

Figure T8 The active fault scarp near Palmi (Calabria, Italy) (photo E722, April 1996).

marine terrace is 10–15 m high, but its elevation gradually increases westwards, deformed by undulation movements. Its altitude ranges from 5 to 36 m along a distance of less than 10 km (Liew and Lin, 1987).

Rates of vertical neotectonic movements

Because of the various potential tectonic processes, rates of vertical movements in coastal areas have varied much in space and time.

Over the long term (10^6–10^5 yr), structural tectonics is indeed the predominant factor, with rates reaching several millimeters per year reported from tectonically very active areas such as near Ventura, California, where Pleistocene marine terraces have been uplifted at average rates exceeding 8 mm/yr (Lajoie, 1986), with even much higher rates reported from the land interior (about 10–15 mm/yr estimated for the crest of the Ventura anticline during the last few hundred thousand years (Wallace, 1986, p. 7)). At Huon Peninsula, Papua New Guinea, uplift of up to 3 mm/yr during the last 124 ka has been demonstrated by Chappell (1974). Evidence of well-preserved Quaternary shorelines is, however, rare in areas of rapid uplift, because strong erosion rates most often effaced most ancient marine marks. At the same timescale, volcano-isostatic factors can be estimated to have been on the order of a few millimeters per year for subsidence and one order of magnitude less for arching uplift; thermo-isostatic vertical displacement rates usually remain limited to the order of a fraction of millimeters per year: average subsidence rates have been estimated at 0.2 mm/yr during the last 60 Myr in the Marshall Islands and 0.12 mm/yr since the Pliocene in Mururoa Atoll (Tuamotus). Glacio- and hydro-isostatic factors cannot be assessed at the timescale of 10^6–10^5 yr.

During the last 20 kyr, the most important vertical displacements have been of glacio-isostatic origin; near the center of the former Fennnoscandian ice sheet, uplift since the last glacial maximum has been estimated at about 800–850 m, that is, at 40–42 mm/yr on average (Mörner, 1979); since the beginning of the Holocene, uplift has been about 284 m, that is, 28 mm/yr on average, decreasing exponentially from over 50 mm/yr at that time to about 9 mm/yr at the present time (Pirazzoli, 1991). Residual present-day uplift is therefore almost 1 m/century. In Canada, with a marine limit close to 300 m in elevation near the eastern part of Hudson Bay (Andrews, 1989) maximum uplift values are probably of the same order as in Fennoscandia.

Hydro-isostatic displacements are also significant over the timescale of 10^4 yr. They can be estimated on the order of about one-third of the sea-level change, that is, some 40 m in deep-sea areas for an eustatic change of 120 m (i.e., 2 mm/yr on average since the last glacial maximum) and less in shallow coastal areas.

At the timescale of 10^4 yr, structurally tectonic displacements can be in some cases on the same order as glacio-isostatic ones: on the east coast of Taiwan, marine evidence suggests an average uplift rate of about 8 mm/yr since 14 kyr ago near Tulan (Pirazzoli *et al.*, 1993) as well as since 4,000–3,000 kyr ago near Hualien (Konishi *et al.*, 1968) (Figure T9). Because the amount of glacio- and hydro-isostatic displacements has been decreasing exponentially with time after complete melting of the

Figure T9 *In situ* coral reefs about 3 kyr old, uplifted up to 25 m above sea level, are visible (arrows) above a cliff at Hualien, east coast of Taiwan (photo B805, January 1990).

ice caps, movements of structurally tectonic origin have become predominant and easily recognizable in several areas during the last few thousand years. Very rapid rates of vertical displacement of thermo-isostatic origin can be reached in volcanic areas during relatively short periods. The most spectacular example is probably Iwo (Sulphur) Island, a volcano situated 1,200 km south of Tokyo, where over 20 steps of marine terraces occupying nearly the whole area of the island (used as military airstrips by the Japanese during the last world war) have been recognized up to 120 m above sea level. The average uplift rate was here of more than 100 mm/yr over the last several centuries, reaching 200 mm/yr in some parts of the island (Kaizuka, 1992). In the case of the Pozzuoli area mentioned above, average vertical displacement rates have reached even faster peaks, such as the uplift rate of up to 800 mm/yr recorded between 1968 and 1984.

Rates of subsidence in delta areas induced by sediment compaction and sedimento-isostasy is often significant over the long term. At the shelf edge of the eastern Mississippi delta, for example, the long-term averaged subsidence rate since the last glacial period has been estimated to at least 1 mm/yr (Stanley *et al.*, 1996). For the four largest modern depocenters in the Mediterranean, long-term mean lowering rates have been reported of 4–5 mm/yr from the Ebro Delta, >7 mm/yr at the mouth of the Grand Rhône Delta, possibly from 1 to 3 mm/yr from the Po Delta, and 4–5 mm/yr from the Nile Delta (Stanley, 1997). The best-known example of land subsidence produced by sediment compaction, accelerated by man-induced drainage during the last centuries, is indeed that of the Netherlands, where the soil level may be as low as 6 m below sea level. More recently, man-induced land subsidence following oil and natural gas extraction or groundwater exploitation has taken place during the last century in many coastal plains of estuarine, delta, or lagoonal areas: the greatest measured land subsidence (a maximum of 4.6 m) is probably that reported from the Tokyo Lowland region, where an area of about 70 km^2, supporting more than half a million inhabitants, is below mean sea level. Significant amounts of land subsidence have been reported also from the Po Delta (3.2 m), Houston-Galveston (3.0 m from 1906 to 1987), Shanghai (2.7 m), Tianjin (2.5 m), the southwestern part of Taiwan (2.4 m), Taipei (1.9 m), and Bangkok (1.6 m). In several big cities built in coastal plains of developing countries (Jakarta, Hanoi, Haiphong, Rangoon), where detailed measurements are lacking, the probable occurrence of land subsidence is revealed by increasing difficulties in drainage which produce more frequent flooding.

Among the many possible causes of coastal change and evolution, tectonics and neotectonics have been indeed major controlling factors, virtually at all timescales. For the present time, however, a new tectonic cause—man-induced land subsidence—is becoming significant in many densely populated or industrial coastal areas, with short-term impacts which may exceed those of all other natural causes occurring in the same area.

Paolo A. Pirazzoli

Bibliography

American Geological Institute, 1960. *Glossary of Geology and Related Sciences*, 2nd edn. Washington DC: National Academy of Sciences.

Andrews, J.T., 1989. Postglacial emergence and submergence. In Fulton, R.J. (ed.), *Quaternary Geology of Canada and Greenland*. Geology of Canada no. 1. Geological Survey of Canada.

Bloom, A.L., 1971. Glacial-eustatic and isostatic controls of sea level since the last glaciation. In Turekian, K.K. (ed.), *The Late Cenozoic Glacial Ages*. New Haven and London: Yale University Press, pp. 355–379.

Chappell, J., 1974. Geology of coral terraces, Huon Peninsula, New Guinea: a study of Quaternary tectonic movements and sea-level changes. *Geological Society of America Bulletin*, **85**: 553–570.

Crowell, J.C., 1986. Active tectonics along the western continental margin of the conterminous United States. In Wallace, R.E. (panel chairman), *Active Tectonics*. Studies in Geophysics. Washington DC: National Academy Press, pp. 20–29.

Dickinson, W.R., 1998. Geomorphology and geodynamics of the Cook–Austral island-seamount chain in the South Pacific Ocean: implications for hotspots and plumes. *International Geology Review*, **40**: 1039–1075.

Kaizuka, S., 1992. Coastal evolution of a rapidly uplifting volcanic island: Iwo-Jima, western Pacific Ocean. *Quaternary International*, **15/16**: 7–16.

Konishi, K., Omura, A., and Kimura, T., 1968. 234^{U}-230Th dating of some Late Quaternary coralline limestones from southern Taiwan (Formosa). *Geology and Palaeontology of Southeast Asia*, **5**: 211–224.

Lajoie, K.R., 1986. Coastal tectonics. In Wallace, R.E. (panel chairman), *Active Tectonics*. Studies in Geophysics, Washington DC: National Academics Press, pp. 95–124.

Liew, P.M., and Lin, C.F., 1987. Holocene tectonic activity of the Hengchun Peninsula as evidenced by the deformation of marine terraces. *Memoir of the Geological Society of China*, **9**: 241–259.

Mitrovica, J.X., and Peltier, W.R., 1991. On postglacial geoid subsidence over the equatorial oceans. *Journal of Geophysical Research*, **96**(B12): 20053–20071.

Morhange, C., Bourcier, M., Laborel, J., Giallanella, C., Goiran, J.P., Crimaco, L., and Vecchi, L., 1999. New data on historical relative sea level movements in Pozzuoli, Plaegrean Fields, southern Italy. *Physics and Chemistry of the Earth*, A, **24**(4): 349–354.
Mörner, N.A., 1979. The Fennoscandian uplift and Late Cenozoic geodynamics: geological evidence. *Geojournal*, **3**: 287–318.
Mörner, N.-A., 1980. The INQUA Neotectonic Commission. *Bulletin INQUA Neotectonic Commission*, **3**: 1.
Paskoff, R., 1996. *Atlas de Las Formas de Relieve de Chile*. Instituto Geografico Militar de Chile.
Pirazzoli, P.A., 1991. *World Atlas of Holocene Sea-Level Changes*. Amsterdam: Elsevier.
Pirazzoli, P.A., 1995. Tectonic shorelines. In Carter, R.G.W., and Woodroffe, C.D. (eds.), *Coastal Evolution: Late Quater- nary Shoreline Morphodynamics*. Cambridge University Press, pp. 451–476.
Pirazzoli, P.A., Arnold, M., Giresse, P., Hsieh, M.L., and Liew, P.M., 1993. Marine deposits of late glacial times exposed by tectonic uplift on the east coast of Taiwan. *Marine Geology*, **110**: 1–6.
Pirazzoli, P.A., Laborel, J., Saliège, J.F., Erol, O., Kayan, I., and Person, A., 1992. Holocene raised shorelines on the Hatay coasts (Turkey): palaeoecological and tectonic implications. *Marine Geology*, **96**: 295–311.
Pirazzoli, P.A., Laborel, J., and Stiros, S.C., 1996. Earthquake clustering in the Eastern Mediterranean during historical times. *Journal of Geophysical Research*, **101**(B3): 6083–6097.
Pirazzoli, P.A., and Montaggioni, L.F., 1988. Holocene sea-level changes in French Polynesia. *Palaeogeography, Palaeoclimatology, Palaeoecology*, **68**: 153–175.
Pirazzoli, P.A., and Salvat, B., 1992. Ancient shorelines and Quaternary vertical movements on Rurutu and Tubuai (Austral Isles, French Polynesia). *Zeits, Geomorphol.*, **36**: 431–451.
Plafker, G., 1965. Tectonic deformation associated with the 1964 Alaska Earthquake. *Science*, **148**(3678): 1675–1687.
Stanley, D.J., 1997. Mediterranean deltas: subsidence as a majpr control of relative sea-level rise. In Briand, F., and Maldonado, A. (eds.), *Transformations and Evolution of the Mediterranean Coastline*. Monaco, CIESM Science Series No, 3, pp. 35–62.
Stanley, D.J., and Warne, A.G., 1994. Worldwide initiation of Holocene marine deltas by deceleration of sea-level rise. *Science*, **265**: 228–231.
Stanley, D.J., Warne, A.G., and Dunbar, J.B., 1996. Eastern Mississippi delta: late Wisconsin unconformity, overlying transgressive facies, sea level and subsidence. *Engineering Geology*, **45**: 359–381.
Stewart, I.S., and Hancock, P.L., 1994. Neotectonics. In Hancock, P.L. (ed.), *Continental Deformation*. Tarrytown: Pergamon Press, pp. 370–409.
Valensise, G., and Ward, S.N., 1991. Long-term uplift of the Santa Cruz coastline in response to repeated earthquakes along the San Andreas Fault. *Bulletin of the Seismological Society of America*, **81**: 1–11.
Vita-Finzi, C., 1986. *Recent Earth Movements: An Introduction to Neotectonics*. London, Academic Press, 226 p.
Wallace, R.E. (panel chairman), 1986. *Active Tectonics*. Studies in Geophysics, Washington DC: National Academy Press.

Cross-references

Changing Sea Levels
Coastal Subsidence
Coastline Changes
Faulted Coasts
Isostasy
Physical Models
Seismic Displacement
Submerging Coasts
Uplift Coasts

THALASSOSTATIC TERRACES

A term derived from Greek roots, the expression "Thalassostatic" obtains its prefix from "thalassic," having to do with the sea, and its suffix "static" referring to its equilibrium state. A fluvial terrace is said to be thalassostatic when it is the product of a former sea level that caused flood-stage sediment load to accumulate as an alluvial surface. Its seaward limit is usually determined by a beachridge or beachridge plain.

The actual term itself was coined by Zeuner (1945), who saw that an estuary created by downcutting during a Pleistocene low sea-level state would become aggraded during a following interglacial phase. Even earlier, Ramsay (1931) pointed out that alluvial terraces of this sort would provide clearer evidence of former eustatic levels than strictly marine terraces that are more liable to erosion or other factors. Clayton (in Fairbridge, 1968, p. 142) mentions the actualistic evidence of modern aggradation behind dams. This filling will extend only a short distancer upstream where the gradient is changed. Zeuner believed that gravel terraces of the lower Thames in England must be solifluction products of glacial-phase stream loading, thus having nothing to do with thalassostatic events.

In the classic work of de Lamothe (1918) on the Somme Valley in northern France, the concept of eustatic control was first presented, but erroneously extended to the entire fluvial terrace system. The Dutch worker, Brouwer (1956) considered the question of thalassostatic terraces in general and showed that only the lowest sectors were appropriately so-named. Earlier, Pleistocene examples would be largely obliterated by erosion.

In the world's major deltas like the Mississippi and the Rhine, a quasi-stable tectonic fulcrum develops, with subsidence on the seaward side, uplift on the landward side. Thus the fluvial terraces are stacked progressively lower and in stratigraphic sequence downstream. On the Mississippi the fulcrum area is above Baton Rouge, on the Rhine about at Nijmegen. However, the exact fulcrum point shifts up- or downstream depending upon the current eustatic level. A problem develops, however, with these large rivers in their propensity to major flood events, their discharge being greatly amplified by the large area and multiple sources of their drainage basins. Such events are likely to overtop levees in the deltaic regions. These so-called "avulsion" breaks are commonly matched by the drying out of other distributary channels, so that a rather complex stratigraphy develops (Törnquist, 1994; Berendsen, 1995). Nevertheless, a strongly cyclical pattern is discernible.

Where an extensive data source has been provided by detailed geophysical transects, as in the Mississippi delta, it can be seen that layers of Holocene transgressive facies are alternating with regressive events when channels deepen and their outlets shift seawards. In contrast, the transgressive intervals are marked by sandy chenier ridges (sand from longshore sources) that form distinctive markers in the otherwise muddy or peaty deltaic facies. Compaction of these water-saturated substrates leads to areas of accelerated subsidence, as shown by very variable tide-gauge evidence, that can be misleading for the delta as a whole, which has a very modest subsidence rate. With each transgression the thalassostatic terrace building shifts upstream, but with the renewed downcutting after avulsion an entirely new delta-lobe is likely to evolve (Lowrie and Hamiter, 1995).

A pioneering attempt to relate thalassostatic eustasy to the rivers of Borneo (Kalimantan) was made by Smit Sibinga (1953). A detailed glacial-age drainage network extends over much of the Sunda Shelf according to the submarine mapping by Molengraaff and referred to as the "Molengraaff River System" by Umbgrove (1947). With each successive eustatic rise thalassostatic terrace deposits must have lined these former waterways. The eustatic curve is not a sine-wave, however, but is a strongly fluctuating one, so that there are extended intervals of pause or reversal, during which there would have been accelerated thalassostatic sedimentation.

Best known and well dated of these interruptions was the Younger Dryas event which lasted approximately 1,000 yr, 11,740–10,740 cal. BP, with abrupt transitions, both at its commencement and terminations. Eustatic sea level at that time fell to about 30 m below present mean sea level. Its coastline is particularly well marked by the presence, in the warmer latitudes and along the more stable coasts, of an abrupt line of reefs or distinctive beach deposits. A wave-resistant beachrock facies is a distinctive marker where carbonate sands permit rapid lithification. In some sectors carbonate sands led to the development of littoral dunes, which also become rapidly lithified to form "eolianites." Today these eolianites form reefs that are well known to fishermen (e.g., "Twelve Fathom Reefs" of Western Australia).

Other sea-level fluctuations were marked in the deep ocean by the "Heinrich events" (at 3–7 kyr intervals) which mark iceberg distribution in the N.W. Atlantic. Heinrich events are marked by climatic extremes, both positive and negative, thus raising the chicken-and-egg problem. Analogous eustatic fluctuations are indicated by the thalassostatic data. Sudden warming of the water around Greenland, such as occurred in AD 1912, accelerates iceberg calving (with disastrous effect on RMS *Titanic*). In this case, however, it was not so much climatic, as tidal, because 1912 marked the largest lunisolar tidal extremes in over 500 years (Wood, 2001).

Thalassostatic terrace building (and its erosion) thus relates to movement of relative sea level (RSL). The bed of a river on reaching the sea is nicely adjusted to velocity and bed-load, which dictate the equilibrium water depth at the river mouth, subject to tidal characteristics and seasonal changes such as storminess and monsoonal shifts.

The equilibrium status may be modified by one or more of several processes: (a) Tectonic, which may be sudden as a result of an earthquake, such as the historic shifts in 1855 and 1931, of Wairapa (Wellington) or Napier in New Zealand or the subsidence of parts of Yokohama in Japan. It may also be secular as a result, for example, of glacioisostatic rebound such as observed today at Stockholm in Sweden (at 5 mm/yr) or at Great Whale River in Quebec (at 8 mm/yr). (b) Eustatic and/or climatic; these are related but partly independent, although not easily distinguished. When a dominant wind system changes, it affects also RSL as observed, for example, in the Hudson Bay, in the Gulf of California and in the eastern Gulf of Mexico, all being partly related to the 11 yr sunspot cycle and the 18.6 yr lunar cycle (Fairbridge, 1992). It is also observed on a 6-month basis in the Red Sea and Arabian Sea and Bay of Bengal, a consequence of the Asiatic monsoon reversal. Longer cycles, of the order of 45, 90, and more years are observed in the arctic and subarctic latitudes, relating in part to the polar anticyclone and the magnetic pole in its periodic shifts from the longitude of Greenland to that of eastern Siberia, and back again.

For whatever reason, a fall in RSL leads to a shallowing of the debouching river mouths. Two possibilities ensue. If the fluvial discharge is linked, as it usually is, to climate fluctuation, a decreased discharge coinciding with a shallowing of the stream beds, then the result will be a "siltation" of local harbors. Around the Mediterranean in Roman times (1st century AD) there was widespread siltation and as a result many Roman docks, harbor facilities, and fish tanks had to be re-engineered. Some harbors, as in Turkey and farther afield in India (entrance of the Narbada River) were abandoned altogether, or the city shifted somewhat downstream (Kayan, 1997).

The opposite condition is associated with a systematic rise in RSL. This occurred, for example, in the post-Carolingian (after Charlemagne) global warming that persisted until about AD 1300. River courses were backed up and deepened. Roman-era dock or defensive facilities were drowned. An example of the latter can be seen at Portchester in the south of England.

Rhodes W. Fairbridge

Bibliography

Berendsen, H.J.A., 1995. Holocene fluvial cycles in the Rhine delta? *Journal of Coastal Research*, **17**(Special issue): 103–108.

Brouwer, A., 1956. Thalassostatic terraces and Pleistocene chronology. *Leidse Geologische Mededelingen*, **20**: 22–33.

De Lamothe, Gen., 1918. Les anciennes nappes alluviales et lignes de rivage du bassin de la Somme et leur rapports avec celles de la Méditerranée occidentale. *Bulletin de la Societe Géologique de France*, **18**(4): 3–58.

Fairbridge, R.W., 1968. *The Encyclopedia of Geomorphology*. New York: Van Nostrand Reinhold.

Fairbridge, R.W., 1992. Holocene marine coastal evolution of the United States. *S.E.P.M.. Publication*, **48**: 9–20.

Kayan, I., 1997. Bronze Age regression and change of sedimentation on the Aegean coastal plain of Anatolia (Turkey). In Dalfes, H.N. *et al.* (eds.), *Third Millennium BC Climate Change and Old World Collapse* Berlin: Springer, NATO-ASI series 1, Global Environment Change **49**: 431–450.

Lowrie, A., and Hamiter, R., 1995. Fifth and sixth order eustatic events during Holocene (fourth order) high stand influencing Mississippi delta-lobe switching. *Journal of Coastal Research*, **17**(Special issue).

Ramsay, W., 1931. Changes of sea-level resulting from the increase and decrease of glaciation. *Fennia*, **52**(5): 1–62.

Smit Sibinga, G.L., 1953. Pleistocene eustasy and glacial chronology in Borneo. *Geologie en Mijnbouw*, **15**: 365–383.

Törnquist, T.E., 1994. Middle and Late Holocene avulsion history of the River Rhine (Rhine-Meuse delta, Netherlands). *Geology*, **22**: 711–714.

Umbgrove, J.H.F., 1947. *The Pulse of the Earth*. The Hague: Nyhoff.

Wood, F.J., 2001. *Tidal Dynamics*, Vol. 2. *Journal of Coastal Research*, Special Issue 31.

Zeuner, F.E. (1945). *The Pleistocene Period*. London: Ray Society.

Cross-references

Archaeological Site Location, Effect of Sea-Level Changes
Beach Ridges
Cheniers
Eolian Processes
Late Quaternary Marine Transgression
Offshore Sand Banks and Linear Sand Ridges

THALASSOTHERAPY—See HEALTH BENEFITS

TIDAL CREEKS

Is a tidal creek an estuary? Stamp (1966) in his *Glossary of Geographical Terms* quotes the OED definition of creek as, "*a narrow recess or inlet in the coastline of the sea, or the tidal estuary of a river; an armlet of the sea which runs inland in a comparatively narrow channel ...*." The *Dictionary of Geological Terms* published by the American Geological Institute (1976) defines a creek as, "*a small inlet, narrow bay, or arm of the sea, longer than it is wide, and narrower and extending farther into the land than a cove.*"

Figure T10 Tidal creek in mangroves showing the narrow shallow channel and low hydrodynamic energy conditions.

Geomorphically, the essential characteristics of a tidal creek are that they are relatively long and narrow, are shallow, and exhibit tidal water level fluctuations and weak tidal currents. They are relatively small-scale landforms with a low hydrodynamic energy environment without significant wave action or strong current action (Figure T10). Typically along its banks the tidal creek is well vegetated. The friction from the banks and the length of the creek ensure that the tidal wave within the creek is hyposynchronous (reducing tidal range upstream). Such conditions facilitate sediment deposition, so tidal creeks exhibit active deposition and infilling over a time frame of decades. Tidal creeks may drain to the open coast of an enclosed coastal sea or bay, or may be part of a larger estuary or delta system. They also occur at the distal ends of arms of drowned river valley (ria) harbors where tidal creeks drain sectors of mangrove stands and/or salt marsh. In tropical areas, tidal creeks and distributaries create a dense network of small-scale drainage through the mangrove forests, while a similar reticulated plexus of creeks occurs in temperate salt marsh (Guilcher, 1958). Pethick (1984) notes that salt marsh creek systems have a very high ratio of total creek length to drainage area—on the order of 40 km/km^2. Tidal creeks also occur in sinuous meandering forms draining extensive intertidal mudflats. However, tidal creeks do not tend to occur draining to open coast high-energy sandy beaches.

Often the creeks extend as the headwaters of a large estuary into the surrounding drainage basin, and on occasion carry freshwater discharges and episodic floodwaters from surrounding catchments. Thus the tidal creeks are frequently the mixing zones of fresh and saline tidal water, and exhibit an enhanced vertical salinity structure due to the low-energy conditions. Accordingly, they are also favored zones for flocculation processes, so that deposition of muds is typical and ongoing within tidal creeks. Unfortunately, tidal creeks often tend to become a receptacle for human waste and rubbish, and are frequently reclaimed.

Terry R. Healy

Bibliography

American Geological Institute, 1976. Dictionary of Geological Terms. Garden City, NY: Anchor Press.

Guilcher, A., 1958. *Coastal and Submarine Morphology*. London: Methuen.

Pethick, J., 1984. *An Introduction to Coastal Geomorphology*. London: Edward Arnold.

Stamp, L.D., 1966. Glossary of Geological Terms. London: Longmans.

Cross-references

Estuaries
Mangroves, Ecology
Mangroves, Geomorphology
Muddy Coasts
Ria
Salt Marsh
Tidal Flats
Vegetated Coasts

TIDAL DATUMS

Introduction

A sea-level datum is a surface constructible statistically from sea-level observations that is used as a reference for measuring and describing vertical positions near the earth's surface. At official intervals, the elevations of sea-level datums are revised to reestablish their relationship to a new mean water level. A multitude of sea-level datums are in use. Each has its own advantages and disadvantages with no universally superior choice. Preferences depend on location, purpose, and past practices. Focus here will be on tidal datums defined at different phases of the tide; for example, high water and low water (Figure T11). Discussion of tidal datums illustrates some of the uses, practices, and limitations common to the broader class of vertical datums.

Scope

Details in the operational definitions of tidal datums evolved as concepts matured, needs for precision increased, field methodologies improved, and new legal issues arose. In this discussion of broad tidal datum issues, operational details may be sacrificed to provide an overall understanding of the basics, at least as understood by a coastal morphologist limited in practice to North America. Specific details continue to evolve with advances in measurement technology and changing society needs. Prescriptions of such details are left to official documents.

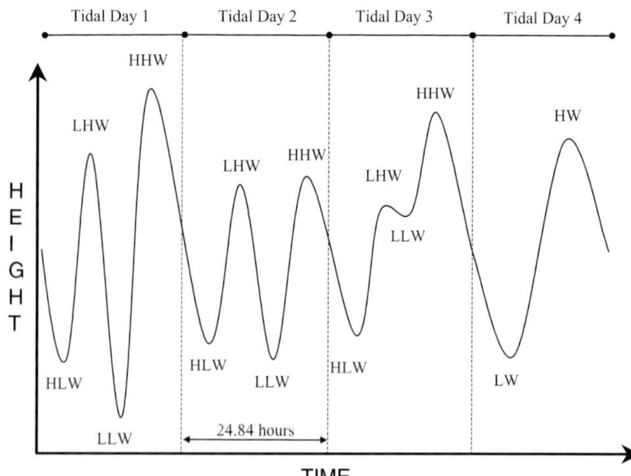

Figure T11 Tide phases labeled on a stylized tide curve. When two high-tide phases can be distinguished in the tide curve during a single tidal day (24.82 h), the higher of the two is called higher high water (HHW) and the other lower high water (LHW). When two low-tide phases are discernable in one day, the higher of the two lows is called higher low water (HLW) and the other lower low water (LLW). When only one high and one low-tide phase appear (e.g., Tidal Day 4), the high phase is called high water (HW) and the low is called low water. The record from sites that usually have only a single high and low per day will periodically transition in and out of multiple-cycles-per-day episodes that can have distinctly different phases like Tidal Day 2 or less distinct phases, more like Tidal Day 3 above.

Historical developments

The astronomical tide causes sea level to vary on daily, monthly, and longer cycles. These deterministic cycles are superimposed on more random water level fluctuations related primarily to weather. Normally, the reference point for repeated measurements should be stable, that is, anchored to the system containing the objects measured. What factors explain widespread adoption of a continually fluctuating water surface as a vertical reference? Early communities developed at sea ports where knowledge of the water depths in navigable channels was vital for the economy and defense. Soil productivity [EBH1] was another fundamental concern related to height above water because of altitudinal climate gradients if not more directly through frequencies of flooding. Referencing elevations to water levels was, therefore, of keen interest to early civilizations. Sea level also provided a common reference elevation applicable widely throughout the world. It could be reestablished with little loss of accuracy near coasts and without the need for long-distance surveys or recover of any destructible marker (e.g., benchmark). Fluctuations of the sea surface (and hence the zero reference) were probably small relative to the precision required to make decisions, especially for civilizations developing around the Mediterranean Sea where the range of tidal fluctuations is small compared with a typical range in the ocean. As coastal property ownership disputes increased, navigation clearances tightened, and tidal fluctuations became recognized (though not yet understood), comparison of elevations with respect to different sets of sea- level observations became more problematic. Later recognition of a broad spectrum of sea-level changes required further adjustments to allow tidal datums to persist as widely used references. Continuing advantages of referencing to sea level, instead of a fixed datum, include: conveniences related to historical precedence, natural differences in human use of property above and below sea level, lateral continuity (at least along coasts), apparent simplicity, lack of geographic bias in definition, and the greater relevance of mean water levels over any truly level surface for most hydrologic and navigation considerations.

Because concerns of navigation and coastal engineering are closely tied to sea level, the reference point in these fields should change with long-term changes in mean sea level. For endeavors in which sea level is irrelevant, a stable nontidal datum may be preferable. For this reason, and to better understand tidal datums themselves, some discussion of nontidal datums is included.

The name of the organization having responsibility for official datum and charting standards within the United States changed with time from the Survey of the Coast, to the Coast Survey, Coast and Geodetic Survey, National Ocean Survey and, now, the National Ocean Service (NOS). The NOS's authoritative references on present and historic coastal boundary determinations are Reed (2000) and Shalowitz (1962, 1964). Kraus and Rosati (1997) summarize procedures for determining shoreline positions for coastal engineering. Prescription for US Army Corps of Engineer surveys of waterways and additional guidance on tidal datums can be found in USACE (2002). All five references can be found on the World Wide Web (www).

Methods of tidal datum determination

Statistically averaging sea-level measurements is the process by which an unsteady surface is fixed to serve as an unambiguous and repeatable datum. There are a number of ways to average sea level, giving rise to a number of different tidal datums. For an unambiguous datum definition, the frequency of the measurements and the time span over which they are to be averaged must be specified. For any tidal datum, the manner of relating the measurements to a clearly defined tidal phase (Figure T11) must also be specified. Breakdown in any one of these fundamental requirements, causes problems as we shall see.

Astronomical cycles and the tidal response

Astronomical tides are the sea's response to gravitational forces of the moon and sun (and to a lesser extent other celestial bodies) that vary over the earth's surface. These forces also vary with time as relative astronomical positions continually change over a wide spectrum of cycles. Procedures to calculate the local elevation of each of six standard tidal datums (upper portion of Table T1) follow similar official algorithms regardless of which of the six is selected. Selection would typically be based on location and purpose, for example, whether it involves resolving legal disputes, promoting navigation, assessing environmental resources, or analyzing geophysical phenomena.

The longest cycle conventionally damped by tidal averaging (the Metonic cycle) has a period of about 18.61 years. The NOS selects a specific 19-year time interval, called the *National Tidal Datum Epoch*, as the official span over which tide measurements are averaged to obtain the current tidal datum elevations throughout North America.

Gauges

The NOS presently maintains about 189 *primary tide gauges* where hourly water levels have been recorded continuously for more than 19 years. *Tidal benchmarks* are physical monuments maintained along the coast and to which tidal datum elevations have been surveyed (see *Geodesy*). Three to five bench marks are tied to NOS tide gauges by precise surveying and monitored to detect possible gauge disturbances.

Elevations are established at a larger set of *secondary gauges* by comparing their shorter tidal records (less than 19 years, but more than one year) with simultaneous measurements from nearby primary stations. Tidal datums can be reestablished by leveling short distances from tidal benchmarks. Numerical modeling of the tide can be used to extend datums longer distances between gauge sites including inland across the continent.

The elevation of water levels (and, therefore, tidal datums) in estuaries and coastal lagoons is typically higher than along adjacent open coasts. This super elevation is primarily due to freshwater flowing into the bay and nonlinear friction in tidal currents. Adjacent to inlets, the transition is smooth, but relatively steep. Therefore, primary gauges are often located away from their immediate vicinity (see *Tide Gauges*).

Table T1 A few North American vertical datums used by coastal engineers and scientists

Acronym	Name and definition	Origin, use or advantages
Tidal datums (based on gauge records)		
MHHW	Mean higher high water Mean of all higher high waters[a, b, c]	
MHW	Mean high water Mean of all high water heights (i.e., HW, HHW, and LWH, see Figure T11)[a, b, c]	Frequently used boundary separating private from state lands. See text. Landward limit for US Corps of Engineers jurisdiction over navigable waters.
MTL	Mean tide level Mean of MHW and MLW datums[a,b,c,d]	More descriptively called half-tide level.
MSL	Mean sea level Mean of hourly water surface heights[c]	Maintained US-wide by the National Geodetic Survey as "the most practicable and stable datum for general engineering," Shalowitz (1962).
MLW	Mean low water Mean of all low water heights (i.e., LW, LLW, and HLWs, see Figure T11)[a, c, d]	Use of the synonymous term, mean low tide, is discouraged by NOS.
MLLW	Mean lower low water Mean of all lower low water heights[a, c, d]	Principal chart datum for all of U.S. since 1981. Used by six states as boundary separating private and state lands.
Orthometric vertical datums (based on topographic surveying)		
NGVD 29	National Geodetic Vertical Datum 1929	Earlier standard geodetic datum. Based on first order survey nets of the United States and Canada fit to MSL at 26 tide gauges.
NAVD 88	North American Vertical Datum 1988	Official vertical datum for all of North America since June 24, 1993. An upgrade of NGVD 1928 based on more measurements over a larger area and independent of any tide gauge measurements.
Three-dimensional datums (based on measurements from space)		
NAD 83	North American Datum 1983	An upgrade using space-based measurements of the previous standard, NAD27.
WGS 84	World Geodetic System 1984	US Defense Mapping Agency equivalent of NOAA's NAD 83.

[a] Each stage of the tide in the time series to be averaged must, by official convention, differ by at least a "0.10 ft" and "2.0 h" from the adjacent measurements in the series. The series usually extend over a specified National Tidal Datum Epoch, but sometimes over a shorter stated duration.

[b] A high water is a maximum in a tide curve reached by a rising tide. The higher of two unequal high waters during the same day (24 h 50 min = the period of the M_1 tidal component) is the higher high water. The other high water is the lower high water.

[c] Definitions from US Department of Commerce Tide and Current Glossary (Hicks, 1989).

[d] A low water is a minimum in the tide curve that is reached by a falling tide. The lower of two unequal low tides during the same day is the lower low water. The other is the higher low water.

Tidal datum elevation algorithms

To calculate the elevations of a particular tidal datum at a primary tide gauge site, select the elevation measurements associated with the chosen tide phase (e.g., high tide) from the set of hourly data that covers the current National Tidal Datum Epoch. Official rules specify which hourly readings are associated with the chosen phase of the astronomical tide, but other factors (primarily weather) contributed to the water level measured at that time. Arguments have been made, especially for regions where the wind is strong, to loosen the required association between astronomical tidal phases and selection of which observations to average. Following the official rules (more details in next major section and in Table T1), a single daily high (or low) water elevation was traditionally selected at stations where the tide is classified as being *diurnal* (displaying a single cycle with one high and low per day, Figure T11, Tidal Day 4); elevations of two high (or low) waters are selected daily at stations classified as *semidiurnal* or *mixed* (displaying two cycles per day, Figure T11, Tidal Days 1, 2, and 3). Sum the selected observations and divide by the number selected. By thus averaging long series of data, the shorter cyclic variations (astronomical tide) and the random (weather-related) variations are damped. Including the full 19th year avoids biasing related to the strong annual cycle. Significant longer term (>19-year) variations arise from astronomical forces, thermal expansion (contraction) of the oceans, crustal movement, glacial melting, etc. These longer-term changes (see *Eustasy* and *Changing Sea Levels*) are compensated for, as needed at irregular intervals, by adopting evermore recent 19-year periods (Epochs) as the National Tidal Datum Epoch. For North America, the current Epoch is 1960–78; previous Epochs include 1941–59 and 1924–42. Agencies responsible for datum definitions elsewhere agree that tidal epochs should span complete 19-year periods, but they do not all use the same 19-year periods.

Spaced-based measurements

After a period of improvement and acceptance, spaced-based measurements derived from the NAVSTAR Global Positioning System (GPS) are now widely used not only as a convenient way to reestablish local datum elevations near benchmarks, but also in defining new coordinate systems (lowest portion of Table T1). Numerical models can introduce tidal datums into these three-dimensional coordinate systems based on idealized reference ellipsoids rather than tidal data. This approach is now quicker and less expensive than collecting tidal measurements or leveling from local tidal datum benchmarks. In the future, direct measurement of sea surface elevations from space may become accurate enough to establish tidal datum elevations essentially continuously along the coast in an iterative blend of observation and tidal dynamic modeling.

Limitations and precautions in applications

Temporal and spatial variations

The difference between successive high and low waters, *tide range*, is neither uniform nor constant. The phases of the moon and the inclination of its orbital plane with respect to the plane of the earth's orbit around the sun (*declination*) are two of the factors contributing to daily variations in tide range on approximately 28-day and 14-day cycles, respectively. In contrast to the elegant theory explaining temporal variations in the astronomical tide, its spatial variations are complex. Tidal dynamics vary with location on the earth, outline of the coast, bathymetry of the seafloor, and such factors as wind, salinity, and river discharge. Each tidal component is characterized by an amplitude and frequency (see *Tides*). The bathymetry and outline of the coast (*basin shape*) modulates sea-level response to the total gravitational force in a manner somewhat analogous to the way the material properties and shape of a poorly tuned musical instrument distorts chords. Some frequencies resonate and are amplified. Others are out of tune with the basin (instrument) and are damped (muffled). Furthermore, frequency-dependent nodes (antinodes) may exist within the basin where responses are maximized (minimized).

Nonuniform height differences between datums

Complexities of tidal propagation cause the mean differences among tidal datums to vary spatially, for example, the difference between MLW and MLLW varies along shore. The mean tide range varies around the world being about 10 m at Anchorage, Alaska; 1 m at San Diego, California; 0.1 m at Galveston, Texas; and less than a centimeter at Chicago, Illinois, on Lake Michigan.

Advent of the www improved the availability and dissemination of the empirically determined elevation differences among tidal datums in the form of tables issued by federal organizations throughout the world (e.g., http://co-ops.nos.noaa.gov/bench. html). Many of these tables also report the highest and lowest observed sea levels. Some sea-level datums (e.g., the British chart datum, Lowest Astronomical Tide) are based on extremes.

Type of tide

The range and timing of the tide goes through cycles as lunar and solar declinations change. They also vary spatially over the earth's surface. The type of tide varies spatially and temporally as well. The type of tide is predominately *semidiurnal* (two similar, semi-sinusoidal cycles per day with little difference between the two lows or the two highs) over most of the US East Coast and *mixed* (two lows and two highs per day that are unequal in elevation) along most of the US West Coast. Along the Gulf of Mexico, the type of tide is not so uniform, but at most locations it is *diurnal* (only one high and one low per day). Figure T11 shows a mixed type during Tidal Day 1 and Tidal Day 3. Tidal Day 2 is semidiurnal and Tidal Day 4 is diurnal. A real tide curve would transition between these types more gradually (see *Tides*).

The most common type of tide throughout the world is semidiurnal. Figure T12 depicts regional variations in type of tide along the north shore of the Gulf of Mexico and a portion of the Atlantic. The basin shapes in Tampa Bay and Charlotte Harbor, on the west coast of Florida, are tuned to diurnal tidal forces. Therefore, the tide type in these estuaries is predominately diurnal even though the open Gulf beaches of west Florida are predominately mixed. *Predominately* means that the type of tide is as designated for more than half of the time at that particular site. Along the Mediterranean Sea, stretches of coast that are predominately semidiurnal alternate with reaches that are predominately mixed. Tides are predominately semidiurnal around most of England, mixed around most of Australia, and diurnal along much of southeast Asia, Japan, and northeast Siberia.

Analysis procedures and datum elevations were established in the Gulf of Mexico prior to a full appreciation of these complex changes in type of tide (Hicks *et al.*, 1988). With the benefit of longer term and spatially denser measurements, it became apparent that some initial datum calculation procedures (as well as inconstancies in terminology) were causing problems. One of the problems was that the MHW and MLW datums had abrupt breaks (*jump discontinuities*) in elevation as one moved alongshore. The global continuity of sea level was a prime feature promoting the widespread adoption of tidal datums. MHW and MLW discontinuities arose because, traditionally, their official definitions included provisions that where the tide was classified as predominately diurnal, only one high (or low) would be included daily in the 19-year series for datum determination. Where the tide was predominately semidiurnal or mixed, both lows and both highs were traditional (and still are) included in these calculations. There are times during the monthly lunar cycle when diurnal forces are small relative to semidiurnal forces. During these times, the tide signal (even at sites where the tide is classified as diurnal) may take on mixed-type characteristics briefly (i.e., exhibit a second daily high and/or low). Likewise at sites that are mixed-type, the second daily high and/or low may become indiscernible for a few days during the month. Tide type is unambiguously classified according to a ratio of amplitude terms for the major tidal constituents. The shape of tide curves vary smoothly along the shore with transition zones between areas with different types of tide. Discontinuities arose

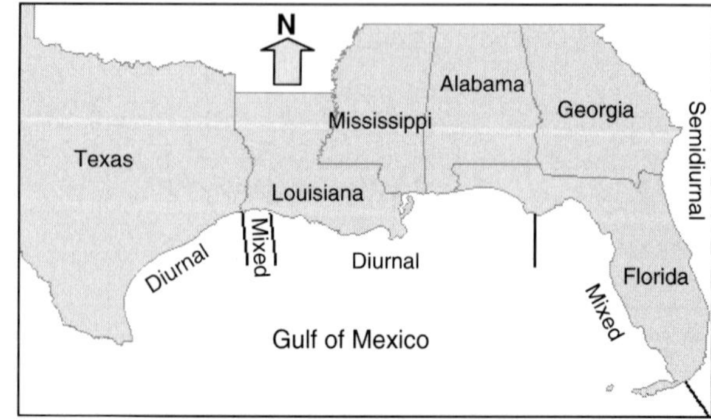

Figure T12 Regional variations in type of tide (modified from Hicks *et al.*, 1988).

from the traditional definitions of MHW and MLW being dependent on tide type. Resulting problems were especially acute in the case of MHW because of its role as a legal state boundary.

Through a series of carefully planned and well-coordinated moves that began as early as the 1970s, the procedures for defining MHW and MLW and the inconsistent naming practices that had crept in at some Gulf sites were changed. Table T1 follows the revised, uniform NOS definitions. Another half dozen modifications were collected into one endeavor that unified analysis, created a single uniform chart datum (MLLW), freed MHW and MLW definitions from dependence on tide type, updated the National Tidal Datum Epoch, moved private–state property boundary seaward along certain reaches, and eliminated all datum discontinuities (Hicks *et al.*, 1988). Though announced much earlier, this comprehensive corrective action officially took effect on November 28, 1980. Complete updating of charts and NOS's Coast Pilot actually proceeded incrementally over the following decade. Tide tables predicting tidal heights for East Coast of North and South America switched chart datum from MLW to MLLW starting with the 1989 volume (making it consistent with NOS tables for other regions). This correction (The National Tidal Datum Convention of 1980) thus simplified standards for tidal datums throughout the Americas.

Nontidal datums

The *geopotential elevation* of tidal datums vary spatially (i.e., the potential energy of a unit mass at a given elevation with respect to each datum varies geographically). For long periods, a tidal datum can deviate on the order of a meter from an *equipotential surface* (surface of constant specific potential energy) due to prevailing atmospheric pressure, temperature, wind, currents, and salinity differences (see *Altimeter Surveys, Coastal Tides and Shelf Circulation*). For example, the geopotential elevation of sea level rises (falls) generally northward along the west (east) coast of the United States. For certain applications (e.g., long-distance pumping, ballistics) it is desirable that the elevation of an equipotential surface be spatially invariant. In such cases, a nontidal datum is clearly required (lower portion of Table T1).

The latest geodetic datum (NAVD 88, middle of Table T1) is not parallel to any tidal datum or to the geodetic datum it replaced (NGVD 29). The NGVD 29 was defined so that it approximated the MSL datum. The new NAVD 88 is on the order of a meter above the MSL datum on the West Coast and 0.1–0.5 m below the MSL datum on the East Coast (Figure T13) and is independent of sea level.

Shorelines

Conceptually, the intersection of any tidal datum with the beach rigorously defines a shoreline. The acronym for this shoreline is obtained by appending an "L" to the acronym for the selected tidal datum. In the United States, the mean high water line (MHWL) was adopted by most states as the boundary between private and public land. Six northeastern states allow private ownership down to the mean lower low water line (MLLWL). In Texas, private ownership has, at least in certain cases, been restricted to being above the mean higher high water line (MHHWL). Federal submerged lands, exclusive fishing zones, and national economic zones are related back to the MLLWL by their legal definitions.

In practice, charted shorelines are sometimes inferred from rectified aerial photographs taken at specific phases of the tide. Such mapped shorelines only approximate the rigorously defined tidal-datum shorelines, but are much easier to establish. Unfortunately, imprecise usage often results in them being labeled MHWL or MLWL. Even less-accurate approximations of tidal-datum shorelines are inferred from the wetted bound on the beach, berm line, and toe of the dune as identified in aerial photographs. These charted shorelines may be unlabeled or mislabeled.

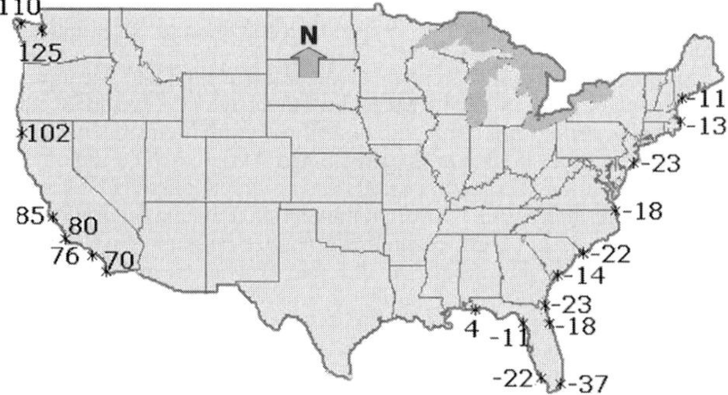

Figure T13 Height in centimeters of MSL datum above NAVD88 (redrawn from USACE, 2002).

Statements like "this chart uses the MLLW datum," do not imply that the shoreline is the MLLWL or any other tidal datum, only that printed soundings are referenced to MLLW. The shoreline could be the water's edge at an arbitrary height of the tide, a high water mark, or follow a geomorphic feature like a berm or toe of dune. Caution is advised not to infer too much from shoreline labels without consulting the description of the procedures used in production of that specific chart. Increased use of digital maps may exacerbate problems because chart developers can no longer anticipate the scale to which the user will take measurements from these zoomable products.

The label "HWL" (for high water line as distinct from MHWL) specifically acknowledges that the shoreline has been located less precisely than tidal-datum shorelines. Nevertheless, the HWL does approximate the MHWL, and is therefore considered "of primary importance in distinguishing the upland from the shore on charts" (Shalowitz, 1964). For practical reasons of determination, however, the surveyed HWL is based on observed physical features in the field (such as berm crest). Thus, the HWL stands above the MHWL by an uncertain distance related to wave runup and surveyor judgment.

Concluding guidance

Tidal datums are listed in Table T1 in the order of decreasing elevation from top to bottom. The vertical offset of orthometric and three-dimensional datums (lower portions of Table T1) from tidal datums varies spatially (e.g., Figure T13). Chart depths usually refer to a tidal datum (MLLW is the official chart datum to which depths are referred in the United States). Elevations on maps usually refer to an orthometric or a three-dimensional datum (NAVD 88 or NAD 83 for older maps in the United States). By convention, elevations are positive above datum; depths are positive below datum (i.e., an object 10 m below the water has an elevation of −10 m and a depth of +10 m).

In the 1990s, proliferation of personal computer software made it easy to convert between reference systems (see *Geographic Information Systems*). Caution is advised, however, when analyzing measurements made to different references. For example, the effectiveness of coastal erosion mitigation is often assessed by analyzing the difference between variable rates of shoreline change after project completion and a series of measurements made before project initiation. Consider a scrabbled mixture of HWLs and MHWLs. Early shorelines exist only as HWLs. The ratio of HWLs to MHWLs would tend to decrease as aerial surveying and GPS technology became pervasive. The horizontal distance separating a HWL from a MHWL is not recoverable after the HWL survey. Because shoreline positions based on tidal and orthometric datums are more precisely defined than the HWL, MHWL positions may appear more reliable (for analyzing shoreline change rates) than the HWLs that only approximate them. For just the years that have both a HWL and a MHWL, it might seem desirable to delete the less precise of the two measurements, but this would tend to bias the post-construction period and produce the appearance of unwarranted erosion reduction (i.e., deletion the HWLs would tend to displace the shoreline landward more frequently during later years). In most such cases, the more precisely located MHWLs might well be ignored to obtain an unbiased (though imprecise) estimate of changes in the rate of shore retreat.

Adoption of a new National Tidal Datum Epoch is pending. Previous updates have raised datums (reduced the elevations of benchmarks above tidal datums) by amounts that varied with the particular epoch transition, gauge location, and datum selection. For the MHW datum, an increase on the order of 0.25 m would not be unusual from one Epoch to the next. In areas of subsidence (Louisiana, the Netherlands, and portions of Texas, see *Coastal Subsidence*) or crustal uplift (e.g., Alaska, see *Changing Sea Levels*), datums could be raised or lowered on the order of a meter. Comparisons of shoreline positions documented during different Epochs can, therefore, be problematic. Adjustments can be made to estimate the shoreline positions for a common Epoch based on the vertical shift in tidal datum between two Epochs and the slope of the beach at these elevations. The adjusted results would only be as good as the knowledge of the beach slope used to translate the vertical shift into a shoreline displacement, but the results could be useful. Beaches (including the dune, berm, and longshore bars) generally migrate upward with the historic long-term rise of sea level. So, comparing unadjusted shorelines, referred to datums that shifted between epochs, can also be valid from the perspective of quantifying how the volume of beach sand changes in concert with sea-level changes. Whether to accept the inter-Epoch shoreline displacement or adjust the shorelines would depend on the time span, magnitude of sea-level change, and knowledge of slope changes. Analysis from both perspectives can be useful. For example, along a 50 km reach of Lake Michigan, the average shore retreat of 17.9 m (over a specific six years during which the mean lake level rose persistently) was decomposed into 14.5 m (at a fixed datum elevation) due to erosion and recession of the profile and 3.4 m due to submergence under a 0.39 m higher mean lake level (Figure 11, Hands, 1979).

For accuracy, charts and tables should clearly state the method of determining elevations and shorelines, coordinate systems, units of measurement, and any coordinate transform procedures employed. To maintain accuracies on the order of 0.1 m or better, be certain of and document which particular Epoch was used in the tidal datum definition.

Edward B. Hands

Bibliography

Hands, E.B., 1979. Changes in rates of shore retreat, Lake Michigan, 1967–76. *CERC TP 79-4*, Vicksburg, MS: US Army Engineer Research and Development Laboratory.

Hicks, S.D., 1989. *Tide and Current Glossary*. Washington, DC: US Department of Commerce.

Hicks, S.D., Hull, W.V., Weir, J.P., Long, E.E., and Hickman, L.E., Jr., 1988. The national tidal datum convention of 1980—the final push. *Proceeding, 3rd Biennial National Ocean Service International Hydrographic Conference*, Washington, DC: National Ocean Service/NOAA. pp. 121–128.

Kraus, N.C., and Rosati, J.D., 1997. Interpretation of shoreline-position data for coastal engineering analysis. *Coastal Engineering Technical Note II-39*, Vicksburg, MS: US Army Engineer Research and Development Laboratory.

Reed, M., 2000. *Shore and Sea Boundaries*, Vol. 3; *The Development of International Maritime Boundary Principles through United States Practice*. Washington, DC: US Department of Commerce.

Shalowitz, A.L., 1962. *Shore and Sea boundaries;* Vol. 1; *Boundary problems Associated with the Submerged Lands Cases and the Submerged Lands Acts*. Washington, DC: US Department of Commerce.

Shalowitz, A.L., 1964. *Shore and Sea Boundaries*; Vol. 2; *Interpretation and use of Coast and Geodetic Survey data*. Washington: US Department of Commerce.

USACE, 2002. *Hydrographic Surveying*, Engineer Manual 1110-2-1003, especially Chapter 5, *Tidal Datums*, Sections 5–7 through 5–9. Washington, DC: US Army Corps of Engineers.

Cross-references

Altimeter Surveys, Coastal Tides and Shelf Circulation
Changing Sea Levels
Coastal Changes, Gradual
Coastal Subsidence
Eustasy
Geodesy
Geographic Information Systems
Sea-Level Change During the Last Millennium
Tide Gauges
Tides

TIDAL ENVIRONMENTS

Definitions

A tidal environment is that part of a marine shore which is regularly submerged and exposed in the course of the rise and fall of the tide. Such environments exhibit particular physical and biological characteristics which, among others, play an important role in coastal dynamics, coastal ecology, coastal protection and engineering works, and integrated coastal zone management.

The coastal area affected by the ocean tides is known as the intertidal or eulittoral zone. Being a long-period wave, the tidal water level oscillates about a mean water level, which usually corresponds to the mean sea level. The vertical distance covered by the tide is known as the tidal range, whereas the part above or below the mean tide level is the tidal amplitude, which can hence have a positive or negative sign. In practice, a number of critical tide levels are distinguished on the basis of longer-term averages. Proceeding from high to low water levels, these are: MEHWS, mean equinoxial high water springs (highest astronomical tide); MHWS, mean high water springs; MHWN, mean high water neaps; MWL, mean water level (commonly mean sea level); MLWN, mean low water neaps; MLWS, mean low water springs; and MELWS, mean equinoxial low water springs (lowest astronomical tide). These mean tide levels are well correlated with various morphological and biological characteristics of tidal environments.

Besides the astronomical modulations, instantaneous tidal elevations can, in the short term, be substantially modified by water-level fluctuations induced by wind and/or wave set-up or set-down. The degree of wave and wind exposure can thus have a substantial influence on the nature of a tidal shore. Where such secondary and irregular fluctuations in coastal water levels are so frequent and strong that they completely mask the tidal signal, the coast is considered to be nontidal (e.g., the Baltic Sea).

In summary, tidal shores are highly variable environments which are not only influenced by the astronomically induced periodic rise and fall of the sea level, but also by numerous secondary processes. In combination, these factors define the local physical nature of a tidal environment (e.g., Davies, 1980; Allen and Pye, 1992; Allen, 1997; French, 1997).

Tidal forcing factors

The tides are essentially produced by the interactive forces exerted on the oceans by the sun and the moon. Since the motions of the sun and the earth–moon system are known with great precision, the tide-generating potential can be mathematically resolved into strictly periodic components. These components vary over time as the position of the sun and the moon relative to the position and orientation of the earth changes. The sum of all the tractive forces at any one time defines the total instantaneous potential. Doodson (1922) computed no less than 390 such harmonic components, of which about 100 are long period, 160 diurnal, 115 semidiurnal, and 14 one-third diurnal. Of these, only seven (i.e., four semidiurnal and three diurnal components) are of practical importance (e.g., McLellan, 1975). Indeed, only four of these are used to define the character of the tides around the globe (Figure T14), in which semidiurnal, diurnal, and two mixed tidal types, comprising a predominantly semidiurnal and a predominantly diurnal one, are distinguished (cf. Table T2). Since the tidal type has an important influence on the physical nature of tidal environments, the global distribution of these four types are illustrated in Figure T15.

As the sun, the earth, and the moon move along their elliptical orbits, they continually change their positions relative to each other. As a result, the total potential defining the height of the astronomical tide is modulated as a function of geographic location in the course of a day, a month, a year, and also on longer timescales. The most prominent tidal period is the fortnightly spring–neap cycle (synodic tide). Thus, spring tides coincide with full moon and new moon, whereas neap tides occur at the corresponding half-moon phases. These tidal ranges are further modulated in the course of a lunar month (anomalistic tides), the highest spring tides occurring when the moon is closest to the earth (perigee), the lowest when the moon is furthest away from the earth (apogee), the difference in distance amounting to roughly 13%. The solar component varies in similar manner between perihelion (currently coinciding with the winter solstice) and aphelion (currently coinciding with the summer solstice), the difference in distance between the two amounting to about 4%.

Another important astronomical feature influencing tidal environments on short timescales is the daily inequality of the tide (declinational tide). This feature results from the inclination of the earth's axis relative to the plane of the ecliptic, and hence the tidal bulge. As the earth rotates around its axis, the position of a geographic locality continually changes relative to the tidal bulge, alternating between a maximum and a minimum tidal elevation every 12.42 h, a feature which affects the elevations of both successive high tides and low tides. On this, the declination of the moon is superimposed which, in turn, causes a progressive change in the daily inequality over time. Thus, with the moon over the equator, successive tides are equal in height (equatorial tides), whereas towards the position of maximum declination successive tides become progressively more unequal in height, dividing into a "large tide" and a "half tide" (tropical tides). The solar tide shows a similar inequality, being zero at the equinoxes (spring and autumn) and largest at the solstices (summer and winter).

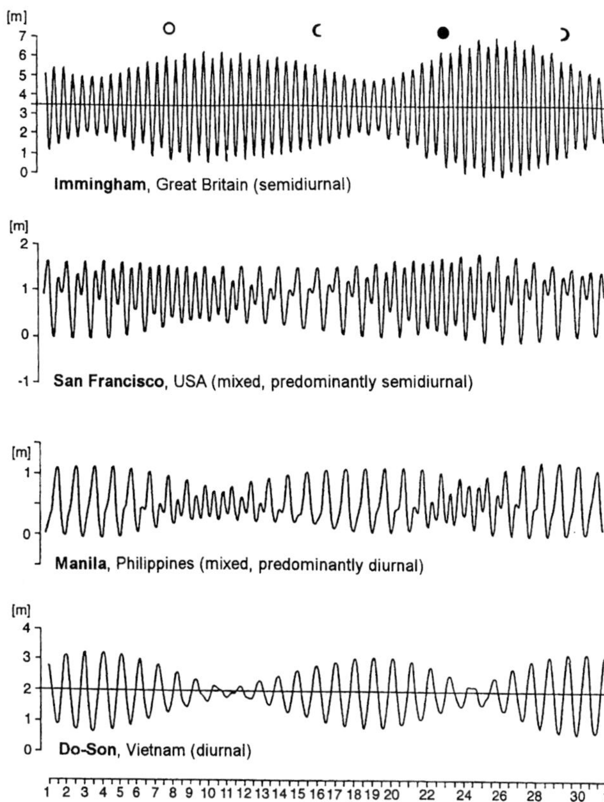

Figure T14 Selected examples of tidal curves from different geographic locations illustrating the main tidal types (adapted from Dietrich *et al.*, 1975). (A) semi-diurnal tides (e.g., Immingham, Great Britain); (B) mixed, predominantly semidiurnal (e.g., San Francisco, USA); (C) mixed, predominantly diurnal (e.g., Manila, Philippines); (D) diurnal (e.g., Do-Son, Vietnam).

The largest astronomical tides, that is, the greatest spring and lowest neap tides, in the course of a year occur at the vernal and autumnal equinoxes when the orbital path of the earth–moon system crosses the plane of the celestial equator (nodal points), that is, when the total potential of the tide-generating forces is largest because of their closest alignment.

In addition to these short-period astronomical cycles, the tide is also modulated by longer period phenomena which need to be considered when analyzing ancient tidal deposits (e.g., Archer *et al.*, 1991; Williams, 1991; Oost *et al.*, 1993; de Boer and Smith, 1994). Among these are the nutational motion or nodal cycle (≈18.6-year cycle) caused by the oscillation of the earth's axis about its mean position, the precessional motion (≈23,000-year cycle) of the earth's axis, the obliquity (≈41,000-year cycle) which defines the angle of inclination of the earth's axis between 22° and 24.8° (currently 23.5°), and the eccentricity (≈100,000-year cycle) controlling the rate of change in the elliptical radius of the earth's orbit around the sun. In addition, it has been shown that the length of the year has decreased from 420 to currently 365 days, while the length of the day has increased from 21 to currently 24 h in the course of the last 500 million years or so (e.g., Williams, 1991).

Besides the geographic variation in tidal type and tidal range resulting from astronomical modulations, the physical nature of a tidal environment is also influenced by numerous secondary factors. Among the more important of these are variations in tidal range with distance from the amphidromic point around which the tidal wave rotates, the Coriolis effect as a function of geographic location, the rate of change in water depth, the coastline configuration (plan shape and slope angle), as well as resonant effects resulting from the shape and depth of a tidal basin. Thus, at the center of an amphidrome the tidal range is considered to be zero, but it progressively increases in height with distance along the axis of the two opposing tidal bulges which rotate around the center offset by 180° or 6 h (tidal phases). The Coriolis force, in turn, forces the tidal wave to rotate clockwise in the Southern Hemisphere and anticlockwise

Table T2 The tidal character (F) as defined on the basis of the ratio between the sum of the lunisolar diurnal (K_1) and principle lunar diurnal (O_1) and the sum of the principle lunar semidiurnal (M_2) and principle solar semidiurnal (S_2) tidal components

$F = K_1 + O_1/M_2 + S_2$	Tidal character
0–0.25	Semidiurnal tides
0.25–1.5	Mixed, predominantly semidiurnal tides
1.5–3.0	Mixed, predominantly diurnal tides
>3.0	Diurnal tides

Source: Dietrich *et al.* (1975).

in the Northern Hemisphere, but this principle can be upset near the Equator. The direction in which the tidal wave propagates along a shore thus depends entirely on the geographic location of the coast. Where tidal waves rotating around neighboring amphidromes meet, the tidal water motion can even be perpendicular to the coast.

In addition, the amplification of the tidal wave and the tidal type are strongly influenced by the configuration of the coastline and the rate of shoaling, the effect of this interaction being illustrated in Figures T15 and T16. With decreasing water depth (h) the length of the tidal wave (λ) progressively decreases proportionally in the form $\lambda \propto h^{0.5}$. When propagating into funnel-shaped estuarine water bodies, the initial increase in tidal range due to convergence of the opposite shores is eventually compensated for by friction along the seabed as a result of which the tidal wave gradually decreases in height. This interplay between friction and convergence is proportionally related in the form $a_0 \propto b^{-0.5} * h^{-0.25}$ where a_0 is the amplitude of the tidal wave (m), b is the width (m), and h is the water depth (m) at a particular location along the estuary. This relationship can take on complicated patterns and in nature three basic modes of tidal wave propagation can be distinguished. Thus, in the case where friction dominates over convergence the tidal range progressively decreases in height up-estuary (hypersynchronic mode). In the opposite case, the tidal range increases in height (hyposynchronic mode), and where friction and convergence are in balance the tidal range remains constant (synchronic mode). Even neighboring estuaries will exhibit quite different modes of tidal wave propagation if they differ in shape and water depth (e.g., Borrego *et al.*, 1995). In many estuarine environments a combination of two or even all three modes can be observed.

In some cases, the entrance channel of an estuary or lagoon is so narrow and shallow that the propagation of the tidal wave is "choked" with the effect that the tidal range is dramatically reduced. This filtering mechanism, expressed by the so-called coefficient of repletion (K), has the numerical form of $K = (T/2a_0\pi) * (A_c/A_b) * [2a_0g/(1 + 2gln^2r^{-4/3})]^{0.5}$ where T is the tidal period, $2a_0$ is the tidal range (a_0 being the tidal amplitude), l is the length of the entrance channel, A_c is the cross-sectional channel area, A_b is the surface area of the water body, r is the hydraulic radius of the channel, g is the gravitational acceleration, and n is the Manning's friction (0.01–0.10 s m$^{-1/3}$). This coefficient of repletion controls reductions in tidal range, phase shifts between ocean and lagoonal tides, non-sinusoidal variations of the lagoonal tides, and flow exchanges between the ocean and the estuarine or lagoonal water bodies (e.g., Kjerfve and Magill, 1989).

A final forcing factor which may affect the behavior of tidal waves in shallow water is resonance, a feature associated with standing waves. Clearly, a standing wave superimposed on a normal tidal wave would dramatically affect the physical nature of a tidal environment. In this context, we distinguish two types. In the case of a half-wave oscillator or seiche, the length of the water body is half the wavelength of the standing wave. The fundamental period (T) of a seiche is defined as $T = 2l/(gh)^{0.5}$ where l is the length of the water body, h is the water depth, and g is the gravitational acceleration. Seiches are particularly common in lakes and marginal seas, where they are forced by wind stress, and hence of less importance in tidal environments. However, quarter-wave oscillators come into operation where the lengths of open-ended elongate gulfs or deep estuaries together with adjacent bays correspond to a quarter of the tidal wavelength ($l = 0.25T * (gh)^{0.5}$). In this case, the period $T = 4l/(gh)^{0.5}$ with notations as above. If this condition is fulfilled, or nearly so, tidal amplification can be quite considerable (e.g., Bay of Fundy, Canada).

Coastal classification by tidal type and range

Besides classifying the world's coastline according to tidal type (Figure T15), coastal tidal environments are also classified on the basis of tidal range, the scheme of Davies (1964, 1980) having been the most

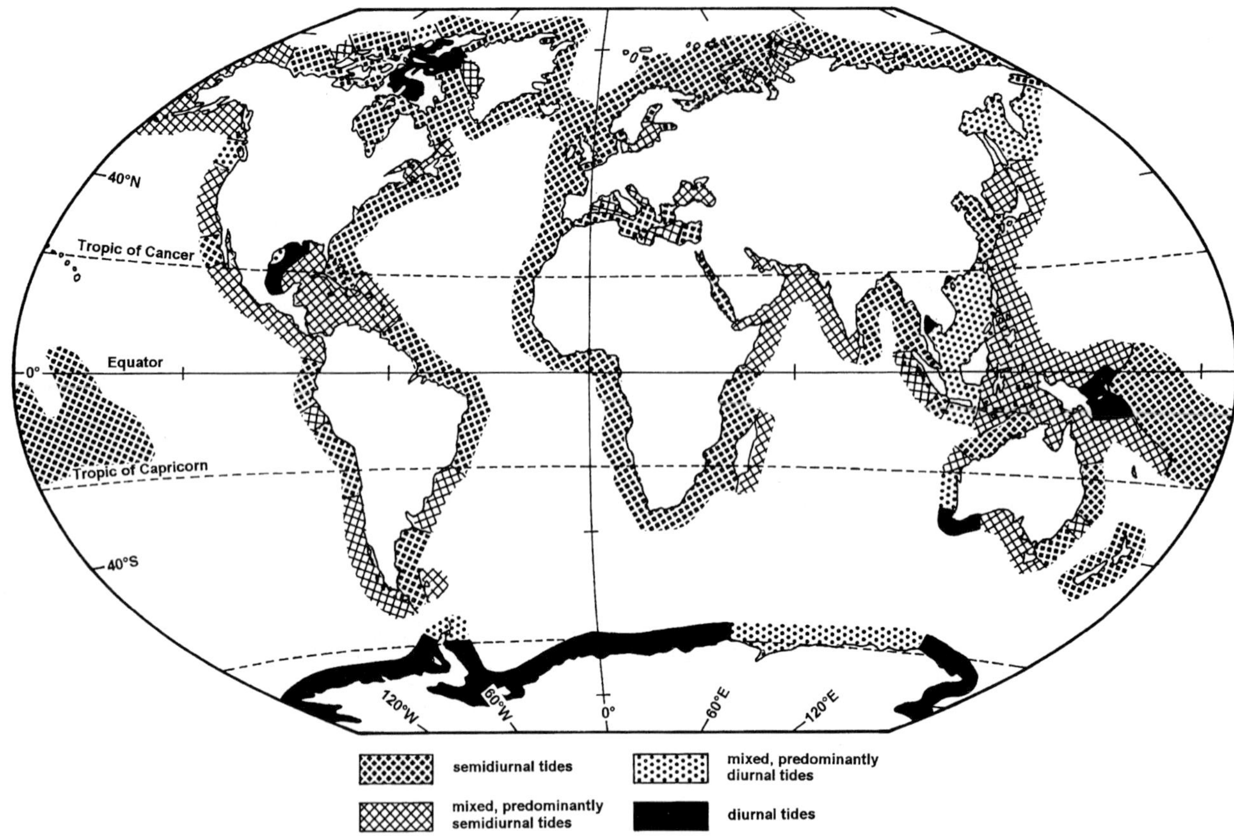

Figure T15 Global distribution of the main tidal types (adapted from Davies, 1980). Note that the transitions between semidiurnal and diurnal tidal types (here represented by so-called mixed tides) are progressive and not abrupt.

Table T3 Contrasting two existing classification schemes of tidal shores on the basis of tidal range

Davies (1964)		Hayes (1979)	
Tidal range (m)	Class name	Tidal range (m)	Class name
<2.0	Microtidal	<1.0	Microtidal
2.0–4.0	Mesotidal	1.0–2.0	Lower mesotidal
>4.0	Macrotidal	2.0–3.5	Upper mesotidal
		3.5–5.5	Lower macrotidal
		>5.5	Upper macrotidal

widely applied to date (Table T3). However, with only three subdivisions, this rather arbitrary approach prevents differentiation where it is most needed, that is, near the lower and upper limits of the potential tidal regime. For example, the Gulf of Mexico coast with an average tidal range of only 0.5 m is very different in character from the west coast of southern Africa where the tidal range averages at 1.6 m, yet both are classified as microtidal. In contrast to this, a more pragmatic classification, comprising five subdivisions (see Table T2), has been proposed by Hayes (1979). This latter scheme takes distinct, process-related geomorphic features into consideration, for example, the upper limit of barrier island occurrence at a tidal range of 3.5 m, which hence marks the transition between upper mesotidal and lower macrotidal regimes in this classification. To contrast the two classification schemes, the global pattern of coastal subdivision using the latter scheme is presented here for the first time (Figure T16).

Rocky versus sandy and muddy tidal environments

The coastlines of the world can be divided into four basic types, namely rocky shores, sandy shores, muddy shores, and bio-shores. In all cases, the character of a particular shore reflects the interaction between the substrate, the local wave climate, the tides, and the biology (e.g., Newell, 1979; Davies, 1980; Raffaeli and Hawkins, 1996). Geographic location, which controls climatic influences and biological species composition (e.g., Chapman, 1974), and the Holocene evolution of a coast (e.g., Bird and Schwartz, 1985) being important additional factors to consider. Rocky shores occupy the smallest overall area because the intertidal zone is commonly narrow as compared with the other shore types. Shore processes have received considerable attention by engineers for constructional purposes (e.g., Horikawa, 1989), whereas biologists have long been intrigued by the distinct faunal and floral zonation patterns along tidal gradients which evidently reflect high degrees of adaptation to the intertidal environment and the overprinting effects of competition between specific organisms (e.g., Lewis, 1972; Raffaeli and Hawkins, 1996). However, the interplay between biological and physical factors in defining zonation patterns is still not well enough understood to allow accurate predictions to be made (e.g., Delafontaine and Flemming, 1989). Different tidal environments are also characterized by different biogeochemical processes due to different climates, substrates, and biological community structures (Alongi, 1998).

The biological zonation pattern observed along the rocky shores of Great Britain as a function of tidal gradient and the degree of exposure to wave action is illustrated in Figure T17 (adapted from Lewis, 1972; Raffaeli and Hawkins, 1996). This basic scheme can be applied to most rocky shores of the world, the only difference being the species composition and distribution, a good example from subtropical Bermuda being shown in Thomas (1985). Important to note here is that the tidal gradient in Figure T17 is relative, expanding or contracting proportional to the tidal range.

In contrast to rocky shores, into which bio-shores can also be included, sandy and muddy tidal shores can attain shore normal extensions of many kilometers (e.g., Davis, 1994; Flemming and Hertweck, 1994). French (1997) distinguishes no less than seven intertidal coastal types based mainly on morphology and facies successions. The final transition between land and sea along sheltered coasts is characterized by sharp boundaries separating different floral zones, the typical pattern observed along the barrier island shore of the southern North Sea being illustrated in Figure T18 (adapted from Streif, 1990).

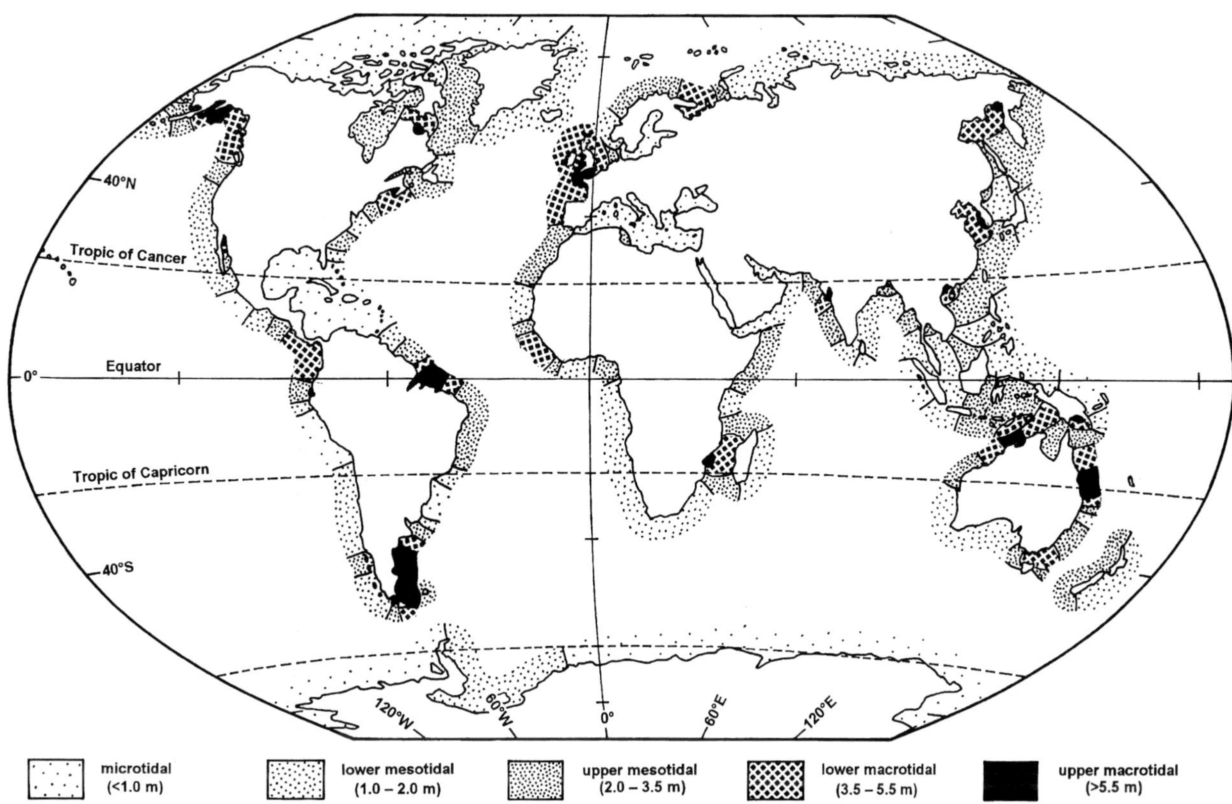

Figure T16 Global distribution of tidal shores based on tidal range according to the classification scheme of Hayes (1979).

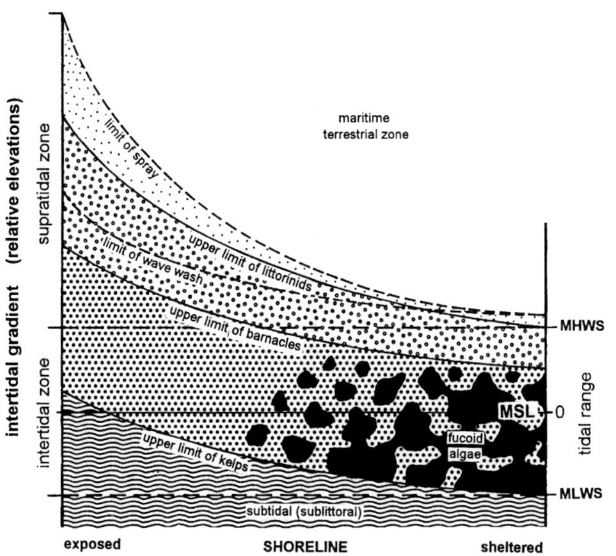

Figure T17 Zonation patterns along rocky tidal shores as a function of exposure to wave action (adapted from Lewis, 1972; Raffaeli and Hawkins, 1996). Note that the scaling is relative to any particular observed tidal gradient.

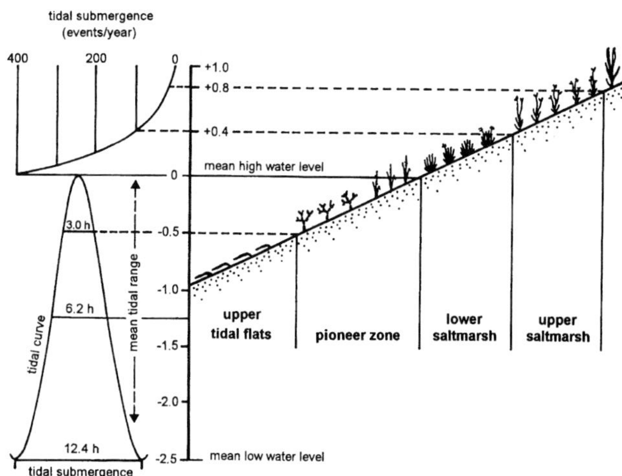

Figure T18 Floral zonation pattern observed at the transition between upper intertidal flats and the terrestrial environment of the Wadden Sea (southern North Sea) as a function of tidal elevation and the frequency of tidal submergence in the course of a year (adapted from Streif, 1990).

In this example the zones are separated by the frequency of tidal submergence in the course of the year which is controlled by the elevation along the tidal gradient. In principle, this basic pattern should be applicable the world over, individual transition levels being dependent on local floral associations, the tidal range, the seasonal wave climate, and the difference in elevation between mean high tide levels at spring and neap tide (e.g., Chapman, 1974; Lugo and Snedaker, 1974). A systematic investigation of the factors controlling the upper and lower limits of occurrence of *Spartina anglica* relative to mean sea level along the coast of the United Kingdom has generated quantitative relationships between the spring tidal range, the fetch available for wave generation and propagation, the area of the tidal basin, and in the case of the upper limit also the geographic location (Gray, 1992). Thus, the lower limit (L_l) is defined as $L_l = -0.805 + 0.366R_s + 0.053F + 0.135 * \log_e A_b$, where R_s is the spring tidal range (m), F is the fetch in the direction of the transect (km), and A_b is the area of the tidal basin (km^2). The upper limit, by contrast, is defined as $L_u = 4.74 + 0.483R_s + 0.068F - 0.199L°$, where $L°$ is the degree North Latitude expressed as a decimal. The correlation coefficients of $r = 0.97$ for L_l and $r = 0.95$ for L_u demonstrate the predictive potential of this approach.

A comprehensive overview of physical and biological processes active along sandy tidal shores is provided in McLachlan and Erasmus (1983).

Outlook

Many features of tidal environments are still poorly understood. Among these are the quasi-periodic, decadal to subdecadal fluctuations in the elevation of mean high tide and mean low tide levels. Being a worldwide phenomenon, one might assume that they result from variations in the astronomical factors defining the tidal potential. A clear correlation, however, is still lacking. As far as sandy tidal environments are concerned, accurate sediment budgets and transport pathways have remained elusive problems whose solution becomes more pressing in view of the predicted acceleration in sea-level rise. The distinction between strictly local features and others of global relevance requires more attention. A number of other unresolved issues have been addressed in the text.

B. W. Flemming

Bibliography

Allen, J.R.L., and Pye, K. (eds.), 1992. *Salt marshes, Morphodynamics, Conservation and Engineering Significance*. Cambridge: University Press.

Allen, P.A., 1997. *Earth Surface Processes*. Oxford: Blackwell Science.

Alongi, D.M., 1998. *Coastal Ecosystem Processes*. Boca Raton: CRC Press.

Archer, A.W., Kvale, E.P., and Johnson, H.R., 1991. Analysis of modern equatorial tidal periodicities as a test of information encoded in ancient tidal rhythmites. In Smith, D.G., Reinson, G.E., Zaitlin, B.A., and Rahmani, R.A. (eds.), Clastic tidal sedimentology. *Canadian Society of Petroleum Geologist, Memoirs*, **16**: 189–196.

Bird, E.C., and Schwartz, M.L. (eds.), 1985. *The World's Coastline*. New York: Van Nostrand Reinhold.

Borrego, J., Morales, J.A., and Pendon, J.G., 1995. Holocene estuarine facies along the mesotidal coast of Huelva, south-western Spain. In Flemming, B.W., and Bartholomä, A. (eds.), *Tidal Signatures in Modern and Ancient Sediments*. Oxford: Blackwell Science. *Special Publication International Association of Sedimentologists*, **24**: 151–170.

Chapman, V.J., 1974. *Salt Marshes and Salt Deserts of the World*, 2nd edn. Lehre (Germany): Cramer.

Davies, J.L., 1964. A morphogenetic approach to world shorelines. *Zeitschrift für Geomorphologie*, **8**: 127–142.

Davies, J.L., 1980. *Geographical Variation in Coastal Development*. Geomorphology Texts 4. London: Longman.

Davis, R.A., Jr. (ed.), 1994. *Geology of Holocene Barrier Island Systems*. Berlin: Springer.

de Boer, P.L., and Smith, D.G. (eds.), 1994. *Orbital Forcing and Cyclic Sequences*. Spec. Publ. Int. Ass. Sediment. No. 19. Oxford: Blackwell Science.

Delafontaine, M.T., and Flemming, B.W., 1989. Physical factors in barnacle community structure: a conceptual model. In Ros, J.D. (ed.), *Topics in Marine Biology. Scientia Marina*, **53**: 405–410.

Dietrich, G., Kalle, K., Krauss, W., and Siedler, G., 1975. *Introductory Oceanography*, 3rd edn. in German. Berlin: Gebr. Borntraeger.

Doodson, A.T., 1922. The harmonic development of the tide-generating potential. *Proceedings of the Royal Society of London*, **A100**: 305–329.

Flemming, B.W., and Hertweck, G. (eds.), 1994. Tidal flats and barrier systems of continental Europe: a selective overview. *Senckenbergiana maritima*, **24**: 1–209.

French, P.W., 1997. *Coastal and Estuarine Management*. London: Routledge.

Gray, A.J., 1992. Salt marsh plant ecology: zonation and succession revisited. In Allen, J.R.L., and Pye, K. (eds.), *Salt Marshes*. Cambridge: Cambridge University Press, pp. 63–79.

Hayes, M.O., 1979. Barrier island morphology as a function of tidal and wave regime. In Leatherman, S.P. (ed.), *Barrier Islands*. New York: Academic Press, pp. 1–27.

Horikawa, 1989. *Nearshore Dynamics and Coastal Processes: Theory, Measurement, and Predictive Models*. Tokyo: University of Tokyo Press.

Kjerfve, B., and Magill, K.E., 1989. Geographic and hydrodynamic characteristics of shallow coastal lagoons. *Marine Geology*, **88**: 197–199.

Lewis, J.R., 1972. *The Ecology of Rocky Shores*. London: The English University Press.

Lugo, A.E., and Snedaker, S.C., 1974. The ecology of mangroves. *Annual Review of the Ecology and Systematics*, **5**: 39–64.

McLachlan, A., and Erasmus, T. (eds.), 1983. *Sandy Beaches as Ecosystems*. Developments in Hydrobiology 19. The Hague: Dr. W. Junk Publishers.

McLellan, H.J., 1975. *Elements of Physical Oceanography*. Oxford: Pergamon Press.

Newell, R.C., 1979. *Biology of Intertidal Animals*. Faversham (UK): Marine Ecological Surveys Ltd.

Oost, A.P., de Haas, H., Ijnsen, F., van den Boogert, J.M., and de Boer, P.L., 1993. The 18.6 year nodal cycle and its impact on tidal sedimentation. *Sedimentary Geology*, **87**: 1–11.

Raffaeli, D., and Hawkins, S., 1996. *Intertidal Ecology*. London: Chapman & Hall.

Streif, H., 1990. *The East Frisian Coast. North Sea, Islands, Wadden Sea, and Marshes* (in German). Berlin: Borntraeger.

Thomas, M.L.H., 1985. Littoral community structure and zonation on the rocky shores of Bermuda. *Bulletin of Marine Science*, **37**: 857–870.

Williams, G.E., 1991. Upper Proterozoic tidal rhythmites, South Australia: sedimentary features, deposition, and implications for the earth's paleorotation. In Smith, D.G., Reinson, G.E., Zaitlin, B.A., and Rahmani, R.A. (eds.), Clastic Tidal Sedimentology. *Canadian Society of Petroleum Geologists, Memoir*, **16**: 161–178.

Cross-references

Barrier Islands
Beach Processes
Bioerosion
Classification of Coasts (see Holocene Coastal Geomorphology)
Estuaries
Littoral
Microtidal Coasts
Rock Coast Processes
Sandy Coasts
Tidal Flats
Tides
Wave-Dominated Coasts

TIDAL INLETS

Introduction

Tidal inlets are found along barrier coastlines throughout the world. They provide a pathway for ships and small boats to travel between the open ocean to sheltered waters. Along many coasts, including much of the East and Gulf Coasts of the United States, the only safe harborages, including some major ports, are found behind barrier islands. The importance of inlets in providing navigation routes to these harbors is demonstrated by the large number of improvements that are performed at the entrance to inlets such as the construction of jetties and breakwaters, dredging of channels, and the operation of sand bypassing facilities.

Diversity in the morphology, hydraulic signature, and sediment transport patterns of tidal inlets attests to the complexity of their processes. The variability in oceanographic, meteorologic, and geologic parameters, such as tidal range, wave energy, sediment supply, storm magnitude, and frequency, freshwater influx, and geologic controls, and the interactions of these factors, are responsible for this wide range in tidal inlet settings.

What is a tidal inlet

A tidal inlet is defined as an opening in the shore through which water penetrates the land thereby proving a connection between the ocean and bays, lagoons, and marsh and tidal creek systems. Tidal currents maintain the main channel of a tidal inlet.

The second half of this definition distinguishes tidal inlets from large, open embayments or passageways along rocky coasts. Tidal currents at inlets are responsible for the continual removal of sediment dumped into the main channel by wave action. Thus, according to this definition, tidal inlets occur along sandy or sand and gravel barrier coastlines, although one side may abut a bedrock headland. Some tidal inlets coincide with the mouths of rivers (estuaries) but in these cases inlet dimensions and sediment transport trends are still governed, to a large extent, by the volume of water exchanged at the inlet mouth and the reversing tidal currents, respectively.

At most inlets over the long term, the volume of water entering the inlet during the flooding tide equals the volume of water leaving the inlet during the ebbing cycle. This volume is referred to as the tidal prism. The tidal prism is a function of the open water area and tidal range in the backbarrier as well as frictional factors, which govern the ease of flow through the inlet.

Inlet morphology

A tidal inlet is specifically the area between two barriers or between the barrier and the adjacent bedrock or glacial headland. Commonly, the recurved ridges of spits, consisting of sand that was transported toward the back-barrier by refracted waves and flood-tidal currents form the sides of the inlet. The deepest part of an inlet, the inlet throat, is located normally where spit accretion of one or both of the bordering barriers constricts the inlet channel to a minimum width and minimum cross- sectional area. Here, tidal currents normally reach their maximum velocity. Commonly, the strength of the currents at the throat causes sand to be removed from the channel floor leaving behind a lag deposit consisting of gravel or shells or in some locations exposed bedrock or indurated sediments.

Tidal deltas

Closely associated with tidal inlets are sand shoals and tidal channels located on the landward and seaward sides of the inlets. Flood-tidal currents deposit sand landward of the inlet forming flood-tidal deltas and ebb-tidal currents deposit sand on the seaward side forming an ebb-tidal delta.

Flood-tidal delta. Their presence or absence, size, and development are related to a region's tidal range, wave energy, sediment supply, and back-barrier setting. Tidal inlets that are backed by a system of tidal channels and salt marsh (mixed-energy coast) usually contain a single horseshoe-shaped flood-tidal delta (i.e., Essex River Inlet, Massachusetts; Figure T19). Contrastingly, inlets that are backed by large shallow bays may contain multiple flood-tidal deltas. Along some microtidal coasts, such as in Rhode Island, flood deltas form at the end of narrow inlet channels cut through the barrier. Changes in the locus of deposition at these deltas produce a multi-lobate morphology resembling a lobate river delta (Boothrdyd *et al.*, 1985). Flood delta size commonly increases as the amount of open water area in the backbarrier increases. In some regions, flood deltas have become colonized and altered by marsh growth, and are no longer recognizable as former flood-tidal deltas. At other sites, portions of flood-tidal deltas are dredged to provide navigable waterways and thus are highly modified.

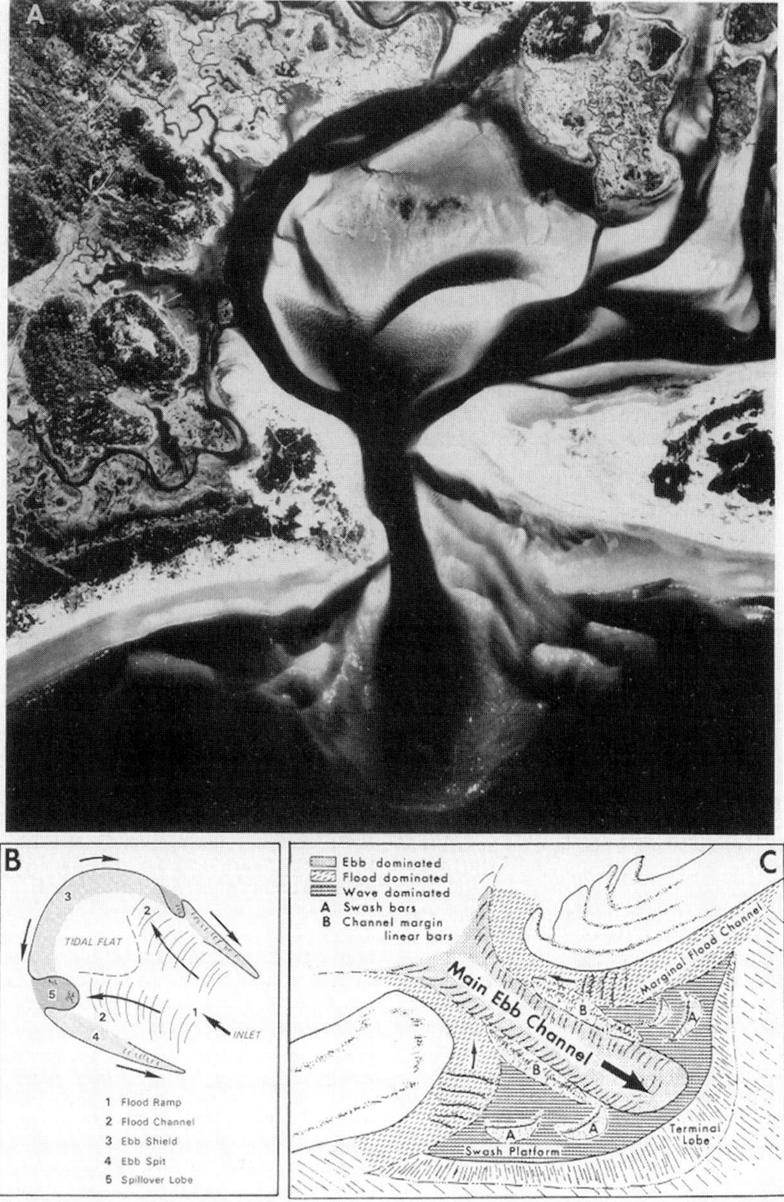

Figure T19 (A) Vertical aerial photograph of Essex River Inlet, Massachusetts with well-developed flood- and ebb-tidal deltas. (B) Flood-tidal delta model (after Hayes, 1975). (C) Ebb-tidal delta model (after Hayes, 1975).

Flood-tidal deltas are best revealed in areas with moderate to large tidal ranges (1.5–3.0 m) because in these regions they are well exposed at low tide. As tidal range decreases, flood deltas become largely subtidal shoals. Most flood-tidal deltas have similar morphologies consisting of the following components (Hayes, 1975, 1979).

1. *Flood ramp.* This is a landward shallowing channel that slopes upward toward the intertidal portion of the delta. Strong flood-tidal currents and landward sand transport in the form of landward-oriented sand-waves dominate the ramp.
2. *Flood channels.* The flood ramp splits into two shallow flood channels. Like the flood ramp, these channels are dominated by flood-tidal currents and flood-oriented sand waves. Sand is delivered through these channels onto the flood delta.
3. *Ebb shield.* It defines the highest and landwardmost part of the flood delta and may be partly covered by marsh vegetation. It shields the rest of the delta from the effects of the ebb-tidal currents.
4. *Ebb spits.* These spits extend from the ebb shield toward the inlet. They form from sand that is eroded from the ebb shield and transported back toward the inlet by ebb-tidal currents.
5. *Spillover lobes.* These are lobes of sand that form where the ebb currents have breached through the ebb spits or ebb shield depositing sand in the interior of the delta.

Through time, some flood-tidal deltas accrete vertically and/or grow in size. This is evidenced by an increase in areal extent of marsh grasses, which require a certain elevation above mean low water to exist. At migrating inlets new flood-tidal deltas are formed as the inlet moves along the coast and encounters new open water areas in the backbarrier. At most stable inlets, however, sand comprising the flood delta is simply recirculated. The transport of sand on flood deltas is controlled by the elevation of the tide and the strength and direction of the tidal currents. During the rising tide, flood currents reach their strongest velocities near high tide when the entire flood-tidal delta is covered by water. Hence, there is a net transport of sand up the flood ramp, through the flood channels and onto the ebb shield. Some of the sand is moved across the ebb shield and into the surrounding tidal channel. During the falling tide, the strongest ebb currents occur near mid to low water. At this time, the ebb shield is out of the water and diverts the currents around the delta. The ebb currents erode sand from the landward face of the ebb shield and transport it along the ebb spits and eventually into the inlet channel where once again it will be moved onto the flood ramp thus completing the sand gyre.

Ebb-tidal delta. This is an accumulation of sand that has been deposited by the ebb-tidal currents and which has been subsequently modified by waves and tidal currents. Ebb deltas exhibit a variety of forms dependent on the relative magnitude of wave and tidal energy of the region as well as geological controls. Along mixed energy coasts, most ebb-tidal deltas contain the same general features including:

1. *Main ebb channel.* This is a seaward shallowing channel that is scoured into the ebb-tidal delta sands. It is dominated by ebb-tidal currents.
2. *Terminal lobe.* Sediment transported out the main ebb channel is deposited in a lobe of sand forming the terminal lobe. The deposit slopes relatively steeply on its seaward side. The outline of the terminal lobe is well defined by breaking waves during storms or periods of large wave swell at low tide.
3. *Swash platform.* This is a broad shallow sand platform located on both sides of the main ebb channel, defining the general extent of the ebb delta.
4. *Channel margin linear bars.* These are bars that border the main ebb channel and sit atop the swash platform. These bars tend to confine the ebb flow and are partially exposed at low tide.
5. *Swash bars.* Waves breaking over the terminal lobe and across the swash platform form arcuate-shaped swash bars that migrate onshore. The bars are usually 50–150 m long, 50 m wide, and 1–2 m in height.
6. *Marginal-flood channels.* These are shallow channels 0–2 m deep at mean low water located between the channel margin linear bars and the onshore beaches. The channels are dominated by flood-tidal currents.

Ebb-tidal delta morphology

The general shape of an ebb-tidal delta and the distribution of its sand bodies tell us about the relative magnitude of different sand transport processes operating at a tidal inlet. Ebb-tidal deltas that are elongate with a main ebb channel and channel margin linear bars that extend far offshore are tide-dominated inlets. Wave-generated sand transport plays a secondary role in modifying delta shape at these inlets. Because most sand movement is in onshore–offshore direction, the ebb-tidal overlaps a relatively small length of inlet shore. This has important implications concerning the extent to which the inlet shore undergoes erosional and depositional changes.

Wave-dominated inlets tend to be small relative to tide-dominated inlets. Their ebb-tidal deltas are driven onshore, close to the inlet mouth by the dominant wave processes. Commonly, the terminal lobe and/or swash bars form a small arc outlying the periphery of the delta. In many cases, the ebb-tidal delta of these inlets is entirely subtidal. In other instances, sand bodies clog the entrance to the inlet leading to the formation of several major and minor tidal channels.

At mixed energy tidal inlets the shape of the delta is the result of tidal and wave processes. These deltas have a well-formed main ebb channel, which is a product of ebb-tidal currents, their swash platform and sand bodies substantially overlap the inlet shore many times the width of the inlet throat due to wave processes and flood-tidal currents.

Ebb-tidal deltas may also be highly asymmetric such that the main ebb channel and its associated sand bodies are positioned primarily along one of the inlet shores. This configuration normally occurs when the major backbarrier channel approaches the inlet at an oblique angle or when preferential accumulation of sand on the updrift side of the ebb delta causes a deflection of the main ebb channel along the downdrift barrier shore.

Tidal inlet formation

The formation of a tidal inlet requires the presence of an embayment and the development of barriers. In coastal plain settings, the embayment or backbarrier was often created through the construction of the barriers themselves, like much of the East Coast of the United States or the Friesian Island coast along the North Sea. In other instances, the embayment was formed due to rising sea level inundating an irregular shore during the late Holocene. The embayed or indented shore may have been a rocky coast such as that of northern New England and California or it may have been an irregular unconsolidated sediment coast such as that of Cape Cod in Massachusetts or parts of the Oregon coast. The flooding of former river valleys has also produced embayments associated with tidal inlet development.

Breaching of a barrier

Rising sea level, exhausted sediment supplies, and human influences have led to thin barriers that are vulnerable to breaching. The breaching process normally occurs during storms after waves have destroyed the foredune ridge and storm waves have overwashed the barrier depositing sand aprons (washovers) along the backside of the barrier. Even though this process may produce a shallow overwash channel, seldom are barriers cut from their seaward side. In most instances, the breaching of a barrier is the result of the storm surge heightening waters in the backbarrier bay. When the level of the ocean tide falls, the elevated bay waters flow across the barrier toward the ocean gradually incising the barrier and cutting a channel. If subsequent tidal exchange between the ocean and bay is able to maintain the channel, a tidal inlet is established. The breaching process is enhanced when offshore winds accompany the falling tide and if an overwash channel is present to facilitate drainage across the barrier (Fisher, 1962). Many tidal inlets that are formed by this process are ephemeral and may exist for less than a year, especially if stable inlets are located nearby. Barriers most susceptible to breaching are long and thin and wave-dominated.

Spit building across a bay

The development of a tidal inlet by spit construction across an embayment usually occurs early in the evolution of a coast. The sediment to form these spits may have come from erosion of the nearby headlands, discharge from rivers, or from the landward movement of sand from inner shelf deposits. Most barriers along the coast of the United States and elsewhere in the world are 3,000–5,000 years old coinciding with a deceleration of rising sea level. It was then that spits began enclosing portions of the irregular rocky coast of New England, the West Coast, parts of Australia, and many other regions of the world. As a spit builds across a bay, the opening to the bay gradually decreases in width and in cross-sectional area. It may also deepen. Coincident with the decrease in size of the opening is a corresponding increase in tidal flow. The tidal prism of the bay remains approximately constant, so as the opening gets smaller, the current velocities must increase. A tidal inlet is formed as the spit reaches a stable configuration.

Drowned river valleys

Tidal inlets have formed at the entrance to drowned river valleys due to the growth of spits and the development of barrier islands which have served to narrow the mouths of the estuaries. It has been shown through stratigraphic studies, particularly along the East Coast of the United States, that in addition to drowned river valleys, many tidal inlets are positioned in paleo river valleys in which there is no river leading to this site today (Halsey, 1979). These are old river courses that were active during the Pleistocene when sea level was lower and they were migrating across the exposed continental shelf. Tidal inlet become situated in these valleys because tidal currents easily remove the sediment filling the valleys.

Tidal inlet migration

Some tidal inlets have been stable since their formation, whereas others have migrated long distances along the shore. In New England and along other glaciated coasts, stable inlets are commonly anchored next to bedrock outcrops or resistant glacial deposits. Along the California coast most tidal inlets have formed by spit construction across an embayment with the inlet becoming stabilized adjacent to a bedrock headland. In coastal plain settings, stable inlets are commonly positioned in former river valleys. One factor that appears to separate migrating inlets from stable is the depth to which the inlet throat has eroded. Deeper inlets are often situated in consolidated sediments that resist erosion. The channels of shallow migrating inlets are eroded into sand.

Although the vast majority of tidal inlets migrate in the direction of dominant longshore transport, there are some inlets that migrate updrift (Aubrey and Speer, 1984). In these cases, the drainage of backbarrier tidal creeks control flow through the inlet. When a major backbarrier tidal channel approaches the inlet at an oblique angle, the ebb-tidal currents coming from this channel are directed toward the margin of the inlet throat. If this is the updrift side of the main channel, then the inlet will migrate in that direction. This is similar to a river where strong currents are focused along the outside of a meander bend causing erosion and channel migration. Inlets that migrate updrift are usually small to moderately sized and occur along coasts with small to moderate net sand longshore transport rates.

Tidal inlet relationships

Tidal inlets throughout the world exhibit several consistent relationships that have allowed coastal engineers and marine geologists to formulate predictive models: (1) Inlet throat cross-sectional area is closely related to tidal prism, and (2) Ebb-tidal delta volume is a function of tidal prism.

Inlet throat area–tidal prism relationship

The size of a tidal inlet is tied closely to the volume of water going through it (Figure T20(A); O'Brien, 1931, 1969). Although inlet size is primarily a function of tidal prism, to a lesser degree inlet cross-sectional area is also affected by the delivery of sand to the inlet channel. For example, at jettied inlets tidal currents can more effectively scour sand from the inlet channel and therefore they maintain a larger throat cross section than would be predicted by the O'Brien Relationship.

Similarly, for a given tidal prism, Gulf Coast inlets have larger throat cross sections than Pacific Coast inlets. This is explained by the fact that wave energy is greater along the West Coast and therefore the delivery of sand to these inlets is higher than at Gulf Coast inlets. Jarrett (1976) has improved the tidal prism-inlet cross-sectional area regression equation for US inlets, separating, into three classes the low-energy Gulf Coast inlets, moderate-energy East Coast inlets, and higher-energy West Coast inlets. Even better correlations are achieved when structured inlets are distinguished from natural inlet.

Variability. It is important to understand that the dimensions of the inlet channel are not static but rather the inlet channel enlarges and contracts slightly over relatively short-time periods (<1 year) in response to changes in tidal prism, variations in wave energy, effects of storms, and other factors. For instance, the inlet tidal prism can vary by more than 30% from neap to spring tides due to increasing tidal ranges. Consequently, the size of the inlet varies as a function of tidal phases. Along the southern Atlantic Coast of the United States water temperatures may fluctuate seasonally by 35–40°F. This and other factors cause the surface coastal waters to expand, raising mean sea level by 30 cm or more. In the summer and early fall, when mean sea level reaches its highest seasonal elevation, spring tides may flood backbarrier surfaces that normally are above tidal inundation. This produces larger tidal prisms, stronger tidal currents, increased channel scour, and larger inlet cross-sectional areas. At some Virginia inlets this condition increases the inlet throat by 5–15% (Byrne *et al.*, 1975). Longer-term (>1 year) changes in the cross section of inlets are related to inlet migration, sedimentation in the backbarrier, morphological changes of the ebb-tidal delta, and human influences.

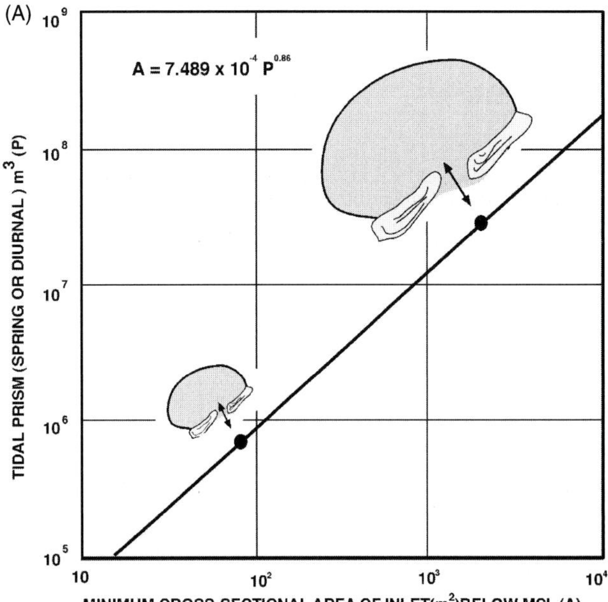

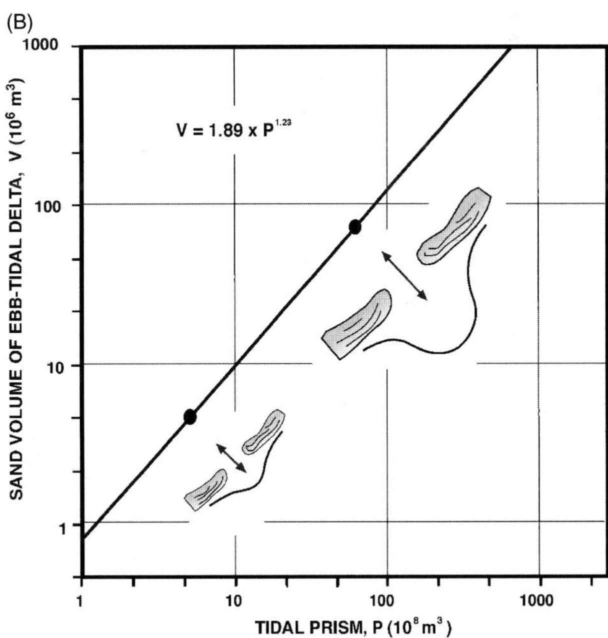

Figure T20 (A) Graph depicting relationship between tidal prism and inlet throat cross-sectional area (data after O'Brien, 1969). (B) Graph showing correspondence between tidal prism and volume of the ebb-tidal delta (data after Walton and Adams, 1976).

Ebb-tidal delta volume–tidal prism relationship

In the mid-1970s, Walton and Adams (1976) showed that the volume of sand contained in the ebb-tidal delta is closely related to the tidal prism (Figure T20(B)). Walton and Adams also showed that the relationship was improved slightly when wave energy was taken into account in a manner similar to Jarrett's divisions. Waves are responsible for transporting sand back onshore thereby reducing the volume of the ebb-tidal delta. Therefore, for a given tidal prism, ebb-tidal deltas along the West Coast contain less sand than do equal sized inlets along the Gulf or East Coast.

Variability. The Walton and Adams Relationship works well for inlets all over the world. However, field studies have shown that the volume of sand comprising ebb-tidal deltas changes through time due to the effects of storms, changes in tidal prism, or processes of inlet sediment bypassing (FitzGerald *et al.*, 1984). When sand is moved past a tidal inlet, it is commonly achieved by large bar complexes migrating from the ebb delta and attaching to the landward inlet shoreline. These large bars may contain more than 300,000 m^3 of sand and represent more than 10% of sediment volume of the ebb-tidal delta (FitzGerald, 1988; Gaudiano and Kana, 2000).

Sand transport patterns

The movement of sand at a tidal inlet is complex due to reversing tidal currents, effects of storms, and interaction with the longshore transport system. The inlet contains short- and long-term reservoirs of sand varying from the relatively small sandwaves flooring the inlet channel that migrate meters each tidal cycle to the large flood-tidal delta shoals where some sand is recirculated but the entire deposit may remain stable for hundreds or even thousands of years. Sand dispersal at tidal inlets is complicated because in addition to the onshore–offshore movement of sand produced by tidal and wave-generated currents, there is constant delivery of sand to the inlet and transport of sand away from the inlet produced by the longshore transport system. In the discussion below the patterns of sand movement at inlets are described including how sand is moved past a tidal inlet.

General sand dispersal trends

The ebb-tidal delta has segregated areas of landward versus seaward sediment transport that are controlled primarily by the way water enters and discharges from the inlet as well as the effects of wave-generated currents. During the ebbing cycle the tidal flow leaving the backbarrier is constricted at the inlet throat causing the currents to accelerate in a seaward direction. Once out of the confines of the inlet, the ebb flow expands laterally and the velocity slows. Sediment in the main ebb channel is transported in a net seaward direction and eventually deposited on the terminal lobe due to this decrease in current velocity. One response to this seaward movement of sand is the formation of ebb-oriented sandwaves having heights of 1–2 m.

In the beginning of the flood cycle, the ocean tide rises while water in the main ebb channel continues to flow seaward as a result of momentum. Due to this phenomenon, water initially enters the inlet through the marginal flood channels, which are the pathways of least resistance. The flood channels are dominated by landward sediment transport and are floored by flood-oriented bedforms. On both sides of the main ebb channel, the swash platform is most affected by landward flow produced by the flood-tidal currents and breaking waves. As waves shoal and break, they generate landward flow, which augments the flood-tidal currents but retards the ebb-tidal currents. The interaction of these forces acts to transport sediment in net landward direction across the swash platform. In summary, at many inlets there is a general trend of seaward sand transport in the main ebb channel, which is countered by landward sand transport in the marginal flood channels and across the swash platform.

Inlet sediment bypassing

Along most open coasts, particularly in coastal plain settings, angular wave approach causes a net movement of sediment, which along much of the East Coast of the United States varies from 100 to 200,000 m^3/yr. The manner whereby sand moves past tidal inlets and is transferred to the downdrift shore is called inlet sediment bypassing. The primary mechanisms of sand bypassing natural inlets include: (l) Stable inlet processes, (2) Ebb-tidal delta breaching, and (3) Inlet migration and spit breaching. One of the end products in all the different mechanisms is the landward migration and attachment of large bar complexes to the inlet shore. Discussion of this topic can be found in FitzGerald (1982) and FitzGerald *et al.* (2000).

Stable inlet processes. This mechanism of sediment bypassing occurs at inlets that do not migrate and whose main ebb channels remain approximately in the same position. Sand enters the inlet by: (1) wave action along the beach, (2) flood-tidal and wave-generated currents through the marginal flood channel, and (3) waves breaking across the channel margin linear bars. Most of the sand that is dumped into the main channel is transported seaward by the dominant ebb-tidal currents and deposited on the terminal lobe.

At lower tidal elevations waves breaking on the terminal lobe transport sand along the periphery of the delta toward the landward beaches in much the same way as sand is moved in the surf and breaker zones along beaches. At higher tidal elevations waves breaking over the terminal lobe create swash bars on both sides of the main ebb channel. The swash bars (50–150 m long, 50 m wide) migrate onshore due to the dominance of landward flow across the swash platform. Eventually, they attach to channel margin linear bars forming large bar complexes. Bar complexes tend to parallel the beach and may be more than a kilometer in length. They are fronted by a slipface (25–33 degrees), which may be up to 3 m in height.

The stacking and coalescence of swash bars to form a bar complex is the result of the bars slowing their onshore migration as they move up the nearshore ramp. As the bars gain a greater intertidal exposure, the wave bores, which cause their migration onshore, act over an increasingly shorter period of the tidal cycle. Thus, their rate of movement onshore decreases. Eventually the entire bar complex migrates onshore and welds to the upper beach. When a bar complex attaches to the downdrift inlet shore, some of this newly accreted sand is then gradually transported by wave action to the downdrift beaches, thus completing the inlet sediment bypassing process. It should be noted that some sand bypasses the inlet independent of the bar complex. In addition, some of the sand comprising the bar reenters the inlet via the marginal flood channel and along the inlet shoreline.

Ebb-tidal delta breaching. This means of sediment bypassing occurs at inlets with a stable throat position, but whose main ebb channels migrate through their ebb-tidal deltas like the wag of a dog's tail. Sand enters the inlet in the same manner as described above for *Stable inlet processes*. However, at these inlets the delivery of sediment by longshore transport produces a preferential accumulation of sand on the updrift side of the ebb-tidal delta. The deposition of this sand causes a deflection of the main ebb channel until it nearly parallels the downdrift inlet shore. This circuitous configuration of the main channel results in inefficient tidal flow through the inlet, ultimately leading to a breaching of a new channel through the ebb-tidal delta. The process normally occurs during spring tides or periods of storm surge when the tidal prism is very large. In this state the ebb discharge piles up water at the entrance to the inlet where the channel bends toward the downdrift inlet shoreline. This causes some of the tidal waters to exit through the marginal flood channel or flow across low regions on the channel margin linear bar. Gradually over several weeks or convulsively during a single large storm, this process cuts a new channel through the ebb delta thereby providing a more direct pathway for tidal exchange through the inlet. As more and more of the tidal prism is diverted through the new main ebb channel, tidal discharge through the former channel decreases causing it to fill with sand.

The sand that was once on the updrift side of the ebb-tidal delta and which is now on the downdrift side of the new main channel is moved onshore by wave-generated and flood-tidal currents. Initially, some of this sand aids in filling the former channel while the rest forms a large bar complex that eventually migrates onshore and attaches to the downdrift inlet shore. The ebb-tidal breaching process results in a large packet of sand bypassing the inlet. Similar to the stable inlets discussed above, some sand bypasses these inlets in a less dramatic fashion, grain by grain on a continual basis.

It is noteworthy that at some tidal inlets the entire main ebb channel is involved in the ebb-tidal delta breaching process, whereas at others just the outer portion of main ebb channel is deflected. In both cases, the end product of the breaching process is a channel realignment that more efficiently conveys water through the inlet, as well as sand being bypassed in the form of a bar.

Inlet migration and spit breaching. A final method of inlet sediment bypassing occurs at migrating inlets. In this situation, an abundant sand supply and a dominant longshore transport direction cause spit building at the end of the barrier. To accommodate spit construction, the inlet migrates by eroding the downdrift barrier shore. Along many coasts as the inlet is displaced further along the downdrift shore, the inlet channel to the backbarrier lengthens retarding the exchange of water between the ocean and backbarrier. This condition leads to large water level differences between the ocean and bay, making the barrier highly susceptible to breaching, particularly during storms. Ultimately, when the barrier spit is breached and a new inlet is formed in a hydraulically more favorable position, the tidal prism is diverted to the new inlet and the old inlet closes. When this happens, the sand comprising the ebb-tidal delta of the former inlet is transported onshore by wave action commonly taking the form of a landward migrating bar complex. It should be noted that when the inlet shifts to a new position along the updrift shore a large quantity of sand, has effectively bypassed the inlet. The frequency of this inlet sediment bypassing process is dependent on inlet size, rate of migration, storm history, and backbarrier dynamics.

Bar complexes. Depending on the size of the inlet, the rate of sand delivery to the inlet, the effects of storms, and other factors, the entire process of bar formation, its landward migration, and its attachment to the downdrift shore may take from 6 to 10 years. The volume of sand bypassed can range from 100,000 to over 1,000,000 m^3. The bulge in the shore that is formed by the attachment of a bar complex is gradually eroded and smoothed as sand is dispersed to the downdrift shore and transported back toward the inlet.

In some instances, a landward migrating bar complex forms a salt water pond as the tips of the arcuate bar weld to the beach stabilizing its onshore movement. Although the general shape of the bar and pond may be modified by overwash and dune building activity, the overall shore morphology is frequently preserved. Lenticular-shaped coastal ponds or marshy swales become diagnostic of bar migration processes and are common features at many inlets.

Tidal inlet effects on adjacent shorelines

In addition to the direct consequences of spit, accretion and inlet migration are the effects of volume changes in the size of ebb-tidal deltas, sand losses to the backbarrier, processes of inlet sediment bypassing, and wave sheltering of the ebb-tidal delta shoals (FitzGerald, 1988).

Number and size of tidal inlets

The degree to which barrier shore are influenced by tidal inlet processes is dependent on their size and number. As the O'Brien Relationship demonstrates, the size or cross-sectional area of an inlet is governed by its tidal prism. This concept can be expanded to include an entire barrier chain in which the size and number of inlets along a chain are primarily dependent on the amount of open water area behind the barrier and the tidal range of the region. In turn, these parameters are a function of other geological and physical oceanographic factors. Wave-dominated, microtidal coasts tend to have long barrier islands and few tidal inlets and mixed energy coasts have short stubby barriers and numerous tidal inlets (Hayes, 1975, 1979). Presumably, the mesotidal conditions produce larger tidal prisms than along microtidal coasts, which necessitate more holes in the barrier chain to let the water into and out of the backbarrier. Many coastlines follow this general trend but there are many exceptions due to the influence of sediment supply, large versus small bay areas, and other geological controls (Davis and Hayes, 1984).

Tidal inlets as sediment sinks

Tidal inlets not only trap sand temporarily on their ebb-tidal deltas, but they also are responsible for the longer-term loss of sediment moved into the backbarrier. At inlets dominated by flood-tidal currents, sand is continuously transported landward enlarging flood-tidal deltas and building bars in tidal creeks. Sand can also be transported into the backbarrier of ebb-dominated tidal inlets during severe storms. During these periods increased wave energy produces greater sand transport to the inlet channel. At the same time the accompanying storm surge increases the water surface slope at the inlet resulting in stronger than normal flood tidal currents. The strength of the flood currents coupled with the high rate of sand delivery to the inlet results in landward sediment transport into the backbarrier. Along the Malpeque barrier system in the Gulf of St. Lawrence, New Brunswick it has been determined that during a 33 year period 90% of the sand transferred to the backbarrier took place at tidal inlets and at former inlet locations along the barrier (Armon, 1979).

Sediment may also be lost at migrating inlets when sand is deposited as channel fill. If the channel scours below the base of the barrier sands, then the beach sand, which fills this channel, will not be replaced entirely by the deposits excavated on the eroding portion of the channel. Because up to 40% of the length of barriers is underlain by tidal inlet fill deposits ranging in thickness from 2 to 10 m (Moslow and Heron, 1978; Moslow and Tye, 1985) this volume represents a large, long-term loss of sand from the coastal sediment budget. Another major process producing sand loss at migrating inlets is associated with the construction of recurved spits that build into the backbarrier. For example, along the East Friesian Islands recurved spit development has caused the lengthening of barriers along this chain by 3–11 km since 1650. During this stage of barrier evolution the large size of the tidal inlets permitted ocean waves to transport large quantities of sand around the end of the barrier forming recurves that extend far into the backbarrier. Due to the size of the recurves and the length of barrier extension, this process has been one of the chief natural mechanisms of bay infilling (FitzGerald and Penland, 1987).

Changes in ebb-tidal delta volume

Ebb-tidal deltas represent huge reservoirs of sand that may be comparable in volume to that of the adjacent barrier islands along mixed-energy coasts (i.e., northern East and West Friesian Islands, Massachusetts, southern New Jersey, Virginia, South Carolina, and Georgia). For instance, the ebb-tidal delta volume of Stono and North Edisto Inlets in South Carolina is 197×10^6 m^3 and the intervening Seabrook-Kiawah Island barrier complex contains 252×10^6 m^3 of sand (Hayes *et al.*, 1976). In this case, the deltas comprise 44% of the sand in the combined inlet-barrier system. The magnitude of sand contained in ebb-tidal deltas suggests that small changes in their volume dramatically affect the sand supply to the landward shorelines.

A transfer of sand from the barrier to the ebb-tidal delta takes place when a new tidal inlet is opened, such as the formation of Ocean City Inlet when Assateague Island, Maryland was breached during the 1933 Hurricane. Initially, the inlet was only 3 m deep and 60 m across but quickly widened to 335 m when it was stabilized with jetties in 1935. Since the inlet formed more than 14 million cubic meters of sand have been deposited on the ebb-tidal delta. Trapping the southerly longshore movement of sand by the north jetty and growth of the ebb-tidal delta have led to serious erosion along the downdrift beaches. The northern end of Assateague Island has been retreating at an average rate of 11 m per year. The rate of erosion lessened when the ebb tidal delta reached an equilibrium volume and the inlet began to bypass sand (Stauble and Cialone, 1996).

In contrast to the cases discussed above, the historical decrease in the inlet tidal prisms along the East Friesian Islands has had a beneficial effect on this barrier coast. From 1650 to 1960 the reclamation of tidal flats and marshlands bordering the German mainland as well as natural processes, such as the building and landward extension of recurved spits, decreased the size of the backbarrier by 80%. In turn, the reduction in bay area decreased the inlet tidal prisms, which led to smaller sized inlets, longer barrier islands, and smaller ebb-tidal deltas. Wave action transported ebb-tidal delta sands onshore as tidal discharge decreased. This process increased the supply of sand to the beaches and aided in lengthening of the barriers (FitzGerald *et al.*, 1984).

Wave sheltering

The shallow character of ebb-tidal deltas provides a natural breakwater for the landward shore. This is especially true during lower tidal elevations when most of the wave energy is dissipated along the terminal lobe. During higher tidal stages intertidal and subtidal bars cause waves to break offshore expending much of their energy before reaching the beaches onshore. The sheltering effect is most pronounced along mixed-energy coast where tidal inlets have well-developed ebb-tidal deltas.

The influence of ebb shoals is particularly well illustrated by the history of Morris Island, South Carolina that forms the southern border of Charleston Harbor. Before human modification, the entrance channel to the harbor paralleled Morris Island and was fronted by an extensive shoal system. In the late 1800s jetties were constructed at the harbor entrance to straighten, deepen, and stabilize the main channel. During the period prior to jetty construction (1849–80) Morris Island had been eroding at an average rate of 3.5 m/yr. After the jetties were in place the shoals eroded and gradually diminished in size, so did the protection they afforded Morris Island, especially during storms. From 1900 to 1973 Morris Island receded 500 m at its northeast end increasing to 1,100 m at its southeast end, a rate three times what it had been prior to jetty construction (FitzGerald, 1988).

Effects of inlet pediment bypassing

Tidal inlets interrupt the wave-induced longshore transport of sediment along the coast, affecting both the supply of sand to the downdrift beaches and the position and mechanisms whereby sand is transferred to the downdrift shores. The effects of these processes are exhibited well along the Copper River Delta barriers in the Gulf of Alaska. From east to west along the barrier chain the width of the tidal inlets increases as does the size of the ebb-tidal deltas (Hayes, 1979). In this case, the width of the inlet can be used as a proxy for the inlet's cross-sectional area. These trends reflect an increase in tidal prism along the chain, which is caused by an increase in bay area from east to west while tidal range remains constant. Also quite noticeable along this coast is the greater downdrift offset of the inlet shore in a westerly direction. This morphology is coincident with an increase in the degree of overlap of the ebb-tidal delta along the downdrift inlet shoreline. The offset of the inlet shore and bulbous shape of the barriers are produced by sand being trapped at the eastern, updrift end of the barrier. The amount of shore progradation (build out) is a function of inlet size and extent of its ebb-tidal delta. What we learn from the sedimentation processes along the Copper River Delta barriers is that tidal inlets can impart a very important signature on the form of the barriers (FitzGerald, 1996).

Drumstick barrier model. In an investigation of barrier islands shores in mixed energy settings throughout the world, Hayes (1979) noted that many barriers exhibit a drumstick barrier island shape. In this model, the meaty portion of the drumstick barrier is attributed to waves bending around the ebb-tidal delta producing a reversal in the longshore transport direction. This process reduces the rate at which sediment bypasses the inlet, resulting in a broad zone of sand accumulation along the updrift end of the barrier. The downdrift, or thin part of the drumstick, is formed through spit accretion. Later studies demonstrated that landward-migrating bar complexes from the ebb-tidal delta determine barrier island morphology and overall shore erosional–depositional trends, particularly in mixed energy settings.

Studies of the Friesian Islands demonstrate that inlet processes exert a strong influence on the dispersal of sand and in doing so dictate barrier form (FitzGerald *et al.*, 1984). In addition to drumsticks, the East Friesian Islands exhibit many other shapes. Inlet sediment bypassing along this barrier chain occurs, in part, through the landward migration of large swash bars (>1 km in length) that deliver up to 300,000 m^3 of sand when they weld to the beach. In fact, it is the position where the bar complexes attach to the shore that determines the form of the barrier along this coast. If the ebb-tidal delta greatly overlaps the downdrift barrier, then the bar complexes may build up the barrier shore some distance from the tidal inlet, forming *humpbacked barriers*. If the downdrift barrier is short and the ebb-tidal delta fronts a large portion of the downdrift barrier, then bar complexes weld to the downdrift end of the barrier forming *downdrift bulbous barriers*.

Human influences

Dramatic changes to inlet beaches can also result from human influences including the obvious consequences of jetty construction that reconfigures an inlet shore. By preventing or greatly reducing an inlet's ability to bypass sand, the updrift beach progrades while the downdrift beach, whose sand supply has been diminished or completely cut off, erodes. There can also be more subtle human impacts that can equally affect inlet shores, especially those associated with changes in inlet tidal prism, sediment supply, and the longshore transport system. Nowhere are these types of impacts better demonstrated than along the central Gulf Coast of Florida where development has resulted in the construction of causeways, extensive backbarrier filling and dredging projects, and the building of numerous engineering structures along the coast. A detailed study of this region by Barnard and Davis (1999) has revealed that since the late-1980s 17 inlets have closed along this coast and at least five closures can be traced to human influences caused primarily by changes in inlet tidal prism. For example, access to several barriers has been achieved through the construction of causeways that extend from the mainland across the shallow bays. Along most of their lengths the causeways are dike-like structures that partition the bays, thereby changing bay areas and inlet tidal prisms. In some instances, tidal prisms were reduced to a critical value causing inlet closure. At these sites, the tidal currents were unable to remove the sand dumped into the inlet channel by wave action. Similarly, when the Intracoastal Waterway was constructed along the central Gulf Coast of Florida in the early 1960s, the dredged waterway served to connect adjacent backbarrier bays thereby changing the volume of water that was exchanged through the connecting inlets. The Intracoastal Waterway lessened the flow going through some inlets while at the same time increased the tidal discharge of others. This resulted in the closure of some inlets and the enlargement of others (Barnard and Davis, 1999).

Duncan M. FitzGerald

Bibliography

Armon, J.W., 1979. Landward sediment transfers in a transgressive barrier island system, Canada. In Leatherman, S.P. (ed.), *Barrier Islands: From the Gulf of St. Lawrence to the Gulf of Mexico*. New York: Academic Press, pp. 65–80.

Aubrey, D.G., and Speer, P.E., 1984. Updrift migration of tidal inlets. *Journal of Geology*, **92**: 531–546.

Barnard, P.L., and Davis, R.A., 1999. Anthropogenic versus natural influences on inlet evolution: West-Central Florida. *Proceedings, Coastal Sediments '99, ASCE*, pp. 1489–1504.

Boothroyd, J.C., Friedrich, N.E., and McGinn, S.R., 1985. Geology of microtidal coastal lagoons, RI. In Oertel, G.F., and Leatherman, S.P. (ed.), *Barrier Islands. Marine Geology*, **63**: 35–76.

Byrne, R.J., Bullock, P., and Taylor, D.G., 1975. Response characteristics of a tidal inlet: a case study. In Cronin, L.E. (ed.), *Estuarine Research*, Vol. 2. New York: Academic Press, pp. 201–216.

Davis, R.A., Jr., and Hayes, M.O., 1984. What is a wave-dominated coast? *Marine Geology*, **60**: 313–329.

Fisher, J.J., 1962. Geomorphic expression of former inlets along the Outer Banks of North Carolina, unpub. Masters thesis. Chapel Hill: University of North Carolina.

FitzGerald, D.M., 1982. Sediment bypassing at mixed energy tidal inlets. *Proceedings 18th Coastal Engineering Conference, ASCE*, pp. 1094–1118.

FitzGerald, D.M., 1988. Shoreline erosional–depositional processes associated with tidal inlets. In Aubrey, D.G., and Weishar, L. (eds.), *Hydrodynamics and Sediment Dynamics of Tidal Inlets.* Berlin: Springer, pp. 186–225.

FitzGerald, D.M., 1996. Geomorphic variability and morphologic and sedimentological controls on tidal inlets. In Mehta, A.J. (ed.), *Understanding Physical Proc.* at Tidal Inlets. *Journal of Coastal Research* (Special Issue), **23**: 47–71.

FitzGerald, D.M., Kraus, N.C., and Hands, E.B., 2000. Natural mechanisms of sediment bypassing at tidal inlets, ERDC/CHL-IV-Vicksburg, MS: US Army Engineer Research and Development Center.

FitzGerald, D.M., and Penland, S., 1987. Backbarrier dynamics of the East Friesian Island. *Journal of Sedimentary Petrology,* **57**: 746–754.

FitzGerald, D.M., Penland, S., and Nummedal, D., 1984. Control of barrier island shape by inlet sediment bypassing: East Friesian Islands, West Germany. *Marine Geology*, **60**: 355–376.

Gaudiano, D.J., and Kana, T.W., 2000. Shoal bypassing in South Carolina tidal inlets: geomorphic variables and empirical predictions for nine mesoscale inlets. *Journal of Coastal Research*, **17**: 280–291.

Halsey, S., 1979. Nexius: a new model of barrier island development. In Leatherman, S.P. (ed.), *Barrier Islands: From the Gulf of St. Lawrence to the Gulf of Mexico*. New York: Academic Press, pp. 185–210.

Hayes, M.O., 1975. Morphology of sand accumulations in estauries. In Cronin, L.E. (ed.), *Estuarine Research*, Vol. 2. New York: Academic Press, pp. 3–22.

Hayes, M.O., 1979. Barrier island morphology as a function of tidal and wave regime. In Leatherman, S.P. (ed.), *Barrier Islands: From the Gulf of St. Lawrence to the Gulf of Mexico*. New York: Academic Press, pp. 1–28.

Hayes, M.O., FitzGerald, D.M., Humes, L.J., and Wilson, S.J., 1976. *Geomorphology of Kiawah Island, South Carolina*, Columbia: Coastal Research Division, University of South Carolina, pp. 80–100.

Jarrett, J.T., 1976. Tidal prism–inlet area relationships. Vicksburg, MS: Us Army Corps of Engineers, Waterways Experiment Station, GITI Report No. 3.

Moslow, T.F., and Heron, S.D., 1978. Relict inlets: preservation and occurrence in the Holocene stratigraphy of southern Core Banks, North Carolina. *Journal of Sedimentary Petrology*, **48**: 1275–1286.

Moslow, T.F., and Tye, R.S., 1985. Recognition and characteristic of Holocene tidal inlet sequences. *Marine Geology*, **63**: 129–151.

O'Brien, M.P., 1931. Estuary tidal prisms related to entrance areas. *Civil Engineering*, **1**: 738–739.

O'Brien, M.P., 1969. Equilibrium flow areas of inlets on sandy coasts. *Journal of Waterways, Harbors, and Coastal Engineering ASCE*, **95**: 43–55.

Stauble, D.K., and Cialone, M.A., 1996. Ebb shoal evolution and sediment management techniques Ocean City Inlet, Maryland. *Procedings 9th National Conference on Beach Nourishment*, St. Petersburg, FL, pp. 209–224.

Walton, T.L., and Adams, W.D., 1976. Capacity of inlet outer bars to store sand. *Proceedings of 15th Coastal Engineering Conference*, ASCE, Honolulu, Hawaii, pp. 1919–1937.

Cross-references

Bars
Barrier Islands
Bypassing at Littoral Drift Barriers
Coasts, Coastlines, Shores, and Shorelines
Longshore Sediment Transport
Shore Protection Structures
Spits
Tidal Prism
Tide-Dominated Coasts
Wave-and-Tide Dominated Coasts
Wave-Dominated Coasts

TIDAL FLATS

Introduction and definition

Tidal flats are low-gradient tidally inundated coastal surfaces. Jackson (1997) defines them as extensive, nearly horizontal, marshy or barren tracts of land alternately covered and uncovered by the tide, and consisting of unconsolidated sediment. Tidal flats may be muddy, sandy, gravelly, or covered in shell pavements, and compositionally they may be underlain by siliciclastic or carbonate sediments. Depending on climate, tidal level, substrate and salinity, tidal flats may be covered biologically in parts by salt marsh, mangroves, sea grass, algal mats, microbial mats, biofilms, as well as mussel beds, oyster beds and reefs, and worm-tube beds and reefs, and inhabited by a burrowing benthos of molluscs, polychaetes, and crustacea.

Tidal flats have been of great interest to sedimentologists and stratigraphers as coastal systems that are readily accessible to sampling and study, and rich in processes and products resulting from oceanographic, sedimentologic, geohydrologic, hydrochemical, and biotic interactions (Ginsburg, 1975; Klein, 1976; Alexander *et al.*, 1998; Black *et al.*, 1998). They contrast with other steeper gradient wave-dominated sedimentary coasts, such as sandy beaches, composed dominantly of sand, and with a relatively limited biota, because with their generally lower energy conditions, and less scope for physical reworking, tidal flats develop a profusion of natural history coastal features. For instance, there are the sedimentologic products of interactions between waves and tides (e.g., cross-laminated sand, ripple-laminated sand, lenticular bedding, flazer bedding, laminated mud, ripple-laminated silt in clay), the products of interactions between sediments and biota (e.g., various burrow forms zoned tidally across the shore, various types of root-structuring, skeletal remains related to tidal levels), the geomorphic products of tides (e.g., tidal runoff on low gradient slopes to form meandering tidal creeks), and the products of hydrochemical interactions with sediments (e.g., dissolution of carbonate by acidic pore water; cemented crusts and their breccia and intraclast derivatives; carbonate nodules; gypsum precipitates; and products of redox reactions such as biologically mediated precipitation of iron sulfide). For stratigraphers and students of sedimentary rocks, identifying tidal flats in the geologic record is often an important step in the reconstruction of paleoenvironments, the location of facies associated with coastlines, and the recognition of such markers in stratigraphic sequences in basin analyses. Tidal flat signatures derived from studies of modern environments provide important analogs in such analyses.

Coastal settings of tidal flats

Tidal flats around the globe occur in a variety of regional geomorphic settings (Table T4 and Figure T21). Since they are surfaces exposed and inundated by tides, they may simply be part of larger coastal systems, that is, the shores of deltas, estuaries, lagoons, gulfs, bays, straits, rias, sounds, and cuspate forelands. Alternatively, they may be the sole coastal form developed along an open coast or broad embayment, or may comprise wholly tidal lagoons leeward of barriers. The best-developed tidal flats occur along estuarine coasts, protected embayments, or barred lagoons, where the shore slopes are gentle due to sediment accretion, and tides are large. Along many coasts, tidal flats are part of prograded shores (Kendall and Skipwith, 1968; Thompson, 1968; Hagan and Logan, 1975; Reineck and Singh, 1980); but in some instances, they may comprise modern sediment veneers on wave or tidally cut unconformities on rock or Pleistocene sediment, or earlier Holocene sediments (Semeniuk, 1981).

Tides and tidal levels

The tidal ranges that expose tidal flats vary globally from less than 1 m to ca. 15 m amplitude, and are diurnal (one tide daily), semidiurnal (two tides daily), or mixed (two tides daily, but with inequality between tide maxima and tide minima across the day). Over a lunar cycle, tides vary from a lower amplitude neap range (during quarter and three-quarter moon phases) to a higher amplitude spring range (during new and full moon phases). Higher than normal tides occur during equinoctial periods, and in response to the Lunar Nodal Periodicity. As a result, and depending on shore slope, tidal flat width may vary from being a narrow coastal strip, to being broad and expansive coastal forms.

Part of the coast emergent during low tide and submerged during high tide is the *intertidal* zone. That part of the coast permanently submerged below the low-water line is the *subtidal* zone. That occurring above the zone of high-tide inundation is the *supratidal* zone. Some authors consider the "supratidal zone" as the zone above the mean high-water line but sometimes under the water during extremely high tides, or even spring tides, but it is preferable to refer to all gently inclined surfaces and terrain *above* the highest tides as supratidal, and to treat all surfaces flooded by both neap, spring, and equinoctial spring tides as intertidal, and to separate these various tidal zones and levels.

Tidal ranges have been classed by Davies (1980) into three groups: *microtidal* <2 m, *mesotidal* 2–4 m, and *macrotidal* >4 m. While this classification has been generally accepted, large tidal ranges >8 m, might be further classed as *extreme macrotidal*. Generally, tides are microtidal along open oceanic coasts, and tend towards macrotidal where tides are semidiurnal. Tidal range amplification also may occur due to bay geometry and coastal constriction. For example, the Bay of Fundy, in Nova Scotia, because of its basin geometry, amplifies the tide from ca. 5.4 m at entrance to the bay to 15 m at its head.

Zones across the tidal flats are best exhibited in macrotidal settings, where there are marked distinctions in slope, sediments, and biology between the interval of spring tidal and neap tidal range. On microtidal flats, these various differences related to tidal levels are less pronounced.

Various levels within a tidal flat, often delineated by biological and/or sediment zones, can be distinguished as follows:

- low tidal flats—exposed by the mean and extreme low spring tides, generally underlain by sand, and vegetation free,
- middle tidal flats—the flats and low gradient slopes centered around mean sea level, exposed and inundated by neap tides; the upper parts of these flats may be vegetated by samphire in temperate latitude areas, and by mangrove in tropical latitudes,
- high tidal flats—inundated by the mean and extreme high spring tides, generally underlain by mud, and vegetated by salt marsh or mangrove, or in more arid settings, vegetation free and salt encrusted (salt flat).

Typical cross sections through some microtidal to extreme macrotidal flats are shown in Figure T22.

Tidal flats and their particle sizes and sediment composition

Tidal flats may be underlain by mud, sand, rock gravel, and shell pavements, or mixtures of these. Often, where all particle sizes are present, there is a zonation of sediment types across the flats, or an interlayering at a specific tidal level, but in many instances, one sediment type may dominate across the entire tidal flat. This partitioning of sediments across the tidal flat lends itself to a classification of tidal flats, or zones within tidal flats, according to particle size. For example, those composed wholly of mud may be termed muddy tidal flats, and those composed wholly of sand are sandy tidal flats. Tidal level zones within the tidal flat may be classed according to substrate, for example, sandy low tidal flats, muddy high tidal flats. A range of possible tidal flat types based on substrate, with field examples, is presented in Table T5.

In regard to sediment composition, two major groups are recognized: *siliciclastic tidal flats*, composed of terrigenous sediments such as quartz sand, quartz silt, and phyllosilicate clay, and *carbonate tidal flats*, composed of carbonate silt and clay, various sand-sized carbonate grains, and products of cementation (e.g., crusts, breccias, intraclasts). These major groups reflect two extremes in settings: an abundant supply of terrigenous sediment to the tidal coast, such as in deltas or estuaries versus a low supply relative to the rate of carbonate sediment production (as along terrigenous sediment starved coasts). From a historical perspective, the majority of earlier investigations of tidal flats were centered on siliciclastic systems, and much information emerged from studies in the North Sea (Reineck, 1972; Evans, 1975). Later,

Table T4 Some well-known tidal flats, and some extreme macrotidal flats, ordered in tidal range

Tidal flat location	Tidal range (m)	Composition	Tidal flat setting
Bay of Fundy (Nova Scotia)	15.0 m Extreme Macrotidal	Siliciclastic	Broad tidal flats of gravel, sand and mud, with local salt marsh, peripheral to an estuarine gulf in a humid temperate climate (Knight and Dalrymple, 1975)
Bay of Mont St Michel (NW France)	15.0 m Extreme Macrotidal	Siliciclastic	Broad tidal flats of sand and mud, with salt marsh, within complex of funnel-shaped estuary in humid temperate climate (Larsonneur, 1975)
King Sound (NW Australia)	11.0 m Extreme Macrotidal	Siliciclastic	Broad tidal flats of sand and mud, with some cheniers, and erosional tidal channels, with mangrove; peripheral to a seasonal estuarine gulf in a semi-arid tropical climate (Semeniuk, 1981)
Roebuck Bay (NW Australia)	10.5 m Extreme Macrotidal	Carbonate	Broad tidal flats along semi-sheltered embayment, dominated by mud; erosional tidal channels, with mangrove; in a semi-arid subtropical climate (Semeniuk, 1993)
Gulf of California (USA)	6–8 to 10 m Macrotidal to extreme macrotidal	Siliciclastic	Broad tidal flats, dominated by mud, with intermittent beach ridges, with salt marsh; part of the Colorado River Delta within a gulf in a semiarid subtropical climate (Thompson, 1968)
The Wash (England)	7.0 m Macrotidal	Siliciclastic	Broad tidal flats of sand and mud, with salt marsh; along the shore of a large embayment in humid temperate climate (Evans, 1975)
Delta of the Klang and Langat Rivers (Malaysia)	4.5 m Macrotidal	Siliciclastic	Compound delta, with insular/peninsular development of tidal flats, traversed by tidal creeks with mangrove; in a humid tropical climate (Coleman *et al.*, 1970)
The Jade and the Dutch Wadden Sea (North Sea)	2.6–4.1 m Mesotidal	Siliciclastic	Broad tidal flats of sand and mud, with salt marsh, leeward of barriers in humid temperate climate (Reineck, 1975)
Niger Delta (Western Africa)	1.0–2.8 m Microtidal	Siliciclastic	Extensive mangrove vegetated tidal flats of mud and sand, developed behind a beach barrier in a delta system in a humid tropical climate (Allen, 1970)
Gascoyne Delta (Western Australia)	2 m Microtidal	Siliciclastic	Local narrow tidal flats of sand and mud, with mangrove, fringing lagoons within a delta in an arid subtropical climate (Johnson, 1982)
Trucial Coast (western Persian Gulf)	2 m Microtidal	Carbonate	Broad tidal flats of carbonate muds and gypsum, with algal mats, salt marsh and mangroves, shoreward of prograding carbonate complex fringing a large gulf in an arid tropical climate (Purser, 1973)
Chesapeake Bay (eastern USA)	1.5–2.1 m Microtidal	Siliciclastic	Narrow to broad tidal flats, with salt marsh, within inlets and along the shore of an estuary in a humid temperate climate
Delmarva Peninsula to Sapelo Island (eastern USA)	<1 m Microtidal	Siliciclastic	Broad protected tidal flats, with salt marsh, leeward of barriers in a humid subtropical climate (Howard *et al.*, 1972; Harrison, 1975)
Shark Bay (Western Australia)	0.5 m Microtidal	Carbonate	Local tidal flats of sand or pelleted mud, with salt marsh and algal mat, shoreward of prograding sea grass banks and hypersaline platforms in large elongate embayments in an arid subtropical climate (Hagan and Logan, 1975)
Andros Island (Bahamas)	0.5 m Microtidal	Carbonate	Broad tidal flats of pelleted mud, with mangrove, salt marsh and algal mat, developed capping the Bahama Bank Carbonate Complex in a humid subtropical climate (Shinn *et al.*, 1969)

as interest in carbonate rocks grew during the 1960s, linked to their petroleum reservoir potential, a range of studies were undertaken in carbonate tidal flats (Shinn *et al.*, 1969; Purser, 1973; Hagan and Logan, 1975; Shinn, 1983).

Generally, regardless of whether the tidal flats are dominantly siliciclastic or carbonate, their sediments commonly contain both siliciclastic *and* carbonate particles. In dominantly siliciclastic settings, there may be minor to moderate carbonate components of shell gravel, shell grit, skeletal sand (e.g., shell fragments, foraminifera), skeletal silt-sized material, and carbonate clay transported to or generated on the flats. Similarly, in dominantly carbonate environments there may be siliciclastic sand, mud, or gravel from oceanic, aeolian, or local erosional sources. The range and origin of mud, sand, and gravel-sized particles comprising tidal flat sediments are noted in Table T6.

Some of the best known siliciclastic tidal flats are along the North Sea coast, for example, the Jade and the Dutch Wadden Sea, The Wash in southeastern England, the Gulf of California, the Bay of Fundy, the compound high-tidal delta of the Klang and Langat Rivers, King Sound in northwestern Australia, Bay of Mont St. Michel in France (Klein, 1963; Thompson, 1968; Allen, 1970; Coleman *et al.*, 1970; Reineck, 1972; Evans, 1975; Larsonneur, 1975; Semeniuk, 1981). With most of these examples, there is a grain size variation across the flats from sand in low tidal zones to mud in high tidal zones, with specific biogenic contributions in particular tidal zones, depending on climate setting and biogeography, and sediment types and sedimentary structures are dominantly the result of physical and biologic processes. With increase upslope in pore water salinity, particularly in semiarid and arid climates, the upper parts of siliciclastic tidal flats may develop carbonate nodules or gypsum crystals, or be salt encrusted.

Carbonate tidal flats generally occur in mid- to low latitude warm climates. The best known are Andros Island of the Bahama Banks (Shinn *et al.*, 1969), the Trucial Coast along the Persian Gulf Coast (Kendall and Skipwith, 1968; Purser, 1973), and Shark Bay in northwestern Australia (Hagan and Logan, 1975). In these examples, there is little or no terrigenous influx from terrestrial sources to dilute the carbonate accumulation contributed by local biogenic and abiotic sources, and hence the sediments are carbonate-rich. There are a range of diagnostic sediments and structures formed on carbonate tidal flats as result of tidal deposition, biogenic contribution and alteration, and primary and secondary effects of cementation. Cementation of sediments, and formation of their (secondary) structural and sedimentary derivatives is an important and common feature on upper parts of carbonate tidal flats. Under conditions of hypersalinity on the higher zones of such tidal flats, precipitation of carbonate-minerals often is prevalent, and in contrast to siliciclastic tidal flats, since there is an abundance of carbonate grains to act as nuclei for interstitial cements, there is a plethora of diagenetic and sedimentary products such as cemented layers and crust development, progressing to surface mounding, formation of compressional polygons and teepees, and then leading to fragmentation, brecciation, and formation of intraclasts. Carbonate tidal flats set in the

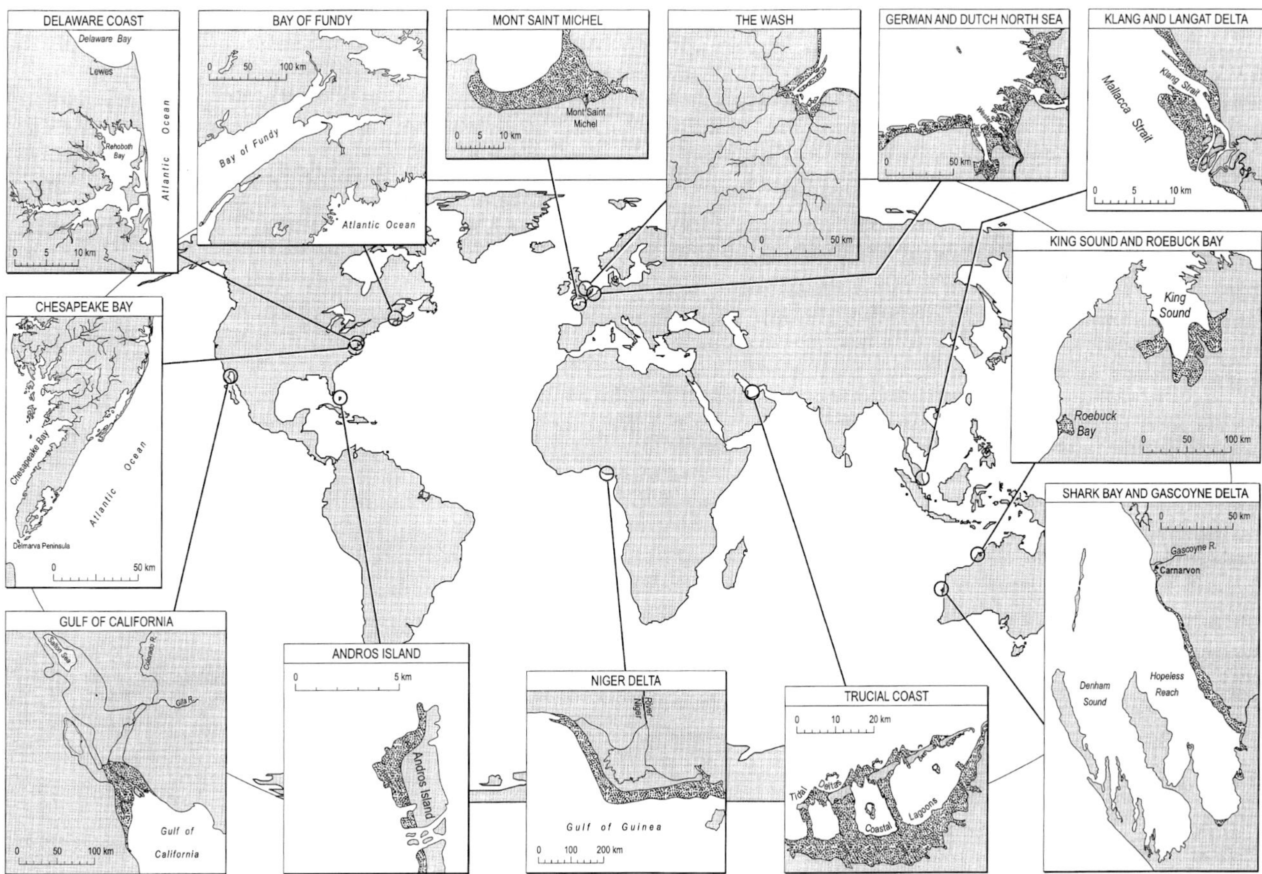

Figure T21 Location, settings, and sizes of various tidal flats around the globe. Table T4 provides more detailed information and references.

more arid climates also develop evaporitic mineral suites such as beds of gypsum nodules, gypsum platey crystals, gypsum mud, halite crusts.

Geomorphic features of tidal flats

While the surface of a tidal flat at the macroscale generally is flat to gently inclined, there may be a range of mesoscale to microscale features therein (Table T7, Figure T23). At the macroscale, the tidal flat may exhibit varying degrees of slope (Figure T22), reflecting the effects of position within either the low and high spring tidal zones, or the neap tidal zone. For example, the low tidal zone may be nearly flat or very gently inclined, the middle tidal flat may be more moderately inclined, and the high tidal flat again may be nearly flat or very gently inclined.

At mesoscale, the geomorphic features of tidal flats may include local cliffs, cheniers, sand waves, shell mounds, skeletal reefs, and gullies, channels and creeks (also called tidal creeks). Cliffs, commonly cut into mud, often separate vegetated and vegetation-free tidal flat zones, but some cliffs are formed due to either the effect of wave energy concentrated at a specific tidal level, or the undercutting of mud through erosion of the underlying sand. Tidal creeks may be ramifying or meandering, with point bars and steep banks. At smaller scales, the surface of tidal flats may be planar and smooth, or hummocky to slightly irregular, or may exhibit linear scours, surface mounding and teepees, desiccation polygons, or mud cracks, produced physically, chemically, or biogenically.

Hydrology

The groundwater hydrology of tidal flats is important for several reasons. Interstitial pore water salinity gradients and moisture gradients, for instance, influence macrophytes (such as mangroves and samphires), microbial mats, and invertebrate biota in relation to their occurrence and zonation. Interstitial pore water salinity and moisture gradients also influence precipitation of evaporitic minerals. Microscale shallow groundwater hydrologic recharges and discharges influence development of sedimentary structures (e.g., seepage zones out of sand mounds to initiate sand erosion, or to initiate hydrochemical exchanges and cementation; formation of bubble sand). The hydrologic functioning of tidal flats additionally can drive geochemical processes that diagenetically modify sediments (e.g., formation of iron sulfide precipitation to form gray sediments, or the oxidation of buried iron sulfide impregnated vegetation to form goethite pseudomorphs).

Tidal flat groundwater levels fluctuate on a diurnal to semidiurnal basis, following the tides, with a dampened effect from mid-tidal levels to upslope. All tidal flat groundwater rises during flood tide, and of course is inundated on high tide. Recharge and discharge, and lateral groundwater flow through tidal flat sediments may be facilitated by specific lithologic layers, or stratigraphic intervals, and at the small scale by burrow and root structures.

Groundwater salinity across tidal flats is commonly zoned, generally with near marine water salinities at about mean sea level, unless the marine waters fronting the tidal flats are hypersaline (e.g., Shark Bay, Australia), grading to hypersaline and extremely hypersaline upslope, and in wet climates becoming fresh where tidal flats interact with terrestrial freshwater. The main source waters for groundwater of tidal flats are marine water, rain, and (through seepage and land overflow) land-derived freshwater. Evaporation, macrophyte transpiration, and increasing infrequency of tidal inundation upslope combine to develop a gradient of increasing salinity across tidal flats. This gradient results in zonation of biota, exemplified by zonation of mangroves, and in zonation of evaporitic minerals and pore water precipitates. Where marine waters are oceanic (ca. 35,000 ppm salinity) and evaporation is extreme, high tidal groundwater may reach 100,000–200,000 ppm salinity, that is, carbonate mineral and gypsum precipitating, but where source waters are already hypersaline, tidal flat groundwater reaches up to ca. 300,000 ppm salinity, resulting in precipitation of halite.

Key processes on tidal flats

Tidal flats are located at the triple junction between land, sea, and atmosphere. In this context, as low-gradient shores, they exhibit a myriad of

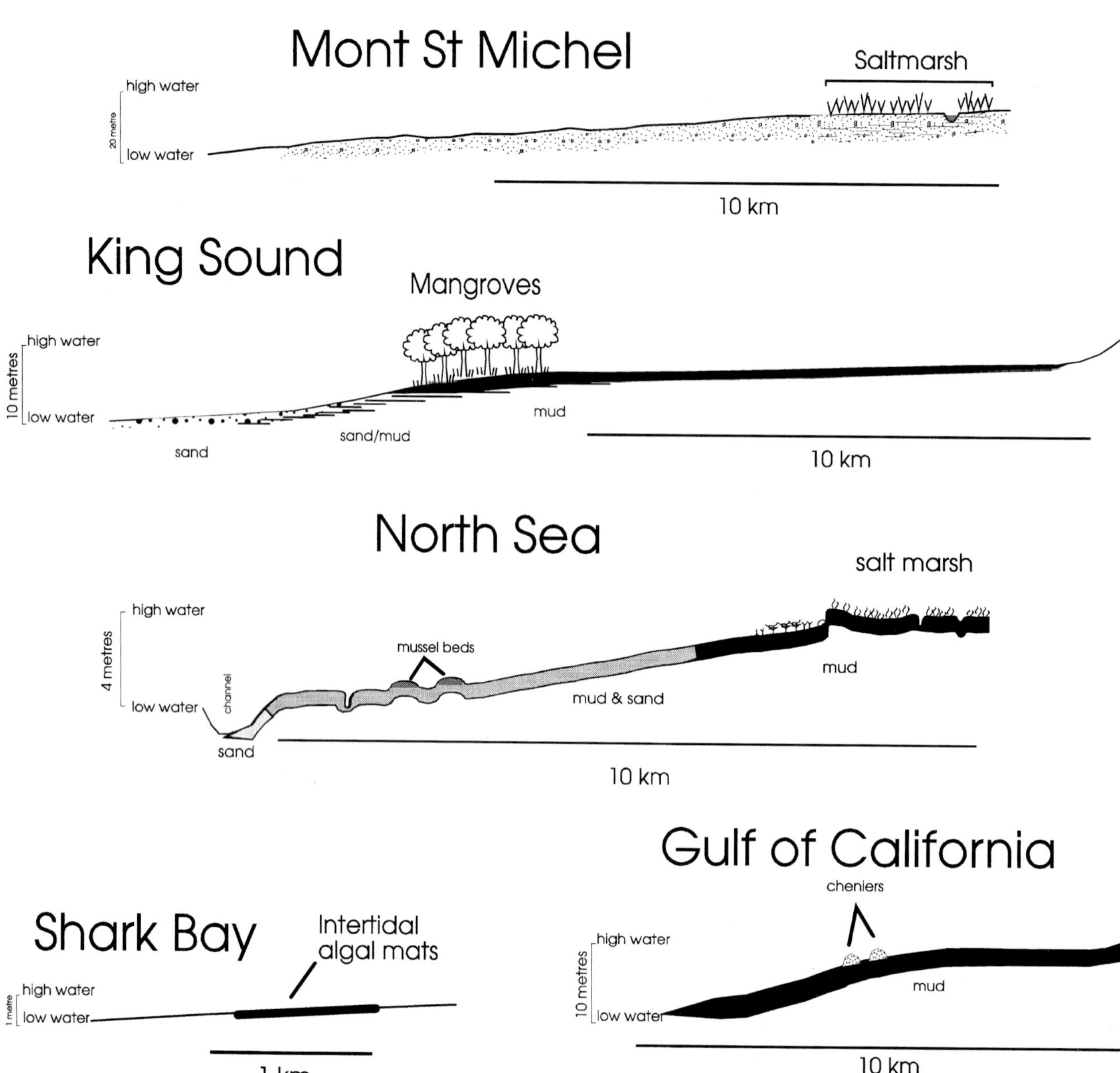

Figure T22 Profiles across various macrotidal to microtidal siliciclastic and carbonate tidal flats (see Table T4) showing nature of the slopes, sediments, vegetation, or morphology.

Table T5 Tidal flat types, according to substrate

Substrate type underlying tidal flat	Terminology	Examples
Whole tidal flats		
Mix of particle sizes across the whole tidal flat, or differentiation not intended	Tidal flat	North Sea, The Wash, Bay of Mont St. Michel, King Sound
Tidal flats wholly underlain mainly by		
mud	Muddy tidal flat	Gulf of California
sand	Sandy tidal flat	
gravel	Gravelly tidal flat	Parts of the Bay of Fundy
shelly pavement across tidal flat	Tidal shell pavement	
Specific zones on tidal flats		
High tidal flats wholly underlain mainly by mud	Muddy high-tidal flat	King Sound
Mid-tidal flats wholly underlain mainly by mud	Muddy mid-tidal flat	King Sound
Mid-tidal flats underlain mainly by mud and sand	Mixed mid-tidal flat	North Sea
Low tidal flats wholly underlain mainly by sand	Sandy low-tidal flat	North Sea; Bay of Mont St. Michel
Shelly pavement on specific zone of tidal flat, e.g., low tidal	Low-tidal shell pavement	Parts of Shark Bay
Crust pavement on specific zone of tidal flat, e.g., high tidal	High-tidal crust pavement	Dampier Archipelago
Breccia pavement on specific zone of tidal flat, e.g., high tidal	High-tidal breccia pavement	Parts of Shark Bay

Table T6 Types of sedimentary particles on tidal flats, and their origin

Sediment particle	Origin (information from various tidal flats)
Clay (<4 μm particle size)	
Phyllosilicate clay (kaolinite, illite, montmorillonite)	Fluvially delivered to the coastal system Reworked Pleistocene coastal deposits Reworking of glacial deposits Aeolian
Calcitic and aragonitic clay	Reworked, comminuted skeletons Precipitated from seawater Disintegrated calcareous algae
Goethite	Fluvially delivered to the coastal system Aeolian
Quartz clay	Aeolian
Amorphous silica	Diatom (*in situ* or transported)
Silt (4–63 μm particle size)	
Quartz, felspar, various silicate minerals	Fluvially delivered to the coastal system Reworked Pleistocene coastal deposits Reworking of glacial deposits Aeolian
Skeletal silt	Reworked and *in situ* comminuted shelly exoskeletons
Amorphous silica	Diatom (in situ or transported)
Sand (63–2,000 μm particle size)	
Quartz, felspar, various silicate minerals, rock fragments	Fluvially delivered to the coastal system Reworked Pleistocene coastal deposits Reworking of glacial deposits Aeolian
Skeletal sand	Comminuted to whole reworked and *in situ* exoskeletons, e.g., shell fragments, foraminifera
Carbonate sand (ooids, pellets)	Generated nearshore and reworked onto tidal flat, and for pellets, carbonate grain destruction by boring algae
Carbonate intraclast sand	Reworking of cemented carbonate crusts
Gravel (>2,000 μm particle size)	
Quartz pebbles, rock fragments	Fluvially delivered to the coastal system Reworked Pleistocene coastal deposits Reworking of glacial deposits
Mud pebbles and cobbles	Eroded tidal mud
Armored mud balls	Mud pebbles and cobbles with adhering gravel and shell
Skeletal gravel	Comminuted to whole, reworked and in situ shell
Carbonate intraclast gravel	Reworking of cemented carbonate crusts

Table T7 Geomorphic features of the tidal flats

Geomorphic, or surface feature	Origin
Microscale surface features (<meter sized)	
Smooth planar surface	Deposition on and erosion of the surface
Linear scours (mm to cm deep)	Tidal erosion
Slightly irregular	Tidal erosion of the surface, and/or bioexcavations by small biota and fish
Hummocky surface	Tidal erosion of the surface, and/or bioexcavations by stingrays, fish, and large burrowing benthos
Mud cracks	Desiccation
Surface moundings grading to teepees and brecciation	Mineral precipitation in surface sediments and resultant surface crust expansion
Mounded surface	Mineral precipitation in surface sediments
Mesoscale surface features (>meter-sized, up to tens of meters long)	
Meandering gullies, channels, creeks, meandering or ramifying	Tidal erosion, with local deposition on point bars
Crust lined, and locally brecciated meandering channels	Tidal erosion, with local deposition on point bars, with mineral precipitation in surface sediments and resultant surface crust expansion
Sand waves	Large bodies of sand developed in low tidal zones
Spits	Shoestring sand and sandy gravel body across tidal flat from local headland, formed by tidal currents and wave action
Cheniers	Isolated shoestring sand and sandy gravel body on tidal flat, variably formed by tidal currents, wave action, and storms/cyclones
Salting cliff	Small cliff, 30–100 cm high, cut in to salt marsh, marking junction between high-tidal salt marsh and vegetation-free mid-tidal to low-tidal flat
Mid-tidal cliff	Small cliff, up to 100 cm high, marking junction between mangrove front at ca. MSL and vegetation-free mid-tidal to low-tidal flat

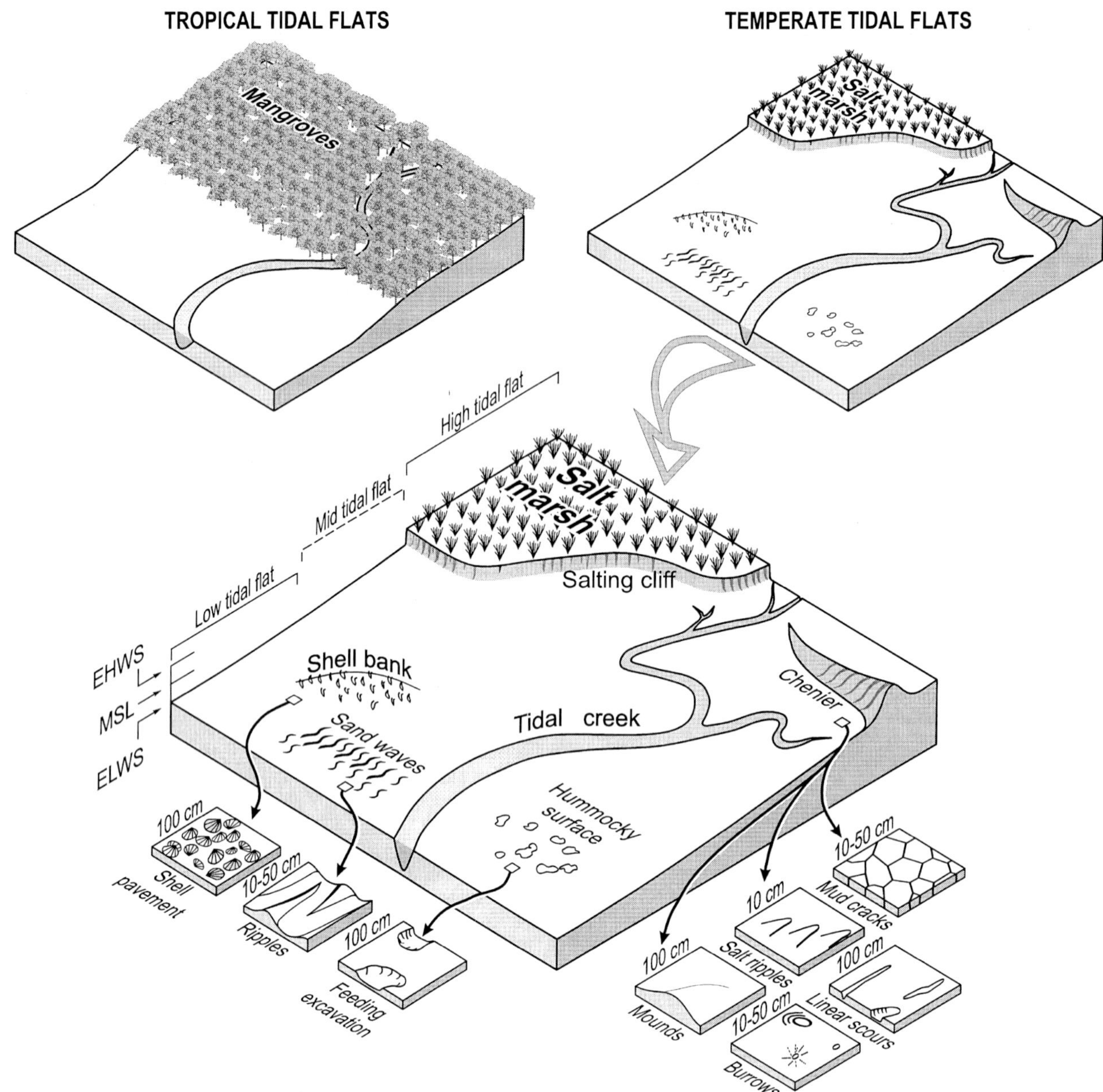

Figure T23 Generalized geomorphology of tidal flats. Macroscale geomorphology of tropical climate tidal flats with mangroves, and temperate climate tidal flats with salt marsh. More detail shown of mesoscale and microscale features of a temperate climate tidal flat.

products resulting from interactive, interrelated and overlapping exogenous and endogenous agents and processes, which include oceanographic, meteorologic, atmospheric, fluvial, hydrologic and hydrochemical, and biological processes (Table T8). These processes commonly are distributed along physicochemical gradients (e.g., tidal, wave, chemical) and operate on a range of basic sediment types such as mud, sand, and shell gravel to develop a complex of geomorphic, sedimentologic, and diagenetic products which are commonly zoned across the tidal flats and are often specific to a coastal setting, sediment setting, climate, and biogeography.

Oceanographic processes involve erosion, transport, and deposition associated with tidal currents, wave action, storms, and cyclones. Meteorologic/atmospheric processes involve evaporation, wind erosion, transport and deposition, rain, ice crystallization, and water temperature fluctuations. Fluvial processes include the delivery of sediment and freshwater to the shore, especially in estuaries and deltas. Hydrologic and hydrochemical processes involve the rising and falling of the water table under the tidal flat, the solution and precipitation of carbonate minerals, evaporite minerals and iron minerals, and redox reactions. Biological processes include: at the largest scale, the accumulation of beds of biogenic material (shell beds, biostromal reefs, and plant material), the modification of tidal current by macrophytes (e.g., mangroves) to induce sedimentation, and the trapping and binding of sediment by vegetation and algal mats; at intermediate scales, the burrow structuring and root-structuring of sediment; and at the smallest scales, the boring of shells, the pelletization of grains by endolithic algae, and the biomediation of the precipitation of minerals such as iron sulfides.

The dynamics of tidal currents is a major factor in the transport and deposition of sediment on tidal flats. Tidal currents transport mud in suspension, and sand by traction. The rise and fall of the tide, with periods of slack water, result in a systematic increasing, decreasing, zero flow, and then reversal of tidal currents. Transport of sand is effected during the main part of the flooding and ebbing tidal cycle when tidal currents are progressing to and regressing from their maximum velocities, with various bedforms developed as the currents systematically increase and then decrease across the tidal cycle resulting in the development of ripples, then megaripples, and then ripples. Deposition of mud is effected during periods of low current velocity and slack water (the times of near-zero to zero current velocities). Mud deposition is accentuated further by fluctuations in water temperature, since cold water and warm water have different viscosity which results in varying mud particle settling velocities (Krogel and Flemming, 1998).

Table T8 Key processes

Some selected processes	Examples of products on the tidal flat
Oceanographic	
Flood and ebb-tidal currents, and slack water	Deposition of mud from suspension, and sand transport by traction; silt ripples, sand ripples, sand waves and megaripples; laminated mud, lenticular bedding, and flazer bedding; scour, and cut-and -fill; tidal creek formation; erosion of mud beds along creek banks and clifflines to form mud ball conglomerate; rolling of mud balls on sandy/gravelly floors to form armored mud balls
Waves	Winnowed sand sheets and shell gravel; rippling; erosion of clifflines cut into mud
Meteorologic/atmospheric	
Wind (erosion, transport, deposition)	Sand transport and fall-out deposition onto tidal flats; formation of adhesion ripples
Evaporation minerals	Mud desiccation and cracking; increasing pore water salinity of groundwater; precipitation of minerals
Rain, ice crystallization	Rain imprints; ice crystal imprints; cryogenic disruption
Water temperature variation	Mud deposition from suspension; mortality of benthos
Groundwater hydrologic/hydrochemical	
Rising and falling of tide	Wetting and drying to form desiccated sediment; development of bubble sand
Solution/precipitation of carbonate minerals	Shell voids, other vughs; cemented crusts, teepees
Evaporite mineral precipitation	Beds of nodular to platey crystalline gypsum; precipitation of gypsum disrupting primary structures
Iron mineral precipitation	Staining of sediment to dark gray with iron sulfide; staining of sediments to orange-brown with iron oxides
Biological	
Accumulation of shell beds and biostromal reefs	Shelly sediments and coquinas, and skeletal biostromes
Plant detritus accumulation	Organic rich sediments, peat
Sediment trapping and binding by vegetation and algal mats	Root-structured sediment; algal-laminated sediment
Feeding/foraging by nekton	Pocked, excavated and hummocky surfaces
Burrowing by benthos	Burrow-structured to fully bioturbated sediment

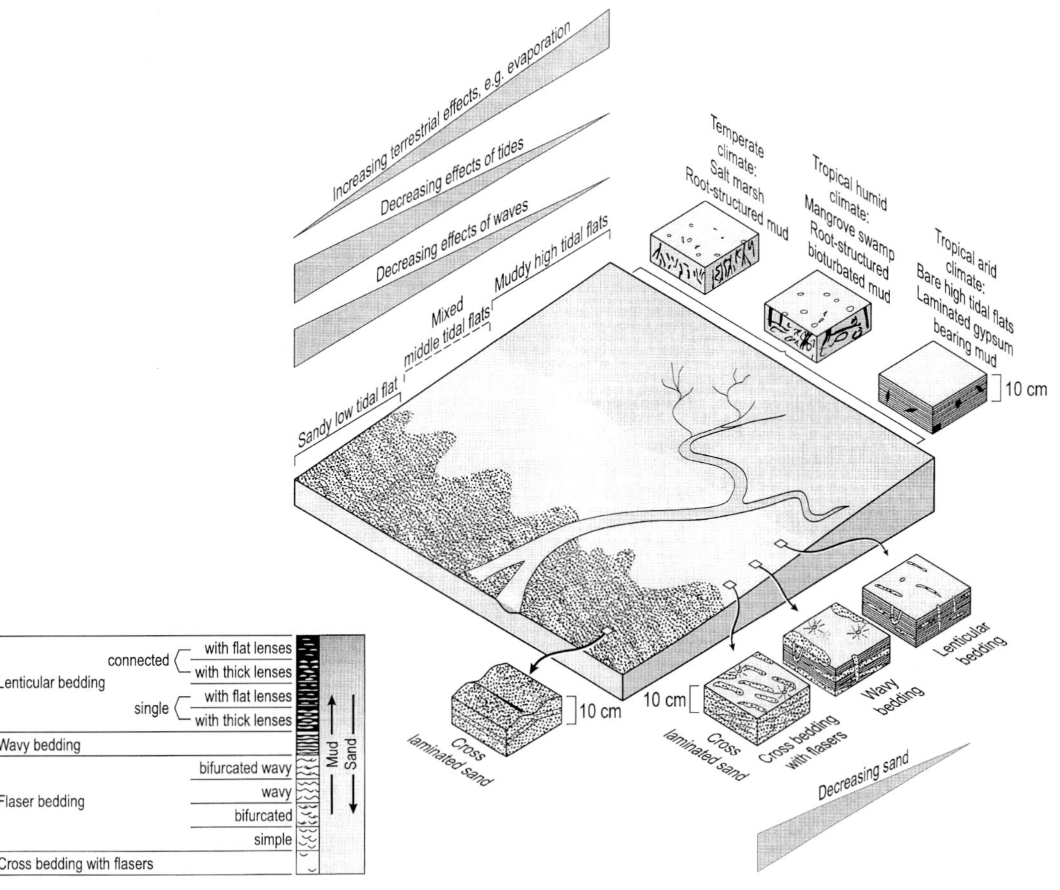

Figure T24 Generalized sedimentology of a typical tidal flat, relating sediment types to oceanographic and terrestrial processes, and inset detail of some sediment types in relation to facies setting and position on tidal flat. Also shown is the systematic variation in bedding types as the sand to mud ratio changes (modified after Reineck and Singh, 1980).

The basic tidal flat sediment types are strongly related to their position on the tidal flat slope as a result of the interplay of tidal action and waves (Figure T24). If there is sand and mud on the tidal flat, tidal and wave processes result in a partitioning of particle sizes: generally, sand dominates the low tidal flats, mixed sand and mud occur on mid-tidal flats, and mud on the high tidal flats. If biogenic activity is not intense enough to bioturbate sediments, the sand and mud on mid-tidal flats are separated in layers and laminae to develop flazer bedding, lenticular bedding, wavy bedding, or interlaminated sand and mud. With more intense burrowing, mid-tidal flats are thoroughly mixed muddy sand with bioturbation structures.

Accumulation of mud in the upper parts of tidal slopes is a general product of tidal processes. There are three main reasons for this. First, with scour lag and settling lag, mud is progressively transported up the tidal slope to accumulate ultimately at the level of the highest tide (Postma, 1961). Scour lag and settling lag processes are particularly accentuated if the tidal flood and ebb are asymmetric. Second, mid- to lower tidal slope environments, more constantly under water, are subject to more continuous and intense wave and tidal current reworking, and hence any mud settled there is prone to remobilization. In contrast, high tidal areas generally are inundated by tides on slackening water, with low to nil current, and any wave trains arriving here are dampened by translation across the tidal flat floor. Hence, there is less scope for reworking. Third, high tidal zones in many regions are vegetated by mangrove or salt marsh, which function in current baffling, and sediment trapping and binding.

The interplay of tidal currents with mud and sand results in an interesting and geologically important range of bedforms and sedimentary structures. Sand transported during flood and ebb tide develops ripple bedforms with internal cross-lamination. Mud deposited during slack water blankets these ripples, or preferentially settles in inter-ripple troughs. Tidal currents, during the ensuing flood or ebb tide, rework the mud layers to diminish their thickness, leave inter-ripple lenses of mud, and remobilize sand forming ripples to bury the mud layers and to form more ripple cross-lamination. Ongoing deposition and burial of these bedforms results in mud-dominated bedding with scattered sand lenses (lenticular bedding), to subequal mud and sand (wavy bedding), to sand-dominated bedding with thin inter-ripple mud lenses (flazer bedding). Thus depending on the proportion of mud to sand, which is a function of the location of the sediment type on the tidal flat, or the inherent proportion of mud to sand regionally, the sedimentary structures produced by the processes described above range from mud-dominated lenticular bedding to flazer bedding (Figure T24).

Sedimentology, sedimentary structures, and stratigraphic sequences

Sediment bedforms, surface features, and near-surface features on tidal flats are produced by oceanographic, other physical, biotic, and hydrochemical processes. Wave action and tides and winnowing result in ripples, megaripples, sand waves, plane sand beds, linear scours, plane mud beds, and gravel pavements. A range of other physical processes result in mud cracks, air escape holes, bubble structures. Biological activity results in burrow-pocked surfaces, animal tracks, crab burrow workings, vesicular structures, crab balls, accumulation of shell banks and shell gravel. Chemical and physical processes combine to develop sheets of gypsum mush and nodules, platey gypsum pavements, carbonate crusts, and breccia pavements.

Sedimentary structures deriving from burial of the sediment bedforms, and the surface and near-surface features include cross-bedding and cross-lamination, herring bone cross-lamination, sand ripple cross-lamination, silt ripple cross-lamination, lenticular bedding, flazer bedding, laminated mud, sand dykes, mud dykes, bubble sand, vesicular mud, root-structuring, vertical burrows to labyrinthoid burrow networks, shell laminae and beds, shell reefs, silt and sand balls, bioturbation and swirl structures, breccias, nodular gypsum beds, platey gypsum beds, and teepee structures.

Key sediments, diagnostic of their formative processes, occur in different parts of the tidal flat. For example, mangrove-vegetated muddy tidal flats develop root-structured and bioturbated (shelly) mud, and crustacean-dominated mixed tidal flats develop burrow structured interbedded sand and mud varying to bioturbated muddy sand. Some examples of sediments and the processes involved in their development from siliciclastic tidal flats are noted in Table T9.

Table T9 Examples of sediments in their setting, and processes in their development

Environment	Main processes	Resulting sediment(s)
Siliciclastic sediment settings		
Mangrove or salt marsh vegetated high-tidal mudflat	Mud accumulation; root-structuring, bioburbation; shell contribution; groundwater alteration	Gray bioturbated root-structured (shelly) mud
Algal mat covered high-tidal mudflat	Mud accumulation; binding; trapping; redox reactions; cracking	Laminated mud; desiccated laminated mud
Bare high-tidal mudflat	Mud accumulation; surface shear; cracking of mud; reworking of desiccation polygons; gypsum precipitation	Laminated mud; desiccated mud; mud chip breccia; gypseous mud
Burrow-pocked midtidal mudflat	Mud accumulation; surface shear; benthic fauna burrowing	Burrow-structured laminated mud; bioturbated mud
Mollusc inhabited mid-tidal mudflat	Mud accumulation; surface shear; accumulation of shell winnowing to concentrate shells	Laminated shelly mud; shell gravel bed
Mid-tidal mudflat with sand ripples	Mud accumulation; surface shear; traction transport of sand	Flaser bedding
Megarippled low- to mid-tidal sand flat	Traction transport of sand; air trapped by rise and fall of tide	Cross-laminated sand; bubble sand
Mid-tidal burrow-pocked sand flat	Traction transport of sand; benthic fauna burrowing	Burrow-structured cross-laminated sand; bioturbated sand
Carbonate sediment settings		
High-tidal breccia pavement	Mud accumulation; carbonate cementation; root-structuring; groundwater alteration	Limestone breccia sheet
High-tidal algal mat covered mudflat	Mud accumulation; binding; trapping; redox reactions; cracking	Laminated mud; desiccated laminated lime mud
High-tidal bare mudflat	Mud accumulation; surface shear; cracking of mud; reworking of cracks; gypsum precipitation	Laminated lime mud; desiccated lime mud; mud chip breccia; gypseous lime mud laminated gypsum; gypsum nodule bed
Mid-tidal burrow-pocked mudflat	Mud accumulation; surface shear; benthic fauna burrowing	Burrow-structured laminated mud; bioturbated mud

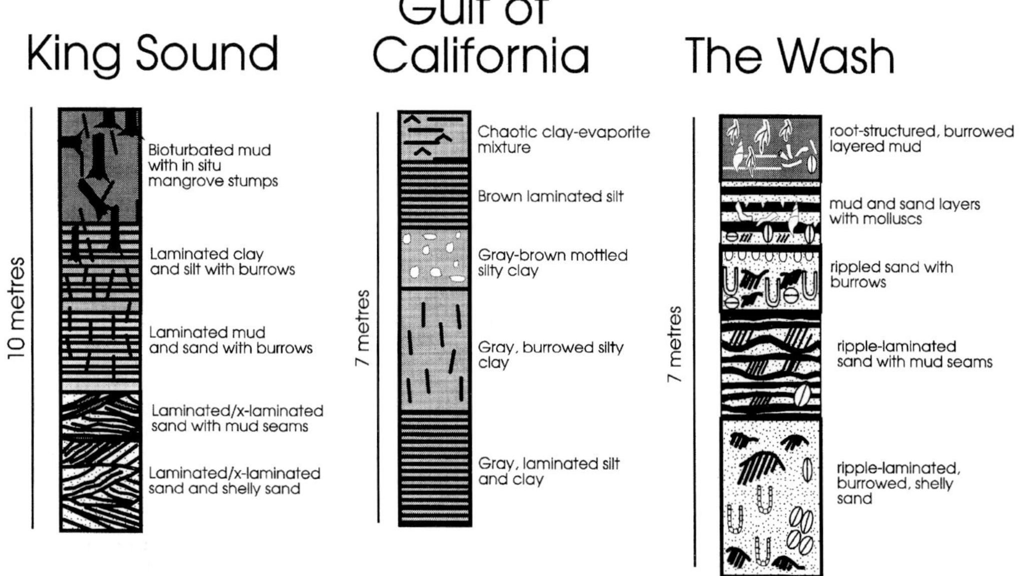

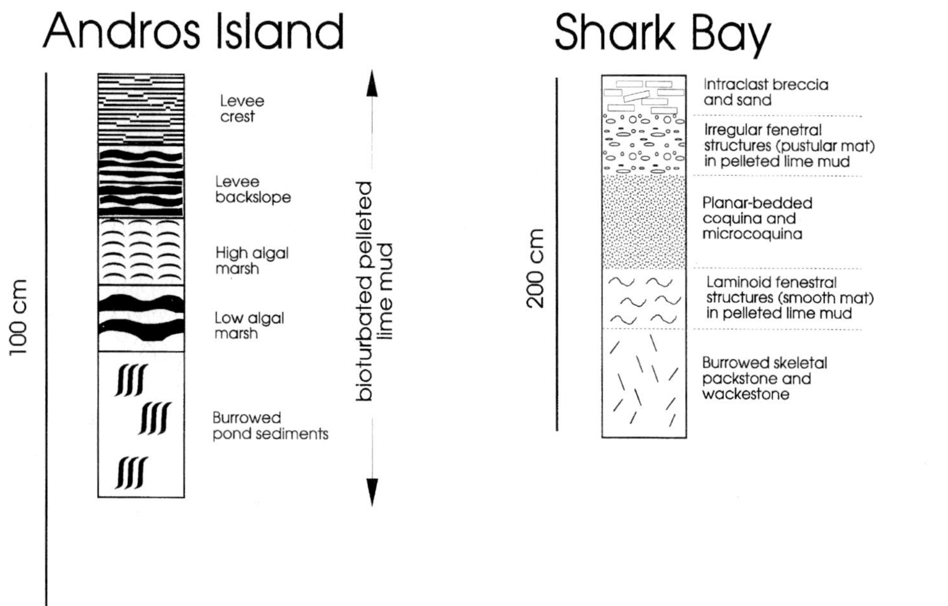

Figure T25 Stratigraphic sequences from various tidal flats (see Table T4). For the macrotidal siliciclastic settings there is comparison for three sequences: a tropical semiarid mangrove-vegetated tidal flat, shoaling from sand to mud, with burrows, root-structures, and *in situ* mangrove stumps; a subtropical semiarid vegetation-free tidal flat dominated by mud, with local burrows, and evaporitic mineral structures; and a temperate humid salt marsh vegetated tidal flat, shoaling from sand to mud, with burrows, root-structures, and shell. The microtidal carbonate sequences compare the structures of a subtropical humid tidal flat with that of a subtropical arid tidal flat.

With progradation, siliciclastic tidal flats develop characteristic stratigraphic sequences. A range of stratigraphic sequences are shown in Figure T25, from various macrotidal to microtidal settings, from flats that are mud dominated, to sand to mud sequences, from temperate to tropical climates. Some examples of sediments from carbonate tidal flats, and the processes involved in their development, are noted in Table T9. With progradation, carbonate sediment tidal flats also develop characteristic stratigraphic sequences, some of which are shown in Figure T25.

Table T10 Key biota of the tidal flats

Biota	Occurrence and function
Mangroves	Tropical tidal flats; massive primary production in the mid-upper tidal zone; detritus sustains biota in adjoining tidal zones (Tomlinson, 1986)
Samphires (salt marsh)	Temperate to tropical tidal flats; primary production in the mid-upper tidal zone; detritus sustains biota in adjoining tidal zones (Chapman, 1977; Beeftink *et al.*, 1985)
Spartina (salt marsh)	Temperate tidal flats; primary production in the mid-upper tidal zone; detritus sustains biota in adjoining tidal zones (Chapman, 1977)
Algal mats and stromatolites	Tropical tidal flats; primary production in the mid-upper tidal zone (Kendall and Skipwith, 1968; Ginsburg and Hardie, 1975)
Molluscs, polychaetes, crustacea	These invertebrates present generally on all tidal flats, although species diversity may decrease towards the temperate regions; molluscs, polychaetes, and crustacea are primary and secondary consumers, and sustain higher level trophic feeders (Dankers *et al.*, 1983; Knox, 1986)
Fish and avifauna	On tidal flats; fish and avifauna generally are primary and secondary consumers, and sustain higher level trophic feeders, and in many instances are the highest trophic level in the region (Knox, 1986; Sylva, 1975; Owen and Black, 1990)

Some key biota of tidal flats

The well-known biota of tidal flats include mangroves, salt marsh, algal mats and stromatolites, polychaetes, molluscs, crustacea, resident fishes and invading nektonic fishes, and avifauna (Table T10). Biogeography and climate, substrate and hydrochemistry are major factors that determine what biota inhabit tidal flats. Species abundance and zonation at site-specific level is determined by physicochemical and biological conditions. For example, with macrophytes, at a global scale, mangroves dominate mid- to upper tidal flats in tropical climates and are replaced by salt marsh in temperate climates. With increased salinity, the upper tidal interval may be inhabited by algal mats and stromatolites. Diversity of flora and fauna is linked to climate setting, with high species richness and abundance in tropical areas, and relatively lower species richness in temperate areas.

Primary production within specific parts of the tidal flat, for example, from mangroves and salt marsh, often drives the ecosystems of tidal flats. With mangroves and salt marshes, these macrophytes fix nutrients and carbon on the mid- to upper tidal flats, supporting the local resident fauna, and the export of detritus sustains benthic biota of polychaetes, molluscs, and crustacea elsewhere on the mid- to low-tidal flats. The biologically rich tidal flat environments also support nekton and avifauna. Fish and other nekton invade the tidal zone for feeding on the high tide, and the avifauna invade the tidal flats at low tide.

Tidal flats typically are biologically zoned. For any benthic group, such as polychaetes, molluscs, or crustacea, there is species zonation across the flats related to frequency of inundation, substrate, pore water salinity, inter-species competition and predation pressure, among other factors. Macrophytes (mangrove and salt marsh) also exhibit zonation, as related to groundwater salinity, substrates, and elevation of habitat above mean sea level. Many of the benthos are burrowing forms, and the macrophytes have diagnostic root structures, and hence sedimentologically, zonation of the biota results in facies and tidal level specific signatures across tidal flats: sand-constructed *Arenicola* burrows, for instance, are diagnostic of low-tidal sand flats, vertical to u-shaped to labyrinthoid crustacean burrows in a root-structure-free mud are diagnostic of mid- to low tidal flats, coarse root-structured substrates and associated faunal burrows are diagnostic of mangrove vegetated high tidal flats, while fine root-structured substrates are diagnostic of salt marsh vegetated high tidal flats. Some diagnostic biogenic structures, and biofacies related to tidal assemblages, are often signatures for specific tidal levels and lithofacies within a given region.

V. Semeniuk

Bibliography

Alexander, C.R., Davis, R.A., and Henry, V.J. (eds.), 1998. *Tidalites: Processes and Products.* SEPM (Society for Sedimentary Geology), Special Publications, 61.

Allen, J.R.L., 1970. Sediments of the modern Niger Delta: a summary and review. In Morgan, J.P. (ed.), *Deltaic Sedimentation: Modern and Ancient.* Society of Economic Paleontologists and Mineralogists, Special Publication No. 15, pp. 138–151.

Beeftink, W.G., Rozema, J., and Huiskes, A.H.L. (eds.), 1985. *Ecology of Coastal Vegetation.* Dordrecht: Dr W. Junk Publishers.

Black, K.S., Paterson, D.M., and Cramp, A. (eds.), 1998. *Sedimentary Processes in the Intertidal Zone.* Geological Society, London, Special Publications, 139.

Chapman, V.J. (ed.), 1977. *Ecosystems of the World, Vol 1, Wet Coastal Ecosystems.* Amsterdam: Elsevier Scientific Publishing Co.

Coleman, J.M., Gagliano, S.M., and Smith, W.G., 1970. Sedimentation in a Malaysian high tide tropical delta. In Morgan, J.P. (ed.), *Deltaic Sedimentation: Modern and Ancient.* Society of Economic Paleontologists and Mineralogists, Special Publication No. 15, pp. 185–197.

Dankers, N., Wolff, W.J., and Zijlstra, J.J. (eds.), 1983. Fishes and fisheries of the Wadden Sea, Report 5. In Wolff, W.J. (ed.), *Ecology of the Wadden Sea, Vol 2.* Rotterdam: A. A. Balkema.

Davies, J.L., 1980. *Geographical Variation in Coastal Development*, 2nd edn. London: Longman.

Evans, G., 1975. Intertidal flat deposits of the wash, western margin of the North Sea. In Ginsburg, R.N. (ed.), *Tidal Deposits: A Casebook of Recent Examples and Fossil Counterparts.* Berlin: Springer-Verlag, pp. 13–20.

Ginsburg, R.N. (ed.), 1975. *Tidal Deposits: A Casebook of Recent Examples and Fossil Counterparts.* Berlin: Springer-Verlag.

Ginsberg, R.N., and Hardie, L.A., 1975. Tidal and storm deposits, northwestern Andros Island, Bahamas. In Ginsburg, R.N. (ed.), *Tidal Deposits: A Casebook of Recent Examples and Fossil Counterparts.* Berlin: Springer-Verlag, pp. 201–208.

Hagan, G.M., and Logan, B.W., 1975. Prograding tidal-flat sequences: Hutchinson Embayment, Shark Bay, Western Australia. In Ginsburg, R.N. (ed.), *Tidal Deposits: A Casebook of Recent Examples and Fossil Counterparts.* Berlin: Springer-Verlag, pp. 215–221.

Harrison, S.C., 1975. Tidal flat complex, Delmarva Peninsula, Virginia. In Ginsburg, R.N. (ed.), *Tidal Deposits: A Casebook of Recent Examples and Fossil Counterparts.* Berlin: Springer-Verlag, pp. 31–38.

Howard, J.D., Frey, R.W., and Reineck, H.-E., 1972. Georgina Coastal Region, Sapelo Island, U.S.A.: Sedimentology and biology. VIII Conclusions. *Senckenbergiana marit.* **4**: 217–222.

Jackson, J.A., 1997. *Glossary of Geology*, 4th edn. Alexandria: American Geological Institute.

Johnson, D.P., 1982. Sedimentary facies in an arid zone delta: Gascoyne delta, western Australia. *Journal of Sedimentary Petrology*, **52**: 547–563.

Kendall, C.G.St.C., and Skipwith, Sir Patrick A. d'E., 1968. Recent algal mats of a Persian Gulf Lagoon. *Journal of Sedimentary Petrology*, **38**: 1040–1058.

Klein, G. deV., 1963. Bay of Fundy intertidal zone sediments. *Journal of Sedimentary Petrology*, **33**(4): 844–854.

Klein, G. deV. (ed.), 1976. *Holocene Tidal Sedimentation.* Stroudsburg: Dowden, Hutchinson and Ross, Inc.

Knight, K.J., and Dalrymple, R.W., 1975. Intertidal sediments from the South Shore of Cobequid Bay, Bay of Fundy, Nova Scotia, Canada. In Ginsburg, R.N. (ed.), *Tidal Deposits: A Casebook of Recent Examples and Fossil Counterparts.* Berlin: Springer-Verlag, pp. 47–55.

Knox, G.A., 1986. *Estuarine Ecosystems: A Systems Approach*, Vol. II. Boca Raton: CRC Press, Inc.

Krogel, F., and Flemming, B., 1998. Evidence for temperature-adjusted sediment distributions in the back-barrier tidal flats of the East Frisian Wadden Sea (Southern North Sea). In Alexander, C.R., Davis, R.A., and Henry, Vernon, J. (eds.), *Tidalites: Processes and Products*. Tulsa: Society for Sedimentary Geology.

Larsonneur, C., 1975. Tidal deposits, Mont Saint-Michele Bay, France. In Ginsburg, Robert N. (ed.), *Tidal Deposits: A Casebook of Recent Examples and Fossil Counterparts*. Berlin: Springer-Verlag, pp. 21–30.

Owen, M., and Black, J.M., 1990. *Waterfowl Ecology*. New York: Chapman and Hall.

Postma, H., 1961. Transport and accumulation of suspended matter in the Dutch Wadden Sea. *Netherlands Journal of Sea Research*, **1**(1,2): 148–190.

Purser, B.H. (ed.), 1973. *The Persian Gulf: Holocene Carbonate Sedimentation and Diagenesis in a Shallow Epicontinental Sea*. Berlin: Springer-Verlag.

Reineck, H.-E., 1972. Tidal flats. In Rigby, J.K., and Hamblin, W.K. (eds.), *Recognition of Ancient Sedimentary Environments*. Society of Economic Paleontologists and Mineralogists, Special Publication No. 16, pp. 146–159.

Reineck, H.-E., 1975. German North Sea Tidal Flats. In Ginsburg, Robert, N. (ed.), *Tidal Deposits: A Casebook of Recent Examples and Fossil Counterparts*. Berlin: Springer-Verlag.

Reineck, H.-E., and Singh, I.B., 1980. *Depositional Sedimentary Environments*, 2nd edn. Berlin: Springer-Verlag.

Semeniuk, V., 1981. Sedimentology and the stratigraphic sequence of a tropical tidal flat, north-western Australia. *Sedimentary Geology*, **29**: 195–221.

Semeniuk, V., 1993. The mangrove systems of Western Australia—1993 Presidential Address. *Journal Royal Society W.A.*, **76**: 99–122.

Shinn, E.A., 1983. Tidal flats. In Scholle, Peter, A., Bedout, Don, G., and Moore, Clyde, H. (eds.), *Carbonate Depositional Environments. Tulsa*. The American Association of Petroleum Geologists, pp. 171–210.

Shinn, E.A., Lloyd, R.M., and Ginsburg, R.N., 1969. Anatomy of a modern carbonate tidal-flat, Andros Island, Bahamas. *Journal of Sedimentary Petrology*, **39**(3): 1202–1228.

Sylva, de, D.P., 1975. Nektonic food webs in estuaries. In Cronin, L.E. (ed.), *Estuarine Research, Vol. 1, Chemistry, Biology, and the Estuarine System*. New York: Academic Press, Inc.

Thompson, R.W., 1968. *Tidal Flat Sedimentation on the Colorada River Delta, Northwestern Gulf of California*. Boulder: Geological Society of America Memoir 107.

Tomlinson, P.B., 1986. *The Botany of Mangroves*. Cambridge: Cambridge University Press.

Cross-references

Bay Beaches
Coastal Sedimentary Facies
Endogenic and Exogenic Factors
Hydrology of Coastal Zone
Mangroves, Ecology
Mangroves, Geomorphology
Muddy Coasts
Ripple Marks
Salt Marsh
Tidal Creeks
Tidal Flats, Open Ocean Coasts
Tides
Vegetated Coasts

TIDAL FLATS, OPEN OCEAN COASTS

Definitions and distribution

Although tidal-flat deposits have been studied extensively over several decades, most studies were focused on embayment and estuary tidal flats, where wave energy is typically low. In contrast, studies of open-coast tidal flats, which differ significantly from the embayment and estuary tidal flats, are scarce. This entry describes the characteristics of open-coast tidal flat in comparison with the well-documented embayment and estuary tidal flats.

A tidal flat is generally defined as an extensive, nearly horizontal, marshy or barren tract of land that is alternately covered and uncovered by the tide, and consisting of unconsolidated sediment, mostly mud and sand (Bates and Jackson, 1980). The tidal flat is also often referred to as an intertidal zone, although in some general discussions, it may include subtidal and supratidal zones. In the following discussion, the term tidal flat strictly refers to the intertidal zone, lying between low-spring tide and high-spring tide levels, as defined in the Glossary of Geology (Bates and Jackson, 1980). A tidal flat generally shows a zonation related to the duration of submergence, which is reflected in apparent differences in sedimentary characteristics. Therefore, based on characteristics of the sedimentary structures and the general trend of lamina thickness, a tidal flat is often divided into upper, middle, and lower intertidal zones (Reineck and Singh, 1980). Although the transitions between the zones are gradual and somewhat subjective, the division provides convenience in describing sedimentary characteristics.

Tidal currents are generally considered to be the dominant driving force for sediment movement on tidal flats, whereas wave-driven sediment motion is often regarded to be minimal and is often neglected. Over 70% of locations classified as tidal flat occur in wave-sheltered areas, such as bays, estuaries, lagoons, and behind spits or barriers, while the remainder occur along open coasts, the majority of which are characterized by low wave conditions (Eisma, 1998). Although high wave energy is often considered to be unfavorable for the development of gentle tidal flats (Boggs, 1995), they can nonetheless develop rather extensively along open coasts given a large tidal range and tremendous sediment supply. Examples of this type of open-coast tidal flats are found in the vicinity of the large river mouths along the Chinese coast (Figure T26; Chen, 1998; Shi and Chen, 1996).

In comparison with extensively studied wave-sheltered tidal flats, open-coast tidal flats are characterized by: (1) facing an open ocean or sea; and (2) flooding and ebbing tidal currents are not confined and/or regulated by tidal channels. In other words, large tidal channels are generally absent along open-coast tidal flats, especially those along the Chinese coasts.

Hydrodynamics and sediment dynamics

Most of the published studies on modern tidal flat deposition have concentrated on wave-sheltered areas in Europe and North America (e.g., Klein, 1976; Boersma and Terwindt, 1981; Dalrymple *et al.*, 1991; Allen and Duffy, 1998; Eisma, 1998). Not surprisingly, therefore, sedimentary characteristics associated with tidal currents, especially those flowing through tidal channels, have been studied extensively, with little attention being paid to sediment motions driven by waves. Generally speaking,

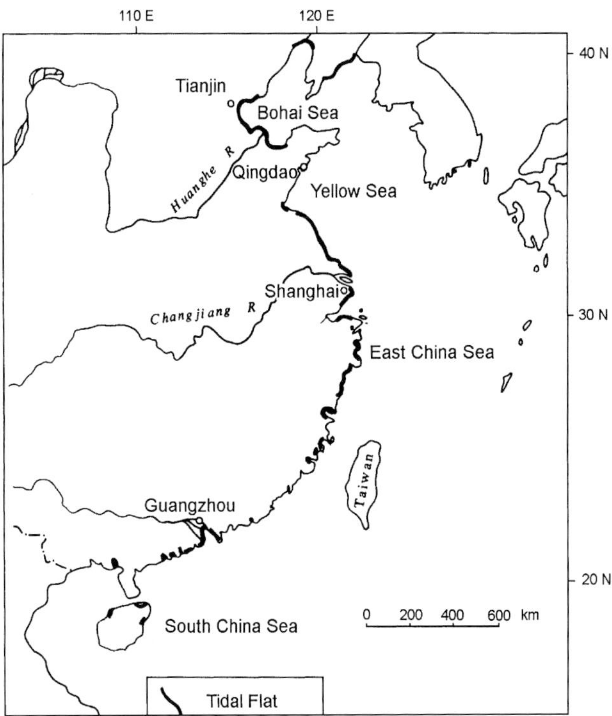

Figure T26 Distribution of open-coast tidal flats in China.

open-coast tidal flats are rather poorly studied and documented, and as a result, the characteristics of hydrodynamics and sediment dynamics presented subsequently are largely based on the studies along the Chinese coasts.

Along open-coast tidal flats, tidal channels are generally absent and, therefore, do not significantly regulate flood and ebb currents. Sedimentary features, such as mega-ripples and dunes, associated with bedform migration in the tidal channels are rare to non-existent. Wave energy is significantly dissipated over the extensive and gentle muddy flats during normal weather. However, during storms, due to the lack of wave shelter, open-coast tidal flats are vulnerable to the impact of storm waves, which may induce substantial reworking and redeposition of normal weather tidal deposits. The tidal flats, especially the middle and upper intertidal zones, are much sandier during the storm seasons than during the calm-weather seasons.

Sediment dynamics along open-coast tidal flats carry strong regional characteristics. The sediment dynamics along the Chinese open-coast tidal flats are significantly influenced by the tremendous sediment supplies from the adjacent large rivers, for example, the Changjiang river (Figure T26). Most open tidal flats along the Chinese coast are accretional owing to the abundant fine (silt and clay) riverine sediment supply. Coastal erosion is typically caused by starvation of sediment supply due to switching of river mouths, artificial flood controls, or drainage withdrawal. Mean sediment grain size along the Chinese open-coast tidal flat ranges from 4 to 8 ϕ (0.063–0.004 mm). The fine nature of the sediment also contributes to the absence of the mega-ripples in addition to the general lack of tidal channels. The width of the tidal flats ranges from 3 to 4 km with a maximum of approximately 8 km. The average slope of the tidal flat is typically 1 : 1,000 with a maximum of 1 : 200 and a minimum of 1 : 5,000.

Characteristics of sedimentary structures

The sequence of sedimentary structure variation from upper to lower intertidal zones as described by Reineck and Singh (1980) was also observed on open-coast tidal flats. The upper intertidal zone is characterized by relatively finer sediment with thicker muddy laminae. Lenticular bedding is common in the upper intertidal zone. The lower intertidal zone is characteristic of coarser sediment and thicker sandy laminae. Flaser bedding is common in the lower intertidal zone. Wavy bedding is common in the middle intertidal zone.

It is generally accepted that four laminae may theoretically be deposited during one tidal cycle (Allen, 1985). Two sandy laminae may be formed during flood and ebb phases, and two muddy laminae deposited during high and low tide slack water. It was also found that thicker sand laminae correspond to relatively higher-energy events during spring tides, while thinner sand laminae correspond to relatively low-energy events during neap tides (Boersma and Terwindt, 1981; Allen, 1985). Time-series analysis of laminae thickness has been applied to quantify the paleo-tide periodicities and paleo-sedimentation rates in both modern and ancient tidal deposits (Yang and Nio, 1985; Kvale *et al.*, 1989; Tessier and Gigot, 1989; Kuecher *et al.*, 1990; Kvale and Archer, 1990; Tessier, 1993; Miller and Eriksson, 1997), and an approximate 14-day periodicity related to neap–spring cycles has been identified in most of these studies. In the above studies, deposition and erosion induced by waves have largely been neglected. Along open-coast tidal flats, four laminae were rarely observed to have been deposited during one tidal cycle due to their poor preservation potential. Over a continuous daily observation of eight months, the preservation of all four laminae after one tidal cycle was only recorded twice (Li *et al.*, 1965).

Two different grouping patterns of sandy and muddy laminae were distinguished on open-coast tidal flat deposits (Figure T27). Groups with generally thicker sandy laminae than adjacent groups, are termed sand-dominated layers (1), while groups with generally thinner sandy laminae than adjacent groups, are referred to as mud-dominated layers (2). Although determination of the exact boundaries between sand- and mud-dominated layers was somewhat subjective, the overall differences between adjacent sand- and mud-dominated layers were apparent (Figure T27). The thickness and number of sandy and muddy laminae in each sand- or mud-dominated layer were not necessarily identical. Detailed description of sand- and mud-dominated layers can be found in Li *et al.* (2000).

Daily, monthly, and yearly sedimentation monitoring along a Chinese open-coast tidal flat near the Changjiang river mouth indicated that the mud-dominated layers described above correspond to calm-weather deposition, while the sand-dominated layers are related to high-energy storm events. These findings are contrary to the neap–spring-cycle interpretation of lamina-thickness variations. The above interpretation of event-related lamina-thickness variation incorporated the much more significant influence of waves, especially storm waves, to the open-coast tidal flat deposits.

High-energy storm events along the east-central Chinese coasts are mainly driven by the typhoon passages typically during the months of August to November, and to a lesser extend the passages of winter cold fronts.

Sedimentation rate and preservation potential

The correspondence of thick–thin variations of tidal laminae with the neap–spring tidal cycles has provided a promising tool to study the paleo-tide periodicities and the paleo-sedimentation rate (Tessier and Gigot, 1989; Miller and Eriksson, 1997). Time-series analysis of the tidal bundle thickness variation has been applied successfully in tidal channel deposits (Yang and Nio, 1985). However, direct application of this neap–spring analysis is not valid along open-coast tidal flats due to the non-negligible and largely random impact of wave-induced erosion and sedimentation. Sedimentation rate and preservation potential vary from region to region, and are strongly influenced by sediment supplies and regional hydrodynamics. In the following, a case study from east-central China is discussed as an example.

Figure T27 Different grouping patterns of tidal bedding. 1 Denotes a sand-dominated layer, and 2 denotes a mud-dominated layer.

Realizing that 100% preservation potential may be far from realistic on an open-coast tidal flat, Li *et al.* (2000) conducted an intensive, time-series *in situ* monitoring of the sedimentation rate and preservation potential. Their methodology included two aspects to understand and quantify the deposition and preservation of tidal bedding. The first aspect emphasized *in situ* observations on the modern tidal flat. Selected points were visited daily during low tide to examine deposition and erosion of the previous two tidal cycles. This daily observation had been conducted for 17 days over a neap–spring tidal cycle. Seasonal sedimentation and erosion observations across an intertidal profile were conducted over a four-month period extending from calm-weather season to storm season. The second aspect of the study focused on examination of the vertical characteristics of tidal laminae and bedding, in terms of the lamina numbers and lamina-thickness variations. Knowledge gained from the *in situ* observations was applied to interpret the vertical distribution of tidal bedding and the preservation potential of individual sandy and muddy laminae in cores and trenches.

Short-term observation was conducted using two thin plastic plates (40 cm long, 40 cm wide, and 2 mm thick). The plates were placed in the transition area between middle and upper intertidal zones. The thin plates were placed flush with the average sediment surface, and their surfaces were sanded to increase the roughness. Plate 1 was left in place for a period of 17 days and one observation was made at the end of the experiment. The objective for monitoring plate 1 was to obtain short-term information without daily disturbance. Deposits on plate 2 were measured daily, or every two days in cases of poor weather conditions. The objective of the plate 2 experiment was to quantify daily sedimentation rate and the number of laminae formed.

The daily monitoring spanned one neap–spring tidal cycle. No apparent trend of lamina-thickness variation was observed from the daily plate experiments. The daily sedimentation rates were rather uniform ranging 17–22 mm/day, except two abnormal values of 12 and 45 mm/day, both measured during spring tides. The calculated 17-day deposition was 378 mm with 50 sandy and muddy laminae, while 12 laminae with a total thickness of 75 mm were measured on the 17-day plate. Thus, the uninterrupted 17-day sedimentation was 20% of the cumulative daily deposition in terms of thickness (75 mm versus 378 mm), or 24% in terms of total number of laminae (12 versus 50). Furthermore, comparison between the results from the 17-day monitoring (12 laminae) and the theoretical estimate (4 laminae per tidal cycle for 33 tidal cycles over 17 days) of 132 laminae indicates that only approximately 9% of the laminae were preserved, even during a short period of one neap–spring tidal cycle. Continuous daily observation of tidal lamina number and thickness, conducted over three neap–spring cycles in 1999, yielded similar preservation potential.

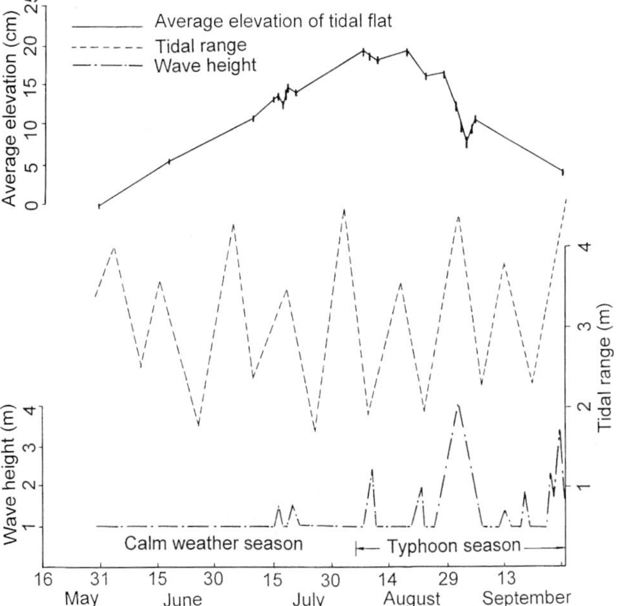

Figure T28 Average elevation change at Donghai Farm on the southern flank of the Yangtze delta, neap–spring tidal cycles and wave heights during a 4-month study, 1992. Wave heights lower than 1 m were neglected.

Seasonal monitoring spanned four months with two months at the end of a calm-weather season and two months at the beginning of the following storm season. Sedimentation and/or erosion were measured at 35 locations relative to a series of graduated rods. A total of 22 measurements were conducted during the four-month period. Overall, the intertidal zone was accreting during the studied calm-weather season (Figure T28), as indicated by the increasing of the average elevation at all the rods. During the studied storm season, net erosion (elevation decrease) was measured and the flat was covered by a sandy lamina that was much thicker than those deposited during the calm-weather season. Sharp elevation decrease was usually measured directly after the storms, apparently indicating the erosion caused by storm waves. For the convenience of discussion, the calm- and storm-weather season was divided somewhat subjectively by the first significant typhoon impact in the study area. The frequent observations throughout the four-month period indicated that the tidal flat was generally muddier in the calm-weather season than during the stormy season. The changes of average elevation were related not to the neap–spring tidal cycles, but closely to high wave events (Figure T28). Deposition of a relatively thicker sandy lamina was directly related to the high-energy wave events induced by the passage of a typhoon, instead of during spring-tide conditions.

Long-term sedimentation rate and preservation potential were determined through counting the number of laminae and mud- and sand-dominated layers in core sections that were deposited over 100 years. Details are described in Li *et al.* (2000). The centennial sedimentation rate was found to be on the order of 4 cm per year. It is worth emphasizing that the high sedimentation rate is closely related to the tremendous sediment supply from the adjacent river. Such a high sedimentation rate should not be expected at locations without overwhelming sediment supply. Over the 100-year period, the preservation potential of individual lamina, including both calm-weather and storm deposits, was found to be on the order of 0.2%, which was 45 times smaller than the 9% estimated for a short-term of a neap–spring cycle. It is expected that the preservation potential decreases as temporal interval increases. The 100-year preservation potential of storm-induced sand-dominated layers was estimated to be of the order of 10%, 50 times higher than the overall potential of 0.2%.

Summary

Waves, especially high storm waves, have a significant influence on sedimentation and preservation of intertidal deposits along the open-coast tidal flats. The thickness variation of sandy laminae on an open-coast tidal flat is related not to neap–spring tidal cycles, but directly to storm activities. The mud-dominated layers containing thinner sandy laminae were deposited during calm-weather conditions, while the sand-dominated layers containing relatively thicker sandy laminae were deposited during storm seasons. In other words, the thick–thin variation of sandy laminae may reflect a much longer cycle of calm-weather and storm seasons, instead of the fortnightly neap–spring tidal cycles as suggested from studies on wave-sheltered tidal flats.

One hundred percent preservation of both the number and thickness of individual laminae in tidal flat deposits, which has often been assumed in the interpretation of time-series analysis of laminae-thickness variation, was found to be unrealistic along the studied open-coast tidal flat. Preservation potential decreases as timescale increases. During one neap–spring tidal cycle under calm-weather conditions, the preservation potential of individual lamina was approximately 9%. However, over a period of 100 years, the preservation rate of individual lamina decreased to about 0.2%. The preservation rate of storm-induced, sand-dominated layers during the 100-year period was found to be on the order of 10%, much higher than the 0.2% of the individual sandy and muddy lamina. Storm deposits have a much higher potential for being preserved due to the higher energy level at which they were deposited.

Congxian Li, Ping Wang and Daidu Fan

Bibliography

Allen, J.R.L. 1985. *Principles of Physical Sedimentology*. London-Boston-Sydney: George Allen & Unwin.

Allen, J.R.L., and Duffy, M.J., 1998. Temporal and spatial depositional patterns in the Severn Estuary, southwestern Britain: intertidal studies at spring–neap and seasonal scales, 1991–1993. *Marine Geology*, **146**: 147–171.

Bates, R.L., and Jackson, J.A. (eds.), 1980. *Glossary of Geology*. Alexandria, VA: American Geological Institute.

Boersma, J.R., and Terwindt, J.H.J., 1981. Neap–spring tide sequences of intertidal shoal deposits in a mesotidal estuary. *Sedimentology*, **28**: 151–170.

Boggs, S., 1995. *Principles of Sedimentology and Stratigraphy*. Upper Saddle River, NJ: Prentice Hall.
Chen, X., 1998. Changjian (Yangtze) river delta, China. *Journal of Coastal Research*, **14**: 838–858.
Dalrymple, R.W., Makino, Y., and Zaitlin, B.A., 1991. Temporal and spatial patterns of rhythmite deposition on mud flat sedimentation in the macrotidal Cobequid Bay–Salmon River estuary, Bay of Fundy, Canada. In Smith, D.G., Reinson, G.E., Zaitlin, A., and Rahmani, R.A. (eds.), *Clastic Tidal Sedimentation*. Canadian Society of Petroleum. *Geology, Memoir* 16, 137–160.
Eisma, D., 1998. *Intertidal Deposits—River Mouths, Tidal Flats, and Coastal Lagoons*. Boca Raton, FL: CRC Press.
Klein, G. deV., 1976. *Holocene Tidal Sedimentation*. Stroudesburg: Dowden, Hutchinson & Ross.
Kuecher, G.J., Woodland, B.G., and Broadhurst, F.M., 1990. Evidence of deposition from individual tides and of tidal cycles from the Francies Creek Shale. *Sedimentary Geology*, **68**: 211–221.
Kvale, E.P., Archer, A.W., and Johnson, H.R., 1989. Daily, monthly, and yearly tidal cycles within laminated siltstones of the Masfield Formation of Indiana. *Geology*, **17**: 365–368.
Kvale, E.P., and Archer, A.W., 1990. Tidal deposits associated with low sulfur coals, Brazil FM (Lower Pennsylvanian), Indiana. *Journal of Sedimentary Petrology*, **60**: 563–574.
Li, C., Yang, X., Zhuang, Z., Zian, Q., and Wu, S., 1965. Formation and evolution of the intertidal mudflat. *Journal of Shangdong College of Oceanography*, **2**: 21–31 (in Chinese with Russian abstract).
Li, C., Wang, P., Fan, D., Dang, B., and Li, T., 2000. Open-coast intertidal deposits and the preservation potential of individual lamina: a case study from east-central China. *Sedimentology*, **47**: 1039–1051.
Miller, D.J., and Eriksson, K.A., 1997. Late Mississippian prodeltaic rhythmites in the Appalachian Basin: A hierarchical record of tidal and climatic periodicities. *Journal of Sedimentary Research*, **67**: 653–660.
Reineck, H.E., and Singh, I.B., 1980. *Depositional Sedimentary Environments*. New York: Springer-Verlag.
Shi, Z., and Chen, J., 1996. Morphodynamics and sediment dynamics on intertidal mudflats in China (1961–94). *Continental Shelf Research*, **16**: 1909–1926.
Tessier, B., 1993. Upper intertidal rhythmites in the Mont-Saint-Michel Bay (NW France): perspectives for paleo-reconstruction. *Marine Geology*, **110**: 355–367.
Tessier, B., and Gigot, P., 1989. A vertical record of different tidal cyclicities: an example from the Miocene Marine Molasse of Digne. *Sedimentology*, **36**: 767–776.
Yang, C., and Nio, S., 1985. The estimation of palaeohydrodynamic processes from subtidal deposits using time-series analysis methods. *Sedimentology*, **32**: 41–57.

Cross-references

Asia, Eastern, Coastal Geomorphology
Coastal Lakes and Lagoons
Deltas
Tidal Environments
Tidal Flats
Tides
Waves

TIDAL POWER

Tidal energy

Tidal energy is derived from the earth's inherent force, the earth's rotation within the disturbing field of moon and sun. In relatively shallow seas friction dissipates almost all the energy. Though tides can be predicted very accurately, they are not in phase with moon and sun movements, the tide waves being distorted as landmasses, narrow passages and shallow depth areas impede and/or influence their progression. (Charlier, 1982). The tidal current is the rotary current accompanying the turning tide crest in an open ocean. It becomes a reversing current, nearshore, moving in and out as flood and ebb currents. Both of these can be harnessed to produce mechanical and/or electrical power (Charlier and Justus, 1993).

Major disadvantages of tidal power electricity generation is that tides are linked to the lunar rather than to the solar cycle, and vary in range throughout the year due to their periodic components. These negative aspects can be partially overcome by ingenious engineering and retiming of use of potential energy accumulated at low-demand periods. Such retiming had even been suggested several decades ago using retaining basins, compressed-air, or hydrogen—even electrolytically produced—usable as fuel when tide and peak-power demand are not synchronic (Gilbert, 1982). Storing has been proposed in exhausted deposits, abandoned mines, and artificial cavities.

Of the 3,000 million kilowatts equivalent dissipated by tidal energy, one billion (10^9) develop in shallow seas. But only a fraction of this "power" can be captured. The geographical site must be suitable from engineering and economics viewpoints and the "usable head" has to be high (5 m or more). The latter requirement has drastically changed with the development of low, and ultra-low, head turbines, greatly increasing the number of potential sites; thus the 200 million kW theoretically harnessable 20 years ago has sizably increased. The economic geographical factor has also been altered as a result of improvement in transmission possibilities; indeed, national grid systems and high voltage transmission lines have minimized the problem of distance between generating plant and consumer (e.g., in Canada) and the problem of protection (insulation) against extreme temperature amplitudes was resolved more than a decade ago by the Soviets. Finally, where the emphasis has been placed on huge, even gigantic (e.g., Chausey Islands, France) projects, small plants are often favored currently, an advantage for developing countries or isolated areas (Suriname, Half Moon Cove [Maine, USA], China).

The plant

The tide mill (gm) and the tidal power plant are similar. The latter has an electric generator. The plant's major components are a dam which houses the powerhouse, a retaining basin, and a link to the electricity grid. Construction of a barrage or dam is necessary; earthen dams have been used in construction of Chinese plants. The dam consists additionally of dikes connecting with the natural embayment and a sluiceway. A passage way for fishes is now provided. A reasonable tidal range is required, although quite small ones will suffice today. Estuaries and gulfs in shallow areas are privileged sites. A closed basin is thus created which fills and empties daily, sometimes more than once, depending on the local tidal regime. Equipped with sluices and turbines placed in the barrage, the system retains the water entering at flood tide and releases it at ebb tide.

Generally turbines produce electricity as the water flows out of the retention basin (or "pool") but reversible blade (the so-called *bulb*) turbines, originally invented by Harza and later developed by French engineers (Hollenstein and Soland, 1982), can produce electricity both as the water enters and when it exits the basin. Practice—based upon 30 years of operation of the Rance River, France, plant—has, however, shown that entering water generation is perhaps not always economically profitable. By judiciously selecting tide gates opening and closing times, power generation can be synchronized with peak-demand periods, even if these do not coincide with tide peaks. The hydraulic head can be increased by reversing power units, temporarily turning the turbine-generator into a pump-motor; the bulb turbine precisely regulates flow in both directions and acts as turbine and pump. Pumping, however, consumes electricity.

A so-called "site value coefficient" for plants (k) was calculated by Robert Gibrat based on dam length and natural energy (k = L/NE), however, these are not the only "factors" involved in site selection; considered should be tidal range, potential pool size, ratio of basin aperture section to basin surface, length of dam, basin characteristics (such as geometric shape and surface, opening, widening shape), geological structure, lithology and petrology, foundation soil quality, gradient of resistance layer, probability of silting, rate of sedimentation, climate, market distance, competitiveness of conventional and/or alternative (e.g., nuclear, aeolean) power sources. Some economists maintain that today tidal power is competitive and may even be less expensive than fossil fuel and nuclear generation. Calculations made by Voyer and Penel (1957) have been revised (Charlier, 1998). All the factors are seldom, if ever, simultaneously favorable, which, with the high capital investment for large plants, probably accounts for the small number of plants constructed (Charlier and Justus, 1993).

Types of plants

Plants are single- or multiple-basins systems, a tide-powered air storage scheme, tide-powered hydrostatic pump scheme or Gorlov setup. Single-basin plants function either as a one-way operation (ebb generation only), two-way operation (generation at both ebb and flood flow), two-way operation with pump-turbines generation (excess used to pump water in storage reservoir[s]), or high-tide pumped storage (non-tide-connected power production). Multiple-basin schemes are either double-pool (station placed in-between basin, filling one basin at flood, emptying the other at ebb), both basins pumping (pumping high pool up and low pool down, pumping at off-peak), pool-to-pool dam pumping, pool-to-pool dam pumping combined with pump-turbines, tide-boosted pumped storage (basins may be of different sizes and in different places), or variations of the latter.

The tide-powered air storage scheme differs in that it uses the tidal energy to drive air turbo-compressors and stores the compressed air. In the tide-powered hydrostatic pump a propeller turbine is linked to a pump and the tidal energy is converted in a flow of high-pressure oil which turns a Pelton wheel coupled to an alternator. The Gorlov proposal involves a thin plastic barrier hermetically anchored to bottom and bay sides supported by a bay-spanning cable. The tidal energy would be converted into power by an air motor piston at ebb with direct generation or compressed air storage (Gorlov, 1982). Finally, one technology would anchor in line a series of floating turbine and generator units along the flow of the tidal current (Charlier and Justus, 1993). This approach, less onerous probably than a "traditional" tidal power plant, would perhaps be attractive for small local or regional schemes.

Reversible blade (bulb) turbines have been placed in the Rance River (France) and the Kislogubskaia (Russian Federation) single-basin plants. The Russian plant was built in modules, which considerably reduced construction costs and avoided construction of expensive cofferdams. The turbines can be used as low-head pumps.

Operating plants

If the literature on tidal power plants is extensive (cf. Charlier, 1982, 1998, 2003; Charlier and Justus, 1993), it is not easy to always determine where plants were built (Figure T29). Little is known of a plant built in the mid-1920s in Suriname (then Dutch Guyana) or in Boston Bay (end of 19th century, and dismantled because of harbor extension). Information on China's post-World War II "more than one hundred" plants is scant and even in the country itself is poor, though a few papers have been recently published (Ch'iu Hou-Ts'ung, 1958). Occasionally construction of plants is announced, as in Korea or India (cf. Sharma, 1982; Song, 1987), but proves premature. Three major plants have been built and are operating: the Rance River (France), the Kislogubskaia (Russia), and the Annapolis-Royal (Nova Scotia, Canada) plants. The first one is large and dates from 1956, the second, a pilot plant, is much smaller and was heralded as a forerunner of more ambitious ones to come, and the third one is an experimental and pilot installation.

The Rance River Plant

Twenty-four bulb groups of 10,000 kW were placed in horizontal hydraulic ducts entirely surrounded by water at the Rance plant. The cost of the cofferdams represented a third of the building cost, an expense which can now be dispensed with as was in the Russian plant and to some extent in the Canadian plant. The cost of the project ran about US$100 million (in 1966-$).

Started in 1993 and completed in 1966, the 53 m wide and about 390 m long plant protrudes 15 m above water, with foundations 10 m below sea level, accommodates a four-lane roadway (eliminating a ferry) and has substantially contributed to the economic development of the former lethargic region besides furnishing 544 MW h of usable power. The mean net annual energy production exceeds slightly 500 GW h and the mean capacity throughout the year is 65 MW. It has six modes of operation. The operation policy has aimed at optimizing the value of the energy generated, reaching maximum profit, instead of eyeing to generate the maximum amount of energy. Generation by overemptying the basin (by pumping) has been done only rarely during the last decade.

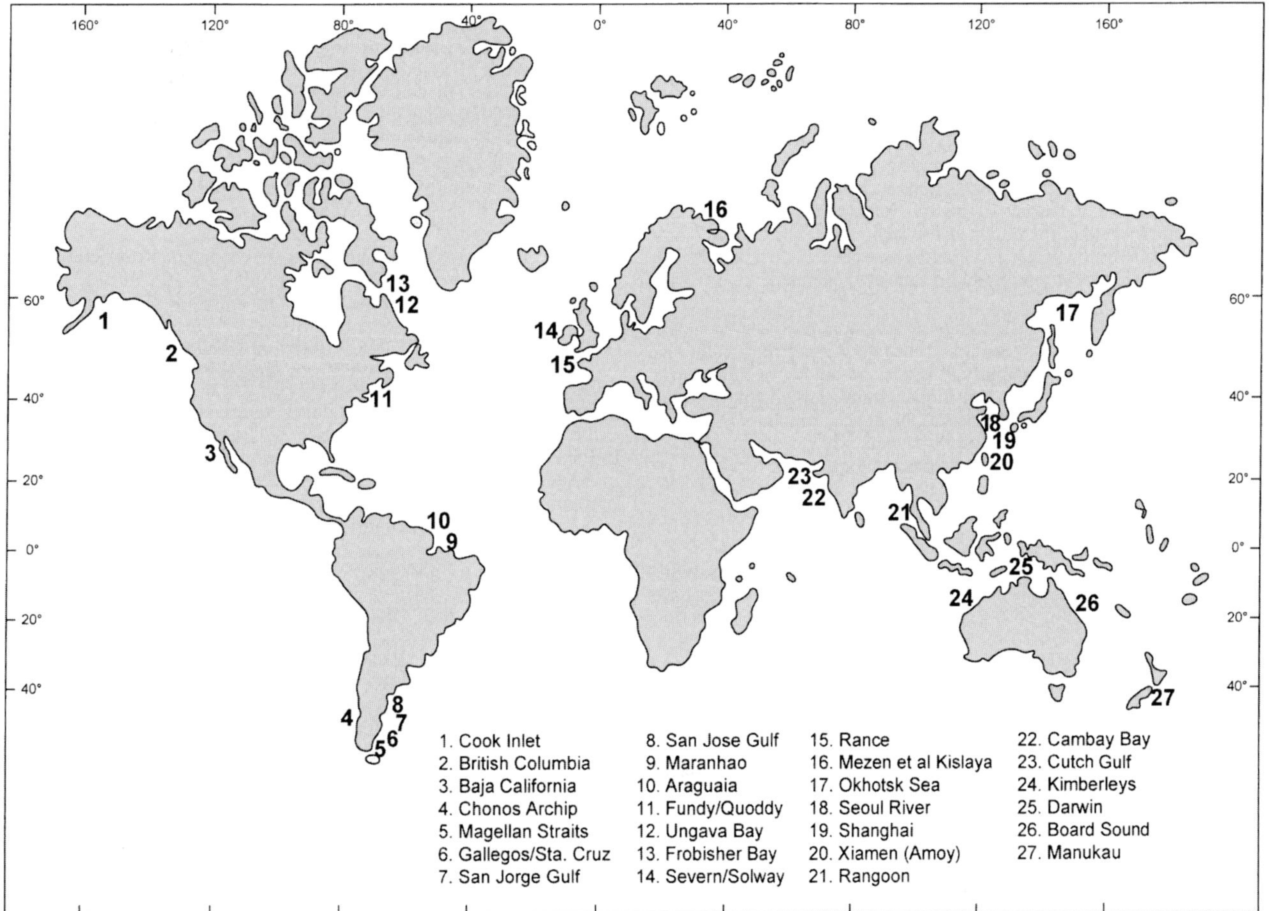

Figure T29 Location of major tidal power plants (after Charlier, 1993, with permission from Elsevier Science).

Environmental aspects at the Rance Plant. Environmental impact, at the French Rance River Plant, has been extremely mild causing only the disappearance of one species of fish. No major biological modifications have occurred. New species have appeared. Sandbanks have disappeared, high speed currents have appeared near sluices and also near the powerhouse where sudden surges have been observed. Tidal ranges have been reduced from 13.5 m to 12.8 m and minima increased. The fishing industry has not been affected contrarily to fears, and tourism has increased.

The Russian plant

The Kislaya Bay plant is located near Mezen on the Arctic. The small Russian Arctic coastal bays hold a 3.2 billion kW potential. Completed in 1968, the 18 m long, 36 m wide, 15 m high, with a 15.2 m head, plant was built by towing preconstructed concrete caissons to the site and then sinking them into place. A narrow passage connects bay and sea; no long dikes were necessary as the powerhouse itself closes the basin. Transmission lines existed close to the site. The power set capacity is 400 kW. The reversible hydraulic turbine is coupled with a synchronous generator, with the help of a multiplicator.

Russians are still considering a variety of schemes and stress that by harnessing tides in the Sea of Okhotsk bays, some 174 billion kW h could be produced.

The Canadian plant

An experimental plant was constructed in the early 1980s at Annapolis-Royal (Nova Scotia) in the Bay of Fundy whose "electricity potential" is huge; studies on Fundy tidal power go back to at least 1944. It is here that the first tide mill was built in North America (1607) using tidal and river water. A low-head STRAFLO turbine has been installed in the new plant whose generator and turbine form a coupled unit without driving shaft, particularly compact, and thus cutting powerhouse costs (Douma and Stewart, 1982). The costs amounted to CA$46 million (approximately US$44 at the time). Situated on Hog's Island in the middle of an intake canal on the lower reaches of the Annapolis River where tidal ranges reach 7 m, it took advantage of an existing dam protecting farm land. The Chinese, similarly, have commonly used existing (often earthen) dams to install tidal plants.

The turbine is a single effect one and generates electricity only during discharge of water from the retaining pool. In the event that the pool level would be below sea level, the turbine's water passage would sluice seawater into the basin during each tidal cycle. A cofferdam of local material was used to construct, in the dry, the powerhouse. Turbine and generator are located in the center of the turbine pit. The dam is 225 m long, 60 m wide at low water level, and 18 m wide at the crest. A fish pathway is provided. Fifty million kilowatts are generated.

Environmental impact. The Canadian project was designed to assess both the operational characteristics and the reliability of the Straflo turbine, and, of course, the possibility of implementing major tidal power utilization in eastern Canada. The impact study showed that interactions can be mitigated and that physical, biological, and human impact would be minimal and residual impacts small, while economic outfall would be considerable both for the short- and long-terms.

The Chinese plants

Information is difficult to secure, even in China. Claims of a large number of plants—as many as 105—and their successful operations have been repeatedly made, and so have plans to undertake new construction (Ch'iu, 1958). There are also records of the "temporary" abandon of some plants due to siltation.

Chinese engineers acknowledged a problem with siltation whose extent proves difficult to assess, but it has been said that some tidal power plants were put out of commission because of sediments accumulation. A comparative study on devising siltproof systems was carried out using data from the Baishakou tidal power plant and envisioned environmental protection to control sedimentation in the tallwater channel and reservoir (Xhikui Zu, 1992); at about the same time other studies explored the optimum patterns for double-effect single-basin plants (Shuyu Want *et al.*, 1991) and calculated (modelization) of the optimum tidal energy at any location (Lee Kwang Soo *et al.*, 1994). Zhikui Zhu presented in 1992 an analysis of data pertaining to the Baishakou tidal power station. It describes measures apparently taken to control sedimentation in the tallway channel and the reservoir area. Comprehensive management is outlined, including mechanical sandproofing methods and environmental protection.

The interest of Chinese researchers in tidal power plants has not waned, nor has it been somnolent as in Europe and the United States, due undoubtedly to the abundance and generally low cost of fossil fuel. An exploratory study and modelization were recently conducted (Zhuang Ji, 1991).

A review article on Chinese activity in tidal power utilization covering the 1950–90 period was published in the somewhat unfamiliar [to westerners] *Collection of Oceanographic Works*. The information rather contradicts other releases which talk about a hundred small tidal power plants. Guixiang Li (1991) mentions eight small power stations and discusses their tangible economic and social benefits. He further discloses that "many more" small and medium size, even one or two large, plants "will be built by the year 2000" along the coasts of China's mainland and the coastal islands. The publication is rather difficult to obtain, except perhaps through the good offices of the University of Karlsruhe (Germany).

Other geographical locations

No accurate information is available about the two-basin plant that functioned in Boston harbor at the turn of the preceding century until it was dismantled to make room for the port's extension, nor of the Van Bemmelen plant in Suriname. The tidal current plant in Iceland seems to have left no traces either. Plans were afoot in the 1990s to build a tidal power plant in [South] Korea (e.g., Garolim Bay) but came to a halt for political reasons (recognition of North Korea by France whose Sogreah was to build the facility) (Song, 1987). Other sites which have been considered, and for which periodically plans re-surface are the Kimberleys in Australia, San Jorge Gulf in Argentina (Aisiks and Zyngierman, 1984), and the Severn River in Great Britain (Severn Barrage Committee, 1986). There are of course numerous other suitable sites worldwide (Charlier and Justus, 1993), and studies were conducted for some of them, for instance for India (Sharma, 1982).

Tidal stream's rapid currents have been recently examined as sources of power, for example, in the Orkney and Shetland islands (Bryden *et al.*, 1993–1995), perhaps using Darrieus turbines (Khio *et al.*, 1996), and conversion of kinetic to electrical energy with a barrage in New York's East River (Birman, 1994). In Maine and South Carolina experiments proved that using tidal stream power with speeds up to 1.5 knots cuts the cost of seeding rafts with a wooden scoop on the bow that directs the following water to an enclosed compartment containing upwelling units and is less expensive that using other sources of power (Hadley, 1994; Rhode *et al.*, 1994).

Roger H. Charlier

Bibliography

Aisiks, E.G., and Zyngierman, I., 1984. The San Jose Gulf tidal power plant, Argentina. *Proceedings of ECOR '84 & 1st Altern. Ener. Argent. Conf.* **II**, pp. 1–9.

Birman, G., 1994. Prospects of tidal energy generation. *Bulletin of N. Y. Academy of Science*, pp. 147–149.

Bryden, I.G. *et al.*, 1993–1994. Tidal stream power for the Orkneys and Shetlands? *Underwater Technology*, **19**(4): 7–11.

Bryden, I.G. *et al.*, 1995. An assessment of tidal streams as energy source in the Orkneys and Shetlands. *Underwater Technology*, **21**(2): 21–29.

Ch'iu Hou-Ts'ung, 1958. The building of the Shamen TPP. *Tien Chi-Ju Tung-Hsin*, **9**: 52–56.

Charlier, R.H., 1982. *Tidal energy*. New York: Van Nostrand-Rheinhold.

Charlier, R.H., 1998. *Re-invention or aggorniamento? Tidal power at 30 years. Renewable and Sustainable Energy Review*, **I**(4): 271–289.

Charlier, R.H., 2003. Justainable cogeneration from the tides: bibliography. *Renewable & Sustainable Energy Reviews*, **7**(3): 215–247.

Charlier, R.H., and Justus, J.R., 1993. *Ocean Energies. Environmental, Economic and Technological Aspects of Alternative Power Sources*. Amsterdam, London, New York, Tokyo: Elsevier Science.

Douma, A., and Stewart, G.D., 1982. Annapolis Straflo turbine will demonstrate Bay of Fundy tidal power concept. *Hydro Power Modern Power Systems*, **I**: 53–65.

Gilbert, R., 1982. Retiming with hydrogen. *Proceedings of International Conference on New Approaches to Tidal Power*, New Bedford N.S. **3**: 1–4.

Gorlov, A.M., 1982. Hydropneumatic approach to harnessing tidal power. *Proceedings of International Conference New Approaches to Tidal Power (Bedford Institute of Oceanography Dartmouth, NS)* **5**: 5–11.
Guixiang Li, 1991. Prospects for the resources on the tidal energy development in China. *Collection of Oceanographic Works—Haiyang Wenji*, **14**(1): 128–134.
Hadley, N.H. *et al.*, 1994. Performance of a tidal powered upwelling nursery system for juvenile clams in South Carolina. *Journal of Shellfish Research*, **13**(1): 285.
Hollenstein, M., and Soland, W., 1982. The bulb turbine for the Rance power station. *Escher-Wyss News*, 54/55.
Khio, S., Shiono, M., and Suzuki, K., 1996. the power generation from tidal current by Darrierus turnines. *Renewable Energy*, **9**(1/4): 1242–1245.
Lee Kwang Soo *et al.*, 1994. A simple analytical model for the design of the tidal power scheme. *Ocean Research*, **16**(2): 111–124.
Rhode, R.J. *et al.*, 1994. Coast analysis of a tidal powered upwelling nursery for juvenile clams in South Carolina. *Journal of Shellfish Research*, **13**(1): 286.
Severn Barrage Committee, 1986. *Tidal Power from the Severn*. London: Telford.
Sharma, H.R., 1982. India embarks on tidal power. *Water Power & Dam Construction*, **34**(6): 32.
Shuyu Wang, Xigi Su, and Zhiyu Jin, 1991. Exploratory study on the optimum pattern of the tidal power plant. *The Ocean Engineer/Haiyand Gongcheny*, **9**(2): 82–90.
Song, W.O., 1987. Reassessment of Garolim tidal power project. *Ocean Research [Korea]*, **9**(1/2): 29–33.
Voyer, M., and Penel, M., 1957. Les calculs de la production d'une usine marémotrice. *La Houille Blanche (Conférence Quatrièmes Journées de l'Hydraulique)* **II**: 472–485.
Zhikui Zhu, 1992. Comparison of siltation protection measures in a Chinese tidal power station. In Larsen, P., and Eisenhauer, N. (ed.), *Proceedings of 5th International Symposium On River Sedimentation, Karlsruhe 6/10 April 1992. Sediment Management*, Vol. 2, pp. 847–852.
Zhuang Ji, 1991. Exploring study of optimum patterns of the tidal power plant. *The Ocean Engineer—Haiyang Gongcheng*, **9**(2): 82–90.

Cross-references

Engineering Applications of Coastal Geomorphology
Microtidal Coasts
Tidal Environments
Tidal Prism
Tide-Dominated Coasts
Tide Mill
Tides
Wave Power

TIDAL PRISM

The tidal prism is the amount of water that flows into and out of an estuary or bay with the flood and ebb of the tide, excluding any contribution from freshwater inflows. For this reason, it is often reported as the volume of the incoming tide, and the contribution of river inflow calculated from the difference of the ebb and flood volumes. The tidal prism can be determined from hydrographic charts, where the volume in the estuary between low water and high water is calculated from sounding data. The tidal prism can also be determined by measuring the amount of water flowing into an estuary using a technique known as a tidal gauging. This is normally undertaken at a narrow section in an estuary of regular shape, and at a time when river inputs are low. Measurements of current velocity and water depth are made continuously at various points throughout the section using current meters suspended from a boat or bridge, along with measurements of water (tide) level in the section over the tide. The tidal prism is computed as the product change in cross-section area and mean velocity in section, integrated over half the tidal cycle. Today a vessel mounted ADCP (acoustic doppler current profiler) which returns depth, tidal change, and current at many points in the water column as the vessel travels back-and-forth across the section greatly simplifies and improves the accuracy of a tidal gaugings.

The tidal prism is an important metric for an estuary. It is an indicator of the hydrodynamic processes operating in an estuary. In shallow estuaries where the tidal prism forms a large proportion of the water in the estuary at high tide, tidal processes dominate and flushing is good. In deep estuaries with relatively small tidal prisms density-driven flows and river inputs play a greater role in the hydrodynamics. In estuaries where there are semidiurnal tides and a large difference in spring and neap tidal ranges, the amount of water flowing into the estuary (the tidal prism) on a spring tide may be double that on a neap tide and tidal currents and sand transport increase along with this. Empirical relationships developed between tidal prism and various estuary parameters help with the conceptual understanding of processes and provide a simple tool for easy calculations of stable channel design (e.g., Bruun, 1990).

The tidal prism is related to inlet dimensions, the amount of sand stored in tidal deltas, and has been used to describe inlet stability. The empirical relationship between the gorge cross-sectional area and tidal prism of a tidal inlet has been described in numerous studies (O'Brien, 1931; Jarrett, 1976; Hume and Herdendorf, 1992) by:

$$A = c\Omega^n,$$

where A is the gorge cross-sectional area (m^2), Ω is the tidal prism (m^3) and c and n are constants. Inlet gorges that are stable conform to the relationship and there is considered to be a balance between the inlet geometry and tidal flow through the gorge. Those lying off the line are out of equilibrium and characterized by either scour or deposition. A similar relationship has been found to exist between the ebb tidal delta sand volume, which increases with increasing tidal prism. In this situation, the volume of the sand body also increases with decreasing wave energy and as the angle that the ebb jet makes with the beach on the adjacent barrier shore increases (Walton and Adams, 1976; Hicks and Hume, 1996). On coasts where there is little wave energy and estuaries have large tidal prisms the delta will be elongated offshore under the influence of the ebb tidal jet. Where estuaries with small tidal prisms occur on coasts with high wave energy the ebb tidal sand body will be more flattened against the shore. A–Ω relationships have been used to quantify the morphological stability of tidal inlets. Bruun and Gerritsen (1960) showed that the size of the tidal gorge is one of the main factors determining the ability of flow to transport sediment through the entrance. Inlet gorges that are morphologically stable (i.e., have the ability to return to their original configuration after a disturbance) conform to the relationship because there is a balance between tidal flow and littoral drift to the gorge, so the inlet stays open. They demonstrated that when the ratio of the tidal prism to total (gross) annual littoral drift delivered to the inlet from the ocean is in excess of 300 the inlet has a high degree of stability. In cases where the ratio is less than 100 there will be a low degree of stability and entrance bars will shallow and navigation difficult. A–Ω relationships like those for tidal inlets on sandy shores also hold for many different estuary types ranging from lagoon to river mouth and even to large coastal embayments, and have been used to characterize and classify inlets (Hume and Herdendorf, 1993).

Terry M. Hume

Bibliography

Bruun, P., 1990. *Port Engineering*. Houston, TX: Gulf Publishing Co.
Bruun, P., and Gerritsen, F., 1960. *Stability of Coastal Inlets*. Amsterdam: North Holland.
Hicks, D.M., and Hume T.M., 1996. Morphology and size of ebb tidal deltas at natural inlets on open-sea and pocket bay coasts, North Island, New Zealand. *Journal of Coastal Research*, **12**: 47–63.
Hume, T.M., and Herdendorf, C.E., 1992. Factors controlling tidal inlet characteristics on low drift coasts. *Journal of Coastal Research*, **8**: 355–375.
Hume, T.M., and Herdendorf, C.E., 1993. On the use of empirical stability relationships for characterising inlets. *Journal of Coastal Research*, **9**: 413–422.
Jarrett, J.T., 1976. Tidal prism-inlet area relationships. Vicksburg, MS: US Army Corps of Engineers, Coastal Engineering Research Center and Waterways Experimental Station, *GITI Report No. 3*.
O'Brien, M.P., 1931. Estuary tidal prisms related to entrance areas. *Civil Engineer*, **1**: 738–739.
Walton, T.L., Jr., and Adams, W.D., 1976. Capacity of inlet outer bars to store sand. In *Proceedings, 15th International Conference on Coastal Engineering*. American Society of Civil Engineers, New York, pp. 1919–1937.

Cross-references

Estuaries
Instrumentation (see Beach and Nearshore Instrumentation)

Tidal Environments
Tidal Flats
Tidal Power
Tide-Dominated Coasts
Tide Gauges
Tides

TIDE-DOMINATED COASTS

As first enunciated by Price (1955), and later elaborated upon by others (e.g., Hayes, 1975), tides play an important role in defining the geomorphology of depositional shores, namely coastal and deltaic plains. On such shores, it is the ratio of tidal energy, usually dictated by tidal range, to wave energy, a function of average wave height, that determine the morphology of the coast, with sediment supply being an important modifier near major river mouths. As a generalization, coasts with small tidal ranges (microtidal ≤2 m (Davies, 1964)) are dominated by wave energy. Such coasts were termed *wave-dominated coasts* by Price (1955). On the other hand, coasts with large tidal ranges (macrotidal ≥4 m) are typically dominated by tidal energy, and were hence termed *tide-dominated coasts* by Price. The effectiveness of wave energy diminishes (i.e., waves cannot break in a concentrated area for a long period of time), and tidal current energy increases as the vertical tidal range increases. Exceptions to this generalization occur where waves are so small that even small tides generate adequate energy to shape the coast. An example of a *tide-dominated coast* with a relatively small tidal range occurs at the head of the embayed coastline of northwest Florida (Hayes, 1979). Coasts with intermediate wave energy and tidal ranges (typically mesotidal = 2–4 m) were termed *mixed-energy coasts* by Hayes (1979).

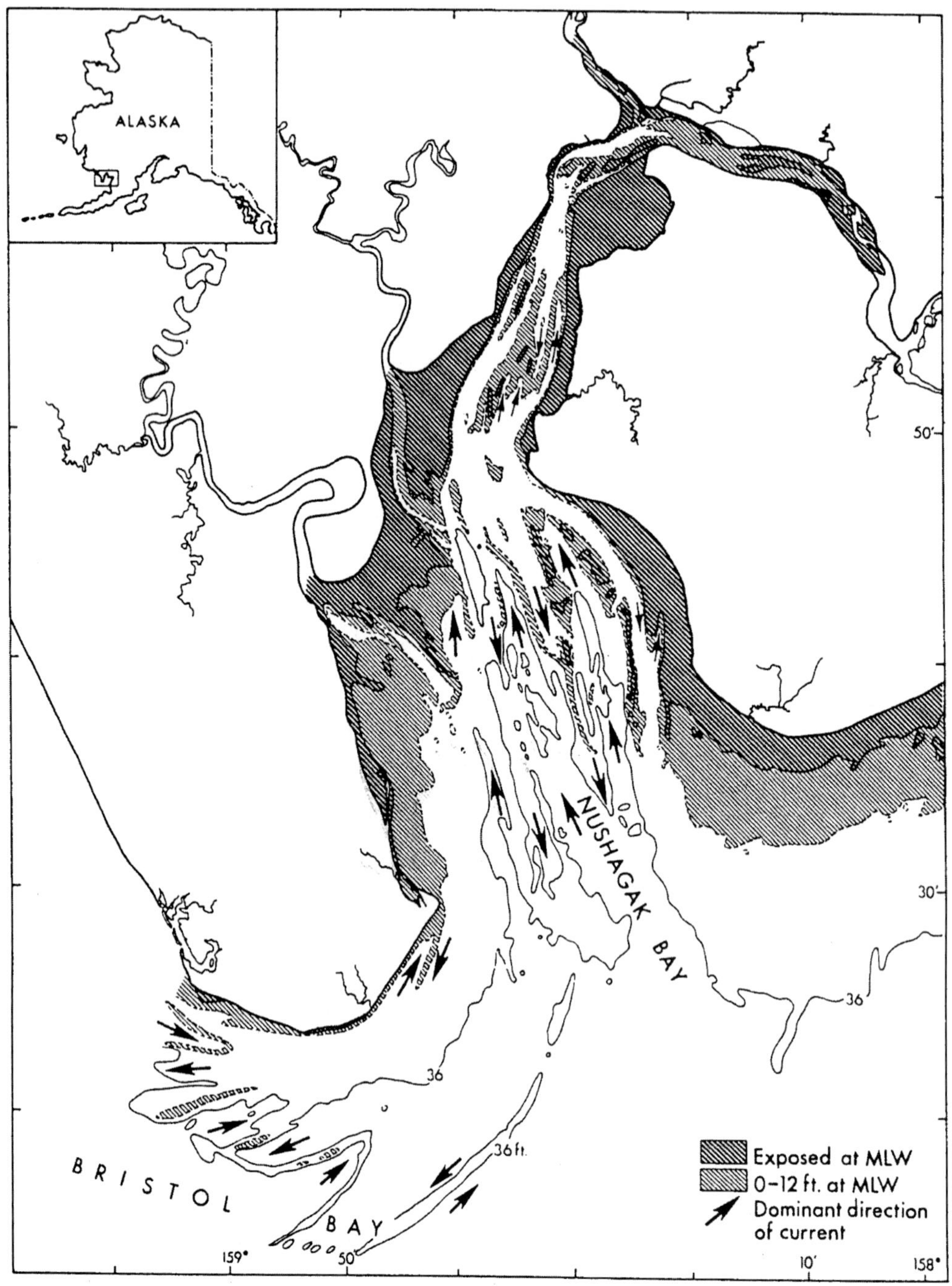

Figure T30 Nushagak Bay, Alaska, a tide-dominated embayment (tidal range = 5.5 m). Note lineation of shoals parallel with tidal currents.

Tide-dominated, non-deltaic coasts

The bathymetry of a typical tide-dominated, non-deltaic shore, Nushagak Bay, Alaska, is illustrated in Figure T30. The coastal morphology of major river mouths on tide-dominated coasts are most commonly open-mouthed estuaries, as shown in Figure T30. Between major rivers on these types of coasts, the shore is occupied by extensive salt marshes and tidal flats. Barrier islands are completely missing, because wave action is not focused enough at a single topographic level to build a barrier island and tidal currents are strong enough to disperse the sand to offshore regions. Examples of this type of coast occur in northwest Australia (Coleman and Wright, 1975), western Korea (Kim *et al.*, 1999), the Bay of Bengal, northern end of the Gulf of California (Thompson, 1968), the Wash, England (Evans, 1979), and many other depositional coasts with tidal ranges greater than 4 m. Generally speaking, the sediment distribution patterns on coasts of this type are exactly opposite to those on wave-dominated coasts, inasmuch as finest sediments occur on mudflats and in wetlands of the upper intertidal zone and coarsest sediments occur lower in the intertidal zone and offshore where tidal currents are strongest.

Studies by Evans (1975) documented the characteristics of the sediments of the tidal flats of the Wash, an indention in the coastline of the east coast of England, which has a tidal range of 7.0 m. The seaward portions of the flats are made up of complex sand bodies covered by sand waves, whereas the upper part is fringed by a salt marsh composed of fine-grained sediments. Studies of the tidal flats of the Bay of Fundy, by Knight and Dalrymple (1975), Dalrymple *et al.* (1990), Yeo and Risk (1981), and numerous others, describe the sedimentology, sediment transport dynamics, and potential stratigraphy of this intertidal zone, which has the largest tidal range in the world.

Figure T31 gives a hypothetical prograding stratigraphic sequence for the intertidal zone of a tide-dominated coast based on the references cited above and other sources.

Tide-dominated deltas

On tide-dominated coasts where the rivers have enough sediment load to have filled the antecedent lowstand valley and build a bulge in the shore (within the time frame of the present highstand), the resulting river deltas are typically referred to as *tide-dominated deltas* (Fisher *et al.*, 1969; Galloway, 1975). However, the distinction between such deltas and estuaries is somewhat obscure. Tide-dominated deltas usually are composed of a series of funnel-shaped water bodies (estuaries?) at multiple river mouths with a series of shore perpendicular sand ridges that extend offshore of the river mouths. Broad tidal flats and marsh or mangrove wetlands occur along the shore of the embayments. A generalized model of a tide-dominated delta is given in Figure T32.

The tide-dominated Ord River Delta, described by Coleman and Wright (1979), is located on the coast of northwest Australia were the tidal range varies between 3.8 and 6.6 m. Tidal currents are oriented primarily in an onshore and offshore direction within an embayment which composes the major mouth of the Ord River. Linear sand ridges, which range in relief from 10 to 22 m and average 2 km in length, occur in the most seaward portion of the embayment. These sand ridges are formed and shaped primarily by tidal currents, which dominate over wave energy effects. A composite prograding stratigraphic column for this delta shows a fining upward sequence resulting from the progradation of muddy upper intertidal flats and marsh sediments over the sand bodies of the lower intertidal and shallow subtidal regions.

Tide-dominated estuaries

The concept of tide and wave dominance has also been applied to estuaries by Dalrymple *et al.* (1992). However, they state that estuaries are unlike many other coastal systems, because they are "geologically ephemeral." If the rate of sediment supply is sufficient to eventually fill the lowstand valley within which the estuary is located, the filled valley then becomes a delta. The estuaries they term wave-dominated are composed of a sand body complex (barrier/tidal inlet) at the entrance, a muddy central basin, and a bayhead delta system. *Tide-dominated estuaries*, on the other hand, contains elongate sand bars and sandy tidal flats at the entrance and complex tidal channel/ wetland habitats further inland.

Miles O. Hayes

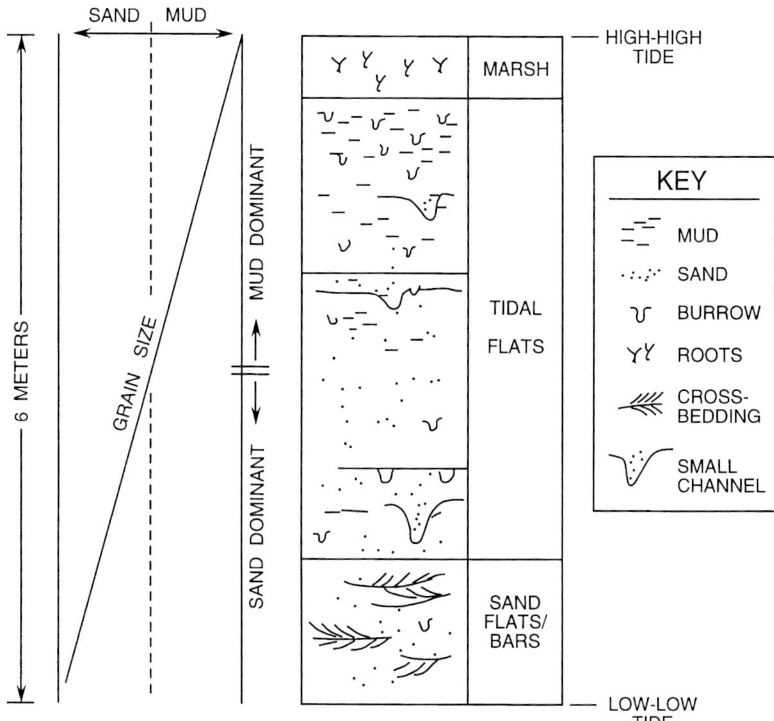

Figure T31 Hypothetical regressive sequence for the intertidal zone of a prograding macrotidal (tide-dominated) shoreline in a non-deltaic setting.

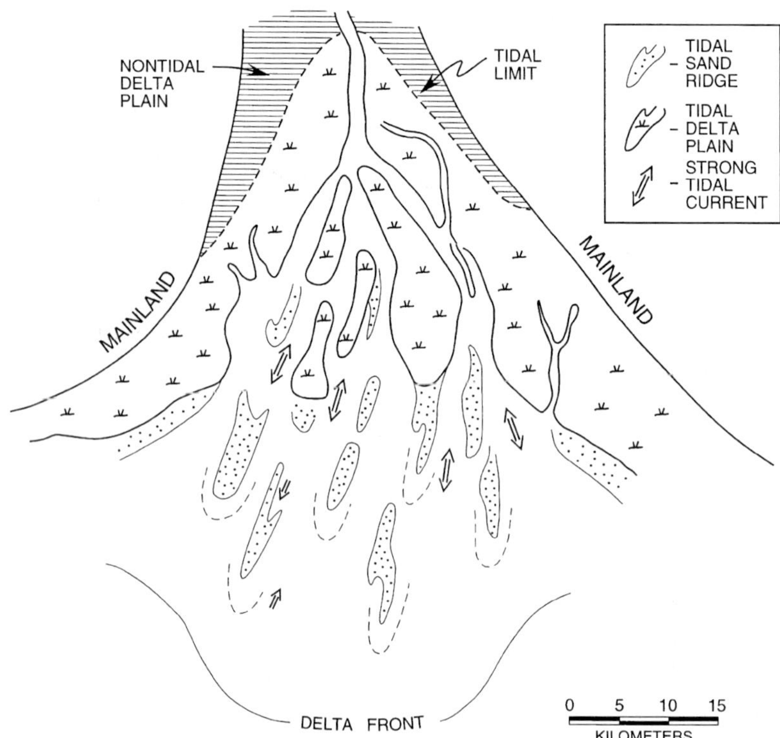

Figure T32 Schematic sketch in plan view of a tide-dominated delta. Note the presence of tidal sand ridges at the offshore entrances to the delta complex.

Bibliography

Coleman, J.M., and Wright, L.D., 1979. Sedimentation in an arid macro-tidal alluvial river system: Ord River, Western Australia. *Journal of Geology*, **85**: 621–642.

Dalrymple, R.W. *et al.*, 1990. Dynamics and facies model of a macrotidal sand-bar complex, Cobequid Bay-Salmon River Estuary (Bay of Fundy). *Sedimentology*, **37**: 577–612.

Dalrymple, R.W., Zaitlin, B.R., and Boyd, R., 1992. Estuarine facies models: conceptual basis and stratigraphic implications. *Journal of Sedimentary Petrology*, **63**: 1130–1146.

Davies, J.L., 1964. A morphogenic approach to world shorelines. *Zeitschrift fur Geomorphologie*, **8**: 27–42.

Evans, G., 1975. Intertidal flat deposits of the Wash, western margin of the North Sea. In Ginsburg, R.N. (ed.), *Tidal Deposits*. Springer-Verlag: New York, pp. 13–20.

Fisher, W.L. *et al.*, 1969. *Delta Systems in the Exploration for Oil and Gas, a Research Colloquium*. Texas University Bureau of Economic Geology.

Galloway, W.E., 1975. Process framework for describing the morphological and stratigraphic evolution of deltaic depositional systems. In Broussard, M.L. (ed.), *Deltas*. 2nd edn. Houston, TX: Houston Geological Society, pp. 87–98.

Hayes, M.O., 1975. Morphology of sand accumulation in estuaries. In Cronin, L.E. (ed.), *Estuarine Research*, Vol. 2. New York: Academic Press, pp. 3–22.

Hayes, M.O., 1979. Barrier island morphology as a function of tidal and wave regime. In Leatherman, S. P. (ed.), *Barrier Islands from the Gulf of St. Lawrence to the Gulf of Mexico*. New York: Academic Press, pp. 1–27.

Kim, Y.H. *et al.*, 1999. Holocene transgressive stratigraphy of a macrotidal flat in the southeastern Yellow Sea: Gomso Bay, Korea. *Journal of Sedimentary Research*, **69**: 328–337.

Knight, R.J., and Dalrymple, R.W., 1975. Intertidal sediments from the south shore of Cobequid Bay, Bay of Fundy, Nova Scotia, Canada. In Ginsburg, R.N. (ed.), *Tidal Deposits*. New York: Springer-Verlag, pp. 47–56.

Price, W.A., 1955. *Development of Shorelines and Coasts*: Department of Ocean. Project 63, Texas A&M University.

Thompson, R.W., 1968. *Tidal Flat Sedimentation on the Colorado River Delta, Northwestern Gulf of California*. Geological Society of America Memoir 107.

Yeo, R.K., and Risk, M.J., 1981. The sedimentology, stratigraphy, and preservation of intertidal deposits in the Minas Basin system, Bay of Fundy. *Journal of Sedimentary Petrology*, **51**: 245–260.

Cross-references

Barrier Islands
Deltas
Estuaries
Tidal Environments
Tides
Wave-and-Tide Dominated Coasts
Wave-Dominated Coasts

TIDE GAUGES

Historical origin

Tide gauges have a relative long history. At some places, systematic sea level observations have been performed and recorded since the late 17th or early 18th century. This is the case in Amsterdam, for instance, since 1682, in Liverpool since 1768, or in Stockholm since 1774.

The first devices were simply graduated rods, usually called tide poles, placed at locations where the instantaneous water level height of the sea could be read off at any time by an observer. Most of such measurements were undertaken in or at the entrance of harbors. They were restricted to observations of high and low water levels, as well as the time of their occurrences (Cartwright, 1999).

Automatic recording devices appeared only in the 1830s, although basic instructions were already detailed and published in an Italian journal in 1675. An extract was transcribed in the "Journal des Sçavans" of April 22, 1675 (Observatoire de Paris, 1675). These were mechanical gauges, equipped with float, wires, counterweights, clock, pen, paper chart recorder, and stilling well. They provided the first complete tidal curves which could be examined in detail and digitized for further analysis.

Tide or sea-level gauges?

Whatever technique is employed, the basic quantity provided by tide gauges is an instantaneous height difference between the level of the sea surface and the level of a fixed point on the adjacent land. The vertical reference point is a nearby benchmark which can be observed by traditional surveying techniques. Thence, tide gauges not only record ocean tides but also a large variety of sea-level signals that can be caused by variations in atmospheric pressure, density, currents, continental ice melt... as well as vertical motions of the land upon which the measurement instrument is located, due to tectonic changes, isostatic adjustments, volcanism inflation, sediment consolidation, pier subsidence, etc. Records of such devices are indicative of what are called relative sea-level changes (Pugh, 1987). The recorded processes have characteristic timescales from several minutes to centuries.

Therefore, tide gauge data is valuable information to a wide range of activities over a variety of timescales, for scientific research as well as for many practical applications. For instance, data have uses: for navigation and ship traffic (guidance, tidal timetables...), for coastal engineering design (dike building, dredging works...), for statistics of extreme levels over long periods, for studies of upwelling and fisheries, for hydrographic surveys (sound charts, chart datum...), for storm-surge predictions and alert, for analysis of the risk of flooding and coastal protection, for input or validation of ocean circulation models, for long-term trend of sea-level variations, for global change research and monitoring and, of course, for tidal analysis, prediction and validation of tidal models. In spite of the numerous applications, the historical and conventional term of tide gauge still prevails for such devices, although sea-level gauge would be more adequate.

Technical evolution

The most common type of gauge in use around the world still consists of the float and stilling well system which was devised more than a century ago. In this system, the height measurement of a floating gauge is taken by measuring, on a reduced scale, the length of the wire holding a counter-balanced float. The float sits on the surface of the water inside a well. The vertical movement of the float is transmitted and reduced in scale through a more or less sophisticated system of wires, pulleys, and counterweights to a pen. A continuous record of water height against time is obtained in this way as a drawn curve on paper, the paper being in motion at a fixed speed in a normal direction to the pen displacement. The stilling well is a vertical tube long enough to cover any possible range of tide. It prevents the float from drifting in the presence of currents or winds. The well is designed to provide a mechanical filtering of short-period oscillations due to waves by restricting the flow of water into and out of the well.

Traditional mechanical float devices are progressively replaced by new technologies. Modern type gauges are mainly based either on the measurement of the subsurface pressure or on the measurement of the time of flight of a pulse, acoustic or radar.

The principle of a pressure system is the measurement of the hydrostatic pressure of the water column above a fixed point below the lowest expected tide level. The conversion of that hydrostatic pressure into a sea-level equivalent height is performed according to the law: $\boldsymbol{h} = (\boldsymbol{p} - \boldsymbol{p}_a)/(\rho g)$ by measuring or assuming values of water density ρ, air pressure $\boldsymbol{p}_a$, and local acceleration due to gravity g. Pressure sensors usually exploit strain gauge or ceramic technology. Water pressure translates then into changes in resistance or capacitance in the pressure element. The resulting signal is normally a specific frequency which is converted into physical units of pressure.

The acoustic or radar systems determine the vertical distance from a transducer, located above the sea surface, to the water by measuring the elapsed time of a pulse that is emitted, reflected, and returned back to the sensor. The distance to the water is then derived from the velocity of the type of wave considered (sound or radar).

Whereas technical description may be provided by manufacturers, IOC manuals (1985, 1993, 1997) are a helpful source of information. Valuable detailed information can be found there on each type of gauge, their respective advantages, drawbacks, performances, and limitations, as well as advice on operational methods and environmental conditions of use.

A critical part of the tide gauge system is probably the benchmark on land as it provides the fundamental zero point or datum to which the values of sea level are referred. This benchmark is extremely important as it serves to build useful long-term sea-level time series, even if parts of the time series were obtained from different gauges and different benchmarks (as long as they were geodetically connected). Tide gauge benchmarks are ultimately the source of the long-term coherence and stability of the measurements. It is therefore common sense to preserve the datum by installing and connecting a set of 5–10 benchmarks within a few hundred meters of the tide gauge. Usually, one of them is arbitrarily called the tide gauge benchmark, the "most" stable, the "most" secure or the closest, although all of them are representive of the datum. Even though the station is equipped with the most modern equipment, long-term instrumental drift and local stability surveying should be performed at least annually.

In the era of modern communication technologies, data in electronic form is essential as it can be retrieved immediately from a gauge to a data center by automated modem dial-up or satellite transmissions. Data can then easily be made accessible worldwide via the Internet. Paper chart recorders are no longer acceptable. They require slow labor-intensive digitization and are cost-effective. Moreover, they contain many sources of inaccuracy and are inadequate for certain applications.

Scientific applications

Many scientific applications other than the natural tidal research and modeling benefit from tide gauge records. For instance, tide gauge data are used to establish vertical reference systems on land and on sea in order to define the height and depth datums. The belief, about a century ago, that the average level of the sea was constant over long periods of time led to define the concept of geoid and, subsequently, to establish the origin of the leveling networks on "mean sea level." Typically, countries chose one tide gauge station to compute this quantity over an arbitrary time period: in France, for example, the datum was determined at Marseilles from continuous tide gauge records performed during the period 1885–97; in Britain, the Ordnance Datum was determined at Newlyn from records extending from May 1915 to April 1921. However, mean sea level varies from place to place and at one specific place over time. Today, mean sea level at Marseilles is about 11 cm above the local 1885–97 datum, whereas it is about 0.2 m above the Ordnance Datum at Newlyn. Thus, the datums no longer represent the "real" average of the sea level at these sites.

Tide gauge data are also used to establish the datums to which the depths are referred on nautical charts and above which tide predictions are provided for practical purposes. Since 1996, the International Hydrographic Organization has recommended that Lowest Astronomical Tide be adopted as the International Chart Datum. This datum is defined as the lowest tide level which can be predicted in average meteorological conditions and in any combination of astronomical conditions.

Owing to the rhythmic nature of the tide, the components of its oscillations can be represented by a series of sinusoidal curves, mainly depending on the relative positions of the moon and the sun. In each place, the major tide components are determined empirically from hourly records during at least 29 days (a moon cycle), by the use of harmonic analysis. Because astronomical orbits are known, tides can be predicted at any time. Random deviations from the predicted tide can be due to changes in air pressure or wind (sea surges), currents, changes in water density, discharge (especially at a river mouth) and can be analyzed after filtering the astronomical tide. The occurrence of short-lived oscillations in random deviations, following a sea surge, may correspond to seiches. Finally, if periodic, short-lived and random components are removed by filtering processes, and if the series investigated are long enough, long-term trends will appear.

Tide gauges have been carefully studied for indications of recent global sea-level rise. They include land movement and sea-level movement and depend, therefore, on local or regional tectonics as well as global climate change effects (eustasy). By analyzing the difference in trend between the records in two stations, local and regional components can be easily revealed, though their interpretation may be not univocal. In order to keep the statistical accuracy of a trend estimation below ± 0.5 mm/yr, almost continuous records of at least 40 to 60 years long are usually necessary. At the present time, less than 300 stations in the world can provide such long records. Their geographical distribution is unfortunately very uneven, most of the stations being located in the Northern Hemisphere, with a great majority on both sides of the North Atlantic and few or no data in very wide coastal areas. Since the late 1980s, international projects like GLOSS have made deserving efforts to improve the existing tide-gauge network, but the duration of the new records is still too short to assess long-term trends. In spite of such globally unrepresentative distribution, of the difficulty in separating the various components of relative sea-level change, and of the fact that long-term trends show a great spatial variability, several authors have attempted an estimation of the recent global (average) sea-level rise from tide-gauge records, with various approaches (Emery and Aubrey, 1991; Pirazzoli, 1996). The results obtained are quite variable: a "global" sea-level rise between 0.5 and 3.0 mm/yr is inferred by various authors, with several estimates around 1 mm/yr; higher rates are obtained when tide-gauge trends are "corrected" using isostatic models; however, other authors infer that an accurate global sea-level trend is indeterminable of the basis of tide-gauge data alone. This precludes for the moment a more precise quantification of the recent global rise in sea level only with tide gauges.

Perspectives: synergy with space techniques

Recent advances in space geodesy and gravity measurements allow consideration of monitoring of land movements in a global geocentric reference frame rather than in a local one. Repeated or continuous precise positioning of tide gauge benchmarks by geodetic techniques like GPS or DORIS, over periods of a decade or so, will enable vertical crustal movement to be determined and, subsequently, provide a possible discrimination between eustatic sea-level rise and land subsidence within tide gauge records (IOC, 1997; Neilan *et al.*, 1998). Moreover, tide gauge data will then be expressed in the same global geodetic reference frame as satellite altimeter observations and can therefore be directly compared and combined with the altimetric sea levels, providing at last a more reliable and accurate estimate of sea-level variations.

Probably the most significant improvement in sea-level research comes from satellite radar altimetry. However, altimetry cannot provide detailed local high-frequency sea-level information due to the specific sampling pattern. Tide gauges continue to possess important attributes for certain applications, like continuity with historic measurements, high accuracy, continuous sampling, and ability to record at the coast at a relatively low cost.

Guy Woppelmann and Paolo Antonio Pirazzoli

Bibliography

Cartwright, D.E., 1999. *Tides—a scientific history*. Cambridge: Cambridge University Press.

Emery, K.O., and Aubrey, D.G., 1991. *Sea Levels. Land Levels, and Tide Gauges*. New York: Springer Verlag.

IOC, 1985 and 1994. *Manual on Sea-Level Measurement and Interpretation*. Intergovernmental Oceanographic Commission Manuals and Guides No. 14, Vol. 1 (1985) and Vol. 2 (1994).

IOC, 1993. *Joint LAPSO-IOC workshop on sea level measurements and quality control*. Intergovernmental Oceanographic Commission, Workshop Report No. 81.

IOC, 1997. *Global Sea Level Observing System (GLOSS) Implementation Plan 199*. Intergovernmental Oceanographic Commission, Technical Series No. 50.

Observatoire de Paris (ed.), 1675. *Extrait du Journal d'Italie contenant quelques avertissements pour observer les différens périodes de la marée; avec la description and la figure de la machine dont il est parlé. Journal des Sçavans, du 22 avril 1675*. Tome 2176, pp. 65–67.

Neilan, R.E., Van Scoy, P.A., and Woodworth, P.L. (eds.), 1998. Workshop on methods, for monitoring sea level: GPS and tide gauge benchmark monitoring. GPS altimeter calibration. In *Proceedings of the Workshop Organized by the IGS and PSMSL*, Pasadena, March 17–18, 1997.

Pirazzoli, P.A., 1996 . *Sea-Level Changes. The Last 20000 Years*. Chichester: John Wiley & Sons.

Pugh, D.T., 1987. *Tides, Surges and Mean Sea Level: A Handbook for Engineers and Scientists*. Chichester: John Wiley & Sons.

Cross-references

Altimeter Surveys, Coastal Tides and Shelf Circulation
Changing Sea Levels
Coastal Climate
Coastal Currents
Geodesy
Global Positioning Systems
Greenhouse Effect and Global Warming
Storm Surge
Submerging Coasts
Tectonics and Neotectonics
Tidal Datums
Tides
Uplift Coasts

TIDE MILL

Tide Mills (*moulins à marée, molinos de mar, Gezeitenmoehlen, getijdenmolens*) dotted the coasts and estuaries for several centuries until more efficient, but not less costly, alternatives all but wiped them off the map. A few are making a comeback. Some were still functioning during World War II in England and Wales, but most were derelict or their buildings transformed and used for other purposes (Wailes, 1941). History records show that tide mills once functioned on the rivers Thames (London Bridge), Tiber (in besieged Rome), and Danube (Charlier and Menanteau, 1998). A mill stood at the entrance of Dover Harbor in (1000) according to the *Domesday Book*.

Tide Mills consisted of a dam with sluices, a retaining basin, and a float or a water wheel and transformed the energy of running water into mechanical power to run flour-mills, saw-mills, even breweries, and as late as 1880 to pump sewage. They apparently were also put to work in the *polder-works*.

Tide Mills require sites with tidal amplitude—and thus a tidal current—though even a tidal creek may do. They are the forerunners of today's tidal power plant in the same manner as the windmill is the precursor of the contemporary air turbines. The tide mill is in fact a conventional water mill using the tidal current as its source of power, occasionally both ebb and flood tide, but most used only the ebb current as the retaining basin filled at flood time, emptied at ebb time. A few mills used a proportion of freshwater from the stream in addition to seawater.

The most common idea was the "float method" by which the incoming water raised a floating mass which, as it fell down to original position, provided "work;" another approach included a shaft-mounted rotating paddle wheel activated by ebb and flood, with power transmitted by the shaft; finally somewhat more sophisticated types would let air contained in a metal or concrete conduit be compressed by the incoming tide, thus furnishing compressed air power, an idea which has resurfaced today in proposed schemes (Gorlov, 1982). The fourth system, even more elaborate, dams part of the sea (bay, gulf) and this pool fills up at incoming tide; the water when released at low tide passes through turbines to flow back to sea or to another basin. The latter scheme appeared in the 19th century and with its wheel rotating at as much as 150 rpm, had a much higher yield.

Tidal mills were quite numerous in England, Wales, The Netherlands, Brittany, and Spain. The Europeans brought them to the United States, seemingly on both coasts (Creek, 1952) and Canada. There are some claims that they were not uncommon along the South China Sea coast. Besides the touristic value of restored mills, as near Southampton (Ewing Mill), Plougastel (Brittany), and in Massachusetts (Chatham Spice Mill), tide mills are making a timid come back, in an improved version, to provide power in remote areas.

Roger H. Charlier

Bibliography

Charlier, R.H. and Menanteau, L., 1998. The saga of tide mills. *Renewable and Sustainable Energy Review*, **1**(3): 1–39.

Creek, H., 1952. Tidal mill near Boston. *Civil Engineering* **22**: 840–841.

Gorlov, A.M., 1982. Hydropneumatic approach to harnessing tidal power. New approach. *Proceedings of Tidal Power Conference [New Bedford, NS, Canada]*, **4**: 1–5.

Wailes, R., 1941. Tide mills in England and Wales. *Junior Institute of Engineering, Journal and Record of Transactions*, **51**: 91–114.

Cross-references

Polders
Tidal Creeks
Tidal Power
Tidal Prism
Tides

TIDES

Introduction

Tides are the periodic motion of the waters of the sea caused by the changing gravitational effects of the moon and the sun as they change position relative to the rotating earth. The tides in the oceans are actually very long waves hundred or thousands of miles long. Although produced by astronomical forces, their behavior in the oceans and connected bays (and the size of resulting water level oscillations) is determined by hydrodynamics (i.e., the physics of the water movement).

The vertical rise and fall of the water surface is usually referred to as the *tide*, while the accompanying horizontal movement is referred to as the *tidal current*, with the tidal flow into a bay called the *flood* and the flow out of a bay called the *ebb*. For most areas of the earth the rise and fall, and flood and ebb, occur twice a day (referred to as a *semidiurnal* tide), but in some areas there may only be one *high water* and one *low water* per day (referred to as a *diurnal* tide). In many areas, there are two high waters and two low waters per day, but one high water and/or one low water is a different height than the other (referred to as a *mixed* tide).

The tide is only one phenomenon that produces variations in water level and currents. Such variations can also be caused by changes in the wind, atmospheric pressure, river discharge, and water density (due to changes in salinity and temperature), but they are not periodic like the astronomical tide and are not nearly as predictable, being associated with weather. Nontidal water level changes caused by changes in the wind and barometric pressure are usually referred to as *storm surges* (*q.v.*). The term *sea level* is generally used for longer-period, slower changes in water level. *Mean sea level* is the average of water level measurements over some time period (such as a day, a month, or a year), which averages out shorter-term oscillations like the tide.

Water level is the height of the water surface above some reference level, called a *datum*. A datum for a particular waterway is generally defined as an average height of a particular stage of the tide. For example, *chart datum* on a nautical chart in the United States is defined as the mean lower low water (MLLW) at each location. (*Lower low water* is the lower of the two low waters that occur each day, and MLLW is the average of all the lower low waters over some time period, usually at least a year). Depth soundings on a nautical chart are the depths below the chart datum, and the predicted tidal heights found in Tide Tables are the heights above the chart datum. Adding the two together gives the total water depth at that moment in time. These tidal datums also provide the legal definition of *marine boundaries*. MLLW, for example, is the dividing line between federal territorial seas and state submerged lands, and mean high water (MHW) is the dividing line between state tidelands and private uplands. Tidal datums at a particular tide gauge are referenced to the land through geodetic leveling to a number of *benchmarks*, which are brass markers driven into solid rock or other permanent structures. Tidal datums can change over decades if the land subsides (or rises due to glacial rebound) or if relative sea-level rises due to other effects.

The tide dominates our thinking about changes in water level, not only because it usually causes the largest changes (except during storms), but also because it is very predictable (especially in comparison to how well we can predict the weather). After analyzing only a month's worth of water level measurements from a tide gauge, we can predict the tide quite accurately (for that location) for years into the future. This high predictability is due to the tide's periodic nature and our very precise knowledge of its astronomical forcing. The earth–moon orbit, the revolution of the earth around the sun, and the rotation of the earth on its axis involve periodic motions with fixed and precisely known time periods. Tidal energy is found at the same frequencies that describe these astronomical motions.

To fully understand and predict the tides one must understand both its astronomical forcing and the hydrodynamics of the oceans and bays. While it is the astronomical forcing of the tide that is the basis for the tide's predictability, it is the hydrodynamics of the tide that is responsible for the size of the *tidal range* (i.e., the height difference from low water to high water), the timing of high and low waters, and the *type of tide* (i.e., semidiurnal, mixed, or diurnal). It is the length, width, and depth of the bay or river (and of any adjoining waterways) that control the hydrodynamics. In shallow waterways, the hydrodynamics also transfers tidal energy to new frequencies, and distorts the shape of the tide curve away from perfect sine curves. These same shallow-water processes also lead to interactions between the tide and nontidal phenomena such as storm surge and river discharge.

The largest tidal ranges occur in shallow coastal waters, in particular, at the ends of certain bays and along coasts with very wide continental shelves. The increase in tidal range and tidal current speeds that one sees in the shallow waters of bays, rivers, and straits can go to dramatic extremes if the circumstances are right. Tidal ranges reach 15 m (50 ft) in Minas Basin in the Bay of Fundy. Tidal ranges greater than 12 m occur at the northern end of Cook Inlet near Anchorage in Alaska, in the Magellan Strait in Chile, in the Gulf of Cambay in India, along the Gulf of St. Malo portion of the French coast bordering the English Channel, in the Severn River in England, and along the open coast of southern Argentina. In a few rivers, a portion of the tide wave propagates up the river as a tumultuous wall of water, called a *tidal bore*. The largest tidal bores are found in the Tsientang River near Hanchow, China, and in the Amazon River, Brazil, where at certain times they can reach 7.5 m in height and travel up the river at a speed of 7 m/s. Smaller bores occur in the Meghna River in India, in the Peticodiac River at the end of the Bay of Fundy, in Turnagain Arm near Anchorage, and in the Severn River in England. Tidal current speeds greater than 7.5 m/s occur in Seymour Narrows, between Vancouver Island and the mainland of British Columbia, Canada. Tidal currents of 5 m/s are found in South Inian Pass in southeast Alaska and in Kanmon Strait, Japan. In some narrow or shallow straits, the tidal currents create dangerous whirlpools or *maelstroms*. Most famous is the whirlpool in the Strait of Messina (between Sicily and the southern tip of the Italian mainland), which Homer depicted in his *Odyssey* as the second of two monsters, Scylla and Charybdis, faced by Ulysses.

The generation of tides

The tides are caused by both the moon and the sun, but the moon though smaller has roughly twice the effect because it is much closer to the earth than the sun. Although the moon appears to orbit around the earth, the earth and moon both actually revolve around a common point, which, because the earth is much more massive than the moon, is inside the earth, but not at the earth's center (see Figure T33). At the

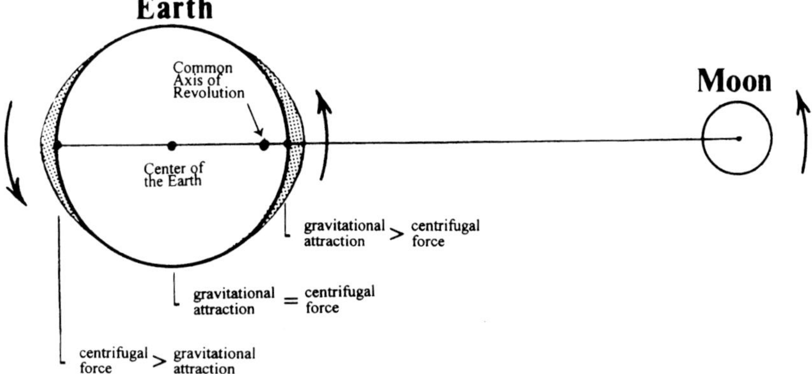

Figure T33 The earth–moon system (viewed from above the North Pole) revolving around a common axis (inside the earth). The earth is shown with a hypothetical ocean covering the whole earth (with no continents) and *two* bulges, resulting from the imbalances of gravitational and centrifugal forces.

center of the earth there is a balance between the gravitational attraction (trying to pull the earth and moon together) and the centrifugal force (trying to push the earth and moon apart). At a location on the earth's surface closest to the moon, the gravitational attraction of the moon is greater than the centrifugal force. On the opposite side of the earth, facing away from the moon, the centrifugal force is greater than the moon's gravitational attraction. Figure T33 shows a hypothetical ocean (covering the whole earth with no continents) with *two* bulges, one facing the moon and one facing away from the moon, that result from the two imbalances of gravitational and centrifugal forces. However, if we look at the side of the earth facing the moon, the force vertically upward from the earth toward the moon is so small compared with the earth's gravitational force that it could not cause the bulges. If we move away from the equator to another point on the earth that is not directly under the moon, we see that the attractive force is still pointing toward the moon, but is no longer perfectly vertical relative to the earth (see Figure T34). At this point, the force toward the moon can be separated in a vertical component of the force and a horizontal component, the latter one being parallel to the earth's surface. This horizontal force, though small, has nothing opposing it, and so it can move the water in the ocean. One can see from Figure T34 that all the horizontal components shown tend to move the water into a bulge centered around the point that is directly under the moon. Similarly, on the other side of the earth another bulge results.

One can easily envision the earth rotating under these bulges in this hypothetical ocean that covers the entire earth. In one complete rotation in one day there will be two high tides (when under a bulge) and two low tides (when halfway between bulges), and thus one entire tidal cycle would be completed in approximately half a day (actually 12.42 h, for reasons to be explained below). However, this is an extreme simplification (called the *equilibrium tide*) used merely to show how the tide generating forces change as the earth rotates. Not only are the continents left out, but this assumes that the oceans respond instantly to the tide-generating force, which they do not.

Now consider the addition of continents and look at one of the oceans, with a bay connected to it. The tide-generating forces are too small to cause a tide directly in a small body of water like a bay. Only in a large ocean are the cumulative effects of the tide-generating forces throughout the ocean large enough to produce a tide. What is actually generated is a very long wave with a small amplitude, on the order of a half a meter or less (see Figure T35). However, when this wave reaches the reduced depths of the continental shelf, there is a partial reflection of the wave, and the part of the wave that continues toward the coast is increased in amplitude. At the coast another reflection further increases the height of this long wave, now reaching at least a meter along most coasts. When the wave moves up into a bay there can be even more amplification depending on the depth, length, and width of the bay, with tidal ranges reaching 5, 10, or even 15 m for bays with the right dimensions.

How large the tide range is depends on how close the natural period of free oscillation of the basin is to the period of the tide-generating force. If the natural period of the basin is same as the period of the tide-generating force, then the energy from the tidal forcing will be input in the same direction as the water is already moving and the resulting tide range will be larger. This is called *resonance*. The natural period of a basin, T_n, is approximately equal to $2L/(gD)^{1/2}$, where L is the length of the basin, D is the depth, and g is the acceleration due to gravity. The Atlantic Ocean is too wide for there to be resonance (its 19-h natural period being much longer than the 12.42-h tidal period). The largest tide ranges in the world are in shallower basins with just the right length and depth combination to have natural periods close to the tidal period.

Astronomical considerations

Because it is a forced oscillating system, the tide will oscillate with the periods determined by the relative motions of the earth, moon, and sun. There are many different periods involved due to the complex nature of the orbit of the moon around the earth and of the orbit of the earth around the sun. However, astronomers have very precisely determined all of these periods. To predict the tide at a specific location for any time in the future one must simply analyze water level data from that location to determine the amplitude and phase associated with each of the important tidal periods.

If the earth–moon orbit and the earth–sun orbit were circular and in the plane of the earth's equator, there would only be two tidal frequencies, and only thus two semidiurnal constituents, M_2 and S_2 (defined below), would be needed to make tidal predictions. However, the orbital motions are more complicated. Distances between the moon and earth and between the earth and sun vary with time (the latter changes over a 20,942-year period and so is of no concern, except perhaps in paleoclimatology). Orbital planes are at angles relative to the earth's equatorial plane and these angles also vary with time. All these motions modulate the tidal forces, so that tidal energy shows up at many more frequencies than at just M_2 and S_2. The changing angles of the orbital planes also means that the moon and sun will not always be directly over the earth's equator, thus making the two tidal bulges (on the opposite sides of the earth) asymmetric with respect to the axis of rotation, which introduces diurnal tidal frequencies. The fundamental periods in the motions of the earth, moon, and sun are shown in Table T11. Table T12 shows how key tidal constituents are derived from these fundamental frequencies. (The *Doodson numbers* used in many tidal papers are a shorthand that indicates which of the six frequencies, ω_L, ω_1 through ω_5 from Table T11, are used to produce a particular constituent.) To make this a little clearer, we will describe the origins of a few of the more important tidal frequencies.

The moon orbits around the earth in the same approximate direction as the rotation of the earth, so that one lunar day (i.e., one complete rotation of the earth *with respect to the moon*) is 24.8412 h long ($1/\omega_L$, see Table T12). There are two tidal high water bulges on the earth, so the period of the largest semidiurnal lunar harmonic constituent, M_2, is half a lunar day, or 12.4206 h.

The earth turns under the sun exactly once every solar day, which leads to the main solar semidiurnal tidal constituent, S_2, with a period of 12.0000 h. Because the sun is so much farther from earth than the moon (and the tidal force is inversely proportional to the cube of the distance), S_2 is much smaller in size than M_2. When the moon and sun are in alignment (at new and full moons) their tidal forces work together to create *spring tides* with larger tidal ranges, while they work against

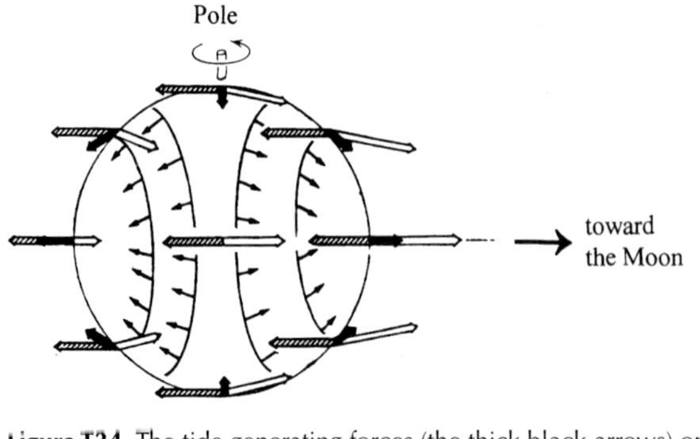

North Pole

toward the Moon

Figure T34 The tide generating forces (the thick black arrows) on the earth resulting from the difference between gravitational attraction (the open arrows) and centrifugal force (the hatched arrows). The small blacks arrows are the *horizontal* components of the tide generating forces, which tend to move the water into the two bulges shown in Figure T33.

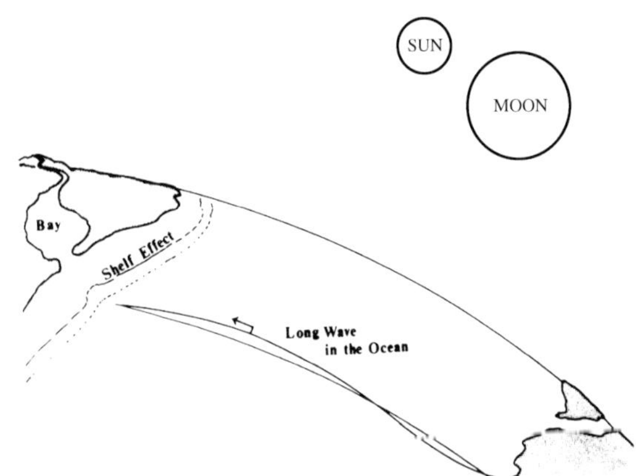

Figure T35 The tide-generating forces caused by the moon and sun produce a very long wave of relatively small amplitude in the ocean. When this long wave reaches the continental shelf, then the coast, and finally propagates up a bay, it is amplified by an amount that depends on the length and depth of each of the basins.

Table T11 Fundamental periods in the motions of the earth, moon, and sun (after Parker *et al.*, 1999)

Description	Period (mean solar units)	Frequency (1/period)
Sidereal day (one rotation wrt vernal equinox)	23.9344 h	Ω
Mean solar day (one rotation wrt to the sun)	24.0000 h	ω_S
Mean lunar day (one rotation wrt to the moon)	24.8412 h	ω_L
Period of lunar declination (tropical month)	27.3216 days	ω_1
Period of solar declination (tropical year)	365.2422 days	ω_2
Period of lunar perigee	8.847 years	ω_3
Period of lunar node	18.613 years	ω_4
Period of perihelion	20,940 years	ω_5

Table T12 Tidal constituents and their origins (after Parker *et al.*, 1999)

Symbol	Description	Period	Speed (°/h)	Derived from	Coefficient C
Semidiurnal tides					
K_2^L	declinational to M_2	11.967 h	30.0821373	$2\omega_L + 2\omega_1\ (=2\Omega)$	0.0768
K_2^S	declinational to S_2	11.967 h	30.0821373	$2\omega_S + 2\omega_2\ (=2\Omega)$	0.0365
S_2	principal solar	12.000 h	30.0000000	$2\omega_S$	0.4299
M_2	principal lunar	12.421 h	28.9841042	$2\omega_L$	0.9081
N_2	elliptical to M_2	12.658 h	28.4397295	$2\omega_L - (\omega_1 - \omega_3)$	0.1739
L_2	elliptical to M_2	12.192 h	29.5284789	$2\omega_L + (\omega_1 - \omega_3)$	0.0257
Diurnal tides					
K_1^L	declinational to O_1	23.934 h	15.0410686	$(\omega_L - \omega_1) + 2\omega_1\ (=\Omega)$	0.3623
K_1^S	declinational to P_1	23.934 h	15.0410686	$(\omega_S - \omega_2) + 2\omega_2\ (=\Omega)$	0.1682
P_1	principal solar	24.066 h	14.9589314	$(\omega_S - \omega_2)$	0.1755
O_1	principal lunar	25.819 h	13.9430356	$(\omega_L - \omega_1)$	0.3769
Q_1	elliptical to O_1	26.868 h	13.3986609	$(\omega_L - \omega_1) - (\omega_1 - \omega_3)$	0.0722
Long-period tides					
Mf	declinational to M_0	13.661 days	1.0980331	$2\omega_1$	0.1564
Mm	elliptical to M_0	27.555 days	0.5443747	$(\omega_1 - \omega_3)$	0.0825
Ssa	declinational to S_0	182.621 days	0.0821373	$2\omega_2$	0.0729

The speed is another form of frequency. M_0 and S_0 represent constant lunar and constant solar forces. The coefficient, C, gives a global measure of each constituent's relative portion of the tidal potential (i.e., it ignores a latitudinal variation that is different for each species).

each other at first and third quarters to create *neap tides* with smaller ranges (see Figure T36).

The earth–moon orbit is elliptical, so that the distance between them varies over a 27.5546-day period ($1/[\omega_1 - \omega_3]$, see Table T12), from perigee (the moon closest to the earth, and so a stronger tidal force) to apogee (the moon farthest from the earth, and so a weaker tidal force) and back to perigee. This modulates the lunar tidal force. This modulation of M_2 shows up in a spectra (of water level or current data) as a line to the left of the line for M_2, this lower frequency line representing a second lunar harmonic constituent, N_2, whose period is 12.6583 h. The stronger *perigean* tidal force will occur when M_2 and N_2 come into phase (leading to larger tidal ranges), while the weaker *apogean* tidal force will occur when M_2 and N_2 are exactly out of phase (leading to smaller tidal ranges). We can use Figure T36 to illustrate this, if, in that figure, we replace S_2 with N_2, spring tide with perigean tide, and neap tide with apogean tide. The difference is that with the M_2 plus S_2 case there really are two distinct effects being added, but in the case of the changing distance between the moon and earth, this directly varies the amplitude of the tide; and N_2 is merely a convenient way (in combination with M_2) to represent this variation of amplitude. There are several times a year when lunar perigee is reasonably close in time to new or full moon to produce the largest tidal ranges of the year, called *perigean spring tides*.

The plane of the moon's orbit around the earth is at an angle to the plane through the earth's equator. Thus, as the earth rotates under the moon, there will be times of the month when the moon is north of the equator (Northern Declination), over the equator (Equatorial Declination), and south of the equator (Southern Declination). When the moon is north or south of the equator, one of the tidal bulges is more north of the equator and one is more south, so that at a particular location on the earth there will either be only one high water per day (a diurnal tide), or, if there are two, they will be of different heights (the difference being the *diurnal inequality*) (see Figure T37). The diurnal lunar tidal forces resulting from lunar declination are represented by two tidal constituents, O_1 and K_1, with periods of 25.8193 and 23.9345 h, since they must cancel each other out every 13.66 days ($1/2\omega_1$, see Table T12), at the times when the moon is over the equator. (The maximum angle between the plane of the moon's orbit and the earth's equator varies from 18.3° to 28.5° over a 18.6-year period; see below for further discussion of this *nodal* cycle.) The sum of the O_1 and K_1 frequencies is equal to the M_2 frequency, so that the time of the diurnal high water does not change with respect to the times of the two semidiurnal high waters.

The plane of the earth's orbit around the sun (called the *ecliptic*) is also at an angle to the plane through the earth's equator. Around December 21st the sun is furthest south of the equator (December solstice) and around June 21st it is furthest north of the equator (June solstice), the angle between the ecliptic and the equator reaching 23.5° in each case. December solstice marks the beginning of winter in the Northern Hemisphere and the beginning of summer in the Southern Hemisphere, and *vice versa* for June solstice. Around March 21st the sun is over the equator (vernal equinox) and again around September 21st (autumnal equinox). This movement of the sun north and south of the equator also leads to diurnal tidal constituents, in this case P_1 with a period of 24.0658 h ($1/(\omega_S - \omega_2)$, see Table T12), and another K_1. Thus, the K_1 used for tidal prediction has both lunar and solar parts. P_1 and the solar part of K_1 cancel each other out every 182 days, at vernal and autumnal equinoxes.

As mentioned above, the angle between the plane through the moon's orbit and the plane through the equator varies over a 18.6-year period ($1/\omega_4$, see Table T12). This is referred to a *lunar nodal regression* because the intersection of the moon's orbital plane with the ecliptic, called the ascending lunar node, regresses backwards along the ecliptic over this 18.6-year period. Lunar distance also varies with time because the longitude of the lunar perigee rotates with an 8.85-year period ($1/\omega_3$, see Table T12). The spectral splitting due to these long-period effects can also be represented by harmonic constituents, called "satellite" constituents, but harmonic analyses using them must use data series that

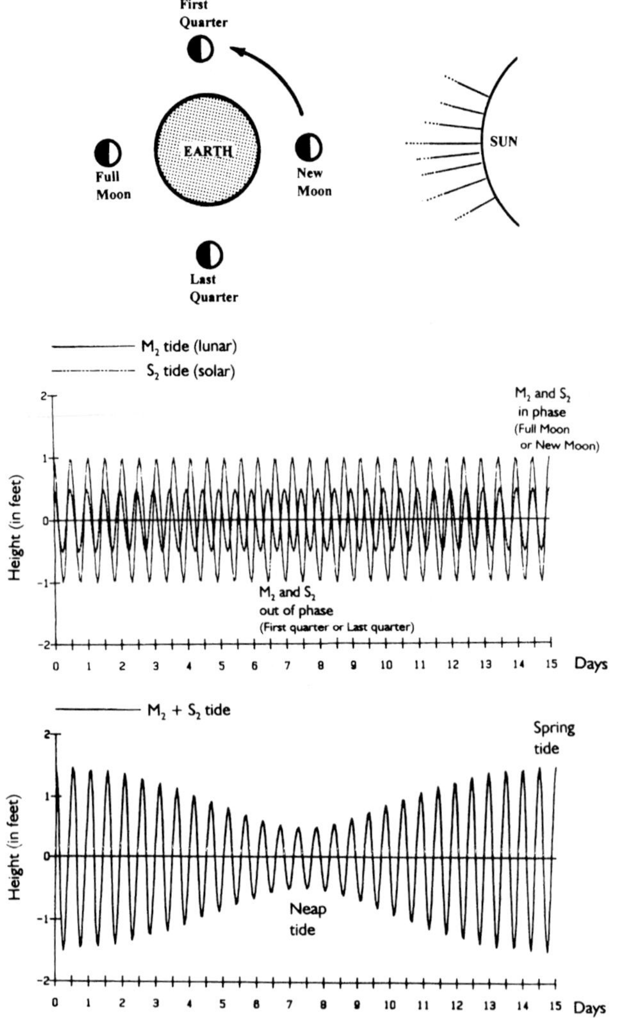

Figure T36 The combined effect of the moon and sun varies throughout the month. When the moon and sun are working with each other (at full moon and new moon) one sees the largest tidal ranges (*spring tides*). At First Quarter and Last Quarter the moon and sun work against each other resulting in smaller tidal ranges (*neap tides*).

are 18.6 years long. Traditionally the nodal effects have been handled in a form that directly represents the modulation of each lunar tidal constituent. The amplitude of each modulation is called the *node factor, f*, and the phase is called the *equilibrium argument, u. f* and *u* are regarded as constants for the period of analysis (or prediction) and are obtained from astronomically calculated tables (e.g. Tables 14 and 15 in Schureman, 1958). The largest variation in f over a 18.6-year period is found in O_1, which varies ±18%, and in K_1, which varies ±11%. The variation of f for M_2 is ±4%, but since M_2 is much larger than O_1 and K_1 in many locations, it may only be the 4% variation with which one is usually concerned. There is no direct nodal effect on solar constituents such as S_2 or P_1.

Tidal analysis and prediction

Knowing only the periods of these and other smaller tidal constituents, one can analyze a data series of water level observations from a particular location. The result of such a *harmonic analysis* is an *amplitude* and *phase* for each of these tidal constituents, which represent how large each effect is *at that particular location*, and when in time the peak of each effect will take place (such as relative to when the moon passes over (transits) that location). The hydrodynamics affect both the amplitude and phase (timing) of each tidal constituent, but we really do not need to know the details of how it happened, only that it did happen. Hydrodynamics in shallow-water areas will also produce additional tidal constituents not seen in deeper water (discussed in the next section) that

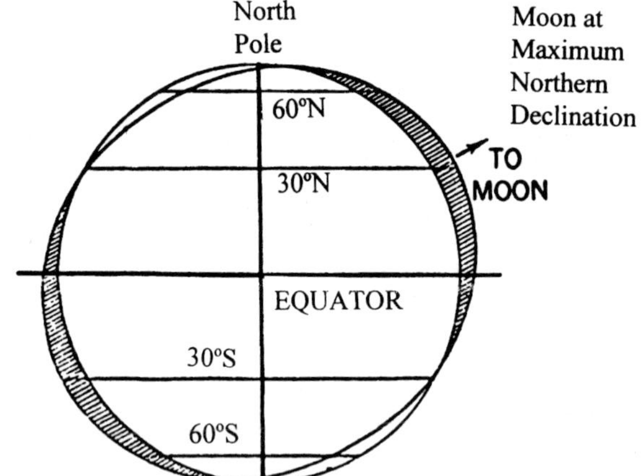

Figure T37 When the moon is at maximum declination north or south of the equator the tidal bulge also shifts north or south. When this happens certain locations on the earth would rotate under only one high water "bulge," that is, there will be a diurnal component of the tidal forces. Hydrodynamics will determine which oceans and bays have significant diurnal tides.

can also be easily handled by the harmonic method. As long as the hydrodynamics stay the same (i.e., the depth of the basin stays the same), the tide predictions from these calculated harmonic constants will be accurate. For small bays, shoaling or dredging can change the hydrodynamics and thus change the tide range and times of high and low waters.

Harmonic analysis is still the most common method used for the analysis and prediction of tides and tidal currents. It has not changed much from the theory first developed by Lord Kelvin in 1867 and later refined by George Darwin, except that the Fourier analysis solution technique (e.g., Schureman, 1958) has generally been replaced by a least-squares solution technique (e.g., Foreman and Henry, 1989) that minimizes the squared differences between measurements and computed tidal predictions. Doodson (1921) carried out a complete harmonic development of the tide-generating potential, determining over 400 tidal constituents (most very small), which has been the standard reference work for tidal analysis and prediction. This was recalculated and updated 50 years later by Cartwright and Tayler (1971).

Harmonic analysis is used to produce the predictions at the reference stations in all nationally published Tide and Tidal Current Tables. Nonharmonic analysis, which is simply one of several methods of comparing the tide or tidal current at two locations, is used to calculate the time differences, height differences, velocity ratios, etc. for the thousands of secondary stations in these tables that are referred to the reference stations. Several refinements to the harmonic methods have been developed in recent decades, as well as the nonharmonic/cross-spectral approach of the response method developed by Munk and Cartwright (1966). Whatever the method used, some principles will always apply. For example, the analysis of longer data time series will lead to more accurate predictions because more tidal frequencies can be resolved in a longer series.

Using the harmonic method the tide is represented by the sum of various tidal constituents, each representing one of these frequencies. The height of the tide at any time is typically represented by a formula such as (Schureman, 1958):

$$h = H_0 + \sum f\, H \cos[at + (V_0 + u) - \kappa],$$

where

h	=	height of tide at any time t
H_0	=	mean height of water level above the datum used for prediction
H	=	mean amplitude of any constituent A
f	=	node factor
a	=	speed of constituent A (i.e., its frequency)
t	=	time, reckoned from some initial epoch (such as the beginning of the year of predictions)

$(V_0 + u) =$ value of equilibrium argument of constituent A at $t = 0$.

κ = epoch (phase) of constituent A.

The *"speed" of a constituent* is merely its frequency, but it has been traditionally given in terms of degrees per solar hour, where 360° is one complete cycle. Thus, M_2, which has a period of 12.42 h and a frequency of 1.932 cycles per day, has a speed of 28.984104°/h.

For tidal current predictions the same equation is used twice, once for each of two orthogonal components (e.g., major and minor axes of flow), which when combined will give the speed and direction of the flow. The major and minor component for each tidal constituent can be combined to produce a *tidal constituent ellipse*, which shows what the constituent flow would be for each instant in a constituent cycle.

The more tidal constituents that can be calculated the more accurate the tidal prediction will be. The number of constituents that can be calculated depends on the length of the data series. The length of a data series needed to resolve two tidal constituents is inversely proportional to the difference in the frequencies. Table T13 lists the 37 typically most important tidal constituents, listed in order of length of time needed to resolve that constituent from a nearby larger constituent. One sees natural groupings near 15 days, 29 days, 6 months, and one year.

If one has only 15 days of data, for example, then a harmonic analysis will provide values for the major constituents M_2, S_2, K_1, and O_1, plus a few higher harmonics and a couple of less important constituents. However, these calculated values will also include energy from the constituents that could not be separated out in only 15 days. For example, the M_2 will include energy from N_2 (which could have been resolved from M_2 if there had been 29 days of data). This N_2 contribution could make the M_2 value calculated from 15 days of data larger than it should be, or smaller than it should be, depending on when the data were measured. Likewise, K_1 will include the effects of P_1 (which could have been resolved from K_1 if there were 6 months of data). If we harmonically analyzed successive 15-day periods, we would see the amplitude of K_1 slowly vary over a 6-month period because of the influence of P_1. The key to an accurate tidal prediction is determining amplitudes and phases for the most tidal constituents that can be calculated with a given length data time series.

To carry out a tide prediction in the era before the above equation could be programmed on a computer (as it is done routinely today), large machines were built with gears and pulleys connected by a wire to a pen. Each tidal constituent had a different size rotating gear and a pin and yoke system connected to a pulley (see Figure T38). The pin and yoke system turned the rotating motion of the gear into a vertical up and down motion of the pulley, which moved the wire over it and thus moved the pen up and down on a roll of moving paper. The wire ran over a number of pulleys so all the constituent effects could be added together. The first tide predicting machine was a wooden model built for Kelvin in 1872, but later models were huge brass machines with dozens of finely made gears and pulleys. The first one built in the United States was by William Ferrel in 1885. Prior to the use of harmonic analysis there were other less sophisticated methods based on recognized relationships between the tides and the movements of the moon and sun. For example, for a particular place, high tide might occur a certain number of hours after the moon was directly overhead, and the highest (spring) tide might occur a certain number of days after full moon or after new moon. In many of the early maritime nations, tide prediction schemes were treasured family secrets passed on to the next generation. The two earliest tide tables discovered were for the tidal bore in the Tsientang River in China in 1056 and for London Bridge in England in the early 1200s.

Table T13 Thirty-seven tidal constituents that can be calculated from a one-year series, listed in order of length of data time series needed to resolve each constituent from a nearby larger constituent, and size

Constant	Speed (°/h)	Origin	Days needed to separate	From	Amplitude at Trenton, NJ (feet)
M_2	28.984104	Lunar	–	–	3.547
M_4	57.968208	Shallow-water	0.5	M_2	0.517
M_6	86.952313	Shallow-water	0.5	M_4	0.266
M_8	115.936417	Shallow-water	0.5	M_6	0.120
K_1	15.041069	Lunisolar	1.1	M_2	0.349
S_6	90.000000	Shallow-water	4.9	M_6	0.005
S_4	60.000000	Shallow-water	7.4	M_4	0.005
O_1	13.943036	Lunar	13.7	K_1	0.288
MK_3	44.025173	Shallow-water	13.7	$2MK_3$	0.120
$2MK_3$	42.927140	Shallow-water	13.7	MK_3	0.116
OO_1	16.139102	Lunar	13.7	K_1	0.030
$2Q_1$	12.854286	Lunar	13.8	O_1	0.028
S_2	30.000000	Solar	14.8	M_2	0.461
MS_4	58.984104	Shallow-water	14.8	M_4	0.148
$2SM_2$	31.015896	Shallow-water	14.8	S_2	0.025
M_3	43.476156	Lunar	27.3	MK_3	0.034
M_1	14.492052	Lunar	27.3	K_1	0.027
N_2	28.439730	Lunar	27.6	M_2	0.553
MN_4	57.423834	Shallow-water	27.6	M_4	0.176
Mm	0.544375	Lunar[a]	27.6	Mm	0.124
Q_1	13.398661	Lunar	27.6	O_1	0.022
J_1	15.585443	Lunar	27.6	K_1	0.016
$2MN_2/L_2$	29.528479	Shallow-water/lunar	31.8	S_2	0.409
$2MS_2/\mu_2$	27.968208	Shallow-water/lunar	31.8	N_2	0.219
MSf	1.015896	Lunar[a]	182.6	Mf	0.186
Mf	1.098033	Lunar[a]	182.6	MSf	0.132
P_1	14.958931	Solar	182.6	K_1	0.110
K_2	30.082137	Lunisolar	182.6	S_2	0.094
ν_2	28.512583	Lunar	205.9	N_2	0.210
λ_2	29.455625	Lunar	205.9	$2MN_2$	0.102
$2NM_2/2N_2$	27.895355	Shallow-water/lunar	205.9	$2MS_2$	0.042
ρ_1	13.471514	Lunar	205.9	Q_1	0.013
Sa	0.041069	Solar[a]	365.2	Ssa	0.430
Ssa	0.082137	Solar[a]	365.2	Sa	0.169
S_1	15.000000	Solar	365.2	K_1	0.062
T_2	29.958933	Solar	365.3	S_2	0.056
R_2	30.041067	Solar	365.3	S_2	0.028

Amplitudes from a 1981 analysis of water level data (from Trenton, NJ, on the Delaware River) are provided as an example (after Parker *et al.*, 1999).

[a] Values are determined predominantly by long-term meteorological effects and thus vary from year to year.

Hydrodynamic considerations in coastal waters

When the very long tide wave generated in the ocean reaches the shallower water of the continental shelf and the even shallower water of the bays and rivers, it is slowed up, amplified, modulated, and distorted by a number of hydrodynamic mechanisms. The long wave enters and propagates up a river as a *progressive wave*, that is, the crest of the wave (high water) moves progressively up the river, as does the trough of the wave (low water) (see Figure T39). In such a progressive tide wave the flood current (the tidal current flowing up the river) is fastest at approximately the same time as high water, and the ebb current (the tidal current flowing down the river) is fastest at approximately the same time as low water. Slack water (the time of no current) occurs approximately halfway between the times of high water and low water.

If there is nothing in a river to impede or stop the tide wave (like a dam or rapids or a sudden decrease in width), it will continue to travel up the river until bottom friction wears it down. However, if the width of the river decreases as the tide wave moves upriver, then the tidal range will be increased, because the same energy is being forced through a smaller opening. If the depth of the river decreases there is a similar but less dramatic amplifying effect generally outweighed by the increased frictional energy loss.

The greatest amplification of a tide wave usually occurs in a bay (or in a river with a dam). In this case, the tide wave is reflected at the head of the bay and travels back down the waterway toward the ocean. This *reflected* wave is not observable by someone on the shore because it is superimposed on the next incoming tide wave that is propagating up the bay, and it is the combination of the two waves that one observes. The resulting combined wave is called a *standing wave*, because the high and low waters do *not* progress up the bay or river. The water simply moves up and down everywhere at the same time (see Figure T40), with the greatest tidal range at the head of the bay. With a standing wave, the tidal range decreases as one moves from the head of the bay toward the ocean entrance, and, if the bay is long enough, reaches a minimum at one location (called a *node*) and then starts increasing again. This node occurs at one-fourth of a tidal wavelength from the head of the bay (see Figure T40). (In a progressive wave high water comes half a wavelength before low water, so if a high water travels a distance equal to one-fourth of a tidal wavelength up the bay to the head, where it is reflected, and then travels one-fourth of a wavelength back down the bay, it will have gone half a wavelength and so coincide with low water of the next incoming wave, and the two will cancel each other out at that location, producing a very small tidal range.) For a standing wave high waters occur at the same time everywhere on one side of the node, which is the same time as low waters occur on the other side of the node. The strongest tidal currents occur when water level is near mean tide level, approximately halfway between the times of high water and low water. At the times of high water and low water there is no flow (slack water). The water flows into the bay, stopping the inward flow at high water, reverses direction, flows out of the bay until low water, at which time it reverses again and starts flowing into the bay again.

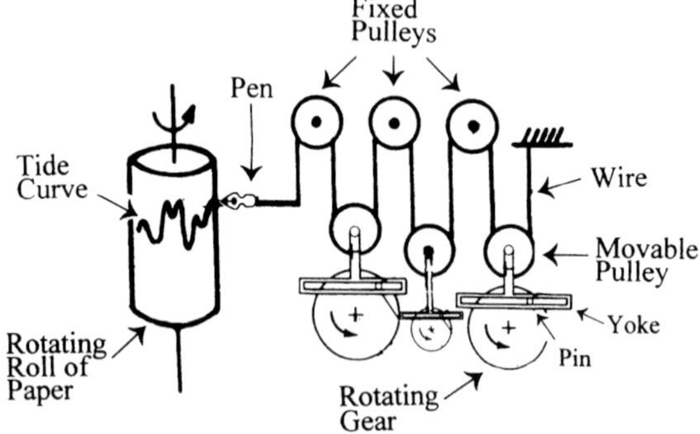

Figure T38 A schematic of an early tide prediction machine. Each gear and pulley combination represented one tidal constituent. The wire running over every pulley summed the motions and moved a pen on a moving roll of paper to draw a tide curve.

When length of a bay is exactly one-fourth of a tidal wavelength, then resonance occurs, which creates the largest tides possible. When the water in the bay is forced to move up and down by the tide at the entrance, it will freely oscillate (slosh up and down) with a natural period that depends directly on its length and inversely on the square root of its depth. If the basin has the right combination of length and depth so that the natural period is the same as the tidal period, then the oscillation inside the bay will be synchronized with the oscillation at the entrance due to the ocean tide. In other words, the next ocean tide will be raising the water level in the bay at the same time that it would already be rising due to its natural oscillation (stimulated by the previous ocean tide wave), so that both are working together, thus making the tidal range inside higher.

Most bays actually fall in between the extremes of pure progressive wave and pure standing wave described above, because bottom friction reduces the amplitude of the tide wave as it travels. Thus, the reflected wave will always be smaller than the incoming wave, especially near the bay entrance, and the combination of the two frictionally damped progressive waves will not be a pure standing wave. There will be no point of zero tidal range, but only an area of minimum tidal range. There will be some progression of high waters and low waters up the bay, and maximum flood or ebb currents will not occur exactly half way between high water and low water. A basin one-fourth of a wavelength long will still produce the largest possible tidal range at the head of the bay, but friction keeps that tidal range much smaller than it would be without friction.

In some bays, the very high tidal range at the head of the bay is due to a combination of both a narrowing width and a near resonant situation (due to the right length and depth). The highest tidal ranges may involve several amplifications, the bay being perhaps connected to a gulf with perhaps a wide continental shelf beyond that, with amplifications of the tide wave occurring in each basin. This is the case with the Bay of Fundy tides, the tide wave being already amplified by the Gulf of Maine and the continental shelf prior to entering the Bay of Fundy.

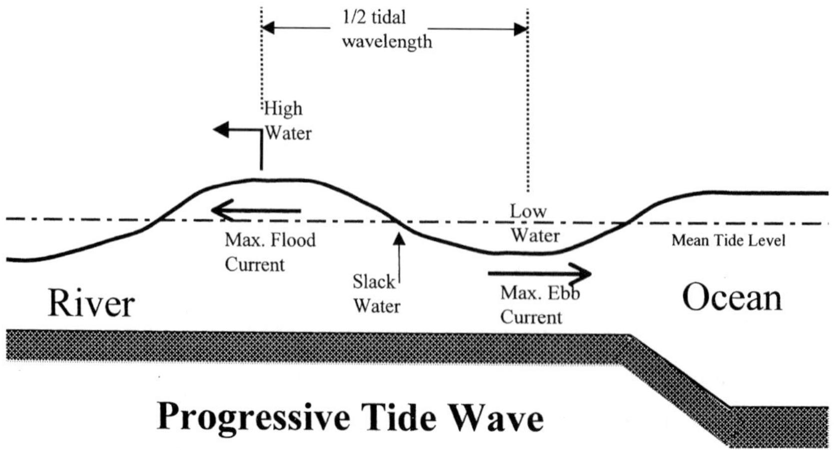

Figure T39 The tide propagating up a river as a *progressive wave*. High water occurs later as one moves upstream. The tidal wavelength is typically on the order of hundreds of miles.

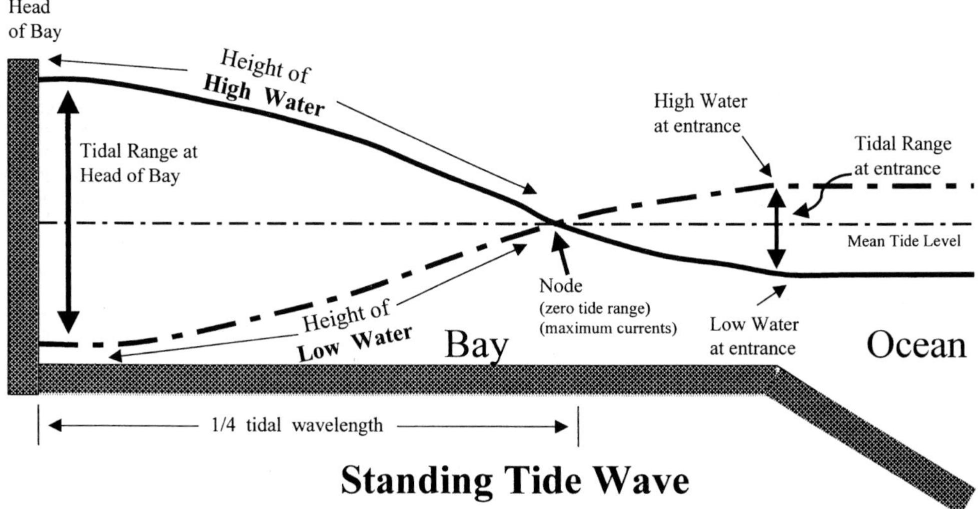

Figure T40 The tide in a bay as a *standing wave* (the water level is shown for two extremes, high water and low water). High water occurs at approximately the same time everywhere on one side of the node (the point of zero tidal range). This is an idealized case assuming there is no bottom friction effect. With friction the tidal range at the node is not zero and the times of high water do progress slightly up the bay.

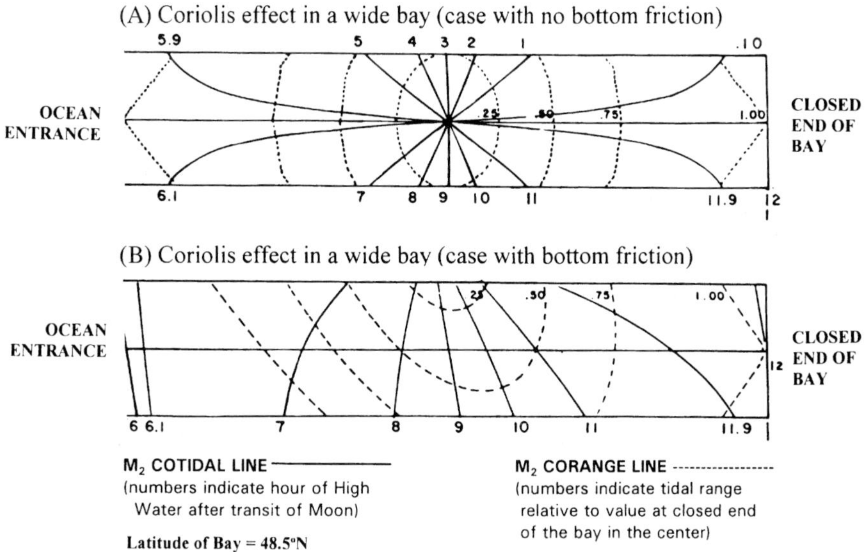

Figure T41 The effect of Coriolis force on M_2 tidal range (corange lines) and the times of high water (cotidal lines) for an idealized rectangular bay. The top figure assumes no bottom friction effects, and a single point of zero tidal range in the middle of the bay results. The bottom figure, which includes friction, is more realistic, and there is no point with zero tidal range.

If a bay is wide enough one also sees larger tidal ranges on the right side of the bay (looking up the bay) due to the Coriolis effect. Figure T41(A) shows lines of locations with the same time of high water (cotidal lines) and lines of constant tidal range in an idealized rectangular basin for the case where the effect of bottom friction is ignored. A single point of zero tidal range occurs in the center of the bay. A more realistic case, including the damping effect of bottom friction, is shown in Figure T41(B), with the point of no tidal range disappearing onto land. (The effect of Coriolis can also be seen in Figure T42, where the shape of the M_2/K_1 lines near Victoria are caused by the point of zero M_2 tidal range having moved to the left on land.)

Huge tidal ranges are not restricted to bays. If the continental shelf is the right combination of depth and width, a near resonant situation can also result. This is the reason for the 12 m tidal ranges along the coast of southern Argentina. The continental shelf there is over 1,000 km (600 miles) wide, and includes the Falkland Islands near the edge of the shelf (where the tidal range only reaches 2 m. The distance from the Argentinean coast to the edge of the shelf is fairly close to one-fourth of a tidal wavelength for that depth of water. Essentially, that wide shelf has a natural period of oscillation that is fairly close to the tidal period.

The largest tidal currents in bays tend to be near the entrances. Maximum tidal current speeds are zero at the head of the bay (since there is no place for the water to flow). As one moves down the bay toward the ocean the maximum flood and ebb tidal current speeds increase, with the greatest speeds occurring at the entrance, or, if the bay is long enough, at the area of smallest tidal range (the nodal area). However, if the width of the bay decreases at any point, the current speeds will be increased in that narrow region (since the same volume of water is being forced to flow through a smaller cross section, it must flow faster). This can be especially dramatic if there is a sudden decrease in width and depth. The largest tidal currents are found in narrow straits in which the tides at either end have different ranges or times of

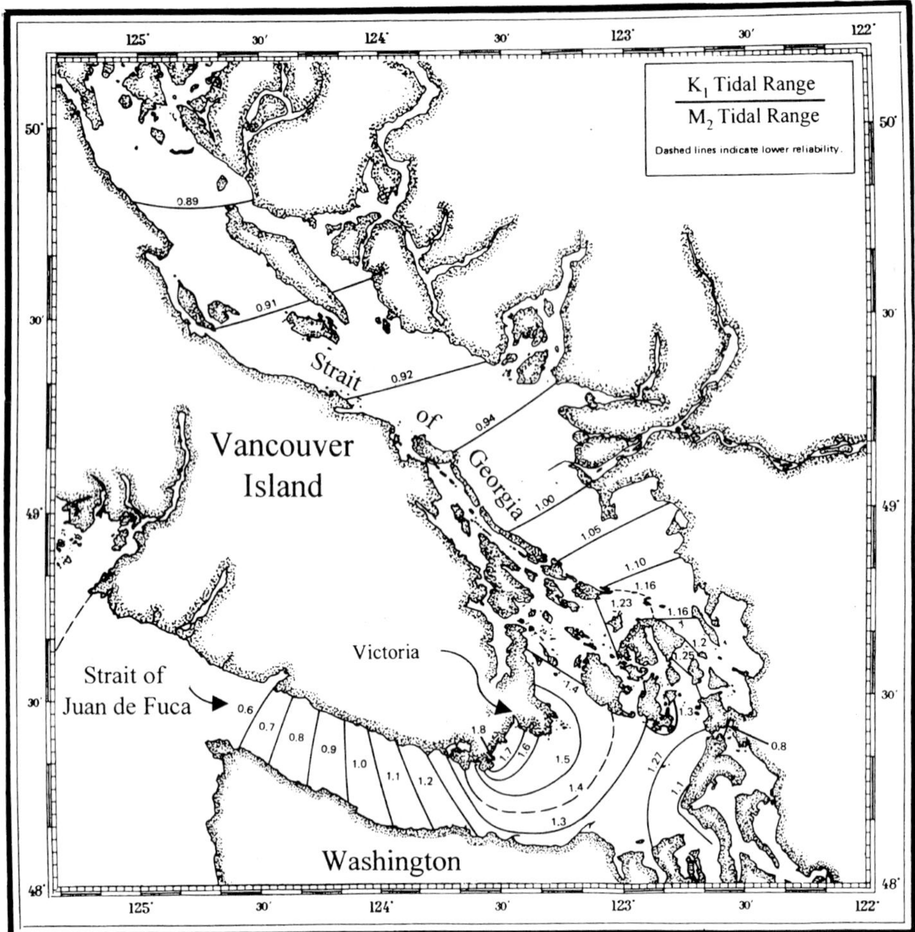

Figure T42 The ratio of the largest diurnal component of the tide (K_1) to the largest semidiurnal component of the tide (M_2) along the length of the Strait of Georgia-Strait of Juan de Fuca. The highest ratio of diurnal range to semidiurnal range occurs near Victoria, British Columbia, because that is the area of the semidiurnal node (minimum M_2 tidal range), which is one-fourth of a semidiurnal tidal wavelength from the northern end of the Strait of Georgia (see text).

high water. Where a strait suddenly becomes very narrow or where it bends, eddies and whirlpools can be formed as the result of the sheltering effect of the land and the inertia of the coastal flow.

The dimensions of a basin can also determine the size of the diurnal tidal signal compared with the usually dominant semidiurnal tidal signal. A particular bay could have a natural period of oscillation that is closer to the diurnal tidal period (approximately 24.84 h) than to the semidiurnal period, thus amplifying the diurnal forcing at the entrance to the bay more than the semidiurnal signal. Depending on the size of the diurnal signal at the entrance the result could be a mixed tide or a diurnal tide. At such locations (such as parts of the Gulf of Mexico) the tide will be diurnal near times of maximum lunar declination, but will be semidiurnal near times when the moon is over the equator.

The wavelength, L, of a tide wave in a bay depends on the depth of the water, D, and on the tidal period, T, according to $L = T(gD)^{1/2}$ (if we ignore frictional effects). The shallower the bay the shorter the wavelength. The longer the tidal period the longer the wavelength. A diurnal tidal component has a wavelength twice as long as a semidiurnal tidal component since its period is twice as long. When a waterway is shallow enough and long enough so that more than one-fourth of a semidiurnal wavelength fits in the waterway, there will be a nodal area with a very small semidiurnal tidal range. This will be an area where the diurnal tide could dominate, since the diurnal tide would still be large at the semidiurnal nodal area (its the diurnal node being twice as far from the head of the bay). Thus near the head of the waterway the tide could be semidiurnal, but near the semidiurnal nodal area the tide could be mixed or even diurnal. This is the case near Victoria, British Columbia, at the southeastern end of Vancouver Island (see Figure T42). At that location along the Strait of Georgia-Strait of Juan de Fuca waterway, the semidiurnal tidal component decreases to a minimum, but the diurnal component does not, and so the tide becomes diurnal (while at the northern end of the Strait of Georgia the tide is semidiurnal).

Whether due to a basin size conducive to amplifying the diurnal signal or due to the existence of a semidiurnal nodal area (leaving the diurnal signal as the dominant one), there are numerous areas around the world with strong diurnal tides—places like Norton Sound in Alaska near the Bering Strait, and various (but not all) locations in the Philippines, New Guinea, and the islands of Indonesia. In southern China, at Beihai, and at Do Son, Vietnam, the diurnal signal is very dominant, with tidal ranges that reach 5 m and 3 m, respectively (near times of maximum southern declination of the moon); the tide remains diurnal even near times when the moon is over the equator.

The primary effect of shallow water on the tide that we have discussed so far is that it shortens the tidal wavelength down to the same order of magnitude as the lengths of bays and river basins, thus bringing the dynamic situation closer to resonance and increasing the tidal ranges. (Or, one can also look at it from the point of view of the shallower depths increasing the natural periods of these bays and rivers (which are very small basins compared to the ocean) to be closer to the tidal period.) However, very shallow water can have other effects on the tide, for example, distorting the shape of the tide wave, that is, making it very asymmetric, so that its rise and fall (and its flood and ebb) are no longer equal (see the second curve in Figure T43). The tide can then no longer be described by a simple sine wave (the first curve in Figure T43). In some cases such distortion leads to double high waters or double low waters (see third curve in Figure T43). The extreme case of distortion is a tidal bore (the fourth curve in Figure T43).

Shallow water distorts the tide through several mechanisms. The speed, C, at which a long tide wave travels depends on the depth of the water, D, according to the formula $C = (gD)^{1/2}$. When depth of the

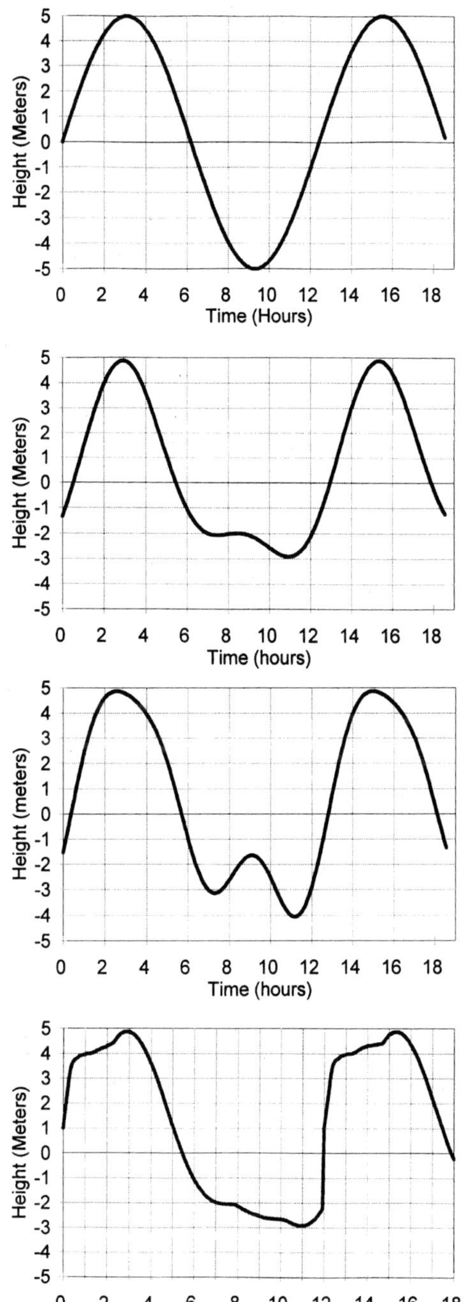

Figure T43 Typical tide curves (i.e., tidal height plotted with respect to time), over one-and-half tidal cycles for an area with no shallow-water effects (top panel) and for three areas with different degrees of distortion caused by the shallow water. In the third panel a double low water occurs. The fourth panel shows the almost instantaneous rise in water level due to the passage of a tidal bore.

water is much greater than the tidal range, the speed of the crest of a tide wave and the speed of the trough are virtually the same, since the tide wave itself has only a very small effect on the total water depth. But in the shallow water where the depth is not much greater than the tidal range, the total water depth under the crest is significantly larger than the total water depth under the trough. In this case, the crest of the wave (high water) travels faster than the trough of the wave (low water). If the tide wave travels far enough the crest begins to catch up with the trough ahead of it (which is falling behind the crest ahead of it). The shape of the tide wave then begins to look like the second curve in Figure T43, with a more rapid rise to high water and a slower fall to low water. In terms of harmonic constituents, this distortion transfers energy from M_2 into a constituent called M_4, with half the period.

Another shallow-water distorting mechanism is caused by bottom friction, which can have both asymmetric and symmetric effects. The asymmetric effect (similar to that just discussed and represented in Figure T43) results because friction has a greater effect in shallow water than in deepwater (there being less water to have to slow down), and so it slows down the trough more than the crest, contributing to the distortion of the tide wave and the generation of M_4. A symmetric effect results because energy loss due to friction is proportional to the square of the current speed. This means that there will be much more energy loss during times of maximum flood and maximum ebb than near slacks. This results in the generation of another higher harmonic, M_6, with a period of one-third that of M_2. This effect, combined with the asymmetric effect, can lead to double high or low waters (see third curve in Figure T43). Higher harmonic tidal constituents like M_4 and M_6 are referred to as *overtides*.

The above symmetric frictional effect also causes the interaction of two tidal constituents, for example, M_2 and N_2. M_2 and N_2 go in and out of phase over a 27.6-day cycle (perigee to apogee to perigee). In this case, the greatest energy loss occurs when M_2 and N_2 are in phase and producing the strongest tidal currents, and the lowest energy loss occurs 13.8 days later with M_2 and N_2 are out of phase and producing the weakest tidal currents. The increased energy loss when M_2 and N_2 are in phase is greater than the decreased energy loss when they are out of phase, and the result is that each constituent will be smaller than if it existed without the other present. There is a 27.6-day modulation of this energy loss from M_2 and N_2, which produces two new *compound* tidal constituents called $2MN_2$ and $2NM_2$. (Similarly, the above asymmetric mechanisms also cause interactions between constituents, producing higher frequency constituents such as MN_4 from M_2 and N_2.)

Friction, of course, dissipates energy from the entire wave and slowly wears the entire wave down. However, if, as the wave propagates up the river, the river's width is decreasing significantly, this can keep the amplitude of the wave high in spite of the friction. Thus, the tide wave can continue to travel up a narrowing river, getting more and more distorted in shape. A further distortion can be caused by the river flow interacting with the tide (see below). In the extreme case, the distortion from all these effects can lead to the creation of a tidal bore (see fourth curve in Figure T43).

In a river there will also be the river current itself (resulting from fresh water flowing downhill) added onto the tidal current, the result being a faster and longer lasting ebb current and a slower shorter flood current phase. Far up a river where the river flow is faster than the strongest tidal current, the flow of water will always be downstream, but the speed of flow will oscillate, flowing the fastest downstream at the time of maximum ebb for the tidal current and flowing the slowest downstream at the time of maximum flood for the tidal current. This is a simple addition to the tide, but the river flow also interacts with the tide and distorts it through interaction caused by bottom friction. As just mentioned, energy loss due to friction is proportional to the square of the total current speed. When the river current, flowing in the same direction as the ebb current, creates a larger combined ebb current, there is a greatly increased energy loss. Likewise, when the river current flowing opposite to the flood current reduces the total speed, the energy loss is greatly reduced. This not only has an asymmetric effect that distorts the tide (causing a faster rise to high water, delaying the time of low water, and contributing to M_4), but it also further wears down the entire wave because the increased energy loss during ebb is larger than the decreased energy loss during flood.

Another type of shallow-water effect causes interactions between tide and storm surge (generated by the wind) that have periods longer than tidal periods. In this case, when the water level is raised by an onshore wind, that increases the water depth and changes the tidal dynamics, usually increasing the tidal range. When an offshore wind lowers the water level, decreasing the water depth, the result is usually a decreased tidal range. Knowing that river discharge and storm surge can modify the tide, it is important when harmonically analyzing water level data to make sure that these data were not taken only during such meteorological events. These shallow-water effects that distort and modulate the tide and cause interactions with storm surge and river discharge are called *nonlinear* effects because the mechanisms that produce these effects are represented by several *nonlinear* terms in the equations of motions used to model the tidal hydrodynamics.

We have shown how tidal prediction can be accomplished quite accurately by harmonic analysis of a water level data time series knowing only the astronomical frequencies involved, and ignoring hydrodynamics. Accurate inclusion of the additional tidal constituents produced by nonlinear shallow-water effects into the harmonic prediction method does involve some knowledge of hydrodynamics (especially when deciding the formula for each constituent's node factor). However, recently tidal prediction has begun to involve the use of sophisticated hydrodynamic numerical models, which have two important advantages over statistical techniques. First, such models can provide tide and tidal current predictions at locations where there are no water level data, thus providing predictions at hundreds or thousands of locations in an area, as opposed to the few locations where statistical methods can provide them (limited by needing data to analyze). Second, such models handle very nicely the nonlinear interaction between the tide and nontidal phenomena like storm surge and river discharge.

Bruce Parker

Bibliography

Cartwright, D.E., 1999. *Tides: A Scientific History*. Cambridge: Cambridge University Press, UK, 292 pages.

Cartwright, D.E., and Tayler, R.J., 1971. New computations of the tide-generating potential. *Geophysical Journal of the Royal Astronomical Society*, **23**: 45–74.

Doodson, A.T., 1921. The harmonic development of the tide-generating potential. *Proceedings of the Royal Society A*, **100**: 305–329.

Foreman, M.G.G., and Henry, R.F., 1989. The harmonic analysis of tidal model time series. *Advances in Water Resources*, **12**(3): 109–120.

Munk, W.H., and Cartwright, D.E., 1966. Tidal spectroscopy and prediction. *Philosophical Transactions of the Royal Society A*, **259**: 533–581.

Parker, B.B. (ed.), 1991. *Tidal Hydrodynamics*. New York: John Wiley & Sons.

Parker, B.B., Davies, A.M., and Xing, J., 1999. Tidal height and current prediction. In Mooers, C.N.K. (ed.), *Coastal Ocean Prediction*. Coastal and Estuarine Studies, American Geophysical Union. **56**, 277–327,

Platzman, George, 1971. Ocean tides and related waves. *Lectures in Applied Math*, **14**: 239–291.

Pugh, D.T., 1987. *Tides, Surges, and Mean Sea Level*. New York: John Wiley & Sons.

Schureman, P., 1958. *Manual of Harmonic Analysis and Prediction of Tides*. S.P. 98 (Revised 1940 edn with corrections. Washington, DC: Coast and Geodetic Survey, US Department of Commerce.

Cross-references

Altimeter Surveys, Coastal Tides and Shelf Circulation
Coastal Currents
Meteorologic Effects on Coasts
Sea-Level Changes During the Last Millennium
Sea-Level Datum
Storm Surges
Tidal Environments
Tidal Inlets
Tidal Prism
Tide Gauges

TIME SERIES MODELING

Fundamental concepts of coastal time series

There has been a considerable expansion in the coastal database in the past two decades. Discrete and continuous data are collected from the coast and its interacting hydrodynamic, biological, morphological, sedimentological, and other associated subsystems in order to make better decisions on coastal management and sustainable development. While the quality and length of the collected coastal data vary greatly, many datasets fall within the domain of time series analysis. Any time series can be considered as a time (or space) ordered sequence of realizations (or observations) of a variable of interest. The set of observations generated sequentially in time could be either continuous or discrete. In statistical terms, a time series is a realization or sample function from a certain stochastic or random process. The time series to be analyzed can be considered as a particular realization, produced by the underlying probability mechanism of the system (i.e., coastal) under consideration. The observed values of a stochastic process are generally considered as a realization of the stochastic process. Time series analysis can consider a class of stochastic processes, called stationary processes, which assume that the process is in a particular state of statistical equilibrium (Box *et al.*, 1994). In the analysis of stationary stochastic processes it is worthwhile to note that "stationary processes generally arise from any 'stable' system which has achieved a 'steady-state' mode of operation" (Priestley, 1981, p. 14), whereas a nonstationary series is regarded as having properties which change with time.

Since in time series analysis, inferences can be made from a realization to the generating process, it is necessary to consider the statistical properties of the time series. Essentially, the properties of a time series can be obtained from a single realization over a time interval or based on several realizations at a particular time. The properties based on a time interval of a single realization are referred to as time-averaged properties. The properties from several realizations at a given time are considered as the ensemble properties. "Since different sections of a time series resemble each other only in their average properties, it is necessary to describe these series by probability laws or models" (Jenkins and Watts, 1969, p. 2).

Coastal time series modeling

Since stochastic processes deal with systems which develop in accordance with probabilistic laws, stochastic models can be used to gain insights on the spatial and temporal behavior of the coastal system (Lakhan, 1982; Lakhan and Trenhaile, 1989). Time series modeling of the many natural phenomena occurring in the coastal environment, which appear to behave in random or probabilistic ways, is essential for understanding not only the operating processes, but also for many coastal applications. Many examples of the use and applications of probabilistic and time series models in coastal studies have been provided by Guedes Soares (2000). Even the modeling of one sea-state parameter has many applications in coastal and offshore engineering. For example, modeling a time series of significant wave height at a location is a useful complement to long-term probabilistic models that describe the wave climate in different areas (Cunha and Guedes Soares, 1999).

The complex nature of the coastal system makes it necessary to develop models which can explain both the deterministic and random features of the time series for any coastal process (Lakhan, 1989). For example, sea-states can be visualized as random and unpredictable. However, there is some constancy of a sea-state for a short duration of several minutes to fractions of minutes, and therefore it is possible to assume stationarity for such short wave records. Conversely, a wave record of long duration will not exhibit stationarity because the significant wave height and other statistics are known to vary with respect to time (Goda, 2000). To characterize and predict the probabilistic nature of the sea surface a time series model can be developed to represent the sea surface elevation as a nonstationary stochastic process, with the sea-state considered to be a Gaussian, statistically stationary stochastic process in short time periods. In modeling stationary or nonstationary time series, coastal and allied researchers have utilized either the frequency domain or the time domain approach.

Approaches to time series analysis

Given the fact that there are various terminology, theoretical and practical aspects of time series analysis, this short review will introduce only the fundamental concepts for coastal and allied researchers. Hannan *et al.* (1985) pointed out that since its inception the theory and practice of the analysis of time series has followed two lines. One of these proceeds from the Fourier transformation of the data and the other from a parametric representation of the temporal relationships. The two approaches to time series analysis, commonly referred to as the frequency domain approach (or spectral analysis approach) and the time domain approach, have been discussed in several books (e.g., Anderson, 1975; Otnes and Enochson, 1976; Gottman, 1981; Kendall and Ord, 1990; Wei, 1990; Bendat and Piersol, 1993; Harvey, 1993; Hamilton, 1994; Brockwell and Davis, 1996; Pollock, 1999; Shumway and Stoffer, 2000). In brief, the frequency domain approach uses spectral functions to study the nonparametric decomposition of a time series into its different frequency components. The time domain approach concentrates on the use of parametric models to model some future value of a time series as a parametric function of the current and past values.

While the frequency and time domain approaches can be used to provide different insights into the nature of the actual time series it should, however, be pointed out that both approaches are mathematically equivalent. According to Gottman (1981), the two approaches are linked by the famous theorem called the Wiener-Khintchine Theorem which provides a shuttle between the frequency and time domains. The Wiener-Khintchine Theorem shows that there is a one-to-one relationship between the autocovariance function of a stationary process and its spectral density function (Pollock, 1999). Knowing the correlation structure in the time domain corresponds to knowing the form of the spectrum in the frequency domain. Essentially, the autocorrelation function and the spectrum function form a Fourier transform pair (Kendall and Ord, 1990).

Although the two approaches are complementary rather than competitive (Harvey, 1993), there are situations when one approach is more appropriate to use than the other. According to Shumway and Stoffer (2000) the two approaches may produce similar answers for long series, but the comparative performance over short samples is better done in the time domain. Given this observation, brief remarks and selected coastal applications for both approaches will, therefore, be provided. Emphasis will then be placed on modeling in the time domain because there are well-established techniques for model selection, identification, and estimation.

The frequency domain approach

Extensive discussions on the frequency domain, or spectral approach, to time series analysis can be found in several books (e.g., Jenkins and Watts, 1969; Rayner, 1971; Kanasewich, 1973; Koopmans, 1974; Brillinger, 1975; Bloomfield, 1976; Priestley, 1981; Brillinger and Krishnaiah, 1983; Brigham, 1988; Brockwell and Davis, 1996; Fuller, 1996; Ramanathan, 1998; Pollock, 1999). According to Jenkins and Watts (1969, p.16), "spectral analysis brings together two very important theoretical approaches, the statistical analysis of time series and the methods of Fourier analysis." Since, in the frequency domain approach, Fourier transforms play a very important role (Brillinger and Krishnaiah, 1983), some brief remarks will be made on Fourier's methods which form the basis of all spectral analysis (Rayner, 1971).

Time series spectral analysis can be traced to Jean Baptiste Joseph de Fourier (1768–1830) who made the claim in 1807 that an arbitrary function defined on a finite interval could be represented as a infinite summation of cosine and sine functions (see Lasser, 1996). Many mathematicians have worked on the development of the techniques of Fourier analysis, and contemporary books (e.g., Körner, 1988; Lasser, 1996; Ramanathan, 1998; Howell, 2001) have presented various aspects of the mathematics of Fourier analysis. Without discussing the details of Fourier techniques it is worthwhile to note that one notable early investigation which focused on analyzing time series in the frequency domain was Schuster (1898) who employed the technique of periodogram analysis. The early underlying model expressed the series as a weighted sum of perfectly regular periodic components upon which a random component was superimposed. While much of the theory of spectral analysis of random processes focused on stationary processes, in recent years, "a new form of spectral analysis has been developed which, while not accommodating all nonstationary processes, does however enable us to treat a fairly large class of such processes in a unified theory which includes stationary processes as a special case" (Priestley, 1981, p. 17).

The Fourier analysis of stochastic processes can provide a representation of an infinite sequence in terms of an infinity of trigonometric functions whose frequencies range continuously. The underlying stochastic process can be represented by the Fourier integral. This can be attained by describing the stochastic processes which generate the weighting functions. There are two weighting processes, associated respectively with the sine and cosine functions; and the function that defines their common variance is the so-called spectral distribution function whose derivative is the spectral density function or the "spectrum" (Pollock, 1999). For practical purposes, the spectral density function can be referred to as the power spectrum. "Since the power spectrum is the Fourier cosine transform of the autocovariance function, knowledge of the autocovariance function is mathematically equivalent to knowledge of the spectrum, and vice-versa" (Box *et al.*, 1994, p. 39). The power spectrum provides significantly more information about the time series than simply the total power. The estimated spectrum can be used to obtain insights about the mechanism that generated the data (Koopmans, 1974).

The power spectrum plays a very important role in the analysis of coastal time series. By considering the probability distribution of the sea surface as nearly Gaussian, a good approximate description is provided by the covariance function, the Fourier transform of which is referred to as the wave spectrum. In coastal studies, the wave spectrum has physical meaning because it can be demonstrated to be the density function specifying the distribution of energy over wave components with different wave number vectors and frequencies. Its integral over all wave components is proportional to the total wave energy per unit area (see Komen *et al.*, 1994). It should be noted that the components of the spectrum are the squares of the wave amplitude at each frequency, which are related to wave energy. Since the covariance function can be related directly to the energy spectrum or the wave spectrum, the covariance function is computed, and then its Fourier transform is calculated to obtain the power spectrum (Dean and Dalrymple, 1984).

With the implementation of the Fast Fourier Transform (FFT) algorithm of Cooley and Tukey (1965), many researchers have followed the contributions of some significant early studies (e.g., Pierson and Marks, 1952; Munk *et al.*, 1959; Hassleman *et al.*, 1963) and utilized spectral concepts for either the analysis or the modeling and prediction of ocean waves. Simplified discussions on the use of spectral techniques for the analysis and modeling of the spectrum of ocean waves have been provided by Dean and Dalrymple (1984), Tucker (1991), and Goda (2000). Timely reviews on the importance and applications of spectral concepts to the description and modeling of water waves have been presented by several authors, among them Cardone (1974), Cardone and Ross (1979), Komen *et al.* (1994), and Cardone and Resio (1998). Recent advances in the development of spectral wave modeling for various application purposes have been presented by several researchers (e.g., Holthuijsen *et al.*, 1993; Rivero *et al.*, 1998; Booij *et al.*, 1999; Ris *et al.*, 1999; Monbaliu *et al.*, 2000; Schneggenburger *et al.*, 2000).

In the frequency domain perspective, some researchers (e.g., Guedes Soares and Ferreira, 1995; Lakhan, 1998) have modeled parameters such as significant wave height. A Fourier representation described the periodic components of time series of significant wave height values from different coastal locations. In addition, Hegge and Masselink (1996) also demonstrated the usefulness of spectral techniques to model data from topographic profiles. The results of the spectral analysis represented the amount of variance of the time series as a function of frequency. With spectral modeling, coastal researchers are obtaining greater insights on the physical characteristics of the generating mechanisms of coastal time series.

The time domain approach

The time domain approach, which focuses on the contribution of parametric models for single series or models for two or more causally related series, can be traced to the classical theory of correlation. Details on modeling and forecasting in the time domain can be found in several books (e.g., Box and Jenkins, 1970, 1976; Kendall, 1973; Anderson, 1975; Gottman, 1981; Abraham and Ledolter, 1983; Hoff, 1983; Pandit and Wu, 1983; Pankratz, 1983; Vandaele, 1983; Chatfield, 1984; Montgomery *et al.*, 1990; Bowermand and O'Connell, 1993; Harvey, 1993; Box *et al.*, 1994; Hamilton, 1994; Hipel and McLeod, 1994; Armstrong, 2001; Pourahmadi, 2001). The time domain models that originated with Yule (1927) and Slutsky (1937) provided a strong foundation for time series analysis in the time domain. Yule (1927) pioneered the concept of autoregression and his autoregressive model was generalized by Walker (1931) to allow for dependence on more than two previous values. Wold (1938) followed the practice of Yule and Walker and plotted correlation coefficients against their lags. According to Pourahmadi (2001, p. 35), "the term correlogram was coined by Wold (1938, p. 7) as a substitute to the Schuster's periodogram and has been used effectively ever since as a means of identifying model or appropriate probabilistic description of time series data." The early autoregressive (AR) model of Yule (1927) and moving average (MA) models of Walker (1931) and Slutsky (1937) were not widely used because of the lack of appropriate methods for identifying, fitting, and checking these models (Jenkins, 1979). AR and MA types of models were combined into the mixed autoregressive moving average (ARMA) model.

ARMA models have their foundation in the work of Box and Jenkins (1970, 1976) who utilized concepts from mathematical statistics and classical probability theory. Box and Jenkins also extended ARMA models to include certain types of nonstationary series, and proposed an entire family of models, called autoregressive integrated moving average (ARIMA) models, which can be applied to practical problems in several disciplines. Besides ARMA and ARIMA models, Box and Jenkins (1976) presented other models, including transfer function noise (TFN) models and seasonal autoregressive integrated moving average (SARIMA) models. "The ARMA and ARIMA classes can provide very useful descriptions for a wide range of time series data and the Box-Jenkins approach (as it has become known) has been used extensively by time series practitioners in a wide variety of fields" (Priestley, 1997, p. 16). In the field of coastal research, several autoregressive models have been utilized for data on sea-state parameters. For example, Lakhan (1981) and Lakhan and LaValle (1986) modeled the correlation structure of wave heights and wave periods, and then parameterized stochastic models with autocorrelated distributed variates. Spanos (1983) used ARMA processes to simulate individual waves in short-term periods. Scheffner and Borgman (1992) also simulated individual waves but accounted for long-term variability. Several other studies (e.g., Guedes Soares and Ferreira, 1996; Cunha and Guedes Soares, 1999; Guedes Soares and Cunha, 2000) have modeled time series of sea-state parameters with autoregressive models. Besides sea-state parameters, researchers (e.g., Walton, 1999; LaValle *et al.*, 2001) have also modeled shoreline changes with time series techniques. The study by LaValle *et al.* (2001) utilized Box–Jenkins modeling procedures to identify models which best described a time series (1978–94) of beach and shoreline data. By following the Box–Jenkins model construction approach described below, it was found that a spatial model described the variation for beach net sediment flux, and a space–time autoregressive model provided the best fit for data on spatial–temporal variations of shoreline retreat. Here, it must be mentioned that ARMA models can be generalized to include spatial location (Cressie, 1993).

Time domain modeling—the Box–Jenkins approach

The success of the Box–Jenkins modeling approach in coastal and other studies can be attributed to the fact that Box–Jenkins models have several advantages over traditional models because there is a large class of models, there is a systematic approach to model identification, and the validity of models can be verified (Hoff, 1983). While a publication of this kind cannot provide technical details on the construction of Box–Jenkins models, it is, nevertheless, worthwhile to note that no matter what type of stochastic model is to be fitted to a given dataset it is recommended that the identification, estimation, and diagnostic check stages of model construction be followed (Box and Jenkins, 1976). By following Box and Jenkins, univariate models for any coastal time series can be constructed by using the iterative approach presented in Figure T44.

Consideration stage

In the consideration stage, it is necessary to be cognizant of all standard time series models. Consideration must be given to the various families of stochastic models, which can be fitted to the coastal time series. It is of paramount importance to consider those models, which on the basis of theory, practical experience, understanding of the problem, and the published literature, have the potential to fit the observed data.

Identification stage

At the beginning of the identification stage it is best to ascertain the subclasses of models that hold greater promise for adequately modeling the coastal time series. The first step is to obtain a graphical plot of the data because a plot of the time series can demonstrate some of the essential mathematical characteristics of the data. From the plot, it can be determined whether the series contains a trend, outliers, seasonality, nonconstant variances, and other non-normal and nonstationary phenomena. This knowledge allows for possible data transformations. Differencing and variance stabilizing transformations can be used. The normal procedure is to apply variance stabilizing transformations before taking differences because differencing may create some negative values. To stabilize the variance the Box–Cox power transformation can be used. For a series with nonconstant variance a logarithmic transformation can be employed (Wei, 1990). In modeling time series of significant wave height data, Guedes Soares and Ferreira (1996) tried both the Box–Cox transformation and the logarithmic transformation. Following transformation to stationarity it is necessary to compute and examine the sample autocorrelation function (ACF) and the sample partial autocorrelation coefficient (PACF) of the original series to determine the need for further differencing.

After performing the necessary differencing, the ACF and PACF are computed for the properly transformed and differenced series. The autocorrelations and partial autocorrelations of a series are considered principal tools for identifying the correct parameters to include in a Box–Jenkins ARIMA model. Autocorrelations are statistical measures computed from the time series data. An autocorrelation measures how strongly time series values at a specified number of periods apart are correlated to each other over time. The number of periods apart is called the lag. The rule of thumb is that the maximum number of lags should not exceed one-fifth of the number of observations. The partial autocorrelation is similar to autocorrelation, except that when calculating it, the (auto)correlations with all the elements within the lag are partialled out (Box and Jenkins, 1976). In univariate Box–Jenkins modeling, it is normal practice to produce correlograms depicting the autocorrelation coefficients plotted against the lag intervals. According to Harvey (1993), the correlogram is the basic tool of analysis in the time domain. An inspection of the correlogram provides important information as to whether the series exhibits a pattern of serial autocorrelation which can be modeled by a particular stochastic process or whether the series is random. Figure T45 is a correlogram of significant wave height data which shows that the data values are not independent of each other. The series is not random because the autocorrelations should be near zero for randomness. By understanding the association between the ACF and PACF and their corresponding processes, a tentative model can be identified. In the case of Figure T45, with the autocorrelation function decaying exponentially, and knowing that the PACF has a distinct spike at lag 1, it is possible to specify an AR(1) model. It should, however, be stressed that several other models could fit the data. The primary goal is to identify a model with the smallest number of estimated parameters.

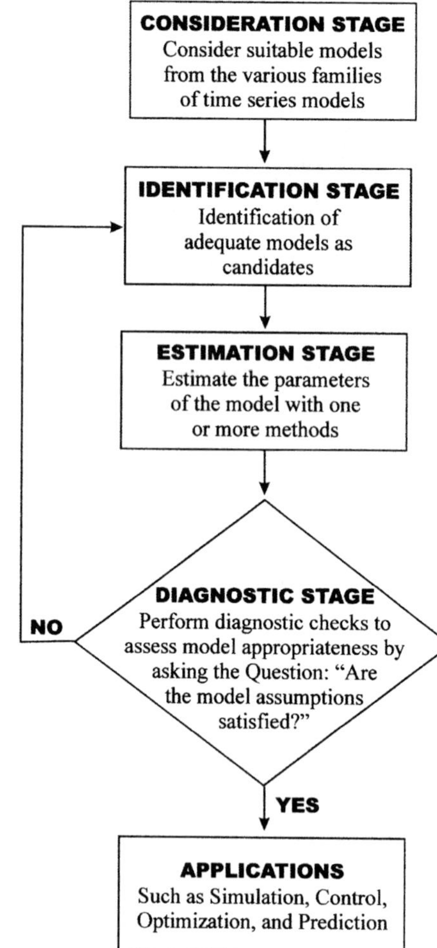

Figure T44 Stages in the iterative approach to model construction (modified from Box *et al.* (1994, p. 17)).

Parameter estimation

Subsequent to identifying an adequate model for fitting to a particular series, the next step is to estimate the parameters in the model. Hoff (1983) outlined several objectives in estimating a model, among them obtaining fitted values that are nearly identical to the original series values, obtaining residuals that are not correlated to each other, and using a minimum number of parameters as necessary. A good model incorporates the smallest number of estimated parameters which are needed to fit the patterns in the data. To estimate parameters such as the mean of the series, AR parameters, MA, and other parameters for an identified ARMA model, several estimation procedures can be utilized. Details in estimation theory and estimation procedures can be found in several publications (e.g., Kruskal and Tanur, 1978; Sachs, 1984; Mendel, 1987;

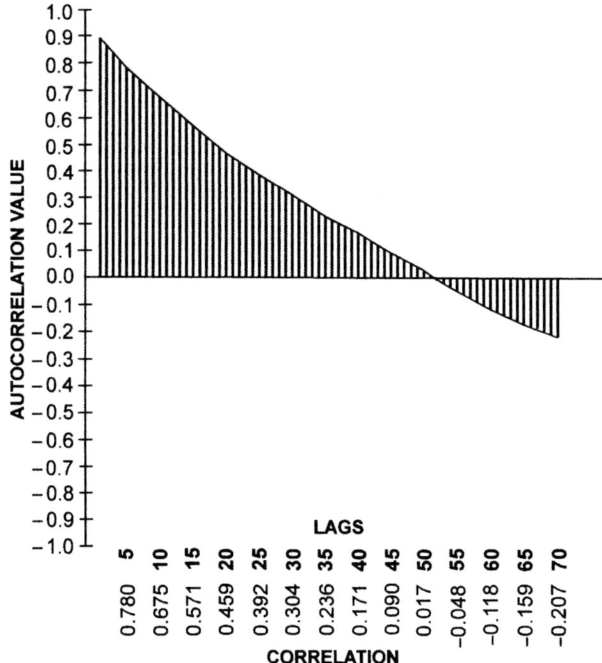

Figure T45 Correlogram of significant wave height (from Lakhan, 1998).

Kotz and Johnson, 1988; Box *et al.*, 1994). For the time series modeling of coastal data, the method of moments, and the maximum likelihood method are widely used approaches to parameter estimation.

Diagnostic checking

Once the parameters of the identified model have been estimated, the next phase is to perform diagnostic checks to determine whether certain assumptions about the model can be verified. Two assumptions to be checked are usually the normality of residuals of the model, and independence. To ascertain whether the residuals are white noise, the residuals from the estimated model are used to calculate the autocorrelation coefficients. Ideally, the residual autocorrelation function for a properly constructed ARIMA model will have autocorrelation coefficients that are all statistically zero. If the residuals are autocorrelated they are not white noise, and this requires formulating another model with residuals that are consistent with the independence assumption (Pankratz, 1983).

The portmanteau lack of fit test is usually applied for testing the independence of a time series. The portmanteau lack of fit test, originally proposed by Box and Pierce (1970), was improved by Ljung and Box (1978). Applications of the portmanteau test for physical time series can be found in several studies (e.g., Hipel and McLeod, 1977; Salas *et al.*, 1980; Hipel and McLeod, 1994). The test statistic is the modified Q statistic which uses all the residual autocorrelations as a set to check the joint null hypothesis. The Q statistic approximately follows a chi-square distribution, with degrees of freedom equal to the total number of lags used minus the number of model parameters and their associated probability values. Recent studies (e.g., LaValle *et al.*, 2000, 2001) have demonstrated that the Q statistic is appropriate for determining the goodness-of-fit of autoregressive models fitted to data on water levels and beach and shoreline changes.

Applications

When a model is accepted it could be used for various applications, among them filtering and control, simulation and optimization, and prediction. A model that fails one or more diagnostic checks is rejected. To construct a good model it becomes necessary to return to the identification stage, and repeat the iterative process of identification, estimation, and diagnostic checking.

Selection of the best model

In coastal time series analysis, it is possible to have several appropriate models that can be used to represent a given dataset. To solve the problem of choosing the best model from the various adequate models, several model determination procedures and model selection criteria have been proposed (e.g., Stone, 1979; Hannan, 1980). Selection criteria are different from the model identification methods discussed above because when there are several adequate models for a given dataset the selection criterion is normally based on summary statistics from residuals computed from a fitted model or on forecast errors calculated from the out-sample forecasts (Wei, 1990).

Some well-known model selection criteria based on residuals are the Akaike Information Criterion (AIC) of Akaike (1974), Parzen Criterion for Autoregressive Transfer (CAT) Functions of Parzen (1977), and the Schwartz's Bayesian Criterion (SBC) of Schwartz (1978). Of the various selection criteria, the AIC is widely used in time series model fitting because it increases the speed, flexibility, accuracy, and simplicity involved in choosing the "best" model. In addition, the AIC is useful for application to many different kinds of time series (Hipel, 1981; Hipel and McLeod, 1994), and facilitates the selection of a parsimonious model that, at the same time, provides a good statistical fit to the data being modeled.

V. Chris Lakhan

Bibliography

Abraham, B., and Ledolter, J., 1983. *Statistical Methods for Forecasting*. New York: John Wiley and Sons, Inc.

Akaike, H., 1974. A new look at the statistical model identification. *IEEE Transactions on Automatic Control*, **19**: 716–723.

Anderson, O.D., 1975. *Time-Series Analysis and Forecasting: The Box–Jenkins Approach*. London: Butterworths.

Armstrong, J.S., 2001. *Principles of Forecasting: A Handbook for Researchers and Practitioners*. Boston, MA: Kluwer Academic Press.

Bendat, J.S., and Piersol, A.G., 1993. *Engineering Applications of Correlation and Spectral Analysis*, 2nd edn. New York: John Wiley and Sons, Inc.

Bloomfield, P., 1976. *Fourier Analysis of Time Series: An Introduction*. New York: John Wiley and Sons, Inc.

Booij, B., Ris, R.C., and Holthuijsen, L.H., 1999. A third-generation wave model for coastal regions: 1. Model description and validation. *Journal of Geophysical Research*, **104**(C4): 7649–7666.

Bowermand, B.L., and O'Connell, R.T., 1993. *Time Series and Forecasting: An Applied Approach*, 3rd edn. N. Scituate, MA: Duxbury Press.

Box, G.E.P., and Jenkins, G.M., 1970. *Time Series Analysis: Forecasting and Control*. San Francisco, CA: Holden-Day.

Box, G.E.P., and Jenkins, G.M., 1976. *Time Series Analysis Forecasting and Control*, 2nd edn. San Francisco, CA: Holden-Day.

Box, G.E.P., and Pierce, D.A., 1970. Distribution of the residual autocorrelations in autoregressive integrated moving average models. *Journal of the American Statistical Association*, **65**: 1509–1526.

Box, G.E.P., Jenkins, G.M., and Reinsel, G.C., 1994. *Time Series Analysis. Forecasting and Control*, 3rd edn. Englewood Cliffs, NJ: Prentice-Hall, Inc.

Brigham, E.O., 1988. *The Fast Fourier Transform and its Applications*. Englewood Cliffs, NJ: Prentice Hall.

Brillinger, D.R., 1975. *Time Series: Data Analysis and Theory*. New York: Holt, Rinehart, and Winston.

Brillinger, D.R., and Krishnaiah, P.R. (eds.), 1983. *Time Series in the Frequency Domain. Handbook of Statistics*, Vol. 3. Amsterdam, The Netherlands: Elsevier Science Publishers.

Brockwell, P.J., and Davis, R.A., 1996. *Introduction to Time Series and Forecasting*. New York: Springer.

Cardone, V.J., 1974. Ocean wave predictions: two decades of progress and future prospects. In Society of Naval Architects and Marine Engineers. *Seakeeping 1953–1973/Sponsored by Panel H-7 (Seakeeping Characteristics) at Webb Institute of Naval Architecture, Glen Cove, New York, October 28–29, 1973*. pp. 5–18.

Cardone, V.J., and Resio, D.T., 1998. An assessment of wave modelling technology. *Proceedings of the 5th International Workshop on Wave Hindcasting and Forecasting*. Melbourne, FL, pp. 468–495.

Cardone, V.J., and Ross, D.B., 1979. State-of-the-Art wave prediction methods and data requirements. In Earle, M.D., and Malahoff, A. (eds.), *Ocean Wave Climate*. New York: Plenum Press, pp. 61–91.

Chatfield, C., 1984. *The Analysis of Time Series: An Introduction*. London and New York: Chapman and Hall.

Cooley, J.W., and Tukey, J.W., 1965. An algorithm for the machine calculation of complex Fourier Series. *Mathematics of Computation*, **19**: 297–301.

Cressie, N.A.C., 1993. *Statistics for Spatial Data*. New York: John Wiley and Sons, Ltd.

Cunha, C., and Guedes Soares, C., 1999. On the choice of data transformation for modelling time series of significant wave height. *Ocean Engineering*, **26**: 489–506.

Dean, R.G., and Dalrymple, R.A., 1984. *Water Wave Mechanics for Engineers and Scientists*. Englewood Cliffs, NJ: Prentice-Hall, Inc.

Fuller, W.A., 1996. *Introduction to Statistical Time Series*, 2nd edn. New York: John Wiley and Sons, Inc.

Goda, Y., 2000. *Random Seas and Design of Maritime Structures*. Singapore: World Scientific Publishing Co.

Gottman, J.M., 1981. *Time-Series Analysis*. Cambridge, England: Cambridge University Press.

Guedes Soares, C., 2000. Probabilistic based models for coastal studies. *Coastal Engineering*, **40**: 279–283.

Guedes Soares, C., and Cunha, C., 2000. Bivariate autoregressive models for the time series of significant wave height and mean period. *Coastal Engineering*, **40**: 297–311.

Guedes Soares, C., and Ferreira, A.M., 1995. Analysis of the seasonality in nonstationary time of significant wave height. In Spanos, P.D. (ed.), *Computational Stochastic Mechanics*. New York: Balkema, pp. 501–521.

Guedes Soares, C., and Ferreira, A.M., 1996. Representation of non-stationary time series of significant wave height with autoregressive models. *Probabilistic Engineering Mechanics*, **11**: 139–148.

Hamilton, J.D., 1994. *Time Series Analysis*. Princeton, NJ: Princeton University Press.

Hannan, E.J., 1980. The estimation of the order of an ARMA process. *The Annals of Statistics*, **8**: 1071–1081.

Hannan, E.J., Krishnaiah, P.R., and Rao, M.M. (eds.), 1985. *Time Series in the Time Domain. Handbook of Statistics*, Vol. 5. Amsterdam, The Netherlands: Elsevier Science Publishers.

Harvey, A.C., 1993. *Time Series Models*, 2nd edn. Cambridge, MA: The MIT Press.

Hassleman, K., Munk, W., and MacDonald, G., 1963. Bispectrum of ocean waves. In Rosenblatt, M. (ed.), *Time Series Analysis*. New York: John Wiley and Sons, Inc., pp. 125–139.

Hegge, B.J., and Masselink, G., 1996. Spectral analysis of geomorphic time series: auto-spectrum. *Earth Surface Processes and Landforms*, **21**: 1021–1040.

Hipel, K.W., 1981. Geophysical model discrimination using the Akaike information criterion. *IEEE Transactions on Automatic Control*, **AC-26**(2): 358–378.

Hipel, K.W., and McLeod, A.I., 1977. Advances in Box–Jenkins modeling. 1. Model construction. *Journal of Water Resources Research*, **13**(3): 567–575.

Hipel, K.W., and McLeod, A.I., 1994. *Time Series Modelling of Water Resources and Environmental Systems*. Amsterdam, The Netherlands: Elsevier Science Publishers.

Hoff, J.C., 1983. *A Practical Guide to Box–Jenkins Forecasting*. Belmont, CA: Wadsworth, Inc.

Holthuijsen, L.H., Booij, N., and Ris, R.C., 1993. A spectral model for the coastal zone. *Proceedings of the 2nd International Symposium on Ocean Wave Measurement and Analysis*. New Orleans, LA, pp. 630–641.

Howell, K.B., 2001. *Principles of Fourier Analysis*. Boca Raton, FL: Chapman & Hall/CRC.

Jenkins, G.M., 1979. Practical experiences with modelling and forecasting time series. In Anderson, O.D. (ed.), *Forecasting*. Amsterdam, The Netherlands: North-Holland Publishing Company, pp. 43–166.

Jenkins, G.M., and Watts, D.G., 1969. *Spectral Analysis and its Applications*. San Francisco, CA: Holden-Day, Inc.

Kanasewich, E.R., 1973. *Time Series Analysis in Geophysics*. Edmonton, AB: University of Alberta Press.

Kendall, M., 1973. *Time-Series*. New York: Hafner Press.

Kendall, M., and Ord, J.K., 1990. *Time Series*, 3rd edn. New York: Oxford University Press.

Komen, G.J., Cavaleri, L., Donelan, M., Hasselmann, K., Hasselmann, S., and Janssen, P.A.E.M., 1994. *Dynamics and Modelling of Ocean Waves*. Cambridge: Cambridge University Press.

Koopmans, L.H., 1974. *The Spectral Analysis of Time Series*. New York: Academic Press, Inc.

Körner, T.W., 1988. *Fourier Analysis*. Cambridge, England: Cambridge University Press.

Kotz, S., and Johnson, N.L. (eds.), 1988. *Encyclopedia of Statistical Sciences*, Vols 1 to 9. New York: John Wiley and Sons, Inc.

Kruskal, W.H., and Tanur, J.M., 1978. *International Encyclopedia of Statistics*, Vols 1 and 2. New York: The Free Press.

Lakhan, V.C., 1981. Parameterizing wave heights in simulation models with autocorrelated Rayleigh distributed variates. *Journal of the International Association of Mathematical Geology*, **13**(4): 345–350.

Lakhan, V.C., 1982. *Stochastic simulation of wave action on concave-shaped nearshore profiles*. Ph.D. thesis. Toronto, ON: University of Toronto.

Lakhan, V.C., 1989. Modeling and simulation of the coastal system. In Lakhan, V.C, and Trenhaile, A.S. (eds.), *Applications in Coastal Modeling*. Amsterdam, The Netherlands: Elsevier Science Publishers, pp. 17–41.

Lakhan, V.C., 1998. Modeling waves and sediments associated with mudbank formation along the Guyana coast. Berbice, Guyana: Canada Caribbean Research Group, *Technical Report No. 38*.

Lakhan, V.C., and LaValle, P.D., 1986. Development and testing of a stochastic model to simulate nearshore profile changes. *Studies in Marine and Coastal Geography*, Halifax, NS: Saint Mary's University, pp. 61–81.

Lakhan, V.C., and Trenhaile, A.S., 1989. Models and the coastal system. In Lakhan, V.C, and Trenhaile, A.S. (eds.), *Applications in Coastal Modeling*. Amsterdam, The Netherlands: Elsevier Science Publishers, pp. 1–16.

Lasser, R., 1996. *Introduction to Fourier Series*. New York: Marcel Dekker, Inc.

LaValle, P.D., Lakhan, V.C., and Trenhaile, A.S., 2000. Short term fluctuations of Lake Erie water levels and the El Niño/Southern Oscillation. *The Great Lakes Geographer*, **7**(1): 1–8.

LaValle, P.D., Lakhan, V.C., and Trenhaile, A.S., 2001. Space-time series modelling of beach and shoreline data. *Environmental Modelling and Software*, **16**: 299–307.

Ljung, G.M., and Box, G.E.P., 1978. On a measure of lack of fit in time series models. *Biometrika*, **65**: 297–303.

Mendel, J.M., 1987. *Lessons in Digital Estimation Theory*. Englewood Cliffs, NJ: Prentice-Hall.

Monbaliu, J., Padilla-Hernández, R., Hargreaves, J.C., Carretero Labiach, J.C., Luo, W., Sclavo, M., and Günther, H., 2000. The spectral wave model, WAM, adapted for applications with high spatial resolution. *Coastal Engineering*, **41**: 41–62.

Montgomery, D.C., Johnson, L.A., and Gardiner, J.S., 1990. *Forecasting and Time Series Analysis*. 2nd edn. New York: McGraw-Hill, Inc.

Munk, W.H., Snodgrass, F.E., and Tucker, M.J., 1959. Spectra of low frequency ocean waves. *Bulletin Scripps Institution of Oceanography*, **7** (4): 283–362.

Otnes, R.K., and Enochson, L.D., 1976. *Applied Time Series Analysis*. New York: John Wiley and Sons, Inc.

Pandit, S.M., and Wu, S-M., 1983. *Time Series and System Analysis with Applications*. New York: John Wiley and Sons, Inc.

Pankratz, A., 1983. *Forecasting with Univariate Box–Jenkins Models*. New York: John Wiley and Sons, Inc.

Parzen, E., 1977. Multiple time series modeling: determining the order of approximating autoregressive schemes. In Krishnaiah, P. (ed.), *Multivariate Analysis IV*. Amsterdam, The Netherlands: North-Holland, pp. 283–295.

Pierson, W.J., and Marks, W., 1952. The power spectrum analysis of ocean wave records. *Transactions of American Geophysical Union*, **33**: 834–844.

Pollock, D.S.G., 1999. *A Handbook of Time-Series Analysis, Signal Processing and Dynamics*. London: Academic Press.

Pourahmadi, M., 2001. *Foundations of Time Series Analysis and Prediction Theory*. New York: John Wiley and Sons, Inc.

Priestley, M.B., 1981. *Spectral Analysis and Time Series*, Vols 1 and 2. New York: Academic Press.

Priestley, M.B., 1997. A short history of time series. In Subba Rao, T., Priestley, M.B., and Lessi, O. (eds.), *Applications of Time Series Analysis in Astronomy and Meteorology*. London: Chapman & Hall, pp. 3–23.

Ramanathan, J., 1998. *Methods of Applied Fourier Analysis*. Boston, MA: Birkhäuser.

Rayner, J.N., 1971. *An Introduction to Spectral Analysis*. London: Pion Limited.

Ris, R.C., Holthuijsen, L.H., and Booij, B., 1999. A third-generation wave model for coastal regions: 2. Verification. *Journal of Geophysical Research*, **104**(C4): 7667–7681.

Rivero, F.J., Arcilla, A.S., and Carci, E., 1998. An analysis of diffraction in spectral wave models. In Edge, B.L., and Hemsley, J.M. (eds.), *Ocean Wave Measurement and Analysis*, Vol. 2. Reston, VA: American Society of Civil Engineers, pp. 431–445.

Sachs, L., 1984. *Applied Statistics. A Handbook of Techniques*, 2nd edn. New York: Springer-Verlag.

Salas, J.D., Delleur, J.W., Yevjevich V., and Lane, W.L., 1980. *Applied Modeling of Hydrologic Time Series*. Littleton, CO: Water Resources Publications Ltd.
Scheffner, N.W., and Borgman, L.E., 1992. Stochastic time series representation of wave data. *Journal of Waterway, Port, Coastal and Ocean Engineering*, ASCE, **118**(4): 1955–2012.
Schneggenburger, C., Günther, H., and Rosenthal, W., 2000. Spectral wave modelling with non-linear dissipation: validation and applications in a coastal tidal environment. *Coastal Engineering*, **41**: 201–235.
Schuster, A., 1898. On the investigation of hidden periodicities with application to a supposed twenty-six day period of meteorological phenomena. *Terrestrial Magnetism*, **3**: 13–41.
Schwartz, G., 1978. Estimating the dimension of a model. *The Annals of Statistics*, **6**: 461–464.
Shumway, R.H., and Stoffer, D.S., 2000. *Time Series Analysis and Its Application*. New York: Springer-Verlag New York, Inc.
Slutsky, E., 1937. The summation of random causes as the source of cyclical processes. *Econometrica*, **5**: 105–146.
Spanos, P.D., 1983. ARMA algorithms for ocean wave modelling. *Journal of Energy Resources Technology* ASME, **105**: 300–309.
Stone, M., 1979. Comments on model selection criteria of Akaike and Schwartz. *Journal of Royal Statistical Society*, Series B, **41**: 276–278.
Tucker, M.J., 1991. *Waves in Ocean Engineering*. New York: Ellis Horwood Limited.
Vandaele, W., 1983. *Applied Time Series and Box–Jenkins Models*. New York: Academic Press.
Walker, G., 1931. On periodicity in series of related terms. *Proceedings of the Royal Society*, A **131**: 195–215.
Walton, T.L., Jr., 1999. Shoreline rhythmic pattern analysis. *Journal of Coastal Research*, **15**(2): 379–387.
Wei, W.W.S., 1990. *Time Series Analysis. Univariate and Multivariate Methods*. Redwood City, CA: Addison-Wesley Publishing Company, Inc.
Wold, H., 1938. *A Study in the Analysis of Stationary Time-Series*. Uppsala, Sweden: Almquist & Wiksell.
Yule, G.U., 1927. On a method of investigating periodicities in distributed series, with special reference to Wölfer's sunspot numbers. *Philosophical Transactions of the Royal Society*, A, **226**: 267–298.

Cross-references

Coastal Modeling and Simulation
Numerical Modeling
Simple Beach and Surf Zone Models
Surf Modeling
Wave Climate

TOPIC CATEGORIES—See APPENDIX 6

TORS

The word "tor," Celtic in origin, is used generally in the British Isles to denote a rather tall rock column (Cunningham, 1968; Jackson, 1997). Linton (1955) was the first to propose it as a scientific term in describing the tors at Dartmoor, Devonshire, England, now considered the type area for the feature (Palmer and Neilson, 1962). Early hypotheses for the origin of tors invoked deep weathering along joint planes in granite, with subsequent removal of the loose material leaving exposed columns. Alternate possibilities outlined by Cunningham (1968) include differential erosion during scarp recession and relict subaerial prominences formed in the Tertiary. Since tors have been found worldwide, often in granite but also in other igneous, sedimentary, and metamorphic rocks, it is appropriate to consider Palmer and Neilson's (1962) pronouncement that "It is not possible to offer a definition that will encompass the many landforms to which the name "tor" has been given." With such wide distribution and varied lithology it is arguable that the most scenic among all of these are coastal tors, as can be seen in the Seychelles and Virgin Islands.

Seychelles

The Seychelles, along with Madagascar, were displaced in a northeasterly direction away from the African landmass during the early formation, in the Jurassic, of the Indian Ocean (Brathwaite, 1984). As such, 42 of the 116 islands comprising the Seychelles Archipelago are the world's only mid-ocean islands composed of granitic rocks (Cílek, 1978). Grey and pink amphibolitic granite, of late Precambrian age, is spectacularly displayed on the northern island of La Digue and at Mahe as discussed by Wagle and Hashimi (1990).

La Digue Island has long been known for its world-class resorts featuring tropical flora, white-sand beaches, swimming in crystal clear waters, snorkeling among coral reefs and magnificent scenic views. Anse source d'argent beach, located on the island, is the site of several tors, chief among them that is pictured in Figure T46. The size of the pink tor can be judged when compared with the heads of the three swimmers seen in the mid-foreground. Graphic too is the weathered "fluting" described as typical of tors by Linton (1955) in his pioneering work, and by Brathwaite (1984) for a tor on Mahe.

Figure T46 The tor at Anse Source D'argent, La Digue Island, Seychelles, Indian Ocean; often considered to be the most beautiful beach in the world (Photo courtesy of New Adventures).

Figure T47 Grotto at the base of the tors at The Baths, Virgin Gorda, British Virgin Islands (Photo, M. Schwartz).

Virgin Islands

The Virgin Islands, located in the northeastern Caribbean along the leading edge of the Caribbean plate, are composed of Mesozoic and lower Tertiary deformed island-arc terrane. Much of the northeastern British Virgin Islands region is underlain by the Virgin Gorda granitic pluton or batholith, which was intruded into the surrounding country rock in mid- to later Eocene time (Mattson *et al.*, 1990). Weathered exposures in tonalitic rocks at the southern end of the island of Virgin Gorda reveal huge boulders now located upon the beaches (Weaver, 1962) in a park system managed by the B.V.I. National Parks Trust.

The origin of these boulders has been described by Ratté (1986) in the classic Linton (1955) style for tors of deep weathering along joints followed by removal of the rotted material. A favorite with tourists, the site is called "The Baths," not because of the underlying batholith as a geologist would imagine, but for the salt water pools in the grotto at the base of the tors (Figure T47). Here one may walk, and crawl, along a trail between boulders that range up to 4 m in height.

Maurice Schwartz

Bibliography

Brathwaite, C.J.R., 1984. Geology of the Seychelles, In Stoddart, D.R. (ed.), Biogeography and ecology of the Seychelle Islands. The Hague: W. Junk, pp. 17–38.

Cílek, V., 1978. Geological investigations on the beaches of Mahe in the Seychelles Archipelago. *Casopsis pro Mineralogii a Geologii*, **23**(2): 149–157.

Cunningham, F.F., 1968. The significance of Caribbean evidence in the elucidation of tors. *Caribbean Journal of Science*, **8**: 187–197.

Jackson, J.A. (ed.), 1997. Glossary of Geology, 4th edn. Alexandria, VA: American Geological Institute.

Linton, D.L., 1955. The problem of tors. *Geographical Journal*, **121**(4): 470–486.

Mattson, P., Draper, G., and Lewis, J.F., 1990. Puerto Rico and the Virgin Islands. In Dengo, G., and Case, J.E. (eds.), *The Caribbean Region, The Geology of North America*, H, Boulder: Geological Society of America, pp. 112–120.

Palmer, J., and Neilson, R.A., 1962. The origin of granite tors on Dartmoor, Devonshire. *Proceedings of the Yorkshire Geological Society*, **33**(3): 315–339.

Ratté, C.A., 1986. *The Story of the Boulders*. Burlington, Vermont: Queen City Printers.

Wagle, B.G., and Hashimi, N.H., 1990. Coastal geomorphology of Mahe Island, Seychelles. *International Journal of Remote Sensing*, **11**(2): 281–287.

Weaver, J.D., 1962. Notes on some erosional features in Virgin Gorda, B. Virgin Islands. *Caribbean Journal of Science*, **2**(4): 159–167.

Cross-references

Boulder Beaches
Caribbean Islands, Coastal Ecology and Geomorphology
Coastal Hoodoos
Indian Ocean Islands, Coastal Ecology and Geomorphology
Tourism, Criteria for Coastal Sites
Weathering in the Coastal Zone

TOURISM AND COASTAL DEVELOPMENT

Coastal tourism is a process involving tourists and the people and places they visit. It is more specifically defined as tourism brought to bear on the coastal environment and its natural and cultural resources. Most coastal zone tourism takes place along the shore and in the water immediately adjacent to the shoreline. Coastal tourism activities occur outdoors and indoors as recreation, sport and play, and as leisure and business (Miller and Ditton, 1986). As with other human endeavors in the coastal zone associated with development, tourism is viewed positively by some for the opportunities it creates. Others condemn coastal tourism for its unacceptable consequences.

Coastal tourism destinations fall all along an urban–rural continuum (see *Demography of Coastal Populations, q.v.*). At one end of the scale are major cities and ports (Hong Kong, Venice, New York, Rio de Janeiro, and Sydney come to mind) known for their cultural, historical, and economic significance. At the other end of the continuum are the relatively isolated and pristine coastlines found around the world that are valued for their natural beauty, flora, and fauna. Of course, many coastal tourism destinations offer rich mixtures of cultural, historical, social, environmental, and other values to visitors.

Coastal tourism technologies of travel include both those which carry tourists from their homeland (e.g., airplanes, ships, cars, buses, and trains) and which are regarded by travelers as mere means to the end of arriving at destinations, and those which transport tourists at coastal destinations but which become part of the touristic experience (e.g., cruise ships, personal watercraft, sailboats, dive boats, motorcycles, bicycles, and forms of animal transportation). Again, some transportation technologies can, depending on the circumstance, be important for being both convenient and for being interesting or pleasing.

In a manner of speaking, all tourism is a matter of supply and demand. With this perspective, coastal tourism is a business for those who make a living by developing accommodations and attractions, and by providing touristic and recreational products and services. Competing marketing programs of a multifaceted industry alert tourists and would-be travelers to coastal tourism amenities. Today, tourists travel to the coastal zone for parts of a day, for weekends, for short vacations, and for prolonged stays. Depending on the circumstances, they may travel alone, with family, or in groups. Some coastal tourism is organized for a special purpose such as ecotourism, adventure tourism, scientific tourism, and dive tourism. Coastal tourism accommodations range from small residences and camping sites rented out as opportunities arise, to single bed-and-breakfast and hotel rooms, to luxury suites in resort enclaves.

Many coastal tourism activities count as a business for those in the tourism industry and as an experience for tourists. Scuba diving, for example, provides an excellent example of how advances in technology have provided foundations for business and have facilitated touristic access to the marine environment. Other coastal activities that have a business aspect (involving, for example, guides and instructors, or special equipment) include recreational and sport fishing, boating, sailing and parasailing, and whale and bird watching. Then too, there are many forms of coastal tourism—swimming and body surfing, snorkeling, beachcombing, hiking and rock climbing, sketching and painting, photographing, sightseeing—that are "free," or for which costs to providers are recovered indirectly through taxes, or are incorporated in standard hotel or accommodation billing practices. In recent years, windsurfing, body-boarding, wake-boarding, kite-surfing in addition to surfing (*q.v.*), have reached new levels of popularity in the coastal zone.

At the same time that coastal tourism fosters economic relationships between industry producers and tourist consumers, the process has shown itself to be an enormously potent force in transforming the natural environment and the lives of people who are neither part of the business of tourism nor a member of the community of tourists.

Coastal tourism is inherently controversial. The coastal zone is a scarce resource prized not only by those who engage in and profit by tourism, but also by those with personal residences near the sea, and those who find employment in fishing, aquaculture (*q.v.*), maritime shipping, nuclear energy, and national defense, among other industries. Congestion and competition in the coastal zone frames the characterization and the resolution of tourism issues. Coastal tourism problems and opportunities are therefore properly debated as "multiple-use" or "multiple-value" conflicts.

Origins of tourism

Although the early Greeks and Romans were known to enjoy the seashore for leisure purposes, coastal tourism has its roots in the Grand Tour traditions beginning with the Renaissance. As an educational institution, the Grand Tour offered young men first-hand exposure to European courts, customs, and prominent cities and ports. Not surprisingly, the Grand Tourists and their "bear-leader" tutors mixed education with pleasure. By the start of the 18th century, these tourists had begun to develop an aesthetic vocabulary that allowed them to more fully appreciate not only the "beautiful" in nature, but the "picturesque" and the "sublime" as well. Especially popular with the Grand Tourists were seascape paintings by Claude Lorrain (1600–82), Salvator Rosa (1615–73), and Gaspard Poussin (1615–75), depicting storms, shipwrecks, harbors, rocky coastlines, and ruins.

The mid-18th century European "discovery" of the seashore for spa and medicinal purposes in England gave rise to early forms of coastal tourism. In the first half of the 19th century, coastal resorts saw a faster rate of population increase than manufacturing towns and by the mid-19th century the medicinal beach was replaced by the pleasure beach. In 1841, the London to Brighton railway was opened, and in the same year Thomas Cook began a legendary career by promoting his first group excursion (Manning-Sanders, 1951; Hern, 1967; Corbin, 1994). Since that time, beaches, atolls, islands, and harbors around the world have supported coastal tourism (see *Beach use and Behaviors, q.v.*).

Magnitude of coastal tourism

Although there are no standardized practices for reporting tourism statistics within the coastal zone, it is not difficult to see how tourism has a major coastal aspect. More than 70% of the earth is covered by water, and only several dozen out of well over 200 nations in the world lack coastlines (Miller and Auyong, 1991a).

International trends

World Tourism Organization statistics confirm that tourism is the world's largest industry as measured by the number of people involved and by economic impacts. From 1945 to 2000, international arrivals increased from 25 million to a record 699 million (WTO, 1996 and 2001). Between 1970 and 1990, tourism grew by nearly 300% (United Nations Environment Programme, 1992).

By the year 2020, it is estimated that international tourist arrivals will reach over 1.56 billion. Statistical estimations for total tourist arrivals by region show that in 2020 the top three receiving regions will be Europe (717 million tourists), east Asia and the Pacific (397 million tourists), and the Americas (282 million tourists), followed by Africa, the Middle East, and south Asia (WTO, 2001).

The World Travel and Tourism Council (WTTC, 1995) reports that between 1980 and 1989, economic expenditures on international travel (excluding transportation) doubled to \$209 billion, rising one-third faster than the world gross national product. Additionally, travel and tourism generated an estimated \$3.4 trillion in gross output in 1995—creating employment for 211 million people, and producing nearly 11% of the world gross domestic product. This growth reflects investments of \$694 billion in new facilities and equipment, and contributions of more than \$637 billion to global tax revenues (WTTC, 1995).

In 2000, tourism generated total international receipts of US\$ 476 billion (WTO, 2001). Of the world's top 15 tourism destination countries in 2000, 12 were countries having coastlines (WTO, 2001). In 2000, the cruise ship industry expected to host over 6.5 million passengers which would represent a 1,200% increase in number of passengers since 1966 (Godsman, 2000).

Sun, beautiful beaches, and warm ocean waters have become standard vacation requirements for many tourists. Of those visiting the Caribbean 49% do so for the beaches, while 28% are primarily interested in sightseeing, and 17% in water sports (Waters, 2001). The Pacific region (which includes Australia) has enjoyed a healthy annual growth rate of nearly 4% since the mid-1990s, though arrivals represent only 1.4% of the world's inbound travelers. Of the Pacific territories, French Polynesia is the most dependent on travel with 78% of the nation's GDP coming from tourism. Tourism in the African region has been growing at a faster rate than for many other parts of the world (Waters, 2001).

US trends

The Travel Industry Association of America (TIA, 2001) reports that tourism is the nation's largest services export industry, the third largest retail industry, and one of America's largest employers. TIA (2001; see also WTTC, 1995) has tabulated that travel and tourism in the United States alone has an impact exceeding \$541 billion a year in expenditures which includes spending by US resident and international travelers within the United States on travel-related expenses (i.e., transportation, lodging, meals, entertainment, and recreation, as well as international, fares on US flag air carriers). This generates more than 17.5 million jobs, and fuels the largest trade surplus of any industry, totaling nearly \$25 billion in 1999. Between 1986 and 1996, international visitation to the United States grew by 78% and expenditures by foreign visitors grew by 223%. It is estimated that more than 90% of foreign-tourist spending occurs in coastal states (US Travel and Tourism Administration, 1994).

US coastlines are popular sites for tourism and recreational activities. Coastal beaches, wetlands, fisheries, aesthetic landscapes, and the human-designed facilities and attractions in the touristic hinterland combine in an endless list of inviting opportunities for visitors, local residents, and entrepreneurs. The major recreational elements of coastal tourism are visiting beaches, swimming, snorkeling and scuba diving, boating, fishing, surfing, and wildlife watching.

It is important to note that coastal tourism and recreation activities often overlap and are not always confined to the marine and coastal environment. For example, diving, fishing, and whale watching are often done while boating, surfing, swimming, and bird watching are usually done while visiting beaches and coastlines; and not all recreational boats are used exclusively in marine and coastal waters.

In the United States, beaches are the leading foreign and domestic tourist destinations (Houston, 1996). In 1995, coastal states made up 11 of the top 15 destinations for overseas travelers visiting the United States (Waters, 1997). A 1999 survey that measured travelers' satisfaction with their visits revealed that Hawaii, Alaska, California, and Florida—all coastal states—were the top four "most liked" destinations in the United States (Volgenau, 2000).

Tourism and recreation are highly significant economic activities in the US coastal zone. By one estimate, approximately 180 million people visit the coast for recreational purposes, and 85% of tourist-related revenues are generated by coastal states (Houston, 1996). Overall, beach tourism and recreation have been estimated to contribute \$170 billion annually to the US economy (Houston, 1995). Coastal states receive 85% of all tourist-related revenues in the United States (Houston, 1995). Coastal districts (defined in terms of state congressional districts) received more than \$185 billion in tourism expenditures in 1997 (TIA, 1998). In addition, it has been estimated that US beaches and marine waters support 28.3 million jobs (Environmental Protection Agency, 1995).

Comprehensive and time-series statistics measuring employment, and the economic and social value of coastal tourism and recreation in the United States are not available. Quantitative and reliable data measuring involvement in specific coastal recreation and tourism activities in the United States are limited (and often proprietary). Nonetheless, many small and unconnected studies have been conducted on specific tourism topics and destinations in the coastal zone.

Several boating and fishing statistics provide some idea of the economic and social importance of coastal tourism. In 1998, according to the US Coast Guard, registered boats numbered 12.3 million, with 10 coastal and Great Lakes states (Michigan, California, Florida, Minnesota,

Texas, Wisconsin, New York, Ohio, South Carolina, and Illinois) accounting for nearly half of them (National Marine Manufactures Association, 2000). In 1999, 77.8 million people participated in recreational boating and recreational boaters spent nearly $23 million on related products and services (National Marine Manufacturers Association, 2000). Between 1991 and 1996 the number of Americans (age 16 and older) who participated in recreational saltwater fishing increased 5.6% from 8.9 million to 9.4 million (Cordell *et al.*, 1997).

Coastal tourism systems

Coastal tourism systems involve interactions between people and place in destinations that include small communities and villages, self-contained resorts, and cosmopolitan cities. From a sociological perspective, coastal tourism systems have three kinds of actors: (1) *tourism brokers*, (2) *tourism locals*, and (3) *tourists* (Miller and Auyong, 1998a).

A "broker-local-tourist" (BLT) model of a coastal tourism system is displayed in Figure T48 (see, Miller and Auyong, 1991a, 1998b). Tourism brokers consist of persons who in one way or another pay professional attention to tourism. Main subcategories include (1) private sector brokers who are part of the tourism industry, (2) public sector brokers at various levels of government who study, regulate, and plan tourism, and (3) social movement brokers in nongovernmental, nonprofit, and environmental organizations who address tourism issues. Tourism brokers of these and other types do not necessarily agree on the kind of tourism that is "best" for coastal tourism systems. Indeed, broker–broker conflict is as common as cooperation. Tourism locals consist of persons who reside in the general region a coastal tourism destination, but do not derive an income from tourism or engage in its management and regulation. Finally, tourists consist of persons of domestic and international origin who travel for relatively short periods of time for business, recreation, and educational purposes before returning home.

Motivation

From the times polite society planned their Grand Tour itineraries through Europe to Rome and other Italian destinations in the 18th and 19th centuries, all who have participated in or witnessed the growth of tourism have pondered the motivations of those fortunate enough to travel. While there are many psychological, social psychological, and social concepts and frameworks for accounting for tourism, only several are mentioned here.

First and looking to the motives of tourists, Miller and Ditton (1986, p. 11) suggest that the fundamental promise of travel "lies in its promise of *contrast*." In elaboration, these authors show that individual trips and vacations allow opportunities for contrast or personal change along three dimensions. *Recreational tourism* as engaged in by the athlete or escapist has a restorative purpose, and provides for change in the physiological or emotional state of the tourist. *Educational tourism* as pursued by the student has a philosophical purpose and provides a basis for change in the intellectual and artistic understanding of the tourist. *Instrumental tourism* as involving entrepreneurs, reformers, and pilgrims exhibits an economic, political, or religious purpose and leads to change in business, network, or moral opportunities available to the tourist. With this framework, a trip by one tourist to, say, a South Pacific island might be experienced as highly recreational, mildly educational, and not at all instrumental. Those accompanying such a tourist could, of course, experience the trip with different weightings along the three dimensions of touristic contrast.

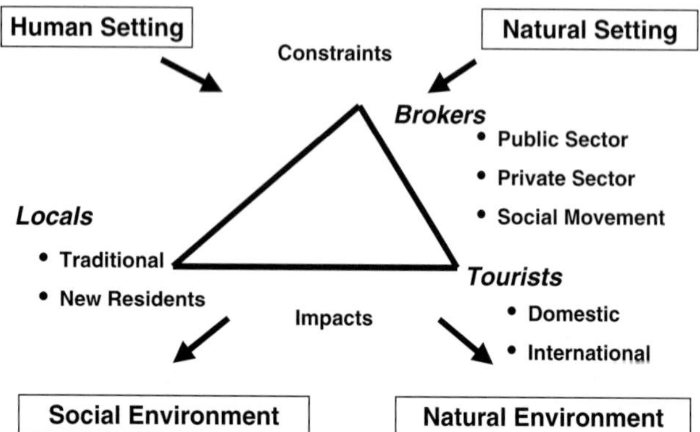

Figure T48 Broker-local-tourist (BLT) model of coastal Tourism (adapted from Miller and Auyong, 1991a, p. 75).

A second way of considering the motivation of tourists emphasizes their intention to experience a psychological state of *challenge* that Csikszentmihalyi (1975, 1990) terms *flow* or *optimal experience*. When in a state of flow—as one might be when surfing, sailing, scuba diving, or even engaging in stimulating conversation—the tourist has found a fine match between his or her abilities and the physical, intellectual, or social challenge at hand. Flow, then, is a state of mind between boredom and anxiety. More fully:

> [f]low denotes the wholistic sensation present when we act with total involvement. It is the kind of feeling after which one nostalgically says: 'That was fun,' or 'That was enjoyable' (Csikszentmihalyi, 1975, p. 43).

According to Csikszentmihalyi (1975) the flow experience is engaged in for its own sake and is marked by (1) a merging of action and awareness, (2) a centering of attention on a limited stimulus field, (3) a feeling variously described as "loss of ego," "self-forgetfulness," "loss of self-consciousness," and even "transcendence of individuality," and "fusion with the world," (4) a feeling of control over one's actions and the environment, (5) coherent, noncontradictory demands for action, and clear unambiguous feedback, and (6) its autotelic [from Greek *auto* = self and *telos* = goal, purpose] nature.

It is often remarked that people who travel together gradually develop a kind of touristic solidarity. By seeing and doing the same things, by sharing emotions and reactions, by facing a common set of logistic obstacles, and even by jointly creating a set of "story lines" with which they might talk about a trip with others, tourists are brought together through the small and multiple secular rituals of travel. In acknowledging this ritual potential, a third perspective on touristic motivation stresses the passionate *commitment* that some tourists exude in performing their favorite coastal touristic activity.

In a series of sociological studies of amateurs, volunteers, and hobbyists in sports, science, and the arts, Stebbins (1992, p. 3) noted intense levels of personal involvement and high levels of technical competence, and coined the term *serious leisure* to describe commitment that was tantamount to professionalism:

> [S]erious leisure can be defined as the systematic pursuit of an amateur, hobbyist, or volunteer activity that is sufficiently substantial and interesting for the participant to find a career there in the acquisition and expression of its special skills and knowledge.

In the realm of coastal tourism, tourists who pursue serious leisure are omnipresent as evidenced by scuba divers, sailors, whale watchers, amateur naturalists and marine conservationists, and the like.

System dynamics

Coastal tourism systems change in size and character over time. To understand and ultimately to predict these changes, and also to plan for desired societal and environmental outcomes, analysis must focus on the behavior of components of the system. In this regard, two processes merit attention.

First, population dynamics of the BLT model should be monitored. It is not unusual for individuals in the system to change statuses. This can occur as, for example, tourists who visit a coastal destination decide to stay and take on a residence, either as a broker of some kind (e.g., as a scuba dive shop entrepreneur or restaurant owner), or as a local (e.g., as a lawyer or teacher). Other transformations in status take place as locals change occupations and become private sector brokers by engaging in a tourism business, or become public sector tourism brokers by finding government employment that concerns tourism. Of course, locals and brokers take on the role of a tourist when they vacation on travel of their own.

Second, power dynamics of the BLT model should be taken into account. Tourism is often examined as a product of the aggregate decisions of individual tourists. The relationship forged between the tourist and the local is accordingly depicted as socioeconomic in nature; tourists and locals interact as "guests" and "hosts" or as consumers and producers. Where power relationships are perceived to exist (e.g., as between First World tourists and Third and Fourth World locals), it is argued that these reveal the colonial and imperialistic leverage tourists have over those whom they visit. From this perspective, tourism systems are controlled and determined—often in unfortunate ways—by the behavior of tourists.

While there certainly are many instances in which tourists have exercised their influence to selfish and inappropriate advantage in coastal tourism settings, a narrow concentration on the power of tourists can result in analysts missing the power of tourism brokers. As Cheong and Miller (2000) have pointed out, tourists are frequently vulnerable to the power and control of brokers and locals. This is the case when, for example, tourists abide by laws and regulations of public sector brokers, and when they follow the advice and instructions of private sector brokers such as tour guides and travel agents.

Tourism development

Coastal tourism development in the coastal zone has become a constant since the end of World War II. Well-known examples are found on the coastlines and islands of Europe, North and South America, Africa, and Asia (Miller and Auyong, 1991b, 1998a; Conlin and Baum, 1995; Lockhart and Drakakis-Smith, 1997). Tourism development necessarily leads to changes in society and the environment of some kind. While conclusions about the "appropriateness," "success," "inappropriateness," or "failure" of coastal tourism development projects vary to a degree with the political and economic orientations, aesthetic standards, and environmental philosophies of analysts and observers, there is no question about the power of tourism development to quickly effect dramatic change.

From a societal point of view, tourism development promises better quality of life. In theory, poverty is alleviated through the creation of new jobs. Personal income and taxes derived from tourism then fosters better health, education, and other social services. In practice, these goals are only sometimes met. In many cases, failures of political institutions have led to unfair distributions of tourism-generated revenues and to problems of environmental justice.

As noted above, the tourism process provides incentives for locals to become tourism brokers. The lives, then, of both locals and new brokers are changed by coastal development. In some instances changes in the community that are derivative of tourism are undeniably positive. In other cases, the effect is negative. In a study of tourism in a Mexican coastal community, McGoodwin (1986) has identified a *tourist impact syndrome* which identifies the possible cultural costs to tourism system locals as including: (1) loss of political and economic autonomy (including loss of real property), (2) loss of folklore and related cultural institutions, (3) social disorganization (including radical changes in value orientations and in norms regarding social relations; heightened desire for material objects; changes in norms regarding work, sexual behavior, and drug use; promotion of illusory life aspirations; and loss of parental control and of respect for elders), and (4) hostility towards tourists (e.g., thievery, hustling, verbal aggression).

From an environmental perspective, tourism development is often seen to promise degradation of ecosystems (see *Human Impacts on Coasts, q.v.*). This, of course, is unavoidable with the building of airports, ports, road systems, hotels, resorts, and other facilities. This said, tourism development can also provide financial support for the protection of the marine environment and endangered species, as for example, in the creation of underwater and marine parks (*q.v.*) and protected areas.

Over the last decades, there has been growing recognition of the social and environmental trade-offs of tourism and also of the unintended consequences and economic externalities of tourism development (e.g., see Mathieson and Wall, 1982; Edwards, 1988; Pearce, 1989; Clark, 1996; Orams, 1999). With this, coastal tourism development is increasingly designed, debated, and evaluated against the ideal of *sustainable development*. Two prominent statements on this important concept follow:

> Economic growth always brings risk of environmental damage, as it puts increased pressure on environmental resources. But policy makers guided by the concept of *sustainable development* will necessarily work to assure that growing economies remain firmly attached to their ecological roots and that these roots are protected and nurtured so that they may support growth over the long term. (World Commission on Environment and Development, 1987, p. 40)
>
> (*Sustainable development* means) improving the capacity to convert a constant level of physical resource use to the increased satisfaction of human needs. (World Conservation Union, the United Nations Environment Programme, and the World Wide Fund for Nature, 1990, p. 10)

Toward the resolution of coastal tourism issues

Coastal tourism has been seen to be responsible for both positive and negative impacts to the natural and social environment (see Figure T48). The impacts of coastal tourism on the social environment involve social, cultural, political, and economic issues. On the positive side, coastal tourism can foster community pride, improved quality of life and new job opportunities; on the negative side, coastal tourism can lead to problems of overcrowding, social displacement, and crime. Impacts on the natural environment are often biological, physical, and ecological in nature. Increased protection and conservation of many areas and species has been a positive result of coastal tourism; nevertheless problems of erosion, pollution, and loss of species diversity occur far too frequently. It should not surprise that many coastal tourism issues simultaneously affect the social and natural environment.

The viability of coastal tourism systems and the natural environment in which they occur is very much dependent on human behavior. The resolution of coastal tourism issues can arise from the work of tourism system brokers and also from the individual decisions of locals and tourists.

Of the multiple ways available to society to control human conduct, three broker-driven mechanisms are prominent in the coastal tourism context. These mechanisms are *tourism management*, *tourism planning*, and *tourism education*.

Tourism management, planning, and education are crucial to the sustainable evolution of a touristic destination. It is therefore imperative that each are administered in such a way as to provide for the social and economic needs of the community, while at the same time ensuring that environmentally sensitive areas and ecologically important habitats are identified and excluded from tourism pressure. It is also recognized that tourism management, planning, and education are necessarily not only for scientific purposes and to conserve the environment for the benefit of residents, but also for the protection of long-term investments in tourism infrastructure, attractions, facilities, services, and marketing programs. It deserves to be noted that coastal tourism management, planning, and education programs are often designed and implemented by the same agencies and organizations. This overlap is often desirable and is found in some instances of larger efforts of government to promote integrated coastal zone management (see *Coastal Zone Management, q.v.*; see, also, Clark, 1996; Cicin-Sain and Knecht, 1998).

The manner in which a country, region, or community chooses to conduct touristic management, planning, and education activities is framed by societal (political, economic, etc.) and environmental (geography, natural resource, etc.) constraints. In the long run, the wisest course of action is to balance environmental, business, management, and social concerns so that tourism development is recognized as a potentially dangerous, but also potentially valuable and responsible course of action.

Coastal tourism management

Very generally, management concerns the actions of an executive decision-making entity in accordance with overarching goals of the larger enterprise in which it is housed. Although resorts, hotels, restaurants, transportation businesses, and many other firms in the coastal tourism sector do make decisions in strategic and professional ways, management, as the term is employed here, points to work engaged in or sponsored by public sector brokers to address problems and opportunities of coastal tourism.

Throughout the world, coastal tourism is managed by regulatory entities in accordance with the structure and procedures of the prevailing political system. In the United States, coastal tourism management is undertaken by federal agencies, and by regional, state, and local authorities at other levels of government. These executive entities rely on the two other branches of government for guidance. Thus, legislatures provide the mandate for tourism management in the design of laws, and the judicial branch of government interprets law as it applies to regulatory actions and the behavior of constituencies and tourists.

In the United States the growing importance of coastal tourism to the Nation was communicated to the general public with publication of *Year of the Ocean: Discussion Papers* in 1998. In a major chapter in this document, "Coastal Recreation and Tourism" is defined as embracing:

> the full range of tourism, leisure, and recreationally oriented activities that take place in the coastal zone and the offshore waters. These include coastal tourism development (hotels, resorts, restaurants, food industry, vacation homes, second homes, *etc.*), and the infrastructure supporting coastal development (retail businesses, marinas, fishing tackle stores, dive shops, fishing piers, recreational

boating harbors, beaches, recreational fishing facilities, and the like). Also included is ecotourism and recreational activities such as recreational boating, cruises, swimming, recreational fishing, snorkeling, and diving. Coastal tourism and recreation … likewise includes the public and private programs affecting all the aforementioned activities (US Federal Agencies, 1998, p. F-2).

In stressing the need to coordinate federal coastal tourism policies and programs, the chapter discusses governmental management and planning, management of clean water, and healthy coastal ecosystems, management of coastal hazards, and beach restoration programs.

At the federal level, the United States does not have a "Department of Tourism" with regulatory authority. Instead, touristic, recreational, and leisure activities are regulated by a host of executive agencies in accordance with a suite of legal mandates. Prominent in the control of coastal tourism and recreation are the National Park Service, the National Marine Fisheries Service, the Fish and Wildlife Service, and the US Army Corps of Engineers (Table T14). These agencies oversee many programs that conserve and protect natural resources and the environment while also fostering public access to the shore and to the marine environment.

At the state-level and at the local-level of government, coastal tourism is managed through the regulatory efforts of a variety of departments (e.g., departments of fish and wildlife, and park departments) with mandates that resemble those of federal counterparts. It is also common for city governments to cooperate with chambers of commerce in the development of infrastructural and business practice standards.

In practice, then, coastal tourism management is conducted by public sector brokers at all levels of government, by private sector brokers in businesses, and by some NGOs and environmental and social movement brokers. An array of tourism management tools (e.g., licensing regulations, zoning rules, tourist quotas, time and areas restrictions, and carrying capacity and limits of acceptable change regulations) have been used successfully throughout the Pacific, in the Caribbean, in the Atlantic and elsewhere (see Pearce, 1989; Miller and Auyong, 1991b, 1998; Conlin and Baum, 1995; Lockhart and Drakakis-Smith, 1997; Orams, 1999). The concepts of *zoning* and *carrying capacity* deserve further attention due to their particular applicability to the management of tourism in coastal areas.

Two very significant parks utilize zoning as a means of tourism management. The Great Barrier Reef Marine Park in Australia is a multiple-use protected area. With zoning, conflicting uses are physically separated. The range of protection in the park varies from virtually no protection to zones where human activity is conditionally permitted. The adoption of this zoning scheme allows the park authority, in association with interested members of the public and with other agencies, to develop and apply a tourism strategy for the entire Great Barrier Reef Marine Park. Zoning ensures that the Reef will not become overpopulated with tourist and other structures, but also allows for careful development in areas which are suitable for that purpose. The Galapagos Islands National Park in Ecuador also employs zoning strategies. The park is effectively managed with intensive use, extensive use, and scientific use (off limits to all but a few visitors) zones.

Table T14 US federal statutes and actions influencing coastal tourism (selected)

- Antiquities Act (1906)
- National Park Service Organic Act (1916)
- Fish and Wildlife Coordination Act (1934)
- The Wilderness Act (1964)
- National Sea Grant College Program Act (1966)
- National Historic Preservation Act (1966)
- Executive Reorganization Plan Number 4 and Executive Order 11564 establishing the National Oceanic and Atmospheric Administration (1970)
- Coastal Zone Management Act (1972)
- National Marine Sanctuaries Act (1972)
- Fishery Conservation and Management Act (1976; renamed Magnuson-Stevens Fishery Conservation and Management Act)
- Archaeological Resources Protection Act (1979)
- Fish and Wildlife Conservation Act (1980)
- Presidential Proclamation 5030 establishing a 200-mile Exclusive Economic Zone (1983)
- The Recreational Boating Safety Act (1986)
- Abandoned Shipwreck Act (1987)
- Executive Order 13158 establishing a national system of Marine Protected Areas (2000)

Carrying capacity (*q.v.*) regulations illustrate the "precautionary principal" method of natural resource management and are highly regarded by practitioners of tourism management. Coastal tourism managers who seek to determine the appropriate level of use that can be sustained by the natural resources of an area are well aware that carrying capacities and use-intensity limits of tourism destinations are dynamic, and depend greatly on the biological and ecological processes of natural resources.

Coastal area carrying capacity can be evaluated in four ways (Sowman, 1987). Physical carrying capacity is concerned with the maximum number of "use units" (e.g., people, vehicles, boats) which can be physically accommodated in an area. Economic carrying capacity relates to situations where a resource is simultaneously utilized for outdoor recreation and economic activity. Ecological carrying capacity (sometimes referred to also as physical, bio-physical, or environmental carrying capacity) is concerned with the maximum level of recreational use that can be accommodated by an area or an ecosystem before an unacceptable or irreversible decline in ecological values occur. Social carrying capacity (also referred to as perceptual, psychological, or behavioral capacity) is concerned with the visitor's perception of the presence (or absence) of others simultaneously utilizing the resource of an area. This concept is concerned with the effect of crowding on the enjoyment and appreciation of the recreation site or experience.

The *limits of acceptable change* (LAC) framework developed by Stankey *et al.* (1985) enables managers to move beyond calculation of carrying capacity figures to address actions needed for management goals. This approach concentrates on establishing measurable limits to human-induced changes in the natural and social setting of parks and protected areas, and on identifying appropriate management strategies to maintain and/or restore desired conditions (Stankey *et al.*, 1985). Knowledge of the natural (physical, biological) setting is combined with knowledge of the human (social, political) setting in order to define appropriate future conditions.

The LAC method employs nine steps as follows: (1) identification of area concerns and issues, (2) definition and description of opportunity classes, (3) selection of indicators of resource and social conditions, (4) inventory of resource and social conditions, (5) specification of standards for resource and social indicators, (6) identification of alternative opportunity class allocations, (7) identification of management actions for each alternative, (8) evaluation and selection of an alternative, and (9) implementation of actions and monitoring of conditions. To date, the LAC system has proved to be a valuable tourism management tool in several wilderness areas in the United States and has direct applications to coastal areas as well.

Coastal tourism planning

Planning, broadly conceived, entails the consideration of a range of actions likely to contribute to the attainment of organizational goals. In some instances, overarching goals are well known in advance and planning professionals concentrate on the means that will ensure these ends. In other situations, the determination of goals requires prolonged deliberation.

Coastal tourism planning is often integrated with other resource analyses in the development of coastal area or region. Planners take into account not only visitation rates and statistics, but also the fact that tourists increasingly insist that destinations be high-quality and pollution-free, as well as inherently interesting. Therefore, it is in both the public and private sector brokers' interest to implement a planning strategy for tourism. The goals and policies of government agencies and businesses are, however, frequently different from one other and may even be in direct conflict. To minimize and even prevent disruptions and loss of time, communication between tourism brokers is crucial. Planning also leads to equitable distributions of coastal tourism benefits.

The success or failure of a tourism project frequently hinges on the conditions of natural amenities in the surrounding environment. This is especially true for tropical environments found, for example, on Pacific and Caribbean islands. Parks and natural resource areas, scenic vistas, archaeological and historic sites, and coral reefs are all touted tourism attractions. Marketing strategies for coastal, marine, and island tourism especially promote destinations for being close to white sand beaches. However, development of permanent structures for tourism near beaches often exacerbates beach erosion, property damage, and requires construction of shore protection structures. If touristic facilities are to be sited near beaches, proper planning is essential for the protection of the coastal zone and private property.

In many numerous coastal and island states located in the Mediterranean, Caribbean, and the Pacific where tourism is a major economic force, major national-level departments of government shape coastal tourism through the design of investment incentives and international joint venture opportunities. In nations such as Mexico and Costa Rica, these activities are linked to the preparation of strategic tourism plans.

In the United States—and with notable exceptions such as those provided by the National Park Service—very little coastal tourism planning takes place in the federal government. At the state-level, it is commonplace for departments of tourism to promote tourism. While many states have experienced great success in attracting tourists with advertising strategies, most state departments of tourism have yet to augment the marketing of tourism with the monitoring and assessment of coastal tourism's effects on the environment and quality of life. At the local-level, many city governments have utilized their planning departments to recommend approaches to issues having to do with public use of the coastline and natural resources, the revitalization of waterfronts, and zoning appropriate to resort and marina development.

Within the private sector, coastal tourism planning is an established professional specialty. Firms of all sizes develop coastal tourism plans tailored by expert consultants to the needs of developer clients. Increasingly, social movement brokers are being seen to engage in professional coastal tourism planning.

Coastal tourism planning falls into two main categories, depending on whether the project in question is driven by a preservation or a development ethic. Preservation goals predominate in the planning of recreational areas, in national park and marine protected area planning, and in planning that is part of natural resource management. The development framework has found application in seaside resort and theme park planning, in condominium time-share planning, and in varieties of coastal city planning. There are many examples worldwide of coastal tourism zones, replete with both preservation and development projects, that extend from major cities. The Costa Brava in Spain, the French Riviera, the Yucatan Peninsula, the East Coast of Australia, and the coastlines of the United States and many Polynesian islands illustrate mixed planning.

Landscape architecture and urban planning are important in shaping coastal tourism. Both fields have public and private applications, and design and planning aspects. Both tailor products and services to preservation and development goals and, accordingly, address biophysical and social and economic objectives.

Landscape architects design parks and gardens, resort and hotel facilities, marinas and waterfronts, plazas and squares, and transportation corridors providing access to coastal touristic destinations. Urban planners design circulation facilities, city districts and spaces, and produce master planning and site design products.

Planning activities and products of landscape architects include resource management plans, environmental analyses, and multidisciplinary feasibility studies, and needs assessment and community structure plans. In overlapping ways, urban planners produce tourism policy plans, functional plans, and environmental assessments.

Because coastal tourism planning efforts are attuned to local conditions, constituencies, and financial constraints, there is no single planning process for guaranteeing success. This said, most professional planning endeavors share a general structure. Grenier *et al.* (1993) suggest a three-phase tourism planning process. With this, a first "Front-end Planning" phase encompasses scoping (entailing a statement of project philosophy, pre-assessments of key issues and themes, and formulation of objectives) and research (involving data collection and analyses supporting cultural, institutional, and environmental profiles; site reconnaissance; eco-determinant mapping; and analyses of constraints and opportunities). A second "Project Planning" phase is focused on refinement of project objectives, design and evaluation of alternative development plan concepts, and selection and approval of the preferred development plan concept. A third and final "Project Management" phase concerns activities of implementation, monitoring and evaluation, and refinement.

In summary, coastal tourism planning has been fostered by public sector brokers at all governmental levels, by consultants among other private sector brokers, and by an impressive range of nongovernmental and environmental organizations in roles these have taken on as social movement tourism brokers. Coastal tourism planning practitioners have developed an array of planning methodologies (e.g., comprehensive land-use planning, integrated coastal zone planning, and strategic and special use planning) and have utilized these throughout the world, in many instances by cooperating with tourism brokers with management expertise (see, Gunn, 1988; Pearce, 1989; Miller and Auyong, 1991a, 1998a; Conlin and Baum, 1995; Lockhart and Drakakis-Smith, 1997; Orams, 1999; Hadley, 2001).

Coastal tourism education

The two mechanisms for the control of human behavior in coastal tourism systems discussed above—management and planning—are similar to one another in that the tourism experts who analyze coastal tourism situations channel their recommendations upward to regulatory and planning authorities. These tourism brokers then implement policies and plans downward, influencing tourism businesses, tourists, and locals.

A third mechanism concerns coastal tourism education and communication. Although education about coastal science and environmental issues is effectively transmitted in classrooms, discussion here focuses on education and outreach in nontraditional settings and how people learn through the experience of being tourists or learn in the course of daily life. Guided tours, museums, brochures, public lectures, newspapers, and signage are but a few of the devices that figure importantly in the educational processes linked to coastal tourism.

In a manner of speaking, tourism education contrasts with management and planning in that the first clients of analysts are not managers and planners in positions of authority, but tourists and locals. Whereas managers achieve goals through policies and regulations and planners depend on plans, coastal tourism educators succeed when people take personal initiative to change their own behavior because they have been taught something. Tourism education, then, is a process in which analyst brokers direct their ideas outward to people involved in tourism. Tourism educators and communicators do not evaluate success or failure at attaining their goals with studies of "enforcement" or "compliance." This is so because successful education motivates individuals by persuasion, not coercion.

By definition and referring to Figure T48, coastal tourism educators are public sector, private sector, and social movement brokers. These brokers design products and strategies to educate people and through this to change human behavior in coastal tourism systems. While educator brokers seek to impart their message to tourists and to locals, they also educate one another as, for example, when a nongovernmental organization (NGO) educates public sector and private sector brokers.

Efforts to resolve problems and opportunities of coastal tourism through education are steadily growing throughout the world. Tourism brokers who are advancing this promising agenda are benefiting from the work of educators who have focused on environmental and sustainability issues. Monroe (1999) has characterized successful environmental education and communication projects as having features that allow for: (1) empowerment of local communities and use of their expertise, (2) attention to scientific, social, economic, political, and cultural topics, (3) identification of a variety of stakeholders and integration of them into the process, (4) advancement of an environmental ethic as well as assistance to residents in developing decision-making skills, (5) development of a gender component, (6) flexibility in project design (including realistic timetables), and (7) project evaluation.

Coastal tourism brokers (e.g., those in government or in NGOs) that provide international aid in developing and poverty-plagued states have also benefited from the cross-cultural advice of Brazilian educator and philosopher, Paulo Freire. Freire has contended that the education process has for too long been regarded as a "delivery service" from the scientific and technological elite of the Western World to those suffering in the Third World. Freire's (1999, p. 61, emphasis added) solution lies in education projects that emphasize collaborations between experts and clients at all stages of the process:

> Through dialogue, the teacher-of-the-students and students-of-the-teacher cease to exist and a new term emerges: teacher-student with student-teachers. The teacher is no longer merely the-one-who-teaches, but one who is himself taught in dialogue with the students, who in turn while being taught also teach. *They become jointly responsible for a process in which all learn.*

Few would disagree with the proposition that coastal tourism education has great potential to enhance the quality of tourism for tourists and locals, and to also protect the environment through responsible human conduct. The importance of education (and of overlapping fields such as communication, journalism, and environmental and science reporting by the media) is recognized by virtually all marine scientists and researchers (see Pearce, 1989; Miller and Auyong, 1991b, 1998; Conlin and Baum, 1995; Lockhart and Drakakis-Smith, 1997; Orams, 1999). Still, many opportunities to integrate coastal tourism education with the mechanisms of management and planning have been missed.

Challenges ahead

Over the last several centuries, the world's coastlines have been substantially transformed to support recreational and touristic pursuits. In some cases, coastal tourism dominates the skyline. In others, tourism is one of many industries. As coastlines become more populated and accessible, it is ever more clear that, however, beneficial coastal tourism is to the tourist, it is neither a panacea that will invigorate any local economy, nor a pollution that will necessarily ruin environments and corrupt cultural traditions and values. Coastal tourism is a process amenable to management, planning, and education. Sustainable coastal tourism obliges humanity to have respect for other life forms and the environment, while it affords opportunities for people to learn, recreate, and reach their potential as individuals through travel.

Because the stakes are high and because mistakes can be virtually irreversible, societal resolution of pressing tourism and coastal development issues requires imagination as well as sustained scientific and policy attention. Work to be done falls in the areas of research, and tourism broker and individual responsibility.

Tourism research

Researchers in government, academe, and in the private and social movement sectors constitute a first group of practitioners whose work induces positive change in coastal tourism systems. Fundamental questions about physical, biological, ecological, social, cultural, economic, demographic, and political processes of coastal tourism are posed and answered in assessments, impact statements, profiles, and other products of natural, biological, and social scientists. With reference to the condition of the environment and society, the possibilities of coastal tourism development raise not only the question "What is?" but also questions about "What is ethical?," "What is fair?," and "What is beautiful?" As a result, analyses by professionals with backgrounds in the humanities and arts have proven to be useful in complementing those of scientists.

The need to formally study tourism is recognized more than ever in academe. Tourism research methods are under continual development in such fields as public affairs, business and marketing, architecture, urban planning and design, political science, sociology, geography, cultural anthropology, marine affairs, and environmental studies (e.g., see Gunn, 1979; Murphy, 1985; Ritchie and Goeldner, 1987; McIntosh and Goeldner, 1990).

Tourism broker responsibilities

A second professional group made up of the different types of coastal tourism brokers will be counted upon heavily in the future to cooperate with one another. This can occur, for example, when investors and developers in the private sector coordinate goals and activities with those of government agencies and NGOs to make sustainable tourism a reality. Another kind of cooperation calls for tourism brokers to work effectively with government, business, and nongovernmental organizations in other economic sectors. Better understandings of tourism-fishery interactions, tourism-aquaculture interactions, and tourism-ocean shipping interactions can lead to an improvement on single-sector governance with partially (or, under ideal conditions, fully) integrated coastal management.

In the aftermath of the terrorism attack on the World Trade Center in 2001, the responsibilities of tourism brokers have been enlarged. Brokers now must function not only as stewards of the coastal environment, businessmen, and representatives of constituencies, but also as protectors of residential and traveling publics.

Uncertainties generated by the terrorism of 2001 will change the ways in which coastal tourism is conducted in the United States and elsewhere. It has long been known that too much tourism can be bad by when it leads to degraded ecosystems and undesirable changes in quality of life. Now it is apparent that too little tourism can put entire coastal economies at risk. Declines in coastal tourism can create serious social problems in the same way declines in fishery resources can threaten livelihoods. Ultimately, coastal tourism and recreation destinations negatively affected by security-related changes in itineraries will become sustainable only to the extent brokers make tourism safe.

Individual responsibilities

The discussion above has concerned the proactive roles researchers and brokers can play in promoting sustainable tourism and coastal development. To this must be added a comment about the personal responsibilities of tourists and locals to contribute toward sustainability in the coastal tourism systems which they visit or in which they live.

To a certain extent, the social role of the ethical tourist can be formulated to resemble that of the good citizen. Good citizens learn from their families and schools to reach their potential in society while knowing how to behave in socially appropriate ways. Using this template, tourists would be expected to behave in ecologically and culturally appropriate ways in the course of their domestic and international travel. Ecotourism and ethnic tourism are two forms of tourism that have emerged to stress this self-conscious orientation. Many tourism brokers in business, in government, and in non-governmental organizations are now promoting the development of "best practices" and "tourism guidelines" to this end.

It is obvious that there are many benefits of coastal travel that accrue to the tourist. These are found in recreational, aesthetic, and educational activities. In exchange, the ethical tourist will strive to behave in a culturally and environmentally responsible manner. As this occurs, locals are given an added incentive to orient their conduct to the same ends. Improvements in the behavior of tourists and locals toward one another and toward the coastal environment will assist tourism providers and managers as they do their part to monitor and control tourism, and improve the tourism experience for all involved.

Concluding remarks

Coastal tourism has demonstrated its considerable power to influence the fundamental configurations of coastlines and the social structures these support. Coastal tourism is sometimes found to be unfortunate in every respect. Coastal tourism can, however, be designed to improve the lives of tourists and those who are part of the tourism industry, conserve natural resources and protect the environment, and not offend locals. For this to occur, coastal tourism brokers—in government, business, and non-governmental organizations—will need to cooperate to insure that tourism is sustainable and safe.

It will also be necessary for tourists and locals to adopt "best practices" that underwrite cross-cultural communication and respect for the environment. In the eyes of many, it is time for all to abide by a coastal system *tourism ethic*. Such an ethic might reasonably incorporate Aldo Leopold's (1949, p. 224–225) famous caution about natural resource use based solely on economic self-interest:

> [a] thing is right when it tends to preserve the integrity, stability, and beauty of the biotic community. It is wrong when it tends otherwise.

Through the implementation of responsible management, planning, and education policies—together with the diffusion of a tourism ethic—tourism and coastal development can be shaped to reflect the best tendencies of humanity.

Marc L. Miller and Nina P. Hadley

Bibliography

Cheong, S.-M., and Miller, M.L., 2000. Power and tourism: a Foucauldian observation. *Annals of Tourism Research*, **27**(2): 371–390.

Cicin-Sain, B., and Knecht, R.W., 1998. *Integrated Coastal and Ocean Management: Concepts and Principles*. Washington, DC: Island Press. (with the assistance of D. Jang and G.W. Fisk).

Clark, J., 1996. *Coastal Zone Management Handbook*. Boca Raton, FL: CRC Press.

Conlin, M.V., and Baum, T. (eds.), 1995. *Island Tourism: Management Principles and Practice*. New York: John Wiley & Sons.

Corbin, A., 1994 [1988]. *The Lure of the Sea: The Discovery of the Seaside 1750–1840*. (translated by J. Phelps) New York: Penguin Books.

Cordell, H.K., Teasley, J., Super, G., Bergstrom, J.C., and McDonald, B., 1997. *Outdoor Recreation in the United States: Results from the National Survey on Recreation and the Environment*. Athens, GA: US Forest Service and the Department of Agriculture and Applied Economics, University of Georgia.

Csikszentmihalyi, M., 1975. Play and intrinsic rewards. *Journal of Humanistic Psychology*, **15**(3): 41–63.

Csikszentmihalyi, M., 1990. *Flow: The Psychology of Optimal Experience*. New York: Harper Perennial.

Edwards, F., (ed.), 1988. *Environmentally Sound Tourism in the Caribbean*. Calgary: The University of Calgary Press.

Environmental Protection Agency (EPA), 1995. *National Demand for Water Based Recreation*. Washington, DC: EPA.

Freire, P., 1999 [1970]. *Pedagogy of the Oppressed*, Revised edn. New York: Continuum.

Godsman, J., 2000. *2001 Outlook for the Cruise Industry*. White paper presented at the 2000 TIA [Travel Industry Association] Marketing Outlook Forum. (October 25–28, 2000) New York: *National Geographic Traveler*.
Grenier, D., Kaae, B.C., Miller, M.L., and Mobley, R.W., 1993. Ecotourism, landscape architecture and urban planning. *Landscape and Urban Planning*, **25**: 1–16.
Gunn, C.A., 1979. *Tourism Planning* (2nd edition). New York: Crane, Russak.
Gunn, C.A., 1988. *Vacationscape: Designing Tourist Regions*. New York: Van Nostrand Reinhold.
Hadley, N., 2001. Cooperative tourism management of midway atoll national wildlife refuge: planning, assessment, and strategy. *Tourism*, **49**(3): 189–202.
Hern, A., 1967. *The Seaside Holiday: The History of the English Seaside Resort*. London: The Cresset Press.
Houston, J.R., 1995. Beach nourishment. *Coastal Forum, Shore and Beach*, **64**(1): 21–24.
Houston, J.R., 1996. International tourism and US beaches. *Coastal Forum, Shore and Beach*, **64**(2): 3–4.
Leopold, A., 1949. *A Sand County Almanac and Sketches Here and There*. New York: Oxford University Press.
Lockhart, D.G., and Drakakis-Smith, D. (eds.), 1997. *Island Tourism: Trends and Prospects*. New York: Pinter.
Manning-Sanders, R., 1951. *Seaside England*. London: B.T. Batsford.
Mathieson, A., and Wall, G., 1982. *Tourism: Economic, Physical, and Social Impacts*. New York: John Wiley & Sons.
McGoodwin, J.R., 1986. The tourism-impact syndrome in developing coastal communities: a Mexican case. *Coastal Zone Management Journal*, **14**(1/2): 131–146.
McIntosh, R.W., and Goeldner, C.R., 1990. *Tourism: Principles, Practices, Philosophies*, 6th edn. New York: John Wiley & Sons.
Miller, M.L., and Auyong, J. 1991a. Coastal zone tourism: a potent force affecting environment and society. *Marine Policy*, **15**: 75–99.
Miller, M.L., and Auyong, J. (eds.) 1991b. *Proceedings of the 1990 Congress on Coastal and Marine Tourism: A Symposium and Workshop on Balancing Conservation and Economic Development (Volumes I and II)*. Newport, OR: National Coastal Resources Research and Development Institute.
Miller, M.L., and Auyong, J. (eds.), 1998a. *Proceedings of the 1996 World Congress on Coastal and Marine Tourism: Experiences in Management and Development*. Seattle: Washington Sea Grant Program and the School of Marine Affairs, University of Washington and Oregon Sea Grant College Program, Oregon State University.
Miller, M.L., and Auyong, J., 1998b. Remarks on tourism terminologies: Anti-tourism, mass tourism, and alternative tourism. In Miller, M.L., and Auyong, J. (eds.), *Proceedings of the 1996 World Congress on Coastal and Marine Tourism: Experiences in Management and Development*. Washington Sea Grant Program and the School of Marine Affairs, University of Washington and Oregon Sea Grant College Program, Oregon State University, pp. 1–24.
Miller, M.L., and Ditton, R., 1986. Travel, tourism, and marine affairs. *Coastal Zone Management Journal*, **14**(1/2): 1–19.
Monroe, M.C. (ed.), 1999. *What Works: A Guide to Environment Education and Communication Projects for Practitioners and Donors*. Gabriola Island, British Columbia: New Society Publishers.
Murphy, P.E., 1985. *Tourism: A Community Approach*. New York: Methuen.
National Marine Manufacturers Association, 2000. *Boating 2000: Facts and Figures at a Glance*. Chicago, IL: NMMA.
Orams, M. 1999. *Marine Tourism: Development, Impacts and Management*. New York: Routledge.
Pearce, D. 1989. *Tourist Development*, 2nd edn. New York: John Wiley & Sons.
Ritchie, J.R., and Goeldner, C.R., (eds.), 1987. *Travel, Tourism, and Hospitality Research: A Handbook for Managers and Researchers*. New York: John Wiley & Sons.
Sowman, M., 1987. A procedure for assessing recreational carrying capacity for coastal resort areas. *Landscape and Urban Planning*, **14**(4): 331–344.
Stankey, G., Cole, D., Lucas, R., Peterson, M., and Fissell, S., 1985. The Limits of Acceptable Change (LAC) system for wilderness planning. Seattle: US Department of Agriculture, *Technical Report INT-176*.
Stebbins, R.A., 1992. *Amateurs, Professionals, and Serious Leisure*. Montreal: McGill-Queen's University Press.
Travel Industry Association of America (TIA), 1998. *Travel and Tourism Congressional District Economic Impact Study*. Washington, DC: TIA and US Department of Commerce.
Travel Industry Association of America (TIA), 2001. *Impact of Travel on State Economies 1999*. Washington, DC: TIA.
United Nations Environment Programme (UNEP), 1992. Sustainable tourism development. *UNEP Industry and Environment*, **15**(3–4): 1–96.
US Federal Agencies (with ocean-related programs), 1998. *Year of the Ocean Discussion* Papers. Washington, DC: US Department of Commerce (NOAA).
US Travel and Tourism Administration, 1994. *Impact of International Visitor Spending On State Economies*. Washington, DC: US Department of Commerce.
Volgenau, G., 2000. The Plog survey: Our favorite places. *The Seattle Times*. Travel Section K, pg. 6 (Sunday, 16 January).
Waters, S. (ed.), 1997. *Travel Industry World Yearbook—The Big Picture 1996–97*, Vol. 40. Spencertown, NY: Travel Industry Publishing Inc.
Waters, S. (ed.), 2001. *Travel Industry World Yearbook—The Big Picture 2001*, Vol. 44. Spencertown, NY: Travel Industry Publishing Inc.
World Commission on Environment and Development, 1987. *Our Common Future ("The Brundtland Report")*. New York: Oxford University Press.
World Conservation Union (IUCN), the United Nations Environment Programme (UNEP), and the World Wide Fund for Nature (WWF), 1990. *Caring for the World: A Strategy for Sustainability*. (Second Draft). Gland: IUCN.
World Tourism Organization (WTO), 1996. *Compendium of Travel 1989–1996*. Madrid: WTO.
World Tourism Organization (WTO), 2001. *Tourism Highlights 2001*. Madrid: WTO.
World Travel and Tourism Council (WTTC), 1995. Agenda 21 for the Travel and Tourism Industry: Towards Environmentally Sustainable Development. London: WTTC, *Progress Report No. 1*.

Cross-references

Aquaculture
Beach Use and Behaviors
Carrying Capacity in Coastal Areas
Coastal Zone Management
Demography of Coastal Populations
Economic Value of Beaches
Environmental Quality
Human Impacts on Coasts
Marine Parks
Surfing
Tourism, Criteria for Coastal Sites

TOURISM, CRITERIA FOR COASTAL SITES

Historically, seaside resorts date back to Roman times with a string of resorts along the Campanian littoral on the northern shore of the Bay of Naples (Turner and Ash, 1976, p. 24). The modern seaside resorts had their origin in England and their early growth was attributed to the therapeutic value of seawater drinking and bathing. From the mid-19th century, the era of railway saw the rapid growth of more seaside resorts, spreading to western Europe (Romeril, 1984; Walton, 1997). Partly as the result of the railway ending at about right angles to the coast, the European coastal resort has a basic T-shaped morphology, although factors varied. Usually, coastal resort morphology is dependent on site characteristics, tourist elements, and other urban functions (Pearce, 1995, pp. 136–140; Nordstrom, 2000, pp. 10–13).

This entry is concerned with such physical site factors as the "emphasis … on aspects of the physical implications of site selection such as coastal erosion rather than resort form … underlies the potential which this approach has to complement the more traditional resort morphologies" (Pearce, 1995, p. 137). This approach is increasingly significant, as developing countries seek out appropriate beaches and islands for tourism development. On a worldwide scale, coastal tourism (see *Tourism and Coastal Development*) in the traditional Mediterranean area is complemented by the Caribbean area, South Pacific area, and Southeast Asia. Islands, especially *small islands (q.v.)*, are actively sought for tourism development. To the usual three "Ss" (sun, sea, and sand) for coastal tourism, one could add two more "Ss" (sunrise and sunset) if one were on a small island.

Coastal site criteria

Coastal tourism development involves the identification of a suitable site, developing the site by clearing, and providing access, accommodation, recreation facilities, and services for the tourists (Ahmad, 1982). Site criteria for coastal tourism are more than just association with white sand beaches and coral reefs. An analysis of the site criteria includes a wide variety of physical and other factors such as land-use, ownership, etc. The objective is to assess the site opportunities and constraints which have a significant implication for the resort entrance, backdrop, views, beaches and swimming areas, buildable area, vegetation, boating tours, fishing and diving opportunities, etc. In addition, various site planning considerations and development standards, including architectural, landscaping, and engineering design, have to be considered (Inskeep, 1991, pp. 303–335).

Planners and architects usually acknowledge that two powerful forces make coastal tourism distinctive and have a bearing on site criteria. The first is the special amalgam where land meets the sea and the coastal site is a junction of landscape and seascape. The second is the linear character of the coast in which the natural resources for tourism are arranged in a linear fashion (Gunn, 1988, p. 87).

The linearity of the coast presents special challenges for planning and tourism design. Generally, four zones with different characteristics can be identified with the more important criteria and implications for tourism development given in Table T15 (Gunn, 1988, p. 88; Mieczkowski, 1990, pp. 243–246). On the seaward side is the marine zone or neretic zone which is the ecological zone from the continental shelf to the beach. This area contains the marine life, reefs, and sandbars and is suitable for a number of marine-based activities. The beach consisting of the foreshore and backshore is the most important of the four zones. It is used for many activities, especially if it is sufficiently wide and sandy. More specific requirements identified for beach resort development in this zone include beach protection and beach capacities (Baud-Bovy and Lawson, 1998, pp. 71–72). The third zone is the shoreland which includes the dunes and is the area for camping, hiking, and other accommodation, food, shopping, and other service businesses. Its serves as the visual linkage between land and sea. The most landward zone is the vicinage or hinterland which provides the setting for tourist business and vacation homes. It is often the zone of population and supporting services and is enhanced by variations in topography and vegetation. Its nearness and access to the sea is more important than the visual linkage.

An example of the application of the linear zones is the planning of development for southern Thai beaches facing the Andaman Sea (Tourism Authority of Thailand, 1989). Three zones for development with 12 categories of landuse have been identified. The first two zones have specific widths. The beachfront area extends 300 m away from the beach; the interior area is 700 m wide and the hinterland is landward of the interior area.

As the coast is a basic component in coastal tourism, any potential resort site can be assessed by an initial understanding of its coastal geomorphology. As coasts differ widely, each type of site has implications for coastal tourism. For example, the potential of a specific coastal type or landform, such as, coral reef, coastal dune, sand spit, river-mouth barrier, rocky headland, can be known. Also, specific advantages and disadvantages of each landform, coastal type, or ecosystem have their influence on the choice of resort sites and also influence the pattern and development of resorts. Coastal geomorphology also helps to reduce negative impacts, protect valuable habitats and provide valuable information for subsequent alteration to the coastal environment, for example, changes to the drainage and water bodies (Wong, 1999, 2000).

More important and often underestimated, geomorphology also takes note of the seasonal factor which can have a marked impact on the coast. For example, in Southeast Asia and Indian Ocean islands, a strong seasonality prevails in the coastal environment. Beaches undergo accretion/erosion cycles and during the onshore wind season, wave action can reach a higher level or further inland. With reversals in wind, wave action and currents, sand movement varies in the offshore–onshore

Table T15 Major criteria in coastal zones for tourism development

Zone[a]	Factor	Comments
Sub-tidal to offshore (Marine or neretic zone)	• Climate • Waves • Tides • Currents • Water temperature; clarity of water • Biodiversity, e.g., marine life, corals, seaweeds	• Physical conditions determine type, extent and seasonality of many recreational activities, e.g., swimming, water skiing, surfing, sailing, boating, travel to nearby island • Biodiversity presents additional attractions for recreational use, e.g., snorkeling, scuba diving, fishing • Free of pollution
Intertidal-nearshore (Beach)	• Beaches: width, gradient, material size, color • Risks from tidal movements • Potential erosion • Public access • Shore platforms: width, access • Wetlands: extent, access	• Physical properties of beach and coast influence type and extent of recreational activities • Beach capacity as useful management tool • Soft protection measures are preferred if need arises • Environmental guidelines, especially for wetlands
Backshore (Shoreland)	• Area • Views • Geomorphology (cliffs, dunes, wetlands) • Coastal vegetation • Microclimate • Scope for improvement • Access, e.g., roads	• Location of various tourist accommodation and service businesses • Proper setback, conditions for use • Preserve view (visual linkage is important) • Maximize specific advantages of geomorphological features • Dunes normally preserved as defense line and for selective uses • Good views from cliffs, headlands • Phased development; minimize degradation • Improvement, e.g., drainage
Onshore (Hinterland or vicinage)	• Topography • Vegetation • Existing development, e.g,, population, supporting services	• Provides setting or backdrop • Separate planning zone • Access to sea is important

[a] Terms in parentheses are used by Gunn, 1988 and Mieczkowski, 1990.
Sources: Compiled from Baud-Bovy and Lawson, 1998; Gunn, 1988; Mieczkowski, 1990; Viles and Spencer, 1995; Wong, 1991.

direction and alongshore direction. The seasonal site features include beach morphological changes, a backshore with two berms, nearshore topographic changes, changing stream mouths, flooding, shifts in beach vegetation belts, etc. In particular, the analysis of the seasonal factor provides some idea of the potential hazards in various coastal zones. The seasonality arising from waves, wind, and tides, and the potential erosion and pollution are among the coastal hazards considered in environmental planning for site development (Beer, 1990, pp. 63–64).

Besides geomorphological criteria associated with tourist sites, other criteria in planning and development of resorts also have a strong physical base. These include adequate access, *setbacks* (*q.v.*), and EIA (environment impact assessment) before construction and carrying capacity. Pearce and Kirk (1986) have suggested specific types of carrying capacity that are closely associated with the linear zones of the coast. One recent development has been the application of a single criterion in the form of an ecolabel to assess the suitability of resort beaches. An example is the Blue Flag award for European beach resorts that focuses on *water quality* (*q.v.*), beach management, and safety (see *life saving and beach safety*) (UNEP/WTO/FEEE 1996).

Coastal site classification

Of various types of *classification of coasts* (*q.v.*) (Bird, 2000) none is suitable for determining site criteria for coastal tourism. Classifications with an emphasis on coastal processes can be useful for resort sites focusing on recreation, such as *surfing* (*q.v.*). For example, Bird (1993) identified various types of surf reflecting geomorphological and oceanic factors. Since the mid-1980s, the beach morphodynamics model with its identification and explanation of beach hazards, such as steep beaches and rip currents, is particularly useful for resort beaches in Australia (Short, 1999).

Generally, physical or morphological types of coastal classification are more useful for recreation/tourism planning, development, and management. For recreational purposes in the temperate countries, coastal landforms are classified as sand and shingle beaches, tidal forms (mudflats, salt marshes), estuaries, cliffs, and shore platforms (Pickering, 1996). Defert (1966 in Mieczkowski, 1990, pp. 247–248) provided one of the earliest classifications for coastal resort development, in which four types of coasts were identified:

1. *Oceanic: continuous and linear*: Straight oceanic beaches with tourism facilities following straight sandy coasts.
2. *Oceanic: discontinuous and concentrated*: Bays alternate with promontories and peninsulas with tourism located in the bays.
3. *Mediterranean: continuous and linear*: Wide and gentle beaches prevail as a result of the absence of tides or very small tides.
4. *Mediterranean: discontinuous and concentrated*: This consists of two types (1) wide bays bordered by promontories or capes, and (2) coves with small beaches.

In a situation where the site is clearly restricted or limited to one coastal type, it is possible to have further categorization in terms of criteria other than physical. For example, in the Ko Samui/Surat Thani region located in the Gulf of Thailand, the beaches are further identified for tourism development as follows: (1) conservation beaches, where tourism activities are not allowed; (2) nature-oriented development beaches where activities and services are permitted to a certain extent; and (3) progressive development beaches where activities and services can be developed to meet international standard (Tourism Authority of Thailand, 1985).

For the east coast of Peninsular Malaysia which is exposed to the northeast monsoon, resort sites are identified for tourism development with a strong consideration given to the seasonal factor. Four major types of sites are identified on the mainland coasts (Tourist Development Corporation, 1979).

1. Beach front site. This is backed by relatively flat land, sandy soils, and coconut groves and exposed to the sea.
2. Site oriented to both ocean and river or brackish lagoon. The topography is generally flat or rolling gently with coconuts and other vegetation.
3. Site situated adjacent to a substantial headland promontory with sufficient flat land for development. The beach may be interrupted by large boulders. Hill and weather patterns influence architectural design and site plan.
4. Site on hillside or hilltop with panoramic views located on headlands adjacent to good beaches. Weather and wind can be either greater or sheltered depending on the position of the development area. More constraints are placed on resort design.

Table T16 Types of resort sites on tropical coasts

Coastal type	Form/feature	Resort sites
A. Rock coast	1. Cliff 2. High headland	1. Good view; generally exposed; adequate setback necessary 2. Good view; stability is crucial; can be sheltered or exposed. Requires compact design solutions
B. Mainland beach Coast	1. Linear 2. Crescentic bay	1. Large area available; can be exposed 2. Can be at head of bay but has higher wave energy; decreasing wave energy towards the limbs of bay
	3. J-shaped bay (zetaform) 4. U-shaped bay 5. Cuspate foreland 6. Spit	3. Sheltered in upcoast curve; increasing exposure to downcoast straight sector 4. Can be at head of bay; often decreasing sand toward the limbs of bay 5. Only on large foreland; need to determine stability of convexity 6. Only on large stable spits; river can be integrated into design; adequate setback from channel to avoid flooding
	7. River mouth associated with above	7. To be avoided because of rapid changes and seasonal closure of mouths, especially of small streams
C. Barrier coast	With/without dunes	Can be on barrier; but behind active dunes. Maintain seaward line of dunes as buffer zone. Lagoon can be integrated into design
D. Small island	As in A and B, where applicable.	Wider choice on many types of beaches on sheltered side. Limited beaches on exposed side. Access is important, especially during seasonal weather
E. Coral island	1. With/without lagoon 2. Cay	1. Adequate setback required avoid destruction of reefs 2. Adequate setback required. Strong seasonal changes; no dredging or structures interfering with sand movement
F. Mangrove coast		On piles to minimize impact on ecosystem. Tidal range can be critical factor in accessibility and jetty length
G. Developed coast	1. Original sandy coast 2. Original low rock coast	1. Sufficient setback from protection measures. Beach nourishment required 2. Selective removal or rock to create bays, artificial beaches. Beach nourishment required. Also raised sandy platform with coastal protection

Sources: Revised from Wong, 1990, 1991, 1999, 2000.

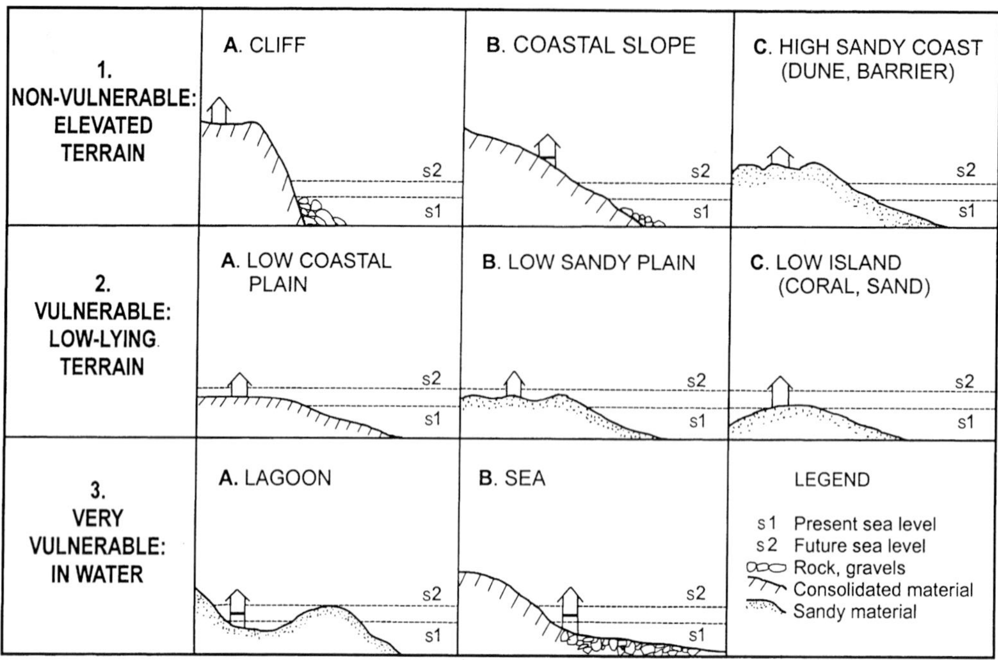

Figure T49 Types of resort sites affected by a sea level rise (redrawn from Teh, 2000).

These four types of sites can also be found on the islands which may provide a change of wind and sun orientation not possible with the mainland coastal site. However, access to islands is an important factor in construction and operation.

Based mainly on field examples from Southeast Asia and Indian Ocean islands a classification of resort sites is proposed for tropical coasts (Table T16). Where possible, the type of coast or coastal form is followed by an identification of specific suitable sites. Mainland *sandy coasts (q.v)* are the most important for resort development, with various suitable sites depending on the planform and degree of protection offered by rock headlands. Small islands, especially *coral reef islands (q.v)*, are also favored for resort sites but they have fragile environments. Mangrove coasts hold a potential although the vegetation is being cleared cut down for other uses. Dunes are limited on tropical coasts and lagoons are actively used for local fishing. Stream mouths can be integrated into the resort site but requires adequate setback from possible flooding. Except for selective headlands, rock coasts (see *rocky coasts*) are underutilized for resorts. High rock coasts offer good views but require careful design and engineering. Low limestone coasts can be altered variously for resort sites, as evident on Mactan Island, where the modification can be limited or minimal (e.g., short seawalls, stone bunds), localized (e.g., groins, breakwaters, artificial islands) or effective (e.g., artificial beaches, selective excavation of rock to form small bays) (Wong, 1999).

With a *sea-level rise (q.v.)* in the future, many resorts on low-lying sandy plains or beach-ridge plains will be threatened as the coastline is cut back by the rising sea. For small islands dependent on tourism, the situation can be serious. Perhaps the most badly hit would be the island resorts of the Maldives, where the height of cays are within a couple of meters of the sea level and the entire island resort industry can be wiped out (Domroes, 2001). Mauritius has projected its loss of tourist beaches and in the major tourist area of Flic-en-Flac, an estimated 26,000 m^2 of beaches could be flooded by a meter sea-level rise (Mauritius National Climate Committee, 1998, p. 45). The response to the threat of rising sea level can be adaptive strategies classified as (managed) retreat, accommodation, and protection. Although various hard structural and soft structural options are considered for protection, new methodologies are being sought to estimate coastal vulnerability and resilience to the sea-level rise. For some low islands in the Pacific, coastal types can be used to assess vulnerability if further information is available. For example, while the upper shore can be of sand or shingle or a mixture of both, the lower shore can be a conglomerate platform, ramp outcrop, beachrock, mangroves, or seawall. Such details would be useful for assessing the vulnerability of a resort coast (Mimura and Harasawa, 2000).

Except for some island resorts, the vulnerability of resort sites to a sea-level rise has not been fully appreciated. In Malaysia, Teh (2000) has classified coastal resort sites according to their vulnerability to inundation arising from a sea-level rise. The sites can be (1) non-vulnerable (on a coastal slope or a high beach-ridge plain), (2) vulnerable (on a low coastal plain, low beach-ridge plain, or low island of coral, sand, or mud), and (3) highly vulnerable (in a lagoon or built over the sea). An additional coastal type (cliff) is added to this classification to cover other coastal types in Southeast Asia and the Indian Ocean islands (Figure T49). Compared with the classification of coasts for resort sites, this classification emphasizes the sea level relative to the two-dimensional coastal profile and much of the variety of coastal types and landforms is consequently made redundant. Nevertheless, it is an initial step in providing some idea of the vulnerability of resort sites to a sea-level rise.

Conclusion

Although many factors have to be considered in coastal sites for tourism, geomorphology remains basic as long as the coast is a necessary resource for tourism. Site analysis has to assess various types of coasts and its various zones. As coastal tourism caters to a widening demand, coastal sites for resorts are also being assessed for their suitability for other demands such as golf courses, marinas, oyster culture, and other related types of development. In the future, resort sites are likely to extend beyond the usual sandy coast and rock headland and include more adaptive use of the mangrove coast which has a potential for coastal ecotourism. Coastal sites are also likely to incorporate better technology not only for resort construction and infrastructure but also for coastal protection against erosion and beach management.

P.P. Wong

Bibliography

Ahmad, Y.F., 1982. *Environmental Guidelines for Coastal Tourism*. Nairobi: UNEP Environmental Management Guidelines, No. 6.

Baud-Bovy, M., and Lawson, F., 1998. *Tourism and Recreation: Handbook of Planning and Design*. Oxford: Architectural Press.

Beer, A.R., 1990. *Environmental Planning for Site Development*. London: E & FN Spon.

Bird, E., 1993. Geomorphological aspects of surfing in Victoria, Australia. In Wong, P.P. (ed.), *Tourism vs Environment: the Case for Coastal Areas*. Dordrecht: Kluwer Academic Press, pp. 11–18.

Bird, E., 2000. *Coastal Geomorphology: An Introduction*. Chichester: John Wiley, Chichester.

Domores, M., 2001. Conceptualizing state-controlled resort islands for an environment-friendly development of tourism: the Maldivian experience. *Singapore Journal of Tropical Geography*, **22**: 122–137.

Gunn, C.A., 1988. *Vacationscape: Designing Tourist Regions*, 2nd edn. New York: Van Nostrand Reinhold.
Inskeep, E., 1991. *Tourism Planning : An Integrated and Sustainable Development Approach*. New York: Van Nostrand Reinhold.
Mauritius National Climate Committee, 1998. *A Climate Change Action Plan*. Mauritius.
Mieczkowski, Z., 1990. *World Trends in Tourism and Recreation*. New York: Peter Lang.
Mimura, N., and Harasawa, H. (eds.), 2000. *Data Book of Sea-Level Rise*. Ibaraki: Centre for Global Environmental Research.
Nordstrom, K.F., 2000. *Beaches and Dunes of Developed Coasts*. Cambridge: Cambridge University Press.
Pearce, D., 1995. *Tourism Today: A Geographical Analysis*. 2nd edn. Harlow: Longman Scientific & Technical.
Pearce, D.G., and Kirk, R.M., 1986. Carrying capacities for coastal tourism. *UNEP Industry and Environment*, **9**(1): 3–7.
Pickering, H., 1996. Limitations for coastal recreation. In Goodhead, T., and Johnson, D. (eds.), *Coastal Recreation Management*. London: E & FN Spon, pp. 69–91.
Romeril, M., 1984. Coastal tourism—the experience of Great Britain. *UNEP Industry and Environment*, **7**(1): 4–7.
Short, A.D., 1999. Beach hazards and safety. In Short, A.D. (ed.), *Handbook of Beach and Shoreface Morphodynamics*. Chichester: John Wiley, pp. 293–304.
Teh, T.S., 2000. Sea level rise implications for coastal and island resorts. In Teh, T.S. (ed.), *Islands of Malaysia: Issues and Challenges*. Kuala Lumpur: University of Malaya, pp. 297–317.
Tourism Authority of Thailand, 1985. *Master Plan for Tourism Development of Ko Samui/Surat Thani*. Bangkok: Thailand Institute of Scientific and Technological Research. Unpublished Agency Report.
Tourism Authority of Thailand, 1989. *The Study on Potential Tourism Development for the Southern Region of Thailand. Tourism Data : Phuket, Surat Thani/Ko Samui, Songkhla/Hat Yai. Final Report, Vol. 3*. Tokyo: Japan International Cooperation Agency. Unpublished Agency Report.
Tourist Development Corporation, Malaysia, 1979. *Malaysia's East Coast: Master Plan Study for the Development of Tourism in the areas of Merang-Besut-Dalam Phu, Rompin-Endau-Mersing*. Los Angeles: Pannell Kerr Foster. Unpublished Agency Report.
Turner, L., and Ash, J., 1976. *The Golden Hordes: International Tourism and the Pleasure Periphery*. New York: St. Martin's Press.
UNEP/WTO/FEEE (United Nations Environment Programme/World Tourism Organization/Foundation for Environmental Education in Europe), 1996. *Awards for Improving the Coastal Environment: the Example of the Blue Flag*. Paris: UNEP.
Viles, H., and Spencer, T., 1995. *Coastal Problems: Geomorphology, Ecology and Society at the Coast*. London: Edward Arnold.
Walton, J.K., 1997. The seaside resorts of Western Europe, 1950–1939. In Fisher, S. (ed.), *Recreation and the Sea*. Exeter: University of Exeter Press, pp. 36–56.
Wong, P.P., 1990. The geomorphological basis of beach resort sites—some Malaysian examples. *Ocean and Shoreline Management*, **13**: 127–147.
Wong, P.P., 1991. *Coastal Tourism in Southeast Asia*. Manila: International Center for Living Aquatic Resources Management.
Wong, P.P., 1999. Adaptive use of a rock coast for tourism—Mactan Island, Philippines. *Tourism Geographies*, **1**: 226–243.
Wong, P.P., 2000. Coastal tourism in Southeast Asia: research from the environmental perspective. In Chon, K.S. (ed.), *Tourism in Southeast Asia: A New Direction*. New York: Haworth Hospitality Press, pp. 107–121.

Cross-references

Bay Beaches
Coral Reef Islands
Classification of Coasts (see Holocene Coastal Geomorphology)
Cliffed Coasts
Headland-Bay Beach
Human Impact on Coasts
Indian Ocean Islands, Coastal Ecology and Geomorphology
Lifesaving and Beach Safety
Natural Hazards
Pacific Ocean Islands, Coastal Geomorphology
Rating Beaches
Rock Coasts Processes
Sandy Coasts
Sea-Level Rise, Effect
Setbacks
Shore Protection Structures
Small Islands
Surfing
Tourism and Coastal Development
Water Quality

TRACERS

Introduction

Tracers are essentially sediment particles that can be easily identified within a large mass of grains having different characteristics. The concept is widely used in sedimentary petrography, where particular mineral assemblages are inherited from the characteristics of the provenance basin. Another approach is to use sediment from the environment to be studied and tag it using an artificial agent (e.g., paint or radioactivity).

Since the early stages of research on sediment transport, it became evident that natural and artificial tracers had a high potential, both for qualitative and quantitative assessments. The first examples of tracing experiments using artificial materials date back to the beginning of the 20th century, with the experiments of Richardson at Chesil Beach (UK) in 1902 and the reports of the Royal Commission on Coast Erosion published in 1907 (in Kidson and Carr, 1971). Although the last 20 years have seen the development of alternative methodologies for measuring sediment transport (see entry on *Instrumentation*), it can be noted that tracers are the only technique that can be applied at a broad range of temporal and spatial scales. Although both the sandy and the coarser fractions have been the object of tracer studies in sedimentary research, sand size material is generally used. A comprehensive review of tracer studies using pebbles can be found in Kidson and Carr (1971), while recent field studies and applications are described in Cooper *et al.* (1996).

Historical and technical development of the tracer technique

Fluorescent tracers

From the 1950s onwards, a group of investigators in the former Soviet Union started to use sands marked with fluorescent paints on a large scale (Zenkovitch, 1960; Zenkovitch and Boldyrev, 1965). The methodology consisted in marking sand grains with a fine film of a colloid (agar-agar) containing fluorescent material. The film was resistant to water, chemical reactions, and mechanical abrasion. The marked sand was then dried and injected on the beach. Thereafter, at regular time intervals (e.g., from minutes to days), samples were collected at fixed distances from the injection point. The research team also experimented using different colors of marked grains, to be injected at different water depths in the nearshore or for different grain sizes to assess differential transport. Almost at the same time, Portuguese investigators were testing the technique using a slightly different approach. Abecassis *et al.* (1962) tried initially to use heavy minerals and grain size trends as indicators of transport, but with scarce results, and therefore decided to employ sand marking either with fluorescent paints or using artificially tagged radioactive sands. The first method did not give good results, since natural luminescence was present in the sands. The methodology employed using fluorescent tracers was similar to that of the Russian team and they injected and collected samples on the foreshore.

In the following years, there was a boom in the use of fluorescent tracers for the study of sediment transport and investigators in the United States started to experiment with this technique. Yasso (1965) studied selective transport of different grain sizes in New Jersey; he used different colors of fluorescent paint, using a mixture of acrylic lacquer and beetle resin. Ingle (1966) undertook a study on beaches in southern California where sand samples were collected along profiles orthogonal to the coastline spacing from the foreshore to the inshore areas. He tested five marking techniques that are described in his book, probably the only published manual on the topic. Boon (1968) undertook one of the few studies that concentrated on carbonate beaches, in this case in the Bahamas, Florida. Three sand populations were considered, with different grain sizes and shapes, and marked using different colors, using fluorescent lacquer. Boon (1970) also applied his expertise on a beach in Virginia, where sand was collected from the local dunes and sieved through a series of meshes to obtain a population with a range of

0.29–0.59 mm. The sand was painted in large plastic bags, where it was mixed with acrylic lacquer, toluene, and beetle-resin.

In the 1970s, the application of tracers to measure longshore drift (see entry on *Longshore Sediment Transport*) became widespread. Empirical measurement of longshore transport to calibrate mathematical predictors for engineering studies was needed. Komar and Inman (1970) carried out several experiments on the western coast of the United States to build a database that eventually lead to their famous transport predictor. Another example of tracer application within the context of a coastal engineering project (see entry on *Engineering Applications of Coastal Geomorphology*) is that of Allen and Nordstrom (1977). In a multidisciplinary study for the assessment of beach form changes within a groin field (see entry on *Shore Protection Structures*) in New Jersey, the authors studied sediment exchange between the foreshore, the surf zone, and the bar. One of the largest experiments ever carried out on beaches is that of Chapman and Smith (1977) on the Gold Coast of Australia. Fifty tons of sand were marked and continuously injected using a dredge for about three weeks. Because the sand had a high degree of natural luminescence, the use of fluorescent paints was ruled out and a blue dye was used. Radioactive tracers were also excluded because of the large quantities involved and the subsequent environmental impact.

The fluorescent marking technique has become of interest to many European researchers in the 1990s, with the development of experiments in France (Corbau *et al.*, 1994; Pedreros *et al.*, 1996), Belgium (Voulgaris *et al.*, 1998), and Portugal (Taborda *et al.*, 1994; Ciavola *et al.*, 1997a, 1998). These experiments introduced several improvements to the older methodology, in particular regarding the automatic detection of marked sand grains and the study of tracer advection in three dimensions.

Radioactive tracers

Despite the fact that radioactive tracers were at first very popular, they never became widely used, since the marking technique is complex and the method has a considerable environmental impact. Trying to compare results obtained using at the same time fluorescent and radioactive tracers, Abecassis *et al.* (1962) had already experimented with isotopes ^{110}Ag and ^{32}P, obtaining the best results with the latter. Two noteworthy field experiments using radioactive tracers were carried out by Heathershaw and Carr (1977) to assess tidally induced transport in the Severn Estuary (UK). The tracer was produced by neutron irradiation of Scandium (^{46}Sc) into glass. The glass was then sieved to obtain a population with different size classes. The detector consisted of a sodium iodide (NaI) crystal, doped with Thallium (TI) and optically coupled to a photomultiplier tube. This assembly, together with electronic components, was housed in a waterproof brass case. Measurements were made by lowering the counter on the seabed and readings were transmitted to a ship.

Clearly, the radioactive method offers many advantages such as the small quantities required, the possibility of tagging several grain size populations, the fact that the hydraulic properties of the grains are left untouched. However, although weak radioactive isotopes are used, it is most unlikely that nowadays an application for this method could be approved in the context of an environmental impact assessment.

Essential concepts for applying the fluorescent tracer method

Applicability and assumptions

The method must fulfill some basic assumptions: the marked sands should have a hydraulic behavior comparable to the unmarked ones; *advection* of the tracers should be prevalent over *diffusion* and *dispersion*; the transport system must be in equilibrium. Madsen (1987) presents an exhaustive mathematical description of the terms advection, diffusion, and dispersion.

It is no surprise that many investigators have used this method for studying sediment transport. The usage of fluorescent sands is simple, marking can be done easily and rapidly, and there is no impact on human health and the environment. Different colors can be used to tag various grain fractions, to study differential transport, and the sensitivity of the technique is on the order of 1 ppm (Ingle, 1966). The most interesting experiment on differential transport is that described by Komar (1977) on a reflective shoreface (see entry on *Reflective Beaches*) in Baja California, Mexico. A single injection of tracer took place and collection of over 200 samples was undertaken along a grid extending more than 200 m alongshore. The samples were sieved into different size fractions and the number of marked grains within each class was annotated. Later Blackley and Heathershaw (1982) revisited the issue and, in agreement with the previous author, reached the conclusion that different grain size transport was related to the transport mechanism, for example, the relationship between bedload and suspended load at a point on the foreshore. Allen and Nordstrom (1977) even used different colors to examine spatial changes of sediment exchange between the foreshore and the bar, albeit in a qualitative way.

Preparation of the tracers

Regarding this first stage of the method, an essential factor is the quantity of tracer sand to be produced. For large quantities (e.g., 1,000 kg or more), industrial preparation is probably the only feasible method, trying to use sediments with a mean grain size and density comparable to those of the natural environment. In any case, the best balance between handling during tagging and percentage of recovery during the experiment should be met. If sand is collected from the local beach, a composite sample could be obtained by mixing equal subsamples collected on the lower-, mid-, and upper-shoreface. It is important to carry out grain size analyses (see entry on *Beach Sediment Characteristics*) of each sub-environment and of the composite distribution, to compare the effects of mixing. Care should be taken to avoid local sedimentary effects due, for example, to the presence of large-scale bedforms or moribund beach cusps (see entry on *Rhythmic Patterns*).

The sand is then washed with freshwater to remove the salt and dried. Regarding the marking procedure, a number of different types of resins and paints can be used. The main requirement for the paint is to be resistant to chemical and mechanical abrasion caused by seawater. Initially, substances such as seaweed glue, bone-glue, gum, and starch were used but many authors later switched to acrylic paints, which offer better resistance to abrasion. If natural luminescence of the sediment is a problem, the solution of Chapman and Smith (1977), a blue dye dissolved in methylated spirit, can be adopted. Knoth and Nummedal (1977) marked the sand using a mixture of fluorescent paint, resin, and organic solvent. In recent experiments, fluorescent paint soluble in toluene has been widely used (Corbau *et al.*, 1994; Ciavola *et al.*, 1997a, 1998).

The next step is to dry the sand, minimizing the effects of aggregation. As far as the paint is maintained as thin as possible, aggregation should generally be negligible, especially if the sand is rapidly dried in the open air under sunlight. Even in the case of aggregate formation, if the sand is sieved with appropriate meshes after tagging, the mode of the populations should remain the same.

Injection of the tracer into the transport system

Injection of the sand into the transport system takes place according to the type of study being carried out. As Madsen (1987) concluded, it is possible to adopt three different techniques to analyze tracer advection. The *Time Integrated Method* (TIM) has an Eulerian nature, whereby a known quantity of tracer is released at a point and variations in tracer concentration are monitored throughout time at a location downdrift. The *Continuous Injection Method* (CIM) is similar to the previous one, with the exception that injection of the tracer is continuous at a known rate. The *Spatial Integration Method* (SIM) involves a Lagrangian approach, since it allows monitoring of tracer movement both in space and in time. Other advantages of the SIM over the previous methods is that the velocity of transport is computed referring it to the *centroid* (or center of mass) of the tracer cloud and that it is possible to calculate with confidence a recovery rate.

In the case of continuous injection on the submerged beach outside the breakers, dumping from a vessel is a possible method, while in the case of working in the breakers this may be unfeasible in terms of safety of the boat. For this reason, researchers using the TIM and CIM methods preferred injecting the tracer from several points along a transect orthogonal to the beach (Yasso, 1965; Ingle, 1966; Boon, 1970). Other experimentalists, mainly using the SIM method, placed the tracer at low tide into a shallow trench dug on the beach face, after washing it with liquid soap to avoid grain floating. There is no standard for the size and depth of the trench: the most important requirement is that all the tracer grains are removed almost instantaneously as soon as the site is covered by water. A delay in removal or burial of the tracer at the trench site could invalidate the basic assumptions of the method.

Sediment sampling

Despite the fact that many authors only worked at surface level (e.g., Ingle, 1966; Corbau *et al.*, 1994; Taborda *et al.*, 1994) it is preferable to collect shallow cores to assess the three-dimensional (3-D) advection of the tracer (e.g., Boon, 1970; Kraus *et al.*, 1982; Ciavola *et al.*, 1997a, 1998). PVC cores or other simple coring techniques are suitable for this

Figure T50 The SADAM system of Pinto *et al.* (1994) for *in situ* mapping of surface dispersal of fluorescent grains. It consists of an ATV, a dark camera with UV light, and computer with image processing software.

purpose. Inman *et al.* (1980) collected their samples according to the type of analysis to be performed: the first technique was employing a grab sampler, especially designed to collect only the first two centimeters of the surface layer and was used to carry out a spatially distributed sampling to apply the SIM. The second method was using corers that were collecting samples along a cross-shore profile at constant time intervals, for the TIM method. Duane and James (1980) adopted a similar approach for the CIM technique. Sampling was frequent along four profiles, to obtain a description of changes in concentration with time at each station. Samples were collected using a hand-held sloop, built on broad runners, that was penetrating the sand column no more than 1 cm, to avoid sampling at depths where the sand concentration was already in equilibrium. If sampling takes place on submerged sites, vaseline coated cards like those of Ingle (1966) can be used.

In case it is decided to assess surface distribution, to increase the significance of the results from the cores, it is possible to use automatic detection techniques like the system of Pinto *et al.* (1994), where a computer system in installed on an ATV (All Terrain Vehicle), to perform rapid assessments using image analysis techniques (Figure T50).

Dyed grain detection

Counting of the grains in the samples under a UV lamp can be tedious and involve several days of work. Whenever the number of tracer grains in a sample is high, automatic detection or indirect detection methods can be helpful, once the considerable calibration difficulties are overcome. In the study of Chapman and Smith (1977), where the dye was not fluorescent, counting took place following two different methodologies: photography and subsequent enlargement; affixing the samples to long strips of pressure-sensitive adhesive tape and passing the strips under a binocular microscope at an enlargement of 30–40 times to detect the grains. In both cases, the counting was undertaken using an automatic image analyzer.

At the very beginning of tracer research, Zenkovitch (1960) and Zenkovitch and Boldyrev (1965) examined the samples in the laboratory using a luminoscope. Such a piece of equipment used a UV source that stimulated fluorescence from the grains, similarly to the photometer that Yasso (1966) developed to count the number of marked grains directly on the beach. The spectrofluorimetric of Farinato and Kraus (1981) was instead based on a more original approach: the technique consisted in washing the sand samples with a solvent able to dissolve the resin and the paint covering the grains. The fluorescence of the effluent can then be measured using a spectrophotometer: the measured intensity is proportional to the number of tagged grains, the solution transmission, and the uniformity of the paint cover. The calibration of the intensity coefficient versus the number of marked grains per gram has to be undertaken according to the degree of marking, the grain size distribution, the chemical properties of the solvent and of the resin. For the method to work, all sand grains should be almost equi-dimensional: certainly in the case of a uni-modal, well-sorted sand that can be true, but as sorting becomes worse, the method looses significance. Besides, the technique assumes that the paint cover of the grains is uniform; probably this might be possible if industrially done, but since most authors mark the sands themselves to save money and time, especially when small quantities are involved, variations in color thickness must be expected.

Controversies and gaps in current knowledge

A simple conceptual model of longshore transport describes the system as two layered. The upper layer is where movement of the tracer occurs as advection longshore, while in the lower layer the sand is not activated. In order to calculate the volume of transported sand, it is necessary to calculate the *mixing depth* (see entry on *Depth of Disturbance*), that is considered equivalent to the thickness of the moving layer described above. Within this layer, grains are subjected to vertical and lateral movements caused by direct wave action and by the superimposed longshore current. There are different methods in the literature for determining the thickness of the mixing layer. While in many experiments it was simply determined in the field by digging one or more control holes, filled with marked sand, a different methodology was generally used in parallel with the previous one, considering as mixing depth the interval within the beach cores where 80% of the total mass recovered in each sample was observed (Kraus *et al.*, 1982; Kraus, 1985; Ciavola *et al.*, 1997b, 1998).

Workers using a detector at surface level may be confronted with the problem of assessing indirectly the burial depth of the tracer. Heathershaw and Carr (1977) estimated it using the *tracer balance method*, assuming that a reduction in the recovery rate of the tracer is

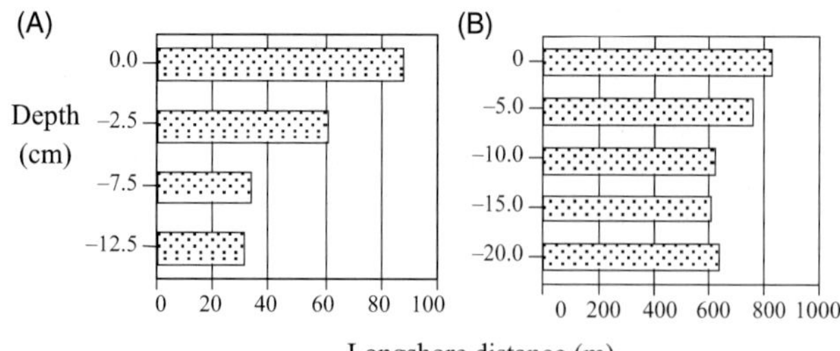

Figure T51 Vertical patterns of advection of tracer clouds from shallow beach cores: data in figure (A) was collected by Ciavola *et al.* (1997a); data in figure (B) was collected by Ciavola *et al.* (1998).

simply due to burial, and that burial can be estimated using a response function typical of the detector used. The method assumes that the tracer is uniformly distributed down to the depth of maximum burial; several field evidences points out that this might not always be true, unless total mixing is achieved, which implies that the tracer had enough time to reach equilibrium conditions. From the diagrams in Figure T51 (data from Ciavola *et al.*, 1997a, 1998), it is possible to notice that the displacement of the center of mass of the tracers, and consequently its velocity, decreases with depth, becoming almost constant in proximity of the maximum depth of activation. When there is an anomalous variation with depth in the speed of displacement, this is probably related to tracer burial. In this case, it might be necessary to reject some of the core samples: simple surface studies could therefore lead to an overestimation of the speed of displacement of the center of the tracer cloud, ultimately causing a miscalculation of the volume of sediment being transported. In experiments where the tracer sampling was extended to several days (e.g., Heathershaw and Carr, 1977; Ciavola *et al.*, 1997a), it was found that as time passed by, total mixing was taking place, with the whole sediment column moving almost at the same speed.

Another limitation of the method is the calculation of a significant *percentage of recovery* of the tracer. Recent experiments (Sherman *et al.*, 1994; Ciavola *et al.*, 1998) used grain size statistics of the tagged sediment to derive approximations for concentrations, based upon assumptions of uniform density and diameter. Estimates of total tracer recovery were made by extrapolating the point concentrations to representative control volumes, by constructing surface polygons and using tracer-depth distributions obtained from recovery depths.

Conclusions

Recently, many investigators throughout the world have been revisiting the tracer technique, especially for field measurements of longshore transport. *Tracer advection* can be measured on a 3-D basis; it is considered essential that investigators adopt such an approach if they want to obtain meaningful observations.

Most field experiments have been confined to small- to medium-scale beach studies, but tracers can be employed successfully to study transport along sandbanks and on continental shelves. In a similar fashion, the timescale of these experiments tends to be short (hours to days), because of the small quantities of tracer injected. In the past, long-term monitoring was undertaken using radioactive tracers, where a small quantity (e.g., less than 1 kg), provides enough material for a monthly exercise; however, possible negative impacts on the natural environment rule out this method. In the case of fluorescent tracers, only the usage of thousands of kilos of sand will ensure a significant recovery rate at the end of a long-term experiment. Large quantities mixed with sand during a beach recharge scheme (see entry on *Beach Nourishment*) could, for example, provide coastal managers with an independent method of assessing *where the sand is going* during monitoring activities, in addition to data for the calibration of longshore drift predictors, to decide whether further intervention is needed.

The main limitations of the method are the time and the effort required for data collection and analysis. Recent applications of automatic analyzer, particularly using image analysis methods, suffered from the cost and availability of high-resolution systems. With the recent boom of digital cameras, the power of image analysis has increased in parallel with a decrease in the costs involved, therefore a rapid development of automatic methods for grain detection and counting is likely to happen in the years to come.

Paolo Ciavola

Bibliography

Abecassis, F., Matias, M.F., Reis de Carvalho, J.J., and Vera-Cruz, D., 1962. Methods of determining sand and silt movement along the coast, in estuaries and in maritime rivers. *Laboratório Nacional de Engenharia Civil Technical Paper No. 186*. Lisbon, Portugal: Ministério das Obras Públicas.

Allen, J.R., and Nordstrom, K.F., 1977. Beach form changes in the lee of groins at Sandy Hook, New Jersey. In *Proceedings of the Coastal Sediments '77*, American Society of Civil Engineers, pp. 33–47.

Blackley, M.W.L., and Heathershaw, A.D., 1982. Wave and tidal current sorting of sand on a wide surf zone beach. *Marine Geology*, **49**: 345–356.

Boon, J.D., 1968. Trend surface analysis of sand tracer distributions on a carbonate beach, Bimini, B.W.I. *Journal of Geology*, **76**: 71–87.

Boon, J.D., 1970. Quantitative analysis of beach sand movement, Virginia Beach, Virginia. *Sedimentology*, **17**: 85–103.

Chapman, D.M., and Smith, A.W., 1977. Methodology of a large scale sand tracer experiment. In *Proceedings of the 3rd Australian Conference on Coastal and Ocean Engineering*, pp. 185–189.

Ciavola, P., Taborda, R., Ferreira, Ó., and Dias, J.A., 1997a. Field measurements of longshore sand transport and control processes on a steep meso-tidal beach in Portugal. *Journal of Coastal Research*, **13**: 1119–1129.

Ciavola, P., Taborda, R., Ferreira, Ó., and Dias, J.A., 1997b. Field observations of sand-mixing depths on steep beaches. *Marine Geology*, **141**: 147–156.

Ciavola, P., Dias, N., Ferreira, Ó., Taborda, R., and Dias, J.M.A., 1998. Fluorescent sands for measurements of longshore transport rates: a case study from Praia de Faro in southern Portugal. *Geo-Marine Letters*, **18**: 49–57.

Cooper, N.J., King, D.M., and Hooke, J.M., 1996. Collaborative research studies at Elmer Beach, West Sussex, UK. In Taussik, J., and Mitchell, J. (eds.), *Partnership in Coastal Zone Management*. Cardigan, UK: Samara Publishing Limited, pp. 369–376.

Corbau, C., Howa, H., Tessier, B., de Resseguier, A., and Chamley, H., 1994. Evaluation du transport sédimentaire sur une plage macrotidale par traçage fluorescent, Dunkerque Est, France. *Compte, Rendus Académie des Sciences de Paris*, **319**: 1503–1509.

Duane, D.B., and James, W.R., 1980. Littoral transport in the surf zone elucidated by an eulerian sediment tracer experiment. *Journal of Sedimentary Petrology*, **50**: 929–942.

Farinato, R.S., and Kraus, N.C., 1981. Spectrofluorometric determination of sand tracer concentrations. *Journal of Sedimentary Petrology*, **51**: 663–665.

Heathershaw, A.D., and Carr, A.P., 1977. Measurements of sediment transport rates using radioactive tracers. In *Proceedings of Coastal Sediments '77*, American Society of Civil Engineers, pp. 399–416.

Ingle, J.C., 1966. *The Movement of Beach Sand*. Amsterdam: Elsevier.

Inman, D.L., Zampol, J.A., White, T.E., Hanes, D.M., Waldorf, B.W., and Kastens, K.A., 1980. Field measurements of sand motion in the surf zone. In *Proceedings of the 17th International Coastal Engineering Conference*, American Society of Civil Engineers, pp. 1215–1234.

Kidson, C., and Carr, A.P., 1971. Marking beach materials for tracing experiments. In Steers, J.A. (ed.), *Introduction to Coastline Development*. Basingstoke, UK: MacMillan, pp. 69–93.

Knoth, J.S., and Nummedal, D., 1977. Longshore sediment transport using fluorescent tracers. In *Proceedings of Coastal Sediments '77*, American Society of Civil Engineers, pp. 383–398.

Komar, P.D., 1977. Selective longshore transport rates of different grain-size fractions within a beach. *Journal of Sedimentary Petrology*, **47**: 1444–1453.

Komar, P.D., and Inman, D.L., 1970. Longshore sand transport on beaches. *Journal of Geophysical Research*, **75**: 5514–5527.

Kraus, N.C., 1985. Field experiments on vertical mixing of sand in the surf zone. *Journal of Sedimentary Petrology*, **55**: 3–14.

Kraus, N.C., Isobe, M., Igarashi, H., Sasaki, T.O., and Horikawa, K., 1982. Field experiments on longshore sand transport in the surf zone. In *Proceedings of the 18th Coastal Engineering Conference*, American Society of Civil Engineers, pp. 970–988.

Madsen, O.S., 1987. Use of tracers in sediment transport studies. In *Proceedings of Coastal Sediments '87*, American Society of Civil Engineers, pp. 424–435.

Pedreros, R., Howa, H.L., and Michel, D., 1996. Applications of grain size trend analysis for the determination of sediment transport pathways in intertidal areas. *Marine Geology*, **135**: 35–49.

Pinto, J.R.C., Dias, J.M.A., Fernandes, S.P., Ferreira, Ó., Silva, A.V., and Taborda, R., 1994. Automatic system for tagged sand detection. *Gaia*, **8**: 161–164.

Sherman, D.J., Nordstrom, K.F., Jackson, N.L., and Allen, J.R., 1994. Sediment mixing-depths on a low-energy reflective beach. *Journal of Coastal Research*, **10**: 297–305.

Taborda, R., Ferreira, Ó., Dias, J.M.A., and Moita, P., 1994. Field observations of longshore sand transport in a high energy environment. In de Carvalho, S., and Gomes, V. (eds.), *Proceedings of Littoral 94*. Lisbon, Portugal: EUROCOAST Portugal, pp. 479–487.

Voulgaris, G., Simmonds, D., Michel, D., Howa, H., Collins, M.B., and Huntley, D.A., 1998. Measuring and modelling sediment transport on a macrotidal ridge and runnel beach: an intercomparison. *Journal of Coastal Research*, **14**: 315–330.

Yasso, W.E., 1965. Fluorescent tracer particle determination of the size-velocity relation for the foreshore sediment transport, Sandy Hook, New Jersey. *Journal of Sedimentary Petrology*, **34**: 989–993.

Yasso, W.E., 1966. Formulation and use of fluorescent tracers coatings in sediment transport studies. *Sedimentology*, **6**: 287–301.

Zenkovitch, V.P., 1960. Fluorescent substances as tracers for studying the movements of sand on the sea bed. *The Dock and Harbour Authority*, **40**: 280–283.

Zenkovitch, V.P., and Boldyrev, V.L., 1965. Alongshore sediment streams and methods of their study. In *Proceedings of the 11th Congress of the International Association for Hydraulic Research*, Volume 5. Leningrad, USSR: International Association for Hydraulic Research, pp. 139–148.

Cross-references

Beach and Nearshore Instrumentation
Beach Nourishment
Beach Sediment Characteristics
Depth of Disturbance
Engineering Applications of Coastal Geomorphology
Longshore Sediment Transport
Reflective Beaches
Rhythmic Patterns
Shore Protection Structures

TROTTOIRS

The term "trottoir," French for pavement, is a very imprecise one in coastal geomorphology, referring to a range of intertidal landforms formed by a variety of processes. At their simplest, they form a bioconstructional overhang perhaps a meter wide attached to steep rocky shores at about mid-tide level. Elsewhere, on less steeply sloping coastlines, erosional platforms tens of meters wide may be covered by a similar variety of calcareous algae, vermitids, and other bioconstructional organisms which, on the outer edge form a distinctive rim where wave agitation encourages more rapid accumulation. These forms are particularly prominent over calcareous substrates and in the tropics, where they cover platforms cut into raised reef limestones, may merge with the algal terraces of high-energy reef margins (Emery, 1962). In deeper water beyond the trottoirs, mushroom-shaped bosses constructed of the same organisms, may rise to the sea surface where their morphology may resemble that of coral micro-atolls.

Processes

Because of the range of morphologies which have been termed trottoirs, there are probably a number of processes involved in their formation. The simpler intertidal overhang or corniche may be entirely the result of bioconstruction. However, the protection of the coating of algae, vermitids, and other organisms give the underlying rock surface in comparison to the zones above and below this mid-tide level, where mechanical, chemical, and bio-erosional processes may be acting, can make the protuberance more prominent and give it a bedrock core. On wider rimmed platforms the paradox is that the bioconstructional layer overlies a platform which is obviously erosional which has led to contradictory hypotheses to explain the landforms' origin as either constructional or erosional (see e.g., Emery, 1962). Probably, both processes occur, either at different times or even simultaneously as many bio-erosional organisms including blue-green algae and sipunculid worms may be associated with or beneath the veneer of rock-building biota.

Organisms involved

The constructive or protective activity of many organisms is involved in the formation of trottoirs. The most ubiquitous are calcareous algae (e.g., *Neogoniolithon notarisii, Lithothamnium* sp., *Lithophyllum tortuosum, Tenarea tortuasa*), vermetid worms (e.g., *Dendropoma petraeum, Vermetus triqueter, Vermetus nigricans, Pomatocerus caeruleus*), Sabellariae (*Sabellaria kaiparaensis, Sabellaria vulgaris, Galeolaria caespitosa, Phragmatopoma iapidosa, Vermilia* sp., and oysters (*Crassostrea amasa, Crassostrea glomerata, Saxostriea* sp.) the exact combination of species depending on climate, exposure, and the environmental factors (Kelletat, 1989, 1995).

Location

Although the overall distribution, limiting factors, and importance in coastal processes are not fully investigated, trottoirs, *sensu lato* are found from northern Scotland to the tropics (Kelletat, 1989). A very wide range of forms has been described from the Mediterranean (e.g., Crete, Kelletat and Zimmerman, 1991). Rimmed platform types, are more common in warmer waters and have been described from the southern Mediterranean, the Caribbean, Micronesia, and Australia (e.g., Kelletat, 1989, 1995).

Rimmed platforms are best developed over limestone substrate and may occur closer to high tide levels than the corniche type of trottoirs, though still permanently wetted by spray even at low tide. Low to moderate tidal ranges therefore appear to be most appropriate for the development of all types of trottoirs. Moderate wave activity also appears favorable. Where too great, the amount of overhang may be limited by mechanical breakage.

Rate of growth

Trottoirs can accumulate rapidly, in tropical waters their growth rates being only a little less than those of coral reefs (Kelletat, 1989). In Crete, trottoirs of algal reefs have grown to a width of 10 m since the coastline was last uplifted 1,530 years ago representing lateral growth rates of more than 6 mm/yr. Marginal extension rates of more than 13 mm/yr and increase in thickness of 0.8 mm/yr have been reported from algal reefs in St. Croix (Adey and Vassar, 1975). Given sufficient stability, sea-level trottoirs have the ability to build significant coastal features.

David Hopley

Bibliography

Adey, W.H., and Vassar, J.M., 1975. Colonization, succession and growth rates of tropical crustose coralline algae (Rhodophyta, Cryptonemiales). *Phycologia*, **14**: 55–69.

Emery, K.O., 1962. Marine geology of Guam. *US Geological Survey Professional Paper* 403-B.

Kelletat, D., 1989. Zonality of rocky shores. In Bird, E.C.F., and Kelletat, D. (eds.), *Zonality of Coastal Geomorphology and Ecology*. *Essener Geographische Arbeiten*, **18**: 1–29.

Kelletat, D., 1995. Atlas of coastal geomorphology and zonality. *Journal of Coastal Research*, **13** (Special issue): 286pp.

Kelletat, D., and Zimmerman, L., 1991. Verbreitung und Formtypen rezenter und subrezenter organischer Gesteinsbildungen an den Küsten Kretas. *Essener Geographische Arbeiten*, **23**: 163p.

Cross-references

Algal Rims
Bioconstruction
Bioherms and Biostromes
Notches

TSUNAMIS

Introduction

The word "Tsunami" is derived from the Japanese meaning "Harbor Wave." Tsunamis are often described as tidal waves but this view is incorrect since they have nothing to do with tides. Tsunamis are generated by offshore earthquakes, submarine slides, and occasionally by subaerial landslides that enter water bodies. In each case, large-scale displacement of water takes place as a result of submarine sediment slides or from earthquakes that induce faulting of the seabed. The initial water movement is often characterized by a rapid drawdown and a lowering of the sea surface at the coast as the water moves into the area of seabed displacement. Thereafter, large kinematic waves are propagated outwards from the zone of seabed disturbance. The waves travel across the ocean at very high velocities, often in excess of 450 km/h, and possess very long wavelengths and periods. At the coast, the tsunami flood level (runup) is partly a function of the dimensions of the propagated waves but is greatly influenced by the topography and bathymetry of the coastal zone and as such the waves can reach considerable elevations causing widespread destruction and loss of life.

Historical accounts of tsunamis

The majority of tsunamis occur around the Pacific Ocean but many are also known from other areas. The frequency of Pacific tsunamis is due to the high occurrence of severe earthquakes under or close to the seabed as a result of the subduction of oceanic crust adjacent to continental margins. Such geological processes are a characteristic feature of Japan where there is a long history of devastating earthquakes and tsunamis. Between 1596 and 1938 the Japanese islands were struck by no less than 15 major tsunamis. One of the worst of these took place on June 15, 1896 as a result of a large submarine earthquake on the ocean floor 150 km offshore. Since the epicenter of the earthquake was located beneath the ocean floor, the inhabitants on nearby coasts, although they knew that an earthquake had taken place, were unaware that a major tsunami had been generated. The following account provides a vivid description of what followed. (Myles, 1985).

> This great Sanriku tsunami came on a festival day when the townspeople were enjoying a holiday. Twenty minutes after the first shock, the sea was seen to recede, while a little past eight in the evening, noises like that of a rainstorm were heard. The tsunami was now on them—a wall of water some tens of feet high—and the holiday revellers, before they realised the awful situation, were swept away and drowned ... the fishermen, who at the time were some distance out at sea, and had noticed nothing unusual, were on their way home the next morning, amazed to find the sea for miles strewn with house wreckage and floating corpses.

According to the Japanese Government, 10,617 houses were swept away, 2,456 houses were partly demolished, 27,122 persons killed, and 9,247 persons injured. Practically, every coastal town and village in the provinces of Mino and Owari on the Sanriku coast of Japan was destroyed.

One of the most severe earthquakes and tsunamis took place on the January 11, 1693 in eastern Sicily. This disastrous earthquake resulted in the loss of life of ca. 70,000 victims. A tsunami occurred at Catania and also at Augusta. According to reports, there were three withdrawals of the sea and three major waves. Similarly severe earthquakes and tsunamis took place in Calabria on February 5 and 6, 1783. At this time, a tremendous earthquake occurred in this area associated with five very strong quakes. Considerable stretches of the coastline of Calabria were badly affected by a tsunami and the sea was reported to have receded and then inundated the shore with recessions and inundations repeated at least three times at intervals of about 10 min. At Messina harbor, quays and buildings were flooded and in one area the sea withdrew for more than 7 m leaving the sea bottom dry and a lot of fish on the beach. Then suddenly the water came back surpassing the limit previously reached and flooding the coast. Local tsunamis were also produced as a result of a large earthquake-induced rockfall into the sea. The tsunami of February 6 was particularly disastrous because of the very high number of victims where many, frightened by the earthquake shocks, escaped to the beach and were drowned by waves which reached the roofs of harbor buildings. In excess of 1,500 people were drowned and the tsunami flood level was estimated to have been between 6 and 9 m. Tinti and Maramai (1996) observed that in one area the tsunami was associated with deposition of "... some sand on the ground."

A very strong earthquake and tsunami took place in the Messina Straits, Italy, on December 28, 1908. The towns of Messina and Reggio di Calabria were completely destroyed together with many neighboring villages. The area of destruction was about 6,000 km^2 and more than 60,000 people died. The earthquake produced a violent tsunami in the Straits of Messina that caused severe damage and many victims. In all places the first observed movement was the withdrawal of the sea (in some places by about 200 m). Thereafter coastal flooding took place in association with at least three large waves. According to Tinti and Maramai (1996) the tsunami lasted many hours and reached its maximum intensity along parts of the Calabrian coast and on the coast of Sicily. In some localities, the biggest wave was the first while at others it was the second. Tsunami runup was observed to decrease for increasing distances away from the epicenter but in the Messina Straits this was obscured by the effects of wave resonance. At Messina the wave height reached 3 m, numerous boats were damaged, harbor quays were destroyed, walls collapsed, and several boats were transported onshore. In certain areas the maximum level reached by the tsunami was in excess of 10 m above contemporary sea levels, resulting in the destruction of many buildings and considerable drowning (Tinti and Maramai, 1996).

Probably, the most destructive tsunami in Europe during historical times took place on November 1, 1755. An earthquake took place offshore ca. 200 km WSW of Cape St. Vincent, on the Gorringe Bank on the seafloor west of Portugal and attained a magnitude estimated at 8.5 Ms. The epicenter of the earthquake was in an area along the Azores–Gibraltar plate boundary that forms the western part of the lithosphere boundary between the Eurasian and African plates (Moreira, 1985). The eastern section of the Azores–Gibraltar plate boundary (which includes the Gorringe Bank) is a zone of active plate compression and in this area faults tend to have a large source component that results in high-magnitude and deep-seated tsunamigenic earthquakes).

The considerable destruction that took place in Lisbon, in addition to widespread fires, was mostly attributable to three tsunami waves estimated to be between 5 and 13 m high that took the lives of 60,000 people in Portugal alone. There are also numerous reports of tsunami flooding and fatalities on a large-scale along the Algarve coast and on the coastline of Morocco. In England, contemporary observations by Borlase (1755, 1758) describe the arrival of the tsunami in Mounts Bay, Cornwall. Borlase noted

> ... the first and second refluxes were not so violent as the 3rd and 4th (tsunami waves) at which time the sea was as rapid as that of a mill-stream descending to an undershot wheel and the rebounds of the sea continued in their full-fury for fully 2 hours ... alternatively rising and falling, each retreat and advance nearly of the space of 10 minutes until 5 and a half hours after it began.

Reconstructed tidal changes for this day for the Isles of Scilly show that the time of high tide coincided approximately with the arrival of the first tsunami wave some 5 h after the first shocks were reported on the Portuguese coast. There are no known reports of the progress of the tsunami northeast along the Channel but it is reasoned here that the coastal flooding effects must have been considerable. There is some evidence to indicate that the 1,755 Lisbon tsunami was not solely caused by a seabed fault. Recently, a large turbidite/submarine slide complex has been identified on the seafloor adjacent to the Gorringe Bank and tentatively dated to AD 1755. This discovery raises the possibility that the tsunami was partly generated by an earthquake-triggered fault on the seabed and partly by submarine sediment slumping.

Geological evidence for tsunamis in prehistory

Historical accounts of former tsunamis have particular value since they can provide information on the frequency and magnitude of events for the time period for which historical records are available. For example, in Italy the oldest historical record of a tsunami having taken place is for the AD 79 eruption of Vesuvius. Since numerous tsunamis have taken place along the coastline of Italy since then, historical accounts are a particularly valuable archive that can be used to estimate tsunami (and earthquake) recurrence. In recent years, however, geological investigations have been used to identify former tsunamis that took place in prehistory. This has proved possible owing to the recognition that, in many cases, tsunamis deposit sediment in the coastal zone (Dawson and Shi, 2000). Identification of such sediment layers in coastal sediment sequences has led to a different perspective on the past frequency and magnitude of tsunami events in different coastal regions.

Geological investigations of former tsunamis is a relatively new research area. The recognition that many tsunamis deposit sediment in the coastal zone has only become an accepted idea during the last 5–10 years. Discussion of this concept has been accompanied by a proliferation of academic papers that have described a range of sediments that have been attributed to a series of former tsunamis (Figure T52).

Unlike storm surges, tsunami runup across the coastal zone is frequently associated with the rapid lateral translation of water and suspended sediment. Thus, tsunami deposits can be used to provide an indirect record of former offshore earthquakes and underwater landslides. It is exceptionally difficult, however, if not impossible, to

Figure T52 Sheet of sand and overlying boulders, Boco do Rio, Algarve, Portugal deposited by tsunami that accompanied the Great Lisbon earthquake of November 1, AD 1755.

differentiate tsunami deposits attributable to former submarine slides or offshore earthquakes. In particular areas of the world, especially in areas of an active plate motion where an offshore earthquake has taken place, it may be a gross oversimplification to attribute the triggering mechanism solely to earthquake-induced seabed faulting. Frequently, an offshore earthquake may also generate local submarine slides thus leading to complex patterns of tsunami flooding at the coast. In other areas (e.g., Hawaiian Islands, Norwegian Sea) submarine sediment slides may be the dominant mechanism of tsunami generation.

Tsunami deposits are distinctive. They are frequently associated with the deposition of continuous and discontinuous sediment sheets across large areas of the coastal zone. Frequently they consist of deposits of sand containing isolated boulders. On occasion, such boulders exhibit evidence of having been transported inland from the nearshore zone. In addition, microfossil assemblages of diatoms and foraminifera contained within sand sheets may provide information of onshore transport of sediment from deeper water.

Field observations of tsunami flooding usually describe the rapid lateral translation of water across the coastal zone. Frequently, the lateral water motion associated with runup is influenced by local wave resonance. Thus, the tsunami waves as they strike the coast are unlike waves associated with storm surges since not only are they associated with considerably greater wavelengths and wave periods, but they are essentially constructive as they move inland across the coastal zone. The rapid water velocities (provided that there is an adequate supply of sediment in the nearshore zone), are in most cases associated with the transport of a variety of grain size ranging from silt to boulders. Unlike storm surges individual tsunami waves reach a point of zero water velocity prior to backwash flow. At this point, large volumes of sediment may be deposited out of the water column onto the ground surface. Young and Bryant (1992) have made reference to isolated boulders in tsunami deposits in Southeast Australia. In that area, thicknesses of massive sands and silts include occasional isolated boulders, described by Young and Bryant (1992) as "boulder floats".

One of the most awkward problems in reconstructing chronologies of former tsunamis for different areas of the world is how to be able to distinguish tsunami deposits from sediments deposited as a result of hurricane-induced storm surges. For example, in coastal Alabama, USA, a series of hurricanes during historical time have resulted in the deposition of multiple sand layers in low-lying coastal wetlands. While it is accepted that storm surges result in the deposition of discrete sedimentary units, it is argued that tsunamis in contrast to storm surges, generally result in deposition of sediment sheets, often continuous over relatively wide areas and considerable distances inland. For example, sediment sheets in the Algarve, Portugal, associated with the Lisbon earthquake tsunami of 1,755 occur in excess of 1 km inland. In addition, it may be the case that tsunami deposits contain distinctive microfossil assemblages that can be differentiated from those produced by storm surges (Dawson and Shi, 2000).

Tsunamis and submarine slides

Along the coasts of the northern North Sea, Norwegian Sea, and northeastern Atlantic Ocean, a very prominent sand layer, first thought to have been deposited by a storm surge, has more recently been attributed to a large tsunami ca. 7,100 ^{14}C years ago (Figure T53). This event was generated in the Norwegian Sea as a result of the Second Storegga submarine slide. The widespread deposit is now regarded as a marker horizon against which to compare the age of related deposits and with which to more closely define patterns of land uplift.

The detail with which the tsunami is known is impressive. Several studies have examined the sedimentology of the deposit, including its particle size, microfossil content, and even the time of year it occurred. The diatom ecology of the layer has been examined and over 100 ^{14}C dates on biogenic material both within the layer and from adjacent horizons have been obtained.

Geological investigations of these tsunami deposits on the northern and eastern coastlines of Scotland as well as in uplifted lake basins along the west coast of Norway provide evidence of minimum tsunami runup. In eastern Scotland, the minimum value of runup associated with this tsunami is on the order of 4–6 m above contemporary high water mark. However, this value as stated above, should be treated with caution since tsunami flooding to higher elevations may have taken place yet did not leave a sedimentary record. Harbitz (1992) attempted to develop a numerical model of the Second Storegga submarine slide. He noted that the scale of tsunami runup along the Scottish and Norwegian coastlines very much depended upon the average landslide velocity that was used into the model. For example, he noted that an average slide velocity of 20 m/s resulted in runup values onto adjacent coastlines of between 1 and 2 m. By contrast, a modeled landslide velocity of 50 m/s resulted in runup values of between 5 and 14 m, values significantly in excess of the estimates for adjacent coastlines based on geological data. Harbitz (1992) concluded that a landslide velocity of 30 m/s provided the closest

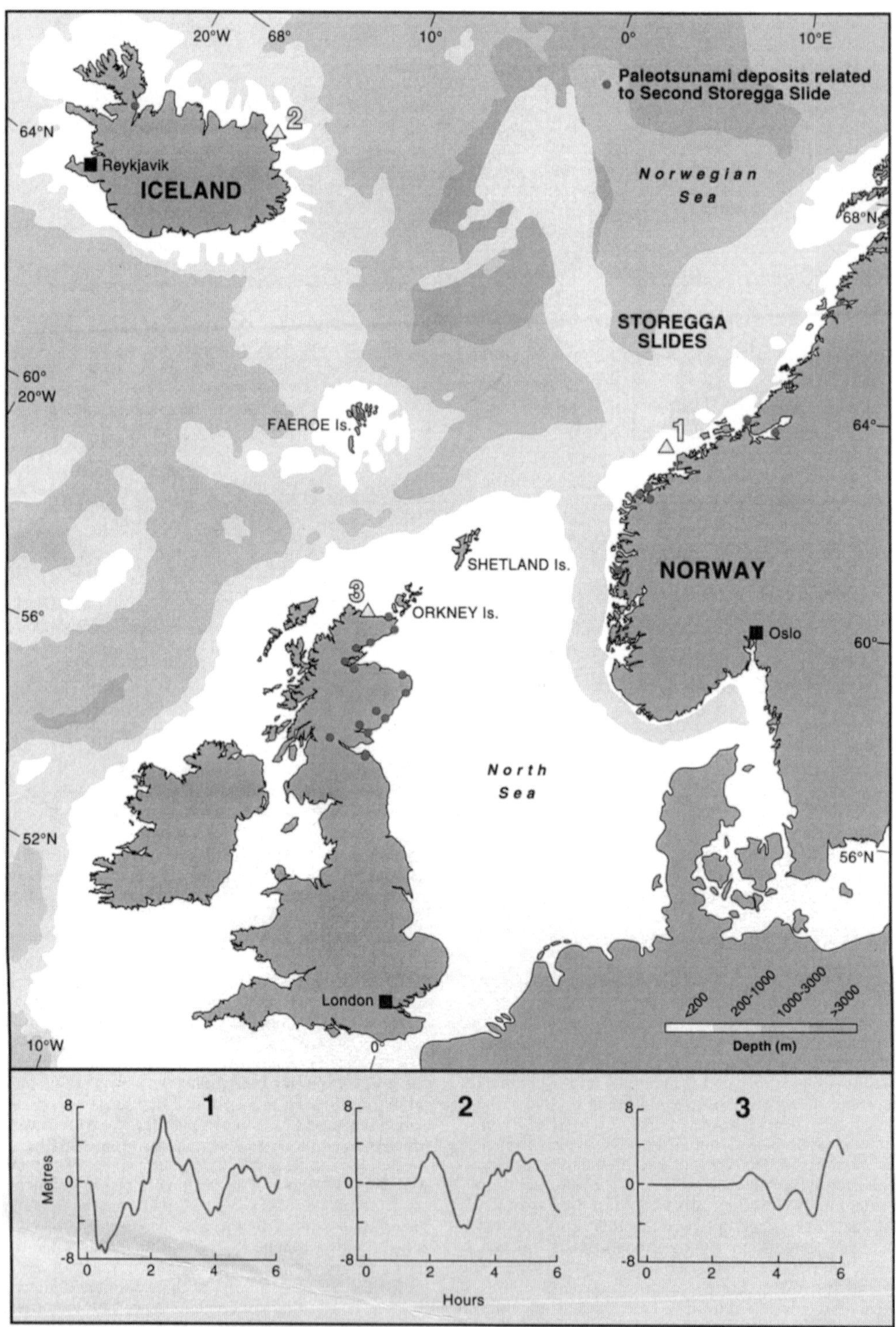

Figure T53 Map of the location of the Second Storegga slide, believed to have generated a tsunami about 7,100 ^{14}C years BP, with sites where the tsunami deposits have been found (see Dawson and Shi, 2000).

approximation to the estimated runup values based on geological data. However, the weakness in this argument is that the geological data only provide minimum estimates of likely flood runup and therefore the related numerical models of the same tsunami will always underestimate the likely average value of the submarine slide velocity.

The occurrence of this tsunami is unusual since it appears to have been generated by one of the worlds largest submarine sediment slides rather than by an earthquake. This serves to demonstrate that severe tsunamis can be generated by submarine slides in aseismic areas where there are considerable thicknesses of unconsolidated sediments on the seafloor.

Giant submarine slides and their potential to generate tsunamis are not restricted to the Norwegian Sea. Recently, Nisbet and Piper (1998) recognized a giant submarine slide occupying the majority of the seafloor of the western Mediterranean. The slide appears to have been generated in deepwater adjacent to western Sardinia and radiometric dates appear to indicate that it took place during the last glaciation of the Northern Hemisphere (probably ca. 20–30,000 years BP). At present, there is no geological evidence that this submarine slide generated a large tsunami. That such a large tsunami took place, however, can hardly be doubted. However, paleoenvironmental reconstruction of this event appeared to indicate that the slide took place at a time when sea level in the western Mediterranean may have been at ca. −100 m below present and hence any geological record of the tsunami having taken place may lie below present sea level. Submarine slides and offshore earthquakes are not mutually exclusive, however, in their capacity to generate tsunami. For example, recent investigations have shown that the islands of Amorgos and Astipalea in the southern Aegean Sea were subject to severe tsunami flooding caused by an offshore earthquake in 1956 that simultaneously generated a major sediment slump and the two processes together acted to generate complex patterns of tsunami flooding.

Tsunami warning systems

As a result of the severe damage and loss of life caused by tsunamis, attempts have been made to develop warning systems designed to alert the public in advance of the arrival of individual tsunamis. Tsunami warning systems did not exist prior to the Aleutian Trench earthquake and tsunami of April 1, 1946. However, the destruction of the Scotch Cape Lighthouse on Uminak Island and the devastation of Hilo, Hawaii by this tsunami eventually led to vociferous calls for the establishment of a tsunami warning system designed to protect life and property.

The network was developed by the US Coast and Geodetic Survey and has its center of operations on the Island of Oahu, Hawaii. The center, known as the Pacific Tsunami Warning Center (PTWC) became operational in 1948 and was linked to over 30 seismographic stations throughout the Pacific Basin. These provide data on Pacific earthquakes whose magnitude and epicenters make them tsunamigenic (capable of producing tsunamis). When such an earthquake has taken place, the PTWC issues a tsunami watch to all receiving stations. The PTWC is also linked to over 50 tide-gauge stations located throughout the Pacific. Any tsunami that has been generated by an earthquake is automatically recorded in the tide-gauges closest to the epicenter. If a tsunami has been detected then the tsunami watch is upgraded to a tsunami warning. At this stage, the estimated times of arrival of the first waves are computed for all stations across the Pacific. Once the PTWC has issued a tsunami warning and has provided information on the times of arrival of the first waves, it is the responsibility of the local police, military, and civil defense agencies to decide on whether or not particular areas should be evacuated.

The accuracy of PTWC tsunami warnings is well-illustrated by the famous Chilean earthquake and tsunami that took place on May 21, 1960. Once the earthquake epicenter had been calculated and a brief study of local tide-gauge data had been completed, it was estimated that the velocity of the tsunami was 710 km/h and that the tsunami would reach the Hawaiian islands 14 h and 56 min after its generation off the Chilean coast. The prediction was that the first wave would strike Hilo at 9.57 p.m.—it arrived 1 min late.

In areas where the coast is located close to the epicenter of a tsunamigenic earthquake, the time that elapses between the generation of the tsunami and its arrival on the coast is often frighteningly short. For example, the first tsunami waves that struck the Chilean coast on May 21, 1960 arrived only 15 min after the main earthquake shock. Similarly, there was only a relatively short time interval between the March 27, 1964 earthquake in Alaska and the arrival of tsunami waves at the coast. In both cases, loss of life was due to the fact that the tsunamis struck coastlines long before any PTWC warning could be given.

As a result of these two great tragedies, it was realized that it was important to establish regional warning systems for Alaska (the Alaskan Regional Warning System—ARWS) and the Hawaiian Islands. In 1967, a regional center was established in Alaska at Palmer, Anchorage. The center makes use of numerous automated seismographic and tide stations and is capable of issuing tsunami watches and warnings within seconds of a particular earthquake. A regional tsunami warning center for the Hawaiian Islands was later established in 1975.

Tsunami hazards and paleoseismicity

In Europe, due to the much lower frequency of tsunamis there are almost no tsunami warning systems in place. The only one that exists has been built by Portuguese authorities who have installed seismometers and wave recorders west of Portugal with the aim of providing advance warning of any future tsunami similar to that which caused such devastation in AD 1755.

In response to the perceived hazard posed by tsunamis to European coastlines, the European Union funded a major research initiative in 1992 concerned with European tsunami risk. The Project "Genesis and Impact of Tsunamis on European Coasts (GITECs)" completed in 1998, had two principal objectives. The first was to study tsunami generation mechanisms in Europe, for example, those caused by earthquakes and by submarine or coastal landslides. The second objective is to evaluate the tsunami hazard in European seas in order to reduce tsunami risk in Europe. As part of this research effort, a unified catalog of historic and prehistoric European tsunamis has been prepared. In addition, it has proved possible to estimate tsunami frequency for selected coastal regions over long (geological) timescales. In this respect, attempts are also being made to assess tsunami hazard and the likely impact of such tsunamis for particular coastal areas. Detailed investigations have also been made to develop numerical model simulations of major European tsunamis. The project has also attempted to compare the development and implementation of two new European tsunami warning systems.

Summary

In addition to the importance of paleotsunami research in terms of coastal hazard assessment, the calculation of long-term tsunami frequency for particular coastal regions provides valuable data on the past frequency of offshore earthquake activity. In certain parts of the world (e.g., Japan), the historic frequency of offshore earthquakes and tsunamis has been so high that preparedness for future earthquakes and tsunamis is of the highest priority. By contrast, it is now known that while earthquakes have induced minor tsunamis that have struck the Portuguese coastline during the last 3,000 years, there has been only one major earthquake and one tsunami during this period that have been highly destructive. In areas where the past occurrence of tsunamis is extremely rare, a difficult problem is presented to politicians and engineers since many complex decisions have to be made in establishing appropriate coastal defenses for catastrophic marine flooding that may only take place once in several thousand years. This problem becomes more acute in areas where nuclear power plants are located close to present sea level. Sadly for most areas of the world, coastal populations have no protection from tsunamis and rely on chance that none will ever take place.

Alastair Dawson

Bibliography

Borlase, W., 1755. Letter to the Rev Charles Lytteton. *Philosophical Transaction of the Royal Society of London*, **49**: 373–378.

Borlase, W., 1758. *Observations on the Islands of Scilly*. (Reprinted 1966, Frank Graham, Newcastle upon Tyne.)

Dawson, A.G., and Shi, S., 2000. Tsunami deposits. *Pure and Applied Geophysics*, **157**: 875–897.

Harbitz, C.B., 1992. Model simulations of tsunami generated by the Storegga Slides. *Marine Geology*, **105**: 1–21.

Moreira, V.F., 1985. Seismotectonics of Portugal and its adjacent area in the Atlantic. *Tectonophysics*, **117**: 85–96.

Myles, D., 1985. *The Great Waves, Tsunami*. New York: McGraw Hill.

Nisbet, E.G., and Piper, D.J.W., 1998. Giant submarine landslides. *Nature*, **392**: 329–330.

Tinti, S., and Maramai, A., 1996. Catalogue of tsunamis generated in Italy and in Côte d'Azur, France: a step towards a unified catalogue of tsunamis in Europe. *Annali Di Geofisica*, **XXXIX**(6): 1253–1299.

Young, R.W., and Bryant, T., 1992. Catastrophic wave erosion on the southeastern coast of Australia: impact of the Lanai Tsunami ca. 105ka. *Geology*, **20**: 199–202.

Cross-references

Coastal Changes, Rapid
Coastal Sedimentary Facies
Global Vulnerability Analysis
Mass Wasting
Natural Hazards
Seismic Displacement
Storm Surge
Waves

U

UPLIFT COASTS

From the late 1800s the early geologists and geomorphologists such as W.M. Davis identified coasts of "emergence" (uplift coasts) and "submergence" (drowned coasts) relative to modern sea level. The idea was promulgated further by Johnson (1919) and Cotton (1974). In a more modern sense we now recognize continental margins as "passive" or "active" (subject to ongoing tectonic processes). Uplift coasts are typically associated with the latter.

Three primary mechanisms of coastal uplift may be identified:

(1) Contemporaneous subduction processes at active plate margins are particularly evident around the circumference of the Pacific Ocean. Above the subduction zone the coastline is subjected to numerous tectonic and seismic processes, but particularly uplift and lateral

Figure U1 The uplift shoreline at Wellington, New Zealand, which emerged ~2 m after the 1855 earthquake. Note the degraded cliff and uplifted shore platform, now grassed (photo: T. Healy).

Figure U2 The tectonically upthrust emerged coastal plain, which was previously the Ahururi lagoon (remnant in the foreground), as viewed in 1981, some 50 years after the uplift of ~2 m associated with the 1931 earthquake at Napier, New Zealand. Note the abandoned stacks, and degraded cliffs. This is in an active subduction zone, with the hills in the background subject to continuing tectonic uplift (photo: T. Healy).

(wrench) faulting, as for example along the northern California coast. Active plate marginal processes creating uplift coasts also occur at a medium regional scale along the Pacific South American coast, New Zealand, New Guinea, Indonesia, and Japan. A type site for an active subduction coast undergoing uplift is the East Coast province of New Zealand.

(2) Isostatic rebound from the melting of continental ice sheets, whose load depressed the earth's crust, provides a second mechanism for uplift of the coast. Resulting uplift coasts are particularly evident over large coastal regions in the northern hemisphere, in particular Alaska, Canada, and northeastern USA; Russia, Scandinavia, and northern Europe. The latter include uplift coasts of the ancient rock terrains of Scandinavia and the British Isles, as well as the modern Pleistocene glacial tills of the north German plain.

(3) Diastrophism—tectonic movements of the earth's crust, including both epeirogenic (regional uplift without significant deformation) and orogenic (mountain building with deformation)—can produce sectors of uplift coasts. For example, block faulting producing horst and graben structures occur along the South Australian coast (Twidale, 1968), while block faulting influences much of the central coast of China, as well as the coast of Chile and Japan.

Seibold and Berger (1993) illustrate how throughout geological time, a major sedimentary wedge formed around the perimeter of the continents. Upon uplift, it is these sedimentary wedge deposits which undergo erosion from marine action at the coast. Subsequently when they are uplifted beyond the marine influences, subaerial denudation over time erodes out and masks the original specific coastal geomorphic features. However, epeirogenic uplift may also occur on continental margins where there is no sedimentary wedge, so that granitic rocks of continental origin characterize the coastal geomorphology subject to the uplift. On a larger spatial scale, some uplift coasts also present emerged "continental shelves" and continental slopes, and possibly even the "continental rise."

When considering uplift coasts, one needs also to bear in mind the timescale of its formation: (1) Recent emergence of the coastline leaves a distinctive imprint on the terrain from the previously active marine processes. Landforms that result include raised terraces, often in flights as in New Guinea, abandoned strand lines as in Scandinavia, coastal plains and abandoned cliffs, shore platforms, and stacks. Examples of the latter include the uplifted shore platforms from historical earthquakes such as the 1855 Wellington event which raised the shoreline by ~2 m (Figure U1), and the coastal plain at Napier which, prior to the 1931 earthquake which destroyed the City of Napier, was a large shallow lagoon, but is now a coastal plain (Figure U2). (2) As time of emergence progresses, normal subaerial weathering and denudation processes take over, acting on the emerged features. Thus the cliffs become degraded, and stream incision occurs on the raised terraces. Eventually all that is identifiable in the landscape is the break of slope indicating location of a previous shoreline.

Terry R. Healy

Bibliography

Cotton, C.A., 1974. *Bold Coasts*. Wellington: Reed.

Johnson, D.W., 1919. *Shore Processes and Shoreline Development*. New York: Wiley.

Seibold, E., and Berger, W., 1993. *The Sea Floor. An Introduction to Marine Geology*, 2nd edn. Berlin: Springer-Verlag.

Twidale, C.R., 1968. *Geomorphology*. Melbourne: Thomas Nelson (Australia).

Cross-references

Coastal Changes, Gradual
Coastal Changes, Rapid
Coral Reefs, Emerged
Faulted Coasts
Isostasy
Marine Terraces
Shore Platforms
Submerging Coasts
Tectonics and Neotectonics

V

VEGETATED COASTS

Vegetated coasts are those where rooted vascular plants are a persistent feature of the coastal landscape. This can include dunes, salt marshes, sea-grass beds, gravel barriers, and rocky coastlines. The type of vegetation, as well as its importance in shaping the coastal environment, varies according to climate, sedimentary deposits, and wave and tidal energy regimes. These variables control both the type of vegetation that can survive, and the physical context within which the plants interact with sedimentary and geomorphic processes to shape coastal landscapes. Vegetation not only exists in many coastal settings, it is an important biogeomorphic agent and as such is an integral link between the ecological and landform dynamics of coastal systems. While the presence of vegetation in dunes or on salt marshes has long been recognized as an essential characteristic of those environments, the importance of plant growth forms and their adaptations to stressful physicochemical conditions in shaping coastal landforms has only more recently received attention. For example, Bauer and Sherman (1999, p. 73) describe foredunes as being "geomorphologically conditioned by the germination, colonization, and succession of vegetation assemblages characteristic of coastal environments." Clearly, understanding the dynamics of vegetated coasts requires not only knowledge of the prevailing physical and sedimentological conditions but insights into vegetative form and function.

Vegetation at the coast

Ecological role

Coastal ecosystems are widely recognized as some of the most productive on earth. Carter (1988) summarizes the net productivity of coastal ecosystems and notes that, with the exception of coral reefs, highest levels are associated with vegetated coastal systems including subtidal seaweeds and sea grasses, and intertidal marshes and mangroves. The ecological importance of vegetation to coastal systems was recognized by Teal (1962) who suggested that plant detritus, after undergoing microbial decomposition and enrichment, was the basis of the salt marsh food chain. While more recent work has shown that much of the detritus in mangroves and salt marshes is directly consumed by benthic scavengers, and that benthic microalgae and phytoplankton provide important high-quality food sources for higher trophic levels, the importance of vegetation as structure and refuge for secondary producers and consumers, especially nekton, is widely accepted (see Weinstein and Kreeger (2000) for review of current understanding of tidal marsh ecology).

The high productivity of some vegetated coastal systems can result in a large amount of plant debris in the coastal zone. Where high tides and wave action raft this debris onto beaches, the ecological processes can be similar to those described above for marshes. Alongi (1998) described how detached algae and sea grasses can form stacks up to 2 m high on beaches in western Australia, and that the main utilization pathway for the detritus is via colonizing microbes and surf-zone amphipods.

Adapting to stress

Rapid and regular fluctuations in water level, along with salinities varying from fresh to hypersaline, are among a number of stresses facing coastal vegetation. In some environments these combine with direct attack from waves, human disturbances, and grazing pressure to produce conditions where only specialized plant growth forms can survive. Table V1 illustrates the variety and combination of stresses facing coastal plants. The presence of vegetation in these various coastal environments is testament to their ability to adapt and cope with such harsh conditions. Duke (1992, p. 65) notes of mangroves that the "combination of morphological and physiological adaptations seen in this diverse and unique group of plants have no equal." The prop roots of *Rhizophora* spp. and the pneumatophores of *Avicennia* spp. are two of a number of ways in which different mangrove species deal with one of these stresses—waterlogging and soil anoxia. Another of the more commonly known adaptations of coastal plants is that way in which the common dune grass *Ammophilia arenaria* rolls its leaves under dry conditions to limit the leaf area exposed and reduce water losses. Even lower plants, such as seaweeds, show adaptations such as the way they cope with extreme water losses during low tide and can rapidly rehydrate and resume photosynthesis soon after resubmergence.

The adaptations made by coastal plants to survive in these conditions are frequently structural. Plants that are exposed to excessive flooding and anoxic soils conditions respond by developing air spaces, or aerenchyma, in their root and stem tissue allowing oxygen to diffuse from the aerial parts of the plant to the roots. Common responses to high salt concentrations include barriers to salt uptake within the root system, such as in the mangrove *Avicennia*, and the excretion of unwanted salts through specialized salt glands in the leaves.

Some plants which can tolerate stressful conditions are so specialized that they are readily outcompeted as conditions ameliorate. For example, in laboratory experimental studies, Adams (1963) found that *Spartina alterniflora* tolerates a wide range of salinities. However, in the natural environment it is restricted to the lowest parts of salt marshes where salinity and waterlogging stress are highest. Bertness (1991) showed that at higher elevations in New England salt marshes, *S. alterniflora* is outcompeted by other species such as *Spartina patens*. Importantly, Bertness (1991) concludes that while the lower intertidal limit of growth for individual marsh species is set by their tolerance for physical stresses, the upper intertidal limits are set by plant competition.

Table V1 Examples of the physical stresses on vegetation in various coastal environments

Stress/environment	Rocky shore	Gravel barriers	Sand dunes	Sea grasses	Salt marshes	Mangroves
Salinity	■	■	■	■	■	■
Waterlogging			■		■	■
Water deficits	■	■	■		■	■
Lack of anchorage		■	■			
Lack of nutrients		■	■			
High wave activity	■	■				
High temperature fluctuations	■	■	■			■
Grazing	■		■	■	■	
Human disturbances		■	■	■	■	■

A similar type of situation seems to occur with the beachgrass *Ammophilia*. According to Packham and Willis (1997) it is capable of withstanding sand accretion rates of approximately 1 m/year, but measures of vegetative vigor usually decline with lower burial rates and in more stable parts of coastal dune systems it is commonly replaced by other species. This has been attributed to competitive interactions among plants, but Disraeli (1984) showed that the growth of *Ammophilia* was lower in areas of shallower burial in the absence of competitors. This phenomena has variously been attributed to the lesser ability of older roots to uptake nutrients and the effects of soil-borne pathogens and parasites which are ameliorated by the delivery of new sand.

The net result of these adaptation and interactions is that many vegetated coasts exhibit clear zonation of plant species in response to stress gradients (e.g., in tidal inundation, salinity, or wave energy).

Sensitivity to disturbances

Most coastal environments are subjected to periodic physical disturbance by storm waves, storm surges, or other events that alter, at least in the short term, the stability of the substrate. In addition to such episodic events, vegetated coasts can be sensitive to more regular erosional or depositional events that result in progressive environmental change. For gravel barrier systems, Scott (1963) identified five stability classes each characterized by different plants. Very unstable areas were devoid of vegetation, while gravel beaches stable between spring and fall were characterized by annuals. Increasing stability led to the presence of short- and then long-lived perennial plants and areas which had been stable for longer periods became covered with heath-like vegetation. The response of the vegetative community in this case appears to be to the frequency of physical disturbance. However, the importance of organic matter within gravels both for holding moisture and for preventing seeds from sinking too deeply for establishment and germination (Packham and Willis, 1997) implies that this is not simply a response to surficial disturbance of the plants. With increased gravel stability, the accumulation of organic matter can proceed providing an environment more hospitable to a wider array of plants. Many smaller annual plants in gravel systems rely on rain and dew as a source of moisture, while larger perennials can penetrate deeper and although unlikely to be able to access the water table may be able to reach freshwater lenses held in smaller gravel barriers.

While physical disturbances associated with storm events or floods usually impact coastal systems at the scale of kilometers, coastal plants can be sensitive to natural disturbances at smaller spatial scales. The most common of these in coastal salt marshes are ice and wrack. In northern latitudes, seasonal ice cover and ice rafting of vegetation, debris, and substrate, as well as scouring by ice blocks is an almost annual occurrence. Bertness (1999) characterizes salt marshes in these areas as in a continual state of recovery—both vegetatively and morphologically. The deposition of wrack—mats of dead plant material rafted to high marsh areas by high tides or storms—can lead to the burial of underlying marsh vegetation. In north Norfolk marshes, Pethick (1974) attributed the increased density of shallow marsh pans on high marshes to wrack. Once vegetation dies, soil salinity increases in the bare areas on the high, infrequently flooded marsh. This salinity can prevent the recolonization by marsh plants and the pans become permanent features of the marsh surface. Bertness (1999) describes how when such bare areas can be colonized by more salt tolerant species, these "fugitives" can ameliorate soil conditions facilitating the invasion of the site by the dominant plant species. Thus the persistence of the disturbance "feature" in the coastal system depends upon the harshness of the physical conditions and the availability of plants that can tolerate such conditions.

Vegetation as biogeomorphic agent

The role of plants in coastal biogeomorphology is largely constructive—acting to enhance deposition or protect surfaces from erosive forces. The vascular plants inhabiting soft sediment substrates on the Atlantic Coast of the United States have been termed "bioengineers" by Bertness (1999) because of their ability to stabilize substrates and enhance sedimentation. The role of salt marsh plants in contributing to the vertical growth of coastal marshes, both directly through the accumulation of organic substrate and indirectly by baffling tidal flows and enhancing sediment deposition has been well documented (Reed, 1995). Similar direct and indirect effects of vegetation appear to operate on cobble beaches in New England, where Bertness (1999) describes how, where *S. alterniflora* has colonized areas of the high intertidal, it ameliorates summer heating effects by shading, and stablizes the cobbles which would otherwise be subject to dislodgement and rolling during winter storms. The *Spartina* also "buffers" the effect of waves and provides a calmer environment where other salt-tolerant plants can survive. The recruitment of these plants is also facilitated by the *Spartina* as it traps waterborne seeds.

Perhaps one of the clearest examples of a vegetated coast where the vegetation plays a crucial role in the formation and development of geomorphology is the beach–foredune environment. Hesp (1984) asserts that foredunes develop on the seaward most vegetated sand surface behind active beaches and identifies four types of incipient foredune formation, each distinguished by their relationship to specific vegetative forms or structures. As Hesp also claims that beach ridges are actually relict foredunes, he implies an overriding role for vegetation, in various forms, for the initiation of coastal dune systems, clearly stating (p. 88) that he "believes that biologic processes and pan-aerodynamic interactions assume foremost importance in influencing initial morphologic variation on foredunes." Furthermore, Carter and Wilson (1990) attribute the role of vegetation in the initial stages of foredune development to the development of complex three-dimensional fluid flow over dunes at Magilligan Point, northern Ireland. They note that gaps in vegetation during foredune initiation can result in gullies or "low saddles" that cross the foredune. These features were observed to funnel airflow through the dune, sometimes creating aeolian mounds on the leeward side, and occasionally resulting in blowouts.

Intertidal vegetation interacts with waterflows slightly differently than dune grass with airflows in that the vegetation can be either completely or partially submerged by the waterflow depending upon the tide. For instance, sea grasses have been shown to strongly influence flows over the bed by reducing turbulence within the canopy and the development of a "stratificiation" above the canopy. Living in seawater means sea grasses do not require the structure support as terrestrial vegetation and their growth forms are frequently extremely flexible. This leads to deflection and compression of the plant canopy under strong flows, reducing friction. Differences in sea-grass morphology, from small shallow rooted to larger strap-bladed forms, alter the specific interaction between the plants and the flow but in most cases the presence of the vegetation significantly increases the mean threshold velocity for sediment motion when compared to adjacent bare sands.

The role of vegetative bioengineering in coastal dynamics is sometimes more readily seen when disturbance to the vegetation results in dramatic system change. Bertness (1999) reports a dramatic change from sandy and muddy subtidal substrates to coarse cobble bottoms after a wasting disease destroyed much of the eelgrass beds on the northern Atlantic shores of the United States. In addition to this physical change, the ecological consequences were substantial with a population crash in the bay scallop which relies on eelgrass not only as a food source but to provide structures for young recruits to attach to.

In dune environments, the loss of vegetative cover due to human disturbances such as trampling and all-terrain vehicle (ATV) use are frequently viewed as leading to the degradation of the dune system and prompting a need for management action. However, Thomas (1999) cautions that patchy development of blowouts is a natural feature of most dune systems and that although disturbance of the vegetation can undoubtedly lead to increased turbulence this should not simply trigger efforts at revegetation. Rather a detailed understanding of the vegetation surface conditions must be used to project the implications of the apparent disturbance and plan any remedial action. This call for a more holistic recognition of vegetated coastal environments as biogeomorphic systems recognizes that management actions must appreciate the natural dynamics of the system as well as its form and surficial characteristics.

Future issues for vegetated coasts

The importance of plant growth and survival to the sustainability of both landforms and ecosystems of vegetated coasts means that environmental changes that alter vegetative structure can have important implications. Two factors that threaten vegetated coasts are the massive increases in coastal populations, and the associated development pressures, and future climate changes.

The coastal population of the United States is currently growing faster than the nation's population as a whole, a trend that is projected to continue. By 2010, 75% of the US population is expected to live within 50 miles of the coast. Nicholls and Small (2002) identify 23% of the global population as living in coastal areas and note that most of the near-coastal population does not live in large cities but in smaller towns and in rural areas. The growing coastal population, increasing economic development, and expanding coastal communities exert pressure on vegetated coasts through recreational pressures, direct damage due to filling or development, and lowered water tables. Pinder and Witherick (1990) point to the loss of coastal wetlands associated with growth of ports and urbanization in the second half of the 20th century, and dunes and other intertidal vegetated habitats have been similarly consumed. The societal need for development space at the coast, especially for industries and port facilities that are directly functionally related to the coastal ocean, combined with the diminishing area of land available and the recognition of the ecological, recreational, and esthetic value of vegetated coasts, has led to increasing examination of the potential for creating new land specifically to support development. However, while this approach may relieve development pressure on existing systems, it has consequences for coastal sedimentary and hydrodynamics. Arens *et al.* (2001) examined the potential environmental impacts of offshore island development in the Netherlands and noted effects associated with construction and to the marine ecosystem based on the footprint of the island and the direct loss of habitat. Coastal dunes and higher plants were largely unaffected and Arens *et al.* projected only local alterations in the sediment budgets and wave climates that support the adjacent vegetated coasts.

While vegetated coasts are often characterized by plants adapted to living in stressful conditions, changes in climate may increase stresses beyond their tolerance range resulting in shifts to nonvegetated coastal systems and potentially dramatic geomorphic transitions. Scavia *et al.* (2002) note that projected increases in sea level are unlikely to have near-term catastrophic impacts on coastal wetlands and mangroves, but when combined with other stresses such as alterations to local hydrology or sediment movement patterns associated with development pressures, the long-term consequences may be severe. For vegetated coastal forms to survive they must increase elevation in place (e.g., accrete soil), or migrate inland to keep pace with the rising sea. However, coastal structures and river management alter sediment supplies, and thus limit vertical building, and developed coastlines limit migration. Management approaches for vegetated coasts during climate change must be focused on alleviating the stresses imposed by human alteration of coastal systems, and allowing transitions in form and ecology to cope with the changing sea level, temperature, and precipitation regimes. Vegetated coasts are among our most valued, and highly pressured systems. By their very nature they can survive change and harsh conditions. Our management actions must allow the natural cycles that have sustained them in the past to continue long into the future.

Denise J. Reed

Bibliography

Adams, D.A., 1963. Factors influencing vascular plant zonation in North Carolina salt marshes. *Ecology*, **44**: 445–456.

Alongi, D.M., 1998. *Coastal Ecosystem Processes*. Boca Raton: CRC Press.

Arens, S.M., Jungerius, P.D., and van der Meulen, F., 2001. Coastal dunes. In Warren, A., and French, J.R. (eds.), *Habitat Conservation: Managing the Physical Environment*. Chichester: John Wiley & Sons, pp. 229–272.

Bauer, B.O., and Sherman, D.J., 1999. Coastal dune dynamics: problems and prospects. In Goudie, A.S., Livingstone, I., and Stokes, S. (eds.), *Aeolian Environments, Sediments and Landforms*. Chichester: John Wiley & Sons, pp. 71–104.

Bertness, M.D., 1991. Zonation of *Spartina patens* and *Spartina alterniflora* in a New England salt marsh. *Ecology*, **72**: 138–148.

Bertness, M.D., 1999. *The Ecology of Atlantic Shores*. Sunderland, Massachusetts: Sinauer Associates.

Carter, R.W.G., 1988. *Coastal Environments: An Introduction to the Physical, Ecological and Cultural Systems of Coastlines*. London: Academic Press.

Carter, R.W.G., and Wilson, P., 1990. The geomorphological, ecological and pedological development of coastal foredunes at Magilligan Point, northern Ireland. In Nordstrom, K.F., Psuty, N., and Carter, R.W.G. (eds.), *Coastal Dunes: Form and Process*. Chichester: John Wiley & Sons, pp. 129–158.

Disraeli, D.J., 1984. The effects of sand deposits on the growth and morphology of *Ammophilia brevigulata*. *Journal of Ecology*, **72**: 145–154.

Duke, N.C., 1992. Mangrove floristics and biogeography. In Robertson, A.I., and Alongi, D.M. (eds.), *Tropical Mangrove Ecosystems*. Washington, DC: American Geophysical Union, pp. 63–100.

Hesp, P.A., 1984. Foredune formation in southeast Australia. In Thom, B.G. (ed.), *Coastal Geomorphology in Australia*. Sydney: Academic Press, pp. 69–97.

Nicholls, R.J., and Small, C., 2002. Improved estimates of coastal population and exposure to hazards released. *EOS*, **83**: 301–305.

Packham, J.R., and Willis, A.J., 1997. *Ecology of Dunes, Salt Marshes and Shingle*. London: Chapman Hall.

Pethick, J.S., 1974. The distribution of salt pans on tidal salt marshes. *Journal of Biogeography*, **1**: 57–62.

Pinder, D.A., and Witherick, M.E., 1990. Port industrialization, urbanization and wetland loss. In Williams, M. (ed.), *Wetlands: A Threatened Landscape*. Oxford: Basil Blackwell, pp. 234–266.

Reed, D.J., 1995. The response of coastal marshes to sea-level rise: survival or submergence? *Earth Surface Processes and Landforms*, **20**: 39–48.

Scavia, D., Field, J.C., Boesch, D.F., Buddemeier, R.W., Burkett, V., Cayan, D.R., Fogarty, M., Harwell, M.A., Howarth, R.W., Mason, C., Reed, D.J., Royer, T.C., Sallenger, A.H., and Titus, J.G., 2002. Climate change impacts on US coastal and marine ecosystems. *Estuaries*, **25**: 149–164.

Scott, G.A.M., 1963. The ecology of shingle beach plants. *Journal of Ecology*, **51**: 517–527.

Teal, J.M., 1962. Energy flow in a salt marsh ecosystem in Georgia. *Ecology*, **43**: 614–624.

Thomas, D.S.G., 1999. Coastal and continental dune management in the twenty-first century. In Goudie, A.S., Livingstone, I., and Stokes, S. (eds.), *Aeolian Environments, Sediments and Landforms*. Chichester: John Wiley & Sons, pp. 105–127.

Weinstein, M.P., and Kreeger, D.A., 2000. *Concepts and Controversies in Tidal Marsh Ecology*. Dordrecht: Kluwer Academic.

Cross-references

Bioengineered Shore Protection
Dune ridges
Mangroves, Coastal Ecology
Salt Marsh
Tidal Creeks
Tidal Environments
Wetlands

VIBRACORE

Vibracoring is a state-of-the-art sediment sampling methodology for retrieving continuous, undisturbed cores. Also referred to as *vibrocoring*, this mechanical drilling technique is used to collect core samples (referred to as either *vibracores* or *vibrocores*) from unconsolidated, loosely compacted, or semi-lithified materials by driving a tube with a vibrating device. Large, heavy-duty vibracorers can work in water up to 5,000 m deep and can retrieve core samples up to 13 m in length. In coastal shallow-water environments where use of heavy equipment is limited by trafficability and ground support on land and high-energy conditions alongshore, cores are shorter, usually about 6 m in length.

The coastal/marine environment presents specialized conditions that are not encountered on *terra firma* or in deeper oceanic waters. Because many onshore areas are characterized by coastal wetlands with extensive areas of organic soils in marshy or swampy conditions, tidal sand and muds, or lacustrine and lagoonal facies, beach sands, or chenier-type materials, access for conventional drilling equipment and personnel is often limited, if not by biophysical conditions, then at least economically. Financial considerations are often limiting when costs of sediment retrieval are very high on a per-sample basis. Large multinational corporations involved in petroleum exploration can go almost anywhere to penetrate the most inhospitable environments while sparing no expense to obtain conventional long-drill cores. Most coastal/marine research, however, operates on a reduced cost basis that must be efficient and cost-effective. There are also numerous subaqueous environments that are normally hostile to the positioning of conventional drilling equipment and so alternative sediment-sampling methods are sought. Retrieval of shallow-water sediments using conventional cylindrical coring methods (e.g., those that employ gravity or piston cores) is limited by shallow water depth as well as the nature and habit of the materials requiring undisturbed collection. Resistance of the sediment restricts penetration of gravity cores and core-lengths typically obtained are less than 3–5 m, depending upon the firmness of the sediment.

Sediment cores obtained from a vibracore system are invaluable because they permit direct, detailed examination of composition and layering in sequences of subsurface sediment (Lanesky *et al.*, 1979; Watson and Krupa, 1984). Examination of material sequences in vibracores provides information regarding the history of depositional environments (e.g., Brooks *et al.*, 1995) and the physical processes that were operative during sedimentation. Vibracores and subsamples derived from them find almost limitless applications in scientific geological research, engineering and geotechnical pre- and re-design, and environmental investigations. The versatility of the vibracoring technique is illustrated by its utilization in sampling riverine sediments, lacustrine deposits, organic accumulations in marshes and bogs, as well as nearshore shallow water coring. Although examples of applications are legion, of primary interest to most coastal specialists are results that are relevant to the scientific study of coastal zone morphodynamics, evolutionary sequences, and exploratory searches for beach-quality sand that is suitable for beach renourishment.

The utility of vibracores, as seen in these few examples from a huge literature, has been demonstrated in: (1) subsurface exploratory sampling for geotechnical purposes (e.g., Meisburger, 1990; Larson *et al.*, 1997), (2) determination of depositional stratigraphy and geomorphological history in marine, estuarine, fresh water, or wetland environments including marshes, swamps, and peat bogs (e.g., Snowden *et al.*, 1977; Amos, 1978; Kraft *et al.*, 1979; Stevenson *et al.*, 1986; Morang *et al.*, 1993; Kirby *et al.*, 1994), (3) glacioeustatic sea-level changes (e.g., Gehrels, 1994; Harvey *et al.*, 1999), (4) integrating studies of barrier island evolution, bars, lithofacies, and sedimento-stratigraphy (e.g., Davis and Kuhn, 1985; Davis *et al.*, 1993; Brooks *et al.*, 1995), (5) offshore sand searches to locate beach-quality sands for beach restoration (e.g., Finkl *et al.*, 1997; Freedenberg *et al.*, 2000), (6) environmental studies of pollution or contamination by hydrocarbons or heavy metals (e.g., Varekamp, 1991), (7) investigations to determine seafloor environments or bottom types (e.g., Barnhardt *et al.*, 1998; Knebel and Poppe, 2000), (8) deltaic and sedimentary shelf processes (e.g., Delaune *et al.*, 1983; Brooks *et al.*, 1995; Levitan *et al.*, 2000; Toldo *et al.*, 2000), (8) collection of samples for chemical, biological, and physical analyses (e.g., Kadlec and Robbins, 1984; Gehrels, 1994), (9) verification of seismic stratigraphy or seismic stratigraphic sequences (e.g., Shipp *et al.*, 1991; Knebel and Poppe, 2000), (10) preliminary investigations for the purpose of determining the presence and positions of paleoshorelines (e.g., Gayes and Bokuniewicz, 1991; Shipp *et al.*, 1991), and (7) stratigraphic and mineral surveys (e.g., Barusseau *et al.*, 1988; Brooks *et al.*, 1995). Use of vibracores to substantiate seismic interpretations (e.g., sidescan sonar, Uniboom seismic reflection—sub-bottom boomer profiles) is a major application that integrates geological and geophysical methodologies to advantage in many studies. Marine or subaerial unconformities, seismic reflectors, stratigraphic units, internal structures of morphodynamic units, sediment textural properties, and even geomorphic features can be identified in vibracores. Although vibracores find use in many diverse applications, the overall setup of coring systems is an important consideration.

Early attempts to obtain (undisturbed) samples in coastal/marine environments

Retrieval of marine sediments is nearly as old as the science of oceanography itself. Initial interest focused on getting an idea of what kinds of sediments were deposited on the seafloor but, then, as technological advancements assisted the desire for more complete information, new methods of bottom sampling were developed. The purposes of sampling marine sediments varied among professionals as the pioneering biologist, petrographer, sedimentologist, and civil engineer had different demands. Reviews of some historical efforts in bottom sampling are summarized, for example, by Trask (1939), Sanders (1960), Rossfelder and Marshall (1967), and Watson and Krupa (1984) who note the pros and cons of numerous platforms, rigs, and devices such as the Strøm coring tube, Twenhofel coring tube, Varney pile-driver sampler, Trask suction sampler, Renn sampler, Piggot coring apparatus, Kudinov vibro-piston core sampler, etc. These early devices suffered from a variety of shortcomings that were eventually overcome by improved sampling methods. Modern methods of coring are not without problems, but they have been reduced to the greatest possible extent to satisfy the purpose of a particular analysis.

Terrestrial deposits in coastal marshes and on coastal plains were initially sampled using hand augers, which are still in use today for special applications (see discussion in Soil Survey Division Staff, 1993). Screw or worm augers do not provide undisturbed samples, but they can bring up materials from several meters depth by adding extra lengths to the shaft. Barrel augers, core augers, bucket augers, on the other hand, have a cylinder or barrel to hold the soil, which is forced into the barrel by cutting lips at the lower end. The upper end of the cylinder is attached to a length of pipe with a crosspiece for turning by hand at the top. Although both ends of the cylinder are open, the soil generally packs so that it stays in place while the auger is removed from the hole. Barrel augers disturb the soil less than screw augers. Soil structure, porosity, consistence, and color can be better observed. Barrel augers work well in loose or sandy soils and in compact soils. They are not well suited for use in wet or clayey soils, though an open-sided barrel is available that works well. They also work poorly in stony and gravelly soils. Barrel augers bore more slowly than screw augers or probes, but they are easy to pull from the hole.

The Dutch auger is a modified barrel auger having two connected straps with lips. The cylinder is about 5–10 cm in diameter. The Dutch mud auger works well in moist or wet soils of moderately fine or fine texture, but poorly in other moist or wet soils and in all dry soil.

Although soil augers are simple in design and somewhat crude in appearance, considerable skill is required to use them effectively and safely. They must be pulled from the soil by using a technique that puts stress on the leg muscles, rather than the back muscles, to avoid serious back injury. Twisting the auger firmly while pulling takes advantage of the inclined plane of the screw to break the soil loose. A pair of pipe wrenches is needed to add and remove lengths of shafts and bits.

Examinations of deep deposits of peat are made with special tubelike samplers. A peat sampler designed by the Macaulay Institute for Soil Research (Aberdeen, Scotland), for example, takes a relatively undisturbed volume that can be used for measurement of bulk density. The Davis peat sampler, consists of 10 or more sections of steel rods, each 60–120 cm in length, and a cylinder of brass or Duralumin, approximately 35 cm long with an inside diameter of about 1.9 cm. The cylinder has a plunger, cone-shaped, at the lower end and a spring catch near the upper end. The sampler is pressed into the peat until the desired depth for taking the sample is needed. Then the spring catch is released, allowing the plunger to be withdrawn from the cylinder. With the plunger withdrawn, forcing it further downward fills the cylinder. The cylinder protects the sample from contamination and preserves its structure when the sample is removed. With this instrument, one can avoid the error of thinking that firm bottom has been hit when actually a buried log is encountered.

Probes consist of a small-bore tube that has a tempered sharp cutting edge slightly smaller in bore but larger in outside diameter than the barrel. Approximately one-third of the tube is cut away above the cutting edges so that the soil can be observed and removed. Probes are about 2.5 cm in diameter and about 20–40 cm in length. The tube is attached to a shaft with a "T" handle at the opposite end. Adding or removing sections can vary shaft length. Probes can be used to examine the soil to a depth of 2 m. Rubber or plastic mallets can be used to drive the tube into the soil; a pair of pipe wrenches is needed to add and remove lengths of shaft.

Probes work well in moist, medium textured soils that are free of gravel, stones, and dense layers. Under these conditions, the soil can be examined faster than with an auger. Probes are very difficult to use in dry, dense, or poorly graded soil, and in soil containing gravel or stones. Probes disturb the soil less than augers, but they retrieve less soil for examination. Probes are light and easily carried, and they pull from the hole more easily than screw augers. Use of a soil probe is the fastest method to collect samples of surface layers for analysis. Probes used with power equipment have wide applications in surveys of surficial sediments. Coastal specialists are increasing turning to mechanical sampling devices as the need increases for longer cores of undisturbed samples. Vibracores meet these needs in many different types of environmental situations, but there are still limitations to collection of coastal/marine sediments. The essentials of vibracoring systems, which supplement hand sampling in terrestrial coastal environments and partially replace other types of marine bottom samplers, are briefly summarized in what follows.

Vibratory coring systems

These relatively simple devices, referred to as vibracorers, consist of three essential components: frame, core barrel, and vibrator (Hoyt and Demarest, 1981). The frame, which allows the corer to stand free on the seafloor, consists of a quadrapod or tripod arrangement with legs attached to a vertical beam that in turn supports and guides the core barrel and vibrator (Figure V1). The core barrel and vibrator slide on the beam for coring and retraction. The core barrel assembly consists of 7.5 or 10 cm diameter (thin walled) aluminum pipe fitted with a cutter head and core catcher (which holds the sediment inside the barrel when it is withdrawn from the sediments), and a plastic tube inner liner that contains the core material (cf. Figure V11). There are many different kinds of innovative assemblies that can include, for example, an electronic penetrometer to record time and depth of penetration of the core pipe into the sediments (Smith, 1992). Deploying and retrieving the corer usually requires a hydraulic crane (Figure V2), A-frame, or similar winch or hoisting equipment with a lifting capacity of at least 10–15 tons (Meisburger, 1990). Shallow protected waters may allow use of small barges, often of simple construction from polystyrene blocks wrapped in heavy-duty industrial plastic and encased in a wooden frame, mounted with a superstructure from which the vibracoring device can be deployed (e.g., Wright *et al.*, 1999). A marine Kiel Vibracorer can, for example, be adapted for deployment from a well in the barge.

Smaller, lightweight, hand-carried units are used to advantage in coastal environments on marshland or in very shallow water. Figure V3, for example, shows a small backpack power unit that is deployed in shallow water to obtain marsh sediments. Once driven into the marsh sediments, the core is retrieved by various methods for pullout but in this case a hand-operated come-along is used to extract the core (Figure V4). After the core is extracted, it can be transported to the laboratory or split in the field as shown here (Figure V5). Because there was easy access to a grassy work area near the sample site, the core barrel was carried to dry land and split using a skill saw. Splitting in the field provides opportunity for immediate inspection of the core and assessment of environmental conditions, as shown in Figure V6. Other advantages accrue from splitting a core in the field because it can be immediately determined whether the core is short due to loss of sediment or compaction, or whether there are other abnormalities such as coarse materials that plug the core causing gaps in sediment retrieval, etc. The sampling program can be modified on the basis of what is observed in the recovered materials. This flexibility in the field is important because deployment and setup are often significant costs in sample retrieval.

Figure V1 A 10-m vibracore (Alpine rig) being raised from the Gulf of Mexico seafloor offshore from Louisiana. The quadrapod stand stabilizes the vibracore frame, which sits on the seabed and allows the pneumatic vibrahead to slide downwards with core penetration. The entire apparatus is lifted aboard the research vessel using a hydraulic crane that reaches over the stern (photo courtesy of Coastal Planning & Engineering, Boca Raton, Florida).

Equipment, design, and function

The main consideration in developing vibracoring systems, often referred to as the "Rule of Deployment," that most researchers learn by experience is that "the cost of an operation is related to the size of the

Figure V2 Same vibracore assembly unit shown in Figure V1 being laid down on the deck. The long core inside the 10-cm-diameter core barrel will provide valuable undisturbed sediment data from the seabed on the continental shelf off Louisiana. Operations such as this one require experienced crew to handle these large vibracoring units (photo courtesy of Coastal Planning & Engineering, Boca Raton, Florida).

Figure V3 Small, portable vibracore. A backpack power unit operates this pneumatic vibracore; the core barrel is stabilized by hand as the unit penetrates the marsh sediments. The lightweight aluminum A-frame is carried to the site and waits for use in retrieval of the core (photo courtesy of Steve Krupa, South Florida Water Management District, West Palm Beach, Florida).

Figure V4 Retrieval of vibracore using a portable aluminum A-frame. The lightweight A-frame has pads on the feet so the unit does not sink into the sediment when load is applied. One leg of the A-frame has steps so the driller can extract the core barrel with a hand-operated come along. The A-frame must be placed vertically over the core barrel to ensure easy extraction without flexure or bending of the core. The A-frame collapses into a compact unit for transporting (Photo courtesy of Steve Krupa, South Florida Water Management District, West Palm Beach, Florida).

Figure V5 Splitting a vibracore. As shown here, the core barrel is laid in a wooden frame and an electric circular saw uses the frame as a guide to cut the flight lengthwise. The upper half is then rotated away from the frame and placed on the ground for observation and sampling (Photo courtesy of Steve Krupa, South Florida Water Management District, West Palm Beach, Florida).

Figure V6 Split core. After cutting the core barrel lengthwise, the contents are visually inspected for a range of physical parameters such as sediment grain size, color, stratigraphic discontinuities, coarse fragments, etc. Shown here is a fine-grained marsh sediment from the south coast of Long Island, New York (photo courtesy of Steve Krupa, South Florida Water Management District, West Palm Beach, Florida).

vessel which is related to the size of the draw works which is related to whatever hangs at the end of the cable." These concerns are especially important to boggy onshore sites where access is limited and in shallow-water work where vessel size and equipment load is restricted. Realizing the need to minimize the weight of individual parts while maximizing the overall force/weight ratio, it becomes obvious to select equipment that can be handled with limited manpower from all-terrain vehicles, small vessels, and even inflatable barges (e.g., Hoyt and Demarest, 1981). A variety of vibracore units are commercially available. Some are small, lightweight, and portable, whereas others are large heavy units that can only be deployed from large vessels. There are various types of bottom-standing rigid frames that are ordinarily used for stabilizing and guiding vibracores (cf. Figures V1 and V2 for heavy-duty marine examples and Figures V3 and V4 for light-duty terrestrial setups).

The principle behind a vibracore is the development of high-frequency, low-amplitude vibration that is transferred from the vibracore head (vibrahead) down through the attached barrel or core tube (Figure V7).

Figure V7 Detail of the parallel mount of electric vibracore heads, showing extension of the core barrel below the vibrahead (photo courtesy of Coastal Planning & Engineering, Boca Raton, Florida).

Figure V8 The 7.5-cm-diameter core barrel is attached to the electric vibrahead assembly prior to being lowered overboard into the Atlantic Ocean as part of an offshore sand search near Miami, Florida. Although the unit is of medium size, the weight of the vibracore assembly requires crane winching from the deck (photo courtesy of Coastal Planning & Engineering, Boca Raton, Florida).

The vibrator generates sufficient vibrations by means of pneumatic/hydraulic/electric motors that are sealed in a submersible housing. Vibrations combined with instrument weight drive the core barrel into the sediment/substrate. The vibrational energy delivered to the core tube (Figure V8; cf. Figures V1 and V2) induces vertical penetration by temporarily displacing sediment particles to overcome frontal resistance and wall friction. The technique is thus most efficient in water-saturated sediments because vibration increases the pore pressure along the wall of the core tube by generating a thin layer of liquefaction. As the vibrating tube penetrates the sediment, it displaces the bedded particles on both sides of the wall. This results in the collection of a largely undisturbed core of sediment within the vibrocore tube. Although useful in water-saturated subaerial environments, such as in marshes, underwater sediments present the optimum medium of application. Starting in the 1950s, vibracoring became increasingly accepted as an efficient and useful method for collecting underwater core samples. The technique was initially slow to gain acceptance because vibrators could not be easily outfitted for underwater use; the problem was soon overcome and a variety of vibrators are now available. Figure V9 shows a typical medium-sized vibracore assembly being lowered over the side of a research vessel.

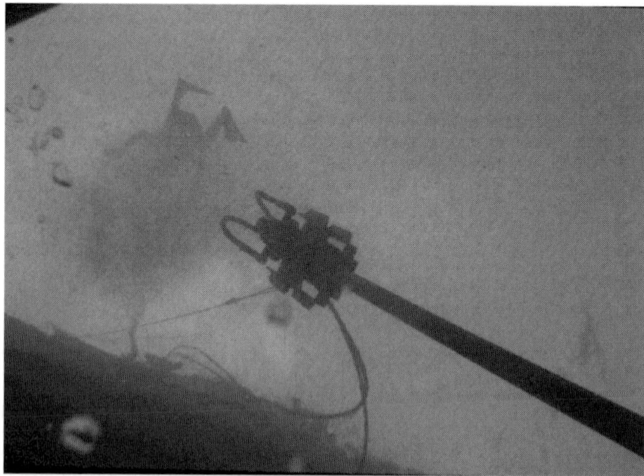

Figure V9 Diver's view of vibracore being lowered over the side of the research vessel. The dark area in the upper left of the photo is the bottom of the vessel's hull. Cables to power the vibrahead and a lifting cable are visible to the left of the diver near the water surface (photo courtesy of Coastal Planning & Engineering, Boca Raton, Florida).

Among the three basic types of vibrators (i.e., pneumatic, hydraulic, and electric), pneumatic piston vibrators were preferred in early research because they worked underwater and did not involve the undersea use of electrical current. Although still deployed in many surveys, this type of vibracoring setup requires an air compressor and the hoses that sometimes become an impediment in swift or choppy waters. Hydraulic vibrators use fluid flows in a closed circuit in balance with the ambient environment. A hydraulic power plant and an umbilical hose are required, presenting similar drawbacks to pneumatic vibracorers. All things considered, including the force/weight ratio from the power source to the vibrahead, electric vibracorers become an attractive choice, particularly when the power source is already part of the vessel system. Many researchers conclude that electric vibracores are good choices for work in shallow waters viz. those characterized by surf or spray zones. There are two main types of electric vibracorers, electromagnetic vibrators and rotating-eccentric vibrators. The second type is often preferred because the dynamics of the overall spring-like system soil–coretube–vibrohead, the contra-rotating eccentric vibrators mounted in parallel (cf. Figure V7), does not require mechanical linkage or special gear for establishing synchronized motion for delivering oscillatory force along a vertical axis. Many combinations of vibracore assembly are available for a variety of environmental conditions. Lightweight parallel-mount twin vibrators that, for example, run at 8,000 V/m on standard 110 V AC current, can be operated with a 5 kW camping-type generator. Heavier models often run at 2,800–3,400 V/m on 3-phase, 50 or 60 Hz, 220–440 V power sources. There are also vibracorers, operating in the medium-frequency range, based on a single vibrator that delivers an oscillatory force to the vertical axis, also often operating on 3-phase 220–440 V. Other variations of basic setups include a "vibro-torsional" operating mode where the two contra-rotating vibrators are mounted along the same axis instead of being in parallel. This coaxial configuration or arrangement synchronizes spontaneously, as in a parallel mount, while adding a horizontal oscillatory torque of small amplitude to the main vertical oscillation. Above the medium range, there is a domain of vibracoring known as "resonant drive" with frequencies on the order of 200–360 Hz (12,000–22,000 V/m) and amplitudes down to a fraction of a millimeter. In this resonant drive mode, the core tube vibrates like a musical string with stationary nodes and antinodes that help to efficiently overcome wall friction, but these units lack force and amplitude for frontal penetration. There is thus a range of vibracore types that can be deployed to specific use.

Advantages and disadvantages of vibracoring

Classical problems associated with coring in general include, for example, flow of external sediment into the core barrel due to increased or reduced stress below the core nose, wall friction, sediment deflection below the cutter, shear failures, thinning or compression of softer strata, thixotropic liquefaction, and textural rearrangement (e.g., Rossfelder and Marshall, 1967; Smith, 1984; Crusius and Anderson, 1991; Morton and White, 1997). Withdrawal disturbances may take place in response to decrease in hydrostatic pressure below the sample, increase in pressure over the sample, lack of wall friction, and adhesion between the sample and the core wall. The sampler may inadvertently be tipped over and dragged along the bottom because of improper retrieval. Even though modern sampling methods have been overcome, there are still sampling problems associated with state-of-the-art core retrieval using vibracore samplers. Some of these difficulties are briefly outlined as precautionary observations for a system of bottom sediment sampling that is generally reliable (e.g., Blomqvist, 1985).

Vibracorers have the advantages of simple construction and easy mobilization, but they are sometimes unwieldy in congested commercial areas such as harbors, and their cost may be beyond the budget of small consultants or universities (Larson *et al.*, 1997). Vibracores are not very effective in either very compact (dense) or very loose sediment; they are not suitable for sediments containing large clasts that can block penetration of the core barrel.

While common vibratory corers are capable of penetrating 6 m or more in unconsolidated sediment, actual performance depends on the nature of the sub-bottom sediment. Under unfavorable circumstances (viz. rough seas, rocky bottom) very little sediment may be recovered. Limited recovery may be primarily due to lack of penetration of the core barrel, blockage of the bit by rock, or loss of sediment during recovery. Core penetration is measured both visually and with an electronic penetrometer. Core recovery is determined by measuring the total length of sediment in the core barrel immediately following retrieval, and the percent recovery calculated as follows (Brooks *et al.*, 1999):

$$\% \text{ Recovery} = (\text{length recovered/length penetrated}) \times 100$$

In certain instances, penetration refusal is met before full penetration of sediment is achieved (Anders and Hansen, 1990; Morton and White, 1997). In such cases the vibracore is removed, and the short core is extracted and stored. A new core liner is installed and the vibracore is again deployed. The core barrel is hydraulically jetted down to the depth of the refusal and then the regular vibracoring is resumed to the targeted depth. In other cases, refusal means that the vibracore cannot be driven any further into the sediment and the borehole is terminated. Coarse fragments such as diamictite, scree, lag, or carbonate/coral rubble may, for example, be larger than the core barrel or the deposit may be so densely packed, if containing smaller diameter clasts, that vibracoring into these materials becomes problematic. Core barrels sometimes get stuck in the sediment and removal requires patience, strength, and some ingenuity.

Vibratory corers are capable of penetrating up to 12 m of unconsolidated sediment, but actual performance depends on the nature of the sub-bottom material. Under unfavorable conditions, however, less than 1 m may be recovered; no coring device has been developed that eliminates the potential for core shortening. There are, however, practical solutions to problems associated with vibracore sample loss or compaction (Smith, 1992). Core shortening may result from several factors that include, for example, physical compaction, sediment thinning, or sediment bypassing (Morton and White, 1997). Limited recovery occurs in response to lack of penetration where stiff clays, gravel, and hard-packed fine to very fine sand are usually most difficult to penetrate. "Freezing" of material in the core liner, which is an age-old coring problem due to skin friction before full penetration is reached (e.g., Rossfelder and Marshall, 1967), stops new material from entering the sampler while the core barrel continues to penetrate; this process may result in exclusion of underlying sediments so that some strata are bypassed and not recovered in the core barrel. Lubrication of the inner wall of the core barrel can reduce friction and prevent plugging as additional sediments enter the core barrel. Choice of lubricant to reduce friction is recommended as long as the chemical additives do not interfere with chemical analyses planned for the cores (Morton and White, 1997). Compaction and loss of material during recovery can also cause discrepancy between penetration and recovery, but occurs less frequently. It is sometimes not possible to correct for shortening when it is not certain whether the shrinkage was due to dewatering or loss of core out of the bottom (Wright *et al.*, 1999). Harvey *et al.* (1999) found, for example, that compaction may range from negligible in clay-rich cores up to 40% for organic-rich mangrove-dominated cores. The coring technique, including setup and deploying the corer, coring and recovery is quite rapid compared with standard soil boring operations. Usually, a 6-m-long core can be obtained in a matter of minutes under ideal conditions.

In some areas where vibracoring is difficult, rough ideas of sediment composition can be obtained by alternative sampling procedures. Wash borings, for example, can sometimes provide estimates of grain size and composition by flushing out sediments using compressed air (Figure V10). Jet probes (*q.v.*) also provide rough estimates of bottom sediment types but, like wash borings, they do not provide undisturbed samples that are recoverable for further offsite analysis.

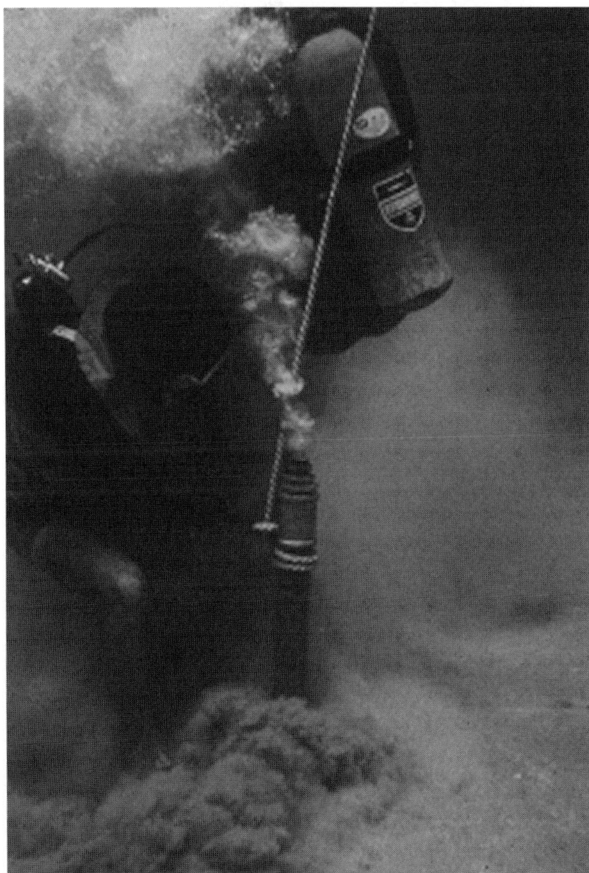

Figure V10 A "wash boring" that is operated by jetting down a pipe within a pipe at intervals of 1–2 m and "washing" up a sample with a blast of compressed air between the pipes. This exploratory technique was previously used to provide preliminary information on the quality of sand in a potential borrow area before engaging more expensive vibracoring operations. Although surface sand samples can be easily obtained, it is necessary to know what type of material lies at depth. This method of sampling provided an inexpensive way to sample sediment at depth, which would be "washed up" the pipe and trapped in cloth bags that were used to catch the samples. The heavier materials (gravel, etc.) could not be lifted in the pipe, but divers could "hear" gravel and rocks banging against the casing to establish the presence of unusable coarse materials. Fines could not be trapped because the material would pass through the cloth bags used to trap sediment samples; the "cloudiness" of the water, however, provided experienced divers with a rough indication of how much silt plus clay was present in the sediment. The penetration of the pipe was similar to jet probes that are now used in order to test the depth of unconsolidated sediment. Although crude, this method provided useful information for refining sand searches. The limitations of disintegrated wash borings compared to undisturbed vibracores emphasize the value of cores that can be subsampled, analyzed, and archived. Wash borings today find specialized applications by most researchers (photo courtesy of Coastal Planning & Engineering, Boca Raton, Florida).

Core logging and sample analysis

During the field data collection phase in large surveys, a preliminary analysis of the cores and samples from the cores is made on a daily basis to obtain information for making advantageous modifications to the survey plan. In addition, when the specific site surveys of high-potential borrow sources are undertaken immediately following the general survey; the preliminary analysis must suffice for selection of these sites, and needs to be as complete as possible. The scope of preliminary core and sample analyses in the field is limited by: (1) only partial visual and physical access to the cores for preliminary logging and sampling; (2) the type and extent of sample analyses that are possible in the field, which in many respects are not comparable to laboratory analysis; and (3) the field analysis for each core that must be completed in a limited time frame in order to keep pace with the progress of the survey.

After each core is taken, the liner containing the cored materials is removed and replaced by a fresh liner. Liners may be clear acrylic plastic tubes that allow observation through the wall. Heavy scratching by granular material or silt and clay particles may smear the inner wall obscuring the contents. Where the cored material is visible, logging and selection of samples can be made. Some corers use aluminum or opaque plastic tubing as core barrels without liners and use the core barrel themselves as the containers using a fresh core barrel for each core run.

Access to cores is best obtained by splitting the cores lengthwise to expose the cored section (Figures V6 and V11). Prior to sampling or disturbance, the core materials are usually photographed next to a scale. Prior to sampling for laboratory analysis, cores are examined visually to determine pertinent characteristics such as size distribution and composition in terms of relatively broad categories. For this purpose, a hand lens and size comparison charts are useful aids. In addition, samples that appear on visual examination to be possibly suitable as fill material should be further analyzed to obtain data on their size distribution characteristics by more accurate means than visual inspection. This can be done using small-diameter sieves to separate small samples into appropriate size fractions, which are then weighed to determine the percent weight of each size fraction. Minimal equipment needed for this procedure is a drying oven, small-diameter sieves covering the Wentworth sand size ranges at $\frac{1}{2}$ phi intervals, and a small top-loading electronic balance with a precision of at least 0.01 g (Meisburger, 1990).

Figure V11 Split core liners from vibracore that penetrated sandy bottom sediments in an inter-reefal area offshore from Miami, Florida. The vibracores are split longitudinally, one half is retained for archival purposes and the other is logged and analyzed in the laboratory for grain-size parameters, organic content, coarse fragments (e.g., loose shell hash, pieces of coral, chunks of coquina), mineral composition (e.g., silicates versus carbonates), etc. Note: Core barrels are generally cut into sections that are about 2 m in length. Depending on the length of core obtained, there may be a small section left over that would be difficult to split lengthwise. These small leftover sections also tend to be disturbed because they are at the end of the core. The two PVC ends shown here are used to store "bit samples" because the core-cutting bit does not have a liner. Typically, when a core is being recovered, the material in the bit falls out because it is below the core catcher. Sometimes the sample stays in the bit and it is saved as shown in the photo. The two piled samples sitting on paper in the center of the photograph are plugs of rock, cut from the subsurface, that were lodged in the bit (photo courtesy of Coastal Planning & Engineering, Boca Raton, Florida).

Usually, cores are opened in the laboratory, logged, photographed, and portions are then sampled for various purposes. The presence of carbonate shelly debris or shell layers, bioturbation clasts and infills, gravels, (de)oxygenated surfaces, color changes, and other notable features are recorded prior to sample removal (Figures V12 and V13). Notations are also made for the presence of epifauna, abraded foraminifera tests, coproliths, or any other unusual biological feature such as insect carapaces and appendages. Molluscan assemblages (including microfossils) are, for example, often removed from the core prior to subsampling for other types of analyses (e.g., Barusseau *et al.*, 1988; Wright *et al.*, 1999). Standard laboratory sediment analysis may be conducted following the sieving methods proposed by Wentworth (1929), by using a rapid sediment analyzer (sand fraction) (e.g., Schlee, 1966), or a Coulter Counter (silt and clay fractions) (e.g., Shideler, 1976). In addition to standard particle-size analyses, there are numerous other kinds of analysis that can be performed on core materials, depending on their nature and composition. Organic-rich deposits such as peat, for example, can be vibracored taking care to make corrections for compaction (usually –10%) by making stratigraphic comparisons with an uncompacted Eijkelkamp core from the same site (Gehrels, 1994). Detrital plant fragments are often selected for dating by the accelerator mass spectrometer (AMS) ^{14}C method (e.g., Gehrels and Belknap, 1993), in preference to bulk samples that can be contaminated by humic acids and younger roots (Belknap *et al.*, 1989). Methods for foraminifera sampling and sample preparation are similar to those described by Scott and Medioli (1980).

Bottom sampling for coastal sand searchers

Samples of seafloor sediments are required for numerous purposes, the least ambitious of which is to gain a rough idea of sediment types. Different kinds of sampling devices are usually geared for the collection of fine or coarse-grained surficial sediments or continuous cores to specified depths. There are a variety of grab-type samplers of different sizes and design that are used for obtaining surficial samples. Most consist of a set of opposing, articulated, scoop-shaped jaws that are lowered to the bottom in an open position and closed by some mechanism. In this process, a sample is retrieved between the closed jaws. Many grab samplers can be deployed by hand while others require lifting gear. Dredge samplers, which can be dragged a short distance along the bottom to dredge up a sample, are sometimes used in place of grab samplers that are subject to losing finer material during recovery when shells or gravel prevent complete closure of grab samplers. Jet probes (*q.v.*) and wash bores are sometimes used on a reconnaissance basis to gain general information on sediment types. While obtaining surficial samples is helpful, it is of limited value because vertical projection of surface data is highly unreliable. Also, the expense of running tracklines for the sole purpose of sampling surficial sediments is not economically justified by the value of the data obtained (Meisburger, 1990).

Direct sampling of sub-bottom materials is essential for borrow source identification and evaluation. This is usually accomplished by means of a continuous coring apparatus that can obtain cores 7–13 m in length (cf. Figure V1). In the types of sediment usually encountered in borrow site exploration, gravity corers are not suitable for obtaining cores of the requisite length, and some form of powered corers must be used. In most cases, vibrator-driven coring devices have been used for this purpose (Meisburger and Williams, 1981).

Collection of surface samples is not useful for most geological and geotechnical applications because depth parameters are required for various purposes. Although surficial samples (e.g., those obtained from grab samplers or jet probes) may provide rough ideas about recent submarine processes, they do not provide at-depth information about sediment thickness, structure, composition, and stratigraphic information. Jet probes (*q.v.*) provide useful information on a reconnaissance basis because large areas of seafloor can be covered relatively quickly at modest costs. Even though jet probes provide a rough idea of sediment types, specialized information is required to prove out potential borrow areas. Because vibracores provide the kind of detailed information that is required, vibracore surveys are often conducted in tandem with geophysical investigations. Geophysical surveys and geotechnical work is thus able to definitely establish deposit geometry, and the quality and quantity of sand. Vibracores are thus commonly deployed in the final stages of sand search investigations, that is, after evaluation of all available bathymetric, seismic, and jet probe data (e.g., Finkl *et al.*, 1997). Closely spaced (e.g., about 300 m apart) vibracores thus help to minimize uncertainty factors related to sediment quantification and type, thereby maximizing confidence in selection of offshore borrows. Figure V14 shows an east–west, cross-shore seismic reflection profile

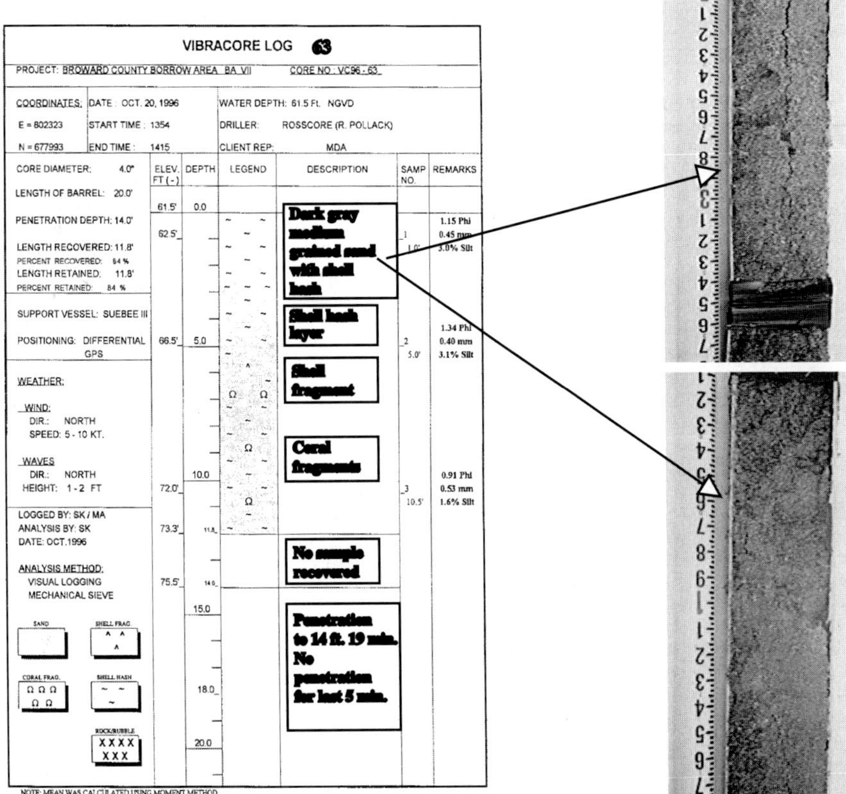

Figure V12 Example of a vibracore log showing uniform sands. This standardized summary format brings together many different kinds of data that are used to help characterize the materials contained in the cores. In addition to photographs of the core section (right side of diagram) with numbers inverted for easy determination of depth, the core log provides visual description, results of mechanical sieve analyses, notable features of the cored materials, penetration rates, and depth at refusal. The arrows point to relatively uniform dark gray, medium-grained sand with sparse inclusions of shell hash (courtesy of Coastal Planning & Engineering, Boca Raton, Florida).

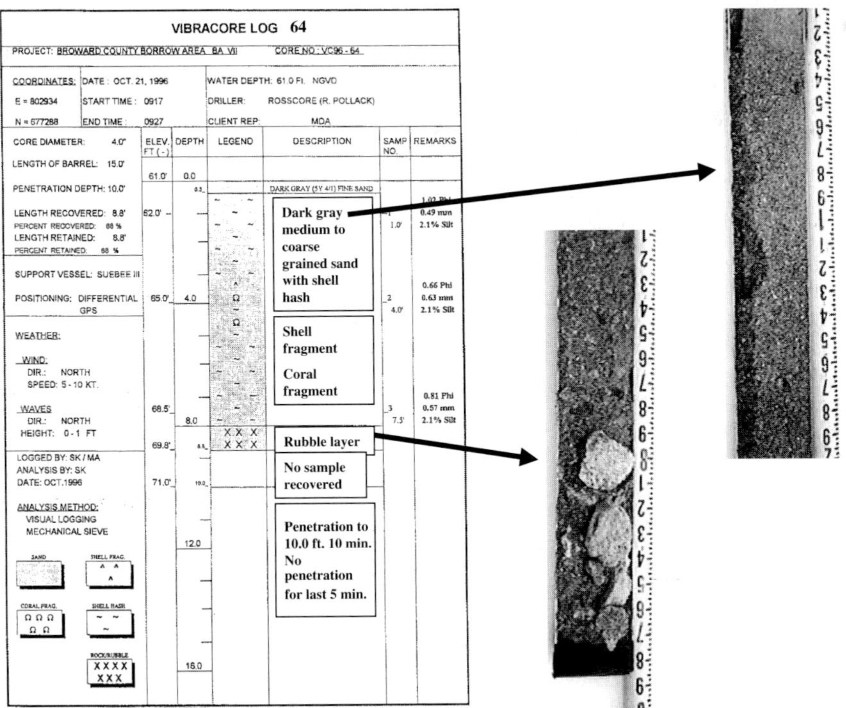

Figure V13 Example of a vibracore log showing a marked discontinuity in a basal rubble layer. Photographs of the split core (right side of diagram) show medium to coarse-grained sand interspersed with shell hash inclusions that terminate in a very coarse carbonate rubble layer. Sediment loss and refusal of the core barrel occur in the rubble layer that cannot be penetrated. The value of composite vibracore logs, as displayed here, is immediately apparent because they red flag potential problems when dredging for beach-quality sands that will be mined offshore and pumped onshore for beach replenishment (courtesy of Coastal Planning & Engineering, Boca Raton, Florida).

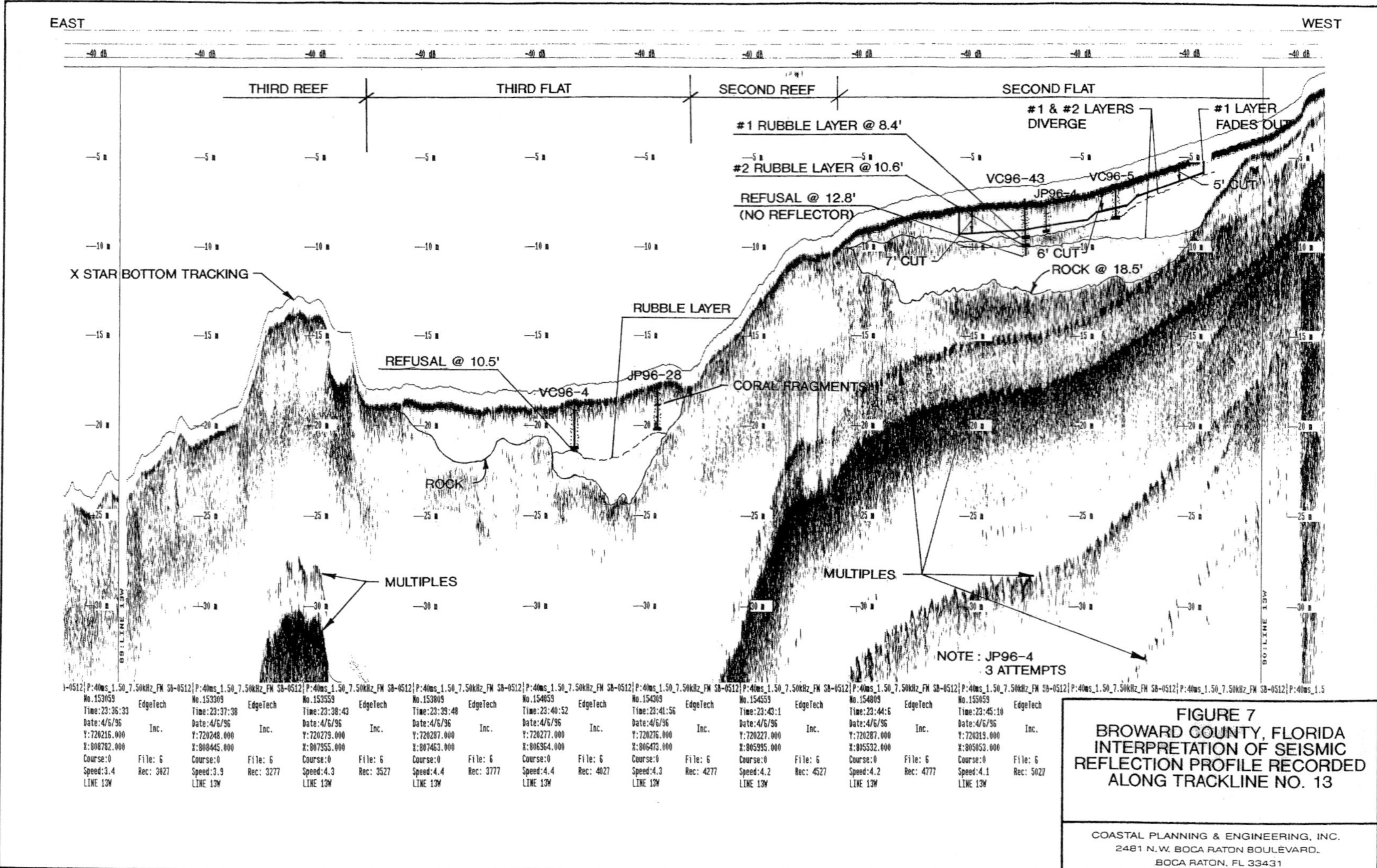

Figure V14 Cross-shore seismic reflection profile and vibracore locations, southeast coast of Florida in Broward County. This north to south view transects the shore-parallel Florida Reef Tract (so that the east or left side of the diagram leads to deeper offshore water). Depressions between parabathic coral reefs fronting the southeast Florida Peninsula contain sedimentary deposits that range from areas of silt plus clay to coarse rubble accumulations, but most areas supply sandy deposits. The second sand flat lies at about 5 m depth whereas the deeper third flat lies under approximately 17 m of water. The vibracores groundtruth the seismic reflection record by verifying the thickness of inter-reefal sandy sediments. Note the presence of carbonate rubble layers at depth in the second and third sand flats (shore is to the west, right side of diagram). Information obtained from the vibracores is essential to complete interpretation of seismic data. The presence of carbonate rubble and coral fragments delimits dredging operations to areas within inter-reefal sand flats that are devoid of materials that are unsuitable for beach renourishment. Offshore sand searches often rely on vibracore data for detailed analysis of sedimentary deposits and interpretation of geophysical surveys (courtesy of Coastal Planning & Engineering, Boca Raton, Florida).

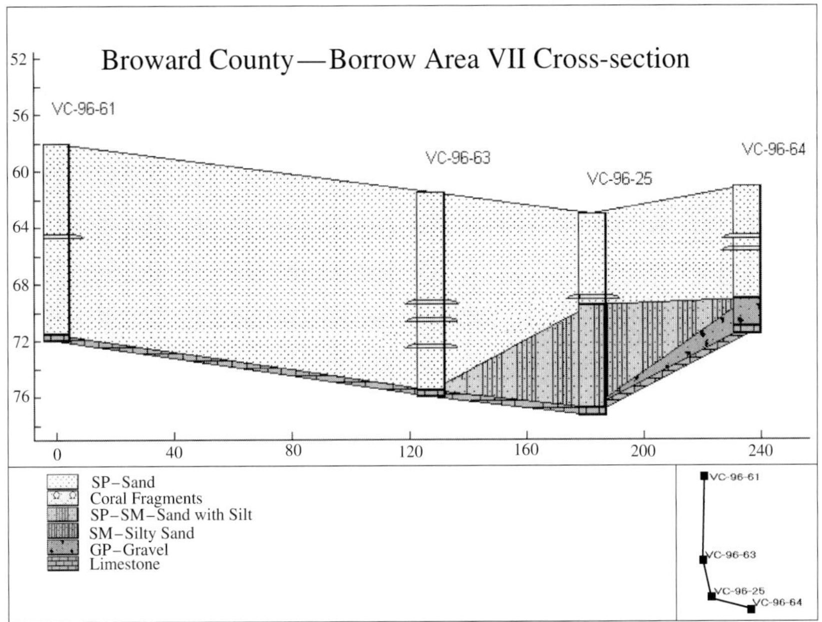

Figure V15 Example of a typical cross-section constructed from vibracore logs for a borrow area offshore southeast Florida in Broward County. The section shows the presence of inter-reefal sands overlying a limestone base (left side of section) and highlights unsuitable fine-grained (silty) materials and coarse gravels (right side of section) as in core VC-96-25. Fine-grained and coarse-grained materials are unsuitable for beach renourishment projects and must be identified prior to dredging so they are not placed on the beach. Because such materials are not compatible with native beach sands, their presence reduces the potential of borrow areas and becomes an important factor in estimating offshore reserves of extractable beach-quality sands (courtesy of Coastal Planning & Engineering, Boca Raton, Florida).

(based on a CHIRP X-Star sonar) on the inner continental shelf of southeast Florida in Broward County. The interreted and annotated sub-bottom profile shows the sedimentary cover that overlies and partially infills inter-reefal troughs, identified as the second and third flats. Even though the seismic survey results can identify rock surfaces of the coral reefs (and buried reef), some carbonate rubble layers are not shown on the trace if they do not present a reflector. Examination of vibracores precisely locates the depth and thickness of coral fragments present in the sandy inter-reefal infills. Summary cross-sections, as shown in the example of Figure V15, although based on seismic and geotechnical data, are confidently constructed using closely spaced vibracore logs. As shown in Figure V15, materials that are unsuitable for beach renourishment (i.e., should not be dredged) are indicated (by a warning red color in the original diagram) as sand with silt, silty sand, gravel, coral fragments, and limestone. Vibracore locations are noted in this fence diagram by a simple coded annotation such as VC-96-61, which indicates a vibracore borehole (VC) that was collected in 1996 (96) as borehole number 61 in a sequence. The strategic role of vibracores as verification or seatruthing of geophysical survey data and as key indicators of actual sedimentary conditions in their own right is patently obvious.

Conclusion

Vibracores are specialized sampling procedures for obtaining continuous, undisturbed cores. Although limited by the maximum length of retrievable core (about 7–10 m for most purposes), vibracores find application in many different kinds of coastal studies where undisturbed samples need to be collected from surficial sediments on land or under water. Vibracores can provide invaluable physical, chemical, and biological information that is otherwise unobtainable. Vibracoring systems range from portable, inexpensive setups to more complicated assemblies that require elaborate service platforms on ships or barges fitted with powerful hoisting equipment. Vibracores are an essential component of multifaceted surveys, such as offshore sand searches that attempt to locate potential borrow areas containing large volumes of beach-quality sands. Although vibracores have a number of limitations that preclude their use under a range of conditions, they are deployed to advantage in many different kinds of scientific and engineering applications. In many respects, vibracores are the unsung heroes of coastal research that depends on acquisition of information posited in the sedimentary record. Collection of unbiased reference information concerning vibracores is somewhat difficult because few papers focus on the vibrocoring methodology *per se*. Vibracoring is a sample collection technique that usually only finds mention in the "methods" section of research reports, often as a backup to broader topics such as geophysical or engineering survey. Nevertheless, information obtained from vibracores should not be minimized, as ancillary when in fact in many cases it is primary, even though it is often collected last.

Charles W. Finkl and Syed M. Khalil

Bibliography

Amos, C.L., 1978. The postglacial evolution of the Minas Basin, Nova Scotia. A sedimentological interpretation. *Journal of Sedimentary Petrology*, **48**: 965–982.

Anders, F.J., and Hanson, M., 1990. Beach and borrow site sediment investigation for a beach nourishment at Ocean City, Maryland. US Army Corps of Engineers, *Technical Report*, CERC-90-5, 77 pp.

Barnhardt, W.A., Kelley, J.T., Dickson, S.M., and Belknap, D.F., 1998. Mapping the Gulf of Maine with side-scan sonar: a new bottom-type classification for complex seafloors. *Journal of Coastal Research*, **14**: 646–659.

Barusseau, J.P., Giresse, P., and Manongo, L., 1988. Genesis of a Holocene phosphate placer deposit on the continental shelf of Congo. *Journal of Coastal Research*, **4**(3): 369–379.

Belknap, D.F., Shipp, R.C., Struckenrath, R., Kelley, J.T., and Borns, H.W., Jr., 1989. Holocene sea-level change in coastal Maine. In Anderson, W.A., and Borns, H.W., Jr. (eds.), Neotectonics of Maine. *Maine Geological Survey Bulletin, 40*: 85–105.

Blomqvist, S., 1985. Reliability of core sampling of soft bottom sediment—an *in situ* study. *Sedimentology*, **32**: 605–612.

Brooks, G.R., Kindinger, J.L., Penland, S., Williams, S.J., and McBride, R.A., 1995. East Louisiana continental shelf sedimentation: a product of delta reworking. *Journal of Coastal Research*, **11**(4): 1026–1036.

Brooks, G.R., Doyle, L.J., Suthard, B.C., and DeWitt, N.T., 1999. Inner west-central Florida continental shelf: sedimentary facies and facies associations. US Department of the Interior, *USGS Open File Report* 98–796 (USGS, Eckerd College, and University of South Florida, 1999).

Crusius, J., and Anderson, R.F., 1991. Core compression and surficial sediment loss of lake sediments of high porosity caused by gravity coring. *Limnology and Oceanography*, **36**: 1021–1031.

Davis, R.A., and Kuhn, B.J., 1985. Origin and development of Anclote Key, west-peninsular Florida. *Marine Geology*, **63**(1985): 153–171.

Davis, R.A., Klay, J., and Jewell, P., 1993. Sedimentology and stratigraphy of tidal sand ridges southwest Florida inner shelf. *Journal of Sedimentary Petrology*, **63**(1): 91–104.

Delaune, R.D., Bauman, R.H., and Gosselink, J.G., 1983. Relationships among vertical accretion, coastal submergence, and erosion in a Louisiana Gulf coast marsh. *Journal of Sedimentary Petrology*, **53**: 147–157.

Finkl, C.W., Khalil, S.M., and Andrews., J.L., 1997. Offshore sand sources for beach replenishment: potential borrows on the continental shelf of the Eastern Gulf of Mexico. *Marine Georesources and Geotechnology*, **15**: 155–173.

Freedenberg, H., Hoenstine, R., and Dabous, A., 2000. Preliminary identification of sand resources in federal waters along the central Florida east coast. *Proceedings 2000 National Conference on Beach Preservation Technology* (February 2–4, 2000, Melbourne, Florida). Tallahassee: Florida Shore and Beach Preservation Association, pp. 247–257.

Gayes, P.T., and Bokuniewicz, H.J., 1991. Estuarine paleoshorelines in Long Island Sound, New York. *Journal of Coastal Research*, Special Issue No. **11**: 39–54.

Gehrels, W.R., 1994. Determining relative sea-level change from salt-marsh foraminifera and plant zones on the coast of Maine, U.S.A. *Journal of Coastal Research*, **10**(4): 990–1009.

Gehrels, W.R., and Belknap, D.F., 1993. Neotectonic history of eastern Maine evaluated from historic sea-level data and ^{14}C dates on salt-marsh peats. *Geology*, **21**: 615–618.

Harvey, N., Barnett, E.J., Bourman, R.P., and Belperio, A.P., 1999. Holocene sea-level change at Port Pirie, South Australia: a contribution to global sea-level rise estimates from tide gauges. *Journal of Coastal Research*, **15**(3): 607–615.

Hoyt, W.H., and Demarest, J.M., III, 1981. Vibracoring in coastal environments: a description of equipment and procedures. Newark, Delaware: University of Delaware Sea Grant College Program, DEL-SG-01-81, pp. 20–31.

Kadlec, R.H., and Robbins, J.A., 1984. Sedimentation and sediment accretion in Michigan coastal wetlands. *Chemical Geology*, **44**: 119–130.

Kirby, R., Hobbs, C.H., and Mehta, A.J., 1994. Shallow stratigraphy of Lake Okeechobee, Florida: a preliminary reconnaissance. *Journal of Coastal Research*, **10**(2): 339–350.

Knebel, H.J., and Poppe, L.J., 2000. Sea-floor environments within Long Island Sound: a regional overview. *Journal of Coastal Research*, **16**(3): 533–550.

Kraft, J.C., Allen, E.A., and Belknap, D.F., 1979. Processes and morphological evolution of an estuarine and coastal barrier system. In Leatherman, S.P. (ed.), *Barrier Islands*. New York: Academic, pp. 149–183.

Lanesky, D.E., Logan, B.W., Brown, R.G., and Hine, A.C., 1979. A new approach to portable vibracoring underwater and on land. *Journal of Sedimentary Petrology*, **49**: 654–657.

Larson, R., Morang, A., and Gorman, L., 1997. Monitoring the coastal environment; Part II, Sediment sampling and geotechnical methods. *Journal of Coastal Research*, **13**(2): 308–330.

Levitan, M.A., Kuptsov, V.M., Romankevich, E.A., and Kondratenko, A.V., 2000. Some indication for late Quaternary Pechora River discharge: results of vibrocore studies in the southeastern Pechora Sea. *International Journal of Earth Sciences*, **89**: 533–540.

Meisburger, E.P., 1990. Exploration and sampling methods for borrow areas. Vicksburg, Mississippi, *Technical Report*, CERC-90-18.

Meisburger, E.P., and Williams, S.J., 1981. Use of vibratory coring samplers for sediment surveys. Vicksburg, Mississippi, *Technical Report*, CERC Coastal Engineering Technical Aid No. 81–9.

Morang, A., Mossa, J., and Larson, R.J., 1993. Technologies for assessing the geologic and geomorphic history of coasts. US Army Corps of Engineers, *Technical Report*, CERC-93-5, 140 pp.

Morton, R.A., and White, W.A., 1997. Characteristics of and corrections for core shortening in unconsolidated sediments. *Journal of Coastal Research*, **13**(3): 761–769.

Rossfelder, A.M., and Marshall, N.F., 1967. Obtaining large, undisturbed, and oriented samples in deep water. In Richards, A.F. (ed.) *Marine Geotechnique*. Urbana, Illinois: University of Illinois Press, pp. 243–251.

Sanders, J.E., 1960. Kudinov vibro-piston core sampler; Russian solution to underwater sand-coring problem. *International Geological Review*, **2**: 174–178.

Schlee, J., 1966. A modified Woods Hole rapid sediment analyzer. *Journal of Sedimentary Petrology*, **30**: 403–413.

Scott, D.B., and Medioli, F.S., 1980. Quantitative studies of marsh foraminiferal distributions in Nova Scotia: implications for sea-level studies. *Cushman Foundation for Foraminiferal Research Special Publication No. 17*, 57 pp.

Shideler, G.L., 1976. A comparison of electronic particle counting and pipet techniques in routine mud analysis. *Journal of Sedimentary Petrology*, **46**: 1017–1025.

Shipp, R.C., Belknap, D.F., and Kelley, J.T., 1991. Seismic-stratigraphic and geomorphic evidence for a post-glacial sea-level lowstand in the northern Gulf of Main. *Journal of Coastal Research*, **7**(2): 341–364.

Smith, D.G., 1984. Vibracoring fluvial and deltaic sediments: tips on improving penetration recorvery. *Journal of Sedimentary Petrology*, **54**(2): 660–663.

Smith, D.G., 1992. Vibracoring: recent innovations. *Journal of Paleolimnology*, **7**: 137–143.

Snowden, J.O., Simmons, W.B., Traughber, E.B., and Stephens, R.W., 1977. Differential subsidence of marshland peat as a geologic hazard in the greater New Orleans area, Louisiana. *Transactions Gulf Coast Association of Geological Societies*, **27**: 169–179.

Soil Survey Division Staff, 1993. *Soil Survey Manual*. Washington, DC: US Department of Agriculture Handbook No. 18, 437 p.

Stevenson, J.C., Ward, L.G., and Kearney, M.S., 1986. Vertical accretion rates in marshes with varying rates of sea-level rise. In Wolfe, D.A. (ed.), *Estuarine Variability*. New York: Academic, pp. 241–259.

Toldo, E.E., Jr., Dillenburg, S.R., Corrêa, C.S., and Almeida, L.E.S.B., 2000. Holocene sedimentation in Lagoa dos Patos Lagoon, Rio Grande do Sul, Brazil. *Journal of Coastal Research*, **16**(3): 816–822.

Trask, P.D., 1939. *Recent Marine Sediments: A Symposium*. Tulsa, Oklahoma: The Society of Economic Paleontologists and Mineralogists (SEPM), Special Publication No. 4. [Unabridged and corrected edition of 1955 SEPM publication, republished in 1968.]

Varekamp, J.C., 1991. Trace element geochemistry and pollution history of mudflat and marsh sediments from the Connecticut coastline. *Journal of Coastal Research*, Special Issue No. **11**: 105–123.

Watson,I., and Krupa, S., 1984. Marine drilling exploration—technical and environmental criteria for rig selection. *Litoralia*, **1**(1): 65–82.

Wentworth, C.K., 1929. Method for computing mechanical composition of sediments. *Geological Society of America Bulletin*, **40**: 771–790.

Wright, C.I., Cooper, J. A.G., and Kilburn, R.N., 1999. Mid Holocene paleoenvironments from Lake Nhlange, northern Kwazulu-Natal, South Africa. *Journal of Coastal Research*, **15**(4): 991–1001.

Cross-references

Beach and Nearshore Instrumentation
Beach Stratigraphy
Coastal Sedimentary Facies
Coastal Soils
Jet Probes
Mining of Coastal Materials
Monitoring, Coastal Geomorphology
Nearshore Geomorphological Mapping
Offshore Sand Sheets
Sequence Stratigraphy
Shoreface

VORTICITY

Vorticity is the tendency for spin or rotation in a fluid (i.e., vortex flow). As such it is a vector and can be separated into components of spin about the vertical axis and either or both of the horizontal axes. Eddies observed as water moves past obstacles such as bridge pilings is an example of relative vorticity flow about the vertical axis.

Figure V16 shows a Cartesian coordinate system with the x-axis pointing eastward, the y-axis pointing northward, the z-axis pointing upward in the opposite direction to the gravity vector $\vec{g}$. The x–y plane is parallel to a level surface, and relative vorticity (ζ) about the z-axis is defined as $\zeta = (\partial v/\partial x) - (\partial u/\partial y)$. In this equation $+v$ is the flow in the northward direction and $+u$ is the flow in the eastward direction, relative to the fluid in which the eddy is imbedded. Components of relative vorticity can be defined about the other two axes, but the most important one in mesoscale oceanography and meteorology is the vertical component ζ.

To interpret ζ, imagine a disk rotating cyclonically (counterclockwise in the northern hemisphere) about a spindle (*q.v.* Figure V16). Along the x-axis, the farther from the center of rotation, the larger will be the positive v-component of velocity. Similarly along the y-axis, the negative u-component increases as distance from the center increases. On such a disk, both $\partial v/\partial x$ and $-\partial u/\partial y$ are positive quantities, thus defining positive relative vorticity as cyclonic circulation.

Since the ocean and the atmosphere are imbedded on a rotating planet, there is another component of vorticity called planetary vorticity. Planetary vorticity is the familiar Coriolis parameter $f = 2\Omega \bullet \sin \phi$, where Ω is the rotation rate of Earth, 7.2921×10^{-5} radians per second [or 360° per sidereal day], and ϕ is latitude. The sum of relative and planetary vorticity, $\zeta_A \equiv \zeta + f$, is known as the absolute vorticity.

Absolute vorticity, or more precisely the timechange of absolute vorticity, $d\zeta_A/dt$, is related to vertical motion in geophysical fluids. Hydrodynamicists have shown that when following along the path of a parcel of fluid, the so-called Lagrangian perspective, that $d\zeta_A/dt = -\zeta_A D_h$. In this equation D_h is the horizontal divergence of flow where $D_h = (\partial u/\partial x) + (\partial v/\partial y)$. This very important expression shows that in the ocean changes of absolute vorticity are related to upwelling and downwelling, because if a surface flow is divergent, water from below must upwell to replace the water that is moving away.

As an example of the relationship between absolute vorticity time-change and vertical motion, consider the Gulf Steam as it meanders eastward in the offing of Cape Hatteras, North Carolina. For simplicity, think of the meanders as horizontal waves and that changes in f are small. As a parcel of fluid moves from the crest of a meander where ζ_A is negative toward the trough where ζ_A is positive, the time-change $d\zeta_A/dt$ is positive and thus D_h is negative. Negative divergence is convergence, and thus there is an area of downwelling water between the meander crest and trough. Similarly, as the parcel moves from the trough to the next meander crest, $d\zeta_A/dt$ is negative and an area of upwelling is observed.

When the ocean is regarded as a barotropic fluid and D_h is integrated from the surface to the bottom where the water is H deep, yet another form of vorticity can be written: $(\zeta + f)/H =$ constant. This ratio is known as potential vorticity, and it is a conservative property of the ocean and the atmosphere. This very important equation shows that geophysical fluids change direction when either its latitude changes or when the water depth H changes.

Two well-known observed examples of conservation of potential vorticity are found in the Atlantic Ocean (Figure V17). First, as water from the southern hemisphere flows across the equator in the Brazil Current, the Coriolis parameter f changes from negative to positive. If H is constant, then ζ must become negative in order to conserve the sum $\zeta + f$. A negative relative vorticity is anticyclonic flow, and thus the water must turn eastward as it passes north of the equator. This is the formation of the Brazil Current Retroflection that eventually leads to the North Equatorial Countercurrent in the Atlantic Ocean.

A second example is found in the Gulf Stream off Charleston, South Carolina. As the Gulf Stream flows northward it encounters a shallow bottom topography feature known as the Charleston Bump. In order to conserve potential vorticity, with f approximately constant, a decreasing H must be balanced by a negative ζ. As with the Retroflection region, negative relative vorticity ζ implies that the Gulf Stream must turn anticyclonically offshore and flow eastward. However, H increases quickly as the current enters the deep water of the continental slope, and ζ changes again, causing the Gulf Stream to meander cyclonically onshore. Thus conservation of potential vorticity initiates meanders in the Gulf Stream that are amplified downstream off Cape Hatteras.

Further reading on this subject may be found in the following bibliography.

George A. Maul

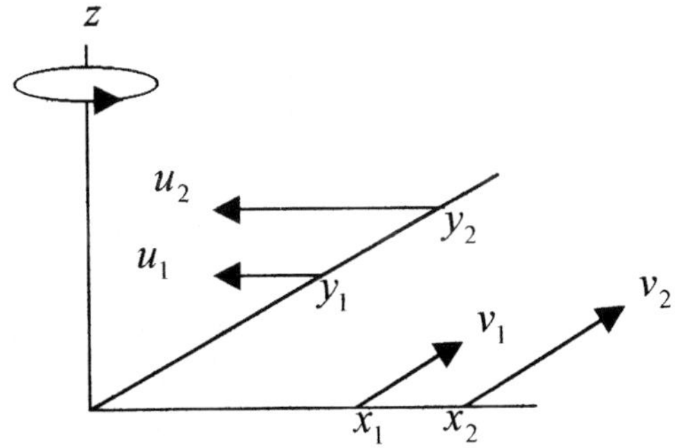

Figure V16 Illustration of relative vorticity ζ in anticlockwise (cyclonic in the northern hemisphere) rotation about the z-axis at a point $x = 0$, $y = 0$.

Bibliography

Knauss, J.A., 1997. *Introduction to Physical Oceanography*, 2nd edn. Upper Saddle River, NJ: Prentice Hall.

Neuman, G., and Pierson, W.J., Jr., 1966. *Principles of Physical Oceanography*, Englewood Cliffs, NJ: Prentice Hall.

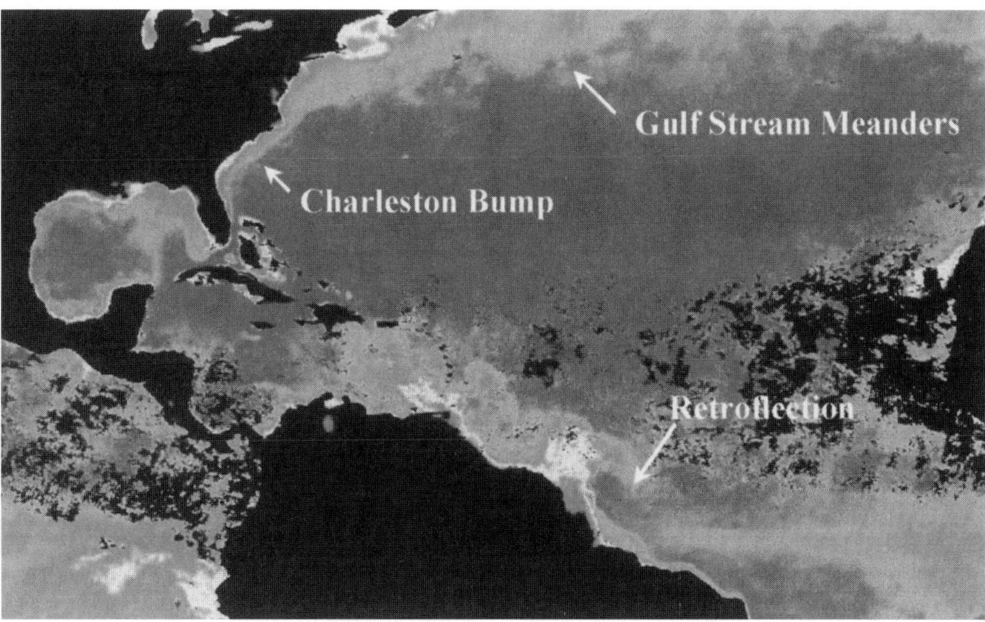

Figure V17 SeaWIFS image of the Intra-Americas Sea showing two regions of potential vorticity conservation: the Brazil Current Retroflection Region off the mouth of the Amazon River, and the Charleston Bump area of the Gulf Stream.

Ocean Circulation, 2nd edn. The Open University. Oxford: Butterworth-Heinemann, 2001.

Cross-references

Coastal Currents
Coastal Upwelling and Downwelling
Pressure Gradient Force
Remote Sensing of Coastal Environments

VOLCANIC COASTS

Volcanism, the ejection of molten magma from the earth's interior onto the surface, is petrologically classified into three types, namely, basic, intermediate, or acidic, depending upon the proportion of silica in the volcanic ejecta. Basic volcanic rocks contain a high proportion of ferromagnesian minerals (Fe + Mg) relative to silica (SiO_2), and are thus relatively dense, heavy, and dark colored, and tend to be ejected as relatively nonexplosive and fluid lava streams. Acid volcanic ejecta contains a much higher proportion of silica, is accordingly relatively light and light colored, more viscous, and is ejected explosively. Intermediate volcanic ejecta products are also intermediate in composition and explosive properties. These properties determine the rock hardness and influence the susceptibility of the rocks to physical and chemical weathering.

Basalts are the major basic volcanic deposits. These are typically extruded as fluid lava streams, such as at Kiluea, on the main island of Hawaii (Macdonald *et al.*, 1983). The fluidity of basalt lava normally allows it to exude interstitial gases easily so that the eruptions are relatively explosive-free, and the flowing lava streams follow pre-existing valleys, forming lava pools or sheets in low-lying topography. Often the fluid lava solidifies as massive thick sheets on land, and as a result of the slow cooling, the basalt fractures into hexagonally jointed vertical columns. When fluid basalt is extruded into water the surface cooling is much faster, typically resulting in closely jointed "pillow lavas." Basalt extruded from a single vent often forms a concave upward volcanic cone due to minor "fire fountaining" of pyroclastic deposits occurring around the vent, and resulting in scoria cone formation. However, little volcanic ash (tephra) is ejected.

Andesite lavas tend to possess greater viscosity, with which is associated greater explosiveness as the lava is ejected. Andesitic eruptions can form thick sheets but when ejected from a single vent tend to be explosive and accompanied by ejection of large volcanic "bombs" into the atmosphere, along with scoriaceous material, and finer tephra material. When ejected from a single vent, the volcano tends to form a mound around the vent. Frequently, however, much of the volcano form is built up from the intermittent eruption of volcanic lava chunks, bombs, and tuff. If these materials become fluidized with waters, they flow as mudflows or *lahars* down the volcano slopes, ending up as irregular hummocky topography around the lower slopes of the volcano (Cotton, 1944). Frequent lahar activity may form a ring plain of laharic deposits—Mt Taranaki in New Zealand is a classic example.

Acid, or *rhyolitic* volcanism, produces gas-rich and viscous lava (Cotton, 1944), and typically the volcanic ejecta is explosively extruded as viscous rhyolitic or dacitic lava. In extreme cases of Pelean type eruptions, massive burning clouds of molten incandescent pumiceous and fine ash ejecta, exuding gases, are explosively erupted and reach high into the atmosphere. The explosion cloud of incandescent particles eventually collapses, creating *nuées ardentes*, consisting of hot glowing clouds of self-lubricating gas-exuding pumice and ash, which may flow at great speed across the landscape. When this material pools in topographic lows and cools relatively slowly, the fragmented rock clasts liquefy, weld together, and solidify as *ignimbrite* sheets—but typically of less density and hardness than the andesitic and basaltic flows. Acid volcanic eruptions often result in considerable ejection of pumice deposits, which may blanket the landscape and are subsequently easily eroded to provide considerable sediment input from the rivers to the coastal littoral zone.

Apart from these "fundamental" types of volcanic deposits, volcanic breccias, and conglomerates can be deposited into the marine environment. The resulting rocks possess a fine tuffaceous matrix of mixed marine and volcanic origin separating the larger breccia clasts of basalt or andesite.

The various volcanic lithologies subjected to coastal marine processes exhibit a wide variety of rock hardness, composition, jointing pattern, as well as existing in a variety of wave energy, tidal range, and climatic (temperature) regime environmental conditions. As expected, a concomitant variety of coastal volcanic landforms result.

Erosion processes and geomorphology of volcanic rocky coasts

Erosion processes along rocky coasts are primarily physical, subaerial, and biological. Physical processes include the destructive power of breaking waves, and the consequent hydraulic and pneumatic pressures exerted when broken wave bores are forced along joint planes. Abrasion by the sediments carried by the waves can occur at the base of cliffs and on shore platforms surfaces, but for coastal outcropping volcanic rocks, is generally a minor process. Subaerial processes include weathering occurring within the atmosphere. Particularly potent, especially for fine grained tuffaceous volcanic rocks, is the effect of constant wetting and drying in the intertidal and supratidal zones, a process controlled by the level of permanent pore space saturation of the rock, termed "water level weathering" by Wentworth (1938). Above the level of pore water saturation, subaerial weathering and chemical oxidation predominates. Volcanic breccias can illustrate the importance of subaerial weathering and wetting drying process in shore platform evolution. Wetting and drying is particularly effective on the fine-grained tuffaceous matrix, so that in areas of low wave energy a near-horizontal, high-tide level bench-type shore platform evolves (Figure V18). On the other hand, where the breccia outcrops and is subjected to high-energy waves, a higher-level irregular notch forms as a result of a higher level of rock pore space saturation, but no horizontal bench occurs.

Biological weathering is typically not an important geomorphic process for volcanic deposits, although it is for certain other coasts such as limestone.

Figure V18 A high-tide bench-type shore platform with undercut notch cut into a cliff of volcanic breccia. The sub-horizontal platform planed at the high-tide level is due to "water level weathering" (photo: T. Healy).

Because volcanic rocks are often massive, indurated, hard, and with coarse jointing patterns, they are typically resistant to the various erosion processes operating at the coast, especially the subaerial and biological processes. Paradoxically a major function of marine action (i.e., the physical power of the breaking waves) is to erode along joint planes and fissures, leading to destruction of the platforms and cliffs, and removing the weathering debris from the foot of the cliffs, thereby maintaining a steep cliff profile. The weathered products from the marine erosion of the volcanic rocks are typically of boulder and cobble size, which are often transported by the waves to create coarse boulder and gravel beaches nearby (Figure V19).

Where unweathered basalts or andesites outcrop at the coast, these hard, dense rocks, are resistant to weathering by marine processes, especially where they possess a low intensity of jointing, that is, the resulting geomorphology is of terminally steep cliffs, perhaps exhibiting incipiently formed narrow high-level benches and topographically irregular platforms. Caves are frequent where the breaking waves have forced along the vertical joints. A classic case is the well-known "Giants Causeway" in Ireland. For tropical coasts, chemical weathering is much more intense, and basalts may be weathered to clays; in such cases broader shore platforms can evolve.

For rhyolite flows, the more viscous lava tends to form "plastered layer mounds" or domes, and these are also relatively resistant to erosion by the sea. Over Holocene time since sea level has been at its approximate present level, the sea has often eroded jagged irregular cliffs, as at Mayor Island, New Zealand. Ignimbrites are likewise typically massive with few joints, and thus form high cliffs, as along New Zealand's Coromandel coast. Compared to basalt and andesite, the rhyolites and ignimbrites are less dense and susceptible to greater weathering rates. Some ignimbrites with high calcium content may evolve coastal geomorphology similar to tropical limestones (Figure V20) with undercut notches and sub-horizontal, subaerially weathered shore platform surfaces planed to the level of rock pore space saturation.

Erosion of coastal lahar deposits along New Zealand's high-energy Taranaki coast, results in lahar cliffs, irregular shore platforms, often littered with boulders, small embayments with boulder and cobble beaches, and rugged offshore reefs surmounting submarine platforms. Erosion of the lahar deposits also produce large quantities of titanomagnetite "black" ironsand (Figure V21) which further updrift produce high dunes mined as an iron ore deposit at Taharoa.

Figure V19 Erosion of an andesite cone in a high-wave energy environment, resulting in steep cliffs, and a coarse clastic beach of boulders surmounting a subaqueous shore platform (photo: T. Healy).

Other volcanic landforms at the coastline

Often specialized volcanic landforms occur coincidentally in the presently active coastal environment. Thus, large-scale collapsed volcanic vents, or *calderas*, such as Santorini in the Greek archipelago (Figure V22), or Banks Peninsula of New Zealand, have become modified by stream erosion, and flooded by the sea, forming enclosed harbors (Cotton, 1942). Scoriaceous tuff rings and *maars* formed from relatively small-scale phreatic explosions, are relatively easily eroded and the sea can erode through to form a horseshoe-shaped harbor (Searle, 1964).

Active volcanism in the coastal zone can lead to some spectacular visual effects. When lava flows into the sea "pillow lavas" are formed, along with steam and phreatic eruptions, from the rapid cooling of the lava skin in contact with the water, creating closely packed ellipsoidal masses with radial jointing.

Collapsing *nuée ardente* pumice clouds flowing into the sea are a spectacular site and are even believed capable of causing tsunamis from the rapid displacement of the seawater. Likewise a major eruption in the sea can also cause tsunamis (Dudley and Lee, 1988; Seibold, 1995). In most cases a flank of an active volcano is suddenly uplifted or depressed,

Figure V20 Gently sloping shore platform (foreground) with a veneer of sediment, passing to a high-tide level, sub-horizontal bench and notch cut into ignimbrite in a low-wave-energy environment. The high-tide bench and notches are reminiscent of tropical limestone cut features (photo: T. Healy).

Figure V21 Boulder beach surmounting a broad intertidal platform, formed from erosion of low lahar deposits, Taranaki, New Zealand. In the foreground are remnants of a deposit of pure black titanomagnetite "ironsand," also an erosion product of the andesitic lahars (photo: T. Healy).

Figure V22 View into the caldera at the town of Thera on Santorini, Greece (photo: M. Schwartz).

thereby generating the tsunami. But for the case of the island of Krakatoa in 1883, the actual violent explosion itself produced waves as high as 40 m crashing ashore in Java and Sumatra, and killing some 36,000 people.

Sediments originating from coastal volcanic rocks

A range of sediments are derived from weathering and erosion of volcanic deposits in the coastal zone.

- Basalts and andesites rarely weather to sand, but rather to cobbles pebbles and boulders. An unusual example of a specific beach material originating from basalt flows is at Kalapana beach, Hawai, which comprises black glass fragments from the nearby active Kilauea volcano lava tongue entering the sea (Macdonald *et al.*, 1983).
- Iron-rich heavy minerals weathered from andesitic and rhyolitic deposits may introduce a specific concentration of beach sediments, occasionally in vast volumes. The black ironsand deposits of titanomagnetite on the west coast beaches and dunes of New Zealand's North Island, result from weathering of the Taranaki andesitic laharic deposits which form the cliffs around Taranaki. But heavy minerals more typically occur as trace rather than bulk minerals; thus they can be used as tracers for the direction of littoral drift and the provenance of the beach sands (Komar, 1998).
- Acid volcanism, typically producing softer deposits of pumiceous tephras and ignimbrites, may produce large suites of minerals for sandy beach deposits. Pumiceous tephra, for example, is easily broken down in the fluvial and marine environments to its constituent minerals such as quartz, feldspars, obsidian (volcanic glass), and heavy minerals (including augite, hypersthene, hornblende, cummingtonite, and titanomagnetite). New Zealand's Bay of Plenty Holocene dune ridge barrier systems provide an excellent example of sand deposits related exclusively to acid volcanic provenance from the nearby Taupo Volcanic Zone.
- Clays weathered from acid volcanic products include allophane, while volcanic glass weathers in very short time to the montmorillonitic type clay, smectite, in shallow marine deposits.
- In areas of coastal dune deposits blanketed by a series of tephras, the sequence of tephras can be used for dating the coastal progradation and evolution. The classic paper illustrating this method is by Pullar and Selby (1971).

Terry R. Healy

Bibliography

Cotton, C.A., 1942. *Geomorphology An Introduction to the Study of Landforms*. Christchurch: Whitcomb & Tombs Ltd.

Cotton, C.A., 1944. *Volcanoes as Landscape Forms*. Christchurch: Whitcomb & Tombs Ltd.

Dudley, W.C., and Lee, M., 1988. *Tsunami*. Honolulu: University of Hawaii Press.

Komar, P.D., 1998. *Beach Processes and Sedimentation*, 2nd edn. New Jersey: Prentice Hall.

Macdonald, G.A., Abbott, A.T., and Peterson, F.L., 1983. *Volcanoes in the Sea. The Geology of Hawaii*, 2nd edn. Honolulu: University of Hawaii Press.

Pullar, W.A., and Selby, M.J., 1971. Coastal progradation of Rangitaiki Plans, New Zealand. *New Zealand Journal of Science*, **14**: 419–434.

Searle, E.J., 1964. *City of Volcanoes*. Auckland: Pauls.

Seibold, E., 1995. *Entfasselte Erde. Vom Umgang mit Naturkatastrophen*. Stuttgart: Deutsche Verlags-Anstalt.

Wentworth, C.K., 1938. Marine bench-forming processes: water-level weathering. *Journal of Geomorphology*, **1**: 6–32.

Cross-references

Cliffed Coasts
Gravel Beaches
Shore Platforms
Weathering in the Coastal Zone

WASHOVER EFFECTS

Although there is extensive geomorphological literature on barrier islands (e.g., Schwartz, 1973) and a similar volume of work on overwash processes (e.g., Leatherman, 1981), studies of the barrier islands associated with present and past positions of the Mississippi Delta (Penland *et al.*, 1988), where there is accelerated subsidence and an absence of a sufficient supply of sand, have demonstrated the critical role of sand dunes in the evolutionary sequence (Ritchie and Penland, 1990a; Ritchie, 1993). This model may have wider applications as evidence accumulates for greater coastal submergence as a consequence of a possible global rise in sea level (Bird, 1993).

In addition to the normal processes of coastal dune development associated with beach-dune exchanges, barrier island dunes and sub-eolian sand terraces have three characteristics which relate to the typical evolution of barrier islands; these are: rapid change, the importance of position (i.e., at the center or at the flanks of the island, Ritchie and Penland, 1990b) and the significance of overwash processes. (Other factors such as sand supply and patterns of vegetation are equally important but are not considered further here insofar as they are not unique to barrier island dunes, although according to Goldsmith (1985), "The most important contribution that humans can make toward the preservation of barrier islands is to prevent damage to the dune vegetation.") Similarly, the evolution of transgressive or regressive barrier islands need not be described. Nevertheless, for reasons such as a rapid reduction in sand supply or tectonic subsidence, a relative rise in sea level will, typically, enhance the frequency and therefore the significance of washover processes (Viles and Spencer, 1995).

The model barrier island (Figure W1) will retreat landwards and migrate alongshore depending, primarily, on the strength and direction of coastal currents. Hydraulic processes at tidal inlets also control the shape and rate of change. Larger beaches occur at the ends of the island and are nourished by alongshore transport and from accretionary berms and sand bars which are formed by flood and ebb tidal flows. This pattern of sand movement along the beach toward the flanks of the barriers produce large, sometimes multiple accretionary dunes of several types. In contrast, central areas, unless there is abundant sand supply from the beach or exceptional relict features such as old sand hills or beach ridges (cheniers in Louisiana, Ritchie, 1972), are typically eroding. Thus, in the central part of the barrier dunes are often little more than narrow, low, ephemeral accretions along the coastal edge.

Where overwash processes occur, usually as tongue-shaped penetrations leading to a depositional fan, a surface of bare sand is produced which, on drying, provides a local source for wind transport with the sand being trapped at the line of vegetation along the perimeter of the sand flat. In time this surface will revegetate to become a terrace which may have some additional wind-blown sand accumulations from the beach or from local coastal edge erosion and other redepositional processes. Although it is often a fine distinction, the terminology which has been developed consists of "washover terrace" if low and flat and "dune terrace" if higher and undulating. Dune terraces might also originate as thin spreads of sand which form the backslope of most types of coastal dunes. From numerous surveys of barrier island dune systems, including investigations of the rate of change, a generalized model of dune types has been constructed (Figure W2) for the Louisiana barrier islands.

The rate, frequency, and penetration of overwash events is highly variable and are functions of hurricane, storm, and subtropical frontal passages. The preexisting height of the dune barrier and the width of the beach are the prime defenses against overwash. Thus, the high dunes at the ends of the barrier may be surmounted once every 5 to 10 years whereas the low coastal terraces may be crossed by surges several times every year (Ritchie and Penland, 1990a). Overwash processes transfer sand from the beach and dunes inland thereby contributing to landwards retreat. If these processes are dominated by extreme events such as the passage of a hurricane or a severe storm, consequential topographic orientations can be at different angles to the orientations which are produced by normal eolian processes. At the north end of the Chandeleur Islands, for example, the biggest, dune ridges run inland at 90° to the coastline, being a combination of post overwash residual features and later eolian deposition. It is often possible on the barrier islands of coastal Louisiana to detect relict washover features including lagoon and bay deposits which can be correlated with known hurricane events over the last 20 to 30 years.

Where severe storm events also produce significant elevation of sea levels which are normally accompanied by very strong storm wave effects, barrier island dunes, of all types, may be eliminated completely and the mass of sand which is stored in these features is translated landwards as sheets and spreads some of which cross the entire width of the island and extend as delta-like deposits into the back—barrier lagoon or bay—and as such are lost to the active coastal beach–dune zone. If these back-barrier spreads of sand are laid down during exceptional high water events (and this is normally true), when the water level subsides they will be colonized by vegetation and remain as visible features until the next extreme event. Complete planation of the barrier island can occur at the penultimate stage of the barrier island cycle (Penland *et al.*, 1988) but in younger, more sand-rich barrier islands, the higher sand dunes, normally at the downdrift end of the island survive, albeit severely eroded and often with local washover penetrations through preexisting low and weaker sections. Accordingly, the importance of dune and similar forms to the maintenance and evolution of barrier islands cannot be overemphasized but these dunes (unlike coastal dunes in non-barrier coastlines or in higher latitudes or in areas lacking frequent

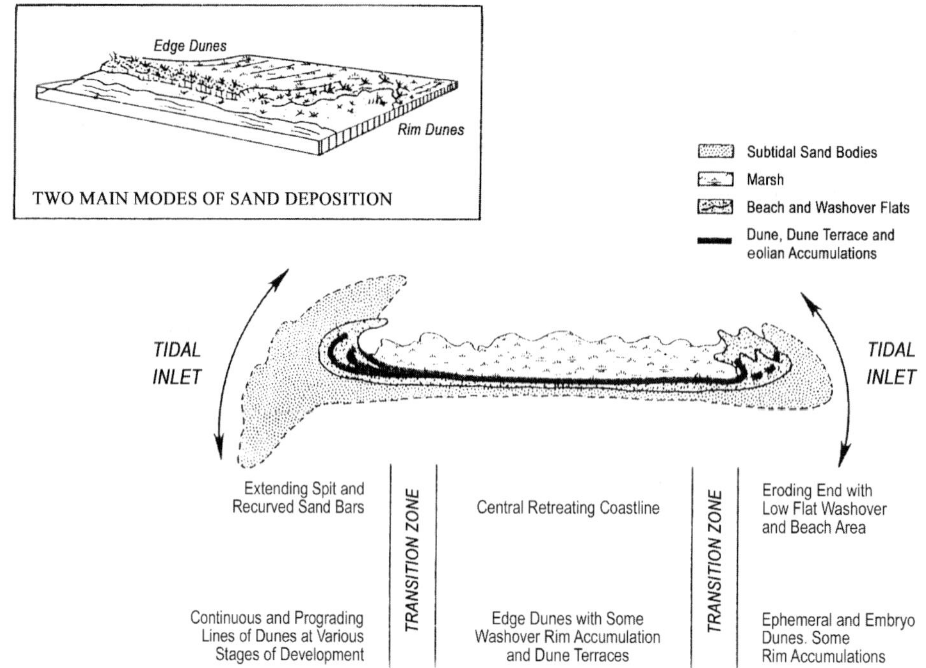

Figure W1 Generalized location of eolian accumulation forms.

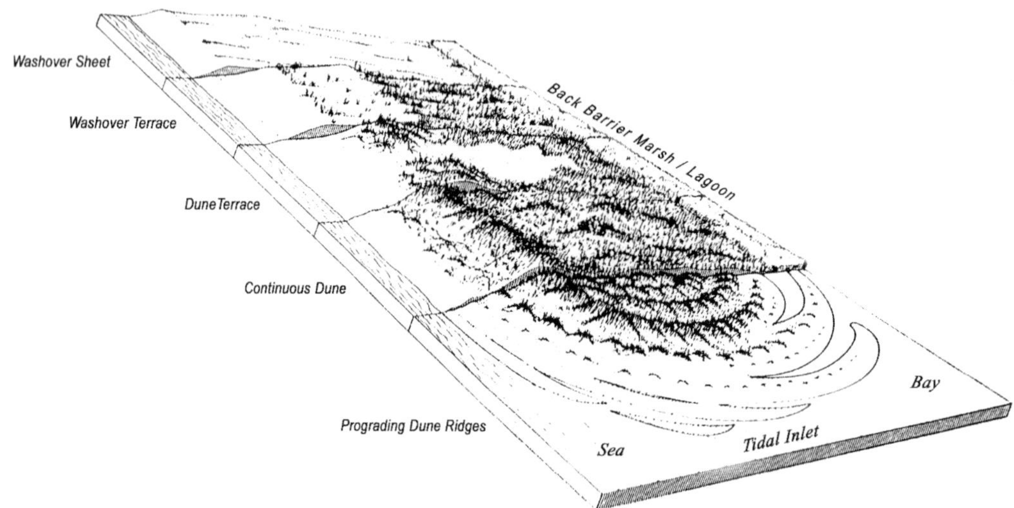

Figure W2 Dune types associated with barrier islands in south Louisiana.

periods of significantly elevated sea levels) change, with great rapidity, not solely as a consequence of eolian processes but also as a consequence of powerful episodic effects of overwash surges. In most locations, however, it is possible to detect cycles of washover activity, with the 7- to 10-year return period of hurricanes of the Louisiana coast being the shortest (Ritchie and Penland, 1990a). It is therefore necessary to conclude that eolian coastal dune processes need to be considered alongside overwash processes in regions of rapid barrier island submergence which, if the predictions of global rises in sea level are correct, might be found, increasingly, in areas other than the margins of large deltaic systems.

Extensive surveys of all the barrier islands of Louisiana were undertaken by the Louisiana Geological Survey and published in four regional volumes, that is, Isles Dernieres, 1989; Plaquemines, 1990; Chandeleur Islands, 1992; Lafourche Barrier System, 1995.

William Ritchie

Bibliography

Bird, E.C.F., 1993. *Submerging Coasts*. Chichester: John Wiley.

Goldsmith, V., 1985. Coastal dunes. In Davis, R.A. (ed.), *Coastal Sedimentary Environments*. New York: Springer-Verlag, pp. 303–378.

Leatherman, S.P. (ed.), 1981. *Overwash Processes, Benchmark Papers in Geology*, Volume 58. Stroudsburg: Hutchinson Ross.

Penland, S.P., Boyd, R., and Suter, J.R., 1988. The transgressive depositional systems of the Mississippi delta plain: a model for barrier shoreline and shelf sand development. Journal of Sedimentary Petrology, **58**: 932–949.

Ritchie, W., 1972. A preliminary study of the distribution and morphology of the Caminada/Moreau sand ridges (Louisiana). *Southeastern Geology*, **14**(2): 113–125.

Ritchie, W., 1993. A classification of the coastal dunes of Louisiana. In Laska, S., and Puffer, A. (eds.). *Coastlines of the Gulf of Mexico*. American Society of Civil Engineers, pp. 18–28.

Ritchie, W., and Penland, S.P., 1990a. Aeolian sand bodies of the south Louisiana coast. In Nordstrom, K.F., Psuty, V.P., and Carter, R.W.G. (eds.), *Coastal Dunes*. Chichester: John Wiley, pp. 105–127.
Ritchie, W., and Penland, S.P., 1990b. Coastline erosion and washover penetration along the Bayon Lafourche barrier coastline between 1978 and 1985, with special reference to hurricane impacts. O'Dell Memorial Monograph. Volume 23. University of Aberdeen, 23pp.
Schwartz, M.L. (ed.), 1973. Barrier Islands. *Benchmark Papers in Geology*, Volume 9. Stroudsburg: Dowrden, Hutchinson and Ross.
Viles, H., and Spencer, T., 1995. *Coastal Problems*. London: Arnold.

Cross-references

Barrier Islands
Beach Processes
Beach Ridges
Changing Sea Levels
Cheniers
Dunes and Dune Ridges
Eolian Processes
Sea-Level Changes During the Last Millennium
Storm Surge
Tidal Inlets

WARFARE—See COASTAL WARFARE

WATER QUALITY

There are around 177 nations open to an ocean, sea, or gulf and around another 30 coastal semi-sovereign states have the legal power to manage their own natural resources and land. Coastal waters are used for a variety of purposes such as food production, hydroelectricity generation, recreational activities, as a transport medium, and as a repository for sewage and industrial waste. For some of these purposes, primarily, food production and recreational activities, the quality of the water is vitally important. The water environment and associated habitats such as coral reefs and beaches may offer significant financial benefit to the associated communities and these opportunities provide the motivating factors for creating a management program.

Water, however clean, is an alien environment to man and thus it can pose hazards to human health even when it is of pristine quality. It is therefore necessary to implement effective management policies in order to minimize and reduce the health consequences of all anthropogenic activities related to water use. In addition to human health implications of poor water quality, poor land use practices, and excessive domestic and agricultural pollution have lead to ecological effects such as the decline of localized coral reef ecosystems (Huber and Jameson, 2000).

Factors affecting the quality of water

Water quality is a multi-faced concept. In addition to microbiological factors, monitoring of coastal waters used for recreational or shellfish harvesting purposes includes the observation of turbidity or transparency, the presence of floating objects, foams on the surface and odor. Apparently clean waters—that is, those that appear clear, may in fact be polluted with high bacterial concentrations or toxic substances.

Coastal waters have traditionally been considered as the ultimate sink for the by-products of human activities. The key sources of pollutants affecting coastal water quality are riverine inputs of domestic, agricultural, and industrial effluents and direct sewage discharges from the local population. Discharges may be regular, through long and short sea outfalls and irregular through storm water and overflow outfalls, and unregulated private discharges. It is the point sources of pollution due to domestic sewage discharges that cause most human health concern. Discharge of sewage to coastal waters exert a variable polluting effect that is dependent on the quantity and composition of the effluent and on the capacity of the receiving waters to accept that effluent. Thus, enclosed, low-volume, slowly flushed water systems will be affected by sewage discharges more quickly than water bodies that are subject to fast change and recharge.

Pollution originating from water-based recreation such as leisure crafts, jet skis, etc. includes sanitation discharges, environmentally toxic effects of antifouling paints and general debris. Estuarine water will be particularly vulnerable to pollution due to the enclosed nature of the system and the subsequent accumulation of pollutants. Controls on sanitation discharges from pleasure craft are dependent on suitable holding tanks and port reception facilities (see entry on *Marine Debris*).

Water quality may also be adversely affected by fuel spillages and discharges from pleasure craft. Oil, petrol, and diesel may be spilt at filling barges and bilge waters may be discharged, adding to pollution of the coastal water. In addition to the fuels themselves, the emissions from the engines may be harmful and the oil and gasoline mix in two-stroke fuels has been implicated in the contamination of fish and shellfish products.

Antifouling paints applied to boats and coastal installations to prevent the subsequent attachment and growth of organisms may have an undesirable effect on water quality, especially where the area is closed or where the vessels or installations are concentrated.

Aesthetic pollution of water quality stems from a variety of sources such as recreational users, leisure boats, fishermen, etc. It has been estimated that 70–80% of marine debris comes from land-based sources (Pond and Rees, 1994) (see entry on *Marine Debris*).

Environmental factors such as rainfall or sunshine may also significantly influence the microbiological quality of water. Heavy rains can cause sewage overflows thus bypassing sewage treatment facilities, increasing surface water runoff and directly influencing microbiological water quality. In contrast, as the intensity of sunlight increases some microorganisms (e.g., *Escherichia coli*) die-off rapidly.

Salinity is another factor that has been shown to affect the die-off of microorganisms. Hanes and Fragala (1967) suggest that enterococci, for example, die-off more rapidly in saline water than in freshwaters.

Eutrophication (the enhancement of the natural processes of biological production, caused by increases in levels of nutrients) has been recognized as a growing concern since the 1950s. The most visible indicator of coastal eutrophication is extensive blooms of phytoplankton and/or benthic macroalgae. In many regions, eutrophication in large water bodies has been characterized by blooms of planktonic species such as *Phaeocystis* (the North Sea; Davidson and Marchant, 1992), *Noctiluca miliaris* (the Black Sea; Mee, 1992), and mucus aggregates (the northern Adriatic Sea; Stachowitsch *et al.*, 1990). Some of the algae are toxic and may cause fish kills while others are aesthetically "nuisance" algae, causing spoiling of beaches, offensive odors, and slimy water. Many cause secondary blooms of undesirable fauna, such as the ctenophore *Mnemiopsis leidyi* (the Black Sea; Mee, 1992) which has reached bloom biomass densities of 1 kg m^{-2}.

In Australia, a large number of estuaries, bays, and coastal lakes have begun to experience algal blooms in the last 30 years. In coastal marine environments many toxic species of dinoflagellates, diatoms, nanoflagellates, and cyanobacteria occur, and have led tohuman health impacts after consumption of shellfish and fish. In Tasmania and Victoria, the closure of shellfish beds has resulted from blooms of algae suspected to have been introduced into Australia in ballast water (Hallegraeff and Bolch, 1992).

In 1995–96, nearly 400 manatees (about 20% of the Florida population) died from exposure to red tide, a toxic algae bloom that occurs naturally in the Gulf of Mexico, and to which manatees have been exposed for many years. It is thought that the blooms were more concentrated due to increasing levels of pollution in Florida's coastal waters, and that pollution factors may also reduce the manatees natural level of resistance to disease.

Continuing construction and development in coastal areas, loss of wetland habitat, which filters surface water runoff (which had previously reduced pollution to coastal waters), and the resulting contamination and loss of the coastal grasses which are the manatee's food supply, all contribute to forcing manatees into smaller ranges.

Most of the problems affecting humans associated with toxin-containing aquatic cyanobacteria have involved freshwater planktonic species. Severe illness due to direct dermal contact with cyanobacterial mats has been reported in tropical marine bathing sites.

Management of nutrient loading to coastal environments, and any subsequent problems of eutrophication, is based on the reduction of nutrient-rich effluents from the land or better dispersion of existing discharges. This may be achieved by the control of soil erosion, changes in the use or nature of fertilizers, reuse rather than discharge of nutrient-rich effluents, diversion of discharges into less-sensitive or better-flushed environments, engineering works to improve flushing, and nutrient removal from effluents.

Human health implications of poor water quality

Epidemiology is the scientific study of disease patterns in time and space (Kay and Dufour, 1999). Epidemiological studies have played an important role in providing information to characterize risks associated with swimming in waters of poor quality and of particular relevance, those risks associated with swimming in feces-contaminated recreational waters. Although the water environment is of a highly variable nature, a number of credible studies have been carried out since 1953 (see Kay and Dufour, 1999 for a review), which have shown an association between a variety of symptoms and exposure to fecally polluted recreational waters.

Coastal waters generally contain a mixture of pathogenic and nonpathogenic microbes derived from sewage effluents, the population using the water for recreational purposes, industrial activities, and farming and wildlife, in addition to indigenous microorganisms. If an infective dose of pathogens colonizes a suitable growth site in the body of a bather there is the possibility of disease.

There are a broad variety of illnesses that have been associated with swimming in marine and fresh recreational waters. *E. coli*, Leptospira, Norwalk virus, and adenovirus 3 are some of the microbes that have been linked to swimming-associated disease outbreaks. A number of outbreaks of *Giardia* and Cryptosporidium have been shown to occur in very small, shallow bodies of water that are generally frequented by children. Epidemiological investigations of the outbreaks found that the source was usually the bathers themselves, most likely children (WHO, 1999).

Enteric illness, such as self-limiting gastroenteritis, is the most frequently reported adverse health outcome investigated in recreational users of contaminated bathing waters and it is now accepted that there is an association between gastrointestinal symptoms and indicator-bacteria concentrations in recreational water. Most of the literature suggests a causal relationship between increasing recreational exposure to fecal contamination and frequency of gastroenteritis (WHO, 1999). Higher symptom rates have been reported usually in the lower age groups (Cabelli, 1983; Pike, 1994). Children are particularly at risk from waterborne diseases compared to other age groups due to weak body defenses, susceptibility, and inadequate or no understanding of how to avoid hazards. Children have a natural curiosity and disregard for safety, and thus may play in or drink contaminated water. As their mobility increases they expose themselves to more hazards thus increasing risk of infection.

Escherichia coli is a well-known indicator of fecal pollution. *E. coli* are capable of causing many diseases and is a major cause of gastroenteritis: Enterotoxigenic *E. coli* cause a dehydrating diarrhea in children, enteroinvasive *E. coli* can cause fever and watery mucoid diarrhea in infants; enterohemorrhagic *E. coli* is associated with hemorrhagic colitis, some cases progressing to hemolytic uremic syndrome, particularly in children. Many waterborne outbreaks of *E. coli* are known—some associated with drinking water and others with recreational water.

Other associations, although fewer, have been reported for other health outcomes, such as ear infections and respiratory symptoms. There is less evidence for the link between recreational water use and more serious diseases although it is biologically plausible. Cholera, for example, can be transmitted through water and food. Environmental sources of the causative agent include sewage, sea, and surface waters. Direct person-to-person transmission is uncommon.

Giardiasis is a common disease throughout the world, affecting all age groups, although the highest incidence is in children (Hunter, 1997). Waterborne outbreaks are particularly common and a few outbreaks have been reported from recreational waters.

Meningitis, hepatitis A, typhoid fever, and poliomyelitis could also be acquired through bathing but it is difficult to determine unequivocally.

Limited evidence suggests that local populations bathing in sewage-contaminated water may suffer less illness than visiting populations, probably due to the influence of the immune status (WHO, 1999). It has also been suggested that viruses could be transferred from the water to the air and therefore nonswimmers could be at a certain risk (Feachem *et al.*, 1982).

Water quality standards

Coastal water quality standards relate primarily to recreational waters and water used for shellfish production. The implementation of water quality standards have had some successes in promoting cleanups, increasing public awareness, contributing to informed personal choice, and to a public health benefit.

Water quality is primarily measured against standards for microbiological parameters and present regulatory schemes for the microbiological quality of recreational waters are primarily or exclusively based on compliance with fecal indicator counts. A variety of water quality standards exist throughout the world (Table W1). The World Health Organization (WHO) have recently developed Guidelines for Safe Recreational Water Quality Environments (WHO, 1998). To comply with standards of quality authorities must monitor the water a minimum number of times during the defined bathing season. There is considerable concern about the financial cost of monitoring water quality, especially in the light of the precision with which the monitoring effort assesses the risk to the health of water users and the effectiveness with which it supports decision-makers to protect public health.

In addition to microbiological hazards there are a number of other diverse physical and chemical pollutants that should be addressed in monitoring programs where they are known or suspected to be locally significant. It is important that standards, monitoring and implementation enable preventative and remedial actions that will prevent health effects arising from such hazards.

Water quality determination

The wide variety of factors affecting the quality of coastal water (together with the vagaries associated with monitoring) makes it inappropriate to compare the quality between different coastal locations. However, individual studies have highlighted particular problems in various regions. For example, Ozkoc *et al.* (1997) investigated the specific issues affecting the quality of the Black Sea. Land-based pollutants were considered the main pollution source in this case.

Jensen *et al.* (1997) investigated the sources of pollution and other impacts for the Mediterranean Sea, the Gulf of Suez and the Red Sea, and Gulf of Aqaba. The study found that industrial wastewater and domestic sewage from residential and tourist areas were a major problem in the Mediterranean Sea. Oil pollution from refineries, ships, and offshore oil platforms were considered the main pollutants for the Gulf of Suez and sewage, oil pollution, landfilling, dredging, and siltation the main contributors to poor water quality in the Red Sea and the Gulf of Aqaba.

Standard methods exist for the determination of the microbiological parameters that identify the quality of the water according to current legislation and methods (see e.g., ISO, 1992, 1997). Many waterborne pathogens are difficult to detect and/or quantify and instead present regulatory schemes for the microbiological quality of recreational waters are primarily based on percentage compliance with fecal indicator organisms (organisms that are always found in feces and which are easy to detect on simple bacteriological media) to show the potential presence of organisms that cause gastrointestinal disease (Table W1). Fecal streptococci are suggested as the recommended indicator for saltwater and either fecal streptococci or *E. coli* can be used for monitoring freshwaters. Sulfite-reducing clostridia may be suitable as indicator organisms for parasitic protozoa and viruses from sewage-impacted waters (Ferguson *et al.*, 1996).

Other indicator organisms for sewage, include the bacteriophages to *Bacteroides fragilis* HSP40, somatic coliphages, F-specific RNA phages, and fecal sterols. The *B. fragilis* phages appear to survive in the same way as human enteric viruses against a variety of environmental conditions, but they may exist in lower numbers than various coliphages and it is thought that only between 1% and 5% of humans excrete these phages and therefore they may be unsuitable indicators in areas of small communities.

Sanitary plastics can act as an immediate indicator of sewage contamination. However, the presence of sanitary plastics may reflect old sewage contamination and may not be a good indicator of human health threats.

No one indicator organism is suitable for representing all the issues associated with fecal contamination of recreational water.

Monitoring only for microbiological parameters does not provide the most accurate and feasible index of health risk. There are a number of acknowledged constraints associated with the current standards and guidelines:

- Management actions are retrospective and can only be instigated after users have been exposed to the hazard.
- The risk to human health is mainly from human excreta, the traditional indicators of which may derive from other sources.
- There is poor inter-laboratory and international comparability of microbiological analytical data.
- Recreational waters are generally classified as safe or unsafe, but there is no gradient of severity.

It is suggested that the measure of a microbiological indicator of fecal contamination be combined with an inspection-based assessment of the susceptibility of an area to direct influence from human fecal contamination. This allows regulators to identify the possible source of the

Table W1 Microbiological quality of water guidelines/standards per 100 ml (adapted from WHO, 1999)

	Shellfish harvesting		Primary contact recreation		
Country	Total coliforms	Fecal coliforms	Total coliforms	Fecal coliforms	Other
Brazil		100% < 100	80% < 1,000[i]	80% < 1,000[i]	
Columbia			1,000	200	
Cuba			1,000[a]	200[a]	
				90% < 400	
EEC, Europe			80% < 500 (guide)	80% < 100 (guide)	Fecal streptococci 100 (guide)
			95% < 10,000 (mandatory)	95% < 2,000 (mandatory)	Salmonella 0/L (mandatory) Enteroviruses 0 PFU/L (mandatory) Enterococci 90% < 100
Ecuador			1,000	200	
France			<2,000	<500	Fecal streptococci < 100
Israel			80% < 1,000[c]		
Japan	70		1,000		
Mexico	70[b]		80% < 1,000[e]		
	90% < 230		100% < 10,000[g]		
Peru	80% < 1,000	80% < 200	80% < 5,000[e]	80% < 1,000[h]	
	0	100% < 1,000			
Poland					*E. coli* < 1,000
Puerto Rico	70[d]			200	
	80% < 230			80% < 400	
United States, California	70[b]		80% < 1,000[e,f] 100% < 10,000[g]	200[a,f] 90% < 400[h]	
United States, USEPA		14[a] 90% < 43			Enterococci 35[a] (marine) 33[a] (fresh) *E. coli* 126[a] (fresh)
Former USSR					*E. coli* < 100
UNEP/WHO		80% < 10		50% < 100[j]	
		100% < 100		90% < 1,000[j]	
Uruguay				<500[j]	
<1,000[k]					
Venezuela	70[a]	14[a]	90% < 1,000	90% < 200	
	90% < 230	90% < 43	100% < 5,000	100% < 400	
Yugoslavia			2,000		

[a] Logarithmic average for a period of 30 days of at least five samples.
[b] Monthly average.
[c,d] Minimum 10 samples per month.
[e] Period of 30 days.
[f] Within a zone bounded by the coastline and a distance of 1,000 ft from the coastline or the 30-foot depth contour, whichever is further from the coastline.
[g] Not a sample taken during the verification period of 48 h should exceed 10,000/100 ml.
[h] Period of 60 days.
[i] Satisfactory waters, samples obtained in each of the preceding weeks.
[j] Geometric mean of at least five samples.
[k] Not to be exceeded in at least five samples.

contamination and since sources other than human fecal contamination are generally less of a risk to human health, it is possible to reflect this modified risk. Knowing the possible sources of contamination and their likely influence upon water use provides a fast and robust means to increase the reliability of the overall assessment. This could lead to a classification system based on relative risk as suggested by the WHO and United States Environmental Protection Agency (USEPA) (WHO, 1999).

Environmental health indicators

The WHO has recently developed a set of environmental health indicators to aid policy makers in their roles in public and environmental health policies (WHO, 2000) (see entry on *Environmental Quality*). This included coastal waters used for recreation. To evaluate the relevance of these indicators for widescale implementation a pilot testing has been initiated in several countries in the European region of the WHO.

Conveying information on water quality to the public

The primary reason for monitoring water quality and for informing the public is for the protection of public health. The public is unlikely to want to know the technical details of sample treatment in the laboratory but wish to know only whether the water is safe or not. It is essential that any information relating to water quality is presented to the public in a clear, unambiguous, and easily understood way.

Raising awareness and enhancing the capacity for informed personal choice is an important factor in ensuring the safe use of coastal water environments. It acts directly (by allowing users to choose an area which is known to be safe), and indirectly (the exercise of preference for safer environments will encourage investments in improvements). In order for it to be successful it is essential that the public is generally aware and that the information is available and easily understood.

Currently, a number of schemes exist which convey information to the public. Unfortunately, as yet the public is often provided with information on the previous bathing seasons water quality rather than current results. For example, in the European Community, the EC Directive on the quality of bathing waters requires that Member States submit to the Commission a comprehensive report on their bathing water and the most significant characteristics. The Commission then publishes this information by means of a report just before the beginning of the next bathing season. Real-time information services do exist—generally provided by local authorities for individual beaches.

Other schemes to inform the public of water quality exist; for example, award schemes are common—in Europe the best known is European Blue Flag scheme. Many countries have national equivalents. Human health is not always the prime concern of these schemes. The quality of recreational water is often used in publicity to attract visitors to the area. Award schemes are often used as incentive programs to involve all parties concerned in participating in optimizing beach safety, water quality, and education activities (see entry on *Rating Beaches*).

Pollution abatement and water quality

Water quality is often affected by pollution due to sewage and industrial discharges, combined sewer overflows, and urban runoff. Pollution abatement is therefore a key part of coastal zone management aimed at minimizing both health risks to bathers and ecological impacts. The main action to reduce sewage pollution requires a large investment in sewage treatment and discharge. Pollution abatement measures for sewage may be grouped into three alternatives: wastewater treatment; dispersion through sea outfalls; discharge to non-surface waters (reuse). The traditional measures are often prohibitively expensive and therefore in some cases in view of costs of control, it may be preferable for integrated beach zone management to focus on alternative options such as restricting beach use or warning the public of the potential health risks during and after risk events (see entry on *Coastal Zone Management*).

Summary

Exposure to fecal pollution through contaminated waters leads to detectable health effects and there is clear evidence of a dose-response relationship linking fecal pollution with enteric and non-enteric illnesses. Evidence also indicates that health effects linked to pollution occur at levels of fecal indicator bacteria which are found in recreational waters globally and which may be below the legal standards in many parts of the world. In order to manage coastal water quality, national authorities should take account of a variety of factors, including social and economic factors, some of which are conflicting, in order to ensure safe waters.

Katherine Pond

Bibliography

Cabelli, V., 1983. *Health Effects Criteria for Marine Recreational Waters*. EPA-600/1-80-031, Cincinnati: US Environmental Protection Agency.

Davidson, A.T., and Marchant, H.J., 1992. The biology and ecology of *Phaeocystis*. *Progress in Phycological Research*, **8**: 1–45.

Feacham, R., Garelick, H., and Slade, J., 1982. Enteroviruses in the environment. *World Health Forum*, **3**: 170–80.

Ferguson, C.M., Coote, B.G., Ashbolt, N.J., and Stevenson, I.M., 1996. Relationships between indicators, pathogens and water quality in an estuarine system. *Water Research*, **30**(9): 2045–2054.

Hallegraeff, G.M., and Bolch, C.J., 1992. Transport of diatom and dinoflagellate resting spores in ships' ballast water: implications for plankton biogeography and aquaculture. *Journal of Plankton Research*, **14**: 1067–1084.

Hanes, N.B., and Fragala, R., 1967. Effect of seawater concentration on the survival of indicator bacteria. *Journal of Water Pollution Control Federation*, **39**: 97.

Huber, R.M., and Jameson, S.C., 2000. Integrated coastal zone management in the tropical Americas and the role of decision support models. In Gustavson, K., Huber, R.M., and Ruitenbeek, K. (eds.), *Integrated Coastal Zone Management of Coral Reefs: Decision Support Modeling*. Washington, DC: The World Bank, pp. 3–14.

Hunter, P.R. (ed.), 1997. *Waterborne Disease, Epidemiology and Ecology*. New York: John Wiley & Sons.

ISO, 1992. *Water Quality—Sampling—Part 9: Guidance on Sampling from Marine Waters*. ISO 5667-9, Geneva: International Organization for Standardization.

ISO, 1997. *Water Quality—Detection and Enumeration of Escherichia coli and Coliform Bacteria—Part 1: Membrane Filtration Method*. ISO/DIS 9308-1, Geneva: International Organization for Standardization.

Jensen, A., El Zarka, M., and Povlsen, E., 1997. A strategy for an Egyptian coastal monitoring program. In Ozhan, E. (ed.), *Proceedings of the Third International Conference on the Mediterranean Coastal Environment, MEDCOAST 97*, November 11–14. Qawra, Malta: pp. 367–376.

Kay, D., and Dufour, A., 1999. Epidemiology. In Bartram, J., and Rees, G. (eds.), *Monitoring Bathing Waters*. London, New York: E & FN Spon, pp. 313–329.

Mee, L.D., 1992. The Black Sea in crisis: a need for concerted international action. *Ambio*, **21**: 278–286.

Ozkoc, H.B., Saricaoglu, K., Yuzbasi, N., Balkaya, N., Buyukgungor, H., and Onar, A.N., 1997. Investigation of land-based pollution parameters in the surface waters of the Black Sea. In Ozhan, E. (ed.), *Proceedings of the Third International Conference on the Mediterranean Coastal Environment, MEDCOAST 97*, Novem- ber 11–14. Qawra, Malta: pp. 315–326.

Pike, E.B., 1994. *Health Effects of Sea Bathing (WMI 9021)—Phase III: Final Report to the Department of the Environment*. Report No: DoE 3412/2, pub. Medmenham, UK: Water Research Centre plc, pp. 1–38.

Pond, K., and Rees, G., 1994. *Norwich Union Coastwatch UK 1994 Survey Report*. Hampshire, UK: Farnborough College of Technology.

Stachowitsch, M., Fanuko, N., and Richter, M., 1990. Mucus aggregates in the Adriatic Sea: an overview of stages and occurrences. *Marine Ecology*, **11**: 327–350.

WHO, 1998. *Guidelines for Safe Recreational Water Environments: Coastal and Freshwaters. Draft for Consultation*. Geneva: World Health Organization.

WHO, 1999. *Health-Based Monitoring of Recreational Waters: The Feasibility of a New Approach (the 'Annapolis Protocol')*. Geneva: World Health Organization.

WHO, 2000. *Environmental Health Indicators: Development of a Methodology for the WHO European Region*. Interim Report, 6 November 2000. Bilthoven: World Health Organization Regional Office for Europe.

Cross-references

Coastal Zone Management
Environmental Quality
Human Impact on Coasts
Marine Debris—Onshore, Offshore, Seafloor Litter
Rating Beaches

WAVE AND TIDE-DOMINATED COASTS

Introduction

Wave and tide-dominated coasts may be defined as coasts whose morphology is shaped by short- to long-term background hydrodynamic conditions generated jointly by waves and tides. Few of the world's coasts are devoid of wave action (Davis and Hayes, 1984), such that waves are generally considered as the dominant agent of coastal change at short (order of hours to weeks) to medium (order of months to years) timescales. The importance of tides in determining coastal processes and morphology has long been overlooked, mainly because of lack of studies that properly evaluate their role, notably in settings deemed as wave-dominated. One probable reason for this is that by far the largest number of process studies on the open coastline has concentrated on beaches in areas of low tidal range where the tidal process signature is extremely weak or negligible. This situation has been changing in recent years, especially with regards to beach studies carried out in areas with large tidal ranges, which increasingly recognize the importance of tidal modulation of the hydrodynamics, sediment transport patterns, and resultant morphology (Short, 1991; Masselink and Short, 1993; Levoy *et al.*, 2000). Identification of the joint influence of waves and tides is important in terms of understanding the physical processes involved in coastal change and its implications for coastal management.

From first principles, tidal forces are felt universally on all coasts, although their effect is modulated by position. However, waves are variable in time and space. As a result, waves might mute, modulate, obliterate, amplify, or dominate the tidal signal. Coasts where waves are absent or where waves obliterate the tidal signal are the extremes previously recognized as tidal- or wave-dominated coasts. Inevitably, the rest must be mixed wave and tide-dominated coasts. Between the wave-dominated coast and the tide-dominated coast therefore exists a wide range of mixed wave–tide-dominated coastal types that comprises a considerable proportion, if not the majority of the world's coastline. Any tidal influence in shaping the coastal morphology occurs directly through tidal currents and their asymmetry, through enhancement of tidal currents due to shoreline configurations, as in the case of estuaries and inlets, through tidal interactions with waves, and indirectly through the effects the large vertical and horizontal tidal translations have on wave processes.

Wave and tide-dominated coastal settings and characteristics

Between the wave- and tide-dominated coastal extremes is a broad spectrum of wave and tide-dominated coasts. These range from settings with high wave energy and perceptible tidal energy associated with a low tidal range, to settings with low wave energy and large tidal ranges. While tides show a regular predictable temporal cycle, wave conditions generally exhibit large temporal variability expressed by irregular, but short-term (order of days to weeks), to seasonal variations in energy and period. As a result, the wave and tide-dominated spectrum is not constant in power level and is probably very irregular, characterized by a few key spikes associated with certain coastal types. Examples are the stubby barrier coasts with frequent tidal inlets associated with low- to moderate-energy waves and moderate tidal ranges (Hayes, 1979), and beaches with very large tidal ranges and low- to moderate-wave energy whose shorefaces are characterized by storm wave and tidally molded sand dunes and ridges. Further variability occurs in wave and tide-dominated settings where wind forcing plays an additional important role in modulating the power level.

In the high wave energy–low tidal range part of the wave and tide-dominated spectrum, strong tidal currents are only generated where local bathymetric/topographic factors lead to tidal strengthening and slackening. This situation may occur in both fixed and migrating tidal inlets linked to estuaries or lagoons. A case of wave and tide-dominated coast that is often overlooked is that of alongshore migrating inlet systems associated with barriers (see entries on *Tidal Inlets* and *Barriers*). Such migrating inlet systems, especially active in areas with low tidal ranges favorable to barrier development, are maintained by the tidal flux. The tidal inlets may capture and store significant amounts of sand transported alongshore by wave-induced longshore drift as tidal inlet fill and flood tidal delta deposits. It is noteworthy that the sediment record of barrier coasts comprising inlets may exhibit important longshore sequences of wave-deposited barrier sand that overlies such tidal deposits (Hayes, 1980; Reinson, 1984). Another situation where this may occur is in the vicinity of headlands where sand banks may accumulate as a result of tidal eddies (Pingree, 1978; Ferentinos and Collins, 1980). It must be noted, however, that this headland–tidal current relationship may span the whole spectrum, depending on tidal range and wave-energy conditions. The high wave-energy end of the wave and tide-dominated spectrum may comprise a spatially narrow or limited zone of tidal influence vis-à-vis a spatially wide but temporally fluctuating zone of wave influence.

At the other end of the spectrum are coasts with low net wave energy and strong tidal action. Coasts subject to low wave action are found where fetch or wave energy conservation conditions are such that significant waves are either not generated regularly or are not energetic enough as to have the dominant impact on the coast. The former situation may be found in protected settings such as large bays and epicontinental seas. Wave conservation depends on nearshore morphology which largely determines frictional energy dissipation and energy spread through refraction and diffraction. The highly periodic nature of wave generation or extreme wave attenuation may result in open fine/cohesive clastic-dominated coasts whose process signature and morphology are dominantly shaped by tidal currents. However, for this to happen, these currents need to be sufficiently strong to mobilize or entrain sediments. This end of the wave and tide-dominated spectrum may comprise, in contrast to the high wave-energy end, a spatially narrow but widely temporally fluctuating zone of wave influence vis-à-vis a spatially wide zone of tidal influence whose temporal fluctuations are regular, being hinged on the lunar cycle. Swell wave attenuation may be expected on coasts fronted by wide, shallow continental shelves that are themselves favorable to tidal range amplification (Clarke and Battisti, 1981).

Schematically, the wave–tide relationship may be presented in terms of a spatial band wherein the tidal influence tends to increase in the offshore zone and the wave influence toward the beach. As wave energy increases, waves would tend to mute the tidal signal and even dominate. This relationship would however, vary both spatially and temporally as a function of the temporal fluctuations in wave energy and in tidal power during the lunar cycle.

On modally low wave-energy coasts, especially in protected wave environments, the tidal signal would tend to dominate over a wider area of nearshore zone. This tidal domination would be maintained up to the high-tide shoreline in the periodic absence of waves, or in areas where attenuation of swell wave energy is complete up to the high-tide shoreline, or would diminish toward the mid- to high-tide part of the beach where wave energy still attains the high-tide shoreline. This mixed energy regime may also be expressed by depth variations in protected settings characterized by short-period waves, with wave action being efficient over the shallower coastal and nearshore deposits, and tidal action in the deeper areas.

Wave and tide-dominated coasts may vary from sandy beaches to mudflats and mangrove-colonized tidal flats (see entries on *Tidal Environments, Tidal Flats, Mangroves, Muddy Coasts*, and *Salt Marsh*). Within the moderate to high wave and high tidal energy end of the spectrum, beaches may exhibit characteristics typical of their counterparts in settings with low tidal ranges. However, tidal modulation of their wave hydrodynamics and sediment transport patterns still occurs, together with significant tidal current activity offshore. Examples of this are Cable Beach in Western Australia (Wright *et al.*, 1982) and beaches on the Aquitaine coast of France. Low- to moderate-energy beaches in settings with very large tidal ranges (above 8 m at mean spring waters) such as those of Normandy, France (Levoy *et al.*, 2000) show even stronger tidal modulation of their morphology and dynamics, essentially through large fluctuations in water level due to the important tidal excursion. Ridge and runnel beaches occupy the modally low to moderate wave energy part of the spectrum in settings with short-period waves and moderate to large tidal ranges.

The topography of the nearshore zone or inner shoreface on wave and tide-dominated coasts may be very regular and akin to that of wave-dominated coasts on open coasts exposed to moderate-to high-energy waves, such as the Aquitaine coast. As tidal action increases, this topography may, where abundant loose sediment is available, be characterized by numerous banks and ridges (see entry on *Offshore Banks and Linear Ridges*) that are typical of a tide-dominated process signature, such as in the eastern English Channel and the Huanghai (Yellow Sea).

Muddy wave and tide-dominated coasts may be considered as being close to the tide-dominated end of the spectrum of coastal types and are generally found in the vicinity of major river mouths with large tidal ranges and that drain catchments with fine-grained sediments such as the Amazon, Huanghe, Yangtze, Ganges–Brahmaputra, and Mekong. They may also be found in tropical areas where several smaller high-discharge rivers form coalescing estuaries along significant stretches of coasts with moderate to large tidal ranges, as on the Guinea coasts of West Africa (Anthony, 1991, 1996). On these coasts, significant swell wave-energy dissipation may occur over muddy inner shorefaces and over alongshore migrating mud banks. The longshore transport of particulate mud and mud banks on these coasts is generally assured by strong tidal currents aided by synoptic wind-forced currents. In all such settings, periodic storm wave activity over the muddy shores may lead to the reworking of disseminated shells, sand, or gravel into distinct ridges called cheniers (see entry on *Cheniers*).

Examples of wave and tide-generated processes and interactions

The fundamental process feature of wave and tide-dominated coasts which differentiates them from pure wave or tide-dominated coasts is the way the waves and the tides interact to give distinct process signatures, sediment transport patterns, and coastal morphologies.

Depending on the position in the wave and tide-dominated spectrum, the inner shoreface on moderate- to high-energy swell wave coasts may show a regular bathymetry related to long-term planing by waves, with tidal and wind-induced currents aiding in the transport of sediment that becomes suspended by the high energy waves. On low- to moderate-wave-energy coasts with strong tidal action, the inner shoreface may exhibit either an irregular, poorly reworked topography comprising inherited drowned features, or may be reworked, when loose sand is abundant, into tidal current ridges. Off high sediment discharge river mouths, the inner shoreface may be draped with mud that may be in equilibrium with waves and tidal and wind-induced currents. On sand-rich shorefaces, the organization of tidal current ridges may be hinged on a long-term balance reflecting the joint action of strong background tidal currents and the imprint of periodic storm waves. The linear current ridges and dunes are generally oriented more or less parallel to the shore in response to the longshore tidal currents. From detailed monitoring of sand dunes off the Belgian coast, Van Lancker (1999) showed a vertical pattern of wave-and-tide reworking. Dune mobility in shallow depths (<5 m), whatever the distance offshore, was due mainly to wave action while deeper water mobility was caused by tides. In these shallow epicontinental seas of the eastern English Channel and southern North Sea, storm waves tend to drive dunes and ridges inshore, a process that sometimes tends to be countered by the longshore tidal currents. This may result in stretching and eventually division of the ridges (Tessier *et al.*, 1999). Ridges that do get close inshore may eventually become attached to the beach, leading to significant accretion (Anthony, 2000) and onshore feeding of dunes. On coastal sectors where the morphodynamic balance is such that the ridges do not move inshore, this may eventually deprive beaches and dunes of sand in spite of the abundant nearshore sand stocks.

The nearshore residual tidal current signature may become strengthened or weakened by synoptic winds, thus supporting the differentiation of tidal asymmetry patterns that determine medium- to long (order of tens to hundreds of years)-term sediment transport. These ebb- or flood-dominated flows lead to well-defined sand transport pathways such as those that run along both the French and English coasts in the eastern Channel and Dover Straits (Beck *et al.*, 1991; Grochowski *et al.*, 1993). As a result of various hydrodynamic conditions, notably orientation of the synoptic wind field and Coriolis deflection of the water mass, tides are larger on the French coast and the tide-dominated sand transport pathway on this coast has been much more active than that on the English side. It has been suggested that this sand-rich pathway has fed the large Holocene dune fields on the French coast via storm reworking of the nearshore sand stocks, and shoreward transport by winds over the wide dissipative beaches (Anthony, 2000).

The relationship between tides and waves has been most readily invoked in explaining the influence of large tidal ranges on beaches (Davies, 1980), and on beach morphodynamics (Wright *et al.*, 1982; Carter, 1988; Short, 1991; Masselink and Short, 1993; Masselink and Turner, 1999; Levoy *et al.*, 2000). Generally, when sediment supply conditions are favorable, coasts with large tidal ranges adjust, over the long term, to the important vertical tidal excursion by having a significant intertidal volume of sand or gravel that necessitates relatively high wave-energy levels in order for large-scale morphological changes to occur. In northern France, ridge and runnel beaches with mean spring tidal ranges of 5–9 m have intertidal volumes of up to 1,200 m^3/m of coastline, compared to the moderate volumes of beaches in settings with low tidal ranges (generally less than a tenth of this volume). As a result, whatever the background wave-energy level, the rates of sediment transport and beach morphological change are retarded. The greater the tidal range and the lower the modal wave energy, the more retarded are these changes. Daily wave reworking of beach morphology in such cases may become limited to minor changes in bedforms that are themselves strongly hinged on the tidal cycle. In storm wave settings, such as those of northwestern Europe, change becomes significant only during storms. This effect may extend to the shoreface. From hydrodynamic data collected on several megatidal beaches (beaches with tidal ranges exceeding 8 m), Levoy *et al.* (2000) identified significant beach and shoreface tidal modulation of wave heights, with wave heights being lower at low tide than at high tide. The effect is exacerbated over shallow offshore areas where the tidal fluctuation induces corresponding changes in water depth.

The large beach volume and the important horizontal tidal translation also imply a reduction in overall beach gradient. A common feature of the tidal influence on beaches with large tidal ranges is that the various wave zones migrate rapidly across this wide, low-gradient profile during the tide, resulting in significant cross-shore variations in the hydrodynamics and resulting morphology. The wave influence is toned down because the sum of energy spent per unit area over time is much less than on beaches with low tidal ranges. While entrainment thresholds are the same, the volume of sediment transport is thus much less on beaches with large tidal ranges. The beach sediment volume and slope being equal, the larger the tidal range the greater the variations in morphodynamic behavior between the lower beach and the upper beach. On megatidal beaches in northern France, the morphodynamic domains may range from extremely dissipative at low tide on the lower beach to moderately reflective on the upper beach at high tide. At high tide, the extreme lower beach is subject to a combination of shoaling waves and strong longshore tidal currents. On similar megatidal gravel beaches fronted by sand flats in northern France, the upper beach is extremely reflective and is dominated by subharmonic gravity wave motions while the lower beach is highly dissipative and shows infragravity edge wave motions and strong, tidally induced longshore currents. Where wind, sand size, and supply conditions are favorable, wide, low-gradient beaches associated with large tidal ranges may provide significant surfaces for eolian reworking of beach sand into coastal dune fields. These wide beaches are also sometimes characterized by strong groundwater table fluctuations that may actively affect beach face hydrodynamic conditions and stability (Turner, 1993; Masselink and Turner, 1999).

On wave and tide-dominated beaches, waves induce sediment suspension while transport is by strong tidal currents (Davidson *et al.*, 1993; Masselink and Pattiaratchi, 2000), although the most active tidal phase appears to differ on different beaches, depending on the effects of friction and on whether the tidal wave is progressive or standing. Tidal flow asymmetry on such beaches may determine preferential sediment transport patterns and directions. On some French sandy beaches in the eastern English Channel, the progressive tidal wave is associated with strong longshore currents at high water that are highly effective in terms of sand transport because they coincide with higher high-tide waves that lead to suspension of fine sand. Such currents are always directed northwards or eastwards, and are therefore very important in terms of the net long-term bedload and suspended sediment (including pollutants) transport in these directions. Low-tide waves and tidal currents on the lower beaches are generally much weaker, probably because the shallower water results in both enhanced wave dissipation and tidal retardation by increased bed friction, while tidal current strength is generally greater seaward in deeper water.

Concluding remarks: management aspects of wave and tide-dominated coasts

One rationale for recognizing distinct wave and tide-dominated coasts is that such coasts may be characterized by specific management aspects that require an understanding of the mixed process regime and sediment transport and accumulation patterns. The commonly wide beaches in settings with large tidal ranges may be utilized for a wide range of activities (notably space-consuming leisure activities such as speed-sailing) that may be mutually exclusive and whose practice may therefore require set regulations and space allotments in highly frequented zones. The distinctive coastal profile of such areas may require the clear recognition in management strategies of the specific cross-shore and longshore dynamics due to both wave action and tidal currents. The classic management notion of coastal sediment cells (e.g., Bray *et al.*, 1995), generally based on longshore wave-energy gradients when coasts with low tidal ranges are considered, is not tenable on wave and tide-dominated coasts because of the significant modulation of the energy spectrum by tidal currents. Sediment cell definition criteria need to integrate the tidal dimension in sediment transport when transferred to wave and tide-dominated coasts. Wave and tidal sediment transport may work over wide low-gradient intertidal and subtidal profiles and this dimension needs to be taken into account in the design of coastal structures such as groins. Port breakwaters, for instance, are much more easily bypassed by tidal transport of sediment than on simple wave-dominated coasts, thus requiring constant dredging operations downdrift.

In areas where the nearshore zone is characterized by tidal sand banks and ridges, such as the English Channel and southern North Sea, changes in the volume and positions of these features in time may lead to navigation hazards as well as exposure or destabilization of cables and pipelines. These banks and ridges are also important in dissipating storm waves and are therefore significant in terms of shorefront protection, while being also a potential source of sand for beaches and dunes, as well as of dredged aggregate.

Although it has become clear that tides play a significant role in settings associated with waves, thus warranting the recognition of mixed, wave and tide-dominated coasts, more studies are needed to elucidate the various ways tides interact with, and modulate waves, and overall sediment transport.

Edward J. Anthony

Bibliography

Anthony, E.J., 1991. Coastal progradation in response to variations in sediment supply, wave energy and tidal range: examples from Sierra Leone, West Africa. *Géodynamique*, **6**: 57–70.

Anthony, E.J., 1996. Evolution of estuarine shoreline systems in Sierra Leone. In Nordstrom, K., and Roman, I. (eds.), *Estuarine Shores: Evolution, Environments and Human Alterations*. Chichester: Wiley, pp. 39–61.

Anthony, E.J., 2000. Marine sand supply and Holocene coastal sedimentation in northern France between the Seine estuary and Belgium. In Pye, K., and Allen, J.R.L. (eds.), *Coastal and Estuarine Environments—Sedimentology, Geomorphology and Geoarchaeology*. Special Publications of the Geological Society of London, 175, pp. 87–97.

Beck, C., Clabaut, P., Dewez, S., Vicaire, O., Chamley, H., Augris, C., Hoslin, R., and Caillot, A., 1991. Sand bodies and sand transport paths at the English Channel–North Sea border: morphology, dynamics and radioactive tracing. *Oceanologica Acta*, **11**: 111–121.

Bray, M.J., Carter, D.J., and Hooke, J.M., 1995. Littoral cell definition and budgets for central southern England. *Journal of Coastal Research*, **11**: 381–399.

Carter, R.W.G., 1988. *Coastal Environments*. London: Academic Press.
Clarke, A.J., and Battisti, D.S., 1981. The effect of continental shelves on tides. *Deep-Sea Research*, **28**: 665–682.
Davidson, M.A., Russell, P.E., Huntley, D.A., and Hardisty, J., 1993. Tidal asymmetry in suspended sand transport on a macrotidal intermediate beach. *Marine Geology*, **110**: 333–353.
Davies, J.L., 1980. *Geographical Variation in Coastal Development*, 2nd edn. London: Longmans.
Davis R.A., Jr., and Hayes, M.O., 1984. What is a wave-dominated coast? *Marine Geology*, **60**: 313–329.
Ferentinos, G., and Collins, M.B., 1980. Effects of shoreline irregularities on a rectilinear tidal current and their significance in sedimentation processes. *Journal of Sedimentary Petrology*, **50**: 1081–1094.
Grochowski, N.T.L., Collins, M.B., Boxall, S.R., Salomon, J.C., Breton, M., and Lafite, R., 1993. Sediment transport pathways in the eastern English Channel. *Oceanologica Acta*, **16**: 531–537.
Hayes, M.O., 1979. Barrier island morphology as a function of tidal and wave regime. In Leatherman, S.P. (ed.), *Barrier Islands*. New York: Academic Press, pp. 1–27.
Hayes, M.O., 1980. General morphology and sediment patterns in tidal inlets. *Sedimentary Geology*, **26**: 139–156.
Levoy, F., Anthony, E.J., Monfort, O., and Larsonneur, C., 2000. The morphodynamics of megatidal beaches in Normandy, France. *Marine Geology*, **171**: 39–59.
Masselink, G., and Pattiaratchi, C., 2000. Tidal asymmetry in sediment resuspension on a macrotidal beach in northwestern Australia. *Marine Geology*, **163**: 257–274.
Masselink, G., and Short, A.D., 1993. The effect of tide range on beach morphodynamics and morphology: a conceptual beach model. *Journal of Coastal Research*, **9**: 785–800.
Masselink, G., and Turner, I.L., 1999. The effect of tides on beach morphodynamics. In Short, A.D. (ed.), *Handbook of Beach and Shoreface Morphodynamics*. Chichester: Wiley, pp. 204–229.
Pingree, R.D., 1978. The formation of the Shambles and other banks by tidal stirring of the seas. *Journal of the Marine Biological Association of the U.K.*, **58**: 211–226.
Reinson, G.E., 1984. Barrier island and associated strand-plain systems. In Walker, R.G. (ed.), *Facies Models*, 2nd edn. Geoscience Canada, pp. 119–140.
Short, A.D., 1991. Macro-meso tidal beach morphodynamics—an overview. *Journal of Coastal Research*, **7**: 417–436.
Tessier, B., Corbau, C., Chamley, H., and Auffret, J.P., 1999. Internal structure of shoreface banks revealed by high-resolution seismic reflection in a macrotidal environment (Dunkerque area, northern France). *Journal of Coastal Research*, **15**: 593–606.
Turner, I.L., 1993. Water table outcropping on macro-tidal beaches, a simulation model. *Marine Geology*, **115**: 227–238.
Van Lancker, V., 1999. Sediment and morphodynamics of a siliciclastic near coastal area, in relation to hydrodynamical and meteorological conditions: Belgian continental shelf. Unpublished Ph.D. thesis, University of Ghent.
Wright, L.D., Nielsen, P., Short, A.D., and Green, M.O., 1982. Morphodynamics of a macrotidal beach. *Marine Geology*, **50**: 97–128.

Cross-references

Barrier Islands
Beach Processes
Dissipative Beaches
Shelf Processes
Tide-Dominated Coasts
Tides
Wave–Current Interaction
Wave-Dominated Coasts

WAVE CLIMATE

Ocean waves may be defined as periodic undulations about an interfacial reference level, and as such can move horizontally, vertically, or at angles inclined to the horizontal. In the simplest sense, waves transport energy without transporting mass. Sound waves, electromagnetic waves, and sand waves, amongst others, are also part of the oceanic realm, but are not dealt with herein.

Most human experience is with those ocean waves visually described as vertical undulations about the air–sea interface. These are generally known as *gravity* waves because gravity is the restoring force. Oceanographers also recognize internal waves, Rossby waves, tidal waves, solitons, Kelvin waves, and many other complex periodic motions. Table W2 is a brief summary of the commonly recognized wave types in the marine environment. This article however, will focus on those waves generally known as *sea* and *swell*, because most of the wave energy in the ocean is contained in such gravity waves—of which sea and swell are the major types.

Waves are characterized by wavelength, height, period, and steepness. Wavelength (L) is the horizontal distance from crest to crest or trough to trough. Wave height (H) is the vertical distance from crest to trough, and steepness is the ratio H/L. When steepness exceeds 1/7, waves tend to break as is seen routinely at a beach. Period is the time for two successive crests or troughs to pass a fixed point, the reciprocal of which is frequency. Significant wave height ($H_{1/3}$) is the height of the highest 1/3 of the waves in a large number of waves as estimated by an observer. The human eye seems to pick $H_{1/3}$ as the observed wave height.

The effect of the wind on water is to produce capillary waves (which are seen as ripples), ultragravity waves (which are seen in sunglint), sea (which are the shorter-period gravity waves), and swell (which are the longer-period gravity waves). Tsunami waves and seiches are examples of infragravity waves, storm surges such as caused by tropical storms are generally thought of as long-period waves, and transtidal waves are the realm of Kelvin waves and Rossby waves. The classification boundaries in Table W2 should not be held too rigid as many waves extend across a broad range of periods. The general category of *wind-waves* includes capillary and ultragravity waves, sea, and swell. *Sea* is the result of the direct action of wind on the ocean, and *swell* is caused by a nonlinear transfer of energy from shorter-period waves toward longer-period waves.

Table W3 is an adaptation of the Beaufort Scale using verbiage from the World Meteorological Organization supplemented by the classic "seaman's" description of wind and sea. This scale is the basis of much of the data oceanographers have regarding winds and waves. It is at best qualitative and depends on the experience and training of the observer for its accuracy. Scientists who study wave climates over extended time need to appreciate the limitations of such data for trend analysis and climate change.

A quantitative theory for wind-waves from first principles does not exist. The sea surface is rarely at rest, and thus almost always presents an uneven surface for the wind to act upon. Most likely, gustiness (turbulence) in air moving over still water has sufficient spatial inhomogeneity in speed to cause capillary waves. Once the sea surface is roughened, small pressure differences on the windward and leeward sides of wave crests transfers more energy from the air to the water, and the waves grow.

Stress (τ) between two fluids or between fluid layers is a force per unit area. At the air–sea interface it is classically given by $\tau = \rho_{\text{air}} C_{\text{D}} v^2$, where ρ_{air} is the density of air, C_{D} is the drag coefficient, and v^2 is the square of the wind velocity. Surface friction (F_{r}), a force per unit volume, is the vertical gradient of stress $F_{\text{r}} = \partial\tau/\partial z$, and is related to the square of the wind speed. Since wind velocity is a vector, stress and friction too are vector quantities. Interestingly, linear wave theory is approached by assuming that wave-motion is frictionless.

Based on numerous observational experiments, oceanographers and ocean engineers have developed parametric relationships between wind speed (i.e., τ), fetch, and duration leading to quantitative values of wave height and wave period. The maximum wave height and period from a given wind speed, blowing over a minimum distance (fetch) and for a minimum time (duration) leads to the concept of a fully developed sea. Table W4, calculated from parameters in the *Shore Protection Manual*

Table W2 Classification of ocean waves

Name	Period	Disturbing force	Restoring force
Capillary waves	<0.1 s	Wind	Surface tension
Ultragravity waves	0.1–1 s	Wind	Gravity
Gravity waves	1–30 s	Wind	Gravity
Infragravity waves	0.5–5 min	Wind	Gravity
Long period waves	0.1–12 h	Storms/earthquakes	Gravity/coriolis
Tidal waves	12–25 h	Gravitation	Gravity/coriolis
Transtidal waves	>1 day	Land–Air–Sea coupling	Gravity/coriolis

Table W3 Beaufort wind scale

Beaufort number	Wind description	Wind speed ($m\ s^{-1}$)	Wave height (m)	Wave description
0	Calm	0–0.2	0	Calm, mirror-like
1	Light air	0.3–1.5	<0.05	Glassy, scale-like
2	Light breeze	1.6–3.3	<0.1	Rippled, no white caps
3	Gentle breeze	3.4–5.4	0.1–0.5	Wavelets, few white caps
4	Moderate breeze	5.5–7.9	0.6–1.2	Slight, numerous white caps
5	Fresh breeze	8.0–10.7	1.2–2.4	Moderate, some spray
6	Strong breeze	10.8–13.8	2.5–4.0	Rough, white caps everywhere
7	Near gale	13.9–17.1	4.1–6.0	Very rough, streaky foam forms
8	Gale	17.2–20.7	4.1–6.0	Very rough, marked foam streaks
9	Strong gale	20.8–24.4	4.1–6.0	Very rough, dense foam streaks
10	Storm	24.5–28.4	6.1–9.1	High, visibility reduced
11	Violent storm	28.5–32.6	9.2–13.7	Very high, foam everywhere
12	Hurricane	>32.7	>13.8	Phenomenal, sea completely white

Table W4 Characteristics of a fully developed sea

Wind speed ($m\ s^{-1}$) at 10 m	Fetch (km)	Duration (h)	Significant wave height (m)	Period (s)
5	60	10	0.6	4
10	235	20	2.5	8
15	530	30	6	12
20	940	41	10	17
25	1,475	51	16	21
30	2,125	61	22	25

of 1984, summarizes typical values of fetch and duration for generation of the maximum deep-water significant wave height ($H_{1/3}$) and the longest period in a fully developed sea.

The higher the wind speed, the longer must be the duration and fetch for the sea to develop fully. Wave height and period in a fully developed sea will not increase if the minimum fetch and duration required is exceeded. If either the minimum fetch or duration required for a fully developed sea is not met, the resulting waves will have smaller heights and shorter periods. A comparison between Tables W3 and W4 shows that the two approaches are not fully compatible, but there is general agreement between the seaman's view of wave height and the ocean engineer's.

Energy (E) in sea and swell is a function of the wave height squared (H^2) and the density of the water (ρ_{water}), and is given by $E = 1/8\rho_{water}gH^2$; wave power $P = E \cdot c_g$ is the product of wave energy and the wave's group velocity (c_g). From linear wave theory, the *group velocity* of a shallow-water wave (a wave whose length L is 20 times or greater than the water depth Z) is $c_g = c = \sqrt{gZ}$ and for a deep-water wave (a wave whose length is less than one-fourth the water depth) is $c_g = 1/2c = 1/2\sqrt{(L \cdot g)/2\pi}$. In these expressions, g is the acceleration of gravity, 9.8 $m\ s^{-2}$, and c is the phase speed or *celerity* of the wave. In the expressions for wave energy (E) and power (P), the units are energy per unit wave area and energy per unit of wave front along a crest or trough, respectively. Longer waves have more energy and power for a given height than shorter waves.

As waves transit from the deep sea into shallow water, their period remains the same, but their height increases. This can be seen by equating the shallow-water wave power in one water depth (Z_1) with that in another water depth (Z_2), and solving for the height (H) changes. The resulting equation,

$$\sqrt{\frac{Z_1}{Z_2}} = \frac{H_2^2}{H_1^2},$$

quantifies the growth in the height of a wave as it enters shallower water. The underlying physical reason for this is conservation of energy and power in shoaling wave systems. Another consequence of wave period remaining constant and energy (per unit area) conservation is the steepening of deep-water waves if they encounter an ocean current flowing in opposition to the direction of the waves. Marine meteorologists often use the term "... and slightly higher in the Gulf Steam." to warn mariners of increased risk due to wave–current interaction.

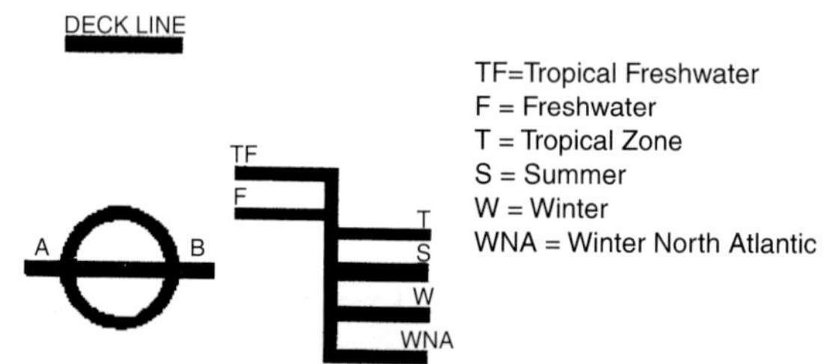

Figure W3 Load line markings for merchant vessels plying the oceans. A vessel loading in tropical freshwater may add cargo until the water is at the TF marking. Then as the ship proceeds out to sea, the greater buoyancy of saltwater over freshwater will cause the ship to ride higher and add additional freeboard or margin of safety.

Risk to shipping and marine operations has been of great concern to sailors, marine transportation companies, and insurance firms for well over 100 years (*q.v.* Table W3). In the late 19th century, Samuel Plimsoll championed the use of load lines to guide mariners in safety. An example of load lines (or Plimsoll marks as they are often called) is given in Figure W3. The A–B designates the American Bureau of Shipping as the issuing agency. The other terms are marks that show how deep a ship may be loaded in different environments. The marking "WNA" designates "Winter North Atlantic" and requires the vessel operator to allow the maximum freeboard (safety margin) for a ship operating in that environment. From well before Plimsoll's time, the Atlantic Ocean in winter was known to have one of the most dangerous wave climates on earth.

Surface winds over the ocean are to first approximation given by the geostrophic equation $\rho_{air}f\nu_g = \partial p/\partial x$, where the Coriolis parameter $f = 1.459 \cdot 10^{-4} \sin \phi$, p is the sea-level air pressure, ϕ is latitude, and x is the horizontal distance. The term $\partial p/\partial x$ is known as the pressure gradient and is a force per unit volume. Over the area of the WNA, ∂p is the sea level pressure difference typically chosen as that at Iceland minus that at the Azores; ∂x is the horizontal distance between Iceland and the Azores. The deep Icelandic Low minus the Bermuda–Azores High gives rise to the large pressure gradients that cause geostrophic winds to exceed $\nu_g = 25\ m\ s^{-1}$. Given that these winds often blow most of the winter and across thousands of kilometers of fetch, a glance at Table W4 will convince the reader that $H_{1/3} = 16$ m waves with >20 s periods is why WNA requires the greatest caution for the mariner operating in those seas.

Until the late 20th century, knowledge of conditions at sea such as winds, waves, and currents, were from reports by the vessels plying those waters using the codes shown in Table W3. American navigators credit a US Naval Officer, Matthew Fontaine Maury, with bringing together much of the information that underlies scientific understanding of winds, temperatures, sea, and swell. Those thousands of individual reports, collated and digested, have resulted in numerous atlases and pilot charts depicting average conditions to be expected. Since ships infrequently visit certain areas (such as the southeast Pacific Ocean), or when shipping routes change (such as after the opening of the Panama Canal in 1914), in certain areas the data in these atlases and pilot charts are nonuniform, very sparse, or even nonexistent. Satellite oceanography has changed much of this lack of data homogeneity, both spatially and temporally.

Figure W4 was developed from the satellite altimeters flown during the last decades of the 20th century. The upper panel depicts the average wave heights as measured from the altimeter return-pulse-spreading due to surface roughness. While the WNA is obviously an area of large wave heights, it is the region of the Southern Ocean where Earth's greatest waves are observed. Here, the sea-level atmospheric pressure gradients $\partial p/\partial x$ are large, and the fetch is essentially infinite as there is only one significant land barrier: the Drake Passage between South America and Antarctica.

In low latitudes, Figure W4 shows the lower sea states associated with the gentle Trade Winds. While the mean wave heights in the tropics and subtropics is on average small, these are the same regions whose weather is punctuated by tropical storms, hurricanes, typhoons, and cyclones. Sea states in these extreme storms are very high because the winds are oftentimes stronger than anything listed in Table W4, and also because the circular air flow at the sea surface leads to a long fetch. In addition, such storms create a solitary wave known as the storm surge, which can easily raise sea level at least 5 m—on top of which is the sea and swell!

The lower panel in Figure W4 depicts the variability of sea state about the mean shown in the upper panel. With the large seasonal wind changes in high northern latitudes come the largest changes in wave climate. Again the wisdom of Samuel Plimsoll's "WNA" designation warns the mariner to take due precautions. In the Southern Ocean, the variability is much lower than in the North Pacific Ocean or the North Atlantic Ocean; the enormous ice-mass of Antarctica provides the thermal inertia for fairly constant and large sea-level pressure gradients $\partial p/\partial x$ and hence large geostrophic winds v_g and resulting waves $H_{1/3}$. The northwestern Indian Ocean (lower panel, Figure W4) shows the large variability in sea states from the reversing monsoonal winds of western India, with a smaller effect caused by the Southeast Asian Monsoon seen in the vicinity of the Philippines.

Earth's wave climate is not static. Although the general picture presented herein is stochastically stable, the same satellite data that allows Figure W4 to be compiled, has shown decadal trends in certain areas. Notably, the wave climate of the Norwegian Sea seems to be changing toward higher sea states. This change in wave height is correlated with a deepening of the Icelandic–Azores atmospheric sea-level pressure gradient—a variable known as the North Atlantic Oscillation. Whether this increase in $H_{1/3}$ is long term or cyclical is unknown, but it would be prudent to continue to monitor sea and swell both from spacecraft and oceanographic observatories. Only through the concerted efforts of scientists such as Matthew Fontaine Maury will oceanographers and ocean engineers be able to make statements with certainty about wave climate change.

The coastal science community is embarking on the design of an integrated and sustained ocean observing system. Conceptually, an Integrated Ocean Observing System (IOOS) must include measurements, modeling, and forecasts of numerous variables including wave height, direction, and period. Modern issues in coastal zone management require improved forecasts of natural hazards for Earth's burgeoning coastal populations, and wave energy will be of paramount importance not only for the immediate need in emergency management, but in the design of future infrastructure. The "S" in IOOS will require

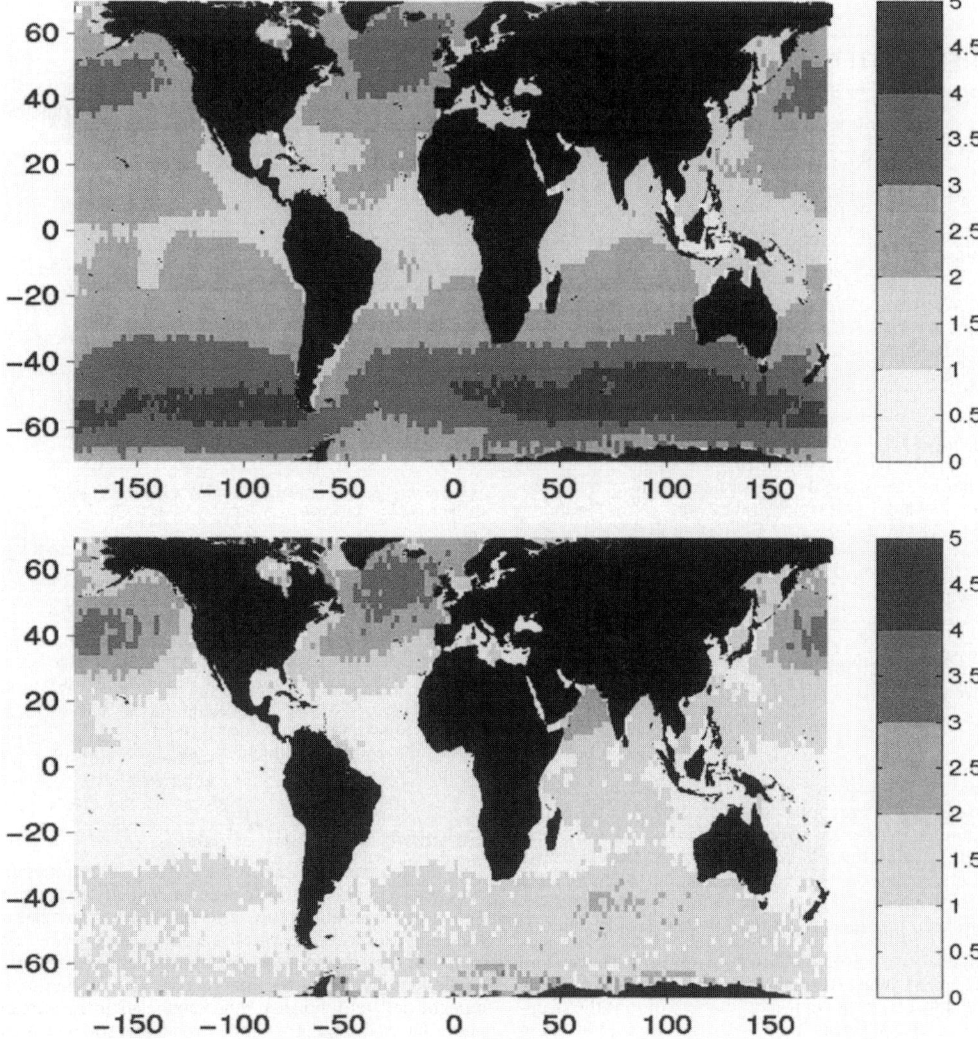

Figure W4 Average wave height (upper panel) and variability about the average (lower panel) in meters. Dr. David Woolf processed these data at the Southampton Oceanography Centre in the United Kingdom from satellite altimeter data for the period 1985–97. Figure used by permission.

effective integration of science, engineering, and management—an intellectual environment where wave climate is a dynamic, not a static, concept. Ocean waves are one component of the earth system, and it will be from a systems-approach that all components will be incorporated into effective decision-making.

Additional reading on this topic may be found in the following bibliography.

George Maul

Bibliography

Kinsman, B., 1965. *Wind Waves—Their Generation and Propagation on the Ocean Surface*. Englewood Cliffs: Prentice-Hall Publishers.

Shore Protection Manual, 1984. U.S. Department of the Army, Corps of Engineers. Washington, DC: U.S. Government Printing Office.

Woolf, D.K., Challenor, P.G., and Cotton, P.D., 2002. Variability and predictability of the North Atlantic wave climate. *Journal of Geophysical Research*, **107**(10–3): 9.

Cross-references

Beaufort Wind Scale
Coastal Wind Effects
Meteorologic Effects on Coasts
Storm Surge
Surf Zone Processes
Wave Hindcasting
Waves

WAVE–CURRENT INTERACTION

Flows observed in shallow marine environments are frequently combinations of currents and waves. Currents, with variations on timescales of hours or days, are principally tidal, wind-driven, or related to larger-scale patterns of ocean circulation. Wave-generated oscillatory flows, with periods on the order of seconds to tens of seconds are produced by surface gravity waves, though lower frequency internal waves can also be present. When wave-generated oscillatory motion of the water column extends to the seabed, wave–current interactions lead to increases in bed shear stress and the bottom roughness affecting bottom-boundary-layer currents. These effects are important for understanding ocean circulation and sediment transport in coastal environments.

Currents and wave-driven flows interact with the seabed to produce boundary layers where frictional effects are important. The ubiquitous presence of currents in the ocean results in a turbulent bottom boundary layer with a thickness on the order of meters to a few tens of meters. Waves produce a boundary layer only when wave oscillatory motion extends to the seafloor, a condition that is satisfied when water depth is less than half the wavelength of surface gravity waves. The boundary layer produced by waves in the ocean is most often turbulent, with a thickness of the order of 0.1 m. The disparity in wave- and current-boundary-layer thickness has several important consequences, including wave-generated bed shear stresses that are generally higher than current-generated stresses, the confinement of combined wave–current turbulence to the region of the wave boundary layer, and a large gradient in shear stress and turbulent mixing near the top of the wave boundary layer when wave shear velocities are significantly larger than current shear velocities.

Theory

Analyses of wave–current interactions are commonly carried out on timescales long enough to average over many individual waves and short enough that the currents can be considered quasi-steady. Hourly averaged currents and wave conditions are often used to characterize effects of wave–current interaction. Near the seafloor, the velocity profile associated with the currents is logarithmic under steady, uniform flow conditions

$$u_c(z) = \frac{u_{*c}}{\kappa} \ln \frac{z}{z_{0c}},$$

where u_c is current velocity, u_{*c} is current shear velocity, κ is von Karman's constant (0.41), z is elevation above the seabed and z_{0c} is a roughness parameter. In the absence of waves, z_{0c} is related to the physical roughness of the bed. When waves are present, an apparent roughness associated with the presence of the wave boundary layer dominates the roughness parameter z_{0c}. The applicability of the equation above is limited to near-bed depths greater than the height of the wave boundary layer, δ_w. Wave boundary layer height, assuming turbulent flow, can be estimated as $\delta_w = u_{*w}/\omega$, where u_{*w} is wave shear velocity (discussed below), $\omega = 2\pi/T$, and T is wave period.

The velocity of wave-generated water motion varies throughout the wave period. The wave-generated velocity just above the seabed (at the top of the wave boundary layer), termed the bottom or near-bed wave orbital velocity, u_b, can be related to surface wave conditions and water depth using linear wave theory. For monochromatic waves

$$u_b = u_{bm} \cos \omega t, \quad u_{bm} = \frac{\pi H}{T \sin h(kh)},$$

where u_{bm} is maximum near-bed orbital velocity, t is time, H is wave height, $k = 2\pi/L$, L is wavelength, and h is water depth. When a spectrum of waves is present, as is typical of coastal environments, u_{bm} is most accurately determined by applying the above equation to each frequency band in the full wave spectrum to determine the average or significant maximum near-bed orbital velocity (e.g., Madsen, 1994); significant orbital velocity is analogous to significant wave height.

The time-varying wave orbital velocity at the top of the wave boundary layer results in a time-varying bed shear stress. It is common practice in calculations of wave–current interaction to characterize wave-generated bed shear stress, and associated turbulent mixing within the wave boundary layer, in terms of the maximum value of near-bed wave orbital velocity, u_{bm}. Wave shear velocity, u_{*w}, or bed shear stress, τ_{bw}, can be related to u_{bm} through the equations for an oscillatory boundary layer (e.g., Smith, 1977; Grant and Madsen, 1979) or a wave friction factor, f_w:

$$\tau_{bw} = \frac{\rho}{2} f_w u_{bm}^2,$$

where ρ is fluid density. Wave friction factor is a function of wave orbital amplitude ($a = H/[2\sin h(kh)]$) and bed roughness, k_s. A number of empirical relationships for f_w are available, such as:

$$f_w = 0.04\left(\frac{a}{k_s}\right)^{-1/4} \quad \text{for } \frac{a}{k_s} > 50,$$

(Fredsøe and Deigaard, 1992). Because of the dependence of u_{bm} on water depth, τ_{bw} decreases with increasing depth for a given set of wave conditions.

When waves and currents both contribute to near-bed flow, the velocity profile of the currents within the wave boundary layer can be expressed as

$$u_c(z) = \frac{u_{*c}^2}{\kappa u_{*cw}} \ln \frac{z}{z_{0c}}, \quad z < \delta_w,$$

where u_{*cw} is the combined wave–current shear velocity and z_0 is the hydrodynamic roughness related to the physical roughness of the bed. The combined flow shear velocity, u_{*cw}, depends on the shear velocities associated with the current, u_{*c}, and with the waves, u_{*w}. The wave–current shear velocity can be found from the combined-flow bed shear stress as,

$$\tau_{bcw} = [(\tau_{bw} + \tau_{bc} \cos \varphi)^2 + \tau_{bc}^2 \sin \varphi^{1/2}] = \rho u_{*cw}^2$$

where φ is the angle between the direction of wave propagation and the direction of the currents, and τ_{bc}, τ_{bw}, and τ_{bcw} are the current, wave, and combined wave–current bed shear stresses, respectively. In this case, wave boundary layer thickness and turbulent mixing within the wave-boundary-layer scale with u_{*cw} rather than u_{*w}.

Solutions and applications

Solving the equations given above for $u_c(z)$ requires specification of z_0 and matching the profiles for current velocity above and below the wave boundary layer at the top of the wave boundary layer. Often, particularly in shallow water, wave-boundary-layer flow is fully rough, so that z_0 is just proportional to the physical roughness length of the bed, k_s. Wave-formed ripples on the bed typically dominate bottom roughness in shallow, sandy coastal environments, though grain roughness and saltating grains also contribute to bottom roughness, particularly when ripples are very small or absent. Current shear velocity, u_{*c}, which also must be specified when calculating the velocity profile $u_c(z)$, is seldom known in advance. In practice, a measured value of current velocity at some level near the bed is specified, and an iterative calculation is performed to determine the value of u_{*c} that is consistent with the measured velocity given the bottom roughness parameter. The equations for $u_c(z)$ provided here assume a linear eddy viscosity of the form $\nu_e = \kappa u_* z$; within the wave boundary layer $u_* = u_{*cw}$ while above it $u_* = u_{*c}$. This form of eddy viscosity, while yielding a simple analytical solution for $u_c(z)$, is discontinuous at the top of the wave boundary layer. Other formulations for ν_e exist that are continuous across the top of the wave boundary layer (e.g., Wiberg and Smith, 1983), but most of these require a numerical solution for $u_c(z)$. Higher-order closures have also been used.

Solving the equations for current velocity within and above the wave boundary layer yields a characteristic profile exhibiting an inflection near the top of the wave boundary layer (Figure W5). Current velocity within the wave boundary layer is more uniform than is the profile above it owing to the larger shear velocity (u_{*cw}) in the wave boundary layer. The zero-velocity intercept of a semilogarithmic line drawn through the velocity profile above the top of the wave boundary layer gives the apparent roughness felt by the currents above the wave boundary layer (Figure W5). The equations for $u_c(z)$ given above can be rearranged to solve for the apparent roughness, z_{0c}, giving

$$z_{0c} = \delta_w \left(\frac{z_0}{\delta_w} \right)^{\lambda}, \quad \lambda = \frac{u_{*c}}{u_{*cw}}.$$

Resulting values of apparent roughness are often several orders of magnitude larger than the roughness parameter associated with the physical roughness of the seabed.

Field observations of bottom boundary flow motivated the initial formulation of the theory of wave–current interaction, but the first comprehensive wave–current interaction models (Smith, 1977; Grant and Madsen, 1979) were developed before there were adequate data available to test them. Since that time, many field studies of near-bed waves and currents have been carried out in wave-dominated coastal environments around the world. Some laboratory studies have also been performed. These have demonstrated the importance of waves in increasing bed shear stress above values generated by currents alone. Under storm conditions, waves can dominate the combined wave– current bed shear stress across most or all of the continental shelf. Calculated and measured values of current shear velocity have been shown to agree reasonably well in a number of studies. However, calculated values depend on bed roughness, a term that is difficult to predict because of its dependence on bedform height and spacing, which vary in ways that are not fully understood. Nonuniform flows and stratified flows also complicate the application of wave–current interaction models to field data.

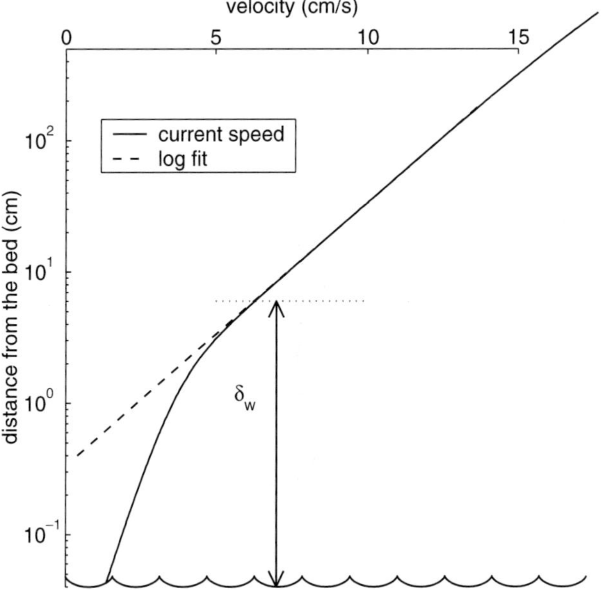

Figure W5 Profile of near-bed current illustrating the effect of wave–current interaction. Profile slope changes near the top of the wave boundary layer (δ_w, indicated by the horizontal dotted line) owing to differences in shear velocity above and below δ_w when waves and currents are both present. The slope of a semi-logarithmic line fit to the profile above δ_w is related to current shear velocity, u_{*c}, and the zero-velocity intercept of the line is the apparent bottom roughness felt by the current, z_{0c}. The profile shown in the figure is cut off above the level, z_0, at which the velocity reaches zero.

Increases in bed shear stress, turbulent mixing within the wave boundary layer, and apparent bottom roughness associated with wave–current interaction at the seabed have implications for a variety of processes. For example, observations from a number of continental shelves have concluded that bottom drag coefficients for near-bed currents can be increased by up to an order of magnitude larger when waves are present. This is important for studies of wind-driven coastal ocean circulation. Effects of wave–current interaction on bottom drag have been incorporated into models of ocean circulation applied to the coastal ocean.

Sediment transport in coastal environments is strongly affected by wave–current interactions. Coastal currents in many locations are insufficient to mobilize sediment outside of the nearshore zone. In these regions, it is only when moderate to high wave conditions are present that bed shear stresses are sufficient to mobilize sediment at the seabed.

Even at depths of 50–100 m, mean winter combined wave–current shear velocities can be considerably larger than current shear velocity. The wave boundary layer, because of its high shear velocity, is able to maintain both higher concentrations of sediment and coarser sediment in suspension than can the overlying current boundary layer. When differences between u_{*cw} and u_{*c} are large, gradients in suspended sediment concentrations near the top of the wave boundary layer can cause significant stratification which inhibits turbulent mixing and modifies near-bed flow and sediment transport.

Patricia L. Wiberg

Bibliography

Fredsøe, J., and Deigaard, R., 1992. *Mechanics of Coastal Sediment Transport*. Singapore: World Scientific.

Grant, W.D., and Madsen, O.S., 1979. Combined wave and current interaction with a rough bottom. *Journal of Geophysical Research*, **84**: 1797–1808.

Madsen, O.S., 1994. Spectral wave–current bottom boundary layer flows. In *Coastal Engineering '94, Proceedings of the 24th International Conference*. Kobe, Japan: Coastal Engineering Research Council, American Society of Civil Engineers, pp. 384–397.

Smith, J.D., 1977. Modeling of sediment transport on continental shelves. In Goldberg, E.D., McCave, I.N., O'Brien, J.J., and Steele, J.H. (eds.), *The Sea, Volume 6, Marine Modeling*. New York: Wiley, pp. 539–577.

Wiberg, P.L., and Smith, J.D., 1983. A comparison of field data and theoretical models for wave–current interaction at the bed on the continental shelf. *Continental Shelf Research*, **2**: 126–136.

Cross-references

Coastal Currents
Coastal Upwelling and Downwelling
Cohesive Sediment Transport
Rip Currents
Ripple Marks
Surf Modeling
Wave-Dominated Coasts
Waves

WAVE-DOMINATED COASTS

Price (1955) used the term *wave-dominated coasts* to describe the morphology of depositional shores where consistent, relatively large waves, with their associated strong wave-generated currents, have produced a smoothed shore of sandy sediments. Where such shores are building seaward (prograding), sand beaches associated with long barrier islands and prograding beach ridges are the dominant intertidal habitats. As a generalization, coasts with small tidal ranges (microtidal = <2 m (Davies, 1964)) are dominated by wave energy, thus most wave-dominated coasts are microtidal. On the other hand, coasts with large tidal ranges (macrotidal = >4 m) are typically dominated by tidal energy, and were hence termed *tide-dominated coasts* by Price. Coasts with intermediate wave energy and tidal ranges (typically mesotidal = 2–4 m) were termed *mixed-energy coasts* by Hayes (1979).

Wave-dominated non-deltaic coasts

Between major rivers on wave-dominated coasts, the shore is typically occupied by long, uninterrupted barrier islands. Inasmuch as wave-dominated barrier islands occur most commonly on microtidal coasts, the barrier islands are typically backed by open-water bays and lagoons. The sediment patterns of barrier islands show a simple gradation of grain size from coarse to fine away from shore as a result of decreasing bottom agitation by waves with increasing water depth.

Barrier islands occur primarily on depositional coasts on the trailing edges of continental plates, or on enclosed tideless seas such as the Baltic and the Mediterranean. Barrier islands are the dominant coastal type along the Atlantic and Gulf coasts of the United States, with those on the Gulf Coast of Texas, the Outer Banks of North Carolina, and Northeast Florida being wave-dominated. Tidal inlets are widely spaced on wave-dominated barrier islands (e.g., over 40 km apart in North Carolina). See Davis (1994) for a detailed synthesis of the barrier islands of the United States.

Most barrier islands have formed within the last 4,000–5,000 years during a near stillstand in sea level. Two types of barrier islands may be present—those that consistently migrate landward (retrograding) and those that build seaward (prograding). How an island builds depends upon the ratio of relative sea-level change to sediment supply. Where the sediment supply is inadequate or sea level is rising rapidly, the island retreats landward. Sediment supply for an island can be diminished through several natural mechanisms and by dams and jetties.

Retrograding barrier islands are composed of coalescing washover fans and terraces that are overtopped at high tides, usually several times a year. Stratigraphically, a relatively thin wedge of sand and shell from the washover terrace overlies back-barrier sediments, which are typically composed of muddy sediments deposited in the lagoons behind the islands. Prograding barrier islands are composed of multiple beach ridges. Stratigraphically, they consist of sand 8–10 m thick which has prograded over offshore muds.

Wave-dominated deltas

On wave-dominated coasts, where the rivers have enough sediment load to have filled the antecedent lowstand valley and build a bulge in the shore (within the time frame of the present highstand), the resulting river deltas are typically referred to as *wave-dominated deltas* (Fisher, *et al.*, 1969; Galloway, 1975). A generalized model of a wave-dominated delta is given in Figure W6. Note the arcuate shape of the delta, which is comprised mostly of prograded beach ridges. The prograded sand sheet associated with such deltas is typically thick because of the massive shorefaces that are created by the large waves.

The depositional patterns of sand at wave-dominated river mouths is illustrated in Figure W7 (modified after Wright, 1977). On river mouths where waves approach the shore with a parallel orientation (normal wave incidence), the crest of the submerged river-mouth bar is arcuate in shape and projects straight offshore with the intertidal portion of the bar being covered with swash bars that migrate more or less straight onto the shore. With oblique wave incidence, the river mouth, as well as the submerged river-mouth bar, is diverted in a downdrift direction.

The São Francisco delta on the coast of southern Brazil, which bears a close resemblance to the general model given in Figure W6, is a spectacular example of a wave-dominated delta. The average tidal range is 1.9 m, which is on the border between micro- and mesotidal conditions. However, the offshore slope is steep, and waves are exceptionally large for deltaic shores. According to Coleman and Wright (1975, p. 131), more wave energy is expended in 10 h on the São Francisco delta front than in 365 days on the Mississippi delta shore. The result is a smooth-faced delta plain dominated by beach ridges and coastal dunes. The composite stratigraphic column for this delta shows a thick sequence of sandy beach, dune, and upper shoreface sediments in the upper half of the stratigraphic column.

Wave-dominated estuaries

The concept of tide and wave dominance has also been applied to estuaries by Dalrymple *et al.* (1992). However, they state that estuaries are unlike many other coastal systems, because they are "geologically ephemeral." If the rate of sediment supply is sufficient to eventually fill the lowstand valley within which the estuary is located, the filled valley then becomes a delta. The estuaries they term wave-dominated are composed of a sand body complex (barrier/tidal inlet) at the entrance, a muddy central basin, and a bayhead delta system. Their general model for this type of estuary is given in Figure W8.

Miles O. Hayes

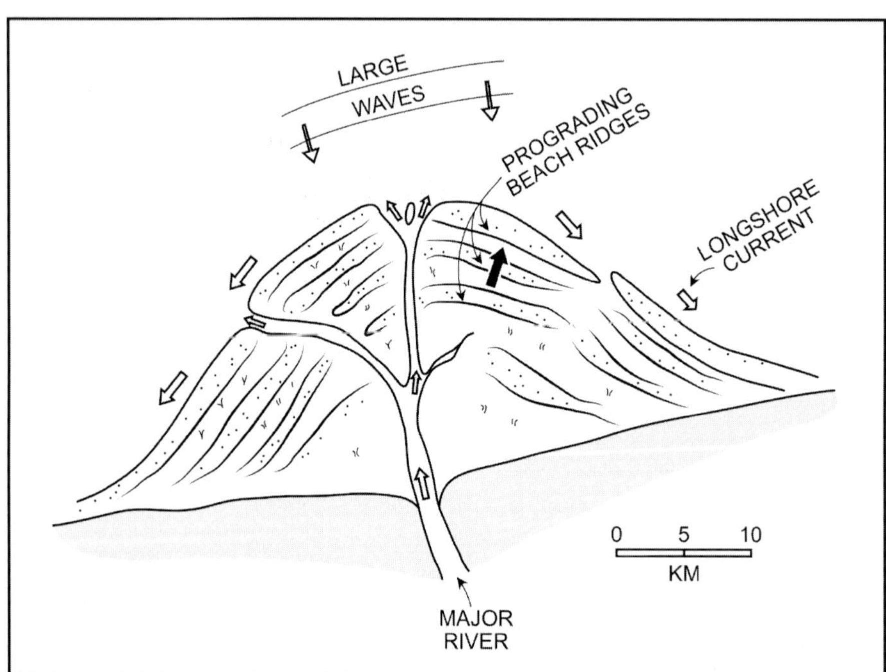

Figure W6 Schematic sketch in plan view of a wave-dominated delta. Note dominance of prograding beach ridges throughout the delta plain surface.

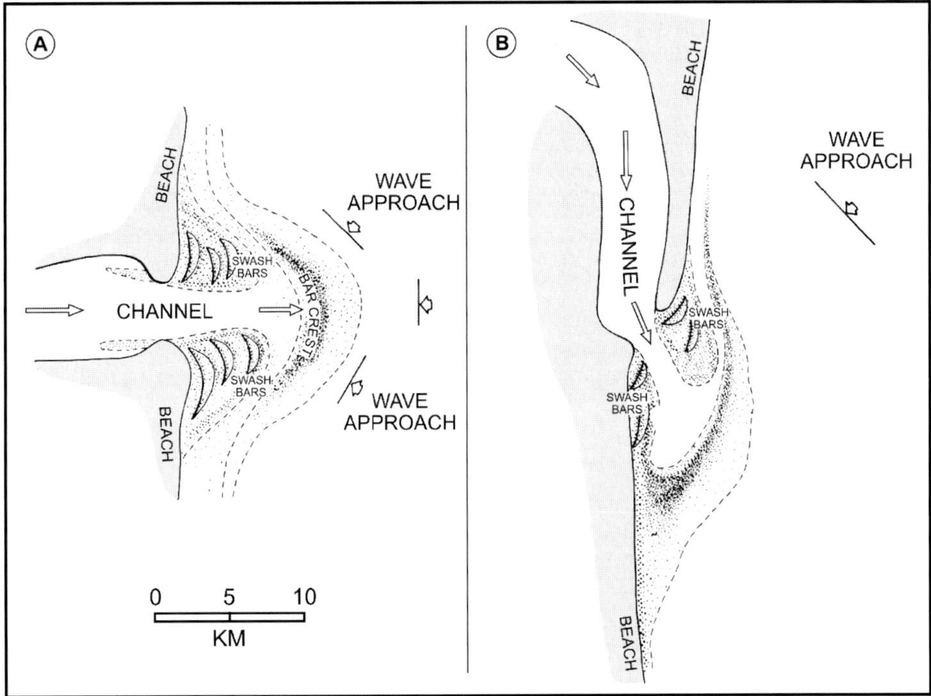

Figure W7 Depositional patterns associated with wave-dominated river mouths: (A) normal wave incidence; (B) oblique wave incidence (after Wright, 1977).

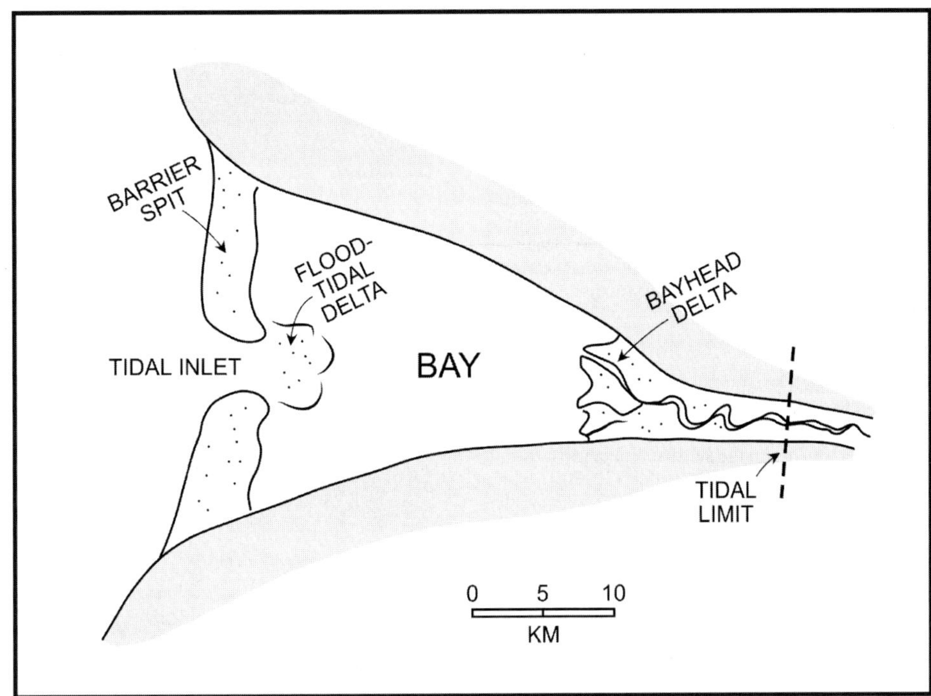

Figure W8 Schematic sketch in plan view of a wave-dominated estuary as described by Dalrymple *et al.* (1992).

Bibliography

Coleman, J.M., and Wright, L.D., 1975. Modern river deltas: variability of processes and sand bodies: In Broussard, M.L. (ed.), *Deltas*, 2nd edn. Houston, TX: Houston Geological Society, pp. 99–150.

Dalrymple, R.W., Zaitlin, B.R., and Boyd, R., 1992. Estuarine facies models: conceptual basis and stratigraphic implications. *Journal of Sedimentary Petrology*, **62**: 1030–1046.

Davies, J.L., 1964. A morphogenic approach to world shorelines. *Zeitschrift für Geomorphologie*, **8**: 27–42.

Davis, R.A., Jr. (ed.), 1994. *Geology of Holocene Barrier Island Systems*. Berlin: Springer-Verlag.

Fisher, W.L. *et al.*, 1969. *Delta Systems in the Exploration for Oil and Gas, a Research Colloquium*. Austin, University of Texas, Bureau of Economic Geology.

Galloway, W.E., 1975. Process framework for describing the morphological and stratigraphic evolution of deltaic depositional systems. In Broussard, M.L. (ed.), *Deltas*, 2nd edn. Houston, TX: Houston Geological Society, pp. 87–98.

Hayes, M.O., 1979. Barrier island morphology as a function of tidal and wave regime. In Leatherman, S. P. (ed.), *Barrier Islands from the Gulf of Mexico to the Gulf of St. Lawrence*. New York: Academic Press, pp. 1–28.

Price, W.A., 1955. *Development of Shorelines and Coasts*. Department of Oceanography, Project 63. College Station, TX: A&M University.

Wright, L.D., 1977. Sediment transport and deposition at river mouths: a synthesis. *Geological Society of America Bulletin*, 88: 857–868.

Cross-references

Barrier Islands
Bars
Coastal Inlets
Deltas
Estuaries
Tide-Dominated Coasts
Wave and Tide-Dominated Coasts

WAVE ENVIRONMENTS

Introduction

Wave environments are those area of the seas and oceans that have similar wave sources and climatologies. The environments are closely related to wave generating systems, namely the global and regional wind and cyclonic regimes, and second to the shape and orientation of the oceans and their surrounding coastlines.

Waves are the most characteristic surface feature of the globe, with a presence over 71% of the surface. They are also the most transitory feature as they continually form, travel, and ultimately, after a few minutes to days, dissipate their prodigious energy. As ephemeral as waves are, they are ever present, moving across the seas and great oceans, and at the coastline supplying on the order of 2.5×10^9 kW energy each year (Inman and Brush, 1973) to build, maintain, rearrange, and erode the coastal systems. This represents over half the energy in the coastal zone.

The seas, oceans, and water column play a passive role in the formation and movement of waves. While waves do move water, their source of energy and movement comes from external sources. For ocean or gravity waves that source is wind, particularly the great zonal westerly and trade wind systems, including cyclones in all their forms.

Wave sources

Ocean waves are entirely dependent on wind for their formation, and on a global scale can be defined by the location of the zonal wind systems. In the high latitudes are the polar easterlies (>70° latitude), in the mid-latitudes the strong westerlies (30–70°), in the subtropics the most extensive, though moderate wind system, the trade winds (30–70°), while straddling the equator are the quieter doldrums (10°N–10°S). These idealized wind systems and their ocean impacts are however, substantially modified by continents and oceans, seasonal shifts, latitude and in high-latitudes sea ice. In addition, cyclones (tropical, east coast, and mid-latitude) while more limited in extent generate the world's biggest waves. At a regional scale the monsoons and local sea breeze wind systems all generate waves and have important impacts on their regional wave environments. Finally, the Coriolis effect plays a major role in redirecting waves as they travel across the oceans veering wave direction to the right in the Northern Hemisphere and left in the Southern Hemisphere. An overview of wave environments must therefore take all these factors into account.

Wave generation

Waves are generated by the wind blowing over a stretch of water (estuary, lake, sea, or ocean). Wave height and period will both increase with increasing wind velocity and duration, together with length of fetch, the length of ocean over which the wind blows, and uniformity of wind direction. Water must also be deep (>150 m) otherwise the higher waves will shoal and break. The largest waves are therefore produced in those parts of the world's deep oceans where strong, unidirectional winds, blow for long periods (days) over long stretches of ocean (1,000 km). The single most important factor is wind speed, with wave energy increasing exponentially with increasing wind speed.

Most of the worlds wind systems the polar easterlies, trades, and doldrums are low to moderate velocity winds with wind speed rarely exceeding 12.5 m s^{-1}, and while they might blow for long periods, over long section of oceans, particularly the trade winds, they are not capable of producing high waves, with trade wind waves less than 3 m 90% of the year and less than 2 m 50% of the year (Young and Holland, 1996). In addition, the polar eastleries blow over land in Antarctica, and for much of the year, over sea ice in the Arctic.

Wave sources

The world's largest and most influential wave environments is that produced under the westerly wind stream between 30° and 70° latitude, but focused on 50–60°N and S. In addition to the strong flow of westerly winds are the mid-latitude cyclones. At any time between 6 and 15 cyclones are embedded in this westerly stream, encircling the globe at these latitudes. The cyclones form year round over the Southern Ocean and more in the winter season over the North Pacific and North Atlantic oceans. The cyclones and their associated winds exceed 15 m s^{-1} 10% of the year, and 10 m s^{-1} 50% of the year. They generate waves 2–3 m high 90% of the time across the Southern Ocean, and in excess of 5–6 m 10% of the time in both hemispheres (Figure W9; Young and Holland, 1996).

Secondary cyclonic sources are the East Coast Cyclones that form off the east coasts centered around 25–35°N and S. They occur in the United

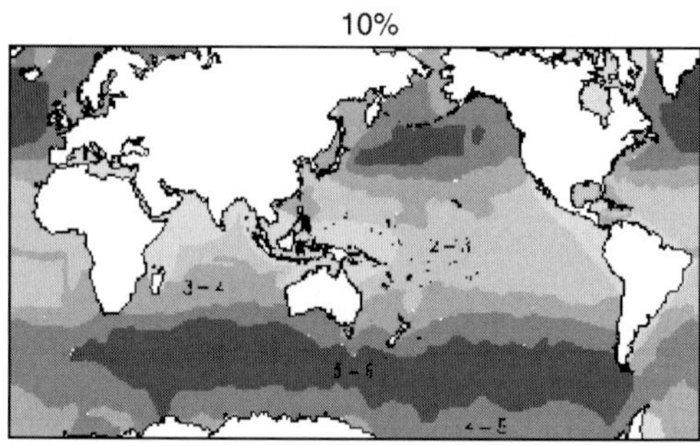

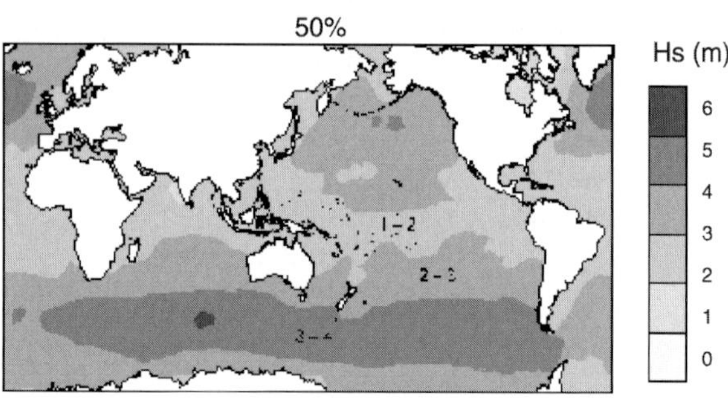

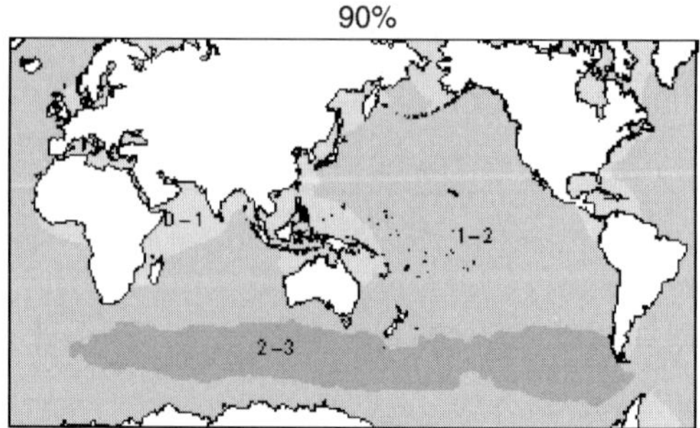

Figure W9 Global values for significant wave height which will be exceeded 10% (top), 50% (middle), and 90% (lower) of the time (from Short, 1999, based on Young and Holland, 1996 with permission of Elsevier).

States (the northeast storms), southeast Australia (east coast cyclones), south Brazil, and South Africa. In Australia they form on average 10 times a year in all months with an early winter maximum and regularly generate waves in excess of 10 m, with waves recorded up to 19 m. Tropical cyclones (hurricanes, typhoons) are generated in summer months between 5° and 15°N and S toward the warmer western side of the tropical oceans. While they can generate both high seas and associated storm surges in affected regions (10–30°N & S), their low and seasonal frequency (several per year in each region), and variable source area and erratic trajectory, means that they do not have a significant impact on longer-term wave climates.

Important regional winds include the reversing monsoons of India and southeast Asia, northern Australia and central Africa, and the local sea breezes that can affect most of the world's coast, though more so in the mid- to lower latitudes.

All the above wind and cyclonic systems vary spatially, shifting with the seasons, and temporally in terms of occurrence and intensity. Some such as polar easterlies and trade winds are relatively uniform in direction and velocity and blow over large areas of oceans. All the cyclonic sources are more variable in their occurrence, trajectory, and wind direction and speed. Both the monsoons and tropical cyclones are seasonally dependent, and like weather and climate all have a high degree of variability, both in terms of location, trajectory, frequency, duration, and intensity.

Davies (1964, 1980) was the first to put some sense into this seemingly wide range of wave sources and types. He developed a morphogenetic classification of the world's wave environments, based on their wind sources, their latitudinal location, and the nature and direction of wave travel.

Global wave environments

Davies identified four major deepwater wave environments the high-energy storm wave of the upper mid-latitudes, the west coast swell, the east coast swell, and finally low-energy protected coasts of both polar and tropical regions (Table W5 and Figure W10).

Storm wave

The most important and most energetic wave environment is the storm wave environment located between 40° and 60°N and S, under the circumpolar west wind belt and its migratory mid-latitude cyclones or subpolar lows. This is a region where all the ingredients for high waves are found. Winds exceed 15 m s^{-1} 10%, 10 m s^{-1} 50%, and 5 m s^{-1} 90% of the time. While they rotate around, the cyclones at any given latitude tend to dominate from a relatively uniform direction (south through west). In the Northern Hemisphere there is the wide fetch of the North Pacific and North Atlantic oceans, while in the Southern Hemisphere there is the world's only continuous stretch of ocean the circumpolar Southern Ocean, providing in principle unlimited fetch, all across deep oceans with few impediments to wave generation or movement, such as islands or reefs.

Storm wave environments are exposed to persistent strong westerly winds, year round in the Southern Hemisphere, and in the winter season in the Northern Hemisphere. They produce the world's highest waves averaging over 2–3 m 90% of the year and 5–6 m 10% of the year (Young and Holland, 1996), with extreme waves reaching several meters. Periods are 10–14 s, however, being a *sea* environment in the area of generation, waves are high, short, steep, and highly variable in both size, shape, and direction (Table W5). Sailors often use the term confused seas when referring to such conditions, with white caps and "rouge" or "freak" breaking waves characteristic of such environments. Wave direction is predominately to the west, with the Coriolis effect deflecting waves equatorward in both hemispheres.

West coast swell

West coast swell refers to the long persistent swell that reaches the west coast of many of the world's mid- to low-latitude continents, including the west coast of the Americas, Africa, Australia, and New Zealand. The *swell*, unlike the storm seas, refers to waves that have left the area of wave generation, the storm wave environment, or the wind has stopped blowing. When this happens the sea waves quickly reform into lower, longer, faster, more uniform swell waves, uniform in size, shape, and direction of travel. Once transformed into swell they are able to travel across deep oceans for thousands of kilometers with minimal loss of energy (Figure W11). The waves are however, affected by the Coriolis effect, which turns them equator-ward causing the initially westerly waves to arrive from the northwest in the Northern Hemisphere, and southwest in the Southern Hemisphere. The swell is higher in the higher latitudes, slowly decreasing toward the equator. They are characterized by moderate to high (2–3 m 50%), long

Table W5 Characteristics of major world deepwater wave environments (from Short, 1999)

	Latitude/ location	Source—wave type	Seasonality	Deepwater height	Period	Direction
Storm wave	40–60°N & S S & W facing coasts	Mid-latitude cyclones—sea (roaring 40s raging 50s)	Northern Hemisphere—winter Southern Hemisphere—year round	High (2–5 m)	Long (10–14 s)	Westerly to south westerly
West coast swell	40–0°N & S west-facing coasts	Mid-latitude cyclones—swell	Northern Hemisphere—winter Southern Hemisphere—year round	Moderate–high (1.5–3 m)	Long (12–14 s)	Northern Hemisphere—NW Southern Hemisphere—SW
East coast swell	40–0°N & S east-facing coasts	Mid-latitude cyclones—swell	Northern Hemisphere—winter Southern Hemisphere—year round	Moderate (1–2 m)	Moderate–long (8–12 s) coasts	Northern Hemisphere—NE Southern Hemisphere—SE
East coast cyclones	25–35°N & S	East coast cyclones—sea/swell	Year round winter max	High (2–5 m)	Moderate–long (8–12 s)	Northern Hemisphere—E Southern Hemisphere—E
Trade winds	25–0°N & S exposed coasts	Trades—sea	Trades in winter	Moderate trades (0.5–1.5 m)	Trades—6–9 s	Parallel winds
Monsoons Tropical cyclones		Monsoons—sea Tropical cyclones—sea	Monsoons and tropical cyclones in summer	Low monsoons (0.5–1 m) TC—high	monsoons—4–6 s TC—short	NE & SE trades NW & SW monsoons TC—variable
Protected coasts	70–90° N & S pole facing coasts	Polar easterlies (+ sea ice)	Summer open water season	Low (0.5 m max)	Short (3–5 s)	NE
	Tropics 10–0°N & S	Doldrums	Year round	Low (<0.5 m)	short (<5 s)	variable

N, north; S, south; E, east; W, west; NW, northwest; SW, southwest; NE, northeast; SE, southeast; TC, tropical cyclones.

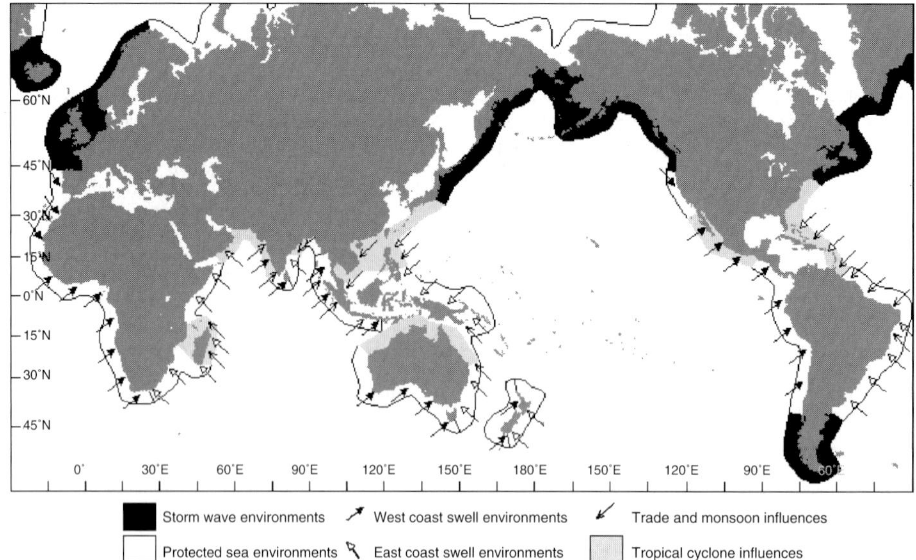

Figure W10 The world's major wave environments (adapted from Davies, 1980, and reprinted by permission of Pearson Education Limited). Note storm wave and west coast swell environments operate year round in Southern Hemisphere and during winter season in Northern Hemisphere.

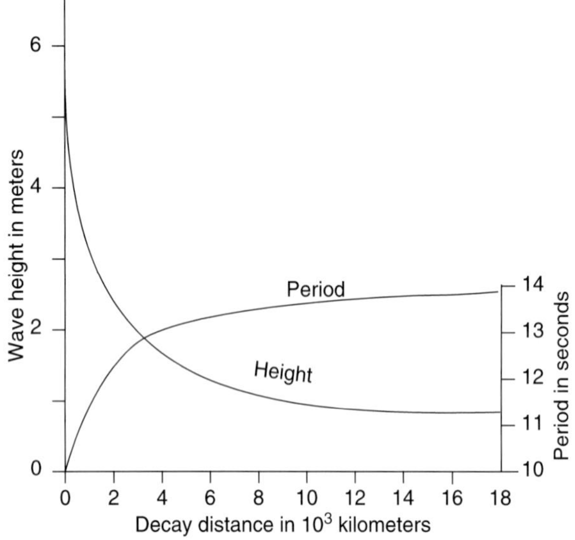

Figure W11 Transformation of sea (left) to swell (from Davies, 1980, and reprinted by permission of Pearson Education Limited).

period, uniform swell, with the period of higher swell associated with major cyclones, and lower swell in between, which in the Southern Hemisphere arrives on average 350 days a year. It is however, distinctly seasonal in the Northern Hemisphere (Table W5) where it delivers the big winter surf to Hawaii and the California coast. The higher winter swell and lower summer swell along the California coast, gave rise to the winter cut-summer fill beach cycle, which while applicable to California, has limited application in other wave environments.

East coast swell

East coast swell refers to swell generated in the storm wave environment, that travel equatorward, but has to undergo more refraction to reach the east coast of the continents below 40°N and S. Because of the initial trajectory and greater refraction required to reach the coast the swell is lower and arrives less frequently. On the southeast coast of Australia it arrives 200 days, as opposed to 350 days on the south coast, and averages 1.5 m down from 2.5 m on the south coast. It still provides a relatively persistent (60%) low to moderate swell. In addition, the mid- to low-latitude east coasts are exposed to additional wave regimes, which produce the world's most complex wave environment (Short and Trenaman, 1992). These include periodic east coast cyclones centerd on 40–25°, persistent trade winds peaking between 20° and 5°, summer tropical cyclones—hurricanes originating between 15 and 5° and summer sea breezes. All these wind regimes produce largely sea waves, ranging from lower frequency but high wave conditions associated with the cyclones, to moderate short waves with the trade winds, to low, short waves with the sea breeze. In addition, all these waves can be generated while swell arrives from the storm wave environment, resulting at times two wave trains arriving from different sources and directions. East coast swell environments also undergo an equatorward transition from predominately easterly swell conditions in the higher latitudes (40°), to increasing influence from first east coast cyclones (40–25°), strong trade winds influence between 20° and 0°, together with periodic tropical cyclones between 30° and 10°.

Protected wave environments

Protected wave environments refer to ocean and sea environments protected from ocean swell. They occur in both the tropics where they are protected by location in the doldrums, distance from major wave sources, and physically by coral reefs, island archipelagos, and smaller seas, as well as polar locations protected by sea ice from the low velocity polar easterlies. Around the Antarctic continent the Coriolis effect sends most of the westerly tending storm waves equatorward, ensuring further protection of the polar continent. All protected locations are devoid of swell, relying on local winds for generation of short, low seas, interspersed with long calms.

Other influences

Sea ice can form over the ocean surface as around Antarctica and across the Arctic Ocean, and seasonally along the shores of high latitude coasts such as Labrador and the Bering Sea. The ice both prevents the formation of waves, and at the shore prevents waves reaching the shore.

Islands and reefs

Islands and reefs wherever they occur in the mid-ocean or fringing the coastline will cause wave attenuation and refraction, leading to a reduction in wave height and changes in wave direction.

Breaker wave environment

All the above refers to deepwater sea and ocean wave environments. As waves move from deep to shallow water surrounding the continents and coastlines they undergo a predictable transformation from deep to shallow water waves. In doing so they may undergo wave attenuation as they shoal across the continental shelf and nearshore zone, wave refraction over variable submarine topography, and ultimately wave breaking. The height of the breaker wave is therefore not only a function of the deepwater wave environment, which will determine the maximum possible wave height,

Ocean Wave Generation, Transformation and Breaking

Wave type	Breaking	Shoaling	Swell	Sea
Environment	Shallow water–surf zone	Inner continental shelf	Deep water >>100m	Deep water>>100m Long fetch= sea/ocean surface Wind velocity waves Wind duration waves Wind direction = wave direction
Distance travelled	100m	1km	100s–1000s km	100s–1000s km
Time required	seconds	minutes	hours to days	hours to days
Wave profile	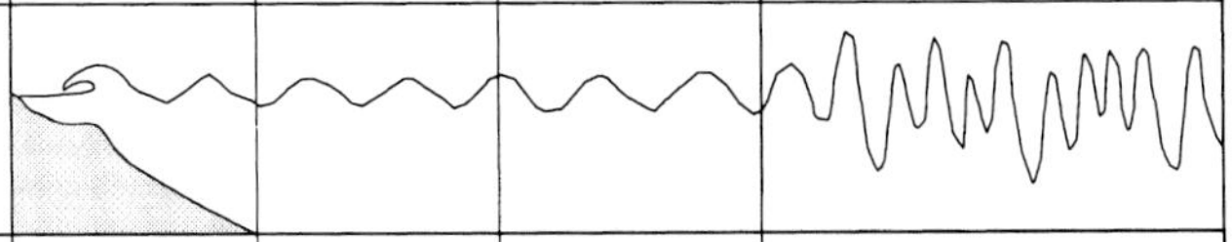			
Water depth	1.5 x water depth	<100m	>>100m	>>100m
Wave character	wave breaks wave bore swash	higher shorter steeper same speed	regular lower longer flatter faster	variable height high short steep slow
Example; height (m) period (s) length (m) speed (km/hr)	2.5–3m 12s 50–0m 15–0	2–2.5m 12s 220–50m 66–15	2m 12s 220m 66 km/h	3–5m 6–8s 50–100m 33–45 km/h
distance travelled (km/day)			1600 km/day	800–1100 km/day

Figure W12 Ocean waves begin life as "sea waves" produced by strong winds blowing over the surface of the deep ocean. When they leave the area of wave generation they transform into lower, longer, faster, and more regular "swell" which can travel for hundreds to thousands of kilometers. As waves reach shallow water they undergo a process called "wave shoaling" which causes them to slow, shorten, steepen, and finally break. This figure provides information on the characteristics of each type of wave (from Short, 1993).

but also the shoaling processes, which can reduce wave height from between 0% and 100% (Figure W12).

Andrew D. Short

Bibliography

Davies, J.L., 1964. A morphogenic approach to world shorelines. *Zeitschrift fur Geomorphologie*, **8**, Mortensen Sonderheft, 127–142.

Davies, J.L., 1980. *Geographical Variation in Coastal Development*, 2nd edn., London: Longman.

Inman, D.L., and Brush, B.M., 1973. The coastal challenge. *Science*, **181**: 20–32.

Short, A.D., 1993. *Beaches of the New South Wales Coast*. Sydney: Australian Beach Safety and Management Program.

Short, A.D., (ed.), 1999. *Handbook of Beach and Shoreface Morphodynamics*. Chichester: John Wiley and Sons.

Short, A.D., and Trenaman, N.L., 1992. Wave climate of the Sydney region, an energetic and highly variable ocean wave regime. *Australian Journal of Marine and Freshwater Research*, **43**: 765–791.

Young, I.R., and Holland, G.J., 1996. *Atlas of the Oceans, Wind and Wave Climate*. New York: Elsevier.

Cross-references

Beaches
Climate Patterns in the Coastal Zone
Coastal Climate
Ice-Bordered Coasts
Monitoring, Coastal Geomorphology
Sea Breeze Effects
Wave Climate
Wave-Dominated Coasts
Wave Hindcasting
Waves

WAVE FOCUSING

Swell waves traversing the inner shelf and shoreface undergo shoaling and refraction processes around seafloor topographic highs, such as offshore reefs or transverse sand ridges. These seafloor topographic irregularities act as a "bathymetric lens" (Speranski and Calliari, 2001) inducing refraction processes such that the wave orthogonals become compressed in the lee (shoreward) of the seafloor topographic high, that is, energy along the wave front becomes concentrated, resulting in a localized greater wave heights (Healy, 1987; Calliari *et al.*, 1998). Away from the zones of wave energy concentration the refraction may spread the wave orthogonals, dispersing the wave energy, and resulting in lower wave heights. As a result of this process, localized zones of higher wave energy may be concentrated or *focused* at the coastline (Finkl and Bruun, 1998). This feature of enhanced

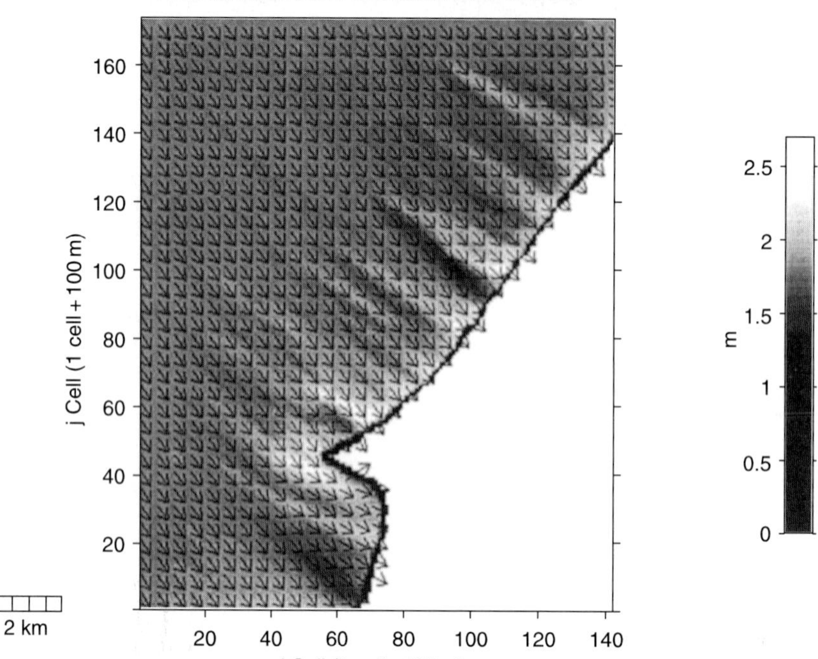

Figure W13 Wave refraction simulation commencing with deepwater wave height H_o = 1.63 m, and period T = 10.64 s. Note the general focusing of energy around the headland where the light patches show wave heights reach 2.5 m, and at various focused sectors further along the coast where waves reach 2.5 m at breaking. In between the major wave-energy focus zones are sectors of lower wave heights (*Source*: H. Easton, personal communication).

wave heights is well-known at headlands, but is not often recognized along sandy beach coastlines (Figure W13). Relatively short swell waves (T ~ 7 s) undergo little refraction when crossing the shoreface, but longer period waves (T ≥ 10 s) may undergo marked refraction and focusing in water depths less than 50–60 m.

Some remarkable erosion and geomorphic effects may result from wave energy focusing along a sandy coastline (Finkl and Bruun, 1998). Sectors of beach subject to wave focusing exhibit localized higher breaking waves, resulting in enhanced wave set-up and run up, and thus are the zones where accelerated dune erosion or dune overwash occurs during episodic storm events. Local variation in littoral drift gradients result. Healy (1987) demonstrated from several examples of numerical wave refraction simulation that wave energy focusing exacerbated rates of erosion along sectors of frontal dune on generally long straight beaches. These sectors of "accelerated" erosion created "arcuate duneline embayments" of order 200–500 m long and 10–20 m amplitude which are cut into the dune face (Stephens *et al.*, 1999). Similar effects have been reported along the coast of Brazil (Speranski and Calliari, 2001). Wave focusing has also been demonstrated as a fundamental cause of barrier spit breaching and breakthrough during a storm, and was identified as a factor in coastal hazard zone assessment (Healy, 1987).

On the inner shelf or shoreface seafloor, the zones of focused wave energy (higher waves) are subjected to greater bottom orbital velocities, and therefore potentially greater scour. These are the zones where side-scan sonar surveys demonstrate that large-scale ripples (often termed *"megaripples"* or *"coarse grained ripples"*) are formed in zones of otherwise fine sands (Healy *et al.*, 1991; Bradshaw *et al.*, 1994).

Terry R. Healy

Bibliography

Bradshaw, B.E., Healy, T., Nelson, C.S., Dell, P.M., and de Lange, W., 1994. Holocene sediment lithofacies on a storm dominated, back-arc shelf margin: the east Coromandel coast, New Zealand. *Marine Geology*, **119**: 75–78.

Calliari, L.J, Speranski, N.S., and Boukareva, I.I., 1998. Stable focus of wave rays as a reason of local erosion at the southern Brazilian coast. *Journal of Coastal Research*, **26**(Special issue): 19–23.

Finkl, C.W., and Bruun, P., 1998. Potentials for manipulating wave action via port and coastal engineering. *Journal of Coastal Research*, **26**(Special issue): 1–10.

Healy, T., 1987. The importance of wave focusing in the coastal erosion and sedimentation process. *Coastal Sediments '87*, American Society of Civil Engineers, pp. 1472–1485.

Healy, T., Harms, C., and de Lange, W., 1991. Dredge spoil and inner shelf investigations off Tauranga Harbour, Bay of Plenty, New Zealand. *Coastal Sediments '91*, American Society of Civil Engineers, pp. 2037–2051.

Speranski, N., and Calliari, L., 2001, Bathymetric lenses and localized coastal erosion in southern Brazil. *Journal of Coastal Research*, **34** (Special Issue): 209–215.

Stephens, S.A., Healy, T.R., Black, K.P., and de Lange, W.P., 1999. Arcuate duneline embayments, infragravity signals, rip currents, and wave refraction at Waihi Beach, New Zealand. *Journal of Coastal Research*, **15** (3): 823–829.

Cross-references

Beach Erosion
Littoral Drift Gradient
Wave Refraction Diagram
Waves

WAVE HINDCASTING

Definition

Wave hindcasts refer to the predictions of wind waves on the water surface for a past event. Wave nowcasts and forecasts similarly refer to the predictions in real time and in the future, respectively. But the relations or models used for predictions for a past, present, or future event are the same. The wave parameters of interest are wave height and period, and the required wind parameters for predictions are wind speed (U) and duration (t). Wind speed should represent an average, typically over a timescale of 1–15 min. Since wind waves represent an irregular undulated water surface comprising a multitude of superimposed wave frequencies, rather than a monochromatic wave, its parameters are best described in statistical terms. These terms are significant wave height (H_s) and significant wave period (T_s), which represent the average of the highest one-third of the parameters. However, in spectral-based computations, H_{mo} is used to represent wave height related to the total energy density as given by the zero-th moment of the wave spectrum. H_s is slightly larger than H_{mo} on most occasions, but they are equivalent for deepwater waves (Goda, 1974; Thompson and Vincent, 1985; Sorensen, 1993). Depth (d) and fetch (F) are the two water body variables required to compute wave parameters. Fetch is defined as a region of the water surface over which wind blows with speeds and directions that vary within a specified limit. The accepted limit for wind speed variation is 2.5 m/s (or 5 knots), and the same for wind direction is 45° (U.S. Army Coastal Engineering Research Center, 1984). In lakes and coastal water bodies, the fetch is often limited by land boundaries. The limiting lengths of fetch and duration give rise to three wave-generating conditions for a particular wind speed. A *fetch-limited* condition applies when the wind duration exceeds the wave travel time over the fetch. When the opposite happens, a *duration-limited* condition applies. If both the fetch and the wind duration are sufficiently large, a *fully arisen sea* develops for a particular wind speed. Depth of the water body is only important for shallow water wave hindcasts.

Wind waves or seas are distinct from swells. Swells are smooth undulations representing decayed and dispersed wind waves that have traveled out of their generating area, and are no longer subject to wind input. A typical ocean wave climate during a windstorm comprises both wind waves and swells.

Wind waves are generated by a complex interaction between the blowing wind and the water surface with the transfer of energy from the former to the later. Two wind-wave generation theories, both based on resonance phenomenon, were pioneered by Phillips (1957) and Miles (1957). The model proposed by Phillips (1957, 1960) is based on a resonant interaction between the forward moving turbulent pressure fluctuations and the generated free waves that propagate at the same speed as the pressure fluctuations. The pressure fluctuates with a varying magnitude and frequency, and is generated by turbulent eddies in the wind field. Phillips' model gives rise to a linear growth of the wave spectrum in time, and best explains the initiation and beginning stages of wave generation. A shear flow model proposed by Miles (1957) provides for a useful mechanism of momentum transfer. It considers the resonant interaction between the wave-induced pressure fluctuations and the free surface waves. An elaborate review of the wave generation processes can be found in an article by Janssen (1994).

In the following sections three methods of wave hindcasting, empirical relations, spectral models, and numerical models are discussed, with a brief history of the development of wave hindcasting methods.

Wave hindcasting methods

History

Probably, the earliest wave hindcasting method started with the pioneering works of Admiral Sir Francis Beaufort of the British Navy in 1805. His devised scale is popularly known as the Beaufort Wind Scale, and gives a qualitative description of winds and visible wave features. The modern history of the quantitative methods of wave hindcasting dates back to World War II. To forecast conditions during landings with amphibious vehicles, H.V. Sverdrup and W.H. Munk (1947) developed methods that related wind speeds to the higher and more distinct oscillations on the sea surface. The waves were characterized using the terms *significant wave height* and *significant wave period*. Statistical analysis of waves, later showed that these characteristic terms were equivalent to the average of the highest one-third of the wave parameters. More works followed with compilations of wave data, statistical analyses and derivation of refined relations (Bretschneider, 1952; Darbyshire, 1952; and Longuet-Higgins, 1952). Neumann (1953) derived wave spectral relations relating significant wave heights and periods. This was followed by the works of Pierson *et al.* (1955) that used Neuman spectra to derive graphical methods for engineering applications. Further developments are covered in the following sections and interested readers could consult Sylvester (1974) for more on historical developments.

Empirical relations

Using a simple wave energy growth concept and an early prediction method developed by Sverdrup and Munk (1947), Bretschneider (1952, 1958) refined the method by an improved calibration with a large amount of field data. It is known as the SMB (Sverdrup–Munk–Bretschneider) method. Deepwater (depth/wavelength >0.5) wave parameters that can be read graphically to establish a relationship between dimensionless significant wave heights and periods with dimensionless fetch and windstorm duration. In functional relationship, this means,

$$\frac{gH_s}{U^2} \text{ and } \frac{gT_s}{2\pi U} = f\left[\frac{gF}{U^2}, \frac{gt}{U}\right], \qquad \text{(Eq. 1)}$$

where U is wind speed, F is fetch, t is windstorm duration, g is acceleration due to gravity, and H_s and T_s are significant wave height and period, respectively. With known wind speed, fetch, and duration, H_s and T_s could be read from graphs. In deepwater, the celerity of wave is $C = gT/2\pi$, and the ratio C/U is known as wave age, an important parameter defining the wave growth. The deepwater graphical relations were later presented as equations by Bretschneider (1970).

For a circular wind field such as a hurricane, Bretschneider (1957) proposed relations based on an analysis of 13 hurricanes in the Atlantic Ocean off the US East Coast. The relations are given as follows (U.S. Army Coastal Engineering Research Center, 1984):

$$H_{sp} = 16.5e^{0.01R\Delta P}\left[1 + \frac{208\alpha V_F}{\sqrt{U_R}}\right] \qquad \text{(Eq. 2)}$$

and

$$T_{sp} = 8.6e^{0.005R\Delta P}\left[1 + \frac{104\alpha V_F}{\sqrt{U_R}}\right], \qquad \text{(Eq. 3)}$$

where H_{sp} (feet) and T_{sp} (seconds) are peak significant deepwater wave heights and periods, respectively. R is distance (nautical miles) from the center out to the point of maximum wind velocity, ΔP is pressure difference (inches of mercury) from the center to the periphery of the hurricane, V_F is forward speed (knots) of the hurricane, U_R is maximum wind speed (knots) at R, and α is a correction factor based on the hurricane speed and is equal to 1 for a slow-moving hurricane. The calculated H_{sp} and T_{sp} develop in the vicinity of the point of maximum wind velocity.

Spectral models

Based on empirical fits to measured waves, three one-dimensional deepwater spectral models have been proposed. These three models assume wave growth in the direction of the wind. The Bretschneider Spectrum (Bretschneider, 1959) shows a period spectrum as a function of wind speed, fetch, and wind duration. The spectrum can be used to hindcast wave height and period, as also mentioned earlier for significant wave parameters. Pierson and Moskowitz (1964) proposed a wave frequency spectrum based on analysis of wave and wind records from British weather ships operating in the North Atlantic. Their analysis considered fully arisen sea for wind speeds (estimated at an elevation of 19.5 m) ranging from 20 to 40 knots. The wave frequency spectrum known as Pierson–Moskowitz Spectrum, representing only the fully arisen sea provides wave frequency as a function of wind speed. The JONSWAP (Joint North Sea Wave Project) Spectrum (Hasselmann *et al.*, 1973) is developed from wave and wind measurements with sufficient wind duration and provides a fetch limited spectrum. The spectrum gives a relation between wave frequency, and wind speed, and fetch. The *Shore Protection Manual* (U.S. Army Coastal Engineering Research Center, 1984) recommended a parametric method based on the JONSWAP spectrum for deepwater wave prediction, and replaced the earlier recommended SMB method. According to this method, the following relations are used for both fetch and duration limited conditions.

$$\frac{gH_{mo}}{U_A^2} = 0.0016\sqrt{\frac{gF}{U_A^2}}, \qquad \text{(Eq. 4)}$$

$$\frac{gT_p}{U_A} = 0.286\left(\frac{gF}{U_A^2}\right)^{1/3}, \qquad \text{(Eq. 5)}$$

and

$$\frac{gt_d}{U_A} = 68.8\left(\frac{gF}{U_A^2}\right)^{2/3}. \qquad \text{(Eq. 6)}$$

The adjusted wind speed, U_A is given by

$$U_A = 0.71U_{10}^{1.23} \qquad \text{(Eq. 7)}$$

where U_{10} is wind speed at 10 m above mean sea level, T_p is peak wave period, and other notations are as before. The significant wave period T_s is estimated from T_p as, $T_s = 0.95T_p$. If the actual wind duration (t) is greater than the calculated limiting wind duration (t_d), a fetch-limited case applies. A duration limited case applies if the converse is true, and an effective fetch is computed using equation 6 by replacing t_d with the actual duration of the storm, t. With the known effective fetch, equations 4 and 5 are used to estimate H_{mo} and T_p.

For fully arisen sea when $gt/U_A \geq 71{,}000$ and $gF/U_A^2 \geq 22{,}800$, the deepwater wave parameters are estimated as,

$$\frac{gH_{mo}}{U_A^2} = 0.243 \quad \text{(Eq. 8)}$$

and

$$\frac{gT_p}{U_A} = 8.13. \quad \text{(Eq. 9)}$$

If wind speed (U_z) is measured at a height z above mean sea level, the velocity at 10 m (U_{10}) can be estimated using one-seventh power law profile,

$$U_{10} = U_z\left(\frac{10}{z}\right)^{1/7}. \quad \text{(Eq. 10)}$$

JONSWAP spectrum can also be used to estimate wave parameters resulting from a circular wind field such as a hurricane using a method developed by Young (1988). An equivalent hurricane fetch is first estimated using hurricane forward velocity and maximum wind velocity. This is then used to estimate wave parameters using the above equations.

For shallow water (depth/wavelength <0.5) applications, U.S. Army Coastal Engineering Research Center (1984) recommended the use of the following equations pertaining to the JONSWAP spectrum.

$$\frac{gH_{mo}}{U_A^2} = 0.283 \tan h\left[0.53\left(\frac{gd}{U_A^2}\right)^{3/4}\right] \tan h\left[\frac{0.00565(gF/U_A^2)^{1/2}}{\tan h\left\{0.53(gd/U_A^2)^{3/8}\right\}}\right], \quad \text{(Eq. 11)}$$

$$\frac{gT_s}{U_A} = 7.54 \tan h\left[0.833\left(\frac{gd}{U_A^2}\right)^{3/8}\right] \tan h\left[\frac{0.0379(gF/U_A^2)^{1/2}}{\tan h\left\{0.833(gd/U_A^2)^{3/8}\right\}}\right], \quad \text{(Eq. 12)}$$

$$\frac{gt_d}{U_A} = 537\left(\frac{gT_s}{U_A}\right)^{7/3}, \quad \text{(Eq. 13)}$$

where d is water depth. However, shallow water wave parameters can best be estimated from deepwater waves using wave transformation techniques.

Numerical models

The above wave hindcasting methods do not give directional distribution of the wave field. The directional spectra are basically one-dimensional spectra corrected by a dimensionless directional spreading function that depends on the wave frequency and direction. The directional spectra can best be hindcasted by numerical models. Numerical models are used to forecast, nowcast, and hindcast waves or wave climates. The models are based on numerical integration of the spectral energy balance equation over a spatial grid. The general expressions for the source functions (Hasselmann, 1962) are the wind energy input, the nonlinear transfer of energy from one frequency to another by wave–wave interaction, and the energy dissipation by wave breaking, turbulence, and bottom friction. These source functions are then allowed to result growth of the wave spectrum as a function of time and space.

The SWAMP (Sea Wave Modeling Project) Group—a group of 10 institutes from Germany, Italy, Japan, the Netherlands, Norway, United Kingdom, and the United States—has examined and tested 10 numerical models on a common platform and has published their findings in 1985 (The SWAMP Group, 1985). They classified the existing models into three groups: Decoupled Propagation (DP) models, Coupled Hybrid (CH) models, and Coupled Discrete (CD) models. DP is a first generation wave model in which the spectral components evolve independent of each other in accordance with the linear input of source function. CH models work on the basis of a hybrid combination of parametrical windsea models with the standard discrete spectral representation for the swell components. The limitations of DP and CH models are eliminated in the CD models, which are based on discrete spectral representation for the entire spectrum of windsea and swell. WAVEWATCH III (Tolman, 1999) is a third generation model of this type developed at NOAA/NCEP in the spirit of WAM (Wave Modeling Group) model (WAMDIG [Wave Model Development and Implementation Group], 1988; Komen *et al.*, 1994). It represents a further development of the original WAVEWATCH I model developed at Delft University of Technology (Tolman, 1991).

The best reference texts for wave hindcasting are *Shore Protection Manual* by U.S. Army Coastal Engineering Research Center (1984), *Basic Wave Mechanics for Coastal and Ocean Engineers* by Sorensen (1993), *Basic Coastal Engineering* by Sorensen (1997), *Coastal Engineering: An Introduction to Ocean Engineering* by Horikawa (1978), and *Dynamics and Modelling of Ocean Waves* by Komen *et al.* (1994). A brief review of wave prediction methods (Bishop and Donelan, 1989) can also be found in chapter 4 of *Applications in Coastal Modeling* by Lakhan and Trenhaile (1989).

Dilip K. Barua

Bibliography

Bishop, C.T., and Donelan, M.A., 1989. Wave prediction models. In Lakhan, V.C., and Trenhalie, A.S. (eds.), *Applications in Coastal Modelling*, Amsterdam: Elsevier.

Bretschneider, C.L., 1952. Revised wave forecasting relationship. *Proceedings of the 2nd Conference on Coastal Engineering*. Council on Wave Research, University of California, pp. 1–5.

Bretschneider, C.L., 1957. Hurricane design wave practices. *Journal of Waterways and Harbors Division*, American Society of Civil Engineers, pp. 1–33.

Bretschneider, C.L., 1958. Revisions in wave forecasting: deep and shallow water. *Proceedings of the 6th Conference on Coastal Engineering*. Council on Wave Research, University of California, Berkeley, pp. 1–18.

Bretschneider, C.L., 1959. Wave variability and wave spectra for wind-generated gravity waves. Washington, DC: US Army Beach Erosion Board. *Technical Memorandum* 118.

Bretschneider, C.L., 1970. Forecasting relations for wave generation. *Look Lab Hawaii*, **1** (3): 31–34.

Darbyshire, J., 1952. The generation of waves by wind. *Proceedings of the Royal Society Series A*, **215**: 299–328.

Goda, Y., 1974. Estimation of wave statistics from spectral information. *Proceedings of the Ocean Waves Measurement and Analysis Conference*. American Society of Civil Engineers, New Orleans, pp. 320–337.

Hasselmann, K., 1962. On the non-linear energy transfer in a gravity-wave spectrum, part 1: general theory. *Journal of Fluid Mechanics*, **12**: 481.

Hasselmann, K., Barnett, T.P., Bouws, E., Carlson, H., Cartwright, D.E., Enke, K., Ewing, J.A., Gienapp, H., Hasselmann, D.E., Kruseman, P., Meerburg, A., Muller, P., Olbers, D.J., Richter, K., Sell, W., and Walden, H., 1973. Measurement of wind-wave growth and swell decay during the Joint North Sea Wave Project (JONSWAP). Report. Hamburg: German Hydrographic Institute.

Horikawa, K., 1978. *Coastal Engineering: An Introduction to Ocean Engineering*. New York: Wiley.

Janssen, P.A.E.M., 1994. Wave growth by wind. In Komen *et al.* (eds.), *Dynamics and Modelling of Ocean Waves*. Cambridge: Cambridge University Press.

Komen, G.J., Cavaleri, L., Donelan, M., Hasselmann, K., Hasselmann, S., and Janssen, P.A.E.M., 1994. Dynamics and modelling of ocean waves. Cambridge: Cambridge University Press.

Lakhan, V.C., and Trenhaile, A.S., 1989. *Applications in Coastal Modeling*. Oceanography Series 49, Amsterdam: Elsevier

Longuet-Higgins, M.S., 1952. On the statistical distribution of the heights of sea waves. *Journal of Marine Research*, **11**: 245–266.

Miles, J.W., 1957. On the generation of surface waves by shear flows. *Journal of Fluid Mechanics* **3**: 185–204.

Neumann, G., 1953. On ocean wave spectra and a new method of forecasting wind-generated sea. Beach Erosion Board, *U.S. Army Technical Memorandum*, 43.

Phillips, O.M., 1957. On the generation of waves by turbulent winds. *Journal of Fluid Mechanics*, **2**: 417–445.

Phillips, O.M., 1960. On the dynamics of unsteady gravity waves of finite amplitude, 1, the elementary interactions. *Journal of Fluid Mechanics*, **9**: 193–217.

Pierson, W.J., and Moskowitz, L., 1964. A proposed spectral form for fully developed wind seas based on the similarity theory of S.A. Kitaigorodskii. *Journal of Geophysical Research*, **69**: 5181–5190.

Pierson, W.J., Neumann, G., and James, R.W., 1955. Practical methods for observing and forecasting ocean waves by means of wave spectra and statistics. H.O. Pub 603, US Navy Hydrographic Office.

Sorensen, R.M., 1993. Basic wave mechanics for coastal and ocean engineers. New York: John Wiley & Sons, Inc..

Sorensen, R.M., 1997. *Basic Coastal Engineering*. New York: 301 Chapman & Hall.

Sverdrup, H.U., and Munk, W.H., 1947. *Wind, Sea and Swell: Theory of Relations for Forecasting*. Publication 601. Washington, DC: U.S. Navy Hydrographic Office.

Sylvester, R., 1974. Coastal Engineering 1—Generation, Propagation and Influence of Waves. Amsterdam: Elsevier Scientific Publishing Company.

The SWAMP Group, 1985. *Ocean Wave Modeling*. New York: Plenum Press.

Thompson, E.F., and Vincent, C.L., 1985. Significant wave height for shallow water design. *Journal of Waterway Port, Coastal, and Ocean Engineering Division,* **111**(5): 828–842.

Tolman, H.L., 1991. A third-generation model for wind waves on slowly varying, unsteady and inhomogeneous depths and currents. *Journal of Physical Oceanography*, **21**: 782–797.

Tolman, H.L., 1999. User manual and system documentation of WAVEWATCH III version 1.18. NOAA/NWS/NCEP/OMB Technical Note, 166, 110pp.

U.S. Army Coastal Engineering Research Center, 1984. *Shore Protection Manual*, 4th edn. Washington, DC: U.S. Government Printing Office.

WAMDIG, 1988. The WAM model—a third generation ocean prediction model. *Journal of Physical Oceanography*, **18**: 1775–1810.

Young, I.R., 1988. Parametric hurricane wave prediction model. *Journal of Waterway, Port, Coastal, and Ocean Engineering Division*. American Society of Civil Engineers, September, 637–652.

Cross-references

Beaufort Wind Scale
Coastal Warfare
Coastal Modeling and Simulation
Numerical Modeling
Time Series Modeling
Wave Climate
Wave Environments
Waves

WAVE POWER

Of the "implemented" systems to extract energy from the ocean, power from the waves is perhaps the one that has received most attention during the 20th century, from Pacifica (California) to Royan (France). The number of patents taken out, and of systems proposed, is impressive. Projects are being examined in several locations in Asia and the Pacific Ocean, in Canada and in Europe. Waves are considered a "small energy source" that can be profitably tapped by industrialized and less-developed countries alike.

Extraction systems utilize either the vertical rise and fall of successive waves in order to build water- or air-pressure to activate turbines, or take advantage of the to-and-fro, or rolling, motions of waves by vanes or cams which rotate turbines. Still another approach concentrates incoming waves in a converging channel, thus allowing the buildup of a head of water, which then makes it possible to operate a turbine. Wave refraction effects can be used to focus wave energy and they have played an expanding role in current wave-energy conversion thinking. However, in shallow waters the amount of such energy available is reduced due to shoaling and seafloor friction.

Most recently proposed schemes show economic feasibility. Japanese and British researchers have been in the forefront of research, though the Japanese have been closer to implementation, and Scandinavians have overtaken the British.

Conversion devices

Systems involve a movable body, an oscillating column, or a diaphragm and the "object" that moves floats is anchored on the seabed. Attenuator type devices in a two-dimensional case will provide only a 50% efficiency, unless dispersion waves strike the object only on one side. A superior apparatus must have a high absorption efficiency for a wide range of wave frequencies. Most currently considered primary wave converters are linear, two- or three-dimensioned. Among the latter are those systems known as Kaimei, Flexible Bag, and National Electronic Laboratory's Oscillating Water Column (NEL's OWC). Converters involving a movable body use vertical, rotational, lateral, coupled movable bodies or a raft or float. They include, respectively, the Point Absorber, Salter Duck (Figure W14), Wave Power Water Turbine, Bristol Cylinder, and Cockerell Raft. There are systems that double as breakwaters, such as the Pendulum and NEL's OWC. Some apparatus are moored, fixed on the shore or on the seabed. The Focusing Wave Energy System is in a somewhat separate category.

Based on *uses* energy conversion devices can be grouped into four "classes:" propulsion, buoy power supply, offshore plants, shore-based power plants. Another grouping considers rather conversion *methods:* utilization of the rise and fall sequence, of the rolling motion, or convergence to create a hydraulic head. A "*physics principles*" classification recognizes: intervention in wave orbits, utilization of the pressure field, acceleration devices, and use of the horizontal transport from breaking waves. A much older (1892) classification was based on *mechanical concepts*: motors operated by the rise and fall of a float, by to-and-fro wave movement, by the varying slope of wave surface, and by impetus of waves rolling up a beach. With about 40 different *systems* proposed over time still another grouping is possible: surface profile variations of traveling deepwater waves, subsurface pressure variations, subsurface fluid particle motion, and naturally or artificially induced unidirectional motion of fluid particles in a breaking wave.

Systems may intervene in the wave orbits, utilize the pressure field, be accelerative or utilize the water mass horizontal displacement. They may involve flaps and paddles; heaving, pitching and rolling bodies, pressure devices, rotating outriggers, pneumatic or cavity resonators, or be waves focusing, surging, or mixed systems.

Performance of systems

Satisfactory performance and economically sustainable results appear to have been achieved by several systems. Charlier and Justus (1993) list an air turbine driven by water oscillation in a vertical borehole (Royan, France, Gironde River estuary), floats activated by horizontal and vertical motion attached to a pier (Atlantic City, New Jersey and Pacifica, California—USA), a Savonius rotor operating pump (Monaco, Musée Océanographique), pump operated by a rising and falling heavy float (Monaco), low-head hydroelectric plant supplied from a forebay with converging channels (Pointe Pescade and Sidi-Ferruch, Algeria), air turbine buoys (Japan, United States, United Kingdom), air turbine generator (Osaka, Japan), Kamei barge with compressed-air chambers (Japan), hydraulic pumping over pliable strips in a concrete trough (Boston, MA, USA), autobailer bilge pump (Sweden), and a sea-lens concentration scheme (Norway). Still other schemes which either function or show promise are utilized in bouys and desalination plants.

The machine must be able to amplify the relatively low water head, have a broad range of response, be nonresonant, and have the capacity of responding to smaller waves, while simultaneously be able to withstand the effects of large storm waves.

Recent developments

Just before the dawn of the 19th century a patent was taken out in France to turn wave energy into mechanical energy by using a float. Several wave motors were patented in the 19th and 20th centuries which supplied electrical energy for buoys, pumps, lighthouses, and even small residences. Duckers (1989), Scheer and Gandhi (1994), Charlier and Justus (1993), Ross (1985, 1987), and Salter (1975a,b, 1976) among others, discussed the more recent devices tested and put into service. OWCs are perhaps the most promising scheme; indeed, being robust and utilizing relatively conventional, proven technology.

In 1990, Wave Energy Inc. took out patents contracts (hydraulic method) on the heels of several successful tests carried out at Scripps Institution of Oceanography for a wave pump and energy extraction system. Norwave conducted a feasibility study for a 5 MW plant on Java. Sixteen prototypes and ten "future prototypes" were listed by Duckers for the period 1965–92; several were operational for a while, but eventually abandoned; they were mostly installed in Japan, with one each in Great Britain, Denmark, Norway, and India, or planned for Japan, Scotland, the United States, among other locations. Eire has concentrated on devices and so have Korea and China; a small power

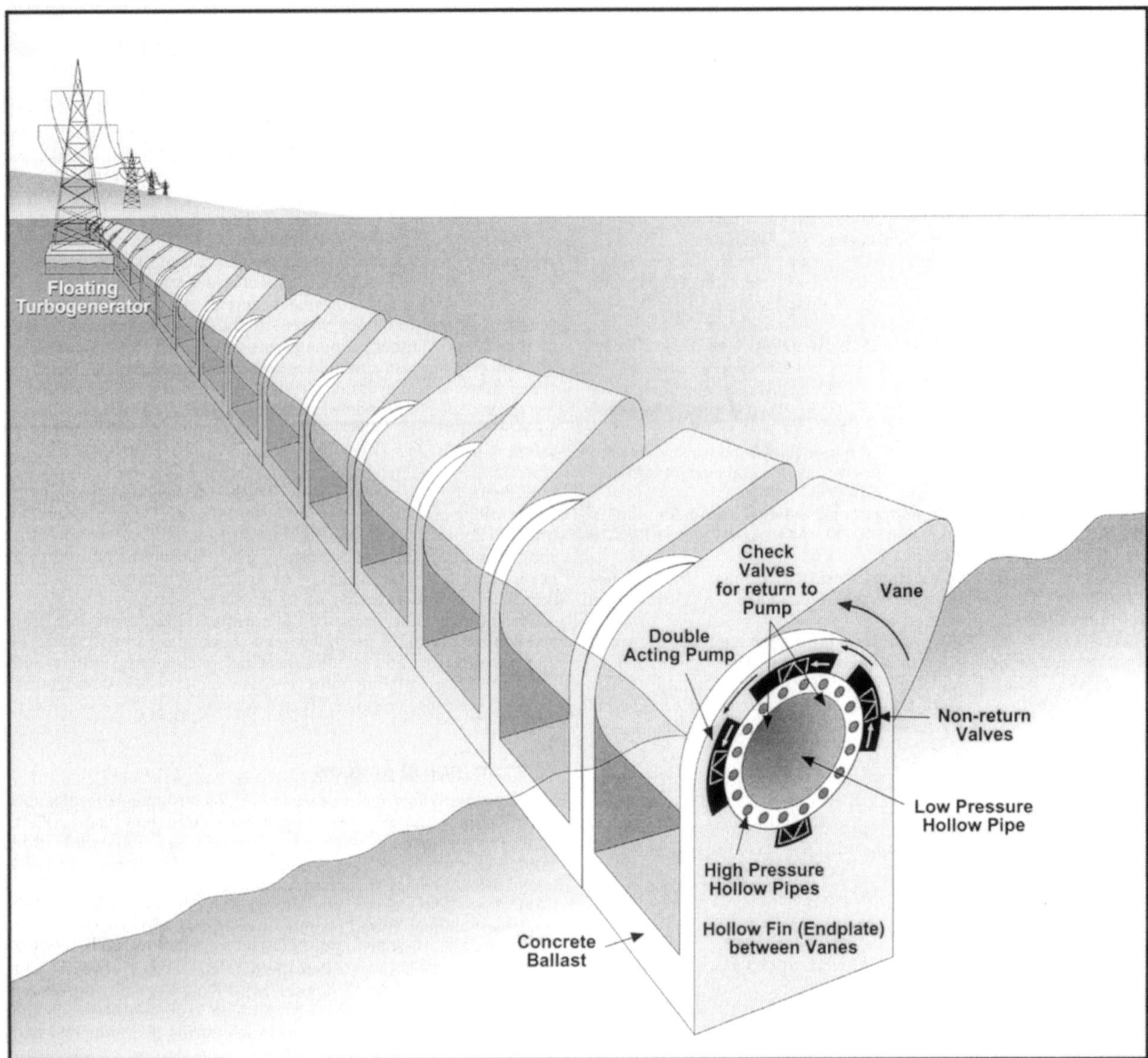

Figure W14 Schematic view of "Salter's Duck" (from Charlier and Justus, 1993, with permission from Elsevier Science).

plant (8 KW) was installed in the Pearl River estuary around 1991. China's interest in wave power has not diminished (Zhi, 1993; You and Zhi, 1995) and India has conducted surveys and tests (Sivaramakrishnan, 1992; Raju and Ravindram, 1996). Ocean Power Technology (USA) developed in 1995–96 a power system using the piezoelectric effect to generate electricity from waves; the bending of laminated sheets suspended from rafts secured on the seafloor generate electricity. The European Union supported in 1994 the construction of a full-scale 2 MW converter of the OWC type in Scotland (You and Zhi, 1995).

Mighty Whale is a Japanese constructed offshore floating wave power device, 50 m long and 30 m wide, the world's largest. It was moored in late 1998 off Gokasho Bay. Resembling a whale, it transforms wave motion into high velocity oscillating airflow, achieved near the "mouth," by three chambers; waves enter and exit through these ducts at their top, with air turbines driving electric power generators (one each of 10 and 50 kW and two of 30 kW). Electricity drives the air compressors.

Environmental impact

Wave energy is clean, safe, environment friendly, but its harnessing affects beaches, marine life, fishing, shipping, coastal tourism and recreation, and should be viewed in the overall picture of integrated coastal management.

Though limited, wave-energy utilization has thus some environmental consequences, biological, physical, and social. Marine organisms will be moderately affected, but may pose biofouling problems for the system. Fisheries will be impacted, differently according to species, though floating structures may act as a fish reef. More severe impact on fisheries is caused by seabed fixed apparatus.

Longshore currents may be rerouted, tidal currents influenced, particularly with apparatus fixed on the seabed, and coastlines may consequently be modified. Navigational problems may arise. Wave absorbers alter wave patterns and drift patterns of sand, thus possibly causing coastal erosion.

Production of electricity at an affordable cost, particularly in power deprived developing countries areas, may lead to implantation of new industries, and/or improvement of existing ones. A social and developmental "plus" may flow forth.

The future

If wave energy is to make a significant contribution to national energy demand, sustained research is needed into its application to the offshore production of hydrogen. Wave power has also been proposed for use in communications and spacecraft propulsion (Korde, 1990) and continues to be eyed for desalination plants (Crerar and Pritchard, 1991). In some countries, for example, Norway, successful plants have been abandoned because of the availability of huge oil reserves; only when these will have dwindled, will new enthusiasm for alternative sources of energy be rekindled there.

Further reading on this subject may be found in the following bibliography.

Roger H. Charlier

Bibliography

Charlier, R.H., and Justus, J.R., 1993. *Ocean Energies. Environmental, Economic and Technological Aspects of Alternative Power Sources*. Amsterdam, London, New York, Tokyo: Elsevier.

Crerar, A.J., and Pritchard, C.L., 1991. Wave powered desalination. Experimental and methodological modelling. In *Proceedings of the 12th International Symposium Desalination and Water Reuse (Malta)*, Amsterdam: Elsevier, pp. 391–398.

Duckers, L. (ed.), 1989. Wave energy devices. *Proceedings of the Conference of Whitefriars Monastery, Coventry, November 1989*. London: Solar Energy Society.

Korde, U.A., 1990. Study of a wave device for possible application in communication and spacecraft propulsion. *Ocean Engineering*, **17** (6): 587–599.

Raju, V.S., and Ravindram, M., 1996. Wave energy: power and progress in India. *Renew. Energy*, **9** (1/4): 339–345.

Ross, D., 1985. *Wave Energy*. London: Graham and Trotman.

Ross, D., 1987. First wave power contracts (Norway). *Electrical Reviews*, **220**(2): 16–17.

Salter, S.H., 1975a. Characteristics of wave power devices. *Nature*, **254**, 504–506.

Salter, S.H., 1975b. *The Environmental Aspects of Wave Power*. Edinburgh: University of Edinburgh.

Salter, S.H., 1976. The architecture of nodding duck wave power generators. *The Naval Architect (London)*, 21–24.

Scheer, H., and Gandhi, M., (ed.), 1994. *The Yearbook of Renewable Energy*. London: James & James.

Sivaramakrishnan, T.T., 1992. Wave power over the Indian seas during southwest monsoon. *Energy*, **17**(6): 625–627.

You Yage, and Zhi Yu, 1995. Wave loads and protective techniques of an on-shore wave power device. *Chinese Ocean Engineering*, **914**: 455–464.

Zhi Yu, 1993. The development of ocean wave power generation technique. *The Ocean Engineering/Haiyang Gongcheng*, **11**(1): 86–93.

Cross-references

Desalination
Engineering Applications of Coastal Geomorphology
Geohydraulic Research Centers
Tidal Power
Tide Mill
Wave-Dominated Coasts
Wave Environments
Waves

WAVE REFRACTION DIAGRAMS

Wave refraction diagrams are used to illustrate and predict the refraction of waves approaching the shoreline. They are an invaluable tool for understanding coastal morphology and processes, and their construction is practically standard practice in coastal engineering applications. Amongst others, wave refraction diagrams can be used to compute alongshore variations in wave energy and breaker angles, and subsequently longshore sediment transport rates and directions (see *Longshore Sediment Transport*). Wave refraction diagrams are also used to indicate the stability of embayed beaches (see *Headland Bay Beach*). A good summary and comparison of techniques to construct wave refraction diagrams is given in Wiegel (1964).

Theoretical background

The theoretical background to the wave refraction process is discussed in the section on *Waves*. Briefly, wave refraction is the bending of wave rays that occurs in intermediate and shallow water depth. As a result of wave refraction, wave crests become increasingly lined up with the bottom contours, thereby reducing the angle of wave approach. The refraction of waves is similar to the bending of light rays and the change in direction is related to the change in the wave celerity C through the same Snell's law

$$\frac{\sin\alpha_1}{C_1} = \frac{\sin\alpha_2}{C_2} = \text{constant}, \qquad \text{(Eq. 1)}$$

where α_1 is the angle of a wave front with the depth contour h_1 over which the wave is passing, α_2 is a similar angle measured as the wave front passes over the second depth contour h_2, C_1 is the wave celerity at the depth of the first contour, and C_2 is the wave celerity at the second contour (refer to Figure 7a in the section on *Waves*).

Wave refraction for coasts with parallel bottom contours

Changes in wave direction due to wave refraction on a straight coast with parallel bottom contours can be determined analytically using

$$\sin\alpha = \frac{C}{C_o}\sin\alpha_o, \qquad \text{(Eq. 2)}$$

where α is the angle between wave crest and depth contour at an arbitrary depth, α_o is the angle between wave crest and depth contour in deepwater, C is the wave celerity at an arbitrary depth, and C_o is the deepwater wave celerity. Computing sin α for different water depths simply involves solving equation 2. The relevant equations to do this computation are given in the section on *Waves*. Alternatively, a nomograph can be used to determine changes in wave height and direction due to refraction (Figure W15). For example, for a deepwater wave angle α_o of 60° and a deepwater wave length L_o of 100 m, the wave angle α in 10 m water depth is 38° and the refraction coefficient K_r is 0.79. Table W6 lists an example of wave refraction (and shoaling) over a bathymetry with parallel depth contours. It can be seen that wave shoaling results in an increase in wave height with decreasing depth ($K_s > 1$), whereas wave refraction reduces the wave height ($K_r < 1$).

Wave refraction for coasts with complex bottom contours

In reality, parallel depth contours rarely exist and in this case the wave refraction pattern has to be determined manually. Refraction diagrams for such conditions can be constructed in two ways. The first, known as the "wave-front method," is essentially a map showing the wave crests at a given time or the successive positions of a particular wave crest as it propagates shoreward (Johnson *et al.*, 1948). Wave rays or orthogonals can then be drawn perpendicular to the wave crests to indicate the direction of wave travel. In the second method, known as the "orthogonal method," the wave rays are drawn directly (Arthur *et al.*, 1952). The orthogonal method is considered the most practical, quickest, and widely used technique at present and will be discussed here.

A number of assumptions underly the orthogonal technique: (1) wave energy between wave rays remains constant, that is, there is no leakage

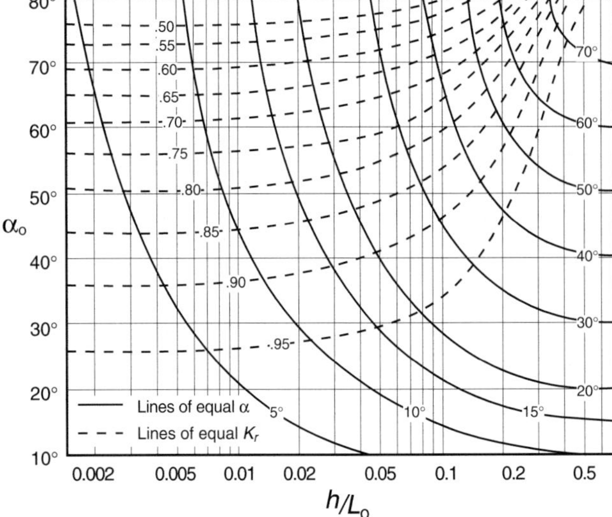

Figure W15 Change in wave direction and height due to refraction on a straight coast with parallel bottom contours (after CERC, 1984) (h = water depth, L_o = deepwater wave length, α_o = deepwater wave angle, α = wave angle, K_r = refraction coefficient).

of wave energy perpendicular to the direction of wave travel; (2) the direction of wave advance is parallel to the wave rays; (3) the celerity of a wave of a given period depends only on the water depth at that location; (4) changes in seabed topography approaching a coastline are gradual; (5) waves are long-crested, constant period, small amplitude, coastline and monochromatic; and (6) other factors that may affect wave refraction, such as currents, winds, and reflected waves are ignored.

Before constructing the wave rays, a tracing paper overlay is placed on the bathymetric chart. The shoreline and the bottom contours are then traced with all bottom irregularities within a local spatial scale of five wave lengths smoothed out. Mid-contour lines are then drawn between the existing contour lines. Subsequently, a set of evenly spaced and parallel wave rays from the chosen direction of wave approach are plotted up to the bottom contour closest to $0.5L_o$ and extended to the first mid-contour line. The number of wave rays should be selected to sufficiently cover the coastline. The closer the spacing between the rays, the greater the resolution when estimating wave energy at the coastline. Finally, for each mid-contour line the C_1/C_2 ratio should be computed, where C_1 represents the wave celerity of the deeper depth contour and C_2 represents the wave celerity of the shallower depth contour.

Table W6 Wave transformation characteristics for an incident wave with a deepwater wave height of 1 m, a wave period of 8 s and a deepwater wave angle of 45°

h (m)	L (m)	C (m/s)	n (−)	α (°)	K_s (−)	K_r (−)	H (m)
100	99.9	12.5	0.50	45	1	1	1
50	99.4	12.4	0.51	45	0.99	1	0.99
30	95.9	12.0	0.58	43	0.95	0.98	0.93
20	88.7	11.1	0.67	39	0.92	0.95	0.88
15	81.8	10.2	0.73	35	0.91	0.93	0.85
10	70.9	8.9	0.81	30	0.93	0.90	0.84
5	53.1	6.6	0.90	22	1.02	0.87	0.89
3	42.0	5.3	0.94	17	1.13	0.86	0.97
2	34.7	4.3	0.96	14	1.23	0.85	1.05

h, water depth; L, wave length; C, wave celerity; n, coefficient; α, wave angle; K_s, shoaling coefficient; K_r, refraction coefficient; H, wave height.

Starting with any one wave ray the following steps are taken in extending the wave ray from offshore to the coastline (Figure W16).

1. Extend the incoming wave ray to the mid-contour and determine the intersection point P.
2. Construct a tangent line to the mid-contour at point P.
3. Measure the angle α_1 between the tangent line and wave crest 1 (normal to wave ray 1) using a protractor.
4. Compute the angle α_2 between the tangent line and wave crest 2 using equation 1 and the C_1/C_2 ratio.
5. Using a protractor, construct wave ray 2 which is the line normal to wave crest 2 going through the intersection point.
6. Repeat the procedure for successive contour intervals and all incoming orthogonals.

Manually measuring and plotting wave angles using a protractor can be quite cumbersome. Templates have been constructed that preclude the manual measurement of wave angles. The template shown in Figure W17 is most commonly used and will be described here. The template should be constructed such that the length of the template is about 25 cm long and should be printed or copied on transparency paper. The following steps are taken in extending the wave ray from offshore to the coastline (Figure W18).

1. Extend the incoming wave ray to the mid-contour and determine the intersection point P.
2. Construct a tangent line to the mid-contour at point P.
3. Place the line on the template labelled "wave ray" along the incoming wave ray with the point representing C_1/C_2 1.0 at point P.
4. Rotate the template about the turning point R until the C_1/C_2 value corresponding to the contour interval being crossed intersects the tangent to the mid-contour at point S. The refracted wave ray now lies in the direction of the turned wave ray on the template.
5. Construct a line parallel to the turned wave ray on the template from P representing the refracted wave ray.
6. Repeat the procedure for successive contour intervals and all incoming wave rays.

Strictly speaking, point P is not the correct intersection point between the incoming and refracted wave ray. The "true" intersection point P′ is found by moving parallel to the turned wave ray until the distance AP′ is equal to the distance BP′ (Figure W18). This ensures that the distances travelled by the incoming and refracted wave rays within the contour interval C_1–C_2 are equal. However, if the change in depth between the

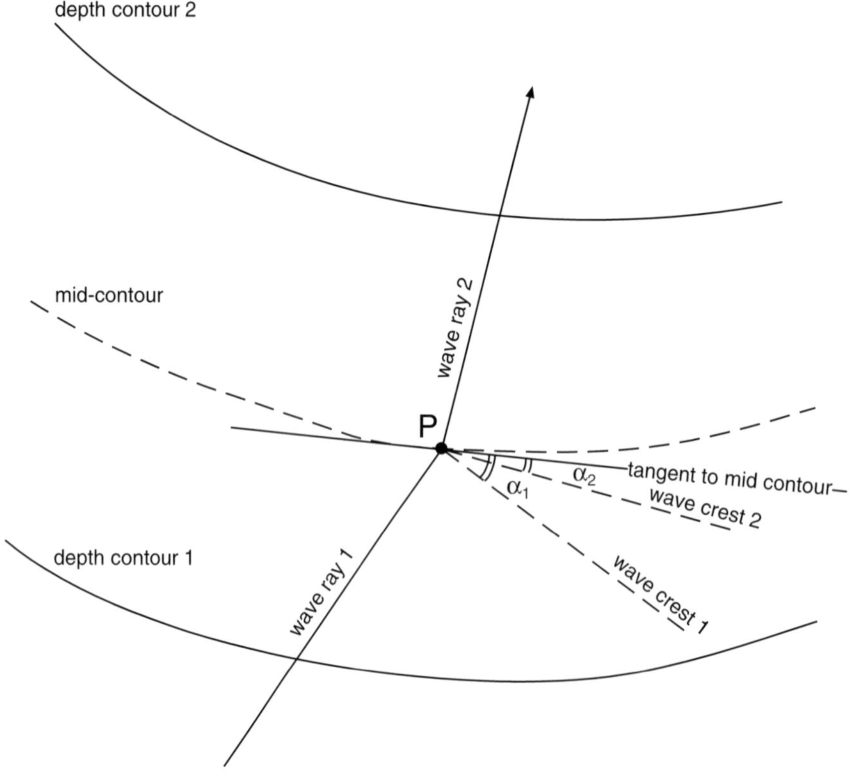

Figure W16 Construction of wave rays using a protractor (α_1 = wave angle of incoming wave, α_2 = wave angle of refracted wave).

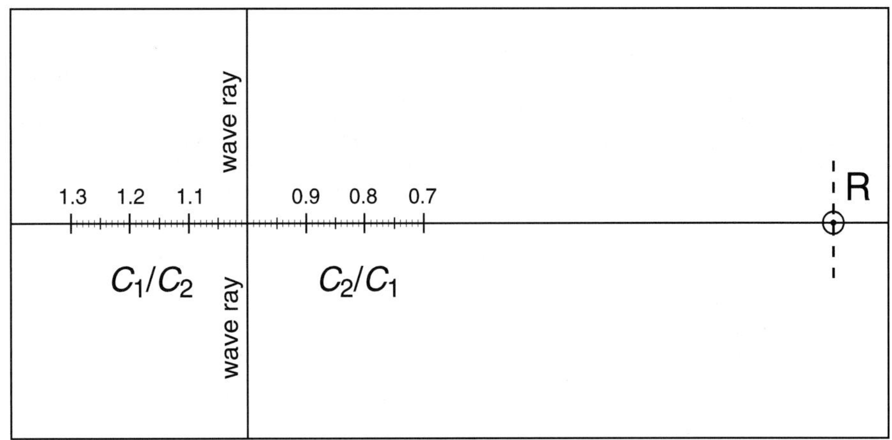

Figure W17 Refraction template (after CERC, 1984) (C_1 wave celerity at contour 1, C_2 wave celerity at contour 2).

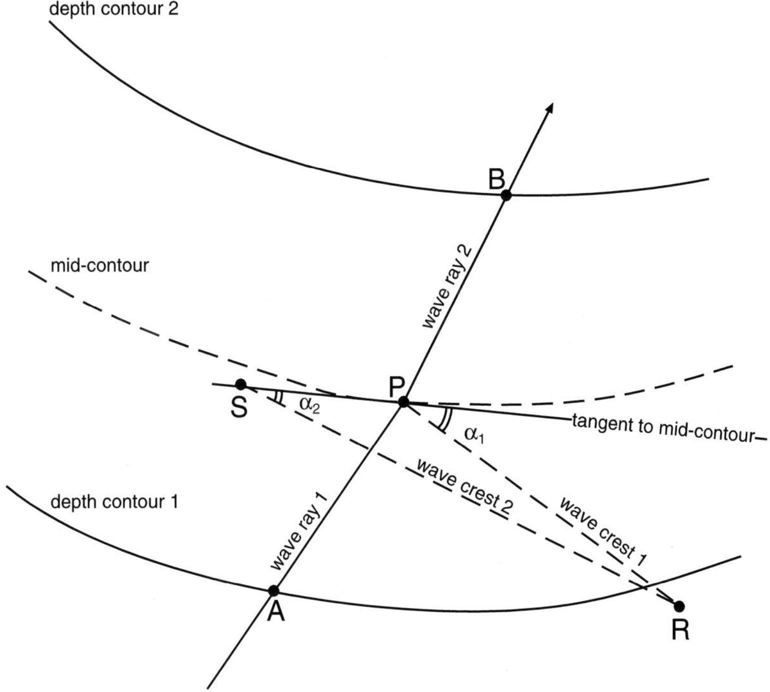

Figure W18 Construction of wave rays using a template (α_1 wave angle of the incoming wave, α_2 = wave angle of refracted wave).

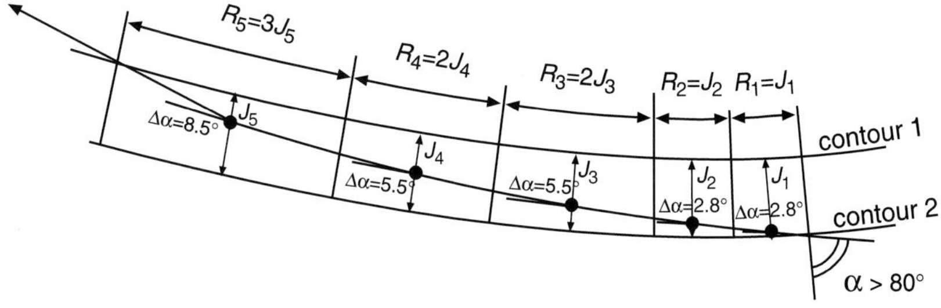

Figure W19 Construction of wave rays using the *R*/*J* method when the wave angle α is larger than 80° (after CERC, 1984). The wave ray has been constructed assuming $C_1/C_2 = 0.95$ (R = distance along wave ray, J = distance between contours, $\Delta\alpha$ = turning angle of the wave ray).

two contours is small and $C_1/C_2 < 1.2$, P and P′ will almost coincide and it is permissable to construct the refracted wave ray from P rather than P′.

For wave angles larger than 80°, the orthogonal method is no longer reliable and an alternative method of constructing wave rays must be used. The *R*/*J* method is generally used for this purpose and consists of the following steps.

1. Divide the contour interval to be crossed (from h_1 to h_2) into segments by transverse lines drawn perpendicular to the contours (Figure W19). The spacing R of the transverse lines is arbitrarily set at some small number times the distance J between the contours.
2. Determine the turning angle of the wave ray $\Delta\alpha$ from the *R*/*J* nomograph (Figure W20).

3. Turn the wave ray by the angle $\Delta\alpha$ and extend it to the middle of the segment.
4. Repeat the procedure for each segment in the sequence.

Once the wave ray crosses a new contour and the wave angle α is less than 80° the *R*/*J* method should be replaced by the orthogonal method.

Both the orthogonal and the *R*/*J* method can be used to construct orthogonals from shallow to deepwater. In this case, the same procedure can be employed, except that the ratio C_1/C_2 (>1) is replaced by C_2/C_1 (<1).

The variation in wave height due to wave refraction can be determined based on the principle that the transported energy between two adjacent wave orthogonals is conserved. By measuring the width between two orthogonals at depth h and in deepwater the refraction coefficient K_r is given by

$$K_r = \left(\frac{b_o}{b}\right)^{0.5}, \qquad \text{(Eq. 3)}$$

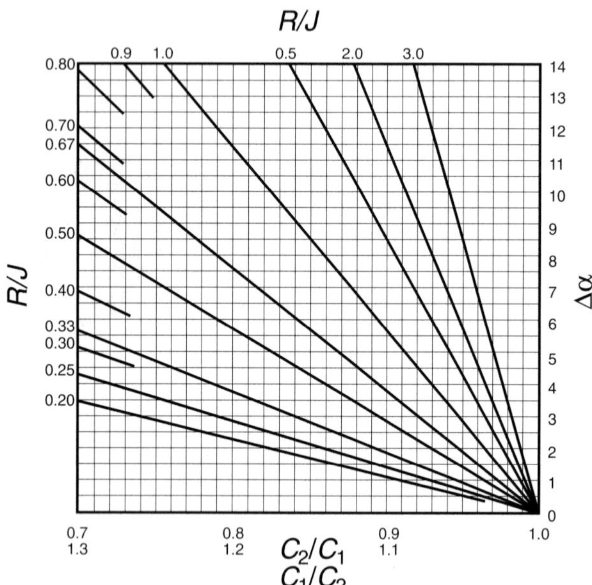

Figure W20 Nomograph to determine the turning angle of the wave ray $\Delta\alpha$ from the ratios *R*/*J* and C_1/C_2 (after CERC, 1984) (R = distance along wave ray, J = distance between contours, C_1 = wave celerity at contour 1, C_2 = wave celerity at contour 2).

where b and b_0 are the width between two wave rays at depth h and in deepwater, respectively.

An example of a wave refraction diagram in shown in Figure W21 which indicates the refraction pattern associated with submarine canyons. Note that the spacing of the wave rays is not constant; where more detail is required, such as in the vicinity of the canyons, the wave rays are spaced closer together. Strong focusing of the wave rays, known as wave convergence, occurs at the headland. At this location, $K_r > 1$ and relatively energetic wave conditions will prevail. Spreading of the wave rays, known as wave divergence, takes place at the heads of the canyons. As a result, $K_r > 1$ and the coastline landward of the canyons will experience relatively calm wave conditions.

Numerical wave refraction models

In the past, wave refraction diagrams were invariable constructed manually. Currently, numerical models are often used to construct wave refraction diagrams (e.g., Dalrymple, 1988). Such models, which often include the effects of wave diffraction, are much faster and more accurate than the manual method, but require considerable technical expertise and preparationary work (e.g., digitising the bathymetry). Rather than yielding orthogonals, the numerical analyses obtain solutions for wave direction and wave height for a finite number of grid cells that cover the area of interest. The obvious advantage of numerical wave refraction models is that once the model has been set up and found to perform satisfactory, a large number of scenarios (different incident wave direction and period) can be investigated with relative ease.

Gerhard Masselink

Bibliography

Arthur, R.S., Munk, W.H., and Isaacs, J.D., 1952. The direct construction of wave rays. *American Geophysical Union Transactions*, **33**: 855–865.

CERC, 1984. *Shore Protection Manual*. Coastal Engineering Research Center, Waterway Experiment Station, Corps of Engineers. Vicksburg, Mississippi, USA.

Dalrymple, R.A., 1988. A model for the refraction of water waves. *Journal of Waterway, Port, Coastal, and Ocean Engineering*, ASCE, **114**: 423–435.

Johnson, J.W., O'Brien, M.P., and Isaacs, J.D., 1948. Graphical construction of wave refraction diagrams. *U.S. Navy Hydrological Office Publication*, No. 605.

Munk, W.H., and Traylor, M.A., 1947. Refraction of ocean waves: a process linking underwater topography to beach erosion. *Journal of Geology*, **55**: 1–26.

Wiegel, R.L., 1964. *Oceanographical Engineering*. Englewood Cliffs: Prentice-Hall.

Cross-references

Engineering Applications of Coastal Geomorphology
Headland-Bay Beach

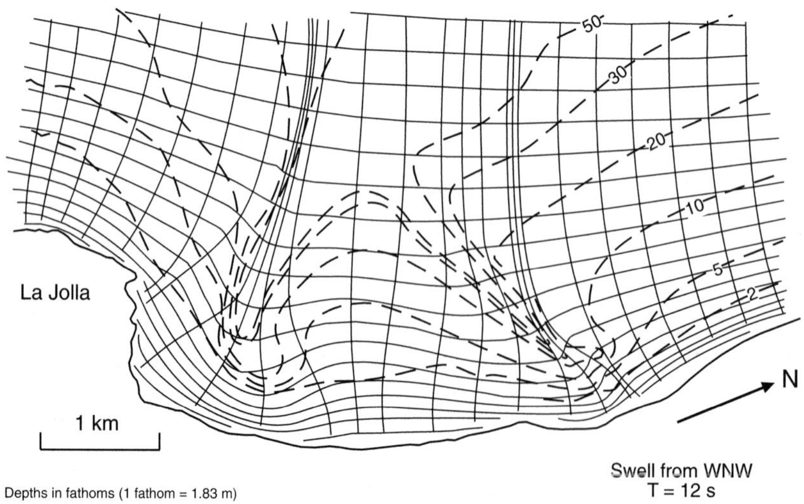

Figure W21 Wave refraction over submarine canyons and along the headland at La Jolla, CA (after Munk and Traylor, 1947).

Longshore Sediment Transport
Numerical Modeling
Wave Focusing
Waves

WAVES

In most coastal regions, waves provide the dominant source of energy to the nearshore. Although part of the incident wave energy is converted into heat or sound, or is reflected at the shore, a significant proportion of the energy is used for sediment transport and ensuing morphological change. An understanding of wave processes is therefore of fundamental importance when investigating nearshore sediment transport processes and morphology.

The characteristics of natural waves have received much attention in the oceanographic and coastal engineering literature. The classic book by Kinsman (1965) summarizes much of what was known about ocean waves in the early 1960s and remains authoritative. In the last three decades, advances in instrument design and data handling have considerably improved our knowledge of natural wave processes as described in Massel (1996). A succinct, but comprehensive overview of our current understanding of wave processes is provided by Komar (1998).

Wave analysis and statistics

An important characteristic of natural waves is that they are often highly irregular and made up of a range of wave heights and periods (Figure W22(A)). Hence, statistical techniques are required to properly describe the sea state in quantitative terms. Wave-by-wave analysis is one of the most common approaches to describe irregular waves. This technique consists of identifying the individual waves in the wave record using the zero-downcrossing method (IAHR, 1989) and determining representative wave parameters from the subset of wave heights and periods (Figures W22(B), (C)). An alternative method to describe the properties of irregular waves is spectral analysis (Figure W22(D)). The wave spectrum plots wave energy (variance) as a function of frequency (inverse of period) and can readily be used to identify the dominant wave frequencies in the wave record. In addition, the wave spectrum is useful for determining the partitioning of wave energy over distinct frequency bands (e.g., wind and swell waves). The directional wave spectrum can also be computed and plots spectral energy as a function of wave direction and frequency.

The two most common wave parameters to describe an irregular wave field are the significant wave height H_s and the peak wave period T_p. The significant wave height is sometimes denoted by $H_{1/3}$ and represents the average wave height of one-third of the highest waves in a wave record. The significant wave height approximately corresponds to visual estimates of wave heights. The peak period is derived from the wave spectrum and is defined as the wave period associated with the maximum wave energy. Other frequently used wave parameters to be derived using wave-by-wave analysis include $\overline{H}$ (mean wave height), H_{rms} (root mean square wave height), $H_{1/10}$ (average wave height of one-tenth of the highest waves), H_{max} (maximum wave height), T_z (mean wave period), T_s (mean wave period associated with H_s), $T_{1/10}$ (mean wave period associated with $H_{1/10}$) and T_{max} (maximum wave period associated with H_{max}).

Longuet-Higgins (1952) suggested that the probability distribution of wave heights in an irregular wave field can be described by the Rayleigh distribution (Figure W23)

$$P(H) = \frac{2H}{H^2_{rms}} \exp\left[-\left(\frac{H}{H_{rms}}\right)^2\right]. \quad \text{(Eq. 1)}$$

According to equation 1, the probability that a particular wave height has a value in the range $H + \delta H$ is given by the product $P(H)\delta$H. Thus, the probability of occurrence in a certain class range increases for increasing width of the class ranges. The probability that a particular wave height exceeds a prescribed value H_* is given by

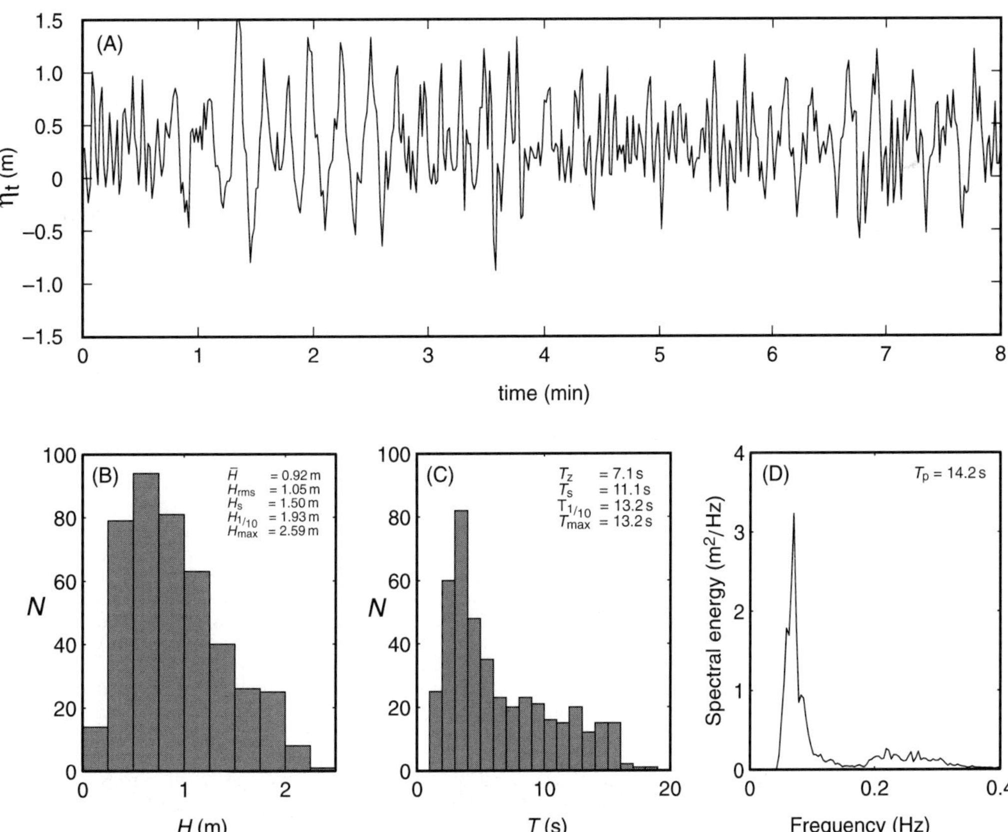

Figure W22 Analysis of 1 h of irregular offshore wave data: (A) time series of 8-min section of demeaned water surface elevation η_t; (B) frequency distribution of wave heights *H*; (C) frequency distribution of wave periods *T*; and (D) wave spectrum. The wave field represents a combination of wind waves and swell as indicated by the bimodal wave spectrum. The data were collected in 48 m water depth off the coast of Perth, Western Australia.

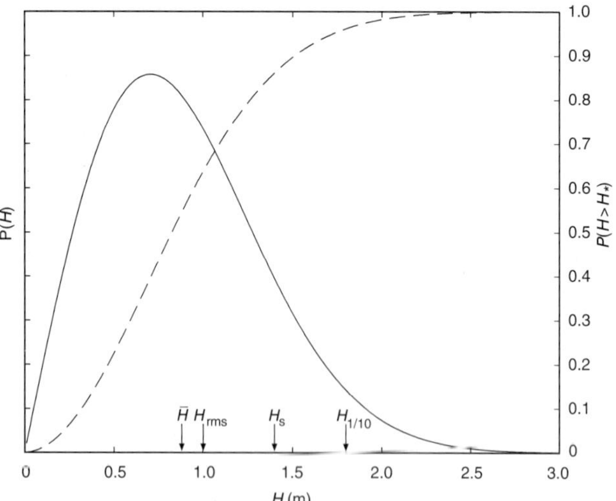

Figure W23 Relative (solid line; equation 1) and cumulative (dashed line; equation 2) probability density functions for wave heights according to the Rayleigh distribution for a root mean square wave height of $H_{rms} = 1$ m.

$$P(H > H_*) = \exp\left[-\left(\frac{H_*}{H_{rms}}\right)^2\right]. \qquad \text{(Eq. 2)}$$

Based on the Rayleigh distribution, various wave height measures can be determined from the standard deviation of the water surface elevation record σ_η

$$\bar{H} = 2.505\sigma_\eta \qquad \text{(Eq. 3a)}$$

$$H_{rms} = 2.828\sigma_\eta \qquad \text{(Eq. 3b)}$$

$$H_s = 4.004\sigma_\eta \qquad \text{(Eq. 3c)}$$

$$H_{1/10} = 5.090\sigma_\eta. \qquad \text{(Eq. 3d)}$$

Equation 3 enables direct determination of representative wave height parameters from the wave record (through the standard deviation) without the need to conduct wave-by-wave analysis.

The periods of individual waves generally show a narrower distribution than that of the wave heights and are in the range of 0.5–2 times the mean wave period T_z. The smallest waves in a record often have the shortest wave periods and site-specific correlations between the wave height and period can be derived empirically. For example, wave records from the North Sea suggest that $T \approx 4H^{0.4}$. One of the main problems with determining the characteristic wave period of a wave field is that natural waves generally consist of more than one wave field (refer to Figure W22), rendering a single wave period statistic not very useful.

An important application of statistical analysis of wave conditions is the estimation of extreme wave conditions on the basis of short-term wave records. This involves choosing a suitable probability distribution to fit to the available data and then extrapolating to obtain the probability of occurrence of extreme wave conditions. The probability distribution that is commonly used for this is a lognormal, Gumbel or Weibull distribution (Massel, 1996).

Wave generation

Waves are generated by wind acting on a water surface. In physical terms, the formation of waves constitutes a transfer of energy from wind to waves. At present, the generally accepted theory to account for the growth of wind waves is the combined Miles–Phillips mechanism. The mechanism incorporates two distinct phases of energy transfer from wind to waves: (1) an initial, linear growth phase that accounts for the formation of waves on a calm water surface (Phillips, 1957); and (2) an ongoing, exponential growth of the waves due the interactive coupling between wind and waves (Miles, 1957). An additional process that occurs within a developing wave field is the transfer of energy from high to low frequencies (Stewart, 1967; Longuet-Higgins, 1969). This process may occur when short-period waves steepen and break on the crests of the long-period waves. Since the long-period waves are traveling faster in deepwater than the short-period waves, they sweep up the energy and momentum contained within the short-period waves that are being overtaken. Thus, in a developing wave field both the wave energy level and the dominant wave period progressively increase. Wave growth is not infinite and when waves reach their limiting steepness ($H/L = 1/7$ in deepwater, where H is wave height and L is wavelength) they break in the form of white caps. An equilibrium can eventually be achieved whereby the energy losses by wave breaking are balanced by the addition of new energy being transferred from the wind to the waves. Such an equilibrium wave field is referred to as a fully arisen sea.

Wave conditions can be predicted as a function of wind speed, wind duration, and fetch length. This technique is known as wave forecasting when predicted wind data are used, or wave hindcasting when historical wind data are used. There are two main approaches to wave prediction: (1) the significant wave approach; and (2) the wave spectrum approach. The former approach is relatively simple and aims to predict wave statistics such as significant wave height and spectral wave period from wind conditions (CERC, 1984). The latter approach is more sophisticated and characterizes the predicted wave field by their spectra. One of the most common spectral formulations to parameterize the growth of a wave field is the JONSWAP spectrum which is based on extensive field measurements conducted in the North Sea (Hasselmann *et al.*, 1976).

Wave theories

Wave theories are mathematical formulations that predict the change in wave properties, such as water particle velocity, wave height, and wave energy, with depth. A basic assumption of linear and nonlinear wave theory is irrotational flow, that is, bed friction is neglected and there are no internal shear stresses. Under this assumption, the full momentum equations, the Navier–Stokes equations, reduce to the Euler equation of motion that contains only terms for the water particle velocities and local pressures. A second relation that must be included in the analyses of wave motion is the continuity equation to ensure conservation of mass. To provide an analytical solution to these equations, additional simplifying assumptions are required. The greater the number of assumptions, the simpler the resulting equations, but the greater the departure from the actual wave motion one is trying to describe. A large number of wave theories have been developed with different levels of complexity and accuracy.

Airy (or linear) wave theory (Airy, 1845) gives the least complicated expressions for wave motion and is therefore the most widely used wave theory. The simplicity of Airy wave theory is attained by making a number of assumptions. Most significantly, it is assumed that the wave amplitude is negligibly small compared with water depth. This is not a problem for waves in deepwater, but the theory is less suitable for waves in shallow water. Stokes wave theory (Stokes, 1847) was developed to overcome this main limitation of Airy wave theory and is a theory for waves of finite height (i.e., wave height is not negligible). Stokes wave theory is essentially Airy wave theory with nonlinear terms added to the equations. The additional terms have the effect of enhancing the crest amplitude and subtracting from the trough amplitude. The resulting wave profile is characterized by narrow wave crests and wide wave troughs which provides a more realistic wave shape of waves as they enter shallow water. When waves are in very shallow water, their crests peak up and are separated by wide, flat troughs, very much resembling solitary waves. Although solitary waves are not oscillatory waves and do not have a wave length or period, solitary wave theory (Boussinesq, 1872) can be used to determine properties of waves very close to breaking. The most sophisticated wave theory available is cnoidal wave theory (Korteweg and de Vries, 1895) and has potentially the widest range of application. In deepwater, the cnoidal wave solution reduces to the Airy wave theory, whereas in shallow water the wave period and length become infinite such that the cnoidal waves become equivalent to solitary waves. Cnoidal wave theory provides the most accurate description of the wave profile but is also the most cumbersome to use. A final wave theory that deserves mention is stream function wave theory (Dean, 1965) which resolves water motions associated with waves numerically without the need for simplifying assumptions required to obtain analytical solutions.

Selection of the appropriate wave theory depends upon the intended application. Figure W24 suggests the fields of application in terms of the ratios H/h and h/L. In the construction of this graph, the widest possible regions are given to the simpler wave theories. For example, the difficult cnoidal wave theory is given only a restricted region of application within its potentially much greater field. An important consideration is that Airy wave theory predicts water motions that are symmetrical, that is, onshore velocities are identical to offshore velocities. The nonlinear wave theories, on the other hand, predict water flows that are distinctly asymmetric in shallow water depths. This flow asymmetry is consistently directed landward whereby the onshore stroke of the wave motion is stronger, but of shorter duration than the offshore stroke. An accurate description of the flow asymmetry under waves is of

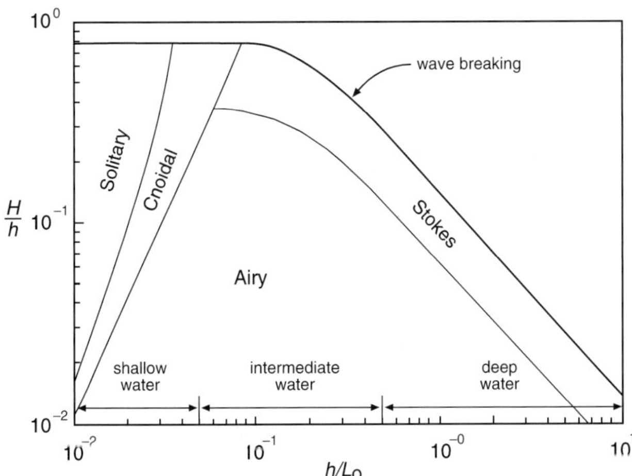

Figure W24 The areas of applications of the various wave theories as a function of the ratios H/h and h/L_o, where preference is given to the simpler theories such as Airy wave theory (after Komar, 1998).

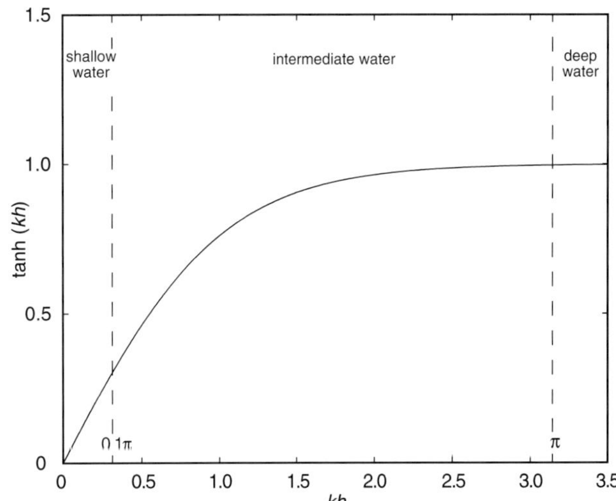

Figure W25 Properties of the tan h function.

fundamental importance for modeling wave-driven sediment transport processes (see *Cross-Shore Sediment Transport*).

Airy wave theory

Airy wave theory is still the most widely used wave theory in oceanographic research and important wave processes such as shoaling and refraction are adequately described by treating ocean waves as Airy waves. According to Airy wave theory the water surface elevation $\eta(x, t)$ is described by

$$\eta(x, t) = \frac{H}{2}\cos(kx - \sigma t), \qquad \text{(Eq. 4)}$$

where x is the coordinate axis in the direction of wave advance, t is time, H is the wave height, $k = 2\pi/L$ is the wave number (L is the wave length), and $\sigma = 2\pi/T$ is the radian frequency (T is the wave period). An important relationship to emerge from Airy wave theory is the dispersion equation which expresses the functional relationship between the wave period and the wavelength,

$$\sigma^2 = gk \tan h(kh), \qquad \text{(Eq. 5)}$$

which can be rewritten as

$$L = \frac{g}{2\pi}T^2 \tan h\left(\frac{2\pi h}{L}\right), \qquad \text{(Eq. 6)}$$

where h is the water depth and g is the gravitational acceleration. From equation 6 the wave celerity C can be derived:

$$C = \frac{L}{T} = \frac{g}{2\pi}T \tan h\left(\frac{2\pi h}{L}\right). \qquad \text{(Eq. 7)}$$

The wave celerity is also referred to as the wave phase velocity and represents the propagation speed of individual waves.

The dispersion equation can not be solved explicitly because it contains L on either side of the equation and therefore has to be solved numerically (iteratively). However, a convenient function which is accurate to 0.1% is given by Hunt (1979)

$$(kh)^2 = y^2 + \frac{y}{1 + 0.666y + 0.355y^2 + 0.161y^3 + 0.0632y^4 + 0.0218y^5 + 0.00654y^6} \qquad \text{(Eq. 8)}$$

where

$$y = \frac{\sigma^2 h}{g} = \frac{4\pi^2 h}{gT^2} = 4.03\frac{h}{T^2}. \qquad \text{(Eq. 9)}$$

At present, the use of computers makes computations involving the general dispersion relationship a relatively painless task, especially if equations 8 and 9 are used. In the past, however, further simplications to the dispersion relationship were desirable and these were made possible due to the characteristics of the tanh function (Figure W25). If $kh = 2\pi h/L$ approaches zero, $\tanh(kh) \approx kh$ and equations 6 and 7 reduce to

$$L_s = T\sqrt{gh} \qquad \text{(Eq. 10)}$$

and

$$C_s = \sqrt{gh}. \qquad \text{(Eq. 11)}$$

Equations 10 and 11 are known as the shallow-water approximations, hence the use of the subscript "s," and are valid for $kh < 0.1\pi$ (or $h/L < 0.05$). If, on the other hand, $kh = 2\pi h/L$ becomes large, $\tan h(kh) \approx 1$ and equations 6 and 7 reduce to

$$L_o = \frac{gT^2}{2\pi} \qquad \text{(Eq. 12)}$$

and

$$C_o = \frac{gT}{2\pi}. \qquad \text{(Eq. 13)}$$

Equations 12 and 13 are known as the deepwater approximations and are valid for $kh > \pi$ (or $h/L > 0.5$). Commonly, the subscript "o" is used to denote deepwater conditions and is therefore used here. Intermediate depth conditions prevail for $0.05 < h/L < 0.5$ and here the general equations 6 and 7 should be used because the errors in the deepwater and shallow-water approximations will exceed 5%.

Waves have potential and kinetic energy associated with the deformation of the water surface and the orbital motion of the water particles, respectively. According to Airy wave theory, the two energy forms are equal and the energy density E (in Newtons per unit area) is given by

$$E = \frac{1}{8}\rho g H^2, \qquad \text{(Eq. 14)}$$

where ρ is the density of water. The rate at which the energy density is carried along by the moving waves is the wave energy flux P and is given by

$$P = ECn \qquad \text{(Eq. 15)}$$

where

$$n = \frac{1}{2}\left[1 + \frac{2kh}{\sin h(2kh)}\right]. \qquad \text{(Eq. 16)}$$

The velocity Cn is the speed at which the energy density is carried along and is referred to as the group velocity C_g since it is the movement of groups of waves as distinguished from individual waves that travel according to the wave celerity C. In deepwater $n \approx 1/2$ but n increases in value as the waves travel into water of intermediate depth, becoming $n = 1$ in shallow water. The implication is that in deepwater, individual waves travel at twice the speed as the wave groups, whereas in shallow water, individual waves propagate at the same speed as the wave groups.

Airy wave theory also provides a description of the movement of water particles associated with wave motion (Figure W26). In deepwater, water particles move in a circular path with the radius of the circles decreasing with increasing depth beneath the water surface. The wave motion of deepwater waves is not felt at the seabed. In intermediate water depths the water particles follow an elliptical path with the ellipses

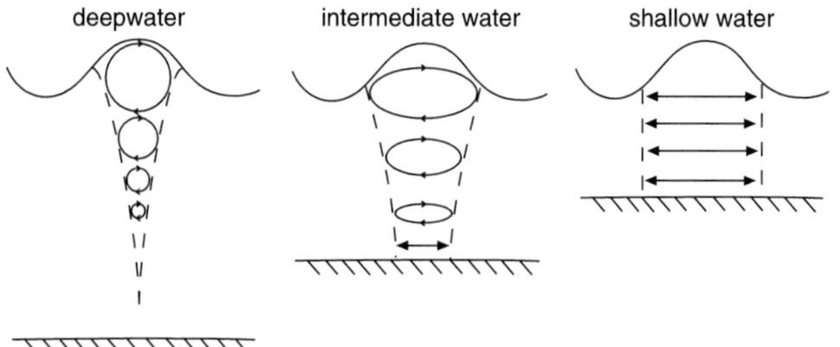

Figure W26 Description of the orbital motion of Airy waves in deep, intermediate, and shallow water.

becoming flatter and smaller as the seabed is approached. At the bottom, the water particles undergo a to-and-fro motion with a horizontal extent of the water motion d_0

$$d_0 = \frac{H}{\sin h(kh)} \quad \text{(Eq. 17)}$$

and a maximum near-bed velocity u_m

$$u_m = \frac{\pi H}{T \sin h(kh)}. \quad \text{(Eq. 18)}$$

In shallow water, Airy wave theory predicts that all water motions consist of to-and-fro horizontal movements uniform with depth with

$$d_0 = \frac{HT}{2\pi}\sqrt{\frac{g}{h}} = \frac{H}{kh} \quad \text{(Eq. 19)}$$

and

$$u_m = \frac{H}{T}\sqrt{\frac{g}{h}}. \quad \text{(Eq. 20)}$$

Parameters d_0 and u_m are important with respect to the bed morphology (wave ripples) and sediment transport processes. It should be emphasized that nonlinear wave theories such as Stokes theory provide a better description of water particle motions in shallow water than Airy wave theory.

Wave dispersion

A developing wave field is characterized by a broad-banded wave spectrum, indicating that a large range of wave periods are represented in the wave field. Such waves are referred to as sea. According to Airy wave theory, the wave celerity in deepwater C_o increases with wave period T (equation 13). In a broad-banded wave field, therefore, waves propagate at a range of wave celerities with the longer-period waves traveling faster than the short-period waves. Given sufficient time, the longer-period waves will outrun and leave behind the shorter-period waves. This sorting of the waves by period is termed wave dispersion and results in the transformation of broad-banded and confused sea into regular swell. The longer the distance of travel from the area of wave generation, the more effective the wave sorting process and the narrower the wave spectrum becomes. If a wave field is generated by an offshore storm, wave disperion causes the longest-period waves within this storm-wave field to reach the coast first. These first waves, the so-called storm-forerunners, will then be followed by waves of progressively decreasing shorter periods.

Wave groups

Ocean waves often occur in successive groups of higher or lower waves. Using linear wave theory, Longuet-Higgins and Stewart (1964) demonstrated that a small lowering of the mean water level accompanies groups of high waves, whereas a small rise in mean water level occurs under groups of small waves. This leads to the development of the group-bound long wave which has the same wave length and period as the incident wave group, but is 180° out of phase with the group. Offshore-directed currents of the long wave are maximum under groups of high waves and may constitute an important mechanism of offshore transport under shoaling waves (see *Shelf Processes*). Wave groups and the group-bound long wave are also of interest through their link with infragravity wave motion in the surf zone (see *Surf Zone Processes*).

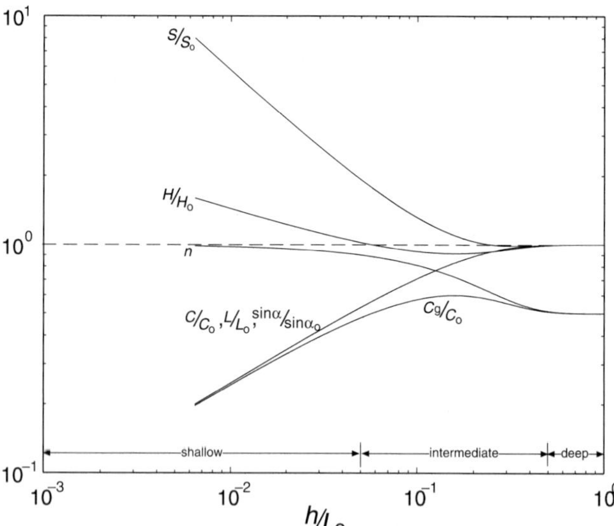

Figure W27 Shoaling transformations for Airy waves as a function of the ratio h/L_o.

Wave shoaling

As waves propagate from deep to shallow water they undergo a number of transformations which can be predicted using Airy wave theory. Some of these variations in wave properties have been plotted in Figure W27 and indicate that both the wavelength L and the wave celerity C systematically decrease with decreasing water depth. The wave period is the only variable that remains constant. Similarly, there are variations in n and hence changes in the wave group velocity C_g as the depth decreases.

The variation in the heights of the shoaling waves can be calculated from a consideration of the energy flux. Assuming energy losses due to bed friction and reflection can be ignored, the wave energy flux $P = ECn$ remains constant during wave propagation. This can be expressed as:

$$P = E_2C_2n_2 = E_1C_1n_1 = \text{constant}, \quad \text{(Eq. 21)}$$

where the subscripts "1" and "2" indicate two different locations along the path of wave travel ($h_1 > h_2$). Inserting $E = 1/8\rho g H^2$ in equation 21 yields

$$H_2 = \left(\frac{C_1n_1}{C_2n_2}\right)^{0.5} H_1. \quad \text{(Eq. 22)}$$

The ratio of the local wave height H (i.e., H_2) to the deepwater wave height H_o (i.e., H_1) can easily be derived from equation 22

$$\frac{H}{H_o} = \left(\frac{1}{2n}\frac{C_o}{C}\right)^{0.5} = K_s, \quad \text{(Eq. 23)}$$

where K_s is referred to as the shoaling coefficient. Equation 23 is plotted in Figure W27 and it can be seen that during wave shoaling, the wave height initially decreases slightly while entering intermediate water depth followed by a rapid increase.

The wave steepness $S = H/L$ also varies in the shoaling waves. Because the increase in H during shoaling is accompanied by a decrease in L, the wave steepness shows a dramatic increase during shoaling, especially over the shallow water region (Figure W27). It is the steepening of the waves that ultimately results in wave breaking.

Wave refraction

When a wave approaches the coast with its crest at an angle to the bottom contours, the water depth will vary along the wave crest. If the wave is in intermediate or shallow water, the wave celerity C will also vary along the wave crest with the deeper part of the wave propagating at a faster rate than the shallow part of the wave (equations 7 and 11). This results in a rotation of the wave crest with respect to the bottom contours, or in other words a bending of the wave rays or wave orthogonals. This processes is known as wave refraction (see *Wave Refraction Diagrams*) and is of great relevance to nearshore currents, sediment transport, and morphology (see *Beach Processes*, *Headland Bay Beach*, *Longshore Sediment Transport*, and *Surf Zone Processes*).

The refraction of waves is similar to the bending of light rays and the change in direction is related to the change in the wave celerity C through the same Snell's law

$$\frac{\sin\alpha_1}{C_1} = \frac{\sin\alpha_2}{C_2} = \text{constant}, \qquad \text{(Eq. 24)}$$

where α refers to the angle between the wave crest and the bottom contours and the subscripts "1" and "2" are used to indicate two different locations along the path of wave travel ($h_1 > h_2$) (Figure W28(A)). For a straight coast with parallel bottom contours, the angle at a given depth can be related to the angle of wave approach of the wave in deepwater

$$\sin\alpha = \frac{C}{C_o}\sin\alpha_o. \qquad \text{(Eq. 25)}$$

The expression shows that as the wave celerity decreases in shallow water, the angle made by the wave with the bottom contour also decreases.

The refractive bending of the wave rays also causes the wave rays to spread out, that is, the distance between rays increases as the waves are being refracted. If b_1 and b_2 are the spacing of the rays at two consecutive depths then the energy flux between the wave rays at these two depths should be constant:

$$P = E_1C_1n_1b_1 = E_2C_2n_2b_2 = \text{constant}. \qquad \text{(Eq. 26)}$$

Inserting $E = 1/8\rho gH^2$ in equation 26 then yields

$$H_2 = \left(\frac{n_1C_1}{n_2C_2}\right)^{0.5}\left(\frac{b_1}{b_2}\right)^{0.5} H_1. \qquad \text{(Eq. 27)}$$

For straight coasts with parallel contours, simple geometry considerations give

$$\frac{b_1}{b_2} = \frac{\cos\alpha_1}{\cos\alpha_2}. \qquad \text{(Eq. 28)}$$

The ratio of the local wave height H to the deepwater wave height H_o can easily be derived from equation 27

$$\frac{H}{H_o} = \left(\frac{1}{2n}\frac{C_o}{C}\right)^{0.5}\left(\frac{b_o}{b}\right)^{0.5} = K_sK_r\,, \qquad \text{(Eq. 29)}$$

where K_r is referred to as the refraction coefficient.

Irregular bottom topography can cause waves to be refracted in a complex way and produce significant variations in wave height and energies along the coast (Figure W28(B)). Waves diverge over relatively deepwater (e.g., depression in the seafloor), resulting in a spreading of the wave rays ($K_r < 1$) and a decrease in wave energy and wave height. In contrast, wave rays converge over relatively shallow water (e.g., shoal on the seafloor), resulting in a decrease in the spacing of the wave rays ($K_r > 1$) and an increase in wave energy and wave height.

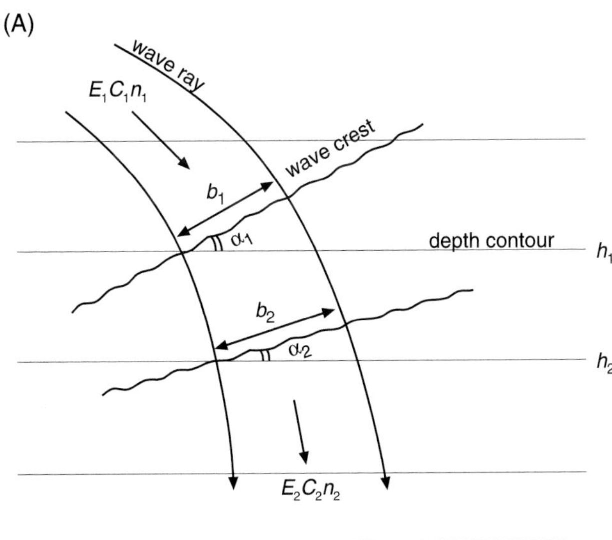

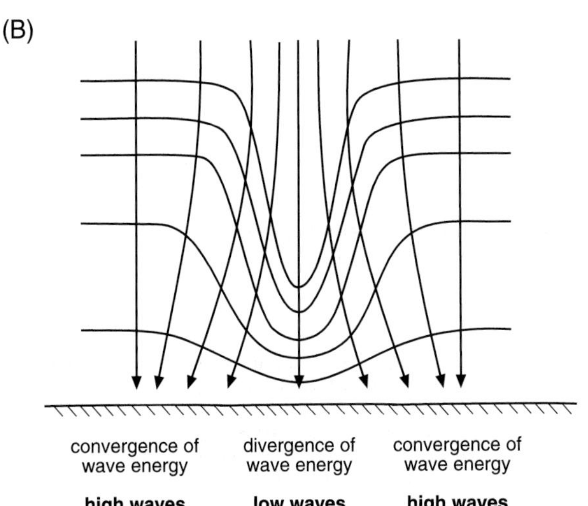

Figure W28 Wave refraction in case of: (A) parallel bottom contours; and (B) complex bathymetry.

Wave diffraction

Wave diffraction is the process of wave energy transfer in a direction other than the wave propagation direction and occurs irrespective of water depth. As such, diffraction is fundamentally different from wave refraction, although both processes often operate in concert. Wave diffraction occurs when an otherwise regular train of waves encounters an impermeable structure, for example, an island. A wave shadow zone will be created behind the obstacle, but wave diffraction will cause wave energy to leak into the shadow zone. Wave diffraction also enables wave energy to enter into narrowly confined bays and harbors.

Wave breaking

At some point during wave transformation, the water depth becomes too shallow for a stable waveform to exist and the waves break (Figure W29). The mechanism of wave breaking is that the horizontal velocities of the water particles in the wave crest exceed the phase velocity of the wave. Consequently, the water particles leave the waveform, resulting in a disintegration of the wave into bubbles and foam. The breaking wave height is related to the water depth by the breaker index γ according to

$$H_b = \gamma h_b\,, \qquad \text{(Eq. 30)}$$

where H_b is the height of the breaking wave and h_b is the mean water depth at the point of breaking. When the height-to-depth ratio of the wave exceeds the breaker index, breaking will occur. Solitary wave theory prescribes that $\gamma = 0.78$ and this provides a good first-order approximation.

Komar and Gaughan (1972) obtained a useful expression of the breaker height as a function of the deepwater wave height and steepness by applying linear wave theory (and ignoring energy losses by bed friction) to evaluate the wave energy flux from deepwater up to the breakpoint and using $\gamma = 0.78$,

$$\frac{H_b}{H_o} = \frac{0.563}{(H_o/L_o)^{1/5}}. \qquad \text{(Eq. 31)}$$

The value 0.563 was determined empirically from laboratory and field data. Due to wave shoaling, the height of the breakers always exceeds that of the deepwater waves and the ratio H_b/H_o increases with decreasing wave steepness H_o/L_o.

A continuum of breaker shapes occur in nature. However, three main types of breakers are commonly recognized: spilling, plunging, and surging (Galvin, 1968). Spilling breakers are characterized by a gradual peaking of the wave until the crest becomes unstable, resulting in a gentle forward spilling of the crest. Plunging breakers are distinguished by the shoreward face of the wave becoming vertical, curling over, and plunging forward and downward as an intact mass of water. In surging breakers,

Figure W29 Plunging breaker.

the front face and crest of the wave remain relatively smooth and the wave slides directly up the beach without breaking. Spilling and plunging breakers occur in the surf zone, whereas surging breakers are restricted to the swash zone. The transition from one breaker type to another is gradual and in natural surf zones, a mixture of breaker types is often present.

The breaker type principally depends on the beach gradient and wave steepness, which may be expressed by the surf similarity parameter

$$\xi_b = \frac{\tan\beta}{\sqrt{H_b/L_o}}, \qquad \text{(Eq. 32)}$$

where tan β is the beach gradient and the subscripts "b" and "o" indicate breaker and deepwater conditions, respectively. Spilling breakers occur for $\xi_b < 0.4$, plunging breakers occur when $\xi_b = 0.4$–2 and surging breakers occur for $\xi_b > 2$ (Battjes, 1974).

Gerhard Masselink

Bibliography

Airy, G.B., 1845. On tides and waves. *Encyclopaedia Metropolitana*, Volume 5. London: B. Fellowes.

Battjes, J.A., 1974. Surf similarity. *Proceedings of the 14th International Conference on Coastal Engineering*, ASCE, pp. 466–480.

Boussinesq, J., 1872. Théorie des ondes et de remous qui se propagent le long d'un canal rectangulaire horizontal, en communiquant au liquide contenu dans ce danal des vitesses sensiblement paralleles de la surface au fond. *Journal de Mathematique Pures et Appliquées*, **17**: 55–108.

CERC, 1984. *Shore Protection Manual.* Vicksburg: Coastal Engineering Research Center, Waterway Experiment Station, Corps of Engineers. Vicksburg, Mississippi, USA.

Dean, R.G., 1965. Stream function representation of nonlinear ocean waves. *Journal of Geophysical Research*, **70**: 4561–4572.

Galvin, C.J., 1968. Breaker type classification on three laboratory beaches. *Journal of Geophysical Research*, **73**: 3651–3659.

Hasselmann, K., Ross, D.B., Muller, P., and Sell, W., 1976. A parametric wave prediction model. *Journal of Physical Oceanography*, **6**: 200–228.

Hunt, J.N., 1979. Direct solution of wave dispersion equation. *Journal of Waterways, Ports, Coastal Oceans Division*, **105**: 457–459.

IAHR, 1989. List of sea-state parameters. *Journal of Waterway, Port, Coastal, and Ocean Engineering*, **115**: 793–808.

Kinsman, B., 1965. *Wind Waves*. Englewood Cliffs: Prentice-Hall.

Komar, P.D., 1998. *Beach Processes and Sedimentation*. Upper Saddle River, NJ: Prentice Hall.

Komar, P.D., and Gaughan, M.K., 1972. Airy wave theory and breaker height prediction. *Proceedings of the 13th International Conference on Coastal Engineering*, ASCE, pp 405–418.

Korteweg, D.J., and de Vries, G., 1895. On the change of form of long waves advancing in a rectangular canal, and on a new type of long stationary wave. *Philosophical Magazine, Series 5*, **39**: 422–443.

Longuet-Higgins, M.S., 1952. On the statistical distribution of the height of sea waves. *Journal of Marine Research*, **11**: 245–266.

Longuet-Higgins, M.S., 1969. A nonlinear mechanism for generation of sea waves. *Proceedings of the Royal Society of London, Series A*, **311**: 371–389.

Longuet-Higgins, M.S., and Stewart, R.W., 1964. Radiation stresses in water waves—a physical discussion with applications. *Deep-Sea Research*, **11**: 529–562.

Massel, S.R., 1996. *Ocean Surface Waves: Their Physics and Prediction.* Singapore: World Scientific.

Miles, J., 1957. On the generation of surface waves by shear flows. *Journal of Fluid Mechanics*, **3**: 185–204.

Phillips, O.M., 1957. On the generation of wind waves by turbulent wind. *Journal of Fluid Mechanics*, **2**: 417–445.

Stewart, R.W., 1967. Mechanics of the air–sea interface. *Physics of Fluids*, **10** (Supplement): 547–555.

Stokes, G.G., 1847. On the theory of oscillatory waves. *Transactions Cambridge Philosophical Society*, **8**: 441.

Cross-references

Beach Processes
Cross-Shore Sediment Transport
Headland-Bay Beach
Longshore Sediment Transport
Nearshore Wave Measurement
Surf Zone Processes
Wave Climate
Wave Environments
Wave Focusing
Wave Hindcasting
Wave Refraction Diagrams

WEATHERING IN THE COASTAL ZONE

Weathering, anywhere on the earth's surface, is a destructive process whereby preexisting rocks, boulders, pebbles, or smaller particles are altered in composition, texture, color, and/or hardness with little or no transport from their original *in situ* positions, except as mediated by chemical solutions. In soil science, it is the initial step in a long and complex process (Merrill, 1897/1921; Carroll, 1962; Barshad, 1972; Yatsu, 1988; Birkeland, 1999). As a process in physics, this rock decay follows the Second Law of Thermodynamics, as positive entropy, in which systematic ordering of chemical elements is progressively broken down. In a mineral petrological sense, the Bowen reaction series is reversed (Colman and Dethier, 1986).

The *agencies of weathering* are multiple, often operating in a synergistic way, and rarely in total independence. The net product contributes to the

"*Mantle*" or "*Regolith*," the term coined by Merrill (1897/1921, p. 299) for unconsolidated material without specifying genesis (Fairbridge, 1968; Gale and Hoare, 1991). The agents of weathering are threefold: (1) *Physical weathering*, such as thermoclastic processes ("thermoclasty"), involving expansion and contraction, operating both in low latitudes (notably the hot deserts) and mid- to high-latitudes (due to freeze and thaw). There is also "Fretting," a process of weathering by wetting and drying (especially by sea spray). The scale ranges from granular-sized disintegration ("granulation") to the splitting of giant boulders, 3–5 m in diameter. (2) *Chemical weathering*, operating principally in the humid climates, both warm-wet and cool-wet types. The weathered material is termed *Saprolite* (literally "rotted rock"). Chemical susceptibility depends partly on the nature of the solvent (the pH of the ground or seawater, acidic or alkaline), and on the minerals themselves, the solubility of which may be positively or inversely related to temperature, for example, the calcite or aragonite of limestone dissolves more freely as the temperature falls, but the K-feldspars of granite dissolve more readily with a rise in temperature and at low pH, but quartz (SiO_2) dissolves faster at a high pH. The progressive nature and stage is referred to in terms of "*Weathering Maturity*" (Garner, 1974, p. 269). Under warm freshwater conditions the ultimate mineral product of granitoid weathering is *kaolinite* (which has a group formula of $Al_2Si_2O_5(OH_4)$ and has the lowest cation exchange capacity). In the warm coastal zone the littoral belt is often marked by mangrove swamps. There, the decaying leaves and organic debris forces the normally high pH seawater to an acidic range (pH 4–6), creating mud-filled pools. This acidification favors removal of SiO_2 and concentration of *bauxite* (ideally $Al_2O_3 \cdot 2H_2O$; but more often a mixture of aluminous laterite). Commercial bauxites on the seaward side of the Guyana Shield and of the West-Australian Shield appear to mark former (pre-Quaternary) eustatically high sea levels. (3) *Biological weathering* relies largely on the presence of liquid water (at equable temperatures), and is thus inhibited by the hot-dry climates of deserts and by the cold-dry climates of the very high latitudes. The synergism of plant roots is particularly significant, for example, the thermoclastic crack in a mineral is exploited by root penetration. But the physical expansion of the root opens the crack still further, accelerating the disintegration process. Contributing also is the chemical (biochemical) role of decaying organic matter, and root "exudation" or granulation. Mineral diagenesis that involves hydration leads to expansion and thus further accessibility for groundwater (and roots). Animals play an important role, especially in limestones, which often favor the roots of plants or algae that crabs, gastropods, bivalves, and ecninoids like to eat. "Lithophagous" mollusca do not actually "eat" the limestone, but mechanically bore into it for the algae living in it. The byssus of *Mytilus*, and related bivalves, is an anchoring device, but is often torn away to expose fresh rock to weathering.

In the coastal zone, weathering is an essential forerunner to the various processes of *Erosion* (*q.v.*) which shifts or displaces the original material that has been made softer or more "erodable" by the loosening or softening effects of weathering. The process may be (1) subaerial, or (2) marine in its action, but for most rock material the rates involved are generally about five or six orders of magnitude less for the latter. The boundary between these two realms is accentuated by the action of tides, waves, and spray, and as a rule-of-thumb the level of mean low tide may be taken as the lower limit of subaerial processes. As an exception to this rule may be the occasion of a tropical hurricane on a coral-reef coast, when a meteorological blocking leads to a dead calm with three days or so of steady rain creating an aerated freshwater layer up to about 60 cm deep, effectively killing the surface reef.

Normally speaking, the low-tide boundary means that coastal rocks are made systematically more erodable, while their submarine extensions are differentially preserved. Depending upon the lithology, there is often a littoral bench or platform produced, the loose weathering products having been removed by mass wasting, wave action, or floating ice.

Lithology of the coastal belt plays a critical role in the effectiveness, style, and rate of weathering. Bearing in mind that since the last glacial phase most of the world's coastlines in the more stable regions have only been exposed to marine forces for about 6,000 yr, the rate of weathering is often critical. Important for this rate three lithic categories may be noted:

1. Massive plutonic rocks of granitic, dioritic, and gabbroic types, together with deeply crystallized metamorphics of granitoid types.
2. Feldspathic volcanics, pyroclastics, and derivative sandstones, including some graywackes.
3. Limestones, beachrocks, calcareous eolianites or calcarenites, algal and coral breccia.

These three categories need to be taken in conjunction with the long-term geological history of a given coastline, in other words, its paleoclimatic inheritance. Among the three categories listed above, rates of weathering are difficult to determine but certainly vary greatly with latitude in view of the physical processes introduced (Loughnan, 1969; Ollier, 1969a; Colman and Dethier, 1986; Birkeland, 1999).

Shore platforms develop mainly in the (2) and (3) categories (Stephenson, 2001). Important also is the role of inheritance, former climates and eustasy, and also the role of sea ice which develops *well before* the glacier buildup causing sea-level fall (Fairbridge, 1971, 1977).

For category (1), the massive plutonic rocks, weathering of a chemical nature is most evident in equatorial and temperate latitudes, where it penetrates the rock along major joint planes which in uncomplicated structural examples tend to produce cube-shaped blocks. Sharp corners and intersections became gradually rounded off until a so-called "woolsack" is created. Around the fractured borders of Gondwanaland, much of this weathering began in the early or mid-Mesozoic (200–150 Myr ago). In the tropics and subtropics the repeated eustatic fall of sea level during the Quaternary glaciations led to brief episodes of torrential precipitation and giant mudflows, which transported the unweathered crystalline blocks to the coast, where recent and present-day wave action is removing the weathered "rind," exposing totally unweathered, rounded boulders (Fairbridge and Finkl, 1984). Fine examples can be seen around Africa, Brazil, and Australia. Where the massive crystalline rocks are still *in situ*, for example, in the Recherche Archipelago, off S.W. Australia, the smooth surfaces of unweathered material plunges directly from the coastal hills well far below present mean sea level (MSL). Little trace of an intertidal notch is seen, and present rate of intertidal coast retreat (over the last 6,000 yr) is almost zero.

In the same category of massive plutonic rocks, but in the intermediate latitudes of the high-pressure belts, a strong alternation of wet and dry seasons prevails. During the Quaternary glacial stages, however, total aridity often prevailed, with seasons of cold-fog ("Atacama Desert" type). Also seen in the Namib desert of SW Africa, weathering pits develop on various scales (Goudie and Migon, 1997). The process is called "exudation." This is also the setting for a peculiar geomorphic form of pitting, *tafoni* (Fairbridge, 1968). Its classic setting is on the island of Corsica in the Mediterranean; the mountains there were not glaciated but subjected to severe frost action and cold dry winds blowing off the Alps. Fine examples may be seen on the south coasts of Australia, South Africa, and South America. The mechanism appears to be a process of weathering pit development. On the shady side of the outcrop, the daily dew precipitation leads to chemical weathering of feldspar minerals. Wind action removes the insolubles but diffusion and evaporation leads to the strengthening of a hard rim. The pit progressively deepens in size to cave where a human being can stand up in it.

Yet a third type of weathering of the first category appears on the coasts of the high latitudes, where alternate wet-dry, freeze-melt conditions apply. These have been studied particularly in Quebec and eastern Canada (Dionne and Brodeur, 1988); as well as in the southern Baltic where the clays become winnowed to concentrate the boulders. Winter freeze-up combined with tides and current action creates a mixture of frost debris, till clays and boulders that the French call *glaciel*. Freeze-thaw, combined with shore-ice pushed by currents and tides, is also seen on the coasts of Norway, where a prominent feature, the *strandflat* (*q.v.*) has long been a problem, a product of multiple cycles, involving both glacio-isostasy and glacio-eustasy.

The second weathering lithic type is that dominated by feldspathic minerals, which are susceptible to slow dissolution in acid groundwater (high levels of CO_2). This form of weathering is most apparent around volcanic centers that have erupted with the andesitic-rhyolitic lava suites, that are typical of the circum-Pacific "Ring of Fire." Weathering platforms are typically swept clean by wave action, and in places display the very evocative patterns of truncated pillow lavas.

Arkosic sandstones (i.e., rich in K-feldspars) are particularly common in the *flysch*-type facies and graywackes that are a feature of the great orogenic belts near former plate boundaries. These rock types are often heavily folded, but in coastal sectors they may be dramatically transected by contemporary (or at least Quaternary) marine erosion. That erosion involves a differential truncation made possible by the subaerial "preparation" by chemical weathering (Stephenson, 2000). The role of subaerial weathering in the development of the Kaikoura Peninsula shore platforms on the south island of New Zealand has been studied in depth by Stephenson and Kirk (2000), who have comprehensively studied the roles of both animal and plant types as well as their interactions with the results of wave energy. A useful review of the whole topic of shore platforms was issued separately by Stephenson (2000), with the question of platform width (Stephenson, 2001).

Finally, there are the limestones, along with beach rock, calcareous eolianites (calcarenites), coral and algal reef rock, and their breccias (Fairbridge, 1968; Miller and Mason, 1994). These are susceptible to solution in CO_2 rich (acid) waters, but seawater is alkaline (pH ca. 8), so at first sight this would not appear to be an effective process.

The explanation lies in the development of biochemical microenvironments. Tide pools disclose an extraordinary range of pH (5–10) in 24 h periods. During hours of darkness, CO_2 production and cooling cause the pH to fall into the acid range and the pool margins become undercut (helped by gastropoda). The next morning, the pool warms and solar evaporation shifts the pH quickly to the alkaline range. A white precipitate (aragonite) is often seen, and the pure spaces in the country rock are tightly cemented. The limestone shores are commonly marked by nearly horizontal intertidal platforms with undercut notches (and possibly overhanging visors) on the landward side (Fairbridge, 1968, p. 653). Tidal action is constantly saturating this undercut which is populated by unicellular blue-green algae which bore into the rock to depths up to 2–3 mm. The platform, for its part, is differentially hardened by interstitial precipitation of $CaCO_3$, in a plane approximating mean low tide (higher on exposed points). A limestone shore platform is thus a product of differential biochemical weathering and lithic hardening. An outer rim of algal construction (the so-called "Lithothamnion Rim" often forms a massive protective barrier on the seaward edge of the platform. Measurements of Holocene platforms suggest their landward growth is often at a rate of >1 mm/yr. When measuring Holocene changes of relative sea level it is found the limestone shores provide the clearest evidence, even to cm-precision in favored places.

As to the *agents of weathering*, to say mechanical, chemical, and biological is simplistic, because in most settings there is synergistic interplay, so that there is commonly a highly complex sequence of reactions. In many standard works on coastal science, the subject is not a even listed in the index. (Nevertheless, it gets 20 citations in Fairbridge, 1968.) It became a topic of debate during the mid- to late-19th century, particularly in Britain, when the physiographic questions of landscape came under consideration. The prominent action of marine (wave) erosion in western Europe tended to overwhelm the associated agencies and was favored by major physical geographers of the 19th century (e.g., Von Richthofen, 1870–72). The role of weathering particularly chemical weathering, was able to gain ground in some circles (e.g., Whitaker, 1867), and notably in New Zealand during the next century (Bartrum, 1926). In the South Pacific, the "Old Hat" phenomenon, observed by Dana in the previous century on the US Exploring Expedition, was a critical factor, although little noticed in the Northern Hemisphere. Thus, it was mainly in Australia and New Zealand during the 20th century that weathering became recognized as the key agency in coastal erosion.

Anomalies abound in coastal weathering situations. Disharmonic climatogenetic features are commonly sources of controversial interpretations. A classic example is at the Giant's Causeway in Northern Ireland, to be seen repeated on the opposing shores of Scotland (Fingel's Cave on Staffa), where the same Cenozoic basalts display spectacular columnar jointing. Paradoxically, the boulder beaches nearby are not characterized by sharply edged pentagonal or hexagonal blocks, but by well-rounded boulders. Their source is evident in the deeply weathered tropical soil that forms a transition zone above the fresh basalt, showing the progressive weathering of the columns into spheroidal "core stones" analogous to those of Devon's well-known granite tors.

Differential hardness of the parent rock material is the explanation for the sorting and concentration of the Upper Cretaceous chert (flint) horizons of the Chalk, as notably developed along the English Channel coast of Dorset where it forms a massive boulder beach, the Chesil Bank. The soft chalk has been lost to weathering. Another process involving Upper Cretaceous chalk is observed on the Baltic coastal cliffs of eastern Germany and Poland. Frozen spray in winter constitutes a powerful weathering agent.

Salt spray is effective in all latitudes, leading to "haloclasty", or what E.S. Hills called "fretting," besides simple wetting drying, that is a grain-by-grain disintegration of the coastal rocks, most typical of sandstones. Due to inhomogeneities to the rock, cavities develop and the result is a network or lace-pattern, that is generally known as *honeycomb* or *alveolar weathering* (see numerous illustrations in Fairbridge, 1968). As in the case of tafoni, noted earlier, extra-large examples are attributed to a dew-plus-chemical action that can operate far inland from the coastal zone (e.g., in the mid-Sahara), but the typical spray-activated alveolar weathering is most typical of the salt-crystal expansion in the supratidal zone. Ice-crystal expansion takes its place in the high latitudes (e.g., the "dry valleys" of Antarctica). Yet another agency in the coastal zone is biological where crabs and gastropods loosen mineral grains as they browse on algae.

Landsliding in the coastal zone is particularly evident in the "*Zone of Weathering*," label introduced by Ruxton and Berry (1957) after a survey undertaken in the territory of Hong Kong which displayed weathering on a scale unfamiliar to eyes accustomed to the landscapes of NW Europe or eastern North America where repeated Quaternary glaciations and periglacial activity has greatly modified the warm-wet products of what is "normal" weathering in much of the world far removed from the direct influences of glaciation over the last 1–2 Myr. In the deeply dissected coastal areas of Hong Kong, likewise in the Seychelles and eastern Brazil, particularly in granite or granitoid rocks, the zone of weathering is commonly 50–80 m thick. Deforestation and other human activities have helped to accelerate landscape modification by massive landsliding. The latter, while not a weathering process itself, may be greatly enhanced by the preparation and lubrication provided. Even in a temperate zone like the south coast of England, landsliding of coastal cliffs is particularly prevalent, thanks to the action of groundwater in the porous stratified formations.

A combination of deep weathering and landsliding is also observed in the region of Rio de Janeiro on the coast of Brazil, but here it has been accelerated by differential uplift following the plate-tectonic separation from Africa. Granite plutons form "sugar-loaf" mountains (inselbergs) that in a relative sense have grown progressively higher through time, aided by eustatic fluctuations and secular lowerings of the water table (Bremer, 1999).

Weathering profiles exposed in deeply dissected volcanic rocks such as in Fiji or in New Guinea (Ollier, 1969b) may frequently exceed 100 m. However, if the volcanic necks are traced eastwards into the Louisiade Archipelago, they are often seen to have been completely isolated from their weathered mantle and the waves beat against a vertical wall of unweathered igneous rock.

An unusual volcanic landform results from the weathering of a porous pyroclastic lithology that is rich in K-feldspars. The result is a so-called "volcano-karst" which includes weathering pits, caverns, and gullies reminiscent of limestone karst.

Important for shore platform development is the "*Weathering Front*," a concept introduced in Australia by Mabbutt (1961), and now adopted worldwide for the interface between fresh and weathered rock, to replace an earlier term "basal surface." In crystalline rocks it represents the ultimate saturation depth of groundwater, and therefore of atmospheric gases such as CO_2 and O_2. In porous, sedimentary sequences its definition is less straightforward, and in artesian basins the term has little use. On the contrary, in the ancient cratons it is very instructive, especially where such features reach the present-day coast.

For the Gondwanaland cratons such weathering fronts may be traced back many million years in time to the initiation of subaerial exposure. Over broad areas this zero date was in the Late Permian with the retreat of the last Late Paleozoic ice sheets about 250 Myr ago. A comparable date in the regions of the Laurasian glaciations would be only about 6,000 yr ago, but nevertheless in some exceptional spots, for example, in eastern Quebec pre-Cretaceous weathering fronts are known. In extensive areas of South America, South Africa, Antarctica, and Tasmania the last major defining events were the basaltic lava and diabase intrusions that ended in the Early Jurassic about 200 Myr ago. In the Indian peninsula the same type of eruption, marked the Cretaceous/Tertiary boundary, was 65 Myr ago.

In each region the "geomorphic clock" started ticking at a different point in time. With the far-reaching taphrogeny that marked the breakup of Gondwanaland came isostatic adjustments and drainage system reorientations. Rivers do not rapidly abrade their beds in fresh crystalline rocks, but where the "bedrock" has been preconditioned by a deep weathering front, that stream bed will be rapidly incised.

One might contrast the interior Guyana Shield in South America (Garner, 1974) and its limited weathering of sandstones, with the deep weathering front of the West Australian Craton where it is often situated at a depth of about 100–120 m (Finkl and Fairbridge, 1979). In this case, the boundary can be directly inspected by taking one of the gold mining elevators in the region of Kalgoorlie. The mean surface elevation of the craton is about 400–500 m, but the highest eustatic level of the Late Cretaceous was nearly 300 m, so there has been progressive lowering of the streambeds in the major rivers, but only near the cratonic margins. A mean rate of cratonic lowering over much of Gondwanaland has been 10–100 cm Myr. This snail's pace constitutes the "primary weathering cycle" on Gondwanaland. Only in the water-saturated stream beds approaching the grade dictated by sea level is chemical weathering feasible. And that MSL has fluctuated markedly, but shown a secular fall during the last 65 Myr.

Where the deeply weathered margin of the Deccan Plateau in India can be seen at the present-day coast, for example, around Goa and Kerala, the unsilicified latosol is terraced by marine erosion. The soft paleosols in the intertidal and supratidal zones there is subject to a *secondary weathering cycle* under its exposure to sea spray, tidal action, and biological agencies, further supplemented by the mechanical role of wave action. This exemplifies the predominant role of weathering, on every scale, in predicting the erosion of coastlines.

Rhodes W. Fairbridge

Bibliography

Barshad, I., 1972. Weathering—chemical. In Fairbridge, R.W. (ed.), *Encyclopedia of Geochemistry and Environmental Sciences*. New York: Van Nostrand Reinhold, pp. 1264–1269.
Bartrum, J.A., 1926. Abnormal shore platforms. *Journal of Geology*, **34**: 793–806.
Birkeland, P., 1999. *Soils and Geomorphology*. (Formerly: *Pedology, Weathering and Geomorphological Research.*) New York and Oxford: Oxford University Press.
Bremer, H., 1999. *Die Tropen*. Berlin and Stuttgart: Gebr. Borntraeger.
Carroll, D., 1962. Rainwater as a chemical agent of geological processes—a review. *U.S. Geological Survey*, Water-supply paper, 1535-G, 18p.
Colman, S.M., and Dethier, D.P. (eds.), 1986. *Rates of Chemical Weathering of Rocks and Minerals*. New York: Academic Press.
Dionne, J.C., and Brodeur, D., 1988. Frost weathering and ice action in shore platforms development, with particular reference to Quebec, Canada. *Zeitschrift fur Geomorphologie, Supplement Band*. **71**: 117–130.
Fairbridge, R.W. (ed.), 1968. *The Encyclopedia of Geochemistry and Environmental Science*. New York: Van Nostrand Reinhold Co., 1321p.
Fairbridge, R.W., 1971. Quaternary shoreline problems at INQUA, 1969. *Quaternaria* (Rome), **15**: 1–18.
Fairbridge, R.W., 1977. Rates of sea-ice erosion of Quaternary littoral platforms. *Studia Geologica Polonica*, **52**: 135–141.
Fairbridge, R.W., and Finkl, C.W., Jr., 1984. Tropical stone lines and podzolized sand plains Paleoclimatic indicators for weathered cratons. *Quaternary Science Reviews*, **3**: 41–71.
Finkl, C.W., Jr., and Fairbridge, R.W., 1979. Paleogeographic evolution of a rifted cratonic margin. S.W. Australia. *Paleogeography, Paleoclimatology, Paleoecology*, **26** (3–4): 221–252.
Gale, S.J., and Hoare, P.G., 1991. *Quaternary Sediments: Petrographic Methods for the Study of Unlithified Rocks*. New York: Wiley (Halsted).
Garner, H.F., 1974. *The Origin of Landscapes*. New York: Oxford University Press, 734p.
Goudie, A.S., and Migón, P., 1997. Weathering pits in the Spitzkoppe area, central Namib Desert. *Zeitschrift Geomorphologie, NF*, **41** (4): 417–444.
Loughnan, F.C., 1969. *Chemical Weathering of the Silicate Minerals*. Amsterdam: Elsevier, 154p.
Mabbutt, J.A., 1961. "Basal surface" or "weathering front." *Proceedings, Geological Association London*, **72**: 357–358.
Merill, G.P., 1897/1921. *Rocks, Rock Weathering, and Soils*. New York, London: Macmillan, 411p.
Miller, W.F., and Mason, T.R., 1994. Erosional featurse of coastal beachrock and aeolianite outcrops in Natal and Zululand, South Africa. *Journal of Coastal Research*, **10**: 374–394.
Ollier, C.D., 1969a. *Weathering*. Edinburgh: Oliver and Boyd, 304p.
Ollier, C., 1969b. *Volcanoes*. Cambridge, MA, and London: MIT Press, 177p.
Robinson, D.A., and Williams, R.B.D. (eds.), 1994. *Rock Weathering and Landform Evolution*. New York: John Wiley.
Ruxton, B.P., and Berry, L.R., 1957. Weathering of granite and associated erosional features in Hong Kong. *Geological Society of America Bulletin*, **68**: 1263–1292.
Stephenson, W.J., 2000. Shore platforms: a neglected coastal feature? *Progress in Physical Geography*, **24**(3): 311–327.
Stephenson, W.J., 2001. Shore platform width—a fundamental problem. *Zeitschrift fur Geomorphologie, NS*, **45**(4): 511–527.
Stephenson, W.J., and Kirk, R.M., 2000. Development of shore platforms on Kaikoura Peninsula, South Island, New Zealand, II. The role of subaerial weathering. *Geomorphology* (Elsevier), **32**: 43–56.
Twaidale, C.R., 1971. *Stsuctural Landforms*. Cambridge, MA and London: MIT Press, 247p.
Von Richthofen, F., 1870/72. "*Führer für Forschungsreisende*." Hannover, Jänecke, 734p.
Whitaker, W., 1867. On subaerial denudation, and on cliffs and escarpments of the Chalk and Lower Tertiary Beds. *Geolgical Magazine*, **4**: 447–454; 483–493.
Yatsu, E., 1988. *The Nature of Weathering*. Tokyo: Sozo-Sha Publishers (Maruzen Co.), 624p.

Cross-references

Bioerosion
Cliffs, Erosion Rates
Ice-Bordered Coasts
Marine Terraces
Notches
Rock Coast Processes
Shore Platforms
Standflats
Tafoni
Tidal Flats
Tors

WETLANDS

Wetlands are neither terrestrial nor aquatic systems and at the coast form part of a continuous gradient between upland and ocean. The most obvious characteristic of all wetlands is continuous, seasonal, or periodic standing water or saturated soil. However, the increasing understanding of wetlands as important ecosystems and the parallel recognition of the threats posed to them by urban development, drainage for agriculture, and other human pressures has led to the need for clearer definitions and classifications. Conservation and management measures focused at wetlands often involve restriction or regulation of activities at the site, and so the need for "delineation" or defining the boundaries of individual wetlands has become of concern to landowners, managers, and political leaders.

While much of the ongoing debate over "what is a wetland?" encompasses a range of environments from peat bogs to vernal pools to cypress swamps, the focus here will be on wetlands found in coastal areas—where the "wet" aspect of the wetland is specifically associated with tidal action, but not necessarily saline tidal waters. These would be classified as part of the marine, estuarine, and riverine-tidal systems under the classification methodology of Cowardin *et al.* (1979) or the marine and coastal section of the wetland classification approach adopted by the Ramsar Convention Bureau (Scott and Jones, 1995). There are many excellent texts which describe the range of processes and characteristics of wetlands around the world including Mitsch and Gosselink (2000), Finlayson and van der Valk (1995), and Mitsch (1994).

Here we will discuss the definitions that have been used to delineate wetlands and how these may be used relative to coastal wetlands, assess the functions and values of various types of coastal wetlands, and evaluate some of the current threats to coastal wetlands and the management measures that are needed to appropriately respond.

Definitions

Although most definitions encompass three fundamental characteristics of wetlands—the presence of water at or near the surface, soil conditions different from those in upland areas, and vegetation dominated by hydrophytes—the diversity of wetland types means that the combination of these three elements into a meaningful definition has been challenging. The purpose of such a definition should be to readily distinguish wetlands from other ecosystems. In some cases the definitions used have emphasized characteristics of the wetland that serve particular purposes or interests. An example is an early definition, the first in the United States, by the US Fish and Wildlife Service (Shaw and Fredine, 1956):

> The term "wetlands" ... refers to lowlands covered with shallow and sometimes temporary or intermittent waters. They are referred to by such names as marshes, swamps, bogs, wet meadows, potholes, sloughs, and river-overflow lands.

This definition goes on to encompass shallow lakes and ponds but not deepwater areas and permanent streams. Also excluded are areas that are flooded on such a temporary basis that the area does not support "moist-soil" vegetation. This on the basis that they were of little value to the wildlife species of concern at the time.

Under the Convention on Wetlands of International Importance Especially as Waterfowl Habitat, a global environmental treaty signed in Ramsar, Iran in 1971 and commonly known as the Ramsar Convention, "wetlands" were defined as:

> ... areas of marsh, fen, peatland or water, whether natural or artificial, permanent or temporary, with water that is static or flowing, fresh, brackish or salt, including areas of marine water the depth of which at low tide does not exceed six metres.

The Ramsar Convention also provides that wetlands "may incorporate riparian and coastal zones adjacent to the wetlands, and islands or bodies of marine water deeper than six metres at low tide lying within the wetlands." This very broad definition encompassed many areas, such as rice fields, that although technically wetlands, were unlikely to be considered of conservation value under Ramsar. While this definition still officially defines wetlands under the treaty, the Contracting Parties have adopted a classification system based loosely on that developed in the United States in the mid-1970s. The Cowardin Classification (Cowardin *et al.*, 1979) divided wetlands into systems, subsystems, classes, and subclasses in combination with a series of modifiers that describe water regime, water chemistry (e.g., salinity), and soil. The definition of wetlands used by Cowardin *et al.* (1979) includes descriptions of vegetation, hydrology, and soil:

> Wetlands are lands transitional between terrestrial and aquatic systems where the water table is usually at or near the surface or the land is covered by shallow water .... Wetlands must have one or more of the following three attributes: (1) at least periodically, the land supports predominantly hydrophytes, (2) the substrate is predominantly undrained, hydric soil, and (3) the substrate is nonsoil and is saturated with water or covered by shallow water at some time during the growing season each year.

In an attempt to convey the ecosystem concept of wetlands, and to ensure that the essential role of hydrology as a driving force behind the physical, chemical, and biological features of the system is retained, the US National Research Council Committee on Characterization of Wetlands (NRC, 1995) developed a broad reference definition of a wetland:

> A wetland is an ecosystem that depends on constant or recurrent, shallow inundation or saturation at or near the surface of the substrate. The minimum essential characteristics of a wetland are recurrent, sustained inundation or saturation at or near the surface and the presence of physical, chemical and biological features reflective of recurrent, sustained inundation or saturation. Common diagnostic features of wetlands are hydric soils and hydrophytic vegetation. These features will be present except where specific physicochemical, biotic or anthropogenic factors have removed them or prevented their development.

At the coast, the flood and ebb of the tide will usually be the controlling factors in determining soil inundation or saturation. Importantly, while many coastal wetlands support halophytic vegetation, the presence of salt is not an essential feature of a coastal wetland. They also include areas further inland and the landward limit can be defined by the limit of tidal fluctuation. According to Field *et al.* (1991) there are over 4.5 million ha of coastal wetlands in the conterminous United States of which only 16% are salt marshes. The remainder are forested and shrub-scrub wetlands and fresh marsh areas. There may be some uncertainty in determining the boundary between tidal and nontidal wetlands and Mitsch and Gosselink (2000) note that tidal freshwater marshes can be described as intermediate in the continuum from coastal salt marshes to freshwater marshes. Similarly, freshwater woody wetlands or swamps, can occur in fresh nontidal waters and in areas influenced by the tide. This tidal–inland boundary has received less attention than that between wetlands and their adjacent uplands, likely because there is no jurisdictional difference between wetlands on either side of it. However, in some areas lands flooded by the tide are in public ownership by statute and it is likely, as this "tidal" boundary changes with sea-level rise, that criteria for the discernment of tidal versus nontidal wetlands in the coastal zone may be required.

Functions and values of coastal wetlands

Coastal wetlands are valuable ecosystems. By providing refuge and forage opportunities for fishes and macrocrustaceans, swamps, marshes, and mangroves around the world are the basis of many communities' economic livelihoods. Shallow ponds and seed producing vegetation provide overwintering habitat for millions of migratory waterfowl, and the structure of trees and forests is vital to songbirds and waders alike. The role of wetlands in uptaking nutrients and reducing loading to the coastal ocean is widely recognized, as is their value for protecting local communities from flooding—either by damping storm surges from the ocean or providing storage for riverine flood-waters. The economic value of tidal marshes and mangroves has been estimated at \$9,990/ha/yr in a recent assessment of ecosystem services (Costanza *et al.*, 1997).

According to Mitsch and Gosselink (2000) coastal wetlands include tidal salt marshes, tidal freshwater marshes, and mangrove swamps. Tidal swamps will also be considered here as they are extensive at the heads of estuaries and in coastal plain environments where tides influence water levels at great distances from the coastline.

Salt marshes

While saline marshes around much of the United States are dominated by *Spartina alterniflora*, regional variations include *Salicornia virginica* and *Spartina foliosa* in California and southern Oregon, to more diverse assemblages including *Distichlis spicata, Triglochin* spp., and *Plantago maritima* in the Pacific Northwest. In the west, these native halophytes are being supplanted at low tidal elevations by invasive *S. alterniflora* which can grow vigorously on open mudflats unavailable to the native species. Salinity typically varies between 15 and 33 ppt depending on coastal physiography and the level of freshwater inputs to the coastal ocean. Coastal salt marshes are highly productive systems. Tides, nutrients, and regular flushing offset the stresses associated with salinity, fluctuating temperatures, and alternate wetting and drying of the substrate. The detritus derived from this primary productivity supports higher trophic levels both via direct consumption and through microbial degradation which enhances the food value of the detritus. By providing shelter and refuge from predation, salt marsh margins are important nursery habitats for many fisheries species. For more detail see *Salt Marsh*.

Mangroves

Mangroves, comprising approximately 60 species of trees and shrubs, are the predominant form of vegetation in the intertidal zone of tropical estuaries, lagoons, and sheltered coastlines. They are largely confined to latitudes between 30°N and S of the equator, with a few notable exceptions that may be explained by the occurrence of warm ocean currents or by the presence of relict populations of more poleward past distributions (Duke, 1992). According to one recent estimate, the total area of mangroves in the world is 181,000 km^2 (Spalding *et al.*, 1997). While mangrove trees are notable for their characteristic adaptations to anoxic soil environments, such as prop roots and pneumatophores (see *Vegetated Coasts*), their ecological function is very similar to coastal salt marshes. For more detail see *Mangroves*.

Tidal freshwater marshes

Inland from salt marshes but still close enough to the coast to experience tidal fluctuations in water level are wetlands dominated by a variety of grasses and annual and perennial broadleaved aquatic plants. This wetland type is found most commonly in the mid- and south-Atlantic coasts and in Texas and Louisiana. No known plant species appears exclusively in tidal freshwater areas. Most marshes are dominated by a combination of annuals and perennials most of which also occur in inland freshwater wetlands. Plant diversity is much higher than in salt marshes and on much of the Atlantic coast of the United States there are substantial seasonal changes in plant coverage (Odum *et al.*, 1984). In the late winter and early spring there is a period of almost bare mud with some remaining stubble from previous years growth, followed by coverage with broad leaved plants in the late spring. By late summer the marshes are dominated by grasses and herbaceous plants. The soil conditions and hydrology of these marshes are very dependent on local conditions; although Odum *et al.* (1984) note that they are most likely fine mineral soils or peats.

In the Mississippi River delta plain a particular type of tidal freshwater marsh has developed in which, rather than being rooted in sediment, the plants live in an organic mat composed of live roots and decomposing litter. The level of this mat rises and falls with the tide. These marshes are termed "flotant." While Mitsch and Gosselink (2000) noted similar floating marshes in the Danube Delta and in lakes around the world, in coastal Louisiana O'Neil (1949) reported 100,000 ha of freshwater and brackish floating wetlands. Flotant is described by Sasser *et al.* (1994) as "wetlands of emergent vegetation with a mat of live roots and associated dead and decomposing organic material and mineral sediments, that moves vertically as ambient water levels rise and fall." Sasser *et al.* (1994) suggests that the floating marshes of the Mississippi delta plain have most likely developed in fresh marshes removed from riverine sediment inputs. There are several possible explanations for the formation of floating marshes. O'Neil (1949) suggests that floating marshes form when natural "attached" organic marshes are subjected to subsidence or sea-level rise and the buoyant organic mat is subjected to increasing upward tension. It eventually breaks free from its mineral substrate and floats (Figure W30(A)). Alternatively, the floating mat could grow and expand out into open water areas from existing marshes (Figure W30(B)).

Tidal freshwater marshes support a variety of waterfowl and animals from alligators to nutria. Their high primary production is only partly consumed directly with most being available to consumers through the detrital foodchain where benthic invertebrates play an important role. Mitsch and Gosselink (2000, p. 279) note that "of all wetland habitats, coastal freshwater marshes may support the largest and most diverse populations of birds." These include wading birds, dabbling ducks, birds of prey, and shorebirds, and the structural diversity of the wetland vegetation interspersed with shallow ponds is an important factor in their use of these wetlands.

Freshwater forested wetlands

Coastal wetland forests frequently have an extremely diverse flora of tree, shrub, and herbs. They can be roughly divided into deepwater swamps, in the southeastern United States dominated by bald cypress (*Taxodium distichum*) and tupelo gum (*Nyssa aquatic*), with a red maple (*Acer rubrum*) and buttonbush (*Cephalanthus occidentalis*) understory; and seasonally flooded bottomland hardwood forests dominated by several oak species (*Quercus* spp.), green ash (*Fraxinus pennsylvanica* var. *lanceolata*), and other hardwood species.

The character of coastal wetland forests is very sensitive to water level conditions, influenced in many areas by the tidal and gradual changes in sea level; as well as seasonal changes in riverine water levels at the heads of estuaries. Increases in freshwater discharge during floods are unlikely to result in any deleterious effects to freshwater forested wetlands as long as seasonal low-water periods allow for substrate drainage and forest regeneration. Bald cypress regenerates well in swamps where the substrate is moist and competing species are unable to cope with flooding. Seedlings must experience dry periods long enough to allow growth and survival of future flooding. However, prolonged flooding, associated with sea-level rise or a major change in riverine discharge regime, can result in changes. DeLaune *et al.* (1987) recognized gradients in seedling survival for both *T. distichum* and *Quercus lyrata* along an elevation gradient in a Louisiana coastal swamp. Increased flooding will thus reduce the area of appropriate elevation for seedling survival, and the findings of Rheinhardt and Hershner (1992) suggest that increased saturation of substrate will alter species composition in freshwater swamps. They found *Fraxinus* spp. and *Nyssa sylvatica* in wetter areas compared to *A. rubrum* and *Liquidambar styraciflua* within swamps of the Virginia coastal plain.

Sensitivities, threats, and management approaches

Coastal wetlands are undoubtedly some of the most valuable ecosystems in the world and also some of the most threatened. Climate change and variability compound existing stresses from human activities such as dredging and/or filling for development, navigation or mineral extraction, altered salinity and water quality resulting from watershed management and non-point source contamination, and the direct pressures of increasing numbers of people living and recreating in close proximity.

The survival of coastal wetlands under conditions of altered climate really depends primarily on the ability of the plants to survive. The biogeomorphic processes which control the wetland landscape may produce gradual changes, for example, from marsh to mangrove swamp, as long as the environmental thresholds which control plant survival are not crossed. The challenge in predicting coastal wetland response, either the nature of the gradual change, or when and where the thresholds are exceeded, depends upon interactions among the responses to various climate forcing. The very low gradients of coastal wetlands make them particularly susceptible to changes in water level and any adjustments in the hydrodynamic forcing from both ocean and watershed. While submergence associated with sea-level rise is frequently considered a serious threat to coastal systems, their inherent adaptation to coastal hydrology makes them perhaps less sensitive than terrestrial ecosystems

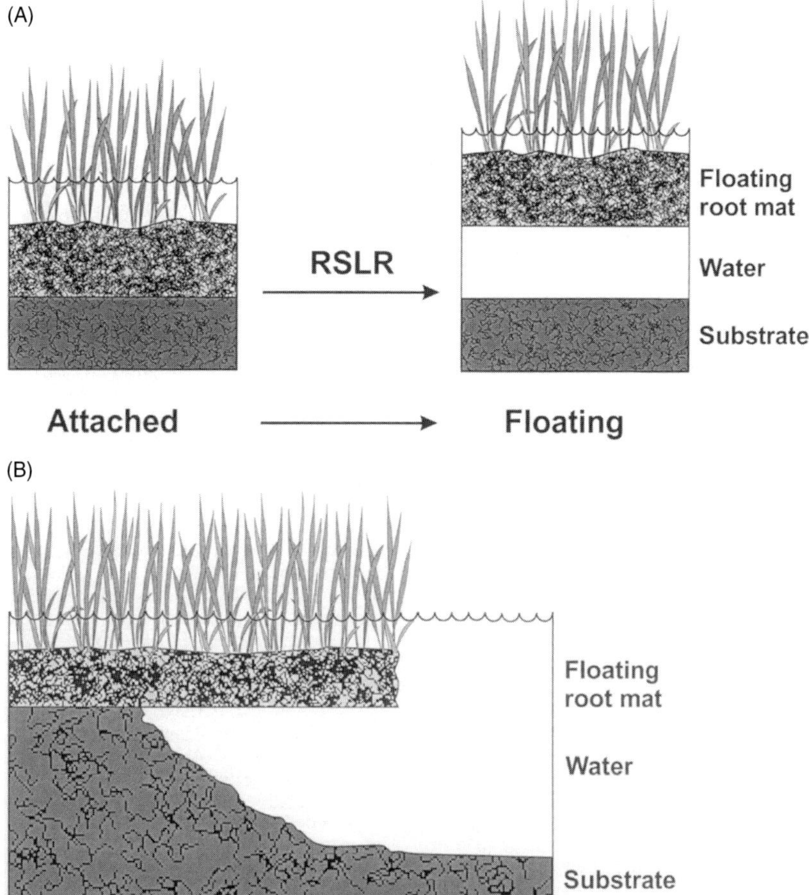

Figure W30 Alternative views on the formation of "flotant" marshes in Louisiana. (A) The process described by O'Neil (1949) involving relative sea-level rise (RSLR). (B) The process described by Russell (1942) involving the growth of marsh into open water from an existing stand.

subjected to frequent flooding by the encroaching sea. An examination of the response of various coastal wetland environments to climate change forcing illustrates the resilient nature of these systems and the complex biogeomorphic interactions that sustain them.

Increased freshwater is generally beneficial to coastal wetlands and a decrease may result in salinity stress for some communities. However, a gradual change does not necessarily result in plant death and wetland loss, and salt and brackish marshes may replace freshwater marshes and swamps. For example, the variation in salinities found by Visser *et al.* (1998) in marshes dominated by *Spartina patens* was so wide (average salinity 10.4 ± 5.8 ppt), and overlapped with the zone dominated by *S. alterniflora* (average 17.5 ± 5.9 ppt) that it seems likely that an increase of more than 1–2 ppt in salinity would be required to result in a significant change in habitat types in the more saline marshes. Decreases in salinity will likely increase the productivity of salt marsh plants as osmotic stresses are decreased. However, the salinity data of Visser *et al.* (1998) suggest that decreases in salinity of more than 5 ppt might be required for *S. alterniflora* to be replaced by more brackish marsh species.

Most coastal wetlands already cope with gradual sea-level rise. During the last 100 years, globally averaged sea level has risen at about 1–2 mm/yr. In most instances, these changes have not produced dramatic changes in coastal wetlands and accretionary processes have maintained wetland elevation and hydroperiod. Increased sediment delivery to coastal systems, resulting from future increased frequency of high intensity rainfall events, enhances sustainability by providing for vertical accumulation of substrate. Catastrophic sediment delivery could increase elevation sufficiently to convert coastal wetlands to areas of upland vegetation. However, unless the net effect is greater than the cumulative effects of sea-level rise, the effect should be seen as a type of dynamic equilibrium where episodic inputs counter long-term but low-level rise in sea level. Indeed it appears that many coastal wetlands in areas of high subsidence—and thus high relative sea-level rise—maintain their elevation through the episodic input of sediments associated with moderate-low frequency storms such as frontal passages and hurricanes (Reed, 1989; Cahoon *et al.*, 1995).

Coastal storms can bring more than beneficial sediments. The resilience of coastal wetlands to storm effects likely varies with salinity. Fresher systems are more susceptible to saline incursions during storms. For example, in South Carolina the storm surge caused by Hurricane Hugo flooded some low-lying coastal forests greater than 3 m deep with saltwater, and the surge carried saline water (> 20 ppt) along waterways up to several hundred kilometers inland (Blood *et al.*, 1991). Elevated soil salinity levels lingered for months after Hugo, and had adverse affects on tree survival and growth, as well as on nutrient cycling processes. The effect of tropical cyclone activity on forested wetlands can be structural as well as hydrologic. Low salinity and freshwater forested wetlands differ in their susceptibility to tropical storm wind impacts, with damage to forest types generally greatest to pines, least to swamp, with hardwoods intermediate (Conner, 1998). In the Atchafalaya River basin in south Louisiana, Hurricane Andrew caused little damage to cypress-tupelo communities but caused significant damage and mortality to bottomland hardwood forests (Doyle *et al.*, 1995). However, in some areas in the vicinity of Cape Cod, MA, USA, Atlantic white cedar wetlands were more severely affected by blowdowns from Hurricane Bob than adjacent upland forests, a fact attributed to their relatively shallow root systems (Valiela *et al.*, 1998). The long-term effect of storms on coastal wetlands depends upon the recovery time and the return interval of the storms. Increased frequency, if not intensity, of storm impact may therefore result in irreversible damage to forested systems or fragile organic substrates. For minerogenic salt marsh soils, however, storms can provide inputs of sediment essential to vertical sustainability.

It is important to recognize that these various responses may interact in ways that are difficult to predict, and that climatic and other stress factors and their interactions may result in a host of indirect effects on coastal wetlands, such as changes in patterns of herbivory or the rise in importance of alien species.

Management measures adopted to remediate loss of coastal wetlands must emphasize future sustainability of existing systems rather than re-creation of lost, impaired, or threatened habitats. The fisheries productivity, storm protection, and avian habitat provided by wetlands is dependent upon their landscape configuration, their internal structure, and their linkages with other coastal and marine resources. Such architecture is challenging to recreate and likely impossible to recreate given the other changes occurring at our coasts. Thus, management and planning strategies that provide for sustainable vertical accumulation, biodiversity, and linkages, both internally and with watershed and coastal ocean, are essential to wetland survival in the face of climate variability and long-term change. Climate change will continue to occur and it seems unlikely that development pressures will diminish in the near future (Nicholls and Small, 2002). Appropriate planning and management to facilitate change and adjustment of the coastal wetland landscape, rather than total loss of this valuable habitat, must be implemented to ensure the wetlands have a fighting chance (see *Coastal Zone Management*).

Summary and conclusions

Coastal wetlands occur at the transition from terrestrial to aquatic ecosystems and on the topographic gradient between upland and sea. Their definition on the ground has been a topic of much discussion in recent decades but importantly, this discussion has clearly identified fundamental wetland characteristics that vary in nature from place to place but which can be used to successfully distinguish wetlands from adjacent lands or waters. At the coast, important gradients in salinity and tidal inundation control the specific ecology and geomorphology of wetlands. When combined with climatic factors these drivers result in environments as different as low salt marshes to cypress-tupelo swamps. Coastal wetlands are undoubtedly among the most productive and valuable ecosystems on earth.

The vital role of vegetation in the form and function of coastal wetlands makes them particularly susceptible to pressures associated with urbanization and industrial development at the coast (see *Vegetated Coasts*). However, coastal wetlands are able to withstand disturbances and stresses associated with climate variability and in most cases future changes in climate will result in transitions and adjustments between and within coastal wetlands rather than the catastrophic loss sometimes projected. Most mangrove and freshwater forested wetland plant species are millions of years old, and therefore have survived through numerous and very large-scale changes in climate and sea level, including several ice ages and changes in sea level on the order of 100 or more meters. Survival under such circumstances generally depended on the ability to migrate, rather than on the persistence of individual wetlands. Individual coastal wetlands today can either accumulate vertically to maintain their position in the coastal zone or migrate as in the past. Both can occur—where conditions are favorable.

Denise J. Reed

Bibliography

Blood, E.R., Anderson, P., Smith, P.A., Nybro, C., and Ginsberg, K.A., 1991. Effects of Hurricane Hugo on coastal soil solution chemistry in South Carolina. *Biotropica*, **23**: 348–355.

Cahoon, D.R., Reed, D.J., Day, J.W., Steyer, G.D., Boumanns, R.M., Lynch, J.C., McNally, D., and Latif, N., 1995. The influence of Hurricane Andrew on sediment distribution in Louisiana coastal marshes. *Journal of Coastal Research*, **18** (Special issue): 280–294.

Conner, W.H., 1998. Impacts of hurricanes on forests of the Atlantic and Gulf Coasts, U.S.A. In Laderman, A.D. (ed.), *Coastally Restricted Forests*. New York: Oxford University Press, pp. 271–277.

Costanza, R., d'Arge, R., de Groot, R., Farber, S., Grasso, M., Hannon, B., Limburg, K., Naeem, S., O'Neil, R.V., Paruelo, J., Raskin, R.G., Sutton, P., and van den Belt, M., 1997. The value of the world's ecosystem services and natural capital. *Nature*, **387**: 253–260.

Cowardin, L.M., Carter, V., Golet, F.C., and LaRoe, E.T., 1979. *Classification of Wetlands and Deepwater Habitats of the United States*. FWS/OBS-79/31. Washington, DC: U.S. Fish and Wildlife Service.

DeLaune, R.D., Patrick, W.H., and Pezeshki, S.R., 1987. Foreseeable flooding and death of coastal wetland forests. *Environmental Conservation*, **14**: 129–133.

Doyle, T.W., Smith, T.J., III, and Robblee, M.B., 1995. Wind damage effects of Hurricane Andrew on mangrove communities along the southwest coast of Florida, USA. *Journal of Coastal Research*, **21** (Special issue): 159–168.

Duke, N.C., 1992. Mangrove floristics and biogeography. In Robertson, A.I., Alongi, D.M. (eds.), *Tropical Mangrove Ecosystems*. Washington, DC: American Geophysical Union, pp. 63–100.

Field, D.W., Reyer, A.J., Genovese, P.V., and Shearer, B.D., 1991. *Coastal Wetlands of the United States*. Washington, DC: National Atmospheric and Oceanic Administration and U.S. Fish and Wildlife Service.

Finlayson, C.M., and van der Valk, A.G., 1995. *Classification and Inventory of the World's Wetlands*. Dordrecht: Kluwer Academic.

Mitsch, W.J., 1994. *Global Wetlands: Old World and New*. Amsterdam: Elsevier.

Mitsch, W.J., and Gosselink, J.G., 2000. *Wetlands*. New York: John Wiley and Sons.

Nicholls, R.J., and Small, C., 2002. Improved estimates of coastal population and exposure to hazards released. *EOS*, **83**: 301–305.
NRC (National Research Council), 1995. *Wetlands: Characteristics and Boundaries*. Washington, DC: National Academy Press.
Odum, W.E., Smith, T.J., Hoover, J.K., and McIvor, C.C., 1984. *The Ecology of Tidal Fresh Marshes of the United States East Coast: A Community Profile*. FWS/OBS-83/17. Washington, DC: U.S. Fish and Wildlife Service.
O'Neil, T., 1949. *The Muskrat in the Louisiana Coastal Marshes*. New Orleans, LA: Wildlife and Fisheries Commission.
Reed, D.J., 1989. Patterns of sediment deposition in subsiding coastal salt marshes, Terrebonne Bay, Louisiana: the role of winter storms. *Estuaries*, **12**: 222–227.
Rheinhardt, R.D., and Hershner, C., 1992. The relationship of belowground hydrology to canopy composition in five tidal freshwater swamps. *Wetlands*, **12**: 208–216.
Russell, R.J., 1942. Geomorphology of the Rhone Delta. Annals of the Association of American Geographers, **32**: 147–254.
Sasser, C.E., Swenson, E.M., Evers, D.E., Visser, J.M., Holm, G.O., and Gosselink, J.G., 1994. *Floating Marshes in the Barataria and Terrebonne Basins, Louisiana*. Baton Rouge: Louisiana State University.
Scott, D.A., and Jones, T.A., 1995. Classification and inventory of wetlands: a global overview. *Vegetatio*, **118**: 3–16.
Shaw, S.P., and Fredine, C.G., 1956. *Wetlands of the United States: Their Extent, and their Values for Waterfowl and Other Wildlife*. Circular 39. Washington, DC: U.S. Fish and Wildlife Service.
Spalding, M., Blasco, F., and Field, C., 1997. *World Mangrove Atlas*. Okinawa, Japan: International Society for Mangrove Ecosystems.
Valiela, I., Peckol, P., D'Avanzo, C., Kremer, J., Hersh, D., Foreman, K., Lajtha, K., Seely, B., Geyer, W.R., Isaji, T., and Crawford, R., 1998. Ecological effects of major storms on coastal watersheds and coastal waters: Hurricane Bob on Cape Cod. *Journal of Coastal Research*, **14**: 218–238.
Visser, J.M., Sasser, C.E., Chabreck, R.H., and Linscombe, R.G., 1998. Marsh vegetation types of the Mississippi River deltaic plain. *Estuaries*, **21**: 818–828.

Cross-references

Coastal Zone Management
Conservation of Coastal Sites
Deltaic Ecology
Deltas
Greenhouse Effect and Global Warming
Mangroves, Coastal Ecology
Salt Marsh
Sea-Level Rise, Effect
Storm Surge
Vegetated Coasts
Wetlands Restoration

WETLAND RESTORATION

Wetlands are usually thought of as vegetated habitat on land that is flooded for at least part of each year. However, specific definitions vary. For example, the Canadian Wetland Registry defines wetlands as "land having the water table at, near, or above the land surface or which is saturated for a long enough period to promote wetland or aquatic processes as indicated by hydric soils, hydrophytic vegetation, and various kinds of biological activity which are adapted to the wet environment" (Mitsch and Gosselink, 1993, p. 26). In contrast, the Convention on Wetlands of International Importance Especially as Waterfowl Habitat (also known as the Ramsar Convention) defines wetlands as "areas of marsh, fen, peatland or water, whether natural or artificial, permanent or temporary, with water that is static or flowing, fresh, brackish, or salt including areas of marine water, the depth of which at low tide does not exceed 6 meters" (Mitsch and Gosselink, 1993, p. 26). While the first definition includes only vegetated habitats, the second definition includes nonvegetated habitats such as mudflats and shallow reefs. In a narrow sense, then, wetlands include salt marshes, mangrove forests, swamps, bogs, fens, marshes, muskegs, playas, moors, vernal pools, wet meadows, and similar areas, but in a broader sense wetlands also include sea grass beds, mudflats, coral reefs, shallow lakes, and shallow rivers (see bogs, mangroves, salt marsh, and wetlands).

Restoration can be defined as "the process of repairing damage caused by humans to the diversity and dynamics of indigenous ecosystems" (Jackson *et al.*, 1995, p. 71). Thus, wetland restoration is the repair of anthropogenic damage to habitat that is saturated or flooded by shallow water for at least part of the year. Related terms include construction, creation, enhancement, mitigation, and rehabilitation (Table W7). Approaches to restoration include activities such as planting hydrophytes, control of invasive plants, grazing control, removal of fill, removal of structures that restrict water flow, and placement of dredged material to provide elevations suitable for wetland establishment.

Why restore wetlands?

In some areas, such as California in the United States and Java in Indonesia, more than 90% of the historic area of wetlands has been lost (Dahl, 1990; Crisman and Streever, 1996). Although estimates are influenced by many factors, most estimates suggest that about 50% of the world's wetlands have disappeared in the past few hundred years. Recognition of these losses, along with recognition of the economic and ecological value of wetlands and a general increase in environmental awareness, has inspired interest in wetland restoration. This interest is reflected in community action, incentive measures, and laws that require restoration under some circumstances.

Community action is often catalyzed by the efforts of nonprofit organizations, including relatively small organizations that operate locally or within a single country, such as the Yadfon ("Raindrop") Association in Thailand, and larger organizations that operate internationally, such as the Mangrove Action Plan and Wetlands International (Quarto, 1999). In some cases, community action manifests itself in the efforts of individuals undertaking restoration on their own property. For example, Kym Denver and his family have made extensive efforts to restore wetlands on their 1,200-ha farm near the mouth of the Murray River in South Australia (Denver, 1999) (Figure W31).

In some nations, government-backed incentive plans provide the impetus for restoration. These incentive plans usually involve direct payments or tax relief for landowners participating in restoration programs. For example, in southwest England the Environment Agency has paid about £450 per ha per year to landowners for participation in the Levels and Moors Strategy, which seeks to restore and maintain wetland habitat (Jenkins and Sturdy, 1999). Similarly, the Wetland Reserve Program in the United States pays farmers to restore wetlands on marginal croplands (Gaddis and Cubbage, 1998). For government projects, policies, and funding packages provide incentives for restoration. Examples include the Commonwealth Wetlands Policy in Australia, which encourages government agencies to consider restoration opportunities, and the Coastal Wetlands Planning, Protection, and Restoration Act in the United States, which provides millions of dollars for restoration of wetlands. At the international level, Resolution VII.17 of the Ramsar Convention encourages signatory nations to incorporate wetland restoration as an element of national planning.

Some nations also have laws that can require wetland restoration to compensate (or "mitigate") for destruction of wetlands that occurs as the result of development. In New South Wales, Australia, *State Environmental Planning Policy Number 14—Coastal Wetlands* can require restoration as compensation for the destruction of wetlands. Similarly, Section 404 of the Clean Water Act in the United States can require restoration to compensate for wetland losses caused by filling of wetlands.

Approaches to wetland restoration

There are dozens of approaches to wetland restoration, ranging from planting hydrophytic vegetation to constructing substantial earthworks to using innovations that set the stage for natural processes to restore wetlands. Often, different approaches are used together, in a single project. In all cases, restoration should begin with consideration of the impacts that caused degradation followed by development of specific objectives for the restoration project. In addition to ecological knowledge, successful restoration requires an understanding of hydrology and engineering as well as coordination among stakeholders that may include landowners, funding sources, nonprofit organizations, government agencies, and the local community. Although many handbooks provide guidelines for restoration, these guidelines are neither universally applicable nor definitive—every restoration project is to some degree unique, and every project requires some level of innovation and creativity.

One approach to restoration involves little more than planting. In the tropics, thousands of acres of mangroves have been planted, leading to

Table W7 Definitions of terms related to wetland restoration

Constructed wetland: Increasingly used to refer to wastewater or stormwater treatment wetlands, but occasionally used interchangeably with "created wetland."
Created wetland: Usually refers to a wetland built in an area that did not previously support a wetland.
Enhancement: An attempt to improve a specific wetland function, such as the ability of the wetland to support ducks.
Mitigation: Literally refers to the reduction of some impact, but is commonly used to describe a wetland that was restored or created to replace a wetland destroyed by development, especially under Section 404 of the US Clean Water Act.
Rehabilitation: Sometimes used to refer collectively to restoration, creation, and enhancement, but also used to refer to improving aesthetic qualities of a wetland.

Figure W31 Australian Prime Minister John Howard (right) awards Kym Denver the 1998 National Landcare Primary Producer Award, in recognition of Denver's efforts to restore wetlands on his family's property in South Australia (from Streever, 1999; with permission of Kluwer Academic Publishers).

reestablishment of mangrove forest that had been cut for production of charcoal or for building materials (see mangrove restoration). When vegetation has been removed from a wetland but hydrological conditions remain unchanged, planting may be all that is required to meet restoration objectives. Often, however, construction of dikes or other activities change hydrological conditions, rendering planting alone insufficient as a means of restoration.

If hydrological conditions have been changed, restoration should address the hydrological change. Complete restoration of hydrology is not always possible because of constraints related to urban infrastructure, flood control, and other factors, but partial restoration of hydrology can contribute substantially to wetland restoration (Streever, 1998). For example, industrial and agricultural development has significantly affected the entire Hunter River Estuary in New South Wales, Australia, but replacement of small-diameter culverts with bridges in selected locations has led to increased tidal flushing and subsequent increases in wetland area (Streever and Genders, 1997). Similarly, in Florida, USA, dikes constructed as part of a mosquito-control program led to severe degradation of wetland habitat, but breaching of dikes to allow some tidal exchange led to rapid recovery of vegetation and fish populations (Brockmeyer *et al.*, 1997). However, even if hydrological conditions can be restored, impacts resulting from drainage may be difficult to overcome. In some coastal areas that have been temporarily drained for farming or other purposes, oxidation and subsidence can lower elevations or oxidation of naturally occurring pyrite can result in soil acidification so severe that substrates will not support plants even after the land is reflooded (White *et al.*, 1997).

Removal of fill has been undertaken at several locations to restore wetlands. In the Gog-Le-Hi-Te Wetland, in the Puyallup River estuary in Washington, USA, removal of about 55,000 m^3 of dredged river sediments and solid waste contributed to partial restoration of a 2.2-ha area (Simenstad and Thom, 1996). Other examples include the Tourle Street Bridge site near Newcastle, New South Wales, Australia, where slag from the iron-smelting industry was removed to restore intertidal wetlands (Day *et al.*, 1999), the Sweetwater River Wetlands Complex near San Diego, California, USA, where fill was removed to mitigate for wetland losses resulting from highway construction (Pacific Estuarine Research Laboratory, 1990), and numerous sites on the Alaskan North Slope, where gravel fill has been removed to promote recovery of tundra near the Beaufort Sea (Jorgenson and Joyce, 1994) (Figure W32). In some cases, natural habitat is excavated to create wetlands. For example, on the Grand Bay-Bangs Lake Estuary in Mississippi, USA, about 10 ha of salt marsh was created by excavating sediment from a pine forest (LaSalle, 1996).

In contrast to fill removal for wetland restoration, sediment dredged from navigation channels is sometimes placed in shallow water to restore or create vegetated wetlands (Streever, 2000). Typically, hydraulic dredges remove sediment from a navigation channel and pump it through a pipeline to the wetland project site. Fine sediments must be contained within earthen berms or other structures to allow consolidation, while sandy sediments can be placed on shallow bay bottoms without confining structures. In the 1970s, when technology for dredged material wetlands was in the early stages of development, success was typically defined by establishment of vegetation, but in recent years increased emphasis has been placed on maximizing all aspects of similarity between natural and dredged material wetlands. For example, in the Atchafalaya Delta, Louisiana, USA, attempts have been made to build dredged material marshes with geomorphology similar to that of naturally accreting deltaic marshes (Figure W33).

Restoration sometimes requires removal of contaminants, including contaminants from oil spills and industrial development. In some cases, contaminants enter groundwater. For example, the Jauá Lake wetlands, a system of freshwater wetlands sitting between an extensive stretch of sand dunes and the Atlantic Ocean in Bahia, Brazil, were contaminated when domestic and industrial waste infiltrated into groundwater, and restoration required removal of contaminated groundwater (da Silva *et al.*, 1999) (Figure W34).

Figure W32 Bill Streever points to peat exposed by removing gravel from an abandoned work pad near the Beaufort Sea on the North Slope of Alaska. Gravel removal is the first step in tundra rehabilitation.

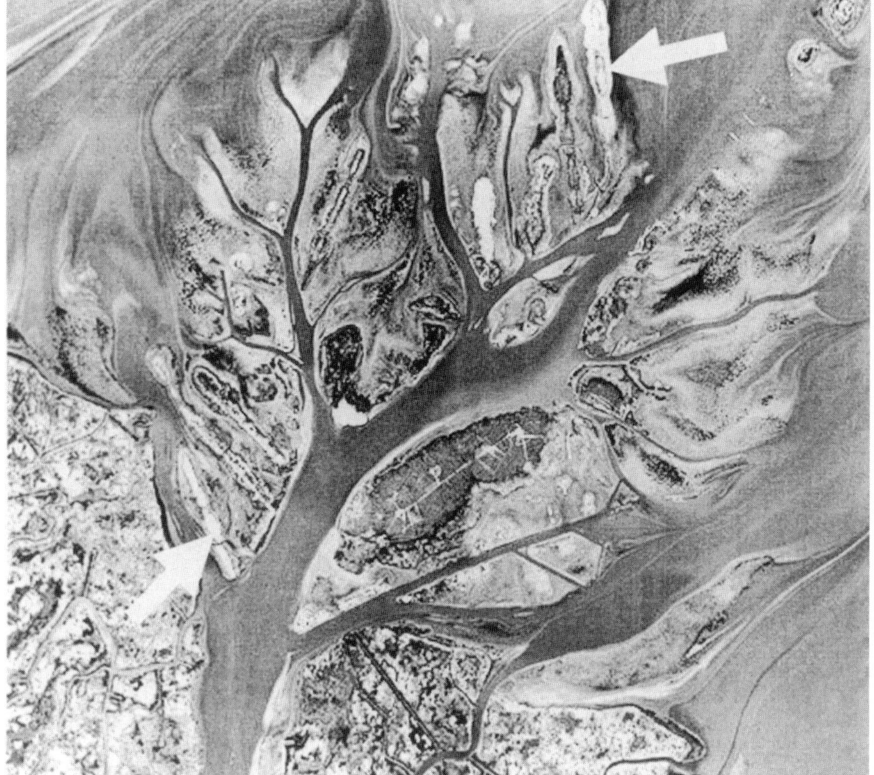

Figure W33 Arrows point to recently created dredged material wetlands built to mimic the natural geomorphology of the Atchafalaya Delta, Louisiana, USA.

For large-scale restoration, costs can prohibit extensive on-the-ground activities. This reality has prompted development of methods that rely on natural processes to restore wetlands. For example, in Louisiana, USA, natural levees have been intentionally breached to facilitate deposition of Mississippi River sediments on shallow bay bottoms, leading to development of vegetated wetlands at costs as low as US$54 per ha by mimicking the natural delta-building process of crevasse-splay formation (Turner and Boyer, 1997) (Figure W35).

Key research issues

As the number of restoration projects has increased over the past few decades, the amount of research on restoration has also increased, but the majority of this research has been recorded in reports with limited distribution. Most research consists of site-specific attempts to track wetland development, but some research has attempted to assess restoration efforts by looking at collections (or "populations") of restored wetlands. Most research, whether it is site-specific or more

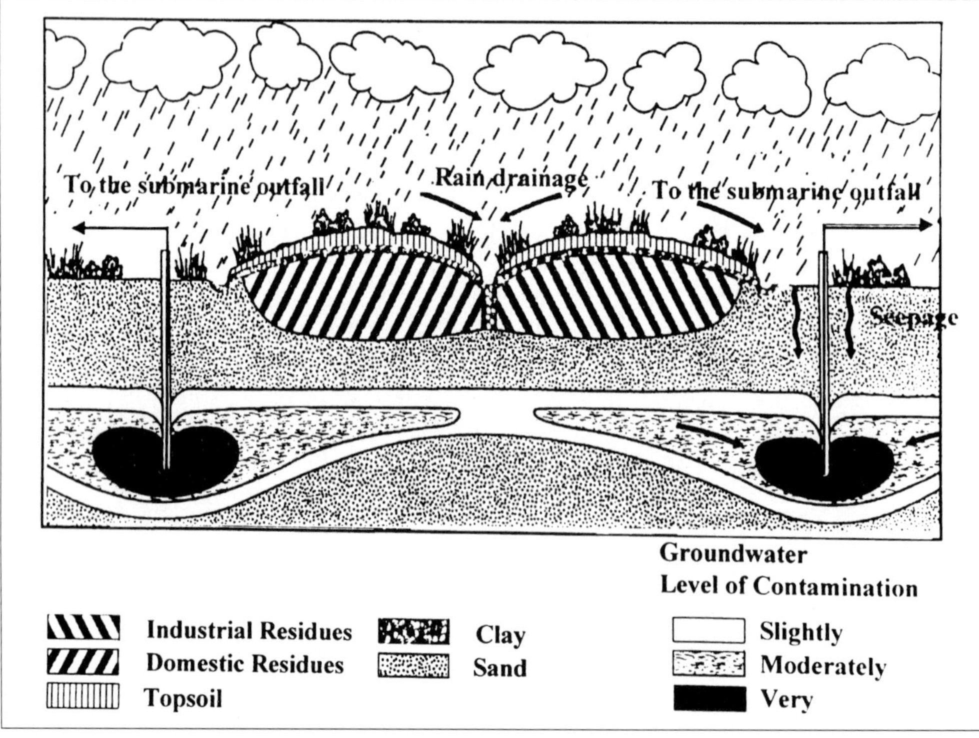

Figure W34 The Jauá Lake wetlands in Brazil (top) were degraded by spills that contaminated the groundwater, leading to a need for restoration that included groundwater withdrawal (bottom) (from Streever, 1999; with permission of Kluwer Academic Publishers).

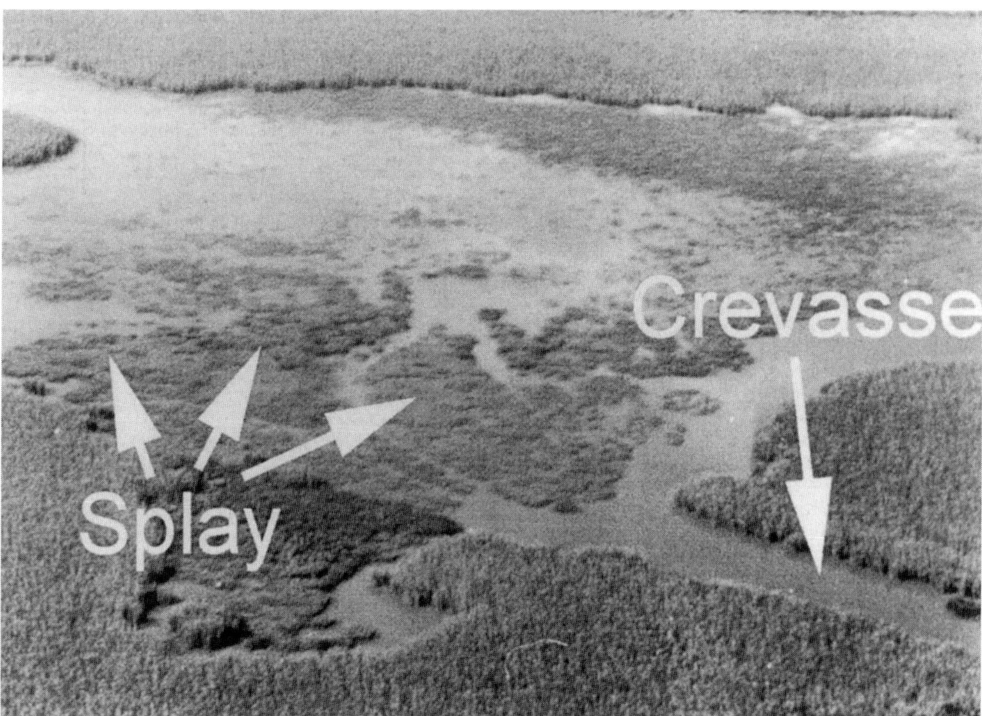

Figure W35 Crevasse splay restoration project near the mouth of the Mississippi River.

broad, addresses two questions. First, how do restored wetlands compare to natural wetlands? And second, do restored wetlands follow a developmental performance curve, or trajectory, leading to increased similarity with natural wetlands over time? In addition to addressing these questions, restoration research has contributed to improved methods of plant establishment, development of engineering alternatives for shore protection, a better understanding of relationships among physical and biological components of wetlands, and other topics. Despite the amount of research that has been completed, useful broad generalizations about wetland restoration remain elusive.

Studies comparing natural and restored wetlands often suffer from pseudoreplication (in the sense of Hurlbert (1984)), in which conclusions about all natural and restored wetlands are generalized from data based on replicate samples collected from a single natural reference wetland and a single restored wetland. However, over time data have accumulated from hundreds of studies comparing pairs of natural and restored wetlands. Cumulative results from these studies and from studies of populations of wetlands clearly indicate that restored wetlands often do not have all of the same structural and functional attributes found in nearby natural wetlands, even many years after restoration efforts have been completed. That is, some attributes of restored wetlands, such as percentage cover by vegetation and abundance of invertebrates, may be indistinguishable from those of natural reference wetlands, but other attributes of restored wetlands, such as soil organic content and rates of nutrient cycling, may be clearly different from those of natural reference wetlands. Furthermore, some differences may persist indefinitely. Zedler (1999) noted that similarity between restored and natural wetlands depends on the degree of degradation at the beginning of the restoration project as well as the degree of effort invested in restoration.

When variables such as abundance of invertebrates or organic content in soils are plotted against time, the resulting curve is sometimes called a "performance curve" or "trajectory." However, because the term "trajectory" implies a specific type of curve (i.e., the curve described by a projectile moving through space, or a curve that passes through a set of points at a constant angle), the phrase "performance curve" may be more appropriate for describing plots that illustrate wetland development over time. Performance curves can be plotted in at least two ways: as measured values plotted against time, or as relativized values plotted against time. Relativized values are measured values that have been standardized against values found at a natural reference site or sites, usually by dividing the measured value from the restoration site by the measured value of the natural reference site or sites. Relativized values help account for intra- and inter-annual variability in measured values. Because there are few long-term data sets for individual restored wetlands, some researchers have endeavored to analyze data from multiple restored wetlands of different ages, making it possible to plot performance curves across time but adding the confounding factor of site-to-site variability. Simenstad and Thom (1996) described five types of performance curves, including curves that (1) converge with the mean found for reference wetlands, (2) approach and surpass the mean found for reference wetlands, (3) develop toward the mean found for reference wetlands but then stabilize at a level lower than that of reference wetlands, (4) follow a sigmoidal trend that reaches an asymptote at the mean found for reference wetlands, and (5) decline from a level higher than that of reference wetlands to a level below the mean of reference wetlands. In keeping with results of studies comparing natural and restored wetlands, currently available research clearly shows that different wetland attributes can have different performance curves even within an individual wetland. For most restored wetlands that have been closely examined, performance curves for at least some attributes have not reached levels found in natural reference wetlands even after many years of development.

The future of wetland restoration

Research illustrating the shortcomings of restoration approaches may eventually lead to a reduced willingness to trade natural wetlands for restored wetlands, as is frequently allowed under legislation such as Section 404 of the Clean Water Act in the United States. However, wetland restoration has a short history, and current trends indicate that interest in restoring wetlands will continue to increase dramatically around the world. Research and experience will continue to contribute to improvements in restoration approaches, and innovative techniques are likely to reduce costs per area. The scale of some restoration efforts is also likely to increase, as exemplified by efforts to restore parts of the Florida Everglades and Louisiana's coastal wetlands in the United States. Techniques for wetland restoration used in wealthy nations are increasingly exported and adapted for use in the developing world. Although it is unlikely that restoration will ever lead to complete replacement of wetlands destroyed over the past two centuries, it is becoming increasingly likely that restoration efforts will complement preservation efforts and one day stem the ongoing loss of wetlands around the globe.

William Streever

Bibliography

Brockmeyer, R.E., Jr., Rey, J.R., Virnstein, R.W., Gilmore, R.G., and Earnest, L., 1997. Rehabilitation of impounded estuarine wetlands by hydrological reconnection to the Indian River Lagoon, Florida (USA). *Wetlands Ecology and Management*, **4**: 93–109.

Crisman, T.L., and Streever, W.J., 1996. The legacy and future of tropical limnology. In Schiemer, F., and Boland, K.T. (eds.), *Perspectives in Tropical Limnology*, Amsterdam: SPB Academic Publishing, pp. 27–42.

da Silva, E.M., Barros, A.F., Navarro, M.F.T., Mota, M.F.V., Cotsifis, P., and Chastinet, C.B.A., 1999. Rehabilitation following industrial contamination: Jauá Lake, a coastal wetland in Camaçari, Bahia, Brazil. In Streever, W.J. (ed.), *An International Perspective on Wetland Rehabilitation*. Dordrecht, The Netherlands: Kluwer Academic Publishers, pp. 197–203.

Dahl, T.E., 1990. *Wetlands Losses in the United States, 1780s to 1980s*. Washington, DC: US Department of the Interior, Fish and Wildlife Service.

Day, S., Streever, W.J., and Watts, J.J., 1999. An experimental assessment of slag as a substrate for mangrove rehabilitation, *Restoration Ecology*, **7**: 139–144.

Denver, K., 1999. Rehabilitating Wyndgate: bringing back wetlands on a family property in South Australia. In Streever, W.J. (ed.), *An International Perspective on Wetland Rehabilitation*. Dordrecht, The Netherlands: Kluwer Academic Publishers, pp. 107–111.

Gaddis, D.A., and Cubbage, F.W., 1998. Wetlands regulation: development and current practices. In Messina, M.G., and Conner, W.H. (eds.), *Southern Forested Wetlands; Ecology and Management*. Boca Raton, Florida: Lewis Publishers, pp. 49–84.

Hurlbert, S.H., 1984. Pseudoreplication and the design of ecological field experiments. *Ecological Monographs*, **54**: 187–211.

Jackson, L.L., Loupoukhine, N., and Hillyard, D., 1995. Ecological restoration: a definition and comments. *Restoration Ecology*, **3**: 71–75.

Jenkins, A.L., and Sturdy, I.D., 1999. Putting policy into practice in South West England. In Streever, W.J. (ed.), *An International Perspective on Wetland Rehabilitation*. Dordrecht, The Netherlands: Kluwer Academic Publishers, pp. 81–89.

Jorgenson, M.T., and Joyce, M.R., 1994. Six strategies for rehabilitating land disturbed by oil development in arctic Alaska. *Arctic*, **47**: 374–390.

LaSalle, M.W., 1996. *Assessing the Functional Level of a Constructed Intertidal Marsh in Mississippi*. (Wetlands Research Program Technical Report WRP-RE-15). Vicksburg, MS: Waterways Experiment Station.

Mitsch, W.J., and Gosselink, J.G., 1993. *Wetlands*. New York: Van Nostrand Reinhold.

Pacific Estuarine Research Laboratory, 1990. *A Manual for Assessing Restored and Natural Coastal Wetlands with Examples from Southern California*. (California Sea Grant Report T-CSGCP-021). La Jolla, California: California Sea Grant College.

Quarto, A., 1999. Local community involvement in mangrove rehabilitation: Thailand's Yadfon. In Streever, W.J. (ed.), *An International Perspective on Wetland Rehabilitation*. Dordrecht, The Netherlands: Kluwer Academic Publishers, pp. 139–142.

Simenstad, C.A., and Thom, R.M., 1996. Functional equivalency trajectories of the restored Gog-Le-Hi-Te estuarine wetland. *Ecological Applications*, **6**: 38–56.

Streever, W.J., 1998. Kooragang wetland rehabilitation project: opportunities and constraints in an urban wetland rehabilitation project. *Urban Ecosystems*, **2**: 205–218.

Streever, W.J. (ed.), 1999. *An International Perspective on Wetland Rehabilitation*. Dordrecht: Kluwer.

Streever, W.J., 2000. *Spartina alterniflora* marshes on dredged material: a critical review of the ongoing debate over success. *Wetlands Ecology and Management*, **8**: 295–316.

Streever, W.J., and Genders, A.J., 1997. The effect of improved tidal flushing and competitive interactions at the boundary between salt marsh and pasture. *Estuaries*, **20**: 804–815.

Turner, R.E., and Boyer, M.E., 1997. Mississippi River diversions, coastal wetland restoration/creation and an economy of scale. *Ecological Engineering*, **8**: 117–128.

White, I., Melville, M.D., Wilson, B.P., and Sammut, J., 1997. Reducing acidic discharges from coastal wetlands in eastern Australia. *Wetlands Ecology and Management*, **5**: 55–72.

Zedler, J.B., 1999. The ecological restoration spectrum. In Streever, W.J. (ed.), *An International Perspective on Wetland Rehabilitation*, Dordrecht, The Netherlands: Kluwer Academic Publishers, pp. 301–318.

Cross-references

Bogs
Human Impacts on Coasts
Hydrology of Coastal Zone
Managed Retreat
Mangroves, Coastal Ecology
Oil spills
Polders
Reclamation
Salt Marsh
Shore Protection Structures
Tidal Creeks
Wetlands

APPENDIX 1: CONVERSION TABLES

Metric to English Units—Equivalents of Length
1 micron (μ) = 0.001 millimeter (mm) = 0.00004 inch (in.)
1 mm = 0.1 centimeter (cm) = 0.03937 in.
1000 mm = 100 cm = 1 meter (m) = 39.37 in. = 3.2808 feet (ft)
1 m = 0.001 kilometer (km) = 1.0936 yard (yd)
1000 m = 1 km = 0.62137 mile (mi)
12 in. = 1 ft = 0.3048 m
1 cm = 0.39370 in. = 0.032808 ft
1 km = 10^5 cm = 0.62137 mi
1 fathom = 6 ft = 1.8288 m
1 nautical mile = 1.85325 km
1 in. = 2.54001 cm
1 ft = 30.480 cm
1 statute mile = 1.60935 km = 5280 ft
1 astronomical unit = 1.496×10^8 km = 92,957,000 mi
1 light year = 9.460×10^{12} km = 5.878×10^{12} mi
1 parsec = 3.085×10^{13} km = 1.917×10^{13} mi

Square Measures	
1 square foot = 0.00002295684 acre = 929.0 cm^2	
1 acre = 43,560 ft^2 = 0.0015625 mi^2	
1 yd^2 = 0.836127 m^2	
1 hectare = 2.471054 acre	
1 mi^2 (statute) = 640 acres = 2.5900 km^2	
1 cm^2 = 0.1550 $in.^2$ = 0.0010764 ft^2	
1 km^2 = 10^{10} cm^2 = 0.3861 mi^2	
1 mm^2 = 0.00155 $in.^2$	1 $in.^2$ = 6.452 cm^2
1 m^2 = 10.764 ft^2	1 ft^2 = 0.09290 m^2
1 km^2 = 0.3861 mi^2	1 mi^2 = 2.5900 km^2

Cubic Measures

1 gal (UK) = 4.5461 liters = 1.200956 gal (US)
1 liter = 0.219969 gal (UK) = 0.264173 gal (US)
1 gal (US) = 3.7854 liters = 0.832670 gal (UK)

1 cc = 0.0610 cu. in. = 0.000035314 cu. ft
1 cu in. = 16.387 cc
1 cu ft = 28317 cc

1 mm^3 = 0.000061 $in.^3$	1 $in.^3$ = 16.387 cm^3 (cc)
1 cm^3 (cc) = 0.0610 $in.^3$	1 ft^3 = 0.028317 m^3
1 m^3 = 35.315 ft^3	1 mi^3 = 4.1681 km^3
1 km^3 = 0.239911 mi^3	

Statute Miles to Nautical Miles to Kilometers

Statute	Nautical	Kilometers	Statute	Nautical	Kilometers
1/4	0.22	0.40	9	7.82	14.48
1/2	0.43	0.80	10	8.68	16.10
3/4	0.65	1.21	20	17.36	32.20
1	0.87	1.61	30	26.05	48.30
2	1.74	3.22	40	34.74	64.35
3	2.61	4.84	50	43.42	80.45
4	3.48	6.45	60	52.10	96.55
5	4.35	8.05	70	61.00	113.00
6	5.22	9.65	80	69.60	129.00
7	6.08	11.27	90	78.16	145.00
8	6.96	12.90	100	87.00	161.00

Fathoms to Feet to Meters

Fathoms	Feet	Meters	Fathoms	Feet	Meters
1/4	1.5	0.5	6 1/2	39.0	11.9
1/2	3.0	0.9	6 3/4	40.5	12.3
3/4	4.5	1.4	7	42.0	12.8
1	6.0	1.8	8	48.0	14.6
1 1/4	7.5	2.3	9	54.0	16.5
1 1/2	9.0	2.7	10	60.0	18.3
1 3/4	10.5	3.2	11	66.0	20.1
2	12.0	3.7	12	72.0	21.9
2 1/4	13.5	4.1	13	78.0	23.8
2 1/2	15.0	4.6	14	84.0	25.6
2 3/4	16.5	5.0	15	90.0	27.4
3	18.0	5.5	16	96.0	29.3
3 1/4	19.5	5.9	17	102.0	31.1
3 1/2	21.0	6.4	18	108.0	32.9
3 3/4	22.5	6.9	19	114.0	34.7
4	24.0	7.3	20	120.0	36.6
4 1/4	25.5	7.8	30	180.0	54.9
4 1/2	27.0	8.2	40	240.0	73.2
4 3/4	28.5	8.7	50	300.0	91.4
5	30.0	9.1	60	360.0	109.7
5 1/4	31.5	9.6	70	420.0	128.0
5 1/2	33.0	10.1	80	480.0	146.3
5 3/4	34.5	10.5	90	540.0	164.6
6	36.0	11.0	100	600.0	182.9
6 1/4	37.5	11.4			

APPENDIX 2: JOURNALS

Professional journals, periodicals, gazetteers, government publications, and trade magazines are a primary means of communicating information among coastal scientists. These information sources are usually provided to users as hard copy on paper stock, but there is an increasing trend toward dual publication with electronic versions becoming more widely available. Consideration of the topic in terms of content, availability, and cost is, however, not as simple as it might first appear. There have been many recent changes in the media of communication. Most senior researchers are familiar with paper copy volumes that were searched via perusal of tables of contents in various research publications (e.g., *Current Contents*) or by tracking down references cited in research papers. Prior to the last decade, journals were generally inexpensive enough for most professionals to subscribe on a personal basis. Many researchers built up a personal research library comprised by books and journals, the latter supplying the most recent cutting edge information. These private libraries were often conveniently located at a university or home office. Notes kept on index cards or bookmarks often signaled important articles or passages in journals. The coastal researcher was conversant with the literature and it often was mostly at hand for immediate perusal in the office or research laboratory.

Sometimes professional organizations publish lists of citations in specialized fields for specific time frames, as a means to facilitate or expedite the compilation and perusal of the coastal literature. Examples include supplements to the *Annotated Bibliography of Quaternary Shorelines* published by the Academy of Natural Sciences of Philadelphia (e.g., Special Publication No. 6, 1965; Special Publication No., 10, 1970; Special Publication No. 11, 1974), GeoAbstracts (East Anglia, UK) (*Third Supplement*, 1979), and *Journal of Coastal Research* (*Fourth Supplement*, 1986). The *International Bibliography of Coastal Geomorphology* (Sherman, 1992; Kelletat, 1996), published by the *Journal of Coastal Research* and the International Geographical Union (Commission on Coastal Systems), is another example of efforts to summarize a great literature of coastal research in comprehensive bibliographic lists for specialized fields. Such efforts are noteworthy in themselves because they are not common. When available in electronic format on a CD, for example, the search capabilities are enormously increased and the research value of the compendium is substantially increased over paper copy.

The convenience and familiarity of yesteryear is but a memory of the past. Research procedures and access to information was changed forever with the advent of the modern personal computer. With the development of online databases and electronic libraries, it became possible to conduct global searches via the Internet without leaving the office. With advances in modern technology came many innovations that assisted the document search process, but there were also downsides. With declining numbers of subscribers, journal costs increase to the point where it becomes impossible for individuals to maintain complete personal libraries of hard (paper) copies. Publishers, responding to declining subscriberships thus offer electronic copies of individual articles or entire issues. Most journals now provide access to free abstracts of papers in an effort to encourage purchase of the article for a royalty or access fee. There are also service bureaus and search services that provide electronic summaries of search results that list papers in a wide range of journals. Not only has the shelf space in the professorial office shrunk, but libraries as well often do not subscribe to hard copies of journals because they rely on a central source for interlibrary loan or use electronic services to order requested papers. This is an increasingly electronic age where much research is conducted in front of a computer screen.

There are pros and cons in these new venues for dissemination of information related to coastal research. There are many physical and psychological disadvantages to spending long hours in front of computer screens. Back radiation from monitors, eyestrain, carpal tunnel syndrome, and lack of physical exercise are but a few of the drawbacks. Positive aspects focus on access to vast amounts of information not heretofore possible to the average person. But, again there are pros and cons to the flood of data offered by computer searches. Search results require much sifting of data for relevancy and computer searchers often miss critical, non-mainline sources of information. It is a trade-off, computer searchers versus perusal of personal hard copies or visits to the stacks in libraries.

Journals, the basic resource for armchair research, cover all aspects of coastal research but inclusion of all relevant or pertinent journals becomes problematic depending on the definition of coastal research (see Introduction; Management) and specific fields of interest. Even more basic is consideration of what is "coastal" from a definitional point of view, at least as far as reporting research results is concerned. Greater diffusivity of the concept of what constitutes a *coastal topic* relates to viewpoints, perspectives, and orientation of study problems. The focus or content of journals is relatively clear-cut for the basic sciences (e.g., biology, geology) but becomes blurred in applications (e.g., coastal engineering, ecology, geophysics, geochemistry, geotechnique, biogeochemistry, hydrophysics, hydrochemistry, hydrobiology, sedimentology, remote sensing), and extremely broad ranging in the management subfields (e.g., coastal environments; marine pollution; use of maritime and marine resources; legal, political, social, and economic analyses). Although studies in the pure sciences (e.g., physics, chemistry) can be very specific, coastal research is mostly conducted from the purview of some application or for greater understanding of coastal shape or configuration due to wave action (e.g., coastal morphodynamics). The scope of coastal science is thus potentially vast and the possibilities of research materials are almost limitless. As an example of the wide range of source materials for coastal research, Table A2.1 is the result of a partial and edited printout of holdings in the coastal library at NOAA's Coastal Services Center (Charleston, South Carolina). This eclectic assemblage of journals represents one perspective of what constitutes a useful public research library in the coastal sciences. Even though data fields in the compilation are sometimes incomplete, the list illustrates a broad interest base for a national coastal service center. Fortunately, the central core of coastal science clumps together lines of research that focus on biological and geological principles, engineering practices, and to a lesser extent various management issues and mathematical modeling. The objectives, goals, and interests of the supporting organization thus restrain the universe of serials retained in any library devoted to coastal research. If nothing else, the list in Table A2.1 emphasizes the wide range of subject matter that must be considered in a government service center library. Some holdings in the library are mainline professional journals, as listed in Table A2.2,

Table A2.1 List of some journal holdings by the Coastal Services Center (NOAA, National Ocean Service, Charleston, South Carolina), illustrating the diverse range of information sources that are required to support coastal research interests

Journal title	ISSN, Website or e-mail contact	Frequency of publication	Publisher
Alabama NPS Newsletter	jmiddle@trojan.troyst.edu[1]	Quarterly	Troy State University
Ambio	0044-7447	Monthly	New York: Pergamon Press, Inc.
Amicus Journal	0276-7201	Quarterly	Washington, DC: National Resources Defense Council
ARC News	1064-6108	Quarterly	Redlands, California: Environmental Systems Research Institute, Inc.
ASFPM News & Views	larry@floods.org	Bi-monthly	Association of State Floodplain Managers
Backscatter	1205-6766	Tiannual	Bedford, Nova Scotia, Canada: Atlantic Center for Remote Sensing of Oceans
Bulletin of Marine Science	0007-4977	Bi-monthly	Miami, Florida: University of Miami
Calypso Log	CL-8756-6354	Bi-monthly	Chesapeake, Virginia: The Cousteau Society, Inc.
Climatological Data	0027-0296	Monthly	Washington, DC: NOAA
Coastal Connection	Not available[2]	Semi-annually	Washington, DC: Center for Marine Conservation
Coastal Guardian	Not available	Bi-monthly	Charleston, South Carolina: South Carolina Coastal League
Coastal Heritage	annette.dunmeyer@scseagrant.org	Quarterly	Charleston, South Carolina Sea Grant Consortium
Coastal Management	0892-0753	Quarterly	Washington, DC: Taylor & Francis
Coastal Review	nccf@nccoast.org	Quarterly	Newport, North Carolina: North Carolina Coastal Federation
Coastal Zone Management	Not available	Bi-monthly	Washington, DC: Nautilus Press, Inc.
Coastal Zone Management Journal	0892-0753	Quarterly	Washington, DC: Taylor & Francis, Inc.
Coastlines	ed_coast@horsleywitten.com	Quarterly	Barnstable, Massachusetts: Horsley & Witten, Inc.
Communicator: Official Publication for Communicators of the National Sea Grant program	Not available	Quarterly	Charleston, South Carolina: South Carolina Sea Grant Consortium
Conservation Issues	Not available	Bi-monthly	Washington, DC: World Wildlife Foundation
Continental Shelf Research	0278-4343	Monthly	New York: Elsevier
Coral Reefs: Journal of the International Society for Reef Studies	0722-4028	Quarterly	New York: Springer-Verlag
Corporate Meetings & Incentives	0745-1636	Monthly	Maynard, Massachusetts: Adams Laux Co.
Current: The Journal of Marine Education	0889-5546	Monthly	Pacific Grove, California: National Marine Educators Association (NMEA)
Deep Sea Research Part I	0967-0637	Monthly	New York: Pergamon Press, Inc.
Deep Sea Research Part II. Topical Studies in Oceanography	0967-0645	Monthly	New York: Pergamon Press, Inc.
Dolphin Log	cousteau@cousteausociety.org	Bi-monthly	New York: The Cousteau Society
Environmental Magazine	1046-8021	Bi-monthly	Westport, Connecticut: Earth Action Network, Inc.
Earth Island Journal	1041-0406	Quarterly	San Francisco, California: Earth Island Institute
Earthviews	eisinfo@unep.org	Quarterly	New York: United National Environment Program
Ecologist	0261-3131	Bi-monthly	Sturminster Newton, Dorset, United Kingdom: Ecosystems Limited
Economist	0013-0613	Bi-monthly	London: The Economist Newspaper Limited
Ecosystem Health	1076-2825	Quarterly	Cambridge, Massachusetts: Blackwell Science, Inc.
EOM	editorial@eomonline.com	Monthly	EOM, Inc.
EPA Watershed Events	840-N-00-002	Quarterly	Washington, DC: Environmental Protection Agency
Estuaries	0160-8347	Quarterly	Port Republic, Maryland: Estuarine Research Federation
Estuarine, Coastal and Shelf Science	0272-7714	Quarterly	Sidcup, Kent, England: Harcourt Brace & Co.
Estuarine Research Federation Newsletter	webmaster@erf.org	Quarterly	Port Republic, Maryland: Estuarine Research Federation
Federal Computer Week	0893-052X	Weekly	Falls Church, Virginia: FCW Group
Federal Employees News Digest	1065-0970	Weekly	Reston, Virginia: Federal Employees News Digest, Inc.

Fish & Wildlife Reference Service Newsletter	kris_lamontagne@fws.gov	Quarterly	Washington, DC: US Fish and Wildlife Service
Fisheries	0363-2415	Monthly	Bethesda, Maryland: American Fisheries Society
Fishery Bulletin	0090-0656	Bi-monthly	Silver Spring, Maryland: NOAA National Marine Fisheries Service
Geo Info Systems	1015-9858	Monthly	Riverton, New Jersey: ADVANSTAR
GeoTimes	0016-8556	Monthly	Alexandria, Virginia: American Geological Institute
Georgia Sound	Not available	Quarterly	Georgia Department of Natural Resources
GIS World	0897-5507	Monthly	Fort Collins, Colorado: GIS World, Inc.
Global Change	editor@globalchange.org	Quarterly	Pacific Institute for Studies in Development, Environment, and Security
Governing	0894-3842	Monthly	Washington, DC: Congressional Quarterly, Inc.
GPS World	1048-5104	Monthly	Cleveland, Ohio: Advanstar Communications
Great Lakes United Newsletter	Not available	Monthly	Buffalo, New York: Great Lakes Limited, Buffalo State College
Gulf Estuarine Research Society	cynthia.moncreiff@usm.edu	Quarterly	Ocean Springer, Mississippi: Gulf Estuarine Research Society
Hourly Precipitation Data	0364-636X	Monthly	Nashville, North Carolina: NOAA
IEEE Transactions on Geoscience and Remote Sensing	0196-2892	Monthly	New York: IEEE
Intercoast Network	noelle@gso.uri.edu	Quarterly	Narragansett, Rhode Island University of Rhode Island Coastal Research Center
International Journal of Geographical Information Systems	1365-8816	Bi-monthly	Bristol, Pennsylvania: Taylor & Francis
International Journal of Remote Sensing	0143-1161	Quarterly	Bristol, Pennsylvania: Taylor & Francis
Interstate Certified Shellfish Shippers List	0364-7048	Monthly	US Food and Drug Administration
ISPRS	0924-2716	Monthly	New York: Elsevier
Journal of Coastal Research	0749-0208	Quarterly	West Palm Beach, Florida: Coastal Education & Research Foundation (CERF)
Journal of Geophysical Research: Oceans	0148-0227	Monthly	Washington, DC: American Geophysical Union
Journal of Marine Science	0022-2402	Bi-monthly	New Haven, Connecticut: Sears Foundation for Marine Research
Journal of Physical Oceanography	0022-3670	Monthly	Boston, Massachusetts: American Meteorological Society
Journal of Shellfish Research	webmaster@shellfish.org	Bi-monthly	Oxford, Maryland: National Shellfisheries Association
Journal of the American Planning Association	japaweb@pdx.edu	Monthly	Washington DC: American Planning Association
Landlines	seanc@lincolninst.edu	Bi-monthly	Lincoln Institute of Land Policy
Limnology and Oceanography	0024-3590	Monthly	Lawrence, Kansas: American Society of Limnology and Oceanography
Louisiana Coast Lines	crdinfo@dnr.state.la.us	Irregular	Baton Rouge: Louisiana Department of Natural Resources
Marine Conservation News	cmc@dccmc.org	Quarterly	Washington, DC: Center for Marine Conservation
Marine Geodesy	0149-0419	Monthly	Bristol, Pennsylvania: Taylor & Francis
Marine Mammal News	Not available	Monthly	Washington, DC: Nautilus Press
Meeting Manager	8750-7218	Monthly	Dallas, Texas: Meeting Planners International
Meeting News	0145-630X	18 times per year	New York: Miller Freeman, Inc.
Météo France	Not available	Monthly	Paris, France: CNRM
Mote News	Not available	Quarterly	St. Petersburg, Florida: Mote Marine Laboratory
National Waterline	Not available	Monthly	NWRA Publications
Natural Hazards Observer	sylvia.dane@colorado.edu	Bi-monthly	Natural Hazards Center
Nature	0028-0836	Monthly	Riverton, New Jersey: Macmillan Magazines
Naval Meteorological & Oceanographic Command	lammonsg.chl@cnmoc.navy.mil	Bi-monthly	US Navy
NCMC Marine Bulletin	christine@savethefish.org	Quarterly	Leesburg, Virginia: National Coalition for Marine Conservation, Inc.
New Waves	0897-5094	Quarterly	Texas Water Resources Institute, Texas Agricultural Experiment Station

Table A2.1 *(Continued)*

Journal title	ISSN, Website or e-mail contact	Frequency of publication	Publisher
NOAA in the News	Not available	Bi-weekly	NOAA
NOAA Report	Not available	Monthly	NOAA Public Affairs
Nonpoint Source Pollution News-Notes	jtaggart@erols.com	Bi-monthly	Alexandria, Virginia: Terrene Institute
Nor'Easter	allard@gso.uri.edu	Quarterly	Narragansett, Rhode Island: University of Rhode Island Northeast Sea Grant
Ocean Alert	Not available	Quarterly	Earth Island Institute
Ocean and Coastal Management	0964-5691	Quarterly	New York: Elsevier
Ocean News & Technology	1082-6106	Bi-monthly	Palm City, Florida: TSC Holding Group
Oceanography	1042-8275	Monthly	Washington, DC: Oceanography Society
Oceanus	oceanusmag@whoi.edu	Semi-annual	Woods Hole, Massachusetts: WHOI Publication Services
Ohio Coastal Management Program	Not on Website	Quarterly	Ohio Department of Natural Resources
PE&RS: Photogrammetric Engineering and Remote Sensing	0099-1112	Monthly	Bethesda, Maryland: American Society for Photogrammetry and Remote Sensing
People, Land & Water	toni_dufficy@nps.gov	Monthly	Washington, DC: US Department of Interior
Photogrammetry & Remote Sensing	0924-2716	Bi-monthly	Amsterdam: Elsevier
Planning	0001-2610	Monthly	Chicago, Illinois: American Planning Association
Publish!	0897-6007	Monthly	San Francisco, California: IDG, Inc.
Remote Sensing of the Environment	0034-4257	Monthly	New York: Elsevier
Research Technology Management	0895-6308	Bi-monthly	Washington, DC: Industrial Research Institute, Inc.
Runoff Report	Not available	Bi-monthly	Terrene Institute
Science	0036-8075	Semi-weekly	Washington, DC: American Association for the Advancement of Science
Science News	0036-8423	Weekly	Washington, DC: Science News Services, Inc.
Scientific American	0036-8733	Monthly	New York: Scientific American, Inc.
Sea Grant in Brief	gfrederick@ucsd.edu	Bi-monthly	Sea Grant California
Sea Technology	0093-3651	Monthly	Arlington, Virginia: Compass Publications, Inc.
South Carolina River News	farr@water.dnr.state.sc.us	Quarterly	SC River Conservation Program
South Carolina Wildlife	0038-3198	Bi-monthly	Columbia: South Carolina Department of Natural Resources.
Space News	1046-6940	Bi-weekly	Springfield, Virginia: Army Times Publishing Company
Spotlight	Not available	Quarterly	Toulouse, France: SPOT Image Corporation
Storm Data	0039-1972	Monthly	NOAA
Successful Meetings	0148-4052	Quarterly	New York: Bill Communications, Inc.
Sun Expert	1053-9239	Monthly	Brookline, Massachusetts: BPA International
Systems Administration	1061-2688	Monthly	San Francisco, California: Miller Freeman, Inc.
Texas Shores	Phone: (409) 862 3767	Quarterly	College Station, Texas: Texas A&M University Sea Grant
Tidelines	bsga@island.net	Quarterly	BCSGA
Training & Development	1055-9760	Monthly	Alexandria, Virginia: American Society for Training & Development
Trends in Ecology and Evolution	1069-5347	Monthly	New York: Elsevier
Twineline	cruickshank.3@osu.edu	Quarterly	Columbus, Ohio: Ohio State University Sea Grant
UNIX Review	0742-3136	Monthly	San Francisco, California: Miller Freeman, Inc.
Urban Land	0042-0891	Monthly	Washington, DC: Urban Land Institute
URISA	1045-8077	Quarterly	Madison, Wisconsin: URISA
Visual Developer	1053-6205	Bi-monthly	Scottsdale, Arizona: Coriolis Group, Inc.
Volunteer Monitor	Not available	Irregular	San Francisco, California: The Volunteer Monitor
Water Monitor	Not available	Irregular	Water/Office of Wetlands, Oceans and Watersheds
Wetland Journal	0277-5212	Quarterly	Lawrence, Kansas: Society of Wetland Scientists
World Climate News	ipa@gateway.wmo.ch	Semi-annual	World Meteorological Organization

[1] Information concerning publications with no ISSN number can be obtained from the Website or e-mail address noted.

[2] Based on computer searches during the compilation of this table, it was not possible in some cases to locate ISSN numbers or a Website contact. Further information can be obtained from the NOAA Coastal Services Center Library at www.csc.noaa.gov/library/. Conventional contact information for the center is as follows: 2234 South Hobson Avenue, Charleston, SC 29405-2413.

Table A2.2 Select list of primary international, national, and regional coastal science journals, based on relevance of subject matter and orientation of readership, published on a regular basis in English

Journal	Content, scope, or society affiliation	Publisher	ISSN	1st issue	Frequency/cost	Description and notes
Primary sources of academic and scientific coastal research information						
Bulletin of Marine Science	International journal: marine science	Rosensteil School of Marine and Atmospheric Science (Miami, Florida)	0007-4977	1951	6/year $225 Institution $85 personal	All aspects of marine science are included: marine biology, biological oceanography, fisheries, marine affairs, applied marine physics, marine geology and geophysics, marine and atmospheric chemistry, and meteorology and physical oceanography.
Coastal Engineering	International journal: engineering applications	Elsevier (The Netherlands)	0378-3839	1977	12/year $1207 Institution	Combines practical application with modern technological and scientific achievements in fundamental studies and case histories for coastal, harbor and offshore engineering: studies on waves and currents; coastal morphology; estuary hydraulics; harbor and offshore structures.
Coastal Engineering Journal	International journal: coastal, harbor, and offshore engineering	World Scientific (USA)	0578-5634	1959	4/year $292 Institution	Deals with waves and currents, sediment motion and morphodynamics, predictive methods for environmental processes, hard and soft technologies related to coastal zone development, shore protection, and prevention and mitigation of coastal disasters.
Coastal Management	International journal: marine environment, resources, law, and society	Taylor & Francis (USA)	0892-0753	1973	4/year $335 Institution	Explores the technical, legal, political, social, and policy issues surrounding the utilization of valuable and unique coastal environments and resources.
Continental Shelf Research	International journal	Elsevier Science (The Netherlands)	0278-4343	1982	15/year $216 Institution	Presents research results in physical oceanography, chemistry, ecology, sedimentology, and applied aspects of continental shelf research.
Coral Reefs	International Coral Reef Society	Springer-Verlag (Germany)	0722-4028	1982	4/year $468 Institution	Covers reef structure and morphology, biogeochemical cycles, behavioral ecology, sedimentology, and evolutionary ecology of the reef biota.
Estuarine, Coastal and Shelf Science	Estuarine, Coastal, and Shelf Science Society	Academic Press (United Kingdom)	0272-7714	1973	12/year $1524 Institution	Devoted to the analysis of saline water phenomena ranging from the outer edge of the continental shelf to the upper limits of the tidal zone.
Estuaries	International journal: estuarine and coastal waters	Estuarine Research Federation (USA)	0160-8347	1978	6/year $320 Institution	Includes papers on any aspect of research on physical, chemical, geological, or biological systems, as well as management of those systems, at the interface between the land and the sea.
Geo-Marine Letters	International journal: marine geology	Springer-Verlag	0276-0460	1981	4/year $496 Institution	Publishes papers dealing with all marine geological aspects including: (1) marine geology, (2) marine geophysics, (3) marine geochemistry, (4) marine geotechnique, (5) environmental problems where geo-marine studies play a role, (6) new techniques or modifications, and (7) applications.
Journal of Coastal Conservation	International conservation	European Union of Coastal Conservation (Sweden)	1400-0350	1995	4/year $200 Institution	A scientific journal for integrated research and management of the coastal zone. Emphasis is on natural resources and their sustainable use in the context of past and future social and economic developments.

Table A2.2 *(Continued)*

Journal	Content, scope, or society affiliation	Publisher	ISSN	1st issue	Frequency/cost	Description and notes
Journal of Coastal Research	International forum for the littoral sciences	The Coastal Education & Research Foundation (USA)	07049-0208	1985	4/year $145 Institution $75 Personal	Dedicated to all aspects of coastal (marine) research. Encourages dissemination of knowledge and understanding of the coastal zone by promoting cooperation and communication between specialists in different disciplines.
Marine Geology	International journal: marine geology, geochemistry, and geophysics	Elsevier Science	0025-3227	1964	28/year $2768 Institution $167 Personal	Original research and comprehensive reviews in the field of marine geology, geochemistry, and geophysics.
Marine Georesources and Geotechnology	International journal: seabed resources, methods of exploration and extraction	Taylor & Francis	1064-119X	1993	4/year $275 Institution $146 Personal	Devoted to all scientific and engineering aspects of seafloor sediment and rocks.
Marine Mammal Science	International journal: marine mammals	Society for Marine Mammalogy (USA)	0824-0469	1985	4/year $18 each issue	Presents original research and observations on marine mammals, their evolution, form, function, husbandry, health, populations, and ecological relationships.
Marine Policy	International journal: ocean law and policy	Elsevier	0308-597X	1977	6/year $723 Institution	Offers researchers, analysts and policy makers a unique combination of legal, political, social, and economic analysis.
Marine Pollution Bulletin	International journal: pollution in estuaries, seas, and oceans	Elsevier	0025-325X	1970	6/year $942 Institution	Concerned with the rational use of maritime and marine resources in estuaries, the seas and oceans. Documents marine pollution and new forms of measurement and analysis. Includes research reports on effluent disposal and pollution control.
Ocean and Coastal Management	International journal: all aspects of ocean and coastal management	Elsevier Science	0964-5691	1973	12/year $1138 Institution $107 Personal	Covers all aspects of ocean and coastal management at local, regional, national, and international levels.
Ocean Engineering	International journal: research and development	Elsevier Science	0029-8018	1968	8/year $1798 Institution $223 Personal	Covers the design and building of ocean structures; submarine soil mechanics; coastal engineering; ocean energy; underwater instrumentation, marine resources and other related issues.
Shore & Beach	American Shore & Beach Preservation Association	ASBPA (USA)	0037-4237	1926	4/year $250 Institution	Deals with shore protection measures, coastal engineering, beach nourishment, methods and techniques, and beach management.
Secondary sources of academic and scientific information related to coastal research						
Acta Oceanologica Sinica	Chinese Society of Oceanography	China Ocean Press	0253-4193	1979	4/year $420 Institution	Publishes scholarly papers on marine science and technology, including physics, chemistry, engineering, remote sensing, and instrumentation. (Text in English)
ASLO Bulletin	American Society of Limnology and Oceanography	ASLO (USA)	0024-3590	1990	3/year $75 Members	Informs members of society events and provides a forum to discuss issues.
Australian	Commonwealth	CSIRO (Australia)	0067-1940	1950	8/year	Presents research in physical oceanography,

Marine and Freshwater Research	Scientific & Industrial Research Organization				$650 Institution $165 Personal	marine chemistry, marine and estuarine biology, and limnology.
Chinese Journal of Oceanology and Limnology	National research journal	Science Press (China)	0254-4059	1982	4/year	Covers hydrophysics, hydrochemistry, hydrobiology, geomorphology, apparatus research and manufacture, comprehensive reviews, and academic activities.
Helgoland Marine Research	Alfred-Wegener-Institute for Polar and Marine Research	Springer-Verlag	1438-387X	1946	4/year $169 Institution	Publishes original research papers, invited reviews, and comments on any aspect and level of the biology of marine and brackish water organisms.
ICES Journal of Marine Science	International Council for the Exploration of the Sea	Academic Press	1054-3139	1926	4/year $527 Institution	Contains original papers within the broad field of marine and fisheries science. References subjects including ecology, population studies, plankton research, and physical and chemical oceanography.
Indian Journal of Marine Science	Indian National Science Academy	CSIR (India)	0379-5136	1972	4/year $100 Personal	Publishes mainly papers and scientific articles in coastal and all fields of marine sciences about India, Oceania, and Red Sea regions.
Journal of Ecology	International journal: marine and terrestrial ecology	British Ecological Society (England)	0022-0477	1996	6/year $550 Institution	Publishes original research papers on all aspects of ecology of plants (including algae) in both aquatic and terrestrial ecosystems.
Journal of Geophysical Research—Oceans	International journal: geophysical research in the oceans	American Geophysical Union (USA)	0148-0227	0148-0227	12/year $120 Members	Covers physical, biological, and chemical oceanography.
Journal of Physical Oceanography	American Meteorological Society	American Meteorological Society (USA)	0022-3670	1971	12/year $445	Publishes research related to the physics of the ocean and of the processes operating at its boundaries.
Journal of Waterway, Port, Coastal and Ocean Engineering	American Society of Civil Engineers	American Society of Civil Engineers (USA)	0733-950X	1956	6/year $220 Institution $74 Personal	Presents information regarding the engineering aspects of dredging, floods, ice, pollution, sediment transport, and tidal wave action that affect shorelines, waterways, and harbors.
Marine Environmental Research	International Journal	Elsevier	0141-1136	1978	12/year 1196.00 Institution	Publishes original research papers on chemical, physical, and biological interactions in the oceans and coastal waters.
Marine Technology Society Journal	International journal: ocean and marine engineering, science, and policy	Marine Technology Society (USA)	0025-3324	1963	4/year $50 Membership	International interdisciplinary journal devoted to the exchange of information in ocean and marine engineering, science, and policy.
Netherlands Journal of Sea Research	Netherlands Institute for Sea Research	Netherlands Institute for Sea Research (The Netherlands)	0077-7579	1961	8/year $453 Institution $112.00 Personal	Contains papers on marine research with emphasis on non-applied aspects and contributions to the understanding of the functioning of marine ecosystems, including abiotic systems.
Oceanologica Acta	Review of European Oceanology	Gauthier-Villars (France)	0399-1784	1978	6/year $386 Institution	Presents results of works in all sections of oceanography and from all parts of the oceans and their adjacent estuaries and brackish water systems.
Photogrammetric Engineering and Remote Sensing	American Society for Photogrammetry & Remote Sensing	Potomac Publishing Services (USA)	Not listed	1996	12/year $160 Institution	Commonly referred to as *PE&RS*, the journal for imaging and geospatial information science and technology, is the flagship publication of the Society.

Table A2.2 (*Continued*)

Journal	Content, scope, or society affiliation	Publisher	ISSN	1st issue*	Frequency/cost	Description and notes
Remote Sensing of Environment	International journal: multidisciplinary remote sensing	Elsevier	0034-4257	1969	12/year $1896 Institution $115 Personal	Serves the remote sensing community with the publication of scientific and technical results on theory, experiments, and applications of remote sensing of earth resources and environment.
Tertiary sources of information that are occasionally relevant						
Acta Hydrochimica et Hydrobiologica	European focused journal	John Wiley & Sons, Ltd. (United Kingdom)	0323-4320	1972	6/year $698 Institution	Reports on the latest from all areas of water and environmental research.
Aquatic Conservation: Marine and Freshwater Ecosystems	International journal	John Wiley & Sons, Ltd.	1099-0755	1990	6/year $630 Institution	Dedicated to publishing original papers that relate specifically to freshwater, brackish, or marine habitats and encouraging work that spans these ecosystems.
Earth Surface Processes and Landforms	British Geomorphological Research Group	John Wiley & Sons	0197-9337	1976	9/year $1920 Institution	Contains important research papers on all aspects of geomorphology interpreted in its widest sense, including both pure and applied.
Environmental Pollution	International journal	Elsevier	0269-7491	1970	12/year $3111 Institution $98 Personal	Addresses issues relevant to the nature, distribution, and ecological effects of all types and forms of chemical pollutants in air, soil, and water.
Geology	Geological Society of America	GSA (USA)	0091-7613	1973	12/year	Topical scientific papers on all earth science disciplines worldwide.
Hydrobiologia	International journal	Kluwer Academic Publishers (The Netherlands)	0018-8158	1973	12/year $7216 Institution	A wide range of papers is published, including ecology, physiology, biogeography, methodology, and taxonomy.
International Journal of Marine and Coastal Law	International journal	Kluwer Law International (The Netherlands)	0927-3522	1985	4/year $377 Institution	Addresses all aspects of marine (maritime) and coastal law.
International Journal of Remote Sensing	International journal for remote sensing	Taylor & Francis	0143-1161	1980	18/year $2567 Institution	Is concerned with the science and technology of remote sensing and the applications of remotely sensed data in all major disciplines. Principal topics include: data collection; analysis, interpretation and display; surveying from space, air, and water platforms; sensors; image processing; use of remotely sensed data; economic surveys and cost–benefit analyses.
Journal of Geology	International journal: geological sciences	University of Chicago Press (USA)	0022-1376	1893	6/year $113 Institution $48 Personal	Publishes original contributions dealing with any aspect of geology including space science. Contributions should have wide appeal to geologists, present new concepts, and/or derive new geological insights through the use of new approaches and methods.
Journal of Great Lakes Research	International Association for Great Lakes Research	IAGLR (USA)	Not listed	1974	4/year $100.00 Institution $70 Personal	A eclectic mix of papers from various scientific, management, and policy perspectives all focused on a single topic: large lakes of the world.

Journal of Sea Research	International journal	Elsevier	1385-1101	1997	6/year $453 Institution $112 Personal	An international and multidisciplinary periodical on marine research, with an emphasis on marine ecosystems, including both biotic and abiotic aspects of all types of marine and estuarine systems, benthic as well as pelagic.
Journal of Marine Science and Technology	International journal	Springer-Verlag	0948-4280	1996	4/year $180 Institution	Provides a forum for the discussion of current issues in marine science and technology by publishing original, full-length, refereed contributions on research and/or developments in this field.
Journal of Marine Systems	International journal	Elsevier	0924-7963	1990	12/year $1789 Institution $102 Personal	Provides a medium of exchange for those engaged in marine research where there exists an interplay between geology, chemistry, biology, and physics.
Journal of Oceanography	Oceanographic Society of Japan	Kluwer Academic Publishers	0916-8370	1942	6/year $479 Institution $288 Personal	This bimonthly journal is published in English, in order to promote the research in pure and applied oceanography internationally.
Journal of Sedimentary Research	International journal: sediments and sedimentary rocks	Society for Sedimentary Geology (USA)	1073-130X	1926	6/year $195/year Nonmembers and institutions	Includes research papers, perspectives, and methods in any area of sedimentary geology.
Marine Biology	International journal: life in coastal and ocean waters	Springer-Verlag	0025-3162	Not listed	6/year $4028 Institution	Publishes articles in plankton research, experimental biology, molecular biology, biochemistry, physiology and behavior, biosystem research, evolution, theoretical biology related to the marine environment, methods and others.
Marine Chemistry	International journal	Elsevier	0304-4203	1973	12/year $1670 Institution $282 Personal	An international medium for the publication of original studies and occasional reviews in the field of chemistry in the marine environment, with emphasis on the dynamic approach.
Marine and Freshwater Research	Regional journal: Australia Academy of Sciences	Commonwealth Scientific & Industrial Research Organization (Australia)	1323-1650	1949	8/year $550 Institution $150 Personal	The broad subject scope imparts a particular strength in publishing multidisciplinary papers such as those in biogeochemistry.
Marine Geophysical Researches	International journal: study of the earth beneath the sea	Kluwer	0025-3235	1970	$583/volume Institution	Emphasizes the description and analysis of structures that can be investigated with geophysical methods only, and the study of the physical processes that led to the origin of these structures. Includes fundamental studies and case histories coastal, harbor, and offshore engineering: waves and currents; coastal morphology; estuary hydraulics; harbor and offshore structures.

Table A2.2 *(Continued)*

Journal	Content, scope, or society affiliation	Publisher	ISSN	1st issue*	Frequency/cost	Description and notes
Progress in Oceanography	International journal	Elsevier	0079-6611	1963	8/year $1962 Institution $162 Personal	Publishes the longer, more comprehensive papers that most oceanographers feel are necessary, on occasion, to do justice to their work.
Sea Technology	News magazine	Compass Publications (USA)		1965	12/year $35 in USA $115 Foreign	Provides current information on the worldwide marine/ocean industry through state-of-the-art and application articles.
Sedimentary Geology	International journal: pure and applied sedimentology	Elsevier Science	0037-0738	1967	24/year $2347 Institutional $156 personal	Provides a forum for the publication of research across the entire spectrum, from analytical techniques to regional or geodynamical aspects of sedimentary systems and basin analysis.
Sedimentology	International Association of Sedimentologists	Blackwell Scientific Publications (United Kingdom)	0037-0746	1952	6/year $835 Institution	Promote the study of sedimentology, interchange of research and international cooperation between sedimentologists.
Zeitschrift für Geomorphologie	International journal: geomorphology	K. Triltsch, Wuzburg (Germany)	0044-2798	1957	12/year	International journal for geomorphology.

(1) Prices are quoted on an annual basis, unless otherwise noted, for institutions. Personal rates are given for some journals, if available on the journal Website. Because some journals offer reduced personal rates for members of socities, these rates are not always readily available.

(2) Journal prices are quoted in US$ on a 2000–2001 volume cost basis, unless otherwise noted. Most journals are published in volume-years, one volume to one year, but some journals produce more than one volume per year. The price noted thus includes variable numbers of issues per volume on an annual basis.

(3) This list of journals is based on Internet-based computer searches initiated by general and specific titles, key words, or subject areas. Additional information was secured from the publisher's Website and used to fill out comments concerning scope and orientation, the ISSN number, date of first issue, frequency and cost, and ancillary comments or notes about the journal.

but there are also a large number of "soft" information sources that are required for effective coastal zone management. Even though some of these tertiary information sources have a definite corporate bias (e.g., *ARC News*, *Calypso Log*, *Geo Info Systems*, *GIS World*, *GPS World*, *Meeting News*, *Publish!*, *Spotlight*, *Successful Meetings*, *Sun Expert*, *UNIX Review*, *Visual Developer*), they provide practical information to coastal researchers who require expertise in diverse technical and socioeconomic fields to supplement their professional or academic expertise.

There are thus many different possibilities for the organization of a journal list. One approach might be to list journals alphabetically without comment or consideration of frequency of use, relative importance, by coastal researchers. The choice of what journal should be listed in a "coastal" list thus becomes subjective and some criteria must be applied for selective inclusion, in preference to personal impressions or bias. Another approach might be to organize the list in terms of frequency of citation in articles relating to the coast. Such an effort would be tedious to do by hand, but could be accomplished by a commercial citation indexing service. Criteria for selection would be difficult to set up and would, of course, again be subjective. Another approach would be to rely on the expertise of seasoned (longtime) researchers who could lend professional opinions to the selection. This approach is clearly subjective but no approach identified here can be regarded as unambiguous. The approach that I have adopted is to use the *Journal of Coastal Research* (JCR) as a baseline that is fairly representative of the central tendency of scientific research in the coastal zone. Although the JCR considers all aspects of coastal research, experience has shown over the years that most papers feature coastal aspects of biology (incl. ecology), geology (incl. geomorphology, hydrology, sedimentology), physical geography (incl. classification and mapping, GIS/LIS/MIS), littoral oceanography, hydrography, coastal hydraulics, environmental (resource) management, engineering, and remote sensing. Policy and legal aspects of are most often appropriately considered in other journals (e.g., *Journal of Coastal Conservation*, *Coastal Management*, *Marine Policy*, *Coastal and Ocean Management*). Even though the basis of the journal selection process is biased, it is hoped that it represents a fair appraisal of primary sources. Another limitation is language; this list features journals where papers are primarily published in English.

Although definition of the term "coastal" is specifically avoided here, the reader is referred to topics in this volume that provide insight into the scope and dimension of the subject area (see types of coasts, management and engineering topics, ecology of specific geographic regions). The list of professional (academic) journals presented in Table A2.2 is not comprehensive in the sense that it contains all journals that may contain papers related to some aspect of the coast. Journals listed here will direct neophytes to a spectrum of coastal journals and as such it serves as an introduction. This is a starting point that will get the interested coastal researcher into the literature. Professionals who have been conducting specialized coastal research for many years will know of more obscure sources that are useful in narrow lines of inquiry. The list points to journals that essentially deal with coastal topics on a regular basis or those that occasionally contain relevant articles. Journals listed in Table A2.2 are broadly organized into three categories that indicate whether the journal provides primary, secondary or tertiary support to the average coastal researcher. Some journals are dedicated to coastal research whereas others might have a different primary focus (e.g., oceanography, ecology, marine mammals, marine technology, ocean law and policy, seabed resources, fisheries science, meteorology, engineering) but often include coastal-based subject matter. The secondary journals are mostly based in the marine realm whereas those considered to be tertiary sources are primarily focused on other disciplines where the subject matter might occur in the coastal zone or be related to coastal processes in some way (e.g., journals dealing with biological, geological, or engineering topics). The listings in Table A2.2 are thus not an endorsement of journal contents or a reflection of citability or credibility. It is benign in the sense that it tries to steer interested readers to sources most likely to contain information related to the coast.

Most of the journals listed in Table A2.2 have international audiences in mind, but the scope of coverage may be limited to biophysical or geographic subdivisions of the coastal zone, define rather specific subject areas, or narrowly focus on legal or conceptual issues viz. *Continental Shelf Research, Coral Reefs, Estuaries, Helgoland Marine Research, International Journal of Marine and Coastal Law, Journal of Great Lakes Research, Marine Biology, Marine Chemistry, Netherlands Journal of Sea Research, International Journal of Remote Sensing*. On the other hand, some journals offer a wide-ranging approach to many different considerations in the coastal zone, for example, *Bulletin of Marine Science, Journal of Coastal Research, Journal of Marine Science and Technology, Geo-Marine Letters, Ocean and Coastal Management*. The scope and interests of journals covering the coastal zone are thus focused and yet multidisciplinary in approach.

Coastal science, as a discrete subject area, is a relatively young discipline. Recognition of diverse subject areas as a coherent corpus is evidenced in the journals themselves, the oldest continuously surviving primary journal being *Shore & Beach* (established in 1926). Some of the other older primary journals include, for example, *Australian Marine and Freshwater Research* (1950); *Bulletin of Marine Science* (1951); *Coastal Engineering Journal* (1959); *Journal of Waterway, Port, Coastal and Ocean Engineering* (1956); and *Marine Geology* (1964). A recent primary player is the *Journal of Coastal Conservation* (1995), the official publication of the European Union of Coastal Conservation. Papers of interest to coastal researchers appear in other venerable journals, listed in Table A2.2 as secondary or tertiary journals, which have broad-ranging interests that overlap or subsume the coastal zone as aspects of it, as for example, *Journal of Geology* (1893); *Journal of Oceanography* (1942); *Journal of Sedimentary Research* (1926); and *Sedimentology* (1952). A relative newcomer to the field is the *Journal of Coastal Research* (1984) which covers a wide range of topics that include but are not limited to geology, biology, geomorphology (physical geography), climate, littoral oceanography, hydrography, hydraulics, environmental (resource) management, engineering, and remote sensing. The *Journal of Coastal Research* invites contributions dealing with theory, methodology, techniques, and field or applied topic studies on interdisciplinary issues within the broad subject areas listed above.

To assist the researcher in gaining access to the journals, the list in Table A2.2 highlights critical information such as publisher, ISSN, and frequency of publication. A brief description of journal content has been abstracted from the journal itself, a journal or professional society Website, or from *Ulrichs International Periodicals Directory* (New Providence, New Jersey: R.R. Bowker).

Charles W. Finkl

Bibliography

Kelletat, D.H., 1996. International Bibliography of Coastal Geomorphology, 1991–1994. *Journal of Coastal Research*, Special Issue No. 19, 203 pp.

Richards, H.G., Gallagher, W.B., and Colquhoun, D.J., 1986. Annotated Bibliography of Quaternary Shorelines, Fourth Supplement 1978–1983. *Journal of Coastal Research*, Special Issue No. 2, 88 pp.

Sherman, D.J., 1992. International Bibliography of Coastal Geomorphology, 1987–1992. *Journal of Coastal Research*, Special Issue No. 16, 146 pp.

APPENDIX 3: ORGANIZATIONS

In the last two decades there has been an explosion of organizations dealing with the coast. Most coastal researchers probably initially think in terms of scientific groups that consider academic, theoretical, or practical applications of scientific principles or engineering practices. With coastal populations increasing worldwide (e.g., Culliton *et al.*, 1992), there have been a number of movements that attempt to organize political and public opinion in favor of limiting overexploitation of coastal resources (see discussions in Clark, 1996). Disregard for the coast is not a new trend, but public awareness is increasing as coastal living space becomes more limited and use of coastal environments becomes more belligerent. In many ways, the coastal zone is becoming a battleground (Finkl, 1997) for different interest groups. In many places, coastal environments have been totally replaced by urbanization and industrialization, seriously degraded by overuse and effluent disposal, or are presently threatened by various facets of human action. There are now many conservation groups interested in organizing various types of resistance to degradation of coastal and marine resources.

Today the coastal zone is seen from many divergent points of view ranging from commercial and industrial use (e.g., ports and harbors, petroleum tank farms, transportation facilities, free trade zones), tourism, urban, and suburban living space, waste disposal sites, land reclamation (e.g., drainage of coastal wetlands), parks and reserves (including marine parks), military installations, and finally as political and military battlespace (e.g., strategic choke points, sites for amphibious landings, approachways for invasion from the sea). With all of these diverse, and often conflicting, uses of coastal space, it would be a disservice, to perception of the problems and potential remedies, to list only organizations of interest to academic researchers. Indeed, many of the most exciting and important considerations of the coast reside in nontraditional organizations that have as agendas the protection and preservation of coasts through avenues of public information campaigns, education, and private or nongovernmental organizations (NGO). Response to issues of coastal use is, unfortunately, largely not met by public agencies responsible for the care and protection of coasts. Leaders of coastal protection are thus often found in nonauthoritarian organizations that do not pander to political whims or inaction. On the other side of the coin, there are many examples of government agencies that have taken appropriate steps to better understand coastal environments and the natural processes responsible for them. By and large, however, the response by governments worldwide is too little too late, except possibly in North America and Western Europe, and Oceana.

Presentation of the organizations dealing with coastal issues requires organization because their number is legion. A mere decade ago if one were to conduct a computer search on the Internet by requesting a search based on the word "coastal," the response would have been minimal. Today, it is a different story. An Internet search using the word "coastal" as a key term would return literally hundreds of hits (e.g., web pages containing that word or access to other sources of information). With such vast amounts of information now readily available, it is clear that many organizations are involved with coasts in a way that was unimaginable in the recent past.

Before the widespread use of the Internet, it was possible to keep track of coastal-marine researchers through printed directories such as the *International Directory of Marine Scientists* (UNESCO, 1983), *Orbis Geographicus* (IGU, 1992), or to locate major repositories of information as listed, for example, in the *International Directory of Marine Science Libraries and Information Centers* (IAMSLIC, 1987). Organizations could be gleaned from these kinds of source materials (i.e., from their contained institutional lists) but there was no easy way to compile lists of coastal organizations. With the development of digital information systems, access changed and large amounts of information became available and retrievable to coastal researchers almost instantly. Because there is no comprehensive published list of coastal organizations worldwide, these lists are conceived as part of an attempt to indicate the range of organizations, public and private, that provide services to researchers of the coastal zone. Addresses, phone and fax numbers, e-mail addresses or websites are provided when available in the following lists. Statements of purpose, objectives or goals, mottoes or slogans, or advertizing self-images of various organizations are listed in table format as examples. The information provided here is not to be inclusive, but it is comprehensive and global in perspective. If there are organizations that are not listed in the following tables (1–16), it is an oversight resulting from the vagaries of computer searches on the net. All reasonable efforts were made to include as many organizations as possible but, of course, no list can be complete in a subject area where groups come together for a period of time and then disperse or where there are name changes, as so frequently happens in government with changes in political parties. In any case, the tables presented here are submitted in good faith in the belief that they represent a fair and unbiased summary of what can be found on the Internet. Access to these organizations is thus possible through traditional means by mail, telephone, or fax. The Internet offers new possibilities for rapid personal interaction and a means of rapidly acquiring information without much effort on the part of the searcher. The collection and availability of data more or less dictated organizational groupings, which turn out to be quite subjective. Mostly, the assembly of organizations was by continental or subcontinental geographic regions but sometimes a hemispheric approach was more convenient whereas some other organizations are essentially global in focus. It is thus anticipated that most readers will have interests in specific geographic regions rather than an organizational focus.

Table A3.1, for example, lists some of the main North American agencies that are maintained by federal governments. In some cases it is not entirely logical nor convenient to list subdepartments and so parent organizations are indicated viz. the National Science Foundation (NSF), US Department of the Navy (USN), National Oceanographic and Atmospheric Administration (NOAA), US Army Corps of Engineers (USACE). For these larger organizations, some of the larger coastal programs and projects are indicated. In the United States there are 33 state coastal zone management programs (Table A3.2). Alaska and Hawaii probably come readily to mind as programs separated from the conterminous United States but territories and commonwealth programs are also included in the number of "state" programs, viz. Guam, Northern Mariana Islands, Puerto Rico, and the Virgin Islands. Indiana, Michigan, Ohio, Pennsylvania, and Wisconsin have coastal management programs for management of Great Lakes shores. Other organizations having responsibility for management of freshwater and marine shores include government agencies and quasi-government bodies that feature concerted efforts to manage wildlife (e.g., Florida Fish and Wildlife Conservation Commission), protect natural coastal resources (e.g., Columbia River Estuary Study Task Force, San Francisco Bay Conservation and Development Commission), or restore coastal areas (e.g., Office of Coastal Restoration Indiana, Plan to Restore America's Everglades) (Table A3.3).

Some worldwide government agencies and programs related to the coastal zone are listed in Table A3.4. Some entries here appear to be national, and they are, but affiliated agencies also work overseas in a related capacity (e.g., Coastcare or the Coast Protection Board in Australia) as for example seen in the Australian Antarctic Division (AAD) as part of the Australian Government's Department of the Environment and Heritage. The European Union for Coastal Conservation (EUCC) attempts to ensure wise use and protection of Baltic Sea coasts and makes concerted efforts to ensure cooperation between government and non-government organizations viz. local communities, scientific organizations, and individuals involved in conservation of the coastal zone. The EUCC fosters participation in intergovernmental movements connected with coastal conservation and management as well as the integration of science and management in the coastal zone. Another example listed in Table A3.4 is the Intergovernmental Oceanographic Commission (IOC) of UNESCO that promotes marine scientific investigations and related ocean services. The Eastern African Coastal Area Management effort strives to assist East African countries to implement and coordinate coastal management activities as follow-up to the Arusha Resolution and the Seychelles Statement on Integrated Coastal Zone Management.

The main North American research and educational institutions that deal with coastal zone management and related topics are listed in Table A3.5. This sort of table is problematic because it is impossible to include every institution that deals with coasts. This table is thus the result of a computer search that turned up more than 60 organizations that conduct coastal research or instruction in coastal zone management related topics. Even though the list is not complete, it provides entry into the sphere of coastal zone research and education in North America. Some organizations, although they are part of state or private university systems, can be obscure (e.g., the Center for Coastal and Land-Margin Research (CCALMR) is part of the Oregon Graduate Institute of Science and Technology, Oregon Health and Science University, Beaverton, Oregon) or well known as part of major research centers of long-standing excellence (e.g., Center for Coastal Studies, Scripps Institution of Oceanography, University of California, La Jolla, California; Department of Earth, Atmospheric, and Planetary Sciences, Massachusetts Institute of Technology, Cambridge, Massachusetts). Although some organizations do not grant academic degrees as from colleges and universities, they do provide useful short courses, study groups and projects, applied research, and in-service training as, for example, seen at the US Army Coastal and Hydraulics Laboratory (Engineer Research and Development Center, Vicksburg, Mississippi). Other organizations function as clearinghouses for digital information (e.g., Oregon Coastal Geospatial Clearinghouse) or as providers of overarching regional research guidance and facilitation (e.g., Regional Association of Research on the Gulf of Maine) as associations of institutions or organizations of organizations (e.g., Coastal Resources Research Network based at Dalhousie University, funded by the International Development Research Centre, and working with partners in South East Asia).

Similar comments could be made with regard to Table A3.6, which lists some of the main European research and educational institutions that deal with the coastal zone. For convenience, the institutions here are listed in alpha-order by host country. The list contains organizations of international renown (e.g., Danish Hydraulics Institute, Institut Oceanographique, Delft Hydraulics Institute, Netherlands Institute for Sea Research, Cambridge Coastal Research Unit, Plymouth Marine Laboratory, Proudman Oceanographic Laboratory) as well as those that are perhaps less well known (e.g., Center for Marine Research, Zagreb, Croatia; Sandgerdi Marine Centre, Sandgerdi, Iceland). In either case, these organizations enjoy "hits" from web searches due to their websites. Last among the tables listing the main research and educational institutions that deal with the coastal zone is Table A3.7, which takes a hemispheric perspective in Australia, Brazil, New Zealand, and South Africa. Additional short listings are included for Argentina, Chile, southern India, and the Philippines.

Proprietary consulting companies provide useful services to both public and private sectors in the coastal zone (Table A3.8). These organizations are mostly "for profit" but they often provide information to the public in the form of competitive bidding for projects, reports, public debates and journal publishing, and participation in public meetings and forums. Nevertheless, these kinds of organizations provide many useful functions in the coastal zone that are not available from governmental organizations. Again, this list is not comprehensive and represents the kinds of results that can be obtained from simple computer searches based on critical key words. Most of the consulting companies deal with environmental issues and coastal engineering, although some consultants are lawyers (e.g., Conservation Law Foundation) or lobbyists (e.g., The Coastal Advocate). A related list contains coastal consulting companies based in the United States but with branches or additional offices overseas (Table A3.9). These are listed in alpha-order with indications of overseas location where the companies are active. Some consulting companies listed in Table A3.10 are truly worldwide in scope and operations. Clearly, there are many other consulting companies that deal with coastal conservation, shore protection, dredging, and so on, but they do not show up in this computer search as the main focus was not engineering, dredging, etc.

Other kinds of organizations active in the coastal zone that have not yet been considered include those that are comprised by professional societies, either governmental or private, that can include societies, federations, coalitions, campaigns, associations, not-for-profit corporations, foundations, alliances, clubs, projects, etc. These organizations listed in Table A3.11, include a broad range of interests in the coastal zone. Primary here are groups that have interest in protecting or preserving coastal resources such as beaches (e.g., American Shore and Beach Preservation Association, Florida Shore and Beach Preservation Association), habitat (e.g., American Littoral Society; Canadian Ocean Habitat Protection Society, Newellton, Nova Scotia, Canada), rivers and estuaries (e.g., Clean Annapolis River Project, Annapolis Royal, Nova Scotia, Canada; Clean Water Fund, Washington, DC), and historic cultural sites (e.g., Coastal America, Washington, DC). Some groups appeal to professional coastal researchers and academics rather than lay people and they attempt to foster study and research of coastal problems and issues (e.g., American Geological Institute, American Society of Civil Engineers, American Society of Limnology and Oceanography, The Coastal Education & Research Foundation, the Coastal Research and Education Society of Long Island, Marine Technology Society). Different examples are found in other kinds of organizations that fight coastal erosion (e.g., Beach Erosion Authority for Clean Oceans and Nourishment, Ventura, California), or strive to revitalize ocean resources (e.g., American Oceans Campaign) or which focus on maintaining sustainable tourism (e.g., Bay of Fundy.com, Chance Harbour, New Brunswick, Canada). There are thus many different orientations and the scope of effort is great in professional organizations that have the wellbeing of the coastal zone at heart. Those organizations listed in Table A3.11 are but a few of the many different kinds of organizations that can be accessed via the Internet.

Table A3.12 continues the previous line of inquiry but the organizations noted here are for coastal regions beyond the shores of North America. Some of these organizations have a global outlook (e.g., Estuarine and Coastal Sciences Association, Greenpeace International, Seas at Risk, Surfers Against Sewage, United Nations Environment Programme) while others have a national perspective (e.g., Asociación Oceánica de Panamá, Australian Coral Reef Society) or regional charge (e.g., European Artificial Reef Research Network, European Coastal Association for Science and Technology). These interest groups outside North America have similar orientations and thrusts of purpose as those listed in Table A3.11.

Asian groups that focus their attention on coastal research in general and marine science specifically are listed in Table A3.13. This regional listing is given as an example of concern for the coast in one of the most densely populated regions on earth. Oceanographic organizations and institutes of marine science conduct most of the coastal effort here. In a similar vein, Brazilian coastal organizations are listed separately in Table A3.14. There is concern in Brazil for maintaining clean coasts and there are numerous state divisions of IBAMA, the Brazilian environmental protection agency. Fisheries research institutes are listed here as well because of their concern for maintaining good coastal environmental quality that will support coastal fisheries, which in turn provide employment for small villages and towns along the coast.

By way of an example of another very specialized list, Table A3.15 lists coastal resource management companies in Hawaii. The island is clearly a Mecca for coastal consulting companies due to its strategic geographic location. Companies based here have ready access to North America, especially the West Coast, and to the eastern Asian continent and Oceana. Full contact information is provided for most of these companies based in Hawaii. Some of these companies are also listed in Tables A3.11, A3.12 or A3.13, depending on the location of their home office or branches. Others are not listed in these previous three tables because they did not show up in the computer search for those listings. The list in Table A3.15 is particularly comprehensive and complete and is a good example of an effort to provide useful information for a specific region, in this case the State of Hawaii. A smaller complimentary list is found in Table A3.16, which summarizes coastal resource management companies in Hawaii based on nonprofits.

Table A3.1 Primary North American agencies and programs related to the coastal zone, as maintained by federal governments in the United States and Canada

Canadian Hydrographic Service (Canada)
615, Booth Street
Ottawa, Ontario K1A 0E6
Phone: (613) 995 4413
Fax: (613) 947 4369

Canadian Ice Service (Canada)
373, Sussex Drive
Block E, Third Floor
Ottawa, Ontario K1A 0H3

Coastal America (USA)
Coastal America Reporters Building
300, 7th Street, SW Suite 680
WA 20250
Phone: (202) 401 9928
Fax: (202) 401 9821
E-mail: wanda.brown@usda.gov

A Decade of Commitment to Protecting, Preserving, and Restoring America's Coastal Heritage

Department of Fisheries & Oceans—Maritimes Region (Canada)
P.O. Box 1035
Dartmouth, Nova Scotia B2Y 4T3
Phone: 1 800 782 3058
E-mail: info@dfo-mpo.gc.ca

Environment Canada (Canada)
45, Alderney Drive
Dartmouth, Nova Scotia B2Y 2N6
Phone: (902) 426 7231
Fax: (902) 426 6348
E-mail: 15th.reception@ec.gc.ca

EPA (Environmental Protection Agency)—Office of Water (USA)
Deals with coastal waters and related aspects, has sub-offices of wetlands, oceans, and watersheds.

Contacts—Headquarters:

Rick Hoffmann
USEPA, 1200 Pennsylvania Ave. NW-4305
WA 20460
Phone: 202 260 0642
Fax: 202 260 9830
E-mail: hoffmann.rick@epa.gov

Charles Kovatch
USEPA, 1200 Pennsylvania Ave. NW-4305
WA 20460
Phone: 202 260 3754
Fax: 202 260 9830
E-mail: kovatch.charles@epa.gov

Promotes a watershed approach to manage, protect, and restore the water resources and aquatic ecosystems of our marine and fresh waters. This strategy is based on the premise that water quality and ecosystem problems are best solved at the watershed level and that local citizens play an integral role in achieving clean water goals. Through its many programs, OWOW provides technical and financial assistance and develops regulations and guidance to support the watershed approach.

Fisheries & Oceans Canada (Canada)
Communications Branch
200, Kent Street
13th Floor, Station 13228
Ottawa, Ontario K1A 0E6
Phone: (613) 993 0999
Fax: (613) 990 1866
TDD: (613) 941 6517

The CCG will ensure the safe and environmentally responsible use of Canada's waters, support understanding and management of oceans resources, facilitate the use of our waters for shipping, recreation and fishing, and provide marine expertise in support of Canada's domestic and international interests.

Fleet Numerical Meteorology and Oceanography Center (US Department of the Navy)
Commanding Officer: cdo@fnmoc.navy.mil
Executive Officer: cdo@fnmoc.navy.mil
Command Master Chief: fn-cmc@fnmoc.navy.mil
Public Affairs: fn-pao@fnmoc.navy.mil

National Marine Fisheries Service (NMFS) (USA)
NOAA Fisheries
1315, East West Highway
SSMC3, Silver Spring, MD 20910

The NMFS administers NOAA's programs which support the domestic and international conservation and management of living marine resources.

National Ocean Service (NOAA, National Oceanography & Atmospheric Administration) (USA)
National Ocean Service
1305, East West Highway
Silver Spring, MD 20910
Phone: (301) 713 3074

As the nation's principal advocate for coastal and ocean stewardship, the National Ocean Service develops the national foundation for coastal and ocean science, management, response, restoration, and navigation. The National Ocean Service maintains its leadership role in coastal stewardship by bridging the gap between science, management, and public policy.

Naval Meteorology and Oceanography Command (US Department of the Navy)
1100, Balch Boulevard
Stennis Space Center MS 39529
Phone: (228) 688 4187
Fax: (228) 688 5743
E-mail: pao@cnmoc.navy.mil

The Naval Meteorology and Oceanography Command's mission is to collect, interpret and apply global data and information for safety at sea, strategic and tactical warfare, and weapons system design, development, and deployment. The command provides meteorological, oceanographic, and geospatial information and services to increase the effectiveness of our Navy in both peacetime and in war.

Naval Oceanographic Office (US Department of the Navy)
N24 Customer Service Division
1002, Balch Boulevard
Stennis Space Center, MS 39522-5001
Phone: (228) 688 5661/5216/5382
DSN: 828 5661/5216/5382
Fax: (228) 688 4688
E-mail: wsc@navo.navy.mil

Naval Research Laboratory (US Department of the Navy)
4555, Overlook Ave., SW
WA 20375

The Naval Research Laboratory (NRL) is the Navy's corporate laboratory. NRL conducts a broadly based multidisciplinary program of scientific research and advanced technological development directed toward maritime applications of new and improved materials, techniques, equipment, system, and ocean, atmospheric, and space sciences and related technologies.

NOAA Coastal Service Center (USA)
2234, South Hobson Avenue
Charleston, SC 29405-2413
Phone: 843 740 1200
Fax: 843 740 1224

The mission of the NOAA Coastal Services Center is to foster and sustain the environmental and economic wellbeing of the nation's coast by linking people, information, and technology.

Table A3.1 *Continued*

NOAA Coral Health Program (USA)

Located at NOAA's
Atlantic Oceanographic and Meteorological Laboratory
4301, Rickenbacker Causeway
Miami, FL 33149-1026

The mission of the Coral Health and Monitoring Program is to provide services to help improve and sustain coral reef health throughout the world.

NOAA Office of Public Affairs (USA)

14th Street & Constitution Avenue, NW
Room 6013
WA 20230
Phone: (202) 482 6090
Fax: (202) 482 3154

NSF (National Science Foundation) Division of Ocean Sciences (USA)

4201, Wilson Boulevard
Room 725, Arlington, VA 22230
Phone: 703 292 8580
Fax: 703 292 9085

The Division of Ocean Sciences (OCE) supports basic research and education to further understanding of all aspects of the global oceans and their interactions with the earth and the atmosphere. The division also offers opportunities to participate in global change research programs and other focus programs.

Topographic Engineering Center (US Department of the Army)

Tammy Scroggins, CEERD-TG-S
Alexandria, VA
Phone: (703) 428-6902
E-mail: tscroggi@tec.army.mil

TEC's Mission is to provide the Warfighter with a superior knowledge of the battlefield, and support the Nation's civil and environmental initiatives through research, development, and the application of expertise in the topographic and related sciences.

US Army Corps of Engineers (USACE) (USA)

Headquarters:
441, G. Street, NW
WA 20314
Phone: 202-761-0008

US Coast Guard (USA)

Headquarters:
2100, Second Street, SW
WA 20593-0001
Website: www.uscg.mil

US Fish & Wildlife Service (USA)

Information about the Coastal and National Coastal Wetlands Conservation Grant Programs, contact:

US Fish and Wildlife Service, Branch of Habitat Restoration
4401, N. Fairfax Drive, Arlington, VA 22203
Phone: (703) 358 2201
Fax: (703) 358 2232

Information about the Coastal Barrier Program, contact:

US Fish and Wildlife Service, Branch of Federal Activities
4401, N. Fairfax Drive, Arlington, VA 22203
Phone: (703) 358 2183
Fax: (703) 358 2232

The US Fish and Wildlife Service is working to conserve coastal resources to benefit present and future generations. Three programs form the core of these coastal conservation efforts which are: The Coastal Program, the Coastal Wetlands Conservation Grants, and the Coastal Barriers Resources System.

Table A3.2 US state coastal zone management programs

Alabama Coastal Area Management Program

ADECA, Coastal Programs Office
1208, Main Street
Daphne, AL 36526
Phone: (334) 626 0042
Fax: (334) 626 3503
E-mail: buddy.sullivan@noaa.gov

Alabama's coastal program balances coastal activities to ensure that the environment on which Alabama business depends remains healthy for generations to come. Alabama's coast faces wetlands loss; coastal erosion; residential, commercial, port, and industrial development; population growth; and nonpoint source pollution problems.

Alaska Coastal Management Program

Division of Governmental Coordination
P.O. Box 110030
Juneau, AK 99811-0030
Phone: (907) 465 3562

With key industries like timber harvesting, oil and gas development, mining, and seafood processing, Alaska's coast faces pressures from resource development and subsistence use of resources. The coastal program balances these uses and the needs of the environment for the long-term health of the state's coastal industries.

California Coastal Management Program

California Coastal Commission
45, Fremont Street, Suite 2000
San Francisco, CA 94105
Phone: (415) 904 5200

For California's extensive coast, resource management and conservation means minimizing the impact of port and residential development, oil transportation, and runoff pollution. To deal with coastal problems, the program oversees almost all activities on the coast, from adding a deck to a private home to building a new refinery.

Connecticut Coastal Management Program

Connecticut DEP
Office of Long Island Sound Programs
79, Elm Street
Hartford, CT 06106-5127
Phone: (860) 424 3034

Connecticut's coastal management program addresses water quality issues including runoff pollution, habitat protection and restoration, public access to the coast, and the use of public lands and waters. The state manages its coastal resources by applying coastal policies to land uses, by overseeing activity in fishery habitats and coastal waters, and by demanding consistency between local, state, and federal actions.

Delaware Coastal Management Program

Delaware DNREC
89, Kings Highway
Dover, DE 19903
Phone: (302) 739 3451

To keep Delaware's coast healthy and productive, the state coastal program monitors activities in the coastal zone. Major challenges are runoff pollution and cumulative and secondary impacts of population growth and urban development.

Florida Coastal Management Program

The Department of Community Affairs
Florida Coastal Management Program
2555, Shumard Oak Boulevard, Tallahassee, FL 32399-2100
Phone: (850) 922 5438
Fax: (850) 487 2899

The Florida Coastal Management Program (FCMP) coordinates among local, state, and federal entities involved in coastal

Table A3.2 *Continued*

management activities. In addition to working with DCA's programs, the FCMP coordinates among the eight state agencies, five water management districts, and local governments that have responsibilities for coastal management under the federally approved Florida Coastal Management Plan. The FCMP also develops partnerships with local communities to actively solve problems related to coastal development.

Georgia Coastal Management Program

Georgia DNR
Coastal Resources Division
One Conservation Way, Suite 300
Brunswick, GA 31520-8687
Phone: (912) 264 7218

Georgia's major coastal issues include pollution, a rapidly growing coastal population, and erosion on the state's developed barrier islands. To deal with these and other issues, the state has developed a federally approved coastal zone management program.

Guam Coastal Management Program

Bureau of Planning
Coastal Management Program
P.O. Box 2950
Agana, Guam 96910
Phone: (617) 472 4201

Coastal hazards, public access, urban growth, and wetlands degradation are some of the key issues for Guam's coastal management program. In an effort to combat coastal problems, the program cooperates with other territory agencies that require permits for coastal activities.

Hawaii Coastal Management Program

Hawaii Office of Planning
Hawaii Coastal Zone Management Program
Hawaii Department of Business, Economic Development, and Tourism
P.O. Box 2359, Honolulu, HI 96804
Phone: (808) 587 2846

With no point in Hawaii more than 29 miles from the shore, almost any activity that accurs inland will impact Hawaii's coastal and ocean resources. Hawaii's coastal program balances the needs for economic growth, a clean environment on which that growth depends, and a vibrant local culture that reflects Hawaii's uniqueness.

Indiana Coastal Coordination Program

Indiana DNR
Division of Water
402, W. Washington
Room W264
Indianapolis, IN 46204-2748
Phone: (317) 233 0132

Indiana's most challenging coastal issues include public access to the shore, beach closures, water quality, brownfields dredging, shoreline erosion, and preservation of natural areas.

Louisiana Coastal Resources Program

Louisiana DNR
Coastal Management Division
P.O. Box 44487
Baton Rouge, LA 70804
Phone: (225) 342 7591

To ensure the environment on which its industry depends is healthy for generations, Louisiana's coastal program must turn around declining fishery habitats, such as wetlands, and reduce erosion. The coastal resources program works with parishes to design programs which resolve conflicting local uses of the coast.

Maine Coastal Management Program

Maine State Planning Office
Coastal Programs
State House Sta. 38
Augusta, ME 04333
Phone: (207) 287 3261

Maine's coastal managers contend with challenges of population growth, water quality, public access, and the impacts of development. The coastal management program manages activities in or on wetlands, flood plains, sand dunes, and other coastal resources.

Maryland Coastal Zone Management Program

Maryland DNR
Coastal Zone Management Division
580, Taylor Avenue
Annapolis, MD 21401
Phone: (410) 974 2784

Maryland's coastal program encourages sound economic development and minimizes the impact people have on vital coastal resources, such as fisheries.

Massachusetts Coastal Zone Management Program

Massachusetts Office of Coastal Zone Management
251, Causeway Street, Suite 900
Boston, MA 02114-2119
Phone: (617) 626 1200
E-mail: mczm@state.ma.us

Because of the beauty and bounty of the coast, many different interests compete for use of coastal resources. Massachusetts' coastal program must balance the competing demands of dredging and dredge material disposal, coastal erosion, runoff pollution, public access, ocean resource management, port revitalization, and harbor planning.

Michigan Coastal Zone Management Program

Michigan DNR
Land & Water Manangement Division
P.O. Box 30028
Lansing MI 48909
Phone: (517) 373 1950

With coasts on four Great Lakes, Michigan has the world's largest freshwater coastline. The state uses coastal management to encourage responsible growth and development along the coast, improve public access to the coast, and aid winter navigation. The program manages coastal activities such as shipwreck salvaging, building piers and marinas, development, and changes to the coast. The program is working creatively to find solutions to the loss of agricultural land and wildlife habitat to sprawling development.

Minnesota Coastal Program

Minnesota DNR, Division of Waters
1201, E. Highway 2
Grand Rapids, MN 55744
Phone: (218) 327 4417

Minnesota is considering participation in the federal Coastal Zone Management program through the program development process. The 4-year process is addressing concerns and evaluating benefits of the program. Local issues which the program could help to address include: shoreline erosion, inadequate sewage and stormwater systems, local watershed and land use planning, habitat restoration, waterfront revitalization, and water access.

Mississippi Coastal Program

Mississippi Department of Marine Resources
1141, Baywiew Avenue
Suite 101
Biloxi, MS 39530
Phone: (228) 374 5000
Fax: (228) 374 5008

Wetlands preservation and restoration is a key issue for Mississippi coastal management, as are dockside gambling and casinos. Construction, public access, land acquisition, and fisheries are some of the coastal activities supervised by the state agencies watching out for the state's coast.

New Hampshire Coastal Program

NH Office of State Planning
Coastal Program Office
2-1/2 Beacon Street
Concord, NH 03301
Phone: (603) 271 2155

Table A3.2 *Continued*

New Hampshire's coastal program strives to protect and improve water quality, restore fishery habitat (wetlands), and balance coastal resource use. The program monitors coastal and estuarine waters to identify the runoff pollution in the state. As a result, some of the clam flats in the Hampton-Seabrook estuary have been conditionally reopened for recreational clamming.

New Jersey Coastal Management Program

New Jersey DEP
Office of Coastal Planning and Program Coordination
401, East State Street, Box 418
Trenton, NJ 08625
Phone: (609) 777 3251

A densely populated coast brings challenges of water quality, overdevelopment, coastal hazards, and runoff pollution to the forefront of coastal management in New Jersey. Barnegat Bay, for example, loaded with fish and plant life, faces intense fishing, recreational uses, and polluted runoff.

New York Coastal Management Program

New York Department of State
Division of Coastal Resources
41, State Street
Albany, NY 12231
Phone: (518) 474 6000

New York's coast faces issues such as wildlife habitat protection and coastal hazards such as floods and erosion. The coastal management program does not directly restrict any activities on the coast, but it does work with other state agencies to make sure the permits they issue comply with coastal management efforts.

North Carolina Coastal Management Program

North Carolina DENR
Division of Coastal Management
1638, Mail Service Center
Raleigh, NC 27699-1638
Phone: (919) 733 2293

Wetlands loss, coastal hazards, and the impacts of population growth and development are among the pressures confronted by the North Carolina coastal management program. The state ensures responsible development and the use of the coast by overseeing coastal activities. Setback laws keep property out of harm's way during storms, and a prohibition on erosion structures keeps the beaches, vital for tourism, from starving.

Northern Mariana Islands Coastal Resources Management Program

Coastal Resources Management Office
Office of the Governor
2nd floor, Morgen Building
San Jose Saipan, Mariana Islands 96950
Phone: (670) 234 6623

Solid waste disposal and water pollution are two of the major threats to the Northern Marianas' coasts. To combat these threats, the program oversees activities along the shoreline, in lagoons and reefs, in wetlands and mangrove swamps, and for port and industrial activities.

Ohio Coastal Management Program

Ohio DNR
Office of Real Estate and Land Management
1952, Belcher Drive
Building C-4
Columbus, OH 43224
Phone: (614) 265 6413

Managing activities in erosion-prone areas and restoring and enhancing coastal marshes are the major challenges facing the Ohio coast. Coastal managers in 1997 completed a program for federal approval to deal with these and other coastal issues.

Oregon Coastal Management Program

Coastal and Ocean Management Program
Oregon Department of Land Conservation and Development
800, NE Oregon Street #18
Portland, OR 97232
Phone: (503) 731 4065

Oregon is nationally recognized as a leader in coastal ocean planning. Two major initiatives for Oregon's coastal managers are mitigating coastal hazards and managing Pacific Ocean resources. Oregon's waters, which extend 3 miles from the coast, include intertidal areas, offshore rocks, and reefs—wildlife habitat susceptible to damage from human recreation. Local governments oversee activity along the state's coast by following local land use plans that are consistent with statewide goals for the coast.

Pennsylvania Coastal Zone Management Program

Pennsylvania DEP
Coastal Zone Management Program
P.O. Box 8555
Harrisburg, PA 17105
Phone: (717) 787 5259

Pennsylvania's Lake Erie and Delaware Estuary coasts face threats from runoff pollution, shoreline erosion, bluff recession, and wetlands loss. The state oversees construction and other activities that alter wetlands. The coastal program also offers free site analysis and recommendation services to coastal property owners. The services include site inspections and recommendations on surface and groundwater control, stabilizing bluffs, and the use of vegetation to stabilize loose soil.

Puerto Rico Coastal Management Program

Puerto Rico DNER
Bureau of Reserves, Refuges, and Coastal Resources
Department of Natural and Environmental Resources
Pda. 3-1/2, Ave. Munoz Rivera
Puerta de Tierra, Box 9066600
San Juan, Puerto Rico 00906-6600
Phone: (787) 721 7593

Puerto Rico's coastal program confronts the challenges of sedimentation, erosion, coastal hazards, and illegal use of the island's maritime zone (its shoreline, territorial waters, and submerged lands).

Rhode Island Coastal Management Program

Coastal Resources Management Council
4808, Tower Hill Road
Stedman Building
Wakefield, RI 02879
Phone: (401) 783 3370
Fax: (401) 783 3767

The Coastal Resources Management Council is an environmental regulatory and management agency responsible for the preservation, protection, development and where possible the restoration of the coastal areas of the state.

South Carolina Coastal Management Program

South Carolina DHEC
Office of Ocean and Coastal Resource Management
1362, McMillan Avenue
Suite 400
Charleston, SC 29405
Phone: (843) 744 5838

South Carolina's coastal program protects marine resources from declining water quality, protects fish habitats such as wetlands, and reduces the risk to coastal property from storms and other hazards. To meet these challenges, the program oversees wetlands filling and commercial and residential construction, including docks and piers. The program also sets construction back a safe distance from the ocean.

Texas Coastal Management Program

Texas GLO
Coastal Division
1700, North Congress Street
Austin Building
Austin, TX 78701
Phone: (512) 463 5054

Texas coastal managers are confronting issues of dredging, erosion, beach access, and wetlands and dune protection. The coastal resources

Table A3.2 *Continued*

program seeks to balance commercial and recreational activity with preservation of its unique coastal resources.

Virgin Islands Coastal Zone Management Program

Virgin Islands Department of Planning & Natural Resources
Coastal Zone Management Program
Cyril E. King Airport
2nd Floor
St. Thomas, US Virgin Islands 00802
Phone: (809) 774 3320
Fax: (809) 775 5706

To keep the islands beautiful and a source of pride and productivity into the next generation, the Virgin Islands emphasizes the importance of healthy terrestrial and coastal resources as development takes place. The territory's coastal program oversees construction and other coastal uses throughout the territory to meet these goals.

Virginia Coastal Resources Management Program

Virginia Coastal Program
Virginia Department Of Environmental Quality
629, East Main Street, 6th Floor
Richmond, VA 23219
Phone: (804) 698 4320

Virginia's coastal zone encompasses the eastern third of the state including the Chesapeake Bay and its tributary rivers, part of the Albemarle-Pamlico watershed, and the Atlantic coast with its vast barrier island lagoon system. The Virginia Coastal Resources Management Program supports its coastal residents and industries, and the plants and animals that rely on these coastal habitats.

Washington Coastal Zone Management Program

Washington Department of Ecology
P.O. Box 47600
Olympia, WA 98504-7600
Phone: (360) 407 6000

Wetlands degradation, population growth, and coastal erosion and flooding are among the major challenges facing the Washington coastal management program. To counter these problems, the program oversees most activities on the state's shoreline except agriculture and activities related to single-family homes.

Wisconsin Coastal Management Program

Wisconsin DOA
Wisconsin Coastal Management Program
P.O. Box 7868
Madison, WI 53707
Phone: (608) 267 7982

To balance competing uses of its Great Lakes coast, Wisconsin's coastal management program encourages wetlands protection and awareness; solutions to runoff pollution, primarily from agriculture; greater public access to the shoreline; solutions to erosion; and resolving water quality threats from failing septic systems.

Table A3.3 Other US state agencies, programs and organizations within the coastal zone (cf. Tables A3.1 and A3.2)

California Coastal Commission

Headquarters:
45, Fremont Street, Suite 2000
San Francisco, CA, 94105-2219
Phone: (415) 904 5200
Fax: (415) 904 5400

The California Coastal Commission was established by voter initiative in 1972 (Proposition 20) and made permanent by the Legislature in 1976 (the Coastal Act). The primary mission of the Commission, as the lead agency responsible for carrying out California's federally approved coastal management program, is to plan for and regulate land and water uses in the coastal zone consistent with the policies of the Coastal Act.

California Coastal Conservancy

1330, Broadway
11th Floor
Oakland, CA 94612
Phone: (510) 286 1015
Fax: (510) 286 0470

Works to preserve, improve, and restore public access and natural resources along the California coast and on San Francisco Bay.

California Department of Boating and Waterways

2000, Evergreen, Suite 100
Sacramento, CA 95815-3888
Phone: (888) 326 2822
(916) 263 4326
Fax: (916) 263 0648

Protect[s] significant natural resources through its programs to provide for public access to the waterways, and promot[es] recreational boating safety. These programs include funding and designing the construction and improvement of boating facilities, beach erosion control, aquatic weed control, boating safety education, and supporting and training local boating law enforcement officers.

California Department of Fish and Game

1416, Ninth Street
Sacramento, CA 95814
Phone: (916) 653 7664
Fax: (916) 653 1856

The Mission of the Department of Fish and Game is to manage California's diverse fish, wildlife, and plant resources, and the habitats upon which they depend, for their ecological values and for their use and enjoyment by the public.

Columbia River Estuary Study Taskforce

750, Commercial Street, Room 205
Astoria, OR 97103
Phone: (503) 325 0435
Fax: (503) 325 0459 (please call first)
E-mail: crest@columbiaestuary.org

CREST is Council of Governments that includes the local counties, cities, and port districts surrounding the Columbia River Estuary in both Oregon and Washington. CREST is not a regulatory agency. It is a regional organization providing a forum for members to identify and discuss issues of regional importance; to monitor and comment on governmental activities related to the development and management of the natural, economic, and human resources of the Columbia River Estuary; and to improve communication and cooperation between member governments.

Delaware Department of Natural Resources and Environmental Control

Delaware Department of Natural Resources and
Environmental Control
89, Kings Hwy
Dover, DE 19901
Phone: (302) 739 4403

Protecting Delaware's Environment for Future Generations.

Florida Department of Environmental Protection

DEP Office of Ombudsman
3900, Commonwealth Blvd. M.S. 49
Tallahassee, FL 32399

Florida Fish and Wildlife Conservation Commission

Our mission is the managing fish and wildlife resources for their long-term wellbeing and the benefit of people.

Northwest Region:
Lt. Col. Louie Roberson, Regional Director
3911, Hwy. 2321
Panama City, FL 32409-1658
Phone: (850) 265 3676

Table A3.3 *Continued*

North Central Region (formerly Northeast Region):
Lt. Col. Julie L. Jones, Regional Director
Route 7, Box 440
Lake City, FL 32055-8713
Phone:(904) 758 0525

Northeast Region (formerly Central Region):
Dennis David, Regional Director
1239, SW 10th Street
Ocala, FL 34474-2797
Phone: (352) 732 1225

Southwest Region (formerly South Region):
Greg Holder, Regional Director
3900, Drane Field Road
Lakeland, FL 33811-1299
Phone: (863) 648 3203

South Region (formerly Everglades Region):
Mark Robson, Regional Director
8535, Northlake Boulevard
West Palm Beach, FL 33412
Phone: (561) 625 5122

Florida Public Interest Research Group
704, West Madison Street
Tallahassee, FL 32304

Michigan Department of Environmental Quality

P.O. Box 30458
Lansing, MI 48909-7958
E-mail: deq-lwm-webmaster@state.mi.us

The core programs of the Land and Water Management Division protect Michigan's sensitive natural resources and the public trust at the land/water interface. Development and construction activities are regulated on the Great Lakes, on inland lakes and streams, floodplains, wetlands, and sand dunes to minimize environmental disruption and to protect the public health and safety. Information and technical assistance are also provided to the public and to the private sector. The programs promote wise management and use of the State's natural resources for present and future generations.

Minnesota's Lake Superior Coastal Program

DNR Information Center
500, Lafayette Road
St. Paul, MN 55155-4040
Phone: 651 296 6157 or 888 MINNDNR
TTY: 651 296 5484 or 800 657 3929
For DNR Info: info@dnr.state.mn.us

The goal of Minnesota's Lake Superior Coastal Program is to preserve, protect, develop, and where possible restore or enhance coastal resources along Minnesota's North Shore of Lake Superior.

Mississippi Department of Marine Resources

1141, Bayview Avenue, Suite 101
Biloxi MS 39530
Phone: 228 374 5022/5254
E-mail: webmaster@dmr.state.ms.us

New York Division of Coastal Resources

Division of Coastal Resources
George Stafford, Director of Coastal Resources
41, State Street, Albany, NY 12231-0001
Phone: (518) 474 6000
Fax: (518) 473 2464
E-mail: coastal@dos.state.ny.us

The Division of Coastal Resources is responsible for administering New York State's Coastal Management Program, adopted in 1982 under the Waterfront Revitalization of Coastal Area and Inland Waterways. In voluntary partnership with local governments, the Coastal Management Program seeks to meet the needs of coastal residents and visitors, while striving to advance economic development opportunities and protect our natural coastal resources.

Office of Coastal Restoration Indiana

P.O. Box 44487
Baton Rouge, LA 70804-4487

Oregon Department of fish and Wildlife

2501, SW First Avenue
P.O. Box 59
Portland, OR 97207
Phone: 503 872 5268

San Francisco Bay Conservation and Development Commission

50, California Street
Suite 2600, San Francisco CA 94111
Phone: (415) 352 3600
Fax: (415) 352 3606
E-mail: info@bcdc.ca.gov

BCDC is the California coastal management agency responsible for the San Francisco Bay-Delta portion of the coastal zone.

South Florida Water Management District

3301, Gun Club Road
West Palm Beach, FL 33416-4680
Phone: (561) 686 8800; 1 800 432 2045 (Florida Only)

The Plan to Restore America's Everglades

US Army Corps of Engineers:
Jacksonville District
P.O. Box 4970, Jacksonville, FL 32232-0019
400, West Bay Street (map/directions)
Jacksonville, FL 32235
Phone: 904 232 2235; 800 291 9405
Fax: 904 232 2237

South Florida Water Management District
3301, Gun Club Road (map/directions)
P.O. Box 24680, West Palm Beach, FL 33416-4680
Phone: 561 686 8800
877 Glades1 (1 877 452 3371) In Florida
Fax: 561 682 6010

This plan is a work in progress. Technical studies and more detailed designs, involving several pilot projects will come next. We will continue to involve the public throughout the process of implementing the Comprehensive Plan. Find out more about the plan and help us improve it. Help us to better involve your community in this effort. Please contact us for information.

WA State Department of Ecology

Mailing Address: P.O. Box 47600, Olympia WA 98504-7600
Physical Address: 300, Desmond Drive, Lacey WA 98503

Shorelands and environmental assistance program: Works to protect and enhance Washington's shorelands, wetlands and other land resources through education, technical assistance, and collaborative environmental management.

Table A3.4 A brief overview of worldwide government agencies and programs related to the coastal zone

Australian Antarctic Division (Australia)

Headquarters:
Australian Antarctic Division
Channel Highway
Kingston Tasmania 7050
Phone: +613 6232 3209
Fax: +613 6232 3288
E-mail: information@aad.gov.au

Administering Australia's wide-ranging activities in Antarctic and subantarctic regions is the responsibility of the Australian Antarctic Division (AAD), a part of the Australian Government's Department of the Environment and Heritage.

Table A3.4 *Continued*

Coastcare (Australia)

Marine and Water Division
Environment Australia, G.P.O. Box 787
Canberra Act 2601
Phone: (02) 6274 1967
Fax: (02) 6274 1006

Coastcare is a national program that encourages community involvement in the protection, management and rehabilitation of our coastal and marine environments. The program assists local communities to form partnerships with local land managers to undertake projects that aim to improve and protect our coastal and marine habitats.

Coast Care (South Africa Coastal Information Center) (South Africa)

Zain Jumat, Project Manager
Coastal Management Office
Department of Environmental Affairs and Tourism
Private Bag X2, Roggebaai 8012
Phone: (+27 21) 402 3030
Fax: (+27 21) 418 2582
E-mail: mzjumat@mcm.wcape.gov.za

The Coastal Management Office of the Department of Environmental Affairs and Tourism recognizes that the more the people are informed about our coast, the easier it will be to protect it and ensure that its development is to the benefit of current and future generations. To this end, it established a special programme, CoastCARE, to assist in the education and exchange of information about coastal issues.

Coast Protection Board (Australia)

Office for Coast and Marine
Department for Environment and Heritage
6th Floor, Australis House
77, Grenfell Street
Adelaide SA 5000, Australia

The Coast Protection Board is specifically responsible under the Coast Protection Act (1972).

Danish EPA (Denmark)

29, Strandgade, DK-1401 København K
Phone: +45 32 66 01 00
Fax: +45 32 66 04 79
E-mail: mst@mst.dk

The Danish Environmental Protection Agency (EPA) spheres of activity are concentrated on preventing and combating water, soil, and air pollution. The Agency belongs under the Danish Ministry of Environment and Energy and some 425 employees.

Environmental Protection Agency (Ireland)

P.O. Box 3000
Johnstown Castle Estate, Co.
Wexford, Ireland
Phone: +353 53 60600
Fax: +353 53 60699

Mission: To promote and implement the highest practicable standards of environmental protection and management which embrace the principles of sustainable and balanced development.

European Union for Coastal Conservation (Poland)

Europejska Unia Ochrony Wybrzeza—Polska ul. Wąska 13
71-415 Szczecin, Poland
Phone/fax: (0 91) 421 00 97
E-mail: szakow@sus.univ.szczecin.pl

Established at 1996 to realize the following most important aims: strengthening the importance of Baltic Sea coasts and solving problems connected with wise using and protection; creating platforms for cooperation between G.O., NGOs, local communities, scientific organizations and individuals involved in conservation problems of coastal zone; participation in international movements connected with coastal conservation and management; integration science and management in the coastal zone.

Federal Waterways Engineering and Research Institute, Coastal Division (Germany)

BAW—Dienststelle Hamburg
Wedeler Landstraße 157
D-22559 Hamburg
Phone: [+49] 40 81908 0
Fax: [+49] 40 81908 373
E-mail: postmaster@hamburg.baw.de

The department Hydraulic Engineering in Coastal Areas is consultant for the coastal offices in the Federal Waterways and Shipping Administration in the field of hydraulic, soils, and foundation engineering. It carries out field measurements, laboratory investigations and theoretical studies, operates physical and numerical models of the large German estuaries and does standardization work. It is also engaged in research and development projects.

Forum Skagerrak (Sweden)

Pege Schelander, Project Leader
BOSAM, Box 305,
S-451 18 Uddevalla
Phone: +46 522 15980
Fax: +46 522 511796
E-mail: pege.schelander@bosam.se

Forum Skagerrak is a common initiative of the regions surrounding Skagerrak, in Denmark, Norway, and Sweden, to find solutions to prioritized environmental problems where cooperation can lead to effective measures.

Gerência Executiva do Ibama no Distrito Federal (Brazil)

Eulália Arlete Machado de Carvalho
SAS, Qd. 05, Lote 05, Bl. "H", 1° Andar
70.070 000, Brasília/DF.
Phone: 225 1686, 223 6155, 323 1150/1132/9962 7834
Fax: 321 6964; R: 367 5150

Inter-Agency Committee on Marine Science and Technology (United Kingdom)

Southampton Oceanography Centre, European Way,
Empress Dock,
Southampton SO14 3ZH
Phone: 023 8059 6611

The IACMST maintains an overview of marine activities across Government. It encourages links between Government and the national marine community, the wider application of marine science and technology, optimum use of major UK marine facilities, training and education, and international links.

Instituto de Conservacao da Natureza (Portugal)

Rua da Lapa, 73
1200 701 Lisboa
Phone: (351) 213938900/3974044
Fax: (351) 213938901/3901048
E-mail: icn@icn.pt

O ICN é o instituto responsável pelas actividades nacionais nos domínios da conservação da natureza e da gestão das áreas protegidas.

Intergovernmental Oceanographic Commission (IOC) of UNESCO (France)

The IOC is composed of its *Member States*, an *Assembly*, an *Executive Council* and a *Secretariat*. The Secretariat is based in Paris, France. Additionally the IOC has a number of Subsidiary Bodies.
Service des Ressources humaines
213, rue La Fayette
75480 Paris Cedex 10, France
Phone: (1) 48 03 76 77
Website: ioc.unesco.org

Or contact the UNESCO office at your country.

The work of the IOC, over the three decades since its inception, has focused on promoting marine scientific investigations and related ocean services, with a view of learning more about the nature and resources of the oceans. This has laid the foundation toward an expanded role of the IOC in meeting new challenges.

Table A3.4 *Continued*

Marine and Coastal Management (South Africa)

P.O. Box X2
Roggebaai, 8012
Cape Town
Phone: (27 21) 4023111
Fax: (27 21) 252920

To provide for responsible custodianship of South Africa's marine and coastal resources and ecosystems for the benefit of current and future generations of South Africans.

Marine and Water Division Environment Australia (Australia)

G.P.O. Box 787, Canberra Act 2601
Phone: (02) 6274 1967
Fax: (02) 6274 1006

MMA—Ministerio do Meio Ambiente (Brasil)

Esplanada dos ministerios bloco B, do andar 5 ao 9
Cep 70-066-900
Brasilia, DF

National Coastal Management Office (South Africa)

The Coastal Management Office
Department of Environmental Affairs & Tourism
Private Bag X2, Roggebaai 8012
Phone: 021 402 3228
Fax: 021 418 2582
E-mail: czm@sfri.wcape.gov.za

The Coastal Management Office of the Department of Environmental Affairs & Tourism acts as the government's national coordinating coastal management agency, empowering coastal users, decisionmakers and the public to manage the coastal zone and its resources wisely, in order to ensure its continued wellbeing.

National Oceans Office (Australia)

Veronica Sakell, Director
National Oceans Office
Phone: (03) 6221 5001
E-mail: office@oceans.gov.au

The National Oceans Office is the lead Commonwealth agency for implementing Australia's Oceans Policy.

National Institute of Ocean Technology (India)

National Institute of Ocean Technology (NIOT)
Velacherry–Tambaram Main Road
Narayanapuram
Chennai, Tamil Nadu 601 302
Phone: 91 44 2460063/2460064/2460066/2460067
Fax: 91 44 2460645

Telex The Department of Ocean Development, Government of India in coordination with Indian Institute of Technology (IIT) Madras has established the National Institute of Ocean Technology (NIOT). The institute's main aim is technology and development.

Natural Heritage Services (Finland)

Metsähallitus
Director, Natural Heritage Services
Rauno Väisänen, Ph.D.
Vernissakatu 4, P.O. Box 94, 01301 Vantaa
Phone: +358 205 64 4386
Fax: +358 205 64 4350
E-mail: rauno.vaisanen@metsa.fi

Scottish Coastal Forum (Scotland)

Martyn Cox, Coastal Project Officer, Scottish Coastal Forum
1 J—South, Victoria Quay, Edinburgh, EH6 6QQ
Phone: 0131 244 1540
Fax: 0131 244 4071
E-mail: coastalforum@scotland.gov.uk
martyn.cox@scotland.gov

Secretariat for Eastern African Coastal Area Management (Mozambique)

874, Av. Amílcar Cabral, 1st floor
Caixa Postal 4220
Maputo
Phone: (258) 1 300641/2
Fax: (258) 1 300638
E-mail: seacam@virconn.com

Has as main objective to assist the Eastern African coastal countries to implement and coordinate coastal management activities following up on the Arusha Resolution and the Seychelles Statement on Integrated Coastal Zone Management.

Servicio de Oceanografía, Hidrografía y Meteorología de la Armada (Uruguay)

Servicio de Oceanografía, Hidrografía y Meteorología de la Armada
Capurro 980, Casilla de Correos 1381
Montevideo, R.O.del Uruguay
Phone: (598 2) 309 3861/309 3775
Fax: (598 2) 309 9220

SRH—Secretaria de Recursos Hídricos (Brasil)

SGAN, Qd. 601
Lote 01 Ed. Codevasf, 4° Andar
CEP: 70.830-901, Brasília/DF

Standing Conference on Problems Associated with the Coastline (United Kingdom)

SCOPAC
C/o Isle of Wight Council
County Hall, Newport
Isle of Wight PO30 1UD

SCOPAC works to promote sustainable shoreline management, and to facilitate the duties and responsibilities of local authorities and other organizations managing the coastal zone of central southern England.

Swedish Environmental Protection Agency (Sweden)

Swedish Environmental Protection Agency (Naturvårdsverket), SE-106 48 Stockholm
Phone: +46 8 698 10 00
Fax: +46 8 20 29 25
E-mail: natur@environ.se

A governmental authority which coordinates and promotes environmental work.

Table A3.5 Overview of some main North American research and educational institutes (academic units) that deal with the coastal zone

Acadia Centre For Estuarine Research (Canada)

Acadia University
Wolfville, Nova Scotia B0P 1X0
Phone: (902) 585 1113
Fax: (902) 585 1054

The primary objective of the Centre is to focus research attention on the estuaries and nearshore coastal waters of Eastern Canada, with emphasis on the estuarine systems of the Bay of Fundy and the hydrographically related Gulf of Maine and Georges Bank.

Bay of Fundy Marine Resource Centre (Canada)

Bay of Fundy Marine Resource Centre
P.O. Box 273
Cornwallis Park, Nova Scotia CANADA B0S 1C0
Phone: 902 638 3044
Fax: 902 638 3284

Table A3.5 *Continued*

Website: http://www.bfmrc.ns.ca/

The Bay of Fundy Marine Resource Centre is a community-based nonprofit, nongovernmental organization offering services, facilities, and technical support to all sectors of the marine economy and ecosystem. The MRC was established in 1997 through the efforts of the Western Valley Development Authority and the Fundy Fixed Gear Council in order to give the Digby and Annapolis region the capacity to take on a greater role in the integrated management of its coastal resources.

Belle W. Baruch Institute (USA)

Belle W. Baruch Institute for
Marine Biology and Coastal Research
University of South Carolina
Columbia, SC 29208
Phone: (803) 777 5288
Fax: (803) 777 3935
E-mail: fletcher@biol.sc.edu

The Belle W. Baruch Institute conducts basic and applied research in marine and coastal environments—research that addresses the critical need for knowledge and improved understanding of these essential ecosystems.

Bodega Marine Laboratory (USA)

2099, Westside Road
P.O. Box 247
Bodega Bay, CA 94923-0247
Phone: (707) 875 2009

California Environmental Education Interagency Network (USA)

Gray Davis, Governor 2000, State of California.
Consortium of environmental educators representing California agencies with oversight responsibilities to protect California's environment.

Center for Alaskan Coastal Studies (USA)

P.O. Box 225, Homer, AK 996003
Phone: 907 235 6667
E-mail: cacs@xyz.net

The Center for Alaskan Coastal Studies' mission is to foster responsible interaction with our natural surroundings, and to generate knowledge of the marine and coastal ecosystems of Kachemak Bay through education and research programs.

Center for Applied Coastal Research, University of Delaware (USA)

Ocean Engineering Laboratory
University of Delaware
Newark, DE 19716
Phone: 302 831 6531
Fax: 302 831 1228
E-mail: rad@udel.edu

This interdisciplinary center provides a focal point for research in coastal processes and coastal engineering. Members of the center are coastal engineers, coastal geologists and oceanographers, primarily from the University of Delaware and the Middle-Atlantic region.

Center for Coastal and Land-Margin Research, Oregon Health & Science University (USA)

Department of Environmental Science and Engineering
OGI School of Science and Engineering
Oregon Health & Science University
20000, NW Walker Rd. Beaverton, OR 97006
Phone: (503) 690 1147
Fax: (503) 690 1273

The Center for Coastal and Land-Margin Research (CCALMR) of the Oregon Graduate Institute of Science & Technology addresses, through advances in scientific understanding, technology, and education, society's need to manage increased development and manipulation of coasts and landmargins while preserving and enhancing their environmental integrity, and protecting human populations from natural and man-made hazards.

CCS—Center for Coastal Studies (USA)

Center for Coastal Studies
59, Commercial St., Provincetown, MA 02657
E-mail: ccs@coastalstudies.org

A private nonprofit organization for research, conservation, and education in the coastal and marine environments. For 25-years, CCS has worked to increase our understanding and protection of coastal and marine environments.

Center for Coastal Studies (CCS), Scripps Institution of Oceanography (USA)

Center for Coastal Studies
Scripps Institution of Oceanography
University of California, San Diego
La Jolla, CA 92093-0209
Phone: (858) 534 4333
Fax: (858) 534 0300

The Center for Coastal Studies (CCS) is a research division of Scripps Institution of Oceanography (SIO), University of California, San Diego (UCSD). Located adjacent to the SIO pier, the Center engages in worldwide scholarly studies of the coastal environment, the development of data acquisition systems and research instrumentation, and advising on coastal protection and sediment management.

Centre for Earth and Ocean Research, University of Victoria (Canada)

Petch Building, Room 169
P.O. Box 3055, Victoria, British Columbia, V8W 3P6
Phone: (250) 721 8848
Fax: (250) 472 4100

The Centre for Earth and Ocean Research (CEOR) was established in 1987 to initiate, foster, promote, and coordinate research in earth, ocean, and atmospheric sciences at the University of Victoria and to engage in collaborative projects and programs with other institutions and agencies.

Center for Marine Science, University of NC at Wilmington (USA)

5600, Marvin K. Moss Lane
Wilmington, NC, 28409
019.052.2300

CMS General Information:
Nancy Stevens
Phone: 910 962 2300
E-mail: stevensn@uncwil.edu

Dedicated to providing an environment that fosters a multidisciplinary approach to questions in basic marine research. The mission of the center is to promote basic and applied research in the fields of oceanography, coastal and wetland studies, marine biomedical and environmental physiology, and marine biotechnology and aquaculture.

Center for the Study of Marine Policy, University of Delaware (USA)

301, Robinson Hall, University of Delaware, Newark, DE 19716
Phone: 1 302 831 8086
Fax: 1 302 831 3668
E-mail: bcs@udel.edu

The Center conducts a broad range of research and policy studies emphasizing the application of policy analysis and other analytical tools to the management of ocean and coastal areas on national, regional, and global scales. A major emphasis of the Center has been on integrated coastal and ocean management, particularly on the development of governance approaches that move beyond present single-sector approaches toward multiple-use management regimes.

CEROS—National Defense Center of Excellence for Research in Ocean Sciences (USA)

73-4460, Queen Kaahumanu Highway, Suite 111
Kailua-Kona, HI 96740
Phone: (808) 327 4310
Fax: (808) 327 4320
E-mail: info@ceros.org

CEROS' mission is to support the Department of Defense technology requirements; encourage leading edge R&D in ocean sciences and technology in Hawaii; foster use of ocean R&D facilities in Hawaii; provide an interface between specialized small businesses with expertise in ocean-related R&D and DoD users of advanced technology and

Table A3.5 *Continued*

develop avenues to ocean science expertise and facilities at the University of Hawaii (UH).

Chesapeake Bay Research Consortium (USA)

Chesapeake Research Consortium
645, Contees Wharf Rd.
Edgewater, MD 21037
Phone: 410 798 1283
Fax: 410 798 0816

The Scientific and Technical Advisory Committee (STAC) provides scientific and technical guidance to the Chesapeake Bay Program on measures to restore and protect the Chesapeake Bay. As an advisory committee, STAC reports quarterly to the Implementation Committee and annually to the Executive Council.

Coastal and Hydraulics Laboratory (USA)

Coastal and Hydraulics Laboratory
Engineer Research and Development Center
3909, Halls Ferry Road, Vicksburg, MS 39180
Phone: (601) 634 3111

Coastal Morphodynamics Laboratory, Louisiana State University (USA)

Coastal Studies Institute
336, Howe-Russell Geoscience Complex
Louisiana State University, LA 70803
Fax: 225 388 2520

The Coastal Morphodynamics Laboratory (CML) was founded in 1991 to facilitate graduate student and faculty research in coastal morphodynamics. The CML offers a wide range of state-of-the-art field, laboratory equipment and computers for research in coastal processes including wave hydrodynamics, hurricane impacts, sediment transport, beach and nearshore profile measurements, GIS/RS, mapping, database, and sedimentology.

Coastal Ocean Observation Lab, Rutgers University (USA)

Rutgers University
Institute of Marine and Coastal Sciences
Coastal Ocean Observation Lab
71, Dudley Road
New Brunswick, NJ 08901-8521
Phone: (732) 932 6555
Fax: (732) 932 1821

COOL's research focus is on New Jersey coastal waters, primarily within an area we call "LEO-15." Since 1996 we have been acquiring data from above and below the ocean. Below is a table of contents for the site, beginning with the real-time ocean data that we know you will love if you are planning on spending any time in or on the ocean.

Coastal Resources Research Network (Canada)

1321, Edward Street
Halifax, Nova Scotia B3H 3H5
Phone: (902) 494 1842
Fax: (902) 494 1216

The Coastal Resources Research Network (CoRR) supports researchers in developing countries in their efforts to research and promote Community Based Coastal Resources Management (CBCRM). The Network is based at Dalhousie University, is funded by the International Development Research Centre (IDRC, Canada) and is primarily working with partners in South East Asia.

Coasts Under Stress (Canada)

Dr. Rosemary E. Ommer, Director
Phone: (403) 220 7238
Fax: (403) 282 7822
E-mail: cus@mun.ca

The Calgary Institute for the Humanities
2500, University Drive NW
Calgary, Alberta T2N 1N4

Analyzes the Impact of Social and Environmental Restructuring on Environmental and Human Health in Canada.

Our goal is to identify the important ways in which changes in society and the environment in coastal British Columbia and coastal Newfoundland and Labrador have affected, or will affect, the health of people, their communities and the environment over the long run.

Columbia River Estuary Study Taskforce (USA)

750, Commercial Street, Room 205
Astoria, OR 97103
Phone: (503) 325 0435
Fax: (503) 325 0459 (please call first)
E-mail: crest@columbiaestuary.org

CREST is Council of Governments that includes the local counties, cities, and port districts surrounding the Columbia River Estuary in both Oregon and Washington. CREST is not a regulatory agency. It is a regional organization providing a forum for members to identify and discuss issues of regional importance; to monitor and comment on governmental activities related to the development and management of the natural, economic, and human resources of the Columbia River Estuary; and to improve communication and cooperation between member governments.

Département D'océanographie (Canada)

UQAR 300, allée des Ursulines
Rimouski, Québec G5L 3A1
Phone: (418) 724 1770
Fax: (418) 724 1842
E-mail : dep_oceano@uqar.uquebec.ca

Department of Earth and Ocean Sciences (Canada)

The University of British Columbia
6339, Stores Road, Vancouver, British Columbia, V6T 1Z4
Phone: 604 822 2449
Fax: 604 822 6088

Department of Marine Sciences, University of Georgia (USA)

Department of Marine Sciences
Marine Sciences Bldg
University of Georgia
Athens, GA 30602-3636
Phone: (706) 542 7671

An interdisciplinary department of biological, chemical, and physical oceanography, with special emphasis on coastal and estuarine processes.

Department of Oceanography at Dalhousie University (Canada)

1355, Oxford Street
Halifax, Nova Scotia, B3H 4J1
Phone: 902 494 3557
Fax: 902 494 3877
E-mail: oceanography@dal.ca

Dalhousie is a world leader in oceanographic research with state-of-the-art facilities. The Department of Oceanography pursues specialized and interdisciplinary research and consists of 23 faculty members, 40 graduate students and 41 research support staff.

Department of Ocean Engineering, University of Rhode Island (USA)

College of Engineering
Department of Ocean Engineering
217, Sheets Building, Narragansett Bay Campus
Narragansett, RI 02882
Phone: 401 874 6139
Fax: 401 874 6837
E-mail: spaulding@oce.uri.edu

The Department of Ocean Engineering provides a challenging and diverse intellectual environment offering academic programs leading to B.S., M.S., and Ph.D. degrees.

Department of Ocean Engineering, Florida Atlantic University (USA)

Department of Ocean Engineering
Florida Atlantic University
777, Glades Road

Table A3.5 *Continued*

P.O. Box 3091
Boca Raton, FL 33431
Phone: 561 297 3000; 954 236 1000

The Department's mission is to provide an outstanding academic environment and offer unique programs in engineering education, research, and technology development. Undergraduate Ocean Engineering students benefit from a traditional university campus life on the Boca Raton campus followed by a senior year spent entirely at the SeaTech complex.

Department of Geography and Geology

777, Glades Road
P.O. Box 3091
Boca Raton, FL 33431
Phone: 561 297 3000; 954 236 1000

Duke University Marine Laboratory (USA)

Nicholas School of the Environment
135, Duke Marine Lab Road
Beaufort, NC 28516-9721
Phone: 252 504 7503
Fax: 252 504 7648
Website: www.env.duke.edu/marinelab/

Earth, Atmospheric, and Planetary Sciences, Massachusetts Institute of Technology (USA)

Program in Atmospheres, Oceans and Climate
Bldg. 54, Room 1524
Department of Earth, Atmospheric, and Planetary Sciences
Massachusetts Institute of Technology
77, Massachusetts Avenue
Cambridge, MA 02139-4307
Fax: (617) 253 4464

Field Research Facility, US Army Corps of Engineers (USA)

USACE, Field Research Facility
1261, Duck Road
Kitty Hawk, NC 27949-4472
Phone: 252/261 3511
Fax: 252/261 4432

Open since 1977, the FRF is internationally recognized for its coastal studies. Instruments at the facility constantly record the changing waves, winds, tides, and currents. Central to the facility is a 560-m-long (1840 ft) pier and unique specialized equipment like the LARC, CRAB, and SIS.

Fisheries and Marine Institute of Memorial University of Newfoundland (Canada)

Fisheries and Marine Institute of Memorial University of Newfoundland
P.O. Box 4920
St. John's, Newfoundland A1C 5R3
Phone: (709) 778 0200; 1 (800) 563 5799
Fax: (709) 778 0346
Website: http://www.mi.mun.ca

Florida Caribbean Science Center (USA)

7920, NW 71st Street
Gainesville, FL 32653
Tel.: 352 378 8181
Fax: 352 378 4956

Florida Center for Environmental Studies (USA)

Florida Center for Environmental Studies
Florida Atlantic University
Northern Palm Beach Campus
3932, RCA Boulevard
Palm Beach Gardens, FL 33410
Phone: (561) 691 8554
Fax: (561) 691 8540

The center acts as a facilitator and coordinator of research and training related to the environment and as a locus for environmental information. Grounding its activities in the Florida subtropical environment, its mandate encompasses global tropical and subtropical environments, especially the issues and problems of water-dominated ecosystems.

Florida Institute of Oceanography (USA)

830, First Street South
St. Petersburg, FL 33701
Phone: (727) 553 1100
Fax: (727) 553 1109

The Florida Institute of Oceanography (FIO) was established by the State University System (SUS) to support and enhance Florida's coastal marine science, oceanography and related management programs through education, research, and public outreach.

Florida Institute of Technology (USA)

Florida Institute of Technology
College of Engineering,
Department of Marine and Environmental Systems
150, West University Boulevard
Melbourne, FL 32901-6975
Phone: (321) 674 8096
Fax: (321) 674 7212
E-mail: dmes@marine.fit.edu

Our mission is to integrate oceanography, ocean engineering, environmental science, meteorology, and related academic concentrations into interdisciplinary knowledge-based optimal solutions to vital contemporary issues through education, research, and service.

Florida Marine Research Institute (USA)

Education & Information Program
100, Eighth Avenue SE
St. Petersburg, FL 33701-5095

Through effective research and technical knowledge, we provide timely information and guidance to protect, conserve, and manage Florida's marine and coastal resources.

Florida State University, Department of Oceanography (USA)

Department of Oceanography
329 OSB, West Call Street
Florida State University
Tallahassee, FL 32306-4320
Phone: (850) 644 6700
Fax: (850) 644 2581

Geological Survey of Canada (Atlantic) (Canada)

Jacob Verhoef, Director
Geological Survey of Canada (Atlantic)
Bedford Institute of Oceanography
1, Challenger Drive, P.O. Box 1006
Dartmouth, Nova Scotia B2Y 4A2
Phone: (902) 426 3448
Fax: (902) 426 1466

Marine Resources Geoscience (Don McAlpine)
Phone: (902) 426 2730
Fax: (902) 426 4465

Marine Environmental Geoscience (Dick Pickrill)
Phone: (902) 426 5387
Fax: (902) 426 4104

GSC Atlantic is the principle marine geoscience facility in Canada a division of the Geological Survey of Canada (GSC) co-located at the Bedford Institute of Oceanography, with the Department of Fisheries and Oceans, a part of an Atlantic Provinces marine research and technology community centerd in the Halifax-Dartmouth metropolitan region of Nova Scotia.

Harbor Branch Oceanographic Institution (USA)

5600, US 1 North
Fort Pierce, FL 34946
Phone: (561) 465 2400; (800) 333 4264
Fax: (561) 465 2446
E-mail: webmaster@hboi.edu

Table A3.5 *Continued*

Hatfield Marine Science Center, Oregon State University (USA)

2030, SE Marine Science Dr
Newport, OR 97365
Phone: 541 867 0100
Fax: 541 867 0138
E-mail: hmsc@hmsc.orst.edu

Huntsman Marine Science Centre (Canada)

Dr. Mark J. Costello, Executive Director
Phone: 506 529 1200
Fax: 506 529 1224
E-mail: costello@huntsmanmarine.ca
Reception: huntsman@huntsmanmarine.ca

Through research and education, the Huntsman Marine Science Centre will enhance knowledge and provide the leadership necessary to achieve sustainable development and effective management of the coastal environment.

Institute of Ocean Sciences (Canada)

P.O. Box 6000
9860, West Saanich Road
Sidney, British Columbia, V8L 4B2

The Institute of Ocean Sciences is the departments center for research on the coastal waters of British Columbia, the North Pacific Ocean, the western Canadian Arctic and the navigable fresh waters east to the Manitoba/Saskatchewan border.

International Oceans Institute of Canada (Canada)

International Oceans Institute of Canada
Dalhousie University
1226, LeMarchant Street
Halifax Nova Scotia B3H 3P7
Phone: 1 902 494 3879
Fax: 1 902 494 1334
E-mail: ioic@dal.ca

The International Oceans Institute of Canada is a nongovernmental organization dedicated to promoting and supporting the sustainable and rational use, management, and regulation of ocean and coastal resources, as well as the protection and conservation of the marine environment in Canada and internationally.

Kalakaua Marine Education Center (USA)

200, W. Kawili St., Hilo, HI 96720-4091
E-mail for general information: kmec@hawaii.edu

As a leader in undergraduate marine science education in the Pacific, our mission is to offer high-quality experiential undergraduate education in marine sciences relevant to the needs of the Pacific region.

Marine Law Institute (USA)

University of Maine School of Law
246, Deering Avenue, Portland, ME 04102

The Marine Law Institute is the research and public service component of the Ocean and Coastal Law Program and is the only law school-affiliated marine policy research program in the Northeast. MLI has dedicated its program of legal and policy research to the analysis of ocean and coastal resource issues for the express purpose of improving management practices and public understanding.

Maryland Sea Grant, University of Maryland (USA)

0112, Skinner Hall
College Park MD 20742
Phone: (301) 405 6371
Fax: (301) 314 9581
E-mail: mdsg@mdsg.umd.edu
Website: http://www.mdsg.umd.edu/

The Maryland Sea Grant College supports innovative marine research and education, with a special focus on the Chesapeake Bay. With funding from the National Oceanic and Atmospheric Administration and the State of Maryland, Sea Grant-supported research targets practical problems, with the aim of promoting wise decisionmaking.

Mote Marine Laboratory (USA)

1600, Ken Thompson Parkway
Sarasota, FL 34236
Phone: (941) 388 4441

NSU Oceanographic Center (USA)

NSU (Nova Southeastern University) Oceanographic Center
8000, North Ocean Drive
Dania Beach, FL 33004
Phone: (800) 39 OCEAN; (954)262 3600
Website: http://www.nova.edu/ocean
webmaster@mako.ocean.nova.edu

Ocean Engineering Program, Texas A&M University (USA)

Ocean Engineering Program
Department of Civil Engineering
Texas A&M University
College Station, TX 77843-3136

Ocean Engineering Studies (USA)

Prof. Spyros A. Kinnas
Department of Civil Engineering (ECJ 8.604)
The University of Texas at Austin
Austin, TX 78712

The University of Texas at Austin is committed to providing a top-quality education to highly qualified students who wish to pursue graduate studies in the department of Civil Engineering with a focus on the field of Ocean Engineering. The ocean engineer of the present and future must not only have a strong grasp of the principles in the related fields, but also be capable of using and or developing sophisticated computational tools for the design and assessment of engineered or natural systems.

Oregon Coast Geospatial Clearinghouse (USA)

Department of Geosciences
104, Wilkinson Hall
Oregon State University
Corvallis, OR 97331-5506
Phone: 541 737 1229

Pacific Marine Environmental Laboratory (USA)

NOAA R/PMEL
7600, Sand Point Way NE
Seattle, WA 98115-6239
Phone: 206 526 6239
Fax: 206 526 6815

Regional Association for Research on the Gulf of Maine (USA)

Eugenia F. Braasch, Executive Director
RARGOM
Dartmouth College
8000, Cummings Hall
Hanover, NH 03755
Phone: 603 646 3480
Fax: 603 646 3856
E-mail: braasch@dartmouth.edu

The Regional Association for Research on the Gulf of Maine is an association of institutions which have active research interests in the Gulf of Maine and its watershed. The Association was founded in 1991 and is presently housed at Dartmouth College. The basic missions of the Association are to advocate and facilitate a coherent program of regional research; to promote scientific quality; and to provide a communication vehicle among scientists and the public.

School of the Coast and Environment, Louisiana State University (USA)

(Formerly CCEER, Center for Coastal, Energy, and Environmental Resources)
Louisiana State University
E302, Howe-Russell, Baton Rouge, LA 70803
Phone: (225) 388 6316

The School of the Coast and Environment exists to provide knowledge, technology, and human resources for successful management of natural resources and resolution of environmental

Table A3.5 *Continued*

issues important to Louisiana, the Gulf of Mexico region, and comparable areas throughout the nation and the world.

School of Marine Sciences (USA)

214, Libby Hall
University of Maine
Orono, ME 04469-5741
Phone: (207) 581 4381
Fax: (207) 581 4388
E-mail: davidt@maine.edu

The University of Maine's School of Marine Sciences (SMS) is the center of excellence for marine education in Maine.

School of Ocean and Earth Science and Technology (USA)

School of Ocean and Earth Science and Technology
University of Hawaii
1680, East-West Road, POST 802
Honolulu, HI 96822

SOEST brings together in a single-focused ocean, earth sciences and technology group, academic departments, research institutes, federal cooperative programs, and support facilities of the highest quality in the nation to meet challenges in the ocean and earth sciences. Scientists of SOEST intend to understand the subtle and complex interrelations of the seas, the atmosphere, and the earth in order to learn how to preserve the quality of our lives.

School of Oceanography, University of Washington (USA)

School of Oceanography
Box 357940
University of Washington
Seattle, WA 98195-7940
Phone: (206) 543 5060

The School of Oceanography is a national leader in oceanographic research and instruction of graduate and undergraduate students. With its roots in the UW Oceanographic Laboratories founded in 1930 and directed by Professor Thomas G. Thompson, the School was organized formally in 1951.

Scripps Institution of Oceanography, University of California San Diego (USA)

UC San Diego
9500, Gilman Drive
La Jolla, CA 92093
Phone: (858) 534 2839
(858) 534 5306

Mission: To seek, teach, and communicate deep scientific understanding of the oceans, atmosphere, earth, and other planets for the benefit of society and the environment.

Skidaway Institute of Oceanography (USA)

10, Ocean Science Circle
Savannah, GA 31411
Phone: (912) 598 2453
Fax: (912) 598 2310

SkIO is an autonomous research unit of the University System of Georgia. The mission of the Institute is to provide the State of Georgia with a nationally and internationally recognized center of excellence in marine science.

US Geological Survey, Coastal Marine and Geology Program (USA)

US Geological Survey
Center for Coastal Geology and Regional Marine Studies
600, Fourth Street South
St. Petersburg, FL 33701-4846

The Center investigates geologic processes related to societal problems arising in coastal and marine environments including natural hazards, resources, and environmental change. Increased understanding of these topics will provide the basis for predicting future coastal erosion, the fate of wetlands, accumulation of sediments, sediment transport and stability, circulation, movement of pollution through aqueous environments, and the locations of economically valuable hard minerals.

USM College of Marine Sciences (USA)

University of Southern Mississippi
703, East Beach Drive (39564)
P.O. Box 7000
Ocean Springs, MS 39566-7000
Phone: (228) 872 4200
Fax: (228) 872 4204

Vero Beach Marine Laboratory (VBML) (USA)

Division of Marine and Environmental Systems
Florida Institute of Technology
150, West University Boulevard
Melbourne, FL 32901-6988
Phone: (407) 674 7273
E-mail: harris@marine.fit.edu

The Vero Beach Marine Laboratory (VBML) is a field laboratory established in 1981 in support of marine science research and education for the academic programs and research institutes of Florida Institute of Technology.

Virginia Institute of Marine Science (USA)

P.O. Box 1346 (mailing)
Rt. 1208, Greate Road (shipping)
Gloucester Point, VA 23062-1346
Phone: 804 684 7000
Fax: 804 684 7097

Chartered in 1940, the School of Marine Science/Virginia Institute of Marine Science (SMS/VIMS), has a tripartite mission of research, education, and advisory service in marine science.

Woods Hole Oceanographic Institution (USA)

Woods Hole Oceanographic Institution
Information Office
Co-op Building, MS #16
Woods Hole, MA 02543
Phone: (508) 548 1400
Fax: (508) 457 2034
E-mail: information@whoi.edu

Research Departments and Centers at Whoi
Applied Ocean Physics & Engineering; Biology; Geology & Geophysics; Marine Chemistry & Geochemistry; Physical Oceanography; Marine Policy Center; Rinehart Coastal Research Center; Cooperative Institute for Climate & Ocean Research.
WHOI is dedicated to research and higher education at the frontiers of ocean science. Its primary mission is to develop and effectively communicate a fundamental understanding of the processes and characteristics governing how the oceans function and how they interact with earth as a whole.

Table A3.6 Overview of some of the main European research and educational institutes (academic units) that deal with the coastal zone

BELGIUM

Flanders Hydraulics Institute

Flanders Hydraulics is a research institute of the Waterways and Marine Affairs Administration of the Department of Environment and Infrastructure of the Ministry of the Flemish Community.

Flanders Hydraulics

Berchemlei 115
B-2140 Antwerp
Phone: ++32 3224 60 35
Fax: ++32 3224 60 36
E-mail: watlab@lin.vlaanderen.be

Laboratory of Oceanology

University of Liège
Sart Tilman B6
B-4000 Liège
Fax: +32 4 3663325

Table A3.6 *Continued*

Renard Centre of Marine Geology
Universiteit Gent
Geologisch Instituut
Krijgslaan 281 S8, B-9000 Gent
Phone: +32 (0)9 264 45 94
Fax: +32 (0)9 264 49 67

CROATIA

Center For Marine Research
Bijenicka c. 54
HR-10000 Zagreb
Phone: ++385 1 425 808
Fax: ++385 1 420 437

Hydrographic Institute of the Republic of Croatia (HHI)
21000 Split
Zrinsko-Frankopanska 161
Phone: +385 (0)21 361 840/344 433
Fax: + 385 (0)21 347 208/347 242
Telex: 26 270 HIRH RH
E-mail: dhi-office@dhi.tel.hr

Hydrographic Institute of the Republic of Croatia (HHI) carries out scientific research, development, and professional works with regard to safety of navigation in the Adriatic, hydrographic–geodetic survey of the Adriatic, marine geodesy, design, and production of maps and charts, as well as nautical publications and aids, oceanographic research, submarine geology research, and finally publishing and printing activities.

DENMARK

Department of Earth Sciences Marine Geology Program
University of Aarhus
C.F. Møllers Allé 110
8000 Århus C
Phone: +45 8942 2899
Fax: +45 86139248

DHI—Danish Hydraulics Institute
DHI Water & Environment
Agern Allé 11
DK-2970 Hørsholm
Phone: +45 4516 9200
Fax: +45 4516 9292
E-mail: dhi@dhi.dk

Institute of excellence in coastal zone deals mainly with shoreline management, sedimentation in harbors and navigation channels, tidal inlet stabilization, dredging and reclamation, storm surges and coastal flooding, environmental impact assessment studies, ports and hydraulic Structures, survey and monitoring, ecology and water quality.

National Environmental Research Institute—Coastal Zone Ecology Department of Coastal Zone Ecology
Grenåvej 12
DK-8410 Rønde
Phone: +45 89 20 17 00
Fax: +45 89 20 15 14

The department of Coastal Zone Ecology undertakes research focusing on the ecology and population dynamics of vertebrate species, and is responsible for national monitoring of species and habitats in terrestrial and marine coastal areas.

The National Environmental Research Institute (NERI), Denmark
National Environmental Research Institute
Frederiksborgvej 399
4000 Roskilde
Phone: 45+ 46 30 12 00
Fax: 45+ 46 30 11 14
E-mail: dmu@dmu.dk

The National Environmental Research Institute (NERI) is a research institute in the Ministry of Environment and Energy, Denmark.

FRANCE

Institut Universitaire Européen de la Mer—France
Technopole BREST-IROISE
Place Nicolas Copernic
29280 Plouzane
Phone: 02 98 49 86 00

L'IUEM est d'abord un pôle pluridisciplinaire de recherche et d'observation dont l'objectif général est l'étude et la modélisation du système couplé atmosphère-océan-géosphère-biosphère de la planète Terre.

Institut Océanographique
Institut océanographique
195, rue Saint-Jacques, F-75005 Paris
Phone: +33 1 44 32 10 70
Fax: +33 1 40 51 73 16

L'Institut océanographique est une fondation française de droit privée, reconnue d'utilité publique et non subventionnée par l'État. Elle a été créée en 1906 par Albert Ier, prince de Monaco.

La Fondation est un organisme autonome, indépendant, doté de la personnalité civile et juridique. Elle regroupe l'établissement de Paris, où est fixé son siège social, et le Musée de Monaco. Chacun des deux éléments a son directeur.

Observatoire Océanologique
BP.44-66651
Banyuls sur Mer Cedex
Phone: 04 68 88 73 00
Fax: 04 68 88 16 99

Observatoire Océanologique
Observatoire Océanologique de Villefranche-sur-Mer
F-06234 Villefranche-sur-Mer Cedex
Dernière modification le 29.06.01

GERMANY

Centre for Marine and Climate Research
Bundesstraße 55
D-20146 Hamburg
Phone: +49 40 42838 4523/5
Fax: +49 40 42838 5235

Center for Tropical Marine Ecology, University of Bremen
Zentrum für Marine Tropenökologie
Fahrenheitstraße 6
D-28 359 Bremen
Phone: +49 +421 23800 21
Fax: +49 +421 23800 30

ZMT takes an integrated interdisciplinary approach to ecosystem research and coastal management in tropical coastal areas.

Coastal Research Laboratory, Christian Albrechts University, Kiel
Otto-Hahn-Platz 3
D-24118 Kiel
Phone: +49 431 880 2851
Fax: +49 431 880 7303
E-mail: info@corelab.uni-kiel.de

The Coastal Research Laboratory (Corelab) at the University of Kiel is a research and teaching unit established to foster research in coastal environments. The Laboratory is jointly coordinated by the Institute of Geosciences in Kiel and the Research and Technology Centre West Coast on the North Sea coast.

Forschungs Zentrum Kueste (Coastal Research Center)
Forschungszentrum Küste
Merkurstraße 11
D-30419 Hannover
Phone: (0511) 762 92 27
Fax: (0511) 762 92 19
E-mail: office@fzk.uni-hannover.de

Universities of Hannover and Braunschweig, and home of the Grossen Wellen Canal.

Table A3.6 *Continued*

Terramare Research Centre
Schleusenstraße 1
D-26382 Wilhelmshaven
Phone: +49 4421/944 0
Fax: +49 4421/944 199

Center for research on shallow seas, coastal zones, and the marine environment.

ICELAND

Marine Research Institute
Skulagata 4
P.O. Box 1390
121 Reykjavik
Phone: +354 552 0240
Fax: +354 562 3790
E-mail: librarian@hafro.is

Sandgerði Marine Centre
Sandgerdi Marine Centre, Gardvegi 1
IS-245 Sandgerdi

The essential objective of the TMR LSF-programme is to provide researchers or research teams throughout the Member States of the Community and the Associated States with access to facilities in Europe that are important for high-quality research and to complement national efforts in respective field.

IRELAND

Coastal Resources Centre
Coastal Resources Center
Environment Research Institute
Old Pres Building, Western Road
University College, Cork
Phone: +353 (0)21 4904129

The Coastal Resources Centre (CRC) is a multidisciplinary group within University College, Cork. As an integral part of the overall Environment Research Institute (ERI), the CRC serves as a critical source of expertise dedicated to ocean and coastal research and resource studies.

Coastal Studies Research Group at the University of Ulster
School of Environmental Studies
University of Ulster at Coleraine
Coleraine, County Londonderry
BT52 1SA, Northern Ireland
Phone: 00 44 (0)28 70324428
Fax: 00 44 (0)28 70 324911

The Coastal Studies Research Group (CSRG) was formed in 1991 and comprises ca. 40 Academic Staff, Research Officers and PhD/MRes Students. The group examines various aspects of coastal environments, from the physical processes to the human impacts on today's coastline.

Coastal Zone Institute
Coastal Zone Institute
Munster Institute
University College Cork
Website: hmrc@ucc.ie

The location of the Coastal Zone Institute (CZI), established under the aegis of University College Cork (UCC), arises from a strong traditional base of research and expertise in coastal studies.

Irish Marine Institute
The Irish Marine Institute
80, Harcourt Street
Dublin 2
Phone: +353 1 476 6500
Fax: +353 1 478 4988

Principal function: to undertake, to coordinate, to promote, and to assist in marine research and development, and to provide such services related to marine research and development, that in the opinion of the Institute will promote economic development and create employment and protect the environment.

Sherkin Island Marine Station
Sherkin Island
County Cork
Phone: +353 28 20187
Fax: +353 28 20407
Website: sherkinmarine@eircom.net

ITALY

Euro-Mediterranean Centre on Insular Coastal Dynamics
Foundation for International Studies
University of Malta
St. Paul Street
Valletta VLT07, Malta
Phone: +356 240746
Fax: +356 230551/245764

Istituto Nazionale di Oceanografia e di Geofisica Sperimentale
Borgo Grotta Gigante 42/C
34016 Sgonico, Trieste
Phone: 39 (40) 21401
Fax: 39 (40) 327307
Telex: 460329 OGS I
E-mail: webmaster@ogs.trieste.it

Tethys Research Institute
Venice Natural History Museum
Santa Croce 1730, 30135 Venezia
Phone: +39 0412750206
Fax: +39 041721000
E-mail: tethys@tethys.org

The Tethys Research Institute, founded in 1986, is a nonprofit NGO dedicated to the preservation of the marine environment. It focuses on marine animals and particularly on cetaceans inhabiting the Mediterranean Sea, and aims at protecting its biodiversity by promoting the adoption of a precautionary approach for the management of natural resources.

NETHERLANDS

Delft Hydraulics institute
Visiting address:
Rotterdamseweg 185
2629 HD Delft
Postal address:
P.O. Box 177
2600 MH Delft
Phone: +31 (0)15 285 8585
Fax: +31 (0)15 285 8582

Founded in 1927, WL|Delft Hydraulics is an independent consulting and research institute located in the Netherlands. WL|Delft Hydraulics has a long-standing reputation for excellence in hydrology, hydraulics, morphology, water quality, and ecology. Construction and design matters related to offshore, coasts, harbors, estuaries, rivers and canals, and industry.

Netherlands Centre for Coastal Research
The Centre is housed at WL|Delft Hydraulics
P.O. Box 177
2600 MH Delft
Phone: +31 15 2858577
Fax: +31 15 2858582
E-mail: ad.vanos@wldelft.nl

RIVO Netherlands Institute for Fisheries Research
Visitors address:
Haringkade 1
IJmuiden

Correspondence:
P.O. Box 68 1970
AB IJmuiden

Table A3.6 *Continued*

Phone: +31 255 56 46 46
Fax: +31 255 56 46 44

Biological, technical, technological, environmental hygiene, and quality research.

The Department of Marine Biology, University of Copenhagen

University of Groningen
Biological Center
Department of Marine Biology
Kerklaan 30, P.O. Box 14
9750 AA Haren (Gn)
Phone: 050 3632259
Fax: 050 3632261
E-mail: g.van.roon-ter.horst@biol.rug.nl

The Institute for Marine and Atmospheric Research Utrecht (IMAU)

IMAU Secretariat
P.O. Box 80005
NL-3508 TA Utrecht
Phone: (+31/0) 030 253 3275
Fax (+31/0) 030 254 3163
E-mail: imau@phys.uu.nl

The IMAU is an interfaculty University research institute established on September 24, 1991. It is composed of the Meteorology and Physical Oceanography Department of the faculty of Physics and Astronomy and the section Coastal research of the Physical Geography Department of the faculty of Geographical Sciences.

The Netherlands Working Group of International Wetland Experts
E-mail: mailto:info@wiw.nl

The Netherlands Working group of International Wetland-experts (WIW) wants to provide a platform for free exchange of information and opinions concerning activities carried out by Dutch institutions or persons in wetlands outside our own country. The WIW secretariat is supported by WWF, the Netherlands.

The Netherlands Institute for Sea Research

The visitor's address of NIOZ is:
Nederlands Instituut voor Onderzoek der Zee
Landsdiep 4 't Horntje
Texel

The mailing address of NIOZ is:
Nederlands Instituut voor Onderzoek der Zee
P.O. Box 59
NL-1790 AB Den Burg, Texel
Phone: (+31) (0)222 369300
Fax: (+31) (0)222 319674

The Netherlands Institute for Sea Research (NIOZ) was founded in July 1876 as the Marine Zoological Station and is presently one of the major European oceanographic institutes. NIOZ is a research institute under the Netherlands Organization for Scientific Research (NWO). Its mission is to pursue multidisciplinary marine research related to phenomena and mechanisms in coastal and shelf seas as well as the open ocean and involves close cooperation between physicists, chemists, geologists, and biologists.

NORWAY

Coastal and Ocean Engineering

Coastal and Ocean Engineering
NO-7465 Trondheim
Phone: +47 73 59 23 38
Fax: +47 73 59 23 76
E-mail: coastal.request@fish.sintef.no

The activities of the department center on harbor, coastal, and ocean engineering, oceanography, marine environmental modeling, and maritime IT.

Institutt for fiskeri- og marinbiologi (IFM), University of Bergen

Institutt for fiskeri- og marinbiologi
Thormøhlensgt. 55
5020 Bergen
Phone: +47 55 58 44 00
Fax: +47 55 58 44 50

Institutt for fiskeri- og marinbiologi (IFM) holder til i Høyteknologisenteret i Bergen. Instituttet har som formål å drive forskning og gi undervisning innen fagområdene marinbiologi. fiskeribiologi og akvakultur inkl. Fiskehelse.

The Institute of Marine Research, Ministry of Fisheries

Institute of Marine Research—Havforskningsinstituttet
Postboks 1870
Nordnes
5817 Bergen
Phone: +47 55 23 85 00
Fax: +47 55 23 85 31
E-mail: havforskningsinstituttet@imr.no

The Institute of Marine Research is Norway's largest research institution in the fields of marine resources, marine environment, and aquaculture. With over 500 employees the Institute is among the largest in the world in this area and is an international leader in several areas of research. The Institute of Marine Research is the research arm of the Ministry of Fisheries.

POLAND

Institute of Oceanology, Polish Academy of Sciences
Powstancow Warszawy 55
P.O. Box 68
81 712 Sopot
Phone: (+48 58) 551 72 81; (+48 58) 550 32 32
Fax: (+48 58) 551 21 30

Polish Academy of Sciences, Institute of Hydroengineering

Koscierska 7
P.O. Box 61
80 953 Gdansk
Phone: +48(0)58 552 20 11
+48(0)58 552 39 03
Fax: +48(0)58 552 42 11
E-mail: sekr@ibwpan.gda.pl

The Institute of Hydroengineering was established in 1953 as a research institution belonging to the Polish Academy of Sciences. The Institute's research activities cover the basic problems of inland and maritime hydroengineering, geotechnics, and geomechanics, as well as other disciplines related to environmental engineering.

University of Szczecin Institute of Marine Sciences

University of Szczecin
Institute of Marine Sciences
Wąska S. 13, 71-415 Szczecin
Phone: (+48 91) 422 64 11 ext. 236
Fax: (0 91) 455 31 20

PORTUGAL

IMAR—Institute of Marine Research

"Centro Interdisciplinar de Coimbra"
c/o Department of Zoology
University of Coimbra
3000 Coimbra
Phone: +351 39 836386
Fax: +351 39 823603
E-mail: imar@ci.uc.pt

National Laboratory of Civil Engineering, Hydraulic Department

Av. do Brasil, 101
1700-066 Lisboa
Phone: +351 21 8443000
Fax: +351 21 8443016

LNEC's Hydraulics Department (DH), created in 1949, carries out research activity on water and the environment. Special areas of interest are: hydrology and river hydraulics, groundwater, hydraulic structures, estuaries and inlets, maritime works, water supply, and sewages.

Table A3.6 *Continued*

Secção Biologia Marinha e Oceanografia

Departmento de Biologia, Universidade da Madeira
Praça do Município, 9000 Funchal/Madeira
Phone: +351 91 233012

Unidade de Ciências e Tecnologias dos Recursos Aquáticos

Universidade do Algarve
Campus de Gambelas
8000 117 Faro

This department is located on the Gambelas Campus, about 10 km from the center of Faro, South Portugal, Europe. UCTRA runs courses and carries out research in the area of aquatic resources.

SPAIN

Canary-Island Institute of Marine Sciences

Office of Research and Universitie
Address:
Aptdo. 56
Telde 35200, Las Palmas
Phone: 34 28 132900, 34 28 132904
Fax: 34 28 132908

The Council of Education, Culture and Sports of Canary-Island Government Taliarte, Telde, Grand Canaria, Canary Islands, Spain

Institut de Ciències del Mar CIMA

Passeig Marítim de la Barceloneta, 37–49
E-08003 Barcelona
Phone: +34 93 230 95 00
Fax: +34 93 230 95 55

Founded in 1951 as the Institute of Fisheries Research, it belongs to the *Consejo Superior de Investigaciones Científicas* (CSIC), within its natural resources area. Its main objective is the multidisciplinary study of the sea, through research projects focusing on different aspects of the marine environments and ecosystems.

Instituto Español de Oceanografía

Instituto Español de Oceanografía
Servicio de Coordinación y Publicaciones
Avda. del Brasil, 31 28020 Madrid
Phone: +34 914 17 54 11
Fax: +34 915 974 770
E-mail: biblioteca@md.ieo.es

En la actualidad el IEO es un Organismo autónomo con personalidad jurídica y patrimonio propios, que depende orgánicamente del Ministerio de Ciencia y Tecnología a través de la Secretaría General de Política Científica, y está clasificada como un Organismo Público de Investigación según la Ley de Fomento y Coordinación General de la Investigación Científica y Técnica de 14 de abril de 1986, siendo su campo de actividad el estudio de la mar y sus recursos por lo que actúa como asesor de la Administración.

International Centre for Coastal Resources Research

Jordi Girona, 1-3
Edif. D-1
08034 Barcelona
Phone: +34 93 280 6400
Fax: +34 93 280 60 19

The International Centre for Coastal Resources Research (CIIRC) is a coordination center for interdisciplinary applied coastal resources research created by the *Generalitat de Catalunya* (*Departament de Política Territorial i Obres Públiques, Comissionat d'Universitats i Recerca, Departament de Medi Ambient* and *Departament d'Agricultura, Ramaderia i Pesca*), the *Universitat Politècnica de Catalunya* (UPC) and the International Federation of Institutes for Advanced Study (IFIAS), with support of the United Nations Environment Programme (UNEP).

Laboratori d'Enginyeria Marítima (LIM/UPC)

Laboratori d'Enginyeria Marítima (LIM/UPC)
Universitat Politècnica de Catalunya
Jordi Girona, 1-3, Campus Nord-UPC, Edif. D-1
08034 Barcelona
Phone: +34 93 401 64 68
Fax: +34 93 401 18 61
E-mail: info.lim@upc.es

The *Laboratori d'Enginyeria Marítima* (LIM/UPC) is a Research Centre within the *Departament d'Enginyeria Hidràulica, Marítima i Ambiental* (*E.T.S. Eng. Camins, Canals i Ports de Barcelona*) of the *Universitat Politècnica de Catalunya* (UPC) in Barcelona. LIM/UPC is thus a nonprofit public Research Centre, with the sole aim of generating and transferring technology in the field of Maritime Engineering and Ocean Sciences.

Planificación y Gestión de Zonas Costeras

Area de ordenación del litoral
Centro andaluz superior de estudios marinos, (c.a.s.e.m.)
Campus universitario-polígono rio san pedro
Puerto real, 11510 cádiz
Phone: 34 956 015546/34 956 016091
Fax: 34 956 015501/34 956 016040
E-mail: juan.barragan@uca.es

SWEDEN

Göteborg University Marine Research

Göteborg University Marine Research Center
Box 460
405 30 Göteborg
Phone: +46 31 772 2295
Fax: +46 31 772 2785

Göteborg University Marine Research Center was established in 1989 by the Government. It belongs to Göteborg University.

Umeå Marina Forskningscentrum

Norrbyn
910 20 Hörnefors
Phone: 090 786 79 74 (kansli)

UNITED KINGDOM

British Marine Life Study Society

14, Corbyn Crescent
Shoreham-By-Sea
Sussex BN43 6PQ
Phone: 01273 465433
E-mail: bmlss@compuserve.com

Cambridge Coastal Research Unit

Department of Geography
University of Cambridge
Downing Place
Cambridge CB2 3EN
Phone: +44 (0)1223 339775; +44 (0)1223 333350
Fax: +44 (0)1223 355674
E-mail: geog-CCRU@lists.cam.ac.uk

Providing highest quality scientific research to underpin sustainable coastal management.

CEFAS

CEFAS Lowestoft Laboratory
Pakefield Road
Lowestoft, Suffolk NR33 OHT
Phone: +44 (0) 1502 562244
Fax: +44 (0) 1502 513865
E-mail: marketing@cefas.co.uk

CEFAS is a scientific research and advisory center for fisheries management and environmental protection. We provide contract research, consultancy, advice, and training in fisheries science and management, marine environmental protection, aquaculture, and fish and shellfish disease and hygiene to a large number of public and private sector clients around the world.

Centre for Coastal Conservation and Education

Centre for Coastal Conservation and Education
School of Conservation Sciences
Bournemouth University

Table A3.6 *Continued*

Poole, Dorset BH12 5BB
Phone: +44 (0)1202 59 53 52
Fax: +44 (0)1202 59 52 55

Promoting care and understanding of our oceans through science, research, and education. The Centre for Coastal Conservation and Education was established in January 1998 at Bournemouth University in recognition of the School of Conservation Science's growing reputation and expertise in coastal conservation, management, and education. Under the Directorship of Dr. Carolyn Heeps, the Centre provides a focus for a wide range of scientific, research, and educational activities and opportunities.

CCM—Centre for Coastal Management

Ridley Building, University of Newcastle
Newcastle upon Tyne, NE1 7RU
Phone: +44 (0)191 222 5607
Fax: +44 (0)191 222 5095
E-mail: enquiries@sustainablecoasts.com

The mission of CCM is to promote coastal management through the coupling of fundamental and applied research in coastal systems with advice on practical and policy issues.

Centre for the Economics and Management of Aquatic Resources (CEMARE)

Department of Economics
University of Portsmouth
Milton Campus, Locksway Road
Portsmouth PO4 8JF
Phone: +44 (0) 23 9284 4082
Fax: +44 (0) 23 9284 4037
E-mail:christopher.martin@port.ac.uk

European Artificial Reef Research Network

Dr. Antony Jensen
School of Ocean and Earth Science
University of Southampton, Southampton Oceanography Centre
European Way, Southampton SO14 3ZH
Phone: +44 1703 593428
Fax: +44 1703 596642
E-mail: a.jensen@soc.soton.ac.uk

The European Artificial Reef Research network (EARRN) formed in May 1995 with funding from the European Commission (EC) AIR programme. The 51 scientists from 36 laboratories that formed EARRN were all active in artificial reef research and the network has provided recommendations for the direction of future research to the EC. The formal EC funding has finished but EARRN continues its activities using Internet, e-mail (discussion groups) and "meetings of opportunity" to continue its work; expanding its membership beyond the original 51 members. Building on the professional relationships and friendships developed between 1995 and 1998 members will undoubtedly prepare proposals for Framework V. EC-funded collaborative research between members will be the next goal for EARRN.

Institute of Marine Studies, University of Plymouth

Drake Circus, Plymouth
Devon PL4 8AA
Phone: +44 (0) 1752 232470
Fax: +44 (0) 1752 232472
E-mail: alc@plymouth.ac.uk

Plymouth Marine Laboratory

Plymouth Marine Laboratory
Prospect Place, The Hoe
Plymouth, England PL1 3DH
Phone: +44 (0)1752 633100
Fax: +44 (0)1752 633101

The Plymouth Marine Laboratory undertakes fundamental and strategic research to underpin the marine requirements of the United Kingdom. The PML executes its mission through key partnerships and collaborations with organizations throughout the world.

POL—Proudman Oceanographic Laboratory

Bidston Observatory
Birkenhead CH43 7RA
Phone: +44(0)151 653 8633
Fax: +44(0)151 653 6269

POL's scientific research focuses on oceanography encompassing global sealevels and geodesy, numerical modeling of continental shelf seas and coastal sediment processes. This research alongside activities of surveying, monitoring, data management, and forecasting provides strategic support for the wider mission of the Natural Environment Research Council.

Scott Polar Research Institute

University of Cambridge
Lensfield Road
Cambridge CB2 1ER, England
Phone: SPRI Switchboard: 01223 336540; +44 1223 336540
Fax: 01223 336549; +44 1223 336549

Welcome to the website of The Scott Polar Research Institute (SPRI), the oldest international research center in the world covering both the Arctic and Antarctic regions. The Scott Polar Research Institute is part of the Faculty of Earth Sciences and Geography in the School of Physical Sciences of the University of Cambridge.

Southampton Oceanography Centre

University of Southampton
Waterfront Campus, European Way
Southampton SO14 3ZH
Phone: 023 8059 6666
Fax: 023 8059 6667

The Southampton Oceanography Centre (SOC) opened in 1996 and is a £49 million development creating a center for some 450 research scientists, lecturing and support staff as well as 600 undergraduate and postgraduate students. With a turnover of around £20 million per annum, the new centre's objective is clear and ambitious: to play a strategic role in global interdisciplinary marine and earth sciences.

OTHER EUROPEAN COUNTRIES

Coastal Research Center For Environment Conservation (CRCEC) (Romania)

Mail Address:
O.P. 54
P.O. Box 3
Bucharest

Visiting Address:
Bd. Nicolae Balcescu
No. 1, Floor IV, Room 404
Phone: +40 1 314 35 08; +40 1 220 45 18
Fax: +40 1 315 30 74
E-mail: emilves@geo.unibuc.ro

CRCEC is a coastal research center affiliated to the University of Bucharest, Department of Geography. Its mission is to promote research and conservation of the coastal environment with a special emphasis on the Romanian Black Sea Coast, and providing education and training in coastal science.

Department of Oceanology, Moscow State University (Russia)

Vorobijvy Gory
Moscow 119899
Phone: (095) 939 2215
Fax: (095) 932 8836
E-mail: ocean@ocean.geogr.msu.su

Administrator:
Dr. Arkhipkin Victor Semenovich
E-mail: arkhip@ocean.geogr.msu.su

Institute of Marine Biology of Crete (Greece)
P.O. Box 2214,
GR 71003, Iraklio,

Table A3.6 *Continued*

Crete
Phone: +30 81 0346860

The IMBC is one of Europe's newest institutions, founded in 1987 as an independent research organization by the then Greek Ministry of Industry, Energy, Research, and Technology, under the direct supervision of the General Secretariat for Research and Technology.

Merentutkimuslaitos—Finnish Institute of Marine Research (Finland)

Lyypekinkuja 3 A
P.O. Box 33
FIN-00931 Helsinki
Phone: + 358 9 613 941
E-mail: forename.surname@fimr.fi

Table A3.7 Overview of some of the main Southern Hemisphere research and educational institutes (academic units) that deal with the coastal zone

AUSTRALIA

Australian Institute of Marine Science

Australian Institute of Marine Science
PMB No. 3, Townsville MC,
Queensland 4810
Phone: +61 7 4753 4444
Fax: +61 7 4772 5852

The Australian Institute of Marine Science (AIMS) was established by the Commonwealth government in 1972 to generate the knowledge needed for the sustainable use and protection of the marine environment, through innovative world-class scientific and technological research.

Centre for Marine Studies

The University of Queensland
Brisbane, Queensland 4072
Phone: 61 7 3365 4333
Fax: 61 7 3365 4755
E-mail: cms@uq.edu.au
Director: Professor Ove Hoegh-Guldberg
E-mail: oveh@uq.edu.au

Coastal CRC—Cooperative Research for Coastal Zone Estuary and Waterway Management

Brisbane (Centre Office):
Indooroopilly Sciences Centre
80, Meiers Road
Indooroopilly, Queensland 4068
Phone: +61 7 3362 9399
Fax: +61 7 3362 9372
E-mail: roger.shaw@dnr.qld.gov.au

The Coastal CRC provides decisionmaking tools and knowledge necessary for the effective management and ecosystem health of Australia's coastal zone, estuaries, and waterways.

Lincoln Marine Science Centre

Physical Address:
Hindmarsh Street, Kirton Point
Port Lincoln, South Australia
Postal Address:
P.O. Box 2023
Port Lincoln, SA 5606

Lincoln Marine Science Centre (or LMSC) has been established to support research and tertiary level education in marine science. Located on the shore of Boston Bay in Port Lincoln, South Australia, it is situated in the heart of a region with a temperate climate, clean waters, abundant marine life and many developing aquaculture industries.

Manly Hydraulics Laboratory

Manly Hydraulics Laboratory
110B King Street
Manly Vale NSW 2093, Sydney
Phone: +61 2 9949 0200
Fax: +61 2 9948 6185
Mr. Tony Bolton
E-mail: tbolton @mhl.nsw.gov.au

Manly Hydraulics Laboratory (MHL) provides specialist services in the area of water, coastal, and environmental solutions.

Queensland Government Hydraulics Laboratory

27, Quinlan Street
Deagon, Queensland 4017
Phone: +61 7 3869 9500
Fax: +61 7 3869 9501

The Queensland Government Hydraulics Laboratory (QGHL) is a premier facility for evaluating hydraulic and coastal structures before their construction, and investigating coastal processes by using scale models. Since 1975, the Laboratory has provided extensive services for governments and private organizations in Australia and overseas.

School of Environmental Science and Management

School of Environmental Science & Management
Southern Cross University
P.O. Box 157
Lismore, NSW 2480

School of Environmental Science & Management

Southern Cross University
Military Road
Lismore NSW 2480

Our School focuses on the challenges facing Australia—conservation of marine resources, restoring land and water quality, coastal management, wildlife conservation, and sustainable forest and fisheries.

Sedimentary, Marine & Environmental Geoscience Research Group

School of Earth Sciences
James Cook University
Townsville, Qld 4811
Phone: +61 7 4781 4536
Fax: +61 7 4725 1501
E-mail: earth.sciences@jcu.edu.au

University of Sydney, Coastal Studies Unit

Coastal Studies Unit, Division of Geography School of Geosciences
Sydney University Madsen Building F09
Sydney NSW 2006
Phone: +61 2 9351 2886
Fax: +61 2 9351 3644

The Unit is responsible for research and communication on problems related to the geomorphology and management of coastal environments.

Water Research Laboratory

King Street
Manly Vale
Sydney NSW 2093
Phone: +61 2 9949 4488
Fax: +61 2 9949 4188
E-mail: office@wrl.unsw.edu.au

BRAZIL

Centro de Estudos Do Mar, UFPR, Brasil

Av. Beira-mar s/n, Caixa Postal 02
Pontal do Sul
Pontal do Parana, PR
CEP: 83255-000

Table A3.7 *Continued*

Phone: +55 41 4551333
Fax: +55 41 4551105

CTTMar, Centro de Ciências Tecnológicas da Terra e do Mar

Universidade do Vale do Itajai—UNIVALI
Centro de Ciencias Tecnologicas da Terra e do Mar—CTTMAR
Curso de Oceanografia
P.O. Box 360 Itajai, SC
CEP: 88202-302
E-mail: www.cttmar.univali.br

INPE—Instituto Nacional de Pesquisas Espaciais

Av. dos Astronautas,1.758
Jd. Granja
São José dos Campos, SP
CEP: 12227-010

Phone: 55-12-3945-6000

Instituto Oceanográfico (IO), Brasil

Praça do Oceanográfico
191, Cidade Universitária
São Paulo
CEP: 05508-900
Phone: (011) 3818 6501
Fax: (011) 3032 3092
E-mail: io@edu.usp.br

Laboratório de Estudos Costeiros (LEC), UFBA

Laboratório de Estudos Costeiros
Instituto de Geociências, UFBA
Campus Ondina
Salvador, Bahia 40210-340

Phone: +55 71 332 0550/237 0408
Fax: +55 71 247 3004

O Laboratório de Estudos Costeiros do CPGG/UFBA foi criado no ano de 1995, com a missão institucional de integrar, estimular e dar suporte aos estudos na Zona Costeira do Estado da Bahia e da Região Nordeste do Brasil.

Núcleo de Educação e Monitoramento Ambiental (NEMA)

CGC 911 00 909/ 0001-77
Rua Maria Araújo
450, Cassino, Rio Grande
Rio Grande do Sul
CEP: 96207-480

Phone: 0532 362420
Fax: 0532 361435
E-mail: nema@super.furg.br

A Principal finalidade do NEMA é a harmonização da relação homem-ambiente para a melhoria da qualidade ambiental e de vida.

Programa Train-Sea-Coast, Brasil

Avenida Itália, km 8
Campus Carreiros
Caixa Postal 474
Rio Grande, RS
CEP: 96.201-900
Fax: (0532) 33 6560

NEW ZEALAND

Cawthron Institute

98, Halifax Street East
Nelson

Phone: (+64) 03 548 2319
Fax: (+64) 03 546 9464
E-mail: info@cawthron.org.nz

Cawthron Institute is a private, independent, not-for-profit research center which has been operating for more than 75-years. Our fundamental purpose is to benefit the region and the nation through science and technology.

National Institute of Water and Atmospheric Research (NIWA)

Private Bag 999 40
269, Khyber Pass Road
Newmarket, Auckland

Phone: +64 9 375 2090
Fax: +64 9 375 2091

Established in 1992 as one of nine New Zealand Crown Research Institutes (CRIs), NIWA's mission is to provide a scientific basis for the sustainable management of New Zealand's atmospheric, marine and freshwater systems, and associated resources.

SOUTH AFRICA

Department of Oceanography, University of Cape Town

RW James Building
9, University Avenue

Phone: (021) 650 3278
Fax: (021) 650 3979

University of Cape Town's Centre for Marine Studies

Centre for Marine Studies

University of Cape Town
Private Bag
Rondebosch 7701

University of Cape Town's center of marine expertise, controlled by a Board and a Steering Committee and run by a full-time manager. Consultancy service which draws on the tremendous resource of highly skilled specialists in marine and coastal sciences among its teaching and research staff to provide multidisciplinary marine expertise to the broader community.

OTHER SOUTHERN HEMISPHERE COUNTRIES

Center of Excellence in Coastal Resources Management (The Philippines)

Silliman University
6200 Dumaguete City

Phone: (63 35) 225 6711/225 6855
Fax: (63 35) 225 4608
E-mail: admsucrm@mozcom.com

Centro de Investigaciones del Mar y la Atmósfera (Argentina)

CIMA/CONICET-UBA
Pabellón II-2do. piso
Ciudad Universitaria
(1428) Buenos Aires

Phone: (54)(1) 787 2693; (54)(1) 781 5020/29 Int. 388
Fax: (54)(1) 788 3572
E-mail: webmaster@at1.fcen.uba.ar

El Programa Regional de Oceanografía Física y Clima (Chile)

Universidad de Concepción (Cabina 7)
Casilla 160-C
Concepción 3

Phone: (+56) 41 203585
Fax: (+56) 41 239900

National Institute of Oceanography (NIO) (India)

2000, Dona Paula
Goa 403 004

Fax: 91 (832) 223340
E-mail: ocean@darya.nio.org

NIO, the premier oceanographic institution in India, was founded in 1966 on completion of the International Indian Ocean Expedition (1962–65). From an initial emphasis on marine biology, NIO's research and development activities have evolved in the last 35 years to include almost all major branches of coastal and high seas oceanography.

Table A3.8 Some North American consultant companies

Acqua Engineering Inc. (Canada)
Otavio Sayao, Ph.D. P.Eng.
Acqua Engineering Inc.
4496, Credit Pointe Dr.
Mississauga, Ontario L5M 3M2
Phone: (905) 821 2985
Fax: (905) 821 9617
E-mail: osayao@acqua.on.ca
Acqua Engineering Inc., of Ontario, Canada is a consulting company established in 1995 to provide expert professional services in the fields of port and waterways, shoreline management, coastal engineering, environmental hydraulic, and construction reviews.

ADAMA Engineering Inc. (USA)
33, The Horseshoe, Covered Bridge Farms
Newark, DE 19711-2066
Phone: (302) 368 3197
Fax: (302) 731 1001

Anchor Environmental L.L.C. (USA)
Tom Schadt
1411, 4th Avenue, Suite 1210
Seattle, WA 98101
Phone: 206 287 9130
Fax: 206 287 9131
Anchor is an environmental science and engineering firm whose expertise and focus is shoreline projects, addressing issues in sediment management, environmental review, natural resources, and waterway, coastal, and geotechnical engineering. It has offices in Seattle (WA), Long Beach, Oakland and San Francisco (CA), and College station (TX).

Andrews, Miller & Associates, Inc. (USA)
401, Academy St.
Cambridge, MD 21613
Phone: 410 228 7117
Complete coastal & civil engineering services.

Applied Coastal Modeling (USA)
Consulting Services
Jon M. Hubertz, Ph.D.
2733, Deborah Drive
Punta Gorda, FL 33950-8182
Phone/Fax: 941 505 4079
A resource for the expert application of numerical models to coastal problems, analysis of data, display, and interpretation of results.

Applied Coastal Research and Engineering, Inc. (USA)
766, Falmouth Rd.
Building A, Unit 1-C
Mashpee, MA 02649
Phone: (508) 539 3737
E-mail: info@appliedcoastal.com
Applied Coastal Research and Engineering, Inc. focuses on developing and implementing scientifically defensible solutions to problems in the marine environment.

Applied Fluids Engineering, Inc. (USA)
Private Mail Box #237
5710 E, 7th Street
Long Beach, CA 90803
E-mail: phil.watts@appliedfluids.com
A consulting firm specializing in: air and water motion, suspension dynamics, rapid phase change, and vertebrate locomotion.

Aqua Solutions (Canada)
Judy Sullivan, P.Eng.
3405, Greenwood Road
Greenwood, Ontario L0H 1H0
Phone/fax: 905 428 3365
E-mail: judy.sullivan@sympatico.ca
Coastal & river engineering coastal zone management planning, environmental hazard management & planning, flooding & erosion evaluation, project management guidelines & criteria development.

ASL Environmental Sciences (Canada)
1986, Mills Road
Sidney, BC V8L 5Y3
Phone: 1 877 656 0177
When it comes to physical aquatic measurement problems, whether it is wave, ice, current or flow, ASL is uniquely qualified to meet the challenge.

AXYS Environmental Consulting Ltd. (Canada)
Head Office:
600-555, Fourth Ave. SW
Calgary, Alberta, T2P 3E7
Phone: (403) 269 5150
Fax: (403) 269 5245
As one of Canada's leading environmental consulting firms, we have consistently applied innovative and precise science in achieving balanced solutions. Since 1974, our professional staff has been offering clients a range of experience and knowledge to meet the diverse needs of each project.—Robert H. Seager, President.

Baker Coastal Services (USA)
Alexandria, VA
Phone: 703 960 8800
Annapolis, MD
Phone: 410 571 8706
Virginia Beach, VA
Phone: 804 468 8243
Elmsford, NY
Phone: 914 333 5300
Tampa, FL
Phone: 813 2897546
Deals with port & harbor engineering, coastal engineering, coastal zone management, dredging strategies & disposal designs, marina planning & design, and hydrografic surveying/digital mapping.

Cammaert Consultants—Newfoundland (Canada)
Dr. Gus Cammaert
Cammaert Consultants
1, Winter Place
St. John's, Newfoundland A1B 1J5
Phone: (709) 738 3581
Fax: (709) 738 3588
Cammaert Consultants provides specialized consulting and testing services in the field of coastal engineering for the fishing, aquaculture, and boating interests. The company offers design and analysis of coastal facilities and prediction of wave and ice climates.

Canadian Hydraulic Centre (Canada)
Ottawa, Ontario, K1A OR6
Phone: (613) 993 2417
Fax: (613) 952 7679
For all physical & numerical modeling and analysis needs in the general fields of hydraulics.

Cashin Associates, P.C. (USA)
1200, Veterams Memorial Hwy,
Hauppauge, NY, 11788
Phone: (516) 348 7600
50, Tice Blvd,Woodcliff lake, NJ 07675
601, Brickell Key Drive, Miami, FL 33131
Engineering and environmental consulting.

Coastal Engineering Company, Inc. (USA)
260, Cranberry Highway
Orleans, MA 02653
Phone: 508 255 6511 (Orleans); 508 778 9600 (Hyannis); 508 487 9600 (Provincetown)
Fax: 508 255 6700
E-mail: info@ceccapecod.com

Table A3.8 *Continued*

The Mission of Coastal Engineering Company is to help our clients achieve their goals. We do this by understanding our clients' needs, understanding the issues that impact their projects, and by providing appropriate consulting, engineering, and surveying solutions.

Coastal Engineering Consultants, Inc. (CEC) (USA)

Lee County Office
17595, S. Tamiami Trail, #102
Fort Myers, FL 33908
Phone: (941) 590 9900
Fax: (941) 590 9909

Charlote County Office
20020, Veterans Blv. #12
Port Charlote, FL 33948
Phone: (941) 743 6611
Fax: (941) 743 6694

Founded in 1977, Coastal Engineering Consultants, Inc. (CEC) is a team of experienced professionals who possess a sound understanding of the disciplines of Engineering and Geology. Our staff of engineers, geologists, environmental specialists, and planners design workable alternatives to produce timely, cost-effective results that are in harmony with natural ecosystems.

CIS—Coastal Information Services (USA)

12932, Victory Church Road
Raleigh, NC 27613
Phone: (919) 676 8684

Deals with beach erosion and beach nourishments, storms and storm damages, inlet dynamics and stabilization, marine and coastal litigation.

Coastal Planning & Design, Inc. (USA)

Coastal Planning & Design, Inc.
849, Cormier Road
Green Bay, WI 54304
Phone: (920) 499 6006
Fax: (920) 499 6116
E-mail: greenbay@coastalplanning.com

Coastal Planning & Design, Inc. specializes in providing professional engineering services for waterfront development, lake and harbor restoration, shore protection, erosion control, stormwater management, flood studies, and wetland delineations.

Coastal Planning & Engineering, Inc. (CPE) (USA)

2481, NW Boca Raton Blvd.
Boca Raton, FL 33431
Phone: (561) 391 8102
Fax: (561) 391 9116
E-mail: cpeboca@aol.com

Coastal Planning & Engineering, Inc. (CPE) is a coastal engineering firm that provides services in coastal engineering, coastal planning, coastal surveying, environmental science, and regulatory permitting. Established in 1984, CPE has an office in Boca Raton, Florida.

Coastal Resource Management

Box 133
Franktown, VA 23354
Phone: (804) 442 5640
Fax: (804) 787 4039

Coastal Resource Management is a diverse company, with environmental and steel fabrication departments. The Environmental department specializes in alternative forms of shoreline erosion control, wetlands delineation, design and creation, soils delineations, water sampling, and marsh and beach vegetation.

Coastal Systems (USA)

Coastal Systems—USA
464, South Dixie Highway
Coral Gables, FL 33146
Phone: 305 661 3655
Fax: 305 661 1914

Coastal Systems International strives to be customer focused, identifying and resolving clients' needs throughout the design process.

Coastal Zone.com (USA)

Coastal Zone.com
P.O. Box 359
Solomons, MD 20688
Phone: (571) 212 9587

Coastal ~ Watershed ~ Environmental; Planning & Management; Constructed Wetlands & Sustainable Development.

Conservation Law Foundation (USA)

62, Summer Street
Boston, MA 02110-1016
Phone: (617) 350 0990
Fax: (617) 350 4030

The Conservation Law Foundation is the largest regional environmental advocacy organization in the United States. We are based in New England, where our attorneys, scientists, economists, and policy experts work on the most significant threats to the natural environment of the region, and to the health of its residents.

David A. Lienhart, FGS (USA)

7229, Longfield Drive
Cincinatti, OH 45243 2209
Phone: 513 561 7049
E-mail: lenhart@ix.netcom.com
The US armourstore expert.

Emerald Ocean Engineering (USA)

107, Ariola Dr.
Pensacola Beach, FL 32561
Toll Free (877) 932 9111
Fax: (850) 932 9111
E-mail: bigwave@emeraldoe.com

First Coastal Corporation (USA)

First Coastal Corporation
4, Arthur Street
P.O. Box 1212
Westhampton Beach, NY 11978
Phone: 631 288 2271
Fax: 631 288 8949
E-mail: mail@firstcoastal.net

Long Island coastal incorporation firm.

Foster Wheeler Environmental Corporation (USA)

Headquarters:
1000, The American Road
Morris Plains, NJ 07950
Phone: (973) 630 8000
Fax: (973) 630 8025
E-mail: webmaster@fwenc.com

Foster Wheeler Environmental Corporation is a leading environmental consulting, engineering and remediation firm employing more than 2,900 dedicated professionals in 24 US offices and 19 international locations. We provide our clients with a full range of traditional and innovative services, that are delivered cost-effectively, timely and in compliance with applicable regulations and requirements.

Gahagan & Bryant Associates (USA)

3802, W. Bay to Bay Blvd
Suite B-22
P.O. Box 18505
Tampa, FL 33679

Dredging consultants, coastal engineering, beach erosion, and hydrographic surveys.

Table A3.8 *Continued*

Halltech Environmental Inc. (Canada)
Exploration Outfitters/Halltech Atmospheric Systems
503, Imperial Rd. N.
Unit #4, Guelph
Ontario N1H 6T9
Phone: (519) 766 4568
Fax: (519) 766 0729
E-mail: sales@htex.com

Holmberg Technologies, Inc. (USA)
1800, Second St. Suite 714
Sarasota, FL 34236
Phone: (941) 351 1144
E-mail: info@erosion.com
Breakthrough in Beach Restoration Technology. Patented beach restoration technology reverses erosion without the addition of artificial fill, and without causing adverse side-effects to adjacent shorelines.

HydroQual, Inc. (USA)
One Lethbridge Plaza
Mahwah, NJ 07430
Phone: (201) 529 5151
Fax: (201) 529 5728
HydroQual, Inc. is an environmental engineering and science firm that combines the latest scientific research with sound engineering principles to solve environmental problems. Established in 1980 and with a staff of over 100 employees, HydroQual's range of services addresses issues dealing with water quality, Total Maximum Daily Load (TMDL) analyses, floatables pollution, ecological risk assessment, watershed management, marine circulation, thermal discharge plume and mixing zone analyses, water and wastewater treatment, hazardous waste management, and permitting, to name a few.

Langley and McDonald, P.C. (USA)
Virginia Beach
Phone: (757) 473 2000
Williamsburg
Phone: (757) 253 2975
Erosion control studies, hydrographic surveys, dredging, beach replenishment, marinas, bulkheads, terminals.

Moffat & Nichol Engineers (USA)
Headquarters:
Moffatt & Nichol Engineers
320, Golden Shore, Suite 300
Long Beach, CA 90802
Phone: (562) 590 6500
Fax: (562) 590 6512
E-mail: jbauer@moffattnichol.com
We are a leading multidisciplinary any engineering firm providing integrated services from concept through planning and design to construction support for a diverse array of projects for public, corporate, and private clients.

Noble Consultants (USA)
San Francisco Bay Area (Marin County)
359, Bel Marin Keys Blvd., Suite 9
Novato, CA 94949-5637
Phone: (415) 884 0727
Fax: (415) 884 0735
E-mail: noble@nobleconsultants.com
A civil, coastal & harbor engineering firm that specializes in the investigation studies, regulatory permitting, planning, field surveying/ monitoring, engineering design, construction management and inspection services for coastal, beach restoration, waterfront structures, dredging and marine/harbor projects.

Ocean and Coastal Consultants, Inc. (USA)
Main Office:
35, Corporate Drive
Trumbull, CT 06611
Phone: 203 268 5007
Fax: 203 268 8821
Ocean and Coastal Consultants, Inc. is a consulting firm founded in order to provide the private and public sector with unique expertise for solving problems in the offshore and coastal environments.

Olko Engineering (USA)
136, West 21st Street
New York, NY 10011
Phone: 212 645 9898
Bulkheads, piers, marinas, beach stabilization, site development structures, and expert witness testimony.

Olssen Associates Inc. (USA)
4438, Herschel Street
Jacksonville, FL 32210
Phone: (904) 387 6114
Fax: (904) 384 7368
We are a coastal engineering firm located in Jacksonville, Florida specializing in the study, design, permitting, and management of projects located in coastal, insular, and estuarine environments.

Philip Williams & Associates, Ltd. (USA)
San Francisco Bay Area Office
770, Tamalpais Drive Suite 401
Corte Madera, CA 94925
Phone: 415 945 0600
Fax: 415 945 0606
E-mail: sfo@pwa-ltd.com
Our professional services include all important aspects of hydraulic engineering and environmental hydrology, from field data collection and analysis to sophisticated hydrodynamic computer modeling. It has offices in San Francisco, Seattle, and Portland.

Rock Products Consultants (USA)
7229, Longfield Drive
Cincinnati, OH 45243-2209
Phone: 513 561 7094
Specialists in the assessment of suitability of armor/stone riprap sources for rubble mound breakwaters and shore protection structures.

Scientific Marine Services, Inc. (SMS) (USA)
Main office:
101, State Place, Suite N
Escondido, CA 92009
Phone: (760) 737 3505
Scientific Marine Services, Inc. (SMS) provides specialized technical consulting services and custom-engineered products to the marine and offshore industries applying advanced methods and technologies. The multidisciplinary nature of the solutions to projects in the marine environment demands expertise in a wide range of fields.

Shiner Moseley and Associates Inc. (USA)
Headquarters:
555, N. Carancahua, Suite 1650
Corpus Christi, TX 78478
Phone: (361) 857 2211
Fax: (361) 857 7234
E-mail: mail@shinermoseley.com
Shiner, Moseley, and Associates is a civil, structural, and marine engineering and consulting firm with offices in Houston and Corpus Christi. SMA has multiple successful projects locally and nationally that incorporate innovative techniques to reach definite and implementable solutions with outstanding results.

Shoreplan Engineering Limited (Canada)
298, Belsize Drive
Toronto, Ontario M4S 1M8
Phone: (416) 487 4756
E-mail: splan@ican.net

Table A3.8 *Continued*

Smith Warner International Limited
Phone: (876) 978 8950; (876) 978 7415
Fax: (876) 978 0685
E-mail: daysmith@infochan.com
pwarner@infochan.com

Smith Warner International Limited is a dynamic company focusing on the coastal and marine environments. Since incorporation in 1995, we have undertaken a number of projects throughout the Caribbean for a variety of clients, including individuals, private sector developers, industry and international funding agencies.

The Coastal Advocate (USA)
2101, Central Ave.
P.O. Box 475
Ship Bottom, NJ 08008
Phone: (609) 361 0550
Fax: (800) 901 0550

Our mission is simply to provide the very best professional voice for you who live, work and invest at the Jersey shore. As a registered lobbying firm in Washington and Trenton we take your cause to the decisionmakers, and we are effective. Please join us, for your home and your coast.

The Sand Web Systems (USA)
100, Aviation Drive South, Suite 202
Naples, FL 34104
Phone: 941 403 7107
520, People street
Corpus Christ, TX 78401
Phone: 888 818 5325

The sand web systems harnesses nature's own energy to reclaim eroded beaches.

Table A3.9 Coastal consultant companies, with US main offices and branches in other countries

Baird & Associates Coastal Engineers (Canada, USA & Chile)
Canada: W.F. Baird & Associates Coastal Engineers
1145, Hunt Club Rd., Suite 1
Ottawa, Ontario K1V 0Y3
Phone: (613) 731 8900
Fax: (613)731 9778

USA: W.F. Baird & Associates Ltd.
2981, Yarmouth Greenway
Madison, WI 53711
Phone: (608) 273 0592
Fax: (608) 273 2010

Chile: Atria Baird Consultores S.A.
Fidel Oteiza 1953, oficina 602
Providencia, Santiago de Chile
Phone: (56 2) 341 4833
Fax: (56 2) 204 6094

Innovation Excellence & Service Oceans, Lakes & Rivers.

Coastal Systems (USA and South America)
USA: 464, South Dixie Highway
Coral Gables, FL 33146
Phone: (305) 661 3655
Fax: (305) 661 1914

South America
Buenos Aires, Argentina
Phone: (011) 54 11 4149 8685
Fax: (011) 54 11 4751 1323

Coastal Systems International, Inc. (Coastal Systems) is a professional consulting engineering firm specializing exclusively in projects within the coastal and marine environment.

Collins Engineers, Inc. (CEI) (All over USA and Ireland)
US Headquarters:
300, West Washington, Suite 600
Chicago, IL 60606-1217
Phone: (312) 704 9300; (877) 346 3234
Fax: (312) 704 9320
E-mail: ilmgr@collinsengr.com

International (Ireland):
Regus House Block 4
Harcourt Centre
Harcourt Road
Dublin
Phone: (01) 417 4339; (877) 346 3234
Fax: (01) 402 9590
E-mail: ilmgr@collinsengr.com

Collins Engineers, Inc. (CEI) is a civil, structural, and water resources engineering firm established in 1979 to provide engineering services to various private and public clients. The initial expertise of the firm was in the areas of structural and transportation analysis, design, and underwater engineering.

Taylor Engineering (US, Mexico, and Argentina)
Headquarters:
9000, Cypress Green Drive
Suite 200
Jacksonville, FL 32256
Phone: 904 731 7040
Fax: 904 731 9847

Founded in Jacksonville, Florida in 1983, Taylor Engineering began as a coastal engineering consulting company. While we continue to devote much of our energy to coastal work, we have expanded our services to include dredging and dredged material management, hydrology and hydraulics, environmental services, and construction support services.

Table A3.10 Coastal zone consultantancy companies worldwide

Ecological Consultancy Services Ltd. (Ireland)
17, Rathfarnham Road
Terenure, Dublin 6W
Phone: 00 353 1 4903237
Fax: 00 353 1 4925694
E-mail: ecoserve@ecoserve.ie

A company providing technical environmental services, including impact and nature conservation assessment, ecotoxicology, monitoring, evaluation and authoritative analysis, interpretation and management of computerized data, with specialization in marine and freshwater system.

Fugro Group (Worldwide)
Headquarters:
Fugro N.V., Veurse Achterweg 10
P.O. Box 41, 2260
AA Leidschendam, The Netherlands
Phone: +31 (0) 70 3111422
Fax: +31 (0) 70 320 2703
E-mail: holding@fugro.nl

Pro Natura (Germany)
Göteborg
Träringen 66

Table A3.10 *Continued*

S-416 79 Göteborg
Phone: + 46 31 14 24 80
Fax: + 46 31 14 24 80
E-mail: pro.natura@pro-natura.net

The preservation of biodiversity and the development of methods for sustainable use of natural resources will be among the most important issues in the environmental field for a long time to come. Pro Natura has been working many years with these tasks. Our main fields of activity are biological inventories, management plans, education, research, and investigations in Swedish terrestrial ecosystems.

Allan Willians Coastal Engineering and Consultant Services (UK)

104, Thurstaston Road
Thurstaston Wirral
CH61 0HG
Phone: +44 (0)151 648 8896; 07771 697403
Fax: +44 (0) 151 648 8896
E-mail: alan.wil@virgin.net

Alan Williams is a Chartered Coastal Engineer based in the UK with over 20 years experience in the field of Coastal & Maritime Engineering.

Hidrosfera Consultoria Ambiental (Brazil)

Rua Agenor de Oliveira Costa, 255
Cassino, Rio Grande/RS
Caixa Postal 1011-96200 972
Phone: +55 (53) 236 5655
Fax: +55 (53) 236 5655

A HIDROSFERA é uma empresa voltada à prestação de serviços técnicos, especializada em diagnosticar e avaliar problemas relacionados à Oceanografia e Meio Ambiente.

HR Wallingford (UK)

Howbery Park, Wallingford,
Oxfordshire OX10 8BA
Phone: +44 (0) 1491 835381
Fax: +44 (0) 1491 832233
E-mail: hrinfo@hrwallingford.co.uk

HR Wallingford is an independent research and consultancy organization specializing in civil engineering hydraulics and the water environment.

lkyon Hydraulic Consultancy & Research (The Netherlands)

Postal address:
P.O. Box 248, 8300 AE
Emmeloord

Visiting address:
Voorsterweg 28
8316 PT Marknesse
E-mail: info@alkyon.nl

An independent Dutch company, founded in 1996 by a group of experts and consultants with an extensive record in coastal and offshore hydraulic engineering and research, aiming to bring fit-for-purpose advice and services of a high quality onto the market at competitive price.

Nouel Engineering Consultants (Venezuela)

Grupo Nouel, C.E. La Pirámide,
Piso 1, Ofic. 106, Urb. Prado Humboldt.
Apdo Postal: 80680, Caracas
Phone: (582) 979 8111; (582) 979 6311
Fax: (582) 979 5427

Nouel Engineering Consultants, since 1952, has been the leader in Venezuela in the areas of ports, terminals and marine consulting engineering, embracing, and also in other areas like industrial facilities, environmental engineering, and public infrastructure.

OCEANOR—Oceanographic Company of Norway (Norway)

Pir-Senteret, N-7462
Trondheim
Phone: +47 73 54 52 00
Fax: +47 73 54 52 01
E-mail: oceanor@oceanor.no

OCEANOR is a high-tech company specializing in delivering integrated real-time environmental monitoring and information systems for oceans, rivers, lakes, groundwater, and soil.

The Coastline Surveys Limited (UK)

Headquarters:
Bridgend Farmhouse
Bridgend Stonehouse
Gloucestershire, GL10 2AX
Phone: +44 01453 826772
Fax: +44 01453 826762
Marine Operations:
Unit 17 & 18

Frampton on Severn Industrial Park
Bridge Road, Frampton on Severn
Gloucestershire, GL2 7HE
Phone: +44 01452 740941
Fax: +44 01452 740811

Coastline Surveys Ltd (CSL) Group provides a professional independent marine data acquisition, interpretation and consultancy service worldwide from operating bases in England.

W.S. Ocean Systems Ltd. (UK)

Omni Business Centre,
Omega Park, Alton,
Hampshire, GU34 2QD
Phone: +44 (0) 1420 541555
Fax: +44 (0) 1420 541499
E-mail: info@wsocean.com

Table A3.11 North American professional societies, nongovernmental, and nonprofit organizations active in the coastal zone

American Coastal Coalition (USA)

American Coastal Coalition
5460, Beaujolais Lane
Fort Myers, FL 33919
Phone: (941) 489 2616
Fax: (941) 489 9917
E-mail: kategapr@cs.com

The American Coastal Coalition is a national membership organization composed of governmental entities, government officials, business people, academics, national and regional interest groups and advocacy organizations, property owners' associations, individual coastal community residents, and others. It has been organized to serve as the voice of the nation's coastal communities in Washington, DC. ACC's goals are to preserve the role of the federal government in shore protection; support policies and programs which promote travel and tourism to coastal regions of the United States; promote the preservation, protection, and restoration of sandy beaches along America's coastline, including the Great lakes; foster public understanding of the importance of well-maintained beaches to the national economy and to national disaster protection policy; and support these objectives in an environmentally and fiscally sound manner.

American Fisheries Society (USA)

5410, Grosvenor Lane
Bethesda, MD 20814
Phone: (301) 897 8616
Fax: (301) 897 8096
E-mail: main@fisheries.org

The mission of the American Fisheries Society is to improve the conservation and sustainability of fishery resources and aquatic ecosystems by advancing fisheries and aquatic science and promoting the development of fisheries professionals.

American Geological Institute (USA)

American Geological Institute
4220, King Street

Table A3.11 *Continued*

Alexandria, VA 22302-1502

The American Geological Institute is a nonprofit federation of 37 geoscientific and professional associations that represent more than 100,000 geologists, geophysicists, and other earth scientists. Founded in 1948, AGI provides information services to geoscientists, serves as a voice of shared interests in our profession, plays a major role in strengthening geoscience education, and strives to increase public awareness of the vital role the geosciences play in mankind's use of resources and interaction with the environment.

American Littoral Society (USA)

Building 18, Sandy Hook
Highlands, NJ 07732
Phone: (732) 291 0055

The American Littoral Society (ALS) is a national, not-for-profit, membership organization, dedicated to the environmental wellbeing of coastal habitat.

American Oceans Campaign (USA)

600, Pennsylvania Ave SE, Suite 210
Washington, DC 20003
Phone: (202) 544 3526
Fax: (202) 544 5625
E-mail: info@americanoceans.org

American Oceans Campaign is working in Washington (DC), Los Angeles (CA), and in coastal communities across the country to revitalize the nation's oceans and coastal waters. AOC has two primary goals: restore and protect ocean habitats and ensure clean, safe beach water.

American Society of Civil Engineers (USA)

1801, Alexander Bell Drive
Reston, VA 20191
Phone: (800) 548 2723

Founded in 1852, the American Society of Civil Engineers (ASCE) represents more than 123,000 members of the civil engineering profession worldwide, and is America's oldest national engineering society. ASCE's vision is to position engineers as global leaders building a better quality of life.

American Society of Limnology and Oceanography (USA)

ASLO Business Office
5400, Bosque Boulevard, Suite 680
Waco, TX 76710-4446
Phone: (254) 399 9635; 1 800 929 2756
Fax: 254 776 3767
E-mail: business@aslo.org

The purposes of ASLO are to promote the interests of limnology, oceanography and related sciences, to foster the exchange of information across the range of aquatic science, and to further investigations dealing with these subjects. ASLO is best known for its journal, *Limnology and Oceanography* (L&O), its interdisciplinary meetings, and its special symposia.

America Shore & Beach Preservation Association (USA)

Gregori Wodell, President
1724, Indian Way
Oakland, CA 94611
Phone: (510) 339 2818
Fax: (510) 339 6710
E-mail: president@asbpa.org

ASBPA dedicated to the sound, far-sighted and economical development and preservation of the shore of our oceans, lakes and rivers which will aid in placing their benefits within the reach of the largest possible number of people in accordance with the ideals of a democratic nation.

Association of Coastal Engineers (USA)

P.O. Box 7800
Alexandria, VA 22307

Founded in 1999 to promote excellence in coastal engineering practice, education, and research. The Association of Coastal Engineers is a professional organization dedicated to the advancement of excellence in education, research, and the practice of coastal engineering. The Association recognizes Coastal Engineering as the skills, knowledge, expertise, and theory associated with purposeful engineering intervention in the coastal system.

Atlantic Coastal Action Program Saint John (Canada)

76, Germain Street,
P.O. Box 6878, Station A
Saint John
New Brunswick
E2L 4S3
Phone: (506) 652 2227
Fax: (506) 633 2184
E-mail: acapsj@fundy.net

ACAP Saint John is a nonprofit community-based environmental management and research organization. In this capacity, we represent all of the interests in the Saint John community and work with the stakeholders to move toward better management of our local environment.

Atlantic Coastal Watch (USA)

Sustainable Development Institute
3121, South St., NW
Washington, DC 20007
Phone: (202) 338 1017
E-mail: susdev@igc.org
Website: www.susdev.org

Bay of Fundy.com (Canada)

P.O. Box 243
Chance Harbour
New Brunswick E5J 2B8
Phone: (506) 659 2044
E-mail: bof@nbnet.nb.ca

Bay of Fundy.com is distinctive because it is dedicated to sustainable tourism—tourism that enables visitors to experience and enjoy the natural and cultural attractions of Fundy in ways that do not exhaust the resource and which generate income for those who are stewards of the region.

Beach Erosion Authority for Clean Oceans and Nourishment (USA)

800, South Victoria Avenue
Room L1050
Ventura, CA 93009

105, East Anapamu St
Suite 201, County Counsel
Santa Barbara, CA 93101
E-mail: staff@beacon.dst.ca.us

The Beach Erosion Authority for Clean Oceans and Nourishment (BEACON) is a California Joint Powers agency established to deal with coastal erosion and beach problems on the Central Coast of California. The agencies making up BEACON are Santa Barbara and Ventura Counties and the cities of Port Hueneme, Oxnard, San Buenaventura, Carpinteria, and Santa Barbara.

California Shore and Beach Preservation Association (USA)

250, W. Wardlow Road
P.O. Box 7707
Long Beach, CA 90807
Phone: (310) 426 9551 Ext. 294
Fax: (310) 424 7489
E-mail: cwebb@moffattnichol.com

State chapter of the American Shore and beach preservations association. CSBPA is an educational and professional association with members from government, academics, coastal engineering, and other professions, as well as property owners and individuals and groups interested in the coast of California.

California Coastal Coalition (CalCoast) (USA)

1133, Second Street, Suite G
Encinitas, CA 92024
Phone: (760) 944 3564
Fax: (760) 944 7852
E-mail: steveaceti@calcoast.org

The California Coastal Coalition (CalCoast) is a nonprofit advocacy group comprising 28 coastal cities; six counties; AMBAG, BEACON, SANDAG, and SCAG; along with business associations and allied

Table A3.11 *Continued*

groups committed to restoring California's shoreline through sand replenishment, increasing the flow of natural sediment, wetlands recovery, and improved water quality. CalCoast was the cosponsor, with the CA Shore and Beach Preservation Association, of the CA Public Beach Restoration Act (AB 64-Ducheny) which was signed into law in October, 1999.

Canadian Ocean Habitat Protection Society (Canada)

Box 13, Newellton,
Nova Scotia
BOW 1PO
E-mail: dkpjones@klis.com

Exploring, understanding, protecting, and restoring eastern Canada's incredible northern coral forests & those fisheries that can coexist with them.

Caribbean Conservation Corporation (USA)

4424, NW 13th St., Suite #A1
Gainesville, FL 32609
Phone: 1 800 678 7853; (352) 373 6441
Fax: (352) 375 2449
E-mail: ccc@cccturtle.org

Caribbean Conservation Corporation is a not-for-profit 501(c)(3) membership organization based in Gainesville, Florida. CCC was founded in 1959 by Mr. Joshua B. Powers in response to renowned ecologist Dr. Archie Carr's award-winning book, The Windward Road, which first alerted the world to the plight of sea turtles. Since its founding, CCC has been dedicated to the conservation of sea turtles and related marine and coastal wildlife through research, training, advocacy, education, and protection of natural areas.

CERF—Coastal Education and Research Foundation, Inc (USA)

P.O. Box 210187
Royal Palm Beach, FL 33411

CERF is a nonprofit corporation dedicated to the advancement of the coastal sciences. The foundation is devoted to the multidisciplinary study of complex problems of the coastal zone. The purpose of the foundation is to help translate and interpret coastal issues for the public and to assist professional research and public information programs. CERF is the publisher of the international *Journal of Coastal Research* (JCR), a coastal-marine science research journal that deals with all aspects of the coastal zone.

Charlotte Marine Research Team

Phone: (813) 571 9750; 626 5478

The Charlotte Marine Research Team (CMRT) is a nonprofit corporation established in 1999 and composed of volunteer citizens.

Clean Annapolis River Project (Canada)

P.O. Box 395,
Annapolis Royal Nova Scotia
BOS 1A0
Phone: (902) 532 7533
E-mail: carp@fox.nstn.ca

The Clean Annapolis River Project (CARP) is a charitable, community-owned corporation created to work with the community and interested organizations to foster conservation, restoration, and sustainable use of the freshwater and marine ecosystems of southwestern Nova Scotia's Annapolis River and its watershed.

Clean Water Fund (CWF) (USA)

4455, Connecticut Ave.
NW, Suite A300-16
Washington, DC 20008
Phone: (202) 895 0432
Fax: (202) 895 0438
E-mail:cwf@cleanwater.org

Clean Water Fund (CWF), a national 501(c)3 nonprofit, brings diverse communities together to work for changes that improve our lives, promoting sensible solutions for people and the environment.

Clean Water Network (USA)

1200, New York Avenue
NW, Suite 400,
Washington DC 20005
Phone: (202) 289 2395
Fax: (202) 289 1060
E-mail: cfw@cleanwater.org

The Clean Water Network (CWN) is an alliance of more than 1,000 organizations that endorse its platform paper, the National Agenda for Clean Water, which outlines the need for strong clean water safeguards to protect human health and the environment.

Coalition to Restore Coastal Louisiana (USA)

200, Lafayette Street
Baton Rouge, LA 70801
Phone: 188-LA-COAST

Early in 1988, the Coalition to Restore Coastal Louisiana was incorporated to address and advocate for the restoration and preservation of the only great delta ecosystem in North America—the Mississippi River Delta. A land of extraordinary riches, the area supplies a large portion of our nation's commercial fish landings—oysters, blue crabs, menhaden, and shrimp, to name a few.

Coast Alliance

600, Pennsylvania Ave.
SE, Suite 340
Washington, DC 20003
Phone: (202) 546 9554
E-mail: coast@coastalliance.org

The Coast Alliance is a nonprofit organization, formed in 1979 by a number of groups and individuals concerned about the effects of unprecedented development pressure and pollution on the coasts.

Coastal America (USA)

300, 7th Street
SW, Suite 680
Washington, DC 20250
Phone: (202) 401 9928
Fax: (202) 401 9821
E-mail: wanda.brown@usda.gov

A decade of commitment to protecting, preserving, and restoring America's coastal heritage.

Coastal Conservancy (USA)

1330, Broadway, 11th Floor
Oakland, CA 94612
Phone: (510) 286 1015
Fax: (510) 286 0470

The Coastal Conservancy acts with others to preserve, protect, and restore the resources of the California Coast. Our vision is of a beautiful, restored, and accessible coastline.

Coastal Conservation Association (USA)

4801, Woodway, Suite 220W
Houston, TX 77056
Phone: (713) 626 4234
Telex: (800) 201 FISH
E-mail: ccantl@joincca.org

CCA is a national organization dedicated to the conservation and preservation of marine resources. Coastal Conservation Association (CCA) is an organization of strong state chapters comprising of avid recreational fishermen who have banded together to address conservation issues nationally and within their respective states.

Coastal Research and Education Society of Long Island, Inc. (USA)

Southampton College of Long Island University
Campus Box 1764
239, Montauk Highway
Southampton, NY 11968

Coastal Society (USA)

P.O. Box 2081
Glouster, MA 01930

Table A3.11 *Continued*

CMC—Center for Marine Conservation (USA)

1725, DeSales Street,
NW, Suite 600
Washington, DC 20036
Phone: (202) 429 5609
Fax: (202) 872 0619
E-mail: cmc@dccmc.org

The mission of the Center for Marine Conservation is to protect ocean ecosystems and conserve the global abundance and diversity of marine wildlife.

Dr.Beach.org (USA)

International Hurricane Center
Florida International University
University Park Campus
Miami, FL 33199
Phone: (305) 348 1607
E-mail: leatherm@fiu.edu

Earth Island Institute (USA)

300, Broadway, Suite 28
San Francisco, CA 94133
Phone: (415) 788 3666
Fax: (415) 788 7324

Earth Island Institute (EII), founded in 1982 by veteran environmentalist David Brower, fosters the efforts of creative individuals by providing organizational support in developing projects for the conservation, preservation, and restoration of the global environment. EII provides activists the freedom to develop program ideas, supported by services to help them pursue those ideas, with a minimum of bureaucracy.

Fisheries and Oceans Canada (Canada)

Station 12E239,
200, Kent Street
Ottawa, Ontario K1A 0E6
Phone: (613) 990 6840
Fax: (613) 952 6802
Website: oceanscanada.com

Florida Defenders of the Environment (USA)

4424, NW 13th Street, Ste. C-8
Gainesville, FL 32609
Phone: (352) 378 8465

Florida Defenders of the Environment was founded in 1969 to fight construction of the Cross-Florida Barge Canal. We succeeded, and have been trying ever since to repair the damage caused during initial construction stages of the barge canal-to remove Rodman (Kirkpatrick) Dam and restore the Ocklawaha River.

Florida Engineering Society (USA)

125, S. Gadsden St.
Tallahassee, FL 32301
Phone: (850) 224 7121
E-mail: fes@fleng.org

Florida Oceanographic Society (USA)

890, NE Ocean Boulevard,
Hutchinson Island
Stuart, FL
Phone: (561) 225 0505

Florida Oceanographic Society is a non-profit 501C(3) organization established in 1964, and is "Dedicated to the preservation and enhancement of Florida's Coastal Ecosystems, through education, research, personal stewardship and fun!"

Florida Shore & Beach Preservation Association, Inc. (USA)

2952, Wellington Circle
Tallahassee, FL 32308
Phone: (850) 906 9227
Fax: (850) 906 9228

The Florida Shore & Beach Preservation Association is a very different kind of state association, which is the source of our effectiveness. On the one hand, we function as a "league of cities and counties" on beach and coastal issues. Most coastal cities and counties are members. This gives us enormous clout in the Florida Legislature. On the other hand, FSBPA represents more than 1,000 private citizens concerned about beach preservation. We provide these citizens with a forum and a strong voice. This public–private partnership is vital to get local beach projects off the ground. FSBPA is a nonprofit corporation with offices in Tallahassee close to the state capital. We enjoy 501 (c) (3) tax exempt status with the I.R.S. as an educational organization.

Marine Technology Society (USA)

1828, L Street, NW #906
Washington, DC, 20036-5104m
Phone: (202) 775 5966
Fax: (202) 429 9417
E-mail: mtspubs@aol.com

As a professional society, we are constantly striving to rise to the challenges and changes within our ocean professions. We do this by ongoing examination of the services we offer our members, by increasing our cooperative efforts with other societies, by expanding our international presence, and by fostering education of the public and our youth.

National Parks Conservation Association (USA)

1300 19th St.
NW, Suite 300,
Washington, DC 20036

Since 1919, the National Parks Conservation Association has been the sole voice of the American people in the fight to safeguard the scenic beauty, wildlife, and historical and cultural treasures of the largest and most diverse park system in the world.

North Carolina Coastal Federation (USA)

North Carolina Coastal Federation
3609, Highway 24 (Ocean)
Newport, NC 28570
Phone: (252) 393 8185; 800 232 6210
Fax: (252) 393 7508

"Citizens working together for a healthy coastal environment" best summarizes the North Carolina Coastal Federation's mission. NCCF is a nonprofit, tax exempt organization which seeks to protect and restore the state's coastal environment, culture, and economy through citizen involvement in the management of coastal resources.

Ocean Conservation Society (USA)

P.O. Box 12860
Marina del Rey, CA 90295
Phone: (310) 822 5205
Fax: (310) 822 5729
E-mail: info@oceanconservation.org

Ocean Conservation Society is a 501 (c) (3) nonprofit corporation engaged in marine biology research, conservation education, public outreach, and the protection of our oceans.

Oceanic Society (USA)

Headquarters:
Fort Mason Center
Building E
San Francisco, CA 94123

The Oceanic Society is a nonprofit, membership organization founded in 1969 to protect the marine environment.

Ocean Voice International (Canada)

P.O. Box 37026,
3332, McCarthy Road
Ottawa, ON K1V 0W0
Phone: (613) 721 4541
Fax: (613) 721 4562

Ocean Voice International is a nonprofit membership-based marine environmental organization dedicated to the harmony of humankind, sea and its life.

Table A3.11 *Continued*

Oregon Coastal Zone Management Association (OCZMA) (USA)

P.O. Box 1033
Newport, OR 97365
Phone: (541) 265 8918

Pensacola Gulf CoastKeepers, Inc. (USA)

811, W. Garden Street
Pensacola, FL 32501
Phone: (850) 429 8422
E-mail: gckeepermoore@cs.com

The Pensacola Gulf CoastKeepers, Inc. is an environmental group whose members are citizens in the Northwest Florida Gulfcoast area who are dedicated to protecting the waterways of this area. The CoastKeepers Mission is to take care of the Gulf Coast and its watershed because it takes care of us. As soon as you are born, you are part owner of the water. You! Not some institution.

Save Our Shores (USA)

222, East Cliff Dr.
Suite # 5A
Santa Cruz, CA 95062
Phone: (831) 462 5660

Since 1978, the volunteers and sponsors of Save Our Shores have defended the health of our coastal environment. Balancing a tough proenvironment stance with a reputation for creative problem solving. We build and sustain cooperative partnerships with government, business, and the public.

Save San Francisco Bay Association (USA)

1600, Broadway Suite 300
Oakland, CA 94612
Phone: (510) 452 9261
Fax: (510) 452 9266
E-mail: savebay@savesfbay.org
Website: www.savesfbay.org

Save San Francisco Bay Association seeks to preserve, restore, and protect the San Francisco Bay and Sacramento/San Joaquin Delta Estuary as a healthy and biologically diverse ecosystem essential to the wellbeing of the human and natural communities it sustains.

Save The Bay (USA)

Headquarters:
434, Smith Street
Providence, RI, 02908
Phone: (401) 273 7153; 1 800 NARRBAY

Save The Bay's Narragansett BayStation
Seamen's Church Institute on Bowen's Wharf
18, Market St
Newport, RI
Phone: (401) 324 6020; (401) 324 6021
E-mail: savebay@savebay.org

The mission of Save The Bay is to ensure that the environmental quality of Narragansett Bay and its watershed is restored and protected from the harmful effects of human activity. Save The Bay seeks carefully planned use of the Bay and its watershed to allow the natural system to function normally and healthfully, both now and for the future.

Save the Manatee Club
Save the Manatee Club
500, N. Maitland Ave.
Maitland, FL 32751, USA
Phone: 1 800 432 JOIN (5646); (407) 539 0990
E-mail: membership@savethemanatee.org
education@savethemanatee.org

Save the Sound Inc. (USA)

185, Magee Ave
Stamford, CT 06902
Phone: (203) 327 9786
Fax: (203) 967 2677
E-mail: membership@savethesound.org

Save the Sound is a bi-state, nonprofit membership organization dedicated to the restoration, protection, and appreciation of Long Island Sound and its watershed through education, research, and advocacy.

SeaWeb (USA)

1731, Connecticut Ave.
NW, 4th Floor
Washington, DC 20009
Phone: (202) 483 9570

SeaWeb is a project designed to raise awareness of the world ocean and the life within it. The ocean plays a critical role in our everyday life and in the future of our planet. We believe that as more people understand this and begin to appreciate the earth as a water planet, they will take actions to conserve the ocean and the web of life it supports.

St. John's Harbour ACAP (Canada)

6, Bruce Street
Mount Pearl
Newfoundland A1N 4T3
Phone: (709) 747 4973
Fax: (709) 772 6309

St. John's Harbour ACAP, Inc. was founded in early 1993 by a group of citizen and government stakeholders. It is a nonprofit organization with a mandate to implement a community directed, concensus-based Comprehensive Environmental Management Plan (CEMP)for the Harbour and its related environs. This CEMP represents a major step toward fulfillment of that mandate.

Surfrider Foundation (USA)

122, S. El Camino Real #67
San Clemente, CA 92672
Phone: (949) 492 8170
Fax: (949) 492 8142

The Surfrider Foundation is a nonprofit organization dedicated to protecting our oceans, waves, and beaches.

Tampa BayWatch, Inc. (USA)

8401, Ninth Street North,
Suite 230-B
St. Petersburg, FL 33702
Phone: (727) 896 5320
Fax: (727) 896 5325

Tampa BayWatch, Inc. is a nonprofit stewardship program dedicated exclusively to the charitable and scientific purpose of protecting and restoring the marine and wetland environments of the Tampa Bay estuary, the largest open water estuary in the State of Florida.

The American Littoral Society (USA)

28, West 9th Road,
Broad Channel, NY 11693
Phone: (718) 634 6467
E-mail: donriepe@aol.com

We are an environmental organization concerned about issues that affect the littoral zone: that area on the beach between low and high tide. The American Littoral Society (ALS) is a national, nonprofit, public-interest organization comprising over 6,000 professional and amateur naturalists, with headquarters in Sandy Hook, New Jersey.

The Chesapeake Bay Foundation

Headquarters:
Philip Merrill Environmental Center
6, Herndon Avenue
Annapolis, MD 21403
Phone: (410) 268 8816; (410) 269 0481 (from Baltimore); (301) 261 2350 (from D.C. metro)

As you wander through this site, you will see concrete examples of how CBF is helping to restore the tapestry that is the Bay. Our efforts are organized under the headings of "resource protection and restoration" and "environmental education."

Table A3.11 *Continued*

The Coastal Society (USA)

The Coastal Society
P.O. Box 25408,
Alexandria, VA 22313-5408
Phone: (703) 768 1599
Fax: (703) 768 1598
E-mail: coastalsoc@aol.com

The Coastal Society is an organization of private sector, academic, and government professionals and students dedicated to actively addressing emerging coastal issues by fostering dialogue, forging partnerships, and promoting communication and education.

The Fishermen and Scientists Research Society (Canada)

P.O. Box 25125 (Canda)
Halifax, Nova Scotia B3M 4H4
Phone: (902) 876 1160
Fax: (902) 876 1321
Website: www.fsrs.ns.ca

The Fishermen and Scientists Research Society (FSRS), a nonprofit organization, is an active partnership between fishermen and scientists. The objective of this partnership is to establish and maintain a network of fishermen and scientific personnel that are concerned with the long-term sustainability of the marine fishing industry in the Atlantic Region.

The Marine Technology Society (USA)

1828, L Street
NW, #906
Washington, DC 20036-5104
Phone: (202) 775 5966
Fax: (202) 429 9417
E-mail: mtspubs@aol.com

From its inception in the early 1960s, the Marine Technology Society has embraced a charter of inclusiveness. We support all the components of the ocean community: marine sciences, engineering, academia, industry, and government. The core objectives of our society remain valuable in today's rapidly changing world: we are dedicated to the development, sharing, and education of information and ideas.

The Sea Turtle Restoration Project (USA)

P.O. Box 400,
Forest Knolls, CA 94933
Phone (415) 488 0370

The Sea Turtle Restoration Project fights to protect endangered sea turtles in ways that make cultural and economic sense to the communities that share the beaches and waters with these gentle creatures. With offices in California and Costa Rica, STRP has been leading the international fight to protect sea turtle populations worldwide.

The Wildlife Conservation Society (USA)

2300, Southern Boulevard
Bronx, NY 10460
Phone: (718) 220 5100

Since 1895, WCS has worked from our Bronx Zoo headquarters to save wildlife and wild lands throughout the world.

World Aquaculture Society (WAS) (USA)

143, J. M. Parker Coliseum
Louisiana State University
Baton Rouge, LA 70803
Phone: +1 225 388 3137
Fax: +1 225 388 3493

The World Aquaculture Society (WAS) is an international nonprofit society with over 4,000 members in 94 countries. Founded in 1970, the primary focus of WAS is to improve communication and information exchange within the diverse global aquaculture community.

1000 Friends of Florida (USA)

926, East Park Avenue
P.O. Box 5948
Tallahassee, FL 32314-5948
Phone: (850) 222 6277
Fax: (850) 222 1117

Table A3.12 Worldwide societies, nongovernmental, and nonprofit organizations active in the coastal zone

AINCO—Interocean

Costa Rica 11 (1, A-26),
28016 Madrid, Spain
Phone: 34 91 350 4394
Fax: 34 91 350 2414

AINCO7-Interocean is an independent nonprofit organization founded in 1985, with the principal goal of developing Oceanographic and Environmental Sciences in Spain and the rest of the world.

Asociación Oceánica de Panamá

6-2305 El Dorado, Panamá,
República de Panamá

6-3998, El Dorado, Panamá,
República de Panamá
Phone: (507) 226 2020; (507) 260 8265

La Asociación Oceánica de Panamá (AOP), es la organización no gubernamental (ONG) panameña sin fines de lucro que desde su fundación en enero de 1991 se dedica a la investigación, protección y conservación del medio ambiente marino panameño. Esta integrada en su mayoría por jóvenes buzos y personas que se sienten atraídas por la belleza de nuestros océanos y que han decidido unir esfuerzos para proteger y preservar nuestros recursos marinos.

Association of Marine Scientists

Room B19, New Academic Complex
University of Mauritius
Réduit, Mauritius
Phone: 4541041 Ext: 1409
Fax: 3952005
E-mail: ams.mauri@intnet.mu

The Association of Marine Scientists (AMS) is a nongovernmental organization which aims at providing a common platform for marine scientists in Mauritius to interact with each other. AMS also acts as the local chapter of the Western Indian Ocean Marine Science Association (WIOMSA).

Australian Coral Reef Society

66, Oogar Street
Alexandra Headland
Queensland 4572, Australia
Phone: (07) 5443 6565
E-mail: acrs@jcu.edu.au

The ACRS plays a key role by promoting scientific research on Australian coral reefs. It is a forum for discussion and information transfer among scientists, management agencies, and reef-based industries that are committed to ecological sustainability. Because it is not aligned to any vested interests, the Society's views are sought by government policymakers, conservationists and all those interested in coral reefs who need impartial and expert advice.

Australian Marine Sciences Association (AMSA)

P.O. Box 902
Toowong
Queensland 4066, Australia
E-mail: amsa@uq.edu.au

Table A3.12 *Continued*

The Australian Marine Sciences Association (AMSA) is a national nonprofit organization dedicated to promoting marine science and coordinating discussion and debate of marine issues in Australia.

Australian Meteorological and Oceanographic Society (AMOS)

Phone: +61 (0)2 9296 1618
Fax: +61 (0)2 9296 1657
E-mail: m.speer@bom.gov.au
Website: www.amos.org.au/sydney

The Australian Meteorological and Oceanographic Society (AMOS) is an independent Australian society that supports and fosters interest in meteorology and oceanography through publications, meetings, courses, conferences, grants, and prizes. It also represents the views of its members to various institutions and the public.

Cape Nature Conservation (CNC)

Phone: + 27 21 426 0723
Fax: + 27 21 426 4266
E-mail: cncinfo@cape-town.org

Cape Nature Conservation (CNC) is concerned with the conservation of our natural environment within the western Cape, South Africa (see our mission statement below). This area includes the fynbos biome—one of the six plant kingdoms of the world. Cape Nature Conservation manages nature reserves and wilderness areas in the Western Cape, and invites you to experience and revel in their pristine natural beauty.

Conference of Peripheral Maritime Regions of Europe

CRPM, 6 Rue Saint-Martin 35700
Rennes, France
Phone: +33 2 99 35 40 50
Fax: +33 2 99 35 09 19

CERM—Consortium for Estuarine Research and Management

University of Port Elizabeth
P.O. Box 1600
Port Elizabeth, South Africa
Phone: +27 (0) 41 5042877
Fax: +27 (0) 41 5832317
E-mail: btagcb@upe.ac.za

Organizations of South African scientists collaborate in promoting the wise management of estuarine systems trough joint participation in direct research, training, and technology transfer.

Dorset Coast Forum

C/o Dorset County Council,
County Hall
Dorchester, Dorset, DT1 1XJ, UK
Phone: 01 305 225 132

The Dorset Coast Forum was established in 1995 to look at the long-term strategic issues facing the Dorset coast. The overriding aim of the Forum is to promote a sustainable approach to the management, use, and development of Dorset's coastal zone, which will ensure that its inherent natural and cultural qualities are maintained and enhanced for the benefit of future generations.

ESPO (European Sea Ports Organisation)

Avenue Michel-Ange, 68
B-1000 Brussels
Phone: 32 2 736 34 63
Fax: 32 2 736 63 25
E-mail: mail@espo.be

ESPO's mission is twofold. It aims at influencing public policy in the European Union and to achieve a safe, efficient, and environmentally sustainable European Port sector, operating as a key element of a transport industry where free and undistorted market conditions prevail, as far as practicable.

Estuarine and Coastal Sciences Association

Department of Biological Sciences
University of Hull
Hull HU6 7RX, UK
E-mail: t.c.telfer@stir.ac.uk

ECSA is an academic organization, with a worldwide membership, which promotes research and study of all aspects of estuarine and coastal regions. The Association was founded in 1971, as the Estuarine and Brackish-Water Biological Association, to promote production and dissemination of scientific knowledge and understanding of estuaries and coastal waters, in order to encourage resource management for the public benefit.

European Aquaculture Society

Slijkensesteenweg 4
B-8400 Oostende, Belgium
Phone: +32 59 32 38 59
Fax: +32 59 32 10 05
E-mail: eas@aquaculture

The European Aquaculture Society (EAS) was established on April 30, 1976 as an international, nonprofit association, with the principal objective of being the European forum for contacts and information exchange between all actors within the aquaculture industry.

European Artificial Reef Research Network

Dr. Antony Jensen
Southampton Oceanography Centre
European Way
Southampton SO14 3ZH, UK
Phone: +44 1703 593428
Fax: +44 1703 596642
E-mail: a.jensen@soc.soton.ac.uk

Artificial reef research programs exist in Italy, Spain, Portugal, the UK, the Netherlands, France, Greece, Norway, Israel, Monaco, Russia, Poland, Turkey, and Finland. Denmark has an interest in artificial reefs, although no structures have yet been placed. European reef research is varied; from biofiltration through habitat protection to fishery enhancement.

European Coastal Association for Science and Technology

Jeanette Owen
Department of Maritime Studies and International Transport
Cardiff University
P.O. Box 907
Cardiff CF1 3YP, UK
Phone: +00 44 2920 874271
Fax: +00 44 2920 874301
E-mail: eurocoast@cardiff.ac.uk
owenJ4@cardiff.ac.uk

EUROCOAST was established in 1989 as an association of scientists, engineers, and decisionmakers within the European community.

European Union for Coastal Conservation

P.O. Box 11232
2301 EE Leiden
The Netherlands
Phone: +31 71 5122900
Fax: +31 71 5124069
E-mail: admin@eucc.nl

EUCC is dedicated to the integrity and natural diversity of the coastal heritage and to ecologically sustainable development. EUCC is the largest coastal network in Europe, with 750 members and member organizations in 40 countries, 14 active National Branches, and 7 professional offices.

EUCC bridges the gap between scientists, ecologists, conservation site managers, planners, and policy makers, especially at an international level.

Fiskardo's Nautical and Environmental Club and Ionian Sea Research Centre

28084 Fiskardo
Kephalonia, Greece
Phone: 00 30 (0) 674 41182
Fax: 00 30 (0) 674 41182

Table A3.12 *Continued*

E-mail: fnec@otenet.gr
fnec@otenet.gr

FNEC is a nonprofit making and nongovernmental organization conducting environmental activities in Kephalonia. FNEC'S main aims are to provide an informal environmental information center, promote environmental education & conservation and encourage youth exchanges for community development. The Ionian Sea Research Centre (ISRC) is an independent scientific body within the club FNEC carrying out marine research, recording marine sightings and promoting marine conservation. The ISRC possesses a number of permits from the Ministry of Fisheries and Forestry to carry out research in the Ionian Sea.

Fundação sos Mata Atlântica

Rua Manoel da Nóbrega, 456
CEP 04001-001
São Paulo/SP, Brazil
Phone: (0XX11) 3887 1195
Fax: (0XX11) 3885 1680
E-mail: smata@ax.apc.org
Website: www.sosmatatlantica.org.br

A Fundação SOS Mata Atlântica é uma entidade privada, sem vínculos partidários ou religiosos e sem fins lucrativos. Seus principais objetivos são defender os remanescentes da Mata Atlântica, valorizar a identidade física e cultural das comunidades humanas que os habitam e conservar os riquíssimos patrimônios natural, histórico e cultural dessas regiões, buscando o seu desenvolvimento sustentado.

Greenpeace International

Kiezersgracht 176
1016 DW Amsterdam
The Netherlands
Phone: +31 20 523 6222

I Love the Ocean—Philippines

Blue Seas, Inc., 5/F CIFC Towers
J. Luna cor. Humabon Sts.
North Reclamation Area
Cebu City, Philippines
Phone: (32) 232 1821 22 412 0487 89
Fax: (32) 232 1825

The "I Love the Ocean" Movement is a recognition of the need to approach the problem of sustainable coastal resource use from all angles. It recognizes that all individuals have a stake in what happens to our marine resources, and that each individual, from the president of the country to the company executive to the nameless man on the street, has an important role to play in saving these precious gifts of nature.

Instituto Ecológico Aqualung

Rua do Russel, 300, grupo 401
Glória, Rio de Janeiro, RJ, Brazil, CEP: 22210-010
Phone: (021) 558 3428; (021) 558 3429
Fax: (021) 558 3419; (021) 558 1233
E-mail: instaqua@uol.com.br

O instituto ecológico aqualung é hoje uma das maiores e mais atuantes entidades preservacionistas brasileiras. Atuando na área de preservação e educação ambiental, criando e implantando ações concretas de forma a arrecadar maiores recursos para patrocinar diversas entidades e organizações ambientalistas, divulgando a informação e o conhecimento sobre as causas ecológicas, criando publicações sobre o meio ambiente e a fauna marinha e lançando campanhas de conscientização da importância de se preservar o meio ambiente, o instituto ecológico aqualung vem desenvolvendo um trabalho moderno e eficiente de inestimável valor para as gerações futuras—a preservação do nosso planeta.

KIMO, UK

Mr. Rick Nickerson, KIMO Secretariat
Shetland Islands Council, Infrastructure Services
Grantfield, Lerwick
Shetland ZE1 0NT UK
Phone: +44 01595 744800
Fax: +44 01595 695887

KIMO is an international association of Local Authorities, which was formally founded in Esbjerg, Denmark, in August 1990 to work toward cleaning up pollution in the North Sea. It has over 100 members in 8 countries including the United Kingdom, Norway, Sweden, Denmark, the Faeroes Islands, and the Netherlands, the Republic of Ireland with associate members in Germany. KIMO's primary objective is the cleaning up of the existing pollution in Northern Seas and coastal waters, of preventing future pollution and of working to preserve and enhance them and to leave them in a fit and healthy state for the wellbeing of future generation.

Marine Conservation Society, UK

9, Gloucester Rd Ross-on-Wye
Herefordshire HR9 5BU, UK
E-mail: mcsuk@mcmail.com

Beach watch and shark education program.

National Committee on Coastal and Ocean Engineering, Australia

University of New South Wales
King St, Manly Vale
NSW 2093, Australia
E-mail: r.cox@unsw.edu.au

The NCCOE was formed in 1971 and is an honorary group of specialist engineering professionals, whose objective is to advance the science and art of coastal and ocean engineering throughout the general engineering profession and the community.

Norwegian Marine Fauna—Underwater Wildlife

Frank Emil Moen
Ufôrfjellveien 11
4370 Egersund, Norway
Phone: +47 51491351

Pictures and descriptions (only norwegian) of different arthropoda (mostly crustacean), cnidarians and 77 fishes. You will find five different thumbnails galleries including pictures of fish, sponges, cold water corals, and more.

River Ocean Research and Education

113-117 Queens Road
Brighton, BN1 3XG, England
Phone: +44 (0)1273 234032
Fax: +44 (0)1273 234033

Our mission is to work in partnership with others to encourage stewardship for, and sustainable use of, all areas of the water environment.

SAMS—The Scottish Association for Marine Science

Prof. Graham Shimmield, Director
SAMS Dunstaffnage Marine Laboratory
Oban, Argyll PA34 4AD
Phone: 01631 559 000
Fax: 01631 559 001
E-mail: mail@dml.ac.uk

The Scottish Association for Marine Science (SAMS) promotes marine research and education in Scotland.

Seas At Risk

Drieharingstraat 25
NL-3511 BH Utrecht
The Netherlands
Phone: +31 30 670 1291
Fax: +31 30 670 1292

Seas At Risk is an independent nongovernmental federation of national and international environmental organizations concerned with the protection and restoration of the marine environment.

Stichting De Noordzee, The Netherlands

North Sea Foundation
Drieharingstraat 25
3511 BH Utrecht
The Netherlands

Table A3.12 *Continued*

Phone: 030 2340016
Fax: 030 2302830
E-mail: info@noordzee.nl

Lid Stichting Waterpakt, Participant organization Seas at Risk. NGO for protection and sustainable use of the North Sea (in Dutch).

Stichting Duinbehoud, The Netherlands

Postbus 664
2300, AR Leiden
The Netherlands
Phone: 071 5160490
Fax: 071 5160499

Coastal Dune Conservation; De Stichting Duinbehoud is een landelijke organisatie voor de bescherming van de duinen. Zij komt op voor de belangen van de natuur langs de Nederlandse kust en voor de belangen van mensen die willen genieten van de natuur.

Surfers Against Sewage

Wheal Kitty Workshops
Agnes, Cornwall
England TR5 0RD
VAT No. 557 6758 85
Phone: +44 (0) 1872 553001
Fax: +44 (0) 1872 552615

Surfers Against Sewage exist because "Everyone Needs Protecting." Everyone has the right to a clean, safe water environment, and the right to enjoy that environment, without fear of getting ill.

TAMAR

Rua Antonio Atanázio, 273
Itaguá, Ubatuba-SP, Brazil
CEP: 11680-000
Phone: (012) 432 6202
Praia do Forte (Base Mãe)
Caixa Postal 2219
CEP: 40-210-970, Salvador-BA, Brazil
Phone: (071) 876 1113; (071) 876 1045; (071) 824 1193 (Arembepe); (071) 374 0201 (Camping Ecológico de Itapuã)

Ao longo dos seus 20 anos, o Tamar-Ibama foi aperfeiçoando sua forma de trabalhar, buscando sempre soluções criativas para preservar as tartarugas marinhas. Da Bahia, Espírito Santo e Sergipe, foi se espalhando pelo litoral brasileiro e ilhas oceânicas.

United Nations Environment Program

United Nations Avenue, Gigiri
PO Box 30552,
Nairobi, Kenya
Phone: (254 2) 621234
Fax: (254 2) 624489/90

Wetlands International—Home

Wetlands International—Africa, Europe, Middle East
P.O. Box 7002
Droevendaalsesteeg 3A
6700 CA Wageningen
The Netherlands
Phone: +31 317 478884
Fax: +31 317 478885
E-mail: post@wetlands.agro.nl

Wetlands International is the leading nonprofit organization dedicated solely to the crucial work of wetland conservation and sustainable management. Well-established networks of expert and close partnerships with key organizations provide Wetlands International with the essential tools for catalyzing conservation activities worldwide. Our activities are based on sound science and carried out in over 120 countries around the world.

Table A3.13. Asian institutions for coastal research and marine sciences

Coastal Fisheries Laboratory, National Research Institute of Fisheries Science

6-31-1 Nagai, Yokosuka
Kanagawa 238-0316, Japan
Phone: 81 468 56 2887
Fax: 81 468 57 3075

College of Oceanography and Environmental Science

Xiamen University (361005)
Xiamen, Fujian
The People's Republic of China
Phone: 86 592 2183065
Fax: 86 592 2183064
Email: coe@jingxian.xmu.edu.cn

JAMSTEC—Japanese Marine Science and Technology Center

Headquarters:
2-15, Natsushima-Cho
Yokosuka City
Kanagawa 237-0061, Japan

National Oceanographic Research Institute (NORI), Korea

7ga Hang-dong
Jung-gu
Incheon, Korea
Phone: 82 32 885 3827
Fax: 82 32 885 3829

NORI under the Ministry of Maritime Affairs and Fisheries has carried out hydrographic survey and oceanographic observation covering its national jurisdiction, and has established a database for the production of nautical charts, publications, and other oceanographic data.

National Research Institute of Fisheries Science

2-12-4 Fukuura
Kanazawa, Yokohama
Kanagawa 236-8648, Japan
Phone: 81 45 788 7615
Fax: 81 45 788 500

Ocean Research Institute, The University of Tokyo

1-15-1, Minamidai, Nakano-ku
Tokyo 164-8639, Japan
Phone: 03 5351 6342
Fax:03 5351 6836

Oceanographic Society of Japan

Dr. Michio J. Kishi
Faculty of Fisheries
Hokkaido University
Hakodate
Hokkaido 041-8611, Japan

South China Sea Institute of Oceanology

164, West Xingang Road
Guangzhou 510301, China
Phone: +86 20 84451335
Fax: +86 20 84452672
Website: www.scsio.ac.cn

SCSIO is well known at home and abroad with its systematic and comprehensive data collection and research achievements on the South China Sea and adjacent tropical waters. In fields of marine geology, geophysics, hydrology, meteorology, biology, chemistry, and physics, SCSIO has accomplished 458 projects, among which 192 won prizes.

Table A3.13 *Continued*

Currently SCSIO is undertaking a series of major and key research projects and continuing to play an important role in marine sciences and development in China.

Tokai University Department of Oceanography

424-8610
3-20-1, Orido
Shimizu
Shizuoka, Japan

Tropical Marine Science Institute, Singapore

14, Kent Ridge Road S (119223)
Singapore
Phone: 7749656
Fax: 7749654

The Tropical Marine Science Institute is a new institution formed within the National University of Singapore. Through active collaboration with academic, government, and industrial sectors, TMSI aims to play a strong role in promoting integrated marine science, in R&D, as well as to establish itself as a regional and international education and training center.

Table A3.14 Brazilian IBAMA state divisions, together with fisheries research institutes and coastal protected areas maintained by IBAMA, the Brazilian national environmental agency, from north to south going down the coastline

Área de Proteção Ambiental Cairuçu

Rua Dotor Geraldo N° 11
Praça Da Matriz
Paraty, RJ
CEP: 23970-000
Phone: (0243) 712051 (Patrimônio Histórico)

Área de Proteção Ambiental de Anhatomirim

Caixa Postal N° 660
Florianópolis, SC
CEP: 88020 302
Phone: (048) 234 4293
Fax: 234 1580

Área de Proteção Ambiental Guapimirim

Estrada Do Contorno da Baía de Guanabara
Br 493, Km 13
Guapimirim, Magé, RJ
CEP: 25910 000
Phone: (021) 747 7160; (021) 221 682

Centro ce Pesquisa e Extensão Pesqueira do Norte, Cepnor

Avenida Tancredo Neves, S/N0.
Terra Firme, Belém, Pa
CEP: 66077-530
Phone: (091) 246 1237

Centro de Pesquisa do Rio Grande—Ceperg

Rua Visconde De Paranaguá,
S/N° Entreposto De Pesca
Rio Grande, RS
CEP: 96200-190
Phone: (0532) 32 6285; (0532) 32 6990
Fax: (0532) 32 6990

Centro de Pesquisa e Extensão Pesqueira das Regiões Sudeste e Sul—Cepsul

Av. Ministro Victor Konder
SNo, Centro
Itajaí, SC
CEP: 88301-280
Phone: (047) 48 6058
Fax: (047) 48 6058

Centro de Pesquisa e Treinamento em Aqüicultura—Cepta

Rodovia Brigadeiro Faria Lima
S/No, Km 65
Pirassununga, SP
Cep:13630-000
Phone: (0195) 65 1299; (0195) 65 1075
Fax: 565 1075; 565 1318

Centro Nacional de Conservação e Manejo de Tartarugas Marinhas

Centro Tamar—ibama
Largo dos Aflitos, S/No, Ed. Ceres
Ministério da Agricultura
Térreo, Salas 3 e 4
Salvador, Ba
CEP: 40060-030
Phone: (071) 321 3174
Fax: 321 3174

Estação de Piscicultura de Jequié

Estrada Da Barragem De Pedra
Km 14,
Jequié, Ba
CEP: 45200-000

Estação Ecológica de Tamoios

Mambucaba
Angras dos Reis, RJ
CEP: 23908-000
Phone: (0243) 43 4455 Ramal: 307

Estação Ecológica do Taim

Taim Rio Grande
Rio Grande, RS
CEP: 96211-000
Phone: (053) 503, 3151

Gerência Executiva do Ibama no Estado da Bahia/Salvador

Jose Guilherme Da Mota
Avenida Juracy Magalhaes Junior,
N° 608, Rio Vermelho
Salvador, Ba
CEP: 41930 080
Phone: Pabx (071) 345 7322/240 79
Fax: 240 7913; 248 9427

Gerência Executiva do Ibama no Estado de Santa Catarina

Luiz Hamilton Martins
Avenida Mauro Ramos, No 187, Centro
Caixa Postal 660
Florianópolis, SC
CEP: 88020-301
Phone: (048) 224 6202; (048) 224 9549; (048) 223 3465; (048) 224 6077
Fax: 224 6077

Gerência Executiva do Ibama no Estado de São Paulo/São Paulo

Wilson Almeida Lima
Alameda Tietê, No 637
Jardim Cerqueira Cesar
São Paulo, SP
Cep: 01417-020
Phone: (011) 3083 1300; (011) 3081 8752; (011) 3088 0227
Fax: (011) 3081 8599

Gerência Executiva do Ibama no Estado do Ceará/Fortaleza

Romeu Aldigueri De Arruda Coelho
Rua Visconde Do Rio Branco, No 3.900
Tauapé, Fortaleza, Ce

Table A3.14 *Continued*

CEP: 60055-172
Phone: (085) 272 7950; (085) 272 9081
Fax: 227 9081; 272 9386

Gerência Executiva do Ibama no Estado do Espírito Santo—Supes/ES
José Olímpio Vargas
Avenida Marechal Mascarenhas De Moraes,
No 2.487, Bento Ferreira
Vitória, Es
CEP: 29052-121
Phone: (027) 225 8510; (027) 324 1811; (027) 222 4777
Fax: (027) 324 1837

Gerência Executiva do Ibama no Estado do Maranhão/São Luiz
Antonio Myses Da Silva Nrto
Avenida Jaime Tavares
No 25, Centro, São Luiz, Ma
CEP: 65025-470
Phone: (098) 221, 2776; (098) 221 2125
Fax: 231 4332

Gerência Executiva do Ibama no Estado do Pará/Belém
Selma Bara Melgaço, Avenida Conselheiro Furtado,
No 1303, Batista Campos, Belém, Pa
CEP: 66035-350
Phone: (091) 241 2621; (091) 224 5899
Fax: 223 1299

Gerência Executiva do Ibama no Estado do Paraná/Curitiba
Luis Antonio Mota Nunes De Melo
Rua Brigadeiro Franco, No 1.733
Caixa Postal N° 691
Curitiba, PR, CEP: 80420-200
Phone: (041) 222 7488; (041) 322 5125
Fax: 225 7588

Gerência Executiva do Ibama no Estado do Piauí/Teresina Delcleciano Guedes Ferreira
Avenida Homero Castelo Branco,
N° 2.240, Jockey Club Teresina, Pi
CEP: 64048-400
Phone: (086) 232 1142; (086) 232 1652
Fax: 232 5323

Gerência Executiva do Ibama no Estado do Rio De Janeiro/RJ
Carlos Henrique Abreu Mendes
Praça 15 Novembro, No 42
8° Andar, Centro
Rio De Janeiro, RJ
CEP: 20010-010
Phone: (021) 224 6214; (021) 224 6463; (021) 3506 1734; (021) 3506 1735
Fax: 221 4911

Gerência Executiva do Ibama no Estado do Rio Grande do Norte/Natal
Francisco pondofe cavalcanti
rua alexandrino de alencar,
N° 1.399, tirol
Natal, RN
CEP: 59015-350
Phone: (084) 201 5840; (084) 201 4335; (084) 201 4230; (084) 201 4068 (084) 985 9393
Fax: 201 4422

Gerência Executiva do Ibama no Estado do Rio Grande do Sul
Rodney Ritter Morgado
Rua Miguel Teixeira
No 126, Cidade Baixa
Porto Alegre, RS
CEP: 90050-250
Phone: (051) 226 0002; (051) 225 2594; (051) 225 2144; (051) 228 7186; (051) 228 7290
Fax: 226 6392

Parque Nacional da Lagoa do Peixe
Praça Luiz Martins, No 30
Mostardas, RS
CEP: 96270-000
Phone: (051) 673 1464

Parque Nacional Marinho dos Abrolhos
Rua Praia Do Kitongo, S/No
Caravelas, Ba
CEP: 45900-000
Phone: (073) 297 1111

Porto de Paranaguá
Armazen 8-A
Paranaguá, PR
Phone (041) 423 2566
Fax: (041) 423 2566

Table A3.15 List of coastal resource management companies in Hawaii, based on major private firms

AECOS Laboratory of Hawaii, Inc.
75-5586, Olioli Rd., #207
Kailua-Kona 96740
Phone: (808) 329 8411
Fax: (808) 329 6343
E-mail: kklein9@juno.com
Provides analytical and consulting environmental services.

AECOS Laboratory of Hawaii, Inc.
Hawaii Aquaculture Company
P.O. Box 61970
Honolulu 96839
Phone: (808) 956 8286
Fax: (808) 956 4483
E-mail: small113@aol.com
Aquaculture consulting, research, and development.

AECOS Laboratory of Hawaii, Inc
TerraSystems, Inc.
2800, Woodlawn Dr.,
Suite 264
Honolulu 96822
Phone: (808) 539 3745
Fax: (808) 539 3746
E-mail: office@terrasys.com
Website: www.terrasys.com
TSI provides data collection and professional services for coastal environmental assessment,. monitoring and GIS mapping of coastal/agricultural/urban areas via satellite, high-resolution airborne spectral imagery and other remote sensing techniques.

AECOS, Inc.
970, N. Kalaheo Ave.,
Suite C-311, Kailua, 96734
Phone: (808) 254 5884
Fax: (808) 254 3029
E-mail: aecos@pixi.com
Website: www.wco.com/~aecos/aecosinc.html
AECOS provides analytical and consulting environmental services through its analytical and biological laboratory.

Table A3.15 *Continued*

American Deepwater Engineering, Ltd.
Pier 14, First Floor
Honolulu 96817
Phone: (808) 545 5190
Fax: (808) 545 1988
E-mail: ademta@gte.net
American Deepwater Engineering utilizes manned submersibles for deep ocean marine research and sub-sea construction. State of the art design, enhanced by a working depth of up to 2000 ft, provides the submarines with a wide range of applications far beyond the scope of both commercial diving and ROV operations.

American Marine Services Group
Pier 14, First Floor
Honolulu 96817
Phone: (808) 545 5190
Fax: (808) 538 1703
E-mail: amsg@amsghq.com
Website: amsghq.com

American Ocean Systems, Ltd.
626, Poipu Drive
Honolulu 96825
Phone: (808) 394 0361
Fax: (808) 394 2191
E-mail: kim@deeperthanthat.com
Website: www.deeperthanthat.com
Marine services company specializing in surveys, consulting, and commercial diving.

Analytical Laboratories of Hawaii
1320, Aalapapa
Lanikai, 96734
Phone: (808) 235 1395
Fax: (808) 247 1910
E-mail: miles@interpac.net
ALH provides services in living marine resource management including environmental monitoring and assessment for long-term trend and impact analysis. Analytical services include GIS output.

Applied Analysis Incorporated
P.O. Box 10631
Honolulu 96816
Phone: (808) 737 3033
Fax: (808) 735 6553
E-mail: rowland@aloha.com
Website: homepages.infoseek.com/~rpmhawaii/
Aquaculture systems analysis, environmental impact analysis, proposal development, and project management.

Applied Technology Corporation
1441, Kapiolani Blvd.,
Suite 810
Honolulu 96814
Phone: (808) 973 1800
Fax: (808) 973 1808
E-mail: atc@lava.net
Applied Technology Corporation specializes in the design, construction, and supervision of reinforced, precast, and prestressed concrete offshore vessels and land structures including buildings, bridges, waterfront facilities, and towers.

Aquacultural Concepts
P.O. Box 560
Waimanalo, 96795
Phone: (808) 259 5042
Fax: (808) 259 8049
E-mail: shleser@aloha.net
Aquaculture consultants, research & development, coastal planning.

Aquaculture Technology, Inc.
455, Anolani Street
Honolulu 96821
Phone: (808) 377 5087
E-mail: aquatech@akamai.ceros.org

Aquasearch, Inc.
73-4460, Queen Kaahumanu Hwy., #110
Kailua-Kona, 96740
Phone: (808) 326 9301
Fax: (808) 326 9401
E-mail: mhuntley@kona.net
Website: www.aquasearch.com
A marine biotechnology company, and a global leader in photobioreactor technology. Industrial photobioreactors are a new platform technology that unlocks the commercial potential of microscopic plants-30,000 species of unexploited microalgae.

Aquasense, Inc.
2800, Woodlawn Dr.,
Suite 156,
Honolulu 96822
Phone: (808) 539 3988
Fax: (808) 539 3719

Aquatic Farms
49-139, Kamehameha Hwy.
Kaneohe, 96744
Phone: (808) 239 2929
Fax: (808) 239 8436
E-mail: officeafl@aol.com
Aquaculture and fisheries feasibility studies, environmental impact statements, design, management and training, genetic improvement, and research in the development of molecular markers for aquaculture, fisheries, and coastal environmental resource management.

Aquatic Sciences Corporation
864, S. Beretania St.
Honolulu 96813
Phone: (808) 537 9971 Ext. 17
Fax: (808) 524 6313
E-mail: dhirota@aloha.net
Website: planet-hawaii.com/dhirota
Oceanography, aquaculture, marine biology.

Bartholomew Aquatic Services
31, Kua Place
Lahaina, Hawaii 96761
Phone: (808) 667 5319
Fax: (808) 667 6934
E-mail: edbarth@maui.net
Aquaculture consulting (manageable systems for the production of food and ornamental species).

Bay Pacific Consulting
1919, Hunnewell St.
Honolulu 96822
Phone: (808) 947 1523
Fax: (808) 941 0180
E-mail: bay@hula.net

Belt Collins Hawaii
680, Ala Moana Blvd.,
1st Floor
Honolulu 96813
Phone: (808) 521 5361
Fax: (808) 538 7819
E-mail: hawaii@beltcollins.com
Website: www.beltcollins.com
Environmental, planning, civil engineering, surveying, and golf course landscape architecture services. Develop plans and conduct investigations, sampling, analysis, and design. Headquartered in Hawaii, with offices on the mainland and overseas.

Table A3.15 *Continued*

Black Pearls, Inc.

Black Pearls, Inc.
P.O. Box 525
Holualoa 96725
Phone: (808) 325 6516
Fax: (808) 325 3425
E-mail: nasims@aloha.net

Develops hatchery techniques for black-lip pearl oysters and uses OTEC water for land-based black pearl culture. Consulting joint-ventures and project management of South Pacific black pearl farms.

Carlsmith Ball

Pacific Tower, Suite 2200
1001, Bishop St.
Honolulu 96813
Phone: (808) 523 2500
Fax: (808) 523 0842
E-mail: dhr@carlsmith.com
gsumida@carlsmith.com
Website: www.carlsmith.com

Specializes in Admiralty & Maritime Law in Hawaii and the Pacific, including Law of the Sea matters, and represents marine interests including insurance ship owners/charters cargo marine financing tour operators and developers of ocean resources, including fishing and ocean and deep sea resources.

CH2M Hill

1585, Kapiolani Blvd.,
Suite 1420
Honolulu 96814
Phone: (808) 943 1133
Fax: (808) 941 8225
Website: www.ch2m.com

Feasibility studies, research, and design of ocean structures, environmental investigations, and planning.

Common Heritage Corp.

4921, Waa St.
Honolulu 96821
Phone: (808) 623 6666
Fax: (808) 377 1530
E-mail: craven@aloha.com
Website: www.aloha.com/~craven

Deep ocean water systems.

Consulting & Research Services, Inc.

721, S. Alu Rd.
Wailuku 96793
Phone: (808) 242 2954
Fax: (808) 242 2954
E-mail: jbhcrs@gte.net

Coastal engineering, dredging engineering, remediation of contaminated sediments, preparation of Environmental Impact Statements, litigation expert witness.

CSL Co.

1914, University Avenue, #104
Honolulu 96822
Phone: (800) 815 9823
Fax: (808) 946 4334
E-mail: winston@lava.net

Cost-effective analysis of ocean farming, agriculture cold technology, and compelling research into the benefits to Ke Ahole Point development. We are biotech research funded, but are open to outside jointventures. Computer resources like laptops, pentiums, and Macintosh clones are also available.

Dames & Moore

615, Piikoi St.,
Suite 900
Honolulu 96814
Phone: (808) 593 1116
Fax: (808) 593 1198
E-mail: hongk@dames.com
Website: www.dames.com

Consultants, engineering ocean & coastal, engineering maritime services, coastal and/or marine planning, marine policy, resource management.

Dashiell, Eugene P. Planning Services

1314, S. King St.,
Suite 951
Honolulu 96814
Phone: (808) 593 8330
Fax: (808) 593 8330
E-mail: dashiell@lava.net
Website: www.lava.net/environmental-planning/

Plans, EA/EIS, permits, watershed management, historic preservation, economic, infrastructure, harbors, ports, airports, GIS, natural resource plans, stream restoration, recreation & parks, golf course impacts, storm water runoff, CZM, public involvement, community-based planning.

Detection Limit Technology, Inc.

1051, Keolu Dr.,
Suite 204
Kailua, 96734
Phone: (808) 263 2364
Fax: (808) 263 2578
E-mail: cschoen@dlimit.com

Fiber-optic remote sensing including CO_2, pH, and temperature.

Dillingham Construction Pacific, Ltd.

614, Kapahulu Ave.
Honolulu 96815
Phone: (808) 735 3356
Fax: (808) 735 7416
E-mail: dredging@pixi.com
Website: www.dillinghamconstruction.com

Construction of docks, wharves, bridges, fishing piers, dredging, shoreline protection.

ELS

1524, Halekoa Drive
Honolulu 96821
Phone: (808) 734 4751
E-mail: sheae@ewc.hawaii.edu

Environmental science/resource management consulting specializing in: ocean and coastal resources; climate and global change research, and societal applications; scientific planning and evaluation; and communication, education, training and outreach.

EnterOcean Group, Inc.
3375, Koapaka,
Suite G310
Honolulu 96819
Phone: (808) 836 3330
Fax: (808) 836 3330
E-mail: laidlaw@enterocean.com

Environmental Assessment Co.

1820, Kihi St.
Honolulu 96821

Environmental Laboratory of the Pacific

930, Mapunapuna St.,
Suite 100
Honolulu 96819
Phone: (808) 831 3090
Fax: (808) 831 3098
E-mail: elpacific2@aol.com

Table A3.15 *Continued*

EL. Pacific is an environmental laboratory dedicated to the preservation of the environment in Hawaii and the Pacific Basin. EL Pacific provides environmental analysis for a wide variety of constituents in soil, water, oil, hazardous wastes, and other matrixes.

Gaffney, Rick & Associates, Inc.
73-1062, Ahikawa St.
Kailua-Kona 96740
Phone: (808) 325 5000
Fax: (808) 325 7023
E-mail: captrick@kona.net
Rick Gaffney & Associates is a 15-year-old consulting firm, specializing in ocean recreation development planning. Experience in Hawaii, Japan, China, Australia, New Zealand, Fiji, Micronesia, Tahiti, Samoa, Guam, etc.

Garcia and Associates
729, Emily St., Suite B
Honolulu 96813
Phone: (808) 597 8865
Fax: (808) 597 8864
E-mail: laurat@gandahi.mhs.compuserve.com

Gateway Technologies Intl., Inc.
Roger Webb
1188, Bishop St.,
Suite 3406
Honolulu 96813
Phone: (808) 537 5522
Fax: (808) 537 1596
E-mail: gtiirw@aol.com

Geolabs Inc.
2006, Kalihi St.
Honolulu 96819
Phone: (808) 841 5064
Fax: (808) 847 1749
E-mail: geolabs@gte.net
Geotechnical consulting and groundwater monitoring services.

Global Ocean Consultants, Inc.
1130A, Mano Dr.
Kula 96790
Phone: (808) 878 2929
Fax: (808) 878 3511
E-mail: goc@maui.net
Undertake appraisals of coastal and offshore fishery resources, prepare business plans/feasibility studies on the establishment of processing facilities, management, and marketing operations.

Hamnett, Michael P. & Associates
47-655, Hui Kelu St.,
No. 5, Kaneohe 96744
Phone: (808) 239 6213
Fax: (808) 239 2510
Conducts policy research and planning in the following areas: coastal zone, ocean, and environmental management, fisheries economics, foreign aid policy, disaster preparedness and mitigation, and Pacific Island development.

Hawaiian Marine Enterprises
P.O. Box 301
Kahuku 96731
Phone: (808) 293 1230
Fax: (808) 293 0059
E-mail: limu@lava.net
Aquaculture of edible seaweed ("limu ogo"), other marine algae, and ornamental fish.

High Health Aquaculture.
P.O. Box 1095
Kurtistown 96760
Phone: (808) 982 9163
Fax: (808) 982 9163

INALAB, Inc.
3615, Harding Ave.,
Suite 308
Honolulu 96816
Phone: (808) 735 0422
Fax: (808) 735 0047
E-mail: mark_hagadone@msn.com
Website: www.gtesupersite.com/inalab
Analysis of sea water and other ocean materials and structures for composition and wide variety of contaminants.
Consultation re-ocean chemistry. Isolation of drugs, etc. from marine organisms.

Innovations Hawaii
P.O. Box 17097
Honolulu 96817
Phone: (808) 847 4732
Fax: (808) 847 4732
E-mail: ruediger@lava.net

Innovative Technical Solutions, Inc.
2800, Woodlawn Drive,
Suite 192
Honolulu 96822
Phone: (808) 539 3660
Fax: (808) 539 3670

Integrated Environmental Technologies, Inc. (INTECH)
1200, College Walk,
Suite 203
Honolulu 96817
Phone: (808) 531 8330
Fax: (808) 531 8374
E-mail: jackharmon@compuserve.com
Advanced systems & technology, environmental monitoring, deep ocean engineering.

Inter-Pacific Ocean Consulting Group Ltd.
611, Hahaione Street
Honolulu 96825
Phone: (808) 395 6112
Fax: (808) 395 6112

Kona Cold Lobsters
P.O. Box 3314
Kailua-Kona 96745
Phone: (808) 329 4332
Fax: (808) 326 2882
Closed-cycle OTEC, marine tropical reef fish reproduction, lobster aquaculture, and marine mammal rehabilitation.

KRP Information Services
1314, South King Street
Suite 951
Honolulu 96814
Phone: (808) 593 8331
Fax: (808) 593 8330
E-mail: krplan@hgea.org
Environmental planning, coastal zone planning and management; SMA permits, CZM consistency; water quality certification.

Marine Analytical Specialists (MAS)
1738, Laukahi St.
Honolulu 96821-1358
Phone: (808) 373 5129
Fax: (808) 373 5129
E-mail: walsht001@hawaii.rr.com
With over 25 years of research & consulting expertise, MAS provides accurate & precise water quality analyses approved by the EPA & DOH. Analyses of all types of water samples are offered for Environmental Monitoring, Wastewater Treatment & Reuse, Golf Course Management and Marine & Coastal Research.

Table A3.15 *Continued*

Marine Analytical Specialists (MAS)

P.O. Box 6882
Kamuela 96743
Phone: (808) 885 6354
Fax: (808) 885 6474
E-mail: dtarnas@aloha.net
carolyn@aloha.net

Private consultant, with local and international experience, specializing in coastal and marine resources management, planning, research, and education.

Marine and Coastal Solutions International

P.O. Box 1206
Kailua 96734
Phone: (808) 259 8871
Fax: (808) 259 8238
E-mail: makai@makai.com

Ocean engineering, design and analysis: pipelines, submarine cable installation, seawater air conditioning, aquaculture, marine vehicles, surveying, construction management.

Marine Minerals Technology Center Associates

2179, Makiki Heights Drive
Honolulu 96822
Phone: (808) 959 1237
Fax:(808) 959 1237
E-mail: mcruick@aol.com

Consultants in economic development of marine mineral resources. Over 200 person years of experience in all disciplines.

Marine Research Consultants

4467, Sierra Dr.
Honolulu 96816
Phone: (808) 734 4009
Fax: (808) 732 1813
E-mail: sdolar@aloha.net

McCorriston Miho & Miller

5, Waterfront Plaza, 4th Floor
500 Ala Moana Blvd.
Honolulu 96813
Phone: (808) 529 7300
Fax: (808) 524 8293
E-mail: m4hawaii@aol.com

Legal support for maritime commercial ventures and international and admiralty law, including Law of the Sea, shipping, and ocean resource development (OTEC, aquaculture, etc.).

McCorriston Miho & Miller

705, Nunu St.
Kailua, 96734
Phone: (808) 254 0203
Fax: (808) 254 3029

Molokai Sea Farms

P.O. Box 560
Kaunakakai 96748
Phone: (808) 553 3547
E-mail: shrimp@aloha.net

Hatchery—*Penaeus vannamei* nauplii and postlarvae, disease-free broodstock development *Clarius fucus, Oreochromis mossambica; Growout-P. vannamei, Chanos chanos, C. fucus, O. mossambica*; and Consulting.

Navatek Ltd.

841, Bishop St.,
Suite 1880
Honolulu 96813
Phone: (808) 531 7001 Ext. 18
Fax: (808) 523 7668
E-mail: pacmar@aloha.net

High-tech ship design and construction.

Neptune Technologies, Inc.

1200, College Walk
Suite 203
Honolulu 96817
Phone: (808) 531 8330
Fax: (808) 531 8374
E-mail: jackh@ceros.org

Supporting ocean engineering, science, technology, and related activities.

Noda, Edward K. & Associates Inc.

615, Piikoi St.,
Suite 300
Honolulu 96814
Phone: (808) 591 8553
Fax: (808) 593 8551
E-mail: enoda@hawaii.edu

Coastal and ocean engineering, specializing in oceanographic field surveys, computer modeling of ocean processes, design criteria development,and environmental impact analysis.

Ocean Engineering Consultants

2250-A, Noah St.
Honolulu 96819
Phone: (808) 735 2775
Fax: (808) 737 3201

Ocean Innovators

P.O. Box 88121
Honolulu 96830
Phone: (808) 533 6434
Fax: (808) 537 5607
E-mail:ocean_innovators@worldnet.att.net

Development of specialized ocean equipment for marine surveys and the ocean recreation field.

Ocean Marine Associates

58-274A, Kamehameha Hwy
Haleiwa 96712
Phone: (808) 638 7469
Fax: (808) 638 8564
E-mail: omahawaii@eudoramail.com

Consulting in: commerical diving, offshore and marine construction.

Oceanic Companies, Inc.

1287, Kalani Street
Suite 203
Honolulu 96817
Phone: (808) 843 8300
Fax: (808) 843 2800
E-mail: oci@aloha.net
Website: www.hawaiibiz.com/oceanic

Engineering and construction for power and processing facilities.

Oceanic Imaging Consultants

Manoa Innovation Center
2800, Woodlawn Dr.,
Suite 270A
Honolulu 96822
Phone: (808) 539 3708
Fax: (808) 539 3710
E-mail: info@oicinc.com
Website: www.oicinc.com

Provides software and systems for acquiring, real-time processing, and mosaicking of sidescan sonar, bathymetry, sub-bottom, and navigation data; and consulting services pertaining to the acquisition and analysis of Marine Geophysical Data for seafloor mapping applications.

Table A3.15 *Continued*

Oceanit Laboratories, Inc.
1100, Alakea St.,
31st Floor
Honolulu 96813
Phone: (808) 531 3017
Fax: (808) 531 3177
E-mail: oceanit@oceanit.com
Website: www.oceanit.com

Specializes in environmental and coastal engineering services and research & development. Special capability is problem solving. Services include marine environmental surveys and assessments, environmental documents, physical hydraulic modeling, physical oceanography, defense systems, data acquisition systems, etc.

Oceantek, Inc.
41-945, Kalanianaole Highway
Waimanalo, 96795
Phone: (808) 259 9102
Fax: (808) 259 0809
E-mail: oceantek@lava.net

Oceantronics, Inc.
711, Nimitz Hwy.,
Pier 24, Honolulu 96817
Phone: (808) 522 5600
Fax: (808) 522 5222
E-mail: oceantronics@worldnet.att.net
Website: www2.11net.com/fritz

Marine electronics, marine safety-lifesaving, land mobile radio, manufacturer of GPS products and radiobuoys, electronic charting installation, FCC and GMDSS inspections.

OffCoast, Inc.
146, Hekili St.
Suite 101
Kailua, 96734
Phone: (808) 261 1999
Fax: (808) 261 2074
E-mail: info@offcoast.com

Specializes in offshore and coastal engineering, structural and hydrodynamic design, and analysis of very large floating structures; ocean engineering software development; platform motions, wave refraction & diffraction, oil-spill spreading, circulation.

Ogden Environmental and Energy Services Co.
680, Iwilei Rd.,
Suite 660
Honolulu 96817
Phone: (808) 545 2462
Fax: (808) 528 5379
E-mail: dlhazelwood@oees.com
Website: www.ogden.com

Full-service environmental consulting, specializing in impact assessment, GIS services, archaeology services and contamination assessment and cleanup.

Pacific Aquatic Environmental
758, Kapahulu Avenue,
Suite 227
Honolulu 96816
Phone: (808) 942 7618
Fax: (808) 942 7618

Pacific Environmental Technologies
963, Kaahue Street
Honolulu 96825
Phone: (808) 395 1005
Fax: (808) 395 1005
E-mail: garym@soest.hawaii.edu

Pace Tech designs and builds advanced geochemical sensors for *in situ* monitoring of the deepsea, coastal marine or terrestrial aqueous environment, including radon, methane, and helium gases, and dissolved ions. PET also provides systems integration, data storage/telemetry and long-term deployments.

Pacific Island Technology, Inc.
2800, Woodlawn Drive,
Suite 113
Honolulu 96822
Phone: (808) 539 3636
Fax: (808) 539 3637
E-mail: khorton@pitec.com
Website: www.pitec.com

High-tech research and development in airborne and satellite remote sensing and ground-based geophysics. Instrumentation development, field surveys, ordnance location, GPS, environmental evaluation.

Pacific Marine Contracting, Inc.
65-1235, Opelo Rd,
Suite #9
Kamuela 96743
Phone: (808) 885 5426
Fax: (808) 885 5298
E-mail: pmchutch@aol.com

General Engineering Contractor specializing in: composite marine fender systems, commercial diving, hazardous waste removal and marine construction.

Pacific Planktonics
73-998, Ahikawa Street
Kailua-Kona, 96740-9407
Phone: (808) 325 1761
Fax: By arrangement only
E-mail: kraul@konacoast.net
Website: www.nethawaii.net/~kraul

Consultant in mahimahi (*Coryphaena hippurus*) and other marine fish aquaculture, including troubleshooting, staff training, facility designs, and project start-up. Experience in Hawaii, Tahiti, Australia, and Greece.

PacMar, Inc.
3615, Harding Ave.,
Suite 409
Honolulu 96816
Phone: (808) 735 2602
Fax: (808) 734 2315
E-mail: pacusa@pixi.com

A Hawaii-based international management consulting firm. Regional focus: Asia-Pacific. Project activities: Enterprise development. Includes CRM impacts into business planning of appropriate land and marine-based businesses. Without viable economic alternatives, abuse of coastal resources is likely to prevail regardless of CRM policies.

PBR Hawaii
Pacific Tower, Suite 650
1001, Bishop St.
Honolulu 96813
Phone: (808) 521 5631
Fax: (808) 523 1402
E-mail: pbrhi@aloha.net

PBR Hawaii provides environmental planning, permit preparation, and processing services for inland and coastal zone development projects throughout Hawaii and the Pacific Basin. Professional services also include preparation of master plans, site plans, and landscape architectural plans.

Plan Pacific, Inc.
737, Bishop St.,
Suite 1520
Honolulu 96813
Phone: (808) 521 9418
Fax: (808) 521 9468

Table A3.15 *Continued*

E-mail: planpac@aol.com

Specialist in land use and policy planning analysis related to coastal zone management, especially shoreline and beach protection, harbor and marina facilities, and endangered species habitats.

Raytheon Systems Company

2828, Paa St.,
Suite 3005
Honolulu 96819
Phone: (808) 396 3309
Fax: (808) 396 5368
E-mail: dick_porter@mukilteo.hac.com

Underwater sound system research and development.

Rescue Technologies Corporation

99-1350, Koaha Place
Aiea 96701
Phone: (808) 395 1688
Fax: (808) 395 4470
E-mail: seerescue@aol.com
Website: www.seerescue.com

Military-approved emergency signaling technologies.

Royal Hawaiian Sea Farms, Inc.

P.O. Box 3167
Kailua-Kona 96745
Phone: (808) 329 5468
Fax: (808) 329 5468
E-mail: limu@ilhawaii.net

Commercial seaweed aquaculture farm and research facility. Conducting research on sea cucumbers, edible seaweeds, limpets, warm-water abalone and marine tropicals.

SAIC-AMSEC

Naval Magazine Lualualei
442-A, Everest St.
Waianae 96792
Phone: (808) 668 8816
Fax: (808) 668 8360

SAIC-AMSEC is a technical services firm supporting the US Navy in the electronics, command and control, combat systems and hull mechanical and electrical, (HM&E) disciplines.

Sandwich Islands Analysis Co.

95-306, Kaloapau St.,
Suite 105, Mililani, 96789 1258
Phone: (808) 473 2638
Fax: (808) 625 7297
E-mail: siac@hawaii.rr.com

Naval operations research consulting.

Science & Technology International (STI)

733, Bishop St.,
31st Floor
Honolulu 96813
Phone: (808) 540 4700
Fax: (808) 540 4850
E-mail: nick@sti.hawaii.com
Website: www.sti-hawaii.com

Sensors, software, and services for visible/near-infrared hyperspectral data acquisition, image processing, analysis, and archiving for maritime and general environmental monitoring and mapping of effluents, coral, vegetation, hydrology, etc.

Sea Engineering, Inc.

Makai Research Pier
41-202, Kalanianaole Hwy,
Suite 8
Waimanalo 96795
Phone: (808) 259 7966
Fax: (808) 259 8143
E-mail: seaeng@lava.net

Ocean and coastal engineering in Hawaii and the Pacific Basin; numerical modeling; hydrographic and geophysical surveys; marine construction and diving services.

Seatech Contracting, Inc.

P.O. Box 2115
Kailua-Kona 96745
Phone: (808) 325 2020
Fax: (808) 325 2022

Research and development of underwater pipelines and pumping stations, anchoring systems, shoreline crossings and pier structures along with the construction and installation of all the above.

SEATRAC, Ltd.

2870, Von Hamm Pl.
Honolulu 96813
Phone: (808) 537 2112
Fax: (808) 528 2113

Project management and financing for commercial ventures utilizing ocean resources including OTEC, wave energy conversion, cold water cooling, aquaculture and ocean mining.

SSFM Engineers, Inc.

501, Sumner Street,
Suite 502
Honolulu 96817 5304
Phone: (808) 531 1308
Fax: (808) 521 7348
E-mail: projects@ssfm.com

Structural Solutions

98-030, Hekaha St.,
Suite 20
Aiea 96701
Phone: (808) 488 0655
Fax: (808) 488 1655
E-mail: ssi@lava.net
Website: www.cablecad.com

Structural and mechanical engineering; expertise in finite element modeling of marine structures; low-cost composite pressure hulls; design software for cables, ropes, and flexible pipe.

Submersible Systems Development

P.O. Box 3954
Honolulu 96812
Phone: (808) 988 5988
Fax: (808) 988 1206
Website: www.ricklaney.com/ssd

Boat/ship building & repairs; engineering, ocean & coastal; naval architecture & small craft design.

Synthetic Technology Corp. (SYNTEK)

3615, Harding Ave.,
Suite 308
Honolulu 96816
Phone: (808) 735 0422
Fax: (808) 735 0047
E-mail: mark_hagadone@msn.com
Website: www.gtesupersite.com/inalab

SYNTEK focuses on the development of fine biochemical products, produced entirely in Hawaii and sold to an ever-expanding global market. Research and developmental strategy targets natural marine biochemical products/metabolites never before synthesized or isolated.

Table A3.15 *Continued*

TCI-Hawaii, Inc.

Philomene A. Verlaan, of Counsel
P.O. Box 235766
Honolulu 96823
Phone: (808) 944 5116
Fax: (808) 235 5161
E-mail: paverlaan@gn.apc.org

Marine research & development, marine resource economic studies and business evaluations, project management, ocean law and policy.

Tetra Tech E M Inc.

2828, Paa St.,
Suite 3080
Honolulu 96819
Phone: (808) 831 6600
Fax: (808) 836 1689
E-mail: farrisj@ttemi.com

Provides oceanographic science services including site characterization studies for ocean disposal of dredged material, modeling of oceanographic processes, marine toxicology, and sediment geochemistry.

Textron Systems Kauai

P.O. Box 730, Waimea 96796
Phone: (808) 338 0080
Fax: (808) 338 8301
E-mail: rmullins@systems.textron.com

Tongg, Clarke & McCelvey

2752, Woodlawn Drive,
Suite 5-211
Honolulu 96822
Phone: (808) 521 2908
Fax: (808) 528 2854
E-mail: tcm@tcmhawaii.com
Website: www.tcmhawaii.com

Landscape architects with experience in coastal resort projects throughout the Pacific basin. Services begin with the preparation of landscape master plans and continue through the design and implementation of award-winning landscape architectural projects.

Towill, R.M. Corp.

420, Waiakamilo Rd.,
Suite 411, Honolulu 96817
Phone: (808) 842 1133
Fax: (808) 842 1937
E-mail: rmtowill@i-one.com

Consultants—engineering, civil environmental & coastal.

TYCO Submarine Systems, Ltd.

1001, Sand Island Pkwy.
Honolulu 96819
Phone: (808) 845 0687
Fax: (808) 845 2719

Coastal surveying & cartography; communications & navigation.

Wilson Okamoto & Associates, Inc.

1907, S. Beretania St.,
4th Floor
Honolulu 96826
Phone: (808) 946 2277
Fax: (808) 946 2253
E-mail: woa@aloha.net

Master planning; marine resource management planning; environmental impact assessment; civil, structural and architectural design.

Xamanek Researches

P.O. Box 131
Pukalani 96788
Phone: (808) 572 8900
Fax: (808) 572 8900
E-mail: xamanek@juno.com

Table A3.16 List of coastal resources management companies in Hawaii, based on nonprofit organizations

Earthtrust

25, Kaneohe Bay Dr.,
Suite 205
Kailua 96734
Phone: (808) 254 2866
Fax: (808) 254 6409
E-mail: earthtrust@aloha.net
earthtrust.org

International wildlife conservation organization based in Hawaii dedicated to preserving whales, dolphins, tigers, rhinos, and endangered species through direct intervention, education, research, and political action.

Institute for Pacific Marine Research

1118, Maunawili Rd.
Kailua 96734
Phone: (808) 262 0284
Fax: (808) 261 7820

Promotion/research of sustainable fisheries internationally. Information bank on destructive and nonsustainable fisheries, wildlife kills.

KeKua'aina Hanuana Hou

HC-01 Box 741
Kaunakakai 96748
Phone: (808) 558 8393
Fax: (808) 558 8453

Aquaculture research.

Nature Conservancy, The Pacific Region

1116, Smith St.,
Suite 201
Honolulu 96817
Phone: (808) 537 4508
Fax: (808) 545 2019

Ocean Law & Policy Institute

Pacific Forum—CSIS
Pauahi Tower, Suite 1150
1001, Bishop St.
Honolulu 96813
Phone: (808) 521 6745
Fax: (808) 599 8690
E-mail: pacforum@lava.net
linpaul@aloha.net

Oceanic Institute

Makapuu Point
41-202, Kalanianaole Hwy
Waimanalo 96795
Phone: (808) 259 7951
Fax: (808) 259 5971
E-mail: oi@oceanicinstitute.org
Website: www.oceanicinstitute.org

The Oceanic Institute is a private nonprofit $20 million facility for aquaculture and marine biotechnology research. Activities include development and transfer of ocean environment research and management and sustainable aquaculture technologies to clients worldwide.

Table A3.16 *Continued*

Pacific Basin Development Council 711, Kapiolani Blvd., Suite 1075 Honolulu 96813 Phone: (808) 596 7229 Fax: (808) 596 7249 E-mail: jnorris@elele.peacesat.hawaii.edu Website: www.pixi.com/~pbdc/index.html	Scientific research of saltwater gamefish, with special emphasis on marlin, tuna, and related species.
Pacific International Center for High Technology Research (PICHTR) 2800, Woodlawn Dr., Suite 180 Honolulu 96822 Phone: (808) 539 3900 Fax: (808) 539 3892 E-mail: vega@htdc.org Renewable energy: OTEC, wind, PV, and hybrid system design construction and operation; engineering, ocean & coastal; research & development.	**Pacific Whale Foundation** 101, N. Kihei Rd. Kihei 96753 Phone: (808) 879 8860 Fax: (808) 879 2615 E-mail: info@pacificwhale.org Website: www.pacificwhale.org Scientific study of marine mammals and their ocean environment; public education outreach and marine conservation programs 1 800 942 5311.
Pacific Ocean Research Foundation 74-381 Kealakehe Pkwy., #E Kailua-Kona 96740 Phone: (808) 329 6105 Fax: (808) 329 1148	**PACON International, Inc.** P.O. Box 11568 Honolulu 96828-0568 Phone: (808) 956 6163 Fax: (808) 956 2580 E-mail: pacon@soest.hawaii.edu Website: www.eng.hawaii.edu/~pmp/paconwww Organizes the biennial Pacific Congress on Marine Science and Technology; promotes ocean education and ocean research; marine policy.

A great deal of information about coastal organizations is provided in these 16 tables. The information that is highlighted here could have been organized in many different ways. Due to the large number of organizations, it was convenient to present lists in a format that was similar to an Internet search result. The table thus compiled required minimal efforts at reworking the lists into usable summaries. Other methods could have been used to obtain this kind of information but due to time constraints and the great effort of hand searches in books, the results would not have been as comprehensive. Perusal of these 16 tables would suggest, even to a novice researcher first approaching the coastal realm of endeavor, that an enormous amount of information from a wide variety of coastal organizations that have interest in coastal topics is now available. This list can now serve as a "heads up" for academic coastal researchers who should now be aware of many brethren of the same ilk that they were not aware of previously. Besides university researchers, there are now many private nonprofit organizations and for-profit consulting companies that conduct research in the coastal zone. The past two or three decades have witnessed enormous increases in coastal populations, expansion of industry and commercial facilities into coastal zones, and tourism and recreation dependent on clean/safe environments and adequate support infrastructure. These coastal uses mostly conflict with one another and governmental oversight has not kept pace with the rapid urban–industrial–commercial development and associated environmental degradation. The large numbers of organizations that have concern for coastal issues evidences the expansion of private and public interest groups as well as service organizations into the coastal zone. It is hoped that these lists of coastal organizations indicate the importance and level of interest in this special part of the world, the thin interface between land and water that is coming under increasing threat from over use. The organizations listed here exist, in large part, to better understand natural coastal systems and to protect, preserve and manage coastal/marine resources.

Charles W. Finkl

Bibliography

Clark, J.R., 1996. *Coastal Zone Management Handbook*. Boca Raton, Florida: Lewis.

Culliton, T.J., McDonough, J.J., Remer, D.G., and Lott, D.M., 1992. *Building Along America's Coasts: 20 Years of Building Permits, 1970–1989*. Rockville, Maryland: NOAA, Strategic Environmental Assessments Division.

Finkl, C.W., 1997. The coastal zone as new battlespace from the purview of an academic graduate program in environmental science. In Miller, M.C., and Cogan, J. (eds.), *Conference Proceedings Coastal Zone 97* (Boston, Massachusetts), Vol. II, pp. 584–586.

IAMSLIC (International Association of Marine Science Libraries and Information Centers), 1987. International Directory of Marine Science Libraries and Information Centers. Woods Hole, Massachusetts: Woods Hole Oceanographic Institution.

IGU (International Geographical Union Staff), 1992. *Orbis Geographicus* (*World Directory of Geography*) 7th edn. Wiesbaden, Germany: F. Steiner Verlag.

UNESCO, 1983. *International Directory of Marine Scientist* 3rd edn. Paris, France: UNESCO (Intergovernmental Oceanographic Commission) and FAO (Rome Italy).

APPENDIX 4: DATABASES

A database can be defined as a collection of interrelated data stored together with minimum redundancy to serve multiple applications. A common and controlled approach is used in adding new data or in modifying and retrieving the existing data within the database.

Proper coastal data management necessitates a comprehensive knowledge of the coastal environment, rapid access to required information, effective communication, and decision support tools. The development of a coastal data management system for multiple use spans a broad range of fields. In the last decades considerable changes have occurred in the methodology of storage and retrieval of data. In this presentation, information on bibliographic as well as numeric/textual databases relevant to coastal geomorphology has been included in tabular form. Databases cover a broad spectrum of related subjects like coastal environment and pollution aspects, coastline, elevation data, landscape and landform images, shore zone morphology, erosion, grain size, bathymetry, topography, hydrology, terrain, oil spills, coral morphology, coastal changes, coastal zone management, etc.

With increasing utilization of the coastal zone for various purpose the need for an Integrated Coastal Management (ICM) system has become necessary. The management of data requirements and compilation of coastal inventories are often among the first steps in ICM. In addition to contributing to the understanding of the environment and potential impacts of development, the compilation of an inventory also ensures a common information base for all decisionmakers. Once the problems of data access are overcome, building a coastal zone inventory involves constructing an intelligent and cohesive mosaic by integrating numerous separate data sources.

However, upon closer scrutiny, it is often found that the databases are incompatible and inconsistent, which hinders the process of inventory building. Differences in data collection equipment and protocol, scale, projection, temporal and attribute resolution, terminology and nomenclature, data gaps, problem solving approaches, and quality control standards, especially data validity, tolerances of error, and uncertainty need to be resolved before integrated databases are compiled. Problems of data inconsistency can supersede those of data availability, accuracy, geographic coverage, large data volume, and lack of structure. Traditional coastal inventories have been compiled as atlases and more recently digital coastal information systems. Further developments have culminated in the development of large digital atlases such as the North Sea Project Database and UK Digital Marine Atlas.

In recent times the more useful way of having a database in GIS (Geographic Information System) format has proved to have manifold support. GIS has emerged as a valuable media for coastal database management, most notably for the ability to store and analyze the diverse amount of data required for coastal management in an integrated format. The tremendous capabilities of GIS, such as infinite precision, the capability to change scale, the capability to combine and overlay data from various data sources, and the production of high-quality cartographic displays; which are most alluring, are also most deceptive. The ease with which one can create maps with GIS and the inexperience of many users in cartographic design can lead to the production of ineffective and misleading maps.

Advances in many technological fields have been well incorporated in development of coastal databases. Traditional digital atlases have been augmented with multimedia video, remote sensing images, photographic images, and hypertext medium. Traditional databases have been converted into CD-ROM databases and/or WEB databases.

Data management technology has advanced from tapes to index files, to hierarchical databases, to network database, to relational databases, to object databases, to object-related database, and finally web databases. Coastal databases have also grown through these stages and hence such databases are found in one of the above stages.

With advances in communication technology, local databases are now accessible through LAN (local area network), WAN (wide area network), Intranet, Internet (WEB databases), etc. Similarly, improvements have occurred in user's interface with database from character interface, Graphic User Interface, Object-oriented UI, Hypermedia, Hypertext mark-up language, and so on. Depending upon the need to resolve the coastal conflicts, using GIS technology, the databases have grown to Decision Support System (DSS), to Multi-criteria DSS, to Spatial DSS, to Collaborative DSS, to expert system, and so on.

Since database creation involves data from various sources, the issue of data quality is considered one of the main areas requiring attention while building databases. Of paramount importance is the need to assess the suitability and correctness of a data set to be incorporated in the database. Unfortunately the drive to build database applications often neglects the fabric of the data. Too often efforts are focused on "feeding the beast" and minimal regards to "what the beast is being fed."

Information on representative databases relevant to coastal geomorphology are presented in tabular form under:

Table A4.1: Bibliographic coastal databases;
Table A4.2: Numeric/textual coastal databases;
Table A4.3: Developer/provider's address.

Pravin D. Kunte

Acknowledgments

The author thanks Dr. E. Desa, Director, NIO, for permitting publication of this article as NIO contribution No. 3558.

Cross-references

Geographic Information Systems
Global Positioning Systems
Monitoring, Coastal Ecology
Monitoring, Coastal Geomorphology
Organizations (see Appendix 3)

Table A4.1 Bibliographic databases relevant to coastal geomorphology

No.	Database title	Provider code[a]	Subject coverage	Period[b]	Geographic coverage	Output mode	Access code	Hardware used
B1	Environmental Abstracts	P7	Publications reporting on all aspects of environment including management planning, technology, coastal geomorphology, biology	1984	Global	C	Fe	IBM PC
B2	Many (Numeric and Bibliographic)	P4	Geology, geomorphology, and hydrology	1986	Argentina	F, P	Re	IBM PC/AT
B3	GEOSCAN	P10	Geosciences including geomorphology	1840	Canada	O, F, P	Fr	HP 3000
B4	CGRG Internet Bibliography	P10	Geomorphology	1940	Canada	O, F, P,	Fr	Internet site
B5	GEOLINE	P9	Geoscience, natural resources including coastal sciences	1970	Worldwide	O, F, P	Fe	IBM PC
B6	INDOCEAN	P15	Oceanography including coastal geomorphology of Indian Ocean	1989	India	P, O, F,	Re, Fr	IBM PC
B7	MORPHO	P23	Bibliographic information on geomorphological articles	–	Japanese articles	O, F, P	Fr	Internet site
B8	GeoRef	P24	Bibliographic information	1785	Worldwide	O, C, Web	Fe	Internet search
B9	GeoBase	P24	Bibliographic information on geosciences including geomorphology	1960	Worldwide	O, C, Web	Fr	Internet site
B10	Indian National Bibliography	P16	Science, arts, and social sciences	–	India	P, F	Re, Fe	–
B11	Indian Science Abstracts	P11	Science and technology including marine field	–	India	F, P, C	Re, Fr	–
B12	Current Contents Search	P12	Earth sciences, applied sciences	1990	Worldwide	O, F, P, C	Fr	–
B13	National Union Catalogue of Science	P11	Science and technology including oceanography	–	India	P, F, O, C	Re, Fe	–
B14	ENVIROLINE	P3	Environmental science including coastal pollution	1971	Worldwide	O, C	Fe	–
B15	Environmental Bibliography	P8	Coastal environment, water resources	1971	Worldwide	O, C, P	Fe	PC based
B16	Pollution Abstracts	P2	Environmental sciences, coastal pollution	1970	Worldwide	O, C, P	Fe	PC based
B17	Georchive on CD-ROM	P14	Earth sciences, geology, geophysics, meteorology	1974	Global	O, C, P	$1400	MS-DOS
B18	Aquatic Sciences and Fisheries Abstracts	P2	Oceanography, geology, geophysics	1978	Worldwide	O, C, P	Fe	MS-DOS
B19	Science Citation Index	P12	Sciences including marine geology	1974	Worldwide	O, C, P	Fe	MS-DOS
B20	General Science Index	P17	Sciences and technology including oceanography	1978	Worldwide	O, C, P, T	Fe	MS-DOS
B21	Oceanic Abstracts	P2	Oceanography, geology, and geophysics	1964	Worldwide	O, C, P	Fe	MS-DOS
B22	Streamline	P36	Hydrology, geomorphology, and environmental geology	–	Australia	O, T, P	Fe	FACOM M380, IBM

B23	AESIS, Australian Earth Sciences Information System	P37	Earth sciences including coastal geomorphology	1975	Australia & Papua New Guinea	O, P, reports, microfilms	Fr, Fe	IBM computers
B24	GIS Bibliography	P38	Publications in GIS and Land Information System areas	1980	International	O, P	Fr	–
B25	Titles of Index and databases at National Library	P40	Web accessible products, CD-ROM, reference material, other products, links to site, etc.	–	Australia	O, P	Fr	Website
B26	West CAT-	P42	Western Michigan University On-line Library Catalog	–	Michigan University	O, P	Fr	Website
B27	The CGRG Bibliography of Canadian Geomorphology—a searchable database	P43	Citations related to the fields of aeolian, applied, coastal, fluvial, glacial, hillslope, karst, periglacial, permafrost, and offshore geomorphology	–	Canadian	O, P, T	Fr	On-line
B28	International Bibliography of Geographic Information Systems (GIS) for Coastal Managers Coastal Information Directory	P44	An international compilation of documented geographic information system (GIS) applications in the field of coastal management (1) system design, (2) coastal processes, (3) natural resource management, (4) coastal planning, (5) disaster management, (6) water quality management, and (7) coastal geology	1997	Global	O, P, T	Fr	Website
B29	Melvyl Catalog database	P45	Abstracting and indexing databases, the Melvyl Catalog and California Periodicals Database, and other catalogs	–	Global	O, P, T	Fr	Website

[a] Provider code (providers/ developers code) is given in Table A4.3.
[b] Single year means from that year uptill present.
Abbreviations used: (Access/cost) Op = Open access, Fe = Fees, Fr = Free, Re = Restricted; (Output mode) F = Floppy, T = Tape, P = Print, O = On-line, C = CD-ROM.

Table A4.2 Numeric/textual databases relevant to costal geomorphology

No.	Database title	Provider code[a]	Subject coverage	Period[b]	Geographic coverage	Output mode	Access code	Hardware used
N1	Canadian Geomorphology Images Database	P43	Landscapes and landforms	1970	Canada	O	Fr	Internet site
N2	Canadian Landform Databases	P43	Coastal, fluvial, aoline, etc. landforms	–	Canada	O	Fr	Internet site
N3	The Virtual Geomorphology Database	P21	Geomorphology including coastal	1995	Worldwide	O	Fr	Internet site
N4	Coastal Relief Gridded database	P22	Coastal elevation data and geomorphology	–	US coast	O, C	Fe	IBM PC, Internet site
N5	Coastal Information System, Canada	P22	Geomorphic features, shore zone morphology, types of material, survey, and sample data.	1981	Beaufort Newfoundland, Nova Scotia, Canada	O, F, P GIS, Oracle	Fr	Unix, IBM PC Internet site
N6	South Pacific Marine Geoscience Database	P5	Coastal geology and geophysics	–	South Pacific	F, P	Fr	–
N7	Database on Coastal Erosion and Littoral Environment	P1	Coastal erosion and environment	1987–91	Europe	T, F, P	–	–
N8	Byma Martin Channel Coastal Morphology and Sedimentology	P6	Coastal geomorphology, geology, and contamination	1986	Arctic islands	F, P	O, Fr	IBM PC
N9	Sediment File Grain Size Data	P10	Geomorphology and grain size	1984	West offshore of Canada	F, P	Re, Fr	COMPAQ-386
N10	Alaskan Marine Contaminants Database	P38	Environment, coastal pollution, and contaminants	1955–90	Alaska, USA	C, P	Re, Fr	MS-DOS
N11	Global Relief Data on CD-ROM	P13	Bathymetry, topography, gravity, and coastline	–	Global	C, P, F	$101	MS-DOS, Macintosh
N12	Oregon's Dynamic Estuary Management Information System DEMIS	P18	Geomorphology, hydrology, terrain, vegetation, population, political boundaries, and infrastructure	1996	Local Oregon State USA	C, P, F, O, GIS based	Fr	MS-DOS
N13	Marine Resource Inventory Survey, Fogo Island	P19	Coastal infrastructure; coastal geomorphology; coastline classification; and environmental features	1996	Coastline— Fogo Island, Newfoundland	C, P, F, O GIS based	Fr	MS-DOS
N14	IAG-GEOMORPHLIST Master Directory	P20	Directory of geomorphologists and their specialization, contact address, and e-mail address		Global	C, P, F, O	Fr	Internet site
N15	NOAA's Medium Resolution Digital Vector Shoreline	P25	Digital shoreline data set	–	World shoreline	C, O, I, GIS based	Fr	Internet site

N16	GSHHS—A Global Self-consistent, Hierarchical, High-resolution Shoreline Database	P26	Digital shoreline	–	Global shoreline	C, O, I, GIS based	Fr	Internet site
N17	ECNC Database: SAXIFRAGA	P27	Landscape slides including coastal geomorphology	–	European	O, I	Fe	Internet site (for browsing)
N18	An Inventory of Australian Estuaries and Enclosed Marine Waters-Database	P58	Catchment area, runoff rate, rainfall, etc.	1989	Victoria	O, I, GIS based	Fr	Record file, Internet site
N19	Placentia Bay Database on Oil Spill Response	P29	Oil spill and geomorphology	–	Placentia Bay	O, I, GIS based	Fr	Internet site for searching
N20	Australian Coastal Atlas	P30	All coastal parameters including geomorphology	–	Australia	O, I, GIS based	Fr	Internet site browsing
N21	Marine & Coastal Data Directory	P41	Blue Pages describing website that provides linkages, information, other sites, and atlases and other material available on Internet	–	Australia	Online	Fr	Internet site
N22	Interim Marine and Coastal Regionalization of Australia (IMCRA)	P41	–	–	Australia	O, I	Fr	Internet site
N23	KeySHORE Coastal Database (SHOREBASE)	P31	Survey, Profiles, and Ground Modeling, GIS, AutoCAD Photos	1995	England	O, I, GIS based	Fe	Access software, IBM
N24	The Canadian Great lakes Shoreline database	P32	Geomorphology, video, aerial photos, and shore protection	1992–95	Canada	O, I, GIS based	Fr	Arc Info, SPANS 5.2
N25	The Coastal Engineering Data Retrieval System (CEDRS)	P33	Wave climatology, observed and computed wave parameters of US shore	–	US coasts	O, I, GIS based	Fr	PC based system, access database
N26	Coastal Resources Database	P34	All parameters required for integrated coastal zone management	–	East Africa	O, I GIS based	Fr	
N27	ReefBase: a Global Database on Coral Reefs and their Resources	P35	Coral reefs and their resources. Designed as a repository of available information on coral reefs	1995	Global	C, O, I	Fe	Website
N28	Distributed Ocean Data System	P66	A DODS server is a Web server with a set of CGI scripts that are specific to the format of the dataset it serves	–	Global	I, O	Fe	Internet site

Table A4.2 Continued

No.	Database title	Provider code[a]	Subject coverage	Period[b]	Geographic coverage	Output mode	Access code	Hardware used
N29	Coastal Relief Model, East Coast, Vol. 1 & 2, 1999	P3	Gridded 3 arc-second bathymetric and USGS topographic data	–	East Coast of the US	C, I, O	Fe	IBM PC
N30	Global AVHRR Derived Land Climatology	P13	Climatological data from NOAA's Advanced Very High Resolution Radiometer (AVHRR) for coast as well	1997	Global	C, I, O	Fees	–
N31	Global Ecosystems Database—Disc A & B	P13	Data sets focusing on global causes and effects of greenhouse warming	1994 & 1997	Global	C, I, O	Fe	–
N32	Global View Project	P13	Coastal change analysis, global ecosystems, and terrain-base are all included	1994	Global	C, I, O	Fe	–
N33	Multi-beam for US East Coast, 1999	P13	Full resolution multi-beam data for the east coast of the US	1999	US East Coast	C. I, O	Fe	PC based
N34	TerrainBase (Global Digital Terrain Model)	P13	A global land and marine 5-min Digital Terrain Model (DTM)	1994	Global	C, I, O	Fe	PC based
N35	ACZISC—Atlantic Coastal Zone Database Directory	P40	Lists and describes 608 databases of relevance to the integrated management and sustainable development of the coastal zone of Atlantic Canada	1994–98	Canada	WEB	Fr	PC based
N36	NatMIS—National Marine Information System	P41	Distributed Network Concept, whereby marine agencies are linked dynamically on the Internet.	–	Australia	WEB	Fr	Internet site
N37	Digital Shoreline of South Carolina	P44	The most current and accurate depiction of South Carolina's coastline in various formats for use in different GIS software packages	–	S. Carolina coast	C, I, O	Fe	WEB server based
N38	South Carolina's Coasts: A Remote Sensing Perspective	P44	It contains land cover and shoreline data; software, and information about South Carolina's coastal management program	–	S. Carolina land and coast	GIS supported CD-ROM	Fe	WEB server based
N39	New Technologies for Coastal Mapping	P44	It provides data regarding high-resolution coastal topography and hurricane-induced coastal ocean impacts. Laser beach mapping data can be used to profile current beaches and shorelines	–	International	GIS based CD-ROM		PC, Macintosh
N40	Yakutat Bay, Alaska	P44	It contains the land cover and change data were derived from Landsat Thematic Mapper (TM) satellite scenes, as per Coastal Change Analysis Program (C-CAP) protocols	1986–93	Yakutat Bay, Alaska	GIS based CD-ROM	Fe	PC, Macintosh

N41	San Francisco Bay, California	P44	The land cover and change data were derived from Landsat Thematic Mapper (TM) satellite scenes from 1986 and 1993, as per Coastal Change Analysis Program (C-CAP) protocols	1986–93	San Francisco Bay	GIS supported CD-ROM	Fe	On PC, Macintosh
N42	Alabama Coastal Hazards Assessment	P44	CD product supports the development of a local Coastal Erosion and Hazard Mitigation Plan	–	Alabama coast	GIS supported CD-ROM	Fe	PC, Macintosh
N43	Cal-OCEAN—The California Ocean and Coastal Environmental Access Network	P46	Cal OCEAN is a web-based virtual library for the discovery of and access to ocean and coastal data and information from a wide variety of sources and in a range of types and formats	–	Global	WEB database	O, Fr, P	On Internet
N44	Global Mangrove Database and Information System (GLOMIS)	P47	Contents information helpful to promote research and development aimed at the conservation and sustainable production of mangrove ecosystems throughout the world	1997	Global	–	O, Fr	PC

[a] Provider code (providers/developers code) is given in Table A4.3.
[b] Single year means from that year uptill present.
Abbreviations used: (Access/cost) Op = Open access, Fe = Fees, Fr = Free, Re = Restricted; (Output mode) F = Floppy, T = Tape, P = Print, O = On-line, C = CD-ROM.

Table A4.3 Addresses of database providers/developers

Code no.	Name and address	Database no. (in Tables A4.1 and A4.2)
P1	BRGM, Domaine Luminy, Int. Dep. Avenue De Concyr, B.P. 6009 45060 Orleans Cedex, France	N7
P2	Cambridge Scientific Abstracts 7200, Wisconsin Avenue, Bethesda MD 20814-4823, USA	B16, 18, 21
P3	Cent. Environ. Information Inc. 33, South Washington Street Rochester, NY 14608, USA	B14
P4	Coastal Geol. Res. Cent. Caille De Carreo 722 7600 Mar Del Plata, Argentina	B2
P5	Dep. Geol., Australian Natl. Univ. Canberra, A.C.T. 2601 Australia	N6
P6	Dep. Indian Affairs and Northern Development Northern Resources and Economic Development Branch, Ottawa Ont, KIA OH4, Canada	N8
P7	CSC Library's Database Collection Coastal Service Center, NOAA E-mail: library@csc.noaa.gov	B1
P8	Environmental Studies Inst. Santa Barbara, CA, USA	B15
P9	Federal Inst. Goes. Nat. Resour. Stillweg 2, P.O. Box 510153 D-3000 Hannover, 51, Germany	B5
P10	Earth Science Information Center Natural Resources, Canada. 615, Booth St., Room 121 Ottawa, ON, Canada K1A 0E9 E-mail: esic@nrcan.gc.ca	B3, N9
P11	Indian National Scientific Documentation Center (INSDOC) 14, Satsang Vihar Marg Off S.J.S. Sansanwal Marg Special Information Area New Delhi 110067, India	B11, 13
P12	Institute for Scientific Information 3501, Market Street Philadelphia, PA 19104, USA	B12, 19
P13	NOAA/NGDC Mail Code E/GC3 325, Broadway, Boulder, CO USA 80303 E-mail: rwarnken@ngdc.noaa.gov	N29, 30, 31, 32, 33, 34
P14	National Information Services Cooperation, USA	B17
P15	National Institute of Oceanography Dona Paula, Goa 403 004, India	B6
P16	National Reference Library Belvedere Road, Calcutta West Bengal, India	B10
P17	The H.W. Wilson Company 950, University Avenue Bronx, NY 10452, USA	B20
P18	Oregon Ocean Coastal Program Dept. of Land Conservation and Development 800, NE Oregon Street, #18 Portland, Oregon 97232, USA E-mail: tanya.haddad@state.or.us	N12
P19	Bernadette Dwyer Fogo Island Cooperative Society Ltd. P.O. Box 70 Seldom, Fogo Island, NF A0G 3Z0 E-mail: andersont@dfo-mpo.gc.ca	N13
P20	International Association of Geomorphologists Montana State University, Boneman Department of Earth Sciences, Bozeman	N14

Table A4.3 Addresses of database providers/developers

Code no.	Name and address	Database no. (in Tables A4.1 and A4.2)
P21	Maintained by Zbigniew Zwolinski (Adam Mickiewicz University, Poznan)	N3
P22	Bob Taylor Natural Resources Canada Geological Survey of Canada (Atlantic) Bedford Institute of Oceanography P.O. Box 1006, Artmouth, NS B2Y 4A2, Canada E-mail: taylor@agc.bio.ns.ca	N4, 5
P23	The Japanese Geomorphological Union (JGU) and operated by the Data Processing Center of Kyoto University. Kobashi, Sumiji, Faculty of Agriculture, Kyoto Univ. Kitashirakawa, Sakyo-Ku, Kyoto 606, Japan E-mail: c53814@sakura.kudpc.kyoto-u.ac.jp	B7
P24	Kay Yost American Geological Institute 4220, King Street Alexandria, VA 22302-1502 E-mail: kyost@agiweb.org	B8, 9
P25	NOAA/Office of Ocean Resources Conservation & Assessment Maintained by: cmoore@ngdc.noaa.gov	N15
P26	Dr. Walter H.F. Smith NOAA Laboratory for Satellite Altimetry, National Oceanographic Data Center Silver Spring, MD, USA E-mail: walter@amos.grdl.noaa.gov	N16
P27	European Center for Nature Conservation ECNC, P.O. Box 1352, 5004 BJ Tilburg The Netherlands Jan van der Straaten (saxifraga@ecnc.nl)	N17
P28	Saenger, Dr. P., and Bucher, D. (1989), Citation: An Inventory of Australian Estuaries and Enclosed Marine Waters—Database. Citation: ANPWS Unpublished Consultancy Report	N18
P29	Placentia Bay Newfoundland, Canada	N19
P30	Department of Environment and Heritage G.P.O. Box 787, Canberra, ACT, 2601, Australia	N20
P31	Redditch, B98 9PA, England E-mail: post@keyterra-firma.com	N23
P32	Environment Canada by Geomatics International Inc. in support of the International Joint Commission Water Levels Reference Study Water Issues Division of Environment Canada	N24
P33	CERC, Coastal Engineering Research Center E-mail: webmaster@cerc.wes.army.mil	N25
P34	Lieven Bydekerke, Mwangi Theuri GIS Analyst, Project Assistant The Division of Environmental Information Assessment and Early warning UNEP, P.O. Box 30552 Nairobi, Kenya E-mail: lieven.bydekerke@unep.org	N26
P35	Reefbase Project International Center for Living Aquatic Resources Management M.C.P.O. Box 2631, 0718 Makati City, Philippines E-mail: reefbase@cgnet.com	N27
P36	Cameron-Stephen, Sally Streamline Coordination Unit Dept. of Primary Industries and Energy P.O. Box 858, Canberra A-C-T 2601 Australia	B22
P37	D.A. Tellis Information Center/Library 63, Conyngham Street, Glenside 5065 South Australia, Australia	B23

Table A4.3 Continued

Code no.	Name and address	Database no. (in Tables A4.1 and A4.2)
P38	Australian Key Center in Land Information Studies St. Lucia 4067, Qld, Australia	B24
P39	Claudette LeBlanc ACZISC Secretariat, Oceans Institute of Canada 1226, Le Marchant Street Halifax, NS Canada B3H 3P7 E-mail: leblancc@fox.nstn.ca	N35
P40	National Library of Australia Canberra, ACT 2600 Australia	B25
P41	Environmental Resources Information Network [ERIN] Department of the Environment Sport and Territories G.P.O. Box 787 Canberra ACT 2601, Australia E-mail: davidc@erin.gov.au	N21, 22, 36
P42	University Library Western Michigan University Michigan, USA E-mail: westcat-voyager@wmich.edu	B26
P43	The Canadian Geomorphology Research Group (CGRG) at International Association of Geomorphology Congress in Hamilton Ontario, Canada	B4, N1, 2
P44	Dr. Dan Smith, University of Victoria National Oceanic and Atmospheric Administration NOAA Coastal Services Center E-mail comments to: csc@csc.noaa.gov	N38, 39, 40, 41, 42, 43
P45	California Digital Library University of California CA, USA	B29
P46	The California Environmental Resources Evaluation System (CERES) California, CA, USA	N43
P47	Dr. Marta Vannucci, Vice-President International Society for Mangrove Ecosystems (ISME) c/o College of Agriculture, University of the Ryukyus Nishihara, Okinawa 903-0129, Japan E-mail: mangrove@ii-okinawa.ne.jp	N44

APPENDIX 5: GLOSSARY OF COASTAL GEOMORPHOLOGY

Abrasion - The wearing away of a rock surface by friction as a result of the impact of wind, wave, current, or ice action, particularly when these agents are armed with rock fragments (notably sand and gravel): sometimes termed *corrasion*. Abrasion of a rocky shore can form ramps or platforms, or excavate furrows or potholes, while abrasion of a cliff can result in the cutting of caves, clefts, and crevices. Submarine abrasion takes place on the nearshore seafloor, diminishing offshore as water depth increases; it becomes imperceptible at depths greater than half the wave length (Bradley, 1958). Rock fragments (such as quartz grains) can also be worn and pitted by abrasion when they are impacted against each other, or against a rock surface: this is known as mechanical abrasion, and is distinguished from etching produced by chemical solution (Corrosion, *q.v.*) (Margolis, 1968).

Abrasion notch - An elongated cliff-base hollow (typically 1–2 m high and up to 3 m recessed) cut out by abrasion, usually where Breaking Waves (*q.v.*) are armed with rock fragments.

Abrasion platform - A smooth, seaward-sloping surface formed by abrasion, extending across a rocky shore and often continuing below low tide level as a broad, very gently sloping surface (plain of marine erosion) formed by long-continued abrasion (Johnson, 1916). The intertidal section is typically 50–100 m wide, increasing with tide range.

Abrasion ramp - A smooth, seaward-sloping segment formed by abrasion on a rocky shore, usually a few meters wide, close to the cliff base.

Accretion - A gradual or intermittent natural process of deposition of sediment by wind, wave or current action, or by rivers, glaciers, solifluction or mass movement, resulting in the natural raising or extension of a land area.

Aeolian calcarenite, Aeolianite (Eolianite) - See Dune Calcarenite.

Abrasion notch at Cape Liptrap, Victoria, Australia.

Intertidal abrasion platform cut in Chalk near Birling Gap, Sussex, England.

Abrasion ramp at cliff base, North Island, New Zealand.

Aggradation - The raising of a land surface by deposition (vertical accretion) of sediment, as on a beach, dune, mudflat, marsh, coastal plain, or delta.

Algal mat - A carpet of blue-green algae (cyanophytes) that stabilize intertidal sediments and precipitate carbonates on the sheltered shores of hypersaline areas, found in the Arabian Gulf and the Red Sea, Shark Bay in Western Australia, the Bahamas, and Texas coast lagoons. See Stromatolite.

Algal rampart, Algal reef - A structure typically 10–20 m, occasionally up to 100 m wide, formed in shore and nearshore waters by the growth of calcareous algae (notably *Lithothamnion, Lithophyllum*, or *Porolithon* spp.).

Algal ridge - A structure built by calcareous algae on the surface of a reef or shore platform.

Algal rim - A ridge built on the more exposed (seaward) margins of coral reefs or shore platforms by encrusting calcareous algae, such as *Lithothamnion* or *Porolithon*; they are usually no more than 20 cm high. Networks of ridges constructed by algae may enclose shallow pools on coral reef flats and shore platforms.

Artificial beach - A beach emplaced by human action, as where sand brought from the land, or alongshore or offshore sources is dumped on the shore, strictly where there was no natural beach, but the term is often used where a natural beach depleted by erosion is restored or renourished (see Beach Nourishment).

Algal ridges on shore platform at Port Hedland, Western Australia.

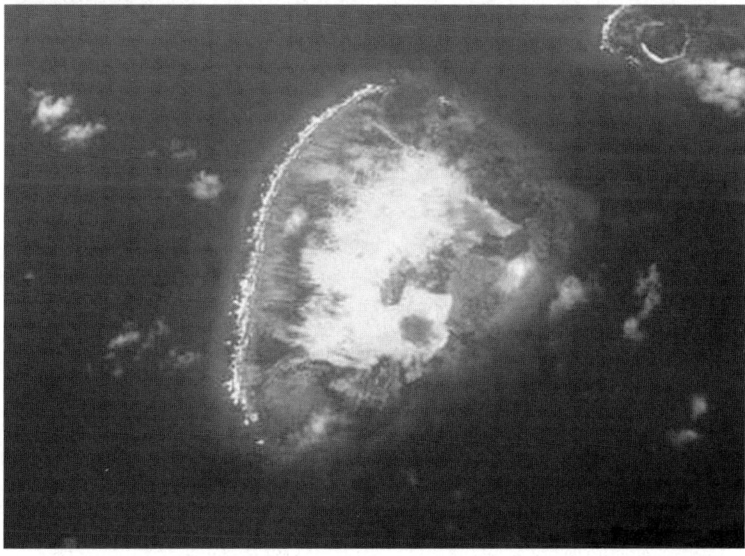

Amirante Atoll, Seychelles, Indian Ocean.

Atoll - A ring-shaped coral reef structure (ranging from <1 km to >100 km in diameter), partly or wholly enclosing a lagoon (typically 30–100 m deep and several kilometers wide); the outer (seaward) slopes plunge steeply to the deep ocean floor (oceanic atoll) or to the continental shelf (shelf atoll). The term comes from the Maldives, where "atolu" are government districts, each being a circular reef enclosing a lagoon. Atolls with a central (noncoral) island, such as Bora Bora, are termed "almost atolls"; those with ring-shaped reefs enclosing remnants of an earlier atoll, as in Houtman Abrolhos off the coast of Western Australia, "compound atolls." Some compound atolls take the form of chains of small atolls (diameter about a kilometer), known as Faros, as in the Maldive Islands, Indian Ocean. Horsehoe reefs form where isolated reef platforms (patch reefs) have marginal spurs that have grown to leeward.

Attrition - The wearing down of rock fragments by friction when they are mobilized and thrown against a rock surface, or ground against other rock fragments by wind action, waves, or currents.

Avulsion - The separation of an area of land when a river or estuary channel suddenly changes its course, usually during floods, or when a meander is breached. The term has been used obscurely, mainly by lawyers, to describe rapid erosion of the shore by waves. Another interpretation may be the transference of land along a coast as a result of rapid erosion on one sector and nearby related accretion, but this is not strictly avulsion, the placing of an area of pre-existing land on the opposite side of an earlier channel as the result of river migration.

Backshore - The coastal fringe lying above (i.e., landward of) the normal high tide line, but occasionally inundated by exceptionally high tides or storm surges. Also known as the supralittoral or supratidal zone.

Backwash - The seaward (return) flow of water after the swash (uprush) produced when a wave breaks on the shore.

Bank - A localized broad elevation of the seafloor, smooth in profile, usually submerged even at the lowest tides, and composed of unconsolidated sediment (usually sand, sometimes glacial drift), subject to movement by waves or current action. Typically a few meters wide, but the term is used for larger features, such as the Dogger Bank in the North Sea or the Grand Bank of Newfoundland. See also Bar (shoal). A banner bank trails in the lee of an island or headland.

Bar - An elongated bank, ridge or mound of sediment (sand or gravel) a few meters wide, deposited and shaped by waves and currents, and submerged at least at high tide (i.e., can be partly or wholly exposed at low tide) (Shepard, 1952). Bars commonly form off beaches, but also

Sand bars on the southeast coast of Port Phillip Bay, Australia.

occur off river mouths and lagoon entrances. A break-point bar is a concentration of sediment formed by breaking waves where material that is being carried shoreward meets that withdrawn from the beach by backwash. Longshore bars are formed parallel to a beach, and at least partly exposed at low tide, while oblique bars run at an angle to the beach, and transverse bars at right angles to it. Bars may also occur in cuspate, lunate, looped, crescentic, reticulate, or chevron patterns. Multiple parallel bars and troughs form on gently sloping sandy shores where spilling waves break and re-form: they are exposed at low tide (see Ridge and Runnel). Trailing bars (banner banks) occur in the lee of headlands or islands, and become flying bars when they are disconnected (cf. trailing and flying spits, which differ in that they extend above high tide level). It should be noted that before Shepard defined bars, the terms such as offshore bar, bay bar, and looped bar (Johnson, 1919) were often used to describe features that would now be termed barriers.

Barchan - A dune form, an isolated, mobile crescentic mound of bare sand, typically up to 30 m high, with a steep advancing leeward slope and a gentler concave windward slope, flanked by lateral horns curving downwind and spaced at up to 350 m. Commonly found in deserts, barchans also form on extensive areas of bare drifting coastal dune sand, and on unvegetated backshores.

Barrier (Coastal barrier) - An elongated ridge of deposited sediment (sand, pebbles, occasionally boulders) that has been built up by wave action above high tide level along the coast or across an embayment (bay barrier): differing from a bar, which is submerged at least at high tide (Shepard, 1952). Typically a barrier is backed by a lagoon or swamp that separates it from the mainland or from earlier barriers, but this is not essential: some barriers abut older land surfaces. A barrier may be from a few meters to more than a kilometer wide and up to 100 km long, attached to the mainland at one end (see Barrier Spit) or both ends, or interrupted by tidal inlets (see Barrier Island).

Short (1999, p. 307): "A barrier is defined as a shore-parallel sub-aerial and sub-aqueous accumulation of detrital sediment (sand/boulders) formed by waves, tides and aeolian processes."

Barrier beach - A single, narrow (usually <200 m) elongated ridge built parallel to the coast, without surmounting dunes.

Barrier island - A barrier segment bordered by transverse gaps (tidal inlets, lagoon entrances, river outlets) which may be migratory and subject to closure; it usually bears beach ridges, dunes and associated swamps and minor lagoons. Barrier islands are typically 0.5–5.0 km wide, 1–100 km long and 6–100 m high.

Barrier lagoon - A lagoon extending roughly parallel to the coastline, behind a barrier or reef. See Coastal Lagoon.

Barrier reef - An elongated, narrow coral reef built up offshore and parallel to the coast, from which it is separated by a broad lagoon (which may be several kilometers wide), typically with extensive areas too deep for coral growth.

Barrier spit - A barrier attached at one end to the mainland, with or without recurves, and backed by a bay, lagoon or marshland (swamp).

Bay - A general term for a wide (typically >1 km) coastal re-entrant between two headlands, its seaward boundary generally wider than the extent of landward penetration. A small bay is termed a cove, a large bay a gulf. The term bay is also used in the United States for coastal waters largely or entirely cut off from the sea by spits and barrier islands (e.g., Netarts Bay on the Oregon coast) which elsewhere are called coastal lagoons, and for the ovoid depressions of uncertain origin that contain lakes or swamps on the coastal plain of South Carolina.

Beach - An accumulation on the shore of generally loose, unconsolidated sediment, ranging in size from very fine sand up to pebbles, cobbles, and occasionally boulders; often also containing shelly material. Conventionally, shores with silt or clay sediment are not regarded as beaches (but in some languages the terms beach and shore overlap, so that silt shores and clay shores may be translated as silt beaches and clay beaches). A distinction can sometimes be made between an upper

Coastal sand barrier at Marlo, southeastern Australia.

Barrier reef with storm-tossed boulders.

Barrier spit, Lake Onoke, North Island, New Zealand.

beach, often swash-built, around high tide level, and a gentler lower beach exposed as the tide falls.

Short (1999, p. 3) defined a beach as "a wave-deposited accumulation of sediment lying between modal wave base and the upper swash limit where wave base is the maximum depth at which waves can transport beach material shoreward." However, since wave base is conventionally where the water depth is half the wave length, this limit can be in water at least 50 m deep and perhaps several kilometers offshore. Most accounts of beaches admit their extension below lowest tide level and for some distance (usually not stated) beyond the breaker zone.

Beach berm - An ephemeral flat or landward-sloping step or terrace built on a beach face by swash action. On shingle beaches there may be several such berms, resulting from successive episodes of swash action at successively lower tidal levels. A berm built at the top of a beach may persist to become a Beach Ridge (*q.v.*).

Beach budget - The quantified gains and losses of sediment from a defined beach sector.

Beach compartment - A beach occupying a sector of coast bounded by rocky reefs, promontories or artificial structures such as breakwaters. See Coastal Sediment Compartment.

Beach cusps - Regular successions of half-saucer (crescentic) depressions opening seaward between cuspate points in the swash zone on the upper beach face, the points being of coarser material (shingle or coarse sand) than the intervening hollows. Also known as Swash Cusps.

Swash-built sand berm on the Ninety Mile Beach, southeastern Australia.

Shingle beach cusps at Ringstead, Dorset, England.

Beach drift (Beach drifting) - The zigzag movement of beach material along the shore in the swash zone when waves arrive at an angle to the shoreline, producing an oblique swash, followed by a Backwash (*q.v.*) that descends orthogonally from the beach face. The same waves generate a longshore current which carries sediment along the coast in the same direction in the nearshore zone. See Longshore Drifting.

Beach, drift-aligned (drift-dominated) - A beach with an orientation determined by the drifting of sediment alongshore in response to waves arriving at an angle to the coastline: they are aligned at an angle to the dominant direction of wave approach, with alignments parallel to the line of maximum longshore sediment flow, generated by obliquely incident (typically 40–50°) waves. Contrast Swash-Dominated Beach (*q.v.*), aligned parallel to the dominant waves.

Beach face - The seaward slope of a beach between the low tide line and the upper limit of wave swash.

Beach gravel - An American term describing beach sediment coarser than sand, ranging up to cobble size, and often (but not always) well-rounded: the British term shingle is similar.

Beach lobe - Roughly triangular or lobate protrusions from a beach that form and may migrate alongshore downdrift in response to oblique wave action.

Beach nourishment - The natural or artificial supply of sand or gravel to a beach. Also termed Beach Restoration, Beach Fill, and Beach Renourishment (*qq.v.*).

Beach, profile of equilibrium - A beach profile (concave, steepening upward past the high tide line and declining seaward to below the low tide line) that represents the attainment of an equilibrium with incident waves, whereby sediment gains and losses are balanced (Johnson, 1919). A cyclic equilibrium is attained where a succession of beach profiles corresponds with changing wave regimes through a Cut-and-fill (*q.v.*) sequence, and a dynamic equilibrium where the beach profile is maintained even though the beach may be prograding or retreating as the result of erosion.

Beach ridge - An elongated low ridge of beach material (sand, gravel, or shells) piled up above high tide level by swash action. Many beach ridges are surmounted by wind-deposited sand, which may develop into a Foredune (*q.v.*), but a beach ridge is strictly a feature built and shaped by wave action. Sandy beach ridges are typically 5–50 m wide, measured from bordering troughs (Swales, *q.v.*), shingle beaches usually narrower.

Beach ridge plain - A series of beach ridges formed successively, parallel, or roughly parallel to the coastline, and separated by elongated hollows or swales, each ridge marking a former position of the prograding coastline.

Sandy beach lobe at Lang Lang, Westernport Bay, Victoria, Australia.

Storm-piled shingle beach ridge at Redcliff Point, Dorset, England.

Shingle beach ridge plain at Hoed, Denmark.

Beach rock exposed on a sandy shore at Gili Bidara, Lombok, Indonesia.

Beach rock - A sandstone layer within a beach, formed by interstitial precipitation of carbonates (calcite or aragonite) to cement the beach sand in the zone of fluctuating groundwater levels, usually where the sediments are strongly calcareous, as on shelly, calcarenite, or coralline beaches (Stoddart and Cann, 1965). Where the material cemented consists of rounded pebbles the term beach conglomerate is used, while cementation of angular gravel yields beach breccia.

Beach scarp - A steep or vertical cliff cut in the beach face by large waves during a storm or tsunami.

Beach state - The state of a beach may be reflective (with a high proportion of wave energy reflected from beach face), dissipative (wave energy diminished through breaker and surf zones), or intermediate between these two.

Beach, swash-aligned (swash-dominated) - A beach formed where incoming waves are refracted into curved patterns that anticipate, and on arriving fit, its curved outline (Davies, 1980), so that the beach is parallel to incoming wave crests (particularly refracted ocean swell).

Beach system - Beaches and the processes at work on them, with adjustments in morphology in response to energy inputs from waves, currents, tides, and winds.

Bench - A flat or gently sloping rock ledge, terrace or platform, typically 5–50 m wide, but sometimes much wider, backed and fronted by steeper slopes. A structural bench coincides with the upper surfaces of a hard rock outcrop, formed where weaker overlying material has been removed by erosion, an erosional bench has been planed across tilted or

Scarp cut by storm waves, Ninety Mile Beach, southeastern Australia.

Structural benches on Bay Cliff, Wonboyn, New South Wales, Australia.

folded rocks by erosion. A bench between high and low tide levels is called a Shore Platform (*q.v.*).

Berm - See Beach Berm.

Beveled cliff - See Slope-over-wall Coast.

Bioconstruction - Formed by organisms (e.g., algae, corals, vegetation) that trap sediment, including the products of their own decay (e.g., shells, peat).

Biodeposition - A little used term for sedimentation generated by organisms, such as coral reefs and algal rims, shelly deposits or peat.

Bioerosion - Removal of rock material by the physical and chemical processes associated with the activities and metabolism of plants and animals that inhabit the shore, especially on limestone coasts (Healy, 1968).

Bioherm - A mound, bank, or (biogenic) reef built by sedentary organisms, such as corals, algae, foraminifera, molluscs or gastropods, or a reef built in the nearshore zone by oysters, mussels, gastropods or worms (serpulid and sabellariid reefs). Vermetid reefs are built by long, coiled gastropods.

Blowhole - A hole or fissure in the roof of a cave on a rocky coast, usually steeply sloping or vertical, through which geyser-like fountains of water and spray are forced by the intermittent release of compressed air trapped in a cave by incoming waves.

Blowout - A small saucer-shaped hollow or trough excavated in vegetated dunes by wind action, with an advancing nose of sand spilling downwind. See Parabolic Dune.

Blue hole - A submerged sinkhole in a coral reef, generally formed by solution when the reef was subaerially exposed during a low-sea-level-stage.

Bluff - A bold, steep, sometimes rounded coastal slope on which soil and vegetation conceal, or largely conceal, the underlying rock formations, in contrast with a cliff in which these formations are exposed. Bluffs may be termed abandoned, degraded, or fossil cliffs. The term seems to have originated in North America for a headland that was rounded rather than cliffed. There is confusion when the term bluff has been used as a place name for a feature that is actually a cliff. Bluffs may form where cliffs stop retreating (as when basal marine erosion is halted by emergence, the Accretion (*q.v.*) of a protective beach, or artificial protective structures such as seawalls or rock ramparts): they may then become degraded by subaerial processes to gentler slopes on which a soil forms and vegetation establishes. Bluffs are generally relatively stable, in comparison with receding cliffs, but the steep forested slopes on the Pacific coast of Oregon and Washington have receded as the result of occasional slumping.

Bombora - Australian term for a bank or reef that causes local raising and steepening of waves moving through shallow water, producing wave fronts that are good for surfing.

Boulder - A rock particle with a diameter exceeding 256 mm (or about 25 cm). See Granulometry.

Breakaway - A fracture at the top of a cliff behind an area where a rock mass has subsided or slid seaward, leaving a small scarp.

Beveled cliff on glacial drift over vertical Chalk, Flamborough Head, Yorkshire, England.

Reef formed by calcareous tubeworms (*Galeolaria*) at Beaumaris, Victoria, Australia.

Blowhole at Quobba, west coast, Western Australia.

Bluffs on the southeast coast of Phillip Island, Victoria, Australia.

Breakaway on cliff top near Lyme Regis, Dorset, England.

Breakers, Breaking waves - When waves break they produce (1) surging breakers, which are low and gentle waves until they sweep up a relatively steep beach, (2) plunging breakers, with fronts that curve over and crash (producing little swash but a strong backwash), (3) collapsing breakers that subside as they move toward the shore, and (4) spilling breakers, which are short and high, and produce foaming surf as the swash runs up a beach of gentle gradient (Galvin, 1972). Alternatively, there are constructive breakers (which wash sediment up on to the beach) and destructive breakers (which cause beach erosion), depending on whether the swash and backwash achieve net shoreward or seaward movement of beach material.

Break-point bar - A bar built in the zone where waves break, deposition occurring where shoreward drifting beneath incoming waves meets seaward drifting by backwash, leaving a parallel trough to landward.

Breakwater - An artificial structure built into the sea, often curved, and designed to impede wave action so as to shelter a harbor or protect a stretch of coastline. The terms jetty and pier are sometimes used as synonyms.

Bund - A term used mainly in India and southeast Asia for an artificial embankment, usually of earth or gravel, built along the coastline or the banks of a river or estuary.

Cala, Calanque - A deep steep-sided marine inlet on a limestone coast: calas are found in the Balearic Islands and calanques on the coast of Provence.

Calcified seaweed, grit - Coarse angular sand or gravel produced by *Lithothamnion calcareum*, which grows extensively on some seafloor areas, as in the western part of the English Channel.

Calcilutite - A fine-grained (silt-sized) calcareous formation with an admixture of clay, often coherent enough to stand as a vertical cliff. The Port Campbell Limestone formation (Miocene) in Victoria, Australia, contains calcilutite deposits that stand in spectacular vertical cliffs.

Calcirudite - A calcareous conglomerate or breccia, consisting of broken or worn fragments of coral, shells, or limestone cemented by precipitated carbonates, often occurring in layers in Dune Calcarenite (*q. v.*).

Calcrete (Caliche) - A hard rock calcareous formation cemented by precipitated carbonates; a limestone or calcareous duricrust usually

Calanque at Wied-il-Ghazri, Gozo, Malta.

Coral cay in the Maldive Islands, Indian Ocean.

formed on or immediately below the land surface in an arid or semiarid environment by precipitation of carbonates derived from groundwater that moved upward to an evaporation level. Calcrete can occur as a horizon within dune stratigraphy (usually formed on an old land surface subsequently buried by younger dune sand) or as a caprock (carbonates having been delivered in rising groundwater and precipitated near the drying surface). Known as kunkar in Australia. See Dune Calcarenite.

Can - A fan of beach material washed through a low permeable coastal barrier by Storm Waves (Storm Surges) (*qq.v.*) to form a delta-like projection into a lagoon or on to backing marshland. There are good examples behind the shingle barrier of Chesil Beach, Dorset.

Cape - A large, often rounded coastal protrusion, located where the coastline intersects a range of mountains, hills, or a plateau, usually where a drainage divide reaches the coast. However, some capes are low-lying (e.g., Cape Canaveral (Kennedy) and others on the American Atlantic coast).

Cay - A small low-lying depositional island of coralline sand or gravel (shingle cay) built up just above high tide level by wave action on a reef flat, usually toward the lee side. In the Caribbean a cay is termed a key.

Chenier - A long, low-lying narrow strip of sand, often shelly and typically up to 3 m high and 40–400 m wide, deposited in the form of wave-built beach ridge on a swampy (peat and clay), deltaic, or alluvial coastal plain by wave action. Many cheniers contain shelly sand and gravel. In Louisiana such ridges often have vegetation dominated by oak trees (chêne, hence chênière), emphasizing the contrast with the adjacent peat or clay terrain (Russell and Howe, 1935). Cheniers on the wide coastal plain east of Darwin, Australia, have pandanus palms: had they been originally studied here they may have been called pandaniers. Most cheniers have been deposited at the limit of storm surge swash, and may be termed transgressive, but some may have formed along the coast during a brief phase of raised sea level, followed by progradation that leaves them stranded inland: these may be termed regresssive.

A chenier plain is a coastal plain with several cheniers scattered across it, usually parallel or subparallel to the coastline. Cheniers may pass laterally into beach ridges where the fine-grained substrate becomes sandy or gravelly.

Chine - A deep and narrow ravine cut into soft rock on a cliffed coast by a stream descending steeply to the shore, often over waterfalls. A term used in the Isle of Wight and on the Hampshire coast in southern England.

Clastic - Consisting of broken and transported fragments of pre-existing rock. The term Coarse clastic beach has sometimes been applied to a beach of gravel or Shingle (*q.v.*).

Clay - A sediment consisting of particles with a diameter smaller than 1/256 (about 0.004) mm.

Cliff - A steep (usually >40°, often vertical and sometimes overhanging) coastal slope cut into (and thus exposing) rock formations,

Sandy chenier within mangroves, Karembé, New Caledonia.

Whale Chine, Isle of Wight, England.

produced by basal marine erosion (undercutting), but occasionally by faulting or earlier fluvial or glacial erosion. Cliffs cut in unconsolidated formations are sometimes known as Earth Cliffs (*q.v.*), and there are also Ice Cliffs (*q.v.*) at the seaward terminations of glaciers and ice sheets. Cliffs rising to 100–500 m above sea level are termed high cliffs, while those exceeding 500 m (as in Peru and Western Ireland) are termed megacliffs (Guilcher, 1966): the cliff at Enniberg on the north coast of Vidoy (Faerö Islands) is 725 m high. Coastal cliffs are generally receding as the result of marine erosion at their base, accompanied by subaerial erosion of the cliff face.

Cliff fall (Rock fall) - The collapse of the face of a rocky cliff into an apron of debris. On some cliffs columns of rock bordered by vertical joints may topple on to the shore.

Clifflet - See Microcliff.

Cliff-top dunes - Usually found where sand blown from a beach moved up and over a cliff, but the link between the beach and the dune has been removed by erosion, exposing the cliff; occasionally the dunes have arrived from inland (Jennings, 1967).

Coast - A zone of varying width where the land meets the sea, and where the lithosphere, hydrosphere, and atmosphere meet and interact. Some use the term as a synonym for shore, but generally it is taken to include the land margin, extending inland to the limit of penetration of marine processes or to the first major change in landform features, such as a rising slope, and the nearshore zone, out at least to the line where waves break. A wider definition includes the whole zone between the highest and lowest coastlines related to sea-level changes during the Quaternary (*q.v.*).

Cliff fall and scar on Chalk coast, North Foreland, Kent, England.

In American literature the coast is sometimes elaborated to the coastal zone.

Coastal barrier - See Barrier.

Coastal dimensions - Coastal morphology may be classified as follows:

First-order features - about 1000 km long, 100 km wide, and 10 km high (e.g., continental coasts, related to global tectonics).
Second-order features are about 100 km long, 10 km wide, and 1 km high (e.g., deltas, fiords).
Third-order features about 10 km long, 1 km wide, and 100 m high (e.g., coastal barriers).
Fourth-order features are about 1 km long, 100 m wide, and 10 m high (e.g., foredunes).
Fifth-order features are about 100 m long, 10 m wide, and 1 m high (e.g., beach berms, shore platforms, sand bars).
Sixth-order features are about 10 m long, 1 m wide, and 10 cm high (e.g., beach cusps).
Seventh-order features are about 1 m long, 10 cm wide, and 1 cm high (e.g., current ripples).

In each case the dimension given should be regarded as being within a range of from 50% to five times the figure given (i.e., Sixth-order features 5–50 m long, 0.5–5 m wide, and 5–50 cm high).

Coastal dunes - Hills, banks, or ridges of aeolian (wind-blown) sand backing present or former coastlines.

Coastal erratics - Large shore rocks that came from outcrops elsewhere, delivered to their present position on the coast or on the shore, usually by glaciers or icebergs.

Coastal gorge - See Geo.

Coastal lagoon (Barrier lagoon) - A shallow, often brackish (estuarine) water body formed where a coastal inlet, river mouth or embayment has been enclosed (or almost enclosed) behind a depositional barrier or barrier spit. Known as an étang in France, a haff in Germany, an estero in Portugal, a liman on the Black Sea coast and a pond in New England. A distinction is made between coastal lagoons enclosed by depositional barriers and lagoons backed by coral atolls, or standing between a barrier reef and the mainland coast.

Coastal land reclamation - The making of new land by enclosing or filling shore and nearshore areas. An alternative term land claim has been suggested, because this is not strictly reclaiming, but land claim risks confusion with territorial claims by political groups.

Coastal landslide - The movement of a mass of rock, earth or debris down a coastal slope. Also known as a landslip (Lyell, 1833). Where the coastal rocks are fine-grained, the downslope movement may be in the form of a mudflow lubricated by exuded groundwater or surface runoff after heavy rain or melting snow or ice.

Coastal plain - Low-lying coastal terrain, usually of depositional origin, but occasionally consisting of planed-off rock outcrops (see Strandflat), between the coastline and rising ground in the hinterland.

Coastal sediment compartment - A sector of coast within which sediment is largely or completely confined, delimited by rocky reefs, promontories, tidal inlets, river mouths, or artificial structures such as breakwaters (Davies, 1974). See Beach Compartment.

Coastal tors - Outcrops of harder or more massive rock protruding as buttresses (coastal tors) from a coastal slope.

Coastal waters - A general term for the sea area adjacent to the coast, comprising the nearshore and offshore zones. The seaward limit is usually indefinite, and an arbitrary distance (such as 3 nautical miles) has been used in Law of the Sea schedules.

Coastline - The edge of the land at the limit of normal high spring tides; the subaerial land margin, often marked by the seaward boundary of terrestrial vegetation. On cliffed coasts it is taken as the cliff foot at high spring tide level. Use of the term Shoreline (*q.v.*) as a synonym for Coast or Coastline is vague and misleading, and should be avoided—shorelines move to and fro as the tides rise and fall, whereas coastlines

Giant's Rock, a coastal erratic on the shore near Porthleven, Cornwall, England.

Coastal landslide at Blackgang, Isle of Wight, England.

Coastal tor (buttress) at Trewalvas Head, Cornwall, England.

are related to the high-spring-tide shoreline, and thus submerged only in exceptional circumstances (e.g., Storm Surges, *q.v.*).

Cobble - A rock particle with a diameter between 64 and 256 mm. See Granulometry.

Colk - A relatively deep circular or oval depression in the seafloor (or on the floor of an estuary or lagoon) excavated and kept clear of sedimentation by locally strong current action (including the siphoning that can occur beneath ice). Also known as a tidal colk or scour hole. Originally a pothole on a river bed.

Compound spit - A recurved spit with several recurves on the inner shore marking former terminations.

Concentric (Contraction) ridges - A series of small parallel beach ridges formed successively on the shore of a contracting shallow lake or lagoon by intermittent deposition along the margins of a prograding salt marsh, as described from Lake Reeve, Victoria, Australia, by Jenkin (1966).

Contraposed coast - A discordant coast developed on previously concealed rock formations where marine erosion has removed bordering and overlying weaker deposits (Clapp, 1913). Examples of contraposed coasts include sectors where Archaean gneiss has been exhumed from a Pleistocene dune calcarenite cover on the west coast of Eyre Peninsula, South Australia, and where Palaeozoic rocks have been exposed by the removal of a fringe of glacial drift on the coasts of the Lleyn Peninsula in North Wales. In southwest England the slope-over-wall profiles formed where a mantle of Pleistocene periglacial Head has been undercut by marine erosion to expose Devonian rocks could also be classified as contraposed.

Coral garden - An open structure built by branching (e.g., staghorn) corals.

Coral rampart - A depositional ridge of coral fragments built near the windward margin of a coral reef by storm waves or surges.

Coral reefs - A structure built in the sea by corals, which form the reef framework, together with algae and other organisms (such as algae, molluscs, crinoids, bryozoans) and precipitated carbonates to make rock formations sufficiently resistant to withstand normal wave action.

Corrosion - Dissolving of rock or minerals by chemical action in water (rain, sea, or spray).

Currents - Movements of water generated by one or more of the following: density currents occur where water of higher specific gravity (colder or more saline) moves to displace water of lower specific gravity; discharge (fluvial jet) currents occur where a river flows into the sea; ebb and flow (flood) currents are generated by falling and rising tides; wave-generated currents such as the longshore currents that develop when waves arrive at an angle to the shoreline; Rip Currents (*q.v.*) flow back into the sea through breaking waves at intervals along the shore; wind-generated currents flow in the direction of the wind; and ocean currents are slow mass movements of water in response to variations in water temperature and salinity, atmospheric pressure and wind stress.

Cuspate bar - A triangular depositional bank of sand or shingle, submerged, at least at high tide, extending out from the coast with straight or concave shores that meet in a seaward point, and enclosing a depression occupied by a lagoon at low tide.

Cuspate barrier - Similar to a cuspate bar, except that it has been built above high tide level.

Cuspate foreland - A triangular depositional area of sand or shingle with straight or concave shores extending out to a seaward point, with multiple beach ridges marking stages in progradation on one or both flanks. Cuspate forelands may migrate by erosion of one flank and the drifting of beach material to accrete on the other. Known in Britain as a ness (Dungeness and Duddon Ness).

Cuspate spit - A triangular spit that projects from the coast with straight or concave shores extending out to a seaward point. Often formed in the lee of an islet, stack, reef or shoal, or on the shores of a bay or lagoon as a result of convergent wave refraction. Formerly known as a cuspate bar, but this term is not applicable to a feature built above high tide level (see Bar).

Cut-and-fill - The cyclic sequence of changes on a beach profile resulting from erosion by storm waves (cut) and subsequent restoration by constructive waves (fill).

Delta - A depositional landform produced by sedimentation at and around the mouth of a river. Deltas usually protrude from the coast, and are typically triangular, named from the Greek letter "delta" applied to the Nile delta by Herodotus in ancient times, but they may alternatively be Digitate (finger-like, as in the Mississippi delta), Cuspate (Tiber delta), Arcuate or rounded (Niger delta), or Lobate (Rhône delta), in shape.

Coral garden, Great Barrier Reef, Australia.

Dungeness: a cuspate foreland in southeastern England.

Cuspate spit, Moreton Island, Queensland, Australia.

Dissipative beach - A beach where wave energy is greatly diminished as spilling breakers move in through shallow nearshore water.

Distributary - One of a number of channels formed in a delta region where the river branches downstream.

Drift-aligned beach - See Beach, Drift-Aligned.

Drowned valley-mouth - See Ria.

Dune - A mound or ridge of unconsolidated wind-blown sediment, usually sand but occasionally silt or clay (where the source area is a dry mudflat: crescent-shaped silt or clay dunes on the lee shore of a lake or lagoon are known as lunettes).

Dune calcarenite - A generally consolidated aeolian sandstone lithified by the cementation or partial cementation of dune sand by secondary internal precipitation of carbonates from groundwater. The proportions of carbonate vary, but are typically at least 50%, the dune sand being calcareous or partly noncalcareous (quartzose sandstones with less than 50% carbonate are termed quartz-arenites). Dune calcarenite is usually of Pleistocene age, and may be overlain by unconsolidated Holocene dune topography. Dune calcarenite is also known as Aeolian Calcarenite (*q. v.*), Aeolianite, Calcareous aeolianite, Dune limestone or Dune sandstone; American spelling: eolianite.

Dune swale - A hollow within dune topography, especially between parallel dune ridges. Wet swales (often with marsh vegetation) are termed *slacks* and deeper hollows may be seasonally or permanently occupied by *dune lakes*.

Earth cliff - A term used by May (1977) for a cliff cut in soft rock formations (sand, clay, and chalk) and subject to rapid recession and instability leading to recurrent mass movements.

Dune calcarenite cliff, Jubilee Point, Victoria, Australia.

Emerged beach, Falmouth Bay, Cornwall.

Edge waves - Standing oscillations that develop at right angles to the coastline as the result of resonance between waves approaching the shore and waves reflected from it.

Emerged beach - A beach that stands above the level at which it originally formed, on or behind the present shore. Known in the earlier literature as a Raised Beach (*q.v.*), when it was thought that upward land movement was necessary to produce such a feature, but it can also be formed as a result of a lowering of sea level, or some combination of land and sea-level change that has left the sea at a relatively lower level.

Emerged shore platform - A shore platform that stands above the level at which it originally formed (often with an emerged beach), on a coast where the land has been uplifted, sea level has fallen, or some combination of land and sea-level change that has left the sea at a relatively lower level.

Emergence - A rise in the level of the land relative to the sea, achieved by actual uplift of the land, a lowering of sea level, or some combination of land and sea-level change that leaves the sea at a relatively lower level. Also known as a negative change (or fall) in base level. An emerged coast (or feature) is one that stands at a higher level relative to the sea than when it originally formed; an emerging coast is one actually rising relative to sea level.

Epeirogenic movement - Upward or downward tectonic movements of large (continental) land masses.

Emerged shore platform (raised in 1855 earthquake), Wellington, New Zealand.

Escarpment cliff (bluff) - A steep coastal slope cut across rock formations that are horizontal, or dipping landward.

Estuary - The seaward end of a river, opening toward the sea, typically through a funnel-shaped inlet, and usually subject to tidal movements and incursions of salt water from the sea.

Estuary threshold - A bank of inwashed sand or gravel at the mouth of an estuary.

Eustasy (Eustatic movements) - Worldwide movements of sea level resulting from changes in the volume of water in the ocean basins. Such changes have occurred as a result of the waxing and waning of the Earth's glaciers, ice sheets, and snowfields (glacio-eustatic movements), but similar changes have taken place as a result of modification of the shape and capacity of the ocean basins by deposition of sediment (sedimento-eustatic movements), submarine vulcanicity (volcanic-eustatic movements) or tectonic deformation (tectono-eustatic movements). There are also *steric changes* due to expansion or contraction of the oceans with rising or falling temperatures.

Fault coast - A steep or cliffed coast produced by faulting, where the seaward slope coincides with the plane of the fault, along which the land has been raised. Some coasts were initiated as fault coasts, but have been cut back by marine erosion and now stand landward of the fault.

Fault-line coast - A steep or cliffed coast following a fault line, where the seaward slope has been formed by differential erosion of rock formations juxtaposed by prior faulting.

Ferricrete - A sedimentary rock formation indurated by the precipitation of iron oxides or other iron compounds, usually derived from percolating groundwater.

Fetch - The distance of open water across which the wind generates waves approaching a coastline from a particular direction.

Fiard (Fjard) - An inlet formed by marine submergence of a river valley (or wide, shallow valley excavated by ice movement) incised in low-lying glaciated rocky terrain. Known in Scotland as a firth.

Fiord (Fjord) - A long, deep steep-sided marine inlet formed by submergence of part of a valley (U-shaped glacial trough) previously shaped by a glacier and incised into coastal uplands. Known in Scotland as a sea loch.

Escarpment cliff, Ballard Down, Dorset, England.

A fiard near Hjortholm, Denmark.

Fiord at Milford Sound, South Island, New Zealand.

Fitting boulders - Interlocking accumulations of blocks and boulders on the shore, resulting from the jostling of the rocks by wave agitation and their consequent abrasion until they have become mutually worn into a complex three-dimensional jigsaw pattern.

Flandrian marine transgression - The sea-level rise that began about 18,000 years ago, in the Late Pleistocene, and continued into the Holocene. This transgression was originally defined in Flanders, northeast France and southern Belgium (Dubois, 1924). See Late Quaternary Marine Transgression (*q. v.*).

Foredune - A ridge of wind-blown sand at the back of a beach, parallel to the coastline, and retained by vegetation. Some foredunes originate as a result of wind deposition of sand on a beach ridge (*q. v.*), but others may begin to form along an upper beach strandline that persists long enough for plants to germinate and intiate sand trapping. The terms primary dune and frontal dune are synonyms for foredune in Australia.

Foreshore - The shore zone, between high and low tide lines. In some countries there is confusion because the term has been used for the fringe of the land, which is really the backshore.

Fringing reef - A structure built adjacent to the coast by coral and associated organisms. A distinction may be made between a shore reef, built out as a prograded terrace from the coastline, and an attached reef, which originated a short distance offshore and became linked to the coastline by subsequent deposition. Some fringing reefs decline gently landward, passing beneath a shallow (ca. 1 m) lagoon or moat, sometimes called a boat channel.

Frost shattering - The disintegration of rock surfaces, notably on cliffs and shore outcrops, as a result of repeated expansion and contraction of the rock by freezing and thawing, particularly where contained moisture forms ice crystals as temperature ranges through 0–4°C. Also termed thermal abrasion.

Foredune at Seaspray, Ninety Mile Beach, Victoria, Australia.

Fringing reef on the southeast coast of Bali, Indonesia.

Gat - A channel cut or maintained across a bar by waves or (more often) currents.

Geo - A Scottish term for a long, deep, and narrow steep-sided inlet (coastal gorge or chasm) cut into rocky cliffs by marine erosion, usually along structural planes of division (joints, faults, bedding planes, cleavage planes). Known as a zawn or gut in Cornwall. A distinction may be made between geos, excavated entirely by marine erosion, and gorges that are the mouths of deeply incised steep-sided valleys invaded by the sea during the later stages of the Flandrian Marine Transgression (*q.v.*).

Geological column - The sequence of rock formations arranged by age (my = million years):

Geological structure - See Structure.

Giant waves - Exceptionally high waves generated by major disturbances (earthquakes, landslides, volcanic eruptions) on the coast or seafloor. See Tsunami.

Grain Size Categories - The Wentworth scale of particle diameters. The ø scale is based on the negative logarithm (to base 2) of the particle diameter in millimetres [$ø = \log_2 d$], so that coarser particles have negative values.

An alternative is a decimal scale centered on 2 mm, in which sand ranges from 0.2 to 2.0 mm, but this does not correspond well with generally perceived grain size categories. The sand range excludes sediment that would be generally classified as fine to very fine sand, and the coarser (>2.0 mm) and finer (<0.2 mm) divisions do not match widely accepted categories of pebbles and cobbles or of silt and clay.

Quaternary Period	Holocene (Recent)	10,000 years
	Pleistocene Epoch	2.3 my
Tertiary Period	Pliocene Epoch	5 my
	Miocene Epoch	23 my
	Oligocene Epoch	36 my
	Eocene Epoch	53 my
	Palaeocene Epoch	65 my
Mesozoic Era	Cretaceous Period	144 my
	Jurassic Period	213 my
	Triassic Period	248 my
Palaeozoic Era	Permian Period	290 my
	Carboniferous Period	360 my
	Devonian Period	405 my
	Silurian Period	436 my
	Ordovician Period	510 my
	Cambrian Period	560 my
Pre-Cambrian Era		

Wentworth scale category	Particle diameter (mm)	ø scale
Boulders	>256	below −8ø
Cobbles	64–256	−6ø to −8ø
Pebbles	4–64	−2ø to −6ø
Granules	2–4	−1ø to −2ø
Very coarse sand	1–2	0ø to −1ø
Coarse sand	1/2–1	1ø to 0ø
Medium sand	1/4–1/2	2ø to 1ø
Fine sand	1/8–1/4	3ø to 2ø
Very fine sand	1/16–1/8	4ø to 3ø
Silt	1/256–1/16	8ø to 4ø
Clay	<1/256	above 8ø

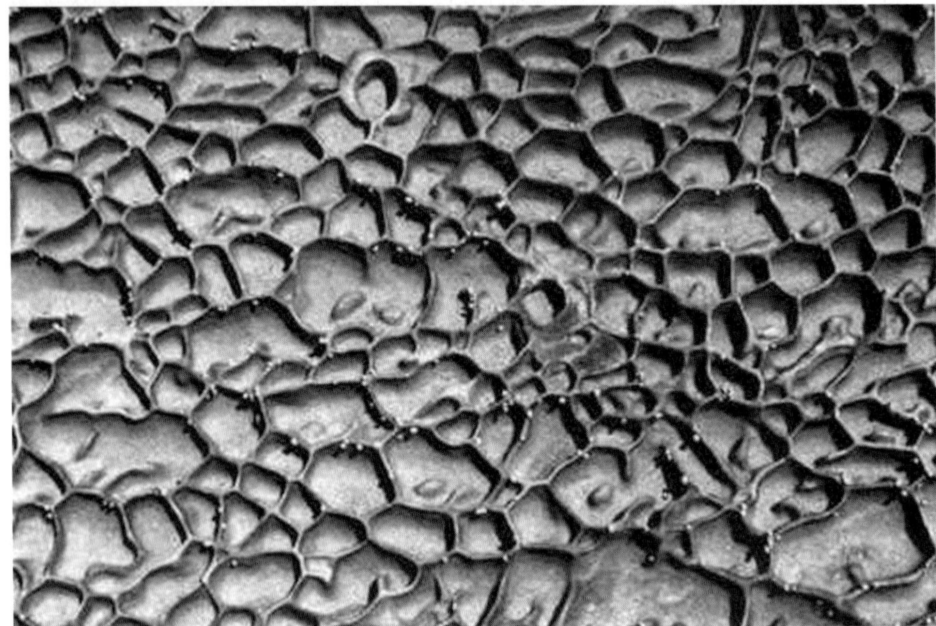

Honeycomb weathering on Cretaceous sandstone, Otways coast, Victoria, Australia.

Granule - A rock particle with a diameter between 2 and 4 mm. See Granulometry.

Granulometry - The measurement and classification of the size of sediment grains.

Groyne - A wall built out at right angles from the coastline, intended to intercept drifting beach material. American spelling: groin.

Gulch - A deep and narrow channel cut by abrasion into a cliff or shore platform or across a rocky shore, often extending below low tide level.

Hairpin dune - An elongated, narrow blowout or parabolic dune with parallel trailing arms, well developed in northeastern Tasmania.

Hanging valley - A valley truncated by cliff recession, so that the stream pours out as a coastal waterfall.

Headland - A high protrusion of the land into the sea, usually cliffed. Generally smaller than a Cape or Promontory (*qq.v.*).

Holocene epoch - The last of the geological epochs, which began about 10,000 years ago.

Holocene marine transgression - See Late Quaternary Marine Transgression or Flandrian Marine Transgression.

Honeycomb weathering (Alveolar weathering) - A process producing an intricate pattern of small cell-like cavities on a rock surface, often penetrating a harder (or indurated) crust.

Humate - A consolidated and indurated sandrock formed within a sandy formation (Dunes, Beach Ridges, *qq.v.*) by interstitial precipitation of iron oxides (derived from the thin coating of quartz sand grains by ferruginous material, responsible for their initial yellow or brown color, mobilized by percolating groundwater) and precipitation of organic matter (washed down from surface soil and decaying vegetation) within the zone where the water table rises and falls (seasonally or irregularly). Also known as coffee rock or hardpan.

Hutberge - A high, steep-sided hill rising abruptly from a Strandflat (*q.v.*), formed as a residual island in a shallow sea subject to intensive freeze–thaw processes accompanying tidal alternations. Similar in form, but not origin, to Old Hat Islands (*q.v.*).

Hydro-isostasy - Vertical movements of a coast and continental shelf in response to loading and unloading of water as sea levels rise and fall.

Cliffs cut in humate, Chatham Islands, New Zealand.

Ice coast/Ice cliffs - Steep cliffs up to 50 m high, formed where glaciers or ice sheets (ice shelves) reach the sea, and from which floes and icebergs become separated during summer melting. Where the ice is afloat the cliff is called an ice front; where it is grounded, an ice wall.

Ice rafting - The delivery of material (rocks, beach sediments, marsh fragments, driftwood) to the shore on or within layers of ice (often in icebergs detached from a glacier front), to be deposited when the ice melts in summer. See also Coastal Erratics.

Induration - The hardening of a rock surface by precipitation of material (carbonates, iron compounds, silica) from groundwater exudations, forming a resistant crust (case-hardening).

Intermediate beach - A beach state transitional between a dissipative and a Reflective Beach (*q.v.*).

Isostasy - An equilibrium between an area of the Earth's crust floating and the underlying plastic mantle, whereby areas loaded with sediment, volcanic deposits or ice subside, and areas unloaded (e.g., when an ice cover melts) rise (isostatic rebound).

Jetty - A solid structure built out more or less at right angles to the coastline or on either side of a river mouth or lagoon entrance. The terms breakwater and pier are sometimes used as synonyms.

Karst coast - A coast with landforms shaped by solution processes, notably on limestone, where the dissolving of the rock leads to the formation of surface depressions, sinkholes, caves, and underground drainage.

Klint - A term used for a cliff (generally in limestone) in the Baltic region. In some countries, notably Estonia, it describes an active cliff, but more generally it indicates an inland bluff that was an active cliff when sea level was higher.

Landlocked - An area of water (usually a bay or lagoon) surrounded or nearly enclosed by land.

Late Quaternary marine transgression - The worldwide rise of sea level that began about 18,000 years ago, when sea stood between 100 and 140 m below its present level, and the continental shelves were subaerially exposed, as the result of global warming and the release of water from melting glaciers, ice sheets, and snow fields. It came to an end about 6,000 years ago, but the history of sea-level change has been complicated on many coasts by accompanying and continuing uplift or depression of coastal land margins. See Flandrian Marine Transgression. Sometimes called the Holocene marine transgression, but the term Late Quaternary Marine Transgression (*q.v.*) is more accurate as the sea-level rise began late in Pleistocene times and culminated within the past 10,000 years (Holocene).

Lateral grading - A gradual change in the caliber of beach sediment along the shore. Grading is a condition; sorting and attrition are processes that may lead to a beach becoming graded.

Longshore current - The flow of water along the shore or nearshore as the result of oblique waves, often augmented by wind-driven and tidal currents.

Longshore (Littoral) drifting - The movement of beach sediment along the shore (and Nearshore, *q.v.*) by waves arriving at an angle to the coastline (Beach Drifting, *q.v.*) and by currents generated by such waves (nearshore drifting). Also known as Shore Drift (*q.v.*).

Lobate (Looped) bar, barrier - A depositional feature curving out from the coastline and back again, in such a way as to enclose a lagoon or swamp. A lobate bar is submerged at least at high tide, and a lobate barrier has been built up above high tide level.

Log-spiral beach - See Zetaform Beach.

Low islands (Low wooded islands) - Small low-lying depositional islands of coralline sediment, comprising a leeward sand Cay (*q.v.*), mainly of sand, a shingle rampart on the windward side, and an intervening depression occupied by a lagoon with mangroves.

Machair - A low flat or hummocky plain of calcareous sand, generally formed on the landward side of a coastal dune, as on the coasts of Scotland and Ireland.

Macrotidal - Where mean spring tide range is 4–6 m.

Mangrove - A type or community of halophytic trees and shrubs that can grow on shores sheltered from strong wave action (bays, inlets, delta, estuary, and lagoon shores) in the intertidal zone, subject to regular or frequent submergence of their root systems by brackish or seawater. Mangroves are found mainly on tropical and subtropical coasts, but extend locally into temperate latitudes.

Low Isles, low wooded islands off Port Douglas, Queensland, Australia.

Megaripples in the Bay of Fundy, Canada.

Megaripple - A large form or Ripple (*q.v.*) or sand wave with an amplitude of 0.1–1.0 m and length between 1 and 100 m. See Ridge and runnel.

Megatidal - Where mean spring tide range exceeds 6 m.

Mesa - A flat-topped, steep-sided residual hill. The term has also been used to describe small upstanding flat-topped rocks on a shore platform.

Mesotidal - Where mean spring tide range is between 2 and 4 m.

Microatoll - A circular organic reef structure (diameter 1–6 m) consisting of a raised rim built by coral (usually porites), algae and other organisms, surrounding a shallow depression on the shore or on a coral reef.

Microcliff - A small cliff, generally less than a meter high, found on the seaward margins of salt marsh and mangrove terraces, and sometimes in intertidal mudflats. Also known as a Clifflet (*q.v.*).

Microtidal - Where mean spring tide range is less than 2 m.

Mud - A sticky fine-grained sediment (silt, clay, sometimes with organic matter).

Mudflat - A relatively level unvegetated area of fine sediment on the shore, especially in sheltered inlets, estuaries, or tidal lagoons. Intertidal

Microatoll, White Lady Bay, Magnetic Island, Queensland, Australia.

Mushroom Rock on Pleistocene dune calcarenite, Sorrento, Victoria, Australia.

mudflats are commonly exposed seaward or salt marshes of mangroves at low tide, and supratidal mudflats (Tanns in West Africa) occur landward of salt marshes or mangroves on arid or semiarid coasts.

Mudrock - A massive or blocky rock composed of indurated fine-grained sediment (silt and clay), also termed siltstone or claystone, differing from a shale in being nonfissile.

Multicausality - Where similar coastal landforms (e.g., coastal barriers, cuspate forelands, laterally graded beaches) may be produced in different ways (Schwartz, 1971).

Muricate weathering - the intricate pitting of coastal rock surfaces in the spray zone, especially on limestones, calcareous sandstones, and dune calcarenites, as a result of physical processes (wetting and drying), chemical processes (corrosion by sea spray, i.e., aerated sea water; growth of salt crystals), and biological processes (corrosion by algae and other organisms; scraping and browsing by shore fauna).

Mushroom rock - A table-like stack that has been marginally undercut by abrasion, solution, or bioerosion so that it is surrounded by a notch and visor, and stands above a narrower pedestal.

Natural arch - A tunnel extending through a headland, island or stack, beneath a connecting bridge, usually formed where a cave has been enlarged by abrasion or solution. Also known as a Sea Arch (*q.v.*).

Natural arch, La Manne Porte, Etretat, France.

Old Hat island, Taylor Island, Paihia, New Zealand.

Neap tide - A diminished tide range when the sun and the moon are at right angles in relation to the earth, so that their gravitational effects are not combined.

Nearshore - The shallow water zone between the low tide line and the line where waves begin to break; a zone that migrates to and fro as the tide rises and falls.

Nearshore water circulation - An association of processes that operates as wave move through the nearshore zone to break on the Shore or Beach, generating Swash and Backwash (*qq.v.*).

Neotectonic - Movements that have produced existing landforms (e.g., fault scarps) and associated structures, generally within Quaternary (*q.v.*) Times.

Notch (Nip) - A narrow hollow excavated along the base of a cliff near high tide level by abrasion, solution or bioerosion, often with an overhanging rock visor (protruding ledge of rock). See Abrasion Notch, Solution Notch.

Old Hat island - A stack or islet surrounded by a flat or gently sloping shore platform, formed where much of a soft or weathered rock formation has been removed by erosion above a harder horizontal intertidal rock layer, which persists as a structural bench. Originally described from North Island, New Zealand.

Open coast - A coast unprotected by islands, promontories or reefs, and so exposed to the full force of wave action.

Overwash (Washover) - The washing of sediment over the crest of a beach or coastal barrier by exceptionally strong wave swash to form a depositional fan on the landward side. See Washover Fan.

Paired spits - Spits on either side of a coastal inlet, river mouth or lagoon entrance, or protruding toward each other between two islands or between the mainland and an offshore island. They have developed either by convergent longshore drifting or the breaching of a former coastal barrier.

Pan - A shallow steep-sided and flat-floored natural depression on a shore platform or in a salt marsh (see Salt Pan). A soil pan is a crust or subsurface horizon of compacted or indurated sediment (see Humate).

Parabolic dune - A dune with an advancing convex nose of spilling sand and trailing (roughly parallel) arms of partly vegetated sand on either side of an elongated low corridor formed or maintained by deflation. Typically its axial length is more than three times its mean width.

Paired spits, Shoal Inlet, Victoria, Australia.

Parabolic dune near Lakes Entrance, Victoria, Australia.

Parallel dunes near Robe, South Australia.

Parallel dunes - A succession of foredunes developed on a prograding coast (usually surmounting Beach Ridges, *q. v.*).

Pebble - A rock particle with a diameter between 4 and 64 mm. See Granulometry.

Periglacial (Paraglacial) - Processes and environments found beyond the limits of glaciation, typified by freezing and thawing, and frequent accumulation and melting of snow.

Pier - An open structure on multiple supports, usually designed to permit ships to berth: beneath it waves, currents and drifting sediment pass almost unimpeded. The term Jetty (*q. v.*) is sometimes used as a synonym. Occasionally these terms are used to describe solid stone structures.

Pit, Pitting - superficial indentations (typically up to 5 mm in diameter), etched on a rock surface by Weathering or Abrasion (*qq. v.*). See Muricate Weathering.

Plunging cliff - A steep or vertical cliff that descends into deep water inshore without any intervening Shore Platform, Rocky Shore, or Beach (*qq. v.*).

Point - A small protrusion of the land into the sea, usually sharp, tapering and low-lying, but sometimes a high headland (Hartland Point, Devonshire).

Progradation - The building seaward of a coastline by deposition of sediment, as on a beach or a dune, or where a marsh or mangrove shoreline advances.

Promontory - A coastal protrusion or headland, high and bordered by cliffs or bluffs, usually smaller than a Cape (*q. v.*).

Quaternary - The geological period which began about 2.3 million years ago. It comprises the Pleistocene epoch (2.3 million years ago to 10,000 years ago) and the Holocene or Recent epoch (the last 10,000 years).

Raised beach - A beach that has been uplifted by tectonic movements to stand above the level at which it originally formed, on or behind the present shore. See Emerged Beach.

Plunging cliff near Milford Sound, South Island, New Zealand.

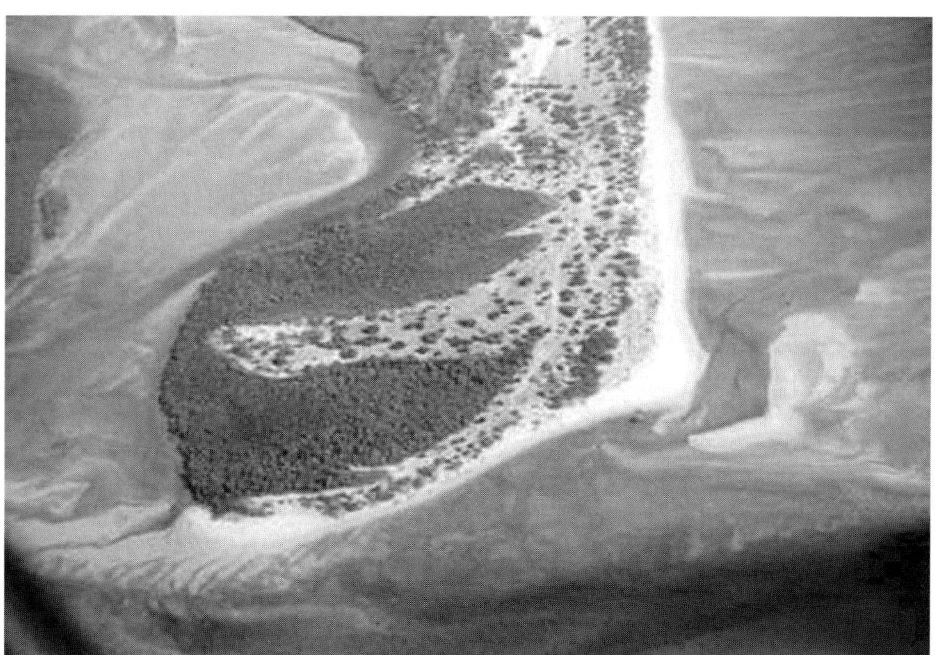

Recurved spit near Mackay, Queensland, Australia.

Rampart - A slight rise or wall-like ridge up to 2 m high toward the seaward edge of a shore platform or reef, or toward the crest of a cliff. Some ramparts are residual rock features that have escaped weathering and erosion; others are constructional, formed by the growth of organisms (see Algal Rim) or depositional, as on the coaming found on some cliff crests. They may consist of boulders, cobbles, pebbles or sand, generally angular and often cemented. See Coral Rampart.

Recurved spit - A spit that ends in a landward hook or recurve. (e.g., Sandy Hook, New Jersey).

Reef - A bank, ridge or mound with a rocky structure, either an eroded rock formation or built by organisms such as corals and algae, usually irregular in outline (but often flat-topped), and not moved by waves or current action.

Reflective beach - A beach where wave energy is partly reflected seaward as plunging breakers move in through relatively deep nearshore water.

Relict features - Features that developed under different environmental conditions (climate, vegetation, sea level) in the past, and have persisted in the present coastal landscape, having not yet been destroyed by modern processes.

Reliction - The exposure of land as a result of seafloor emergence during a slow or gradual withdrawal of the relative fall in sea level, as around the Caspian Sea between 1930 and 1977.

Ria - A long, narrow, often branching inlet formed by marine submergence of parts of a river valley that had previously been incised to a lower sea level: a drowned valley-mouth. Cotton (1956) separated ria *sensu lato*, thus defined, from ria *sensu stricto*, where the inlet runs parallel to geological outcrops that run at right angles to the coastline, as in the Rias of Galicia, northwest Spain, which Richthofen (1886) quoted in the original definition. Some of the Galician rias are wide and deep, and the valleys that have been drowned may have been shaped partly by tectonic subsidence or by the recession of bordering scarps.

Ridge and runnel - Several subdued bars and troughs running parallel or nearly parallel to the coastline and exposed at low tide on a sandy shore. Also known as low and ball.

Ria (Aber Benoît) on the Brittany coast, France.

Ridge and runnel on the shore at Saltburn, Yorkshire.

Rip channel - A channel cut by the seaward flow of a Rip Current (*q.v.*) across the nearshore zone, usually through nearshore bars.

Rip current - A strong narrow current (up to two knots) flowing seaward through breakers at right angles or an oblique angle to the coastline.

Ripple marks - Undulations (ridges generally less than 10 cm high, and troughs a few centimeters long between successive crests, often exposed at low tide on the foreshore), formed on a sandy foreshore and nearshore area by waves and/or currents. Some are parallel, others form intersecting, sometimes rhomboidal patterns. They may be symmetrical in cross-section, but are usually asymmetrical with steeper slopes in the direction of wave or current flow and elongated at right angles to this direction.

Rocky shore - An irregular, rugged rocky area between high and low tide where shore platforms have failed to develop, or have been intricately dissected.

Round hole - An enlarged blowhole on coastal slopes above cliffs that have been penetrated by caves, notably in southwest England.

Sabkha - See Sebkha.

Salt marsh - a flat or gently sloping vegetated wetland in the upper intertidal zone on sheltered parts of the coast (estuaries, inlets, lagoon shores). Often in the form of a depositional terrace, periodically submerged, with halophytic grasses, herbs, and shrubs; dissected by tidal creeks, and may contain enclosed Salt Pans (*q.v.*).

Salt pan - An enclosed bare depression within a salt marsh, apt to dry out, leaving an algal or saline crust. The term is also used for shallow artificial basins in which seawater is trapped and concentrated by evaporation to brine, which crystallizes into salt for harvesting.

Salt weathering - Disintegration or decomposition of a rock surface by stress caused by the growth of salt crystallizing from sea spray, resulting in the formation of pits, cavities, or shallow basins.

Sand - A sediment consisting of rock particles with a diameter between 0.125 and 2.0 mm. Subdivisions are very coarse sand (1–2 mm), coarse sand (1/2 or 0.5–1 mm), medium sand (1/4 or 0.25–1/2 or 0.5 mm), fine sand (1/8 or 0.125–1/4 or 0.25 mm) and very fine sand (1/16 or 0.0625–1/8 or 0.125 mm). See Granulometry.

Sandflat - A relatively level unvegetated sandy intertidal area.

Rip currents (arrowed) on Woolamai Beach, Phillip Island, Victoria, Australia.

Round Hole at The Lion's Den, Lizard Point, Cornwall, England.

Seagrass - A marine grass that grows in the intertidal and shallow subtidal zones.

Sea level - The level at which the sea stands against the coast, conventionally taken as mean sea level, the arithmetic mean of the calm sea surface (excluding waves and oscillations related to winds and atmospheric pressure variations) measured at hourly intervals over at least 18.6 years.

Sebkha (Sabkha) - A low-lying very gently sloping saline area above normal high tide level on an arid coast, subject to occasional flooding by the sea or rain water, and prolonged phases of evaporation and dessication, resulting in hypersaline conditions. Some sebkhas have the branching form of rias. Some carry Salt Marshes (*q.v.*).

Sediment budget - The relative proportions of inputs, outputs and storage in a coastal sediment system (e.g., a Beach, *q.v.*).

Segmentation - The division of an elongated lagoon or bay into a chain of smaller, oval, or circular lagoons by the growth and coalescence of cuspate spits and forelands and the erosion of intervening embayments, usually with narrow connecting channels (Price, 1947; Zenkovich, 1959).

Seiche - Alternating, diminishing fluctuations of water level after a rapid change in atmospheric pressure or where strong wind action has built up water level downwind and lowered it upwind in a landlocked bay, inlet, or coastal lagoon.

Shale - A fine-grained stratified or laminated sedimentary rock which breaks readily into thin layers. Contrast Mudrock (*q.v.*).

Sharm - A long narrow marine inlet on an arid coast, usually at the mouth of a wadi (a generally dry valley in a coastal upland).

Shingle - A British term for Beach Gravel (*q.v.*), a coarse, loose deposit which may range from granules through pebbles to cobbles and small boulders, generally well-rounded particles that vary in shape from roughly spherical through ovoid to flat or platy. Shingle beaches are sometimes termed coarse Clastic Beaches, Clastic (*q.v.*) meaning consisting of rock fragments.

Shore - The zone between the water's edge (Shoreline, *q.v.*) at high and low tide. Sometimes referred to as tidelands.

Sebkha on the coast of King Sound, Western Australia.

Shore drift - See Longshore drift, which is more accurate because shore drift could include movements landward or seaward across the intertidal zone.

Shoreline - The water's edge, moving to and fro as the tides rise and fall, so that there is a low tide shoreline, a mid-tide shoreline, and a high tide shoreline. The term has been used as a synonym for Coastline (*q. v.*), but it is useful to maintain a distinction between the two terms, taking Coastline as equivalent to the high tide shoreline. Where the tide range is large and the shore profile gently sloping there is much variation in the position of the various shorelines.

In the United States the term Shoreline (*q. v.*) is defined legally as mean high water (MHW), as shown on nautical charts produced by the National Oceanic and Atmospheric Administration (NOAA), while shorelines at other levels are called lines (e.g., the mean lower low water line) which is a private-property seaward boundary in some eastern states. It should be noted that the American shoreline, thus defined, is submerged at high spring tides, and is not the margin of normally dry land.

Shore platform - A flat or gently sloping smooth or relatively smooth rock surface formed in the zone between high and low tide levels. The term is preferred to wave-cut platform, except on some soft rock outcrops where the platform has been literally cut by wave action alone. Some shore platforms are sub-horizontal, and may be *high tide shore platforms*, submerged only at the highest tides, or *low tide shore platforms*, exposed only at low tide. Others are seaward sloping intertidal platforms, extending down to below low tide level. Some are structural (coinciding with the upper surface of a seaward sloping hard rock formation), others erosional (cut across outcropping structures).

The term Beach Platform (*q. v.*) is not acceptable as an alternative to Shore Platform (*q. v.*) because of the risk of confusion with depositional terraces (e.g., Beach Berm, *q. v.*).

Shore pothole - A circular or oval depression scoured in a rocky shore platform by abrasion where sand or pebbles are circulated by wave motion (Wentworth, 1944). Similar features may occur on a limestone coast where soil pipes formed beneath an ancient soil (palaeosol) are exhumed and excavated by wave scour.

Shore potholes formed by abrasion, Pearce's Beach, Rye, Victoria, Australia.

Significant wave height - The average height of the highest one-third of waves measured over a 20-min observation period. As the number of waves within 20 min varies (240 5-s waves, but only 80 15-s waves) it may be preferable to measure the highest 33 of a set of 99 successive waves.

Silt - A sediment consisting of particles with a diameter between 1/256 (about 0.004) and 1/16 (0.0625) mm.

Skerry - A low, rugged rocky reef or scatter of reefs, generally intertidal but sometimes extending above high tide level, off a hard rock coast, particularly in Scandinavia and Scotland. Usually there are many skerries; the Norwegian term skjergaard has been wrongly translated as "skerry-guard" when in fact it means the area of calm water in the lee of skerries.

Slope-over-wall profile - A steep coast on which an upper, sloping facet descends to a steeper, often vertical, basal cliff. Such profiles are found on steep coasts where the rock formations dip seaward, on soft formations where a subaerial slope is recurrently undercut by marine erosion, and where a weak formation (such as glacial drift) forms a slope above a cliff in more resistant rock. On hard rock coasts that were periglaciated in Pleistocene times, as in southwest England, the upper slope is mantled by frost-shattered solifluction talus deposits (termed head) and the basal rocky cliff has been exposed by later marine erosion. Some slope-over-wall coasts have beveled slopes of uniform seaward gradient (Beveled Cliffs, *q. v.*), but many are convex (hogsback coasts) and some concave. Also known as two-storied cliff.

Solution notch - An elongated cliff-base hollow cut out near high tide level by solution (sometimes also bioerosion), often with an overhanging rock visor (protruding ledge of rock).

Solution pan - A flat-floored basin developed on a limestone shore platform by solution, bordered by a microcliff or a constructional rim of algae.

Solution pipe - A cylindrical depression formed on a limestone surface by the dissolving of carbonates, often beneath soil and around a root system. There may be no surface depression, but soil has often subsided into a deepening hollow, the margins of which may become case-hardened by secondary precipitation of the leached carbonates. On Dune Calcarenite (*q. v.*) coasts solution pipes (with encircling calcrete) may be exposed by wind erosion (as structures sometimes confused with petrified forests) or as shore platforms are cut, and the soil washed out to leave a protruding vase-shaped structure.

Sound - A narrow marine channel or strait between an island and the mainland, or between two islands, or between two land areas. A Strait (*q. v.*).

Spit - A finger-like ridge or embankment of beach material built up above high tide level and diverging from the land at one end (proximal) to terminate (distal end) usually in one or more recurves or hooks curving landward. See Recurved Spit.

Spring tide - An augmented tide range when the sun and the moon are in alignment with the earth, so that their gravitational effects are combined.

Squeaking sand - Beach or dune sand that emits sounds (squeaks, squawks, shrieks, sings, barks, booms, roars, or whistles) when walked upon, as the result of the mutual impact of sand grains. Sometimes called musical sand.

Stack - An isolated upstanding steep-sided rock pillar, column or pinnacle rising from the shore, a shore platform, or the seafloor close to a cliffed coast.

Stillstand - A condition of stability when the relative levels of land and sea remain the same for a prolonged period (several centuries), either because there has been no change in land or sea level, or because these levels have risen or fallen the same amount.

Slope-over-wall profile (periglaciated slope), Dodman Point, Cornwall, England.

Solution notch on a limestone coast, Baron, Java, Indonesia.

Stack at Sentinel Rock, Port Campbell, Victoria, Australia.

Storm surge - A temporary abnormal rise of sea level on a coast, as when an exceptionally high tide (often with sea level raised by low atmospheric pressure) is accompanied by strong wave action generated by an onshore gale (cyclone, hurricane, typhoon).

Storm wave - A short, steep high wave (wave period generally <10 s) generated by strong winds.

Strait - A narrow passage of water between two land areas. A Sound (*q. v.*).

Strand - See Shore or Beach.

Strandflat - An extensive shallow submarine and low-lying emerged coastal platform (up to 65 km wide and typically with transverse gradients of up to 10° and many surmounting mounds and hillocks that were formerly stacks and islands), sharply contrasted with a high and rugged hinterland.

Strandline - See Shoreline.

Stromatolite - A low calcareous sedimentary domal structure, formed in shallow water where a mat or assemblage of blue-green algae have trapped sediment and precipitated calcium carbonate. See Algal Mat.

Structure, Structural features - The assemblage of rock formations (their arrangement, disposition, nature, and form) upon which erosional processes are, or have been, at work in shaping a landscape or coastline.

Subdelta lobe - Portion of a delta formed around the mouth of a distributary channel.

Submergence - A fall in the level of the land relative to the sea, achieved by actual subsidence of the land, a raising of sea level, or some combination of land and sea-level change that leaves the sea at a relatively higher level. A submerged coast (or feature) is one that stands at a lower level relative to the sea than when it originally formed; a submerging coast is one actually subsiding relative to sea level.

Swale - An elongated hollow or low-lying area between dune ridges, usually running parallel to the coastline.

Swamp encroachment - The advance of sediment-trapping swamp vegetation (reedswamp, salt marsh, mangroves) into shallow nearshore water, notably in sheltered bays, inlets and coastal lagoons, resulting in progradation of the land.

Swash - The rapid flow (uprush) of a breaking wave up the beach face.

Swash bar - A nearshore bar built parallel to the shore by wave swash, with a steeper landward slope advancing into a shallow lagoon.

Swash (Swash-backwash) cusps - See Beach Cusps.

Swash-dominated beach - A beach shaped primarily by the action of swash generated by breaking waves that arrive parallel to the shoreline, so that the beach has an alignment at right angles to the dominant direction of wave approach. Contrast with Drift-Aligned Beach (*q. v.*).

Swashway - A low sector of a barrier through which storm waves or surges flow. Sand or gravel swept through a swashway is deposited as a Washover Fan (*q. v.*) on the inner shore of the barrier.

Swatchway - A channel cut and maintained across a bar or shoal by wave and current scour. A term used in a novel by Erskine Childers (1903).

Swash zone - The zone regularly covered and uncovered as breaking waves generate swash and backwash.

Sweep zone - The zone (vertical plane) between high (often summer) and low (often winter) beach profiles, surveyed at right angles to the coastline, and indicating the extent of loss and gain resulting from cut and fill sequences.

Swell - Long, low waves (wave period typically 12–16 s) generated by distant storms and transmitted across oceans and seas.

Tafoni - Large cavities excavated in a rocky cliff by weathering, notably where the rock surface has been indurated and the underlying material is softer.

Tectonic - Crustal movements that produce geological structures, notably folding and faulting.

Terrigenous sediment - Sediment derived from the land surface and delivered to the coast by runoff, landslides, rivers, glaciers, or wind action, in contrast with marine sediment supplied to the coast from the seafloor.

Threshold - A shallow area near the seaward end of a drowned valley. In fiords it is often a rocky sill (sometimes partly depositional), but in estuaries and inlets on high wave energy coasts it is commonly a bank of inwashed sand, prograding landward.

Tidal bore - A large turbulent wave that develops a steep advancing water slope (up to 5 m high) that moves rapidly (5–8 m/s) into a long

Swamp encroachment (reedswamp) on the shore of Lake Wellington, Victoria, Australia.

Tafoni on granite coast, Victor Harbour, South Australia.

shallow funnel-shaped inlet or estuary during a rising spring tide on a macrotidal or megatidal coast. Known as the mascaret on the Gironde estuary; eagre on the Trent and Humber; pororocas on the Amazon.

Tidal current - A current generated by the rise or fall of the tide.

Tidal delta - An intertidal or subtidal bar or shoal, typically triangular or lobate, formed on the inner (flood delta) and often also the outer (ebb delta) sides of a tidal inlet or gap between barrier islands. Usually of sand, sometimes gravel, with diverging channel systems.

Tidal divide - An intertidal isthmus that emerges between an island and the mainland, or between neighboring islands, as the tide falls. There are usually tidal creeks heading away on either side and converging downstream. Sometimes called a tidal watershed, but this is inappropriate as the term watershed can also indicate a catchment area.

Tidal environment - A Macrotidal (*q.v.*) environment where mean spring tide range is between 4 and 6 m; a Mesotidal (*q.v.*) environment where mean spring tide range is between 2 and 4 m, and a Microtidal (*q.v.*) environment where mean spring tide range is less than 2 m. The term Megatidal (*q.v.*) is used where mean spring tide range is more than 6 m.

Tidal flat - A depositional feature, usually with a very gentle intertidal slope, consisting of sand (Sandflat, *q.v.*) or finer sediment (Mudflats, *q.v.*).

Tidal prism - The volume of seawater that flows in and out of an estuary or lagoon as the result of the rise and fall of the tide; a process known as tidal ventilation. It is computed as the difference between high and low tide volume, or as mean tide range times the mid-tide area of the basin.

Tide-dominated morphology - Landforms such as intertidal Mudflats (*q.v.*), salt marsh, and mangrove swamp, shaped primarily by tides and tidal currents, rather than by wave action. Most intertidal mudflats and sandflats, and most salt marshes and mangrove swamps, have to some extent been shaped by occasional wave action (notably producing shoreward drifting of sediment). It is suggested that tide-dominated shore morphology occurs where significant wave height is less than 20 cm at high tide.

Tide range - Mean tide range is the difference between mean high and mean low tides; neap tide range the difference between mean high neap and mean low neap tides; spring tide range the difference between mean high spring and mean low spring tides.

Tombolo - A Ridge or Barrier (*qq.v.*) of sand (occasionally Shingle, *q.v.*) built above high tide level in such a way as to link a former island to the mainland, or unite two islands. Sometimes this Italian term (which actually means the sand dune on top of the barrier) is taken to include the island as well as the depositional linking feature. Where the connecting deposit is submerged at high tide the terms tie-bar or tombolino are used. The term usually indicates a constructional feature, but similar forms can develop as a result of erosion of weak or unconsolidated material in the lee of a Headland or Breakwater (*qq.v.*).

Trailing spit - A spit that runs out from the lee shore of a headland or island. Also termed arrow spit or comet tail.

Transgressive barrier - A barrier that is migrating landward as a result of overwash and the movement of dune sand by onshore winds, encroaching on the lagoon or swamp that lies behind it.

Transgressive dune - A dune migrating away from the prevailing wind, some of which may assume barchan form, with lateral arms pushed leeward, as in desert dunes.

Trottoir - A constructional bench formed by the growth of algae at about mid-tide level on cliffed coasts (notably limestones), particularly in the Mediterranean. In some places the algae have colonized a pre-existing rock ledge. Trottoirs have also been described from the Mariana Islands in the western Pacific. The mearls of Brittany are similar features, built by *Lithophyllum*.

Tsunami - Long seismic sea waves, generated by a major disturbance within an ocean basin (mainly due to earthquakes, but sometimes explosive volcanic eruptions or submarine landslides). They are of subdued form in deep water, but on entering shallow nearshore areas their height increases greatly, and can exceed 30 m when they break over the coastline.

Undertow - Laminar basal flow of water seaward beneath incoming waves; a continuation of Backwash (*q.v.*) from the shore into the nearshore and offshore areas. However, seaward flow of water generally takes the form of segregated Rip currents (*q.v.*) rather than laminar basal flow.

Washover fan (Overwash fan) - A fan of beach material washed over low sectors of a coastal barrier by Storm waves (Storm surges) (*qq.v.*) to form a delta-like projection into a lagoon or on to backing marshland. See Overwash. In Texas the term washaround has been used for features produced by overwash on either side of a residual island or mound.

Wave diffraction - Increased curvature in the pattern of waves moving through a natural or artificial entrance to a bay or lagoon because of frictional retardation by bordering Headlands or Breakwaters (*qq.v.*).

Wave energy environments - High wave energy coasts are those exposed to strong ocean swell and large storms waves, with significant wave (Breaker, *q.v.*) heights of more than 1 m. Moderate wave energy coasts are where significant wave height is 0.3–1 m, and low wave energy coasts (where wave energy is limited by headlands, islands or reefs, or where there is extensive shallow water offshore), have significant wave heights of less than 0.3 m. The term zero wave energy is applied where wave action is completely absent.

Wave quarrying - Excavation or displacement of rock masses by wave impact.

Tombolo at Broulee Island, New South Wales, Australia.

Trailing spit, Halifax Island, Queensland, Australia.

Wave quarrying has disintegrated the shore platform at Tessellated Pavement, Tasmania.

Wave refraction from seawall, Black Rock, Victoria, Australia.

Wave reflection - Where waves breaking against a rock formation, steep beach or artificial structure return seaward.

Wave refraction - Changes in the pattern of waves moving toward a coast because of frictional retardation by the shallowing seafloor and bordering headlands; waves in deep water have parallel crests, but as they move into a gulf or bay these become curved, and may anticipate and fit the outline of a beach or a cliffed coastline cut in soft rock formations.

Wave sluicing - the washing away by waves of material disintegrated from the surface of a shore platform by Weathering (*q.v.*) or erosion.

Waves - Undulations produced on the sea surface by disturbance, generally the frictional drag of wind action, but see Giant Waves (*q.v.*). Wave height is the vertical distance between adjacent crests and troughs, wave length the horizontal distance between successive wave crests, wave period the time taken by successive crests to pass a fixed point, wave steepness the ratio of wave height to wave length, and wave velocity the speed at which a wave crest moves forward. Wave energy is taken as length multiplied by the square root of the height.

Weathering - The *in situ* disintegration or decomposition of rock surfaces exposed to the atmosphere and the upper part of an outcropping rock formation as the result of physical, chemical, and biological processes. It is usually indicated by changes in color, texture, composition, coherence, or firmness. Excludes erosion and induration. Residual rocks more resistant to erosion may produce strange forms (columns, pinnacles, pillow or cannon-ball structures, pillars and pedestals, sometimes termed hoodoo rocks).

Wetland - a general term describing swamps, bogs, marshes, and shallow (up to 5 m) lagoons and lakes. In coastal environments these include salt marshes, mangrove swamps, reedswamp, rush swamp, and seagrass beds.

Wind resultant - an expression of the long-term effect of winds of varying strength, duration, and direction, usually developed as a vector diagram from which a directional resultant may be determined.

Zetaform beach - an asymmetrical beach between headlands, where the outline is shaped by waves arriving obliquely, and refracted round

Wave refraction in Seven Mile Bay, Tasmania.

the headlands. Also known as a Log-spiral (*q.v.*) or half-heart beach, and sometimes as a Headland (*q.v.*) bay beach (but this term does not indicate that the beach outline is asymmetrical).

 Eric Bird

Bibliography

Bradley, W., 1958. Submarine abrasion and wave-cut platforms. *Bulletin of the Geological Society of America*, **69**: 967–974.
Childers, E.R., 1903. *The Riddle of the Sands*, xii. London: Sidgwick and Jackson.
Clapp, C.H., 1913. Contraposed shoreline. *Journal of Geology*, **21**: 537.
Cotton, C.A., 1956. Rias sensu stricto and sensu lato. *Geographical Journal*, **122**: 360–364.
Davies, J.L., 1974. The coastal sediment compartment, *Australian Geographical Studies*, **12**: 139–151.
Davies, J.L., 1980. *Geographical Variation in Coastal Development*. London: Longman, 130 pp.
Dubois, G., 1924. *Recherches sur les terrains quaternaires du Nord de la France*. University of Lille, France: Lille.
Galvin, C.J., 1972. Waves breaking in shallow water. In Meyer, R. (ed.), *Waves on Beaches*. London: Academic Press, 413–455.
Guilcher, A., 1966. Les grandes falaises et megafalaises des cotes sud-ouest et ouest d. Irlande. *Annals de Géographie*, **75**: 26–38.
Healy, T.R., 1968. Bioerosion on shore platforms. *Earth Science Journal*, **2**: 26–37.
Jenkin, J.J., 1966. *The Geomorphology and Upper Cainozoic Geology of South-east Gippsland*. Geological Survey of Victoria, Memoir, 27: 94.
Jennings, J.N., 1967. Cliff top dunes. *Australian Geographical Studies*, **5**: 40–49.
Johnson, D.W., 1916. Plains, planes and peneplanes. *Geographical Review*, 443–447.
Johnson, D.W., 1919. *Shore processes and shoreline development*. Wiley, New York.
Lyell, C., 1833. *Principles of Geology*, 2, xiv. London: J. Murray.
Margolis, S.V., 1968. Electron microscopy of chemical solution and mechanical abrasion features on quartz sand grains. *Sedimentary Geology*, 2: 243–256.
May, V.J., 1977. Earth Cliffs. In Barnes, R.S.K. (ed.), *The Coastline*. Chichester: John Wiley & Sons, pp. 215–236.
Price, W.A., 1947. Equilbrium of form and forces in tidal basins on the coast of Texas and Louisiana. *Bulletin of the American Association of Petroleum Geologists*, **31**: 1619–1663.
Richthofen, F. von, 1886. *Führer für Forschungreisende*. Hanover: Jänecke, 308–309.
Russell, R.J., and Howe, H.V., 1935. Cheniers of southwestern Louisiana. *Geographical Review*, **25**: 449–461.
Schwartz, M.L., 1971. The multiple causality of barrier islands. *Journal of Geology*, **79**: 91–93.
Shepard, F.P., 1952. Revised nomenclature for coastal depositional features. *Bulletin of the American Association of Petroleum Geologists*, **36**: 1802–1812.
Short, A.D. (ed.), 1999. *Handbook of Beach and Shoreface Morphodynamics*. Chichester: Wiley.
Stoddart, D.R., and Cann, J.R., 1965. Nature and origin of beach rock. *Journal of Sedimentary Petrology*, **25**: 243–247.
Wentworth, C.K., 1944. Potholes, pits and pans: subaerial and marine. *Journal of Geology*, **52**: 117–130.
Zenkovich, V.P., 1959. On the genesis of cuspate spits along lagoon shores. *Journal of Geology*, **76**: 169–177.

APPENDIX 6: TOPIC CATEGORIES

Atmosphere and oceanography

Altimeter Surveys, Coastal Tides, and Shelf Circulation
Asteroid-Impact Coasts
Beaufort Wind Scale
Changing Sea Levels
Climate Patterns in the Coastal Zone
Coastal Climate
Coastal Currents
Coastal Lakes and Lagoons
Coastal Temperature Trends
Coastal Upwelling and Downwelling
Coastal Wind Effects
El Niño–Southern Oscillation
Endogenic and Exogenic Factors
Energy and Sediment Budgets of the Global Coastal Zone
Eolian Processes
Estuaries
Geographical Coastal Zonality
Greenhouse Effect and Global Warming
Ingression, Regression, and Transgression
Late Quaternary Marine Transgression
Littoral
Meteorological Effects on Coasts
Microtidal Coasts
Natural Hazards
Nearshore Geomorphological Mapping
Nearshore Sediment Transport Measurement
Nearshore Wave Measurement
New Zealand, Coastal Geomorphology and Oceanography
Oil Spills
Oil Spills, High-Energy Coasts
Offshore Sand Banks and Linear Sand Ridges
Offshore Sand Sheets
Pressure Gradient Force
Rip Currents
Scour and Burial of Objects in Shallow Water
Sea-Level Changes During the Last Millenium
Sea-Level Rise, Effect
Sediment Suspension by Waves
Shelf Processes
Shoreface
Storm Surge
Submarine Groundwater Discharge
Submerged Coasts
Surf Zone Processes
Tidal Creeks
Tidal Datums
Tidal Environments
Tidal Inlets
Tidal Flats
Tidal Flats, Open Ocean Coasts
Tidal Prism
Tide-Dominated Coasts
Tides
Tsunamis
Vorticity
Water Quality
Wave and Tide-Dominated Coasts
Wave Climate
Wave–Current Interaction
Wave-Dominated Coasts
Wave Environments
Wave Focusing
Wave Hindcasting
Wave Power
Wave Refraction Diagrams
Waves
Weathering in the Coastal Zone

Ecology

Africa, Coastal Ecology
Algal Rims
Antarctica, Coastal Ecology and Geomorphology
Aquaculture
Arctic, Coastal Ecology
Asia, Eastern, Coastal Ecology
Asia, Middle East, Coastal Ecology and Geomorphology
Atlantic Ocean Islands, Coastal Ecology
Atolls
Australia, Coastal Ecology
Bioconstruction
Bioengineered Shore Protection
Bioerosion
Biogeomorphology
Bioherms and Biostromes
Black and Caspian Seas, Coastal Ecology and Geomorphology
Bogs
Carbonate Sandy Beaches
Caribbean Islands, Coastal Ecology and Geomorphology
Coral Reef Coasts
Coral Reef Islands
Coral Reefs
Deltaic Ecology
Driftwood
Europe, Coastal Ecology

History, Coastal Ecology
Indian Ocean Coasts, Coastal Ecology
Indian Ocean Islands, Coastal Ecology and Geomorphology
Mangroves, Ecology
Middle America, Coastal Ecology and Geomorphology
Monitoring Coastal Ecology
New Zealand, Coastal Ecology
North America, Coastal Ecology
Pacific Ocean Islands, Coastal Ecology
Peat
Reefs, non-coral
Salt Marsh
Sea-Level Indicators, Biologic
Sea-Level Indicators, Biological in Depositional Sequence
South America, Coastal Ecology
Vegetated Coasts
Wetlands
Wetlands Restoration

Engineering and technology

Airborne Laser Terrain Mapping (ALTM) and Light Detection and Ranging (LIDAR)
Altimeter Surveys, Coastal Tides, and Shelf Circulation
Artificial Islands
Beach and Nearshore Instrumentation
Beach Drain
Beach Nourishment
Bioengineered Shore Protection
Bypassing at Littoral Drift Barriers
Capping of Contaminated Coastal Areas
Coastal Modeling and Simulation
Dams, Effect on Coasts
Desalination
Dikes
Dredging of Coastal Environments
Engineering Applications of Coastal Geomorphology
Geochronology
Geographic Information Systems
Geohydraulic Research Centers
Geotextile Applications
Global Positioning Systems
Ground Penetrating Radar
History, Coastal Protection
Jet Probes
Managed Retreat
Mangroves, Remote Sensing
Mapping Shores and Coastal Terrain
Mining of Coastal Materials
Modeling of platforms, terraces, and Coastal Evolution
Monitoring, Coastal Ecology
Monitoring, Coastal Geomorphology
Navigation Structures
Nearshore Geomorphological Mapping
Nearshore Sediment Transport Measurement
Nearshore Wave Measurement
Numerical Modeling
Photogrammetry
Physical Models
Polders
Profiling, Beach
RADARSAT-2
Reclamation
Remote Sensing of: Coastal Environments
Remote Sensing, Wetlands Classification
Shore Protection Structures
Simple Beach and Surf Zone Models
Surf Modeling
Synthetic Aperture Radar Systems
Tidal Power
Tide Gauges
Tide Mill
Time Series Modeling
Tracers
Vibracore
Wave Hindcasting
Wave Power
Wave Refraction Diagrams

Geomorphology

Accretion and Erosion Waves on Beaches
Africa, Coastal Geomorphology
Alluvial Plain Coasts
Antarctica, Coastal Ecology and Geomorphology
Archaeological Site Location, Effect of Sea Level Changes
Arctic, Coastal Geomorphology
Asia, Eastern, Coastal Geomorphology
Asia, Middle East, Coastal Ecology and Geomorphology
Asteroid-Impact Coasts
Atlantic Ocean Islands, Coastal Geomorphology
Atolls
Australia, Coastal Geomorphology
Barrier
Barrier Islands
Bars
Bay Beaches
Beach Erosion
Beach Features
Beach Nourishment
Beach Processes
Beach Profile
Beach Ridges
Beach Sediment Characteristics
Beach Stratigraphy
Beachrock
Biogeomorphology
Black and Caspian Seas, Coastal Ecology and Geomorphology
Boulder Barricades
Boulder Beaches
Boulder Pavement
Carbonate Sandy Beaches
Caribbean Islands, Coastal Ecology and Geomorphology
Cays
Chalk Coasts
Changing Sea Levels
Cheniers
Cliffed Coasts
Cliffs, Erosion Rates
Cliffs, Lithology versus Erosion Rates
Coastal Changes, Gradual
Coastal Changes, Rapid
Coastal Hoodoos
Coastal Sedimentary Facies
Coastal Soils
Coastal Subsidence
Coastline Changes
Coasts, Coastlines, Shores, and Shorelines
Cohesive Sediment Transport
Continental Shelves
Coral Reefs, Emerged
Cross-Shore Sediment Transport
Cross-Shore Variation of Grain Size on Beaches
Cuspate Forelands
Dalmatian Coasts
Deltas
Depth of Closure
Depth of Disturbance
Desert Coasts
Dissipative Beaches
Drift and Swash Alignments
Dune Ridges
Dynamic Equilibrium of Beaches
Energy and Sediment Budgets of the Global Coastal Zone
Engineering Applications of Coastal Geomorphology
Eolianite
Eolian Processes
Erosion: Historical Analysis and Forecasting
Erosion Processes
Eustasy
Europe, Coastal Geomorphology
Faulted Coasts
Geodesy
Geographical Coastal Zonality
Glaciated Coasts
Glossary of Coastal Geomorphology
Gravel Barriers
Gravel Beaches

Gross Transport
Headland-Bay Beach
History, Coastal Geomorphology
Holocene Coastal Geomorphology
Holocene Epoch
Honeycomb Weathering
Hydrology of Coastal the Zone
Ice-Bordered Coasts
Indian Ocean Coasts, Coastal Geomorphology
Indian Ocean Islands, Coastal Ecology and Geomorphology
Ingression, Regression, Transgression
Isostasy
Karst Coasts
Klint
Late Quaternary Marine Transgression
Littoral
Littoral Cells
Littoral Drift Gradient
Longshore Sediment Transport
Machair
Mangroves, Geomorphology
Mapping Shores and Coastal Terrain
Marine Terraces
Mass Wasting
Middle America, Coastal Ecology and Geomorphology
Modes and Patterns of Shoreline Change
Monitoring Coastal Geomorphology
Muddy Coasts
Natural Hazards
Nearshore Geomorphological Mapping
Net Transport
New Zealand, Coastal Geomorphology and Oceanography
North America, Coastal Geomorphology
Notches
Pacific Ocean Islands, Coastal Geomorphology
Paleocoastlines
Paraglacial Coasts
Placer Deposits
Pleistocene Epoch
Pluvial Lake Shore Deposits
Profiling, Beach
Reflective Beaches
Rhythmic Patterns
Ria
Ripple Marks
Rock Coast Processes
Sandy Coasts
Sea-Level Indicators, Geomorphic
Sediment Budget
Sedimentary Basins
Seismic Displacement
Sequence Stratigraphy
Sharm Coasts
Shore Platforms
Shoreface
South America, Coastal Geomorphology
Spits
Strandflats
Submerged Coasts
Submerging Coasts
Tafoni
Tectonics and Neotectonics
Thalassostatic Terraces
Tidal Creeks
Tidal Flats
Tidal Flats, Open Ocean Coasts
Tide-Dominated Coasts
Tors
Trottoirs
Uplift Coasts
Volcanic Coasts
Washover Effects
Wave- and Tide-Dominated Coasts
Wave-Dominated Coasts
Weathering in the Coastal Zone

Human activities

Archaeological Site Location, Effect of Sea-Level Changes
Archaeology
Beach Use and Behaviors
Carrying Capacity in Coastal Areas
Cleaning Beaches
Coastal Boundaries
Coastal Warfare
Coastal Wells
Coastal Zone Management
Conservation of Coastal Sites
Databases
Demography of Coastal Populations
Developed Coasts
Economic Value of Beaches
Environmental Quality
Estuaries, Anthropogenic Impacts
Global Vunerability Analysis
Health Benefits
Human Impact on Coasts
Journals
Lifesaving and Beach Safety
Managed Retreat
Marine Debris—Onshore, Offshore, Seafloor Litter
Marine Parks
Organizations
Rating Beaches
Sand Rights
Setbacks
Shell Middens
Small Islands
Surfing
Tourism and Coastal Development
Tourism, Criteria for Coastal Sites
Water Quality

Maurice Schwartz

Index

F

Q

R